RUDOLF WALTHER

POLYTECHNICAL DICTIONARY

POLYTECHNICAL DICTIONARY

GERMAN-ENGLISH

Edited by

Ing. RUDOLF WALTHER

Containing about 100,000 terms

PERGAMON PRESS

Oxford · London · Edinburgh · New York
Toronto · Sydney · Paris · Braunschweig

VEB VERLAG TECHNIK BERLIN

Pergamon Press Ltd.
Headington Hill Hall, Oxford
4 & 5 Fitzroy Square, London W. 1

Pergamon Press (Scotland) Ltd.
2 & 3 Teviot Place, Edinburgh 1

Pergamon Press Inc.
44-01 21st Street, Long Island City, New York 11101

Pergamon of Canada, Ltd.
207 Queen's Quay West, Toronto, Ontario

Pergamon Press (Aust.) Pty. Ltd.
Rushcutters Bay, Sydney, N. S. W.

Pergamon Press S.A.R.L.
24 rue des Écoles, Paris 5e

Vieweg & Sohn G.m.b.H.
Burgplatz 1, Braunschweig

Library of Congress Catalog Card No. 67-25817

Printed in Germany

08 003525 6

PREFACE

This new German-English Polytechnical Dictionary is a companion to the previously published English-German volume. Both dictionaries present a comprehensive aid for everyone interested in using scientific and technological English and German.

The subject matter and standard for this German-English volume correspond to that established in the first volume. This, of course, does not mean that it is simply a "mirror image" of the latter. All sub-sections have been re-considered and adapted to the requirements of the language change-over. Numerous new terms were added to bring the dictionary up to date, and a few fields were included for the first time, e.g. production engineering, nuclear physics, nuclear engineering, polygraphics, agricultural engineering, ship building and navigation. Important monolingual sources, including "Meyers Neues Lexikon" in eight volumes published by VEB Bibliographisches Institut, Leipzig (1961 to 1964), were evaluated, thereby ensuring both accuracy of terminology and a high degree of completeness.

The editor and publishers hope that this new dictionary will contribute to an exchange of information and will be welcomed by users to the same extent as was the English-German volume.

Suggestions and hints which could lead to an improvement of this dictionary would be appreciated, and they should be forwarded to VEB Verlag Technik, 102 Berlin, Oranienburger Straße 13/14.

Both the editor and publishers thank all institutions and individuals who contributed to the compilation of this book.

Publishers Editor

LIST OF COLLABORATORS

Ernst-Georg Berends, Berlin; *Luise Berthold*, Karl-Marx-Stadt; Dr.-Ing. *Georg Brack*, Merseburg; *Willy Eberl*, Berlin; Dr.-Ing. *Kurt Fiedler*, Leipzig; Dipl. phil. *Horst Görner*, Berlin; *Arno Gräf*, Berlin; Dipl.-Math. *Werner Heidrich*, Berlin; Ing. *Horst Hummelke*, Berlin; Dipl. rer. hort. *Madeleine Karrer*, Berlin; *Ruth Körner*, Dresden; Dipl.-Ing. *Ulrich Lagowitz*, Berlin; *Herbert Lasch*, Berlin; Dipl.-Ök. *Herbert Liebscher*, Leipzig; *Fritz Mühler*. Karl-Marx-Stadt; *Wolfgang Müller*, Leipzig; Dipl.-Chem. Dr. *René Otto*, Leipzig; Dipl.-Ing. *Manfred Röhnisch*, Berlin; Dipl.-Gwl. *Helmut Schmidt*, Berlin; *Hans-Joachim Schuch*, Jena; Dipl.-Math. *Ralf Sube*, Berlin; *Peter Wennrich*, Berlin; Dipl.-Ing. oec. *Günter Wohlfahrt*, Berlin; Dipl.-Chem. *Gottfried Zimmermann*, Leipzig.

BENUTZUNGSHINWEISE · DIRECTIONS FOR USE

1. BEISPIELE FÜR DIE ALPHABETISCHE ORDNUNG
EXAMPLES OF THE ALPHABETICAL ARRANGEMENT

Hilfsanode
Hilfsantenne
Hilfselektrode
Hilfsenergie
  H. / mit
  H. / mit hydraulischer
  H. / mit pneumatischer
  H. / ohne
Hilfsenergieversorgung
Hilfsregler
Hilfsruder
Himmelsmechanik
hinauftransformieren
hindurchgehen
  durch einen Stoff h.
hindurchsickern/durch die Filtermasse
Hinterachsantrieb

Kochung
  K. ohne Umwälzsystem
Kochversuch
Kode
  K. / ohne
  dicht gepackter K.
  fehlererkennender K.
  fehlerverbessernder K.
  selbstprüfender K.
Kodein
Kodeumwandler
kodieren
Kodierer
Kodierung
  numerische K.
Koeffizient
  K. der gegenseitigen Induktion
Koeffizientenpotentiometer

2. BEDEUTUNG DER ZEICHEN
SIGNS USED IN THE DICTIONARY

/ — Korrektionsfaktor / quantentheoretischer = quantentheoretischer Korrektionsfaktor

( ) — charge exchange (transfer) = charge exchange *or* charge transfer
bearing (sliding, running) surface = bearing surface *or* sliding surface *or* running surface
isobaric (constant-pressure) line = isobaric line *or* constant-pressure line

[ ] — ball[-type] valve = ball valve *or* ball-type valve
charge[d particle] carrier = charge carrier *or* charged particle carrier
Liefer- und Gütebestimmung [/technische] = Liefer- und Gütebestimmung *oder* technische Liefer- und Gütebestimmung

( ) — Die Klammern enthalten Erklärungen
These brackets are used to include explanations

ABKÜRZUNGEN · ABBREVIATIONS

ac acoustics / Akustik

aero aeronautics / Flugwesen

ag agricultural engineering / Landtechnik

Am Americanism / Amerikanismus

astr astronomy / Astronomie

aut automotive engineering / Kraftfahrzeugtechnik

build building and surveying / Bauwesen

ch chemistry, chemical engineering / Chemie, chemische Technik

cont control engineering / Regelungstechnik

dat data processing; office machines / Datenverarbeitung; Bürotechnik

ec economics / Ökonomie

el electrical engineering, electronics / Elektrotechnik, Elektronik

el.m. electric machines, transformers / Elektromaschinenbau

f feminine / Femininum

ff fluid flow / Strömungslehre

geol geology / Geologie

leath leather / Leder

m masculine / Maskulinum

mach machine building / Maschinenbau

math mathematics / Mathematik

mech mechanics / Mechanik

met metallurgy, materials testing, foundry practice / Metallurgie, Werkstoffprüfung, Gießereiwesen

mil military engineering / Militärtechnik

min mineralogy / Mineralogie

mine mining, mine engineering / Bergbau, Bergbautechnik

n neuter / Neutrum

nucl nuclear physics, nuclear engineering / Kernphysik, Kerntechnik

opt optics / Optik, optische Geräte

ph physics / Physik

pl plural / Plural

plast plastics / Plast

print typography / Typografie

refr refrigerating engineering / Kältetechnik

rock rockets / Raketentechnik

ry railway / Eisenbahnwesen

s. see / siehe

s.a. see also / siehe auch

ship ship construction / Schiffbautechnik, Schiffahrt

telec telecommunications / Fernmeldetechnik

telev television / Fernsehtechnik

text textile industry / Textiltechnik

therm thermodynamics / Wärmelehre

A

a-Ader *f* A-wire, positive wire, tip-wire

A-Betrieb *m (el)* class-A operation

A-Bombe *f* atom[ic] bomb, A-bomb

A-Horizont *m (geol)* A-horizon, zone of removal

A-Mast *m* A-tower, A-type pole

A-Verstärker *m* class-A amplifier

A0-Wellen *fpl* type A0 waves

A1-Wellen *fpl* type A1 waves

A2-Wellen *fpl* type A2 waves, modulated keyed continuous waves

A3-Wellen *fpl* type A3 waves

A4-Wellen *fpl* type A4 waves, facsimile waves

Aalkorb *m* eel basket

Aaltreppe *f* eel ladder

Aas *n* fleshings

Aasseite *f (leath)* flesh side

AB-Betrieb *m (el)* AB method

AB-Verstärker *m* class-AB amplifier

abackern *(ag)* to plough off

Abakafaser *f* abaca [fibre]

Abakus *m* abacus *(1. calculating frame; 2. upper member of capital of column)*

Abampere *n (el)* abampere

abändern to alter; to modify; to change

Abänderung *f* alteration; modification; change

A. des Frachtvertrags modification of the contract of carriage

Abandon *m* abandonment *(marine insurance)*

Abandonerklärung *f* notice of abandonment

Abandonnement *n*, **Abandonnierung** *f s.* Abandon

abankern to unmoor *(a vessel)*

abarbeiten 1. to come off, to refloat, to get afloat, to get away, to get off, to bring off *(a stranded vessel)*; 2. to work off; to remove

Abart *f* variety, modification

abätzen to corrode off; to etch

abbacken *(ship)* to clear the messtable, to clear away

abbaken *(ship)* to set beacons

Abbakung *f (ship)* beaconing

abbalgen to flay; to skin

Abbau *m* 1. mining; exploitation; working; quarrying *(stone)*; extraction; stoping; getting; winning; 2. removal; demounting; dismantling; 3. decomposition; degeneration; degradation; 4. panel, stope, face, working; breaking ground; 5. reduction *(wages; staff)*; 6. *(el)* collapse, dying *(of fields)*

A. der Bande *(ph)* destruction of band

A. durch Seilförderung *(mine)* extraction by means of rope

A. mit Bergeversatz *(mine)* mining with filling

A. mit Rahmenzimmerung *(mine)* square-set method (stoping)

A. mit Rahmenzimmerung und Bergeversatz *(mine)* square-set-and-fill

A. von Erzschweben *(mine)* pillar caving

A. von Schallvorgängen reduction of harmonic content in sound analysis

belegter A. *(mine)* manned face

harmonischer A. *(mine)* harmonic system of working

schwebender A. *(mine)* working to the rise

unsystematischer A. *(mine) (Am)* coyoting, *(Am)* gophering

unvollständiger A. *(mine)* partial extraction

versetzter A. *(mine)* filled stope

vollständiger A. *(mine)* full extraction

Abbauabteilung *f (mine)* working panel

Abbaublock *m (mine)* block

abbauen 1. to mine; to quarry *(stone)*; to win; to get; to extract; to break down; to excavate; to exploit; 2. to remove; to dismantle; to demount; 3. *(ch)* to decompose; to degradate; to degenerate; 4. to reduce *(wages; staff)*

Abbauen *n s.* Abbau

Abbaufeld *n (mine)* 1. panel; 2. district

Abbaufördermittel *n* face conveyor

Abbaufortschritt *m* advance rate, face advance

A. in die Teufe average mining rate vertically

Abbaufront *f* coal face

stempelfreie A. *(mine)* prop-free-face

Abbaugeschwindigkeit *f (mine)* face advance

Abbauhammer *m (mine)* pneumatic pick

Abbauhammerbetrieb *m* pneumatic-pick face

Abbauhammerstreb *m* pneumatic-pick face

Abbaukammer *f* / **offene** *(mine)* open stope

Abbaukonzentration *f (mine)* concentration of working

Abbaumatte *f (mine)* mat

abbäumen 1. *(text)* to let off; 2. to shore up, to boom off, to breast off *(a vessel)*

Abbäumen *n (text)* let-off motion

Abbauprodukt *n* reduction product

Abbaurechte *npl (mine)* mining concessions

Abbaurevier *n* working panel

Abbaurichtung *f* direction of advance (mining)

Abbauscheibe *f (mine)* slice

Abbauschema *n (mine)* working scheme

Abbauschild *m (mine)* shield [support]

Abbauspieß *m (mine)* pneumatic pointed chisel

Abbaustrecke *f (mine)* board; *(Am)* entry, *(Am)* butt entry, panel entry, gate (stall) road, roadway, stope drift

Abbauverfahren *n* mining (working) method

A. mit Versatz filled stoping

kombiniertes A. combined mining method

abbauwürdig mineable, workable

Abbe-Beleuchtungsapparat *m* Abbe illuminating device

Abbe-Längenmeßgerät *n* Abbe vertical metroscope
Abbe-Prisma *n* Abbe prism
Abbe-Refraktometer *n* Abbe refractometer
Abbe-Spektrometer *n* Abbe spectrometer
abbeizen to pickle
Abbeizmittel *n* pickling [agent]; paint (varnish) remover
abbiegen to bend [off] *(reinforcing bars)*; to deflect
Abbiegung *f* bend *(reinforcement)*
Abbild *n* image, picture
thermisches A. thermal dissipation curve
abbilden 1. *(math)* to map; 2. *(opt)* to form an image; to image; to depict; to make an image (a picture)
Abbilden *n* / **konformes** *(math)* conformal mapping
abbildend *(ph)* image-forming
Abbildtheorie *f (ph)* theory of representation
Abbildung *f* 1. *(math)* mapping; 2. *(opt)* image; depiction; projection; imaging
elektronische A. electronic imaging
flächentreue A. equal-area projection
konforme A. conformal mapping
latente A. latent image formation
mikroskopische A. microscopic depiction
punktförmige A. *(ph)* point image
stigmatische A. stigmatic depiction
winkeltreue A. conformal mapping
Abbildungsfehler *m* optical aberration; image defect
Abbildungsgesetz *n* **[von Abbe]** *(opt)* [Abbe's] law of imagery
Abbildungsgleichung *f* imaging equation
Abbildungsgüte *f* definition (quality) of image
Abbildungsmagnet *m* focusing magnet
Abbildungsmaßstab *m* image scale
Abbildungsoptik *f* imaging (focusing) optics
Abbildungsspule *f* focusing coil
Abbindebeginn *m (build)* initial set
Abbindebeschleuniger *m (build)* accelerator, accelerating agent
Abbindedauer *f (build)* setting time; curing time
Abbindedraht *m (el)* binding wire, tie-wire
Abbindefähigkeit *f (build)* setting quality
Abbindegarn *n* twine, binding thread
Abbindegeschwindigkeit *f (build)* rate of setting; rate of cure
Abbindeklemme *f (el)* bonding clip
abbinden 1. to harden, to set *(cement)*; to cure *(plastic)*; 2. to join; to frame *(wood)*
ein Kabel a. to tie off, to tie up, to lace [a cable]
eng a. to interlace tightly
Abbinden *n* 1. set[ting], hardening *(cement)*; curing *(plastic)*; 2. joining, framing *(wood)*; 3. tying *(cable)*
vorzeitiges A. false set *(cement)*
Abbindeplatz *m (build)* timber yard
Abbindeprüfung *f (build)* setting test
Abbindetemperatur *f (build)* setting temperature, temperature of setting
Abbindeverhalten *n (build)* setting behaviour
Abbindeverzögerer *m (build)* retarder, retarding agent, dope
Abbindevorgang *m (build)* [mechanism of] setting
Abbindewärme *f (build)* setting heat, heat of setting
Abbindezeit *f* setting time *(cement)*; assembly time *(glueing)*
abblasen 1. to blow off; 2. to blow down *(blast furnace; steam boiler)*; 3. to quench *(cold)*; 4. to blast *(sand)*
Abblasen *n* 1. blowing [off] *(steam)*; 2. blowing down *(blast furnace; boiler)*; 3. *(cold)* quenching; 4. blasting *(sand)*
A. mit kalten Gasen gas quenching
Abblaseventil *n* exhaust (blow-out) valve; bleeder
Abblasöffnung *f* vent hole *(pneumatics)*
abblättern 1. to flake [off], to peel [off], to slab, to spall; to scale off; 2. to exfoliate
Abblättern *n* 1. flaking [off]; slabbing [off]; spalling; scaling off; 2. exfoliating
Abblätterung *f* exfoliation (*s. a.* Abblättern)
Abblendeinrichtung *f* anti-dazzle device
abblenden 1. to dim *(cinema)*; 2. to stop down, to reduce the aperture of a lens by means of the stops *(camera)*; 3. to dip [the headlamp]; to dim [lights]
Abblenden *n* 1. dimming *(cinema)*; 2. stopping down; reducing the aperture of a lens by means of the stops *(camera)*; 3. dipping [the headlamps]; dimming *(lights)*
A. beider Scheinwerfer double-dipping
Abblendfaden *m* dimming (dip) filament
Abblendfußschalter *m* pedal-operated dip switch
Abblendhaube *f* anti-dazzle cap
Abblendkappe *f* headlight shield
Abblendlampe *f* double-dipping headlamp, passlamp
Abblendlicht *n* passing (anti-dazzle, dimmed) light
Abblendschalter *m* dimming (dimmer, anti-dazzle, beam-deflecting) switch, dipper
Abblendvorschriften *fpl* anti-dazzle regulations
abborken to bark
abböschen to slope; to scarp *(steep or vertical slope)*
Abbrand *m* 1. melting loss by oxidation; 2. roasting residue; 3. burn-up *(reactor)*; waste; 4. black ends
relativer A. *(nucl)* burn-up fraction

spezifischer A. *(nucl)* specific burn-up
Abbrandschale *f* **[am Funkenstativ]** *(opt)* scale-collecting cup
Abbrandtiefe *f (nucl)* specific burn-up
abbrausen to rinse; to spray
abbrechen 1. *(build)* to demolish; to pull down; 2. to break off; to discontinue
Abbrechfehler *m (dat)* truncation error
abbremsen to brake; to slow down *(moderator)*; to run the engine on the chocks *(aeroplane)*
Abbremsung *f* braking; retardation; slowing-down, moderation, thermalization; stopping
A. des Eigengewichts braking of the tare
Abbremsungszeit *f* relaxation time *(electronics)*
Abbremszeit *f* slowing-down time
abbrennen 1. to burn up; 2. to gr[e]ave, to bream, to burn off, to sweal *(the bottom of a ship)*; 3. *(ch)* to deflagrate; to calcine; 4. *s.* abbeizen
Abbrennen *n* burning [up]
A. der Hochofengase excess-gas burning
Abbrennkontakt *m (el)* arcing tip (contact)
Abbrennschweißen *n* flash welding
Abbrennstumpfschweißen *n* flash butt welding
Abbrennstumpfschweißnaht *f* flash butt weld
Abbrennstumpfschweißung *f* flash butt welding
abbringen to bring off, to get off, to get away, to get afloat, to refloat *(a stranded vessel)*
abbröckeln to spall, to crumble *(cutting edge)*
Abbruch *m* 1. breaking down *(house)*; demolition; pulling down; 2. rupture; breaking-off; discontinuance; 3. dismounting, breaking-off *(vessel)*
Abbruchfehler *m (dat)* truncation error
Abbruchhammer *m* demolition pick (hammer)
Abbruchholz *n* old timber
Abbruchwerft *f* breaking up yard
Abbruchwerkzeug *n* demolition tool
Abbruchzone *f (geol)* faulted zone
abbrühen to scald; to boil out
Abbuchung *f* deduction
abbuffen *(leath)* to buff
Abbügeletikett *n (print)* hot transfer label
Abcoulomb *n* abcoulomb
abdachen 1. to slope; 2. to unroof
Abdachung *f* 1. [direction of] slope; 2. scarp, escarpment
abdämmen to dam; to seal off
Abdämmung *f* damming
Abdampf *m* exhaust (waste) steam
abdampfen to vaporize, to evaporate
abdämpfen 1. to damp down; 2. to absorb
Abdampfinjektor *m* exhaust steam injector
Abdampfleitung *f (Am)* vent pipe
Abdampfprobe *f* gum test
Abdampfrückstand *m* residue from evaporation
Abdampfschale *f (ch)* evaporating dish; porcelain dish
Abdampfstutzen *m* exhaust connecting branch
Abdampfturbine *f* exhaust (waste) steam turbine
Abdampfverwertung *f* utilization of waste steam
Abdampfverwertungsanlage *f* exhaust-steam utilizing plant
Abdampfvorwärmer *m* exhaust-steam preheater
abdarren to [kiln-]dry
Abdeckasphalt *m* masking asphalt
Abdeckblech *n* flashing; cover plate (sheet)
Abdeckblende *f* shutter mask
abdecken 1. to cover; to mask; to blanket *(vessel)*; 2. to uncover; to unroof, to untile *(roof)*
Abdecken *n* **der Schleifsteinoberfläche** blinding of the stone surface
A. des Leitapparates covering the guide-blade disk
Abdecker *m* flayer
Abdeckerei *f* flaying-house, knackery, flaying-yard
Abdeckerhaut *f* renderer
Abdeckhaube *f* dust hood; covering hood
Abdecklack *m* masking lacquer, covering varnish; paint for cocoonization
Abdeckleiste *f* cover fillet (strip)
Abdeckmatte *f* curing mat *(concrete)*
Abdeckplatte *f* cover plate (slab); covering plate *(chimney)*
Abdeckschablone *f* mask
Abdeckschicht *f* resist coating
Abdeckstein *m* cover brick, capstone [for cables]
Abdeckstreifen *m* masking tape
Abdecktusche *f* masking ink
Abdeckung *f* 1. *(build)* cover[ing], capping *(cable)*; coping; 2. uncovering; untiling *(roof)*, unroofing; 3. blanket *(melt)*
A. des Stapels roof of pile *(wood)*
Abdeckungsfrequenz *f* blanketing frequency
abdestillieren to distil off, to retort
abdichten to seal [off]; to pack; to [make] water-proof; to [make] damp-proof; to caulk; to make tight; to stop; to stuff
Abdichten *n* sealing off; making tight; water-proofing; damp-proofing; caulking
abdichtend antidribble (*comp. also* abdichten)
Abdichtung *f* seal[ing]; packing; damp-proofing, water-proofing; caulking
A. durch Kupfereinlage packing with copper
A. ortsfester Teile static seal
statische A. static seal
Abdichtungsmittel *n* sealant
Abdichtungsplatten *fpl* **am Schleifer / seitliche** finger bars on the grinder *(papermaking)*
Abdichtwachs *n* sealing wax
Abdrängkraft *f* thrust force *(tool)*

Abdrängung *f (rock)* drift
Abdreheinrichtung *f* truing device *(grinding)*
abdrehen 1. to true *(grinding)*; to turn [off] *(lathe)*; 2. *(ship)* to turn off, to turn to port, to turn to starboard *(course)*; 3. to retread *(tyres)*; 4. to shut off *(faucet)*
  Faß a. to clean off the cask
Abdrehen *n* **der Schleifscheibe** truing of the grinding wheel
Abdrift *f* drift
abdrosseln 1. *(el)* to fit (load) with chokes, to block by chokes; 2. to throttle [down] *(motor)*; to stall
Abdruck *m* replica; reprint, copy
Abdruckblech *n* covering knife *(knitting)*
Abdrücken *n* pressure test; forcing-off
Abdrücklokomotive *f* hump locomotive
Abdrückschraube *f (mach)* forcing screw
Abdrucktechnik *f (opt)* replica technique
Abdrückungen *fpl (mine)* pressure partings
abdunkeln *(opt)* to darken, to dim
abdunsten to evaporate
Abendfarbe *f (text)* evening shade
Abendrot *n* sunset glow
aberkennen / die Patentfähigkeit to deny the patentability
Abernathyit *m* abernathyite *(uranium mineral)*
Aberration *f* aberration
  astronomische A. astronomic aberration
  chromatische A. chromatic aberration
  planetarische A. planet aberration
  sphärische A. spherical aberration
aberrationsfrei without aberration
Aberrationswinkel *m* angle of aberration
aberregen *(el)* to de-energize; to de-excite
Aberregung *f* de-energization; de-excitation
Abessinierbrunnen *m (build)* Abyssinian well
Abessinierpumpe *f (build)* Abyssinian pump
abfackeln to burn; to bleed off; to flare *(gas)*
Abfackeln *n* **der Hochofengase** excess-gas burning
abfaden *(ship)* to fathom
Abfahrauftrag *m* order of departure
abfahren 1. to leave; to start; to depart; to sail, to set sail, to put to sea *(ship)*; to set out; 2. to trace *(copying)*
Abfahrt *f* 1. departure *(train)*; *(ship)* sailing, leaving, start; setting-out; 2. *(build)* interchange; exit *(motorways)*
Abfahrtsbahnsteig *m* departure platform
Abfahrtshafen *m* port of sailing (departure); *(Am)* port of clearance
Abfahrtstag *m (ship)* sailing day (date)
Abfahrtszeit *f* departure time
Abfall *m* 1. waste [matter]; loss; refuse [matter]; scrap *(metal)*; garbage *(kitchen)*; chips; trash *(residue of peel)*; husk *(residue of cocoons)*; cuttings *(tailoring)*; 2. drop *(pressure, temperature, voltage)*; fall; 3. decay *(electronics)*; release *(relay)*
  exponentieller A. exponential decay
  geringaktiver A. low-level waste
  heißer A. *(nucl)* hot waste
  hochaktiver A. hot waste
  inaktiver A. cold waste
  radioaktiver A. radioactive (atomic) waste
Abfallabscheider *m* duster; waste extractor
Abfallauge *f* **[beim Natronzellstoffprozeß]** green liquor
Abfallbehälter *m* waste bin
Abfallbeseitigung *f* waste (refuse) disposal
  dauernde (endgültige) A. permanent [waste] disposal
Abfallbeseitigungsanlage *f* disposal plant
Abfalldauer *f* releasing (fall, drop-out, decay) time
Abfälle *mpl* waste [material], scraps
  flüssige radioaktive A. radioactive effluents
  wieder aufgearbeitete A. *(plast)* reground material
abfallen 1. to drop *(pressure, temperature, voltage)*; to fall; to drop out *(relay)*; to decrease; to decay; to restore; to release; 2. to slant, to drop, to rake; 3. *(ship)* to fall (drop) astern
  auf Null a. to decrease (drop) to zero
  schräg a. to taper off, to slant
abfallend / steil steeply sloping [downward], dropping sharply *(curve)*
Abfallgummi *m* waste rubber
Abfallhalde *f* waste heap
Abfallholz *n* waste wood
Abfallkammer *f* waste chamber
Abfallklauber *m (text)* waste picker
Abfallklopfwolf *m (text)* waste shaker (beating willow)
Abfallkohle *f* waste coal
Abfallkoks *m* scrap (waste) coke
Abfallkonzentrierung *f (nucl)* waste treatment
Abfallkübel *m* dust bin
Abfallösung *f* liquid waste; waste solution
Abfallpapier *n* offcut, waste paper
Abfallprahm *m* refuse (garbage) lighter, dump (garbage) scow
Abfallprodukt *n* [useless] by-product; waste product
Abfallraum *m* waste chamber
Abfallreinigung *f* waste cleaning
Abfallreinigungsmaschine *f* waste cleaner (cleaning machine)
Abfallreinigungstrommel *f* waste shaker
  A. mit Staubhaube waste shaker with dust hood
Abfallreißer *m* waste breaker
Abfallrohr *n* downcomer *(blast furnace)*; waste-pipe; rain-pipe; gutter-pipe

barometrisches A. barometric discharge pipe
Abfallrost *m* dust bars
Abfallsäure *f* residuary acid
Abfallschaum *m* froth *(paper)*
Abfallschute *f s.* Abfallprahm
Abfallsicherheitsfaktor *m* safety factor for drop-out *(relay)*
Abfallsichter *m* waste sorter
Abfallstoffe *mpl* waste matter; refuse [matter]
Abfallstrom *m (el)* release (drop-out) current
Abfallverwertung *f* waste utilization
Abfallverzögerung *f (el)* releasing delay
Abfallwärme *f* waste heat
Abfallwert *m* drop-out value *(relay)*
Abfallzeit *f* releasing time *(relay)*; decay time *(pulse)*; fall time
A. der Rückflanke trailing edge fall time
Abfallzellstoff *m* waste pulp
Abfallzellstoffaufbereitung *f* waste pulp preparation
Abfaltung *f (geol)* downfolding
abfangen 1. to support *(building)*, to underpin, to shore; 2. to right *(aeroplane)*; 3. to intercept; to catch; to capture *(neutrons)*
eine Nachricht a. to intercept a message
Abfangen *n (aero)* hold-off; righting; flattening out
hartes A. sudden pull-up
Abfangjäger *m* interceptor [aircraft]
Abfanglinie *f* minimum line of interception
Abfangradius *m (aero)* pull-out radius
Abfangrakete *f* anti-missile missile
Abfangstab *m s.* Absorberstab
Abfangung *f* 1. *(build)* shoring, underpinning; 2. interception
Abfangwahrscheinlichkeit *f* probability of interception
Abfarad *n* abfarad
abfärben to stain; to bleed
abfasen to chamfer, to bevel, to cant, to edge-machine
Abfasen *n* chamfering, bevelling, edge trimming
Abfasmeißel *m* chamfering tool
Abfasung *f* chamfer
abfedern to spring, to lift [from]
Abfederung *f* 1. springing; spring suspension; 2. lifting
A. von Maschinen introduction of [the flexible springs of] a vibration absorber *(between the machines and the foundation)*
abfeilen to file off
abfendern to fend off, to breast off *(vessel)*
Abferkelbucht *f* farrowing pen
abferkeln to farrow
Abferkeln *n* farrowing
Abferkelstall *m* pig farrowing house
abfertigen to register *(luggage)*; to dispatch *(luggage; vessel)*; to clear [goods through customs]
Abfertigung *f* **von Gepäck** forwarding of luggage
durchgehende A. through registration
Abfertigungsart *f* method of registration (forwarding)
Abfertigungsgebäude *n* terminal building *(airport)*
Abfertigungsgebühren *fpl* terminal charges
Abfett *n (leath)* skimming
abfeuern to discharge
abfiltern to filter [off], to remove by filtration, to filtrate off
abflachen to truncate; to flatten
abflächen to spot-face *(metal cutting)*
Abflächen *n* spot facing
Abflächmesser *n* spot facing cutter
Abfläch- und Zentriermaschine *f* face milling and centring machine
Abflachung *f* truncating, truncation; flattening; *(math)* platikurtosis
abflammen *(text)* to gas, to singe
abflauen to abate, to calm down, to die away *(wind)*
abflecken to stain
abfliegen to take off, to start
abfließen to flow off; to leak off
Abfließen *n s.* Abfluß
abflößen to float down a river
abfluchten to align; to range into line
Abflug *m* start, take-off
Abfluggeschwindigkeit *f* take-off speed
Abflugdeck *n* flight (flying-off) deck
Abflugkurs *m* heading
Abflugpeilung *f (aero)* outbound bearing
Abflugschneise *f* departure corridor
Abfluß *m* 1. drain; run-off *(furnace)*; 2. discharge; flowing off; leakage *(neutrons)*
Abflußgebiet *n* gathering-ground, drainage area (basin)
Abflußgeschwindigkeit *f* 1. velocity of flow; 2. leakage rate
Abflußgraben *m* draining (drainage) ditch
Abflußgrube *f* drainage pit
Abflußhahn *m* discharge cock
Abflußkanal *m* port, discharge pocket *(pump)*
Abflußleitung *f* discharge piping; waste pipe
Abflußmenge *f* outflow *(fluid flow)*; run-off; flow
jährliche A. annual flow
Abflußrinne *f* drainage channel; gutter[ing]
Abflußrohr *n* waste pipe; drain pipe; discharge pipe; down-pipe
Abflußrohrschelle *f* holderbat
Abflußverbindung *f* drain connection
Abfohlbox *f* foaling box
abfohlen to foal
Abfohlen *n* foaling
Abfragebetrieb *m (telec)* direct trunking
Abfrageimpuls *m* interrogation pulse *(magnetic core memory)*

Abfrageklinke *f* *(telec)* answering (calling) jack
abfragen to accept a call; to answer; to interrogate *(magnetic core memory)*
Abfrageregister *n* / **assoziatives (zuordnendes)** associative interrogation register *(for recovery of stored data)*
Abfrageschalter *m* answering (listening) key
Abfragesender *m* interrogator
Abfragestöpsel *m* *(telec)* answering plug
Abfragetaste *f* *(telec)* speaking key; answering button
Abfragewicklung *f* interrogate winding *(magnetic core memory)*
abfräsen to mill (cut) off
abfühlen to scan
Abfuhr *f* removal; delivery; unloading *(work)*; dissipation *(heat)*
abführen / Wärme to dissipate heat
Abführlattentuch *n* *(text)* delivery lattice
Abführmittel *n* purgative, aperient
Abfuhrrinne *f* unloading chute *(machine)*
Abfuhrrollgang *m* live-roller table
Abführrutsche *f* unloading (delivery, exit, outlet) chute
Abführtablette *f* decongestant tablet, purgative (cathartic) pill
Abführtisch *m* delivery table
Abführung *f* unloading *(work)*
  A. der Luft aus der Dampfrohrleitung elimination of air from the steam pipes
  A. der Reibungswärme friction heat dissipation
Abführungseinrichtung *f* unloading device *(work)*
Abführungszeichen *n* close quote *(sign)*
abfüllen to fill; to bottle
Abfüllen *n* transfusing, filling out
Abfüllmaschine *f* filling machine, filler
Abfüllraum *m* filling room; bottlery
Abfüllschlauchhahn *m* nozzle trigger
Abfüllwagen *m* *(Am)* gasoline service truck
Abfunkeffekt *m* effect of sparking off
Abfunkmaschine *f* spark eroder
abfüttern to line
Abgabe *f* delivery; output *(power)*; dissipation *(heat)*
Abgabefläche *f* electron-emitting area
Abgabeleistung *f* *(telec)* operating speed
Abgabeseite *f* output side *(power)*
Abgabespannung *f* discharge voltage *(of cells)*
Abgang *m* 1. waste; 2. junction; 3. departure
  fetthaltiger A. *(text)* oily waste
Abgänge *mpl* dirt, discard, refuse, rejects, tailings
Abgangsbahnhof *m* departure station
Abgangsflughafen *m* aerodrome of departure
Abgangshafen *m* port of sailing (departure); *(Am)* port of clearance
Abgangskabel *n* outgoing cable
Abgangsmast *m* end (distributing) pole
Abgangsschalter *m* outgoing feeder switch
Abgangsverwaltung *f* originating (departure) administration
Abgarung *f* carbonization
Abgas *n* exhaust (waste, burnt, spent, flue) gas, gaseous products of combustion, foul gas; fume
Abgasanalysator *m* exhaust-gas analyzer
abgasen to blow down (back)
Abgasfackel *f* surplus-gas burner
Abgasgeräuschdämpfer *m* *(aut)* exhaust silencer
Abgasheizung *f* exhaust-operated air heating
Abgaskanal *m* waste-gas flue
Abgaskessel *m* waste-heat boiler, exhaust-gas boiler
Abgasleitung *f* gas-exhaust pipe, exhaust lead (line); *(mine)* devil gas line
Abgasprozeß *m* blow-down process *(pulp)*
Abgasprüfgerät *n* exhaust-gas analyzer
Abgasrohr *n* exhaust[-gas] pipe
Abgasstrom *m* exhaust-gas flow
Abgasturbine *f* exhaust-gas turbine
Abgasturboaufladung *f* *(aut)* exhaust-turbo supercharging
Abgasturbogebläse *n* exhaust-turbo-driven blower; exhaust-type turbo-blower
Abgasturbolader *m* exhaust-gas turbo-driven supercharger
Abgasverwertung *f* exhaust-gas utilization
Abgasvorwärmer *m* economizer, waste gas feed heater
  ausschaltbarer A. economizer with by-pass flue
  Greenscher A. Green's economizer
abgeben to deliver; to yield; to dispense; to release; to emit; to radiate; to give off
  Elektronen a. to lose electrons
  Energie a. to release (lose) energy
  Impulse a. to deliver pulses
abgedeckt covered *(way)*
abgedichtet sealed
abgefast bevelled
abgefedert sprung
abgeflacht 1. *(math)* platikurtic; 2. flattened
  seitlich a. flat-sided
abgeglichen aligned; adjusted; balanced
  falsch a. misaligned
abgegrenzt defined
abgegriffen tapped *(voltage)*; worn; worn-out *(cloth)*
abgehend outgoing
abgekantet canted, bevelled; folded *(sheet metal)*
abgekürzt abbreviated
abgelagert aged
abgeleitet derived *(function)*; deduced
abgenutzt worn [out]

abgeplattet 1. *(math)* platikurtic; 2. flattened, oblate
abgerichtet / mit Profilrolle crush-formed
abgerundet rounded; radiused *(corner)*
abgeschält bark-bared
abgeschaltet disconnected; interrupted *(circuit)*
abgescheuert chafed *(yarn)*
abgeschirmt screened; clad; shielded
mit Blei a. lead-shielded
abgeschlossen / hermetisch hermetically sealed
luftdicht a. hermetically sealed
Abgeschlossenheit *f (math)* closure
abgeschrägt battered, splayed, bevelled, canted, raked
abgeschreckt quenched *(steel)*
in Öl a. oil-quenched
in Wasser a. water-quenched
abgesenkt lowered
abgesetzt offset *(tool)*, shouldered, stepped; recessed
im Durchmesser a. stepped
abgesichert fused
Abgesiebtes *n* screenings
abgespannt guyed *(with rope)*
abgestimmt tuned *(grid)*; pitched
auf Resonanz a. tuned to resonance
mit zwei Einzelkreisen a. double-tuned
nicht a. untuned
scharf a. sharply tuned
unscharf a. flat tuned
zweifach a. double-tuned
abgestuft stepped; graded
nach Größe a. graded in size
abgestumpft blunt
abgetragen worn-out *(cloth)*
abgetreppt stepped
abgewandelt modified
abgewickelt rolled out; wound off
abgießen to decant; to pour [off]
abglänzen *(text)* to remove the lustre
Abgleich *m* balance; balancing, setting up, gauging
A. auf Tonlosigkeit silent balance
A. der Durchlaßkurve *(telev)* tight alignment
optischer A. optical balance
Abgleichanzeiger *m* balance detector
Abgleichbereich *m* balance range *(radar)*
Abgleichbesteck *n* alignment tools
abgleichen to adjust; *(el)* to balance; to set up; to gauge; to level; to make even; to compensate; to parfocalize *(microscopy)*; to equilibrate *(photoelectric cell)*
auf Null a. to restore to zero
Rechenverstärker a. to balance an operational amplifier
Verstärkerstufen a. to line up amplifier stages
Abgleichfehler *m* unbalance; alignment error
Abgleichkondensator *m* balance capacitor, trimmer
Abgleichlänge *f (opt)* parfocalizing length
Abgleichmotor *m* balancing motor
Abgleichpotentiometer *n* balancing potentiometer
Abgleichprüfer *m* balancing tester
Abgleichprüfung *f* balance check *(analogue computer)*
Abgleichpunkt *m* balance point
Abgleichschalter *m* balancing switch
Abgleichschaltung *f* balancing connection (circuit)
Abgleichsender *m* alignment generator
Abgleichung *f* balance; balancing
Abgleichwerkzeug *n* alignment tool
Abgleichwiderstand *m* adjustable resistor; balancing (compensating) resistance
abgleiten to glide (slip, slide) off
Abgleitgeschwindigkeit *f* creep rate
abgraben to excavate *(open-cast mine)*
abgraten to [de]burr, to flash
warm a. to hot trim
Abgraten *n* deburring; snagging *(grinding)*
Abgratmaschine *f* deburring (burr removing) machine, flash trimmer
Abgratmeißel *m* burring chisel
Abgratpresse *f* trimming press
Abgratschere *f* trimming shears
abgreifen to tap [off]; to take off *(voltage)*; to select
Abgreifer *m* tapper; tapping switch
Abgreifschalter *m* tapper; tapping switch
Abgriff *m* tap[ping]; pick-off, pick-up
Abgriffschelle *f (el)* sliding clip
Abgriffspule *f* pick-up coil
Abguß *m* cast[ing]; mould
Abgußstreichmaschine *f* cast-coater *(papermaking)*
abhaaren to unhair
abhacken to chop off, to cut off
abhaken to unhook
Abhaltebaum *m (ship)* [spur] shore, breasting-off spar, fender
abhalten / einen Lehrgang to conduct a course
von Land a. to keep off [from the shore]
Abhang *m* slope
abhängen to unhook; to detach
den Hörer a. to remove the receiver
abhängig von dependent on; a function of; interlocked
gegenseitig a. interdependent
Abhängigkeit *f* dependence; relationship
A. von / in vs., versus; as a function of; with respect to
funktionelle A. functional dependence
gegenseitige A. interdependence, correlation
quadratische A. square law dependence

spektrale A. spectral dependence
Abhängigkeitsgebiet *n* dependency area
Abhängigkeitsschaltung *f* interlocking control; safety interlock device
A. / in interconnected, interlocked
A. / mit [electrically] interlocked
Abhängigkeitsverschluß *m* reverse interlocking
Abhäsion *f* abhesion
abhaspelfähig easily reeled, reelable
Abhaspelmaschine *f (text)* reeling machine
abhaspeln *(text)* to reel [off]
Abhaspeln *n* **der Seide** silk reeling
abhauen *s.* abschroten
Abhauen *n (mine)* dip heading
abhäuten to flay
Abhebebewegung *f* relief motion *(tool)*
Abhebeeinrichtung *f* tool relieving unit
Abhebefallschirm *m* lift-off parachute
Abhebegeschwindigkeit *f* get-away speed *(aeroplane)*
abheben to rise, to relieve, to raise, to lift *(tool)*; to lift off *(telephone)*; to unstick, to get away *(aeroplane)*; to lift clear off; to curl up; to skim *(gliding boat)*
Abhebepunkt *m* shrink point *(continuous casting)*
abhebern to siphon
Abhebestift *m* guide pin *(flask)*
Abhebevorrichtung *f* lifting device
Abhebung *f* lift *(tool)*; *(telev)* superelevation
abheften to file
Abhenry *n* abhenry
Abhieblinie *f (mine)* offset line
Abhitze *f* waste heat
abhobeln *(wood)* to jack-down, to plane off
Abhobeln *n (wood)* planing off
Abholdienst *m* collection service
abholzen to deforest, to cut down trees
Abhördienst *m* tapping service *(telecommunications)*
abhören to intercept; to tap the wires
Abhörlautsprecher *m* pilot (monitoring) loudspeaker
Abhörraum *m* monitoring room
Abhörsicherheit *f (telec)* safety from interception
Abietinsäure *f* abietic (sylvic) acid
abirren to deviate; *(opt)* to aberrate
Abirrung *f* aberration *(of light)*
abisolieren to strip, to skin, to bare [a wire]
Abisoliermaschine *f* wire stripper; wire stripping machine
Abisolierpinzette *f* stripping tweezers
abkalben to calve
Abkalben *n* calving
Abkämmesser *n (ag)* retainer
abkanten 1. to bevel; 2. to bend
Abkantmaschine *f* bending machine; folding machine
Abkantpresse *f* bending-off press
abkappen to cap, to cut off, to top; to clip, to limit *(amplitude)*
abketteln to hook up *(hosiery)*; to close off *(knitting)*
abkippen to fall off, to pitch down; to dump, to tip
Abkippstelle *f* tip (dump) site
abklappbar tilting, tiltable; hinged
abklären to clarify, to filter, to decant
Abklärgefäß *n (ch)* decanter
abklemmen 1. to disconnect; 2. *(ph)* to squeeze off, to pinch off
Abklingbehälter *m (nucl)* decontamination vessel
Abklingcharakteristik *f* decay characteristic
Abklingdauer *f* dying-out time
abklingen to abate, to fade, to be damped out, to decay, to die out
Abklingen *n* damping, dying out [of oscillation]; decay[ing], fading away; tailing off
A. der Lumineszenz decay of luminescence
A. der Schwingung dying-out of the oscillation
Abklingkonstante *f* damping coefficient (constant, factor), decay constant
Abklingkurve *f* decay curve (characteristic)
Abklinglänge *f* attenuation length
Abklingrate *f* decay rate
Abklingraum *m (nucl)* depot, storage facility (space)
Abklingverhältnis *n* subsidence ratio
Abklingzeit *f* decay (dying-out, fade-out, recovery) time; die away time, dying-down time; *(nucl)* cooling period
Abklopfapparat *m* rapper *(moulding)*
abklopfen 1. to knock off; to scale; to beat off; 2. to feel
Abklopfen *n* **der Firste** *(mine)* drumming
abkneifen to pinch (nip) off
abknöpfen to unbutton
abknüpfen to untie
Abkochbad *n* degumming (boiling-off, steeping) bath *(silk)*; boiling bath *(jute)*
abkochen to decoct; to boil off, to degum *(silk)*; to boil *(jute)*
Abkochung *f (ch)* decoction
Abkochverlust *m* loss of weight by boiling off *(silk)*
abkommen to deviate, to come away *(course)*
von der Startbahn a. to swing off
Abkömmling *m (ch)* derivative, derivate
abköpfen to cap, to cut off, to top
abkratzen to scrape off
abkühlen to cool; to refrigerate; to chill; to quench *(steel)*
Abkühlflüssigkeit *f* quenching liquid
Abkühlmittel *n* cooling agent

Abkühlung *f* cooling; hold-up *(of radioactive substances)*
Abkühlungsfläche *f* cooling surface
Abkühlungsgeschwindigkeit *f* cooling rate, rate of cooling
  kritische A. critical cooling rate, critical rate of cooling
  thermometrische A. thermometrical rate of cooling
Abkühlungsgesetz *n* law of cooling *(Newton)*
Abkühlungsgröße *f* loss of heat per square centimetre
Abkühlungskurve *f* cooling curve; recalescence curve
Abkühlungsverlauf *m* cooling curve
Abkühlungszeit *f* cooling period
Abkühlverlust *m* loss due to condensation
abkuppeln to uncouple *(wagon)*; to disengage; to disconnect
abkürzen to shorten; to abbreviate; to abridge
Abkürzsäge *f* bucking saw
Abkürzung *f* abbreviation
Abladeanlage *f* unloading installation
Abladekran *m* unloading crane
abladen to unload, to discharge; to dump
Abladen *n* unloading, discharging; dumping
Abladeplatz *m* spoil (dumping) ground
Ablader *m* shipper, consigner
Abladestelle *f* place of deposit, dumping place *(excavator)*
Ablagebock *m* **für Papier** skid for paper
Ablagebrett *n* **/ hinteres** *(aut)* rear parcel shelf
ablagern 1. to deposit; to settle; to sedimentate; 2. to season *(timber)*; to age
Ablagerung *f* 1. deposit; deposition; sediment; sedimentation; deposition (settling) on ground; alluviation; settlings; 2. seasoning *(timber)*; aging
  innere A. *(nucl)* internal deposition
  selektive A. *(nucl)* selective localization
Ablagerungsperiode *f* period of deposition
Ablagerungsriß *m* deposit crack
Ablagerungsstruktur *f* deposition structure
Ablagerungszone *f* sedimentation zone, zone of sedimentation (deposition)
Ablaktation *f* ablactation
Ablammen *n* lambing
ablängen to cut (saw) off to length, to face to length
Ablängen *n* **[/ maßgenaues]** sawing-off to [dead] length [accuracy], cutting-off to [dead] length [accuracy]
Ablängsäge *f* bucking saw
Ablaß *m* drain[age] *(oil)*; run-off; outlet
ablassen to drain [off] *(fluid)*; to vent *(air)*; to discharge
  Dampf a. to exhaust steam
  einen Kessel a. to empty a boiler
  Luft a. to let the tyres down, to deflate
Ablaßhahn *m* drain cock
Ablaßpfropfen *m* drain plug
Ablaßschraube *f* drain plug (screw)
Ablaßtrichter *m* emptying funnel
Ablaßventil *n* drain valve
Ablation *f* ablation, surface melting *(of glaciers)*
Ablauf *m* 1. drainage; outlet; run-off; discharge; waste pipe; 2. sequence [of operations]; 3. *(ship)* launch[ing]
Ablaufarm *m (text)* take-off arm
Ablaufbahn *f* 1. *(ship)* ways, standing (fixed, launching, running) ways; 2. inrun, approach *(ski-jumping)*
  gekrümmte A. *(ship)* cambered standing ways
Ablaufbahngründung *f (ship)* ground ways
Ablaufbahnneigung *f (ship)* declivity (inclination) of standing ways
Ablaufberg *m* hump in a gravity yard
Ablaufberglokomotive *f* hump locomotive
ablaufen to run down; to return *(dial)*; to drain *(liquid)*
  a. lassen 1. to drain off *(liquid)*; 2. to launch *(ship)*
Ablaufgeschwindigkeit *f* 1. *(ry)* shunting speed; 2. *(ship)* launching speed
Ablaufgestell *n (ch)* draining rack
Ablaufgewicht *n* launching weight *(ship)*
Ablaufgleis *n (ry)* falling track
Ablaufhaspel *f* running-off reel
Ablaufmenge *f* run-off; flow
  jährliche A. annual flow
Ablaufplan *m* work (construction) schedule, progress chart
Ablaufplanung *f* scheduling
Ablaufrangierbetrieb *m* gravity shunting, *(Am)* gravity switching
Ablaufrinne *f* gutter
Ablaufrohr *n* drain (discharge) pipe
Ablaufschlitten *m (ship)* running (sliding) ways, runner
Ablaufspule *f* wind-off spool; creel bobbin
Ablaufstellwerk *n* marshalling yard control office
Ablaufsteuerung *f* sequence (sequential) control
Ablauftrichter *m* discharging funnel
Ablaufwasser *n* discharged water
Ablaufwinkel *m* angle of lapping
Ablaufzeit *f* time for running out
Ablauge *f* waste liquor; black (spent) liquor
Ablaugentrockensubstanz *f (text)* dry substance of the spent liquor
abläuten *(telec)* to ring off
ableeren to press off *(knitting)*
Ableerschloß *n* press off cam *(knitting)*
ablegen *(text)* to doff; *(off)* to file; *(print)* to distribute; *(ship)* to cast off, to set sail, to depart; to eject *(punch card machinery)*

Ableger *m* 1. *(print)* distributor; 2. *(ag)* layer; new swarm
Ablegesatz *m (print)* dead matter
Ablegeschlitten *m* distributor carriage
Ablegestange *f* **der Setzmaschine** distributing (distributor) bar
Ablegevorrichtung *f* delivery mechanism
ablehren to gauge
Ablehren *n* **des Messerkopfes** gauging the cutter block
ableichtern to lighten, to lighter *(vessel)*
Ableichterung *f* lightening *(vessel)*
ableitbar deducible; derivable; differentiable
Ableitbarkeit *f* deducibility; derivability; differentiability
ableiten 1. to drain [off]; to bleed off *(liquid)*; to carry off; 2. to conduct; to leak *(current)*; 3. *(math)* to deduct; to derive; 4. to divert
  den Vorschub a. von to take the feed from
Ableiter *m (el)* arrester; charge eliminator; differentiator
  A. mit Mehrfachfunkenstrecke *(el)* multigap arrester
Ableiterfunkenstrecke *f (el)* arrester gap
Ableiterstrom *m* discharge current
Ableitkondensator *m* by-pass capacitor
Ableitkreis *m (el)* derived circuit
Ableitstrom *m (el)* current leakage, leakage current; discharge current
Ableitung *f* 1. draining-off, bleeding-off *(liquid)*; 2. conduction; leak[age] *(current)*; grounding, earthing; downlead *(aerial)*; 3. *(math)* derivation; derivative; differentiation
  A. der Wärme abstraction (dissipation) of heat
  dielektrische A. dielectric leakance (conductance)
  partielle A. *(math)* partial derivative
  substantielle A. derivative following the fluid
  totale A. *(math)* total derivative
  zweite A. *(math)* flexion
Ableitungsdämpfung *f* leakage attenuation; shunt loss
Ableitungsdraht *m* shunt wire
Ableitungsverlust *m* leakage loss
Ableitungswiderstand *m* leakage resistance
Ableitvermögen *n (el)* discharge capacity *(of arrester)*
Ableitweg *m (el)* current leakage path
Ableitwiderstand *m* bleeder [resistance]
Ablenkamplitude *f* maximum deflection
Ablenkbarkeit *f* deviability
Ablenkblech *n* deflector plate
Ablenkelektrode *f* deflecting electrode, deflecting (deflection, deflector) plate
Ablenkempfindlichkeit *f* deflection sensitivity, sensitivity of deflection
ablenken to deflect; to divert; to deviate; to sweep *(oscillograph)*; to scan
Ablenkfaktor *m* deflection factor (coefficient)
Ablenkfeld *n* deflection field
Ablenkfrequenz *f* sweep frequency
Ablenkgenerator *m* deflection generator; scanning generator; time-base generator
Ablenkgeradlinigkeit *f* sweep linearity
Ablenkinduktivität *f* deflecting inductance
Ablenkjoch *n (telev)* deflection (scanning) yoke
Ablenkkeil *m* whipstock
Ablenkkreis *m* / **getriggerter monostabiler** triggered start-stop sweep circuit
Ablenkplatte *f* deflecting (deflection, deflector) plate; baffle plate; splash baffle
Ablenkplattenpaar *n* / **schirmnahes** screen-side deflection plates
Ablenkprisma *n* deflecting prism
Ablenkrolle *f* steering idler
Ablenkschaltung *f* electronic sweep circuit
Ablenkscheibe *f* deflecting sheave
Ablenkseil *n* damper, damping (rubbing) rope, rubber *(shaft hoisting)*
Ablenkspannung *f* deflection voltage, sweep[ing] voltage
Ablenkspannungen *fpl* sweeps; X-potentials *(horizontal)*, Y-potentials *(vertical)*
Ablenkspule *f* deflector, deflection (sweeping) coil, scanning yoke
  magnetische A. magnetic yoke
Ablenksystem *n* sweeping system
Ablenkung *f* deflection; deviation; scanning
  asymmetrische A. asymmetrical deflection
  symmetrische A. symmetrical deflection
  unsymmetrische A. asymmetrical deflection
Ablenkungsdruck *m* deflection pressure *(turbine)*
Ablenkungseffekt *m* deflecting effect
Ablenkungselektrode *f* deflecting electrode
Ablenkungsempfindlichkeit *f* deflection sensitivity, sensitivity of deflection
Ablenkungsgröße *f* magnitude of deflection *(turbine)*
Ablenkungskondensator *m* deflecting capacitor
Ablenkungskraft *f* deflecting force
Ablenkungslinie *f* deviation line
Ablenkungsmagnetfeld *n* deflecting magnetic field
Ablenkungsoszillator *m* sweep oscillator
Ablenkungsprisma *n* deviation prism
Ablenkungsspannung *f* deflecting voltage (potential)
  symmetrische A. symmetric deflection potential
Ablenkungsweite *f (telev)* sweep amplitude
Ablenkungswinkel *m* angle of deflection (deviation)
Ablenkverzeichnung *f* deflection distortion
Ablenkweite *f* trace interval
Ablenkwinkel *m* angle of deflection (deviation)
Ablenkzeit *f* deflection time

ablesbar readable
direkt a. capable of direct reading
leicht a. easy to read
Ableseeinrichtung *f* read-out device
Ablesefehler *m* reading error, error in reading
Ablesefenster *n* [reading window] screen
Ablesefernrohr *n* reading telescope
Ablesegenauigkeit *f* accuracy of reading
Ablesemikroskop *n* reading microscope
Ablesemöglichkeit *f* reading, readability
ablesen to read off, to take readings
Ablesen *n* reading off
Ableserichtung *f* direction of reading off *(pattern)*
Ableserolle *f* indexed roller
Ablesescheibe *f* indexing disk
Ablesestrich *m* reading line, graduation mark
Ablesezeit *f* access time
Ablesgenauigkeit *f* reading accuracy
Ablesgerät *n* indicator
Ablesung *f* reading
direkte A. direct reading
unmittelbare A. direct reading
Ableuchtlampe *f* inspection lamp
Ablieferungshafen *m* port of delivery
Ablieferungskasten *m* delivery box
Ablieferungsnachweis *m* advice of delivery
Ablieferungswalzen *fpl* delivery rollers
ablöschen to slake *(lime)*; to quench *(coke)*
Ablöschen *n* slaking *(lime)*; quenching, watering *(coke)*; decalescence
Ablösearbeit *f (nucl)* work function
ablösen 1. *(mine)* to slab; to delaminate *(layers)*; to strip *(colour)*; to loosen; to peel; 2. to spell; to relieve the watch *(ship)*
sich a. to detach; to loosen; to come off *(chip)*
Ablösen *n* **der Strömung** *(rock)* flow separation
Ablösung *f* flow separation *(fluid flow)*; stripping; liberation; separation; scaling
A. der Schichten *(mine)* bed separation
A. von Elektronen detachment of electrons, electron detachment
einfache A. simple-detached layer
zweifache A. double-detached layer
Ablösungsfläche *f (geol)* cleat
Ablösungspunkt *m* separation point *(fluid flow)*
Ablösungsversuch *m* stripping test
Ablösungswiderstand *m* stripping resistance
abloten to plumb
Abloten *n* plumbing
ablöten to unsolder
Abluft *f* discharge (exhaust, vitiated, outgoing) air
Abluftejektor *m* exhaust ventilator
ablüften *(text)* to [expose to] air, to hang in the air
Ablüfterkopf *m* exhaust ventilator
Abluftfilter *n* waste air filter
Abluftkanal *m* air-outlet conduit
Abluftstutzen *m* ventilating outlet
abmanteln to bare *(cable)*
Abmaß *n* 1. dimensional variation *(tolerance)*; 2. allowance
oberes A. upper deviation
oberes und unteres A. limits
unteres A. lower deviation
zulässiges A. permissible deviation
abmatten to dim, to dull, to deaden
abmeißeln to chisel off; to chip
Abmelden *n* **eines Zuges** announcing the departure of a train
Abmeldesignal *n* train starting signal
abmeßbar measurable
Abmeßbarkeit *f* measurability
abmessen 1. to measure; to gauge; 2. to batch; to proportion
Abmeßkasten *m* batch (gauge) box
Abmessung *f* 1. dimension; 2. measuring
gesamte A. overall dimensions
Abmessungen *fpl* dimensions; data; size
Abminderung *f* reduction
abmontierbar demountable; detachable; removable
abmontieren to detach; to remove; to take off; to dismantle; to demount
abmustern *(ship)* to pay off, to discharge *(crew)*
Abmusterung *f (ship)* discharge, signing off, paying off *(crew)*
Abnäher *m (text)* gusset, wedge-like fold
Abnahme *f* 1. acceptance *(machine)*; 2. demounting; removal; taking-off; 3. drop; decrease; fall; reduction
A. der Spannung fall in tension; voltage drop
A. der Stromstärke fall of current
A. der Variablen decrement of variables
A. des Werkstücks unloading the work
exponentielle A. exponential decrease
lineare A. linear decrement
abnahmebereit ready for delivery
Abnahmeelektrode *f* output (collector) electrode
Abnahmekoeffizient *m* coefficient of reduction
Abnahmekontrolle *f* inspection control
Abnahmelehre *f* inspection gauge
Abnahmemessung *f* final (trial) measurement
Abnahmeprobefahrt *f (ship)* acceptance trial-trip
Abnahmeprotokoll *n* test certificate
Abnahmeprüfung *f* acceptance test
Abnahmetoleranz *f* acceptance tolerance
Abnahmeversuch *m* acceptance test
Abnahmevorschrift *f* quality specification
Abnarbeisen *n (leath)* buffing slicker
abnehmbar detachable; demountable; removable

abnehmen 1. to detach; to demount; to remove; to take off; to take up *(receiver)*; 2. to accept; 3. to collect *(current)*; 4. *(text)* to doff *(card)*; to cast off, to narrow *(meshes)*; 5. to decrease; to drop
den Antrieb a. von to take the drive from
eine Spannung a. to extract a voltage
einen Span a. to take a cut, to lift a chip, to remove metal (material)
Abnehmen *n* **von Werkstücken** unloading components
Abnehmer *m* 1. customer; 2. *(text)* doffer; 3. catcher *(rolling mill)*; 4. pick-up; collector
Abnehmerbeschlag *m (text)* doffer covering
Abnehmerkamm *m (text)* doffer (doffing, noil stripping) comb
schwingender A. oscillating evener comb
Abnehmerkratze *f (text)* doffer
Abnehmerleiste *f (text)* stripper bar
Abnehmerspule *f* pick-up coil
Abnehmersystem *n (text)* doffer system
abnieten to unrivet
abnorm anomalous
abnormal gekrümmt *(math)* allokurtic
abnutschen to filter by suction
abnutzbar wearable
Abnutzbarkeit *f* abrasion resistance; wearing properties
abnutzen [/ sich] to wear [out]
Abnutzung *f* wear [and tear]; abrasion
Abnutzungsaufwand *m* depreciation cost
Abnutzungsbeständigkeit *f* resistance to wear; abrasion resistance
Abnutzungsfläche *f* wearing surface *(e.g. of the rail)*
Abnutzungsgeschwindigkeit *f* speed of wear
Abnutzungsversuch *m* rattler (attrition) test; abrasion test
Abnutzungswiderstand *m* resistance to wear; resistance to abrasion
Abohm *n* abohm
abölen *(leath)* to oil off
Abonnement *n* subscription
Abonnementsauftrag *m* subscription order
Abonnementsfernsehen *n* subscription television
Abonnent *m* subscriber
Abperleffekt *m (text)* water repellency
Abpfeilern *n (mine)* pillaring
abpflocken to peg out
abpicken / einen Kessel to scale a boiler
abplanen to face *(metal cutting)*
abplatten to flatten
Abplattung *f* oblateness; ellipticity
Abplattungstheorie *f* theory of ellipticity
abplatzen to peel off, to flake, to slab, to spall [off]
Abprall *m* rebound, ricochet
abprallen to rebound
Abpraller *m* ricochet
Abprallgeschwindigkeit *f* velocity of rebound
Abprallpunkt *m* point of rebound
Abprallweite *f* length of ricochet
Abprallwinkel *m* angle of ricochet (rebound)
Abprallwirkung *f* ricochet effect
abpressen to press out (off); to squeeze out
Abpressen *n* **des Kessels** hydraulic boiler test
abprotzen to unlimber
abpuffern *(ch)* to buffer
abpumpen to pump out, to evacuate
abputzen 1. to clean; to polish; to wipe; 2. to plaster
Abquetscheffekt *m (text)* squeezing effect
abquetschen to squeeze out; to pinch off
Abquetschen *n* squeezing [off]
Abquetschfoulard *m (text)* squeezing mangle
Abquetschmaschine *f (text)* squeezing machine
Abquetschrand *m (plast)* shear edge
Abquetschwalze *f (text)* squeezing roller; *(ch)* squeeze roll
abrammen to ram, to stamp
abrauchen *(ch)* to evaporate
Abraum *m* rubbish; trash; waste; spoil; barren rock; overburden, overlay; cap rock, [mine] cover
Abraumband *n* overburden disposal conveyor belt
Abraumbeseitigung *f* removal of overburden
Abraumbewegung *f* overburden handling (moving, removal), spoil moving, waste removal (stripping)
abräumen to clear away; to strip; *(mine)* to uncover
Abraumförderbrücke *f* overburden conveyor bridge
Abraumförderwagen *m* wagon for the removal of overburden
Abraumhalde *f* overburden dump
Abraumkippe *f* overburden dump, spoil bank
Abraumsalz *n* abraum salt, saline deposit
Abraumschichten *fpl* [mine] cover
Abraumseilbahn *f* overburden ropeway
Abregung *f* de-excitation
abreiben to rub down; to abrade; *(text)* to crock
Abreibfestigkeit *f* abrasion resistance
abreichern *(nucl)* to deplete, to exhaust
Abreicherung *f (nucl)* depletion
Abreißbogen *m (el)* intermittent arc
Abreißbogenerzeuger *m* intermittent arc producer
abreißen 1. to pull (break) down; to demolish *(building)*; 2. to tear off; 3. *(text)* to rupture, to break *(thread)*; 4. *(mine)* to break *(roof)*; 5. to burble, to break away *(fluid flow)*
Abreißen *n* break [of spark]

A. der Funkverbindung loss of radio contact
A. der Strömung *(rock)* flow separation
A. des Schmierfilms breaking of the oil film
Abreißfunke *m* discharge spark, break[ing] spark; touch spark
Abreißfunkenstrecke *f* breaking spark gap
Abreißkarte *f* stub card
Abreißkontakt *m* disconnecting (arcing) contact
Abreißrolle *f* counter reel
Abreißschaltung *f* break circuit
Abreißvorgang *m* *(rock)* break-off phenomenon
Abreißwalze *f* detaching roller *(combing machine)*
belederte A. leather detaching roller *(combing machine)*
Abreißzündung *f* coil (make-and-break) ignition
Abreutersieb *n* *(ag)* grubbing screen
Abrichtarbeit *f* wheel-dressing operation
Abrichtdiamant *m* diamond wheel dresser; dressing diamond
Abrichteinrichtung *f* dressing fixture, grinding wheel dresser
abrichten 1. to level; to plane; 2. to true, to dress *(grinding wheel)*
mit Profilrolle a. to crush-form, to dress with formed crusher roll
Abrichten *n* **mit Profilrolle** crush[-form] dressing, crush forming
Abrichter *m* dresser *(grinding)*; smooth planer
A. mit Flatterrädchen star-wheel dresser
Abrichthobel *m* planer
Abrichthobelmaschine *f* head straightening machine
Abrichtmaschine *f* smooth planer
Abrichtstift *m* abrasive-stick dresser
Abrichtwerkzeug *n* [grinding wheel] dresser, dressing tool
Abrieb *m* 1. abrasion, wear, attrition; 2. *(mine)* attrition, degradation; degradation product; breeze
Abriebbildung *f* *(mine)* attrition, degradation
abriebfest abrasive (attrition) resistant, non-abrasive
Abriebfestigkeit *f* abrasion resistance, non-abrasive quality, non-abrasiveness
Abriebprüfung *f* abrasive test
abriegeln to bolt
abrinden to bark
Abriß *m* 1. summary; abstract; 2. *(build)* demolition, pulling down; 3. break
Abrißpunkt *m* *(aut)* firing point
Abroll-Abrichtgerät *n* wheel crushing attachment
abrollen 1. to uncoil; to unwind; to unreel; 2. to taxi off *(aeroplane)*
a. auf to roll along
abrösten to roast
Abrücksignal *n* humping signal
abrufen *(telec)* to ring off
abrunden to round, to radius; *(math)* to round off
Abrunden *n* radiusing *(external radius)*; *(math)* rounding-off
Abrundung *f* external radius
A. der Ecke corner radius
A. der Eintrittskante rounding off the inlet *(turbine)*
Abrundungsfehler *m* error in rounding off, rounding error
Abrundungshalbmesser *m* nose radius *(tool)*; radius of curvature *(turbine nozzle)*
abrüsten to take down the scaffold *(building)*; to dismantle; to lay up, to put out of commission *(vessel)*
abrutschen to slide; to slip
über eine Tragfläche a. *(aero)* to fall off
absacken 1. to pancake *(aeroplane)*; 2. to sink; to settle; to subside *(building)*; 3. to bag
nach achtern a. to fall astern
Absackmaschine *f* bagging (bag packing) machine
Absackwaage *f* sack-weighing and packing machine
absägen to saw off
Absalzung *f* salting-out
absanden to sandblast
Absanden *n* sandblasting, sanding
Absättigung *f* **[der Kernkräfte]** saturation [of the nuclear forces]
Absatz *m* 1. ledge, terrace, bench, berm; offset; heel, toe; 2. shoulder *(turning)*; 3. deposit[ion]; 4. paragraph; 5. sale
rechtwinkliger A. square shoulder *(turning)*
Absatzbecken *n* settling tank (basin)
absatzfähig marketable
Absatzferkel *n* weaner
Absatzkosten *pl* cost of marketing
Absatzleder *n* heel leather
Absatzmöglichkeit *f* marketing opportunity
Absatzsägeblatt *n* tenon saw blade
Absatzschott *n* *(ship)* end bulkhead, break bulkhead *(a transverse bulkhead at the end of a superstructure)*
Absatzvorbaumaschine *f* *(leath)* heel building machine
absatzweise at intervals, intermittent[ly], cyclic; in steps
absäuern to acidify, to sour; *(build)* to acid wash
Absäuerung *f* acid wash; acidification
Absäuerungsbad *n* acidifying bath
Absaugelektrode *f* extracting electrode
absaugen to exhaust, to draw off; to evaporate, to pump out
Absaugen *n* extraction, exhaustion, suction
A. der Rauchgase drawing off the flue gases
A. von Elektronen electron collection

Absauger *m* exhaust fan
Absaugpumpe *f* vacuum pump
Absaugspannung *f* extraction potential, draining-off voltage
abschaben to scrape off; to abrade
abschälen *(text)* to peel off *(stalk)*; to strip off *(hemp)*
abschaltbar disconnectable; disengageable
Abschaltbatterie *f* cut-off battery
abschalten to switch off, to cut [off], to de-energize, to break (*comp.* ausschalten)
  den Reaktor a. to shut down the reactor
Abschalten *n* **des Reaktors** reactor shut-down
  A. eines Transistors cut-off of a transistor
Abschaltfunke *m* break spark
Abschaltklinke *f* break jack
Abschaltleistung *f* breaking capacity
Abschaltlichtbogen *m* breakdown arc
Abschaltmagnet *m* disconnecting magnet
Abschaltorgane *npl* shut-off mechanism; scram mechanism *(reactor)*
Abschaltrelais *n* cut-off relay
Abschaltstab *m* shut-off rod; emergency shut-off rod; scram rod *(reactor)*
Abschaltstrom *m* breaking current
Abschaltung *f* disconnection, switching-off
  A. aus Sicherheitsgründen safety shut-down *(reactor)*
  A. eines Kreises circuit opening
  elektrische A. radio signal cut-off
Abschaltverfahren *n* combustion cut-off method
Abschaltzeit *f* shut-down time *(reactor)*
abschärfen to chamfer, to scarf
Abschärfmaschine *f (leath)* slicing machine
Abschattung *f (opt)* adumbration
Abschattungsverlust *m (telec)* shadow effect, mountain effect
abschätzen to evaluate; to excess; to estimate
Abschätzung *f* estimation; estimate; evaluation; *(math)* appraisal
Abschaum *m* scum, dross
abschäumen to scum, to skim off; to dross
Abschäumlöffel *m* skimmer
abscheidbar precipitable, separable
abscheiden to precipitate, to separate; to exude
Abscheider *m* precipitator, separator; trap, interceptor
  elektromagnetischer A. electro-magnetic separator
Abscheidung *f* precipitation; deposit; removal
  elektrolytische A. electrodeposition
  elektrostatische A. electrostatic precipitation
Abscheidungsspannung *f* decomposition voltage
Abscherbolzen *m* shearing bolt, shear pin
abscheren to shear off
Abscheren *n* shearing off, shear cutting
Abscherfestigkeit *f* shearing resistance (strength)
Abschergeschwindigkeit *f* velocity of shear
Abscherstift *m* shear pin
Abscherung *f* shearing off; shear
Abscherwinkel *m* shear angle
Abscherwirkung *f* shearing action
Abschiebevorrichtung *f* stripping device *(rolling)*
Abschirmanschluß *m (el)* shield terminal
Abschirmbecher *m (el)* screening (shielding) can
abschirmen *(el)* to screen [off]; to shield
Abschirmfaktor *m* screening factor, degree of screening
Abschirmgeflecht *n (el)* shielding harness
Abschirmgehäuse *n* screening box *(cover)*
Abschirmhaube *f* shielding can, screening box (can)
Abschirmkabel *n* screened cable
Abschirmkammer *f* shielding castle
Abschirmkonstante *f* screening constant
Abschirmmaterial *n* shield[ing] material
Abschirmschicht *f* shield layer
Abschirmschlauch *m* screening tubing
  geklöppelter A. shield braid
Abschirmspule *f* field neutralizing coil
Abschirmstrom *m* screening current
Abschirmung *f* 1. shield[ing]; screen[ing]; screening effect; 2. [screening] can; 3. radio shielding; electron screening
  A. gegen Gamma-Strahlung gamma[-ray] shield
  biologische A. *(nucl)* biological shield
  geflochtene leitende A. braided conducting covering
  magnetische A. magnetic shield
Abschirmungsklemme *f* shield terminal
Abschirmungskonstante *f* screening constant
Abschirmungsträger *m* shield base
Abschirmungszahl *f* screening number (constant)
Abschirmwicklung *f* shield winding
Abschirmwirkung *f* screening (guarding) effect
Abschirmzahl *f* screening number (constant)
abschirren to unharness
abschlacken to remove the slag [from a melt]; to separate the dross
Abschlackung *f* removal of slag
Abschlag *m* 1. *(text) (Am)* cast-off cam; 2. *(mine)* pull, round of shots
Abschlagen *n (text)* backing-off
Abschlagexzenter *m (text)* knocking-over cam
Abschlagkamm *m* stripping comb *(spinning)*; knocking-over comb *(knitting)*
Abschlagkammsteg *m* knockover verge wall *(knitting)*
Abschlagkante *f* / **mittlere** centre knocking over surface *(holding down sinker)*
  obere A. upper knocking over surface *(holding down sinker)*
  untere A. lower knocking over surface *(holding down sinker)*

Abschlagplatine *f (text)* holding down sinker, knock-off sinker, knockover bit
Abschlagpunkt *m (text)* knockover point
Abschlagschloß *n (text)* knockover cam
Abschlagslänge *f (mine)* advance per round, depth of round (pull)
Abschlagstellung *f (Am)* cast-off position *(knitting machine)*
Abschlagstiefe *f (mine)* advance per round, depth of round (pull)
Abschlagvorrichtung *f (text)* backing-off motion
Abschlagzahn *m (text)* comb, tooth, verge bit
eingesetzter A. inserted tooth
abschlämmen to elutriate
Abschlämmhahn *m* sludge drain cock
abschleifen to grind off; to abrade; to smooth; to sandpaper
Abschleiffestigkeit *f* abrasion resistance
Abschleifversuch *m* abrasion (attrition) test
Abschleppdienst *m (aut)* recovery service
abschleppen to drag, to haul off; to tow off *(vessel)*
Abschleppkran *m* wrecking crane
Abschleppkupplung *f* tow-bar coupling
Abschleppseil *n* tow rope
Abschleppstange *f* towing rod
Abschleppwagen *m* wrecking car (truck); breakdown vehicle
abschleudern to centrifuge; to catapult
abschließen 1. to seal; to inclose; to shut off; 2. *(el)* to terminate; 3. to lock *(door)*; to finish
elektrisch a. to terminate
abschließend mit flush with
Abschliff *m* 1. abrasion; 2. depth of cut in grinding
Abschluß *m* 1. seal; 2. *(el)* termination *(resistance)*; 3. *(nucl)* filling *(shell)*; 4. [conclusion of] contract
A. des Beförderungsvertrages conclusion of the contract of carriage
angepaßter A. *(el)* matched termination
luftdichter A. air-tight seal
oberer A. des Platinenringes *(text)* cap of [the] sinker
reflexionsfreier A. *(el)* matched termination; non-reflecting termination *(waveguides)*
Abschlußbedingung *f* terminal condition
Abschlußbetrag *m* ledger total
Abschlußfehler *m* closing error
Abschlußgeschwindigkeit *f* closing speed
Abschlußimpedanz *f* terminal (terminating) impedance
Abschlußkabel *n* termination cable
Abschlußkappe *f* cap end, protective (cover) cap
Abschlußkasten *m* terminal box
Abschlußmuffe *f* terminal box
vergossene A. pot head jointing sleeve
Abschlußplatte *f* closing plate
Abschlußprüfung *f* final test
Abschlußscheinwiderstand *m* terminating impedance
Abschlußschott *n* end-bulkhead
Abschlußspindel *f* stop valve spindle *(turbine)*
Abschlußwiderstand *m (el)* 1. terminating impedance; 2. terminating resistor
angepaßter A. matched (non-reflecting) termination
Abschlußzeit *f* shut-down time
Abschmelzbrenner *m* tipping torch, sealing-off burner *(valve production)*
Abschmelzdauer *f* fusing time
Abschmelzelektrode *f* fusible arc-welding electrode
abschmelzen to fuse; to melt off; to smelt; to seal *(valve)*
Abschmelzgeschwindigkeit *f* rate of deposition *(welding)*
Abschmelzkonstante *f* fusion coefficient
Abschmelzschweißen *n s.* Abbrennschweißen
Abschmelzsicherung *f* safety (fusible) cut-out, [safety] fuse
Abschmelzstrom *m* fusing (blowing) current
Abschmelzstromstärke *f* fusing (blowing) current
Abschmelzung *f* / **luftdichte** air-tight seal *(of electron tubes)*
Abschmiereinrichtung *f* lubricating equipment
abschmieren 1. to lubricate; 2. to set off
Abschmierfett *n* [lubricating] grease
Abschmiergerät *n* lubricating equipment
Abschminkkrem *f* cleansing cream
abschmirgeln to abrade with emery
Abschmutzbogen *m (print)* set-off sheet
Abschmutzrolle *f (print)* anti-set-off roller
Abschmutzwalze *f (print)* anti-set-off roller
Abschnappkupplung *f (aero)* impulse starter
Abschneideenergie *f* cut-off energy
Abschneidefrequenz *f* cut-off frequency
abschneiden to cut off; *(text)* to doff; to chop *(waves)*
Abschneiden *n* **der Dampfeinströmung** cutting-off of the steam admission
A. einer Welle chopping of a wave
Abschneideröhre *f* threshold valve; clipper valve
Abschneideschaltung *f* clipper
Abschnitt *m* section; segment; sector; cut-off [piece] *(sawing)*; *(math)* intercept
Abschnittgestänge *n* transposition pole *(overhead lines)*
Abschnittkarte *f* stub card *(punched)*
Abschnittlänge *f* cut-off length *(sawing)*
abschnüren 1. to pinch off; 2. to impound *(water)*; 3. *(ship)* to line out, to lay off (down) *(lines on the mould-loft floor)*

Abschnüren *n (ship)* lofting, laying off *(on the mould-loft floor)*
Abschnürung *f* pinch-off
Abschnürungsbereich *m* pinch-off region
abschopfen to crop; to chop *(the ends of billets)*
abschöpfen to skim [off]; to scum
abschotten *(ship)* to bulkhead off, to bound by bulkheads
Abschottung *f* screening *(boiler)*; bulkheading *(vessel)*
abschrägen to bevel; to chamfer; to scarf; to taper; to slope; to rake; to cant; to splay; to batter
Abschräghobel *m* shooting plane
Abschrägmaschine *f* bevelling machine
Abschrägung *f* 1. chamfering; bevelling; scarfing; tapering; sloping; raking; canting; splaying; battering; 2. bevel; chamfer; slope; taper
Abschrägungswinkel *m* chamfer angle
abschrauben to screw off; to unscrew; to loosen
Abschrauben *n* screwing-off, unscrewing, loosening
Abschreckalterung *f* quench ag[e]ing
Abschreckaustenit *m* retained austenite
Abschreckbad *n* quenching bath
Abschreckbehälter *m* quenching tank
abschrecken to quench *(hardening)*; to chill *(casting)*
Abschrecken *n* quenching; chilling; decalescence
  A. in Blei lead-quenching
  A. in Wasser water-quenching
Abschreckflüssigkeit *f* quenching liquid
Abschreckhärtetiefe *f* hardness penetration by quenching
Abschreckmittel *n* quenchant *(hardening)*
Abschreckspannung *f* quenching stress
Abschrecktiefe *f* depth of chill
Abschreckturm *m* quenching tower
Abschreckversuch *m* quenching test
abschreiben to depreciate
Abschreibung *f* depreciation
Abschrift *f* / **beglaubigte** certified copy
  rechtsgültige A. legalized copy
abschroten to cut off with a chisel; to chisel off
abschruppen *(wood)* to jack up
abschürfen *(build)* to skim; to scratch
abschurren to scuff
abschwächen to attenuate; to reduce; to diminish; to soften; to tone down *(photography)*
Abschwächer *m* reductor, reducer; attenuator *(waveguide)*
  festeingestellter A. pad [in a waveguide]
  nichtabsorbierender A. reactive attenuator
  reaktiver A. reactive attenuator
Abschwächung *f* attenuation; damp[ing]; fading; reduction; fall
Abschwächungsglied *n (cont)* attenuator
Abschwächungswiderstand *m* reducing resistor; reducing resistance
abschwarten to square *(timber)*
abschwefeln to calcine *(ore)*; to desulphurate
abschwemmen to elutriate, to erode, to wash away, to float off, to rinse
abseifen *(text)* to soap
abseigern to liquate
abseihen to strain [off], to filter
Abseite *f* backing
absengen to singe
absenkeln to plumb
absenken to sink *(shaft)*; to lower, to bring down; to dip; to move in, to insert *(rods)*; to depress, to reduce *(flux)*; to settle *(rock)*
  einen Schacht a. to sink [down] a shaft
Absenkungsfaktor *m (nucl)* disadvantage factor
Absenkungsgeschwindigkeit *f (rock)* settling speed
Absenkungskurve *f* depression curve
Absetzbecken *n* 1. *(build)* sedimentation tank (basin), settling tank (basin); draining tank, drainer; 2. *(mine)* clarifying basin, slush pit
Absetzbehälter *m s.* Absetzbecken 1.
Absetzbütte *f* drainer *(paper)*
Absetzdock *n (ship)* depositing dock
absetzen 1. to settle; to deposit; 2. to wean *(animal)*; 3. to shove off, to push off *(a boat)*; to prick, to set up, to lay off *(a course in the chart)*
  a. lassen to decant, to allow to deposit
  Besteck a. to plot, to prick the chart
  eine Schicht a. *(geol)* to abut against
  Kurs a. to lay the course
  sich a. to settle; to precipitate; to disengage
Absetzen *n* 1. settling, settlement, sedimentation; 2. weaning *(animal)*
  A. des Bestecks *(ship)* plotting, prickling the chart
  A. des Schleifschlammes settling of the grinding sludge
Absetzer *m* overburden spreader; *(Am)* stacker
Absetzerkippe *f (mine) (Am)* stacker dump
Absetzgefäß *n* sediment (settling) vessel; clarifier; *(ch)* entrainment separator
Absetzgeschwindigkeit *f* settling rate
Absetzglas *n* settling glass
Absetzkasten *m* settling tank
Absetzklärung *f* sedimentation
Absetzplatz *m (ship)* repair berth
Absetzprüfung *f* sedimentation analysis (test)
Absetzraum *m* sedimentation chamber *(Imhoff tank)*
Absetztank *m* settling tank; thickener
Absetzverfahren *n* sedimentation method (process)
Absetzversuch *m* sedimentation test
Absetzzeit *f* settling rate

absichern to fuse, to protect by fuse
absieben to screen, to separate by screening, to sieve; to filter out
absieden to decoct, to boil up
absinken auf to drop to, to settle down to
Absinken *n* **der Intensität** decrease of intensity
Absitzenlassen *n* sedimentation
absolut absolute
Absolutbestimmung *f* absolute determination
Absolutbewegung *f* absolute motion
Absolutgeschwindigkeit *f* absolute velocity
Absolutglied *n (math)* absolute term
Absolutkolorimeter *n* absolute colorimeter
Absolutmessung *f* absolute measurement
Absolutrechnung *f* absolute calculus
Absolutschwelle *f* absolute threshold
Absoluttheorie *f* atomic absolute theory
Absolutwert *m* absolute value
  A. der Kraft absolute value of the force
Absolutwertelement *n* absolute-value element *(analogue computer)*
Absolutzählung *f* absolute counting
absondern to isolate; to segregate; to separate
Absonderung *f* elimination; segregation; abstraction; secretion; isolation
Absonderungsfläche *f (geol)* sheet
Absonderungsfuge *f (geol)* joint
Absorbens *n* absorbent, absorbing substance (agent)
Absorber *m* 1. absorber; 2. plant for the absorption [of a gas]; 3. *(nucl)* absorber, poison
  abbrennbarer A. *(nucl)* burnable poison
Absorberstab *m (nucl)* absorbing rod; control rod
absorbierbar absorbable
absorbieren to absorb; to attenuate; to capture
Absorption *f* absorption; attenuation; capture
  A. durch Paarbildung *(ph)* pair-production absorption
  A. von geladenen Teilchen absorption of charged particles
  A. von Schwingungen absorption of vibrations
  akustische A. acoustic absorption
  atomare A. atomic absorption
  dielektrische A. dielectric absorption
  kontinuierliche A. continuous absorption
  molekulare A. molecular absorption
  selektive A. selective absorption
  unvollständige A. incomplete absorption
  vollständige A. complete absorption
Absorptionsapparat *m* absorber, absorption apparatus
Absorptionsbande *f* absorption band
Absorptionsdämpfung *f* attenuation due to absorption
Absorptionsdruckfarbe *f* pressure-set ink
Absorptionsfähigkeit *f* absorbency, absorbability, absorptivity, absorption (absorbing) capacity, absorptive power; degree of absorption
Absorptionsfalle *f* absorption trap
Absorptionsfilter *n* absorption filter
Absorptionsfläche *f* surface of absorption
Absorptionsflasche *f* absorption bottle
Absorptionsfrequenzmesser *m* absorption-type frequency meter
Absorptionsgefäß *n* absorption vessel (cell), absorber
Absorptionsgesetz *n* law of absorption
  optisches A. law of optical absorption
Absorptionsgrad *m* absorption factor, absorptance
Absorptionsinfrarotanalysator *m (cont)* negative filtering infrared analyzer
Absorptionskälteanlage *f s.* Absorptionskältemaschine
Absorptionskältemaschine *f* absorption-type refrigeration machine
Absorptionskante *f* absorption edge (limit)
Absorptionskeil *m* absorption (absorbing) wedge
Absorptionskoeffizient *m* coefficient of absorption, absorption coefficient (factor)
  atomarer A. atomic absorption coefficient
  kontinuierlicher A. continuous absorption coefficient
  mittlerer A. mean absorption coefficient
Absorptionskreis *m* absorption (absorbing) circuit
Absorptionskühlschrank *m* absorption-type refrigerator
Absorptionskurve *f* absorption curve; transmission curve *(radiobiology)*
Absorptionsküvette *f* absorption cell
Absorptionslinie *f* absorption line (band)
Absorptionsmittel *n* absorbent
Absorptionsmodulation *f* absorption modulation
Absorptionsökonometer *n* absorption econometer
Absorptionsquerschnitt *m (nucl)* absorption cross section
Absorptionsregelung *f* absorption control
Absorptionsrohr *n* absorption tube
Absorptionsschwund *m* fading by absorption
Absorptionsspektrum *n* absorption (dark-line) spectrum
Absorptionssprung *m* absorption discontinuity
Absorptionsstrom *m* absorption current
Absorptionsturm *m* absorption tower
Absorptionsveredlung *f (text)* absorbent finish
Absorptionsvermögen *n* absorptivity, absorptance, absorptive power (capacity)
Absorptionswärme *f* heat of absorption
Absorptionsweglänge *f (nucl)* absorption mean free path
Absorptionswellenmesser *m (el)* absorption wavemeter; absorption-type frequency meter
abspalten to spall *(stone)*; to split off; to cleave; to crack *(oil)*; to separate

Fluor a. *(ch)* to defluorinate
Abspalten *n* **von Stromkreisen** segregation of circuits
Abspaltung *f* spalling *(stone)*; splitting-off; cleavage *(crystal)*; cracking *(oil)*; separation; elimination; segregation *(of circuits)*
A. der Azetylgruppe deacetylation
abspanen to remove *(chips)*
Abspannanker *m (el)* pole guy
Abspannbindung *f* **der Leitung** *(el)* termination of wire
Abspannbund *m* dead-end tie, termination, *(Am)* dead-tie *(overhead line)*
Abspanndraht *m* span (guy, stretching, stay) wire, back stay
abspannen to guy *(pole)*; to unload, to unclamp; to terminate
Abspannen *n* guying *(pole)*; unclamping, unloading
Abspanner *m* step-down transformer; anchor tower
Abspanngestänge *n* terminal (storm guyed) pole
Abspanngittermast *m* anchor tower; angle support
Abspannholzmast *m* anchor pole
Abspannisolator *m* terminal (strain, terminal strain, shackle) insulator
Abspannkette *f* insulator chain, string of insulators
Abspannklemme *f* anchor (straining, terminal) clamp
Abspannmast *m* straining (dead-end, anchor, supporting) tower, anchor (end, terminal) pole
Abspannpfahl *m* dead man, span pole
Abspannpflock *m* anchor log
Abspannseil *n* span (anchoring) rope; aerial support; guy [cable], guy rope
Abspannstempel *m (mine)* anchor (stell) prop
Abspannstütze *f (el)* terminal spindle (bracket)
Abspannung *f* rigging; guying; [aerial] support
A. der Leitungen termination of wires on terminal poles
absperren to shut off *(steam)*; to seal; to close; to lock; to throttle; to isolate
Absperrflüssigkeit *f* sealing liquid
Absperrfurnier *n* crossband *(wood)*
Absperrhahn *m* stopcock, cut-off cock
Absperrkreis *m (ph)* antiresonant circuit
Absperrlage *f* crossband *(wood)*
Absperrorgan *n* shut-off device
Absperrschieber *m* shut-off valve; gate valve
Absperrspindel *f* stop valve spindle
Absperrung *f* / **kreuzweise** crossbanding *(wood)*
Absperrventil *m* blocking (shut-off, stop) valve
kolbenbetätigtes A. plunger-operated shut-off valve
selbsttätig schließendes A. automatic stop valve
Absperrvorrichtung *f* stopping device
selbsttätige A. automatic stopping device
Abspieldose *f* gramophone pickup
Abspieleinrichtung *f* playback unit
Abspielen *n* playback
unmittelbares A. nach der Aufnahme direct playback
absplittern *(build)* to spall, to splinter off
Absplitterung *f* spalling; spallation; chipping
Absplitthebel *m* column splits *(tabulator)*
Absprengabsteller *m (text)* cloth fall-out detector
absprengen 1. to spall *(stone)*; 2. to crack *(glass)*; 3. to strut (truss) a beam
Absprengen *n (text)* pressing off *(knitting)*
A. des Stoffes cloth fall-out *(knitting)*
Absprengsystem *n (text)* press-off feeder *(knitting)*
Absprengteil *n (text)* press-off part *(knitting)*
Absprengung *f* spalling *(stone)*
abspringen 1. to abandon the aeroplane; to descend; 2. to rebound; 3. to jump off
abspritzen to wash [by spraying], to cleanse
Absprungtisch *m* take-off, platform *(ski-jumping)*
abspulen 1. *(text)* to reel off, to unspool, to wind off; 2. *(el)* to uncoil, to run off; to wind off, to unwind
abspülen to rinse, to scour, to cleanse
Abspulmaschine *f (text)* reeler, reeling frame
Abstand *m* distance; spacing, space; clearance; offing *(ship)*; *(regular)* interval; spacing *(radar)*
A. / **in gleichem** equidistantly, equally spaced
A. / **in ungleichem** unevenly spaced
A. der Buckel *(el)* spacing of the humps *(filters)*
A. der Niveaus *(ph)* spacing of levels
A. der Resonanzspitzen *(el)* spacing of the humps *(filters)*
A. Magnetkopf-Beleg *(off)* head-to-coating separation
A. Strahlungsquelle-Haut *(nucl)* source-skin distance
A. vom Pfeifpunkt singing margin, margin of stability
A. von Mitte zu Mitte centre-to-centre spacing
A. zwischen den Atomen interatomic distance
A. zwischen den Elektroden interelectrode distance, electrode spacing
A. zwischen den Linsen *(opt)* inter-lens distance
als Winkelmaß angegebener A. der Teilung angular space distance
fester A. fixed distance
interatomarer A. interatomic distance (spacing)
lichter A. clear distance (width)
zwischenatomarer A. *s.* interatomarer A.
Abstandgeber *m* range transmission unit *(radar)*

Abstandhalter *m* spacer, spacing stay (block), distance piece; bar spacer *(reinforced concrete)*; spacing clip *(overhead conductors)*; cage, separator *(bearing)*
Abstandkode *m* space code
Abstandring *m s.* Abstandsring
Abstandsbestimmung *f* determination of distance *(navigation)*
Abstandsgenauigkeit *f* spacing accuracy *(bores)*
Abstandsgesetz *n* / **quadratisches** *(nucl)* inverse square law
abstandsgleich equidistant; equally spaced
Abstandshülse *f* distance tube
Abstandskurzschluß *m* kilometric (short line) fault
Abstandsleiste *f* spacing strip
Abstandsmesser *m* telemeter, distance meter
Abstandsplatte *f* spacing plate
Abstandsring *m* distance ring; spacing collar, spacer *(milling machine)*; marker range ring *(radar)*
Abstandsschelle *f* spacing clamp (clip), spacer
Abstandsunterscheidung *f* range discrimination *(radar)*
Abstaubband *n* dusting band
abstäuben to dust
Abstaubwalze *f* dusting roller
abstechen 1. to cut off; to part off [to length] *(metal cutting)*; 2. to tap *(furnace)*; to teem
Abstechen *n* 1. cutting-off, parting-off, single-point cut-off *(metal cutting)*; 2. tapping *(furnace)*
abstechend contrasting *(shade)*
Abstechmaschine *f* cutting-off machine
Abstechmeißel *m* parting-off tool
Abstechpflug *m* paring plough
Abstechschaufel *f* turf spade (spud)
Abstechwerkzeug *n* cutting-off tool
abstecken 1. *(text)* to pin; 2. *(build)* to peg out, to set out, to stake out
Absteckkette *f* surveyor's chain
Absteckleine *f* tracing cord, marking line
Absteckung *f* *(build)* pegging out, setting out, staking out; *(mine)* alignment
abstehen lassen to kill *(melt)*
abstehend projecting
Absteifbrett *n* poling board
absteifen to prop [up]; to shore; to brace *(ribbing)*; to strut *(pole)*
Absteifung *f* *(build)* shoring, propping; stiffening, bracing *(ribbing)*; strutting *(pole)*
absteigend descending
Abstellbahnhof *m* storage sidings
Abstelleinrichtung *f* stop-motion device
abstellen to stop *(machinery)*; to disconnect; to switch off; to shut down
Abstellgleis *n* railway siding, siding track; shunt line
Abstellgrenzzeichen *n* shunting limit signal
Abstellhebel *m* stop motion lever
Abstellnadel *f* *(text)* detector wire, drop (wire) pin
Abstellraum *m* store, storage room
Abstellring *m* annular ring *(circular knitting machine)*
Abstellventil *n* stop valve
Abstellvorrichtung *f* locking device, stop mechanism, stop motion
abstemmen to chisel off
abstempeln to stamp, to mark; *(mine)* to prop; to stake down
Absterbezeit *f* / **mittlere** *(nucl)* median lethal time, MLT
Abstich *m* tapping *(furnace)*; run-off
Abstichbühne *f* tapping platform *(furnace)*
Abstichloch *n* tap hole *(furnace)*; bunghole *(barrel)*
  eingefrorenes A. hard tap *(furnace)*
Abstichofen *m* drossing oven
Abstichöffnung *f* tapping door, tap-hole, metal notch
Abstichpfanne *f* tap ladle
Abstichrinne *f* tapping spout
Abstieg *m* descent
Abstiegswinkel *m* angle of descent
Abstimmanzeige *f* indication of tuning
Abstimmanzeiger *m* [visual] tuning indicator
Abstimmanzeigeröhre *f* magic eye, tuning eye; tuning (electron-ray) indicator tube
Abstimmauge *n* magic (tuning) eye; electron eye
abstimmbar tunable
Abstimmbarkeit *f* tunability
Abstimmbereich *m* tuning range
Abstimmeinheit *f* tuner
Abstimmelement *n* tuning element
Abstimmempfindlichkeit *f* tuning sensitivity
  thermische A. thermal tuning sensitivity
abstimmen to tune
  auf Maximum a. *(el)* to peak
  zeitlich a. to time
Abstimmen *n* tuning
  zeitliches A. timing
Abstimmfähigkeit *f* tuning property
Abstimmhilfe *f* aid to tuning
Abstimminduktivität *f* tuning inductance
Abstimmittel *n* tuning unit, tuner; tuning circuits
Abstimmkapazität *f* tuning capacitance
Abstimmknopf *m* tuning control (knob)
Abstimmkondensator *m* tuning capacitor
Abstimmkörper *m* slug tuner *(waveguides)*
Abstimmkreis *m* tuned circuit
Abstimmkreisglied *n* tuned circuit element
Abstimmkurve *f* selectivity characteristic
Abstimmpeilfehler *m* plain tuning error
Abstimmpunkt *m* tuning point
Abstimmschärfe *f* selectivity, clearness (sharpness) of tuning

Abstimmschieber *m* tuning slide
Abstimmskala *f* tuning dial
Abstimmspule *f* tuning coil
Abstimmstichleitung *f* stub tuner *(waveguides)*
Abstimmung *f* tuning *(radio)*; timing *(time)*
A. der Tonhöhe sound (note) tuning
A. der Tonhöhe und Welle sound and wave tuning
A. durch veränderlichen Widerstand resistance tuning
A. von Hand manual tuning
falsche A. mistuning
grobe A. coarse tuning
kapazitive A. capacitive tuning
magnetische A. permeability tuning
optische A. visual tuning
scharfe A. sharp tuning
schlechte A. mistuning
thermische A. thermal tuning
unscharfe A. imperfect (flat) tuning
Abstimmungsregelung *f* tuning control
Abstimmungsschärfe *f* sharpness of tuning
Abstimmvorrichtung *f* tuner, tuning device (arrangement)
abstoppen to stop, to record a period by using a stop-watch
abstoßen *(ph)* to repel, to repulse; *(ry)* to push off, to throw *(shunting)*
Fett a. to sleek off *(leather)*
Narben a. to degrain, to peel off *(leather)*
abstoßend repulsive
Abstoßmaschine *f (leath)* grease jack
Abstoßung *f* repulsion
A. von Elektronen repulsion of electrons
elektrische A. electric repulsion
elektrostatische A. electrostatic repulsion
magnetische A. magnetic repulsion
Abstoßungskraft *f* repelling force (power), repulsive force (power)
Abstoßungspotential *n* repulsion potential
abstrahlen to emit; to radiate [into space]
Abstrahlfläche *f* emitting surface
Abstrahlung *f* emission; radiation
senkrechte A. vertical radiation
Abstrahlungswinkel *m* angle of radiation, radiation angle
Abstrahlvorrichtung *f* dissipative system
Abstreichbohle *f* screeding (levelling) beam
abstreichen to strike off; to level, to screed; to scrape off
Abstreicher *m* plough *(on conveyor belts etc.)*; scraper [rake]; skimmer
Abstreichlineal *n* levelling board *(moulding)*
Abstreichlöffel *m (ch)* scummer
Abstreichmesser *n* doctor blade
Abstreichwalze *f* evener cylinder
abstreifen 1. to wipe, to sweep clear off; 2. to scrape off *(oil)*; 3. to strip *(mould)*
Abstreifen *n* 1. peeling; 2. *(nucl)* stripping
Abstreifer *m* stripper *(distilling)*; deflector, plough, scraper; rake; wiper
A. für Späne wiper [for sweeping swarf clear off bed slides], bed wiper
Abstreiferform *f* stripper plate mould
Abstreifkamm *m* collecting comb, [noil] stripping comb *(combing machine)*
Abstreifmesser *n* mote (scraping) knife
Abstreifreaktion *f (nucl)* stripping reaction
Abstreifring *m* scraper ring
Abstreifschaber *m* deckle doctor
Abstreifwalze *f* doctor roll
A. mit zurücktretenden Stiften *(text)* evener roller with receding spikes
Abstrich *m* scum *(lead)*
abströmen to flow away (off)
a. lassen to bleed off; to drain
Abströmgeschwindigkeit *f (ship)* velocity of efflux
Abströmregler *m* overflow (dead-end pressure) regulator
Abströmung *f* outflow, efflux
Abströmventil *n* relief valve
abstufen 1. to step; 2. to grade; 3. to truncate; 4. to shade, to tint
Abstufung *f* 1. stepping; 2. graduation; 3. shading
feine A. fine graduation
abstumpfen 1. to dull *(tool)*; 2. *(ch)* to neutralize
Abstumpfen *n* 1. dulling *(tool)*; 2. *(ch)* neutralizing
Abstumpfungsfase *f* wear land
Abstumpfungsfläche *f* truncating face *(crystal)*
Abstumpfungswirkung *f (ph)* dulling action
abstürzen / brennend *(aero)* to go down in flames
abstützen to support; to back up; to prop; to shore; to reinforce
Abstützung *f* support; backing up; propping; shoring; reinforcement; shoring-up *(deck)*
Absuchen *n* **auf Fehler** perching
A. mittels Radar searchlighting *(radar)*
Absud *m (ch)* lixivium, decoction
Absüßen *n* **[des Filterkuchens]** *(ch)* sweetening off, desugarizing
Abszisse *f* abscissa
Abszissenachse *f* abscissa axis
abtafeln *(text)* to take off
abtakeln to unrig, to clear off, to strip, to cast off *(a vessel)*
Abtastblende *f* scanning diaphragm
Abtastelektrode *f* exploring electrode
abtasten 1. to scan, to sample; 2. to sense; 3. to read *(memory)*; 4. to trace *(machining)*; 5. to pick off *(radar)*
Abtasten *n* **/ fotoelektrisches** photoelectric scanning

zeilenweises A. linear scanning
Abtaster *m* scanner, scanning aerial; sampler
Abtastfeinheit *f* fineness of scanning, scanning density
Abtastfläche *f* scanning field
Abtastfolge *f* scanning cycle; scanning sequence
Abtastfrequenz *f* sampling rate
Abtastgeschwindigkeit *f* pick-up velocity; scanning rate; velocity of scanning
Abtastglied *n* sampler
Abtastimpuls *m* sampling pulse
Abtastkreis *m* sweep circuit *(radar)*
Abtastlichtfleck *m* scanning (exploring) spot
Abtastlichtstrahl *m* scanning beam (pencil) of light
Abtastmoment *m* sampling instant
Abtastöffnung *f* scanning aperture
Abtastoptik *f* scanning lens
Abtastoszillograf *m* sampling oscilloscope
Abtastperiode *f* sampling interval
Abtastrate *f* sampling rate
Abtastregelung *f* sampled-date control [system]
Abtastregler *m* periodic controller
Abtastrichtung *f* direction of scanning
Abtaströhre *f* scanning (camera, pick-up) tube
Abtastschreiber *m* scanning recorder
Abtastspannung *f* scanning voltage
Abtastspule *f* pick-up coil
Abtaststift *m* 1. *(el)* transmitting stylus; 2. tracer *(machining)*
Abtaststrahl *m* scanning beam
Abtastsystem *n* sampled-data system
auswechselbares A. pick-up cartridge
Abtasttheorem *n* sampling theorem
Abtastung *f* 1. scanning; sensing; sampling *(by pulses)*; 2. tracing *(machining)*
A. mit Geschwindigkeitsmodulation variable-speed scanning
A. mit Zeilensprung progressive scanning
elektronische A. sampling
fortlaufende A. progressive scanning
fotoelektrische A. photoelectric scanning
kompensierte A. compensated scan
mechanische A. mechanical scanning
punktförmige A. spot-type scanning
Abtastverfahren *n* tracer method *(machining)*
Abtastverlust *m* scanning loss
Abtastvorgang *m* scanning action
Abtastvorrichtung *f* 1. sampler; scanning device, scanner; 2. tracing device *(machining)*
Abtastvorschub *m (telec)* scanning traverse
Abtastwinkel *m* scanning angle
Abtastzeile *f* scanning line
Abtastzeit *f* scan time
Abtastzeitpunkt *m* sampling instant
Abtastzyklus *m* sampling (scanning) cycle
Abtauen *n* defrosting *(windscreen)*
Abteil *n* compartment
A. für Zugpersonal *(ry)* crew compartment
abteilen to divide; to partition; to classify
Abteilung *f* 1. department *(factory)*; 2. compartment *(of a vessel)*; 3. division
Abteilungsleiter *m* head of department
Abteilungsstrecke *f (mine)* section road
Abteufarbeit *f (mine)* sinking [operation]
Abteufarbeiter *m (mine)* sinker
Abteufausrüstung *f* sinking plant
Abteufbühnenwinde *f* sinking scaffold winch
abteufen to sink *(a shaft)*
Abteufen *n* sinking *(a shaft)*
gleichzeitiges A. und Einrichten *n* concurrent shaft sinking and equipping
Abteuffördergerüst *n* sinking head gear (frame)
Abteufördermaschine *f* sinking engine (hoist)
Abteuffortschritt *m* sinking advance (rate)
Abteufgerüst *n* sinking head frame (gear)
Abteufgreifer *m* lasher, lashing unit, sinking grab
Abteufkübel *m* hoppit, kibble
Abteuftrum *m* sinking compartment
Abteufung *f* sinking *(a shaft)*
abtönen to shade; to soften; to tint
Abtönfarbe *f* tint, tinting colour
Abtönungsfarbstoff *m* shading dyestuff
Abtönvermögen *n* tinting power
abtorfen to dig peat
Abtrag *m* spoil; excavation
abtragen 1. to remove soil; to abrade; to erode; *(mine)* to uncover; 2. to mark off *(drawing)*
Abtragen *n* / **chemisches** chemical milling
Abtragung *f* abrasion; erosion; attrition; soil removal
abtreiben 1. to trim; to distill; to scrub; to cupel *(noble metal)*; to sag, to drift away *(wind)*
Abtreibepfahl *m (mine)* plank *(piling)*
Abtreiber *m* stripper, [stripping] still
Abtreibeverfahren *n (mine)* piling
Abtreibezimmerung *f (mine)* forepoling
Abtreibung *f* distillation; cupellation *(noble metal)*
abtrennen to isolate, to separate; to detach; to part off *(machining)*
Abtrennen *n* isolation, separation; parting off *(machining)*
Abtrennrelais *n* cut-off relay
Abtrennschalter *m* splitting key
Abtrennung *f* isolation, separation
A. der Grenzschicht boundary layer separation
abtreppen to step
Abtreppung *f* benching, stepping; offset
Abtretung *f* abandonment *(insurance term)*
Abtrieb *m* driven side; output
Abtriebsdrehmoment *n* output torque
Abtriebsdrehzahl *f* output speed

Abtriebsglied *n* driven member (link), output member
Abtriebsrad *n* follower, driven gear
Abtriebsseite *f* **eines Motors** back of a motor
Abtriebswelle *f* output shaft
Abtrift *f (ship)* drift, leeway
Abtriftanzeiger *m* drift indicator
Abtriftberichtigung *f* drift correction
Abtriftgeschwindigkeit *f* drift velocity
Abtriftkorrektur *f* drift correction
Abtriftmesser *m* drift indicator
Abtriftrechner *m* drift computer
Abtriftwinkel *m* drift angle
abtrommeln to run off *(cable)*; to pay out *(wire)*
Abtropfbehälter *m* drainer
Abtropfblech *n* drip pan
Abtropfbrett *n* drain[ing] board
abtropfen to trickle down, to drop off, to drip off; to drain *(paper)*
Abtropfpfanne *f* drainer
abtrudeln to go into a spin, to come down in a spin
abtun *(mine)* to ignite, to fire
Abtun *n (mine)* firing, lighting, igniting
Abutilonfaser *f* abutilon fibre
Abvolt *n* abvolt
Abwälzbewegung *f* relative rolling motion between the work and the cutter *(hobbing)*
abwälzend aufeinander rolling together
Abwälzschleifen *n* **[von Stirnrädern]** [spur gear] generation grinding
Abwälzverfahren *n s.* Wälzverfahren
Abwanderung *f* **der Frequenz** frequency drift
Abwärme *f* distilled heat; waste heat
Abwärmekraftmaschine *f* waste heat engine
Abwärmeverlust *m* waste heat loss
Abwärmeverwertung *f* waste heat utilization
Abwärtsbewegung *f* down movement; downward travel *(machine)*
Abwärtsförderung *f* downward conveying
Abwärtshub *m* downstroke
Abwärtstransformator *m* step-down transformer
Abwärtstransformierung *f* step-down transformation
Abwärtswandler *m s.* Abwärtstransformator
abwaschbar washable
Abwaschbecken *n* sink
abwaschen to wash, to rinse, to cleanse, to scour
Abwasser *n* waste (foul) water, sewage; back water; white water *(papermaking)*
Abwässer *npl* sewage
  gewerbliche A. commercial sewage
  radioaktive A. radioactive effluents
  unbehandelte A. crude sewage
Abwässerabführung *f* sewage disposal
Abwasseranlagen *fpl* sewerage works
Abwasserbearbeitung *f* treatment of waste water
Abwasserbehälter *m* sewage tank
Abwasserbehandlung *f* sewage treatment
Abwasserbelüftung *f* bio-aeration
Abwasserberieselung *f* sewage irrigation
Abwasserbeseitigung *f* sewage disposal; disposal of radioactive waste water
Abwasserchlorung *f* sewage chlorination
Abwasserfallrohr *n* soil pipe
Abwasserfilter *n* sewage filter
Abwasserkanal *m* sewer
Abwässerkanalisation *f* sewerage
Abwasserkanalreinigung *f* sewer cleaning
Abwasserklärtrichter *m* save-all *(papermaking)*
Abwässerklärung *f* clarification of sewage, sewage purification
Abwasserlandbehandlung *f* sewage irrigation
Abwasserleitung *f* sewer, disposal (sewage pipe) line
Abwasserprüfgerät *n* effluent tester
Abwasserreinigung *f* sewage purification (treatment)
Abwassertechnik *f* sewage engineering
Abwasserwesen *n* sewerage
abwechseln to alternate
abwechselnd alternate
abweichen to deviate, to deflect *(from the course)*; to swerve; to vary *(from tolerance)*
Abweichung *f* 1. deviation; 2. error; 3. variation *(from tolerance)*; 4. *(math)* variance; 5. drift *(course)*; 6. aberration; 7. deflection *(ray)*; 8. declination *(compass needle)*
  A. des Nullpunkts zero error
  A. von der elektrischen Neutralität deviation from electric neutrality
  A. von der Lotrechten *(mine)* deviation, drift *(boring)*
  chromatische A. der sphärischen Aberration chromatic variation of spherical aberration
  mittlere A. average deviation
  mittlere quadratische A. standard (mean square) deviation
  normale A. standard deviation
  periodische A. periodic deviation
  prozentuale A. percentage deviation
  sphärische A. *(opt)* spherical aberration
  zulässige A. allowable (permissible) variation
Abweichungsanzeiger *m* deviation indicator
Abweichungsquadrat *n* squared deviation
Abweichungsverhältnis *n* deviation ratio
Abweichungswinkel *m* angle of divergence; angle of deviation (deflection); magnetic azimuth (declination) *(compass needle)*
Abweiser *m (ship)* defence boom, fender
Abweiseranode *f* beam-forming plate
abwelken *(leath)* to sam

Abwelkmaschine *f (leath)* samming machine
Abwelkpresse *f (leath)* drying press
abwerfbar droppable
abwerfen 1. to disconnect *(load)*; 2. *(mine)* to abandon [permanently]; 3. to drop; 4. to release *(bomb)*
den Schutzkegel a. *(rock)* to drop off the nose-cone
Abwerfen *n* **der Last** *(el)* disconnection of load
Abwesenheitszeit *f* absence time
bezahlte A. attendance time
Abwetter *pl (mine)* return air
abwettern to weather *(a ship)*
abwickelbar *(math)* developable *(surface)*
Abwickelbarkeit *f (math)* developability *(surface)*
Abwickelmaschine *f* untwisting machine
abwickeln 1. *(math)* to develop, to roll out on a plane *(geometry)*; 2. to uncoil *(wire)*; to wind off *(cable)*, to unwind
Abwickelspule *f* feed reel (spool)
Abwickelwalze *f* lap roller
Abwickler *m (ship)* developer, loftsman
Abwicklung *f* 1. *(math)* development, rolling out on a plane *(geometry)*; 2. uncoiling *(wire)*; winding-off *(cable)*, unwinding
A. des Schaufelrades diagram of the blade wheel *(turbine)*
A. einer Fläche development of a surface
A. einer Kurve unwinding of a curve
Abwicklungskurve *f (math)* evolvent, involute
abwiegen to weigh
Abwind *m* downward [air] current, downwash
Abwinden *n (text)* backing-off
Abwindevorrichtung *f (text)* backing-off motion
abwinkeln to snake *(conveyor)*
abwischen to wipe off; to dust
Abwrackbetrieb *m (ship)* scrapping (demolition, dismantling) works
abwracken to scrap, to break up, to dismantle *(a vessel)*
Abwracken *n* ship-breaking
Abwracker *m* ship-breaker
Abwrackung *f* ship-breaking
Abwurf *m* jettisoning; dropping
Abwurfbehälter *m* jettisonable tank
Abwurfende *n* delivery end
Abwurfhöhe *f* dumping height
Abwurfkontrolle *f* dump check *(computer)*
Abwurfparabel *f* discharge trajectory
Abwurframpe *f* coke wharf
abwürgen to stall *(engine)*
Abyssalregion *f (geol)* abyssal region
abzählbar countable, denumerable, enumerable
nicht a. indenumerable
Abzählbarkeit *f* countability, denumerability, enumerability
abzählen to denumerate
Abzählung *f* denumeration

Abzapfbrunnen *m* bleeder well
abzapfen to draw off; to bleed off; to drain; to tap
Gas aus der Lagerstätte a. to drain gas from the reservoir
Abzapfhahn *m* drawing-off cock
Abzapfventil *n* bleeder
abzäumen to unbridle
Abzehrung *f* / **äußere** external corrosion *(boiler)*
innere A. internal corrosion *(boiler)*
rillenförmige A. grooving *(boiler)*
Abzehrungen *fpl* / **vereinzelte grübchenförmige** isolated pitting *(boiler)*
zahllose grübchenförmige A. honeycombing *(boiler)*
zusammenfließende grübchenförmige A. confluent pitting *(boiler)*
abzeichnen to copy
Abziehbild *n* decalcomania [picture], metachromotype
Abziehbildverfahren *n* metachromotype process
Abziehbohle *f* screeding beam (board)
Abziehemulsion *f* stripping emulsion (film)
abziehen 1. *(text)* to doff *(spinning)*; to degum *(silk)*; to take off *(material)*; 2. to hone, to oilstone *(tool)*; to smooth; to sharpen; 3. to level, to screed; 4. to take proofs; 5. *(ch)* to distill, to elutriate; 6. to seal off; 7. to deduct; 8. to slag off
Beton a. to screed
Schlacke a. to seperate the dross, to remove the cinder
Abziehen *n* **der Farbe** *(text)* removal of colour, stripping
A. der Schleifscheibe dressing the grinding wheel
A. mit Ölstein [oil-]stoning; sharpening; smoothing
A. von Förderwagen auf der Hängebank *(mine)* banking [out], uncaging
Abzieher *m (text)* doffer
Abziehfarbe *f* cocoon
Abziehfeder *f* pull-off spring
Abziehfilm *m* stripping emulsion (film)
Abziehformkasten *m* snap flask
Abziehlack *m* strippable film paint
Abziehlehre *f* screeding template
Abziehmesser *n* finishing knife, back iron
Abziehmesserkasten *m* drawer for dressing knives
Abziehmittel *npl* dye removers
Abziehpflug *m* draining plough
Abziehpresse *f* proof press
Abziehrohr *n* drain pipe
Abziehstein *m* oilstone, whetstone; hone
Abziehvorrichtung *f* puller, extractor
Abzug *m* 1. proof *(printing)*; 2. copy *(photography)*; print; proof; 3. trigger *(rifle)*; 4.

outlet; vent; fume hood; exhaust hood; eduction; 5. deduction; 6. doffing *(spinning)*; 7. flaying
im Maschinengestell gelagerter A. base-supported take-up *(knitting machine)*
korrigierter A. *(print)* clean proof
Abzugsband *n* drawing-off band
Abzugsbügel *m* trigger guard
Abzugsgraben *m* irrigation ditch, draining channel (trench)
Abzugsgrube *f* sink hole (trap)
Abzugshebel *m* firing lever
Abzugskanal *m* turf drain; flue, exhaust duct
Abzugsrohr *n* flue [pipe], exhaust (vent) pipe
Abzugsschrank *m* fume chamber (hood)
Abzugsspannung *f* drain voltage
Abzugswalze *f* take down roller
Abzugswiderstand *m* drain voltage
Abzugswinkel *m* angle of take off *(of web from felt)*, take down angle
abzundern to scale off
Abzweig *m (el)* branch; tap; subcircuit
Abzweigbahnhof *m* junction station
Abzweigbefehl *m* branch order *(programming of computers)*
Abzweigdose *f* junction (branch, distribution, joint) box, access fitting
abzweigen to branch, to tee, to tap off, to shunt
Abzweigfeld *n* branching panel
Abzweigfilter *n* shunt-arm electric wave filter
Abzweiggestänge *n* pole with cross arms
Abzweigkabel *n* branch (stub) cable
Abzweigkasten *m* distributing (branching) box; joint box, cable chamber
Abzweigklemme *f* tee-joint, T-joint, branch terminal
Abzweigleitung *f* branch (junction, side) line, branch conductor; shunt line *(hydraulics)*
Abzweigmuffe *f* T-joint, joint box; Y-joint
Abzweigrohr *n* branch pipe
Abzweigschalter *m* branch switch
Abzweigschaltung *f* branching connection
Abzweigspule *f* tapped coil
Abzweigstecker *m* socket-outlet adaptor, distribution plug
Abzweigstelle *f* railway siding
Abzweigstromkreis *m* subcircuit, branch circuit
Abzweigung *f* branch[ing]; junction; tapping; *(ry)* turn-out; bifurcation
Abzweigungen *fpl* **von Leitungen** *(el)* taps
Abzweigungspunkt *m* tapping point
Abzweigungsstromkreis *m* derived circuit
abzwicken to pinch of *(wire)*
Achat *m* agate
Achatmörser *m* agate mortar
Achatpolierstein *m* agate burnisher
Achroit *m* achroite
Achromasie *f* achromatism, achromaticity, degree of achromatic correction
A. der Bildgröße achromatism of magnification
A. der Bildlage achromatism of position
A. des Bildortes achromatism of position
Achromat *m (n)* achromat, achromatic lens
achromatisch achromatic
achromatisieren to achromatize
Achromatisierungskurve *f* achromatization curve
Achromatismus *m s.* Achromasie
Achsabstand *m* 1. *(ph)* [centre-to-]centre distance; 2. axle base *(vehicle)*
Achsantrieb *m* axle drive
beid[er]seitiger A. bilateral transmission
Achsbolzen *m* front-axle pivot pin
Achsbolzenbuchse *f* king-pin bush
Achsbruch *m* breaking of [an] axle
Achsbüchse *f* axle-bearing, journal-box, axle box
Achsbund *m* axle-collar
Achsdruck *m* axle load; load on axle
Achsdynamo *m* axle-driven generator
Achse *f* 1. *(math, ph)* axis; 2. axle *(vehicle)*
elektrische A. electric axis, X-axis *(of crystals)*
fliegende A. floating axle
freie A. free axis
gekröpfte A. cranked axle
gerade A. straight axle
geradlinige A. rectilinear axis
geteilte A. divided axle
gezogene A. solid-drawn axle
kristallografische A. crystallographic axis
magnetische A. magnetic axis
mechanische A. mechanical axis, Y-axis *(of crystals)*
neutrale A. neutral axis *(statics)*
optische A. optical axis, Z-axis *(of crystals)*
polare A. polar axis
verschiebbare A. sliding axle
Achsel *f (print)* shoulder *(of a letter)*
Achselkeil *m (text)* key
Achsenabschnittsform *f (math)* intercept form
Achsenabweichung *f* amount of off-centre
Achsendrehmaschine *f* axle lathe
Achsendrehung *f* axial rotation
Achsenebene *f* axial plane *(of crystals)*
Achsenfertigdrehmaschine *f* axle finishing lathe
Achsenfett *n* axle grease
Achsengelenk *n* pin joint
Achsenkraft *f* axial force
Achsenkreuz *n* system of coordinates
Achsenlibelle *f* axis level
Achsenneigung *f* inclination of an axis
Achsenöl *n* axle oil
Achsenquantenzahl *f* beam quantum number
Achsenregler *m* shaft governor
Achsenring *m* ring of [an] axle

Achsenschälmaschine *f* axle peeling lathe
Achsenschnittform *f (math)* intercept form
Achsenschnittverfahren *n (opt)* axial section method
Achsenschruppdrehmaschine *f* axle roughing lathe
Achsensehprüfscheibe *f* axis test disk *(for astigmatic eyes)*
achsensymmetrisch axially symmetric
Achsenverhältnis *n* axial ratio
Achsenwinkel *m* shaft angle *(bevel gear)*
Achsfeder *f* axle spring
Achsfolge *f* axle sequence (arrangement)
Achsgabelsteg *m* axle guide stay, *(Am)* pedestal binder
Achsgehäuse *n* axle box
Achshals *m* journal
Achshalter *m* axle guard
Achskilometer *n* axle-kilometre
Achskreuzpunkt *m* crossed-axes point *(gear shaving)*
Achskreuzschaben *n* [rotary] crossed-axes shaving
Achskreuzwinkel *m* crossed-axes angle *(gear shaving)*
Achslager *n* axle bearing (journal); journal-box
Achslagerbügel *m* axle-box stirrup, stirrup of the axle box
Achslagerführung *f* axle-box guide (slide), pedestal frame, hornblock patches
Achslagergehäuse *n* axle box
Achslagergleitplatte *f* axle-box guide plate
Achslagerschale *f* axle-box [brass] bearing
Achslagersteg *m* axle guide stay, *(Am)* pedestal binder
Achslagerstellkeil *m (ry)* axle-box key (wedge), adjusting key (wedge)
Achslast *f* load on an axle; axle load
Achsmaß *n* centre-to-centre distance
Achsmotor *m* 1. direct-drive motor; 2. direct-drive engine
Achsmutter *f* axle nut; collar nut
achsparallel axially parallel
Achsprüflehre *f* axle-testing gauge
Achsrohr *n* axle tube
Achsschenkel *m* axle journal (stub), stub axle, steering knuckle
Achsschenkelbolzen *m* steering swivel pin; swivel (king) pin
Achsschenkelbuchse *f* steering swivel bush
Achsschenkelbund *m* bearing (axle journal) collar
Achsschenkeldrehmaschine *f* journal lathe
Achsschenkellenkung *f* king-pin steering assembly
Achsschenkelschleifmaschine *f* axle journal grinding machine
Achsspiel *n* clearance of the axle
Achsstand *m* axle base
Achsstummel *m* axle stub (journal)
Achssturz *m* axle dip (camber)
Achsstütze *f* axle bracket
Achsteilung *f* axial pitch *(gear)*
Achsverlängerungsstück *n* spindle coupling
Achswelle *f* axle [driving] shaft
Achszähleinrichtung *f (ry)* wheel counting device, axle counter
Achszapfen *m* journal *(bearing)*; swivel pin, axle journal; king pin
Achszapfensturz *m* king-pin inclination
achtbasisch octobasic
achtbindig *(text)* eight-harness weave
Achteck *n* octagon
achteckig octagonal
Achtel *n* eighth
Achtelektrodenröhre *f* octode
Achter *m* 1. eight [oared shell] *(sporting boat)*; 2. *(telec)* double-phantom circuit
achteraus astern, sternward *(navigation)*
Achterbrassen *fpl (ship)* after braces
Achterbrücke *f* warping (after, docking) bridge *(of a vessel)*
Achtercharakteristik *f (el)* figure-[of-]eight pattern *(dipoles)*
Achterdeck *n* quarter (after) deck *(of a vessel)*
Achterdiagramm *n s.* Achtercharakteristik
Achterfigur *f s.* Achtercharakteristik
Achterhütte *f (ship)* poop
Achterklüse *f* cat hole *(of a vessel)*
Achterkreis *m (telec)* double-phantom circuit
achterlastig *(ship)* [down] by the stern
Achterlastigkeit *f (ship)* trim by the stern
Achterlaterne *f* poop-lantern *(of a vessel)*
achterlich abaft *(navigation)*
achterlicher als dwars abaft the beam
Achterliek *n (ship)* [after]leech
Achtermast *m (ship)* mizen-mast, after mast
achtern abaft, astern, aft[er]
Achternaht *f (text)* figure-of-eight suture
Achterpiek *f (ship)* after peak
Achterraum *m (ship)* after hold
Achterschale *f (nucl)* octet (eight-lamina) shell
Achterschaltung *f (el)* double-phantom circuit; octuple circuit
Achterschiff *n* 1. after body, [after] quarter; 2. after-end arrangement, all-aft arrangement ship *(all superstructure aft)*
Achtersegel *npl* after sails
Achterspanten *npl* after frames
Achterspule *f* octagonal coil; eight-film reel
Achtersteven *m* stern post (frame); back stern-post
Achtertanker *m* all-aft tanker *(all superstructure aft)*
Achtertelegrafie *f* supherphantom telegraphy
Achterwicklung *f* figure-eight winding *(of coils)*
achtfach eightfold; octuple

Achtfachuntersetzer *m (nucl)* scale of eight circuit
achtflächig octahedral
Achtflächner *m* octahedron
Achtganggetriebe *n* eight-speed transmission
achtgliedrig eight-membered
Achtkant *n* octagon
Achtkantstahl *m* octagon bar steel
achtmotorig eight-engined
Achtpolröhre *f* octode
Achtschloßflachstrickmaschine *f* eight-lock flat knitting machine
Achtschloßmaschine *f (text)* eight-cam machine, eight-lock machine
Achtspindelautomat *m* eight-spindle automatic
Achtstundentag *m* 8-hr. working day
Achtundvierzigflächner *m* hexakisoctahedron, hexoctahedron *(crystal)*
achtwertig octavalent
Achtwertigkeit *f* octavalency
achtzählig eightfold
Acker *m* field; acre; farm land
Ackerbau *m* farming; tillage
Ackerboden *m* arable land; cultivated soil
Ackerbürste *f* [spring-tine] weeder
Ackeregge *f* tilling harrow
Ackerfurche *f* furrow
Ackergare *f* tilth
Ackergeräte *npl* farming (agricultural) implements, farm (agricultural) tools
Ackerland *n* tillage
absolutes A. arable land
ackern to plough, to till the soil
Ackerschiene *f* toolbar, tilling beam
Ackerschleife *f s.* Ackerschleppe
Ackerschleppe *f* field drag, clod crusher
Ackerschlepper *m* farm tractor
Ackerschlichte *f s.* Ackerschleppe *f*
Ackerscholle *f* clod
Ackerwagen *m* field cart (wagon)
Ackerwalze *f* land roller
Acornröhre *f s.* Eichelröhre
AD-Umsetzer *m (dat)* analog-to-digital converter
Adamin *m (min)* adamite
Adaptation *f* adaptation
Adaptationsbreite *f* range of adaptation
Adapter *m* adapter
Adapterkabel *n* extension cable (lead)
adaptieren to adapt
Adaptionsleuchtdichte *f* adaption brightness
Adaptometer *n* adaptometer
Adcock-Antenne *f* Adcock aerial
Addend *m (math)* addend
addieren to add
Addierer *m (dat)* adder
symbolischer A. half adder
Addierimpuls *m* add pulse
Addiermaschine *f* **/ schreibende (druckende)** adding listing machine
Addierschaltung *f* adding circuit
analoge A. Kirchhoff adder
Addierstufe *f* counter stage
Addier- und Subtrahierzählkette *f* add-and-subtract counting chain
Addierwerk *n* adder
Addierwerkzwischenspeicher *m* adder accumulator
Addition *f* addition
logische A. logical addition *(without carry)*
Additionsduplex *n (telec)* incremental duplex
Additionsmaschine *f* adding machine
Additionspolymerisation *f* addition polymerization
Additionsregel *f s.* Additionstheorem
Additionssatz *m s.* Additionstheorem
Additionsschaltung *f* summing circuit
Additionsstelle *f (dat)* mixing (summing) point
Additionstheorem *n* addition theorem
Additionsverbindung *f* addition compound
Additionszeichen *n* plus sign
additiv additive
Additivität *f* **der Bindungsenergie** *(nucl)* additivity of binding energy
A. der Varianzen additivity of variances
Adelsvorschub *m s.* Adelszone
Adelszone *f (geol)* ore-shoot, pay-streak
Adenosintriphosphorsäure *f* adenosine triphosphoric acid
Ader *f* 1. *(el)* core *(of cable)*; wire; conductor; 2. *(mine)* vein
Aderbruch *m* cable fault
Adergneis *m (geol)* vein gneiss
Aderhaut *f* choroid membrane *(of the eye)*
Aderkreuzung *f* crossing of wires
Adermin *n (ch)* adermin, pyridoxin
adern to grain
Adern *fpl* **auf einem Gußstück** veins on a casting
Adernbündel *n* group of conductors
Adernlage *f* arrangement of cable conductors
Adernpaar *n (el)* pair [of cores]
Adernzählfolge *f* numbering of cable cores
Aderpaar *n* pair of wires
Aderschleifenwiderstand *m* conductor loop resistance
Aderverlagerung *f* decentralization of cores
Adhäsion *f* adhesion, adherence
Adhäsionsfähigkeit *f* adhesiveness
Adhäsionskraft *f* adhesive force (power)
Adhäsionsvermögen *n* adhesiveness, adhesive power
Adiabate *f* adiabatic curve (line)
Adiabatenprinzip *n* adiabatic principle
adiabatisch adiabatic
Adipinsäure *f* adipic acid

adjektiv adjective *(dye)*
adjungieren *(math)* to adjoin
Adjustage *f* finishing shop
Adjustiertank *m* reserve buoyancy tank
Admiralitätsgericht *n* Admiralty Court
Admiralitätsknoten *m (ship)* admiralty knot
Admiralitätskonstante *f (ship)* admiralty (displacement) coefficient
Admiralitätsmessing *n* admiralty metal (brass)
Admissionsarbeit *f* work done during admission
Admissionsdruck *m* admission pressure
Admissionsdruckkurve *f* admission pressure curve
Admittanz *f* admittance
Admittanz-T-Glied *n* shunt T
Adrenalin *n* epinephrine, adrenaline
Adressat *m* addressee
Adresse *f* address *(computer)*
  absolute A. absolute address
  indirekte A. indirect address
  relative A. relative address
  symbolische A. symbolic (floating) address
Adressenaufruf *m* address selection *(computer)*
Adressendekodierung *f* address decoding
Adressenregister *n* address (index) register
Adressenteil *m* address code *(instruction word)*
Adressenumsteuerung *f* address modification *(digital computer)*
Adressenwerk *n* address unit *(computer)*
Adressiermaschine *f* addressing machine
  programmierbare A. programmable addressing machine
Adriabindung *f* **mit ungleichen Flottungen** *(text)* diagonal rib with irregular floats
Adsorbat *n* adsorbate, adsorbed substance
Adsorbens *n* adsorbent
adsorbierbar adsorbable
Adsorbierbarkeit *f* adsorbability
adsorbieren to adsorb
Adsorption *f* adsorption
  physikalische A. physical adsorption
  reversible A. physical adsorption
Adsorptionsanalyse *f* / **chromatografische** chromatographic analysis
Adsorptionschromatografie *f* adsorption chromatography
Adsorptionsfähigkeit *f* adsorbability, adsorptive power (capacity), degree of adsorption
Adsorptionsfiltration *f* adsorption filtration
Adsorptionsgesetz *n* adsorption law
Adsorptionsisotherme *f* adsorption isotherm
Adsorptionskohle *f s.* Aktivkohle
Adsorptionsmittel *n* adsorbent
Adsorptionsprodukt *n s.* Adsorbat
Adsorptionssäule *f* column of adsorbent
Adsorptionsschicht *f* adsorbent bed
Adsorptionsthermostat *m* adsorption thermostat
Adsorptionsvermögen *n* adsorption power
Adsorptionswärme *f* heat of adsorption
Adsorptiv *n* adsorbed substance
Adular *m (min)* adularia
Adventurinfeldspat *m (min)* adventurine feldspar
Advektion *f* advection
Advektionsnebel *m* advection fog
aerob aerobic
Aerodynamik *f* aerodynamics
aerodynamisch aerodynamic
Aerokartograf *m* aerocartograph
Aerologie *f* aerology
Aeromechanik *f* aeromechanics
Aeronautik *f* aeronautics
Aerosol *n* aerosol
  radioaktives A. radioactive aerosol
Aerosolfilter *n* aerosol (aerial dust) filter
Aero-Sprühmittel *n (ag)* aerial (low-volume aircraft) spray
Aerostatik *f* aerostatics
Aerotriangulation *f (opt)* aerial triangulation
affin affine
Affinade *f* affinated sugar
Affinität *f* [chemical] affinity
Affinitätsachse *f* axis of affinity
Affinitätskurve *f* affinity curve
Affinitätsrest *m* residual affinity
Afterheu *n* aftermath
Afterkristall *m* pseudomorph, pseudomorphous crystal
Afterschleuse *f* swelling sluice
AG-Beschleuniger *m (nucl)* strong-focusing accelerator, alternating-gradient accelerator
AG-Fokussierung *f (nucl)* strong (alternating-gradient) focusing
AG-Maschine *f s.* AG-Beschleuniger
AG-Synchrotron *n (nucl)* strong-focusing synchrotron, alternating-gradient synchrotron
Agar-Agar *m (n)* agar-agar
Age *n s.* Fermi-Alter
Agens *n* agent
  chelatbildendes A. chelating agent
  Kettenabbruch bewirkendes A. chain stopper (terminator)
Age-Theorie *f (nucl)* age theory
Agglomerat *n* agglomerate; agglomerated cake
Agglomeration *f* agglomeration
Agglomeratkuchen *m* sinter cake; agglomerated cake
Agglomerieranlage *f* agglomerating plant
agglomerieren to agglomerate
Aggregat *n* set *(engineering)*; compound machinery
  hydraulisches A. hydraulic package[-type unit]
Aggregatgleichung *f* missile equation
Aggregation *f* aggregation
Aggregationskinematik *f* kinetics of aggregation

Aggregatlagerwinkel *m* missile angle
Aggregatpolarisation *f* aggregative polarization
Aggregatrückstoß *m* aggregate recoil
Aggregatsteuerung *f* missile steering control
Aggregatwinkel *m* guided missile angle
Aggregatzustand *m* state of aggregation
  fester A. solid state
  flüssiger A. liquid state
  gasförmiger A. gaseous state
Aggregatzustandsänderung *f* change of state
aggressiv corrosive; aggressive
Aggressivität *f* aggressive action, aggressiveness, aggressivity
Agone *f* agonic line (curve)
Agraffenmaschine *f (leath)* hook setting machine
Agrartechnik *f* agricultural engineering (technology)
Agrarwissenschaften *fpl* agricultural (agrarian) science
Agrikultur *f* agriculture, farming
Agrikulturchemie *f* agricultural chemistry
Agrikulturphysik *f* agricultural physics
Agronom *m* agronomist
Agronomie *f* agronomy
Agrotechnik *f* agricultural technology
Ahle *f* bodkin; awl, pricker; broach, reamer
Ahming *f (ship)* draught-marks
ähnlich similar *(dynamically, geometrically)*
  geometrisch a. geometrically similar
Ähnlichkeit *f* similarity, similitude
  dynamische Ä. dynamical similarity (similitude)
  mechanische Ä. mechanical similarity (similitude)
  thermodynamische Ä. thermodynamic similarity (similitude)
Ähnlichkeitsgesetz *n* similarity principle (law), law of similarity (similitude)
Ähnlichkeitsmechanik *f* mechanics of similitude
Ähnlichkeitstransformation *f (math)* similarity transformation
Ahorn *m* maple
Ahornholz *n* maple [wood]
Ahornsirup *m* maple syrup
Ahornzucker *m* maple sugar
Ährenheber *m* pick-up reel
Ährenkurve *f (math)* Cotes's spiral
Ährenlesen *n* gleaning
Ährenmäher *m* header
Ährensieb *n* upper (capes, chob) riddle
Ailanthusspinner *m* ailanthus silkworm
Ajax-Ofen *m* Ajax-[Wyatt] furnace
Akarizid *n* acaricide
Akaroidharz *n* acaroid resin
Akausalität *f (nucl)* non-causality
Akazie *f* acacia
Akaziengummi *m* acacia gum

Akkommodation *f* accommodation
  elektrische A. electrical accommodation
Akkommodationsbreite *f* amplitude of accommodation
Akkommodationsruhe *f* relaxed state of accommodation
Akkommodationsvermögen *n* power of accommodation
akkommodieren to accommodate
Akkord *m* 1. contract; 2. *(ac)* chord
  A. / in by contract
  alterierter A. chromatic chord
  dissonanter A. discord
  konsonanter A. concord
Akkordabteilung *f* rate-fixing office
Akkordarbeit *f* piece-work, job (contract) work
Akkordlohn *m* piece[-rate] wages
Akkordsatz *m* piece-per-hour rate
Akku *m s.* Akkumulator
Akkumulation *f* accumulation
Akkumulationsgebiet *n* accumulation region
Akkumulator *m* accumulator [battery], storage (secondary) battery
  A. mit elastischer Trennblase flexible bag-type accumulator
  A. mit gelatiniertem Elektrolyt unspillable accumulator
  A. mit Gummisack rubber sag-type accumulator
  A. mit kegelförmiger Speicherblase conical bag accumulator
  A. ohne Trennkolben pistonless-type of accumulator
  alkalischer A. alkaline accumulator (storage battery)
  druckgasbelasteter A. mit Trennelement separator type accumulator
  pneumatischer A. compressed gas accumulator, air[-hydraulic] accumulator
  säuredichter A. nonspillable accumulator
Akkumulatorenanlage *f* accumulator plant
Akkumulatorenantrieb *m* battery drive
  A. / mit battery powered
Akkumulatorenbatterie *f* storage battery, accumulator
Akkumulatorenelement *n* accumulator (storage) cell
Akkumulatorenladegerät *n* accumulator charging set
Akkumulatorenplatte *f* storage-battery plate
Akkumulatorensäure *f* accumulator acid
Akkumulatorenschalter *m* battery regulating switch
Akkumulatorenzelle *f* storage cell
Akkumulatorenzugförderung *f* battery electric traction
Akkumulatorladegleichrichter *m* rectifier for battery-charging

Akkumulatorlokomotive *f* storage-battery locomotive
Akkumulatorplattenblock *m* subassembly of positive and negative accumulator plates
Akkumulatorplattensatz *m* set of unipolar accumulator plates
Akkumulatorsäure *f* electrolyte; storage-battery acid, accumulator acid
Akkumulatorzelle *f* accumulator (storage-battery) cell
akkumulieren to accumulate
Akonitsäure *f* aconitic (citridic) acid
Akridinfarbstoff *m* acridine dye
Akrylat *n* acrylate
Akrylfaser *f* acrylic fibre
Akrylharz *n* acrylic resin
Akrylharzkunststoff *m* acrylic plastic
Akrylnitril *n* acrylonitrile
Akrylsäure *f* acrylic acid
Akt *m* process; event
Akte *f* file, record
Aktend-Alarm *m* reel-end signal *(cinema)*
Aktendeckel *m* folder, document cover
aktenkundig notorious
Aktensuchmaschine *f* file searching machine
Aktentasche *f* portfolio
Aktentaschenleder *n* portfolio (bag) leather
Aktenvernichter *m* shredding machine
Aktenzeichen *n* serial symbol (number)
Aktiengesellschaft *f* joint stock company
Aktinid *n (nucl)* actinide
Aktiniden *npl* actinide series
Aktinität *f* actinism
Aktinium *n* actinium
Aktinium[zerfalls]reihe *f* actinium series
aktinoelektrisch actinoelectric
Aktinolith *m (min)* actinolite
Aktinometer *n* actinometer
Aktinon *n* actinon
Aktinouran *n* actinium uranium, actinouranium, uranium-235
Aktinouranfamilie *f s.* Aktinium[zerfalls]reihe
Aktionsbereich *m* operation range; steaming range (radius)
Aktionsprinzip *n* principle of action *(turbine)*
Aktionsrad *n* impulse wheel *(turbine)*
Aktionsradius *m* efficient range; reach; radius of operation (action), operating range *(of vessels)*
Aktionsreaktionsturbine *f* impulse reaction turbine
Aktionsturbine *f* action (impulse) turbine
Aktiuran *n s.* Aktinouran
aktiv [radio]active, radio-
Aktivator *m* 1. sensitizer; 2. activating agent, activator
Aktivatoratom *n* activator atom
Aktivhobel *m* activated plough
A. mit Selbstauslösung auto-percussive plough
aktivieren to activate; to sensitize; to radioactivate
Aktivierung *f* sensitization; [radio]activation
Aktivierungsanalyse *f* activation analysis
aktivierungsanalytisch by activation analysis
Aktivierungsausbeute *f* activation yield
Aktivierungsenergie *f* activation energy
A. infolge einer Verunreinigung impurity activation energy *(semiconductor)*
Aktivierungsimpuls *m* sensitizing pulse
Aktivierungsquerschnitt *m* activation cross section
Aktivierungssonde *f* activation foil
Aktivierungsspannung *f* activation overvoltage
Aktivität *f* [radio]activity
absolute A. absolute activity
induzierte A. induced [radio]activity
optische A. optical activity
relative A. relative activity
spezifische A. specific [radio]activity
Aktivitätsabfall *m* activity decay
Aktivitätsanalyse *f* activity analysis *(economical statistics)*
Aktivitätseinheit *f* activity unit
aktivitätsfrei *(nucl)* cold
Aktivitätskoeffizient *m* activity coefficient
Aktivitätskurve *f* activity curve
Aktivitätsnachweis *m* radioactivity detection
Aktivitätsniveau *n* level of activity, activity level
Aktivitätsverteilung *f* activity distribution
Aktivkohle *f* activated coal; absorbent carbon
Aktivruder *n (ship)* active rudder
Aktivschlamm *m (build)* activated sludge
Aktivschlammverfahren *n* activated sludge process
Aktivwert *m* asset
Akustik *f* acoustics
Akustikplatte *f* acoustic board
akustisch acoustic[al]
Akzeptor *m* acceptor
Akzeptordichte *f* acceptor density
Akzeptorniveau *n* acceptor level
Akzeptorstörstelle *f* acceptor impurity *(semiconductor)*
Akzeptorverunreinigung *f s.* Akzeptorstörstelle
Akzidenzarbeit *f* jobbing work
Akzidenzdruck *m* job printing
Alabaster *m* alabaster
Alabastergips *m* alabastrite, gypseous alabaster
Alarmanlage *f* alarm circuit (set)
Alarmapparat *m* alarm apparatus
Alarmeinrichtung *f* **mit Arbeitsstrom** open circuit alarm system
Alarmglocke *f* alarm bell; call bell
Alarmpfeife *f* alarm whistle

Alarmsignal *n* danger signal
Alarmverriegelung *f* alarm locking
Alarmvorrichtung *f* alarm device
Alaun *m* alum
Alaunerde *f* aluminous earth, alumina
Alaungerberei *f* alum tanning
Alaungerbung *f* alum tanning
alaunhaltig aluminiferous, aluminous
Alaunleder *n* alum leather
Alaunschiefer *m* alum shale
Alaunstein *m* aluminous limestone, alunite, alumstone
Albedo *f* albedo; reflection coefficient
Aldehyd *m* aldehyde
Aldehydgerbung *f* aldehyde tanning
Aldehydharz *n* aldehyde resin
Aldolkondensation *f* aldol condensation
Aldoxim *n* aldoxime
aleatorisch *(dat)* random, stochastic
Alexanderson-Maschine *f (el.m.)* Alexanderson alternator
Alfa *f* esparto, [h]alfa
Alfapapier *n* esparto paper
Algebra *f* algebra
  A. der Logik Boolean algebra
  Boolesche A. Boolean algebra
algebraisch algebraic
Algenbekämpfung *f* algae control
Algenvernichtungsmittel *n* algaecide
Algenvertilger *m* algaecide
Alginatfaser *f* alginic fibre
Alginsäure *f* alginic acid
Algorithmus *m* algorithm
Alhidade *f* alidade
aliphatisch aliphatic
Alizarinfarbstoff *m* alizarin dye[stuff]
alizyklisch alicyclic
Alkali *n* alkali; alkaline salt
Alkaliaufschluß *m* fusion with alkali
alkalibeständig alkali-proof
Alkalibeständigkeit *f* resistance to alkalis
Alkalibestimmung *f* alkalimetry
Alkalibildner *m* alkaline
Alkaliechtheit *f* fastness to alkalis
Alkalielement *n* caustic soda cell, alkaline accumulator cell
Alkalihalogenid *n* alkali halide
Alkalimetall *n* alkali metal
Alkalimetrie *f* alkalimetry
alkalisch alkaline, alkalinous
Alkalisieranlage *f* causticizing plant *(paper-making)*
Alkalisierung *f* causticizing process
Alkalisilikat *n* alkali silicate
Alkalisprödigkeit *f* caustic embrittlement
Alkalität *f* alkalinity, alkalescence
Alkaliverfahren *n* alkaline process
Alkalizellulose *f* alkali cellulose
Alkaloid *n* alkaloid
Alkan *n* alkane
Alken *n* alkene
Alkin *n* alkine
Alkohol *m* alcohol
  absoluter A. pure alcohol
  denaturierter A. methylated spirit, denatured alcohol
  dreiwertiger A. trihydric alcohol
  fünfwertiger A. pentahydric alcohol
  mehrwertiger A. polyhydric alcohol
  primärer A. primary alcohol
  sekundärer A. secondary alcohol
  vergällter A. methylated spirit, denatured alcohol
  vierwertiger A. tetrahydric alcohol
Alkoholat *n* alcoholate
Alkoholauszug *m* alcoholic extract
Alkoholbestimmung *f* alcoholometry
alkoholfrei non-alcoholic
Alkoholgehalt *m* percentage of alcohol
  A. im Blut blood alcohol content, level of alcohol in [the] blood
Alkoholgemisch *n* alcohol-blended fuel
alkoholisch alcoholic
alkohollöslich soluble in alcohol
Alkoholometer *n* alcoholometer
Alkoholometrie *f* alcoholometry
Alkoholtreibstoff *m* alcohol-blended fuel
Alkoholyse *f* alcoholysis
Alkoven *m* alcove
Alkydharz *n* alkyd resin
Alkylat *n (ch)* alkylate
Alkylgruppe *f (ch)* alkyl group
alle Mann an Deck! *(ship)* all hands on deck!
  a. Maschinen äußerste Kraft voraus! *(ship)* all ahead full!
  a. Maschinen äußerste Kraft zurück! *(ship)* all back emergency!
Alleinerfinder *m* sole inventor
Alleingerbung *f* self-tannage
Alleininhaber *m* sole owner
Alleinrecht *n* exclusive right
Allesbläser *m (ag)* general-purpose blower
Allesförderer *m* general-purpose elevator
Alleskleber *m* all-purpose adhesive
Alleslader *m* general-purpose loader
Allgebrauchslampe *f* general-purpose lamp
Allgemeinbildung *f* general education (knowledge)
Allglasröhre *f* all-glass valve
Alligatorzahnringdübel *m (build)* alligator connector
Allobar *n (nucl)* allobar
allochromatisch allochromatic
allomorph allomorphic
allotrop allotropic
Allotropie *f* allotropy

Allpaß *m s.* Allpaßfilter
Allpaßeigenschaft *f* all-pass property
Allpaßfilter *n* all-pass filter
Allpaßschaltung *f* all-pass circuit
Allradantrieb *m* all-wheel drive
Allradtraktor *m* four-wheel drive tractor
Allrichtungsfunkfeuer *n* omnidirectional range
allseitig polydimensional
Allseitigkeit *f* polydimensionality *(of pressure)*
Allstromgerät *n* all-mains set, ac-dc set
Allstromradiogerät *n* universal-mains radio set
Allstromröhre *f* ac-dc valve
Allstromspeisung *f* ac-dc power supply
Allverstärker *m* universal amplifier
Allwellenantenne *f* multi-band aerial
Allwellenempfänger *m* multirange (all-wave) receiver
Allylalkohol *m* allyl alcohol
Allylen *n (ch)* propine, allylene
Allylsenföl *n (ch)* allyl (artificial) mustard oil, allyl isothiocyanate
Allylumlagerung *f (ch)* allylic rearrangement
Allzweckdiode *f* general-purpose diode
Allzweckgerät *n* general-purpose tool
Allzweckmaschine *f* all-purpose machine
Allzweckpflug *m* general-purpose plough
Allzweckrechner *m* general-purpose computer
Allzwecktanker *m* all-purpose tanker
Aloxidverfahren *n* electrolytic oxidation process
Alpaka *n* 1. *(text)* alpaca [wool]; alpaca fabric; 2. *(met)* German silver; pakfong
Alpakafaser *f* alpaca hair
Alpha-Absorption *f* alpha-particle absorption
alpha-aktiv alpha [radio]active, alpha-emitting
Alpha-Aktivität *f* alpha [radio]activity
Alpha-Bindungsenergie *f* alpha-particle binding energy
Alpha-Bronze *f* alpha bronze
Alpha-Dosimeter *n* alpha survey meter
Alpha-Eisen *n* alpha-iron, ferrite
Alpha-Emission *f* alpha-particle emission
Alpha-Kammer *f* alpha-chamber, alpha counter tube
Alpha-Phase *f* alpha phase
Alpha-Spektrum *n* alpha-particle spectrum
Alpha-Strahlen *mpl* alpha-rays, alpha radiation
Alpha-Strahlendosimetrie *f* alpha-ray dosimetry
Alpha-Strahler *m* alpha[-ray] emitter, alpha radiator
Alpha-Strahlung *f* alpha radiation (rays)
Alpha-Teilchen *n* alpha particle (ray)
Alpha-Umwandlung *f* alpha decay (disintegration)
Alpha-Zähler *m* alpha counter (meter)
Alpha-Zerfall *m* alpha decay (disintegration)
Alpha-Zerfallsenergie *f* alpha disintegration energy
Alphabetlocher *m* alphabetical punch
Alphabetlochprüfer *m* alphabetical verifier
Alphabetmischvorrichtung *f* alphabetical collating device
Alphatron *n* alphatron
Alphina-Teilchen *n (nucl)* alphina particle
Altan *m (build)* balcony
Altazimuth *m (n)* altazimuth
Alteisen *n* iron scrap, scrap iron
Alter *n* age
  absolutes A. absolute [geological] age
  geologisches A. geological age
Alter Mann *m (mine)* goaf, gob, caved area (goaf, waste), waste area
altern to season, to age
Altern *n* seasoning, ag[e]ing
Alternanz *f* biennial (alternate) bearing
Alternativdruckregler *m* non-bleed pressure regulator, pressure regulator with three-way valve
Alternativhypothese *f* alternative hypothesis
Alternativsteuerung *f* three-way valve *(pneumatics)*
Altersbestimmung *f (nucl)* dating, age determination (measurement)
  A. nach dem ^{14}C-Gehalt radiocarbon dating
Altersgleichung *f* age equation
Altersklasse *f* age class
alterssichtig presbyopic
Alterssichtigkeit *f* presbyopia
Alterstheorie *f* / **Fermische** *(nucl)* age theory
Alterung *f* ag[e]ing, seasoning
  A. eines Dauermagneten weakening of a permanent magnet
  beschleunigte A. accelerated ag[e]ing
  künstliche A. artificial ag[e]ing
Alterungsbehandlung *f* ag[e]ing treatment
alterungsbeständig ag[e]ing resistant, non-ag[e]ing
Alterungsbeständigkeit *f* durability, ag[e]ing resistance
Alterungsdiagramm *n* performance curve
Alterungsprüfung *f* ag[e]ing (storage) test
  beschleunigte A. accelerated ag[e]ing (storage) test
Alterungsriß *m* season (ag[e]ing) crack
Alterungsschutzmittel *n* age-protecting agent
Alterungsverfahren *n* ag[e]ing process
Altgrad *m* degree
Altgummi *m* waste (scrap) rubber
Altholz *n* old growth
Altmaterial *n* used material
Altmetall *n* scrap metal
Altminute *f* minute
Altmungo *m (text)* mungo from old rags
Altöl *n* used (waste) oil
Altpapier *n* waste paper

Altpapieraufbereitungsanlage *f* waste paper dressing plant
Altsand *m* floor (used) sand *(moulding)*
Altsekunde *f* second
Altwolle *f* reused wool
Altwuchs *m* old growth
Alu *n s.* Aluminium
Aluminat *n* aluminate
aluminieren to calorize; to aluminize
Aluminium *n* aluminium *m*; *(Am)* aluminum
Aluminiumbronze *f* aluminium bronze
Aluminiumfolie *f* leaf aluminium, aluminium foil
aluminiumhaltig aluminiferous
Aluminiumhülle *f* aluminium can (cladding)
Aluminiumoxid *n* alumina, oxide of aluminium
Aluminiumoxidschneide *f* aluminium-oxide tool tip
Aluminiumspritzguß *m* aluminium-die-casting
Aluminiumtriäthyl *n* triethyl aluminium
Aluminiumwalzstraße *f* aluminium-rolling mill
Aluminiumwalzwerk *n* aluminium-rolling mill
Aluminothermie *f* aluminothermic process
aluminothermisch aluminothermic
Aluminothermitschweißung *f* alumino-thermic welding
Amalgam *n* amalgam
Amalgamation *f* amalgamation
Amalgambad *n* amalgamating solution
amalgamieren to amalgamate
Amalgamzersetzer *m* denuder
Amateur *m* amateur, fan
Amateurastrokamera *f* amateur astro-camera
Amateurband *n* amateur [frequency] band
Amateurfernrohr *n* amateur telescope
Amateurfilm *m* amateur film
Amateurfrequenzband *n* amateur [frequency] band
Amateurfunker *m* amateur operator
Amateursender *m* amateur (home-made) transmitter
Amateurspiegelteleskop *n* amateur reflecting telescope
Amboß *m* anvil
Amboßbahn *f* anvil face
Amboßkontakt *m (el)* buffer contact
Ambrettenmoschus *m* musk ambrette
Ameisensäure *f* formic acid
Amerizium *n* americium
Amethyst *m* amethyst
Amici-Prisma *n* Amici (direct-vision) prism
Amid *n* amide
Amidogruppe *f* amido group
Amin *n* amine
Aminoessigsäure *f* glycine, glycocoll
Aminogruppe *f* amino group
Aminophenol *n* aminophenol
Aminoplast *m* amino plastic
Aminosäure *f* amino acid
 essentielle A. essential amino acid
Ammeter *n* ammeter, ampere meter
Ammoniak *n* ammonia
Ammoniakabtreiber *m* ammonia still
Ammoniakanlage *f* ammonia plant
Ammoniakauslaßventil *n* ammonia outlet valve
Ammoniakdampf *m* ammonia fume (vapour)
Ammoniakentfernung *f* separating of ammonia
Ammoniakfabrik *f* ammonia plant
Ammoniakkühlschlange *f* ammonia refrigerating coil
Ammoniaksynthese *f* ammonia synthesis, Haber process
Ammoniaksynthesegas *n* ammonia synthesis gas
Ammoniakverarbeitung *f* working-up of ammonia
Ammoniakverdichter *m* ammonia compressor
Ammoniakwäscher *m* ammonia scrubber (washer)
Ammoniakwasser *n* gas liquor, ammoniacal water
Ammonium *n* ammonium
Ammoniumbase *f* ammonium base
Ammoniumchlorid *n* chloride of ammonium, salmiac
Ammoniumsuperphosphat *n* ammoniated superphosphate
amorph amorphous
Ampere *n* ampere
Amperemeter *n* ampere meter, ammeter
 A. mit Thermoumformer thermocouple ammeter
 aperiodisches A. dead-beat ammeter
 elektromagnetisches A. moving-iron ammeter
Amperesekunde *f* ampere-second
Amperestunde *f* ampere-hour
Amperestundenzähler *m* ampere-hour meter
Amperewindung *f* ampere-turn
Amperewindungszahl *f* number of ampere-turns
Amperezahl *f* amperage
Amperometrie *f* amperometry
Amphibienflugzeug *n* amphibian plane
Amphibienpanzer *m* amphibious tank
Amphibolit *m (min)* amphibolite
Ampholyt *m (ch)* ampholyte
amphoter amphoteric
Amplidyne *f* amplidyne generator
Amplitude *f* amplitude; *(math)* departure
 A. einer symmetrischen Wechselgröße amplitude of a symmetrical alternating quantity
 abnehmende A. decreasing amplitude
 konstante A. constant amplitude
 optimale A. optimum amplitude
Amplitudenabbildung *f* amplitude presentation
Amplitudenabhängigkeit *f* amplitude response
Amplitudenabnahme *f* decrement; reduction in amplitude

Amplitudenanalysator *m s.* Impulshöhenanalysator
Amplitudenänderung *f* amplitude variation
Amplitudenanstieg *m* amplitude increase
Amplitudenaufschaukelung *f* amplitude increase
Amplitudenaussteuerung *f* amplitude range
Amplitudenbedingung *f* amplitude relationship
Amplitudenbegrenzer *m* peak limiter, clipper
Amplitudenbegrenzung *f* amplitude limiting
  selbsttätige A. automatic amplitude control
Amplitudenbegrenzungskreis *m* clipper circuit
Amplitudencharakteristik *f* amplitude characteristic
Amplitudendichte *f* amplitude density
Amplitudenentzerrung *f* correction of amplitudes
Amplitudenfrequenzgang *m* amplitude response (of frequency characteristic)
Amplitudenfunktion *f* amplitude function
Amplitudenhub *m* amplitude swing
Amplitudenkennlinie *f* amplitude characteristic
Amplitudenmodulation *f* amplitude modulation
amplitudenmoduliert amplitude-modulated
Amplitudenobjekt *n* amplitude specimen
Amplitudenrand *m* gain margin
Amplitudenregelung *f* amplitude control
Amplitudenschnittfrequenz *f* gain-crossover frequency
Amplitudenschrift *f* amplitude trace
Amplitudenschwankung *f* amplitude fluctuation (swing)
Amplitudenschwund *m* amplitude fading
Amplitudensieb *n* amplitude separator (discriminator)
Amplitudensteuerung *f* amplitude control
Amplituden- und Phasengang *m* / **logarithmischer** *(cont)* Bode-diagram; log-magnitude and phase diagram
Amplituden- und Phasengitter *n (opt)* amplitude and phase grating
Amplitudenverhalten *n* amplitude response
Amplitudenverzerrung *f* amplitude distortion
Amplitudenvorrat *m* amplitude (gain) margin
Amplitudenwahl *f* amplitude selection
Amplitudenwelle *f* wave of the amplitude
Ampulle *f* ampoule, vial
Amt *n (telec)* office; station; exchange
  bemanntes A. attended station
  betriebsführendes A. controlling exchange
  ferngespeistes A. dependent station, *(Am)* auxiliary station
  fernspeisendes A. power-feeding station
  halbautomatisches A. semi-automatic exchange
  regelndes A. control station
  unbemanntes A. unattended station
Amtsanruf *m* exchange call
Amtsanrufschlüssel *m* exchange call key
Amtsbatterie *f* exchange battery
Amtsfreizeichen *n* dial[ling] tone
Amtskabel *n* office cable
Amtskennziffer *f* exchange code
Amtsleitung *f* [direct] exchange line
Amtsverbindungsleitung *f* junction line (circuit)
  ankommende A. incoming junction line
Amtswähler *m* junction selector
Amtszeichen *n* dial[ling] tone
Amwind-Segelfläche *f* all plain sail area
Amylalkohol *m* amyl alcohol
anaerob anaerobic
Anaglyph *m (opt)* anaglyph
Anaglyphenverfahren *n* anaglyphic method
anallagmatisch *(math)* anallagmatic
analog analogous, analogue, *(Am)* analog
Analog-Digital-Konverter *m* analogue-[to-]-digital converter
Analog-Digital-Umsetzer *m* analogue-[to-]-digital converter
Analog-Digital-Wandler *m* analogue-[to-]-digital converter
Analogaufzeichnung *f* analogue recording
Analogberechnung *f* analogue computation
Analogdaten *pl* analogue data
Analogfrequenzteiler *m* analogue frequency divider
Analogie *f* analogy
Analogiemodell *n* simulator
Analogierechenmaschine *f s.* Analogrechner
Analogierechner *m s.* Analogrechner
Analogon *n* analog[ue]
Analogrechenautomat *m s.* Analogrechner
Analogrechenmaschine *f s.* Analogrechner
Analogrechner *m* analogue computer (computing machine)
  repetierender A. repetitive analogue computer
Analogsichtgerät *n* analogue display unit
Analogvervielfacher *m* analogue multiplier
Analysator *m* analyzer
  magnetischer A. magnetic analyzer, analyzing magnet
Analyse *f* analysis
  anorganische A. inorganic analysis
  elektrolytische A. electro-analysis
  gravimetrische A. gravimetric analysis
  momentane A. instantaneous sampling
  organische A. organic analysis
  qualitative A. qualitative analysis
  quantitative A. quantitative analysis
  radiometrische A. radiometric analysis
  schlußfolgernde A. inferential analysis
  spektrografische A. spectrographic analysis
  technisch-wirtschaftliche A. cost and method analysis

thermische A. thermoanalysis, thermal analysis
volumetrische A. volumetric analysis
Analysenbefund *m* analytical result
Analysenbereich *m* analyzing range
Analysenmeßgerät *n* analyzer, *(Am)* end point analyzer
Analysenmeßtechnik *f* chemical composition measurements
Analysenprobe *f* analytical sample
Analysentrichter *m* analysis funnel
Analysenwaage *f* chemical balance
analysierbar analyzable
analysieren to analyze
Analysis *f* mathematical analysis
höhere A. calculus
Analytiker *m* analyst, analytical chemist
analytisch analytical
Anamorphot *m* anamorphote lens
anamorphotisch anamorphote, anamorphotic
Anästhetikum *n* anesthetic
anästhetisch anesthetic
Anastigmasie *f* anastigmatism
Anastigmat *m* anastigmat
anastigmatisch anastigmatic
Anastigmatismus *m* anastigmatism
Anbau *m* 1. *(build)* extension; annex; addition; attachment; 2. *(ag)* growing; cultivation
Anbaublock *m* attachment support
Anbaudrehpflug *m* mounted one-way plough
Anbaudrillmaschine *f* tractor-mounted drill
anbauen 1. to extend [a building], to build an extension; 2. *(ag)* to grow; to cultivate; 3. to attach; to fit
Anbaufläche *f (ag)* cultivated acreage
Anbauflansch *m* mounting flange
Anbaugerät *n (ag)* [tractor-]mounted implement; *(build)* attachment
Anbaugrubber *m* tractor cultivator
Anbauhobel *m (mine)* attachment plough
Anbaukopfstück *n* attachment head piece
Anbaumähwerk *n* mounted mower
Anbaumotor *m* flanged (shell-type) motor
Anbaunebelmaschine *f* mounted fogger
Anbaupflug *m* mounted plough
Anbauplan *m* cropping plan
Anbauraupe *f* semi-tracklaying tractor
Anbaurechen *m* tractor-mounted hay rake
Anbauspritze *f (ag)* tractor-mounted spray
Anbausystem *n* suspension system
Anbauvielfachgerät *n* mounted toolbar
Anblasen *n* blowing-in *(blast furnace)*
anblatten to halve *(carpentry)*
anbohren 1. to start a hole, to spot-drill *(metal)*; 2. to scuttle *(vessel)*; 3. *(mine)* to collar
Anbohren *n* 1. spot-drilling, starting the (a) hole *(metal)*; 2. scuttling *(vessel)*; 3. *(mine)* collaring
Anbohrer *m* spotting drill
Anbohrkopf *m* trepanning [bar cutter] head
Anbohrmaschine *f* centring lathe
anböschen to slope
anbrennen to light; to kindle
anbringen to fix, to fit; to attach; to mount
gelenkig a. to mount pivotically
Anbringen *n* **des Bestecks** *(ship)* fixing positions
Andalusit *m (min)* andalusite
andauern to continue; to be permanent[ly present]
ändern to change, to alter; to vary; to rectify; to transform; to modify
die Frequenz ä. to change (alter) the frequency
Kurs ä. nach (auf) to alter (change) the course to
maßstäblich ä. to scale
stufenlos ä. to vary steplessly
Vorzeichen ä. to change sign
Anderthalbdecker *m* sesquiplane
Anderthalbseitenbandverfahren *n (telev)* sesqui-side-band system
Änderung *f* change; alteration; variation; transformation; modification
Ä. der Aufschlußzahl change in lignin content *(papermaking)*
Ä. der Größenordnung scaling
Ä. der Lage des Stößels *(mach)* alteration of the position of the ram
Ä. der Steigung *(math)* flexion
Ä. des Leitungsweges route deviation
Ä. des Reiseweges change of route
schrittweise Ä. gradual variation
thermodynamische Ä. thermodynamic transformation
umgekehrte Ä. inverse transformation
Änderungsgeschwindigkeit *f* rate of change
andicken to thicken
Andicken *n* thickening
Andradit *m (min)* andradite
Andrehapparat *m (text)* twisting-in frame
andrehen 1. to crank; to start an engine; 2. to join by twisting *(spinning mill)*; 3. to switch on
Andrehen *n* 1. cranking; 2. *(text)* twisting-in; 3. switching-on
A. der Kette *(text)* chain twisting
Andreher[in *f***]** *m (text)* twister; piecer *(spinning mill)*
Andrehklaue *f* cranking jaw, starting dog
Andrehkurbel *f* crank; crank handle, starting crank
Andrehkurbelarm *m* starting crank arm
Andrehkurbelgriff *m* starting crank handle
Andrehkurbelklaue *f* starting crank dog
Andrehkurbelwelle *f* starting crankshaft
Andrehmotor *m (el.m.)* pony motor
Andruck *m* 1. *(print)* proof; 2. *(mine)* thrust; pressure
andrucken to proof

andrücken an to force against; to press on
Andruckmagnet *m* pressure solenoid
Andruckpresse *f (print)* proof-press
Andruckwalze *f* back pressure roller
aneinanderfügen to join (fit) together
  stumpf a. to butt-joint
aneinanderhaften to cling together
aneinanderpassend mating *(surfaces)*
Aneinanderreiben *n* **der Schnüre** *(text)* chafing of the cords
Anemograf *m* anemograph
Anemologie *f* anemology
Anemometer *n* anemometer, wind gauge (speed indicator)
Anerkennung *f* certification; recognition; acknowledgement; acceptance
Aneroidbarometer *n* aneroid barometer
Anethol *n* anethol, anis camphor
Aneurin *n* aneurin[e], thiamin[e]
anfachen to excite *(oscillations)*
Anfachung *f* excitation *(of oscillation)*
Anfahrbeschleunigung *f* [starting] acceleration
Anfahrdrehmoment *n* starting torque
anfahren 1. to start *(machine)*; to start up *(a new plant)*; to take off *(vehicle)*; 2. *(mine)* to approach, to meet *(man winding)*
Anfahren *n* starting *(machine)*; start-up *(of reactor)*; take-off *(vehicle)*
Anfahrendgeschwindigkeit *f* speed at end of rheostatic starting period
Anfahrleistung *f* starting power
Anfahrperiode *f* start-up period *(of reactor)*
Anfahrschwierigkeit *f* launching difficulty
Anfahrstellung *f* starting position
Anfahrstrecke *f* starting distance
Anfahrstrom *m* starting current
Anfahrstück *n* dummy bar
Anfahrstufe *f* preliminary stage
Anfahrt *f* 1. conveyance; transport; 2. approach; *(mine)* descent; 3. start; take-off *(vehicle)*
Anfahrverlust *m* accelerating loss
Anfahrweg *m* starting distance
Anfahrwiderstand *m* starting resistance
Anfahrwirbel *m* starting (initial) vortex
Anfahrzeit *f* preoperation period, initiation time
Anfahrzugkraft *f* starting force
Anfall *m* yield; output
Anfälligkeit *f* **für Korrosion** corrodibility
Anfang *m* start, beginning; origin; initiation
anfangen to begin, to start; to initiate
Anfänger *m* 1. beginner; learner; 2. *(mine)* starter *(drill bits)*
anfänglich initial[ly]
Anfangsaktivität *f* initial activity
Anfangsanreicherung *f (nucl)* initial enrichment
Anfangsbahn *f (rock)* initial trajectory
Anfangsbedingung *f* initial condition
Anfangsbelastung *f* initial loading
Anfangsboje *f* key-buoy
Anfangsdruck *m* initial pressure; admission pressure
Anfangsdurchmesser *m (mine)* starting diameter *(bore hole)*
Anfangserhärtung *f* initial hardening
Anfangserregungsgeschwindigkeit *f (el)* initial voltage response of an exciter
Anfangsfestigkeit *f* initial strength
Anfangsförderung *f* initial rating of well
Anfangsgehalt *n* initial salary
Anfangsgeschwindigkeit *f* initial velocity (speed)
Anfangskapazität *f (el)* initial capacity
Anfangskonzentration *f (nucl)* initial concentration; initial enrichment; input concentration
Anfangsleistung *f* initial rating
Anfangsmetazentrum *n* initial metacentre
Anfangspermeabilität *f* magnetic initial permeability
Anfangsreihe *f* starting course *(knitting)*
Anfangsrichtung *f* initial direction
Anfangsspannung *f* 1. *(el)* initial voltage; 2. initial stress *(statics)*
  entgegengesetzte A. *(el)* initial inverse voltage
Anfangsstabilität *f* initial stability
Anfangsstrom *m* initial current
Anfangstemperatur *f* initial temperature
Anfangstonne *f* key-buoy
Anfangstraglast *f* setting load *(support)*
Anfangsvorspannung *f* initial bias
Anfangswert *m* initial value
Anfangswirbel *m (rock)* initial (starting) vortex
Anfangszustand *m* initial state
Anfärbbarkeit *f* affinity for dyes (dyestuffs), dye affinity
Anfärben *n* **des Papierstoffes** dyeing of the stock
Anfärbevermögen *n* colouring strength
Anfärbung *f* **/ egale** *(text)* dyeing in equal depth
anfasen to chamfer; to debur
Anfasen *n* chamfering; deburring
Anfeilhärteprüfung *f* file scratch test, file hardness testing
anfetten to grease
anfeuchten 1. *(text)* to damp; 2. *(mine)* to moisten, to sprinkle; 3. to temper *(sand)*
Anfeuchtmaschine *f* damping machine
Anfeuchtwalze *f* damping roller
Anfeuern *n* **des Kessels** firing-up the boiler
anflächen to spot face *(metal cutting)*
anflanschen to flange
anfliegen to approach; to land
Anflug *m* approach
  hindernisfreier A. unobstructed approach
Anflugbefeuerungsanlage *f* approach lighting system
Anflugbeginn *m* initial approach
Anflugfeuer *n* approach light

Anflugfrequenz *f* approach frequency
Anfluggrundlinie *f* extended centre line, principal course of approach
Anflughilfe *f* approach aid
Anflugrichtungsfeuer *n* approach light
Anflugrichtungsfunkbake *f* runway-localizing beacon
anfordern to require; to ask for; to claim; to order
Anforderung *f* **von Leerwagen** requisition for empty wagons
anfressen to pit; to corrode
Anfressen *n* **der Fasen des Bohrers** grabbing, binding, hooking into the work, seizing
Anfressung *f* corrosion; erosion
Anfuhr *f* cartage; delivery; supply
A. der Kohlen conveyance of coal
anfüllen to fill; to charge; to prime *(pump)*
Angabe *f* information; indication; specification
A. der Leistung capacity rating
detaillierte A. specification
Angaben *fpl* data; specification
fehlende A. missed values
angeben to state; to indicate; to specify; to rate *(capacity)*
angebogen closed *(spring)*
Angebot *n* **und Nachfrage** *f* supply and demand
angefault rotten
angeflanscht flange-mounted, flanged
angefräst milled
angegossen cast-on, cast integral
Angel *f* 1. hinge; pivot; 2. tang *(file)*; 3. fishing rod
angelassen / blau blue-finished *(steel)*
angelegt applied *(voltage)*
angelernt semi-skilled *(worker)*
Angelhaken *m* fish hook
Angelleine *f* fishing line
Angelleinenfahrzeug *n (ship)* hooker
angeln to fish, to hook, to catch fish by hook
Angelrute *f* fishing rod
angemacht ready[-made]
angemeldet filed; pending *(patent)*
angeordnet / drehbar pivoted
feststehend a. [arranged] stationary
sinnfällig a. *(mach)* directional *(controls)*
angerben *(leath)* to set the grain
Angerbfleck *m (leath)* tanning stain
angereichert enriched *(melt)*
angerostet slightly rusty
Angerufener *m (telec)* called party
angeschlossen connected
gelenkig a. pin-jointed, pin-connected, hinged
Angestellter *m* [salaried] employee
angesteuert / durch Sprache *(el)* speech-excited
angestoßen werden to receive impulses *(computer)*
angetrieben / durch Motor motor-driven
angewandt applied; practical
Angewende *n (ag)* headland
angewendet auf applied to
angezapft / in der Mitte *(el)* centre-tapped
angießen to cast on
angleichen an to match to
Angorawolle *f* Angora wool
Angoraziegenhaar *n* mohair
angreifen to attack, to affect; to pit; to take hold *(rolling)*; to act *(force)*; to engage *(tool)*; to corrode
chemisch a. to attack chemically
angrenzen to abut; to border [on]; to join
angrenzend *(build)* abutting; bordering [on]; joining; adjacent
Angriff *m* 1. attack; 2. application *(of a force)*
A. durch Korrosion attack by corrosion
chemischer A. attack by chemical action
Angriffspunkt *m* **einer Kraft** point of application of a force
Angriffswinkel *m* angle of attack *(coal cutting)*
Anguß *m* runner *(foundry)*
Angußfarbe *f* engobe *(ceramics)*
Angußmasse *f* engobe *(ceramics)*
anhaftend adherent, adhesive
anhaken 1. to hook on; 2. to mark off; to check
Anhaltdauer *f* **des Tisches** *(mach)* period of table dwell
Anhaltebremsung *f* stopping brake
anhalten 1. to stop; to stay; to bring to rest; 2. to continue
Anhalten *n* stopping, stoppage; shut-down *(reactor)*
anhaltend continuous
Anhaltevorrichtung *f* stopping device *(automatic recording)*
Anhang *m* appendix
A. zu einem Gesetz schedule to a law
Anhänge *mpl (ship)* appendages *(shaft bossings, bilge keel, etc.)*
Anhängeetikett *n* tie-on label, tag
Anhängefeldspritze *f (ag)* field crop trailed sprayer
Anhängegerät *n* trailed implement
Anhängelast *f (aut)* towed load
anhängen *(math)* to affix *(index)*; to annex; to join; to trail *(vehicle)*
den Hörer a. *(telec)* to put back, to restore *(receiver)*
Anhängepflug *m* trailer plough
Anhänger *m* 1. trailer *(vehicle)*; 2. *(text)* tag
A. für Bootstransport boat trailer (carrier)
A. für Kabeltransport *(Am)* reel cable trailer
schleppergezogener A. tractor trailer
Anhängerbremskraftregler *m (aut)* trailer-brake pressure regulator
Anhängerbremsventil *n* trailer brake-valve

Anhängerkupplung *f* / **selbsttätige** automatic trailer coupling
Anhängerschiff *n* trailer ship
Anhängeschiene *f* drag beam
Anhängeschild *n* tag
Anhängespritze *f (ag)* tractor-drawn spray
Anhängestraßenaufreißer *m* towed-type road ripper, trailer-type road ripper
Anhängestraßenhobel *m* towed-type grader, trailer-type grader
Anhängestraßenwalze *f* towed-type road roller, trailer-type road roller
Anhängevibrationswalze *f* towed vibratory roller
Anhängevorrichtung *f* hitch
Anhängewagen *m* trailer
anharmonisch non-harmonic
anhauen *(mine)* to open *(seam)*
Anhauen *n (ag)* preliminary mowing
Anhäufeln *n (ag)* ridging, *(Am)* hilling up
anhäufen / sich to accumulate, to pile up
Anhäufung *f* accumulation, aggregation; cluster; agglomeration, conglomeration
Anhäufungszeichen *n (telec)* group busy signal
Anhebeeinrichtung *f* elevator *(broaching)*; jack *(vehicle)*
anheben to elevate; to lift; to raise; to jack *(vehicle)*; to hoist; *(nucl)* to rise, to withdraw, to draw out, to move out *(rods)*; *(el)* to boost
Anhebeschlitten *m* elevating (broach-handling) slide
Anhebevorrichtung *f* 1. lifting (elevating) device; hoisting device; jack; 2. *(therm)* easing (testing) lever
anheften an to pin to
anheizen to warm up; to heat up; to set fire
Anheizen *n* warm-up; heating-up
Anheizstrom *m (el)* stand-by current
Anheizverzögerung *f* warming-up delay
Anheizzeit *f (el)* heating-up period, warm-up period (time), heating time
anheuern to hire, to sign, to engage *(sailors)*
Anhieb *m (mine)* first cut
Anhöhe *f* rise
Anhub *m* first lift; fetching
Anhydrid *n* anhydride
Anhydrit *m (min)* anhydrite; cube spar
Anilin *n* aniline, aminobenzene
Anilinblau *n* aniline blue, azuline
Anilindruck *m* flexographic (aniline) printing
Anilindruckmaschine *f* flexographic [printing] press
Anilineindruckwerk *n* aniline imprinting unit
Anilinfarbstoff *m* aniline dyestuff
Anilinrot *n* aniline red
Anilinschwarz *n* aniline black
Anilinviolett *n* aniline violet
animalisieren to animalize *(fibres)*
Animalisierung *f* animalization *(fibres)*
Animpfen *n* inoculation; seeding of crystallization
Anion *n* anion
Anionenaustausch *m* anion exchange
Anionenaustauscher *m* anionite
Anis *m* anise
Anisaldehyd *m* aubepine, anisic aldehyde
Anisoklinalfalte *f (geol)* anisoclinal fold
Anisöl *n* aniseed oil, oil of anise (aniseed)
anisomer anisomeric
anisometrisch anisometric
anisotrop anisotropic
Anisotropie *f* anisotropy
  A. im Gitterbau anisotropy due to crystalline lattice structure
Anisotropieanteile *mpl* anisotropic components
Anissäure *f* anisic acid
Ankaufspreis *m* cost price, prime cost
ankeimen to chit, to sprout
Anker *m* 1. anchor *(vessel)*; 2. armature *(motor)*; 3. guy *(pole)*; stay *(boiler)*; tie; *(mine)* rock bolt; roof bolt
  A. mit Hammerkopf foundation bolt with hammerhead
  A. mit Keilverschluß foundation bolt secured with a cotter
  A. mit Stabwicklung bar-wound armature
  gefädelter A. tunnel-wound armature
  genuteter A. slotted core armature
  gepolter A. polarized armature
  gewickelter A. wire-wound armature
  glatter A. smooth core armature
  oberständiger A. overtype armature
  polarisierter A. polarized armature
  tragender A. riding anchor
  unklarer A. foul anchor
  unterständiger A. undertype armature
  vielpoliger A. multipole armature
  zweipoliger A. two-pole armature
Ankerabfall *m* release of the armature
Ankeranschlag *m (telec)* armature stop
Ankeraufwindebaum *m (ship)* fish boom
Ankerausbau *m (mine)* roof (strata, rock) bolting
Ankerbalken *m (mine)* anchor beam
Ankerball *m (ship)* black (anchor) ball
Ankerbandage *f* armature band
Ankerbaum *m (ship)* fish boom
Ankerblech *n* armature lamination
Ankerblechschnitte *mpl* armature punchings
Ankerblock *m* dead man, stay block
Ankerbohrmaschine *f (mine) (Am)* roofbolter
Ankerbohr- und -einbringmaschine *f* roof pinner
Ankerbohrung *f* armature bore
Ankerboje *f* beacon, mooring (cable) buoy
Ankerbolzen *m* 1. anchor bolt; fishtail bolt; lewis bolt; holding-down bolt; 2. *(mine)* roof (rock) bolt

Ankerbuchse *f* armature hub
Ankerdraht *m* stay wire
Ankerdrehrichtung *f* sense of armature rotation
Ankerdruckplatte *f* armature end plate
Ankerfahne *f* armature end connection
Ankerfederlagerung *f* armature-reed support
Ankerfeld *n* armature field
Ankerflügel *m (ship)* fluke of the anchor, anchor fluke (blade)
Ankerformspule *f* diamond (former-wound) coil
Ankerfuß *m* stay block
Ankergebühr *f* groundage *(vessel)*
Ankergegenfeld *n* opposing magnetic field of armature
Ankergeschirr *n (ship)* ground tackle
Ankergrund *m* 1. *(ship)* holding (anchor) ground, anchorage *(vessel)*; 2. *(build)* roadstead, anchorage
schlechter A. *(ship)* foul ground (bottom), bad holding ground
Ankergut *n (ship)* ground tackle
Ankerhalteschraube *f* armature holding screw
Ankerhand *f (ship)* fluke of the anchor, anchor fluke (blade)
Ankerhohlwellenantrieb *m (el.m.)* hollow shaft motor drive
Ankerhub *m* armature stroke
Ankerit *m (min)* ankerite
Ankerkatt *n (ship)* cat tackle
Ankerkern *m* armature core
Ankerkette *f* [chain] cable, anchor cable
Ankerkettenendglied *n* anchor chain bitter ends
Ankerkettenhiever *m* cable jack
Ankerkettenkneifer *m* cable compressor
Ankerkettenpoller *m* chain bitt
Ankerkettenschäkel *m* joining shackle
Ankerkettenslipper *m* Senhouse slip
Ankerkettenstopper *m* slip (cable, chain, riding) stopper, riding chock, controller
Ankerklotz *m (el)* stay block
Ankerklüse *f (ship)* [anchor] hawse, hawse hole (pipe)
Ankerklüsenband *n* hawse-hook, hawse band
Ankerklüsenwulst *m* hawse bolster
Ankerkörper *m (el.m.)* armature body
Ankerkraftfluß *m* armature flux
Ankerkühlkanal *m* armature duct
Ankerkühlschlitz *m* armature duct
Ankerkurzschlußbremse *f* armature short circuiting brake
Ankerlager *n* armature bearing
Ankerlaterne *f* riding (anchor) lantern
Ankerleichter *m* mooring lighter, chain boat
Ankerleiter *m (el)* armature conductor
Ankerlicht *n* riding (anchor) light
Ankerloch *n* foundation bolt hole
Ankerlochbohrmaschine *f (mine) (Am)* roof bolter
Ankerlochbohr- und -einbringmaschine *f* roof pinner
Ankerluftspalt *m* armature gap
Ankermanöver *n* anchoring manoeuvre
Ankermischer *m* anchor agitator
Ankermutter *f* tie nut
ankern to anchor, to drop anchor
Ankernabe *f* armature hub (quill)
Ankernabenstern *m* armature spider
Ankernocken *m* boss (lug) for foundation bolt
Ankernut *f* armature slot
Ankerpaket *n* armature lamination
Ankerpfahl *m* stay block
Ankerpfeiler *m* anchorage pier
Ankerplatte *f* base shoe, anchor plate, foundation washer; foundation plate
Ankerplatz *m* [tying-up] berth, anchoring ground, place of mooring, anchorage
offener A. open berth
Ankerpol *m* armature pole
Ankerprellung *f (el.m.)* armature chatter
Ankerreaktanz *f (el.m.)* armature reactance
Ankerreibfeder *f (el)* anti-chatter armature reed
Ankerritzel *n* armature pinion
Ankerrückwirkung *f* armature reaction
Ankerrührer *m* horseshoe mixer
Ankerschaft *m (ship)* shank, anchor shank (head)
Ankerschäkel *m (ship)* anchor (bending) shackle, jew's harp
Ankerschaufel *f (ship)* fluke of the anchor, anchor blade
Ankerschiene *f* tie bar
Ankerschnur *f* mooring line
Ankerschraube *f* anchor bolt; fishtail bolt; lewis bolt; foundation bolt
Ankerseil *n* guy [cable], guy rope
Ankerslipper *m (ship)* tumbler
Ankerspanner *m* strainer
Ankerspannschloß *n* turnbuckle
Ankerspannschraube *f* stay tightener
Ankerspannung *f* armature voltage
Ankerspannungsgleichrichter *m* armature-voltage rectifier
Ankerspiel *n (el)* armature play
Ankerspill *n (ship)* [capstan] windlass
Ankerspule *f* armature coil
Ankerstab *m* 1. armature conductor *(motor)*; 2. *(build)* tie bar
Ankerstopper *m (ship)* riding stopper
Ankerstreufluß *m* armature stray flux
Ankerstreuung *f* armature leakage
Ankerstrom *m* armature current
Ankerstromkreis *m* armature circuit
Ankertasche *f (ship)* anchor pocket
Ankerwelle *f* armature shaft (spindle)
Ankerwicklung *f* armature winding
eingängige A. singly re-entrant armature winding

offene A. open-coil armature winding
Ankerwiderstand *m* armature resistance
Ankerwinde *f* windlass; gipsy, *(Am)* gypsy *(small winch)*
Ankerzweig *m (el)* path [of armature winding]
anklammern to attach; to fasten with cramps
ankleben to agglutinate; to glue on; to paste on
Ankleidezimmer *n* dressing room
anklemmen *(el)* to secure under terminals; to clamp
Anklopfmaschine *f (leath)* pounding-up machine
anknipsen to snap on; to switch on
anknoten to knot
anknüpfen to tie [piece] ends
Anknüpfmaschine *f (text)* tying-in machine, knotter
ankohlen to char
ankommen 1. to arrive; 2. to come in *(call)*
ankommend incoming *(call)*
Ankoppelkreis *m (el)* coupling circuit
ankoppeln to couple
Ankopplung *f* 1. *(el)* coupling; 2. probe-to-specimen contact *(materials testing)*
ankörnen to punch, to centre punch [scribed lines]
Ankörnen *n* [centre] punching
Ankreis *m (math)* excircle; escribed circle *(triangle)*
ankreuzen to check, to mark
Ankündigungsabstand *m* warning distance
Ankündigungssignal *n* warning signal
Ankunft *f* arrival *(at a station)*
Ankunftsbahnsteig *m* arrival platform
Ankunftsgleis *n* arrival track
Ankunftshafen *m* port of arrival (entry)
Ankupplung *f* coupling *(of the locomotive)*
Anlage *f* 1. plant; installation; lay; field; establishment; rig; equipment; 2. bearing surface; 3. building; erection
A. über Tage *(mine)* surface establishment (workings)
A. unter Tage *(mine)* underground establishment (workings)
A. von Freileitungen erection of overhead lines
elektrische A. electric plant (installation)
gefährdete A. exposed installation
großtechnische A. industrial plant
provisorische A. temporary installation
wetterfeste A. weatherproof installation
Anlagefläche *f* bearing face (surface); locating surface; attachment surface
A. der Dichtungsringe face of packing rings
Anlagekontrolleinrichtung *f* misfeed detector
Anlagekosten *pl* initial cost; cost of installation (construction, maintenance), purchasing cost
Anlagemaß *n* backfocal length *(photography)*
Anlagemittel *pl* investment fund
Anlagenausrüstung *f* systems instrumentation
Anlagenschema *n* flow diagram *(process control)*
Anlagenteile *npl* / **unterirdische** below-ground structure
anlagern to adsorb; to add; to trap; to capture
Anlagerung *f* trapping; capture; addition, additive action; cumulative action; adsorption *(of gas)*
A. von Bromwasserstoff hydrobromination
A. von Chlorwasserstoff hydrochlorination
A. von Elektronen electron attachment
Anlagerungsprozeß *m (nucl)* nuclear transmutation without emission of particles; addition reaction
Anlagerungsreaktion *f* combination (addition) reaction (*s.a.* Einfangreaktion)
Anlagerungsverbindung *f* addition compound
Anlagewinkel *m* feeding angle
Anlandebrücke *f* landing bridge
anlanden to land
Anlandeplatz *m* landing place, boat (ship) landing
Anlängen *n* **der Kette** *(text)* joining the warp threads
Anlaßakkumulator *m* starter battery
Anlaßbatterie *f* starter battery
Anlaßbeständigkeit *f (met)* retention of hardness
Anlaßdauer *f* duration of tempering *(steel)*
Anlaßdruckknopf *m* starter button
Anlaßeinspritzanlage *f* priming fuel injector
Anlaßeinspritzpumpe *f* primer pump
Anlaßeinspritzung *f* priming
anlassen 1. to start *(motor)*; to start up *(reactor)*; to crank; 2. to temper *(steel)*
a. mittels Handkurbel to start by manual cranking
Anlassen *n* 1. starting *(motor)*; 2. tempering, drawing *(steel)*
A. in Öl oil tempering *(steel)*
A. mit Hilfsphase capacitor starting
A. mit Spartransformator auto-transformer starting
A. mit Widerständen rheostatic starting
asynchrones A. asynchronous starting
Anlasser *m* starter *(engine; motor)*; starting motor; igniter
A. mit Tippbetrieb inching (jogging) starter
automatischer A. self-starter
Anlasserdruckknopf *m* starter push-button
Anlasserleitung *f* starter cable
Anlassermotor *m* pony motor
Anlasserzahnkranz *m* starter ring
Anlaßfarbe *f* 1. temper colour *(steel)*; 2. *(text)* graining colour
Anlaßgemisch *n* starting mixture
Anlaßhärte *f* retention of hardness, hot hardness *(steel)*
Anlaßhärtung *f* artificial aging

Anlaßhäufigkeit *f* starting abundance; starting frequency
Anlaßhebel *m (ship)* starting ram *(launching)*
Anlaßkondensator *m* starting capacitor
Anlaßkraftstoff *m* primer, priming fuel
Anlaßleistung *f* starting capacity
Anlaßluftflasche *f* starting-air cylinder; starting-air bottle
Anlaßmittel *n (met)* tempering medium
Anlaßmotor *m* cranking (starting) motor, starter
Anlaßofen *m (met)* tempering (drawing) furnace
Anlaßöl *n* tempering oil
Anlaßrelais *n* start[ing] relay
Anlaßschalter *m* starting switch; switch starter
Anlaßschaltung *f* starting connection
Anlaßschieber *m* starting gate valve
Anlaßschütz *n* accelerating contactor, contactor starter
Anlaßsprödigkeit *f (met)* temper brittleness
Anlaßstromstoß *m* starting impulse
Anlaßtemperatur *f (met)* tempering (drawing) temperature
Anlaßtransformator *m* compensator (transformer) starter, starting transformer
Anlaßventil *n* starting valve
Anlaßvorrichtung *f* starting gear
Anlaßwalze *f* drum starter
Anlaßwicklung *f* starting winding
Anlaßwiderstand *m* starting resistor, rheostat starter
Anlaßzündkerze *f* igniter plug
Anlaßzündspule *f* booster coil
Anlauf *m* 1. start[ing]; [moments after] start[ing]-up; starting from rest; 2. approach; 3. *(build)* batter
schneller A. *(el)* rapid starting
Anlaufbahn *f* inrun, approach *(ski-jumping)*
Anlaufbereich *m (nucl)* counter range
Anlaufbetrieb *m* starting operation
Anlaufdrehmoment *n* starting torque
anlaufen 1. *(ship)* to touch (call) at a port, to make [for a] port; 2. to be started up; to start; 3. to tarnish *(metal surface)*; 4. *(mine)* to start up *(face)*
a. lassen 1. to start; to get started; 2. to sull coat *(wire)*
Anlauffarbe *f* temper colour
Anlaufhafen *m* port of call
Anlaufherzstück *n* **der Weiche** run-over frog
Anlauflänge *f* first loading coil section
Anlaufmoment *n* starting torque
Anlaufperiode *f* warming-up period
Anlaufreibung *f* friction when starting
Anlaufschritt *m* start signal
Anlaufspannung *f* initial voltage
Anlaufspule *f* starting coil
Anlaufstrecke *f* starting distance
Anlaufstrom *m* initial velocity current; initial current flow; residual current *(tube)*
Anlaufstromgebiet *n* residual current region *(electron valve)*
Anlaufwiderstand *m* starting resistance *(of a valve)*
Anlaufzeitkonstante *f* acceleration constant
Anlaufzustand *m* residual current state
Anlegebrett *n* feed board
Anlegebrücke *f* landing pier, wharf quay
Anlegegoniometer *n* contact goniometer
Anlegekante *f* lay edge
Anlegemarke *f* feeding mark, lay mark
Anlegemaschine *f* spreader, spreading machine
anlegen 1. to accost, to come to shore, to go alongside, to land, to moor *(a ship into a port)*; to board *(another ship)*; 2. to set the course by the compass; 3. *(print)* to feed; 4. *(mine)* to bind
a. an *(el)* to tap on to
eine Spannung a. to introduce (impress, inject) a voltage, to apply a tension, to put (place) a voltage across
eine Spannung an einen Widerstand a. to put a voltage across a resistance
Anlegen *n* 1. application *(of a voltage)*; 2. binding *(miners)*; 3. *(text)* spreading; 4. *(print)* feeding
Anlegeplatz *m (ship)* landing place
Anleger *m (print)* feeder
Anlegerahmen *m* focal plane frame; register frame
Anlegerfloß *n s.* Anlegerponton
Anlegerponton *m* dummy barge, ferry bridge
Anlegestelle *f* tying-up berth, landing place
Anlegetisch *m (print)* feed table
Anlegewandler *m* split-core type transformer
Anlegewinkelmesser *m* bevel protractor
anleimen to glue on
Anleimmaschine *f* gluing machine
anleiten to instruct, to guide
Anleitung *f* instruction
Anlenkbolzen *m* knuckle pin
anlenken to hinge
anlernen to instruct
Anlerner *m* trainee
Anlernling *m s.* Anlerner
Anlernzeit *f* training time
anleuchten to illuminate, to light up
anliefern to supply, to deliver
Anlieferung *f* supply, delivery
anliegend adjacent
Anliegerstraße *f* service road
anlöten to solder to, to join by soldering
anlüften to lift *(tool)*
Anlüfthebel *m* easing (testing) lever
anmachen to mix *(concrete, mortar etc.)*
Anmachwasser *n (build)* mixing water
Anmähen *n* preliminary mowing

anmelden to book *(call)*; to register
ein Patent a. to apply for a patent
Anmeldezeit *f* filing time *(of calls)*
Anmeldung *f* **eines Fernsprechanschlusses** application for telephone service
anmustern to sign on, to engage, to enrol *(seamen)*
Annabergit *m (min)* annabergite, nickel bloom
annageln to nail
Annageln *n* nailing
annähen to sew [on]
Annähen *n* sewing[-on]
annähern to approach
Annähern *n* approach
annähernd approximate[ly]
Annäherung *f* approximation
A. von Hangendem und Liegendem convergence *(rock mechanics)*
Annäherungsbeleuchtung *f* approach signal lighting
Annäherungsverfahren *n* approximation (trial-and-error) method
Annäherungsverschluß *m* approach locking
Annahme *f* supposition, assumption; acceptance
Annahmebunker *m* intake hopper
Annahmekennlinie *f* operating characteristic
Annahmekontrolle *f* **durch Stichproben** acceptance sampling
Annahmekriterium *n* acceptance criterion
Annahmesperre *f* stopping of traffic
Annahmeverweigerung *f* refusal to take delivery
Annahmewahrscheinlichkeit *f* probability of acceptance
Annahmezahl *f* acceptance number
annehmen to assume; to accept
annieten to fix by riveting
Annihilation *f* annihilation
Annihilationsspektrum *n* annihilation spectrum
Annihilationsstrahlung *f* annihilation radiation
annullieren to cancel
Anode *f* anode, positive pole (terminal), *(Am)* plate
kugelförmige A. spherical anode
massive A. solid (heavy) anode
ringförmige A. annular anode
Anodenabstand *m* anode spacing
Anodenabstimmkreis *m* anode (plate) tuned circuit
Anodenanschluß *m* anode lead
Anodenarm *m* anode arm
Anodenausgleichsdrossel *f* anode balancing coil
Anodenbasisschaltung *f* cathode follower
Anodenbasisverstärker *m* anode base amplifier, cathode follower, cathode-coupled circuit
Anodenbatterie *f* anode battery, B-battery, high tension battery, plate supply (battery)
Anodenbelastung *f* anode dissipation, *(Am)* plate drop
zugelassene A. anode rating
Anodenbelastungswiderstand *m* anode load, *(Am)* plate load
Anodenbeschneidung *f* anode clipping
Anodenbleche *npl* anode sheet metals
Anodenbrumm *m* anode hum
Anodenbügel *m* anode clamp
Anodenclip *m* top cap connector
Anodendrossel *f* anode reactor
Anodendurchführung *f* anode leading-in wire
Anodendurchgriff *m* inverse amplification factor
Anodenentkoppelwiderstand *m* anode decoupling resistor
Anodenfall *m* anode drop (fall)
Anodenfleck *m* anode spot
Anodenflügel *m* anode fin
Anodengitterkapazität *f* anode grid-plate capacitance
Anodengleichrichter *m* anode [bend] detector, *(Am)* plate [power] detector
Anodengleichrichtung *f* anode [bend] detection; anode rectification
Anodengleichspannung *f* d-c anode voltage
Anodengleichstrom *m* d-c anode current
Anodenglimmlicht *n* anode glow
Anodenhülse *f* anode shield
Anodenkappenanschluß *m* top cap connector
Anodenkennlinie *f* anode bend
Anodenkennlinien[feld]schar *f* anode family
Anodenklemme *f* positive terminal
Anodenknick *m* anode bend
Anodenkorb *m* anode cage
Anodenkreis *m* anode circuit
abgestimmter A. tuned anode circuit
Anodenkreisspule *f* anode coil
Anodenkreistastung *f* anode circuit keying
Anodenleitung *f (Am)* plate conduction
doppelt herausgeführte A. double-ended plate
Anodenlicht *n* anode light
Anodenmodulation *f* anode modulation, *(Am)* plate modulation
Anodenplatte *f* anode plate
Anodenpolarisation *f* anodic polarization
Anodenreaktion *f* anodic reaction
Anodenrest *m* anode scrap
Anodenrohr *n* anode arm
Anodenrückkopplung *f* anode feedback
Anodenrückwirkung *f* anode reaction
Anodenruhestrom *m* feed (anode rest) current; steady anode current
Anodenschlamm *m* anode slime (mud)
Anodenschutznetz *n* anode guard net
Anodenschutzrohr *n* anode shield
Anodenschutzwiderstand *m* anode feed resistance, *(Am)* plate-dropping resistor

Anodenschwamm *m* anode sponge *(lead)*
Anodenschwingkreis *m* anode tank
Anodenschwingkreiskapazität *f* anode tank capacity
Anodenspannung *f* anode voltage
Anodenspannungsabblockkondensator *m* anode return by-pass
Anodenspannungsabfall *m* valve drop; anode potential drop
Anodenspannungsamplitude *f* anode voltage swing
Anodenspannungsgegenkopplung *f* anode voltage degeneration
Anodenspannungsmodulation *f* choke (constant current) modulation
Anodenspannungsquelle *f* anode voltage supply
Anodenspannungsschwankung *f* anode voltage fluctuation
Anodenspannungssiebkondensator *m* anode supply by-pass capacitor
Anodenspannungstastung *f* anode keying
Anodenspannungsumformer *m* anode converter
Anodensperrkreis *m* anode rejector circuit
Anodenspitzenspannung *f* peak anode voltage
Anodenspulenabgriff *m* anode tapping point
Anodenstrahl *m* anode ray
Anodenstrom *m* anode current
  maximaler A. maximum peak anode current
Anodenstrom-Anodenspannungskennlinie *f* anode-current-anode-voltage characteristic
Anodenstrom-Gitterspannungskennlinie *f* anode-current-grid-voltage characteristic
Anodenstrommodulation *f* anode current modulation
Anodenstromsparschaltung *f* anode current economy circuit
Anodenstromtastung *f* anode modulation
Anodenstromversorgung *f* anode power supply
Anodenstumpf *m* anode butt
Anodentastung *f* anode keying
Anodenverlust *m* / **größter** maximum anode dissipation
Anodenverlustleistung *f* anode dissipation, *(Am)* plate drop
Anodenverlustleistungsgrenze *f* anode dissipation limit
Anodenwechselspannung *f* a-c anode voltage
Anodenwiderstand *m* anode load, *(Am)* plate load; anode load resistor, *(Am)* plate load resistor
Anodenwinkel *m* target angle *(X-ray tube)*
Anodenwirkungsgrad *m* anode efficiency, *(Am)* plate efficiency
Anodenzerstäubung *f* anode sputtering
Anodenzündspannung *f* anode breakdown voltage
Anodenzündung *f* anode firing
anodisch anodic
Anolyt *m* *(el)* anolyte
anomal anomalous
Anomalie *f* 1. *(math)* amplitude; argument; 2. *(ph)* anomaly
anordnen to arrange; to lay out; to specify; to dispose
  gemeinsam a. to group
Anordnung *f* arrangement; lay-out *(machinery)*; order; array; configuration; assembly, assemblage, assembling *(of the fuel)*
  A. der Atome arrangement of [the] atoms, atomic arrangement (configuration)
  A. der Bauelemente component lay-out
  A. der Raketenstufen staging
  A. der Strickschlösser *(text)* camming
  bistabile A. bistable arrangement
  gemeinsame A. centralized grouping
  gemeinsame A. der Schaltelemente central control grouping
  geradlinige A. *(mine)* in-line system *(support)*
  kristalline A. crystalline array
  kritische A. *(nucl)* critical assembly (facility)
  logische A. logical configuration
  räumliche A. spatial arrangement; physical lay-out
  schwenkbare A. swivelling arrangement
  unterkritische A. *(nucl)* subcritical assembly (facility)
  zentrale A. centralized grouping
  zentrale A. der Schaltelemente central control grouping
  zufällige A. randomization
Anordnungsgütegrad *m* **des Propellers** relative rotative efficiency of the propeller
anorganisch inorganic
anormal abnormal
anoxydieren to preoxidize
Anpaßblindleitung *f s.* Anpaßstichleitung
anpassen to fit; to match; to adjust; to adapt *(a sub-program in digital computers)*
Anpaßkörper *m* slug *(waveguide)*
Anpaßstichleitung *f* matching stub *(aerials)*
Anpaßtransformator *m* adapting (matching, work head) transformer
Anpaßübertrager *m* matching transformer; balance converter *(aerial)*
Anpassung *f* matching; adaptation
Anpassungsblende *f* matching plate *(waveguide)*
Anpassungsdämpfung *f* matching (non-reflection) attenuation
anpassungsfähig flexible, adaptable
Anpassungsfähigkeit *f* flexibility; adaptability; compatibility
Anpassungsfehler *m* reflection coefficient (factor); mismatching
Anpassungsglied *n* *(el)* matching section
Anpassungskoeffizient *m* return current coefficient

Anpassungskreis *m* matching circuit
Anpassungsleitung *f* matching line (section, stub)
Anpassungsschaltung *f* matching connection
Anpassungsstab *m (nucl)* shim rod
Anpassungsstift *m* matching pillar
Anpassungstransformator *m* matching transformer
Anpassungsverhältnis *n (el)* impedance matching condition
Anpassungsvermögen *n* flexibility, adaptability
Anpassungswiderstand *m* matching pad (impedance)
Anpassungswirkungsgrad *m* matching efficiency
anpeilen to bear, to take a bearing
Anpeilung *f* bearing; taking a bearing; direction finding
Anpfahl *m (mine)* head board
Anpflanzung *f* plantation
anprallen to collide; to impinge
Anpreßbalken *m* clamping bar
Anpreßdruck *m* contact (surface) pressure
Anpreßkraft *f* contact force
Anpreßplatte *f* pressure plate *(clutch)*
Anpressungsdruck *m* contact pressure; coupling pressure
Anrauhung *f (build)* skid-proofing *(floor)*
Anregbarkeit *f* excitability
Anregelzeit *f* rise time
anregen to excite, to energize; to agitate; to stimulate; to power
Anreger *m* activator
Anregung *f* excitation; agitation; stimulation
  A. durch elektromagnetische Strahlen radiation excitation
  A. durch Licht light excitation
  A. eines Gases excitation of a gas
  A. hoher Intensität high intensity excitation
  fotoelektrische A. photoexcitation, photoelectric excitation
  konstante A. steady excitation
  schwache A. low-power excitation *(laser)*
  stufenweise A. step-by-step excitation *(of a gas)*
Anregungsband *n* excitation band
Anregungsbedingung *f* excitation condition
Anregungsenergie *f* excitation energy
Anregungsfähigkeit *f* excitability
Anregungsfrequenz *f* excitation frequency
Anregungsfunktion *f* excitation function
Anregungsimpuls *m* excitation pulse; pumping pulse *(of optical maser)*
Anregungskurve *f* excitation curve
Anregungsleuchten *n* excitation luminescence
Anregungsmaximum *n* excitation peak
Anregungsmittel *n* stimulant
Anregungspotential *n* excitation potential
Anregungsquerschnitt *m (nucl)* excitation cross section
Anregungsspannung *f* excitation potential (voltage), potential (voltage) of excitation
Anregungsstoß *m* excitation collision
Anregungswahrscheinlichkeit *f* probability of excitation
Anregungswelle *f* excitation wave
Anregungszahl *f* excitation number
Anregungszeit *f* excitation time
Anregungszustand *m* excitation (excited) state
anreiben to grind *(printing ink, dyestuff)*
anreichern *(met)* to concentrate; to enrich
  mit Kohlenstoff a. to carburize
  mit Sauerstoff a. to oxygenate
Anreicherung *f (met)* concentration, increasing the metal content of an ore; enrichment; accumulation; beneficiation
  grobe A. ragging
Anreicherungsanlage *f* enrichment plant
Anreicherungsfaktor *m* enrichment factor
Anreicherungsgebiet *n* accumulation region
Anreicherungsgrad *m* enrichment factor
Anreicherungshorizont *m (geol)* B-horizon, zone of concentration
Anreicherungsprozeß *m* accumulative process; concentration process
Anreicherungsschicht *f* enriched layer; barrier layer *(semiconductor)*
Anreißarbeit *f* marking-off operation
Anreißbezugsfläche *f* setting up datum
anreißen to mark; *(Am)* to lay out *(work in metal cutting)*; to scribe
Anreißen *n* marking; *(Am)* lay out *(work in metal cutting)*; scribing; lining out, setting out *(work)*
Anreißer *m* marker, liner, *(Am)* layout man
Anreißflüssigkeit *f* marking-out fluid, *(Am)* layout fluid
Anreißhaken *m (ship)* scrieve hook *(mould-loft tool)*
Anreißkörner *m* marking-out punch, *(Am)* layout punch; prick punch
Anreißnadel *f* scriber
Anreißplatte *f* marking-out table, marking-off table, surface (bench) plate, *(Am)* laying-out plate
Anreißverfahren *n* setting-out process
  optisches A. optical tracing method
Anreißwerkzeug *n* marking-off tool, *(Am)* layout tool
Anreißwinkel *m* flat square
Anreißzeug *n* marking-off tool, *(Am)* layout tool
Anreiz *m* incentive
  materieller A. incentive, material inducement, profit motive, stimuli
Anrichte *f* kitchen counter, sideboard
Anrichteraum *m (ship)* pantry
Anriß *m* 1. marking; 2. incipient crack *(fatigue)*; surface crack

Anrißlinie *f* scribed line
anrollen to taxi *(aeroplane)*
anrosten to corrode [by rusting]
anrösten to preroast
Anruf *m* call; telephone call; calling, *(Am)* ringing
  A. an alle general call
  A. im Fernverkehr trunk call
  drahtloser A. wireless call
  dringender A. express call
  gebührenfreier A. free call
  selektiver A. selective ringing
  unerledigter A. call on hand
Anrufanzeiger *m* call indicator
Anrufbestätigungszeichen *n* call-confirmation signal
Anrufbetrieb *m* signal junction working, *(Am)* straight forward trunking method
anrufen to call [up], to ring up
  einen Teilnehmer a. to call up a subscriber
Anrufer *m* calling party
Anruffeststeller *m* call detector *(telephony)*
Anruffrequenz *f* calling frequency
Anrufhilfsrelais *n* auxiliary calling relay
Anrufklappe *f* drop[-indicator] shutter, calling indicator (drop) *(telephony)*
Anruflampe *f* calling lamp, *(Am)* line lamp
Anrufordner *m* allotter, traffic distributor
Anrufsuchen *n* call finding (hunting)
Anrufsucher *m* call (line) finder
  erster A. *(Am)* primary line switch
Anrufverteiler *m* hunting switch, allotter for calls, call distributor
Anrufzeichen *n* call[ing] signal
Anrüsten *n* udder massage
Ansage *f* announcement *(radio)*
ansagen to announce *(radio)*
Ansager *m* announcer *(radio)*
ansammeln to accumulate
Ansammlung *f* accumulation
Ansatz *m* 1. shoulder *(turning)*; lug; tongue; neck *(shaft)*; 2. *(math)* statement, setup, ansatz; 3. *(met)* charge, batch, run; 4. scale *(of boiler)*
Ansatzbad *n* starting bath
Ansatzbohrung *f* stepped hole, hole with shoulder[s]
Ansatzdurchgangsbohrung *f* stepped through hole, through-hole with shoulder[s]
Ansätze *mpl* **eines Schachtofens** accretions of a shaft furnace
Ansatzkegel *m* tapered end
Ansatzpunkt *m* *(mine)* 1. point of collaring *(borehole)*, 2. placing
Ansatzstück *n* throat *(waveguide)*
ansäuern to acidify, to acidulate
Ansäuern *n s.* Ansäuerung
Ansäuerung *f* acidification, acidulation
Ansaugdruck *m* inlet pressure
ansaugen to prime, to draw in
Ansaugen *n* suction, induction *(engine)*; drawing-in
Ansauggeräuschdämpfer *m* intake silencer
Ansaughub *m* intake (induction) stroke
Ansaugkanal *m* suction port
Ansaugkrümmer *m* inlet elbow
Ansaugleistung *f* suction capacity
Ansaugleitung *f* inlet (induction) manifold; suction line (main)
Ansaugluft *f* induction air; inlet air
Ansaugluftfilter *n* induction air filter
Ansaugluftvorwärmer *m* carburetter-air heater
Ansaugrohr *n* intake (induction) pipe
Ansaugstutzen *m* intake connection (manifold), suction connection
Ansaugtakt *m* suction stroke
Ansaugung *f* suction
Ansaugventil *n* intake valve
Ansaugvolumen *n* intake volume
Anschaffungskosten *pl* prime (first, initial, installation) cost
Anschäften *n* grafting
anschalten *s.* einschalten
Anschaltklinke *f* operator's [telephone] jack
Anschaltleitung *f* transfer circuit
anschärfen to scarf; *(mine)* to sharpen *(drill bit)*
Anschärfen *n* scarfing; *(mine)* sharpening *(drill bit)*
Anschauungsmittel *n* demonstration material
anschirren *(text)* to warp; *(ag)* to harness; to yoke
Anschirrhaken *m (text)* twisting hook
Anschlag *m* 1. stop, trip [dog], dog *(movement of machine)*; 2. flanning *(masonry)*; 3. *(mine)* plat; 4. firing position *(shooting)*
  fester A. fixed (positive) stop
  A. kniend aufgelegt present kneeling with support
  A. kniend freihändig present kneeling without support
  A. liegend aufgelegt present lying with support
  A. liegend freihändig present lying without support
  selbstauslösender A. automatic stop
  A. sitzend present sitting
  A. stehend aufgelegt present standing with support
  A. stehend freihändig present standing without support
  verstellbarer A. adjustable dog
Anschlagbacke *f* fence
anschlagbetätigt dog-actuated
Anschlagbolzen *m* stop pin
Anschlagbühne *f* / **obere** *(mine)* apex; bank head *(braking incline)*

anschlagen *(mine)* to attach, to fix *(rope)*; to bounce (strike) against
Anschlagen *n* 1. beating-up *(weaving)*; 2. slinging
Anschläger *m (mine)* banksman, bellman, bottom cager, bottomer, cage tendent, cager, hanger-on, hitcher, lander, shifter
anschlaggesteuert dog-controlled
Anschlagkette *f* slinging chain, chain slings
Anschlagknagge *f* stop dog
Anschlagleiste *f* stop bar
A. zum Kopieren master template
Anschlagmittel *n* lifting tackle, hoisting gear
Anschlagnocke *f* dog
verstellbare A. adjustable trip dog
Anschlagreihe *f* commencing course *(knitting)*
Anschlagrolle *f* tappet roller
Anschlagschiene *f (ry)* guard rail
Anschlagwalze *f* multiple position stop
Anschlagwinkel *m* try square
anschleifen to grind; to resharpen
den Schaufelwinkel a. to give the blades an angle by grinding *(turbine)*
Anschleifer *m* rigger
anschließen to join, to connect
an die Steckdose a. to plug into connection
Anschliff *m* 1. grind; ground surface; 2. polished and etched surface; 3. resharpening *(tool)*
fasenartiger A. ridge
treppenförmiger A. des Schneidkopfes stepping of the drill point
Anschluß *m* 1. contact; connection; junction; 2. connector
A. an die Druckleitung delivery connection *(hydraulics)*
A. für zweiten Lautsprecher external speaker jack *(radio)*
A. von hinten rear connection
rückseitiger A. rear connection
verdrosselter A. impedance bond
Anschlußbahn *f* connecting railway
Anschlußbereich *m* exchange area *(telephone)*
Anschlußbereichsgrenze *f* exchange boundary
Anschlußbrett *n (el)* connection board
Anschlußbuchse *f* connecting socket
Anschlußdichte *f* subscriber line density
Anschlußdose *f* junction (connection) box, wall socket
Anschlußdraht *m* connecting (terminal) wire; pigtail *(lead)*
A. am Bauelement pigtail at the component
Anschlußdrähtchen *n* pigtail *(lead)*
Anschlußfahne *f (el)* lug; tag
Anschlußfeld *n* terminal board
Anschlußgebühr *f* private siding charges
Anschlußgerät *n* connecting set
Anschlußgleis *n (ry)* siding
Anschlußkabel *n* outlet (connection, connecting) cable; subscriber's (distribution) cable
Anschlußkasten *m* terminal box
Anschlußklemme *f* [connecting] terminal; binding post; outlet terminal
zugentlastete A. outrigger clamp
Anschlußklemmenleiste *f s.* Anschlußleiste
Anschlußkonsole *f* terminal bracket
Anschlußleiste *f (el)* connecting (connection) block, terminal strip
rückwärtige A. rear terminal block
Anschlußleitung *f* connecting line (wire)
Anschlußlitze *f* flexible pigtail
Anschlußmaß *n* assembly dimension
Anschlußmessung *f* **/ astronomische** field astronomical observation, astronomical measurement by conjunction
Anschlußmuffe *f (el)* cable jointing sleeve, junction box
Anschlußnetz *n* subscriber line network
Anschlußnummer *f* directory number
Anschlußort *m* 1. *(el)* wiring point, point of connection; 2. secondary port *(a port for which tidal differences from a standard port have been calculated and published in tide tables)*
Anschlußöse *f (el)* eyelet, *(Am)* terminal lug
Anschlußplan *m (el)* connection diagram, wiring layout (diagram)
Anschlußpunkt *m (el)* junction, point of connection, post
Anschlußrinne *f* ramp pan *(conveyor)*
Anschlußrohr *n* joining pipe
Anschlußschiene *f (el)* connecting (connection, junction) bar
Anschlußschnur *f (el)* flexible cord (lead), *(sl)* flex
A. für Fernsprechgehäuse instrument cord
Anschlußschraube *f* binding post
Anschlußstreifen *m* connecting strip
Anschlußstück *n* connecting piece; union, nipple *(pipe)*; junction block *(cable)*
Anschlußstutzen *m* connecting branch
Anschlußtemperatur *f* stud temperature *(semiconductor)*
Anschlußverteilung *f (el)* subdistribution
Anschlußwert *m (el)* connected (installed) load
Anschlußzug *m* connection train; boat train *(channel)*
anschmelzen to join (adhere) by casting
anschmieden to forge on
anschmiegen / sich einer Geraden to conform to a straight line
anschmoren to scorch
anschnallen to strap
Anschnallgurt *m* safety belt
anschneiden 1. *(mach)* to begin to cut, to enter the cut; to originate [a hole]; to start

the drill into the workpiece *(metal cutting)*; 2. *(mine)* to jib in *(seam)*

Anschnitt *m* 1. start of the cut *(metal cutting)*; 2. bevel lead, chamfer *(reamer)*; starting taper *(hand reamer)*; 3. [pouring] gate *(casting)*; 4. shelf *(civil engineering)*

doppelter A. double bevel *(reamer)*

kurzer A. slight bevel *(reamer)*

Anschnittlänge *f* bevel lead length, *(Am)* chamfer length *(reamer)*

Anschnittschneide *f* leading edge, cutting edge at the end, *(Am)* chamfer cutting edge *(reamer)*

Anschnittseite *f* approach side *(milling cutter)*; entering (starting) end *(broach)*

Anschnittweg *m* **des Fräsers** approach of cutter, length of the approach cut in milling

Anschnittwinkel *m* complementary angle of the bevel lead angle

Anschnittzahn *m* bumper tooth

anschnüren *(text)* to tie up

anschrauben to bolt to, to screw on

anschreiben *(math)* to escribe

anschuhen *(el)* to shoe *(poles)*

anschütten to fill

Anschüttung *f* fill[ing]

anschweißen to weld to

Anschwemmung *f (ch)* silting, washing, deposit

anschwingen *(el)* to start *(oscillator)*

Anschwingsteilheit *f* oscillation build-up transconductance

Anschwingstrom *m* transit state current, pre-oscillation (starting) current

Anschwingzeit *f* build-up period (time) *(oscillation)*

anschwöden *(leath)* to daub with ashes and lime

ansegeln to make (put into) a port

Ansegelungsboje *f* making buoy

Ansegelungsfeuer *n* making (landfall) light, entrance light of a harbour

Ansegelungstonne *f* sea (making) buoy, entrance buoy of a harbour

anseilen to rope

ansenken to spot-face

Ansenken *n* spot-facing

ansetzen 1. *(ch)* to prepare, to mix; 2. *(mine)* to collar *(borehole)*; to place; 3. to deposit; 4. to apply *(a spanner)*

zur Landung a. to come into land

Ansetzen *n* accretion *(crystals)*

A. von Bohrlöchern *(mine)* placing of holes

A. von Kristallen accretion of crystals

Ansetzkamera *f* attachment camera

Ansicht *f* view *(drawing)*, elevation

A. von hinten rear-side view, view looking forward

A. von innen looking outboard

Ansieden *n* immersion plating

anspannen to tighten; to stress; *(ag)* to harness

anspinnen to attach (join, piece) a thread

Ansprechbereich *m* range of sensitivity, sensitive region

Ansprechdauer *f (ph)* reacting duration; *(el)* response time [of relay], operating time, pre-arcing time *(fuse)*; transit time *(armature)*; *(Am)* answering time

Ansprechempfindlichkeit *f* response sensitivity

ansprechen *(el)* to react, to actuate; to respond; to pick up; to trip; to pull up *(magnet)*; to blow, to fuse

Ansprechen *n* response; actuating, operation *(relay)*

verzögertes A. delay-action

zufälliges A. random operation

ansprechfähig *(el)* responsive

Ansprechfrequenz *f* operating frequency

Ansprechgeschwindigkeit *f* speed of response, responding speed, rapidity of action

A. der Regelstrecke process reaction rate

Ansprechglied *n* current-responsive element

Ansprechpegel *m* operating level

Ansprechschwelle *f* threshold of sensitivity (response)

Ansprechsicherheitsfaktor *m* safety factor for pick-up

Ansprechspannung *f* minimum operating voltage, pick-up voltage, sealing voltage; response voltage

Ansprechspule *f* driving coil

Ansprechstrom *m* minimum operating (working) current, actuating (pick-up, release, sealing) current; fusing current

Ansprechvermögen *n* sensitivity, sensitiveness; efficiency

Ansprechverzögerung *f* impulse inertia *(of pulses)*; releasing delay *(of relay)*, delayed action; *(cont)* response latency

Ansprechverzug *m* make-time *(switch)*

Ansprechwahrscheinlichkeit *f* **[des Zählrohrs]** counter (counting) efficiency, efficiency of counter (counting)

Ansprechwert *m* minimum operating value, pick-up value

Ansprechzeit *f* 1. *(el)* operating (response) time; 2. time for putting into operation *(brakes)*

A. eines Zeichenempfängers receiver response time

Ansprechzeitverhältnis *n* speed ratio of a fuse

ansprengen to wring *(gauge block)*

anspülen to wash ashore

anstauchen to upset; to squeeze; to head *(bolt)*

anstechen to open the furnace

Anstecken *n (mine)* piling *(shaft sinking)*

gewöhnliches A. piling

schräges A. inclined piling

senkrechtes A. vertical piling

Ansteckhahn *m* tap *(beer)*
Ansteckraupe *f* semi-tracklaying tractor
Ansteckverfahren *n (mine)* piling
anstehend *(mine)* solid
Anstehendes *n (mine)* bedrock
ansteigen to rake, to slope, to slant; to rise *(temperature, pressure)*; to ascend
Anstellgeschwindigkeit *f* screw-down speed *(rolling)*
Anstellung *f* 1. screw-down movement *(rolling)*; 2. position, job, employment
Anstellventil *n* start valve
Anstellwalze *f* balanced roll
Anstellwinkel *m* angle of incidence; *(Am)* angle of attack *(aeroplane)*
  A. der Düse nozzle inclination (angle)
ansteuern / einen Hafen to make for a port
Ansteuerungsfeuer *n* making light, entrance light of a harbour; approach beacon
Ansteuerungsfunkfeuer *n* localizer
Ansteuerungstonne *f* entrance buoy of a harbour, making (approach) buoy; fairway (mid-channel) buoy
Anstichquerschnitt *m* leading pass section
Anstieg *m* rise *(pressure, temperature)*; slope *(electron tube)*
  A. einer Kurve slope of a curve
  exponentieller A. exponential rising
  kurzzeitiger A. transient rise
  unkontrollierter A. uncontrolled rise
Anstiegsantwort *f (dat)* ramp response
Anstiegsdauer *f* time of rise
Anstiegsflanke *f* forward edge [of a pulse]
Anstiegsfunktion *f (cont)* ramp function
Anstiegskurve *f* growth curve
Anstiegsspitze *f (el)* top end of ramp
Anstiegssteilheit *f* rate of rise
Anstiegszeit *f* rise (transition) time, time response (of rise)
  A. der Vorderflanke eines Impulses leading edge pulse time
  A. eines Signales build-up time
Anstiegwinkel *m* angle of climb
anstiften to peg
Anstoß *m* impulse; impact; impulsion
Anstoßatom *n* knocked-on atom
Anstoßelektron *n* impact electron
anstoßen *(build)* to butt, to abut; *(text)* to plank *(felt)*
anstoßend abutting; adjoining; impulsive *(force)*
Anstoßprozeß *m (nucl)* knock-on process; impact acceleration process
anstrahlen to [floot]light
anstreichen to paint, to coat [with paint]; to brush
Anstreicher *m* painter
Anstreifen *n* **der Schaufeln** touching of the blades *(turbine)*
Anstrich *m* coating of paint, paint coat; coating
  anwuchsverhindernder A. anti-fouling paint
  säurefester A. acid-resisting paint
Anstrichfläche *f* painting surface
Anstrichstoff *m* paint
Anstrichsystem *n* system of painting coats
Anstrichtechnik *f* painting technique[s]
Anstrichverfahren *n* brush method
anstricken to foot *(hose)*
Anströmgeschwindigkeit *f* local velocity of current, speed of advance
Anströmung *f* / **ungleichförmige** non-uniform inflow
Anströmungswinkel *m* angle of attack
anstücken *(text)* to piece [on]; to lengthen
Anstückung *f* lengthening
Anteil *m* portion; quota; share
Anteilsberechnung *f* division statement
Antenne *f* aerial, *(Am)* antenna; array *(aerial)*
  A. eindeutiger Peilrichtung sense aerial
  A. mit gestaffelten Strahlern echelon aerial
  A. mit schwenkbarer Charakteristik array with variable polar diagram
  A. mit starker Richtwirkung highly directional aerial
  A. zur Zielverfolgung und Datenaufnahme tracking and data acquisition aerial
  abgeschirmte A. screened aerial
  abgestimmte A. tuned aerial
  aperiodische A. aperiodic (untuned) aerial
  auf Oberwellen erregte A. harmonic[al] aerial
  belastete A. loaded aerial
  dachförmige A. roof-shaped aerial
  dielektrische A. dielectric aerial
  direkt gespeiste A. directly fed aerial
  einfache A. plain aerial
  eingebaute A. build-in aerial
  einseitig gerichtete A. unidirectional aerial
  einstellbare A. steerable aerial
  freihängende A. trailing aerial
  geräuscharme A. anti-interference aerial
  gerichtete A. directional aerial
  geschichtete A. laminated aerial
  geschlossene A. closed aerial
  gespeiste A. driven (active) aerial
  kapazitiv beschwerte A. top-capacitor aerial, top-loaded aerial
  künstliche A. artificial (phantom, dummy) aerial
  mechanisch bewegte A. mechanically nodded aerial
  mehrdrähtige A. multiple-wire aerial
  mittelbar gespeiste A. parasitic (indirectly fed, passive) aerial
  nachteffektfreie A. spaced frame aerial
  ortsveränderliche A. portable (mobile) aerial
  rundstrahlende A. omni-directional aerial

schwundmindernde A. fading reduction aerial, fade reducing aerial, anti-fading aerial
störungsarme A. anti-interference aerial
strahlungsgekoppelte A. passive (indirectly fed, parasitic) aerial
unbelastete A. unloaded aerial
ungerichtete A. equi-radial aerial; omni-directional aerial, non-directional aerial
verstimmte A. dumb (non-resonating) aerial
wirksame A. active aerial
zusammenlegbare A. collapsible aerial

Antennenabdeckung *f* radome
Antennenabgleichelement *n* earth equalizer conductor
Antennenabgleichmittel *n* aerial matching and compensating means
Antennenableitung *f* down lead
Antennenabschalter *m* lightning switch
Antennenabschwächer *m* aerial (strong signal) attenuator
Antennenabstimmittel *n* aerial tuning equipment
Antennenabstimmung *f* aerial tuning
Antennenanordnung *f* arrangement of the aerial
gebündelte A. stacked aerial (dipol array)
Antennenanpassungskreis *m* aerial matching (coupling) circuit
Antennenanschluß *m* aerial connection
Antennenaufhängung *f* aerial suspension
Antennenbauvorschriften *fpl* regulations pertaining to the installation of aerials
Antennenbuchse *f* aerial jack
Antennencharakteristik *f* radiation pattern (characteristic)
Antennendämpfung *f* aerial damping, damping of the aerial
Antennendraht *m* aerial wire
Antennendrehgeschwindigkeit *f* aerial rotation speed
Antennendurchführung *f* leading-through of the aerial
Antennendurchführungsisolator *m* down-lead insulator
Antennendurchhang *m* aerial dip
Antenneneffekt *m* [residual] aerial effect
Antenneneigenfrequenz *f* natural frequency of an aerial
Antenneneigenschwingung *f* natural frequency of an aerial
Antenneneinführung *f* aerial lead-in
Antenneneinführungsisolator *m* lead-in insulator
Antenneneingangswiderstand *m* aerial feed impedance
Antennenelement *n* elementary aerial
strahlungsgekoppeltes A. parasitic element
Antennenerdleitung *f* earth system of an aerial
Antennenersatzwiderstand *m* dummy aerial resistor
Antennenfläche *f* / **effektive** effective area of aerial
Antennenfuß *m* aerial base
Antennengewinn *m* aerial gain, gain of aerial
Antennengittermast *m* lattice mast; derrick-style aerial mast
Antennenhaspel *f* aerial reel (winch)
Antennenhöhe *f* / **wirksame** effective height of aerial
Antenneninduktivität *f* aerial inductance
Antennenkabel *n* aerial feeder
Antennenkapazität *f* aerial capacitance
Antennenkoppelleitung *f* aerial coupling feeder
Antennenkreis *m* aerial circuit
Antennenkreistastung *f* aerial circuit keying
Antennenleistung *f* aerial power
Antennenlitze *f* [stranded] aerial wire
Antennenmast *m* aerial mast
abgespannter A. guyed aerial mast
freistehender A. self-supporting aerial mast
freitragender A. self-supporting aerial mast

Antennenmastverstärker *m* masthead amplifier
Antennenmodulation *f* absorption modulation
Antennenmodulationsdrossel *f* high power choke modulator
Antennenneigungsanzeiger *m* aerial tilt indicator
Antennenöffnung *f* aerial aperture
Antennenrauschen *n* aerial noise
Antennenreihe *f* array [of aerials]
Antennenrichtdiagramm *n* aerial directivity pattern
Antennenrichtungsgewinn *m* directive gain
Antennenrolle *f* aerial reel
Antennenschalter *m* aerial [change-over] switch
Antennenscheinwiderstand *m* aerial impedance
Antennensockel *m* aerial base
Antennenspannung *f* aerial voltage
Antennenspeiseleitung *f* aerial feeder
Antennenspeisescheinwiderstand *m* aerial feed impedance
Antennenspeisung *f* aerial feeding
Antennenspiegel *m* aerial reflector
Antennenspule *f* aerial coil
Antennenstab *m* aerial rod
Antennenstrahlung *f* radiation of the aerial
Antennenstrom *m* aerial current
Antennenstützpunkt *m* aerial support
Antennensystem *n* aerial array
Antennentransformator *m* jigger
Antennenturm *m* aerial tower
Antennenüberspannungsschutz *m* aerial overload protection
Antennenübertrager *m* aerial transformer
Antennenumschalter *m* aerial change-over switch
Antennenverkürzungsfaktor *m* aerial shortening coefficient

Antennenverkürzungskondensator *m* aerial shortening capacitor
Antennenverlängerungsspule *f* aerial loading coil
Antennenverlust *m* aerial loss
Antennenverlustdämpfung *f* aerial loss damping
Antennenverstärker *m* aerial booster
Antennenverstärkereinschub *m* aerial amplifier insert
Antennenverstärkung *f* aerial gain
Antennenwand *f* curtain array of aerials
Antennenwirkung *f* aerial effect
Antennenwirkungsgrad *m* aerial efficiency
Antennenzug *m* aerial strain
Anthrachinon *n* anthraquinone
Anthrachinonderivat anthraquinone derivative
Anthrachinonfarbstoff *m* anthraquinone dyestuff
Anthrachinonküpenfarbstoff *m* anthraquinone vat dye[stuff]
Anthranilsäure *f* anthranilic acid
Anthrazen *n* anthracene
Anthrazenöl *n* anthracene oil
Anthrazenzählrohr *n (nucl)* anthracene counter
Anthrazit *m* anthracite [coal], hard coal
minderwertiger A. lambskin
Anthrazitbildung *f* anthracitization
Anthrazitgrus *m* / **gesiebter** barley, bird's eye
Anthrazitrohfeinkohle *f* culm coal
Antiabschirmungsfaktor *m* antishielding factor
Antiballonvorrichtung *f* / **halbringförmige** *(text)* open antiballooning device
Antibiotikum *n* antibiotic
Antichlor *n* antichlorine, sodium hyposulphite
Antidröhnmasse *f* antinoise compound
Antielektron *n* positron, anti-electron
Antiferroelektrika *npl* antiferroelectrics
Antiferroelektrizität *f* antiferroelectricity
Antiferromagnetikum *n* antiferromagnet
Antiferromagnetismus *m* antiferromagnetism
Anti-fouling-Anstrich *m* antifouling [paint]
Antifriktionsmetall *n* antifriction metal
Antihyperon *n (nucl)* anti-hyperon
Antikatalysator *m* anticatalyst
Antikatode *f* anticathode
Antikdruckpapier *n* antique printing paper
antiklastisch *(math)* anticlastic
Antiklopfmittel *n* antiknock component
Antiklopfzusatzmittel *n* antiknock fuel dope
Antikoinzidenz *f* anticoincidence
Antikoinzidenzanalysator *m* anticoincidence analyzer
Antikoinzidenzschaltung *f* anticoincidence circuit
Antikoinzidenzstufe *f* anticoincidence circuit
Antikoinzidenzzähler *m* anticoincidence counter
Antikörper *m* antibody, anti-enzyme
Antilichthofschicht *f* antihalation backing
Antiloch *n* antihole
Antilogarithmus *m (math)* antilogarithm
Antilopenleder *n* antelope leather
antimagnetisch non-magnetic
Antimaterie *f* anti-matter
Antimeson *n (nucl)* anti-meson
Antimon *n* antimony
Antimonat *n* antimonate
Antimonblei *n* antimonial (antimon[i]ous) lead, hard lead
Antimonblende *f (min)* pyrostibnite, kermesite
Antimonblüte *f* antimony bloom; valentinite
Antimonerz *n* antimony ore
Antimonfahlerz *n* tetrahedrite
Antimonglanz *m* antimonite, stibnite
Antimonit *m (min)* antimonite
Antimonmetall *n* metallic antimony
Antimonpentasulfid *n* antimony pentasulphide, antimonic sulphide
Antimonpentoxid *n* antimonic anhydride, antimony (stibic) pentoxide
Antimonsäure *f* antimonic acid
Antimonsilberblende *f* pyrargyrite
Antimontrioxid *n* antimonous anhydride (oxide), antimony trioxide
Antimon(III)-verbindung *f* antimonous (stibious) compound
Antimon(V)-verbindung *f* antimonic (stibic) compound
Antimonvergiftung *f* stibialism, antimony poisoning
Antimonwasserstoff *m* stibine, hydride of antimony
Antimonyl *n* antimonyl
Antineutrino *n* anti-neutrino
Antineutron *n (nucl)* antineutron
Antinukleon *n* antinucleon
Antioxydans *n s.* Antioxydationsmittel
Antioxydationsmittel *n* oxidation inhibitor, antioxidant agent
antiparallel anti-parallel, inversely parallel
Antiparallelkurbel *f* nonparallel equal crank
Antiparallelschaltung *f* anti-parallel connection (coupling)
Antipartikel *f* antiparticle
Antiphase *f* antiphase
Antiproton *n* anti-proton
Antipyrinsalizylat *n* salipyrine
Antiqua *f* Roman type
Antiraketenrakete *f* antimissile missile
Antireflexschutzschicht *f* antihalation backing
antireflexvergütet coated
Antiresonanz *f* antiresonance, parallel impedance
Antischaummittel *n* antifoaming agent
Antischleiermittel *n* anti-fog agent
Antischmiermittel *n* antilubricating agent
Antiseptikum *n* antiseptic
antiseptisch antiseptic, germicidal

Antiskorbutmittel *n* antiscorbutic
Antistatikgerät *n (nucl)* static eliminator
antisymmetrisch anti-symmetric
Antiteilchen *n* antiparticle
Antithese *f* antithesis
Antitranspirationsmittel *npl* antiperspirants
Antivalenzbahn *f* antibonding orbital
Antivalenzelektron *n* antibonding electron
Antivibrationssockel *m (el)* anti-vibration cap
Antrag *m* application
Antragsformular *n* application form
Antransport *m* supply, delivery
antreiben to drive
Antrieb *m* 1. drive *(machine)*; propulsion *(vessel; aeroplane)*; 2. *(ph)* impulse; change of momentum
  A. des Stellwerkes operating gear of the interlocking plant
  A. mit Kernenergie *s.* Kernenergieantrieb
  diesel-elektrischer A. oil-electric drive
  formschlüssiger A. positive drive
  hydraulischer A. hydraulic drive
  kraftschlüssiger A. non-positive drive
  turboelektrischer A. turbo-electric propulsion (drive)
  zwangsläufiger A. positive drive
Antriebsaggregat *n* drive unit
Antriebsanlage *f* propelling plant
Antriebsart *f* kind of drive; system of powering *(vessel)*
Antriebsbahn *f (rock)* powered-flight phase (trajectory), power phase, propulsion branch
Antriebsdampfmaschine *f* driving steam engine
Antriebsdauer *f (rock)* duration of thrust
Antriebsdrehmoment *n* driving torque
Antriebsdrehzahl *f* driving speed
Antriebselement *n* drive element
Antriebsflansch *m* drive flange
Antriebsflug *m (rock)* powered flight
Antriebsgelenk *n (aut)* universal joint
Antriebsgelenkgehäuse *n* universal-joint housing
Antriebsglied *n* driving (input) member, driver *(kinematics)*
Antriebshebel *m* actuating arm
Antriebskegelrad *n* bevel drive pinion
Antriebskopf *m* **mit Rädergetriebe** geared head *(machine tool)*
Antriebskraft *f* driving force
Antriebskupplung *f* driving clutch
Antriebsmaschinen *fpl* driving machinery
Antriebsmotor *m* driving motor
Antriebsorgan *n* drive element
Antriebsrad *n* driver, driving gear
  A. für den Schloßmantel cam driving gear *(knitting machine)*
Antriebsreaktor *m* propulsion reactor
Antriebsregelung *f* drive control
Antriebsritzel *n* driving pinion
Antriebsrolle *f* driving drum
Antriebsseil *n* driving string (rope); pulley cord
Antriebsseite *f* 1. *(mach)* driving end; 2. *(mine)* drivehead
  A. eines Generators back of a generator
Antriebsspindel *f* driving screw
Antriebsstation *f (mine)* driving drum, drivehead
Antriebsstirnrad *n* reduction spur pinion
Antriebsstrahl *m* propulsive jet
Antriebsteil *m* propulsion portion
Antriebstrommel *f* drive (driving) drum
Antriebswelle *f* main (drive) shaft
  A. mit Antriebszahnrad transmission shaft with pinion
Antriebswellenstumpf *m* drive-shaft spigot (stub), input shaft extension
Antriebswirkungsgrad *m* propulsive efficiency
Antrittstufe *f* bottom step
Antwort *f* reply; response
Antwortbake *f* responder beacon, racon, transponder; radar beacon
Antwortflagge *f* code pennant (flag)
Antwortgerät *n* radar beacon *(navigation)*
Antwortimpuls *m* reply (response) pulse
Antwortimpulsgenerator *m* reply pulse generator
Antwortklinke *f* answering jack
Antwortkode *m* reply code
Antwortsender *m* responder, radar beacon transmitter
Antwortstöpsel *m* answering plug
Antworttelegraf *m* repeating telegraph
Antwortwimpel *m* answering pennant
anvisieren to sight; to take aim
Anvulkanisation *f* scorching
anwachsen to increase [to], to rise, to swell; *(ag)* to grow to
anwählen *(telec)* to dial
Anwaltsgebühren *fpl* attorney's fees
anwärmen to heat up; to preheat
  eine Trockenkammer a. to warm up a kiln
Anwärmer *m* feed-water heater
Anwärmperiode *f* heating-up period *(of the steam engine)*
anwendbar applicable
  a. sein für to apply to
anwenden to apply
Anwendung *f* application
  A. des niedrigsten Tarifsatzes application of the lowest tariff
  friedliche A. der Kernenergie peaceful use of nuclear energy
Anwendungsbereich *m* field (range) of application; adaptability; score of work *(machine tool)*
Anwendungsgebiet *n* field of use (application)
Anwendungsmöglichkeit *f* applicability
Anwerfdrehmoment *n* cranking torque

Anwerfdrehzahl *f (aut)* cranking speed
anwerfen to crank
Anwerfen *n* cranking, hand swinging
Anwuchs *m* barnacles, scruff, marine fouling *(ship's hull)*
Anwurfmotor *m (el.m.)* non-starting motor
Anwurfschalter *m* motor-starting switch
Anzahl *f* number; quantity
  A. der Bildzeilen *(telev)* number of scanning lines
Anzapfdampf *m* bled steam
Anzapfdampfmaschine *f* bleeder-type steam engine
Anzapfdrossel *f (el)* tapped variable inductor; tapped choke
anzapfen to tap; to bleed
Anzapfschütz *n (el)* tapping contactor
Anzapfstelle *f (el)* tap connection
Anzapftransformator *m* split (tapped voltage) transformer
Anzapfung *f* tap[ping]
Anzapfungsumschalter *m* tap changer
Anzapfungsumschaltung *f* tap-changing
Anzapfwiderstand *m* tapped resistor
anzeichnen to mark
Anzeichnen *n* **/ optisches** optical marking
Anzeige *f* display; indication *(device)*; reading; response
  A. / mit automatischer self-indicating
  A. einer Geschwindigkeitsbeschränkung restricted speed aspect
  A. von beweglichen Zielen mit Festzielunterdrückung moving target indication *(radar)*
  direkte A. direct indication
  eindeutige A. unambiguous reading
  einmalige A. single-shot reading
  elektrische A. electrical indication
  horizontale A. in-line display *(electronic counters)*
  kontinuierliche A. continuous display
  lineare A. linear display
  mehrstellige A. multi-digit presentation
  numerische A. numerical indication
  senkrechte A. columnar read-out *(electronic counters)*
  sichtbare A. visual presentation
  waagerechte A. in-line read-out *(electronic counters)*
Anzeigeauflösung *f* read-out resolution
Anzeigebereich *m* indicating range
Anzeigeeinrichtung *f* display; read-out device; read-out circuit
Anzeigefehler *m* indication error
Anzeigefenster *n* indicator window
Anzeigefolge *f* display period *(electronic counters)*
Anzeigegerät *n* indicator, indicating instrument, display (presentation) unit
Anzeigeinstrument *n* indicating (recording) instrument, indicator; meter; reading instrument
Anzeigelampe *f* tell-tale light, warning lamp, indicating light; pilot bulb
anzeigen to indicate; to display; to register; to read
Anzeigenchiffre *f* box number
Anzeigenpapier *n* weddings
Anzeiger *m* indicator
Anzeigeröhre *f* visual indicator valve
Anzeigeschaltung *f* indicator circuit
Anzeigetafel *f* indicator panel
Anzeigeübermittlung *f* transmission of indication
Anzeigeunsicherheit *f* reading accuracy
Anzeigeverstärker *m* directly indicating amplifier
Anzeigevorrichtung *f* indicating device
anzetteln *(text)* to warp
anziehen 1. to attract; 2. to draw in; to tighten *(nut)*; to draw home; 3. to pull-up *(relay)*; 4. to set *(cement)*; 5. to put on *(brake)*; 6. to rise *(prices)*
  zu fest a. to overtighten
  zu locker a. to undertighten
  zu straff a. to overtighten
Anziehen *n* tightening [nuts]
Anziehung *f* attraction
  elektrische A. electrostatic attraction
  gegenseitige A. mutual attraction
Anziehungsbeschleunigung *f* acceleration due to attraction
Anziehungskraft *f* force of attraction
  magnetische A. magnetic pull (attraction)
Anziehungskräfte *fpl* **/ molekulare** forces of molecular attraction
Anziehungspotential *n* attraction potential
Anziehvorrichtung *f* tightening device
Anzug *m (build)* batter
Anzugschraube *f* draw-in bolt *(milling spindle)*
Anzugsdrehmoment *n* starting torque
Anzugsgewinde / mit with tapped shank; tapped-end
Anzugsmoment *n* starting (locked-rotor) torque
Anzugsstange *f* drawbar *(milling spindle)*
Anzugsstrom *m* pick-up (starting) current
Anzugsvermögen *n* get-away power
Anzugszeit *f* lap-time
anzünden to kindle; to light; to ignite
Apatit *m (min)* apatite
aperiodisch aperiodic; dead-beat
  a. ausschwingend dead-beat
  a. gedämpft dead-beat
  a. gesteuert aperiodically controlled
Aperiodizität *f* aperiodicity
Apertometer *n (opt)* apertometer
Apertur *f (opt)* aperture
  numerische A. numerical aperture
Aperturbegrenzung *f* aperture limitation

Aperturblende *f* aperture plate (stop, diaphragm)
Aperturkorrektur *f* aperture correction
Aperturstrahler *m* radiating aperture
apfelförmig pomiform
Äpfelsäure *f* hydroxysuccinic (malic) acid
Apfelsinenschaleneffekt *m (plast)* orange peel effect
Apfelsinenschalenöl *n* orange peel oil, essence of orange
Aplanat *m (opt)* aplanatic [objective]
aplanatisch *(opt)* aplanatic
Apochromat *m (n) (opt)* apochromatic [objective]
apochromatisch *(opt)* apochromatic
Apogäum *n* apogee
Apotheke *f* pharmacy, apothecary's shop; dispensary *(on board of a ship)*
Apotheker *m* pharmacist, pharmaceutical chemist, apothecary
Apothekergewicht *n* officinal weight
Apparat *m* apparatus; device; appliance; unit; set; instrument; accessory; equipment
  Kippscher A. Kipp's apparatus
Apparategabel *f (el)* switch-cradle, receiver rest
Apparateklemme *f (el)* binding post
Apparatfärberei *f (text)* machine dyeing
Apparatur *f* **zur Arsenbestimmung** arsenic apparatus
  A. zur Erforschung der oberen Schichten der Atmosphäre upper atmosphere instrumentation
Apparatwecker *m* station ringer
Apparatzuleitung *f* instrument lead
Appartement *n* apartment
Appartementhaus *n* apartment block (building)
Applikation *f* appliqué *(laid-on work)*
Applikatorenmethode *f* administration method
Appretbrechmaschine *f s.* Appreturbrechmaschine
appretieren *(text)* to finish; to dress; to season *(leather)*
Appretiermaschine *f (text)* dressing frame; seasoning machine *(leather)*
Appretierung *f (text)* finish[ing]; dressing
  einseitige A. back-filling finish
Appretkocher *m (text)* finish boiler
Appretur *f (text)* finish, [texturing] finishing
  daunendichte A. downproof finish
  erste A. bottom season
  wasserbeständige A. waterproof finish
Appreturbrechmaschine *f* finish breaking machine
Appreturechtheit *f* fastness to finishing
Appreturfoulard *m* finishing padding machine
Appreturkörper *m* loading agents on fillers *(paper)*
Appreturmasse *f* finishing compound; sizing compound (preparation)
Appreturöl *n* textile oil
Aprikosenöl *n* apricot seed oil
Apsidenlinie *f* apse line, line of apsides
Apsis *f* apsis
Aquadag *n (nucl)* aquadag
Aquädukt *m* aqueduct
Aquamarin *m* aquamarine
Aquarellfarbe *f* water colour
Aquariumreaktor *m s.* Schwimmbadreaktor
äquidistant equidistant
Äquidistante *f* equidistant
äquimolekular equimolecular
Äquipartitionstheorem *n* principle of equipartition of energy
Äquipotentialfläche *f* equipotential surface
Äquipotentialkatode *f* unipotential cathode
Äquipotentiallinie *f* equipotential line
äquivalent equivalent
Äquivalent *n* equivalent
  kalorisches Ä. thermic equivalent
Äquivalentbrennweite *f* equivalent focal length
Äquivalentgesetz *n* equivalence principle, equivalence law
Äquivalentgewicht *n* equivalent weight
Äquivalentleitfähigkeit *f* equivalent conductivity
Äquivalenz *f* equivalence
  Ä. von Energie und Masse equivalence of energy and mass
Äquivalenzgesetz *n* equivalence law
  Einsteinsches Ä. Einstein's law of photochemical equivalence, photochemical equivalence law
Äquivalenzkonzentration *f* equivalent concentration
Äquivalenztheorem *n* equivalence theorem
Aquosystem *n (ch)* aquo system
Ar *n (m)* are *(1 a = 100 m^2)*
Arabinose *f* pectinose, arabinose, pectin sugar
Arabit *n* arabitol
Arabonsäure *f* arabonic (arabitic) acid
Aragonit *m* aragonite
Aräometer *n* hydrometer, areometer
Arbeit *f* 1. work *(mechanics; economics)*; 2. job; labour; task
  A. im Schichtlohn work by the day
  A. im Schwerefeld work in field of gravity, work in gravitational field
  A. in Wechselschichten rota system
  A. pro Hub work done by stroke
  elektrische A. electric work
  indizierte A. indicated work
  mechanische A. mechanical work
  schöpferische A. creative work
  unproduktive A. deadwork; non-productive work
  verfügbare A. available energy *(turbine)*
  virtuelle A. virtual work
arbeiten 1. to work; 2. to operate *(relay)*; 3. to contract and expand *(building materials)*

auf eine Belastung a. *(el)* to work into a load
auf einen hohen Außenwiderstand a. *(el)* to work through a high resistance
auf einen Verstärker a. *(el)* to drive an amplifier
auf Frequenzen a. *(el)* to operate on frequencies
in einer Schicht a. to work a single shift
in zwei Schichten a. to be employed on two shifts
mit Gegenstrom a. to work with counter flow *(steam engine)*
mit Gleichstrom a. to work with parallel flow *(steam engine)*
über einen hohen Widerstand a. *(el)* to work through a high resistance

Arbeiten *n* 1. working; 2. operation; 3. internal action *(concrete)*
richtungsabhängiges A. directional operation
unnötiges A. unnecessary operation
unterbliebenes A. missing operation
zweifelhaftes A. doubtful operation

Arbeiter *m* operative, operator; worker; labourer
angelernter A. semi-skilled worker
gelernter A. skilled worker
ungelernter A. labourer

Arbeiterin *f* female (woman) worker
Arbeiterzug *m* workmen's train
Arbeitsablauf *m* 1. [machining] cycle; 2. sequence of operations
halbautomatischer A. semi-automatic cycle
selbsttätiger A. automatic cycle

Arbeitsablaufkarte *f* flow sheet
Arbeitsaufnahme *f* absorption of work
Arbeitsauftrag *m* work order (specification)
Arbeitsaufwand *m* energy expended; productive labour; expenditure of labour; amount of work given out
Arbeitsaustausch *m* interchange of work
Arbeitsbeanspruchung *f* working stress
Arbeitsbedarf *m* amount of work required
Arbeitsbedingungen *fpl* working conditions
Arbeitsbeispiel *n* case study, typical working example; application
Arbeitsbekleidung *f* work clothing
Arbeitsbelastung *f* working load
Arbeitsbereich *m* field (scope) of work; work capacity
Arbeitsbewegung *f* / **kreisförmige** rotation of work; rotating movement
Arbeitsboot *n* work boat, working boat
Arbeitsbreite *f* working (operating) width
Arbeitsbühne *f* [working] platform
Arbeitsbummelei *f* absenteeism
Arbeitsdampf *m* working steam
Arbeitsdiagramm *n* diagram of work
Arbeitsdruck *m* operating (actuating, effective) pressure
A. beim Sägen feed pressure in sawing
normaler A. normal operating pressure

Arbeitseichkreis *m* working standard (reference system)
Arbeitseinheit *f* 1. *(ph)* unit of work, work unit; 2. *(ag)* labour unit
Arbeitseinsatz *m* arrangement of work; [engagement in] practical work
Arbeitselektrode *f* operational electrode
Arbeitsersparnis *f* labour saving
Arbeitsfaktor *m* *(nucl)* total exposure time per week
Arbeitsfläche *f* 1. machined (final) surface *(metal cutting)*; 2. effective area
A. beim Hobeln planed surface
A. beim Räumen broached surface

Arbeitsflüssigkeit *f* operating (working) fluid *(hydraulics)*
Arbeitsfolge *f* operation sequence, sequence of operations
Arbeitsfolgeeinstellung *f* sequencing
Arbeitsfrequenz *f* working frequency
Arbeitsfuge *f* construction (working) joint *(concrete)*
Arbeitsgang *m* operation; cycle; procedure; process of manufacture
A. / in einem in one operation, in a single pass

Arbeitsgebiet *n* field of activity
Arbeitsgenauigkeit *f* working accuracy
Arbeitsgeschwindigkeitsstufen *fpl* working speeds
Arbeitsgewinn *m* gain of work
Arbeitsgruppe *f* team, gang
Arbeitsgüte *f* / **erforderliche** limits of accuracy required
Arbeitshub *m* 1. working stroke; cutting stroke *(metal cutting)*; 2. fire (expansion, power) stroke, expansion *(engine)*
Arbeitshublänge *f* cutting-stroke length *(metal cutting)*
Arbeitskaliber *n* live pass
Arbeitskanal *m* *(nucl)* fuel channel; exposure hole
Arbeitskarte *f* routing card *(workpiece)*
Arbeitskenngröße *f* performance characteristic
Arbeitskennlinie *f* dynamic characteristic, operating curve
Arbeitskolben *m* power (servo) piston
Arbeitskontakt *m* make-contact unit, operating (open-circuit, normally open auxiliary) contact
Arbeitskopf *m* cutting head
Arbeitskosten *pl* operation cost
Arbeitskräfte *fpl* manpower, labour [force]; workers, workmen
angelernte A. semi-skilled workers
gelernte A. skilled workers
ungelernte A. unskilled labourers

Arbeitskräftemangel *m* labour shortage

Arbeitskraftstunde *f* manpower-hour
Arbeitslehre *f* shop gauge
Arbeitsleistung *f* 1. labour efficiency; 2. output; production
Arbeitsleuchte *f* work light
Arbeitslohn *m* wages; earnings
Arbeitsmaschine *f* machine [doing mechanical work]
Arbeitsmaßstab *m* scale of work
Arbeitsmedium *n* operation (working) medium *(hydraulics)*
Arbeitsöffnung *f* rabbling hole *(furnace)*
Arbeitsordnung *f* factory regulation
arbeitsparend labour-saving
Arbeitspferd *n* farm horse
Arbeitsplan *m* 1. working scheme; 2. instruction sheet
Arbeitsplanung *f* production scheduling
Arbeitsplattform *f* working platform
Arbeitsplatzbeleuchtung *f* local lighting
Arbeitsproduktivität *f* labour productivity
Arbeitsprozeß *m* cycle of work *(steam engine)*
Arbeitsprüfung *f* service test
Arbeitspunkt *m* operating point *(electron tube)*
  statischer A. quiescent point
Arbeitsraum *m* steam space *(cylinder)*
Arbeitsrecht *n* labour laws, industrial and factory law
Arbeitsschacht *m s.* Kabelschacht
Arbeitsschutz *m* labour safety [measures]; protection of labour
Arbeitsschutzanordnungen *fpl* labour safety regulations
Arbeitsseite *f* 1. working side; 2. slave end *(of manipulator)*
Arbeitsspannung *f* working (load, terminal, closed-circuit) voltage
arbeitssparend labour-saving
Arbeitsspeicher *m* working storage *(computer)*
Arbeitsspiel *n* work cycle; duty cycle *(e.g. digging cycle in excavating)*
Arbeitsspindel *f* [main] work-driving spindle, [main] work spindle
Arbeitsspindeldrehzahl *f* work-spindle speed
Arbeitsstelle *f* work station *(automatic machine)*
Arbeitsstellung *f* 1. *(el)* on-position; 2. working position *(machine tool)*
Arbeitsstich *m* live pass
Arbeitsstoß *m (mine)* breast
Arbeitsstrom *m* operating (working) current; load current
Arbeitsstromalarmgerät *n* open-circuit alarm device
Arbeitsstromauslöser *m* operating-current release
Arbeitsstrombetrieb *m* open-circuit operation (working)
Arbeitsstromkontakt *m* open contact
Arbeitsstromschaltung *f* circuit-closing connection
Arbeitsstück *n* workpiece
  rohes A. blank
Arbeitsstudie *f* work (method) study, job analysis
Arbeitsstufe *f* stage of operation; stage *(turbine)*
Arbeitsstunde *f* working-hour, man-hour
Arbeitstakt *m* 1. [working] cycle time, [working] cycle *(machining)*; 2. power (expansion) stroke *(engine)*
  automatischer A. automatic cycle
Arbeitsteilung *f* division of labour
Arbeitstemperatur *f* operating temperature
Arbeitstempo *n* working tempo
Arbeitstisch *m* work table, *(Am)* work platen *(machine tool)*
Arbeitstrum *m* tight side *(belt)*
Arbeitsüberschuß *m* excess of work
Arbeitsunfall *m* accident at work, occupational accident
Arbeitsunfallrente *f* industrial injuries disablement benefit
arbeitsuntauglich unfit for work
Arbeitsunterbrechung *f* work interruption
Arbeitsvereinfachung *f* work simplification
Arbeitsverfahren *n* working method, [manufacturing] process
Arbeitsverlust *m* loss of work *(steam engine)*
Arbeitsvermögen *n* **des Reglers** power of the governor
Arbeitsversuch *m* trial operation
Arbeitsvertrag *m* contract of employment
Arbeitsvorbereiter *m* planner, planning engineer
Arbeitsvorbereitung *f* routing; job planning
Arbeitsvorgang *m* operation
Arbeitswalze *f* 1. *(text)* worker; 2. work roll *(rolling)*
Arbeitsweg *m* work length *(machine)*
Arbeitsweise *f* procedure, technique; performance
Arbeitswelle *f* working (service) wave
Arbeitswert *m* mechanical equivalent
  A. der Wärme mechanical equivalent of heat
  A. eines Kilogrammes Dampf work-equivalent of one kilogramme of steam
Arbeitswiderstand *m* dynamic resistance
Arbeitszeichnung *f* working drawing
Arbeitszeit *f* working hours; operating time
Arbeitszeitstudie *f* time study
Arbeitszimmer *n* study
Arbeitszug *m* work train
Arbeitszustand *m* marking *(telegraphy)*
Arbeitszyklus *m* work[ing] cycle
Arbeitszylinder *m* operating (working) cylinder
  eingegossener A. cast-in working cylinder
  einseitig beaufschlagter A. single-acting cylinder
  hydraulischer A. hydraulic cylinder

wechselseitig beaufschlagter A. double-acting cylinder
Arcatombogen *m* shielded arc
Arcatomschweißung *f* atomic hydrogen arc welding
Arcatomverfahren *n* atomic-hydrogen welding process
archimedisch Archimedean
Architekt *m* architect
Architektur *f* architecture
Architekturvermessung *f* architectural survey
Archiv *n* record office
Ardometer *n* ardometer
Argandbrenner *m* Argand burner
Argentit *m (min)* argentite
Argentometrie *f (ch)* argentometry
Argon *n* argon
Argonalter *n (nucl)* argon age
Argon-arc-Schweißen *n* argon-arc welding
Argument *n (math)* argument; amplitude
Arithmetik *f* arithmetic
arithmetisch arithmetic
Arkade *f* arcade
Arkansasstein *m* Arkansas stone
Arkus *m* arc
Arkuskosinus *m* arc cosine
Arkuskotangens *m* arc cotangent
Arkussinus *m* arc sine
Arkustangens *m* arc tangent
Arm *m* arm, spider *(gear)*; lever; crank; bearer, support
arm low-grade *(ore)*; poor; lean
Armatur *f* instrument; element; detail; armature; fitting[s]
feine A. fine (small) fittings *(boiler)*
grobe A. rough (large) fittings *(boiler)*
Armaturenbrett *n* instrument board (panel); dashboard
Armbanddosimeter *n (nucl)* hand dosimeter
Armco-Eisen *n* Armco iron
Armeereaktor *m* military[-type nuclear] reactor
Ärmeleinnähmaschine *f* sleeving machine
Ärmelsaum *m* sleeve hem
Ärmelschoner *m* sleeve protector
Armerz *n* poor ore
Armgas *n* lean (residue) gas
armieren to reinforce *(concrete)*; to armour; to sheathe *(cable)*
Armierung *f* armouring; sheathing *(cable)*; reinforcement *(concrete)*; braid *(cord)*
Armlehne *f* arm rest
Armozement *m* reinforced cement-mortar, ferrocement
Arnszinken *m (ag)* spring-loaded tine, spring-mounted tine
Aroma *n* aroma, flavour, fragrance
Aromaten *pl (ch)* aromatics
aromatisch aromatic
Aromatisierung *f* aromatization
Aron-Schaltung *f (el)* Aron measuring circuit
Aron-Zähler *m (el)* pendulum (Aron) meter
Arretierbolzen *m* locking bolt, stop pin
arretieren to arrest; to block; to impede
Arretierhebel *m* catch, stop-lever
Arretierstift *m* locating pin
Arretierung *f* detent; catch
Arretierungsdraht *m* lock wire
Arsanilsäure *f* arsanilic (atoxylic) acid
Arschleder *n (mine)* arse leather, back-skin, leather apron
Arsen *n* arsenic
Arsenal *n* ordnance
Arsenäscher *m (leath)* arsenic lime
Arsenat *m* arsenate
Arsenglas *n* arsenic glass
arsenhaltig arsenical
Arsenid *n* arsenide
Arsenikblüte *f (min)* arsenolite
Arsenit *n* arsenite
Arsenkies *m* arsenopyrite, mispickel
Arsennachweis *m* estimation of arsenic
Arsenolith *m (min)* arsenolite
Arsenopyrit *m (min)* arsenopyrite, mispickel
Arsensäure *f* arsenic acid
Arsentrioxid *n* arsenic [trioxide], anhydrid of arsenious acid
Arsen(III)-verbindung *f* arsenious compound
Arsen(V)-verbindung *f* arsenic compound
Arsenwasserstoff *m* arsenic hydride, arsine, arseniuretted hydrogen
Arsin *n* arsine
Arsinat *n* arsinate
Art *f* 1. kind; species, variety; sort; class; 2. method; manner; way; process; pattern
A. des Gutes description of good
Arten *fpl* **der Passungen** grades of fits
artfremd of different nature *(e.g. ceramics)*
Artillerie *f* artillery
schwere A. medium artillery; *(ship)* primary artillery
Artilleriebekämpfung *f* counterbattery fire
Artilleriebeobachter *m* spotter, artillery observer
Artilleriebeobachtungsflugzeug *n* artillery spotting aircraft
Artilleriebeschuß *m* artillery bombardment
Artillerieerkundung *f* artillery reconnaissance
Artilleriemeßtrupp *m* artillery survey detachment
Artilleriepunkt *m* locating (trigonometric) point
Artilleriepunkttafel *f* artillery point-designation grid
Artillerieschießplatz *m* artillery range
Artillerieschlepper *m* artillery prime mover
Artilleriesperrfeuer *n* artillery barrage
Artillerievermessung *f* artillery survey

Artilleriezugmaschine *f* artillery prime mover
Artikel *m* 1. article; 2. item, material
Artikelkarte *f* stock card
Aryl *n (ch)* aryl
arylieren *(ch)* to arylate
Arylierung *f (ch)* arylation
Arylierungsmittel *n* arylating agent
Arylierungsprodukt *n* arylate
Arylrest *m* aryl residue
Arzneimittel *n* medicinal substance, medicine, remedy
Arzneimittelbuch *n* pharmacopeia, dispensatory
Arzneimittelkunde *f* pharmacology
Arzneimittelsynthese *f* drug synthesis
Arzneipflanze *f* medicinal plant
Asbest *m* asbestos
asbestartig asbestic, asbestoid, amianthine
Asbestbaupappe *f* asbestos wallboard
Asbestbeton *m* asbestos cement
Asbestbetonplatten *fpl* asbestos cement sheeting
Asbestdichtung *f* asbestos joint (packing)
Asbestdrahtnetz *n* asbestos wire gauze
Asbestfaser *f* asbestos fibre
Asbestfeuerschutzanzug *m* asbestos fire-proof suit
Asbestgarn *n* asbestos yarn
Asbestgewebe *n* asbestos textile
Asbest-Graphit-Dichtungsring *m* asbestos-graphite gasket
Asbesthandschuhe *mpl* asbestos gloves
Asbestisolierungspapier *n* asbestos electrical isolation paper
Asbestkarton *m* asbestos board
Asbestpackung *f* asbestos packing
Asbestpappe *f* asbestos board
Asbestplatte *f* asbestos wallboard
Asbestring *m* asbestos ring
Asbestschiefer *m* asbestos slate
Asbestschnur *f* asbestos cord (twine)
  geklöppelte A. braided asbestos cord
Asbestzement *m* asbestos cement
Asbestzementpappe *f* asbestos cement board
Asbestzementplatte *f* asbestos cement sheet
Asbestzementschiefer *m* asbestos shingle
Asbestzementwelltafel *f* corrugated asbestos sheet
Asbestzopf *m* asbestos gasket
Asbestzwischenlage *f* asbestos layer
Asbolan *m (min)* asbolane, asbolite
Asche *f* cinders, ash[es]
  gebundene A. fixed (inherent) ash
  radioaktive A. radioactive ash
Ascheauswerfer *m* ash ejector
Aschebildner *m* ash constituent
Ascheejektor *m* ash ejector
aschefrei ash-free, a.f.
Aschegehalt *m* ash content
Aschekasten *m* ash pan
Aschekeller *m* ash cellar
Aschenabfuhrkanal *m* ash chute
Aschenauswerfer *m* ash ejector
Aschenbahn *f* cinder [athletics] track, cinder path *(running track)*
Aschenfall *m* ash pit
Aschenfallklappe *f* ash pit damper; ash pit drop door
Aschenprahm *m* sullage lighter, ash boat (lighter)
Aschenraum *m* ash-pit
Äscher *m (leath)* slake-lime pit; [chalk] lime
  milder Ä. mellow lime
Äscherfaß *n* liming tub
Äscherfehler *m* lime fault
Äschergang *m* shift [liming]
Äschergrube *f* lime-pit
Äscherkalk *m* slack lime
äschern to soak in lime, to lime
Äschern *n* liming
Äscherwerkstatt *f* lime yard
Aschesack *m* ash pocket
Ascheschieber *m* ash stop
Ascheschmelztemperatur *f* ash fusion temperature
Ascheschütte *f* ash-shoot
Ascheträger *m* ash component
asche- und wasserfrei dry ash-free, d.a.f.
Aschewinde *f* ash hoist
Aschfall *m* ash-pit
Aschprahm *m* sullage lighter, ash boat (lighter)
Asepsis *f* asepsis
Asparagin *n (ch)* asparagine
Asparaginsäure *f* asparagic (aminosuccinic) acid
Aspe *f* trembling poplar
Asphalt *m* asphalt
Asphaltauskleidung *f* asphaltic lining
Asphaltbelag *m (build)* asphalting, asphalt pavement
Asphaltbeton *m* bituminous concrete, asphalt[ic] concrete
Asphaltbitumenpappe *f* asphalted felt
Asphaltgestein *n* rock asphalt
asphalthaltig asphaltic, bituminiferous
asphaltieren to asphalt, to spread asphalt
Asphaltierung *f* asphalting
Asphaltkitt *m* asphalt[ic] cement
Asphaltkohle *f* pitch coal
Asphaltkopierverfahren *n* asphalt process
Asphaltlack *m* asphalt varnish
Asphaltmastix *m* bituminous (asphalt) mastic, mastic asphalt
Asphaltstraße *f* bituminous road
asphärisch aspherical
Aspirant *m* doctorial candidate
Aspirateur *m* aspirator; winnowing machine

Aspirin *n* acetosalicylic acid
Assistent *m* assistant
Assoziation *f* association
assoziativ *(math)* associative
Assoziativitätsgesetz *n* associativity law
Ast *m* 1. knot; 2. branch
  durchgehender A. passing knot
  eingewachsener A. encased (living, live, firm) knot
  gesunder A. live knot
  grober A. gross knot
  herausgefallener A. loose (dropped out) knot
  hohler A. hollow knot
  vorstehender A. protruding knot
Astabilität *f (cont)* instability
Astasie *f* astaticity
Astatin *n (ch)* astatine
astatisch astatic
Astaxt *f* lopping axe
Astbohrer *m* knot borer
Äste *mpl* limbwood
Ästehackmaschine *f* knot chipper
Asterismus *m* asterism
Asteroide *f (math)* astroid, tetracuspid
Astfänger *m* sliver screen, knotter
  rotierender A. rotary sliver screen
  schüttelnder A. shaking sliver screen
astfrei knotless, free from knots
Astholz *n* lopwood
Astigkeit *f* knottiness
astigmatisch astigmatic
Astigmatismus *m* astigmatism; spot-shape distortion
  A. schiefer Bündel astigmatism of oblique pencils
Astigmeter *n* astigmeter
Astknorren *m* knorl, knub, knag
Astknoten *m* branch knot
Astloch *n* knot hole, open knot
astlos knotless
ASTM-Prüfmaschine *f* ASTM test rig
Astrachan *m (n)* astrakhan
astrein free from knots
Astrofotografie *f* astrophotography
Astrograf *m* astrograph
Astroide *f* astroid, tetracuspid
Astrokamera *f* astrographic camera
Astrokompaß *m* astro-compass
Astrolenkung *f* celestial (star-tracking) guidance
Astrometrie *f* astrometry
astrometrisch astrometric
Astronaut *m* astronaut, space traveller
Astronautik *f* astronautics
Astronautin *f* female astronaut
Astronavigation *f* astronavigation; celestial navigation
Astronom *m* astronomer
Astronomie *f* astronomy
  nautische A. celo-navigation
astronomisch astronomical
Astrophysik *f* astrophysics
Astsäge *f* pruning saw
  zweischneidige A. pruning saw with teeth on both ends
Astschere *f* tree pruner
Aststummel *m* dead knot
Ästuarium *n* estuary
Ästung *f* pruning
Asymmetrie *f* asymmetry
  A. der Verteilung skewness of distribution
asymmetrisch asymmetric
Asymptote *f* asymptote
Asymptotenkegel *m* asymptotic cone
asymptotisch asymptotic
asynchron asynchronous, non-synchronous, out-of-step
Asynchronbetrieb *m* variable cycle operation *(computer)*
Asynchrongenerator *m* induction generator, asynchronous alternator
Asynchronmaschine *f* non-synchronous machine
Asynchronmotor *m* asynchronous (non-synchronous) motor
  A. mit Anlauf- und Betriebswicklung double-deck motor
  A. mit Repulsionsanlauf repulsion-start induction motor
  A. mit Schleifringläufer slip-ring induction motor
  phasenkompensierter A. all-watt motor
  polumschaltbarer A. change-pole induction motor
Asynchronphasenschieber *m* asynchronous phase compensator
Asynchronrechner *m* asynchronous computer
Atakamit *m (min)* atacamite
Atelier *n* studio
Atelierkamera *f* studio camera
Atemgift *n* fumigant
Atemschutzgerät *n* respiratory protection apparatus
Atemschutzmaske *f* respirator
Atemwege *mpl* respiratory tract
Äthan *n* ethane
Äthanol *n* ethanol
Äthen *n* ethene
Äther *m* ether
  gemischter Ä. complex (compound) ether
ätherisch etherial; essential *(oil)*, volatile
Ätherlibelle *f (opt)* spirit level
atherman athermanous
Äthertheorie *f* **des Lichtes** ether theory of light
Äthinylierung *f (ch)* ethynylation, ethinylation
Äthoxylinharz *n* ethoxyline resin
Äthyl *n* ethyl

Äthylalkohol *m* ethyl alcohol
Äthylat *n* ethylate, ethoxide
Äthylazetat *n* ethyl acetate
Äthylchlorid *n* ethyl chloride
Äthylen *n* ethylene
Äthylenbindung *f* ethylenic (olefinic) bond
Äthylenglykol *n* ethylene glycol
Äthylenoxid *n* ethylene (ethene) epoxide
Äthylierungsprodukt *n* ethylate
Athylmerkaptan *n* ethyl hydrosulphide (mercaptane)
Äthylschwefelsäure *f* vinic acid, ethylhydrogen sulphate
Äthylsenföl *n* ethyl mustard oil
Atlas *m* / **doppelbindiger** *(text)* double-stitched satin
Atlasband *n* satin ribbon
Atlasbindung *f* *(text)* satin weave
Atlaserz *n* *(min)* false emerald, malachite
Atlasglanz *m* satin lustre
Atlasgrat *m* *(text)* satin rib
Atlasspat *m* *(min)* satin spar, calcite
Atmosphäre *f* 1. atmosphere; 2. *(ch)* medium
  geschichtete A. stratified atmosphere
  physikalische A. normal atmosphere
Atmosphärendruck *m* atmospheric pressure
Atmosphärentiefe *f* atmospheric depth
atmosphärisch atmospheric
Atmungsgerät *n* respirator
Atom *n* atom
  angeregtes A. excited atom
  gebundenes A. bonded atom
  geladenes A. charged atom
  heißes A. hot atom
  ionisiertes A. ionized atom
  komplexes A. complex atom
  markiertes A. labelled (tagged) atom
  nacktes A. stripped atom
  substituiertes A. substitutional atom
Atomabfall *m* atomic waste
Atomabstand *m* interatomic distance, atomic spacing
  A. im Molekül atomic distance in a molecule
Atomanordnung *f* arrangement of atoms
Atomanregung *f* atomic excitation
Atomantrieb *m* nuclear propulsion
atomar atomic[al]
Atomasche *f* *s.* radioaktive Asche
Atombatterie *f* atomic [energy storage] battery, radioactive battery
Atombau *m* atom[ic] structure
Atombeschuß *m* atomic bombardment
Atombewegung *f* atomic motion
Atombindung *f* atomic linkage (bond), covalence, covalent bond (link), electron linkage
Atombombe *f* atom[ic] bomb, A-bomb
  nominelle A. nominal atomic bomb
Atombombenangriff *m* atomic bomb attack
Atombombenexplosion *f* atomic blast (bomb burst), atomic [bomb] explosion
Atombombenträger *m* atomic bomber
Atombunker *m* atomic [bomb] shelter
Atomchemie *f* atomic chemistry
Atomdichte *f* atomic density
atomdispers atomically dispersed
Atomdrehung *f* atomic rotation
Atomdurchmesser *m* atomic diameter
Atomebenen *fpl* atomic planes
Atomeigenstrahlung *f* characteristic atom radiation
Atomeisbrecher *m* nuclear-powered icebreaker
Atomelektron *n* atomic (atom-bound) electron
Atomenergie *f* *s.* Kernenergie
Atomenergieantrieb *m* *s.* Kernenergieantrieb
Atomenergiegesetz *n* Atomic Energy Law; *(Am)* Atomic Energy Act
Atomflugzeug *n* nuclear [-powered] plane
Atomformfaktor *m* atom[ic] form factor, atomic scattering factor
Atomforschung *f* atomic research; nuclear research
Atomforschungszentrum *n* atomic research centre (establishment)
Atomfriedhof *m* *(nucl)* burial ground, graveyard, dumping ground
atomgetrieben nuclear-powered
Atomgewicht *n* atomic weight
Atomgewichtsskala *f* atomic weight scale
Atomgitter *n* atomic [space] lattice
Atomhülle *f* atomic shell
atomig atomic
Atomion *n* ionized atom, atomic ion
Atomionisierung *f* atomic ionization
Atomistik *f* atomistics
Atomkern *m* [atomic] nucleus
Atomkernenergie *f* *s.* Kernenergie
Atomkernmodell *n* *s.* Kernmodell
Atomkernspaltung *f* *s.* Kernspaltung
Atomkonstante *f* atomic constant
Atomkraft *f* *s.* Kernenergie
Atomkraftantrieb *m* *s.* Kernenergieantrieb
Atomkraftwerk *n* atomic (nuclear) power station; atomic energy plant, nuclear power plant
Atomkrieg *m* nuclear war[fare], atomic war[fare]
Atomladung *f* atomic charge
Atommeiler *m* *s.* Kernreaktor
Atommodell *n* model of the atom, atom[ic] model
  Rutherford-Bohrsches A. Rutherford-Bohr atom model
  Thomsonsches A. Thomson's atom model
Atommoment *n* atomic moment
Atommotor *m* *s.* Kernenergieantrieb
Atommüll *m* *s.* Atomabfall
Atomniveau *n* atomic [energy] level
Atomnummer *f* atomic number
Atomofen *m* *s.* Kernreaktor

Atomphysik *f* atomic physics
Atomphysiker *m* atomic physicist
Atompolyeder *n* atomic polyhedron
Atomradius *m* atomic radius
Atomrakete *f* nuclear[-powered] rocket
Atomreaktor *m s.* Kernreaktor
Atomrumpf *m* atomic trunk (hull, residue, core, torso)
Atomschale *f* atomic electron shell
Atomschiff *n* nuclear-propelled ship
Atomschmelzwärme *f* atomic heat of fusion
Atomschwächungskoeffizient *m* atomic attenuation coefficient
Atomspektrum *n* atomic spectrum
Atomstoß *m* atomic collision
Atomstrahlen *mpl* atomic rays
Atomstrahlmethode *f* atomic beam method
Atomstrahlung *f* atomic radiation
Atomstruktur *f* atom[ic] structure
Atomsuszeptibilität *f* atomic susceptibility
Atomtheorie *f* atomic theory
Atomuhr *f* atomic clock
Atomumlagerung *f* atomic displacement *(in the lattice)*, discomposition, knocking-out
Atomumwandlung *f* atomic transmutation (transformation)
Atomunterseeboot *n* atomic (nuclear-powered) submarine
Atomverhältnis *n* atomic ratio
Atomverkettung *f* atom bonding (linkage)
Atomverteilung *f* atom distribution
Atomvolumen *n* atomic volume
Atomwaffe *f s.* Kernwaffe
Atomwärme *f* atomic heat
Atomwertigkeit *f* atomic valence
Atomwirtschaft *f s.* Kerntechnik
Atomwissenschaftler *m* atomic scientist
Atomzeitalter *n* atomic (nuclear) age
Atomzerfall *m* disintegration of atom, atomic disintegration (decay)
Atomzertrümmerung *f s.* 1. Spallation; 2. Kernumwandlung
Atomzustand *m* atomic state
Atramentieren *n* atrament phosphating
Atropin *n* atropine
Attenuator *m* attenuator, fader
Attributprüfung *f* inspection by attributes, attribute sampling (inspection)
Ätzalkali *n* caustic alkali
  Ä. in Schuppenform caustic in flakes, flake caustic
Ätzalkalischmelze *f* caustic fusion
Ätzdauer *f* etching time
Ätzdruck *m (text)* discharge print
ätzen to etch *(metals)*; to corrode; to cauterize; *(ch)* to macerate; *(text)* to discharge
  galvanisch ä. to electro-etch
Ätzen *n* etching *(metal)* (*comp.* ätzen)
  automatisches Ä. automatic etching
  elektrolytisches Ä. electrolytic etching
  maschinelles Ä. machine-etching
ätzend caustic, corroding, mordant
Ätzfigur *f* etch (etched, etching) figure *(crystal)*
Ätzgrube *f* etching pit *(crystal)*
Ätzgrund *m* etching ground
Ätzkali *n* caustic potash
Ätzkalk *m* caustic lime, quicklime
Ätzkraft *f (ch)* causticity
Ätzlauge *f (ch)* caustic lye
Ätzlösung *f* etching solution *(metal)*
Ätzmaschine *f* etching machine, etcher
Ätzmittel *n* 1. etchant *(metal)*; 2. *(ch)* etch, corroding agent; caustic, mordant
Ätzmuster *n (text)* discharge pattern
Ätznäpfchen *n* gravure cell
Ätznatron *n* caustic soda, sodium hydroxide
Ätznatronlösung *f* solution of caustic soda
ätzpolieren to attack-polish
Ätzspitze *f (text)* burnt-out lace
Ätzstickerei *f* burnt-out embroidery
Ätzstift *m* caustic (corrosive) stick
Ätztiefe *f* etching depth *(metal)*
Ätz- und Retuschiertisch *m* etching and retouching desk
Ätzung *f* 1. engraving; 2. etching; corrosion
  elektrolytische Ä. electro-engraving
Ätzverfahren *n* etching process
Ätzwanne *f* etching basin (sink)
Ätzwirkung *f* etching action; causticity
Ätzzeit *f* etching time
Audiogramm *n* audiogram
Audiometer *n* audiometer
Audion *n (el)* grid detector, audion
  selbstschwingendes A. self interference audion
Audionempfänger *m* audion receiver
Audiongleichrichter *m* grid-leak detector
Audiongleichrichtung *f* grid-leak detection (rectification)
Audionverstärker *m* amplifying detector
Auerbrenner *m* Auer (Welsbach) burner
aufarbeiten 1. to repair, to rework, to rebuild, to regenerate, to reprocess, to recondition; 2. to use up, to finish
Aufarbeitung *f* reprocessing; regeneration; treatment, treating, processing; reconditioning; rebuilding; repairing, overhauling, repairs
Aufarbeitungsanlage *f* reprocessing plant
Aufarbeitungsverluste *mpl* reprocessing loss
Aufbau *m* 1. erection; building; assembly *(machine)*; 2. *(ch)* synthesis; 3. arrangement; construction; structure; 4. establishment *(voltage)*; increase; 5. superstructure *(vessel)*; 6. body[work], coachwork; superstructure *(body or body and cab where integral construction is not adopted)*

A. an der Schneide chip build-up
A. der Materie structure of matter
A. einer Schaltung construction of a circuit
A. nach Baukastensystem standardized unit construction, unit construction on the building-block principle, design[ing] on the building-block principle, unitized construction
bordfreier A. *(ship)* board-free erection
freistehender A. *(ship)* detached superstructure
geschlossener A. packed construction
kompakter A. compact design
konstruktiver A. mechanical construction
über die ganze Schiffslänge durchlaufender A. complete superstructure

Aufbauanordnung *f* mounting arrangement
Aufbauchung *f* fattening
Aufbaudeck *n* superstructure deck
Aufbaueinheit *f* unit, assembly *(machine)*
aufbauen 1. to erect; to build up; to construct; to assemble; to rig up *(scaffolding)*; to set up; 2. *(el)* to make [a field]; 3. *(ch)* to synthesize, to compose
Aufbaufaktor *m (nucl)* build-up factor
Aufbaufeld *n* mounting panel
Aufbauinstrument *n* surface mounted instrument
Aufbaumaschine *f* unit-type machine, building-block machine
Aufbäumen *n (text)* beaming
A. der Flammrohre durch ungleichmäßige Ausdehnung hogging of the flues
A. des Werkzeugs lift

Aufbäummaschine *f* beaming machine
Aufbäumung *f* deflection
Aufbaupflegemittel *n (aut)* body cleaning compound
Aufbaureaktion *f s.* Fusionsreaktion
Aufbauschiff *n* **/ durchlaufendes** complete superstructure
Aufbauschmelze *f* build-up heat
Aufbauschneide *f* built-up [cutting] edge
Aufbauspant *n (ship)* superstructure frame
Aufbaustoffe *mpl (ch)* builders
Aufbauten *mpl* **/ durchlaufende** complete superstructure *(vessel)*
Aufbauzeit *f (el)* ionization time; formation time *(of spark discharge)*
Aufbereitbarkeit *f (met)* washability
aufbereiten 1. to rework; to recover; to reclaim *(oil)*; to repulp *(paper)*; 2. to treat, to purify *(water)*; to prepare; to dress *(ore)*; to wash *(coal)*; to refine, to upgrade; to process *(seed)*
Aufbereitung *f* 1. reworking; recovery; reclamation *(oil)*; repulping *(paper)*; 2. treating, treatment, purification *(water)*; preparation; dressing *(ore)*; washing *(coal)*; refining; upgrading; [treating] processing *(seed)*; benefication
Aufbereitungsanlage *f* 1. *(met)* cleaning plant, concentrator, mill *(ore)*; 2. *(mine)* preparation plant; coal washery; 3. *(text)* screen room
Aufbereitungsgut *n* product of dressing *(ore)*
Aufbereitungskette *f* seed processing system
Aufbereitungskonzentrat *n* preparation concentrate
Aufbereitungsverlust *m* loss in cleaning, dressing loss
Aufbeton *m* concrete topping (screed)
aufbewahren to store, to stock, to keep; to preserve
Aufbewahrungsgebühr *f* cloak room charges, *(Am)* check-room charges
aufbiegen to bend up *(reinforcing bars)*; to turn up; to deflect
Aufbiegung *f* bend *(reinforcement)*
aufbinden to untie; to loosen
aufblasbar inflatable
aufblasen to inflate; to blow up
Aufblasestelle *f* hot spot
Aufblätterung *f* **der Schichten** *(mine)* bed separation
aufblitzen to flash up
aufbocken to jack up
aufbohren to bore *(metal)*; *(mine)* to enlarge a borehole
Aufbohren *n* boring *(metal)*; *(mine)* enlarging a borehole
A. mit einschneidigem Werkzeug single-point boring

Aufbohrer *m* **für vorgegossene Löcher** core drill
Aufbohrleistung *f* boring capacity
Aufbohrmaschine *f* boring machine
aufbrausen to effervesce
aufbrechen 1. to force (break) open; 2. to start
aufbringen 1. *(ship)* to capture, to bring in a prize; 2. to apply; to evaporate; to mount; to deposit
a. auf to apply to, to impose on

Aufbringung *f* application *(force)*; deposition
aufbuchsen to bush on
Aufbuchsen *n* bushing on
Aufbügelfolie *f* hot transfer foil
aufbügeln to iron
Aufdampfanlage *f* deposition apparatus
aufdampfen to vapour-deposit, to evaporate
Aufdampfen *n* evaporation coating, vacuum metalling; vapour deposition
Aufdampfschicht *f* evaporated film
aufdecken to uncover
Aufdeckmasche *f* eyelet stitch *(knitting)*
Aufdeckmaschine *f* eyelet machine *(knitting)*
Aufdeckmuster *n* eyelet in pelerine design *(knitting)*
Aufdeckplatine *f* eyelet jack *(knitting)*

aufdocken to dock up
aufdornen 1. to enlarge, to expand; to pierce; 2. to put on a mandrel; 3. to drift *(rivet holes)*
aufdrehen 1. to [screw] open; 2. *(text)* to untie; 3. *(ship)* to luff into the wind; 4. to advance *(potentiometer)*
Aufdruck *m* 1. [im]print, stamp; 2. total liquid pressure against a submerged plane area
aufdrucken to [im]print
aufdrücken / eine Spannung *(el)* to impress a voltage
Aufdüsen *n (text)* flame spraying
Aufeinanderfolge *f* succession
aufeinanderfolgend consecutive, successive
aufeinandergeschichtet stacked one on top of another
aufeinanderliegend mating *(surfaces)*
Aufeinandertürmen *n* **der Impulse** pile-up of pulses
aufentern *(ship)* to go aloft
Aufenthalt *m* stop, halt
Aufenthaltsbescheinigung *f (ry)* notice giving authority to stop (stay)
Auferregung *f (el)* build-up of a self-excited generator
auffädeln to thread
auffahren 1. *(mine)* to drive, to head; 2. to go aground *(vessel)*
Auffahren *n* **der Weiche** trailing the points, forcing open the points
A. im breiten Blick *(mine)* shortwall working
A. von Flözstrecken *(mine)* driving
Auffahrleistung *f (mine)* driving (heading) performance
Auffahrt *f* driveway; approach ramp (embankment), access ramp, ramp lane
Auf-Fahrt-Bleiben *n* **der Signale** signals remaining at "line clear"
Auffahrtsklappe *f* gangway *(of a ferry)*
Auffahrtsrampe *f* ramp lane, approach ramp (embankment), access ramp
Auffahrtsystem *n* drive-on system *(cargo vessel)*
Auffahrungsarbeit *f (mine)* drifting
Auffallwinkel *m* angle of incidence, glancing angle
Auffaltung *f (geol)* anticlinal fold
Auffanganode *f* collecting (gathering) anode, collector plate
Auffangausbeute *f (nucl)* collection efficiency
Auffangbecken *n (build)* catch basin
Auffangelektrode *f* collecting electrode; collector plate
auffangen to trap; to collect; to catch
Auffangen *n* collection
Auffänger *m* target; trap; catcher
Auffängerelektrode *f* collector; bucket; target electrode
Auffängerelement *n* target element
Auffängerfläche *f* target area
Auffängermaterial *n* target material
Auffängerplatte *f* collector
Auffängertasche *f* target pocket
Auffanggefäß *n (nucl)* collecting tank
Auffangquerschnitt *m* absorption cross section
Auffangtrog *m* catch basin
Auffangvorrichtung *f* trap
auffärben to redye, to dye afresh
auffasern to unravel
Auffaßbereich *m* pick-up range *(radar)*
aufflammen to deflagrate, to flame up, to burst into flame
aufforsten to reforest, to afforest, to restock
Aufforstung *f* reforestation, afforestation
auffrischen *(ch)* to regenerate; to revivify *(moulding sand)*
Aufführseil *n* rope carrier *(papermaking machine)*
auffüllen 1. to refill, to replenish; to recharge *(furnace)*; 2. to fill up
Auffüllen *n* refilling, replenishing, replenishment; recharging; filling
Auffüllung *f* recharging; refilling *(furnace)*
Aufgabe *f* 1. task, purpose, function; 2. feeding *(work)*
Aufgabeamt *n (telec)* office of origin
Aufgabebecherwerk *n* feed[ing] elevator
Aufgabebehälter *m* feed hopper
Aufgabegut *n* [feed] material
Aufgabengröße *f (cont)* desired value; ultimately controlled variable
Aufgabensammlung *f* collection of problems
Aufgabentafel *f* prepatch panel, problem board *(analogue computer)*
Aufgabenvariable *f* problem variable *(analogue computer)*
Aufgabenveränderliche *f* problem variable *(analogue computer)*
Aufgabenwert *m* desired value
umgeformter A. *(Am)* reference input
Aufgabeplattenband *n* apron feeder
Aufgabeschurre *f* feed chute
Aufgabestelle *f (telec)* station of origin
Aufgabetrichter *m* feed hopper *(machine tool)*; *(build)* feed funnel; tremie *(concrete)*
Aufgabewagen *m* feeder car
aufgebaut built-on
aufgeben 1. to charge; to feed; to forward *(goods)*; 2. *(mine)* to abandon, to give up; 3. to discontinue
Aufgebung *f* abandonment *(insurance term)*
aufgedampft vapour-deposited
aufgehängt suspended
federnd a. spring-suspended
frei a. freely suspended
gelenkig a. pivoted
aufgehend / nicht *(math)* aliquant

ohne Rest a. *(math)* aliquot
aufgekeilt auf keyed to, feathered to
aufgeladen charged
aufgelassen abandoned, disused *(mine)*
aufgelötet soldered on; brazed on
aufgenietet riveted to
aufgerauht roughened, hacked
aufgeschraubt auf bolted to
aufgeschrumpft shrunk on; fitted by shrinking
aufgeschüttet filled-in
aufgesetzt auf seated on
aufgesteckt auf seated on
aufgetakelt rigged up
aufgichten *(met)* to feed
Aufgießen *n* extrusion laminating
Aufglasurfarbe *f* glazing colour
Aufgleisung *f (ry)* rerailing
Aufgleisungskran *m* heavy breakdown crane
Aufgleisungsschuh *m* rerailing ramp
aufgliedern 1. to classify; 2. to break down; 3. to subdivide
Aufgliederung *f* 1. classification; 2. breakdown; 3. subdivision
Aufgrundlaufen *n* running aground *(of a vessel)*
Aufgrundliegen *n* lying aground *(vessel)*
Aufguß *m* infuse
aufhaken to unhook
aufhalten to stop; to delay; to trap
Aufhängeband *n* suspension band
Aufhängebügel *m* suspension hook
Aufhängedraht *m* catenary (suspending, messenger, suspension) wire
Aufhängegestänge *n* suspension rods
Aufhängeisolator *m* suspension insulator
aufhängen 1. to suspend; 2. to hang out *(washing)*
kardanisch a. to hang from gimbals
Aufhängeöse *f* suspension eye
A. der Zugfeder extension spring eye
Aufhängepunkt *m* fulcrum of suspension, suspension point; centre of suspension *(conductors)*
Aufhängevorrichtung *f* suspension device
Aufhängung *f* suspension; linkage
A. am Kesseltragkranz suspension by means of a segment of angle iron riveted on the shell crown
A. der Wiege secondary suspension of bolster beam
A. durch Hängestangen suspension by means of rods
A. eines Fahrzeuges suspension of a vehicle
bifilare A. bifilar suspension
halbstarre A. semi-rigid suspension
kardanische A. gimbal suspension, gimbal [type mounting]
rhombische A. rhombic suspension
starre A. rigid suspension
umgekehrte A. inverted suspension *(pendulum)*

aufhaspeln to reel up (in), to unspool, to wind off
Aufhauen *n (mine)* raise, rise [heading]
Aufhäufeln *n (ag)* ridging, *(Am)* hilling up
aufhäufen to pile up; to store
aufheben 1. to neutralize, to equalize; 2. *(math)* to cancel out; 3. to disconnect a call *(telephone)*; 4. to lift, to raise; 5. to preserve; to store; to keep
sich gegenseitig a. 1. *(math)* to annul each other; 2. to cancel each other *(voltage)*
Aufheben *n* **der Sperrung** unblocking
Aufhebung *f* cancellation
aufheißen to hoist up, to pull up
aufheizen to heat up
Aufheizperiode *f* heating-up period
Aufheizzeit *f* heating-up time (period)
aufhellen to tint; to lighten *(paint etc)*; to clarify; *(text)* to brighten
mit Säure a. *(text)* to tone down *(acid)*
Aufhellen *n s.* Aufhellung
Aufheller *m (ch)* lightener; fining agent; *(text)* brightening agent; bleaching agent
optischer A. optical dye (bleaching agent)
Aufhellimpuls *m* bright-up pulse
Aufhellung *f* tinting; lightening *(paint)*; clarification; *(text)* brightening; bleaching; toning down *(acid)*
optische A. optical bleaching
Aufhellungssteuerung *f* Z-axis modulation
Aufhellungsvermögen *n* tinting power
Aufhellungswert *m* tinting value
Aufholleine *f (ship)* tripping (tricing) line
aufhören to stop, to end
aufkämmen to cog *(carpentry)*
Aufkämmung *f* cogging *(carpentry)*
Aufkaschieren *n* laminating
aufkeilen auf to key to
Aufkeilwinkel *m* angle of advance
aufkimmen to raise *(ship's bottom)*
Aufkimmung *f* rise of floor (bottom), dead rise *(ship's bottom)*
starke A. sharp rise of floor
aufklappbar hinged; folding
aufklappen to fold upwards; to turn up
aufklaren 1. to clear up; 2. to brighten
Aufklärer *m* scout[er] *(vessel)*
Aufklärung *f (mil)* reconnaissance
Aufklärungsradar *n* search[tracking] radar
aufklauen to birdsmouth *(carpentry)*
Aufklauung *f* birdsmouthing *(carpentry)*
aufkleben to paste on, to glue on
aufklettern to ride on the teeth *(chain)*
Aufklettern *n* **der Kette auf Zahnflanken** riding out too near the ends of the teeth, riding of the links on the teeth
Aufklotzung *f (ship)* dead wood, [shell] chock, blocking, back shore

aufkohlen to carburize, to increase the carbon content; to cement *(iron)*
Aufkohlen *n* carburization; cementation
A. mit Roheisen carburization by pig iron
Aufkohlungsmittel *n* recarburizing agent
aufkommen to come up (in); to enter a port
aufkratzen *(text)* to nap
aufkrausen *(leath)* to grain up
aufladbar *(el)* rechargeable
Aufladedruck *m (aut)* supercharge pressure
Aufladegebläse *n* supercharger, booster
Aufladegerät *n* loader
Auflademotor *m* forced-induction engine
aufladen 1. to load [on]; 2. to [re]charge *(accumulator)*; 3. to supercharge, to boost *(engine)*
sich a. *(el)* to charge up
Aufladen *n* 1. loading [on]; 2. [re]charging *(accumulator)*; 3. supercharging, boosting *(engine)*
A. der Strahlungsquelle *(nucl)* source loading
elektrostatisches A. von Textilfasern electrostatic charge of textile fibres
Aufladestrom *m* charging current
Aufladung *f* 1. supercharging, boosting *(engine)*; 2. [re]charging *(accumulator)*; electrification
A. / mit supercharged *(engine)*
A. / ohne unsupercharged *(engine)*
statische A. static electrification
Auflage *f* 1. support; rest; bearing surface; seating; 2. edition *(book)*
gute A. haben to be well supported
ohne A. unsupported
volle A. haben to be fully supported (seated)
Auflagedruck *m* support pressure (reaction)
Auflagefläche *f* locating (bearing, mounting) surface; working face
Auflagekraft *f* bearing reaction
Auflagepunkt *m* point of support (contact), resting point; contact centre *(of a tyre)*
Auflager *n* support, bearing
festes A. fixed bearing
Auflagerbedingungen *fpl* support conditions
Auflagerblock *m* bearing pad
Auflagerdruck *m* support (bearing) pressure
Auflagerfläche *f* bearing (supporting) surface
Auflagergegendruck *m* counter-boost pressure
Auflagerkraft *f* supporting force, support reaction
auflagern auf to bear on; to be supported by
Auflagerplatte *f* base (supporting, bed, bearing) plate
Auflagerreaktion *f* support reaction
Auflagerschräge *f* [tapered] haunch *(reinforced concrete)*
Auflagerung *f* support[ing]
A. des Kessels supporting of the boiler
Auflagerungsstelle *f* point of support
Auflagerverschiebung *f* displacement of support
Auflagewerkstoff *m* cladding material
Auflanger *m (ship)* futtock *(second frame to cover the joints of main frame)*
auflassen *(mine)* to abandon, to give up
Auflast *f* live (superimposed, applied) load, surcharge
Auflaufbacke *f* leading shoe *(brake)*
Auflaufbremse *f* loading (overrunning) brake
auflaufen 1. *(ag)* to sprout; 2. *(ship)* to run aground, to take the ground, to strand
Auflaufen *n* **der Tide** rise (inflow) of the tide
auflegen 1. to replace *(receiver)*; to abandon a call *(telephony)*; 2. to lay up, to put up *(vessel)*; 3. to load *(workpiece)*
a. auf to place on
Auflegerschießen *n (mine)* plaster-shooting, *(Am)* mud-capping
Auflegezeit *f* lay-up time (days) *(of a vessel)*
aufleuchten to light [up]; to flash; to be illuminated; to go on *(signal light)*
Auflicht *n (opt)* incident light
Auflichtbeleuchtung *f* episcopic illumination
Auflichtkondensor *m (opt)* incident-light condenser, epi-condenser
Auflichtmikroskop *n* incident-light microscope
aufliegen 1. to lie up *(ship)*; 2. to be supported; to seat
a. auf to bear on, to be supported by
aufliegend / frei freely supported
auflockern 1. to loosen [up], to bulk; 2. *(text)* to open; 3. to fluff *(foundry sand)*
Auflockern *n* 1. loosening, bulking; 2. *(text)* opening; 3. fluffing *(foundry sand)*
Auflockerung *f* bulking; loosening
Auflockerungsfaktor *m* scarifying factor
auflösbar soluble; *(opt)* resolvable
Auflösbarkeit *f* solubility; *(opt)* resolvability
Auflösebehälter *m (text)* dissolving chest
Auflöseholländer *m* breaker (broke) beater
auflösen 1. *(math)* to solve; 2. *(opt)* to resolve; 3. *(ch)* to dissolve; to liquefy; to decompose
Auflösen *n* **der Fahrstraße** release of route locking, route release
A. eines Zuges splitting up a train
Auflösung *f* 1. *(ch)* dissolution; 2. *(math, ch)* solution; 3. *(opt, ch, telec)* resolution; *(opt)* resolving power; decomposition *(light)*
elektrolytische A. electro-dissolution
räumliche A. spatial resolution
schlechte A. poor resolution
selbsttätige A. automatic route release
spektrale A. spectral resolution
Auflösungsfähigkeit *f* resolution capability
Auflösungsfunktion *f (nucl)* resolution function
Auflösungsgeschwindigkeit *f* velocity of dissolution, dissolving rate (speed)

Auflösungsgrenze *f* resolution (resolving) limit
Auflösungsmittel *n* solvent
Auflösungstest *m (opt)* test for resolving power
Auflösungsverlust *m* loss of resolution
Auflösungsvermögen *n* resolution power *(data storage and reproduction)*; *(opt)* resolving power; *(ch)* dissolving power
  begrenztes A. *(cont)* granularity
  zeitliches A. *s.* Zeitauflösung
Auflösungswärme *f* heat of solution
Auflösungszeit *f* resolving time *(minimal pulse spacing)*; *(ch)* dissolving time
auflöten to solder (brace) on
aufmachen to open; to undo *(lace)*
  Dampf a. to raise (get up) steam
aufmagnetisieren to remagnetize
Aufmaß *n* admeasure, admeasurement
aufmodulieren to modulate on (upon)
aufmontieren auf to fasten to; to fix to; to mount on
aufnähen to sew on
Aufnahme *f* 1. *(ph, ch)* absorption; taking-up; admission; uptake, intake *(radiobiology)*; 2. input *(power, energy)*; 3. *(mach)* reception; collet *(tool)*; piloting portion *(broach)*; accommodation *(workpiece)*; receiving, accomodating; 4. *(el)* receptacle *(plug)*; 5. exposure *(camera)*; photograph; 6. recording *(tape)*; plotting *(graph)*; 7. survey
  mikrofotografische A. microphotograph
Aufnahmebehälter *m* receiving hopper
Aufnahmebohrung *f* socket
  A. der Spindel socket of the spindle
Aufnahmebunker *m* receiving bunker
Aufnahmedorn *m* mounting mandrel *(machine tool)*; guide bar, horn *(broaching keyways)*
Aufnahmeentzerrung *f* pre-equalization
aufnahmefähig *(ch)* absorbable
Aufnahmefähigkeit *f* 1. *(ch)* absorbability, absorbing capacity; 2. [receiving] capacity *(container; reception)*
  A. des Spanraums chip-carrying capacity *(broach)*
  A. eines Sammlers ampere-hour capacity
  magnetische A. susceptibility
Aufnahmefaktor *m* uptake factor *(radiobiology)*
Aufnahmegefäß *n* receptacle
Aufnahmegerät *n* input unit *(part of computer)*
Aufnahmegeschwindigkeit *f* uptake rate *(radiobiology)*
Aufnahmekegel *m* mounting (receiving) taper
Aufnahmekopf *m* recording magnetic head
Aufnahmeliste *f* record of cuttings
Aufnahmematerial *n* film stock *(photography)*
Aufnahmeobjektiv *n* taking lens
Aufnahmeröhre *f (telev)* camera (pick-up) tube
Aufnahmesperre *f* blocking of the recording; safety button *(tape recorder)*
Aufnahmespitze *f* work holding centre *(machine tool)*
Aufnahmestromkreis *m* input circuit
Aufnahmestutzen *m* intake connection (manifold) *(hydraulics)*
Aufnahmeteil *n* input unit *(part of computer)*
Aufnahmevermögen *n* 1. *(ch)* absorptivity, absorption (absorptive) capacity, absorptive power; 2. receiving capacity *(container, reception)*
Aufnahmevorrichtung *f* phonograph recorder
Aufnahmewagen *m* television car
Aufnahmezähler *m* exposure counter
Aufnähmuster *n* appliqué pattern
aufnehmen 1. to absorb *(moisture; pressure)*; to adsorb; to take up; to permit; to admit *(flowing liquid)*; to seat, to accommodate *(workpiece)*; to house; to allocate; to carry *(chips)*; to receive *(tool)*; to take; 2. to photograph; 3. to record *(tape)*; 4. to survey; 5. to resist *(pressure)*; 6. to accept *(signal)*
  den Kurs wieder a. to retrace the course
  ein Zeichen a. to receive (accept) a signal
  eine Kennlinie a. to plot a characteristic
  eine Lehre a. to start an apprenticeship
  einen Schaft a. to take a shank
  Farbstoffe a. to take dyes
  Leistung a. to take watts
  Wärme a. to absorb heat
Aufnehmer *m (therm)* receiver
Aufnehmerdampf *m (therm)* receiver steam
Aufnehmerdampfmaschine *f* steam engine with receiver
Aufnehmerheizung *f (therm)* heating of receiver
Aufnehmertrommel *f* pick-up drum
Aufpackung *f (ship)* making up, packing, fillings, stopping up
aufpfropfen to graft
Aufpfropfen *n* grafting
Aufpinseln *n* brushing
aufpolstern to upholster
Aufprall *m* impingement; impact; bound
Aufprallkorrosion *f* impingement corrosion
Aufpressen *n* **der Radkörper** *(ry)* pressing-on of the wheels [to the axles]
  A. des Rades pressing-on of the wheel *(turbine)*
aufpumpbar inflatable
aufpumpen to inflate
Aufpunkt *m (ph)* test point *(point for which potential is determined)*
Aufputzmontage *f* surface installation (mounting), over-plaster installation
Aufputzschalter *m* switch for surface mounting, surface switch
Aufputzsteckdose *f* surface socket
aufquellen to swell up
Aufradlinie *f s.* Epizykloide

Aufrahmen *n* *(text)* tentering
aufrauhen to rough up; to hack, to roughen
die Schleifscheibe a. to sharpen the grinding wheel
Aufrauhmaschine *f* roughing-up machine; *(wood)* machine for roughing veneer
Aufrauhungsprofil *n* roughened profile
Aufräumarbeit *f* clearance operation
aufräumen to clear, to tidy up
Aufräumung *f* clearance, clearing
aufrecht upright
aufrechterhalten to maintain
aufreiben to ream *(metal cutting)*
Aufreiben *n* reaming *(metal cutting)*
aufreihen to sequence
Aufreißband *n* tab
aufreißen 1. to plot; to scribe; to lay out; 2. to scarify *(road)*; to rip; to crack
Aufreißen *n* 1. plotting; scribing; laying-out; 2. scarification *(road)*; ripping; 3. cracking *(of the canning)*
Aufreißer *m* ripper, scarifier
Aufreißkamm *m* scarifier *(road)*
aufrichten to right; to erect
Aufrichten *n* **der tiefliegenden Stöße** *(ry)* rectifying of low joints
A. eines Konverters turning up of a converter
Aufrichteprisma *n* erecting prism
Aufrichtungsprisma *n* erecting prism
Aufrichtungsvermögen *n* righting capacity
Aufriß *m* front view *(technical drawing)*
Aufrißfläche *f* **des Schornsteins** vertical section of the chimney
aufrollen to coil *(wire)*; to reel; to curl *(chip)*
sich in einer festen Spirale a. to coil up in a tight spiral
Aufrollen *n* **des Spans** chip curling
Aufruf *m* *(dat)* call-in
Aufrufzahl *f* *(dat)* call number
aufrühren *(ch)* to agitate, to stir
aufrunden *(dat)* to round off, to truncate
Aufsammelfeldhäcksler *m* pick-up loader *(forage harvester)*
aufsammeln to pick up
Aufsammelpresse *f* clearing baler, baling press
Aufsammelschneidegebläse *n* *(ag)* pick-up cutter-blower
Aufsatteldruck *m* *(aut)* fifth-wheel load
Aufsattellast *f* *(aut)* fifth-wheel load
Aufsatzbacke *f* false jaw *(chuck)*
Aufsatzrahmen *m* filling frame *(moulding)*
aufsaugen to soak, to absorb, to resorb
Aufsaugen *n* absorption
aufsaugend absorbing
Aufsaugung *f* absorption; suction; imbibition
aufschablonieren to stencil on
aufschalten 1. to lock on *(radar)*; *(rock)* to mix signals
Aufschaltmotor *m* *(rock)* signal mixing motor
aufschaukeln to amplify, to build up *(waves)*
Aufschaukeln *n* amplitude increase *(of a vibration)*; build-up, increase of amplification *(of waves)*
Aufschaukelung *f* **des Signals** resonant rise of a signal
Aufschaukelungsgeschwindigkeit *f* speed of resonant rise
aufschäumen 1. to foam [up], to froth; to effervesce; 2. *(text)* to foam onto; 3. to bubble *(melt)*
Aufschäumen *n* 1. *(ch)* frothing, foaming-up; effervescing; 2. *(text)* foaming onto; 3. bubbling *(melt)*
Aufschäumtechnik *f* frothing technique *(laminating)*
aufschichten to stack; to pile up
aufschieben to delay, to postpone; to push open
Aufschieben *n* *(mine)* caging, decking, on-setting
Aufschiebung *f* *(geol)* thrust [fault]
aufschießen to luff into the wind *(navigation)*
Aufschlag *m* 1. impact; 2. extra charge
Aufschlagbarkeit *f* *(wood)* defibratability
Aufschlagen *n* 1. striking the ground *(aeroplane)*; 2. *(text)* disintegration
Aufschlaggerät *n* defibrator
Aufschlaghaspel *f* **für den Äscher** lime reel
Aufschlagständer *m* *(ry)* supporting post
Aufschlagzünder *m* percussive fuse
aufschlämmen to suspend; to triturate
Aufschlämmung *f* [liquid] suspension; dredge
A. in Wasser suspension in water
Aufschleppdock *n* slip dock
Aufschleppe *f* railway dry dock; [railway] slip
Aufschlepphelling *f* slipway, patent (railway, hauling) slip, marine railway, railway dry-dock
Aufschleppwagen *m* railway (slipway) cradle
aufschließen 1. to unlock; 2. *(mine)* to open out (up), to develop; 3. *(met)* to disintegrate, to pulverize; 4. to digest *(Bayer process)*
Rückstände a. *(ch)* to fuse residues
Stärke a. *(ch)* to hydrolize starch
Aufschließen *n* **von Bauxit** digesting bauxite
Aufschlitzen *n* *(leath)* ripping cut
Aufschluß *m* 1. *(mine)* opening-up, development; 2. digestion *(Bayer process)*; pulping *(paper)*; 3. disintegration, pulverizing
Aufschlußarbeiten *fpl* initial work *(opencast mining)*
Aufschlußgrad *m* extent of delignification *(wood pulp)*
A. des Zellstoffs pulp hardness
Aufschlußmischung *f* fusion mixture
Aufschlußverfahren *n* fusion process
Aufschneiden *n* *(leath)* ripping cut

A. der Florschläuche cutting of the pile tubes
A. von Weichen *(ry)* forcing (bursting) open the points
aufschnüren to unlace
Aufschnüren *n* **des Schiffskörpers** ship lofting
aufschobern to cock
aufschrauben auf to bolt on, to screw on [upon]
Aufschraubfräser *m* screw-on cutter
Aufschraubverlängerung *f* screwed-on extension
aufschreiben to write down; to record *(device)*
aufschrumpfen to shrink on
Aufschrumpfen *n* shrinking-on
A. unter Vakuum snap-back forming, vacuum snap-back
Aufschrumpfsitz *m* shrinkage fit
aufschütten to fill, to make up *(ground)*; to pile *(goods)*
Aufschüttung *f* fill[ing], filled (made-up) ground; embankment
Aufschweißbiegeprobe *f* bead bend specimen
Aufschweißbiegeversuch *m* bead bend test
aufschweißen 1. to weld on; 2. to open with a welding torch
Aufschweißlegierung *f* hard-facing alloy; filler metal
aufschwimmen to float
Aufschwingfaktor *m* rise factor
Aufseher *m* supervisor
aufsetzen to put on; to attach
ein Faß a. to hoop a cask, to mount the staves
Aufsetzen *n* **eines Flickens [auf das Kesselblech]** patching
A. von Werkstücken loading components
Aufsetzvorrichtung *f* *(mine)* bearing-up stop, cage seats, catch, keeps, kep, landing chair (dogs)
Aufsicht *f* control *(airport)*; supervision
aufsintern to sinter-fuse
aufsitzen auf to be seated on
aufspachteln to trowel; to float
aufspalten to split [up], to break up, to crack; to cleave
in optisch aktive Komponenten a. *(ch)* to resolve in optically active components
Aufspalten *n* **der Spektrallinien** splitting-up of the spectrum lines
Aufspaltung *f* splitting [up]; cleaving; *(nucl)* fission; *(ch)* dissociation
A. der Atome dissociation of atoms
A. der Versetzungen splitting of dislocations
Aufspaltungsfaktor *m* *(nucl)* [Landé] *g*-factor, interval (splitting) factor
Aufspannbock *m* planer jack
Aufspanndorn *m* mounting mandrel
aufspannen 1. *(mach)* to chuck, to clamp, to mount *(work)*; 2. to spread out
Aufspannen *n* 1. *(mach)* chucking, clamping, mounting *(work)*; 2. spreading out; 3. *(text)* tentering
Aufspanner *m* step-up transformer
Aufspannfläche *f* **des Tisches** table working surface, work clamping area
Aufspannfutter *n* chuck
Aufspanngerät *n* holding tool
Aufspannplatte *f* face plate of chuck; magnetic chuck of rectangular pattern
Aufspanntisch *m* clamping table, worktable, work-holding table
Aufspannung *f* clamping; set-up
A. / mit einer at one setting
Aufspannvorrichtung *f* [clamping] fixture, clamping device, work-holding device (fixture)
Aufspannwerkzeug *n* holding tool
Aufspannwinkel *m* angle plate
Aufspannzeit *f* clamping (loading) time; setting-up time
aufspeichern to store up *(energy)*; to stock; to accumulate
aufspleißen to fan out *(cable)*
Aufspleißen *n* fan-out *(cable)*
aufspritzen to spray
aufsprühen to spray
Aufsprühstreichmaschine *f* spray coater
Aufsprungbahn *f* alighting ground, landing hill *(ski-jumping)*
aufspulen to reel up, to coil, to wind up (on reels)
Aufspülung *f* hydraulic fill
aufstampfen to ram up *(mould)*
Aufstand *m* *(build)* upstand; cresting
Aufstandspfahl *m* *(build)* end-bearing pile
aufstapeln to stack, to pile up
Aufsteckantenne *f* plug-in aerial
Aufsteckblech *n* spill-plate
Aufsteckbrett *n* creel board
Aufsteckdorn *m* stub arbor
Aufsteckeinheit *f* *(el)* plug-in device
aufstecken to slip over; to mount on; *(text)* to creel *(bobbins)*
Aufstecken *n* slipping-over; mounting-on
Aufsteckfilter *n* *(opt)* push-on filter
Aufsteckfräser *m* arbor (hole-type) cutter, arbor-type mill
Aufsteckgatter *n* *(text)* creel
Aufsteckglas *n* *(opt)* attachment lens
Aufsteckgrundreibahle *f* rose shell reamer, shell-type reamer
Aufsteckhalter *m* accessory shoe *(camera)*
Aufsteckkappe *f* slip-on cap
Aufsteckrahmen *m* *(text)* creel frame
Aufsteckreibahle *f* shell reamer
Aufsteckspindel *f* *(text)* skewer
Aufsteckspule *f* *(el)* plug-in coil
Aufsteckwerkzeug *n* shell-type tool

Aufsteckwinkelfräser *m (Am)* stub angular cutter
aufsteigen to rise *(gases)*; to ascend; to take to the air, to start *(aeroplane)*
Aufsteigungsunterschied *m* ascensional difference
aufstellen 1. *(mach)* to erect; to assemble; to allocate; to mount; *(ag)* to stack *(sheaves)*; 2. *(math)* to set up *(equation)*
  Standards a. to establish standards
Aufstellen *n* **von Garben** stacking [of] sheaves
  A. von Maschinen erection of machines
Aufstellung *f* installation; assembly; *(mach)* erection; *(el)* mounting position
  erschütterungsfreie A. anti-vibration mounting
Aufstellungsfläche *f* storage space
  benötigte A. floor space occupied *(machine)*
Aufstellungskosten *pl* cost of installation (erection)
aufstemmen to open by means of a chisel
aufsteppen to quilt on
aufsticken 1. *(met)* to nitride *(surface)*; 2. *(text)* to embroider on
Aufstickung *f (met)* nitration
Aufstieg *m* rise, ascent, climb; *(build)* step
Aufstiegbahn *f* trajectory of climb, climbing path, ascent trajectory *(of rockets)*
Aufstiegkanal *m* cable chute (shaft)
Aufstoßen *n* running on *(knitting)*
Aufstoßnadelbarre *f* points bar *(knitting)*
Aufstoßstellung *f* running-on position *(knitting)*
Aufstrandholen *n (ship)* beaching, stranding
aufstreichen to brush on, to apply by brush
aufstreifen to slip on
aufstreuen to dust on *(mould)*; to spread over
Aufstrichmasse *f* coating material
Aufstromklassierer *m* upward-current classifier
aufstützen to support
Aufsuchen *n* prospecting; search
auftakeln to rig up *(a vessel)*
auftanken to refuel
Auftankung *f* refuelling
Auftastgenerator *m* gate generator
Auftastimpuls *m* gate (gating) pulse
Auftastimpulskreis *m* gate
Auftastkreis *m* interval-selector circuit
Auftastschaltung *f* gate circuit
auftauchen to surface, to emerge, to break surface *(U-boat)*
auftauen to thaw
Auftauen *n* thawing
Auftausalz *n* de-icing salt
aufteilen to distribute; to split up; to break down; to subdivide; to separate
Aufteilungsmuffe *f* distribution cable connector
Auftrag *m* 1. order; 2. fill[ing], filled (made-up) ground; coat[ing]
auftragen 1. to coat; to spread; to apply *(material)*; 2. to plot *(graph)*
Auftragen *n* / **beiderseitiges** double spreading
Auftragmaterial *n* fill; coating material
Auftragmetall *n* deposit (build-up) metal; coating metal
Auftragnehmer *m* contractor
Auftragschweißen *n* deposit (building-up) welding, hard [sur]facing
Auftragsdienst *m (telec)* absent subscriber service, *(sl)* absentee service
Auftragsforschung *f* sponsored [industrial] research
Auftragsnummer *f* job number
Auftragszusammenstellung *f* order picking
Auftragung *f* 1. deposition; coating; spreading; application, applying *(material)*; 2. plotting *(graph)*
Auftragwalze *f* inking roller; applicator roll; casting roll; spreader roll
Auftreffelektrode *f* target electrode
auftreffen to strike; to impinge; to be incident on; to encounter
  auf einen Kontakt a. to strike a contact
auftreffend incident *(sound)*
Auftreffgeschwindigkeit *f* impact velocity (speed)
Auftreffpunkt *m* incidence point
Auftreffwinkel *m* angle of impact
Auftreffwucht *f* energy of impact
auftreiben to force on; to blow on
auftrennen to sever; to rip [up]
Auftrenntechnik *f (telec)* sectional toll switching
Auftreten *n* occurrence; appearance
Auftrieb *m* buoyancy *(liquid)*; lifting power *(air)*; upward pressure
  thermischer A. thermal lift
Auftriebsbeiwert *m* lift coefficient
Auftriebsbelastung *f* lift loading
Auftriebsdichtemesser *m* float-type densitometer
Auftriebserhöhung *f (aero)* augmentation of lift
Auftriebskammer *f (ship)* balance chamber
Auftriebsklappe *f (aero)* lift flap
Auftriebskraft *f* buoyancy; lift component
Auftriebskurve *f (ship)* curve of buoyancy
Auftriebsregelung *f (aero)* lift control
Auftriebsverlust *m (aero)* loss of lift
Auftriebszahl *f (aero)* coefficient of lift
Auftriebszentrum *n* centre of buoyancy
Auftrommelmaschine *f* coil winder
auftrommeln to reel in (up)
Auf- und Abbewegung *f* 1. *(mach)* reciprocating motion in vertical direction; 2. *(ship)* heave
Auf- und Abfahrtsvorrichtung *f* drive-on-drive-off layout *(of a vessel)*
Aufwallung *f* ebullition

aufwältigen *(mine)* to re-open
Aufwältigung *f (mine)* re-opening
aufwalzen to roller-coat
Aufwalzvorrichtung *f* roll coater
Aufwand *m* 1. expenditure; expense; 2. power input
 optischer A. optical outlay
aufwärmen to reheat
aufwärtsbewegen to ascend, to move upwards
Aufwärtsbewegung *f* 1. upstroke, up-movement *(machine part)*; 2. *(ship)* scend *(lifting or heaving upward in a seaway)*
Aufwärtskomponente *f* upward component
Aufwärtstransformator *m* step-up transformer
aufweichen to soften
aufweiten to expand; to widen; to stretch; to enlarge *(borehole)*
 die Düsen a. to expand the nozzle section *(turbine)*
Aufweiten *n* expanding; widening; stretching; enlarging *(borehole)*
Aufweiteversuch *m* expansion (bulging) test
Aufweitewalzwerk *n* expanding (becking) mill *(tubes)*
Aufweitung *f* bell mouthing *(bore)*; opening-out
Aufwickelfriktion *f* take-up friction *(photography)*
Aufwickelkassette *f* take-up cassette *(photography)*
aufwickeln to wind up; to reel; to take up
 sich wendelförmig a. to curl out in the form of a helix *(chip)*
Aufwickeln *n* winding on
 A. auf Kopse winding into cop form
Aufwickelseite *f* reeling end
Aufwickelspule *f* 1. *(text)* winding bobbin; 2. take-up spool *(photography)*
Aufwind *m* ascending current
aufwinden to wind on; to hoist
Aufwinder *m (text)* winding faller *(selfactor)*
Aufwinderdraht *m (text)* winding faller wire *(selfactor)*
Aufwindespannung *f* **des Garns** winding tension of the yarn
Aufzehrung *f* 1. consumption; 2. clean-up *(discharge tubes)*
Aufzeichnen *n* recording; plotting *(graph)*; charting
 thermoplastisches A. thermoplastic recording
Aufzeichnung *f* record[ing]; plot[ting]
 A. auf Landkarten mapping
 A. in Seitenschrift lateral recording
 A. von Fernsehsendungen television recording
 grafische A. graphic recording
 sichtbare A. visible record
 thermoplastische A. thermoplastic recording
Aufzeichnungsamplitude *f* amplitude of write-out
Aufzeichnungsdichte *f* recording density
Aufzeichnungsfläche *f* plotting area
Aufzeichnungsfrequenz *f* recording frequency
Aufzeichnungsgerät *n* recorder; plotter
Aufzeichnungsgeschwindigkeit *f* writing (record) speed
Aufzeichnungskopf *m* recording head
Aufzeichnungsskala *f* logging scale
Aufzeichnungsstrukturen *fpl* head-surface configuration, recording structures *(magnetic tape technics)*
aufziehen to wind [up] *(clock)*
 auf Pappe a. to mount on cardboard
 warm a. to shrink on *(wheel tyre)*
Aufziehen *n* 1. winding-up; 2. *(text)* chaining
 A. des Radreifens tyring [a wheel], pressing on a tyre
Aufziehgeschwindigkeit *f (text)* rate of absorption
Aufziehkarton *m* mounting board
Aufziehvorrichtung *f* **für Kratzenbänder** *(text)* card mounting machine
Aufzischen *n* fizzing
Aufzuchtbatterie *f (ag)* tier brooder
Aufzug *m* 1. elevator; lift; hoist; 2. packing, dressing
Aufzugbogen *m* packing sheet
Aufzugleine *f* rip (release) cord *(parachute)*
Aufzugschacht *m* elevator well (shaft)
Aufzugseil *n* hoisting rope
Aufzugskasten *m (build)* skip
Aufzugskübel *m (build)* lifting vat, skip
Aufzugswinde *f* hoisting winch
Auf-Zu-Regler *m* on-off control
aufzwicken to last a shoe
Augapfel *m* eye ball
Auge *n* eye; eyelet; lug; boss
 akkommodationsloses A. unaccommodated (fixed) eye
 magisches A. magic eye, tuning (cathode ray) indicator, electron ray tube
 unbewaffnetes A. *(opt)* unaided eye
Augenabschnitt *m (opt)* segment of the eye
Augenabstand *m (opt)* interocular (interpupillary) distance
Augenabstandsmesser *m* interpupillary distance gauge
Augenarztbesteck *n* oculist's diagnostic set
augenblicklich instantaneous; momentary
Augenblicksfrequenz *f (el)* instantaneous frequency
Augenblicksleistung *f* instantaneous power
Augenblicksspannung *f* instantaneous voltage
Augenblickswert *m* instantaneous value
 A. des Stromes instantaneous current
Augenblickswerte *mpl* / **flüchtige** elusive transients
Augenbrauenstift *m* eyebrow pencil

Augendraht *m* mail wire
Augenlager *n (mach)* solid journal bearing
Augenoptik *f* ophthalmic optics
Augenoptiker *m* ophthalmic optician
Augenpunkt *m (math)* centre of vision; point of sight, visual point
Augenschraube *f (mach)* eyebolt
Augenschutz *m* eye protection, eyeshield
Augenspiegel *m* ophthalmoscope
einfacher A. simplified ophthalmoscope
Auger-Effekt *m* Auger effect
Auger-Schauer *m (nucl)* extensive (giant air, Auger) shower
Augmentationsschiff *n* augmentation vessel
Aula *f* assembly hall
Auramin *n* auramin[e] *(bright yellow ketonimine dye)*
Aurat *n (ch)* aurate
Aureole *f* aureole
Auripigment *n (min)* orpiment
Auroralinie *f (ph)* auroral line
aus off, stop *(switch position)*
ausarbeiten to elaborate, to prepare, to work out
Ausästen *n* trimming, pruning
Ausäthern *n* etherizing, shaking out with ether
ausbaggern to dredge *(in water)*; to excavate *(on land)*
Ausbaggerung *f* dredging *(in water)*; excavation *(on land)*
ausbalancieren to compensate; to equilibrate; to [counter]balance
das Ventil a. to balance the valve
Ausbau *m* 1. *(build)* interior work, internal finish, finishing operations, completion, finishes and services; 2. *(mine)* lining, support, timbering, tubbing; 3. dismantling *(machine part)*, demounting
A. / ohne *(mine)* unlined
A. im Abbau *(mine)* face support
endgültiger A. *(mine)* permanent support
ingenieurtechnischer A. building services
mechanisierter A. *(mine)* powered support
schreitender A. *(mine)* walking (self-advancing roof, self-moving roof) support
vorläufiger A. *(mine)* provisional (temporary) support
wasserdichter A. *(mine)* coffering
Ausbauabschnitt *m* planning period, stage of extension
Ausbauanker *m (mine)* rock (roof) bolt
Ausbaubereichsplan *m* extension survey map
Ausbaubühne *f* 1. tubbing platform *(shaft)*; 2. support scaffold *(drift)*
ausbauchen 1. to bulge, to buckle; 2. to expand *(tool)*
Ausbauchen *n* 1. bulging *(rubber)*; 2. expanding *(tool)*
Ausbauchung *f* flare *(of bow or sides of a vessel)*
ausbauen 1. *(build)* to finish internally, to complete; 2. to demount, to dismantle *(machine part)*; 3. *(mine)* to crib, to line, to support, to timber
Ausbaufeld *n (mine)* track
Ausbaugeschwindigkeit *f* design speed
Ausbaugeviert *n (mine)* frame
Ausbaugewerke *npl* finishing trades
Ausbauhauptnetzplan *m* extension main network map
Ausbaukabelkanalnetzplan *m* extension conduit network map
Ausbaurahmen *m (mine)* timber set
Ausbauregel *f (mine)* support rule
Ausbauschema *n (mine)* support pattern
Ausbauverbindungsnetzplan *m* extension junction network map
Ausbauverzweigungsnetzplan *m* extension distribution network map
ausbeißen *(geol)* to come out to the day, to crop out
ausbessern to repair, to mend; to fettle *(lining of furnace)*, to recondition
Ausbessern *n* **eines Ofens** patching of furnace
Ausbesserung *f* **/ kleine** minor repairs
Ausbesserungsdock *n* repair dock
Ausbesserungsgleis *n* repair track; *(Am)* rip track
Ausbesserungswerk *n* repair shop
Ausbesserungswerkstatt *f* repair shop
ausbeulen to bulge; to buckle
Ausbeulen *n* bulging; buckling *(outward)*
Ausbeute *f* output; yield; produce; efficiency
A. an angereichertem Material *(nucl)* enriched output
A. des abgereicherten Materials *(nucl)* stripped output
A. je Stufe *(nucl)* fractional yield
Ausbeute-Masse-Kurve *f (nucl)* yield-mass curve
ausbeuten to exploit; to mine
Ausbeutung *f* 1. exploitation; 2. mining
ausbiegen to bulge, to bow out; to bend; to deflect
Ausbiegung *f* bulging; bend[ing]; deflection
ausbilden 1. to train *(apprentices)*; to instruct; 2. to form
Ausbildung *f* 1. training, education; 2. formation
konstruktive A. design
Ausbildungsleiter *m (Am)* training director
Ausbildungsmethoden *fpl* training methods
Ausbildungspersonal *n* training staff
Ausbildungsplan *m* training scheme *(apprentices)*
Ausbildungsreaktor *m* training reactor
Ausbildungswerkstätte *f* training workshop
ausbinden *(print)* to tie up

Ausbindeschnur *f (print)* page cord
Ausbiß *m (geol)* apex, crop[ping], outcrop
Ausblasebütte *f* blow pit *(papermaking)*
ausblasen to blow out *(borehole)*; to blow down *(furnace)*
  Dampf a. to blow off (out) steam, to exhaust steam
Ausbläser *m (mine)* blow[n]-out shot
Ausblaseventil *n* blow-off valve
ausblatten 1. to jag, to calk; to notch; 2. *(ag)* to remove side shoots and sick leaves
ausbleiben to fail *(ignition)*
ausbleichen to fade *(shade)*
ausblenden to fade down (out); *(opt)* to block out
Ausblendstufe *f (el)* double limiter; gate
Ausblendung *f* 1. collimation *(radiation)*; 2. fade-out
Ausblendungsverhältnis *n* front-to-back ratio *(aerials)*
Ausblendwirkung *f* suppression ratio *(aerials)*
ausblühen to effloresce, to bloom
Ausblühung *f* efflorescence, bloom
ausbluten *(text)* to bleed
Ausbluten *n (text)* bleeding
Ausbohren *n* 1. drilling *(from the solid)*; 2. boring *(enlarging)*
Ausbohrmaschine *f* boring machine
ausbooten to disembark
Ausbrand *m s.* Abbrand
ausbrechen 1. to break out; to chip *(cutting edge)*; 2. *(ag)* to pinch; 3. to swerve in taking off *(aeroplane)*
Ausbrechen *n* 1. breaking out; chipping *(cutting edge)*; 2. *(ag)* pinching; 3. swerving in taking-off *(aeroplane)*
  A. des Kornes tearing loose, release [of abrasive grains]
Ausbreitemaschine *f* expander
ausbreiten to spread [out]; to extend; to expand; to unfold *(cloth)*; to propagate *(waves)*
Ausbreiter *m* expander, spreader
Ausbreitmaß *n* slump
Ausbreitprobe *f s.* Ausbreitversuch
Ausbreitung *f* spreading [out]; propagation *(waves)*; expansion; unfolding *(cloth)*
Ausbreitungsart *f* mode of propagation
Ausbreitungsbedingung *f* propagation condition
Ausbreitungscharakteristik *f* propagation pattern
Ausbreitungsfaktor *m* propagation factor
Ausbreitungsform *f* mode of propagation
Ausbreitungsformel *f* radio-transmission formula
Ausbreitungsgeschwindigkeit *f* velocity (speed) of propagation, propagation (spread) velocity
Ausbreitungskonstante *f* propagation coefficient (constant)
Ausbreitungsmaß *n* propagation constant
Ausbreitungsrichtung *f* direction of propagation
Ausbreitungsträger *m* medium of propagation
Ausbreitungsverlust *m* propagation loss
Ausbreitungsvermögen *n* 1. diffusivity; 2. spreading power *(of concrete)*
Ausbreitungsweg *m* propagation path
Ausbreitungswiderstand *m* earthing resistance, *(Am)* grounding resistance
Ausbreitversuch *m (build)* slump (flow) test
Ausbrenneffekt *m (text)* burning-out effect
ausbrennen 1. to burn out; 2. to cleanse by fumigation
ausbringen to yield; to produce
Ausbringen *n* yield
Ausbringung *f* output, production [rate]; capacity
Ausbröckeln *n* crumbling *(tool edge)*
Ausbruch *m* 1. chipping *(cutting edge)*; 2. outburst, eruption; 3. excavation; 4. *(mine)* backbreak
Ausbruchdurchmesser *m (mine)* excavated diameter
Ausbruchleistung *f (mine)* cubic yardage
Ausbruchquerschnitt *m (mine)* excavated cross-section
ausbüchsen to bush
Ausbüchsen *n* bushing
ausbuchten to bulge, to buckle, to bow out
ausdehnbar expandable; extensible; ductile *(material)*
Ausdehnbarkeit *f* expansibleness; extensibility, extendibility; ductility *(material)*
ausdehnen to expand; to stretch; to extend; to dilate
Ausdehnung *f* expansion *(engine)*; dilatation; distension; extension; elongation; stretch
  adiabatische A. adiabatic expansion
  kubische A. volume expansion
  räumliche A. volume expansion
  spezifische räumliche A. specific volumetric dilatation
  thermische A. thermal expansion
Ausdehnungsfuge *f* expansion joint
Ausdehnungsgefäß *n* 1. expansion tank *(hot-water system)*; 2. conservator [chamber] *(of transformers)*
Ausdehnungskoeffizient *m* coefficient of expansion
  kubischer A. coefficient of cubical expansion
  linearer A. coefficient of linear expansion
  räumlicher A. coefficient of cubical expansion
  relativer A. coefficient of apparent expansion
  thermischer A. coefficient of thermal expansion
Ausdehnungsmesser *m* dilatometer; extensometer
Ausdehnungsraum *m* expansion chamber
Ausdehnungsrohr *n* expansion pipe

Ausdehnungsspannung *f* expansion stress
Ausdehnungstakt *m* expansion stroke
Ausdehnungsthermometer *n* filled-system thermometer
Ausdehnungsvermögen *n* expansivity; dilatability
ausdocken to undock, to float out *(a vessel)*
ausdrehen 1. to turn (switch) off; 2. to bore [in a lathe], to turn internal surfaces
Ausdrehen *n* turning internal surfaces, boring in [a] lathe
Ausdrehmeißel *m* boring (internal turning) tool
Ausdrehwerkzeug *n* / **fliegendes** fly cutter in stub bar
ausdreschen to thresh out
Ausdruck *m (math)* expression; term
  algebraischer A. algebraic expression
  analytischer A. analytical expression
  arithmetischer A. arithmetical expression
  mehrgliedriger A. polynomial
  quantitativer A. quantitative expression
ausdrückbar expressible
Ausdruckdauer *f* print-out period
ausdrucken to print [out]
Ausdrucken *n* logging; print-out
ausdrücken to express; to press (force) out
Ausdruckfolge *f (dat)* logging cycle
Ausdrückgleis *n* ram track
Ausdrückmaschine *f* pushing machine, pusher
Ausdrückseite *f* pusher side, ram side
Ausdrückstange *f* ram bar (rack)
Ausdrückstift *m (mach)* ejector plate
Ausdrucksweise *f* / **technische** style of technical writing
Ausdruckzyklus *m (dat)* logging cycle
ausdünnen *(ag)* to thin [out]
Ausdünngerät *n (ag)* down-the-row thinner
Ausdünnmaschine *f (ag)* thinner
Ausdünnung *f (ag)* thinning [out]
auseinander apart
auseinanderlaufen to diverge; to spread *(colour)*
Auseinanderlaufen *n* divergence; spreading *(colour)*
auseinandernehmen to dismantle; to take apart; to dismember
Auseinandernehmen *n* dismantling; taking apart; dismembering
Auseinanderspreizen *n* expansion; expanding
Auseinanderziehen *n* **der Kette** *(text)* drawing apart of the warp
ausfachen to shelve
Ausfachung *f* infilling
Ausfachungstafel *f* infill panel
Ausfahrbühne *f* larry car platform
ausfahren 1. *(nucl)* to discharge, to unload, to withdraw, to draw out, to rise, to move out, to extract *(the rods, the fuel etc.)*; 2. *(mine)* to come up, to walk out; 3. *(mach)* to move out; 4. *(aero)* to extend *(landing gear)*
  eine Kurve a. to round a curve
Ausfahren *n* **des Tisches** moving out the table
Ausfahrgleis *n* departure track (line)
Ausfahrsignal *n* starting signal
Ausfahrt *f* 1. opening out, outlet *(of a harbour)*; departure; 2. *(mine)* ascent; 3. exit, interchange
Ausfall *m* 1. deficiency; loss; 2. stoppage *(factory)*; breakdown, failure; 3. *(nucl)* fall-out; 4. *(ship)* flare *(of bow or sides)*; rake *(of the stern)*
ausfällen to precipitate
ausfallend emergent *(ray)*
Ausfallerwartung *f* probability of failure
Ausfallmuster *n* outturn sheet
Ausfallrate *f* failure rate
Ausfallrutsche *f* billet chute *(rolling mill)*
Ausfallstraße *f* exit (arterial) road
Ausfallswinkel *m* angle of emergence *(ray)*
Ausfällung *f* precipitation; deposit; flocculation
Ausfällungsmittel *n* precipitant
Ausfallverteilung *f* mortality distribution
Ausfallzeit *f* down (lost) time
ausfärben to dye completely
ausfertigen to finish
Ausfeuereinrichtung *f* sparking-out equipment
ausfeuern to spark out
Ausfeuerzeit *f* sparking-out period *(grinding)*
ausfiltern / **eine Frequenz** to reject (trap out) a frequency
ausflammen to flame out
Ausfleckung *f* spotting out
ausfließen to flow out; to issue; to discharge
Ausfließzeit *f* outflow time
ausflocken to separate as a flocculent, to flocculate, to coagulate
Ausflockung *f* flocculation, coagulation
ausfluchten to align
Ausfluchten *n* alignment
Ausflugsschiff *n* excursion vessel
Ausfluß *m* discharge, issue, outflow; leakage; outlet
Ausflußdruck *m* open pressure
Ausflußdüse *f* nozzle, orifice
Ausflußgeschwindigkeit *f* velocity of efflux
Ausflußkoeffizient *m* outflow coefficient *(turbine)*; coefficient of discharge
Ausflußmenge *f* quantity discharged *(turbine)*; efflux
Ausflußrichtung *f* direction of outflow (exhaust) *(turbine)*
Ausflußspektrum *n* leakage spectrum
Ausflußviskosimeter *n* efflux viscosimeter
Ausflußwiderstand *m* back pressure; exhaust (outflow) resistance *(turbine)*

Ausflußzeit *f* time of efflux; exhaust period; outflow period *(turbine)*
Ausflußziffer *f* [efflux] velocity coefficient
ausfräsen to mill out
Ausfressung *f* scouring *(lining of furnace)*; erosion; pitting
Ausfrierverfahren *n* freeze-out technique
ausfugen to joint [brickwork]
Ausfuhr *f* export
ausführen 1. to perform, to accomplish; 2. to construct; 3. to export
eine hin- und hergehende Bewegung a. to reciprocate
einen Befehl a. 1. *(dat)* to obey; 2. to carry out an order
Ausfuhrschein *m* export certificate
Ausführung *f* 1. construction; constructional features; 2. model; style; type; design; version
geschlossene A. packaged construction
korrosionsbeständige A. corrosion-resistant finish
maßgerechte A. accurate workmanship (design)
nicht geschlossene A. *(el)* unpackaged construction
Ausführungsart *f* version, construction, type
Ausführungsbestimmung *f* code of practice
Ausführungszeichnung *f* working drawing
Ausfuhrverbot *n* embargo
ausfunken to spark machine
a. lassen [ohne Zustellung] to allow to spark out [with the same setting]
Ausfunken *n* sparking out
ausfüttern to line *(furnace etc)*; to pad
Ausfütterung *f* lining; padding
Ausgabe *f* output *(computer)*; read-out; output unit; edition *(book)*; issue *(ticket)*
A. eines Fahrausweises issue of a ticket
einbändige A. one-volume edition
zweibändige A. two-volume edition
Ausgabeeinrichtung *f (dat)* read-out device
Ausgabefenster *n* serving (service) hatch
Ausgabegerät *n* output unit *(in computer)*; output equipment *(separated from computer)*
Ausgabeinformation *f (dat)* output
Ausgabespeicher *m (dat)* external memory; output block
Ausgabeteil *n (dat)* output unit
Ausgabevorrichtung *f (dat)* display
Ausgabewerk *n* output device (element) *(computer)*
Ausgang *m* exit; outlet; outflow; receiving end; output *(power)*
unmodulierter A. continuous wave output
Ausgangsbandführung *f* delivery tape guide
Ausgangsbuchse *f* exit hub *(punch card machinery)*
Ausgangsdrehzahl *f* 1. output speed; 2. initial speed
Ausgangsdrossel *f* output inductance coil
Ausgangselektrode *f* output electrode
Ausgangselement *n* parent [element]
Ausgangserzeugnis *n* initial product
Ausgangsfeld *n (el)* parent field
Ausgangsfläche *f* starting surface
Ausgangsfunktion *f* original function
Ausgangsgleichung *f* initial equation
Ausgangsgröße *f (dat)* output [variable]
Ausgangshub *m* output swing
Ausgangsimpedanz *f* output impedance
Ausgangsimpuls *m* output pulse
Ausgangsinformation *f* output *(computer)*
Ausgangskabel *n* outgoing cable
Ausgangskapazität *f* output capacitance *(of an electronic valve)*
Ausgangskern *m (nucl)* original (parent) nucleus
Ausgangsklemme *f (el)* outlet terminal, terminal at the output side
Ausgangsklemmen *fpl* **des Zählers** consumer's terminals
Ausgangskontrolle *f* inspection of end-product
Ausgangsleistung *f* power output
effektive A. effective power output
momentane A. instantaneous power output
unverzerrte A. undistorted power output
Ausgangsleistungsmesser *m* output meter
Ausgangsleitwert *m* output admittance *(of electronic valves)*
Ausgangslinie *f* parent [mass] peak
Ausgangsmetall *n* parent metal
Ausgangsmoment *n* output torque
Ausgangsplatz *m (telec)* outgoing position
Ausgangsprodukt *n* starting material, feed stock; reactant; raw material
Ausgangspunkt *m* base, starting point
Ausgangsquerschnitt *m* starting section
Ausgangsscheinwiderstand *m s.* Ausgangsimpedanz
Ausgangsschiffsform *f* parent hull
Ausgangsseite *f* 1. *(mach)* exit side; output side; 2. *(el)* receiving end
Ausgangssignal *n (dat)* output
Ausgangsspannung *f* output voltage
Ausgangsspannungsteiler *m* output attenuator (voltage-divider)
Ausgangsspeicher *m (dat)* output block
Ausgangsstellung *f* original (starting, initial) position
Ausgangssteuerung *f* regulation of output
Ausgangsstrom *m* output current
Ausgangsteil *n* blank
Ausgangsübertrager *m (el)* output transformer
Ausgangswerkstoff *m* raw material
Ausgangswert *m* original value
Ausgangszellulose *f* basic cellulose *(papermaking)*

Ausgarzeit *f* killing period *(steel)*
Ausgasung *f* 1. *(mine)* gas emission; 2. fumigation *(of a vessel)*
ausgeben / optisch *(dat)* to display
ausgefleischt / stark *(leath)* scooped
ausgeglichen balanced; equalized
ausgeglüht annealed
ausgehärtet precipitation-hardened
ausgehen 1. to go off *(light)*; 2. to run short *(material)*
Ausgehendes *n* *(geol)* crop[ping], outburst, outcrop
ausgeizen *(ag)* to pluck off suckers *(vineyard)*
Ausgeizen *n* *(ag)* suckering *(vineyard)*
ausgekleidet / basisch lined with a basic refractory *(furnace)*
ausgekragt cantilevered, corbelled
ausgekuppelt disengaged, out of mesh
ausgelaufen worn-out *(bearing)*
Ausgelaugtes *n* pulp *(sugar)*
ausgelernt apprenticed
ausgelöst / durch Sprache speech-excited
ausgerückt out of mesh, thrown out
ausgerundet filleted, rounded
ausgerüstet / mit Radar radar-equipped
ausgeschaltet sein to be switched off, to be disconnected; to be disengaged
ausgesetzt exposed to, subjected to
Ausgesiebtes *n* screenings, siftings
ausgespart recessed
  beiderseitig a. recessed two sides
  einseitig a. recessed on one side
ausgespitzt point-thinned *(drill)*
ausgewuchtet balanced
ausgießen 1. to pour out; 2. to line; 3. to seal *(cable)*; 4. to remetal *(bearing)*
Ausgleich *m* compensation, balancing, balance, equalization; neutralization; adjustment; *(cont)* inherent regulation
  A. / mit selbsttätigem self-compensating
  A. der Abnutzung compensation for wear
  A. der Unwuchten balancing
  A. des Querruders aileron balance
  A. im Verteilergetriebe *(aut)* transfer-case differential
  optischer A. optical compensation
Ausgleichapparat *m* balancing apparatus
  statischer A. static balancing apparatus
Ausgleicharm *m* balance arm
Ausgleichbecken *n* *(build)* balancing (compensating) reservoir
Ausgleichbeton *m* levelling screed
Ausgleichbogen *m* **/ aufgehängter** suspended expansion bend
Ausgleichbrückenschaltung *f* *(el)* balanced bridge transition
Ausgleichdruck *m* equalizing pressure
Ausgleichdüse *f* *(aut)* compensating jet
Ausgleicheinrichtung *f* **für verschiedene Fräserdurchmesser** cutter diameter compensation unit *(punch tape control)*
ausgleichen 1. to compensate [for] *(wear)*; to cancel out; to equalize; to make up *(losses)*; 2. *(build)* to trim, to level, to even out, to smooth
  die Zentrifugalkraft a. to neutralize the centrifugal force
  durch Gegengewicht a. to counterbalance
  Durchhang a. to take up slack
  Lochmittenunterschiede a. to align holes
  eine Phasenverschiebung a. to back off a phase shift
  Verschleiß a. to take up wear
Ausgleichen *n* 1. balancing; 2. *(build)* levelling, trimming, smoothing, evening out
Ausgleicher *m* *(therm)* expansion (compensating) pipe
Ausgleichfarbe *f* complementary colour
Ausgleichfrequenzgang *m* *(el)* transient response
Ausgleichgehäuse *n* *(aut)* differential casing
Ausgleichgetriebe *n* differential gear [unit]
Ausgleichkanal *m* *(therm)* equalization passage
Ausgleichkegelrad *n* differential bevel (spider) pinion
Ausgleichkreis *m* balancing circuit; corrector circuit
Ausgleichmasse *f* balance mass
Ausgleichmoment *n* balancing moment
Ausgleichradachse *f* differential pinion shaft
Ausgleichrinne *f* adjusting pan
Ausgleichrohr *n* *(therm)* expansion (compensating) pipe
Ausgleichsbatterie *f* buffer (balancing) battery
Ausgleichsbehälter *m* header tank
Ausgleichsbunker *m* surge bin (hopper)
Ausgleichschaltung *f* *(el)* equalizing connection, correction (corrective, signal-shaping) network
Ausgleichschicht *f* *(build)* levelling course (layer)
Ausgleichschlangenrohr *n* coiled expansion pipe
Ausgleichschleife *f* loop expansion pipe
  A. mit drehbarem Flansch expansion loop with rotary flange, compensating loop with swivel joint
  A. mit Gelenkrohren expansion loop with spherical joint
Ausgleichsdrossel *f* *(el)* interphase reactor, balance coil
Ausgleichseinrichtung *f* compensator, compensation device *(jig borer)*
Ausgleichsentwickler *m* compensation developer *(photography)*
Ausgleichsgewicht *n* *(mach)* balance weight, counterbalance [weight]; *(mine)* balance bob
Ausgleichsgrube *f* *(met)* soaking pit
Ausgleichsimpuls *m* equalizing pulse

Ausgleichskapazität *f (el)* balancing capacitance
Ausgleichskolben *m* dummy piston
korrespondierender A. corresponding dummy piston *(turbine)*
Ausgleichskondensator *m (el)* balancing capacitor, neutralizer
Ausgleichskreis *m (el)* compensated circuit
Ausgleichskupplung *f* flexible coupling
Ausgleichsladung *f* equalizing charge *(battery)*
Ausgleichsleitung *f* extension wire *(thermocouple)*
Ausgleichslineal *n* correction bar (strip)
Ausgleichsmaschinensatz *m (el.m.)* balancer set
Ausgleichsnetzwerk *n (el)* correcting network
Ausgleichsperre *f* differential lock
Ausgleichsrechnung *f* calculus of observations
Ausgleichsregler *m (el)* balancer-booster
Ausgleichsrohrverbindung *f* expansion pipe joint
Ausgleichsruder *n (ship)* balanced rudder
Ausgleichsschalter *m (el)* equalizer switch
Ausgleichsschaltung *f (el)* compensating circuit, correcting (balancing) network
Ausgleichsschlaufe *f* control take-up *(conveyor)*
Ausgleichsspannung *f* equalizing (compensating) voltage; transient voltage
Ausgleichsspule *f* compound (compensating) coil
Ausgleichsstrom *m* balance current; transient current
A. bei Ausschaltung circuit breaking transient current
Ausgleichsstromkreis *m* compensating circuit
Ausgleichstern *m (aut)* differential spider
Ausgleichstirnrad *n (aut)* differential spur gear
Ausgleichstopfbüchse *f* expansion stuffing box
Ausgleichstransformator *m* hybrid coil (transformer); compensator (alternating current, static) balancer, balancer-transformer
Ausgleichsübertrager *m s.* Ausgleichstransformator
Ausgleichsverbindung *f (el)* equipotential connection
Ausgleichsvorgang *m (el)* transient state
Ausgleichsvorrichtung *f s.* Ausdehnungsrohr
Ausgleichswellrohr *n (therm)* corrugated expansion pipe
Ausgleichswiderstand *m (el)* balancing resistance
Ausgleichszeit *f characteristic time, related to the approximation of a step function response by a dead time and a reaction rate*
Ausgleichszustellung *f* equalizing
Ausgleichteilen *n* differential indexing *(machining)*
Ausgleichübertrager *m s.* Ausgleichstransformator
Ausgleichung *f (therm)* balancing
dynamische A. dynamic balancing
statische A. static balancing
Ausgleichventil *n* pressure sequence valve, make-up valve
Ausgleichvorrichtung *f (therm)* compensating device
Ausgleichwahrscheinlichkeit *f* initial (a priori) probability
Ausgleichwicklung *f (el)* compensating winding
Ausgleichzeit *f* soaking time *(furnace)*
Ausgleichzustand *m* equilibrium state
ausgleiten to slip, to skid
ausglühen to anneal; to roast
Ausglühen *n* full annealing
ausgraben to excavate; to dig [out]
Ausguck *m (ship)* lookout, crow's nest
Ausguckmann *m* lookout man
Ausgußbecken *n* sink
Ausgußmasse *f (el)* insulating (sealing) compound *(for cables)*
Ausgußrinne *f* spout
aushalten 1. to withstand, to resist, to stand; 2. *(mine)* to sort out, to clean, to pick *(by hand)*
rauhen Betrieb a. to stand rough work
Aushängebogen *m* advance (specimen) sheet, proof-sheet
Aushänger *m* advance sheet
Aushangfahrplan *m* time-table poster
aushärtbar age-hardenable
warm a. hardenable by heat treatment
aushärten 1. *(plast)* to cure; to harden permanently; 2. *(met)* to age-harden
Aushärtung *f* 1. *(plast)* curing, hardening permanently; 2. *(met)* age-hardening
Aushebegabel *f* lifting fork
ausheben to withdraw *(drill)*; to dig, to excavate *(ditch)*
Ausheben *n* **[von Gräben]** excavating (excavation of) ditches, trench digging
Ausheber *m* lifter
Aushebewalze *f* fancy roller
Ausheilung *f* annealing *(of radiation damage)*
Aushilfskraftwerk *n* emergency power station
Aushobeln *n* routing
aushöhlen to hollow out
Aushöhlung *f* hollow; cavitation; groove
Aushub *m* excavated soil, excavation, spoil
Aushubboden *m* spoil, excavated earth
Aushubböschung *f* excavation slope
Aushubmasse *f* spoil, excavated earth
auskehlen to fillet, to groove, to recess; *(wood)* to chamfer
Auskehlen *n* recessing, filleting, grooving
Auskehlfräser *m* moulding cutter
Auskehlung *f* [internal] groove; recess
ringförmige A. spool, annular groove
auskeilen to edge away; to die out; to wedge out

auskellen to ladle [out]
auskernen to stone *(fruit)*
Auskesseln *n* **eines Bohrlochs** *(mine)* chambering, springing, squibbing
auskippen to dump [out]
ausklammern *(math)* to factor out
Ausklarierung *f* clearance
ausklauben *(mine)* to sort out
auskleiden to line *(furnace)*; to coat; to plate
Auskleidung *f* lining *(furnace, bearing)*; coating
feuerfeste A. fireproof lining
säurefeste A. acidproof lining
Ausklinkeinrichtung *f* tripping mechanism *(machine)*
ausklinken 1. to unlatch; to trip out; to release; to disengage; to notch
ein Klischee a. to mortise
Ausklinkung *f* 1. unlatching; release; disengagement; 2. notch[ing]
ausklopfen to scale *(boiler)*
ausknicken to buckle
Ausknicken *n* buckling
auskolken 1. to crater, to cup, to erode, to pit *(tool)*; 2. *(build)* to undermine, to scour, to underwash
Auskolkung *f* 1. [surface] cratering *(wear on the tool face)*, crater wear, eroding, erosion, pitting; 2. *(build)* scour[ing], undermining, underwashing
Auskopierpapier *n* printing out paper
Auskoppeldämpfung *f* coupling attenuation
Auskoppelraum *m (el)* catcher [space], output resonator (gap)
Auskoppelspalt *m (el)* output gap
Auskopplungsfeld *n (el)* catcher; output resonator *(clystron)*
Auskopplungsglied *n s.* Auskopplungsfeld
auskragen to project, to protrude, to corbel, to cantilever
auskragend projecting
Auskragung *f* projection
Auskratzen *n* **von Mauerfugen** raking out
auskristallisieren 1. to recrystallize; 2. to sugar *(sugar industry)*
Auskunft *f* information
Auskunftsbüro *n* enquiry office, information bureau
Auskunftsgespräch *n (telec)* inquiry call
auskuppelbar disconnectable
auskuppeln to declutch, to disengage the clutch, to step on the clutch pedal
ausladen 1. *(build)* to overhang, to project, to corbel, to cantilever; 2. to unload; to discharge; to empty
Ausladen *n* 1. *(build)* overhang[ing]; projection; projecting; 2. unloading; discharge; discharging
ausladend projecting, jutting out
Ausladerampe *f* unloading platform
Ausladestelle *f* unloading yard
Ausladung *f* 1. *(build)* projection; overhang; 2. *(mach)* throat clearance (depth, distance), depth of gap; 3. outreach *(derrick)*; swinging radius, length of jib *(crane)*; *(ag)* working (operating) radius
Ausladungshafen *m* port of discharge
Auslandsauftrag *m* foreign order
Auslaß *m* outlet; discharge
Auslaßhub *m (aut)* exhaust stroke
Auslaßkante *f (therm)* exhaust edge
Auslaßöffnung *f* outlet
Auslaßschlitz *m (aut)* exhaust port
Auslaßsteuerung *f (aut)* exhaust-gear mechanism
Auslaßstutzen *m* exhaust connecting branch; outlet connection
Auslaßüberdeckung *f (therm)* exhaust (inside) lap
negative A. negative inside lap
Auslassung *f* 1. out, omission; 2. *(dat)* skip
Auslaßventil *n* delivery valve; exhaust valve
Auslaßventilgehäuse *n* exhaust valve chest
Auslaßverlust *m* outlet loss
auslasten to balance; to equalize; to distribute weight
Auslastung *f* 1. weight distribution; 2. duty; rate of utilization
Auslastungsgrad *m* rate of utilization
Auslauf *m* 1. coasting, run down; stopping *(motor)*; 2. outrun, finish *(ski-jumping)*; 3. *(ag)* pen; outlet; exercise yard
A. / beim at stopping *(machine)*
Auslaufbahn *f* outrun, finish *(ski-jumping)*
Auslaufbauwerk *n* outfall works
auslaufen 1. *(ship)* to sail, to make (set) sail, to make off, to run out, to put (proceed) to sea, to clear the port; 2. to shallow out *(curves)*; to narrow; 3. to bleed *(colour, ink)*; 4. to flow out; to leak; 5. *(mine)* to reach the boundary *(face)*, to spill out *(coal)*; 6. to come to rest, to stop; 7. to wear out *(bearing)*
Auslaufen *n* **des Schiffes** starting, departure; outgoing vessel, setting sail (vessel)
A. nicht erstarrter Blöcke bleeding of ingots
Auslaufenergie *f* deceleration energy
Ausläufer *m (ag)* off-shoot, runner, sprig
Auslaufrinne *f* shoot *(rolling mill)*
Auslaufseite *f* delivery side *(rolling mill)*
A. der Walzen outgoing side of rolls
Auslaufventil *n* drain cock
Auslaufverfahren *n (el.m.)* retardation method
Auslaufzeit *f* run out time *(liquid)*; time for running out *(machine)*
Auslaugbottich *m (leath)* leach pit
auslaugen to leach; to extract; to macerate
Auslaugeturm *m* diffusor *(sugar industry)*
Auslaugung *f* extraction; lixiviation; leaching
Auslaugungshorizont *m (geol)* A-horizon, zone of removal

Auslaugungszone *f (geol)* leached zone
ausleeren to empty, to drain; to clear
Auslegearm *m* radial arm *(radial drill)*
auslegen 1. to line, to dress; to deliver *(sheets)*; 2. to design; to rate *(el. machinery)*; 3. to lay [out] *(cable)*
eine Schaltung optimal a. to optimize a circuit
Ausleger *m* boom, jib, cantilever *(crane)*; bracket, side arm, arm *(radial drill)*; cross arm; outrigger *(gig)*
drehbarer A. swing jib
freitragender A. cantilever
Auslegerbohrmaschine *f s.* Radialbohrmaschine
Auslegerboot *n* outrigger [boat]
Auslegerbrücke *f* cantilever bridge
Auslegerdachbinder *m* cantilevered roof truss
Auslegerechen *m* delivery gate
Auslegerfeder *f* cantilever spring
Auslegerförderband *n* stacker belt
Auslegergerüst *n* fan-guard, projecting scaffold
Auslegerklemmung *f* locking arrangement securing sleeve to pillar *(radial drill)*
Auslegerkopf *m* jib point, *(Am)* boom point
Auslegerkran *m* jib crane
Auslegermast *m* bracket pole
Auslegerrüstung *f* projecting scaffold, fanguard
Auslegerseil *n* jib cable, *(Am)* boom cable
Auslegerspitze *f* jib point, *(Am)* boom point
Auslegersystem *n (print)* delivery system
Auslegerträger *m* cantilever girder
Auslegerwinde *f* jib hoist
Auslegestapel *m (print)* delivery pile
Auslegetisch *m (print)* delivery table
ausleihbar loanable
auslenken 1. to sweep; to deflect, to displace *(vibration)*; 2. *(rock)* to correct deviation, to compensate
Auslenkung *f* 1. excursion; deflection; 2. *(rock)* correction of deviation; 3. displacement
Auslenkungsstrecke *f (mine)* secondary (subsidiary) drift
auslernen to complete the apprenticeship
Ausleseband *n* sorting conveyor
auslesen 1. to read-out *(memory)*; 2. to classify; to sort [out]; to grade; to separate
Auslesespeicher *m* read-out memory, read-only memory *(for automatic machine translation)*
Auslesevorrichtung *f* **mit Datenvorgabe** datebiased read-out device
ausleuchten to [fully] illuminate
Ausleuchtung *f* illumination
A. einer Fläche *(opt)* luminous coverage
auslichten to thin; *(ag)* to clear
ausliefern to deliver; to supply
auslitern to dispense, to meter [by the litre]
auslöschen to extinguish, to quench
Auslöschmagnet *m* obliteration magnet; cleaning magnet
Auslöschung *f* extinction *(waves, arc)*, quenching
völlige A. benachbarter Störer infinite adjacent channel rejection *(receiver)*
Auslöschungslage *f* extinction position *(of a Nicol prism)*
Auslösebalken *m (mine)* chock release
Auslösebereich *m (nucl)* Geiger [counting] region
Auslöseeinrichtung *f* / **federgespannte** *(mach)* spring load-and-fire mechanism
Auslöseimpuls *m* release (triggering, trigger, initiating, pilot) pulse; main bang *(radar)*
Auslöseimpulsverstärker *m* gate amplifier
Auslöseklinke *f (telec)* release pawl
Auslösekolben *m* impulse piston
auslösen *(mach)* to drop out of gear (mesh), to disengage, to trip out; to trigger; *(el)* to deenergize, to release, to trip
Auslöser *m* 1. *(mach)* detent; trigger; tripping device; 2. *(el)* cut-out; 3. release *(camera)*
Auslöserelais *n* release relay
Auslöseschalter *m* release (trip) switch
Auslösesicherung *f* tripping fuse
Auslösesignal *n* trigger signal; start order
Auslösespule *f* trip coil
Auslösestab *m* **der Setzmaschine** key rod, trigger bar
Auslösestrom *m (el)* release current, tripping current
Auslösetriode *f* trigger triode
Auslöseventil *n* release valve
Auslösevermögen *n* trigger performance
Auslösevorrichtung *f* trigger; releasing device
Auslösezähler *m s.* Auslösezählrohr
Auslösezählrohr *n (nucl)* release (Geiger-Müller) counter
Auslösezeichen *n* release signal; initiating signal
Auslösezeit *f* releasing time; time of liberation *(electronics)*
Auslösung *f* trip-out *(machine)*, release *(worm, camera)*; knock-off
A. eines künstlichen Fehlers *(el)* fault throwing
begrenzt verzögerte A. *(el)* definite time-lag tripping
direkte A. *(telev)* direct trip
einmalige A. non-recurrent triggering
mechanische A. mechanical tripping
mittelbare A. indirect tripping [circuit control]
selbsttätige A. automatic tripping (cut-out)
thermische A. thermal circuit-breaking
unmittelbare A. direct tripping
verzögerte A. time-lag circuit-breaking
vorzeitige A. premature release
Auslösungseinrichtung *f* trip device, kickout

Auslösungsventil *n* unloading valve

Auslösungsverzögerung *f* releasing delay

Auslösungszeichen *n (telec)* release guard signal

ausloten to fathom, to sound

Ausmahlung *f* beating of pulp *(papermaking)*

Ausmaß *n* extent

ausmauern to line *(furnace)*

Ausmauerung *f* [brick]lining *(furnace)*; masonry *(reactor)*; *(mine)* bricklining, bricking

saure A. acid lining

ausmessen to measure; to check the dimensions

ausmisten *(ag)* to muck out

ausmitten to centre

ausmittig eccentric

ausmörteln *(build)* to grout [up]

Ausmündung *f* opening *(turbine)*

Ausmusterung *f* taking out of service

ausnähen *(text)* to darn, to mend

Ausnahmefall *m* exceptional case

Ausnehmerglas *n (text)* pick glass

Ausnehmung *f (mach)* recess

ausnutzen to utilize; to exploit

Ausnutzungsfaktor *m* utilization (exploitation) factor; commercial efficiency

A. des Wicklungsraumes *(el)* space factor of winding

Ausnutzungsgrad *m* utilization factor

Ausnutzungskoeffizient *m* utilization coefficient

auspacken to unpack

Auspendler *m (build)* commuter

auspflanzen to transplant

auspressen *(build)* to inject, to pressure-grout; to dry by pressing *(paper)*; to force out

Auspressen *n* forcing out *(fit)*

Auspreßgeschwindigkeit *f* speed of discharge *(man-made textiles)*

Auspreßmörtel *m* grout

Auspreßverfahren *n* pressure grouting

Auspuff *m* exhaust; escape; eduction

Auspuffbetrieb *m* non-condensing working *(steam engine)*

Auspuffdampfmaschine *f* non-condensing steam engine

auspuffen to exhaust; to escape

Auspuffflammendämpfer *m* exhaust-flame damper

Auspuffgas *n* exhaust gas (fume)

Auspuffgasturbine *f* exhaust gas turbine

Auspuffhauptschalldämpfer *m* main exhaust silencer

Auspuffhub *m* exhaust (scavenge) stroke

Auspuffklappe *f* muffler cut out

Auspuffknallen *n* backfiring

Auspuffkrümmer *m* exhaust manifold (stack)

Auspuffleitung *f* exhaust pipe (piping)

Auspuffmaschine *f* non-condensing engine

Auspufföffnung *f* exhaust port

Auspuffrohr *n* exhaust pipe (stack)

Auspuffschalldämpfer *m* exhaust silencer (muffler)

Auspuffstutzen *m* exhaust head (stack)

Auspufftakt *m s.* Auspuffhub

Auspufftopf *m* exhaust silencer (muffler)

Auspuffventil *n* exhaust valve

Auspuffventilabschlußdeckel *m* exhaust-valve lid *(turbine)*

Auspuffvorschalldämpfer *m* pre-expansion chamber

auspumpen to evacuate, to exhaust; to pump out, to suck out

Ausputzen *n* cleaning; finishing *(shoe)*

Ausradieren *n* rubbing out

ausrechnen to calculate, to compute, to figure out

Ausrechnen *n* calculation; computation

Ausreckeisen *n (leath)* setting slicker

ausrecken *(leath)* to flat the grain; *(text)* to lengthen, to stretch

Ausreckmaschine *f (leath)* setting-out machine

Ausrecktisch *m (leath)* striking-out table

Ausregelzeit *f* settling time; correction time

Ausreibebürste *f (leath)* finishing brush

ausreiben to ream *(metal cutting)*

Ausreiben *n* reaming out *(metal cutting)*

Ausreißen *n* **der Schwellenschrauben** extraction of sleeper-screws

Ausreißer *m* wild shot *(shooting)*

ausreutern to grub

ausrichten 1. to line up; to align; to straighten *(sheet)*; 2. to orient; 3. *(mine)* to develop, to open out (up); 4. to train *(conveyor)*

Ausrichten *n* 1. lining-up; aligning, alignment *(shaft)*; straightening *(sheet)*; fairing *(laying down the ship's lines)*; 2. orientation; 3. *(mine)* development, opening-up; 4. training *(conveyor)*

A. der Höhenlage levelling

A. der Linienführung [re-]lining

ausrichtend / sich selbst self-aligning *(bearing)*

Ausrichtung *f* 1. lining-up; alignment; levelling-up; straightening *(sheet)*; 2. orientation; 3. *(mine)* development, opening-up

schlechte A. misalignment

Ausrichtungsbau *m (mine)* development opening

Ausrichtungsberge *pl (mine)* development rock

Ausrichtungsstrecke *f (mine)* development opening

ausroden to stub up

ausrollen to coast

Ausrückeinrichtung *f* cut-out device

ausrücken to disengage, to disconnect, to declutch, to throw out *(lever)*; to cut out

Ausrückgabel *f* clutch release yoke

Ausrückhebel *m* clutch release lever

Ausrücklager *n* clutch release bearing

Ausrückmuffe *f* withdrawal collar, disengaging sleeve
Ausrückplatte *f* clutch release plate
Ausrückvorrichtung *f* shifting device
Ausrückwelle *f* clutch release shaft
ausrunden 1. *(build)* to ease; 2. *(mach)* to radius, to fillet
Ausrunden *n* 1. *(build)* easing; 2. *(mach)* radiusing, filleting
Ausrundung *f* 1. *(build)* easement curve; 2. *(mach)* internal radius, fillet
ausrüsten 1. to equip, to furnish; to fit [out]; 2. *(build)* to strike formwork; 3. *(text)* to finish
Ausrüstung *f* 1. outfit, equipment, installation; fittings *(lamp)*; 2. paper coating, adding surface sizing *(paper)*; 3. *(text)* finish; finishing
  A. für gleislosen Betrieb *(mine)* off-track equipment
  A. mit Sinterhartmetallwerkzeugen carbide tooling
  antistatische A. *(text)* antistatic finish
  elektrische A. electrical equipment
  fäulniswidrige A. rotproof finish; antifouling finish
  flammfeste A. uninflammable finish
  knitterarme A. crease-resistant finish
  knitterbeständige A. wrinkle-resistant finish
  nichtbeständige A. nondurable finish
  quellfeste A. swelling-resistant finish
  wasserabweisende A. waterproofing
Ausrüstungsbecken *n s.* Ausrüstungsdock
Ausrüstungsdock *n (ship)* fitting berth, fitting-out berth (basin)
Ausrüstungshilfsmittel *n (text)* finishing auxiliary
Ausrüstungskai *m* outfitting quay, fitting-out shore
Ausrüstungssaal *m* finishing room
Aussaat *f (ag)* seeding, sowing
aussäen *(ag)* to seed, to sow
aussägen to saw out
Aussagenkalkül *m (dat)* sentential (propositional) calculus
aussalzen *(ch)* to salt out
Aussalzen *n (ch)* salting-out, salt displacement
ausschaben to scrape
ausschachten to excavate
Ausschachter *m* navvy
Ausschachtung *f* excavation, pit
ausschalen to strip, to strike formwork, to demould
Ausschalen *n* [form] stripping, striking formwork, demoulding
Ausschaltdauer *f* break-time, breaking period, cleaving time
ausschalten 1. *(el)* to switch off; to disconnect; to cut-off; to de-energize; to open a circuit; 2. *(mach)* to disengage; to throw out; to trip out
Ausschalter *m* cut-out, circuit breaker, disconnecting (single-throw) switch
  A. mit Sicherung fuse cut-out
  einfacher A. single-throw switch, one-way switch, contact breaker
Ausschaltknopf *m* stop button
Ausschaltleistung *f* breaking capacity
Ausschaltlichtbogen *m* break arc
Ausschaltspitzenstrom *m* cut-off current [of a fuse]
Ausschaltstellung *f* off-position
Ausschaltstoß *m* circuit breaking transient current
Ausschaltstrom *m* breaking current
Ausschalttaste *f* release-to-break switch, circuit-breaking key
Ausschalttor *n (cont)* normally closed gate
Ausschaltung *f* circuit-breaking, circuit-opening, disconnection, switching off, cutting out
Ausschaltvermögen *n* breaking capacity
  A. des einzelnen Kontaktes contact current breaking capacity
Ausschaltverriegelung *f* switch-off interlock
Ausschaltverzug *m* opening-time
ausschärfen to bevel *(sheet metal)*
ausschaufeln to shovel out
ausscheiden *(ch)* to precipitate, to separate; to exude, to exudate
Ausscheidung *f* 1. *(math)* elimination; 2. segregation, separation, isolation; 3. precipitation, deposit
Ausscheidungsfolge *f (geol)* succession
Ausscheidungshärtung *f* precipitation hardening
Ausscheidungsprodukt *n* excretion product
Ausscheidungsspiel *n* play-off
Ausscheren *n* **[aus dem Flugverband]** peeling off
ausschieben to exhaust *(internal combustion engine)*
Ausschieben *n* exhaust *(internal combustion engine)*
ausschießen *(build)* to render *(first coat)*
  Ballast a. to unballast
ausschiffen to disembark, to put on shore, to land *(persons)*; to unload, to discharge *(cargo)*; to put to sea *(vessel)*
Ausschiffung *f* disembarkation, debarkation *(persons)*; unloading, discharge *(cargo)*; landing
Ausschiffungshafen *m* port of debarkation
ausschirren to unharness *(horse)*
ausschlachten to salvage; to cannibalize *(motor car)*
ausschlacken to draw the slags; to clear from dross
Ausschlag *m* deflection; departure; amplitude; reading *(of a meter)*, response *(of an instru-*

ment); eccentricity; arm (ratio) of eccentricity
A. des Schiebers aus der Mittellage distance of the valve from the centre position
ausschlagen 1. to deflect; to deviate; 2. *(metal)* to hammer, to beat out
über die Skala hinaus a. to go off scale
ausschlaggebend determining, decisive
Ausschlagkompensator *m* deflection potentiometer
Ausschlagrückgang *m* dip
Ausschlagweite *f* amplitude
Ausschlagwinkel *m* angle of deflection (oscillation)
ausschlämmen *(ch)* to elutriate, to levigate; to clear from mud
ausschleifen to grind internally
Ausschleifen *n* **des Zahngrundes** grinding of the tooth gullet
A. zylindrischer Bohrungen internal grinding of cylindrical holes
ausschleppen *(ship)* to tow out, to haul out
ausschleudern 1. *(ch)* to centrifuge; 2. *(text)* to hydro-extract; 3. *(mine)* to whizz out *(water)*
ausschleusen 1. to lock out, to tow a ship out of a sluice; 2. *(nucl)* to extract, to deflect *(accelerator)*
Ausschleusen *n* 1. locking out, towing a ship out of a sluice; 2. *(nucl)* deflection, extraction *(accelerator)*
Ausschleusvorrichtung *f (nucl)* beam extractor
ausschließen *(print)* to justify, to space
Ausschließen *n (print)* justification, spacing
ausschließend exclusive
Ausschließungsprinzip *n (ph)* [Pauli] exclusion principle
Ausschluß *m (print)* blank space, spaces
großer A. quads
Ausschlußtaste *f* spacing key
Ausschlußtrommel *f* justifying scale
ausschmelzen to liquate, to melt, to fuse
Ausschmelzverfahren *n (met)* cire perdu process *(casting)*
ausschmieden to draw out, to hammer, to thin
Ausschneidbohrer *m* hole cutter
Ausschneidearbeit *f (met)* cut-out operation; blanking operation
ausschneiden to cut [out]
Ausschneiden *n* finish of the cut *(in metal cutting process)*; blanking *(sheet metal)*
Ausschneidfräsen *n* cut-out milling *(of sheets)*
Ausschnitt *m* cut[-out]; ingate *(moulding)*
A. des Mittelstücks opening in the distance piece
A. des Ringes opening of the ring
Ausschnittende *n* log end
ausschreiben to invite tenders; to announce
Ausschubtakt *m* exhaust stroke *(engine)*
Ausschuß *m* scrap [parts], scrap work, rejects *(metal working)*; *(text)* refuse, waste [stuff], broke; spoil sheets *(paper)*
Ausschußaufbereitung *f* broke handling
Ausschußbogen *m* spoil [sheet], waste sheet
Ausschußbütte *f* broke pit
ausschußfrei scrapless *(metal working)*
Ausschußgrenze *f* rejection limit
Ausschußlehre *f* not-go gauge
Ausschußpapier *n* defective paper
Ausschußprozentsatz *m* **/ zulässiger** allowable percent defective
Ausschußquote *f* scrap percentage (rate)
A. der Gutlage allowable percent defective
Ausschußseite *f* not-go end [of gauge], scrap side
Ausschußteil *n* scrap component
Ausschußware *f* damaged goods; cull timber
Ausschütteln *n* **mit Äther** *(ch)* etherizing, shaking out with ether
ausschütten 1. to dump *(excavator)*; 2. to pour out *(liquid)*
Ausschütthöhe *f* dumping height
Ausschüttweite *f* dumping reach
ausschwefeln to sulphur
Ausschwefeln *n* sulphuration
ausschwemmen 1. *(ch)* to scour, to sluice; 2. *(text)* to flush out
ausschwenkbar swing-out
ausschwingen 1. *(ph)* to die out, to decay *(vibration)*; 2. to swing out, to turn out *(davit)*
Ausschwingen *n* decay, dying-away, dying-out
ausschwingend / aperiodisch dead-beat
Ausschwingstrom *m* decay current
Ausschwingungsverzerrung *f* facsimile transient distortion
Ausschwingvorgang *m* decay (dying-out) process
Ausschwingzeit *f* dying-down time, time of decay
ausschwitzen *(ch)* to bleed, to exude, to exudate
Aussehen *n* appearance
ausseigern *(ch, met)* to liquate, to reduce by liquation; to segregate
Außenabmessung *f* external dimension
Außenabteiler *m (ag)* [binder] divider
Außenachtersteven *m (ship)* false post
Außenanschluß *m (el)* external connection
Außenanschlußkappe *f* external connection cap
Außenansicht *f* exterior view; *(ship)* outboard profile
Außenanstrich *m* exterior painting; outside coating
Außenantenne *f* exterior (open) aerial

Außenarbeiten *fpl (build)* exterior work
Außenarmatur *f* outdoor fitting
Außenaufnahme *f (telev)* field pick-up, remote
Außenbackenbremse *f (aut)* outside (external-contracting) shoe brake
Außenbahn *f (nucl)* outermost orbit
Außenballast *m* outside ballast
Außenbelag *m* outer coating
Außenbeleuchtung *f* outdoor lighting (illumination)
Außenbeplankung *f (ship)* outside planking; berthing *(sheer-strake)*
Außenbeplattung *f (ship)* shell (outside) plating
Außenbereich *m* external district
Außenbezirk *m* outskirts, suburb
Außenblech *n* outside panel
Außenbordlöschung *f (ship)* discharging overside, overside delivery
Außenbordmotorboot *n* outboard motor-boat
außenbords outboard
Außenbordventil *n (ship)* sea valve
aussenden *(ph)* to emit; to transmit; to throw a beam
Aussendung *f* / **ungewollte** spurious radiation
Außendock *n* open basin
Außendrehen *n* turning external diameters
Außendruck *m* external pressure
Außendurchmesser *m* outside (outer) diameter
  A. eines Gewindes major diameter of a thread
  A. über der Isolierung dielectric outer diameter *(cables)*
Außenelektrode *f* external electrode
Außenelektron *n* outer electron
Außenfedertaster *m* outside spring-joint caliper
Außenfeuerung *f (therm)* extension furnace, furnace in front of the boiler
Außenfläche *f* [external] surface
Außenführung *f* **der Zahnkette** side guide *(silent chain)*
Außenfurnier *n* outer ply
Außenganghaus *n* gallery (balcony access) block
Außengewinde *n* external (male) thread
Außengewindedrehmeißel *m* external threading tool
Außengewindeschleifen *n* external thread grinding
Außengewindeschneidarbeit *f* external threading operation
Außengewindeschneiden *n* external threading; external thread cutting
Außengitter *n (el)* exterior (outer) grid
Außenhafen *m* outer harbour (port), outport
Außenhandbremse *f (aut)* outside hand brake
Außenhandel *m* foreign trade
  seewärtiger A. oversea trade
Außenhaupt *n* outer (lower) gates *(of a sluice)*
Außenhaut *f* 1. *(aero)* [outer] skin; 2. *(ship)* outside planking, shell (skin) plating; hull *(yacht)*
  abgewickelte A. expanded (developed) shell plating
  gewellte A. corrugated shell plating
Außenhautbeplattung *f* shell plating
Außenhautfüllstück *n (ship)* shell chock
Außenhautlängsspant *n* shell longitudinal
Außenhautpforte *f (ship)* shell door
Außenhautplatte *f* outer plate *(stern frame)*
Außenhautreibung *f* skin friction
Außenhautwinkel *m* shell attachment (angle) *(of a frame)*
Außenisolator *m* outdoor insulator
Außenkabel *n* outside (external) cable
Außenkante *f* outside edge
aussenken to counterbore
Aussenken *n* counterboring
Außenkontaktsockel *m (el)* external contact base, base with external contacts
Außenkorrosion *f* external corrosion
Außenkurzgewinde *n* external short thread
Außenlager *n* tail (outboard) bearing
Außenleiter *m* external (outer) conductor *(of cable)*; leg of a circuit, phase conductor
Außenliek *n* afterleech
Außenlinie *f* contour
Außenlufttemperatur *f* outdoor air
Außenlunker *m* surface blowhole
Außenmantel *m* 1. outer surface [of cylinder]; 2. covering [of a cable]
Außenmaß *n* external dimension
Außenmauer *f* external (exterior) wall
Außenmessung *f* external measurement
Außenmorsekegel *m* Morse taper on the outer surface
Außennebenstelle *f (telec) (Am)* off-premises extension station
Außenpolgenerator *m* overhung-type alternator
Außenpolgestell *n* external pole frame
Außenpolschwungradgenerator *m* flywheel alternator with external poles
Außenprofil *n* external profile
Außenpropeller *m* wing propeller, outer screws
Außenputz *m* rendering
Außenrad *n* external gear
Außenraum *m* outer (cosmic) space
Außenräumen *n* external surface broaching
Außenraumklima *n* open-air conditions
Außenräummaschine *f* external (surface) broaching machine
Außenräumwerkzeug *n* external (surface) broach
  mit Hartmetallschneiden bestücktes A. carbide surface broach
Außenrauschen *n* atmospheric noise
Außenreportage *f* [television] outside broadcast, *(Am)* field pick-up, remote

Außenring *m* outer ring (race) *(bearing)*; outer member *(free-wheel)*
Außenrundläppen *n* external cylindrical lapping
Außenrundschleifen *n* externally grinding cylindrical parts
Außenrundschleifmaschine *f* external cylindrical grinding machine
Außenrüttler *m* external vibrator
Außensammler *m (rock)* external exhaust collector ring
Außenschale *f* outer[most] (valence) shell *(of electrons)*
Außenschere *f* exterior cutter *(loom)*
Außenschicht *f* outer layer
Außenschott *n (ship)* wing bulkhead
Außenschrauben *fpl (ship)* outer screws, wing propeller
Außenseite *f* outside
Außenspeicher *m (dat)* external memory
Außenstempel *m (mine)* lower prop
Außensteven *m (ship)* false stem
Außenstrom *m* external current
Außentaster *m* outside caliper[s]
  A. mit Schraubenscharnier firm joint outside caliper, stiff jointed outside caliper
  A. mit Spannfeder spring[-joint] outside caliper
Außenteil *n* female part; outer piece (member) *(fit)*
Außentemperatur *f* ambient (outdoor, outside) temperature
Außenthermometer *n (rock)* external thermometer
Außentransformator *m* free-air transformer
Außentreppe *f* fliers
Außentrittstuhl *m (text)* outside treading loom
Außentrittvorrichtung *f (text)* outside treading motion
Außenverzahnung *f* external teeth
  A. / mit external-toothed
Außenwand *f* exterior (external) wall
Außenwandplatte *f* facade (external wall) panel
Außenwange *f* outer stringer *(stairs)*
Außenwiderstand *m* load resistance
Außenzentrierung *f* major-diameter fit
Außenzug *m (therm)* outer (external) flue
außerachsig eccentrical
außeratmosphärisch extra-atmospheric
außeraxial off axis
Außerbetriebsetzen *n (el)* putting out of action; stopping
außergalaktisch extra-galactic
außerhalb der Erdatmosphäre extraterrestrial, beyond the earth atmosphere
  a. der Klammern *(math)* unbracketed
  a. des Kerns extranuclear
Außermitteauffuttern *n* chucking eccentrically
Außermittebohrung *f* eccentric bore
Außermittedrehen *n* turning off centre, eccentric turning
Außermittemaß *n* eccentricity
Außermittespannfutter *n* eccentric chuck
außermittig eccentric, off-centre
Außermittigkeit *f* eccentricity
außerordentlich extraordinary
Außertrittfallen *n (el)* falling out of phase (step, synchronism)
Außertrittfallmoment *n (el)* pull-out torque *(of a synchronous motor)*
Aussetzbetrieb *m (el)* intermittent operation, periodic duty
aussetzen 1. to interrupt *(a motion)*; to fail; 2. to lower, to launch, to heave out, to drop out *(boat)*; 3. to expose, to subject to; 4. *(leath)* to flat the grain
  die Segel mit den Spieren a. to boom out the sails
  ein Boot achtern a. to drop a boat astern
  Wagen aus einem Zug a. to take coaches off a train
Aussetzen *n* 1. *(el)* failure; interruption; 2. lowering, launching *(boat)*; 3. exposing
  A. der Schwingungen failure of oscillation
aussetzend intermittent; interrupted
Aussetzer *m (dat)* skip
Aussetzvorrichtung *f* launching arrangement, boat lowering device
Aussichtsfernrohr *n* terrestrial telescope
Aussichtspavillon *m* gazebo
Aussichtsturm *m* belvedere, viewing tower
Aussichtswagen *m* observation car, *(Am)* dome car
aussieben 1. to screen; 2. *(el)* to filter out; to select *(frequency)*
  der Größe nach a. to screen in sizes
aussondern to segregate, to separate
aussortieren to sort, to eliminate
ausspänen to clear the chips
Ausspänen *n* chip relief, swarf removal
ausspannen to unclamp *(work)*
Ausspannen *n* unclamping *(work)*
  A. des Werkzeugs tool releasing (unclamping)
aussparen to block out *(in concrete construction)*; to leave a recess *(in brickwork)*; to recess *(metal cutting)*
Aussparung *f* recess, pocket, embrasure, blockout
  kegelige A. tapered recess
aussperren to lock out
Ausspinnen *n* final (fine) spinning
Ausspitzarbeit *f* thinning operation *(drill)*
ausspitzen to thin, to point *(drill)*
Ausspitzen *n* **[des Spiralbohrers]** drill pointing, web (point) thinning, thinning the point of the drill

Ausspitzvorrichtung *f* web thinning attachment *(drill)*
ausspülen to rinse [out], to wash out; to scavenge *(engine, gases)*; to scour, to underwash
Ausspülung *f* rinsing; underwashing, scour[ing]; scavenging *(engine)*
ausstatten to furnish [with], to equip [with], to fit [out]
Ausstattung *f* equipment; outfit
A. mit Betriebskontrollgeräten instrumentation
Ausstattungsgegenstände *mpl* furnishings, fitment, equipment
Ausstattungspapier *n* fancy paper
Ausstechmeißel *m* tool bit [for very large holes in thin metal]
Ausstecken *n (ag)* pricking out
aussteifen to brace; to reinforce
Aussteifung *f* bracing; stiffening
Aussteifungsträger *m* diaphragm beam
aussteigen to get out; to leave [the bus]; to abandon the aeroplane
Aussteller *m* exhibitor; drawer *(bill)*
Ausstellfenster *n (aut)* ventilator window
„Aus"-Stellung *f* off-position *(of relays)*
Ausstellung *f* exhibition; exposition
Ausstellungsfläche *f* exhibition area
überdeckte A. hall (under-roof) exhibition space
Ausstellungsgebäude *n* exhibition building
Ausstellungsgelände *n* exhibition grounds
Ausstellungshalle *f* exhibition hall
Ausstellungsraum *m* showroom
Ausstellungsschiff *n* fair ship
ausstemmen to mortise *(joinery)*; to chisel out
Aussteuerbereich *m s.* Aussteuerungsbereich
aussteuern to control; to modulate
Aussteuerung *f* drive; driving; tone control *(broadcast transmitter)*
Aussteuerungsbereich *m* range of modulation (uniform control), workable control range; full scale *(analogue computer)*
Aussteuerungsgrad *m* depth of modulation; phase control factor *(rectifier)*
Aussteuerungskontrolle *f* recording level indication *(tape recorder)*
Aussteuerungsmesser *m* tone control measuring set, volume (level) indicator
Aussteuerungsmeßgerät *n* **für Sprache** speech voltmeter
Ausstieg *m* exit hatch
ausstopfen to stuff
Ausstoß *m* 1. output, rate of production; 2. *(mach)* ejection
ausstoßen 1. to turn out; to produce; to manufacture; 2. to expel, to eject *(part)*; 3. *(text)* to strip *(card)*; 4. *(leath)* to flat the grain; 5. to exhaust *(gases)*
Ausstoßer *m* ejector
Ausstoßhebel *m* ejector lever
ausstrahlen to radiate *(into space)*; to emit *(radiation)*
Ausstrahlung *f* radiation; emission *(of rays)*
Ausstrahlungsverlust *m* radiation loss *(heat)*
Ausstrahlungsvermögen *n* emissivity, emittance, emissive (emitting, radiating) power
Ausstrahlungswinkel *m* angle of radiation
ausstraken *(ship)* to fair, to sheer *(a line)*
Ausstrecke *f (text)* finisher frame
ausstreichen *(geol)* to come out to the day, to crop out
Ausstrich *m* 1. *(mine)* outcrop; 2. smear *(microscope)*
Ausströmarbeit *f (therm)* work done during exhaust
ausströmen to flow out [from], to issue [from]; to escape *(gases)*
Ausströmgeschwindigkeit *f* outlet flow velocity; [nozzle] exhaust velocity, exit (jet, efflux) velocity
Ausströmhals *m* exhaust passage *(turbine)*
Ausströmung *f* outflow; effusion; exhaust
ringförmige A. annular outflow *(turbine)*
Ausströmungsarbeit *f (therm)* work done during exhaust
Ausströmungsenergie *f (therm)* energy of [out]flow
Ausströmungsgeschwindigkeit *f s.* Ausströmgeschwindigkeit
Ausströmungslinie *f (therm)* exhaust curve (line)
austasten to blank [out]; to gate; to zero
Austastimpuls *m* blanking pulse
Austastlücke *f* blanking interval
Austastpegel *m (el)* blanking (pedestal) level
Austastsignal *n (el)* blanking pulse (signal), pedestal signal, *(Am)* blanking signal
Austastübertrager *m* blocking transformer
Austastung *f* blanking, gating
Austastzeit *f* gate time
Austauchen *n (ship)* lifting at the bow
Austausch *m* 1. *(nucl)* exchange; mass transfer; *(math)* substitution; 2. *(ch)* metathesis, exchange, interchange
austauschbar interchangeable, exchangeable
Austauschbarkeit *f* interchangeability; *(math)* permutability
Austauschbau *m* interchangeable manufacture
Austauscheffekt *m* interchange effect
austauschen to exchange; to substitute
die Röhre a. to interchange the valve
Austauschenergie *f* exchange energy
Austauscher *m* exchanger; *(nucl)* ion exchanger, ionite
Austauscherharz *n* exchange resin

Austauscherkolonne *f (nucl)* ion-exchange column
Austauschgeschwindigkeit *f* rate of exchange
Austauschkraft *f (ph)* exchange force *(Heisenberg)*
Austauschpotential *n* exchange potential
Austauschprogramm *n* alternate programme *(computer)*
Austauschreaktion *f (ch)* replacement (exchange) reaction
Austauschreaktor *m* exchange reactor
Austauschschiff *n* cartel ship
Austauschstoff *m* substitute [material]
Austauschstoß *m* exchange collision
Austauschstreuung *f (nucl)* charge-exchange scattering
Austauschvorgang *m* interaction
Austauschwechselwirkung *f* exchange interaction
Austauschwerkstoff *m* substitute material
Austauschzahl *f* / **theoretische** *(nucl)* number of transfer units, NTU
Austauschzeit *f* exchange time
A. der Energie time of energy exchange, energy-exchange time
Austenit *m* austenite
austenitisch austenitic
Austenitisierung *f* austenitizing
Austenitkorngröße *f* austenitic grain size
Austenitzerfall *m* austenite dissociation
austonnen *(ship)* to buoy
Austrag *m* 1. drag-out *(galvanics)*; 2. discharge, discharging, unloading
austreiben 1. *(ag)* to drive out; 2. to drift out *(wedge)*; 3. to flash out the gas; to deaerate
Austreiben *n* 1. *(ag)* driving-out; 2. drifting out *(wedge)*
A. des Bohrers removing (removal) of the drill
Austreiber *m* taper drift, *(Am)* taper[ed] key
Austreiberlappen *m* tang
Austreiberschlitz *m* tang slot
Austreiblappen *m* tang
austreten to emerge [from] *(liquid)*; to break loose [from] *(electrons)*
austretend emergent *(ray)*
Austriebsexzenter *m* clearing cam *(knitting)*
Austriebsstellung *f* clearing position *(knitting)*
Austriebsteil *n* clearing element *(knitting)*
Austritt *m* exit, outcome, outflow, outlet; leakage *(oil)*
A. aus der Düse outlet from the nozzle *(turbine)*
Austrittdampfspannung *f* back pressure *(turbine)*
Austrittgeschwindigkeit *f s.* Austrittsgeschwindigkeit
Austrittkammer *f* exhaust-chamber *(turbine)*
Austrittkante *f* outlet edge *(turbine)*
Austrittsarbeit *f* work function, electron affinity
äußere A. outer work function
fotoelektrische A. photoelectric (electronic) work function
innere A. inner work function
wahre A. true work function
Austrittsdosis *f (nucl)* exit dose
Austrittsdreieck *n* outlet diagram (triangle) *(turbine)*
Austrittsdurchmesser *m (rock)* exit diameter
A. der Düse nozzle exit diameter
Austrittseite *f* delivery side *(pump)*; exit side *(furnace)*
Austrittsentfernung *f* escape distance
Austrittsgefälle *n* exit gradient
Austrittsgeschwindigkeit *f* 1. velocity of discharge *(liquid)*; 2. [nozzle] exhaust velocity, exit (jet, efflux) velocity; 3. emission velocity *(of electrons)*; escape velocity
absolute A. absolute terminal velocity
Austrittskammer *f* discharge (exhaust) port
Austrittskanal *m* outlet port *(pump)*
Austrittsöffnung *f* outlet, outcome, portal
Austrittspotential *n (el)* exit potential; work function
Austrittspupille *f* exit pupil
Austrittsrichtung *f* emergent direction, direction of emission
Austrittsspalt *m* exit slit
Austrittsstelle *f* point of escape (exit)
Austrittsstrahl *m* exit beam
Austrittsstutzen *m* outlet piping connection, pressure connection *(pump)*
Austrittstemperatur *f* outlet temperature
Austrittstiefe *f* depth of escape *(of electrons)*
Austrittswinkel *m* angle of emergence (emission)
Austrittverlust *m* outlet loss *(turbine)*
Austrittwinkel *m* angle of outlet *(turbine)*
austrocknen to dry out; to dehumidify; to desiccate; to exsiccate, to season *(wood)*
Austrocknung *f* drying out; dehumidifying; desiccation; exsiccation; seasoning *(wood)*
ausüben to exercise, to practise
Druck a. to exert pressure to
ausvauen to Vee-groove
„Aus"-Vorrang *m (el)* OFF-priority, OFF-precedence *(switchgear)*
Auswahl *f* selection; choice; sampling *(statistics)*
Auswahleinheit *f* feed control unit *(collator)*
auswählen to select, to choose; to sample *(statistics)*
Auswähler *m* selector
Auswahlimpuls *m* selection pulse
Auswahlregel *f* selection rule
Auswahlreihe *f* preferred series, series of preferred tolerances

auswalzen to roll out
Auswandererschiff *n* emigrant ship
auswaschbar washable; removable
Auswaschbarkeit *f* washability
auswaschen *(ag)* to leach, to wash out; to underwash; to scour
Auswaschmaschine *f* scouring machine
Auswaschstein *m* scouring stone
Auswaschtafel *f* scouring table
Auswaschung *f* 1. *(ag)* leaching; 2. underwashing; scouring
Auswaschverlust *m (text)* washing-out loss
auswechselbar interchangeable; replaceable; exchangeable
auswechseln to interchange; to replace; to substitute
Auswechselobjektiv *n* interchangeable lens
Auswechs[e]lung *f* trimming *(carpentry)*
ausweichen to give way [to]; to turn aside
Ausweichflughafen *m* emergency airport
Ausweichgleis *n* siding, passing track
Ausweichlandeplatz *m* dispersal (emergency) landing field
Ausweichliegeplatz *m (ship)* lay-by
Ausweichmanöver *n (ship)* emergency turn
Ausweichstelle *f* passing place
ausweiten to expand; to extend; to widen
Auswerfeinrichtung *f* ejection fixture, ejector mechanism
auswerfen to eject
  die Gußform a. to open the mould
Auswerfer *m* ejector
Auswerfvorrichtung *f* work ejector mechanism
Auswerteimpulsgenerator *m* strobing pulse generator
auswerten to analyze, to evaluate; to utilize
Auswerter *m* analyst
Auswertetisch *m* plotting table
Auswertung *f* evaluation; utilization
  punktweise A. point-by-point evaluation
auswiegen to level; to balance
auswintern *(ag)* to winter, to be lifted by frost
auswittern to weather
Auswitterung *f* weathering; seasoning; efflorescence
Auswuchs *m* burl
Auswuchtbock *m* balancing stand
Auswuchtdorn *m* balancing arbor
auswuchten to balance; to counterbalance
Auswuchtmaschine *f* balancing machine
Auswuchtscheibe *f* balancing washer
Auswuchtstein *m* balancing weight
Auswuchtung *f* balance; balancing
  dynamische A. running balance
Auswurf *m* spittings *(converter)*
Auswurfklappe *f (text)* grid door *(opener)*
auszacken to indent
Auszackmaschine *f (leath)* scalloping machine
Auszählung *f* counting out; count[ing]
auszeichnen 1. to trace; 2. to mark out; 3. to display
Auszeichnungsschrift *f* display type
ausziehbar 1. telescopic; 2. extractable
ausziehen 1. to draw (pull) out; 2. to extract; 3. to abstract
Ausziehen *n* / **automatisches** *(dat)* auto abstracting [technique]
Auszieher *m* drawing-out device *(rolling mill)*
Ausziehleiter *f* extending ladder
Ausziehschacht *m (mine)* upcast shaft
Ausziehtubus *m (opt)* draw tube
Ausziehtusche *f* drawing ink, China (Indian) ink
Ausziehversuch *m* pull-out test
Ausziehvorrichtung *f* drawing device
Ausziehwalze *f* drawing roller
Ausziehwetterstrom *m (mine)* upcast ventilation
auszimmern *(mine)* to tube; to prop the shaft
Auszug *m* 1. *(ch)* lixivium, extract; 2. abstract *(scientific journal)*; auto-abstract; 3. extension *(camera)*
Auszwirnen *n* twisting at the head
Auto *n* autocar, [motor]car, automobile
Autoanhänger *m* automobile trailer
Autoanlasser *m* starter
Autoantenne *f* car (automobile) aerial
Autoausstellung *f* motor[car] show
Autobahn *f* expressway, express highway, motorway, freeway, autobahn
  A. mit zwei Fahrbahnen dual (double) carriageway [motor] road
Autobahnmeisterei *f* motorway breakdown and repair shop
Autobatterie *f* automobile (car) battery
Autobus *m* [auto]bus, motorbus, omnibus
Autobushaltestelle *f* bus stop
Autobushof *m* bus depot (terminal)
Autochromverfahren *n (print)* autochrome system
Autodeck *n* car deck *(of a vessel)*
Autodrehkran *m* truck slewing crane
Autodyn *n* autodyne
Autodyndetektor *m* autodyne detector
Autodynempfang *m* autodyne reception
Autoempfänger *m* car radio (receiver)
Autoentstörung *f (el)* car shielding
Autofähre *f* auto-ferry, car-ferry
Autofahrer *m* car driver
Autofrachter *m* car carrier
Auto-Freilichtkino *n* drive-in cinema, *(Am)* ozoner
Autogarage *f* automobile garage
autogen autogenous
Autogenschweißen *n s.* Gasschweißen
Autohupe *f* motor-car horn
Autoindustrie *f* automotive (car) industry
Autoionisation *f* autoionization

Autokatalyse *f* autocatalysis
Autoklav *m* autoclave
Autoklavbehandlung *f* autoclave curing, autoclaving, highpressure steam curing *(concrete products)*
Autoklavgips *m* autoclaved gypsum
Autokollimation *f* autocollimation
Autokollimationsfernrohr *n* autocollimator
Autokollimationskamera *f* autocollimation camera
Autokran *m* truck[-mounted] crane
Autoleder *n* motor-car upholstery leather
Automat *m* 1. automatic [machine]; automatic screw machine; single-spindle screw machine; multi-spindle automatic; 2. slotting machine; 3. automaton; *(Am)* robot; 4. *(el)* tripping fuse, automatic cut-out; 5. vending machine
A. mit Endzuständen finite automaton
erkennender A. *(dat)* perceiving automaton (computer, robot)
lernender A. learning automaton (machine)
schließender A. reasoning computer
urteilender A. reasoning computer
wahrnehmender A. *s.* erkennender A.
Automatenfernrohr *n* coin-automatic telescope
Automatenstahl *m* free cutting steel
Automatenwebstuhl *m* automatic loom
Automation *f s.* Automatisierung
automatisch automatic[al]; self-acting
automatisieren to automate
Automatisierung *f* automation
Automobil *n* automobile, autocar, [motor]car
Automobilausstellung *f* motor[car] show
Automobilismus *n* automobilism, motoring
Automobilkarosserie *f* [motor]car body
Automobilkupplung *f* automobile clutch
Automobilreifenkord *m* automobile tyre cloth
automorph *(math)* automorphic
Automorphismus *m (math)* automorphism
Autopilot *m* autopilot, automatic pilot (direction finder)
Autoradio *n* automobile radio
Autoradiografie *f* radioautography, *(Am)* autoradiography
Autoradiogramm *n* radioautograph, *(Am)* autoradiograph
Autoreifen *m* automobile [pneumatic] tyre
Autorennbahn *f* car racing track
Autorennen *n* car racing
Autoreparaturwerkstatt *f* motor (car) repair shop
Autorkorrektur *f* author's proof
Autorotation *f* auto-rotation, tailspin
Autoscheinwerfer *m* automobile headlamp
Autoschleppstart *m* car launching
Autoschlosser *m* engine mechanic
Autoschütter *m* dumper, front dipper
Autotransduktor *m (el)* auto-transductor
Autotransformator *m* self-transformer, autotransformer
Autotransportschiff *n* car carrier
A. mit Rampenauffahrt drive-on car carrier
Autotypie *f* autotypy, half-tone etching
Autotypieraster *m* half-tone screen
Autoxydation *f* autoxidation
Autunit *m (min)* autunite
Auxochrom *n (ch)* auxochrome
Aventurinfeldspat *m* sunstone, aventurine fel[d]spar
Aventurinquarz *m* aventurine [quartz]
Avertin *n* avertin
Aviso *m* advice boat
Avitaminose *f* hypovitaminosis, avitaminosis
Avivage *f* avivage; aftertreatment; *(text)* brightening
avivierecht *(text)* fast to brightening
avivieren to aftertreat; *(text)* to brighten, to clear the shade
Awningdeckschiff *n* awning deck vessel, awning-decked vessel
axial axial[ly]; endwise
Axialbeanspruchung *f* axial stress
Axialbelastung *f* axial load
Axialbewegung *f* end (axial) movement
Axialdruck *m* end (axial) pressure, end (axial) thrust
Axialdruckausgleich *m* balance for end thrust
Axialgebläse *n* axial compressor
Axialität *f* alignment, linear accuracy
Axialkolbenhydromotor *m* axial cylinder-type hydraulic motor
Axialkolbenpumpe *f* axial cylinder type of piston (plunger) pump, axial piston pump
A. mit Sitzventilen axial piston pump with seated valves
A. mit Taumelscheibe swash plate pump
A. mit umlaufendem Zylinderkörper axial piston pump with rotary valves
Axialkolbenregelpumpe *f* axial type of multiple-piston variable displacement pump
Axialkomponente *f* axial component *(turbine)*
Axialkompressor *m* gas turbine
Axialkraft *f* axial force
Axiallager *n* journal (end-thrust) bearing
Axiallast *f* axial (concentric) load
Axiallüfter *m* axial fan
Axialpumpe *f* axial pump
Axialschub *m* axial (end) thrust
Axialspannvorrichtung *f* axial stretching device
Axialspiel *n* end play, end clearance, amount of axial freedom
Axialturbine *f* axial-flow turbine
Axialverdichter *m* axial-flow compressor
Axialverfahren *n* **der Schneckenradherstellung** worm wheel hobbing by the tangential-feed method

Axinit *m (min)* axinite
Axiom *n* axiom
Axonometrie *f* axonometry
axonometrisch axonometric
Axt *f* ax[e]; adze
Axtschneide *f* axe blade
Axtstiel *m* axe handle
Azelainsäure *f* azelaic (anchoic) acid
azeotrop azeotropic
Azetal *n (ch)* acetal
Azetaldehyd *m* acetaldehyde
Azetat *n* acetate
Azetatfaser *f* acetate (staple) fibre
Azetatfilm *m* acetate film
Azetatseide *f* acetate rayon
Azetatträger *m* acetate base
Azetessigester *m* acetoacetic (diacetic) ester
Azetessigsäure *f* acetoacetic (diacetic) acid
Azeton *n* acetone
azetonlöslich acetone-soluble
Azetophenon *n* acetophenone
Azetylen *n* acetylene
Azetylenbrenner *m* acetylene burner
Azetylenentwickler *m* acetylene (carbide-to-water gas) generator
Azetylengas *n* acetylene gas
Azetylenmonokarbonsäure *f* propiolic acid
azetylieren to acetylate
Azetylieren *n* acetylation
Azetylierungsmittel *n* acetylating agent
Azetylsalizylsäure *f* acetosalicylic acid
Azetylzahl *f* acetyl[ation] number
Azetylzellulose *f* cellulose acetate
Azid *n* hydrazoate, azide
Azidimetrie *f* acidimetry
Azidität *f* acidity
Aziditätsgrad *m* acidity degree, degree of acidity
azidophil acidophil
Azimut *m* 1. *(opt)* azimuth; 2. radio bearing *(radar)*; 3. *(math)* argument, amplitude
azimutal azimuthal
Azimutfehler *m* bearing error *(radar)*
Azimutkreisel *m* azimuth gyro
Azimutpeilvorrichtung *f* azimuth indicating goniometer
Azimutverzerrung *f* azimuth distortion *(radar)*
Azimutwinkel *m* azimuth angle
Azobenzol *n* azo benzene
Azofarbstoff *m* azo dye[stuff], azoic dye[stuff]
Azotometer *n* azotometer
Azowalkrot *n* milling azo red
Azurblau *n* azure-blue
Azurit *m (min)* azurite
azyklisch acyclic[al]
Azylgruppe *f* acyl group
azylieren *(ch)* to acylate
Azylierung *f (ch)* acylation

B

b-Ader *f (telec)* B-wire
B-Horizont *m (geol)* B-horizon, zone of concentration
B-Modulation *f (el)* class-B modulation
B-Schirm *m* B-scope *(radar)*
B-Signal *n (telev)* picture signal
B-Verstärker *m (el)* class-B amplifier
Babbeln *n (telec)* babble *(unintelligible crosstalk)*
Babbitmetall *n* babbit metal
Babelquarz *m (min)* Babylonian quartz
Babotrichter *m* Babo flask holder
Babyausstattung *f* layette
Babyzelle *f (Am)* penlite cell
Bachgeröll *n* rolled pebbles
Bachwacke *f* river boulder
Bachwackenmauerwerk *n* masonry from river boulders
Back *f* forecastle *(of a vessel)*
  feste B. topgallant forecastle
  gewölbte B. turtle-back forecastle
  halbe B. monkey forecastle
  hohe B. topgallant forecastle
  niedrige B. monkey forecastle
  offene B. open-end forecastle
  versenkte B. sunk-forecastle *(a forecastle partly above and partly below the upper or weather deck)*
back *(ship)* backward, aback, abaft
Backbord *n* port side
  B. / an *(ship)* aport
  hart B. hard-a-port
Backbordlaterne *f* port lantern (lamp, light)
Backbordlicht *n* port light
Backbordmaschine *f* port engine
Backbordruder *n* port rudder
Backbordseite *f* port side
  B. / zur aport
Backdeck *n* forecastle deck (head)
Backdeckkreuzer *m* raised deck cruiser *(motor-boat)*
Backe *f s.* Backen *m*
backen to bake *(cores)*; to cake *(coal)*
Backen *n* baking *(cores)*; caking *(coal)*
Backen *m* jaw *(chuck, vice)*; block, shoe *(brake)*
Backenbrecher *m* jaw breaker (crusher)
Backenbremse *f* block (shoe) brake
Backenfutter *n* jaw chuck
Backenhobel *m* fence plane
Backenschiene *f (ry)* guard rail, stock-rail of a switch
Backenschmiege *f* oblique cut of a hip rafter *(carpentry)*
Bäckerei *f* bakery
Bäckereimaschine *f* bakery (bakehouse) machine
Backfähigkeit *f (mine)* agglutinating value, caking capacity (power, property); bakability,

baking capacity *(moulding)*; cementing property
Background *m s.* Untergrund
backhalsen *(ship)* to boxhaul
Backhefe *f* baker's yeast
backholen to back, to haul aback *(a vessel)*
Backkohle *f* caking (fat, bituminous) coal
Backlackdraht *m* heat-bonding [enamelled] wire
Backmuffel *f* baking muffle
Backmulde *f* kneading trough
Backobst *n* dried fruit
Backofen *m* bakery oven; baking oven
elektrischer B. electric baking oven
Backprüfung *f* adhesive test *(of enamelled wire)*
Backpulver *n* baking powder
Backschaft *f* mess[mates] *(a group of seamen forming a mess)*
Backschott *n (ship)* forecastle bulkhead
B. der versenkten Back break bulkhead of sunk forecastle
Backsgeschirr *n* mess trap, [mess] kid
Backspiere *f* swing (boat, lower, riding) boom *(for boats)*
Backsreling *f (ship)* forecastle rail
backstags *(ship)* abaft the beam, off the quarter
Backstein *m s.* Ziegel
backtrimmen to trim aback
Backtrog *m* kneading trough
Backvermögen *n s.* Backfähigkeit
Bad *n* 1. bath; 2. dip, dye; 3. cell, tank
altes B. *(text)* standing bath
elektrolytisches B. electrolytic bath
galvanisches B. fused electrolyte, galvanic tank (bath)
Badbewegung *f* circulation of electrolyte
Badeanstalt *f* bath[ing] house
Badekleidung *f* bathing clothes
Badematte *f* bathing mat
Bademode *f* bathing fashion
Badeofen *m* bath [room] stove, bath [room] oven
Bäderschiff *n* excursion vessel
Badesalz *n* bath salt
Badeseife *f* bath soap
Badevorleger *m* bathing mat
Badewanne *f* bath[ing] tub
Badewannenkondensator *m (el)* bath tub capacitor
Badoberfläche *f (met)* top of bath
Badofen *m (met)* bath furnace
Badspannung *f* bath (tank, cell) voltage; electrolytic bath potential
Badstrecke *f (text)* bath travel *(yarn)*
Badstrom *m* bath current *(electrodeposition)*
Badstromregler *m* tank current regulator *(for electrolysis)*
Bagasse *f* [sugar cane] bagasse, crushed sugar cane
Bagger *m* 1. dredger *(excavating under water)*; 2. dredger excavator *(on land)*; power shovel; 3. digger
B. mit Heckschlitz stern-well dredger
B. mit Mittelschlitz centre-well dredger
Baggereimer *m* 1. dredger bucket; 2. excavator bucket
Baggerführer *m* 1. dredger operator; 2. excavator operator
Baggergreifer *m* excavator grab
Baggergut *n* 1. dredged material; 2. excavated material; spoil material
Baggergutladeraum *m* hopper
Baggerkorb *m* grab [bucket]
Baggerlöffel *m* dipper, shovel; excavator bucket
baggern 1. to dredge; 2. to excavate; 3. to trench *(ditches)*
Baggerplanum *n* dredger track level; excavator track level
Baggerponton *m* dredging pontoon
Baggerprahm *m* hopper [barge], mud lighter, dredging boat
Baggerpumpe *f* dredging pump
Baggersand *m* dredged river sand
Baggerschaufel *f* bucket of a dredger
Baggerschleppkabel *n* dredger trailing cable
Baggerschneidkopf *m* dredging cutter
Baggerschute *f* dredger barge
Baggerschüttstelle *f* spoil (dumping) ground
Baggerstelle *f* digging site (point)
Baggerung *f* 1. dredging *(under water)*; 2. excavation *(on land)*
Baggerwand *f* working face
Bahn *f* 1. *(ph)* path, track *(of a body)*; trajectory, orbit *(of electrons)*; *(astr)* path, orbit *(of celestial bodies)*; 2. guide way; guide rail; 3. face *(of an anvil)*; 4. [continuous] web *(paper, board)*; breadth, width *(of fabric)*; sheet[ing]; 5. *s.a.* Eisenbahn
beschichtete B. coated web *(of fabric)*
Bohrsche B. Bohr orbit
geradlinige B. straight-line path (trajectory)
gestörte B. disturbed orbit
elliptische B. elliptical orbit *(of a planet)*
hyperbolische B. hyperbolic orbit *(of a planet)*
mittlere B. mean orbit
momentane B. instantaneous orbit
parabelnahe B. nearly-parabolic orbit *(of a planet)*
parabolische B. parabolic orbit *(of a planet)*
stabile B. stable orbit
übergebende B. transferor railway
ungestörte B. undisturbed orbit
Bahnabschnitt *m (nucl)* track segment
Bahnachse *f (rock)* flight axis
Bahnbeschleunigung *f* acceleration along the path
Bahnbetriebswagenwerk *n* car maintenance and overhaul shop

Bahnbetriebswerk *n* railway repair and servicing shop
Bahnbewegung *f* orbital motion
Bahnböschung *f* slope of the embankment
Bahndrehimpuls *m (nucl)* orbital angular momentum, orbital moment [of momentum]
Bahndruck *m* pressure normal to blade surface *(turbine)*
Bahnebene *f* orbit[al] plane, plane of the orbit, trajectory plane
Bahneigentum *n* railway property
Bahneinengung *f* path contraction
Bahneinlenkung *f (rock)* return to standard path
Bahnelektron *n* orbital electron
Bahnelektroneinfang *m* orbital-electron capture
Bahnelement *n* 1. *(ph)* path element; 2. *(rock)* element of the orbit, orbital element
Bahnentrockner *m (ch)* tenter drier
Bahnfehler *m (rock)* trajectory error
Bahngelände *n* railway territory
Bahngeschwindigkeit *f (rock)* [flight] path speed, velocity in trajectory, satellite velocity; *(nucl)* orbital velocity, velocity of motion along the path
Bahnhof *m* railway station
Bahnhofsanlagen *fpl* station premises
Bahnhofsvorsteher *m* stationmaster, yardmaster
Bahnimpuls *m* orbital moment (*s. a.* Bahndrehimpuls)
Bahnkontraktion *f* path contraction
Bahnkörper *m* road (track) bed; right of way
Bahnkorrektur *f (rock)* trajectory correction
Bahnkraft *f (rock)* path force
Bahnkreis *m (mach)* crank-circle
Bahnlage *f* web position
Bahnlänge *f* 1. orbit circumference; 2. table *(sheet)*
Bahnmagnetismus *m* orbital magnetism (magnetic moment)
Bahnmeister *m* district permanent way inspector
Bahnmeisterei *f* permanent way district
Bahnmoment *n (nucl)* orbital moment (*s. a.* Bahndrehimpuls)
  magnetisches B. orbital magnetic moment
Bahnmotor *m* railway (traction) motor
Bahnneigung *f (rock)* path angle; inclination of path
Bahnnormale *f (rock)* perpendicular to the trajectory plane
bahnparallel *(rock)* parallel to the path
Bahnpolizei *f* railway police
Bahnräumer *m (ag)* divider
Bahnrennrad *n* track racing cycle
Bahnschranken *fpl* level-crossing gates
Bahnschutzpfeiler *m* railway protective pillar
Bahnschwingung *f* trajectory oscillation
bahnsenkrecht *(rock)* at right angles to flight path, normal to the path
Bahnspannung *f (text)* web tension
Bahnspur *f* track (trace) of trajectory; orbital track
Bahnstabilität *f* orbit stability
Bahnsteig *m* platform
Bahnsteigkante *f* edge of platform
Bahnsteigkarte *f* platform ticket
Bahnsteigschaffner *m* platform inspector
Bahnsteigsperre *f* platform barrier
Bahnsteuerung *f* continuous-path control *(machine tool)*
Bahnstromversorgung *f* traction [power] supply
Bahntangente *f (rock)* tangent to the trajectory
Bahntransformator *m* traction transformer
Bahnüberführung *f* overbridge
  offene B. open draw *(papermaking)*
Bahnübergang *m* railway-crossing
  niveaugleicher B. [railway] level crossing
Bahnunterführung *f* underbridge
Bahnunterhaltungsarbeiter *m* platelayer, *(Am)* trackman
Bahnunterhaltungsrotte *f* track maintenance gang
Bahnverfolgung *f (rock)* path tracking
  fotografische B. photographic tracking
  optische B. optical tracking
Bahnverlauf *m* form of trajectory
Bahnvermessung *f* path tracking, survey of the trajectory
Bahnwinkel *m* path angle
Bahnwinkeländerung *f* variation in path angle
Bahnzahl *f (nucl)* track population
Bahnzustand *m* orbital state
Bai *f* bay, creek
Bainit *m (min)* bainite
Bajonettfassung *f* bayonet holder (socket); bayonet mount
Bajonettrahmen *m* bayonet frame
Bajonettsockel *m (el)* bayonet base (cap)
Bajonettverbindung *f* twist lock connection
Bajonettverschluß *m* bayonet fixing (catch, joint)
Bake *f* beacon; radio beacon
  akustische B. talking beacon
  unbemannte B. unattended beacon
Bakelit *n* bakelite
Bakelitgehäuse *n* moulded bakelite jacket
Bakenboje *f* beacon[-surmounted] buoy, beacon fire
Bakengeld *n* beaconage, buoy dues
Bakentonne *f* beacon[-surmounted] buoy, topmark buoy
Bakterienbeize *f (leath)* bacterial bate
Bakterizidausrüstung *f (text)* bactericidal (antibacterial) finish

Balance *f* balance; equilibrium
Balanceruder *n* balanced rudder
Balanceschweberuder *n* spade (balanced underhung) rudder
Balancierdampfmaschine *f* beam steam engine
Balancierverhältnis *n* balance ratio *(balanced rudder)*
Balata *f* balata
Baldachin *m* 1. baldachin, canopy; 2. ceiling plate *(of the lamp)*
Baldachinstrebe *f* centre section strut
Baldrianöl *n* valerian oil
Balg *m* bellows
Balgendurchflußmesser *m* aneroid flowmeter; bellow flowmeter
Balgenkamera *f* bellows camera
Balgenmantel *m* corrugated cable sheathing
Balgennaheinstellgerät *n* bellows for close-ups
Balgfedermanometer *n (cont)* bellow pressure gauge
Balken *m* 1. *(build)* beam; girder; joist (*comp.* Träger); 2. *(telev)* bar
  B. auf zwei Stützen simple (simply supported) beam
  B. mit behauenen Kanten canted beam
  B. mit Schwalbenschwanz dovetailed beam
  B. mit Seitenverstärkungen fished beam
  abgefaster B. canted beam
  baumkantiger B. rough-edged beam, dull-edged beam
  eingespannter B. fixed beam
  kurzer B. stub beam
  schräger B. raker beam
  verdübelter B. flitched (dowelled, keyed) beam
  verdübelter B. mit Schwalbenschwanz dovetailed beam
  verlaschter B. fishing beam
  verzahnter B. joggle beam
  waldkantiger B. dull-edged beam
  zusammengesetzter B. built-up beam
Balken *mpl* **ohne Herz** flitches
  in der Mitte aufgetrennte B. flitches
Balkenanker *m* beam bar (tie)
Balkenauflage *f* joist bearing *(on wall)*
Balkenbiege- und -richtmaschine *f* beam bending and straightening machine
Balkenbrücke *f* beam bridge
Balkenbucht *f (ship)* camber, camber (round, crown, crop) of beam
Balkendiagramm *n* bar chart
Balkenenden *npl* timber ends
Balkenfloß *n* loft raft
Balkengenerator *m (telev)* bar generator
Balkenherdglühofen *m* walking-beam annealing furnace
Balkenherdofen *m* rocker bar-type furnace
Balkenholm *m (aero)* straight spar
Balkenkiel *m* bar (vertical iron-plate) keel
Balkenkopf *m (wood)* beam head; joist end
Balkenlage *f* beams
Balkenmuster *n (telev)* bar pattern
Balkenpumpe *f* beam pump
Balkenrost *m (ship)* gridiron
Balkenschleppe *f (ag)* field drag
Balkenschmiege *f* bevel of a beam
Balkensperre *f* [harbour] boom, boom defence
Balkenstoß *m* scarfed joint
Balkenstringer *m (ship)* beam stringer
Balkenvorsteven *m* bar stem
Balkenwaage *f* beam[-type]scale, equal-armed balance
Balkenweger *m (ship)* [beam] clamp, shelf piece
Balkon *m* balcony
Balkweger *m (ship)* [beam] clamp, shelf piece
Ballast *m* ballast; stiffening; kentledge
  fester B. stationary (permanent, solid) ballast
  fliegender B. shifting ballast
Ballastbäume *mpl* stiffening booms, ballast logs *(lashed anlongside to prevent capsizing)*
Ballasteinrichtung *f* ballasting device
Ballasteisen *n* [iron] kentledge
Ballastfahrt *f* ballast trip (passage, voyage)
Ballastgehalt *m* inerts content
Ballastgüter *npl* kentledge goods
Ballasthochtank *m* deep ballast tank
Ballastladelinie *f* ballast load-line, light water line
Ballastladung *f* dead freight
ballastreich dirty
Ballastwasserlinie *f* ballast water-line
Ballastwiderstand *m (el)* 1. ballast (loading) resistance; 2. ballast resistor
Ballastwurf *m* jettison
Ballempfang *m (el)* ball (rebroadcasting, relay) reception
Ballempfänger *m (el)* ball (rebroadcast) receiver, re-receiver
Ballempfangssystem *n* radio-relay system
Ballen *m* bale; balling; ball; bundle
  B. einer Walze roll barrel
Ballenbeförderungssystem *n* bale handling system
Ballenbrecher *m (text)* bale breaker
  B. mit Pedalvorrichtung *(text)* pedal bale breaker
Ballendraht *m* baling wire
Balleneinlage *f* bottom filler *(shoe)*
Ballenförderer *m* bale handler (transporter, conveyor)
Ballengriff *m (mach)* ball handle
Ballenkarre *f* bale truck
Ballenlader *m* bale loader
Ballenladung *f (ship)* bale cargo
Ballenlänge *f* surface (body) length *(of a roll)*
Ballenleistung *f* baling capacity
Ballenoberfläche *f* cylinder surface *(rolling mill)*

Ballenpresse *f* baler, baling press
Ballenschlitten *m* bale sledge
Ballenschurre *f* bale chute
Ballenstapel *m (text)* stack of bales
Ballenstropp *m (ship)* bale sling
Ballenverlader *m* bale loader
Ballhammer *m* ball peen [hammer]
ballig crowned; cambered
Balligdrehen *n* spherical turning
Balligkeit *f* convexity; crown dimension
Balligschliff *m* crowned (convex) grinding
Balligtragen *n* localized tooth bearing
balligtragend localized bearing, bearing with localized tooth
Balligverzahnung *f* crowned gears
Ballistik *f* ballistics
ballistisch ballistic
Ballon *m* 1. *(aero)* balloon; 2. demijohn; carboy; 3. *(el)* bulb *(of valves)*
bemannter B. manned balloon
unbemannter B. unmanned balloon, balloon without crew
Ballonantenne *f* balloon aerial
Ballonaufstieg *m* balloon ascent (ascension)
Ballonbereifung *f* balloon tyres
Ballonbildung *f* ballooning of the yarn *(spinning)*
Ballonett *n* ballonet
Ballonflieger *m* flying balloon jib, balloon jib topsail *(yacht)*
Ballonklüver *m* balloon jib *(yacht)*
Ballonkontrollring *m (text)* balloon control ring
Ballonreifen *m* balloon tyre
Ballonsegel *n* balloon jib *(yacht)*
Ballonsonde *f* recording (sounding, test) balloon
Ballonsperre *f* balloon barrage
Ballonstart *m s.* Ballonaufstieg
Ballonstoff *m* balloon fabric
Ballontheodolit *m* balloon theodolite
ballsenden to rebroadcast
Ballsender *m* relay broadcast station, radio repeating station, rebroadcast (reradiating) transmitter
Ballung *f* bunching *(of valves)*; package, packing *(of components)*; agglomeration; concentration
optimale B. optimum bunching
überkritische B. overbunching
unterkritische B. underbunching
Ballungsgebiet *n (build)* conurbation
Balmer-Serie *f* Balmer series
Balmer-Sprung *m* Balmer discontinuity
Balsaholz *n* balsawood
Balsamtanne *f* balsam fir
Bambus *m* bamboo
Bambusrohr *n* bamboo [stem]
Bambuszucker *m* tabasheer, tabashir
Bananenbuchse *f (el)* banana jack
Bananendoppelstecker *m (el)* U-plug
Bananenschiff *n* banana (fruit) carrier
Bananenstecker *m* banana pin (plug), spring contact plug; pin-plug
Band *n* 1. *(text)* ribbon; sliver (spinning); 2. tape *(of magnetic recorder)*; 3. belt[ing], band *(conveyor)*; 4. strip *(steel, sheet)*; hoop; 5. band *(architecture)*; 6. brace; 7. *s.* Bande
B. zum Öffnen einer Verpackung pull strip (tape)
besetztes B. *(nucl)* filled band
endlos umlaufendes B. endless moving belt
gummiertes B. adhesive tape
kaltgewalztes B. cold-rolled strip
magisches B. magic band valve
nichtbesetztes B. *(nucl)* unoccupied band
schmales B. narrow band
thermoplastisches B. thermoplastic tape, thermotape
verbotenes B. energy gap *(semiconductor)*
Bandablage *f* tape file
Bandabsetzer *m (Am)* boom stacker
Bandabstand *m* energy gap, band gap *(semiconductor)*
Bandabtastung *f* tape sensing
Bandachat *m (min)* ribbon (banded) agate
Bandage *f* 1. bandage, dressing; 2. shroud ring, shrouding *(turbine)*
Bandagen *fpl* handwraps *(boxing)*
Bandagenblock *m* tyre ingot *(rolling mill)*
Bandagendraht *m* binding wire
Bandagenglühofen *m* tyre heating furnace *(rolling mill)*
Bandagenindustrie *f (text)* surgical cloth industry
Bandagenverfahren *n* bandage process
Bandagenverluste *mpl* band (binding wire) losses
Bandagenwalzwerk *n* strip rolling mill
Bandantrieb *m* tape drive
Bandaufliegezeit *f* belt life
Bandaufnahme *f* tape record; tape recording
Bandaufnahmegerät *n* tape recorder
Bandbebauung *f* ribbon development
Bandbecherwerk *n* belt elevator
Bandbefehl *m (Am)* bootstrap *(computer)*
Bandbegrenzung *f (telev)* band limiting
Bandbelag *m* band (belt) cover
Bandbeschicker *m* belt feeder
Bandbeschneidung *f* band-width limitation
Bandblechwalzwerk *n* broad strip mill
Bandbohrer *m* brad-awl
Bandbreite *f (telec)* band-width; spread
B. bei Verstärkung gain-band merit
belegte B. occupied band-width
effektive B. effective band-width
spektrale B. spectral band-width

Bandbreitebeschneidung *f* reduction of bandwidth
Bandbremse *f* band brake
Bandbruch *m* tape breakage
Bandbrücke *f* belt conveyor bridge
Bändchen *n* small strand *(spinning)*
Bändchengalvanometer *n* tape galvanometer
Bändchenlautsprecher *m* ribbon loudspeaker
Bändchenmikrofon *n* tape (velocity) microphone; ribbon microphone
Bändchenstickerei *f* braid embroidery
Bandchronograf *m* tape chronograph
Banddehner *m (telec)* speech inverter
Banddehnung *f (telec)* band spread[ing]
Banddüngung *f (ag)* in-row fertilization, sideband placement *(of fertilizer)*
Bande *f (ph)* [molecular] band
  aufgelötete B. resolved band
  breite B. broad band
  kurzwellige B. short-wavelength band
  langwellige B. long-wavelength band
Bandeinlegeschlitz *m (el)* drop-in tape loading slot
Bandeisenlehre *f* rolling mill gauge
Bandenfolge *f* band progression *(spectrum)*
Bandenkante *f s.* Bandenkopf
Bandenkopf *m* band head *(spectrum)*
Bandenserie *f* band series
Bandenspektrum *n* band spectrum
Bandensystem *n* band system
Banderder *m* earthing strip, strip-type earthing conductor
Bänderjaspis *m (min)* striped jasper
Banderole *f* bill band
Bänderton *m (geol)* varve[d] clay
Bandfilter *n* band-pass filter
Bandfilterkondensator *m* band-pass capacitor
Bandfilterkopplung *f (el)* filter coupling
Bandfilterresonanzkurve *f (el)* band-pass response
Bandflechtmaschine *f (text)* braid-plaiting machine
Bandförderer *m* belt (band) conveyor
Bandführung *f* tape guide; tape guidance
Bandführungsrolle *f* guide pulley *(band saw)*
Bandgenerator *m (el)* belt-type generator, Van-de-Graaff generator
Bandgeschwindigkeit *f* tape speed *(of magnetic tape recorder)*
Bandgeschwindigkeitsumschalter *m* speed selector
bandgesteuert tape-controlled *(machine tool)*
Bandgetriebe *n* pulley gear drive
Bandglühofen *m s.* Bandagenglühofen
Bandhaken *m* belt fastener
Bandjaspis *m (min)* striped jasper
Bandkabel *n* tape (ribbon, flat) cable
Bandkondensator *m* tape capacitor
Bandkupplung *f* band expansion clutch
Bandlauf *m* tape run *(recorder)*
Bandlautsprecher *m* ribbon-type dynamic speaker
Bandlebensdauer *f* belt life
Bandleiter *m (el)* strip conductor
Bandleitung *f (el)* ribbon conductor (line), stripline
Bandlesegerät *n* tape reading unit
Bandleser *m* tape reader
Bandmagnetscheider *m* magnetic separating belt
Bandmaß *n* [measuring] tape, [steel] tape for measuring, flexible rule
Bandmesserspaltmaschine *f (leath)* bandknife splitting machine
Bandmischer *m* ribbon mixer
Bandmittenfrequenz *f* midband (centre) frequency
Bandpaß *m (el)* band-pass filter
Bandrauschen *n* tape background noise *(recorder)*
Bandrauschzahl *f* average noise factor
Bandreinigungsrolle *f* belt cleaner
Bandriß *m* belt breakage *(conveyor)*; turn tape *(tape recorder)*
Bandrolle *f* [roller] idler, roller
Bandrost *m* endless grate
Bandsäge *f* band (endless) saw
Bandsägeblatt *n* band-saw blade, endless saw-blade
Bandscheider *m* belt separator
Bandschleife *f* loop take-up *(conveyor)*
Bandschleifen *n* abrasive-belt grinding
Bandschleifmaschine *f* belt grinding (sanding) machine, abrasive-belt machine
Bandschnecke *f* ribbon screw
Bandschnellader *m* belt delivery shovel loader
Bandschreiber *m* tape (strip-chart) recorder
Bandseil *n* band rope, flat cable (rope)
Bandsieb *n* endless-belt screen
Bandspan *m* ribbon[like] chip
Bandspannhaken *mpl* belt grips
Bandspannvorrichtung *f* belt take-up, belt tensioning device
Bandspeicher *m* tape store *(for data)*
Bandspeiser *m* strip feeder
Bandsperre *f (el)* band-rejection filter, band-elimination filter, filter stop band
Bandsperrfilter *n* band-stop filter
Bandspule *f* tape spool *(recorder)*; ribbon coil
Bandstahl *m* strip (hoop) steel
Bandstahlarmierung *f s.* Bandstahlbewehrung
Bandstahlbewehrung *f* ribbon (tape) steel armouring *(of cables)*
Bandstahlhaspel *f* hoop steel spool
Bandstahlstraße *f* band mill train
Bandsteuersystem *n* tape control system *(machine)*

Bandsteuerung *f* tape control *(machine)*
Bandtraggerüst *n* belt conveyor structure
Bandtrockner *m* belt (conveyor) drier, endless belt conveyor drier, conveyor drying machine
Bandtrocknung *f* belt drying
Bandtrum *m* strand
Bandüberlappung *f* band overlap
Bandübertragung *f (text)* ribbon feed
B. mit Kreuzung cross ribbon feed
B. ohne Kreuzung straight ribbon feed
Bandüberwachung *f (el)* band-watching
Bandverbinder *m* belt joint
Bandverbreiterung *f* band broadening
Bandverschiebung *f* band shift
Bandverzerrung *f* band distortion
Bandvorschub *m* tape feed
Bahnwächter *m* automatic belt-control device
Bandwalze *f* belt drum
Bandwalzwerk *n* strip rolling mill
Bandwärter *m* conveyor attendant
Bandweberei *f* ribbon weaving
Bandwebstuhl *m* ribbon loom
Bandwickelmaschine *f (text)* sliver lap machine, derby doubler
Bandwickelvorbereitung *f* comber lap preparation
Bandwickler *m* sliver lap machine
Bandzählwerk *n* tape footage counter *(magnetic tape recorder)*
Bandzeitschreiber *m* tape chronograph
Bandzug *m* 1. train conveyor; 2. tape tension; tension of magnetic band
Banjoachse *f (aut)* banjo axle
Bank *f* 1. bench, seat; 2. bank *(money; clay, sand)*; 3. *(geol)* stratum, seam, layer; 4. contact bank (panel)
optische B. optical bench
Bankazinn *n* Banca tin
banken to swift *(a ship)*
Bankett *n* berm *(civil engineering)*
Bankfiliale *f* branch bank
Bankgewerbe *n* banking industry
Banknotenpapier *n* bond paper
bankrecht *(geol)* normal to the stratification
Banse *f (ag)* bay [of a barn]
Bar *n* bar
bar for cash; in cash
Bär *m* tup, hammer, monkey *(piling plant)*; ram[mer]
Baracke *f* hut; shanty
Barbitursäure *f* barbituric acid
Barboteur *m (ch)* bubbler
Barchent *m (text)* fustian
Barchentrauhmaschine *f* top gig
barieren to bariate
barisch baric
Barium *n* barium
Bariumbeton *m* barium concrete
Bariumgetter *m* barium getter
bariumhaltig baric
Bariumkarbonat *n* barium carbonate
Bariumkatode *f* barium cathode
Bariumoxid *n* barium oxide, [calcined] baryta
Bariumtitanat *n* barium titanate
Bark *f* barque, bark
Barkantine *f (ship)* barkentine, barkentyne, barquentine
Barkasse *f* launch, long-boat, quarter boat
Barkassenheck *n* square stern
Barkentine *f s.* Barkantine
Barkhausen-Effekt *m* Barkhausen effect
Barkhausen-Kurz-Schwingungen *fpl* Barkhausen-Kurz oscillations
Barkhausen-Sprung *m* Barkhausen jump (effect)
Barkschoner *m* barkentine, barkentyne, barquentine
Barkunen *pl* luffing (hinged, radial) davits
Barlow-Rad *n* Barlow's wheel
Barn *n* barn, b
Barnett-Effekt *m* Barnett effect
Barograf *m* barograph; recording barometer
Barometer *n* barometer
Barometerablesung *f* barometer reading
Barometerdruck *m* barometric (barometer) pressure
Barometereffekt *m* barometer effect
Barometersäule *f* barometric column
Barometerschwankung *f* barometrical variation
Barometerstand *m* barometer height; barometric pressure; barometric reading
Barometrie *f* barometry
barometrisch barometric[al]
barotrop barotropic
Barotropiefeld *n (ph)* barotropic field
Barras *m (wood)* barras
Barrehafen *m* bar port
Barrel *n* barrel *(a measure used in English fish trade = 13,608 kg)*
Barrelotse *m* bar pilot
Barren *m* 1. bar; ingot; pig; bullion *(noble metal)*; billet; 2. parallel bars
Barrenedelmetall *n* bullion bar
Barrenform *f* ingot mould
Barrensilber *n* silver ingot
Barrenzimmerung *f (mine)* crown-bar timbering
Barrenzink *n* bar zinc
Barretiefe *f* bar draught
Barretonne *f* bar buoy
Barriere *f (el)* barrier *(high voltage insulation)*
Barscheck *m* cash cheque
Bart *m* burr
Baryon *n (nucl)* baryon, barion
Baryonenzahl *f* baryon number
Baryt *m (min)* barytes, barite, heavy spar
Barytbeton *m* baryte (barite) concrete

Barytwasser *n* baryta water
Barytweiß *n* permanent white, baryta white
baryzentrisch barycentric
Basalt *m* basalt
Basalteisenerz *n* basaltic iron ore
Basaltlava *f* basaltic lava
Basaltpflasterstein *m* basalt paving sett
Basalttuff *m* basaltic debris; trap tuff
Basaltwacke *f* basaltic wacke
Basaltwolle *f* basalt (rock) wool *(mineral wool)*
Base *f (ch)* base
  dreisäurige B. triacid base
  einsäurige B. monoacid base
  viersäurige B. tetracid base
  zweisäurige B. biacidic base
Basenaustausch *m* exchange of bases, base exchange
Basenaustauschanlage *f* base exchange equipment
basenbildend basifying, base-forming
Basenbildner *m* basifier
Basengrad *m* basicity factor
Basenstärke *f* basicity
Basenüberschuß *m* excess base
Basenumtausch *m s.* Basenaustausch
Basis *f* 1. *(build)* basis; base; 2. base *(logarithm)*; radix *(of a number system)*; 3. base line *(surveying)*
  B. natürlicher Logarithmen base of natural logarithms
  offene B. open base *(transistor)*
  veränderliche B. variable base length
Basisanschluß *m (el)* base contact (terminal)
Basisbreite *f* base width
basisch basic
Basisebene *f* basal plane *(crystal)*
Basiselement *n* base element *(vector)*
Basisemitterschaltung *f* base-emitter circuit
Basisemitterübergang *m* base-emitter junction
Basisfläche *f* base surface; base
Basishafen *m* standard port
Basiskomplementkode *m (dat)* radix complement code
Basiskontakt *m* base contact
Basiskreis *m* base circuit
Basislatte *f (Opt)* subtense bar
Basislinie *f* base line; base
Basisring *m* base ring *(transistor)*
Basisschaltung *f* grounded-base circuit (configuration) *(semiconductor)*
Basisschicht *f* base layer *(transistor)*
Basisstrom *m* base current
Basisvektor *m* base vector
Basiswiderstand *m* base resistance
  äußerer B. external base resistance
  innerer B. internal base resistance
Basiswinkel *m* base angle
Basiszeitraum *m* base period
Basiszone *f* base region
Basizität *f* basicity
Basizitätsverhältnis *n* slag ratio
Basketballkorb *m* basket ball gasket (goal)
Basketballständer *m* backstop
Baskülbrücke *f* bascule bridge
Basküle *f* bascule
Basrelief *n* bas relief
Baßabsorption *f* bass absorption
Baßabstimmung *f* bass tuning
Baßanhebung *f* low note accentuation, bass compensation (boost)
Basselissestuhl *m (text)* low warp loom
Baßentzerrung *f* bass corrector (compensator)
Bassin *n* reservoir, basin, tank, pool
Bassinkondensator *m (therm)* submerged condenser
Baßverdeckung *f* bass masking
Baßwiedergabe *f* bass response
Bast *m (wood)* inner bark, bast
Bastardbahnen *fpl (ch)* hybrid orbitals
Bastardfeile *f* bastard file
Bastfaser *f* bast fibre
Bastler *m* home constructor; amateur constructor
Bastpapier *n* bark paper
Bastseide *f* raw (unscoured) silk
Bathometer *n s.* Bathymeter
Bathymeter *n* bathymeter, bathometer
Bathyscaph *m (n)* bathyscaph *(deep-diving boat)*
Batik *f* batik *(dyeing)*
Batikdruck *m* batik printing
Batikeffekt *m* batik effect
Batikfärberei *f* multicolour dyeing; batik dyeing
Batikreserve *f* batik resist
Batist *m* batiste
Batschemulsion *f (text)* batching medium
batschen *(text)* to batch
Batschen *n* **im Fach** *(text)* batching in bins
Batschfach *n* batching bin
Batschmaschine *f* batching machine
Batschöl *n* batching oil
Batschreife *f* batching ripeness
Batterie *f* 1. *(el)* battery, primary cell; 2. vertical multi-mould *(for precasting concrete)*
  entladene B. run-down battery
  ortsfeste B. stationary battery
  vergossene B. sealed-in battery
Batterieanschlußklemme *f* battery connecting terminal
Batterieausfall *m* battery failure
Batterieausgangsleistung *f* battery output
batteriebetrieben battery-operated, battery-powered
Batteriedecke *f* battery top *(coking)*
Batterieempfänger *m* battery-operated receiver
Batterieendkopf *m* buttress of battery *(coking)*
Batterieentlüftungsventil *n* battery vent valve
Batterieersatzgerät *n* battery eliminator

Batteriefertigung *f* vertical multi-moulding *(concrete)*
Batterieform *f* 1. vertical multi-mould *(for precasting concrete)*; 2. bank furnace *(coking)*
batteriegespeist battery-powered, battery-operated
Batteriekarton *m* jacket, case
Batteriekasten *m* accumulator (battery) container, *(Am)* battery box
Batteriekessel *m* battery boiler
Batterieklemme *f* battery terminal, terminal pillar (post)
Batterieladeeinrichtung *f* battery charging set
Batterieladegerät *n* accumulator (battery) charging set, battery charger
Batterieladevorschriften *fpl* battery charge regulations
Batteriepolklemme *f* terminal post
Batterieprüfer *m* battery tester
Batterieröhre *f* battery valve
Batteriesäureheber *m* battery syringe
Batterieschalter *m* battery switch
Batteriespannband *n* battery strap
Batteriestromverbrauch *m* battery drain
Batterieträger *m* battery mounting (carrier)
Batterietrog *m* tray
Batterieumschalter *m* battery reversing switch
Batterieverfahren *n* vertical multi-moulding *(concrete)*
Batteriezellenverbindung *f* intercell connector
Batteriezuführung *f* battery lead
Batteriezündung *f* battery-coil ignition
Bau *m* building; construction; structure; *(mine)* support unit
repräsentativer B. prestige-type building
Bauablaufplan *m* construction (work) schedule, progress chart
Bauabschnitt *m* stage [of construction]
Bauabteilung *f (mine)* panel
Bauakustik *f* architectural acoustics
Bauarbeiter *m* building worker (operative), builder
Bauart *f* design, type, construction; constructional features
rahmenlose B. frameless (integral, chassisless) construction
Bauaufnahme *f* survey of a building
Bauaufsicht *f* supervision of works
Bauaufwand *m* cost of construction
Bauausführung *f* [building] construction
Baubeschläge *mpl* builders' hardware
Baubesteck *n* table of the dimensions and scantlings, list of scantlings *(of a ship)*
Baubestimmungen *fpl* building regulations
Baubetrieb *m* contracting company (firm), [building] contractor
Baubude *f* workmen shelter
Baubüro *n* site (job) office
Bauch *m* 1. antinodal point, antinode *(wave)*; 2. bottom, bilge *(of a ship)*
Bauchfreiheit *f s.* Bodenfreiheit
bauchig barrel-shaped
Bauchstag *n (ship)* belly stay
Bauch-Transformator *m* Bauch-transformer, earthing auto-transformer, neutral compensator, neutralizer
Baud *n* baud *(unit of speed of telegraphic code transmission)*
Baudichte *f (el)* packing density
Baudock *n* building basin (dock)
Baudurchführung *f* [building] construction
Baueinheit *f (mach)* standard [element] component, basic unit, building block; constructional unit
Bauelement *n (mach)* component; *(build)* structural unit (member, element, component, part); *(el)* unit; module
angepaßtes B. matched component
bewickeltes B. wound component
dichtes B. sealed component
elektronisches B. electronic component
integriertes B. integrated component
punktförmiges B. dot component
widerstandsfähiges B. rugged unit
Bauelemente *npl* components; *(Am)* hardware *(computer)*
bauen to build; to construct
Bauen *n* / **industrielles** industrialized building, industrialized construction technique, building by industrialized methods
ländliches B. rural construction
landwirtschaftliches B. agricultural building, farm-building construction
Bauernhaus *n* farmhouse
Bauernhof *m* farmyard
Baufachmann *m* builder; building expert
baufällig dilapidated
Baufehler *m* building fault
Baufeld *n (mine)* working panel
Baufirma *f* contracting firm (company), [building] contractor
Bauflucht *f* building (property, frontage) line
Bauform *f* design
gedrungene B. compact design
Baufristenplan *m* progress chart, construction (work) schedule
Bauführer *m* assistant supervisor
Baugelände *n* building works area
Baugenehmigung *f* [planning] building permission
Baugenossenschaft *f* building society
Baugerät *n* construction equipment (plant), builders' (contractors') plant
Baugerüst *n* scaffold[ing]
Baugesetzgebung *f* building legislation
Baugewerbe *n* building trade

Bauglas *n* constructional glass
Bauglied *n* structural unit (member, component, part)
Baugröße *f (mach)* size
Baugrube *f* foundation pit, excavation
Baugrund *m* subsoil, foundation soil
Baugrunderkundung *f* investigation of foundation conditions
Baugrundmechanik *f* soil mechanics
Baugrundstabilisierung *f* soil (ground) stabilization
Baugrundverbesserung *f* soil (ground) stabilization, soil (ground) solidification
chemische B. soil cementation *(by injection of chemicals)*
Baugrundverdichtung *f* compaction of soil, soil compaction
Baugrundverfestigung *f* / **chemische** *s.* chemische Baugrundverbesserung
Baugruppe *f* element, component *(machine tool)*; group of parts, sub-assembly; package[d] unit
elektronische B. module
kompakte B. package
Bauhandwerk *n* building trade
Bauhelling *f (ship)* slipway
Bauhilfsarbeiter *m* builder's labourer
Bauhof *m* contractor's yard; timber yard
Bauhöhe *f* height of construction; head room; *(mine)* lift; overall height *(ship)*
flache B. *(mine)* projection of face length in the direction of full dip, inclined depth
seigere B. *(mine)* vertical projection of face length
Bauholz *n* structural (building) timber, *(Am)* lumber
Bauindustrie *f* building industry
Bauingenieur *m* civil engineer
Bauingenieurwesen *n* civil engineering
Baujahr *n* building year; year of manufacture
Baukastenanlage *f (el)* unit-composed plant, block-building plant, single-unit plant, combined substation installation, unit-combined plant
Baukastenmöbel *n* modular furniture
Baukastenprinzip *n* module principle of construction, unit principle; building-block principle; unit assembly (construction) principle; *(build)* modular concept, sectional system
Baukastensystem *n* unit assembly (construction) principle, building-block system; modular building (unit) system, modular design and construction method
Baukeramik *f* structural clay product
Bauklammer *f* clamp[-iron], cramp, dog
Bauklempner *m* building plumber
Baukonstruktion *f* building construction, structural design
Baukonstruktionslehre *f* structural theory
Baukörper *m* structure, building
Baukosten *pl* cost of construction
Baukunst *f* architecture
Baulänge *f* constructional (overall) length
optische B. opto-constructional length
streichende B. *(mine)* life of face *(in terms of distance)*
Bauleiter *m* site engineer (supervisor), field superintendent
Baum *m* 1. tree; 2. *(text)* beam; 3. *(ship)* boom, derrick
Baumantenne *f* fishbone aerial
Baumaschinen *fpl* construction equipment (machinery), builder's (contractor's) plant
Baumaschinenindustrie *f* construction machinery and equipment industry
Baumaxt *f* felling axe; cleaver
Baumeister *m* master builder
Baumfock *f (ship)* boom fore-sail
Baumgalgen *m (ship)* beam gallows
Baumgrenze *f* tree limit
Baumgrundschleppnetz *n (ship)* beam trawl [-net]
Baumhacke *f* felling axe; cleaver
Baumkante *f* wane, natural (bed) bevel, waney corner, dull (rough) edge
baumkantig rough-edged *(wood)*
Baumkratzer *m* bark scraper
Baumkurre *f (ship)* beam trawl
Bäummaschine *f (text)* beamer, beaming machine
B. für gewickelte Ketten ball warp beaming machine
Baummesser *n* pruning knife (hook)
Baumodell *n* built[-up] model
Baumpfahl *m* tree prop, stay
Baumsäge *f* tree (pruning, bow) saw
Baumschären *n (text)* beam-warping
Baumschere *f* pruning shears
Baumschnitt *m* pruning
Baumschule *f* [tree] nursery
Baumsetzmaschine *f* tree planter
Baumstamm *m* trunk
Bäumstuhl *m (text)* beaming frame
Baumstumpf *m* stock, stub, stump, snag
Baumtoppnant *f (ship)* boom (quarter) lift
Baumumleger *m* tree puller
Baumwollabfälle *mpl* cotton waste
Baumwollabfallspinnerei *f* cotton waste spinning
Baumwollanbau *m* cotton growing
Baumwollanbaufläche *f* cotton growing area
Baumwollballen *m* cotton bale
Baumwollballenpreßanlage *f* cotton baling plant
Baumwollbaum *m* kapok tree
baumwollbesponnen cotton-covered *(wire)*

Baumwollblüte *f* flower of cotton plant
Baumwollbuntspinnerei *f* spinning of coloured cotton yarn
Baumwolle *f* cotton
animalisierte B. animalized cotton
außergewöhnlich weiße B. blue cotton
fleckige B. stained cotton
kurzstapelige B. short-staple cotton
langstapelige B. long-staple cotton
merzerisierte B. mercerized cotton
Baumwollegreniermaschine *f* mechanical cotton gin
Baumwollentkörnungsmaschine *f* mechanical cotton gin
Baumwollernte *f* cotton crop
Baumwollfarbstoff *m* cotton dye[stuff]
Baumwollfaser *f* cotton fibre
Baumwollfrucht *f* **/ ausgereifte** mature cotton pod
Baumwollgarn *n* cotton yarn
gekämmtes B. combed cotton yarn
Baumwollgewebe *n* cotton fabric (goods, linen, textiles)
reines B. all-cotton cloth
Baumwollinters *pl* cotton linters
Baumwollkräuseldruckgewebe *n* crinkle crêpe
Baumwollpackung *f* cotton packing *(hydraulics)*
Baumwollpflanze *f* cotton plant
Baumwollpflücken *n* cotton picking
Baumwollplüsch *m* cotton shag
Baumwollsaatkuchen *m* cotton cake
Baumwollsamen *m* cottonseed
Baumwollsamenkapselschalen *fpl* cottonseed hulls
Baumwollsamenöl *n* cottonseed oil
Baumwollschnur *f* cotton cord
Baumwollseidenkabel *n* silk and cotton covered cable
Baumwollselfaktor *m* [cotton] mule
Baumwollspeiser *m* cotton [automatic] feeder
Baumwollspinner[in *f***]** *m* cotton spinner
Baumwollspinnerei *f* 1. cotton spinning; 2. cotton mill
Baumwollstapel *m* cotton staple
Baumwollstengel *m* cotton stalk
Baumwollstrauch *m* cotton plant
Baumwolltara *f* bale tara
Baumwolltrockenfilz *m* drier felt
Baumwollumspinnung / mit doppelter double-cotton covered
baumwollumsponnen cotton-covered
Baumwollvliesstoff *m* non-woven cotton
Baumwollweberei *f* cotton weaving
Baumzucht *f* arboriculture
Baunivellier *n* builder's level
Baunummer *f* yard number *(of a ship)*
Bauordnung *f* building code
Baupappe *f* felt
Bauplan *m (el)* component layout
Bauplatte *f* building board
Bauplatz *m* 1. *(build)* construction (building, job, work) site, site; 2. building berth *(for ships)*
Baurecht *n* building legislation
Bauschaltplan *m (el)* wiring diagram (scheme)
Bauschein *m* builder's certificate
bauschen to bulge, to bag *(thread)*
Bauschgarn *n* textured (bulk) yarn
Bausohle *f (mine)* working horizon
Bauspant *n (ship)* constructional section, frame
Bauspantenriß *m* [framing] body plan, framing plan
Baustahl *m* mild (structural) steel
hochwertiger B. high-tensile steel
legierter B. structural alloy steel
Baustahlgewebe *n* steel [mesh] fabric, welded fabric
Baustange *f* pole
Baustapel *m* building block
Baustatik *f* structural [engineering] statics, structural analysis
Baustein *m* 1. building stone; 2. *(cont)* building block, module, modular (packaged) unit, subunit, add-on unit, standard component, building-block module; block
B. für Mikrominiaturschaltung micro-miniature circuit module
B. mit gedruckter Schaltung printed circuit unit
elektronischer B. electronic component
standardisierter B. standard modular unit, standardized building block
Bausteinbauweise *f (el)* modular construction
Bausteintechnik *f (el)* modular construction
Baustelle *f* building site, [construction] site; erection site *(steel erection)*
B. / auf der on-site
B. / nicht auf der off-site
Baustellenbeton *m* site (job-mixed) concrete
Baustellenbüro *n* site (job) office
Baustelleneinrichtung *f* site organization
Baustellenlabor *n* field laboratory
Baustellenmontage *f* on-site assembly (erection), site assembly (erection)
Baustellenniet *m* field (site) rivet
Baustellenorganisation *f* site organization
Baustellenprüfung *f* field (site) test
Baustellenschweißung *f* site (field) welding
Baustellensilo *m* site hopper, bulk cement silo
Baustellenüberwachung *f* field (site) supervision
Baustellenverbindung *f* field (site) connection
Baustellenvermessung *f* site survey
Baustellenvorfertigung *f* on-site precasting
Baustoff *m* building material
Baustoffe *mpl* **/ feuerfeste** refractories
Bautagebuch *n* job diary (record)

Bautechnik *f* constructional engineering
Bautechniker *m* building site technician (engineer)
Bauteil *n* 1. *(mach)* element, basic unit; 2. [structural] member *(e.g. ship)*, structural unit (component)
vorgefertigtes B. pre-constructed unit, prefabricated unit
Bauteile *npl* / **längslaufende** fore-and-afters
Bauten *mpl* / **gesellschaftliche** public buildings
unterirdische B. below-ground structure
Bautenschutzmittel *n* proofing material for buildings
Bautischler *m* joiner
Bautischlerei *f* joinery, joiner's shop
Bauunternehmen *n* contracting company (firm), [building] contractor
Bauvorhaben *n* project
Bauvorschrift *f* / **Lloyds** *(ship)* Lloyd's Rules and Regulations
Bauvorschriften *fpl (build)* building regulations
Bauweise *f* 1. *(mach)* design; 2. *(build)* [system of] construction
industrielle B. industrialized building, industrialized construction technique, industrialized building method
monolithische B. monolithic construction
wickellose B. *(el)* non-lined construction system
Bauwerft *f* [building] yard
Bauwerk *n* structure, building
Bauwerksabdichtung *f* waterproofing of buildings
Bauwerksachse *f* centre line
bauwürdig / **bedingt** *(geol)* marginal
Bauwürdigkeitsgrenze *f (geol)* payability limit
Bauxit *m* alumina hydrate, bauxite
Bauzaun *m* hoarding
Bauzeichnung *f* construction drawing
Bauzeit *f* 1. construction period; 2. *(ship)* building time *(on the stocks)*
Bauzug *m* work train
Bayer-Verfahren *n* Bayer process
Baykogarn *n* metal-covered cotton yarn
Bayöl *n* bay oil
Beachtet Funkstille! Observe silent period!
Beachtung *f* consideration; observance
Beackerung *f* tilling
Beanspruchbarkeit *f* stressability; permissible (allowable) stress
beanspruchen to load, to stress, to impose stress (a load) on
beansprucht loaded, stressed
Beanspruchung *f* load, stress; duty
beaufschlagte B. applied stress
elektrische B. voltage stress, [di]electric stress
hohe B. severe duty *(valve)*
mechanische B. mechanical stress
plastische B. superelastic stress
rauhe B. rough usage
starke B. hard wear
thermische B. thermal stress
überelastische B. superelastic stress
zulässige B. permissible stress
beanstandenswert objectionable
Beanstandung *f* rejection; complaint
bearbeitbar workable; machinable
Bearbeitbarkeit *f* 1. workability *(metal working)*; 2. machinability *(metal cutting)*, cutability
bearbeiten to work; to machine, to cut *(metal)*; to process, to treat
allseitig b. to machine all over
auf dem Baum b. *(leath)* to work over the beam
den Grund b. *(mach)* to bottom [the bore]
spanend b. to cut, to machine
Bearbeiten *n* **auf Maß** *(mach)* sizing
B. mit hohen Geschwindigkeiten *(mach)* high-speed machining
spanendes B. machining
Bearbeitung *f* machining; working; processing; tooling
B. der Informationen data recovery (processing)
B. in Querrichtung *(mach)* cross operation
B. mit Rundvorschub *(mach)* circular work
spanende B. cutting, machining
steinmetzmäßige B. tooling
Bearbeitungseinheit *f (mach)* machining unit
Bearbeitungsfolge *f* sequence of operations
Bearbeitungsgenauigkeit *f (mach)* machining accuracy
Bearbeitungsgrad *m* workability
Bearbeitungsmethode *f* manufacturing method; machining method
spanlose B. chipless machining method
Bearbeitungsplan *m* operation sheet
Bearbeitungsrichtwerte *mpl* **beim Spanen** machining data
Bearbeitungsriefe *f* ridge left by the tool
Bearbeitungsstation *f (mach)* working (machining) station
Bearbeitungszeit *f* machining (working) time; processing time
Bearbeitungszugabe *f* machining (working) allowance
Bearbeitungszuschlag *m* machining allowance
Beaufortskale *f* Beaufort wind scale
beaufschlagen to admit
Beaufschlagung *f* admission *(turbine)*
äußere B. external discharge
axiale B. axial admission
gemischte B. mixed admission
innere B. internal discharge
partielle B. *s.* teilweise B.
radiale B. radial admission
tangentiale B. tangential admission

teilweise B. partial admission
volle B. full admission
Beaufschlagungsdurchmesser *m* admission diameter *(turbine)*
Beaufschlagungsebene *f* plane of admission *(turbine)*
Beaufschlagungsgebiet *n* zone of admission *(turbine)*
Bebakung *f* beaconing *(navigation)*
bebaut built-up
Bebauung *f (build)* development; *(ag)* cultivation
Bebauungsdichte *f* density
Bebauungsplan *m* master plan
beben to tremble, to quake; to vibrate, to oscillate
Beben *n* 1. trembling, quake; vibration, oscillation; 2. earthquake
Beblasung *f* / **magnetische** *(el)* magnetic blowout
bebunkern to bunker *(a vessel)*
Bebunkerungsschiff *n* bunkering boat
Becher *m* bucket *(conveyor)*; cup
Becherförderanlage *f s.* Becherförderer
Becherförderer *m* bucket conveyor; bowl feeder
Becherglas *n* beaker, jar
Becherkondensator *m* encased (potted, box-type) capacitor
Becherwerk *n* [chain] bucket conveyor, bucket elevator
Beck-Bogenlampe *f* Beck arc lamp
Becken *n* pool; basin; reservoir; tank
Beckmann-Thermometer *n* Beckmann thermometer
Becquerel-Effekt *m* photovoltaic effect, Becquerel [photo]effect
Becquerel-Strahlen *mpl* Becquerel rays
bedachen to roof
bedampfen to steam cure *(concrete)*; to [vapour-]coat; to evaporate
Bedampfen *n* [vacuum] deposition
Bedampfung *f* steam curing *(concrete)*; evaporating
Bedarf *m* want, need; demand, requirement
Bedarfsausbesserung *f* casual repairs
Bedarfsfahrpläne *mpl* spare train paths
bedecken to coat, to cover; to overcast *(sky)*
Bedeutung *f* meaning
bedeutungslos insignificant
bedienen to operate; to attend; to service
Bedienung *f* operation; operator attention *(machine)*; attendance
B. der Feuerung feeding (stoking) of the fire
Bedienungsanleitung *f* operating (working) instructions
Bedienungsanweisung *f* operating instructions, instruction[s] to operator; operator's manual, servicing information, instruction book
Bedienungsbrücke *f* operations area
Bedienungseinrichtung *f* control (operating) device
Bedienungselement *n* control element
Bedienungselemente *npl* controls
Bedienungsfehler *m* error in operation
Bedienungsgriff *m* operating handle *(machine)*, master handle *(manipulator)*
Bedienungshandbuch *n* operator's manual
Bedienungshebel *m* control (operating) lever
Bedienungsknopf *m* operator's control; push-button
Bedienungskomfort *m* operational comfort
Bedienungskosten *pl* cost of attendance
Bedienungsmann *m* operator
Bedienungsperson *f* operator
Bedienungspult *n* control desk (console)
Bedienungsseite *f* operating side *(machine)*; working side *(furnace)*
Bedienungsspannung *f* operating voltage
Bedienungsteil *n* control element
Bedienungsvorschrift *f* working instructions
bedingt *(dat)* conditional
zweifach b. biconditional *(e.g. equivalence)*
Bedingung *f* condition
B. am Leitungsende terminal condition
Braggsche B. *(nucl)* Bragg condition
Bedingungsgleichung *f* equation of condition
bedingungslos unconditional
bedrahten *(el)* to wire
Bedrahtung *f (el)* wiring
Bedrahtungsplan *m* wiring diagram
bedruckbar printable
Bedruckbarkeit *f* printability
Bedruckbarkeitsprüfgerät *n* printability tester (test printer)
bedrucken to print
beeinflussen to influence, to affect
Beeinflussung *f* influence
B. des Hangenden *(mine)* roof control
gegenseitige B. mutual influence; interaction; interference
Beeinflussungsdämpfung *f* interaction loss
beeinträchtigen to impair; to interfere with
Beerenobst *n* soft fruit
Beet *n* bed; land
beeteln *(text)* to beetle
Beetlekalander *m (text)* beetle calender
Beetpflug *m* furrow plough
befähigen to enable
Befähigung *f* qualification, capability
befahrbar navigable *(waters)*; passable *(road)*; runnable
Befahrbarkeit *f* navigability *(waters)*; *(mine)* accessibility
B. des Kessels accessibility of the interior and exterior of the boiler
befahren to navigate, to sail along; to use *(road)*; to pass

den Schornstein b. to climb up the chimney
Befahren *n* **des Kessels** inspection of the boiler
B. des Schornsteins climbing up the chimney
befahren able-bodied *(sailor)*
Befehl *m* instruction, order *(computer)*; command
fiktiver B. dummy order
gespeicherter B. presetting control *(railway signalling)*
symbolischer B. pseudo instruction
Befehlsantrieb *m (el)* makeshift drive
Befehlsauslösung *f* command action
Befehlsbibliothek *f (dat)* library
Befehlscode *m s.* Befehlskode
Befehlsfolge *f (dat)* control sequence
kodierte B. coding
relative B. relative coding
Befehlsfolgeregister *n* sequence control register *(computer)*
Befehlsimpuls *m* command pulse
Befehlskode *m* order (instruction) code
Befehlsmindestdauer *f* minimum command time
Befehlsnummer *f* operating number
Befehlsschalter *m* command (control, auxiliary) switch
Befehlsschema *n (dat)* flow-chart
Befehlsschleife *f (dat)* loop of instructions
Befehlswort *n* instruction word
Befehlszähler *m* programme (instruction) counter, sequence control register
Befehlszeile *f* coding line *(digital computer)*
befestigen to attach [to], to fasten [to], to fix [to], to secure [to], to apply [to]; to pave *(roads)*
gelenkig b. to hinge
mit Kette b. to chain on
befestigt / außermittig eccentrically mounted
gelenkig b. hinged
starr b. rigidly attached [to]
Befestigung *f* attachment, mounting, fixing, fastening; paving, pavement *(roads)*
B. der Schleifscheibe mounting of the grinding wheel
stoßsichere B. shock mounting
Befestigungseinrichtung *f* **einer Katodenstrahlröhre** cathode-ray tube harness
Befestigungsflansch *m* mounting flange
Befestigungsgewinde *n* fastening screw thread
Befestigungskegel *m* mounting cone, reception taper
Befestigungskerbe *f* blade groove *(turbine)*
Befestigungslappen *m* fastening lug *(turbine)*
Befestigungsloch *n* clearance hole
Befestigungsmittel *n* fastener; *(build)* fixing means (accessories); means for mounting purposes
Befestigungsmutter *f* fixing nut
B. der Schleifscheibe spindle-end nut
Befestigungsplatte *f* bracket plate; *(ship)* keep plate
Befestigungsring *m* fastening ring
zweiteiliger B. two-part fastening ring *(turbine)*
Befestigungsschelle *f* mounting bracket; fixing clip
Befestigungsschraube *f* fastening screw (bolt), set screw (bolt); threaded fastener
Befestigungsstift *m* securing (steady) pin
Befestigungsstiftschraube *f* holding-down stud
befeuchten to damp[en], to moisten, to humidify
Befeuchter *m* humidifier
Befeuchtung *f* humidification; moistening; damp[en]ing
Befeuchtungsanlage *f* humidifying installation
Befeuchtungskammer *f* moistening chamber
befeuern to beacon, to light
Befeuerung *f* [group of] lights; beacons; navigation lights; lighting
befindlich / im Eingriff in contact, intermeshing, meshing into each other, engaging
im Wasser b. afloat *(vessel)*
beflechten to braid
Beflechtung *f* braiding
beflocken to flock
Beflocken *n* flocking
Beflockung *f* / **elektrostatische** electrostatical flocking
befördern to haul, to convey, to transport, to ship, to handle
Befördern *n* **in senkrechter (schräger) Richtung** lifting
Beförderung *f* transport, conveying, conveyance, haulage, shipment, handling
kombinierte B. combined transport
palettisierte B. palletized traffic
Beförderungsbedingungen *fpl* conditions of carriage
Beförderungsfrist *f* period allowed for conveyance (carriage)
Beförderungsfristberechnung *f* calculation of transit periods
Beförderungshindernis *n* circumstances preventing carriage
Beförderungsirrtum *m* wrong routing
Beförderungskosten *pl* carriage charges
Beförderungsmittel *npl* transport facilities, means of transport (conveyance)
Beförderungspflicht *f* obligation to carry
Beförderungspreis *m* fare *(passenger)*; rate
Beförderungssteuer *f* tax on transport
Beförderungstuch *n* / **endloses** *(text)* creeper
Beförderungsvertrag *m* contract of carriage
Beförderungsweg *m* route of travel *(passengers)*; forwarding route *(goods)*
billigster B. cheapest (most favourable) route
befrachten to charter *(ship)*

Befrachter *m (ship)* charterer
Befrachtung *f (ship)* affreightment, chartering
Befrachtungskommission *f* charter commission
Befrachtungsmakler *m* charter (clearing) broker
Befrachtungsvertrag *m* charter party, contract of affreightment
befreien to liberate; to free; to release; to set free
befugt entitled
Begasungsmittel *n (ag)* gassing preparation, fumigant
begegnen to meet [with]; to fall in *(with another vessel)*
begehen to patrol *(electrical lines)*
Begehkanal *m (mine)* gas gallery
begichten to charge the furnace; to burden
Begichtung 1. charging operation; burdening; 2. batch; charge
B. von Hand barrow charging
Begichtungsanlage *f* charging floor
Begichtungsbühne *f* charging floor
Begichtungseinrichtung *f* charging device
Begichtungskran *m* charging crane
Begichtungswagen *m* charging barrow
Begiensegel *n* jigger sail
begießen to water, to sprinkle; to moisten
Beginn *m* commencement; beginning; start; initiation; origin
B. des Dampfaustritts beginning of the exhaust [of steam]
B. des Stromflusses *(el)* initiation of current
beginnen to start; to begin; to commence; to originate, to initiate
Beginnzeichen *n* answer signal, *(Am)* off-hook signal, start signal
Beglaubigung *f* certification
begleichen to settle
Begleichung *f* **der Rechnung** settlement of account
Begleitbrief *m* covering letter
Begleiteffekt *m* disturbance *(radar)*
Begleitelement *n (met)* impurity
Begleiter *m* 1. guide, convoyer; 2. companion *(of ores)*
Begleiterscheinung *f* attendant phenomenon
Begleitflöz *n* accompanying bed
Begleitgestein *n* country rock
Begleitjagdflugzeug *n* escort fighter
Begleitkultur *f (ag)* companion crop
Begleitmatrix *f (math)* companion matrix
Begleitmetall *n* foreign-metal impurity
Begleitmineral *n* accompanying (secondary, subordinate) mineral
Begleitort *n (mine)* back entry, parallel gallery
Begleitpanzer *m* accompanying tank
Begleitpapiere *npl* accompanying documents
Begleitschiff *n* escort vessel, tender convoy
B. für U-Boote submarine depot ship
Begleitstrecke *f (mine)* back entry, parallel heading (road)
begradigen to straighten out
begrenzen to limit; to terminate; to clip *(signals)*
Begrenzer *m* limiter, clipper, peak chopper, clipping circuit
doppelseitiger B. slicer, double-end limiter; clipper limiter
doppeltwirkender B. *s.* doppelseitiger B.
Begrenzerdiode *f* limiter diode
Begrenzerkennlinie *f* limiter characteristic
Begrenzerkreis *m* clipper, clipping circuit
Begrenzerschaltung *f* limiting (suppressor) circuit, suppressor *(circuit)*
Begrenzerstufe *f* limiting stage
Begrenzerwirkung *f* limiter action
begrenzt durch bounded by
Begrenzung *f* limitation; limit; termination; clipping *(signals)*
B. durch die Erdkrümmung horizon limitation
B. durch Gitterstromeinsatz cut-off limiting
Begrenzungsanschlag *m* back stop
Begrenzungsbake *f (aero)* boundary beacon
Begrenzungsdrossel *f (Am) (el)* [current] limiting coil; current limiting reactor
Begrenzungsebene *f* boundary plane
Begrenzungsfläche *f (math)* facet; boundary surface *(of a solid)*
Begrenzungslicht *n* side light *(motor car)*
Begrenzungslinie *f* boundary line
B. der festen Anlagen obstruction gauge limit
B. für Pufferteller gauge for buffer heads
B. für Wagen vehicle gauge
Begrenzungsplanke *f (rock)* margin plank
Begrenzungsring *m* outer ring *(turbine)*
Begrenzungsschott *n (ship)* boundary (end) bulkhead
Begrenzungsverstärker *m* peak limiter; limiting amplifier
Begrenzungswiderstand *m (el)* feed impedance; [current] limiting resistor
Begriff *m* term; notion; idea; concept
Begriffsbestimmung *f* definition, meaning
begrünen to plant
Begrünung *f* planting
Begußmasse *f* engobe *(ceramics)*
Begutachter *m* expert
Begutachtung *f* expert opinion
Behälter *m* container; bin; reservoir; tank; case; receptacle; receiver
flexibler B. *(ship)* flexible barge (container) *(a long flexible tube with tapering ends to carry oil and other liquids lighter than water)*
gemauerter B. brick reservoir
kastenförmiger B. rectangular tank
kugelförmiger B. sphere-shaped container

unterirdischer B. underground tank
zusammenlegbarer B. collapsible container
Behälterbau *m* tank construction
Behälterfassungsvermögen *n* [container] capacity, tankage capacity
Behälterhaut *f* tank skin *(of rockets)*
Behälterkuppel *f* tank dome
Behälterlader *m (mach)* hopper [feeder]
Behälterpappe *f* container board
Behälterschiff *n* container ship, container vessel
Behälterspitze *f (rock)* front section *(containing instrument and pressure tank compartments)*
Behälterverkehr *m* container system
behandelbar treatable; processable
behandeln to treat; to process; to work; to handle
anodisch b. to anodize
Behandlung *f* treatment; handling; processing; working; manipulation
B. der Güter handling of goods
ladedienstliche B. handling [of goods]
rauhe B. hard usage
Behandlungsweise *f* method (manner) of treatment, procedure
beharren to persist; to remain; to stay
Beharren *n* persistence
Beharrungsgeschwindigkeit *f (el)* balancing speed
beharrungsgesteuert inertia-controlled
Beharrungskraft *f* inertia
Beharrungsmasse *f* inertia mass
Beharrungsregler *m* inertia governor
Beharrungsvermögen *n* inertia
Beharrungswirkung *f* inertia effect
Beharrungszustand *m* state of resistance (inertia), equilibrium state (condition); *(el)* steady state condition (regime)
wiederkehrender B. periodic state of resistance
behauen to dress *(stone)*; to trim; to adze, to chisel off *(wood)*
behäufeln *(ag)* to hill up, to earth up
beheben to remove; to cure; to clear; to eliminate
beheizen to heat; to fire
beheizt heated; fired
Beheizung *f* heating
Beheizungsart *f* heating system, method of heating
Behelfsanker *m* jury anchor
Behelfsanlage *f* makeshift installation
Behelfsantenne *f* emergency (makeshift) aerial
Behelfsbauten *pl* temporary buildings (structures)
Behelfskabel *n* shift cable
Behelfskonstruktion *f (build)* temporary structure; *(met)* makeshift design
Behelfsmast *m* jury mast *(of a vessel)*
Behelfsruder *n* jury rudder
Beherrschung *f* **des Hangenden** *(mine)* roof (strata) control
Behördenschiff *n* government vessel
beibehalten to maintain, to retain
Beiboot *n* tender, ship's boat; cock boat *(of a yacht)*
beidäugig binocular
Beidraht *m (el)* supplementary earth wire
beidrehen to tack *(aeroplane)*; to heave to, to lay to, to bring to
beidseitig on both sides, both-way
Beifahrer *m* co-driver, assistant driver
beigeklappt flapped back *(flight)*
Beiholer *m (ship)* bull rope
Beil *n* axe; hatchet
Beiladung *f* additional load
Beilage *f* inset
Beilbrief *m (ship)* builder's certificate, bill of sale
Beilegering *m* packing [ring]
Beilegescheibe *f* shim, packing washer
beiliegend enclosed *(letter)*
Beiluft *f* admixed air
beimengen to admix, to add
Beimengung *f* addition, admixture, additive
beimischen to admix, to add
Beimischung *f* addition, admixture, additive
Bein *n (mine)* leg, post *(prop)*
Beinling *m* leg *(hose)*
Beinschiene *f* pad *(icehockey)*
Beinschwarz *n* bone black
Beinwolle *f (text)* crutchings
beiordnen to coordinate; to assign
Beipaß *m* by-pass
Beischiff *n* tender
Beißkeil *m* wedge grip
Beistellbetrag *m* amount of feed
beistellen [von Hand] to feed [by hand]
Beistellhandrad *n* **der Schleifscheibe** hand wheel for wheel feed
Beistellung *f* **der Schleifscheibe** wheel (lateral) feed
Beitel *m* chisel *(for cutting wood)*
Beiwagen *m* 1. side car *(motor-cycle)*; 2. trailer *(tram)*
Beiwert *m* coefficient; factor
Beizahl *f s.* Beiwert
Beizanlage *f (met)* pickling plant
Beizbad *n* 1. *(text)* mordanting bath; 2. *(met)* pickling bath
elektrolytisches B. electrolytical pickling bath
Beizbrüchigkeit *f* acid brittleness
Beize *f (met)* pickle; mordant *(dye)*; stain *(wood)*; corrosive; caustic
beizen *(met)* to pickle; *(text)* to taw; to mordant *(dye)*; to stain *(wood)*; *(leath)* to bate; to corrode; *(ag)* to dress, to disinfect; to bran
elektrolytisch b. to electrobrighten

beizend *(met)* pickling; caustic, mordant; staining; corroding
Beizenfarbstoff *m* mordant dye, adjective dye-[stuff]
Beizer *m (ag)* dresser
Beizhaspel *f (leath)* puering paddle
Beizkorb *m (met)* pickling crate
Beizmittel *n* pickling medium; mordant; *(ag)* dressing agent, dresser, disinfectant
chemisches B. *(ag)* chemical dresser, disinfectant
Beizsprödigkeit *f* acid (pickle, hydrogen) brittleness
Beizung *f (ag)* dressing, disinfection; *(met)* pickling; staining *(wood)*
katodische B. cathode pickling
Beizzusatz *m* pickling compound, addition agent *(electrodeposition)*; inhibitor
Bekämpfung *f* control; abatement
bekiesen to grit; to gravel *(roof)*
Beklebezettel *m* label, placard, poster, form
Bekleidungsfabrik *f* clothing factory
Bekleidungsindustrie *f* clothing industry
Bekleidungsleder *n* clothing leather
Bekleidungssektor *m* apparel field
Beklöppelung *f (el)* concentric[al] braiding
bekohlen to coal, to take in coal
Bekohlung *f* coaling *(locomotive)*
Bekohlungsanlage *f* coaling (coal-handling) plant, facilities for coaling
Bekohlungsbühne *f* coaling stage
Bekohlungsschiff *n* coaling ship
bekommen / frei to get free *(a stranded vessel)*
Bel *n* bel
Beladefrist *f* period allowed for loading
beladen to load [on]; to charge; to lade; to burden; to affreight *(ship)*
Beladen *n* **des Reaktors** reactor loading (charging)
Belader *m* shipper
Beladestelle *f* loading station
Beladung *f* load[ing]; stem *(of a ship to a certain time)*
B. unter Last *(nucl)* hot[-reactor] loading
volle B. full (total) load
Beladungsmaschine *f* charging machine
Beladungsprüfgerät *n (ship)* loading calculator
Belag *m* 1. lining; coat[ing]; cover[ing]; plating; flooring; surface; 2. *(el)* characteristic *(transmission line)*
B. eines Kondensators plate of a capacitor
säurefester B. acid lining
Belastbarkeit *f* stressability; load capacity; load-carrying capacity; wattage, load factor; power rating
B. der Wähler *(telec)* selector carrying capacity
B. eines Trägers load-carrying capacity of a beam
belasten to load; to stress; to weight
das Ventil b. to load the valve
einen Kreis b. *(el)* to impose a load on a circuit
belastet *(el)* loaded; active
exzentrisch b. excentrically loaded
stark b. heavily loaded
Belastung *f* 1. load[ing]; 2. *(el.m.)* demand; coil loading; 3. *s.* Strahlenbelastung
B. am Tage day load *(power station)*
B. außerhalb der Spitzenzeit *(el)* off-peak load
B. durch Wechselstromwiderstand reactive load
B. durch Wirkwiderstand non-reactive load
B. im Seegang *(ship)* wave load
B. in Richtung der Längsachse longitudinal (transverse) load
ausgeglichene B. balanced loading
aussetzende B. intermittent load
bewegliche B. mobile (traffic) load
dynamische B. dynamic load[ing], impulse loading
einseitig freitragende B. cantilever load
gleichförmige B. continuous loading
höchstzulässige B. safety (maximum safe) load
induktive B. inductance (lagging, coil, reactive) load
kapazitive B. capacitive (leading) load
mittlere B. average (mean) load
ohmsche B. resistive (non-reactive) load
rein ohmsche B. pure[ly] resistive load
ruhende B. static loading
schwellende B. alternating load
schwingende B. dynamic stress
statische B. static loading
symmetrische B. balanced load
unsymmetrische B. unbalanced load
volle B. total (full) load
wechselnde B. alternating load; live load
zulässige B. allowable (safe) load
Belastungsabschnitt *m* loading section
Belastungsanzeiger *m* load indicator
Belastungsbereich *m* load range
Belastungsbild *n* load scheme, loading diagram
Belastungsebene *f* plane of loading
Belastungsfähigkeit *f (el)* carrying (load) capacity; *(ship)* loading capacity
Belastungsfaktor *m (el)* capacity factor (coefficient), load (utilization, output) factor
Belastungsfehler *m* load[ing] error *(analogue computer)*
Belastungsgrenze *f* load limit, permissible (maximum) load
Belastungsimpedanz *f* load impedance

Belastungskennlinie *f* load characteristic
B. für reine Blindlast zero power-factor characteristic *(synchronous machine)*
Belastungskoeffizient *m* load factor
Belastungskörper *m* weight-giving substance *(turbine)*
Belastungskreis *m (el)* load circuit
Belastungskurve *f* load line; working line *(radio)*
Belastungsprobe *f* bearing (load) test
Belastungsschwankung *f* load fluctuation (variation)
Belastungsspannung *f (el)* on-load voltage; *(ph)* load voltage
Belastungsspitze *f (el)* peak load
Belastungsspule *f* loading inductance (coil)
Belastungsvermögen *n* load capacity
Belastungsverteilung *f* load distribution
Belastungswiderstand *m* load (terminal) resistance; load rheostat
Belastungszeit *f* loading time
Belastungszeitverhältnis *n* duty factor
belaufen auf / sich to amount to
beleben to activate *(flotation)*
Belebtschlamm *m* activated sludge *(sewage treatment)*
Belebtschlammbecken *n* activated sludge chamber
Belebtschlammverfahren *n* activated sludge process; biological waste treatment
Belebung *f* activation
Belebungsmittel *n* restorative
Belebungsverfahren *n s.* Belebtschlammverfahren
beledern to face with leather
Belederung *f* facing with leather
Beleg *m* document; voucher
Belegablage *f* document file
automatische B. automatic document file
Beleganalyse *f* check analysis
belegen 1. to occupy *(area)*; 2. to man *(a mine)*; 3. to line; to cover
mit Folien b. to foliate
mit Platten b. to tile, to flag
Spiegel b. to silver mirrors
Belegen *n* **der Krempeln mit Kratzenbändern** *(text)* card mounting
Belegexemplar *n* file (voucher) copy
Belegklampe *f (ship)* cavil, cavel, belaying cleat
Belegnagel *m (ship)* belaying pin
Belegschaft *f* staff; crew; personnel; number of employees
Belegschaftsschutz *m* personnel protection
belegt 1. occupied *(railway track)*; 2. faced, lined
Belegung *f* 1. distribution (number) of personnel; 2. *(telec)* seizure; holding; 3. covering; coating; plating; lining
B. der Kühlfläche deposit of the cooling surface
B. eines Kondensators charge of a capacitor
Belegungsdauer *f (telec)* holding time
Belegungsdichte *f* surface density
Belegungsfaktor *m* occupancy factor *(dosimetry)*
Belegungsplan *m (Am)* cable assignment record
Belegungsverteilung *f* distribution of occupancies
Belegungswahrscheinlichkeit *f (telec)* utilization factor
Belegungszahl *f* number of occupancies, telephone traffic unit
Belegungszeichen *n (telec)* seizing signal
beleuchten to illuminate; to light
Beleuchterbrücke *f* light bridge *(theatre)*
Beleuchtung *f* illuminating, illumination; lighting
diffuse B. diffuse lighting (illumination)
direkte B. direct lighting
gleichförmige B. general diffused lighting
halbdirekte B. semi-direct lighting
halbindirekte B. semi-indirect lighting
indirekte B. indirect (concealed) lighting
ungerichtete B. undirectional lighting
vorwiegend direkte B. semi-direct lighting
vorwiegend indirekte B. semi-indirect lighting
Beleuchtungsanlage *f* lighting installation
Beleuchtungsapparat *m* substage
Abbescher B. Abbe substage
Beleuchtungsbühne *f (telev)* studio light boards
Beleuchtungsdichte *f* irradiancy
Beleuchtungseinheit *f* unit of illumination
Beleuchtungsfenster *n* lighting aperture
Beleuchtungsgerät *n* illuminating apparatus; illuminator
Beleuchtungsindustrie *f* lighting industry
Beleuchtungsingenieur *m* illumination engineer
Beleuchtungskörper *m* lighting fixture (fitting)
Beleuchtungsleitung *f* lighting feeder
Beleuchtungsmesser *m* illumination meter, illuminometer, luxmeter
Beleuchtungsstärke *f* illumination [intensity], intensity of illumination, illuminance
Beleuchtungsstrahlengang *m* illuminating path of rays
Beleuchtungstechnik *f* illuminating (lighting) engineering
beleuchtungstechnisch lighting-engineering
Beleuchtungsvorschriften *fpl* lighting regulations
belichten to expose; to subject to the action of actinic rays
Belichtung *f* exposure [to light]
natürliche B. daylighting, natural lighting
Belichtungsautomat *m (print)* electronic exposure control, light integrating timer

Belichtungsmesser *m* exposure meter
fotoelektrischer B. photoelectric exposure meter
Belichtungsspielraum *m* exposure latitude
Belichtungsuhr *f* exposure timer
Belichtungszeit *f* exposure speed (time), [camera] exposure
beliebig arbitrary; *(math)* any
beliefern to supply; to feed
Belladonnawurzel *f* deadly nightshade root, belladonna root
Belleville-Kessel *m* Belleville boiler
belüften to ventilate; to air; to aerate; to fan
Belüftung *f* aeration; ventilation; airing
künstliche B. induced ventilation
Belüftungsanlage *f* 1. system of ventilation, ventilation [plant]; 2. *(ag)* warm-air drier, dried-air drier
Belüftungsbecken *n* aeration tank
Belüftungshaube *f* air scoop
Belüftungsleitung *f* air intake
Belüftungsmesser *m* air meter
Belüftungsrohr *n* *(rock)* breather pipe, ram charger
Belüftungssilo *m* *(ag)* ventilated silo
Belüftungssystem *n* ventilation system
bemängeln to find fault with; to criticize
Bemängelungsanzeige *f* advice (notice) of irregularity
bemannen to man; to rig
bemannt / leicht light-handed *(vessel)*
ungenügend b. short-handed *(vessel)*
Bemannung *f* manning
bemaßen to dimension
Bemaßung *f* dimensioning; indication of dimensions
Bemaßungsregeln *fpl* [standard] rules for the dimensioning of drawings
bemasten to mast *(vessel)*
Bemastung *f* masting, sparring *(vessel)*
bemessen to design, to dimension, to determine safe dimensions
genau b. closely measured
Bemesserung *f* filling *(roller mill)*
doppelt gesicherte B. double safety-locking filling
Bemessung *f* dimensioning, design, determination of [safe] cross-sectional dimensions
Bemessungstafel *f* design table (chart)
Bemessungsverfahren *n* design method
bemustern to sample; to decorate
Bemustern *n* sampling
benachbart adjacent; adjoining; neighbouring
Benadelung *f* **des Nadelstabes** *(text)* needling of the comb strip *(combing machine)*
benennen to denominate; to term; to designate
Benennung *f* designation
B. der Frequenzen nomenclature of frequencies
benetzbar wettable
Benetzbarkeit *f* wettability
Benetzbarkeitsprüfung *f* wettability test
benetzen to wet; to moisten; to damp
Benetzung *f* wetting; moistening; damping
B. durch Lösungsmittel lyosorption
Benetzungsfähigkeit *f* wetting (moistening) power
Benetzungsmittel *n* wetting agent
Benetzungsspannung *f* wetting potential
Benetzungsvermögen *n* wetting (moistening) power
Benetzungswärme *f* *(ch)* heat of wetting
Bensonkessel *m* Benson boiler
Bentonit[ton] *m* bentonite [clay]
benutzbar usable, available; utilizable
benutzen to use, to employ; to utilize
Benutzer *m* user
Benutzung *f* use, utilization
B. der Gleise in beiden Richtungen either direction working, two-way working
gegenseitige B. der Güterwagen reciprocal use of wagons
Benutzungsgebühr *f* wayleave charges *(telephony)*
Benzaldehyd *m* benzaldehyde
Benzidin *n* benzidine
Benzin *n* petrol; *(Am)* gas[oline]; benzine *(petroleum hydrocarbon)*
B. mit definierten Siedegrenzen end point petrol
bleifreies B. unleaded petrol
gebleites B. doped petrol
gekracktes B. cracked petrol
geruchfreies B. sweet petrol
hochoktaniges B. high-octane petrol
klopffestes B. high-octane petrol
mit Zusätzen versehenes B. doped petrol
Benzinabscheider *m* petrol trap
Benzindüse *f* petrol jet
Benzineinfüllstutzen *m* petrol filler
Benzineinspritzanlage *f* petrol-injection system
Benzineinspritzmotor *m* petrol-injection engine
Benzineinspritzung *f* petrol injection
Benzineinspritzverfahren *n* petrol-injection system
Benzinfaß *n* petrol barrel
Benzinkanister *m* petrol can
Benzinlager *n* petrol dump (storage)
Benzinmotor *m* petrol engine
Benzinpumpe *f* petrol pump
Benzinrückstand *m* petrol combustion product
Benzinseifen *fpl* petroleum soaps
Benzinstandsanzeiger *m* petrol gauge
Benzintank *m* fuel (petrol) tank
Benzinuhr *f* petrol gauge

Benzinverbrauch *m* petrol consumption
Benzinversorgung *f* petrol supply
Benzinzapfstelle *f* filling station
Benzoat *n* benzoate
Benzoe *f* benzoin
Benzoeharz *n* styrax benzoin
Benzoemilchsäure *f* benzolactic acid
Benzoesalz *n* benzoate
Benzoesäure *f* benzoic acid
Benzol *n* 1. *(ch)* benzene; 2. benzol *(industry)*
Benzolanlage *f* benzol plant
Benzolfabrik *f* benzol recovery plant
Benzolgewinnung *f* benzol recovery
Benzolkern *m* benzene nucleus
Benzolring *m* benzene ring
Benzolvorprodukt *n* primary benzol
Benzolwaschanlage *f* benzol washing plant
Benzolwäsche *f* benzene recovery plant
Benzolwäscher *m* benzol scrubber
benzoylieren to benzoylate
Benzylalkohol *m* benzyl alcohol
Benzylgrün *n* fast green
Benzylzellulose *f* benzyl cellulose
beobachten to observe; to watch; to view
Beobachtersystem *n* laboratory system
Beobachtung *f* observation; viewing
B. im Hellfeld *(opt)* bright-field observation
monokulare B. *(opt)* monocular viewing
Beobachtungsbereich *m* detection range *(radar)*
Beobachtungsbühne *f* observation platform
Beobachtungsdeck *n* observation deck
Beobachtungsfehler *m* observational error, error of observation
Beobachtungsfenster *n* eye-sight, viewing window, observation port
Beobachtungsfernrohr *n* observation telescope
Beobachtungsplatz *m* observation point
Beobachtungspunkt *m* observation point
Beobachtungsreihe *f* set of observations, observed series
Beobachtungsspalt *m* eye slit
Beobachtungsstand *m* observation post
Beobachtungsstation *f* observing station
Beobachtungsstelle *f* observation point
Beobachtungstechnik *f* observing technique
Beobachtungsturm *m* look-out (watch) tower
Beobachtungsuhr *f* hack (deck) watch
Beobachtungswinkel *m* viewing angle
bepflanzen to plant
Bepflanzung *f* planting
bepichen *(wood)* to pitch
beplanken *(wood)* to coat
Beplattung *f* sheeting; plating
Bequemlichkeit *f* comfort
Berandung *f* edge; boundary, limit[ation]; interface
Berandungspunkt *m* end point
berasen *(ag)* to sod

Beratungsausschuß *m* advisory committee
Beratungsdienst *m* / **technischer** technical advisory service
beräumen to clear; to trim; to scale *(mine)*
Beräumung *f* clearing
berechenbar computable; calculable
Berechenbarkeit *f* calculability, computability
berechnen to compute; to calculate; to design, to analyze
Berechnung *f* computation; figuring; calculation; analysis, design
B. nach der Elastizitätstheorie elastic design
baustatische B. *(build)* structural (design) calculation, statical analysis
falsche B. miscalculation
n-freie B. ultimate strength method (design), load-factor method, ultimate load method, limit (ultimate load) design
statische B. static test, statical analysis
Berechnungsannahmen *fpl* design assumptions
Berechnungsbreite *f (ship)* moulded breadth
Berechnungsfehler *m* mistake (error) in computation
Berechnungsformel *f* formula of computation
Berechnungsgrundlage *f* base of calculation
Berechnungslänge *f* designed length
Berechnungstiefe *f* designed draught
Berechnungstiefgang *m* designed draught
Berechnungsverfahren *n* computation (calculation) method; design method
Berechtigter *m* entitled party (person)
Bereederer *m* outfitter
Beregnung *f (ag)* sprinkling, spray irrigation
Beregnungsanlage *f* 1. *(ag)* overhead spraying plant, spraying aggregate; 2. sprinkler installation *(fire-fighting equipment)*
Beregnungsprüfung *f (text)* rain test
Bereich *m* 1. area, region; 2. interval; sphere; zone, field; 3. range; measuring range
B. begrenzter Proportionalität region of limited proportionality
B. guten Empfangs good service area
B. hoher Energien high-energy region
B. niedriger Energien low-energy region
akustischer B. audible range
ausnutzbarer B. aerodynamischen Auftriebs *(rock)* effective depth of atmosphere for aerodynamic lift
äußerer B. outside region
einfach zusammenhängender B. *(math)* simply connected region
empfindlicher B. sensitive volume (region) *(radiobiology)*; range of sensivity
erweiterter B. extended range
geeichter B. calibrated range
geradliniger B. linear range
großer B. wide range
koaxialer B. coaxial region

kontrollierter B. controlled area
kurzwelliger B. short wavelength coverage (region)
langwelliger B. long wavelength coverage (region)
mehrfach zusammenhängender B. *(math)* multiply connected region
oberflächenaktiver B. active region of the surface
sichtbarer B. visible region *(of the spectrum)*
stabiler B. stable range
strahlenempfindlicher B. sensitive volume (region) *(radiobiology)*
ultravioletter B. ultraviolet region
wirksamer B. einer Durchlaßkurve actual response curve

Bereichsautomatik *f* autoranging
Bereichseinengung *f* range suppression
Bereichserweiterung *f* range extension
Bereichsfaktor *m* multiplying factor
Bereichsplan *m* survey map
Bereichsumschalter *m* range switch (selector), band switch
Bereichsumschaltung *f* range (band) switching
Bereichswähler *m* *(telec)* band selector
bereifen 1. to tyre; 2. to hoop, to band *(a cask)*; 3. to frost
Bereifung *f* 1. tyres; 2. formation of hoar frost
bereinigen to clear
bereißen *(mine)* to bar down
Bereißen *n* *(mine)* barring down, scaling
Bereisungsboot *n* inspector's (inspection) launch
Bereisungsschiff *n* ranger vessel, surveying ship
B. für magnetische Messungen magnetic-survey vessel

Bereitschaft *f* stand-by; readyness
ständige B. *(telec)* continuous attention [method]; *(Am)* utility operating method
technische B. [auf dem Flugplatz] refuelling and maintenance duty squad

Bereitschaftsbehälter *m* everready case
Bereitschaftsdienst *m* emergency (reserve) duties
Bereitschaftslokomotive *f* stand-by (reserve, emergency) locomotive
Bereitschaftszeichen *n* proceed-to-dial signal
bereitstellen to place at disposal, to make available
Bereitstellungsfrist *f* *(ry)* period within which available
Berg *m* 1. mountain; 2. *(mine)* incline[d raise]
Bergahorn *m* rock (greater) maple
Bergakademie *f* college (school) of mining
Bergarbeiter *m* mineworker, miner
Bergarbeitergewerkschaft *f* miner's union
Bergarbeitersiedlung *f* miner's camp
Bergassessor junior inspector
Bergaufrutsche *f* uphill shaker
Bergaufsichtsbehörde *f* Mines Inspectorate Organization
Bergbahn *f* mountain railway
Bergbau *m* mining
Bergbaubetrieb *m* mine; colliery *(coal)*
Bergbaubezirk *m* mining district
Bergbauindustrie *f* mining industry
Bergbaupacht *f* mining lease
Bergbausenkung *f* mining subsidence
Bergbautechnik *f* mining engineering
Bergbauzweig *m* mining branch
Bergbehörde *f* Mines Inspectorate
oberste B. Supreme Mining Authority

Berge *pl* *(mine)* dirt, discard, refuse, shale, tailings
Bergeablagerung *f* tailing disposal
Bergeanfall *m* make of refuse
Bergeaustrag *m* de-shaler, refuse extractor
Bergebrecher *m* stone breaker
Bergedamm *m* pack
Bergefalleitung *f* dirt descending pipe, drop stowing pipeline
Bergegeld *n* *(ship)* salvage [money], civil salvage
Bergehalde *f* bing, dirt (pit) heap, dump, muck stack, refuse tip
bergehaltig dirty
Bergeiche *f* rock-oak
Bergeisen *n* gad, picker
Bergekippe *f* waste tip
Bergeklein *n* debris
Bergelohn *m* *(ship)* [compensation paid for] salvage
Bergemassen *fpl* rubbles
Bergemauer *f* gateside (roadside) pack
Bergemittel *n* dirt band (bed), drift band
bergen to salvage *(ship)*; to recover *(goods)*; to furl *(sails)*
Berger *m* salvor, wrecker *(ship)*
Bergerippe *f* pack
Bergerolle *f* waste chute
Bergetrum *m* chute compartment *(raise)*
Bergeversatz *m* stowage *(by machine)*; packing *(by hand)*
Bergfahrt *f* 1. upstream navigation, upstream trip; 2. *(mine)* uphill run *(coal cutter)*
Bergfeste *f* 1. barrier pillar; 2. rock pentice *(shaft sinking)*
B. an der Markscheide boundary pillar

Bergganggetriebe *n* hill gear
Berggelb *n* *(min)* iron ochre
Berggesetz *n* law of mines
Berggesetzgebung *f* mining legislation
Berggipfel *m* mountain summit
Berggrün *n* mountain green
Bergholz *n* *(ship)* wale
Bergingenieur *m* mining engineer
Berginvalide *m* disabled miner
Bergius-Verfahren *n* Bergius process

Bergkiefer *f* mountain pine
Bergkletterschuh *m* climbing boot
Bergkristall *m* rock (mountain) crystal, quartz
Bergkupfer *n* native copper
Bergmann *m* miner, mineworker, hewer, collier
Bergmannrohr *n (el)* slip (Bergmann) conduit
Bergmannslampe *f* miner's lamp
Bergmilch *f (min)* rock-milk
Bergöl *n* rock-oil
Bergpech *n* asphalt, bitumen
Bergrat *m* senior inspector
Bergrettungsdienst *m* mountain rescue service
Bergrüster *f* mountain elm
Bergschaden *m* damage due to mining, mining (subsidence) damage
Bergschuh *m* mountaineering boot
Bergschuhbenagelung *f* nailing of climbing boots
Bergschuhsteigeisen *n* crampons of climbing boots
Bergschule *f* mining school
Bergsee *m* mountain lake
Bergseife *f* mountain (rock) soap
Bergsenkung *f* mining subsidence
Bergstadt *f* mining town
Bergsteiger *m* mountain climber
Bergsteigerausrüstung *f* mountaineering outfit, alpinist's equipment
Bergsteigereishaken *m* ice piton
Bergsteigerkarabiner *m* mountaineering snap-ring
Bergstrom *m* torrent, mountain stream
Bergstütze *f* sprag
Bergteer *m* mineral tar, maltha; bitumen
Bergulme *f s.* Bergrüster
Bergung *f* salvage; recovery
  B. der Aufzeichnungen data recovery
  automatische B. automatic recovery
Bergungsarbeiten *fpl* salvage works
Bergungsdienst *m* recovery service
Bergungsfahrzeug *n* recovery vehicle; salvage vessel
Bergungsfallschirm *m* recovery parachute
Bergungsgeräte *npl* salvage appliances
Bergungsschiff *n* salvage ship (vessel), recovery vessel
Bergungsschlepper *m* salvage (wrecking) tug
Bergverkehr *m* upstream traffic
Bergvermessung *f* mine surveying
Bergwachs *n* earth wax, ozokerite
bergwärts upstream, up the river
Bergwerk *n* mine
Bergwerksabgabe *f* contribution (impost) of mines, royalty
Bergwerksaktien *fpl* mining shares (stocks)
Bergwerksausrüstung *f* mining equipment
Bergwerksindustrie *f* mining industry
Bergwerksunternehmen *n* mining concern
Bergwesen *n* mining
Bergwirtschaft *f* mineral industry
Bergzinn *n (min)* vein-tin
Bericht *m* report; information; notice; message
berichtigen 1. to correct, to rectify *(value)*; 2. to adjust *(setting)*, to set right; 3. *(ry)* to straighten out, to dress, to rectify; to correct (amend) a charge
Berichtigung *f* 1. correction, rectification; adjustment; 2. *(ry)* rectification, amendment *(charge)*
  B. der Nulleinstellung zero clearing
Berichtigungsbeiwert *m* corrective factor
Berichtigungsfaktor *m* correction factor
Berichtigungslinse *f* corrective (correction) lens, collimator
Berichtigungstaste *f* reset key
Berichtigungszeichen *n* correcting signal
berieseln to sprinkle; to spray; to wet
  mit Öl b. to flood with oil
Berieselung *f* sprinkling; subsurface watering (irrigation)
Berieselungsanlage *f* sprinkler installation; irrigation system; cooling stack *(in cooling towers)*
Berieselungsapparat *m* sprinkler
Berieselungskühler *m* trickle (spray) cooler; washer, scrubber
Berieselungsturm *m* spray tower
Berkelium *n* berkelium
Berliner Blau *n* Berlin (Prussian) blue
Berme *f* berm, bench; ledge, terrace
Bermudatakelung *f (ship)* Bermuda rig
Bernstein *m* amber
Bernsteinsäure *f* succinic acid
Berohrung *f (mine)* tube-works, tubing
Berstdruck *m* bursting pressure
Berstdruckprüfer *m* bursting strength tester
bersten to burst; to rupture
Bersten *n* **der Schleifscheibe** rupture (flying apart) of the grinding wheel
Berstfestigkeit *f* bursting strength
Berstfestigkeitsprüfgerät *n* mullen tester
Berstindex *m* burst ratio *(paper)*
Berstscheibe *f* rupture-disk, shear-disk
Berstwiderstand *m* bursting strength *(paper)*
berücksichtigen to take into consideration
Berücksichtigung *f* consideration
Beruf *m* occupation; trade, craft; profession
Berufsausbildung *f* vocational (craft) training
  technische B. engineering craft training
Berufsausbildungswettbewerb *m* vocational training competition
Berufsberater *m (Am)* guidance counsellor
Berufsberatung *f* vocational guidance service
Berufsbild *n* specification of the skills required for a job, job specification
Berufsdosis *f (nucl)* occupational radiation dosage

Berufskleidung *f* professional clothing
Berufskrankheit *f* occupational disease
Berufspraktikum *n* outside practical experience *(student)*; works training
Berufsschule *f* trade (apprentice training, vocational) school
Berufswettbewerb *m* apprentice competition
beruhigen to kill *(steel)*, to deoxidize; to stabilize; to steady; to quiet; to smooth
beruhigt dead *(melt)*
Beruhigung *f* 1. damping; smoothing; 2. desoxidation; quieting *(melt)*
Beruhigungsbecken *n* stilling (absorption) basin
Beruhigungsglied *n* smoothing section
Beruhigungskapazität *f* smoothing capacitance
Beruhigungskondensator *m* smoothing capacitor
Beruhigungsmittel *n* 1. killing agent *(steel)*; 2. sedative, tranquillizer
Beruhigungswiderstand *m* steadying (smoothing) resistance
Beruhigungszeit *f* response time; damping period; setting time
Beruhigungszone *f* recalming zone *(rock pressure)*
berühren to touch, to be in contact with
  einen Kontakt b. to strike (engage with) a contact
Berührung *f* 1. contact; touch *(conductors)*; 2. *(math)* osculation; tangency
  ohne B. mit clear of, without contact with, contactless
  metallische B. metallic (metal-to-metal) contact
  zufällige B. accidental contact
Berührungsebene *f* tangent[ial] plane
Berührungselektrizität *f* contact electricity
Berührungsfläche *f* contact area *(surface finish)*; zone (area) of contact; arc of contact *(grinding wheel)*; *(math)* interface; transfer area *(distillation)*
berührungsfrei non-contacting, contactless
Berührungsgefahr *f* risk (hazard) of electric shock, risk (hazard) of electric contact
Berührungsgift *n* contact poison (insecticide)
Berührungsheizfläche *f* contact heating surface
Berührungskorrosion *f* contact corrosion
Berührungslinie *f* line of contact; tangent[ial] line, tangent
berührungslos non-contacting
Berührungspotential *n* contact potential
Berührungspunkt *m* point of contact, contact (touching) point
  B. von (an) Kreisen point of tangency
Berührungsschutzfassung *f* shock-proof lamp socket
Berührungsschutzkapazität *f* capacitance for shock protection
Berührungsschutzkondensator *m* shock-protection capacitor
berührungsschutzsicher shock-proof
berührungssicher shock-proof
Berührungsspannung *f* contact voltage (potential), shock-hazard voltage
Berührungsspannungsschutz *m* protection against electric shock hazard
Berührungsstelle *f* point (place) of contact
  B. der Walzen *(text)* bite, nip
Berührungswiderstand *m* contact resistance
Berührungswinkel *m* angle of contact
Berührungszeichen *n* contact (touch) signal
Berührungszeit *f* duration of contact
Berührungszwillinge *mpl* juxtaposition twins *(crystals)*
berußen to soot
Beryll *m* beryl
Beryllium *n* beryllium
Berylliumbronze *f* beryllium bronze
Berylliumerde *f* beryllia
Berylliumfenster *n* beryllium window
berylliumgebremst beryllium-moderated *(reactor)*
Berylliumkupfer *n s.* Berylliumbronze
Berylliumoxid *n* glucina, beryllium oxide
Berylliumreaktor *m* beryllium-moderated reactor
Besamungsstation *f* *(ag)* artificial insemination station
Besan *m* spanker, miz[z]en
Besanewer *m* sailing (ketch-rigged) barge
Besanmast *m* jigger mast
Besansegel *n* jigger sail
Besatz *m* 1. *(mine)* stemming [material] *(shot hole)*; 2. *(text)* trimmings and edgings
Besatzartikelindustrie *f* trimmings' industry
Besatzmaterial *n* *(mine)* stemming
Besatzpatrone *f* *(mine)* tamping bag
Besatzstein *m* filler brick *(open-hearth furnace)*
Besatzung *f* crew; manning
Besatzungsräume *f* crew accommodation *(ship)*
besäumen to square [up]
Besäumgatter *n* side saw blade frame
Besäumkreissäge *f* edger *(circular saw)*
Besäummaschine *f s.* Besäumsäge
Besäumsäge *f* circular saw for edging boards, saw for parallel sawing, siding machine
Besäumschnitt *m* slab cut
besäumt square edged; plain edged
beschädigen to damage; to mar *(surface)*; to impair; to injure
beschädigt defective, damaged
Beschädigung *f* damage; defect
  unsichtbare B. hidden (concealed) damage
  zufällige B. accidental damage
beschaffen to supply, to provide
Beschaffenheit *f* condition; constitution; quality
  neutrale B. des Schmiermittels neutral constitution of the lubricant

beschäftigen to employ, to engage
Beschäftigungsfaktor *m (nucl)* total time of exposure per week
beschallen to expose to ultrasonic waves, to irradiate acoustically
Beschallung *f* exposure to ultrasonic waves; acoustic irradiation; ultrasonic checking
beschalten *(el)* to wire
Beschaltung *f* wiring; cabling; connecting-up
Beschaltungsplan *m* wiring scheme (plan)
Beschattung *f* shading
Beschauprüfung *f* visual inspection
Bescheid *m* / **amtlicher** official notice
  patentamtlicher B. patent-office notification
bescheinen to shine upon *(sun)*
bescheinigen to certify; to attest
beschichten to coat
Beschichten *n* coating
Beschichtung *f* coating
Beschichtungskalander *m* lining calender
beschicken 1. to load *(machine tools)*; to feed; 2. to charge *(furnace)*; 3. *(text)* to prepare *(bath)*, to set; 4. to fire, to stoke
  mit der Hand b. to stoke by hand
Beschickung *f* 1. loading *(machine tool)*; load; feeding; feed; 2. charging *(furnace)*; charge; supply; burden; 3. firing, stoking
  abwechselnde B. alternate firing
  satzweise B. charging by batches
  unterbrochene B. intermittent firing
  ununterbrochene B. continuous firing
Beschickungsaufzug *m* feeder skip hoist
Beschickungsbunker *m* feed tank
Beschickungsdiagramm *n* stoking diagram
Beschickungseinrichtung *f* feeding device
Beschickungskran *m* charging crane
Beschickungsmulde *f* feeding trough; charging box
Beschickungsöffnung *f* charging door
Beschickungsperiode *f* firing period
Beschickungstrichter *m (mach)* [feed] hopper
Beschickungswaage *f* feeder scale
Beschickungswagen *m* hopper car
Beschickungsweise *f* method of firing
Beschickungszone *f* feed section (zone)
beschießen to bombard *(elements)*
Beschießung *f* bombarding *(elements)*
  B. mit Elektronen electron bombardment
  B. mit Kernen nuclear bombardment
beschiffbar navigable
beschiffen to navigate
Beschlag *m (text)* clothing *(tambour)*
Beschläge *mpl* ironmongery, ironwork, hardware, fittings
beschlagen 1. to mist [up] *(glass, screen)*; 2. to shoe *(horse)*
Beschlagen *n* 1. misting [up] *(glass, screen)*; age-coating *(lamp bulb)*; 2. shoing *(horse)*
Beschlagzeising *n (ship)* gasket
beschleunigen *(ph)* to accelerate; *(ch)* to activate; to promote
Beschleuniger *m (ph)* [electron] accelerator; *(ch)* activator; promoter
  zyklischer B. *s.* Ringbeschleuniger
beschleunigt fast *(operation of furnace)*
Beschleunigung *f (ph)* acceleration, speeding-up; *(ch)* activation, promotion
  generalisierte B. generalized acceleration
  geradlinige B. rectilinear acceleration
  gleichförmige B. uniform acceleration
  relative B. relative acceleration
  resultierende B. resultant acceleration
  ungleichförmige B. nonuniform acceleration
Beschleunigungsanlage *f (nucl)* accelerating unit
Beschleunigungsanode *f* second anode (gun), accelerator anode
Beschleunigungsarbeit *f* work of acceleration
Beschleunigungsdruck *m* acceleration pressure
Beschleunigungsdüse *f* acceleration nozzle *(turbine)*
Beschleunigungselektrode *f* accelerating electrode
Beschleunigungsfehler *m* acceleration error (misalignment) *(servomechanism)*; ballistic deflection error *(gyroscope)*
Beschleunigungsfeld *n* accelerating field
Beschleunigungsfläche *f* acceleration area
Beschleunigungshöhe *f* head required to accelerate the liquid
Beschleunigungskammer *f (nucl)* accelerating chamber
Beschleunigungskomponente *f* component of acceleration
Beschleunigungskonstante *f* acceleration error (misalignment) coefficient *(servomechanism)*
Beschleunigungskraft *f* accelerative (accelerating) force
Beschleunigungskurve *f* acceleration curve
Beschleunigungslinse *f* accelerating electronic lens *(electron optics)*
Beschleunigungsmesser *m* accelerometer
Beschleunigungspotential *n* acceleration potential
Beschleunigungspumpe *f* accelerating pump
Beschleunigungsraum *m* acceleration space *(of electrode)*
Beschleunigungsrohr *n (nucl)* accelerating tube
Beschleunigungsspalt *m (nucl)* accelerating slit (gap)
Beschleunigungsspannung *f* accelerating voltage
Beschleunigungsstoß *m (ph)* accelerating impact
Beschleunigungsverluste *mpl (el)* accelerating losses
Beschleunigungsvermögen *n (aut)* throttle response

Beschleunigungsvorrichtung *f* accelerator
Beschneideabfälle *mpl* cuttings
Beschneidelinie *f* trimming (cutting) line
Beschneidemaschine *f* 1. cutting machine, trimmer *(paper)*; 2. *(el)* clipping machine
beschneiden 1. to trim, to cut *(paper)*; 2. *(el)* to clip, to chop *(amplitude)*; to cut off *(frequency)*
Beschneiden *n* 1. trimming, cutting *(paper)*; 2. *(el)* peak chopping *(of amplitude)*; cut-off *(of frequency)*
Beschneider *m* trimmer *(paper)*
Beschneidung *f s.* Beschneiden
Beschnitt *m* trim[ming]
Beschnittkante *f* fore-edge
Beschnittrand *m* trimming edge
beschottern to ballast, to macadamize
Beschottern *n* ballasting
Beschotterung *f* ballasting
beschränken to limit [to]; to restrict, to confine
beschränkt limited; *(math)* bounded
Beschränkung *f* constraint, restraint *(e.g. linear programming)*
beschreiben to describe; to specify
beschreibend descriptive
Beschreibung *f* specification *(patent)*
Beschreibungsfunktion *f* describing function; harmonic balance
beschriften to letter *(technical drawing)*
Beschriftung *f* lettering *(technical drawing)*; legend *(instruments)*; engraving *(of an escutcheon)*
Beschriftungsschild *n* tab
Beschriftungsstelle *f* **[am Spiralbohrer]** mark recess
Beschuß *m* bombardment
Beschwerdestelle *f* complaint section
beschweren to load, to fill, to weight *(textiles)*
den Dampf b. to add weight to the steam *(turbine)*
Beschwerungsappretur *f (text)* filling finish
Beschwerungsmittel *n* weighting (loading) agent, extender
Beschwerungsstoff *m* heavy material, medium solids
Besegelung *f* sails, sail area
beseitigen to dispose of; to remove; to eliminate; to clear *(rubble)*
eine Störung b. to clear a fault
Beseitigung *f* disposal; removal; elimination
B. des Abraums *(mine)* uncovering
B. von Störungen debugging
hydromechanisierte B. des Abraums hydraulic stripping
Besenstiel *m* broom stick
Besenverankerung *f* fan anchorage *(prestressed concrete)*
besetzen to tamp *(fill a charged shot hole with stemming material)*; to occupy
besetzt occupied; *(ph)* closed, completed, filled *(shell)*; occupied *(state)*; busy *(telephone)*
Besetztanzeige *f* **/ optische** visual busy signal
Besetztfall *m* engaged condition *(telephone)*
Besetztflackerzeichen *n (telec)* engaged (busy-flash) signal
Besetztkennzeichen *n s.* Besetztflackerzeichen
Besetztlampe *f* [visual] engaged lamp, visual busy lamp
Besetztprüfung *f* **mit Flackerzeichen** flashing *(telephone)*
optische B. visual engaged test
Besetztsignal *n* blocking signal *(telephony)*
Besetztzeichen *n* engaged (busy) signal, busy-back tone
Besetzung *f* manning *(of a ship)*; watch *(radio)*
Besetzungszahl *f* absolute frequency; class frequency
Besitz *m* **/ gemeinsamer** co-ownership
Besonnung *f* sunlighting, insolation
bespannen to cover
Bespannung *f* 1. covering [material]; clothing; 2. coupling *(locomotive)*
Bespannungsstoff *m* covering material
Bespantung *f* [ship's] framing
bespritzen to sprinkle
besprühen to spray
bespulen *(el)* to load with coils
bespult *(el)* coil-loaded
mittelschwer b. medium heavy loaded
Bespulung *f (el)* coil load, coil (Pupin, series) loading, pupinization
B. für Musikübertragung programme circuit loading
besonders leichte B. extra-light loading
gleichmäßige B. continuous loading
mittelschwere B. medium loading
Bespülung *f* air scavenging *(hollow insulators of circuit breakers)*
Bessemerbirne *f* acid Bessemer converter
Bessemerei *f* Bessemer foundry [for acid steel]
Bessemerflußstahl *m* acid Bessemer low carbon steel
Bessemerroheisen *n* pig iron for acid process
Bessemerschmelze *f* Bessemer blow [in acid process]
Bessemerstahl *m* acid converter steel
Bessemerverfahren *n* acid converter process
Bestand *m* stand, stock
beständig continuous; stable
b. gegen resistant to, stable to
b. gegen Frost und Tauwasser freeze-thaw-stable *(lacquer)*
b. gegen normales Waschen *(text)* launder-proof
Beständigkeit *f* continuity; stability; durability; resistance

B. gegen Zerdrücken *(text)* crush resistance *(fabric)*
Bestandsalter *n* age of stand, age-class
Bestandsaufnahme *f* inventory [list], stock taking
Bestandsdichte *f* density of growth
Bestandskarte *f* balance card *(e.g. punched card)*
Bestandslücke *f (ag)* blank
Bestandsnachweis *m* stock inventory
Bestandswert *m* estimated value of a crop
Bestandteil *m* component; constituent; ingredient
Bestandteile *mpl* accessories *(coal)*
akzessorische B. auxiliary accessories
brennbare B. combustible matter
flüchtige B. volatile matter
Bestätigung *f* **eines Signals** acknowledgement of a signal
bestäuben *(ag)* to pollinate; *(opt)* to sputter
Bestäuben *n (opt)* sputtering
Bestäubung *f (ag)* pollination
Besteck *n* 1. set of instruments (tools); 2. determination of the position; reckoning [position]; ship's position
gegißtes B. dead reckoning
Besteckrechnung *f* ship account, reckoning, ship's (dead) reckoning
bestehen aus to consist of
eine Prüfung b. to pass an examination
eine Prüfung nicht b. to fail to pass an examination
bestellen 1. to order; 2. *(ag)* to till *(field)*
Bestellerrisiko *n* consumer's risk
Bestellformular *n* order form
Bestellnummer *f* order number
Bestellung *f* 1. order; 2. *(ag)* tillage *(field)*
Bestellungsabteilung *f* purchasing department
Bestellungsplan *m (ag)* cropping plan
bestimmen to determine; to analyze
bestimmt determinate *(equation)*
statisch b. statically determinate
Bestimmung *f* determination; analysis
Bestimmungen *fpl* / **gesetzliche** statutory requirements
Bestimmungsbahnhof *m* destination (receiving) station
Bestimmungsflughafen *m* aerodrome of destination
Bestimmungshafen *m* port of destination (delivery), treaty (entrepôt) port
Bestimmungsort *m* destination
Bestockung *f (ag)* tillering
bestoßen to abut *(wood)*
Bestoßhobel *m* edging plane
bestrahlen to irradiate; to expose to rays (radiation); to bombard
bestrahlt [radiation-]exposed; irradiated; spent
Bestrahlung *f* irradiation; exposure [to rays]; bombardment; radiation treatment (processing)
äußere B. external exposure (irradiation)
berufliche B. occupational exposure (irradiation)
berufsbedingte B. occupational exposure (irradiation)
innere B. internal exposure (irradiation)
interstitielle B. interstitial irradiation
intrakavitäre B. intracavitary irradiation
kurzzeitige B. acute exposure
nichtberufliche B. non-occupational exposure
protrahierte B. protracted (long continued) irradiation
ständige B. chronic exposure
ultraviolette B. ultraviolet irradiation
zufällige B. accidental exposure
Bestrahlungsanlage *f (nucl)* irradiation plant
Bestrahlungsapertur *f* beam aperture
Bestrahlungsdosis *f (nucl)* radiation (exposure) dose (dosage)
Bestrahlungsdosisleistung *f* exposure dose rate
Bestrahlungsgeometrie *f (nucl)* geometry of irradiation
Bestrahlungskammer *f* radiation chamber, exposure cell (cavity)
Bestrahlungskanal *m* beam (exposure) hole
Bestrahlungsreaktor *m* reactor for irradiation
Bestrahlungsstärke *f* irradiation, irradiance, radiation intensity
Bestrahlungstubus *m (nucl)* [X-ray therapy] applicator, localizer, treatment cone
Bestrahlungsverhältnis *n* / **optimales** *(nucl)* advantage factor
Bestrahlungszeit *f* irradiation (exposure) time
Bestrahlungszone *f* radiation zone
bestreichen to spread; to coat with
mit Gummilösung b. to gum
Bestreichungswinkel *m* maximum traverse
bestücken to tip *(cutting tools)*
Bestücken *n* **von Schneidwerkzeugen** tipping of cutting tools
bestückt mit Hartmetall carbide-tipped
Bestückung *f* complement *(tube)*; tipping *(cutting edge)*
B. mit Bauelementen insertion
B. mit Transistoren transistorization
Bestwirkungsgrad *m (ph)* optimum efficiency
Bestzeitprogramm *n (dat)* optimally coded programme
Beta-Absorptionsdickenmesser *m* beta absorption gauge
beta-aktiv beta-[radio]active, beta-emitting
Beta-Aktivität *f* beta-[radio]activity
Beta-Dickenmesser *m* beta-thickness gauge
Beta-Elektron *n* beta electron
Beta-Emission *f* beta-ray emission, beta radiation
Beta-Kammer *f* beta counter
Beta-Linie *f* beta line

Beta-Messing *n* beta-brass
Beta-Phase *f* beta-phase
beta-radioaktiv beta-[radio]active
Beta-Reflexion *f s.* Beta-Rückstreuung
Beta-Reflexionsdickenmesser *m* beta-reflection gauge
Beta-Rückstreudickenmesser *m* beta-reflection gauge
Beta-Rückstreuung *f* beta backscattering (reflection)
Beta-Schwefel *m* beta-monoclinic sulphur
Beta-Spektrometer *n* beta-ray spectrometer
Beta-Spektrum *n* beta-ray spectrum
beta-stabil beta-stable
Beta-Strahl *m* beta ray (particle)
Beta-Strahldosimetrie *f* beta-ray dosimetry
Beta-Strahlenbündel *n* beta-particle beam
Beta-Strahler *m* beta-emitter, beta-radiator
Beta-Strahlung *f* beta radiation, beta-ray emission
Beta-Teilchen *n* beta particle (ray)
Beta-Umwandlung *f* beta decay (disintegration)
Beta-Zerfall *m* beta decay (disintegration)
  doppelter B. double-beta decay (disintegration)
Betakelung *f (ship)* rigging
betanken to refill, to [re]fuel
Betankung *f* [re]fuelling
Betankungseinrichtungen *fpl* fuelling facilities
Betankungskupplung *f* filler-neck coupling
Betankungsstutzen *m* filler neck
Betankungsventil *n* [oxygen] filler valve
betätigen to actuate, to operate; to control
betätigt / indirekt remotely controlled
  zwangsläufig b. positively actuated
Betätigung *f* operation; actuation, actuating *(switch)*
  B. mit Kraftantrieb power operation
  B. von Hand manual (hand) actuation
  automatische B. automatic actuation
  magnetische B. solenoid operation
  willkürliche B. independent (trip-free) actuation
Betätigungselement *n (mach)* operating element (control)
Betätigungsgriff *m* master handle *(manipulator)*
Betätigungshebel *m (mach)* operating lever
Betätigungsimpuls *m* activating impulse
Betätigungsnocke *f* actuating cam
Betätigungsschalter *m* actuating (remote-control, mechanically operated, trip) switch
Betätigungsseil *n* control wire
Betätigungsspannung *f* auxiliary (operation, control) voltage
Betätigungsstrom *m* actuating current
Betätigungszylinder *m* operating cylinder
Betatron *n* betatron
Betatronbahn *f* betatron orbit
Betatronbeschleunigung *f* betatron acceleration
betäubend anesthetic, anaesthetic
Betäubung *f (el)* black-out effect *(of receiver)*
Betäubungsmittel *n* narcotic, anesthetic
Bethe-Effekt *m (ph)* Lamb shift
Bethe[-Weizsäcker]-Zyklus *m (nucl)* Bethe (carbon-nitrogen) cycle
Beting *m (f) (ship)* bitt
Beton *m* concrete
  B. mit gerissener Zugzone cracked concrete
  bewehrter B. reinforced concrete
  erdfeuchter B. *s.* schwachplastischer B.
  fetter B. rich concrete (mix)
  feuerfester B. refractory concrete
  hitzebeständiger B. refractory concrete
  hochplastischer B. high-slump concrete (mix)
  magerer B. lean concrete (mix)
  monolithischer B. site (poured-in-place, cast-in-place, cast-in-situ, in-situ) concrete
  nagelbarer B. nailing concrete
  schwachplastischer B. stiff (low-slump) concrete, stiff (low w/c) mix
  steifer B. harsh concrete (*s.a.* schwachplastischer B.)
  unbewehrter B. plain (non-reinforced) concrete
Betonabschirmung *f (nucl)* concrete shield
Betonabwasserrohr *n* concrete sewer pipe
Betonanlieferung *f* concrete delivery
Betonarbeiter *m* concretor
Betonaufbereitung *f* concrete fabrication
Betonaufbereitungsanlage *f* mix[ing] plant
  zentrale B. central mix[ing] plant
Betonaufbruchhammer *m* concrete breaker
Betonaufzug *m* concrete hoist
Betonausbau *m (mine)* concrete lining
Betonauskleidung *f* concrete lining
Betonbau *m* concrete construction (engineering)
Betonbaustein *m* concrete brick
Betonbewehrung *f* concrete reinforcement
Betonblock *m* concrete block
Betonbohle *f* concrete plank
Betonbrecher *m* concrete breaker
Betondachstein *m* concrete roofing tile
Betondecke *f* concrete floor
Betondeckung *f* concrete cover
Betondichtungsmittel *n* concrete densifier (waterproofing compound)
Betondruckfestigkeit *f* concrete compressive strength
Betondruckrohr *n* concrete pressure pipe
Betoneinbau *m s.* Betoneinbringung
Betoneinbringung *f* concrete placement (placing), concreting
Betoneisenbahnschwelle *f* concrete sleeper
Betoneisenbieger *m* rod bender
Betonemulsion *f* concrete emulsion
Betonerzeugnis *n* concrete product

Betonestrich *m* concrete floor
Betonfacharbeiter *m* concretor
Betonfahrbahn *f* concrete decking
Betonfertiger *m* concrete finisher
Betonfertigteil *n* precast concrete [unit], precast unit, concrete product
Betonfestigkeit *f* concrete strength
Betonförderleitung *f* concrete discharge pipe
Betonformstahl *m* deformed bar
Betonformsteinausbau *m* concrete-block walling
Betonfüllung *f* concrete filling
Betonfundament *n* concrete foundation
Betongefüge *n* concrete texture
Betongießrinne *f* concrete chute
Betongleitfertiger *m* concrete extruding machine, slip-forming machine
Betongüte *f* grade (quality) of concrete
Betonhärtungsmittel *n* concrete hardener
Betonierabschnitt *m* concreting section
Betonieranlage *f* concreting plant
Betonierbett *n* casting bed
Betonierbrücke *f* concrete placing gantry
Betonierbühne *f* concrete placing platform, concreting stage
betonieren to concrete, to place concrete
Betonieren *n* concreting
Betonierer *m* concretor
Betonierfläche *f* casting yard
Betonierplatz *m* casting yard
betoniert cast
Betoniertrichter *m* tremie
Betonierungstemperatur *f* placing temperature
Betonkaminkühler *m (therm)* concrete chimney cooler
Betonkies *m* concrete gravel
Betonkippkarren *m* concrete buggy (tip cart)
Betonkriechen *n* creep of concrete
Betonmantel *m* concrete lining; concrete envelope
Betonmischer *m* concrete mixer (mixing machine)
  kontinuierlich arbeitender B. continuous [concrete] mixer
Betonmischmaschine *f s.* Betonmischer
Betonmischplatz *m* concrete mixing site
Betonmischturm *m* concrete mixing tower
Betonmischung *f* concrete mix[ture]
  fette B. rich mix
Betonmischwaage *f* concrete batcher scale
Betonnung *f* buoyage *(navigation)*
  B. und Befeuerung *f* navigational aids
Betonpfahl *m* concrete pile
Betonplattenpresse *f* concrete tile press
Betonpumpe *f* pumpcrete machine; concrete pump
Betonrohr *n* cement (concrete) pipe
Betonrüttler *m* concrete vibrating equipment
Betonsäge *f* concrete cutter (saw)
Betonschiff *n* reinforced concrete vessel, concrete ship
Betonschiffskörper *m* concrete hull
Betonschild *m (nucl)* concrete shield
Betonschleifmaschine *f* concrete grinder
Betonschutz *m (nucl)* concrete shield
Betonskelett *n* concrete frame
Betonspannung *f* concrete stress
Betonspritzmaschine *f* concrete [placing] gun, jetcrete gun
Betonstahl *m* reinforcing steel, steel for concrete reinforcement
Betonstahlgewebe *n* reinforcing mesh, mesh reinforcement
Betonstein *m* concrete block
Betonsteinherstellung *f* block making
Betonstraßenbau *m* concrete road construction
Betonstraßenfertiger *m* concrete road finisher (paver)
Betontransportfahrzeug *n* concrete hauler (hauling unit)
Betontrichter *m* concrete placement funnel
Betontunnel *m (nucl)* canyon
betonummantelt concrete-encased *(steelwork)*
Betonung *f (el)* emphasis
Betonverdichtung *f* concrete compaction
Betonverflüssiger *m* plasticizer, plastifying agent
Betonversuch *m* concrete test
Betonverteiler *m* concrete spreader (distributor)
Betonwalzverfahren *n* concrete-rolling technique, continuous rolling method
Betonwandschale *f* concrete wall shell
Betonwerk *n* concrete factory, precasting plant
Betonwerkstein *m* cast (artificial) stone
Betonzusammensetzung *f* concrete composition
Betonzusatz *m* concrete admixture
Betonzuschlag *m* concrete aggregate
Betonzuschlagstoff *m* concrete aggregate
Betrachten *n* examination; considering, consideration; *(opt)* viewing
  pseudoskopisches B. pseudoscopic viewing
Betrag *m* amount; rate
  absoluter B. absolute value
betragen / durchschnittlich to average
Betragsfeld *n (cont)* amount field
betreiben to run; to operate; to conduct
  Bruchbau b. *(mine)* to cave
  eine Anlage b. to operate a plant
  eine Funkstelle b. to operate a station
  mit 220 Volt b. to operate at 220 volts
Betrieb *m* 1. factory *(manufacturing)*; mill *(textile, steel)*; 2. operation; service
  B. / außer out of operation (service, use, order); disabled
  B. / in in running order, working; in service, in use
  B. mit erzwungener Erregung (Stromform) constrained current operation, forced excitation

B. mit freier Stromform *(el)* free current operation
B. mit natürlicher Erregung *(el)* natural excitation
B. mit Trägerübertragung *(el)* transmitted-carrier operation
aussetzender B. *(el)* intermittent duty
aussetzender B. mit gleichbleibender Belastung intermittent service
diskontinuierlicher B. *(cont)* batch process
durchlaufender B. mit aussetzender Belastung *(el)* continuously-running duty with intermittent loading
fehlerhafter B. *(el)* faulty operation
gesteuerter B. *(el)* controlled operation
halbautomatischer B. *(el)* semi-automatic service
kurzzeitiger B. mit gleichbleibender Belastung *(el)* short-time duty
periodisch aussetzender B. mit gleichbleibender Belastung *(el)* intermittent periodic duty
rauher B. rough usage
störungsfreier B. trouble-free service
unbeaufsichtigter B. unattended operation
Betriebsabrechnung *f* cost account
Betriebsabrechnungsbogen *m* cost accounting statement, process cost account, cost summary card
Betriebsabteilung *f* production department
Betriebsaggregat *n* plant unit
Betriebsanlage *f* plant; industrial plant
Betriebsanleitung *f* operating instructions, instruction leaflet, operator's manual, instruction book *(for operator)*
Betriebsanweisung *f s.* Betriebsanleitung
Betriebsarchiv *n* plant record
Betriebsart *f* mode [of operation], duty, method (system) of working, method (system) of operation
Betriebsartenschalter *m* mode (wafer) switch
Betriebsartenstecker-Programmierung *f* pinboard programming
Betriebsartenumschalter *m s.* Betriebsartenschalter
Betriebsaufwand *m* operating expenditure
Betriebsausgaben *fpl* operating expenditure
Betriebsausrüstung *f* factory equipment
Betriebsausstattung *f* factory equipment
Betriebsbeanspruchung *f* service load
Betriebsbedingung *f* service (working, operating) condition
Betriebsbelastung *f* work-load *(radiobiology)*
Betriebsbereitschaft *f* readiness for operation, stand-by
Betriebsbesichtigung *f* works visit
Betriebschemiker *m* plant chemist
Betriebsdampf *m* working steam
Betriebsdampfmaschine *f* driving steam engine
Betriebsdämpfung *f (el)* insertion loss, overall [attenuation] loss, operative attenuation
Betriebsdaten *pl* data; performance, operating characteristics
Betriebsdauer *f* time of operation, working time
Betriebsdefizit *n* operating deficit
Betriebsdrehzahl *f* normal (working, operating, service) speed
Betriebsdruck *m* working (design, system) pressure
Betriebseingriffswinkel *m* operating pressure angle
Betriebseinnahmen *fpl* operating receipts
Betriebseinrichtung *f* operating facilities
Betriebsempfänger *m* communications (commercial) receiver
Betriebserde *f (el)* earth bus
Betriebserfahrung *f* operating (works) experience; technical know-how
Betriebsergebnis *n* operating result
Betriebsertrag *m* net operating profit
betriebsfähig ready for service, in running order
Betriebsfaktor *m* / **effektiver** *(dat)* operation ratio
Betriebsforschung *f* operational research
Betriebsfrequenz *f (el)* operating (power) frequency
günstigste B. optimum working frequency
höchstmögliche B. maximum usable frequency
Betriebsführer *m* **unter Tage** *(mine)* undermanager
Betriebsführung *f* operating [management], management
Betriebsfunkspruch *m* service (routine) radio message
Betriebsfunkstelle *f* service radio station
Betriebsgas *n* fuel gas
Betriebsgemeinkosten *pl* general operating cost
Betriebsgerät *n* operational apparatus
Betriebsgüte *f* grade of service
Betriebsingenieur *m* manufacturing (works, plant maintenance) engineer
Betriebskennlinie *f* operation (operating) characteristic
Betriebskennzahl *f s.* Betriebsdaten
Betriebskennziffern *fpl* operating figures
Betriebskontrollabteilung *f (cont)* instrument department
Betriebskontrollanlage *f (dat)* data-logging system
Betriebskontrolle *f* 1. instrument department; 2. process measurements
Betriebskontrolleinrichtung *f* data logger
Betriebskonzentration *f* concentration of workings
Betriebskosten *pl* operating expenses (cost); running cost *(ship)*

Betriebslaboratorium *n* works (industrial) laboratory
Betriebslänge *f* length of line operated
Betriebslautsprecheranlage *f* factory public address system
Betriebsleistung *f* 1. manufacturing efficiency; 2. operating performance
Betriebsleitung *f* factory management
betriebsmäßig commercial; industrial
Betriebsmerkmale *npl* operating characteristics
Betriebsmeßinstrument *n* service instrument
Betriebsordnung *f* operating regulations
Betriebspoliklinik *f* factory medical centre
Betriebspraxis *f* manufacturing practice
Betriebsprüfgerät *n* field monitoring unit
Betriebspunkt *m* working place
Betriebsregime *n* duty, dynamics; operating condition
Betriebsschaltung *f* service connection
Betriebsschmelze *f* production cast
Betriebsschule *f* works school
Betriebsschutz *m (Am)* plant security
Betriebssender *m* commercial transmitting station
Betriebssetzmaschine *f* commercial jig
Betriebssicherheit *f* 1. reliability in service, safety (reliability) of operation; 2. factory security
Betriebssicherung *f* active fuse
Betriebsspannung *f* working (service, operating, operational) voltage
  gleichmäßige B. steady voltage
  obere B. highest system voltage
Betriebssteilheit *f* **[einer Verstärkerröhre]** mutual conductance slope
Betriebsstellbereich *m (cont)* turndown *(valve)*
Betriebsstellung *f* operating (running) position
Betriebsstofflager *n* petrol and lubricant bulk-storage plant
Betriebsstörung *f* breakdown, shut-down, stoppage; traffic disturbance, breakdown of service
Betriebsstrom *m* operating (running, working) current
Betriebsstromkreis *m* service circuit
Betriebsstunden *fpl* working hours
Betriebsstundenzähler *m* engine-hour indicator
Betriebstemperatur *f* operating temperature; working temperature
Betriebstiefgang *m* service draught *(of vessels)*
betriebstüchtig serviceable
Betriebsübertragungsmaß *n (telec)* propagation constant
Betriebsüberwachung *f* process monitoring
Betriebsüberwachungsanlage *f* monitoring system
Betriebs- und Verkehrsdienst *m (ry)* operating department
betriebsunfähig unserviceable
Betriebsunfall *m* factory (occupational) accident, industrial injury
Betriebsunkosten *pl* / **allgemeine** factory overhead cost
Betriebsunterbrechung *f* operation interruption
Betriebsverhalten *n* performance; behaviour
  B. des Motors engine performance
Betriebsverhältnisse *npl* service conditions
Betriebsverzerrungsgrad *m* degree of distortion in service
Betriebswagenwerk *n (ry)* shop for light repairs to rolling stock
Betriebswälzkreisdurchmesser *m (mach)* operating pitch diameter
betriebswarm warmed up
Betriebswelle *f* signal wave
Betriebswellenlänge *f* operating wavelength
Betriebswert *m* efficiency factor
Betriebswiderstände *mpl* source and load resistances
Betriebswirtschaftler *m* manufacturing engineer; economist
Betriebszahlen *fpl* operating figures
Betriebszeit *f* working hours
Betriebszusammenfassung *f* concentration of workings
Betriebszuschnitt *m* mine layout
Betriebszuschuß *m* equipment bounty; equipment subsidy *(for shipowner)*
Betriebszustand *m* working (operational, service) condition
Betriebszuverlässigkeit *f* reliability in service, reliability of operation
Bett *n* 1. bed *(machine tool)*; 2. *(build)* bedding; 3. *(mine)* compartment *(jigger)*
  gekröpftes B. gap bed
Bettbarchent *m (text)* bed ticking
Bettbauart / **in** *(mach)* bed-type
Bettdecke *f* / **gesteppte** quilt
  wollene B. blanket
Bettenstation *f* ward unit
Bettführungsbahn *f (mach)* bed way[s] (track, slide)
Bettkarte *f (ry)* sleeping-berth supplement
Bettnische *f* alcove
Bettschlitten *m (mach)* carriage; saddle *(lathe, slotter)*
Bettung *f* ballast; bedding, foundation
  elastische B. elastic foundation (bedding)
Bettungshöhe *f* depth of ballast
Bettungskrone *f* crown (top) of ballast
Bettungsprofil *n* ballast profile
Bettungsräumungsmaschine *f* ballast scarifier, cribbing machine, *(Am)* ballast cribber
Bettungsreinigungsmaschine *f* ballast cleaner (screening machine)
Bettungsschicht *f* bedding course, underlay
Bettwange *f (mach)* cheek

Bettzahnstange *f (mach)* bed rack
beuchecht *(text)* fast to kier-boiling
Beuchen *n (text)* kier-boiling
Beuchfaß *n (text)* kier
beugen 1. to bend; to deflect; to crank; 2. to diffract *(ray)*
Beugung *f (ph)* diffraction
  B. des Lichtes diffraction of light
  Fraunhofersche B. Fraunhofer diffraction
Beugungsbild *n* diffraction image (pattern)
Beugungsdiagramm *n* diffraction pattern
Beugungserscheinungen *fpl* diffraction fringes
Beugungsfigur *f* diffraction image (pattern)
Beugungsgitter *n* diffraction grating
Beugungsraum *m* / **Fresnelscher** Fresnel diffraction fringe
Beugungsring *m* diffraction ring
Beugungsscheibchen *n s.* Beugungsscheibe
Beugungsscheibe *f* diffraction[al] disk
Beugungsspektrograf *m* diffraction spectrograph
Beugungsspektrum *n* diffraction spectrum
Beugungsstreifen *m* diffraction fringe
Beugungsstreuung *f* diffraction scattering
Beugungsverlust *m* diffraction loss
Beugungsvermögen *n* diffraction ability
Beugungswinkel *m* diffraction angle, angle of diffraction
Beugungszentrum *n* centre of diffraction
Beugungszone *f* diffraction zone
Beule *f* bulge *(sheet metal)*
beulen to buckle *(plate, shell)*
Beulen *n* buckling
Beulfestigkeit *f* strength to buckling, buckling strength
Beullast *f* collapsing (buckling) load
Beulsicherheit *f* safety against buckling
Beulspannung *f* buckling stress
Beutel *m* bag; sack
Beutelelement *n (el)* sack cell (element)
Beutelnetz *n* seine net, purse seine *(fishery)*
Beutelverschließmaschine *f* bag sealing machine
Bevatron *n (nucl)* bevatron
Bevölkerungsschutz *m* protection of [the] population
Bevölkerungsstatistik *f* demography
bevollmächtigt authorized
bevorzugen to prefer
bevorzugt preferred *(crystallography)*
  nicht b. random *(crystallography)*
bewalden to afforest
bewaldet arboraceous
bewässern to irrigate
Bewässerung *f* irrigation
Bewässerungsanlage *f* irrigation plant
Bewässerungskanal *m* irrigation canal
Bewässerungsrohrsystem *n* irrigation pipe system
Bewässerungsspeicherung *f* irrigation storage
bewegbar movable
bewegen to move
  b. gegen to move past
  b. um to move about
  b. vorbei an to move past
  sich b. to move, to travel
  sich zwangsläufig b. to make a constrained movement
beweglich movable; portable
Beweglichkeit *f* mobility, manoeuvrability; flexibility
Bewegung *f* movement, motion; swirling *(melt)*; travel *(sliding machine part)*
  B. in einer Ebene movement in one plane
  auf- und abwärtsgehende B. reciprocating movement in vertical direction, up and down movement
  beschleunigte B. accelerated movement (motion)
  Brownsche B. Brownian movement (motion)
  drehende B. rotary motion
  ebene B. uniplanar movement, movement in one plane
  epizyklische B. epicycloidal motion
  erzwungene B. forced motion
  fortschreitende B. translation
  gegenseitige B. relative movement
  geradlinige B. rectilinear motion
  gleichförmig beschleunigte B. uniformly accelerated motion
  gleichförmig verzögerte B. uniformly retarded (decelerated) motion
  gleichförmige B. uniform motion
  harmonische B. harmonic motion
  hin- und hergehende B. reciprocating motion
  hypozyklische B. hypocycloidal motion
  krummlinige B. curvilinear motion (movement), movement in a curve
  laminare B. laminar motion
  oszillierende B. oscillatory motion
  periodische B. periodic motion
  räumliche B. movement in space
  regellose B. irregular (disordered) motion
  resultierende B. resultant motion
  rückläufige B. retrogressive (retrograde, reverse) motion
  scheinbare B. apparent motion; relative movement
  schwingende B. oscillating movement; oscillatory motion
  sinusförmige B. sinusoidal motion
  sphärische B. motion about a fixed point
  stationäre B. stationary motion
  ungeordnete B. disordered (irregular) motion
  ungleichförmige B. irregular (non-uniform) movement

unregelmäßige B. irregular movement
veränderliche B. variable movement
verborgene B. concealed movement
verzögerte B. retarded movement
Bewegungsablauf *m* sequence of motions
Bewegungsänderung *f* change in motion
Bewegungsbegrenzung *f* restraint [of movements]
Bewegungselektrizitätslehre *f* electrokinetics
Bewegungsfolge *f* order of movements
Bewegungsform *f* form of motion, motion pattern
Bewegungsfreiheit *f* freedom of motion (movement)
Bewegungsfuge *f (build)* expansion (movement) joint
Bewegungsgeometrie *f* geometry of kinematics
Bewegungsgeschwindigkeit *f* rate of traverse
B. des Satelliten satellite velocity
Bewegungsgesetz *n* law of motion
Bewegungsgewinde *n* power-transmission screw thread, thread for [the] transmitting [of] motions
Bewegungsgleichung *f* equation of motion
Bewegungsgröße *f* quantity of motion, [linear] momentum
B. der Mischung product of mass and velocity of the mixture *(turbine)*
Bewegungsimpedanz *f* motional impedance
freie B. free motional impedance
Bewegungskomponente *f* component motion
Bewegungskontrollpult *n* flight progress board
Bewegungslehre *f* kinematics, theory of motion
geometrische B. geometry of kinematics
bewegungslos motionless
bewegungsmäßig kinematic
Bewegungspause *f (mach)* dwell, period (time) of dwell
B. bei Umsteuerung *(mach)* reversal dwell
Bewegungsreibung *f* kinetic friction
Bewegungsrichtung *f* movement direction, direction of movement
Bewegungsrichtungsumkehr *f (mach)* reversal of [the] travel direction
Bewegungsschraube *f* screw
Bewegungsstudie *f* motion study *(work cycle)*
Bewegungsübertragung *f* motion transmission, transmission of motion
Bewegungsumkehr *f (mach)* reversal of movement (motion, travel); return of motion (movement)
B. der Spindel beim Gewindebohren tapping reversal
Bewegungsunschärfe *f* [movement] blur
Bewegungswiderstand *m* resistance of motion
Bewegungszwang *m* constrained movement
bewehren to reinforce *(concrete)*; to armour, to sheath *(cable)*
bewehrt reinforced *(concrete)*; armoured, sheathed *(cable)*
doppelt b. doubly reinforced
einachsig b. singly (one-way) reinforced
kreuzweise b. two-way reinforced
schlaff b. normally (conventionally) reinforced
Bewehrung *f* reinforcement *(concrete)*; armouring, sheathing *(cable)*
B. im Bereich negativer Momente negative reinforcement
einachsige B. one-way reinforcement
konstruktive B. nominal reinforcement
kreuzweise B. two-way reinforcement
schlaffe B. non-tensioned reinforcement, non-prestressed reinforcement
Bewehrungsanteil *m (build)* percentage of reinforcement
Bewehrungseisen *n* reinforcing steel
Bewehrungsgrenze *f* reinforcing limit
Bewehrungskorb *m* reinforcement cage
Bewehrungsmatte *f* reinforcement (bar) mat, mesh (steel fabric) reinforcement
Bewehrungsstab *m* reinforcing rod (bar)
B. mit Endhaken hooked bar (rod)
aufgebogener B. bent-up bar (rod)
glatter B. plain bar (rod), smooth bar (rod)
Bewehrungsstahl *m* reinforcing steel
Bewehrungsstoß *m* bar splice
Beweis *m (math)* proof, demonstration; evidence
Beweisantritt *m* introduction of evidence
Beweisaufnahme *f* hearing of witnesses
Beweisbarkeit *f (math)* demonstrability
beweisen *(math)* to prove
Beweisstück *n* exhibit; evidence
bewerben / sich to apply for
Bewerber *m* applicant
Bewerbung *f* application
bewerfen / mit Mörtel to plaster
bewerkstelligen to accomplish
Bewertungsfaktor *m* **eines Stromes** current weighting factor
Bewertungsfilter *n* weighting network
Bewetterung *f (mine)* ventilation
blasende B. blowing (forcing) ventilation
fallende B. descensional ventilation
gegenläufige B. antitropal ventilation
gleichlaufende B. homotropal ventilation
künstliche B. forced (mechanical) ventilation
natürliche B. natural ventilation
saugende B. exhaust ventilation
steigende B. ascensional ventilation
Bewetterungsmeßgerät *n (text)* weatherometer
bewickeln to wind, to wrap, to whip
Bewilligung *f* licence
bewirken to effect; to cause
Bewitterungsversuch *m* weathering
Bewoid-Harzleim *m (text)* bewoid size

bewölkt overcast, cloudy
Bewölkung *f* cloudiness, overcast
Bewölkungsgrad *m* degree (extend) of cloudiness
Bewölkungskarte *f* cloud map
Bewölkungsmesser *m* nephoscope
Bewölkungszunahme *f* thickening of clouds
Bewuchs *m* [marine] growth, [marine] fouling
bezahlen to pay; to settle
  die Rechnung b. to settle an account
Bezahlung *f* payment; settlement
  B. nach dem Gewicht payment by weight basis
bezeichnen to mark; to notate; to denote; to designate; to specify; to indicate
Bezeichnung *f* marking; notation; designation; specification
  B. der Funkwellen designation of radio-electric waves
  B. der Güter description of goods
  falsche B. misnomer
Bezeichnungsstreifen *m* designation strip *(cable)*
bezetteln to affix a label, to label
Bezettelung *f* labelling, *(Am)* tagging
Beziehung *f* relation; relationship
  in B. zueinander interrelated with each other
  Bragg-Graysche B. *(nucl)* Bragg-Gray relation
Beziehungsgleichung *f* regression equation
Beziehungspunkt *m* point of reference
Bezirk *m* area; district; zone
Bezirkskabel *n (telec)* trunk cable
Bezirksnetz *n (telec)* trunk network
Bezirkssender *m* district transmitter, regional broadcasting station
Bezug *m* 1. covering; 2. *(ph, math)* reference
bezüglich with reference to
Bezugsdämpfung *f* reference (relative, loudness volume, volume, transmission) equivalent
  B. eines Übertragungssystems volume (effective transmission, loudness volume) equivalent
  B. je Längeneinheit *(telec)* volume loss per unit length
Bezugsdaten *pl* reference data
Bezugsebene *f* reference (datum) plane
Bezugselektrode *f* reference electrode
Bezugsfläche *f* reference [sur]face, datum [sur]face
Bezugsform *f* model, master form *(machining)*
Bezugsfrequenz *f* reference (standard) frequency
Bezugshafen *m* standard port
Bezugshöhe *f* datum level
Bezugsinstrument *n* reference instrument
Bezugskraftstoff *m* reference fuel
Bezugskreis *m* reference circuit
  fiktiver B. hypothetical reference circuit, nominal maximum circuit
Bezugslänge *f* reference length; *(text)* gauge distance
Bezugslautstärke *f* reference volume
Bezugslinie *f* reference (datum) line
  mittlere B. mean reference line
Bezugsniveau *n* level of reference
Bezugsnormal *n* reference standard
Bezugspegel *m* reference level (volume)
  B. Null zero reference level
Bezugspunkt *m* reference point (mark)
Bezugsrauschen *n* reference noise
Bezugsrauschpegel *m s.* Bezugsrauschen
Bezugsrauschwert *m (Am)* reference noise
Bezugssprechkopf *m* reference standard for recording sound head
Bezugsstellung *f* reference position
Bezugsstoff *m (text)* upholstery fabric
Bezugsstromkreis *m* reference circuit
Bezugsstück *n* master [component], [reference] model; sample workpiece
Bezugstemperatur *f* standard (reference) temperature
Bezugsverbindung *f* / **fiktive** hypothetical reference circuit
Bezugswert *m* reference (standard) value
BF_3-Kammer *f (nucl)* BF_3-chamber, boron-filled chamber
biaxial biaxial
Biberpelz *m* beaver fur
Biberschwanz *m* plain [roofing] tile
Biberschwanzantenne *f* fanned-beam aerial
Biberschwanzheck *n (ship)* turtle-backed stern
Bibliothek *f* library
Bibliothekar[in *f*] *m* librarian
Bibliotheksprogramm *n (dat)* library [sub]routine
bichrom bichromatic
bidestilliert *(ch)* redistilled
biegbar bendable
Biegbarkeit *f* bendability, flexure
Biegeachse *f* neutral axis
biegebeansprucht subjected to bending
Biegebeanspruchung *f* flexural load; bending stress
Biegebelastung *f* flexural load
Biegeblech *n* bending plate
Biegebock *m* bending horse
Biegedorn *m* bending mandrel
Biegedruckzone *f* moment compression zone
Biegefähigkeit *f* bendability; bend test ductility
Biegefestigkeit *f* flexural (bending) strength
Biegehalbmesser *m* bending radius
Biegemaschine *f* rod (bar) bender, rod (bar) bending machine
Biegemodell *n* bending block
Biegemoment *n* flexural (bending) moment
biegen to bend; to crank

Biegeplan *m* 1. *(build)* bending schedule; 2. *(ship)* frame-bending slabs
Biegeplatz *m* steel bending yard
Biegepresse *f* bending brake
Biegeprobe *f* bending specimen
Biegeprüfung *f* bending test
Biegerichtpresse *f* bending and straightening press
Biegering *m* bending ring
Biegerohrdurchflußmesser *m* elbow flowmeter
Biegeschlagversuch *m* resilience (impact) bending test
Biegeschwellfestigkeit *f* pulsating bending fatigue limit
Biegeschwingung *f* bending vibration
Biegeschwingungsfestigkeit *f* repeated bending stress strength, bending fatigue resistance
Biegespannung *f* flexural (bending) stress
Biegestab *m* flexural member (bar)
biegesteif rigid; not subject to permanent deformation by bending
 b. angeschlossen (verbunden) rigid-jointed
Biegesteifigkeit *f* flexural rigidity
Biegetisch *m* rod (bar) bender
Biegetrommel *f* bending drum
Biegeversuch *m* bending test
Biegewalze *f* bending roll
Biegewechselfestigkeit *f* reverse bending strength
Biegewechselzahl *f* [number of] bending cycles
Biegewelle *f* flexible shaft
Biegezugfestigkeit *f* tensile bending strength
Biegezugspannung *f* tensile bending stress
biegsam flexible; bendable
Biegsamkeit *f* flexibility; bendability
Biegung *f* 1. flexure, bending; 2. curve *(road)*; curvature
 B. des anbindenden Fadens *(text)* curvature of the stitching thread
 B. im elastischen Bereich elastic bending
 B. mit Längskraft combined bending and axial load (stress), direct stress and bending
 B. ohne Längskraft simple bending
 reine B. simple bending
 schiefe B. bending in two planes
Biegungsabstand *m* bending distance *(bearded needle)*
Biegungselastizität *f* flexional elasticity
Biegungsfeder *f* / **gewundene** helical torsion spring
Biegungsfläche *f* deflection surface
Biegungshöhe *f* bending height *(knitting)*
Biegungslinie *f* curve of deflection, line of flexure, bending line
Biegungspfeil *m* pitch of deflection sag
Biegungsring *m* bending ring
Biegungsschwingung *f* flexural mode of vibration *(e.g. quartz)*
Biegungswelle *f* flexural wave
Biegungswiderstand *m* resistance to bending
Biegungswiderstandsmoment *n* section modulus of bending
Biegungszentrum *n* centre of flexure
Biene *f* bee
Bienenhalter *m* bee-keeper, bee-master, apiarist
Bienenhaltung *f* bee-keeping, apiculture
Bienenkorb *m* beehive, skep
Bienenkorbofen *m* beehive oven
Bienenmaske *f* bee-veil
Bienenschwarm *m* swarm of bees
Bienenstock *m* beehive
Bienenwachs *n* beeswax
Bienenzucht *f* bee-keeping, apiculture
Bienenzüchter *m* bee-keeper, bee-master, apiarist
Bier *n* beer
Bierbrauerei *f* brewery
Bierdeckel *m* beer mat
Bierhefe *f* yeast, barm
Bierwürze *f* wort
Biesenschnellnäher *m* *(text)* high-speed cording machine
bifilar bifilar
Bifilarantenne *f* two-wire aerial
Bifilaraufhängung *f* bifilar suspension
Bifilardraht *m* bifilar wire
Bifilarwicklung *f* bifilar winding
Bifilarwiderstand *m* non-inductive resistance
bifokal bifocal
biharmonisch biharmonic
Bikarbonathärte *f* permanent hardness
bikonkav biconcave, double-concave
Bikonkavlinse *f* biconcave (double-concave) lens
bikonvex biconvex, double-convex
Bikonvexlinse *f* biconvex (double-convex) lens
Bilanzbogen *m* balance sheet
Bilanzlogger *m* balance logger
Bild *n* 1. *(opt)* image; 2. *(telev)* picture, frame; 3. *(math)* graph; 4. pattern; oscillogram
 B. im Ton vision on sound
 abgehacktes B. *(telev)* truncated picture
 abgeschnittenes B. *(telev)* truncated picture
 aufrecht stehendes B. erect image
 aufrechtes B. *(opt)* erect image
 bewegtes B. moving picture
 detailreiches B. *(telev)* high-contrast picture (image)
 dünnes B. *(telev)* soft picture
 durchgezeichnetes B. *(telev)* deep dimension picture
 feststehendes B. still picture
 fotografisches B. photograph
 gerastertes B. *(telev)* half-tone process image
 gespeichertes B. *(telev)* latent electronic image
 geteiltes B. split image

hartes B. *(telev)* hard image
hochzeiliges B. *(telev)* hard (high-contrast) picture
imaginäres B. *(opt)* virtual image
kontrastarmes B. *(telev)* flat picture; picture without contrast
lichtschwaches B. *(telev)* picture of low brightness
negatives B. reversed (negative) image
optisch entworfenes B. image
projiziertes B. picture
reelles B. *(opt)* real image
scharf eingestelltes B. sharply focused picture
scharfes B. sharp (high-definition) picture
stehendes B. still image
stillstehendes B. stationary pattern *(discharge valve)*
umgekehrtes B. reversed (inverted) image
unscharfes B. blurred (non-sharp, unfocused) image
unverzeichnetes B. undistorted picture
vergrößertes B. enlarged image
verkehrtes B. *(opt)* inverted (reversed) image
verkleinertes B. reduced image
verwaschenes B. blurred (non-sharp, unfocused) image
virtuelles B. virtual image
weiches B. soft (weak, uncontrasty) picture
welliges B. *(telev)* ragged picture

Bildablenkschaltung *f* vertical time-base generator; frame time-base circuit
Bildablenkspule *f* frame coil
Bildablenkstufe *f* scanning stage
Bildabschattung *f* shading *(camera tube)*
Bildabtaster *m* scanning device, scanner
Bildabtaströhre *f* scanning tube
Bildabtastscheibe *f* disk scanner
Bildabtastung *f (telev)* scanning
B. mit hoher Anfangszerlegungsgeschwindigkeit initial high velocity scan
Bildauflösung *f* picture resolution (definition)
vertikale B. vertical definition
Bildaufnahme *f* vision pick-up
Bildaufnahmeröhre *f* camera (pick-up) tube, vision pick-up tube
Bildaufzeichnung *f* display, picture tracing
Bildausfall *m* / **sektorweiser** blind sector *(radar)*
Bildausgangstransformator *m* frame output transformer
Bildauslösung *f* video trigger
Bildausschnitt *m* image sector
vergrößerter B. close-up sectional view
Bildaustastsignal *n* blanked television signal
Bildaustastsynchronsignal *n* composite video signal
Bildaustastung *f* **am Ende eines Rasters** frame suppression
Bildbandbreite *f* vision bandwidth
Bildbandführung *f* film-strip transport
Bildbegrenzer *m* vertical shaper
Bildbreiteregelung *f* width (line amplitude) control
Bilddarstellung *f* display *(radar)*; picture presentation
vergrößerte B. close-up [view]
Bilddauer *f* picture duration; *(Am)* frame rate
Bilddetail *n* picture detail
Bildebene *f* image (picture) plane; perspective plane
Bildebnungseinrichtung *f (opt)* image-flattening device
Bildeinstellspule *f* picture control coil; *(Am)* frame coil
Bildeinstellung *f* framing
horizontale B. horizontal centring
Bildelektrode *f* image plate
Bildelement *n* picture element (point)
Bildempfänger *m* picture (vision) receiver; photographic receiver
bilden 1. to form; to shape; 2. to educate; 3. to generate, to produce, to create
eine Schneide b. to take (form) an edge
Bildendstufe *f* video output stage
Bildendverstärker *m* image power amplifier
Bilderform *f (print)* half-tone (picture) forme
Bilderrahmentechnik *f* picture-frame technique
Bildfahrplan *m (ry)* train diagram, graphic timetable
theoretischer B. theoretical graph of train running
Bildfang *m* hold control
Bildfangregler *m* hold control
Bildfangröhre *f* vision pick-up tube
Farnsworthsche B. Farnsworth image dissector tube
Bildfehler *m* distortion of the image; picture defect
Bildfeinheit *f* definition of picture
Bildfeld *n* picture area; image field
Bildfeldwölbung *f* curvature of the field
Bildfeldzerlegung *f* image dissection; scanning
Bildfenster *n* picture gate; aperture
Bildfläche *f* image area; picture field; image plane
Bildfolge *f* series of single pictures
Bildfolgeregler *m* picture-interval regulator
Bildformat *n* picture ratio; frame size
Bildfrequenz *f* picture frequency; *(Am)* field frequency
Bildfunk *m* facsimile radio
Bildfunkanlage *f* facsimile apparatus
Bildfunkdienst *m* radio facsimile service
Bildfunkgerät *n* wireless picture telegraph
Bildfunksender *m* picture (facsimile) transmitter
Bildfunktelegramm *n* radio photo
Bildgeber *m* picture transmitter

Bildgeometrie *f* picture geometry
Bildgeräusch *n* frame noise
Bildgewebe *n* fancy cloth, figured fabric
Bildgleichlaufimpuls *m* frame synchronizing impulse
Bildgleichrichtung *f* video detection
Bildgüte *f* image (picture) quality
Bildhälftenverschiebung *f* out of frame
Bildhauer *m* sculptor
Bildhauerin *f* sculptress
Bildhauerkunst *f* sculpture
Bildhauerton *m* plastic clay
Bildhelligkeit *f* brightness of image, picture brightness
Bildhöhe *f* picture (frame, image) height
Bildingenieur *m* video engineer
Bildkante *f* picture border
Bildkippfrequenz *f* image sweep frequency
Bildkippgenerator *m* vertical time-base generator
Bildkippspannung *f* frame sweep voltage
Bildkonservierung *f* sticking
Bildkontrast *m* image contrast
Bildkontrolle *f* image control
Bildkontrollgerät *n* picture monitor
Bildkontrollröhre *f* monitor tube
Bildleistungsverstärker *m* image power amplifier
Bildlupe *f (telev)* magnifier
Bildmaske *f* framing mask
Bildmeßverfahren *n* photographic recorder system
Bildmischpult *n* video mixer
Bildmitteneinstellung *f* vertical centring
Bildmodulation *f* vision (image) modulation
Bildmodulationsfrequenz *f* image frequency
Bildnachleuchten *n* afterglow of picture
Bildnerei *f* sculpture
Bildpaar *n* stereogram, stereograph, stereopair
Bildplan *m* controlled mosaic
Bildpunkt *m (telev)* picture point (element)
Bildpunktfrequenz *f* video frequency
Bildpunktzahl *f* number of picture elements
Bildraster *m (print)* half-tone screen, grating; *(telev)* frame; raster; field
Bildreaktor *m* image reactor
Bildrechner *m* pictorial computer
Bildregler *m* vertical hold
Bildröhre *f* picture tube
  B. für Direktbetrachtung direct vision tube
  B. mit Metallhals metal cone kinescope
Bildröhrenmaske *f* mask
Bildröhrenprüfer *m* picture-tube emission tester
bildsam plastic
Bildsamkeit *f* plasticity
Bildschärfe *f* image definition, picture sharpness; contrast
Bildschaukeln *n* image drift
Bildschirm *m (el)* [viewing] screen; cathode ray tube screen; display scope *(radar)*
  B. mit aufgedampfter Aluminiumschicht aluminized picture-tube screen
  aluminisierter B. aluminized picture-tube screen
Bildschirmfläche *f* display area
Bildschirmhelligkeit *f* brightness [of the fluorescent screen]
Bildschreiber *m* facsimile recorder
Bildschrumpfung *f* image contraction
Bildseitenverschiebung *f* lateral-shift control
Bildsendeleistung *f (telev)* vision (picture) transmitter power
Bildsender *m* facsimile broadcast transmitter; visual (video) transmitter
Bildsignal *n (telev)* picture (video) signal
Bildsondenröhre *f* image dissector
Bildspaltung *f* image splitting
Bildspannung *f* image (picture) voltage
Bildspeicherröhre *f* image storing tube, storage camera tube
Bildsperre *f* picture trap
Bildsperrschwingtransformator *m* frame blocking oscillator transformer
Bildsprung *m* break of vision
Bildstand *m* framing condition
Bildstatistik *f* pictorial statistics
Bildsynchronisation *f* synchronizing of images
Bildsynchronisierfrequenz *f* field frequency
Bildsynchronisierlücke *f* blank bar
Bildtanzen *n* jumping of picture
Bildtelegraf *m* facsimile telegraph, phototelegraphic apparatus
Bildtelegrafie *f* facsimile (picture) telegraphy
Bildtelegrafiegerät *n* wirephoto equipment
Bildtelegramm *n* photo-telegram; facsimile telegram
Bildträger *m (telev)* picture (vision, visual, image) carrier
Bildträgerfrequenz *f* visual frequency
Bildtriangulation *f (opt)* triangulation from photographs
Bildüberblendeinrichtung *f* fade-in-and-out control
Bildübertrager *m* picture transformer
Bildübertragung *f* facsimile transmission, transmission of pictures
  indirekte B. indirect transmission of pictures
Bildübertragungsdauer *f* picture transmission period
Bildübertragungsgerät *n* picture telegraph apparatus
Bildung *f* 1. generation, production, formation; 2. education
  fotochemische B. photochemical formation
  spontane B. spontaneous generation
Bildungsenergie *f* energy of formation

Bildungsgeschwindigkeit *f* velocity of formation
B. der Ionen velocity of formation of ions
Bildungsquerschnitt *m* formation cross section
Bildungsrate *f* birth-rate, formation rate
Bildungsreaktion *f* reaction of formation
Bildungswärme *f* heat of formation, formation heat
atomare B. atomic (nuclear) heat of fission
Bildungswasser *n* natural water
Bildungswesen *n* / **technisches** engineering education
Bildunschärfe *f* lack of picture definition
Bildunterschrift *f* caption
Bildverdopplung *f* *(telev)* ringing
Bildverriegelung *f* / **horizontale und vertikale** picture lock
Bildverschiebung *f* image shift
Bildverstärker *m* video-frequency amplifier, vision (video) amplifier
Bildverstärkung *f* video amplification
Bildverstellung *f* framing
Bildvervielfachung *f* split image
Bildverzerrung *f* distortion of the image, field distortion
glockenähnliche B. cobs *(radar)*
Bildwand *f* projection screen
Bildwandler *m* image converter
Bildwandlerröhre *f* image [converter] tube
Bildwechsel *m* *(telev)* repetition of pictures
Bildwechselfrequenz *f* image frequency, frame repetition rate, *(Am)* frame frequency
Bildwechselimpuls *m* frame pulse
Bildweichheit *f* *(telev)* bloom
Bildwerfer *m* projector
Bildwerferlampe *f* projection lamp
Bildwerferraum *m* projection room (box)
Bildwiedergaberöhre *f* picture tube
Bildwinkel *m* angle of image
toter B. clip position, dead sector
Bildzahl *f* **je Sekunde** image frequency
Bildzählwerk *n* frame counter
Bildzeile *f* *(telev)* horizontal (scanning) line, horizontal scanning strip
Bildzeitbasis *f* *(telev)* image output
Bildzerleger *m* image dissector (analyzer); picture scanner
Bildzerlegung *f* analyzing of picture; scanning
Bildzerreißung *f* *(telev)* tearing
Bildzusammendrückung *f* *(el)* picture compression
Bilge *f* *(ship)* bilge
Bilgeabdeckung *f* limber boards
Bilgebrunnen *m* bilge well
Bilgedecke *f* limber board
Bilgehahn *m* bilge cock
Bilgeleitung *f* bilge main (pipe-line)
Bilgelenzbrunnen *m* bilge [drain] well
Bilgelenzeinrichtung *f* direct bilge suctions
Bilgelenzpumpe *f* direct bilge pump
Bilgepumpe *f* bilge pump
Bilgetank *m* / **begehbarer** conduit bilge tank *(on ocean-going ships)*
Bilgewasser *n* bilge-water
Bilgewasserentöler *m* oil (oily-bilge) separator
Bilgewegerung *f* limber boards, bilge ceiling
bilinear bilinear
Bilinearform *f* bilinear form *(polynomial)*
Billard *n* billiards
Billardqueue *n* [billiard] cue
Billardstock *m* [billiard] cue
Billen *pl* *(ship)* buttocks
Billetlochzange *f* [conductor's] ticket punch
Billitonit *m* billitonite
Biluxlampe *f* *(aut)* bilux (double-dipping, anti-dazzle) lamp
Bimetall *n* bimetal
Bimetallauslöser *m* bimetallic trip (releaser)
Bimetallfeder *f* bimetal spring
bimetallisch bimetallic
Bimetallmeßinstrument *n* thermo-bimetallic instrument
Bimetallplatte *f* bimetal plate
Bimetallregler *m* *(el)* bimetallic thermostat (thermostatic strip)
Bimetallstreifen *m* bimetallic strip
Bimetallstrommesser *m* bimetallic ammeter
Bimetalltemperaturfühler *m* bimetallic thermostat
bimolekular bimolecular
Bims *m* pumice
Bimsbeton *m* pumice concrete
bimsen to pumice; to fluff
Bimskies *m* pumice gravel
Bimsmaschine *f* fluffing machine
Bimsstein *m* pumice [stone]
Bimswalze *f* fluffing wheel
binär binary
Binäraddierer *m* binary adder
Binäraddierglied *n* binary adder
Binäraddierstufe *f* binary adder
Binärdarstellung *f* binary notation *(of numbers)*
Binär-Dezimalumsetzung *f* binary-to-decimal conversion
Binärelement *n* binary cell
Binärentscheidungsprogramm *n* binary-decision programme
Binärkode *m* *(dat)* binary code
fortschreitender B. progressive binary code
reflektierter B. Gray code
Binärkodeübersetzer *m* binary conversion equipment
Binärkodewandler *m* binary-code converter
binärkodiert binary-coded
Binärkomma *n* *(dat)* binary point
Binärschreibweise *f(dat)* binary notation
Binärschritt *m* *(dat)* binary digit, bit

Binärspeicherelement *n* two-state device
Binärstellenzahl *f* / **äquivalente** *(dat)* equivalent binary digit
Binäruntersetzer *m* binary (scale-of-two) counter
Binärverschlüsselung *f* binary coding
Binärzahl *f* binary number
Binärzähler *m* binary (scale-of-two) counter, radix 2-counter
Binärzählstufe *f* binary circuit, scale-of-two circuit
Binärzählwerk *n* binary counter
Binärziffer *f* [binary] bit
Binauraleffekt *m* binaural effect
Bindedraht *m (ag)* baling (binding) wire; *(build)* tie (splicing, lashing) wire
Bindefaden *m (text)* backing (binding) yarn
Bindefadenfutter *n* single yarn backing; fleecy fabric
Bindegarn *n* 1. *(text)* binder yarn; 2. *(ag)* binder twine; 3. lacing twine *(for cables)*
Bindeglied *n* connecting link
Bindekette *f (text)* binder warp
Bindekraft *f* adhesive force; adhesiveness; bonding strength
Bindelatte *f* joined lath
Bindemäher *m* [combined reaper and] binder
Bindemittel *n* binding medium, binding agent *(grinding wheel)*; binder; fixing agent; agglutinate; matrix; cement; agglomerant *(briquetting)*
  anorganisches B. cementing agent
  hydraulisches B. hydraulic binder
Bindemitteldosierung *f* binder batching
Bindemittelhaut *f* binder film
Bindemittelsilo *m* bulk cement silo
binden to bind *(sheaves; books)*; to tie, to bond *(chemical bond)*; to cement; to fix *(chemicals)*; to hoop *(a cask)*
  ein Faß b. to hoop a cask, to mount the staves
  fest b. to tie tightly
  Garben b. to tie (bind) sheaves
Bindepunkt *m* binding point; interlacing point
  gezeichneter B. raiser
Binder *m* 1. *(ag)* [combined reaper and] binder; 2. [roof] truss; girder; 3. header, bonder, bondstone *(brick)*; 4. cement; binder; matrix; agglomerant
Binderbarte *f* broad axe
Bindereif *m* binding hoop *(for casks)*
Bindereife *f (ag)* binding ripeness
Binderholz *n* barrel stave
Binderiemen *m* belt lace
Binderschicht *f* heading (bonding) course *(brick work)*
Bindersparren *m* binding rafter
Bindeschuß *m (text)* binder pick
Bindestrich *m* hyphen
Bindetechnik *f (text)* bonding technique
Bindetuch *n (ag)* binding (binder) cloth
Bindevorrichtung *f (ag)* binding apparatus
Bindewert *m* binding value
Bindfaden *m* twine
Bindfadenentferner *m* string catcher
bindig cohesive
Bindung *f* 1. [chemical] bond; linkage; 2. tying *(overhead line)*; 3. *(text)* weave; fabric structure
  B. der Oberware *(text)* top cloth weave
  B. der Schleifscheibe bond of the wheel
  B. der Unterware *(text)* bottom cloth weave
  B. für Doubles *(text)* backed cloth weave
  abgeleitete B. *(text)* derived weave
  anorganische B. inorganic (vitrified) bond
  doppelte B. *(text)* double tie
  elastische B. elastic bond *(grinding wheel)*
  gerippte B. *(text)* corded weave
  gratartige B. *(text)* twill interlacing
  heteropolare B. *(ch)* ionic linkage, electrovalence
  homöopolare B. *(ch)* atomic bond, covalence
  intermolekulare B. *(ch)* molecular association
  keramische B. vitrified (porcelain) bond *(grinding wheel)*
  kunstharzhaltige B. synthetic resin bonding, resinoid bonding
  naturharzhaltige B. resin bonding; rubber bonding *(grinding wheel)*
  organische B. organic bond *(grinding wheel)*
  polare B. polar bond
  schwer ausspringende B. *(text)* difficult intersection
  starre B. rigid bond (linkage)
  unpolare B. non-polar (covalent) bond
  zusammengesetzte B. *(text)* combination weave
Bindungsanalyse *f (text)* weave analysis
Bindungsart *f* construction of weave
Bindungselektronen *npl* shared electrons
Bindungsenergie *f* binding (bond) energy
  B. des Kerns nuclear binding energy, nucleus bond energy
  B. des Neutrons neutron binding energy
Bindungsfähigkeit *f* bondability
Bindungsfehler *m* bonding defect
Bindungsfestigkeit *f* bonding strength
Bindungsform *f* form of bond (linkage)
Bindungshärte *f* **der Schleifscheibe** strength of [the] bond of the [grinding] wheel
Bindungslehre *f (text)* theory of interlacing
Bindungsrapport *m (text)* weaving repeat
Bindungswärme *f* combination heat
Bindungswechsel *m* change of weave
Binnenachtersteven *m (ship)* inner [stern-]post, sternson, stern rider
binnenbords *(ship)* inboard
Binnendruck *m* intrinsic pressure
Binnenfahrt *f (ship)* inland navigation (sailing)

Binnenfahrzeug *n (ship)* inland vessel
Binnengang *m (ship)* inside strake
Binnengewässer *n* inland water
Binnenhafen *m* basin [of a port], inland harbour (port), inner harbour (port)
Binnenhaupt *n* upper gates *(sluice)*
Binnenjolle *f* open inland centre-boarder, centre-board racer, dinghy
Binnenkiel *m (ship)* keelson
Binnenklüver *m (ship)* inner jib
Binnenleichter *m* / **flachgehender** *(ship)* gundalow
Binnenmeer *n* inland sea
Binnenschiff *n* inland (river) vessel, home trade ship
Binnenschiffahrt *f* internal (interior, inland, inshore) navigation, inland trade
Binnenschiffahrtskanal *m* navigation canal
Binnenschiffahrtsnetz *n* network of inland waterways
Binnenschiffahrtsweg *m* inland waterway
Binnenschiffer *m* waterman, boatman
Binnensee *m* inland lake
Binnenseekahn *m* inland lake scow
Binnentarif *m* internal tariff
Binnenverkehr *m* internal (domestic) traffic
Binnenvordersteven *m (ship)* stem rider, apron
Binnenwasserstraße *f* inland waterway
Binnenwasserstraßenordnung *f (Am)* Inland Rules
Binnenzollamt *n* inland customs house
Binode *f (el)* binode
binokular binocular
Binokularperiskop *n* binocular periscope
Binom *n* binomial
Binomialformel *f* binomial formula
Binomialkoeffizient *m* binomial coefficient
Binomialpapier *n* binomial probability paper
Binomialreihe *f* binomial series
Binomialtorder *m (el)* binomial twist
Binomialverteilung *f* binomial distribution
Binomialwinkelsystem *n (el)* binomial corner
Binomialzerlegung *f* binomial expansion
binomisch binomial
Binormale *f* binormal
Binse *f* rush
Binsenmatte *f* rush mat
Biobatterie *f* bio-battery
Biochemie *f* biochemistry
biochemisch biochemical
bioelektrisch bioelectric
Bioelektrizität *f* bioelectricity
Bioenergiewandler *m* bio-power unit
Biokatalysator *m* biocatalyst
biologisch biological
Biolumineszenz *f* bioluminescence
Bionik *f* bionics
Biophysik *f* biophysics
Biose *f* disaccharose
Biosolarelement *n* bio-solar cell
Biosynthese *f* biosynthesis
Bioverstärker *m (dat)* bioamplifier
bipolar bipolar
Bipolarkoordinate *f* bipolar coordinate
Bipolarsystem *n* bipolar electrode system
Bipyramide *f* bipyramid
biquadratisch biquadratic
Biquinärkode *m (dat)* biquinary code
Biquinärverschlüsselung *f (dat)* biquinary coding
Birke *f* birch
Birkenholz *n* birch wood
Birkenrindenöl *n* birch oil
Birkenteer *m* birch tar
Birne *f* 1. *(met)* converter; 2. punching bag *(boxing)*; 3. [incandescent] bulb
birnenförmig pear-shaped
Bisamratte *f* musk rat, musquash
Bisamrattenfell *n* musquash
Bisektrix *f* bisectrix
Biskuitporzellan *n* biscuit porcelain
Bismit *m (min)* bismite *(bismuth ochre)*
Bismuthin *m (min)* bismuthinite, bismuth glance
bistabil bistable
Bisulfit *n* hydrosulphite, bisulphite
Bit *n (dat)* bit, binary digit
  höchstes B. most significant digit
Bitterholz *n* quassia
Bittermandelöl *n* bitter almond oil
Bittermandelölgrün *n* malachite green
Bittersalz *n* magnesium sulphate, bitter salt, Epsom salts
Bitterspat *m* bitter spar
Bitterstoff *m* bitter principle
Bitumen *n* bitumen
  dampfraffiniertes B. steam refined bitumen
  langsamabbindendes B. slow-curing bitumen
  schnellabbindendes B. rapid-curing bitumen
Bitumendachpappe *f* bituminous (asphalt) roofing felt
Bitumendecke *f* asphaltic bitumen pavement
Bitumenfarbe *f* asphaltic bitumen paint
bitumengetränkt bitumen-impregnated
bitumenhaltig bituminous; bituminiferous
Bitumenkanone *f* bitumen gun
Bitumenkohle *f* bituminous coal
Bitumenlack *m* bituminous varnish
Bitumenpappe *f* bituminous (asphalt) felt
Bitumenpumpe *f* asphaltic bitumen pump
bituminös bituminous
Biuretreaktion *f* biuret reaction
bivariabel bivariant
Bivibrator *m* bivibrator, bistable multivibrator
bizirkular bicircular
bizyklisch *(ch)* bicyclic
blähen to swell; to inflate

Blähgrad *m* swelling capacity
Blähschieferton *m* expanded slate (shale)
Blähton *m* expanded clay
Blähvermögen *n* swelling capacity
Blanchiereisen *n* whitening slicker
blanchieren to whiten
Blanchiermaschine *f* whitening machine
Blanchierspäne *mpl* whitenings
Blanchierstahl *m* whitening steel
blank bright *(metal)*; *(el)* bare, naked, uninsulated *(wire)*
Blankdraht *m* blank [metallic welding] wire; naked (bare) wire
Blankdrahtleitung *f* open wire line
Blanket *n s.* Brutmantel
Blankglühen *n* bright annealing
Blankglühofen *m* scaling (bright-annealing, non-oxidizing annealing, wash heating) furnace
Blankhammer *m (ag)* scythe whetting hammer
Blankhärteöl *n* non-oxidizing annealing oil
Blankkupfer *n* bare copper
Blankleder *n* harness leather
Blankofahrausweis *m* blank-to-blank ticket, blank coupon
Blankstahl *m* cold-drawn steel; bright steel
Blankverdrahtung *f* bare[-wire] wiring, *(Am)* strapping
Blankwalzöl *n* straight mineral oil
Blankziehen *n* bright-drawing
Blasdruck *m* blowing pressure
Blase *f* bubble *(water)*; void, blowhole, blister *(metal)*; matrass *(distilling)*; *(ch)* still
Blasebalg *m* [smith's] bellows
Blaseeinrichtung *f* / **magnetische** magnetic blow-out
Blasenbahn *f* **[des Torpedos]** wake of the torpedo
blasenbildend blister-forming, vesicant
Blasenbildung *f* cavitation; blistering
Blasendestillation *f* batch distillation
Blasengewebe *n* blister fabric
Blasenkammer *f* bubble chamber
Blasenkavitation *f* bubble cavitation
Blasenkupfer *n* blister copper
Blasenloch *n* pit hole
Blasenraum *m* cavity
Blasensieden *n (nucl)* bulk boiling
Blasenstahl *m* blister steel
Blasenströmungsmesser *m* bubble flow meter
Blasenverdampfung *f (nucl)* bulk boiling
Blasenzähler *m* bubble counter
Blasform *f* blow die (mould); tuyere
Blasformebene *f* tuyere level
blasig blown *(casting)*
Blasigkeit *f* blistering *(coated papers)*
Blaslunker *m* blow hole
Blasmagnet *m* arc deflector, blowing magnet, magnetic blow-out
Blasofen *m* wind furnace
Blasrohr *n* muffle
Blasspule *f* blow-out coil
Blasversatz *m (mine)* pneumatic stowing
Blasversatzmaschine *f* pneumatic stowing machine
  B. mit angebautem Bergebrecher crusher stower
Blatt *n* 1. leaf *(tree)*; 2. blade *(saw, file; oar, propeller)*; 3. sheet *(paper)*; foil; lamella; 4. halved (half) joint *(carpentry)*; 5. *(text)* vamp; reed *(loom)*
  B. der Kratze *(text)* card sheet
  fliegendes B. loose reed
  gerades B. half (halved) joint *(carpentry)*
  schräges B. oblique scarf with butt ends *(carpentry)*
Blattaluminium *n* aluminium foil, leaf aluminium
Blattanschlag *m (text)* beat of the reed
Blattbildung *f* formation of sheet
Blattbinden *n (text)* reed binding
Blattbreite *f* blade width *(sawblade)*; *(text)* reed space *(loom)*
Blättchen *n* lamina
blättchenförmig lamellar
Blattdichte *f (text)* set of the reed
Blattdrucker *m* page printer *(telegraphy)*
Blatteinstellung *f (text)* set of warp in the reed
Blattelektroskop *n* leaf electroscope; gold-leaf electroscope; aluminium-leaf electroscope
Blättererz *n* nagyagite
Blätterkohle *f* laminated (leaf) coal
Blättertellurerz *n* nagyagite
Blätterton *m* foliated clay
Blattfaser *f* leaf fibre
Blattfeder *f* leaf spring
  einfache B. single-leaf spring
  geschichtete B. laminated (leaf) spring
Blattfederhammer *m* spring hammer
Blattfederventil *n* laminated spring valve
Blattfläche *f* blade area *(airscrew)*
Blattfliegereinrichtung *f (text)* loose reed mechanism
Blattfuge *f* groove of lay for the reed
Blattgold *n* gold foil (leaf), beaten gold
  echtes B. foliated (leaf) gold
Blattgröße *f* sheet size, size of drawing paper *(technical drawing)*
Blatthalter *m* blade holder *(saw)*
Blatthöhe *f (text)* depth of reed
Blattkohle *f* slate coal
Blattkupfer *n* copper sheet
Blattkurve *f (math)* folium
Blattleiste *f (text)* reed baulk
Blattmeißel *m* drag (scraper) bit
Blattnumerierung *f (text)* reed counting

Blattrost *m* grate consisting of thin bars
Blattrücken *m* back *(saw)*
  geschweifter B. skew back blade *(saw)*
Blattschreiber *m* page printer
Blattstechen *n (text)* drawing in the reed, reeding
Blattstecher *m (text)* reed hook
Blattstechmaschine *f (text)* denting machine
Blattstreifen *m* reed mark *(cloth)*
Blatttragfeder *f* suspension spring
Blattvergoldung *f* gilding with gold leaf, gold blocking
Blattvorderseite *f* blade face
Blattzapfen *m* scarf tenon, foot of the spoke
Blattzinn *n* tin foil
Blau *n* blue [colour]
  Berliner B. Berlin (Prussian) blue
Blaublech *n* blue steel plate
Blaubrenne *f* blue dip
Blaubruch *m* blue brittleness
blaubrüchig blue-short
Blaubrüchigkeit *f* blue brittleness (shortness)
Bläuen *n* blueing
Blauerde *f (min)* blue earth
Blauerz *n* brown ore
Blaufäule *f* blue stain *(timber)*
Blaufäulepilz *m* blue-sap fungus
Blaufuchs *m* blue fox
blauglühen to open-anneal
Blauglühen *n* open annealing
Blaupause *f* blue print
Blaupausen *n* blue printing
Blaupauspapier *n* blue-print paper
Blaupilz *m* blue-sap fungus *(timber)*
Blausäure *f* hydrocyanic (prussic) acid
Blauspat *m (min)* lazulite
Blausprödigkeit *f* blue (ferrite) brittleness
blaustichig blue tinged (stained) *(wood)*
Blauwärme *f* blue heat range
Blech *n* plate; metal sheeting *(sheets)*; sheet metal
  B. zweiter Wahl mender
  buckliges B. jumped sheet
  dickes B. plate
  doppelseitig plattiertes B. sandwich metal
  dünnes B. thin sheet
  kunststoffplattiertes B. skin-plate
Blechabfall *m* plate scrap
Blechausstechbohrer *m* fly cutter [for holes in thin metal]
Blechbearbeitung *f* sheet metal working
Blechbearbeitungsbetrieb *m* plate shop
Blechbearbeitungsmaschine *f* sheet metal working machine
Blechbearbeitungsmaschinen *fpl* sheet metal machinery
Blechbiegemaschine *f* plate-bending machine
Blechbohrer *m* sheet drill, hole cutter
Blechbördelmaschine *f* sheet bordering machine, plate flanging machine
Blechbüchse *f* tin canister (box), tin [can]
Blechdicke *f* gauge of a sheet
Blechdickenmesser *m* sheet-iron gauge
Blechdoppler *m* plate doubler (doubling machine)
Blechdose *f s.* Blechbüchse
Blechdruck *m (print)* tin-printing
Blechdruckfarbe *f* tin-printing ink
Blechdruckmaschine *f* tin-printing machine
Blechduo *n* two-high plate mill
Blechfalzmaschine *f* seaming machine
Blechfangschale *f* **für Tropföl** sheet metal coolant tray
Blechformmaschine *f* sheet metal forming (working) machine
Blechfundament *n* sheet-iron foundation
Blechglühofen *m* plate heating furnace
Blechhalter *m* blank holder, hold down *(deep drawing)*
Blechhandschere *f* sheet metal workers' hand shears, snips
Blechhaube *f* sheet-iron casing *(chimney)*
Blechhülse *f* sheet metal tube
Blechkantenhobelmaschine *f* edge planer, plate edge planing machine
Blechkantennachformarbeit *f* plate edge profiling [work]
Blechkastenträger *m* box plate girder
Blechkern *m (el)* laminated core
Blechlehre *f* plate gauge; sheet metal gauge
Blechpaket *n* laminated sheet[-iron] plate; sheet pack; stacked sheets
Blechradscheibe *f* wheel disk stamped out of a plate
Blechrichtmaschine *f* sheet straightening machine
Blechrichtpresse *f* plate straightening press
Blechschablone *f* sheet metal templet
Blechschere *f* snip; plate cutting (shearing) machine, plate shears (cutter)
Blechschließung *f* clamping together [of] the plates *(rivets)*
Blechschornstein *m* iron plate chimney
Blechschraube *f* sheet metal screw
Blechschrott *m* sheet (plate) scrap
Blechstanzpresse *f* sheet stamping press
Blechstraße *f* / **schwere** plate mill
Blechteil *n* sheet metal part
Blechträger *m* [steel] plate girder
Blechverkleidung *f* sheet metal lagging
Blechwalzstraße *f* sheet rolling mill
Blechwalzwerk *n* sheet mill
Blei *n* lead
  arsensaures B. lead arsenate
Bleiabgang *m* lead dross (scoria)
Bleiabschirmung *f (nucl)* lead screen (shielding)

Bleiabstrich *m* lead-scum
Bleiakkumulator *m* lead accumulator (storage battery, cell)
Bleialter *n (nucl)* lead age
Bleiantimonglanz *m (min)* zinkenite
Bleiäquivalent *n (nucl)* lead equivalent
Bleiarbeit *f* lead-smelting; plumbing
Bleiarsenat *n* lead arsenate
Bleiarsenglanz *m (min)* arsenical lead
bleiartig plumbeous
Bleiasche *f* lead dross (suboxide)
Bleiauskleidung *f* lead lining
Bleiazetat *n s.* Bleizucker
Bleiazid *n* lead azide
Bleibad *n* lead bath
Bleibadhärtung *f* lead hardening
Bleibadofen *m* lead bath furnace
Bleibadpatentierung *f* lead-bath patenting
Bleibarren *m* lead pig
bleiben to remain; to stay; to be permanent *(strain)*
  am Apparat b. *(telec)* to hold on, *(sl)* to hang on
  auf Empfang b. to stand-by
  außerhalb [der Dreimeilenzone] b. to keep the offing *(vessel)*
  in der Leitung b. *(telec)* to hold on (the line), *(sl)* to hang on
  in der Luft b. to remain aloft *(aeroplane)*
  in Landsicht b. to keep the land aboard *(vessel)*
  unter Land b. to keep to the shore *(vessel)*
bleibend permanent, plastic *(strain)*
Bleibenzin *n (aut)* leaded petrol
Bleibergwerk *n* lead mine
Bleiblech *n* sheet lead; lead sheet
Bleiblock *m* lead pig
Bleiblockausbauchung *f* bulging of lead blocks *(used for testing explosives)*
Bleichanstalt *f* bleaching works (home), bleachery
Bleichartikel *mpl* bleached goods
Bleichbarkeit *f* bleachability
Bleiche *f* 1. bleaching; 2. bleaching ground
  chemische B. chemical bleaching
  natürliche B. grass bleaching
  optische B. optical bleaching
bleichecht fast to bleaching
Bleichechtheit *f* fastness to bleaching
bleichen to bleach
Bleichen *n s.* Bleiche 1.
Bleicherde *f* active (bleaching, fuller's) earth, podsol, discolouring (percolation) clay
  aktivierte B. activated earth
Bleicherdebehandlung *f* clay treatment
Bleichflotte *f* bleaching bath; bleaching liquor
Bleichhilfsmittel *n* bleaching assistant
Bleichholländer *m* bleaching engine *(paper)*
Bleichkalk *m* bleaching lime
Bleichlauge *f* bleaching liquid, chlorate liquor
Bleichlaugenbereitung *f* bleach manufacture
Bleichlaugenzubereitungsanlage *f* bleach mixing plant
Bleichlorid *n* lead chloride
Bleichmittel *n* bleach[ing agent]
Bleichmittelbedarf *m* bleaching requirements
Bleichplatz *m* bleaching ground
Bleichpulver *n* bleaching powder
Bleichromat *n* lead chromate
Bleichstabilisator *m* bleaching stabilizer
Bleichton *m* active (bleaching) earth, active (bleaching) clay
Bleichturm *m* bleaching tower
Bleichverfahren *n* bleaching process
Bleichware *f* bleached goods
Bleidach *n* lead covering, plumb roofing
Bleidichtung *f* lead joint
Bleidioxid *n* lead dioxide
Bleieisenstein *m* iron-lead matte
Bleiempfindlichkeit *f* lead susceptibility; lead response *(fuel)*
Bleierz *n* lead ore
Bleiessig *m* lead subacetate
Bleifarbe *f* lead paint
Bleifassung / mit leaded; connected by leaden cames
Bleifolie *f* lead foil
bleifrei unleaded
Bleigelb *n* massicot, yellow lead
Bleigießer *m* plumber
Bleiglanz *m* lead glance, galena
Bleiglanzdetektor *m* galena detector
Bleiglas *n* lead glass
Bleiglasur *f* lead glazing
Bleiglasziegel *m* lead-glass brick
Bleiglätte *f* litharge, massicot
Bleigleichwert *m (nucl)* lead equivalent
Bleigrube *f* lead mine
Bleigummischürze *f* body (lead-rubber, protective) apron
Bleiguß *m* lead casting
bleihaltig plumbiferous
Bleihülle *f* lead sheath
Bleihütte *f* lead smelting plant
Bleikabel *n* lead-covered cable
  B. mit Korrosionsschutz served lead-covered cable
Bleikammer *f* lead chamber
Bleikammerverfahren *n* lead chamber process
Bleikelle *f* lead ladle
Bleikrätze *f* lead dross
Bleikristall *m* lead crystal
Bleilagermetall *n* lead-base bearing metal
Bleilagerstätte *f* lead deposit
Bleilasur *f (min)* linarite
Bleilegierung *f* lead alloy
Bleilot *n* lead solder
Bleimantel *m* lead sheath (sheathing, coating)

Bleimantelkabel *n* lead-covered cable, lead-sheathed cable
Bleimine *f* lead *(plumbago stick)*
Bleimuffe *f* lead sleeve
Bleioriginal *n* original pencilled drawing
Bleioxid *n* lead oxide
Bleipflaster *n* lead plaster, diachylon
Bleiraffinerie *f* lead refinery
Bleirohr *n* lead tube (pipe)
Bleisammler *m* lead accumulator (storage battery)
überlaufsicherer B. *(el)* unspillable lead acid rechargeable battery
Bleischachtofen *m* lead blast-furnace
Bleischaum *m* lead dross (ashes)
Bleischwamm *m* lead sponge, mossy lead
Bleiskizze *f* pencil sketch
Bleispiegel *m* specular galena
Bleisteg *m (print)* lead furniture
Bleistein *m* matte of lead, leady matte
Bleistift *m* lead pencil
Bleistifthärteversuch *m (el)* pencil hardness test
Bleistiftholz *n* pencil cedar
Bleistiftspitzmaschine *f* pencil sharpener (sharpening machine)
Bleistiftzeichnung *f* pencil drawing
Bleistrahlenschutzschicht *f* lead shielding
Bleitetraäthyl *n* tetraethyl lead
Bleiumhüllung *f* lead sheathing *(cable)*
Bleiunterlage *f* lead base
Blei(II)-verbindung *f* plumbous compound
Blei(IV)-verbindung *f* plumbic compound
Bleivergiftung *f* lead poisoning
Bleiverschluß *m* seal *(wagons)*
Bleiwalzwerk *n* lead mill
Bleiweiß *n* white lead
Bleiwolframat *n* lead tungstate
Bleizeichnung *f* pencilled drawing
Bleiziegel *m* lead brick
Bleizucker *m* lead acetate, sugar (acetate) of lead
Blendbogen *m* blind arch
Blenddeckel *m* false cover
Blende *f* 1. stop; *(opt)* aperture; [matching] diaphragm; orifice; 2. blende *(ore)*
zylindrische B. *(Am)* hat orifice
blenden to dazzle *(motor car)*
Blendendurchflußmengenmesser *m* orifice plate flow meter
Blendeneinstellung *f* stop adjustment; iris control
Blendenöffnung *f* aperture
Blendenraum *m* / **feldfreier** field grid space
Blendenring *m* aperture plate
Blendentheorie *f* theory of stops
Blendholz *n* facing board
Blendlicht *n* glare
Blendschutz *m* anti-glare device *(motor car)*
Blendschutzglas *n* anti-dazzle glass
Blendschutzlampe *f* anti-dazzle lamp
Blendschutzscheibe *f* anti-glare screen, anti-dazzle glass *(motor car)*
Blendung *f* glare
akustische B. aural dazzle (dazzling)
Blendungswinkel *m* angle of glare
Blendwirkung *f* glare effect
Blickebene *f* visual plane; plane of vision
Blickfeld *n (opt)* field of vision (view), range of vision
Blickrichtung *f* line of vision; visual line
Blicksilber *n* lightened (refined) silver
Blickwinkel *m* angle of view
blind blind *(mine)*; *(build)* false; blank
Blindanflugverfahren *n* blind approach procedure
Blindband *m (print)* dummy
Blindbelegung *f* dummy connection
Blindboden *m* false floor
Blinddeckel *m* false cover
Blindendruck *m* Braille printing
Blindenergie *f* wattless (reactive) energy
Blindenschrift *f* Braille
Blindflansch *m* 1. blank flange; 2. *(mine)* blind shaft
Blindflug *m* blind (instrument) flight; blind (instrument) flying
Blindgänger *m* dud
Blindgußzeile *f* blank slug
Blindhärteversuch *m* blank hardness test
Blindholz *n* ordinary (base) wood; inferior wood
Blindkaliber *n* dead pass
Blindkochung *f* blank cook
Blindkomponente *f* reactive (quadrature, idle, wattless) component
Blindlandeanlage *f* blind landing aid
Blindlandebake *f* instrument landing approach beacon
Blindlandeeinrichtung *f* radio approach system
Blindlandefeuer *n* blind approach beacon
Blindlandung *f* blind (instrument) landing
Blindlast *f (el)* reactive load
Blindlastfaktor *m* capacity factor
Blindlastregler *m* reactive load regulator
Blindlasttransduktor *m* transductor reactor
Blindlastzählung *f* reactive metering
Blindlauf *m* racing *(propeller)*
Blindleistung *f (el)* reactive (blind, wattless) power, reactive factor (volt-amperes), susceptance
B. der gegenläufigen Komponente negative-sequence reactive power
B. des mitläufigen Systems positive-sequence reactive power
Blindleistungseinheit *f* unit of reactance output
Blindleistungsfaktor *m* reactive factor
Blindleistungsgenerator *m* reactance oscillator

Blindleistungsmaschine *f* reactive power generator
Blindleistungsmesser *m* varmeter, idle current wattmeter, wattless-power meter
Blindleistungsrelais *n* reactive power relay
Blindleitung *f (el)* non-dissipative stub; variable reactance line
 B. zur Anpassung non-dissipative stub
Blindleitwert *m* susceptance
Blindmaterial *n (print)* furniture, spacing (blank) material
Blindnut *f* empty slot
Blindort *n (mine)* blind drift, dead end, dean, dummy road
Blindortversatz *m* dummy-road packing
Blindpermeabilität *f (el)* reactive permeability
Blindprägung *f (print)* blind blocking, relief embossing
Blindprobe *f* blank *(analytical chemistry)*
Blindröhre *f* reactance valve
Blindschacht *m (mine)* staple shaft
 tonnlägiger B. sub-incline shaft
Blindschaltbild *n (el)* mimic diagram, diagram board, graphic panel
Blindschwanz *m (el)* stub
Blindsegment *n* idle segment
Blindspannung *f* reactive voltage
Blindstecker *m* dummy plug
Blindstich *m* blind stitch *(sewing machine)*; blind pass *(rolling)*
Blindstöpsel *m* dummy plug
Blindstrecke *f (mine)* blind drift
Blindstrom *m (el)* reactance (idle, wattless) current
Blindstromamperemeter *n* *s.* Blindstrommesser
Blindstromkompensation *f* power-factor compensation
Blindstromkondensator *m* phase advancer (advancing capacitor)
Blindstrommesser *m* idle current meter
Blindstromzähler *m* reactive current meter, wattless component meter
Blindstutzen *m* dummy plug
Blindverbrauchszähler *m (el)* reactive energy meter, var-hour meter, wattless component meter
Blindversuch *m (ch)* blank trial (experiment); *(nucl)* cold run
Blindwalze *f* dummy (dead) roll
Blindwiderstand *m (el)* reactance, reactive impedance
 B. für das Nullsystem zero-sequence reactance
 B. für die gegenläufige Komponente negative-sequence reactance
 akustischer B. acoustical reactance
 gemeinsamer B. common reactive element
 induktiver B. inductance, magnetic (inductive, positive) reactance
 kapazitiver B. capacitance, condensive (capacitive, negative) reactance
 mechanischer B. mechanical reactance
 nichtlinearer B. non-linear reactance
Blindwiderstandsröhre *f* reactance valve
blinken to flash; to indicate the direction; to blink; to flicker
Blinker *m* flashing-light direction indicator, flashing direction indicator lamp *(motor car)*
Blinkfeuer *n* flashing light
Blinkgeber *m* flasher unit
Blinkgerät *n* blinker light; flash-light; signal lamp
Blinkkennfeuer *n* group flashing light
Blinkleuchte *f* direction-indicator lamp
Blinklicht *n (ship)* intermittent (flashing) light
Blinklichtanlage *f* flashing direction indicating equipment
Blinkmotor *m* flasher motor
Blinkschaltung *f* ratchet circuit
Blinkscheuche *f (ag)* bird scarer
Blinkschlußleuchte *f* combined flasher and tail lamp
Blinksignal *n* flashing (flash-light) signal
Blinkzeichengeber *m* flasher
Blitz *m* lightning; flash
Blitzableiter *m* lightning conductor (rod)
Blitzbahn *f* track (path) of lightning
Blitzentladung *f* lightning discharge
Blitzfeuer *n* quick flashing light
 unterbrochenes B. occulting quick flashing light
Blitzgerät *n* flash-light unit *(photography)*
Blitzgespräch *n* flash *(priority use of trunk telephone)*; lightning call
Blitzlampe *f* photo-flash lamp
Blitzleuchte *f* flash gun
Blitzlicht *n* flash-light, magnesium light
Blitzlichtaufnahme *f* flash-light photograph
Blitzlichtkontakt *m* X-contact
Blitzlichtlampe *f* photo-flash lamp
Blitzlichtpulver *n* flash-light powder
Blitzrate *f* flashing rate
Blitzröhre *f* flash tube *(gas discharge)*
Blitzschlag *m* lightning stroke
Blitzschutz *m* lightning protective system; lightning protection (proofing)
Blitzschutzerde *f* protective earth
Blitzschutzrelais *n* lightning arrester relay
Blitzschutzschalter *m* lightning switch
Blitzschutzspirale *f* inductance spiral
Blitzschutzvorrichtung *f* lightning protector
Blitzstrom *m* lightning current
Blob *n (nucl)* blob
Bloch-Funktion *f (ph)* Bloch function
Bloch-Welle *f (ph)* Bloch wave

Block *m* 1. *(mine, build, ry)* block; 2. *(dat)* block *(some machine words carried as a whole)*; 3. pad *(paper)*; 4. *(met)* ingot; bloom; bar; slab
ausgelaufener B. bled ingot
unberuhigter (unberuhigt vergossener) B. rimming ingot
vorgewalzter B. rough rolled ingot
Blockabschnitt *m (ry)* block section
Blockabstechmaschine *f* ingot parting machine
Blockabstreifer *m (met)* ingot stripper
Blockade *f (print)* turned letter
Blockadresse *f* block address *(programming)*
Blockanlage *f (ry)* block signal post
Blockaufsatz *m* hob top
Blockausdrücker *m* ingot pusher
Blockauszieher *m* ingot withdrawing device
Blockbandsäge *f* band saw for logs
Blockbauweise *f* large[-sized] block construction
Blockblei *n* pig lead
Blockbruchbau *m (mine)* block (undercut) caving
Blockdiagramm *n (cont)* block diagram
Blockdruck *m (text)* block print
Blockdrücker *m* furnace pusher
Blockeinsetzkran *m* ingot charging crane
Blockeinsetzwagen *m* ingot charging bogie
Blockende *n* log end
unteres B. butt on an ingot
Blockfett *n* briquetted grease
Blockform *f* ingot mould
Blockfundament *n* foundation block, pad foundation
Blockgatter *n* saw mill with one saw
Blockgerüst *n* trestle work
Blockgreifer *m (met)* ingot gripper
Blockgußstück *n* en bloc casting
Blockhaus *n* log house (cabin)
Blockheftung *f* block stitching
Blockhobelmaschine *f* ingot planing machine
Blockierdraht *m* inhibit wire
blockieren to block; *(el)* to interlock; to jam *(road; passage)*
Blockierschaltung *f* *(dat)* clamping circuit
Blockierung *f* blocking; locking; *(el)* interlock[ing] *(machine)*
Blockkaliber *n* roughing (blooming, cogging) pass *(rolling)*
Blockkasten *m* monobloc container [of storage battery]
Blockkipper *m* ingot tipper
Blockklemme *f (el)* terminal block
Blockkokille *f* ingot mould
Blockkondensator *m* block[ing] capacitor
Blockkonstruktion *f* unit construction
Blocklocher *m (off)* gang punch
Blockmaß *n* log measure
Blockmetall *n* ingot metal
Blockpolymerisation *f* bulk polymerization
Blockreifen *m* block tyre
Blockrundmaß *n* log measure
Blocksäge *f* pit (log) saw
Blockschaltbild *n (el)* block diagram [of circuit]; symbolic circuit
Blockschaum *m* ingot scum
Blockscheibe *f* sheave
Blockschere *f* bloom shears, bar-cutting machine
Blockschiff *n* hulk, raft, dismantled ship
Blockschrift *f* block (capital) letters
Blockschrott *m* heavy scrap
Blockschwinger *m* shock-excited oscillator
Blockseigerung *f* major (mass) segregation, *(Am)* ingotism
Blocksignal *n (ry)* block signal
Blocksilber *n* silver bullion
Blockstein *m* building block
Blockstelle *f (ry)* block post
Blockstoßofen *m* pusher-type billet heating furnace
Blockstraße *f (met)* cogging mill train
Blockstrecke *f (ry)* block
geblockte B. closed block
Blockstromkreis *m* clearing circuit
Blockstufe *f* massive tread
Blocktransportvorrichtung *f* ingot conveying device
Blocktrio *n* three-high blooming mill
Blockübersicht *f* block record
Blockverband *m* English bond *(brickwork)*
Blockwagen *m* ingot chariot (buggy)
Blockwalze *f* blooming (cogging) roll
Blockwalzwerk *n* blooming (cogging) mill
Blockwand *f* log wall
Blockware *f* / **unbesäumte** logs sawn through and through
blockweise in full cut logs
Blockzange *f* ingot tongs (dog)
Blockzinn *n* block tin
Blockzurichterei *f* ingot dressing shop
Blöße *f (leath)* pelt
Blötwolle *f* carrion (skinners') wool
Blühen *n (text)* blooming
Blume *f* 1. flower; 2. *(text)* bloom *(bath)*
Blumendraht *m* binding wire
Blumenschere *f* flower gatherer
Blumenseite *f (leath)* outside of the skin
Blumentopf *m* flower pot
Blumenzwiebelheber *m* bulb lifter
Blumenzwiebelpflanzmaschine *f* bulb planter
Blutbild *n* blood picture (count)
Blutbilduntersuchung *f* blood count
Blutdruckmessung *f* blood pressure measurement
Blütenfarbstoff *m* colouring material of flowers
Blütenöl *n* flower oil
Blütenstand *m* inflorescence
Blutersatzmittel *n* plasma expander
Blutkörperchenzählung *f* blood count

Blutlaugensalz *n* prussiate of potash
gelbes B. potassium ferrocyanide, yellow prussiate of potash
rotes B. potassium ferricyanide, red prussiate of potash
Blutmehl *n* dried blood, blood meal
Blutprobe *f* sample of blood
Blutsenkung *f* sedimentation test
Blutstein *m s.* Hämatit
blutstillend styptic
BMSR-Technik *f* process instrumentation and control engineering
Bö *f* gust; squall, flurry
schwarze B. black squall
Bob *m* bobsleigh, bobsled
Bobbahn *f* bob run
Bobfahrer *m* bobman
Bobine *f* bobbin *(spool)*
Bobinenfördermaschine *f* *(mine)* reel-drum hoist
Bobinetmaschine *f* *(text)* bobbinet frame
Bock *m* 1. bent, trestle; horse *(scaffolding)*; stand; 2. vaulting buck
Bockbrücke *f* trestle bridge
Bockgestänge *n* A-type pole
Bockkran *m* gantry crane (hoist), portal (frame) crane
Bocklagermotor *m* pedestal-type motor
Bockleiter *f* double-ladder
Bockprahm *m* shear hulk
Bockwolle *f* ram's wool
Bodediagramm *n* *(cont)* log-magnitude and phase diagram, Bode-diagram
Boden *m* 1. ground; bottom; plate, stage *(column)*; floor; 2. soil; earth; 3. attic; garret; loft
B. einer Rektifizierkolonne rectifying tray
abgeknickter B. *(ship)* edge bottom
angeschwemmter B. alluvial deposits
aufgeschütteter B. filled (made-up) ground
bewachsener B. foul bottom *(of a vessel)*
bindiger B. cohesive soil
doppelwandiger B. double (hollow-cast) cover
einfacher B. single bottom
einwandiger ebener B. single flat cover
einwandiger konischer B. single conical cover
gewachsener B. unmade ground
neutraler B. neutral bottom
nichtbindiger B. cohesionless (non-cohesive, granular) soil
rolliger B. granular soil
steiniger B. stony ground
theoretischer B. theoretical stage (plate) *(distillation)*
Bodenabdeckplatte *f* ground plate
Bodenabdichtung *f* earth (soil) waterproofing
Bodenabfluß *m* soil pipe
Bodenabrichtmaschine *f* head dressing machine
Bodenabstand *m* ground clearance; road clearance *(vehicle)*
Bodenanalyse *f* soil analysis
Bodenanlage *f* ground installation[s] (equipment)
Bodenanteil *m* 1. soil fraction; 2. *(ag)* share of soil *(put into a co-operative farm)*
Bodenantwortstation *f* responder beacon
Bodenauftrag *m* filled ground, filling
Bodenaushub *m* excavation
Bodenbeanspruchung *f* ground pressure
Bodenbearbeitung *f* *(ag)* soil cultivation, cultivation of the soil; tilling of land
oberflächliche B. surface (shallow) cultivation
Bodenbearbeitungsgerät *n* *(ag)* implement used to till the land *(or to prepare seed beds)*
Bodenbearbeitungsmaschine *f* *(ag)* machine used to till the land *(or to prepare seed beds)*
Bodenbedeckung *f* *(ag)* mulch, carpet, ground cover
Bodenbegasung *f* *(ag)* soil fumigation
Bodenbelag *m* *(build)* floor covering, flooring; *(ship)* bottom covering
rutschfester B. non-skid flooring
Bodenbeplankung *f* *(ship)* bottom planking
Bodenbeplattung *f* *(ship)* bottom plating
Bodenbeschaffenheit *f* *(ag)* soil condition; nature of the soil
Bodenbeschlag *m* *(ship)* bottom sheathing
Bodenbeutel *m* flat-bottom bag
Bodenbewässerung *f* soil irrigation
Bodenbewegung *f* earth displacement (movement), earthmoving
Bodenblech *n* bottom plate
Boden-Boden-Rakete *f* surface-surface-missile, SSM
Boden-Bord-Verbindung *f* *(el)* point-to-ship communication
Bodenbrett *n* foot (bottom) board *(in a boat)*; head piece
Bodendämpfanlage *f* *(ag)* soil sterilization plant
Bodendämpfgeräte *npl* *(ag)* soil sterilizing equipment
Bodendämpfung *f* *(ag)* soil sterilization
Bodendeckel *m* bottom (back cylinder) cover
Bodendesinfektion *f* soil disinfection
Bodendesinfektionsmittel *n* soil disinfectant
Bodendruck *m* ground pressure; soil pressure
Bodendübelpresse *f* head compressor
Bodenecho *n* ground return *(radar)*
Bodenelektrode *f* bottom electrode
Bodenentleerer *m* bottom dump truck; bottom discharge waggon
Bodenentleerung *f* bottom discharge, bottom-dumping
Bodenentleerungsventil *n* bottom drain valve
Bodenentseuchung *f* soil disinfection

Bodenentwässerung *f* [soil] drainage
Bodenerkundung *f* soil survey
Bodenerosion *f* soil erosion
Bodenerschütterung *f* concussion of the ground
Bodenertrag *m (ag)* soil produce (yields)
Bodenfauna *f* soil fauna
Bodenfestigung *f* soil consolidation (stabilization, solidification)
  chemische B. soil cementation
Bodenfeuchtigkeit *f* soil (ground) humidity, soil moisture, humidity of soil
Bodenfeuchtigkeitsmessung *f* soil humidity measurement
Bodenfläche *f* floor space
Bodenflora *f* soil flora
Bodenförderung *f* earthmoving
Bodenform *f (met)* open sand mould
Bodenfräse *f* rotary cultivator, motor hoe; rotary hoe, rotavator
  selbstfahrende B. rotary motor hoe
Bodenfreiheit *f* ground (road) clearance
Bodenfrost *m* ground frost
Bodenfruchtbarkeit *f* soil fertility
Bodenfügemaschine *f* head jointing machine
Bodenfunkanlage *f* ground radio installation
Bodenfunkfeuer *n* ground radio beacon
Bodenfunkstelle *f* ground station
Bodenfunkverkehr *m* ground-to-ground radio traffic
Bodengang *m (ship)* bottom strake
Bodengare *f* tilth
bodengebunden soil-bound
Bodengeologie *f* agrogeology
Bodengeschwindigkeit *f* ground speed
bodengesteuert ground-controlled
Bodenhebung *f* uplift
Bodenhobel *m* grader, scraper
  B. mit Eigenantrieb tractor-scraper
Bodenhobelmaschine *f* head planing machine
Bodenhorizont *m* soil horizon
Bodenimpfstoff *m* soil inoculant
Bodenimpfung *f* soil inoculation (injection)
Bodeninjektor *m* soil injector
Bodenkarte *f* soil map
Bodenklappe *f* 1. bottom-dump hatch; 2. *(aero)* bottom door; trap-door; 3. *(ship)* hopper door, false bottom
Bodenklassifizierung *f* soil classification
Bodenkolonne *f* plate column *(distillation)*
Bodenkontakt *m (el)* contact plate [of lamp]
Bodenkunde *f* soil science; pedology
Bodenlängsspant *n* bottom longitudinal
Bodenlängsträger *m (aut)* bottom runner
Bodenlockerung *f* tormenting of soil
Bodenlockerungsgerät *n* soil tormentor
Boden-Luft-Rakete *f* surface-air rocket (missile), SAM
Bodenmatte *f* floor mat
Bodenmechanik *f* soil mechanics
Bodenmeißel *m (ag)* ground chisel
Bodenmesser *n* bottom blade *(coal plough)*
Bodenmischer *m* soil mixer (mixing machine); rotary soil mixer
Bodenmüdigkeit *f* soil exhaustion; soil sickness
Bodennadel *f* plug pin *(bessemer converter)*
Bodennährstoff *m* soil nutrient
Bodennavigation *f* ground navigation
Bodennebel *m* ground fog
Bodennutzung *f* soil utilization
Bodenöffnung *f* bottom opening
Bodenpacker *m (ag)* furrow presser
Bodenpeiler *m* ground direction finder
Bodenpeilstelle *f* ground direction-finder station
Bodenpeilsystem *n* ground-controlled approach *(radar)*
Bodenpersonal *n* ground crew
Bodenphysik *f* soil physics
Bodenplatte *f* bottom plate; bottom cover *(instrument)*
Bodenpresse *f* head compressor
Bodenpressung *f* ground pressure
Bodenprobe *f* soil sample
  gestörte B. remoulded sample
  ungestörte B. undisturbed sample
Bodenprobeentnahme *f* soil sampling
Bodenprofil *n* profile (cross section) of soil
Bodenprüfung *f* soil test
Bodenrahmen *m (aut)* floor-frame
Bodenreaktion *f* soil reaction
Bodenreflexionsfaktor *m* ground reflection factor
Bodenreform *f* land reform
Bodenrille *f* groove
Bodenring *m* foundation ring, fire box casing *(locomotive)*
Bodenrundung *f (ship)* bilge
Bodensatz *m* deposit; sediment; residuum
Bodensatzbildung *f* settling, settlement
Bodensau *f (met)* furnace sow, salamander, bear
Bodenschädling *m* soil pest
Bodenschicht *f (geol)* stratum
Bodenschwelle *f* ground plate
Bodenschwert *n* bottom guide-plate *(coal plough)*
Bodenseil *n (el)* earth continuity conductor *(overhead systems)*
Bodensenkung *f* subsidence
Bodensonde *f* soil probe
Bodenspant *n (ship)* bottom frame, floor plate
Bodenspantwinkel *m* bottom angle of floor plate
Bodenstabilisierung *f* soil stabilization (solidification)
Bodenstation *f* ground station *(radar)*
Bodenstein *m* [hearth] bottom; flagstone *(furnace)*
Bodenstrahl *m s.* Bodenwelle
Bodenstruktur *f* soil structure

Bodenstück *n* 1. breech *(gun)*; 2. *(ship)* floor, ground (naval) futtock
Bodentank *m (ship)* water bottom, bottom tank
Bodentemperatur *f* soil temperature
Bodentextur *f* soil texture
Bodenüberwachungsradar *n* ground surveillance radar
Boden- und Deckelstanzmaschine *f* bottom and lid stamping machine
Bodenuntersuchung *f* soil investigation
Boden-Unterwasser-Rakete *f* surface-underwater missile, SUM
Bodenventil *n* sea valve (cock), drain hole, bottom valve
Bodenverbesserung *f (ag)* land improvement, amelioration; *(build)* soil stabilization
Bodenverdichtung *f (ag)* soil compression; *(build)* soil compaction (densification)
  künstliche B. artificial soil compression
Bodenverfestigung *f* soil stabilization (solidification, consolidation)
  chemische B. soil cementation
Bodenverkrustung *f* cap
Bodenvermörtelung *f* soil stabilization (cementation)
Bodenverseuchung *f* contamination of the ground
Bodenwanne *f (mil)* belly turret, ventral bola (turret)
Bodenwasser *n* ground water
Bodenwegerung *f (ship)* bottom (floor) ceiling; dumping boards
Bodenwelle *f* 1. surface wave; ground wave, direct ray; 2. *(geol)* undulation of ground
Bodenwert *m (nucl)* height equivalent to a theoretical stage (plate), HETS, HETP
Bodenwiderstand *m* soil resistance
Bodenwirkungsgrad *m (nucl)* [Murphree] plate efficiency, plate efficiency [factor]
Bodenwrange *f (ship)* floor [plate], frame floor
  durchlaufende B. continuous floor
  durchschnittene B. intercostal floor
  hohe B. rising (deep) floor
  niedrige B. shallow floor
  offene B. skeleton floor
Bodenzahl *f (ch)* number of plates *(distillation)*
  theoretische B. number of theoretical plates, NTP
Boden-zu-Boden-Zeit *f* floor-to-floor time *(machining)*
Boden-zu-Bord-Anruf *m* ground-to-air calling
Bodenzünder *m* base fuse
bodmen to head a cask
Böenfront *f* squall front
Böenmesser *m* gust meter
Böenschreiber *m* anemograph; gust recorder
Bögchen *n (text)* small bend *(bearded needle)*
Bögchenlänge *f* length of small bend *(bearded needle)*

Bogen *m* 1. *(el, math)* arc; 2. *(build)* arch; bow; curve; curvature; bend; 3. sheet *(paper)*
  B. mit Zugband bowstring (tied) arch
  beiderseits eingespannter B. arch built-in at both ends
  biegesteifer B. rigid arch
  eingespannter B. fixed-end arch
  elliptischer B. elliptical arch
  flacher B. depressed (diminished) arch
  gemauerter B. brick arch
  gotischer B. Gothic arch
  scheitrechter B. camber (straight, flat, jack) arch
  schiefer B. skew arch
  schiefwinkliger B. skew arch
Bogenableger *m* lay boy *(paper)*
Bogenanleger *m (print)* [sheet] feeder
bogenartig arcuate
Bogenausbau *m (mine)* arch girder set
Bogenbeschneiden *n* sheet trimming
Bogenbeschneider *m* sheet cutter
Bogenbinder *m (build)* arched girder
Bogenbrücke *f* arch[ed] bridge
  B. mit Zugband tied-arch bridge
Bogendiagramm *n* arc plot
Bogendurchlaß *m* arch[ed] culvert
Bogenentladung *f (el)* arc discharge
Bogenentladungsröhre *f* [discharge] arc tube
Bogenfachwerk *n* arch[ed] truss
  B. mit Zugband bowstring (tied-arch) truss
bogenförmig arched, arch-like; bent; crooked; curved; arch-shaped
Bogenführung *f* / **zweiseitige** double turning *(paper)*
Bogenführungstrommel *f* conveyor roller (cylinder) *(paper)*
Bogenfunke *m* arc spark
Bogengang *m* arcade, archway
Bogengewichtsmauer *f* arched gravity dam
Bogenglattstoßmaschine *f* jogger
Bogengröße *f* size of sheet
Bogenhobel *m* compass plane
Bogenhöhe *f (text)* height of bend *(bearded needle)*
Bogenkatode *f* arc cathode
Bogenkontrolle *f* sheet control
Bogenkorrektur *f* proof in sheets
Bogenlager *n* arch bearing
Bogenlampe *f* arc lamp
  sich selbstregulierende B. automatic feed arc lamp
Bogenlampenkohle *f* arc lamp carbon
Bogenlänge *f (el)* arc length
Bogenläufigkeit *f* **[eines Fahrzeuges]** suitability for running in curves, running through curves, taking of curves
Bogenleibung *f (build)* intrados

Bogenlöschung *f* / **magnetische** *(el)* magnetic blow-out
Bogenmaß *n* circular measure
Bogenminute *f* minute of [an] arc
Bogenpappe *f* sheet board
Bogenprojektion *f* arc projection
Bogenrohr *n* bent pipe
Bogenrotationsmaschine *f* sheet-fed rotary press
Bogenrücken *m (build)* extrados
Bogensäge *f* web (coping, bow-type) saw
Bogenscheitel *m (build)* crown, vertex, top
Bogenschenkel *m (build)* haunch
Bogenschießen *n* archery
Bogenschubkurbel *f* oscillating crank mechanism
Bogenschüttelmaschine *f* jogger
Bogenschütze *m* archer, bowman
Bogensehne *f* bow string
Bogensekunde *f* second of an arc
Bogensignatur *f* [folio] signature
Bogenspannung *f* arc voltage
Bogenspektrum *n* arc spectrum
Bogensperrmauer *f* arch[ed] dam
Bogenstaumauer *f* arch[ed] dam
Bogenstreichmaschine *f* sheet coater
Bogenstück *n* bend *(fitting)*; *(el)* conduit bend; bridge piece; *(ship)* propeller arch *(stern frame)*
Bogenübergabe *f* sheet transfer
Bogenüberschlag *m (el)* arcing-over
Bogenverlangsamungsmechanismus *m* sheet slow-down mechanism
Bogenverlust *m (el)* arc-drop loss
Bogenvorschub *m* / **schuppenförmiger** sheet-by-sheet feeder
Bogenwinkelstück *n* conduit elbow
Bogenzähler *m* sheet counter
Bogenzirkel *m* bow compass; wing divider
Bogenzuführung *f* sheet feeding
  fließende B. stream feeding
  pneumatische B. suction (pneumatic) feeding
Bogenzündung *f* arc initiation
Bogenzwickel *m* spandrel
Bohle *f* [square] plank; deal; structural timber
Bohlenbelag *m* planking
Bohlenfußboden *m* plank floor
Bohlenholz *n* plank timbers
Bohlensäge *f* cleave (plank) saw
Bohlenweg *m* plank way, plank road
Bohlwerk *n* bulwark *(harbour)*
Bohnenenthülser *m* bean sheller
Bohnenerntemaschine *f* bean harvester
Bohnenmähmaschine *f* bean harvester
Bohnenschober *m* bean stack
Bohnergerät *n* floor polisher
Bohnermaschine *f* electric floor polisher
bohnern to wax, to polish
Bohnerwachs *n* floor wax (polish), rubbing wax
Bohrabteilung *f* drilling department
Bohranlage *f (mine)* drilling (boring) rig
  fortlaufend kernende B. continuous core drilling rig
Bohrapparat *m* drilling rig
Bohrarbeit *f* drilling (boring) operation
Bohrarbeiter *m (mine)* drill runner
Bohrarm *m* boring head
Bohrautomat *m* automatic drilling (boring) machine
Bohrbarkeit *f* drillability; borability
Bohrbereich *m* drilling range
Bohrbericht *m* boring journal
Bohrbild *n* hole pattern (layout) *(metal)*
Bohrbock *m* boring trestle
Bohrbrunnen *m* bore (drilled) well
Bohrbuchse *f* drill bush[ing], jig bush[ing] *(metal)*
Bohrbühne *f* derrick platform
Bohrdrittel *n (mine)* drilling shift
Bohrdruck *m* drilling (boring) pressure
Bohrdurchmesser *m* 1. drilling capacity; 2. diameter drilled
  größter B. aus dem Vollen drill capacity from solid
Bohreinheit *f* drill (boring) unit *(metal)*
Bohreinrichtung *f* drilling attachment, boring head *(metal)*
bohren 1. to drill *(from the solid)*; to bore *(metal)*; 2. to hole *(slates, roofing)*
  abwärts b. to bore down
  aufwärts b. to bore from below
  Gewinde b. to tap
  kegelig b. to taper bore
  Lehren b. to jig bore
  lehrenhaltig b. to drill true
Bohren *n* drilling *(from the solid)*; boring
  B. aus dem Vollen drilling from the solid
  B. kleiner Lose batch boring (drilling) *(metal)*
  B. mit Diamantkronen diamond boring (drilling)
  B. mit heißen Abgasen rocket exhaust drilling
  B. mit Schrotkrone shot boring (drilling)
  B. mittels Stoßwellen shock wave drilling
  drehendes B. rotary drilling
  drehend-schlagendes B. rotary-percussive drilling
  einmännisches B. *(mine)* single-hand drilling
  gerichtetes B. controlled (directional) drilling
  pennsylvanisches B. Pennsylvanian drilling system
  schlagendes B. percussive drilling
  stoßendes B. mit Schlagbaum spring-pole drilling
Bohrer *m* drill
  B. mit beidseitig abgesetztem Schneidenteil double-end drill
  B. mit enger Nutensteigung (Spiralsteigung) high-helix drill

B. mit genutetem Schaft grooved-shank drill
B. mit geschlitzter Querschneide drill with split point grind
B. mit Innenanzugsgewinde threaded-shank drill
B. mit kleinem Durchmesser small-size drill
B. mit schlanker Spirale low-helix drill
B. mit verschiedenen Spanwinkeln drill with increase twist
B. mit verschiedener Steigung der Nuten drill with increase twist
B. mit weitem Drall low-helix drill
engspiraliger B. high-helix drill
gewundener B. twisted drill
Bohrerachse *f* axis of [the] drill, drill axis
Bohrerausspitzen *n* drill thinning
Bohreraustreiber *m* drill drift
Bohreraustritt *m* break-through of the drill
Bohrerdrehachse *f* axis of drill rotation, rotating axis of the drill
Bohrerdurchmesser *m* drill diameter
Bohrerfestigkeit *f* drill strength
Bohrerführungsbuchse *f* drill guide bush
Bohrerkern *m* web (core) of the drill
Bohrerlänge *f* overall length [of the drill]
Bohrerlippe *f (Am)* rib
Bohrernut *f* drill flute
Bohrerrohling *m* drill blank
Bohrerrücklauf *m* drill return
Bohrerschleifmaschine *f* drill grinding machine
Bohrerseele *f s.* Bohrerkern
Bohrerspitze *f* drill point
Bohrerspitzenwinkel *m* drill point angle, included angle of the drill tip, included cutting angle
Bohrerstandzeit *f* drill life *(time between regrinds)*
Bohrerstarrheit *f* drill rigidity
Bohrersteifigkeit *f* drill stiffness
Bohrerumdrehung *f* drill revolution, revolution of the drill
Bohrerverlängerung *f* drill extension
Bohrervorschub *m* drill feed
Bohrervorschubgröße *f* drill feed rate
Bohrerwechsel *m* drill changing
Bohrfortschritt *m* drilling rate (speed), penetration rate (advance)
Bohr-, Fräs- und Gewindeschneidmaschine *f* drilling, boring, milling, and tapping machine
Bohrfutter *n* drill chuck
Bohrfutterkegel *m* taper sleeve for [the] drill chuck
Bohrgarnitur *f* drilling string, drill stem with bit
Bohrgerät *n* drilling appliance
Bohrgerüst *n (mine)* drill mounting
Bohrgeschwindigkeit *f (mine)* drilling rate (speed), penetration advance (rate)
Bohrgestänge *n* drill rod[s], boring rod[s]
aus Einzelstangen bestehendes B. sectional drill rod
Bohrgestängezug *m* drill pipe string
Bohrgestell *n (mine)* drill mounting
Bohrhammer *m* hammer-drill, rock drill, drifter, stoper, sinker
B. mit automatischem Vorschub automatic-feed drill
leichter B. jack hammer
leichter B. mit automatischer Bohrstütze autostoper
Bohrhammerstütze *f (mine)* feed leg
Bohrhauer *m (mine)* drill man (runner), driller
Bohrhülse *f* [spindle] sleeve *(drilling machine)*
Bohrkasten *m* box [drill] jig *(metal)*
Bohrkern *m* boring core, plug *(trepanning)*
Bohrklein *n (mine)* debris, drill cuttings
Bohrknarre *f* ratchet brace
Bohrknecht *m (mine)* air (feed) leg
Bohrkopf *m* boring head
B. mit mehreren Meißeln multiple-cutter boring head
B. mit zwei Bohrmeißeln boring head employing two cutters
dreimeßriger B. three-cutter boring head
Bohrkopfstange *f* multiple-cutter boring tool
Bohrkörner *m* centre punch
Bohrkrone *f* core drill bit
Bohrkronenabschlagvorrichtung *f (mine)* detaching device for taper joint drill bits
Bohrlager *n* boring collar
Bohrleistung *f* drilling (boring) capacity *(metal)*; *(mine)* drill footage
Bohrloch *n (mine)* bore [hole], boring, drill-hole; shot-hole *(blasting)*
abwärts gerichtetes B. down (water) hole
aufwärts gerichtetes B. dry (up) hole, upper
ausgekesseltes B. sprung hole
enges B. slim hole
falsch angesetztes B. misplaced hole
gekesseltes B. chambered hole
gekrümmtes B. curved hole
geneigtes B. angle hole
nicht gekommenes B. missed hole
söhliges B. dry hole
waagerechtes B. beim Wandsprengen toe hole
Bohrlochabstand *m (mine)* hole space (spacing)
Bohrlochkristall *m* well-type crystal
Bohrlochmund *m* hole collar
Bohrlochneigungsmessungen *fpl* surveying of bore holes
Bohrlochpfeife *f (mine)* blown-out hole
Bohrlochprobe *f* bore hole sample
Bohrlochsohle *f* hole bottom
Bohrlochtiefstes *n* hole bottom
Bohrlochuntersuchung *f* **mit Gammastrahlung** bore-hole logging by gamma-rays

Bohrlöffel *m* sludger
Bohrmannschaft *f* drilling crew
Bohrmaschine *f* drilling (boring) machine
B. mit Antriebskopf mit Rädergetriebe geared drilling machine
B. mit selbsttätiger Positionierung automatic positioning boring machine
freistehende B. vertical drilling machine
ortsveränderliche B. *s.* transportable B.
tragbare B. *s.* transportable B.
transportable B. portable drilling machine
Bohrmaschinentisch *m* drilling (boring) machine table
Bohrmehl *n* fine drillings
Bohrmeißel *m* boring bit, single-point boring tool, boring-bar cutter
Bohrmesser *n* boring-bar cutter, boring bit; cutting blade
beiderseitig arbeitendes B. double-ended boring-bar cutter
doppelseitig arbeitendes B. double-ended boring-bar cutter
Bohrmotor *m* driving motor [of drilling machine]
Bohröl *n* drilling oil, oil for drilling
Bohrpfahl *m* bored pile
Bohrplatte *f* plate jig
Bohrprogramm *n* drilling campaign
Bohrratsche *f* ratchet brace
Bohrrechteck *n* area of drilling
größtes B. maximum area of drilling
Bohrreibung *f* pivoting friction
Bohrring *m* hollow mill
Bohrrohr *n* 1. trepanning bar, drill tube *(metal cutting)*; 2. *(mine)* casing
Bohrrohrfangkeil *m* drill pipe slip
Bohrsäule *f* drill column
Bohrschablone *f* boring jig; drilling template
Bohrschema *n (mine)* drilling pattern
Bohrschlamm *m* rotary mud
Bohrschlammaufbereitung *f* mud conditioning
Bohrschlitten *m* drilling saddle *(radial drill)*, drilling head
Bohrschneide *f* bit
Bohrschrämkopf *m* rotary head
Bohrschrämlader *m* boring miner *(cutter loader)*
Bohrschute *f* drilling barge
Bohrschwengel *m* walking beam
Bohrseil *n* drilling cable (line)
Bohrsetzstock *m* boring stay
Bohrspan *m* boring (drilling) chip
Bohrspindel *f* drill[ing-machine] spindle
Bohrspindeldrehzahl *f* drilling (boring) spindle speed
Bohrspindelhülse *f* drill spindle sleeve
Bohrspindelschlitten *m* drilling head
Bohrspülungsmittel *n* drilling mud
Bohrspülungszusatzmittel *npl* drilling mud additives
Bohrstand *m* drilling platform
Bohrständer *m* upright carrying the boring spindle
Bohrstange *f* boring bar *(metal cutting)*; *(mine)* drill collar (rod, stem)
B. mit Schneide *(mine) (Am)* bit
Bohrstangenhalter *m* boring-bar holder
Bohrstangenmesser *n* boring bar cutter
zweischneidiges B. double-ended boring-bar cutter
Bohrstation *f* / **schwimmende** floating drilling platform
Bohrstellung *f* drilling (boring) position
Bohrstraße *f* transfer line for drilling
Bohrstück *n* drill bit
Bohrstufentiefe *f* depth of the drilling step
Bohrstütze *f* / **hydraulische** *(mine)* hydraulically operated support leg, hydro-leg
Bohrsupport *m* boring head *(planer)*
Bohrtiefe *f* drill (drilling, boring) depth, depth of bore (drilling)
Bohrtiefenanzeiger *m* drill depth indicator
Bohrtiefenbegrenzung *f* depth control device, feed depth gauge *(drilling machine)*
Bohrtiefengenauigkeit *f* depth accuracy of a bore
Bohrtiefenkontrolle *f* [hole] depth control
Bohrtiefenmeßeinrichtung *f* depth measuring device
Bohrtiefenrundskala *f* depth dial
Bohrtiefenskala *f* graduated depth gauge
Bohrtisch *m* drill table *(drilling machine)*
Bohrtrommel *f* bull wheel
Bohrtrübe *f* drilling fluid (mud)
Bohrturm *m* drill[ing] tower, derrick
Bohr- und Fräswerk *n* horizontal boring and milling machine
Bohrung *f* bore [hole]; [drilled] hole
B. geringer Tiefe short-length bore
B. mit Gewinde tapped hole
B. ohne Gewinde plain hole
frei ausfließende B. flowing well
kegelige B. taper bore (bored hole)
verlaufene B. off-centre hole
vorgegossene B. cored hole
vorgeschmiedete B. punched hole
Bohrungsabmessungen *fpl* hole size; dimensions of the bore
Bohrungsabstand *m* hole spacing
Bohrungsachse *f* axis of the hole (bore), bore axis
Bohrungsdurchbruch *m* hole exit
Bohrungsdurchmesser *m* bore diameter (dia, size), hole diameter (dia, size), diameter of [the] bore
Bohrungsdurchmessertoleranz *f* bore diameter limits
Bohrungsgrund *m* bottom of [the, a] drilled hole

Bohrungsmantel *m* inner surface of a cylinder
Bohrungsmeßgerät *n* bore meter
Bohrungsmitte *f* centre point of bore
Bohrungsmittenabstand *m* hole centre distance
Bohrungstiefe *f* bore (hole) depth, depth of hole drilled (bored)
Bohrungstoleranz *f* bore tolerance
Bohrungswand[ung] *f* bore (hole) wall, wall of the bore (hole)
Bohrvorrichtung *f* drill jig
Bohrvorschub *m* drilling (boring) feed
Bohrvorschubbereich *m* range of drilling (boring) feed
Bohrvorschubgröße *f* rate of drill[ing] feed
Bohrwagen *m* drill carriage, wagon drill, *(Am)* jumbo
Bohrwerk *n* boring mill
Bohrwurm *m* marine borer
Bohrzeit *f* drilling time
  reine B. actual drilling time
Bohrzeug *n* drilling string, drill stem with bit
Boje *f* buoy *(sea-mark)*
Bojeanker *m* mooring anchor
Bojenboot *n* buoy boat
Bojenleger *m* buoy-laying vessel
Bol *m* bole
Bollwerk *n* bulwark *(in harbours)*
Bolometer *n* bolometer
Bolometerabschluß *m* thermistor mount *(of waveguides)*
Bolometerbrücke *f* bolometer bridge
Bolometerdetektor *m* bolometer detector
bolometrisch bolometric
Boltzmann-Gleichgewicht *n* Boltzmann equilibrium
Boltzmann-Gleichung *f* Boltzmann (transport) equation
Boltzmann-Statistik *f* Boltzmann statistics
Bolus *m* bole
Bolzen *m* 1. *(mach)* bolt; pin; dowel [pin]; pintle; 2. *(mine)* [pit] prop, strut *(support)*; corner post, studdle *(cribbing)*
  versenkter B. countersunk bolt
Bolzengelenk *n* pinned hinge, pin[ned] joint
Bolzengewinde *n* bolt (external) thread
Bolzengewindeschneidmaschine *f* bolt-thread cutting machine
Bolzenkopf *m* bolt head
Bolzenloch *n* eye; bolt hole
Bolzenlochbohrmaschine *f* stud inserting machine
Bolzenschaftfräsmaschine *f* bolt milling machine
Bolzenschießgerät *n* *(build)* stud gun, automatic stud driver, cartridge hammer
Bolzenschneider *m* bolt clipper
Bolzenschrotausbau *m* *(mine)* [open] cribbing
Bolzenschrotzimmerung *f* *s.* Bolzenschrotausbau
Bolzenschußapparat *m* *(ag)* humane stunner
Bolzenverbindung *f* bolted union (joint)
Bombagestoff *m* *(text)* lining
Bombardement *n* bombardement
Bombe *f* bomb
  Berthelotsche B. Berthelot's calorimeter
  kalorimetrische B. calorimetric bomb
Bombenabwurfautomat *m* automatic bomb release
Bombenabwurfhebel *m* bomb release lever
Bombenabwurfvorrichtung *f* bomb-dropping device
Bombenangriff *m* bombing raid
bombenbeschädigt bomb-damaged
Bombenflugzeug *n* bombing plane, bomber
Bombenkern *m* bomb core
Bombenpunktwurf *m* pinpoint (pin-point) bombing
Bombenreihenabwurf *m* stick-bombing
Bombenrichtgerät *n* bomb-aiming mechanism (device)
Bombenrohr *n* *(ch)* sealed (bomb) tube
Bombenschacht *m* bomb bay
bombensicher bomb-proof
Bombentrichter *m* bomb crater
Bombenvisier *n* *s.* Bombenzielgerät
Bombenwurf *m* release of bombs; bombing
Bombenzielgerät *n* bomb sight (equipment)
Bomber *m* bomber, bombing plane
bombieren to chase, to emboss
bondern to bonderize *(surface)*
Boostergebläse *n* booster fan
Boostersteuerung *f* boost (power assisted) control
Boot *n* boat
  B. aus glasfaserverstärktem Harz resin glass boat
  B. aus Segeltuch canvas boat
  aufblasbares B. bubble boat
  flachgehendes B. shoal-draft boat
  in Einzelkonstruktion gebautes B. custom[-built] boat
  kräftiges B. lively boat
  offenes B. open boat, undecked boat
  schnittiges B. rakish boat
  selbstaufrichtendes B. self-righting boat
  starkes B. lively boat
  starkgebautes B. husky boat
  wendiges B. handy boat
Bootsabdeckung *f* boat cover
Bootsanker *m* / **kleiner** killick
Bootsausrüstung *f* boat's outfit; boat equipment *(of ship)*
Bootsaussetzvorrichtung *f* launching arrangement *(for life-boats)*
Bootsbau *m* boat-building

Bootsbezug *m* boat's cover, spray hood
Bootsdavit *m (n)* boat davit
Bootsdeck *n* boat deck
Bootsdolle *f* thole [pin]
Bootsducht *f* thwart *(of a boat)*
Bootsfaß *n* keg, water breaker
Bootsführer *m* waterman, boatman, boat driver, coxwain, cockswain
Bootsgalgen *m* skid beam, boat skid (beam)
Bootsgast *m* oarsman
Bootsgruppe *f* nest *(two or more small vessels moored together)*
Bootshafen *m* boat harbour (haven, basin); *(Am)* marina, sally
Bootshaken *m* setting pole, boat hook
Bootshaus *n* boat-house
Bootshaut *f* rubber skin *(folding canoe)*
Bootsheißmaschine *f* boat hoist
Bootsklampe *f* boat cradle
  klappbare B. shifting chock
Bootslack *m* [marine] spar varnish, boat varnish
Bootsleine *f* tow-rope
Bootsmann *m* boatswain, bosun
Bootsmotor *m* boat motor
Bootsreling *f* boat rail
Bootsriemen *m* oar
Bootsrucksack *m* folding canoe bag
Bootsschleppanhänger *m* boat trailer (carrier)
Bootsschwert *n* centre-board
Bootssteg *m* landing stage, boat-walk
Bootstrap-Integrator *m (dat) (Am)* bootstrap integrator
Bootswagen *m* boat trolley
Bootswerft *f* boatbuilding yard, boat yard
Bootswinde *f* boat winch
Bootswinsch *f* boat winch
Bootszurring *m* gripe lashing
Bor *n* boron
Boran *n* borane, boron hydride
Boräquivalent *n* boron equivalent
Borat *n* borate
Borax *m* borax
Boraxkalk *m* calcium borate
Boraxperle *f* borax bead
Boraxsee *m* borax lake
Borazit *m (min)* boracite
Bord *m* 1. board *(ship)*; 2. restraining flange *(roller bearing)*
  B. / an aboard
  B. / über overboard
  B. an Bord side by side *(vessels)*
Bordabfangradar *n* airborne intercept radar
Bordabtastgerät *n* airborne scanner *(radar)*
Bordanlage *f* airborne system
Bordanlasser *m (aero)* cockpit starter
Bordantenne *f* ship's aerial
Bordaufzug *m* marine elevator, ship's lift
Bordbetriebsbedingungen *fpl (ship)* service condition on board
Bord-Bord-Verbindung *f* ship-to-ship communication
Bord-Bord-Verkehr *m* air-to-air communication
Bord-Bord-Waffe *f* air-to-air weapon
Borddienst *m* service aboard (afloat)
borddienstfähig fit for service afloat
Bordeauxbrühe *f* Bordeaux mixture
Bördelmaschine *f* flanging machine *(sheet metal)*
Bördelmutter *f* knurled nut
bördeln to flange; to bead
Bördelnaht *f* flanged-edge type of weld
Bördelriß *m* **durch schlechtes Umbördeln** fracture at the flange due to bad flanging
Bördelschweißnaht *f* flange-welding seam
Bördelwalze *f* flanging roller
Bordempfangsgerät *n* airplane receiving set
borden to board *(a ship)*
Bordflak *f* ship-based anti-aircraft artillery
Bordflugzeug *n* ship's (carrier-borne, shipboard) plane, shipborne craft
Bordfunkanlage *f* aircraft radio equipment
Bordfunker *m* wireless (radio) operator
Bordfunkgerät *n* aircraft radio
Bordfunkstelle *f* 1. ship radio station; 2. aircraft radio room; airplane radio station
bordgeführt *(mach)* flange-guided
Bordkante *f* kerb, curb
Bordkran *m* deck crane
Bordküche *f* galley
Bord-Land-Verbindung *f* ship-to-shore communication
Bordleiste *f (ship)* wale
Bordpeiler *m* 1. wireless (radio, ship's) direction finder; 2. airborne direction finder
Bordpeilstelle *f* ship direction finding station
Bordradar *n* airborne radar
Bordrakete *f* airborne (aircraft) rocket
Bordregistriergerät *n* missile-mounted recording instrument
Bordsammelschiene *f* missile-mounted terminal board
Bordschwelle *f (build)* curb, kerb
Bordseite *f* ship's side (board)
Bordsender *m* airborne transmitter
Bordstein *m* curbstone, kerbstone
Bordsteinfühler *m* kerb feeler
Bordstoß *m* edge joint
Bordtelefon *n* aircraft interphone
Borduhr *f* ship's time
Bordürenmuster *n (text)* pattern for trimmings, edge pattern
Bordwaffen *fpl* aircraft armament
Bordwand *f* ship's side (wall)
  eingezogene B. turnover
Bordwerkzeug *n (aero)* service tool-kit

Bordzielanfluggerät *n* aircraft homing device
Borfluorwasserstoffsäure *f* fluoboric acid
Borgmechanismus *m (dat)* borrow mechanism *(subtraction)*
Borid *n* [metallic] boride
Borionisationskammer *f (nucl)* boron[-filled ionization] chamber
Borkammer *f (nucl)* boron chamber
Borkarbid *n* boron carbide
Borke *f* bark *(wood)*
Borkenkrepp *m (text)* bark crêpe
Borneokampfer *m* Borneo camphor
Borneol *n* borneol, Borneo camphor
Bornit *m* bornite
Bornitrid *n* boron nitride
Borsalbe *f* boric-acid ointment
Borsäure *f* boric acid
Borsäureregelung *f* boric acid control *(reactor)*
Borstab *m (nucl)* boron rod
Borstahl *m (nucl)* boron steel
Borste *f* bristle
Bort *m* bort, boart
Bortenweberei *f* weaving of trimmings
Bortrifluorid *n* borofluoride, boron trifluoride
Bortrifluoridionisationskammer *f s.* BF_3-Kammer
Bortrifluoridkammer *f s.* BF_3-Kammer
Bortrioxid *n* boron trioxide
Borwasserstoff *m* boron hydride
Borzählrohr *n (nucl)* boron[-filled] counter tube
Böschungsausbruch *m (mine)* backbreak
Böschungsbefestigung *f* slope stabilization
Böschungshobel *m (build)* slope grader
Böschungsmauer *f* retaining wall; toe wall
Böschungsneigung *f* inclination of slope; slope of repose
Böschungsprofil *n* profile of slope
Böschungsschutz *m* slope protection
Böschungsverkleidung *f* revetment
Böschungswinkel *m* slope [gradient], slope angle; angle of repose
  natürlicher B. angle of repose
Bose-Einstein-Statistik *f* Bose-Einstein statistics
Bose-Teilchen *n s.* Boson
Boson *n* boson
Bossenwerk *n (build)* bossage
Bostonpresse *f (print)* Boston (lever) press
Böttcher *m* cooper, barrel maker
Böttcherei *f* cooperage
Böttcherholz *n* cask wood
Böttcherzange *f* cooper's dog
Bottich *m* vat; vessel; tank; beck; back; tub; tun
Bottichgärung *f* tun (tub) fermentation
  offene B. fermentation in open tubs
Boucherieverfahren *n* Boucherie process
Bouclégarn *n* bouclé yarn
Bouclestoff *m* bouclé
Bougie-Kerze *f (el)* decimal candle *(unit)*
Bouilleurkessel *m* elephant (French) boiler
Bourdonmanometer *n* Bourdon tube, Bourdon pressure gauge
Bournonit *m* bournonite
Bowdenzug *m* Bowden wire (cable)
Box *f* pen, compartment
Boxermotor *m* opposed piston (cylinder) engine, horizontally opposed engine
Boxkalf *n* box calf
Boxpalette *f* box pallet
Brache *f (ag)* fallow
Brachistochrone *f* brachistochrone
Brachland *n (ag)* fallow
Brachyachse *f* brachyaxis, brachydiagonal
Brachydiagonale *f* brachydiagonal, brachyaxis
Brachzeit *f (mach)* idle time
Brackelsbergofen *m* Brackelsberg furnace
Brackett-Serie *f* Bracket series
Brackwasser *n* brackish water
Bradford-Brecher *m* Bradford breaker
Bradford-Spinnverfahren *n* Bradford spinning
Bramme *f* slab; slab ingot; plate slab
Brammenblock *m* slab ingot
Brammendrehvorrichtung *f* slab turning device
Brammenform *f* ingot slab mould
Brammenkaliber *n* slabbing pass
Brammenschere *f* slab shears
Brammentiefofen *m* slab heating furnace
Brammenwalzwerk *n* slabbing (slab cogging) mill
Brand *m* fire; burning
Brandbekämpfung *f* fire-fighting
Brandbombe *f* incendiary bomb
Brandbombenbündel *n* cluster of incendiary bombs
Branddamm *m (mine)* dam, air-tight barrier against fire; break; fire-wall
Brände *mpl (leath)* brands
Brandeisen *n (leath)* brand iron
Brandflasche *f* incendiary bottle
Brandgefahr *f* fire hazard
Brandgiebel *m* fire-proof wall
Brandgold *n* refined gold
Brandgraben *m* fire lane
Brandgranate *f* incendiary shell
Brandmauer *f* fire[-division] wall, party (fire-resisting) wall
Brandmunition *f* incendiary ammunition
Brandriß *m* crazing *(mould)*
Brandsalbe *f* burns ointment
Brandsatz *m* powder train
Brandschaden *m* fire damage
Brandschiefer *m* bituminous (carbonaceous) shale, schist slate, black bat
Brandschott *n* fire-resisting bulkhead

Brandschutz *m* fire protection, precautions against fire
Brandschutzmaßnahme *f* fire precaution
Brandschutzpolizei *f* fire brigade
Brandsohle *f* insole *(shoe)*
Brandsohlenheftmaschine *f* insole tacking machine
Brandsohlenrißmaschine *f* insole chanelling machine
Brandsohlleder *n* insole leather
Brandspant *n* fire wall, bulkhead dividing engine compartment from fuselage
Brandstiftung *f* arson
Brandung *f* breakers, surf, surges, seething of waves
Brandungsboot *n* surf-boat
Brandversuch *m* fire test
Brandzeichen *n* brand
Brannerit *m* brannerite *(uranium mineral)*
Branntelektrode *f*/ **amorphe** amorphous carbon electrode
Branntkalk *m* quicklime, anhydrous (fired) lime
Branntwein *m* brandy, spirits
Branntweinbrennerei *f* brandy distillery, distillery of spirits
Brasilettoholz *n* braziletto, Jamaica wood
Brasilholz *n* Brazil wood
Brasilienholz *n* Brazil wood
Brasilnußbaum *m* Brazil-nut tree
Brasse *f (ship)* brace
Bratofen *m* roasting (baking) oven
Bratpfanne *f*/ **elektrische** electric frier
Bratröhre *f s.* Bratofen
Braubottich *m* brewing tun (vat, cask)
Brauchbarkeit *f* serviceability
Brauchwasser *n* process water, water for industrial use
brauen to brew
Brauer *m* brewer
Brauerei *f* brewery
Brauereimaschine *f* brewing machine
Brauerpech *n* brewer's pitch
Brauhaus *n* brewery, brew-house
Braumalz *n* brewing malt
Braumeister *m* brew-master
braun brown
Braunbeizen *n* browning, bronzing
Braunbleierz *n* pyromorphite
Brauneisenerz *n* limonite
Brauneisenstein *m* limonite
bräunen 1. *(met)* to burnish; 2. to brown
Braunfäule *f* brown rot *(wood)*
Braunit *m (min)* braunite
Braunkohle *f* lignite, brown (soft) coal
  bituminöse B. bituminous lignite
  lignitische B. woody (ligneous) brown coal
  ölreiche B. bituminous lignite
Braunkohlenbergbau *m* lignite mining
Braunkohlenbrikett *n* lignite briquette
Braunkohlenhalbkoks *m* partially carbonized lignite
braunkohlenhaltig lignitic, lignitiferous
Braunkohlenhochtemperaturkoks *m* lignite high-temperature coke
Braunkohlenkokerei *f* lignite coke plant
Braunkohlenkoks *m* lignite coke
Braunkohlenlager *n* lignite field (deposit)
Braunkohlensandstein *m* brown-coal grit
Braunkohlenschiefer *m* schistous lignite
Braunkohlenschwelerei *f* lignite carbonization plant
Braunkohlenschwelkoks *m* brown coal low-temperature coke
Braunkohlenschwelung *f* carbonization of lignite
Braunkohlenstaub *m* powdered brown coal
Braunkohlentagebau *m* open-cast lignite mine
Braunkohlenteer *m* lignite tar
Braunkohlenvorkommen *n* lignite deposit, deposit of brown coal
Braunschliff *m* brown mechanical woodpulp
Braunschliffpappe *f* brown mechanical pulp board
Braunstein *m* pyrolusite
Brauprozeß *m* brewing process
Brause *f* shower bath
Brausekabine *f* shower stall
brausen *(ch)* to effervesce
Brausenische *f* shower stall
Brausepulver *n* effervescent powder
Brausesieb *n* rinsing (spraying) screen
Brauwesen *n* brewing industry
Breccie *f s.* Brekzie
Brechanlage *f* [stone-]crushing plant, [stone-] breaking plant, rock crushing (breaking) plant
Brechbacke *f* crusher jaw
Brecheisen *n* wrecking bar
brechen 1. to break; to fracture *(material)*; 2. to crush, to reduce in size; 3. to break the corner, to bevel; 4. to beat *(flax)*; 5. to refract *(sound, light)*; 6. to quarry *(stone)*; to mine
  über die Poop b. *(ship)* to be pooped *(wave)*
Brechen *n* **der Kanten** breaking the corners, bevelling
Brecher *m* 1. [rock] crusher; 2. *(ch, text)* breaker
Brecher *mpl* breakers *(sea)*
Brechgut *n* crushing material
Brechkegel *m* crushing cone
Brechkoks *m* coke nuts, crushed (cut) coke
Brechkraft *f (opt)* refractive power
Brechmaschine *f* breaker *(hemp)*
Brechmittel *n* emetic
brechreizerregend emetic
Brechstange *f* wrecking bar, crowbar
  kleine B. jemmy, jimmy

Brech- und Siebanlage *f* crushing and screening plant
Brechung *f (opt, ac)* refraction
B. des Schalls refraction of the sound
B. von Kraftlinien refraction of lines of forces
akustische B. acoustic refraction
Brechungsachse *f* axis of refraction
Brechungsbrennfläche *f* focal surface of refraction, diacaustic surface
Brechungsebene *f* plane of refraction
Brechungseinheit *f* unit of refractivity
Brechungsexponent *m* refractive index, refraction coefficient
Brechungsgesetz *n (ph)* law of refraction
Brechungsgrenzwinkel *m* critical angle
Brechungsindex *m* refractive (refraction) index, index of refraction
Brechungsmesser *m* refractometer
Brechungsquotient *m* index of refraction
Brechungsschicht *f* refracting layer
Brechungsspektrum *n* refraction spectrum
Brechungsverhältnis *n s.* Brechungszahl
Brechungswinkel *m* angle of refraction, refraction (refracting) angle
Brechungszahl *f* refractive (refraction) index, index of refraction
absolute B. absolute index of refraction
Brechwalze *f* crushing roll *(rock)*; *(text)* toothed roller
Brechwalzwerk *n* crushing mill
Brechwasserplatte *f* manger plate
Brechweinstein *m* tartar emetic
Breeder *m s.* Brutreaktor
Breguetspiralfeder *f* Bréguet spring
Brei *m* pulp *(papermaking)*, paste, slurry
breiartig pulpy, pasty
breit wide; broad
Breitauffahren *n (mine)* shortwall working
Breitband *n* wide band *(rolling; telecommunications)*
Breitbandantenne *f* wide-band aerial, broad-band aerial
Breitbandempfang *m* wide-band reception
Breitbandklirren *n* wide-band distortion
Breitbandleistungserzeugung *f* wide-band power generation
Breitbandoszillograf *m* wide-band oscilloscope
Breitbandrichtfunkübertragung *f* wide-band microwave relay
Breitbandstraße *f* wide strip mill
Breitbandübertragung *f* broad-band transmission
Breitbandverstärker *m* wide-band amplifier
Breitbandwalzwerk *n* broad strip mill
Breitbeil *n* broad axe
Breitdüngerstreuer *m (ag)* manure (fertilizer) broadcaster
Breite *f* width; breadth; latitude *(geography)*
B. auf Spanten *(ship)* moulded breadth (beam), beam (breadth) moulded
B. der Stoßschicht width of shock zone
B. der Stoßwelle width of shock wave
B. des Impulsspektrums pulse bandwidth
B. eines Frequenzbandes width of a band
B. über alles extreme breadth, beam (breadth) over all
B. über Planken *(ship)* moulded beam (breadth), breadth over planks
B. über Scheuerleisten *(ship)* beam (breadth) over fenders, extreme breadth
gemallte B. *(ship)* moulded beam, beam moulded
geografische B. geographical latitude
größte B. maximum beam (breadth), extreme breadth
lichte B. clear width
Breiteisen *n* semi-finished flat *(rolling)*
breiten to flatten *(rolling mill)*
Breiteneffekt *m (nucl)* latitude effect
Breiteneingang *m (text)* shrinkage in width
Breiteneinstellung *f* setting in width
Breitengrad *m* degree of latitude
Breitenkrümmung *f* transverse warping
Breitenmaßstab *m* scale of breadths
Breitenrapport *m (text)* horizontal repeat
Breitenträgheitsmoment *n* lateral moment of inertia
Breitfilm *m* wide film
breitflanschig wide-flanged
Breitflanschprofil *n* H-section
Breitflanschträger *m* wide flange beam (girder), H-beam
Breitfräsarbeit *f* width-milling operation
Breitfußschiene *f* wide-base rail
Breithalten *n* **der Ware** stretching of the cloth
Breithalter *m* stretcher; temple *(loom)*
Breithalterschere *f* temple blade
Breitholz *n* half round wood
Breitsämaschine *f* broadcast sower, seed broadcaster
Breitschlichten *n* broad cut finishing, wide finishing *(metal cutting)*
Breitschlichthobelmeißel *m* flat finishing planing tool
Breitschlichtmeißel *m* broad finishing tool, *(Am)* broad nose finish tool, square-nosed tool
Breitschnitt *m* square edge *(shoe)*
Breitschnittholz *n* broad timber
breitseifen *(text)* to soap in full width
Breitspülen *n (text)* full width rinsing
Breitspur *f (ry)* broad (wide) gauge
Breitstahl *m* paring chisel
Breitstrahler *m (el)* spread beam; wide angle lighting fitting
breitstreuen to spread; to broadcast

Breitwandfilm *m* wide-screen film
Breitwender *m (ag)* tedder
Brekzie *f (geol)* breccia
Bremsabnutzung *f (ry)* wear of brakes
Bremsanker *m (ship)* drag
Bremsanlage *f* braking system, brake assembly
hydromechanische B. hydro-mechanical brake system
Bremsanschlag *m* brake buffer
Bremsäquivalent *n (nucl)* stopping equivalent, equivalent stopping power
Bremsaudion *n* electron-oscillating detector
Bremsausgleich *m* brake compensator
Bremsausgleichhebel *m* brake-compensating lever
Bremsausgleichwelle *f* brake-compensating shaft
Bremsausrüstung *f (ry)* brake equipment
Bremsbacke *f* brake shoe (block, clip)
Bremsbackenhalter *m* brake-shoe holder
Bremsbackenlager *n* brake-shoe pin bushing
Bremsbackenlagerbolzen *m* brake-shoe mounting bolt
Bremsbackenrückholfeder *f* brake-shoe return spring
Bremsbackenrückzugfeder *f* brake-shoe check spring
Bremsband *n* brake band (strap)
Bremsbelag *m* brake lining
Bremsbelastung *f* load on a brake
Bremsberg *m (mine)* brake (braking, self-acting) incline, dook, haulage (gravity, go-devil, self-acting) plane
eintrümiger B. mit Gegengewicht back balance, jig plane
Bremsbergführer *m* incline attendant
Bremsberghaspel *f* lowering winch
Bremsbetätigung *f* brake operation
Bremsblock *m (nucl)* moderating block
Bremsbüchse *f* brake bushing
Bremsdauer *f* braking period
Bremsdichte *f* slowing-down density
Bremsdreieck *n (ry)* brake triangle, triangular brake beam
Bremsdruck *m* brake pressure
Bremsdruckleitung *f* brake pressure line
Bremsdruckventil *n* brake pressure valve
Bremsdynamometer *n* brake dynamometer
Bremse *f* brake
doppeltwirkende B. double-acting brake
druckluftbetätigte hydraulische B. air-operated hydraulic brake
hydraulische B. hydraulic brake
saugluftbetätigte hydraulische B. vacuum-operated hydraulic brake
selbsttätige B. automatic brake
Bremseinrichtung *f* brake
Bremseinstellvorrichtung *f* brake-adjusting mechanism
Bremselektrode *f* decelerating (retarding) electrode
bremsen to brake, to apply the brake[s]; to decelerate
Bremsenleistung *f* brake performance
Bremsentlüftung *f* bleeding of the brake
Bremser *m (ry)* brakeman, brake tightener
Bremserhaus *n* brakeman's cabin, *(Am)* brakeman's caboose
Bremsfähigkeit *f* moderating (slowing-down) power
Bremsfallschirm *m* deceleration (brake) parachute
Bremsfeld *n* retarding field
Bremsfeldaudion *n* reverse-field detector, retarding field detector
Bremsfeldoszillator *m* positive-grid oscillator, negative-transconductance oscillator
Bremsfeldröhre *f* brake-field valve, retarding-field tube
Bremsfeldröhrenoszillator *m* transitron oscillator
Bremsfeldschaltung *f* Barkhausen-Kurz circuit
Bremsfeldschwingung *f* retarding-field oscillation
Bremsfläche *f* braking surface
Bremsflüssigkeit *f* brake fluid
Bremsflüssigkeitsbehälter *m* brake fluid tank
Bremsförderer *m* retarding conveyor
Bremsfußhebel *m* brake pedal
Bremsgestänge *n* braking leverage (linkage, rigging), brake leverage (linkage, rigging)
Bremsgestängesteller *m* brake rod adjuster
Bremsgestängeträger *m* brake-gear support, brake rigging supporting device
Bremsgewicht *n* brake weight
Bremsgitter *n (el)* suppressor grid
Bremsgittermodulation *f* suppressor modulation
Bremsgleichung *f (nucl)* slowing-down equation
Bremshandhebel *m* brake hand lever
Bremshauptzylinder *m* master cylinder
Bremshebel *m* brake lever
Bremshebelführung *f* brake-lever guide
Bremshundertstel *n (ry)* effective (percentage of) braking power
Bremsketten *fpl* launching drags, drag chain
Bremsklaue *f* brake-shoe holder
Bremsklotz *m* brake shoe
Bremskolben *m* brake piston
Bremskraft *f* braking force
B. am Radumfang braking effort *(el. traction)*
B. bei Gefällebremsung holding braking effort *(el. traction)*
Bremskraftregler *m* brake pressure regulator
Bremskupplung *f* brake coupling
Bremskurve *f (rock)* braking curve
Bremsleistung *f* brake horsepower, BHP

Bremsleitung *f* brake line
Bremsleuchte *f* brake light, stop lamp (light)
Bremslicht *n* stop light
Bremslichtdrehschalter *m* stop-light rotating switch
Bremslichtöldruckschalter *m* hydraulic stop-light switch
Bremslichtzugschalter *m* stop-light pull switch
Bremsluftbehälter *m* compressed-air brake cylinder
Bremsluftbohrung *f* air bleeder
Bremslüfter *m* brake magnet (lifter), brake lifting magnet
Bremsluftschraube *f* fan brake, static airscrew
Bremsmagnet *m* braking magnet [of a meter], magnetic brake
Bremsmanschette *f* brake sealing cup
Bremsmedium *n s.* Bremsmittel
Bremsmittel *n (nucl)* moderator, moderating medium
Bremsmoment *n* braking torque *(of a motor)*
Bremsnocken *m* brake cam
Bremsnockenhebel *m* brake cam lever
Bremsnockenlager *n* brake cam bushing
Bremsnockenwelle *f* brake camshaft
Bremsnutzung *f (nucl)* resonance escape probability
Bremspedal *n* brake pedal
Bremspferdestärke *f* brake horsepower, BHP
Bremspotential *n* stopping (retarding) potential
Bremsprobe *f s.* Bremsprüfung
Bremsprüfung *f* brake trial (test)
Brems-PS *s.* Bremspferdestärke
Bremsquerschnitt *m (nucl)* stopping (slowing-down) cross section, atomic stopping power
Bremsrakete *f* retro-rocket, retrograde (retro-active, retardation, forward-facing, reverse-thrust) rocket, reverse jato
Bremsregulierlager *n* brake-adjusting bearing
Bremsschalter *m* braking controller
Bremsschaltung *f* braking circuit
Bremsscheibe *f* brake pulley
Bremsschild *m (ship)* shield *(launching)*
Bremsschlauch *m* [air] brake hose
Bremsschlußkennzeichenleuchte *f* combined stop-tail-number plate lamp
Bremsschlußleuchte *f* stop and tail lamp, stop-tail lamp
Bremsschlußlicht *n* stop and tail light
Bremsschlußnummernlicht *n* stop-tail-licence plate lamp
Bremsschuh *m* brake shoe
Bremsschuhhalter *m* brake shoe-holder
Bremsschutzwiderstand *m* protective resistance for rheostatic braking
Bremsschwund *m* brake fading
Bremssegment *n* brake-lever sector, brake quadrant
Bremsseil *n* brake wire (cable)
Bremsseilzug *m* brake cable assembly
Bremsspalt *m* retarding slit
Bremsspannung *f* retarding-field potential; stopping potential
Bremsspektrum *n* bremsspectrum, retardation spectrum
Bremssperrklinke *f* brake ratchet
Bremsspur *f* tyre marks; braking impression
Bremsstoff *m* moderator
Bremsstrahl *m* retardation (decelerating) jet
Bremsstrahlung *f* 1. bremsstrahlung; 2. braking radiation
Bremsstrahlungsquelle *f* bremsstrahlung source
Bremsstrecke *f* stopping (braking) distance
Bremsstrom *m* braking current
Bremsstützlager *n* brake-support bearing
Bremssubstanz *f* moderator
Bremstheorie *f (nucl)* slowing-down theory
Bremsträger *m* brake anchor pin (plate)
Bremstrommel *f* brake drum
Bremstrommelnabe *f* brake drum hub
Bremstrommelrückwand *f* brake backing plate
Bremsung *f* braking; deceleration; retardation; slowing down; stopping; moderation; thermalization; delay
  B. von Elektronen deceleration (retardation) of electrons
  elektrische B. electric braking
  elektromagnetische B. electro-magnetic braking
Bremsventil *n* brake valve
Bremsverhältnis *n* moderating ratio
Bremsvermögen *n* 1. braking power; 2. *(nucl)* stopping (moderating, slowing-down) power
Bremsversuch *m* brake test
Bremsverzögerung *f* deceleration under the influence of braking
Bremswahrscheinlichkeit *f (nucl)* slowing-down probability
Bremswärme *f* braking heat
Bremsweg *m* braking distance
  B. bis zum Stillstand stopping distance
Bremswelle *f* brake toggle, brake [cross] shaft
Bremswiderstand *m* braking resistance, retarding torque
Bremswirkung *f* slowing-down effect; braking effect; retardation
Bremszahl *f* stopping number
Bremszaum *m* Prony brake; brake dynamometer
Bremszeit *f* moderation time
Bremszugstange *f* connecting rod of brake-lever, brake pull rod
Bremszwischenwelle *f* brake cross shaft
Bremszylinder *m* brake (vacuum) cylinder; dashpot
Brennachse *f (opt)* focal axis

brennbar combustible; inflammable; burnable *(ceramics)*
Brennbarkeit *f* combustibility
Brennbarkeitsversuch *m* burning test *(fibres)*
Brennbarkeitsverzögerer *m* *(text)* fire retardant agent
Brennbock *m* *(text)* crabbing jack
Brenndauer *f* 1. duration of combustion; combustion time; 2. burning time, life *(valve)*
Brennebene *f* *(opt)* focal plane
  hintere B. rear focal plane
  vordere B. front focal plane
Brennelement *n* fuel element
  abgebranntes B. spent fuel element
  nacktes B. uncanned fuel element
  plattenförmiges B. plate-type fuel element
  stabförmiges B. rod-type fuel element, fuel rod
  verbrauchtes B. spent fuel element
Brennelementefabrik *f* fuel fabrication plant
Brennelement[en]hülle *f* *(nucl)* can[ning], cladding, fuel element can (cladding, jacket, sheath), fuel can (cladding, jacket, sheath)
Brennelementhüllendefekt *m* *(nucl)* fuel element failure
Brennelementhüllenschaden *m* *(nucl)* fuel element failure
brennen 1. to burn *(fuels, bricks)*; 2. to cauterize *(with hot iron)*; 3. to fire *(pottery)*; to bake *(ceramics)*; 4. to calcinate *(lime)*, to calcine; to roast *(siderite)*
  Ziegel b. to burn bricks *(in a kiln)*
Brennen *n* **der Schleifscheibe** burning of the grinding wheel
Brenner *m* 1. blowpipe *(welding)*; *(Am)* torch; 2. cutting blowpipe (torch), cutter; 3. burner *(Bunsen)*; 4. distiller
Brennerdüse *f* *(rock)* burner nozzle
Brennerei *f* distillery
Brennereinbruch *m* *(mine)* burn (shatter) cut, Michigan-cut
Brennereinstellung *f* combustion adjustment
Brennerspitze *f* burner tip
Brennerüberwachung *f* combustion control
Brennerzange *f* burner pliers
Brennfläche *f* focal (caustic) surface
  brechende B. diacaustic surface
  reflektierende B. catacaustic surface
Brennfleck *m* *(opt)* focal spot; cathode spot *(mercury rectifier)*
  bandförmiger B. strip focus
Brennflotte *f* *(text)* crabbing liquor
Brenngas *n* combustible (heating, fuel) gas
Brenngas-Luft-Flamme *f* air-acetylene flame
Brenngas-Sauerstoff-Flamme *f* oxy-acetylene flame
brenngehärtet flame-hardened
Brenngeschwindigkeit *f* burning rate, rate of combustion (burning)
  stündliche B. rate of combustion per hour
Brennglas *n* burning glas
Brennhärtemaschine *f* flame-hardening machine
Brennhärten *n* flame hardening
Brennholz *n* firewood, cordwood; fuel wood; stove wood; split wood
Brennholzspalterei *f* firewood factory
Brennkammer *f* combustion chamber *(gas turbine; heating)*
Brennkammereinsatz *m* *(rock)* combustion chamber liner
Brennkammerfestigkeit *f* *(rock)* strength of the combustion unit
Brennkammerkühlrohr *n* *(rock)* combustion unit cooling line
Brennkammerleistung *f* *(rock)* performance of combustion unit
Brennkapsel *f* saggar, sagger
Brennkegel *m* melting (pyrometric, fusible, Seger) cone
Brennkopf *m* *(rock)* igniter chamber
Brennkraftmaschine *f* *s.* Verbrennungskraftmaschine
Brennlinie *f* caustic line (curve), focal line (curve); catacaustic curve *(mirror)*; diacaustic curve *(lens)*
Brennmaterial *n* fuel, combustible material
Brennofen *m* [firing] kiln; calcinator; roasting furnace
Brennöl *n* fuel (burning) oil
Brennpunkt *m* *(opt)* focal point; fire point
  B. / mit gemeinsamem confocal
Brennpunkte *mpl* **der Ellipse** foci of the ellipse
brennputzen to deseam
Brennputzen *n* flame scarfing (chipping, descaling), *(Am)* torch deseaming
Brennrahmen *m* testing screen
Brennraum *m* combustion space
Brennreisig *n* faggot wood for fuel
Brennrohr *n* combustion cowl
Brennschluß *m* *(rock)* cut-off, combustion (propellant) cut-off, burn-out, combustion termination, termination of combustion
  B. der zweiten Raketenstufe second-stage burn-out
Brennschlußanlage *f* combustion cut-off system
Brennschlußbodenanlage *f* cut-off ground installation
Brennschlußempfänger *m* cut-off evaluation (signal) receiver
Brennschlußempfängerwagen *m* cut-off evaluation truck
Brennschlußfrequenz *f* combustion cut-off frequency
Brennschlußgabe *f* cut-off signal transmission
Brennschlußgeschwindigkeit *f* cut-off (burn-out) velocity, burn-out speed

Brennschlußgewicht *n* weight at combustion termination, burn-out weight
Brennschlußgröße *f* cut-off value
Brennschlußhöhe *f* burn-out altitude
Brennschlußkommando *n* combustion cut-off signal
Brennschlußsignal *n* combustion cut-off signal
Brennschlußsignalanlage *f* cut-off ground installation
Brennschlußstelle *f* combustion cut-off signal site
Brennschlußstreuung *f* variation in combustion cut-off point
Brennschlußvermessung *f* cut-off control
Brennschlußweite *f* abscissa of point of cut-off
Brennschlußwinkel *m* cut-off angle
Brennschlußzeit *f* burn-out time
Brennschneiden *n* flame cutting
Brennschwindung *f* firing shrinkage, shrinkage in the process of firing
Brennspannung *f (el)* operating (working) voltage, running voltage [of an arc]
Brennspiritus *m* methylated spirit
Brennstand *m (rock)* static (combustion) test stand
Brennstaubfeuerung *f* pulverized coal firing
Brennstelle *f (el)* power consuming point, lighting outlet
Brennstoff *m* fuel; fuel material *(reactor)*
  B. auf Anilinbasis *(rock)* aniline base fuel
  fester B. solid fuel
  flüssiger B. liquid fuel
  keramischer B. *(nucl)* ceramic fuel [material]
  metallischer B. *(nucl)* metallic fuel
  minderwertiger B. lower-grade fuel
  rauchloser B. smokeless fuel
Brennstoffanordnung *f (nucl)* fuel assembly
Brennstoffäquivalent *n* / **thermokinetisches** *(rock)* thermokinetic fuel equivalent
Brennstoffaufarbeitung *f* fuel reconditioning (regeneration, reprocessing)
Brennstoffaufarbeitungsanlage *f* fuel reprocessing plant
Brennstoffausnutzung *f* fuel utilization
Brennstoffbatterie *f* fuel battery
Brennstoffbehälter *m* fuel reservoir (tank, container)
Brennstoffdurchsatz *m* fuel throughput
Brennstoffdüse *f* fuel injection nozzle
Brennstoffeinsatz *m* fuel charge (inventory)
Brennstoffeinspritzer *m* fuel injector
Brennstoffelement *n (nucl)* fuel element; *(el)* gas cell
Brennstoffersparnis *f* saving of fuel
Brennstoffgemisch *n* fuel mixture
Brennstoffgitter *n* [reactor] lattice
Brennstoffhauptventil *n (rock)* main propellant valve
Brennstoffilter *n* fuel strainer
Brennstoffkanal *m (nucl)* fuel channel
Brennstoffkassette *f (nucl)* fuel assembly
Brennstoffknappheit *f* fuel shortage
Brennstofflager *n* fuel dump
Brennstofflagerung *f* fuel storage
Brennstoffleitung *f* fuel supply line
Brennstoffnadelventil *n* **im Vergaser** carburetter needle valve
Brennstoffnutzung *f* fuel utilization
Brennstofförderpumpe *f* fuel feed pump
Brennstoffpumpe *f* fuel pump
Brennstoffraum *m* fuel tank (space)
Brennstoffspareinrichtung *f* fuel economizer
Brennstoffspeiseeinrichtung *f* fuel feeder
Brennstoffstab *m (nucl)* fuel rod
Brennstoffsuspension *f (nucl)* fuel suspension
  nasse B. slurry fuel
Brennstofftagestank *m* service fuel tank
Brennstofftechnik *f* fuel engineering
Brennstoffübernahme *f* fuelling
Brennstoffüberwacher *m* fuel watcher
Brennstoffumhüllung *f (nucl)* fuel sheathing
Brennstoffumladung *f* fuel reloading
Brennstoffumsetzung *f* fuel reloading
Brennstoffverbrauch *m* fuel consumption
Brennstoffverbrauchsmesser *m* fuel-flow gauge
Brennstoffvorventil *n (rock)* first propellant valve
Brennstoffwechsel *m (nucl)* refuelling
Brennstoffwiederaufbereitungsanlage *f* *s.* Brennstoffaufarbeitungsanlage
Brennstoffzerstäubersystem *n* fuel-atomizer system
Brennstoffzuführung *f* fuel feed
Brennstoffzyklus *m (nucl)* fuel cycle
Brenntemperatur *f* baking (firing) temperature; temperature of calcination
Brennweite *f (opt)* focal length
Brennwert *m* heating power
Brennzeit *f* combustion (burning) period; duration of combustion; *(el)* conducting period; [rocket] firing duration, firing (burning) time
Brenzkatechin *n* pyrocatechin, catechol
Brenztraubensäure *f* pyruvic acid
Brenztraubensäurealdehyd *m* pyruvaldehyde
Brett *n* 1. *(build)* deal; board; plank; batten; 2. panel *(controls)*
  gekehltes B. moulded board
  gezinktes B. dovetailed board
  vierzölliges B. quarter stuff
  vollkantiges B. flatted plank
brettartig *(text)* boardy *(feel)*
Brettbinder *m* built-up truss
Brettchen *npl* planchettes
Brettchenhobelmaschine *f* thicknessing machine for boards

Bretter *npl* boards; timber
gespundete B. matched boards
schmale B. planchettes
verleimte B. laminated timber
Bretterdach *n* board roof
Bretterlager *n* timber store
Brettersäge *f* pit saw
Bretterschnitt *m* board cut
Bretterstoß *m* stack of boards
Brettertür *f* batten door; ledged door
Bretterverschlag *m* 1. partition boards; 2. timber-lined shed
Bretterwagen *m* timber carriage
Bretterwand *f* batten wall; board[ed] partition
Bretterzaun *m* close-boarded fence, boarding, planking
Brettsäge *f* long (pit) saw
Brettsägefeile *f* long saw file
Brettschaltung *f* breadboard [assembly], baseboard layout (setup)
Brettverkleidung *f* boarding
Brettverschalung *f* sheathing
Brief *m* letter
Briefhülle *f* correspondence envelope
Briefpapier *n* note paper
Brieftaube *f* carrier pigeon
Briefumschlag *m* envelope
Briefumschlagfutterseide *f* envelope lining
Briefumschlagmaschine *f* envelope making machine
Briefwaage *f* letter balance
Brigade *f* team, gang, crew
Brigadier *m* foreman, job captain, team leader; *(mine) (Am)* face boss
Brigg *f (ship)* brig
Brikett *n* briquette
Brikettdruckfestigkeit *f* briquette crushing strength
Brikettfabrik *f* briquetting factory
brikettieren to briquette
brikettierfähig briquettable
Brikettierfähigkeit *f* briquettability
Brikettierkohle *f* briquetting coal
Brikettierung *f* briquetting
Brikettierungseigenschaft *f* briquetting quality, briquettability
Brikettpresse *f* briquetting press
Brillant *m* brilliant
Brillantgrün *n* brilliant green
Brillantsucher *m* brilliant view-finder *(photography)*
Brillanz *f* brilliancy
Brille *f* 1. *(opt)* spectacles, glasses; goggles; 2. *(mine)* socket, back of shot-hole; 3. gland *(stuffing box)*; 4. dividing cam *(knitting machine)*; 5. *(mach)* centre rest
B. der Stopfbüchse stuffing-box gland
Brillenanpassung *f* fitting of spectacles
Brillenfutter *n* flange of gland, gland lining (bush) *(stuffing box)*
Brillengestell *n* spectacle frame
Brillengestellmaß *n* spectacle frame gauge
Brillenglas *n* spectacle lens
achsensymmetrisches B. spherical spectacle lens
astigmatisches B. cylindrical spectacle lens
neutrales B. non-corrective lens
Brillenglaskondensor *m* spectacle-lens condenser
Brillenofen *m* spectacle furnace
Brinellhärte *f* Brinell hardness
Brinellhärteprüfung *f* Brinell hardness test
Brinellhärtezahl *f* Brinell hardness number
Brinellkugeldruckprüfung *f* Brinell ball hardness test
Brinellprüfung *f* Brinell test
bringen to bring
auf eine Umlaufbahn b. *(rock)* to bring into orbit
auf Fertigmaß b. to finish to size
auf Gleichlauf b. to gang; to synchronize
aus dem Eingriff b. to disengage, to drop out of mesh
aus dem Gleichgewicht b. to unbalance
aus der Phase b. to outphase
außer Eingriff b. to disengage, to drop out of mesh
das Raumschiff aus dem Bereich der Erdanziehung b. to make the space ship escape terrestrial gravity
den Satelliten auf seine Bahn b. to place the satellite into its orbit
in Eingriff b. *(mach)* to put into mesh
in Phase b. to phase
in Umlauf b. to issue *(coins)*
wieder auf Null b. to restore to zero
zu Tal b. *(mine)* to flit
zum Stehen b. to bring to rest
brisant highly-explosive
Brisanz *f* shattering power, violence, disruptive power *(of explosives)*
Brisanzgranate *f* high-explosive shell
Brisanzsprengstoff *m* high-explosive
Brise *f* breeze
leichte B. flurry, gentle breeze
Briseur *m (text)* licker-in
Bristolpapier *n* Bristol (ivory) paper
Britanniametall *n* Britannia metal
bröckelnd crumbly; brittle; friable
Brockengespenst *n* spectre of the Brocken
Bröckler *m (ag)* husker shredder
bröcklig crumbly; brittle; friable
Brodeln *n* bubble, boiling noise *(radio)*
Brodelstörung *f* alternating-current hum, grinders *(radio)*
Brom *n* bromine
Bromabspaltung *f* debromination

Bromazeton *n* bromoacetone
Brombenzol *n* bromobenzene
Bromide *n* bromide
Bromidemulsion *f* bromide emulsion
bromieren to brominate
Bromierung *f* bromination
Bromoform *n* bromoform
Bromöldruck *m* bromoil process; bromoil print
Bromphenolblau *n* bromophenol blue
Bromsäure *f* bromic acid
Bromsilber *n* silver bromide
Bromsilberdruck *m* bromide print; bromide printing [process]
Bromsilberpapier *n* silver bromide paper
Bromsilberverfahren *n* bromide process
Bromvergiftung *f* bromism
Bromwasserstoff *m* hydrogen bromide
Bromwasserstoffsäure *f* hydrobromic acid
Bromzahl *f* bromine value (number)
Bromzählrohr *n* *(nucl)* bromic counter
Bronze *f* bronze
Bronzebezug *m s.* Bronzebuchse
Bronzebuchse *f* bronze-box
Bronzedruckfarbe *f* bronze (metallic) ink
Bronzegießerei *f* bronze foundry
Bronzeglanz *m* *(text)* bronziness
Bronzelack *m* bronze varnish
Bronzepigment *n* bronze powder
Bronzepulver *n* bronze powder
Bronzeunterdruckfarbe *f* bronzing medium
Bronzezeit *f* *(geol)* brazen age
bronzieren to bronze
Bronzier- und Abstaubmaschine *f* bronzing and dusting machine
Bronzit *m* bronzite
Brookit *m* brookite
broschieren *(text)* to brocade
Broschierlade *f* *(text)* embroidery lathe
Broschierstuhl *m* *(text)* swivel loom
Broschierwürfel *m* *(leath)* fulling cube
Broschur *f* stitching
Broschüre *f* booklet, brochure, paper-bound book, folder
Broschürenheftmaschine *f* booklet stitcher
Broschüren- und Blockdrahtheftmaschine *f* brochure and pad wire stitcher
Brotgärung *f* panary fermentation
Brotröster *m* / **elektrischer** electric toaster
Bruce-Antenne *f* Bruce aerial, Grecian[-type] aerial
Bruch *m* 1. *(math)* fraction; 2. fracture *(material, bone)*; rupture; 3. failure; crack; *(text)* break[age]; breach; fault; 4. *(mine)* fall *(face)*; fold
  echter B. *(math)* proper fraction
  faseriger B. *(met)* fibrous fracture
  feinkörniger B. *(met)* fine-grained fracture
  gemeiner B. *(math)* vulgar fraction
  gemischter B. *(math)* mixed number
  gitterartiger B. *(met)* lattice fracture
  glänzender B. *(met)* splendent (glossy) fracture
  grauer körniger B. *(met)* grey granular fracture
  grober B. *(met)* granular fracture
  interkristalliner B. *(met)* intercrystalline fracture
  körniger B. *(met)* granular fracture
  lamellarer B. *(met)* laminated fracture
  muscheliger B. *(min)* conchoidal fracture
  nadelartiger B. *(met)* acicular fracture
  schieferartiger B. splintery fracture
  unechter B. *(math)* improper fraction
  verformungsloser B. *(met)* non-deformed fracture
  winkliger B. *(met)* slip fracture *(tensile test)*
Bruchaussehen *n* appearance of fracture
Bruchbau *m* *(mine)* caving [method]
Bruchbauverfahren *n* *(mine)* caving [method]
Bruchbeanspruchung *f* breaking stress
Bruchbelastung *f* breaking (fracturing) load, ultimate stress
Bruchbild *n* [appearance of] fracture
Bruchbildung *f* fracturing, rupturing
Bruchdehnung *f* strain at failure, failure strain; breaking elongation
Bruchebene *f* plane of fraction
Bruchfalte *f* *(geol)* faulted overfold, fold-fault
Bruchfeld *n* *(mine)* caved area (goaf, waste)
bruchfest unbreakable, fracture-proof
Bruchfestigkeit *f* ultimate (rupture, crushing, breaking) strength
Bruchfestigkeitsgrenze *f* breaking-down limit (point)
Bruchfläche *f* [area of] fracture, fracture surface
  faserige B. fibrous fracture
  feinkörnige B. smooth (even) fracture
  frische B. freshly fractured surface
  grobkörnige B. bright-crystalline fracture
  splitterige B. barked (splintery) fracture
Bruchflächenbeschaffenheit *f* surface texture *(of aggregates)*
Bruchgefahr *f* risk of breakage
Bruchgefüge *n* grain structure
  ungleichmäßiges B. irregular fracture
Bruchgestein *n* quarry rocks
Bruchgleichung *f* fractional equation
Bruchgrenze *f* 1. *(met)* ultimate stress limit; 2. *(mine)* breaking limit
brüchig brittle; fragile; short
Brüchigkeit *f* brittleness; fragility; shortness
  interkristalline B. cleavage brittleness
Bruchkante *f* *(mine)* breaking edge *(roof)*
Bruchkantenstempel *m* *(mine)* chock
Bruchlandung *f* crash-landing
Bruchlast *f* ultimate (rupture, crushing, failure, collapse) load; break load *(text; insulator)*

Bruchlinie *f* 1. break line *(technical drawing)*; 2. line (envelope) of rupture; 3. *(geol)* fault line, line of faulting
Bruchlinientheorie *f (build)* yield line theory
Bruchlochwicklung *f (el)* fractional-slot winding
Bruchmassen *fpl* rubbles
Bruchmodul *m* modulus of rupture
Bruchmoment *n* ultimate moment
Bruchneigung *f* tendency of fracture
Bruchpilot *m* hard-luck pilot
Bruchprobe *f* specimen for strength test
Bruchprüfung *f* breaking (breakdown) test
Bruchpunkt *m (plast)* brittle (breaking) point
Bruchrechnung *f* fractional operation[s]
Bruchsattel *m (geol)* fault saddle
bruchsicher break-proof; shatterproof; resistant to fracture (rupture)
Bruchsicherheit *f* safety against fracture
Bruchspalte *f (geol)* break, crack
Bruchspaltenbildung *f (geol)* rifting
Bruchspan *m* discontinuous (segmental) chip
Bruchspannung *f* ultimate stress; failure (crushing) stress
Bruchstein *m* quarry stone
Bruchsteinmauerwerk *n* rubble masonry (work)
Bruchsteinunterlage *f* rubble bed
Bruchstelle *f* spot of rupture
Bruchstempel *m (mine)* breaker prop
Bruchstempelreihe *f* breaker row
Bruchstrich *m (math)* fractional line
Bruchstück *n* broken log
Bruchteil *m* fraction
Bruchwinkel *m* angle of break
Bruchzone *f (geol)* fracture zone
Brucit *m* brucite
Brücke *f* bridge
  B. mit Gleitkontakt slide bridge *(measuring)*
  B. mit untenliegender Fahrbahn trough bridge
  akustische B. acoustic bridge
  fliegende B. *(ship)* fore and aft bridge
  schiefe B. skew bridge
  schiefwinklige B. skew bridge
  selbstabgleichende B. *(el)* self-balancing bridge
  Wheatstonesche B. resistance bridge
  Wiensche B. Wien bridge
Brückenabgleich *m* bridge balance *(measuring)*
Brückenanordnung *f (el)* bridge circuit
Brückenarm *m* bridge (ratio) arm
Brückenauffahrt *f* bridge approach
Brückenbau *m* bridge engineering (construction)
Brückenbauingenieur *m* bridge engineer
Brückenbildung *f* arching
  B. zwischen Kettenmolekülen *(ch)* chain bridging
Brückenbindung *f (ch)* bridge bond
Brückenduplexschaltung *f* bridge duplex connection
Brückenendschott *n (ship)* bulkhead at after end of bridge
Brückenfähre *f* ferry bridge
Brückenfahrstand *m (ship)* bridge control
Brückenflügel *m* bridge leaf
Brückenförderer *m* piggyback conveyor
Brückenfrontschott *n (ship)* bulkhead at front of bridge
Brückengegensprechsystem *n (telec)* bridge duplex system
Brückengeländer *n* bridge railing
Brückengeld *n* [bridge] toll, [bridge] due
Brückengleichrichter *m* bridge[-connected] rectifier
Brückenhaus *n (ship)* bridge house
Brückenkleid *n (ship)* dodger
Brückenkontakt *m* bridge contact
Brückenkörper *m* bridge wall *(furnace)*
Brückenkran *m* bridge (overhead, travelling) crane
Brückenlager *n* bridge bearing
Brückenlegepanzer *m* bridge-laying tank
Brückenmischer *m* 1. mixer-spreader *(highway construction)*; 2. *(el)* balanced mixer
Brückenmodulationsmesser *m* differential modulation meter
Brückennock *n (f) (ship)* wing of a bridge, wing of the navigating bridge, bridge wing (sponson)
Brückenpanzerkampfwagen *m* bridge-carrying tank
Brückenpfeiler *m* bridge pillar
Brückenschaltung *f* bridge [network], bridge circuit (connection, transition)
Brückenschaltungsverstärker *m* bridge limiter *(analogue computer)*
Brückenspannung *f* bridge voltage
Brückenspeisung *f* feeding by bridges
brückenstabilisiert *(el)* bridge-stabilized
Brückenstanze *f (leath)* revolution press
Brückensternschaltung *f* bridged T-connection, bridged Y-connection
Brückensteuerung *f* bridge (remote) control *(of vessels)*
Brückenstück *n* bridge piece *(of vessels)*
Brückentafel *f* bridge deck
Brückenübertrager *m* hybrid (differential) transformer
Brückenwiderstand *m (el)* bridge resistance
Brückenzünder *m* low-tension electric detonator
Brückenzweig *m (el)* bridge branch (arm)
Brüden *pl* vapour[s]
Brüdenabzug *m* vapour escape (hood)
brühen to scald, to blanch
Brühe *f* / **Bordelaiser** Bordeau mixture
Brühezufuhr *f (ag)* spray feed

Brühpumpe *f (leath)* tan liquor pump
Brüllen *n* **des Motors** roaring of the engine
Brumm *m* hum *(radio)*
Brummabstand *m* signal-to-hum ratio
Brummeinstreuung *f* hum pick-up
brummen to hum *(radio)*; to vibrate *(piston)*
Brummen *n* hum[ming] *(radio)*; vibration *(piston)*
B. des Kolbens vibration of the piston
Brummfaktor *m* ripple ratio
Brummfilter *n* hum eliminator
brummfrei hum-free
Brummschleife *f* hum pick-up
Brummspannung *f* hum (ripple) voltage
Brummspannungsverhältnis *n* ripple ratio, percent ripple voltage
Brummton *m* hum [tone]
Brummverzerrung *f* ripple effect
brünieren to burnish, to bronze, to black [finish], to brown, to blue
Brunnen *m* well
artesischer B. artesian well
Dortmunder B. Dortmund tank
Brunnenbau *m* well construction
Brunnenbohren *n* well drilling (boring)
Brunnendeck *n* well deck *(of a vessel)*
Brunnengründung *f* well foundation
Brunnenring *m* well lining [precast concrete] segment
Brust *f* 1. front *(blast furnace)*; 2. *(text)* chest *(needle)*; 3. *(mine)* breast
Brustbaum *m* breast beam *(weaving)*
Brustbohrer *m s.* Brustbohrmaschine
Brustbohrmaschine *f* breast drill
Brusthöhe *f* breast height
Brustholz *n* waler, waling
Brustleier *f s.* Brustbohrmaschine
Brustmikrofon *n* [breast] transmitter, chest set
Brustschild *m* pusher-blade *(bulldozer)*
B. / mit of the breast type *(drill)*
Brustschleifen *n* face sharpening
Brüstung *f* parapet [wall], balustrade
Brüstungsmauer *f* parapet [wall]
Brustwalze *f* breast roller
Brutanstalt *f* hatchery
Brutapparat *m* incubator
Brutausbeute *f (nucl)* breeding processing efficiency
brütbar *(nucl)* fertile
brüten 1. *(ag)* to hatch; to incubate; 2. *(nucl)* to breed *(reactor)*
Brüten *n* 1. *(ag)* hatching; incubation; 2. *(nucl)* breeding
Brüter *m (nucl)* breeder reactor
schneller B. fast breeder [reactor]
thermischer B. thermal breeder [reactor]
brutfähig *(nucl)* fertile
Brutfaktor *m s.* 1. Brutverhältnis; 2. Konversionsverhältnis
Brutgewinn *m* breeding gain
Brutmantel *m* [breeding] blanket
Brutmaschine *f (ag)* incubator
elektrische B. electric incubator
Brutmaterial *n (nucl)* breeding (fertile) material
Brutreaktor *m (nucl)* breeder [reactor]
schneller B. fast breeder [reactor]
thermischer B. thermal breeder [reactor]
Brutstoff *m (nucl)* fertile material
Bruttoleistung *f* gross output
Bruttolohn *m* gross wage
Bruttomasse *f* gross weight
Bruttoraumgehalt *m (ship)* gross tonnage closed in space
Bruttotonnenkilometer *npl* gross tonnage-kilometres
Bruttovermessung *f (ship)* gross tonnage (measurement)
Brutverhältnis *n (nucl)* breeding ratio
Brutzeit *f (ag)* breeding season; incubation period
Brutzone *f (nucl)* [breeding] blanket
Brutzyklus *m (nucl)* breeding cycle
Bruzin *n (ch)* brucine
Buch *n* book
Schwarzes B. *(ship)* black book *(Lloyd)*
Buchbesprechung *f* book review
Buchbinder *m* bookbinder
Buchbinderei *f* 1. bookbinding; 2. bookbinding establishment
Buchbinderklebestreifen *m* binding tape
Buchbinderleder *n* bookbinding leather
Buchbinderleinen *n* book cloth
Buchbindermusselin *m* binding cloth (muslin)
Buchbinderpappe *f* binder's board
Buchbinderpresse *f* screw (binding) press
Buchblock *m* book block
Buchblockfälzelmaschine *f* book backbone lining-up machine
Buchdecke *f* case, book cover
Buchdeckenanbiegemaschine *f* book-case bending machine
Buchdeckenmaschine *f* case making machine
Buchdruck *m* letterpress-printing
Buchdrucker *m* [letterpress] printer, typographer, pressman
Buchdruckfarbe *f* printing ink
Buchdruckpapier *n* book paper
Buchdruckplatte *f* / **flexible** flexible relief plate
Buchdruckpresse *f* [printing] press, letterpress
Buchdruckrotationsmaschine *f* rotary letterpress
Buchdruckschnellpresse *f* high speed printing machine [for letterpress work]
Buche *f* beech
Bucheckernöl *n* beechnut oil
Bucheinhängemaschine *f* casing[-in] machine
Buchenfrostspanner *m* northern winter moth

Buchenholz *n* beech
Buchenkohle *f* beech charcoal
Buchenspinner *m* pale tussock moth
Buchfahrplan *m* time-table
Buchfalzeinbrennmaschine *f* joint setting machine
Buchführung *f* **mit Lochkartenmaschinen** punched card machine accounting
Buchholz-Relais *n* *(el)* Buchholz [protective] relay, gas pressure relay
Buchholz-Schutz *m* *(el)* Buchholz protection (*s. a.* Buchholz-Relais)
Buchlager *n* book depository
Büchnertrichter *m* Buchner funnel
Buchpapier *n s.* Buchdruckpapier
Buchrückenleimapparat *m* spine gluing apparatus
Buchrückenrundemaschine *f* back rounding machine
Buchsbaum *m* boxwood
Buchsbaumholz *n* boxwood
Buchschnitt *m* fore edge [of book]
Buchse *f* bush[ing]; collar; socket; sleeve; hub
  kegelige B. taper bushing
Büchse *f* pot, tin; *(Am)* can
Buchsenauszieher *m* bush extractor
Buchsenbrett *n* socket strip
Büchsenfilter *n* canister filter
Büchsenfleisch *n* tinned meat, *(Am)* canned meat
Buchsenkette *f* bush roller chain
Buchsenlager *n* bush-bearing
Büchsenmilch *f* tinned (evaporated) milk
Buchsenrollenkette *f* bush roller chain
Buchsenventil *n* sleeve valve
Buchstabe *m* type, letter, character
  fetter B. extrabold letter
  großer B. capital letter
  halbfetter B. bold letter
  kleiner B. lower-case [letter]
  kurzer B. short letter *(without ascender and/or descender)*
  langer B. tall letter *(with ascender and/or descender)*
  magerer B. light letter
  unterschnittener B. kerned letter
  verfischter B. wrong fount
buchstabengekennzeichnet lettered; marked by letters
Buchstabengießmaschine *f* single-type casting machine
Buchstabengruppe *f (dat)* word
Buchstabeninversionssignal *n* letter-shift signal *(telegraphy)*
Buchstabenrechnung *f* algebra, algebraic calculus
Buchstabenreihe *f* letters case *(telegraphy)*
Buchstabentelegraf *m* A.B.C. telegraph
Buchstabentelegrafie *f* alphabetic (printing) telegraphy
Buchstabenumschaltung *f* letter shift
Buchstabenwechsel *m* letter shift *(telegraphy)*
Bucht *f* 1. *(telec)* bay; 2. inlet, bay, embayment, bight
  kleine B. *(ship)* inlet
Buchtschäkel *m* *(ship)* harp-shape shackle, jew's harp
Buchungsautomat *m* automatic accounting machine
Buchungsbeleg *m* journal voucher
Buchungsmaschine *f* accounting (book-keeping) machine
Buckel *m* 1. hump; 2. boss
Buckelblech *n* buckled plate
Buckelschweißung *f* projection welding
Buckling *n*, **Bucklung** *f s.* Flußwölbung
Buckskinkurbelwebstuhl *m* buckskin crank loom
Büffelleim *m* leather glue
buffen to velour-finish
Buffetwagen *m* buffet-coach, buffet-car
Bug *m* 1. bow, head, prow; 2. angle brace, strut *(carpentry)*
  scharfer B. lean bow
  senkrechter B. plumb bow
  springender B. flare-out bow
  voller B. bluff bow
  völliger B. bluff bow
Buganker *m* bower [anchor]
Bugankergeschirr *n* head gear
Bugankerklüse *f* bow hawse-hole
Bugantennenverkleidung *f* nose radome *(aeroplane)*
Bugaufklotzung *f* bow chock
Bügel *m* 1. bow; frame *(saw)*; 2. stirrup, tie, link *(reinforced concrete)*; 3. *(ship)* strap, bridle, hoop, bow, bale
  B. des Mannlochverschlusses manhole dog
Bügelabstand *m* pitch of links *(reinforced concrete)*
Bügelechtheit *f* *(text)* fastness to pressing (ironing)
Bügeleisen *n* flat-iron
  elektrisches B. electric iron
Bügelelektrode *f* bow-type electrode
Bügelfalte *f* crease
Bügelfleck *m* *(text)* greenish stain during ironing
bügelfrei non-iron
Bügelmaschine *f* / **elektrische** electric ironing machine
Bügelmeßschraube *f* external (outside) micrometer
bügeln to press, to iron
Bügeln *n* pressing, ironing
Bügelpresse *f* pressing machine

Bügelsäge *f (met)* hacksaw; web (framed cross-cut) saw; *(ag)* bow tree saw
handbetätigte B. hand hacksaw
Bügelsägeautomat *m* automatic hacksawing machine
Bügelsägeblatt *n* hacksaw blade
Bügelsägemaschine *f* hacksaw machine, power hacksaw
Bügelstromabnehmer *m* bow trolley (current collector)
Bugfender *m* bow (stem) fender, wedges, whiskers *(for boats)*
Bugfestmacher *m* bow (head) fast
Bugfigur *f* figure-head
Bugflagge *f* jack
Bugfüllstück *n* eiking, ek[e]ing
Buggeschütz *n* bow gun
Buggmaschine *f* folding machine
Bugklüse *f (ship)* bow hawse-hole
runde B. bow-chock mooring pipe
buglastig bow-heavy
Buglastigkeit *f* trim by the bow
Bugpforte *f* bow-port, raft-port *(opening in the bow of a ship)*
Bugpropeller *m* bow (forward) propeller
Bugradfahrwerk *n (aero)* nose [wheel] landing gear
Bugraum *m (ship)* bow compartment
Bugschanzkleid *n (ship)* bow bulwark, *(Am)* buffalo
Bugsee *f* head sea, bow wave
Bugsierdampfer *m* steam tug
Bugsierdienst *m* tug service
bugsieren to track, to tow, to warp
Bugsieren *n* towage, towing
Bugsierer *m* tower
Bugsierlohn *m* tug charge, towage
Bugspriet *m (ship)* bowsprit
einzuziehender B. running bowsprit
fester B. standing bowsprit
Bugsprietklampen *fpl (ship)* bees of a bowsprit
Bugsprietwanten *pl (ship)* bowsprit-shrouds
Bugstag *n (ship)* bowsprit shroud
Bugstrahlmanövriersystem *n* bow jet manoeuvring system
Bugstrahlruder *n* lateral thrust units
Bugstrahlruderantrieb *m* bow thruster drive
Bug- und Heckvertäuung *f (ship)* fore-and-aft moorings
Bugverstärkung *f* ice-lining, ice-doubling
Bugwelle *f* bow wave
Buhne *f* dike, dam, break water, quay, wharf, groyne, *(Am)* groin; bush weir, jetty
Bühne *f* 1. platform *(charging)*; 2. stage
Bühnenarbeiter *m* scene shifter
Bühnenbeleuchtung *f* stage lighting
Bühnenwagen *m* wagon *(theatre)*
Bühnloch *n (mine)* hitch
Buise *f* buss *(ship for fish-transport)*
Bulbbug *m (ship)* bulb bow
Bulbwinkel *m (ship)* angle bulb, bulb angle
Buleine *f (ship)* bowline
Bulin *f (ship)* bowline
Bulkfrachter *m* / **umgebauter** jumboized break-bulk freighter
verlängerter B. jumboized break-bulk freighter
Bulkladung *f (ship)* bulk cargo
Bullauge *n* side-light, bull's eye, portlight, *(Am)* porthole
Bulldozer *m* [bull]dozer
Bullenhütte *f* bull pen
Bumboot *n* bumboat
Buna *n (m)* buna
Bunagummi *m* buna rubber
Bund *m* 1. *(mach)* collar, shoulder *(shaft)*; flange; 2. band, cord *(book)*; 3. federation, union
Bundaxt *f* mortise cleaner, blocking axe
Bundbalken *m* joining beam
Bünde *mpl* / **echte** raised cords
Bündel *n* 1. sheaf; pack[et], bunch; cluster; bale; 2. *(math, opt)* beam; pencil; 3. *(telec)* group of lines
breites B. *(opt)* broad beam
enges B. *(opt)* narrow beam
inneres B. internal (circulating) beam
schmales B. *(opt)* narrow beam
umlaufendes B. circulating (internal) beam
Bündelbreite *f* beam width *(radar)*
Bündelendröhre *f (el)* beam-power valve
Bündelkabel *n* bank cable
Bündelleiter *m* bunch conductor *(overhead lines)*
bündeln 1. to tighten; to cluster; to bunch; 2. *(opt)* to focus; to phase
Bündelpfeiler *m* annulated (clustered) column
Bündelpotential *n* electron-stream potential
Bündelpresse *f* baling press
Bündelschweißung *f* multiple electrode welding
Bündeltetrode *f* beam tetrode *(radio)*
Bündelung *f (ph)* convergence; focus[s]ing; concentration; narrowing; beaming *(rays)*
dynamische B. dynamic convergence
horizontale B. horizontal directivity
Bündelungselektrode *f* focus[s]ing electrode, beam-forming plate
Bündelungspotential *n* electron-stream potential
Bündelungsröhre *f (ph)* buncher
Bündelungsschärfe *f* sharpness of directivity
Bündelungssteilröhre *f (ph)* buncher
Bündelweite *f* beam width *(rays)*
Bundgatter *n* gang saw, [multiple blade] frame saw
Bundgewicht *n* coil weight *(rolling, drawing)*
bündig fair[-faced] *(masonry)*; flush
b. mit flush with

b. mit der Oberfläche [eingebaut] flush mounted
bundlos shoulderless, flangeless
Bundseite *f* exterior face *(carpentry)*
Bundstahl *m* faggot steel
Bundsteg *m* binding margin, gutter-stick *(of books)*
Bundzeichen *n* jointing mark *(carpentry)*
Bunker *m (ship, build)* bunker; silo; *(mach)* bin; hopper
Bunkerabzugsrinne *f* bin discharge conveyor
Bunkeranlage *f* stock bin
Bunkerauslauf *m* bunker outlet
Bunker-C-Öl *n* bunker fuel oil
Bunkerdeckel *m* bunker ring, cover of the bunker
Bunkerentleerungswagen *m* bunker discharge (evacuating) wagon
Bunkerfüllstandszeiger *m* bin level indicator
Bunkerkohle *f* bunker (steam) coal, bunkers
Bunkerluke *f* scuttle hatch
bunkern to bunker, to coal, to fuel
Bunkerschieber *m* bunker sluice (slide), coal door, sliding door for bunker
Bunkerschiff *n* bunkering boat (vessel)
Bunkerschott *n* bunker bulkhead
Bunkerstandsanzeiger *m* bin level indicator, material-level indicator
Bunkertasche *f* bunker pocket
Bunkerzug *m* bunker (mucking, slusher) train, trainloader
Bünn *f* [fish] well, life-fish well
Bünnschiff *n* well boat (smack), wet boat *(fishing boat)*
Bunsenbrenner *m* Bunsen burner
Bunsenflamme *f* Bunsen flame
bunt varicoloured, many-coloured
Buntätzdruck *m (text)* coloured discharge print
Buntätze *f (text)* colour discharge
Buntbleichartikel *mpl (text)* coloured bleach goods
Buntbleiche *f (text)* bleaching of coloured goods
Buntbleierz *n* pyromorphite
Buntfärberei *f* multicolour dyeing
Buntheit *f (telev)* chrominance
Buntkette *f (text)* multicoloured warp
Buntkupferkies *m* bornite, peacock ore
Buntmetall *n* non-ferrous metal
Buntmetallfräsen *n* non-ferrous milling
Buntpapier *n* glossy (fancy, tinted) paper
Buntpapiermaschine *f* colouring machine
Buntregler *m (telev)* colour control
Buntreserve *f* **mit Küpenfarbenüberdruck** *(text)* coloured resist under vat printing
Buntreservierung *f* **von Anilinschwarz** *(text)* coloured resist under aniline black
Buntspinnerei *f* spinning of coloured yarn[s]
Buntwaschmittel *n* detergent for the coloured wash
Buntweberei *f* colour weaving
Bürde *f (el)* burden
Bürette *f (ch)* burette
Bürettenhalter *m* burette holder
Bürgersteig *m* footpath, footway, pavement
Bürgschaft *f* security; bail
Büroarbeit *f* paper (office) work
Bürogebäude *n* office building
Büroklammer *f* letter clip
Büroleim *m* mucilage
Büromaschine *f* office machine
Büromöbel *npl* office furniture
Bürooffsetdruckmaschine *f* offset duplicator
Bürorechenmaschine *f* accounting machine; business machine; desk-calculator
Bürste *f* brush
lamellierte B. *(el)* laminated brush
versetzt angeordnete B. *(el)* staggered brush
bürsten to brush
Bürsten *n (met)* power brush finishing, Tampico brushing *(polishing)*
Bürstenabhebeeinrichtung *f (el)* brush lifter
Bürstenabzug *m (print)* brush (galley, copy) proof
Bürstenandruckfeder *f (el)* brush spring
Bürstenanschlußkabel *n (el)* pigtail
Bürstenarm *m* trailer *(telephone)*
Bürstenbrücke *f (el)* brush-rocker, brush yoke
Bürstenfeuchtwerk *n* brush-type sheet remoistener
Bürstengestell *n (el)* brush gear, rocker
Bürstenhalter *m (el)* brush holder
schräger B. reaction (trailing) brush holder
Bürstenhalterkasten *m* brush box
Bürstenhalterpatrone *f* cartridge-type brush holder
Bürstenholz *n* brush wood
Bürstenkante *f* / **ablaufende** *(el)* trailing edge
auflaufende B. *(el)* entering edge
Bürstenkontakt *m (el)* brush contact
Bürstenmotor *m* collector motor
Bürstenputz *m* dinging
Bürstensatz *m* set of brushes
Bürstenschalter *m* laminated brush switch
Bürstenstreichmaschine *f* brush coater
Bürstenstreichverfahren *n* brush spreading
Bürstenstromabnehmer *m* brush [contact]
Bürstenträger *m (el)* wiper [shaft], selector rod; brush carriage *(punched card machine)*
Bürstenverschiebung *f (el)* brush shift (displacement)
Bürstenverstellmotor *m* brush-shifting motor
Bürstenvorverstellung *f (el)* forward lead of the brushes

Bürstenwähler *m* brush selector, trip rod *(telephony)*
Bürstenwalze *f* brush roll, circular (rotary) brush
Bürstenwelle *f (el)* wiper shaft
Bürstenzerstäuber *m* brush atomizer
Bürstenzungenöffner *m* brush latch opener
Bürstfärbung *f (leath)* staining, brush-dyeing
Bürstmaschine *f (text)* brushing machine
Bürstwalze *f* brushing roller
Bus *m* bus
Busch *m* bush
Buschegge *f (ag)* bush harrow
Büschel *n (math)* pencil; family
Büschelentladung *f* brush discharge
Büschelstruktur *f* cluster structure
Buschholz *n* underwood
Büse *f* buss *(ship for fish-transport)*
Bussole *f (opt)* compass
Bussolenrichtkreis *m* compass aiming circle (director)
Bussolenzug *m* compass traverse
Butadien *n* butadiene
Butan *n* butane
Butanol *n* butanol
Bütte *f* butt *(wine)*; [pulp-]vat; trough; mould; tub; bucket
Büttenkarton *m* deckle-edged board, feather-edged board
Büttenleimung *f* tub sizing
Büttenofen *m* pot furnace
Büttenpapier *n* vat (hand-made, mould-made) paper
Büttenpresse *f* vat (pulp) press
Büttenrand *m* deckle edge
Büttenrandpapier *n* deckle-edged paper
Butter *f* butter
Butterbrotpapier *n* food wrapper, greaseproof paper
Butterfaß *n* churn[ing barrel], churning vat
Butterfertiger *m* working churn, combined churn and butter worker, butter-making machine
Butterknetmaschine *f* butter blending machine
Buttermaschine *f* butter churn
Buttermilch *f* buttermilk
buttern to churn, to make butter
Buttersäure *f* butyric acid
Buttersäureester *m* butyric ester
Buttersäuregärung *f* butyric fermentation
Butterstößel *m* churn dasher (staff)
Butterworth-Tankreinigungsanlage *f* Butterworth tank cleaning system
Butylalkohol *m* butyl alcohol
Butylazetat *n* butyl acetate
Butylen *n* butylene
Butylkautschuk *m* butyl rubber
Byte byte *(quantity from 1 to 8 bits)*

C

c-Ader *f (el)* C-wire
C-Darstellung *f* C display
c-Draht *m* C-wire
C-Gehalt *m* carbon content
C-Haken *m* C-hook
C-Horizont *m (geol)* C-horizon
C-Karte *f* c chart, control chart for number of defects per unit
c-Kontaktarm *m (telec)* test brush
c-Leiter *m* C-wire
C-Rad *n* Curtis wheel
C-Schirm *m* C-scope
C-Stahl *m* carbon steel
C-Verstärker *m* class-C amplifier
C-Verstärkung *f* class-C valve-operation
CAB-Schiff *n* captured air bubble vessel, CAB vessel
Cachou *n* catechu
Caisson *m* caisson
Caissonkrankheit *f* caisson disease, bends
Caissonverfahren *n (mine)* pneumatic sinking method
Californium *n* californium
Callan-Element *n (el)* Callan cell
Calutron *n (nucl)* calutron *(California University Cyclotron)*
Calyx-Bohrgerät *n* Calyx drill
Cambridge-Walze *f (ag)* Cambridge roller
Campbell-Brücke *f (el)* Campbell bridge
Campecheholz *n* campeachy wood
Campingplatz *m* camping site
Candela *f (el)* candela, cd
Candelillawachs *n* candelilla wax
Carbamidharz *n* carbamide (urea) resin
Cardoxschießen *n (mine)* Cardox shotfiring
Cargocaire-Anlage *f (ship)* cargocaire installation *(dehumidifying)*
Carnallit *m (min)* carnallite
Carnaubawachs *n* carnauba wax
Carnotit *m (min)* carnotite
Carraramarmor *m* Carrara marble
Cassegrain-Gitterspektrograf *m* Cassegrain grating spectrograph
Cassegrain-Spiegelteleskop *n* Cassegrain reflecting telescope
Celsiusgrad *m* centigrade [degree], degree centigrade
Celsiusskale *f* centigrade scale
Čerenkov-Effekt *m (nucl)* Čerenkov effect
Čerenkov-Strahlung *f (nucl)* Čerenkov radiation
Čerenkov-Zähler *m (nucl)* Čerenkov counter
Cermet *n* metal-ceramic, cer[a]met
Cetanzahl *f* cetane number
Cetenzahl *f* cetene number
Chabasit *m (min)* chabazite
chagrinieren to shagreen

Chagrinierrolle *f (leath)* embossing roller
chagriniert shagreened
Chagrinnarben *m (leath)* pebble grain
Chalkanthit *m (min)* chalcanthite
Chalkopyrit *m* chalcopyrite
Chalkosin *m (min)* chalcocite
Changeanteffekt *m* shot (iridescent) effect
Changier-Spannrahmentrockenmaschine *f* jig stenter
Charakteristik *f* characteristic [curve]
  C. der nichtlinearen Verzerrung harmonic distortion characteristic
  C. der spektralen Empfindlichkeit spectral sensitivity characteristic
  fächerförmige C. fan marker
  fallende C. negative resistance characteristic
  negative C. negative resistance characteristic
  statische C. static characteristic *(of transductors)*; transfer curve
charakteristisch characteristic
Charge *f* [furnace] charge; batch; heat; melt; load
Chargenbetrieb *m* batch process
Chargenmischer *m (build)* batch mixer
Chargenprozeß *m* batch process
Chargensilo *m* batch holding hopper
Chargenzusammensetzung *f* composition of batch
Chargierbühne *f* charging floor
chargieren to charge, to load
Chargierkran *m* [furnace] charging crane
Chargierlöffel *m* charging spoon (peel)
Chargiermaschine *f* charging machine
Chargiermulde *f* charging box
Chargieröffnung *f* charging door
Chargiertrichter *m* feed hopper
Chargierwagen *m* charging barrow
Charter *m* **eines Schiffes** bare boat (pole, hull) charter *(without a crew)*
Charterer *m* charterer *(ship)*
Chartergebühr *f* charter money
chartern *(ship)* to charter
Charterpartie *f* charter party
Charterung *f* chartering, affreightment
chasen *(text)* to chase
Chassis *n* 1. chassis, base frame; 2. colour trough *(fabric printing)*
Chassisanordnung *f* chassis layout
Chassisantenne *f* under-car aerial
Chassisaufbau *m* chassis mounting
Chassisauflage *f* chassis support
Chassisbaukastensystem *n* modular chassis system
Chassismontage *f* chassis mounting
Chaulmugraöl *n* chaulmoogra oil
Cheddit *m* cheddite *(explosive)*
Chefingenieur *m* chief engineer
Chefkonstrukteur *m* chief designer
Chelatbildung *f (ch)* chelation, chelate formation
Chelatverbindung *f (ch)* chelate compound
Chemie *f* chemistry
  analytische C. analytical chemistry
  angewandte C. applied chemistry
  anorganische C. inorganic chemistry
  gerichtliche C. forensic chemistry
  heiße C. hot atom chemistry
  organische C. organic chemistry
  pharmazeutische C. pharmaceutical chemistry
  physikalische C. physical chemistry
  physiologische C. physiological chemistry
  reine C. pure chemistry
  technische C. chemical engineering
  theoretische C. theoretical chemistry
Chemieanlagen *fpl* chemical processing plant (equipment)
Chemieanlagenbau *m* chemical plant construction
Chemiefaser *f* synthetic (man-made, artificial) fibre
Chemieingenieur *m* chemical engineer
Chemieseide *f* continuous filament yarn; rayon
Chemigraf *m* block maker, process engraver
Chemigrafie *f* process engraving, photoengraving
Chemikalien *fpl* chemicals
chemikalienfest chemical resistant
Chemikalienflasche *f* reagent bottle
Chemikalienregenerierung *f* chemical recovery
Chemiker *m* chemist
Chemilumineszenz *f* chemiluminescence
chemisch chemical
Chemisorption *f* chemisorption
Chemotherapie *f* chemotherapy, chemotherapeutics
Chenille *f* chenille
Chenillegarn *n* chenille yarn
Chevillieren *n* chevilling *(silk)*
chevilliert *(text)* chevilled
Cheviot *m* cheviot
Chevretteleder *n* chevrette [leather]
Chiffre *f* cipher, code
Chiffreanzeige *f* keyed (box number) advertisement
chiffrieren to [en]code, to cipher
Chiffriernummer *f* code number
Chiffrierschlüssel *m (telec)* key; code
Chilesalpeter *m* Chile saltpetre (nitre)
Chinafahrer *m* chinaman *(ship)*
Chinagras *n* ramie, China grass
Chinaldin *n (ch)* quinaldine
Chinarinde *f* cinchona [bark]
Chinasäure *f* quinic (chinic) acid
Chinhydron *n (ch)* quinhydrone
Chinhydronelektrode *f* quinhydrone electrode
Chinhydronhalbzelle *f* quinhydrone half-cell

Chinidin *n (ch)* quinidine
Chinin *n* quinine
Chinizarin *n* quinizarine
chinoid *(ch)* qui[no]noid
Chinolin *n* quinoline
Chinon *n* quinone
Chinoxalin *n* quinoxaline
Chintz *m (text)* chintz
Chintzausrüstung *f* chintz finish
Chintzdruck *m* chintz printing
Chintzeffekt *m* chintz effect
Chireixmodulation *f (el)* Chireix (outphasing) modulation
Chitosamin *n (ch)* glucosamine
Chloanthit *m (min)* c[h]loanthite
Chlor *n* chlorine
Chlorabspaltung *f* dechlorination
Chloral *n* chloral
Chloramin *n* chloramine
Chlorat *n* chlorate
Chlorbenzol *n* chlorobenzene
Chlorbleiche *f* chlorine bleaching
Chlorbleispat *m* phosgenite
Chlorbromsilberfotopapier *n* gaslight paper
Chlorechtheit *f* chlorine fastness
chloren to chlorinate
Chloressigsäure *f* chloracetic acid
Chlorgehalt *m* chlorinity; chlorine content
chlorhaltig chlorous
Chlorid *n* chloride
Chlorieranlage *f* chlorination plant
chlorieren to chlorinate, to chloridize
chloriert chlorinated
Chlorierung *f* chlorination; chlorination stage
C. unter Aufspaltung der Kohlenwasserstoffkette chlorinolysis
Chlorierungsgefäß *n* chlorinator
Chlorierungsmittel *n* chlorinating agent
Chlorit *m (min)* chlorite *(green earth)*
Chloritbildung *f* chloritization
Chloritschiefer *m* chlorite schist
Chlorkalk *m* chlorinated lime; chloride of lime; bleaching powder
Chlorkalklösung *f* bleaching liquid
Chlorkautschuk *m* chlorinated rubber
Chlorknallgasreaktion *f* chlorine hydrogen reaction
Chlormethylierung *f* chloro-methylation
Chlormolekül *n* chlorine molecule
Chloroform *n* chloroform
Chlorophyll *n* chlorophyl[l]
Chlorophyllzersetzung *f* **durch Licht** protolysis
Chloropren *n* chloroprene
Chloroprenkautschuk *m* polychloroprene
Chlorplatinsäure *f* chloroplatinic acid
Chlorrückhaltevermögen *n* chlorine retention
Chlorsäure *f* chloric acid
Chlorsilberpapier *n* gelatino-chloride paper
Chlorstörstellen *fpl* Cl impurities *(semiconductor)*
Chlorsulfonsäure *f* chloro-sulfonic acid
Chlor[su]peroxidbleiche *f* chlorine-peroxide bleach
Chlorung *f* chlorination
Chlorwasser *n* chlorine water
Chlorwasserstoff *m* hydrogen chloride, hydrochloric acid
Choleinsäure *f* choleic acid
Cholesterin *n* cholesterine
Cholesterol *n* cholesterol
Cholin *n* choline
Cholsäure *f* cholic acid
Chopper *m* mechanical [velocity] selector, chopper
Chorbrett *n (text)* lower hole board
Chrom *n* chromium
Chromalaun *m* chrome-alum
Chromat *n* chromate
Chromatätze *f* chromate discharge
chromatisch chromatic *(aberration)*
chromat[is]ieren to chromize
Chromatografie *f* chromatography
Chromatografieschreiber *m* chromatography recorder
chromatografisch chromatographic
Chromatogramm *n* chromatogram
Chromazofarbstoff *m* chrome azo dye
Chrombeizenfarbstoff *m* chrome mordant dye
chromdiffundieren to chromize *(to form a high chromium content surface layer by diffusion)*
Chromeisenerz *n* chrome iron ore, chromite
Chromeisenstein *m* chrome iron ore
Chromfarbe *f s.* Chromfarbstoff
Chromfarbenteig *m* chrome-dye printing paste
Chromfarbstoff *m* chrome dye
Chromfilter *n* chromium-layer filter
chromgar chrome-tanned
Chromgelatine *f* bichromated gelatine
Chromgelb *n* chrome yellow
Chromgerbeextrakt *m* chrome tanning extract
Chromgerber *m* chrome tanner
Chromgerberei *f* chrome tannage
Chromgerbung *f* chrome tannage (tanning)
Chromgrün *n* chrome green
chromhaltig chromiferous
chromieren to chrom[al]ize
Chromier[ungs]farbstoff *m s.* Chromfarbstoff
Chromit *m (min)* chromite
Chromitstein *m* chromite (chrome) brick
Chromkomplexfarbstoff *m* chrome complex dye
Chromleder *n* chrome[-tanned] leather
Chrommagnesitstein *m* chrome magnesite brick
Chrommolybdänstahl *m* chromium-molybdenum steel
Chromnickeldraht *m* nichrome wire

Chromnickelstahl *m* nichrome (chromium-nickel) steel
Chromoersatzkarton *m* imitation chromo-board
Chromokarton *m* chromo-board
Chromolithografie *f* chromolithography
Chromopapier *n* chromo paper, coated stock
Chromophor *m* chromophore
Chromosomenaberration *f* chromosome aberration
Chromosomenschaden *m* chromosome aberration
Chromosphäre *f* chromosphere
Chromrot *n* chrome red
chromsauer chromic
Chromsäure *f* chromic acid
Chromsäureanhydrid *n* chromic anhydride
Chromsäureelement *n* bichromate (chromic acid) cell
Chromschwefelsäure *f* chromatosulphuric acid
Chromsohlleder *n* chrome sole leather
Chromsohlspaltleder *n* chrome sole leather splits, flexible chrome splits
Chromstahl *m* chrome steel
nichtrostender reiner C. straight chrome stainless steel
Chrom(II)-verbindung *f* chromous compound
Chrom(III)-verbindung *f* chromium compound
Chrom(VI)-verbindung *f* chromic compound
Chromwolframstahl *m* chrome-tungsten steel
Chromylchlorid *n* chromyl chloride, chromic oxychloride
Chromzinnober *m* Austrian cinnabar; chrome red
Chronograf *m* chronograph
Chronometer *n* chronometer
Chronometerhemmung *f* chronometer escapement
Chronometrie *f* chronometry
Chronotron *n* chronotron *(pulse technique)*
Chrysoberyll *m (min)* chrysoberyl
Chrysoidin *n (ch)* chrysoidine
Chrysokoll *m (min)* chrysocolla
Chrysopras *m (min)* chrysoprase
Chrysotil *m (min)* chrysotile
Chrysotilasbest *m s.* Chrysotil
cis-Anlagerung *f (ch)* cis-addition
cis-Form *f (ch)* cis-isomer
cis-Stellung *f (ch)* cis-situation
cis-trans-Isomerie *f (ch)* cis-trans-isomerism
cis-Verbindung *f (ch)* cis-compound
Claisen-Kolben *m* Claisen flask
Clarit *m (min)* clarain
Claude-Verfahren *n* Claude process
Clausius-Rankine-Prozeß *m* Rankine cycle
Clean-up-Effekt *m (nucl)* clean-up effect
Climaxkessel *m (therm)* Climax boiler
Clusius-Dickel-Trennrohr *n* Clusius column
Cluster *m (nucl)* cluster
Clusterzählung *f* cluster count[ing]
cm-Gerät *n (el)* microwave set
cm-Welle *f (el)* microwave
CO_2-Schweißen *n* carbon-dioxide welding
Cockcroft-Walton-Generator *m (nucl)* Cockcroft-Walton accelerator
Cockpitschiebehaube *f (aero)* sliding cockpit enclosure
Code *m s.* Kode
Codleine *f* cod[end] line, draw (bull) rope, poop line
Coelostat *m (opt)* coelostat
Coffinit *m* coffinite *(uranium mineral)*
Colemanit *m (min)* colemanite
Colorimeter *n s.* Kolorimeter
Colorimetrie *f s.* Kolorimetrie
Compiler *m (dat)* compiler
Compoundkern *m (nucl)* compound (intermediate) nucleus
Compoundmaschine *f* compound engine; *(el)* compound-wound machine
Compton-Absorption *f* Compton absorption
Compton-Effekt *m* Compton effect
Compton-Elektron *n* Compton electron
Compton-Rückstoß *m* Compton recoil
Compton-Stoß *m* Compton collision
Compton-Streuung *f* Compton scattering
Compton-Verschiebung *f* Compton shift
Compton-Wellenlänge *f* Compton wavelength
Confinement *n* confinement *(of plasma)*
Conradson-Test *m* Conradson carbon-residue test *(for lubricating oils)*
Conradson-Zahl *f* Conradson carbon value
Container *m s.* Kontainer
Containment *n (nucl)* [reactor] containment
Coolidge-Röhre *f* Coolidge[-type ray] tube
Copyright-Vermerk *m* copyright imprint
Cor *m s.* Spaltzone
Cordierit *m (min)* cordierite, iolite
Corioliskraft *f* Coriolis force
Corliss-Maschine *f* Corliss engine
Corliss-Steuerung *f* Corliss valve gear
Corliss-Ventil *n* Corliss valve
Cornealschale *f (opt)* corneal lens
Cornwallkessel *m* Cornish boiler
Cosinus *m s.* Kosinus
Costabirne *f (ship)* Costa propulsion bulb
Cottonmaschine *f (text)* full-fashioned hosiery machine
Cotton-Mouton-Effekt *m* Cotton-Mouton-effect
Cottonöl *n* cotton[seed] oil
Cottonstrumpf *m* full-fashioned hose
Cottrellanlage *f* Cottrell electrical precipitator
Coudé-Gitterspektrograf *m* Coudé grating spectrograph
Coudé-Refraktor *m* Coudé refractor
Coulomb *n* coulomb
Coulomb-Anregung *f* Coulomb excitation

Coulomb-Energie *f* Coulomb energy
Coulomb-Feld *n* Coulomb field
Coulomb-Kraft *f* Coulomb force
Coulomb-Streuung *f* Coulomb scattering
Coulometer *n* coulometer
Coulometrie *f* coulometry
Couturier *m (text)* apparel designer
Covellin *m (min)* covellite
Cowper *m* Cowper stove
CPM-Verfahren *n* critical path method
Crackanlage *f* cracking plant
Cracken *n* cracking
Creed-Fernschreiber *m* Creed printing telegraph system
Cremonaplan *m* Cremona (Maxwell) diagram
Croning-Verfahren *n* C-process, shell process
Crosby-Schaltung *f* Crosby phase modulation circuit
Cross-Verfahren *n* moment distribution method, Hardy Cross method
Croupon *m (leath)* butt
Cryosar *s.* Kryosar
Cryotron *n s.* Kryotron
Cumberlandverfahren *n (el)* cathodic protection
Cuprit *m (min)* cuprite
Curie *n* curie
Curie-Gesetz *n* Curie's law
Curie-Konstante *f* Curie constant
Curie-Punkt *m* Curie point
Curie-Temperatur *f* Curie temperature
Curietherapie *f s.* Radiumtherapie
Curium *n* curium
Curtis-Rad *n* Curtis wheel
cuttern to cut
CW-Laser *m* c.w. laser, continuous wave laser
Cyclotron *n s.* Zyklotron
Cymol *n (ch)* cymene, cymol
Cymophan *m (min)* cymophane

D

D-Anteil *m* derivative action (control factor)
D-Darstellung *f* D-display *(radar)*
D-Einfluß *m* derivative action
D-Elektrode *f* dee, duant
D-Glied *n* derivative element, lead network
D-Neutron *n* D neutron
D-Regelung *f* derivative control
D-Schirm *m* D-scope *(radar)*
D-Sichtgerät *n* D-indicator *(radar)*
D-Verhalten *n* derivative action
D-Wirkung *f* derivative action
D-Zustand *m (nucl)* D-state
DA-Umsetzer *m* digital-to-analogue converter
Dach *n* 1. *(mine)* back, [hanging] wall, roof; 2. *(build)* roof; 3. *(el)* top *(of pulses)*
  abgebundenes D. framed roof
  flach geneigtes D. shallow pitched roof
  gebrochenes D. mansard (curb) roof
  halbkreisförmiges D. compass roof
  V-förmig nach innen geneigtes D. pent (double lean-to) roof
Dachabspanngestänge *n* roof standard
Dachabspannmast *m* roof standard
Dachantenne *f* overhouse (roof) aerial
Dachaufbau *m* penthouse
Dachausschalung *f* ashlering
Dachaussteigluke *f* trap door on roof
Dachbalken *m* roof beam (tree)
Dachbalkenlage *f* plate
Dachbaugruppe *f* roof-panel assembly
Dachbeplankung *f* roof panel
Dachbinder *m* roof truss
Dachboden *m* loft, attic, garret
Dachbodenantenne *f* loft aerial
Dachdecker *m* roofer
Dachdeckung *f s.* Dacheindeckung
Dacheindeckung *f* roof covering (decking), roofing
Dachentwässerung *f* roof drainage
Dachfenster *n* roof-light, skylight; dormer window
Dachfilz *m* roof felt
Dachfirst *m* roof ridge
Dachform *f* roof type (profile)
Dachgarten *m* roof garden
Dachgaupe *f* dormer *(of a roof)*
Dachgeschoß *n* attic, loft; garret story
Dachgrat *m* arris
Dachhaken *m* roof hook
Dachhammer *m* sax; slate ax
Dachkammer *f* garret; attic
Dachkantprisma *n* pentagonal (ridge) prism
Dachkapazität *f* top loading capacity *(aerial)*
Dachkonstruktion *f* roof structure
Dachlaterne *f* roof lantern, roof-light, lantern light
Dachlatte *f* tile (roof) batten
Dachlüfter *m* roof ventilator
Dachluke *f* loading hatch
Dachnagel *m s.* Dachpappnagel
Dachneigung *f* roof pitch, slope of roof
Dachpappe *f* roofing felt
  besandete D. mineral surface roofing paper
  unbesandete D. smooth roofing paper
Dachpappnagel *m* roofing nail
Dachpfanne *f* Flemish tile, pantile
Dachpfette *f* purlin
Dachplatte *f* roof slab (panel)
Dachrahmen *m* roof frame
Dachrinne *f* gutter, eaves
Dachrinnenkessel *m* rainwater head
Dachrost *m (therm)* roof-shaped grate
Dachrostfeuerung *f* furnace with ridge grate

Dachsbeil *n* adz[e]
Dachschalung *f* roof boarding (boards), roofing; *(Am)* [wood] sheathing
Dachschiefer *m* roofing slate
Dachschindel *f* shingle, shide
Dachschräge *f* ramp-off
  negative D. *(el)* pulse droop, sag
  positive D. *(el)* [pulse] tilt
Dachsparren *m* *(build)* rafter; *(ry)* roof rib (rafter)
Dachspriegel *m* *(ry)* roof bow (stick)
Dachständer *m* roof standard (pole)
Dachstein *m* roof[ing] tile, pantile
Dachstube *f* garret, attic
Dachstuhl *m* roof truss, framework
Dachstütze *f* roof-type supporting insulator, roof insulator spindle *(aerial)*
Dachtragwerk *n* roof structure
Dachtraufe *f* eaves
Dachverband *m* roof truss, principal
Dachwippe *f* lever *(rolling mill)*
Dachziegel *m* roof[ing] tile, pantile
Dahlander-Schaltung *f* *(el)* Dahlander pole-changing connection
Dalbe *f* dolphin
Dalbenbake *f* *(ship)* dolphin beacon
Dalbenpfahl *m* dolphin pile
d'Alembert-Kraft *f* d'Alembert auxiliary force
Dallrohr *n* Dall tube *(flowmeter)*
Damastbindung *f* *(text)* damask weave
Damastleinen *n* damask linen
Damastpapier *n* linen paper
Damastwebstuhl *m* damask loom
Damaszenerstahl *m* damask steel
damaszieren to damascene *(steel)*
Damenfahrrad *n* lady's [bi]cycle
Damenkonfektionsstoffe *mpl* ladies' dress goods
Damenunterwäsche *f* lingerie
Damm *m* dam *(stream)*; ridge; bank *(for water)*; barrier [wall]; embankment; dike; mole; levee; causeway; *(mine)* stoping
Dammar[harz] *n* dammar [resin]
Dämmaterial *n* insulating material
Dämmbrett *n* sleeker *(moulding)*
Dämmeigenschaft *f* insulation property (quality)
dämmen to insulate, to dam up
Dämmerlicht *n* dusk; twilight
Dämmerungseffekt *m* night effect (error)
Dämmerungsfrequenz *f* *(telec)* transition frequency
Dämmerungsschalter *m* twilight switch *(street lighting)*
Dämmerungssehen *n* mesopic (twilight) vision
Dämmerungssehleistung *f* visual capacity in twilight
Dammhobel *m* *(mine)* fast-end plough
Dammhöhe *f* embankment height; dam height
Dammkrone *f* crest of dam
Dammkultur *f* *(ag)* ridge crop
Dämmpappe *f* ceiling board
Dämmplatte *f* insulating slab (board)
Dämmschicht *f* *(build)* insulating course (layer)
Dämmstoff *m* insulating material, insulant
Dammstraße *f* causeway
Dammtür *f* *(mine)* dam door
Dämmung *f* [sound and heat] insulation
Dämmzahl *f* [sound] damping factor
Dampf *m* steam *(water)*; vapour; fume
  gesättigter D. saturated steam
  indirekter D. latent steam
  kondensierter D. condensed steam
  nasser D. wet steam
  satter D. saturated steam
  trockener D. dry steam
  überhitzter D. superheated steam
  ungesättigter D. surcharged steam
Dampfabblasrohr *n* steam blow-off pipe, waste-steam pipe, exhaust-steam pipe, waste (blow-off) pipe of safety valve
Dampfabgabe *f* steam utilization *(for heating purposes)*
Dampfabgangsventil *n* eduction valve
Dampfablaßhahn *m* blow cock
Dampfabscheider *m* steam separator
Dampfabsperrhahn *m* steam stop cock
Dampfabsperrorgan *n* steam shutting-off device
Dampfabsperrschieber *m* steam gate (parallel slide stop) valve
Dampfabsperrung *f* steam stop-valve; cutting off steam
Dampfabsperrventil *n* steam stop valve
dampfangetrieben steam-driven
Dämpfanlage *f* *(ag)* steamer installation (plant), sterilization plant
Dämpfapparat *m* ager *(printing cloth)*
Dampfarbeit *f* work done by the steam
  indizierte D. indicated work done by the steam
Dampfaschwinde *f* steam ash-hoist
Dampfausdehnung *f* expansion of steam
Dampfauslaß *m* steam outlet
Dampfauslaßrohr *n* exhaust-steam pipe, waste-steam pipe
Dampfauslaßseite *f* steam exhaust side, exhaust port side
Dampfauslegerkran *m* steam jib crane
Dampfausnutzung *f* utilization of steam
Dampfausströmkanal *m* exhaust passage
Dampfausströmperiode *f* exhaust period
Dampfausströmung *f* steam exhaust, emission of steam
Dampfausströmungsöffnung *f* exhaust port
Dampfausströmungsrohr *n* steam exhaust pipe
Dampfaustritt *m* escape of the steam
Dampfaustrittskanal *m* exhaust-steam passage
Dampfaustrittsöffnung *f* steam (exhaust) port

Dampfaustrittsrohr *n* exhaust-steam pipe
Dampfbad *n* steam bath
Dampfbagger *m* steam excavator; steam dredger
Dampfbehandlung *f* *(text)* ag[e]ing; *(build)* steam curing *(concrete)*
Dampfbetrieb *m* steam service; driving by steam
Dampfbildung *f* formation of steam
stürmische D. violent formation of steam
Dampfblase *f* steam (vapour) bubble; steam void
Dampfblasenbildung *f* **in der Kraftstoffleitung** fuel vapour lock
Dampfbläser *m* steam ejector
Dampfbohranlage *f* steam drilling rig
Dampfboot *n* steam boat
Dampfbremse *f* steam brake
Dämpfbürstmaschine *f* steaming and brushing machine
Dampfchloren *n* *(text)* steam chemicking
Dampfdehnung *f* steam expansion
Dampfdestillation *f* steam distillation
Dampfdiagramm *n* steam diagram
dampfdicht steamproof, steamtight
Dampfdichte *f* density of steam
Dampfdichtemesser *m* condenser gauge
Dampfdichtungsring *m* oil ring
Dampfdom *m* *(ry)* steam dome
Dampfdrosselung *f* steam throttling
Dampfdruck *m* vapour pressure *(petrol)*; steam tension
statischer D. static steam pressure
Dampfdruckdiagramm *n* vapour [pressure] diagram
Dampfdruckerniedrigung *f* lowering of vapour pressure
Dampfdruckfläche *f* steam-pressure area
Dampfdruckkammer *f* autoclave
Dampfdruckkurve *f* vapour [pressure] diagram
Dampfdruckminderer *m* *s.* Dampfdruckverminderer
Dampfdruckminder[ungs]ventil *n* steam [pressure] reducing valve
Dampfdruckschaubild *n* steam indicator diagram
Dampfdrucksterilisator *m* autoclave; apparatus for sterilization by steam at high pressure
Dampfdrucktopf *m* steam digester
Dampfdruckverminderer *m* steam-pressure reducer
selbsttätiger D. automatic steam-pressure reducer
Dampfdüse *f* steam nozzle
Dampfdynamo *m* steam dynamo
Dampfeinlaßbüchse *f* steam chest *(turbine)*
Dampfeinlaßhahn *m* admission steam cock
Dampfeinlaßseite *f* steam admission (port) side
Dampfeinlaßventil *n* steam inlet (supply) valve; admission steam valve; controlling valve
Dampfeinströmlinie *f* steam admission line
Dampfeinströmung *f* steam admission
Dampfeinströmungsrohr *n* steam admission pipe
Dampfeintritt *m* steam admission, admission of steam
Dampfeintrittskanal *m* [admission] steam port
Dampfeintrittsrohr *n* supply steam pipe
Dampfeintritts- und -austrittskanal *m* steam admission and exhaust port
Dampfeintrittsventil *n* admission steam valve
dampfen to steam
dämpfen 1. to damp *(vibration)*; *(el)* to attenuate; to absorb; 2. to steam; to expose to the action of steam; to age *(printing cloth)*; 3. to bank *(blast furnace)*; 4. to muffle *(sound)*; to cushion
Dampfentnahme *f* extraction of steam, bleeding
Dampfentnahmeventil *n* steam supply valve
Dampfentöler *m* grease (oil) separator
Dampfentweichung *f* escape of steam, steam leaking
Dampfentwicklung *f* production (generation) of steam, getting up steam
Dämpfentwicklung *f* *(text)* age-development
Dämpfentwicklungsverfahren *n* *(text)* steam developing process
Dampfer *m* steamer, steamship
Dämpfer *m* antivibrator, dampener *(noise)*; muffler; digester; *(text)* [steam] ager
dampferhärtet steam cured *(concrete)*
Dampferhärtung *f* steam-curing *(concrete)*
Dampferkai *m* steam wharf
Dampferlicht *n* steaming light, masthead lantern
Dampfersparnis *f* saving of steam, steam saving *(economy)*
D. durch Kondensation saving of steam due to condensation
Dampferweg *m* [shipping] lane
Dämpferwicklung *f* *(el)* amortisseur (damper, damping) winding
unterteilte D. *(el)* open amortisseur winding
Dampferzeuger *m* steam generator (generating unit)
Dampferzeugung *f* steam generation
Dampferzeugungsanlage *f* steam-generating plant
Dampffeuchtemesser *m* hygronom
Dampffeuchtigkeit *f* wetness of steam
Dampffeuerlöschanlage *f* steam fire-extinguishing plant
Dampffeuerlöschleitung *f* steam smothering line
Dampffeuerlöschung *f* steam smothering
Dampffeuerspritze *f* steam syringe
dampfförmig vapourous
Dampffüllapparat *m* steam packer
Dampfgeschwindigkeit *f* velocity of steam
Dampfgewicht *n* weight of steam

Dampfgrenzkurve *f* steam-limit curve
Dampfhahn *m* steam cock
Dampfhammer *m* steam hammer
Dampfhärtekessel *m* steam curing vessel
Dampfheizschlange *f* steam heating coil
Dampfheizung *f* steam heating
Dampfheizungsanlage *f* steam heating installation
Dampfhemd *n* cylinder (steam) jacket
Dampfherstellungskosten *pl* cost of steam production
Dampfheuler *m* steam hailer
dämpfig werden to ret *(hay)*
Dampfkammer *f* 1. steam chamber; 2. curing chamber (room) *(concrete)*
Dampfkanal *m* steam port (passage)
Dampfkasten *m* steam chest
  D. zum Holzdämpfen steam box, wood boiler
Dämpfkasten *m (text)* steam ager
Dampfkessel *m* steam boiler
  feststehender D. stationary boiler
  liegender D. horizontal boiler
  ortsfester D. stationary boiler
  stationärer D. stationary boiler
  stehender D. vertical boiler
  transportabler D. portable boiler
Dampfkesselbetrieb *m* steam boiler working
Dampfkesselmanometer *n* steam gauge
Dampfkesselüberwachung *f* boiler inspection
Dampfkesseluntersuchung *f* steam boiler inspection
Dampfkesselwartung *f* maintenance (management) of steam boilers
Dampfklärung *f* clearing [sugar] by steam
Dampfkolben *m* steam piston
Dämpfkolonne *f (ag)* steamer (sterilization) column
Dampfkondensator *m* steam condenser
Dampfkraft *f* steam power
Dampfkraftmaschine *f* steam power engine
Dampfkraftwerk *n* steam power plant
Dampfkran *m* steam crane
Dampfleistung *f* produced quantity of steam
  spezifische D. evaporation per sq.m. of heating surface
Dampfleitung *f* steam line (conduit, pipe)
Dampfleitungsrohr *n* steam [conduit] pipe
Dampflogger *m* steam drifter (logger)
Dampflokomotive *f* steam locomotive
Dampfluftpumpe *f* steam airpump
Dampfmantel *m* steam (cylinder) jacket
Dampfmantelrohr *n* steam-jacketed pipe
Dampfmaschine *f* steam engine
  D. mit Expansion expansion steam engine
  D. mit Expansion und Kondensation expansion and condensing steam engine
  D. mit fester Expansion steam engine with fixed expansion
  D. mit kombinierten Dämpfen binary vapour engine
  D. mit Schiebersteuerung steam engine with slide valve
  D. mit schwingenden Kolben steam engine with oscillating cylinders
  D. mit Umsteuerung reversing steam engine
  D. mit Ventilsteuerung steam engine with drop valve gear
  D. mit Wärmeaustausch reheat[er steam] engine
  D. ohne Expansion steam engine working without expansion
  D. ohne Kondensation non-condensing steam engine
  atmosphärische D. atmospheric steam engine
  bewegliche D. portable steam engine
  direktwirkende D. direct-acting steam engine
  doppeltwirkende D. double-acting steam engine
  einfachwirkende D. single-acting steam engine
  einkurblige D. single-crank steam engine
  einzylindrige D. single-cylinder steam engine
  freistehende D. independent (self-contained) steam engine
  kombinierte liegende und stehende D. combined horizontal and vertical steam engine
  kurzhubige D. short-stroke steam engine
  langhubige D. long-stroke steam engine
  langsamlaufende D. slow-speed steam engine
  liegende D. horizontal steam engine
  mehrkurblige D. multi-crank steam engine
  ortsfeste D. stationary steam engine
  rotierende D. rotary steam engine
  schnellaufende D. high-speed steam engine
  schrägliegende D. inclined steam engine
  schwingende D. steam engine with oscillating cylinders
  stehende D. upright (vertical) steam engine
Dämpfmaschine *f (ag)* steamer unit, soil sterilizer
Dampfmaschinenanlage *f* steam engine plant
Dampfmaschinensteuerung *f* steam engine control gear
Dampfmaschinenteil *n* steam engine part
Dampfmenge *f* amount (quantity) of steam
Dampfmotor *m* steam motor
Dampfnässe *f* wetness of steam
Dampfniederschlaggefäß *n* receiver for condensed water
Dampföffnung *f* steam port
Dampföler *m* steam lubricator
Dampfpfeife *f* steam whistle, hooter
Dampfpflug *m* steam tackle plough
Dampfpinasse *f* steam pinnace
Dampfpinne *f* steam tiller
Dampfplattentrockner *m* steam platen drier

Dampfprahm *m* steam lighter
Dampfpumpe *f* steam pump; donkey pump *(steam)*
direktwirkende D. direct-acting steam pump
unmittelbar angetriebene D. direct-acting steam pump
Dampframme *f* steam pile driver
Dampfraum *m* steam space
Dampfreduktionsventil *n s.* Dampfreduzierventil
Dampfreduzierventil *n* steam pressure reducing valve
Dampfreibungsarbeit *f* steam friction work
Dampfreibungsverlust *m* steam friction loss
Dampfreiniger *m* steam purifier
Dampfreinigungssieb *n* steam strainer
Dampfrohr *n* steam pipe
Dampfrohrüberhitzer *m* superheater with U-tubes placed horizontally underneath the boiler
Dampfrudermaschine *f* steam steering gear
Dampfsägewerk *n* steam saw mill
Dampfsammelrohr *n* dry (dome steam) pipe
Dampfsammler *m* steam accumulator (collector, dome); dry (separating) drum; receiver *(Niclausse boiler)*
Dampfschieber *m* steam slide valve
Dampfschiff *n* steamship, steamboat, steam vessel
Dampfschiffahrt *f* steam navigation
Dampfschiffsmaschinenanlage *f* / **kompakte** integrated marine steam power plant
Dampfschlag *m* beat of the blast pipe
Dampfschlange *f* steam coil
Dampfschlauch *m* steam hose
Dampfschleierfeuerung *f* furnace with steam jets above grate, steam-fan furnace
Dampfschleife *f* automatic return of water
Dampfschlepper *m* steam tug
Dampfschmierapparat *m* steam lubricating apparatus
Dampfschmierung *f* steam lubrication
Dampfschwitze *f (leath)* steam sweat
Dampfsirene *f* steam siren
Dampfspannung *f* steam tension, tension of vapour
Dampfspannungsthermometer *n* vapour-actuated thermometer
Dampfspeicher *m* steam accumulator
Dampfspeisepumpe *f* feed (donkey) pump
Dampfsperre *f* vapour barrier
Dampfspill *n* steam capstan
Dampfspritze *f* steam syringe
Dampfsteuerapparat *m* steam steering gear
Dampfstrahl *m* steam jet
Dampfstrahlbrenner *m* steam jet burner
Dampfstrahlgebläse *n* steam [jet] blower
Dampfstrahlluftsauger *m* steam jet air-ejector
Dampfstrahlpumpe *f* steam jet pump, steam injector
Dampfstrahlreiniger *m* steam cleaner
Dampfstrahlunterwindfeuerung *f* forced draught furnace with steam jet blower underneath the grate
Dampfstrahlzerstäuber *m* steam jet sprayer (atomizer)
Dampfstunde *f* steam hour
Dampfstutzen *m* steam connection
Dampftabelle *f* steam table
Dampftafel *f s.* Dampftabelle
Dampfteil *m* steam end *(pump)*
Dampftrockenfarbe *f* steam-set ink
Dampftrockenheit *f* dryness of steam
Dampftrockentrommel *f (text)* steam-heated can
Dampftrockenzylinder *m (text)* steam-heated can
Dampftrockner *m* steam drier (drying apparatus)
Dampfturbine *f* steam turbine
Dampfturbinenkraftwerk *n* steam turbine power station
Dampfüberführung *f* conveyance of steam
Dampfüberhitzer *m* steam superheater
Dampfüberhitzung *f* steam superheating
nukleare D. nuclear steam superheating
Dampfumformer *m* steam transformer (converter)
Dampfumsteuerung *f* steam reversing gear
Dämpfung *f* 1. absorption; attenuation; damping; fading; transmission loss; dissipation; 2. shock-absorber; built-in damping
D. der Bodenwelle *(el)* ground absorption
D. der harmonischen Gesamtverzerrung *(el)* total harmonic ratio
D. der hohen Frequenzen de-emphasis *(receiver)*
D. durch Stöße collision[al] damping
D. eines Schwingkreises damping of a circuit
akustische D. acoustic damping
elektromagnetische D. electromagnetic damping
geometrische D. geometrical attenuation
konjugiert-komplexe D. conjugate attenuation coefficient
kritische D. critical damping *(measuring)*
räumliche D. attenuation
relative D. damping ratio
selektive D. gain sensitivity control
spezifische D. attenuation constant
überkritische D. overdamping *(measuring)*
zeitliche D. damping
Dämpfungsaufteilung *f (el)* allocation of attenuation
Dämpfungsausgleich *m* attenuation correction (compensation)

Dämpfungsbereich *m* attenuation range, range of attenuation
Dämpfungsdekrement *n* damping factor
Dämpfungsdiode *f* damping diode
Dämpfungsentzerrer *m* attenuation equalizer (compensator)
Dämpfungsentzerrung *f* attenuation equalization (correction)
Dämpfungsfähigkeit *f* cushioning capacity
Dämpfungsfaktor *m* damping coefficient (factor), attenuation (decay) factor
Dämpfungsfehler *m* attenuation error
Dämpfungsflosse *f (ship)* damping fin
Dämpfungsflügel *m* vane
Dämpfungsflüssigkeit *f* damping fluid
Dämpfungsfunktion *f* damping function
Dämpfungsgang *m (el)* attenuation response (frequency characteristic), frequency response of attenuation
Dämpfungsglied *n* attenuator [pad]; pad; resistance pad
 festes D. *(el)* pad
 koaxiales D. coaxial pad
 regelbares D. variolosser
 scheibenförmiges D. disk attenuator
Dämpfungsglimmer *m* damping mica *(valve manufacture)*
Dämpfungsgrad *m* attenuation ratio
Dämpfungskonstante *f s.* Dämpfungsfaktor
Dämpfungslänge *f* attenuation distance
Dämpfungsleitung *f* attenuator
Dämpfungsmagnet *m* damping magnet
Dämpfungsmaschine *f* regenerative braking motor *(of Ward-Leonard-connection)*
Dämpfungsmaß *n* attenuation constant (equivalent)
Dämpfungsmaterial *n* sound absorbing material
Dämpfungsmesser *m* transmission [efficiency] measuring set, decible meter
Dämpfungsmessung *f (el)* attenuation measurement (test)
Dämpfungsmodul *m s.* Dämpfungsfaktor
Dämpfungsmoment *n* damping torque
Dämpfungsplan *m* transmission plan
Dämpfungsrolle *f* impact idler
Dämpfungsschalter *m* muting switch
Dämpfungsschaltung *f* attenuation network, anti-sidetone circuit
Dämpfungsverbreiterung *f* broadening by damping *(spectrum)*
Dämpfungsverhältnis *n* damping coefficient
Dämpfungsverlauf *m* attenuation characteristic (curve)
Dämpfungsverminderung *f* damping reduction
Dämpfungsversteilerung *f* attenuation increase
Dämpfungsverzerrung *f* attenuation (frequency) distortion
Dämpfungswiderstand *m* damping resistance
Dämpfungswirkung *f* damping action
Dämpfungszahl *f s.* Dämpfungsfaktor
Dämpfungszeit *f* attenuation time
Dämpfungsziffer *f s.* Dämpfungsfaktor
Dämpfungszylinder *m* dashpot
Dampfventil *n* steam valve
Dampfverbrauch *m* steam consumption
 erreichter D. actual steam consumption
Dampfverbrauchsmesser *m* steam consumption meter
Dampfversuch *m* steam trial
Dampfverteilung *f* steam distribution
Dampfverteilungskanal *m* steam distribution passage *(turbine)*
Dampfverteilungsrohr *n* steam distributing pipe
Dampfvorrat *m* steam reserve
Dampfvorwärmer *m* feed heater
Dampfvulkanisator *m* steam vulcanizer
Dampfwalze *f* steam roller
Dampfwärme *f* heat of steam
Dampfwasserheber *m* pulsometer
Dampfwasserrückleiter *m* apparatus for returning water for condensation to boiler; automatic receiver and pump
Dampfweg *m* steam passage (port); port hole
Dampfwinde *f* steam winch
Dampfwirkung *f* / **einseitige** steam acting on one side
Dampfzähler *m* steam meter (counter)
Dampfzufluß *m* steam supply
Dampfzufuhr *f* steam supply
 regelmäßig eingestellte D. controlled steam supply
Dampfzuführung *f s.* Dampfzufuhr
Dampfzugförderung *f* steam traction
Dampfzuleitung *f* steam supply
Dampfzuleitungsrohr *n* steam supply pipe
Dampfzustand *m* vapour state
Dampfzutritt *m* steam admission (supply)
Dampfzylinder *m* steam cylinder
Dampfzylinderöl *n* steam cylinder oil
Dampfzylinderschmierapparat *m* steam-cylinder lubricator
Dandywebstuhl *m* dandy-loom
Daniell-Element *n* Daniell cell
dann und nur dann, wenn *(math)* if and only if
Darmsaite *f* gut, catgut string *(tennis racket)*
Darrbühne *f* drying room (kiln)
Darre *f* 1. drying stove *(foundry)*; 2. *(ag)* drier; drying house; 3. kiln *(malt)*
darren to dry; to torrefy *(copper)*
Darrgewicht *n* kiln-dry weight
Darrkupfer *n* liquated copper
Darrmalz *n* kiln-dried malt
Darrofen *m* kiln
darstellen to display, to represent, to obtain, to plot; *(ch)* to prepare, to synthesize

digital d. *(dat)* to digitize
grafisch d. to represent graphically, to graph
schematisch d. to represent diagrammatically
ziffernartig d. *(dat)* to digitize
darstellend descriptive *(geometry)*
Darstellung *f* 1. description; 2. representation; display *(radar)*; notation *(data)*; diagram; figure; 3. *(ch)* preparation, synthesis
D. der Aufgabenlösung *(dat)* display of solution
D. im Schnitt sectional view
D. in auseinandergezogener Anordnung exploded view
bildliche D. diagram
bildliche D. der Bindungen *(text)* sketch of interlacing
biquinäre D. *(dat)* biquinary notation
elektronische D. electronic display
freihändige D. free-hand sketching (drawing, representation)
grafische D. graphical representation; graph[ic illustration]; plot[ting]; diagram
halblogarithmische D. semi-logarithmic plot
isometrische D. isometric drawing
lineare D. linear display
maschineneigene D. *(dat)* hardware representation
schematische D. schematic diagram (representation)
sichtbare D. visual representation
vektorielle D. vectorial display
Darstellungsgerät *n* plotter
Darstellungsmatrix *f* chart matrix
Darstellungsschalter *m* presentation selector *(radar)*
Darstellungsweise *f* representation *(technical drawing)*
amerikanische D. American representation, [third-angle] orthographic representation
deutsche D. German representation *(technical drawing)*
Dasselbeule *f (leath)* warble lump *(fault)*
Dasymeter *n* dasymeter
Daten *pl* data, information *(numerical control)*
technische D. technical data
Datenabtastsystem *n* data sampling system
Datenaufzeichnung *f* data logging
Datenauswahl *f* selection of data
Datenauswertung *f* data handling
Datenbearbeitungssystem *n* data handling system
Datenbit *n* data bit *(zero or one in the binary notation)*
Datenblock *m* data block
Datenbuchungsmaschine *f* **für Umsätze** data transaction posting unit
Datendarstellung *f* data representation
Datendruckwerk *n* data printer
Datenerfassung *f* data gathering; data logging; data recording
Datenfluß *m* data flow
Datenflußplan *m* data flow chart
Datengeber *m* data generator
Datengeschwindigkeit *f* data rate
Datengewinnung *f* data acquisition
Datenkanal *m* data channel
Datenmaterial *n* data material
Datenreduktion *f* data reduction
parallele D. on-line data reduction
Datenregistriergerät *n* data logger
Datenregistrierung *f* data recording
Datensammlung *f* data collection
Datenschreiber *m* data logger
Datensende- und -empfangsgerät *n* data-transceiver, data transmitter-receiver
Datensortiergerät *n* data selector [unit]
Datenspeicher *m* data storage unit
Datenspeicherung *f* data storage
Datenübertragung *f* data transmission
fehlerfreie D. error-free data transmission
Datenübertragungssystem *n* data link (transmission) system
Datenumsetzer *m* data translator
datenverarbeitend data (information) processing
Datenverarbeiter *m* [data] processor
Datenverarbeitung *f* data processing (handling)
automatische D. automatic data processing
elektronische D. electronic data processing
ganzheitliche D. integrated data processing
kaufmännische D. commercial data processing
unabhängig betriebene D. off-line operation
Datenverarbeitungsanlage *f* data processing equipment
digitale D. digital computer
Datenverarbeitungsmaschine *f* / **elektronische** electronic data-processing machine (system)
Datenverarbeitungssystem *n* data processing system
Datenverschlüßler *m* data-coding unit
Datenverstärker *m* data amplifier
Datenwortformat *n* data word format *(digital computer)*
Datenzusammenfassung *f* data gathering (reduction)
Datierung *f* dating, age determination (measurement)
Datumdrucker *m* date printer
Datumsgrenze *f* date line
Daube *f* stave
Daubenabkürzsage *f* stave cross-cut saw
Daubenabrichtmaschine *f* stave dressing machine
Daubenkürzsäge *f* double-stave cross-cut saw
Daubholz *n* stave (cask) wood

Dauer *f* period; duration *(of time)*; continuance
D. des Antriebsfluges *(rock)* duration of powered flight
D. des freien Ausflusses flowing life *(borehole)*
Dauerappretur *f (text)* permanent finish
Daueraufzeichnung *f* permanent record
Dauerausrüstung *f (text)* durable finish
Dauerbeanspruchung *f* repetition of stress
Dauerbelastung *f* continuous ratings (load); sustained loading; steady (permanent) load
Dauerbestrahlung *f (nucl)* protracted (long-continued) irradiation; chronic exposure
Dauerbetrieb *m* continuous repetition work; continuous use (work, operation, duty)
D. mit gleichbleibender Belastung continuous duty
D. mit periodisch veränderlicher Belastung periodic duty
Dauerbiegefestigkeit *f* bending stress fatigue limit
Dauerbiegeversuch *m* fatigue bending test
Dauerbogenerzeuger *m* permanent-arc producer
Dauerbrandbogenlampe *f* enclosed-arc lamp
Dauerbrandofen *m* / **eiserner** slow-burning iron stove
Dauerdipol *m* permanent dipole
Dauereinsatz *m* permanent installation
Dauerelektrode *f* continuous (self-baking) electrode
selbstbrennende D. self-baking electrode
Dauerentladung *f* continuous (permanent) discharge
Dauererdschluß *m* continuous (permanent) earth
Dauerfalte *f (text)* permanent crease
Dauerfestigkeit *f* fatigue strength
Dauerfestigkeitsschaubild *n* stress-number curve
Dauerfeuer *n (mil)* continuous fire
Dauerflug *m* sustained flight
Dauerform *f* permanent mould
Dauerfrostboden *m* permafrost
Dauergeschwindigkeit *f (aut)* cruising speed
Dauerglanzappretur *f (text)* permanent sheen finish
Dauergleichgewicht *n* secular equilibrium
radioaktives D. secular radioactive equilibrium
Dauerhaftigkeit *f* durability
Dauerimplantation *f (nucl)* permanent implant
Dauerkragen *m (text)* everclean collar
Dauerkultur *f (ag)* perennial crop
Dauerkurzschluß *m* sustained short-circuit
Dauerkurzschlußstrom *m* sustained short-circuit current
Dauerladung *f* continuous charge
Dauerlast *f* 1. *(el)* continuous (steady) load; 2. *(build)* sustained load
Dauerleistung *f* continuous rating (power output); constant power
mittlere D. mean output *(radio)*
Dauermagnet *m* permanent magnet
Dauermagnetfutter *n* permanent-magnet chuck
Dauermagnetwerkstoff *m* permanent-magnet material
dauernd permanent
Dauerrechnen *n* long-time computation
Dauerriß *m* fatigue crack
Dauerschaden *m* permanent damage
Dauerschlagbiegefestigkeit *f* repeated impact bending strength
Dauerschlagbiegeversuch *m* repeated impact bending test
Dauerschlagfestigkeit *f* impact fatigue limit
Dauerschlagprüfmaschine *f* impact fatigue testing machine
Dauerschlagversuch *m* impact fatigue test, repeated impact test
Dauerschlagwerk *n* repeated transverse stress testing machine
Dauerschlagzugversuch *m* repeated impact tensile test
Dauerschmierlager *n* self-lubricating bearing
Dauerschwingung *f* maintained vibration
dauerschwingungsbeansprucht fatigue-loaded
Dauerschwingversuch *m* fatigue test
Dauerspannung *f* constant voltage
Dauerspeicher *m* permanent memory
Dauerstandfestigkeit *f* creep strength
Dauerstandverhalten *n* creep behaviour
Dauerstandversuch *m* creep test
Dauerstrich *m* continuous dash signal; *(telec)* long (continuous) dash; continuous wave, c.w.
reiner D. pure continuous wave
Dauerstrichmagnetron *n* constant current magnetron
Dauerstrichmodulation *f* continuous-wave modulation
Dauerstrichsender *m* continuous-signal transmitter
Dauerstrichverfahren *n* continuous-wave method *(navigation)*
Dauerstrom *m (el)* continuous (constant) current
Dauertauchversuch *m* permanent immersion test
Dauerton *m* continuous sound (tone), permanent note, steady (sustained) sound, sustained audio signal, steady hum *(radio)*
Dauertonsektor *m* on-course zone *(radio navigation)*
Dauerverdrehversuch *m* repeated torsion test
Dauerversuch *m* 1. extended time test, long-duration test; 2. fatigue test; endurance test
Dauerwahl *f (telec)* continuous hunting

Dauerwechseldrehversuch *m* alternating torsion fatigue test
Dauerwechselschlagwerk *n* alternating impact machine
Dauerzugfestigkeit *f* tensile fatigue strength
Dauerzugversuch *m* repeated direct stress test
  D. bei ruhender Belastung static tension endurance test
Daumen *m* thumb; lift, cog, tappet, cam
Daumenantrieb *m* cam gear
Daumenrad *n* tappet wheel
Daumenregel *f* thumb rule
Daumensteuerung *f* cam gear
Daumenwelle *f* camshaft
Daunen *fpl* down
Davit *m (n) (ship)* davit
Davitboot *n* davit craft
Dazit *m (min)* dacite
Deaktivierung *f* deactivation
Deakzentuierung *f* de-emphasis *(radio)*
Dealkylierung *f* dealkylation
Debitorenauszug *m* accounts receivable
Deblockierung *f (el)* unblocking
de-Broglie-Wellenlänge *f* de Broglie wavelength
Debye-Radius *m* Debye shielding distance
Debye-Scherrer-Aufnahme *f* powder pattern (photograph)
Debye-Scherrer-Verfahren *n* Debye-Scherrer method, Debye-Sherrer method
Debye-Term *m* Debye term
Decca-Kette *f* decca chain
Decca-Navigationsverfahren *n* decca navigation
Dechenit *m (min)* dechenite
dechiffrieren to decipher, to decode
Dechiffrierung *f* decoding
Dechsel *f* adze
Dechselfläche *f (ry)* adzed surface *(sleeper)*
dechseln *(wood)* to adze, to dub
Deck *n* deck, top-side
  D. / über overdeck, top-side; aloft
  D. / unter below deck, alow
  freies D. exposed (free) deck
  loses D. preventer deck
  oberes D. upper deck
  rundes D. turtle[-back] deck
  tiefgesetztes D. drop deck
  unterstes D. lowermost deck
  wetterdichtes D. weathertight deck
Deckanstrich *m* top coat, finish[ing] coat
Deckbalken *m* transverse deck beam
Deckballasttank *m* **/ seitlicher** topside tank *(of vessels)*
Deckbewegung *f (ph)* symmetry operation
Deckblatt *n* wrapper *(cigar)*
Deckbrandsohle *f* sock lining *(shoe)*
Deckbrett *n* facing board
Deckdruck *m (text)* blotch print
Decke *f* 1. *(text)* blanket; tarpaulin; wagon sheet; 2. tyre cover, outer casing (shoe) *(tyre)*; 3. *(mine)* roof; 4. *(build)* ceiling, floor; road surface; pavement; covering; topping; surfacing
  elastische D. flexible pavement *(road)*
  getäfelte D. panelled ceiling
  Kleinesche D. *(build)* Kleine floor
  unterhängte D. architectural (suspended, hung, counter, false, dropped) ceiling
Deckel *m* cover; lid; cap
  D. zur Hinterachsbrücke *(aut)* rear axle casing cover
  gerippter D. ribbed cover
  nach außen gewölbter D. dished cover
Deckelausputz *m (text)* flat strips (waste)
Deckelbeschlag *m (text)* flat covering
Deckelflansch *m* cover (blank) flange
Deckelheizung *f* heating (jacketing) of the cylinder cover
Deckelkarde *f (text)* flat card
Deckelplatte *f* flat bearing cap
Deckelputzapparat *m (text)* flat stripping apparatus
Deckelraste *f* lid arrester *(tape)*
Deckelriemen *m* deckle strap
Deckelring *m* follower [plate], junk ring
Deckelsatinage *f* board-glaze process *(paper)*
Deckelschraube *f* cylinder[-cover] bolt
  verbohrte D. tap bolt secured by drilling
Deckelschraubenkreis *m* bolt circle of junk ring
decken [mit Schiefer] to slate
  sich d. *(math)* to coincide with, to be congruent with
Deckenaufhängung *f* **für Jacquardmaschine** mounting of the Jacquard machine on a suspended frame
Deckenauslaß *m* lighting outlet
Deckenbalken *m* floor beam, floor (ceiling) joist, joist
deckend opaque *(paint)*
  sich d. *(math)* congruent
Deckendruckmaschine *f* blanket printing machine
Deckeneinbau *m* surfacing, paving
Deckeneinbauzug *m* paving train
Deckenfertiger *m* road finisher (finishing machine), spreader-finisher
Deckenfüllung *f* non-conducting material for covering the top of the boiler
Deckengewölbe *n* roof
Deckenhaken *m* ceiling hook
Deckenheizung *f* [radiant] ceiling heating
Deckenlampe *f* dome lamp
Deckenleuchte *f* ceiling fitting
Deckenplatte *f* floor panel (slab)

Deckenputz *m* ceiling plaster
Deckenrosette *f* ceiling rose
Deckenschalter *m* ceiling switch
Deckenschauer *m (nucl)* ceiling shower
Deckenstrahlungsheizung *f* radiant ceiling heating
Deckenträger *m* floor beam (joist), joist
Deckenventilator *m* ceiling fan
Deckfaden *m (text)* covering (plating, face) thread
Deckfadenfutter[stoff *m*] *n* fleecy fabric
Deckfähigkeit *f* hiding (covering) power, opacity *(paint)*
Deckfaktor *m (text)* cover factor
Deckfarbe *f (build)* finishing paint; *(text)* covering colour; masking ink; opaque pigment, body colour
Deckfenster *n* skylight
Deckfrucht *f (ag)* cover crop
Deckfuge *f (ship)* deck seam
Deckgebirge *n (mine)* burden, *(Am)* capping [mass], [mine] cover, overburden, *(Am)* rock capping
Deckglas *n* cover glass *(microscope)*
Deckglaskorrektion *f* correction for covered specimens
Deckhausboot *n* house boat
Deckkraft *f* hiding (covering) power, opacity *(paint)*; coverage
Decklack *m* top coat of paint
Decklasche *f* cover plate
Deckleiste *f* cover strip (fillet)
Deckmaschine *f* narrowing (tickler) machine *(hose)*
Deckmetall *n* metal cladding; electroplate
Deckmuster *n* stitch transfer design
Decknadel *f (text)* covering needle
Deckoperation *f (ph)* symmetry operation
Deckpapier *n* cover paper (board); liner
Deckpeilung *f (ship)* transit bearing
Deckplatte *f* cover plate; top cover *(instrument)*; lid
Deckpolitur *f* finishing coat of varnish
Deckring *m* end-ring *(seal)*
Decks *npl* / **übereinanderliegende** superimposed decks
Decksaufbau *m* [deck] erection
  alleinstehender D. detached superstructure
Decksbalken *m* beam
Decksbalkenträger *m* deck stringer
Decksbalkenweger *m* deck-beam clamp
Decksbelag *m* decking, deck sheathing (composition, covering), deck-flat
Decksbelastung *f (ship)* deck load
Decksbeleuchtung *f* illumination (lighting) of deck
Decksbeplankung *f* deck planking
Decksbeplattung *f* deck plating
Decksbreite *f* beam, breadth of deck
Deckscheibe *f* shroud plate
Deckschicht *f* topping, finishing (top) layer, covering, surfacing; overburden, overlap
  feste D. strong lining *(paper)*
Deckschichten *fpl s.* Deckgebirge
Deckseinrichtungen *fpl* deck fittings
Decksfahrstand *m* deck control
Decksfenster *n* patent glass
Decksfläche *f (ship)* deck area, airing space
Deckshaus *n* deck house [forecastle]
Deckshaussüll *n* house coaming
Deckshilfsmaschinen *fpl* deck machinery (auxiliaries)
Deckshöhe *f* deck height, height between decks
Deckskajüte *f* deck cabin *(boat)*
Decksklüse *f* chain locker
Decksladung *f* deck cargo (load)
Deckslängsträger *m* deck strongback
Deckslast *f* deck cargo (load)
Deckslinie *f* / **gesetzliche** statutory deck-line
Decksloch *n* scuttle hatch
Deckslog *n* taffrail (deck) log
Decksnaht *f* deck seam
Deckspersonal *n* deck hands
Decksruderstopper *m* deck stops
Decksseitenstütze *f* deck quarter pillar
Deckssprung *m* sheer [of the deck]
Deckssprunglinie *f* sheer line
Deckssteuerstand *m* deck control station
Decksstrak *n* beam (sheer) line
Decksstringer *m* deck stringer
Decksstütze *f* deck pillar, stanchion
  seitliche D. quarter pillar
Deckstein *m* cap jewel, end stone *(watch)*
Decksuhr *f* hack watch
Decksunterzug *m* deck girder; runner; riser *(wooden ship)*
Decksverladung *f* shipment on deck
Deckswrange *f* deck transom (hook)
Deckung *f* coincidence; congruence
  D. der Auslagen reimbursement of the outlay
  einfache D. single coverage [of the roof]
deckungsgleich *(math)* congruent
Deckungssignal *n (ry)* covering (protection) signal
Deckvermögen *n* hiding (covering) power *(paint)*, opacity
Deckweiß *n* zinc white
decodieren *s.* dekodieren
Deduzierbarkeit *f (math)* deducibility
Dee *n (nucl)* dee, duant
Deemphasis *f* de-emphasis
defekt defective, faulty
Defekt *m* defect, imperfection; deficiency
Defektbogen *m (print)* oversheet, faulty sheet
Defektelektron *n* defect electron, electron hole
Defektelektronenbewegung *f* hole motion

Defektelektronendichte *f* hole density
Defektelektronenleitfähigkeit *f* hole conductivity *(semiconductor)*
Defektelektronenleitung *f* hole conduction *(semiconductor)*
Defekthalbleiter *m* defect (p-type) semiconductor, hole conductor
Defektleitfähigkeit *f* hole conductivity
Defektleitung *f* hole (defect, p-type) conduction
Defektoskop *n* flaw detector
Defektoskopie *f* radiography, inspection by means of rays
Defibrator *m* defibrator
defibrinieren to defibrinate
Definitionsbereich *m* *(math)* domain of definition
Deflagration *f* deflagration
deflagrieren to deflagrate
Deflektor *m* deflector
Defokussierung *f* defocusing
  D. bei Ablenkung deflection defocusing
Deformation *f* deformation
  D. der Wicklung winding deformation
Deformationsmaß *n* measure of deformation
Deformationspotential *n* deformation potential
Deformationstensor *m* deformation (strain) tensor
Deformationsverlust *m* deformation loss
Deformierbarkeit *f* deformability
deformieren to deform
Deformierung *f* deformation
Degradation *f* *(nucl)* degradation
Degras *n* moellen, degras
degrassieren to fat liquor
degummieren to degum *(silk)*
dehnbar strainable; yielding; extensible; ductile; elastic
Dehnbarkeit *f* strainability, extensibility, expandability, stretchability, elasticity
dehnen to strain, to expand, to stretch
Dehner *m* [automatic volume-]expander
Dehnfuge *f* contraction (expansion) joint
Dehnmeßstreifen *m* resistance strain gauge
Dehnpreßpassung *f* expansion [press] fit
Dehnreibahle *f* expanding reamer
Dehnung *f* strain; extension, elongation, dilatation, distension, expansion, stretch
  D. von magnetischen Feldlinien stretching of magnetic field lines
  bleibende D. permanent (plastic) strain *(material)*; *(text)* permanent stretch
  elastische D. elastic strain; *(text)* elastic (recoverable) stretch
Dehnungsarbeit *f* *(therm)* work done during expansion
Dehnungsfestigkeit *f* *(ph)* tensile strength
Dehnungsfuge *f s.* Dehnfuge
Dehnungsmesser *m* extensometer, strain gauge
  mechanischer D. strain gauge
Dehnungsmeßgerät *n* strain gauge
Dehnungsmeßstreifen *m* wire (foil) strain gauge
Dehnungsprüfapparat *m* apparatus for testing the extensibility
Dehnungsriß *m* expansion crack
Dehnungsrohr *n* expansion (compensating) pipe
Dehnungsschlaufe *f* expansion bend *(pipe)*
Dehnungsschwingung *f* dilatation mode
Dehnungsspannung *f* tensile stress
Dehnungswelle *f* dilational wave
Dehnungszahl *f* coefficient of expansion
Dehnvermögen *n* ductility
Dehnweg *m* stretch, elongation
dehydratisieren to dehydrate
Dehydratisieren *n* dehydration
dehydrieren to dehydrogenize
Dehydrieren *n* dehydrogenation
Dehydrogenase *f* dehydrogenase
Deich *m* dike, dyke; levee
Deicheinbruch *m* dike burst
Deichsel *f* 1. adz[e]; 2. *(ag)* pole shaft
Deichselschere *f* [pole] cleave
Deionisationsschalter *m* deion (quenched arc) circuit breaker
Deionisierungsgitter *n* deionizing grid
Dekade *f* decade
Dekadendrehschalter *m* rotary decade switch
Dekadenimpulsgeber *m* decade pulse generator
Dekadenkontakt *m* *(telec)* decade contact
Dekadenmeßbrücke *f* decade bridge
Dekadenschalter *m* decade switch
Dekadenschaltung *f* decade connection
Dekadenstufe *f* decade step
Dekadenteiler *m* decade divider
Dekadenuntersetzer *m* decade scaler, decascaler, scale of ten circuit
Dekadenvervielfacher *m* decade multiplier
Dekadenwiderstand *m* 1. decade resistor; 2. decade resistance
Dekadenzähler *m* *(nucl)* decade counter; decade scaler, decascaler, scale of ten circuit
Dekadenzählröhre *f* decade counter tube
Dekaeder *n* decahedron
Dekalin *n* decalin, decahydronaphthalene
Dekameter *n* oscillometer
Dekan *n* *(ch)* decane
Dekantierapparat *m* *(therm)* settling apparatus
dekantieren to decant, to pour off
Dekantieren *n* decantation
Dekantiergefäß *n* decanter, decanting glass
Dekantierventil *n* decanting valve
Dekarboxylierung *f* *(ch)* decarboxylation
dekatieren to decatize; to hot press; to remove the gloss
Dekatiermaschine *f* decating machine
Dekatiertuch *n* decating blanket

Dekatur *f* decatizing
Dekaturechtheit *f* fastness to decatizing
Deklination *f* declination
Deklinationsachse *f* declination axis
Deklinationskreis *m* declination circle
Deklinationsnadel *f* declinometer; declination needle
Deklinationswinkel *m* angle of declination
Deklinograf *m* declinograph
Deklinometer *n* declinometer
Dekodiereinrichtung *f* decoding device
dekodieren to decode, to decipher
Dekodierer *m* decoder
Dekodiermatrix *f* decoding matrix
Dekodierung *f* decoding; code translation
Dekontamination *f* [radioactive] decontamination
Dekontaminationsmittel *n* decontaminating agent, decontamination substance
Dekontaminationsverhältnis *n* decontamination ratio
Dekontaminierung *f s.* Dekontamination
Dekor *m* scenery *(television)*; pattern
Dekorateur *m* decorator
Dekorationspapier *n* fancy paper
dekorieren to dress *(shop-window)*; to decorate
Dekrement *n* decrement
  logarithmisches D. logarithmic decrement
dekrepitieren to decrepitate
Dekupiersäge *f* scroll saw
Deleaturzeichen *n* deletion mark
Delon-Gleichrichter *m (el)* Delon rectifier
Delon-Schaltung *f* Delon rectifier connection, voltage doubling connection
Delta *n* delta
Delta-Anpassung *f* delta matching *(of aerials)*
Delta-Antenne *f* delta-matched impedance aerial
Delta-Elektron *n* delta-electron, knocked-on electron, delta-ray
Delta-Metall *n* delta metal
Delta-Modulation *f* delta modulation
Delta-Schaltung *f* delta connection (circuit)
Delta-Strahl *m* delta ray, delta-electron, knocked-on electron
Delta-Teilchen *n* delta ray
deltaförmig deltoid
Deltoid *n* deltoid *(geometry)*
Deltoiddodekaeder *n* deltoid dodecahedron
Deltoidikositetraeder *n* deltoid icositetrahedron, trapezohedron
Demarkationspotential *n* demarcation potential
Dematerialisation *f* annihilation radiation (of matter), dematerialization
Demodulation *f* demodulation, detection
Demodulationsverfahren *n* demodulation method
Demodulator *m* demodulator
  D. für Frequenzmodulation frequency discriminator
Demodulatorröhre *f* detecting (detector) valve
Demodulatorschaltung *f* demodulator circuit
demodulieren to demodulate, to detect
Demonstration *f* demonstration
Demonstrationsauge *n (opt)* demonstration eye
Demonstrationsokular *n (opt)* demonstration eyepiece
Demonstrationsschaltbild *n* working diagram
Demontage *f s.* Demontierung
demontieren to dismantle, to dismount
Demontierung *f* dismantling, dismounting
Demulgator *m* defoaming agent, antifroth
demulgieren to de-emulsify, to demulsify
Demulgierung *f* de-emulsification, demulsification
denaturieren to denature
Denaturierung *f* denaturation
Denaturierungsmittel *n* denaturant
Dendrit *m* tree-like crystal, dendrite
Dendritenbildung *f* dendrite formation
Dendritenverästelung *f* branching dendrites
dendritisch dendritic
Dendrolith *m (geol)* dendrolite
Dendrometer *n* dendrograph
Dengelamboß *m* scythe anvil
Dengelhammer *m* scythe hammer
dengeln to sharpen, to beat out a scythe
Dengelstein *m* scythe stone
Denier *n* fineness *(silk)*
Denitrifikationsbakterien *fpl* denitrifying bacteria
Densitometer *n* densitometer
Densometer *n* densitometer
Depeschenboot *n* advice boat
Dephlegmator *m* dephlegmator
Deplacementsschwerpunkt *m (ship)* centre of buoyancy
Deplistor *m* deplistor *(semiconducting device)*
Depolarisation *f* depolarization
Depolarisationsgrad *m* degree of depolarization
Depolarisationskorrektur *f* depolarization correction
Depolarisierungsfaktor *m* depolarizing factor
Depolymerisation *f* depolymerization
Depot *n* depot, storage facility (space)
Depression *f* depression
Depressionsbahn *f (rock)* path of the depression
Depressionsfront *f* front of depression *(meteorology)*
Depressionswinkel *m* angle of depression
Deputatkohle *f* miners' coal (issues)
Derbgehalt *m* solid contents of stacked wood
Derbholz *n* compact wood
Derbmasse *f* solid contents of stacked wood
Derbstangen *fpl* compact bars *(wood)*

Deri-Motor *m (el)* Deri[-type] motor, Deri repulsion motor
Derivat *n* derivative
Derrick *m* derrick [crane]
Derrickkran *m* / **abgespannter** guy derrick
Desaktivierung *f* desactivation
Desaminase *f (ch)* desaminase
Desaminierung *f (ch)* desamination
Desensibilisator *m* desensitizer
desensibilisieren to desensitize
Desensibilisierung *f* desensitization
Desinfektion *f* disinfection
Desinfektionsmittel *n* disinfectant
desinfizieren to disinfect
desodorieren to deodorize
Desodorierungsmittel *n* deodorant
desodorisieren to deodorize
desorbieren to desorb
Desorption *f* desorption
Desoxydation *f* deoxidation
Desoxydationslegierung *f* deoxidation alloy
desoxydieren to deoxidize; to oxidize off
Dessauer-Schaltung *f* Dessauer series connection
Dessertwein *m* dessert (sweet) wine
Dessin *n* / **modisches** fashion design
Dessinateur *m* designer
Dessindruck *m* pattern printing
Destillat *n* distillate
Destillation *f* distillation, distilling
  D. im Vakuum low-pressure distillation
  D. über Adsorptionsmittel contact distillation
  azeotrope D. azeotropic distillation
  extraktive D. extractive distillation
  fraktionierte D. fractional distillation
  stufenweise D. *s.* fraktionierte D.
  trockene D. dry (destructive) distillation
Destillationsanlage *f* still
Destillationsapparat[ur *f*] *m* still, distilling apparatus
Destillationsaufsatz *m* distillation head, ball-top attachment
Destillationsbenzin *n* straight-run petrol (gasoline)
Destillationsblase *f* batch still
Destillationskolben *m* distillation flask
Destillationskolonne *f* column still
Destillationsnachlauf *m* distillation tail, tail run, last runnings
Destillationspunkte *mpl* running points
Destillationsverhalten *n* distillation characteristics
Destillationsverlauf *m* distillation range
Destillationsvorlage *f* run-down tank, receiver
Destillationsvorlauf *m* first (fore) runnings
Destillierapparat *m* still, distilling apparatus
destillierbar distillable
Destillierblase *f* still, distilling vessel, alembic
destillieren to distil, to rectify, to fractionate
Destillierkolben *m* distillation flask
Detachieren *n (text)* removal of stains
Detachiermittel *n (text)* spot removing agent
Detail *n* detail
Detailauswertung *f (opt)* detailing
  linienweise D. detailing in lines
Detailbild *n* detail (split) picture
detailreich *(telev)* of high definition
Detailvermessung *f* / **polare** detailing by polar coordinates
Detailzeichnung *f* detail drawing
Detektor *m* detector
  D. für langsame Neutronen slow neutron detector
  D. für schnelle Neutronen fast neutron detector
  D. für thermische Neutronen thermal [neutron] detector
  fest eingestellter D. permanent crystal detector
  lichtempfindlicher D. light-sensitive detector
  quadratischer D. square-law detector
  schneller D. fast detector
  zweiter D. audio detector
Detektorapparat *m* detector receiver (receiving set)
Detektordrähtchen *n* cat's whisker
Detektorempfänger *m* detector (crystal) receiver
Detektorglied *n* crystal receiver
Detektorröhre *f* detector (rectification) valve
Detergens *n* detergent
Determinante *f* determinant
Determinantenteiler *m (math)* determinantal divisor *(matrix)*
Detonation *f* detonation
Detonationsgeschwindigkeit *f* quickness, velocity of detonation *(explosive)*
Detonationsherd *m* explosion centre
Detonationsübertragungsfähigkeit *f* propagating power *(explosive)*
detonieren to detonate
Deuterium *n* deuterium, heavy hydrogen
Deuteriumkern *m* deuteron
Deuteriumoxid *n* heavy water, deuteroxide, deuterium oxide
Deuteron *n* deuteron
Deuteronenbündel *n* deuteron beam
Deuteroprisma *n (ph)* deuteroprism, prism of the second order
Deutung *f* interpretation
  atomistische D. atomistic interpretation
  physikalische D. physical interpretation
  statistische D. statistical interpretation
Deviation *f* deviation
Deviationsboje *f* swinging-buoy

Deviationstabelle *f* calibration card (table)
Deviometer *n* deviometer
Devonzeit *f (geol)* devonian period
Dewargefäß *n* Dewar flask (vessel)
Dextran *n (ch)* dextran
Dextrin *n* dextrin
Dextrose *f* dextrose, grape sugar
Dezibel *n* decibel
Dezibelmeßgerät *n* decibel meter
Dezibelstufe *f* decibel step
Dezimalablesung *f* decimal reading
dezimalbinär *(dat)* decimal binary
Dezimalbinärumsetzung *f (dat)* decimal-to-binary conversion
Dezimalbinärumwandlung *f s.* Dezimalbinärumsetzung
Dezimalbruch *m* decimal fraction
Dezimalbruchrechnung *f* decimal system of calculations
Dezimaldarstellung *f* decimal notation
 binär verschlüsselte D. binary coded decimal notation
 biquinär verschlüsselte D. biquinary coded decimal notation
 mit 0 und 1 verschlüsselte D. 0-L coded decimal notation *(L = binary one)*
 verschlüsselte D. coded decimal notation
Dezimalkomma *n* decimal point
Dezimalkonvertierung *f* decimal conversion
Dezimalpunktautomatik *f* automatic indication of decimal point
Dezimalschreibweise *f s.* Dezimaldarstellung
Dezimalstelle *f* decimal place
Dezimaluntersetzer *m s.* Dekadenuntersetzer
Dezimalziffer *f* decimal digit
Dezimeterwelle *f* decimetric (decimetre) wave
Dezimeterwellenleistungsmesser *m* VHF power meter
Dezimeterwellentechnik *f* ultra-high frequency technique
Dezimeterwellenverstärker *m* amplifier for microwaves
Dia *n* diapositive
Dia-Abtastung *f* slide scanning
Diagnoseprogramm *n (dat)* diagnostic routine
diagonal diagonal
Diagonalastigmatismus *m* diagonal astigmatism
Diagonale *f* diagonal
Diagonalköper *m (text)* upright twill
Diagonalmatrix *f* diagonal matrix
Diagonalschaben *n* diagonal shaving
Diagonalschichtung *f (geol)* current-bedding
Diagonalschneider *m* angle-cutting machine
Diagonalschnitt *m* diagonal grain *(wood)*
Diagonalspeisung *f* diagonal feed
Diagonalstab *m* diagonal member (brace, strut, tie, bar), angle brace
Diagonalstrebe *f* diagonal brace (strut)
Diagonalverband *m* raking bond *(brickwork)*; diagonal bracing
Diagonalverrippung *f* **über volle Bettlänge** diagonal cross ribbing over full length of bed
Diagonalverspannung *f* cross bracing
Diagonalzeile *f* slanting line
Diagramm *n* diagram, chart, graph, plot
 Boulvinsches D. Boulvin diagram
 fortlaufendes D. continuous diagram
 rankinisiertes D. Rankine diagram
Diagrammfläche *f* area of the diagram
Diagrammhöhe *f* / **mittlere** mean height of diagram
Diagrammkarte *f* cartogram
Diagrammkurve *f* diagram curve
Diagrammlänge *f* length of diagram
Diagrammlinie *f* line of diagram
Diagrammpapier *n* indicator card
Diagrammpunkt *m* point of diagram
Diagrammsatz *m* series of diagrams
Diagrammschleife *f* diagram loop
Diagrammtransportwerk *n* chart drive
Diagrammvoreilwinkel *m* angle of lead in diagram
diakritisch diacritical
Dialysator *m* dialyzer
Dialyse *f* dialysis
dialysieren to dialyze
Dialysierzelle *f* dialyzer
diamagnetisch diamagnetic
Diamagnetismus *m* diamagnetism
Diamant *m* diamond
Diamantabrichter *m* diamond wheel dresser
Diamantabzieheinrichtung *f* diamond dressing device
diamantbestückt diamond-tipped
Diamantbohren *n* diamond boring (drilling)
Diamantbohrer *m* diamond drill
Diamantbohrkrone *f* diamond drill bit
Diamantbohrmaschine *f* diamond drill
Diamantgitter *n* atomic structure of diamond, diamond structure
diamanthaltig diamondiferous
diamanthart adamantine
Diamantkrone *f* diamond drill bit
Diamantpyramide *f* diamond pyramid
Diamantpyramidenhärte *f* diamond pyramid hardness
Diamantschleifscheibe *f* diamond [grinding] wheel
Diamantspitze *f* diamond point
Diamantsplitter *m* diamond particle
Diamantstaub *m* diamond dust (powder)
Diamantwerkzeug *n* diamond tool
diametral diametrical; diametral
Diaminfarbstoff *m* diamine dye
Diaphragma *n* septum, diaphragm

Diapositiv *n* diapositive
Diapositivabtastung *f* lantern slide scanning
Diapositivprojektion *f* slide projection
Diaprojektion *f* slide projection
Diaspor *m (min)* diaspore
Diastase *f (ch)* diastase
diatherman diathermanous
Diäthyläther *m* diethyl ether
Diäthylenglykol *n* diethylene glycol
Diatomeenerde *f* diatomaceous (infusorial) earth
Diazoechtfarbstoff *m* diazo fast dye
Diazofarbe *f s.*Diazofarbstoff
Diazofarbstoff *m* diazo dye
Diazoniumsalz *n* diazonium salt
diazotieren to diazotize
Diazotierung *f* diazotization
Diazotypie *f* diazotype
Diazoverbindung *f* diazo compound
Dibbelmaschine *f (ag)* dibbler, spacing drill
dibbeln *(ag)* to dibble
Dichroismus *m* dichroism
Dichroit *m* dichroite, cordierite
dicht leakproof; dense; close *(fabric)*
d. am Land inshore
d. unter Land close inshore
Dichte *f* 1. density; 2. compactness; 3. tightness; 4. specific gravity
D. fotografischer Schichten density of photographic deposits
bezogene D. relative density
optische D. optical density
räumliche D. der elektrischen Ladung spatial residual charge density
relative D. relative density
spektrale D. spectral concentration
Dichteabfall *m* slope of density
Dichteänderung *f* density change
Dichteanreicherung *f* gravity concentration
Dichtekurve *f* density curve
Dichtematrix *f* density matrix
Dichtemesser *m* densitometer
Dichtemeßgerät *n s.* Dichtemesser
dichten to pack, to seal, to caulk; to damp-proof, to water-proof
ein Leck d. to stop a leak
Dichteplan *m* density map
Dichteschrift *f* variable density recording *(sounding pictures)*
Dichteschwankung *f* **in Schallfeldern** fluctuation of intensity in sound fields
Dichteverhältnis *n (ph)* relative density
Dichteverteilung *f* density distribution
Dichtezahl *f (ph)* relative density
Dichtfläche *f* packing (gasket) surface
Dichtheit *f* 1. *(math)* denseness; 2. compactness
Dichtnietung *f* close riveting
Dichtprüfung *f* proofing; leak[age] test; pressure build-up test

Dichtring *m* gasket
Dichtung *f* sealing, packing, caulking; damp-proofing, water-proofing
bewegliche D. floating (running) packing
festliegende D. fixed packing
rotierende D. rotary seal
statische D. static seal
Dichtungsband *n* caulking strip
Dichtungsbinde *f* packing bandage *(of cable)*
Dichtungsbuchse *f* liner (bush) of diaphragm
Dichtungsdraht *m* seal wire
Dichtungsfenster *n* seal [in a waveguide]
Dichtungsfett *n* joint grease
Dichtungsfläche *f* packing surface, seal face
Dichtungshaut *f* damp-proof membrane
Dichtungskegel *m* packing cone
zweiteiliger D. two-part packing cone
Dichtungskitt *m* luting agent
Dichtungslager *n* stuffing box bearing
Dichtungslänge *f* length of packing
Dichtungsleiste *f* face of joint
Dichtungsmanschette *f* gasket
Dichtungsmasse *f* sealing (caulking, jointing) compound
Dichtungsmaterial *n* sealing (packing) material; cementing agent
Dichtungsmittel *n* sealing medium, cement[ing agent]; densifier, waterproofing compound *(concrete admixture)*
D. gegen Flüssigkeitsdurchtritt anti-leak cement
Dichtungsring *m* packing (seal) ring, ring gasket
D. für den Flansch joint ring [for flanged joint]
defekter D. blown gasket
elastischer D. elastic (spring) packing ring
gewellter D. corrugated packing ring
glatter D. smooth packing ring
Dichtungsringe *mpl* / **ineinandergreifende** compression packing
Dichtungsringlage *f* set of packing rings
Dichtungsscheibe *f* sealing washer
lederne D. leather washer
Dichtungsschnur *f* packing cord
Dichtungsschürze *f* blanket
Dichtungsstreifen *m* caulking strip, weather strip *(door)*
Dichtungsstrick *m* gaskin
Dichtungswerkstoff *m* seal (packing) material
Dichtungszopf *m* packing cord
Dichtungszusatz *m* densifier, waterproofing compound *(concrete)*
dick thick
dickbankig *(geol)* heavy bedded
Dicke *f* thickness; diameter; gauge; consistency *(of fluid)*
D. des Hohlraums thickness of cavity

Dickenhobelmaschine *f* thicknessing machine
Dickenlehre *f* thickness gauge
Dickenmeßeinrichtung *f* / **berührungslose** non-contacting thickness gauge
Dickenmesser *m* thickness gauge; caliper profile recorder *(paper, board)*
Dickenmeßgerät *n* thickness gauge
Dickenmessung *f* thickness measurement (gauging)
berührungslose D. non-contact thickness measuring (gauging)
Dickenschwingung *f* width (thickness) vibration *(of crystal)*
dickflüssig viscous, consistent
dickflüssiger machen (werden) to thicken
Dickflüssigkeit *f* consistency *(oil)*
Dickicht *n* thicket
Dickit *m (min)* dickite
Dicksaft *m* syrup, thick juice
Dicksaftkocher *m* thick juice blow-up *(sugar industry)*
Dicksaftsaturation *f* thick juice carbonatation
Dickschlamm *m* slurry
Dickstoffbleiche *f* high density bleaching *(paper)*
Dickstoffpumpe *f* slush pump
Dickte *f* ply
Dickteer *m* crude tar, thick tar
Dickteerabscheider *m* thick-tar decanter
Dicktenlehre *f* feeler gauge
Dicktenschablone *f* feeler gauge
Dicktrübe *f* regenerated dense-medium
Dickungsmittel *n* thickener
dickwandig thick-walled
Diele *f* 1. hall, lobby, vestibule; 2. deal, plank, board
Dielektrikum *n* dielectric
festes D. solid dielectric
ideales D. perfect (ideal) dielectric
kriechstromfestes D. non-tracking dielectric
künstliches D. artificial dielectric
schlechtes D. poor dielectric
verlustbehaftetes D. imperfect dielectric
verlustfreies (verlustloses) D. perfect (no-loss, loss-free) dielectric
Dielektrikumtester *m* dielectric test set
dielektrisch dielectric
Dielektrizität *f* dielectricity
Dielektrizitätskonstante *f* dielectric constant, permittance
absolute D. [absolute] permittivity
relative D. relative permittivity
Dielenbalken *m* boarding joist *(carpentry)*
Dielenfußboden *m* boarded floor
Dielung *f* boarding, planking
Dieme *f* hay rick
Dien *n (ch)* diene
dienen to serve, to be of use, to be utilized [for]
Dienst *m* service; duty
durchgehender D. continuous service
meteorologischer D. weather service
Dienstabfrageklinke *f (telec)* order-wire jack
Dienstabteil *n* service compartment
Dienstanruflampe *f (telec)* order-wire lamp
Dienstanschluß *m* service telephone
Dienstfahrausweis *m* duty pass
Dienstgeschwindigkeit *f* service (commercial) speed
Dienstgespräch *n (telec)* service call
Dienstgewicht *n* weight in working order
Dienstgipfelhöhe *f* service ceiling
Dienstgutbeförderung *f (ry)* traffic conveyed on railway service, non revenue earning traffic
Dienstgutverkehr *m (ry)* traffic on railway service
Dienstleistung *f* service
Dienstleitung *f (telec)* service circuit
Dienstreise *f* journey on duty
Dienstunfähigkeit *f* incapability of service
dienstuntauglich ineligible
Dienstvernachlässigung *f* carelessness while on duty
Dienstvorschrift *f* regulations governing conditions of service, service regulations
Dienstwähler *m* discriminating selector
Dienstzug *m* service train
Diesel *m s.* 1. Dieselmotor; 2. Dieselöl
D. für Überwasserfahrt surface Diesel-engine *(of a submarine)*
Dieselaggregat *n* Diesel-driven generating set
Dieselantrieb *m* Diesel drive
Dieselbetrieb *m (ry)* Diesel traction
dieselelektrisch Diesel-electric
Dieselfahrzeug *n* Diesel engine vehicle
Dieselhorst-Martin-Verseilung *f (el)* multiple twin formation
dieselhydraulisch Diesel-hydraulic
Dieselkarren *m* Diesel truck
Dieselkraftstoff *m* Diesel fuel
Diesellokführer *m* Diesel engine driver
Diesellokomotive *f* Diesel locomotive
Dieselmotor *m* Diesel engine
D. mit Turboaufladung turbo-charged Diesel engine
direkt umsteuerbarer D. directly reversible Diesel engine
kompressorloser D. compressorless Diesel engine
turboaufgeladener D. turbo-charged Diesel engine
Dieselmotorschiff *n* Diesel motorship
Dieselöl *n* Diesel oil, *(ship)* fuel oil
Dieselschweißaggregat *n* engine-driven welding set
Dieselstraßenfahrzeug *n* Diesel-engined road vehicle

Dieseltraktor *m* Diesel tractor
Differential *n* 1. *(math)* differential; 2. differential [gear unit] *(motor vehicle)*
totales D. *(math)* total differential
Differentialachse *f* differential pinion shaft
Differentialanalysator *m* differential analyzer
digitaler D. digital differential analyzer, DDA
Differentialanteil *m* derivative action (control factor)
Differentialantrieb *m* differential drive
Differentialantriebskegelrad *n* differential axle drive bevel gear
großes D. axle-drive bevel gear, differential master gear, crown wheel
kleines D. differential (axle-drive) bevel pinion
Differentialbalgendurchflußmesser *m* bellows differential flowmeter
Differentialbandbremse *f* differential band brake
Differentialbewegung *f* differential motion
Differentialbogenlampe *f* differential arc lamp
Differentialbremse *f* differential brake
Differentialbrücke *f (el)* differential bridge
Differentialempfänger *m* differential receiver *(synchro)*
Differentialerdungsdrossel *f* drainage coil
Differentialerregung *f* differential excitation
Differentialflaschenzug *m* differential pulley block
Differentialflyer *m (text)* flyer frame with differential motion
Differentialgalvanometer *n* differential galvanometer
Differentialgeber *m* differential transmitter *(synchro)*
Differentialgehäuse *n* differential casing
Differentialgeometrie *f* differential geometry
Differentialgetriebe *n* differential mechanism (gear)
Differentialgleichung *f* differential equation
D. der Bewegung differential equation of the motion
D. der schwingenden Saite differential equation of string vibrations
gewöhnliche D. ordinary differential equation
partielle D. partial differential equation
Differentialgleichungslöser *m* differential analyzer
digitaler D. digital differential analyzer, incremental digital computer
Differentialionisationskammer *f* differential ionization chamber
Differentialkammer *f s.* Differentialionisationskammer
Differentialkolben *m* differential piston *(circuit breakers with low oil content)*
Differentialkolbenmanometer *n* differential piston gauge
Differentiallagerkasten *m* differential carrier
Differentialmanometer *n* differential manometer (pressure gauge)
Differentialmelder *m* differential transmitter (fire-alarm)
Differentialmeßbrücke *f* differential bridge
Differentialmessung *f* differential measurement
Differentialoperator *m* differential operator
Differentialpermeabilität *f* magnetic differential permeability
Differentialquotient *m (math)* derivative, differential coefficient
partieller D. partial derivative
Differentialrechnung *f* differential calculus
Differentialregler *m* differential governor
Differentialrelais *n* differential (balanced current) relay
Differentialschaltung *f* differential (duplex) connection
Differentialschutzwandler *m* biasing (differential protective) transformer
Differentialseitenwelle *f* drive (axle) shaft
Differentialspannungsmesser *m* differential voltmeter
Differentialsperre *f* differential lock
Differentialstern *m* differential star piece; differential spider
Differentialstirnrad *n* differential spur gear
Differentialstrommesser *m* differential current meter
Differentialteilen *n* differential indexing
Differentialthermoanalyse *f* differential thermal analysis
Differentialthermopyrosynthese *f* differential thermal pyrosynthesis
Differentialtransformator *m* hybrid coil, differential transformer
Differentialübertrager *m* differential transformer, hybrid coil
Differentialverstärker *m* difference (differential) amplifier
Differentialwicklung *f (el)* differential (counter-acting) winding
Differentialwirkung *f* derivative action
Differentialzähler *m* differential counter
Differentialzeit *f* derivative control factor
Differentialzylinder *m* differential cylinder
Differentiation *f (math)* differentiation
Differentiationsschaltung *f (dat)* differentiator
Differenz *f* difference
D. zwischen Spannungsmaximum und -minimum peak-to-peak voltage
Differenzdiagramm *n (therm)* differential diagram
Differenzdruck *m* differential pressure

Differenzdruckdurchflußmesser *m (Am)* differential flowmeter
Differenzdruckmanometer *n* flowmeter manometer
Differenzdruckventil *n* differential pressure valve
Differenzengleichung *f (math)* difference equation
Differenzenverfahren *n* method of differences
Differenzfrequenz *f* difference frequency
differenzierbar differentiable
Differenzierbarkeit *f* differentiability
differenzieren *(math)* to derive *(function)*, to differentiate
Differenzierglied *n* differentiator, differentiating circuit; advance element
Differenzierkreis *m* peaking circuit
Differenzierkreisel *m* rate gyro
Differenzierschaltung *f* differentiating circuit, peaker
Differenzindikator *m* differential indicator
Differenzmesser *m* differential measuring instrument
Differenzschaltung *f* differential connection (wiring)
Differenzstrom *m* differential current
Differenztemperaturregler *m* differential thermostat
Differenzton *m* difference tone
Differenzträgerverfahren *n* intercarrier-sound system
Differenzverstärker *m* difference (differential) amplifier, comparator
Differenzzugmesser *m* *(therm)* differential draught gauge
Diffraktion *f (ph)* diffraction
Diffraktionsgitter *n* diffraction grating
Diffraktionsmikroskopie *f* diffraction microscopy
Diffraktionsstreuung *f* diffraction scattering
diffundieren to diffuse
diffus diffuse
Diffusat *n* diffusate
Diffuseur *m (ch)* diffusion apparatus; blow pit
Diffusion *f* diffusion
  D. der magnetischen Feldlinien diffusion of magnetic field lines
Diffusionsatmosphäre *f* diffusion atmosphere
Diffusionsfläche *f* diffusion area
Diffusionsgeschwindigkeit *f* rate (speed) of diffusion, diffusion rate
Diffusionsgleichung *f* diffusion equation
Diffusionsgrenze *f* diffusion limit
Diffusionskammer *f* diffusion chamber
Diffusionskapazität *f* diffusion capacitance, hole storage effect *(semiconductor)*
Diffusionskoeffizient *m* diffusion coefficient, coefficient of diffusion
Diffusionskonstante *f* diffusion constant
Diffusionskühlung *f* diffusion cooling
Diffusionslänge *f* diffusion length
diffusionslegiert alloy-diffused
Diffusionslegierungstechnik *f* alloy diffusion technique
Diffusionsmethode *f* diffusion method
Diffusionsnäherung *f* diffusion approximation
Diffusionsnebelkammer *f* diffusion cloud chamber
Diffusionspotential *n* diffusion potential
Diffusionspumpe *f* diffusion pump
Diffusionsquerschnitt *m* diffusion cross section
Diffusionsschicht *f* diffused junction *(semiconductor)*
Diffusionsspannung *f* potential difference, diffusion voltage
Diffusionssprung *m* diffusional jog
Diffusionsstrom *m* flow of diffusing molecules, diffusion current
Diffusionstransistor *m* diffused transistor
Diffusionstrennkolonne *f* diffusion separating column
Diffusionsverfahren *n* diffusion process
Diffusionsverlust *m* loss by diffusion
Diffusionsvermögen *n* diffusibility
Diffusionsweg *m* diffusion path
Diffusionszone *f* diffusion zone
Diffusität *f (ac)* diffusivity
Diffusor *m* diffuser
digerieren *(ch)* to digest
digital digital
Digitalablesung *f* digital read-out
Digital-Analogumsetzer *m* digital-to-analogue converter
Digital-Analogwandler *m* digital-to-analogue converter
Digitalanzeige *f* digital display (indication)
Digitalausgang *m* digital output
Digitalbaustein *m* digital module
Digitaldarstellung *f* digitizing; digital display
Digitaldatenverarbeiter *m* digital data processor
Digitaldifferentialanalysator *m* digital differential analyzer
Digitaldrehgeber *m* digital shaft encoder
Digitaldrucker *m* digital printer
Digitalfeldrechner *m* digital field computer
Digitalmodul *m* digital module
Digitalrechenautomat *m* digital computer
Digitalrechenmaschine *f* digital computer
Digitalrechner *m* digital computer
Digitalrechnertechnik *f* digital computing system
Digitalschaltung *f* digital circuit
Digitalteilung *f* digital division
Digitalübertragung *f* digital transmission
Digitalumsetzer *m* digitizer

Digitalvoltmeter *n* digital voltmeter
Digitalwandler *m* digital converter
Digitron *n* digitron
Dikaliumtartrat *n* dipotassium tartrate
Dikarbonsäure *f* dicarboxylic acid
Diketon *n (ch)* diketone
Diktafon *n* dictaphone
Diktiergerät *n* dictating machine
Dilatation *f* dilatation
Dilatometer *n* dilatometer
  D. zur Bestimmung der kubischen Ausdehnung volume dilatometer
Dimension *f* dimension
dimensional dimensional
dimensionieren to dimension; to calculate dimensions
Dimensionierung *f* dimensioning, proportioning, sizing; calculation of [safe] dimensions
  binomische D. binominal dimensioning
  geometrische D. geometric dimensioning
  optimale D. optimum design
Dimensionsbeständigkeit *f* dimensional stability
Dimensionsholz *n* dimension timber
dimensionslos nondimensional, dimensionless
Dimensionsstockholz *n* dimension stock
dimer *(ch)* dimeric
Dimerisation *f* dimerization
Dimethylanilin *n* dimethylaniline
Dimethylglyoxim *n* dimethylglyoxime
dimetrisch dimetric
dimorph dimorphic, dimorphous
Dimorphie *f* dimorphism
DIN-Empfindlichkeit *f* DIN photographic film speed factor
DIN-Grad *m s.* DIN-Empfindlichkeit
Dinasstein *m* Dinas brick
Dinghi *n* jolly boat, dinghy
Diode *f* diode
  reihengeschaltete D. series diode
  vorgespannte D. biased diode
Diodenabschnitt *m* diode section
Diodenadmittanz *f* diode admittance
Diodenaußenwiderstand *m* diode load resistance
Diodenbegrenzer *m* diode limiter
Diodenbegrenzung *f* diode limiting
Diodenerwärmung *f* diode heating
Diodengleichrichter *m* diode detector
Diodengleichrichtung *f* diode detection (rectification)
Diodenheizfaden *m* diode filament
Diodenkennlinie *f* diode characteristic
Diodenmischer *m* diode mixer
Diodenpumpe *f* diode pump
Diodenrauschen *n* diode noise
Diodenschaltung *f* diode circuit
Diodenspitzenspannung *f* diode peak voltage
Diodenspitzenstrom *m* diode peak current
Diodenstromeinsatzpunkt *m* diode-current starting point
Diodentorschaltung *f* diode gate
Diodenverstärker *m* diode parametric amplifier
Diodenzähler *m* diode counter
Diolefin *n* diolefine
diophantisch diophantine
Diopsid *m (min)* diopside
Dioptas *m (min)* dioptase, emerald copper
Diopter *n (opt)* collimator
Dioptrie *f (opt)* diopter *(unit)*
Dioptrienfernrohr *n* attachment telescope with dioptre scale
Dioptrienteilung *f* dioptre scale
Dioxid *n* dioxide
Dipeptid *n* dipeptide
Dipeptidase *f* dipeptidase
Diphenyl *n* diphenyl
Diphenylamin *n* diphenylamine
Diphosphorsäure *f* pyrophosphoric acid
Diplom *n* 1. diploma *(document given by university testifying to a degree taken)*; 2. university degree in *(e.g. technology)*; *(no real equivalent is available; comparable roughly in level to master degree)*
Diplomagronom *m* diplomaed agronomist; university graduate of agronomy
Diplombetriebswirtschaftler *m* university graduate of production economics
Diplomgärtner *m* diplomaed horticulturist; university graduate of horticulture
Diplomingenieur *m* professional (academically trained) engineer; university graduate of technology
Diplomlandwirt *m* diplomaed agriculturist; university graduate of agriculture
Dipol *m* dipole
  abgestimmter D. tuned dipole
  elektrischer D. electric dipole
  gefalteter D. folded dipole
  Hertzscher D. Hertzian doublet (oscillator), half-wave doublet
  in der Mitte erregter D. centre-fed dipole
  rotierender D. rotating dipole
  spannungsgespeister D. voltage-fed dipole
  stromgespeister D. current-fed dipole
Dipolachse *f* dipole axis
Dipolanordnung *f* dipole array *(of aerials)*
Dipolanregung *f* dipole excitation
Dipolantenne *f* dipole [aerial]
Dipolbildung *f* dipole formation
Dipolfeld *n* dipole field
Dipolimpedanz *f* dipole impedance
Dipolkörper *m* dipole solid
Dipolmolekül *n* dipole molecule
Dipolmoment *n* dipole moment
Dipolreihe *f* [col]linear array
Dipolschicht *f* dipole (electric double) layer

Dipolschwingung *f* dipole oscillation
Dipolstrahlung *f* dipole radiation
magnetische D. magnetic dipole radiation
Dipolübergang *m* dipole transition
Dipolzeile *f* [col]linear array
Dipolzeilenantenne *f* [col]linear array
Dirac-Elektron *n* Dirac electron
Dirac-Gleichung *f* Dirac equation
Dirac-Teilchen *n* Dirac particle
Direktablesegenauigkeit *f* direct reading accuracy
Direktablesung *f* direct reading
Direktanzeige *f* direct indication
Direktaufnahme *f* contact image *(X-ray)*
Direktbeobachtung *f* direct viewing
Direktbestrahlung *f* direct exposure
Direktbetrieb *m* direct operation
Direktdruck *m* direct print
Direktdruckwerk *n* direct printer
Direktfarbe *f* direct (substantive) dye; flushed colour
Direktfarbstoff *m* direct (substantive) dyestuff
Direktfärbung *f* substantive dyeing
Direktionskraft *f* directing force
Direktkühlung *f* direct cooling
Direktmessung *f* direct measurement
Direktor *m* director
Direktorenelement *n* director
Direktschreiber *m* direct writing recorder
Direktschreibung *f* direct recording
Direktsendung *f* on-the-scene broadcast
Direktteilkopf *m* single plate [dividing] head
Direktteilscheibe *f* direct indexing plate, rapid index plate
Direktverbindung *f* *(telec)* direct connection; hot line
Direktvergrößerung *f* screen magnification *(electron microscope)*
Direktwahlsystem *n* *(telec)* step-by-step-system
direktwirkend self-operated; direct-acting
direktziehend substantive *(dye)*
Dirk *m* *(ship)* boom lift
Disaccharose *f* disaccharose
Disadvantagefaktor *m* *(nucl)* disadvantage factor
Disazofarbstoff *m* disazo dye[stuff]
Dischwefelsäure *f* disulphuric (pyrosulphuric) acid
Disgregationsarbeit *f* *(therm)* work done during change of state
Disgregationsvolumen *n* *(therm)* variation of volume by change of physical condition
Disilan *n* silicoethane, disilane
Disjunktion *f* disjunction *(logical operator)*
Diskenmeißel *m* disk bit
diskontinuierlich discontinuous
Diskontinuität *f* discontinuity
Diskontinuitätsfläche *f* surface of discontinuity
Diskontinuum *n* discontinuum
Diskordanz *f* *(geol)* unconformity
Diskriminante *f* discriminant
Diskriminationsfaktor *m* common-mode rejection quotient
Diskriminationspegel *m* discrimination level
Diskriminator *m* discriminator
Diskriminatorausgang *m* discriminator output
Diskriminatorgleichrichtung *f* discriminator detection
Diskriminierpegel *m* discrimination level
Diskriminierschwelle *f* discrimination level
Diskus *m* discus, disk
Diskuswerfen *n* throwing the discus, discus throw[ing]
Diskuswurf *m s.* Diskuswerfen
Dislokation *f* dislocation
Dismutation *f* dismutation
Disparität *f* disparity
dispergieren to disperse
Dispergier[ungs]mittel *n* dispersing agent
dispers disperse, dispersal
Dispersion *f* dispersion
D. des Lichtes dispersion of light
anomale D. anomalous dispersion
räumliche D. spatial dispersion
Dispersionsbeziehung *f* dispersion relation
Dispersionsbrennelement *n* *(nucl)* dispersion-type [fuel] element
Dispersionsfarbstoff *m* dispersal colour, disperse dye[stuff]
Dispersionsgesetz *n* dispersion relation
Dispersionshärtung *f* dispersion hardening, hardening by precipitation
Dispersionskurve *f* dispersion curve
Dispersionsmittel *n* dispersant, dispersion medium
Dispersionsprisma *n* dispersing prism
dispersionsstark highly dispersive
Dispersionstheorie *f* theory of dispersion, dispersion theory
Dispersionsvermögen *n* dispersive power
Dissipation *f* dissipation
D. der Energie energy dissipation
Dissoziation *f* dissociation
elektrolytische D. electrolytic dissociation
stufenweise D. dissociation in stages
Dissoziationsenergie *f* energy of dissociation, dissociation energy
Dissoziationsgleichgewicht *n* dissociation equilibrium
Dissoziationsgleichung *f* dissociation equation
Dissoziationsgrad *m* degree of dissociation
Dissoziationsgrenze *f* limit of dissociation
Dissoziationskonstante *f* dissociation constant
Dissoziationspotential *n* dissociation potential
Dissoziationsspannung *f* dissociation potential
Dissoziationswärme *f* heat of dissociation

dissoziieren to dissociate
dissoziiert dissociated
Distanz *f* distance
Distanzbüchse *f* spacer [bushing]
Distanzhalter *m* spacer, separator
Distanzplatte *f* spacing plate
Distanzrelais *n* distance relay
Distanzring *m* spacer
Distanzschutz *m (el)* distance protection
D. mit Stufenkennlinie stepped curve distance-time protection
Distanzstück *n* distance (spacing) piece; spacer
Disthen *m (min)* disthene, kyanite
Distickstoffoxid *n* nitrous oxide
Distickstoffpentoxid *n* nitrogen pentoxide, nitric anhydride
Distickstofftrioxid *n* nitrogen trioxide, nitrous anhydride
distributiv *(math)* distributive
Distributivgesetz *n* distributive law [of algebra]
ditetragonal ditetragonal *(crystal)*
Dithiokohlensäure *f* dithiocarbonic acid
Dithionat *n* hyposulphate, dithionate
Dithionit *n* hyposulphite, dithionite
Dithionsäure *f* dithionic acid
divergent divergent *(series)*
Divergenz *f (math)* divergence
Divergenzverlust *m* divergence loss
Divergenzwinkel *m* angle of divergence
Diversity-Empfang *m* diversity reception
Diversity-Empfänger *m* diversity receiver
Dividend *m* dividend
dividieren to divide
Dividierer *m* divider *(analogue computer)*
Dividiertrieb *m (dat)* mechanical dividing device
Divis *n* hyphen
Division *f* division
Divisionsgetriebe *n (dat)* mechanical dividing device
Divisionszeichen *n* division sign
Divisor *m* divisor
Dizyan *n* cyanogen
DK-Meter *n (cont)* oscillometer
dm-Welle *f* decimetric wave
DME-Abfragesender *m* distance-measuring interrogator
Docht *m* wick
D. einer Bogenlampenelektrode core
Dochtgewebe *n* wicks
Dochtkohle *f* cored carbon
Dochtschmierung *f* wick-feed lubrication
Dock *n* dock
D. mit eigener Kraftzentrale self-contained dock
selbstdockendes D. self-docking dock
Dockarbeiter *m* docker; dock hand *(unskilled)*
Dockboden *m* bottom of dock
Dockbrücke *f* warping (docker) bridge, afterbridge
Dockcaisson *m* dock caisson
Dockdrempel *m s.* Dockhaupt
Docke *f s.* Dockenspindel
docken to [dry] dock *(a vessel)*
Docken *n* dockage, [dry-]docking
Dockenspindel *f* mandrel
Dockenstock *m* mandrel
Dockgebühr *f* berthing dues, berthage, dockage, dock fee
Dockgeld *n* berthing dues, berthage, dockage, dock fee
Dockhaupt *n* dock sill (head, entrance)
Dockkiel *m* docking keel
Dockponton *m* dock pontoon
Dockschiff *n* lifting vessel
Dockschlepper *m* dock[ing] tug
Dockschleuse *f* [tide] lock
Dockseitenkasten *m* dock wing-structure, side walls, wing structure
Dockseitenstütze *f* spur (dock side) shore
Docksohle *f* apron, dock floor
Dockstütze *f* dock shore
Docksüll *m (n) s.* Dockhaupt
Docktor *n* dock gate
Dockverschlußponton *m* dock pontoon
Dodekaeder *n* dodecahedron
dodekaedrisch dodecahedral
Dokumentalist *m* abstractor
Dokumentarfilm *m* documentary film
Dokumentenfilm *m* document film
dolieren to fluff *(leather)*
Doliermaschine *f* buffing (fluffing) machine
Dolierrad *n* fluffing wheel
Dollbord *m (ship)* inwale, gunwale; washboard
Dolle *f* thole pin
Dollenboot *n* inrigger
Dollenrennboot *n* inrigged racing boat
Dollring *m* button *(rowing)*
Dolmetscheranlage *f* interpreter installation
Dolomit *m* dolomite
Dolomitgestein *n* dolomite [rock]
Dolomitmarmor *m* dolomite marble
Dolomitstein *m* dolomite brick
Dolomitzustellung *f (met)* dolomite lining
Dom *m* dome
Domäne *f* domain *(ferromagnetism)*
Domänenbreite *f* domain width
Domänenstruktur *f* domain structure
dominant [pre]dominant
Donator *m* donor *(semiconductor)*
Donatordichte *f* donor density
Donatorenwanderung *f* donor migration *(semiconductor)*
Donatorion *n* donor ion
Donatorkonzentration *f* donor concentration
Donatorniveau *n* donor level

Donatorplatz *m* donor site
Donatorterm *m* donor level
Donatortiefe *f* donor depth
Donatorverunreinigung *f* donor impurity
Donatorwanderung *f* donor migration
Donatorzentrum *n* donor centre
Donatorzustand *m* donor state
Dongolaleder *n* Dongola leather
Donkeymann *m (ship)* donkeyman; *(Am)* water tender
Donnanpotential *n* Donnan potential *(electrobiology)*
Dopen *n* doping *(semiconductors)*
Doppel-J-Naht *f* double-J butt weld
Doppel-T-Anker *m* two-pole armature; shuttle (drum) armature
Doppel-T-Filter *n* parallel-T network
Doppel-T-Netzwerk *n* parallel-T network
Doppel-T-Profil *n* I-section
Doppel-T-Querschnitt *m* I-section
Doppel-T-Schaltung *f* parallel-T network
Doppel-T-Stahl *m* H-steel
Doppel-T-Träger *m* flanged beam (girder); I-beam
Doppel-U-Naht *f* double-U butt weld
Doppelableger *m* double distributor
Doppelabnehmerkrempel *f* double-doffer card
Doppelader *f* pair of cores
  abgeschirmte D. screened pair
  gekreuzte D. crossed pair
doppeladrig bifilar
Doppelanker *m (el)* double armature
Doppelankerrelais *n* double-armature relay
Doppelantenne *f* twin aerial
Doppelarbeitskontakt *m* double-make contact
Doppelarmkreuz *n* double set of arms *(flywheel)*
doppelatomig biatomic, diatomic
Doppelausschalter *m* double cutout
Doppelbackenbremse *f* double-shoe brake
Doppelbahnablauf *m (ship)* double-way launching, two-way launching
Doppelbalkenbett *n* double-beam frame
Doppelband *n* double sliver
Doppelbandschleifmaschine *f* dual-belt sander
Doppelbasisdiode *f* unijunction transistor
Doppelbegrenzerkreis *m* dual-limiter circuit
Doppelbeil *n* double chopper
Doppelbelichtung *f* double exposure *(photography)*
Doppelbelichtungssperre *f* double-exposure prevention
Doppelbemannung *f* double crew
Doppelbereifung *f* dual tyres
Doppelbild *n (telev)* echo images
Doppelbildokular *n* double-image ocular
Doppelbindung *f (ch)* double bond (linking), ethylenic (olefinic) bond
  endständige D. end double bond
  konjugierte D. conjugated double bond
  kumulierte D. cumulative (adjacent) double bond
  semipolare D. semipolar double bond
Doppelblattruder *n (ship)* double-plate rudder
Doppelboden *m (ship)* double bottom, *(Am)* water bottom
  D. mit Bilge edge bottom
  D. mit offenen Bodenwrangen open-floor double bottom
Doppelbodenbeplattung *f* tank-top plating, double-bottom plating *(of a ship)*
Doppelbodendecke *f* tank top, top (crown) of tank
Doppelbodentank *m* double-bottom tank, *(Am)* water bottom
Doppelbogen *m* return (double) bend, syphon *(piping)*
Doppelbogenanlage *f (print)* two-up feeding
Doppelbogenkontrolleinrichtung *f* double-sheet detector
Doppelboot *n* catamaran
doppelbrechend birefringent, double-refracting
Doppelbrechung *f* birefringence, double refraction
Doppelbruch *m (math)* complex (compound) fraction
Doppelbrücke *f* 1. *(build)* twin bridge; 2. *(el)* Kelvin (Thomson, double) bridge
Doppelbrückenschaltung *f* bridge duplex connection
Doppelbuchse *f* duplicate sockets
Doppelbuchstabe *m* ligature
Doppelbündelleiter *m* double-bunch conductor
Doppelbundgatter *n* double-log frame
Doppeldampfraumkessel *m* boiler with double steam space
Doppeldecker *m* 1. biplane; 2. *(print)* two-magazine composing machine; 3. (build) double-deck screen
  mehrstieliger D. multi-strut biplane, multi-bay biplane
Doppeldeckersieb *n* double-deck screen
Doppeldeckung *f* double coverage *(roof)*
Doppeldiode *f* duodiode, twin-diode, double diode
Doppeldipolantenne *f* double-dipole aerial
Doppeldraht *m* twin wire
Doppeldrahtzwirnmaschine *f* double-twisting machine
Doppeldrehkondensator *m* two-gang capacitor
Doppeldrehtischpresse *f* double-revolver press
Doppeldreieckschaltung *f* double-delta connection
Doppeldrucksystem *n* / **hydraulisches** dual-pressure hydraulic system
Doppelduostraße *f* four high rolling mill
Doppelduowalzwerk *n* double two-housing mill
Doppeldurchbrucharbeit *f (text)* cut work

Doppeldüsenvergaser *m* *(aut)* double-jet carburetter
Doppeleinseitenband *n* *(el)* independent sideband
Doppelempfang *m* dual (double) reception
Doppelendball *m* punch[ing] ball *(boxing)*
Doppelender *m* double-bowed boat, double-ended vessel *(ship)*; double ender *(ship or boiler)*
Doppelenderkessel *m* double-ended boiler
D. mit gemeinsamer Flammkammer double-ended boiler with common combustion chamber
D. mit getrennten Flammkammern double-ended boiler with separate combustion chambers
Doppelenderschiffskessel *m* double-ended Scotch marine boiler
Doppelerdschluß *m* polyphase earth, double-earth fault
Doppelfadenlampe *f* double-filament lamp
Doppelfahrdraht *m* twin contact wires
Doppelfeinbank *f s.* Doppelfeinflyer
Doppelfeinflyer *m* jack (fine roving) frame
Doppelfeldmotor *m* double-field induction motor
Doppelfeldumformer *m* double-field converter
Doppelfeldwicklung *f* double-field winding
Doppelfenster *n* double window
Doppelflammofen *m* double-reverberatory furnace
Doppelflammrohrkessel *m* Lancashire (double-flue) boiler
Doppelflankenmodulation *f* double-edge modulation
doppelflanschig double-flanged
Doppelflor *m* double pile
Doppelflorzweikrempelsatz *m* two-doffer two-card set
Doppelfokussierung *f* double focusing
Doppelführung *f* **eines Zuges** duplication of a train, provision of a relief train
Doppelgabelläufer *m* *(text)* double-armed traveller
doppelgängig double-start *(thread)*, double-threaded
Doppelgegenbetrieb *m* *(el)* quadruplex system
Doppelgegenschreiben *n* quadruplex telegraphy
Doppelgegensprechsatz *m* quadruplex system
Doppelgegentaktgleichrichter *m* ring-type modulator *(carrier frequency)*
Doppelgelenk *n* constant-velocity joint
Doppelgeschränk *n* furnace front with double door
Doppelgespann *n* *(ag)* double team
Doppelgewebe *n* double cloth
D. mit Ausfüllungsfäden double cloth with stuffing threads
D. mit bauschiger Stelle baggy fabric
Doppelgewebebindung *f* double-cloth weave
Doppelgewindeschraube *f* double-threaded screw
Doppelgitter *n* double grid
doppelgleisig *(ry)* double-track
Doppelglockenisolator *m* double-shed insulator, double-petticoat insulator
Doppelhacke *f* double chopper
Doppelhaus *n* semi-detached house
Doppelhebel *m* crank lever
Doppelhechelfeldstrecke *f* gill box with double set of fallers, intersecting gill box (drawing frame)
Doppelhiebfeile *f* double-cut file
doppelhiebig double-cut *(file)*
Doppelhitzdrahtspannungsmesser *m* compensated hot-wire voltmeter
Doppelhobel *m* German jack plane with double iron
Doppelhobeleinrichtung *f* double-plough installation with additional drive in the face
Doppelhobeleisen *n* **mit langer Schraube** long screw double-plane iron
Doppelhobelmaschine *f* duplex planing machine
doppelhöckerig *(el)* double humped
Doppelhöckerkurve *f (el)* double-humped (peaked) response
Doppelholländer *m* duplex beating engine
Doppelhub *m* [work] cycle, double stroke
Doppelhübe *mpl* **je min** speed cycles per min.
Doppelhubjacquardmaschine *f* double-lift Jacquard machine
Doppelimpuls *m* double pulse
Doppelimpulsgenerator *m* double-pulse generator
Doppelindex *m* double-subscript
Doppelindikator *m* double indicator
Doppelinduktor *m* dual magneto
Doppelinjektor *m* double-tube injector, injector with two sets of cones (nozzles)
Doppelintegral *n* double integral
Doppelionisationskammer *f* double-ionization chamber
Doppelisolator *m* duplex insulator
Doppelkabine *f* double[-berth] cabin, two-berth cabin
Doppelkäfiganker *m* double [squirrel]-cage armature
Doppelkäfiginduktionsmotor *m* double squirrel-cage induction motor
Doppelkäfigläufer *m* double squirrel-cage [rotor]
Doppelkäfigmotor *m* double squirrel-cage motor
Doppelkäfigwicklung *f* double squirrel-cage winding
Doppelkappenisolator *m* double-cap insulator
Doppelkarde *f (text)* double card
Doppelkassette *f* twin magazine
Doppelkassettensystem *n* cartridge-to-cartridge film feed
Doppelkegelfeder *f* double-cone spring

Doppelkegelkupplung *f* double-cone friction clutch
Doppelkehlnaht *f* double-fillet weld
Doppelkeil *m* folding wedges
Doppelkeilanker *m (mine)* sliding-wedge [roof-] bolt
Doppelkernrohr *n* double-tube core-barrel
D. mit feststehendem Innenrohr rigid double-tube core-barrel
D. mit mitdrehendem Innenrohr swivel double-tube core-barrel
Doppelkette *f* double-strand chain
endlos umlaufende D. pair of endless chains
Doppelkettenkratzförderer *m* double-chain scraper conveyor
Doppelkettenstich *m (text)* double in-and-out stitch
Doppelketten[web]stuhl *m* double-warp loom
Doppelkohlemikrofon *n* push-pull carbon microphone
Doppelkoinzidenzspektrometer *n* double-coincidence spectrometer
Doppelkoje *f* double-deck bunk
Doppelkolbenzylinder *m* opposed thrust cylinder
Doppelkollektormotor *m* double-commutator motor
Doppelkommutatormotor *m* double-commutator motor
Doppelkondensator *m* twin capacitor
Doppelkontakt *m* double contact; change-over contact
Doppelkontrollkarte *f* joint chart
Doppelkonus *m* double cone
Doppelkonusantenne *f* biconical aerial
Doppelkonusstrahler *m* biconical radiator *(aerial)*
Doppelkopfschiene *f* double-head[ed] rail, bullhead rail
Doppelkreisrichtcharakteristik *f (el)* bilateral characteristic
Doppelkreuzpoller *m (ship)* double bollard with a cross-piece, double-head riding bitt
Doppelkristall *m* twin crystal
Doppelkrümmer *m* return bend *(piping)*
Doppelkurbel *f* / **gleichschenklige** isosceles drag link
rotierende (umlaufende) D. knee joint, rotating double crank
Doppelkurbelgetriebe *n* / **durchschlagendes** knee joint, rotating double crank
Doppelkurbelmechanismus *m* drag-link mechanism
Doppelkurbelwiderstand *m* / **gleichläufiger** *(el)* double-resistance box
Doppelladepfosten *mpl (ship)* pair masts, double-king posts
Doppellasche *f* double clip
Doppelleiste *f* double tabbing *(of mill saw)*
Doppelleiter *m* twin conductor
Doppelleitung *f (el)* two-wire line, twin [line]; loop circuit
abgeschirmte D. shielded pair *(line)*
asymmetrische D. asymmetrical two-wire system
konzentrische D. coaxial pair *(line)*
symmetrische D. balanced twin *(line)*
unabgeschirmte D. open-wire twin *(line)*
verdrillte D. twisted pair *(line)*
Doppellitze *f* stranded two-wire lead
verdrillte D. twisted pair
Doppelmaschine *f* outsole stitcher
Doppelmast *m (el)* H-pole
Doppelmesserschalter *m* double-bladed knife switch
Doppelmetalldraht *m* bimetallic wire
Doppelmikrofon *n* double-button microphone (transmitter)
Doppelmischer *m* double-type mixer
Doppelmodulation *f* double (dual) modulation
Doppelmotor *m* twin-engine
Doppelmustereinrichtung *f* double pattern attachment
doppeln to double *(spinning; sheet metal)*, to fold *(sheet metal)*
Doppeln *n* **von Blechen** folding of sheets
Doppelnadel *f* / **astatische** astatic needles
Doppelnadelstabstrecke *f (text)* intersecting [gill drawing] frame
Doppelnadeltrieurzylinder *m (ag)* dual-needle cylinder
Doppelnadelwalzenstrecke *f (text)* double-head porcupine drawing
Doppelnadler *m (text)* gill box with double set of fallers, intersecting gill box
Doppelnahtmaschine *f* two-needle vamping flat bed sewing machine *(shoe)*
Doppelneutron *n* dineutron, bineutron
Doppelnippel *m* double nipple
Doppelnutmotor *m* double squirrel-cage motor
Doppelofen *m* pair-oven
Doppelölabscheider *m* duplex oil trap
Doppeloxid *n* binoxide
Doppelparallelkurbel *f* double-parallel crank
Doppelparallelkurbelmechanismus *m* double-parallel crank mechanism
Doppelpart *m (n)* bight
Doppelpedalsteuerung *f* double-pedal control system
Doppelpeilung *f* cross-bearing
Doppelpendel *n* double pendulum
Doppelpfeilverzahnung *f* triple helical teeth
Doppelphantomkreis *m (telec)* double-phantom circuit
Doppelplan *m* double-sampling plan, double scheme
Doppelplanrost *m* double-horizontal grate
Doppelplattenkiel *m* double-plate keel

Doppelplattenruder *n* side plate rudder, double-plate rudder
doppelpolig bipolar, double polar
Doppelpoller *m (ship)* double bollard, double-head bitt (bitton, bollard)
Doppelpotentiometer *n* dual-operated potentiometer
Doppelprisma *n* double prism
Doppelproton *n (nucl)* diproton
Doppelpuddelofen *m (met)* double furnace
Doppelpumpe *f* double pump
Doppelpunkt *m (therm)* instantaneous centre
D. einer Kurve crunode
Doppelrad *n* twin wheel
Doppelreaktor *m s.* Zweizweckreaktor
Doppelreifen *m* dual tyre
Doppelreihe *f (math)* double series
Doppelreihenbefeuerung *f* double-row lighting
Doppelreihenentwicklung *f (math)* double-series expansion
Doppelreihenmotor *m* double-tandem engine
doppelreihig double-row
Doppelrelais *n* double-armature relay
doppelrichtend bidirectional
Doppelriemenantrieb *m* two-belt drive
Doppelröhre *f (el)* double (duplex) valve
Doppelröhrenverstärker *m* two-valve amplifier
Doppelrohrverflüssiger *m* double-pipe condenser
Doppelrollenkette *f* two-strand roller chain
Doppelruder *n* twin rudder
Doppelrumpfboot *n* catamaran, two-hull boat
Doppelrumpfflugzeug *n* twin-boom aircraft
Doppelrundtisch-Flächenschleifmaschine *f* **senkrechter Bauart** vertical-spindle grinding machine with twin rotary tables
Doppelsägeblattanordnung *f (mine)* double saw-tooth system *(support)*
Doppelsalz *n* double salt
Doppelsammelkontakt *m* master number double contact *(telephony)*
Doppelsamt *m* double velvet
Doppelsamtstuhl *m* double-velvet loom
Doppelsaugrohr *n* double-throat Venturi tube
Doppelsäule *f* twin (double) column
Doppelschacht *m (mine)* twin shaft
Doppelschalter *m* duplex (twin-pole, double-pole) switch
Doppelschaufel *f* double bucket *(turbine)*
symmetrische D. symmetrical double bucket
Doppelscheibenegge *f* tandem disk harrow
Doppelscheibenschleifmühle *f* double-disk attrition mill
Doppelschicht *f* double layer
Doppelschichtfilm *m* double-coated film
Doppelschichtkorbwicklung *f (el)* chain winding
Doppelschiebermotor *m* double-sleeve valve engine
Doppelschloßflachstrickmaschine *f* twin-cam flat knitting machine
Doppelschlüssel *m* double-ended spanner
Doppelschlußerregung *f* compound excitation
Doppelschlußgenerator *m* compound[-wound] generator
Doppelschlußmotor *m* compound[-wound] motor
Doppelschlußwicklung *f* cumulative compound winding
Doppelschmiege *f* T-bevel
Doppelschneckenmischer *m* twin-worm mixer
Doppelschrägzähne *mpl* double-helical teeth
Doppelschrämarm *m (mine)* twin-jib
Doppelschrämausleger *m (mine)* twin-jib
Doppelschrauben *fpl (ship)* twin screws (propeller)
Doppelschraubenfahrgastschiff *n* twin-screw passenger ship
Doppelschraubenschiff *n* twin-screw vessel
Doppelschraubenschlüssel *m* double open-end wrench, double-head wrench
Doppelschraubstock *m* double-screw vice
Doppelschuß *m (text)* double picks *(fault)*
Doppelschwad *m (ag)* windrow
Doppelschwelle *f* twin-sleeper *(track)*
Doppelschwinggetriebe *n* double-swing mechanism
Doppelschwingung *f* double vibration
doppelseitig bilateral; double-end, double-faced
Doppelsicherheitsventil *n* double safety valve
D. mit Signalvorrichtung double safety valve with alarm
Doppelsilikat *n* bisilicate
doppelsinnig bidirectional
Doppelsitzregelventil *n* double-ported control valve, double-seated control valve
Doppelsitzventil *n* double-seat valve
Doppelskala *f* double (two-range) scale
Doppelspalt *m* double slit
Doppelspaltoszillator *m* floating drift oscillator
Doppelspannreck *n* double horizontal bars
Doppelspant *n (ship)* double frame, frame and futtock
Doppelsperrklinke *f* pair of pawls *(telephony)*
Doppelspindelfräsmaschine *f* **mit feststehendem Tisch** two-spindle bed-type miller
Doppelspindelgesenkfräsmaschine *f* two-spindle die sinking machine
Doppelspindelplanfräsmaschine *f* duplex manufacturing-type milling machine, duplex-head milling machine, two-spindle bed-type miller
Doppelspindelschraubstock *m* double-screw vice
Doppelspindelsenkrechtfräseinrichtung *f* two-spindle vertical milling attachment
Doppelspinnrad *n* double-spinning wheel
Doppelspitze *f (math)* tacnode

Doppelsprechbetrieb *m* diplex operation *(telephony)*
Doppelsprechschaltung *f* diplex phone connection
Doppelspule *f* twin (double, compound, two-storey) coil
Doppelspur *f* double track
Doppelspurbetrieb *m* dual-trace operation
Doppelspurtonbandgerät *n* half-track recorder
Doppelstabläufer *m (el)* double-cage rotor (armature)
Doppelstander *m (ship)* burgee, split flag
Doppelständer *m* double standard (column)
Doppelständerbauart *f* double-column construction
Doppelständerdrehwerk *n s.* Zweiständerkarusselldrehmaschine *f*
Doppelständerhobelmaschine *f* double-column planing machine, standard (two-housing) planer
Doppelstatordrehkondensator *m* split-stator variable capacitor
Doppelstecker *m (el)* socket-outlet adapter; double (two-way) plug
Doppelstern *m* double (twin) star
Doppelsternkabel *n* quad-pair cable
Doppelsternschaltung *f* duplex-star connection
Doppelsternverseilung *f* quad-pair formation *(cables)*
Doppelsteuerung *f* dual (double) control
Doppelstichprobenverfahren *n* double sampling
Doppelstockbrücke *f* double-deck bridge
Doppelstockwagen *m* double decker
Doppelstößelräummaschine *f* duplex-type broaching machine
Doppelstrahlröhre *f* double-stream amplifier
Doppelstreb *m (mine)* double-unit face
Doppelstreuung *f* double scattering
Doppelstrich *m* double dash
Doppelstrom *m* double current *(telegraphy)*
Doppelstrombetrieb *m* double-current working, transmission by double current
Doppelstromsystem *n* double-current signalling (system)
Doppelstromturbotriebwerk *n* turbojet fan
Doppelstromübertragungsverfahren *n* double-current transmission, *(Am)* polar direct-current system
Doppelstuhl *m* double-cloth weaving loom
Doppelstütze *f* 1. *(el)* twin (double) insulator spindle, double petticoat; 2. *(build)* double (twin) column
doppelt dual, double
Doppeltarif *m* two-part tariff
Doppeltarifzähler *m* double-tariff meter, two-rate meter
Doppelteig *m (text)* double-strength paste
doppeltgeladen *(ph)* double-charged
doppeltgelascht with cover plates each side
doppeltgerichtet both way
doppeltkonisch tapered on both sides
doppeltlogarithmisch *(math)* full-logarithmic, log-log
doppeltmagisch doubly magic
Doppeltonfarbe *f* double-tone ink
Doppeltontastung *f (telec)* two-tone keying
Doppeltontelegrafie *f* two-tone telegraphy
Doppeltonverzerrung *f* intermodulation distortion
Doppeltragdrahtaufhängung *f* double-catenary suspension
Doppeltransformator *m* double transformer
Doppeltreibstoff *m* bipropellant [fuel]
doppeltrichterförmig converging-diverging
Doppeltriode *f* duotriode, twin-triode
Doppeltrommelkrempel *f* double card
doppeltseidenisoliert double-silk covered
Doppeltulpennaht *f* double-U butt weld
Doppelturbine *f* / **gegenläufige** double turbine with the blade wheels running in opposite directions
doppeltwirkend double-acting
Doppelüberlagerungsempfang *m* double-superheterodyne reception
Doppelüberlagerungsempfänger *m* double-(dual-) heterodyne receiver
Doppel- und Fehllochfeststellung *f* double punch and blank column detection
Doppeluntersetzungsgetriebe *n* double-reduction gear
Doppelverhältnis *n (math)* cross ratio
Doppelvierer *m* four scull *(boat)*
Doppelwalzenkessel *m* double-cylindrical boiler
Doppelwalzenschrämlader *m (mine)* double-drum cutter loader
doppelwandig double-walled
Doppelwandung *f* jacketed (double) wall
Doppelweggleichrichter *m* full-wave rectifier, bi-phase rectifier
Doppelweiche *f (ry)* double turnout
Doppelwelle *f* full wave
Doppelwendel *f* coiled-coil filament
Doppelwendellampe *f* coiled-coil lamp
Doppelwerkstückträger *m* twin work-holding fixture
Doppelwinkelprisma *n* double-prismatic square
Doppelwirkung *f* double action
Doppelwurzel *f* double root
Doppelzackenschrift *f (ac)* bilateral area track
Doppelzapfen *m* double-pin
Doppelzeichen *n* eccentricity signals, split response *(radar)*
Doppelzellenschalter *m* double-battery switch
Doppelzünder *m* double-action fuse
Doppelzweckmaschine *f* dual-purpose machine
Doppelzweier *m* double scull *(boat)*
Doppelzylinderkessel *m* double-cylindrical boiler

Doppelzylinderstrickmaschine *f* double-cylinder knitting machine
Doppelzylinderstrumpfautomat *m* automatic double-cylinder hosiery knitting machine
Döpper *m* rivet set, punch-type dolly, snap
Doppler *m* duplicating punch, duplicator, reproducing puncher *(punched card)*
Doppler-Breite *f* Doppler width
Doppler-Effekt *m* Doppler effect
Doppler-Radar *n* Doppler radar
Doppler-Verbreiterung *f* Doppler broadening
Doppler-Verschiebung *f* Doppler shift
Dopplung *f* 1. lamination; 2. *(dat)* double precision
Dorn *m* arbor; mandrel; punch; plug; bolt
fliegend angeordneter D. stub arbor
freitragender D. stub arbor
kurzer D. stub arbor
zentrischer D. centring mandrel (pin) *(turbine)*
Dorndurchmesser *m* arbor diameter
dornen to indent; to expand; to pierce
Dornhubwagen *m* ram-type lift truck
Dornmutter *f* arbor nut
Dornpresse *f* piercing (mandrel, broaching) press
Dornstützlager *n* arbor supporting bracket
Dorntraglager *n* arbor bracket
Dornwalze *f* mandrel roll
Dorr-Eindicker *m* Dorr thickener
Dorr-Klassierer *m* Dorr classifier
Dörrgemüse *n* dried vegetables
Dörrobst *n* dried fruit
Dörrschrank *m* / **elektrischer** electric dehydrator
Dose *f* box *(barometer)*; *(Am)* can, tin; cell
Dosenbarometer *n* aneroid barometer
Dosenlibelle *f* box level
Dosenrelais *n* box relay
Dosenschalter *m* **für Montage unter Putz** recessed switch
Dosenspinnmaschine *f* can spinning frame
Dosieranlage *f* batching plant
Dosierband *n* feed-regulating conveyor
Dosierbehälter *m* dosing tank
dosieren to meter out *(hydraulics)*; to dose; to time; to mix; to proportion; *(build)* to batch; to gauge
nach Masseteilen d. to weight-batch
nach Raumteilen d. to volume-batch
Dosieren *n* metering [out]; dosage; proportioning; *(build)* batching
Dosierkasten *m* batch (gauge) box
Dosierkiste *f s.* Dosierkasten
Dosierpumpe *f* proportionating (controlled volume) pump
Dosiertrichter *m* scale hopper
Dosierung *f* *(build)* batching, proportioning; dosage; feed regulation *(flotation)*; metering-out *(hydraulics)*
D. nach Masseteilen weight-batching
D. nach Raumteilen volumetric batching
Dosiervorrichtung *f* *(build)* batcher; automatic reagent feeder *(flotation)*
Dosierwaage *f* *(build)* weigh-batcher; bag-fitting scale
Dosierzähler *m* batching counter
Dosimeter *n* dose meter, dosimeter
chemisches D. chemical dose meter
integrierendes D. integrating instrument (dose meter)
kolorimetrisches D. colourimetric dosimeter
Dosimetrie *f* dosimetry, dosage measurement
speichernde D. accumulative dosimetry
Dosis *f* dosage *(rays)*, dose
D. der beruflichen Bestrahlung occupational [radiation] dose
D. der inneren Bestrahlung internal dose
absorbierte D. *s.* Energiedosis
akkumulierte D. accumulated dose
höchstzulässige D. maximum permissible dose (*Am*: exposure)
innere D. internal dose
integrale D. integral dose
maximal zulässige D. maximum permissible dose, MPD
physikalische D. physical dose
spezifische D. specific dose
subletale D. sublethal dose
tödliche D. lethal (fatal) dose, LD
unterschwellige D. subthreshold dose
verträgliche D. permissible (sub-tolerance) dose, sub-tolerance
zulässige D. permissible dose
Dosisäquivalent *n* dose equivalent
Dosisaufbaufaktor *m* dose build-up factor
Dosiseffektkurve *f* dose-effect curve
Dosisfraktionierung *f* dose fractionation
Dosiskonstante *f* specific gamma-ray constant (emission)
D. eines Radionuklids specific gamma-ray output of a radionuclide
Dosisleistung *f* dose (dosage) rate
höchstzulässige D. maximum permissible constant dose rate
lokale D. local dose rate
Dosisleistungsmesser *m* dose rate meter
Dosismeßbereich *m* dosage-measuring range
Dosismesser *m* dose meter, dosimeter
Dosismessung *f* dosage measurement, dosimetry
Dosisprotrahierung *f* dose protraction, protraction of dose
Dosisrate *f s.* Dosisleistung
Dosiszuwachsfaktor *m* dose build-up factor
Dotierung *f* 1. doping *(semiconductor)*; 2. dosage
Dotierungsatom *n* doping atom
Dotierungsbereich *m* doping range

Dotierungsmittel *n* doping agent, dopant
Dotierungstechnik *f* doping technique
Douglastanne *f* Douglas fir, Oregon pine
Drachen *m* kite
Drachenantenne *f* kite aerial
Drachenballonslup *m (ship)* kite balloon sloop
Drachenboot *n* dragon [boat]
Drachenholm *m (aero)* kite spar
Drachenseeanker *m* kite drag
Dragee *n* dragée
Dragganker *m* grapnel, [kite] drag
Draggen *m s.* Dragganker
Draht *m* wire
D. mit Seidenumflechtung silk-covered wire
abgeschirmter D. screened (shielded) wire
abisolierter D. bared (stripped) wire
baumwollisolierter D. cotton-covered wire
besponnener D. braided (covered) wire
bewehrter D. armoured wire
biegungsfrei gezogener D. dead-drawn wire
blanker D. uninsulated (bare, naked) wire
blankgeglühter D. bright annealed wire
bleibender D. *(text)* permanent twist
brüchiger D. cracked wire
dicker D. heavy wire
doppeltseidenisolierter D. double-silk covered wire
dünner D. fine wire
einfach umklöppelter D. single-covered wire
emaillierter D. enamel-insulated wire
feuerverzinkter D. hot-galvanized wire
feuerverzinnter D. hot-tinned wire, tin-coated wire
galvanisierter D. galvanized wire
geerdeter D. earthed wire
gelackter D. varnished wire
gelitzter D. stranded wire
gewalzter D. rolled wire
gezogener D. drawn wire
gummiisolierter D. rubber[-covered] wire
isolierter D. covered (insulated) wire
kupferplattierter D. copper-plated wire
lötbarer D. solderable wire
papierisolierter D. paper-covered wire
seideumsponnener D. silk-covered wire
umwickelter D. taped wire
verdrillter D. twisted wire
verkupferter D. copper-clad wire
verseilter D. stranded (twisted) wire
verzinkter D. galvanized wire
verzinnter D. tinned wire
vorgezogener D. process wire
zweimal gezogener D. two-draft wire
Drahtabfall *m* scrap wire
Drahtangaben *fpl* wire particulars
Drahtanker *m (el.m.)* wire-wound armature
Drahtanschluß *m* wire lead
Drahtantenne *f* / **aufgehängte** supported wire aerial
Drahtaufnahme *f* magnetic wire recording
Drahtauge *n* wire eye
Drahtauslösung *f* wire release
Drahtbarren *m* wire bar
Drahtberührung *f* contact on line
Drahtbruch *m* rupture of wire, wire breakage (break), disconnection
Drahtbruchausschalter *m* conductor break cut-out
Drahtbruchrelais *n* line break relay
Drahtbruchschutzschaltung *f* conductor-break protective system, wire-break disconnecting system, line-fault protective circuit
Drahtbund *m* binding for aerial line
Drahtbündel *n* wire bundle
Drahtbürste *f* wire brush
Drahtdaten *pl* wire particulars
Drahtdrehwiderstand *m* wire-wound [volume] control
Drahtdrucker *m s.* Nadeldrucker
Drähte *mpl* / **korrespondierende** corresponding wires
kurzgeschlossene D. shorted wires
verwickelte D. entangled wires
Drahteinlage *f* wire core *(sealing)*
Drahtemaille *f* wire enamel
drahten to wire, to cable
Drahtende *n* wire lead
isoliertes D. skinner *(cable)*
Drahtfernsehsystem *n* jeep [television system]
Drahtfernsprechen *n* wire telephony
Drahtfunk *m* wired wireless, carrier current transmission, line radio, carrier telegraphy (transmission); *(Am)* wire program distribution, program distribution system
hochfrequenter D. carrier transmission, radio distribution system, carrier [frequency] wire broadcasting
niederfrequenter D. audio frequency distribution, audio-frequency wire broadcasting
Drahtfunkendschaltung *f* carrier terminal circuit
Drahtfunknetz *n* distribution network
Drahtgaze *f* wire gauze
Drahtgeflecht *n* wire cloth (screen, netting)
drahtgenäht wire-sewn
Drahtgewebe *n* wire cloth (fabric, mesh, netting)
Drahtgitter *n* wire grating *(waveguides)*
Drahtglas *n* wire[d] glass
Drahtglasschutzhülse *f* water gauge protector of armoured glass
Drahtglasschutzmantel *m s.* Drahtglasschutzhülse
Drahtglühen *n* wire annealing
Drahthakenverbindung *f* wire hook lacing *(belt)*
Drahtheftmaschine *f* [wire-]stitcher

Drahtheftung *f* wire-stitching, stapling
Drahtheizkörper *m* wire-wound heating element
Drahthindernis *n s.* Drahtverhau
Drahtkern *m* bead wires *(tyres)*; wire core
Drahtklemme *f* wire grip
Drahtkorb *m* wire crate
Drahtkreuzung *f* transposition of wires; wire crossing
Drahtlehre *f* wire gauge
Drahtliste *f* wire list *(design automation)*
Drahtlitzenschaft *m* wire heald shaft
drahtlos wireless
Drahtmaschenschutz *m* wire mesh guard
Drahtmeldung *f* wire report
Drahtmeßbrücke *f* slide wire bridge
Drahtnachricht *f* cablegram
Drahtnetz *n* metal gauze, wire screen
Drahtnetzelektrode *f* wire gauze electrode
Drahtnetzhülse *f* wire-netting protector
Drahtpanzerkabel *n* wire-armoured cable
Drahtpatentierofen *m* wire patenting furnace
Drahtpotentiometer *n* wire-wound potentiometer
Drahtquerschnitt *m* wire cross-section
Drahtregelwiderstand *m* wire-wound rheostat
Drahtrundfunk *m* wire broadcasting, wired radio
Drahtrundspruch *m* diffusion by wire
Drahtschlinge *f* wire sling (strap) *(cargo-handling gear)*
Drahtseil *n* wire rope
Drahtseilbahn *f* aerial cableway (ropeway), telpher railway
Drahtseilschwebebahn *f* overhead-cable railway
Drahtseiltrieb *m* steel-rope drive
Drahtsicherung *f (el)* wire fuse
Drahtspanner *m* wire stretcher *(overhead line)*
Drahtspeichenrad *n* wire-spoked wheel
Drahtspule *f* wire reel (bobbin)
Drahtstift *m* sprig, wire nail
Drahtstraße *f* wire mill *(rolling)*
Drahtstropp *m* wire sling (strop) *(cargo-handling gear)*
Drahtstück *n* / **kurzes** short length (piece) of wire
Drahttelefonie *f* wire telephony
Drahttelegrafie *f* line telegraphy
Drahttongerät *n* wire recorder
Drahtübertragungsweg *m (el)* metallic circuit
Drahtverbindung *f* wiring connection; wire communication
Drahtverhau *m* wire entanglement
Drahtverteilung *f (text)* distribution of twist
Drahtwalzwerk *n* looping (rod-rolling) mill
Drahtweberei *f* wire cloth weaving
Drahtwiderstand *m (el)* wire-wound resistor
  D. mit Abgriffschelle adjustable dropper
Drahtzaun *m* wire fence
Drahtziehbank *f* wire drawing bench
Drahtzieheisen *n* gauge-plate, wire-drawing plate
Drahtziehen *n* wire drawing
Drahtzieherei *f* wire drawing mill
Drahtziehmaschine *f* wire drawing machine
Drahtziehstein *m* wire drawing plate
Drahtzug *m* 1. wire drawing; 2. wire transmission (gearing)
  D. für Signale und Weichen *(ry)* signal and point wire
Drahtzugspannwerk *n (ry)* wire compensating device
Drahtzugstellwerk *n (ry)* locking frame with wire gear
Drainage *f s.* Dränage
Draisine *f (ry)* inspection trolley, track motor car
Drall *m* 1. *(mech)* angular momentum; *(ph)* spin *(electron)*; 2. helix; twist; 3. lay *(rope)*
  enger D. *s.* kurzer D.
  flacher D. *s.* kurzer D.
  kurzer D. *(35 bis 45°)* quick helix *(drill) (30° to 40°)*
  normaler D. medium helix *(drill)*
Drallachse *f (ph)* longitudinal (roll) axis
Drallbohrer *m s.* Spiralbohrer
Drallbrennkammer *f (rock)* swirl combustion chamber
Dralldüse *f (rock)* twist (swirl) nozzle
drallförmig helical
drallgenutet with helical flutes, spiral-fluted
Drallmoment *n (el)* roll (spin) momentum
Drallnut *f* helical flute (groove)
Drallrichtung *f* hand of [the] helix
Drallsteigungswinkel *m* helix angle [of the flutes], flute helix angle
Drallwinkel *m* helix angle, angle of the [tooth] helix
  D. / mit großem high-helix
Dränage *f* drainage
dränen to drain
Drängelampe *f* speeding up lamp *(telephony)*
  D. im Fernverkehr trunk congestion lamp
Drängeräte *npl* draining tools
Drängraben *m* draining ditch (channel, trench)
Dränleitung *f* drain pipe
Dränmaschine *f* draining machine
Dränmaschinen *fpl* drainage machinery
Dränpflug *m* draining plough
Dränrohr *n* drain pipe (tile)
Dränrohrstrang *m* subsurface [pipe] drain, draining channel
Dränspachtel *m* drain scoop
Dränspaten *m* ditching spade
Dränstrang *m s.* Dränrohrstrang
Dränung *f* [artificial] drainage; subsurface pipe drainage

Draufbohrer *m* brace
Draufholz *n* top wood
Draufsicht *f* top view
draußen off, outside *(at sea)*
Dravit *m* dravite *(tourmaline)*
Drechselbank *f* woodworking (wood turning) lathe
drechseln to turn wood
aus freier Hand d. to turn on the chuck lathe
Drechslerarbeit *f* turnery
Drechslerbank *f* wood turning lathe
Drechslerei *f* turnery
Drechslermeißel *m* turning chisel
Drechslerrohr *n* turning gouge
Dregganker *m s.* Dragganker
Drehachse *f* rotation axis, axis of rotation (revolution); fulcrum
D. des Stangenstromabnehmers trolley-pivot
augenblickliche D. der Bewegung instantaneous axis of rotation
permanente D. permanent axis of rotation
Drehanker *m* revolving (rotating, rotary) armature
Drehankermagnet *m* rotary armature magnet
Drehanodenröhre *f* rotating-anode tube *(X-ray)*
Drehantenne *f* rotating aerial; scanning aerial, scanner [unit] *(radar)*
Drehantennenmotor *m* scanner turning motor
Drehantrieb *m* rotary actuator *(hydraulics)*
Dreharbeit *f* turning operation
Drehausleger *m* hinged cantilever
Drehautomat *m* automatic [screw machine]
Drehbahnteiler *m (ag)* corkscrew divider
Drehbakensender *m* rotating beacon transmitter
Drehbandkolonne *f (ch)* rotating strip column
Drehbank *f* 1. wood turning lathe; 2. *s.* Drehmaschine
drehbar rotatable; swivelling
Drehbarkeit *f* rotatability
Drehbeanspruchung *f s.* Torsionsbeanspruchung
Drehbereich *m* 1. angle of slew *(crane)*; 2. control span
Drehbeschleunigung *f* circular (rotary) acceleration
Drehbeschleunigungsaufschaltung *f (rock)* mixing ratio for angular acceleration signal
Drehbewegung *f* rotary motion
Drehbohren *n* rotary drilling
Dreh-, Bohr- und Abstechmaschine *f* parting off, turning, and boring lathe
Drehbolzen *m* fulcrum pin, pivot
Drehbrücke *f* swing (pivot, turning) bridge; turntable
Drehbuch *n* scenario, script
Drehbuchautor *m* shooting script writer
Drehbühne *f* revolving stage *(theatre)*
Drehdavit *m (n) (ship)* round-bar davit
Drehdolle *f* swivel lock *(rowing)*
Drehdorn *m* lathe mandrel
Drehdrossel *f (el)* variometer
Drehdurchmesser *m* turning diameter, swing
größter D. maximum swing
Drehebene *f* plane of rotation (revolution)
Dreheisen *n (el)* moving iron
Dreheiseninstrument *n* moving-iron instrument, soft-iron instrument, moving-magnet instrument
D. mit Magnet permanent-magnet moving-iron instrument
Dreheisenmeßinstrument *n* moving-iron instrument, soft-iron instrument
Dreheisenmeßwerk *n* moving-iron instrument
Dreheisenspannungsmesser *m* moving-iron voltmeter
Dreheisenstrommesser *m* moving-iron ammeter
Drehelektrode *f* rotating electrode
drehen 1. to rotate; to turn, to swing; to haul, to back *(wind)*; to swivel, to slew *(crane)*; to shift *(phases)*; 2. to turn *(metal cutting)*
die Nummernscheibe d. to dial *(telephony)*
die Polarisationsebene d. to rotate the plane of polarization
einen Regler d. to rotate (turn) a control
sich d. um to rotate about
sich einmal um die Erde d. to complete a revolution around the earth
Dreher *m* turner, [skilled] lathe operator
Dreherbindung *f* leno (doup, gauze) weave
Drehergewebe *n (text)* leno, gauze
Dreherkette *f (text)* crossing warp
Dreherlehrling *m* turner apprentice
Drehfahnenrelais *n* vane-type relay
Drehfeder *f* torsion spring
gerade D. torsion bar (rod) spring
Drehfeld *n (el)* revolving (rotating, rotary) field
elliptisches D. elliptical field
magnetisches D. rotating magnetic field
Drehfeldadmittanz *f* cyclic admittance
Drehfeldempfänger *m* selsyn; synchro motor
Drehfeldgeber *m* rotary inductor, synchro (selsyn) generator
Drehfeldhysterese *f* rotary hysteresis
Drehfeldhysteresis *f s.* Drehfeldhysterese
Drehfeldimpedanz *f* cyclic impedance
Drehfeldinstrument *n* induction (rotating field) instrument
Drehfeldleistungsmesser *m* induction wattmeter
Drehfeldreaktanz *f* cyclic reactance
Drehfeldrichtungsanzeiger *m* phase-sequence indicator

Drehfeldumformer *m* transformer generator
Drehfeldwattmeter *n s.* Drehfeldleistungsmesser
Drehfenster *n* swivel (pivoting) window, opening side light
Drehfestigkeit *f* torsion[al] strength
Drehfeuer *n* rotating beacon, revolving light *(navigation)*
Drehfilter *n* revolving (rotating) filter
Drehflammofen *m* revolving reverberatory furnace
Drehflansch *m* rotary flange
Drehflügelflugzeug *n* rotary-wing aircraft
Drehflügelmotor *m* vane-type actuator, *(Am)* rotary ram
Drehflügelschraube *f* feathering (reversible, variable pitch) propeller
Drehflügelzähler *m* rotating vane flowmeter
Drehfunkenstrecke *f (el)* rotary discharger
Drehfunkfeuer *n* revolving (rotating, rotating loop) radio beacon *(navigation)*
Drehgelenk *n* swivel joint
Drehgenauigkeit *f* turning accuracy *(lathe)*
Drehgeschwindigkeit *f* 1. *(ph)* angular velocity; velocity (speed) of rotation, rotational (rotative) speed; 2. turning speed
Drehgestell *n* bogie, *(Am)* truck
vorderes D. einer Lokomotive leading bogie (truck) of a locomotive
Drehgestellrahmen *m* bogie frame
Drehgestellwange *f* bogie sole bar, bogie frame member
Drehgrubber *m* rotary cultivator
Drehhacke *f* rotary hoe
Drehhaken *m* heel tool
Drehhalbautomat *m* automatic lathe, semiautomatic screw machine
Drehherdofen *m* rotating hearth furnace
Drehherz *n* lathe (driving) dog, lathe [driving] carrier
gebogenes D. bent-tail[ed] carrier, bent-tail[ed] dog
Drehimpuls *m* angular momentum; moment of momentum, spin
Drehimpulsachse *f* axis of angular momentum (spin)
Drehimpulsdurchflußmesser *m* angular-momentum flowmeter
Drehimpulsoperator *m* angular momentum operator
Drehimpulssatz *m* law of moment of momentum
Drehkastenwechsel *m (text)* circular box motion
Drehklappe *f* swivel (revolving) damper
Drehklinke *f* turning pawl
Drehknopf *m* [adjusting] knob
Drehknopfausschalter *m* rotary (turn) switch
Drehko *m s.* Drehkondensator
Drehkolben *m* rotary jack (piston)
Drehkolbenlader *m (aut)* displacement supercharger
Drehkolbenmotor *m* rotary [piston] engine
Drehkolbenpumpe *f* rotary [piston] pump
Drehkolbenschieber *m* rotary valve based on circular cylindrical surface, circular cylindrical valve
Drehkolbenumsteuerschieber *m* barrel selector
Drehkolbenventil *n* cylindrical rotary valve
Drehkolbenverdichter *m* rotary piston compressor
Drehkolbenzähler *m* rotating lobe meter
Drehkondensator *m* variable capacitor
frequenzgerader (frequenzgleicher) D. straight-line frequency capacitor
kapazitätsgerader D. plain capacitor
kombinierter D. compound variable capacitor
Drehkörper *m (ph)* body (solid) of revolution
Drehkraft *f* torque; rotary power; *(therm)* tangential pressure
Drehkraftdiagrammlinie *f* crank effort diagram, tangential pressure diagram
Drehkran *m* rotary (slewing, revolving, swing, jib) crane
Drehkranz *m* slewing ring
Drehkreis *m (ship)* turning circle
Drehkreisdurchmesser *m* diameter of the turning circle, final diameter
Drehkreisversuch *m* manoeuvring (circle) test
Drehkreuz *n* capstan; turnstile
Drehkreuzantenne *f* turnstile aerial
Drehkristallmethode *f (ph)* rotating-crystal method
Drehkü[c]ken *n* cone plug
Drehkü[c]kenhahn *m* cone plug
Drehkupplung *f (el)* rotating joint
Drehlänge *f* turning length *(lathe)*
größte D. turning length, distance between centres
Drehlaufkatze *f* slewing trolley
Drehmagnet *m* rotary (moving) magnet
Drehmagnetgalvanometer *n* moving-magnet galvanometer
Drehmagnetinstrument *n* moving-magnet instrument
Drehmantel *m s.* Mantelrohr
Drehmaschine *f* 1. lathe; turning machine *(metal cutting)*; 2. barring engine
D. mit Flachtischrevolverkopf flat-turret lathe
D. mit Kastenfuß cabinet lathe
Drehmaschinenbett *n* lathe bed
Drehmaschinenleitspindel *f* lathe lead screw
Drehmaschinenspanndorn *m* lathe mandrel
Drehmaschinenspitze *f* lathe centre
Drehmaschinenspitzenhöhe *f* height of the lathe centres

Drehmaschinensupport *m* lathe carriage
Drehmeißel *m* lathe (turning) tool
 abgesetzter D. offset lathe tool
 gebogener D. bent lathe tool
 gerader D. straight lathe tool
Drehmelder *m* synchro, selsyn
Drehmelderanzeigeempfänger *m* synchro indicator
Drehmelderdifferentialempfänger *m* synchro differential receiver
Drehmelderdifferentialgeber *m* synchro differential transmitter
Drehmelderempfänger *m* synchro repeater (receiver)
Drehmeldergeber *m* synchro transmitter
Drehmelderkoordinatenwandler *m* synchro resolver
Drehmelderrückmelder *m* synchro control transformer
Drehmeldersteuerempfänger *m* synchro control transformer
Drehmoment *n* torque; starting torque; turning moment
 D. bei festgebremstem Rotor static (locked-rotor) torque *(of a motor)*
 D. in bezug auf eine Achse moment of momentum about an axis
 bezogenes D. torque-weight ratio *(of a meter)*
Drehmomentabgleich *m* torque adjustment
Drehmomentbegrenzer *m* torque-limiting device
Drehmomentdiagramm *n* torque diagram
Drehmomentenkontrolle *f* torque control
Drehmomentenschlüssel *m* torque wrench
Drehmomentenverstärker *m* torque amplifier
Drehmomentfluß *m* torque flux
Drehmomentgefälle *n* torque gradient
Drehmomentmesser *m* torquemeter
Drehmomentverstärker *m* torque amplifier
Drehmomentwandler *m* torque converter
 hydraulischer D. hydraulic torque converter
Drehofen *m* rotary kiln (furnace), revolving kiln (furnace)
Drehoszillograf *m s.* Drehspiegeloszillograf
Drehpendel *n* ring (torsion) pendulum
Drehpendelbewegung *f* oscillatory motion
Drehpfanne *f* centre casting, pivot bearing (socket)
 D. der Drehscheibe bearing socket of [the] turntable
 obere D. upper centre casting
 untere D. lower centre casting
Drehpfannenbolzen *m* king (centre) pin
Drehpflug *m* reversible plough
Drehphase *f* rotary phase
Drehphasenschieber *m (el)* rotary phase changer, induction phase shifter
Drehpotentiometer *n* rotary potentiometer
Drehpuddelofen *m* rotary puddling furnace
Drehpunkt *m (ph)* centre of rotation; fulcrum *(lever)*; pivot
 fester D. fixed pivot (fulcrum)
Drehradius *m* radius of gyration
Drehrahmenantenne *f (el)* rotating loop aerial
Drehrahmenpeiler *m* frame (loop) direction finder
Drehraster *m* circular screen
Drehrauchschieber *m* revolving (swivel) damper
Drehregler *m* rotary-type regulator; induction regulator
Drehrichtung *f* direction (sense) of rotation; *(text)* direction of twist
Drehrichtungsanzeiger *m (el)* phase-sequence indicator
Drehrichtungssinn *m* direction (sense) of rotation
Drehrichtungsumschalter *m* reversing controller
Drehrohrofen *m* cylindrical rotary (revolving, rotating) kiln, cylindrical rotary (revolving, rotating) drier, rotary tubular kiln; *(met)* revolving cylindrical furnace
Drehrohrsinterofen *m* rotary sintering kiln
Drehrost *m* revolving (rotary) grate
Drehrostgenerator *m* revolving grate [gas] producer
Drehschalter *m* revolving (rotary, turn) switch; snap switch
Drehscheibe *f* 1. *(ry)* turntable; 2. *(mach)* swivel [head plate] *(shaper)*
 D. mit Gradeinteilung *(mach)* graduated swivel base
Drehscheibenabschwächer *m (el)* rotary attenuator
Drehscheibenlafette *f (mil)* turning platform mounting
Drehscheibenverschluß *m (opt)* rotating disk shutter
Drehschemel *m* swivelling (pivoted) bolster
Drehschemellenkung *f* fifth-wheel steering
Drehschemelwagen *m* wagon with radial bolster
Drehschieber *m* rotary slide valve
 flacher D. rotary disk valve
Drehschlagbohren *n (mine)* rotary-percussive drilling, rotary-percussion drilling
Drehschlagbohrgerät *n (mine)* rotative and percussive drilling machine
Drehschritt *m* rotary step *(telephony)*
Drehschwingung *f* torsional vibration
 Brownsche D. Brownian rotation
Drehschwingungsdämpfer *m* torsional damper
Drehsieb *n* revolving (rotary) screen
Drehsinn *m* direction (sense) of rotation
 D. / im positiven counter-clockwise
Drehspan *m* turning
Drehspiegeloszillograf *m* rotating-mirror oscillograph
Drehspindel *f* lathe spindle

Drehspulaufhängung *f* moving-coil support
Drehspule *f (el)* moving coil
Drehspulenschnellschreiber *m* high-speed moving-coil recorder
Drehspulenvariometer *n* rotating-coil variometer
Drehspulgalvanometer *n* moving-coil galvanometer
Drehspulinstrument *n* moving-coil instrument; permanent-magnet moving-coil instrument
Drehspulleistungsmesser *m* / **eisenloser** coreless wattmeter
Drehspuloszillograf *m* moving-coil oscillograph
Drehspulrelais *n* moving-coil relay, magnetoelectric relay
Drehspulschnellschreiber *m* high-speed moving-coil recorder *(telegraphy)*
Drehspulspannungsmesser *m* moving-coil voltmeter
Drehspulspiegelgalvanometer *n* moving-coil mirror (reflecting) galvanometer
Drehspulstrommesser *m* moving-coil ammeter
Drehspultragkonstruktion *f* moving-coil support
Drehspulvariometer *n* rotating-coil variometer
Drehspulvibrationsgalvanometer *n* moving-coil vibration galvanometer
Drehspulzeigergalvanometer *n* moving-coil pointer galvanometer
Drehstab *m* torsion bar
Drehstabaufhängung *f* torsion bar suspension
Drehstabfeder *f* torsion bar (rod) spring
Drehstabfederung *f* torsion bar springing
Drehstabstabilisator *m* torsion bar stabilizer
Drehstahl *m s.* Drehmeißel
Drehstangenverschluß *m* espagnolette bolt
Drehstift *m* fulcrum pin
Drehstrom *m* three-phase current
Drehstromanlasser *m* three-phase starter
Drehstromasynchronmotor *m* three-phase induction motor
Drehstrombrückenschaltung *f* three-phase bridge-connection, three-phase bridge-circuit
Drehstromdreileiternetz *n* three-phase three-wire circuit
Drehstromdreileiterzähler *m* three-phase three-wire meter
Drehstromgenerator *m* three-phase generator
Drehstromkreis *m* three-phase circuit
Drehstrommagnet *m* three-phase magnet
Drehstrommotor *m* three-phase motor
Drehstromnebenschlußkommutatormotor *m* polyphase shunt conduction motor
Drehstromnetz *n* threephase (three-phase, polyphase, multiphase) system
  D. mit Nulleiter three-phase four-wire system
  D. ohne Nulleiter three-phase three-wire circuit
  symmetrisches D. symmetrical three-phase system
Drehstromschalter *m* three-phase switch
Drehstromsiebenleiternetz *n* seven-wire three-phase system
Drehstromsystem *n* three-phase system
  offenes D. six-wire three-phase system, three-phase six-wire system
Drehstromtransformator *m* three-phase transformer
Drehstromvierleiteranlage *f* three-phase four-wire system
Drehstromwicklung *f* polyphase winding
Drehstromzähler *m* three-phase meter
Drehsupport *m* swivel head *(turning)*
Drehsymmetrie *f (ph)* rotation[al] symmetry
Drehteil *n (mach)* swivel base (plate), head swivel *(shaper)*
Drehtellerspeiser *m* rotary table feeder
Drehtisch *m (mach)* rotary (revolving) table; turntable; *(opt)* rotary stage
Drehtransformator *m* phase[-shifting] transformer, three-phase induction regulator, induction phase shifter
Drehtrenner *m* rotary (pivoted-layer) isolator, rotary disconnector (isolating switch)
Drehtrennschalter *m s.* Drehtrenner
Drehtrommelröstofen *m* rotary calcining kiln
Drehtür *f* revolving door
Drehturm *m (mil)* rotary turret
Drehumformer *m* dynamotor, motor-generator
Drehung *f* rotation; twist[ing]; revolution; torsion
  D. der Polarisationsebene rotation of the plane of polarization
  D. der Spindeln während der Wagenausfahrt *(text)* spindle turn during outward running of carriage
  D. entgegen[gesetzt] dem Uhrzeigersinn anti-clockwise rotation
  D. im vollen Kreis full-circle rotation
  magnetische D. magnetooptic rotation
  molekulare D. der Polarisationsebene molecular rotation of the plane of polarization
  optische D. optical rotation
  rechtsläufige D. clockwise rotation
Drehungsachse *f* rotation axis
Drehungselastizität *f* torsional elasticity
Drehungsgeschwindigkeit *f* velocity (speed) of rotation, rotational (rotative) speed
Drehungsschwingung *f* torsional vibration (oscillation)
Drehungswiderstand *m* rotational resistance *(turbine)*
Drehungswinkel *m* 1. *(ph)* angle of rotation, rotation angle; 2. angle of twist *(yarn)*
Drehvariometer *n (el)* [rotating-coil] variometer
Drehvermögen *n* rotary power
  optisches D. optical rotary power

Drehvorrichtung *f* barring gear
Drehwaage *f* torsion balance
Drehwähler *m* uniselector; rotary preselecting line switch
Drehwählersystem *n* rotary [automatic telephone] system, rotary dial system
Drehwerk *n* turning gear
  D. mit Seitensupport vertical turret lathe [with side head]
Drehwerkstück *n* turned component; component to be turned
Drehwiderstand *m (el)* [rotary] rheostat, variable resistor
Drehwinkel *m* angle of rotation (tilt)
Drehwuchs *m* spiral grain, twisted growth *(wood)*
drehwüchsig spiral-grained *(wood)*
Drehzahl *f* speed [of rotation in rpm]; rotational speed
  D. / mit hoher high-speed
  gestufte D. stepped (variable) speed
  kritische D. für die Auferregung (Selbsterregung) *(el)* critical build-up speed
  stufenlos veränderliche D. steplessly variable speed
Drehzahlabfall *m* speed drop
Drehzahländerung *f* variation in speed, speed variation (control)
  stufenlose D. stepless speed changing
Drehzahlanstieg *m* speed rise
Drehzahlbegrenzer *m* governor
Drehzahlbereich *m* range of speeds, speed range
drehzahleinstellbar adjustable-speed
Drehzahleinstellung *f* speed adjustment
Drehzahlgrenze *f* speed limit
Drehzahlgruppe *f* speed range
Drehzahlmesser *m (mach)* speedometer, tachometer; speed (revolution) counter; speed indicator
  elektrischer D. magneto speed indicator
Drehzahlregelbereich *m* range of speed regulation
Drehzahlregelung *f* speed regulation (governing, control)
  stufenlose D. infinitely variable speed control
Drehzahlregelwiderstand *m* speed regulating rheostat
Drehzahlregler *m* speed governor (controller)
Drehzahlreihe *f* group of speeds
Drehzahlstufung *f* speed graduation
Drehzahlveränderung *f* **durch Polumschaltung** pole changing control
Drehzahlverstellmotor *m* speed changer *(engine governor)*
Drehzahlvorwahl *f* speed [pre]selection
Drehzahlvorwählung *f* preselection of speeds
Drehzahlwähler *m* speed [pre]selector
Drehzahlwechsel *m* speed change
Drehzahn *m* cutter bit *(not ground)*
Drehzapfen *m* central pivot, trunnion, spigot; gudgeon; swivel (king) pin
Drehzapfenabstand *m* distance between bogie (*Am:* truck) centre pins
Drehzapfenlager *n* trunnion bearing
Drei-Exzeß-Kode *m s.* Drei-Überschuß-Kode
Drei-Überschuß-Kode *m (dat)* excess-three-code
dreiachsig 1. *(ph)* triaxial; 2. *(aut)* six-wheel
Dreiadressensystem *n (dat)* three-address system
dreiadrig three-core, three-wire
Dreiarmflansch *m* three-armed flange
Dreiarmnabe *f* triple-sector clutch hub
Dreiäscherverfahren *n (leath)* three-pit system
Dreibackenbohrfutter *n* drill chuck with three jaws
Dreibackenfutter *n* three-jaw chuck
Dreibahnenbett *n* triple-track bed
dreibahnig three-way, triple-track
Dreibaum *m* tripod
Dreibein *n* tripod
Dreibeinkran *m* three-leg crane, shear legs (leg derrick)
Dreibeinlafette *f (mil)* tripod mounting
Dreibeinmast *m* tripod mast (tower)
dreibettig three-bed
dreibindig three-leaf
Dreiblattruder *n* three-plate rudder
Dreibleimantelkabel *n* separately lead-covered cable
Dreibockreuter *m (ag)* tripod
Dreibockstützer *m* three-leg insulator, three-pin [supporting] insulator
Dreidecker *m* 1. triplane *(aeroplane)*; 2. *s.* Dreideckschiff
Dreideckschiff *n* three-deck[ed] vessel, three-decker
dreidimensional three-dimensional
Dreidrahtmethode *f* **des Gewindemessens** three-wire method of thread measurement
Dreiebenenschalter *m (el)* three-storeyed switch
Dreieck *n* triangle
  gleichschenkliges D. isosceles triangle
  gleichseitiges D. equilateral triangle
  kongruentes D. congruent triangle
  nautisches D. *(opt)* nautical triangle
  Pascalsches D. Pascal's triangle
  rechtwinkliges D. right[-angled] triangle
  sphärisches D. spherical triangle
  spitzwinkliges D. acute triangle
  stumpfwinkliges D. obtuse triangle
  ungleichseitiges D. scalene triangle
Dreieckantenne *f* triangular aerial
Dreieckaufhängung *f* three-point suspension
Dreieckbinder *m* pitched truss
Dreiecke *npl* / **ähnliche** similar triangles
Dreieckfachwerk *n* pitched truss
dreieckgeschaltet *(el)* delta-connected

dreieckig triangular
Dreieckimpuls *m* triangular [transient] pulse, triangle pulse
Dreieckkettenfahrleitung *f* double-catenary suspension line
Dreiecklast *f* triangular load
Dreieckplatte *f* triangular insert *(single-point tool)*
Dreieckrückstrahler *m* triangular reflector
Dreiecksanordnung *f (mine)* diagonal (triangular) system *(support)*
Dreieckschalter *m (el)* delta switch
Dreieckschaltung *f (el)* delta connection
Dreieckschwingung *f* triangular oscillation
Dreieckseinbruch *m (mine)* draw-cut
Dreieckslenkeraufhängung *f (aut)* wishbone suspension
Dreiecksmessung *f* trigonometry
Dreiecksnetz *n* network of triangles, triangulation system *(surveying)*
Dreieckspannung *f* delta (mesh) voltage
Dreieckspunkte *mpl* triangulation points, control points [of known position and elevation]
Dreiecksquerschnitt *m* triangular cross section
Dreiecksternschaltung *f* delta-star connection
Dreiecksterntransformation *f* star-mesh transformation
Dreieckstufe *f* spandrel step
Dreieckverband *m* triangular (diagonal) bracing, triangular (diagonal) tieing
Dreieckwelle *f* triangular (triangle) wave
Dreieckwicklung *f* delta winding
Dreieckzahlen *fpl* triangular numbers
Dreieckzickzackschaltung *f* delta-zigzag-connection
Dreielektrodenquarz *m* piezo-electric coupler
Dreielektrodenröhre *f* three-electrode valve
Dreielementenantenne *f* three-element aerial
Dreierbündel *n (el)* triple-bunch conductor
Dreierspaltung *f s.* Dreifachspaltung
Dreierstern *m (nucl)* three-pronged star
Dreierstoß *m (ph)* triple (threefold) collision
Dreietagenofen *m* three-storeyed furnace
dreifach threefold, triple, treble
Dreifachbandfilter *n (el)* treble (three-section) band-pass filter
Dreifachbindung *f (ch)* triple bond
Dreifachdrehkondensator *m* three-gang capacitor
Dreifachexpansionsdampfmaschine *f* triple expansion steam engine
  D. mit geteiltem Niederdruck four-cylinder triple expansion steam engine
Dreifachexpansionsmaschine *f* triple expansion engine
Dreifachkabel *n* three-conductor cable, three-core cable
Dreifachkette *f* triplex (triple-strand) chain
Dreifachkoinzidenz *f (nucl)* triple coincidence
Dreifachmodulation *f* triple (treble) modulation
Dreifachpoller *m (ship)* triple bollard
Dreifachröhre *f* three-in-one valve, three-unit valve
Dreifachrollenkette *f* triple roller chain
Dreifachschalter *m* three-point switch
Dreifachschieber *m* open centre-type slide valve
Dreifachschnur *f* triple cord
Dreifachspaltung *f (nucl)* ternary fission, tripartition
Dreifachstecker *m* three-way [plug] adaptor; three-pin plug
Dreifachtarifzähler *m* three-rate meter
Dreifachumschalter *m* three-point switch, three-way switch, triple-throw switch
Dreifachverstärker *m* three-stage amplifier
Dreifachverzweigungsmuffe *f* trifurcating box
dreifachwirkend triple-action, triple-acting *(press)*
Dreifachzeilensprungabtastung *f* triple interlaced scanning
Dreifadenlampe *f* three-filament lamp
Dreifarbenbildröhre *f* tricolour (three-colour) picture tube
Dreifarbendruck *m* three-colour [process of] printing
Dreifarbenröhre *f* tricolour kinescope
Dreifarbenverfahren *n (telev)* three-colour method
  additives D. der Farbenfotografie additive process of colour photography
  subtraktives D. der Farbenfotografie subtractive process of colour photography; subtractive colour process
dreifarbig three-coloured
Dreifeldrahmen *m (build)* three-bay frame, three-span frame
dreifeldrig *(build)* three-span, three-bay
Dreifingerregel *f (el)* right-hand rule
Dreiflächenruder *n* three-plate rudder, three-blade rudder, three-body rudder
Dreiflammrohrkessel *m* triple flue boiler
Dreiflügelmeißel *m* **mit Basaltdüseneinlagen** *(mine)* basalt nozzle inserted three-wing chisel
dreiflüglig three-bladed *(propeller)*
Dreifurchenpflug *m* three-furrow plough
Dreifuß *m* tripod
Dreiganggetriebe *n* three-speed transmission
dreigängig three-start *(thread)*, triple[-threaded]
Dreigelenkbogen *m* three-hinged arch, three-pin arch
Dreigelenkrahmen *m* three-hinged frame
Dreigitterröhre *f* three-grid valve, pentode
dreigleisig triple-tracked
dreigliedrig *(math)* three-term; ternary
Dreigutscheidung *f* three-product separation
Dreihalskolben *m* three-necked bottle
Dreiinselschiff *n* three-island [type] vessel

Dreikantbohrer *m* three-cornered drill
Dreikantfeile *f* triangular (three-square) file
dreikantig three-edged, three-cornered, three-square
Dreikantleiste *f* angle fillet, chamfer strip
Dreikantlitzenförderseil *n* triangular strand winding rope
Dreikegelrollenmeißel *m* tricone bit
dreikernig trinuclear
Dreiklang *m (ac)* triad
Dreikörperproblem *n (nucl)* three-body problem, three-particle problem
Dreikrempelsatz *m (text)* three-card set
dreikurbelig three-throw
Dreilagenkarton *m* **/ minderwertiger** filled pulp-board
dreilagig three-coat *(plaster)*
dreilappig tri-lobe
Dreileiterkabel *n* triplex (three-conductor, three-core) cable
Dreileitermaschine *f* three-wire generator
Dreileiternetz *n (el)* three-wire system
Dreileitersystem *n (el)* three-wire system
dreilippig three-lip
Dreilochspeicherkern *m* three-hole memory core *(for data storage)*
Dreimastgaffelschoner *m* tern (three-mast fore and aft) schooner
Dreimastschoner *m* three-mast schooner
Dreimeilengrenze *f* three-mile limit, marginal sea
dreimeßrig three-cutter *(milling)*
Dreimomentengleichung *f* three-moment theorem
  Clapeyronsche D. Clapeyron's theorem
Dreimomentensatz *m* theorem of three moments
Dreiphasenkohlebogenlampe *f* three-phase arc lamp
Dreiphasenmotor *m* three-phase motor
Dreiphasenstrom *m* three-phase current
Dreiphasensystem *n (el)* three-phase system
Dreiphasenwechselstrom *m* three-phase alternating current
Dreiplattenstoß *m* junction of three plates
dreipolig triple-pole, three-pole; three-pin
Dreipolröhre *f* triode, three-electrode valve
Dreipolstecker *m* three-pin plug, three-pole plug
Dreiprismenspektrograf *m* three-prism spectrograph
dreiprofilig three-ribbed, triple-edge
Dreipunktaufhängung *f* three-point suspension
Dreipunktkupplung *f* three-point linkage
Dreipunktlandung *f* three-point landing
Dreipunktregelung *f* three-step action
Dreipunktschaltung *f* autodyne connection, three-point circuit (connection)
  kapazitive D. Colpitt's circuit
Dreipunktwirkung *f (cont)* three-step action
Dreiquartier *n* three-quarter bat *(brick)*
Dreirad *n* tricycle; tricar
Dreiräderblock *m (mach)* sliding triple gear
Dreiradwalze *f* three-wheeled roller
Dreiring *m (ch)* three-membered ring
Dreiringverbindung *f (ch)* tricyclic compound
Dreiröhrenempfänger *m* three-valve receiver
Dreiröhrenverstärker *m* three-valve amplifier
Dreirollenmeißel *m* Hughes bit
Dreirumpfboot *n* tri-maran, three-hull boat
Dreisatz *m (math)* rule of three
Dreischichtenbetrieb *m* three-shift working
Dreischichtenfilm *m* tri-pack film
Dreischichtenplatte *f* sandwich panel
Dreischichtenverbundplatte *f* three-layer board
dreischichtig sandwich; in three layers
dreischiffig three-bay, three-span
Dreischleifenwicklung *f* triplex winding
Dreischneidenbohrer *m* three-flute drill
Dreischneider *m* three-knife trimmer, bock trimming machine
Dreischraubenschiff *n* triple-screw vessel
Dreischußbindung *f* three-weft binding
Dreisitzer *m* three-seater
Dreispiegelhaftglas *n (opt)* three-mirror contact lens
Dreispindelfräsmaschine *f* triplex milling machine
dreispindelig *(mach)* triple-spindle; triple-screw *(pump)*
Dreispindelmaschine *f* triple-spindle machine
Dreispindelschraubenpumpe *f* three-screw type of pump
dreispurig three-lane, three-track
dreißigflächig triacontahedral
dreistellig three-place, three-figure
Dreistiftstecker *m* three-pin plug
dreistöckig three-stor[e]y
Dreistofflegierung *f* ternary alloy
Dreistoffsystem *n* ternary system *(alloy)*; three-component system, triangular classification chart
Dreistufenfilter *n* three-step filter
Dreistufenrakete *f* three-stage rocket, three-step rocket
  D. ohne Stabilisierungsflächen finless three-stage rocket
dreistufig three-step, in three steps
Dreitafelprojektion *f* three-plane projection
Dreitankschiff *n* twin-bulkhead tanker
dreiteilen to trisect
dreiteilig tripartite, three-piece
Dreiteilung *f* trisection *(angle)*
Dreiviertelziegel *m* three-quarter bat
Dreivoltmeterleistungsmessung *f* three-voltmeter method

Dreivoltmeterverfahren *n* three-voltmeter method

Dreiwalzengerüst *n* three-high rolling stand

Dreiwalzenkalander *m* three-roller calender

Dreiwalzenstreckwerk *n* set of three pairs of drawing rollers

Dreiwegebohrmaschine *f* three-way drilling machine

Dreiwegehahn *m* three-way cock

Dreiwegeschieber *m* three-way valve

Dreiwegestück *n* three-way pipe

Dreiwegschalter *m* three-point switch

dreiwertig *(ch)* trivalent, tervalent; trihydric *(alcohol)*

Dreiwertigkeit *f (ch)* trivalence, tervalence

Dreiwickler *m s.* Dreiwicklungstransformator

Dreiwicklungstransformator *m* three-winding transformer

Dreizackereignis *n s.* Dreierstern

Dreizimmerwohnung *f* three-room[ed] flat

Dreizinkengrubber *m* scuffler

Dreizylindermotor *m* three-cylinder engine

Drell *m s.* Drillich

Drempel *m* sill

Drempelwand *f* jamb wall

dreschen *(ag)* to thresh, to thrash

Dreschflegel *m* flail

Dreschkorb *m (ag)* concave

Dreschmaschine *f* threshing machine, thresher

Dreschmühle *f* threshing mill

Dreschsatz *m* threshing aggregate

Dreschtenne *f* threshing floor

Dreschtrommel *f* [threshing] drum

Dreschtrommelkorb *m* threshing drum concave

Dreschwerk *n* threshing gear (machinery)

Dreschzylinder *m* threshing cylinder

Dressiergerüst *n* temper mill, skin pass

Dressierwalze *f* cold roll

Dressierwalzwerk *n* temper mill, skin pass

Drift *f* drift

Driftbeweglichkeit *f* drift mobility

Driftbewegung *f* drift [motion]

driften to drift

Driftfehler *m* drift (phasing) error *(computer)*

Driftgeschwindigkeit *f* drift velocity

driftkompensiert drift-corrected

Driftkomponente *f* drift component

Driftkorrektion *f* drift correction

Driftlänge *f* drift length

Driftnebeln *n (ag)* drift-fogging

Driftröhre *f* drift tube

Driftspruhen *n (ag)* drift-spraying

Driftstörung *f* drift disturbance

Drifttransistor *m* drift transistor

Drillbohrer *m* Archimedian drill

Drilldocke *f* drill chuck

drillen to drill, to sow in drills

Drillfestigkeit *f s.* Torsionsfestigkeit

Drillich *m* ticking

Drilling *m* three-barrel rifle; three-barrelled gun

Drillingsdampfmaschine *f* triple (three-) cylinder steam engine

Drillingskristall *m* triplet crystal

Drillingsmaschine *f* triple engine

Drillingsturm *m (ship)* triple turret

Drillingswalzwerk *n* trio mill

Drillkasten *m* seeding box, hopper

Drillknicken *n* torsional buckling

Drillmaschine *f (ag)* drill [machine]

Drillmoment *n* twisting (torsional) moment

Drillreihe *f (ag)* drilled row

Drillsaat *f* drilled seed

Drillschar *n* [drill] coulter

Drillung *f* torsion, twist

Drillwiderstand *m* torsional resistance

Dringlichkeitsfrequenz *f* urgency frequency

Dringlichkeitsgrad *m* degree of priority

Dringlichkeitszeichen *n* urgency signal

Drittel *n (mine)* shift; *(math)* third
  mittleres D. middle third

Drittelspunkt *m* third point

„Drive-on-drive-off"-Frachter *m* „drive-on-drive-off" carrier

Drossel *f* 1. choke; restrictor *(pneumatics)*; throttle; 2. *(el)* reactor, inductance (reactive) coil; choke [of a waveguide]; ballast [of fluorescent lamps]
  D. mit Eisenkern iron-cored inductor
  D. mit Luftkern air-cored inductor
  D. mit stellbarer Anzapfung tapped variable inductor
  D. ohne Eisenkern air-core choke
  eisenlose D. air-cored coil
  nichtlineare D. non-linear coil
  regelbare D. adjustable inductor
  trägersperrende D. carrier isolating choke coil
  veränderliche D. variable-area nozzle
  vorgeschaltete D. series inductance, ballast
  vormagnetisierte D. [magnetic] biased reactor

Drosseldurchflußmesser *m* head flowmeter

Drosselflansch *m* choke connector *(waveguides)*

Drosselflanschverbindung *f* choke coupling *(waveguides)*

drosselgekoppelt impedance-coupled

Drosselkette *f* low-pass filter

Drosselklappe *f* choke control *(air)*; butterfly valve; *(Am)* damper

Drosselklappenachse *f* butterfly throttle spindle

Drosselklappenhebel *m* throttle control lever

Drosselklappenventil *n* butterfly valve

Drosselklappenwelle *f* throttle valve shaft

Drosselkolben *m* choke piston *(waveguides)*

Drosselkolbenschieber *m* spool-type throttling valve

Drosselkopplung *f* resistance inductance coupling
Drosselkreis *m* reactance (rejector) circuit
Drosselkupplung *f* choke coupling *(waveguides)*, resistance inductance coupling
Drosselkurve *f* throttling curve *(turbine)*
Drosselmodulation *f* saturated iron core modulation, modulation due to iron saturation
drosseln 1. to throttle *(engine)*; 2. *(el)* to choke [out]; to restrict *(flow)*
Drosselregelung *f* throttle governing, governing by throttling the steam supply
Drosselrelais *n* high-impedance relay
Drosselring *m (aut)* throttle ring
Drosselschaltung *f* suppressor [circuit]
Drosselscheibe *f s.* Stauscheibe
Drosselspannung *f* choking voltage
Drosselspeisung *f* choke feed
Drosselspinnerei *f* continuous spinning
Drosselspule *f* choke (impedance, choking) coil, self-inductor, reactor
Drosselstoß *m* choke joint *(waveguides)*
Drosseltransformator *m* constant-current transformer
Drosselturbine *f* throttling turbine
Drosselung *f* 1. throttling *(fluid flow)*; 2. *(el)* choking
Drosselventil *n* choke (regulator, restriction, butterfly) valve; throttle-type valve, volume-control valve
  D. mit Stellschraube adjustable screw restriction valve
Drosselverbinder *m* impedance bond
Drosselverstärker *m* impedance coupled amplifier
Drosselweg *m* throttling path *(turbine)*
Drosselwiderstand *m (el)* choking resistance
Drosselwirkung *f (el)* choking effect
Droussierwolf *m* willow
Druck *m* 1. pressure; thrust; compression; 2. print[ing]
  D. der Umgebung ambient pressure
  D. im Kavitationsraum cavity pressure
  absoluter D. absolute pressure
  atmosphärischer D. atmospheric pressure
  außermittiger D. eccentric action of force
  beiderseitiger D. *(text)* duplex print
  drehender D. tangential pressure
  elektrostatischer D. electrostatic printing
  großflächiger D. *(text)* blotch print
  hydrostatischer D. hydrostatic pressure
  kritischer D. critical pressure
  osmotischer D. osmotic pressure
  partieller D. partial pressure
  spezifischer D. specific pressure
  statischer D. static pressure
  verminderter D. reduced pressure
  zweiseitiger D. *(text)* duplex print
Druckabfall *m* pressure drop
druckabhängig pressure-dependent, pressure-controlled
Druckamplitude *f* pressure amplitude
Druckanfang *m* leading edge
Druckanstieg *m* pressure rise, rise in pressure
Druckanzeigegerät *n* pressure indicating device
Druckanzeiger *m* pressure gauge
Druckanzug *m* [high-]pressure suit
Druckaufbau *m* pressure build-up
Druckaufnahmefähigkeit *f* load-carrying properties
Druckaufteilung *f* subdivision of the drop of pressure
Druckausgleich *m* pressure compensation
  D. / mit pressure-balanced
  D. / ohne unbalanced
Druckausgleicheinrichtung *f (therm)* pressure compensation device
druckausgleichend pressure-equalizing
Druckausgleichschalter *m (el)* pressure compensating circuit breaker
Druckausgleichventil *n* pressure compensating valve
Druckautomat *m* automatic platen
Drückbank *f* [metal] spinning lathe
druckbeansprucht subjected to compression
Druckbeanspruchung *f* compressive stress
Druckbehälter *m* pressure vessel (tank)
Druckbelastung *f* compressive load
Druckbestäuber *m (print)* anti-set-off spray
druckbetätigt pressure-operated
Druckbewehrung *f* compression (compressive) reinforcement
Druckblende *f* / **automatische** *(opt)* automatic pressure diaphragm
Druckbuchstabe *m* printed character
Druckchronograf *m* printing-type chronograph
druckdicht pressure-sealed, pressure-tight; pressurized
Druckdichte *f* pressure tightness
Druckdose *f (cont)* pressure capsule
Druckeinheit *f* unit of pressure
Druckeinstellung *f* pressure adjustment
Druckelektrizität *f* piezoelectricity
Druckelektrode *f* resistance welding electrode
Druckempfang *m (telec)* printing (typing) reception
Druckempfindlichkeit *f* pressure sensitivity
drucken to print
  im Schwunge d. to print on the fly
drücken 1. to force, to push, to press; to depress; 2. to spin [metal]
  eine Taste d. to depress a key
Drucken *n* printing
  D. auf beiden Seiten *(text)* duplex printing
  D. der Summe *s.* Summenschreibung
Drücken *n* forcing, pressing, pushing, depressing

Druckende *n* leave
druckentlastet balanced
Druckentlastung *f* pressure relief
Druckentnahmebohrung *f* pressure tap
Drucker *m* printer
Drücker *m* 1. *(el)* push[er], push button; bell-push; 2. depressant, depressing agent *(flotation)*
Druckerballen *m* leather (printer's) ball, tampon
Druckerei *f* printing office
Druckerempfangslocher *m* *(telec)* printing reperforator
Druckerhitzung *f* heating under pressure
Druckerschwärze *f* printing ink *(black)*
Druckerzeuger *m* pressure generator
Druckerzeugung *f* generation of pressure
  D. mittels Kettenübertragung lever and chain weighting *(drawing)*
Druckfarbe *f* printing ink (pigment)
  D. auf Wasserbasis water-base ink
  magnetische D. magnetic ink
Druckfeder *f* helical compression spring
Druckfehler *m* misprint, erratum, printer's error
Druckfeld *n* pressure field
druckfest pressure-proof, resistant to pressure
Druckfestigkeit *f* compressive (crushing) strength
  D. bei behinderter Querdehnung confined compressive strength
Druckfeuerung *f* supercharged furnace
Druckfinger *m* strut
Druckfläche *f* driving (leading) face *(propeller)*; working face
  D. eines Propellerflügels face of propeller blade
Druckflasche *f* pressure bottle
Druckflüssigkeit *f* pressure (hydraulic, operational) fluid, pressure (hydraulic, operational) liquid
Druckfolge *f* printing sequence
Druckform *f* forme, printing block
Druckformherstellung *f* forme preparation
Druckfortpflanzung *f* pressure transmission (propagation), transmission of pressure
druckgasbelastet gas-loaded
Druckgasflasche *f* gas cylinder
Druckgasförderverfahren *n* gas lift
Druckgaskabel *n* pressurized (gas pressure, gas-filled) cable
Druckgasschalter *m* gas-blast circuit breaker
Druckgefälle *n* pressure gradient; drop of pressure *(turbine)*
Druckgefäß *n* pressure vessel (tank)
druckgeschmiert pressure-lubricated
Druckgeschwindigkeit *f* printing speed
druckgesteuert pressure-controlled; pressure-responsive
Druckgewölbe *n* *(mine)* pressure (voussoir) arch
Druckglied *n* compression member
Druckgradientenfeld *n* field of pressure gradient
Druckgurt *m* compression chord (boom) *(trusses)*
Druckguß *m* die casting
Druckhebewinde *f* hydraulic jack
Druckhelm *m* pneumatic helmet
Druckhöhe *f* pressure head
  geodätische D. static [pressure] head
Druckhöhenmesser *m* pressure altimeter
Druckhöhenverlust *m* loss of pressure head
Druckholz *n* compression wood
Druckionisation *f* pressure ionization
Druckkabel *n* compression (non-draining) cable
Druckkabine *f* pressure cabin
Druckkalander *m* swissing calender
Druckkammer *f* pressure chamber; delivery chamber *(pump)*
Druckkammerlautsprecher *m* pneumatic (pressure-chamber) loudspeaker
Druckkammerschalter *m* *(el)* pressure chute (chamber) circuit breaker
Druckknopf *m* 1. push button; 2. snap-fastener, patent-fastener
  D. für Höhenverstellung *(mach)* raise-lower push button
Druckknopfabstimmung *f* push-button tuning *(radio)*
Druckknopfabstimmvorrichtung *f* push-button tuner *(radio)*
druckknopfgesteuert *(mach)* push-button controlled
Druckknopfschalter *m* push-button switch (key)
Druckknopfschalttafel *f* push-button panel
Druckknopfschaltung *f* push-button control
Druckknopfsteuerung *f* push-button control
Druckknopftaster *m* push-button key
Druckkocher *m* pressure digester, autoclave
Druckkoeffizient *m* **der Reaktivität** pressure coefficient of reactivity
Druckkondensator *m* pressure-type capacitor
Druckkontakt *m* pressure (butt) contact
Druckkörper *m* pressure hull *(submarine)*
Druckkraft *f* compressive force, thrust; pressure
Druckkurve *f* / **umkehrbare** *(therm)* equilibrium (reversible pressure) curve
Drucklagen *fpl* *(mine)* pressure partings
Drucklagerwelle *f* thrust shaft
Druckleiste *f* thrust strip
Druckleitung *f* pressure (delivery, discharge) pipeline *(hydraulics)*
Drucklitze *f* lowering heald
drucklos at zero pressure
Druckluft *f* compressed air
Druckluftakkumulator *m* hydro-pneumatic accumulator
Druckluftantrieb *m* pneumatic (compressed-air) drive
Druckluftbehälter *m* compressed-air reservoir

Druckluftbesatzgerät *n (mine)* pneumatic stemmer
Druckluftbespülung *f* compressed-air scavenging *(hollow insulators of circuit-breakers)*
druckluftbetätigt pneumatically operated, air-actuated, air-operated
Druckluftbetätigung *f* pneumatic control (actuation, operation), air control (actuation, operation)
Druckluftbohrer *m* air (pneumatic) drill
Druckluftbremse *f* [compressed] air brake, pneumatic brake
Druckluftdifferentialkolbenakkumulator *m* differential accumulator, air accumulator employing a differential piston
Drucklufterzeuger *m* air compressor, compressed-air generation plant
Druckluftfeinanzeiger *m s.* Druckluftfeinzeiger
Druckluftfeinzeiger *m* [Solex] pneumatic micrometer
Druckluftfeuerung *f* furnace with forced draught
Druckluftfilter *n* compressed-air cleaner
Druckluftfilzführung *f* pneumatic felt guide
Druckluftflasche *f* [compressed] air bottle
Druckluftflüssigkeitsspeicher *m* **ohne Trennelement** non-separator type air-hydraulic accumulator
Druckluftförderer *m* pneumatic (airflow) conveyor
Druckluftförderung *f* pneumatic conveying
Drucklufthammer *m* pneumatic (air) hammer
Drucklufthandbohrmaschine *f* compressed-air motor hand drill, pneumatically operated drilling machine of the portable type
Druckluftkolbenakkumulator *m* air-loaded accumulator with piston
Druckluftleistungsschalter *m* **mit Selbstbeblasung** auto-pneumatic circuit-breaker
Druckluftleitung *f* compressed-air line
Druckluftmanometer *n* compressed-air gauge
Druckluftmotor *m* air (pneumatic) motor
Druckluftregler *m* pneumatic governor
Druckluftschalter *m* air-blast [circuit] breaker, compressed-air circuit breaker, air-pressure circuit breaker
Druckluftschaltung *f* pneumatic change
Druckluftschaltzylinder *m* compressed-air shift cylinder
Druckluftschießen *n (mine)* Armstrong blasting
Druckluftschlauch *m* compressed air hose
Druckluftschnellschalter *m* rapid (quick-operating) air-blast circuit breaker
Druckluftschütz *n* **mit Magnetsteuerung** electro-pneumatic contactor
Druckluftschutzanzug *m* pneumatic frogsuit
Druckluftsenkkasten *m* pneumatic (air) caisson
Druckluftspannung *f* pneumatic (air) chucking *(workpiece)*
Druckluftspeicher *m* air[-hydraulic] accumulator
D. mit Gummiluftsack flexible bag-type accumulator
Druckluftstampfer *m* air (pneumatic) tamper
Druckluftstellwerk *n* electro-pneumatic signal box
Druckluftsteuerschalter *m* electro-pneumatic controller
Druckluftsteuerung *f* pneumatic control
Drucklüftung *f* forced ventilation
Druckluftventil *n* / **elektrisch betätigtes** electro-pneumatic valve
Druckluftversorgung *f* compressed-air supply
Druckluftverteiler *m* compressed-air distributor
Druckluftwerkzeug *n* pneumatic tool
Druckmagnet *m* printer magnet *(telegraphy)*
Druckmaschine *f* printing machine
Druckmaßstab *m (therm)* scale of pressure
Druckmelder *m* pressure indicator
Druckmeßdose *f* pressure (load) cell
Druckmesser *m* pressure gauge
D. für hohes Vakuum high-vacuum gauge
Druckmeßgerät *n* pressure gauge
Druckmessung *f* pressure measurement
Druckmikrofon *n* pressure-actuated microphone
Druckminderer *m* pressure reducing valve
Druckminderung *f* pressure reducing
Druckminderventil *n* pressure relief (reducing) valve
Druckmittel *n* pressure (fluid) medium *(hydraulics)*, operation medium
Druckmittelpunkt *m* centre of pressure
Druckmittelpunktkoeffizient *m* centre of pressure coefficient
Druckmittler *m* pressure transmitting seal; *(Am)* chemical attachment
Drucköl *n* pressure (hydraulic) oil
Druckölantrieb *m* [oil-]hydraulic drive
Druckölpumpe *f* hydraulic (pressure) pump
Druckölregler *m* oil pressure governor
Druckölschaltung *f* hydraulic gear change
Druckölschmierung *f* pressure-feed oil lubrication
Druckölung *f* forced-feed lubrication
Druckpapier *n* printing paper
Druckpaste *f* printing paste
Druckpfahl *m* bearing pile
Druckplattennase *f* pressure-plate lug
Druckprobe *f* compression specimen
Druckprüfung *f* 1. compression test *(strength)*; crushing test; 2. pressure test
Druckpumpe *f* force (pressing, plunger) pump
Druckpunkt *m* 1. centre of pressure; 2. trigger slag
Druckraumabschluß *m* pressure seal
Druckreaktor *m* pressure-type reactor
Druckreduzierventil *n* pressure relief valve

Druckregelventil *n* pressure control (regulating) valve

Druckregler *m* pressure governor (regulator); manostat

Druckrelais *n* pressure relay

Druckring *m* reaction ring, thrust collar

Druckrohr *n* pressure pipe

  D. der Speisepumpe feed delivery pipe

Druckröhrenreaktor *m* [pressure-]tube reactor

Druckrohrleitung *f* pressure pipe line

Druckrohrreibung *f* discharge pipe friction *(hydraulics)*

Druckrohrstutzen *m* pressure pipe tube

Druckrolle *f* 1. *(ag)* roller wheel; 2. pressure (pad) roll *(film)*

Drückrolle *f* [metal] spinning roller

Drucksaal *m* pressroom

Drucksache *f* printed matter

Druckschalter *m* pressure switch

Druckscheidung *f* decantation under pressure

Druckschieferung *f (geol)* flow cleavage

Druckschienenkontakt *m* electrical depression bar

Druckschlauch *m* pressure hose

Druckschlechten *pl (geol)* thrust cleavage

Druckschmierapparat *m* high-pressure grease gun

Druckschmierkopf *m* pressure grease cup

Druckschmierpresse *f* grease gun

Druckschmierung *f* pressure (force) feed lubrication

Druckschott *n* pressure bulkhead

Druckschraube *f* clamping nut; adjusting screw

Druckschreiber *m* manograph, recording manometer

Druckschrift *f* block letters

Druckschrumpfen *n (text)* compressive shrinking

Druckschwankung *f* pressure variation (fluctuation)

Druckseite *f* discharge (delivery) side *(pump)*; working (driving, blade) face *(propeller)*

Druckseitenkavitation *f* face cavitation

Drucksintern *n* pressure sintering

Druckspannung *f* compressive (compression) stress

Druckspeichereinrichtung *f* pressure storage device

Druckspüler *m* valve tap

Druckstab *m* compression bar *(reinforced concrete)*; compression member *(truss)*; strut

Druckstange *f* pusher beam *(rack)*; plunger rod *(brake)*

Druckstauwirkung *f* boost action

Drucksteigerung *f* increase of pressure

Druckstein *m* litho stone

Drucksteuerung *f* pressure control

  D. / mit automatischer with automatic pressure [control]

Druckstock *m* [printing] block, plate, cut, engraving

Druckstufe *f* difference in pressure; pressure stage *(turbine)*

Druckstufenscheidewand *f* diaphragm, guide blade disk

Druckstutzen *m* pressure connection

Drucktank *m* high-pressure vessel

Drucktankgenerator *m* pressure-type Van de Graaff accelerator

Drucktaste *f* push button, press button; press-button switch; tab; key

  betätigte D. activated push button

Drucktastenaggregat *n* press-button block; key unit

Drucktastenauslöser *m* tab

Drucktastenempfänger *m* push-button-type radio set

Drucktastenschalter *m* press-button selector

Drucktelegraf *m* printing telegraph

Drucktelegrafenapparat *m* printing telegraph

Drucktränkungsverfahren *n* pressure method

Drucktuch *n* [cylinder] blanket; back cloth

Druckturbine *f* impulse (action) turbine

  D. mit Druckabstufung impulse (action) turbine with pressure stages, multistage action (impulse) turbine

  D. mit Geschwindigkeitsstufen impulse turbine with velocity stages

  einstufige D. single-stage impulse turbine

  mehrstufige D. multistage action (impulse) turbine, impulse (action) turbine with pressure stages

Drucktype *f* printing (letter) type

Druckübertragung *f* pressure transmission

Druckumlaufschmierung *f* pressure circulating [system of] lubrication, lubrication of the pressure circulating type

druckunabhängig pressure-independent

Druckunterschied *m* difference in pressure, pressure difference

Druckventil *n* delivery (pressure, discharge) valve

Druckventilfeder *f* delivery valve spring

Druckverfahren *n* printing process

Druckvergasung *f* gasification under pressure, high-pressure gasification

Druckverlust *m* pressure loss; loss of head

Druckvermerk *m* imprint

Druckverminderung *f* depression

Druckverschiebung *f* pressure shift *(spectral lines)*

Druckverstärker *m* booster *(piping)*

Druckversuch *m* compression (crushing) test

  D. mit behinderter Querdehnung confined compression test *(soil mechanics)*

  D. mit unbehinderter Querdehnung unconfined compression test *(soil mechanics)*

Druckverteilung *f* pressure distribution
Druckverteilungskurve *f* curve of pressure distribution
Druckvolumendiagramm *n* pressure-volume diagram, p v diagram
Druckwaage *f* pressure maintaining valve
Druckwächter *m* manostat
Druckwagen *m* printing carriage
Druckwalzenstreichmaschine *f* print-roll coater
Druckwasser *n* pressurized water
Druckwasserreaktor *m* pressurized water reactor, PWR
Druckwassertest *m (nucl)* autoclave test
Druckwelle *f* shock wave
Druckwerksplittung *f* splitting of printer *(tabulator)*
Drückwerkzeug *n* [metal] spinning tool
Druckwindkessel *m* air vessel, delivery air chamber
Druckwindreinigung *f (ag)* compressed air cleaning
Druckzone *f* compressive (compression) zone
Druckzug *m (therm)* forced draught
Druckzugschalter *m* push-pull switch
Druckzugventilator *m* forced draught blower
Druckzwiebel *f* pressure bulb *(soil mechanics)*
Druckzylinder *m* printing (impression) cylinder
Drusch *m (ag)* threshing
Druschvorrichtung *f* threshing aggregate
Druse *f (geol)* vug
drusig *(geol)* pockety
dual *(dat)* binary, dual
Dualbruch *m* binary fraction
Dualitätsprinzip *n* duality principle
Dualkode *m (dat)* binary code
Dualnetzwerk *n* dual network
Dualschaltung *f* dual circuit
Dualuntersetzer *m s.* binärer Untersetzer
Dualzahl *f* binary code
Dualzähler *m s.* binärer Untersetzer
Dualziffer *f* binary digit
Duant *m (nucl)* dee, duant
Dübel *m* dowel, nog, plug, key; connector; dowel pin *(wood)*; dook
  runder D. round pin
Dübelbohrer *m* brad-awl
Dübelmaschine *f* peg runner
dübeln to dowel
Dübelstein *m* fixing brick (block)
Dublett *n (nucl)* doublet, duplet
Dublettaufspaltung *f* doublet splitting
Dublette *f* double
Dublettzustand *m* doublet state
dublieren to double
Dubliermaschine *f* doubler, doubling winder (winding frame)
dubliert doubled
Dublierung *f* doubling
Ducht *f (ship)* thwart
Dückdalben *m* mooring post, dolphin, pile moorings
Duftstoff *m* odour compound, perfume
Duktor *m* duct[or] roller
Duktorkasten *m* duct[or] box
Duktorlineal *n* duct[or] blade
Dumper *m* dumper
Düne *f* dune
Dung *m* stable manure
Dungbahn *f* overhead manure carrier
Dungeinleger *m* skim coulter
Düngekalk *m* manuring (fertilizing) lime
Düngelanze *f* root application fertilizer dispenser
Düngemittel *n* fertilizer, manure
Düngemittelfabrik *f* fertilizer plant
düngen to fertilize, to manure
Dünger *m* fertilizer, manure
Düngerdrillhebel *m* fertilizer drill-lever
Düngerkasten *m* fertilizer hopper
Düngermühle *f* fertilizer grinder
Düngerstreuanhänger *m* trailer spreader
Düngerstreuen *n* spreading (distributing) manure
Düngerstreuer *m* fertilizer distributor, manure spreader
  schleppergezogener D. tractor-mounted fertilizer distributor
Düngerstreumaschine *f s.* Düngerstreuer
Düngerstreuwanne *f* hand fertilizer distributor
Düngesalz *n* saline manure
Düngevermögen *n* fertilizing capacity
Düngewert *m* manurial value
Düngewirkung *f* manurial effect
Dunggabel *f* dung-fork
Dunkeladaptation *f* dark adaptation
Dunkelemission *f* dark emission
Dunkelentladung *f* dark discharge
Dunkelfeld *n* dark field (ground)
Dunkelfeldbeleuchtung *f* dark-ground illumination
Dunkelfeldbeobachtung *f* dark-field observation
Dunkelfeldkondensor *m* dark-field condenser
dunkelgelb dark straw *(temper colour)*
Dunkelimpuls *m* dark pulse
Dunkelkammer *f* dark-room, darkroom
Dunkelkammerbeleuchtung *f* dark-room illumination
Dunkelleitfähigkeit *f* dark conductivity
Dunkelleitung *f* dark conduction
Dunkelnebel *m* dark nebula
Dunkelperiode *f* dark period
Dunkelpunkt *m* blackout marker *(radar)*
Dunkelraum *m* dark space
  Astonscher D. Aston dark space
  Crookesscher D. Crookes dark space
  Faradayscher D. Faraday dark space
  Hittorfscher D. Hittorf dark space

Dunkelraumkamera *f* darkroom camera
Dunkelreaktion *f* dark reaction
Dunkelrotglut *f* dull red heat
Dunkelschalter *m* dimmer switch
Dunkelschriftröhre *f* dark-trace tube
Dunkelsteuerung *f* retrace blanking
Dunkelsteuerzeichen *n* blanking (blank out) signal
Dunkelstrahlung *f* dark (obscure) radiation
Dunkelstrom *m* dark current
D. einer lichtelektrischen Zelle dark current of [a] photoelectric cell
Dunkelstromimpuls *m* dark pulse
Dunkelwiderstand *m* dark resistivity
dünn 1. thin; slender; 2. dilute *(fluid)*; 3. sheer *(fabric)*
Dünnbeil *n* chip axe
Dünndruckpapier *n* Bible (India, thin printing) paper
Dünnfilm *m* thin film
Dünnfilmkondensator *m* thin film capacitor
Dünnfilmmikroelektronik *f* thin film microelectronics
Dünnfilmschaltung *f* thin film circuit
Dünnfilmtechnik *f* thin film technique (technology)
Dünnfilmtransistor *m* thin film transistor
dünnflüssig light-bodied, thin-bodied, thin-fluid
Dünnsaft *m* clarified juice
Dünnschichtchromatografie *f* thin-layer chromatography
Dünnschichthalbleiter *m* thin-film semiconductor
Dünnschichtspeicher *m* thin film memory
magnetischer D. magnetic film memory
Dünnschichttechnik *f* thin film technique
Dünnschichtverdampfer *m* thin layer evaporator
Dünnschliff *m* thin ground (polished) section
Dünnschnitt *m* thin cut section
Dünnstoffbleiche *f* low consistence bleaching *(paper)*
Dünntrübe *f* dilute medium
dünnwandig thin-walled
Dunst *m* fume, vapour
Dünung *f* swell [waves]
Duodiode *f* duodiode, double diode
Duoplasmatron-Ionenquelle *f* *(nucl)* duoplasmatron ion source
Duoschaltung *f* *(el)* lead-lag connection
Duowalzen *fpl* duo rolls
Duowalzwerk *n* duo (two high rolling) mill
Duplexbetrieb *m* duplex operation
Duplexbremse *f* duplex brake
Duplexdruckapparat *m* duplex printing machine
Duplexfilmdruckmaschine *f* duplex screen printer
Duplexnachbildung *f* duplex artificial circuit
Duplexpappe *f* two-ply board
Duplexpumpe *f* duplex pump
Duplexradzylinder *m* duplex wheel cylinder
Duplexsystem *n* *(telec)* duplex system
D. mit Brückenschaltung bridge duplex system
D. mit Differentialschaltung differential duplex system
Duplexverfahren *n* duplex process *(steel)*
Duplikatdruckform *f* duplicate plate
Duplikatvergleich *m* duplication check
Duplopapier *n* double paper
Düppel *m* window *(radar)*
Düppelung *f* flasher dropping *(radar)*
Dupuiskessel *m* Dupuis boiler
Dural[uminium] *n* duraluminium
Duraluminiumlegierung *f* duraluminium alloy
durchbauen *s.* durchörtern
Durchbauen *n* *(mine)* back ripping, holing, rerip *(roadway)*
durchbiegen to deflect; to sag
Durchbiegung *f* deflection; sag[ging]
D. der Kolbenstange sag of [the] piston rod
D. durch eigene Schwerkraft natural sag
Durchbiegungsdiagramm *n* load-deflection diagram
Durchbiegungsfestigkeit *f* cross-breaking strength
Durchbiegungskontakt *m* *(ry)* treadle
Durchbiegungsmesser *m* deflectometer
Durchbildung *f* / **konstruktive** design
Durchblasen *n* bubbling, sparging
D. am Kolben piston-ring blow-by
Durchblasreiniger *m* cleansing blower
durchblättern to slip *(paper)*
durchbohren to through-drill
durchbrennen *(el)* to blow *(fuse)*, to fuse; to burn out *(lamp)*
durchbrochen pierced, perforated
Durchbruch *m* break-through, opening; breakout *(melting furnace)*; *(el)* disruptive discharge, breakdown, rupture
D. der Isolierung dielectric breakdown
dielektrischer D. dielectric breakdown
unvollständiger D. partial breakdown
Durchbrucharbeit *f* *(text)* openwork *(embroidery)*, cagework
Durchbruchfeldstärke *f* *(el)* breakdown field strength
Durchbruchköper *m* open-texture twill
Durchbruchskennlinie *f* *(el)* breakdown characteristic
Durchbruchspannung *f* *(el)* breakdown (Zener) voltage, potential at breakdown
Durchbruchswiderstand *m* differential Zener resistance *(semiconductor)*

durchdrehen 1. to crank *(engine)*; 2. to flow over *(dial)*
den Motor von Hand d. to crank the engine by hand
Durchdrehkontakt *m* overflow (eleventh step) contact *(telephony)*
Durchdrehmotor *m* barring motor
Durchdrehsender *m* wideband sweep transmitter
durchdringbar permeable
Durchdringbarkeit *f* permeability, penetrability
durchdringen 1. *(math)* to intersect *(technical drawing)*; 2. *(ph)* to penetrate
sich d. to interpenetrate
durchdringend hard, penetrating, high-energy *(radiation)*
Durchdringung *f* 1. *(math)* intersection *(technical drawing)*; 2. *(ph)* permeation, penetration
D. des Potentialwalls tunnelling [through the barrier], barrier penetration
D. von Strahlen penetration of rays
Durchdringungsfähigkeit *f* penetration capacity
Durchdringungskomponente *f* penetrating (hard) component *(cosmic rays)*
Durchdringungslinie *f* intersection [line] *(technical drawing)*
Durchdringungsvermögen *n* penetrating (penetration) power *(X-rays)*
Durchdringungswahrscheinlichkeit *f* [potential barrier-]penetration probability
Durchdringungswinkel *m* angle of intersection *(technical drawing)*
Durchdringungszwilling *m* interpenetration twin *(crystal)*
Durchdruck *m* screen printing
durchdrücken to force through
durchfahren 1. to pass; 2. *(mine)* to hole
Durchfahren *n* 1. passing; 2. *(mine)* holing
Durchfahrschalter *m* non-stop switch
Durchfahrt *f* passing, passage
Durchfahrtshöhe *f* **einer Brücke** deadroom (clearance) of a bridge
Durchfahrtsilo *m* pit
Durchfahrtsöffnung *f* fairway span *(bridge)*
durchfallen to pass *(sieve)*
Durchfärben *n* / **gleichmäßiges** *(text)* dyeing in equal depth
durchfedern to deflect; to spring
Durchfederung *f* [spring] deflection
durchfließen to flow through; to pass *(current)*
Durchfluß *m* flow; passage *(current)*
Durchflußanzeiger *m* flow indicator
Durchflußelektrode *f* flow-type electrode
Durchflußgeschwindigkeit *f* velocity of flow, rate of flow
Durchflußkennlinie *f* flow characteristic
Durchflußkoeffizient *m* flow coefficient
Durchflußkühlung *f* straight-through cooling
Durchflußküvette *f* flow-through type cell
Durchflußmenge *f* rate of flow, flow volume (capacity); quantity passing (passed) *(turbine)*
sekundliche D. quantity passed per second *(turbine)*
Durchflußmengenmesser *m* flowmeter
Durchflußmesser *m* **mit Pitot-Rohr** Pitot-tube flowmeter
gyrometrischer D. gyro flowmeter
induktiver D. electromagnetic flowmeter
thermischer D. thermal (boundary layer) flowmeter
Durchflußmethode *f* transpiration method
Durchflußmischer *m* flow (pipeline) mixer
Durchflußproportionalzählrohr *n* flow proportional counter
Durchflußregler *m* flow regulator
Durchflußrohr *n* fluid-carrying duct
Durchflußzahl *f* discharge coefficient *(flowmeter)*
Durchflußzähler *m* volumetric flowmeter, flow counter
Durchflußzählrohr *n* flow counter
durchforsten to log, to thin
Durchforstung *f* selective logging
Durchforstungsholz *n* thinnings
Durchfracht *f (ship)* through (transit) freight
Durchführbarkeit *f* practicability
durchführen 1. to thread; to guide through, to pass through; 2. to carry out, to perform, to execute, to accomplish
d. durch to insert through
einen Draht d. to thread a wire [through]
Durchfuhrgut *n* transit (bonded) goods
Durchführung *f (el)* [lead-through] bushing, pin, lead-in [wire]
Durchführungshülse *f* grommet
Durchführungsisolator *m (el)* insulated bushing, bushing (feed-through, wall-tube) insulator
Durchführungsklemme *f (el)* bushing clamp
Durchführungskondensator *m* duct (feed-through) capacitor
Durchführungsperle *f* beading
Durchführungsstromwandler *m* bushing-type current transformer
Durchführungstülle *f* grommet
Durchführungswandler *m* bushing transformer
Durchgang *m* 1. *(build)* thoroughfare; 2. *(mach)* pass[age]; 3. penetration, transmission *(radiation)*; 4. runaway *(reactor)*
D. / **in einem** in one pass
D. des Räumwerkzeugs broaching pass
D. des Werkstücks passage of the workpiece
D. durch das Rad passage through the wheel *(turbine)*
hoher D. high-altitude passage
niedriger D. low-altitude passage
Durchgangsamt *n (telec)* transit exchange, *(Am)* via office

Durchgangsbahnhof *m* intermediate station

Durchgangsbohrung *f* through hole (bore)

Durchgangsdienst *m* / **halbautomatischer** *(telec)* semi-automatic tandem working

Durchgangsdrehzahl *f* runaway speed

Durchgangsfernamt *n* zone centre, through trunk exchange

Durchgangsfernplatz *m* through switching position

Durchgangsfernschrank *m* through switching board

Durchgangsgeschwindigkeit *f* velocity through ports

Durchgangsgüterzug *m* through goods train

Durchgangshafen *m* port of transit, intermediate port

Durchgangshöhe *f* clearance height

Durchgangslage *f* position of passage

Durchgangsleistung *f (el)* transforming capacity, throughput *(e.g. transformer)*

Durchgangsleitung *f* tandem trunk, through (via) circuit *(telephony)*

Durchgangsmeßkopf *m* through probe

Durchgangsmuffe *f* straight joint, joint sleeve, through joint

Durchgangsplatz *m (telec)* through position

Durchgangsquerschnitt *m* section of passage

Durchgangsregelventil *n* globe body control valve

Durchgangsschleifen *n* through grinding

Durchgangsstück *n* / **aufklappbares** hinged section

Durchgangsventil *n* straight-way valve

automatisches D. unloading valve

Durchgangsverbindung *f (telec)* through connection

Durchgangsverfahren *n* through-feed process *(grinding)*

Durchgangsverkehr *m* through traffic; *(Am)* via traffic *(telephony)*

Durchgangsverlust *m* transmission loss

Durchgangswagen *m (ry)* corridor carriage

Durchgangswahl *f (telec)* tandem dialling

Durchgangswähler *m (telec)* tandem selector

Durchgangswellenmesser *m* transmission frequency meter

durchgeben 1. to transmit; 2. to filter

durchgehen 1. to pass; 2. to run away

Durchgehen *n* **der Dampfmaschine** racing of the steam engine

durchgehend throughout; continuous

durchgerostet eaten through by rusting

durchgießen to pour through, to filter

durchglühen to soak

Durchgriff *m (el)* reciprocal of amplification, reciprocal amplification factor

Durchgriffsänderung *f* variation of inverse amplification factor *(electron tube)*

Durchgriffskapazität *f* grid-to-plate capacity

Durchgriffsmeßbrücke *f* Miller bridge

Durchhang *m* sag[ging]

loser D. der Leitung slack

durchhängend slack; sagging

Durchhangsprüfung *f* sag test *(overhead line)*

Durchhangsregelung *f* regulation of wires (cables)

Durchhängung *f* loop formation *(rolling)*

durchhärten to harden throughout (fully)

Durchhärten *n s.* Durchhärtung

Durchhärtung *f* through hardening

Durchhieb *m* 1. *(mine)* cut-through, dog hole, holing[-through], cross-entry; 2. improvement cutting, trimming

durchkontrollieren to screen

Durchkreuzungszwilling *m* cruciform twin group *(crystal)*

Durchlaß *m* 1. passage; 2. *(el)* pass-band; 3. culvert

Durchlaßbandbreite *f* filter pass-band

Durchlaßbereich *m* pass-band width [of a filter], pass (acceptance) band, transmission range

Durchlaßcharakteristik *f* transmission characteristic

durchlassen to let through *(rays)*; to transmit

Wärme d. to transmit heat

Durchlaßfaktor *m* coefficient of transmission

Durchlaßfrequenzbereich *m* filter transmission band

durchlässig transparent; penetrable, permeable; porous, spongy

stark d. high-transmission *(light)*

Durchlässigkeit *f* permeability; penetrability; transmittance, transmission factor (ratio), transmissivity

D. der Potentialschwelle transparency of the potential barrier

magnetische D. permeability

Durchlässigkeitsbereich *m* transmission range

D. eines Filters filter range

Durchlässigkeitsgrad *m* transmitting factor

Durchlässigkeitskurve *f* transmission curve

spektrale D. spectral transmission curve

Durchlässigkeitsprüfung *f* penetration test

Durchlässigkeitszahl *f* coefficient of permeability

Durchlaßkanal *m* culvert

Durchlaßkennlinie *f* pass-band characteristic

Durchlaßkreis *m (el)* acceptor [circuit]

Durchlaßkurve *f s.* Durchlaßkennlinie

Durchlaßperiode *f* conducting period

Durchlaßrichtung *f* forward (conduction) direction

Durchlaßspannung *f* forward voltage

Durchlaßspitzenstrom *m* peak forward current

Durchlaßstrom *m* forward current
Durchlaßstromstoß *m* surge forward current
Durchlassung *f* transmission
 diffuse D. diffuse transmission *(light)*
 gemischte D. mixed transmission *(light)*
 gerichtete D. direct transmission *(light)*
 gestreute D. diffuse transmission *(light)*
Durchlaßwiderstand *m* forward resistance
Durchlauf *m* run; passage
Durchlaufbalken *m* continuous beam
Durchlaufbewehrung *f* continuity steel (reinforcement)
durchlaufen 1. to pass; 2. to leak
 ein Minimum d. to pass through a point
 eine Hysteresisschleife d. to go through a hysteresis loop
 rückwärts d. to retrace
durchlaufend full length; continuous
Durchlauferhitzer *m* geyser, flow heater
Durchlauffräs- und -zentriermaschine *f* continuous-type milling and centring machine
Durchlaufofen *m* continuous furnace
 D. zum Normalglühen continuous normalizing furnace
Durchlaufplatte *f* continuous slab
Durchlaufrahmen *m* continuous (multiple-span, multiple-bay) frame
Durchlaufreaktor *m* straight-through reactor
Durchlaufschmierung *f* once-through lubrication, all-loss lubrication, total-loss lubrication, non-circulating lubrication, non-recovery lubrication; gravity lubrication
Durchlaufträger *m* continuous beam (girder)
Durchlauftrockner *m* continuous vertical (grain) drier
Durchlaufwaage *f* continuous grain weigher
Durchlaufzeit *f* time of passage
Durchlaufzentrifuge *f* concurrent [flow-through] centrifuge
Durchleuchtung *f* fluoroscopy; radioscopy; X-raying; transmission of light
Durchleuchtungsgerät *n* fluoroscope, roentgenoscope
Durchlicht *n* transmitted light
Durchlichtung *f* improvement cutting *(forest)*
durchlüften to air, to ventilate
Durchmagnetisierung *f* through magnetization
Durchmesser *m* diameter; size *(bore, reamer)*
 äußerer D. outer (overall) diameter
 innerer D. inner (inside) diameter
Durchmesserbelastungsgrad *m* diameter load coefficient *(propeller)*
Durchmesserspannung *f* diametral voltage
Durchmesserteilung *f* diametral pitch
Durchmessertoleranz *f* diametrical tolerance
Durchmesserverjüngung *f* longitudinal clearance, *(Am)* longitudinal relief, back taper *(twist drill)*
Durchmesserwicklung *f (el)* diametral (full-pitch) winding
durchmischen to mix thoroughly
durchmoduliert fully modulated
Durchnähmaschine *f* sole sewing machine *(shoe)*
Durchnähzwirn *m* blake thread *(shoe)*
durchörtern *(mine)* to hole
Durchörterung *f (mine)* intersection, passage, holing [through]
Durchperlungselektrode *f* bubbling-type electrode
Durchprojektion *f (telev)* rear projection
Durchprüfung *f* check-out, *(Am)* leap-frog test *(computer)*
Durchquetschprüfung *f* cut-through test [of winding-wire insulation]
durchrauhen to burl, to nap
Durchreichefenster *n* serving (service) hatch
durchrühren to rabble
durchsacken 1. *(aero)* to pancake; 2. to sag; to drop *(a vessel)*
Durchsacken *n* **einer Tragfläche / plötzliches** *(aero)* wing dropping
Durchsackung *f (ship)* sagging
Durchsatz *m* throughput *(furnace)*
Durchsatzgeschwindigkeit *f* throughput (driving) rate
Durchsatzzeit *f* time of passage
durchschalten to connect through
Durchschalterelais *n* connecting relay, *(Am)* cut-through relay
Durchschalthebel *m* block switch
Durchschaltung *f* connecting through
Durchschaltversuch *m* test run
 nasser D. wet test run
durchscheinen to show through *(printing ink)*
durchscheinend translucent; transparent
durchschießen *(print)* to interleave
durchschiffen to sail (ship) through, to navigate, to cross
Durchschlag *m* 1. *(el)* puncture, [disruptive] break-down, rupture *(insulation)*; 2. carbon copy; 3. *(mine)* holing
 D. des Isolierstoffes puncture of the insulation
 dielektrischer D. dielectric breakdown
durchschlagbar exposed to puncture
durchschlagen 1. *(el)* to puncture, to break down, *(Am)* to spark through; 2. to perforate, to punch; 3. to come through, to strike through *(printing ink)*
Durchschläger *m* round punch
durchschlagfest *(el)* puncture-proof
durchschlägig werden to thirl
Durchschlaglage *f* position of passage *(crank)*
Durchschlagmeßtisch *m* **[für Gleichrichterröhren]** breakdown bench

Durchschlagpapier *n* carbon paper
Durchschlagsfestigkeit *f* dielectric (disruptive) strength
spezifische D. specific dielectric strength
durchschlagsicher puncture-proof *(insulator)*
Durchschlagsicherung *f* fusible cut-out, puncture cut-out [with diaphragm]
Durchschlagskraft *f* penetration
Durchschlagsprobe *f* breakdown test
Durchschlagsspannung *f (el)* disruptive (breakdown, puncture) voltage
Durchschlagsspannungsprüfung *f* flash test
Durchschlagsverzögerung *f (el)* time to puncture, impulse inertia
Durchschleifen *n* **mittels Trennscheibe** abrasive cutting
durchschleusen to lock, to tow through sluices *(vessels)*
Durchschleusung *f* passing a vessel through a lock
Durchschlupf *m* average outgoing quality
maximaler D. average outgoing quality limit
Durchschlupflinie *f* average outgoing quality curve
Durchschnitt *m* average; mean; intersection; meet
Durchschnittsausgangsleistung *f* / **größte** *(el)* maximum average output power
Durchschnittsbestimmung *f* average determination
Durchschnittsbetrag *m* average amount
Durchschnittsdrehzahl *f* average speed in rpm
Durchschnittsgeschwindigkeit *f* average speed
Durchschnittskrümmung *f (aero)* mean camber
Durchschnittslast *f* average load
Durchschnittsleistung *f (mach)* average performance; average output
Durchschnittsmenge *f (math)* meet
Durchschnittsquadrat *n* mean square
Durchschnittsstamm *m* average stem
Durchschnittsstundenlohn *m* average hourly earning
Durchschnittswert *m* average value
Durchschreibeformular *n* carbonized (carbon-backed) form
Durchschreibpapier *n* copy paper
Durchschuß *m (print)* [interlinear] space, lead
Durchschußbogen *m* interleaving sheet
Durchschußkasten *m* lead case (box)
Durchschußpapier *n* interleave (interleaving) paper
Durchschwingen *n* overshoot
durchseihen to strain, to filter
Durchsicht *f* 1. overhaul; 2. look through *(structural appearance of a sheet)*
durchsichtig transparent
d. optisch isotrop isotropic transparent
Durchsichtigkeit *f* transparency, diaphanousness, diaphaneity
Durchsichtsfrontkatode *f* transparent front cathode
durchsickern to percolate, to leak through
durchsieben to screen, to sieve
Durchsiebung *f* screening, sieving
durchspülen to flush
durchstarten to touch and go, to go-around
Durchstarthöhe *f* missed approach altitude
durchstecken durch to thread through
Durchsteckschraube *f* through bolt
Durchsteckschraubenverbindung *f* bolted assembly
Durchsteckwandler *m* single-turn current transformer, bar-type current transformer
D. ohne Primärleiter winding-type current transformer
durchsteuern to fully modulate
Durchstich *m* cut
durchstimmbar continuously variable (tunable)
Durchstimmbarkeit *f* continuous-tuning feature
Durchstimmbereich *m* tuning range
Durchstimmkurve *f* tuning curve
durchstoßen *(aero)* to come down through the overcast; to punch *(sheet metal)*
den Wasserstand d. to clean out the water-gauge passages
Durchstoßen *n* **der Wasserstandshähne** purging (blowing-through) of the water gauge cocks
Durchstoßofen *m* continuous discharge furnace
Durchstoßpunkt *m (math)* piercing point
Durchstrahlung *f* radioscopy; transirradiation; transmission
Durchstrahlungsanlage *f* radiological unit
Durchstrahlungsexperiment *n* transmission [beam] experiment
Durchstrahlungsmikroskop *n* irradiation microscope
Durchströmung *f* seepage
Durchtränken *n* **[von Hölzern] mit Kreosot** creosoting
Durchtreiber *m* round punch
durchtreten to break through *(drill)*
Durchtritt *m* **der Bohrerspitze** break-through of the drill point
Durchtrittquerschnitt *m* section of passage
Durchtunnelung *f* tunnelling *(of electrons)*
Durchtunnelungsstrom *m* tunnelling current
durchverbinden to connect through, to put through
Durchwachsung *f* [inter]penetration *(crystals)*
Durchwahl *f* through dialling *(telephony)*
durchwählen to dial through
durchwärmen 1. to warm thoroughly; 2. to soak
Durchwirbelung *f* **der Heizgase** whirling (eddying) of the heating gases

Durchzieherwicklung *f (el)* pull-through winding
Durchziehglühofen *m* continuous annealing furnace
Durchziehverfahren *n* stripping process
Durchzugschaltung *f* progressive system of gear shifting
Durchzugskraft *f* [table] pull of the planer, [table] pull
Durchzugslaufwerk *n* film pulling mechanism
Durchzündung *f (el)* arc-through
durchzwängen to force through
dürftig poor *(e.g. quality)*
Durit *m (min)* durain
Duroplast *m* thermoset[ting plastic]
duroplastisch thermosetting
Dürre *f* drought
dürreresistent xerophytic
Dürrkessel *m* Dürr boiler
Dusche *f* shower bath
Duschkabine *f* shower stall
Duschnische *f* shower stall
Duschraum *m* shower bath
Düse *f* nozzle; spout *(waveguide)*; tuyere *(furnace)*
  D. mit rechteckiger Mündung nozzle with rectangular outlet
  glatte D. nozzle with smooth walls
  große D. high-speed nozzle
  kleine D. low-speed nozzle
  mehrteilige D. nozzle consisting of several pieces *(turbine)*
  nicht nachtropfende D. anti-drip nozzle
  quadratische D. square nozzle
  ringförmige D. ring (annular) nozzle
  runde D. circular nozzle
  sich verengende D. contracting nozzle
  zylindrische D. straight cylindrical nozzle
düsenähnlich nozzlelike
Düsenantrieb *m* jet propulsion
Düsenbefestigung *f* fastening of nozzles *(turbine)*
Düsenbeiwert *m* nozzle coefficient
Düsenblock *m* nozzle block
Düsenbohrung *f* nozzle bore
Düsendecke *f* oven deck pad
Düseneinsatz *m* set of nozzles, nozzle ring segment *(turbine)*
Düsenfärbung *f* spun-dyeing
Düsenflugzeug *n* jet-propelled aircraft
Düsengarnitur *f* nozzle fittings *(turbine)*
Düsengruppe *f* group of nozzles *(turbine)*
Düsenhalter *m* nozzle holder
düsenhochdruckgeschmiert mist-lubricated
Düsenhochdruckkühlung *f* mist cooling
Düsenkammer *f* nozzle box
Düsenkanal *m* nozzle channel *(turbine)*; die channel *(extruder)*
Düsenkasten *m* nozzle box
Düsenklappe *f* nozzle flap
Düsenkörper *m* nozzle head (body)
Düsenkraftstoff *m* jet fuel
Düsenkühler *m (aero)* nozzle radiator
Düsenlackierverfahren *n* gasket coating
Düsenmündung *f* nozzle tip
Düsennadel *f* injector needle, pintle
Düsenneigung *f* nozzle inclination (angle) *(turbine)*
Düsenöffnungsdruck *m* release pressure of the nozzle
Düsenpropeller *m* shrouded propeller
Düsenquerschnitt *m* nozzle cross-section
  engster D. smallest cross-section of nozzle *(turbine)*
Düsenregelung *f* governing by cutting out nozzles *(turbine)*
Düsenring *m* nozzle ring *(turbine)*, nozzle-like shroud ring *(propeller)*
Düsensatz *m* nozzle ring segment, set of nozzles *(turbine)*
Düsenschiff *n* jet-propelled vessel
Düsenspannung *f* nozzle pressure *(turbine)*
Düsenstellung *f* position of nozzle *(turbine)*
Düsenstock *m* pen stock
Düsentrockner *m* jet drier
Düsenvergaser *m* jet carburetter
Düsenverlust *m* nozzle loss *(turbine)*
Düsenverschmutzung *f* nozzle fouling
Düsenwandung *f* wall of nozzle
Düsenwirkungsgrad *m* nozzle efficiency
Düsenzapfen *m* pintle
Düsenzerstäuber *m* nozzle sprayer (atomizer)
Düsenzunge *f* nozzle flap *(turbine)*
  federnde D. elastic nozzle flap
Düsenzwischenboden *m* nozzle diaphragm
Duvetine *m (text)* duvetyn
dwars thwartships, athwartship
  d. ab abeam [off], abreast
Dwarsbalken *m* cross-beam
Dwarsleine *f* breast rope
Dwarssee *f* beam sea
Dwarswind *m* beam (side) wind
Dwight-Lloyd-Sinterverfahren *n (met)* blast sintering process
Dyade *f* dyad
dyadisch dyadic
Dyn *n* dyne
Dynamik *f* dynamics
  D. des Kurbeltriebes dynamics of the crank mechanism
  D. plastischer Massen plasticodynamics
Dynamikbereich *m* volume range
Dynamikdehner *m* [automatic] volume expander
Dynamikentzerrer *m* dynamic expander
Dynamikregler *m* compandor
Dynamiksteigerung *f* volume expansion
dynamisch dynamic
Dynamit *n* dynamite

Dynamitpatrone *f* dynamite cartridge
Dynamo *m* dynamo
fremderregter D. separately-excited dynamo
selbsterregter D. self-excited dynamo
Dynamoanker *m* dynamo armature
Dynamoblech *n* transformer steel; sheet lamination, dynamo sheet
dynamoelektrisch dynamo-electric
Dynamometer *n* dynamometer
Dynamostahl *m* dynamo steel
Dynatron *n* dynatron
Dynatronoszillator *m* dynatron oscillator
Dynatronwirkung *f* dynatron effect
Dynode *f* dynode
Dynodensystem *n* dynode system
Dyotron *n* dyotron
Dysprosium *n* dysprosium

E

E-Bogen *m (el)* E plane bend *(waveguides)*
E-Entladung *f* E-discharge
E-Fassung *f (el)* Edison lamp-holder
E-Modul *m* modulus of elasticity
E-Schirmbild *n* E-display
E-Schweißen *n* electric welding
Eau de Javelle *n (f)* eau de Javel[le], Javel water
Ebbe *f* ebb[-tide], low-tide
E. und Flut *f* ebb and flow, flux and reflux
Ebbe- und Flutmesser *m* tide gauge
eben flat; plane; smooth
Ebenbild *n* replica
Ebene *f* 1. plane; 2. *(el)* tier [of winding]; array *(aerial)*
euklidische E. Euclidean plane
hyperbolische E. hyperbolic plane
invariante E. invariant plane
komplexe E. complex plane
schiefe E. inclined plane
unveränderliche E. invariable plane
Ebenenbündel *n* bundle (sheaf) of planes
Ebenenbüschel *n* axial pencil, pencil of planes
Ebenenschar *f* family of plane surfaces
Ebenführung *f* flat guide
Ebenheit *f* flatness; evenness; smoothness
Ebenheitsgrad *m* degree of flatness
Ebenholz *n* ebony
ebnen to level; to flatten; to smooth
Ebonit *m* ebonite, hard rubber
Ebonitgehäuse *n* ebonite case
Ebullioskopie *f* ebullioscopy
Eccles-Jordan-Schalter *m (dat)* Eccles-Jordan trigger
Eccles-Jordan-Schaltung *f (dat)* bistable multivibrator, Eccles-Jordan-circuit
Echelettegitter *n (ph)* echelette grating
Echo *n* echo
gegenläufiges E. *(telec)* talker echo
künstliches E. plume, feather *(radar)*
mitläufiges E. *(telec)* listener echo
Echoanzeige *f* blip *(radar)*
Echobild *n* echo pattern
Echodämpfung *f* active [balance] return loss
Echoempfänger *m* echo receiver
Echofalle *f* echo trap
Echograf *m* depth recorder, echograph
Echohohlraumresonator *m* echo box *(radar)*
Echoimpuls *m* reflected pulse; blip *(radar)*
Echolaufzeit *f (el)* echo transmission time *(radar)*, echo interval
Echolot *n* echo sounder (sounding device), depth sounder (finder); echo sounding
elektrisches E. electric depth finder
elektroakustisches E. sound-range altimeter
elektronisches E. fathometer
Echolotanlage *f* echo-sounding gear
Echolotmeßeinrichtung *f* echometer sounding device
Echolotung *f* reflection (echo) sounding; echo depth sounding
Echomesser *m* echo-attenuation measuring set
Echomikrofon *n* echo microphone
Echoortung *f* echo ranging
Echopaare *npl (ph)* paired echoes
Echoraum *m* reverberation chamber
Echosignal *n* blip *(radar)*
Echospannung *f* return voltage
Echosperre *f* echo suppressor
stetig arbeitende E. valve-type echo suppressor, rectifier-type echo suppressor
unstetig arbeitende E. relay-type echo suppressor
Echosperrung *f* echo elimination (cancellation)
Echostrom *m* return current
Echounterdrückung *f s.* Echosperrung
Echoweg *m* echo path
Echowelle *f* echo (reflected) wave
Echozacken *m* radar flips *(radar)*; *(ph)* echo amplitude
echt fast *(dye)*; real *(silk)*; pure *(gold)*; genuine, authentic *(signature)*; natural
Echtbeizenfarbstoff *m* fast mordant dyestuff
Echtbordeaux *n (ch)* fast milling red
Echtfarbe *f* fast colour
echtfärben to dye fast
Echtfarbstoff *m* fast dyestuff
Echtgelb *n* fast yellow
Echtheit *f* fastness *(dye)*; purity, trueness
Echtheitsanforderungen *fpl* fastness requirements *(dye)*
Echtheitsanspruch *m* fastness demanded *(dye)*
Echtheitsgrad *m* fastness rating *(dye)*
Echtheitsmesser *m* fadeometer *(dye)*
Echtheitswert *m* fastness rating *(dye)*
Echtwiderstand *m (el)* effective (true) resistance

Echtzeit *f* real time
Echtzeitbetrieb *m (dat)* real-time simulation
Echtzeitrechner *m* real-time computer
Eckaussteifung *f* corner bracing
Ecke *f (math)* vertex; corner
  E. / mit ausgerundeter round-cornered
  E. / mit umgebogener corner-up
  untere E. bottom corner
Eckenabschneidemaschine *f* corner-cutting machine
Eckenbohrmaschine *f* close-quarter drill [for use in corners]
Eckenheftklammern *fpl* corner staples
Eckenheftmaschine *f* corner stitching machine
Eckenheftung *f* corner stitching
Eckenschneidemaschine *f* corner cutting machine
Eckenschneider *m* corner cutting machine
Eckenstoßfänger *m* corner bumper
Eckfrequenz *f* edge (break, corner) frequency
Eckhaut *f (leath)* edge hide
eckig angular; cornered; square *(bracket)*
Ecknaht *f* edge fillet weld *(corner joint)*
Eckofen *m* end oven
Eckpfosten *m* angle post
Eckplatz *m (ry)* corner seat
Eckpolster *n* corner pillow *(boxing)*
Eckring *m* stiffening ring
Eckschutzschiene *f* angle iron
Eckständer *m* angle post
Eckstein *m* quoin, angle stone
Eckstiel *m* angle post
Eckstoß *m* corner joint *(welding)*
Ecküberblattung *f* **mit geradem Schnitt** square corner halving
  E. mit schrägem Schnitt bevelled corner halving
Eckventil *n* angle-body valve
Eckverbindung *f* corner joint
Eckversteifung *f* corner bracing; stiffening of the corner
Eckverzapfung *f* corner tenon jointing
Eckwasserzeichen *n* corner mark, edge watermark *(paper)*
Edeleanu-Verfahren *n (ch)* Edeleanu process
Edelerz *n* rich ore
Edelgas *n* inert (inactive, noble, rare) gas
Edelgasgleichrichter *m* rare-gas rectifier
Edelgaskonfiguration *f (nucl)* rare-gas configuration
Edelgaskristall *m* rare-gas crystal
Edelgas-Laser *m* noble gas laser
Edelgaslichtbogenschweißen *n* inert-gas arc welding
Edelgassicherung *f (el)* inert-gas arrester (safety fuse)
Edelguß *m* high-strength cast iron
Edelholz *n* luxury wood
Edelkoks *m* refined coke
Edelmetall *n* precious (noble) metal
Edelmetallbarren *m* bullion
Edelmetalleiter *m* precious metal conductor
Edelmetallschmelzofen *m* bullion furnace
Edelrost *m* patina
Edelstahl *m* high-quality steel, high-grade steel
Edelstein *m* precious stone, gem [stone]
Edelsteinimitation *f* 1. making artificial gems; 2. artificial gem (jewel)
Edelsteinlager *n* jewel bearing
Edelzellstoff *m* refined pulp
Edison-Akkumulator *m* Edison cell
Edison-Effekt *m* Edison effect
Edison-Fassung *f* screwed-type socket, Edison screw holder
Edison-Gewinde *n* Edison screw
Edison-Sockel *m* Edison screw holder (cap)
Efeulinie *f (math)* cissoid
Effekt *m* effect
  akustoelektrischer E. acoustoelectric effect
  äußerer E. extrinsic effect
  äußerer lichtelektrischer E. photoemissive (external photoelectric) effect
  elektrokalorischer E. electrocaloric effect
  fotodielektrischer E. photodielectric effect
  fotoelektrischer E. photoeffect, photoelectric effect
  glühelektrischer E. thermionic emission
  gyromagnetischer E. gyromagnetic effect
  innerer lichtelektrischer E. photoconductive effect
  kalorischer E. calorific (heating) effect
  lichtelektrischer E. *(nucl)* photoeffect, photoelectric effect
  piezoelektrischer E. piezoelectric effect
  reziproker piezoelektrischer E. converse piezoelectric effect
  stroboskopischer E. stroboscopic effect
  thermoelektrischer E. thermoelectric (Seebeck) effect
Effektbeleuchtung *f* fancy lighting
Effektbogenlampe *f* flaming arc lamp
Effektgarn *n* effect (fancy) yarn
effektiv efficient, effective; useful
Effektivität *f* efficiency, effectiveness
Effektivleistung *f* actual power
Effektivmasse *f* **des Elektrons** effective electron mass
Effektivspannung *f* 1. *(el)* virtual (effective, r.m.s.) voltage; 2. true stress *(statics)*
Effektivstrommesser *m* current square meter
Effektivwert *m (el)* virtual (root mean square) value
  E. der Spannung *(el)* voltage-effective; r.m.s. voltage
Effektkohlenbogenlampe *f* flame cored carbon lamp

Effektlichtbogen *m* flaming arc
Effektschermaschine *f (text)* fancy cutting machine
Effektzwirn *m* effect (fancy) twist
Effektzwirnen *n* fancy twisting
Effektzwirnmaschine *f* fancy doubling frame
Effusiometer *n* effusiometer
Effusion *f* effusion
Egalfärbung *f (text)* level (even) dyeing
Egalisieren *n (text)* level dyeing
egalisierend / gut *(text)* ready levelling
Egalisierrakel *f (text)* levelling doctor
Egge *f* harrow
schwere E. drag harrow
eggen to harrow
Eggenbalken *m* drag beam, harrow bar
Eggenrahmen *m* harrow frame
Eggenschlitten *m* harrow sledge
Eggenzahn *m* spike, tine
Eggenzinke *f* spike, tine
Egoutteur *m* dandy roll
EGR-Anlage *f* electrical precipitator *(gas cleaner)*
egrenieren to gin
Egrenieren *n* ginning
Egreniermaschine *f* gin
EH-Verfahren *n* firecracker welding [process]
Eibenholz *n* yew [wood]
Eichblock *m* calibration block
Eiche *f* oak
Eichelektrode *f (ac)* electrostatic actuator
Eichelröhre *f* acorn valve
eichen 1. to calibrate; 2. to gauge, to measure *(a vessel)*
Eichenholz *n* oak [wood]
Eichenlohe *f* oak bark tan
Eichenlohgerbung *f* oak bark tannage
Eichenrinde *f* oak bark
Eichfrequenz *f* calibration (standard) frequency
Eichgeber *m* standard signal generator
Eichgenauigkeit *f* calibration accuracy, precision of calibration
Eichgenerator *m* fixed range mark generator
Eichinvarianz *f* gauge invariance
Eichkondensator *m* calibration capacitor
Eichkraftstoff *m* standard (reference) fuel
Eichkreis *m* calibration circuit
Eichkurve *f* calibrating plot, calibration curve
Eichleitung *f* attenuator, standard [transmission] line, calibrating line
regelbare E. variable attenuator
Eichmarke *f* calibration mark
Eichmarkengenerator *m* timing-wave generator
Eichmarkenoszillator *m* ringing oscillator
Eichmaß *n* calibration standard, gauge
Eichmikrofon *n* standard microphone (transmitter)
Eichnormal *n* etalon; standard gauge
Eichpräparat *n* reference source; standard; etalon; calibrating preparation
Eichsatz *m* standard gauge block
Eichschein *m* certificate of measurement *(vessel)*
Eichsender *m* standard frequency generator
Eichspannung *f* calibration voltage
Eichstempel *m (mine)* calibrated prop
Eichstrich *m* graduation, calibration (gauge) mark
Eichteiler *m* calibrating attenuator
Eichton *m* reference tone
Eichtransformation *f* gauge transformation
Eichung *f* 1. calibration; 2. measurement *(ship)*
E. der Zeitbasis *(el)* time base calibration
elektrostatische E. electrostatic calibration
Eichverfahren *n* calibration procedure
Eichwiderstand *m* 1. calibration resistor; 2. standard resistance
Eidechse *f* electric truck
Eierbrikett *n* egg coal
Eierhandgranate *f* egg[-shaped] hand grenade
Eierisolator *m* globe strain insulator, egg[-type] insulator
Eiersortiermaschine *f* egg grader
Eigelb *n* yolk
Eigenabfall *m* self decay
Eigenabsorption *f* self-absorption, intrinsic absorption
Eigenantrieb *m* self-propulsion *(of a vessel)*
Eigenbedarfsleistung *f* generating station auxiliary power
eigenbelüftet self-cooled *(machine)*
Eigenbewegung *f (ph)* proper motion
wahre E. proper motion *(radar)*
Eigendichte *f* inherent (natural) density
Eigendiffusion *f* self-diffusion
Eigendrehimpuls *m* spin
Eigendrehung *f* autorotation, autogyration
Eigenenergie *f* self-energy, intrinsic energy
eigenerregt self-excited, self-oscillatory
Eigenerregung *f* self-excitation
Eigenfehler *m* inherent ambiguity *(of an instrument)*
Eigenfeld *n* proper field, self-field
Eigenfestigkeit *f* inherent stability
Eigenfilter *n* inherent filter
Eigenfilterung *f* inherent filtration, self-filtering; self-shielding
Eigenfotoemission *f* intrinsic photoemission
Eigenfotoleitung *f* intrinsic photoconduction
Eigenfrequenz *f* natural frequency
E. des Sperrkreises antiresonance frequency
Eigenfunktion *f* eigenfunction
gerade E. even eigenfunction
ungerade E. odd eigenfunction
Eigengenauigkeit *f* intrinsic accuracy
Eigengeräusch *n* self-noise, inherent noise

Eigengeschwindigkeit *f* true airspeed *(aircraft)*; proper speed *(ship)*
Eigengewicht *n s.* Eigenmasse
Eigengewichtsabbremsung *f (ry)* braking of the tare
Eigengröße *f* proper variable
Eigenhalbleiter *m* intrinsic semiconductor
Eigenheim *n* owner-occupied home
Eigenheit *f* property, peculiarity, characteristic
Eigenimpedanz *f (el)* self-impedance
Eigenionisation *f* autoionization
Eigenkapazität *f* natural capacity, self-capacitance *(aerial)*
Eigenkühlung *f* natural cooling *(electron tubes)*
Eigenladung *f* self-charge; space charge
Eigenlast *f* dead load, self-load
Eigenleiter *m* intrinsic conductor
Eigenleitfähigkeit *f* intrinsic conductivity
Eigenleitung *f* intrinsic conduction (conductivity) *(semiconductor)*
Eigenmasse *f* dead weight; ship's (light) weight; permanent weight; tare [weight]
Eigenmodulation *f* self-modulation
Eigenpeilung *f* self-bearing, homing, direction finding on board, pinpointing *(radar)*
  gerichtete E. directional homing
Eigenrauschen *n* inherent (basic, granular, internal, background) noise, noise background
Eigenreibung *f* **des Reglers** governor friction
Eigenresonanz *f* natural resonance
Eigenschaft *f* property; condition
  elektrooptische E. electro-optical property
  ferroelektrische E. ferroelectric property
  physikalische E. physical property
  wasserabweisende E. water repellency
Eigenschaften *fpl* / **durch Störstellenleitung bedingte** *(el)* extrinsic properties
Eigenschwingung *f* natural vibration
Eigenschwingungsdauer *f* characteristic time, natural period of vibration
Eigenschwingungszahl *f* number of natural oscillations
eigensicher intrinsically safe
Eigensicherheit *f* intrinsic safety
Eigenspannbeton *m* self-stressing concrete, self-stressed concrete
Eigenspannung *f* 1. self-contained stress; 2. *(el)* natural voltage
Eigenspektrum *n* fundamental mode spectrum
Eigenstabilisierung *f* self-stabilization
Eigenstabilität *f* inherent stability
Eigensteuerung *f* automatic control
Eigenstrahlung *f* characteristic radiation; self-radiation, natural radiation, fluorescent radiation
Eigenstreuung *f* self-scattering
Eigentum *n* ownership; property
Eigentumsbahn *f* owning administration
Eigentumsmerkmal *n* owner's (ownership) mark
Eigentumsrecht *n* ownership
Eigenüberlagerungsempfang *m* autodyne reception
Eigenvektor *m* proper vector, eigen-vector
Eigenverbrauch *m* internal consumption
  E. des Zählers meter losses
Eigenvergrößerung *f (opt)* factorial magnification
Eigenverzerrungsgrad *m* degree of inherent distortion
Eigenviskosität *f* internal viscosity
Eigenwellenlänge *f* natural wavelength
Eigenwert *m* proper value; eigenvalue; mode
Eigenwiderstand *m (el)* inherent resistance, resistivity
Eigenzeit *f (el)* opening-time, proper time
Eigenzeitkonstante *f (el)* residual time constant
Eigenzustand *m* eigenstate
Eignungsprüfung *f* aptitude (ability) test
  technische E. technical aptitude test
Eiisolator *m (el)* egg (globe strain) insulator
Eilbewegung *f* rapid traverse (motion)
Eilfracht *f* express (fast) freight
Eilgang *m* rapid (quick, fast) traverse
Eilgangbewegung *f* rapid traversing motion
Eilganggeschwindigkeit *f* rapid traverse rate
Eilgangschaltung *f* rapid traverse controls
Eilgut *n* fast (accelerated) goods traffic, *(Am)* fast freight traffic
Eilgüterzug *m* express parcels train, *(Am)* express
Eilgutkarte *f* waybill, invoice
Eilrohrpost *f* rapid transfer rabbit
Eilrücklauf *m* rapid return stroke, rapid idle movement, quick return *(machine)*
Eilrücklaufgetriebe *n* quick-return mechanism
Eilschiff *n* fast steamer
Eilvorlauf *m* rapid approach (advance) *(machine)*
Eilzug *m* express train
Eimer *m* bucket
Eimerbagger *m* 1. bucket dredger; 2. bucket excavator
eimerförmig bucket-shaped
Eimerkette *f* bucket chain, chain of buckets
Eimerkettenbagger *m* 1. bucket-ladder dredger *(under water)*; 2. bucket-ladder excavator
Eimerkettenförderer *m* bucket conveyor
Eimerkettengrabenbagger *m* ladder-type trenching machine
Eimerkettennaßbagger *m* bucket dredger
Eimerkettenschwimmbagger *m* bucket-ladder dredger
Eimerkettentrockenbagger *m* bucket-ladder excavator
Eimerleiter *m* bucket-ladder
Eimermelkgerät *n* portable bucket machine
Eimerspritze *f* stirrup spray

Eimerwerk *n* bucket conveyor
Einabteilungsschiff *n* one-compartment ship
einachsig *(math)* uniaxial, unidirectional; single-axle *(vehicle)*
Einachsschlepper *m* garden tractor
Einachstraktor *m* garden (walking) tractor
Einachswagen *m* / **schleppergezogener** two-wheeled tractor trailer
Einadreßbefehl *m (dat)* one-address instruction
Einadreßsystem *n (dat)* single-address system
einadrig single-core, containing one core
Einankerfrequenzwandler *m* single-unit frequency converter
Einankerumformer *m (el.m.)* synchronous (single-armature, rotary) converter; *(Am)* genemotor
  asynchroner E. *(el)* binary converter
einarbeiten 1. to sink, to recess *(machining)*; 2. to learn; to train
  Keilnuten e. to keyseat
Einarbeitungszuschlag *m* learner allowance
einatomig mon[o]atomic
Einatomigkeit *f* monoatomicity
einäugig monocular, one-eyed
Ein-Aus-Schalter *m* single-throw circuit-breaker
Ein-Aus-Tastung *f (telec)* on-off keying
Ein-Aus-Zeit *f (el)* on-off period
Einbackenbremse *f* single-block brake
Einbadätzung *f (print)* single-solution etching, single-bath etching, one-bath etching
Einbadfärbung *f* one-bath dyeing
Einbadgerbung *f* one-bath process
Einbahnstraße *f* one-way road (street)
Einbahnverkehr *m* one-way traffic
Einband *m* cover *(book)*
  flexibler E. flexible (limp) cover
Einbandgewebe *n* cloth, bookbinding fabric
einbasisch monobasic
Einbau *m* mounting, assembly, installation; inserting; placing, placement *(concrete)*
Einbauantenne *f* built-in aerial
Einbauchung *f* inward bulging
einbauen to install, to insert, to build in, to fit; to place *(concrete)*
  einen Widerstand e. to insert a resistor
Einbauinstrument *n* panel instrument
Einbaum *m* dug-out [canoe], log canoe
Einbaumikroskop *n* installation microscope
Einbaumotor *m* built-in motor
Einbaumotorboot *n* inside motorboat
Einbaurefraktometer *n* pan refractometer
Einbaureibahle *f* block-type reamer
Einbauschalter *m* recessed (built-in, flush-type) switch
Einbauschicht *f* lift *(concrete)*
Einbauschrank *m* fitted cupboard
Einbausicherungshalter *m* panel mounting fuseholder
einbetonieren to concrete in, to encase (embed) in concrete
einbetten to embed, to merge, to introduce
einbettig single-bed *(machine)*
Einbettkammer *f* single[-berth] cabin
Einbettung *f* [em]bedding
Einbettungsmasse *f* embedding material
Einbeulung *f* inward bulging
einbinden 1. *(build)* to bond in, to fix in; 2. *(text)* to enmesh *(knitting)*
einblatten to adze
Einblatten *n s.* Einblattung
Einblattruder *n* single-plate rudder
Einblattung *f* joining by rabbets (scarfs)
einblenden *(el)* to fade in
Einblenden *n (telev)* cross-fading
einbooten to embark
Einbootung *f* embarkation
Einbrand *m* penetration *(welding)*
Einbrandtiefe *f* depth of penetration
Einbrandzone *f* penetration zone
einbrennen to stove, to bake *(varnish)*; *(leath)* to hot-stuff
Einbrennen *n* baking *(varnish)*; burning in; poisoning *(electronics)*
Einbrennfarbe *f* ceramic colour
Einbrennlack *m* baking enamel (varnish)
Einbrennmaschine *f (text)* crabbing machine
Einbrennofen *m* burning-in stove; baking oven
Einbrenntemperatur *f* baking temperature
einbringen to place *(concrete)*; to charge *(furnace, shot hole)*; to fix *(reinforcement)*
  Besatz e. *(mine)* to tamp
  Beton e. to concrete, to place concrete
Einbringen *n* placement, placing *(concrete)*; [steel] fixing *(reinforcement)*; charging *(furnace, blasthole)*
Einbruch *m (mine)* 1. holing *(coal)*; 2. cut *(stone)*
  stufenförmiger E. bench-cut, stope cut
Einbruchbohrmaschine *f (mine)* sumper drill
Einbruchsalarmanlage *f* burglar alarm
Einbruchsmeldevorrichtung *f* / **fotoelektrische** photo-electric intrusion detector
Einbruchschuß *m (mine)* centre (cut) shot, sumper
einbühnen *(mine)* to hitch
eindämmen to stem; to dyke; *(mine)* to seal off
Eindampfanlage *f* evaporating plant
eindampfen to evaporate [to dryness]; to boil down; to concentrate
Eindampfer *m* evaporator
Eindampfung *f* evaporation
eindecken to roof, to tile
  neu e. to retile, to rethatch
  weich e. to thatch
Eindecker *m* 1. single-decker *(ship)*; 2. monoplane

freitragender E. cantilever monoplane
Eindeckersieb *n* single-deck screen
Eindeckschiff *n* single-deck vessel
Eindeckung *f* / **weiche** thatching
Eindeckungsmaterial *n* roofing materials, roof covers
eindeutig *(math)* unique; unambiguous; univocal
Eindeutigkeit *f (math)* uniqueness
Eindickanlage *f* concentration plant *(papermaking)*
Eindickbütte *f* thickening chest
eindicken to thicken, to concentrate; to evaporate, to boil down; to body *(oil)*
Eindicken *n* thickening, concentration; evaporation; bodying *(oil)*
Eindicker *m* thickener
Eindickung *f* thickening, concentration; bodying *(oil)*
Eindickzyklon *m* cyclone thickener
eindocken to [dry-]dock, to float a vessel in, to come into dock *(a vessel)*
Eindockung *f* [dry-]docking, floating in *(of a vessel)*
Eindrahtlampe *f* single-filament bulb
Eindrahtleitung *f* single-wire circuit
Eindrahtwellenleiter *m* single-wire transmission line
Eindrehung *f* **des Scheibenrandes** groove under the rim of the disk *(turbine)*
eindringen to ingress *(fluid)*; to intrude, to penetrate; to indent *(hardness)*
Eindringen *n* entry; ingress; penetration
Eindringgerät *n* penetrometer
Eindringhärte *f* indentation hardness
Eindringtiefe *f* penetration depth, depth of penetration
E. des Stromes penetration of current
Eindringvermögen *n* penetrating power
Eindruck *m* impression
eindrücken in to press (force) into
Eindrücken *n (mine)* spiring
E. der Rohre bulging-in of the tubes
Eindruckfläche *f* area of indentation *(hardness test)*
Eindruckhärte *f* indentation hardness
Eindruckkalotte *f* indentation cup
Eindruckreaktor *m (nucl)* one-cycle reactor
Eindruckschmierung *f* one-shot lubrication
Eindrucktiefe *f* indentation *(Brinell)*
Eindrucktiefenmesser *m* penetrometer
Eindruckzentralschmierung *f* one-shot lubrication
eindunsten to stew down
eindüsig single-jet
einebnen *(build)* to level
eineindeutig *(math)* one[-to]-one; biunique
Einelektronenbindung *f* one-electron bond
Einenderkessel *m* single-ended boiler
Einenderschiffskessel *m* single-ended Scotch marine boiler
einengen to reduce; to neck down *(beam)*; *(ch)* to boil down, to concentrate, to evaporate
Einengung *f (math)* striction
Einer *m* single [scull], racing skiff; *(Am)* single shell
Einer-Auslegerboot *n* single outrigger shell
Einerkanadier *m* canoe C-1
Einerziffer *f* digit in the units place
Einetagenofen *m* one-storey furnace
einfach simple; single; plain
Einfachbeschleuniger *m s.* Linearbeschleuniger
Einfachbeschleunigung *f* single acceleration
Einfachbetrieb *m (el)* simplex operation
Einfachbiegeversuch *m* single-bend test
einfachbrechend unirefringent
Einfachdampfraumkessel *m* boiler with single-steam space
Einfachdiode *f* single diode
Einfachecho *n* single echo
Einfacherdschluß *m (el)* single earth
Einfachform *f* single-cavity mould
Einfachfräsmaschine *f* plain milling machine
Einfachgenerator *m (telec)* neutral current generator
Einfachhechelfeldstrecke *f (text)* gill box
Einfachimpuls *m* single pulse
Einfachkehlnaht *f* single-fillet weld
Einfachkernrohr *n* single-tube core-barrel
Einfachkessel *m* single boiler
Einfachkette *f* 1. single[-strand] chain; 2. *(text)* single warp
Einfachkoinzidenzspektrometer *n* single-coincidence spectrometer
Einfachkontakt *m* single contact
Einfachleitung *f* simplex circuit
Einfachlinienschreiber *m* one-pen recorder
Einfachmeißelhalter *m* single-tool post (holder)
Einfachplan *m* single-sampling plan, single scheme
Einfachprozeß *m* single-action procedure
Einfachpunktschweißgerät *n* single-spot welder
Einfachriemen *m* single-ply belt
Einfachrollenkette *f* single-strand roller chain
Einfachrollenkettentrieb *m* single-strand roller-chain drive
Einfachschraubenschlüssel *m* single open-end wrench
Einfachsignal *n* single-component signal
Einfachstichprobenverfahren *n* single sampling
Einfachstreuung *f* single scattering
Einfachstrombetrieb *m (telec)* transmission by simplex current, single-current working
Einfachstromübertragungsverfahren *n (telec)* single-current transmission, *(Am)* inverse neutral direct current system

Einfachwendel *f* single-coil filament
Einfachwendellampe *f* single-coil lamp
einfachwirkend single-acting
Einfachzeichen *n* single-component signal
einfädeln to thread
Einfädelung *f* / **selbsttätige** automatic threading
Einfadenaufhängung *f* unifilar suspension
Einfadenelektrometer *n* unifilar electrometer
Einfadenprojektionselektrometer *n* unifilar projection electrometer
einfahren 1. *(mine)* to go down; 2. to raise, to retract *(landing gear)*; 3. to move in *(machine table)*; 4. to insert *(rods)*; 5. to position; 6. to run in *(car)*
Einfahren *n* **der Lokomotive** initial operation of locomotive
  E. des Fahrgestells *(aero)* retraction (raising) of the landing gear
  E. des Tisches moving in of the table *(machine)*
  E. in festgelegte Position automatic positioning
Einfahrsignal *n (ry)* home (square) signal
Einfahrt *f* 1. *(mine)* descent; 2. *(ship)* inlet, access canal; entrance; gateway
Einfahrtshafen *m* port of entry
Einfall *m* incidence
  axialer E. axial incidence
  schräger E. oblique incidence
  senkrechter E. normal incidence
  tangentialer E. tangential incidence
einfallen to collapse
Einfallen *n* 1. *(mine)* dip, gradient, pitch; 2. incidence *(rays)*
  geringes E. *(geol)* flat
einfallend incident *(rays)*
Einfallende *f (geol)* dip
Einfallsdosis *f (nucl)* incident (field) dose
Einfallsebene *f* incidence plane, plane of incidence
Einfallslot *n* normal at the point of incidence
Einfallsrichtung *f* incident direction
Einfallstelle *f (plast)* sunk spot, sink mark
Einfallstrahl *m* incident ray
Einfallswelle *f* incident wave
Einfallswinkel *m* incidence angle, angle of incidence *(light)*
Einfalzung *f (wood)* joining by rabbets (scarfs); rabbet joint
Einfang *m* capture *(neutrons)*; trapping; absorption
  nichtproduktiver E. nonproductive capture
  parasitärer E. parasitic capture
  radiogoner E. *s.* Strahlungseinfang
  spaltungsloser E. non-fission capture
  strahlender E. *s.* Strahlungseinfang
einfangen to collect *(electrons)*; to bracket, to align
Einfang-Gamma-Quanten *npl (nucl)* capture gamma radiation, capture gamma-rays
Einfangkonstante *f* capture constant
Einfangquerschnitt *m (nucl)* capture cross section
Einfangreaktion *f (nucl)* capture reaction
Einfangwahrscheinlichkeit *f (nucl)* capture probability
Einfangweglänge *f (nucl)* capture mean free path, mean free path for capture
Einfangwirkungsquerschnitt *m s.* Einfangquerschnitt
Einfangzeit *f* trapping time
einfärben to ink; to stain
Einfärben *n* **von Präparaten** staining of specimens *(microscopy)*
Einfarbenübertragung *f (telev)* black and white transmission
einfarbig single-coloured; monochromatic; unicolour
einfassen to edge, to border; to line; to enclose with; to mount, to set *(precious stone)*
Einfeldbrücke *f* simple (single-span) bridge
Einfeldrahmen *m (build)* single-bay frame
Einfeldträger *m* simple (simply support) beam
einfetten to grease, to lubricate
Einflammrohrkessel *m* Cornish boiler
Einflankenmodulation *f* single-edge modulation
einfließen to flow in
Einfließen *n* inflow
einfluchten to align, to run together; to flush; to range out
Einflugbake *f* approach beacon
Einflugerlaubnis *f* permit for entry by air
Einfluggenehmigung *f* traffic control clearance
Einflugpunkt *m* point of entry
Einflugzeichen *n* marker
Einflugzeichenbake *f* approach marker beacon
Einflugzeichensender *m* marker (landing beam) beacon
Einfluß *m* 1. influence; 2. [neutron] importance
Einflußfaktor *m* / **subjektiver** human error factor
Einflußfeld *n* influence surface
Einflußfläche *f* influence surface
Einflußfunktion *f* importance function
Einflußlinie *f* influence line
Einflußzahl *f* influence coefficient
einfonturig single-unit
Einformfräser *m* single-formed cutter
Einfrequenzempfänger *m* selective receiver
Einfrequenzsignal *n (telec)* simple signal
Einfriedung *f* fencing, boundary fence (wall); *(ag)* enclosure
Einfriedungsmauer *f* boundary wall
einfrieren to freeze in
einfügen to assemble *(programming of computers)*; to insert; to place between; to fit in, to join
Einfügen *n s.* Einfügung

Einfügung *f* joining by rabbets (scarfs); rabbet joint
Einfügungsdämpfung *f (el)* insertion loss (attenuation)
Einfügungsgewinn *m* insertion gain
Einfügungsverlust *m* insertion loss
Einfügungsverstärkung *f* insertion gain
Einfuhr *f* import
Einfuhrbeschränkung *f* quota restriction on the import
einführen 1. to import *(trade)*; 2. to introduce; to enter, to insert
Einfuhrkosten *pl* inward charges
Einführstab *m* shed stick
Einführtuch *n (text)* feeding apron
Einführung *f* introduction; insertion
Einführungsisolator *m* bushing (lead-in) insulator, insulated bushing
Einführungslehrgang *m* instructional course
Einführungspfeife *f (el)* porcelain (inlet, lead-in) funnel
Einführungsstutzen *m* inlet nipple
Einführungstülle *f* bushing nipple
Einführungsvorlesung *f* introductory lecture
Einfuhrzoll *m* import duty
einfüllen to fill in, to pour in; to feed; to charge *(furnace)*
Einfüllkappe *f* filler cap
Einfüllöffnung *f* filler inlet (hole)
Einfüllstutzen *m* filler neck
Einfülltrichter *m* feed hopper; filling funnel
Einfüllverschluß *m* filler cap
Einfurchenpflug *m* single-furrow plough
Eingabe *f* 1. input *(data)*; input unit; entry; 2. feeding, filling in
Eingabegerät *n* input equipment
Eingabeprogramm *n* input routine *(computer)*
Eingabespeicher *m* input memory (storage, store) *(computer)*
Eingabewerk *n* input device (element) *(computer)*
Eingang *m* 1. *(el)* input end; 2. entrance; 3. entry, receipt
eingängig single-start, single-threaded *(thread, worm)*
Eingangsadmittanz *f* input admittance
Eingangsbuchse *f (el)* entry hub
Eingangsbuchstabe *m* input letter
Eingangsdampfer *m* import steamer
Eingangsdämpfung *f (el)* input damping
Eingangsdrehzahl *f* input speed
Eingangsgröße *f* input quantity
Eingangshalle *f* porch, vestibule, lobby
Eingangsimpedanz *f* input impedance
Eingangsinformation *f (dat)* input [signal]
Eingangskapazität *f* input capacitance *(of an electronic valve)*
Eingangsklemme *f (el)* input terminal
Eingangskondensator *m* input capacitor
Eingangskontrolle *f* incoming-material control, receiving inspection
Eingangskreis *m (el)* input circuit
Eingangsleistung *f (el)* power input, input power
 E. des Belastungskreises load-circuit power input
Eingangsleitung *f (el)* incoming circuit
Eingangsleitwert *m (el)* input admittance
Eingangsmoment *n* input torque
Eingangsprüfung *f* initial check
Eingangsrauschen *n* noise input
Eingangsresonator *m* input resonator
Eingangsschaltung *f* input connection
Eingangsscheinleitwert *m* driving-point admittance
Eingangsseite *f (el)* input side (end)
Eingangssiebschaltung *f* capacitor input filter
Eingangssignal *n* input [signal]
Eingangsspannung *f* input voltage; primary voltage
Eingangsspeicher *m (dat)* input block
Eingangsstromkreis *m* input circuit
Eingangsstufe *f* input stage
Eingangsübertrager *m* input transformer
Eingangsverstärker *m* input amplifier
Eingangswähler *m* incoming selector
Eingangswiderstand *m (el)* input resistance (impedance)
Eingangswirkleitwert *m* input conductance
Eingangszeitkonstante *f* input time constant
eingebaut built-in, inbuilt
Eingebeeinrichtung *f* **für Werkstücke** feeding device
eingeben to enter *(information)*; to feed *(information; workpieces)*
eingebettet in embedded in
 e. zwischen interbedded between
eingeeist ice-bound *(vessel)*
eingefroren frozen
eingegliedert on-line
 nicht e. off-line
eingegossen cast-in
eingehen to shrink
 eine Verbindung e. mit to form a compound with
eingelassen sunk; flush; recessed
eingelegt inlaid
eingepaßt fitted-in
eingepreßt moulded-in
eingerastet engaged
eingereicht filed *(patent)*
eingerichtet adapted for, furnished for
eingerissen ragged
eingerückt thrown in, engaged
eingeschlossen confined *(liquid)*; enclosed
eingeschossig single-storey

eingeschrieben registered *(letter)*; *(math)* inscribed
eingespannt restrained, fixed; clamped
teilweise e. partially restrained
voll e. fully restrained
eingesprengt *(geol)* disseminated
eingestellt adjusted
vorher e. preadjusted, pre-set
eingetragen filed *(patent)*
eingewebt loom-figured
eingießen to pour in
die Schaufel e. to cast the blade *(turbine)*
eingipflig *(math)* unimodal
eingleisig single-tracked, single-track
eingliedrig single-term
eingraben to dig in
eingravieren to engrave [on]
eingreifen to engage; to mate *(gear)*; to force *(programming)*
eingrenzen to limit; to localize
eine Störung e. to locate a fault
Eingrenzung *f* **von Fehlerquellen** localization of error sources
Eingriff *m* engagement, mesh
E. sein mit / im to engage (mesh) with
Eingriffbedienung *f* single-lever operation
Eingriffsbogen *m* arc of action *(gear)*
Eingriffsdauer *f s.* Überdeckungsgrad
Eingriffsflankenspiel *n* backlash determined in the direction of the line of action *(gear)*
Eingriffslinie *f* line of action *(gear)*
Eingriffspunkt *m* point of contact *(gear)*
Eingriffsstelle *f* meshing point, point of gear intermesh, point of [tooth] engagement
Eingriffsstrecke *f* length of action *(gear)*
Eingriffsteilungsmeßgerät *n* base-pitch tester
Eingriffswinkel *m* angle of approach *(miller)*
Eingriffswinkel α_0 *m* pressure angle at pitch point
Eingriffszeit *f* period of contact
Eingruppentheorie *f (nucl)* one-group theory
Einguß *m* down-gate, ingate *(founding)*
angeschnittener E. ingate with side channels
Eingußabschneider *m* sprue cutter *(founding)*
Eingußkasten *m* feed box
Eingußloch *n* grouting hole *(turbine)*
Eingußrohr *n* central gate *(founding)*
Eingußsumpf *m* runner basin
Eingußtrichter *m* sprue *(founding)*; gate of mould *(founding)*; trumpet
Eingußverschluß *m* filler cap
Eingußwalzenstreichmaschine *f* gate-roll coater *(paper)*
einhaken to catch; to hook in; to bite, to dig in *(tool)*
Einhaken *n* **des Bohrers** digging-in [of the drill], hooking into the work
einhalten to hold, to maintain *(tolerance)*; to observe *(rule)*
enge Toleranzen e. to hold close tolerances
Einhängedraht *m* slinging wire
Einhängegestell *n* plating rack
einhängen 1. to hang, to place on the hinges; 2. to abandon a call *(telephony)*; 3. *(mine)* to lower *(shaft winding)*
Einhängeträger *m* suspended beam
Einhebel *m* single lever, mono-lever
Einhebelbedienung *f* single-lever control (operation)
Einhebelbetätigung *f* single-lever operation (actuation)
Einhebeleinschaltung *f* single-lever engagement
Einhebelschalter *m* single-lever switch
Einhebelschaltung *f* single-lever control, mono-lever control
E. / mit controlled by (through) a single lever
Einhebelsteuerung *f* single-lever control
Einhebelverriegelung *f* single-lever locking
Einhebelwahl *f* single-crank (lever) selection
Einheit *f (ph)* unit; item
E. des elektrischen Verschiebungsstromes unit of electric displacement current
abgeleitete E. derived unit
astronomische E. astronomical unit
atomphysikalische E. atomic unit
elektromagnetische E. electromagnetic unit
elektrostatische E. electrostatic unit
fehlerhafte E. defective [unit]
fotometrische E. photometric unit
imaginäre E. imaginary unit
magnetostatische E. magnetostatic unit
mechanische E. mechanical unit
natürliche E. *(ph)* natural unit
physikalische E. physical unit
radioaktive E. unit of radioactivity
spektroskopische E. spectroscopic unit
Einheitenkode *m (el)* equal-length multi-unit code
Einheitensystem *n* system of units
einheitlich standard; standardized; uniform[ly]
Einheitlichkeit *f* uniformity
Einheitsbaustein *m* standard modular unit
Einheitsbohrung *f* basic hole *(fit)*
Einheitsbrennstoffverbrauch *m* specific propellant consumption
Einheitsgröße *f* standard size
Einheitsgüterwagen *m* standard-type wagon
Einheitshobel *m (mine)* slow (standard) plough
Einheitsklasse *f* one-design class *(yacht)*
Einheitsladung *f (el)* unit charge
Einheitsmatrix *f* identity matrix
Einheitspol *m* unit [magnetic] pole
magnetischer E. unit magnetic pole
Einheitsquelle *f* unit source
Einheitsröhre *f* unit tube

Einheitsröhrenfassung *f* standard base
Einheitssärad *n* standard seed barrel
Einheitssätabelle *f* standard seeding (sowing) table
Einheitsschnurpaar *n* universal cord circuit, *(Am)* universal cord pair
Einheitssprung *m* unit step function
Einheitstarif *m* unit fee charge
Einheitswelle *f* basic shaft *(fit)*
einhiebig single-cut
Einhubjacquardmaschine *f* **für Offenfach** *(text)* single-lift Jacquard machine for open shedding
einhüllen *(math)* to envelop
Einhüllende *f (math)* envelope
Einhüllungswirkungsgrad *m* efficiency of the non-conducting covering
einhülsen to can; to encase; to encapsulate
einimpfen to inoculate
einkalken *(leath)* to steep in lime water
Einkammeranhängersteuerventil *n* single-chamber trailer control valve
Einkammerbremszylinder *m* single-chamber brake cylinder
Einkammerkessel *m* water tube boiler with one header
E. mit Doppelrohren single-header boiler with circulating water tubes
E. mit einfachen Wasserrohren single-header boiler with plain water tubes
Einkanalanalysator *m* single-channel analyzer
Einkanaldiskriminator *m* single-channel discriminator
Einkanalkoinzidenzspektrometrie *f* single-channel coincidence spectrometry
einkaufen to purchase, to buy; to shop
Einkäufer *m* purchasing executive
Einkaufsabteilung *f* purchasing department
Einkaufsdienst *m* purchasing department
Einkaufshalle *f* supermarket
Einkaufszentrum *n* shopping centre
einkehlen to channel
einkerben to notch; to indent; to nick; to score; to serrate
Einkerbung *f* notch; indentation; serration
Einkesselschalter *m* single-tank switch
Einkettenförderer *m* single-chain conveyor
Einkitten *n* cementing in *(grinding wheel)*
Einklassenschiff *n* one-class vessel
einkleben to paste in
einklemmen to jam
einklinken to catch; to pawl
Einknopfabstimmung *f* ganged tuning, single-dial control
Einknopfbedienung *f* single-dial control
Einkolbenzähler *m* reciprocating piston meter
einkommen to come in *(vessel)*; to make (enter a) port
Einkoppelstrecke *f* buncher space *(radio)*
Einkornbeton *m* single-sized aggregate concrete, no-fines concrete
Einkreisempfänger *m* single-circuit receiver
Einkreistriebwerk *n* single-spool engine
Einkristall *m* single crystal
Einkristallspektrometer *n* single-crystal spectrometer
Einkristallzüchtung *f* single-crystal growth
einkuppeln to engage the clutch, to throw [the clutch] into gear, to release the clutch pedal
Einlage *f* insert[ion]; deposit *(money)*
Einlagerung *f* incorporation *(impurities)*
linsenförmige E. *(geol)* lenticular intercalation
Einlagerungsmischkristall *m* solid solution
Einlagerungsraum *m (nucl)* storage space
einlagig single-layer *(coil)*
Einlaß *m* intake, inlet; admission
Einlaßbauwerk *n* intake (inlet) structure
Einlaßdosenschalter *m* recessed switch
Einlaßdruck *m* admission pressure
einlassen to admit
Einlaßgeschwindigkeit *f* inlet velocity *(turbine)*
Einlaßhub *m* induction stroke; charging period
Einlaßkanal *m* steam (inlet, induction) port
Einlaßkante *f* admission edge
Einlaßkrümmer *m* induction (inlet) manifold, air inlet
Einlaßöffnung *f* intake
Einlaßquerschnitt *m* admission section
Einlaßrohr *n* induction (suction) pipe, admission tube
Einlaßschalter *m* recessed switch
Einlaßschieber *m* admission valve
Einlaßschlitz *m (aut)* admission port
Einlaßstutzen *m* inlet piping connection, inlet connecting branch
Einlaßtemperatur *f* inlet temperature
Einlaßüberdeckung *f (therm)* outside (steam) lap
Einlaßventil *n* admission (intake, suction) valve
Einlaßventilgehäuse *n* inlet valve chest
Einlauf *m* inlet, intake; downgate *(founding)*
Einlaufbauwerk *n* intake (inlet) structure
einlaufen 1. to wear in *(bearing)*; 2. *(text)* to shrink; 3. *(ship)* to come in, to run in, to put in; to make (enter a) port; 4. to run in *(machine)*
in einen Hafen e. to enter a port
Einlaufen *n* 1. wearing-in *(bearing)*; 2. *(text)* shrinkage
E. ins Dock docking
Einlaufhafen *m* port of entry (arrival)
Einlaufkante *f* / **scharfe** sharp inlet edge
Einlaufkokille *f* git mould
Einlaufkontrolle *f (text)* shrinkage control
Einlauflänge *f* length of entrance [line]
Einlaufschacht *m* gully

Einlaufseite *f* **[der Erntemaschine]** feeding (charging) side
Einlaufstutzen *m* inlet connection
Einlaufwinkel *m* spout angle *(chipping)*
Einlaufzeit *f* run-in period, break-in period *(machine)*
Einlegearbeit *f* intarsia, marquetry; inlaid work; inlaying
Einlegekeil *m* sunk key
einlegen to put in, to insert
Band e. to insert the tape
den Kern e. to place the core *(moulding)*
Filme e. to load films
Einlegen *n* loading *(work, films)*
E. der Kerne laying-in of cores *(moulding)*
Einleger *m* charger *(threshing)*
Einlegevorrichtung *f* loading device *(work)*
Einleistmaschine *f* lasting machine
einleiten to trigger, to start; to pass into; to initiate
Einleiterkabel *n* single[-wire] cable
Einleiterwandler *m* single-turn transformer
Einleitung *f* **[eines Vorganges]** *(el)* commencement
einlenken *(rock)* to guide into line (path) of sight
Einlenkgerät *n* *(rock)* device for guidance to the line of sight
Einlenkgesetz *n* *(rock)* formula for guidance to the line of sight
Einlenkrechner *m* *(rock)* computer for guidance
Einlenkung *f* *(rock)* guidance to the line of sight
Einlenkvorgang *m* *(rock)* guidance to the line of sight
Einlesemaschine *f* reading[-in] machine
einlesen to read in
Einlinienspektrum *n* spectrum of single lines, single-line spectrum
Einlippenbohrer *m* single-fluted drill, single-lip drill
Einlippentieflochbohrer *m* single-fluted drill for deep-hole drilling; single-lip gun drill
einlippig single-lip, single-flute
Einlötelektrolytkondensator *m* tubular electrolytic capacitor
einloten to plumb
einlöten to solder in
einlotsen to pilot in, to take into a port
Einmachglas *n* preserve jar (bottle)
einmaischen to sparge, to mash in
Einmaischschnecke *f* mash screw feeder
Einmaleinskörper *m* *(dat)* multiplier
Einmannbedienung *f* one-man control
Einmannkammer *f* single-berth (bed) cabin
Einmannruderboot *n* single-banked boat
Einmannsäge *f* one-man cross-cut saw
Einmanntorpedo *m* human torpedo
Einmaster *m* single sticker, one-masted ship
einmauern to immure, to embed; to brick in
Einmauerung *f* bricking-in
einmäulig single-ended *(gauge)*
einmieten *(ag)* to pit, to clamp
Einmitten *n* centring
einmotorig single-engined
Einmündung *f* junction *(road)*
Einnadelflachmaschine *f* single-needle flat bed sewing machine *(shoe)*
Einnahmekontrolle *f* audit office
Einnahmen *fpl* revenue, receipts, takings
einnehmen to occupy *(space)*; to receive, to take, to collect *(money)*; *(ship)* to fuel; to coal; to shorten sail
einnivellieren to bone in
einordnen to classify; to file
Einordnung *f* classification; filing
einorten to position to, to locate
Einortung *f* location, positioning
E. des Tisches table positioning
Einortungsgenauigkeit *f* accuracy of positioning *(boring)*
einpacken to wrap; to pack up
einparametrig one-parameter
einpassen to fit in; to register
Einpaßstelle *f* fitting strip
Einpendler *m* commuter *(town planning)*
Einperldrossel *f* *differential pressure regulator for bubbler-gauge*
einpflanzen to plant
einpfropfen to inoculate, to ingraft
Einphasenanlasser *m* single-phase starter
Einphasenbahnsystem *n* single-phase traction system
Einphasendosimeter *n* *(nucl)* single-phase dosimeter
Einphasengenerator *m* single-phase generator
Einphasengleichrichter *m* single-phase rectifier
Einphasenkommutatormotor *m* **mit Kompensationswicklung** neutralized series motor
Einphasenreaktor *m* single-phase reactor
Einphasenreihenschlußmotor *m* **mit Kompensationswicklung** single-phase commutator motor with series compensating winding
Einphasenstrom *m* single-phase current
Einphasenwechselstrommotor *m* single-phase alternating-current motor
Einphasenzähler *m* single-phase meter
einphasig single-phase, monophase
einplanieren to level
Einplattenruder *n* single-plate rudder, centre-plate rudder
einpolig single-pole, unipolar, single-polar
einprägen to impress
einpressen to force in, to press into place *(part)*; to inject, to grout in *(cement, mortar)*

Einpressen *n* **der Düse** forcing-in of the nozzle *(turbine)*
E. des Arbeitszylinders driving-in of the working barrel
Einpreßmörtel *m* intrusion mortar (grout)
Einpreßsonde *f* injection well
Einpreßverfahren *n* pressure grouting, grout injection *(cement)*
einprofilig single-rib[bed] *(thread grinding)*
Einprofilschleifscheibe *f* single-edge wheel
einpudern to powder
Einpudern *n* **einer Form** powdering (facing) of a mould
Einquadrantmultiplizierer *m* one-quadrant multiplier *(analogue computer)*
Einradhacke *f (ag)* push hoe
einrammen to drive [in] *(piles)*
Einrastbolzen *m* drop-in pin
einrasten to engage; to insert [the pin]; to click *(switch)*
in ein Loch e. to enter a hole
Einrastfeder *f* click-in spring
Einraumschiff *n* one-compartment vessel
Einregelzeit *f (el)* attack time
Einreiber *m (build)* casement fastener (turn)
Einreibmittel *n* rubbing agent
einreihen to sequence
Einreihen *n* drawing-in *(weaving)*
einreihig single-row *(bearing)*
Einreißwiderstand *m* resistance to the starting of a tear *(paper)*
Einrichtebogen *m (print)* register sheet
einrichten 1. to arrange; to furnish; 2. to set up *(machine tool)*; 3. to position *(table)*
Einrichten *n* setting-up *(machine tool)*
E. einer Maschine machine setting
Einrichter *m* [machine-]setter, tool setter; *(Am)* setup man
Einrichtezeit *f* setting-up time *(machine)*
Einrichtung *f* 1. furnishment *(house)*; equipment *(shop)*; 2. installation; introduction; establishment; arrangement; 3. *(mach)* device; attachment; apparatus; unit; 4. *(mach)* positioning *(table)*; setting-up *(machine tool)*
E. eines Teilnehmeranschlusses installation of a subscriber's telephone
E. für Balligverzahnung special equipment for hobbing crowned gears
E. für Tonfrequenzfernwahl voice-frequency selecting system
E. für Viehtransportschiff cattle fittings
E. zum Einschleifen von Spanbrechernuten chip-breaker groove grinding unit
E. zur selbsttätigen Spanabfuhr automatic chip disposal unit, chip conveyor
elektrische E. electrical equipment
Einrichtungen *fpl* facilities
Einrichtungsbetrieb *m (cont)* single-ended action
Einrichtungsschott *n (ship)* accommodation bulkhead
Einrichtungswandler *m (el)* unidirectional (unilateral) transducer
Einröhrenanordnung *f (el)* single-valve arrangement
Einröhrenempfänger *m* single-valve receiver
Einröhrenverstärker *m* single-valve amplifier
einrollen to roll up
Einrollverfahren *n* [grinding] wheel crushing
einrosten to rust in, to grow rusty
einrücken to engage; to clutch; to throw in *(lever)*; to throw into gear
Zeilen e. to indent
Einrücken *n* engagement, throwing-in
Einrückhebel *m* engaging lever
einrüsten to scaffold
Eins *f* unit[y]
flüchtige E. elusive one *(in electromechanical calculators)*
einsalben to anoint, to grease
einsalzen to salt, to brine
Einsattelung *f* dip *(between two maxima)*
Einsatz *m* 1. application; use; utilization; service; 2. insert *(part)*; 3. charge *(furnace)*; 4. case *(hardening)*; 5. *(nucl)* charge, inventory; inset
Einsatzanlage *f* field installation
Einsatz-Ausstoß-Analyse *f (dat)* input-output analysis
Einsatzbacke *f* false jaw *(chuck)*
einsatzbereit ready for service
Einsatzbuchse *f* renewable bushing
Einsatzdeckel *m* inserted cover
Einsatzdosenschalter *m* recessed switch
Einsatzeffekt *m* threshold effect
Einsatzenergie *f s.* Schwellenenergie
einsatzgehärtet case-hardened *(steel)*
Einsatzgitter *n (nucl)* inset
einsatzhärten to case harden
Einsatzhärteschicht *f* case
Einsatzhärtung *f* case hardening *(steel)*
Einsatzhülse *f* taper socket *(drill)*
Einsatzkohle *f* consumption coal
Einsatzkontrolle *f (telev)* continuity control
Einsatzmaterial *n* raw material
Einsatzmeißel *m* bit, insert *(machine tool)*
keramischer E. ceramic bit
Einsatzmesser *n* inserted blade *(cutter)*
Einsatzmittel *n* case-hardening compound
Einsatzofen *m* carburizing furnace
elektrischer E. electric carburizing furnace
Einsatzöffnung *f* pot hole *(crucible furnace)*
Einsatzpulver *n* carburizing agent
Einsatzring *m* cone chucking ring
Einsatzschalter *m* recessed switch
Einsatzsicherung *f* switch-fuse

Einsatzspannung *f* cut-off voltage *(of an electronic valve)*; Geiger threshold
Einsatzstahl *m* case-hardening steel
Einsatzstopfbüchse *f* inserted (loose) stuffing box
Einsatztage *mpl* days in use
Einsatzteil *n* mechanism
Einsatztiefe *f* case depth *(hardening)*
Einsatztopf *m* case-hardening pot
Einsatzzeit *f* operating time
Einsatzzylinder *m* cylinder liner
einsäuern to ensilage *(fodder)*
Einsäuerung *f* ensilage *(fodder)*
einsaugen *(ch)* to resorb, to imbibe
einsäumen to hem, to border, to edge, to trim
einschaben to fit [in] by scraping
einschalen *(build)* to form, to shutter
Einschalt-Ausschalt-Zeit *f (el)* make-break time
Einschaltblatt *n* inset
einschalten 1. to switch on, to turn on, to snap on, to make (build up) a circuit, to connect, to close; 2. to start, to put into operation
  den Stromkreis e. to close a circuit
  in ein Gespräch e. *(telec)* to plug in, to tap into
  sich in eine Leitung e. *(telec)* to listen in on a circuit
  wieder e. to bring back into circuit
Einschalter *m* closing (single-throw) switch, circuit closer
  einfacher E. single-way switch
Einschaltfolge *f* switching-on sequence
Einschaltknopf *m* starter button *(machine)*
Einschaltleistung *f* making capacity
Einschaltlichtbogen *m* make arc
Einschaltrelais *n* starting relay
Einschaltspule *f* closing coil
Einschaltstellung *f* on-position
Einschaltstoß *m* make impulse
Einschaltstrom *m* making (starting) current
Einschaltstromkreis *m* completing circuit
Einschaltstromstoß *m* cut-in current rush, current inrush
Einschalttaste *f* push-to-make switch
Einschalttor *n* normally opened gate
Einschaltung *f (el)* circuit closing; starting
Einschaltvermögen *n* making capacity
Einschaltverzug *m (el)* make-time
Einschaltvorgang *m* circuit closing operation
Einschaltzeit *f* make time; starting delay
Einschaltzustand *m (el)* closed circuit condition
Einscharpflug *m* single-furrow plough
Einscheibe *f* single pulley
Einscheibenantrieb *m* single-pulley drive
Einscheibenkupplung *f* single-plate [disk-type friction] clutch
Einscheibenläppmaschine *f* **für Flachläppen** single-lap machine for flat work
Einscheibenschar *n (ag)* single-disk coulter
Einscheibentrockenkupplung *f* single-dry plate clutch, single-plate dry-type clutch
einscheibig single-wheel *(grinding machine)*
Einschiebekarte *f (off)* plug-in card
Einschiebestütze *f* transposition bracket
Einschienenbahn *f* mono-railway, monorail railway
Einschienenförderer *m* monorail transporter
Einschienenkatze *f* monorail trolley
Einschienenkran *m* monorail crane
Einschießbogen *m (print)* set-off sheet, slip[-sheet]
einschießen 1. *(mil)* to find the target; 2. *(nucl)* to inject
Einschießen *n* **der Schnellschlußstäbe** *(nucl)* scramming
einschiffen [/ sich] to ship; to embark, to go on board, to take ship
einschirren to harness, to gear
Einschlag *m* point of impact *(shell)*
Einschlagbürste *f* dabbing brush *(wool comb)*
einschlagen to wrap; to drive in *(stake, nail)*
Einschlaglupe *f* folding magnifier
  anastigmatische E. anastigmatic folding magnifier
  aplanatische E. aplanatic folding magnifier
Einschlagwecker *m* single-stroke bell, one-stroke bell
Einschlagwinkel *m* lock angle *(wheels)*
einschlämmen to jet, to flush
einschleifen 1. to grind in; 2. to loop in *(circuit)*
einschleppen to tow in, to haul in *(vessel)*
einschleusen to lock in, to tow into sluices *(vessel)*
Einschleusen *n (nucl)* injection
einschließen 1. to comprise; to include; 2. *(ch)* to confine *(liquid)*; 3. to lock in; 4. to clear the loop *(knitting)*
Einschließung *f* confinement *(of plasma)*
Einschließungssatz *m (math)* inclusion theorem
einschlitzen to mortise *(joinery)*
Einschluß *m* inclusion, occlusion; entrapping
  fester nichtmetallischer E. solid non-metallic inclusion
  harter E. hard spot
  nichtmetallischer E. non-metallic inclusion
Einschlußstellung *f (text)* clearing position *(latch needle)*
Einschlußverbindung *f (ch)* occlusion compound, adduct
Einschmelzelektrode *f* fused-in electrode
einschmelzen 1. to remelt; to melt down; 2. to seal [in], to fuse in
  in eine Glaswand e. to seal through glass
Einschmelzen *n* 1. remelting; melting-in; 2. sealing [in]
Einschmelzkohlenstoffgehalt *m* melt-down carbon

Einschmelzmaschine *f* sealing-in machine *(valve)*
Einschmelzperle *f* beading
Einschmelzschlacke *f* melt-down slag
Einschmelzstelle *f* seal *(electrode connection)*
Einschmelzung *f* sealing-in joint
Einschmelzversuch *m* melt-down test
einschnappen to catch
einschneiden to cut in; to notch; to tap
einschneidig single-point, single-edged *(tool)*
Einschnitt *m* 1. indentation *(metal)*, notch, kerf, nick; 2. runner *(moulding)*
E. in Erdreich cutting *(open excavation)*
einschnittig single-cut, single-point *(tool)*; single-shear *(shearing test)*
Einschnittkette *f (text)* ground warp
Einschnittschuß *m (text)* binding pick
Einschnüreffekt *m (ph)* pinch effect
einschnüren to restrict, to reduce *(flow)*; to neck down *(tensile test)*
Einschnürung *f* constriction, restriction, reduction; necking, reduction of area, formation of a neck *(tensile test)*
E. der Düse nozzle contraction *(turbine)*
Einschobern *n* piling under cover
einschränken to restrict; to restrain *(movement)*; to reduce *(production)*
einschrauben to screw [into]
Einschraubenantrieb *m* single-screw propulsion
Einschraubenschiff *n* single-screw vessel
Einschraubensteven *m* single-screw stern frame
Einschraubverlängerung *f* screwed-in extension
einschreiben *(math)* to inscribe; to register *(letter)*
Einschreibung *f (dat)* entry
einschrumpfen to shrink
Einschrumpfverbindung *f* expansion fit
Einschub *m* withdrawable (plug-in) unit
Einschubdecke *f (build)* false ceiling
Einschubeinheit *f (off)* plug-in unit
Einschubrahmen *m* rack-mounting frame
Einschurwolle *f* single-clip wool
Einschuß *m* 1. *(nucl)* injection; 2. *(text)* weft
Einschußoptik *f* injection optics
Einschußspannung *f* injection voltage
Einschußvorrichtung *f* injector
Einschüttbehälter *m* feed box [pre-cleaner]
einschütten to pour in *(liquid)*
Einschüttkasten *m* feeding box *(workpieces)*
einschwenken *(mine)* to sump in *(jib)*
einschwimmen *(build)* to float into position
Einschwingdauer *f* build-up time
Einschwingen *n* **[beim Ausschalten]** *(el)* circuit-breaking transient
Einschwingerscheinung *f s.* Einschwingvorgang
Einschwingfrequenz *f* transient frequency
Einschwingkurve *f* transient characteristic (response)
Einschwingspannung *f* transient voltage
Einschwingstrom *m* balance (transient) current
Einschwingung *f (el)* building-up
E. beim Einschalten make transient
Einschwingvorgang *m (el)* transient [state], transient phenomenon, initial transient, building-up process (transient), transient condition (effect)
Einschwingzacke *f (el)* pre-shoot
Einschwingzeit *f (el)* transient time (period); building-up time
Einschwingzustand *m s.* Einschwingvorgang
Einseilförderung *f* direct (main) rope haulage
Einseilgreifer *m* single-rope grab bucket
Einseitenband *n (el)* single sideband
Einseitenbandempfänger *m* single-sideband receiver
Einseitenbandübertragung *f* single-sideband transmission
Einseitenbetrieb *m (cont)* single-ended action
einseitig unilateral; on one direction; in one direction
Einsetzbühne *f* charging floor
einsetzen 1. to insert; to instal[l]; 2. to case harden; 3. *(met)* to charge
e. an Stelle von to substitute for
Einsetzen *n* 1. insertion, installation; 2. case hardening; 3. *(met)* charging; 4. *(math)* substitution
Einsetzlöffel *m* charging spoon (peel)
Einsetzmulde *f* charging box
Einsetzseite *f* charging side *(furnace)*
einsickern to infiltrate; to soak; to leak
Einsickerung *f* infiltration
einsilieren *(ag)* to ensilage
einsinken to cave in; to subside; to sink; *(mine)* to yield
Einsinkweg *m (mine)* yield
Einsinkwiderstand *m (mine)* resistance to yield *(prop)*
Einskuller *m* racing skiff
Einspannbacke *f* chuck jaw
einspannen 1. to chuck, to clamp, to hold *(workpiece)*; 2. to fix, to restrain *(statics)*
Einspannfutter *n* chuck jaws
Einspannmoment *n* fixing (fixed-end) moment *(statics)*
Einspannteil *n* holding portion *(broach)*
Einspannung *f* 1. clamping; chucking *(work)*; 2. fixation, restraint *(statics)*
teilweise E. partial fixation
Einspannvorrichtung *f* clamping (chucking, holding) device
Einsparung *f* saving in cost, economies; cost reduction
einspeisen to feed; to supply
Einspeisezelle *f (el)* feeding cubicle
Einspindelanordnung *f* single-spindle arrangement

Einspindelautomat *m* single-spindle automatic [screw machine]
E. für Stangenarbeit single-spindle automatic bar machine (auto)
E. mit verschiebbarem Spindelstock single-spindle sliding-head type automatic screw machine
Einspindelbauart *f* single-spindle type (model, construction, design)
Einspindelbohrmaschine *f* single-spindle drilling machine
Einspindelmaschine *f* single-spindle machine
Einspindelplanfräsmaschine *f (Am)* single-spindle manufacturing type milling machine, single-spindle fixed-bed type milling machine, simple [manufacturing type] milling machine
einspindlig single-spindle
Einsprengling *m* phenocryst
Einsprengmaschine *f* sprinkling machine
einspringend reentrant
Einspritzdauer *f (aut)* injection period
Einspritzdruck *m* injection pressure
Einspritzdüse *f* [fuel-]injection nozzle, atomizing nozzle; discharge nozzle
einspritzen to inject *(fuel)*; to prime
Anlaßkraftstoff e. to prime
Einspritzen *n* injection; priming
E. von Anlaßkraftstoff engine priming
Einspritzkondensation *f* condensation by injection
Einspritzkondensator *m* jet condenser
Einspritzkühlung *f* cooling by injection
Einspritzmotor *m* fuel-injection engine
Einspritzöffnung *f* inlet (injection) port
Einspritzpumpe *f* [fuel] injection pump
Einspritzpumpenoberteil *n* injection pump upper housing
Einspritzrohr *n* injection pipe
Einspritzung *f* injection
direkte E. *(aut)* solid injection
kompressorlose E. solid injection
Einspritzventil *n* injection (atomizer) valve
Einspritzventilfilter *n* injector filter
Einspritzverstelleinrichtung *f* injection timing gear (device)
Einspritzversteller *m* injection advance device
Einspritzverstellung *f* injection timing *(diesel)*
Einspritzverzögerung *f* injection lag
Einspritzwasser *n* injection water
Einspruchsfrist *f* period for entering opposition
einspülen *(build)* to jet, to flush
Einspulentransformator *m s.* Spartransformator
Einspülverfahren *n (build)* [water] jetting, flushing
einspurig single-line, single-track
einstampfen 1. *(build)* to ram; 2. to pulp *(paper)*
Einständerausführung *f* open-side (single-column) design

Einständerbauart *f* open-side (single-column) construction
Einständerblechkant[en]hobelmaschine *f* open-side plate planing machine
Einständerdrehwerk *n s.* Einständerkarusselldrehmaschine
Einständerexzenterpresse *f* gap frame eccentric press
Einständerfräsmaschine *f* open-side milling machine
Einständerhobelmaschine *f* open-side planing machine
Einständerkarusselldrehmaschine *f* single-column vertical [turning and] boring mill
Einständerlangfräsmaschine *f* single-column plano-milling machine
Einständermaschine *f* single-column (open-sided) type of machine
Einstärkenglas *n (opt)* single-vision lens
Einstauben *n* **mit Gesteinsstaub** *(mine)* stone dusting
Einstauber *m (mine)* stone duster
einstechen *(mach)* to recess; to plunge
Einstechen *n (mach)* recessing; plunging
Einstechmaschine *f* welt sewing machine *(shoe)*
Einstechschleifen *n* plunge[-cut] grinding
spitzenloses E. infeed-method of centreless grinding
Einstechschleifmaschine *f* plunge grinding machine
Einstechverfahren *n* infeed process (method)
E. beim spitzenlosen Schleifen infeed method of centreless grinding
Einsteck[bohr]buchse *f* slip renewable bushing
einstecken to plug in; to insert
Einsteckende *n* shank end *(boring)*
Einsteckhörer *m* insert earphone
Einsteckhülse *f* / **kegelige** taper socket *(drill)*
Einsteckkurbel *f* removable crank
Einsteckmeißel *m* [toolholder] bit, inserted tool
Einsteckschlüssel *m* face spanner
Einsteckspindel *f* loose inserted peg
einsteigen to get in, to enter
Einsteigloch *n* manhole
Einsteigöffnung *f* manhole
Einsteigschacht *m* manhole
Einstein-de-Haas-Effekt *m* Einstein-de-Haas effect, gyromagnetic effect
Einsteinium *n* einsteinium
Einstellanschlag *m* adjustable stop
einstellbar adjustable
in Stufen e. step-by-step adjustable
stetig e. infinitely adjustable
stufenlos e. steplessly adjustable
Einstellbarkeit *f* adjustability
Einstellbereich *m* setting range
Einstelldichte *f* **der Kette** *(text)* set of warp
Einstelldrehmoment *n* adjustment torque

Einstelldrossel *f* variable-area nozzle
Einstellehre *f* setting gauge
einstellen 1. to stop, to suspend, to cease; 2. to garage *(car)*; to shelve; 3. to tune in *(radio)*; to adjust, to set *(instrument)*; 4. to hire; 5. to align; to margin *(relay)*
auf Null e. to set to zero
den Regler e. to adjust the governor
den Verkehr e. to suspend (stop) traffic
die Arbeit e. to lay off; to strike
die Scheinwerfer e. to focus the headlamps
die Spätzündung e. to retard the ignition
Frühzündung e. to advance the ignition timing
sich e. auf to position itself on
wieder e. to reset
Einstellen *n* setting, set-up; adjusting, adjustment *(instrument)*
Einsteller *m* adjuster
Einstellfehler *m* adjusting (setting) error; positioning error; faulty focusing control
E. des Bildpunktes *(telev)* faulty spot control
Einstellgenauigkeit *f* setting (adjustment) accuracy; positioning accuracy *(boring)*
Einstellgeschwindigkeit *f (el)* response time, responding speed *(instrument)*; *(mach)* positioning time *(boring)*
Einstellglied *n (el)* setting element
Einstellhebel *m* **für die Meißelabhebung** tool-lifter engagement lever
Einstellhülse *f* adjusting sleeve
einstellig single-figure, one-figure, one-place
Einstellknopf *m* pointer knob
gerändelter E. serrated knob, knurl
Einstellmarke *f* timing mark; setting mark
Einstellmaß *n* slip gauge
Einstellmikroskop *n* focusing microscope
Einstellmoment *n* controlling torque
Einstellnormal *n* reference gauge, master
Einstellpotentiometer *n* set potentiometer
Einstellregler *m* preset control
E. mit Rändelknopf edge-type volume control
ungekapselter E. open (skeleton) preset, open-type control
Einstellscheibe *f* dial
Einstellschraube *f* adjusting screw; setscrew
E. des Platinenexzenters *(text)* jack cam adjusting screw
Einstellspitze *f* adjusting point
Einstellung *f* setting; adjustment
E. der Bildhöhe *(telev)* height control
E. der Bohrlage locating of a hole
E. der Nullinienhöhe position control
E. des Arbeitsablaufs cycle setting
E. des Kardendeckels *(text)* flat setting *(card)*
E. des Stößelspiels *(aut)* tapped adjustment
E. des Zeitpunktes timing
automatische E. self-adjustment
falsche E. maladjustment
sehr genaue E. exact hair line setting
Einstellungsgenauigkeit *f s.* Einstellgenauigkeit
Einstellungsvertrag *m* registration contract *(for privately owned wagons)*
Einstellvorrichtung *f* **für den Nennlastabgleich** full load meter adjustment
Einstellwert *m* setting
Einstellwiderstand *m* adjustable resistor
Einstellwinkel *m* plan angle *(tool)*; *(mine)* tip working angle; *(aero)* rigging angle of incidence
Einstellzeit *f* response time *(electrical measurement)*; setting time; setting-up time; positioning time *(boring)*; settling time
einstemmen to mortise *(joinery)*
die Schaufel e. to caulk in the blade *(turbine)*
Einsteuern *n* **der Rakete auf ihre Flugbahn** placing of the rocket into its orbit
Einstich *m* recess *(machining)*
E. am Spiralbohrer mark recess, neck
Einstichkasten *m* hole for planting the pole *(pole jump)*
Einstichlänge *f* recess (neck) length *(twist drill)*
Einstieg *m* manhole
Einstiegluke *f* entrance hatch
einstöckig single-storey
Einstöpselschnur *f (el)* single-ended cord-circuit
Einstoßer *m* pusher
Einstrahleinrichtung *f (opt)* single-beam equipment
einstrahlig single-jet
Einstrahlung *f* irradiation
Einstrahlungsgebiet *n* irradiated region; radiation zone
Einstrahlungswinkel *m* angle of incidence
Einstrahlungszahl *f* coefficient of absorption of radiant heat
Einstrangverfahren *n* single-strand process
Einstreichfeile *f* screw head file
Einstrich *m* barring, bunton, divider
einströmen to enter, to flow in, to ingress, to run in, to stream in
Einströmen *n* / **plötzliches** inrush
Einströmgeschwindigkeit *f* admission velocity [of steam]
Einströmmenge *f* amount of inflow *(liquid)*
Einströmstelle *f* / **scharfkantige** sharp inlet edge
Einstromsystem *n* single-current telegraph system
Einstromtaste *f* single-current key *(telegraphy)*
Einströmungsperiode *f* period of admission
einstroppen to strap [a load]
Einstückspannvorrichtung *f* single-station fixture
einstufen to grade
Einstufenätzen *n (print)* powderless etching
Einstufengleitboot *n* single-step hydroplane, monoplane
Einstufenmahlwerk *n* single-stage grinding mill

Einstufenrückführung *f* *(nucl)* single-stage recycle
Einstufenturbine *f* single-stage turbine
einstufig single-stage *(pump)*
Einsturz *m* collapse; falling-in, [down]fall, decay; caving-in, subsiding, sinking
einstürzen to fall in, to collapse
einsumpfen *(build)* to wet slake *(lime)*
Eintafelprojektion *f* *(math)* one-plane projection
Eintagstide *f* diurnal tide
Eintaktverstärker *m* *(el)* single-ended amplifier
eintauchen to immerse; to steep; to dip; to submerge; *(nucl)* to move in, to insert *(the rods)*
Eintauchkühlmittel *n* flood coolant
Eintauchplattierung *f* immersion plating
Eintauchpumpe *f* submersible pump, *(Am)* submergible pump
Eintauchrefraktometer *n* dipping refractometer
Eintauchtiefe *f* depth of immersion
Eintauchtrommel *f* dipping drum
Eintauchzählrohr *n* *(nucl)* dipping counter
Einteilchenübergang *m* single-particle transition
einteilen to divide; to classify; to grade *(size)*; to graduate *(degrees)*; to calibrate
einteilig one-piece; solid *(tool)*
Einteilung *f* division; classification; graduation; grading *(size)*; calibration
Eintonbetrieb *m* *(telec)* single-tone working
Eintonmodulation *f* single-tone modulation
Eintonverfahren *n* *(telec)* single-tone keying
Eintrag *m* drag-in; furnish; weft
eintragen to apply *(a force)*; to plot *(diagram)*
  eine Peilung e. to lay off a bearing
  sich in ein Register e. to make an entry in a register, to enter in a register
Eintragen *n* picking motion *(weaving)*
  E. einer Peilung laying off of a bearing
Einträgerkran *m* single-girder crane
Eintragung *f* **in das Handelsregister** certificate of incorporation
eintreffen / verspätet *(ry)* to arrive late
Eintreffen *n* **des Impulses** pulse arrival
eintreiben to drive home (in) *(wedge)*
Eintreiben *n* driving-in *(wedge)*
eintreten to emerge [into] *(fluid)*
  in die Lehre e. to enter an apprenticeship
  in die Leitung e. *(telec)* to answer on the circuit
  in eine Verbindung e. *(telec)* to enter a circuit
Eintritt *m* intake *(fluid)*, entrance, inlet
Eintrittsarbeit *f* work done during admission
Eintrittsdampfspannung *f* *(therm)* initial (admission) pressure
Eintrittsdreieck *n* inlet diagram (triangle) *(turbine)*
Eintrittsgehäuse *n* inlet casing
Eintrittsgeschwindigkeit *f* admission velocity [of steam]
  absolute E. absolute entrance (inlet) velocity
  relative E. relative entrance velocity [to the moving blade]
Eintrittskammer *f* admission chamber *(turbine)*
Eintrittskanal *m* steam (inlet) port
Eintrittskante *f* inlet edge *(turbine)*
Eintrittsluke *f* *(opt)* entrance port
Eintrittsöffnung *f* intake port
Eintrittsquerschnitt *m* admission section
Eintrittsschleuse *f* air lock
Eintrittsspalt *m* *(opt)* entrance slit; receiving slit
Eintrittsspannung *f* admission potential
Eintrittsstelle *f* point of entry *(electrons)*
Eintrittsstutzen *m* air intake casing, inlet connecting branch
Eintrittstemperatur *f* inlet temperature
Eintrittsvorlage *f* inlet header
Eintrittswinkel *m* inlet angle *(turbine plate)*
Ein- und Ausfahren *n* in-and-out travel *(machine)*
Ein- und Ausschalter *m* on-off switch; make-and-break key
Einwaage *f* [weight-in] quantity
einwalken to mill, to full
einwalzen to roll in
Einwandererschiff *n* immigration ship
Einwegautomat *m* *(dat)* one-way automaton
Einwegbehälter *m* one-way packing (container)
Einwegemaschine *f* single[-way] machine
Einweggleichrichter *m* one-way rectifier, half-wave rectifier
Einweggleichrichterschaltung *f* half-wave rectifier circuit
Einweggleichrichtung *f* single-wave rectification, half-wave rectification
Einwegofen *m* uniflow (single-way) furnace
Einwegrutsche *f* one-way chute
Einwegschalter *m* single-way switch
Einwegschaltung *f* single-way connection (circuit)
Einwegverbindung *f* *s.* Einwegschaltung
Einwegverstärker *m* one-way amplifier
Einwegwähler *m* uniselector
Einweichbottich *m* steeping bowl
einweichen to steep, to soak
Einweichgewicht *n* *(leath)* soaking weight
Einweisung *f* *(aero)* vectoring
Einweisungsradar *n* zone position indicator
Einwellenbetrieb *m* *(el)* single-wave operation
Einwellenstrom *m* single-wave current, simple harmonic current
Einwellenturbine *f* solid turbine
einwellig monochromatic
einwertig *(ch)* univalent, monovalent, *(math)* single-valued *(function)*
Einwertigkeit *f* *(ch)* univalence, monovalence; *(math)* single-valuedness

einwickeln to wrap
Einwickelpapier *n* wrapping paper
einwirken *(ch)* to attack, to affect; to act [upon] *(force)*
  e. lassen to interact, to allow to react
  auf Nadeln e. to act on needles *(knitting)*
Einwirkung *f* action; influence
  E. von Streufeldern *(el)* stray pick-up
  chemische E. chemical action
  gegenseitige E. interaction
Einwirkungsbereich *m* affected overburden *(rock mechanics)*
Einwirkungsfläche *f* area of extraction *(rock mechanics)*
Einwohnerschaft *f* residential population
Einwohnerzahl *f* residential population
einwölben to turn *(arches)*; to vault
Einwölben *n* arching, vaulting
Einzahnformschlagfräser *m* single-toothed form cutter
einzahnig single-tooth[ed]
Einzahnschlagfräser *m* single-tooth (point) cutter, single-blade[d] fly cutter
einzapfen to mortise; to jag; to notch
einzäunen to fence in
Einzäunung *f* fencing, boundary fence
einzeichnen to plot (diagram)
Ein-Zeit *f (el)* ON-period, make-time
Einzelader *f (el)* single strand
Einzelanfertigung *f* individual (on-off, single-piece, job) production
Einzelanruf *m (telec)* selective ringing
Einzelanschlußleitung *f (el)* individual line
Einzelantrieb *m* individual drive; direct motor drive *(machine)*
  mit E. individually powered, with individual drive
Einzelarbeit *f* one off *(machining)*; output per wheel *(turbine)*
Einzelaufspannung *f* single set-up
Einzelbahn *f (ph)* single path (track)
Einzelbaustein *m* individual unit *(electronics)*
Einzelbestimmung *f (ch)* detection by separate operation; single measurement
Einzelbildaufnahme *f* single-frame exposure
Einzelblech *n* single plate (sheet)
Einzelbuchstabensetzmaschine *f* monotype composing machine
Einzeldavit *m (n)* half davit
Einzeldipol *m* single dipole
Einzeldrahtspeicher *m (dat)* single-wire memory
Einzeldüse *f* single nozzle *(turbine)*
Einzeleinstellung *f* individual setting
Einzelelektron *n* single electron
Einzelempfänger *m* single receiver
Einzelfaden *m* monofilament
Einzelfehler *m* individual error *(gear inspection)*
Einzelfehlerprüfung *f* inspection of individual [gear] errors
Einzelfertigung *f* individual (single-piece, job) production
Einzelfeuer *n (mil)* single-shot firing (fire)
Einzelfundament *n s.* Einzelgründung
Einzelgründung *f (build)* single (individual, independent) footing; foundation pad (block)
Einzelhaus *n* detached house
Einzelkabine *f* single-berth cabin, single-bed cabin
Einzelkornsämaschine *f (ag)* root (single-seed) drill
Einzelkosten *pl* itemized cost
Einzelkraft *f* single force
Einzelkreis *m* individual (single-tuned) circuit
Einzellast *f* concentrated (point) load
  E. in Feldmitte centre (midspan) loading
Einzelleitung *f (el)* single-wire circuit
einzellig unicellular
Einzelöler *m* single lubricator
Einzelplattenruder *n* single-plate rudder, centre-plate rudder
Einzelpostenausstoß *m* single-item ejection
Einzelpotential *n* single potential
Einzelradaufhängung *f* individual front-wheel suspension, independent suspension [on all wheels]
Einzelregelung *f* governing by cutting-out individual nozzles *(turbine)*
Einzelschachtanlage *f (mine)* single colliery
Einzelschützsteuerung *f (el)* individual contactor equipment
Einzelspindelantrieb *m* single-spindle drive
Einzelstecker *m* individual plug
Einzelsteuerung *f* individual control
Einzelstreuung *f (ph)* single scattering
Einzelstück *n* single piece
Einzelstücke *npl* **der Ladung** *(ship)* oddments
Einzelstückfertigung *f* one-off production
Einzelteil *n* one-off part
Einzelteilchen *n (ph)* single particle
Einzelteilchenmodell *n* independent (individual) particle model of nucleus
Einzelteilfehler *m* single-pitch error *(gear)*; single-dividing error *(dividing head)*
Einzelteilfertigung *f* single-part production, jobbing
Einzelteilung *f* tooth-to-tooth spacing
Einzeltreibstoff *m* monofuel, monopropellant, monoreactant
Einzelverlustverfahren *n (el)* loss-summation method
Einzelverzug *m (text)* partial draft
Einzelwert *m* individual value
Einzelwirkungsgrad *m* individual efficiency
Einzelzahl *f* single number
Einzelzahn *m* individual tooth

Einzelzähne *mpl* / **eingesetzte** inserted teeth
einziehbar retractable
Einziehdeckskran *m* *(ship)* level-luffing deck crane
Einziehdraht *m* *(el)* conduit (draw) wire
einziehen to retract; to introduce *(paper web)*
  ein Schiff e. to condemn a vessel
  Fensterscheiben e. to fit window panes
Einziehen *n* *(text)* drawing-in *(weaving)*
  E. einer Zeile *(print)* indentation
Einziehfahrwerk *n* folding undercarriage, retractable landing-gear
Einziehhäkchen *n* *(text)* drawing-in hook
Einziehhaken *m* heddle hook
Einziehhaspel *f* *(text)* feeding-in winch
Einziehmaschine *f* *(text)* reeding[-in] machine
Einziehmesser *n* *(text)* reed hook
Einziehnadel *f* *(text)* heddle (drawing-in) hook, threading needle
Einziehschacht *m* *(mine)* downcast shaft
Einziehstrebe *f* radius rod *(landing gear)*
Einziehung *f* **des Gasstromes** contraction of the gas current
Einziehwalze *f* feeding roller
einzig unique
Einzonenreaktor *m* one-region reactor, single-region reactor
Einzug *m* *(text)* drawing-in *(weaving)*
  E. von vorn nach hinten front-to-back drawing-in
Einzugsbreite *f* *(text)* drawing-in width
Einzugsfehler *m* wrong draft
Einzugsgebiet *n* catchment area, *(Am)* watershed
Einzugsseil *n* *(text)* drawing-in band, taking-in band *(selfactor)*
Einzugsvorrichtung *f* 1. insertion device; 2. *(ag)* retraction device
Einzugszylinder *m* *(text)* feed roller
Einzweckmaschine *f* single-purpose machine
Einzweckreaktor *m* single-purpose reactor
Einzweckwerkzeug *n* single-purpose tool
Einzwirnung *f* twist contraction
Einzylinderdampfmaschine *f* single-cylinder steam engine
Einzylinderdiagramm *n* single-cylinder diagram
Eipulver *n* dry egg
Eis *n* ice
Eisabweiser *m* ice fin *(propeller-protection)*
Eisanker *m* ice drag (anchor)
Eisbahn *f* ice (skating) rink
Eisbelastung *f* ice loading *(of aerials)*
Eisberg *m* iceberg
  schwimmender E. floe berg
Eisbildung *f* ice formation, formation of ice
Eisbildungsanzeiger *m* *(aero)* ice detector
Eisboot *n* ice boat
Eisbrecher *m* 1. ice-breaker; 2. ice guard
  E. mit Kernenergieantrieb nuclear-powered ice-breaker
Eisbrecherleistung *f* ice-breaking power
Eisbrechersteven *m* ice-breaker stem
Eisbug *m* ice-doubling, ice-lining, Baltic (ice-breaker) bow
Eisdopplung *f s.* Eisbug
Eisdraggen *m s.* Eisanker
Eisdruck *m* ice pressure
Eisen *n* iron
  grobkörniges E. open-grained iron
  mattes E. dull iron
  sprödes überhitztes E. overburnt brittle iron
  teigiges E. pasty iron
  verunreinigtes E. contaminated iron
Eisenabscheider *m* magnetic separator
Eisenabstich *m* iron notch *(blast furnace)*
Eisenalaun *m* iron alum
Eisenantimonerz *n* berthierite
eisenartig ferruginous, ironlike
Eisenbahn *f* railway, *(Am)* railroad
Eisenbahnbau *m* railway engineering
Eisenbahnbrücke *f* railway bridge
Eisenbahndamm *m* railway embankment
Eisenbahnfähre *f* railway ferry
Eisenbahnfahrplan *m* railway time-table
Eisenbahnfahrzeug *n* railway vehicle
Eisenbahnflak *f* railway anti-aircraft artillery
Eisenbahnknotenpunkt *m* railway junction
Eisenbahnlafette *f* railway mount
Eisenbahnlinie *f* railway line
Eisenbahnnetz *n* railway system (network)
Eisenbahnoberbau *m* railway superstructure
Eisenbahnpostwagen *m* railway mail coach (car)
Eisenbahnreparaturwerkstätte *f* railway (loco) repair shop
Eisenbahnschaffner *m* railway conductor
Eisenbahnschiene *f* rail
Eisenbahnschotterung *f* railway ballasting
Eisenbahnschwelle *f* tie, sleeper
Eisenbahnsignal *n* railway signal
Eisenbahntechnik *f* railway engineering
Eisenbahnüberführung *f* railway bridge crossing
Eisenbahnunterbau *m* railway subgrade
Eisenbahnverkehr *m* railway traffic
Eisenbahnverkehrsordnung *f* regulations concerning carriage by rail
Eisenbahnwaggon *m* railway wagon
Eisenbahnwerkstätten *fpl* rail workshops
Eisenbahnzusammenstoß *m* rail crash
Eisenballast *m* *(ship)* iron kentledge
Eisenbeton *m s.* Stahlbeton
Eisenbetonschiff *n s.* Stahlbetonschiff
Eisenbewehrung *f* iron armouring (sheathing)
Eisenbieger *m* *(build)* steel fixer (reinforcing worker)
Eisenblätterkern *m* laminated-iron core *(ignition coil)*

Eisenblechkern *m* laminated-iron core *(ignition coil)*
Eisenbrand *m (el)* core welding *(transformer)*
Eisendraht *m* iron wire
Eisendrahtkratzenbeschlag *m (text)* iron wire card clothing
Eisendrossel *f (el)* iron-cored choke coil
Eisendrossel[spule] *f (el)* saturable reactor
Eisenerde *f* ferruginous earth
Eisenerz *n* iron ore
Eisenflechter *m (build)* steel fixer (reinforcing worker)
Eisengallustinte *f* ferrogallic ink
eisengesättigt iron-saturated
Eisengießer *m* iron founder
Eisengießerei *f* iron foundry
Eisenglanz *m* iron glance, haematite, specular iron ore
Eisenglimmer *m* micaceous iron ore
Eisenglimmerschiefer *m (min)* itabirite
eisenhaltig ferruginous, ferriferous; chalybeate *(water)*
Eisenhochofen *m* iron blast furnace
Eisenholz *n* iron wood
Eisenhütte *f* iron mill
Eisenhüttenchemie *f* metallurgical chemistry, chemical metallurgy
Eisenhüttenkunde *f* ferrous metallurgy
Eisenhüttenmann *m* steel metallurgist
Eisenhüttenwesen *n* metallurgy of iron and steel
Eisenkarbid *n* cementite, iron carbide
Eisenkern *m (el)* iron core
  E. / mit core-type *(coil)*
  E. / ohne air-core *(coil)*
Eisenkernabstimmung *f* slug tuning
Eisenkerninduktivität *f* iron-core inductance
Eisenkernspule *f* iron-cored coil
Eisenkerntransformator *m* iron-core transformer
Eisenkernverluste *mpl* core losses
Eisenkiel *m (ship)* iron keel
Eisenkies *m* iron pyrite
Eisen-Kohlenstoff-Diagramm *n* iron-carbon diagram
  metastabiles E. metastable diagram
  stabiles E. stable diagram
Eisenkohlenstofflegierung *f* iron-carbon alloy
Eisenkorn *n* ferrite grain
Eisenkreis *m (el)* ferromagnetic circuit
Eisenlängsschnitt *m (ship)* construction (steel) plan
Eisenlegierung *f* iron alloy
Eisenluppe *f* iron loop
Eisenmanganerz *n* ferriferous manganese ore
Eisenmeßgerät *n* permeameter
Eisenmeteorit *m* iron meteorite
Eisennadelinstrument *n* permanent-magnet moving-iron instrument
Eisennickelakkumulator *m* Edison accumulator, Edison storage battery, iron-nickel accumulator
Eisennickelkies *m* pentlandite, nicopyrite
Eisenoolith *m* oolitic iron ore
Eisenoxalat *n* iron oxalate
Eisenoxid *n* ferric oxide; ferrous oxide
Eisenoxidfarbe *f* iron-oxide paint
Eisenpecherz *n (min)* pitticite
Eisenplan *m s.* Eisenlängsschnitt
Eisenportlandzement *m* Portland blast furnace cement
Eisenpulver *n* iron powder
Eisenpulverelektrode *f* contact electrode
Eisenpulverkern *m (el)* iron dust core
Eisenpulvertopfkern *m (el)* pot-type iron dust core
Eisenrogenstein *m (min)* oolitic iron ore
Eisensau *f (met)* bear
Eisenschluß *m (el)* magnetic shunt
Eisenschmelze *f* iron melt
Eisenschwamm *m* sponge iron
Eisen-Silizium-Kernblech *n* iron-silicon transformer sheet
Eisenspat *m (min)* siderite, spathic ore
Eisenspiegel *m* specular iron
Eisenspule *f (el)* iron-core coil
Eisensteg *m* iron furniture
Eisenstein *m* iron-stone
Eisensulfatdosimeter *n s.* Ferrosulfatdosimeter
Eisentransformator *m (el.m.)* iron-core transformer
Eisen(II)-verbindung *f* ferrous compound, ferro-compound
Eisen(III)-verbindung *f* ferric compound, ferri-compound
Eisenverkleidungsblech *n* sheet iron cover
Eisenverlust *m* iron loss
Eisenvitriol *n* green vitriol, copperas, iron sulphate
Eisenwaren *fpl* ironmongery, ironwork, hardware
Eisenwasserstoffwiderstand *m* hydrogen iron resistance, barretter
Eisenwolframat *n* iron tungstate
Eisenzinkspat *m* ferruginous calamine
Eisenzunder *m* iron scale
Eisessig *m* glacial acetic acid
Eisfahrt *f (ship)* navigation in ice
Eisfarbe *f* ice colour (dye)
Eisfarbenbuntreserve *f (text)* coloured resist with azoic dye
Eisgang *m* floating ice
Eisgrenze *f* edge of the ice
Eisgürtel *m* ice fender
Eishaut *f (ship)* ice doubling, ice-lining, bow-grace
Eishockeyschläger *m* [icehockey] stick
Eiskalorimeter *n* ice calorimeter
eisklüftig frost cleft *(wood)*

Eiskristall *m* ice crystal
Eisläufer *m* skater
Eismeer *n* polar sea
Eisnot sein / in to be enclosed by drifting ice
Eispickel *m* ice axe (pick)
Eisräumung *f* ice removal
Eisschneider *m* ice cutter
Eisscholle *f* ice flow
Eisschrank *m* ice box
Eisschranke *f* ice barrier
Eisschutz *m (ship)* bow-grace, ice protection
Eissegeln *n* ice-boat sailing, skate sailing, ice yachting
Eissegelschlitten *m* ice boat
Eissicherung *f* anti-ice measures
Eissporn *m* 1. *(ship)* ice cutter, ice spur, creeper; 2. *(mil)* ice space *(gun)*
Eistrift *f* ice drill
Eisverstärkung *f* ice strengthening; strengthening for navigation in ice; ice trebling
Eiswachschiff *n* ice patrol vessel
Eiswasser *n* ice water
Eiszeit *f (geol)* [great] ice age
Eiweiß *n* albumen; albumin
Eiweißabbau *m* proteolysis
Eiweißabbauprodukt *n* albumen degradation product
eiweißartig albuminous, proteinaceous
Eiweißfaser *f* protein fibre
Eiweißkörper *m* protein, albuminoid
Eiweißstickstoff *m* albumen nitrogen
Eiweißtrübung *f* protein (albumen) turbidity
Eiweißverbindung *f* albuminate, protein compound
Ejektor *m* ejector
Ekatorührer *m (ch)* portable mixer
Ekliptik *f* ecliptic
Ekonomiser *m* economizer, waste gas feed heater
Ekonomiserheizfläche *f* economizer heating surface
Elaidinsäure *f* elaidic acid
Eläostearinsäure *f* elaeostearic acid
Elast *m* elastic, elastomere
Elastikgarn *n* elastic yarn
elastisch elastic
Elastizität *f* elasticity *(material)*; resilience *(fabric)*
vollkommene E. perfect elasticity
Elastizitätsachse *f* elastic axis *(of crystals)*
Elastizitätsgrenze *f* elastic limit, limit of elasticity
Elastizitätshysterese *f* elastic hysteresis
Elastizitätskoeffizient *m* / **räumlicher** bulk modulus of elasticity
Elastizitätsmesser *m* elastomer
Elastizitätsmodul *m* modulus of elasticity, elastic (Young's) modulus
Elastizitätstheorie *f* elastic theory, theory of elasticity
Elastizitätszahl *f s.* Elastizitätsmodul
Elastomer[es] *n s.* Elast
elastoplastisch elasto-plastic
Elastostatik *f* elastostatics
Elchhaut *f* elk hide
elektrifizieren to electrify
Elektrifizierung *f* electrification
Elektriker *m* electrician
elektrisch electric[al]
elektrisieren to electrize, to faradize, to electrify
Elektrisiermaschine *f* electrostatic machine
Elektrizität *f* electricity
atmosphärische E. atmospheric electricity
dynamische E. dynamic electricity
gleichnamige E. electricity of same sign
negative E. resinous electricity *(electrostatics)*
positive E. vitreous electricity
ungleichnamige E. electricity of opposite sign
Elektrizitätsdurchgang *m* passage of electricity *(through gases)*
Elektrizitätslehre *f* science of electricity
Elektrizitätsmenge *f* quantity of electricity
Elektrizitätsträger *m* carrier of electricity
Elektrizitätsversorgung *f* electricity supply
Elektrizitätsverteilung *f* distribution of electricity
Elektrizitätswerk *n* [electric] power station
Elektrizitätszähler *m* integrating (electric, energy) meter
Elektroakustik *f* electro-acoustics
elektroakustisch electro-acoustic
Elektroanalyse *f* electro-analysis
Elektroanästhesie *f* electrical anesthesia
Elektrobackofen *m* electric baking oven
Elektrobiologie *f* electrobiology
Elektrochemie *f* electrochemistry
elektrochemisch electrochemical
Elektrochirurgie *f* electrosurgery
Elektrochromatografie *f* electrochromatography
Elektrodampfkessel *m* electric (electrode) boiler, electric (electrode) steam raiser
Elektrode *f* electrode
abschmelzbare E. consumable electrode
aktive E. active electrode
blanke E. bare electrode
differente E. active electrode
eisenausbringende E. contact electrode
getauchte E. dip-coated electrode
indifferente E. reference (inert) electrode
kalte E. cold electrode
nackte E. bare electrode
negative E. negative electrode
neutrale E. neutral electrode
nicht abschmelzende E. non-focusing arc welding electrode, non-consumable electrode
positive E. positive electrode
selbstbrennende E. self-baking electrode

umhüllte E. coated (cored, sheathed) electrode
verstellbare E. sliding electrode
verzehrbare E. consumable electrode
Elektrodenabbrand *m* consumption of electrode, spark erosion on the electrodes
Elektrodenabstand *m* interelectrode gap (distance), electrode gap (spacing); spark gap
Elektrodenadmittanz *f* electrode admittance
Elektrodenanordnung *f* arrangement of electrodes, electrode system
bipolare E. bipolar electrode system
konzentrische E. concentric electrode system
unipolare E. unipolar electrode system
Elektrodenanschluß *m* electrode connection
Elektrodenausleger *m* electrode arm
Elektrodenblindleitwert *m* electrode susceptance
Elektrodenblindwiderstand *m* electrode reactance
Elektrodencharakteristik *f* electrode characteristic
Elektrodenfahrsäule *f* electrode mast
Elektrodenfassung *f* electrode clamp
Elektrodenfläche *f* electrode surface
Elektrodenform *f* electrode shape
Elektrodengleichspannung *f* supply voltage [of an electrode]
Elektrodenhalter *m* electrode holder (support)
Elektrodenimpedanz *f* electrode impedance
Elektrodenkapazität *f* interelectrode capacitance
Elektrodenkennlinie *f* electrode characteristic
Elektrodenkitt *m* electrode compound
Elektrodenklemme *f* electrode terminal
Elektrodenkonduktanz *f* electrode conductance
Elektrodenkühllamelle *f* electrode radiator
Elektrodenkühlung *f* electrode cooling
Elektrodenleitung *f* electrode lead
elektrodenlos electrodeless
Elektrodenmantel *m* electrode shell
Elektrodenmehl *n* crushed electrodes
Elektrodennachsteller *m* electrode adjusting gear
Elektrodenpotential *n* electrode potential
dynamisches E. dynamic electrode potential
statisches E. static electrode potential
Elektrodenrauschen *n* electrode noise
Elektrodenreaktanz *f* electrode reactance
Elektrodensalzbadofen *m* electrode salt bath furnace
Elektrodenscheinleitwert *m* electrode admittance
Elektrodenscheinwiderstand *m* electrode impedance
Elektrodensonde *f* electrode probe
Elektrodenspanner *m* electrode prong
Elektrodenspitze *f* electrode point
Elektrodenspitzmaschine *f* electrode shaper
Elektrodensteuerung *f* electrode control
Elektrodensuszeptanz *f* electrode susceptance
Elektrodenträger *m* electrode jib
Elektrodenüberschlag *m* arcing between electrodes
Elektrodenverlust *m* electrode dissipation
Elektrodenverlustleistung *f* electrode dissipation
Elektrodenverstellwerk *n* electrode adjusting gear
Elektrodenvorspannung *f* electrode bias
Elektrodenwirkleitwert *m* electrode conductance
Elektrodenwirkwiderstand *m* electrode resistance
Elektrodesikkation *f* electrodesiccation
Elektrodialyse *f* electrodialysis
Elektrodynamik *f* electrodynamics
elektrodynamisch elektrodynamic[al]
Elektrodynamometer *n* electrodynamometer
Elektroenergie *f* electric energy
Elektroenergiebedarf *m* demand for electric energy
Elektroenergieerzeugung *f* generation of electric energy
Elektroentstaubung *f* electrical precipitation
Elektroerosion *f* electroerosion *(metal working process)*
elektroerosiv electroerosive
Elektrofeile *f* rotary file
Elektrofilter *n* electric separator; electrostatic filter (precipitator)
Elektrofotografie *f* electro-photography
Elektrofunkenbearbeitung *f* spark cutting
Elektrogeräte *npl* electrical appliances
Elektrographit *m (Am)* Acheson graphite
Elektrohandbohrmaschine *f* **mit Brustschild** electric hand drill of the breast type
Elektroherdofen *m* electric-hearth furnace
Elektrohubkarren *m* electric elevating platform truck
Elektroinduktionsofen *m* electric induction furnace
Elektroingenieur *m* electrical engineer
Elektroinstallateur *m* electrician
elektrokalorisch thermo-electric; electrothermal
Elektrokapillarität *f* electrocapillarity
Elektrokapillarkurve *f* electrocapillary curve
Elektrokardiogramm *n* electrocardiogram
Elektrokarren *m* electric [baggage and freight] truck, storage battery truck; elektromobile
Elektrokatze *f* electric trolley
Elektrokauter *m* electrocautery
Elektrokauterisation *f* electrocauterization
Elektrokettenzug *m* electric chain pulley (hoist)
Elektrokinetik *f* electrokinetics
elektrokinetisch electrokinetic
Elektrokoagulation *f* electrocoagulation
Elektrokocher *m* electric boiler
Elektrokochplatte *f* electric boiling plate
Elektrokran *m* electric crane
elektrokratisch electrocratic

Elektrokupplung *f* electric clutch
Elektrolichtbogen *m* electric arc
Elektrolichtbogenrapidschweißung *f* submerged arc welding
Elektrolot *n* electric sounder
Elektrolumineszenz *f* electroluminescence
Elektrolyse *f* electrolysis
Elektrolysebad *n* electrolytic bath
Elektrolysenendlauge *f (ch)* spent electrolyte
Elektrolyseschlamm *m* electrolytic slime
Elektrolyseur *m* electrolyzer
Elektrolysezelle *f* electrolytic cell
Elektrolyt *m* electrolyte
verbrauchter E. foul electrolyte
Elektrolytableiter *m* electrolytic surge (lightning) arrester
Elektrolytaufbereitung *f* purification of electrolyte
Elektrolytblei *n* electrolytic lead
Elektrolyteisen *n* electrolytic iron
Elektrolytgleichrichter *m* electrolytic rectifier (valve), chemical rectifier
elektrolytisch electrolytic[al]
Elektrolytkondensator *m* electrolytic capacitor
E. mit Stiftanschlüssen prong-fitting electrolytic capacitor
feuchter E. semi-dry electrolytic capacitor
gepolter E. polarized electrolytic capacitor
nasser E. wet electrolytic capacitor
ungepolter E. non-polarized electrolytic capacitor
Elektrolytkupfer *n* electrolytic copper
Elektrolytzähler *m* electrolytic meter
Elektrolytzink *n* electrolytic zinc
Elektromagnet *m* electromagnet
elektromagnetisch electromagnetic; solenoid
Elektromagnetismus *m* electromagnetism
Elektromagnetkern *m* pole shank (core)
Elektromagnetkupplung *f* electromagnetic clutch
Elektromagnetlamellenkupplung *f* electromagnetic laminated clutch
Elektromagnetstellantrieb *m (cont)* solenoid actuator
Elektromaschinenbau *m* electrical engineering industry
Elektromechaniker *m* electrician
elektromechanisch electromechanic[al]
Elektromedizin *f* electromedicine
Elektrometallurgie *f* electrometallurgy
elektrometallurgisch electrometallurgic[al]
Elektrometer electrometer
Elektrometerröhre *f* electrometer valve
Elektrometrie *f* electrometry, electrometric titration
elektrometrisch electrometric
Elektromobil *n* electromobile
Elektromotor *m* electric motor, electromotor
E. für Unterwasserfahrt submerged propulsion motor
elektromotorisch electromotive
Elektron *n* electron, negatron
angelagertes E. trapped electron
äußeres E. outer-shell electron
befreites E. liberated electron
eingefangenes E. trapped electron
energiereiches E. energetic electron
freies E. free electron
gebundenes E. bound electron
inneres E. inner-shell electron
langsames E. cold electron
positives E. *s.* Positron
rotierendes E. spinning electron
rückgestreutes E. backscattered electron
schnelles E. fast electron
überschüssiges E. extra electron
unpaar[ig]es E. unpaired (odd) electron
zurückgestreutes E. backscattered electron
Elektronarkose *f* electric sleep
Elektron-Defektelektron-Paar *n* hole-electron pair
elektronegativ electronegative
Elektronenabbremsung *f* electron retardation
Elektronenabgabe *f s.* Elektronenemission
Elektronenablenkung *f* deflection of [the] electrons
Elektronenablösung *f* detachment of electrons, electron detachment (liberation)
E. durch Strahlung photo-detachment *(photoionization)*
Elektronenaffinität *f* electron affinity
Elektronenanlagerung *f* electron attachment (trapping)
Elektronenaufbau *m* electronic structure
Elektronenauslösung *f* electron emission, liberation of electrons
Elektronenaussendung *f s.* Elektronenemission
Elektronenaustausch *m* electron exchange
Elektronenaustrittsarbeit *f* electron[ic] work function
Elektronenbahn *f* electron orbit (path, trajectory)
Elektronenbelegung *f* density of electrons, electron density
Elektronenbeschleuniger *m* electron accelerator
linearer E. linear electron accelerator
Elektronenbeschuß *m* electron bombardment
Elektronenbeugung *f* electron diffraction
Elektronenbeugungsaufnahme *f* electron diffraction pattern
Elektronenbeugungsbild *n* electron diffraction pattern
Elektronenbeugungsgitter *n* electronic diffraction grating
Elektronenbild *n* electron image
Elektronenbindung *f* electron binding

Elektronenblitz *m* electronic flash
Elektronenbombardement *n* cathodic bombardment
Elektronenbündel *n* electron (cathode) beam
Elektronenbündelung *f* electron focusing
Elektronendichte *f* electron density
Elektronendonator *m* electron donor
Elektronendosis *f* electron dose
Elektronendrall *m* electron spin
elektronendurchlässig electron-permeable, electron-transmitting
Elektronendurchstrahlung *f* electron transmission
Elektroneneinbau *m* insertion of electrons
Elektroneneinfang *m* electron capture (trapping)
Elektronenemission *f* emission of electrons, electron emission
  lichtelektrische E. photoelectric emission
  thermische E. thermionic emission
Elektronenemissionsfläche *f* electron-emitting area (surface), electron-emission area (surface)
Elektronenemitter *m* electron emitter
elektronenemittierend electron-emissive
elektronenempfindlich electron-sensitive
Elektronenenergie *f* electron energy
Elektronenentladung *f* electron discharge
Elektronenerzeugung *f* electron production
Elektronenfänger *m* acceptor
Elektronenfleck *m* electron (scanning) spot
Elektronenfluß *m* electron flow (flux)
Elektronenfokussierung *f* electron focusing
Elektronenformel *f (ch)* electronic formula
Elektronengas *n* electron gas
Elektronengehirn *n* electronic brain
elektronengekoppelt electron-coupled
Elektronengleichgewicht *n* electronic equilibrium
Elektronengleichrichter *m* electronic rectifier
Elektronenhalbleiter *m* electronic semiconductor
Elektronenhülle *f* sheath of electrons, electron shell (envelope), outerelectrons
  E. des Atoms atomic [electron] shell, electron shell
Elektronenkanone *f* electron gun
Elektronenkonfiguration *f* electron[ic] configuration
Elektronenkonzentration *f* electron concentration
Elektronenkopplung *f* electron coupling
Elektronenladung *f* charge of an electron, electron charge
  spezifische E. specific electronic charge
Elektronenlaufzeit *f* electron transit (orbit) time
Elektronenlawine *f* electron avalanche (cascade)
Elektronenleerstelle *f* vacant electron site
Elektronenleiter *m* electron conductor
Elektronenleitfähigkeit *f* electron[ic] conductivity, p-type conductivity
Elektronenleitung *f* electron conduction
Elektronen-Lichtquanten-Komponente *f* electromagnetic (electron-photon) component
Elektronenlinearbeschleuniger *m s.* Linearbeschleuniger für Elektronen
Elektronenlinse *f* electron lens
Elektronenmangel *m* electron deficit
Elektronenmasse *f* mass of the electron, electron mass
Elektronenmikroskop *n* electron microscope
  E. mit magnetischen Linsen magnetic [electron] microscope
Elektronenmikroskopie *f* electron microscopy
Elektronenniederschlag *m* electron deposition
Elektronenniveau *n* electron[ic] level
Elektronenoptik *f* electron optics
elektronenoptisch electron-optical
Elektronenorgel *f* pipeless organ
Elektronenort *m* electron position (locus)
Elektronenpaar *n* electron pair
  einsames E. lone pair of electrons
Elektronenpaket *n* bunch of electrons
Elektronen-Photonen-Komponente *f* electromagnet (electron-photon) component
Elektronenphysik *f* electron physics
Elektronenplasma *n* electron plasma (gas)
Elektronenpolarisation *f* electron polarization
Elektronenquelle *f* electron [emission] source
Elektronenradius *m* [classical] electron radius
Elektronenrauschen *n* electron background
Elektronenrechner *m* electronic computer
Elektronenreflektor *m* electron mirror
Elektronenreibung *f* electron friction
Elektronenröhre *f* electron [discharge] tube, electron [discharge] valve
Elektronenröhrenherstellung *f* electron tube manufacture
Elektronenröhrenschalter *m* vacuum tube switch
Elektronenröhrentechnik *f* electron tube engineering
Elektronenschale *f* electron shell
  E. des Atoms atomic electron shell
  abgeschlossene E. completed electron shell
Elektronenschalter *m* electronic (vacuum-tube) switch
Elektronenschar *f* electron swarm
Elektronenschleuder *f* betatron
Elektronenschwarm *m* electron swarm
Elektronensonde *f* electron probe
Elektronenspektrum *n* electron spectrum
Elektronenspin *m* electron spin
Elektronensprung *m* electron transition
Elektronenspur *f* electron track
Elektronenstatistik *f* electron statistics
Elektronensteuerung *f* electronic control
Elektronenstoß *m* electron collision (impact)

Elektronenstoßleuchten *n* ionization glow
Elektronenstrahl *m* electron beam; cathode ray
abtastender E. scanning electron beam
Elektronenstrahlabtaster *m* electron-scanning pencil (brush, beam)
Elektronenstrahlbündelung *f* concentration of the electron beam
Elektronenstrahler *m* electron emitter (gun)
Elektronenstrahlerzeuger *m* electron gun
Elektronenstrahlerzeugungssystem *n* electron gun
Elektronenstrahlgerät *n* electron jet instrument
Elektronenstrahlhochleistungsverstärkerröhre *f* beam power amplifier tube
Elektronenstrahlkonzentration *f* electron-beam focusing
Elektronenstrahlofen *m* electron bombardment furnace
Elektronenstrahloszillograf *m* cathode-ray oscillograph
Elektronenstrahlpunktgröße *f* spot size
Elektronenstrahlpunktsucher *m* spot locator
Elektronenstrahlquerschnitt *m* aperture of the beam
Elektronenstrahlröhre *f* cathode (electron) ray tube, cathode (electron) beam tube
Elektronenstrahlschmelzofen *m* electron-beam melting furnace
Elektronenstrahlschweißung *f* electron-beam welding
Elektronenstrahlstrom *m* cathode-ray current
Elektronenstrahlsystem *n* electron gun
Elektronenstrahlverstärker *m* electron-beam amplifier
Elektronenstreuung *f* electron scattering
Elektronenstrom *m* electronic current (flow)
thermischer E. thermionic current
Elektronenstromdichte *f* density of electron current
Elektronenströmung *f* electron stream
Elektronenstruktur *f* electronic structure
Elektronensynchrotron *n* electron synchrotron
Elektronentemperatur *f* electron temperature
Elektronenterm *m* electron level (term)
Elektronentheorie *f* electron theory
Elektronenträger *m* electron carrier
Elektronenträgheit *f* electron inertia
Elektronenübergang *m* electron transition; electron transfer
strahlungsloser E. non-radiative electron transition
Elektronenübergangswahrscheinlichkeit *f* electron transition probability
Elektronenüberholungsgebiet *n* electron catch-up region, electron over-take region
Elektronenüberschuß *m* excess of electrons
Elektronenverdampfung *f* electron evaporation (emission)
Elektronenverteilung *f* distribution of electrons, electron distribution
Elektronenvervielfacher *m* electron multiplier, multiplier phototube
Elektronenvervielfachung *f* electron multiplication
Elektronenvolt *n* electron volt
Elektronenwelle *f* electron wave
Elektronenwolke *f* electron cloud
Elektronenzähler *m* electron counter
Elektronenzerfall *m s.* Beta-Zerfall
Elektronenzustand *m* electron state
Elektronenzyklotron *n* electron cyclotron
Elektronik *f* electronics
molekulare E. molecular electronics
Elektroniker *m* electronician, electronic technician
Elektronikindustrie *f* electronics industry
Elektronikingenieur *m* electronic engineer
Elektron-Ion-Stoß *m* electron-ion collision
elektronisch electronic
Elektronmetall *n* electron metal *(magnesium alloy)*
Elektron-Positron-Paar *n* electron-positron pair
Elektroofen *m* electric furnace
Elektrooptik *f* electrooptics
Elektroosmose *f* electroosmosis
elektrophil *(ch)* electrophilic
Elektrophorese *f* electrophoresis
Elektrophysik *f* electrophysics
Elektroplattformwagen *m* electric platform truck
elektroplattieren to electroplate
Elektroplattierung *f* electroplating
elektropneumatisch electropneumatic
Elektropolieren *n* electro-polishing
elektropositiv electropositive
Elektroraffination *f (ch)* electrorefining
Elektroroheisen *n* electric pig
Elektroschachtofen *m* electric shaft furnace
Elektroschlepper *m* [industrial] electric tractor
Elektroschock *m* electric shock
Elektroschweißen *n* electrical welding
Elektrosherardisierung *f* electrosherardizing
Elektroskop *n* electroscope
Elektrostahl *m* electric steel
Elektrostahlofen *m* electric steel furnace
Elektrostahlschmelzen *n* electric furnace melting
Elektrostatik *f* electrostatics
elektrostatisch electrostatic[al]
Elektrostriktion *f* electrostriction
Elektrotechnik *f* electrical engineering (technology); electrotechnics
Elektrotechniker *m* electrical engineer
Elektroteerscheider *m* electro detarrer
Elektrotherapie *f* electrotherapy
Elektrothermie *f* electrothermics
Elektrotypie *f* electrotyping

Elektrovalenz *f* electrovalence
Elektrowärmelehre *f* electrothermics
Elektrozaun *m* electric fence
Elektrozaunanlage *f* fencing equipment
Elektrozug *m* electric hoist
Elektrozugmaschine *f* electric tractor
Element *n* 1. *(ch)* element; 2. *(build)* member; 3. *(el)* battery, cell, *(Am)* pile
aktives E. active element
fotoelektrisches E. photoelectric device
galvanisches E. galvanic (voltaic) cell
gasförmiges E. gaseous element
leitendes E. conductive element
monoisotopisches E. monoisotopic element
nasses E. wet cell
nichtmetallisches E. non-metallic element
passives E. passive element
radioaktives E. radioactive element, radioelement
stabiles E. stable element
umkehrbares E. reversible cell
elementar elementary; ultimate
Elementaranalyse *f* elementary (ultimate) analysis
Elementararbeit *f (ph)* elementary work
Elementarbüschel *n* elementary pencil (brush)
Elementardipol *m* elementary dipole
Elementarladung *f* elementary (electronic) charge, charge of the electron
elektrische E. electronic (elementary) charge, charge of the electron
Elementarphysik *f* elementary physics
Elementarquantum *n* elementary quantum
elektrisches E. elementary quantum of electricity, charge of the electron
Elementarreaktion *f (ch)* elementary reaction
Elementarströmung *f (el)* elementary circulation
Elementarteilchen *n* elementary particle
Elementarwelle *f* elementary wave
Elementarzelle *f* elementary lattice cell *(of crystals)*
Elementbehälter *m (el)* battery box
Elementenpaar *n* pair *(kinematics)*
höheres E. higher pair
niederes E. lower pair
Elementprüfung *f* element test *(circuit breakers)*; cell test *(batteries)*
Elementschlamm *m* battery mud (sediment)
Elementtrennung *f (ph)* separation of elements
Elendsviertel *n* blighted area (district), slum
Elevator *m s.* Becherwerk
Elevatortuch *n* elevator (elevating) canvas
Elfeck *n* undecagon
Elfenbein *n* ivory
elfenbeinartig ivory-like
Elfenbeinpapier *n* ivory paper
Elferprobe *f (math)* casting out elevens
Elferrest *m (math)* excess of elevens
Elin-Hafergut-Schweißverfahren *n* firecracker welding *(Hafergut method)*
Elinvar *n* elinvar
Ellbogenrohr *n* elbow pipe
Elle *f* cubit
Ellipse *f* ellipse
Ellipsenlenker *m* / **angenäherter** approximately elliptical motion
genauer E. exactly elliptical motion
umgekehrter E. inverted elliptical motion
Ellipsenzeichner *m* ellipsograph, elliptic trammel
Ellipsenzirkel *m* elliptic trammel
Ellipsograf *m* ellipsograph
Ellipsoid *n* ellipsoid
elliptisch elliptic
Ellira-Schweißung *f* Ellira welding [process], Ellira process of submerged arc welding
Elmsfeuer *n* St. Elmo's fire
Eloxalschicht *f* anodized coating
Eloxalverfahren *n* anodic oxidation, anodizing
eloxieren to anodize
Eloxieren *n* anodic treatment, anodization
Eluat *n* eluate
eluieren to elute
Eluierung *f* elution
Elutionsmittel *n* eluent
Email *n* [vitreous] enamel
Emailfarbe *f* enamel paint
Emailglas *n* enamelled glass
Emaillack *m* enamel varnish
Emaillackdraht *m* enamel-insulated wire
Emaille *f s.* Email
emaillieren to enamel
Emaillierofen *m* enamelling furnace
Emaillierung *f* vitreous enamelling
Emailweiß *n* enamel white, lithopone
Emanation *f* [active] emanation; radioactive gas
aktive E. active emanation
Emanationstechnik *f* emanation method
emanieren to emanate
Emaniermethode *f* emanation method
Emballage *f* packing [material], container
Embargo *n* embargo
Emission *f (ph)* emission
erzwungene E. stimulated emission
feldlose E. zero-field emission
glühelektrische E. thermionic emission
induzierte E. negative (induced) emission
lichtelektrische E. photo-electric emission
negative E. negative (induced) emission
sukzessive E. cascade radiation (emission)
thermische E. thermionic emission, Richardson effect
Emissionsanalyse *f* emission analysis
Emissionsbande *f* emission band
Emissionsfläche *f* emitting surface (area)
Emissionsfotozelle *f* photo-emissive cell

Emissionsfrequenzband *n* frequency band of emission
Emissionskennlinie *f* emission characteristic
Emissionskoeffizient *m* emission coefficient
Emissionskonstante *f* emission constant
Emissionslinie *f* emission line
Emissionsmikroskop *n* emission microscope
Emissionsrichtung *f* direction of emission
Emissionsschicht *f* emitting layer
Emissionsspalt *m* emission slit
Emissionsspaltung *f* emissive fission
Emissionsspektrum *n* emission spectrum
Emissionsstärke *f* emissive power
Emissionsstrom *m* emission current
  feldfreier E. field-free emission current
Emissionsvermögen *n* emissivity, emission (emissive) power
  E. eines Temperaturstrahlers emissivity of a thermal radiator
Emitter *m* emitter
Emitteranschluß *m* emitter contact
Emitter-Basis-Diode *f* emitter-base diode
Emitterelektrode *f* emitter electrode
Emitterfolger *m* emitter follower *(semiconducting device)*
Emitter-Kollektor-Diffusionsstrom *m* emitter-collector diffusion current
Emitterschaltung *f* grounded-emitter circuit
Emitterschicht *f* emitter junction
Emitter-Tor-Abstand *m* emitter-to-gate spacing
emittieren to emit; to eject
  Elektronen e. to eject electrons
Empfang *m* reception
  E. mit Mehrfachantenne spaced aerial reception
  gerichteter E. directional reception
Empfänger *m* 1. consignee, depositary; 2. responder *(radar)*; receiving set, receiver *(radio)*
  E. im Lotsenflugzeug pilot indicator *(radar)*
  fotoelektrischer E. photodetector
  fotovoltaischer E. photovoltaic detector
  selektiver E. selective receiver
  tragbarer E. portable receiver (receiving set)
Empfängerabgleich *m* receiver alignment
Empfängerausfall *m* receiver failure
Empfängerausgang *m* receiver output
Empfängerdämpfung *f* receiver attenuation
Empfängereichung *f* receiver calibration
Empfängereigenstrahlung *f* receiver radiation *(radio)*
Empfängererdanschluß *m* receiver earth
Empfängererdklemme *f* receiver earth terminal
Empfängererholung *f* receiver recovery
Empfängergehäuse *n* [receiver] cabinet *(radio)*
Empfängerlautstärke *f* receiver output volume
Empfängerpotentiometer *n* receiver potentiometer
Empfängerrauschen *n* receiver noise
Empfängerröhre *f* receiver tube
Empfängerseite *f* receiving end
Empfangsanlage *f* receiving installation
Empfangsantenne *f* receiving aerial
Empfangsbahn *f* destination railway
Empfangsbereich *m* service area
Empfangsbestätigungssignal *n* acknowledgement signal *(radio)*
Empfangscharakteristik *f* response curve *(of aerials)*
Empfangsgebäude *n* terminal building *(airport)*
Empfangsgerät *n* receiving set, receiver unit
Empfangsgüte *f* reception quality
Empfangshafen *m* receiving port, port of arrival
Empfangskreis *m* receiving circuit
Empfangsleistung *f* incoming power
Empfangslocher *m* receiving perforator, reperforator
Empfangsseite *f* receiving end
Empfangssperrzelle *f* T-R cell *(waveguides)*
Empfangsstelle *f* receiving station
Empfangssystem *n* **für feste Frequenzkanäle** channelized receiving system
Empfangsverstärker *m* receiving amplifier
empfehlen to recommend
empfindlich sensitive
Empfindlichkeit *f* sensitivity, sensitiveness *(film)*; response *(instrument)*
  E. bei naher Besprechung *(telec)* close-talking sensitivity
  E. bei Vollausschlag full-scale sensitivity
  E. beim relativen Pegel-Null zero relative level sensitivity
  E. des Reglers sensitiveness of the governor
  E. im diffusen Schallfeld random-incidence response
  axiale E. axial response
  fotoelektrische E. photoelectric sensitivity, photosensitivity
  spektrale E. spectral response; *(telev)* colour response
Empfindlichkeitsabnahme *f* sensitivity decrease
Empfindlichkeitsfaktor *m* multiplying power *(of a galvanometer shunt)*
Empfindlichkeitsgewinn *m* gain in sensitivity
Empfindlichkeitsgrenze *f* threshold (limit) of sensitivity, threshold (limit) of response, detection limit
Empfindlichkeitskurve *f* response curve
Empfindlichkeitsmesser *m* sensitometer
Empfindlichkeitsprüfung *f* sensitivity test
Empfindlichkeitsschwelle *f* threshold of sensitivity, sensitivity threshold
Empfindlichkeitszeit *f* s. Ansprechzeit
Empireformat *n (print)* Empire *($6\frac{1}{2} \times 4\frac{1}{4}$ in.)*
empirisch empiric[al]
Empore *f* gallery
Emscherbrunnen *m* Emscher tank

Emulgator *m* emulsifying agent, emulsifier
emulgierbar emulsifiable
Emulgierbarkeit *f* emulsibility
Emulgiereinrichtung *f* emulsifier
emulgieren to emulsify
emulgierfähig emulsifiable
Emulgierfähigkeit *f* emulsifiability
Emulgiermittel *n* emulsifying agent
Emulgierung *f* emulsifying, emulsification
Emulgierungsmittel *n* emulsifying agent
Emulgierzusatz *m* emulsifier
Emulseurkessel *m (therm)* boiler fitted with emulseurs
Emulsion *f* emulsion
E. von Öl in Wasser oil-in-water emulsion
E. von Wasser in Öl water-in-oil emulsion
feinkörnige E. fine grain emulsion
Emulsionsbildung *f* emulsification
Emulsionsmischpolymerisation *f* emulsion copolymerization
Emulsionspaket *n (nucl)* pellicle stack
Emulsionspolymerisation *f* emulsion polymerization
Emulsionsschicht *f* emulsion layer
Emulsionsspritzmittel *n* emulsion sprayer
Emulsionsstapel *m (nucl)* pellicle stack
Emulsionsstern *m s.* Zertrümmerungsstern
Emulsionstechnik *f* [nuclear] emulsion technique, photographic emulsion technique
Emulsionsträger *m* base
Emulsoid *n* emulsoid
enantiomorph enantiomorphous *(crystal)*
enantiotrop enantiotropic
Enantiotropie *f* enantiotropy, enantiotropism
Enargit *m (min)* enargite
Endabnahme *f (mach)* final examination
Endamt *n (telec)* terminal exchange (station), district exchange
Endanflugkurve *f* final procedure turn
Endanschlag *m (mach)* stop for adjusting length of stroke, end (limit) stop
Endauflager *n (build)* abutment; end support
Endauslösung *f* end stop (trip)
Endausschalter *m* limit switch
Endausschlag *m* full range deflection
Enddruck *m* final pressure
Ende *n* end; termination; tip
E. der Meßspitze tip
geschlossenes E. blind end *(cylinder)*
Endechosperre *f* terminal echo suppressor
Endeinrichtungen *fpl* local end
Endeinspannung *f* end fixing (restraint, fixation)
Enderzeugnis *n* finished product
Endfenster *n* end window
Endfernamt *n* terminal trunk exchange
Endfestigkeit *f* final strength
Endformat *n* trimmed size
Endgebühr *f* terminal charge
Endgefüge *n* final structure *(heat treatment)*
Endgeschwindigkeit *f* final velocity (speed), terminal velocity (speed)
Endgestänge *n (el)* end (termination) pole
verankertes E. stayed terminal pole
verstrebtes E. strutted terminal pole
Endhafen *m* terminal [port], final (ultimate) port
Endhaken *m* end hook
Endheizwand *f* final heating wall
Endimpedanz *f (el)* terminal impedance *(of load)*
Endkämmung *f* end cogging *(carpentry)*
Endkapazität *f (el)* terminating capacity, top capacity *(aerial)*
Endkappe *f (el)* end bell
Endkern *m* final nucleus
Endkontakt *m* terminal contact
Endkontrolleinrichtung *f (telev)* master monitor
Endkoppelschleife *f* segment-fed loop *(magnetron)*
Endlagenschalter *m* limit switch
Endlauge *f* final liquor
endlich finite
endlos continuous *(filament)*; endless *(belt)*
Endlosformularführung *f* continuous form guide
Endluftkasten *m* end tank *(life-boat)*
Endmaß *n* end measure, gauge block, block gauge
Endmaßblock *m* size (gauge) block
Endmasse *f (rock)* final mass
Endmaßlehrenbohrmaschine *f* end-measure jig borer
Endmaßsatz *m* set of gauge (size) blocks
Endmast *m* end pole
Endmonitor *m* transmission monitor, *(Am)* master monitor
Endmontage *f* final erection (assembling, assembly)
Endofen *m* end oven
endotherm endothermic
Endpartie *f* finishing end *(papermaking)*
Endpaßstück *n* end filling piece, blade stopper *(turbine)*
Endpentode *f* output pentode
Endprodukt *n* final (finished, end) product
Endpunkt *m* end point
E. der Reaktion end-point of the reaction
Endröhre *f* final stage valve
Endschalter *m* limit switch
Endschott *n* end bulkhead; end plate
E. der versenkten Back break bulkhead of sunk forecastle
Endsortierung *f* final separation *(of paper)*
Endspannung *f* final voltage, *(Am)* cut-off voltage
Endspreize *f (build)* face piece (waling, waler)
Endstabilität *f* final (ultimate) stability
Endstein *m* finishing die *(drawing)*
Endstellung *f* final (end) position
Endstreckenband *n* final drawn sliver

Endstrom *m* final current
Endstück *n* follow-rest grip, rear support *(broach)*
Endstufe *f (el)* output (final) stage
Endstufenmodulation *f* high-level modulation
Endsumme *f* grand total
Endtermin *m* target date, deadline
Endtetrode *f* **mit Elektronenbündelung** beam power tetrode
Endtransistor *m* output transistor
Endtülle *f* bushing nipple, terminal socket
Endübertemperatur *f* final temperature rise
Endumschalter *m* travel-reversing switch
Endvakuum *n* final vacuum
Endverankerung *f* end anchorage *(prestressed concrete)*
Endverkämmung *f* end cogging *(carpentry)*
Endverschluß *m* [cable] sealing end, cable terminal (head)
Endverstärkerstufe *f* final amplifier stage, power-amplifier stage
Endverzug *m* finish[ing] draft
Endverzweiger *m* terminal block (point), distributing block (point), distribution terminal
Endvorspannung *f* final prestress[ing] force *(prestressed concrete)*
Endwert *m* terminal (final) value
Endwinkel *m* **der Schaufel** outlet angle of the blade *(turbine)*
Endzustand *m* final (terminal) state
  thermischer E. final thermal state
Energetik *f* energetics
energetisch high-energy
Energie *f* energy
  E. der Bewegung kinetic energy, energy of motion
  E. der Lage potential energy, energy of position
  E. des Reglers regulating power of the governor
  aufgewandte E. applied energy; input energy
  ausgestrahlte E. radiated energy
  äußere kinetische E. external kinetic energy
  chemische E. chemical energy
  elektrische E. electric energy
  elektromagnetische E. electromagnetic energy
  gebundene E. stored (latent, bound) energy
  gespeicherte E. stored energy
  innere E. intrinsic energy
  innere kinetische E. internal kinetic energy
  kinetische E. kinetic energy
  kinetische E. der Spaltneutronen fission neutron [kinetic] energy
  kritische E. critical energy
  mechanische E. mechanical energy
  potentielle E. potential energy
  potentielle E. der Formänderung potential energy of deformation
  ruhende E. potential energy, energy of position
  spezifische E. specific energy
  thermische E. thermal (heat, calorific) energy; thermal power
  zugeführte E. input energy
Energieabfall *m* energy drop
Energieabgabe *f* energy output; energy release (delivery)
energieabhängig energy-dependent
Energieabnahme *f* energy decrement
Energieabsorption *f* energy absorption
Energieäquivalent *n* energy equivalent
energiearm of low energy, low-energy, non-penetrating, soft
Energieauflösung *f* energy resolution
Energieausbeute *f* energy yield; energy efficiency
Energieausgang *m* energy output
Energieausgleich *m* power balance
Energieausstrahlung *f* radiation of energy
Energieaustausch *m* energy exchange
Energieband *n* energy band
  leeres E. empty band
  verbotenes E. forbidden band
  vollbesetztes E. filled band
Energiebandverbreiterung *f* energy band broadening
Energiebedarf *m* power consumption, energy demand
Energiebereich *m* energy range (region)
Energiebetrag *m* amount of energy
Energiebeziehung *f* energy relation
Energiebilanz *f* energy balance
Energiebreite *f* energy width
Energiedekrement *n* energy decrement
  logarithmisches E. logarithmic energy decrement
  mittleres logarithmisches E. mean logarithmic energy decrement
Energiedichte *f* density of energy, energy density
  räumliche E. density of the total energy
Energiedissipation *f* energy (power) dissipation
Energiedosis *f* absorbed dose
  akkumulierte E. cumulative absorbed dose
  integrale E. integral absorbed dose
Energiedosisleistung *f* absorbed dose rate
Energieeinheit *f* energy unit, unit of energy
Energieerhaltung *f* energy conservation, conservation of energy
Energieerzeugung *f* power production, energy generation
Energiefläche *f* energy surface
Energiefluenz *f* energy fluence
Energiefluß *m* energy flux; radiant power (flux)
Energieflußdichte *f* energy flux density (intensity)
Energieform *f* kind of energy
Energiefortleitung *f* energy transmission
Energiefreisetzung *f* energy release

Energiegewinn *m* gain of energy
Energiegewinnung *f* energy generation (release)
energiegleich equal-energy
Energiegleichgewicht *n* energy balance
Energiegleichverteilungssatz *m* law of equipartition of energy
Energiegruppe *f* energy group
Energie-Impuls-Tensor *m* energy tensor
Energiekonverter *m* energy converter
Energieleitung *f* power [transmission] line, feed-line; conduction of energy
Energielücke *f* energy gap *(of semiconductors)*
Energie-Masse-Äquivalenz *f (nucl)* mass-energy equivalence, equivalence of mass and energy
Energiemenge *f* quantity of energy
Energiemesser *m* ergometer
Energiemulde *f* energy well
Energieniveau *n* energy level
Energieniveaubestimmung *f* energy level determination
Energieprinzip *n* energy conservation principle
Energiequant *n* energy quantum
Energiequelle *f* source of energy
Energiereaktor *m* power reactor
energiereich *(ch)* energized; hard, penetrating, high-energy *(radiation)*
Energie-Reichweite-Beziehung *f s.* Reichweite-Energie-Beziehung
Energiesatz *m* conservation of energy principle; energy theorem
Energieschale *f* energy shell
Energieschema *n* energy-level diagram; term scheme (diagram)
Energiespeicher *m* energy store (storage device)
Energiespektrum *n* energy [level] spectrum
  E. der Reaktorneutronen pile (reactor) spectrum
  E. der Spaltneutronen fission spectrum
Energiestau *m* energy accumulation
Energiestrom *m* energy flow, flow (flux) of energy
Energiestromdichte *f* energy flow per unit area
Energieterm *m* energy term
Energietönung *f* reaction energy; Q value
Energieübertragung *f* energy transfer; power transmission
  lineare E. linear energy transfer, LET
Energieumformung *f* power conversion
Energieumwandlung *f* power conversion, transformation of energy
Energieverbrauch *m* power (energy) consumption
Energieverlust *m* loss of energy, energy loss (decrement)
  logarithmischer E. logarithmic energy decrement
  mittlerer logarithmischer E. mean logarithmic energy decrement
Energieversorgung *f* power supply
Energieverteilung *f* energy distribution, distribution of energy
  spektrale E. spectral energy distribution
Energieverteilungskurve *f* energy distribution characteristic
Energiewandler *m* power converter, transducer
  E. mit Energiequelle active transducer
Energiewirtschaft *f* power economy
Energiezufuhr *f* power feed
Energiezustand *m* energy state, term
  E. des Atomkerns nuclear state, energy state [of the nucleus]
  erlaubter E. allowed energy state
  verbotener E. forbidden energy state
Energiezuwachs *m* gain of energy
Enfleurage *f (ch)* enfleurage
eng narrow; close, closely spaced; tight[ly]
enger werden to narrow
engfügig with narrow joints between courses *(carpentry)*
Enghalsflasche *f* narrow-necked bottle
Engler-Grad *m* Engler degree
Engler-Kolben *m* Engler flask
engmaschig fine mesh [gauge]; close-meshed
Engobe *f* engobe *(ceramics)*
Engpaß *m* bottleneck
engringig close grain, narrow-ringed, with small annular rings *(wood)*
engspiralig high-helix, fast spiral
engtoleriert close-tolerance
Enimin-Form *f (ch)* enimic form
Enol-Form *f (ch)* enol form
Enolisierung *f (ch)* enolization
entaktivieren to deactivate, to decontaminate
Entaktivierung *f* deactivation, decontamination
entartet degenerate
Entartung *f* 1. *(ph)* degeneracy, degeneration *(Boltzmann)*; 2. *(ag)* degeneration
entaschen to remove the ash
Entaschung *f* ash removal
entasten to free from branches
entballasten to unballast *(a vessel)*
entbasten to degum, to boil off *(silk)*
Entbastungsechtheit *f* fastness to degumming
entbenzolen to debenzolize
Entbenzolung *f* benzol extraction, debenzolizing
Entblätterung *f* defoliation
entborken to bark
Entbrummer *m (el)* hum potentiometer (eliminator)
Entbrummkondensator *m* anti-hum capacitor
Entbrummung *f (el)* hum balance
entbündeln to debunch; to defocus *(rays)*
entdämpfen to reduce damping
entdecken to discover
Entdecker *m* discoverer
Entdeckung *f* discovery
entdrillen to untwist *(stranded wire)*

enteisen to defrost, to de-ice *(screen)*; to free from ice *(vessel)*
Enteisenung *f* iron removal *(water purification)*; iron extraction (elimination)
Enteisung *f* defrosting; de-icing *(screen)*; freeing from ice *(vessel)*
Enteisungsanlage *f* de-icer, defroster *(vehicle)*; ice-freeing device *(vessel)*
Enteisungseinrichtung *f s.* Enteisungsanlage
Entelektrisator *m* antistatic device
Entemulgierfähigkeit *f* demulsibility, water miscibility
Entenheck *n (ship)* raised stern
Entenschnabellader *m* duckbill
Enterhaken *m* grapnel, grappling iron (hook)
entern to board, to grapple *(a ship)*
Entfärbelösung *f* de-inking solution
entfärben to decolorize *(glass)*; to de-ink *(waste paper)*
Entfärben *n* decolorizing *(glass)*; de-inking *(preparation of waste paper)*
Entfärbungskohle *f* decolorizing charcoal
Entfärbungsmittel *n* decolorizer *(glass)*
Entfaserungsmaschine *f* rasping machine *(sisal)*
entfernen to remove; to clear
  Glanz e. *(text)* to de-lustre
  Kletten auf mechanischem Wege e. *(text)* to extract burs mechanically
  Ölkohle e. to decarbonize
Entfernen *n* **der Späne** swarf clearance
  E. des Stickstoffs denitration
  E. von Garnverunreinigungen clearing of the yarn
entfernt von clear of
Entfernung *f* 1. distance; 2. *(opt)* range; 3. removal
  E. des Ausschusses *(text)* trash removal
Entfernungsfehler *m* distance error *(radar)*
Entfernungsgesetz *n* square law
Entfernungsmesser *m* range finder *(camera)*; telemeter, distance measurer (gauge)
Entfernungsmeßgerät *n* distance-measuring equipment; range finder
Entfernungsmeßmarke *f* range mark[er] *(radar)*
Entfernungsmessung *f* distance measurement; range finding
  akustische E. phonotelemetry
Entfernungsschätzung *f* distance judgment
Entfernungsstreuung *f* long-distance scatter *(electromagnetic waves)*
Entfernungstafel *f* distance table
entfetten to degrease *(metal)*; to scour, to wash back *(wool)*
Entfettungsmittel *n* degreasant, degreasing agent
Entfettungsvorrichtung *f* fat extraction apparatus *(wool)*
entfeuchten to dehumidify
Entfeuchtung *f* dehumidification
Entfeuchtungsanlage *f* dehumidifying plant
entflammbar inflammable
Entflammbarkeit *f* accendibility, inflammability
Entflammpunkt *m* flash point
entfleischen to green-flesh *(leather)*
Entfleischmaschine *f* fleshing machine
Entfleischung *f* fleshing *(of leather)*
entfluorieren to defluorinate
Entfritter *m* decoherer
Entfrosten *n* defrosting
Entfroster *m* defroster
entfurchen *(ag)* to smooth
entgasen to degas; to deaerate
Entgaser *m* feed deaerator (de-aerator)
Entgasung *f* degassing, degasification, degasifying
Entgasungsanlage *f* gas-freeing equipment *(tanker)*
Entgasungsperiode *f* period of distillation of gases
Entgasungsplatte *f* dead (choke) plate
Entgasungsspeisewasservorwärmer *m* de-aerating feed-water heater
entgegengesetzt opposite
entgeizen *(ag)* to pinch
entgiften to detoxicate, to decontaminate
entgleisen to derail
Entgleisung *f* derailment
Entgleisungsmeldeanlage *f* derailment alarm
Entgleisungsschuh *m* derailer, derailing block
Entgranner *m (ag)* awner and smutter
entgraten to deburr, to burr, to flash trim
Entgraten *n* deburring
Entgratungseinrichtung *f* deburring (burr-removing) device
enthaaren to unhair *(leather)*
Enthaaren *n* unhairing
Enthaarmaschine *f* unhairing machine
Enthalpie *f* enthalpy
enthalten to contain; to comprise
enthärten to soften *(water)*
Enthärtung *f* softening *(water)*
  E. des Speisewassers softening of the feed water
Enthärtungsmittel *n* softener
entharzen to deresinify, to extract the resin
enthäuten to skin
enthornen to de-horn
enthülsen *(ag)* to shell, to husk, to hull; *(nucl)* to decan *(fuel elements)*
Enthülsungsmaschine *f (ag)* seed sheller, shelling machine
Entionisierung *f* deionization
Entionisierungszeit *f* recovery time *(gas discharge tube)*
entkalken to delime, to decalcify
entkeimen to sterilize; to degerminate
Entkeimung *f* sterilization; degermination

Entkernen *n s.* Egrenieren
Entkletten *n (text)* bur extracting, burring; picking *(wool)*
Entklettungsmaschine *f* picker
entkohlen *(met)* to decarburize; to decarbonize, to decarburate
Entkohlung *f (met)* decarburization; decarbonization
Entkohlungsmittel *n (met)* decarburizing agent
entkoppeln to isolate, to decouple *(radio)*
Entkoppelung *f* decoupling, falling out of steps, uncoupling, intermediate coupling; disengagement
Entkopplungsglied *n* stopper circuit
Entkopplungskreis *m* anti-resonant circuit
Entkopplungsschaltung *f* decoupling network
Entkopplungswiderstand *m* plate decoupling resistor
entkörnen *(text)* to gin *(cotton)*
Entkörnung *f (text)* ginning *(cotton)*
Entkörnungsmaschine *f* 1. *(ag)* maize sheller; 2. *(text)* cotton gin
Entkräuseln *n* **des Kammzugs** *(text)* straightening the top
entkupfern to decopperate
Entkupfern *n* removal of copper
entkuppeln to declutch, to unclutch, to uncouple
Entladedauer *f* 1. *(el)* discharge period *(of storage battery)*, time of discharge; 2. time for unloading *(cargo)*
Entladeeinrichtung *f* unloading device
Entladefrist *f* period allowed for unloading, unloading period, time for unloading *(cargo)*
Entladegrenzspannung *f (el) (Am)* cut-off voltage
entladen 1. to unload *(cargo, workpiece)*; to unship *(vessel)*; 2. *(el)* to discharge
Entladerinne *f (mach)* unloading trough
Entladestation *f (mach)* unloading station *(transfer line)*
Entladestrom *m* discharge current
Entladeverzug *m* discharge delay (lag, time-delay)
Entladewiderstand *m (el)* bleeder, discharging resistor
Entladung 1. unloading *(cargo)*; unshipping *(vessel)*; 2. *(el, ph)* discharge; discharging, unloading *(of the reactor)*
aperiodische E. aperiodic (impulsive) discharge
atmosphärische E. atmospheric discharge; atmospherics
dunkle E. silent discharge
elektrische E. electric discharge
funkenähnliche E. spark-like discharge
geschichtete E. globular discharge
halbselbständige E. semi-self-maintained discharge
kalte E. cold electronic discharge
selbständige E. self-maintained discharge, self-sustaining discharge
unselbständige E. non-self-maintained discharge
Entladungsfunke *m* discharge spark
Entladungshafen *m* port of discharge
Entladungskammer *f (nucl)* [gas] discharge chamber
Entladungskanal *m (el)* discharge channel
Entladungslampe *f* discharge lamp
Entladungsplasma *n* [gas] discharge plasma
Entladungsrohr *n* discharge tube
Entladungsröhre *f* discharge tube
Entladungsstrom *m* discharge current
entlang along *(a surface, a road)*
e. der Küste coastwise
entlasten to relieve of load, to unload; to balance *(valve)*
entlastet neutral, relaxed, strainfree *(rock mechanics)*; balanced *(valve)*; relieved of load, unloaded *(statics)*
Entlastung *f* unloading; easing *(of wedges etc.)*; relieving; balancing *(valve)*
Entlastungsbogen *m (build)* relieving arch
Entlastungsdamm *m (mine)* buttress pack
Entlastungskanal *m* by-pass, by-wash
Entlastungskolben *m* dummy piston
Entlastungsventil *n* safety (by-pass, cut-out) valve; auxiliary valve
Entlastungszug *m* relief train
entlauben to defoliate
Entlaubung *f* defoliation
Entlaugerei *f* washing department
entleeren to empty; to evacuate; to shake out *(mould)*
Entleerungshahn *m* drain-cock
Entleerungsöffnung *f* discharge gate
entlüften to vent *(mould)*; to ventilate; to deaerate; to bleed *(brake)*
Entlüften *n* **von Formen** venting of moulds
Entlüfter *m* [air] exhauster; deaerator; uptake ventilator; deculator *(papermaking machine)*
Entlüfterkappe *f (aut)* breather cap
Entlüfterstutzen *m (aut)* breather pipe
Entlüftung *f* deaeration, ventilation; airing; vent
E. der Dampfrohrleitung elimination of air from the steam pipes
Entlüftungshahn *m* air relief (escape) cock, airdrain petcock, air vent
Entlüftungskanal *m* air-bleed duct
Entlüftungsleitung *f* breather line
Entlüftungsloch *n* vent hole
Entlüftungsöffnung *f* vent
Entlüftungsrohr *n* vent (exhaust) pipe, ventilation flue, breather tube
Entlüftungsschraube *f* vent screw *(battery)*; bleed screw *(brake)*

Entlüftungsstopfen *m* bleed plug
Entlüftungsventil *n* escape (air-relief, vent, air-drain) valve
Entlüftungsventilator *m* exhaust fan
Entlüftungsvorrichtung *f* deaerator, breather
entmagnetisieren to demagnetize, to degauss
Entmagnetisierung *f* demagnetization
  E. durch Ummagnetisierung demagnetization by continued reversals
  adiabatische E. adiabatic demagnetization
Entmagnetisierungsfaktor *m* factor of demagnetization
entmasten *(ship)* to dismast
Entmaterialisierung *f* dematerialization
Entmetallisierungsbad *n* liberator tank
entmethylieren to demethylate
entmischen to separate out, to segregate
Entmischung *f* segregation, separation
entmisten *(ag)* to muck out
Entmistungsanlage *f* mechanical gutter cleaner
entmodeln to demodulate *(radio)*
Entnahme *f* drawing *(money)*; bleeding *(fluid)*; taking
  E. von Proben sampling
Entnahmebauwerk *n* intake (inlet) structure
Entnahmeelektrode *f* output electrode
Entnahmeförderband *n* collecting conveyor
Entnahmegrube *f* borrow pit (excavation)
entnehmen to take *(samples)*, to sample; to take out
  Energie e. to remove energy
  Stichproben e. to draw samples
  Strom e. to take current
entpechen to depitch
Entpechen *n* depitching
entphosphoren to dephosphorize, to remove phosphor
Entphosphorung *f* dephosphorization, phosphor removal
entpickeln to depickle *(leather)*
Entplattierung *f* / **elektrolytische** electrolytic stripping
entquellen to deplete *(leather)*
entrahmen to skim, to separate cream
Entrahmer *m* cream separator
entrastern to de-screen
Entrasterung *f* de-screening
entregen to de-energize
Entregungsschalter *m* field break-up switch
entriegeln *(mach)* to unclamp *(turret)*; to unlock
Entriegelung *f* **der Weiche** release of the points
entrinden to [de]bark
entrindet bark-bared
Entrindungstrommel *f* drum barker, barking drum
Entropie *f* entropy
Entropiediagramm *n* entropy diagram
Entropieprinzip *n* entropy principle
Entropiesatz *m s.* Entropieprinzip
Entropietemperaturkurve *f* entropy-temperature curve
entrosten to derust, to remove the rust
Entrosten *n* rust removal, removing the rust, derusting
Entroster *m* rust removing agent
entrostet cleaned from rust, derusted
Entrostungsmittel *n* rust remover
entsalzen to desalinate
Entsalzung *f* desalting; desalination
entsäuern to disacidify, to de-acidify, to free from acid; to neutralize
Entsäuerung *f* de-acidification
Entsäurer *m* neutralizer
Entschädigung *f* compensation; allowance *(wage)*
  E. für Beschädigungen compensation for damage
  E. für Betriebsunfall compensation for accident on duty
Entschädigungsantrag *m* claim for compensation
entschalen 1. *(build)* to strip, to strike formwork, to demould *(concrete)*; 2. to decorticate, to remove the husk (bark)
entschälen to degum *(silk)*
Entschalung *f* 1. *(build)* [form] stripping, striking formwork, demoulding *(concrete)*; 2. decortication, removal of the husk (bark)
Entschälverlust *m* loss of gum in scouring
entschärfen to defuse
entscheiden to decide
Entscheidung *f* decision
Entscheidungselement *n (dat)* decision element
Entscheidungsforschung *f* / **mathematische** *(dat)* operational (operations) research
Entscheidungsverfahren *n (dat)* decision procedure
Entscheidungstreffen *n (dat)* decision making
entschichten to strip, to remove [the] coating
entschlacken to remove the cinder (slag)
Entschlackung *f* slag removal, removal of the slag
entschlammen to desludge
Entschlämmung *f* desliming
entschlichten *(text)* to desize
Entschlichten *n* desizing
Entschlichtungsmittel *n* boil-off agent
entschlüsseln *(dat)* to decipher, to decode
Entschlüsselung *f* decoding
Entschlüßler *m* decoder
entschwefeln to desulphurate, to desulphurize
Entschwefelung *f* desulphurization, sulphur removal
entschweißen to scour, to degrease *(wool)*
Entschweißmaschine *f* backwashing machine
Entschweißungsmittel *n* degreasing agent

Entschweißvorrichtung *f* grease extracting apparatus
Entseuchung *f (ag)* disinfecting, disinfection; *(nucl)* decontamination
Entseuchungsfaktor *m (nucl)* decontamination factor
Entseuchungsgebühr *f* disinfection charge
Entseuchungsmittel *n* disinfectant; *(nucl)* decontaminating agent, decontamination substance
entsichern to unlock
entspannen to relax *(stress)*; to unload
Entspannen *n* stress relieving, removal of stress; unloading; expansion *(engine)*
Entspannstation *f (mach)* unloading station
entspannt neutral, relaxed, strainfree *(rock mechanics)*
Entspann- und Spannstation *f (mach)* loading and unloading station
Entspannung *f* stress relief, relaxation of stress; detensioning; stress-relieving *(rock mechanics)*; decrease in tension
Entspannungsabkühlung *f* cooling due to expansion
Entspannungsmittel *n* wettability agent
Entspannungsort *n (mine)* relief roadway
Entspannungsraum *m* expansion chamber
entspiegeln to demirror
Entspiegelung *f* demirroring
entsprechen to correspond to, to conform to
entsprechend corresponding
entstauben to dedust *(gases)*; *(text)* to dust; to remove the dust
Entstaubung *f* dedusting, dust removal, removal of suspended matter
entstehen to develop, to occur; to result; to be created, to be generated; to rise
Entstehungsort *m* place of origin
Entstördiode *f (telev)* black spotter
Entstördrossel *f (el)* suppressor choke
entstören to suppress interference; to clear faults
Entstörer *m* interference suppressor
Entstörkappe *f* radio shielding cap
Entstörkondensator *m* suppressor capacitor
Entstörung *f* interference (noise) suppression; fault clearance; anti-interference
Entstörungsdienst *m (telec)* fault clearing service
Entstörungskondensator *m* reaction (interference suppression) capacitor
Entstörungswiderstand *m* suppressor resistance
entteeren to detar
Entteerung *f* detarring
Enttrichtern *n* **von Gußstücken** spru[e]ing, degating [of castings]
Enttrübung *f* zero clearing *(bearing)*, fast time control
Enttrübungs- und Abbrausesieb *n* drain-and-spray screen
entwässern to dewater, to drain; to lower the ground water *(opencast mine)*; to drain off water *(pulp)*; *(ch)* to dehydrate, to dephlegm *(spirit, acid)*
Entwässern *n s.* Entwässerung
Entwässerung *f* draining, drainage, dewatering; lowering the ground water *(opencast mine)*; *(ch)* dehydration
E. des Arbeitszylinders draining of the cylinder
E. des Überhitzers drainage of the superheater
Entwässerungsfilz *m* pulp felt *(papermaking)*
Entwässerungsförderband *n* drainer conveyor *(papermaking)*
Entwässerungsgebiet *n (build)* drainage area (district)
Entwässerungsgeschwindigkeit *f* draining rate
Entwässerungsgraben *m (build)* drain, drainage ditch (trench); *(mine)* catch drain
Entwässerungshahn *m* drain cock
Entwässerungsloch *n (build)* weep hole
Entwässerungsmaschine *f* decker *(papermaking)*
Entwässerungsrinne *f* dewatering trough
Entwässerungsschacht *m* drainage pit
Entwässerungsschlitz *m* weep hole
Entwässerungssieb *n* drainage screen
Entwässerungssiebrinne *f* dewatering screening trough
Entwässerungsstutzen *m* trap
Entwässerungstopf *m* steam drier (separator, trap)
Entwässerungsvorrichtung *f* draining device
ENTWEDER-ODER-Schaltung *f (dat)* OR-ELSE circuit
entweichen to escape, to disengage
Luft e. lassen to disinflate
entwerfen to design
entwickeln to develop; *(ch)* to set free, to evolve, to liberate, to incubate; to develop; *(math)* to expand
Entwickler *m* developer
Entwicklerbad *n* developing bath
Entwicklerschale *f* developing dish
Entwicklersubstanz *f* developing agent
Entwicklertank *m* developing tank
Entwicklung *f* development; *(ch)* liberation, incubation; *(math)* expansion
E. in eine Doppelreihe *(math)* double-series expansion
asymptotische E. asymptotic expansion
Bornsche E. Born expansion
oxydative E. development by oxidation
Entwicklungsabteilung *f* development department
Entwicklungsarbeit *f* developmental work

Entwicklungsbad *n* developing bath

Entwicklungsfarbstoff *m* developing dye[stuff], developed dye[stuff]

Entwicklungsfärbung *f* developed dyeing

Entwicklungsingenieur *m* design (development) engineer

Entwicklungskinetik *f* kinetics of development

Entwicklungskoeffizient *m* coefficient of expansion

Entwicklungskurve *f* subsidence development curve

Entwicklungsperiode *f* incubation period

Entwicklungsprogramm *n* development programme

Entwicklungsrichtung *f* trend in design

Entwicklungsstadium *n* development stage

Entwicklungstempo *n* rate of development

Entwicklungstendenz *f* trend in design

Entwicklungs- und Forschungsabteilung *f* development and research department

Entwicklungsverfahren *n* development technique

entwirren to disentangle, to untwist

Entwurf *m* design, project; draft

logischer E. logical design

Entwurfsautomatisierung *f* design automation

Entwurfsingenieur *m* project (design) engineer

Entwurfsskizze *f* draft

Entwurfszeichnung *f* preliminary assembly drawing

entwurzeln to uproot

entzerren *(el)* to correct *(distortion)*; to equalize; to compensate; to rectify

Entzerrer *m* equalizer, compensator, corrector

Entzerrerbaustein *m* component equalizer network *(tape recording)*

Entzerrerkette *f* equalizing network

Entzerrerschaltung *f* correction (corrective) network

Entzerrung *f* correction of distortion, equalization, compensation

Entzerrungsbereich *m* frequency range of equalization

Entzerrungsfilter *n* filter-type equalizer

Entzerrungsgerät *n* rectifier

Entzerrungsgrenze *f* frequency limit of equalization

Entzerrungskette *f* filter-type equalizer

entziehen to withdraw; to extract

Energie e. to remove energy

entzinken to dezinc

Entzinkung *f* dezincing

entzinnen to detin

Entzinnung *f* detinning

entzuckern to desugarize, to desaccharify

entzündbar inflammable

Entzündbarkeit *f* inflammability

entzünden to ignite; to initiate

entzundern to descale, to scale off

Entzunderung *f* descaling, scaling-off

Entzunderungsofen *m* wash heating furnace

Entzunderungswärme *f* wash heat

Entzündung *f* ignition *(engine, mine)*; inflammation

Entzündungspunkt *m* inflammation point

Entzündungstemperatur *f* ignition temperature

Enveloppe *f (math)* envelope

Enzym *n* enzyme

enzymatisch enzymatic, enzymic

Enzymwirkung *f* enzymosis

Ephedrin *n (ch)* ephedrine

Epidiaskop *n* epidiascope

Epidot *m (min)* epidote

Epilationsdosis *f (nucl)* epilation dose

Epiprojektion *f* episcopic projection

Episkop *n* episcope

epithermisch *(nucl)* epithermal

Epizykloide *f* epicycloid

Epoxyd *n* epoxide

Epoxydharz *n* epoxy resin

Epoxydträger *m* epoxy base carrier

Epoxygießharz *n* epoxy casting resin

Epoxyharz *n* epoxy resin

Epsomit *m (min)* epsomite

Epstein-Gerät *n* Epstein hysteresis tester

Epstein-Rahmen *m* Epstein square

Erbium *n* erbium

Erbsenauslesemaschine *f* pea grader (sorting machine)

Erbsendreschmaschine *f* pea viner

Erbsenerntemaschine *f* pea harvester

Erbsengröße *f* pea size

Erbsenreinigungs- und -sortiermaschine *f* pea cleaner and grader

Erbsenschwader *m* pea swather

Erbsen- und Bohnenauslesemaschine *f* pea and bean grader

Erdalkalien *npl* alkaline earths

Erdalkalimetall *n* alkaline earth metal

Erdanschluß *m* earth connection (terminal), *(Am)* ground connection (terminal)

Erdantenne *f* earth (buried) aerial, *(Am)* ground antenna

Erdarbeit *f* groundwork, earthwork

Erdarbeiter *m* navvy

Erdatmosphäre *f* earth's atmosphere

Erdaushub *m* excavation

Erdbahn *f* terrestrial orbit

Erdbau *m* earthwork[s]; soil engineering

Erdbauarbeiten *fpl* earthworks

Erdbeben *n* earthquake

Erdbebengebiet *n* seismic region

Erdbebenkunde *f* seismology

Erdbebenzone *f* seismic zone

Erdbehälter *m* underground tank

Erdbeschleunigung *f* acceleration due to gravity, gravitational acceleration
Erdbeton *m* soil cement
Erdbewegung *f* soil displacement, earthmoving, dirt moving
Erdbohrer *m* earth borer; earth digger
Erdbuchse *f (el)* earthing bush
Erddamm *m* earth [fill] dam, embankment
Erddämpfung *f (ag)* soil sterilization
Erddraht *m (el)* earth wire
Erddruck *m* earth (soil) pressure *(wall)*
aktiver E. active earth pressure
passiver E. passive earth pressure
Erddruckmeßdose *f* soil pressure cell
Erde *f* 1. earth; 2. *(el)* earth, *(Am)* ground; 3. *(build)* earth, soil
künstliche E. *(el)* counterpoise; artificial earth
erdelektrisch geoelectric
Erdelektrizität *f* geoelectricity, terrestrial electricity
erden to earth; to connect, *(Am)* to ground
Erden *fpl* / **seltene** noble (rare) earths, lanthanides
Erder *m (el)* earth connection (lead); earth electrode (plate)
Erdfarben *fpl* mineral colours
Erdfeld *n* / **magnetisches** magnetic field of earth
Erdferne *f* apogee
erdfeucht semi-dry *(concrete)*
erdfrei *(el)* earth-free
Erdgas *n* natural gas
schwefelwasserstoffhaltiges E. foul (sour natural) gas
Erdgeschoß *n* ground floor
Erdhaue *f (ag)* mattock
Erdhobel *m* grader
Erdkabel *n* buried cable; underground cable
Erdkapazität *f (el)* capacity to earth
Erdklemme *f (el)* earth[ing] terminal, earth[ing] connection, *(Am)* ground terminal (clamp)
Erdkobalt *m (min)* wad; asbolane
Erdleiter *m (el)* earth wire (lead)
Erdleitung *f (el)* earth lead (connection); earth termination, conductor connected to earth
Erdmagnetfeld *n* terrestrial magnetic field
Erdmagnetismus *m* earth magnetism, geomagnetism
Erdmassen *fpl* earth
eingeschlämmte E. hydraulic fill
eingespülte E. hydraulic fill
Erdmassenbewegung *f* earthworks
Erdmetalle *npl* earth metals
seltene E. rare-earth metals
Erdmine *f (mil)* land mine
Erdnähe *f* perigee
Erdnußöl *n* arachis (peanut) oil
Erdoberfläche *f* surface of the earth
Erdöl *n* crude petroleum (mineral oil), oil
naphthenbasisches E. naphthenic oil, naphthene-base crude petroleum
paraffinbasisches E. paraffin-base oil
Erdölbitumen *n* oil asphalt
Erdöldestillationskolonne *f* crude flash tower
Erdölraffinerie *f* oil (petroleum) refinery
Erdpech *n* mineral pitch
Erdplatte *f (el)* earth plate
Erdpotential *n (el)* earth potential
Erdpunkt *m (el)* neutral point
Erdrückleitung *f (el)* earth return *(circuit)*
Erdrückleitungskreis *m* earth return circuit
Erdrückstrom *m* line-residual current, earth-return current
Erdrutsch *m* landslip, landslide
Erdsammelschiene *f (el)* earthing bus bar
Erdsatellit *m* / **künstlicher** artificial (man-made) earth satellite
Erdschelle *f (el)* earth terminal, *(Am)* ground clamp
Erdschicht *f* soil stratum
Erdschleife *f (el)* earth circuit
Erdschluß *m* earth fault (leakage), line-to-earth fault, accidental earth
intermittierender E. intermittent earth
satter E. dead earth
schleichender E. earth leakage
unvollkommener E. partial earth
vorübergehender E. temporary earth
Erdschlußanzeiger *m* earth detector, leakage indicator
erdschlußfrei earth-free
Erdschlußlichtbogen *m* arc between line and earth
Erdschlußlöschung *f* earth-fault neutralizing
Erdschlußprüfer *m* earth detector
Erdschlußrelais *n* [earth] leakage relay, residual (zero phase-sequence) relay
Erdschlußschutz *m* earth-fault protection
erdschlußsicher earth-free
Erdschlußspule *f* Petersen (arc-suppressing) coil
Erdschlußstrom *m* earth leakage current, short-circuit current to earth
Erdscholle *f* clod
Erdschüttdamm *m* earth fill dam
Erdseil *n* earthed (earthing) wire
Erdstampfbau *m* pisé de terre
Erdstoff *m* soil
Erdstrom *m* stray (return, vagabond) current
natürlicher E. earth current
Erdstromausgleicher *m* earth-current equalizer
Erdstromkreis *m* earth circuit
Erdstrommeßgerät *n* electrotellurograph
Erdstromrelais *n* residual relay
Erdstromschürfung *f* earth-current prospecting
Erdtaste *f* earthing key *(telephone)*
Erdtelegrafie *f* earth [conduction] telegraphy

Erdtopf *m (ag)* soil block
Erdtopfmaschine *f* soil block-making machine
Erdtopfpresse *f* soil block press
Erdtrabant *m* / **künstlicher** artificial (man-made) earth satellite
Erdübergangswiderstand *m* earth entering resistance
Erdung *f* earth lead (connection); earthing, *(Am)* grounding
E. durch Schutzleiter pilot earth core protection
E. über Widerstand resistance earthing
kapazitive E. capacitance earth
starre E. *(el)* solid earthing, solidly earthed system
Erdungsader *f* earthing core
Erdungsanlage *f* earthing system
Erdungsanschluß *m* earth terminal (connection point)
Erdungsbürste *f* earth return brush
Erdungsdraht *m* earth wire
Erdungsdrossel *f (el)* earthing reactor (inductor), drainage (discharge) coil
Erdungsklemme *f* earth[ing] terminal, earth connecting point
Erdungskontakt *m* earthing contact
Erdungskreis *m* leak circuit
Erdungsleiter *m* earth connection (lead); earth electrode
Erdungsmantel *m* earth sheath
Erdungsmesser *m* earth resistance meter
Erdungspunkt *m (el)* earthing connection, earth connection point
Erdungsschalter *m* earthing switch, earthing isolator
Erdungsschelle *f* earth clip, bonding ring (clip)
Erdungsschraube *f (el) (Am)* bonder
Erdungsseil *n* earth wire
Erdungsspule *f* earth[ing] coil
Erdungsstange *f* earthing hook (switch-pole)
Erdungsverhältnisse *npl* / **schlechte** *(Am)* ground imperfections
Erdungswiderstand *m* earthing resistance; discharging resistor
Erdungsziffer *f* coefficient of earthing
Erdverbindung *f* earth[ing] connection
Erdverlegung *f (el)* underground laying, buried wiring
Erdwachs *n* earth wax, ozokerite
Erdwiderstand *m (build)* passive earth pressure
Erdwiderstandsbohrschürfung *f* electrical bore-hole prospecting
Ereignis *n* event
zufälliges E. random event
Erfahrungsaustausch *m* exchange of experience
Erfahrungswert *m* empirical value
erfassen to log; to detect; to sense
Erfassung *f* coverage; acquisition *(of measuring values)*
tagesfertige E. *(dat)* being up-to-date daily
erfinden to invent
Erfinder *m* inventor
Erfinderschutz *m* protection of inventors
Erfindervergütung *f* award to the inventor
Erfindung *f* invention
Erfindungspatent *n* patent for invention
erforderlich required
erfordern to require
Erforschung *f* exploration, research, investigation
E. des Weltraums space research, exploration of space
Erfrischungswagen *m* buffet-coach, buffet-car
erfüllen / die Prüfbedingungen to meet the specifications of the test
seine Arbeitsnorm e. *(Am)* to meet oneself's quota
Erg *n* erg
ergänzen to complement; to supplement; to complete; to replenish
Ergänzung *f* completion; *(math)* complement *(of an angle to 90°)*; supplement *(of an angle to 180°)*; explement *(of an angle to 360°)*
Ergänzungsarbeit *f* filling-in work *(cards)*
Ergänzungsausrüstung *f* supplementary equipment
Ergänzungsband *m* supplementary volume
Ergänzungsnetzwerk *n (el)* line simulator (building-out network)
Ergänzungsspeicher *m* backing store *(for data)*
Ergänzungswinkel *m* complementary angle
ergeben to provide; to yield; to result
Blaufärbung e. to give blue colour
Ergebnis *n* yield; result
numerisches E. numerical result
vorläufiges E. preliminary result
ergiebig productive; *(mine)* payable; rich; fertile
Ergiebigkeit *f (mine)* payability; abundance; fertility; richness
E. der Druckfarbe coverage of printing ink
E. der Strahlungsquelle *(nucl)* source strength, strength of source
ergodisch ergodic
Ergosterin *n* ergosterol
Ergußgestein *n (geol)* extrusive (igneous) rock
erhaben convex; embossed, relief
erhalten to receive, to obtain; *(ph)* to conserve; to preserve; to maintain
Auftrieb e. to be buoyed up
Erhaltung *f* conservation; preservation; maintenance
E. der Energie conservation of energy
E. der Ladung conservation of charge
E. der Masse conservation of mass
E. des Impulses conservation of momentum

Erhaltungssatz *m* conservation law, law of conservation
E. der Masse law of the conservation of mass
E. des Drehimpulses principle of conservation of angular momentum
Erhärtung *f* hardening; solidification
Erhebung *f* 1. elevation; 2. inquiry
E. der Kurbel angle of crank with the horizontal
erhitzen to heat [up]
Erhitzung *f* heating *(bearing)*, overheating; *(ch)* calefaction
E. der Bremsklötze heating of brake shoes
erhöhen to increase, to raise, to stop up; to elevate
die Produktivität e. to increase the productivity
die Spannung e. to rise the voltage, to boost
die Trennschärfe e. to enhance the selectivity *(radio)*
Erhöhungswinkel *m* elevation angle
erholen to recover *(material)*
Erholung recovery *(material)*
Erholungspause *f* recovery (recreation, relief) period *(oscillator)*
Erholungsvermögen *n* recuperative capacity *(storage battery)*
Erholungszeit *f* recovery time
E. eines Empfängers receiver recovery time *(radio)*
Erholungszyklus *m* / **elektrischer** [electrical] recovery cycle
Eriaseide *f* eria [silk]
Erichsen-Tiefungsversuch *m* Erichsen cupping test
erkalten to cool [down], to chill
Erkennbarkeit *f* detectibility; *(opt)* discernibility
erkennen to discover, to detect, to distinguish, to perceive; *(opt)* to discern
Erkennungsfeuer *n* airway beacon
Erkennungsschaltung *f* *(dat)* recognition circuit
Erkennungssignal *n* distinguishing signal
Erker *m* bay, jutty
Erkerfenster *n* bay window
rundes E. compass (bow) window
Erkundung *f* exploration, reconnaissance
Erkundungsstrecke *f* *(mine)* exploratory drift
erläuternd illustrative
Erle *f* alder
erleichtern to facilitate
erleiden / Schaden to suffer damage
Schiffbruch e. to shipwreck
Erlenholz *n* alder
Erlenmeyerkolben *m* Erlenmeyer flask
erlernen / einen Beruf to acquire a craft
erlöschen to extinguish; to expire *(patent)*
Erlöschen *n* **des Patents** expiration of the patent
ermächtigen to authorize
ermäßigen to reduce
Ermäßigung *f* reduction
ermitteln to detect, to find out; to determine *(value)*
Ermittlung *f* determination *(value)*
Ermittlungsverfahren *n* / **empirisches** trial-and-error method
ermöglichen to enable
Ermüdung *f* fatigue *(materials)*
E. des Bedienungsmannes operator fatigue
ermüdungsbeständig anti-fatigue
Ermüdungsbruch *m* fatigue fracture (failure)
Ermüdungsfestigkeit *f s.* Dauerfestigkeit
Ermüdungsgrenze *f* endurance limit (strength), fatigue limit (strength)
Ermüdungskorrosion *f* corrosion fatigue
Ermüdungsversuch *m* fatigue test
ernähren to feed, to nourish
Ernährung *f* nutrition; feeding, nourishment
ernennen to appoint, to nominate, to designate; to assign
Ernennungsproblem *n* assignment problem
Ernennungsurkunde *f* *(mil)* commission
erneuern 1. to renew; to recondition, to overhaul, to repair; 2. *(el)* to restore; 3. *(build)* to face-lift, to reface
Erneuerung *f* 1. renewal; overhauling, repairing, reconditioning; 2. *(el)* restoration; 3. *(build)* rebuilding, remodeling, reconstruction
E. der Fahrzeuge reconditioning of rolling stock
E. des Ausbaus *(mine)* relining, retimbering
Erneuerungswert *m* renewal value
Ernte *f* harvest; crop, yield
Erntebergungsgeräte *npl* harvesting implements
Erntebergungsmaschinen *fpl* harvesting machinery
Erntefläche *f* *(Am)* harvested cropland
Erntemaschine *f* harvester
ernten to harvest, to gather; to pick
Ernterechen *m* hay sweep
Erntewagen *m* harvest wagon
Eroberung *f* **des kosmischen Raumes** conquering of the space
erodieren to erode
erodierend erosive
eröffnen to open, to set up, to start
Eröffnung *f* **des Auslasses** *(therm)* opening of the exhaust port
E. des Kanals *(therm)* opening of the port
Eröffnungsansprache *f* opening address *(fair)*
Eröffnungstag *m* opening day *(fair)*
Eröffnungsweite *f* *(therm)* width of opening
Erosion *f* erosion
Erosionsschutz *m* erosion control

erproben to try, to test
Erprobung *f* **des Schiffsverhaltens im Seegang** sea-keeping trial
errechnen to calculate
erregen to excite; to energize *(relay)*
Erreger *m* exciter [set]
Erregeranode *f* excitation anode
Erregerfeldwiderstand *m* exciter field rheostat
Erregergleichrichter *m* rectifier exciter
Erregergrad *m* effective field ratio
Erregergruppe *f* exciter set
Erregerkreis *m* exciter circuit
Erregerlichtquelle *f* exciter light-source
Erregermaschine *f* exciter, exciting generator
Erregerspannung *f* exciting voltage
 höchste E. ceiling voltage of an exciter
Erregerspule *f* excitation (exciting) coil; driving coil
Erregerstrom *m* inductor current
Erregerturbine *f* exciter turbine
Erregerverluste *mpl (el.m.)* excitation losses
Erregerwicklung *f* excitation (exciting, induction) winding, field (magnetizing) coil
Erregerwiderstand *m* exciter resistance
Erregung *f* exciting, excitation; energization *(relay)*
 E. mit positiver Rückkopplung self-excitation
 magnetische E. magnetic field strength
 mehrfache E. multiple excitation
 wechselseitige E. reciprocal excitation
Erregungsenergie *f* excitation energy
Erregungsfluß *m* induction flux
Erregungskreis *m* energizing circuit
Erregungsstärke *f* strength of induced field
erreichen to reach; to obtain; to arrive at
 die größte Erdferne e. to reach the apogee
 die größte Erdnähe e. to reach the perigee
 Überschallgeschwindigkeit e. to reach supersonic speed
errichten to erect; to build; to set up; *(math)* to raise *(perpendicular)*
erringen to score *(success)*; to win; to achieve
Errungenschaft *f* achievement, result
Ersatz *m* replacement, substitute, spare
Ersatzader *f (el)* spare conductor
Ersatzanker *m* spare armature
Ersatzantenne *f* auxiliary aerial; phantom aerial, *(Am)* dummy antenna
Ersatzbatterie *f* spare (stand-by) battery
Ersatzempfänger *m (el)* emergency receiver
Ersatzgerät *n* emergency unit; spare equipment
Ersatzkapazität *f (el)* equivalent capacitance
Ersatzkernverfahren *n (math)* substitution kernel procedure
Ersatzkolben *m* spare piston
Ersatzkosten *pl* replacement cost *(rolling stock)*
Ersatzkreis *m (el)* equivalent circuit
Ersatzröhre *f* spare (replacement) valve
 eingebaute E. mounted spare valve
Ersatzschaltung *f* equivalent network (circuit)
 E. eines Quarzes equivalent circuit of a piezoelectric crystal unit
Ersatzschiene *f* spare rail
Ersatzsender *m* spare transmitter
Ersatzsicherung *f* spare fuse
Ersatzsignal *n* substitute (emergency) signal
Ersatzspule *f* replacement bobbin
Ersatzsternschaltung *f* **einer Dreieckschaltung** star connection equivalent to delta connection
Ersatzstirnrad *n* equivalent spur gear
Ersatzstoff *m* substitute material
Ersatzstromkreis *m* equivalent circuit
Ersatzteil *n* spare (replacement) part, replacement, spare
Ersatzteilausstattung *f* service kit
Ersatzteilhaltung *f* stocking of spare parts
Ersatzteilplan *m* layout of spare parts
Ersatzweg *m (telec)* alternate route
Ersatzwiderstand *m (el)* equivalent (substitutional) resistance
 E. für die mechanische Rückwirkung motional impedance
 E. für mechanische Verluste motional resistance
ersaufen *(mine)* to drown, to be flooded
Ersaufen *n (mine)* flooding, drowning
erscheinen to appear; to be published *(book)*
Erscheinung *f* appearance; phenomenon
 äußere E. outward appearance
 kosmische E. cosmical phenomenon
Erscheinungspotential *n* appearance potential
Erschlaffung *f* relaxation
erschließbar developable *(site)*
erschließen to develop *(site)*
Erschließung *f* development *(site)*
 E. technischer und wissenschaftlicher Information information retrieval
Erschmelzung *f* melting
erschöpfen *(mine)* to exhaust *(deposit)*; to deplete *(fuel)*
 das Bad e. *(text)* to exhaust the bath
Erschöpfung *f* **der Batterie** exhaustion of the battery
erschürfen *(mine)* to discover
erschüttern to shatter, to rock; to vibrate
erschüttert werden to be shaken
Erschütterung *f* shock, percussion; vibration; concussion
erschütterungsfrei vibrationless
Erschütterungsmesser *m* vibration measuring instrument
erschütterungssicher shockproof
erschweren to weight *(silk)*; to load
Erschweren *n* loading, weighting *(silk)*
 E. des Gewebes weighting of the cloth
Erschwerung *f* loading, weighting *(silk)*

Erschwerungsappretur *f* weight finishing
ersetzbar replaceable; interchangeable
ersetzen to substitute, to replace, to exchange; to supplant, to displace
  e. durch to replace with
ersoffen *(mine)* flooded
Erstanmeldung *f* original application
erstarren 1. *(ch)* to congeal, to coagulate, to gel; 2. *(met)* to solidify, to freeze
Erstarren *n* **einer Schmelze** freezing (solidification) of a melt
Erstarrung *f* 1. coagulation; 2. *(met)* freezing, solidification; 3. *(build)* set; 4. *(plast)* setting
  wiederholte E. repeated solidification *(semiconductors)*
Erstarrungsbeginn *m* initial set
Erstarrungsbereich *m (met)* solidification (freezing) range
Erstarrungsgestein *n (geol)* igneous (eruptive) rock
Erstarrungsintervall *n (met)* solidification range
Erstarrungspunkt *m* 1. *(ch)* congealing point, chill point *(crystal)*; setting point *(liquid, plastic)*; 2. *(met)* solidifying (freezing) point, point of solidification
Erstarrungstemperatur *f (met)* solidification temperature
erstatten to repay, to refund
Erstattungsantrag *m* claim for a refund
Erstausführung *f* **einer Maschine** prototype machine
ersteigen to climb; to ascend; to mount
ersticken / ein Feuer to smother a fire
erstickend riechend with a choking smell
erstklassig first-rate
Erstlingsdrucke *mpl* incunabula
Erstmagnetisierung *f* initial magnetization
erstrecken in / sich to extend into
  sich e. von to range over from
  sich über die gesamte Länge e. to run the full length
Erststoß *m (nucl)* first collision
Erststoßdosis *f (nucl)* first-collision dose
Erteilung *f* allowance *(patent)*
Erteilungsunterlagen *fpl* documentary record of patents
Ertrag *m* yield
Eruption *f* eruption, flare
Eruptivgang *m (geol)* intrusive vein
Eruptivgestein *n* eruptive (igneous) rock
Eruptivsonde *f* flowing well
Erwachsenenbildung *f* adult education
erwärmen to heat [up], to warm [up]
Erwärmung *f* warming, heating
  aerodynamische E. aerodynamic heating
  dielektrische E. dielectric heating
  kapazitive E. dielectric heating
Erwartungswert *m* expected (expectation) value
Erweichen *n* 1. softening *(edge)*; 2. *(plast)* plasticizing
Erweichung *f* 1. softening *(edge)*; 2. *(plast)* plasticization
Erweichungsbereich *m (plast)* softening range
Erweichungspunkt *m* softening point *(melt)*
Erweichwalze *f* diluting roller *(paper)*
erweisen to prove, to render
erweitern to enlarge; to extend, to expand; to widen
  den Meßbereich e. to extend the range
  den Spannungsbereich e. to enlarge the range of voltage
Erweiterung *f (build)* extension; expansion *(pipe)*; widening
  kegelförmige E. conical widening *(turbine nozzle)*
  konische E. conical widening *(turbine nozzle)*
  radiale E. der Schaufelkanäle radial widening of the passage between the blades *(turbine)*
Erweiterungsbau *m* extension building; [building an] extension
Erweiterungsstufe *f (mine)* reaming stage *(drilling)*
Erweiterungsverhältnis *n* ratio of divergence *(turbine)*
  veränderliches E. variable ratio of divergence
erwerben to acquire, to obtain, to gain
Erwerber *m* **eines Schutzrechtes** transferee (assignee) of a protective right
Erythemdosis *f (nucl)* erythema dose
Erythrin *m (min)* erythrite, cobalt bloom
Erythrit *n (ch)* erythritol
Erythrose *f (ch)* erythrose
Erythrozyt *m* red corpuscle, erythrocyte
Erz *n* ore
  abbauwürdiges E. pay ore
  angereichertes E. dressed ore
  armes E. base (lean) ore
  aufbereitetes E. dressed ore
  bauwürdiges E. pay ore
  bergfeines E. native ore
  eisenarmes E. low-grade iron ore
  fein eingesprengtes E. chatty ore
  hochprozentiges E. rich ore
  kalkiges E. calcareous (limey) ore
  mageres E. poor ore
  mulmiges E. dust (friable) ore
  oxidisches E. oxide ore
  plattiges E. flag ore
  reichhaltiges E. high-grade ore
  selbstgehendes E. self-fluxing ore
  stückiges E. lumpy ore
  sulfidisches E. sulphide ore
  toniges E. argillaceous ore
  vorgerichtetes E. blocked out ore
Erzader *f* ore vein

Erzanreicherungsverfahren *n* ore concentration process
erzarm low in ore
Erzart *f* kind of ore
Erzaufbereitung *f* ore dressing
E. auf trockenem Wege dry process of ore dressing
Erzaufbereitungsanlage *f* ore dressing plant
Erzbergbau *m* ore (metal) mining
Erzbergwerk *n* ore (metalliferous) mine
Erzbrecher *m* ore crusher
Erzbrikettierung *f* ore briquetting
Erzbruchbau *m (mine)* ore caving
Erzbunker *m* ore bunker (bin)
Erzcharge *f* ore charge
erzeugen to produce; to generate
Dampf e. to generate steam
Schwingungen e. to excite oscillations
Erzeugende *f (math)* ruling; generatrix
E. eines Kegels ruling (element) of a cone
Erzeugendenschar *f (math)* regulus
Erzeugnis *n* article; product; make; creation
Erzeugung *f* **von Ladungsträgern** charge production
E. von Magnetfeldern generation of magnetic fields
E. von Metallen production of metals
Erzeugungswälzkreis *m* pitch circle of generation, generating pitch circle *(gear)*
Erzfall *m* ore-shoot, pay-streak
Erzfeste *f (mine)* ore pillar
Erzförderung *f* output of ore; mining of ore
Erzfrischverfahren *n (met)* ore process
Erzführung *f* ore bearing
Erzgang *m* course, lode, vein of metal ore
Erzgehalt *m* grade of ore
Erzgemisch *n* ore mixture
Erz-Getreide-Schiff *n* ore-grain carrier
Erzgicht *f* ore charge (burden), dose of ore
Erzgrube *f* ore (metalliferous) mine
Erziehungsbeihilfe *f* education allowance
erzielen / Einsparungen to obtain economies
Erzklein *n* ore fines, broken ore
Erzkörper *m (geol)* ore body
Erzlager *n (geol)* ore body
Erzlagerstätte *f* ore deposit
Erzlaugung *f* ore leaching
Erzmineral *n* metalliferous mineral
Erzmöller *m* ore burden, stock of ores
Erzmühle *f* edge mill
Erznest *n* ore pocket
Erz-Öl-Schiff *n (ship)* ore-oil carrier (tanker)
Erzpartikel *fpl* fine raggings
Erzpfeiler *m (mine)* ore pillar
Erzpochwerk *n* ore stamp
Erzpuddeln *n* ore puddling
Erzreduktionsanlage *f* ore reduction plant
erzreich rich in ore
Erzrolle *f (mine)* ore pass
Erzrolloch *n (mine)* ore pass
Erzröstofen *m* ore roasting kiln
Erzröstung *f* ore roasting
Erzsatz *m* dose of ore
Erzscheider *m* ore separator
Erzschiff *n* ore carrier
Erzschlamm *m* ore sludge
Erzschmelzofen *m* ore-smelting furnace
Erzsinteranlage *f* ore agglomerating plant
Erzstampfwerk *n* ore stamp
Erztrichter *m* ore funnel
Erztrübe *f* ore pulp
Erzverfahren *n* ore process
Erzverhüttung *f* metallurgical working of ores
Erzvorkommen *n* ore deposit, ore body
Erzwagen *m* ore (mineral) wagon
Erzwäscherei *f* ore washing plant
erzwingen to force *(vibration)*
Erzwingung *f* **der Patentausübung** compulsory working of a patent
erzwungen forced *(vibration)*
Erzzerkleinerungsanlage *f* ore crushing plant
Esaki-Diode *f* high-speed tunnel diode
Esche *f* ash
Eschenholz *n* ash[-wood]
Espagnolettenverschluß *m* espagnolette bolt
Espartozellstoff *m* esparto pulp
Espe *f* asp[en]
Eßbesteck *n (ship)* mess gear
Esse *f* chimney; [chimney] stack, smoke stack; funnel
Essenschieber *m* chimney damper
Essenz *f* essence
Essig *m* vinegar
Essigfabrik *f* vinegar distillery
Essigkonserve *f* pickles
Essigsäure *f* acetic acid
Essigsäureanhydrid *n* acetic anhydride
Essigsäuregärung *f* acetous fermentation, acetification
Eßkohle *f (mine)* steam coal
Ester *m* ester
Esterase *f (ch)* esterase
Esterbildung *f* esterification
Esterharz *n* ester gum
Esterzahl *f* ester number
Estragonöl *n* tarragon oil
Estrich *m (build)* floor, screed
schwimmender E. floating screed
Etage *f* 1. *(build)* stor[e]y, floor; 2. *(el)* tier *(winding)*
Etagenbrücke *f* double-deck bridge
Etagenfilter *n* stage filter
Etagenförderer *m* multi-deck conveyor
Etagenkessel *m* multiple-stage boiler
Etagenmaschine *f* balcony-type printing press
Etagenofen *m* storey furnace

Etagenpresse *f* multiplication (multi-bank, multi-daylight) press
Etagenrost *m* multi-stage grate
Etagenrostfeuerung *f* furnace with multi-stage grate
Etagentrockenmaschine *f* *(text)* storey drying machine
Etagentrockner *m* tier drier
Etagenwohnung *f* flat
Etagenzwirnmaschine *f* twisting frame with serveral tiers, uptwister, up-twister
Etalon *m* etalon
Etalonplatte *f* etalon plate
Etappe *f* 1. stage; 2. *(mil)* base
Etat *m* budget
Eternit *n* *(m)* Eternit *(asbestos cement; trade name)*
Etikett *n* label, tag
Etikettdruck *m* label printing
Etikettenausstanzmaschine *f* label puncher
etikettieren to label
Etmal *n* day's run *(distance sailed by a ship during the 24 hours of a nautical day)*; day's work *(work of computation required in navigating a ship for every 24 hours)*
Euchlorin *n* euchlorine
Eudiometer *n* eudiometer
Eugenol *n* eugenol
Eukalyptol *n* eucalyptole
Eukalyptusbaum *m* eucalyptus
Eukalyptusöl *n* eucalyptus oil
Euklas *m* euclase
euklidisch Euclidian
Eukolloid *n* eucolloid
Eulerbereich *m* Euler range
Europium *n* europium
eustatisch eustatic
Eutektikum *n* eutectic [system]
eutektisch eutectic
eutektoid eutectoid
Eutektoid *n* eutectoid
Euterviertelmelkmaschine *f* *(ag)* udder quarter recording milker
evakuieren to evacuate, to pump out (down), to empty
Evakuieren *n* evacuation
evakuiert evacuated
  schlecht e. gassy *(tube)*
Evakuiervorrichtung *f* evacuating machine, *(Am)* exhausting machine
Evolute *f* *(math)* evolute
Evolvente *f* involute [curve]
Evolventenformprüfmaschine *f* involute profile checking machine
Evolventenformräumwerkzeug *n* involute form broach
Evolventenprofil *n* involute [tooth] profile
Evolventenprüfgerät *n* involute tester
Evolventenrad *n* involute gear
Evolventenverzahnung *f* involute [gear] teeth
Evolventenzahn *m* involute gear tooth
Evolventenzahnrad *n* involute [toothed] gear
Evolventenzahnradfräser *m* involute gear milling cutter
Ewer *m* sailing barge, lighter, wherry [boat]
Ewerführer *m* bargee, lighterman, waterman, wherryman
Examen *n* examination
3-Exzeß-Kode *m* *(dat)* excess-3-code, XS-3 code
Exciton *n* *(ph)* exciton
Excitonenbildung *f* exciton formation
Excitonenerzeugung *f* exciton production
Excitonenniveau *n* exciton level
Excitron *n* *(ph)* excitron
Exergie *f* *(ph)* exergy
Exerziermunition *f* dummy (practice) ammunition
Exhaustor *m* exhaust fan
Exkursion *f* 1. work visit *(students)*; 2. excursion, run *(reactor)*
Exkursionsschiff *n* excursion vessel (ship)
Exoelektronen *npl* *(nucl)* exoelectrons
exoenergetisch exoergic
Exophthalmie *f* *(opt)* exophthalmos
Exophthalmometer *n* *(opt)* exophthalmometer
exotherm exothermic
Expansion *f* expansion; *(ph)* dilatation *(of solid bodies by heat)*
  E. von Flüssigkeiten (Gasen) expansion of liquids (gases)
  adiabatische E. adiabatic expansion
  gemischte E. mixed expansion
  isotherme E. isothermal expansion
  vollständige E. complete expansion
Expansionsapparat *m* *(therm)* expansion gear
Expansionsarbeit *f* work done during expansion *(turbine)*
Expansionsbeginn *m* initial point of expansion
Expansionsdampfmaschine *f* expansion steam engine, steam engine working expansively
Expansionsendspannung *f* *(therm)* terminal pressure, final pressure of expansion
Expansionsexponent *m* *(therm)* exponent of expansion
Expansionsgrad *m* *(therm)* degree of expansion
Expansionshub *m* expansion stroke
Expansionskurve *f* *(therm)* expansion curve
Expansionslinie *f* *(therm)* expansion curve
Expansionsluke *f* *(ship)* expansion hatch
Expansionsmaschine *f* expansion engine
Expansionsperiode *f* period of expansion
Expansionsschalter *m* expansion circuit-breaker
Expansionsstufe *f* expansion stage *(turbine)*
Expansionsventil *n* expansion valve *(refrigeration system)*
Expansionsverflüssigung *f* expansion liquefying
Expansionswelle *f* blast wave

Expedient *m* dispatching (forwarding) clerk
Experiment *n* experiment, test, trial
  kritisches E. *(nucl)* critical experiment
Experimentalbau *m* experimental building
Experimentalhaus *n* experimental house
Experimentalphysik *f* experimental physics
experimentell experimental
experimentieren to experiment, to test, to try
Experimentierkanal *m (nucl)* experimental hole
explodieren to explode, to burst, to detonate
  nicht e. to fail to detonate
Explosion *f* explosion, detonation, blast
Explosionsdruck *m* explosion (blast) pressure
explosionsgeschützt explosion-proof
Explosionsgrenzen *fpl* explosivity limits
Explosionshub *m* firing stroke
Explosionsmotor *m* internal-combustion engine
Explosionsramme *f* frog (internal-combustion) rammer
Explosionsschauer *m (nucl)* burst
explosionssicher explosion-proof
Explosionsumformen *n* explosive forming
explosiv explosive
Explosivumformen *n* explosive forming
Exponent *m* exponent
  E. der Expansionslinie *f (therm)* exponent of expansion
Exponentialexperiment *n* exponential experiment
Exponentialfunktion *f* exponential function
Exponentialgleichung *f* exponential equation
Exponentialreaktor *m* exponential pile (reactor)
Exponentialröhre *f* variable-mu tube
Exponentialtetrode *f* variable-mu tetrode
Exponentialwirkung *f (el)* variable-mu action
exponentiell exponential
exponieren *(nucl)* to expose [to radiation]
Exponierte *m (nucl)* exposed person
Exponierung *f (nucl)* exposure
Export *m* export
Exportauftrag *m* export order
Exportausführung *f* export version
Exporteinnahmen *fpl* earnings from exports
Exporteur *m* exporter
exportieren to export
Exportliefertermin *m* export delivery date
Exportlieferung *f* export delivery
Exportlieferzeit *f* export delivery time (period)
Exposition *f (nucl)* exposure
Expreßgut *n* express parcel (goods)
Exsikkator *m* exsiccator, desiccator
Exsikkatoreinsatz *m* desiccator plate
Extinktion *f (ph)* extinction, attenuation
Extinktionskoeffizient *m* coefficient of extinction
Extinktionsregistriergerät *n* extinction recorder
Extradoppelfeinbank *f (text)* double-jack frame
Extrafeinabstimmung *f* extra-fine tuning
Extrafeinflyer *m* double-jack frame, fine roving frame
extrahieren to extract
Extrakt *m* extract; essence
Extraktgerbung *f (leath)* extract tanning
Extraktion *f* 1. extraction; leaching; 2. removal
  elektrische E. electro-extraction
Extraktionsanlage *f* extraction plant
Extraktionskolben *m* extraction flask
Extraktionskolonne *f* extraction column
Extraktionsmittel *n* extraction agent, extractive; extractant
Extraliegetag *m (ship)* day of demurrage
Extrapolation *f (math)* extrapolation
Extrapolationslänge *f (nucl)* extrapolation length (distance)
extrapolieren to extrapolate
Extrazubehör *n (mach)* extras
Extremalwertregelung *f* optimalizing control
Extrempunkt *m (math)* bend point *(curve)*, apse
Extremwert *m* extreme value
Extremwertkarte *f* control chart for largest and smallest values
Extruder *m (plast)* extruder
extrudieren *(plast)* to mould by extrusion
Exzenter *m* eccentric cam
  E. für Kreuzwindung *(text)* cam for crossing layer
  E. für Schalttrommel *(text)* cam for drum racking *(knitting)*
  der Kurbel vorauseilender E. *(therm)* leading eccentric
Exzenterangriff *m* **beim Ladenvorgang** *(text)* cam stroke as the slay beats up
Exzenterbewegung *f* eccentric motion
Exzenterkeil *m (text)* cam bit *(knitting)*
Exzenterkreis *m* eccentric circle
Exzentermittel *n* centre of eccentric
Exzentermittellage *f* middle position of eccentric
Exzenterpresse *f* eccentric press
Exzenterrad *n* eccentric wheel
Exzenterrolle *f (text)* cam bowl
Exzenterscheibe *f* eccentric sheave (disk)
Exzenterschwingsieb *n* eccentric vibrating screen
Exzenterspeisepumpe *f* eccentric-driven feed pump
Exzenterstellung *f* position of eccentric
Exzenterstück *n* **für Offenfach** *(text)* cam piece for open shed
Exzenterstuhl *m (text)* tapped (treadle) loom
Exzentertotlage *f (therm)* dead point (centre) of eccentric
Exzenterwelle *f* eccentric shaft
Exzenterzapfen *m* eccentric pin
exzentrisch eccentric; off-centre, off-radial
Exzentrizität *f* [amount of] eccentricity
  E. der Drehung eccentricity of rotation *(turbine)*

E. der Ruhe eccentricity of rest *(turbine)*
Exzeß *m* excess; *(math)* kurtosis
Exziton *n* exciton
Exzitron *n* excitron

F

F-Absorption *f* F-absorption
F-Bande *f* F-band
F-Schicht *f* F-layer, Appleton layer
F-Verteilung *f* F distribution
Fabrik *f* 1. works; shop; plant; 2. *(mach)* factory; 3. mill *(spinning, paper)*; 4. distillery *(vinegar)*
Fabrikarbeiter *m* factory (shop, mill) worker
Fabrikat *n* make, manufacture; manufactured article
Fabrikation *f* manufacture, making; production
Fabrikationskosten *pl* manufacturing (production) cost
Fabrikationsnummer *f* serial (stock) number
Fabrikationsprogramm *n* manufacturing programme
Fabrikdirektor *m* factory manager
fabrikfertig factory-finished
Fabrikgebäude *n* factory building
Fabrikhafen *m* private port
Fabrikhalle *f* bay, factory hall, shop
Fabrikschiff *n* factory ship
Fabrikschornstein *m* factory chimney
Fabriktrawler *m* factory trawler, trawler factory ship
Fabrikvorfertigung *f* factory precasting *(concrete)*
fabrizieren to manufacture, to produce
Facette *f* facet[te], bevel, chamfer
Facettenfräsmaschine *f* bevelling machine
Facettenhalter *m* flange
Facettenhobel *m* bevelling machine, shootboard
Facettenspiegel *m* tesselated mirror
Facettensteg *m* plate-clip
facettieren to bevel
Fach *n* 1. subject, field, branch, profession, line; 2. *(text)* shed *(Jacquard)*; 3. *(build)* panel *(truss)*, bay; 4. compartment *(shelf)*; drawer *(table)*; shelf; partition, division *(cupboard)*
  feines F. *(text)* V-shed
  gleichbleibendes F. *(text)* shed of fixed length
  reines offenes F. *(text)* clear shed
  schräges F. *(text)* clear shed, V-shed
  technisches F. engineering subject
Facharbeiter *m* skilled worker, craftsman
Facharbeiterausbildung *f* craft training
Facharbeiterlohn *m* skilled rate of pay
Facharbeitermangel *m* skilled-labour shortage, shortage of skilled workers (labour)
Facharbeiterprüfung *f* qualifying examination
Fachausdruck *m* technical term
Fachbildevorrichtung *f* *(text)* shedding motion
Fachbildung *f* 1. knowledge in a special field; 2. *(text)* shedding
Fachbuch *n* / **technisches** technical book
fachen *(text)* to ply
Fachen *n* *(text)* shedding motion
Fächer *m* fan
Fächerantenne *f* fan (spider-web) aerial
Fächereinbruch *m* *(mine)* fan cut, slabbing-cut, swing-cut
Fächerfenster *n* fan light
Fächermarkierungsbake *f* fan marker beacon
Fächermotor *m* double-V engine
Fächerverankerung *f* fan anchorage *(prestressed concrete)*
Fachexzenter *m* *(text)* shedding tappet (wiper)
Fachgebiet *n* field; subject, branch; profession; line
Fachkönnen *n* knowledge in a special field; workmanship
Fachkräfte *fpl* skilled labour (workers)
Fachkräftemangel *m* shortage of skilled labour
Fachliteratur *f* scientific (technological) literature
  chemische F. chemical literature
  technische F. technical (technological) literature
Fachmann *m* expert, specialist; subject specialist
Fachmaschine *f* *(text)* doubling machine
Fachrichtung *f* special field
Fachschluß *m* *(text)* closing the shed
Fachterminologie *f* / **technische** technological terminology
Fachtreten *n* *(text)* shedding by treadles
Fachwählertastatur *f* classification keyboard *(punched card machinery)*
Fachwerk *n* framework; truss; timbered building
  ebenes F. plane truss
  räumliches F. space frame (truss)
Fachwerkbinder *m* trussed (framed, lattice) girder
  parallelgurtiger F. flat (parallel chord) truss
Fachwerkbogen *m* trussed arch
Fachwerkbrücke *f* truss bridge
Fachwerkhaus *n* timber-framed house, timber framework building
Fachwerkknoten *m* truss joint
Fachwerkkonstruktion *f* lattice-type construction
Fachwerkstab *m* truss member
Fachwerkträger *m* lattice girder, trussed (framed, lattice) beam
Fachwerkträgerbrücke *f* trussed girder bridge
Fachwerkverband *m* cross (diagonal) tying
Fachwort *n* technical term
Fachwortkartei *f* card register of [technical] terms

Fachzeitschrift *f* / **technische** technical magazine (journal, periodical)
Fackelgas *n* fackel-gas
Fackelrohr *n* bleeder pipe, gas bleeder
Fädelband *n* *(el)* drawing hoop, fish tape, *(Am)* pulling-in tape
Fädelmaschine *f* threading machine
Fädelwicklung *f* *(el)* pull-through winding
Faden *m* 1. thread; filament; 2. fathom *(measure)*
  F. der Glühlampe filament of the bulb
  F. des Falzmessers thread of the shaving knife
  gespritzter F. colloidal filament
  gezwirnter F. plied yarn
Fadenabsauganlage *f* thread suction plant
Fadenanzugfeder *f* *(text)* thread take-up spring
Fadenauge *n* *(text)* thread guide
Fadenauslegen *n* putting the yarn out of action *(knitting)*
Fadenbildung *f* formation of the thread
  mangelhafte F. faulty formation of the thread
Fadenbruch *m* thread breakage
Fadenbruchabsauganlage *f* broken-end collection
Fadenbruch- und -zupfabsteller *m* stop motion for yarn breakages and yarn drag *(knitting)*
Fadenbruchzahl *f* end breakage rate
Fadenbuchheftmaschine *f* thread book-stitching machine
Fadendichte *f* *(text)* sett
Fadeneinlegeapparat *m* thread insertion apparatus *(knitting)*
Fadeneinlegen *n* insertion of thread *(knitting)*
Fadenelektrometer *n* filament (thread) electrometer
Fadenführer *m* thread guide *(knitting)*
  regulierbarer F. adjustable thread guide
  schwingender F. traversing thread guide
Fadenführeranschlag *m* end top
Fadenführeraufläufer *m* / **kettengesteuerter** chain-controlled carrier stop *(knitting machine)*
Fadenführergleitschiene *f* carrier slide bar
Fadenführerkette *f* chain of traverse rail
Fadenführerschlauch *m* carrier tube
Fadenführung *f* *(text)* thread guidance
Fadenhalter *m* thread holder
Fadenheftmaschine *f* thread-sewing machine
Fadenheftung *f* thread sewing, thread stitching
Fadeninstrument *n* filament instrument
Fadenkatode *f* filamentary cathode
Fadenkonstruktion *f* *(math)* string construction
Fadenkorrektur *f* stem correction
Fadenkreuz *n* 1. *(text)* lease; 2. cross hairs *(measuring)*
Fadenkreuzung *f* *(text)* crossing of threads
Fadenkreuzvisier *n* cross wire sight
Fadenkreuzvorrichtung *f* *(text)* leasing system
Fadenlagennähwirkmaschine *f* *(text)* stitch-bonding machine for layers of weft and warp threads
Fadenlauf *m* *(text)* course of the thread
Fadenlinie *f* involute curve
Fadenlunker *m* *(met)* coky centre; secondary pipe
Fadenmantel *m* surface of the thread
Fadenmantelfläche *f* **der Spule** winding surface of the bobbin
Fadenmikrometer *n* filar micrometer
Fadennetz *n* reticule, reticle
Fadenöse *f* thread guide
Fadenprobe *f* string test *(sugar)*
Fadenreiniger *m* yarn clearer
Fadenreiter *m* *(text)* dropper
Fadensonde *f* filament probe
Fadenspannung *f* 1. *(el)* filament voltage; 2. *(text)* yarn tension
Fadenstrahl *m* thread (narrow) beam
Fadenthermometer *n* filament thermometer
Fadentransistor *m* filamentary transistor
Fadenüberwachungsgerät *n* yarn stop motion detection installation
Fadenwiderstand *m* filament resistance
Fadenwindung *f* coil of yarn
Fadenzähler *m* thread counter (counting glass), pick-glass, pick counter
Fadenzähllupe *f* simplified thread counter
Fadenzerstäubung *f* disintegration of filament
fadenziehend ropy
Fadenzubringer *m* yarn feed installation
Fadeometer *n* *(text)* fadeometer
Fading *n* fading [effect] *(radio)*
Fadingausgleich *m* automatic volume control
Fadingeffekt *m* fading effect
Fadingregelung *f* automatic volume control
Fagaraseide *f* fagara silk
fähig capable
Fähigkeit *f* capability; ability; capacity
Fahlerz *n* fahlerz, fahlore, grey copper ore
Fahlleder *n* russet upper leather
Fähnchenabschwächer *m* *(el)* guillotine attenuator
Fähnchenwattmeter *n* vane wattmeter
Fahne *f* 1. flag; 2. *(print)* galley proof, proof-sheet, slip-proof; 3. *(el)* lug [of commutator]
Fahnenabzug *m* galley proof, proof-sheet
Fahnenbildung *f* *(telev)* tailing
Fahnenkorrektur *f* reader's proof
Fahnenmagnetron *n* vane magnetron
Fahnenmast *m* flagstaff
Fahnenstange *f* flag pole
Fahnenziehen *n* *(telev)* streaking, pulling on whites, black after white
  F. beim Bildorthikon *(telev)* smudging
Fähranleger *m* ferry bridge, floating ferry
Fahraufhauen *n* *(mine)* service raise

Fahrausweis *m* ticket
 abgelaufener F. expired ticket
 ausgedruckter F. fixed destination ticket
 internationaler F. international ticket
Fahrbahn *f* carriageway; track
 zweispurige F. dual-lane carriageway (road)
Fahrbahnbelag *m* road surface
Fahrbahnmarkierung *f* road marking
Fahrbahnplatte *f* decking, deck slab, [bridge] deck
Fahrbahnseite *f* / **linke** offside *(in countries with right-hand traffic)*
 rechte F. offside *(in countries with left-hand traffic)*
Fahrbahntafel *f* [bridge] deck, deck slab, decking
fahrbar *(ship)* navigable; *(build)* portable; mobile *(crane)*
Fahrbereich *m* operating (cruising) range, steaming range (radius); radius of action
fahrbereit ready to sail *(vessel)*
Fährbett *n* ferry rack
Fährboot *n* ferry [boat], train ferry
Fahrbremse *f* pedal brake
Fährbrücke *f* ferry (trail) bridge
Fahrdamm *m* carriageway
Fahrdienstleiter *m* transport manager, station master
Fahrdraht *m* contact (trolley) wire, bus (contact) line
 doppelter F. double-contact wire system
 einfacher F. single-contact wire system
Fahrdrahtaufhängung *f* catenary suspension
Fahrdrahtkreuzung *f* overhead crossing
Fahrdrahtlokomotive *f* trolley locomotive
Fahrdrahtmast *m* trolley tower
Fahrdrahtspanner *m* trolley wire strainer
Fahrdrahtstoßklemme *f* splicing fitting
Fahrdrahtweiche *f* overhead (trolley) frog
Fähre *f* ferry [boat], wherry
Fahreigenschaft *f* *(aut)* roadability
fahren to drive *(a car)*; to traverse *(machine slide)*; to travel *(in a car)*; to run *(wagon on rails)*; to ride *(lorry, motor cycle)*
 auf falschem Gleis f. to run on the wrong line (track)
 Boot f. to boat, to go in a boat
 im Auto f. to travel by car
 in Kiellinie f. *(ship)* to line ahead
 mit Benzin f. to run on petrol *(car)*
 mit überhöhter Geschwindigkeit f. to speed
 rückwärts f. to back [up]; *(ship)* to go astern
 Zickzackkurs f. to zigzag
Fahren *n* **auf Sicht** operation under shunting regulations
Fahrer *m* driver *(car, truck, bus)*, chauffeur
 regelmäßiger F. *(ship)* trader
Fahrerboden *m* cab floor
Fahrerlaubnis *f* driving licence
Fahrersitz *m* driver's seat
Fahrfeld *n (mine)* travelling track (way), travelway, walking track
Fahrgast *m* passenger
Fahrgastbarkasse *f* passenger launch (boat), runabout
Fahrgastdeck *n* passenger deck
Fahrgasteinrichtung *f* passenger accomodation
Fahrgastlinienschiff *n* cabin liner
Fahrgastmotorschiff *n* passenger motorship
Fahrgastraum *m* passenger compartment
Fahrgastschiff *n* passage boat, passenger ship (liner)
Fahrgastschiffsraum *m* passenger tonnage
Fahrgeld *n* [passenger] fare
Fahrgeschwindigkeit *f* travelling speed *(crane)*; *(ship)* service speed (rate); velocity, speed *(car)*
Fahrgestell *n* base frame, chassis *(car)*; *(ry)* underframe; running gear
Fahrgestellrahmen *m (aut)* chassis frame
Fährhafen *m* ferry terminal
Fahrhauer *m (mine) (Am)* fire boss
Fahrkante *f (ry)* inner side of rail head
Fahrkarte *f* ticket
 F. für Hin- und Rückfahrt return ticket
Fahrkartenausgabemaschine *f* ticket issuing machine
Fahrkartenheft *n* book of tickets
Fahrkartenkarton *m* ticket cardboard
Fahrkartenkontrolle *f* inspection of tickets
Fahrkartenkontrolleur *m* travelling ticket-collector, travelling ticket-inspector
fahrklar machen to make operational *(a vessel)*
Fahrkorb *m* car *(elevator)*
Fahrkran *m* mobile (travelling) crane
Fahrkupplung *f* external clutch
Fahrlehrer *m* driving instructor
Fahrleitung *f* contact line
Fahrleitungsanlage *f* overhead contact system
Fahrleitungsausleger *m* bracket arm
Fahrleitungsmast *m* **[der Straßenbahn]** tramway pole, tram tower
Fahrleitungsschalter *m* line breaker
Fährmann *m* waterman, ferryman; *(Am)* floatman
Fahrmarke *f (mine)* tally
Fahrmotor *m* traction motor
Fährnische *f* ferry rack
Fahrpaß *m (mine)* travelling track (way), walking track (way), travelway
Fahrplan *m* time table
Fahrplanänderung *f* time-table alteration
Fährprahm *m* ferry punt
Fahrrad *n* bicycle
Fahrraddynamo *m* bicycle dynamo
Fahrradluftpumpe *f* bicycle pump

Fahrradweg *m* cycle track
Fahrrinne *f* navigational (ship) channel; fairway
Fahrschacht *m (mine)* travelling shaft
Fahrschalter *m* traction (control) switch, controller
Fahrschaltersteuerung *f* directly controlled equipment
Fahrscheinheft *n* coupon ticket, book of coupons
  internationales F. international sectional coupon-ticket
  zusammengestelltes F. combined coupon ticket, booklet of circular tour tickets
Fährschiff *n* ferry [boat]; car ferry
Fahrschule *f* driving school
Fahrschüler *m* learner-driver
Fahrsilo *m (ag)* clamp
Fahrsperre *f (ry)* [automatically released] train stop, [automatically released] train stopping device
Fahrspiel *n (mine)* trip cycle
Fahrspieldauer *f (mine)* cycle time
Fahrspur *f* [traffic] lane
Fahrspurmarkierung *f* lane marking
Fahrstand *m* control platform (position)
Fahrsteiger *m (mine)* district (specialist) overman
Fahrsteuerung *f* train control
Fahrstift *m* tracing point
Fahrstraßenauflösung *f* release of a route locking, route release
Fahrstraßenfestigung *f (ry)* route locking *(before the signal has been passed)*
Fahrstraßenspeicher *m* route storage arrangement
Fahrstraßenspeicherung *f* pre-setting
Fahrstrecke *f (mine)* travelling track (way), travelway, *(Am)* gangway
Fahrstreckenverriegelung *f* section blocking
Fahrstromrückleitung *f* return circuit
Fahrstromschiene *f* contact collecting bar
Fahrstufe *f (ry)* speed stage, working point, running (regulating) step
Fahrstuhl *m* lift
  F. mit Selbststeuerung self-service lift
Fahrstuhlgerüst *n* lift frame
Fahrstuhlschacht *m* elevator well, lift shaft
Fahrstuhlsteuerschalter *m* lift controller
Fahrstuhlstockwerksdruckknopf *m* landing push-button
Fahrt *f* driving, drive; riding, ride run[ning]; going; journey *(train)*; *(ship)* passage, voyage, cruise; way (navigation) of a ship; tour; *(mine)* ladder
  F. auf falschem Gleis *(ry)* wrong-direction running
  F. auf Sicht *(ry)* cautious running, running under cautions
  F. eines Schiffes navigation, way of a ship
  F. in Kiellinie *(ship)* column movement
  F. in Vielfachsteuerung multiple unit running
  einfache F. single journey
  große F. long (ocean) voyage; ocean going; distant trade
  kleine F. short (home) trade
  lange F. distant trade
  mittlere F. intercoastal navigation
Fahrtbereich *m* reach of trade
Fahrtenkreuzer *m* cruising yacht, cruiser
Fahrtentrum *m (mine)* ladderway, manway compartment
fahrtgestört not under control *(vessel)*
Fahrtgrenze *f* reach (limit) of trade
Fahrtlaternen *fpl (ship)* running lights, navigating lanterns
Fahrtmesser *m s.* Geschwindigkeitsmesser
Fahrtminderungsschalter *m* slowing controller
Fahrtregler *m* speed governor, winding speed regulator
Fahrtrenner *m (el)* horizontally moving isolator
Fahrtreppe *f* escalator
Fahrtrichtung *f* direction of travel
Fahrtrichtungsanzeiger *m* direction indicator [lamp]; trafficator
Fahrtrichtungsschalter *m* reverser
Fahrtroute *f* [shipping] lane
Fahrtrum *m (mine)* ladder road, ladderway
Fahrtstellung *f* clear position (aspect), proceed (off) position
Fahrtstörungslaterne *f (ship)* not-under-command light
Fahrttrimm *m* squat *(of a vessel)*
Fahrtwender *m* reverser
Fahrtwende- und Motortrennschalter *m* disconnecting switch reverser
Fahrtwiderstand *m* head resistance
Fahrüberhauen *n (mine)* service raise
Fahruntüchtigkeit *f* unseaworthiness *(of a vessel)*
Fahrwasser *n* [ship] channel, fairway; sea-lane
Fahrwassertiefe *f* depth of fairway
Fahrwassertonne *f* channel buoy
Fahrweg *m (mine)* dog hole
Fahrwerk *n* alighting (landing) gear; undercarriage
Fahrwerkabfederung *f* landing gear cushioning
Fahrwiderstand *m* resistance to forward motion
  F. in der Ebene und in der Geraden *(ry)* train resistance on level tangent track
  spezifischer F. *(ry)* specific train resistance
Fahrwindbelüftung *f (ry)* roof ventilation
Fahrzeug *n* vehicle *(land)*; *(ship)* vessel; craft *(air)*
  F. ohne Eigenantrieb *(ship)* dumb vessel
  schienenloses F. trackless vehicle

stillgelegtes F. laid-up vehicle
Fahrzeugachse *f* vehicle axle
Fahrzeugantriebsbatterie *f* traction battery
Fahrzeugbremse *f* vehicle brake
Fahrzeugdeck *n* vehicle deck *(of a vessel)*
Fahrzeugheizungsanlage *f* car heating installation
Fahrzeugkolonne *f* caravan of vehicles
Fahrzeugmotor *m* vehicle engine; [electric] traction motor
Fahrzeugpanne *f* breakdown
Fahrzeugpapiere *pl* registration book
Fahrzeugpark *m* fleet of vehicles
Fahrzeugverkehr *m* vehicular traffic
Fahrzeugwinde *f* vehicle winch
Fairbairn-Kessel *m* Fairbairn boiler
Fäkalien *pl* f[a]ecal matter, f[a]eces, excreta
Fäkalstoff *m* fecal substance
Faksimile *n* facsimile
Faksimiledruck *m* facsimile printing
Faksimileübertragung *f* facsimile transmission
Faktis *m* factice, oil rubber, vulcanized oil
Faktor *m* factor
isolierter F. isolated factor
Faktorenversuch *m* factorial experiment
Faktorenzerlegung *f (math)* factoring, factorization
Fakturiermaschine *f* billing (invoicing, accounting) machine
Fakultät *f (math)* factorial
Fall *m* 1. case; 2. *(ph)* fall; drop; 3. *(text)* drape; 4. *(ship)* rake *(mast, funnel)*
freier F. free fall
Fällapparat *m* precipitation apparatus
Fällbad *n* coagulating bath *(fibre)*, precipitant
mehrschichtiges F. *(text)* bath with several layers of liquid
Fallbär *m* tup, monkey, hammer *(piling plant)*
fällbar *(ch)* precipitable
Fällbarkeit *f (ch)* precipitability
Fallbehälter *m (aut)* gravity-feed tank
Fallbeschleunigung *f* acceleration due to gravity, gravitational acceleration
Fallblech *n* fall plate
Fallbügel *m (el)* chopper bar; locking device
Fallbügelinstrument *n (el)* instrument with locking device; relay-type recording instrument
Fallbügelregler *m (Am)* chopper bar controller
Fallbügelrelais *n* hoop drop relay
Fallbügelschreiber *m* chopper bar recorder
Falle *f* trap, catch; entrainment separator
magnetische F. magnetic trap
fallen to fall; to drop; to decrease
f. lassen to drop
den Anker f. lassen to drop the anchor
Fallen *n* 1. fall; 2. *(mine)* dip; 3. *(text)* drape
F. einer Form contraction of a mould
fällen *(ch)* to precipitate; to fell, to cut down *(tree)*; to pass *(sentence)*; *(math)* to drop *(perpendicular)*
fallend falling; *(mine)* underhand
frei f. falling freely
Fallennest *n* trap-nest
Fäller *m* precipitation apparatus
Fallfenster *n* drop window
Fallgesetz *n (ph)* law of falling bodies
Fallgewicht *n* tup, monkey, hammer *(piling plant)*
Fallhammer *m* monkey, tup, hammer *(piling plant)*
Fallhärte *f* impact ball hardness
Fallhärteprüfer *m* hardness drop tester
Fallhärteprüfung *f* drop hardness test
Fallhaspel *f (text)* collapsible reel
Fallhobelmaschine *f* tubular planing machine
Fälligkeitstag *m* due date
Fälligkeitstermin *m* due date
Fallinie *f (geol)* line of dip
Fallkasten *m (text)* drop (rising) box
Fällkasten *m* zinc box
Fallkorb *m* scarf cut, undercut, notch, nick
Fallklappe *f* shutter; trap door; *(el)* drop indicator; drop shutter
Fallklappenrelais *n* drop indicator relay
Fallmasche *f (text)* dropped stitch
Fallmaschine *f* / **Atwoodsche** Atwood's machine
Fallmeißelschiff *n* drop-cutting ship, drop cutter
Fallout *m (nucl)* fall-out
lokaler F. local radioactive fall-out
Fallprobe *f* drop test *(anchor)*
Fällprodukt *n* deposit, precipitate
Fallreep *n* boarding ladder, side-rope, gangway, man-rope
Fallreepspforte *f* entering (gangway) port
Fallreepstau *n* entering-rope, man-rope
Fallreepstreppe *f* entering-ladder, accomodation ladder
Fallrichtung *f (geol)* direction of dip
Fallrohr *n* soil pipe *(water closet)*; gutter pipe; fall (down) pipe, down-spout
Fallrohrkondensator *m* dry condenser
Fallschirm *m* parachute
Fallschirmbergung *f* parachute recovery
Fallschirmseide *f* parachute silk (fabric)
Fallschnecke *f* drop worm [mechanism]
Fallspeisung *f* head feed
Fallstromvergaser *m (aut)* down-draught carburetter
Fallstropp *m (ship)* halliard strap
Falltank *m* header [tank], gravity[-feed] tank
Falltreppe *f (mine)* dropping step
Falltür *f (build)* trap door; *(mine)* guillotine door

Fällung *f (ch)* precipitation; felling, cutting-down *(tree)*

Fällungschromatografie *f* chemichromatography

Fällungsmittel *n (ch)* precipitant, precipitating agent

Fallwassergeschwindigkeit *f* velocity of falling water *(condenser)*

Fallwinkel *m* 1. *(build)* angle of pitch; 2. *(mil)* angle of fall

F. von Schichten *(geol)* angle of bedding

Falschanruf *m (telec)* false ring, *(Am)* wrong number call

Falschdrahtspindel *f (text)* false twisting spindle

fälschen to duff; to adulterate *(victuals)*; to forge *(coin)*; to counterfeit

Falschlochungen *fpl* errors in card punching

Falschluft *f* leak of air, entrained air

falschmünzen to forge *(coins)*

falschphasig misphased, outphased

Falschverbindung *f (telec)* wrong connection

Falschwahl *f (telec)* faulty selection

Falschzeichen *n* phantom signal *(radar)*

Falschzwirnmaschine *f* false twisting frame

Faltbeanspruchung *f* folding stress

Faltboot *n* folding (collapsible) canoe, canvas boat

Faltbroschüre *f* folder, leaflet

Faltdach *n* folding hood, collapsible top

Faltdipol *m* folded dipole

Faltdipolantenne *f* folded dipole aerial

Falte *f* 1. *(mine)* fold *(ground)*; 2. *(build)* V-unit *(folded plates)*; 3. *(text)* crease, crinkle, wrinkle

fälteln to pleat

falten to fold; to pleat *(cloth)*; to double

Faltenbalg *m* bellow; *(ry)* gangway (communication) bellows

Faltenbalglasttrennschalter *m* load-breaking switch with wrinkling bellows

Faltenband *n (mine)* folded belt

Faltenbeständigkeit *f* crease-retention *(of fabrics)*

Faltenbildung *f (telev)* fold-over

Faltenfilter *n (ch)* plaited (folded, fluted) filter

Faltenkette *f* crimp warp

Faltfestigkeit *f* folding endurance

Faltlutte *f (mine)* concertina-type ducting

Faltprospekt *m* folder

Faltschachtel *f* folding box

Faltschachtelkarton *m* folding boxboard

Faltschachtelklebemaschine *f* folding box gluing machine

Faltspriegel *m* folding bow

Faltspur *f (text)* crease mark

Faltstelle *f (text)* crease mark

Falttür *f* accordion (folding) door

Faltung *f* convolution *(integral equation)*

Faltversuch *m* doubling (folding) test

Faltwand *f* accordion partition

Faltwerk *n* hipped plate, prismatic slab (shell)

Faltwerkdach *n* folded-plate roof

Faltwinkel *m* angle of bend

Falz *m* fold; *(build)* rebate; seam *(sheet metal)*; welt; joint; notch

Falzbein *n* folding stick, folder

Falzbock *m (leath)* shaving beam

Falzeisen *n (leath)* shaving knife

falzen 1. to fold *(book)*; 2. to welt, to seam *(sheet)*; to turn up edges; to groove; to shave *(leather)*

Falzfestigkeit *f* folding endurance, bursting strength

Falzfräser *m* rebating (rabbeting) cutter, notching cutter

Falzhaken *m* heel tool

Falzhobel *m* grooving plane, fillister [plane]

einfacher F. standing fillister

stellbarer F. moving fillister

Falzmaschine *f* 1. folding machine *(book)*; 2. *(leath)* shaving machine

Falzmesser *n* folding blade; shaving knife

Falzplatte *f (leath)* beam face

Falzrohr *n* slip conduit

Falzschwert *n* folding blade

Falzstoß *m* seam joint *(soldering)*

Falzstreifen *m* slip-fold

Falztrichter *m* former

Falztrommel *f* folding drum

Falz- und Leimmaschine *f* folding and gluing machine

Falzwiderstand *m* folding endurance *(paper)*

Falzziegel *m* interlocking tile

Fanganode *f* collecting (gathering) anode

Fangarbeit *f* fishing job

Fangband *n* rebound strap

Fangbaum *m* trap tree

Fangbereich *m (el)* lock-in range, collecting zone

Fangbügel *m* safety bow; brake safety strap

Fangdüse *f* receiving nozzle

Fangedamm *m* cofferdam

Fangeinrichtung *f (telec)* interception circuit

Fangelektrode *f* target, collecting electrode

Fänger *m* catcher *(fishing)*

Fangfaden *m (text)* catch thread, catch end at the selvadge

Fanggerät *n* fishing tools

Fanggitter *n* suppressor (collecting) grid

Fangglocke *f* overshot

Fanggraben *m* trap trench

Fanghaken *m (mine)* backstay, crowfoot; catch hook

Fanghülse *f* friction socket

Fangkettenware *f (text)* double-rib warp goods

Fangleine *f* 1. shroud line *(parachute)*; 2. *(ship)* lazy jack

Fangmasche *f (text)* cardigan (tuck) stitch

erhabene F. raised tuck stitch
Fangnetz *n* catch-net
Fangpflanze *f* trap-plant
Fangreise *f* **auf Salzfisch** salt trip
Fangschale *f* ring for the sealing liquid
F. für Tropföl coolant tray
Fangschloß *n (text)* cardigan lock
Fangstäbe *mpl* deflector
Fangstelle *f* trap
Fangstellung *f* tucking position *(knitting)*
Fangstoff *m* gettering agent, getter
Fangteil *m* **eines Hebers** *(text)* cardigan cam
automatisch einstellbarer F. automatic tuck bar *(knitting machine)*
Fangversatz *m (text)* cardigan rack
Fangvorrichtung *f (mine)* safety catch (gear)
Fangwerkzeug *n* fishing tool[s]
Fangzeit *f* fishing season
Farad *n* farad
Faraday-Effekt *m* Faraday effect
Faraday-Käfig *m* Faraday['s] cage
Faraday-Kammer *f* Faraday cell
Faradisation *f* faradization, faradism
faradisieren to faradize
Farbabgabe *f (print)* ink release
Farbabnahme *f (print)* ink taking [of rollers]
Farbabstreichmesser *n* colour doctor
Farbabweichung *f (telev)* chromatic aberration
Farbannahme *f (print)* ink acceptance
Farbanstrich *m* coat of paint
Farbaufnahme *f* 1. *(text)* dye absorption; 2. colour print
farbaufnahmefähig dye absorptive *(fabric)*
Farbaufnahmefähigkeit *f* dye affinity
Farbauftrag *m* / **beiderseitiger** duplex printing
Farbauftragrauschen *n (dat)* inking noise *(pattern recognition)*
Farbauftragwalze *f* colour furnisher
Farbausbeute *f (text)* colour yield
Farbauszug *m* chromatic selection, colour record; colour separation
Farbbalkengenerator *m (telev)* colour bar generator
Farbbalkenmuster *n (telev)* colour bar pattern
Farbband *n* 1. typewriter (copying) ribbon; 2. *(text)* wrong colour of weft
Farbbase *f (text)* dye base
Farbbeständigkeit *f* colour fastness
Farbbild *n* colour picture
farbbildend dye-forming
Farbbildröhre *f* colour picture [television] tube, colour kinescope
F. mit drei Komponenten *(telev)* tricolour tube
Farbbildschirm *m* colour picture screen
Farbdia *n* colour transparency
Farbdichtemeßgerät *n* densitometer
Farbdichtemessung *f* ink density control
Farbdifferenzsignal *n* colour difference signal

Farbe *f* colour, *(Am)* color; paint; pigment; *(text)* dye; [printing] ink
anwuchsverhindernde F. anti-fouling paint (composition)
echte F. fast colour
kurze F. soft (smooth) ink
schnelltrocknende F. *(print)* quickset ink
strenge F. *(print)* short (stiff, tacky) ink
subtraktive F. subtractive colour
Färbeapparat *m* dyeing apparatus *(circulating liquor)*
farbecht colour-fast
Farbechtheit *f* colour fastness, fastness of colour
Färbeflotte *f* dye liquor (bath), dyer's bath
Färbefoulard *m (text)* dyeing pad
Färbegeschwindigkeit *f (text)* rate of dyeing
Färbehülse *f* / **perforierte** *(text)* perforated dyeing tube
Färbekraft *f* tinctorial power, colouring value
Färbemaschine *f (text)* dyeing machine *(circulated material)*
Farbempfindlichkeitskurve *f* sensation curve
Farbempfindung *f* colour perception
färben to dye; to colour
echt f. to dye fast shades
egal f. to dye even
einbadig f. to dye in a single bath
im Garn f. to dye in the yarn
im Garnkörper f. to dye in the package
im Schlauch f. to dye in the hose
im Walkfaß f. *(leath)* to colour in the drum
in der Flocke f. to dye in the loose state, *(Am)* to stock dye
in der Mulde f. *(leath)* to colour in the tray
in der Schlichte f. to dye in the size
in der Walke f. to dye during milling
in der Wolle f. to dye in the loose wool
kochend f. to dye at the boil
nach Muster f. to dye to pattern
neutral f. to dye in a neutral bath
unter der Flotte f. to dye under the surface
Färben *n* **auf der Kontinuekufe** continuous vat dyeing
F. auf Vorbeize dyeing on a mordant
F. im Topfapparat potting
F. im Wickelkörper package dyeing
F. in der Spinnlösung spun dyeing *(man-made fibre)*
F. unter atmosphärischem Druck pressure dyeing
F. von Gewebe ohne vorheriges Waschen bale dyeing
F. von Stückware piece dyeing
Farbenabtaster *m* colour scanner
Farbenbereich *m* colour range
Farbenblindheit *f* colour blindness
Farbenbuchdruck *m* letterpress colour printing
Farbendeckung *f (text)* colour registration

Farbendruck *m* colour printing
farbenempfindlich colour-sensitive
Farbenempfindlichkeit *f* chromatic sensitivity
Farbenfehler *m* chromatic aberration
Farbenfernsehen *n s.* Farbfernsehen
Farbenflimmern *n* colour flicker effect
Farbenfolge *f* colour sequence
Farbenfransen *fpl (el)* colour fringes
Farbengang *m (leath)* colour suspender (pits)
Farbengleichung *f* colour equation
Farbenlängsabweichung *f* longitudinal chromatic aberration
Farbenlehre *f* science (theory) of colours, chromatics
Farbenmaßstabfehler *m* lateral chromatic aberration
Farbenortsfehler *m* chromatic aberration of position
Farbenphasenumtastung *f (el)* colour phase alternation
Farbenraster *m* colour frame
Farbenraum *m (ship)* paint locker
Farbensperre *f (telev)* colour killer
Farbenträgerbezugswert *m* colour carrier reference
Farbentüchtigkeit *f* fitness to distinguish between colours, capacity for differentiating colours
Farbenüberdeckung *f (telev)* colour superimposition
Farben- und Bindungseffekt *m (text)* colour and weave effect
Farbenunterschied *m* **der Vergrößerung** chromatic difference of magnification
Farbenvergrößerungsfehler *m* chromatic difference of magnification
Farbenzeilenflimmern *n (telev)* line crawl (flicker)
Farbenzwischenträger *m (telev)* colour subcarrier
Färber *m* dyer
Färberkrapp *m* madder
Färberwau *m* weld, yellowroot, dyer's rocket
Farbfehler *m (opt)* chromatic aberration
Farbfernsehbildröhre *f* colour picture tube, chromoscope, *(Am)* kinescope
Farbfernsehen *n* colour television
Farbfernsehmethode *f* / **additive** additive colour television system
Farbfilm *m* colour film
Farbfilmdicke *f (print)* ink film thickness
Farbfilter *n* colour filter
Farbflimmern *n* colour flicker
Farbfotografie *f* colour photograph[y]
Farbgebung *f* colouring; dyeing; *(print)* inking
farbgleich isochromatic
Farbhaftgrund *m* paint base
Farbheber *m (print)* ink feed roller
Farbhilfsträger *m (telev)* chrominance subcarrier
Farbholz *n* dyewood
Farbkasten *m (print)* duct, ink fountain (duct)
Farbkastenwalze *f (print)* duct roller
Farbklotz *m (print)* ductor key
Farbkonsistenz *f (print)* ink consistency
Farbkörper *m* pigment, colouring body
Farbkorrektur *f* colour correction
elektronische F. electronic colour correction
Farbkraft *f* colouring strength
Farblack *m* lacquer enamel
farblos colourless
Farbmarkierung *f (el)* colour coding *(components)*
Farbmesser *n (print)* ductor (ink) knife, ductor (ink) blade
Farbmessung *f* colorimetry
Farbmischmaschine *f (print)* ink mixer
Farbmischungsverfahren *n* / **additives** additive colour photography
Farbmusterkopie *f* colour answer (check) print
Farbnebel *m (print)* ink mist
Farbortsfehler *m* chromatic aberration of position
Farbpasser *m (print)* colour register
Farbpaste *f (text)* dye paste
Farbpigment *n (text)* paint pigment; pigment dye
Farbpyrometer *n* colour (colorimetric) pyrometer
Farbrakel *f* colour doctor
Farbraster *m* colour screen
Farbrasterfilm *m* lenticulated film
Farbreibewalzen *fpl* distributing rollers
Farbreibmühle *f* paint grinder mill, colour grinding mill
Farbreinheitseinstellung *f (telev)* purity adjustment
Farbreinheitsregelung *f (telev)* purity control
Farbreiz *m* colour stimulus
Farbreproduktion *f* colour reproduction
Farbretusche *f* colour retouching
farbrichtig orthochromatic
Farbrührwerk *n (print)* ink agitator
Farbsaum *m (ch)* chromatic bleeding
Farbschalter *m (telev)* colour sampler
Farbschaltfrequenz *f* colour sampling frequency (rate)
Farbschriftröhre *f* dark trace tube *(skiagraph)*
Farbsignal *n (telev)* colour signal
zusammengesetztes F. composite colour signal
Farbskala *f* scale of colours, colour range
Farbspaltung *f* ink film splitting
Farbspritzen *n* paint spraying
Farbspritzgerät *n* spray-painting plant
Farbspritzpistole *f* paint spray gun
Farbstäuben *n (print)* ink-misting

Farbstein *m* inking stone, ink slab
Farbstich *m* colour tinge, predominance of one colour
Farbstoff *m* dye[stuff]; colouring agent; pigment
F. in Pulverform dye in powder form
adjektiver F. adjective dye
aufgezogener F. absorbed dye
basischer F. basic dye
direktziehender F. direct dye[stuff]
pulverförmiger F. dye in powder form
substantiver F. direct (substantive) dye
unechter F. fugitive dye
walkfester F. milling dye
Farbstoffaffinität *f* dye affinity
Farbstoffdispergiermittel *n* dye dispersing agent
Farbstofflösemittel *n* dye solvent
Farbsynchronisationsaustastung *f* colour burst pedestal
Farbsynchronisationssignal *n* colour synchronization signal
Farbsynchronisationsstufe *f* colour synchronizer
Farbteig *m* dye paste
Farbteilbild *n* colour frame
Farbtemperatur *f* colour temperature
Farbton *m* tint, shade, tone
abstechender F. contrasting shade
Farbtonregler *m (telev)* chroma control
Farbtreue *f* colour fidelity *(film)*
Farbübertragung *f (print)* ink transfer
Farbüberzug *m* coat of paint
Farbumkehrfilm *m* reversible colour film
Farbumschlag *m* colour change, change of colour
Farbumschlagsbereich *m* **eines Indikators** transition interval of indicator
Färbung *f* colour; *(text)* dyeing
geflammte F. *(text)* shaded dyeing *(fault)*
schattierte F. *(text)* ombré dyeing
Farbvalenz *f* colour stimulus specification
Farbveränderung *f* colour change
Farbvergrößerungsfehler *m* chromatic difference of magnification
Farbverteilung *f* ink distribution
Farbverteilungsmittel *n* dye dispersing agent
Farbwalze *f* 1. *(print)* ductor roller; 2. *(text)* colour furnisher
Farbwanderung *f* migration
Farbwerk *n (print)* inking system (unit)
abfahrbares F. roll-back inking unit
Farbwertverschiebung *f (telev)* colour contamination
Farbwiedergabe *f* colour rendering (rendition)
Farbzerstreuung *f* chromatism
Farnsworth-Röhre *f (el)* image dissector
Fasan *m* pheasant
Faschine *f* faggot, fascine
Faschinenbau *m* faggoting, fascine work
Faschinendamm *m* fascine barrier wall
Fase *f* bevel, chamfer, land *(twist drill)*
F. an der Umfangsschneide top land *(milling cutter)*
fasen to chamfer
Fasen *n* chamfering
Fasenbrett *n* thread board
Fasenbreite *f* land width, size of land, *(Am)* marginal width
Fasenfräser *m* bevelled milling cutter
Fasenhöhe *f* land (rib) clearance, body diameter clearance, depth of body clearance
Fasenschneide *f* leading edge [of the land], cutting edge at the periphery of the [twist] drill
Faser *f* 1. filament; 2. *(text)* fibre; 3. grain *(wood)*
F. für Seilerwaren cordage fibre
anorganische F. inorganic fibre
hochgradig isolierende F. high-insulating fibre
neutrale F. neutral fibre
optische F. optical fibre
synthetische F. synthetic fibre
Faseraffinität *f* affinity to the fibre
Faserangriff *m* damaging of the fibre
Faserasbest *m* amianthus, mineral flax
Faserband *n (text)* sliver
Faserbart *m* tuft *(cotton)*
Faserbaustoffplatte *f* fibre-board
Faserbündel *n* fibre bundle
Faserelektrizität *f* electricity of fibres
Faserfeinheit *f (text)* count of fibre
Faserfilz *m* sheet of pulp
Fasergefüge *n* fibre matrix
faserig fibrous; fibriform
Faserisolierung *f* fibrous insulation
Faserkalk *m* agalite, fibrous limestone, satin spar
Faserkohle *f* fusain, mineral charcoal, mother of coal
Faserkunde *f* technology of fibres
Faserkunstleder *n* fibrous artificial leather
Faserlein *m* fibre flax
fasern to unravel
Faseroptik *f* fibre optics
Faserplatte *f* fibreboard, fibre building board, fibre slab
gehärtete F. tempered hardboard
halbharte F. medium density fibreboard
hartverpreßte F. hardpressed fibreboard
harzgebundene F. resin-bonded fibreboard
heißgepreßte F. hot-pressed fibreboard
luftverfilzte F. air-felted fibreboard
naßverfilzte F. wet-felted fibreboard
nicht vorgepreßte F. non-compressed fibreboard
vergütete F. tempered hardboard

vorgepreßte F. compressed fibreboard
zementgebundene F. cement fibre slab
Faserrichtung / in lengthwise of the grain
Faserrückgewinnung *f* fibre recovery
Fasersättigungspunkt *m* fibre-saturation point
Faserschädigung *f* damaging of the fibre
Faserschicht *f* layer of fibres
Faserschonung *f* fibre preservation
Faserstoff *m* fibrous material
natürlicher F. natural fibre
pflanzlicher F. vegetable fibre
Faserstoffaufschwemmung *f* pulp
Faserstoffisolierung *f* fibrous insulation
Faserstoffriemenscheibe *f* fibre pulley
Faserstoffschicht *f* furnish layer *(board)*
Faserstofftechnik *f* technology of textiles
Faserstoffzusammensetzung *f* fibre furnish
Faserstruktur *f* grain *(of wood)*; fibre structure (pattern)
Faserverlauf *m (met)* fibre flow
Faservlies *n* fibre fleece, non-woven fabric, chopped mat
Faservliesnähwirkmaschine *f* fibre fleece chain-stitch machine
Faservliesstoff *m* bonded fabric
Faserwatte *f* fibre wadding
Faserwerkstoff *m* fibrous material
Faserwickel *m* lap of fibre
Faß *n* barrel; cask; vat; tub; drum
Fassade *f* facade
Fassadenbeleuchtung *f* frontage lighting (illumination)
Fassadenklinker *m* facing clinker
Fassadenplatte *f* facade panel
Fassadenschmuck *m* facade decoration
Faßband *n* hoop
Faßbier *n* keg beer, bulkbeer
Faßbindemaschine *f* cask making machine
Faßbinden *n* hooping
Faßbinder *m* cooper, hooper
Faßbinderarbeit *f* hooping
Faßboden *m* barrel head, head of a cask
Faßbohrer *m* piercer, pod auger
Faßboje *f* cask buoy
Faßdaube *f* barrel stave
Faßdaubenbiegemaschine *f* stave bending machine
Faßdaubenfügemaschine *f* stave jointing machine
fassen 1. *(mach)* to grip; to bite; 2. to contain, to hold; 3. to mount *(gem, lens)*, to set; 4. *(text)* to take *(colour)*, to bite
Fassen *n* **der Walzen** bite (gripping) of rolls
Faßfabrik *f* cask manufacture
Faßgärung *f* cask fermentation
Faßgerbung *f* drum tanning
Faßholz *n* cask wood
kurzes F. short barrel staves
langes F. long barrel staves
Faßkammer *f* primary standard X-ray ionization chamber
Faßkrösemaschine *f* barrel crozing machine
Faßöl *n* packed petroleum
Fassondrehmaschine *f* profiling (copying) lathe
Fassoniermaschine *f s.* Fassondrehmaschine
Fassonrohr *n* adapting (making-up, template) piece
fassonschmieden to swage
Fassonschmieden *n* precision forging
Fassonwalzwerk *n* structural rolling mill
Faßpichmaschine *f* barrel pitching machine
Faßreifen *m* [barrel] hoop
Faßrumpf *m* cask body
Faßstäbe *mpl* shook
Faßtonne *f* cask (keg) buoy
Fassung *f* mount[ing] *(lens)*; setting; *(el)* socket *(bulb)*, lamp holder, [bulb] holder
diebstahlsichere F. locking lamp holder
federnde F. cushioned socket
gefederte F. anti-vibration lamp holder
Fassungsader *f* fixture wire
Fassungsschalter *m* socket switch
Fassungsstecker *m* lamp holder plug
Fassungsvermögen *n* capacity; storage capacity; volumetric capacity; loading capacity; *(ship)* carrying capacity
Faßwicklung *f (el)* barrel (imbricated, basket) winding
Faßzusammensetz- und -reifenaufziehmaschine *f* cask trussing and hoop driving machine
fastkritisch *(nucl)* near-critical
fastmagisch *(nucl)* near-magic
fastprompt *(nucl)* near-prompt
Fata Morgana *f* mirage
faul rotten; decaying
Faulast *m* rotten (unsound) knot
Faulbehälter *m* sludge digester, digestion tank; septic tank
faulbrüchig hot short, burnt
Faulbütte *f* fermenting trough, rotting vat *(paper)*
Fäule *f* rot
faulen to rot, to decay, to decompose; to digest *(sewage treatment)*
faulend rotting, decaying; digesting
nicht f. mould-proof, rot-proof
Faulgas *n* sewage gas
Faulgrube *f* septic tank, cesspool
faulig septic
Faulkammer *f* digestion chamber (tank)
Fäulnis *f* putrescence, putrefaction, decay, rot, decomposition
Fäulnisbakterien *fpl* putrefactive bacteria
fäulnisbeständig rot-proof, rot-resistant, mould-proof

Fäulniserreger *m* septic
Fäulnisgärung *f* putrefactive fermentation
fäulnishindernd antirot, antiseptic, antifouling
Faulprozeß *m* **von Abwasserschlamm** digestion of sewage sludge
Faulraum *m* digestion tank (chamber) *(Imhoff tank)*
Faulschlamm *m* sapropel, sludge
Faulung *f* digestion *(sewage treatment)*
Faustball *m* fist ball
Fäustel *m* *(build)* mallet; sledge
Fausthandschuh *m* mitten
Fausthobel *m* small plane
Faustregel *f* hard-and-fast rule, rule of thumb
Faustsäge *f* hand saw
Fayence *f* faience
FD-Zug *m* long-distance express train
Feather-Analyse *f* *(nucl)* Feather analysis
fechten to fence
Feder *f* 1. spring *(machine element)*; 2. feather [key], parallel key; 3. feather, tongue; 4. pen *(writing)*; 5. feather *(bird)*
  F. mit einstellbarer Spannung adjustable-tension spring
  F. mit Vorspannung compressed spring
  F. und Nut *f* key and slot
Federakkumulator *m* spring-loaded type of accumulator
Federaufhängung *f* spring suspension
Federauflage *f* spring seating
Federausgleichvorrichtung *f* spring balancing device
Federball *m* badminton, shuttlecock
Federbandstahl *m* spring-steel cross-bands
Federbarometer *n* aneroid barometer
Federbein *n* shock absorber
federbelastet spring-loaded; spring-opposed
federbetätigt spring-operated
Federblatt *n* spring leaf
Federbolzen *m* spring bolt
Federbuchse *f* spring bushing
Federbügel *m* spring U-bolt
Federdiagramm *n* load deflection curve for spring
federdicht *(text)* feathertight
Federdraht *m* spring wire
Federdrahtrelais *n* wire-spring relay
Federdruckbüchse *f* spring grease cup
federentlastet spring-balanced
Federgalvanometer *n* spring galvanometer
federgespannt spring-loaded
Federhammer *m* spring hammer
Federhärte *f* spring temper
Federhemmung *f* spring detent escapement
Federinnendurchmesser *m* inside diameter of spring
Federkeil *m* feather joint-tongue
Federkennlinie *f* spring deflection with (caused by) given loads; load deflection curve for spring
Federklammer *f* spring clip
Federklemme *f* *(el)* spring terminal
Federklinke *f* spring switch
Federkonstante *f* spring constant
Federkontakt *m* resilient (spring) contact
Federlade *f* *(text)* flyer lathe
Federlager *n* spring hanger
Federleichtpapier *n* featherweight paper
Federleiste *f* edge connector, multiple connector socket, socket assembly
Federmanometer *n* diaphragm (spring pressure) gauge
Federmaßstab *m* spring scale
Federmotor *m* clock-spring motor
Federn *n* springing
Federnagel *m* spring-grip spike
federnd spring-loaded
Federöse *f* spring loop
Federprüfung *f* spring testing
Federpuffer *m* spring bumper pad
Federregler *m* spring governor
Federrohr *n* expansion (compensating) pipe
Federrückführung / mit spring-returned
Federsatz *m* *(el)* spring bank
Federscheibe *f* spring lock washer
Federschlagstuhl *m* *(text)* spring pick loom
Federschraube *f* spring screw
Federschuh *m* spring saddle
Federsicherheitsventil *n* spring-loaded safety valve
Federspanneisen *n* compressive clamp
Federspannfutter *n* chuck spring collet
Federspannplatte *f* spring tension plate
Federspannung *f* spring tension
Federspeicherantrieb *m* auxiliary power stored by winding-up motor drive *(circuit-breaker)*
Federspeicherbremszylinder *m* spring-brake cylinder
Federsprungbrett *n* [elastic] springboard, high-jumping board
Federstahl *m* spring steel
Federstift *m* spring plunger *(indexing)*
Federtaster *m* spring caliper
Federteller *m* spring plate
Federthermometer *n* bulb (pressure-spring) thermometer
Federträger *m* spring bracket
Federung *f* springing; cushioning
  akustische F. *(el)* acoustic compliance
Federungsfähigkeit *f* cushioning capacity *(tyre)*
Federunterlage *f* spring pad
Federvisier *n* spring leaf sight
Federwaage *f* spring balance
Federweg *m* spring deflection
Federwelle *f* spline shaft
Federwerkstoff *m* spring material

Federwickelmaschine *f* spring-coiling machine
Federwickeln *n* spring winding (coiling)
Federwindung *f* spring coil
Federzahngrubber *m (ag)* spring-tine cultivator, *(Am)* weeder
Federzinken *m (ag)* spring tine
Federzinkengrubber *m (ag)* spring-tine cultivator, *(Am)* weeder
Federzirkel *m* spring divider, bow spring dividers
Fegsel *n (ship)* spillage, sweepings
Fehlabgleichung *f* faulty alignment, misalignment
Fehlanflughöhe *f* missed approach altitude
Fehlanflugverfahren *n* missed approach procedure
fehlanpassen to mismatch *(radio)*
Fehlanpassung *f* mismatch; misalignment
Fehlanruf *m (telec)* false start call
Fehlbezeichnung *f* misnomer
Fehlbogen *m* spoil, waste (imperfect) sheet
Fehldruck *m* misprint
Fehler *m* 1. error, fault; 2. defect, flaw *(material)*
  F. erster Art *(dat)* type I error
  F. im eingeschwungenen Zustand steady-state error
  F. je Einheit defects per unit
  F. zweiter Art *(dat)* type II error
  absoluter F. absolute error
  angesammelter F. accumulative error
  apparativer F. instrumental error
  chromatischer F. *(opt)* chromatic aberration
  kritischer F. critical (major, class 2) defect
  mehrpoliger F. multi-phase fault
  mittlerer F. standard error
  mittlerer quadratischer F. *(math)* standard (root-mean-square) deviation
  programmbedingter F. *(dat)* program-sensitive error
  systematischer F. constant error
  überkritischer F. critical (class 1) defect
  unwesentlicher F. class 4 defect, minor B defect
  wahrscheinlicher F. probable error
  wirklicher F. actual error
  zufälliger F. accidental error
  zulässiger F. permissible (allowable) error
Fehlerabschätzung *f* error estimation
Fehleranteil *m* portion of defectives
  relativer F. des Postens population fraction defective
  relativer F. in der Stichprobe fraction defective of a sample
Fehlerausbreitung *f* propagation of error
Fehlerausbrüche *mpl* bursts of errors
Fehlerausgleich *m* compensation of errors
Fehlerbandbreite *f* error latitude
Fehlerberichtigung *f* / **automatische** automatic error correction *(in computers)*
Fehlerbeseitigung *f (el)* removal of faults, trouble elimination (shooting); failure recovery
Fehlerbestimmung *f* localization of faults, trouble shooting
  F. aus dem Spannungsabfall conductivity test
Fehlerbetrieb *m* faulty switching
Fehlerdämpfung *f (el)* [balance] return loss
  F. in Stromkreisen ohne Verstärker *(el)* passive balance return loss
  aktive F. active [balance] return loss
  passive F. passive [balance] return loss
Fehlerdämpfungskoeffizient *m* return current coefficient
Fehlerdämpfungsmesser *m* unbalance attenuation measuring set; impedance unbalance measuring set, *(Am)* return loss measuring set
Fehlerdämpfungsmessung *f* balance attenuation measurement
Fehlerecho *n* flaw echo
Fehlereingrenzung *f* fault-localization, trouble shooting
Fehlererkennung *f* / **automatische** automatic error detection
Fehlerfortpflanzung *f* error propagation, propagation of error
Fehlerfortpflanzungsgesetz *n* law of the propagation of errors, error propagation law
fehlerfrei error-free *(measurement)*
Fehlerfreiheit *f* accuracy *(in digital computers)*; freedom from faults (errors)
fehlerhaft defective, faulty
Fehlerhäufigkeit *f* frequency of errors, error rate
Fehlerhäufungen *fpl* clustered errors
Fehlerklassifizierung *f* classification of defects
Fehlerkurve *f* error [distribution] curve, error characteristic; probability curve
Fehlernachweisempfindlichkeit *f* flaw detection sensitivity
Fehlerortung *f* fault-location, fault determination
Fehlerprozentkarte *f* control chart for fraction defective
Fehlerprozentsatz *m* percentage of defectives, percent defective
  mittlerer F. average percent defective
  zulässiger F. allowable percent defective (nonconforming)
Fehlerprüfschalter *m* feeler switch
Fehlerquelle *f* source of error, chance of error
Fehlerspannung *f* potential to earth (ground); failure voltage
Fehlerspannungsauslöser *m* leakage [protective] release, fault protective system

Fehlerstelle *f* point of fault *(power line)*, location of the fault, fault position
Fehlerstrecke *f* fault section
Fehlerstrom *m* fault (leakage, spill) current
Fehlerstromschutzschalter *m* fault-current protective switch
Fehlersuche *f* error detection; fault finding (localization), trouble shooting
Fehlersuchprogramm *n* diagnostic routine
Fehlerverteilungsgesetz *n* / **Gaußsches** Gaussian (normal) distribution law
Fehlerverteilungskurve *f* / **Gaußsche** Gauss error distribution curve
Fehlerwiderstand *m* fault resistance
Fehlerzahl *f* number of defects
Fehlfarbe *f* off-colour
Fehlfracht *f* deadfreight
Fehlgewicht *n* weight shortage
Fehlguß *m* spoiled (misrun, waste) casting, offcast
Fehlkonstruktion *f* faulty design
Fehlkorn *n* misplaced size
Fehlladung *f* short delivery
Fehlleimung *f* **bei Kaltverleimung** loose veneer in cold bonding
Fehlloch[ung *f*] *n* blank column *(punched card machine)*
Fehllochungssucher *m* blank column detector *(punched card machinery)*
Fehlmessung *f* faulty measurement
Fehlordnung *f* disarrangement; [lattice] disorder, imperfection, defect *(in crystal lattice)*
  chemische F. chemical defect
  kristalline F. crystalline disorder
Fehlordnungsenergie *f* disorder energy *(of crystals)*
Fehlorientierung *f* misorientation
Fehlpeilung *f* spurious bearing
Fehlphasenunterdrückung *f (el)* inverse suppression
Fehlschaltung *f* faulty switching *(operation)*
Fehlschicht *f* absenteeism, lost shift
Fehlschluß *m* fallacious inference
Fehlschmelze *f* misfit cast
fehlsichtig ametropic, defectively sighted
Fehlsichtigkeit *f* ametropia, visual deficiency
Fehlstart *m* wrong take-off
Fehlstelle *f* defect, point imperfection; lattice defect; *(ag)* blank
Fehlverbindung *f* wrong connection
Fehlwahl *f* faulty selection *(telephone)*
Fehlweisung *f* error in bearing
Fehlwinkel *m* **eines Wandlers** phase angle of a transformer
Fehlzündung *f* backfiring of the engine, misfiring
Feigenbaum *m* fig tree
Feile *f* file
  einhiebige F. single-cut file
Feilen *n* filing
Feilenangel *f* file tang
Feilenblatt *n* file blade
Feilenheft *n* file handle
Feilenhieb *m* cut of file
Feilmaschine *f* filing machine
Feilspan *m* filing
Feilspanbild *n (el)* magnetic figures
Feilspankohärer *m (el)* Branly coherer
Feil- und Sägemaschine *f* filing and sawing machine
Feimdach *n (ag)* hay rick cover
Feime *f* hay rick
Feimenstuhl *m* hay rick stand
fein fine; pure
Feinabgleich *m* fine adjustment
Feinablauf *m* high green syrup *(sugar)*
Feinablesung *f* vernier reading
Feinabstimmknopf *m* slow motion control knob *(radio)*
Feinabstimmkondensator *m* vernier capacitor
Feinabstimmregler *m* fine-tuning control
Feinabstimmung *f* precision (slow, sharp, vernier, incremental, sensitive) tuning
Feinabtastung *f (telev)* close (fine) scanning
Feinanzeige *f* microindication
Feinätzung *f* final etching
Feinbau *m* microstructure
Feinbearbeiten *n* 1. fine finishing; 2. refined abrading
Feinbearbeitung *f* [fine] finishing
  F. von Zahnflanken gear finishing
Feinbearbeitungsdrehmaschine *f* finishing lathe
Feinbearbeitungsverfahren *n* 1. finishing operation; 2. refined abrading process
Feinblech *n* sheet metal
  dünnes F. tagger
Feinbohren *n* fine hole drilling
  F. mit Diamant- oder Hartmetallwerkzeug diamond boring, *(Am)* borizing
Feinbohrmaschine *f* fine boring machine
Feinbrecher *m* fine (secondary) reduction crusher, fine (secondary) crusher
Feindehnungsmessung *f* determination of elongation by extensometer
feindispers highly disperse
Feindrahtzug *m* drawing thin wire
Feindrehmaschine *f* precision lathe
Feineinstelleinrichtung *f* fine setting device
Feineinstellkondensator *m* vernier control capacitor
Feineinstellschraube *f* fine adjusting screw
Feineinstellskala *f* vernier dial
Feineinstellung *f* fine setting (adjustment), vernier setting (adjustment); fine control
Feineisenstraße *f* rod (merchant) mill
Feineisenwalzwerk *n* small section rolling mill

feinempfindlich sensitive
feinen to [re]fine; to deoxidize
Feinerz *n* fine ore, fines
Feinflyer *m (text)* roving frame
Feinförderkohle *f* fine small coal
feinfühlig sensitive *(instrument, drilling)*
Feinfühligkeit *f* sensitivity *(instrument, drilling)*
Feingefüge *n* microstructure
Feingehalt *m* fineness, purity
feingemahlen finely ground
Feingemüse *n* garden vegetables
Feingewinde *n* fine screw thread, fine pitch thread
Feingewindereihe *f* fine screw thread series
Feingold *n* fine gold
Feingut *n* fine fraction
Feinheit *f* fineness
  F. des Baumwollgarns cotton count
  F. des Gewebes count of cloth
  F. des Kammgarns worsted count
  F. des Schusses *(text)* pick count
  F. des Wollgarns woollen count
Feinheitsbestimmung *f* **der Seide** numbering of silk
Feinheitsgrad *m* 1. degree of fineness; 2. *(text)* count, number
Feinheitsnummer *f (text)* thread count
Feinheitszahl *f* fineness modulus *(concrete)*
Feinheitsziffer *f* fineness modulus *(concrete)*
Feinhöhenmesser *m* precision altimeter
Feinhorizontierung *f* accurate levelling
feinjährig with small (narrow) annual rings, narrow-circled *(wood)*
Feinkarde *f (text)* finisher (finishing) card
Feinkies *m* fine (pea) gravel
Feinkohle *f* fines, smalls, fine coal (smudge), duff
  gewaschene F. fine small duff
Feinkorn *n* 1. *(mil)* fine sight (rifle); 2. fine grain
Feinkornausbringen *n* true undersize recovery
Feinkorneisen *n* fine-grained iron
Feinkornfilm *m* fine grain film
feinkörnig fine-grained, close-grained
Feinkornsetzmaschine *f* smalls jig
Feinkornstahl *m* structural steel
Feinkreissäge *f* smooth cutter combination saw
Feinkrempel *f (text)* finisher card, second breaker
  F. für Flachswerg mit Streckkopf flax tow finisher card with drawing head
feinkristallin finely crystalline, compact-grained
Feinkupfer *n* refined copper
feinmahlen to grind finely, to pulverize, to finish
Feinmahlung *f* fine grinding
feinmaschig fine-mesh
Feinmechanik *f* precision mechanics
Feinmechaniker *m* precision toolmaker
Feinmeßeinrichtung *f* fine measuring device
Feinmeßfunkenstrecke *f* micrometer gap
Feinmeßgerät *n* precision measuring instrument *(set)*
Feinmühle *f* finishing machine *(paper)*
Feinnivellement *n* precision levelling
Feinplanierung *f* final grading, fine grading
Feinplanum *n* formation
feinporig fine-pored
feinpulverig fine-powdered
Feinregelung *f* slow-motion control
  kontinuierliche F. fine smooth control
Feinrippenrundstrickmaschine *f* fine rib circular knitting machine
Feinrost *m* fine screen
Feinsäge *f* fine (veneering) saw
  F. mit Messingrücken fine saw with brass back
  F. mit poliertem Stahlrücken fine saw with polished steel back
Feinsämereienreinigungsanlage *f* / **elektromagnetische** *(ag)* electro-magnetic seed cleaner
Feinsand *m* fine sand
Feinschlamm *m (met)* pulp
Feinschleifen *n* **von Ventilen** seating valves
Feinsicherung *f (el)* miniature (heat coil, glow tube, fine-wire) fuse, micro-fuse
  träge F. *(el)* delay-action fuse
Feinsicherungspatrone *f (el)* heat coil
Feinsieb *n* fine screen
Feinspaltrost *m* slotted-hole screen
Feinspan *m* fine segmental chip
Feinspanplatte *f* sliver board
Feinspinnen *n (text)* spinning fine counts *(of cotton etc.)*; final (fine) spinning
Feinspinnmaschine *f* fine spinning frame
Feinsplitt *m* fine chippings
Feinstahlwalzwerk *n* small section rolling mill
Feinstbearbeitungsverfahren *n* precision finishing process, operation producing fine finish; superfinishing [process]
Feinstbohrmaschine *f* precision (diamond) boring machine
Feinstdraht *m* finest wire
Feinstein *m* converter matte *(nickel)*
feinsteinstellbar micrometer-adjustable
Feinsteinstellbohrstange *f* boring bar with micrometer adjustable cutter
Feinstellknopf *m* fine regulating knob
Feinstellschraube *f* slow-motion head
Feinstellskale *f* slow-motion dial
Feinstgut *n* fines
Feinstkohle *f* / **angereicherte** cleaned fines
  gewaschene F. washed fines
Feinstkorn *n* finest grain, fines
Feinstkornbrikett *n* finest grain briquette
Feinstmahlung *f* pulverization
Feinstmahlvorrichtung *f* micronizer

Feinstmühle *f* colloid mill
Feinstoff[e *pl***]** *m* fines *(paper)*
Feinstruktur *f* fine structure; microstructure
Feinstrukturkonstante *f* fine-structure constant
Feinstsand *m* silt
feinstufig fine-step, in fine steps; sensitive *(controls)*
  f. regelbar controlled by fine graduation
Feinstufigkeit *f* notching ratio *(control)*
Feinstziehschleifen *n* superfinishing
Feintaster *m* precision dial gauge
Feinteile *npl* fines
Feintrieb *m* fine adjustment
Feinung *f* refining, finishing
Feinungsofen *m* finishing furnace
Feinungsperiode *f* refining period
Feinvakuum *n* pre-vacuum
Feinverschiebung *f (mach)* inching
Feinverstellung *f* fine adjustment
feinverteilt *(ch)* disperse; finely divided; finely deposited
Feinvorschub *m (mach)* fine feed
Feinvorschubreihe *f* fine-feed series
Feinwaschmittel *n* light duty detergent
Feinzeiger *m* **/ mechanischer** lever-type indicator
Feinzeigermeßschraube *f* indicating micrometer
feinzerkleinern to grind finely, to pulverize
Feinzerkleinerung *f* fine grinding
Feinzerlegung *f (opt)* fine-detail resolution
Feinzeugholländer *m* finisher, beater, beating machine
Feinziehschleifen *n* superfinishing
Feinzinn *n* head tin
Feinzug *m* finishing block *(wire drawing)*
Feinzuschlag *m* fine aggregate *(concrete)*
Feinzustellung *f (mach)* fine (micrometer) adjustment
Feld *n* 1. *(ag, el)* field; 2. *(build)* bay, panel; 3. *(mine)* concession; track; field; 4. *(math)* field, net; 5. land *(rifle barrel)*
  F. gleichen Potentials equipotential field
  F. im Streb *(mine)* track
  äußeres F. external field
  baroklines F. baroclinic field
  barotropes F. barotropic field
  elektrisches F. [electric] field
  elektrodynamisches F. electrodynamic field
  elektrostatisches F. electrostatic field
  entgegengesetztes F. opposing field
  entmagnetisierendes F. demagnetizing field
  erdmagnetisches F. geomagnetic (terrestrial magnetic) field
  erregtes F. induced field
  fremdes F. external field
  gleichnamiges F. field of same polarity
  homogenes F. uniform (homogeneous) field
  induziertes elektrisches F. induced electric field
  inhomogenes F. non-uniform field
  inneres F. internal field
  kanonisch konjugiertes F. canonically conjugated field
  magnetisches F. magnetic field
  magnetohydrostatisches F. magneto-hydrostatic field
  magnetostatisches F. magnetostatic field
  Maxwellsches F. Maxwell field
  metrisches F. metric field
  oszillierendes F. oscillating field
  pulsierendes F. pulsating field
  quasistationäres F. quasistationary field
  quellenfreies F. solenoidal field
  remanentes F. retentive field
  rotierendes F. rotating (rotary) field
  ruhendes F. stationary field
  schwaches F. weak (low) field
  starkes F. strong field
  ungleichnamiges F. field of opposite polarity
  wanderndes F. travelling field
  wirbelfreies F. irrotational (non-rotational) field
Feldabhängigkeit *f* field dependence
Feldamperemeter *n* field ammeter
Feldanbauversuch *m* field trial
Feldarbeiter *m* field labourer, agricultural worker, farm hand
Feldausgleichwicklung *f* field compensating winding
Feldbahngleis *n* light railway track
Feldbahnlokomotive *f* light railway engine
Feldbahnschiene *f* light rail
Feldbau *m* agriculture, farming, tillage
Feldbaubrigade *f* tillage (plant cultivation) team
Feldbild *n* field pattern
Feldbogen *m (el)* high-field-emission arc
Feldbrennofen *m* pot (clamp) kiln
Feldbuch *n* engineer's field book
Felddiagramm *n* field [distribution] diagram *(transformer)*
Felddichte *f* field density
Felddüngungsversuch *m* manuring (fertilization) field trial
Feldeffekt *m* field effect
Feldeffekttransistor *m* field effect transistor
Feldelektronenemission *f* field emission [of electrons], cold emission
Feldelektronenmikroskop *n* field emission microscope
Feldemission *f* field (cold) emission
Feldempfindlichkeit *f* field sensitivity
Feldenergie *f* field energy
Felderregung *f* field excitation
Feldesbreite *f (mine)* track-width
Feldesgrenze *f (mine)* boundary

Feldfeinkorrektion *f* shimming
Feldfernsprecher *m* field telephone
Feldflugplatz *m* combat (advanced) airport, front airfield
feldfrei having no electrical field, field-free, zero-field
Feldfrucht *f* crop
Feldgefälle *n* fringing field
Feldgestell *n* magnet frame
Feldgleichrichter *m* field rectifier
Feldgraswirtschaft *f* [short] ley farming
Feldhäcksler *m* forage harvester
Feldindex *m* [magnetic-]field index
Feldintensität *f* field strength (intensity)
Feldionenemission *f* field-ion emission
Feldlader *m* general purpose crop loader
Feldlinie *f* line of force (flux)
  eingefrorene F. frozen-in field line
  magnetische F. magnetic field line
Feldlinse *f* field lens
Feldmagnet *m* field magnet
Feldmesser *m* land surveyor
Feldmessung *f* land surveying
Feldmitte *f* midspan *(statics)*
Feldmoment *n* midspan moment *(statics)*
Feldort *n (mine)* forehead
Feldpol *m (el.m.)* field pole
Feldregelanlasser *m* rheostatic controller
Feldregler *m* field rheostat
  selbsttätiger F. automatic rheostat
Feldrichtung *f* field direction
Feldröhre *f* tube of force
Feldschwächung *f* field weakening, flux shunting, attenuation of field excitation
  F. durch Anzapfung field weakening by tapping
  F. durch Nebenschluß field shunting
  gemischte F. combined field weakening
  selbsttätige F. automatic field weakening
Feldschwächungseinrichtung *f* field suppressor
Feldschwächungsschalter *m* field weakening switch
Feldschwächungsverhältnis *n* field weakening ratio
Feldspannung *f* inductor (field) voltage
Feldspat *m* fel[d]spar
feldspatartig fel[d]spathic
feldspathaltig fel[d]spathic
Feldspatvertreter *m* fel[d]spathoid
Feldspritze *f* field-crop sprayer
Feldspritzrohr *n* spray boom
Feldspule *f (el)* field (stationary, magnetizing) coil
Feldspulenkasten *m* field spool (coil) box
Feldstärke *f (el)* field strength (intensity)
  ausgestrahlte F. radiated field intensity
  differentielle magnetische F. incremental magnetizing force
  elektrische F. electric field strength (intensity)
  erdmagnetische F. geomagnetic field intensity
  magnetische F. magnetic field strength
Feldstärkeanzeiger *m* field-strength indicator
Feldstärkediagramm *n* field pattern *(of aerials)*
Feldstärkemeßgerät *n* field-strength measuring set
Feldstärkeverlauf *m* field-strength distribution
Feldstecher *m* field glass
Feldstein *m* cobble [stone]
Feldsteller *m s.* Feldregler
Feldsteuerung *f* field[-weakening] control
Feldstrecke *f (mine) (Am)* gangway
Feldstrom *m* field (inductor) current
Feldstromkreis *m* field (shunt) circuit
Feldtheorie *f* field theory
  allgemeine F. unified field theory
  einheitliche F. unified field theory
  klassische F. classical field theory
  nichtlineare F. non-linear field theory
  nichtlokale F. non-local field theory
Feldtransistor *m* field effect transistor
Feldtrenner *m* field break-up switch
Feldtyp *m* mode *(waveguides)*
Feldverdrängung *f* field displacement
Feldverlauf *m* field distribution (changing)
Feldverschiebung *f* field displacement
Feldversuch *m (ag)* field trial
Feldverteilung *f* field distribution (configuration)
  F. der wirksamen Strahlungsfläche field distribution of the aperture
Feldverzerrung *f* distortion of the magnetic flux distribution, field distortion
Feldwaage *f* magnetic balance
Feldweg *m* farm track
Feldwellenwiderstand *m* [characteristic] wave impedance
Feldwicklung *f (el)* field coil (winding)
Feldzahl *f (opt)* field-of-view number
Felge *f* [wheel] rim, felloe
  abnehmbare F. detachable rim
  feste F. fixed rim
Felgenband *n* rim band
Felgenbett *n* [rim] well
Felgenbremse *f* rim brake
Felgenflansch *m* rim flange
Felgenholz *n* felloe wood
Felgenhorn *n* clinch
Felgenrand *m* rim flange (ring)
Felgenring *m* rim ring
  geteilter F. split rim ring
  ungeteilter F. solid rim ring
Fell *n* fell; skin
  verschnittenes F. scored (cut) skin
Fellboot *n* skin boat
Fels *m* rock
  anstehender F. bed-rock
Felsbohrer *m* rock drill

Felsbrecher *m* rock breaker (cutter)
Felsenbohrschiff *n* rock driller
felsitisch *(geol)* felsitic
Felslöffel *m* rock dipper
Felsriff *n* rocky ledge
Felssprengung *f* rock blasting
Felsuntergrund *m* bed-rock
Felswand *f* rock face
femisch *(geol)* femic
Fenchelöl *n* fennel oil
Fenchol *n (ch)* fenchol, fenchyl alcohol
Fender *m (ship)* fender, fend-off
fendern to fend off *(a vessel)*
Fenster *n* window
  bewegliches F. opening light
  feststehendes F. fixed light
  französisches F. French door (window), glazed door
  großflächiges F. picture window
  rundes F. compass window
  vorgetäuschtes F. blank (dead) window
Fensteranordnung *f* fenestration
Fensteraufteilung *f* fenestration
Fensterbrett *n* window board
Fensterbrüstung *f* window sill, breast
Fensterfeststeller *m* casement stay, window fastener (stay)
Fensterflügel *m* [window] casement
Fenstergewände *n* window surrounds
Fenstergriff *m* window lift (handle)
Fenstergurt *m (ry)* window strap
Fensterhaken *m* window fastener (stay), casement stay
Fensterkitt *m* glazing compound, bedding (back, oil) putty
Fensterkurbel *f* window crank
Fensterladen *m* window shutter
Fensterleibung *f* reveal
  abgeschrägte F. embrasure
fensterlos windowless
Fensterputzer *m* window cleaner
Fensterrahmen *m* casement, frame
Fensterrose *f* fan-light
Fensterscheibe *f* window pane
Fenstersprossenmaschine *f* window sash bar making machine
Fenstersturz *m* lintel, window head
Fenstertür *f* French (glazed) door, French window
Fensterverstärker *m* window amplifier
Fensterzählrohr *n (nucl)* end-window counter [tube]
Ferienkurs *m* holiday [training] course
Ferkelwaage *f (ag)* piglet weigher
Ferment *n* ferment, enzyme
Fermentation *f* fermentation, fermenting process; curing
Fermentationsfähigkeit *f* fermentability
fermentativ ferment[at]ive, enzymatic
fermentierbar fermentable, fermentative, fermentescible
fermentieren to ferment; to cure
Fermi-Alter *n (nucl)* [Fermi] age
Fermi-Dirac-Statistik *f* Fermi-Dirac statistics
Fermi-Energie *f* Fermi energy
Fermi-Gas *n* Fermi-gas
Fermi-Kante *f* Fermi limit (level)
Fermi-Näherung *f* Fermi approximation
Fermi-Niveau *n* Fermi level
Fermi-Statistik *f* Fermi statistics
Fermion *n (nucl)* fermion, Fermi-particle
Fermium *n* fermium
fern far; remote; distant
Fernablesung *f* distant reading
Fernabstimmgerät *n* remote tuning device
Fernamt *n* trunk exchange, *(Am)* toll exchange
  zwischenstaatliches F. international terminal office
Fernamtsauskunftsstelle *f* trunk directory inquiry
Fernamtstrennung *f* through clearing
Fernanruf *m* trunk call
Fernanrufrelais *n* long-distance line relay
Fernanschluß *m* trunk subscriber's line
Fernantrieb *m* remote (distance) drive
Fernanzeige *f* remote indication
Fernanzeigevorrichtung *f* remote indicating device
Fernartillerie *f* long-range artillery
Fernaufgabe *f* remote feeder (feeding device)
Fernaufklärung *f (mil)* long-distance reconnaissance
Fernbedienung *f* telecontrol; remote control; remote handling (manipulation)
Fernbedienungsgerät *n* remote handling device; manipulator; long-handed tool
Fernbedienungszange *f* expansion tong
Fernbeladung *f* remote loading *(of a reactor)*
Fernbeobachtung *f* remote viewing
fernbesetzt engaged on trunk call
Fernbestrahlung *f (nucl)* tele-irradiation
fernbetätigt remotely operated
Fernbetätigung *f* remote control (operation, actuation), telecontrol
Fernbild *n* telephoto
Fernbildübertragung *f* telephotography
Ferndienst *m* / **wartezeitloser** *(telec)* demand service
Ferndrehzahlmesser *m* distant-reading tachometer
Ferndrucker *m* printing telegraph
Ferneinleger *m* remote feeder (feeding device)
Ferneinrichtung *f* remote aiming (adjustment)
Ferneinstellung *f* remote aiming (adjustment)
Fernempfang *m (el)* distant (long-distance) reception, distant station reception

Fernempfangsantenne *f* DX aerial
Fernempfangszone *f* secondary coverage (service area) *(radio)*
Fernfahrt *f (ship)* long-distance trip, off-shore navigation
Fernfunkfeuer *n* long-range radio beacon
Ferngas *n* grid (long distance) gas
Ferngasleitung *f* long distance gas pipeline
Ferngasnetz *n* long distance gas grid (system)
Ferngasversorgung *f* long distance gas supply
Ferngeber *m* pick-off, pick-up, teletransmitter
Ferngeschütz *n* long-range gun
Ferngespräch *n* trunk call, long-distance [telephone] call, *(Am)* trunk toll
  F. zu bestimmter Zeit fixed-time call
ferngesteuert remotely controlled
Ferngreifer *m (nucl)* manipulator
Ferngüterzug *m* through goods train
Fernheizung *f* district heating
Fernheizwerk *n* heat[ing] station
Fernhörer *m* telephone (hand) receiver, telephone earphone
  elektrodynamischer F. moving-conductor receiver
  elektromagnetischer F. electromagnetic (moving-iron) receiver
Fernhörerschnur *f* receiver (telephone) cord
Fernkabel *n* trunk (long-distance) cable *(telephony)*
Fernkompaß *m* telecompass, remote-indicating compass
Fernkraftwerk *n* long-distance supply station
Fernladung *f* remote charging *(batteries)*
Fernlastverkehr *m* long-distance road haulage
Fernleitung *f (el)* transmission line; trunk line, *(Am)* toll line *(telephony)*
Fernleitungsendverstärker *m* terminal repeater
Fernleitungskabel *n* long-distance cable
Fernleitungsnetz *n* trunk (long distance) network
Fernleitungsschaltplan *m (Am)* toll switching plan
Fernleitungswähler *m* trunk final selector *(telephony)*
Fernlenkflugzeug *n* remote-controlled aircraft
Fernlenkgeschoß *n* ballistic missile
  interkontinentales F. intercontinental ballistic missile
Fernlenkpult *n* radio-control desk
Fernlenkschiff *n* remote-controlled vessel, distance-controlled vessel
Fernlenkung *f* guidance *(aircraft)*, remote control
Fernlenkwaffe *f* remote-control weapon
Fernlicht *n* full headlight beam, long-distance light
Fernlichtkontrollampe *f* headlight main beam indicator
Fernlichtkontrolleuchte *f* high-beam indicator lamp
Fernmeldeanlage *f* communication facility
Fernmeldedienst *m* telecommunication service
Fernmeldegeheimnis *n* secrecy of telecommunication
Fernmeldeleitung *f* telephone line
Fernmeldeluftkabel *n* aerial telecommunication cable
Fernmeldemeßgeräte *npl* telecommunication test gear
Fernmeldemeßkoffer *m* portable voice frequency measuring set
Fernmeldenetz *n* telecommunication network (system)
Fernmeldetechnik *f* telecommunication engineering
Fernmeldeverkehr *m* telecommunication
  fester F. point-to-point communication
Fernmeldevertrag *m* / **Internationaler** International Telecommunications Convention
Fernmeldeweg *m* telecommunication circuit
Fernmeldewesen *n* telecommunication [engineering]
Fernmeßeinrichtung *f* telemetering device
Fernmessen *n* telemetering
Fernmeßgerät *n* remote meter, telemeter
  F. mit Spannungsübertragung voltage-type telemeter
  strombetätigtes F. current-type telemeter
Fernmeßinstrument *n s.* Fernmeßgerät
Fernmessung *f* telemetering; telemetry, remote metering (indication, measuring)
Fernmessungsinstrument *n* distant-reading instrument
Fernnebensprechen *n* far-end cross talk
Fernnetz *n (el)* long distance plant
Fernobjektiv *n* telephoto lens
Fernregistrierinstrument *n* remote recorder (recording instrument), telerecorder
Fernregistrierung *f* remote registration, telerecording
Fernrohr *n* telescope
  holländisches F. Dutch telescope
  Keplersches F. Kepler telescope
  technisches F. industrial telescope
Fernrohrantrieb *m* telescope drive
Fernrohrbrille *f* telescopic spectacles
Fernrohrlupe *f* telescopic magnifier
Fernrohrstütze *f* telescope support
Fernsatz *m* teletype setting
Fernschalter *m* remote-operated switch, remote control (actuated) switch, tele-switch
Fernschaltung *f* remote control change
Fernschmierer *m* lubricator working from a distance
Fernschnellzug *m* express train
Fernschrank *m* trunk switchboard

Fernschreiben *n* / **drahtloses** broadcast writing
Fernschreiber *m* teletype, teleprinter, teletyper
F. hoher Geschwindigkeit high-speed teleprinter
Fernschreibmaschine *f* teleprinter
Fernschreibnetz *n* teletype network
Fernschreibteilnehmer *m* telex subscriber
Fernsehabtaster *m* scanner
Fernsehansager *m* television announcer
Fernsehantenne *f* television aerial
Fernsehapparat *m* television receiver, televisor, *(sl)* telly
Fernsehaufnahme *f* vision pick-up
Fernsehaufnahmeraum *m* television studio
Fernsehaufnahmeröhre *f* [television] camera tube
Fernsehaufnahmetechnik *f* television studio technique
Fernsehaufnahmewagen *m* television car
Fernsehaufzeichnungen *fpl* teletranscriptions
Fernsehaufzeichnungsgerät *n* video tape recorder
Fernsehband *n* television [broadcast] band
Fernsehbearbeitung *f* television version
Fernsehbild *n* television image (picture)
Fernsehbildaufzeichnung *f* television picture recording
Fernsehbildmustergenerator *m* television test pattern generator
Fernsehbildröhre *f* **mit Sicherheitsglasscheibe** bonded shield picture tube
Fernsehbrücke *f* television relay link
Fernsehbrückenweiche *f* diplexer
Fernsehempfang *m* television reception
Fernsehempfänger *m* television set (receiver), televisor, *(sl)* telly
F. mit Gebührenautomatik pay as you see television
F. mit Grauscheibe black screen television set
F. mit umschaltbarer Zeilennorm switchable television set
fernsehen to televise
Fernsehen *n* television
F. für die Dockfahrt docking television
F. im Dunkeln mittels Infrarotstrahlung noctovision
F. im Kurzschlußverfahren closed circuit television
F. mit geringer Auflösung low-definition television
F. über Kabel wired television system
dreidimensionales F. stereoscopic television
farbiges F. colour television
industrielles F. industrial television
niedrigzeiliges F. low-definition television
plastisches F. stereoscopic television
zeilenfreies F. line-free television
Fernsehepiskop *n* television episcope
Fernseher *m s.* Fernsehempfänger
Fernsehfilm *m* television film, telefilm
Fernsehfilterweiche *f* notch diplexer *(aerial)*
Fernsehfrequenz *f* television frequency
Fernsehfrequenzumsetzer *m* television frequency-transposer
Fernsehfunk *m* television, TV
Fernsehgemeinschaftsempfangsanlage *f* multiple-outlet television system
Fernsehgroßbildprojektor *m* telecine projector
Fernsehgroßprojektor *m* large screen television projector
Fernsehindustrie *f* television (TV) industry
Fernsehingenieur *m* television (TV) engineer
Fernsehkamera *f* television camera, telecamera
Fernsehkanal *m* television channel
Fernsehkanalschalter *m* television turret
Fernsehkino *n* television cinema
Fernsehkontrollempfänger *m* picture monitor
Fernsehnorm *f* television standard
Fernsehnormsignal *n* standard television signal
Fernsehprogramm *n* television programme
Fernsehraster *m* television raster
Fernsehrelaisstrecke *f* television relay link
Fernsehröhre *f* television [picture] tube, kinescope
Fernsehrundfunk *m* television (video) broadcasting
Fernsehsendegerät *n* / **tragbares** walkie-lookie
Fernsehsender *m* television (TV) transmitter, television (TV, broadcast) station
Fernsehsenderöhre *f* camera tube
Fernsehsendung *f* television transmission (broadcast), *(Am)* telecast
Fernsehsignal *n* television signal
Fernsehsprechdienst *m* combined television-telephone service
Fernsehstörung *f* television interference
Fernsehstudio *n* telestudio
Fernsehstudioaufnahme *f* studio pick-up
Fernsehsystem *n* **mit Gebührenautomat** fee-television
F. mit teilweiser Übertragung über eine Telefonleitung phonevision system
Fernsehtechnik *f* television engineering, video technique[s]
Fernsehteilnehmer *m* [tele]viewer
Fernsehtelefonie *f* videotelephony, visiotelephony
Fernsehton *m* television sound
Fernsehtruhe *f* console receiver
Fernsehtuner *m* television tuner
Fernsehturm *m* television tower
Fernsehübertragung *f* television transmission
F. durch Relaisstrecke relay television
Fernsehübertragungsnormen *fpl* television transmission standards
Fernsehumsetzer *m* television transposer

Fernsehverfahren *n* type of television

hochzeiliges F. high-definition television

Fernsehvorführung *f* audio-visual

Fernsehwiedergabe *f* **von Filmaufnahmen** film pick-up

Fernspannungsmesser *m* pilot voltmeter

Fernspeisung *f* remote [power] supply

Fernsprechamt *n* telephone exchange (office), *(Am)* telephone central office

halbautomatisches F. semi-automatic exchange

Fernsprechanlage *f* telephone equipment

private F. private telephone plant

Fernsprechanschluß *m* telephone connection

Fernsprechapparat *m* telephone [set]

F. für Wählbetrieb subscriber's automatic telephone

F. ohne Rückhördämpfung side-tone telephone set

Fernsprechauftragsdienst *m* absent subscriber service

Fernsprechbezugsleistung *f* reference volume (telephonic power)

Fernsprechbuch *n* telephone directory

Fernsprechdichte *f* telephone density

fernsprechen to [tele]phone

Fernsprechen *n* telephony

Fernsprecher *m* telephone [set], phone

batterieloser F. sound-powered telephone

Fernsprechformfaktor *m* telephone interference (influence) factor

F. der Spannung voltage telephone influence factor

Fernsprechfreileitung *f* open-wire telephone wire

Fernsprechhauptanschluß *m* subscriber's main station

Fernsprechinneneinrichtung *f* internal telephone installation

Fernsprechkabel *n* telephone cable

Fernsprechkanal *m* telephone channel

Fernsprechkundendienst *m* absent subscriber's service

Fernsprechleitung *f* telephone circuit (line)

Fernsprechnebenanschluß *m* subscriber's extension station

Fernsprechnetz *n* telephone system

öffentliches F. public telephone system

Fernsprechnummer *f* call number

Fernsprechordnung *f* telephone regulations

Fernsprechringübertrager *m* toroidal repeating coil

Fernsprechschnellverkehr *m* no-delay telephone service

Fernsprechschrank *m* telephone switchboard

Fernsprechsperre *f* withdrawal of the telephone service

Fernsprechstelle *f* telephone station, call office

öffentliche F. public call office

Fernsprechstörung *f* telephone breakdown

Fernsprechsystem *n* telephone system

Fernsprechtarif *m* telephone tariff

Fernsprechtechnik *f* telephone engineering

Fernsprechteilnehmer *m* telephone subscriber, user

Fernsprechteilnehmerverzeichnis *n* telephone directory

Fernsprechübertrager *m* repeating (repeater) coil

Fernsprechverbindung *f* telephone communication

Fernsprechverkehr *m* telephone traffic

Fernsprechvermittlung *f* telephone exchange

Fernsprechvermittlungsstelle *f* telephone exchange

Fernsprechverstärker *m* telephone amplifier

Fernsprechvertrag *m* subscription

vorübergehender F. temporary subscription

Fernsprechwählsystem *n* automatic telephone system

Fernsprechwesen *n* telephony; telephone communications

Fernsprechzelle *f* call box, telephone booth (box, cabin)

öffentliche F. public call office, kiosk booth

Fernsprechzentrale *f* telephone exchange

Fernsprechzone *f* telephone zone

Fernsteueranlage *f* remote-control system

Fernsteuerapparat *m* remote-control apparatus; *(ship)* telemotor steering-gear control *(rudder)*

Fernsteuereinrichtung *f* remote-control facility

Fernsteuerschalter *m* remote-control switch

Fernsteuerung *f* remote (distance, distant) control; telecontrol; supervisory control *(power dispatching systems)*; guidance *(aircraft)*

F. einer Strecke *(ry)* centralized traffic control

Fernsteuerungsanlage *f* remote control, distant control gear, follow-up gear

Fernstromversorgung *f* remote power supply

Fernstudium *n* correspondence course (study), home study course

Fernsynchronisierung *f* *(Am)* step-by-step synchronizing

Fernthermometer *n* distance thermometer, telethermometer

Ferntrennschalter *m* remote-controlled master switch

Fernübertragungssystem *n s.* Drehmelder

Fernüberwachung *f* remote control

Fernverkehr *m* 1. long-distance traffic; 2. long-distance communication

Fernverkehrsstraße *f* [arterial] highway (road), trunk road, thoroughfare

F. mit zwei getrennten Fahrbahnen dual carriageway [road]

Fernvermittlung *f* *(telec)* trunk exchange

Fernvermittlungsklinke *f* trunk junction jack
Fernwahl *f* long-distance selection, junction (trunk) dialling
Fernwahlleitung *f* trunk circuit with dialling facilities, *(Am)* dial toll circuit
Fernwarnung *f* early warning
Fernwattmeter *n* telewattmeter
Fernwirktechnik *f* telecontrol engineering
Fernwirkungskraft *f (nucl)* long-range force
Fernzähler *m* telecounter
Fernzeiger *m* long-distance indicator
Fernzündung *f* distant (remote, long-distance) ignition
Ferraris-Instrument *n (el)* Ferraris wattmeter, shaded-pole instrument; induction meter
Ferraris-Motor *m* Ferraris motor *(two-phase induction motor)*, drag-cup motor
Ferraris-Relais *n* induction relay
Ferraris-Tachodynamo *m* drag-cup tachometer
Ferraris-Tachometermaschine *f* drag-cup tachometer
Ferraris-Zähler *m* induction [motor] meter
Ferrimagnetismus *m* ferrimagnetism
Ferrit *n* ferrite
Ferritantenne *f* ferrite aerial
Ferritbegrenzer *m* ferrite limiter
Ferritblock *m* slab of ferrite
Ferrithof *m* ferrite envelope
ferritisch ferritic
Ferritkern *m* ferrite core
Ferritkernspeicher *m* ferrite memory
Ferritnetzwerk *n (met)* ferrite network
Ferritplatte *f* ferrite slab
Ferritschalter *m* ferrite switch
Ferritstab *m* ferrite rod
Ferritstabantenne *f* ferrite rod aerial
Ferritübertrager *m* ferrite transformer
Ferritwerkstoff *m* ferrite material
Ferrit-Zementit-Gemenge *n* ferrite-carbide aggregate
Ferrit-Zementit-Mischgefüge *n* ferrite-carbide aggregate structure
Ferroaluminium *n* ferro-aluminium
Ferrochrom *n* ferro-chromium
Ferrokarbidstein *m* ferrocarbide brick
Ferrokoks *m* ferro-coke
Ferrolegierung *f* ferro-alloy
ferromagnetisch ferromagnetic
Ferromagnetismus *m* ferromagnetism
Ferromangan *n* ferro-manganese
Ferromolybdän *n* ferro-molybdenum
Ferrosilizium *n* ferro-silicon
Ferrosulfat *n* ferrous sulphate, green vitriol
Ferrosulfatdosimeter *n (nucl)* ferrous sulphate dosimeter
Ferrotitan *n* ferro-titanium
Ferrouran *n* ferro-uranium
Ferrovanadin *n* ferro-vanadium
Ferrovorlegierung *f* ferro-alloy
Ferrowolfram *n* ferro-tungsten
Ferrozement *m* ferro-cement, reinforced cement-mortar
Ferse *f* / **verstärkte** *(text)* plated heel
Fersenheftmaschine *f* seat fastening machine
Fersenmaschine *f* heeling machine
fertig ready; finished; complete[d]
Fertiganstrich *m* last coating of paint
Fertigbauteil *n* prefabricated slab
fertigbearbeiten *(mach)* to finish [machine]
fertigbearbeitet finish-machined
  noch nicht f. unfinished
Fertigbearbeitung *f* finish machining
Fertigbeton *m* ready-mixed concrete
Fertigblasen *n (met)* final blowing
Fertigbohrarbeit *f* finish-boring operation
Fertigbohren *n* finish boring
Fertigbohrkopf *m* boring head for finishing cuts
Fertigbreite *f* finished width
Fertigdrehen *n* finish turning
Fertigdrehmaschine *f* finishing lathe
fertigen to manufacture, to produce
  versuchsweise f. to trial produce
Fertiger *m (build)* finisher, finishing machine
Fertigerzeugnis *n* finished product
Fertigfabrikat *n* finished product
Fertigfeinpolieren *n* finish polishing
Fertigfrischen *n (met)* finish refining
Fertigfrischofen *m* finishing furnace
Fertiggerüst *n* finisher, finishing stand *(rolling)*
Fertiggesenk *n* finish-forging die, finisher
Fertigglühung *f* final annealing
Fertighaus *n* prefabricated house, prefab
Fertighobeln *n* finish planing
Fertigkalibrieren *n* final sizing
Fertigkeit *f* art, skill
Fertigkeiten *fpl* / **technische** technical skills
Fertigmaß *n* final (finish) size; finishing dimension
Fertigprodukt *n* finished product
Fertigräumen *n* finish broaching
Fertigreibahle *f* finishing reamer
  F. für kegelige Bohrungen finishing taper reamer
Fertigreiben *n* **mit Maschinenreibahle** flute reaming, finish reaming with flute reamer
Fertigschlacke *f* refining (white) slag
Fertigschliff *m* finish grinding
Fertigschmelzofen *m* finishing furnace
Fertigschneider *m* plug (third) tap
Fertigsintern *n* final sintering
Fertigspinnmaschine *f* fine spinning frame
fertigstellen to complete, to finish
Fertigstellung *f* completion, finishing
Fertigstellungstermin *m* completion date
Fertigstich *m* shaping pass *(rolling)*

Fertigstoff *m* accepted stock, accepts *(papermaking)*
Fertigstraße *f* finishing train (mill)
Fertigteil *n* *(build)* prefabricated (precast concrete) unit
Fertigung *f* manufacture, [engineering] production; yield, output
  F. begrenzter Stückzahlen limited-quantity production
  F. größerer Mengen bulk production
  F. unter Einhaltung enger Toleranzen close limit production
  diskontinuierliche F. intermittent production
  gesteuerte F. timed process
  satzweise F. batch production
Fertigungsabteilung *f* shop
Fertigungsanlage *f* production equipment
Fertigungsauftrag *m* production order; batch
Fertigungsbahn *f* *(build)* finishing line; precasting lane
Fertigungsbetrieb *m* engineering factory
Fertigungseinheit *f* manufacturing unit
Fertigungsfehler *m* manufacturing error
Fertigungsfolge *f* manufacturing sequence
Fertigungsgemeinkosten *pl* overhead cost
Fertigungsingenieur *m* production engineer
Fertigungskennwert *m* manufacturing characteristic
Fertigungskontrolle *f* production testing (control); process control
Fertigungskosten *pl* manufacturing cost
Fertigungsleistung *f* / **stündliche** rate per hour
Fertigungslinie *f* flow process
  F. für Motoren engine line
Fertigungslos *n* lot, run
  kleines F. short run
Fertigungsmaschine *f* production machine
Fertigungsmenge *f* [rate of] output, yield
  jährliche F. annual rate of output
Fertigungsmethode *f* manufacturing (production) method, manufacturing (production) technique
Fertigungsnummer *f* serial number
Fertigungsplanung *f* production (process) planning
Fertigungsplanungsingenieur *m* process planning engineer
Fertigungsprogramm *n* manufacturing programme (schedule)
Fertigungsprozeß *m* manufacturing process
Fertigungsprüfer *m* works inspector
Fertigungsstraße *f* process (production) line
  verkettete F. link line
Fertigungsstückzahl *f* production rate
Fertigungstechnik *f* manufacturing (production) engineering, metal working manufacturing, manufacturing techniques
Fertigungsverfahren *n* manufacturing process
Fertigwalzgerüst *n* finishing rolls
Fertigwalzstraße *f* finishing train
Fertigwalzwerk *n* finishing mill
Fertigzug *m* finishing pass
Fertigzuschlag *m* *(met)* final addition
Fesselballon *m* captive balloon
fest 1. strong *(material)*; 2. *(ph)* solid; frozen, solidified; 3. fixed; stationary; locked; tied; stuck, jammed; 4. massive *(rock)*; compact; 5. stable *(foundation)*; 6. *(mil)* fortified; 7. permanent; definite; firm; safe; established; steady; 8. fixed *(gauge)*; dead *(centre)*
Festanode *f* heavy anode
Festanodenröhre *f* stationary-anode tube *(X-ray)*
Festanschlag *m* fixed stop
Festbett *n* / **bewegtes** *(mine)* moving bed
  ruhendes F. *(mine)* fixed bed
Festbettverfahren *n* fixed bed process
festbinden to lash
Festbrennen *n* **des Deckels** sticking of the cover [due to overheating]
Festdrossel *f* fixed nozzle
Festecho *n* permanent echo *(radar)*
Festenbau *m* *(mine)* breast stoping
festfahren *(mach)* to get (become) stuck, to stick *(tool)*
Festfahren *n* *(mach)* sticking
Festfeuer *n* fixed light *(navigation)*
Festfrequenz *f* spot frequency
festfressen *(mach)* to jam, to seize
Festfressen *n* *(mach)* seizing, jamming, seizure
festgefahren stuck *(tool)*
Festgehalt *m* solid measure *(wood)*
festgelagert compact *(soil)*
festhalten to hold in place, to retain; to arrest *(moving parts)*; to impede
  ein Bild f. to retain an image
  fotografisch f. to register photographically
Festhalteschaltung *f* *(dat)* hold circuit
Festhaltescheibe *f* *(ry)* stop disk *(signal)*
Festhaltetau *n* becket-rope
festigen to strengthen; to stabilize; to compact
Festigkeit *f* strength *(material)*; rigidity; stability; massivity; compactness
  elektrische F. dielectric strength, voltage proof, insulating capacity
  rechnerische F. *(build)* design strength
festigkeitsbedingt dependent on strength
Festigkeitsberechnung *f* strength calculation
Festigkeitsdeck *n* [main] strength deck
Festigkeitseigenschaft *f* strength property
Festigkeitsgrenze *f* strength limit
Festigkeitslehre *f* theory (science) of the strength of materials
Festigkeitsprüfung *f* strength test
Festigkeitsschott *n* strength (structural) bulkhead
Festigkeitsuntersuchung *f* strength test
festkeilen to key

Festkleben *n* **der Kolbenringe** [piston] ring sticking
Festklemmeinrichtung *f* locking device
festklemmen 1. to clamp [in place], to lock; 2. to jam, to stick
Festklemmen *n* 1. locking, clamping [in place]; 2. jamming, sticking
Festklemmhebel *m* clamping lever
Festklemmung *f (mach)* clamping; lock
Festkomma *n (dat)* fixed point
Festkommabetrieb *m* fixed-point calculation
Festkommadarstellung *f* fixed-point representation
Festkommarechnung *f* fixed-point calculation
festkommen *(ship)* to run (go fast) aground
Festkondensator *m* fixed capacitor
Festkontakt *m* fixed contact
Festkörper *m* solid [body]
 anorganischer F. inorganic solid
 gelöster F. solute
 homogener F. homogeneous solid
 kristalliner F. crystalline solid
Festkörperbauelement *n* solid state component
Festkörperdetektor *m* solid state detector
Festkörperdosimetrie *f s.* speichernde Dosimetrie
Festkörperdruckschaltung *f* solid state printed circuit
Festkörperelektronik *f* solid state electronics
Festkörperelement *n* solid state element
Festkörper-Laser *m* solid state laser
Festkörper-Maser *m* solid state maser
Festkörpermodell *n* model of solid
Festkörperphysik *f* solid state physics, physics of solids
Festkörperschalter *m* solid state switch
Festkörperschaltkreis *m* solid state circuit
Festkörperschaltung *f* solid state circuit
 binäre logische F. binary solid-state logic circuit
Festkörperschaltungstechnik *f* solid state circuits art
Festkörperspeicher *m (dat)* solid state memory
Festkörperstromkreis *m* integrated circuit
Festkörperverstärker *m* solid state amplifier
Festland *n* main land, continent
festlegen / eine Fahrstraße to prepare (set up) a route
 Kurs f. to shape course for
Festmacheboje *f* mooring buoy
Festmacheleine *f* securing (working, mooring) line
festmachen *(ship)* to seize, to moor, to stow; to fix
 f. an to fix to; to make fast to
Festmachepfahl *m* mooring post
Festmachetonne *f* mooring buoy
Festmaß *n* solid measure *(wood)*
festnarbig *(leath)* firm-grained
Festoszillator *m* local oscillator
Festphase *f* fixed phase
Festpunkt *m* observation point; *(opt)* fixed point; *(mach)* fixed datum, bench mark; fulcrum *(pivot)*
Festpunkte *mpl* **bei Triangulation** control points of known position and elevation
Festpunkthöhe *f* reference height
Festpunktnetz *n* observation grid, fixed-point net
Festpunktnivellement *n* fixed-point levelling
Festpunktverdichtung *f* densification of the fixed-point net
Festpunktverfahren *n* fixed-point method
Festrost *m* stationary grate
Festscheibe *f (mach)* fixed pulley
Festsieb *n* stationary screen
festsitzen to stick *(tool)*
Festspannung *f* fixed voltage
Feststecken *n* fitchering, sticking *(drilling)*
feststehend 1. *(mach)* fixed; stationary; 2. still *(picture)*
Feststellbremse *f* 1. *(ry)* locking brake; 2. parking brake *(car)*
feststellen 1. to find out; to spot; to ascertain; 2. to clamp, to lock
 die Fehlerquelle f. to isolate the source of trouble
Feststellschraube *f* binder (clamping) screw, binding (retaining) bolt
Feststellvorrichtung *f* clamping device
Feststoff *m* solid [matter]
Feststoffpartikel *f* solid particle
Festtank *m* integrated tank
Festtreibstoff *m* solid propellant
Festwertregelung *f* regulating system *(control system with a fixed value of the set point)*
Festwertspeicher *m* permanent (non-volatile) memory, permanent (non-volatile) store
Festwiderstand *m* fixed resistor
 F. mit Abgriff semifixed resistor
Festzeichen *n* bench mark
Festzeitgespräch *n* fixed time call
festziehen to tighten
Festziehen *n* tightening
festzurren to lash, to seize; to secure
fett bold, fat-faced *(letter)*; rich; bituminous *(coal)*
Fett *n* 1. fat; 2. grease *(lubrication)*
 gehärtetes F. hydrogenated fat
 graphitiertes F. graphited grease *(lubrication)*
 verseifbares F. *(ch)* saponifiable fat
Fettabscheider *m* grease trap
Fettalkohol *m* fatty (aliphatic) alcohol
fettartig fatty, lipoid
Fettausschlag *m (leath)* fatty spew
Fettbüchse *f* grease cup, greaser

fettdicht grease-proof *(paper)*
Fettdichtigkeit *f* grease (oil) penetration resistance
fetten to grease, to lubricate
Fettfleckfotometer *n* grease-spot photometer
Fettgarleder *n* oil leather
Fettgerbung *f* oil tanning
fettig greasy
Fettkalk *m* fat lime
Fettkohle *f* fat coal
Fettlicker *m* fat liquor
Fettnippel *m* grease nipple
Fettöl *n* fixed oil
Fettpistole *f* grease gun
Fettpresse *f* pressure grease gun
Fettreihe *f (ch)* aliphatic (fatty) series
Fettsäure *f* fatty acid
Fettschmierbüchse *f* grease cup
Fettschmierstelle *f* grease lubricating point
Fettschmierung *f* grease lubrication
F. / mit grease-lubricated
Fettschwanzschaf *n* fat-tailed sheep, broad-tailed sheep
Fettspaltung *f (ch)* lipolysis, fat splitting
Fettspritze *f* grease gun
Fettstift *m* coloured wax pencil, tallow pencil
Fettwachs *n (ch)* adipocere
Fettwechsel *m (mach)* renewal of the grease
Fettwechselperiode *f (mach)* grease renewal period
feucht humid; moist; damp
Feuchte *f* humidity; moisture
Feuchtemesser *m* moisture-content meter
feuchten to wet; to moisten; to damp[en]
Feuchtglätter *m* intermediate calender *(papermaking)*
Feuchthalten *n* **des Frischbetons** wet (water, moist) curing
Feuchtigkeit *f* humidity; moisture; dampness, wetness
absolute F. absolute humidity
normale F. normal humidity
relative F. relative humidity
feuchtigkeitsabstoßend moisture-repellent
Feuchtigkeitsanzeiger *m* hygroscope
Feuchtigkeitsaufnahme *f* moisture absorption, absorption of moisture; moisture pick-up
feuchtigkeitsbeständig damp-proof, moisture-proof
Feuchtigkeitsgehalt *m* humidity (moisture) content; content (amount, percentage) of moisture
handelsüblicher F. commercial humidity content
feuchtigkeitsgeschützt moisture-proof, damp-proof
Feuchtigkeitsgleichgewicht *n* moisture equilibrium
feuchtigkeitsisoliert damp-proof
Feuchtigkeitsmesser *m* moisture-content meter, *(ag)* moisture gauge; hygrometer
Feuchtigkeitsmessung *f* moisture measurement
Feuchtigkeitsrest *m* residue of moisture
Feuchtigkeitsschutz *m* damp-proofing, water-proofing, moisture-proofing
Feuchtigkeitssperre *f* water-proof membrane
Feuchtigkeitssperrung *f* water-proofing
feuchtigkeitsundurchlässig moisture-proof
Feuchtigkeitswächter *m* humidistat
Feuchtigkeitswiederaufnahme *f* moisture regain
Feuchtigkeitszahl *f* moisture index *(soil mechanics)*
Feuchtraumanlage *f (el)* vapour-proof installation, moisture-proof installation
Feuchtstrumpfthermometer *n* wet bulb thermometer
Feuchttemperatur *f* wet bulb temperature
Feuchtung *f* dampening
Feuchtungswanne *f* **am Kalander** calender box *(papermaking)*
Feuchtwalze *f* damping roller
Feuchtwalzenstoff *m* moleskin
Feuchtwanne *f* wetting trough
Feuchtwasser *n* damping solution
Feuchtwerk *n (print)* damping device (system)
Feuer *n* 1. fire; heat; 2. beacon; light *(aviation, navigation)*
festes F. fixed light
unbewachtes F. unattended light
unterbrochenes F. occulting light
Feueralarmanzeige *f* fire alarm
Feueranzünder *m* fire-lighter
Feuerbahn *f (therm)* area covered by the fire
Feuerbake *f* lighted beacon
Feuerball *m* fireball, ball of fire *(of nuclear explosion)*
feuerbeständig fire-proof, fire-resistant; refractory *(material)*
Feuerbeständigkeit *f* fire resistance; refractoriness
Feuerbeständigkeitsprüfung *f* fire-resistant test
Feuerbeton *m* refractory cement (concrete)
Feuerbrücke *f (therm)* fire bridge
F. mit Luftzuführung split bridge
Feuerbuch *n (ship)* Admiralty List of Lights
Feuerbuchsbodenring *m* firebox bottom flange
Feuerbüchse *f* fire box; [locomotive] furnace
Feuerbüchsenkessel *m* Cornish multitubular boiler
stehender F. vertical multitubular boiler with fire box
Feuerbüchsenstrebe *f* firebox stay
Feuerbuchskessel *m* internal-flue boiler
Feuerbuchsmantel *m* firebox shell
Feuerbuchsrohrwand *f* front wall of a firebox
Feuerbuchsrückwand *f* back plate of a firebox
feuerdämmend flame-retardant, fire-retarding

Feuererkennungssystem *n (ship)* fire-detecting system
selbsttätiges F. automatic fire-detecting system
feuerfest fire-resistant, fire-proof; refractory *(material)*; *(ch)* apyrous
Feuerfestigkeit *f* refractoriness, fire-proof quality
Feuergas *n* flue gas
Feuergefahr *f* fire hazard
Feuergeschränk *n* furnace front
Feuergeschwindigkeit *f* rate of fire
Feuergewölbe *n* furnace arch
Feuergrube *f* ash pit
Feuerhaken *m* fire iron (hook)
feuerhemmend fire-retarding, fire-resisting; flame-retarding
Feuerhemmschott *n* fire-retarding bulkhead
Feuerkammer *f* fire box
Feuerleiter *f* fire escape [stair]
Feuerleitgerät *n (mil)* fire-control equipment (director)
Feuerleitstelle *f* fire-control post
Feuerleitung *f* fire control
Feuerleitzentrale *f* fire-control post; fire-control station (tower) *(navy)*
Feuerlöschanlage *f* fire extinguishing plant
Feuerlöschboot *n* fire-monitor; fire float (boat)
Feuerlöschbrause *f* sprinkler, fire extinguishing rose
Feuerlöscheinrichtung *f* fire extinguishing installations
Feuerlöscher *m* fire extinguisher (drencher)
Feuerlöschgerät *n* fire extinguisher (drencher)
Feuerlöschmanöver *n (ship)* fire drill
Feuerlöschmittel *n* fire extinguishing substance
Feuerlöschschlepper *m* fire-fighting tug
Feuerlöschteich *m* fire pond
Feuerluke *f* throat of a furnace *(boiler)*; fire escape *(ship)*
Feuermanöver *n (ship)* fire drill
Feuermeldefernsprecher *m* fire telephone
Feuermeldegerät *n* / **elektronisches** fire detection device
Feuermelder *m* fire alarm
selbsttätiger F. automatic fire alarm
feuern to fire; to burn; to fuel
Feuern *n* stoking, firing
Feuerraumdecke *f* crown of furnace, furnace roof
Feuerrohr[dampf]kessel *m* fire-tube boiler
Feuerschiff *n* light ship, light vessel, floating light
unbemanntes F. light boat
Feuerschiffsmann *m* lightsman
Feuerschirm *m (therm)* brick arch
Feuerschirmschott *n* fire screen bulkhead
Feuerschott *n* fire[-resisting] bulkhead
Feuerschottür *f* fire door
Feuerschutz *m* fire-proofing
Feuerschutzartikel *m* fire-proof article
Feuerschutzstreifen *m* fire-protection strip
Feuerschweißen *n* forge welding
Feuerschwindung *f* fire shrinkage
feuersicher fire-proof
Feuerstätte *f (therm)* space above grate
Feuerstein *m* [lighter-]flint
Feuertage *mpl (mine)* days under fire, firing days
Feuerton *m* refractory clay, fire clay
Feuertür *f* furnace door *(boiler)*; fire door *(bulkhead)*
Feuerturm *m* lighthouse
Feuerung *f* firing; furnace; fuel
F. mit Beschickung von unten underfeed furnace
F. mit Druckluft forced-draught furnace
F. mit Handbeschickung hand firing (stoking)
F. mit Oberluftzuführung furnace with overgrate blower
F. mit umgekehrter Flamme down-draught furnace
F. mit unmittelbarer Heizfläche furnace with direct heating surface
gemischte (kombinierte) F. combined gas and coal furnace
rauchverhütende F. smokeless (smoke-preventing) furnace
rauchverzehrende F. smoke-consuming furnace
Feuerungsauskleidung *f* furnace lining
Feuerungsdecke *f* crown of furnace, furnace roof
Feuerungseinrichtung *f* arrangement of furnace
Feuerungsgewölbe *n* furnace arch
Feuerungsraum *m* fire box
ansteigender F. sloping furnace
Feuerungstechnik *f* fuel engineering
Feuerungswehr *n* furnace bridge
Feuerungswirkungsgrad *m* efficiency of [the] stoking
Feuerungszug *m* heating flue
feuerverbleien to lead coat
Feuervergoldung *f* fire gilding
Feuerversicherung *f* fire insurance
Feuerversicherungspolice *f* fire policy
feuerverzinken to hot-dip galvanize
Feuerverzinkung *f* hot-dip galvanizing (galvanization)
Feuerverzinnung *f* hot tinning
Feuerwehr *f* fire brigade
Feuerwehrmann *m (ship)* fire fighter
Feuerwerk *n* fireworks
Feuerwerker *m* pyrotechnist
Feuerwerkerei *f* pyrotechny
Feuerwerkskörper *m* pyrotechnic composition, firework
Feuerwiderstandsfähigkeit *f* fire resistance

Feuerzugüberhitzer *m* superheater placed in the flue
feurig 1. fiery; 2. *(geol)* igneous
Fiber *f* fibre
Fiberrundbürste *f* circular brush, brushing wheel
Fibrilliereigenschaft *f* fibrillation property
Fibrin *n* fibrin
fibrinhaltig fibrinous
Fibrinmangel *m* hypinosis
Fibrinüberschuß *m* hyperinosis
Fibroin *n* fibroin
Fibrolith *m (min)* fibrolite, sillimanite
Fichte *f* spruce
Fichtenharz *n* fir (pine) resin
Fichtenholz *n* spruce
fiebermildernd antipyretic, febrifugal
Fiedelbohrer *m* fiddle (bow) drill
Fieldistor *m (el)* fieldistor
Fieldkessel *m (therm)* Field boiler
fieren to lower
Fieren *n* lowering
Figur *f* figure; pattern
  Lichtenbergsche F. Lichtenberg figure
  Lissajoussche F. Lissajous' figure
Figuren *fpl* / **polar entsprechende** *(math)* polar reciprocals
  Widmannstättensche F. Widmannstätten structure
Figurendruck *m* figure printing
Figurenharnisch *m (text)* figuring harness
Figurkette *f (text)* pattern warp
Figurschuß *m (text)* figuring weft
Figurstelle *f (text)* figure space
Filetarbeit *f (text)* netting
Filethobel *m* fillet plane
Filetnadel *f* netting needle
Filigran *n* filigree
Film *m* 1. cinematograph film; 2. film; coat
  belichteter F. exposed film
  doppelbeschichteter F. double-coated film
  gegossener F. *(plast)* coated film
  unbelichteter F. unexposed film
Filmabtaster *m* film reading device, film pick-up
Filmarchiv *n* film library
Filmaufnahme *f* **eines Fernsehbildes** kinescope recording
Filmaufnahmespule *f* film take-up spool (reel)
Filmaufzeichnung *f* film record
Filmbearbeitung *f* film processing
Filmdosimeter *n* film dosimeter, dosifilm
Filmdosimetrie *f* film dosimetry
Filmdruck *m* screen printing
Filmeinlegen *n* threading of the film, loading
Filmemulsion *f* film emulsion
Filmentwicklungsgerät *n* film developing device
Filmentwicklungsmaschine *f* film processor, processing machine
Filmfassungsvermögen *n* footage capacity
Filmfenster *n* film gate (window)
Filmkamera *f* movie (film) camera
Filmkanal *m* film track
Filmkassette *f* film magazine, dark-slide
Filmkitt *m* film cement
Filmkittlehre *f* splicer
Filmklebeapparat *m* film splicer
Filmkopiermaschine *f* film copying machine
Filmkorn *n* film grain
Filmlaufwerk *n* film-drive mechanism
Filmmerkring *m* film-speed indicator
Filmplakette *f (nucl)* film badge [meter]
Filmprojektor *m* film projector
Filmrahmen *m* film holder
Filmring *m (nucl)* film ring
Filmriß *m* breaking of the film
Filmrolle *f* film reel
Filmsalat *m* buckling, spaghetti, camera buckle
Filmsaugplatte *f* vacuum back *(process camera)*
Filmschneidetisch *m* cutting bench
Filmschrumpfung *f* film shrinkage
Filmsetzgerät *n* photo-setter, filmsetter
Filmsetzmaschine *f* filmsetting machine, filmsetter, film composer
Filmsieden *n (nucl)* film boiling
Filmspannung *f* film tension
Filmspeicher *m* film memory *(data storage)*
Filmstreifen *m* film strip
Filmtransport *m* film feed
Filmverdampfung *f (nucl)* film boiling
Filmvorführer *m* projectionist
Filmvorführgerät *n* cinema projector
Filmvorschub *m* film feed
Filter *n* filter; strainer
  F. mit Drosseleingang *(el)* choke-input filter
  F. mit induktivem Eingang *(el)* choke-input filter
  F. mit kammförmigem Frequenzspektrum *(el)* comb filter
  abgestimmtes F. *(el)* matched filter
  akustisches F. acoustic filter
  optisches F. optical filter
Filterabschluß *m (el)* filter termination
Filteranlage *f (aut)* filter unit
Filterbecken *n (build)* filtering basin
Filterbett *n* filter (contact) bed
Filterbrei *m* filter pulp
Filterbrunnen *m (build)* filter well
Filterdeckel *m* filter cover
Filterdrossel *f (el)* filter choke
Filtereinlage *f (ch)* filter cone
Filtereinsatz *m* filter element (cartridge)
Filtereinsatztopf *m (aut)* filter-element container
Filtergehäuse *n* filter casing (case)
Filterkerze *f (ch)* candle filter
Filterkette *f (el)* filter chain (ladder)

Filterkies *m* filter gravel
Filterkohle *f* filtering charcoal
Filterkörper *m* filter pad
Filterkreis *m (el)* filtering circuit
Filterkuchen *m* filter cake
filtern to filter
Filterpaket *n* filter element
Filterpapier *n* filter[ing] paper
Filterpatrone *f* filter cartridge
Filterpolarisator *m (opt)* filter polarizer
Filterraum *m* filtration chamber
Filterrohr *n* screen pipe; liner; filtering tube
Filterschicht *f* filter bed *(water purification)*
Filtersperrbereich *m (el)* filter stop (attenuation) band
Filterstoff *m* filtering medium
Filtertank *m* filtering basin
Filtertasche *f (ch)* filter bag (cell)
Filtertopf *m* filter bowl
Filtertuch *n* filter (straining) cloth
Filterung *f s.* Filtrierung
Filtervorwärmer *m* filter feed heater
Filterwirkung *f* filtering action
Filtrat *n* filtrate
Filtration *f* filtration, filtering
Filtrationshilfsmittel *n* filter aid
filtrierbar filterable
Filtrierbarkeit *f* filterability, filtering properties
filtrieren to filtrate, to filter, to strain
Filtriergeschwindigkeit *f* rate of filtration
Filtriergestell *n* filter (funnel) stand
Filtrierhadern *pl* rags for filtering paper
Filtrierpapier *n* filter[ing] paper
aschefreies F. ashless filtering paper
Filtrierung *f* filtration
Filz *m* felt
Filzbildung *f* felting
Filzdichtung *f* felt packing (joint, gasket, washer)
Filzdichtungsring *m* felt packing ring
Filzen *n (mach)* felt-wheel polishing
Filzer *m (text)* felter
Filzfilter *n* felt filter
Filzinstandhalter *m* felt conditioner
Filzkalander *m* felt calender
Filzleitwalze *f* felt guide
Filzmaschine *f* felter
Filzöler *m* felt pad lubricator
Filzpapier *n* felt board
Filzpappe felt board
Filzprägung *f* felt finish
Filzrauhmaschine *f* felt napping machine
Filzreiniger *m* felt conditioner
Filzröllchen *n* felt pad *(magnetophone)*
Filzsauger *m (text)* felt suction roll
Filzsaugwalze *f (text)* felt suction roll
Filzscheibe *f (mach)* [solid] felt wheel for polishing
Filzseite *f* felt side
Filzstiefel *m* felt boot
Filztrockner *m* canvas drier *(papermaking)*
Filztuch *n* milled cloth
Filzunterlegscheibe *f* felt washer
Filzwalze *f* dabber
Filzwalzenbelag *m* clearer board
Fimmel *m (mine)* gad
Findling *m (geol)* erratic block
Finger *m* finger
Fingerbalken *m* finger bar
Fingerdock *n* depositing (finger-type) dock
Fingerfräser *m* slot drill, shank-type slotting end mill, keyseating cutter
Fingerhebelwelle *f* steering-knuckle shaft
Fingerhutionisationskammer *f s.* Fingerhutkammer
Fingerhutkammer *f* thimble [ionization] chamber
Finishdekatiermaschine *f (text)* finishing and decatizing machine
Finsen-Lampe *f* Finsen lamp
Firma *f* firm
Firnis *m* boiled oil, oil varnish
First *m* [roof] ridge
Firstabdeckung *f* ridge capping (covering)
Firstbeschickung *f* top feeding
Firste *f (mine)* back, hanging [wall], roof
Firstenbau *m (mine)* overhand (filled, raise) stope
F. mit abgesetzten Stößen back stoping
F. mit Versatz cut-and-fill stoping
F. mit waagerechter Firste und Bergeversatz horizontal cut-and-fill
Firstendruck *m (mine)* top (roof) pressure
Firstenfestpunkt *m (mine)* roof observation point
Firstengetriebe *n (mine)* spiling
Firstengewölbe *n (mine)* roof arch
Firstenkammerbau *m (mine)* back stoping
Firstenläufer *m (mine)* crown runner
Firstensegment *n (mine)* crown (top) section
Firstenstoß *m (mine)* stope face
F. mit schräggestellter Firste [filled] rill stope, inclined cut-and-fill stope
F. parallel zur Sohlenstrecke flat-back stope
söhliger F. filled flat-back stope, horizontal cut-and-fill stope
Firstenstoßbau *m (mine)* overhand (breast) stoping
F. mit sägeblattartigem Verhieb stepped-face overhand stoping
F. mit Spreizenausbau stulled overhead stoping
Firstpfette *f* [roof] ridge purlin
Firstschar *f* ridge course *(roof tiling)*
Firstschneider *m* roof blade *(coal plough)*
Firstziegel *m* ridge tile
Fisch *m* 1. fish; 2. *(print)* wrong fount
Fischbauchschwelle *f (ry)* dished sleeper

Fischbauchträger *m (build)* fish-bellied girder
Fischblase *f* fish bladder
Fischbrutanstalt *f* fish hatchery
Fischdampfer *m* steam trawler
Fischen *n* **bei Nacht** owling
Fischer *m* fisher[man]
Fischerboot *n* fishing boat (smack), fisherman
Fischereifabrikschiff *n* floating fish factory, trawler factory ship
Fischereifahrt *f* fishing trade
Fischereifahrzeug *n* fishing boat, fisherman, trawler; single boater; box trawler
Fischereifang- und -verarbeitungsschiff *n* factory trawler
Fischereischiff *n* fishing vessel
Fischereischutzboot *n* fishery-protection vessel, fishery-protecting vessel, fishery patrol boat
Fischereiverarbeitungsschiff *n* trawler-factory ship
Fischereiflotte *f* fishing fleet
Fischer-Tropsch-Synthese *f*, **Fischer-Tropsch-Verfahren** *n* Fischer-Tropsch process
Fischfabrikschiff *n* trawler-factory ship, fish-factory ship
Fischfanggerät *n* fishing gear
Fischgalgen *m* trawl [warp] gallow
fischgrätenartig herringbone
Fischgrätenmelkstand *m* herringbone milking parlour
Fischgrätenverband *m* raking bond *(brickwork)*
Fischgründe *mpl* fishing grounds
Fischhaken *m* gaff, fish hook
Fischkonserve *f* tinned (preserved) fish, *(Am)* canned fish
Fischkutter *m* fishing cutter (smack), trawl (welled) smack
  F. mit Bünn welled fishing smack
Fischladeraum *m* fish hold
Fischleim *m* fish glue
Fischleiter *f* fish ladder
Fischmehl *n* fish meal
Fischmehlanlage *f* fish meal plant
Fischmehlfabrik *f* fish meal factory
Fischnetz *n* [fishing] net; trawl [net]
  stehendes F. seine
Fischnetzwinde *f* trawl winch; fish hoist
Fischöl *n* fish oil
Fischortungsausrüstung *f* fish locating equipment
Fischortungsgerät *n* fish locating equipment
Fischpaß *m (build)* fishway, fish pass
Fischplanke *f* king plank
Fischraum *m* fish hold
Fischrechen *m* fish grid (screen)
Fischschwanzmeißel *m* fish-tail bit
Fischschwarm *m* shoal
Fischsortierabteile *npl* fish (deck) ponds
Fischsortierkästen *mpl* deck ponds
Fischteich *m* fish pond
Fischtreppe *f* fish ladder
Fischzucht *f* breeding of fishes; pisciculture
Fischzug *m* haul, draught
Fissionskammer *f s.* Spaltungskammer
Fissium *n (nucl)* fissium
Fitschenbeil *n* plugging chisel
Fittings *pl* fittings
Fitzfaden *m (text)* dividing thread
Fixativ *n* 1. fixative *(colour, photo)*; 2. fixature *(hairdressing)*
Fixierbad *n* fixing bath
fixieren to fix, to set *(colour)*
Fixieren *n* fixing, fixation; heat set
Fixiermittel *n* fixing agent
Fixiersalz *n* fixing salt, hypo
  saures F. acid fixer
Fixierstift *m* steady (locating) pin
Fixierung *f* fixing, fixation; heat set
Fixpunkt *m* bench mark, fixed datum (point)
Fixstern *m* fixed star
Fjordschiff *n* fiordvessel
flach shallow; flat; *(mine)* inclined
Flachamboß *m* flat anvil
Flachbacke *f* flat die *(thread rolling)*
Flachbagger *m* scraper, skimmer
Flachbaggerung *f* surface digging (excavation), shallow digging (excavation)
Flachbahn *f (mach)* flat top guide, flat guideway (track)
Flachbahnanlasser *m* face-plate starter (controller), disk-type starter
Flachbahngeschütz *n* flat-trajectory gun
Flachbahnregler *m* disk-type rheostat
Flachbahnschalter *m* wafer[-type] switch
Flachbahnschaufellader *m* flat action shovel-loader
Flachbahnsteuerschalter *m* face-plate controller
Flachband *n* flat belt
Flachbandkabel *n* balanced twin feeder cable
Flachbesäummaschine *f* breast planer
Flachbettfelge *f* flat-base rim
Flachbeutel *m* flat (satchel) bag
Flachbeutelmaschine *f* flat bag making machine
Flachbinder *m (ag)* flat harvester binder
Flachblock *m* flat-shaped ingot
Flachbodenbunker *m* flat-bottom pin
Flachbodenkolben *m* flat-top piston
flachbodig flat-bottomed *(vessel)*
Flachbrunnen *m* surface well
Flachbrüter *m (ag)* single-tier incubator
Flachdach *n* flat roof
Flachdachziegel *m* plain roofing tile
Flachdecke *f (build)* flat slab
Flachdichtung *f* flat gasket
Flachdraht *m* flat wire
Flachdruck *m* flatbed (surface, planographic) printing

Fläche *f* area; surface; face *(of crystals)*
F. gleichen Potentials equipotential (isopotential) surface
F. gleicher Phase *(el)* equiphase surface, surface of equal phase
F. konstanten Potentials equipotential (isopotential) surface
F. zweiten Grades quadric surface
F. zweiter Ordnung *(math)* quadric surface
abwickelbare F. torse
bearbeitete F. machined surface
beheizte F. generator surface
brechende F. refracting surface
ebene F. plane surface
freie F. *(mine)* free face
geneigte F. scarp; inclined plane
geschlossene F. *(ph)* closed surface
glatte F. smooth surface
isobare F. isobaric surface
isopyknische F. isopycnic surface
isostatische F. isostatic surface
isostere F. isosteric surface
isotherme F. isothermal surface
kaustische F. caustic surface
orientierte F. oriented surface
plane F. flat surface
projizierte F. maximum cross-section *(of rocket hull when viewed head-on)*
rauhe F. rough surface
schädliche F. noxious surface
tragende F. supporting surface
tragende F. des Fundaments bearing surface of the foundation
überdeckte F. hall space, covered area, underroof space, under-hall space *(fair)*
wirksame F. active area *(of electronic valves)*
Flacheisen *n (met)* flat iron (billet) (*s. a.* Flachstahl)
grobes F. flat billet *(rolling)*
flächen *(mach)* to spot-face
Flächen *n (mach)* spot-facing
Flächenaberration *f* zonal aberration
Flächenantenne *f* plane aerial
Flächenanziehung *f* surface attraction
Flächenausdehnung *f* surface expansion; areal extension
Flächenbelastung *f* surface load
Flächenbelegung *f* surface density
Flächenberechnung *f* computation of an area
Flächenberührung *f* area contact *(gears)*
Flächenbeschleunigung *f* surface acceleration
Flächenbettungszahl *f* modulus of the foundation
Flächenblitz *m* sheet lightning
Flächenbrüter *m (ag)* single-tier incubator
Flächendiagonale *f* plane diagonal
Flächendichte *f* surface density
elektrische F. surface density of electric charge
Flächendiode *f* junction diode
Flächendrahthindernis *n* multiple-belt entanglement
Flächendruck *m* 1. surface pressure; 2. blotch print
F. beim Stapellauf launching pressure
Flächeneinheit *f* unit of area
Flächeneinheitsbelastung *f* unit load
Flächenfräsen *n* surface milling
Flächengebilde *n* **/ textiles** nonwoven textile fabric
Flächengeschwindigkeit *f (ph)* surface (areal) velocity
Flächengewicht *n* 1. *s.* Flächendichte; 2. weight per unit area; substance *(paper)*
Flächengewichtsmeßgerät *n* isotopic thickness gauge
Flächengitter *n* plane (two-dimensional) lattice *(of crystals)*
Flächengitterwirkung *f* plane lattice effect *(of crystals)*
flächengleich equal [in area]
Flächengleichrichter *m* surface-contact rectifier *(semiconductor)*
Flächengröße *f* **des Ortsnetzes** *(telec)* area of exchange district
Flächengründung *f* spread foundation (footing), pad foundation
Flächenheizkörper *m* **/ elektrischer** electric panel heater
Flächenhelle *f* surface light intensity
Flächeninhalt *m* area
Flächenintegral *n* surface integral
Flächenkraft *f* surface force
Flächenkrümmung *f* surface curvature
Flächenkühlung *f* cooling by outspread surface
Flächenladung *f (el)* surface charge
Flächenlast *f* area load
Flächenleistung *f* power per unit of surface; *(ag)* area capacity
Flächenlichtstromdichte *f* **der Einstrahlung** luminous flux falling on unit area
Flächenmessung *f* measurement by planimeter
Flächenmuster *n* background pattern, mechanical tint
Flächennutzungsplan *m (build)* zoning map; region utilization plan
Flächenpolarisator *m* sheet polarizer
Flächenpressung *f* unit pressure
Flächenquelle *f* surface (plane) source
Flächenreibungskonstante *f* skin friction
Flächenrieselkühler *m* surface cooler
Flächenschar *f (math)* family of surfaces
Flächenschleifmaschine *f* surface grinding machine

F. mit **Langtisch** surface grinding machine with rectangular table
F. mit Rundtisch surface grinding machine with rotary table
F. mit Topfscheibe surface grinding machine using the face of the cup-type wheel
Flächenschliff *m* surface grinding
Flächenschruppschleifen *n* surface rough grinding
Flächenschwerpunkt *m* centre of gravity of an area
Flächenstromdichte *f* surface current density
Flächentetrode *f* junction tetrode
Flächenträgheitsmoment *n* areal moment of inertia
Flächentragwerk *n* plate (two-dimensional) structure
Flächentransistor *m* junction transistor
flächentreu equal-area *(projection)*, *(math)* area-preserving
Flächenverhältnis *n* disk area ratio
Flächenverteilung *f* surface distribution
Flächenwiderstand *m* skin-friction drag
Flächenwinkel *m* interfacial (plane) angle
Flächenwirbel *f* surface distribution of curl
flächenzentriert face-centred *(lattice)*
Flaches *n (mine)* incline
Flachfeder *f* flat spring
Flachfeile *f* flat file
Flachfeuer *n* flat trajectory fire
Flachflanschdichtung *f* flat flange gasket
Flachflanschverbindung *f* butt joint
Flachformmaschine *f* flatbed press
Flachformrotationsmaschine *f* flatbed rotary press
Flachführung *f (mach)* flat track (slideway, guideway)
Flachführungsbahn *f* flat guideway (way, track, slideway)
Flachgalvano *n* flat electrotype
Flachgewinde *n* square (flat) thread
Flachglas *n* flat (plate) glass
Flachgründung *f* spread foundation (footing)
Flachhacke *f* flat pick
Flachherdmischer *m* flat hearth-type mixer
Flachkabel *n* flat cable
Flachkaliber *n* box pass *(roll)*
Flachkämmaschine *f* rectilinear combing machine
Flachkehlnaht *f* / **eingeebnete** flush fillet weld
Flachkeil *m* flat key; wedge
Flachkeilverschluß *m (mil)* sliding square-wedge breech block
Flachkettenstuhl *m* flat warp loom
Flachkettenwirkmaschine *f* flat warp knitting machine
Flachkiel *m (ship)* flat [plate] keel
Flachknüppel *m* slab billet
Flachkommutator *m* radial (disk) commutator
Flachkopf *m* flat head
Flachkopfgeschoß *n* flat-pointed bullet, flat-nose bullet
Flachkulierstuhl *m* flat knitting frame with loop forming sinkers
Flachkulierwirkmaschine *f* [Cotton's patent] full-fashioned knitting machine, flat weft knitting machine
Flachkupfer *n* flat bar copper
Flachkupplung *f* butt joint
Flachläppen *n* flat lapping [operation], lapping [on] flat surfaces
Flachläppmaschine *f* flat lapping machine
Flachlasche *f (ry)* plain fishplate
Flachlitzenförderseil *n* flattened strand winding rope
Flachlitzenseil *n* flattened strand rope
Flachlöffel *m* skimmer
Flachlöffelbagger *m* skimmer
Flachlöschkammer *f* parallel-plate arc-quenching chamber
Flachmeißel *m* 1. square-nosed tool *(machine tool)*; 2. flat chisel; *(mine)* spudding bit
Flachmembrane *f* flush diaphragm
Flachnähmaschine *f* flatbed sewing machine
Flachpackung *f* flat packing *(seal)*
Flachpalette *f (mach)* flat pallet
Flachplattenpresse *f* flat-platen press
Flachregler *m* drum governor
Flachrelais *n* flat-type relay; side armature relay
Flachriemen *m* flat belt
Flachriementrieb *m* flat belt drive
Flachringanker *m (el)* flat (gramme) ring armature
Flachrundschraube *f* round-head bolt, oval-head bolt
F. mit Vierkantansatz step bolt, oval-head elevator bolt
Flachs *m* flax
gekämmter F. hackled flax
Flachsaufbereitung *f* dressing (preparing) flax
Flachsauger *m* flat box *(papermaking)*
Flachsbrechen *n* flax breaking
Flachsbrechmaschine *f* flax breaker, breaking (stripping) machine
Flachsbrech- und -reinigungsmaschine *f* flax breaking and dressing machine
Flachschaber *m (mach)* flat scraper
Flachschieber *m* flat slide valve, slide valve [based on flat surface]
Flachschiebersteuerventil *n* flat plate control valve
Flachschleifen *n* flat grinding
Flachschliff *m (met)* taper section; *(mach)* flat grinding
Flachschnitt *m* flat cut, flat grain, secant cut

Flachseil *n* band rope, flat cable (rope)
Flachsfaserband *n (text)* flax sliver
Flachshechelgerät *n* flax comber (hackler)
Flachshechelmaschine *f* flax hackling machine
Flachshocke *f* flax stack
Flachsieb *n* flat screen
Flachsockel *m* **bei Metallröhren** *(el)* wafer-type socket
Flachspan *m (wood)* flake
Flachspanplatte *f (wood)* flake board
Flachspirale *f* flat spiral
Flachspulamperemeter *n (el)* attraction-type ammeter
Flachspule *f (el)* slab (disk, pancake) coil
Flachsraufmaschine *f* flax puller
Flachsreinigungsgerät *n* flax dresser
Flachsreißmaschine *f* flax cutting machine
Flachsriffelkamm *m* flax rippling comb
Flachsröste *f* rettery *(flax)*
Flachsschneidemaschine *f* flax cutting machine
Flachsstengel *m* stalk of the flax
Flachsstuhl *m (text)* tipple box
Flachstahl *m* flat steel (bar)
Flachstahlschwelle *f (ry)* flat steel sleeper
Flachstampfer *m* flat rammer
Flachstich *m* flat pass
Flachstrickerei *f* flat frame knitting; plain knitting
Flachstrickmaschine *f* flat knitting machine
zweisystemige F. double-lock flat knitting machine, double-system flat knitting machine
Flachstrickware *f* flat fabric
Flachstromvergaser *m* horizontal-draught carburetter, cross-draught carburetter
Flachstrumpfwirkmaschine *f* [Cotton's patent] full-fashioned knitting machine
Flachsvollerntemaschine *f* complete flax harvester
Flachswergaufbereitung *f* flax tow preparing
Flachtischrevolver *m (mach)* flat turret
Flachtrudeln *n (aero)* flat spin
Flachverbunderregung *f* level-compound excitation *(generator)*
Flachwasser *n* shallow (shoal) water
Flachweger *m (ship)* floor ceiling
Flachwicklung *f (el)* plane winding
Flachwulststahl *m* bulb flat (plate)
Flackereffekt *m* flicker effect
Flackerfrequenz *f* flicker frequency
Flackerlampe *f (telec)* flicker lamp
flackern *(el)* to flash; to flicker
Flackerrelais *n* flashing relay
Flackersignal *n* undulating light-signal
Fladepotential *n* Flade potential
Flader *f* curled spot *(wood)*
fladerig mottled *(wood)*
Fladerschnitt *m* flat cut, secant cut
Flagge *f* flag, colours
Flaggendiskriminierung *f (ship)* flag discrimination
Flaggenführung *f (ship)* setting (carrying) of flags
Flaggengruß *m (ship)* flag dipping (salute)
Flaggenknopf *m (ship)* truck, mast truck (ball)
Flaggensetzen *n (ship)* setting of flags
Flaggensignalisieren *n* **mit zwei Flaggen** *(ship)* semaphore flag-signalling
Flaggenstock *m (ship)* ensign staff *(stern)*
Flaggenunterscheidung *f (ship)* flag discrimination
Flaggenzuschlag *m (ship)* flag discrimination
Flaggleine *f (ship)* flag halliard (line)
Flaggoffizier *m* flag officer
Flaggschiff *n* flag ship
Flak *f* anti-aircraft gun
Flakon *m (n)* phial
flammbeständig *(text)* flame-resistant
Flammbeständigkeit *f (text)* flame resistance
Flammbogen *m* flame (flaming) arc
Flammbogenlampe *f* high intensity carbon arc lamp
Flamme *f* flame
nichtleuchtende F. roaring flame
offene F. naked flame
oxydierende F. oxidizing flame
reduzierende F. carbonizing flame
sensitive F. sensitive flame
Flämmen *n* flame scarfing
Flammenaureole *f* gas cap *(safety lamp)*
Flammenbildung *f (therm)* flame formation
Flammenbüschel *n* cluster of flames
Flammenfärbung *f* flame coloration
Flammenfärbungsreaktion *f* flame reaction
Flammenfortpflanzung *f* flame propagation
Flammenfotometer *n* flame photometer
flammengehärtet flame-hardened
Flammenhärten *n* flame hardening
Flammenlötung *f* torch (blowpipe) brazing
Flammenoberflächenhärten *n* flame hardening
Flammenprüfung *f* flame test
Flammenrückschlag *m (aut)* backfiring
Flammenrückschlagsicherung *f (aut)* flame trap
Flammenschutz *m (aut)* flame trap
Flammenschutzmittel *n* flameproofing agent
Flammenspektrometer *n* flame spectrometer
Flammenspektrum *n* flame spectrum
Flammenstrahl *m* flame jet, jet of flame; rocket blast
Flammenstrahlbohren *n* jet piercing
Flammentzunderung *f* flame descaling
Flammenüberwachung *f* combustion controls
Flammenwächter *m* flame failure alarm, flamestat
Flammenwerfer *m* flame thrower
Flammenwerferfüllwagen *m* recharging trailer for flame throwers

Flammenwerferpanzer *m* flame throwing (thrower) tank
Flammenzug *m* flame flue
Flammgarnfärberei *f* yarn shading
Flammhärteofen *m* tempering flame furnace
Flammkammer *f* fire box
Flammkohle *f* free-burning coal, flaming coal
Flammofen *m* reverboratory furnace
flammofenfrischen to puddle
Flammofenfrischen *n* reverberatory furnace refinery, puddling process
Flammofenkuppel *f* reverberating roof
Flammpunkt *m* flash point
  F. im geschlossenen Tiegel closed cup flash point
  F. im offenen Tiegel open cup flash point
Flammpunktprüfer *m* flash point tester
  F. mit geschlossenem Tiegel closed cup tester
  F. mit offenem Tiegel open cup tester
Flammrohr *n* fire tube, flue
  eingedrücktes F. bulged-in flue tube
Flammrohrdoppelkessel *m* twin Cornish boiler
Flammrohrfeuerung *f* flue furnace
Flammrohrheizfläche *f* flue heating surface
Flammrohrkessel *m* flue boiler
  F. mit glattem Flammrohr boiler with plain flue
  F. mit vorgehenden Heizröhren Cornish multitubular boiler
  stehender F. vertical flue boiler
  zylindrischer F. mit durchschlagender Flamme marine boiler
Flammrohr-Tenbrinkfeuerung *f* Tenbrink flue furnace
Flammrohrüberhitzer *m* furnace-flue superheater
Flammschutzausrüstung *f (text)* fire-resistant finish
Flammschutzimprägnierung *f (text)* fire-resistant finish
flammspritzen *(plast)* to flame-spray
Flammspritzen *n (plast)* flame-spraying
flammwidrig fire-resistant, non-inflammable
Flanke *f* 1. flank; 2. slope; wing; tail; 3. side wall *(of a block)*
  F. des Impulses edge (side) of the pulse
  F. eines Gewindes flank of a thread
  ansteigende F. leading edge *(of a pulse)*
  steile F. steep edge
Flanken *fpl* sides of gear tooth, tooth profiles
Flankendurchmesser *m* effective diameter *(thread)*
Flankenfahrt *f (ry)* taking slantwise, collision at points
Flankenform *f* flank profile
Flankenprofil *n* tooth profile *(gear)*
Flankenspiel *n* backlash *(gear)*
flankenspielfrei backlash-free *(gear)*
Flankensteilheit *f* slope steepness; edge steepness
Flankenwinkel *m* thread angle
  F. des Gewindes profile angle of the thread
Flankenzentrierung *f* side-bearing type of fit, sides-of-teeth fit
Flansch *m* flange
  F. mit Dichtungsrillen flange with packing grooves
  eingezogener F. internal flange
  fester F. fixed flange
  glatter F. plain flange
  weit auskragender F. wide flange
Flanschbauform / in flange-mounted *(motor)*
Flanschdichtung *f* **durch eingelegten konischen Ring** flanged joint with conical ring
  F. durch Feder und Nut flanged joint with tongue and groove
flanschen to flange
Flanschendichtung *f* joint-packing, gasket
Flanschenhülse *f (text)* double-flanged bobbin
Flanschenliderung *f* joint-packing, gasket
Flanschenpackung *f* joint-packing, gasket
Flanschenrohr *n* flanged pipe
Flanschgewindeverbindung *f* screwed flange joint
Flanschkupplung *f* flange coupling
Flanschmotor *m* flange-mounted motor, flange-type motor
Flanschnabe *f* flange hub
Flanschrohrverbindung *f* flange pipe connection
Flanschträger *m* I-beam
Flanschverbindung *f* flanged connection, bolted flanges; flange coupling *(waveguides)*
Flanschverschraubung *f* bolted flanges
Flanschzwischenlage *f* waveguide shim
Fla-Panzer *m* anti-aircraft tank
Fla-Rakete *f* anti-aircraft rocket
Flasche *f* 1. bottle; flask; jar; 2. cylinder *(gas)*; 3. block
  Leydener F. Leyden jar
  Woulfesche F. Woulfe's (pressure) bottle
Fla-Scheinwerfer *m* anti-aircraft searchlight
Flaschenbier *n* bottled beer
Flaschenelement *n* bottle battery (cell)
Flaschengas *n* bottled (compressed) gas, gas from cylinders
Flaschenglas *n* bottle glass
Flaschenhalskokille *f* bottle top mould
Flaschenpost *f* drift bottle, bottle papers
Flaschenrüttler *m* needle (poker, immersion, internal, spud) vibrator
Flaschenspülmaschine *f* bottle rinsing machine
Flaschenverschluß *m* bottle closure
Flaschenzug *m* pulley block (system), chain block, tackle
  F. mit elektrischem Antrieb electric hoist block

flattern 1. to knock *(valve)*; 2. to flutter *(radar)*; 3. to screech; to chatter *(saw)*
Flatterrädchen *n* **zum Abrichten** star wheel *(grinding)*
Flaum *m* down, fluff
flaumig downy
flauschig fleeced
Flavanol *n* flavanol
Flavanon *n* flavanone
Flavon *n* flavone
Flechtarbeit *f* 1. wicker work *(baskets)*; 2. fixing steel *(concrete)*
flechten 1. to twist, to braid; 2. to fix steel *(reinforced concrete)*
Flechtgurtträger *m* braiding machine carrier
Flechtmaschine *f* 1. braiding machine; 2. wiring jig
Flechtschnur *f* shoe braid
Flechtwerk *n* plaiting
Flechtwerkzaun *m* wattle fence
Fleck *m* spot, stain; blotch *(of oil)*
Fleckenbeständigkeit *f (text)* stain resistance
Fleckeneffekt *m* patch effect *(semiconductors)*
Fleckenentferner *m* stain (spot) remover, scouring drops
Fleckenentfernung *f* removal of stains
Fleckenentfernungsmittel *n* spot (stain) removal agent
fleckenfrei stainless
Fleckengebiet *n (el)* patch area
Fleckhelligkeit *f* brightness of the spot
Fleckigkeit *f (telev)* spottiness, blurring
Fleckigwerden *n* spotting
  F. durch Wassertropfen water spotting
Fleckunschärfe *f* **bei Ablenkung** deflection defocusing
Fleckverzerrung *f (el)* spot distortion
Fleier *m (text)* flyer
Fleisch *n* flesh *(leather)*
Fleischbeschauer *m* meat inspector
Fleischbrühe *f* meat broth, beef-tea
Fleischextrakt *m* meat extract, extract of beef
Fleischkonserve *f* tinned meat, *(Am)* canned meat
Fleischmilchsäure *f* sarcolactic acid
Fleischseite *f* flesh side *(leather)*
  glatte F. level flesh
Fleischspalt *m* flesh split *(leather)*
Fleischtransportschiff *n* meat ship
Flensdeck *n* flensing deck *(whale factory ship)*
flensen *(ship)* to flense
Flettner-Rotorschiff *n* Flettner rotorship
Flettner-Ruder *n* Flettner rudder (fin, control surface)
flexibel flexible
Flexodruck *m* flexographic (aniline) printing
Flexodruckfarbe *f* flexographic ink
Flexur *f (geol)* fold, flexure
Flickarbeit *f* repair work; *(met)* patch work
flicken to repair; to patch *(furnace)*
Flickmasse *f* fettling material *(furnace)*
fliegen to fly
  ohne Motor f. to glide
Fliegen *n* flying; aviation
Fliegenfenster *n* fly screen
Fliegenholz *n* quassia
Fliegenkopf *m* turned letter
Flieger *m* sprinter *(cycling)*; aviator, flyer, pilot; glider
Fliegerabwehrgeschütz *n* anti-aircraft gun
Fliegerabwehrkanone *f* anti-aircraft gun
Fliegerabwehrmaschinengewehr *n* anti-aircraft machine gun
Fliegerbekleidung *f* aviation wear
Fliegerhorst *m* air field (base), aerie
Fliegerrettungsboot *n* crash boat
Fliegerschule *f* aviation (flying) school, aerie
Fliegerschutzanzug *m* flying suit
Fliegersichtzeichen *npl* shapes
Fliehgeschwindigkeit *f* escape velocity
Fliehkraft *f* centrifugal force
Fliehkraftabscheider *m* cyclone
Fliehkraftanlasser *m* centrifugal starter
Fliehkraftbandkupplung *f* slip-ring clutch
Fliehkraftklassierer *m* centrifugal classifier
Fliehkraftkupplung *f* centrifugal clutch
Fliehkraftpendel *n* centrifugal-force tachometer
Fliehkraftregler *m* centrifugal governor
Fliehkraftschalter *m* centrifugally operated switch
Fliehkrafttachometer *n* centrifugal-force tachometer
Fliehkraftzerstäuber *m* centrifugal disk atomizer
Fliehkraftzündversteller *m* centrifugal spark advance
Fliehmoment *n* centrifugal moment
Fliese *f* tile
Fliesenbelag *m* tiling
Fliesenleger *m* tiler
Fließarbeit *f* flow-line operations, continuous production
Fließband *n* flow-line; conveyor belt
Fließbandfertigung *f* flow-line production, production-line technique
Fließbandmontage *f* flow-line assembly
Fließbedingung *f* yield condition *(material)*
Fließbereich *m* yield region
Fließbeschichtung *f* flush coating *(paper)*
Fließbett *n* fluidized bed, *(Am)* fluid bed
Fließbettreaktor *m* fluidized-bed reactor
Fließbettverfahren *n* moving-bed process
Fließbild *n* flow sheet, flow chart
Fließbrennstoff *m* liquid fuel
Fließdehnung *f* yield strain

Fließdruckpresse *f* impact extrusion press
fließen to flow; to creep
Fließen *n* flow
beginnendes plastisches F. incipient plastic flow
laminares F. telescopic flow
plastisches F. plastic flow
stationäres plastisches F. steady plastic flow
uneingeschränktes plastisches F. unrestricted plastic flow
unmittelbar bevorstehendes plastisches F. impending plastic flow
fließfähig flowable; fluid
Fließfähigkeit *f* fluidity; flowability; filterability
geringe F. hard flow
gute F. free flowing
Fließfertigung *f* flow production; production-line method
Fließfiguren *fpl* stretcher lines, slip bands
Fließgelenkverfahren *n* *(build)* plastic-hinge method
Fließgeschwindigkeit *f* rate of flow
Fließgleichgewicht *n* transient equilibrium
Fließgrenze *f* yield point, flow limit
Fließlinien *fpl* stretcher lines
Fließmethode *f* flow-line method
Fließmontage *f* line assembly work
Fließpapier *n* blotting paper
fließpressen to extrude
Fließpressen *n* extrusion
Fließpreßverfahren *n* extrusion process
Fließprobe *f* *(build)* flow test; pouring test *(melting)*
Fließreihe *f* flow line
Fließscheide *f* non-slip point *(rolling)*
Fließschmierung *f* fluid (hydrodynamic) film lubrication
Fließspan *m* flow chip
Fließspannung *f* yield stress
Fließstraße *f* flow (production) line
Fließverfahren *n* flow[-line] process
Fließvermögen *n* fluidity, mobility; fusibility
Fließversatz *m* *(mine)* flow (controlled gravity) stowing
Fließvorgang *m* *(met)* yielding phenomenon, mechanics of yielding
Fließwiderstand *m* *(met)* yield resistance
Flimmerfotometer *n* flicker photometer
flimmerfrei flickerless
Flimmerfreiheit *f* freedom from flicker
Flimmerfrequenz *f* flicker frequency
flimmern to flicker
Flimmern *n* flicker[ing]
Flimmerverfahren *n* flicker method
flink quick-acting *(fuse)*
Flintglas *n* flint glass
Flintstein *m* flint
Flip-Flop *n* flip-flop
Flip-Flop-Generator *m* flip-flop generator
Flip-Flop-Schaltung *f* *(dat)* flip-flop circuit *(bistable multivibrator)*, scale-of-two circuit
Flitschfurnier *n* flitch veneer
Flitschholz *n* flitches in the veneer
Flittergold *n* tinsel
Flockdruck *m* *(text)* flock print
Flocke *f* flock *(wool)*; flake *(snow)*
flockegefärbt stock-dyed
Flocken *fpl* flakes, flocks, chips
Flockenbast *m* cottonized bast fibre
Flockenflachs *m* cottonized flax
Flockengraphit *m* flaky graphite
Flockenstoffbindung *f* nap cloth weave
flockig flaky; flocculent
Flockigkeit *f* flakiness
Flockpunktgewebe *n* flock dot fabric
Flocktest *m* flocculation test
Flockung *f* coagulation, flocculation
Flockungsmittel *n* flocculating agent
Flockungspunkt *m* *(ch)* flocculation (cloud) point
Floconnébindung *f* *(text)* nap cloth weave
Flor *m* [cut] pile, web
Florbändchen *n* strip of web
Florbindung *f* pile weave
Florettseide *f* waste silk
Florgewebe *n* pile weave
F. aus Baumwolle cotton pile fabric
Florkette *f* nap warp
Flornoppe *f* / **aufgeschnittene** cut loop
Florpostpapier *n* flimsy (airmail) paper
Florschuß *m* *(text)* pile pick
Florschußschlauch *m* weft pile tube
Florteiler *m* *(text)* web divider
Florteppich *m* pile carpet
Florwebstuhl *m* loom for weaving pile fabrics
Floß *n* raft
aufblasbares F. pneumatic raft
Floßbaumholz *n* balsawood
Floßbrücke *f* floating bridge
Flosse *f* stabilizer, fin; fixed plane
flößen to raft
Flossen *fpl* dashers *(in a churn)*
Flossengeschoß *n* fin-stabilized missile
Flossenkiel *m* fin keel
Flossenplatte *f* fin plate
Flossenverstellgerät *n* fin-adjusting instrument
Flossenverstellung *f* fin adjustment (trimming)
Flößer *m* boomsman, rafter, raftsman, log driver
Floßholz *n* floated (floating, rafted) timber
Flotation *f* floatation
Flotationskonzentrat *n* froth fines
Flotationsmittel *n* floatation agent
Flotationsverfahren *n* washing process
Flotationszelle *f* floatation (froth, frother) cell
Flöte *f* flute
Flotierbarkeit *f* floatability

flotieren to float
flott *(ship)* afloat
Flotte *f* 1. navy; fleet; 2. dye solution
Flottenbegleiter *m* escort sloop
Flottenbegleitschiffe *npl* fleet train
Flottenfärbung *f* bath-dyeing
Flottengeleit *n* convoy
Flottenweg *m* / **doppelter** *(text)* double-liquor flow
Flottenzirkulation *f* / **zweiseitige** *(text)* double-liquor circulation
Flottierung *f* flush *(weaving)*
flottmachen to float, to bring afloat; to come off *(a stranded vessel)*
Flottung *f* float [stitch] *(knitting)*
Flöz *n (geol)* bed, seam
  anstehendes F. existing seam, seam to be worked
  bauwürdiges F. workable seam
  bedingt bauwürdiges F. marginal seam
  hangendes F. overlying seam
  liegendes F. underlying seam
  nichtbauwürdiges F. unworkable seam
  stehendes F. vertical seam
  unbauwürdiges F. unworkable seam
Flözarchiv *n* seam memoirs
Flözausbringen *n* seam output
Flözfaltung *f* roll
Flözmächtigkeit *f* seam thickness
Flözprofil *n* seam section
Flözstrecke *f* gate road
Fluat *n* fluate
Flucht *f* 1. row; 2. play; 3. alignment
  F. / außer out of line, misaligned
fluchten to align; to be aligned
  nicht f. to be misaligned
fluchtend in line; aligned
Fluchtfaktor *m s.* Bremsnutzung
Fluchtgenauigkeit *f* accuracy of alignment
Fluchtgeschwindigkeit *f* escape speed *(of a rocket)*
flüchtig volatile
Flüchtigkeit *f* volatility, fugacity
Fluchtlinie *f* vanishing line; straight line
Fluchtlinientafel *f* alignment chart, nomogram
Fluchtpunkt *m* vanishing point
Fluchtschacht *m* escape shaft
Fluchtstab *m* field rod; ranging rod
Fluchtungsfehler *m* misalignment, alignment error
Fluchtungsfernrohr *n* optical alignment testing telescope
Fluchtungsprüfung *f* alignment testing
Flug *m* 1. flight; gliding; 2. *(text)* fly *(waste)*
  F. im intergalaktischen Raum intergalactic travel
  F. mit Überschallgeschwindigkeit supersonic flight
  F. ohne Motor gliding
  F. über den Wolken over-the-top flight, flying on top
  interplanetarischer F. interplanetary flight (trip)
  interstellarer F. interstellar flight
Flugasche *f* fly ash; flue dust
Flugauswertung *f* flight evaluation
Flugbahn *f* trajectory, path *(missile)*
  F. des Sputniks (Satelliten) orbit of the satellite, satellite orbit
  ansteigende F. arched trajectory
  elliptische F. elliptical orbit
  gestreckte F. flat trajectory (path)
  hindernisfreie F. safe clearance path
  sichtbare F. visible trajectory
  vorausberechnete F. intended (prescribed) trajectory, predetermined [flight] path
Flugbahnbild *n* trajectory chart
Flugbahnebene *f* trajectory plane
Flugbahnfläche *f* plane of the orbit
Flugbahngeschwindigkeit *f* flight-path speed
Flugbahnhöhe *f* / **maximale** trajectory peak
Flugbahnlinie *f* curve of flight path
Flugbahnneigungswinkel *m* flight path angle
Flugbahnüberwachungsverfahren *n* plotted trajectory system
Flugbahnverfolgung *f* flight tracking
Flugbahnwerte *mpl* parameters of the orbit
Flugbahnzeichner *m* flight-path recorder (plotter)
Flugbenzin *n* aviation petrol, *(Am)* aviation gasoline
Flugberechnung *f* flight computation
Flugbereich *m* range of flight
Flugbeschleunigung *f* flight acceleration
Flugbetriebslenkung *f* operational control
Flugbetriebsmeldung *f* flight regularity message
Flugblatt *n* pamphlet, leaflet
Flugboot *n* flying boat
Flugbordsprechanlage *f* aircraft intercom
Flugdauer *f* flight time
Flugdeck *n* landing (flight, flying-off, taking-off) deck
Flügel *m* 1. *(aero)* wing, aerofoil; 2. *(mach)* vane *(e.g. pump)*; blade, fan *(propeller)*; [mud] guard *(car)*; 3. *(build)* leaf *(bridge, gate)*; casement, wing *(window)*; 4. *(text)* beater; flyer; 5. *(geol)* leg, limb
  F. mit Nadelbesatz *(text)* carding beater
  F. mit Preßfinger *(text)* flyer with presser arm
  aufgesetzter F. movable blade
  durchlaufender F. continuous wing
  verstellbarer F. adjustable blade
Flügelast *m* spike knot; wing-shaped knot
Flügelbrecher *m* blade crusher

Flügelbreite *f* blade depth (width) *(propeller)*
Flügeldrehung *f* rotation of the flyer
Flügeldruck *m* blade pressure
Flügeldruckseite *f* blade face
Flügelfenster *n* casement light (window)
Flügelflächenverhältnis *n* blade area ratio *(propeller)*
Flügelform *f* shape of blade, blade outline *(propeller)*
Flügelführungsschlitz *m* vane slot *(vane pump)*
Flügelgarn *n* flyer yarn
Flügelholm *m* spar
Flügelkanone *f* wing cannon
Flügelkante *f* edge of a propeller blade
Flügelkolben *m* vane-type piston
Flügelkopf *m (text)* flyer head
flügellastig *(aero)* lopsided
Flügelmaschinengewehr *n* wing gun
Flügelmauer *f* head (wing) wall *(bridge)*
Flügelmine *f* vaned mine
Flügelmittelstück *n* centre-section wing
Flügelmutter *f* wing (fly) nut
Flügelort *n* lateral road, *(Am)* lateral *(ore mining)*
Flügelprofil *n* blade section *(propeller)*
Flügelpumpe *f* vane pump
Flügelquerschnitt *m* blade section *(propeller)*
Flügelrad *n* impeller *(water pump)*
Flügelradanemometer *n* revolving vane anemometer
Flügelradpropeller *m* vane-screw propeller, vane wheel
Flügelradschiff *n* vane wheeler, vane-wheel ship
Flügelradwassermesser *m* rotary (vane) watermeter
Flügelradzähler *m* rotating vane meter
Flügelrahmen *m* casement frame *(window)*
Flügelscheibe *f (text)* wing disk
Flügelschiene *f (ry)* wing rail
Flügelschott *n* wing bulkhead
Flügelschraube *f* thumb screw
Flügelspill *n (ship)* acron, vane spindle
Flügelspindel *f (text)* flyer spindle
Flügelspinnmaschine *f* flyer, fly machine
Flügelsteigung *f* pitch *(propeller)*
Flügelstreckung *f* aspect ratio
Flügeltank *m* topside tank
Flügelwirkungsgrad *m* blade efficiency
Flügelzellen[flüssigkeits]motor *m* rotary motor of the vane type
Flügelzellenpumpe *f* vane pump
außen beaufschlagte F. rotary-vane type of pump
Flügelzwirnmaschine *f* fly doubler (twister), flyer twisting frame
Flugfolgeradar *n* automatic tracking [radar]
Flugförderband *n* flight conveyor
Flugfunk *m* aeronautical (aircraft) radio
Flugfunkdienst *m* aircraft wireless (radio) service, wireless aircraft service
Flugfunkfeuer *n* aeronautical radio beacon
Flugfunksprechzeugnis *n* general radiotelephony licence
Fluggast *m* [aircraft] passenger
zahlender F. revenue passenger
Fluggeschwindigkeit *f* 1. *(aero)* flying speed; 2. travelling velocity *(e.g. of electrons)*
F. bei Brennschluß burned speed, burn-out (burned, burnt, cut-off) velocity
Flughafen *m* airport
städtischer F. municipal airport
Flughafenbefeuerung *f* airport lighting
Flughafenerkennungszeichen *n* aerodrome identification sign
Flughafenfunkstelle *f* airport radio station
Flughafenkennzeichen *n* aerodrome identification sign
Flughafenleuchtfeuer *n* aerodrome beacon
Flughafenradar *n* airfield surface movement indicator
Flughöhe *f* flying altitude
Flugkoks *m* flue coke
Flugkolbenmaschine *f* free-piston engine
Flugkolbenmotor *m* free-piston[-type] engine
Flugkörper *m* missile
Fluglehrer *m* flying instructor
Flugleitungsturm *m* **für Ausbildungszwecke** mock-up tower
Fluglinie *f* air line
Flugmechanik *f* aerodynamics
Flugmeldedienst *m* aircraft reporting service
Flugmotor *m* aero-engine
Flugmotorenindustrie *f* aero-engine industry
Flugmotorenöl *n* aviation lubricant
Flugnavigation *f* air (aerial) navigation
Flugnavigationshilfe *f* navigational aid
Flugplan *m* flight plan
gültiger F. current flight plan
Flugplatz *m* aerodrome, airfield, airport
Flugplatzansteuerungsfeuer *n* aerodrome proximity light
Flugplatzbefeuerung *f* airport lighting
Flugplatzradar *n* airport radar
Flugregelsystem *n* flight control system
Flugregler *m* autopilot
Flugsand *m* shifting (quick) sand
Flugschneise *f* aerial corridor
Flugschüler *m* pilot trainee; student pilot, flying pupil
Flugsicherungsanlage *f* installation for air-traffic control
Flugsicherungsdienst *m* air-safety service
Flugsicherungskontrollturm *m* aerodrome control tower
Flugsimulation *f* flight simulation

Flugsimulator *m* flight simulator
Flugstaub *m* flue dust; smoke dust
Flugstaubbett *n* diluted phase fluid bed, entrained catalyst
Flugstaubkammer *f* dust chamber
Flugstaubverlust *m* dust loss
Flugstaudruck *m* flight-dynamic pressure
Flugsteig *m* ramp
Flugstrecke *f* airline; distance; air route
Flugstreckenbefeuerung *f* airway (route) lighting
Flugstreckenleuchtfeuer *n* airway light
Flugsturzbremse *f* dive flap
flugtauglich fit for flying duties
Flugtauglichkeit *f* aptitude for flying
Flugtrainer *m* link trainer
Flugtriebwerk *n* reaction propulsion [aircraft] engine
Flugüberwachung *f* air traffic control, supervision of air navigation
Flugüberwachungsgerät *n* flight control instrument
Flugübungsgerät *n* synthetic trainer
Flugunterricht *m* flying tuition
Flugweg *m* track
Flugwegabweichung *f* distance off-course
Flugwegdarstellung *f* track display
Flugwegeinhaltung *f* track keeping
Flugwesen *n* aviation, aeronautics
Flugwetter *n* / **ungünstiges** closed weather
Flugzeit *f* transit (flight) time, time of flight
Flugzeitspektrometer *n* time-of-flight spectrometer
Flugzeug *n* aeroplane, *(Am)* airplane; aircraft
  abfliegendes F. departing aircraft
  drahtlos gesteuertes F. radio-controlled plane
Flugzeugantenne *f* aircraft aerial
Flugzeugaufnahme *f* aerial photograph; airplane prospecting
Flugzeugbau *m* aircraft manufacture
Flugzeugbesatzung *f* crew [of the aeroplane]
Flugzeugbordgerät *n* airborne equipment
Flugzeugbordstation *f* airplane radio set
Flugzeugfabrik *f* aircraft factory
Flugzeugfernsteuerung *f* remote control of airplanes
Flugzeugflattern *n (telev)* aeroplane flutter
Flugzeugführer *m* pilot[-in-command]
Flugzeugfunkeinrichtung *f* aircraft radio equipment
Flugzeugfunkmechaniker *m* aircraft radio mechanic
Flugzeugfunkstelle *f* aircraft station
Flugzeuggerippe *n* [aeroplane] framework
Flugzeughalle *f* hangar
Flugzeugindustrie *f* aircraft industry
Flugzeugkennbuchstabe *m* distinguishing letter [of the aeroplane]
Flugzeugkompaß *m* aircraft compass
Flugzeuglack *m* [air] dope
Flugzeuglandungsinsel *f* langley
Flugzeugmotor *m* aeroplane engine
Flugzeugmotorenaufladung *f* aeroplane-engine supercharging
Flugzeugpeilung *f* bearing of aircraft
Flugzeugposition *f* aircraft position
Flugzeugrakete *f* airborne (aircraft) rocket
  ferngesteuerte F. guided aircraft rocket
Flugzeugreaktor *m* airplane reactor
Flugzeugrechner *m* air data computer
Flugzeugschleuder *f* catapult
Flugzeugsender *m* aircraft transmitter
Flugzeugsicherungsschiff *n* aircraft auxiliary
Flugzeugteil *n* aircraft component
Flugzeugträger *m* aircraft carrier
Flugzeugträgerreaktor *m* carrier vessel reactor
Flugziel *n* actual moving target *(radar)*
Fluokieselsäure *f* fluosilicic (hydrofluorsilicic) acid
Fluor *n* fluorine
Fluoren *n* fluorene
Fluoreszenz *f* fluorescence
  sensibilisierte F. sensitized fluorescence
Fluoreszenzabfall *m* fluorescent decay
Fluoreszenzausbeute *f* fluorescence yield
Fluoreszenzdosimeter *n* fluorescence dosimeter
Fluoreszenzdruckfarbe *f* fluorescent ink
Fluoreszenzenergie *f* fluorescent energy
Fluoreszenzfarbe *f* 1. luminous paint; 2. fluorescent (fluorescence) colour
Fluoreszenzfarbstoff *m* fluorescent dye
Fluoreszenzlicht *n* fluorescent light
Fluoreszenzmesser *m* fluorimeter
Fluoreszenzschirm *m* fluorescent screen
Fluoreszenzspektrum *n* fluorescence spectrum
Fluoreszenzstoff *m* fluorescent substance
Fluoreszenzstrahlung *f* fluorescent radiation
fluoreszieren to fluoresce
fluoreszierend fluorescent
Fluorid *n* fluoride
fluorieren to fluorinate, to fluorate
fluoriert fluorinated
Fluorit *n* fluorite
Fluorkohlenwasserstoff *m* fluoro[hydro]carbon, fluocarbon
Fluorochromierung *f* fluorescent dye staining
Fluorwasserstoff *m* hydrogen fluoride
Flur *m (build)* hall, lobby, vestibule
Flur *f* [arable] land; fields
Flurbereinigung *f* redistribution of land
Flurfördermaschine *f* ground-mounted winder
Flurfördermittel *npl* industrial trucks
Flurgarderobe *f* hall set
flurgesteuert floor-controlled
Flursteuerung *f* floor control
Flurwinde *f* floor-mounted electric hoist

Fluse *f* nap *(in the fabric)*
Fluß *m* 1. river; 2. *(electric or magnetic)* flux
F. der schnellen Neutronen fast [neutron] flux
F. der thermischen Neutronen thermal [neutron] flux
elektrischer F. electric flux
integraler F. flux time; integrated [neutron] flux
magnetischer F. magnetic flux
quellenfreier F. conservative flux
remanenter F. remanent flux
schneller F. fast [neutron] flux
thermischer F. thermal [neutron] flux
Flußabflachung *f (nucl)* [flux] flattening
Flußabsenkung *f (nucl)* flux depression
flußabwärts downstream, down the river
flußaufwärts upstream, up the river
Flußbarkasse *f* river launch
Flußbau *m* river engineering
Flußbett *n* bed of a river, river bed
Flußdampfer *m* river steamer
Flußdepression *f s.* Flußabsenkung
Flußdiagramm *n* flow chart *(information theory)*
Flußdichte *f (ph)* flux density
elektrische F. electric flux density
magnetische F. magnetic flux density
Flußdichtemesser *m* fluxmeter
Flußenge *f* narrow
Flußfahrt *f* river service
Flußfahrzeug *n* river-boat, river-craft
Flußfeld *n* flux field
Flußhafen *m* river port
flüssig liquid; fluid
Flüssig-flüssig-Extraktion *f (nucl)* liquid-liquid extraction
Flüssigfrischen *n* melting refine
Flüssiggastanker *m* liquefied petroleum gas tanker, L.P.G.-tanker; liquefied natural gas tanker, L.N.G.-tanker
Flüssigkeit *f* liquid; fluid
barokline F. baroclinic fluid
barotrope F. barotropic fluid
homogene F. homogeneous fluid
ideale F. perfect fluid
inkompressible F. incompressible fluid
korrodierende F. corrosive liquid
piezotrope F. piezotropic fluid
ruhende F. liquid at rest
schwerelose F. imponderable fluid
verdrängte F. displaced fluid
wirkliche F. real fluid
zähe F. viscous fluid
Flüssigkeitsabdichtung *f* liquid (fluid) seal
Flüssigkeitsabschluß *m* liquid seal; water gland, liquid packing
Flüssigkeitsanlasser *m* liquid starter
Flüssigkeitsantrieb / mit hydraulically driven, fluid-driven
Flüssigkeitsbehälter *m* liquid container (tank), cistern
F. der Hydraulik hydraulic tank
Flüssigkeitsbrandbombe *f* liquid incendiary bomb
Flüssigkeitsdämpfung *f* fluid dashpot action, viscous damping
flüssigkeitsdicht liquid-tight
Flüssigkeitsdrehmomentwandler *m* fluid torque converter
Flüssigkeitsdruckmeßdose *f* hydraulic capsule (load-cell)
Flüssigkeitsdruckmesser *m* liquid-pressure gauge
Flüssigkeitsdurchflußzähler *m* liquid-flow counter
flüssigkeitsdurchlässig leaky
Flüssigkeitselement *n* one-fluid cell
Flüssigkeitsenteisung *f* fluid de-icing
Flüssigkeitsfederthermometer *n* liquid-filled thermal system, liquid-expansion thermometer
Flüssigkeitsfeuerung *f* **mit Rohrzerstäuber** liquid fuel furnace with nozzle spraying
Flüssigkeitsgetriebe *n* fluid transmission (gear, drive), fluid power system, hydraulic drive
F. mit veränderlicher Abtriebsdrehzahl hydraulic variable speed gear
Flüssigkeitsgrad *m* degree of viscosity
Flüssigkeitsgradmesser *m* viscosimeter
Flüssigkeitsgrenzkurve *f* liquid-limit curve
Flüssigkeitskreislauf *m* fluid (hydraulic) circuit
Flüssigkeitskupplung *f* hydraulic coupling (transmission) *(Föttinger)*
Flüssigkeitsküvette *f* liquid cell
Flüssigkeitsmanometer *n* fluid (liquid) manometer, liquid-column gauge
Flüssigkeitsmenge *f* amount of fluid
Flüssigkeitsmeßstab *m* telltale level plug
Flüssigkeitsmotor *m* hydraulic (fluid power) motor
F. mit drehender Abtriebsbewegung rotary hydraulic motor
F. mit konstantem Drehmoment constant torque type of fluid power motor
Flüssigkeitsrakete *f* liquid[-propellant] rocket
Flüssigkeitsraketentriebwerk *n* liquid [rocket] engine, liquid-rocket unit, liquid-propellant unit (rocket), liquid-propellant motor (rocket engine), liquid motor, liquid-fuel rocket engine, liquid-fuel and oxidant motor
Flüssigkeitsreaktor *m* liquid-fuel reactor, liquid-moderated reactor
Flüssigkeitsreibung *f* liquid (fluid) friction
Flüssigkeitsreibungsverlust *m* fluid frictional loss
Flüssigkeitsreiniger *m* liquid purifier

Flüssigkeitsrücklaufbremse *f* hydraulic buffer (recoil brake)
Flüssigkeitssäule *f* head *(of a liquid)*
Flüssigkeitsschicht *f* / **zusammenhängende** fluid film
Flüssigkeitsspeicher *m* accumulator
F. ohne Trennkolben pistonless type of accumulator
druckgasbelasteter F. compressed gas accumulator
druckgasbelasteter (pneumatischer) F. mit elastischer Trennblase rubber sac-type accumulator
pneumohydraulischer F. hydro-pneumatic accumulator
Flüssigkeitsspiegel *m* liquid level, level of liquid
Flüssigkeitsstandregler *m* **ohne Hilfsenergie** altitude valve
Flüssigkeitsstrahl *m* jet of a liquid
Flüssigkeitsstrahlbohren *n* abrasive jet drilling
Flüssigkeitsstrahloszillograf *m* jet oscillograph
Flüssigkeitsströmung *f* fluid flow; flow of liquid
Flüssigkeitsszintillationszähler *m (nucl)* liquid scintillation counter
Flüssigkeitstachometer *n* gyrometer
Flüssigkeitsverdrängung *f* liquid displacement
Flüssigkeitsverteiler *m (aut)* fluid distributor
Flüssigkeitsvolumen *n* volume of liquid
verdrängtes F. displacement, volume of the displaced liquid
Flüssigkeitswiderstand *m (el)* liquid rheostat; liquid resistance
Flüssigkeitszähler *m (nucl)* liquid counter [tube]
Flüssigkeitszählrohr *n (nucl)* liquid counter [tube]
Flüssigmetall *n* liquid metal
Flüssigmetallkühlung *f* liquid-metal cooling *(reactor)*
Flüssigmetallreaktor *m* liquid-metal cooled (coolant, fuel) reactor
Flüssigsuspensionsreaktor *m* fluidized[-fuel] reactor
Flußkabel *n* subaqueous (under-river) cable
Flußkahn *m* river (inland waterway) barge
Flußkonverter *m (nucl)* [neutron] flux converter, doughnut
Flußleitwert *m (el)* forward conductance
Flußlinie *f* line of flux
Flußlinienverteilung *f* flux distribution
Flußlotse *m* river pilot, huffler
Flußmittel *n* flux *(soldering)*
F. beim Löten soldering (brazing) flux
ätzendes F. corrosive (acid) flux
säurebeständiges F. non-corrosive flux
Flußmündung *f* mouth of a river
Flußplan *m* flow chart *(information theory)*
Flußquerschnitt *m* water cross-section of a river
Flußraddampfer *m* river paddle-wheel steamer
Flußregulierung *f* regulation of a river
Flußrichtung *f* direction of flux
Flußsand *m* river sand
Flußsäure *f* fluorhydric (hydrofluoric) acid
Flußschiff *n* river boat (vessel)
Flußschiffahrt *f* river navigation
Flußschiffer *m* waterman, boatman, master of a river boat
Fluß-See-Schiff *n* river-and-seagoing vessel
Flußspat *m* fluorite, fluorspar
Flußstahl *m* mild (low-carbon) steel; steel produced in the fluid condition
Flußumkehr *f* flux reversal *(of the magnetic flux)*
Fluß- und Seebau *m* hydraulic engineering
Flußverkettung *f* / **magnetische** flux interlinking
Flußverteilung *f (ph)* [neutron] flux distribution
Flußwasser *n* river water
Flußwölbung *f (nucl)* buckling
Flußzeit *f s.* integraler Fluß
Flut *f* flood; high water, inundation; deluge
Flutanker *m* flood-anchor
flutbar floodable
Flutbarkeit *f* permeability; floodability
Flutbecken *n* wet (tidal) basin, wet (closed) dock
Flutbett *n* channel
fluten to flow; to flood
Fluten *n* flooding *(of reactor)*
Flutgebiet *n* tidal waters
Fluthafen *m* wet basin (dock), tidal harbour
Flutkai *m* tidal quay
Flutkarte *f* tide-table
Flutlicht *n* floodlight
Flutlichtanlage *f* floodlight plant
Flutlichtbeleuchtung *f* floodlighting
Flutlichtlampe *f* flood lantern, floodlight lamp
Flutlinie *f* high-water mark
Flutmarke *f* high-water mark
Flutmesser *m* tide-gauge; tide-dial
Flutmündung *f* estuary
Flutschleuse *f* tide lock
Flutströmung *f* flood current (stream)
Fluttor *n* tide gate, flood gate
Flutuhr *f* tide gauge (dial)
Flutventil *n* 1. *(ship)* sea cock, flooding valve; 2. air-inlet valve
Flutverfahren *n* **mit einer Alkoholbank** alcohol slug process
Flutwasser *n* flooding water
Flutwelle *f* tidal wave
F. in Untiefen eager
Flutzeit *f* flood-tide
Fluxmeter *n* flux meter
Flyer *m* flyer, speed frame
FM- *s.* Frequenzmodulation

FM-Radartechnik *f* frequency-modulated radar technique
Fock *f* jib; fore sail
Fockmast *m (ship)* fore-mast
Focktakelung *f (ship)* Bermuda rig
Focktoppnant *f (ship)* fore-lift
Fokalebene *f (opt)* focal plane
Fokallänge *f (opt)* focal length
Fokometer *n* focometer
Fokus *m (opt)* focus
primärer F. prime focus
Fokuseinstellung *f* focusing; focus control
Fokus-Haut-Abstand *m (nucl)* focus-to-skin distance; target-skin distance
Fokusregelung *f* focus control
Fokussieraufnahme *f* focusing shot
fokussierbar focusable
Fokussierbarkeit *f* focusability
Fokussierelektrode *f* focusing electrode
fokussieren to focus, to bring into focus
Fokussierfeld *n* focusing field
Fokussierlinse *f* focusing lens
Fokussiermagnet *m* focusing magnet
Fokussierspule *f* focusing coil
Fokussierung *f* focusing
elektrostatische F. electrostatic focusing
magnetische F. magnetic focusing
periodische F. periodic focusing
starke F. strong (alternating-gradient) focusing
Fokussierungselektrode *f* focusing electrode
Fokussierungswirkung *f* focusing action
Fokustiefe *f* depth of focus
Fokusverlagerung *f* focus displacement
Folge *f* sequence, succession; *(math)* progression; sequel *(time)*; consequence *(cause)*
F. der Arbeitsstufen (Arbeitsvorgänge) operation sequence, sequence of operations
arithmetische F. arithmetic progression
geometrische F. geometrical progression
stochastische F. stochastic sequence
Folgeabtastung *f (telev)* sequential scanning
Folgebild *n* travel (tracking) shot
Folgeeinrichtung *f (mach)* follow-up servo
hydraulische F. hydraulic servo mechanism
Folgeelement *n (nucl)* daughter element
Folgefehler *m* / **statischer** static error
Folgefrequenz *f* repetition frequency (rate), recurrence rate
Folgegerät *n* sequential unit
Folgeglied *n* follower
Folgekern *m* product (daughter) nucleus
Folgekolben *m* servo-spool *(hydraulics)*
Folgekontakt *m* sequence[-controlled] contact, make-before-break contact
Folgemaßnahmen *fpl* improvement measures
Folgemotor *m* follow-up motor
folgen *(math)* to follow; to result; to succeed; to track *(radar)*
Folgen *n* tracking *(radar)*
Folgepol *m* consequent (commutating, intermediate) pole, interpole
Folgeprodukt *n (nucl)* daughter, decay product; reaction product, descendant; secondary material, after-product
radioaktives F. radioactive product
Folgeprüfprogramm *n (dat)* sequence checking routine
Folgeprüfung *f* sequential test
Folgereaktion *f* consequent reaction
Folgeregelkreis *m (cont)* servo loop
Folgeregelung *f* differential control
Folgeregler *m* servomechanism
F. mit einer Integration type 1 servomechanism
F. mit zwei Integrationen type 2 servomechanism
folgerichtig consequent[ly]
Folgerung *f* 1. conclusion; deduction; 2. *(logical)* inference; 3. *s. a.* Folgerungssatz
Folgerungssatz *m* corollary
Folgeschalter *m* sequence switch
Folgeschalterkamm *m* sequence switch cam
Folgeschaltung *f* sequencing, sequence control
Folgesteuerung *f* sequence (sequential) control
Folgesteuerungsventil *n* sequence valve
Folgestrom *m* follow current
Folgetestkarte *f* sequential control chart
Folgeumschaltkontakt *m* sequence change-over contact
Folgeverstärker *m* follower-amplifier
Folgewald *m* second growth
Folgewerkzeug *n* **mit Bohr- und Reibwerkzeug** subland drill-reamer
Folgezeiger *m* direction pointer (indicator)
Foliant *m* folio volume
Folie *f* foil *(metal)*; film, plastic sheet, lamina
flache F. flat film
genarbte F. embossed sheet
harte F. rigid sheet
kaschierte F. backed foil; metallized paper
weiche F. flexible sheet
Folienkalander *m* sheeting calender
Folienkondensator *m* foil capacitor
Folienzelt *n* igloo cloche
Folioformat *n* folio
Follikelhormon *n* follicular hormone
Fön *m* hair drier
Fond *m (text)* [back]ground; bottom [shade]
Fondboden *m (aut)* main floor
Förderabgabe *f (mine)* royalty, duty, lot
Förderanlage *f* conveyor equipment, conveying plant; *(mach)* handling equipment; mine hoist
Förderanlagen *fpl* conveying machinery; conveyance plants
Förderaufhauen *n (mine)* chute raise

Förderaufseher *m (mine)* dispatcher, traffic controller
Förderband *n* conveyor belt[ing], conveying belt[ing]; belt (band) conveyor
F. mit Seiltraggerüst *(mine)* rope sideframe belt
endloses F. endless conveyor
Förderbanddosierwaage *f* weigh-feeder belt
Förderbandgerüst *n* conveyor structure
Förderberg *m (mine)* chute raise
Förderbrücke *f* conveyor bridge *(open-cast mining)*
Förderdampfmaschine *f* winding steam engine
Förderdruck *m* delivery (discharge) pressure, discharge-line pressure *(hydraulics)*
F. der Pumpe pump outlet pressure
Fördereinrichtung *f* conveyor; *(mach)* handling (conveying) device; *(mine)* winding gear
Förderer *m* conveyor
kurvengängiger F. curving conveyor
zweisträngiger F. double-strand conveyor
Fördererz *n* crude ore
Förderfeld *n (mine)* conveyor track
Fördergefäß *n* 1. conveyance; 2. skip
Fördergerät *n (mach)* materials handling device
Fördergerüst *n (mine)* pithead, head (hoist) frame, headgear
Fördergeschwindigkeit *f (mach)* handling speed; *(build)* hoisting speed; *(mine)* rope speed
Fördergestell *n (mine)* [drawing] cage, [drawing] frame
Fördergrus *m* slack, rough dross
Fördergurt *m* [conveyor] belt, [conveyor] belting
gewendelter F. inverted (twisted) belt
Fördergut *n* loading material, cargo
Förderhöhe *f (mine)* lift; height of delivery, delivery head *(hydraulics)*
F. einer Pumpe pump head
geodätische F. static head
hydrostatische F. hydrostatic head
manometrische F. total head
statische F. static discharge head
Förderkasten *m* tote box
Förderkette *f* chain-type carrier
Förderkohle *f* raw (uncleaned, run-of-mine, R.O.M.) coal
Förderkolben *m* delivery plunger
Förderkorb *m* car; *(mine)* drawing cage (frame)
Förderkübel *m* 1. skip; 2. hudge, kibble *(shaft sinking)*
Förderleistung *f* 1. *(build)* haulage capacity; discharge capacity *(pump)*; *(mach)* capacity to be conveyed (handled); 2. output
Fördermann *m (mine)* hurrier
Fördermaschine *f* hauling (hoisting) machine; winding machine
F. mit konischer Trommel conical-drum hoist
F. mit zylindrischer Trommel cylindrical-drum hoist
F. mit zylindro-konischer Trommel cylindro-conical drum hoist
Fördermaschinen *fpl* hoisting machinery and conveyors
Fördermaschinenbedienungspult *n* hoist operator's control
Fördermaschinengebäude *n (mine)* hoist room
Fördermaschinenmotor *m (mine)* hoist motor
Fördermaschinist *m* hoist man, winder operator
Fördermedium *n* fluid medium *(hydraulics)*
Fördermenge *f* 1. rate of delivery, capacity, delivery [rate], volumetric delivery *(pump)*; 2. *(mine)* output
F. je Zeiteinheit delivery quantity per unit time *(pump)*
theoretische F. theoretical output *(pump)*
wirkliche F. actual output *(pump)*
Fördermittel *n (mach)* handling device; *(mine)* conveyance
fördern 1. to convey; *(mach)* to handle; to haul; to transfer; 2. to deliver *(pump)*; 3. *(mine)* to hoist, to wind, to draw up, to raise; to mine, to produce; 4. to further, to promote
abwärts f. to lower, to descend
im Tagebau f. to mine by the open-cast method
Förderpumpe *f* fuel [supply] pump, feed pump
Förderrichtung *f* direction of discharge *(pump)*
Förderrichtungsumkehr *f* delivery reversal *(pump)*
Förderrinne *f* conveying chute
Förderschacht *m* lift-well; *(mine)* coal (drawing, winding) shaft
Förderschale *f (mine)* drawing frame (cage)
Förderschaukel *f (mach)* carrier
Förderschicht *f (mine)* production shift
Förderschnecke *f* screw (spiral, worm) conveyor; *(ag)* auger conveyor
Förderseil *n* hoisting rope (cable, line); *(mine)* winding (hoisting, head) rope
Förderseite *f* pressure side *(pump)*
Fördersohle *f (mine)* haulage (tramming) level
Fördersoll *n (mine)* target production, production target
Fördersonde *f* producing well
Förderspeicher *m* storage conveyor (bank) *(transfer machine)*
Förderspieldauer *f* cycle time
Förderstrecke *f (mine)* haulage road (drift), haulageway, *(Am)* haulway
Fördertechnik *f* hoisting and conveying, conveying engineering; materials handling engineering *(manufacturing)*
Fördertour *f* oil string
Fördertrommel *f* calf wheel

Fördertrum *m* cage (hoisting, skip, winding) compartment, cageway
Förderturm *m (mine)* head frame, headgear, winder (winding) tower
Förderüberhauen *n (mine)* chute raise, mill
Förderung *f* 1. conveyance, conveying; transport, haulage; *(mach)* handling; 2. delivery *(pump)*; yield; 3. *(mine)* hoisting, winding, raising, drawing-up; output
F. aus der Vorrichtung *(mine)* development stoping
F. je Mann und Schicht *(mine)* output per man-shift, O.M.S.
F. mit Gegengewicht *(mine)* counterweight hoisting
F. mit geschlossenem Seil *(mine)* endless-rope haulage
F. mit Seilgewichtsausgleich *(mine)* balanced hoisting
F. mit Vorder- und Hinterseil *(mine)* tail-rope haulage
absatzfähige F. *(mine)* saleable output
gebrochene F. *(mine)* stage hoisting
mechanische F. mechanical handling
verwertbare F. *(mine)* disposable (net) output
Förderungsvolumen *n* / **theoretisches** theoretical swept volume *(pump)*
Fördervolumen *n* volumetric delivery *(pump)*
tatsächliches F. actual volumetric delivery
Förderwagen *m* dog, hutch, [mine] car
F. mit Mitnehmer grip car
kleiner F. mine tub, *(Am)* buggy
Förderzug *m (mine)* haulage run *(coal)*; slushing run *(stone)*; trip *(shaft winding)*
Fördeschiff *n* fiordvessel
Forke *f* [pitch] fork
Form *f* 1. form, shape; appearance; fashion; 2. *(build)* form[work], mould; 3. *(met)* mould; die
abgeplattete F. oblate shape
abgewandelte F. modified form
biegesteife F. rigid mould *(prestressed concrete)*
bilineare F. bilinear form
geometrische F. geometrical configuration
geschlossene F. *(math)* closed form
gestreckte F. prolate shape
hexagonale F. hexagonal modification
verlorene F. *(met)* sand mould
Formabrichteinrichtung *f* form truing device *(grinding)*
Formabweichung *f* error of form
Formaldehyd *m* formaldehyde
Formaldehydharz *n* formaldehyde resin
Formalin *n* formalin
Formänderung *f* deformation, change of shape, strain
bleibende F. permanent deformation, set
Formänderungsarbeit *f* work of deformation
Formänderungsenergie *f* energy of deformation
Formänderungsfähigkeit *f* ductility, deformability
Formänderungskomponente *f* component of strain
Formänderungsverfahren *n* deformation method
Formänderungswiderstand *m* forming resistance
Formänderungszustand *m* state of deformation
dreiachsiger F. general case of strain
räumlicher F. general case of strain
Formart *f s.* Aggregatzustand
Format *n* format *(book)*; size; size of sheet
beschnittenes F. trimmed size
breites F. oblong size
Formation *f (geol)* formation
Formatleiste *f* former strip
Formatreglette *f* forme reglet
Formatsteg *m* furniture
Formätzen *n* contour etching, chemical milling
Formbacke *f* profiling roller
formbar formable; *(Am)* workable
Formbarkeit *f* formability; *(Am)* workability
Formbeständigkeit *f* 1. *(plast)* dimensional stability; 2. *(text)* shape retention
F. der Drapierung drape retention
Formbett *n (print)* type-bed
Formblatt *n* form
Formdeplacement *n (ship)* moulded displacement
Formdiamant *m* diamond wheel dresser; profile dressing diamond
Formdrehen *n* form turning
Formdrehmaschine *f* form turning lathe
Formdrehmeißel *m* form[ed] turning tool
Formeisen *n* sectional iron
Formeisenwalzwerk *n* girder and section mill
Formel *f (ch, math)* formula
Boltzmannsche F. Boltzmann equation
chemische F. chemical formula
Fresnelsche F. Fresnel reflection formula
Kardanische F. *(math)* Cardan's solution
Formelsprache *f* symbolic language
formen 1. to shape, to form; to work; *(plast)* to fabricate; 2. to mould *(foundry)*
Formen *n* **ohne Modell** patternless moulding
Formenanschnitt *m* gate of mould
Formenschließen *n (print)* quoining, locking of the forme
Formenschlüssel *m (print)* quoin key
Formenschwärzung *f* **mit Graphit** *(met)* graphite-blacking
Formentransportwagen *m (print)* forme trolley
Formentrockenofen *m* mould drying oven
Formen- und Gesenknachformfräsmaschine *f* mould and die duplicator
Formenzonenring *m* tuyere bolt *(blast furnace)*

Former *m* moulder *(foundry)*
Formerbank *f* snap bench
Formerei *f* 1. moulding; 2. moulding shop
Formerlehm *m* moulding clay
Formermaßstab *m* moulder's rule
Formerstift *m* sprig, moulder pin
Formerwerkzeug *n* moulder's tool
Formfaktor *m (el)* form factor
Formfaktormeßbrücke *f* form factor indicator
formfest inelastic
Formfläche *f* mould face
Formfräsen *n* form milling
Formfräser *m* form[ed milling] cutter, cutter of the former type
einzahniger F. single-toothed form cutter
hinterdrehter F. form-relieved cutter
profilgeschliffener F. form profile cutter, profile-type form cutter
spitzverzahnter F. form-ground cutter
Formfräsmaschine *f* form milling machine
Formgebung *f* shaping, forming; *(Am)* working
spanende F. *s.* spanende Formung
spanlose F. *s.* spanlose Formung
formgenau geometrically true, accurate to geometry
f. machen to true up
Formgenauigkeit *f* truth (accuracy) of shape, geometrical accuracy
formgerecht fully fashioned
Formgesenk *n* shaped die
formgleich identical in shape
Formgußstück *n* casting
Formhälfte / obere top flask *(foundry)*
untere F. drag, bottom-half mould *(foundry)*
formhinterdreht form machine-relieved *(cutter)*
Formiat *n (ch)* formiate, formate
Formierung *f s.* Formung
Formierungsspannung *f (el)* formation voltage
Formkanal *m* pressing canal
Formkasten *m* moulding box, [foundry] flask
F. mit Sandrippe flask with flange
mehrteiliger F. more-part flask
zweiteiliger F. two-part flask
Formkastenklammer *f* moulder's clamp
Formkastenschore *f* flask bar
Formkastentraverse *f* flask bar
formkennzeichnend indicating [a] specific shape, indicating specific shapes
Formkern *m* mould core
Formkopier- und Profilfräsmaschine *f* form copying and profile miller
Formkurve *f* hydrostatic curve *(of a vessel)*
Formläppen *n* lapping specially formed surfaces
Formlegen *n* form setting
Formlehre *f* form [tool] gauge, profile gauge
Formlineal *n* former plate *(grinding wheel)*
formlos 1. shapeless; amorphous; 2. informal *(writing)*
Formlosigkeit *f* shapelessness; amorphousness
Formmaschine *f* 1. moulding machine *(foundry)*; 2. *(text)* forming machine
F. für Besatznudeln *(mine)* stemming extrusion press
Formmaskenverfahren *n* shell moulding process
Formmeißel *m (mach)* form[ing] tool, contouring tool
Formpressen *n* pressure (compression) moulding
Formpreßholz *n* moulded plywood
Formpreßling *m* laminated moulding
Formpreßstoff *m* compression moulding material
Formpuder *m* parting medium *(foundry)*
Formrahmen *m* chase; mould frame *(papermaking)*
Formräumwerkzeug *n* form broach
Formrevolverdrehmaschine *f* forming turret lathe
formrichtig properly formed (shaped), geometrically true
Formsand *m* foundry (moulding) sand
feuchter F. damped moulding sand
Formsandaufbereiter *m* sand muller *(foundry)*
Formsandaufbereitung *f* sand cutting, preparing sand for moulding
Formsandbindemittel *n* binder of moulding sand
Formsandschüttelsiebmaschine *f* reciprocating moulding sand riddle
Formscheibenstahl *m* circular forming tool
Formschleifen *n* form (contour) grinding
Formschlichte *f* dressing
formschlüssig positive, solid, conjugate; closed *(pair)*
Formschmieden *n* precision forging
Formschrämarm *m (mine)* curved jib
Formschwärze *f* moulding blackening
Formschwerpunkt *m* centre of buoyancy *(of a vessel)*
Formschwerpunktkurve *f* curve of [centre of] buoyancy
Formspule *f* preformed coil
Formspulenwicklung *f* preformed winding
Formstabilität *f (ship)* hydrostatical (surface) stability
Formstahl *m* structural steel shape, steel section, sectional steel
schwerer F. heavy sections
Formstahlwalzwerk *n* structural rolling mill
Formstampfer *m* moulding pestle *(foundry)*
Formstein *m (build)* shaped brick
Formstich *m (met)* shaping pass
Formstift *m* moulder's pin *(foundry)*
Formstück *n* [fabricated] shape, form, shaped part; fitting *(piping)*, making-up piece, adapting (template) pipe
Formteil *n* formed part; *(plast)* moulded part, moulding; blank *(forming)*

Formularzuführung *f* form feeding
Form- und Stempelhobelmaschine *f* punch and form shaper, form and punch shaping machine
Formung *f* deformation, forming, formation, shaping *(of pulses)*
spanende F. chip-producing formation method
spanlose F. non-cutting shaping; *(Am)* chipless shaping
Formungenauigkeit *f* geometric inaccuracy
Formungsfähigkeit *f* formability
Formverfahren *n* forming process *(gear)*
Formverlauf *m* run *(of a vessel)*
Formwälzmethode *f* formate cutting process *(gear)*
Formwälzwerkzeug *n* circular broach
Formwandler *m (el)* wave converter
Formwerkzeug *n (plast)* forming die
Formwiderstand *m* form (body) drag *(fluid flow)*; *(ship)* hull (wave-making) resistance
formylieren *(ch)* to formylate
Formzeug *n* forming tools
Formziegel *m* shaped (purpose-made) brick
Formzylinder *m (print)* forme (plate) cylinder
Forscher *m* researcher, research worker, investigator
Forschung *f* research, investigation
angewandte F. applied research
Forschungsabteilung *f* research department
Forschungsanlage *f* research laboratory (installation), investigation plant
Forschungsarbeit *f* research activities (work)
Forschungsbericht *m* research report
Forschungschemiker *m* research chemist
Forschungsergebnis *n* research result
Forschungsergebnisse in der Praxis anwenden to adapt the laboratory information to everyday use
Forschungsgebiet *n* sphere (area) of research
Forschungsgegenstand *m* object of research
Forschungsingenieur *m* research engineer
Forschungsinstitut *n* research institute
Forschungskanal *m s.* Experimentierkanal
Forschungslaboratorium *n* research laboratory
Forschungsmikroskop *n* research microscope
Forschungsmittel *npl* research funds (grants), spending on research
Forschungsprogramm *n* research scheme (programme)
Forschungsrakete *f* research rocket
Forschungsrat *m* research council
Forschungsreaktor *m* research reactor
Forschungssatellit *m* research satellite
Forschungsschiff *n* exploration (research) vessel
Forschungsstand *m* state of the art (research activities)
Forschungstätigkeit *f* research activity
Forschungs- und Entwicklungsabteilung *f* research and development department
Forschungsvorhaben *n* research project, project of research
Forst *m* forest
Forstamt *n* local forestry administration
Forstarbeiter *m* wood worker, woodsman
Forstbenutzung *f* forest utilization
Forstbetrieb *m* forest management
Förster *m* forester
Förstersonde *f* Foerster probe, second harmonic magnetic modulator
Forstgarten *m* seedling enclosure (nursery), nursery for trees
Forstkultur *f* forestry, silviculture
Forstschädlinge *mpl* forest destroyers (pests)
Forstwesen *n (ag)* forestry
Forstwirtschaft *f* forestry
Forstwirtschaftsjahr *n* forest year
Forstwissenschaft *f* science of forestry
fortbewegen to move on (away)
mit der Stange f. *(ship)* to boom-off
fortdauern to continue
fortfahren 1. to go (drive) away; 2. to continue
fortlaufend continuous
fortleiten to conduct
Fortleitungswiderstand *m* resistance to conduction
fortpflanzen / sich to propagate, to be transmitted *(wave)*
Fortpflanzung *f* 1. propagation *(wave)*; 2. *(ag)* reproduction
F. des Lichtes propagation of light
Fortpflanzungsgeschwindigkeit *f* speed of propagation
Fortpflanzungskonstante *f (ph)* propagation constant (coefficient)
Fortpflanzungsrichtung *f* direction of propagation *(wave)*
Fortpflanzungswelle *f* wave of translation
Fortschalteinrichtung *f* continuous-switching device
Fortschaltmagnet *m* stepping magnet *(contactor control)*
Fortschaltrelais *n* accelerating (connecting, cut-through, notching) relay
Fortschaltwerk *n (telec)* stepping mechanism
fortschreiten to advance, to make progress
Fortschritt *m* 1. progress; 2. improvement
Fortschrittsgeschwindigkeit *f* speed of advance *(propeller)*
Fortschrittsgrad *m* degree of advance *(propeller)*
Fotaktor *m* photactor
fotoaktiv photoactive
Fotoamateur *m* photo amateur
Fotoanregung *f* photoexcitation
Fotoapparat *m* camera, photographic apparatus
Fotoausrüstung *f* photographic equipment

Fotobelichtungsmesser *m* photographic exposure meter
Fotoblitz *m* photoflash
Fotochemie *f* photochemistry
Fotochemikalie *f* photographic chemical
fotochemisch photochemical
Fotodetektor *m* photoconductive detector
Fotodiffusion *f* photodiffusion
Fotodiode *f* photodiode
Fotodissoziation *f* photodissociation
Fotodruck *m* photographic print
Fotoeffekt *m* photoeffect
äußerer F. photoemissive (external photoelectric) effect
innerer F. photoconductive (internal photoelectric) effect
selektiver F. selective photoelectric emission
Fotoeinwickelpapier *n* black photo paper
fotoelastisch photoelastic
fotoelektrisch photoelectric
Fotoelektron *n* photoelectron
Fotoelektronenerzeugung *f* photoelectron production
Fotoelektronenvervielfacher *m* photoelectron multiplier, photomultiplier
Fotoelement *n* photovoltaic cell
elektrolytisches F. photochemical cell
Fotoemission *f* photoelectric emission, photoemission
Fotoemitter *m* photoemitter
fotoempfindlich photosensitive
Fotoformer *m* cathode-ray function generator, photoformer *(analogue computer)*
Fotograf *m* photographer
Fotografie *f* photo[graphic picture]
fotografieren to photograph, to shoot
fotografisch photographic
Fotogrammeter *m* photogrammetrist
Fotogrammetrie *f* photogrammetry
terrestrische F. terrestrial photogrammetry
Fotoionisation *f* photoionization
Fotokatalyse *f* photocatalysis
Fotokatode *f* photocathode, photoemitter
Fotokernreaktion *f* photonuclear reaction, [nuclear] photodisintegration
Fotokernspaltung *f s.* Fotospaltung
Fotokopie *f* photoprint, photostat, photostatic (photographic) copy
fotoleitend photoconductive
Fotoleiter *m* photoconductor
Fotoleitfähigkeit *f* photoconductivity
Fotoleitung *f* photoconduction
Fotoleitwert *m* photoconductance
Fotoleitzelle *f* photoconductive cell
Fotolinie *f (nucl)* photopeak
Fotolumineszenz *f* photoluminescence
Fotolyse *f* photolysis
Fotolyseprodukt *n* photolyte
Fotomaschinengewehr *n* gun-camera
Fotometer *n* photometer
integrierendes F. integrating photometer
Fotometeraufsatz *m* photometer head
Fotometerbank *f* photometer bench
Fotometerschirm *m* photometer screen
Fotometrie *f* photometry
physikalische (objektive) F. physical photometry
visuelle (subjektive) F. visual photometry
fotometrisch photometric
Fotomikrografie *f* photomicrography
Fotomikroskopie *f* photomicroscopy
Fotomontage *f* photomontage
Fotomorphose *f* photomorphosis
Fotomultiplier *m s.* Fotovervielfacher
Fotoneutron *n* photoneutron
Fotoneutronenquelle *f* photoneutron source, photosource of neutrons
Fotoobjektiv *n* photolens
lichtstarkes F. fast photolens
Fotopapier *n* photographic paper
Fotopeak *m (nucl)* photopeak
Fotoplatte *f* photoplate, photographic plate
Fotoplattentechnik *f* photographic plate technique
Fotopolymerplatte *f (print)* photopolymer plate
Fotoproton *n* photoproton
Fotosatz *m* photocomposition
Fotoschicht *f* photosensitive surface; photolayer
Fotoschichtspur *f* emulsion (photolayer, photographic) track
Fotospaltung *f* photofission
Fotospannung *f* photovoltage, photopotential
Fotostrom *m* photoelectric current, photocurrent
Fotosynthese *f* photosynthesis
Fototheodolit *m* phototheodolite
Fototräger *m* photocarrier
Fototransistor *m* phototransistor
Fotovervielfacher *m* photomultiplier cell (tube), photomultiplier, photoelectric multiplier
Fotovervielfachercharakteristik *f* multiplier-gain-versus-voltage power law
fotovoltaisch photovoltaic
Fotovolteffekt *m* photovoltaic effect
Fotowiderstand *m* photoconductive cell, photoresistor; photoresistance
Fotozelle *f* 1. photocell, photoelectric cell; 2. photoemissive cell
F. mit Elektronenvervielfacher multiplier phototube
gasgefüllte F. gas phototube
Fotozellenkompensator *m* photoelectric-type potentiometer
Fotozellenverstärker *m* photoelectric cell amplifier

Fotozerfall *m* photodisintegration
Föttingergetriebe *n* Föttinger converter
Foucault-Strom *m* eddy (Foucault) current
Foulard *m* padding machine
Foulardfärberei *f (text)* dyeing on the padder
Foulardierlösung *f (text)* padding liquor
Fourier-Analyse *f* Fourier (harmonic) analysis
Fourier-Entwicklung *f* Fourier expansion
Fourier-Reihe *f* Fourier series
Fournisseurrad *n (text)* feedwheel
Fracht *f* 1. freight, cargo, load; 2. freightage *(cost of conveyance)*
F. in Behältern containerized cargo
F. in Überweisung carriage forward (charges to pay)
Frachtaufzug *m* freight lift
Frachtbrief *m* consignment note, way-bill
internationaler F. international consignment note
Frachtdampfer *m* cargo (freight) steamer, freighter
Frachtewer *m* sailing barge
Frachtkahn *m* barge
Frachtkarte *f* way-bill, invoice
Frachtkosten *pl* cargo rates
Frachtmotorschiff *n* cargo motorship (motor vessel)
Frachtraum *m (ship)* deadweight tonnage
Frachtsatz *m* freight rate
Frachtschiff *n* freighter, cargo vessel
F. in der Linienfahrt cargo liner
Frachtschiffmakler *m* loading broker
Frachttonnage *f (ship)* deadweight (freight) tonnage
Frachttonne *f* freight (shipping) ton, ton of shipping
Fracht- und Fahrgastschiff *n* cargo and passenger vessel, cargo-passenger liner, mixed (intermediate) vessel
Frachtvertrag *m* charter [party]
Frachtvorschuß *m* advance freight
Fragebogen *m* questionnaire
Fraktion *f (ch)* fraction; category
abgereicherte F. depleted (stripped) fraction
Fraktionieraufsatz *m* fractionating column
fraktionieren to fractionate
Fraktionieren *n* fractionation *(dose)*; fractionating; fractional distillation
Fraktionierkolben *m* fractional distillation flask
Fraktionierkolonne *f* fractionating column
fraktioniert fractional *(distillation)*
Fraktionierturm *m* bubble tower
Fraktionierung *f* fractionation; fractional distillation; dose fractionation
Francis-Turbine *f* Francis turbine
Frankaturrechnung *f* recharge note
Frankaturzwang *m* compulsory prepayment of charges
Franklinit *m (min)* franklinite
Fransenwirkung *f (telev)* fringe effect
Franzbranntwein *m* French brandy
Franzium *n* francium
Fräsarbeit *f* milling operation (job)
Fräsbereich *m* cutter field
Fräsblech *n (text)* comb plate
Fräsdorn *m* cutter arbor
Fräsdornmutter *f* cutter arbor nut
Fräsdornring *m* milling machine arbor collar
Fräsdruck *m* milling thrust
Fräsdurchmesser *m* milling diameter
größter F. maximum milling diameter
fräsen to mill
im Umriß f. to contour *(mill)*
Fräsen *n* milling
F. auf genaue Toleranz close-tolerance milling
F. geneigter Schlitze angular gashing
F. im Wälzverfahren hobbing
F. in den drei Ebenen cavity milling
F. in Querrichtung cross milling
F. kreisförmiger Auskehlungen circular recessing
F. mit Hartmetallwerkzeug carbide milling
F. schräger Flächen angular milling
pausenloses F. continuous milling
ununterbrochenes F. continuous milling
Fräser *m* [milling] cutter; mill
F. aus Gußlegierung cast-alloy cutter
F. für Halbkreisprofil concave (convex) milling cutter
F. mit eingesetzten Zähnen built-up type cutter
F. mit gefrästen Zähnen fluted cutter
F. mit grober Zahnteilung coarse-pitch cutter
F. mit Grobverzahnung coarse-tooth cutter
F. mit hinterdrehten Zähnen formed-tooth cutter
F. mit Pfeilverzahnung right-and-left-hand helix cutter, cutter with opposing helices
F. mit Rechtsdrall right-hand helix cutter
F. mit Schaft shank cutter, shank-type mill
F. mit schraubenförmiger Verzahnung cutter with helical teeth
F. zum Aufstecken auf fliegend angeordneten Dorn cutter for stub-arbor mounting
geradverzahnter F. cutter with straight teeth
hartmetallbestückter F. carbide-tipped cutter
keramischer F. ceramic tool for milling
kreuzverzahnter F. alternate helical tooth cutter, *(Am)* alternate tooth milling cutter
linksdrehender F. top-going cutter
mehrzahniger F. multi-toothed milling cutter
rechtsdrehender F. top-coming cutter
zylindrischer F. cylindrical cutter

Fräserabhebung *f* [milling] cutter relief, retraction of the milling head
Fräserachse *f* [milling] cutter axis
Fräserbohrung *f* cutter bore
Fräserdrehrichtung *f* hand of milling cutter rotation
Fräserdrehzahl *f* [milling] cutter r.p.m., milling [cutter] speed
Fräserkörper *m* [milling] cutter body
Fräsernachschliff *m* [milling] cutter sharpening
Fräsersatz *m* [milling] cutter gang
Fräserschaft *m* [milling] cutter shank, end mill shank
Fräserschleifmaschine *f* [milling] cutter grinder, [milling] cutter sharpening machine
Fräserstandzeit *f* [milling] cutter life
Fräserübergangshülse *f* [milling] cutter adapter
Fräserverschleiß *m* [milling] cutter wear
Fräserwelle *f* [milling] cutter spindle
Fräserwiege *f* [milling] cutter cradle
Fräserzahn *m* [milling] cutter tooth
Fräserzahnteilung *f* [milling] cutter pitch
Fräskopf *m* milling [spindle] head; [milling] cutter head
  F. mit eingesetzten Messern cone-type face milling cutter with inserted blades
Fräskörperschlitz *m* body slot
Fräskühlmittel *n* coolant for milling
Fräsleistung *f* milling capacity
Fräsmaschine *f* milling machine, miller
  F. mit automatischem Arbeitsablauf automatic cycle type of milling machine
  F. mit drehbarem Spindelkopf swivel head milling machine
  F. mit schwenkbarem Spindelkopf universal head milling machine
  F. mit verstellbarer Traverse adjustable rail milling machine
Fräsmaschinendorn *m* milling machine arbor
Fräspinole *f* milling quill
Fräsring *m* hollow mill
Frässchablone *f* milling pattern (template)
Frässcheibenlader *m* fluted-disk loader, rotary-disk loader
Fraßgift *n (ag)* stomach poison
Frässpan *m* milling chip
Frässpäne *mpl* millings
Frässpindel *f* milling [cutter] spindle
Frässpindeldrehzahl *f* [milling] cutter spindle speed
Frässpindelhülse *f* [milling] cutter spindle quill
Frässpindelstock *m* milling spindle head, [milling] cutter head
Frässupport *m* milling head *(planer)*
Frästeil *n* part to be milled; milled part
Frästisch *m* milling table
Fräs- und Schabmaschine *f* plate milling and shaving machine
Träsverfahren *n* **zur Schneckenradherstellung** worm-wheel hobbing process
Fräsvorrichtung *f* milling fixture
Fräsvorschub *m* milling feed
Fräsweg *m* cutting (milling) path; milling length
  gesamter F. total table travel *(milling machine)*
Fräszeit *f* cutting time
Fregatte *f* frigate
frei free; off ground *(aeroplane)*; vacant *(position)*; disengaged; open *(space)*; detached *(house)*; clear *(rail-line)*; floating *(in water)*; free of charge
  f. an Bord free on board
  f. Bahnwagen free wag[g]on
  f. beweglich sein to be free to move
Freiantenne *f* outdoor aerial
Freiauslösung *f* trip-free release
Freiausstellungsfläche *f* open-air exhibition space
Freiballon *m* free balloon
Freibauten *pl* outdoor (open-air) plant
Freibergit *m (min)* argentiferous tetrahedrite
freiberuflich free-lance *(artist)*
Freibord *m* freeboard
  F. / mit großem high-freeboarded *(vessel)*
Freibordbestimmung *f* assignment of freeboard
Freibordhöhe *f* freeboard depth
Freibordmarke *f* freeboard marking
Freibordtiefgang *m* freeboard draught
Freibrief *m* charter
Freideck *n* weather (freeboard) deck
Freiexemplar *n* complimentary (free) copy
Freifahrschein *m* free pass
Freifahrtversuch *m (ship)* open water test
Freifallbeschickung *f* gravity feed
Freifallklassierer *m* unhindered-settling classifier
Freifallmischer *m* gravity mixer, rotary (revolving) drum mixer
Freifläche *f* 1. open-air space *(fair)*; 2. flank, surface below the cutting edge; 3. *(mine)* free face
  F. am Umfang diametral clearance *(twist drill)*
  F. an der Schneide flank
  F. an der Reibahle secondary clearance
  gewetzte F. primary clearance *(tool)*
Freiflächengestaltung *f* landscaping, landscape planning
Freiflächenverschleiß *m* [tool] flank wear, [tool] flank abrasion
Freiflächenverschleißfase *f* [tool] flank wear land
freifließend free-running
Freiflugbahn *f* free[-flight] trajectory
Freiformschmieden *n* hammer forging

Freiformschmiedestück *n* smith (hand, hammered, flat-die, blacksmith) forging

Freifräsung *f* **am Umfang** land clearance

Freigabe *f* 1. *(mach)* release; 2. *(ch)* liberation

F. einer Strecke section clearing

F. für den Anflug clearance for approach

F. von Schutzrechten restitution of protective rights

Freigabeschalter *m* clear (release) switch

Freigabezeichen *n* release guard signal

freigeben 1. *(mach)* to release; 2. *(ch)* to liberate; 3. to clear *(computer)*

freigelegt exposed, uncovered

Freihafen *m* free port

Freihandbohren *n* hand boring

Freihandlinie *f* free-hand line

Freihandschleifen *n* off-hand grinding

Freihandschleifmaschine *f* off-hand grinding machine

Freihandzeichnen *n* freehand drawing

Freiheitsgrad *m* degree of freedom

Freikennzeichen *n* ringing tone signal

Freikolbengasturbinenanlage *f* free-piston gas turbine plant

Freikolbenmaschine *f* free-piston engine

Freikolbenmotor *m* free-piston-type engine

Freikolbentreibgaserzeuger *m* free-piston gasifier

Freikolbenturbinentriebwerk *n* **für Kraftfahrzeuge** automotive free piston-turbine engine

Freikolbenverdichter *m* free-piston engine compressor

Freikolbenvergaserturbinenanlage *f* free-piston gas generator turbine

Freilager *n* transit (bonded) warehouse

Freilampe *f (telec)* free-line signal

Freilauf *m* freewheel; freewheeling

Freilaufaußenring *m* freewheel outer ring

Freilaufgehäuse *n* freewheel housing

Freilaufinnenring *m* freewheel inner ring

Freilaufklemmrolle *f* freewheel brake roller

Freilaufklemmrollenkäfig *m* freewheel brake roller cage

Freilaufkupplung *f* freewheeling clutch

Freilaufschaltung *f* freewheel change

Freilaufsperre *f* freewheel lock

Freilaufsystem *n (telev)* free running

freilegen to uncover

Freileitung *f* overhead transmission line, open-wire line

F. auf Holzmasten overhead (open-wire) pole line

elektrische F. overhead electric power line

Freileitungsanlage *f* overhead (open-wire) system

Freileitungsbau *m* open-line construction

Freileitungshängekette *f* overhead-line insulator chain

Freileitungsisolator *m* overhead (line) insulator

Freileitungskabel *n* overhead cable

Freileitungskreis *m* open-wire circuit

Freileitungsmontage *f* erection of overhead lines

Freileitungsnachbildung *f* open-line balancing network

Freileitungsseil *n* stranded conductor for overhead lines

Freiluftanlage *f (el)* outdoor (open-air) plant

Freiluftdosis *f (nucl)* [free] air dose, in-air dose

Freiluftisolator *m* outdoor insulator

Freiluftöldurchführung *f* outdoor oil-bushing

Freiluftschaltanlage *f* outdoor substation

Freiluftschalter *m* outdoor circuit breaker

Freiluftstation *f* outdoor substation

Freilufttransformator *m* outdoor (open-air) transformer

Freiluftumspanner *m* outdoor (open-air) transformer

Freiluftunterwerk *n* outdoor substation

Freiluftversuch *m* open-air test *(corrosion)*

Freimachen *n* **der Leitung** line clearing *(telephony)*

Freimachung *f* **von Elektronen** detachment of electrons

Freischaltung *f* clearing *(telephony)*

freischwimmen to float off *(vessel)*

Freischwingerlautsprecher *m* free swinging loudspeaker

Freischwingsieb *n* flexible-drive screen

freisetzen *(ch)* to set free, to liberate

Freisetzung *f* **von Kernenergie** production of nuclear power

Freisignal *n (ry)* clear signal

Freisitz *m* terrace

freistehend free-standing; detached

Freistrahl *m* free (open) jet

Freistrahlschalter *m* open jet circuit breaker

Freistrahlturbine *f* free jet turbine

Freistrahlwindkanal *m* open wind tunnel

Freiton *m (telec)* audible ringing signal

freitragend self-supporting; cantilevered

Freiträger *m* cantilever [beam], cantilevered (corbel) beam, semi-beam

Freitreppe *f* fliers

Freivorbau *m* cantilever construction (method) *(bridge construction)*

Freiwache *f* off-watch, watch below

Freiwerden *n* **von Wärme** liberation of heat

Freiwinkel *m* angle of clearance (relief), clearance (relief) angle; back-off angle (broach)

F. / ohne unrelieved

F. α **beim Spiralbohrer** angle of lip clearance with workpiece face, lip relief angle

Freizeichen *n* free-line signal

Fremdatom *n* impurity (foreign) atom

Fremdbeimengung *f* impurity addition

Fremdbelüftung *f* forced ventilation

Fremdberge *pl (mine)* imported dirt
Fremdbergeversatz *m (mine)* excess filling
Fremdbestrahlung *f* external exposure (irradiation)
Fremdeisen *n* tramp iron
Fremdelektron *n* stray electron
Fremderregung *f* foreign (external) excitation; *(el.m.)* separate excitation
Fremdfeld *n* interfering (external, extraneous, separate, foreign) field
Fremdgas *n (ch)* foreign gas
Fremdgasbeheizung *f* underfiring from external producers
Fremdkörper *m* foreign matter (substance)
Fremdkörpereinschluß *m* foreign-body enclosure
Fremdleiter *m* extrinsic semiconductor
Fremdlöschung *f* external quenching
Fremdlüftung *f* forced ventilation
Fremdmaterial *n* foreign matter (substance)
Fremdmodulation *f* external modulation
Fremdortung *f* ground position finding
Fremdpeilung *f* ground direction finding
Fremdschicht *f* pollution (contamination) layer
Fremdschichtprüfung *f* pollution test
Fremdschrott *m* external scrap
Fremdsignal *n* external signal
Fremdspannung *f* extraneous voltage; external voltage
Fremdspannungsmesser *m* noise-level meter
Fremdspannungsmessung *f* noise-level measuring
Fremdspeicher *m* external memory
Fremdspeicherung *f* secondary storage
Fremdsteuerung *f* external control
Fremdstoff *m* foreign matter (substance)
Fremdstrom *m* parasitic (external, stray) current
Fremdversatz *m (mine)* imported stowing
Fremdzündung *f* spark (applied) ignition
Frenkel-Defekt *m (ph)* Frenkel defect
Frenkel-Fehlordnung *f (ph)* Frenkel disorder
Frequenz *f* frequency
  F. / von gleicher equifrequent
  F. des Bildsignals video frequency
  F. im Kennlinienknick break frequency
  augenblickliche F. instantaneous frequency
  eingestellte F. set frequency
  genehmigte F. authorized frequency
  gestörte F. interfered frequency
  kritische F. critical frequency
  momentane F. instantaneous frequency
  natürliche F. natural frequency
  negative F. negative frequency
  störende F. interfering (interference causing) frequency
  technische F. industrial (commercial) frequency
  veränderliche F. variable frequency
frequenzabhängig variable with frequency, frequency-dependent
Frequenzabhängigkeit *f* dependency of frequency
Frequenzabstand *m* distance between frequencies, frequency separation (spacing), mode separation
Frequenzabweichung *f* frequency deviation (drift)
  F. beim Anheizen warm-up drift
  F. einer Oszillatorfrequenz oscillator drift
  höchstzulässige F. maximum system deviation
  zulässige F. frequency tolerance
Frequenzanalysator *m* wave analyzer
Frequenzänderung *f* frequency change; frequency variation
Frequenzangabe *f* frequency designation
Frequenzanhebungsfilter *n* emphasizer
Frequenzanzeiger *m* frequency meter (indicator)
Frequenzauflösung *f* frequency discrimination
Frequenzauslenkung *f* frequency swing
Frequenzband *n* frequency band
  schmales F. narrow frequency range (band)
  zugeteiltes F. service band
Frequenzbandausbeute *f* wide-band ratio
Frequenzbandbegrenzung *f* limitation of frequency band
Frequenzbereich *m* frequency range
  kommerzieller F. communication band
Frequenzcharakteristik *f* bandwidth characteristic
Frequenzdemodulator *m* discriminator
  F. mit abgestimmten Einzelkreisen double-tuned circuit-type discriminator
Frequenzdifferenzdetektor *m* frequency difference detector
Frequenzdoppler *m* frequency doubler
Frequenzdrift *f* frequency drift, drift of frequency
Frequenzeinheit *f* unit of frequency
Frequenzeintragung *f* frequency recording
frequenzempfindlich frequency-sensitive
Frequenzentzerrung *f* response equalization
Frequenzerniedrigung *f* frequency division
Frequenzfehler *m* frequency error
Frequenzfernmeßverfahren *n* frequency telemetering
Frequenzfilter *n* separating filter
Frequenzgang *m* frequency response
  F. der Amplitude amplitude-frequency characteristic
  F. der Phase phase-frequency characteristic
  F. des geschlossenen Kreises closed-loop frequency response
  F. in logarithmischer Darstellung Bode-diagram
  linearer F. flat frequency response
Frequenzganganalysator *m* transfer function analyzer

Frequenzgangmesser *m* transfer function analyzer, TFA
Frequenzgemisch *n* frequency composition
Frequenzgenauigkeit *f* frequency accuracy
frequenzgleich straight-line frequency
Frequenzgleichrichter *m* discriminator
Frequenzgleiten *n* frequency change [of a magnetron]
Frequenzhalbierung *f* frequency halving
Frequenzhilfsnormal *n* frequency sub-standard
Frequenzhub *m* frequency deviation (swing)
Frequenzhubdauer *f* sweep duration
Frequenzhubgenerator *m* sweep generator
Frequenzkennlinie *f* frequency response characteristic (curve)
Frequenzkomponente *f* component frequency
Frequenzkonstanthaltung *f* frequency stabilization
Frequenzkonstanz *f* frequency (circuit) stability, constancy (stability) of frequency
  F. über kurze Zeit short-time stability *(of frequency)*
Frequenzkontrollgerät *n* frequency monitor
Frequenzkontrolluhr *f* frequency checking clock
Frequenzkorrekturfilter *n* equalizer
Frequenzkurve *f* frequency response characteristic (curve)
frequenzlinear straight-line frequency
Frequenzmarkengenerator *m* frequency marker generator
Frequenzmarkierungsimpuls *m* frequency marker
Frequenzmesser *m* frequency meter (indicator)
  F. nach dem Absorptionsverfahren absorption-type frequency meter
  F. nach dem Durchgangsverfahren transmission frequency meter
  integrierender F. master frequency meter
Frequenzmeßwandler *m* frequency transfer unit
Frequenzmodler *m* frequency modulator
Frequenzmodulation *f* frequency modulation, FM
  F. mit Hilfsträger sub-carrier frequency modulation
Frequenzmodulatorstufe *f* discriminator stage
Frequenzmultiplex *n* frequency-division multiplex
Frequenznachlauf *m* automatic frequency control
Frequenznormal *n* frequency standard
Frequenzplan *m* group allocation
Frequenzregler *m* frequency controller
Frequenzregulierung *f* frequency control
Frequenzrelais *n* frequency relay
Frequenzschachtelung *f* frequency interlacing (multiplexing, overlap)
Frequenzschnitt *m* crossover of frequencies
Frequenzschwankung *f* frequency variation
Frequenzselektion *f* frequency discrimination
Frequenzsieb *n* filter
Frequenzspeicher *m* frequency memory
Frequenzspektrograf *m* frequency spectrograph, panoramic monitor
Frequenzspektrum *n* compass of frequencies, frequency spectrum
  F. einer Impulsreihe pulse-train spectrum
Frequenzspringen *n* moding
Frequenzsprung *m* frequency jumping
frequenzstabil stable in frequency
Frequenzstabilisierung *f* frequency stabilization
Frequenzsteuerung *f* constant frequency drive
Frequenzstreuung *f* frequency scattering
Frequenzstufe *f* frequency step
Frequenzstufen *fpl* **in gleichem Abstand** equidistant frequency steps
Frequenzteiler *m* frequency divider, dividing network
  integrierender F. integrating divider
Frequenzteilung *f* frequency demultiplication
Frequenztransformator *m* frequency transformer
Frequenztransformierung *f* frequency transformation
Frequenztrennschärfe *f* frequency selectivity
Frequenzüberwachungsbehörde *f* discriminating communications authority
Frequenzumformer *m* frequency changer (converter)
Frequenzumsetzer *m* frequency transposer
Frequenzumsetzung *f* frequency inversion
Frequenzumtastung *f* frequency-shift keying
Frequenzunabhängigkeit *f* independence of frequency
Frequenzverdreifacher *m* frequency tripler
Frequenzverhalten *n* harmonic response
Frequenzverlagerung *f* frequency translation [of a channel]
Frequenzverlauf *m* **der Dämpfung** attenuation frequency characteristic
Frequenzvervielfacher *m* frequency multiplier
Frequenzvervielfachungsklystron *n* frequency multiplier clystron
Frequenzverwerfung *f* shift in frequency, frequency shift
Frequenzverzerrung *f* frequency distortion
Frequenzwahl *f* frequency selection
Frequenzwanderung *f* frequency drift; oscillator drift
Frequenzwandler *m* frequency changer
Frequenzwandlung *f* frequency transformation
Frequenzweiche *f* frequency filter
Frequenzwiedergabe *f* frequency response
Frequenzwobbler *m* wobbulator, sweep generator
Frequenzzeitmodulation *f* frequency time modulation

Frequenzziehen *n* frequency pulling
Frequenzziehwert *m* pulling figure
Frequenzzuteilung *f* frequency allocation
fressen to fret *(corrosion)*; *(mach)* to seize
Fressen *n* fretting *(corrosion)*; *(mach)* seizure
  F. des Kolbens seizing of the piston
Freßschaden *m* fretting damage *(corrosion)*
Friedhof *m* burial (dumping) ground, graveyard *(of radioactive waste)*
Fries *m* frieze
Frigen *n* Freon
Friktion *f* friction; frictional transmission
Friktionsantrieb *m* friction drive
Friktionsglättung *f* friction glazing *(papermaking)*
Friktionskalander *m* friction calender *(papermaking)*
Friktionskupplung *f* friction clutch
Friktionsrad *n* friction wheel
Friktionsregler *m* friction governor
Friktionsscheibe *f* friction disk
frisch fresh *(air, water)*; green *(wood)*
Frischarbeit *f* fining process
Frischband *n* blank (new) tape
Frischbeton *m* fresh (wet, green) concrete, fresh (mixed) batch
Frischbetonmischung *f* wet batch, wet [concrete] mix
Frischbetonwerk *n* ready-mix plant
Frischbirne *f* converting vessel
Frischdampf *m* live steam
frischdampfbeheizt live-steam
Frischdampfstutzen *m* inlet connecting branch
frischen to oxidize, to blow, to fine
Frischen *n* decarburization; [re]fining
Frischerz *n* raw ore
Frischfeuer *n* bloomery, forge hearth, fining forge
Frischfleisch *n* fresh meat
Frischgeschwindigkeit *f* *(met)* rate of carbon drop
Frischhaltepackung *f* fresh-keeping wrapping
Frischherd *m* fining (refinery) hearth
Frischhütte *f* melting finery
Frischlauge *f* fresh liquor
Frischluft *f* fresh air
Frischluftgerät *n* *(mine)* fresh-air respirator (hose mask)
Frischluftheizung *f* fresh-air heater
Frischluftzufuhr *f* fresh-air supply
Frischofen *m* refining furnace
Frischöl *n* fresh (virgin) oil
Frischölschmierung *f* once-through lubrication
Frischperiode *f* oxidizing period
Frischprüfung *f* initial output test *(of battery)*
Frischschlacke *f* refining cinder, oxidizing (first) slag
Frischstahl *m* rough steel
Frischverfahren *n* fining (oxidizing, decarburizing) process
Frischwasser *n* fresh water; make-up water
Frischwassererzeuger *m* evaporator, water condenser
Frischwassertank *m* fresh-water [storage] tank
Frischwetter *pl* *(mine)* fresh air
Frischwirkung *f* *(met)* oxidizing reaction
Frisierhilfsmittel *n* hair-grooming preparation
Fristablauf *m* expiration of the time allowed
Fristausbesserung *f* periodical repairs
Fritte *f* frit, fritted glass filter
fritten to frit, to sinter
Fritter *m* *(el)* coherer
Fritterklemme *f* coherer terminal
Fritterung *f* *(el)* coherence
Fritterwiderstand *m* coherer resistance
Fritterwirkung *f* coherer action
Fritürepfanne *f* / **elektrische** electric deep frier
Front *f* front; face
frontal frontal
Frontalbeleuchtung *f* front lighting
Frontansicht *f* front view
Frontantrieb *m* *(mach)* front drive
  F. / mit front-operated
frontbedient *(mach)* front-operated
Frontbogenauslage *f* front sheet delivery
Frontdrehmaschine *f* frontal (front-operated) lathe
Frontlader *m* front-end loader, tractor (loading) shovel
Frontladergabel *f* buckrake
Frontlenker *m* forward control chassis
Frontmäher *m* front-mounted mower
Frontplatte *f* front panel; top panel; face; facing slab
Frontplattenbeleuchtung *f* faceplate illumination
Frontplattenbeschriftung *f* panel legend
Frontplattenschalter *m* panel switch
Fronträumer *m* [bull]dozer
Frontschaufellader *m* front-end loader
Frontschaufler *m* front-end loader
Frontschild *m* escutcheon plate
Frontschirm *m* front screen
Frontschott *n* end (break) bulkhead *(a transverse bulkhead at the end of a superstructure)*
  F. des erhöhten Quarterdecks break bulkhead of raised quarter-deck
Frontseite *f* front [side]
Frontsitz *m* front seat
Froschbeinwicklung *f* *(el)* frog-leg winding
Fröschchen *n* *(print)* lay mark
Froschklemme *f* *(el)* wire stretcher (grip), draw-tongs, toggle
Frost *m* frost, freezing
frostbeständig frost-proof, frost-resistant
Frostbeständigkeit *f* frost resistance

Frostgrenze *f* frost line

frosthart frost-hardy

Frosthärte *f (ag)* cold (frost) resistance

Frosthebung *f* frost heave

Frostkörper *m (mine)* ice wall

Frostleiste *f (text)* frost rib

Frostresistenz *f (ag)* cold (frost) resistance

Frostriß *m* frost crack *(wood)*

Frostschaden *m* frost damage

Frostschutz *m* frost protection; anti-freeze measure

Frostschutzmittel *n* freezing preventive, anti-freeze, anti-freezing composition

Frostschutzscheibe *f* anti-frost screen

Frostschutzschicht *f* frost blanket (layer), sub-base

frostsicher frost-proof

Frosttiefe *f* depth of frost penetration

Frostwand *f (mine)* ice wall

Frostwetter *n* frost weather

Frottiergewebe *n* terry towelling

Frottierstrecke *f* rubber drawing

Frottiertuch *n* terry towel

Fruchtäther *m* fruit ether (oil), essential oil

fruchtbar fertile, fecund

Fruchtbarkeit *f* fertility, fecundity

Fruchtbarmachung *f* fructification, fertilization

Fruchtessenz *f* fruit essence

Fruchtfolge *f* crop rotation

Fruchtkonserve *f* preserved fruits

Fruchtpresse *f* fruit juicer

Fruchtsaft *m* fruit juice

Frucht[transport]schiff *n* fruit ship (carrier), fruit carrying ship

Fruchtwechsel *m* crop rotation

Frühbeetkasten *m* hotbed, forcing bed

Frühdrucke *mpl* incunabula

Frühfolgen *fpl* **der Bestrahlung** immediate effect of irradiation

Frühgemüse *n* forced vegetables

frühhochfest high-early-strength *(cement)*

Frühholz *n* early wood, springwood

Frühjahrsbestellung *f* spring cultivation

Frühschicht *f* morning (early) shift

Frühsymptom *n* premonitary symptom

Frühzündung *f* pre-ignition *(surface ignition, e.g. due to deposits)*, premature (advanced) ignition

Fruktose *f* fructose

FÜ-System *n (cont)* synchro

Fuchs *m* [chimney] flue; skimmer

unterirdischer F. underground main flue

Fuchsbrücke *f* back bridge wall, flue bridge

Fuchsende *n* flue end

Fuchsin *n* fuchsine

Fuchsit *m (min)* fuchsite

Fuchsmündung *f* mouth of the main flue

Fuchsöffnung *f* flue opening

Fuchspelz *m* fox fur

Fuchsschwanzsäge *f* dovetail (pad) saw

deutsche F. backed pad saw, backsaw

englische F. broken span saw, half rip saw

Fuchssohle *f* foundation for bottom of the main flue

Fuchsverlust *m* flue loss

Fuder *n* cart-load, wagon-load

Fugazität *f (ch)* fugacity

Fuge *f* joint *(brickwork)*; groove *(welding)*

F. auf Fuge straight joint

bündige F. *(build)* flat joint

offene F. drained joint *(large panel construction)*

sichtbare F. face joint

versetzte F. broken joint

Fügebank *f* jointer, jointing plane

Fügefräser *m* joint[ing] cutter

Fügemaschine *f* jointing machine

fügen to join *(welding)*; to mate

Fugenbeton *m* joint concrete

Fugenbewehrung *f* tie bar

Fugendeckleiste *f* profile border

Fugenfläche *f* jointing edge (surface)

Fugenkelle *f* pointing (filling) trowel

Fugenkitt *m* joint cement (mastic); joint-filling compound

Fugenlänge *f* groove length *(welding)*

fugenlos without expansion joint[s] *(rail)*; jointless *(floor)*

Fugenmörtel *m* joint cement

Fugenpressung *f* **in Preßverbindung** radial pressure between adjacent surfaces of a press fit

Fugenputzer *m* mortise cleaner

Fugensäge *f* joint cutter

Fugenschnitt *m* **/ radialer** radial joint

Fugenundichtigkeit *f* joint leakage

Fugenvergußmasse *f* joint sealing (filling) compound

Fugenverstrich *m* jointing *(brickwork)*

Fugenvorbereitung *f* edge preparation

Fugenweite *f* width of joint

Fügezwinge *f* joining press

Fühlelement *n* sensor, sensing element

fühlen to sense; to feel; to trace *(copying)*

Fühler *m* sensor, sensing (detection) element; probe; *(mach)* tracer

temperaturempfindlicher F. temperature-sensitive element

Fühlerelement *n* [temperature] sensing element

fühlergesteuert *(mach)* tracer-controlled

Fühlerkondensator *m* scanning capacitor

Fühlerkopf *m* detector head, sensor, sensing head (probe)

Fühlerlehre *f* feeler (thickness) gauge

Fühlernachformeinrichtung *f* tracing attachment

Fühlernachformsteuerung *f* tracer duplicator control

selbsttätige F. automatic tracer duplicator control
Fühlerschieber *m* tracer control valve
Fühlersteuerung *f (mach)* tracer control
Fühlerstift *m (mach)* stylus
Fühlerstiftdruck *m (mach)* stylus pressure
Fühlerventil *n (mach)* tracing valve
Fühlerwalze *f (mach)* roller follower, tracer roll
Fühlhebel *m s.* Feinzeiger
Fühlnadel *f* selecting pin
Fühlstreifen *m* feeler
Fühlungnahme *f* contact
führen 1. to conduct, to guide, to direct, to carry, to lead, to take to; 2. to drive *(car, tram)*; to fly; to navigate; 3. to pilot *(tool)*; to keep aligned
die Reibahle in der Bohrung f. to pilot the reamer in the hole
ein Gespräch f. to talk *(telephony)*
ein Schiff f. to navigate a vessel
eine Flagge f. *(ship)* to fly a flag
Logbuch f. *(ship)* to keep the log
sich selbst f. to be self-piloting *(tool)*
Tagebuch f. to keep the log
Führer *m* driver *(car, locomotive)*; *(aero)* pilot
Führerbremsventil *n (ry)* driver's control (brake) valve
Führergondel *f* control car *(airship)*
Führerhaus *n (ry)* cab
Führerpult *n (ry)* control board (panel)
Führerraum *m (aero)* cockpit
Führerschein *m* driving licence
internationaler F. international driving permit
Führerscheinentzug *m* driving ban
Führersitz *m (aero)* cockpit
Führersitzverkleidung *f* cockpit-enclosure
Führerstand *m* cab, driver's cabin
doppelseitiger F. double-end control
Führerstandautomat *m* canopy-switch
Führerstandsignal *n* cab signal
Führerstandsteuerung *f* single-end control
Führung *f* guidance; run *(cable)*; guide; *(mach)* guideway, [slide]way
F. am Endstück back-end piloting *(broach)*
F. der Schmelze working the heat
F. des Sägeblattes guidance of the saw blade
F. in Querrichtung cross-slide
prismatische F. *(mach)* prismatic guide way
seitliche F. lateral guide
Führungsachse *f* axis of guide
Führungsantenne *f* director
Führungsbahn *f (mach)* guideway, [slide]way
F. der Planscheibe table track *(vertical boring and turning mill)*
abgedeckte F. *(mach)* covered way
Führungsbahnen *fpl* **am Ständer** column ways
Führungsbahnenschleifmaschine *f* slideway grinder (grinding machine)
Führungsbahnenschutz *m (mach)* guideway cover
Führungsbahnschleifen *n* grinding of slideways, slideway grinding
Führungsbeschleunigung *f* acceleration of translation
Führungsbock *m (rock)* guide casing
Führungsbohrung *f* pilot (starting) hole
Führungsbuchse *f* guide bush[ing]
Führungsfase *f* land; *(Am)* margin *(tool)*; primary clearance, circular land *(reamer)*
Führungsfeld *n* guide field
Führungsfläche *f* **für gleitende Teile** guiding surface for sliding members
Führungsgabel *f (ry)* axle guide (hornplate, hornblock)
Führungsgeschwindigkeit *f* velocity of translation
Führungsgröße *f* command, reference input *(control engineering)*
Führungskäfig *m* retaining member *(needle bearing)*
Führungskamm *m* carriage comb
Führungslager *n* arbor yoke *(milling machine)*
Führungsleiste *f* 1. *(mach)* bearing (rubbing) pad *(D-bit)*; 2. *(build)* guide fillet (plate)
F. beim spitzenlosen Schleifen work guide in centreless grinding
Führungslineal *n* fence; guide plate (rule)
Führungsmast *m* guide mast
Führungsplatte *f* guide plate
Führungsregler *m* master controller *(cascade control system)*
Führungsrille *f* track, groove; groove of ski
Führungsring *m (mil)* driving band *(shell)*, *(mach)* guide ring
freibeweglicher F. floating ring
Führungsrohr *n* guide tube
geschlossenes F. bored out guide
Führungsrolle *f (mach)* contact (guide) roller
Führungsschaft *m (mach)* guide, rear (back) pilot *(tool)*
Führungsscheibchen *n* guiding disk
Führungsschlitten *m* crosshead *(shaft sinking)*
Führungsschlitz *m* selector shaft guide *(telephony)*
Führungsschuh *m* cage (guide) shoe
Führungsstange *f* tail rod *(of piston)*
Führungsstein *m* sliding block
Führungsstift *m (mach)* former pin; locator (guide) pin *(electron tube)*
Führungsstreifen *m* felt seam mark
Führungsstück *n* pilot segment, alignment section *(broach)*

Führungstrommel *f* guide drum

Führungsverhalten *n (dat) response to a variation of the reference input*

Führungszahnrad *n* pilot gear

Führungszapfen *m* pilot pin *(of the counterbore)*

Fuhrunternehmen *n* haulage contractors (contracting firm)

Fuhrwerkswaage *f* weigh-bridge

Füllapparat *m* feeding (filling) apparatus

F. für Hackschnitzel chip packer

Füllartikel *m (print)* filler

Füllboden *m* fill

Fülldruck *m* inflation pressure

Fülleitung *f* air-supply line

Füllelement *n (el)* wet cell

füllen to fill; to charge; to pour in *(liquid)*; to crowd *(excavators)*

Füllen *n* **von oben** top charging

Füller *m* 1. filler; 2. *(mine)* mucker

Fullererde *f* fuller's earth

Füllfaden *m (text)* wadding thread

Füllfaktor *m (el)* space (copper) factor

Füllfederhalter *m* fountain pen

Füllgas *n* filling gas

Füllgasabsaugung *f* gas exhausting during charging

Füllgemisch *n* filling-gas mixture

Füllhalterdosimeter *n (nucl)* pen-type dosimeter

Füllhöhe *f* **der Kohle** *(mine)* height of coal line

Füllhöhenmesser *m* level indicator

Füllholz *n* 1. *(build)* packing piece; 2. *(ship)* making up *(launching)*, filling

Fülling *f (ship)* air course board

Füllkasten *m (mach)* hopper

Füllkette *f (text)* wadding warp

Füllkitt *m* stopping compound

Füllklappe *f* charging door *(heating)*

Füllkoks *m* bed coke

Füllkonstante *f* bulk factor

Füllkörper *mpl* 1. *(ch)* packing material *(absorption, distillation)*, tower packing; 2. *(build)* filler block; filling; filler

Füllkörperkolonne *f (nucl)* packed tower (column)

Füllmagazin *n* battery *(loom)*

Füllmasse *f* 1. *(build)* filling compound (material), filler; 2. massecuite *(sugar)*

Füllmaterial *n* filling medium *(fuses etc.)*

Füllmenge *f* unmixed batch capacity *(concrete mixer)*

Füllmengenanzeiger *m* level detector (indicator)

Füllmittel *n* 1. *(plast)* reinforcing agent (material); 2. *(ch)* diluent; extender *(paper)*; 3. *(text)* filling

Füllnut *f* loading slot

Fülloch *n* charging (filling) hole

Füllort *n (mine)* bottom landing, ingate-plot, inset, pit bottom plat

Füllplatte *f* filler plate; transfer chamber retainer plate

Füllrahmen *m* sand frame *(core)*

Füllraum *m* loading cavity (chamber), material well

Füllrumpf *m* feed box *(briquetting press)*; fuel-receiving chamber

Füllsand *m* backing (filler) sand *(mould)*

Füllschacht *m* charging hole

Füllschachtfeuerung *f* continuous charge furnace, self-feeding furnace

Füllschuß *m (text)* wadding weft

Füllspant *n (ship)* filling frame (timber)

Füllstand *m* level of filling, height of fill

Füllstandsanzeiger *m* level indicator

Füllstandsmesser *m* level indicator (gauge, meter)

F. nach dem Auftriebsprinzip buoyancy probe

manometrischer F. air-trap level meter

Füllstandsmeßgerät *n* liquid level gauge

Füllstandsregelung *f* level control

Füllstandsschalter *m* level switch

Füllstandswächter *m* level controller (monitor)

Füllstandszeiger *m* level indicator

Füllstein *m* filling brick

Füllstoff *m (build, plast, text)* filler; loading material

Füllstoffaufbereitung *f* preparation of filling material

Füllstrich *m* filling (gauge) mark

Fülltasche *f (mine)* loading (measuring) pocket

Fülltrichter *m (mach)* [feed] hopper, loading hopper, bin; charging hopper, feed chute, chute feeder

Fülltrichterzuführer *m (mach)* hopper feeder

Fülluftmesser *m* air-inflation indicator

Füllung *f* filling; *(el, met)* charge; load[ing]; *(build)* packing; cut-off

wirksame F. real admission (cut-off)

Füllungsänderung *f* variation of admission (cut-off)

Füllungsdauer *f* duration of filling

Füllungsgrad *m (therm)* degree of filling (feeding)

Füllungsraum *m* admission space

Füllungsstrecke *f* stroke of admission

Füllungsvergrößerung *f* increase of admission

Füllungsverkleinerung *f* reduction of admission

Füllvorrichtung *f* filling (feeding) device; charging device

Füllwagen *m* charging (larry) car, filling bogie

Fulminat *n* fulminate

Fumarsäure *f* fumaric acid

Fundament *n* foundation, groundwork, footing

F. mit Pfahlgründung pile-supported foundation

Fundamentallösung *f (math)* fundamental solution
Fundamentalpunkt *m* fixed point *(thermometer)*
Fundamentanker *m* foundation bolt
Fundamentaushub *m* foundation excavation
Fundamentbolzen *m* holding-down bolt
Fundamentgrube *f* foundation pit
Fundamenthöhe *f* foundation level
Fundamentklotz *m* foundation block
Fundamentmauer *f* foundation wall
Fundamentplatte *f* 1. *(build)* foundation raft (slab); 2. *(ship)* foundation (bed) plate; 3. oven basement raft
Fundamentrost *m* [foundation] grillage
Fundamentschraube *f* foundation (anchor) bolt
Fundamentsohle *f* foundation (formation) level, final grade
Fundamentträger *m* foundation girder
Fundbohrung *f* discovery well
fundieren to found
Fundierung *f* foundation
Fundus *m* fundus
Funduperipherie *f* periphery of the fundus
fünfatomig pentatomic
Fünfeck *n* pentagon
fünfeckig pentagonal
Fünfelektrodenröhre *f* pentode
fünffach quintuple
Fünffachkette *f* quintuple-strand chain
Fünfgitterröhre *f* heptode, pentagrid valve
Fünfkantrevolverkopf *m* five-sided turret
Fünfkörperproblem *n* five-body problem
Fünfleitersystem *n (el)* five-wire network
Fünfmastgaffelschoner *m* five-masted fore-and-aft schooner
Fünfmastvollschiff *n* five-masted full-rigged vessel
Fünfpolröhre *f* five-electrode valve, pentode
Fünfring *m (ch)* five-membered ring, five-ring compound
Fünfröhrensuper *m* five-valve superhet
Fünfschenkelwandler *m* five-core transformer
Fünfschrittkode *m* five-unit code
Fünfspindelautomat *m* five-spindle automatic
Fünfstiftsteckersockel *m* five-pin base
fünfwertig *(ch)* quinquevalent, pentavalent
Fünfwertigkeit *f (ch)* quinquevalence, pentavalence
fünfzählig five-fold
Fungizid *n* fungicide
Funk *m* radio, wireless
Funkabkürzung *f* radio abbreviation
Funkabschirmung *f* **an der Zündeinrichtung** ignition screening
Funkamateur *m* radio amateur, *(sl)* ham
Funkamateurempfänger *m* ham-receiver
Funkanlage *f* wireless installation
Funkapparat *m* radio set
Funkastronomie *f* radio astronomy
Funkausrüstung *f* wireless (radio) equipment
Funkazimut *m (n)* radio azimuth
Funkbake *f* radio beacon
Funkbastler *m* home constructor
Funkbefehl *m* radio command
Funkbefeuerung *f* system of radio beacons
Funkbeobachtung *f* radio observation
Funkbereitschaft *f* radio alert
Funkbeschickungssender *m* calibration station
Funkbeschickungstabelle *f* calibration chart
Funkbeschränkung *f* limitation of radio traffic
Funkbetrieb *m* radio service
  F. sein / im to be on the air
Funkbetriebsverfahren *n* radio procedure
Funkbetriebszentrale *f* control point
Funkbild *n* radio picture
Funkbildübertragung *f* radiophotography
Funkboje *f* radio buoy
Funkbrief *m* radio letter telegram
Funkdämpfung *f* radio attenuation
Funkdienst *m* radio[-communication] service
  beweglicher F. mobile radio service
  fahrbarer F. mobile radio service
Funkdienstband *n* service-band
Funkdisziplin *f* radio discipline
Funke *m* spark
  elektrischer F. electric spark
  gesteuerter F. controlled spark
  überspringender F. jump spark
Funkecholot *n* ground-clearance indicator
Funkeinrichtung *f* radio equipment
Funkeleffekt *m* flicker effect
Funkelfeuer *n (el)* scintillating (quick-flashing) light
funkeln *(ph)* to scintillate, to sparkle
Funkelrauschen *n* flicker noise *(of electronic valves)*
Funkempfang *m* radio (wireless) reception
Funkempfänger *m* radio (wireless) receiver
funken 1. *(el)* to radio[telegraph]; 2. *(mach, ph)* to spark; 3. to arc [over]
Funkenableiter *m* spark arrester
Funkenabschirmung *f* sparking-plug shield
Funkenanzeiger *m* spark indicator
Funkenbahn *f* path of spark
Funkenbildung *f* sparking
Funkenbombe *f* spark bomb *(hydrospark forming)*
Funkenbrücke *f* spark gap
Funkendämpfung *f* spark damping
Funkenentladung *f* spark discharge
Funkenerosionsmaschine *f* spark eroder (erosion machine)
Funkenerzeuger *m* spark generator
Funkenfänger *m* spark catcher
Funkenfängerschornstein *m (ry)* spark arrester funnel
funkenfrei non-sparking

Funkenfrequenz *f* spark frequency
Funkengarbe *f* ray of sparks
Funkeninduktor *m* spark (induction) coil
  F. mit Hammerunterbrecher hammer spark coil
  Ruhmkorffscher F. Ruhmkorff induction coil
Funkenkammer *f (el)* arc chute
Funkenlänge *f* sparking distance
Funkenleiter *m* spark conductor
Funkenlöscher *m* spark quencher (arrester, killer, absorber)
Funkenlöscherspule *f* spark-suppressor coil
Funkenlöschkondensator *m* spark-quenching capacitor *(ignition)*
Funkenlöschkreis *m* spark quench
Funkenlöschspule *f* blow-out coil
Funkenlöschung *f* spark quenching (extinguishing)
Funkenpotential *n* spark potential
Funkenprüfung *f* spark testing *(grinding)*
Funkenschaltröhre *f* / **gesteuerte** trigatron
Funkenschreiber *m* spark recorder
Funkenschutz *m* spark killer
Funkensender *m* spark transmitter
Funkenspannung *f* sparking voltage (potential)
Funkenspektrum *n* spark spectrum
Funkensprühen *n* sparking
Funkenstativ *n* spark stand
Funkenstrecke *f* spark[ing] gap
  F. mit drei Elektroden three-electrode spark-gap, threefold spark gap
  F. mit Feineinstellung micrometer gap
  feste F. fixed discharger
  gesteuerte F. timed spark gap
  rotierende F. disk (rotary) discharger
Funkenstreckenblitzableiter *m* air protector
Funkentstörung *f* radio noise suppression, elimination of radio interference
Funkenüberschlag *m* spark-over, sparkover
Funkenversetzung *f* angular offset of sparks
Funkenwiderstand *m* spark resistance
Funkenzähler *m* spark counter
Funkenziehen *n* striking of sparks
Funkenzieher *m* spark drawer
Funkenzündspannung *f* sparking voltage
Funkenzündung *f* spark ignition
Funker *m* radio (wireless) operator
Funkerkabine *f* radio cabin
Funkerraum *m* radio cabin
Funkerzeugnis *n* operator's certificate
  F. erster Klasse first-class radio telegraph licence
Funkfehlweisung *f* radio-direction-finder error
Funkfernmeßdaten *pl* telemetering (telemetered) data
Funkfernmessen *n* radiotelemetering
Funkfernschreiben *n* radioteletype
Funkfernsprechbetrieb *m* radiotelephone operation
Funkfernsprechen *n* radiotelephony
Funkfernsteuerung *f* radio telecontrol
Funkfeuer *n* radio beacon
  ungerichtetes F. non-directional [radio] beacon
Funkfeuerempfänger *m* radio-beacon receiver
Funkfeuerkennung *f* radio beacon identification
Funkfrequenz *f* radio frequency
Funkgabel *f* radio terminal (terminating unit)
Funkgast *m* wireless operator
Funkgegensprechen *n* radiotelephony
Funkgerät *n* radio set
  tragbares F. mit Trockenbatteriebetrieb *(Am)* all-dry portable
funkgesteuert radio-controlled
Funkhaus *n* broadcasting (studio) centre
Funkhöhenmesser *m* radio altimeter
Funkhorizont *m* radio horizon
Funkkanal *m* radio channel (link)
Funkkommando *n* radio command
Funkkompaß *m* radio compass
Funkkontakt *m* / **interplanetarischer** interplanetary radio contact
Funkkontrolle *f* band-watching
Funkkurspeilung *f* radio position-line determination
Funklandegerät *n* radio landing equipment
Funkleitsender *m* radio beacon
Funkleitstelle *f* radio control station
Funkleitstrahl *m* radio (localizer) beam
Funklinie *f* radio link (circuit)
  feste F. point-to-point communication
Funklinienverkehr *m* radio communications between two stations
Funkmast *m* radio mast (tower)
Funkmechaniker *m* radioman, radio technician
Funkmeldung *f* radio (wireless) message
  verstümmelte F. garbled message
Funkmeßanlage *f* radar station
Funkmeßeinrichtung *f* radar device
Funkmeßgerät *n* wireless range finder, radar-aid
Funkmeßlandeanlage *f* ground-controlled approach
Funkmeßnavigationsgerät *n* radar-aid to navigation
Funkmeßsender *m* radar transmitter
Funkmeßstörsender *m* radar jammer (jamming station)
Funkmeßsuchgerät *n* search radar set
Funkmeßtechnik *f* radar [engineering]
Funkmeßturm *m* radar tower
Funkmeßverfahren *n* radio range-finding
Funkmutung *f* radio prospecting
Funknachricht *f* radio message
Funknachrichtenkanal *m* radio circuit
Funknavigation *f* radionavigation

Funknetz *n* radio network (system)
Funkortung *f* radio location (determination)
Funkpeildienst *m* radio direction finding service
Funkpeilempfänger *m* direction finding receiver
Funkpeiler *m* radio (wireless) direction finder, radio goniometer
Funkpeilgerät *n* radio direction finder
Funkpeilstation *f* radio compass station
Funkpeilstelle *f* radio direction finding station
Funkpeilung *f* radio direction finding, wireless (D.F.) bearing
  aufgenommene F. observed radio bearing
  wahre F. true radio bearing
Funkpersonal *n* radio operators
Funkprüfgerät *n* radio analyzer
Funkpult *n* radio-control desk
Funkraum *m* radio room
Funkrelais *n* radio relay
Funkruf *m* radio call
  selbsttätiger F. automatic radio alarm
Funkschatten *m* radio shadow (pocket)
Funkschneise *f* radio beacon course
Funkschwund *m* fading, fade-out
Funkselbststeuerung *f* radio-autocontrol
Funksignal *n* radio signal
Funksonde *f* radio sonde
Funksonderdienst *m* special radio service
Funkspektrum *n* radio spectrum
Funksprechanlage *f* radiotelephone system
Funksprechband *n* radiotelephone band
funksprechen to radio[tele]phone
Funksprechen *n* radiotelephony, radiotelephone communication
Funksprechgabel *f* radiotelephone terminal
Funksprechgerät *n* radiotelephone
Funksprechkanal *m* voice communication channel
Funksprechkode *m* radiotelephony code
Funksprechverbindung *f* radiotelephone circuit
Funksprechverkehr *m* wireless telephone traffic
Funksprechzentrale *f* radiotelephony exchange
Funksprechzeugnis *n* radiotelephony certificate
Funkspruch *m* radio message
  unerledigter F. pending message
Funkstation *f* wireless (radio) station
  F. im Sendebetrieb station on the air
Funkstelle *f* wireless (radio) station
  tragbare F. portable radio station
  zuständige F. appropriate station
Funksteuerung *f* guidance by radio, radio control
Funkstille *f* silent period, radio silence
  beachtet F.! observe silent period!
  internationale F. international radio silence
Funkstörmeßgerät *n* radio interference meter
Funkstörung *f* radio disturbance (interference)
Funkstreifenwagen *m* radio patrol car
Funktafel *f* radio-code table
Funktagebuch *n* radio logbook
Funktechnik *f* radio engineering
Funktechniker *m* radio engineer
Funktelefon *n* radio[tele]phone
Funktelefonie *f* radiotelephony
Funktelegrafendienst *m* radiotelegraph service
Funktelegrafie *f* radiotelegraphy
Funktelegramm *n* radiogram
Funktion *f* function
  F. einer Funktion composite function, functional
  abgeleitete F. derived function
  algebraische F. algebraic function
  Besselsche F. Bessel function
  Boolesche F. Boolean function
  einwertige F. single-valued function
  elliptische F. elliptic function
  erzeugende F. generating function
  gerade F. even function
  harmonische F. harmonic function
  hyperbolische F. hyperbolic function
  implizite F. implicit function
  komplementäre F. complementary function
  logische F. logical function
  Peircesche F. dagger (Peirce) function *(logical function)*
  periodische F. periodic function
  rationale F. rational function
  reziproke F. inverse function
  trigonometrische F. cyclometric function
  unstetige F. discontinuous function
  zusammengesetzte F. composite function, functional
Funktionalanalysis *f* functional analysis
Funktionaldeterminante *f* functional determinant
Funktionalgleichung *f* functional equation
Funktionalmatrix *f* functional matrix
Funktionentheorie *f (math)* theory of functions
funktionieren to function, to work, to operate, to be in operation
  nicht f. to be out of order
Funktionsgeber *m* function generator *(analogue computer)*
Funktionsgenauigkeit *f* operating accuracy *(of a machine)*
Funktionsgenerator *m (dat)* function generator
  bivarianter F. bivariant function generator
  kartenlesender F. map-reader function generator
Funktionsgruppe *f* functional unit (module)
Funktionsmöglichkeit *f* option of function
Funktionsmultiplizierer *m* function multiplier *(analogue computer)*
Funktionspotentiometer *n* function potentiometer *(analogue computer)*
Funktionsprüfung *f* functional gauging

Funktionsschalter *m* function switch *(analogue computer)*
Funktionsschema *n (dat)* symbolic circuit
Funktionstabelle *f (dat)* function table
Funktionstafel *f (dat)* function (truth) table
Funktionstisch *m (dat)* plotting board, X-Y recorder
funktionstüchtig serviceable
Funktionstüchtigkeit *f* serviceability, functionality
Funktionsumformer *m (dat)* function generator
Funktionsverstärker *m* operational-type amplifier
Funktionszuordner *m (dat)* function translator
Funktor *m (math)* functor
Funkträgerfrequenz *f* radio-carrier frequency
Funkturm *m* radio (aerial) tower
Funkübertragung *f* radio transmission
Funkübertragungsweg *m* radio circuit
Funküberwachung *f* radio monitoring
Funkverbindung *f* radio circuit
Funkverkehr *m* wireless (radio) communication
F. vom Boden zum Flugzeug ground-to-airplane communication
fester F. point-to-point communication
gerichteter F. directional radio communication
Funkverkehrsformen *fpl* types of radio communication
Funkverkehrskarte *f* radio network chart
Funkverständigung *f* **im Zug** *(ry)* communication by train radio
Funkwache *f* radio watch
Funkwagen *m* radio car
Funkweg *m* radio channel
Funkwellen *fpl* wireless waves
Funkwellenspektrum *n* radio spectrum
Funkwesen *n* radio communication
Funkwetterdienst *m* radio meteorological service
Funkzeitlupenaufnahme *f* high-speed spark photograph
Funkzeitlupenkamera *f* high-speed spark camera
Funkzentrale *f* radio centre, central radio station
Furan *n* furane, furfurane
Furanharz *n* furane resin
Furche *f* 1. *(mach)* ridge; 2. *(ag)* furrow
Furchenbreite *f* furrow-width
Furchenrad *n* furrow wheel
Furchentiefe *f* depth of ploughing
Furchenzieher *m* furrow opener
Furfurol *n* furfural, furfurol
Furnier *n* veneer
gemasertes F. *(wood)* figured veneer
Furnierblatt *n* lamina, sheet of veneer
Furnierblock *m* veneer log
Furnierbock *m* veneering press
Furnierbockspindel *f* gluing clamp spindle
Furnierdecke *f* veneer face
furnieren to veneer
Furnierfügemaschine *f* veneer jointing machine
Furnierhobelmaschine *f* veneer cutting machine
Furnierhochtemperaturtrocknung *f* high-temperature veneer drying
Furnierholz *n* veneer
Furniermesser *n* veneer cutting knife
Furnierplatte *f* ply
Furnierpresse *f* veneering press
Furnierpreßwalzwerk *n* veneering rolls
Furniersäge *f* veneering (slitting, board) saw
Furnierschälmaschine *f* veneer lathe (peeling machine)
Furnierschere *f* veneer clippers
Furnierschneidemaschine *f* veneer cutter (cutting machine)
F. mit geradem Schnitt horizontal veneer machine with straight cut
rotierende F. rotary veneer lathe
furniert / mit Buche beech-veneered
Furnierverkleidung *f* veneer paneling
Fusarcverfahren *n* fusarc process
Fusel *m* fusel; bad spirits
Fuselöl *n* fusel oil
Fusion *f* [nuclear] fusion, fusion process
gesteuerte F. controlled fusion
Fusionsbombe *f* thermonuclear (nuclear fusion) bomb; hydrogen bomb, H-bomb
Fusionsreaktion *f* fusion reaction
Fusionsreaktor *m* thermonuclear reactor
Fusit *m* fusain, mineral charcoal, mother of coal
Fuß *m* 1. *(mach)* foot, leg; 2. foot *(30,48 cm)*; 3. *(text)* butt *(needle)*; 4. *(build)* toe *(retaining wall)*; base, foot
Fußabblendschalter *m* foot dip switch
Fußanlaßschalter *m* foot starter switch
Fußballschuh *m* football boot
Fußbauform / in *(mach)* foot-mounted
fußbetätigt pedal-operated
Fußbetätigung *f* actuation (operation) by foot
Fußboden *m* floor
Fußbodenbelag *m* floor covering, flooring
fugenloser F. composition flooring
schwimmender F. floating floor
Fußbodenfliese *f* flooring tile
Fußbodengräting *f (ship)* grill flooring, floor grating
Fußbodenheizung *f* underfloor heating
elektrische F. electric screed heating
Fußbodenhöhe *f* floor level
Fußbodenkontakt *m* floor contact
Fußbodenlack *m* floor varnish
Fußbodenoberfläche *f* **aus verpreßtem Holz** compressed wood laminate flooring surface
Fußbodenöl *n* floor oil
Fußbodenplatte *f* flooring tile

nichtunterkellerte F. slab on grade
Fußbodenschalter *m* floor contact
Fußbodenwachs *n* floor polish
Fußbremse *f* pedal (foot-operated) brake; brake pedal
Fußeinspannung *f* end fixing (restraint, fixation) *(columns)*
Fußflanke *f* flank *(gear)*
Fußgänger *m* pedestrian
Fußgängerbereich *m* pedestrian precinct
Fußgängerbrücke *f* pedestrian bridge, footbridge
Fußgängerschutzinsel *f* street (pedestrian) refuge
Fußgängerstraße *f* pedestrian mall
Fußgängertunnel *m* pedestrian subway
Fußgängerüberweg *m* pedestrian crossing
Fußgängerunterführung *f* pedestrian subway
Fußgängerverkehr *m* pedestrian traffic
Fußgängerweg *m* pavement, sidewalk
rollender F. travolator
Fußgashebel *m* accelerator [pedal]
Fußhebel *m* pedal, foot lever; treadle
fußhebelbetätigt pedal-controlled, pedal-operated
Fußhöhe *f* dedendum *(gear)*
Fußhydraulik *f* pedal-operated hydraulic mechanism
Fußkreis *m* root (dedendum) circle *(gear)*
Fußkreisdurchmesser *m* root diameter, diameter of the root circle
Fußkreuz *n (text)* chain lease
Fußlager *n* footstep bearing
Fußleiste *f* skirting [board], base-board
Fußmaschine *f* footing machine
Fußmotor *m* foot-mounted motor
Fußpfette *f* inferior purlin, wall plate
Fußplatte *f* base (bed) plate; foot plate *(ski)*
Fußpunktspeisung *f* base energizing *(of aerials)*
Fußschalter *m* foot switch
Fußstarterknopf *m* toe switch
Fußstütze *f* foot rest
Fußteil *n* stilt *(support)*
Fußtritt *m* foot treadle
Fußtritthub *m* pedal lifting mechanism
Fußventil *n* foot valve
Fußweg *m* pavement, *(Am)* sidewalk
Futter *n* 1. *(ag)* fodder; 2. *(mach)* chuck *(clamping)*; 3. *(met, build)* lining *(furnace)*; packing
rechts-links-plattiertes F. plain-plated laid-in fabric
saures F. *(met)* acid lining
Futteral *n* case, sheath
Futterarbeit *f* chuck[ing] work, chucking operation
Futterautomat *m* 1. *(mach)* automatic chucking machine *(turning)*; 2. *(ag)* self-feeder
Futterbacken *m (mach)* chuck jaw *(clamping)*
Futterbahn *f* automatic feed meter
Futterbau *m* forage growing
Futterbedarf *m* fodder requirement
Futterbereitungsmaschine *f* fodder-processing machine
Futterbeutel *m* nose bag
Futterboden *m* forage loft
Futterbrett *n* riser *(stairs)*
Futterdämpfer *m* food cooker, fodder steamer
Futtererntemaschine *f* forage harvester
Futterfaden *m (text)* backing thread *(weaving)*; laid-in yarn *(knitting)*
Futterhalbautomat *m* semi-automatic chucking machine *(turning)*
Futterholz *n* filler, packing piece, infiller block
Futterkalk *m* feeding lime
Futterkammer *f* forage shed
Futterkasten *m* corn bin (chest)
Futterklinge *f* chopper, chopping knife
Futterkonservierung *f* fodder preservation
Futterkrippe *f* crib
Futterlader *m* green-crop loader
Futterleder *n* lining leather
Futterleiste *f* filling rod
Futtermasse *f* lining mass *(furnace)*
Futtermauer *f* revetment wall, brickwork lining
Futtermischer *m* food-mixing machine
Futtermischmaschine *f* food-mixing machine
Futtermühle *f* fodder mill
Futtermulde *f* scoop
Futtermuser *m* fodder (forage) pulper
füttern 1. *(ag)* to feed; 2. *(met, build)* to line
Futterpflanze *f* fodder plant
Futterreißer *m* shredder
Futterrohr *n* casing tube
Futterschneider *m* shredder
Futterschuß *m (text)* backing weft
Futtersilo *m* silage silo
Futterstoff *m (text)* lining fabric
Futterstreuer *m* forage spreader
Futterstück *n* filler
Futterverteilungswagen *m* combined food trolley and weigher
Futterwaage *f* food weigher
Futterwagen *m* food trolley
Futterwirtschaft *f* fodder management
Futterzerreißer *m* shredder
Futterzuführungsapparat *m* feed conveyor

G

g-Atom *n s.* Grammatom
g-Beschleuniger *m* g-accelerator
g-Faktor *m* **[/ Landéscher]** *(nucl)* [Landé] g-factor, interval factor
G-Schirm *m* G-scope *(radar)*

G-Sichtgerät *n* G-indicator *(radar)*
G-Wert *m (nucl)* G-value
Gabardine *f* gabardine
Gabel *f* 1. *(ag)* [pitch] fork; 2. *(mach)* yoke *(milling arbor)*; 3. cradle, receiver (termination) set *(telephony)*
  G. mit Gabelzapfen fork with crosshead pin
  G. mit Stirnzapfenlager fork with bearing for crosshead pin
  G. zur Riemenführung belt guide (fork)
Gabelachse *f* fork axle
Gabelamt *n (telec)* hybrid terminal [office]
Gabelauge *n* eye of fork
Gabeldeichsel *f* shaft
Gabelechosperre *f (el)* terminal echo suppressor
Gabelgelenk *n* knuckle joint
Gabelhebel *m* forked steering arm
Gabelholm *m* fork beam
Gabelkrümmung *f* bend of the fork
Gabelmontierung *f* fork mount
Gabelmuffe *f* trifurcating box, Y-joint
gabeln / sich to bifurcate, to bisect
Gabelöse *f* eye of fork
Gabelpfanne *f (met)* ring ladle carrier, shank ladle
Gabelpunkt *m (el)* split [connection] point
Gabelrohr *n* fork (bifurcated, breeches) pipe, Y-piece
Gabelschalter *m* cradle switch
Gabelschaltung *f* terminating (hybrid) set; hybrid coil termination circuit
Gabelschlüssel *m* open-end wrench
Gabelschußwächter *m (text)* weft-fork motion
  seitlicher G. side weft-fork motion
Gabelstapler *m* fork[-lift] truck *(conveying)*; forked lever *(printing)*
Gabelstiel *m* 1. *(ag)* fork handle; 2. *(aero)* forked strut
Gabelstück *n* clutch-actuating fork
Gabelumschalter *m (el)* cradle switch
Gabelung *f* Y-junction, bisection, bifurcation; *(el)* split connection
Gabelverkehr *m* forked working, two-way communication
Gabelweite *f* / **lichte** fork opening
Gabelwender *m (ag)* tedder
Gabelzapfen *m* crosshead pin, gudgeon
Gadolinium *n* gadolinium
Gaffel *f* gaff
Gaffelklaue *f* gaff jaw, jaw of a gaff
Gaffelschoner *m* [fore-and-aft] schooner
Gaffelschuh *m* gaff jaw, jaw of a gaff
Gaffelsegel *n* gaff sail
Gahnit *m (min)* gahnite, zinc spinel
Gal *n* gal
galaktisch galactic
Galaktometer *n* [ga]lactometer
Galaktonsäure *f* galactonic acid
Galakturonsäure *f* galacturonic acid
Galaxis *f* galaxy
Galeere *f* galley
Galeerenschlag *m* galley stroke
Galerie *f (build)* gallery; balcony; platform
Galettseide *f* waste silk
Galgen *m (ship)* [beam, derrick] gallows
Galion *n* beak, [ship's] head
Galionsfigur *f* figure-head
Galionsreling *f (ship)* headrail, head rail
Gallapfel *m* gall-nut
Gallat *n* gallate
Gallenfarbstoff *m* bile pigment
Gallensäure *f* bile acid
Gallierung *f (text)* harness tie
Gallium *n* gallium
Galloway-Kessel *m* Galloway boiler
Gallseife *f* gall soap
Gallussäure *f* gallic acid
Gallustinte *f* gallotannic ink
Galmei *m* calamine
Galtonpfeife *f* Galton's whistle
galvanisch galvanic
Galvanisierbad *n* plating bath
Galvanisierbottich *m* plating vat
galvanisieren to electrodeposit, to [electro]plate
Galvanisiergehänge *n* plating rack
Galvanisierung *f* electroplating
Galvanisierwerkstatt *f* electroplating shop
Galvano *n* electrotype, *(Am)* electro; shell
  plasthinterfüttertes G. plastic-backed electrotype plate
Galvanochromie *f* electrochemical colouring
Galvanokaustik *f* galvanocautery
Galvanometer *n* galvanometer
  aperiodisches G. deadbeat galvanometer
  ballistisches G. ballistic galvanometer
  elektrostatisches G. electrostatic galvanometer
Galvanometernebenschluß *m* galvanometer shunt
Galvanometernebenwiderstand *m* galvanometer resistance
Galvanometerspule *f* galvanometer coil
Galvanoplastik *f* electrotyping
Galvanostegie *f* electroplating, galvanostegy
Galvanotechnik *f* electrodeposition, electroplating
Galvanotherapie *f* galvanism
Gamaschen *fpl* leggings
gamma-aktiv gamma-[radio]active
Gamma-Aktivität *f* gamma [radio]activity
Gamma-Bestrahlung *f* gamma-ray irradiation
Gamma-Bestrahlungsanlage *f* gamma-ray irradiation plant
Gamma-Breite *f (nucl)* radiation width
Gamma-Defektoskopie *f* gamma-ray testing (inspection)

Gamma-Detektor *m* gamma detector
Gamma-Dosimeter *n* gamma-radiation dose meter
Gamma-Dosis *f* gamma-ray dosage
Gamma-Dosisleistung *f* gamma dosage rate
Gamma-Eisen *n* gamma iron
Gamma-Emission *f* gamma emission
Gamma-Karottage *f* bore-hole logging by gamma-rays
Gamma-Kaskade *f* gamma-ray cascade, cascade (successive) gamma-rays
Gamma-Meßkopf *m* gamma probe
Gamma-Monitor *m* gamma monitor
Gamma-Quant *n* gamma[-ray] quantum
Gamma-Quelle *f* gamma-ray source
Gamma-Rückstreumeßgerät *n* gamma-ray backscattering meter
Gamma-Schild *m* gamma[-ray] shield
Gamma-Schutz *m* gamma[-ray] shield
Gamma-Spektrometer *n* gamma-ray spectrometer
Gamma-Strahl *m* gamma ray
Gamma-Strahldosimetrie *f* gamma-ray dosimetry
Gamma-Strahlenschutz *m* gamma[-ray] shield
Gamma-Strahlentherapie *f* gamma-ray therapy
Gamma-Strahler *m* gamma emitter (radiator)
Gamma-Strahlspektrometer *n* gamma-ray spectrometer
Gamma-Strahlung *f* gamma-radiation, gamma-ray emission
  prompte G. prompt gamma-radiation
Gamma-Strahlungsquelle *f* gamma-ray source
Gamma-Therapie *f* gamma-ray therapy
Gamma-Übergang *m* gamma-ray transition
Gamma-Untergrund *m* gamma-ray background
Gamma-Zählrohr *n* gamma-ray counter
Gamma-Zerfall *m* gamma-decay, gamma disintegration
Gammagrafie *f* gammagraphy, [gamma-]radiography
Gammameter *n* gamma-meter
Gamow-Berg *m (nucl)* Gamow barrier (wall)
Gang *m* 1. *(build)* walkway, gangway, aisle; 2. *(ship)* alleyway, passageway, strake; 3. *(mine)* lode, vein; 4. cycle, sequence; trend; 5. *(mach)* running; progress; routine; gear, speed *(motor car)*
  anliegender G. *(ship)* inner (inside, sunken) strake *(shell plating)*
  erster G. starting (first) gear, first speed
  leicht schaltbarer G. easily shifted gear
  niedriger G. low gear
  schnellster G. top gear
  stoßfreier G. *(mach)* smooth running
  toter G. backlash, lost motion, end play
  toter G. der Lenkung *(aut)* steering play
  zerschlagener G. *(geol)* branch vein
Gangänderung *f* change in speed
Ganganordnung *f* gear-shift pattern
Ganganordnungsschaubild *n* gear-change diagram
Gangart *f (geol)* gangue, matrix, waste rock
Gangausfüllung *f (geol)* infilling, ledge matter
Gänge *mpl* **je Zoll** threads per inch
Gangerz *n* lodestone, vein ore
Gangfüllung *f (geol)* infilling, ledge matter
Ganggestein *n (geol)* gangue, matrix
Ganghöhe *f* lead *(thread)*; pitch of spiral *(spiral column reinforcement)*
  G. der Gitterdrähte grid pitch *(electron valve)*
  G. des Gitters grid pitch
  G. eines Gewindes lead of a thread
gängig free-working *(coal)*
Gängigkeit *f* workability *(coal)*
Gangkreuz *n (text)* chain lease
Gangmasse *f (geol)* attle, gangue material
Gangpolkegel *m* polhode cone
Gangrad *n* escapement wheel *(watch)*
Gangschaltbild *n* gear-shift diagram
Gangschaltschema *n* gear-change diagram
Gangschaltung *f* gear-shift control mechanism
Gangspalte *f* vein fissure
Gangspill *n* **mit Handkurbelantrieb** crank capstan
Gangstreichen *n (geol)* bearing, course of ore, line of lode
Gangunterschied *m (el)* path (phase) difference; retardation *(polarization)*
Gangvorwähler *m* / **elektrischer** electric preselector gear
Gangwähler *m* selector
Gangwählerschaltgabel *f* selector fork
Gangway *f* brow, gangway
Gangzahl *f* number of starts *(thread)*
Gangzug *m (geol)* lode
Gänsefüßchen *npl* inverted commas, quotation marks, quotes
Gänsefußmesser *n (ag)* duckfoot tine (knife)
Gänsefußschar *n* duckfoot share
ganz 1. *(math)* integral, whole *(number)*; 2. in order, intact; 3. complete
Ganzbestrahlung *f s.* Ganzkörperbestrahlung
Ganzholzbauweise *f* all-wood construction
Ganzkammerkessel *m* water tube boiler with single header
Ganzkörperbestrahlung *f* whole-body irradiation, total-body irradiation
Ganzkörperdosis *f* whole-body dose
Ganzkörperzähler *m (nucl)* whole-body counter; human-body counter
Ganzlochwicklung *f (el)* integer-slot winding, integral-pitch winding
Ganzmetallflugzeug *n* all-metal plane
Ganzmetallgehäuse *n* all-metal box (housing)

Ganzrost *m* grate in one section
Ganzstahlkarosserie *f* all-steel body
  selbsttragende G. unit construction all-metal body
Ganzstoff *m* [paper] pulp
Ganzwelle *f* full wave
ganzzahlig integer
Ganzzeug *n* whole stuff *(paper making)*
Ganzzeugholländer *m* general purpose beater
Ganzzug *m* complete train load
Garage *f* garage
Garantie *f* guarantee warrant, surety
Garantieleistung *f* after-sales service
garantieren to guarantee; to warrant
Garantieschein *m* guaranty
Garantieversuch *m* guarantee test
Garantiezeit *f* guarantee period
Garantiezink *n* guarantee zinc
Garbe *f (ag)* sheaf
Garbenbinden *n* sheaf binding
Garbenbinder *m* binder
Garbenmesser *n* band cutter
Garbenschober *m* stack of sheaves
Garbenträger *m* sheaf carrier
Gärbottich *m* fermenter, fermenting vat
Garbrand *m* finishing burn *(ceramics)*
Garderobe *f* cloak-room
Gardine *f* window curtain
Gardinenleiste *f* curtain (pelmet) board
Gare *f (met)* refined state
garen to carbonize
gären to ferment
gärfähig fermentable
Gärfähigkeit *f* fermentability
Gärfaulverfahren *n* fermentation putrefaction
garfrischen to refine
Gärfutter *n* [en]silage
Gärfutterbehälter *m* silage silo
Gärfuttersilo *m* silage silo
Gärkartoffelbehälter *m* potato ensilage pit
Gärkartoffelsilo *m* potato ensilage pit
Gärkolben *m* fermentation flask
Garkupfer *n* refined copper
Garn *n* yarn
  G. aus Glasfaser fibreglass yarn
  G. mit künstlicher Kräuselung man-crimped yarn
  G. ohne Verdrehung torqueless yarn
  hochfestes G. high-tenacity yarn
  kardiertes G. carded yarn
  linksgedrehtes G. crossband-twist thread
  texturiertes G. textralized yarn
Garnabfall *m* yarn waste
Garnette *f* garnett machine
Garnettkrempel *f* garnett clothing card
Garnfachung *f* yarn ply
Garnfärben *n* yarn dyeing
Garnführer *m* faller wire
garngefärbt yarn-dyed
Garngleichheitsprüfer *m* yarn evenness tester
Garnier *n (ship)* dunnage
Garnierholz *n (ship)* dunnage wood
Garnitur *f* fittings; outfit; mountings; armatures, set
Garniturteil *n* fitting
Garnkette *f (text)* yarn warp
  einfädige G. single-yarn warp
Garnkonditionierung *f* yarn conditioning
Garnkörper *m* yarn package
Garnnumerierung *f* ounce (dram) system *(silk)*
Garnnummer *f* yarn count, count of yarn
Garnprüfung *f* yarn testing
Garnsengmaschine *f* yarn singer
Garnstropp *m (ship)* selvagee strop
Garnträger *m* yarn carrier
  perforierter G. perforated tube
Garprobe *f* refining assay
Garrösten *n* finishing roasting
Garschaum *m* dross, refined iron froth, kish
Garschaumgraphit *m* kish graphite
Garschlacke *f* refining slag
Gartenbau *m* horticulture, gardening
Gartengeräte *npl* gardening tools
Gartenmesser *n* pruning knife
Gartenpflanze *f* garden plant
Gartenschere *f* garden shears
Gartenschlauch *m* garden hose
Gartentraktor *m* horticultural tractor
Gartenwalze *f* garden roller
Gärtner *m* [market] gardener
Gärtnerei *f* market garden
Gärung *f* fermentation, fermenting process, zymosis
  alkoholische G. vinous (alcoholic) fermentation
  saure G. sour (acid) fermentation
Gärungschemie *f* zymurgy; zymology
gärungserregend zymogenic, fermentative
gärungshemmend antizymotic
Gärungsmilchsäure *f* sour milk acid
Gärungstechnik *f* zymotechnology
Garungszeit *f (mine)* carbonizing (coking) period, coking time
Gas *n* gas
  absorbiertes G. absorbed gas
  adsorbiertes G. adsorbed gas
  atomares G. atomic gas
  dichtes G. dense gas
  eingeschlossenes G. occluded gas *(electron valve)*
  ideales G. perfect (ideal) gas
  inertes G. inert gas
  ionisiertes G. ionized gas
  quasineutrales G. quasi-neutral gas
  reales G. imperfect (non-ideal) gas
  ungereinigtes G. crude (unpurified) gas

verflüssigtes G. liquefied gas
vollionisiertes G. completely (fully) ionized gas
vollständig ionisiertes G. completely (fully) ionized gas
Gasabsaugpumpe *f* off-gas pump
Gasabscheider *m* gas separator
Gasabsorption *f* absorption of gas
Gasabsorptionsmittel *n* gas-absorbing agent
selektives G. burrel sorbent
Gasaktivität *f* gas activity
Gasanalysator *m* gas analyzer
Gasanalyse *f* gas analysis
Gasanalysegerät *n* gas analyzer
Gasansammlung *f* gas accumulation
Gasanstalt *f* gas works
Gasanzeiger *m* gas-leak indicator
Gasanzünder *m* gas lighter
gasartig gaseous
Gasatmosphäre *f* gaseous atmosphere
Gasaufbereitung *f* gas purification (processing)
Gasaufbereitungsanlage *f* gas processing plant
Gasaufzehrung *f* gas clean-up *(electronic valves)*
Gasausbeute *f* gas make (yield)
Gasausbruch *m* gas escape
plötzlicher G. gas outburst
Gasausdehnungsthermometer *n* gas[-expansion] thermometer
Gasausströmung *f* gas leakage (escape)
Gasaustritt *m* gas leakage
Gasballastschraube *f* surplus gas screw
Gasbehälter *m* gas holder; gasometer
gasbeheizt gas-fired
Gasbeleuchtung *f* gas illumination
Gasbeschaffenheit *f* character (condition) of gas
Gasbeton *m* gas (aerated) concrete
Gasbetrieb *m* gas operation
Gasbildung *f* gas formation
Gasblase *f* gas bubble; gas cavity *(casting)*
Gasbrenner *m* gas burner
Gasbürette *f* gas burette
Gaschromatografie *f* gas chromatography
gasdicht gas-proof, fume-proof
Gasdichte *f* gas density
Gasdichtemesser *m* **nach dem Auftriebsprinzip** buoyancy gas balance
Gasdiffusion *f* gaseous diffusion
Gasdiffusionsanlage *f* gaseous diffusion plant (factory)
Gasdrehgriff *m* throttle twist-grip
Gasdruck *m* gas pressure
Gasdruckkabel *n* gas-pressure cable, pressurized gas cable
Gasdruckregler *m* gas[-pressure] regulator
Gasdurchflußzähler *m* [gas-]flow counter
Gasdurchsatz *m* loading velocity
Gasdüse *f* nozzle
Gasdynamik *f* gas dynamics
Gaseinsatzhärtung *f* gas carburizing
Gaseinschluß *m* 1. gas hole, entrapped gas; 2. occlusion of gas
Gaselektrode *f* bubbling-type electrode
gasen to evolve gas
Gasentartung *f* gas degeneration
Gasentgiftung *f* detoxification of gas
Gasentladung *f* gas[eous] discharge
Gasentladungsableiter *m* gas discharge arrester
Gasentladungskammer *f* gas discharge chamber
Gasentladungslampe *f* gas discharge lamp
Gasentladungsplasma *n* [gas] discharge plasma, plasma of gas discharge
Gasentladungsrohr *n* gaseous discharge tube
Gasentladungsröhre *f* ionic (gas-filled) valve
G. mit Außengitter cathetron
Gasentladungsschalter *m* gas discharge switch
Gasentladungsventil *n* gas-filled rectifier
Gasentladungszählrohr *n* gaseous discharge counter tube
Gasentwicklung *f* evolution (formation) of gases, gas evolution (formation)
Gasentwicklungsflasche *f* gas generating bottle
Gasertrag *m* gas yield
Gaserzeuger *m* *(aut)* gas generator (producer)
Gaserzeugung *f* gasification, gas generation (production)
Gaserzeugungsanlage *f* gas processing (producing) plant
Gasexplosion *f* **im Feuerzug** fluegas explosion
Gasfackel *f* excess gas burner
Gasfalle *f* absorption tube
Gasfang *m* gas seal bell
Gasfeuerung *f* gas-fired furnace; gas firing
Gasflammkohle *f* long-flame gas coal, open burning coal
Gasflammofen *m* gas-reverberatory furnace
Gasflasche *f* gas cylinder (bottle)
Gasfördersonde *f* gas producing well
Gasförderung *f* gas transmission
gasförmig gaseous
gasführend gas-containing
Gasfüllung *f* gas filling
Gasgebläse *n* gas blower
gasgefüllt gas-filled
Gasgehalt *m* content of gas
G. des Wassers gas content of [the] water
Gasgenerator *m* gas generator (producer)
Gasgeruch *m* smell of gas
gasgeschmiert gas-lubricated
Gasgesetz *n* *(ph)* gas law
Gasgruskohle *f* gas slack
Gashahn *m* gas tap
gasieren to gas, to singe
Gasinstallateur *m* gas fitter
Gasinterferometer *n* gas interferometer
Gasionisation *f* gas ionization

Gaskalorimeter *n* gas calorimeter
Gaskammer *f* gas chamber
Gaskammerofen *m* gas chamber kiln
Gaskanal *m* gas duct (flue)
Gaskinetik *f* gas kinetics
Gaskocher *m* gas cooker
Gaskohle *f* gas (light) coal
Gaskoks *m* gas coke
Gaskonstante *f* gas constant
Gaskraftmaschine *f* gas engine
Gaskrieg *m* gas (chemical) warfare
Gaskühlung *f* gas cooling
Gasküvette *f* gas cell
Gaslampe *f* gas lamp
Gas-Laser *m* gas (gaseous state) laser, gaseous optical maser
Gasleitung *f* gas main
Gaslöslichkeit *f* solubility of gas
Gaslötkolben *m* gas-air iron
Gaslunker *m* gas pocket
Gasmaschine *f* gas engine
Gas-Maser *m* gas[eous] maser
Gasmaske *f* gas mask
Gasmaskenkohle *f s.* Aktivkohle
Gasmenge *f* gas volume
Gasmischungskammer *f (therm)* gas mixing chamber
Gasmolekül *n* gaseous molecule
Gasmotor *m* gas engine
Gasnachweis *m* gas detection
Gasnetz *n* gas grid
Gasofen *m* gas oven
Gasöl *n* gas (solar) oil
  zündträges G. low-cetane gas oil
Gasolin *n* gasoline, petroleum ether
Gasometer *m s.* Gasbehälter
Gaspartikel *f* gas particle
Gaspedal *n (aut)* accelerator [pedal]
Gasphase *f* gaseous phase
Gaspipette *f* gas sampling tube, bulb pipette
Gaspore *f* gas pocket
gaspreßgeschweißt gas-pressure welded
Gaspreßschweißen *n* pressure gas welding
Gasprobenventil *n* gas sampling valve
Gasprüfer *m* gas analyzer
Gasquelle *f* gas well
Gasreibung *f* gas friction
Gasreinigung *f* cleaning (purification) of gas; off-gas treatment
Gasreinigungsanlage *f* gas cleaner (cleaning plant)
Gasreinigungsmasse *f* gas purifying medium
Gasrekombination *f (nucl)* gas recombination
Gasrest *m* residual gas *(electronic valves)*
Gasrohr *n* gas pipe
Gasrohrbruch *m* gas main fracture
Gasrückstand *m* residual gas
Gasruß *m* gas (carbon) black
Gassack *m* gas bag
Gassammler *m* gas holder
Gasschieber *m* gas sluice valve
Gasschleuse *f* gas trap
Gasschmelzschweißen *n* gas fusion welding
Gasschutzgerät *n* device for protection against poisonous gases, gas protective apparatus
Gasschweißbrenner *m* autogeneous welding torch
Gasschweißen *n* gas welding
Gassenge *f (text)* gas singeing machine
Gassengemaschine *f (text)* gas singeing machine
Gassengen *n (text)* gas singeing
Gasspaltanlage *f* gas reforming plant
Gasspülung *f* barbotage
Gasspurenanalysator *m* high-sensitive gas analyzer
Gasspürgerät *n* gas detector
Gasstrahl *m* jet of gases
Gasstrom *m* gas flow
Gasstromschalter *m* gas-blast circuit breaker
Gasströmung *f* gas flow
Gasströmungsmesser *m* anemometer
Gasstrumpf *m s.* Glühstrumpf
Gasszintillationszähler *m* gas (gaseous) scintillation counter
Gasteer *m* producer gas tar
Gastheorie *f* gas theory, theory of gases
  kinetische G. kinetic theory of gases
Gastonne *f (ship)* gas buoy
Gastriode *f* gas[-filled] triode, thyratron
Gastrockner *m* acid trap *(of storage batteries)*; gas drier
Gasturbine *f* gas turbine
Gasturbinenreaktor *m* gas turbine reactor
Gasuhr *f* gas meter (counter), *(Am)* wet-test meter
Gasumformung *f* gas converting
Gasunterfeuerung *f* under-fired gas furnace, externally fired gas furnace
Gasuole *f* gas pore
Gasverflüssigung *f* liquefaction of gas[es], gas liquefaction
Gasvergiftung *f* gas poisoning
Gasverstärkung *f* gas amplification
Gasvorkühlung *f* gas primary cooling
  direkte G. direct gas primary cooling
Gaswaage *f* gas balance, dasymeter *(instrument for measuring the density of gases)*
  registrierende G. recording gas balance
Gaswandlung *f* gas converting (convertion)
Gaswäsche *f* gas scrubbing
Gaswäscher *m* scrubber, gas washer
Gaswaschung *f* gas scrubbing
Gaswassergrube *f* liquor well
Gaswechseleigenschaften *fpl* breathing characteristics *(of an engine)*
Gaswerk *n* gas works (processing plant)

Gaszähler *m* gas meter
  nasser G. liquid sealed gas meter
  trockener G. dry gas meter
Gaszählrohr *n* gas[eous] counter
Gaszementieren *n* gas carburizing
Gaszentrifuge *f* gas centrifuge
Gaszentrifugenverfahren *n* gas-centrifuge technique
Gaszustand / im in the gaseous state
Gate *n s.* Torschaltung
Gatsch *m* slack way, crude paraffin
Gatt *n (ship)* stern; gut; eyelet
Gatter *n* 1. lattice, gate; 2. gate [circuit], gating circuit *(switching circuit)*; 3. *(text)* creel; 4. saw gate
  fahrbares G. portable log frame
Gatterführung *f* saw guide
Gattersäge *f* mill (gang, frame) saw
Gatterschaltung *f* gate circuit
Gatterschnitt *m* flitch cut
Gattersteuerspannung *f* gate voltage *(for electronic switching)*
Gatterteilung *f (text)* pitch of the creel
gattieren 1. *(build)* to batch, to proportion, to gauge; 2. *(met)* to burden
Gattierungsanlage *f (build)* batching plant
Gattierungsberechnung *f* calculation of mixture
Gattierungswaage *f* 1. *(build)* weigh-batcher; 2. *(met)* burden balance
Gaufrage *f* embossing; goffering
gaufrieren to emboss, to goffer
Gaufrierkalander *m* embossing (goffering) calender
Gauge *n* gauge *(hosiery)*
Gauß *n* gauss
Gauß-Kurve *f* Gaussian [error] curve
Gauß-Messer *m* gauss meter
Gautschbrett *n* pressing board *(papermaking)*
Gautschbruchbütte *f* couch pit *(papermaking)*
Gautschbruchrückgewinnung *f* couch pit recovery
Gautsche *f* couch *(papermaking)*
gautschen to couch *(papermaking)*
Gautschgegenwalze *f* lump breaker
Gautschpresse *f* couch press
Gautschstuhl *m* couch stool
Gay-Lussac-Turm *m* Gay-Lussac tower
Gaze *f* gauze
Gazefilter *n* gauze filter
Gazekatode *f* gauze cathode
Gazestuhl *m (text)* gauze winding loom
Gazeweberei *f* cross weaving
geadert mottled *(wood)*
Gebälk *n* beams, framework
gebändert *(geol)* banded
Gebäude *n* building
  landwirtschaftliches G. farm building
  öffentliches G. public building
  repräsentatives G. prestige-type building
  stützenfreies G. clear span building
Gebäudeachse *f* centre line
Gebäudeanschluß *m* building (house) connection
Gebäudebreite *f* depth of building
Gebäudeflucht *f* frontage (building) line
Gebäudefundament *n* building foundation
Gebäudeschwingung *f* building vibration
Gebäudeskelett *n* building frame
Gebäudetiefe *f* depth of building
geben to give; to yield; to result in; to transmit, to send *(communications)*
  Aufträge g. to place orders
  Flackerzeichen g. to flash
  Gas g. to accelerate
  Stoßglanz g. *(leath)* to put a gloss
  Vorschub g. to cause the tool to feed, to feed [the tool]
  Vorspannung g. to pre-load, to prestress
Geber *m* transmitter *(e.g. signal transmitter)*; indicator *(e.g. position indicator)*; sender; pick-off, *(Am)* transducer
  elektromechanischer G. actuator
  induktiver G. inductance pick-off
  kapazitiver G. capacitance gauge
Geberbrücke *f* transducer bridge
Geberpotentiometer *n* transmitter potentiometer
Geberwicklung *f* pick-up winding
Gebetempo *n* rate of sending *(communications)*
gebettet supported *(statics)*
Gebevorrichtung *f* transmitter arrangement
Gebeweise *f* manner of keying, *(sl)* fist
Gebiet *n* 1. area, zone, sector, domain; 2. field *(branch of science)*
  G. unsicherer Peilung *(el)* bad bearing sector
  abgetastetes G. scanned area, coverage
  bebautes G. built-up area
  Belowsches G. Below region
  instabiles G. unstable region
  kritisches G. critical zone *(flow)*
gebietsweise sectional
Gebinde *n (text)* lea
Gebirge *n* 1. mountains; 2. *(mine)* rock; ground
  beruhigtes G. settled ground
  druckhaftes G. heavy broken ground
  gebräches G. bad ground
  gutes G. workable ground
  lockeres G. loose ground
  nachfälliges G. caving ground
  nichtstandfestes G. unstable ground
  noch nicht zur Ruhe gekommenes G. unsettled ground
  quellendes G. heaving ground
  zerrüttetes G. fractured ground
Gebirgsanker *m (mine)* rock (roof) bolt
Gebirgsbahn *f* mountain railway

Gebirgsbewegungen *fpl (mine)* rock (strata) movement
gebirgsbildend rock-forming
Gebirgsdruck *m* rock pressure
  normaler G. normal rock pressure
Gebirgsentspannung *f* rock de-stressing
Gebirgsfalte *f* mountain fold
Gebirgsfluß *m* torrent, mountain river
Gebirgskette *f* mountain chain
Gebirgskörper *m* rock mass
Gebirgsmechanik *f* rock mechanics
Gebirgsschlag *m (mine)* bump, rock burst
  G. aus dem Festen extradosal burst
  G. aus der Zerrüttungszone intradosal burst
Gebirgsstörung *f (geol)* disturbance
Gebirgsstraße *f* mountain road
Gebirgsstrecke *f (ry)* mountain railway
Gebirgswand *f* mountain face
Gebläse *n* fan; blower; air pipe *(blast furnace)*; blowpipe; gas blower (pump); *(aero)* supercharger
Gebläsebrenner *m* blowpipe
Gebläsedampfmaschine *f* blowing engine
Gebläseegreniermaschine *f* air-blast gin
Gebläseeinlauf *m* compressor inlet
gebläsegekühlt fan-cooled
Gebläsehäcksler *m (ag)* cutter-blower
Gebläsekühlung *f (aut)* forced-draught cooling
Gebläsekupolofen *m* blast cupola
Gebläseluftkühlung *f (aut)* fan-type air cooling
Gebläsemotor *m* forced-induction engine
Gebläsesand *m* sandblasting sand
Gebläsewind *m* blast
  vorgewärmter G. hot blast
gebräch *(mine)* friable
gebrauchen to use; to apply; to utilize
Gebrauchsanweisung *f* instructions for use, operating instructions
Gebrauchsbrille *f* wearer's spectacles
Gebrauchsdauer *f (mach)* service life; *(ch)* spreadable life
Gebrauchsgrafik *f* advertizing (commercial) art
Gebrauchslage *f* position of application (use)
Gebrauchslast *f* service (working) load
Gebrauchsmikroskop *n* routine-work microscope
Gebrauchsmuster *n* **[/ eingetragenes]** registered design
Gebrauchsmusteranmeldung *f* application for utility patent
Gebrauchsmusterschutz *m* registration of designs (trade marks), patent of utilization
Gebrauchsspannung *f* service voltage
Gebrauchswasser *n* process water
gebremst *(nucl)* moderated
gebrochen *(math)* fractional *(function)*
Gebühr *f* charge, rate, fee
  ermäßigte G. reduced rate
Gebührenansage *f (telec)* rate notification
Gebührenanteil *m* proportion of charge
Gebührenanzeiger *m* tax indicator
Gebührenbefreiung *f* exemption from fees
Gebührenberechnung *f* calculation of charge
Gebühreneinheit *f* tariff unit
Gebührenermäßigung *f* reduction of charges
Gebührenerstattung *f* return of charges
gebührenfrei free of charge; exempt from fees, duty-free
Gebührenmelder *m* tax indicator *(telephony)*
Gebührenminuten *fpl* charged minutes *(telephony)*
gebührenpflichtig liable to charge (duty), chargeable
Gebührentafel *f* table of rates
Gebührenzone *f* exchange area
gebündelt bunched *(energy)*; narrowed, focused, collimated *(beam)*
gebunden attached, linked to; bound
  anorganisch g. vitrified-bonded *(grinding wheel)*
  koordinativ g. *(ch)* coordinately linked
gechlort chlorinated
Gedächtnis *n (dat)* memory, storage
Gedächtniseffekt *m* memory effect
gedämpft attenuated, damped; muffled *(sound)*; *(mach)* cushioned
  aperiodisch g. dead-beat
Gedankenstrich *m* dash, em-dash, em-rule
gediegen native, in pure state; solid
gedimpelt indented
Gedinge *n (mine)* contract work, stint
Gedingearbeiter *m* contract worker
Gedingeart *f* kind of contract
Gedingelohn *m* bargain
gedopt doped
gedrallt helical[ly], with helical flutes
gedrängt packed closely together
  dicht g. compact
gedübelt *(wood)* dowelled
geeignet für adapted for
geerdet earthed, *(Am)* grounded
  induktiv g. reactance earthed
Gee-Verfahren *n (el)* Gee-system
Gefahr *f* danger
Gefährdungsfaktor *m (nucl)* danger coefficient
Gefährdungsspannung *f* danger (excess) voltage
Gefahrenfeuer *n* hazard beacon
Gefahrenklasse *f* dangerous-materials class *(chemicals)*
Gefahrenpunktabstand *m* advance of a signal
Gefahrensignal *n* danger signal
Gefahrenstelle *f* danger point
Gefahrenstrahlung *f* danger radiation
Gefahrenzone *f (nucl)* danger space (zone)
Gefahrenzulage *f* danger money, hazard bonus
Gefälle *n* gradient, pitch, slope; drop, fall
  adiabatisches G. adiabatic gradient

barometrisches G. barometric gradient
Gefällebahnhof *m* station on an incline
Gefällebremse *f (ry)* gradient brake
Gefällebremsung *f* holding brake *(electric traction)*
Gefälledruck *m* pressure due to head of water
Gefälleförderung *f* gravity haulage
Gefällehöhe *f* difference of level
Gefällekondensator *m (therm)* dry condenser
Gefällewechsel *m* change in gradient
gefalzt rebated
gefärbt coloured, dyed
im Garn g. *(text)* yarn-dyed
im Kammzug g. *(text)* top-dyed
in der Masse g. beater-dyed
in der Wolle g. stock-dyed
Gefäß *n* vessel, container, cup
G. mit Bodenentleerung bottom-dumping skip
evakuiertes G. evacuated vessel
Gefäßaufzug *m* skip hoist
Gefäßdruck *m* vessel pressure
Gefäßförderung *f (mine)* skip hoisting
Gefäßinhalt *m* bucket capacity
Gefäßversuch *m (ag)* pot experiment
Gefechtsaufklärer *m (mil)* pathfinder
Gefechtskopf *m* warhead
Gefechtsschießen *n* battle practice
gefedert tongued *(woodwork)*; *(mach)* spring-loaded, cushioned
gefertigt / aus einem Stück solid, one-piece
gefeuert / mechanisch machine-fired
Geflecht *n* 1. *(build)* plaiting; 2. *(text)* braided material
Geflechteinlage *f* braid
Gefluder *n (mine)* flume
Geflügel *n* poultry
Geflügelfarm *f* poultry farm
Geflügelhalter *m* poultry farmer
Geflügelhaltung *f* poultry farming
Geflügelhof *m* poultry farm
Geflügelstall *m* poultry house
Geflügelzüchter *m* poultry breeder
geformt profiled, formed, shaped
im Gesenk g. die-formed
Gefrieranlage *f* refrigerating plant
gefrieren to freeze, to congeal; to frost
Gefrierfleisch *n* frozen meat
gefriergetrocknet freeze-dried
Gefrierkonserve *f* deep frozen pack
Gefrierloch *n (mine)* freezing hole
Gefrierpunkt *m* freezing point
Gefrierpunktmesser *m* cryoscope
Gefrierpunktserniedrigung *f* lowering (depression) of freezing point, freezing point depression
molekulare G. molecular lowering of freezing point, freezing constant
Gefrierschacht *m (mine)* freezing shaft
Gefrierschutzmittel *n* freezing preventive, antifreeze [mixture]
Gefriertrocknung *f* freeze-drying
Gefrierverfahren *n (mine)* freezing method (shaft sinking)
Gefüge *n* structure; grain texture; texture
G. der Schleifscheibe structure of the grinding wheel; grain spacing, spacing of abrasive grains
dichtes G. dense structure
feinkörniges G. fine-grained structure
grobes G. coarse-grained structure
großporiges G. wide spacing
kleinporiges G. close spacing
zeilenförmiges G. bended structure
Gefügeänderung *f* constitutional change
Gefügeaufbau *m* structure, structural constitution
Gefügeausbildung *f* structural shape; formation of the structure
Gefügebestandteil *m* constituent
Gefügebild *n* micrograph
Gefügebildung *f* formation of structure
Gefügegrundmasse *f* / **metallische** metal matrix
Gefügeneubildung *f* structural transformation
Gefügeumwandlung *f* structural transformation
Gefügeuntersuchung *f* examination of structure
Gefühlsgewindebohrmaschine *f* sensitive tapping machine
Gefühlshebel *m* feed handle
Gefühlshebelvorschub *m* sensitive [hand] feed
Gefühlsschraube *f* ratchet stop
gefühlvoll sensitive *(drilling)*
gefüllt / mit Luft air-filled *(cavitation)*
gegabelt forked, fish-tailed
Gegenamperewindung *f* back (counter) ampere turn, demagnetizing turn
Gegenamt *n (telec)* distant terminal (exchange)
Gegenantrieb *m* counterdrive
Gegenbewegung *f* countermovement, countermotion
Gegenbindung *f (text)* counter binding
Gegendrehungskurbeln *fpl* crank moving in opposite directions, crossed parallelogram
Gegendruck *m* back pressure, counterpressure
Gegendruckarbeit *f* work due to back pressure, negative work
Gegendruckkolben *m* dummy piston
Gegendruckturbine *f* reaction (back-pressure) turbine
Gegendruckventil *n* governing (back-pressure, counterpressure) valve
Gegendruckwalze *f* back-pressure roller; counterpressure roller
Gegenecho *n (telec)* talker echo
Gegenelektrode *f* counter (opposite) electrode;

(telev) back-plate electrode; back plate *(microphone)*

gegenelektromotorisch back (counter) electromotive

Gegen-EMK *f* back (counter) e.m.f.

Gegenfarbe *f* complementary colour

Gegenfeder *f* reacting spring

Gegenfeld *n (el)* opposing field

Gegenfeuer *n* passing of the flames to the front of furnace

Gegenflußsystem *n (rock)* return-flow system

Gegengewicht *n* 1. *(mach)* counterweight, counterbalance; 2. counterpoise *(earth)*; 3. balancing capacitance *(aerial)*, capacity earth; earth screen

G. / mit counterbalanced, counterweighted

G. im Bremsberg *(mine)* ground-hog

Gegengift *n* antidote

Gegenhaltedruck *m* counterpressure

Gegenhalten *n* holding-up *(riveting)*

Gegenhalter *m* 1. holding-up hammer, bucking bar, dolly *(riveting)*; 2. back rest (broach); 3. overhanging arm, overarm *(milling machine)*; 4. [planer] poppet *(clamping)*

Gegenhalterschere *f* overarm braces, arbor (arm) brace *(milling machine)*

verstellbare G. adjustable overarm braces

Gegenhalterstift *m* holder sinker

Gegenhalterstütze *f* overarm support *(milling machine)*

Gegenimpedanz *f* negative-sequence field impedance

Gegeninduktion *f* [coefficient of] mutual induction

Gegeninduktivität *f* mutual inductance

Gegenkapazität *f* mutual capacitance

Gegenkathete *f* side opposite

Gegenkatode *f* anticathode

Gegenkeil *m* tightening key; gib; cotter

Gegenkiel *m (ship)* top keel

Gegenkolbenmotor *m* horizontally opposed engine

Gegenkolbenrudermaschine *f* opposed ram steering gear

Gegenkommer *m* meeting vessel, meeter

Gegenkomponenten *fpl* **eines Dreiphasensystems** *(el)* negative sequence components of a system of three-phase quantities

Gegenkompoundierung *f (el)* differential compounding (excitation)

Gegenkompoundwicklung *f (el)* counter-compound winding, differential (reverse) compound winding

Gegenkontakt *m* opposite contact

Gegenkoordinate *f* **eines Dreiphasensystems** negative sequence co-ordinate of a three-phase system

Gegenköper *m (text)* reverse twill

Gegenkoppelspannung *f* degenerative voltage

Gegenkopplung *f (el)* degeneration, degenerative (reversed) feedback, negative retroaction; negative feedback

Gegenkopplungsdemodulator *m* demodulator operating with negative feedback

Gegenkopplungsverstärker *m* negative feedback amplifier

Gegenkopplungsweg *m (el)* feedback loop

Gegenkraft *f* counterforce

elektromotorische G. electromotive counterforce, counter-electromotive force

Gegenkrümmung *f* reverse curve

Gegenkurbel *f* crank with drag link

Gegenkurs *m* counter-course, reciprocal (opposite) course

Gegenlager *n* end support, outboard (boring stay) bearing, outer support bearing *(horizontal boring mill)*

Gegenlast *f* counterweight

Gegenlauffräsen *n* conventional (up-cut) milling, cutting up

Gegenlauffrässchnitt *m* conventional cut

gegenläufig *(moving or running)* in opposite directions

g. zu countercurrent to

Gegenlaufmotor *m* opposed piston-type motor

Gegenlaufstreichmaschine *f* contra coater *(papermaking)*

Gegenlaufturbine *f* turbine with nozzles and blades rotating in opposite directions

Gegenlenker *m (mach)* counter-guide link

Gegenlicht *n* counterlight

Gegenlichtaufnahme *f* against-the-light picture

Gegenlichtblende *f (telev)* flag; lens shade *(photography)*; *(auto)* anti-dazzle screen

Gegenmitsprechen *n* side-to-phantom far-end crosstalk

Gegenmutter *f* lock (check, binding) nut, *(Am)* jam nut

Gegennebensprechen *n* far-end cross-talk, receiving-end cross-talk

Gegenpassat *m* anti-trade winds

Gegenphase *f (el)* anti-phase, reverse (opposite) phase

gegenphasig *(el)* inversely (oppositely) phased, in phase opposition

Gegenplatte *f* guard plate; counter punch *(embossing)*

Gegenpol *m* antipole, counterpole, opposite pole

Gegenpropeller *m* counter-propeller, contra-propeller

Gegenrad *n* mating gear

Gegenreaktanz *f (el)* demagnetizing reactance

Gegenreaktion *f* opposing reaction

Gegenregelung *f* counter-regulation *(turbine)*

Gegenreihenschaltung *f (el)* opposed series connection

Gegenrichtung *f* opposite direction
Gegenring *m* clamping (packing) ring
Gegenruder *n* contra-guide rudder, contra-rudder
Gegenschaltung *f (el)* anti-parallel coupling; connection in opposition
Gegenschaltungsmethode *f* opposition method
Gegenscheinleitwert *m (el)* transadmittance
Gegenschlag *m* / **atomarer** atomic (nuclear) retaliation
Gegenschraube *f* contra-propeller, counter-propeller
Gegenschwingsieb *n* screen with two counter-balanced frames
Gegensee *f* head sea
Gegensegler *m* meeting vessel, meeter
Gegenseil *n* check band *(selfactor)*
Gegenseitigkeit *f (math, ph)* reciprocity
Gegenspannung *f* back-potential, counter-voltage, opposing voltage
Gegenspant *n (ship)* reverse frame
Gegensprechanlage *f* intercommunication system; talk-back circuit
Gegensprechbetrieb *m* exploitation of a duplexed line, duplex operation
Gegensprecheinrichtung *f* **in Brückenschaltung** bridge duplex system
Gegensprechgerät *n* / **tragbares** walkie-talkie
Gegensprechleitung *f (el)* duplex circuit
Gegensprechsatz *m (el)* duplex set
Gegensprechschaltung *f* duplex (talk-back) circuit
Gegensprechtelefonie *f* duplex telephony
Gegensprechverkehr *m* two-way speech (telephone) communication
Gegenstand *m (opt)* object
Gegenständer *m (mach)* end column
Gegenstation *f* opposite (remote) station
Gegenstelle *f* counter terminal, fellow station
Gegenstrom *m* 1. *(el)* countercurrent, back (reversed) current; 2. *(ph)* counterflow
Gegenstrombohranlage *f* counter-flush drilling rig
Gegenstrombremsung *f* regenerative (countercurrent) braking
Gegenstromdestillation *f* countercurrent distillation
Gegenstromextraktion *f* countercurrent extraction
Gegenstromkessel *m* countercurrent boiler
Gegenstromklassierer *m* countercurrent classifier
Gegenstromkolonne *f* countercurrent column
Gegenstromkondensation *f* counterflow condensation
Gegenstromkondensator *m* countercurrent condenser
Gegenstromkühler *m* countercurrent condenser
Gegenstromkühlturm *m* counterflow cooling tower
Gegenstrommethode *f* inverse current method, countercurrent method
Gegenstrommischer *m* countercurrent mixer
Gegenstromprinzip / **nach dem** counter-currently
Gegenstromschutzrelais *n* reverse-current protective relay
Gegenstromtrocknung *f* countercurrent flow drying
Gegenströmung *f* foul tide; countercurrent
Gegenstromverfahren *n* countercurrent principle
Gegenstromzentrifuge *f* countercurrent centrifuge
Gegentaktanordnung *f* push-pull arrangement
Gegentaktbetrieb *m* push-pull action
Gegentaktdemodulator *m* push-pull detector
Gegentaktdetektor *m* push-pull detector
Gegentaktendstufe *f* push-pull output stage
Gegentaktfrequenzverdoppler *m* push-pull frequency doubler
Gegentaktgleichrichter *m* push-pull rectifier, full-wave rectifier
Gegentaktmodulator *m* balanced modulator
Gegentaktschaltung *f* push-pull circuit (connection)
Gegentaktschwingkreis *m* split tank circuit
Gegentakttransformator *m* push-pull transformer
Gegentaktverstärker *m* balanced (push-pull) amplifier
Gegentaktzylinder *m* push-pull cylinder
Gegenüberhöhung *f (ry)* counter cant of the outer rail [at curve]
gegenüberliegend opposed; opposite
Gegenverbunderregung *f (el)* differential excitation
Gegenverbundgenerator *m (el)* differential compound generator
Gegenverbundmotor *m* differential compound [-wound] motor, decompounded motor
Gegenverkehr *m* intercommunication
Gegenwelle *f* countershaft
Gegenwicklung *f (el)* opposing winding
Gegenwind *m* head wind
Gegenwinder *m (text)* counter faller
Gegenwinkel *m* corresponding angle
Gegenwirkleitwert *m (el)* intrinsic transconductance
Gegenwirkung *f* reaction, reverse action
Gegenzelle *f* counter-electromotive force cell *(of batteries)*; regulating cell
 alkalische G. alkaline counter-cell
Gegenzug *m* 1. *(ry)* train running in opposite directions, *(Am)* opposing train; 2. *(mach)* back pull *(wire drawing)*

Gegenzugsäge *f* saw cutting on the draw stroke
Gegenzugstuhl *m (text)* positive lift loom
geglättet / einseitig machine-glazed
gegliedert articulated
gegossen cast
Gehalt *n* salary
Gehalt *m* content
  G. an flüchtigen Bestandteilen volatile content
Gehaltsempfänger *m* salaried employee, salary earner
Gehaltsempfänger *mpl* salaried staff
Gehänge *n* attachment, hanger
gehärtet hardened
Gehäuse *n* housing, case, casing; enclosure; cabinet; jacket; kit
  G. mit Berührungsschutz screening enclosure
  G. mit Mantelkühlung double-cased enclosure
  abgeschirmtes G. screened enclosure, shielded case
  explosionsgeschütztes G. explosion-proof enclosure
  feuchtigkeitsgeschütztes G. moisture-proof enclosure
  freihängendes G. overhung casing *(turbine)*
  geschlossenes G. totally enclosed enclosure
  nichtverschiebbares G. stationary housing *(radial pump)*
  schlagwettergeschütztes G. flameproof enclosure
  schwallwassergeschütztes G. splash-proof enclosure
  schwenkbares G. swivelling block *(pump)*
  staubgeschütztes G. dust-proof enclosure
  strahlwassergeschütztes G. hose-proof enclosure
  tropfwassergeschütztes G. drip-proof enclosure
  wettergeschütztes G. weatherproof enclosure
Gehäuseantenne *f* bult-in aerial
Gehäusebohrung *f* casing bore *(pump)*
Gehäusedeckel *m* cylinder cover, cover of casing *(turbine)*
Gehäuseflansch *m* flange of casing *(turbine)*
Gehäuseführung *f* casing guide *(turbine)*
Gehäuseführungsbolzen *m* casing guide-bolt *(turbine)*
Gehäuserippe *f* casing rib *(turbine)*
Gehäusespindel *f* box peg
Gehäusewand *f* casing wall, wall of casing *(pump, etc)*
geheftet / auf Heftgaze sewn on gauze
  im Falz g. saddle-stitched
Gehege *n* enclosure
Geheimtinte *f* sympathetic ink
Geheimtür *f* jib door

gehen to walk, to go; to depart, to leave; to resign *(office)*, to quit; *(mach)* to run, to work, to function, to operate
  an Bord g. to go on board
  an Deck g. to go above
  an die Boje g. to pick up moorings, to head for a buoy, to go to a buoy
  auf die Reise g. to go upon the voyage
  in See g. to proceed to sea
  über Bord g. to break free, to go by the board
  über Stag g. to go about
  über Steuer g. to go astern
  unter Segel g. to make sail
  zu Bruch g. to fail, to rupture, to collapse
Geh-Gabelstapler *m* pedestrian-type fork truck
Gehöft *n* farm
Gehör *n* hearing
  scharfes G. acute hearing
gehorchen / dem Ruder to answer the helm
Gehörfehler *m* hearing defect
Gehörmesser *m* audiometer
Gehörpeilung *f* acoustic direction finding
Gehörprüfung *f* audiometry
Gehre *f* mitre
Gehrmaß *n* bevel way
Gehrung *f* mitre
Gehrungssäge *f* mitring (mitre-box) saw, saw for mitre cutting
Gehrungsschere *f* bevelling shear
Gehrungsschneidgerät *n* mitre cutting machine
Gehrungsschneidmaschine *f* mitre cutting machine
Gehrungsschnitt *m* mitre (bevel) cut
Gehrungsstoß *m* mitred joint
Gehweg *m* sidewalk, pavement
Gehwegplatte *f* paving flag, flagstone; quarry tile
Geigenbogen *m* violin bow
Geiger-Bereich *m* Geiger [counting] region
Geiger-Gebiet *n* Geiger [counting] region
Geiger-Müller-Zählrohr *n* Geiger-Muller counter (counting) tube, GM counter tube
Geiger-Plateau *n* voltage (Geiger) plateau
Geiger-Schwelle *f* Geiger threshold
Geiger-Zähler *m* geiger, Geiger counter
Geißfuß *m (build)* bird's-mouth
Geißlerröhre *f* Geissler tube
Geist *m (telev)* ghost, double image, echo, fold over
Geisterbild *n* ghost, (double) image
Geitau *n (ship)* brail
geizen *(ag)* to pinch
Geizen *n* pinching
gekapselt enclosed, encapsulated; totally enclosed
gekerbt indented, notched
geklebt glued
Geklirr *n* **[von Zahnrädern]** clashing [of gears]

geklüftet fissured
gekoppelt coupled
 direkt g. directly coupled
 induktiv g. inductively coupled
 kapazitiv g. capacitively coupled
 starr g. rigidly coupled
 überkritisch g. over-coupled
Gekrätz *n (mine)* dross
gekreuzt crossed, cruciform *(edges)*
gekröpft cranked; offset *(tool)*; gap *(machine bed)*
 doppelt g. two-throw *(crankshaft)*
 dreifach g. three-throw *(crankshaft)*
 einfach g. one-throw *(crankshaft)*
 nach rückwärts g. swan necked *(tool)*
gekrümmt *(math)* curved, curvilinear; crooked; *(build)* draped *(prestressing)*; warped
 abnormal g. allokurtic
 doppelt g. doubly-curved
 einfach g. singly-curved
 evolventenförmig g. involute
gekuppelt [inter]linked, [inter]connected, coupled; *(mach)* interlocked *(milling cutter)*
Gel *n* gel
geladen *(el)* charged
 einfach g. *(ph)* singly-charged
 negativ g. negatively charged
 positiv g. positively charged
gelagert pivoted; supported *(beam)*; *(mach)* [running] in bearing[s]
 beweglich g. freely supported
 drehbar g. pivoted
 gelenkig g. hinged
Geländeantrieb *m* four-wheel drive
Geländeaufnahme *f* ground survey
Geländebruch *m* soil failure
Geländebruchuntersuchung *f* circular arc method
Geländefahrzeug *n* cross-country vehicle
Geländegang *m* low range of gears
Geländegängigkeit *f* cross-country capacity (properties) *(of a car)*
Geländehöhe *f* grade (ground) level
Geländeoberkante *f* ground (grade) level
Geländer *n* guard (hand) rail, railing, balustrade, parapet
 G. mit Kettendurchzug *(ship)* chain guard
Geländereifen *m* cross-country tyre, ground-grip tyre, off-the-road tyre
Geländersprosse *f* baluster, banister
Geländerstab *m* baluster, banister
Geländerstütze *f* rail stanchion
Geländewagen *m* cross-country car
Gelatine *f* gelatin[e]
gelatineartig gelatinous, jelly
gelatinefrei gelatine-free
Gelatinerelief *n* gelatine relief
Gelatineschicht *f* gelatine film (coating)
Gelatinetafel *f* cake of gelatine
Gelatinewalze *f* gelatine roller
Gelatinierung *f* gelation
gelaugt in Salzlauge wet brined
 g. und getrocknet in Salzlauge dry brined
Gelbätze *f (text)* yellow discharge
Gelbbleierz *n* wulfenite, molybdate of lead
Gelbbrenne *f* pickling plant
Gelbeisenerz *n* xanthosiderite
Gelbglut *f* yellow heat; bright orange *(annealing)*
Gelbguß *m* yellow brass
gelblich yellowish
gelbstichig yellow-tinged
Gelbstrohstoff *m* mechanical (unbleached, yellow) straw pulp *(papermaking)*
Gelee *n* jelly
gelegen situated
Geleitschiff *n* convoy; escort vessel
Geleitschutz *m* convoy guard
Gelenk *n* hinge, [hinged] joint, pin[ned] hinge, pin[ned] joint
 elastisches G. flexible joint
Gelenkausleger *m* hinged jib
Gelenkbogen *m* articulated arch
 nachgiebiger G. articulated yielding arch
Gelenkbohrmaschine *f* **mit einstellbaren Spindeln** adjustable-spindle drilling machine, universal-joint type drilling machine
Gelenkegge *f (ag)* jointed harrow
Gelenkeinlage *f* waist pad
Gelenkflansch *m* flange with spherical seat
Gelenkführung *f* joint guide
Gelenkgetriebe *n* linkage
gelenkig articulated, jointed, hinged
Gelenkkappe *f* articulated (hinged, link) bar
Gelenkkette *f* link chain
Gelenkkupplung *f* joint coupling
 kardanische G. universal coupling
Gelenkpunkt *m* fulcrum point
Gelenkrohr *n* articulated pipe
Gelenkrohrverbindung *f* swivel pipe coupling
Gelenkrohrwelle *f* tubular shaft
Gelenkrunge *f* articulated (hinged) stanchion
Gelenkschiff *n* pinjoint ship
Gelenkspindel *f* universal-joint spindle
Gelenkspindelbohrmaschine *f* universal-joint multiple drilling machine, multi-spindle drilling machine with universal-joint driven spindles
Gelenkspindelhalter *m (mach)* universal-joint spindle holder
Gelenkstahlkappe *f* hinged steel bar
Gelenkstift *m* wrist pin
Gelenkstiftschlüssel *m* jointed fork spanner
Gelenkstrebe *f* joint strut
Gelenkstück *n* shank *(shoe)*
Gelenkträger *m* cantilevered and suspended beam
Gelenkviereck *n* four-bar linkage

überschlagendes (gekreuztes) G. crossed crank mechanism, crossed link quadrilateral

zylindrisches G. four-bar motion with axes of rotation parallel, linked quadrilateral

Gelenkwelle *f* propeller shaft; articulated shaft; universal shaft *(boring machine)*

G. mit Gleitgelenk propeller shaft with sliding joint

ausziehbare G. shaft having universal joints and telescopic connection

Gelenkwellenrohr *n* drive-shaft tube

Geleucht *n* [miner's] lamp

offenes G. naked light

gelieren to gelatinate, to gelatinize, to gel, to jell

Gelierung *f* gelling *(oil)*

gelocht pierced; punched

gelöscht slaked *(lime)*

gelöst *(ch)* solute

gelten für to be true for, to hold for

Geltungsdauer *f* period of availability *(ticket)*, validity of tickets

Gelzustand *m* gelatiniform state

gemasert *(wood)* mottled, grained

gemäßigt moderate; temperate

gemein *(math)* vulgar *(fraction)*

Gemeindebetrieb *m* municipal factory (service)

Gemeine *pl (print)* lower-case letters

Gemeinkosten *pl* overhead expenses (charges, cost)

Gemeinschaftsanschluß *m* party line, party connection

Gemeinschaftsantenne *f* community (common) aerial

Gemeinschaftsarbeit *f* joint work; team work

Gemeinschaftsbahnhof *m* joint station

Gemeinschaftsfernsehempfang *m* community television

Gemeinschaftsforschung *f* co-operative research

Gemeinschaftsleitung *f* party line

Gemenge *n* 1. mixture; 2. *(ag)* dredge corn

gemessen measured

ohne Rinde g. measured under the bark

über der Rinde g. measured over the bark

gemessert knife-cut *(veneer)*

Gemisch *n (ch)* mixture; conglomerate

azeotropes G. azeotropic mixture

binäres G. binary mixture

fettes G. rich mixture *(fuel)*

kältebeständiges G. antifreezing mixture *(fuel)*

zu fettes G. overrich mixture *(fuel)*

Gemischaufbereitung *f* carburetion *(fuel)*

Gemischregelventil *n* mixture control valve *(fuel)*

Gemischschmierung *f* petroil lubrication

Gemischstärkeprüfgerät *n* exhaust-gas analyzer

Gemischtbauflugzeug *n* composite plane

Gemischtbauweise *f* composite construction *(wood and iron)*

Gemmaschine *f* gemming machine

Gemsleder *n* chamois leather

gemuldet troughed

Gemüsebau *m* vegetable gardening (growing)

Gemüsekonserve *f* canned vegetables

Gemüsepflanzmaschine *f* vegetable planter

Gemüsesetzmaschine *f* vegetable planter

Gemüsesortiermaschine *f* vegetable grader

gemustert *(text)* figured, fancy

genagelt nailed

verdeckt g. secret-nailed

genarbt grained

genau exact; accurate; precise; true; correct; intimate *(knowledge)*

g. innerhalb einer Toleranz von true within [limits of], accurate within

Genaubohren *n* precision boring (drilling)

Genauguß *m* precision casting

Genauigkeit *f* accuracy, precision

G. der Kugelform sphericity

wiederholbare G. *(mach)* repetitive accuracy

Genauigkeitsbohrarbeit *f* precision boring operation

Genauigkeitsbohren *n* **mit Diamant (Hartmetallaufbohrwerkzeug)** diamond boring, *(Am)* borizing

Genauigkeitsbohrmaschine *f* precision boring machine

Genauigkeitsgrad *m* degree (order) of accuracy, degree (order) of precision

Genauigkeitsprüfung *f* precision test

Genauigkeitsschliff *m* precision grind[ing]

Genauschmiedestück *n* precision forging

genehmigen to license; to grant

Genehmigung *f* / **amtliche** licence

genehmigungspflichtig subject to authorization

geneigt inclined *(plane)*, battered, sloping, raked, slanting

in der gleichen Richtung g. *(geol)* monoclinal

Generalauftragnehmer *m (build)* main contractor

Generalbebauungsplan *m (build)* master plan

Generaldirektor *m* director general

Generalgedinge *n (mine)* general contract

Generalnenner *m (math)* lowest common multiple

Generalplan *m* general arrangement [plan] *(vessel)*; general plan

Generalreparatur *f* major overhaul

Generalüberholung *f* general overhaul

Generationenfolge *f* neutron cycle

Generationsdauer *f* generation time

Generator *m* generator; producer

G. in Blockschaltung unit-connected generator

akustischer G. acoustic generator

elektrostatischer G. electrostatic generator (accelerator)
polumschaltbarer G. multi-frequency generator
stoßerregter G. shock-excited oscillator
van de Graaffscher G. Van de Graaff generator (accelerator)
Generatorerde *f* main earth cable connection
Generatorgas *n* producer (generator) gas
Generatorgruppe *f* generator group, grouped generators; generating set
Generatorholz *n* producer gas wood
Generatorkohle *f* producer coal
Generatorreaktanz *f* generator reactance
Generatorsatz *m* generating set
Generatorschutz *m* generator protective system
Generatorspannung *f* generator voltage
Generatorwiderstand *m* generator impedance, internal resistance of a generator
genießbar edible; drinkable
genietet riveted
dreireihig g. triple-riveted
einreihig g. single-riveted
Genmutation *f (nucl)* gene mutation
genormt standardized
Genossenschaft *f* co-operative
Genossenschaftsbauer *m* co-operative farmer
genügen / einer Bedingung to satisfy (fulfil) a condition
Geochemie *f* geochemistry
geochemisch geochemical
Geodäsie *f* surveying, geodesy
höhere G. geodesy, geodetic surveying
niedere G. plane surveying
Geodät *m* land surveyor, geodesist
geodätisch geodetic[al]
geoelektrisch geoelectric
Geoelektrizität *f* terrestrial electricity, geoelectricity
Geograf *m* geographer
Geografie *f* geography
Geoid *n* geoid
Geologe *m* geologist
Geologenhammer *m* geologist's hammer
Geologie *f* geology
tektonische G. structural geology
geologisch geological
Geomechanik *f* geomechanics, rock mechanics
Geometrie *f* geometry
G. der Lage topology
G. der Schneide *(mach)* tool (cutting-edge) geometry
darstellende G. descriptive geometry
ebene G. plane geometry
elliptische G. elliptic geometry
euklidische G. Euclidean geometry
gute G. good geometry
nichteuklidische G. non-Euclidean geometry
schlechte G. poor geometry
sphärische G. spherical geometry
Geometriefaktor *m (nucl)* geometry factor
Geophysik *f* geophysics
geophysikalisch geophysical
geothermisch geothermal
geozentrisch geocentric
Gepäck *n* luggage, *(Am)* baggage
abgefertigtes G. registered luggage
herrenloses G. left (abandoned) luggage
Gepäckabfertigungsgebühr *f* registration charge
Gepäckabteil *n* luggage space
Gepäckannahme *f* luggage office
Gepäckaufbewahrung *f* left-luggage office
Gepäckaufbewahrungsfach *n* **[für Selbstbedienung]** customer-operated luggage locker *(railway station)*
Gepäckaufbewahrungsgebühr *f* left-luggage charge
Gepäckaufbewahrungsschein *m* left-luggage ticket
Gepäckaufzug *m* luggage lift, *(Am)* baggage lift
elektrischer G. *(Am)* electric dumbwaiter
Gepäckausgabe *f* luggage office, *(Am)* baggage office
Gepäckhalter *m* luggage grid
Gepäckhinterlegungsschein *m* left-luggage ticket
Gepäckkarren *m* lift-truck
Gepäcknetz *n* luggage rack (net), parcel rack (net)
Gepäcknetzträger *m* luggage-net bracket
Gepäckraufe *f* luggage rack
Gepäckraum *m* luggage compartment (hold)
Gepäckraumverschluß *m* luggage locker *(car)*
Gepäckrost *m* luggage grid
Gepäckschalter *m* luggage registration office
Gepäckschein *m* luggage waybill
gepackt / dicht closely packed
Gepäckträger *m* porter; carrier *(bicycle)*
Gepäckversicherung *f* luggage insurance
Gepäckwagen *m* luggage van, *(Am)* baggage car
gepanzert / mit Blei *(nucl)* lead-shielded
gepfeilt *(el)* backswept
gepflastert paved
geplant planned
gepulst *(nucl)* pulsed
gepulvert powdered
gerade straight *(line)*; even *(number)*; right *(cone)*
Gerade *f* straight line
geradeaus straight on; straight ahead
Geradeausanflug *m* straight-in approach
Geradeausempfänger *m* straight receiver
Geradeausschaltung *f* straight circuit *(receiver)*
Gerade-Gerade-Kern *m (nucl)* even-even nucleus
Geradengleichung *f* equation of a straight line

Gerade-Ungerade-Kern *m* *(nucl)* even-odd nucleus
Gerade-Ungerade-Kontrolle *f* *(dat)* odd-even check
Gerade-Ungerade-Prüfung *f* *(dat)* odd-even check
geradeverzahnt straight-tooth[ed]
geradflankig straight-profile
Geradförderer *m* in-line machine
Geradführung *f* straight-line motion; slide bar
Geradführungsbewegung *f* straight-line motion
Geradführungslänge *f* length of straight-line motion, stroke of straight-sliding link
Geradführungsmechanismus *m* straight-line mechanism
geradgenutet straight-fluted, with straight flutes (cutting edges)
Geradheit *f* straightness
  G. der Führungsbahn *(mach)* trueness (alignment) of the way
geradkettig *(ch)* straight-chain
geradlinig straight-line[d], rectilinear
  g. hin- und hergehend straight-line reciprocating
Geradlinigkeit *f* rectilinearity, straightness
Geradschubkurbel *f* straight-sliding link
  rotierende G. rotating crank
Geradschubkurvengetriebe *n* curved-wedge mechanism, curved cam surface effecting rectilinear motion
Geradsichtprisma *n* direct-vision prism
Geradspantenschiff *n* straight-framed ship
geradstämmig straight-boled
geradverzahnt straight-tooth[ed]
Geradverzahnung / mit straight-toothed
geradzahlig even-numbered
Geradzahn *m* straight tooth
Geradzahnkegelrad *n* straight[-tooth] bevel gear
Geradzahnkupplung *f* straight-tooth clutch
Geradzahnstirnrad *n* [straight] spur gear
Geraniol *n* *(ch)* geraniol
Gerät *n* device; appliance; *(telec)* set; instrument; unit; *(ag)* implement, tool; equipment; apparatus
  G. mit Vollnetzanschluß all-mains set, all-electric set
  G. zur Aufzeichnung von Gesprächen voice recording equipment, telephone voice recorder
  G. zur Lösung von Polynomen polynomial computer
  G. zur Vektorzerlegung complex plane analyzer
  allklimatisches G. pan-climatic equipment
  batteriebetriebenes G. battery-powered instrument
  elektronenoptisches G. electron optical device
  elektronisches G. electronic device
  geschlossenes G. one-piece unit
  kernphysikalisches G. nucleonic instrument
  miniaturisiertes G. miniaturized unit
  temperaturempfindliches G. temperature-sensing device
Geräte *pl* **/ landwirtschaftliche** farm implements
Geräteanordnung *f* arrangement of equipment
Geräteanschlußschnur *f* flexible cable (lead, cord)
Gerätebau *m* instrument manufacture, equipment production
  wissenschaftlicher G. scientific instrument manufacture
Gerätebauindustrie *f* instrument industry
Geräteelektronik *f* device electronics
Geräteform *f* device geometry
Gerätegenauigkeit *f* instrumental accuracy
Geräteglas *n* soft soda glass
Gerätekasten *m* equipment box
Gerätekontakt *m* device terminal
Geräteleistung *f* **/ theoretische** nominal handling rate *(excavator)*
Gerätenummer *f* serial number
Gerätephysik *f* device physics
Geräteplatte *f* electrical control board
Geräteraum *m* instrument compartment (bay), instrumentation section
Geräteschalter *m* appliance switch
Geräteschnur *f* flex, flexible, connecting cord
Geräteschuppen *m* implement (tool) shed
Gerätespitze *f* *(rock)* front compartment
Gerätesteckdose *f* coupler (set) socket, appliance plug-socket; convenience receptable
Gerätestecker *m* appliance coupler (plug, adaptor), [bush-type] connector, coupler (apparatus) plug
Gerätesteckerschalter *m* appliance plug switch
Gerätesteckvorrichtung *f* coupler
Geräteträger *m* implement carrier tractor
Geräteturnen *n* heavy gymnastics
Geräte- und Maschinenpark *m* equipment fleet, *(Am)* spread
Gerätschaften *pl* apparel *(of a vessel)*
gerauht *(text)* brush-finished; with a rough surface
Geräusch *n* *(ph)* noise, sound
  G. des Übertragungsweges circuit noise
Geräuschabstand *m* signal-to-noise ratio
geräuscharm [near-]silent
Geräuschbegrenzer *m* noise limiter
Geräuschbeseitigung *f* suppression of noise
Geräuschbewertung *f* noise weighting
Geräuschboje *f* acoustic (sonic) buoy
Geräuschdämmung *f* **des Wagenkastens** sound insulation of the body of a coach
geräuschdämpfend silencing; sound absorbing (absorbent)
Geräuschdämpfung *f* noise insulation, soundproofing; silencing

Geräuscheinheit *f* noise unit
Geräuschfilter *n* noise suppressor
geräuschlos noiseless, quiet
Geräuschmesser *m* noise meter
Geräuschpegel *m* noise level
Geräuschspannung *f* [audible] noise voltage
 frequenzbewertete G. psophometric voltage
Geräuschspannungsmesser *m* psophometer, circuit noise meter
Geräuschsperre *f* squelch circuit
Geräuschspiegel *m* noise level
Geräuschunterdrückung *f* noise suppression, squelch
Geräuschverhalten *n* **[der Zahnräder]** [gear] noise characteristics
Geräuschverhältnis *n* noise ratio
geräuschvoll noisy
Gerbanlage *f* tan-yard
gerbbar tannable
gerben to tan
 mit Lohe g. to bark tan
Gerber *m* tanner
Gerberei *f* 1. tanning [process]; 2. tannery
Gerbermesser *n* fleshing knife
Gerberträger *m (build)* cantilevered and suspended beam
Gerberwolle *f* tanner's (glover's) wool
Gerbfaß *n* tanning tumbler
Gerbgrube *f* tanning pit
gerbsauer tannic
Gerbsäure *f* tannic acid
Gerbsäurelösung *f* solution of tannic acid
Gerbsäuremesser *m (leath)* tannometer
Gerbstoff *m* tanning agent (material), tannin
 vegetabilischer G. vegetable tanning material
Gerbstoffbestimmung *f* determination of tanning matter
gerbstoffhaltig tanniferous
Gerbtrommel *f* tanning tumbler
Gerbung *f* tannage, tanning
 G. mit mineralischen Stoffen mineral tannage, tanning done by mineral treatment
 gemischte G. combination tannage
 nichtsaure G. non-acid tannage
 pflanzliche G. vegetable tannage
 sämische G. chamois dressing (tannage)
 vegetabile G. vegetable tannage
Gerbwert *m* tanning value
Gerechtsame *f (mine)* leased territory
gerichtet having direction *(quantity)*; directed *(force; ray)*; *(math)* sensed; oriented *(crystal)*; *(cont)* unidirectional
 einseitig g. unidirectional
Gerichtsbeschluß *m* legal resolution, judicial decree
Gerichtsverhandlung *f* hearing
gerieft corrugated; serrated; fluted
geriffelt serrated; fluted
gerillt fluted
gering poor *(e.g. quality)*; low *(force)*
geringaktiv *(nucl)* low-activity, low-level, warm
geringmächtig *(mine)* thin
gerinnen to congeal, to coagulate; to curdle *(milk)*
Gerinnsel *n (ch)* curds, concretion, coagulum
Gerinnungsmittel *n* coagulant, coagulum, runnet
Gerippe *n* framework
gerippt *(mach)* ribbed; reeded *(coin edge)*
Germanat *n (ch)* germanate
Germanischer Lloyd *m* German Lloyd
Germanium *n* germanium
 eigenleitendes G. intrinsic germanium
 n-leitendes G. n-type germanium
Germaniumdetektor *m* germanium detector (crystal)
Germaniumdiode *f* germanium diode
Germaniumflächendiode *f* germanium junction diode
Germaniumplättchen *n* germanium wafer (slab)
Germaniumsperrschichtgleichrichter *m* germanium junction rectifier
Geröll *n* debris; *(geol)* detritus, gravel
Gerste *f* barley
Gerstenkornbindung *f (text)* barleycorn (huckaback) weave
Gerstenmalz *n* barley malt
geruchlos inodorous, odourless; scentless
Geruchsbeseitiger *m* deodorant
Geruchsbeseitigung *f* deodorizing
Geruchverschluß *m* air-trap, stench-trap, water-trap, siphon trap
gerundet radiused; round
Gerüst *n* scaffold[ing], staging; framing
 fahrbares G. portable staging (scaffolding), jumbo
Gerüstbau *m* scaffolding
Gerüstbaum *m* gin pole
Gerüstbock *m* horse
Gerüststange *f* scaffold pole
Gerüststoffe *mpl (ch)* builders
gesamt total, entire, whole
Gesamtabmessung *f* overall dimension
Gesamtabsorption *f* total absorption
Gesamtabweichung *f* cumulative error
Gesamtaktivität *f (nucl)* total activity
Gesamtanalyse *f* **/ kritische** network analysis
Gesamtanodenspannung *f* final anode voltage
Gesamtansicht *f* full-view illustration
Gesamtausschaltdauer *f* total break time
Gesamtbelegschaft *f* total labour force; miners on colliery books; total staff
Gesamtbelegungszähler *m (telec)* traffic recorder
Gesamtbereich *m* overall range
Gesamtbetrag *m* total [amount]

Gesamtbild *n (telev)* vista shot; general (total) view
Gesamtbodenwirkungsgrad *m (nucl)* [overall] plate efficiency
Gesamtbrechung *f* total refraction
Gesamtbreite *f (nucl)* total [level] width
Gesamtdämpfung *f* total attenuation
Gesamtdiagramm *n* complete diagram
Gesamtdrehimpuls / halbzahliger half-integer total angular momentum
Gesamtdrehimpulsquantenzahl *f* total angular momentum
Gesamtdruck *m* total pressure
Gesamtdurchflutung *f (el)* ampere conductors *(of a winding)*
Gesamteinschaltdauer *f* total make time
Gesamtergebnis *n* total result
Gesamtfahrwiderstand *m* total resistance *(electric traction)*
Gesamtfehler *m* total (accumulated) error
Gesamtfeuchtigkeit *f* total moisture
Gesamtfilter *n* total filter
Gesamtfläche *f* total [floor] area, gross floor area
G. unter der Gaußkurve areas under the normal curve
Gesamtflözmächtigkeit *f (geol)* aggregate thickness
Gesamtflugdauer *f* total flight time
Gesamtflugzeit *f* total flight time
Gesamtförderhöhe *f* total head including internal hydraulic losses
Gesamtfunktion *f* overall function
Gesamtgehalt *m* **an Oberwellen** *(el)* total harmonic content
Gesamtgenauigkeit *f* overall accuracy
Gesamtgradationsverhältnis *n (telev)* overall-contrast ratio
Gesamtgüte *f (el)* Q loaded
Gesamtheit *f* ensemble; assembly; [finite] population *(statistics)*
statistische G. population
Gesamtheitsmittel *n* population mean *(statistics)*
Gesamthöhe *f* overall (total) height; total head *(liquid)*
Gesamtinformationsgehalt *m* gross information content
Gesamtkatalog *m* general catalogue
Gesamtklirrfaktor *m (el)* total harmonic distortion
Gesamtkörperbestrahlung *f s.* Ganzkörperbestrahlung
Gesamtlänge *f* total (overall) length
Gesamtlast *f* total load
Gesamtleistung *f* total power output
Gesamtmaß *n* overall dimension
Gesamtmasse *f* 1. total weight; 2. *(math)* population
Gesamtmeßbereich *m* total measuring range
Gesamtplan *m* **/ logischer** *(dat)* overall logical plan *(of a system)*
Gesamtpropulsionswirkungsgrad *m (ship)* overall propulsive efficiency
Gesamtquantenzahl *f* total quantum number
Gesamtquerschnitt *m* total (overall) cross-section
Gesamtresultierende *f* **[der Schnittkraft]** resultant force
Gesamtschaltstrecke *f (el)* length of break
Gesamtspannung *f* overall (total) voltage
Gesamtspannungsverluste *mpl* total voltage losses
Gesamtstrahlung *f (ph)* total radiation
Gesamtstrahlungspyrometer *n* total radiation pyrometer
Gesamtstreuung *f* process variability *(statistics)*
Gesamtstrom *m* total current
Gesamtstückzeit *f (mach)* cycle time
Gesamtteilfehler *m* total accumulative dividing error *(pitch error, gear)*
Gesamttonnage *f* total laid-up tonnage
Gesamttragfähigkeit *f* deadweight-carrying capacity, deadweight [all told]
Gesamttrennfaktor *m (nucl)* overall separation factor
Gesamt- und Spitzenzähler *m (el)* excess and total meter
Gesamtvergrößerung *f (opt)* total magnification
Gesamtverlust *m* overall (total) loss
Gesamtverstärkung *f (el)* overall amplification (gain)
Gesamtverzerrung *f* total distortion
Gesamtverzug *m (text)* total draft
Gesamtwasserhärte *f* total water hardness, total hardness of water
Gesamtwiderstand *m* total resistance
Gesamtwirkungsgrad *m* total (overall) efficiency; commercial efficiency
G. des Kessels total boiler efficiency
G. des Schiffsantriebes overall propulsive efficiency
Gesamtwirkungsquerschnitt *m (nucl)* total cross section
Gesamtzahl *f* total number
Gesamtzeit *f* total time
Gesamtzuladung *f* deadweight carrying capacity, deadweight [all told]
gesättigt *(ch)* saturated
geschäftlich commercial
Geschäftsanschluß *m* business telephone
geschaltet *(el)* connected; switched
geschichtet laminated; in layers; *(mine)* stratified; racked *(wood)*
Geschicklichkeitsflug *m* proficiency flight
Geschiebe *n* debris
Geschiebemergel *m* boulder clay

Geschirr *n (ag)* harness; gear *(cargo handling)*; dishes; *(text)* heald shaft
Geschirrdraht *m* harness wire
Geschirreinzug *m* / **falscher** *(text)* wrong draft
Geschirrhalter *m (text)* harness frame
Geschirrkette *f (text)* harness chain
Geschirrleder *n* harness leather
Geschirrspülmaschine *f* dishwashing machine, dishwasher
Geschirrspülmittel *n* dishwashing agent
geschleudert centrifugally cast *(concrete pipes)*
geschliffen *(mach)* ground; cut *(glass)*; sharpened *(tool)*
geschlitzt slotted; slitted, split *(reamer body)*
geschlossen totally enclosed *(apparatus)*; confined, closed
geschmacklos tasteless, free from taste
Geschmacksstoff *m* flavouring agent
Geschmacksstoffe *mpl* **für Arzneimittel** medicinal flavours
geschmeidig pliable; flexible; supple *(leather)*
g. machen to make pliable; to supple *(leather)*
Geschmeidigkeit *f* 1. flexibility; ductility; 2. *(build)* plasticity; 3. *(text)* drape
geschmiert lubricated
geschmolzen molten
Geschoß *n* 1. *(mil)* projectile; missile; 2. *(build)* floor, storey
scharfes G. live projectile
Geschoßansetzer *m (mil)* rammer
Geschoßbau *m* multi-storey building
Geschoßflächendichte *f (build)* plot ratio
Geschoßgarbe *f* cone of fire
geschoßhoch *(build)* storey-high
Geschoßhöhe *f (build)* floor-to-floor height
Geschoßkern *m* core of projectile
Geschoßmantel *m* jacket of projectile
Geschränk *n* / **aufgehängtes** suspended furnace front
einfaches G. furnace front with single door
Geschränkrahmen *m* fire door frame
geschraubt screwed; bolted
Geschütz *n* gun
schweres G. heavy gun
Geschützbettung *f* gun base
Geschützbronze *f* gun metal
Geschützführer *m* gun-captain
Geschützlafette *f* gun carriage
Geschützpanzerkampfwagen *m* gun tank
geschützt / gesetzlich protected by patent
Geschützturm *m (ship)* gunhouse
geschwächt weakened
geschweißt welded
Geschwindigkeit *f* speed *(scalar)*; velocity *(vector)*; rate *(of a process)*
G. bei Stundenleistung speed at one-hour rating
G. der Energiefortpflanzung velocity of energy transmission
G. der Entionisierung deionization rate
G. des Kriechganges *(mach)* creeping speed
G. im Apogäum apogean (apogeal) velocity
G. im Perigäum perigean (perigeal) velocity
G. in größter Erdferne apogean (apogeal) velocity
G. in größter Erdnähe perigean (perigeal) velocity
G. über Grund 1. *(aero)* ground speed, velocity of an aircraft relative to the ground; 2. *(ship)* overground speed, speed over the bottom
abnehmende G. retarded (decreasing) speed, retarded (decreasing) velocity
absolute G. absolute velocity
astronautische G. escape velocity, speed of escape
augenblickliche G. instantaneous velocity
charakteristische G. characteristic velocity
durch Reibung verminderte G. velocity reduced by friction
generalisierte G. generalized velocity
gleichförmige G. uniform speed (velocity)
kosmische G. escape velocity, speed of escape
kritische G. critical velocity *(Reynolds)*
lineare G. linear velocity
makroskopische G. macroscopic velocity
mäßige G. easy speed
mittlere G. average speed
mittlere quadratische G. mean square velocity
parabolische G. parabolic velocity
resultierende G. resultant velocity
thermische G. *(nucl)* thermal velocity
ungleichförmige G. nonuniform velocity
wahrscheinlichste G. most probable velocity
Geschwindigkeitsabnahme *f* deceleration, speed reduction; decrement (decrease, drop) of speed
Geschwindigkeitsabstufung *f* velocity staging *(turbine)*
Geschwindigkeitsanalysator *m* velocity analyzer
Geschwindigkeitsänderung *f* / **stufenlose** *(mach)* stepless speed changing
Geschwindigkeitsanzeige *f* **einer Langsamfahrstelle** approached restricted speed aspect
Geschwindigkeitsaufschaltung *f* rate (speed) feedback
Geschwindigkeitsbegrenzer *m* speed-limiting device
Geschwindigkeitsbegrenzung *f* speed limit[ation], speed restriction
Geschwindigkeitsbeschränkung *f* speed limit[ation], speed restriction
Geschwindigkeitsdiagramm *n* velocity diagram (triangle)

Geschwindigkeitsdreieck *n* velocity triangle (diagram)
Geschwindigkeitserzeugung *f* production of velocity *(turbine)*
Geschwindigkeitsfehler *m* velocity error (misalignment)
Geschwindigkeitsfokussierung *f* velocity focusing
Geschwindigkeitshöhe *f* velocity head
Geschwindigkeitskonstante *f* reaction rate, velocity constant
Geschwindigkeitskontrolle *f* speed checking
Geschwindigkeitskurve *f* velocity curve
Geschwindigkeitsmesser *m* tachometer, speed indicator
Geschwindigkeitsmessung *f* measuring (measurement) of the velocity, measuring (measurement) of the speed
optische G. optical method of speed measurement
Geschwindigkeitsminderung *f* deceleration
Geschwindigkeitsminderungsventil *n* deceleration valve
Geschwindigkeitsmodulation *f* velocity modulation
Geschwindigkeitsparallelogramm *n* parallelogram of velocities *(turbine)*
Geschwindigkeitsplan *m* velocity diagram (triangle)
Geschwindigkeitspotential *n* velocity potential
Geschwindigkeitsprobefahrt *f* speed trial trip
Geschwindigkeitsregler *m* speed governor
Geschwindigkeitsschaubild *n* velocity diagram
Geschwindigkeitsselektor *m* *(nucl)* velocity selector
mechanischer G. mechanical [velocity] selector, chopper, mechanical selector of neutrons
Geschwindigkeitsspektrograf *m* magnetic spectrograph
Geschwindigkeitsstufe *f* velocity stage *(turbine)*
Geschwindigkeitsverlust *m* loss of velocity *(turbine)*
Geschwindigkeitsverteilung *f* distribution of velocity, velocity distribution
G. in runden Rohren velocity distribution in circular pipes
Maxwellsche G. Maxwellian distribution of velocity, Maxwell's velocity distribution
Geschwindigkeitsverteilungsgesetz *n* law of distribution of velocities
Fermi-Dirac-Sommerfeldsches G. Fermi-Dirac-Sommerfeld velocity-distribution law
Geschwindigkeitszunahme *f* increment (increase, rise) of speed
Geselle *m* journeyman
Gesellschaft *f* company
G. mit beschränkter Haftung limited liability company
eingetragene G. registered company
Gesellschaftsleitung *f* party line *(telephony)*
Gesellschaftsvermögen *n* partnership assets
Gesellschaftsvertrag *m* partnership agreement
Gesellschaftswissenschaft *f* social science
Gesenk *n* 1. *(mine)* winze; 2. *(met, mach)* die, cavity block
G. mit gebrochener Teilfuge lock die
offenes G. open (plain) die
Gesenkeinsatz *m* die insert
Gesenkfräsen *n* die sinking
Gesenkfräser *m* die-sinking cutter
G. für runde Ausnehmungen cherry
kegeliger G. taper-flute die-sinking cutter
Gesenkfräsmaschine *f* die-sinking machine, die-milling machine
Gesenkgravur *f* die cavity
Gesenkherstellung *f* die making
Gesenkhohlraum *m* die impression
Gesenkpresse *f* die (stamping) press
Gesenkpreßstück *n* die pressing
Gesenkscheibe *f* *(mine)* cross-section of a staple pit
Gesenkschmieden *n* die (drop) forging, swaging
Gesenkschmiedepresse *f* drop forging press
Gesenkschmiederohling *m* dummy
Gesenkschmiedestück *n* drop forging
Gesenkstahl *m* die steel
Gesenk- und Formennachformfräser *m* die and mould copy milling machine
Gesetz *n* law *(science)*
G. der großen Zahlen law of great numbers
G. der konstanten Proportionen law of constant proportions
G. der multiplen Proportionen law of multiple proportions
G. der quadratischen Abnahme inverse square law
G. der thermodynamischen Ähnlichkeit law of thermodynamic similitude
G. der übereinstimmenden Zustände law of corresponding states
G. der Winkelkonstanz law of constant angles
G. der Wirkung und Gegenwirkung law of action and reaction
G. des radioaktiven Zerfalls law of radioactive disintegration (decay)
G. von der Erhaltung der Energie law of conservation of energy
G. von der Erhaltung der Masse law of conservation of mass (matter)
G. von der Erhaltung des Drehimpulses law of conservation of moment of momentum
Ampèresches G. Ampère's law
assoziatives G. associative law

Avogadrosches G. Avogadro's law
Biot-Savartsches G. Biot and Savart's law
Boyle-Mariottesches G. Boyle's law, Mariotte's law
Brewstersches G. Brewster's law
Coulombsches G. Coulomb's law
Curie-Weißsches G. Curie-Weiss' law
Daltonsches G. der multiplen Proportionen Dalton's law of multiple proportions
Daltonsches G. der Partialdrücke Dalton's law of partial pressures
Darcysches G. Darcy's law
distributives G. distributive law, law of distribution
Duane-Huntsches G. Duane-Hunt relation, Duane and Hunt's law
Faradaysches G. Faraday's law
Gay-Lussacsches G. Gay-Lussac's law
Hagen-Poiseuillesches G. Poiseuille formula
Hookesches G. Hooke's law
Joulesches G. Joule's law
Keplersches G. Keplerian law
Kirchhoffsches G. Kirchhoff's law
kommutatives G. der Addition commutative law of addition
Lambertsches G. Lambert's law
Laplacesches G. Laplace's law
Moseleysches G. Moseley's law
Ohmsches G. Ohm's law
Pascalsches G. Pascal's law
Poiseuillesches G. [Hagen-]Poiseuille law
Raoultsches G. Raoult's law
Stefan-Boltzmannsches G. Stefan's law, Stefan-Boltzmann law
Torricellisches G. Toricelli's law
Webersches G. Weber-Fechner law of hearing
Wiedemann-Franzsches G. Wiedemann-Franz law
Gesetzestreue *f* obedience to law
Gesetzesübertretung *f* transgression of a law
Gesichtsfeld *n* field of vision (view), visual field
Gesichtskreis *m* 1. range of vision; 2. [sea] horizon
Gesichtswinkel *m* visual (viewing) angle, angle of sight
gesiebt screened
Gesims *n* moulding, cornice, ledge
Gesimsbrett *n* fascia board
Gesimshöhe *f* cornice level
gesintert cemented, sintered
gespalten knapped *(stone)*; split
Gespann *n* team *(of horses)*; yoke *(of oxen)*
Gespanndrillmaschine *f* horse-drawn drill
Gespannegge *f* horse-drawn harrow
Gespanngrubber *m* horse-drawn cultivator
Gespannguß *m* group casting
Gespannhackmaschine *f* horse hoe
Gespannheurechen *m* horse rake
Gespannmähbinder *m* horse-drawn binder
Gespannpflug *m* horse plough
Gespannplatte *f* bottom pouring plate, group teeming stool
Gespannspritze *f* horse spray
Gespannstäuber *m* horse-drawn duster
gespannt tensioned
straff g. highly tensioned
Gespannzug *m* animal traction
Gespannzuggrubber *m* horse cultivator
Gesperre *n* locking mechanism
Gespräch *n* [telephone] call
G. mit Voranmeldung préavis call
G. zwischen zwei Sprechstellen station-to-station call
abgehendes G. outgoing call
ankommendes G. incoming call
dringendes G. express call
gebührenpflichtiges G. chargeable call
gestörtes G. faulty call
zwischenstaatliches G. international call
Gesprächsanmeldung *f* booking of a call
Gesprächsdauer *f* duration of the (a) call, length of conversation
gebührenpflichtige G. chargeable time
Gesprächseinheit *f* unit call, *(Am)* initial period
Gesprächsgebühr *f* call fee (charge)
Gesprächskontrolluhr *f* call control watch
Gesprächsminuten *fpl* duration of a (the) call in minutes
Gesprächsuhr *f* time recorder, chargeable time indicator
Gesprächszähler *m* call counter, call (counting) meter
Gesprächszähltaste *f* meter key
Gesprächszeit *f* conversation time
Gesprächszeitmesser *m* timing register
gesprungen cracked *(ceramics)*
gespundet tongued and grooved, matched *(woodwork)*, joined by grooves
gestäbt beaded
gestaffelt staggered, in echelon; graded
g. sein *(mach)* to be progressively higher *(teeth)*
Gestalt *f* shape, form; design
gestalten to shape; to design; to sculpture
g. in Holz to sculpture, to carve wood
g. in Stein to sculpture, to chisel stone
g. in Ton to sculpture, to model clay
stromlinienförmig g. to streamline
gestaltlos amorphous, formless; shapeless
Gestaltlosigkeit *f* shapelessness
Gestaltsänderung *f* change of shape, non-dilatational strain

Gestaltsfestigkeit *f* form strength

Gestänge *n* 1. *(mach)* leverage; connecting rod; linkage; 2. *(el)* strut, support; 3. *(mine)* track

fliegendes G. movable track, slide rail

Gestängebremse *f* linkage brake

Gestängedruckluftbremse *f* air-operated linkage brake

Gestängegewicht *n* weight of connecting rod

Gestängerohr *n* drill pipe

Gestängeziehvorrichtung *f* rod puller

gestanzt pierced

gestapelt racked

der Länge nach g. straw-piled

luftig g. spacing piled, piled with sufficient air-space

Gestehungskosten *pl* cost of production, actual cost, manufacturing expenses

Gestehungspreis *m* cost price

Gestein *n* rock

hereingeschossenes G. debris, muck

metamorphes G. metamorphic (metamorphosed) rock

taubes G. attle, dead ground, deads, waste rock

Gesteinsablösen *n (mine)* slab

Gesteinsanker *m (mine)* rock (roof) bolt

Gesteinsarbeit *f (mine)* deadwork

Gesteinsbalken *m (mine)* rock beam

gesteinsbildend rock-forming

Gesteinsbohrer *m* rock drill

Gesteinsbohrmaschine *f* rock drilling machine

Gesteinseigenschaften *fpl* properties of the rock

Gesteinsfestigkeit *f* rock strength

Gesteinsgang *m* dike

zusammengesetzter G. composite dike

Gesteinshauer *m* stone man, cutter

Gesteinsmehl *n* powdered mineral[s], mineral filler

Gesteinsmittel *n* dirt band

Gesteinsplatte *f (mine)* rock beam, slab

Gesteinssprengstoff *m* non-permitted explosive

Gesteinsstaub *m* stone dust

Gesteinsstaubverfahren *n (mine)* stone dusting

Gesteinsstrecke *f* drift

Gestell *n* 1. *(mach)* frame, bed; stand; *(el)* frame *(of a switch gear)*; [apparatus] rack *(telephony)*; 2. fixed link *(kinematics)*; 3. hearth *(blast furnace)*

freitragendes G. self-supporting rack

tragendes G. supporting frame

Gestellausführung *f* rack mounting style, rack mounted version

Gestellbauweise *f* rack mode, structural design

Gestellbelastung *f* hearth load *(blast furnace)*

Gestellbremsberg *m (mine)* balance brow

Gestelleinschub *m* modular rack unit; plug-in unit

Gestellerdung *f* frame earthing

gestellfest fixed *(kinematics)*

Gestellförderung *f (mine)* cage hoisting

Gestellglied *n* fixed bar (link) *(kinematics)*

Gestellmantel *m* hearth jacket *(blast furnace)*

Gestellmotor *m* frame-suspended motor

Gestellpanzer *m* hearth casing *(blast furnace)*

Gestellrahmen *m* mounting frame

Gestellreihe *f (telec)* repeater rack

Gestellsäge *f* frame saw

Gestellschaltpult *n* skeleton-type switchboard

Gestellschlußschutz *m* frame-leakage protection

gesteuert controlled

Gestirn *n* star

Gestirnpeilung *f* bearing by stars

gestört faulty, disturbed, troubled

gestranded aground, stranded

gestreckt flat, straight *(angle)*; elongated

gestrichelt dotted, broken

gestrichen coated; painted

frisch g. wet paint

Gestrick *n* latch needle fabric; knitted fabric

Gestrickschlauch *m* fabric tube

Gestrickseite *f* face side

gestuft stepped, multi-step

geometrisch g. *(mach)* in geometrical progression

Gestüt *n* stud farm

Gesuch *n* application, petition, request

schriftliches G. request in writing

Gesundheitsgefährdung *f* biological (health) hazard

Gesundheitsphysik *f s.* physikalischer Strahlenschutz

gesundheitsschädlich detrimental to health

Gesundheitstechnik *f* sanitary engineering

geteilt divided; split

g. durch divided by

getempert annealed, heat-treated

getränkt impregnated; soaked

Getreide *n* corn, grain

Getreidebehälter *m* corn bin

Getreidebeizung *f* dressing of grain

Getreideeinrichtungen *fpl* grainfittings *(of a vessel)*

Getreideeinstreumittel *n* corn bait

Getreidefeeder *m (ship)* grain feeder

Getreidefeuchtigkeitsmesser *m* grain moisture gauge

Getreideheber *m* grain elevator

Getreideladung *f (ship)* grain cargo

Getreideluke *f* grain (trimming) hatch

Getreidenachfüllschacht *m (ship)* grain feeder

Getreidesämaschine *f* grain-sowing machine

Getreideschacht *m (ship)* grain feeder

Getreideschott *n* grain bulkhead

losnehmbares G. shifting boards

Getreideschotthalter *m* shifting iron *(cargo hold)*

Getreideschottstütze *f* shifting-board stanchion
Getreidesieb *n* grain screen
Getreidesilo *m* grain silo
Getreidespeicher *m* granary, grain storehouse
Getreidestengel *m* corn stalk
Getreidetrockner *m* grain drier
Getreidetrocknungsanlage *f* grain drier (drying plant)
fahrbare G. portable grain drying plant
getrennt separate[d]
Getriebe *n* train of mechanisms *(kinematics)*; *(mach)* transmission; gear box *(motor car)*; drive
G. mit ständigem Eingriff constant-mesh gear
achsversetztes G. hypoid gear
automatisches G. shiftless transmission
einfaches G. simple mechanism
gekapseltes G. enclosed gear
hydraulisches G. fluid drive, hydraulic transmission
stufenlos regelbares G. variable speed transmission unit
zusammengesetztes G. compound mechanism
Getriebebremse *f* gear brake
Getriebeelemente *npl* elements of the pair *(kinematics)*
Getriebeendschalter *m* gear-type limit switch
Getriebeflansch *m* gearbox flange
Getriebegehäuse *n* gear box [casing], gear case
Getriebegehäusedeckel *m* gear box case cap
Getriebekasten *m* gear box (case)
Getriebelader *m* gear-driven supercharger
Getriebelehre *f* kinematics
Getriebemotor *m (el)* [back] geared motor
getrieben driven
Getriebenes *n* driven gear
Getriebepfahl *m (mine)* fore pole, lath, spile
Getriebeschaltschütz *n* propulsion unit contactor
Getriebespiel *n* backlash
Getriebespülapparat *m* gear-flushing unit
Getriebespüler *m* gear-flushing unit
Getriebetunnel *m* transmission tunnel
Getriebeturbine *f* geared turbine
Getriebeverfahren *n s.* Getriebezimmerung
Getriebezimmerung *f (mine)* forepoling, spiling
getrimmt trimmed
schlecht g. out of trim
getrocknet dried
künstlich g. kiln-dried
Getter *m* getter
Getterung *f* gettering *(electron tube)*
Getterverdampfung *f* gettering
Getting-Schaltung *f* Getting circuit
getypt standardized
Geviert *n* 1. *(mine)* crib-ring, cribwork; 2. em-quad *(printing)*
Geviertzimmerung *f (mine)* full (four-piece timber) set
Gewächshaus *n* hot-house, glass-house, greenhouse, conservatory
Gewächshauskultur *f* greenhouse (glass-house) crop
gewähren / einen Patentanspruch to grant a patent claim
Gewaltbruch *m* forced rupture
Gewände *n* jamb
Gewässer *n* **beschränkten Querschnitts** restricted waters
Gewebe *n (text)* fabric; cloth; woven material; tissue *(biology)*
G. in Stückbreite fabric at full width
G. mit Abseite backed cloth
G. mit kurvenförmiger Verzerrung der Kette warp-bowed cloth
G. mit kurvenförmiger Verzerrung des Schusses weft-bowed cloth
appretiertes G. finished cloth
bestrichenes G. coated fabric
feuerbeständiges G. fire-resistant fabric
gemustertes G. fancy cloth, figured woven fabric
glattes G. plain cloth
grobschüssiges G. coarse-weft fabric
gummiertes G. rubberized fabric
irisierendes G. iridescent fabric
kaschiertes G. lined fabric
mehrfaches G. compound cloth
metalldurchwirktes G. metallic cloth
pelzartiges G. fur fabric
Gewebeäquivalent *n* tissue-equivalent [material]
Gewebedichte *f* density of cloth
Gewebedosis *f (nucl)* tissue dosis
Gewebefaltmaschine *f* folding machine
Gewebefehler *m* nap
Gewebefüllstoff *m* fabric filler
Gewebehalbwertschicht *f (nucl)* half-value tissue layer
Gewebeisolierung *f* fabric insulation
Gewebekante *f* / **eingerissene** cut selvedge
Gewebekunststoff *m* fabric base plastic
Gewebelagen *fpl* **des Unterbaus** casing plies *(tyre)*
Gewebeoberseite *f* cloth face
Gewebeprüfapparat *m* cloth tester
Gewebeputzmaschine *f* cloth cleaning machine
Gewebeschaden *m* tissue injury
gewebt woven
Gewehr *n* rifle
Gewehrgranate *f* rifle grenade
Gewehrkolben *m* rifle butt
Gewehrlauf *m* rifle barrel
Gewehrlaufbohrer *m* single-lipped rifle drill, bit for rifle barrel drilling
Gewehrlaufzug *m* rifling groove

Gewehrpanzergranate *f* armour-piercing rifle grenade
Gewehrriemen *m* rifle sling
Gewehrzielfernrohr *n* telescopic rifle sight
gewellt corrugated *(sheet)*
Gewerkschaft *f* trade union
Gewicht *n* 1. weight; 2. *s.* Masse
aufgehängtes G. suspended weight
aufgestecktes G. movable (sliding) weight
spezifisches G. *s.* Dichte
Gewichtfunktion *f (cont)* impulse response, weighting function
Gewichtheben *n* weight-lifting *(sport)*
Gewichtsanalyse *f* gravimetric analysis
Gewichtsbär *m* fall hammer
gewichtsbelastet weight-loaded
Gewichtsbelastung *f* weight-loading
Gewichtsdosierung *f* gravimetric (weight) batching, feed weighing
genaue G. weigh-starved feeding
Gewichtseinheit *f* unit of weight
Gewichtsfracht *f (ship)* deadweight cargo
gewichtslos imponderable
Gewichtsnumerierung *f (text)* numbering by yarn weight
Gewichtsprozent *n* percent in weight, percentage by weight
Gewichtsregler *m* weighted governor
Gewichtsschwerpunkt *m (ship)* centre of gravity
Gewichtsschwund *m* loss in weight
Gewichtsteil *n* part by weight
Gewichtsverlust *m* loss of weight
Gewichtsverminderung *f* decrease of weight
Gewichtsware *f (text)* pound goods
Gewichtswirkung *f* effect of weight
Gewichtszuführung *f (mach)* gravitational flow of parts, gravity feed
Gewichtszunahme *f* weight gain
gewickelt wound
auf Abstand g. space-wound
bifilar g. *(el)* double-wound
fest g. tightly rolled
Gewinde *n* [screw] thread
ohne G. non-screwed
doppelgängiges G. double-start thread
dreigängiges G. triple thread
eingängiges G. single-start thread, single-threaded thread
mehrgängiges G. multiple (multi-start, multi-threaded) thread
metrisches G. metric thread *(International Standard)*
viergängiges G. quadruple thread
Gewindeaußendurchmesser *m* major diameter of a thread
Gewindebezeichnung *f* thread specification
Gewindebohrarbeit *f* thread tapping operation, tapping job
Gewindebohreinrichtung *f* tapping attachment
Gewindebohren *n* [thread] tapping
Gewindebohrer *m* [thread] tap
G. mit gebogenem Schaft bent-shank tap
G. mit geschliffenem Gewinde ground-thread tap, tap with ground thread
G. mit geschnittenem Gewinde cut-thread tap, tap with cut thread
Gewindebohrerauszieher *m* tap extractor
Gewindebohrerbruch *m* tap breakage
Gewindebohrernutmaschine *f* / **automatische** automatic tap fluting machine
Gewindebohrerschleifmaschine *f* tap sharpening machine
Gewindebohrerstandzeit *f* tap life
Gewindebohrmaschine *f* tapping machine
Gewindebohrtiefe *f* tapping depth
Gewindebohrung *f* tapped hole
Gewindebolzen *m* bolt
Gewindebuchse *f* threaded ring
Gewindedrehen *n* cutting of threads on a lathe
Gewindedrehmeißel *m* single-point threading tool
Gewindeeinstechschleifverfahren *n* **mit mehrprofiliger Scheibe** multi-rib plunge-cut method of thread grinding
Gewindeflanke *f* thread flank
Gewindeflankenwinkel *m* thread profile angle
Gewindeform *f* thread form
Gewindeformfräser *m* / **scheibenförmiger** formed-type single-thread milling cutter
gewindeförmig thread-shaped
Gewindefräsen *n* thread milling
Gewindefräser *m* thread-milling cutter
scheibenförmiger G. single-thread milling cutter, single-formed cutter
Gewindeganghöhe *f* lead of a thread
Gewindegrenzdorn *m* thread limit plug gauge
Gewindegrenzlehre *f* screw-thread limit gauge
Gewindegrund *m* root of [a] thread
Gewindegrundbohrung *f* blind tapped hole
Gewindekamm *m* screw-pitch gauge
Gewindekerndurchmesser *m* minor diameter of a thread
Gewindekernloch *n* tapping-size hole
Gewindekernlochbohrer *m* tap drill
Gewindekernlochdurchmesser *m* minor diameter of an internal thread
Gewindekurzzeichen *n* thread symbol
Gewindelängsschleifen *n* thread traverse grinding
Gewindelängsschleifverfahren *n* **mit mehrprofiliger Scheibe** multi-rib traverse method of thread grinding
Gewindelehrdorn *m* **[für Innengewinde]** internal thread gauge
Gewindelehre *f* thread gauge
Gewindeloch *n s.* Gewindebohrung

Gewindemeißel *m* thread tool
Gewindemeßschraube *f* thread ring gauge
Gewindemessung *f* measurement of threads
Gewindemuffe *f* threaded (screwed) sleeve
gewinden to thread
Gewinden *n* threading
Gewindenenndurchmesser *m* nominal (major) diameter of thread
Gewindenut *f* thread groove
Gewindereihe *f* thread series
Gewindering *m* **[mit Zahnkranz]** ring nut
Gewinderohrverbindung *f* threaded pipe connection
Gewinderollen *n* cylindrical die thread rolling
Gewindesackloch *n s.* Gewindegrundbohrung
Gewindeschablone *f* thread template; screw-pitch gauge
Gewindeschaft *m* screwed shank
Gewindeschleifen *n* thread grinding
Gewindeschleifmaschine *f* thread grinding machine
Gewindeschleifscheibe *f* thread grinding wheel
Gewindeschneidarbeit *f* threading (screw-cutting, thread-cutting) operation
Gewindeschneidbohrer *m* tap
Gewindeschneideisen *n* threading die
Gewindeschneiden *n* thread (screw) cutting
Gewindeschneidkopf *m* die head
  selbstöffnender G. self-opening die head
Gewindeschneidschraube *f* self-tapping screw, thread-cutting screw
Gewindeschneidwerkzeug *n* [screw] thread cutting tool
Gewindesockel *m* screw cap *(of a lamp)*
Gewindespindel *f* screw
  G. mit Kugelmutter ball screw
Gewindespitze *f* crest of a thread
Gewindesteigung *f* thread pitch, pitch of a thread
$$\left(\frac{1}{\textit{number of threads per inch}}\right)$$
Gewindestift *m* thread plug, threaded stud
Gewindestrählen *n s.* Gewindestrehlen
Gewindestrehleinrichtung *f* thread-chasing attachment
Gewindestrehlen *n* thread chasing
Gewindestrehler *m* thread chaser
Gewindetiefe *f* depth of thread
Gewindewalzbacken *m* thread rolling die
Gewindewalzen *n* flat-die thread rolling
Gewindewalzmaschine *f* [flat die] thread rolling machine
Gewindewalzstahlbacken *m* flat[-faced] thread rolling die
Gewindewirbeleinrichtung *f* thread whirling attachment
Gewindewirbeln *n* thread whirling
Gewinn *m* gain
  G. an Rausch-Signal-Abstand wide-band improvement
Gewinnanteil *m* **an der Patentverwertung** share of profit from the patent exploitation
Gewinnbarkeit *f* workability *(coal)*
gewinnen 1. to get, to win; 2. to extract; 3. to mine *(coal, ore)*; 4. to quarry *(stone)*
  bergmännisch g. to mine
  Elektronen g. to gain electrons
Gewinnfaktor *m* gain factor
Gewinnsaldo *m* balance profit
Gewinnung *f* 1. getting, winning, exploitation; 2. extraction; 3. mining *(coal, ore)*; 4. quarrying *(stone)*
  G. der Flachsfaser extraction of the flax fibre
  G. von Begleitmetallen recovery of by-metals
  G. von Steinen und Erden quarrying
  hydraulische G. hydraulic mining
  mechanische G. mechanical mining
  rhythmische G. cyclic mining
Gewinnungsabteilung *f (mine)* working panel
Gewinnungsfeld *n (mine)* winning track
Gewinnungsmaschine *f* miner
  vor Kopf angreifende G. buttock machine
Gewinnungsschicht *f (mine)* coal[ing] shift, production shift
Gewinnungsstoß *m (mine)* face, breast, stope
Gewinnungsverfahren *n* extraction process
Gewitter *n* thunderstorm
Gewitterbeeinflussung *f* thunderstorm influence
Gewitterbö *f* thunder-squall
Gewitterstörung *f* atmospherics
Gewitterüberspannung *f* overvoltage of atmospheric origin, lightning surge
Gewölbe *n (build)* vault[ing]; arched crown *(furnace)*
  feuerfestes G. refractory arch
  schiefes G. skew-arched vault
  schiefwinkliges G. skew-arched vault
  verkehrtes G. inverted arch foundation
Gewölbebildung *f* bridging, arching *(e.g. in bulk cement silos)*
Gewölbeeinbau *m* arched construction
Gewölbekern *m (mine)* arch-core
Gewölbescharen *fpl* arch rings
Gewölbewirkung *f* arching effect
Gewölbezwickel *m* spandrel
gewölbt domed, vaulted; dished
geworfen warped
Geyserit *m (min)* geyserite
Gezähe *n (mine)* working gear
Gezähekasten *m* tool box
Gezähekübel *m* bowk *(shaft sinking)*
Gezähewirtschaft *f* working gear economy
gezahnt toothed

Gezeitenfeuer *n* tidal light
Gezeitenkraftwerk *n* ocean tides power station
Gezeitenpegel *m* tide-gauge, tide-dial
Gezeitenstrom *m* tidal stream
gegenlaufender G. foul tide
Gezeitentafeln *fpl* tide tables
Gezeitenwelle *f* tidal wave
gezogen drawn *(wire)*
nahtlos g. seamless drawn *(tube)*
gezwirnt thrown, twisted *(silk)*
gg-Kern *m (nucl)* even-even nucleus
Gibbsit *m s.* Hydrargillit
Gicht *f* top *(blast furnace)*; charge
Gichtaufzug *m* blast furnace elevator, skip bridge
schräger G. inclined lift *(furnace hoist)*
Gichtbühne *f* [blast furnace] charging platform
Gichtenfolge *f* cycle of charges, charging sequence
Gichtenzähler *m* / **registrierender** recording furnace-filling counter
Gichtflamme *f* top flame *(shaft furnace)*
Gichtgas *n* exhaust (blast furnace) gas
Gichtgasfackel *f* excess blast-furnace gas burner
Gichtgasfang *m* blast furnace gas take
Gichtgasleitung *f* rough gas main
Gichtglocke *f* furnace-top bell
Gichtglockenwinde *f* bell operating gear
Gichtkübel *m* skip, charging bucket (basket)
Gichtöffnung *f* top opening
Gichtschwamm *m* tutty, cadmia
Gichtsonde *f* charge (stock) level indicator
Gichtstaub *m* flue (blast furnace) dust
Gichtstein *m* stone rest
Gichttemperatur *f* top temperature *(blast furnace)*
Gichtverschluß *m* throat stopper, bell and hopper *(blast furnace)*
Gichtwinde *f* bell operating gear
Giebel *m* gable
Giebeldach *n* gable roof
Giebelfeld *n* tympanum *(architecture)*
Giebelmauer *f* end wall
Giebelseite *f* flank
Gierbewegung *f (aero, ship)* yawing
Gieren *n* yawing
Giermoment *n* yawing moment
Gierschwingungen *fpl* corkscrewing *(of a vessel)*
Gierwinkel *m* angle of yaw, yaw angle
Gießaufsatz *m* hob top
gießbar castable
Gießbarkeit *f* castability
Gießbett *n* pig bed
Gießbühne *f* teeming platform
gießen 1. to cast, to found, to pour *(foundry)*, to teem; 2. to cast *(composing machine)*; to water *(flowers)*
fallend g. to pour from the top, to cast downhill
steigend g. to pour from the bottom, to cast uphill
Gießen *n* **im Gespann** bottom casting
G. mit Gießlöffel hand ladling
G. mit Kern core casting
G. mit verlorener Gießform investment casting
G. von Hohlkörpern slush casting (moulding)
G. von oben top casting
G. von unten bottom casting
stehendes G. pouring on end
Gießerei *f* foundry
Gießereiflammofen *m* foundry air-furnace
Gießereikoks *m* foundry coke
Gießereikupolofen *m* foundry cupola
Gießereimeister *m* master smelter
Gießereiroheisen *n* foundry pig-iron
Gießereischlichte *f* mould wash
Gießereitechnik *f* foundry technique[s]
Gießfähigkeit *f* pourability
Gießform *f* mould *(foundry)*
feste G. permanent mould
Gießgeschwindigkeit *f* pouring speed, teeming rate
Gießgrube *f* foundry (casting) pit
Gießhalle *f* casting bay
Gießhals *m* pot-throat
Gießharz *n* cast resin
geschichtetes G. laminated cast resin
glasfaserverstärktes G. fibre-glass armoured cast resin
Gießharzschalter *m* cast-resin circuit-breaker
Gießharzwandler *m* cast-resin instrument transformer
Gießkanne *f* watering can
Gießkasten *m* casting box
Gießkastenrahmen *m* deckle frame
Gießkelle *f* casting ladle *(foundry)*
Gießkopf *m* runner head
Gießkran *m* ladle crane
Gießlöffel *m* casting ladle
Gießmarke *f* pin mark
Gießmaschine *f* casting machine
Gießmund *m* pot-mouth, nipple
Gießpfanne *f* pouring (foundry) ladle
G. mit Wagen buggy ladle
Gießpfannenauskleidung *f* ladle lining
Gießpfannenbügel *m* ladle bail
Gießpfannenkran *m* ladle crane
Gießplatz *m* foundry (casting) pit
Gießrad *n* mould disk
Gießrinne *f* runner, feeder, pouring channel (hole)
Gießschlicker *m* castable slip
Gießschnauze *f* pouring lip *(ladle)*
Gießschweißen *n* flow welding

Gießtechnik *f* pouring technique
Gießtemperatur *f* pouring temperature
Gießtisch *m* mould frame
Gießtopf *m* [casting] pot; melting pot, crucible *(composing machine)*
Gießtrichter *m* sprue, pouring gate; centre riser; down gate
Gießtrommel *f* casting wheel
Gießtümpel *m* runner (pouring) basin
Gießverfahren *n* deckle box formation
Gießwagen *m* pouring car
Gießwanne *f* tundish
Gift *n* poison
Giftgas *n* poison[ous] gas
giftig poisonous, toxic *(vapours)*, venomous
Giftigkeit *f* toxicity *(vapours)*
Gig *n (f) (ship)* gig
Gigaelektronenvolt *n* billion electron volt, BeV
Gillfeld *n (text)* set of fallers
Gillspinnmaschine *f (text)* gill
Gillstab *m (text)* faller, gill bar
Gillungsheck *n* counter stern
Gillungsplatte *f (ship)* oxter plate
Gimpe *f* gimp [yarn]
Gipfel *m* summit, top, peak, apex, vertex
Gipfeldürre *f* stag headedness
Gipfelhöhe *f (aero)* [absolute] ceiling
errechnete (theoretische) G. calculated ceiling
Gipfelpunkt *m* **der Flugbahn** apex (summit) of the trajectory
Gipfelwert *m s.* Scheitelwert
Gips *m (min)* gypsum; *(build)* plaster of Paris, gypsum plaster
gebrannter G. plaster of Paris, calcined gypsum
gipsartig gypseous
Gipsdiele *f* plaster (gypsum) board
Gipsestrich *m* plaster floor[ing]
Gipsform *f* plaster mould
gipshaltig gypsiferous
Gipsmodell *n* plaster pattern
Gipsmörtel *m* plaster, stucco
Gipsplatte *f* plaster (gypsum) board, plaster slab
Gipsputz *m* gypsum plaster
Gipsschablone *f* plaster pattern
Gipsschlackenzement *m* slag-sulphate cement
Girodofen *m* Girod furnace
Gischt *m* froth
Gissung *f* dead reckoning
Gitter *n* 1. lattice *(of crystals)*; 2. *(el)* grid *(discharge tube)*; 3. *(opt)* grating; 4. *(math)* net; 5. *(build)* lattice, trellis; 6. *(telev)* gate
G. der Elektronenröhre grid of the electron tube
akustisches G. acoustic grating
ebenes G. plane lattice
engmaschiges G. fine grid
flächenzentriertes G. face-centred lattice *(of crystals)*
freies G. *(el)* disconnected (floating, free) grid
gestörtes G. defect lattice *(of crystals)*
ideales G. perfect (ideal) lattice *(of crystals)*
kubisches G. cubic lattice
kubisch-flächenzentriertes G. face-centred cubic lattice
kubisch-raumzentriertes G. body-centred cubic lattice
lineares G. linear lattice
offenes G. disconnected (floating, free) grid
optisches G. optical grating
periodisches G. periodic lattice
raumzentriertes G. body-centred lattice
wabenartiges G. honeycomb
Gitterableiter *m* grid leak *(resistor)*
Gitterableitung *f* grid return (leak)
Gitterableitwiderstand *m* grid leak resistance
Gitterabschaltung *f* arc suppression, grid extinguishing
Gitterabschirmhaube *f* grid shielding can *(electron tube)*
Gitterabstand *m* lattice pitch (spacing)
Gitteranordnung *f* lattice array; lattice configuration
G. der Kristalle arrangement of the crystals in the lattice
Gitteranschluß *m* grid clip
Gitterarbeitswiderstand *m* internal input resistance
Gitteraufbau *m* lattice structure
Gitteraussteuerung *f* grid excitation
Gitterbasis *f* grounded grid
Gitterbasisschaltung *f* grounded-grid circuit
Gitterbasisverstärker *m* grounded-grid amplifier
Gitterbatterie *f* grid battery, bias cell
Gitterbaufehler *m* lattice defect
Gitterblockierung *f* grid cut-off, negative biasing of grid; grid blocking *(by statical charge)*
Gitterbrummen *n* grid hum
Gitterclip *m* grid clip
Gitterdrossel *f* grid inductance
Gitteremission *f* grid emission
Gitterenergie *f* grid energy
Gitterfehler *m* lattice distortion (imperfection, defect) *(crystal)*
Gitterfehlstelle *f* lattice distortion (imperfection, defect) *(crystal)*
Gitterfenster *n* lattice window
gittergesteuert grid-controlled
Gittergleichrichter *m* bias (audion, grid) rectifier, grid detector
Gittergleichrichtung *f* grid (cumulative) rectification, grid (current) detection

Gittergleichspannung *f* grid supply voltage
Gitterionisationskammer *f* gridded (grid ionization) chamber
Gitterkapazität *f* / **dynamische** effective input capacitance *(electron tube)*
Gitterkappe *f* grid cap (clip), screen cap
Gitterkondensator *m* grid capacitor
Gitterkonstante *f* lattice constant *(of crystals)*
Gitterkonstruktion *f (build)* lattice[d] structure
Gitterkopie *f (ph)* grating replica
Gitterkreis *m* grid circuit (branch)
Gitterkreiskondensator *m* capacitor of the grid circuit
Gitterladebaum *m (ship)* [bar] lattice boom
Gitterleiter *f* hanging (wall) ladder
Gitterleitfähigkeit *f* lattice conductance
Gitterlücke *f* [lattice] vacancy
Gittermast *m* tower, pylon, lattice mast
Gittermastleitung *f* tower line
Gittermauerwerk *n* checker brick work
Gittermodulation *f* grid modulation
Gitternetz *n* meshwork of lines at right angles, grid [system]
Gitterneutralisation *f* grid neutralization
Gitterplatte *f* grid plate *(capacitor)*
Gitterplattensammler *m* Faure accumulator
Gitterplatz *m* atom site, lattice point
Gitterpotential *n* grid potential
Gitterpunkt *m* lattice point *(of crystals)*; *(math)* net point
Gitterpunktmethode *f* method of differences
Gitterrad *n (ag)* cage wheel
Gitterraum *m* **der Laufzeitröhre** field grid space
Gitterrauschwiderstand *m* grid-noise resistance
Gitterreaktor *m* lattice reactor
Gitterrost *m (build)* grating, grillage, grate, grille; *(nucl)* diagrid
Gitterschott *n (ship)* [batten and] space bulkhead, lattice bulkhead
Gitterschritt *m s.* Gitterabstand
Gitterschutzwiderstand *m* grid stopper
Gitterschwanzträger *m (aero)* open tail boom
Gitterschwingung *f* lattice vibration *(of crystals)*
Gitterspannung *f* grid voltage (potential)
  automatische G. automatic grid tension
Gitterspannungskennlinie *f* grid voltage characteristic (curve)
Gitterspannungsschwankung *f* grid sweep (voltage swing)
Gitterspektrograf *m (ph, opt)* [diffraction-]grating spectrograph
Gitterspektrum *n* grating (diffraction) spectrum
Gittersperrkreis *m* grid stopper
Gittersperrspannung *f* cut-off bias (voltage, grid potential)
Gittersperrung *f* arc suppression, grid extinguishing
Gitterspule *f* lattice coil
Gittersteg *m* grid support *(electron tube)*
Gittersteilheit *f* electrode conductance
Gittersteuerleistung *f* grid-driving power
Gittersteuerung *f* grid control
  ruhende G. transformer grid control [of a thyratron]
Gitterstörung *f (ph)* lattice distortion (dislocation) *(crystal)*
Gitterstrich *m* ruling of a lattice
Gitterstrom *m* grid current
  G. im Augenblick der Zündung critical grid current
  positiver G. inverse grid current
Gitterstrombegrenzung *f* grid limiting
Gitterstrom-Gitterspannungs-Kennlinie *f* grid characteristic
Gitterstromgleichrichter *m* power grid detector
Gitterstromkennlinie *f* grid characteristic
Gitterstrommodulation *f* grid-current modulation
Gitterstütze *f* latticed column
Gittertastung *f* grid keying
Gitterträger *m* lattice girder
Gittertragwerk *n* lattice[d] structure
  räumliches G. space lattice
Gitterverschiebespannung *f* biasing grid voltage
Gitterversuch *m* lattice experiment
Gittervorspannbatterie *f* grid bias battery
Gittervorspannung *f* grid bias (priming voltage), bias voltage
  automatische G. self-bias
  feste G. fixed bias
  selbsttätige G. automatic grid bias
Gittervorspannungsbatterie *f* grid bias battery (cell)
Gittervorspannungserzeuger *m* grid bias generator
Gittervorspannungserzeugung *f* / **automatische** self-bias arrangement, self-biasing
Gittervorspannungszusatz *m* C power supply
Gittervorwiderstand *m* biasing (series grid) resistor
Gitterwechselspannung *f* grid alternating current voltage, alternating [current] grid voltage
Gitterwechselstrom *m* grid alternating current
Gitterwerk *n (build)* trellis-work, grating, lattice-work
Gitterwiderstand *m* / **äquivalenter** equivalent grid leak resistance
Gitterzündspannung *f* critical grid voltage
Gitterzündstrom *m* critical grid current
Glacéleder *n* glacé kid
Glacépapier *n* / **farbiges** high-glazed coloured art-paper
Gladiatorenstiefel *m* wrestler's shoe *(wrestling)*
Glanz *m* 1. glaze *(textile finish)*; 2. lustre *(reflexion of light)*; 3. gloss *(smooth surface)*
  erster G. *(leath)* bottom season

letzter G. *(leath)* top season
metallischer G. metallic lustre
seidenartiger G. [silky] lustre
Glanzabkochung *f* *(leath)* decoction of lustre, preparation for seasoning
Glanzblech *n* polished sheet metal
Glanzbrenne *f* bright dip
Glanzdruckfarbe *f* gloss ink
Glanzeisenerz *n* specular iron [ore]
glanzeloxieren to anode-brighten, to electro-brighten
glänzen to glisten
elektrolytisch g. to anode-brighten, to electro-brighten
glänzend glossy, lustrous, bright, polished
Glanzfläche *f* smooth surface *(briquette)*
Glanzfolienkaschierung *f* film lamination
Glanzfutterstoff *m* lustre lining
Glänzigwerden *n* **des Schleifmittelbelags** glazing
Glanzkalander *m* glazing (friction) calender *(paper)*
Glanzkohle *f* bright (glance, glossy-black bituminous) coal, lustrous carbon
Glanzlicht *n* high light
glanzlos dull *(surface)*; lustreless
Glanzmaschine *f* glazing machine, pendulum jigger *(papermaking)*
Glanzmesser *m* glossmeter
Glanzpapier *n* glazed (glossy, enamel) paper
Glanzrinde *f* silver bark
glanzstoßen *(leath)* to put a gloss on
Glanzstoßzylinder *m* glazing roller
Glanzteile *npl* **an Bord** bright work
Glanztinte *f* glossy ink
Glanzverzinkung *f* bright zinc coating
Glanzwinkel *m* *(nucl)* glancing (Bragg) angle
Glanzzusatz *m* *(met)* brightener *(galvanic bath)*
Glas *n* 1. glass; 2. glass, jar
G. zum Glanzstoßen *(leath)* glazing glass
feuerfestes G. heat resisting glass
mundgeblasenes G. hand-made glass
optisches G. optical glass
splittersicheres G. safety glass
UV-durchlässiges G. vita glass
Glasanalyse *f* glass analysis
glasartig vitreous, glassy
Glasauge *n* artificial eye
Glasbad *n* glass bath
Glasballon *m* carboy, demijohn
Glasbaustein *m* glass block (brick)
Glasbearbeitungsmaschine *f* glass machine
Glasbehälter *m* glass container
Glasblasen *n* glass blowing
Glasbläser *m* glass blower
Glasblasmaschine *f* glass blowing machine
Glasbohrer *m* glass drill
glasen to strike the ship's bell
Glasen *npl* bells
Glaser *m* glazier
Glaserdiamant *m* glass cutter
Glaserkammer *f s.* Blasenkammer
Glaserkitt *m* glazing compound; bedding (back, glazier's) putty
gläsern vitreous, glassy
Glasfabrik *f* glass factory (works)
Glasfabrikation *f* glass manufacture
Glasfaden *m* glass thread
Glasfaser *f* glass fibre, fibre glass
glasfaserarm glass-fibre starved
glasfaserbewehrt glass fibre reinforced
Glasfaserkitt *m* polyester-putty
Glasfaserkunststoff *m* fibre glass-reinforced plastic
Glasfaserschichtstoff *m* fibre glass-reinforced laminate
Glasfaserstoff *m* / **textiler** glass fibres (filaments)
Glasfehler *m* glass defect; glass crack
Glasfilter *n* glass filter
Glasfilterfritte *f* fritted glass filter
Glasflasche *f* glass bottle
Glasfliese *f* glass tile
Glasfluß *m* molten glass
Glasfolie *f* glass foil
Glasform *f* glass mould
Glasfritte *f* glass frit
Glasgemenge *n* glass batch
Glasgespinst *n* glass fibres, spun glass
Glasgewebe *n* glass cloth (fabric)
Glashafen *m* glass melting pot
Glashahn *m* glass cock
Glashalbzelle *f* glass electrode
glashart dead hard *(steel)*
Glashütte *f* glass factory (works)
glasieren to glaze
Glasierglas *n* *(leath)* glass slicker
Glasiermaschine *f* *(leath)* glassing jack (machine)
Glasindustrie *f* glass industry
Glasisolator *m* glass insulator
Glaskappenisolator *m* glass petticoat insulator
Glaskolben *m* glass bulb *(lamp)*, glass envelope *(electron tube)*
Glaskörper *m* vitreous body *(eye)*
Glaslinse *f* lens
Glasmacher *m* glass maker
Glasmacherpfeife *f* blowpipe, blowing iron
Glasmaler *m* glass painter
Glasmärbel *f* glass grinding marble
Glasmaßstab *m* [precision] glass scale
Glasmehl *n* glass powder
Glas-Metall-Verschmelzung *f* glass-to-metal seal *(electron tube)*
Glasopal *m* *(min)* hyalite
Glaspapier *n* glass-paper
Glasplatte *f* glass panel (plate)

Glaspresse *f* glass press
Glaspreßling *m* glass blank
Glasproportionalzählrohr *n (nucl)* glass proportional counter
Glasraster *m* glass screen
Glasreife *f (ag)* pasty condition [of maize]
Glasrohrabdichtung *f* packing of the gauge glass
Glasröhrchen *n* glass tube
Glassand *m* glass sand
Glasscheibe *f* glass pane, pane of glass
Glasschleiferei *f* glass-grinding [work]shop
Glasschliff *m* ground glass joint
Glasschmelze *f* molten glass
Glasschneidediamant *m* glazier's diamond
Glasschneider *m* glass cutter
Glasschneiderei *f* glass-cutting workshop
Glassin *n* glassine *(transparent paper)*
Glassplitter *m* glass fragment
Glasstab *m* glass (stirring) rod
Glasstandregler *m* glass level controller
Glasstopfen *m* glass stopper
Glastasche *f* pocket
Glasteilkreis *m* glass dial *(optical dividing head)*
Glastinte *f* glass ink
Glasträne *f* glass tear (drop)
Glastrennwand *f* glass screen
Glastrübung *f* glass muddiness (cloudiness)
Glasur *f* glaze, glazing
  halbleitende G. semiconducting glaze
Glasurbrand *m* glaze baking
Glasverarbeitung *f* working (forming) of glass
Glasveredlung *f* glass decoration
Glasvergoldung *f* glass gilding
Glasvergütung *f* refining of glass
Glasversilberung *f* glass silvering
Glasvlies *n* glass fleece
Glaswatte *f* glass wool
Glaswolle *f* glass wool
Glasziegel *m* glass block (brick, tile)
Glaszustand *m* vitreous state
glatt even, smooth *(surface)*; fair[-faced]; slippery *(road)*
Glättbohle *f* finishing screed (beam), smoothing screed (board), screed[ing] board
Glättbrett *n (build)* float
Glattdeckschiff *n* flush deck vessel
Glättdornstange *f (met)* reeling mandrel rod
Glätte *f* smoothness *(surface)*; slipperiness *(road)*
Glätteisen *n (leath)* slater
Glättemeßgerät *n* smoothness tester
glätten 1. to smooth; 2. *(text)* to calender; 3. *(mach)* to burnish, to polish; to finish; 4. *(el)* to equalize; 5. to slick *(mould)*; 6. *(build)* to finish fair (smooth); to trowel, to float
  Faß g. to clean off the cask
Glatthobel *m* jointer
Glättkalander *m (text)* glazing calender; smoothing roll *(coating)*
Glättkelle *f (build)* finishing (smoothing) trowel
Glättmaschine *f* 1. *(text)* calender; pouncing machine *(hat)*; 2. *(build)* power trowel *(float)*; 3. *(leath)* slating machine
Glattputz *m (build)* fair-faced plaster
Glättscheibe *f (build)* power trowel (float)
Glättstein *m (leath)* slating stone
Glättungskondensator *m* smoothing capacitor
Glättungskreis *m (el)* smoothing circuit
Glättungsschaltung *f (el)* filter circuit
Glattwalze *f (ag)* plain (flat) roller
Glättwalze *f* smoothing (sleeking) roll
  G. der Drehkanne *(text)* coiler calender roll
Glattwalzwerk *n* plain-roll crusher
Glättwalzwerk *n* reeling mill
Glattwasserlandung *f* mirage landing
Glattwebstuhl *m* plain loom
Glättzahn *m (mach)* burnishing button (tooth), burnisher *(broach)*
Glättzahneinsatz *m* burnishing shell *(broach)*
Glaubersalz *n* Glauber's salt
Glaukonit *m (min)* glauconite
gleich *(math)* equal; like *(signs)*
  g. Null sein to equal zero
  g. sein to equal
gleichachsig coaxial
gleicharmig equal-armed, with equal arms *(lever)*
gleichartig homogeneous
gleichbleibend constant, invariable; uniform *(section)*
Gleichdruckbeschaufelung *f* impulse-type blading
Gleichdruckhöhe *f* critical altitude
Gleichdruckturbine *f* action (impulse) turbine
gleichelektronisch isoelectronic
gleichen to equal
gleichfarbig homochromatic, homochrome, equal-coloured
gleichförmig uniform
Gleichförmigkeit *f* uniformity
  G. des Ganges uniformity of running
Gleichförmigkeitsgrad *m* coefficient of cyclic variation, degree of uniformity
gleichgerichtet unidirectional, unidirected, equidirectional, equal in direction
Gleichgewicht *n* equilibrium, balance
  G. der Flüssigkeiten equilibrium of liquids
  G. der Kräfte equilibrium of forces
  baroklines G. baroclinic equilibrium
  barotropes G. barotropic equilibrium
  dauerndes radioaktives G. secular radioactive equilibrium
  detailliertes G. detailed balance
  dynamisches G. transient (dynamic) equilibrium; *(mach)* running balance
  heterogenes G. polyphase equilibrium

indifferentes G. neutral equilibrium
labiles G. unstable equilibrium
laufendes radioaktives G. transient radioactive equilibrium
nicht umkehrbares G. complete equilibrium
radioaktives G. radioactive equilibrium
stabiles G. stable equilibrium
statisches G. static balance
thermisches G. thermal equilibrium
Gleichgewichtsbedingung *f* condition of equilibrium
Gleichgewichtsdiagramm *n* equilibrium diagram
Gleichgewichtsdruckkurve *f* equilibrium (reversible pressure) curve
Gleichgewichtseinstellung *f* establishment of equilibrium
Gleichgewichtsform *f* type (form) of equilibrium
Gleichgewichtsgeschwindigkeit *f* balancing speed
Gleichgewichtsgleichungen *fpl* **einer Flüssigkeit** equations of equilibrium of a fluid
Gleichgewichtskonstante *f* equilibrium constant
Gleichgewichtslage *f* position of equilibrium, equilibrium (balanced) position
Gleichgewichtspotential *n* **einer Elektrode** equilibrium electrode potential
statisches G. der Reaktion equilibrium reaction potential
Gleichgewichtsschaubild *n* equilibrium diagram
Gleichgewichtstemperatur *f* temperature of equilibrium
Gleichgewichtsverhältnis *n* / **dynamisches** dynamic equilibrium ratio
Gleichgewichtsverschiebung *f* displacement of equilibrium
Gleichgewichtszustand *m* equilibrium (balanced) state, state (condition) of equilibrium
gleichgradig stetig *(math)* equicontinuous
gleichgroß sein to be of the same size *(dimensions)*; to be equal in magnitude *(forces)*
Gleichheit *f* equality, identity
Gleichheitszeichen *n* equality sign, sign of equality
Gleichklang *m* accord, consonance, unison
Gleichlast *f* uniform (uniformly distributed) load
gleichlastig *(ship)* on an even keel, on level trim
Gleichlauf *m (el)* synchronism, ganging, alignment; parallelism
mangelnder G. misalignment
Gleichlaufabstimmelement *n* **von zwei Kreisen** two-circuit tuner
Gleichlaufabstimmung *f* alignment of tuned circuits
Gleichlaufeinrichtung *f (mach)* synchronizing mechanism
Gleichlauffehler *m* synchronization fault
Gleichlauffräseinrichtung *f* backlash eliminator (eliminating device), *(Am)* anti-backlash device
Gleichlauffräsen *n* down (climb-cut, down-cut) milling, climb cutting (milling), cutting down
Gleichlauffräsmaschine *f* / **programmgesteuerte** climb-cut programme controlled milling machine
Gleichlauffräsverfahren *n* climbfeed method
gleichläufig moving in the same direction
Gleichlaufimpuls *m* governing (correcting) pulse; *(telev)* sync
Gleichlaufkontrolleinrichtung *f* scrutineer *(telegraphy)*
Gleichlaufkorrektur *f (el)* synchronous correction
Gleichlaufmangel *m (el)* tracking error
Gleichlaufrelais *n* correction relay
Gleichlaufselbststeuerung *f (el)* automatic synchronization control; autoscrutineer
Gleichlaufsteuerung *f* gang control
Gleichlaufsystem *n* synchronous system
Gleichlauftrimmer *m* tracking capacitor
Gleichlaufzeichen *n* synchronizing (correcting) signal
Gleichlegemaschine *f* splint levelling machine
gleichmächtig *(math)* equivalent; equipotent; equinumerous
gleichmäßig uniform; even
g. belastet uniformly loaded
g. beschleunigt uniformly accelerated
Gleichmäßigfärben *n* level dyeing
Gleichmäßigkeitsprüfer *m* evenness tester
gleichnamig *(math)* correspondent
gleichphasig cophasal, in phase
Gleichpolfeldmagnet *m* homopolar field magnet
Gleichpolgenerator *m* homopolar (d.c.) generator; homopolar alternator
gleichpolig homopolar, unipolar
Gleichpolmaschine *f* inductor alternator
Gleichraumverbrennung *f* constant volume combustion
gleichrichten to rectify
Gleichrichter *m* rectifier; demodulator
G. in Graetzschaltung bridge rectifier
G. mit Außengittersteuerung cathetron
G. mit flüssiger [Quecksilber-]Katode mercury-pool rectifier
G. mit Gittersteuerung transrectifier
G. mit Kommutator permutator
G. mit Vakuumhaltung maintained-vacuum rectifier
G. mit Zündstift ignitron
elektrolytischer G. electrolytic (chemical) rectifier
gekapselter G. embedded rectifier
geradliniger G. linear rectifier
gittergesteuerter G. grid controlled rectifier
idealer G. perfect rectifier
linearer G. straight-line detector
mechanischer G. mechanical rectifier

phasenabhängiger G. phase-sensitive rectifier
quadratischer G. square-law rectifier
vorgespannter G. biased rectifier
Gleichrichteranode *f (el.m.)* rectifier anode
Gleichrichterblock *m* rectifier stack
Gleichrichterbrücke *f* bridge rectifier
Gleichrichtergruppe *f* metallic rectifier stack assembly, rectifier equipment
Gleichrichterkennlinie *f* rectification curve (characteristic)
Gleichrichtermeßinstrument *n* rectifier-type instrument
Gleichrichterröhre *f* rectifying valve, rectifier valve (tube), valve detector
direkt geheizte G. filamentary rectifying tube
ungesteuerte G. *(Am)* phanotron
Gleichrichterrückzündung *f* backlash
Gleichrichtersatz *m* rectifier unit
Gleichrichtersäule *f* rectifier pile (stack)
Gleichrichterschaltung *f* rectifier circuit (connection)
spannungsverdoppelnde G. voltage doubling rectifier
Gleichrichterscheibe *f* rectifying washer
Gleichrichterschicht *f* rectifying film
Gleichrichtertransformator *m* rectifier (plate) transformer
Gleichrichterunterwerk *n* rectifier substation
Gleichrichterventil *n* rectifying valve
Gleichrichterwirkung *f* rectifying effect (action), valve action, asymmetrical conductivity *(effect)*
Gleichrichterwirkungsgrad *m* efficiency of rectification
Gleichrichterzelle *f* rectifier (asymmetric) cell
Gleichrichtung *f* rectifying; rectification; demodulation
G. mit hoher Leistung power detection
phasenabhängige G. coherent detection, phase-sensitive demodulation
quadratische G. square-law detection (rectification)
Gleichrichtungscharakteristik *f* rectification characteristic
Gleichrichtungsfaktor *m* rectification factor, degree of current rectification
gleichschenklig isosceles *(triangle)*; with equal legs *(steel)*
Gleichschlagseil *n* Albert (Lang's) lay rope, equal-laid rope
gleichseitig *(math)* equilateral
Gleichseitigkeit *f* equilaterality
gleichsetzen to equate
gleichsinnig in the same direction of rotation
Gleichspannung *f (el)* direct current voltage, d.c. voltage, continuous (constant) voltage, direct potential (voltage)
ideelle G. ideal no-load d.c. voltage
unstabilisierte G. unregulated d.c. voltage
Gleichspannungsanalogrechner *m* d.c. analog[ue] computer
Gleichspannungsbereich *m* d.c. voltage range
Gleichspannungsendstufe *f* d.c. output stage
Gleichspannungskonstanthalter *m* d.c. voltage regulator
Gleichspannungsmesser *m* d.c. voltmeter
Gleichspannungsquelle *f* d.c. supply (source)
Gleichspannungsrechner *m* d.c. computer
Gleichspannungsverstärker *m* d.c. amplifier
Gleichspulenwicklung *f (el)* diamond winding
Gleichstandsdrillgerät *n* root (single-seed) drill
Gleichstrom *m* direct current, d.c., continuous current, c.c.
ausgeglichener G. smoothed d.c.
intermittierender G. intermittent d.c.
pulsierender G. pulsating d.c.
unterbrochener G. intermittent current
wellenförmiger G. ripple current
Gleichstromanteil *m* steady (d.-c.) component
Gleichstromanzeiger *m* d.c. detector
Gleichstrombahnmotor *m* d.c. traction motor
Gleichstrombetrieb *m* d.c. operation
Gleichstrombrücke *f* d.c. bridge
Gleichstrom-EMK *f* direct e.m.f.
Gleichstromgenerator *m* d.-c. generator, dynamo
Gleichstromglied *n* d.-c. component
G. eines Kurzschlußstromes aperiodic component of a short-circuit current
Gleichstromkessel *m (therm)* concurrent boiler
Gleichstromkomponente *f* steady (d.c.) component
Gleichstromkondensation *f (therm)* parallel-flow condensation, parallel-current condensation
Gleichstromkreis *m* d.c. circuit
Gleichstromleitvermögen *n* d.c. conductance
Gleichstromleitwert *m* d.c. conductance
Gleichstromlichtbogen *m* d.c. arc
Gleichstrommaschine *f* d.-c. machine; dynamo
Gleichstrommesser *m* d.-c. ammeter
Gleichstrommeßtransduktor *m* d.c. measuring transductor
Gleichstrommeßwandler *m* d.c. instrument transformer
Gleichstrommotor *m* d.c. [electric] motor
Gleichstromnetz *n* d.c. mains
Gleichstromnetzempfänger *m* d.c. operated receiver
Gleichstrompegel *m (telev)* d.c. level
Gleichstrompotentialdifferenz *f* direct potential difference
Gleichstromreihenschlußmotor *m* direct current series wound motor
Gleichstromruf *m* / **überlagerter** superposed ringing *(telephony)*
Gleichstromschweißgenerator *m* d.c. welding generator

selbstregelnder G. self-regulating d.c. welding generator
Gleichstromschweißmaschine *f* d.c. welding machine
Gleichstromspülung *f (aut)* end-to-end scavenging
Gleichstromtelegrafie *f* d.c. telegraphy
Gleichstromtransformator *m* d.c. transformer
Gleichstromübertragung *f* d.c. transmission
Gleichstromumformer *m* d.c. converter
Gleichstromumkehrmotor *m* d.c. reversible motor
Gleichstromunterbrecher *m* d.c. interrupter
Gleichstromverstärker *m* d.c. amplifier
Gleichstromverstärkung *f* d.c. amplification
Gleichstromwandler *m* d.c. transformer, d.c. measuring transductor
Gleichstromwecker *m* trembler bell, d.c. bell, *(Am)* vibrating bell
Gleichstromwiderstand *m* ohmic (d.c.) resistance; plate impedance *(of a tube)*
Gleichstromzähler *m* d.c. meter, c.c. meter
Gleichstromzeichengabe *f* d.c. signalling
Gleichtaktfeuer *n (ship)* intermittent light
Gleichtaktsignal *n (el)* common-mode signal
Gleichtaktspannung *f* in-phase voltage
Gleichtaktunterdrückung *f* in-phase rejection, common-mode rejection
Gleichung *f* equation
G. dritten Grades cubic [equation]
G. mit einer Unbekannten equation with one unknown
G. vierten Grades quartic equation
G. zweiten Grades quadratic (second order) equation
Bernoullische G. Bernoulli's equation
Boltzmannsche G. Boltzmann equation
Braggsche G. Bragg equation
Clausius-Clapeyronsche G. Clausius-Clapeyron equation
Diracsche G. Dirac equation
empirische G. empirical equation
Eulersche G. Eulerian equation
Gibbs-Helmholtzsche G. Gibbs-Helmholtz equation
Hamiltonsche G. Hamiltonian equation
hydromechanische G. hydromechanical equation
kritische G. critical equation
linearisierte G. linearized equation
Maxwellsche G. Maxwell's equation
modifizierte G. modified equation
Poissonsche G. Poisson's equation
quadratische G. quadratic equation
Richardsonsche G. Richardson's equation
stöchiometrische G. stoichiometric equation
van't Hoffsche G. van't Hoff's equation
Gleichungen *fpl* **/gekoppelte** coupled equations
Maxwellsche G. Maxwell field equations
simultane G. simultaneous equations
Gleichungsansatz *m* setup of an equation
Gleichungssystem *n* set of equations
Gleichverteilung *f* equipartition
Gleichverteilungssatz *m* principle of the equipartition of energy
Gleichwellenbetrieb *m* simultaneous broadcasting
Gleichwellenrundfunk *m* simultaneous (shared-channel) broadcasting
gleichwertig equivalent
gleichwinklig *(math)* equiangular, isogonal
Gleichwinkligkeit *f (math)* isogonality
gleichzeitig simultaneous
Gleichzeitigkeit *f* simultaneity
Gleichzeitigkeitsfaktor *m* simultaneity factor
Gleichzeitigkeitslogik *f* concurrency logic
Gleis *n* [rail] track, line
Gleisabschluß *m* buffer stop
Gleisabschnitt *m* **hinter einem Signal** advance of a signal
G. vor einem Signal approach to a signal
Gleisabstand *m* distance between running lines, midway between tracks, six foot way (side)
Gleisanlage *f* railway trackage
Gleisanschluß *m* cross-over, junction, siding
Gleisbagger *m* rail-mounted elevator
Gleisbau *m* construction of the track
Gleisbauwagen *m* track construction car
Gleisbesetzungsplan *m* chart showing occupation of tracks, track occupation diagram
Gleisbett *n* track bed
Gleisbettung *f* railroad bedding
Gleisbildstellwerk *n* geographical panel
Gleisblock *m* automatic block with track circuits
Gleisbogen *m* curve in the track
Gleisbremse *f* rail-brake, car retarder
Gleisbündel *n* set of tracks, group of lines, fan of sidings
Gleisdeck *n* traindeck *(train ferry)*
Gleisdreieck *n* reversing triangle, Y track
Gleisdrossel *f* impedance bond
Gleise *npl* **/ unabhängig gespeiste** independent tracks
gleisgebunden rail-borne, rail-mounted
Gleisjoch *n* track panel
Gleiskette *f (aut)* track
Gleiskettenantrieb *m* caterpillar drive
Gleiskettenkran *m* crawler crane
Gleiskettenzugmaschine *f* caterpillar tractor
Gleiskreuzung *f* level crossing
Gleislage *f* track bed
gleislos railless, trackless
Gleismagnet *m* track magnet
Gleismeßwagen *m* track recording coach
Gleisoberbau *m* permanent way
Gleisrad *n* track wheel

Gleissperre *f* track stop
Gleissteigung *f* gradient of track
Gleisstopfmaschine *f* packing (tamping) machine, tamper
Gleisstrecke *f* trackage
Gleisstromkreis *m* track circuit
induktiv gespeister G. reactance-fed track circuit
kapazitiv gespeister G. capacitor-fed track circuit
Gleisunterhaltung *f* maintenance of the track
Gleisverbindung *f* / **einfache** single cross-over
Gleisverlegung *f* track laying
Gleisverwerfung *f* track distortion, warping (lateral buckling) of track
Gleiswaage *f* weigh bridge
Gleitbahn *f* glide path; skid [beam] *(boat)*; *(mach)* sliding way, slideway
Gleitbahnabdeckung *f* / **harmonikaähnliche** concertina-type covers of the ways
Gleitband *n* slip band *(crystallography)*
Gleitbandgeber *m* moving tape transmitter
Gleitbau *m* sliding technique, slip forming, slip-forming method
Gleitbewegung *f* sliding motion, slip
Gleitboden *m* gliding (skimming) bottom *(hydroplane)*
Gleitbogen *m* *(mine)* sliding (yielding) roadway arch
Gleitbonderflüssigkeit *f* *(met)* anti-friction bonderizing bath
Gleitboot *n* planing craft, gliding boat, glider
G. mit gestaffeltem Boden grade-bottomed launch
G. mit mehreren Stufen multistep hydroplane
Gleitebene *f* slip plane *(crystal)*
Gleiteigenschaft *f* anti-friction property
Gleiteinsatzspannung *f* flash-over threshold
Gleitelektrode *f* sliding electrode
gleiten to slide, to slip; to chute *(work)*
frei g. auf to be free to slide on
Gleiten *n* slip *(crystallography)*
Gleitentladung *f* *(el)* flash-over [discharge]
gleitfähig slidable
Gleitfähigkeit *f* slidability
Gleitfeder *f* feather [allowing parts to slide in a lengthwise direction], sliding key; *(aut)* spring guide
Gleitfertiger *m* concrete extruding machine, slip-forming machine; slip-form paver *(highway construction)*
Gleitfläche *f* sliding surface *(crystallography)*; working surface *(cylinder)*; rubbing surface *(bearing)*; slip *(in coal)*
Gleitflosse *f* plaining fin
Gleitflug *m* gliding, glide
Gleitflüssigkeit *f* lubricant *(caisson method)*
Gleitfunke *m* *(el)* flash-over discharge spark; creepage spark
Gleitfuß *m* sliding support; *(mine)* sliding block
Gleitgeschwindigkeit *f* sliding velocity *(gearing, crystallography)*; planing speed; skimming (gliding) speed
Gleitkanal *m* cable duct *(prestressed concrete)*
Gleitkappe *f* *(mine)* slide bar
Gleitklotz *m* guide (slide) block
Gleitkomma *n* *(dat)* floating point
Gleitkommaprogramm *n* floating-point routine
Gleitkommarechnung *f* floating-point calculation
Gleitkontakt *m* *(el)* sliding (rubbing, wipe) contact, wiper
Gleitkraft *f* pressure on sleeve
Gleitkreuzgelenk *n* slip universal joint
Gleitkurve *f* sliding curve
Gleitlade *f* *(text)* sliding slay
Gleitlager *n* slide (sliding) bearing; plain (journal) bearing
Gleitlagerwerkstoff *m* material for journal bearing
Gleitmetall *n* anti-friction metal
Gleitmikromanipulator *m* sliding micromanipulator
Gleitmittel *n* lubricant; *(met)* parting compound, mould lubricant
Gleitpassung *f* sliding fit
Gleitplatte *f* guide plate; slipper, shoe
Gleitreibung *f* sliding friction
Gleitring *m* lever collar
versenkt liegender G. inner ring
Gleitringdichtung *f* / **axiale** axial seal ring
Gleitschalung *f* sliding (slip) form, sliding shuttering (formwork)
Gleitschalungsdeckenfertiger *m* slip-form paver
Gleitschicht *f* *(geol)* slickensiding
Gleitschieber *m* adjustable block
Gleitschiene *f* slide bar; rail *(boat)*
seitliche G. side plate
übergreifende G. back plate
Gleitschuh *m* slipper, shoe
Gleitschuhkolben *m* trunk piston
Gleitschuhzapfen *m* slipper pin
Gleitschutz *m* non-skid device
Gleitschutzdecksbelag *m* skidless (non-skid) decking
Gleitschutzeinrichtung *f* anti-slip attachment
Gleitschützen *m* *(text)* sliding shuttle
Gleitschutzkette *f* anti-slip chain, anti-skid chain, non-skid chain
Gleitschutzreifenprofil *n* anti-skid pattern
gleitsicher skid-resistant, non-skid, non-slip
Gleitsitz *m* sliding fit
Gleitspriegel *m* *(aut)* slide bow
Gleitstein *m* slide *(kinematics)*; sliding member *(crank)*; slide ring
Gleitstößel *m* roller (sliding) tappet

Gleitstrahlbake *f* glide path beacon
Gleitstück *n* slipper, shoe
Gleittisch *m (opt)* floating stage; sliding stage
Gleitung *f* slip, gliding *(crystal)*; shear deformation *(strain)*
Gleitverhältnis *n (aero)* lift/drag ratio
Gleitweg *m* glide path
Gleitwegleitstrahl *m* glide path beam
Gleitwegsender *m* glide path transmitter
Gleitwerkstoff *m* sliding material
  gesinterter G. sintered sliding material
Gleitwiderstand *m* shearing resistance
Gleitwinkel *m* flight (glide) path angle
Gleitzapfen *m* journal
Gletscher *m* glacier
Gletscherberg *m* glacier berg
Gletscherkunde *f* glaciology
Gletschermühle *f (geol)* pot-hole
Gletschersee *m* lake ponded up by glacier
Gletschersohle *f* sole of the glacier
Glied *n* 1. joint; limb; bar, link *(kinematics)*; 2. *(math)* term; 3. member *(statics)*; 4. *(el)* component *(e.g. of current)*; 5. link *(chain)*; *(dat)* element
  G. einer radioaktiven Zerfallsreihe [radioactive] chain product
  G. zweiter Ordnung *(math)* second order term
  aktives G. *(dat)* power-operated element
  bewegliches G. movable link *(kinematics)*
  bewegtes G. moving link *(kinematics)*
  differenzierendes G. *(dat)* derivative element, lead network
  gleichnamiges G. similar term
  integrierendes G. *(dat)* integrating element
  rückwirkungsfreies G. *(cont)* irreversible element
  signalformendes G. wave-shaping set *(telegraphy)*
  überflüssiges G. [in Gleichungen] *(dat)* redundant term
  unsymmetrisches G. *(el)* unbalanced component
Gliederegge *f* chain harrow
Gliederheizkörper *m* radiator
Gliederhohlleiter *m* vertebrate waveguide
Gliederkessel *m* sectional boiler
Gliederkette *f* plain coil chain
Gliedermaßstab *m* folding rule
Gliederschleppe *f (ag)* chain drag
Gliederung *f* [sub-]division, grouping
  taktische G. *(mil)* tactical grouping
Gliedervorwärmer *m (therm)* sectional economizer
Gliederwelle *f (aut)* articulated shaft
Gliederzug *m* compound (articulated) train
Glimmanzeigeröhre *f* glow discharge indicator
Glimmbogen *m* glowing arc
glimmen to glow, to smolder
Glimmen *n* glow
Glimmentladung *f* glow discharge; corona discharge
Glimmentladungsröhre *f* glow tube
Glimmentladungsventil *n* glow-discharge rectifier
Glimmer *m* mica
glimmerartig micaceous
Glimmerblättchen *n* mica sheet
Glimmerfensterzählrohr *n* mica end-window counter
Glimmerisolation *f* mica insulation
Glimmerkondensator *m* mica [dielectric] capacitor
Glimmerscheibe *f* mica lamination; deflector *(lamp)*
Glimmerscheinung *f* corona (glow) appearance
Glimmerstanzmaschine *f* mica punching machine *(electron tube manufacture)*
glimmfrei corona-free
Glimmgleichrichter *m* cold-cathode rectifier, neon detector (tube rectifier)
Glimmkatode *f* glow-discharge cathode
Glimmlampe *f* glow[-discharge] tube, neon lamp
  elektrodenlose G. neon induction lamp
Glimmlampenoszillator *m* neon oscillator
Glimmlampensummer *m* glow-discharge tube audio generator
Glimmlicht *n* / **negatives** negative glow
Glimmlichtanzeigeröhre *f* neon indicator
Glimmlichtgleichrichter *m* glow[-lamp] rectifier
Glimmlichtindikator *m* neon indicator
Glimmlichtoszilloskop *n* ondoscope
Glimmlichtspannungsteiler *m* glow-gap divider
Glimmlichtziffernröhre *f* digital neon tube
Glimmröhre *f* neon (glow) tube, glow-discharge lamp
Glimmröhrenscheitelspannungsmesser *m* neon tube crest voltmeter
Glimmröhrenstabilisator *m* glow regulator
Glimmsaum *m* glowing seam
glimmsicher glow-proof
Glimmspannung *f* glow potential
Glimmspannungsteiler *m* glow-discharge voltage divider, glow-gap divider
Glimmstabilisator *m* neon stabilizer
Glimmstreckenstabilisator *m* voltage stabilizing tube
Glimmzählröhre *f* glow counting tube
Glimmzone *f* corona area
Globalrakete *f* global rocket (missile)
Globoidschnecke *f* globoid (Hindley) worm
Globoidschneckenlenkung *f (aut)* hour-glass worm-and-sector steering gear
Globoidschneckentrieb *m* globoidal worm gear
Glocke *f* bell; *(ch)* bubble cap, glass jar
Glockenboden *m* bubble plate (tray)

Glockenbodenkolonne *f* bubble-plate column (tower), bubble-cap column (tower)
Glockenboje *f* bell buoy
Glockenbronze *f* bell metal
glockenförmig bell-shaped
g. aufgeweitet bell-mouthed
Glockengalgen *m (ship)* belfry
Glockenimpuls *m (el)* bell-shaped pulse, sine-squared pulse
Glockenisolator *m (el)* petticoat (dome-shaped, cup) insulator
Glockenkurve *f* normal (bell-shaped) curve
Gaußsche G. *s.* Glockenkurve
Glockenläufer *m (cont)* drag-cup rotor
Glockenmühle *f* 1. *(ch)* bell (gyratory, rotary fine) crusher, cone mill; 2. *(leath)* key-stone mill
Glockenprofilausbau *m (mine)* bell-shaped steel arch support
Glockensignal *n* bell signal
Glockenspinnmaschine *f* cap spinning frame
Glockentonne *f (ship)* bell buoy
Glockenturm *m* campanile
Glockenwascher *m* scrubber with hoods
Glockenwinde *f* bell operating gear
Glockenzählrohr *n* end-window counter [tube], bell counter
Glove-Box *f s.* Handschuhbox
Gloversäure *f* tower acid
Gloverturm *m* Glover tower
Glühanlaßschalter *m* heater-plug starting switch, heater starter switch
Glühbirne *f s.* Glühlampe
Glühelektrode *f* incandescent (hot) electrode
Glühelektron *n* thermion, thermoelectron
Glühelektronenemission *f* thermionic emission, Richardson effect
Glühelektronenentladung *f* thermionic discharge
Glühemission *f* thermionic emission
glühen 1. to glow; 2. *(ch)* to calcine, to roast; 3. to anneal *(heat treatment)*
glühend 1. glowing, burning; 2. hot *(pyrometry)*; 3. red-hot *(iron)*; 4. live *(coal)*
Glüherei *f* annealing installation (plant)
Glühfaden *m* [electric lamp] filament
Glühfadenpyrometer *n* disappearing filament pyrometer
Glühfadenträger *m* filament support
Glühfarbe *f* annealing colour
Glühfestigkeit *f* annealed tensile strength
glühfrischen to malleablize
Glühhaube *f* annealing bell
Glühhitze *f* glowing heat
Glühkatode *f* glow (hot, incandescent, thermionic) cathode
Glühkatodengleichrichter *m* thermionic (hot-cathode) rectifier
Glühkatodenlampe *f* hot-cathode lamp
Glühkatodenquecksilberdampfgleichrichter *m* hot-cathode mercury-vapour rectifier
Glühkatodenröhre *f* thermionic (hot-cathode) valve
Glühkatodenventil *n* thermionic rectifier, hot-cathode valve
glühkatodisch thermionic
Glühkaustik *f (el)* hot-wire cautery
Glühkerze *f* glow (heater) plug
Glühkerzenwiderstand *m* heater-plug resistor
Glühkiste *f (met)* annealing box
Glühkopf *m* hot bulb
Glühkopfmotor *m* hot-bulb engine
Glühkopfzündung *f* hot-bulb ignition, surface combustion
Glühkörper *m* incandescent mantle
Glühlampe *f* incandescence lamp, bulb
sockellose G. capless (footless) lamp
Glühlampenbeleuchtung *f* incandescent lighting
Glühlampenfassung *f* lamp holder
Glühlampenschrank *m (telec)* lamp switchboard
Glühmuffel *f (met)* annealing muffle
Glühofen *m* annealing furnace
Glühprobe *f* heating sample
Glühprüfung *f* heating test
Glührohrprüfung *f* ignition tube test
Glührückstand *m* residue on ignition
Glühschalter *m* heater switch
Glühsonde *f (el)* search filament, hot-filament probe
Glühspirale *f* glow plug filament
Glühstartlampe *f* hot-start lamp
Glühstrumpf *m* gas mantle
Glühtemperatur *f* glowing heat
Glühüberwacher *m (aut)* heater-plug control
Glühzeit *f (met)* annealing cycle (time)
Glühzone *f* zone of incandescence *(generator)*
Glühzünder *m* low-tension electric detonator
Glühzündung *f* ignition by incandescence; *(aut)* premature (surface) ignition
Glukonat *n* gluconate
Glukonsäure *f* gluconic acid
Glukose *f* glucose
glukoseartig saccharoid
Glukuronsäure *f* glucuronic acid
Glutamin *n* glutamine
Glutaminsäure *f* glutamic acid
Glutarat *n* glutarate
Glutarsäure *f* glutaric acid
Glykogen *n* glycogen, animal starch
Glykokoll *n* glycine, glycocoll
Glykol *n* glycol
Glykolsäure *f* glycolic (hydroxyacetic) acid
Glykoproteid *n* glycoprotein
Glykose *f* glucose, glycose
Glykosid *n* glucoside, glycoside
Glykosidbindung *f* glycoside linkage
Glyoxal *n* glyoxal, oxalic aldehyde

Glyoxylsäure *f* glyoxylic acid, glyoxal[ic] acid
Glyptalharz *n* glyptal resin
Glyzerid *n* glyceride
Glyzerin *n (Am)* glycerol
Glyzerinseife *f* transparent soap
GM-Zählrohr *n s.* Geiger-Müller-Zählrohr
Gneis *m* gneiss
gnomonisch gnomonic *(projection)*
Gobelin *m* gobelin tapestry
Goethit *m (min)* goethite
Gold *n* gold
  gediegenes G. native (solid) gold
  kolloidales G. colloidal gold
Goldader *f (geol)* gold vein
Goldamalgam *n* gold amalgam
goldartig goldlike
Goldbarren *m* gold bullion
Goldblattelektroskop *n* gold-leaf electroscope
Goldblech *n* sheet gold
Goldbronze *f* gold bronze
golddotiert gold doped
Golddrahtdiode *f* gold-bonded diode
Goldfolie *f* gold foil
goldführend gold bearing
Goldgang *m (geol)* reef
Goldgehalt *m* 1. gold content; 2. gold standard
Goldgewinnung *f* extraction of gold
Goldglätte *f* gold litharge
Goldglimmer *m* yellow mica
goldhaltig auriferous
Goldkies *m* gold quartz
Goldklumpen *m* gold nugget
Goldkorn *n* regulus of gold
Goldmünze *f* gold coin
goldplattiert gold-plated
Goldpurpur *m* / **Cassiusscher** purple of Cassius
Goldrubinglas *n* gold ruby glass
Goldscheideanstalt *f* gold refinery
Goldscheidung *f* gold parting
Goldschmiedetechnik *f* gold smithing
Goldschnitt *m* gilt edge *(of books)*
Goldseife *f (geol)* gold placer
Goldtombak *m* red brass
Gold-I-verbindung *f* aurous compound
Gold-III-verbindung *f* auric compound
Goldwährung *f* gold standard
Goldwäsche *f* gold washing
Goldzahl *f* gold number
Golf *m* embayment, gulf
Goliathsockel *m (el)* Goliath [Edison screw] cap
Gonadendosis *f (nucl)* gonad dose
Gondel *f* nacelle, car *(airship)*
Gonellatrieb *m* ball-and-disk integrator; wheel-and-disk integrator
Gongboje *f* gong buoy
Gongtonne *f* gong buoy
Goniometer *n* goniometer
Goniometrie *f* goniometry
goniometrisch goniometric
Gonioskop *n* gonioscope
Goochtiegel *m* Gooch crucible
Göpel *m* 1. *(ag)* gin; 2. *(mine)* whim
Gösch *f* jack *(flag)*
Gosse *f* gutter [channel]
Goubau-Leitung *f* single-wire transmission line, guide wire
graben to dig, to trench
Graben *m* ditch, drain, dike, trench
Graben *n* trenching
Grabenaushub *m* trenching, ditching, trench (ditch) digging
Grabenbagger *m* trench excavator (digger)
Grabenfräse *f* trench cutter (cutting machine), wheel-type trenching machine
Grabenpflug *m* drainage plough
Grabenräumer *m* ditch cleaner
Grabenräummaschine *f* ditch and dyke cleaner
Grabenräumung *f* ditch cleaning
Grabenreinigungsgerät *n* ditch cleaner
Grabenverbau *m* trench timbering
Grabenverzimmerung *f* trench timbering
Grabenziehanbaugerät *n* trenching attachment
Grabenziehgerät *n* trencher
Grabenziehmaschine *f* ditching machine
Grabgefäß *n* digging bucket
Grabstein *m* gravestone, tombstone
Grabvorgang *m* crowding, digging
Grabwerkzeug *n* digging tool
Grad *m* order *(equation)*; degree *(scale)*
  G. absolut degree absolute (Kelvin)
  G. Celsius centigrade [degree], degree[s] centigrade
  G. der Abweichung drift percentage
  G. der Entartung *(nucl)* degree of degeneration
  G. der statischen Unbestimmtheit degree of statical indeterminancy
  G. Engler Engler degree
Gradation *f* gradation
Gradationswiedergabe *f* gradation reproduction
Gradbogen *m (math)* protractor
Gradeinteilung *f* graduation (marking) in degrees
  G. / mit graduated, having graduation in degrees
Gradient *m* gradient, slope, incline
Gradientenrelais *n* rate-of-change relay
Gradierwerk *n* cooling tower (stack), saline
Gradnetzmeldeverfahren *n (el)* grid referencing
Graduieren *n* graduation
Graetz-Schaltung *f* Graetz (bridge-connected) rectifier
grafisch graphic[al]
Grahamhemmung *f* Graham escapement
Gramm *n* gram[me] *(mass)*

Grammäquivalent *n* gram-equivalent
Grammatom *n* gram-atom
Grammkalorie *f* gram (small) calorie
Grammolekül *n* gram-molecule, mol[e]
Grammophonnadel *f* gramophone needle
Gramm-Röntgen *n* gram-roentgen *(radiobiology)*
Granalie *f* granule, nodule, pellet
Granalien *fpl* granulated metal
Granat *m (min)* garnet
Granate *f* shell
 blindgeladene G. blank shell
 scharfgeladene G. live shell
Granattrichter *m* shell crater (hole)
Granatwerfer *m* [trench] mortar
Granatwerfertrupp *m* mortar squad
Granatzünder *m* percussion fuse
Granierkalander *m* graining calender
Granit *m* granite
granitähnlich granitoid
granitisch granitic
Granulat *n (ch)* granule, chips
granulieren to granulate, to shot, to corn
Granulierrinne *f* granulating spout
granuliert granulated, with granular structure, shotted
Granuliertrommel *f* pelletizing drum
Graphit *m* graphite
 flockiger G. flaky graphite
Graphitabsorber *m* graphite absorber
Graphitbildung *f* graphitization
Graphit-Bronze-Lager *n* oilless bearing
Graphitbürste *f (el.m.)* carbon-brush
Graphitemulsion *f* graphite (black lead) emulsion
Graphit-Gas-Reaktor *m* gas-cooled graphite [-moderated] reactor
Graphitgitter *n* graphite lattice
graphithaltig graphitiferous
Graphitierbürste *f* black leading brush
Graphitieren *n* graphitization
graphitiert graphite-treated, graphitic
Graphitisieren *n* graphitization
Graphitkugeln *fpl* graphite pebbles
Graphitmoderator *m (nucl)* graphite moderator
Graphit-Natururan-Reaktor *m* natural uranium-graphite reactor
Graphitöl *n* graphite oil
Graphitpyrometer *n* graphite pyrometer
Graphitreaktor *m* graphite (carbon) reactor, graphite (carbon) pile
 gasgekühlter G. gas-cooled graphite-moderated reactor
Graphitsäule *f s.* thermische Säule
Graphitschmelztiegel *m* graphite crucible
Graphitschmierapparat *m* graphite lubricator
Graphitschmiere *f* graphite lubricant
Graphitschmiermittel *n* graphite lubricant
Graphitschmierung *f* graphite lubrication
Graphitstabofen *m* graphite bar furnace
Graphitstift *m* / **elektrisch leitender** electrographic pen
Graphittiegel *m* graphite (black-lead) crucible
Graphit-Uran-Reaktor *m* graphite-uranium reactor, carbon-uranium reactor
Graphitziegel *m* graphite brick
Grasmäher *m* mower
Grasnarbe *f* grass cover
Grassamenreiniger *m* grass seed cleaner
Gras- und Heuheber *m* grass pick-up elevator
Grat *m* 1. *(met)* burr, fin, flash; 2. *(build)* groin *(vaults)*; verge *(roof)*; arris; 3. *(print)* burr
 G. / mit burred
 angeschliffener G. shoulder, ridge
Gratbildung *f (met)* formation of burrs
gratfrei burr-free, free from burrs
Grathobel *m* dovetail (rabbet) plane
Gräting *f (ship)* grating
gratlos finless, burrless
Gratsäge *f* dovetail saw
Gratsparren *m* angle (hip) rafter
Gratziegel *m* bonnet (hip, arris) tile
Graufilter *n* grey filter
Grauguß *m* grey cast iron
Graugußstück *n* grey iron casting
Graukeil *m* [neutral] wedge filter; *(telev)* tone wedge, gradation; key
Graumanganerz *n* manganite
Graupappe *f* grey board, chipboard
 gewalzte G. friction board
Grauschleier *m* grey fog
Grauskale *f* grey scale
Grauskalenspeicherung *f (dat)* grey scale recording, half-tone storage
Grauspießglanz *m* stibnite, antimonite
Grauspießglanzerz *n* grey antimony ore
Graustrahler *m* grey-body radiator
Graustufenaufzeichnung *f s.* Grauskalenspeicherung
Grauton *m* grey-shade
grauwertrichtig actual grey-value
Gravierarbeit *f* engraving operation
gravieren to engrave
Gravierfräser *m* engraving cutter
Graviermaschine *f* engraving machine, engraver
 elektronische G. electronic engraver (engraving machine)
Graviernachformmaschine *f* engraving-type form duplicating machine
Graviernadel *f* engraving needle (stylus)
Gravierstichel *m* engraving needle
Gravierung *f* engraving
Gravimeter *n* gravimeter
Gravimetermessung *f* gravimetric measuring
Gravimetrie *f* gravimetry

gravimetrisch gravimetric
Gravis *m* grave accent
Gravitation *f* gravitation
Gravitationsfeld *n* gravitation[al] field
Gravitationskonstante *f* constant of gravitation, gravitation[al] constant
Gravitationspotential *n* gravitational potential
Gravitationsverschiebung *f* **der Spektrallinien** gravitational shift of spectral lines
Gravur *f* engraving
elektronische G. electronic engraving
Gravurraster *m* engraved screen
Gray-Code *m (dat)* Gray code
Greifbagger *m* 1. grab dredger *(under water)*; 2. grab excavator *(on land)*, grabbing crane
Greifer *m* 1. *(mach)* gripper, claw; 2. *(build)* grab [bucket], clamshell bucket; 3. *(text)* looper *(sewing machine)*
Greiferaufzug *m* grab hoist
Greiferbagger *m* grab excavator
Greiferbrückenkran *m* grabbing bridge crane
rundlaufender G. revolving grabbing bridge crane
Greiferbühne *f* grab platform
Greiferkante *f* gripper edge
Greiferkorb *m* grab [bucket]
Greiferkran *m* clamshell (grab) crane
Greiferkübel *m* grab [bucket]
Greiferlaufkatze *f* grabbing trolley
Greiferrundwebmaschine *f* circular loom with gripper
Greiferschale *f* half-scoop
Greifertrenner *m* grab isolator
Greifkübel *m* grab [bucket]
Greiftiefe *f* grab depth *(of excavator)*
Greifweite *f* grab width *(of excavator)*
Greifzirkel *m* caliper
Greinacherschaltung *f* half-wave voltage doubler, Greinacher doubling circuit
Grendelbalken *m* plough-beam
Grenzausschaltleistung *f s.* Grenzausschaltvermögen
Grenzausschaltvermögen *n (el)* breaking capacity
Grenzbahnhof *m* frontier station, *(Am)* border station
Grenzbeanspruchung *f* limiting load
Grenzbedingung *f* boundary (critical, limiting) condition
Grenzbelastung *f* load limit, limiting load
Grenzbereich *m* threshold region
Grenzdichte *f* limiting density
Grenzdrehzahl *f* maximum speed
Grenzdruck *m* limiting pressure
Grenze *f* limit, boundary, threshold
G. der lichtelektrischen Wirkung wavelength threshold
G. der Nachweisbarkeit limit of detection
G. der statischen Stabilität steady-state stability limit
G. des Hörbereiches limit of audibility
extrapolierte G. extrapolated boundary
größte untere G. *(math)* greatest lower bound
kleinste obere G. *(math)* least upper bound
kurzwellige G. short wavelength limit
langwellige G. long wavelength limit
lichtelektrische G. photo-electric threshold
obere G. top limit
scharfe G. sharp demarcation
zulässige G. limit of acceptability
Grenzempfindlichkeit *f* threshold response (sensitivity), limiting (ultimate) sensitivity
Grenzen *fpl* **/ verengte** compressed limits
Grenzenergie *f* limiting energy; end-point energy
Grenzerwärmung *f* limiting temperature-rise
Grenzfaktor *m* limiting factor
Grenzfall *m* borderline case
Grenzfläche *f* boundary surface; interface, surface of separation
grenzflächenaktiv surface (capillary) active
Grenzflächenenergie *f* interfacial surface energy
Grenzflächenerscheinung *f* interfacial phenomenon
Grenzflächenpotential *n* boundary potential; interface potential
Grenzflächenspannung *f* interfacial tension
Grenzflächenwinkel *m* interfacial angle
Grenzfrequenz *f* 1. edge (cross-over) frequency; 2. limiting (cut-off) frequency
effektive G. effective cut-off frequency
obere G. upper limiting frequency
untere G. threshold frequency
Grenzfunkbake *f* boundary marker
Grenzgeschwindigkeit *f* limiting velocity
Grenzkohlenwasserstoff *m* satisfied hydrocarbon, paraffine
Grenzkorn *n* boundary size
Grenzkurve *f* limit curve of critical state, limiting curve; limit cycle
Grenzlast *f* maximum load
Grenzlehrdorn *m* plug limit gauge
Grenzlehre *f* limit gauge
Grenzleistung *f* power limit (margin), limiting power
zulässige G. operating limit, *(Am)* overload level
Grenzlinie *f* boundary
Grenzmaß *n* limit [of size], limiting dimension, dimensional limit
Grenzpotential *n* limit potential
Grenzprozeß *m* limiting process
Grenzrachenlehre *f* external limit gauge
Grenzrad *n* gear with minimum number of teeth [to avoid undercutting]
Grenzreibung *f (ph)* boundary friction

Grenzschalter *m* [main] limit switch

Grenzschaufel *f* limit blade *(turbine)*

Grenzschicht *f* boundary layer

  laminare G. laminar boundary layer

  turbulente G. turbulent boundary layer

Grenzschichtablösung *f* separation of the boundary layer

Grenzschichtablösungspunkt *m* separation point

Grenzschichtdicke *f* thickness of the boundary layer

Grenzschichttemperatur *f* junction temperature *(semiconductors)*

Grenzschmierung *f* boundary lubrication

Grenzspannung *f* 1. limiting stress; 2. cut-off voltage

Grenzstrahl *m* grenz ray, infra-Roentgen ray

Grenzstrom *m* limiting (marginal) current; minimum fusing-current

  dynamischer G. instantaneous short-circuit current *(of a transformer)*

  thermischer G. rated short-circuit current

Grenztemperatur *f* limiting temperature

Grenztragfähigkeit *f* ultimate bearing capacity

Grenzübertemperatur *f s.* Grenzerwärmung

Grenzwelle *f* intermediate (medium high-frequency) wave

Grenzwellenlänge *f* minimum wavelength, quantum limit; critical wavelength *(hollow waveguide)*; cut-off wavelength *(waveguide)*

  obere G. threshold wavelength

Grenzwert *m* limit, limit[ing] value, maximum permissible value

  G. einer Folge limit of a sequence

Grenzwertprüfung *f* marginal checking

Grenzwertüberwachung *f* 1. limit control, monitoring; 2. off-limit check

Grenzwiderstand *m* critical resistance *(e.g. measuring instrument)*

Grenzwinkel *m* angle of draw, limit (critical) angle

  G. der Reibung limiting angle of friction

  G. der Totalreflexion critical angle of total reflection

  dynamischer G. travelling limit angle

Grenzzähnezahl *f* / **kleinste** minimum number of gear teeth *(which can be generated without undercut)*

Grenzzeiger *m* limit indicator

Grenzzeit *f* time limit *(relay)*

Grenzzone *f* contact zone

Grenzzustand *m* critical (limiting, ultimate) state

Grenzzyklus *m* limit cycle

Greyhound greyhound *(sailing ship)*

Grieß *m (telev)* snow; grass

Griff *m* 1. *(mach)* handle; 2. *(text)* feel, body, hand *(fabric)*

  weicher G. soft hand

Griffangel *f* handle end

Griffelement *n* gripper unit, manual element

Griffläche *f* gripping surface

griffig sharp *(grinding wheel)*; non-slip, non-skid; *(wood)* bulky

  die Schleifscheibe g. machen to sharpen the grinding wheel

Griffigkeit *f* 1. *(text)* grip, hand, feel; 2. non-skid property *(road)*; 3. cutting ability *(grinding wheel)*

Griffkreuz *n* star wheel

Griffmuster *n* grip pattern

Griffsicherung *f* handle-type fuse

Griffzeit *f (mach)* idle (non-productive, handling) time

Grindel *m* plough-beam

grob coarse *(grains)*; rough

Grobabstimmung *f* coarse tuning

Grobabtastung *f (telev)* coarse scanning

Grobätzung *f* macroetching

Grobbank *f (text)* slubbing frame

Grobblech *n* thick (coarse) plate, thick sheet iron

Grobblechlagerkran *m* slab yard crane

Grobdrahtziehbank *f* thick wire drawing bench

Grobeinstellung *f* coarse setting (adjustment, positioning)

Groberzofen *m* coarse ore furnace

grobfasrig coarse-grain[ed]

Grobfeile *f* coarse file

Grobfilter *n* coarse filter

Grobflyer *m* slubbing frame

Grobgefüge *n* macrostructure

Grobgewinde *n* coarse screw thread

Grobgewindereihe *f* coarse screw thread series

Grobgut *n* coarse fraction

grobjährig wide-ringed, with large (wide) annual rings *(wood)*

Grobkapillarwasser *n* water of open capillaries

Grobkeramik *f* heavy ceramics (clay products), burnt clay ware

Grobkies *m* coarse gravel

Grobkohle *f* large (lump) coal

Grobkoks *m* lump coke

Grobkorngefüge *n* coarse-grained structure

Grobkornglühen *n* coarse-grain annealing

grobkörnig coarse-grain[ed]

Grobkornsetzmaschine *f* coarse-coal jig

Grobkornstruktur *f* coarse-grain[ed] structure

Grobkrempel *f* breaker card

grobkristallin coarse-crystalline, coarse-grain[ed]

Groblageeinstellung *f* coarse positioning

Grobmahlen *n* primary grinding

grobnarbig rough-grained

Grobpositionierung *f* coarse positioning

Grobraster *m* coarse screen

Grobrasterätzung *f* coarse screen etching
Grobregelstab *m* shim (coarse control) rod
Grobregelung *f* coarse control (adjustment)
Grobregler *m* coarse control
Grobrost *m* disk screen
Grobsaatrad *n* coarse-seed wheel, large-seed wheel
Grobsand *m* coarse sand
Grobsiebung *f* scalping
Grobsplitt *m* coarse stone chippings
Grobspulen *fpl* bobbins for slubbing and roving
Grobsteuerung *f* coarse control (adjustment)
Grobstoff *m* rejects *(paper making)*
Grobstraße *f* blooming train
Grobstrecke *f (text)* first drawing frame
grobstückig coarse, lumpy
Grobtrieb *m* coarse motion
Grobvorschubreihe *f* coarse-feed series
Grobwalzwerk *n* roughing down mill
Grobwaschmittel *n* heavy-duty detergent
Grobzerkleinerung *f* coarse crushing
Grobzug *m* roughing block *(wire)*
Grobzuschlag[stoffe *mpl***]** *m* coarse aggregate *(concrete)*
groß big *(body)*, bulky; large *(area)*; capital *(letter)*; great *(amount)*
Großanlage *f* big plant
Großantenne *f* giant aerial
Großaufnahme *f* close-up
Großbaustelle *f* large-scale building (construction) site
Großblockbau *m* large[-sized] block construction
Großblockbauweise *f* large[-sized] block construction
Großboot *n* long-boat, launch
Großbuchstabe *m* capital [letter]
Großdrehmaschine *f* large (heavy-duty) lathe
Größe *f* size *(dimensions)*; *(ph)* magnitude, quantity; amount *(mass, number)*
  G. der Serie length of production run
  G. des Hohlraumes size of [the] cavity
  G. des Spanwinkels amount of rake
  G. des Winkels amount of angularity
  abgeleitete G. derived quantity
  bekannte G. *(math)* known quantity
  dimensionslose G. dimensionless quantity
  elektrische G. electrical quantity
  erregende G. actuating quantity *(relay)*
  gegebene G. given quantity
  kritische G. critical size (dimensions)
  natürliche G. natural size
  skalare G. scalar quantity
  unbekannte G. unknown quantity
  unendlich kleine G. *(math)* infinitesimal [quantity]
  vektorielle G. vector quantity
größengleich identical in size
Größengleichung *f* dimensional equation
Größenklasse *f* **der Sterne** class of star magnitudes
Größenordnung *f* order of magnitude
  benannte G. denominational order
Größenunterschied *m* size difference
Größenwerter *m* analog-to-digital converter
Großfeldmessung *f (nucl)* broadbeam measurement
Großflächenbewirtschaftung *f* large-area farming
Großflächendrillmaschine *f* large-area seed drill, large-acreage seed drill
Großflächenzähler *m (nucl)* large-area counter
großflächig large
Großformat *n* big size
Großfunkstelle *f* long-distance radio station
Großfutterbehälter *m* bulk-feed hopper
Großfuttertrichter *m* bulk-feed hopper
Großhäcksler *m* high-capacity cutter-blower
großkalibrig heavy-calibre
Großkampfschiff *n* capital ship
Großkegel *m (print)* large body size
Großkern *m* macronucleus
Großkraftübertragung *f* high-power transmission
Großkraftwerk *n* high-power plant, large-scale power plant
Großkreisbogen *m* arc of great circle of sphere
Großlautsprecher *m* high-power loudspeaker
Großlautsprecheranlage *f* public address system
Großlochbohrer *m* 1. large-hole drilling machine *(stone)*; 2. auger *(coal)*
Großlochbohrmaschine *f* large hole drill
Großmast *m* mainmast
Großplatte *f* large (room-sized) panel
Großplattenbau[weise *f***]** *m* large panel construction (method, system)
großporig having wide spacing
Großraumbüro *n* open plan office
Großraumtiefofen *m* large-space soaking pit
Großraumwagen *m (ry)* large-capacity wagon, high-capacity wagon; *(mine)* high capacity mine car
Großrundfunksender *m* high-power broadcasting station
Großsegel *n* mainsail
Großsender *m* high-power transmitter, large transmitting station
Großsendestation *f* long-distance station
Großstapelauslage *f* deep pile delivery
Großstetigschleifer *m* continuous high-power caterpillar grinder
großstückig coarse, lumpy
Großtafelbau[weise *f***]** *m* large panel construction (system, method)

Großtanker *m* supertanker
großtechnisch industrial, commercial
Größtmaß *n* maximum-limit [dimension]
Größtspiel *n* maximum clearance
Größtwert *m* maximum value
Großversuch *m* full scale test
Großwasserraumkessel *m* large water-tube boiler
Grotesk *f* sanserif, grotesque
Grubber *m* grubber, cultivator, scuffler
Grubberegge *f* drag harrow
Grübchenbildung *f* pitting *(corrosion)*
Grube *f* pit; mine; colliery *(coal)*
  G. für die radioaktiven Abfälle burial (dumping) ground, graveyard, storage well
  G. mit hohem Automatisierungsgrad push-button mine
  schlagwetterfreie G. naked-light mine, non-gassy mine
  schlagwettergefährdete G. gassy (safety-lamp) mine
  tiefe G. deep-level mine
Grubenabwässer *pl* mine waters
Grubenarbeiter *m* mine worker
Grubenbahn *f* mine tramway
Grubenbau *m* mine working, underground excavation
  abgeworfener G. disused working
  streichender G. level course
Grubenbeamter *m* mining official
Grubenbrand *m* mine (underground) fire
Grubenfahrdienstleiter *m* mine train dispatcher
Grubenfeld *n* claim, mining property
grubenfeucht freshly mined *(coal)*; freshly quarried *(sand)*
Grubenfeuer *n* mine fire
Grubenformerei *f* pit moulding
Grubengas *n* fire (mine) damp, marsh gas
Grubengasanzeiger *m* mine gas indicator
Grubengebäude *n* mine workings, underground excavations
Grubengerbung *f* pit tannage
Grubengießen *n* pit casting
Grubenglühen *n* pit annealing
Grubenhobelmaschine *f* pit planer
Grubenholz *n* mine timber, pit prop[s]
Grubenklima *n* mine climate
Grubenlampe *f* miner's lamp
  elektrische G. electric safety lamp
Grubenleuchte *f* mine lamp
Grubenlokomotive *f* mining locomotive
Grubenlüfter *m* mine fan
Grubenmauerung *f* walling in mines
Grubenpersonenbahnhof *m* man-riding station
Grubenraum *m* excavation
Grubenrettungsstation *f* mine rescue station
Grubenrettungswesen *n* mine rescuing
Grubenröste *f (text)* pond retting
Grubensand *m* pit (quarry) sand
Grubenschacht *m* mine shaft
Grubenschwelle *f* pit-sleeper
Grubensicherheit *f* safety in mines
Grubensohle *f* mine floor
Grubenstempel *m* mine (pit) prop
Grubenventilator *m* mine fan
Grubenwarte *f* central mine-control station
Grubenwasser *n* mine water
Grubenwasserhaltung *f* mine drainage
Grubenwehr *f* mine rescue corps
Grubenweite *f* / **äquivalente** equivalent orifice
Grubenwetterkühler *m* colliery air conditioner
Grubenzimmerung *f* timbering [of a mine]
Grudekoks *m* coke of brown coal
Grummet *n* aftermath, second hay
Grummetheu *n* second hay crop
Grumt *n* aftermath, second hay
Grün *n* / **Rinmans** cobalt green
  Scheelesches G. mineral green
  Schweinfurter G. emerald green
Grünanlage *f* greenery; park, lawns and plantings
Grünbleierz *n* pyromorphite, green lead ore
Grund *m* 1. bottom *(borehole, etc)*, ground; root *(thread)*; 2. reason
  G. einer Grundbohrung bottom in a blind hole
  G. eines Gewindes root of a thread
  ankerfester G. good anchorage
  lotbarer G. soundings
  nicht mehr lotbarer G. non-sounding
Grundabmessung *f* plan dimension
Grundabsorption *f* fundamental absorption
Grundanstrich *m* priming (prime, primary) coat, priming
Grundausbildung *f* basic training
  G. in der Produktion basic workshop training
  technische G. basic engineering education
Grundbacke *f* master jaw *(chuck)*
Grundbahn *f* basic mat *(paper)*
Grundbalken *m* ground beam
Grundbau *m* foundation engineering
Grundbaugruppe *f* basic unit
Grundbegriff *m* basic notion (idea)
Grundbereich *m* fundamental range
Grundberührung *f (ship)* grounding, touching of the ground
Grundbestandteil *m* basic constituent
Grundbindung *f* elementary (plain, standard) weave
Grundblatt *n* drawsheet
Grundbohrung *f* blind hole
Grundbruch *m (build)* breach, soil failure, rotational slide
Grundbuchse *f* master bushing, permanent bush fastened in the jig
Grunddünung *f* ground swell

Grundebene *f* basal plane *(crystal)*
Grundeigenschaft *f* fundamental property
Grundeinheit *f* fundamental (basic) unit
gründen to found
Grundentstörung *f (el)* basic interference suppression
Grundfaden *m* back yarn *(knitting)*
Grundfarbe *f* primary (basic) colour
  additive G. additive primary colour
Grundfelge *f* base rim
Grundfertigkeit *f* basic skill
Grundfläche *f* base surface (area); floor area; *(opt)* cardinal plane
Grundform *f* basic form (design)
  G. der Maschine basic machine
Grundfrequenz *f* fundamental frequency
Grundfrequenzband *n* baseband
Grundfrequenzbereich *m* baseband
Grundfunktion *f* basic function
Grundgänger *m* sunken torpedo
Grundgarn *n* back (core) yarn
Grundgebühr *f* fixed charge
Grundgehalt *n* base pay
Grundgelenkviereck *n* basic four-bar linkage *(kinematics)*
Grundgeräusch *n* background (random) noise
Grundgesamtheit *f* universe, total (parent, infinite) population
Grundgeschirr *n (ship)* ground tackle
Grundgeschwindigkeit *f* ground speed *(airplane)*
Grundgesetz *n* fundamental law
Grundgewebe *n* backing cloth, foundation fabric
Grundgitter *n* host lattice
Grundgleichung *f* fundamental equation
Grundgröße *f* fundamental quantity
Grundhelligkeit *f* brightness
Grundhelligkeitsregler *m (telev)* background brightness control
Grundhobel *m* router plane
Grundhobeleisen *n* ground plane iron
Grundieranstrich *m* prime coat
grundieren 1. *(build)* to prime *(paint)*; 2. *(text)* to bottom
Grundierfarbe *f* priming paint, first coat
Grundierlack *m* primer, filler, precoat
Grundiermittel *n* wash primer
Grundierung *f* 1. *(build)* prime (priming, primary) coat; priming; foundation; 2. *(text)* bottoming
Grundkapital *n* stock
Grundkette *f (text)* back warp; ground warp
Grundkonstante *f* fundamental constant
Grundkörper *m* backing material
Grundkosten *pl* real cost
Grundkreis *m* fundamental (base) circle *(gear)*
Grundkristall *m* host crystal
Grundkurs *m* true course *(radar)*
Grundlagenforschung *f* fundamental (basic) research
Grundlager *n* main bearing
Grundlast *f* base (normal) load
grundlegend basic, fundamental
Grundlinie *f* base [line]; bottom line *(character recognition)*
Grundlohn *m* basic wage rate
Grundlösung *f* fundamental solution
Grundmaschinentyp *m* basic machine type
Grundmaß *n* basic size
Grundmasse *f (geol)* ground-mass, matrix
Grundmetall *n* parent metal *(welding)*; base metal *(coating)*
Grundmine *f (mil)* ground mine
Grundmodus *m* dominant mode *(of a waveguide)*
Grundniveau *n* 1. ground level; 2. elementary level
Grundoperation *f* unit operation
Grundpatent *n* master (basic) patent
Grundplatte *f* 1. *(mach)* baseplate, bed [plate], bed-frame *(boiler)*; floor plate *(horizontal boring mill)*; 2. *(el)* mounting plate; 3. *(build)* invert; foundation raft (slab)
  G. mit Kondensationseinrichtung bed plate with condenser
Grundplattenanschluß *m* bed plate junction
Grundplattenmaschine *f* **mit Kreuzbett (kreuzbeweglichem Ständer)** floor-type [horizontal boring] machine with cross-sliding column
Grundprozeß *m (ch)* unit process
Grundpunkt *m (opt)* cardinal point
Grundrahmen *m* bed-plate, bed-frame *(turbine)*; base frame
Grundreaktion *f (ch)* essential reaction
Grundrechnungsart *f* fundamental operation
Grundregel *f* basic rule
Grundreibahle *f* bottoming reamer
Grundring *m* bottom ring, internal (inner) spring ring; neck ring
Grundriß *m* plan (top) view *(technical drawing)*; [ground] plan; layout; *(ship)* half-breadth plan
Grundrißstrecke *f* planimetric distance
Grundrißzeichnung *f* layout drawing
Grundschaltbild *n* basic circuit diagram
Grundschaltung *f* basic circuit (wiring)
Grundschicht *f* primary layer
Grundschieber *m* **einer Langhobelmaschine** saddle of a planer
Grundschiffsform *f* parent hull
Grundschleppnetz *n* trawl [net]
Grundschule *f* primary school, *(Am)* elementary school
Grundschuß *m* back weft; binding pick
Grundschwelle *f (mine)* foot block, sill
Grundschwingung *f* 1. fundamental vibration (oscillation); 2. fundamental mode
Grundsee *f* ground swell

Grundspant *n (ship)* main timber (frame)
Grundsteigung *f* neutral pitch *(air screw)*
Grundstein *m* foundation stone
Grundstellung *f* 1. *(el)* normal position *(switch)*; 2. *(ry)* initial position *(signal)*
Grundstoff *m* element; base, primary material
 chemischer G. element
Grundstraffer *m* drawsheet
Grundstrecke *f (mine)* base (bottom) road, *(Am)* butt-entry, main (mother) gate
Grundstruktur *f* basic structure
Grundstück *n* plot, lot, allotment
Grundstücksmessung *f* real estate survey
Grundstufe *f* initial stage
Grundtyp *m* dominant mode *(of a waveguide)*
Grundüberholung *f* general (major) overhaul
Grundüberzug *m* base coat
Gründung *f* foundation, groundwork, footing
 abgetreppte G. benched foundation
Gründünger *m* green manure
Gründungsbauwerk *n* foundation structure
Gründungspfahl *m* foundation pile
Gründungsplatte *f* foundation raft (slab)
Gründungsrost *m* grillage [foundation]
Gründungssohle *f* formation (foundation) level, final grade
Gründungswanne *f* tank
Gründüngung *f* green manuring
Grundverbesserung *f* land (soil) improvement
Grundverbindung *f* parent substance
Grundviskosität *f* intrinsic viscosity
Grundvorgang *m* basic mechanic
Grundwasser *n* ground water
Grundwassersenkung *f* lowering of the water table
Grundwasserspiegel *m* [ground] water table, [ground] water level
Grundwasserverseuchung *f* underground water contamination
Grundwelle *f (ph)* fundamental wave
Grundwellenblindleistung *f* **der Nullkomponente** *(el)* fundamental zero-sequence
Grundwerkstoff *m* base material (metal), parent (base) metal *(welding)*
Grundwinkel *m* base angle *(bulkhead)*
Grundwissen *n* fundamental knowledge
 technisches G. basic technical knowledge
Grundwissenschaft *f* fundamental science
Grundwolle *f* linters
Grundzahl *f* cardinal number; base, radix
Grundzahlkomplementkode *m* radix complement code
Grundzahlschreibweise *f* radix notation
Grundzapfen *m* fulcrum pin
Grundzeit *f* basic time
Grundzug *m* bottom flue
Grundzustand *m* ground state *(atom, molecule)*; original (basic) state
 G. des Kernes ground state of the nucleus
Grünerbsendreschmaschine *f* mobile viner
Grünfeuer *n* barium nitrate firework
Grünfutter *n* green fodder (crop)
Grünfutteraufladegerät *n* green-crop loader
Grünfuttertrocknungsanlage *f* green-feed drier
Grüngewicht *n* green weight *(hide)*
Grüngürtel *m (build)* green belt
Grüngut *n* green crop, forage
Grünkreuzkampfstoff *m* asphyxiant gas
Grünland *n* grassland
Grünlandverbesserung *f* / **umbruchlose** pasture improvement
Grünlauge *f* green liquor
Grünsandform *f* green sand mould
Grünschnitt *m* summer pruning
Grünseife *f* soap base
Grünspan *m* verdigris, aerugo
grünspanartig aeruginous
Gruppe *f* group
 chromophore G. chromophore
 prosthetische G. prosthetic group
Gruppenantrieb *m* group drive
Gruppenanzeige *f* group indication *(punched card machinery)*
 G. auf Grund der letzten Karte last card indication *(punched card machinery)*
Gruppenanzeigearbeitsgang *m* group indication cycle *(punched card machinery)*
Gruppenanzeigeelimination *f* group indication elimination *(punched card machinery)*
Gruppenbau *m (mine)* simultaneous extraction of several seams
Gruppenbearbeitung *f* group technology (production), Mitrofanov group machining method
Gruppencharakteristik *f* space factor *(aerials)*
Gruppendiffusionsmethode *f (nucl)* group-diffusion method
Gruppenfaktor *m* space factor *(aerials)*
Gruppenfällungsreagenz *n* group precipitant
Gruppenfeuer *n (mil)* volley fire
Gruppenfrequenz *f* group frequency
Gruppengeschwindigkeit *f* group (envelope) velocity
Gruppenkode *m* group code
Gruppenlaufzeit *f* envelope delay, group retardation
Gruppenmelkstand *m* two stalls per unit parlour
Gruppenreaktion *f (ch)* group reaction
Gruppenschalter *m* gang switch
Gruppentheorie *f* group theory
Gruppentransformator *m* bank of transformers
Gruppenunterbrechung *f* control break *(punched card machinery)*
Gruppenverbindungsplan *m (el)* trunking (grading) diagram
Gruppenwähler *m* group (intermediate) selector *(telephony)*

Gruppenwechsel *m* group control change *(punched card machinery)*
gruppenweise in groups
Gruppenzähler *m (dat)* batch counter
Gruppierung *f* grouping
  G. von Motoren grouping of motors, motor combination
Grus *m (mine)* breeze; *(build)* fine gravel
Gruskohle *f* slack [coal]
Guajakharz *n* guaiacic resin
Guajakholz *n* lignum vitae
Guajakol *n* guaiacol
Guanidin *n (ch)* guanidine
Guckloch *n* peep-hole, judas *(in a door)*
gu-Kern *m (nucl)* even-odd nucleus
Gülle *f* liquid manure
Gulonsäure *f* gulonic acid
gültig bis [zu] valid up to
  g. sein für to hold for
Gültigkeitsbereich *m* range of validity
Gültigkeitserklärung *f* validation, declaration of validity
Gummi *m* rubber; caoutchouc
Gummi *n* gum
Gummiabsatz *m* rubber heel
Gummiader *f (el)* rubber-covered wire
Gummiaderschnur *f (el)* rubber-insulated cord
Gummiarabikum *n* gum arabic, acacia (Senegal) gum
gummiartig gummous, gummy
Gummiaufzug *m* rubber dressing
Gummiband *n* elastic [ribbon], rubber tape
  G. mit Stahleinlagen steel cord belting
Gummibelag *m* rubber flooring
gummibereift rubber-tyred
Gummibereifung *f* **mit Luftfüllung** pneumatic tyres
Gummibindung *f* rubber bonding *(grinding wheel)*
Gummibleikabel *n* rubber[-insulated] lead-covered cable
Gummiboot *n* rubber boat
Gummidichtring *m* rubber sealing ring
Gummidichtung *f* packing of natural rubber, rubber gasket (packing, seal)
Gummidichtungsring *m* / **hohler** tubular rubber seal
Gummidrucklack *m* gum printing varnish
Gummidrucktuch *n* rubber blanket
Gummieinlage *f* rubber core *(sealing)*
gummieren to gum, to rubberize
Gummieren *n* **von Geweben** coating of fabrics with rubber
Gummiermaschine *f* gumming machine
Gummiersatzstoff *m* rubber substitute
gummiert rubber-coated, rubberized *(fabric)*; gummed
Gummifeder *f* rubber spring
Gummifederlager *n* rubber-cushioned spring hanger
Gummifender *m* / **aufblasbarer** *(ship)* inflatable fender
Gummiformpressen *n* rubber die process
Gummiformteil *n* rubber moulding
Gummigurt *m* rubber belting *(conveying)*
Gummihandschuh *m* rubber glove
Gummiharz *n* gum resin
Gummihaut *f* rubber skin *(folding canoe)*
Gummihülse *f* rubber bush
gummiisoliert rubber-covered, rubber-insulated
Gummikabel *n* rubber-insulated cable, rubber-covered cable
Gummikern *m* high-stretch rubber core
Gummiklischee *n* rubber block (cut)
Gummikreuzgelenk *n (aut)* rubber universal joint
Gummilinse *f (telev)* zoom lens
Gummilitze *f* rubber-covered braided wire
Gummilösung *f* [solvent] rubber cement, rubber solution
Gummimanschette *f* rubber gaiter
Gummimantel *m* waterproof (India rubber) coat
Gummimantelleitung *f (el)* cab-tyre line
Gummimuffe *f* rubber bush
Gummipuffer *m* bump rubber
Gummipufferziehverfahren *n* rubber die process
Gummiriemen *m* rubber belt
Gummiringdichtung *f* rubber ring gasket (packing)
Gummirohr *n* rubber conduit
Gummisack *m* rubber sac *(hydraulic accumulator)*
Gummisauger *m* suction ball
Gummischeibe *f* rubber disk *(coupling)*
Gummischichtfeder *f* rubber spring *(of all-steel cleaning shoe)*
Gummischlauch *m* rubber tube (tubing) *(tyre)*; rubber nose
Gummischlauchleitung *f (el)* cable with cab-tyre sheathing, cab-type line, rubber-sheathed cable
Gummischlauchmantel *m* cab-tyre sheathing
Gummischnur *f* rubber cord
Gummiseil *n* elastic rope, tow-rope, shock cord
Gummispeicherblase *f* rubber sac *(hydraulic accumulator)*
Gummistempel *m* rubber stamp
Gummistereo *n* rubber stereo
Gummistopfen *m* rubber bung (stopper)
Gummistrangsiebkette *f (ag)* sifting chain with rubber-covered links
Gummituch *n* rubber blanket
Gummituchstreichverfahren *n (leath)* rubber blanket coating
Gummiumspinnmaschine *f* covering machine for rubber threads

Gummiventil *n* rubber valve
Gummiwaren *pl* rubber goods
Gummizwischenlage *f* rubber washer
Gurkeneinlegerei *f* cucumber pickling plant
Gurt *m* chord, boom *(in trusses)*; waler, waling *(in timberwork)*; belt *(conveying)*
Gurtbandförderer *m* flat-belt conveyor
Gurtdurchsenkung *f* belt deflection
Gurtförderer *m* belt (band) conveyor
  ansteigender G. ascending belt conveyor
  fahrbarer G. mobile belt conveyor
  tragbarer G. portable belt conveyor
Gurtgesims *n* cornice
Gurtholz *n (build)* waling, waler, runner
Gurtkastenband *n (mine)* rectangular trough-shaped belt conveyor
Gurtschlinge *f* web sling *(cargo handling)*
Gurtstropp *m* web sling *(cargo handling)*
Gurttrommel *f* belt drum
Gurtungsdeck *n* [main] strength deck
Gurtweberei *f* strap (belt, tape) weaving
Gurtwinkel *m* / **unterer** base angle *(bulkhead)*
Guß *m* 1. pouring, casting, founding; 2. casting[s]
  dünnwandiger G. thin walled castings
  fallender G. top casting
  fehlerhafter G. unsound castings
  massiver G. solid castings
  nicht ausgelaufener G. misrun castings
  siliziumreicher G. silicon-rich castings
  steigender G. bottom (rising) casting
  undichter G. unsound castings
Gußasphalt *m* mastic asphalt
Gußasphaltbelag *m* mastic asphalt flooring
Gußbeton *m* cast concrete
Gußblock *m* cast ingot
Gußbruch *m* scrap castings, cast iron scrap
Gußeisen *n* cast iron
  G. mit Kugelgraphit nodular cast iron
  G. mit Stahlschrottzusatz semi-steel
  G. mit Stahlzusatz toughened cast iron
  feinkörniges G. close-grained cast iron
  graphitreiches G. graphitic cast iron
  legiertes G. alloy cast iron
  meliertes G. mottled pig iron
Gußeisengeschränk *n* cast-iron furnace front
Gußeisenüberhitzer *m* / **kammerloser** *(therm)* cast-iron superheater not placed in a chamber
Gußeisenventil *n* cast-iron valve
Gußeisenvorwärmer *m* cast-iron feed-water heater
Gußeisenwalze *f* cast-iron roll
Gußfehler *m* casting defect
Gußform *f* mould
  getrocknete G. baked mould
Gußformbearbeitung *f* mould machining
Gußgefüge *n* cast structure
Gußgrat *m* cast seam
Gußhaut *f* casting skin, cast-iron scale, crust of scale
Gußkasten *m* casting box
Gußkern *m* core
Gußlegierung *f* casting alloy
Gußmagnet *m* cast magnet
Gußmessing *n* cast brass
Gußnaht *f* cast seam, seam of the (a) casting
Gußputzen *n* cleaning of castings
Gußputzerei *f* foundry cleaning room
Gußputzmaschine *f* snagging machine
Gußputztrommel *f* tumbler, rumble, rattle (shaking) barrel
Gußrinde *f* outer skin
Gußschalttafel *f* cast-iron assembly, enclosed distribution panel
Gußschlacke *f* iron dross
Gußschrott *m* scrap castings, casting (cast-iron) scrap
Gußspannung *f* casting stress
Gußstahl *m* cast steel
Gußstahltiegel *m* skillet
Gußstück *n* casting
  sauberes G. neat casting
Gußteil *n* cast member; casting
Gußtrichter *m* feeding gate, runner head
Gußwarze *f* nipple
Gut *n* 1. merchandise, goods, *(Am)* freight; 2. *(ship)* rigging *(ropes)*; 3. farm
  G. in loser Schüttung goods in bulk, *(Am)* bulk freight
  geborgenes G. *(ship)* salvage *(goods)*
  gefährliches G. dangerous goods
  laufendes G. *(ship)* running rigging *(ropes)*
  verschüttetes G. spillage
  versenktes G. logan, lagan *(any heavy article thrown overboard)*
  volkseigenes G. nationally-owned farm
Gutachten *n* judgement, opinion
  fachliches G. expert judgement
Gutachter *m* expert
Güte *f* quality *(article)*; efficiency
Gütefaktor *m* performance (quality) factor; factor of merit; Q value
  G. einer Spule coil amplification factor
Gütegrad *m* quality [level]
Güteklasse *f* grade; quality
Gütekontrollabteilung *f* [works] inspection department
Gütekontrolle *f* inspection, quality control; inspection department
Gütekontrolleur *m* [works] inspector, quality-control man
Gütekriterium *n* 1. control criterion; 2. performance criterion (index)
Güteprüfung *f* quality inspection
Güter *npl* goods, freight
  leicht verderbliche G. perishables

nicht brennbare G. incombustibles
Güterabfertigung *f* 1. goods department, freight office; clearance of freight (goods)
Güterbahnhof *m* goods station, rail freight yard; freight depot
Gütereinteilung *f* general classification of merchandise
Güterfernverkehr *m* long-haul traffic
Gütergemeinschaft *f* joint property
Güternahverkehr *m* short-haul traffic
Güterschnellverkehr *m* rapid transit of goods
Güterschuppen *m* goods shed
Gütertrennung *f* separation of property
Güterumschlag *m* cargo (freight) handling
Güterverkehr *m* transport, goods traffic, *(Am)* freight traffic
Güterwagen *m* goods wagon, truck, *(Am)* freight car
fehlgeleiteter G. wrongly routed wagon
verschleppter G. wrongly routed wagon
Güterzug *m* goods train, *(Am)* freight train
Güterzugbremse *f* goods brake
Güterzugkilometer *n* goods train-kilometre
Gütesicherung *f* quality assurance *(manufacture)*
Gütevorschrift *f* quality (material) specification
Gütezahl *f* figure of merit, magnification [factor] *(radio)*
gutheißen to approve
Gutlage *f* acceptable quality level *(quality control)*
Gutläufer *m* free runner [wagon]
Gutlehre *f* go gauge
Gut-Schlecht-Prüfung *f (mach)* go-not-go testing
Gutseite *f* go end [of gauge]
Gutstoff *m* accepts, accepted stock, first quality pulp *(paper)*
Guttapercha *f (n)* guttapercha
Gyrationsfrequenz *f* gyro frequency
Gyrobus *m (el)* gyrobus
gyromagnetisch gyromagnetic[al]
Gyrometer *n* gyrometer
Gyrostat *m* gyrostat

H

H-Anker *m (el.m.)* H-armature
H-Bombe *f* hydrogen bomb, H-bomb
H-Gestänge *n (el)* H-pole
H-Kabel *n s.* Höchststädter Kabel
H-Motor *m* H-type engine
H-Waffe *f* hydrogen weapon, H-weapon
H-Welle *f (el)* H-wave
Haar *n* hair
Haarbaum *m (leath)* unhairing beam
Haarbleichmittel *npl* hair bleachers
Haarbüschel *n* tuft
Haardraht *m* finest wire
Haareisen *n (leath)* unhairing knife
Haarentfernungsmittel *n* depilatory, epilatory
Haarfärbemittel *n* hair-dyeing tincture, hair dye
Haarfilz *m* hair felt
Haarhygrometer *n* hair hygrometer
Haarkalkmörtel *m* [lime and] hair mortar
Haarmesser *n (leath)* unhairing knife
Haarnadelglühfaden *m* horseshoe filament
Haarnadelkatode *f* hairpin cathode
Haarnadelkurve *f* hairpin bend (curve)
Haarnadelwendel *f (el)* folded filament
Haarriß *m* hair (capillary) crack, check
H. in Glasuren fire crack
Haarrißbildung *f* crazing *(ceramics)*
Haarröhrchen *n* capillary (small-bore glass) tube
Haarröhrchenwirkung *f* capillary attraction
Haarseite *f (leath)* hair side
Haarspatium *n* hair-space
Haarspieß *m (print)* hair-line
Haarstreckmittel *npl* hair straighteners
Haarstrich *m (print)* serif; hair line *(drawing)*
Haartonikum *n* hair tonic
Haarverformungsmittel *n* wave set
Haarwaschmaschine *f (leath)* burring machine
Haarwasser *n* hair tonic
Habannröhre *f* split-anode magnetron tube
Habenzeichen *n* credit symbol
Haber-Bosch-Verfahren *n* Haber[-Bosch] process
Habilitationsschrift *f* habilitation thesis
Habitat *n* habitat *(animals, plants)*
Habitus *m* habit *(crystal)*
Hackbank *f* chopping block
Hackbeil *n* chopper
Hackblatt *n* [hoe] blade
Hacke *f* 1. *(ag)* hoe; *(build)* pick; 2. *(ship)* heel *(rudder)*
leichte H. für Bemusterungszwecke *(geol)* hand pick
zweizinkige H. forked hoe
hacken *(ag)* to hoe *(field, bed)*; to chop *(meat)*, to mince; *(build)* to pick, to break ground with a pick
Holz h. to chop wood
Hackenleistung *f* coal-getting
Hacker *m (text)* doffer comb; noil stripping comb
Hackerei *f* chipping department
Hackfräse *f* rotary cultivator
Hackfrucht *f* cleaning (fallow) crop
Hackfruchtaufladegerät *n* sugar beet elevator (loader)
Hackmaschine *f* 1. chipper; 2. *(ag)* row hoe
H. für Abfallholz waste-wood chipper
mehrreihige H. multiple row hole
Hackmesser *n* hoe blade; meat chopper
Hackpflug *m* cultivator
Hackrotor *m* disk chipper
Hackschnitzelsilo *m* chip loft
Häcksel *m (n)* chaff

Häckselaggregat *n* cylinder cutter head
Häckselauswerfer *m* chaff ejector
Häcksellader *m* green-crop loader
Häckselmaschine *f* chaff (straw) cutter
Häckseltrommel *f* cutting cylinder
Häcksler *m* forage (chaff) cutter
Hackspänebehälter *m* chip bin
Hacktraktor *m* row-crop tractor
Hadern *pl* rags
Hadernholländer *m* rag engine
Hadernkocher *m* rag boiler
Hadernpapier *n* rag paper
Hadernschneider *m* rag chopper (cutter, cutting machine)
Hadernstoff *m* rag pulp
Hadernwolf *m* breaker
Hafen *m* 1. [sea-]port, harbour; haven *(small boats)*; 2. pot
  geschützter H. snug harbour
  künstlicher H. artificial harbour
  natürlicher H. natural harbour
  offener H. open harbour
  ungeschützter H. open harbour
Hafenabfertigung *f* turn-around
Hafenabgaben *fpl* port (harbour) charges, port (harbour) dues, revenue
Hafenanker *m* mooring (single arm) anchor
Hafenanlagen *fpl* port installations
Hafenarbeiter *m* shipman; docker, dock (hand) labourer; [a]longshoreman, lumper
Hafenausfahrt *f* mouth of a harbour, exit
Hafenbahn *f* dock (port, harbour) railway
Hafenbahnhof *m* maritime railway station
Hafenbarkasse *f* harbour launch (tender), harbour service launch
Hafenbau *m* harbour engineering
Hafenbecken *n* harbour basin, [wet] basin; [wet] dock
Hafenbehörde *f* port (harbour) authority, *(Am)* harbor commission
Hafenbetrieb *m* management of the port, port management, harbour routine
Hafenbruch *m* grog *(glass)*
Hafenbugsierdienst *m* tug service
Hafendamm *m* pier, mole, harbour bar, jetty
Hafendock *n* gate (wet) dock, basin
Hafeneinfahrt *f* mouth of a harbour, harbour entrance, channel
Hafeneinrichtungen *fpl* port facilities (equipment); dock system
Hafeneisbrecher *m* harbour icebreaker
Hafenfeuer *n* harbour (port) light
Hafengebühr *f* harbour dues, port charges (dues), keelage
Hafengelder *npl s.* Hafengebühr
Hafenglas *n* pot glass
Hafeninspektor *m* harbour master, port warden
Hafenkosten *pl* disbursement accounts *(for a vessel)*
Hafenleichter *m* harbour lighter, barge
Hafenliegegebühr *f* groundage
Hafenliegezeit *f* lay days; turn-about
Hafenlotse *m* harbour (dock) pilot
Hafenmacher *m* pot maker
Hafenmeister *m* harbour (berthing) master, port warden
Hafenofen *m* pot furnace
Hafenradar *n* port radar [installation]
Hafenreinigungsbecken *n* scouring basin, sluicing pond
Hafenschlepper *m* harbour (docking) tug
  kleiner H. depot tug
Hafenschleuse *f* gate dock, basin
Hafenschmelze *f* pot melting
Hafensperre *f (ship)* embargo; [harbour] boom
Hafentender *m* harbour tender
Hafenton *m* pot clay
Hafentor *n* pot opening *(glass making)*
Hafentörn *m* turn-about
Hafenumschlag *m* turn-round
Hafenverwaltung *f* port administration
Hafenzoll *m* harbour charges (dues), port charges (dues), revenue
Hafer *m* oats
Haferkasten *m* oat bin
Haferquetsche *f* oat crusher
Haff *n* bay, gulf
Hafnium *n* hafnium
haftbeständig adherent
Haftelektrode *f* sticking electrode
haften to adhere, to stick
Haften *n* adherence, adhesion, adhering
haftend adherent
haftfähig adherent
Haftfähigkeit *f* adhesiveness, adhesive power
Haftfestigkeit *f* adhesion (bond) strength
Haftgläser *npl* contact lenses
Haftgrund *m* base surface, [paint] base
Haftkleber *m* contact (pressure-sensitive) adhesive
Haftkraft *f* adhesion, adhesive power
Haftmagnet *m* magnetic chuck, holding magnet
Haftmine *f* adhesive (magnetic) mine
Haftmittel *n* adhesive (anti-stripping) agent
Haftpflicht *f* liability
Haftpflichtanspruch *m* liability claim
Haftpflichtversicherung *f* third-party insurance
Haftreibung *f* static friction, friction of rest, stiction
Haftrelais *n* magnetically locked relay
Haftschale *f* contact lens
Haftspannung *f* adhesive stress (tension); bond stress
Haftstelle *f* trap *(semiconductor)*
Haftstellendichte *f* trap density

Hafttaste *f* locking push-button, locking key
Haftung *f* 1. adherence, bond, adhesion; 2. liability
atomare H. atomic cohesion
magnetische H. magnetic cohesion
Haftvermögen *n* adhesive capacity (power), adherence, adhesiveness, tackiness
Haftwert *m (ry)* adhesive value, coefficient of adhesion *(rails)*
Hagel *m* hail
hageln to hail
Hagelschaden *m* damage due to hail
Hahn *m* tap; faucet; [cone plug] cock *(hydraulics)*; stopcock
H. am Wasserstandanzeiger gauge-cock, water-gauge cock
Hahnfassung *f* switch lamp-holder
Hahnfett *n* stopcock (tap) grease
Hahngehäuse *n* casing of the cock
Hahnküken *n* [taper] plug *(hydraulics)*
Hahnschalter *m* socket switch
Hahnschlüssel *m* key of cock, cock wrench
Hahnventil *n* [cone plug] cock
H. mit Stellschraube screw-down cock
Häkelgarn *n* crochet yarn
häkeln to crochet
Häkelnadel *f* crochet hook
haken to hook *(drill)*
Haken *m* hook; peg *(clothes)*; clamp, clasp; hook *(knitting machine)*
rundgebogener H. common hook *(latch needle)*
Hakenblatt *n (wood)* scarf and key
Hakenbrett *n* pin rail
Hakenisolator *m (el)* swan-neck insulator
Hakenkamm *m (wood)* scarf and key
Hakenmeißel *m (mach)* hook (recessing) tool, undercutting tool *(lathe, planing)*
Hakennadel *f (text)* spring beard[ed] needle, beard[ed] needle
Hakennadelstrickmaschine *f* spring needle machine
Hakenplatte *f* hooked sole-plate
Hakenschalter *m* gravity (hook) switch
Hakenschlüssel *m* pin spanner
Hakenspindel *f* hooked peg
Hakentasse *f* hydraulic dip
Hakenumschalter *m* hook[-type change-over] switch, *(Am)* switchhook; telephone hook switch
Hakenverbindermaschine *f* belt stitcher
Hakenzugkraft *f* hauling power at [the] draw-hook
Halbachse *f* 1. *(math)* semi-axis; 2. half-axle *(vehicle)*
Halbaddierer *m* half adder
Halbautomat *m (mach)* semi-automatic [machine] *(turning)*
halbautomatisch semi-automatic[al]
Halbazetal *n* semi-acetal
Halbbalanceruder *n* semi-balanced rudder
halbberuhigt semi-killed *(steel)*
Halbbild *n* semi-image
Halbbildabtastung *f* / **erste** first interlace
Halbbildfrequenz *f* frame frequency
Halbbleiche *f* holland finish *(cotton, linen)*
Halbbrache *f* bastard fallow
Halbdeckelkrempel *f (text)* mixed card
Halbdieselmotor *m* semi-Diesel engine
Halbdose *f s.* Duant
Halbdrehautomat *m* semi-automatic lathe
halbdurchlässig semi-permeable, semi-transparent
halbdurchsichtig translucent, semi-transparent
Halbebene *f* half-plane
Halbedelstein *m* semi-precious gem
Halbfabrikat *n* intermediate (semi-finished) product
Halbfeder *f* semi-elliptic spring
Halbfernsteuerung *f* semi-remote control
Halbfertigwaren *fpl* semi-finished goods, semi-products
halbfest semi-solid
Halbfestwertspeicher *m* semi-permanent memory
halbfett medium *(printing)*
Halbflachbettfelge *f* semi-flat-base rim *(tyre)*
halbflächig hemihedral
Halbflächner *m* hemihedron
halbflüssig semi-liquid, semi-fluid
Halbfranzband *m* half-binding
halbfreitragend semi-cantilever *(wing)*
Halbgasfeuerung *f* semi-producer type furnace, half-gas fired furnace
halbgeleimt soft-sized, half-sized
halbgeschmolzen semi-fused
Halbgeschoß *n* entresol, mezzanine
halbgeschränkt half-crossed *(belt drive)*
Halbgewebeband *m* half-cloth
halbglänzend English (egg-shell) finished *(paper)*
Halbglied *n* half-section
halbhart *(met)* half-hard, moderately hard
Halbhohlkehlfräser *m* semi gutter cutter
Halbholz *n* half-round wood; scantling
halbieren 1. *(math)* to bisect *(angle)*; 2. to halve
Halbierungsachse *f* bisectrix axis
Halbierungslinie *f* bisector
Halbinsel *f* peninsula
Halbklappboot *n* semi-collapsible boat, pontoon lifeboat
Halbkoks *m* semi-coke
Halbkoksofen *m* semi-coke oven
halbkontinuierlich semi-continuous *(furnace)*
Halbkreis *m* semicircle
Halbkreis-Beta-Spektrograf *m* semicircular beta-spectrograph

Halbkreiseinschnitt *m* *(mach)* semicircular groove
Halbkreisfokussierung *f* semicircular focusing
Halbkreisform *f* semicircular shape
halbkreisförmig semicircular
Halbkreisfräser *m* radius cutter
  nach außen gewölbter H. convex [milling] cutter
  nach innen gewölbter H. concave [milling] cutter
Halbkreiskrümmer *m* return bend *(pipe)*
Halbkugel *f* hemisphere
halbkugelförmig semi-spherical
Halbkugelventil *n* semi-spherical valve
Halblederband *m* half-calf, quarter leather binding
Halbleinen *n* half-linen
halbleitend semiconductive
Halbleiter *m* semiconductor, semi-conductor
  H. mit Eigenleitfähigkeit intrinsic semiconductor
  anisotroper H. anisotropic semiconductor
  binärer H. binary semiconductor
  gedopter H. impurity semiconductor
  gemischter H. mixed semiconductor
  hochohmiger H. high-resistivity semiconductor
  intermetallischer H. intermetallic semiconductor
  organischer H. organic semiconductor
  reiner H. pure semiconductor
  zusammengesetzter H. compound semiconductor
Halbleiteranordnung *f* semiconductor device
Halbleiterbauelement *n* semiconductor element (device)
Halbleiterbetrieb *m* semiconductor regime
Halbleiterblock *m* semiconductor block
Halbleiterdetektor *m* semiconductor detector
Halbleiterdiode *f* semiconductor diode
Halbleitereigenschaft *f* semiconductor property
Halbleiterelektronik *f* semiconductor electronics
Halbleiterfertigung *f* semiconductor manufacturing
Halbleiterfotodetektor *m* semiconductor photodetector
Halbleiterfotoeffekt *m* photoelectric effect of semiconductors, photoconductive (internal photoelectric) effect
Halbleiterglasur *f* semiconducting glaze
Halbleitergleichrichter *m* semiconductor rectifier
Halbleiterherstellung *f* semiconductor production
Halbleiterindustrie *f* semiconductor industry
Halbleiterkristall *m* semiconducting crystal
Halbleitermaterial *n* semiconductor material
Halbleiterphysik *f* semiconductor physics
Halbleiterplasma *n* semiconductor plasma
Halbleiterplättchen *n* semiconductor wafer
Halbleiterschicht *f* semiconductor layer
Halbleitertechnik *f* semiconductor technique
Halbleiterübergang *m* semiconductor junction
Halbleiterwerkstoff *m* semiconductor material
Halbleitfähigkeit *f* semiconductivity
Halbleitung *f* semiconduction
  amphotere H. amphoteric semiconduction
  lichtelektrische H. photoelectric semiconduction
halblogarithmisch semi-logarithmic[al]
Halbmesser *m* radius
  H. der Eckenrundung corner radius *(tool)*
  äußerer H. outer radius
  innerer H. inner radius
  mittlerer H. mean radius
Halbmesserlehre *f* radius gauge
Halbmetall *n* semi-metal; semi-metallic element
Halbmikromaßstab *m* semi-micro scale
halbmonatlich semi-monthly
halbmondförmig semi-lunar
Halbnaßspinnmaschine *f* demi-sec spinning frame, half-dry spinning frame
Halboffenfach *n* semi-open shed *(Jacquard)*
Halbperiode *f (el)* half-cycle, semi-cycle
Halbperiodendauer *f* alternation
halbpolar semi-polar *(bond)*
Halbportalkran *m* semi-portal crane
Halbporzellan *n* feldspathic stoneware, semi-porcelain
Halbraum *m* hemisphere; half-space, semi-space
  elastischer H. elastic half-space, elastic semi-space
Halbröhrenkessel *m* combination (semi-tubular) boiler
halbrund half-round; semicircular *(two-dimensional)*; half-sperical *(three-dimensional)*
Halbrundfeile *f* half-round file
Halbrundkopf *m* button (spherical) head *(rivet)*
Halbrundschraube *f* round-head bolt, button-head bolt
Halbrundstahl *m* half-round [bar]
Halbsamt *m* terry velvet
Halbsäule *f* pilaster, attached column
Halbschatten *m (opt)* penumbra
Halbschieberegister *n (dat)* half shift register
Halbschnitt *m* half section *(technical drawing)*
Halbschreibimpuls *m* half-write pulse *(magnetic core memory)*
Halbschweberuder *n* semi-balanced [underhung] rudder
halbselbsttätig semi-automatic
Halbspur *f* half track *(magnetic tape)*
Halbstamm *m* half-standard
halbsteil semi-steep
Halbsteinwand *f* half-brick wall
Halbstoffholländer *m* breaker

Halbtagsweide *f strip-grazed pasture where the half daily ration of pasturage is provided for the herd*
Halbtiefbettfelge *f* semi-drop-centre rim *(tyre)*
Halbton *m* 1. half-tone *(photography)*; 2. *(ac)* half tone
Halbtonätzung *f (print)* half-tone etching, autotype; heliography *(copper plate printing)*
Halbtonbild *n* half-tone picture
Halbtonspeicherung *f* grey scale recording, half-tone storage
Halbtonverfahren *n* half-tone process
Halbuniversalradialbohrmaschine *f* semi-universal radial drilling machine
Halb-V-Naht *f* level groove weld, single-bevel butt weld
Halbwählimpuls *m* half-select pulse
Halbwassergas *n* semi-water gas
Halbweggleichrichtung *f* half-wave rectification
Halbwegmarkierungssender *m* middle marker *(radar)*
Halbwelle *f* half-wave
Halbwellenantenne *f* half-wave aerial
Halbwellendipol *m* half-wave dipol
Halbwellengleichrichter *m* half-wave rectifier, one-way rectifier
Halbwellengleichrichtung *f* half-wave rectification
Halbwellenlänge *f* half-wave length
Halbwellrohrkessel *m* boiler with partly corrugated flue
Halbwert[s]breite *f* half-width
Halbwertschicht *f* half-value layer, H.V.L., half-value thickness, H.V.T.
Halbwertspunkt *m* half power point *(radio)*
Halbwert[s]zeit *f (nucl)* half-life [period]
  biologische H. biological half-life
  effektive H. effective half-life
  scheinbare H. apparent half-life
Halbwinkel *m* semi-angle
Halbwolle *f* cotton and wool mixture
Halbwollfärberei *f* union dyeing
Halbwollfarbstoff *m* union dye
halbzahlig half-integer
Halbzeilenimpuls *m (telev)* equalizing pulse
Halbzellstoff *m* semi-chemical pulp
Halbzeug *n (met)* semi-finished product (steel); half stuff, first stuff (stock) *(paper)*
  H. für Röhren tube rounds
Halbzeugaufbereitung *f* macerating *(paper)*
Halbzeugholländer *m* breaker engine, half-stock Hollander beater
Halbzeugholländerwalze *f* breaker drum
Halbziffernverzögerung *f* half-digit delay
Halbzylinder *m* semi-cylinder
halbzylindrisch semi-cylindrical
Halde *f* [stock-]heap, stock[-pile]
Haldenarbeiter *m* heap keeper
Haldenerz *n* dump (waste heap) ore
Haldenkohle *f* stock coal
Haldenlagerung *f* stock-piling
Haldenrutsch *m* dump slip
Halfter *m* halter
halftern to halter
Hall-Effekt *m* Hall effect
Hall-Generator *m* Hall generator
Hall-Koeffizient *m* Hall coefficient
hallend reverberatory
Hallendeck *n* hangar deck *(aircraft carrier)*
Hallenschiff *n* bay *(factory)*
Halloysit *m (min)* halloysite
Hallraum *m* reverberatory (acoustically live) room
Hallwachs-Effekt *m* photoemissive effect
Halm *m* blade *(grass)*; stalk, stem *(cereals)*
Halmabteiler *m (ag)* [binder] divider
Halmfrüchte *fpl* cereals
Halmhöhe *f* standing height
Halmteiler *m (ag)* [binder] divider
Halo *m (astr)* halo
Halobildung *f* halation
Halochromie *f* halochromism
Halogen *n* halogen
Halogenid *n* halide, haloid
  lichtempfindliches H. photohalide
halogenieren to halogenize, to halogenate
halogeniert halogenated
  dreifach h. trihalogenated
Halogenierung *f* halogenation
Halogenkohlenwasserstoff *m* halocarbon
Halogenleckfinder *m (nucl)* halide leak detector
Halogenlöschzusatz *m (nucl)* halogen quenching addition
Halogenquecksilber *n* mercury halide
Halogensilber *n* silver halide
Halogenwasserstoffsäure *f* halogen hydracid
Halogenwirkung *f* halogen effect
Halogenzählrohr *n (nucl)* halogen[-quenched] counter
Hals *m* 1. neck, collar; 2. nose *(of a converter)*; 3. ring *(insulator)*
Halsausdrehung *f (mach)* [neck] recess
halsen / mit backen Segeln *(ship)* to boxhaul
Halslager *n* journal bearing, neck journal
Halslänge *f* neck length *(twist drill)*
Halszapfen *m* neck collar journal, journal with collars
haltbar durable; solid; stable
  h. machen to preserve; to stabilize
Haltbarkeit *f* durability, endurance, stability; service life
Haltbefehl *m* / **bedingter** conditional break-point instruction
Halteanode *f* excitation anode; keep-alive electrode

Haltearm *m* tool frame *(filing and sawing machine)*
Haltebahnhof *m* stopping station
Haltebereich *m* retention (hold) range
Haltedauer *f* *(mach)* dwell period, period of dwell
Halteeinrichtung *f* *(mach)* retaining device; interception circuit *(telephony)*
selbsttätige H. automatic holding device
Halteglied *n* clamping circuit, holding unit
Haltegriff *m* supporting strap; hand rail
Halteklammer *f* retaining clip
Halteklemme *f* support clamp
Haltekraft *f* tractive power
Haltemagnet *m* holding[-on] magnet
halten 1. *(mach)* to carry; 2. to hold, to maintain *(temperature)*; 3. to stop
besetzt h. to keep (hold) watch
Dampf h. to keep steam
den Kurs h. to keep to course
die See h. to keep the sea
frei von Land h. to keep the offing
Position h. to keep station
sich in Hörweite h. to keep within hail
sich unter Land h. to keep the land aboard
Haltepfahl *m* *(ship)* mooring post
Haltepunkt *m* 1. *(met)* transformation (critical) point; 2. stop *(bus)*, halt
H. auf dem Rollfeld *(aero)* taxi-holding position
Haltepunktkurve *f* thermal curve
Halter *m* holder; arbor *(reamer)*; stay; clip; mount
H. für Feinsicherung fuse clip
Halterelais *n* locking (holding) relay
Haltering *m* retaining (carrier) ring
Halterung *f* mount [holding components]
spannungsfreie H. strain-free mounting
Halteseil *n* guy rope
Haltespannung *f* holding voltage
Haltespule *f* holding coil
Haltestelle *f* stop, halt
Haltestift *m* retainer pin
H. am Spulenteller bobbin centre rod, cone pin *(knitting machine)*
H. der Teilscheibe plate stop, stop pin
Haltestrahl *m* holding beam
Haltestrom *m* hold current
größter H. limiting no-damage current
Haltestromkreis *m* holding (locking, stick, trap) circuit
Haltetau *n* *(ship)* retaining rope, becket-rope
Halteventil *n* back-pressure valve
Haltewicklung *f* *(el)* holding winding
Haltewiderstand *m* economy resistance
Haltezeit *f* hold-up time, critical interval; holding time *(hardening)*; *(mach)* dwell *(table, slide)*
H. der Tischumkehrung length of time of table dwell
Haltezustand *m* hold condition *(analog computer)*
Haltlage *f* *(ry)* stop (danger) position *(signal)*
Haltsignal *n* *(ry)* stop (warning) signal
bedingtes H. permissive (stop and proceed) signal
permissives H. *s.* bedingtes H.
Halttaste *f* stop bar
Hämatin *n* haematin
Hämatit *m* haematite, *(Am)* hematite
roter H. Indian red
Hämatiterz *n* haematite ore
Hämatitroheisen *n* haematite pig iron
Hamilton-Funktion *f* Hamilton function
Hamilton-Operator *m* Hamiltonian, Hamilton operator
Hämin *n* haemin
Hammelwolle *f* wether's wool
Hammer *m* hammer
Wagnerscher H. *(telec)* Wagner (hammer) interrupter
hämmerbar malleable
Hämmerbarkeit *f* malleability
Hammerbohrmaschine *f* heavy-duty drilling machine
Hammerbrecher *m* hammer crusher
hammergar tough-pitch *(copper)*
Hammerinduktor *m* hammer spark coil
Hammerlampe *f* **nach Heß** Hess hammer lamp
Hammermühle *f* hammer mill; swing-hammer crusher
sieblose H. screenless hammer mill
hämmern to hammer; to beat
Hammernieten *n* impact riveting
Hammerschlag *m* hammer scale, forge cinder; hammer blow
Hammerschlaglack *m* hammer effect enamel
gebrannter H. hammertone stove enamel
Hammerschmiede *f* tilt mill
Hammerschweißen *n* hammer (forge) welding
Hammersperre *f* hammerlock
Hammersplitthebel *m* hammer split
Hammerspur *f* *(nucl)* hammer track, T-[shaped] track
Hammerstern *m* *(nucl)* hammer track, T-[shaped] track
Hammerunterbrecher *m* *(telec)* trembler, vibrating (hammer) break
Hammerwerfen *n* throwing of the hammer, hammer throwing
Hammerwerk *n* 1. hammer (tilt) mill; 2. *(ac)* tapping machine, impact generator
Hammerwurfgitter *n* [hammer throwing] cage
Hammingkode *m* *(dat)* Hamming-code *(error-detecting code)*
Hämoglobin *n* haemoglobin, *(Am)* hemoglobin
Hand / von manually, by hand

Handabstimmung *f* manual tuning
Handabweiser *m* hand (safety) guard
Handabzug *m* 1. hand pull *(printing)*; 2. rip cord *(parachute)*
Handamt *n* manual (hand-operated) exchange *(telephony)*
Handandrehkurbel *f* cranking handle
Handanlage *f* hand-feed, manual feed *(printing)*
Handanlasser *m* hand starter
Handapparat *m* handset, microtelephone
Handarbeit *f* manual work
Handarbeiten *fpl* needle work
Handarbeitsstunde *f* manpower hour
Handauftragwalze *f* hand inking roller *(printing)*
Handaugenspiegel *m* hand ophthalmoscope
H. mit elektrischer Beleuchtung luminous hand ophthalmoscope
Handauslösung *f* manual release, disengagement by hand
Hand-Automatik-Umschalter *m* manual-automatic switch (station)
Hand-Automatik-Umschaltung *f* manual-automatic transfer
Handballspiel *n* fieldball, handball
handbedient hand-operated, manually operated, manual
Handbedienung *f* hand operation, [manual] manipulation, handling
Handbeschickung *f* hand charging *(furnace)*
handbetätigt hand (manually) operated
Handbetätigung *f* hand (manual) operation, hand (manual) control
mit H. mutually operated
Handbetrieb *m* manual operation
Handbohren *n* hand drilling
Handbohrgerät *n* hand-drilling appliance
Handbohrhammer *m* hand-held drill
Handbohrmaschine *f* hand drill, [hand] drilling unit
elektrische H. portable electro-drill
Handbremse *f* hand brake, parking brake operated by hand lever *(car)*
Handbremshebel *m* hand-brake lever
Handbremszuggriff *m* hand-brake handle
Handbrunnen *m* dug well
Handbuch *n* manual, handbook
H. der Luftfahrt aeronautical information publication
Handbügelsäge *f* hand hacksaw
Handbütten *n* hand-made paper
Handdosimeter *n (nucl)* hand dosimeter
Handdruck *m* block print[ing] *(paper)*
Handdynamo *m* hand generator
Hände *fpl* **/ magische** master-slave manipulator
Handeffekt *m* body effect *(radio)*
Handeinstellung *f (mach)* hand adjustment (setting)
Handel *m* trade, commerce
Handelsabkommen *n* trade deal (agreement)
kurzfristiges H. short-term trade agreement
langfristiges H. long-term trade agreement
Handelsadreßbuch *n* trade directory
Handelsbeziehungen *fpl* commercial relations
Handelsdünger *m* fertilizer
Handelsflagge *f* merchant (mercantile) flag
britische H. The Union Jack
Handelsflotte *f* merchant fleet
Handelsgesellschaft *f* commercial company
Handelsgewicht *n* commercial weight
Handelshafen *m* trading (commercial) port
Handelshölzer *npl* common timbers
Handelskammer *f* Chamber of Commerce
Handelskorrespondenz *f* commercial correspondence
Handelskunde *f* commercial science
Handelslänge *f* standard (commercial) length
Handelsmarine *f* merchant (mercantile) marine
Handelsmonopol *n* commercial monopoly
Handelspartner *m* trade partner
Handelsprokura *f* procuration
Handelsqualität *f* commercial grade
Handelsregister *n* commercial register
Handelssaatgut *n* once-grown seed
Handelsschiff *n* trading (trade, merchant) vessel
Handelsschiffahrt *f* merchant shipping (service), carrying-trade, commercial navigation, liner trade
Handelsschiffstonnage *f* merchant tonnage
Handelssorten *fpl* commercial grades [of coal]
Handelsspanne *f* margin
Handelsstatistik *f* commercial statistics
handelsüblich commercial; customary in the trade
Handels-U-Boot *n* submarine cargo-carrier
Handelsverkehr *m* commercial intercourse
Handelsvertreter *m* trade representative
Handelsvollmacht *f* procuration
Handelsvolumen *n* volume of trade
Handelszuschlag *m* **auf Trockengewicht** conventional allowance *(conditioning)*
Handempfindlichkeit *f* body effect
Handfeinzustellung *f* **der Schleifscheibe** hand fine adjustment for feeding wheel to work, hand fine feed
Handfeuerlöscher *m* portable (hand-operated) fire extinguisher, hand apparatus for fire extinguishing
Handfeuerung *f* hand firing (stoking)
Handfeuerwaffe *f* small arm
Handflachstrickmaschine *f* hand-controlled flat knitting machine
Handflaschenzug *m* hand-operated block
Handflügelpumpe *f* half-rotary hand pump, semi-rotary hand pump
Handformerei *f* hand moulding *(foundry)*

Handfräsmaschine *f* hand miller
handgearbeitet hand-made
handgedruckt *(text)* hand-blocked
handgefertigt hand-made
handgeführt hand-guided
Handgemisch *n* hand mixing
handgenäht hand-sewn
Handgepäck *n* hand luggage
handgeschabt *(mach)* hand-scraped
handgeschmiedet hand-forged
handgeschöpft hand-made *(paper)*
handgesetzt hand-set
handgesteuert hand-controlled, manually controlled
handgestrickt hand-knitted
handgewebt hand-woven
Handgewindebohrer *m* hand tap
Handgranate *f* hand grenade
Handgriff *m* handle; hand-grip
handhaben to handle
Handhabung *f* manipulation, handling
handhabungssicher safely handled
Handhacke *f* garden hoe
Handhammernieten *n* hand riveting
Handhebel *m* hand lever
handhebelbetätigt hand-lever operated
Handhebelbremse *f* hand-lever brake
Handhebelschneider *m* hand-lever cutter, hand-operated guillotine
Handhebelsteuerung *f* hand-lever control
Handhebelvorschub *m* hand-lever feed
Handhebelwelle *f* hand-lever cross shaft
Handhobel *m* hand plane
Handhöheneinstellung *f (mach)* manual elevating control
handig manageable; handy *(boat)*
Handkapazitätseffekt *m* body effect
Handkardätsche *f* hand card
Handkarre *f* wheel barrow
Handkegelreibahle *f* hand taper reamer
  H. zum Fertigreiben finishing hand taper reamer
  H. zum Vorreiben roughing hand taper reamer
Handkratze *f* hand card
Handkreuz *n (mach)* hand capstan, star handle (wheel); pilot wheel
Handkultivator *m* cultivator
Handkultivatorschar *n* prong
Handkurbel *f* hand crank; ball-handle crank; crank handle
Handlampe *f (el)* inspection lamp; hand lamp
Handläppen *n (mach)* hand lapping
Handlauf *m* handrail, hand bar
Handleuchte *f* hand (portable, inspection) lamp
handlich manageable; easily operated; handy
Handlocher *m* keyboard perforator *(telegraphy)*
Handlog *n (ship)* hand (chip) log
Handlotleine *f* hand lead line
Handlupe *f* hand magnifier
Handmelken *n* hand milking
Handmikrofon *n* fist (hand-held) microphone
Handmischung *f* hand mixing
Handmorselampe *f* Aldis lamp
Handmühle *f* quern, hand mill
Handmutternbohrer *m* hand nut tap
Handnachstimmung *f* manual tuning correction
Handprüfzählrohr *n* hand counter
Handpuddeln *n (met)* hand puddling
Handpumpe *f* hand (manual, manually operated) pump
Handpumpenvorrichtung *f* hand primer
Handrad *n (mach)* hand wheel
Handradeinstellung *f* adjustment by hand wheel
Handradsäule *f* standard for [the] hand wheel
Handrechenmaschine *f* hand (desk) calculator
Handrefraktometer *n* hand refractometer
Handregelung *f* manual control; hand regulation
Handreglerschalter *m (cont)* manual-automatic switch
Handreibahle *f* hand reamer
Handruf *m* manual ringing *(telephony)*
Handsäge *f* hand saw
Handsämaschine *f* hand seed drill, fiddle [seed sower]
Handsatz *m* hand-set type matter; hand-setting, hand composition
Handschalter *m* hand (manual) switch
Handschalthebel *m* gear-shift lever
Handschaltung *f* hand (manual) control; hand shifting
Handschere *f* hand shears
Handschlaufe *f* supporting loop
Handschliff *m* off-hand grinding; manual grind
Handschrank *m* manual switchboard *(telephony)*
Handschrapper *m (build)* hand scraper
Handschriftenleser *m* handwriting reader *(automaton)*
Handschuhbox *f* glove (dry) box
Handschuhfach *n* glove compartment
Handschuhkasten *m* glove box
Handschuhleder *n* glove leather
Handschuhmacher *m* glover
Handschweißen *n* manual welding
Handschweißkopf *m* / **pistolenartig geformter** pistol-shaped welding gun
Handschweißung *f* manual welding
Handsignalscheinwerfer *m (ship)* Aldis lamp
Handspaltleuchte *f* hand slit lamp
Handspektroskop *n* hand spectroscope
Handspitze *f* hand made lace
Handspritze *f* hand sprayer
Handstampfer *m* hand rammer (tamper)
Handstart *m* manual start-up

Handsteuerung *f* hand (manual) control; *(ship)* hand gear
handstollen *(leath)* to hand stake
Handstrickwolle *f* / **feste** fingering wool
Handstück *n (min)* hand-specimen
Handtachometer *n* hand tachometer
Handtempo *n* key speed *(communications)*
handtrocken hand-dry
Handtuchhalter *m* towel rack
Handtuchkreppapier *n* toweling
Handtuchpapier *n* toweling
Handübertragung *f* **durch Lochstreifen** manual tape relay
Handuhr *f* hack watch
Handvermittlung *f* manual switchboard; manual telephone system, hand-operated exchange
Handvermittlungssystem *n* manual system *(telephony)*
Handverstellung *f (mach)* hand (manual) adjustment
Handvollversatz *m (mine)* solid hand stowing
Handvorschub *m (mach)* hand feed
Handwaffen *fpl* small arms
Handwagen *m* hand truck *(conveying)*
handwarm lukewarm
Handwaschbecken *n* wash basin
Handweber *m* hand loom weaver
Handwebstuhl *m* hand loom
Handweiche *f* points operated on site
Handweiser *m (mine)* half set
Handwerk *n* trade; craft
Handwerker *m* craftsman
Handwerksberuf *m* craft
Handwerkszeug *n* hand tool
Handwinde *f* hand winch
Handzeichnung *f* hand drawing
Handzentrale *f (telec)* manual exchange
Hanf *m* hemp
Hanfanbau *m* hemp farming (growing)
Hanfbrech- und -schwingmaschine *f* hemp breaker and scutcher
Hanfdarre *f* hemp kiln
Hanfdichtung *f* hemp packing
Hanffaser *f* hemp fibre
Hanfgarn *n* hemp thread
Hanfhechelmaschine *f* hemp hackling machine
Hanföl *n* hemp oil
Hanfpackung *f* hemp packing
Hanfpapier *n* hemp paper
Hanfreibe *f* edge (hemp) softening mill
Hanfriste *f* hemp strick
Hanfschneidemaschine *f* hemp knifing machine
Hanfschwingwerg *n* hemp scutching tow
Hanfseele *f* hemp centre (core); jow, heart of a cable
Hanfseil *n* hemp rope
Hanfstrecke *f* hemp drawing frame
Hanfumwicklung *f* hemp packing
Hanfwerg *n* hemp tow
Hanfwergaufbereitung *f* hemp tow preparing (preparation)
Hanfzerreißmaschine *f* hemp knifing machine
Hanfzopf *m* twisted hemp, hemp twist
Hang *m* slope; *(mine)* batter *(prop)*
Hangar *m* hangar
Hangardeck *n* hangar deck
Hangarwinde *f* hangar winch
Hangaufwind *m* hill-side up-current *(gliding)*
Hängeäscher *m (leath)* suspending vat
Hängebahn *f* suspension (overhead) railway *(traffic)*; telpher *(conveying)*
Hängebank *f (mine)* landing, pit [head] bank, pit-top
Hängebankanschläger *m* bankman
Hängebrücke *f* suspension bridge
  erdverankerte H. shore-anchored suspension bridge
Hängedach *n* suspended (cable) roof
Hängedrahtantenne *f* trailing wire aerial
Hängedruckknopfschalttableau *n (mach)* pendant push-button control panel (unit)
Hängedruckknopftafel *f (mach)* pendant push-button station
Hängegatter *n* central-driven log frame
Hängeisolator *m* suspension insulator
Hängekatze *f* overhead hoist *(conveying)*
Hängekette *f* chain insulator, suspension chain
Hängekiel *m* drop (sliding) keel
Hängeklemme *f* suspension clamp
Hängekompaß *m* hanging compass
Hängekorb *m* cradle
Hängekran *m* overhead (suspension) crane
Hängelade *f (text)* suspended slay
Hängelager *n* bearing supported by hanger
Hängelampe *f* pendant [fitting]
Hängeleiter *f* suspension ladder
Hängeleuchte *f* pendant light, suspended lamp
Hängematte *f* hammock
hängenbleiben to seize *(valve)*; to get stuck, to stick; to adhere; *(mine)* to hang up *(chute)*
Hängenbleiben *n* **der Kolbenringe** ring sticking
hängend suspended, pendant
Hangenddruck *m (mine)* roof pressure
Hangendes *n (mine)* back, hanging [wall], roof; hanger
  gebräches H. irregular (weak) roof
  unmittelbares H. immediate roof
  unsicheres H. weak roof
Hangendpflege *f* roof (strata) control
Hangendriß *m* main roof break
Hangendschrämarm *m* overcutting (turret) jib
Hangendschrämwalze *f* roof drum
Hanger *m (ship)* filling-futtock; [derrick] span
Hängeruder *n* underhung rudder
Hängerüstung *f (build)* suspended scaffold
Hangerwinde *f (ship)* instanter (topping) winch

Hängesäule *f (build)* middle post; queen post
Hängeschalter *m* pendant (suspension) switch
Hängeschelle *f* cable suspender (suspension clamp), conduit supporting clip, messenger clamp
Hängeseil *n (mine)* slack rope
Hängesockel *m (el)* festoon cap
Hängespindel *f* elastic (gravity) spindle
Hängestange *f* suspension rod
Hängesteckdose *f* pendant socket outlet
Hängestromschiene *f* overhead conductor rail
Hängestropp *m (ship)* sling
Hängetisch *m (ship)* suspended table
Hängetür *f (ship)* hinged bulkhead door
Hängewagen *m* **[im Stall]** *(ag)* overhead manure carrier
Hängewerk *n* king[-post] truss, king-post trussed beam
Hangneigung *f* slope
Hangsegelflug *m* ridge soaring
Hangwinkel *m* angle of slope
Hanteln *fpl* dumb-bells
Hantelscheibe *f* disk-loading bar-bell, dump-bell disk
Hantieren *n* manipulation, handling
Hapug-Modulation *f* Hapug (controlled-carrier) modulation
Hardyscheibe *f* flexible disk
Harke *f* rake
harken to rake
Harmonikatür *f* accordion (folding) door
harmonisch harmonic
Harmonische *f* harmonic [component]
  geradzahlige H. even harmonic
  ungeradzahlige H. odd harmonic
Harnanalyse *f* urinalysis
harnartig urinous
Harnisch *m* 1. *(text)* harness *(Jacquard)*; 2. *(geol)* slickenslide
  verkreuzter H. *(text)* crossed harness
Harnischbrett *n (text)* comber (lower hole) board, harness reed
Harnischschnur *f (text)* harness cord
Harnischstab *m (text)* harness bar
Harnischstich *m (text)* straight tie up
Harnsäure *f* uric acid
Harnstoff *m* urea
Harnstoffharz *n* urea resin
Harnstofformaldehydharz *n* urea-formaldehyde resin
harntreibend diuretic
Harpune *f* harpoon; whale-lance; gun-harpoon
Harpunenkanone *f* harpoon (whaling) gun
Harpunenschnur *f* harpoon line
hart hard *(material, water)*; hard, penetrating *(ray)*, high-energy; vigorous *(paper in photography)*
  h. am Wind close by the wind
härtbar hardenable

Härtbarkeit *f* hardenability
Hartblei *n* hard (antimonial) lead
Hartbranntstein *m* engineering brick
Hartbraunkohle *f* [hard] brown coal
Härte *f* hardness *(material, water)*; penetrating power *(X-rays)*
  bleibende H. non-carbonate hardness
  nicht bleibende H. carbonate hardness
  permanente H. *s.* bleibende H.
  temporäre H. temporary (carbonate) hardness *(water)*
  vorübergehende H. *s.* temporäre H.
Härteanlage *f* hardening plant
Härteannahme *f* hardenability
Härtebeschleunigung *f* hardening acceleration
Härtebestimmung *f* hardness grading (measurement)
Härtebildner *m (ch)* hardness causing salt; *(met)* hardening constituent
Härteempfindlichkeit *f* sensitivity to hardening
Härtefixierbad *n* hardening fixing bath
Härteflüssigkeit *f* quenchant
Härtegrad *m* 1. *(met)* degree of hardness; 2. *(mach)* grade *(grinding wheel)*; 3. contrasty grade *(photography)*
  H. des Wassers degree of hardness of the water
Härtemesser *m (met)* hardness indicator
Härtemittel *n* hardener
härten 1. *(met)* to harden; 2. to cure, to set *(cement, plastic)*; 3. to bake
  an der Luft h. to set in air
  im Bleibad h. to lead harden
Härteofen *m* hardening furnace
Härteöl *n* annealing (quenching, steel-hardening) oil
Härteprüfeinrichtung *f s.* Härteprüfgerät
Härteprüfer *m* hardness tester
Härteprüfgerät *n* hardness tester (testing machine)
Härteprüfmaschine *f* hardness tester (testing machine)
Härteprüfung *f* hardness test
Härter *m (met)* hardener
Härterei *f (met)* hardening shop
Härteriß *m* hardening (quenching, hardness) crack
Härtesalz *n* hardening salt
Härteschrumpfung *f* hardening shrinkage
Härteskala *f* hardness scale
  H. nach Mohs Mohs' scale
  Mohssche H. Mohs' hardness scale
Härtespannung *f* quenching (internal hardening) stress
Härtestufe *f* hardness grade
Härtetemperatur *f* hardening temperature
Härtetiefe *f (met)* depth of hardening (case), hardening (case) depth

Härtetiefenuntersuchung *f* *(met)* chill (hardening depth) testing
Härtetiegel *m* hardening crucible
Härte- und Vergüteanlage *f* hardening and heat-treating plant
Härteverzug *m* distortion on hardening
Härtezeit *f* 1. *(met)* hardening (quenching) time; 2. cure time *(plastics)*; 3. bake-out period
Härtezeittest *m* stroke test
Hartfaserplatte *f* hardboard, moulded fiber board
Hartgasableiter *m* *(el)* expulsion-type arrester
Hartgasschalter *m* *(el)* hard-gas circuit breaker, expulsion-type circuit breaker
Hartgewebe *n* fabric reinforced laminate
hartgezogen hard-drawn
hartgezwirnt hard-twisted
Hartgummi *m* ebonite, hard rubber, vulcanite
Hartgummiisolator *m* ebonite insulator
Hartgummiverschluß *m* ebonite plug
Hartguß *m* white cast iron; chilled cast iron
Hartgußrad *n* chilled iron wheel
Hartgußwalze *f* chilled roll
Hartholz *n* hardwood, deciduous (leaf) wood
Hartkocherstofföffner *m* breaker trap *(paper)*
Hartkord *m* kraft cord
Hartlackdraht *m* *(el)* enamelled wire
Härtling *m* *(met)* bear; hard head *(tin-iron alloy)*
Hartlot *n* brazing solder (alloy)
hartlöten to hard solder, to braze
Hartlöten *n* hard soldering, brazing
Hartmanganerz *n* psilomelane, black haematite
Hartmattlack *m* *(Am)* dull clear varnish
Hartmessing *n* hard brass
Hartmetall *n* hard metal (alloy); cemented (sintered metal) carbide
gesintertes H. cemented carbide
Hartmetallaufbohrwerkzeug *n* [cemented] carbide boring tool
Hartmetallauflage *f* carbide facing
hartmetallbestückt [sintered] carbide-tipped, tipped with sintered (cemented) carbide; hard-metal tipped
Hartmetallbestückung *f* cutting alloy tip *(plane)*, carbide tip[s]
Hartmetallegierung *f* hard metal alloy
Hartmetalleinsatz *m* cemented [carbide] insert, carbide insert
Hartmetalleinsatzmeißel *m* [cemented] carbide insert
Hartmetallfräser *m* [cemented] carbide milling cutter, [cemented] carbide cutter
Hartmetallhobelmeißel *m* [cemented] carbide planing tool
Hartmetallmeißel *m* [cemented] carbide tool *(planer, lathe)*
Hartmetallplättchen *n* [cemented] carbide tip
Hartmetallräumzeug *n* [cemented] carbide surface broach
Hartmetallreibahle *f* [cemented] carbide reamer
Hartmetallschleifmaschine *f* [cemented] carbide tool grinding machine, [cemented] carbide grinding machine
Hartmetallschneide *f* [cemented] carbide cutting edge, [cemented] carbide cutting tip, [cemented] carbide edge, [cemented] carbide tip; [cemented] carbide lip *(drill)*, carbide tip bit
Hartmetallschneideinsatz *m* carbide cutting section *(broach)*
Hartmetallschneidenteil *m* carbide cutting section *(broach)*
Hartmetallspiralbohrer *m* [cemented] carbide twist drill
Hartmetallwerkzeug *n* cemented (sintered) carbide tool
Hartmetallwerkzeugschleifmaschine *f* [cemented] carbide tool grinder
Hartpapier *n* kraft [paper]; bakelite (bakelized, synthetic resin, pressed) paper
Hartpapiertafel *f* pressed-paper sheet
Hartpappe *f* hardboard
Hartparaffin *n* paraffin [wax], solid (high-melting) paraffin
Hartplatte *f* hardboard
geprägte H. embossed hardboard
Hartsilber *n* silver nickel
Hartspiritus *m* solid spirit
Hartstrahlen *mpl* hard rays
Härtung *f* 1. *(met)* hardening; 2. cure, setting *(cement, plastics)*
H. des Neutronenspektrums neutron hardening
gestufte H. interrupted quench *(metal)*
vorzeitige H. precure *(plastics)*
hartverchromen to deposit hardchromium
Hartverchromen *n* hard chromium plating
Härtverfahren *n* hardening process
Hartwerden *n* **der Röntgenröhre** hardening of the X-ray tube
Harz *n* resin
drucklos härtendes H. no-pressure resin
säurehärtendes H. acid catalyzed resin
technisches H. industrial resin
undispergiertes H. unground resin
Harzabzapfung *f* resin tapping
harzartig resinous, gummy
harzbildend gum-forming *(hydraulics)*
harzen to resin[ate]
Harzester *m* resin ester, ester gum
harzfrei non-resinous, resin-free
Harzgalle *f* gum pocket *(wood)*
Harzgewinnung *f* resin tapping
harzhaltig resiniferous, resinic, resinous
harzig resinous, resinic, resin-like

Harzkanal *m* resin duct *(wood)*
Harzkleber *m* resin adhesive (cement)
Harzleim *m* resin size
Harzlösung *f* resin milk
Harzmischer *m* resin blender
harzreich highly resinous
Harzreserve *f (text)* resin resist
Harzrückstand *m* gum; gummy deposit *(hydraulics)*
Harzsäure *f* resin (abietic) acid
Harzseife *f* resin soap, resinate
Harzstreifen *m* resin canal *(wood)*
Harztasche *f* pitch pocket *(wood)*
Harzträger *m* resin binder, filler *(plastics)*
Harzverseifung *f* size cutting
Haselnußbaum *m* hazel-nut tree
Haspe *f* hasp
Haspel *f* 1. *(text)* reel *(yarn)*; filature *(silk)*; 2. *(build, mine)* capstan; winch; hoist; windlass; 3. reel *(wire drawing)*
Haspelanlage *f* reeling plant
Haspelberg *m (mine)* engine plane
Haspelfärbeapparat *m* winch dyeing machine
Haspelkufe *f* winch vat
Haspelpfosten *m (mine)* knuckle post
Haspelseil *n* haulage rope
Haspelumfang *m* circumference of reel
Haspelwärter *m* winch-man, haulage man
Haspelzieher *m* incline attendant
Haspelzug *m* reel tension
Haspen *m* hasp
Haubarkeitsalter *n* maturity age *(wood)*
Haube *f (mach)* hood
Haubengeschoß *n* capped shell
Haubenkörper *m* top section *(converter)*
Haubenofen *m* bell-type annealing furnace
Haubenschloß *n* bonnet lock
Haubenstütze *f* bonnet stay
Haubitze *f* howitzer
Haue *f* 1. *(mine)* adz[e], axe; 2. *(ag)* hoe
Hauer *m* clearer, cutter, face man, hewer
Hauerfalz *m* blade fold
Hauerschichten *fpl* coal getting shifts
häufbar cumulative
Häufchensaat *f* dibbling
Häufelgerät *n (ag)* ridger
Häufelkörper *m (ag)* ridging body
Häufelkultur *f* ridge crop
häufeln to earth up, *(Am)* to hill
Häufelpflug *m* potato coverer (ridger), ridging plough
Häufelschar *n (ag)* ridging share, *(Am)* hilling share
häufen to accumulate
Häufigkeit *f* frequency *(statistics)*; abundance
H. der Betätigung *(el)* operating frequency *(switch)*
absolute H. absolute frequency *(statistics)*
relative H. relative frequency *(statistics)*; *(nucl)* relative abundance; abundance ratio
Häufigkeitsdiagramm *n (math)* frequency curve (diagram) *(statistics)*
Häufigkeitskurve *f* error (probability) curve
Häufigkeitspapier *n (math)* probability paper
Häufigkeitspolygon *n* frequency polygon *(statistics)*
Häufigkeitsverteilung *f* frequency distribution (pattern) *(statistics)*
Häufung *f* accumulation
Häufungspunkt *m* accumulation point
Häufungsstelle *f (math)* cluster
Haufwerk *n (mine)* debris, muck
weggeschleudertes H. fly rock *(blasting)*
Haufwerkshaufen *m* muck-pile
Hauptabmessungen *fpl (mach)* specifications, main dimensions; *(ship)* principal dimensions (particulars)
Hauptabsperrhahn *m* main cock
Hauptabsperrventil *n* main stop valve
Hauptachse *f* major axis *(ellipse)*; principal axis *(crystal)*
Hauptamt *n (telec)* central office
hauptamtlich full-time
Hauptanfahrventil *n (ry)* main starting valve
Hauptanschluß *m (telec)* main set (station), subscribers main station
Hauptanschlußdichte *f* main station density
Hauptanschlußleitung *f* direct (exclusive) exchange line
Hauptanschlußrelais *n* network master relay
Hauptanschlußteilnehmer *m (telec) (Am)* individual line subscriber
Hauptantrieb *m (mach)* [main] drive
Hauptantriebskupplung *f* main drive clutch
Hauptantriebsmotor *m* main (spindle) drive motor
Hauptanwendungsgebiet *n* main field of application
Hauptauftragnehmer *m* main contractor
Hauptbahnhof *m* central (main, chief) station
Hauptbalken *m* floor beam
Hauptband *n* master tape *(with recorded data)*
Hauptbandfilter *n (el)* principal channel filter
Hauptbeanspruchung *f* principal stress
Hauptbehälter *m* main reservoir
Hauptbestand *m (wood)* superior stand
Hauptbestandteil *m* chief constituent
Hauptbewegung *f* primary motion
Hauptbohrung *f* main bore
Hauptbremsleitung *f* main brake pipe
Hauptbremszylinder *m* master cylinder
Hauptbuchhalter *m* auditor
Hauptdampfmaschine *f* main steam engine
Hauptdaten *pl (ship)* main (leading) particulars
Hauptdeck *n* main (upper) deck
Hauptdehnung *f* principal strain

Hauptdehnungstrajektorien *fpl* trajectories of principal strain
Hauptdock *n* principal dock
Hauptdüse *f* high-speed nozzle
Hauptebene *f (opt)* principal plane
H. bei Astigmatismus principal meridian
Haupteichkreis *m* telephone transmission reference system
Haupteinfallswinkel *m* angle of principal incidence
Haupteinflugzeichen *n* main entrance signal, middle marker, inner marker signal *(radar)*
Haupteingang *m* main entrance
Haupteinspeisung *f* main feeder
Hauptendausschalter *m* main limit switch
Hauptentladungsstrecke *f* main gap *(tube)*
Haupterreger *m* main exciter
Haupterregermaschine *f* main exciter [set]
Hauptfahrwerk *n* main undercarriage
Hauptfaktor *m* main factor
Hauptförderband *n* main (mother) belt
Hauptförderstrecke *f (mine)* carrying gate, main [road], main haulage drift, trunk road, main transportation tunnel
Hauptgang *m (geol)* champion (master) lode
Hauptgasleitung *f* main gas pipe-line
Hauptgebäude *n* main building
Hauptgetriebe *n* headstock gearing
Hauptgrubenlüfter *m* main fan
Haupthahn *m* main cock
Haupthangendes *n (mine)* main roof
Haupthub *m* main hoist *(conveying)*
Hauptimpuls *m* main pulse
Hauptkabel *n* main (primary) cable
Hauptkabelnetz *n* main cable network
Hauptkammer *f* main chamber
Hauptkanal *m* main sewer *(sewage)*
Hauptkatode *f* guide cathode
Hauptkessel *m* main boiler
Hauptkeule *f* main lobe (beam) *(aerial)*
Hauptkiel *m* main keel
Hauptkompaß *m* master compass
Hauptkontakt *m* main [current-carrying] contact
Hauptkrümmung *f* principal curvature
Hauptkupplung *f* main coupling
Hauptladeraum *m (ship)* main hold
Hauptlage *f* / **Gaußsche** Gauss position
Hauptlager *n* main bearing *(ship, vehicle)*, crankshaft bearing
Hauptlastverteilungszentrale *f (el)* main load-distributing centre
Hauptleitung *f (el)* mains; main [pipe] line, main pipe
Hauptlenker *m (therm)* main link
Hauptlenzleitung *f (ship)* bilge main, main piping (drain)
Hauptlicht *n* key light *(studio)*
Hauptlichtschalter *m* main light switch
Hauptluftbehälter *m* main air reservoir
H. der Saugluftbremse main vacuum brake chamber
Hauptluftleitung *f* through brake pipe
Hauptmarkscheider *m* chief surveyor
Hauptmaschine *f (ship)* main engine
Hauptmast *m* mainmast, main mast
Hauptmaximum *n* major peak
Hauptmerkmal *n* main (essential) feature
Hauptmonitor *m* on-the-air monitor
Hauptmotor *m* main engine *(internal combustion engine)*, main motor *(electric motor)*
Hauptnenner *m* common denominator, lowest common multiple (denominator)
Hauptnormal *n* master standard
Hauptnutzzeit *f* utilization time
Hauptoszillator *m* master oscillator; main oscillator
Hauptpatent *n* original patent
Hauptpol *m* main pole
Hauptprogramm *n* master programme *(computer)*
Hauptpunkte *mpl* cardinal points; *(opt)* principal points
Hauptquantenzahl *f* principal quantum number
Hauptrahmen *m* main frame
Hauptraum *m (ship)* main hold
Hauptrettungsstelle *f* main rescue centre
Hauptrichtstrecke *f (mine)* main drift
Hauptrippe *f (aero)* principal wing rib
Hauptrohr *n* main *(piping)*
Hauptrohrleitung *f* trunk pipe line
Hauptruder *n* main rudder
Hauptrumpf *m (ship)* main hull *(without upper works)*
Hauptsammelschiene *f (el)* collecting (main bus) bar
Hauptsammler *m (build)* main sewer
Hauptsatz *m* **der Thermodynamik** law of thermodynamics
Hauptschalter *m* main (master, power supply) switch
Hauptschalttafel *f* main control board; master (main) panel, main (power) switchboard; main (central) switchboard
Hauptschaltwarte *f* main switch station
Hauptscheinwerfer *m* main head lamp
Hauptschieber *m* slide (parallel slide) junction valve, main gate valve
Hauptschlechte *pl (geol)* back
Hauptschlüssel *m* master key, pass-key
Hauptschlußfeldspule *f (el.m.)* series coil
Hauptschlußmotor *m* series-wound motor
Hauptschlußspule *f (el)* series coil
Hauptschneide *f (mach)* primary (principal, main, working, active) cutting edge, working part of the cutting edge
H. am Spiralbohrer lip, principal cutting edge

Hauptschneiden *fpl* **an der Bohrerspitze** drill-point leading edges
Hauptschneise *f* main ride
Hauptschnitte *mpl* buttocks
Hauptschnittkraft *f* tangential cutting force component
H. **(P_h)** cutting force (F_c)
Hauptsender *m* master transmitter (station)
Hauptsicherung *f* main fuse
Hauptsignal *n* stop signal *(traction)*
Hauptsignallampe *f* master pilot [lamp]
Hauptspannung *f* principal stress
Hauptspannungslinien *fpl* trajectories of principal stresses
Hauptspant *n (ship)* midship frame (section)
Hauptspantfläche *f* area of midship section
benetzte H. wetted midship section
eingetauchte H. immersed area of midship section
Hauptspantkurve *f* midship bend
Hauptspantquerschnitt *m* midship section, dead flat
Hauptspantvölligkeitsgrad *m* midship (maximum) section coefficient, midship (midsection) coefficient
Hauptspeicher *m* primary store
Hauptspeiseleitung *f (el)* main feeder (supply line)
Hauptspektrum *n* principal spectrum
Hauptspiegel *m* main mirror
Hauptspindel *f (mach)* main spindle
Hauptspindelbohrung *f* main spindle bore
Hauptständer *m (mach)* main column
Hauptsteuerorgane *npl (aero)* primary controls
Hauptsteuerpult *n* central [control] desk
Hauptsteuerschalter *m* master controller
Hauptstock *m (text)* headstock
Hauptstopper *m* tumble shore *(launch)*
Hauptstrahlenbündel *n* principal beam
Hauptstrahlrichtung *f (el)* primary transmitting direction
Hauptstrahlungskeule *f* major (main) lobe of radiation *(aerial)*
Hauptstrang *m* main *(piping)*
Hauptstraße *f* main (major) road
Hauptstrecke *f* principal route *(railway)*; *(mine)* main [road]
Hauptstreichen *n (geol)* general trend
Hauptstrom *m* main (maximum, series) current
Hauptstromanlasser *m* series starter
Hauptstromauslösung *f* maximum circuit-breaking
Hauptstrombahn *f* **eines Schaltgerätes** main circuit of a switching device
Hauptstromfilter *n* full-flow filter
Hauptstromgenerator *m* series dynamo, series-wound generator
Hauptstromkontakt *m* main current-carrying contact
Hauptstromkreis *m* main (power) circuit
Hauptstrommotor *m* series (direct current series wound) motor
Hauptstromregelanlasser *m* rheostatic starter
Hauptstromrelais *n* primary (series-connected) relay
Hauptstromsteuerung *f* maximum circuit control
Haupttitel *m* full title
Hauptträgheitsmoment *n* principal moment of inertia
Haupttragseil *n* main carrier cable
Haupttturbine *f* main turbine
Hauptüberwachungslampe *f* master pilot lamp
Hauptuhr *f* master (primary) clock
Hauptumlaufpumpe *f* main circulation pump
Hauptvalenz *f (ch)* chief (principal) valency
Hauptvalenzkette *f (ch)* primary valence chain
Hauptverkehrsader *f* arterial road
Hauptverkehrsstunde *f* busy (rush) hour
Hauptverkehrszeit *f* rush (peak) hours, busy period
Hauptverstärkerfeld *n* main repeater section
Hauptverstärkerstufe *f* main amplifier stage
Hauptverteiler *m (el)* main distributor (distribution panel); *(el)* main distribution frame
Hauptverteilung *f* [main] distribution centre
Hauptverwaltung *f* administration
Hauptverwaltungsgebäude *n* head office [building]
Hauptwelle *f* 1. *(ph)* main (principal) wave; 2. *(mach)* main (driving, line) shaft; 3. *(mine)* band wheel
H. mit Schraubenkeilen *(aut)* main shaft with helical splines
Hauptwellenlager *n* main-shaft bearing
Hauptwerk *n* main works
Hauptwetterstrecke *f (mine)* main airway
Hauptwindleitung *f* blast main *(blast furnace)*
Hauptzugspannung *f* principal tensile stress
schiefe H. diagonal tension
Hauptzumeßdüse *f (aut)* main metering jet
Hauptzyklus *m (dat)* major cycle
Hauptzylinder *m (aut)* master cylinder
Haus *n* house
einstöckiges H. single-storeyed house, bungalow
mehrstöckiges H. multi-storeyed house
zweistöckiges H. two-storeyed house
Hausanschluß *m* service entrance, house connection; private connection *(telephone)*
Hausanschlußkabel *n* consumer's cable
Hausanschlußkasten *m* transfer box
Hausanschlußleitung *f* service line (drop); telephone drop, service conductor, house connection line
äußere H. service entrance conductor

Hausboot *n* houseboat
Hausbrandkohle *f* domestic (household) coal
Hausflagge *f* house flag *(of a vessel)*
Hausgebrauch *m* home use
Hausgrundstück *n* curtilage
Haushaltsabfälle *mpl* domestic waste
Haushaltsabwasser *n* domestic sewage
Haushaltsgerät *n* domestic appliance
Haushaltsinstallation *f* domestic installation
Haushaltswaschmittel *n* household detergent
Haushaltszähler *m (el)* house service meter
Haus-Haus-Beförderung *f* door-to-door transport (conveyance)
Haus-Haus-Verkehr *m* door-to-door service (traffic)
Hausinstallation *f* domestic (house) installation
Hauskorrektor *m* printer's reader
Hauskorrektur *f* first (printer's, reader's) proof
Hausmacherleinen *n* homespun linen
Hausmannit *m* hausmannite, black manganese
Hausschuhe *mpl* slippers
Hausschwamm *m* dry rot, fungus
Haustechnik *f* domestic engineering (services)
Haustelefon *n* interphone, house telephone
Haustier *n* domestic animal
Haustür[öffnungs]kontakt *m* plunger door contact
Hausweberei *f* domestic weaving
Hauszustellung *f* delivery at home (domicile)
Haut *f* 1. *(ship, aero)* skin; 2. hide *(animal)*; 3. membrane
  H. der nassen Jahreszeit wet season hide
  H. der trockenen Jahreszeit dry season hide
  einheimische H. native hide
  halbdurchlässige H. semipermeable membrane
  lufttrockene H. air-dried hide
  oberste H. top hide *(of a pile)*
  semipermeable H. semipermeable membrane
  trocken gesalzene H. dry salted hide
  ungesalzene grüne H. green (unsalted raw) hide
  unterste H. bottom hide *(of a pile)*
Hautanzug *m (nucl)* frogsuit
hautätzend blister-forming, vesicant
Hautaufhellungsmittel *npl* skin lighteners
Hautaufkupferungsverfahren *n* skin coppering process
Hautbelastung *f (nucl)* skin dose
Hautbeschädigung *f* **durch Röntgenstrahlen** X-ray dermatitis
Hautdosis *f* skin dose
Hauteffekt *m* skin effect *(radio)*
Hautelissestuhl *m (text)* high warp loom
Hautgift *n* vesicant
Hautkrem *f* skin cream
Hautnährkrem *f* emollient cream
Hautpflegemittel *n* cosmetic
Hautreinigungsmittel *n* personal washing agent
hautreizend skin-irritating
Hautschutzkrem *f* barrier cream
Hautschutzmittel *n* skin protective
Hautwolle *f* skin wool
Havarie *f* average, damage, accident
  große H. general average
  kleine H. particular average
Havarieabschaltung *f* scram *(reactor)*
Havariestab *m* scramming rod *(reactor)*
Havarieuntersuchung *f* inspection of damage, damage survey
Havarievergütung *f* general average
HD-Öle *npl* heavy duty oils
Heavisideschicht *f (ph)* Heaviside layer
Hebbewegung *f* vertical motion
Hebdrehwähler *m (telec)* two-motion selector
Hebe *f* hoist *(conveying)*
Hebebock *m* lifting jack
Hebebrettbohrung *f* holes in the trap board
Hebebrücke *f* lift bridge
Hebebühne *f* lifting (rising) platform, car lift
Hebedock *n* lifting dock
Hebekran *m* lifting (luffing) crane *(deck crane)*
Hebel *m* lever
  H. mit Kugelgriff ball-handled lever
  H. zum Klemmen des Bohrschlittens drilling head clamping lever
  einarmiger H. one-armed lever
  ökonomischer H. incentive
  zweiarmiger H. two-armed lever
Hebelantrieb *m* rotary actuator
Hebelarm *m* lever arm
  H. der Kraft perpendicular distance of the fulcrum from the line of action of the force
  H. der Kurbel crank radius
Hebelarmkurve *f (ship)* curve of statical stability
hebelbetätigt lever-operated
Hebelblechschere *f* hand lever shears
Hebelbremse *f* hand lever brake
Hebeldraisine *f* lever inspection trolley
Hebeleichter *m (ship)* camel
Hebeleinstellung *f* lever setting
Hebelgesetz *n* principle of the lever
Hebelitze *f* lifting heald
Hebelkontakt *m* electrical depression bar
Hebelschalter *m* lever switch
  H. mit Messerkontakten knife switch
Hebelschaltung *f* lever control
Hebelschere *f* lever shears
Hebelübersetzung *f* leverage, lever transmission
Hebelwaage *f* equal-armed balance
Hebelwirkung *f* lever action; leverage
Hebemagnet *m* 1. lifting magnet; 2. *(telec)* vertical magnet
Hebemittel *n* lifting tackle
heben to raise, to elevate, to lift; to jack *(car)*; to hoist *(conveying)*; *(mine)* to wind *(shaft*

winding), to hoist, to raise; to creep, to heave, to lift; to crown *(floor)*
ein Boot aus dem Wasser h. to hoist out a boat
ein Schiff h. to salve a ship
hebend siphoning *(liquid)*
Hebeponton *m (ship)* lifting pontoon, camel
Hebeprahm *m s.* Hebeponton
Heber *m* 1. *(ch)* siphon, pipette; 2. jack *(car)*; 3. vibrator *(printing)*; 4. clearing (raise) cam *(knitting machine)*
hydraulischer H. [hydraulic] jack
Heberglocke *f* bell-shaped container
Heberleitung *f* siphon pipe
Heberloch *n (mine)* lifter hole
hebern *(ch)* to siphon, to pipette
Heberrohr *n* siphon pipe
Heberschreiber *m* siphon recorder *(telephone)*
Heberschuß *m (mine)* lifter
Hebeschiff *n* lifting vessel
Hebespindel *f (mach)* elevating (elevation) screw
Hebestutzen *m* jack socket
Hebetank *m* camel
Hebe- und Fördergeräte *npl* handling equipment
Hebe- und Senkeinrichtung *f* rise-and-fall mechanism
Hebevorrichtung *f* elevator, lifting device *(conveying)*; *(ship)* lifting gear; purchase
Hebewalze *f* lifting roller
Hebewerk *n* 1. pumping station; 2. *(ship)* lift lock; 3. draw work
Hebewerkstrommelwelle *f* drum shaft
Hebezeug *n* hoisting machine (apparatus, unit, gear), lifting tackle (appliance, gear); jack; *(ship)* gin [tackle], hoist
Hebezeugbau *m* hoisting machinery construction
Hebezeuge *npl* hoisting machinery
Hebmagnet *m* vertical magnet *(telephony)*
Hebschritt *m* vertical step *(telephony)*
Hebsperrfeder *f* vertical lock spring *(telephony)*
heb- und senkbar *(mine)* ranging *(cutter head)*; elevating; vertically movable
Hebung *f* heave; elevation; lifting
kapillare H. capillary elevation
Hechel *f* hackle
Hechelapparat *m* gill
Hechelbrett *n* hackling block
Hechelflachs *m* hackled flax
Hechelhanf *m* hemp comb
Hechelkette *f* hackle chain
Hechelmaschine *f* hackling machine
hecheln to hackle, to gill
Hechelstuhl *m (text)* hackling bench
Hechelwerg *n* hackling tow
Heck *n (aero)* tail; *(ship)* stern, transom, tail
elliptisches H. counter (elliptical) stern
löffelförmiges H. long counter stern
plattes H. flat stern
rundes H. elliptical stern
Heckanker *m* stern anchor
Heckankerklüse *f* stern pipe (hawse hole)
Heckantrieb *m* rear drive *(car)*; *(ship)* transom drive
Heckaufschleppe *f* stern chute *(factory ship)*
Heckbalken *m* transom beam, wing (deck) transom
Heckbodenwrange *f* transom floor (plate)
Heckboot *n* stern boat
Heckbord *n* taffrail
Heckdavit *m (n)* stern davit
Hecke *f* hedge, live fence
Heckenschere *f* hedging shears
Heckenschneidemaschine *f* hedge trimmer (cutter)
Heckflagge *f* stern flag
Heckflosse *f (aut)* tail-fin
Heckgeländer *n* taffrail
Heckkanzel *f* tail turret
Heckklappen *fpl (aut)* rear flaps
Heckknick *m* knuckle
Heckkonstruktion *f* transom
Hecklader *m* rear loader
Hecklastigkeit *f* trim by the stern
Hecklaterne *f* stern light, poop lantern
Heckmaschinengewehr *n* tail machine gun
Heckmotor *m* rear engine, rear (back) motor
H. / mit rear-engined
Heckpropeller *m* after propeller
Heckradar *n* aircraft tail warning radar
Heckraddampfer *m* stern-wheel steamer, sternwheeler
Heckreling *f* taffrail, quarter rail
Heckrelinglog *n* taffrail log
Heckruder *n* stern rudder
Heckscheibe *f* stern sheave *(cable-laying vessel)*
Heckschlitz *m* stern well *(dredger)*; stern recess *(ice-breaker)*
Heckschlitzbagger *m* [open-ended] well-type dredger
Heckschütze *m* tail (rear) gunner
Hecksee *f* stern (quarter) wave
Heckseitenradschiff *n* quarter wheeler
Heckseitenradschubboot *n* quarter-wheel pusher tug
Heckspant *n* after (stern) timber *(wood)*; stern (after) frame *(steel)*; transom frame, after rib
Heckspiegel *m* [stern] transom
Heckstand *m* rear gunner's station, tail-gun position
Heckstütze *f* stern timber; side counter timbers
Hecktank *m* stern tank
Hecktrawler *m* stern trawler
Hecküberhang *m* stern overhang, fantail

Heckwelle *f* stern wave, wave of replacement
Heckzelle *f (rock)* rear section
H. ohne Einbauten tail hull
Hede *f* tow
Hefe *f* yeast
obergärige H. top yeast
untergärige H. bottom yeast
Hefepreßsaft *m* yeast press juice
Heft *n* 1. handle *(e.g. of a file)*; 2. booklet
Heftdraht *m* stitching wire
heften to sew *(book)*, to stitch
auf Bänder h. to sew on tapes
Hefter *m* stitcher; baster
Heftfaden *m* binding thread
Heftgarn *n* basting cotton
Heftgaze *f* stitching gauze
Hefthaken *m* stitching hook
heftig violent
Heftklammer *f* staple, stitching hook
Heftlage *f* section
Heftmaschine *f* stapling machine *(book)*
Heftnadel *f* bookbinder's (stitching) needle
Heftniet *m* tack rivet
Heftnietung *f* tack riveting
Heftpflaster *n* adhesive plaster
Heftpflasterverband *m* adhesive bandage
Heftrand *m* binding edge
Heftschweißung *f* tack weld[ing]
Heftstich *m* basting
Heftzwirn *m* sewing thread
Heilmittel *n* remedy, medicine, drug
Heilpflanze *f* medicinal plant (herb), curative herb
Heimarbeit *f* outwork
Heimarbeiter *m* outworker
Heimatbahnhof *m* home station
Heimathafen *m* home port, port of registry
Heimkehr *f* return home
heimkehren to return home
heimkehrend homeword bound
Heimkino *n* home cinema
Heimreise *f* return voyage, homeward trip
H. / auf der homeward bound
Heimwärtsbau *m (mine)* back work, retreat [mining, working], retreating [working], working home
Heimweberei *f* domestic (home) weaving
Heinisch-Riedl-Schutzschalter *m* earth-leakage trip
Heising-Modulation *f* Heising (choke, constant current) modulation
heiß hot; highly radioactive
Heiß *f (ship)* hoist
Heißbeladung *f* **des Reaktors** hot[-reactor] loading
Heißdampf *m* superheated steam
Heißdampfbetrieb *m* working with superheated steam
Heißdampflokomotive *f* superheated steam locomotive
Heißdampfmaschine *f* engine for superheated steam
Heißdampföl *n* oil suitable for working at high temperature[s]
Heißdampfventil *n* valve for superheated steam
heißen to hoist
Heißfixierung *f (text)* thermofixation
Heißformen *n (text)* heat setting
Heißgasschweißen *n* hot-gas welding, hot-dip welding
Heiß-Kalt-Verfahren *n* 1. *(nucl)* dual-temperature process; 2. hot-and-cold-bath process
heißkritisch *(nucl)* hot critical
Heißkühlung *f* hot cooling
heißlaufen to heat, to get hot *(bearing, axle-box)*
Heißläufer *m (ry)* hot box
Heißleiter *m* thermistor, thermally sensitive resistor
heißlöten to hot-solder
Heißlötstelle *f* hot junction
Heißluft *f* hot air
Heißluftanlaßofen *m* hot-air drawing furnace
Heißluftdusche *f* hot-air apparatus, fan-forced heater
Heißluftfeuerbrücke *f* hot-air bridge
Heißluftmansarde *f (text)* hot flue
Heißluftmaschine *f* hot-air engine
Heißluftmotor *m* hot-air engine
Heißlufttrockenmaschine *f (text)* hot flue
Heißlufttrockenschrank *m* fan oven
Heißlufttrocknung *f* desiccation
Heißlufttrocknungsanlage *f* high-temperature drier
Heißluftturbine *f* hot-air turbine
Heißplattenpresse *f* hot press plate
Heißprägung *f (print)* blind blocking (tooling)
heißpressen to hot-press
Heißreparatur *f* hot repair
Heißrichten *n* hot straightening
Heißsäge *f* hot (warm) saw
Heißschlinge *f (ship)* hoisting sling
heißsiegelfähig heat-sealable *(plastics)*
Heißsiegelverfahren *n* hot-sealing *(plastics)*
Heißstrahltriebwerk *n* thermal jet-propulsion unit, thermal jet engine
Heißstropp *m (ship)* [hoisting] sling
Heißtauchen *n* dip-moulding, hot-dip coating *(plastics)*
Heißteer *m* hot tar
Heißtrockenfarbe *f* heat-set ink
Heißveredlung *f* hot refining *(of paper)*
Heißvulkanisation *f* hot vulcanization
Heißwasserbehälter *m* hot well
Heißwasserspeicher *m* / **elektrischer** thermal storage water heater

Heißwasserspinnen *n* short-ratch spinning
Heißwind *m* hot blast
Heißwinde *f (ship)* hoist, hoisting winch
Heißwindkupolofen *m* hot blast cupola, hot-air cupola furnace
Heißwindofen *m* hot blast furnace
Heißwindschieber *m* hot blast slide valve
Heißwinsch *f (ship)* hoist, hoisting winch
Heizakkumulator *m* filament accumulator
Heizanlage *f* heating plant (equipment)
Heizanschluß *m* filament terminal
Heizapparat *m* heater
Heizband *n* strip heater
Heizbatterie *f* filament (low-tension) battery, A-battery
Heizdampf *m* steam for heating purposes, jacket steam
Heizdecke *f* electric blanket
Heizdraht *m* heating wire, filament
Heizdrossel *f* filament choke
Heizeffekt *m* calorific (heating) effect
Heizeffektmesser *m* heating effect indicator
Heizeinrichtung *f* heating plant
Heizelement *n* resistance element
heizen to heat; to stoke, to fire
Heizenergie *f* filament power
Heizer *m* 1. stoker, fireman; 2. *(el)* heater
Heizerlogis *n* firemen's quarters
Heizerstand *m* stoker's place
Heizfaden *m* filament, heater
Heizfadenanschluß *m* filament terminal
Heizfadenbruch *m* filament break
Heizfadenspannung *f* heater voltage
Heizfadenspeisung *f* filament supply
Heizfadenüberbrückungskondensator *m* heater by-pass capacitor
Heizfadenwiderstand *m* filament resistance
Heizfadenwirkungsgrad *m* filament efficiency
Heizfläche *f* heating surface
Heizflächenbelastung *f (ph)* heat velocity
Heizgas *n* fuel (heating, combustible) gas
Heizgebläse *n* heater fan
Heizgenerator *m* filament generator
Heizgewebe *n* heating fabric
Heizkabel *n* heating cable
Heizkammer *f* fire box
Heizkanal *m* heating flue
Heizkessel *m* 1. boiler; 2. caldron
Heizkesselwagen *m* heating van, wagon with heating apparatus
Heizkissen *n* **[/ elektrisches]** electrically heated pad, hot pad
Heizkörper *m* heater, heating unit; radiator
Heizkraft *f* calorific value
Heizkraftwerk *n* electricity power and heat station
Heizkreis *m* filament circuit
Heizkupplung *f* heating coupling
Heizleistung *f* filament input (power), quantity of heat; filament wattage
Heizleiter *m* heating resistor (conductor)
Heizleiterlegierung *f* heating conductor alloy
Heizleitermaterial *n* conducting wire material
Heizleiterträger *m* element former
Heizleitung *f* heating pipe
Heizlüfter *m* fan-forced heater
Heizmantel *m* heat[ing] jacket
Heizmaß *n* emission efficiency *(cathode)*
Heizofen *m* **[/ elektrischer]** electric fire
Heizöl *n* fuel oil
Heizpatrone *f* heating cartridge
Heizplatte *f* heating panel
Heizraum *m* fire (boiler) room
Heizreaktor *m* [industrial] heat reactor, heat producing reactor
Heizregler *m* heat regulator
Heizrohr *n* heating tube
Heizröhrenkessel *m* multitubular boiler
  stehender H. vertical multitubular boiler
Heizrohrkessel *f* fire (smoke) tube boiler, multitubular boiler
  ausziehbarer H. multitubular boiler with withdrawable fire box and smoke tubes
  einfacher H. single cylinder multitubular boiler
  stehender H. mit Feuerbüchse vertical multitubular boiler with fire box
Heizschlange *f* heating (heater) coil
Heizschlauch *m* heating hose
Heizsonne *f* parabolic heater, electric fire
Heizsonnenreflektor *m* electric fire bowl
Heizspannung *f* heater (filament) voltage
Heizspirale *f* **des Düsenzerstäubers** helical core of the nozzle sprayer
Heizspule *f* heating (heater) coil
Heizstrom *m* heating (filament) current
  zulässiger H. safe filament current
Heizstromkreis *m* [filament] heating circuit
Heiztabelle *f* stoking (firing) record
Heiztafel *f* heating panel
Heizteppich *m* heating carpet
Heiztransformator *m* filament transformer
Heiztrommel *f* drying drum
Heiztür *f* fire door
Heiztürgriff *m* fire door handle
Heiztürrahmen *m* fire door frame
Heiztürverschluß *m* fire door latch
Heizung *f* 1. heating; 2. heating system
  H. der Dampfüberhitzer heating of superheaters
  H. des Zylindermantels heating of the cylinder walls
  induktive H. induction heating
Heizungsanlage *f* [domestic] heating system
Heizungs- und Lüftungsanlage *f* heating and ventilating system

Heizversuch *m* stoking test
Heizwendel *f* heating (heater) coil
Heizwert *m* calorific effect, calorific (heating) value
oberer H. gross (upper) calorific value
unterer H. net calorific value
Heizwertbestimmung *f* calorific value determination, determination of the heating value
Heizwertzahl *f* fuel ratio
Heizwicklung *f* filament (heater) winding
Heizwiderstand *m* heating resistor, filament rheostat
Heizzentrale *f* central heating plant
Heizzug *m* heating flue
Heizzugverschluß *m* heating flue casting
Hektar *n (m)* hectare
Hektarzähler *m* acremeter
Hektograf *m* hectograph
Hektografenplatte *f* hectograph pad
hektografieren to hectograph
Helfenberechnung *f* heald calculation
Helfenschiene *f* ridge bar
Helgen *m* slip[way], building berth (slip, ways)
Helianthin *n (ch)* helianthine
Heliarcverfahren *n* heliarc welding process
Heliogravüre *f* heliogravure
Heliotrop *m* bloodstone, heliotrope *(mineral)*
Helium *n* helium
Heliumalter *n* helium age
Heliumkern *m* helium nucleus
Heliumleckprüfung *f (nucl)* helium leak detection
Heliumlinie *f* helium line
Heliumzyklus *m (nucl)* helium cycle
hell bright; light
Helladaptation *f* light adaptation
Hellanpassung *f (opt)* bright adaptation
Hell-Dunkel-Feldkondensor *m* bright-field / darkground change-over condenser
Hellegatt *n (ship)* store room
Hellempfindlichkeitsgrad *m* **des Auges** brightness response of the eye
Hellfeldbeleuchtung *f* brightfield illumination
Hellfeldkondensor *m (opt)* bright-field condenser
Helligkeit *f* brightness; brilliance; brilliancy; luminosity
H. der Sterne brightness (luminosity) of stars
H. des Bildes brightness of the image
absolute H. absolute brightness
Helligkeitsbereich *m (telev)* brightness range
Helligkeitsflimmern *n* brightness flicker
Helligkeitskontrast *m (telev)* brightness contrast
Helligkeitsmodulation *f* brightness (intensity) modulation *(cathode ray oscillograph)*
Helligkeitsregelung *f* brightness control; dimming control *(stage lighting)*
Helligkeitssignal *n (telev)* luminance signal
Helligkeitssprung *m (telev)* jump in brightness
Helligkeitssteuerung *f* intensity (brilliance) modulation
Helling *f* [launch] slip, [launch] slip way, building berth (slip, ways), stocks for shipbuilding
H. mit Stapelklötzen stocks
überdachte H. roofed-over berth
Hellingbahnneigung *f* slope of the slipway, declivity of the standing ways
Hellingbauzeit *f* building time on the stocks
Hellinggerüst *n* staging
Hellingkran *m* slipway crane
Hellingneigung *f* slope of the slipway
Hellinienspektrum *n* bright-line spectrum
Hellsteuerung *f* bright intensification *(cathode ray oscillograph)*
Helltastung *f* bright up, intensity modulation
Hemdenleger *m* folding table for shirts
Hemianopikerbrille *f* hemianopsy spectacles
Hemieder *n* hemihedron
Hemiedrie *f* hemihedry
Hemikolloid *n* hemicolloid
Hemimorphie *f* hemimorphism
Hemimorphit *m* electric calamine, hemimorphite
hemitrop hemitropic
Hemizellulose *f* hemicellulose
hemizyklisch hemicyclic
hemmen to stop, to arrest; *(ch)* to inhibit; to stem, to dam *(liquid)*; to retard, to check *(movement)*
Hemmkörper *m* inhibitor
Hemmschuh *m* skid; trig
Hemmschuhleger *m* lengthman, *(Am)* trackman
Hemmstoff *m* inhibitor
Hemmung *f* stopping, arresting; braking; escapement *(watch)*
Hempel-Bürette *f* Hempel pipette
Henkelplüsch *m* plush loop fabric
Henry *n* henry
Heparprobe *f (ch)* hepar test
Heptaldehyd *m* heptaldehyde
Heptan *n* heptane
Hepten *n* heptene
Heptode *f* heptode, pentagrid valve
Heptylalkohol *m* heptyl alcohol
herabdrücken to force down
herabfließen to flow down
herabführen to lead down
herabgehen to go down; *(mach)* to change down, to slow feed (speed)
herabhängend pendant
herablassen to lower
herabsetzen to reduce, to minimize, to decrease, to lower
Geschwindigkeit h. to decelerate
herabtransformieren to step down
herabziehen to pull down

heranfahren *(mach)* to approach
heranführen *(el)* to link up to *(line)*; *(mach)* to approach, to bring close to
Heranführen *n* / **schnelles** *(mach)* rapid approach motion
heraufmischen to up-convert *(frequency)*
herauftransformieren to step up
herausdrücken to force out
herausführen / eine Leitung to bring a lead out
herausheben to withdraw, to draw out, to rise, to extract *(e.g. fuel)*
herauslösen to dissolve out
herausquetschen to squeeze out
herausragen to protrude, to project
Herausreißen *n* **eines Nukleons aus dem Kern** *(nucl)* pick-up [reaction]
herausspritzen to jet, to spatter [out]
herausspulen *(el)* to reel out
herausstoßen to knock out
Herausstoßen *n* **des Moderators** *(nucl)* expulsion of the moderator
herauswaschen to wash out; to obtain by scrubbing *(benzol)*
herausziehbar retractable, withdrawable
herausziehen to extract, to move out, to draw out; to unplug; to withdraw
Herausziehen *n* **des Bohrers** retraction of the drill
H. des Moderators *(nucl)* expulsion of the moderator
herb *(ch)* acerb, acrid
Herbizid *n* herbicide, weed-killer
ätzend wirkendes H. contact herbicide with a caustic effect
selektiv wirkendes H. selective herbicide
total wirkendes H. non-selective herbicide
Herbstaussaat *f* autumn seeding (sowing)
Herbstbestellung autumn seed bed preparation and sowing
Herbstfurche *f* unbroken furrow
Herd *m* hearth *(furnace)*; cleaning (concentrating) table; range
ausfahrbarer H. mobile (car-bottom) hearth
offener H. open hearth
Herdbuch *n* herd-book
Herdbuchzucht *f* pedigree breeding
Herddosis *f* tumor (depth) dose
Herde *f* herd *(cattle)*; flock *(sheep)*
Herdenvieh *n* drove cattle
Herdfeuerung *f* hearth furnace
Herdflammofen *m* reverberatory hearth-furnace
Herdformerei *f* hearth (floor) moulding
Herdfrischarbeit *f* refinery operation
Herdfrischeisen *n* knobbled iron
Herdfrischen *n* open hearth refining
Herdfrischprozeß *m* refinery (open-hearth) process
Herdfrischstahl *m* open-hearth steel
Herdfrischverfahren *n* refinery (open-hearth) process
Herdfutter *n* [hearth-]fettling
Herdgewölbe *n* arched roof of the hearth
Herdguß *m* furnace (open-sand) casting
Herdofen *m* hearth-type furnace
Herdprobe *f* assay of silver
Herdrahmen *m* test ring
Herdsohle *f* bed of the hearth, hearth level
Herdtiefofen *m* soaking pit furnace
Herdwagenofen *m* bogie hearth furnace
hereinbrechen to cave in, to collapse
Hereinbrechen *n* **des Hangenden** *(mine)* roof fall
hereingewinnen *(mine)* 1. to get, to mine, to win; 2. to break [down] *(ore)*
Hereingewinnung *f* *(mine)* 1. getting, mining, winning; 2. breaking, ground *(ore)*
hereinwerfen *(mine)* to cave
Hereinziehen *n* **der Reibahle in die Bohrung** drawing-in of the reamer into the hole, screwing-in of the reamer into the hole
hergestellt / in Massenproduktion mass-produced
Heringsboot *n* herring boat
Heringsbuise *f* [herring] buss
Heringskutter *m* herring smack
Heringslogger *m* herring drifter, logger
Herkunft *f* *(ag)* provenance; *(ph, mine)* origin
H. der kosmischen Strahlung origin of cosmic rays
Herkunftshafen *m* port of origin
herleiten to deduce; to derive
hermetisch hermetic[al], air-tight, close
Héroult-Ofen *m* Héroult (direct arc) furnace
Herpolhodiekegel *m* herpolhode cone
herstellen to produce, to manufacture, to make; *(ch)* to prepare, to synthesize
Durchhieb h. *(mine)* to hole
Einbruch h. *(mine)* to cut
eine Verbindung h. to set up a call, to establish a connection
einen Anschluß h. to make a connection
Herstellen *n* **von Bohrungen mit gekrümmter Achse** bent-hole boring
Hersteller *m* maker, manufacturer, producer
Herstellerrisiko *n* producer's risk
Herstellung *f* production; manufacture; *(ch)* preparation
H. eines Anschlusses establishment of a connection *(telephony)*
genaue H. accurate workmanship; precise manufacture
großtechnische H. commercial (industrial) production
Herstellungsdatum *n* date of manufacture
Herstellungsfehler *m* manufacturing error

Herstellungsgang *m* manufacturing method (process)
Herstellungsgenauigkeit *f* grade of accuracy *(e.g. of a machine tool)*
Herstellungskosten *pl* production cost
Herstellungslänge *f* factory length *(e.g. of cables)*
Herstellungsmethode *f* manufacturing method
Herstellungstoleranz *f* manufacturing tolerance
Herstellungstoleranzen *fpl* mill limits *(rolling)*
Herstellungsverfahren *n* manufacturing process
Herstellungsvorschriften *fpl* production specifications
Hertz *n* cycles per second, cps
Hertz-Einheit *f* Hertzian unit
herumdrehen to turn around
herumschalten to index around
herumziehen to swing the plane
herunterfahren *(mach)* to lower; to shut down *(rectifier)*
herunterfieren to lower *(boat)*
Herunterfrischen *n* **durch Erzzugabe / nachträgliches** oreing down
heruntergehen *(aero)* to come down
mit der Fahrt h. to slacken in speed, to slow down *(a vessel)*
heruntergleiten to slip off
herunterrutschen to slip off
heruntertransformieren to step down
herunterwalzen to roll (rough) down; to cog down
hervorstehen to protrude, to project
hervorstehend projecting
Herzcharakteristik *f* cardioid diagram *(aerial)*
herzförmig heart-shaped, cardioid
Herzkurve *f (math)* cardioid
Herzriß *m* heart shake *(of wood)*
Herzschnitt *m* flitch cut *(of wood)*
Herzstück *n* frog *(switch)*
Hessit *m (min)* hessite
heterochrom[atisch] heterochromatic
heterogen heterogeneous, inhomogeneous
Heterogenität *f* heterogeneity, inhomogeneity
heteropolar heteropolar
heterozyklisch heterocyclic
Heterozyklus *m* heterocycle
Heu *n* hay
Heuauflader *m* hay sweep
Heuautomat *m* hay self-feeder
Heuballenpresse *f* hay baler
Heubelüftungsanlage *f* hay barn drier
Heuboden *m* hay-loft, hayloft
Heuer *f* hire, engagement *(marine)*; pay, wages *(salary)*
Heuerbass *m* shipping master
Heuerbeamter *m* shipping master (commissioner)
heuern to hire, to engage, to ship *(sailors)*; to charter, to hire, to rent *(ship)*
Heuernte *f* haymaking; hay harvest (crop)
Heufeim *m* hay rick, hayrick
Heugabel *f* hay fork
Heugewinnung *f* haymaking
Heulader *m* hayloader, hay loader
Heulboje *f* howling (sounding, whistling) buoy
heulen *(ac)* to howl
Heuler *m* howler
Heulton *m* howl
Heultonne *f* sounding (howling, whistling) buoy
Heumehl *n* grass meal
Heupresse *f* hay baler
Heuraufe *f* hay-rack, hayrack
Heurechen *m* hay rake
Heusammelpresse *f* hay baler
Heuscheune *f* hay barn
Heuschober *m* haystack
Heuschuppen *m* hay barn
Heuseil *n* hay-band
Heuseildrehmaschine *f* hay-band twister
Heustapler *m* haystacker
Heu- und Strohgebläse *n* forage blower
Heuwagen *m* hay car
Heuwender *m* swath turner, hay tedder
Heuwerbung *f* haymaking
Hevesy-Paneth-Analyse *f (nucl)* tracer analysis
hexadezimal hexadecimal
Hexaeder *n* hexahedron
Hexafluorokieselsäure *f* silicofluoric acid
Hexagon *n* hexagon
hexagonal hexagonal *(crystals)*
Hexan *n* hexane
Hexen *n* hexene
Hexit *n* hexitol
Hexode *f* hexode, six-electrode valve
Hexose *f* hexose
HF *s.* Hochfrequenz
Hickory *m* hickory
Hickoryholz *n* hickory [wood]
Hiddenit *m (min)* hiddenite
Hieb *m* cut *(file, wood)*; stroke, blow
Hiebfläche *f* cutting area, area to be cut
Hiebfolge *f* felling rotation, succession of cuttings
Hiebsatz *m* annual cut
Hiebteilung *f* spacing of file teeth; number of file teeth per cm.
Hiebumlauf *m* felling rotation
Hieve *f (ship)* sling [load]; heave, lift, hoist
hieven to heave, to hoist, to lift, to haul up
Hievleine *f* heaving line
Hilfe *f* help; aid; support; assistance
H. von / mit by means of
erste H. first aid
Hilfsabstellvorrichtung *f* **mit Fühlerfinger** *(text)* drop-finger auxiliary stop motion
Hilfsabteilung *f* service department
Hilfsachse *f* auxiliary idler shaft

Hilfsamt *n* satellite exchange; sub-exchange *(telephony)*
Hilfsanode *f* auxiliary plate (anode); relieving (supplementary) anode
Hilfsantenne *f* auxiliary aerial; dummy aerial
Hilfsantrieb *m* tail-end drive
Hilfsantriebsrad *n* timing gear
Hilfsarbeiter *m* unskilled worker
Hilfsausgang *m (el)* ancillary output
Hilfsballastpumpe *f (ship)* ballast donkey pump
Hilfsbasis *f* secondary base *(surveying)*
Hilfsbremsleitung *f* auxiliary brake pipe
Hilfsdampfmaschine *f* auxiliary steam engine
Hilfsdruckluftbremse *f* compressed-air servo-brake
Hilfseinrichtung *f (mach)* auxiliary attachment
Hilfselektrode *f* auxiliary (secondary) electrode; guard ring *(of a counter tube)*
Hilfsenergie *f* auxiliary power
  H. / mit power-operated
  H. / mit hydraulischer fluid-powered, liquid-operated
  H. / mit pneumatischer air-powered, pneumatically operated
  H. / ohne self-operated
Hilfsenergieversorgung *f* supply line
Hilfserregermaschine *f (el)* pilot exciter
Hilfsflughafen *m* supplementary aerodrome
Hilfsfunkenstrecke *f* auxiliary spark gap
Hilfsfunktion *f* auxiliary function
Hilfsgleichrichter *m* complementary rectifier
Hilfsgruppenrelais *n (telec)* auxiliary group relay
Hilfsheizer *m* assistant (under) stoker
Hilfshub *m* auxiliary hoist *(conveying)*
Hilfskabel *n* auxiliary (emergency) cable
Hilfskessel *m* donkey (auxiliary) boiler
Hilfskesselschornstein *m* funnel for auxiliary boiler
Hilfskesselwärter *m* donkeyman
Hilfskettenfahrzeug *n* auxiliary track vehicle
Hilfskondensator *m (therm)* auxiliary condenser
Hilfskontakt *m (el)* secondary (arcing, auxiliary) contact
Hilfskreis *m* auxiliary (subsidiary) circuit, *(Am)* fall-back circuit; idler circuit *(parametric amplifiers)*
Hilfskreuzer *m* auxiliary cruiser *(armed merchant cruiser)*
Hilfslüfter *m* auxiliary fan
Hilfsmaschine *f* auxiliary (donkey) engine, auxiliary (donkey) machine
Hilfsmaschinenanlage *f* auxiliary machinery
Hilfsmotor *m* auxiliary motor; booster [motor], servo-motor
Hilfsnormal *n* secondary standard
Hilfsöffnungskontakt *m* normally closed auxiliary contact
Hilfspatent *n* subsidiary patent
Hilfsphasengenerator *m* monocyclic alternator
Hilfspol *m* auxiliary (compensating, intermediate) pole
Hilfspumpe *f* auxiliary (booster, ancillary) pump
Hilfsrahmen *m* auxiliary frame
Hilfsregler *m* safety governor
Hilfsrelais *n* auxiliary (supplementary) relay
Hilfsrohrleitung *f* auxiliary piping
Hilfsruder *n (aero)* servo control surface, tab
Hilfssammelschiene *f (el)* auxiliary bus-bar
Hilfssatz *m (dat)* lemma
Hilfsschacht *m (mine)* materials shaft
Hilfsschalter *m* auxiliary (booster) switch
Hilfsschieber *m* auxiliary sluice (parallel slide, gate) valve
Hilfsschiff *n* auxiliary ship
Hilfsschließkontakt *m* normally open auxiliary contact
Hilfsschuß *m (mine)* relieving shot
Hilfssender *m* satellite transmitter
Hilfssenker *m (text)* auxiliary cam
Hilfssicherheitsvorrichtung *f* back-up safety device
Hilfsspeicher *m* auxiliary memory
Hilfssteiger *m (mine)* assistant deputy
Hilfsstempel *m (mine)* catch prop
Hilfssteuerelement *n* servo-mechanism
Hilfssteuerung / mit servo-controlled
Hilfsstoff *m (ch)* auxiliary, accessory agent
Hilfsstrom *m* auxiliary current
Hilfsstromauslösung *f* independent trip
Hilfsstromkreis *m* auxiliary (ancillary) circuit
Hilfsstudio *n* satellite studio
Hilfstank *m* stand-by tank
Hilfsteilung *f* auxiliary scale of graduation
Hilfstragdraht *m* auxiliary carrier cable, *(Am)* emergency cable
Hilfsträgerkanal *m* stereophonic subchannel
Hilfstragseil *n* auxiliary carrier cable, *(Am)* emergency cable
Hilfstriangulation *f* subsidiary triangulation
Hilfstürstock *m (mine)* false set
Hilfsunterbrechungsstelle *f (el)* auxiliary interrupting contact member
Hilfsventil *n* by-pass valve, auxiliary valve *(to facilitate the opening of the main valve)*
Hilfsverstärker *m* sub-repeater
Hilfsvoramt *n* sub-control station *(telephony)*
Hilfswähler *m* discriminating selector *(telephony)*
Hilfsweg *m (el)* alternative route
Hilfswelle *f* auxiliary shaft
Hilfswiderstand *m* secondary resistance
Hilfszieleinrichtung *f* emergency means for laying the gun
Hilfszug *m* breakdown train
Himmel *m* 1. sky; heaven; 2. *(aut)* inside roof lining

Himmelsglobus *m* celestial globe
Himmelskörper *m* celestial (stellar) body
Himmelsmechanik *f* celestial mechanics
Himmelsrichtungen *fpl* cardinal points
hin und zurück *(ry)* there and back, return *(ticket)*
hinabreichen [zu] to reach down [to]
hinauftransformieren to step up
hindern to hamper, to handicap, to impede; to obstruct, to restrain *(movement)*
Hindernis *n* obstruction; obstacle
Hindernisbefeuerung *f* obstruction lights
Hindernisfeuer *n* obstruction light
Hinderung *f* hindrance; obstruction; restraint *(motion)*
 sterische H. steric hindrance
hindurchgehen to pass through *(liquid, movement)*; to reach through *(part)*
 durch einen Stoff h. to penetrate [through] a material, to traverse a material
hindurchsickern / durch die Filtermasse to trickle through the filtering material
hineinreichen [in] to extend [into]
hineinschlagen to drive home
 h. [in] to hammer [into]
hineinschütten to pour in
hineinstecken to put in, to insert [into]
hineinstoßen in to push in
hineinströmen to flow [into]
hineinziehen to draw [in]
Hinfahrt *f* forward (outward) journey
Hinken *n* **der Dampfmaschine** halting (lagging behind) of the steam engine
hinreichend sufficient
Hinreise *f* forward (outward) journey
hinten *(ship)* abaft; *(mach)* in the rear, at the rear end (side)
Hinterachsantrieb *m* final-drive unit
Hinterachsbrücke *f* rear axle casing
Hinterachse *f* rear axle
Hinterachsenübersetzungsverhältnis *n* back axle ratio
Hinterachsgehäuse *n* rear axle housing
Hinterachsgehäusehälfte *f* rear axle housing section
Hinterachskörper *m* rear axle assembly
Hinterachsrohr *n* rear axle tube
Hinterachsschubstange *f* rear axle radius rod
Hinterachsstrebe *f* rear axle strut (tie rod)
Hinterachstrichter *m* rear axle flared tube
Hinterachsübersetzung *f* final gear reduction in the rear axle
Hinterachswelle *f* half (rear axle) shaft
Hinterachswellenrad *n* differential side gear
Hinterachszugstange *f* rear axle tie rod
Hinterbaum *m* back beam *(loom)*
Hinterbesatz *m (leath)* back strap
Hinterbohren *n* back drilling
Hinterbündelecho *n* back echo *(radar)*
Hinterdeck *n* after (quarter) deck, poop
Hinterdreharbeit *f (mach)* relieving operation
Hinterdreheinrichtung *f (mach)* relieving attachment
hinterdrehen *(mach)* to relieve, to back off
Hinterdrehen *n (mach)* relieving, backing off
Hinterdrehmaschine *f* relieving (back-off) lathe
hinterdreht relieved [by turning], machine-relieved, backed-off
Hinterdrehvorrichtung *f (mach)* relieving attachment
hintereinander [angeordnet] [arranged] in tandem
hintereinandergeschaltet *(el)* connected in series, series connected
hintereinanderschalten *(el)* to connect in series
Hintereinanderschaltung *f (el)* series (serial) connection
 H. der Überhitzerrohre series connection of the superheating tubes
 H. von Druck- und Geschwindigkeitsstufen series arrangement of impulse and velocity stages *(turbine)*
Hinterfeder *f (aut)* rear spring
Hinterfederbock *m* rear spring hanger
Hinterfederstütze *f* rear spring support
Hinterflanke *f* trailing edge *(of the pulse)*
hinterfräsen to relief mill, to relieve by milling
hinterfüllen *(build)* to pack, to fill [in], to backfill
Hinterfüllung *f (build)* backfilling, backfill, infill
Hinterfüllungssand *m* back sand *(foundry)*
Hintergeschirr *n (ship)* after sails
hintergießen *(print)* to back up
Hintergießen *n (print)* backing up
Hintergrund *m* background
 getüpfelter H. speckled background *(radar)*
Hintergrundprojektion *f* background projection
Hinterkappe *f* counter *(shoe)*
Hinterkeule *f* back lobe *(aerial)*
Hinterkippen *n* rear (end) dumping
Hinterkipper *m* rear dumper, rear-dump truck, end-dump truck
Hinterladepforte *f (ship)* stern port
Hinterlader *m* breech-loader
Hinterlappen *m* back lobe *(aerial)*
Hinterlegung *f* 1. consignment of goods deposit; 2. [metal] backing
Hinterlegungsbetrag *m* guarantee
Hintermaschine *f* rear-motor, back-motor
hintermauern to back up
Hintermauerung *f* backing-up
Hintermauerungsstein *m* backing (common) brick
Hintermauerungsziegel *m* backing (common) brick

Hinterpforte *f (ship)* stern port
Hinterpiek *f (ship)* after peak
Hinterplicht *f (ship)* after cuddy
Hinterrad *n* rear wheel
Hinterradantrieb *m* rear-wheel drive, rear-axle drive
Hinterradbremse *f* rear-wheel brake
Hinterradfederung *f* rear-wheel springing
Hinterradgabel *f* back fork
Hinterradlenkung *f* rear-wheel steering
Hinterradwelle *f* half shaft
Hinterraum *m (ship)* after hold *(cargo hold)*
Hinterriemen *m* back strap
Hinterschiff *n* after body (quarter), quarter of a ship
Hinterschiffsform *f* run
Hinterschiffsschnitte *mpl* buttocks
Hinterschiffsspant *n* after rib (frame)
Hinterschiffsspantenriß *n* after-body plan
Hinterschleifeinrichtung *f* relief-grinding attachment, backing-off attachment
hinterschleifen to relieve by grinding
Hinterschleifen *n* relieving by grinding, backing-off
Hinterschleifwinkel *m* relief angle
Hinterschliff *m* relief grinding, backing-off; back-off clearance
  H. / ohne unrelieved *(tool)*
hinterschliffen relief-ground, backed-off *(tool)*
  nicht h. unrelieved *(tool)*
Hinterschliffläche *f* flank [of the drill point]
Hinterschrämarm *m* back-shearing jib
hinterschrämen to back-shear
Hinterseil *n* tail rope *(slusher)*
Hintersteven *m (ship)* back (stern) post, stern frame
  loser H. false post
Hinterstevenbogen *m* propeller arch
Hinterstevenhacke *f* heel piece *(of the stern frame)*
Hinterstevenmittelteil *n* horn timber
Hinterstevennuß *f* stern [frame] boss, stern bossing
Hinterstevensohle *f* sole (heel) piece *(of the stern frame)*
Hinterstich *m (text)* backstitch
Hinterteil *n* **des Rumpfes** *(aero)* after-portion of the fuselage
Hintertreppe *f* backstairs
Hintertür *f* backdoor
Hinterwandsperrschichtzelle *f* back-wall barrier [blocking] layer cell
Hinterwetzwinkel *m* primary clearance angle, radial relief (clearance) *(tool)*
hinüberbringen *(math)* to transpose
hinübermultiplizieren *(math)* to cross-multiply
hin- und herbewegen to reciprocate
Hin- und Herbewegung *f* reciprocation
Hin- und Herbiegeversuch *m* reverse (to-and-fro) bend test
hin- und herfahren to shuttle, to move back and forth
Hin- und Hergang *m* to-and-fro motion
hin- und hergehen to reciprocate
hin- und hergehend reciprocating
hin- und herschwingen to oscillate; to surge back and forth *(electromagnetic wave)*
hin- und hertanzen to wobble from side to side *(drill)*
Hin- und Rückflugschein *m* return ticket
hinzufügen to add
Hippe *f (ag)* pruning knife
Hippursäure *f* hippuric acid
Hirnholz *n* cross-grained wood
Hirnholzplatte *f* cross-grain leaf
Hirnholzverbindung *f* end-to-end-grain joints
Hirnschnitt *m* cross-cutting *(wood)*
Hirnseite *f* cross grain *(wood)*
Hirschhornsalz *n* hartshorn salt, ammonium carbonate
Hirse *f* millet
hissen to hoist, to set *(flags)*
Histamin *n (ch)* histamine
Histidin *n (ch)* histidine
Histochemie *f* histochemistry
Histogramm *n* histogram
Hitzbandinstrument *n (el)* hot-strip instrument
Hitzdaube *f* firing cone
Hitzdraht *m* hot wire
Hitzdrahtamperemeter *n* thermal (hot-wire) ammeter
Hitzdrahtanemometer *n* hot-wire anemometer
Hitzdrahtausbrennung *f* hot-wire cautery
Hitzdrahthöhenmesser *m* hot-wire altitude meter
Hitzdrahtinstrument *n* hot-wire [expansion] instrument, thermoexpansion (electrothermal) instrument
Hitzdrahtmanometer *n* hot-wire pressure gauge
Hitzdrahtrelais *n* thermo-relay
Hitzdrahtspule *f* heat coil
Hitzdrahtvoltmeter *n* hot-wire voltmeter
Hitze *f* heat
hitzebeständig heat-proof, heat-resistant (-resisting), thermostable; refractory, fire-proof; *(met)* stable at red heat
Hitzebeständigkeit *f* heat resistance
Hitzeempfindlichkeit *f* sensibility to heat
Hitzefalte *f (text)* heat crease
Hitzegrenze *f* heat barrier
hitzehärtbar thermosetting, heat-setting
hitzeunbeständig thermolabile
Hitzstromverbrauch *m* filament-current consumption
HM-Plättchen *n* cemented-carbide tip

Hobel *m* plane
Hobelbank *f* **mit Holzschraubstock** carpenter's bench vice
Hobelbankeisen *n* bench dog
hobelbar ploughable *(seams)*
Hobelbarkeit *f* ploughability *(seams)*
Hobelbreite *f* planing width
Hobeleisen *n* plane (cutting) iron
Hobelfahrer *m (mine)* plough motorman
Hobelführer *m (mine)* plough minder
Hobelführung *f (mine)* plough-guide
Hobelgasse *f (mine)* plough track
Hobelgehäuse *n* plane stock
Hobelkamm *m* rack shaped cutter
Hobelkasten *m* plane stock
Hobelkopfschlitten *m* tool (ram) head slide
Hobelkörper *m (mine)* plough body
Hobelmaschine *f* planer, planing machine *(metal cutting)*
  H. für Holzgeflechtspäne wood planing machine for cord fibres
  H. mit beweglichem Hobelwerkzeug travelling head planer
  H. mit Kurbel[schleifen]antrieb crank planer
  H. mit Vielschnittgeschwindigkeit multi-speed planer
Hobelmaschinenschraubstock *m* planer vice
  drehbarer H. planer vice with swivel base
  fester H. planer vice with plain base
Hobelmaschinensupport *m* planer tool head
Hobelmaschinentisch *m* planer platen
Hobelmaschinenwinde *f* planer jack
Hobelmaul *n* opening in the plane face
Hobelmeißel *m* 1. planer (planing) tool; 2. shaper tool
  gebogener H. cranked planing tool
Hobelmeißeleinstellehre *f* planer gauge
Hobelmesser *n* 1. plane iron; 2. *(mine)* plough blade
Hobelmesserschleifmaschine *f* grinding machine for plane irons
hobeln 1. to plane *(metal cutting)*; 2. *(mine)* to plane, to plough *(coal)*
Hobeln *n* 1. planing *(metal cutting)*; 2. *(mine)* planing, ploughing *(coal)*
  H. mit Hartmetallmeißel[n] carbide planing
Hobelnase *f* handle of the plane
Hobelschlitten *m (mine)* plough base
Hobelschwert *n (mine)* plough baseplate
Hobelsohle *f* face of the plane
Hobelspan *m* 1. paring, wood shavings; 2. planing *(metal cutting)*
Hobelspanplatte *f* shaving board
Hobel- und Fügemaschine *f* head planing and jointing machine
Hobel- und Kehlmaschine *f* / **vierseitige** four-cutter planing and moulding machine
Hobler *m* **mit ziehendem Stößel** draw-cut shaper
hoch high
  h. drei cube, third power of
  h. fünf to the fifth power
  h. vier to the fourth power
  h. zwei square, second power of
hochaktiv hot, highly radioactive
hochangereichert *(nucl)* highly enriched, high-grade
Hochantenne *f* overhead aerial
Hochantenneneffekt *m* vertical (capacity) effect
hochauflösend high-resolution, with high resolving power
Hochausbeutezellstoff *m* high yield pulp *(papermaking)*
Hochbahn *f* elevated (overhead) railway
Hochbau *m (build)* building construction; *(el)* overhead construction *(line)*
Hochbauingenieur *m* structural engineer
Hochbaukran *m* building crane
Hochbauschgarn *n* high-bulk yarn
Hochbaustahl *m* structural steel
Hochbaustatiker *m* structural engineer
hochbeansprucht highly stressed
Hochbehälter *m* elevated (head) tank; overhead cistern; sanitary (gravity) tank
Hochbelastbarkeit *f* heavy-load capacity
hochbordig *(ship)* high-freeboarded, high-sided
Hochbordwagen *m (ry)* high-sided goods wagon
hochbrechen *(mine)* to raise
Hochbrechen *n* **eines Schachts** shaft raising
Hochdruck *m* 1. high pressure; 2. relief (letterpress) printing
  indirekter H. indirect letterpress, letterset
Hochdruckbehälter *m* high-pressure vessel
Hochdruckdampfbehandlung *f (build)* autoclave (high-pressure steam) curing, autoclaving
Hochdruckdampferhärtung *f (build)* high-pressure steam curing
Hochdruckdampfkessel *m* high-pressure boiler
Hochdruckdampfmaschine *f* high-pressure steam engine
Hochdruckdeckel *m* high-pressure cover
Hochdruckdiagramm *n* high-pressure diagram
Hochdruckfärberei *f* dyeing under pressure
Hochdruckfett *n* pressure-gun grease
Hochdruckfilterpresse *f* high-pressure ram filters
Hochdruckfließpressen *n* high-pressure flow pressing
Hochdruckgebläse *n* high-pressure blower
Hochdruckhandabsperrventil *n* manual high-pressure valve
Hochdruckhochofen *m* pressurized blast furnace

Hochdruckhydrierung *f* high-pressure hydrogenation
Hochdruckionisationskammer *f* high-pressure ionization chamber
Hochdruckkammer *f* high-pressure chamber *(turbine)*
Hochdruckkessel *m* high-pressure boiler
Hochdruckkurbel *f* high-pressure crank *(steam engine)*
Hochdrucklampe *f (opt)* high-pressure lamp
Hochdruckleitung *f* high-pressure pipe line
Hochdrucklokomotive *f* high-pressure locomotive
Hochdruckmarschturbine *f* high-pressure cruising turbine
Hochdruckphase *f* high-pressure stage
Hochdruckpresse *f (ag)* high-pressure baler, high-density baler
Hochdruckpumpe *f* high-pressure pump
Hochdruckrad *n* high-pressure disk *(turbine)*
Hochdruckreaktor *m* high-pressure reactor
Hochdruckreifen *m (aut)* high-pressure tyre
Hochdrucksammelpresse *f (ag)* high-pressure pick-up baler
Hochdruckschmiermittel *n* extreme-pressure lubricant, E.P. lubricant
Hochdruckschmieröl *n* extreme-pressure lubricating oil
Hochdruckschmierpresse *f (aut)* pressure grease gun
Hochdruckschmierung *f* high-pressure lubrication
Hochdruckspritze *f (ag)* high-pressure sprayer
Hochdruckstufe *f* high-pressure stage *(turbine)*
Hochdruck-Van-de-Graaff-Generator *m (nucl)* pressure-type Van de Graaff accelerator
Hochdruckventil *n (rock)* main steam-plant valve
Hochdruckzählrohr *n* high-pressure counter
Hochdruckzusatz *m* extreme-pressure dope
Hochdruckzylinder *m* high-pressure cylinder *(steam engine)*
Hochdurchforstung *f* thinning in the upper storey
hochempfindlich highly sensitive
hochenergetisch high-energy
Hochenergiephysik *f* high-energy [nuclear] physics
Hochenergieumformen *n* high-energy-rate forming *(metal)*
hochevakuiert highly evacuated
Hochfach *n (text)* upper shed
hochfahren to raise, to run up *(e.g. rectifier)*; to start up *(transmitter)*
Hochfahren *n (el)* running-up *(motor)*, raising *(generator)*; start-up, rise *(of the reactor)*
hochfein first rate
Hochferse *f (text)* high-spliced heel

hochfest high-strength
Hochflußreaktor *m* high-flux reactor
Hochformat *n* upright format, portrait size
hochfrequent high-frequency, h.f., radio-frequency, r.f.
Hochfrequenz *f* high frequency, h.f., radio frequency, r.f.
Hochfrequenzabschirmung *f* high-frequency screen; radio shield
Hochfrequenzausgleich *m* high-frequency compensation
Hochfrequenzausgleichsschwingung *f* high-frequency transient oscillation
Hochfrequenzbauteil *n* radio-frequency component
Hochfrequenzbeschleuniger *m* radio-frequency accelerator
Hochfrequenzdrahtfunk *m* wired broadcasting
Hochfrequenzdrossel *f* high-frequency choke
Hochfrequenzeinstreuung *f* radio-frequency pick-up
Hochfrequenzeisen *n* radio-frequency iron
Hochfrequenzeisenkern *m* dust-iron core
Hochfrequenzelektronik *f* radio-frequency electronics
Hochfrequenzerhitzung *f* radio-frequency heating
Hochfrequenzgenerator *m* high-frequency generator
Hochfrequenzgleichrichter *m* radio-frequency rectifier, demodulator
Hochfrequenzhärtung *f* hardening by radio-frequency current
Hochfrequenzheizung *f* dielectric heating
Hochfrequenzhöhenmesser *m* radio altimeter
Hochfrequenzindikator *m* high-frequency indicator
Hochfrequenzinduktionsofen *m* high-frequency induction furnace
Hochfrequenzisolator *m* high-frequency insulator
Hochfrequenzkabel *n* radio-frequency cable
Hochfrequenzkopplung *f* high-frequency coupling
Hochfrequenzleistung *f* radio-frequency power
Hochfrequenzlitze *f* radio-frequency litz wire
Hochfrequenzlot *n* high-frequency sounder
Hochfrequenzlöten *n* high-frequency induction heat brazing
Hochfrequenzmaschine *f* high-frequency alternator
Hochfrequenzmeßbrücke *f* high-frequency bridge
Hochfrequenzmeßsender *m* radio-frequency signal generator
Hochfrequenzmessung *f* radio-frequency measurement

Hochfrequenzmetallsuchgerät *n (cont)* radio metal locator
Hochfrequenzofen *m* high-frequency [induction] furnace
Hochfrequenzregelpentode *f* variable-mu radio-frequency pentode
Hochfrequenzröhre *f* radio-frequency valve
Hochfrequenzschirm *m* high-frequency screen
Hochfrequenzschweißen *n* radio-frequency welding
Hochfrequenzschweißung *f* radio-frequency welding
Hochfrequenzspannung *f* radio-frequency voltage
Hochfrequenzspektrometer *n* radio-frequency spectrometer
Hochfrequenzsperrkreis *m* radio-frequency selective circuit
Hochfrequenzsprechkanal *m* high-frequency carrier current telephone channel
Hochfrequenzspule *f* radio-frequency coil
Hochfrequenzsteilheit *f* radio-frequency transconductance
Hochfrequenzstörschutzfilter *n* radio-frequency interference filter
Hochfrequenzstörung *f* high-frequency interference
Hochfrequenzstrom *m* radio-frequency current
Hochfrequenzstrommesser *m* radio-frequency ammeter
Hochfrequenztechnik *f* high-frequency engineering, high-voltage engineering
Hochfrequenztelefonie *f* radio (carrier current) telephony
Hochfrequenztransformator *m* radio-frequency transformer
veränderlicher H. variocoupler
Hochfrequenztrocknung *f* high-frequency drying
Hochfrequenzübertrager *m* oscillation transformer
Hochfrequenzübertragung *f* high-frequency transmission
Hochfrequenzvakuumprüfer *m* high-frequency vacuum gauge
Hochfrequenzvergleichsschutz *m* high-frequency differential protective system
Hochfrequenzverhalten *n* high-frequency response
Hochfrequenzverstärker *m* high-frequency amplifier
Hochfrequenzverstärkung *f* high-frequency amplification
Hochfrequenzvorstufe *f* preliminary signal-frequency stage
Hochfußnadel *f (text)* high (long) butt needle
Hochgatter *n* log frame for deep cutting
Hochgebirge *n* high mountains
hochgelegt elevated *(heater)*
hochgekohlt high-carbon *(steel)*
hochgerippt high-fin *(tubing)*
hochgesättigt highly saturated
Hochgeschwindigkeitstank *m (ship)* high-speed basin *(model basin)*
Hochglanzbeschichtung *f* high-gloss lamination
Hochglanzblech *n* high mirror-finished sheet, bright-luster sheet
Hochglanzfarbe *f* high-gloss ink
Hochglanzkaschierung *f* high-gloss lamination
Hochglanzpapier *n* glazed (cast-coated) paper
hochglanzpoliert highly polished, mirror-finished *(metal)*
Hochglanzpolitur *f* mirror finish, high (brilliant) polish
Hochglanzstreichkarton *m* enamelled board
Hochglühen *n* full annealing
hochgradig high-grade
hochhalten / die Dampfspannung to keep up the steam pressure
Hochhaus *n* tall (high-rise) building
hochheben to lift, to raise; to hoist
Hochhubbrammenwalzwerk *n* high-lift slabbing mill
Hochhubdoppelsicherheitsventil *n (therm)* [double] safety valve with high lift
hochinkohlt high-rank
hochionisiert highly ionized, stripped
hochkant edgeways, edge-wise, with edge uppermost, on edge
Hochkantsammelschiene *f (el)* edge-type bus-bar
Hochkantwicklung *f (el)* edge winding
hochkippen to tip up
hochklappen to fold upwards
Hochlauf *m* starting *(electrical machine)*
Hochlaufmoment *n* starting torque
hochlegiert high-alloy[ed]
Hochleistungs- heavy-duty, high-duty; high-power; high-speed
Hochleistungsbohren *n* heavy-duty drilling (boring)
Hochleistungsbohrer *m* super drill
Hochleistungsdrehmaschine *f* heavy-duty lathe
Hochleistungselektrode *f* contact (heavy-duty) electrode
Hochleistungsfernschreiber *m* high-speed teleprinter
Hochleistungsinstrument *n* instrument of high performance
Hochleistungsmaschine *f* heavy-duty machine, high-duty machine
Hochleistungsmotor *m (el)* high-power motor; *(aut)* high-powered engine
Hochleistungsöl *n* heavy-duty oil
Hochleistungsreaktor *m* high-power reactor
Hochleistungsröhre *f* high-performance valve
Hochleistungsrundstrickmaschine *f* / **einfon-**

turige high-speed production plain circular knitting machine
Hochleistungsschalten *n* high-speed switching
Hochleistungsschalter *m* heavy-duty switch
Hochleistungsschnell[arbeits]stahl *m* super high-speed steel
Hochleistungssicherung *f* power (quick-break) fuse
Hochleistungsstreifenlocher *m* high-speed tape punch
Hochleistungstechnik *f* high-power engineering
Hochleistungstransformator *m* high-power transformer
Hochleistungsumformen *n* / **hydroelektrisches** electrohydraulic method of high-velocity forging
Hochleistungsverstärker *m* high-gain amplifier
Hochleistungswalzenfräser *m* helical (heavy-duty) plain milling cutter, cutter of the coarse-tooth spiral type
Hochleistungswellendrehmaschine *f* heavy-duty roll turning lathe
Hochlicht *n* highlight
Hochlöffel *m* crowd (face) shovel *(excavator)*
Hochlöffelbagger *m* crowd (power) shovel, crane navvy
hochmolekular high-molecular
Hochofen *m* blast furnace
H. mit geschlossener Brust blast furnace with closed front
gedämpfter H. banked (damped-down) blast furnace
kaltliegender H. stopped furnace
mit Holzkohle betriebener H. charcoal blast furnace
verstopfter H. gobbed-up blast furnace
Hochofenabstich *m* blast-furnace tapping
Hochofenaufzug *m* blast-furnace skip hoist
Hochofenbegichtung *f* blast-furnace charging
Hochofenbeschickung *f* blast-furnace charging
Hochofenbetrieb *m* blast-furnace practice
Hochofenbrust *f* / **offene** open breast of a blast furnace
Hochofenfutter *n* blast-furnace lining
Hochofengas *n* blast-furnace gas
Hochofengerüst *n* blast-furnace framework
Hochofengestell *n* blast-furnace hearth (crucible)
Hochofengicht *f* mouth (top) of a blast furnace
Hochofengichtverschluß *m* blast-furnace hopper
Hochofenkoks *m* metallurgical (blast-furnace) coke
Hochofenkranz *m* blast-furnace lid
Hochofenpanzer *m* blast-furnace armour (jacket)
Hochofenprofil *n* blast-furnace lines
Hochofenrast *f* boshes of a blast furnace
Hochofenrüstung *f* armour of a blast furnace
Hochofensau *f* blast-furnace bear (salamander)
Hochofenschacht *m* blast-furnace stack
Hochofenschachtpanzer *m* blast-furnace stack casing
Hochofenschaumschlacke *f* [Portland] blast-furnace foamed slag
Hochofenschlacke *f* blast-furnace slag
Hochofensumpf *m* sump in a blast furnace
Hochofenverfahren *n* blast-furnace process
Hochofenwindring *m* blast-furnace bustle pipe
Hochofenzement *m* blast-furnace cement
hochohmig high-resistive, high-impedance
hochpaarig multi-paired *(cable)*
Hochpaß *m (el)* high-pass
Hochpaßfilter *n* high-pass filter
hochprozentig high-percentage
Hochpuffen *n (mine)* boiling up *(floor)*
hochradioaktiv highly radioactive
Hochrahmen *m (aut)* elevated frame
Hochreservoir *n* elevated tank (reservoir)
Hochschnitt *m* high face *(open-cast mining)*
H. / im plain sawn *(wood)*
Hochschnittbagger *m* up-dredge
Hochschnittmähbalken *m* high-level cutter-bar
Hochschulbildung *f* university education
Hochschule *f* university
technische H. technological university
Hochschulingenieur *m* university graduate engineer, academically trained engineer
Hochschulreaktor *m* training reactor
Hochsee *f* high (main) sea, deep sea
Hochseebergungsschlepper *m* deep-sea salvage tug
Hochseedampfer *m* ocean[-going] steamer, sea-going steamer
Hochseefischdampfer *m* deep-sea trawler, sea-going trawler
Hochseefischerei *f* deep-sea fishery, off-shore fishery (fishing)
kleine H. small deep-sea fishing
Hochseefischfang *m* pelagic fishing
Hochseeflotte *f* main (deep-sea) fleet
Hochseehafen *m* deep-sea port
Hochseejacht *f* off-shore yacht
Hochseeregatta *f* off-shore race
Hochseeschiff *n* sea-going vessel, ocean-going ship
Hochseeschiffahrt *f* ocean (deep-sea) navigation
Hochseeschlepper *m* ocean-going tug, sea-going tug
Hochseetrawler *m* deep-sea trawler, sea-going trawler
Hochseewalfang *m* pelagic whaling
hochselektiv sharply selective
hochsiedend high-boiling
Hochsilo *m* tower silo
Hochspannung *f (el)* high tension (voltage)
Hochspannungsbeschleuniger *m* high-voltage accelerator

Hochspannungsdurchführung *f* high-voltage bushing
Hochspannungselektrometer *n (Am)* aislometer
Hochspannungserzeugung *f* high-voltage generation
Hochspannungsgenerator *m* high-voltage generator
Hochspannungsgleichrichter *m* high-voltage rectifier
Hochspannungskabel *n* high-voltage cable
Hochspannungskommutatorgleichrichter *m* transverter
Hochspannungskondensator *m* high-voltage capacitor, high-potential capacitor
Hochspannungskreis *m* high-voltage circuit
Hochspannungsleitung *f* power transmission line
Hochspannungsmast *m* transmission tower (pole)
Hochspannungsnetz *n* high-tension mains, high-voltage system (grid)
Hochspannungsporzellan *n* porcelain for high-voltage application
Hochspannungsprüfung *f* high-voltage test
Hochspannungsschaltanlage *f* high-voltage switch plant
Hochspannungsschalter *m* high-voltage switch
Hochspannungsschaltgerät *n* high-voltage switchgear
Hochspannungsseite *f* high-tension end
hochspannungssicher high-tension protected
Hochspannungstechnik *f* high-voltage engineering
Hochspannungsverteilung *f* primary distribution
Hochspannungsverteilungsnetz *n* network primary distribution system
Hochspannungswicklung *f* high-voltage winding
Hochspant *n (ship)* deep frame
Höchstabrichtbetrag *m (mach)* maximum amount removed in dressing *(grinding)*
Hochstamm *m* standard tree
Hochstauung *f* deep stowage *(cargo)*
Höchstbefrachtung *f* full and down *(maximum load)*
Höchstbelastung *f* maximum [permissible] load
Höchstdruckbereich *m* maximum-pressure range
Höchstdrucklampe *f* maximum-pressure lamp
Höchstdruckschmiermittel *n* extreme-pressure lubricant
Höchstdruckventil *n* pressure-limiting valve
Höchstfrequenz *f* hyper-frequency, extra-high frequency
Höchstfrequenzmassenspektrometer *n* mass synchrometer
Höchstgeschwindigkeit *f* maximum speed
Höchstlast *f* peak load
Höchstleistung *f* maximum (peak) power; available power
verfügbare H. available power; rated capacity
Hochstrahlrohr *n* high-power spraying lance, spray gun
Hochstraße *f* elevated road (highway)
Hochstromgenerator *m* high-current generator
Hochstromkohlebogenlampe *f* high-intensity carbon arc lamp
Hochstromkreis *m* high-current circuit
Hochstromprüfung *f* high-current test
Hochstromschalter *m* heavy-current switch
Hochstromshunt *m* high-current shunt
höchstsiedend boiling above 270 °C
Höchstspannung *f* extra-high voltage (tension)
Höchstspannungsdurchführung *f* extra-high voltage bushing
Höchstspannungsgenerator *m* very high voltage (tension) generator
Höchstspannungskondensator *m* extra-high voltage capacitor
Höchstspannungsmeßwandler *m* extra-high voltage instrument transformer
Höchstspannungsnetz *n* extra-high voltage system, *(Am)* supergrid
Höchstspannungsschalter *m* extra-high voltage circuit-breaker
Höchststrom *m* maximum current
Höchststromauslöser *m* ultimate-limit switch
Höchststromrelais *n* overload relay
Höchststromschalter *m* maximum current circuit-breaker
Höchstvakuum *n* extra-high vacuum
Höchstvakuumröhre *f* extra-high vacuum tube
Höchstverzugsstreckwerk *n (text)* drawing frame for maximum draft
Höchstwert *m* maximum value; peak value
Höchstwertanzeiger *m* maximum impulse indicator
Höchstwertzeiger *m* peak modulation meter
Höchstzug *m* maximum tension *(of a line)*
Höchstzuglast *f* maximum permissible tractive power
höchstzulässig maximum permissible (allowable)
Hochtakelung *f (ship)* Bermuda rig
Hochtank *m* gravity (deep) tank
hochtemperaturbeständig high-temperature [resistant]
Hochtemperaturdestillation *f* high-temperature distillation
Hochtemperaturelement *n* fused electrolyte cell
Hochtemperaturfärbung *f* high-temperature dyeing
Hochtemperaturkoks *m* high-temperature coke
Hochtemperaturreaktor *m* high-temperature reactor

Hochtemperaturverkokung *f* high-temperature coking
Hochtemperaturzone *f* zone of high temperature
Hochton *m* high-pitched tone
Hochtonblende *f (el)* treble control
Hochtonlautsprecher *m* tweeter, treble loudspeaker
Hochtonwiedergabe *f* high-note response
Hochtrainieren *n* scragging *(spring)*
hochtransformieren to step up *(voltage)*
Hoch- und Niederdruckturbine *f* **mit entgegengerichteter Dampfströmung** double-flow turbine
Hoch- und Tieffachjacquardmaschine *f (text)* centre shed Jacquard machine
Hochvakuum *n* high (hard) vacuum
Hochvakuumbedampfung *f (met)* high-vacuum evaporation
Hochvakuumbedampfungsapparatur *f (met)* high-vacuum coating plant
Hochvakuumerzeuger *m* high-vacuum pumping system
Hochvakuumgleichrichter *m* high-vacuum [thermionic] rectifier
Hochvakuumgleichrichterröhre *f* high-vacuum rectifier valve
Hochvakuumlampe *f* hard-vacuum lamp
Hochvakuummeßgerät *n* high-vacuum measuring instrument
Hochvakuumofen *m* high-vacuum [melting] furnace
Hochvakuumöldiffusionspumpe *f* high-vacuum oil diffusion pump
Hochvakuumpumpe *f* high-vacuum pump
Hochvakuumröhre *f* high-vacuum tube
Hochvakuumventil *n* high-vacuum valve
hochveredelt high-grade finished
Hochveredelung *f* high-grade finish[ing]
Hochverzahnung *f* full-depth teeth
Hochverzugsstreckwerk *n (text)* high-draft mechanism
hochviskos highly viscous
Hochvoltbeschleuniger *m* high-voltage accelerator
Hochvolttherapie *f* megavoltage therapy, supervoltage [X-ray] therapy
Hochvoltwicklung *f* high-voltage winding
Hochwasser *n* high tide; flood[ing], high water, floodwater
niedriges H. second high water
Hochwasserablagerung *f* flood plain deposit
Hochwasserdeich *m* flood protection dam
Hochwasserentlastung *f* flood discharge
Hochwasserkatastrophe *f* flood disaster
Hochwassermarke *f* high-water mark
Hochwasserregelung *f* flood control
Hochwasserschaden *m* flood damage
Hochwasserschutz *m* flood protection
Hochwasserschutzdeich *m* flood[-control] levee
Hochwasserstand *m* height of high water, high-water level
Hochwasserstillstand *m* high-water stand
hochwertig high-grade
hochwinden to heave, to lift, to raise up, to hoist
Hochzeit *f* double *(printing)*
Hochzucht *f* pedigree breeding
Hockendrusch *m* heap (sheaf) threshing
Höcker *m* hump *(of a curve)*
unterer H. bottom peak
Höckerfrequenz *f* / **untere** bottom frequency
Höckerspannung *f* peak point voltage
Höckerstrom *m* peak point current
Hockeyschläger *m* hockey stick
Hodograf *m* hodograph, locus
Hof *m* 1. yard; farmstead; 2. *(ph)* halo
Hofwirtschaftsmaschinen und -geräte *pl* mechanical handling and farm transport machinery
Höhe *f* height *(body)*, depth; altitude *(sea level; geometry)*; elevation; peak, summit; *(ship)* siding *(frame)*, offing *(distance)*; head *(liquid)*
H. von / auf der abreast of
H. der größten Erdferne apogee altitude
H. der Schüttung *(therm)* depth of fuel layer, thickness of fire
H. des Apogäums apogee altitude
H. über Normalnull altitude above sea-level
freie H. *(mine)* overwind distance
geometrische H. elevation head
geringe H. low height
hydrostatische H. hydrostatic head
lichte H. clear height, [free] headroom, overhead clearance
lichte H. einer Brücke headroom (clearance) of a bridge
metazentrische H. metacentric height
statische H. static head
Hoheitsgebiet *n* sovereign territory; *(ship)* maritime territory
Hoheitsgewässer *npl* maritime belt, territorial waters
Höhenanhebung *f* high-frequency emphasis
Höhenanzeige *f* altitude indication
Höhenaufnahme *f* levelling
Höhenbestimmung *f* levelling
höhenbeweglich vertically movable
Höhendruckkabine *f* pressurized cabin
höheneinstellbar vertically (height) adjustable
Höheneinstellung *f* 1. *(mach)* vertical adjustment; 2. *(telev)* height control; 3. *(aero)* stabilizer adjustment
Höhenflosse *f (aero)* fixed tail plane
Höhenflugzeug *n* substratoplane
Höhenförderer *m* height elevator
Höhenformel *f* formula for altitude

höhengleich [at the same] level
Höhenindexlibelle *f (opt)* altitude index level
Höhenjagdflugzeug *n* high-altitude fighter
Höhenkabine *f* sealed (pressurized) cabin
Höhenkanal *m* elevated duct *(radio)*
Höhenlehre *f* height gauge
Höhenleitwerk *n (aero)* elevator assembly, horizontal tail surfaces
Höhenlinie *f* contour (level) line
Höhenmaß[stab *m***]** *n* height gauge
Höhenmesser *m* altimeter, altitude gauge
  barometrischer H. aneroid altimeter
  elektrischer H. electric depth finder
  kapazitiver H. electric-capacity altimeter
Höhenmessung *f* height measuring; levelling; measurement of altitude
  barometrische H. barometric levelling (measurement of altitude)
  trigonometrische H. trigonometric height determination
Höhenrakete *f* high-altitude rocket
Höhenreißer *m* vernier height scriber
Höhenrichtfernrohr *n* elevation-tracking telescope
Höhenrichtmaschine *f* elevating mechanism
Höhenrudermaschine *f* elevator servo-motor
Höhenscherbrett *n* extra otter board *(fishery)*
Höhenschichtlinie *f* contour line
Höhenschichtung *f* relief
Höhenschnittpunkt *m* orthocentre
Höhenschreiber *m* altitude recorder
Höhenschritt *m* level, vertical step *(telephony)*
  unbesetzter H. spare level
Höhenschublehre *f* height gauge
Höhenstrahlung *f* cosmic radiation
Höhenstrahlungsintensität *f* cosmic-ray intensity
Höhenumschalter *m* height change-over switch
höhenverkehrt reversed top to bottom
höhenverstellbar *(mach)* vertically adjustable
Höhenverstellung *f* vertical (height) adjustment, elevating control
  H. / mit vertically adjustable, adjustable-height, elevating *(radial drill)*
  H. / ohne non-elevating *(radial drill)*
Höhenwechsel *m* altitude change, change of levels
Höhenzielmarke *f* height marker *(surveying)*
Höhenzuweisung *f* altitude assignment
hohl hollow
Hohlanode *f* tubular (hollow) anode
Hohlbeitel *m* gouge
Hohlblock *m* cavity (hollow) block
Hohlblockstein *m* hollow block
Hohlbohren *n (mach)* trepanning
Hohlbohrer *m (mach)* trepanning cutter; hollow drill; hollow boring bit *(wood)*
Hohlbohrgestänge *n (mine)* hollow drill steel (stem)
Hohlbohrkopf *m (mach)* trepanning [bar cutter] head
Hohlbohrmaschine *f* trepanning lathe *(metal cutting)*
Hohlbohrstahl *m (mine)* hollow drill rod
Hohlbolzen *m* hollow pin
Hohldorn *m (mach)* hollow mandrel
Hohleisen *n* tube (hollowed) iron
Hohlform *f* hollow shape, female form, die
Hohlfräser *m* hollow milling cutter
hohlgebohrt bored through
Hohlgefäß *n* vessel, container
  zylindrisches H. shell
hohlgeschliffen hollow-ground; side-relieved *(metal cutting saw)*
Hohlgewebe *n* tubular fabric
Hohlguß *m* hollow (core) casting
Hohlgußbalken *m* **/ kastenförmiger** box section girder
Hohlgußdeckel *m* double (hollow-cast) cover
Hohlkastenstütze *f (build)* box column
Hohlkastenträger *m (build)* box beam (girder)
Hohlkatode *f* hollow (concave) cathode
Hohlkegel *m* internal taper
Hohlkehle *f* round corner, fillet; *(build)* furrow; channel *(weapon)*
Hohlkehlenkavitation *f* fillet cavitation
Hohlkehlhobel *m* hollowing plane
Hohlkehlnaht *f* concave fillet weld
Hohlkeil *m* saddle key
Hohlkolbenschaltventil *n* hollow-plunger control valve
Hohlkopfgeschoß *n (mil)* hollow-charge projectile
Hohlkörper *m (met)* vessel; hollow part *(machining)*; *(build)* hollow block (brick, tile)
Hohlkörperdecke *f* hollow-tile floor
Hohlkugel *f* hollow sphere
Hohlkugelresonator *m (el)* spherical resonator
Hohlladung *f (mine)* shaped charge *(shotfiring)*
Hohlladungsgeschoß *n (mil)* hollow-charge projectile
Hohlladungsmunition *f* hollow-charge ammunition
Hohlleiter *m (el)* [tubular] waveguide
  H. mit axialer Scheidewand septate waveguide
  H. mit Längssteg ridge waveguide
  angepaßter H. matched waveguide
  geschränkter H. twisted waveguide
  homogener H. uniform waveguide
  rechteckiger H. rectangular waveguide
  runder H. circular waveguide
  sich erweiternder H. tapered waveguide
  tordierter H. twisted waveguide
  zylindrischer H. cylindrical waveguide
Hohlleiterabschluß *m* waveguide terminal
Hohlleiterachse *f* axis of a waveguide
Hohlleiteranpaßglied *n* non-dissipative stub

Hohlleiterblende *f* waveguide shutter
Hohlleiterbolzen *m* waveguide post
Hohlleiterserienverzweigung *f* waveguide hybrid junction
Hohlleitertransformator *m* / **einstellbarer** waveguide tuner
Hohlleiterübergang *m* waveguide junction
Hohlleiterwendelstück *n* twist
Hohlleitung *f* hollow waveguide
Hohlmandrill *m (met)* hollow mandrel
Hohlmaß *n* capacity measure; liquid measure
Hohlmauer *f* cavity wall
  zweischalige H. two-leaf cavity wall
Hohlmauerwerk *n* cavity wall
  zweischaliges H. two-leaf cavity wall
Hohlmeißel *m* gouge
Hohlnadel *f* seed; cannula
Hohlnaht *f* concave weld
Hohlniet *m* hollow rivet
Hohlplatte *f* core[d] panel, core[d] slab
hohlprägen to emboss
Hohlprisma *n* hollow [glass] prism
Hohlrad *n (mach)* internal gear
Hohlraum *m* cavity; void; flaw
  H. eines Gesenks impressions of a die, die cavity
  abgelöster H. *(ship)* separated cavity
  abschwimmender H. *(ship)* transient (travelling) cavity
  anliegender H. *(ship)* attached cavity
  aufblitzender H. *(ship)* flashing cavity
  intermittierender H. *(ship)* intermittent cavity
  stetiger H. *(ship)* steady cavity
  unstetiger H. *(ship)* unsteady cavity
Hohlraumbildung *f* cavitation *(fluid flow)*
Hohlraumfrequenzmesser *m (el)* cavity frequency meter
Hohlraumgeschoß *n* hollow-charge projectile
Hohlraumgitter *n* resonator grid
Hohlraumionisation *f* cavity ionization
Hohlraumionisationskammer *f* cavity ionization chamber
Hohlraumisolation *f* air-space insulation
Hohlraumkammer *f* cavity ionization chamber
Hohlraumresonator *m (nucl)* cavity resonator
Hohlraumschießen *n (mine)* cushioned (air-space) blasting
Hohlraumschwingkreis *m* closed resonator
Hohlraumstrahlung *f* black-body radiation
Hohlraumwellenmesser *m* cavity frequency meter
Hohlringladung *f* hollow-ring charge
Hohlrohr *n* hollow tube
Hohlrohrabschwächer *m* cut-off attenuator
Hohlrohrleiter *m s.* Hohlleiter
Hohlrohrleitung *f s.* 1. Hohlleiter; 2. Hohlleitung
Hohlroststab *m* hollow grate bar
  H. mit Luftkühlung hollow air-cooled grate bar
  H. mit Wasserkühlung hollow water-cooled grate bar
Hohlschliff *m* hollow grinding; side clearance *(metal slitting saw)*
Hohlschraubenverbindung *f* banjo fitting
Hohlseil *n* hollow [stranded] conductor
Hohlseilleitung *f* hollow-conductor line
Hohlsog *m* cavitation *(fluid flow)*
Hohlspiegel *m* concave mirror
Hohlspindel *f (mach)* hollow spindle (mandrel)
Hohlspindeldrehmaschine *f* hollow-mandrel lathe
Hohlspindellagerung / mit carried by a hollow spindle
Hohlstein *m* hollow block (brick, tile)
hohlstielig calamiferous
Hohlstützer *m* hollow supporting insulator
Hohltafel *f* cored panel
Hohltoroid *n s.* Ringröhre
Hohlwalze *f* hollow roll
Hohlwalzwerk *n* rotary piercing mill
Hohlwelle *f* hollow shaft
Hohlwirbel *m* hollow vortex *(fluid flow)*
Hohlziegel *m* cavity brick, hollow clay building block
Hohlzylinder *m* hollow cylinder
Holebug *m (ship)* good (fetching) board
Holländer *m* hollander, beating engine (machine), beater
Holländereinsatz *m* beater filling
Holländereintrag *m* beater loading
Holländerfüllung *f* furnish *(beater)*
Holländergeschirr *n* beater tub
Holländergrundwerk *n* beater [bed-]plate, bed knives
Holländerkropf *m* breasting of the beater
Holländerleim *m* beater glue
Holländermesser *n* beater blade (knife)
Holländerraum *m* beater room
Holländersaal *m* beater house
Holländertrog *m* beater trough (tub)
Holländerwalze *f* beater roll
  bänderlose H. bandless beater roll
Holländerzusatz *m* beater additive
Höllenstein *m* lunar caustic, silver nitrate
Höllensteinstift *m* silver nitrate stick
Holm *m* 1. *(build)* waler, waling *(trench timbering)*; capping (cross) beam, runner; 2. *(aero)* stringer, spar; 3. bar *(parallel bars)*
Holmdurchbiegung *f (aero)* deflection of the spar
Holmfräser *m* spar cutter
Holmium *n* holmium
Holoeder *n* holohedron
holoedrisch holohedral

Holundermarkelektroskop *n* pith-ball electroscope
Holundermarkkügelchen *n* pith ball
Holz *n* wood; timber, *(Am)* lumber
 H. für Streichholzherstellung match wood
 H. geringer Güte inferior wood
 astreines H. clean (clear) wood
 aufbereitetes H. processed wood
 ausländisches H. imported wood
 behauenes H. hewn[-squared] timber
 einheimisches H. native wood
 frisches H. green wood
 gebogenes H. bent wood
 gebündeltes H. packaged timber
 geringwertiges H. low rate timber
 gespundetes H. match board
 gestreiftes H. ribbon-grained wood
 halbrundes H. halfround wood
 inländisches H. home-grown wood
 krummfaseriges H. dwarf wood
 langschäftiges H. log clean grown
 lufttrockenes H. air-seasoned timber, air-dried timber
 maseriges H. spottet wood
 wahnkantiges H. back-sided timber
Holzabfälle *mpl* wood refuse, abatement
Holzabfallplatte *f* wood waste board
Holzabfuhr *f* hauling of wood, log hauls
Holzalkohol *m s.* Holzspiritus
Holzarbeiter *m* wood worker
holzartig ligneous, woody
Holzasbest *m* ligneous asbestos
Holzaufschluß *m* pulping process *(papermaking)*
 alkalischer H. alkaline pulping process
Holzausbau *m* timbering
Holzbalken *m* wooden (timber) beam
Holzbalkendecke *f* wooden (joist, timber beam) floor
Holzbau *m* timber construction (engineering)
Holzbauteile *npl (build)* joinery
holzbearbeitend wood-working
Holzbearbeitungsdrehmaschine *f* wood-working lathe
Holzbearbeitungsmaschine *f* wood-working machine
Holzbehandlung *f* / **vorformende** procession of wood
Holzbeize *f* wood stain
Holzbettung *f* **unter Deckshilfsmaschinei** *(ship)* machinery chock
Holzbiegeform *f* wood bending template
Holzboden *m* 1. wooden floor; 2. *(ship)* scrieve board
Holzbohle *f* deal
Holzbohrarbeiten *fpl* wood boring
Holzbohrautomat *m* automatic wood boring machine
Holzbohren *n* wood drilling
Holzbohrer *m* machine bit for wood, wood boring bit; auger, gimlet
Holzbottich *m* wooden vat
Holzbrett *n* wooden board
Holzbrikettmaschine *f* press-to-logs machine
Holzdeck *n* laid (wood) deck
Holzdecksladung *f* timber deck-load, timber deck-cargo
Holzdeckslast *f* timber deck-load, timber deck-cargo
Holzdielenfußboden *m* boarded floor
Holzdraht *m* match wood
Holzdrahtabschlagmaschine *f* splint chopping machine
Holzdrahthobelmaschine *f* splint shaping machine
Holzdrahttrockentrommel *f* splint drying drum
Holzdrehmaschine *f* wood-turning lathe
Holzdübel *m* timber connector; dowel, nog, plug, key
Holzeinlegearbeit *f* marquetry, intarsia
Holzentrindung *f* / **chemische** chemical debarking
hölzern wooden
Holzessig *m* pyroligneous acid
Holzfahrt *f* timber trade
Holzfällen *n* wood-cutting
Holzfäller *m* wood-cutter, lumberman
Holzfaser *f* grain; wood fibre
Holzfaserbauplatte *f* wood fibre building board
Holzfaserbruch *m* woody (fibrous) fracture
Holzfaserplatte *f* wood fibreboard
Holzfaserstoff *m* lignocellulose
Holzfeile *f* rasping file
Holzfelge *f* wood felloe
Holzfender *m* spar (rod) fender, bumper
Holzfeuerung *f* wood fired furnace
Holzfloß *n* log raft
Holzfräser *m* shaper, milling cutter for wood
holzfrei wood-free
Holzgeist *m* wood spirit
Holzgewebe *n* texture of wood
Holzhammer *m* mallet
Holzhaue *f* adze
Holzhaus *n* wooden house
Holzhobelmaschine *f* wood planer (planing machine)
Holzhof *m* timber wharf
Holzhorde *f* wooden grid (hurdle)
Holzkantenbestoßmaschine *f* wood trimmer
Holzkasten *m (mine)* clog pack, cog, pigsty
Holzkitt *m* joiner's putty
Holzkohle *f* charcoal
Holzkohlenfrischverfahren *n* charcoal hearth process
Holzkohlengasgenerator *m* charcoal gas generator
Holzkohlenroheisen *n* charcoal pig iron

Holzkonservierung *f* wood preserving
Holzkonservierungsmittel *n* wood preservative
Holzkonstruktion *f* timber construction
Holzkübel *m* vat
Holzladefähigkeit *f* timber capacity
Holzladelinie *f* timber load line
Holzlager *n* timber wharf (yard)
Holzlastladelinie *f* timber load line
Holzlehre *f* wooden template *(for the bolt hole)*
Holzmantel *m* wooden lagging
Holzmaserung *f* grain, figure
Holzmaß *n* scantling gauge
Holzmassenaufnahme *f* measurement of standing timber
Holzmast *m* [wooden] pole
Holzmastleitung *f* [wooden] pole line
Holzmehl *n* wood flour
Holznagel *m (build)* wooden peg
Holznagelbinder *m* nailed timber truss
Holznagelmaschine *f* wood pegging machine
Holzöl *n* wood oil
Holzpappe *f* pulp board
  weiße H. mechanical pulp board
Holzpfahl *m* timber pile
Holzpfeiler *m (mine)* pigsty, cog
Holzpflaster *n* wooden (block wood) pavement
Holzpflock *m* nog, peg
Holzplatz *m* timber yard
Holzpolitur *f* wood polish (finish)
Holzputzen *n* preparation of logs
Holzputzerei *f* wood room
Holzreifen *m* wood rim
Holzreifenbiegemaschine *f* wood rim bending machine
Holzriß *m* shake, split, check *(of wood)*
Holzrücker *m* skidder
Holzruß *m* wood soot
Holzschalung *f* timber formwork (shuttering)
Holzschiff *n* wood[en] ship
Holzschindel *f* wooden shingle
Holzschleifer *m* skidder
Holzschleifmaschine *f* stuff grinder
Holzschliff *m* ground wood, wood (mechanical) pulp
  brauner H. brown mechanical wood pulp
Holzschliffbleiche *f* pulp bleaching
Holzschliffpappe *f* wood board
Holzschliffverfahren *n* / **chemisches** chemi-groundwood process
Holzschnitt *m* woodcut, wood engraving
Holzschott *n* wooden bulkhead; solid log bulkhead
  diagonal gebautes H. diagonal bulkhead
Holzschrank *m (mine)* clog pack, cog
Holzschraube *f* wood (drive) screw
Holzschuhherstellung *f* manufacture of wooden shoes
Holzschute *f* / **selbstkippende** *(ship)* self-tipping log barge
Holzschutz *m* timber preservation
Holzschutzbühne *f (mine)* timber pentice
Holzschutzmantel *m* wood lagging
Holzschutzmittel *n* timber preservative
Holzschwamm *m* dry rot
Holzschwelle *f (ry)* timber (wooden) sleeper, *(Am)* tie
Holzspan *m* wood shaving
Holzspant *n* timber; futtock [frame]
Holzspielwarenfabrik *f* wooden toy factory
Holzspiritus *m* wood (methyl) alcohol, methanol
Holzspule *f (text)* wooden bobbin
Holzspurlatte *f* wooden guide
Holzstabmaschine *f* slat machine
Holzstapel *m* wood pile
Holzstift *m* nog, peg
Holzstoff *m* mechanical wood pulp
Holzteer *m* wood tar
Holzteich *m* log pond
Holztransport *m* log handling
Holztransportschiff *n* timber (lumber) carrier
Holzunterklotzung *f (ship)* wooden (machinery) chock
Holzverbindung *f* joining of timbers
Holzverkleidung *f* wood lagging
Holzverkohlung *f* carbonization of wood
Holzverschalung *f* wood lagging; timbering, planking
Holzverschlag *m* wooden partition
Holzverschnitt *m (build)* abatement
Holzverzuckerung *f* wood hydrolysis
Holzwerbung *f* harvesting
Holzwolle *f* wood-wool
Holzwoll-Leichtbauplatte *f* wood-wool slab
Holzwollmaschine *f* wood-wool making machine
Holzwollplatte *f* wood-wool board (slab)
Holzzellstoff *m* lignocellulose, wood cellulose
Holzzementplatte *f* cement fibre slab
Holzzinnerz *n* fibrous cassiterite
Holzzucker *m* xylose, wood sugar
Homodynempfang *m* homodyne reception
Homodynempfänger *m* homodyne receiver
Homoeder *n* homohedron
homogen homogeneous
homogenisieren to homogenize
Homogenität *f* homogeneity
Homogenitätstest *m* test of homogeneity
homolog homologous
Homologes *n (ch)* homologue
homöopolar homopolar
homozentrisch homocentric
homozyklisch homocyclic
Honahle *f* honing tool
  sechssteinige H. honing tool with six sticks

honen to hone
Honen *n* honing
Honigaromaöl *n* honey scent oil
honiggelb honey-yellow
Honigmann-Schachtbohrverfahren *n (mine)* Honigmann method
Honigschleuder *f* honey centrifuge (separator)
Honigsteinsäure *f* mellitic acid
Honigwabenbaustoff *m* honeycomb
Honigwabenspule *f (el)* honeycomb coil
Honmaschine *f* honing machine
Honstein *m* honing stone, abrasive stone for honing
Honsteinhalter *m* honing stone holder
Honvorrichtung *f* honing fixture
Hopfen *m* hop[s]
Hopfendarre *f* hop drying kiln; hop oast
Hopfenpflückmaschine *f* hop picking machine
Hopfenspritze *f* hop sprayer
Hopfenstange *f* hop pole
Hopfentrocknungsgerät *n* hop drier
Hopfenwäscher *m* hop washing machine
Hopper *m (ship)* hopper
Hopperbagger *m* hopper dredger
Hopperklappe *f (ship)* hopper (bottom) door
Hörapparat *m* hearing aid
hörbar audible
Hörbarkeit *f* audibility
Hörbarkeitsbereich *m s.* Hörbereich
Hörbarkeitsfaktor *m* audibility factor
Hörbarkeitsgebiet *n* area of audibility
Hörbarkeitsgrenze *f* limit (threshold) of audibility
  obere H. upper limit of audibility
  untere H. lower limit of audibility
Hörbarkeitszone *f* zone of audibility
Hörbereich *m* range of audibility, audible range
hörbereit sein to keep watch *(receiver)*
Horchdienst *m* interception service *(radio)*
Horchgerät *n* sound locator, listening device (set)
Hordentrockner *m* tray drier
Hordenwäscher *m* hurdle washer, hurdle-type scrubber
Hörempfang *m* aural (audio) reception
Hörempfindlichkeit *f* hearing sensitivity
hören to listen, to hear
Hören *n* listening, hearing
Hörer *m* 1. receiver, earpiece *(of telephone)*; 2. listener
Hörergabel *f* cradle of the receiver
Hörergehäuse *n* receiver case (shell)
Hörerschnur *f* receiver cord
Hörfrequenz *f* audio (audible) frequency
Hörgerät *n* hearing aid
Hörgrenze *f* limit of audibility
  untere H. lower audible limit
Hörhilfe *f* hearing aid
Horizont *m* 1. horizon; 2. cyclorama *(theatre)*
  produzierender H. producing horizon
  scheinbarer H. sensible horizon
  sichtbarer H. apparent (visual) horizon
  terrestrischer H. sensible horizon
  wahrer H. rational (true) horizon
horizontal horizontal
Horizontalablenkplatten *fpl* horizontal-deflection plates, x-plates
Horizontalantenne *f* horizontal aerial
Horizontalauflösung *f (telev)* horizontal definition
Horizontalbandsäge *f* horizontal band saw
  H. mit Saugzone horizontal band saw with suction pick-up
Horizontalbandsägemaschine *f* horizontal band-sawing machine
Horizontalbewegung *f* horizontal movement
Horizontalbohreinheit *f* horizontal [way-type] boring unit
Horizontalbohrwerk *n* horizontal boring mill
Horizontaldach *n* horizontal roof
Horizontaldehnung *f (el)* X-expansion
Horizontale *f* horizontal line
Horizontalebene *f* horizontal plane
Horizontalendpentode *f (telev)* line output pentode
Horizontalgatter *n* horizontal saw (deal) frame
Horizontalhobelmaschine *f* horizontal planing machine
Horizontalimpuls *m* horizontal synchronization pulse
Horizontalkamera *f* horizontal camera
Horizontalkammerofen *m* horizontal chamber oven
Horizontalkraft *f* horizontal force
Horizontalkreisklemme *f* horizontal circle clamp
Horizontalofen *m* oven with horizontal flues
Horizontalregelung *f* horizontal centring
Horizontalschnitt *m* horizontal section
Horizontalschub *m* horizontal thrust
Horizontalstrahl *m* horizontal beam
Horizontalsynchronisation *f* horizontal synchronization
Horizontalturbine *f* horizontal turbine, turbine with horizontal shaft
Horizontalverstärkung *f (el)* X-gain
Horizontalwalze *f (met)* longitudinal roll
Horizontalzeilen *fpl* **je Sekunde** *(el)* horizontal repetition rate
Horizontal-Zweiraum-Reproduktionskamera *f* two-room horizontal process camera
Horizontierung *f* levelling
Hörkapsel *f* earphone
Hörmuschel *f* earpiece, earcap, receiver cap
Horn *n (aut)* horn; flare
hornähnlich hornlike
Hornast *m* spike knot
Hornblende *f (min)* hornblende

Hornblendeasbest *m (min)* amphibole
Hornbüchse *f (mine)* horn (slip) socket
Horndruckknopf *m (aut)* horn button
Horndruckring *m (aut)* horn ring
Hörnerableiter *m* horn[-gap] arrester
Hörnerfunkenstrecke *f (el)* horn gap
Hörnerschalter *m* horn[-gap] switch, horn-break switch
Hörnersicherung *f* horn-break fuse
Hornhautmikroskop *n* corneal microscope
Hornkamm *m* horn comb
Hornkontaktschalter *m* horn-break switch
Hornlöffel *m* horn spoon
Hornschiene *f (ry)* wing rail
Hornsignal *n* horn signal
Hornsilber *n (min)* hornsilver, cerargyrite
Hornstein *m (min)* chert
Hornstrahler *m* horn radiator
Hörsaal *m* auditorium, lecture theatre (room)
Hörschärfe *f* acuity of hearing
Hörschärfemesser *m* audiometer
Hörschwelle *f* threshold of audibility (hearing)
H. des Schalldrucks im Ohr minimum audible pressure
H. im freien Schallfeld minimum audible field
Hörspiel *n* radio drama, wireless play
Hör-Sprech-Kopf *m* record-repeat head, recording-playback head
Hör-Sprech-Schalter *m* talk-listen switch
Hörvermögen *n* power of hearing
Hörverstärker *m* audio amplifier
Hörzone *f* audible region
Hosenboje *f* breeches buoy
Hosenrohr *n* Y-pipe
Hosenrutsche *f* breeches chute (shoot)
Hotflue *f (text)* hot flue
HP-Schale *f (build)* hyperbolic paraboloid shell
Hub *m* 1. *(mach)* stroke *(engine)*, travel; lift *(valve)*; throw *(crankshaft)*; 2. *(el)* sweep width
H. / mit veränderlichem variable-stroke
einfacher H. single stroke
Hubänderung *f (mach)* variation of stroke
Hubanfang *m* beginning of stroke
Hubbalkenofen *m* rocker-bar type furnace
Hubbereich *m (el)* sweep [range], sweep width
Hubbewegung *f (mach)* elevating motion; hoist motion *(conveying)*; elevating, lifting *(excavator)*; stroke
Hub-Bohrungs-Verhältnis *n (aut)* stroke-bore ratio
Hubende *n* end of stroke
Hubgerüst *n* lift frame *(conveying)*
Hubgeschwindigkeit *f* hoisting (lifting) speed *(conveying)*
Hubhöhe *f* lift, hoisting height *(conveying)*; valve lift
H. der Ringbank *(text)* lift of the ring rail
Hubkarren *m* lift truck
Hubkasten *m* drop (rising) box
Hubkolbenmotor *m* motor of the reciprocating piston type
Hubkraft *f* hoisting (lifting) power, hoisting (lifting) capacity
Hublader *m* front (fore) loader
Hublage *f (mach)* position of the [ram] stroke
Hublagenänderung *f* alteration of the position of the stroke
Hublagenverstellung *f* adjustment for position of stroke
Hublänge *f* stroke length, length of stroke
Hublängenänderung *f* alteration of the length of stroke
Hublängeneinstellung *f* setting of the length of stroke
Hublängenskala *f* dial indicating the length of stroke, stroke indicator dial
Hublängenverstellung *f* stroke length adjustment
Hubmagnet *m* lifting (crane) magnet
Hubmitte *f* middle of stroke
Hubmotor *m* elevating motor
Hubplattenverfahren *n (build)* lift-slab method, Youtz-Slick method
Hubplattform *f* lifting (elevating) platform
Hubpotentiometer *n* valve-position potentiometer
Hubrad *n* stroke wheel
Hubraum *m* displacement [of piston], cubic capacity, swept volume
Hubraumdruckleitung *f* pressure line of telescopic jack
Hubraumleistung *f* power per unit of displacement
Hubroller *m* lifting truck
Hubsäge *f* power hacksaw
Hubsägeblatt *n* power-hacksaw blade
Hubscheibe *f* crank (main driving) gear
Hubscheibenrad *n* bull gear (wheel)
Hubschrauber *m* helicopter, rotary-wing aircraft
Hubseil *n* hoisting cable (line, rope)
Hubspindel *f (mach)* elevating (elevation) screw; jack screw
Hubspindelmutter *f (mach)* elevating nut
Hubstapler *m* forklift truck
Hubstellung *f (mach)* position of the stroke
Hubtisch *m* lift table
Hubverhältnis *n* deviation ratio
Hubverlegung *f (mach)* stroke positioning
Hubverminderungsrolle *f* reducing pulley
Hubverstellung *f (aut)* stroke adjustment
Hubvolumen *n (aut)* cubic capacity, swept volume, piston displacement
Hubwagen *m* lift truck *(conveying)*

Hubwange *f* crank web
Hubwechsel *m* reversal of stroke
Hubwechselpunkt *m* dead point (centre)
Hubweg *m* **eines Ventils** *(aut)* valve travel
Hubwerk *n* *(mach)* elevating mechanism; hoisting unit (gear)
Hubzahl *f* *(mach)* number of strokes
Hubzahländerung *f* change in the number of strokes [per minute]
Hubzähler *m* stroke counter
Hubzapfen *m* crank pin
Hubzylinder *m* lift cylinder
Huckepackverkehr *m* rail transport of road trailers
Huf *m* hoof
Hufeisen *n* horseshoe
Hufeisenmagnet *m* horseshoe magnet
Hufmesser *n* hoof cutter
Hühnerauslauf *m* chicken run (coop)
Hühnerhaltung *f* poultry keeping (farming)
Hühnerhaus *n* hen house
Hühnerhof *m* poultry-yard
Hühnerstall *m* hen house
Hühnerzucht *f* poultry breeding
Hukboot *n* hocker *(fishing boat)*
Hulebug *m* *(ship)* good board
Hulk *m(f)* hulk *(unrigged hull, used as a floating depot in a harbour)*
Hülle *f* casing; wrapping; sheath; jacket; coma *(of a comet)*; shell *(of electrons)*
  H. des Brennstoffelements [element] can, canning, cladding, jacket, sheath
  H. des Lichtbogens shell (envelope) of arc
  H. des Satelliten skin of the satellite
  abgeschlossene H. *(math)* closure
  bauliche H. *(build)* carcass, carcassing
Hüllenelektron *n* extranuclear (orbital, planetary) electron
Hüllenintegral *n* integral around a closed contour
Hüllenmaterial *n* can (canning, sheathing) material
Hüllenwerkstoff *m* can (canning, sheathing) material
Hüllkurve *f* envelope curve *(of wave)*
Hüllkurvenverzögerung *f* envelope delay
Hüllpapier *n* wrapping paper
Hüllrohr *n* cable duct *(prestressed concrete)*
Hülse *f* 1. tube, case; 2. *(mach)* socket, sleeve, collar, adapter, adaptor; 3. *(ph)* can; 4. *(mine)* bush; 5. chuck *(drill)*; 6. rind
Hülsenabweiser *m* cartridge-case deflector
Hülsenauszieher *m* cartridge-case extractor
Hülsenbewegung *f* *(mach)* quill traverse
Hülsenbrücke *f* bridge of the bolt cover *(rifle)*
Hülsendipol *m* sleeve dipole *(aerial)*
Hülsenfrüchte *fpl* leguminous plants, legumes; pulse foods
Hülsenfruchtgemenge *n* dredge corn
Hülsenfundament *n* socket foundation (base)
Hülsenherstellungsmaschine *f* tube making machine
Hülsenkette *f* sleeve-type chain
Hülsenklebemaschine *f* tube gluing machine
Hülsenlehre *f* ring gauge
Hülsenpuffer *m* *(ry)* sleeve (plunger) buffer
Hülsenreck *n* stationary horizontal bar
Hülsensack *m* spent ammunition bag
Hülsenschneidemaschine *f* tube cutting machine
Hülsenstoß *m* sleeve splice
Hülsenwickelmaschine *f* tube winding machine
Humifizierung *f* humification
Huminsäure *f* humic acid
Humus *m* / **echter** true humus
Humuskohle *f* humic coal
Hundehaus *n* *(ship)* dog house *(small deck erection)*
Hundehütte *f* dog-house *(furnace)*
Hundekurve *f* *(math)* curve of pursuit
hundertfach hundredfold
Hundertfachuntersetzer *m* *(nucl)* ampliscaler
Hundertfachzähler *m* *(nucl)* ampliscaler
Hundert-Grad-Skale *f* centigrade scale
Hungerharke *f* horse rake
Hunt *m* *(mine)* car, dog
Hupe *f* hooter, horn
  elektrische H. electric hooter (horn)
Hürde *f* hurdle
Hürdenlauf *m* hurdle race, hurdles
Hustenmittel *n* cough remedy
Hut *m* / **eiserner** *(geol)* gossan
Hutappretur *f* hat finishing
Hütchenkondensator *m* cup-shaped capacitor, hat-type capacitor
Huth-Kühn-Schaltung *f* tuned-grid tuned-anode oscillator
Hutmanschette *f* hat leather [packing], inverted cup seal, flange leather
Hutmutter *f* acorn nut
Hütte *f* mill, smelting plant
Hüttenaluminium *n* primary-aluminium pig
Hüttenbims *m* foamed (expanded) blast-furnace slag
Hüttendeck *n* poop deck
Hüttenflur *f* casting floor level *(furnace)*
Hüttenkoks *m* metallurgical (blast-furnace) coke
Hüttenkunde *f* [process] metallurgy
Hüttenmann *m* metallurgist; blast-furnace man
hüttenmännisch metallurgical
Hüttenprozeß *m* metallurgical process
Hüttentechnik *f* metallurgical engineering
Hüttenwerk *n* smelting plant
Hüttenwesen *n* metallurgy, metallurgical engineering
Hüttenzement *m* slag cement
Huygens-Okular *n* Huygens' eyepiece
HV-Naht *f s.* Halb-V-Naht

HV-Schraube *f* high-strength [friction grip] bolt, high-tensile [grip] bolt, high-tensile prestressed bolt, friction grip bolt
HV-Schraubenverbindung *f* friction-grip bolting, high-tensile bolting; high-tensile bolted structural joint
HV-Verbindung *f s.* HV-Schraubenverbindung
Hyalit *m (min)* hyalite
Hyazinth *m* hyacinth, jacinth
Hybridbaustein *m* hybrid unit *(electronics)*
Hybride *f* hybrid; hybrid junction
Hybridekoppler *m (el)* hybrid coupler
Hybridisation *f (ag)* hybridization, cross-breeding
hybridisieren *(ag)* to hybridize, to cross-breed
Hybridrechenelement *n* hybrid computing element
Hybridrechner *m* hybrid computer
Hybridrohöl *n* hybrid-base petroleum, mixed base oil
Hybridschaltung *f* hybrid circuit *(electronics)*
Hybridtransformator *m* hybrid transformer
Hydantoinsäure *f* hydantoic acid
Hydrant *m* hydrant
Hydrargillit *m (min)* gibbsite, hydrargillite
Hydrastoffauflöser *m* hydrapulper *(papermaking)*
Hydrat *n* hydrate
Hydratation *f (build)* hydration *(cement)*, aquation
Hydratationswärme *f* heat of hydration
hydratisieren to hydrate
hydratisiert hydrated
Hydraulik *f* 1. hydraulics; 2. hydraulic transmission, hydraulic circuit [of a machine]
Hydraulikaggregat *n* pump and motor unit, power pack
Hydraulikflüssigkeit *f* hydraulic fluid
Hydraulikingenieur *m* hydraulic engineer
Hydraulikleitung *f* hydraulic main (line)
Hydraulikmedium *n* fluid medium
Hydraulikmotor *m* hydraulic motor
Hydrauliköl *n* hydraulic oil
Hydraulikölleitung *f* fluid power line
Hydraulikpumpe *f* hydraulic pump
  umlaufende H. rotary hydraulic pump
Hydraulikpumpengehäuse *n* hydraulic pump housing
Hydraulikrohrverbindung *f* hydraulic pipe coupling
Hydraulikventil *n* hydraulic valve
Hydraulikventilkörper *m* hydraulic valve body
hydraulisch hydraulic
  h. betätigt hydraulically operated (actuated)
Hydrazid *n* hydrazide
Hydrazin *n* hydrazine
Hydrazobenzol *n* hydrazobenzene
Hydrid *n* hydride
hydrieren to hydrogenize, to hydrogenate
hydriert hydrogenated
Hydrierung *f* hydrogenation
hydroaromatisch hydroaromatic
Hydrochinon *n* hydroquinone
Hydrochlorierung *f* hydrochlorination
Hydrodynamik *f* hydrodynamics
hydrodynamisch hydrodynamic
hydroelektrisch hydroelectric, electrohydraulic
Hydroforming *n* hydroforming process
Hydrogensulfat *n* hydrosulphate
Hydrogensulfit *n* hydrosulphite
Hydrogeologie *f* geohydrology
Hydrokinetik *f* hydrokinetics
Hydrokolbenmotor *m* piston-type motor
Hydrokultur *f* hydroponics, soilless culture
Hydrolyse *f* hydrolysis
hydrolysieren to hydrolyze
Hydrolysierungsfähigkeit *f* power of hydrolysis
Hydromagnetik *f* hydromagnetics
Hydromechanik *f* hydromechanics, fluid mechanics
Hydrometer *n* hydrometer, acid tester
Hydromonitor *m* giant
Hydromotor *m* hydraulic (fluid power) motor
  H. für geradlinige Bewegung linear fluid (hydraulic) motor
  H. für kontinuierlich drehende Bewegung continuous hydraulic motor
  H. mit geradliniger Bewegung non-continuous hydraulic motor
Hydromotorantrieb *m (ship)* jet propulsion
Hydromühle *f* hydro-mill, conical refiner, hydrorefiner
Hydroniumion *n* hydronium ion
hydrophil hydrophilic
Hydroponik *f* hydroponics, soilless culture
Hydroskimmer *m (ship)* hydroskimmer
Hydrosol *n* hydrosol
Hydrostatik *f* hydrostatics
hydrostatisch hydrostatic
Hydroxamsäure *f* hydroxamic acid
Hydroxid *n* hydroxide
Hydroxylgruppe *f* hydroxyl group
Hydroxylion *n* hydroxyl ion
Hydrozimtsäure *f* hydrocinnamic acid
Hygrometer *n* hygrometer
hygroskopisch hygroscopic
Hyperbel *f* hyperbola
  gleichseitige H. equilateral hyperbola
Hyperbelbahn *f* hyperbolic path (orbit)
Hyperbelbewegung *f* hyperbolic motion
Hyperbelfunktion *f* hyperbolic function
Hyperbelortung *f* hyperbolic position finding
Hyperbelrad *n* hyperbolical wheel
hyperbolisch hyperbolic
Hyperboloid *n* hyperboloid
  zweischaliges H. parted hyperboloid
Hyperboloidrad *n* hyperboloid gear

Hyperbolparaboloid *n (build)* hyperbolic paraboloid
hypereutektisch hypereutectic
Hyperfeinstruktur *f* hyperfine structure
Hyperfragment *n (nucl)* hyperfragment
Hyperon *n (nucl)* hyperon
Hypersthen *m* hypersthene
Hypnotikum *n* hypnotic
Hypobromit *n* hypobromite
Hypochlorit *n* hypochlorite
Hypochloritraffination *f* hypochlorite sweetening
Hypoidgetriebe *n* hypoid gearing
Hypoidkegelrad *n* hypoid [bevel] gear
Hypoidöl *n* hypoid oil
Hypoidwälzfräsautomat *m* automatic hypoid generator
Hyponitrit *n* hyponitrite
Hypophosphat *n* hypophosphate
Hypophosphit *n* hypophosphite
Hypotenuse *f* hypotenuse
Hypothese *f* hypothesis
Hypozykloide *f* hypocycloid
  dreispitzige H. tricuspid
  vierspitzige H. astroid, tetracuspid
Hypsometer *n* hypsometer
Hysterese *f* hysteresis
  dielektrische H. dielectric hysteresis
  magnetische H. magnetic hysteresis
Hysteresefaktor *m* hysteresis factor
Hysteresefehler *m* hysteresis error
Hysteresemesser *m* hysteresis meter
Hysteresemotor *m* hysteresis motor
Hystereseschleife *f* hysteresis loop
Hystereseschreiber *m* hysteresigraph
Hystereseverlust *m* hysteresis loss
  differentieller H. incremental hysteresis loss
Hysteresis *f s.* Hysterese
Hysteresiskonstante *f* hysteretic constant
Hysteresismoment *n* hysteretic drag

I

I-Anteil *m* integral action (control factor)
I-Folgeregler *m* type 1 servomechanism
I_2-Folgeregler *m* type 2 servomechanism
I-Glied *n (cont)* integrating element
i-Halbleiter *m* i-type semiconductor
I-Naht *f* open square butt weld
I-Profil *n* I-section
I-Querschnitt *m* I-section
I-Regelung *f* integral control
I-Regler *m* integral [action] controller
I-Stahl *m* I-beam
I-Träger *m* I-beam, I-girder
I-Verhalten *n* integral action
I-Wellrohr *n (cont)* integral bellow
I-Wirkung *f (cont)* integral action
ideal ideal; perfect *(dielectric, fluid)*
Idealbasis *f* ideal basis
Idealbedingung *f* ideal condition
Idealkristall *m* ideal (perfect) crystal
Idealweiß *n (telev)* equal-energy white
identifizieren to identify
Identifizierung *f* identification
Identität *f* identity
Identitätsgesetz *n* identity law
idiochromatisch idiochromatic
idiomorph idiomorphic
idiostatisch idiostatic
Idokras *m* idocrase
Igel *m* 1. *(text)* gill; porcupine; 2. *(mine)* chain hedgehog
Igeln *n* inter-row cultivation
Igelstrecke *f (text)* hedgehog drawing; porcupine drawing frame
Igeltransformator *m* hedgehog transformer
Igeltrommel *f (text)* porcupine cylinder
Ignitron *n* ignitron
Ignitronlokomotive *f* ignitron locomotive
Ikonoskop *n* iconoscope, emitron camera
Ikosaeder *n* icosahedron
Ikositetraeder *n* icositetrahedron
Ilgner-Aggregat *n* Ilgner system
Ilgner-Antrieb *m* Ilgner system
Ilgner-Umformer *m* Ilgner system
Illuminator *m* illuminator
illuminieren to illuminate
Illuminieren *n* **der Ätzung** *(text)* colouring of the discharge
Illustrationsdruckpapier *n* supercalendered (art) paper
Ilmenit *m* ilmenite
Imageikonoskop *n* image iconoscope
Imageorthikon *n* image orthicon
imaginär imaginary
Imaginärteil *m* imaginary component (part)
Imhoffbrunnen *m* Imhoff tank
Imhofftrichter *m* Imhoff cone
Imin *n (ch)* imine
Imitatgarn *n* imitation yarn
Imitatspinnerei *f* spinning of imitation yarn
Imker *m* bee-keeper, bee-master, apiarist
Imkerei *f* bee-keeping
immatrikulieren to enrol, to admit to the university
Immediatanalyse *f* proximate analysis
Immersionsöl *n* immersion oil
Immersionssystem *n* immersion objective
immunisieren to immunize
Immunisierungseffekt *m* immunological effect *(of irradiation)*
Impedanz *f (el)* impedance
  I. bei Parallelresonanz rejector (dynamic) impedance

I. des gegenläufigen Feldes negative-sequence field impedance
akustische I. acoustical impedance
mechanische I. mechanical impedance
negative I. negative impedance
Impedanzanpassung *f* impedance matching
Impedanzanpassungsblende *f* matching plate
Impedanzkorrektor *m* impedance corrector (compensator)
Impedanzkurve *f* impedance characteristic
Impedanzmeßbrücke *f* impedance bridge
Impedanzrelais *n* impedance relay; distance relay
Impedanzschutz *m* impedance protective system; impedance protection
Impedanz-T-Glied *n* series T *(waveguides)*
Impeller *m* impeller
impfen to inoculate
Impfen *n* inoculation, seeding of crystallization; vaccination
Impfkristall *m* seed crystal
Impfschlamm *m* activated sludge *(sewage treatment)*
Impfserum *n* vaccine
Impfstoff *m* vaccine
Impfung *f* inoculation *(crystals)*
Implantationstechnik *f* implantation (interstitial) technique
Implikation *f* [material] implication *(logical operator)*
implizit implicit
implodieren to implode
implosionssicher implosion-proof
Import *m* import
Importholz *n* imported wood
importieren to import
imprägnieren to impregnate, to preserve
Imprägnieren *n* **der Schwellen** impregnation of the sleepers
Imprägnierfähigkeit *f* treating characteristic
Imprägnierfoulard *m (text)* impregnating mangle
Imprägnierharz *n* impregnating resin
Imprägniermittel *n* impregnating matter, saturant, coating varnish, bonding material, stiffening agent
Imprägniertank *m* cable tank
Imprägnierung *f* impregnation, preservation
Imprägnierungsmasse *f* impregnation compound
Imprägnierungsmittel *n* impregnant
Impressum *n* imprint
Imprimatur *n* ready for press
imprimieren to pass for press
Impuls *m* pulse; impulse; count; momentum; mark
I. der Kraft impulse of the force
I. mit langer Laufzeit broad pulse
ankommender I. incoming pulse
ausgesandter I. transmitted pulse
einmaliger I. impulse
einseitig gerichteter I. unidirectional pulse
falscher I. spurious pulse
generalisierter I. generalized momentum
glockenförmiger I. bell-shaped pulse, sine-squared pulse
kurzer I. fast pulse; pip, blip
reflektierter I. reflected pulse
steil ansteigender I. steep-wave-front pulse
steiler I. steep-wave-front pulse
unechter I. spurious count, after pulse
zeitlich definierter I. timed pulse
Impulsabfall *m* pulse decay, tail of the pulse
Impulsabfallzeit *f* pulse decay time
Impulsabstand *m* pulse-to-pulse interval, pulse separation (period, spacing)
Impulsabstiegsversteilerung *f* trailing-edge squaring
Impulsabtrenner *m (el)* pulse clipper
Impulsabtrennung *f* pulse clipping
Impulsamplitude *f* pulse height (amplitude)
effektive I. effective pulse amplitude
Impulsanalysator *m s.* Impulshöhenanalysator
Impulsanregung *f (el)* pulse triggering
Impulsanstiegsversteilerung *f* pulse-front squaring
Impulsanstiegszeit *f* pulse rise time
Impulsantwort *f* impulse response
Impulsart *f* pulse mode
Impulsaufstockung *f* pile-up of pulses
Impulsausleseschaltung *f* impulse separator
Impulsauslösung *f* pulse triggering
Impulsbandbreite *f* pulse bandwidth
Impulsbasis *f* pulse base
Impulsbegrenzer *m* [pulse] clipper, clipping circuit
Impulsbetrieb *m* pulsed operation
Impulsbreite *f* pulse width
Impulsbreitenmodulation *f* pulse duration (width) modulation
Impulsdach *n* horizontal part of the pulse, pulse tilt (top)
Impulsdauer *f* pulse duration, duration of the (a) pulse
mittlere I. mean pulse time
Impulsdemodulator *m* impulse detector
Impulsdichte *f* counting (pulse) rate
Impulsdichtemesser *m* [counting-]rate meter
Impulsecho *n* pulse reply
Impulsentladung *f* pulsed discharge
Impulsentladungsröhre *f* pulse discharge tube
Impulsentschlüsselung *f* decoding *(of pulses)*
Impulsentzerrer *m* pulse regenerator
Impulsentzerrung *f* pulse regeneration (restoration)
Impulserneuerer *m* pulse regenerator
Impulserzeuger *m* pulse generator
Impulserzeugung *f* pulse generation

Impulsfernmeßgerät *n* impulse-type telemeter
Impulsflanke *f* pulse front (edge, flank)
Impulsflankensteilheit *f* pulse steepness (slope), width of transition
Impulsfolge *f* impulse train (repetition)
Impulsfolgefrequenz *f* pulse repetition frequency (rate)
Impulsform *f* pulse shape
Impulsformer *m* pulse-shaping stage (unit), [pulse] shaper
Impulsformerleitung *f* pulse-forming line
Impulsformerstufe *f* pulse-shaping stage, [pulse] shaper
Impulsformung *f* pulse shaping
Impulsfrequenz *f* recurrence (repetition) frequency, recurrence (repetition) rate
Impulsfuß *m* pulse base
Impulsgeber *m* impulse machine (sender)
Impulsgenerator *m* [im]pulse generator, [im]pulser
quarzgesteuerter I. crystal-controlled [im]pulse generator, quartz-controlled [im]pulse generator
Impulsgerät *n* pulser unit
Impulsgleisstromkette *f* pulsating current fed track circuit
Impulshöhe *f* pulse height (amplitude, elevation)
Impulshöhenanalysator *m* pulse height analyzer
Impulshöhenanalyse *f* pulse height analysis; kicksorting technique
Impulshöhenanalysierer *m* kicksorter, pulse height analyzer
Impulshöhendiskriminator *m* pulse height discriminator
Impulshöhenselektor *m* pulse height selector
Impulshöhenspektrum *n* pulse height spectrum
Impulshöhenverteilung *f* pulse height distribution
Impulsinstabilität *f* pulse jitter
Impulsintervall *n* pulse separation
Impulsionisationskammer *f* pulsed ionization chamber
Impulskette *f* pulse-forming network
Impulskode *m* pulse code
Impulskodefernübertragung *f* pulse code telemetering
Impulskolonne *f* pulsed column
Impulskontakt *m* impulse contact
Impulskorrektur *f* pulse stretcher
Impulskreis *m* impulse circuit
Impulskurve *f* pulse curve
Impulslage *f* pulse position
Impulslandeverfahren *n* pulsed-guide path landing system
Impulslängenmodulation *f* pulse-duration modulation
Impulsleistung *f* pulse power
Impulsleistungsverhältnis *n* pulse duty factor
Impulslücke *f* [pulse] space
Impulsmagnetisierung *f* impulse magnetization
Impulsmesser *m* peak indicator
Impulsmodulation *f* pulse modulation
impulsmoduliert pulse modulated, pulsed
Impulsmodus *m* pulse mode
ungewollter I. spurious pulse mode
Impulsneutronenquelle *f* pulsed neutron source
Impulspeiler *m* pulse direction finder
Impulsphasenmodulation *f* pulse-position modulation
Impulsrate *f* counting rate; pulse rate
Impulsraum *m* momentum space
Impulsreaktor *m* pulsed reactor
Impulsregenerierung *f* pulse regeneration
Impulsreihe *f* pulse train (group), train (succession) of pulses
Impulsrelais *n* time pulse relay
Impulssatz *m* law of conservation of momentum
Impulsschalter *m* pulse switch
Impulsschaltung *f* switching circuit
Impulsschreiber *m* pulse recorder
Impulsschwanz *m* pulse tail
Impulsserie *f* pulse train; spike train
Impulssieb *n* impulse separator
Impuls-Spannung-Charakteristik *f* counting response (rate-voltage characteristic), plateau characteristic
Impulsspeicherrelais *n* notching relay
Impulsspektrum *n* pulse spectrum
Impulssperrschaltung *f* pulse gate
Impulsspitze *f* pulse spike
Impulssteuerung *f* pulse control [system]
Impulssynchronisierschaltung *f* clock-pulse gate
impulssynchronisiert impulse governed
Impulstaktgeber *m* clock-pulse generator
Impulstal *n* pulse valley
Impulstasteinrichtung *f* echo box *(radar)*
Impulstastverhältnis *n* pulse duration ratio
Impulstechnik *f* pulse technique
Impulstor *n* pulse gate
Impulstrennung *f* [im]pulse separation
Impulsübergangsfunktion *f* impulse response
Impulsüberlagerung *f* pulse overlap
Impulsüberschneidung *f* pulse overlap
Impulsübertrager *m* impulse repeater
Impuls- und Sprungsignal *n* pulse-and-bar signal
Impulsunterschwingen *n* underthrow distortion
Impulsuntersetzer *m* pulse dividing circuit
I. 1:n count-by-n-circuit
einstufiger I. scale-of-two circuit
Impulsuntersetzerschaltung *f* pulse dividing circuit
Impulsverbesserung *f* pulse restoration
Impulsverbreiterungsschaltung *f* pulse stretching circuit

Impulsverflechtung *f* pulse interleaving
Impulsverformung *f* impulse distortion
Impulsverschleifung *f* degradation of pulse
Impulsverschlüsselung *f* pulse coding
Impulsverstärker *m* pulse amplifier
  linearer I. linear pulse amplifier
Impulsvorwahl *f* preset counts
  I. / mit count-controlled
Impulswahl *f* pulse action; pulse selection
Impulswiederholer *m* pulse responder
Impulszacke *f (el)* spike
Impulszähler *m* [im]pulse counter, [im]pulse summator
Impulszählschaltung *f* counting circuit
Impulszentrale *f* pulser, keyer, synchronizing pulse generator
Impulszyklotron *n* pulsed cyclotron
inaktiv inactive
inaktivieren to inactivate
Inaktivierung *f* deactivation
Inbetriebnahme *f* commissioning, putting into operation, starting, placing into operation (service)
Inbetriebsetzung *f s.* Inbetriebnahme
Inchromieren *n (met)* chromizing
Indaminfarbstoffe *mpl* indamines
Indanthrenfarbstoff *m* indanthrene
Index *m* index
  oberer I. *(math)* superscript
  unterer I. *(math)* subscript
Indexfehler *m* index error
Indexklinke *f* indexing latch
Indexraststift *m* indexing latch, indexing-arm pin
Indexregister *n* B-box, base (index) register *(digital computer)*
Indexstift *m* index pin
Indexzahl *f* index number
Indienststellung *f* commissioning *(of a vessel)*
indifferent indifferent, inert, neutral
Indifferenz *f* immunity *(galvanics)*
Indifferenzzone *f* neutral zone
Indigo *m(n)* indigo
indigoblau indigo-blue
Indigodruck *m* indigo print
Indigokarmin *n* indigo carmine (extract)
Indigolith *m (min)* indigolite, indicolite
Indigopflanze *f* indigo plant, indigofera
Indigotin *n* indigo extract (carmine)
Indigoweiß *n* indigo-white
Indikator *m* tracer; indicator; indicating instrument (recorder); engine gauge
  I. für fortlaufende Diagramme continuous indicator, indicator for continuous diagrams
  I. mit außenliegender Feder indicator with external spring
  I. mit innenliegender Feder indicator with internal spring
  inaktiver I. non-radioactive tracer
  isotoper I. isotopic tracer, tracer isotope
  planimetrierender I. integrating indicator
  radioaktiver I. radioactive tracer, radiotracer
  totalisierender I. integrating indicator
Indikatoranalyse *f* tracer analysis
Indikatoranschluß *m* indicator connection
Indikatoratom *n* tracer [atom]
Indikatorbereich *m* indicator range
Indikatorchemie *f* tracer chemistry
Indikatordiagramm *n* indicator diagram
Indikatorelement *n* tracer element
Indikatorfeder *f* / **doppelt gewundene** double coil spring
  einfach gewundene I. single coil spring
Indikatorhahn *m* indicator cock
Indikatorisotop *n* isotopic tracer, tracer isotope
Indikatorkonzentration *f* tracer concentration
Indikatormethode *f* tracer method
Indikatornocken *m* indicator boss
Indikatorpapier *n* indicator paper
Indikatorschnur *f* indicator cord
Indikatorschraube *f* indicator plug
Indikatorschreibstift *m* indicator pencil
Indikatorverfahren *n* tracer technique (method), labelled atoms method; radioactive tracer method
Indikatorversuch *m* indicator test
Indikatrix *f* indicatrix
indirekt indirect
Indium *n* indium
Indiumchlorid *n* indium chloride
Indiumoxid *n* indium oxide
Individualdosimeter *n (nucl)* personal monitor
Indizes *mpl* **Millersche** Miller[ian] indices
Indizienbeweis *m* circumstantial proof
indizieren to indicate; to index
Indol *n* indole
Indophenol *n* indophenol
Induktanz *f (el)* inductance, inductive reactance
Induktion *f (el)* induction, *(magnetic)* flux density
  elektromagnetische I. electromagnetic induction
  elektrostatische I. electrostatic induction
  gegenseitige I. mutual induction, common reactance
  magnetische I. magnetic induction
  remanente I. residual induction
  zusätzliche I. incremental inductance
induktionsarm low-induction
Induktionsbeschleuniger *m* induction accelerator
Induktionserwärmung *f* induction heating
Induktionsfeld *n* induction field
Induktionsfluß *m* induction flux; magnetic flux
  magnetischer I. magnetic flux
induktionsfrei non-inductive, non-reactive
induktionsgehärtet induction-hardened

Induktionsgenerator *m* induction generator
Induktionsgesetz *n* [Faraday's] law of induction
Induktionshärtung *f* induction hardening
Induktionsheizung *f* induction heating
Induktionskoeffizient *m* factor of induction
Induktionskompaß *m* induction compass
Induktionskreisel *m* induction top
Induktionslötverfahren *n* induction brazing process
Induktionsmaschine *f* induction machine
Induktionsmeßgerät *n* induction meter
Induktionsmeßwerk *n* electrodynamic (induction-type) meter
Induktionsmotor *m* induction motor
  synchronisierter I. synchronous induction motor
Induktionsofen *m* induction furnace
  I. mit Eisenkern core-type induction furnace
  kernloser I. coreless induction furnace
Induktionsschutz *m* anti-inductive protection
Induktionsschweißen *n* induction welding
Induktionsspule *f* induction coil; inductor
Induktionsstörung *f* inductive interference
Induktionsstrom *m* induced current
Induktionstiegelofen *m* / **kernloser** coreless induction furnace
Induktionswirkung *f* inductive action
Induktionszähler *m* electrodynamic (induction motor, Ferraris, induction-type) meter
induktiv inductive
  i. geerdet reactance earthed
  rein i. perfectly inductive
Induktivität *f* inductance
  I. der Zuführungen lead inductance
  gegenseitige I. mutual (common) inductance
  innere I. inner self-inductance
  konzentrierte I. lump inductance
  parallelgeschaltete I. shunt inductor
  regelbare I. [continuously] adjustable inductor
  reine I. pure inductance
  veränderliche I. variable inductance
  vorgeschaltete I. series inductance
Induktivitätsbelag *m* inductance per unit length
Induktivitätsmeßbrücke *f* inductance bridge
Induktivitätsmesser *m* inductance meter
Induktivitätsnormal *n* inductance standard
Induktivitätssteuerung *f* indüctivity control
Induktor *m (el)* inductor
Induktoranruf *m (telec)* magneto (generator) ringing, generator call
Induktorkurbel *f* magneto crank
Induktormaschine *f* inductor (magneto) generator
Induktorruf *m (telec)* magneto (generator) ringing, generator call
Induktorspannung *f* inductor voltage
Induktorstrom *m* inductor (rotor) current
Induktorstromkreis *m* inductor circuit
Induktorwicklung *f* inductor (rotor) winding
Indulin *n* induline
Industrialisation *f* industrialization
industrialisieren to industrialize
Industrialisierung *f* industrialization
Industrie *f* / **chemische** chemical industry
  eisenverarbeitende I. steel-using industry
  elektronische I. electronics industry
  elektrotechnische I. electrical industry
  gerätebauende I. instrument industry
  grobkeramische I. heavy (brick and) clay industry
  metallverarbeitende I. metal processing industry
Industrieabwasser *n* industrial sewage
Industrieanlage *f* industrial plant
Industrieausstellung *f* industrial exhibition
Industriebahn *f* industrial railway, *(Am)* industrial track
Industriebau *m* industrial (factory) building
Industriediamant *m* industrial diamond
Industriedübel *m* timber connector
Industrieformgebung *f* industrial design
Industriefußboden *m* industrial floor[ing]
Industriegebäude *n* industrial (factory) building
Industriehafen *m* industrial harbour, private port
Industrieholz *n* industrial wood
Industriemesse *f* industries fair
Industrieofen *m* industrial furnace
Industriereaktor *m* industrial reactor
Industrieschornstein *m* factory chimney
Industriespektroskop *n* industrial spectroscope
Industrie- und Handelskammer *f* Chamber of Industry and Commerce
Industriezweig *m* branch of industry
induzieren to induce
induziert induced
ineinander übergehen to blend into each other *(shade)*
ineinandergreifen to mesh, to mate
Ineinandergreifen *n* meshing, mating
ineinandergreifend inter-engaging, interlocking
ineinanderpassen to mate
ineinanderschachteln to nest
Ineinanderschachtelung *f* nest
inert *(ch)* inert
Inertstaub *m* stone dust
Infanteriebegleitpanzer *m* infantry tank
Infanterieflugzeug *n* army aircraft
infinitesimal *(math)* infinitesimal
Infinitesimalgröße *f (math)* infinitesimal
Influenz *f (el)* electrostatic induction, influence
Influenzfeld *n* electrostatic field
Influenzkonstante *f* influence (absolute permittivity) constant
Influenzmaschine *f* electrostatic generator, influence (induction, electrostatic) machine

Information *f (el)* information
Informationsaufnahme *f* / **automatische** automatic information pick-up
binäre I. binary information
numerische I. numeric information
Informationsbelag *m* information content
Informationsbeschaffung *f* information retrieval
Informationsbild *n* information pattern
Informationsbit *n* information bit
Informationsdarstellung *f* machine language *(for data processing machines)*
Informationsdichte *f* packing density
Informationseinheit *f* information unit, item
Informationselement *n* bit
Informationsentropie *f* negative entropy, negentropy
Informationsfluß *m* information rate *(computer)*
Informationsgehalt *m* information content
Informationsgeschwindigkeit *f* information rate
Informationsrecherche *f* document (information) searching
Informationsspeicher *m* memory
Informationsspeicherungsdichte *f* information storage density
Informationsstand *m* information booth *(fair)*
Informationstechnik *f* information technique
Informationstheorie *f* information theory
Informationsverarbeitung *f* data (information) processing
Informationsverarbeitungszelle *f* artificial neuron, information processing cell
Informationswort *n* information word
infraakustisch infra-acoustic, sub-audio
Infrafernsprechfrequenz *f* sub-telephone frequency, sub-audio frequency
Infrarot *n* infrared
Infrarotaufklärung *f (mil)* infrared reconnaissance
Infrarotfilter *n* infrared filter
Infrarotfotografie *f* infra-red photography
Infrarotheizung *f* infra-red heating
Infrarotlampe *f* infra-red lamp
Infrarotspektralfotometer *n* infrared spectrophotometer
Infrarotstrahler *m* infra-red radiator
Infrarotstrahlung *f* infra-red radiation
Infrarotstrahlungsheizung *f* heating by infrared radiation
Infrarottrockung *f* infrared drying
Infrarot-Zielsuchgerät *n* infrared homing system
Infraschall *m* infrasound
Infraschallfrequenz *f* infra-acoustic frequency, sub-audio frequency
Infusorienerde *f* infusorial earth, guhr
Ingenieur *m* engineer
I. für Städtebau municipal engineer
I. mit Hochschulbildung professional (academically trained) engineer
beratender I. consulting engineer
Ingenieurbau *m* civil engineering
konstruktiver I. structural engineering
Ingenieurbauten *mpl* [civil] engineering structures
Ingenieurbauwerk *n* [civil] engineering structure
Ingenieurchemiker *m* chemical engineer
Ingenieurhochbau *m* structural engineering
Ingenieurholzbau *m* timber engineering
Ingenieurschule *f* technical college
ingenieurtechnisch engineering
Ingweröl *n* ginger oil
Inhalation *f* inhalation
Inhalt *m* contents; *(ship)* capacity; *(math)* area [of surface]; content, volume, capacity
I. des Hohlraumes cavity content
inhibieren to inhibit
Inhibitor *m (ch)* inhibitor, stopping agent
Inhibitschaltung *f (cont)* inhibition circuit
inhomogen heterogeneous; *(math)* inhomogeneous
Inhomogenität *f* inhomogeneity
Initial[e *f*] *n* initial
Initialsprengstoff *m* initial detonating agent
Initialstrahlung *f* initial radiation of explosion
Initialtemperatur *f* initial temperature
Initialzünder *m* initiator, detonator
initiieren to initiate
Initiierung *f* ignition
Injektion *f s.* Einschuß
Injektionsspritze *f* syringe
Injektionsverfahren *n (build)* pressure grouting *(cement)*
Injektor *m* [feed] injector
I. mit einem Düsensatz single tube injector, injector with one set of cones (nozzles)
I. mit zufließendem Wasser non-lifting injector
nichtsaugender I. non-lifting injector
selbstanziehender I. re-starting injector
wieder ansaugender I. re-starting injector
injizieren to inject
Inklination *f (ph)* dip, inclination
Inklinationsmesser *m* inclinometer
Inklinationsnadel *f* dip needle, inclinometer
Inklinometer *n* inclinometer
inkohärent incoherent
Inkohlung *f* carbonization, mineralization, coalification
inkompressibel incompressible
Inkonstanz *f* instability
I. der Frequenz frequency sliding
Inkorporation *f (nucl)* incorporation
Inkorporieren *n (nucl)* incorporation
Inkorporierung *f (nucl)* incorporation
Inkrafttreten *n* coming into force *(law)*
Inkreis *m* inscribed circle
Inkreismittelpunkt *m (math)* incentre
Inkreisradius *m (math)* apothem, inradius
Inkrement *n* increase, increment

I. der Rollschwingung increment of roll
Inkrustierung *f* incrustation
Inkrustierungsmittel *n* incrustifier, seed protectant
Inkubationszeit *f* incubation period
Inkubator *m* incubator
Inlandschiff *n* inland (home trading) vessel, home trade ship
Inlandsfahrt *f* home (inland) trade
Innenabmessung *f* internal dimension
Innenansicht *f* interior view
Innenanstrich *m* interior (inside) painting
Innenanstrichfarbe *f* interior paint
Innenantenne *f* indoor aerial
Innenarchitektur *f* interior design
Innenausbau *m* interior work, internal finish
Innenbackenbremse *f* internal[-expanding] shoe brake
Innenballast *m (ship)* internal (inside) ballast
Innenbeleuchtung *f* interior lighting
Innenblech *n* inside panel
Innenboden *m* **eines Schiffes** inner-bottom plating [of a vessel], tank-top plating [of a vessel]
Innenbodenwrange *f* floor rider *(wooden ship)*
Innenbordmotor *m* inboard motor
innenbords inboard
Innenböschung *f* inside slope
Innenbremsbacke *f* internal shoe
Innendrehen *n* turning internal surfaces
Innendrehmeißel *m* boring (internal turning) tool
Innendruck *m* internal pressure
Innendurchmesser *m* interior diameter
Innenfedertaster *m* inside spring-joint caliper
Innenfeuerung *f* internal furnace
Innenfläche *f* internal surface
I. der Schienen gauge side of the rails
Innenflächenfräsen *n* internal milling
Innenfokussiereinrichtung *f* inside focusing device
Innenfräsen *n* internal milling
Innenführung *f* centre guide *(silent chain)*
Innengestaltung *f* interior decoration (design)
Innengewinde *n* internal (female) thread
I. / mit tapped, internally threaded
Innengewindemeißel *m* internal screw-cutting tool
Innengewindeschleifen *n* internal thread grinding
Innengewindeschneidarbeit *f* internal threading operation
Innengewindeschneiden *n* cutting internal threads, internal threading
Innengewindestrehler *m* internal-thread chaser
Innenhandbremse *f* inside hand brake
Innenhof *m* patio, garth
Inneninstallation *f* indoor wiring (piping)
Innenkabel *n* inside cable
Innenkante *f* inside edge
Innenkegel *m* internal (female) taper
Innenkiel *m* inner keel, keel batten, lay board, hog piece
Innenklüver *m (ship)* inner jib
Innenkopiereinrichtung *f (mach)* internal copying attachment
Innenkorrosion *f* internal corrosion
Innenkurzgewinde *n* internal short thread
Innenlager *n* inner bearing
Innenläppmaschine *f* internal lapping machine
Innenleiter *m* inside (inner, core) conductor
Innenmaß *n* internal dimension
innenmattiert [internally] frosted *(bulbs)*
Innenmessung *f* internal measurement
Innennachformeinrichtung *f* internal copying attachment
Innenperspektive *f* interior view
Innenpolgenerator *m* rotating-field-type alternator, inner-pole alternator
Innenpolmaschine *f* internal pole machine, inner-pole-type alternator, external armature alternator
Innenpolwechselstromgenerator *m* internal field alternator
Innenprofil *n* internal profile
Innenputz *m* interior plastering
Innenrad *n* internal gear
Innenraum *m* interior space
Innenräumarbeit *f* internal broaching operation
Innenräumen *n* internal broaching
Innenräummaschine *f* internal broaching machine
Innenraumstützer *m* indoor supporting insulator
Innenraumstützisolator *m s.* Innenraumstützer
Innenraumtherapie *f* intracavitary therapy
Innenräumwerkzeug *n* internal broach
Innenring *m* inner ring *(bearing)*; inner member *(free wheel)*
Innenriß *m* internal (shatter) crack
Innenrohr *n* inner column *(radial drill)*
Innenrundläppen *n* hole (internal cylindrical) lapping
Innenrundschleifen *n* internal cylindrical grinding
Innenrundschleifmaschine *f* internal cylindrical grinding machine
Innenrüttler *m* poker (spud, internal, needle, immersion) vibrator
Innensäule *f* inner column *(radial drill)*
Innenschleifeinrichtung *f* internal grinding attachment
Innenschleifmaschine *f* internal grinding machine
Innenschliff *m* internal grinding
Innensechskantschraube *f* hexagonal socket-head bolt
Innenspannen *n (mach)* holding from the inside

Innenspannung *f* internal stress
Innenstempel *m (mine)* upper prop
Innensteven *m (ship)* inner post
Innentaster *m* inside caliper[s]
I. mit Schraubenscharnier firm joint inside caliper
I. mit Spannfeder spring[-joint] inside caliper
Innenteil *n* male part; insert, inner member *(fit)*
Innenthermometer *n* internal thermometer
Innenverkleidung *f* wainscoting, interior lining
Innenverrippung *f* internal ribbing
I. / mit internally ribbed
innenverzahnt internally toothed, internal *(gear)*
Innenverzahnung *f* internal toothing (gearing)
I. / mit internal-toothed
Innenverzahnungsschabemaschine *f* internal gear shaving machine
Innenwand *f* inside (internal) wall
Innenwiderstand *m* source (internal, anode, plate) resistance, source impedance; anode differential resistance
I. einer Röhre resistance of a valve, valve alternating current resistance
Innenzahnrad *n* internal gear, annular toothed wheel
Innenzahnradfräsen *n* internal gear milling
Innenzentrierung *f* minor-diameter fit
Innenzug *m* internal (inner) flue
Inneres *n* interior
innerhalb within *(tolerance)*
innerlich internal
innig *(ch)* intimate
Innung *f* guild
Inosit *m (ch)* inositol
Inradlinie *f s.* Hypozykloide
inseitig inside
insektenvernichtend insecticidal
Insektenvertilgungsmittel *n* insecticide
Insektizid *n* insecticide
systemisches I. systemic insecticide
Inselbildung *f (el)* island effect
Inseldampfer *m* interinsular steamer
Inselpfosten *m (build)* bollard
Inserat *n* advertisement
instabil unstable; decaying; non-equilibrium; excited
Instabilität *f* instability
monotone I. non-oscillatory instability
Installateur *m* fitter
Installation *f* installation; services, utilities *(plumbing)*
festverlegte I. hard (permanent) wiring
Installationsgeschoß *n* mechanical floor
Installationskern *m* plumbing (mechanical, utility, service) core
Installationskosten *pl* cost of erection (installation)
Installationsmaterial *n* wiring material
Installationstechnik *f* domestic engineering
Installationswand *f* plumbing wall
Installationszelle *f* service (utility, plumbing, mechanical) unit
installieren 1. *(build)* to install; 2. *(el)* to wire
instandhalten to keep in order, to maintain
Instandhaltung *f* maintenance, maintaining, entertainment; servicing
Instandhaltungsarbeit *f* maintenance (servicing) work
Instandhaltungskosten *pl* maintenance charges (cost)
Instandhaltungspersonal *n* servicing staff; ship keepers
instandsetzen to repair, to recondition, to reservice
Instandsetzung *f* repair, restoration, reconditioning, reservicing
I. und Unterhaltung *f* repair and maintenance
Instandsetzungsarbeit *f* repair work
laufende I. routine repair work
Instandsetzungsdienst *m (el) (Am)* repair service
instationär unsteady
Institut *n* institute *(research)*
Institutsdirektor *m* head of institute
instruieren to instruct, to order *(data)*
Instruktion *f (dat)* instruction, order
Instrument *n* instrument, device
I. mit Fernanzeige distant reading instrument
I. mit Nebenwiderstand shunted instrument
I. mit Nullpunkt (Nullage) in der Mitte centre zero instrument
I. mit unmittelbarer Ablesung direct reading instrument
I. mit unterdrücktem Nullpunkt suppressed zero instrument
aperiodisch gedämpftes I. aperiodic instrument
astatisches I. astatic instrument
chirurgisches I. surgical instrument
direkt anzeigendes I. direct-reading instrument
eigenständiges I. self-contained instrument
elektrodynamisches I. electrodynamic instrument
elektromagnetisches I. electromagnetic instrument
elektrothermisches I. electro-thermal instrument
gedämpft schwingendes I. damped periodic instrument
integrierendes I. integrating instrument
schreibendes I. recording instrument
spritzwassergeschütztes I. splash-proof instrument
thermisches I. thermal instrument

tragbares I. portable instrument
unmittelbar anzeigendes I. direct-reading instrument
Instrumente *npl* / **nautische** *(ship)* navigational aids (instruments)
Instrumentenanordnung *f* instrument layout
Instrumentenanzeige *f* meter indication (presentation, deflection)
Instrumentenbeleuchtung *f* instrument lighting
Instrumentenbrett *n* instrument board; dashboard
Instrumentenfehler *m* index (instrumental) error
Instrumentenflug *m* blind (instrument) flying
Instrumentengehäuse *n* kit
Instrumentenkonstante *f* meter (instrument) constant, constant of measuring instrument
Instrumentenleuchte *f* instrument panel lamp
Instrumentennebenwiderstand *m* instrument (meter) shunt
Instrumententafel *f* instrument panel; meter panel
Instrumententafelausrüstung *f* instrument panel equipment
Instrumententafelstütze *f* instrument-panel support
Instrumententisch *m* instrument table
Instrumentierung *f* instrumentation
I. des Reaktors *(nucl)* reactor instrumentation
Insulin *n* insulin
Intarsia *f* intarsia, inlay
Intarsiaarbeit *f* inlaying
Intarsiaflachstrickmaschine *f* / **einbettige** intarsia plain knitting machine
Intarsie *f s.* Intarsia
integrabel *(math)* integrable
Integral *n* integral
Abelsches I. *(math)* Abelian integral
bestimmtes I. definite integral
elliptisches I. elliptic integral
unbestimmtes I. indefinite integral
uneigentliches I. improper integral
Integralanteil *m (cont)* integral action
Integralbalg *m* integral bellow *(pneumatic controller)*
Integraldarstellung *f (math)* integral representation
Integraldosis *f (nucl)* integral dose
Integraldrossel *f* integral restriction *(pneumatic controller)*
Integralgleichung *f* integral equation
Integralkern *m (nucl)* [integral] kernel
Integralkosinus *m* cosine integral
Integrallogarithmus *m* logarithm integral
Integralrechnung *f* integral calculus
Integralregelung *f* integral control
Integralregler *m* integral controller
Integralsinus *m* sine integral
Integraltransformation *f* integral transform
Integralwirkung *f (cont)* integral action
Integralzeichen *n* integral sign
Integrand *m* integrand
Integration *f* integration
partielle I. partial integration
Integrationsglied *n* integrator, integrating circuit
Integrationskreis *m* integrating circuit
Integrationsmotor *m (cont)* integrating motor, messmotor
Integrationsweg *m (math)* path of integration
Integrationszeitkonstante *f (nucl)* integration time constant
Integrator *m* [counting-]rate meter; integrator, integrating element
mechanischer I. mechanical integrating device
integrierbar integrable
Integriergetriebe *n* mechanical integrating device *(analogue computer)*
Integrierglied *n (dat)* integrator
Integrierkreisel *m (dat)* gyro integrator, integrating gyroscope
Integriertrieb *m* mechanical integrating device *(analogue computer)*
Integrierverstärker *m* integrating amplifier
Integrodifferentialgleichung *f (math)* integro-differential equation
Intensität *f* intensity; strength
I. der kosmischen Strahlung cosmic-ray intensity
I. einer Lichtquelle luminous intensity of a light source
anomale I. anomalous intensity
spektrale I. spectral intensity
Intensitätsabfall *m* decrease of intensity
Intensitätsanstieg *m* increase of intensity
Intensitätsbegrenzer *m* video-gain control
Intensitätsmesser *m* intensity meter
Intensitätsmessung *f* intensity measurement
Intensitätsschrift *f* variable density tracks
Intensitätsverteilung *f* intensity distribution
Intensivgeflügelhaltung *f* intensive poultry farming
intensivieren to intensify
Intensivweide *f* strip-grazed pasture
Intercarrierverfahren *n (telev)* intercarrier sound system
Interferenz *f* 1. *(opt)* interference; 2. *(el)* parasitic noise, mush
Interferenzbild *n* interference pattern
Interferenzerscheinung *f* interference phenomenon
Interferenzfilter *n* interference filter
Interferenzfleck *m* interference spot
Interferenzgerät *n* interference apparatus
Interferenzkomparator *m* interference comparator, gauge-block interferometer
Interferenzmikroskop *n* interference microscope

Interferenzmikroskopie *f* interference microscopy
Interferenzrefraktor *m* / **Jaminscher** Jamin refractor
Interferenzschwund *m* fading by interference
Interferenzspektrometer *n* interference spectrometer
Interferenzstreifen *m* interference fringe
Interferenzstreifen *mpl* **gleicher Neigung** interference fringes of equal inclination
Interferenzvorgang *m* interference (beating) effect
Interferenzzone *f* nuisance area
interferieren to interfere
Interferometer *n* interferometer
Interferrikum *n s.* Luftspalt
Interglazialzeit *f* interglacial period
Interhalogenverbindung *f* interhalogen compound
Interkombination *f (nucl)* intercombination
Interkombinationslinie *f (ph)* intercombination line
interkontinental intercontinental
Interkostalträger *m (ship)* intercostal girder
interkristallin intercristalline
Interlockgestrick *n* interlock
Interlocknähmaschine *f* interlock sewing machine
intermediär intermediary, intermediate
intermetallisch intermetallic
intermittierend intermittent
Intermodulationsverzerrung *f (el)* intermodulation (I.M.) distortion
intermolekular intermolecular
Interpolation *f (math)* interpolation
  trigonometrische I. trigonometric interpolation
Interpolator *m* interpolating relay
Interpoliereinrichtung *f* interpolator
interpolieren to interpolate
Intervall *n* interval
  abgeschlossenes I. closed interval
  festes I. fixed interval
  kritisches I. *(met)* critical interval (range)
  offenes I. open interval
Intervallregel *f* interval rule *(spectroscopy)*
Intervallschachtelung *f (math)* nest of intervals
intraatomar intraatomic
intrakristallin intracrystalline
intramolekular intramolecular
intranuklear intranuclear
Intrittfallen *n* coming into step *(synchronous machine)*
Intrittfallmoment *n (el)* pull-in torque
intrudieren *(geol)* to intrude
Intrusivgang *m (geol)* dyke
Intrusivgestein *n* intrusive rock
Intrusivkörper *m (geol)* boss
Intrusivlager *n* intrusive sheet, sill
Invar *n* invar
Invardraht *m* invar tape
invariant *(math)* invariant
Invariante *f (math)* invariant
Invarianten *fpl* **des Deformationszustandes** invariants of strain
  I. des Spannungszustandes invariants of stress
Invariantentheorie *f* invariant theory
Invarianz *f (math)* invariance
Inventar *n* stock, inventory
  lebendes I. live stock
  totes I. dead stock
Inventaraufnahme *f* stock taking
Inventarbeitrag *m* stock put into a co-operative farm by a prospective member
Inventur *f* stock taking
Inventurmaschine *f* inventory machine
Inventur- und Materialberechnung *f* inventory and material accounting
invers inverse *(e.g. function)*
Inversion *f* inversion
Inversionsdublett *n (nucl)* inversion doublet
Inversionsschaltung *f* inversion circuit
Inversionstemperatur *f (nucl)* inversion temperature
Invertase *f* invertase
Inverter *m* inverter *(logical switching)*
invertiert inverted
Invertzucker *m* invert sugar
Inzidenz *f (nucl)* incidence
Ion *n* ion
  isoelektronisches I. isoelectronic ion
  negatives I. negative ion
  positives I. positive ion, cation
  wasserstoffähnliches I. hydrogenic ion
Ionenableiter *m* ionic discharge device
Ionenantrieb *m* ion drive
Ionenausbeute *f* ion yield
Ionenaustausch *m* ion exchange, ionic replacement
Ionenaustauschchromatografie *f* ion-exchange chromatography
Ionenaustauscher *m* ion exchanger, ionite
Ionenaustauscherharz *n* ion-exchange resin
Ionenbahn *f* ion path (trajectory)
Ionenbeschleuniger *m* ion accelerator
Ionenbeschuß *m* ion bombardment
Ionenbeweglichkeit *f* ion[ic] mobility, mobility of an ion (of ions)
Ionenbewegung *f* ion movement, ionic migration, migration of ions
Ionenbildung *f* ion formation
Ionenbildungsgeschwindigkeit *f* velocity of ionization
Ionenbindung *f* ionic bond (binding), ion linkage
Ionenbrennfleck *m* ion spot

Ionendichte *f* ion density
Ionendosimeter *n* ionic quantimeter
Ionendosis *f* ion dose, radiological dosis
Ionendrift *f* ion drift
Ionenemission *f* emission of ions
Ionenerzeugung *f* ion production
Ionenfalle *f* ion trap
Ionenfallenmagnet *m* ion trap magnet
Ionenfanggitter *n* ion repeller mesh
Ionengitter *n* ion[ic] lattice
Ionenhalbleiter *m* ionic semiconductor
Ionenhaufen *m* ion cluster
Ionenhydratation *f* hydration of ions
Ionenkonzentration *f* ion concentration (density)
Ionenkristall *m* ionic crystal
Ionenladung *f* ionic charge
Ionenlawine *f* avalanche of ions
Ionenleiter *m* ionic conductor
Ionenleitfähigkeit *f* ionic conductivity
Ionenleitung *f* ion[ic] conduction
Ionenoptik *f* ion optics
Ionenpaar *n* ion pair
Ionenpaket *n* bunched (bunch of) ions
Ionenphysik *f* ion physics
Ionenprodukt *n* ionic product
Ionenpumpe *f* ionic pump
Ionenquelle *f* ion source
Ionenradius *m* ionic radius
Ionenrakete *f* ion rocket
Ionenrauschen *n* ion noise
Ionenrekombination *f* ion recombination
Ionenrelais *n* gas relay
Ionenrohr *n* ion tube
Ionenröhre *f* ionic valve (tube)
Ionenrumpf *m* ion core
Ionensättigung *f* ionic saturation
Ionenschlauch *m* ion hose (layer, sheath)
Ionenspektrum *n* ion spectrum
Ionenstärke *f* ionic strength
Ionenstörstelle *f* ionic impurity
Ionenstoß *m* ion impact
Ionenstrahl *m* ion beam
Ionenstrahlstrom *m* positive ray current
Ionenstrom *m* ion[ic] current, stream of ions
Ionenvalenz *f* ionic valency
Ionenventil *n* ionic (gas-filled) rectifier
Ionenwanderung *f* ion migration
Ionenwind *m* electric wind (breeze)
Ionenwolke *f* ion cloud
Ionenzähler *m* ion counter
Ionenzentrifuge *f* ionic separator
Ionenzustand *m* ionic state
Ionisation *f* ionization
  minimale I. minimum ionization
  spezifische I. specific ionization
  thermische I. thermal ionization
Ionisationsdichte *f* ionization density
Ionisationsdosimeter *n* ionization dose meter
Ionisationsdosis *f* ionization dose
Ionisationsgeschwindigkeit *f* ionization rate
Ionisationsgrad *m* ionization degree, degree of ionization
Ionisationskammer *f* ionization chamber
  I. mit luftäquivalenten Wänden air wall ionization chamber
  dünnwandige I. thin-walled ionization chamber
  gasgefüllte I. gas-filled ionization chamber
  gepanzerte I. shielded ionization chamber
  gewebeäquivalente I. tissue-equivalent [ionization] chamber
  integrierende I. integrating ionization chamber
  kompensierte I. *s.* Kompensationskammer
  luftäquivalente I. air-wall ionization chamber
  luftgefüllte I. air[-filled] ionization chamber
Ionisationskammerstrom *m* ionization chamber current
Ionisationskanal *m* ionization channel
Ionisationskoeffizient *m* / **differentieller** specific ionization coefficient
Ionisationsmanometer *n* ionization manometer (pressure gauge)
Ionisationsmikrofon *n* cathodophone
Ionisationspotential *n* ionization potential
Ionisationsschicht *f* ionized layer
Ionisationsspannung *f* ionization potential
Ionisationsstoß *m* ionization impact (collision, pulse); burst
Ionisationsstrom *m* ionization current
Ionisationsvakuummesser *m* molecular (thermionic) vacuum gauge, vacuum ionization gauge
Ionisationsvakuummeter *n* molecular (ionization) vacuum gauge, thermionic pressure gauge
Ionisationsverlust *m* ionization loss
Ionisationsvermögen *n* ionization power
Ionisationswahrscheinlichkeit *f* ionization probability, probability of ionization
Ionisationsweg *m* ionization path (track)
Ionisationszähler *m* ionization counter
ionisch ionic
ionisierbar ionizable
ionisieren to ionize
ionisierend ionizing
Ionisierung *f* ionization; ionization event (*s. a.* Ionisation)
  thermische I. thermal ionization
Ionisierungsarbeit *f* ionization potential
Ionisierungsenergie *f* ionizing (ionization) energy
Ionisierungsgrad *m* degree of ionization
Ionisierungsquerschnitt *m* ionization cross section
Ionisierungsspannung *f* ionization potential
Ionisierungsstärke *f* specific ionization

Ionisierungsstoß *m* ionizing collision
Ionisierungsvermögen *n* ionization (ionizing) power
Ionisierungszeit *f* ionization time
Ionium *n* ionium
Ioniumalter *n* ionium age
ionogen ionogenic
Ionophorese *f* ion transfer, ionophoresis
Ionosphäre *f* ionosphere
Ionosphärenforschung *f* ionospheric research
Ionosphärenstörung *f* ionospheric disturbance
Ipser-Tiegel *m* plumbago crucible
Iraser *m* iraser *(infrared amplification by stimulated emission of radiation)*
Iridium *n* iridium
Iris *f (opt)* iris
Irisblende *f* iris diaphragm
Irisdruck *m* iridescent printing
Irisieren *n* irization, iridescence
irisierend iridescent
Iron *n* irone
irrational irrational
irreversibel irreversible
Irrstrom *m* stray (sneak) current
Irrtum *m* mistake, error
Irrtumswahrscheinlichkeit *f* risk of erroneous rejection *(statistics)*
Irrungszeichen *n* erasure signal
Irrzeichen *n* phantom signal *(radar)*
I_S-Begrenzer *m (el)* short circuit current-limiting device
I-S-Diagramm *n* Mollier diagram
Isallobare *f* isallobar
Isentrope *f* isentropic
isentropisch isentropic
Isherwoodbauart *f (ship)* isherwood framing
isobar isobar[ic]
Isobar *n s.* Isobare
Isobare *f* isobar, isobaric (constant pressure) line
isobarisch isobaric
  i. bei konstantem Druck isobaric at constant pressure
Isobutylalkohol *m* isobutyl alcohol
Isochore *f* isochore
isochorisch isochoric
  i. bei konstantem Volumen isochoric at constant volume
isochromatisch isochromatic
isochron isochronous
Isochronismus *m* isochronism
Isodiapher *n (nucl)* isodiaphere
isodispers isodisperse
Isodispersion *f* isodispersion
Isodosenschreiber *m (nucl)* isodose recorder
Isodrome *f s.* nachgebende Rückführung
Isodromregler *m s.* PI-Regler
Isodyname *f* isodynamic [line]
isoelektrisch isoelectric
Isogamme *f* isogam
Isogeotherme *f* isogeothermal surface
isogonal isogonal
Isogyre *f* isogyre
Isohaline *f* isohaline
Isoimpulskurve *f* isocount, isopulse curve
Isokline *f* isoclin[al] line
Isokorrosionsdiagramm *n* iso-corrosion diagram
Isolation *f* 1. *(el)* insulation; 2. isolation *(separation)*
  I. gegen Erde insulation against earth
  abgestufte I. graded (co-ordinated) insulation
  reduzierte I. reduced insulation
  schadhafte I. faulty insulation
  schlechte I. low insulation
Isolationsabstand *m* insulation distance
Isolationsbeanspruchung *f* insulation stress
Isolationsbemessung *f* insulation rating
Isolationsbezugsspannung *f* basic insulation level *(reference level)*
Isolationsfehler *m* insulation fault; leakage
Isolationsfestigkeit *f* insulating strength
Isolationsklasse *f* insulation class
Isolationsmasse *f* insulating mass
Isolationsmaterial *n* insulant; insulating material
Isolationsmesser *m (el)* insulation detector
Isolationsmessung *f* testing of insulation
Isolationsprüfer *m* insulation indicator; insulation test finger
Isolationsprüfung *f* insulation test
Isolationsverluste *mpl* insulation (leakage) losses
Isolationsvermögen *n* insulating capacity, dielectric strength
Isolationswiderstand *m* insulation (insulating, dielectric) resistance, insulance
Isolationszusammenbruch *m* insulation breakdown
Isolator *m (el)* insulator
  I. mit Schwanenhalsstütze swan-necked insulator
  durchschlagfester I. puncture-proof insulator
  durchschlagsicherer I. puncture-proof insulator
  kriechstromfester I. non-tracking insulator
  verlustarmer I. low-loss insulator
Isolatordoppelstütze *f* double insulator spindle
Isolatorenklöppel *m* insulator pin
Isolatorkette *f* insulator string (chain)
Isolatorstütze *f* insulator bracket (support)
Isolierband *n* insulating tape
Isolierbauplatte *f* structural insulation board
Isolierbuchse *f* insulating bush
isolieren 1. *(el)* to insulate; 2. *(ch)* to separate, to isolate
  den Kessel i. to cover the top of the boiler
Isolierfähigkeit *f* insulating power
Isolierfarbe *f* impenetrable (insulating) paint

Isolierfilz *m* lining felt *(pipes)*
Isolierflüssigkeit *f* insulating liquid
Isolierhülle *f* covering of non-conducting material
Isolierhülse *f* protective sleeve, slot liner
Isolierkarton *m* insulating cardboard
Isolierklemme *f* insulating clamp, insulated bolt
Isolierkörper *m (aut)* plug insulator
Isolierlack *m* insulating lacquer (varnish)
Isolierlackband *n* varnished insulating tape
Isolierlasche *f* insulating fish-plate
Isolierlukendeckel *m (ship)* insulated (plug) hatch
Isoliermaterial *n* insulating (non-conducting) material, insulating (non-conducting) compound
  imprägniertes I. impregnated insulating material
Isoliermatte *f* insulating mat
Isoliermittel *n* insulating compound, insulant
  verlustloses I. no-loss insulant
Isoliermuffe *f* insulating joint
Isolieröl *n* insulating oil
Isolierpapier *n* insulating paper
Isolierpappe *f* insulating [card]board
Isolierperlen *fpl* insulating beads (spacers)
Isolierplatte *f* insulation board
  halbsteife I. semi-rigid insulation board
  steife I. rigid insulation board
Isolierpreßstoff *m* moulded insulating material
Isolierring *m* insulating ferrule
Isolierrohr *n* insulated conduit, insulating tube (conduit, sleeve)
Isolierrohrbiegezange *f* conduit bending pliers
Isolierrohrmuffe *f (Am)* slip joint
Isolierrohrschelle *f* conduit clip
Isolierrolle *f* reel (knob-type) insulator
Isolierschemel *m* insulating stool
Isolierschicht *f* insulating layer; layer of non-conducting material
Isolierschlauch *m* insulated sleeving; spaghetti insulation
Isolierschnur *f* cord covering
Isolierstoff *m* insulating material, insulant
  elektrischer I. insulating material
Isolierstoß *m* insulated rail joint
isoliert 1. *(el)* insulated; 2. covered; isolated
Isolierteppich *m* insulating mat
Isoliertransformator *m* insulating transformer
Isolierung *f (el)* insulation; lagging *(heat)*
  thermische I. heat thermic insulation
Isolierungswirkungsgrad *m* efficiency of the non-conducting covering
Isoliervermögen *n* insulating capacity (power)
Isolierwandler *m* isolated (insulated) instrument transformer
Isolierzange *f* insulated tongs
isomagnetisch isomagnetic
isomer isomeric
Isomer *n* isomer
Isomeres *n* isomeride
  linksdrehendes I. laevoisomer
Isomerie *f* isomerism
  I. des Atomkerns nuclear isomerism, isomerism of atomic nucleus
  optische I. optical isomerism
isomerisieren to isomerizate
Isomerisierung *f* isomerization
Isometrie *f (math)* isometry, isometric projection
isometrisch *(math)* isometric
isomorph *(ch)* isomorphous; *(math)* isomorphic
Isomorphie *f* isomorphism
Isooktan *n* iso-octane
Isoplere *f (therm)* isopleric (constant-volume) line
Isopren *n* isoprene
Isoprenkautschuk *m* isoprene rubber
Isopropylalkohol *m* isopropyl alcohol
Isospin *m (ph)* isotopic spin
Isostasie *f (ph)* isostasy
isoster isosteric
Isotache *f* curve of equal velocity
isotherm isothermal
Isotherme *f* isothermal curve (line)
Isoton *n (nucl)* isotone
Isotop *n* isotope
  I. mit Neutronendefizit neutron-deficient isotope
  I. mit Neutronenüberschuß neutron-rich isotope
  künstlich erzeugtes I. artificial (man-made) isotope
  künstliches I. artificial (man-made) isotope
  neutronenarmes I. neutron-deficient isotope
  neutronenreiches I. neutron-rich isotope
  radioaktives I. radioisotope, radioactive isotope
  stabiles I. stable isotope
Isotopenanalysator *m* isotope analyzer
Isotopenanalyse *f* isotopic analysis (assay)
isotopenangereichert isotope enriched
Isotopenanreicherung *f* isotope concentration (enrichment)
Isotopenanwendung *f* isotope application
Isotopenaustausch *m* isotopic exchange
Isotopenaustauschreaktion *f* isotopic exchange reaction
Isotopenchemie *f* isotope chemistry
Isotopendosimetrie *f* isotope dosimetry
Isotopengemisch *n* isotopic mixture
  natürliches I. natural isotopic mixture
Isotopengeologie *f* isotopic geology
Isotopengewicht *n s.* Massenwert
Isotopenhäufigkeit *f* isotopic abundance, abundance of isotopes
Isotopenlaboratorium *n* isotopic laboratory
Isotopennummer *f* isotope number
Isotopenproduktion *f* isotope production

Isotopenretention *f* isotope retention
Isotopenspin *m* isotopic spin
Isotopentechnik *f* isotope technology
Isotopentrennanlage *f* isotope separation unit
elektromagnetische I. electromagnetic isotope separation unit
Isotopentrenner *m* isotope separator
Isotopentrennung *f* isotope separation, separation of isotopes
elektrolytische I. separation of isotopes by electrolysis, electrolytic separation of isotopes
elektromagnetische I. separation of isotopes by electromagnetic method, electromagnetic separation of isotopes; calutron separation
magnetische I. magnetic isotope separation
Isotopentrennverfahren *n* / **statistisches** statistical method of isotope separation
Isotopenverdünnung *f* isotope dilution
Isotopenverdünnungsanalyse *f* isotope dilution analysis
Isotopenverhältnis *n* isotopic ratio
Isotopenzusammensetzung *f* isotopic composition
Isotopie *f* isotopism
Isotopieeffekt *m* isotope effect
Isotopieverschiebung *f* isotopy shift
Isotopiezahl *f s.* Neutronenüberschuß
Isotron *n (nucl)* isotron
isotrop isotropic
Isozyanat *n* isocyanate *(plastics)*
isozyklisch isocyclic
Istabmaß *n* actual deviation
Istgröße *f* actual dimension
Istmaß *n* actual size (dimension)
Iststellung *f* actual position
Istvorschub *m* actual feed rate
Istwert *m* control point *(actual value of the controlled variable)*; actual value
Iteration *f (math)* iteration
Iterationsverfahren *n* method of successive approximations
Iterierte *f (math)* iterate
Izodprobe *f* Izod specimen

J

J-Naht *f s.* Jotnaht
Jacht *f* yacht
Jachthafen *m* marina *(for power yachts and launches)*
Jachtheck *n* yacht stern
Jachtsegeln *n* yachting, yacht sailing
Jachtwerft *f* yacht-building yard
Jachtwesen *n* yachting
Jachtwettrennen *n* yacht racing, race of yachts
Jacquardeinrichtung *f* Jacquard attachment
Jacquardhebemesser *n* Jacquard lifting knife
Jacquardkarte *f* Jacquard card
Jacquardkartenbindemaschine *f* Jacquard card lacer
Jacquardkartenschläger *m* Jacquard card puncher
Jacquardkette *f* **mit Mustergliedern** Jacquard pattern chain
Jacquardkopiermaschine *f* Jacquard repeating machine
Jacquardmaschine *f* Jacquard machine
J. für Tief- und Hochstellung top and bottom shedding machine
J. mit feinfeiner Teilung Jacquard machine with extra fine pitch
J. mit geteiltem Messerkasten Jacquard machine with double grifle blocks
Jacquardmuster *n* / **patroniertes** paper notation of a Jacquard pattern
Jacquardnadelbrett *n* Jacquard needle-board
Jacquardnadelgehäuse *n* Jacquard needle-box
Jacquardpatrone *f* Jacquard design chart
Jacquardplatinenboden *m* Jacquard bottom board
Jacquardsamt *m* / **zweispuliger** double-weft Jacquard velvet
Jacquardschnurbrett *n* Jacquard comberboard
Jacquardschnürung *f* Jacquard tie
Jacquardstrickmaschine *f* Jacquard knitting machine
Jacquardstuhl *m* Jacquard loom
Jadeit *m (min)* jadeite
Jagdberechtigung *f* shooting licence
Jagdbombenflugzeug *n* fighter-bomber
Jagdflinte *f s.* Jagdgewehr
Jagdflugzeug *n* pursuit-plane, [fury-]fighter
Jagdgewehr *n* sporting gun, shotgun
Jagdpatrone *f* shotgun cartridge
Jäger *m* 1. hunter, huntsman; 2. *(mil)* sharpshooter; 3. *s.* Jagdflugzeug
Jahresabflußmenge *f* annual flow
Jahresabonnement *n* annual subscription
Jahresbelastungsdiagramm *n* annual load diagram
Jahresbericht *m* annual report
Jahresdosis *f (nucl)* annual (yearly) dose
Jahresgebühr *f* annual rate
Jahresproduktion *f* annual production (rate of output), output per annum
Jahresring *m* annual [growth] ring *(tree)*
Jahreszeit *f* / **lichtarme** season wanting in light
Jährlingswolle *f* yearling's wool
Jakarandaholz *n* jacaranda
Jakobsleiter *f* Jacob's (Jack) ladder
Jalousie *f* blind, louvre, Venetian blind
Jalousieauslaß *m* discharge register, outlet grille
Jalousiedrosselklappe *f* louvre [damper]

Jalousiesichter *m* louvre deduster
Janmaat *m* able-bodied seaman
Jansenschaltung *f (el)* Jansen connection (circuit)
Japaner[karren] *m* tip cart, concrete buggy
Japanlack *m* Japan varnish
Jarowisation *f* vernalization
Jarowisierung *f* vernalization
Jasminöl *n* jasmine oil
jäten to pull out weeds, to weed
Jauche *f* liquid manure
Jaucheanhänger *m* liquid manure cart
Jauchedrill *m* liquid manure distributor
Jauchefaß *n* liquid manure tank
Jauchegrube *f* liquid manure pit
Jauchepumpe *f* liquid manure pump
Jaucheschleuderverteiler *m* liquid manure spreader
Jaucheverteiler *m* liquid manure distributor
Jedermannfunksprechverkehr *m* citizens radio service *(particularly in US)*
Je-Kopf-Produktion *f* per capita production
Jenaer Glas *n* Jena glass
Jennymaschine *f* spinning jenny
Jigger *m (text)* jig
Jitter *m (el)* jitter
Joch *n* 1. yoke *(magnetic)*; 2. *(build)* bent, trestle; 3. *(mine)* barring
  kurzes J. *(mine)* end plate
  langes J. *(mine)* wall plate
Jochbrücke *f* trestle bridge
Jochhöhe *f* yoke height *(of transformers)*
Jochverstellhebel *m* yoke adjuster *(cathode ray tubes)*
Jochwandler *m* cross-bar transformer *(waveguides)*
Jod *n* iodine
Jodat *n* iodate
jodhaltig iodiferous
Jodid *n* iodide
jodieren to iodizate, to iodize, to iodinate, to iodate
Jodloch *n (nucl)* iodine hole
Jodmulde *f (nucl)* iodine hole
Jodoform *n* iodoform
Jodometrie *f* iodometry
Jodsalbe *f* iodine ointment
Jodsäure *f* iodic acid
Jodstärkepapier *n* iodized starch paper
Jodtinktur *f* tincture of iodine
Jodwasserstoff *m* hydrogen iodide
Jodwasserstoffsäure *f* hydroiodic acid
Jodzahl *f* iodine number
Joghurt *m (n)* yogurt, yoghourt, yoghurt
Johannit *m* johannite *(uranium mineral)*
Jolle *f* dinghy, jolly boat, wherry; centre-board sloop, yawl
Jollenkreuzer *m* cruising centre-boarder, jolly-cruiser
Jollensegeln *n* jolly boat sailing
Jonon *n* ionone
Jordankegelbemesserung *f* Jordan plug
Jordanmühle *f* Jordan refiner
Jotnaht *f* single-J butt weld
Joule-Thomson-Effekt *m* Joule-Thomson effect, Joule-Kelvin effect
Journalbeleg *m* journal voucher
Juchtenleder *n* Russian leather
Jungfernfahrt *f* maiden trip (voyage)
Jungfernreise *f* maiden trip (voyage)
jungfräulich 1. virgin *(metal; neutron)*; 2. cold *(reactor)*
Jungpflanzenanzucht *f* nursery gardening
Jungviehaufzucht *f* stock rearing
Jungwuchs *m* young growth
justierbar adjustable
Justiereinrichtung *f* adjuster
justieren to adjust
Justierfehler *m* maladjustment
Justiermutter *f* adjusting (levelling) nut
Justierschraube *f* adjusting screw
Justierstift *m* adjusting pin
Justiertemperatur *f* reference temperature
Justierung *f* adjustment
Justierwiderstand *m* trimming resistor
Jute *f* jute [fibre]
Juteabfallklauber *m* jute waste picker
Jutefaden *m* jute thread
Jutehedegarnspinnerei *f* jute tow spinning
Jutequetschmaschine *f* jute softener
Jutequetschwalze *f* jute softener roller
Juteriste *f* / **zopfförmig zusammengedrehte** twisted jute strick
Juteweberei *f* jute weaving
Jutewebstuhl *f* jute loom
Jutewerggarnspinnerei *f* jute tow spinning
Juxtaposition *f* juxtaposition

K

K-Einfang *m (nucl)* K-electron capture, K-capture
K-Elektron *n* K-electron
K-Faktor *m* gauge factor *(strain gauge)*
K-Lampe *f* krypton lamp
K-Naht *f* double-bevel butt weld
K-Schale *f (nucl)* K-shell
K-Schirm *m* K-scope *(radar)*
K-Strahler *m (nucl)* K-emitter
K-Strahlung *f (nucl)* K-radiation, K-emission
K-Umwandlung *f (nucl)* K-conversion
Kabel *n* cable
  K. mit abgeschirmten Leitern screened conductor cable
  K. mit blankem Bleimantel plain-lead covered cable

K. mit Bleimantel und Umhüllung served lead-covered cable
K. mit geringer Adernzahl small-capacity cable, small-sized cable
K. mit Juteisolierung jute insulated cable
K. mit längsveränderlicher Bespulung taper-loaded cable
K. mit Luftisolation (Luftraumisolierung) air-spaced cable, dry-core cable
K. mit Massivisolierung (massiver Isolierung) solid cable, *(Am)* tight-core cable
K. mit Metallmantel metal-clad cable
K. mit ovalem Mantel oval cable
K. mit Umklöppelung braided cable
abgeschirmtes K. screened cable
achterverseiltes K. quadruple pair cable
armiertes K. armoured (sheathed) cable
asphaltiertes K. bitumen cable
bandeisenbewehrtes K. band-armoured cable
bespultes K. pupinized (loaded) cable
bewehrtes K. armoured (sheathed) cable
biegsames K. flexible cable
blankes K. bare cable
doppeladriges K. bifilar (twin) cable
drahtbewehrtes K. wire-armoured cable
dreiadriges K. triplex (three-core) cable
einadriges K. single-core cable, single-wire cable
feuchtegeschütztes K. moisture barrier cable
flexibles K. *s.* biegsames K.
gemischtpaariges K. combined (composite) cable
gummiisoliertes K. rubber cable
hochpaariges K. large-capacity cable
Höchstädter K. Hoechstaedter (shielded-conductor) cable
induktionsfreies K. anti-induction cable
kapazitätsarmes K. low-capacity cable
koaxiales K. coaxial (concentric) cable
konzentrisches K. *s.* koaxiales K.
mehradriges K. multi-core cable, multi-conductor cable
mehrpaariges K. multipair (multiple-twin) cable
mehrphasiges K. ohne Vierer non-quadded cable
mit Steckern versehenes K. cable fitted with plug and socket connectors
nicht abgeschirmtes K. unshielded cable
nicht bespultes K. unloaded cable
nicht normgerechtes K. non-standardized cable, non-association cable
paarverseiltes K. paired cable
pupinisiertes K. pupinized cable
radial unterteiltes konzentrisches K. septate cable
schirmloses K. unshielded cable
schwer bespultes K. heavy loaded cable
schwer pupinisiertes K. heavy loaded cable
stahldrahtbewehrtes K. steel wire armoured cable
sternverseiltes K. star quad cable
unbespultes K. unloaded cable
unbewehrtes K. unarmoured cable
unterirdisches K. underground cable
verseiltes K. stranded (rope-lay) cable
vieladriges K. multi-core cable
vielpaariges K. multipair (multi-way) cable
vieradriges K. four-core cable
viererverseiltes K. quadded cable
zweiadriges K. bifilar (two-core, double-conductor) cable

Kabelabschluß *m* cable terminal (termination); cable head
Kabelabzweiger *m* cable box
Kabelabzweigmuffe *f* multiple cable joint, cable branch-joint box
Kabelabzweigung *f* cable branching *(box)*
Kabelader *f* cable conductor (core)
Kabeladernpaar *n* cable pair
Kabelanschluß *m* cable terminal (connection)
Kabelanschlußkasten *m* cable [joint] box
Kabelarmatur *f* cable fitting
Kabelarmierungsdraht *m* cable armouring wire
Kabelasphalt *m* cable compound
Kabelaufbau *m* cablc make-up
Kabelaufführungspunkt *m* cable head (lifting) point
Kabelaufhängung *f* 1. cable suspension; 2. *s.* Kabelhalterung
Kabelausgang *m* cable outlet
Kabelauslastung *f* cable fill
Kabelauslegung *f* laying of cables
Kabelbagger *m* slackline cable plant; cableway excavator
Kabelbaum *m (el)* cable harness, fanned (preformed) cable
Kabelbautrupp *m* cable-laying team
Kabelbeilauf *m s.* Kabelfüllmaterial
Kabelbett[ung *f*] *n* packing of cable
Kabelbewehrung *f* cable armouring
Kabelboje *f* cable buoy
Kabelbruch *m* cable break
Kabelbruchstelle *f* point of cable break
Kabelbrücke *f (build)* cable bridge
Kabelbrunnen *m* cable pit
Kabeldampfer *m* cable steamer
Kabeldämpfung *f* cable attenuation
Kabeldichtungsmasse *f* cable sealing compound
Kabeleinführung *f* cable entry (lead-in)
Kabelende *n* cable end
K. mit Schutzkappe sealed cable end
verbleites K. sealed cable end
zugebleites K. *s.* verbleites K.
Kabelendgestell *n* cable support (terminating) rack

Kabelendverschluß *m* cable terminal [box]
Kabelendverstärker *m* line final amplifier *(radio transmitter)*
Kabelendverteiler *m* cable distribution head, cable termination distributor
Kabelfehler *m* cable fault
Kabelfehlerortungsverfahren *n* / **kapazitives** Gott's method
Kabelfehlersuchgerät *n* electrode cable detector, cable fault finder
Kabelfett *n* cable grease
Kabelform *f* cable form
Kabelformbrett *n* forming (lacing) board
Kabelformstein *m* [multiple] cable tile
Kabelführung *f* cable run
Kabelführungsplan *m* cable-layout plan, cabling layout
Kabelfüllmaterial *n* cable filler
Kabelgarn *n* cable yarn
Kabelgarnitur *f* cable accessories
Kabelgeschirr *n* cable gear, wire manifold
K. für die Abschirmung screening harness
Kabelgestell *n* cable rack
Kabelgleitfett *n* cable lubricant (grease)
Kabelgraben *m* cable trough (trench, ditch)
Kabelhalter *m* cable support
Kabelhängeschelle *f* cable suspension clamp
Kabelhochführungsschacht *m* vertical wall-duct [for cables], cable chute
Kabelhülle *f* cable housing (duct)
Kabelkanal *m* cable duct (channel, tunnel), raceway; conduit
einzügiger K. single cable duct
Kabelkanalanlage *f* conduit plant
Kabelkanalformstein *m* conduit brick
Kabelkapazität *f* cable capacity
Kabelkeller *m* underground distribution chamber
Kabelklemme *f* cable terminal (lug); cable clip (clamp)
Kabelkran *m* cable[way] crane
Kabelkrananlage *f* cableway; telpherage
Kabellageplan *m* plan of cable lay-out, cable map
Kabellegegeschirr *n* cable (paying-out) gear
Kabellegen *n* laying out cables, cabling
Kabelleger *m* cable-laying ship, cable layer (ship)
Kabelleiter *m* cable conductor
Kabellitze *f* cable strand
Kabellöter *m* cable jointer
Kabellötstelle *f* cable joint
Kabelmantel *m* cable sheath[ing], cable jacket
Kabelmantelpreßanlage *f* cable-sheath pressing plant
Kabelmantelverbinder *m* **für durchgehende Erdung** cable bond
Kabelmasse *f* cable filler, [sealing] compound
Kabelmast *m* cable post
Kabelmerkstein *m* cable marker
Kabelmeßdienst *m* cable test service
Kabelmesser *n* cable stripper, chipping knife
Kabelmeßgerät *n* wire tester, cable measuring set
Kabelmeßkoffer *m* portable cable measuring set
Kabelmeßwagen *m* cable testing car, testing van, mobile testing unit
Kabelmuffe *f* cable sleeve (joint, box, fitting)
kabeln to cable
Kabelnachbildung *f* artificial balancing cable, imitated line
Kabelnachziehschlauch *m* split cable grid
Kabelnetz *n* cable system (network); cable mains
Kabelortung *f* cable localizing
Kabelöse *f* cable lug (tag)
Kabelpackung *f* rope packing
Kabelpapier *n* cable (insulating) paper
Kabelplan *m* cable lay-out
Kabelplanung *f* proposed cable scheme
Kabelprojekt *n* proposed cable scheme
Kabelreduktionsfaktor *m* cable sheath protection factor
Kabelrinne *f* cable channel
Kabelrohr *n* duct, cable tube
Kabelrolle *f* cable sheave [of a cable layer]
Kabelrost *m* cable rack (shelf, grate), wire-ways
Kabelrückleitung *f* metallic return circuit
Kabelschacht *m* cable shaft (pit, chamber), manhole chimney
Kabelschelle *f* cable clamp
Kabelschiff *n* cable layer, cable[-laying] ship, cable[-laying] vessel
Kabelschrank *m* cable terminal box
Kabelschuh *m* *(el)* cable eye (lug, socket, thimble)
Kabelschutzrohr *n* cable [protective] conduit
Kabelschutzschlauch *m* cable protecting hose
Kabelseele *f* cable core
Kabelspleißung *f* splicing of the cable
Kabelstecker *m* wire plug
Kabelsteckverbindung *f* cable coupling
Kabelstollen *m* cable subway
Kabelstrecke *f* plotting of [the] cable run
Kabelstromwandler *m* cable current transformer
Kabelsuchgerät *n* cable finder (locator, detector)
Kabeltank *m* cable [storage] tank
Kabeltankfassungsvermögen *n* spoil capacity
Kabeltonne *f* cable buoy
Kabeltragschelle *f (el)* carrying clamp
Kabeltragseil *n* messenger cable
Kabeltrasse *f* cable route
Kabeltrommel *f* cable drum (reel)
Kabeltrosse *f* cable-laid hawser
Kabeltunnel *m* gallery, cable subway
Kabelüberführung *f* cable terminal

Kabelübergangskasten *m* cable joint box
vergossener K. pothead (sealed) terminal
Kabelverbinder *m* cable fitting
Kabelverbindungskasten *m* cable joint box
Kabelvergußmasse *f* cable sealing compound
Kabelverlegung *f* running of cables, cable-laying
Kabelverlustwinkel *m* cable power factor
Kabelverseilmaschine *f* cable stranding (twisting) machine, cable jig
Kabelverzweiger *m* cable distribution box, distribution cabinet
Kabelverzweigung *f* cable branching
Kabelwachs *n* cable wax, adhesive pitch
Kabelwagen *m* cable trolley
Kabelweg *m* run of [the] cable
Kabelwickelpapier *n* cable binding paper
Kabelwickeltrommel *f* cable reel
Kabelwiderstand *m* characteristic cable impedance
Kabelziehschlauch *m* cable grip
Kabelziehstrumpf *m* cable grip
Kabelzopf *m* cable fan
Kabelzubehör *n* cable equipment
Kabine *f* cabin (*s. a.* Kajüte); *(aero)* cockpit
Kabinendach *n* canopy
Kabinenkreuzer *m* cabin cruiser
Kabotage *f* coasting trade
Kachel *f* tile
kacheln to tile
Kachelofen *m* tile stove
Kachelung *f* tiling
Kadaverin *n* cadaverine, pentamethylene-diamine
Kadenz *f* cadence
Kadettenschiff *n* cadet ship
Kadettenschulschiff *n* cadets' training ship
Kadmieren *n* cadmium plating
Kadmium *n* cadmium
Kadmiumblende *f* greenockite
Kadmiumelement *n* cadmium cell
Kadmiumgelb *n* cadmium yellow
kadmiumhaltig cadmiferous
Kadmiumlinie *f* / **rote** cadmium red line
Kadmiumregelstab *m (nucl)* cadmium regulator (control rod)
Kadmiumverhältnis *n (nucl)* cadmium ratio
Kaffeegerbsäure *f* caffe-tannic acid
Kaffsieb *n* upper (capes, chob) riddle
Käfig *m* 1. *(ag)* [laying] cage; 2. *(bearing) s.* Abstandshalter; 3. *(lapping) s.* Läppkäfig
Faradayscher K. Faraday cage
Käfiganker *m (el)* squirrel-cage armature
Käfigankermotor *m* squirrel-cage motor
Käfigblitzableiter *m* cage-type lightning conductor
Käfigeffekt *m (nucl)* cage effect
Käfighaltung *f* keeping hens in laying cages
Käfigläufer *m (el)* [squirrel-]cage rotor
Käfigmagnetron *n (el)* squirrel-cage magnetron
Käfigrelais *n* cage relay
Käfigwicklung *f (el)* [squirrel-]cage winding
Kahlappretur *f (text)* napless (pileless) finish
kahlscheren to crop the nap *(of a fabric)*
Kahlschlag *m* char felling
Kahn *m* [small] boat, skiff; [canal] barge
kahnfahren to boat, to go in a boat
Kahnschiffer *m* lighterman, master of a boat, waterman, boatman
Kai *m* quay, wharf
Kaiarbeiter *m* docker, lumper, longshoreman; ship labourer, *(Am)* pierman
Kaibesitzer *m* wharfinger
Kaigebühren *fpl* quay dues, wharf charges, wharfage, berthing [dues]
Kaigeld *n* quayage
Kailaufkran *m* dockside travelling crane
Kaimauer *f* quay wall
Kaimeister *m* berthing master, wharfinger
Kainit *m (min)* kainite
Kaischuppen *m* transit shed
Kajak *m(n)* kayak
Kajüte *f (ship)* cabin, berth; stateroom, saloon
Kajütenjunge *m* mess boy
Kajütenluke *f* companion hatch
Kajütsaufbau *m* raised (deck) cabin
Kajütschiff *n* cabin liner
Kajütskompaß *m* hanging compass
Kajütskreuzer *m* cabin cruiser
Kajütsniedergang *m* companion-way
Kajütstreppe *f* companion ladder
Kakaobaum *m* cacao tree
Kakaobohne *f* cacao, cocoa bean
Kakaobutter *f* cacao butter (nut oil)
Kakaofrucht *f* cacao
Kakodylreaktion *f* cacodyl reaction
Kakodylsäure *f* cacodylic acid
Kalander *m* calender
Kalanderappretur *f* calender finish
Kalanderauftrag *m* calender coating
Kalanderausrüstung *f* calender finish
Kalanderfolie *f* calendered sheet
kalandern to calender
Kalandern *n* calendering
K. mit Kupferwalzen copper finishing
K. mit Zinkwalzen zinc finishing
Kalanderpartie *f* surfacing end
Kalandertücher *npl* calender sheeting
Kalanderwalze *f* calender roll
kalandrieren to calender
Kalandrieren *n* calendering
kalandriert calender-finished
Kalandrierwerk *n* calendering unit
Kälberstall *m* calf pen
Kalbfell *n* calfskin
totgeborenes K. still-born calfskin

Kalbskopfspaltmaschine *f* calf head splitting machine
Kaldaune *f* naked (blank, blind) wall
Kalender *m* calendar; diary
Kalenderblock *m* diary pad, date block
Kalenderblockkarton *m* calendar-pad bond board
Kalenderrückenkarton *m* calendar-back board
Kalfaterer *m* [wood] caulker, [wood] calker
Kalfatereisen *n* bent iron
Kalfaterhammer *m* caulking mallet
kalfatern to caulk, to calk
Kalfaterwerg *n* caulk, calk
Kalfatfloß *n* floating stage
Kali *n* potash
Kalialaun *n* potash alum
Kaliber *n* 1. calibre, bore *(pipe)*; 2. *(met)* roll pass *(rolling)*
  abgesetztes K. *(met)* tapering pass
  geschlossenes K. *(met)* closed groove
  offenes K. *(met)* open pass
Kaliberabnahme *f* reduction of roll passes
Kaliberanstellung *f* reduction of roll passes
Kaliberanzug *m* taper of groove
Kaliberbohrer *m* three-flute drill
Kaliberdorn *m* sizing pin, mandrel
Kaliberdruck *m* reduction of roll passes
Kaliberfolge *f* *(met)* pass sequence
Kaliberöffnung *f* *(met)* groove opening
Kaliberpresse *f* sizing press
Kaliberreibahle *f* sizing reamer
Kaliberverjüngung *f* *(met)* taper of groove
Kaliberverschleiß *m* *(met)* groove wear
Kalibervorrichtung *f* *(met)* end sizing installation
Kaliberwalze *f* groove[d] cylinder (roll)
Kalibrierarbeit *f* sizing work
kalibrieren to calibrate; *(mach)* to bring to size, to [finish to] size; to adjust, to scale; to gauge
  Walzen k. to design rolls
Kalibrierpresse *f* coining press
Kalibrierräumwerkzeug *n* sizing broach
Kalibrierringe *mpl* border
Kalibrierteil *m* finishing section *(broach)*
Kalibrierung *f* calibration; *(mach)* sizing, finishing to size; adjusting, scaling, gauging; *(met)* roll pass design
Kalibrierungsschema *n* *(met)* roll pass design
Kalibrierwalze *f* sizing roll
Kalibrierzahn *m* sizing tooth *(broach)*
Kalidünger *m* potash fertilizer (manure)
Kalifornium *n* californium
Kaliglimmer *m* potash mica
kalihaltig potassic
Kalikobalgen *mpl* calico bellows
Kalilauge *f* caustic potash, [aqueous solution of] potassium hydroxide
Kalisalz *n* potassium salt
Kalischmelze *f* potash fusion
Kalium *n* potassium
Kaliumantimonyltartrat *n* antimonyl potassium tartrate
Kaliumbichromat *n* bichromate of potassium
Kaliumbromid *n* potassium bromide
Kaliumdichromat *n* potassium dichromate
Kaliumferrizyanid *n* potassium ferricyanide
Kaliumferrozyanid *n* potassium ferrocyanide
Kaliumhydroxid *n* potassium hydroxide
Kaliumjodid *n* potassium iodide
Kaliumjodidstärkepapier *n* potassium iodide test paper
Kaliumkarbonat *n* potassium carbonate, potash
Kaliumnitrat *n* potassium nitrate, nitre, saltpetre
Kaliumpermanganat *n* potassium permanganate
Kaliumsulfat *n* potassium sulphate
Kaliumzyanid *n* potassium cyanide
Kalk *m* lime
  fetter K. high calcium lime
  gebrannter K. anhydrous (burnt, fired) lime, quicklime
  gelöschter K. slaked (hydrated) lime
  hydraulischer K. hydraulic lime
  luftgelöschter K. air-slaked lime
  ungelöschter K. quicklime, unslaked lime
Kalkäscher *m* *(leath)* lime-pit
Kalkausblühung *f* *(build)* lime bloom
Kalkbeton *m* lime concrete
Kalkbrei *m* *(leath)* lime cream
Kalkbrennen *n* lime burning
Kalkdrehofen *m* rotary lime kiln
Kalkdünger *m* calcium fertilizer (manure)
Kalkdüngung *f* liming, lime dressing
Kalkelend *n* lime set *(blast furnace)*
kalken to whitewash
Kälken *n* liming *(wire)*
Kalkgrube *f* lime pit
kalkhaltig calciferous, calcareous
Kalkhydrat *n* hydrated lime
kalkig limy, calcareous, calcic
Kalklicht *n* / **Drummondsches** Drummond [lime] light
Kalklöschanlage *f* lime slaking plant
Kalklunker *m* fish-eye *(plaster)*
Kalkmergel *m* calcareous clay
Kalkmilch *f* lime milk, milk of lime, limewash
Kalkmilchprobe *f* lime test for cracks
Kalkmörtel *m* lime mortar
Kalkmühle *f* quicklime mill
Kalkofen *m* lime-kiln
Kalkputz *m* lime plaster
Kalkrührwerk *n* lime agitator
Kalksandstein *m* sand-lime brick, calcium-silicate brick
Kalkschachtofen *m* vertical lime kiln
Kalkschlämme *f* neat lime
Kalkschleier *m* chalk fog, chalky veil
Kalkseife *f* lime (calcium) soap
Kalkspat *m* calcite, calcspar

Kalkspatinterferometer *n* calcite interferometer
Kalkstein *m* limestone
Kalksteinbruch *m* limestone quarry
Kalkstickstoff *m* lime-nitrogen
Kalkstreuer *m (ag)* lime distributor (spreader)
Kalktravertin *m* travertine
Kalktuff *m* calc tufa
Kalktünche *f* lime whiting, lime-wash
Kalkül *m (math)* calculus
Kalkulation *f* cost accounting
Kalkulationsabteilung *f* cost department
Kalkulationsnorm *f* cost standard
Kalkulator *m* cost accountant; quantity surveyor, estimator
Kalkung *f (ag)* liming
Kalkwasser *n* lime water
Kalkzementmörtel *m* cement-lime mortar
Kalkzuschlag *m* adding of lime
Kalmusöl *n* calamus oil
Kalomelelektrode *f* calomel electrode
Kalomelhalbzelle *f* calomel half-cell
Kalorie *f* calorie
  mittlere K. mean calorie
Kalorimeter *n* calorimeter
Kalorimetergefäß *n* calorimeter vessel
Kalorimeterwasser *n* water in the calorimeter
Kalorimetrie *f* calorimetry
kalorimetrieren to measure with a calorimeter
Kalorimetrierung *f* calorimetric measurement
kalorimetrisch calorimetric
kalorisieren *s.* aluminieren
Kalotte *f* cap, spherical surface
Kalottenfläche *f (met)* area of ball impression *(Brinell)*
Kalottensignalleuchte *f* signal (indicating) lamp with spherical cap
kalt cold
  k. abbindend cold-setting
  k. aufpressen to drive on cold
  k. umgeformt cold formed (worked)
  k. vergießen to cold cast
Kaltaushärtung *f* aging at room temperature
Kaltbadtauchverfahren *n* cold immersion method
Kaltband *n* cold rolled strip
Kaltbearbeitung *f* cold working
Kaltbeladung *f* **des Reaktors** cold loading of the reactor
Kaltbiegeprobe *f* cold-bend test specimen
Kaltbiegeversuch *m* cold bend test
kaltblasen to cold-blast
Kaltblasen *n* cold blowing (blasting)
Kaltbonderverfahren *n* cold bonderizing
kaltbrüchig cold short
Kaltbrüchigkeit *f* cold shortness
Kaltdampfzylinder *m* cold steam cylinder
Kälte *f* coldness
Kälteanlage *f* refrigerator
Kältebehandlung *f* cold treatment
kältebeständig antifreezing, nonfreezing, cold resisting
Kältebeständigkeit *f* anti-freezing property (quality)
Kältebrücke *f (build)* thermal (cold) bridge
Kältechemie *f* cryochemistry
Kältediagramm *n* psychrometric chart
Kälteeinheit *f* frigorific unit
kälteerzeugend cryogenic, frigorific
Kälteerzeuger *m* / **elektrischer** electric refrigerator
Kälteerzeugung *f* refrigeration
Kälteerzeugungsmaschine *f* refrigerator
kältefest cold-resisting
Kältefestigkeit *f* strength at low temperatures
Kältefließfähigkeit *f* filterability at low temperatures *(Diesel fuel)*
Kältegebläse *n* cold blast
Kältegemisch *n* freezing mixture
Kalteinsenkpresse *f* hobbing (die typing) press
Kältelagerraum *m* chillroom
Kälteleistung *f* refrigerating capacity
Kaltelektronenemission *f* cold (field, autoelectronic) emission of electrons
Kaltemail *n* cold top enamel
Kältemaschine *f* refrigerating machine, refrigerator
Kältemaschinenraum *m* cooling room
Kältemischung *f* freezing (frigorific) mixture
Kaltemission *f* cold (field, autoelectronic) emission of electrons
Kältemittel *n* refrigerant; cooling agent
Kälteprobe *f* cold test
Kältepunkt *m* freezing (solidifying) point
  K. des Öls solidifying point of oil
Kälteregler *m* cryostat
Kältereinigungsskala *f* low-temperature purification plant
Kälteschutzisolierung *f* insulation (insulator) for cold
Kälteschutzmittel *n* cold protective
Kältetechnik *f* refrigeration engineering
Kältethermometer *n* cold-test thermometer
Kälteträger *m* cooling agent
Kälteverhalten *n* low-temperature characteristics
Kaltfahrbedingungen *fpl* arctic driving conditions
Kaltfärber *m (text)* cold-dyeing colour
Kaltfärbeverfahren *n (text)* cold-bath method
Kaltfederindikator *m* indicator with external spring
Kaltfließpressen *n (met)* cold extrusion
Kaltformbarkeit *f (met)* cold plasticity
Kaltformen *n* cold forming, *(Am)* cold working
Kaltformgebung *f* cold forming, *(Am)* cold working

Kaltformvermögen *n* cold plasticity
Kaltgang *m* cold state *(blast furnace)*
kaltgehärtet hardened by cold working
Kaltgesenkpressen *n* cold swaging
kaltgewickelt cold-wound, cold-coiled *(spring)*
kaltgezogen cold-drawn
kalthärtend cold-setting
Kalthärtung *f* strain hardening, hardening by cold working
Kaltkatode *f* cold cathode
Kaltkreissäge *f* cold circular saw
Kaltleim *m* cold glue
Kaltlötstelle *f* cold junction (solder joint)
Kaltlufttrockner *m* low-temperature drier
Kaltlufttrocknung *f* unheated air drying, low-temperature drying
Kaltnadelradierung *f* dry point engraving
Kaltnaßspinnmaschine *f* cold water spinning frame
Kaltnebel *m* cold fog
Kaltnetzer *m (text)* cold wetting agent
Kaltnetzmittel *n (text)* cold wetting agent
Kaltnieten *n* cold riveting
Kaltpilgern *n* cold pilger rolling
Kaltprägung *f* cold blocking (embossing)
Kaltpreßmasse *f* cold moulding compound
Kaltpreßpappe *f* cold pressboard
Kaltpreßschweißen *n* cold welding
Kaltprofil *n* cold rolled section
kaltprofilieren to cold-form
Kaltreaktivität *f* cold reactivity
Kaltsäge *f* cold-sawing machine
Kaltsägemaschine *f s.* Kaltsäge
Kaltsatz *m* cold type (composition)
Kaltschlagmatrize *f* piercing die
Kaltschlagschmieden *n* cold impact forging
Kaltschliff *m* cold ground *(papermaking)*
Kaltschweißstelle *f* cold lap (shut) *(metal)*, weld mark
Kaltsprödigkeit *f* cold brittleness
Kaltstart *m (el)* cold (instant) start; cold starting *(engine)*
Kaltstarthilfe *f* cold-starting aid
Kaltstartlampe *f* cold-start lamp
Kaltstauchen *n* cold heading
  K. von Schraubenköpfen cold heading of bolts
Kaltstrecken *n* cold stretch *(plastics)*
Kalttiefziehen *n* cold cupping
Kalttrockenfarbe *f* cold-set ink
kaltumformbar cold-workable
Kaltumformung *f* cold deformation (working)
Kaltveredlung *f* cold refining *(of paper)*
kaltverfestigen to workharden
Kaltverfestigung *f* workhardening
Kaltverfestigungsfähigkeit *f* strain-hardenability
kaltverformen to cold-form
Kaltverformung *f* cold deformation
Kaltverstrecken *n* cold drawing *(man-made fibre)*
Kaltverwinden *n* cold twisting
Kaltvulkanisation *f* cold curing
Kaltwalze *f* cold roll
Kaltwalzen *n* cold rolling
Kaltwalzwerk *n* cold reduction mill
Kaltwaschechtheit *f (text)* fastness to cold washing
Kaltwasserröste *f (text)* cold water retting
Kaltwasserschwitze *f (leath)* cold-sweat
Kaltwasserwalke *f (text)* cold water milling
Kaltwell-Lösung *f (ch)* cold-waving lotion
Kaltwind *m* cold blast *(furnace)*
Kaltwindofen *m* cold blast furnace
Kaltziehbank *f* cold drawing bench
kaltziehen to cold-draw
Kaltziehmatrize *f* cold-drawing tool
Kaltzug *m* cold drawing
Kalutron *n s.* Calutron
Kalziferol *n* calciferol
Kalzination *f* calcination
kalzinieren to calcine
Kalzinierung *f* calcination
kalziothermisch calciothermal
Kalzit *m* calcite
Kalzium *n* calcium
Kalziumalginatfaden *m* calcium alginate thread
Kalziumalter *n (nucl)* calcium age
kalziumhaltig calcic
Kalziumkarbid *n* calcium carbide, acetylene stones
Kalziumsilikatstein *m* calcium-silicate brick
Kalziumsulfat *n* 1. calcium sulphate, gypsum; 2. pearl white *(papermaking)*
Kamel *n (ship)* lifting pontoon, camel
Kamelhaar *n* camel hair
Kamera *f* camera
  K. für Großaufnahmen close-up camera
  mikrofotografische K. photomicrographic camera
Kameraansatz *m* camera attachment
Kameraauszug *m* camera extension
Kamerabühne *f (telev)* camera crane
Kameradschaft *f (mine)* butty gang, crew
Kameraeinstellung *f* camera line-up
Kameragehäuse *n* camera body
Kamerakappe *f* lens hood
Kamerakontrollgerät *n* camera monitor
Kameraobjektiv *n* camera lens
Kameraverbindungssender *m* pick-up link transmitter
Kameraverschiebung *f* **nach oben / vertikale** *(telev)* pan up
  K. nach unten / vertikale *(telev)* pan down
Kameraverschluß *m* [camera] shutter
Kamerawagen *m* camera dolly, doll buggy
Kamillenöl *n* camomile oil

Kamin *m* 1. chimney [stack], smoke stack, funnel; 2. fireplace
Kaminkühler *m* chimney cooler
  selbstventilierender K. chimney cooler with natural draught
Kaminkühlung *f* cooling by means of a cooling tower
Kaminsäule *f* chimney shaft
Kaminzug *m* chimney draught (draft)
Kamm *m* 1. *(text)* comb; 2. *(ag)* currant stripper
  fester K. *(text)* fast reed
Kämmaschine *f* *(text)* comb[ing machine], comber
  K. mit unterbrochenem Arbeitsgang comber with intermittent action
  Noblesche K. Noble comb
Kammbuchse *f* *(therm)* brasses of the collar thrust bearing
Kammeißel *m* rack[-form] cutter *(for gear generation)*
kämmen 1. *(text)* to comb; 2. to be in mesh, to mesh, to mate *(gears)*, to gear [together]
Kämmen *n* **des Wergs** *(text)* tow combing
kämmend engaging, mating, in contact *(gears)*
Kammer *f* 1. chamber; stall; 2. *(ship)* cabin, state-room, berth
  K. mit mehreren Platten *(nucl)* multiplate chamber
  K. zum Zerkleinern des hereingeschossenen Erzes *(mine)* bulldozing chamber
  heiße K. *(nucl)* hot cave (cell)
Kammerbau *m s.* Kammerbauverfahren
Kammerbauverfahren *n* *(mine)* [pillar and] chamber working, pillar and stall method, pillar working, room and pillar method, rooming
Kämmereiausputz *m* *(text)* comber strippings
Kammerfeuerung *f* furnace with combustion chamber
Kammerfilterpresse *f* chamber filter-press
Kammerhals *m* *(mine)* neck
Kammerkonstante *f* *(opt)* calibrated focal length
Kammerkristall *m* chamber crystal
Kammerofen *m* chamber (compartment, oven-type) furnace
Kammersäure *f* chamber acid
Kammerschleuse *f* 1. *(ship)* entrance-lock, [tide] lock; 2. star feeder
  einfache K. *(ship)* single lock
Kammerschott *n* *(ship)* partition (joiner) bulkhead
Kammersprengung *f* *(mine)* chamber (coyote-hole) blasting
Kammerstrom *m* ionization chamber current
Kammertiefenmesser *m* chamber-depth indicator
Kammerton *m* *(ac)* standard musical pitch, concert pitch
Kammertrocknung *f* kiln-seasoning, kiln drying
Kammerüberhitzer *m* *(therm)* chamber superheater
Kammerverfahren *n* chamber process *(sulphuric acid)*
Kammerwinkel *m* *(opt)* chamber angle
Kammflug *m* *(text)* comber fly
Kammgarn *n* worsted [yarn]
Kammgarndoppelkrempel *f* double-worsted card
Kammgarngewebe *n* blended worsted
Kammgarnspinnverfahren *n* worsted spinning
Kammgarnstoff *m* worsted [yarn] fabric
Kammgarnwalke *f* worsted milling
Kammlager *n* collar thrust bearing
Kämmling *m* *(text)* noil, combing, comber waste
Kämmlingsband *n* *(text)* comber waste sliver
Kammputz *m* *(build)* combed stucco
Kammreihe *f* row of needles *(comber)*
Kammschlitzbarre *f* *(text)* carriage comb
Kammsegment *n* *(text)* comb segment
Kammspiel *n* *(text)* nip
Kammstab *m* *(text)* gill bar
Kammstromabnehmer *m* *(el)* comb[-type] brush
Kammstuhl *m s.* Kämmaschine
Kämmtrommel *f* *(text)* comb cylinder
Kämmungszahl *f* *(text)* carding number
Kammwalze *f* *(text)* porcupine
Kammwolle *f* worsted [wool], combing wool
Kammzapfen *m* dovetail tenon; thrust journal
Kammzug *m* *(text)* combed sliver, top
  fetthaltiger K. *(text)* oily top
Kammzugband *n* *(text)* combed sliver
Kammzugdruck *m* Vigoureux printing
Kammzugfärbeapparat *m* [illegible] dyeing apparatus
Kammzugplättmaschine *f* lisseuse
Kampanje *f* *(ship)* poop
Kampanjedeck *n* poop deck
Kampfer *m* camphor
Kämpfer *m* *(build)* impost, abutment, springer
Kämpferdruck *m* abutment pressure
  nacheilender K. back abutment pressure
  voreilender K. front abutment pressure
Kämpferdruckzone *f* abutment zone
Kampfergeist *m* camphorated spirits
Kämpferlinie *f* springing [line]
Kampferöl *n* camphor oil
Kampfersäure *f* camphoric acid
Kampferspiritus *m* spirit of camphor
Kämpferstein *m s.* Kämpfer
Kampfhandschuh *m* match glove *(boxing)*
Kampfstoff *m* warfare (toxic) agent
  biologischer K. biological warfare agent
  chemischer K. chemical warfare agent
  radioaktiver K. radioactive contaminant, radiological warfare agent
radiologischer K. *s.* radioaktiver K.
Kanadabalsam *m* Canada balsam

Kanadier *m* [Canadian] canoe
Kanal *m* 1. channel *(natural waterway)*; canal *(artificial waterway)*; 2. communication channel; channel *(frequency tape)*; *(dat) (Am)* trunk; 3. port *(valve)*; duct *(ventilation)*; passage *(turbine)*; conduit
K. der Reaktion channel of the reaction
einzügiger K. single-way duct
mehrzügiger K. multiple-way duct
Kanalabstand *m (el)* channel spacing
Kanalbildung *f* channelling
Kanalbreite *f* 1. channel width; duct width *(troposphere)*; 2. width of passage *(turbine)*
Kanaldampfer *m* cross-channel steamer
Kanaleffekt *m (nucl)* channelling [effect]
Kanaleffektfaktor *m s.* Kanalfaktor
Kanaleröffnung *f (therm)* opening of the port
Kanalfaktor *m (nucl)* channel[ling] effect factor
Kanalfeuchteregler *m* insertion humidity controller
Kanalgas *n* sewer gas
Kanalguß *m* sewer castings
Kanalhafen *m* canal harbour (port)
Kanalisation *f* sewerage [system]
Kanalisationsrohr *n* sewer
kanalisieren to sewer
Kanalisierung *f* canalization *(hydraulic engineering)*
Kanalkante *f* / **steuernde** *(therm)* distributing edge of the steam port
Kanalkennbuchstabe *m (el)* channel identification letter
Kanalofen *m* tunnel kiln (drying oven)
Kanalprofil *n (ship)* channel section
Kanalquerschnitt *m* 1. *(ship)* water cross-section; 2. *(therm)* cross-section of port
freier K. free section of the steam port
voller K. full section of the steam port
Kanalschiff *n* cross-channel vessel
Kanalschleuse *f* [canal] lock
Kanalschluß *m (therm)* closing of the port
Kanalsohle *f* ground of the channel; invert
Kanalstrahl *m* canal ray [particle], anode ray
Kanaltemperaturregler *m* insertion thermostat
Kanaltiefe *f* depth of passage *(turbine)*
Kanaltrennung *f (el)* channel separation
Kanaltrockner *m* tunnel drier
Kanalüberhitzung *f (nucl)* hot channel effect
Kanalumschalter *m* channel transfer switch
Kanalverstärker *m (nucl)* window amplifier
Kanalwähler *m (el)* channel selector
Kanalweite *f* width of the [steam] port
Kaninchenfell *n* rabbit [skin]
Kaninchenhaar *n* rabbit hair
Kanisteröl *n* case oil, packed petroleum
kannelieren to chamfer, to groove
Kannelierfräser *m* moulding cutter
Kannelierung *f* grooving, groove, channelling [flute]
Kannengatter *n (text)* can creel
Kannenkühlbecken *n* in-can-cooling system
Kannenkühler *m* in-can cooler
Kannenmelkanlage *f* direct-to-can milking installation
Kannenmelksystem *n* direct-to-can milking system
Kannenwechsel *m (text)* can changer
Kanone *f* gun
starre K. fixed gun
Kanonenbohrer *m* simple D-bit
Kanonenschießen *n (mine)* burn (shatter) cut
kanonisch canonical
Kante *f* edge; corner; border
abgefaste K. chamfer
abgeschrägte K. chamfered edge
brechende K. *(opt)* refracting edge
eingelegte K. inlaid border
gebrochene K. chamfer, bevelled corner
gefaste K. bevelled corner
zugeschärfte K. feather edge
kanten to edge, to cant, to tilt
einen Block k. to edge a bloom
Kantenabfall *m* taper of surfaces
Kantenausroller *m* selvedge uncurling device
Kantenbearbeitung *f* edge trimming
Kantenbiegung *f* springing *(of wood)*
Kantenbläue *f* blueness on the edges *(of wood)*
Kanteneffekt *m* edge effect
Kantenemission *f (nucl)* edge emission
Kantenfilter *n (aut)* edge-type filter
Kantenlänge *f* edge length, length of the side *(cube)*
Kantenpressung *f (build)* edge pressure
Kantenriß *m* edge fracture (break, crack), lap edge fracture
Kantenschutz *m* edging; stair nosing; angle bead *(in plastering)*
Kantenstecker *m* edge-type connector *(for printed wiring)*
Kantenstoß *m* **des Dampfes** shock at entry, shock of steam striking edge *(turbine)*
Kanten- und Facettenfräsmaschine *f* edge and bevel edge milling machine
Kantholz *n (wood)* squared (rectangular) timber, *(Am)* dressed lumber, dressed timber, scantling
Kantholzboden *m* cant deck
Kantholzkreissäge *f* circular saw for squaring logs
Kantrinne *f* guide channel *(rolling mill)*
Kantspant *n (ship)* cant frame; cant timber
Kantvorrichtung *f* manipulator; tilting (canting) device
Kanu *n* canoe
Kanuheck *n* canoe stern
Kanüle *f* cannula, tubule

Kanzel *f* cockpit
Kanzeleinstiegklappe *f (aero)* nose opening door
Kanzleipapier *n* foolscap (document, ledger) paper
Kaolin *n* kaolin, china clay
Kaolinfüllstoff *m* clay filler
Kaolinit *m (min)* kaolinite
Kaolinsand *m* kaolin sand
Kapazitanz *f (el)* capacitance, capacitive reactance
Kapazität *f* capacity; capacitance; efficiency
K. der Leitungen wiring capacity
K. in Ah ampere-hour capacity *(of storage battery)*
K. je Flächeneinheit unit-area capacitance *(of electrolytic capacitor)*
gegenseitige K. mutual capacitance
konzentrierte K. lump capacitance
tote K. dead capacitance
veränderliche K. variable capacitance
verteilte K. distributed capacitance
wirksame K. effective capacitance
kapazitativ capacitative
kapazitätsarm low-capacity, anti-capacitance
Kapazitätsausgleich *m* capacity balancing
kapazitätsbehaftet capacitive, condensive
Kapazitätsbelag *m* unit-area capacitance *(of electrolytic capacitor)*
Kapazitätsdiode *f* capacitance diode, voltage-variable capacitor; varactor
kapazitätsfrei non-capacitive
kapazitätsgerade straight-line capacitance
kapazitätslinear straight-line capacitance
Kapazitätsmeßbrücke *f* capacity (capacitance) bridge
Kapazitätsmesser *m* capacity meter[ing device], faradmeter
Kapazitätsmessung *f* capacity measurement
Kapazitätsnormal *n* capacitance standard
Kapazitätsprüfung *f* **nach Lagerung** *(el)* delayed test
Kapazitätssymmetrie *f* capacity balance
Kapazitätsübertrag *m* capacitance transfer
Kapazitätsunsymmetrie *f* capacity unbalance
Kapazitätsveränderung *f* / **zeitliche** capacity drift with time
Kapazitätsverhältnis *n* capacity ratio
Kapazitätsverlust *m (el)* deterioration
kapazitiv capacitive
Kapellenform *f (met)* cupel mould
Kapellenofen *m* cupellation (assay) furnace
Kapellenprobe *f* cupel test
Kaperei *f* s. Kaperfahrt
Kaperfahrt *f* privateering
Kaperschiff *n* privateer
kapillar capillary
Kapillaraffinität *f* capillary affinity
kapillaraktiv capillary active
Kapillaraktivität *f* capillary activity
Kapillaraszension *f* elevating (rise) of liquid in [a] capillary tube
Kapillardepression *f* capillary depression
Kapillardruck *m* capillary pressure
Kapillare *f* capillary (small-bore glass) tube
Kapillarelektrometer *n* capillary electrometer
Kapillarität *f* capillarity
Kapillaritätskonstante *f* capillary constant
Kapillarkondensation *f* capillary condensation
Kapillarkraft *f* capillary force
Kapillar-Rheometer *n* caplastometer
Kapillarrohr *n* capillary [tube]
Kapillarsäule *f* capillary column *(gas-chromatography)*
imprägnierte K. coated capillary
Kapillarstrom *m* electro-capillary current
Kapillarströmung *f* capillary flow
Kapillarthermometer *n* filled-system thermometer
Kapillarviskosimeter *n* capillary-tube viscosimeter, capillary-type viscosimeter, friction-type viscosimeter
Kapillarwasser *n* water of capillarity, capillary water
Kapillarwirkung *f* capillary effect (action)
Kapillarzone *f* capillary fringe
Kapitalband *n (print)* headband
Kapitalkonto *n (ec)* stock account
Kapitän *m* captain; master; skipper
K. auf kleiner Fahrt coastwise trade master
Kapitänspatent *n* master's certificate
Kapitell *n* capital (architecture)
Kaplan-Turbine *f* Kaplan water turbine
Kapnoskop *n* smoke gauge
Kapok *m* kapok
Kapokentkörnungsmaschine *f (text)* kapok gin[ning machine], kapok twirling machine
Kappe *f* cover; hood; cap; *(mine)* [roof] bar, cap[-piece], crown-piece, girder, girt, head-piece
gewölbte K. *(mine)* cambered girder *(support)*
kappen to poll, to cut [off]
Kappenanschluß *m* **[an einer Röhre]** *(el)* top cap
Kappenausbau *m (mine)* bar timbering, timber set without posts
Kappengeschoß *n (mil)* capped shell
Kappengewölbe *n (build)* jack vault
Kappenisolator *m* cap-and-pin type insulator
Kappenkopf *m* gib and cotter end, butt (strap) end *(rod)*
Kappenlochmaschine *f* toe-cap perforating machine *(shoe)*
Kappenschleifmaschine *f* counter buffing machine *(shoe)*
Kappensteife *f* puffs and stiffeners *(shoe)*
Kapp-Faktor *m (el)* Kapp coefficient
Kappnaht *f* flat fell seam *(shoe)*

Kapp-Phasenschieber *m (el)* Kapp-type phase advancer
Kappsäge *f* cross-cut saw
Kappschuh *m (mine)* bar shoe
Kapr[in]at *n* caprate
Kaprinsäure *f* capric acid
Kaprolaktam *n* caprolactam
Kapronsäure *f* caproic acid
Kaprylat *n* caprylate, caprilate
Kaprylsäure *f* caprylic acid
Kapsel *f* capsule; can, jacket, casing, box
Kapselfedermanometer *n* capsule-type pressure gauge
Kapselfederzugmesser *m* double-diaphragm draught gauge
Kapselgebläse *n* valveless compressor, enclosed blower
kapseln to can, to encapsulate
Kapselpumpe *f* mercury rotating pump
Kapselschalen *fpl* cottonseed hulls
Kapselton *m* saggar clay
Kaptur *f* capture *(of a ship)*
Karabiner *m* car[a]bine
Karabinerhaken *m* snap hook, carabine swivel
Karamel *m* caramel
Karat *n* carat
Karatgewicht *n* troy-weight
Karätigkeit *f* carat value
Karbamidharz *n* urea formaldehyde resin
Karbaminat *n* carbaminate, carbamate
Karbaminsäure *f* carbamic (carbamino) acid
Karbid *n* 1. carbide (*metal*); 2. calcium carbide
Karbidausschlackung *f* carbide slagging
karbidbildend carbide-forming
Karbideinlagerung *f* / **zeilenförmige** *(met)* carbide band
Karbideinwurfentwickler *m* carbide-to-water gas generator
Karbidhartmetall *n* cemented carbide
Karbidlampe *f* carbide lamp
Karbidstabilisator *m (met)* carbide stabilizing element
Karbidstaub *m* carbide powder
Karbidzeile *f (met)* carbide band
Karbinol *n* carbinol
Karbolineum *n* carbolineum
Karbolsäure *f* carbolic acid, phenol
Karbonat *n* carbonate
Karbonatbildung *f* carbon[at]ation
Karbonathärte *f* carbonate hardness (*water*)
Karbondruckmaschine *f* carbon coating machine
Karbonfarbe *f* carbon black
karbonisieren (*met*) to carbonize, to carbonate, to carburize
Karbonisierfehler *m* carbonizing defect
Karbonisierfleck *m (text)* carbonizing stain
Karbonisierflotte *f (text)* carbonizing bath
Karbonisierkammer *f (text)* carbonizing chamber
Karbonisierlauge *f (text)* carbonizing liquor
Karbonisiermaschine *f (text)* carbonizing machine
Karbonisierofen *m (text)* carbonizing stove
Karbonisiertrommel *f (text)* carbonizing drum
Karboniumion *n* carbonium ion
Karbonrohseide *f* carbon base paper
Karbonsäure *f* carboxylic acid
Karbonyl *n* carbonyl
Karbonyleisen *n* carbonyl iron
Karbonylgruppe *f* carbonyl group
Karbonylierung *f* carbonylating
Karbonylsauerstoff *m* ketonic oxygen, carbonyl oxygen
Karborund *m* carborundum
Karbowidwiderstand *m (el)* rod-type carbon resistor
Karboxylase *f* carboxylase
Karboxylgruppe *f* carboxyl group
karburieren *(met)* to carburize; to carburet
Karburierung *f* carburation; carburetting
Kardangelenk *n* cardan (universal) joint, Hooke's coupling
Kardangelenkwelle *f* universal jointed shaft
Kardanring *m* gimbal
Kardanwelle *f* cardan (propeller) shaft
Kardätsche *f (ag)* horse-brush
Karde *f (text)* card, carding engine (machine); teasel
gemischte K. mixed card
Kardenabfallreinigungsmaschine *f* card waste cleaning machine
Kardenband *n* card[ed] sliver
Kardenbandnummer *f* card sliver count number
Kardenbandzuführung *f* card delivery
Kardenflug *m* card fly
Kardennadelrichter *m* card grinder
Kardenrahmen *m* teasel frame
Kardenrauhmaschine *f* teasel raising machine
Kardenraum *m* card room
Kardenschleifer *m* card grinder
Kardensetzer *m* card fixer (setter)
Kardenstaub *m* carding dust
Kardenvlies *n* card web, fleece coming from the card
Karderie *f* card room
Kardierabfall *m* card waste
Kardierflügel *m* carding arm
Kardierkonstante *f* carding constant
kardiert carded
Kardinalfläche *f* cardinal plane
Kardinalpunkt *m* cardinal point
Kardioide *f* cardioid
Kardioidenkennlinie *f* heart-shaped diagram *(aerial)*
Karierfehler *m* wrong checking pattern

Karierwebstuhl *m* multiple box (shuttle) loom
Karkasse *f* carcass
Karmin *n* carmine
Karnaubawachs *n* carnauba wax
Karnotit *m (min)* carnotite
Karolus-Fernsehverfahren *n* Karolus system
Karosserie *f* coachwork; car body
 selbsttragende K. chassisless construction, monocoque (integral body and chassis) unit
Karosseriebeschlag *m* body fitting
Karosserieblech *n* car body sheet
Karosserieverkleidung *f* / **äußere** outer body shell
Karotin *n* carotene
Karotinoid *n* carotenoid
Karottagegerät *n* bore-hole logging apparatus
karren to cart, to wheel
Karren *m* truck; *(build)* buggy, barrow, cart; *(ag)* cart, wagon; *(mach)* carriage, bed
Karrenantrieb *m* carriage (bed) drive
Karrenbahn *f* carriage (bed) track
Karrenbegichtung *f* barrow charging *(blast furnace)*
Karrenpflug *m* wheel plough
Karrenrücklauf *m* bed-return
 zeitlich verkürzter K. accelerated bed-return
Karrenspritze *f (ag)* wheelbarrow sprayer
Karrenwalze *f (leath)* butt roller
Kartätsche *f* case shot
Karte *f* 1. card; 2. map; 3. chart; 4. ticket
 eingespannte K. *(text)* card fitted in
 fehlerhafte K. faulty card
 meteorologische K. meteorological chart
 topografische K. topographical map
Kartei *f* card file (index)
Karteiabfühler *m* card reader *(punched card machinery)*
Karteikarte *f* filing card
Karteikarton *m* index board
Karteikasten *m* filing cabinet
Karteikerbzange *f* index card punch
Karteisystem *n* card system
Kartellschiff *n* ship with flag of truce
Kartenabtaster *m* card reader
Kartenaufnahme *f* surveying; mapping
Kartenbindemaschine *f (text)* card lacing machine
 automatische K. automatic lacing machine
Kartenbinder *m (text)* card lacer
Kartendoppler *m* card duplicator (reproducing puncher), reproducing (gang) puncher
 K. mit Zeichenabfühlung mark-sensing reproducer punch
Kartendurchlauf *m* passage (run) of cards
Kartenhals *m* throat *(punched card machinery)*
Kartenhaus *n (ship)* chart house
Kartenherstellung *f* mapping
Kartenkette *f (text)* chain of cards
Kartenkopieren *n (text)* card repeating
Kartenkunde *f* cartography
Kartenkurs *m (ship)* chart (real, true) course
Kartenlesestation *f* card reader *(punched card machinery)*
Kartenlochen *n* card cutting (perforating)
Kartenlocher *m* card perforator (punch)
 lesender K. card read punch *(punched card machinery)*
 lochstreifengesteuerter K. tape-to-card converter
Kartenlochlage *f* card hole location
Kartenmaßstab *m* map scale
Kartennetzentwurf *m* map projection
Kartennull *f (ship)* chart datum
Kartenprisma *n (text)* card (Jacquard) cylinder
Kartenrückschlagvorrichtung *f (text)* card reversing motion
Kartenschlagen *n* card cutting (perforating, punching)
Kartenschläger *m (text)* card cutter
Kartenschlagmaschine *f (text)* card punching machine
Kartenschnürer *m (text)* card lacer
Kartenspareinrichtung *f (text)* card saving device
Kartenstanzen *n* card cutting (perforating, punching)
Kartenstanzer *m* card punch
Kartenstapler *m* card stacker
Kartenstau *m* card jam *(in punched card machines)*
Kartenvorschub *m* card feed[ing]
Kartenzähler *m* card counter
Kartenzuführer *m* card feeder
Kartenzuführung *f* card feed[ing]
kartesianisch Cartesian *(diver)*
kartesisch *(math)* Cartesian
kartieren to map
Kartieren *n* mapping
Kartiergerät *n* mapping instrument
Kartiertisch *m* plotting table
Kartierung *f* mapping
Kartoffel *f* potato
Kartoffelanbau *m* potato growing
Kartoffeldämpfanlage *f* potato steamer (boiler)
Kartoffeldämpfer *m* potato steamer (boiler)
Kartoffeldämpfmaschine *f* potato steamer
Kartoffeleinsäuerung *f* potato ensilage
Kartoffeleinsäuerungsbehälter *m* potato ensilage pit
Kartoffelelevator *m* potato elevator
Kartoffelernte *f* potato harvest
Kartoffelerntemaschine *f* potato harvester
Kartoffelhöhenförderer *m* potato elevator
Kartoffelkäfer *m* Colorado beetle, potato beetle
Kartoffelkiepe *f* potato hamper
Kartoffelkombine *f* complete potato harvester

Kartoffelkorb *m* potato basket
Kartoffelkraut *n* [potato] haulm
Kartoffelkrautschläger *m* potato haulm cutter
Kartoffellegemaschine *f* potato planter
Kartoffellesen *n* potato picking
Kartoffelmehl *n* potato flour
Kartoffelmiete *f* potato clamp (pit)
Kartoffelmietendecke *f* potato clamp cover
Kartoffelquetsche *f* potato crusher
Kartoffelrodekörper *m* potato raising body
Kartoffelroden *n* potato digging (lifting)
Kartoffelrodepflug *m* potato plough
Kartoffelroder *m* potato digger
Kartoffelschrägförderer *m* potato elevator
Kartoffelsetzmaschine *f* potato planter
Kartoffelsortiermaschine *f* potato grader (sorter)
Kartoffelstampfer *m* potato masher
Kartoffelstärke *f* potato starch
Kartoffelsteckmaschine *f* potato planter
Kartoffelvollerntemaschine *f* complete potato harvester
Kartoffelwaschmaschine *f* potato washer
Kartoffelzüchtung *f* potato breeding
Kartograf *m* cartographer, map drawer
Kartografie *f* cartography
Kartogramm *n* cartogram
Karton *m* 1. carton *(box)*; 2. boxboard, cardboard, pasteboard
gegautschter K. wet-pressed board
kaschierter K. combined board
mehrlagig gegautschter K. combination board
minderwertiger K. cardboard middles
papierkaschierter K. paper board combination
Kartonagemaschine *f* cardboard box machine
Kartonagen *fpl* [paper and] cardboard containers, [paper and] cardboard, cardboard boxes [box products]
Kartonagenfabrik *f* card[board] mill
Kartonagenglanzpapier *n* box enamels
Kartonagenmaschine *f* cardbox (cardboard) machine
Kartonagepappe *f* box board
Kartonagezuschnitte *mpl* board-cuttings for composing
kartonieren to put in boards
kartoniert bound in paper boards
Kartonmaschine *f* board machine
Kartonpapier *n* cardboard
Kartonpappe *f* boxboard
Kartonrollenschneidmaschine *f* cardboard reel cutter
Kartonschneider *m* cardboard trimmer (cutter)
Kartusche *f* 1. gun cartridge; 2. cartouche
Kartuschenfilter *n* / **hydraulisches** hydraulic depth filter
Karussell *n s.* Karusselldrehmaschine
Karussellaufnahme *f (nucl)* panoramic exposure
Karusselldrehmaschine *f* vertical boring and turning mill
K. mit Seitensupport vertical turret lathe, vertical boring and turning mill with side head
Karussellförderer *m* merry-go-round conveyor
Karussellrevolverdrehmaschine *f* vertical turret lathe
Karweel . . . *s.* Kraweel . . .
Karzinom *n* carcinoma
Karzinotron *n* carcinotron; backward-wave oscillator
kaschieren to coat, to apply coatings, to laminate *(fabrics)*, to back, to line
Kaschieren *n* **mit Schaumstoff** foam backing (lining)
K. mit Schneckenfütterung extrusion coating
Kaschierfolie *f* / **klebefertige** pressure-sensitive adhesive film
Kaschierkalander *m* lining calender
Kaschiermaschine *f* lining (paper mounting) machine, board liner; laminating machine
Kaschierpapier *n* lining paper, liner
Kaschierrohmaterial *n* cardboard middles
Kaschierrohpappe *f* filler board
kaschiert / mit Schaumstoff foam-backed
Kaschierung *f* lining
Kaschierwachs *n* laminating wax
Kaschmirwolle *f* Cashmere hair
Kaschmirziege *f* Cashmere goat
Kaschunuß *f* cashew nut
Käseform *f* cheese mould
Kasein *n* casein
Kaseinfarbe *f* casein paint
Kaseinfaser *f* casein fibre
Kaseinleim *m* casein glue
Käsepresse *f* cheese press
Käsewanne *f* cheese vat
Kaskade *f* cascade [set]
Kaskaden-Gamma-Strahlung *f* cascade (successive) gamma-rays
Kaskadengenerator *m (nucl)* cascade generator; Cockcroft-Walton accelerator
Kaskadenkühler *m* cascade cooler
Kaskadenlichtbogen *m* cascade arc
Kaskadenmotor *m* concatenated motor
Kaskadenregelung *f* cascade (metered) control system, master-and-slave control
Kaskadenschaltung *f (el)* cascade (concatenated) connection
Kaskadenschauer *m (nucl)* cascade [shower]
Kaskadensichter *m* cascade-type deduster
Kaskadenspannungswandler *m* cascade-connected voltage transformer
Kaskadenstrahlung *f* cascade radiation (emission)
Kaskadentrockner *m (ag)* cascade drier
Kaskadenübergang *m* cascade transition
Kaskadenüberschlag *m* cascading *(of insulators)*

Kaskadenübertrag *m (cont)* cascade[d] carry
Kaskadenumformer *m* cascade (motor) converter, motor-generator
Kaskadenverstärker *m* cascade amplifier
Kaskadenwandler *m* cascade instrument transformer
Kaskadenzerfall *m* cascade decay
Kaskodenschaltung *f* cascode circuit
Kaskodenverstärker *m* cascode amplifier
Kaskoversicherung *f* comprehensive insurance
Kasse *f* 1. cash-office; booking (ticket) office; 2. cash register
Kassenabschnitt *m* sales slip
Kassenanweisung *f* bank bill; cash order
Kassenblock *m* cash pad (block), check pad
Kassenblockmaschine *f* cash block making machine
Kassenbuchungsblatt *n* cash tally sheet
Kassenschein *m* bank bill
Kassenzettel *m* sales slip
Kassette *f* 1. cassette; 2. *(opt)* dark-slide; 3. *(nucl)* casket; assembly, assemblage, assembling (*s. a.* Brennstoffkassette)
Kassettendecke *f (build)* coffered (panelled) ceiling
Kassettenplatte *f (build)* coffered (waffle, caisson) slab, coffered (waffle, caisson) panel
Kassiaöl *n* cassia oil
Kassiopeium *n* cassiopeium, lutecium
Kassiterit *m (min)* cassiterite, tinstone
Kastanienbaum *m* chestnut [tree]
Kastanienholz *f* chestnut
Kasten *m* 1. box, case; 2. *(ag)* bin *(feeding)*; 3. drawer; trunk; cabinet *(filing)*; 4. *(mine)* chock; 5. vaulting box
Kastenbalken *m* box beam
Kastenballenbrecher *m (text)* hopper bale breaker
Kastenballenöffner *m (text)* hopper bale opener
K. mit Staubabzug exhaust hopper bale opener
Kastenbett *n* box-like bed
Kastenboden *m* floor-plate of magazine *(rifle)*
Kastenbühne *f* box-sided working platform
Kastendamm *m s.* Kastenfang[e]damm
Kastendämpfer *m (text)* cottage steamer
Kastendock *n* box dock
Kastendrän *m* box drain
Kastendurchlaß *m (build)* box culvert
Kastenfang[e]damm *m (build)* box-type cofferdam
Kastenformerei *f* flask moulding
kastenförmig [ausgebildet] *(mach)* box like
k. ausgeführt *(mach)* of box-section construction
Kastenfuß *m (mach)* cabinet leg; box[-type] pedestal
Kastenglucke *f (ag)* hover, brooder
Kastenglühen *n* flask annealing
Kastenglühofen *m* box annealing furnace
Kastenglühung *f* close annealing
Kastenguß *m* box casting *(process)*; box castings *(pieces)*
Kastenkamera *f* box camera
Kastenkiel *m (ship)* duct keel
Kastenkielschwein *n (ship)* box keelson
Kastenkippwagen *m* box tipping wagon
Kastenrahmen *m* box-section frame
Kastenrinne *f* box (trough) gutter
Kastenrudern *n* practice in the fixed tub
Kastensäule *f (build)* box column
Kastenschute *f* closed barge
Kastenschwelle *f (ry)* box sleeper
Kastensinkscheider *m* dense-medium bath
Kastenspeiser *m* hopper feeder, feeder hopper *(cotton spinning)*
selbsttätiger K. automatic hopper feeder
Kastenständer *m (mach)* box-type column, square-box-section column
Kastenständerausführung *f (mach)* box-column model (design, type)
Kastenständerbohrmaschine *f* [vertical] box column drill
Kastentisch *m (mach)* box[-type] table
Kastenträger *m (build)* box beam (girder)
Kastenvorderkipper *m (met)* box end tipping wagon
Kastenwagen *m* box truck *(conveying)*; [box-type] delivery van *(hauling)*
Kastenzimmerung *f (mine)* boxing
Katakaustik *f* catacaustic
Katalase *f* catalase
Katalog *m* catalogue
katalogisieren to catalogue
Katalysator *m* catalyst, catalyzer
flüssiger K. liquid catalyst
Katalysatorgift *n* catalyst (catalytic) poison, paralyzer
Katalysatorträger *m* catalyst support
Katalyse *f* catalysis
negative K. *(ch)* negative catalysis, reaction inhibition
katalysieren *(ch)* to catalyze
katalysierend *(ch)* catalyzing
katalytisch catalytic *(cracking)*
Katamaran *m* catamaran
Kataphorese *f* cataphoresis
Katapultierbeschläge *mpl (aero)* catapult fittings
Katapultiereinrichtung *f* catapult equipment
Katapultstart *m* catapult launching, catapult take-off
Katapultstartanlage *f* catapult launcher
Kataster *n* cadastre, land register
Katasteraufnahme *f* cadastral survey
Katastergenauigkeit *f* cadastral accuracy

Katathermometer *n* kata-thermometer
Katechin *n* catechin
Katechu *n* catechu
Katenoid *n* catenoid
Katgut *n* catgut
Kathedralglas *n* cathedral (diffusing, configurated) glass
Kathete *f (math)* leg
Kathetometer *n* cathetometer
Kathode *f s.* Katode
Katholyt *m s.* Katolyt
Kation *n* cation
Kationenabspaltung *f* decationizing
Kationenaustauscher *m* cation exchanger, cationite
Kationenwanderung *f* cation migration
Katode *f* cathode
 aktivierte K. activated cathode
 bespritzte K. sprayed cathode
 direkt geheizte K. directly heated cathode, filament-type cathode
 einkristalline K. monocrystalline cathode
 flüssige K. pool cathode
 geschichtete K. coated cathode
 indirekt geheizte K. indirectly heated cathode, heater-type cathode
 kalte K. cold cathode
 komplexe K. complex cathode
 konkave K. concave cathode
 mehrkristalline K. polycrystalline cathode
 pastierte K. coated cathode
 reelle K. actual cathode
 vielschichtige K. multi-layered cathode
 virtuelle K. virtual cathode
 zylinderförmige K. cylindrical cathode
Katodenabbau *m* cathode disintegration
Katodenableiter *m* cathode arrester
Katodenableitung *f* cathode tail
Katodenanheizzeit *f* cathode heating time
Katodenauskoppelung *f* cathode follower output
Katodenbändchen *n* cathode connecting strip
Katodenbasisschaltung *f* grounded cathode circuit
Katodenbasisverstärker *m* grounded cathode amplifier
Katodenbecher *m* focusing cup *(Roentgen tubes)*
Katodendunkelraum *m* cathode dark space
Katodenfall *m* cathode fall (drop), potential fall (drop)
Katodenfallableiter *m* cathode (multigap) arrester
Katodenfläche *f* cathode surface
Katodenfleck *m* cathode spot
Katodenfolger *m* cathode follower
Katodenglimmlicht *n* cathode glow
Katodenhals *m* cathode neck
Katodenhalterung *f* cathode anchor *(electron tube)*
Katodenheizleistung *f* cathode heating power
Katodenimpulsmodulation *f* cathode pulse modulation, c.p.m.
Katodenkondensator *m* cathode by-pass capacitor
Katodenkupfer *n* cathode copper
Katodenleitwert *m* cathode conductance
Katodenlicht *n* cathode glow
Katodenmodulation *f* cathode modulation
Katodenniederschlag *m* cathodic (cathode) deposit
Katodenoberfläche *f* cathode surface
Katodenpolarisation *f* cathodic polarization
Katodenröhrchen *n* cathode sleeve
Katodensaum *m* cathode border
Katodenschicht *f* cathode layer
Katodenschutz *m* cathodic protection *(corrosion)*
Katodenspalt *m* cathode gap
Katodenstrahl *m* cathode ray
Katodenstrahlbündel *n* cathode-ray pencil
Katodenstrahlfernsehen *n* television by electronic methods
Katodenstrahlfunktionsgenerator *m* cathode-ray function generator
Katodenstrahlleuchtfleck *m* cathode-ray spot
Katodenstrahloszillograf *m* cathode-ray oscillograph
Katodenstrahlröhre *f* cathode-ray tube
 K. mit ebener Schirmfläche flat-ended cathode-ray tube
Katodenstrom *m* cathode current
Katodenstromsteilheit *f* rate of rise of cathode current, ratio of cathode current change to grid
Katodenstufe *f* cathode follower
Katodentastung *f* cathode keying
Katodenvergiftung *f* cathode contamination (poisoning)
Katodenverstärker *m* cathode amplifier; anode base amplifier; cathode-coupled circuit; cathode follower
Katodenvorspannung *f* cathode bias
Katodenwiderstand *m* grid bias resistance; cathode resistor
Katodenzerstäuben *n* cathode sputtering (disintegration)
Katodolumineszenz *f* cathodoluminescence
Katodophosphoreszenz *f* cathodophosphorescence
Katolyt *m* catholyte
Katoptrik *f (ph)* catoptrics
katoptrisch catoptric
Kattanker *m (ship)* back[ing] anchor
Kattdavit *m(n) (ship)* cathead, cat davit (crane)
katten to cat *(an anchor)*
Kattläufer *m (ship)* cat fall (rope)
Katt-Takel *n (ship)* cat tackle
Kattun *m* calico; cotton [cloth]; chintz *(furniture)*

Kattunbindung *f* calico weave
Kattundruck *m* cotton print
Kattundruckpapier *n* chintz (calico) paper
Katze *f* trolley, crab, jenny *(conveying)*
Katzenauge *n* cat's eye; [rear red reflex] reflector
Katzfahrbahn *f* trolley track *(conveying)*
Katzfahrschaltung *f* crab connection *(trolley)*
Katzrahmen *m* trolley frame *(conveying)*
Kauf *m* purchase, buying
Kaufabschluß *m* purchase
Kaufbedingung *f* condition concerning the purchase
Käuferwunsch *m* customer requirement
Kauffahrteiflotte *f* merchant (mercantile) fleet, commercial navy (marine)
Kauffahrteischiff *n* merchantman, merchant (trading) vessel, trader
Kaufkraft *f* purchasing power
Kaufpreis *m* purchase price
Kaufschrott *m* purchased scrap
Kaufvertrag *m* contract of sale
Kaufwert *m* market value
Kaukamm *m (mine)* axe
Kaurikopal *m* Kauri copal
Kausche *f (ship)* deadeye; thimble
Kaustifizierbottich *m* smelt-dissolving tank *(papermaking)*
Kaustifizieren *n* causticizing operation
Kaustik *f* caustic curve (surface); caustics
Kaustiktransformator *m* cautery transformer
kaustisch caustic
Kaustizieren *n* causticizing operation
Kautschuk *m* caoutchouc
Kautschuklösung *f* caoutchouc solution *(sealing)*
Kavalierperspektive *f (math)* cavalier perspective
Kaverne *f (nucl)* flaw
Kavitation *f* cavitation
  K. an eintretender Kante leading edge cavitation
  K., die sich vom Körper in den Nachlauf erstreckt trailing cavity
  K. hinter maximaler Profildicke aft cavitation
  K. im Nachlauf wake cavitation
  K. vor maximaler Profildicke forward cavitation
  beginnende K. incipient cavitation
  teilweise K. partial cavitation
  voll entwickelte K. fully developed cavitation
Kavitationsbildung *f* cavity formation
Kavitationstank *m* cavitation (Lightgow) tunnel
Kavitationsverlust *m* cavitation loss
Kederheftmaschine *f* tacking machine *(shoemaking)*
Kegel *m* 1. *(math)* cone; 2. *(mach)* taper, nose *(spindle)*; plug *(valve)*; 3. *(text)* cone *(bobbin)*; 4. *(print)* body [size], point size, type body
  abgestumpfter K. *(math)* truncated cone
  gerader K. *(math)* right cone
  Machscher K. Mach front (stem, cone)
  schiefer K. *(math)* oblique cone
  steiler K. *(mach)* steep (abrupt) taper
  stumpfer K. *(math)* truncated cone
  umbeschriebener K. *(math)* circumcone
Kegelachse *f* axis of the cone
Kegelanode *f* cone anode
Kegelantenne *f* conical aerial
Kegelbohrung *f* taper bore (hole), conical bore (hole)
Kegelbrecher *m* cone crusher
Kegeldreheinrichtung *f* taper turning attachment
Kegeldrehen *n* taper turning
Kegeldruckhärte *f (met)* conical indentation hardness
Kegeldruckprüfung *f (met)* cone-thrust test
Kegeleinsatz *m (mach)* taper adapter
Kegelfeder *f (mach)* conical coil spring
kegelförmig *(math)* conical; cone-shaped; tapered
Kegelfräseinrichtung *f* attachment for taper milling, taper milling attachment
Kegelfräsen *n* taper milling
Kegelfußpfahl *m (build)* caisson pile
Kegelhandreibahle *f* tapered hand reamer
Kegelhärteprüfung *f (met)* cone-hardness test
Kegelhobeln *n* taper planing
Kegelhülse *f (mach)* taper sleeve
kegelig conical, taper
  k. aufreiben to taper-ream
  beiderseits k. having a taper on both sides
Kegeligbohren *n* taper drilling (boring)
Kegeligkeit *f* conicity
Kegelklemmhülse *f (mach)* split taper sleeve *(drill)*, collet-type sleeve, drill driver
Kegelkondensator *m* cone-type capacitor
Kegelkreiselmischer *m* cone impeller
Kegelkupplung *f* cone (conical) clutch
Kegellager *n* cone bearing
Kegellehre *f* taper gauge
Kegellineal *n (mach)* tangent bar
Kegelmessung *f* taper measurement, measurement of cone
Kegelmühle *f* conical refiner, *(Am)* jordan mill
Kegelnabe *f (aut)* bevel hub
Kegelöffner *m* **mit Voröffner** *(text)* vertical opener combined with porcupine opener
Kegelpendel *n* conical pendulum
Kegelpendelregler *m* governor with sleeve
Kegelrad *n* bevel gear
  K. mit Bogenverzahnung curved-tooth type bevel gear
  K. mit geraden Zähnen straight[-tooth] bevel gear
  gerades balligtragendes K. Coniflex gear
Kegelradantrieb *m* bevel-gear drive

Kegelradformfräser *m* bevel-gear formed cutter
Kegelradgehäuse *n* bevel-wheel casing
Kegelradgetriebe *n* bevel-wheel train
Kegelradhobelmaschine *f* bevel-gear planer (planing machine)
Kegelradpaar *n* pair of bevel gears
Kegelradtrieb *m* bevel-wheel drive (transmission)
Kegelradwälzfräsmaschine *f* bevel-gear generator
Kegelradwälzstoßmaschine *f* generating gear shaper for cutting bevel gears
Kegelrefiner *m s.* Kegelmühle
Kegelreibahle *f* taper reamer
K. für Morsekegel *(mach)* socket reamer for Morse tapers
K. für Morsekegel mit Zylinderschaft *(mach)* parallel shank socket reamer for Morse tapers
K. mit Kegelschaft *(mach)* taper shank socket reamer
Kegelreibungskupplung *f* cone friction clutch
Kegelritzel *n* bevel-gear pinion
Kegelrollenlager *n* tapered roller bearing
Kegelschaft *m* taper shank *(tool)*
Kegelschaftbohrer *m* taper-shank drill
Kegelschaftreibahle *f* taper-shank reamer
Kegelschale *f* conical shell
Kegelschälreibahle *f* steep-spiral taper reamer, taper reamer with steep spiral angle
Kegelschleifen *n* taper grinding
Kegelschnitt *m* conic [section]
Kegelschnittleitlinie *f* directrix of a conic
Kegelschraube *f* screw with tapered head; conical screw
Kegelsitzventil *n* bevel seated valve
Kegelstärke *f (print)* point (type) size
Kegelstift *m* taper pin
Kegelstiftbohrung *f* taper-pin hole
Kegelstück *n* wedge
Kegelstuhl *m (text)* draw loom
Kegelstumpf *m* frustum of a cone
Kegeltonne *f (ship)* conical [buoy]
Kegelventil *n* cone (conical, poppet) valve
Kegelwalzenlager *n* taper roller bearing
Kegelwinkel *m* **der Düse** angle of cone of the nozzle *(turbine)*
Kegelzahnrad *n* bevel gear
Kegligkeit *f* conicity
Kehlausrundung *f (build)* fillet
Kehlbalken *m (build)* collar beam
Kehlbalkendach *n* collar[-beam] roof
Kehlbeitel *m* gouge
Kehlblech *n (build)* flashing
Kehlbohle *f (build)* valley board
Kehle *f (build)* valley *(roof)*; fillet; chamfer
kehlen to channel, to groove, to chamfer
Kehlenzahn *m* parrot (briar) tooth
Kehlhobel *m* moulding plane
Kehlhobelmaschine *f* contour shaping machine; wood moulding machine
Kehlkopfmikrofon *n (el)* necklace (throat-type) microphone; laryngophone
Kehlleiste *f (build)* fillet strip
Kehlleistenfabrik *f* wood moulding factory
Kehlmaschine *f* **für kegelige (konische) Leisten** moulding machine for conical strips
Kehlmesser *n* moulding iron
Kehlnaht *f* fillet weld
Kehlsparren *m (build)* valley rafter
Kehlung *f* moulding *(of wood)*
Kehrbewegung *f (therm)* retrogressive movement
Kehre *f* return end *(coal plough)*
Kehricht *m* refuse, litter, sweepings
Kehrmaschine *f* mechanical (rotary, road) sweeper, road sweeping machine
Kehrpflug *m (ag)* turnabout (one-way) plough
Kehrtkurve *f* / **hochgezogene** *(aero)* wing-over
Kehrtstrecke *f* lap drawing frame, ribbon lap machine
Kehrwert *m* reciprocal
Keil *m* 1. *(build, telev)* wedge; 2. *(mach)* [taper] key; spline; 3. *(mine)* nog *(cut)*; 4. *(print)* sliding quoin; space band *(linotype)*; wedge *(monotype)*
keilartig wedge-shaped, feather-edged, tapered
Keilbauch *m (mach)* bottom of key
Keilbefestigung *f (mach)* keying
Keilbeilage *f (mach)* gib
Keilbolzen *m (mach)* wedge bolt
Keileinbruch *m (mine)* V-cut, wedge out
Keilfilter *n* wedge filter
keilförmig wedge-shaped
Keilfotometer *n* wedge photometer
Keilkranz *m (mine)* curb ring, wedging crib
Keillappen *m (mach)* tang
Keilleiste *f (mach)* gib strip
Keilloch *n (mach)* slot for cotter, cotter slot
Keillochfräsmaschine *f* cotter slot milling machine
Keilloch- und Keilnutenfräsmaschine *f* cotter and keyway milling machine
Keilnabe *f (mach)* splined hub
Keilnabenprofil *n* internal splines
Keilnachstelleiste *f (mach)* taper[ed] gib
Keilnase *f (mach)* gib head
Keilnut *f (mach)* keyway
Keilnutenfräsen *n* keyway cutting (milling)
Keilnutenfräser *m* keyseating cutter
Keilnutenräumen *n* keyway broaching
Keilnutenstoßmeißel *m* keyway tool
Keilnutenziehmaschine *f* keyseating machine working on the draw cut principle
Keilnutmaschine *f* keyseater
Keilpaar *n (mach)* pair of keys, angular parallels *(clamping)*

Keilplatte *f* key plate
Keilprofil *n (mach)* splines, spline profile
Keilramme *f* wedge ram *(slipway)*
Keilriemen *m* vee belt, *(Am)* V-belt
Keilriemenscheibe *f* vee belt pulley, *(Am)* V-belt pulley
Keilriementrieb *m* vee belt drive, *(Am)* V-belt drive
Keilring *m* conical clutch
Keilrücken *m (mach)* top of key
Keilschraube *f (mach)* splined bolt
Keilschubgetriebe *n (therm)* wedge effecting rectilinear motion, straight-wedge mechanism
Keilsitz *m (mach)* seat for key
Keilspatien *npl (print)* quoin spaces
Keilspundung *f* vee grooving and tonguing
Keilstein *m (build)* arch stone, voussoir
Keilstück *n* wedge-shaped piece
Keilstücke *npl* taper parallels, angular parallels *(clamping)*
Keiltaste *f (print)* space key
Keiltransistor *m* wedge-type transistor
Keiltreiber *m* cotter driver
Keiltrichter *m (met)* wedge-shaped gate
Keilverankerung *f* wedge anchorage *(prestressed concrete)*
Keilverbindung *f* keyed joint; cottering
  stellbare K. gib and cotter
Keilverschluß *m* 1. *(mil)* wedge breechblock (breech mechanism); 2. cam lock
Keilwelle *f (mach)* splined shaft
Keilwellenaußendurchmesser *m* major diameter [of external splines]
Keilwellenfräser *m* spline[-shaft] milling cutter
Keilwellenfräsmaschine *f* spline-shaft milling machine, spline mill
Keilwelleninnendurchmesser *m* minor diameter [of external splines]
Keilwellenprofil *n* external splines
Keilwinkel *m* [cutting-]wedge angle *(tool)*
  K. β [am Spiralbohrer] lip angle
Keilzapfen *m* tang
Keilziegelstein *m* arc (feather-edged) brick
Keim *m* 1. *(ag)* germ; 2. [crystal] nucleus
Keimapparat *m (ag)* germinating apparatus
Keimbildung *f* 1. *(ag)* germination; 2. nucleation *(crystal)*
keimen to germinate, to sprout
Keimen *n* germination
Keimfähigkeit *f* viability
keimfrei sterile
Keimfreimachung *f* sterilization
Keimkraft *f* germinative vigour
Keimkristall *m* nucleus crystal, seed crystal
Keimprobe *f s.* Keimprüfung
Keimprüfung *f* germinating test
Keimpunkt *m* seeding centre
keimtötend bactericidal, germicidal, sterilizing
Keimvernichtungsmittel *n* germicide
Kelle *f* 1. ladle; 2. trowel
Keller *m* basement, cellar
Kellergeschoß *n (build)* basement [floor]
  zweites K. sub-basement
Kellerspeicher *m* list (push down) store *(data storage)*
Kellerwohnung *f* basement flat
Kellog-Schalter *m (el)* Kellog (tumbler) switch
Kelter *f* fruit press
Kelvin-Effekt *m* Kelvin effect
Kelvin-Grad *m* degree Kelvin, °K
Kelvin-Temperaturskale *f* Kelvin scale
Kenafhanf *m* kenaf
Kennbake *f* identification beacon *(radar)*
Kennbuchstabe *m* recognition letter
Kenndaten *pl* characteristics, characteristic values
Kennelly-Heaviside-Schicht *f* Kennelly-Heaviside layer
Kennfaden *m* tracer [thread], marker (marking) thread, cotton binder *(cable)*
  farbiger K. coloured thread
Kennfeuer *n* identification (character) light, marker-beacon light *(navigation)*
Kennfrequenz *f* assigned (marker) frequency
Kenngröße *f* characteristic [value]
Kennimpuls *m* label
Kennlampe *f* pilot lamp
Kennleitwert *m* indicial admittance
Kennlinie *f* characteristic line (curve)
  K. bei Magnetfeldröhren critical-voltage parabola
  K. der Gesamthelligkeitswiedergabe *(telev)* overall-brightness transfer characteristic
  fallende K. falling characteristic
  langsam auslaufende K. tailed characteristic
  statische K. static characteristic *(e.g. of transductors)*; transfer curve
  unsymmetrische K. asymmetrical characteristic
Kennlinienauslauf *m* / **allmählicher** gradual cut-off
Kennlinienfeld *n* family of characteristics, curve families
Kennliniengebirge *n (el)* characteristic surface
Kennlinienknick *m* bend of the characteristic
  unterer K. lower bend of the characteristic
Kennlinienschar *f* family of characteristics
Kennlinienschreiber *m* [characteristic] curve tracer
Kennlinienverlauf *m* / **ebener** flat response
Kennmelder *m (el)* indicating fuse
Kennummer *f* **eines Signals** date-time group
Kennung *f* identification marking *(radar)*
Kennwert *m* characteristic value; parameter
Kennwertregler *m* value control
Kennwiderstand *m (el)* characteristic impedance

Kennzahl *f* characteristic factor, index; routing code; index, operating figure
Kennzeichen *n* distinguishing mark (sign)
polizeiliches K. licence plate *(automobile)*
Kennzeichenbeleuchtung *f (aut)* licence plate light
hintere K. rear number plate light
vordere K. front number plate light
Kennzeichenlampe *f s.* Kennzeichenleuchte
Kennzeichenleuchte *f (aut)* number-plate illumination lamp
Kennzeichenschild *n (aut)* number plate
kennzeichnen to mark, to designate, to characterize
Kennzeichnung *f* marking
Kennzeichnungssystem *n* marking system
Kennziffer *f (math)* characteristic; *(mach)* specification [of size]; *(telec)* code number, answer-back code
Kentermoment *n (ship)* upsetting moment
kentern to capsize, to upset; to keelover; to turn bottom up
Kenterschloß *n* wedge cut
Keps *pl (mine)* landing dogs
Keramik *f* ceramics
Keramikdurchführung *f* ceramic insulated feed-through
Keramikdurchführungskondensator *m* lead-through ceramic capacitor
Keramikerzeugnis *n* clay product (ware), ceramic product
Keramikkleinkondensator *m* midget ceramic capacitor
Keramikkondensator *m* ceramic capacitor
Keramikkörper *m (el)* ceramic body
Keramikmagnet *m* ceramic magnet
Keramikröhrenfassung *f* low-loss valve holder
Keramikscheibenkondensator *m* disk ceramic capacitor
Keramikschneidplatte *f (mach)* ceramic tip
Keramikwerkstoff *m* ceramic material
keramisch ceramic
Kerargyrit *m (min)* cerargyrite
Keratin *n* keratine
Kerb *m s.* Kerbe
Kerbarm *m (mine)* shear jib
Kerbbiegeprobe *f* notch bending specimen
Kerbbiegeversuch *m* notch bending test
Kerbe *f* notch, slot, groove, kerf, nick
kerbempfindlich notch-sensitive
Kerbempfindlichkeit *f* notch sensitivity
kerben to notch, to slot, to groove
Kerbmeißel *m* shearing blade *(coal plough)*
Kerbsäge *f* trim saw
Kerbschlagbiegeprobe *f* notch-bar impact specimen
Kerbschlagfestigkeit *f* notch impact strength
Kerbschlagversuch *m* notched-bar impact test
Kerbschlagzähigkeit *f* notched-bar impact strength
Kerbschnitt *m* jag *(wood)*
Kerbspannung *f* notch stress
Kerbsprödigkeit *f* notch brittleness
Kerb- und Schrämmaschine *f (mine)* header
Kerbung *f* nicking
kerbverzahnt serrated
Kerbverzahnung *f* serration
Kerbwirkung *f* notch effect
Kerbzähigkeit *f* notched impact value
Kerbzahnnabe *f* serrated hub
Kerbzahnnabenprofil *n* internal serration[s]
Kerbzahnwellenprofil *n* external serration[s]
Kerbzugversuch *m* notched-bar tensile test
Kerma *f* kerma *(kinetic energy released in material, dose unit)*
Kermesit *m (min)* kermesite
Kern *m* 1. core *(foundry; cable)*; 2. [atomic] nucleus; [ionization] centre; [integral] kernel; 3. [reactor] core; core of fuel element; heart; 4. *(build)* middle third *(analysis of walls etc.)*; 5. *(mach)* web *(drill)*; 6. stone *(plum)*
angeregter K. excited nucleus
ausgebohrter K. *(mach)* solid core removed *(trepanning)*
deformierter K. deformed nucleus
doppelt gerader K. *s.* gg-Kern
doppelt ungerader K. *s.* uu-Kern
entarteter K. degenerated nucleus
instabiler K. unstable nucleus
konischer K. *(mach)* chamfered core
lösender K. *(math)* resolvent kernel
magischer K. *(nucl)* magic nucleus
schwerer K. *(nucl)* heavy nucleus
stabiler K. *(nucl)* stable nucleus
versetzter K. misplaced core *(moulding)*
Kernabschirmung *f* screening [of the nucleus], electron screening, screening effect
Kernabstand *m* [inter]nuclear distance, [inter]-nuclear separation
Kernachse *f* epipolar axis
kernaktiv nuclear-active
Kernanlage *f* nuclear power station; nuclear engine
Kernanregung *f* nuclear excitation
Kernantrieb *m s.* Kernenergieantrieb
Kernart *f s.* Nuklid
Kernaufbau *m* nuclear structure
Kernausdrücker *m* core knockout *(moulding)*
Kernausrichtung *f* nuclear orientation (alignment)
Kernbatterie *f* atomic [energy storage] battery, radioactive battery
Kernbauart *f* **in** core-type *(transformers, reactors)*
Kernbaustein *m* nuclear particle, nucleon
Kernbeschleuniger *m* nuclear accelerator
Kernbeschuß *m* nuclear bombardment

Kernbewegung *f* nuclear motion (movement)
Kernbildung *f* nucleation *(crystal)*
Kernbinder *m* core binder
Kernbindung *f* nuclear binding (linkage)
Kernbindungsenergie *f* nuclear binding energy, nucleus bond energy
Kernblasmaschine *f* core blower *(moulding)*
Kernblech *n* core lamination (plate)
Kernbohren *n (mach)* trepanning
Kernbohrer *m (mach)* annular borer, trepanning tool
Kernbohrkopf *m (mach)* trepanning [bar cutter] head
Kernbohrkopfmesser *n (mach)* trepanning head cutting bit
Kernbohrung *f* core drilling; *(mach)* trepanning
Kernbombe *f* atomic (nuclear) bomb
Kernbrennen *n* core baking *(moulding)*
Kernbrennstoff *m* nuclear fuel
  abgebrannter K. spent [nuclear] fuel
  angereicherter K. enriched [nuclear] fuel
  flüssiger K. liquid fuel
  metallischer K. metallic fuel
Kernbruchstück *n* [nuclear] fragment; fission fragment
Kernchemie *f* nuclear chemistry
Kerndichte *f* nuclear density
Kerndicke *f (mach)* web thickness *(twist drill)*
Kerndickenverstärkung *f* web taper *(drill)*
Kerndrehbohren *n (mach)* core boring
Kerndrehung *f* nuclear spin
Kerndurchmesser *m* minor diameter [of a thread]
Kernemulsion *f* nuclear emulsion
Kernen *n (mach)* coring
Kernenergetik *f* nuclear energetics, atomic [energy] industry
Kernenergie *f* nuclear energy; nuclear power
  industriell genutzte K. industrial nuclear power
  nutzbare K. nuclear power
Kernenergieanlage *f* nuclear power plant; nuclear engine
  transportable K. package nuclear power plant
Kernenergieantrieb *m* nuclear propulsion; nuclear engine
  K. / mit nuclear-powered
Kernenergiefreisetzung *f* nuclear power production
Kernenergiegesetz *n* nuclear energy law
kernenergiegetrieben nuclear-powered
Kernenergiegewinnung *f* nuclear power production
Kernenergieprogramm *n* programme for peaceful utilization of atomic energy
Kernenergierakete *f* nuclear-energy rocket, nuclear[-powered] rocket
Kernenergietechnik *f* nuclear energetics
Kernenergiezeitalter *n* atomic age, nuclear era
Kernexplosion *f* fragmentation, nuclear explosion
Kernfaden *m* core thread
Kernfänger *m* core catcher
kernfaul heart-rotten *(wood)*
Kernfeld *n* nuclear (meson) field
Kernfernheizwerk *n* nuclear steam generating plant
Kernfolge *f* suite of cores
Kernform *f* core mould *(foundry)*
Kernformfaktor *m* nuclear form factor
Kernforschung *f* nuclear research
Kernforschungszentrum *n* nuclear (atomic) research centre
Kernfotoeffekt *m* photonuclear reaction, nuclear photoeffect (photoelectric effect)
Kernfotoplatte *f* nuclear plate
Kernfusion *f s.* Fusion
Kernfusionsbombe *f* fusion bomb
Kerngehalt *m* content of nuclei
kerngestreut deflected by nucleus
Kerngewinn *m* core recovery
Kerngießmaschine *f* core shooter
Kernguß *m* cored part[s]
Kernhalogenierung *f* nuclear halogenation
Kernholz *n* heartwood
Kernigkeit *f* fullness *(leather)*; body *(skin, hide)*
kerninaktiv nuclear-inactive
Kerninduktion *f* nuclear induction
Kernisomerie *f* nuclear isomerism, isomerism of the atomic nucleus
Kernkasten *m* core box *(foundry)*
Kernkettenreaktion *f* nuclear chain reaction
Kernkorrosion *f* nuclear corrosion
Kernkraft *f* nuclear force (power)
Kernkraftwerk *n* nuclear power station, *(Am)* nuclear power plant, atomic power (energy) station, *(Am)* atomic power (energy) plant
  industrielles K. industrial atomic station
Kernladung *f* nuclear charge
Kernladungszahl *f* atomic number (charge)
  effektive K. effective atomic charge
Kernloch *n* core-hole
Kernlochbohrer *m (mach)* tap drill
Kernlochdurchmesser *m* **[bei Innengewinden]** minor diameter of an internal thread
kernlos coreless *(furnace)*
Kernmacher *m* coremaker *(foundry)*
Kernmacherei *f* core moulding (making)
Kernmagnet *m* core magnet
kernmagnetisch nuclear magnetic *(resonance)*
Kernmagnetmeßwerk *n* core magnet [measuring] system
Kernmagneton *n* nuclear [Bohr] magneton
Kernmarke *f* core mark *(foundry)*
Kernmasse *f* core mixture *(foundry)*; nuclear mass

Kernmasseneinheit *f* atomic mass unit, amu
Kernmodell *n* nuclear model, model of nucleus
  kollektives K. collective model of nucleus, collective nuclear model; unified model [of nucleus]
  optisches K. optical model of nucleus
  statistisches K. Wigner (uniform, statistical) model of nucleus
  thermodynamisches K. thermodynamic model of nucleus
Kernmoment *n* nuclear moment
Kernnagel *m* sprig, brad; core nail
Kernniveau *n* nuclear energy level
Kernöl *n* core oil *(foundry)*
Kernpapier *n* core paper
Kernpermeabilität *f (el)* core permeability
Kernphysik *f* nuclear physics
  theoretische K. theoretical nuclear physics
Kernphysiker *m* nuclear physicist
Kernplatte *f* nuclear plate
Kernprobe *f* core sample
Kernprozeß *m* nuclear reaction
Kernquerschnitt *m (mach)* web section *(drill)*; *(el)* area (cross-section) of a core *(transformer)*
Kernradius *m* nuclear radius, radius of the nucleus
Kernreaktion *f* nuclear reaction
  gesteuerte K. controlled nuclear reaction
  künstliche K. artificial (induced) nuclear reaction
Kernreaktor *m* [nuclear] reactor, pile
Kernreaktorantrieb *m s.* Kernenergieantrieb
kernrein *s.* nuklearrein
Kernresonanz *f* nuclear resonance
  magnetische K. nuclear magnetic resonance
Kernresonanzspektroskopie *f* NMR (nuclear magnetic resonance) spectroscopy
Kernriß *m* [heart] shake, boxheart checking, star shake, heart check
kernrissig heart-shaky *(wood)*
Kernrohr *n* core barrel
  inneres K. core receiving barrel
Kernrückfeinung *f* core refining
Kernrumpf *m* core of the nucleus
Kernsägezähne *mpl* rip-saw teeth
Kernsand *m* core sand *(foundry)*
Kernsatz *m (mine)* suite of cores
Kernschäle *f* cup shake, shakiness *(wood)*
Kernschalenmodell *n s.* Schalenmodell des Atomkerns
kernschälig ring-shaky *(wood)*
Kernschatten *m* umbra
Kernschrott *m* solid scrap
Kernschuß *m* point-blank shot
Kernseife *f* curd soap
Kernseigerung *f* normal segregation
Kernspaltung *f* nuclear (atomic) fission
Kernspaltungsreaktor *m s.* Spaltungsreaktor
Kernspeicher *m* magnetic core memory (store), ferrite memory
Kernspektroskopie *f* nuclear spectroscopy
Kernspin *m* nuclear spin
Kernspuremulsion *f* nuclear emulsion
Kernspurmeßmikroskop *n* nuclear track measuring microscope
Kernspurplatte *f* nuclear plate
Kernspursuchmikroskop *n* nuclear track finding microscope
Kernstabilität *f* stability of atom nucleus
Kernstärke *f (mach)* web (core) thickness *(twist drill)*
Kernsteigung *f (mach)* web (core) taper
Kernstopfen *m (el)* centre core
Kernstrahlung *f* nuclear (radioactive) radiation
Kernstreuung *f* nuclear scattering
Kernströmung *f* core flow
Kernstütze *f* chaplet *(foundry)*
Kernsubstanz *f* nuclear substance
kernsubstituiert nuclear-substituted
Kernsuszeptibilität *f* susceptibility of magnet core
Kernsystematik *f* nuclear systematics
Kerntechnik *f* nuclear engineering (technology), nucleonics
Kernteilchen *n* nuclear particle, nucleon
Kernteilung *f* nuclear division
Kerntemperatur *f* nuclear temperature
Kernterm *m* nuclear term
Kerntheorie *f* nuclear theory
Kerntransformator *m* core-type transformer
Kerntreibstoff *m* nuclear propellant
kerntrocken thorough-dry
Kerntrockenkasten *m* sagger *(foundry)*
Kerntrockenofen *m* core stove (oven)
Kerntrocknung *f* core baking *(foundry)*
Kernübergang *m* nuclear transition
Kernüberzug *m* core coating *(foundry)*
Kernumwandlung *f* nuclear transformation (transmutation); nuclear disintegration (decay)
  spontane K. spontaneous [nuclear] transformation
Kernverdampfung *f* nuclear (nucleus) evaporation, evaporation of the nucleus
Kernvermehrung *f* nuclear increase
Kernverschmelzung *f* [nuclear] fusion; fusion process
Kernverschmelzungsreaktion *f* fusion [reaction]
Kernverschmelzungsreaktor *m s.* Fusionsreaktor
Kernwaffe *f* nuclear weapon
  taktische K. tactical nuclear weapon
Kernwaffenexplosion *f* atomic blast (explosion, bomb burst)
kernwaffenfrei non-nuclear
Kernwaffentest *m* nuclear test (trial)

Kernwaffenversuchsexplosion *f* nuclear test (trial), atomic explosion trial

Kernwand *f* nuclear membrane

Kernwarze *f* tip of nucleus

Kernwechselwirkung *f* nuclear interaction

Kernwucherung *f* proliferation of the nucleus

Kernzerfall *m* nuclear disintegration (decay)

Kernzersplitterung *f* spallation

Kernzertrümmerung *f* spallation

Kernziehen *n* core lifting

Kernzone *f* nuclear zone

Kernzustand *m* nuclear state, state of the nucleus

angeregter K. excited nuclear state

metastabiler K. metastable nuclear state

Kerosin *n* kerosene

Kerr-Effekt *m* Kerr effect

Kerr-Zelle *f* Kerr cell

Kerze *f* 1. *(opt)* candle; 2. spark plug

neue K. *(opt)* new candle

Kerzenelektrodenabbrand *m* point erosion *(of spark plugs)*

Kerzenkörper *m* plug body

Kerzenstärke *f (opt)* candle power

Kerzenzahl *f (opt)* candle power

Kerzenzündung *f* spark ignition

Kessel *m* 1. boiler; basin, vessel, reservoir; 2. *(mine)* chamber; pocket *(of drill hole)*; 3. *(geol)* caldron

K. für Augenblicksverdampfung flash boiler

K. mit einem Dampfraum boiler with single-steam space

K. mit engen Wasserrohren water tube boiler with small tubes, express type of water tube boiler

K. mit großem Wasserraum boiler with large water space

K. mit Innenfeuerung internal-firebox boiler

K. mit kleinem Wasserraum boiler with small water space

K. mit Unterkessel double-cylindrical boiler

K. mit weiten Wasserrohren water tube boiler with large tubes

K. mit zwei Dampfräumen boiler with double-steam space

amerikanischer K. American type of boiler

kombinierter K. combination boiler

schottischer K. Scotch boiler

stehender K. mit waagerechten Quersiedern vertical cross tube boiler

zusammengesetzter K. combination boiler

zusammengesetzter K. mit Heizrohren combination (semi-tubular) boiler

zylindrischer K. cylindrical boiler

Kesselabblasehahn *m* boiler blow-off cock

Kesselabdeckungsmaterial *n* non-conducting material for covering the top of the boiler

Kesselabklopfwerkzeug *n* boiler-scaling tool

Kesselablaßhahn *m* boiler drain cock

Kesselabnahme *f* boiler acceptance

Kesselabwärme *f* boiler waste heat

Kesselanlage *f* boiler plant

Kesselarmatur *f* boiler fittings

Kesselaufhängung *f* boiler suspension

Kesselausrüstung *f* boiler fittings

Kesselbatterie *f* battery of boilers

Kesselbau *m* boiler construction

Kesselbauanstalt *f* boiler shop

Kesselbauart *f* boiler type

Kesselbauer *m* boiler maker

Kesselbauingenieur *m* boiler maker, [graduated] engineer in the field of boiler construction

Kesselbauwerkstatt *f* boiler shop

Kesselbedienung *f* boiler attendance

Kesselbedienungsmannschaft *f* boiler attendants

Kesselbekleidung *f* boiler casing

Kesselbekohlungsanlage *f* boiler-coaling plant

Kesselbetrieb *m* boiler operation

Kesselblech *n* boiler plate

durchgerostetes K. boiler plate eaten through by corrosion

Kesselblechüberhitzung *f* overheating of the boiler plate

Kesselbock *m* boiler bearer (cradle, saddle)

Kesselboden *m* boiler end, [bulk]head, bottom [of the boiler]

Kesseldampf *m* live steam

Kesseldampfmaschine *f* steam-boiler machine; generator machine

Kesseldekatiervorrichtung *f* decatizing machine

Kesseldom *m (ry)* tank dome *(tank wagon)*

Kesseldruck *m* boiler (working) pressure

Kesselexplosion *f* boiler explosion

Kesselfabrik *f* boiler factory

Kesselfeuerung *f* boiler firing; boiler furnace

Kesselflammrohr *n* boiler flue

Kesselfundament *n* boiler seat[ing]

Kesselgas *n* boiler flue gas

Kesselgebläse *n* boiler blower

Kesselgewindebohrer *m* boiler tap

Kesselgruppe *f* battery of boilers

Kesselhammer *m* boiler-scaling hammer

Kesselhaus *n* boiler house

Kesselhausbetrieb *m* boiler house operation

Kesselheizfläche *f* boiler-heating surface

Kesselhersteller *m* boiler maker

Kesselisolierung *f* boiler lagging

Kesselkohle *f* boiler (steam) coal

Kessellager *n* bed[ding] of the boiler

Kessellagerung *f* boiler setting

Kesselleistung *f* boiler output (duty), rating of the boiler

qualitative K. qualitative boiler output

quantitative K. quantitative boiler output

Kesselluke *f* boiler hatch

Kesselmantel *m* boiler casing (envelope); jacket
Kesselniederschlag *m* boiler scale
Kesselölschalter *m* dead-tank oil circuit-breaker
Kesselpferdestärke *f* boiler horsepower
Kesselplatte *f* boiler plate
Kesselpratze *f* boiler bracket (lug)
Kesselprobierpumpe *f* boiler tester (test pump)
Kesselprüfer *m* boiler inspector
Kesselprüfung *f* boiler test
Kesselprüfungskommission *f* boiler inspection commission
Kesselraum *m* boiler room
Kesselraumgräting *f (ship)* grating over stokehold
Kesselraumschott *n (ship)* stokehold (boiler-room) bulkhead
Kesselregelung *f* boiler control
Kesselreibahle *f* boiler maker's reamer
Kesselreihe *f* / **zusammenarbeitende** battery of boilers
Kesselreinigung *f* boiler cleaning
Kesselreparaturwerkstatt *f* boiler repair shop
Kesselrohr *n* boiler (heating) tube
Kesselrost *m* boiler grate
Kesselschacht *m (ship)* fidley, boiler casing
Kesselschild *n* boiler name plate
Kesselschmied *m* boiler maker
Kesselschüsse *mpl* shell rings
Kesselsohle *f* boiler level
Kesselspannung *f* boiler pressure
Kesselspeisepumpe *f* boiler feeder (feed pump), feed injector
Kesselspeisewasser *n* boiler feed-water
Kesselspeisewassermesser *m* boiler-feed-water meter
Kesselspeisewasserpumpe *f* boiler-water feed pump
Kesselspeisewasservorwärmer *m* feed-water heater
Kesselspeisung *f* boiler feed
Kesselstahl *m* boiler steel
Kesselstein *m* boiler scale, water incrustation, furring
Kesselsteinabklopfer *m* tube cleaner; scaling hammer
pneumatischer K. pneumatic scaling hammer (chipper)
Kesselsteinabkratzer *m* scraper
Kesselsteinablagerung *f* [boiler-]scale deposit
Kesselsteinansatz *m* incrustation of boiler fur
Kesselsteinbeseitigung *f* descaling
Kesselsteinbeseitigungsgerät *n* descaling machine
Kesselsteinbeseitigungsmittel *n* scale solvent, disincrustant
kesselsteinbildend [boiler] scale-forming
Kesselsteinbildung *f* [boiler-]scale formation
Kesselsteinlösemittel *n* boiler-cleansing compound, antiscale composition
Kesselsteinlösung *f* incrustation preventings
Kesselsteinschale *f* scale crust
Kesselsteinsplitter *m* piece of scale
Kesselsteinverhütung *f* prevention of scale formation
Kesselsteinverhütungsmittel *n* antiscale boiler fluid, boiler-antiscaling composition, boiler-anti-incrustant composition
Kesselstirnwand *f* front end plate of boiler
Kesselstuhl *m* boiler support
Kesselstütze *f* boiler support
Kesselteil *m* / **freitragender** part of the boiler not supported
Kesseltrommel *f* boiler drum
Kesseltyp *m* type of boiler
Kesselüberwachungsapparat *m* boiler-supervising apparatus
Kesseluntersuchungsbehörde *f* boiler inspection authorities
Kesselverkleidung *f* boiler jacket
Kesselwagen *m (ry)* tank wagon
Kesselwand *f* boiler shell
Kesselwarte *f* boiler switchboard
Kesselwärter *m* stoker, fireman
Kesselwasserbehandlung *f* boiler-water treatment
Kesselwasserreinigung *f* boiler-water treatment
Kesselwirkungsgrad *m* boiler efficiency
Kesselzubehör *n* boiler accessories
Kesselzug *m* boiler draft (flue pass)
Kesselzugüberhitzer *m* superheater placed in the flue
Keten *n* ketene
Ketoalkohol *m* ketol
Keto-Enol-Tautomerie *f* keto-enolic tautomerism
Ketoform *f* keto-form
Ketogenese *f* ketogenesis
Ketol *n* ketol
Keton *n* ketone
Ketonmoschus *m* ketone musk
Ketonspaltung *f* ketonic cleavage
Ketosäure *f* ketonic acid
Ketoxim *n* ketoxime
Ketsch *f* ketch *(sailing ship)*
Kettablaßvorrichtung *f (text)* negative warp let-off motion
Kettbaum *m (text)* warp beam
Kettbaumablaufgestell *n* warp beam creel
Kettbaumbremse *f* warp beam brake
Kettbaumfärben *n* warp beam dyeing
Kettbaumschären *n* warp beam warping
Kettbaumscheibe *f* warp beam flange
Kettdruck *m (text)* warp printing
Kette *f* 1. chain; 2. *(text)* warp; 3. tread, track *(crawler)*; 4. cascade

K. mit großer Teilung large-pitch chain
einsträngige K. single-strand chain
gebleichte K. *(text)* bleached warp
geschlossene kinematische K. closed kinematic chain
kinematische K. kinematic chain
mehrfarbige K. *(text)* multicoloured warp
zusammengesetzte kinematische K. closed kinematic chain
zwangsläufig geschlossene K. closed chain with constrained movement

Kettelmaschine *f (text)* linking machine, looper
Kettelnaht *f (text)* linking seam
Kettenabbruch *m (ch)* [end] stopping of the chain, chain breaking (interrupting, stopping; termination)
Kettenabbruchreaktion *f* cessation reaction
Kettenabweiser *m* chain stripper
Kettenanknüpfmaschine *f* warp tying machine
Kettenantrieb *m (mach)* chain drive
K. / mit chain-driven
Kettenaufspaltung *f (ch)* splitting of the chain
Kettenauge *n* track lug *(crawler)*
Kettenbahnförderung *f* chain haulage
Kettenbecherwerk *n* chain elevator
Kettenbreite *f* chain width
Kettenbruch *m (math)* continued fraction
Kettenbrücke *f* chain bridge
Kettenbügel *m* chain connector, connecting link
Kettendämpfung *f (el)* iterative attenuation [coefficient]
Kettendichte *f (text)* warp setting
Kettendouble *n (text)* warp-backed fabric
Ketteneinziehstuhl *m (text)* drawing-in frame
Kettenfahrleitung *f* catenary suspension (trolley wire)
polygonale K. polygonal overhead contact system
windschiefe K. inclined overhead contact line
Kettenfahrzeug *n* crawler-type vehicle, track-type vehicle
Kettenfärberei *f* chain dyeing
Kettenfläche *f (math)* catenoid
Kettenförderbahn *f* endless chain haulage
Kettenförderer *m* chain conveyor
endloser K. circulating chain conveyor
Kettenfräse *f* chain cutter moulding machine
Kettengeschwindigkeit *f* chain velocity
Kettengetriebe *n* train of sprockets
kettengewirkt warp-knitted
Kettengleichrichtung *f (telev)* successive detection
Kettenglied *n* 1. chain link; track link *(crawler)*; 2. *(el)* section, network mesh *(filter)*; 3. *(therm)* chain grate link
Kettengreifer *m* chain gripper
Kettenhalogenierung *f* chain halogenation
Kettenimpedanz *f (el)* iterative impedance
Kettenindexziffer *f* chain index number
Kettenisolator *m* suspension insulator
Kettenisomerie *f* chain isomerism
Kettenkammer *f* chain locker
Kettenkasten *m* cable (chain) locker
Kettenklaue *f* housing stopper
Kettenkneifer *m (ship)* compressor
Kettenkrad *n* half-track motorcycle
Kettenkratzerförderer *m* scraper [chain] conveyor
Kettenlänge *f* chain length
Kettenleiter *m* recurrent (ladder, lattice, iterative) network, lattice (recurrent) circuit
Kettenleitrad *n* track idler *(crawler)*
Kettenlinie *f* catenary
Kettenmitnehmer *m* chain jockey
Kettennietung *f* chain-spaced rows of rivets
Kettennuß *f* chain sprocket, sprocket wheel
Kettenphasenmaß *n (el)* iterative attenuation factor
Kettenpoller *m (ship)* chain bitt
Kettenrad *n* sprocket wheel
Kettenraddurchmesser *m* sprocket diameter
Kettenradfräser *m* sprocket-wheel milling cutter
Kettenradzahn *m* sprocket tooth
Kettenrapport *m (text)* repeat of warp threads
Kettenreaktion *f (nucl)* chain reaction; chain fission, fission chain reaction
Kettenrechnung *f (math)* chain method
Kettenregel *f (math)* chain rule
Kettenrippen *fpl (text)* warp cords
Kettenröhrenreiniger *m* chain tube-cleaner
Kettenrolle *f* chain pulley
Kettenrost *m* chain grate
Kettenrostfeuerung *f* chain grate stoker
Kettenrostwagen *m* chain grate carriage
Kettensäge *f* [power] chain saw
Kettenschaltung *f (el)* ladder network; concatenation (consecutive, cascade, chain) connection
Kettenschlepper *m* caterpillar, crawler tractor
Kettenschloß *n* chain connector, connecting link
Kettenschluß *m* closing of the chain
Kettenschmiervorrichtung *f* chain greasing device
Kettenschrämmaschine *f* chain breast machine, chain coal-cutter
Kettenschutz *m* track guard
Kettensequenz *f (ch)* chain sequence
Kettenspanner *m* chain tensioner; track tensioner *(crawler)*
Kettenspindelbremse *f (ry)* chain screw brake
Kettenstab *m* chain grate link
Kettenstern *m* chain sprocket, sprocket wheel
Kettenstopper *m* controller, cable (housing) stopper

Kettenstrang *m* strand of chain
Kettenstrecke *f (text)* chain bar drawing frame
Kettenstromübersetzung *f (el)* iterative current ratio *(four-terminal theory)*
Kettenstropp *m (ship)* sling chain, chain sling (strap)
Kettenstuhl *m* tricot machine
Kettenteilung *f* pitch of the chain
Kettentischhobelmaschine *f* chain-type table planing machine
Kettenträger *m (ch)* chain propagator
Kettentrieb *m* chain drive
Kettentrum *n* chain strand
Kettenübertragungsmaß *n (el)* iterative transfer coefficient, iterative propagation constant
Kettenverstärker *m (el)* distributed amplifier; transmission-line amplifier; chain amplifier
Kettenverzweigung *f (ch)* chain branching
Kettenwechsler *m (text)* warp changer
Kettenwiderstand *m (el)* iterative impedance
Kettenwinkelmaß *n (el)* iterative phase constant, iterative phase-change coefficient
Kettenwirkmaschine *f* warp knitting loom
Kettenwirkungsgrad *m (el)* string efficiency
Kettenzahl *f (math)* chain relative
Kettfaden *m (text)* warp thread
gebrochener K. broken warp thread
Kettfadenbruch *m* broken end (warp thread), warp breakage
Kettfadeneinstellung *f* warp setting
Kettfadenknoten *m* warper's knot
Kettfadenreiter *m* warp drop wire
Kettfadenwächter *m* warp stop motion
Kettflorgewebe *n* warp pile fabric
Kettflorteppich *m* warp pile carpet
Kettflottierung *f (text)* warp float
Kettgarn *n* warp yarn
Kettgarnkötzer *m* warp cop
Kettgarnspulmaschine *f* warp winding frame
Kettgewebe *n* warp loom fabric
Kettkammgarn *n* worsted warp
Kettköper *m* warp twill
Kettnummer *f (text)* warp count
Kettplüsch *m* warp plush
Kettrahmen *m (text)* warp frame
Kettrips *m* warp rib
Kettsamt *m* warp pile fabric
kettscheren *(text)* to warp
Kettschlichten *n (text)* warp sizing
kettschonend *(text)* warp-saving
Kettspannung *f (text)* warp tension
Kettspule *f (text)* warp bobbin
Kettspulmaschine *f (text)* warp winder
Kettstreifen *m (text)* warp strip
Kettvorbereitung *f (text)* warp preparation
Kettwirkware *f* warp-knit fabric
Kettzylinder *m (text)* warp roller
Keule *f* 1. [Indian] club; 2. *(el)* lobe; beam
scharfe K. narrow beam
Keulenbreite *f* spread *(aerials)*
Keulendiagramm *n* lobe pattern *(radar)*
Keulenhalbwertsbreite *f* angular width *(waveguide)*
Keulenschwingen *n* Indian club exercise, club swinging
Keulenwandler *m* door-knob transformer *(waveguide)*
KFZ-Elektriker *m* auto-electrician
KFZ-Technik *f* automobile engineering
Kickstarter *m* kick-starter
Kidgerberei *f* kid leather works
Kiefer *f* pine
Kiefernharz *n* pine resin
Kiefernholz *n* pine [wood]
Kiel *m* 1. calamus *(feather)*; 2. *(ship)* keel, bilge
gekrümmter K. rockered keel
loser K. shoe of a keel, false-keel
Kielablauf *m* single-way launching
Kielaufbuchtung *f* hogging
K. und Durchsenkung *f* hogging and sagging
Kielbank *f* gridiron
Kielblockramme *f* keel-track ram
Kielboot *n* keel boat
Kielbruch *m* hogging
Kielbucht *f* sagging
Kielform *f* contour of the keel
Kielgang *m* garboard strake
Kielgeld *n* keelage
Kielhacke *f* skeg of keel
Kielhöhe *f* keel siding
Kielholung *f* careening
Kieljacht *f* keeler, keel yacht
Kielklotz *m* keel block
Kiellegen *n* laying the keel
Kiellegung *f* keel-laying
Kiellinie *f* keel (middle) line
Kielneigung *f* drag
kieloben turned, keelover
Kielpalle *f* keel block
Kielplanke *f* garboard plank
Kielplatte *f* garboard plate
Kielraum *m* bilge
Kielschuh *m* shoe of a keel, keel shoe
Kielschwein *n* carlin, keelson, kelson; keelson timber
Kielschweinplatte *f* foundation plate
Kielschweinsohle *f* rider keelson
Kielschwert *n* centre-board
Kielschwertjacht *f* centre-boarder, centre-board yacht
Kielsenkung *f* sagging
Kielstapel *m* keel block, stocks
Kielstapelklotz *m* keel block
Kielstapelramme *f* keel-track ram
Kielstapelung *f* cribbing
Kielstütze *f* keel shore; skeg shore

Kieltunnel *m* pipe tunnel, duct keel
Kielwasser *n* dead water, wash, wake, steerage-way
Kielwasserwirbel *m* wake
Kielweger *m* limber strake
Kielwegerungsplanke *f* limber strake, futtock plank
Kienapfel *m* pine cone
Kienöl *n* pine oil
Kies *m* gravel *(sand)*; grit *(metal)*
  ungesiebter K. all-in gravel
Kiesabbrand *m* roasted pyrites, pyrite cinder
kiesartig gravelly, gritty
Kiesasphalt *m* gravel asphalt
Kiesaufbereitungsanlage *f* gravel plant
Kiesaufschüttung *f* gravel fill
Kiesboden *m* gravel soil
Kiesel *m* pebble, flint
Kieselerde *f* silica, silex
Kieselgalmei *m (min)* hemimorphite
Kieselgel *n* silica gel
Kieselgelfett *n* silica gel grease
Kieselgur *f* kieselguhr, infusorial (diatomaceous) earth, diatomite
Kieselkupfer *n (min)* chrysocolla
kieselsauer siliceous
Kieselsäure *f* silicic acid
Kieselstein *m* pebble, flint
Kieserit *m (min)* kieserite
Kiesfilter *n* gravel filter
Kiesgrube *f* gravel pit
kiesig gravelly, gritty
Kieslager *n* gravel deposit
Kiessandfilter *n* **mit abgestuften Korngrößen** graded [gravel] filter
Kiesstraße *f* gravel road
Kiloampere *n* kiloamp[ere]
Kilocurie *n* kilocurie, kCi, kc
Kiloelektronenvolt *n* kilo-electron-volt, keV
Kilogramm *n* kilogramme *(mass)*
Kilohertz *n* kilocycles per second, kc/s, *(Am)* kilohertz
Kilokalorie *f* kilocalorie, large calorie
Kilometer *n* kilometre
Kilometerwelle *f* long wave
Kilometerzähler *m* mileage indicator
Kiloohm *n* kiloohm
Kilopond *n* kilogramme force *(weight)*
Kilovolt *n* kilovolt
Kilowatt *n* kilowatt
Kilowattstunde *f* kilowatt-hour, kWh
Kilowattstundenzähler *m* kilowatt-hour meter
Kimm *f* [visible] horizon; [turn of] bilge
  runde K. easy (rounded) bilge
  scharfe K. hard bilge
  scheinbare K. sensible horizon
  sichtbare K. visible horizon
  wahre K. rational horizon
Kimmbeplankung *f (ship)* bilge planking
Kimme *f* back-sight *(rifle)*, rear sight; vee *(measuring)*
Kimmenendmaß *n* vee-jaw gauge block
Kimmgang *m (ship)* bilge strake
Kimmkiel *m (ship)* rolling (bilge) chock, rolling (bilge) keel
Kimmkielschwein *n (ship)* bilge log (keelson)
Kimmpalle *f (ship)* [mechanical] bilge block
Kimmplatte *f (ship)* bilge plate
Kimmreling *f (ship)* bilge rail
Kimmrundung *f (ship)* turn of bilge
  scharfe K. sharp bilge
Kimmschlitten *m (ship)* bilge ways, mechanical bilge block
Kimmschwert *n (ship)* bilge board
Kimmsente *f (ship)* bilge harpin (diagonal)
Kimmstapel *m (ship)* bilge block
Kimmstringer *m (ship)* bilge log (keelson, stringer)
Kimmstütze *f (ship)* bilge shore; bilge log
Kimmstützplatte *f (ship)* bilge bracket; margin plate bracket; tank-side bracket
Kimmtiefe *f (ship)* dip [of the horizon]
Kimmweger *m (ship)* bilge stringer, chine piece; bilge wale
Kimmwegerung *f (ship)* bilge ceiling (planking), ceiling of the floor heads
Kinderfahrkarte *f* child's ticket
Kindergarten *m* nursery, kindergarten
Kinderkrippe *f* crèche
Kinematik *f* kinematics
kinematisch kinematic
Kinemograf *m* kinemograph
Kinemometer *n* kinemometer
Kinetik *f* kinetics
  chemische K. chemical kinetics
kinetisch kinetic
Kink *f (ship)* kink
Kino *n* cinema
Kinoaufnahmeobjektiv *n* cine recording lens
Kinoorgel *f* cinema organ
Kinoprojektion *f* cine projection
Kinoprojektionsobjektiv *n* cine projection lens
Kinoprojektor *m* cine projector
Kinoröntgenografie *f* cineradiography
Kinostativkopf *m* tilting-turning tripod head
Kinovorführgerät *n* motion picture projector
Kiosk *m* booth, kiosk
Kipp *m* sweep *(electronic tubes)*
Kippablauf *m* / **einmaliger** single sweep
Kippablenkung *f (el)* sweep deflection
Kippachse *f* horizontal (tilting) axis; trunnion, hinge pin
Kippachsenhöhe *f* trunnion axis height
Kippamplitude *f* sweep amplitude
Kippanhänger *m* tipping trailer
Kippaufzug *m* skip hoist

Kippbahn *f (ry)* tipping base
kippbar tiltable
Kippbewegung *f* tilting motion
Kippdämpfer *m* tipping steamer
Kippe *f* tip (dump) site; *(mine) s.* Abraumkippe
Kippeinrichtung *f (mach)* tilting arrangement (mechanism)
kippen to tip, to tilt; to cant; to throw; to buckle sideways *(strength)*
Kippen *n* commutation failure; torque breakdown, tilting, stalling, pull-out; sweep, relaxation
K. eines Konverters turning down a converter
Kippenfuß *m (mine)* spoil pile toe
Kippenstoß *m (mine)* spoil bank wall
Kipper *m* tipper; tip-truck; tip-wagon; *(mine)* tilting car; dumper, dump truck
Kippfahrzeug *n s.* Kipper
Kippfenster *n* balance window
Kippflügelfenster *n* horizontal pivot-hung window
Kippform *f* tilting (tilt-up) mould
Kippfrequenz *f* sweep (relaxation) frequency
Kippfrequenzregelung *f* synchronization (hold) control
Kippgefäß *n* skip *(conveying)*
Kippgefäßaufzug *m* skip hoist
Kippgenerator *m* sweep (relaxation) generator; scanning generator
Kipphebel *m* valve rocker, rocker arm, rocking lever
Kipphebelblock *m* valve-rocker bracket
Kipphebelgehäuse *n* rocker box
Kipphebelrolle *f* valve roller
Kipphebelschalter *m* tumbler switch
Kipphebelsteuerung *f (aero)* rocker gear *(engine)*
Kipphebelwelle *f* rocker[-arm] shaft
Kipphebelzapfen *m* valve-rocker fulcrum pin
Kipphelltastimpuls *m (telev)* sweep bright-up pulse
Kippkarren *m (build)* tip cart, buggy; *(met)* tipping waggon, skip
Kippkastenkarosserie *f* dump body
Kippkraft *f* tilting force
Kippkübel *m* tipping (roll-over) skip; *(mine)* gunboat *(inclined shaft)*
Kippkübelbegichtung *f* skip filling *(blast furnace)*
Kipplager *n* rocker bearing *(bridge)*
Kippleine *f (ship)* shank painter
Kippmischer *m* tilting mixer *(concrete)*
Kippmoment *n* 1. *(mech)* tilting (overturning) moment; 2. *(el)* breakdown torque; stalling (pull-out) torque *(of motors)*; 3. *(ship)* upsetting (tilting, capsizing) moment
relatives K. *(el)* breakdown factor
Kippmulde *f* tipping trough
Kippofen *m* tilting (rocking-type) furnace
tiegelloser K. non-crucible tilting furnace
Kipppfanne *f* lip-poured ladle
Kipppflug *m (ag)* right and left hand plough
Kipprahmen *m* erecting (tip) frame
Kippregel *f* telescopic alidade
Kipprelais *n* throw-over relay; thyratron
Kippriegel *m (aut)* rocker bar
Kipprost *m* dumping (tipping) grate
Kippsägezahn *m (el)* sweep saw tooth
Kippschalter *m* switching key; tumbler (toggle) switch; trigger switch
Kippschaltung *f* sweep (time-base) circuit
bistabile K. bistable (flip-flop) circuit
Kippschaufel *f* shovel
Kippschute *f (ship)* tipper barge
Kippschwingung *f (el)* relaxation (sawtooth) oscillation
Kippschwingungsgenerator *m* relaxation generator, relaxation-type oscillator
Kippschwingungsoszillator *m* relaxation (sweep) oscillator
Kippschwingungsschaltung *f (el)* sweep circuit
Kippsegment *n* tiltable shoe *(segment bearing)*
Kippsegmentlager *n* pivoted segmental thrust bearing
Kippsicherheit *f* safety against overturning
Kippspannung *f (el)* sweep voltage
Kippspule *f (el)* tilting coil
Kippständer *m (aut)* prop stand
Kippstrom *m (el)* sweep (scan, relaxation) current
Kippstromgenerator *m* scan (sweep) current generator
Kippstützer *m (el)* tilting insulator
Kippsupport *m (mach)* swivel head
Kippsystem *n* tilting system
Kipptisch *m (mach)* tilting table
Kipptor *n* glide-over door, up-and-over door
Kipptreppengenerator *m (el)* stair-step generator
Kipptrommel *f* tipping (tilting) drum
Kipptrommelmischer *m* tilting[-drum] mixer, tilt-type mixer
Kippvorrichtung *f (mach)* tipping device; dumping mechanism *(vehicle)*
Kippwagen *m* 1. *(build)* dump truck; 2. *(mine)* cradle (hinged-body, tilting) car; tilting (tipping) truck *(conveying)*
Kippwinkel *m* tilt angle, angle of tilt
Kippzeit *f (el)* transit time, time of relaxation
Kirchengebäude *n* church [building]
Kirchenschiff *n* nave
Kirkendall-Effekt *m* Kirkendall effect
Kirschbaum *m* cherry tree
Kirschholz *n* cherry [wood]
kirschrot cherry-red, bright-red *(temper colour)*
Kissen *n* cushion; pillow; pad *(stamp, lubrication)*

Kissenbezug *m* cushion cover
Kissenschmierung *f* pad lubrication
Kissenverzeichnung *f (telev)* pincushion distortion
Kiste *f* box, case, chest
Kistenausmaße *npl* box (case) dimensions
Kistenbeitel *m* case chisel
Kistenbrecher *m* case opener
Kistenbretter *npl* [box] shook
Kistendeckel *m* lid of the chest (box, case)
Kisteneinsatzhärtung *f (met)* pack-hardening
kistenglühen *(met)* to box (pack) anneal
Kistenholz *n* wood for boxes (cases)
Kistenmaschine *f* box making machine
Kistennagelmaschine *f* box nailing machine
Kistenöffner *m* nail puller
Kistenware *f* box stock
Kitt *m* cement *(for lenses)*; putty *(for windows)*; lute, mastic
kitten to cement *(lenses)*; to putty *(windows)*; to lute
Kittfalz *m* fillister *(window)*
Kittlehre *f* splicing gauge
Klafterholz *n* stacked wood
Klagerecht *n* right of action
Klammer *f* 1. *(build)* cramp, clamp, staple; 2. *(el)* clip, brace; 3. squeeze clamp *(fork truck)*; 4. *(math)* bracket; parenthesis; brace
eckige K. *(math)* square bracket
geschweifte K. *(math)* brace, curly bracket
runde K. parenthesis, round bracket
Klammerdraht *m* stapling wire
Klammerlasche *f* fishplate *(support)*
klammern to cramp, to clamp, to brace
Klammerschlittschuh *m* clamp-on skate
Klang *m* sound
Klanganalysator *m* sound analyzer
Klangbild *n* sound impression
Klangblende *f* tone control, tonalizer
Klangboden *m* soundboard
Klangcharakter *m s.* Klangfarbe
Klangdrossel *f* tone-control choke
Klangentzerrung *f* tone correction
Klangfarbe *f* timbre [of sound], tone quality (colour)
Klangfarbenkorrektur *f* tone correction
Klangfarbenregelung *f* tone control
Klangfarbenregler *m* tone control (regulator), tonalizer
Klangfärbung *f* coloration
Klangfiguren *fpl* /**Chladnische** acoustic pattern (figures), Chladni's acoustic figures
Klangfilter *n* sound filter
Klangfülle *f* fullness (richness) of tone
Klanglein *m (text)* rattle flax
Klangprobe *f* resonance test
K. an der Schleifscheibe sounding out the grinding wheel
Klangregler *m* tone control
Klangsieb *n* sound filter
Klangverzerrung *f* sound distortion
Klappbett *n* folding bed
Klappboden *m* hinged bottom
Klappbrücke *f* bascule (balance) bridge
Klappbügel *m (el)* tilting bracket (stand)
Klappdeckelpumpe *f* snap cover-type pump
Klappe *f (mach)* flap *(planer)*, tool block; lid; *(build)* hatch, gate
klappen to flap, to tilt
Klappenbetätigung *f (aero)* flap actuation
Klappendurchflußmesser *m* airfoil (drag-disk, vane) flowmeter
Klappenkolben *m* bucket
Klappen[luft]pumpe *f* flap-valve pump
Klappenruf *m (telec)* drop calling
Klappenschrank *m (telec)* indicator (magneto) board, drop indicator panel, drop-type switchboard
Klappensitz *m* valve seating
Klappenstellung *f (aero)* deflection setting *(landing)*
Klappenstreifen *m* strip of indicators
Klappenströmungsmesser *m* vane anemometer
Klappenträger *m (mach)* clapper (flapper) box
Klappenträgerklemme *f (mach)* clapper box clamp
Klappenventil *n* check (leaf) valve
Klappenverschluß *m (opt)* flap shutter
Klappfenster *n* hinged window
Klappfußtritt *m* folding step
Klappkamera *f* folding camera
Klappodest *n(m)* hinged platform
Klapprost *m* tipping (dumping) grate
Klappschornstein *m* hinged funnel
Klappsessel *m* folding stool
Klappsitz *m* folding seat; tip-up seat
Klappsporn *m* folding trail-spade *(gun)*
Klappstuhl *m* folding chair
Klapptisch *m* folding table
Klappventil *n* flap valve
Klappverdeck *n* folding hood, collapsible top
Kläranlage *f* sewage disposal (treatment) plant, clarification (sewerage purification) plant, *(Am)* water treatment plant
Klärapparat *m* settling apparatus; clarifier
Klärbad *n (ch)* clearing bath
Klärbecken *n* sedimentation (settling) tank, sedimentation (settling) basin, clearing reservoir
Klärbehälter *m* clarification (sedimentation, filtering, settling) tank
klären *(ch)* to settle, to clarify, to clear, to defecate, to fine, to purify
Klärgas *n* sewage (sewer, digester) gas
Klärgefäß *n* settling tank
Klarglas *n* clear glass

Klärgrube *f* cesspool
Klarkohle *f* slack (small) coal
Klarlack *m* clear varnish (lacquer)
Klarmeldelampe *f* all-clear signal light, all-ready signal light
Klärmittel *n* fining agent, clarifier
Klärprahm *m (ship)* sludge vessel
Klarschriftsortierleser *m* clear-text sorter reader
Klärspitze *f* conical settling tank
Klärteich *m* settling pond
Klartext *m (telec)* text in clear, clear text
K. / in in clear
Klärung *f* **der Abwässer** clarification of sewage
Klärwerk *n* sewage works (treatment plant), clarification plant
Klärzyklon *m* cyclone clarifier
Klasse *f* cell, class *(statistics)*; classification
kristallografische K. crystallographic class
Klassenanzahl *f* number of cells (classes) *(statistics)*
Klassenbesetzung *f* number of variates of a class *(statistics)*
Klassenbreite *f* cell width, class interval *(statistics)*
Klassengrenze *f* cell boundary (limit), class limit *(statistics)*
Klassengröße *f (math)* class interval
Klassenhäufigkeit *f (math)* class frequency
Klassenmitte *f (math)* class mark; mid-point
Klassenpunkt *m (math)* class mark
Klassensumme *f (math)* class sum
klassieren to classify, to sort; to separate; to grade, to size; to screen
Klassieren *n* **der Erze** grading (sizing) of ores
Klassierer *m* classifier
Klassierrüttelsieb *n* grading jigging screen
Klassiersieb *n* classifying (grading) screen
Klassierspitze *f* conical classifier
Klassiertrommel *f* sizing drum
Klassierung *f* classification, sorting; sizing, separating; grading *(wool)*
K. der Erze classification of ores
Klassierzyklon *m* cyclone classifier
Klaubeband *n* picking belt
Klaubeberge *pl* picking (hand-picked) dirt
klauben to clean, to cob, to pick, to sort out by hand
Klauben *n* picking, sorting out by hand, cleaning
Klauber *m* sorter
Klaubetisch *m* picking table
Klaue *f (mach)* jaw, claw; *(build)* bird's mouth *(carpentry)*
Klauenkupplung *f* claw clutch (coupling)
Klauenöl *n* neat's (cattle) foot oil
Klauenrad *n* gear with dog clutch
Klauenschere *f* hoof shears
Klebapparat *m* film splicer
Klebdispersion *f* adhesive dispersion
Klebeband *n* adhesive tape
Klebebeschichtung *f* adhesive backing
Klebebindeautomat *m (print)* automatic adhesive binder
Klebebindung *f (print)* flexible (adhesive, perfect) binding
Klebeblech *n* residual plate
Klebefestigkeit *f* bonding strength
Klebefolie *f* adhesive foil
Klebefuge *f* [adhering] joint
kleben 1. to stick *(relay)*; to adhere; 2. to join with adhesives, to cement, to glue, to paste, to gum; to splice *(film)*
Kleben *n* 1. sticking; adhering, adherence; 2. gumming, glueing, joining with adhesives, cementing, pasting; splicing *(film)*
Klebeplakette *f* window sticker
Kleber *m* adhesive, cement, glue, gum
Klebeschmiege *f* oblique cut of a hip rafter *(carpentry)*
Klebestift *m s.* Klebstift
Klebestreifen *m* sealing tape
Klebeverbindung *f (mach)* glued joint
Klebfähigkeit *f* adhesion, binding property
Klebharz *n* adhesive [resin], resin adhesive
Klebkraft *f* / **elektrostatische** electrostatic retentive force
klebrig sticky, adhesive, adherent, glutinous, gluey, agglutinant, viscous
Klebrigkeit *f* adhesiveness, glueiness, tackiness
K. des Schmiermittels stickiness of the lubricant
Klebsand *m* loamy sand
Klebstift *m* distance piece *(relay)*; residual stud
Klebstoff *m* adhesive, glue, cement, bonding agent, agglutinant
stromleitender K. conductive adhesive
vegetabilischer K. vegetable adhesive
warm abbindender K. hot setting adhesive
Klebstoffauftrag *m* application of adhesive
Klee *m* clover
Kleeblattkreuzung *f (build)* cloverleaf junction
Kleeblattzapfen *m* **[an einer Walze]** wabbler
Kleedrescher *m* clover thresher
Kleereibe *f* velvet band dosser
Kleereiber *m* clover rasper
Kleesalz *n* salt of sorrel
Kleeuntersaat *f* undersown clover
Kleiderablage *f* cloak-room
Kleiderfärberei *f* garment dyeing
Kleiderkammer *f* closet
Kleiderschrank *m* closet, locker, wardrobe
Kleidung *f* apparel, clothes, dress
Kleie *f* bran
Kleinatomkraftwerk *n* low-power atomic energy station (*Am* plant)
Kleinauto *n* small car, minicar
Kleinbahn *f* narrow-gauge railway

Kleinbessemerbirne *f* baby Bessemer converter
Kleinbessemerei *f* baby Bessemer steel plant
Kleinbetrieb *m* small factory; small-scale manufacturing
Kleinbildfotografie *f* microphotography
Kleinbildkamera *f* miniature camera
Kleinbohrungsmeßgerät *n* small-bore gauge
Kleinbohrungsmeßmikroskop *n* small-bore measuring microscope
kleindimensioniert small-sized
Kleineisenzeug *n* hardware, ironmongery, ironwork, fittings
Kleinempfänger *m* miniaturized receiver
Kleinfahrzeug *n* minicar
Kleinfeldmessung *f* narrow-beam measurement
kleinformatig small-size
Kleinfunkstelle *f* field-radio station
Kleinhandel *m* retail business
Kleinkaliberbohren *n* small-diameter drilling
Kleinkalibergewehr *n* small-bore rifle
Kleinkaliberstand *m* miniature range
kleinkalibrig of small bore (calibre)
Kleinkampfwagen *m* tankette
Kleinkernkraftwerk *n* low-power atomic energy station (*Am* plant)
Kleinkraftwagen *m* minicar
Kleinkupolofen *m* dwarf cupola
Kleinlampe *f* miniature bulb
Kleinlokomotive *f* tractor locomotive (engine, unit), light motor tractor; light rail motor tractor *(electrical train)*
Kleinlosfertigung *f* small-lot manufacture
Kleinmaterialien *npl* incidentals
Kleinmelkanlage *f* / **transportable** portable bucket machine
Kleinmotor *m (el)* small horsepower motor
Kleinmünzen *fpl* mirror coinage
Kleinpflaster *n* small-sett paving
Kleinpflasterstein *m* small sett
kleinporig having close spacing
Kleinreaktor *m* low-power reactor
Kleinrechner *m* / **elektronischer** electronic calculator
Kleinreihenfertigung *f* small [quantity] batch production, limited-quantity production, short-run production, small-lot production; *(Am)* small-lot run; batch work
Kleinreihenfräsen *n* small-lot milling
Kleinrohr *n* small-bore pipe *(heating)*
Kleinrohrheizung *f* small-bore [heating] system
Kleinschalter *m* installation switch
Kleinschiffahrt *f* small-ship navigation
Kleinschrämmaschine *f* breast machine
Kleinserie *f* small batch (lot) (*s. a.* Kleinreihenfertigung); *(Am)* small-lot run
Kleinserienbau *m* small-scale series production
Kleinspannung *f* low (extra-low, safe) voltage
Kleinstausführungsart *f* microminiaturization
Kleinstbagger *m* midget excavator
Kleinstbohrer *m* small-hole drill
Kleinstetigschleifer *m* continuous low-power caterpillar grinder *(papermaking)*
Kleinstmaß *n* minimum size (limit)
Kleinstmotor *m* miniature (fractional horsepower, pilot) motor, micro-motor
Kleinstregelwiderstand *m* midget trimmer potentiometer
Kleinstrelais *n* subminiature relay
Kleinstschlepper *m* mini-tractor, midget tractor
Kleinstspiel *n* minimum clearance
Kleinstspitzbohrer *m* flat drill for microscopic holes
Kleinstvorrichtung *f* / **elektronische** microminiature electronic device
Kleinstwagen *m* midget car
Kleinstwert *m* minimum value
Kleinstwohnung *f* flatlet
Kleinstzähnezahl *f* minimum number of [gear] teeth
Kleinsuper *m* midget superhet
Kleintriangulierung *f* subsidiary triangulation
Kleinwasserraumkessel *m* boiler with small water space
Kleister *m* bookbinding (adhesive) paste, glue, size
Kleisterauftragmaschine *f* glue applicator
Klemmbacke *f (mach)* clamping jaw
Klemmbock *m (el)* contact jaw
Klemmbolzen *m* binding bolt
Klemmbrett *n (el)* terminal board (block, plate)
Klemmbuchse *f* locking bushing, fixing bush
Klemmdiode *f* clamping diode
Klemme *f* clip, clamp, post; *(el)* terminal
  K. eines Elementes cell terminal
  negative K. *(el)* negative terminal
  positive K. *(el)* positive terminal
Klemmeinrichtung *f* clamping device
klemmen to clamp *(workpiece)*, to jam *(valve piston)*; to seize
Klemmen *n* **des Kolbens** jamming of the piston
Klemmenanschlußstück *n* terminal connection piece
Klemmenbezeichnung *f* terminal marking
Klemmenblock *m* terminal block (plate)
Klemmenbrett *n* terminal board (block, plate)
Klemmendeckel *m* **eines Zählers** meter terminal cover
Klemmenleiste *f (el)* terminal block (board), connecting strip (block)
Klemmenleistung *f* output power
Klemmenpaar *n* set of terminals
Klemmenplatte *f* connection (terminal) plate
Klemmenprüfung *f* insulation test *(of terminals)*
Klemmenspannung *f* terminal voltage
Klemmenstreifen *m* terminal strip (board)
Klemmenverbinder *m* terminal connector

Klemmenzuleitung *f* terminal lead
Klemmhaken *m* clamping hook
Klemmhalter *m (mach)* clamp-type toolholder
Klemmhebel *m* clamp lever (handle), clamping (binder) lever
Klemmisolator *m* cleat (split knob) insulator
Klemmknebel *m* clamping bar-head
Klemmknopf *m* clamping head
Klemmkupplung *f* compression coupling
Klemmlänge *f* grip; thickness of plates being joined *(rivet)*
Klemmlast *f* setting load *(support)*
Klemmleiste *f (el)* terminal block (board), connecting strip (block)
Klemmplättchen *n* locking plate
Klemmplatte *f (therm)* packing (clamping) ring; adjustable block; ram block (slotter); *(el)* connection (terminal) plate
Klemmring *m* clamp[ing] ring, clip ring; retaining ring; packing ring
Klemmrolle *f* pinch roller *(e.g. for magnetic tape)*
Klemmschaltung *f* clamping circuit
Klemmschlittschuh *m* clamp-on skate
Klemmschraube *f* clamp[ing screw]; *(el)* terminal screw; retaining nut, set-screw
K. für den Klappenträger clapper clamping nut
Klemmspannstock *m* nipper temple
Klemmstange *f* clamping bar
Klemmstreifen *m (el)* terminal strip
Klemmstück *n* **[am Stößel der Waagerechtstoßmaschine]** adjustable block
Klemmung *f* clamping; clamping mechanism, clamp
Klemmwirkung *f (text)* gripping (nipping) effect
Klempner *m* pipe fitter; plumber
Klette *f (text)* bur
verkohlte K. carbonized bur
Klettenbrechwalze *f* bur crusher roller
Klettenentfernung *f* bur crushing
klettenhaltig burry
Klettenwolf *m* mechanical bur crusher
Kletterbühne *f (mine)* raise climber (platform)
Klettereisen *n* climbing iron, pole climber; grappler
Kletterkran *m* climbing crane
klettern to climb
in die Takelage k. *(ship)* to go aloft
Kletterschalung *f (build)* climbing formwork (shuttering)
Kletterseil *n* climbing rope
Kletterstange *f* climbing pole
Klima *n* climate
Klimaanlage *f* air-conditioning plant (system), air-conditioned plant, air conditioner
K. / mit air-conditioned
klimafest tropicalized, suitable for use in different climatic zones
Klimakunde *f* climatology
Klimaregelung *f* air conditioning
Klimaschrank *m* air-conditioning cabinet; climatic test cabinet
Klimaschutz *m* climatic protection
Klimatechnik *f* air conditioning
klimatisch climatic
klimatisieren to air-condition
Klimatisierung *f* air-conditioning
Klimatisierungsanlage *f* air-conditioning system (plant)
Klimatologie *f* climatology
Klinge *f* blade
K. des Schraubenziehers screwdriver blade
Klingel *f* bell
Klingelbatterie *f* ringing battery
Klingeldraht *m* bell (ringing) wire
Klingelknopf *m* bell button (push)
klingeln to ring [the bell]
Klingelnberg-Palloidspiralkegelrad *n* Klingelnberg palloid spiral bevel gear
Klingelschaltung *f* ringing circuit
Klingeltransformator *m* bell transformer
klingen to sound
Klingen *n* microphonics; microphonic effect *(tubes)*
Klingenabziehapparat *m* razor sharpener
Klingfestigkeit *f* insensitivity to microphonics
Klingneigung *f* microphony; microphone effect
Klinikboot *n* clinic boat
Klinke *f* 1. door handle; 2. *(mach)* latch; detent; pawl; 3. *(telec)* pawl; [spring] jack; dog
dreipolige K. three-pole jack
fünfpolige K. five-pole jack
Klinkenbrett *n* jack board
Klinkenbuchse *f* jack bush
Klinkenfeder *f* jack spring; ratchet spring
Klinkenfeld *n* jack panel (field), patch panel
Klinkengriff *m* ratchet handle
Klinkenhülse *f* jack sleeve
Klinkenkupplung *f* pawl coupling
Klinkenrad *n* notched (ratchet) wheel
Klinkenstange *f* ratchet rod
Klinkenstöpsel *m* jack plug
Klinker *m* clinker *(cement)*; clinker (blue) brick
Klinkerbauart *f (ship)* clinker (clinch) built type
Klinkerbeplankung *f (ship)* clinker (lapstrake) planking
Klinkerboot *n* clinker-built boat
Klinkerpflaster *n* clinker pavement
Klinkerstein *m* clinker (blue, engineering) brick, engineering clinker
Klinkerziegel *m s.* Klinkerstein
Klinkhebel *m* lever with pawl
Klipper *m* clipper *(sailing boat)*
Klipperbug *m* clipper (fiddle) bow
Klirranalysator *m* distortion analyzer
Klirren *n* **der Zahnräder** clashing of gears

Klirrfaktor *m* [harmonic] distortion factor, distortion factor *(electron tube)*
Klirrfaktorcharakteristik *f* harmonic distortion characteristic
Klirrfaktormeßbrücke *f* distortion bridge
Klirrfaktormeßgerät *n* distortion [factor] meter
Klirrverzerrung *f* non-linear distortion
Klischee *n (print)* block, cut, plate, engraving, galvano, cliché, electro[type]
ausgeklinktes K. mortise
Klischeeanstalt *f* block makers, process engravers
Klischeebearbeitung *f* block finishing
Klischeebiegeapparat *m* block bending apparatus
Klischeebiegevorrichtung *f* block bender
Klischeefuß *m* block mount (base)
Klischeeprüfer *m* plate gauge
Klischeezink *n* process zinc
klischieren to make blocks
Klischiermaschine *f* engraving machine, engraver
Kloakenprahm *m* sludge vessel, sewerage pram
Kloben *m* 1. log, cloven *(wood)*; 2. block, pulley; 3. vice *(filing)*
Klobsäge *f* veneer saw
klopfbegünstigend pro-knock *(fuel)*
Klopfbremse *f* anti-knock blend (agent), knock suppressor
Klopfdichte *f (nucl)* tap density
Klopfeigenschaften *fpl* knock characteristics (properties)
klopfen 1. to knock, to pink *(fuel)*; 2. to break *(stone)*
Klopfer *m* telegraph sounder
klopffest antiknock, knockless, knockproof *(fuel)*
Klopffestigkeit *f* anti-knock quality (value, rating)
Klopfförderer *m* knock promoter
Klopfgrenze *f* engine detonation limit, incipient knocking
Klopfhammer *m* leather hammer
Klopfneigung *f* knocking tendency *(fuel)*
Klopfprüfmotor *m* C.F.R. engine
Klopfstärke *f* knock intensity *(fuel)*
Klopfwert *m* octane rating (value, number)
Klopfwolf *m (text)* shaker, willow
Klöppel *m* 1. hammer *(of bells)*; 2. pin *(insulator)*; 3. *(text)* bobbin *(lace)*
Klöppelarbeit *f* bobbin lace work
Klöppeldurchmesser *m* hammer diameter *(chain insulators)*
Klöppelkissen *n* lace pillow
klöppeln to make bobbin lace[s]
Klöppeln *n* bobbin lace making
Klöppelspitze *f* bobbin lace
Klosett *n* toilet, closet, lavatory
Klosettpapier *n* toilet paper
Klosettpumpe *f (ship)* sanitary pump
Klosettsitz *m* toilet seat
Klosettspüler *m* flushing cistern
Klosettank *m (ship)* sanitary tank
Klothoide *f (math)* spiral [transition] curve, clothoid
Klotje *f (ship)* truck
Klotz *m* 1. chunk, trunk, block of wood; shoe *(brake; segmental bearing)*; 2. *(ship)* chock; 3. *(text)* pad[ding] liquor
Klotz-Dämpf-Küpendruckverfahren *n* pad-steam vat-print process
Klotzdruck *m* block print
Klotzen *n* padding, pad dyeing
Klotzfärben *n* pad dyeing
Klotzflotte *f* padding liquor
Klotzgrundierfarbe *f* padding ground shade
Klotzlager *n s.* Segmentlager
Klotzmaschine *f* padding machine, padder
Klotzreserveartikel *m* padded resist style
Klotzverfahren *n* padding
Klotzwalze *f* padding roller
Kluft *f* 1. *(geol)* fissure; 2. shake *(timber)*
Kluftfläche *f (geol)* joint plane
Kluftwasser *n (geol)* interstitial water
Klumpen *m* 1. lump; clod *(soil)*; 2. nugget *(gold)*
klumpig lumpy
Kluppe *f* 1. *(text)* stenter hook; 2. die-stock *(screw-cutting)*; 3. filing block
Klüse *f (ship)* hawse [hole, pipe]
Klüsenband *n* hawse hook
Klüsendeckel *m* [hawse] buckler
Klüsengatt *n* hawse hole (pipe)
Klüsenholz *n* hawse piece
Klüsenpfropfen *m (ship)* jackass
Klüsholz *n* hawse timber
Klüssack *m* hawse bag
Klütenwalze *f (ag)* clod-crushing roller
Klüver *m (ship)* jib
Klüverbaum *m* jib boom, jibboom
Klüverkopf *m* jib-head
Klydonograf *m (el)* klydonograph, surge recorder
Klystron *n (el)* klystron
Klystrongenerator *m* klystron generator
Klystronmodulator *m* klystron modulator
Knabbelkohle *f* cobbles
Knack *m* clicking noise, crackling
knacken to click *(switches)*
Knackgeräusch *n* spluttering, [acoustic] click, clicking noise, crackling
Knagge *f* 1. *(mach)* dog; trip, catch; cam; 2. *(build)* cleat, bearing block *(carpentry)*
Knall *m* explosion; report *(shot)*; crack; bang
knallen *(mine)* to crack *(prop)*
Knallgas *n* oxyhydrogen [gas]

Knallkapsel *f* detonator, detonating cartridge, *(Am)* torpedo
Knallquecksilber *n* fulminate of mercury
Knallsäure *f* fulminic acid
Knallscheuche *f* bird scarer
knapp insufficient; short [of]; scarce
Knäpperschießen *n (mine)* blockholing
Knappheit *f* shortage
Knappschaft *f* miner's company
Knappschaftskasse *f* miner's benefit fund
Knarre *f* ratchet
knarren to creak, to rattle
Knarrenbohrer *m* ratchet borer (drill)
knattern to sizzle *(radio)*; to crackle
Knäuel *n* ball
Knäuelwickelmaschine *f* balling machine
Knäuelwickeln *n* balling
Knebel *m (mach)* lock
Knebelgriff *m* locking handle
Knebelschalter *m* jack (rotary, revolving, turn) switch
Knebelstropp *m (ship)* becket [bridle]
Kneifstopper *m (ship)* [cable] compressor
Kneifzange *f* pincers
knetbar kneadable
kneten to knead, to masticate; to pug *(clay)*
Kneter *m* kneader; masticator; pug mill
Knetlegierung *f* wrought alloy
Knetmaschine *f* kneading machine, kneader
Knetmühle *f* kneading mill
Knick *m* 1. kink *(wire, hose)*; 2. bend *(curve)*; 3. knuckle *(stern of a ship)*
  oberer K. der Kennlinie upper bend of the characteristic
  unterer K. der Kennlinie lower bend of the characteristic
Knickbeanspruchung *f* buckling stress
Knickbelastung *f* buckling load
knicken to buckle *(column)*; to break, to split; to fold *(paper)*
knickfest non-buckling
Knickfestigkeit *f* 1. *(mech)* buckling strength (resistance) *(column)*; 2. folding endurance *(paper)*
Knickflügel *m (aero)* cranked (gull) wing
Knickformeln *fpl* / **Eulersche** Euler's formulae for columns
Knickfrequenz *f* break (corner) frequency
Knickgang *m (ship)* knuckle strake
Knickgelenkschaufellader *m* articulated loading shovel
Knicklänge *f* / **freie** unsupported (effective) length of column, unsupported (effective) length of strut
Knicklast *f* crippling load
Knicklinie *f* knuckle line *(stern of a ship)*
Knickpunkt *m (mine)* yield load *(prop)*
Knickschott *n (ship)* recessed bulkhead
Knicksicherheit *f* safety against buckling
Knickspannung *f* critical stress
Knickspant *n (ship)* chine-type frame
Knickspantbauweise *f* chine construction
Knickspantboot *n* chine (dead rise, V-bottom) boat, skipjack, sharpie
Knickstab *m* column
Knickung *f* flaw; cracking, breaking, splitting; buckling [of columns]
  K. des Strahlenganges optical folding
Knickungswinkel *m* angle of curvature
Knickversuch *m* buckling test
Knickwert *m* critical value
Knickwiderstand *m* resistance to buckling
Knickzahl *f* buckling coefficient
Knie *n* angle; elbow, quarter [pipe] bend, knee
Kniegelenk *n* toggle, hinged knee; knuckle joint
Kniegelenkgetriebe *n* toggle linkage
Kniehebel *m* toggle lever
Kniehebelkippschalter *m* toggle switch
Kniehebelmatrizenpresse *f* knuckle joint embossing press
Kniehebelpresse *f* toggle plate press
Kniehebelziehpresse *f* toggle drawing press
Knieholz *n* 1. knee-timber; 2. *(ship)* crook
Kniehose *f* knee breeches
Kniekupplung *f* rotating double crank, knee joint
Knierohr *n* quarter (long sweep) bend, knee (elbow) pipe
Knierostfeuerung *f* furnace with knee-shaped grate bars
Kniestück *n* quarter (long sweep) bend, elbow, knee *(pipe)*
Kniff *m (text)* crease; wrinkle
knipsen 1. to snap[shot] *(photo)*; 2. to clip, to punch *(ticket)*
Knipsschalter *m* snap switch
Knirschen *n* scroop *(silk)*
Knirschgriff *m* scroopy feel *(silk)*
Knitter *m (text)* crease, wrinkle
knitterarm crease-resistant, creaseproof
Knitterarmausrüstung *f* anticrease (crease-resistant) finish
Knitterbeständigkeit *f* wrinkle resistance
knitterecht creaseproof
Knittereigenschaft *f* creasing property
Knittererholung *f* crease recovery, wrinkle-recovery
Knittererholungswinkel *m* crease recovery angle
Knitterfalte *f* crease
knitterfest crease-resistant, creaseproof
Knitterfestigkeit *f* crease resistance
Knittermaschine *f* flexing machine
knitterresistent crease-resistant
Knitterversuch *m* creasing test
Knitterwiderstandsvermögen *n* crease resistance
Knoblauchöl *n* garlic oil

Knochenentfettungsanlage *f* bone-grease-extracting apparatus
Knochenfett *n* bone oil
Knochenkohle *f* bone (animal) charcoal
Knochenleim *m* bone glue
Knochenmühle *f* bone mill
Knochenöl *n* bone oil
Knochensucher *m (nucl)* bone seeker
Knochenteeröl *n* bone oil
knollig lumpy *(sand)*
Knopf *m* 1. *(mach)* button *(pressing)*; knob *(turning)*; 2. *(el)* control (instrument) knob
Knopfannähmaschine *f* button sewing machine
Knopflochausnähmaschine *f* buttonhole sewing machine *(shoe)*
Knopflochautomat *m* automatic buttonhole machine
Knopflochmaschine *f* buttonhole machine
Knopflochmikrofon *n* lapel microphone
Knopflochstich *m* buttonhole stitch
Knopfröhre *f (el)* door-knob tube
Knopfsetzmaschine *f* button fixing machine *(shoe)*
Knopfzelle *f (el)* button cell
Knorren *m (wood)* knot; knar
Knorrholz *n* knotty [fire]wood
knorrig knotted; knaggy *(wood)*
Knötchen *n* 1. *(text)* mote *(fault in fabric)*; 2. *(met)* fisheye
Knoten *n* tieing, knotting
Knoten *m* 1. *(text)* burl; knot; kink *(rope)*; 2. node *(wave)*; 3. *(el)* junction; 4. joint *(statics)*; 5. knot *(sm/h)*
absteigender K. descending node
aufsteigender K. ascending node
Knotenamt *n* centre (tandem) exchange, junction centre, *(Am)* tandem central office *(telephony)*
Knotenamtsbereich *m* tandem area
Knotenamtsbetrieb *m* tandem working
Knotenbahnhof *m* junction station
Knotenblech *n* gusset plate
Knotenfänger *m* 1. knotter, [pulp] strainer, knot catcher *(papermaking)*; 2. *(text)* knot catcher stop motion *(knitting)*
Knotenfestigkeit *f* knot tenacity
Knotenfläche *f (ph)* nodal plane (surface)
Knotengarnzwirnmaschine *f* slub yarn doubling frame
Knotenlinie *f (ph)* nodal line
Knotenpunkt *m* 1. *(build)* nodal (junction) point; truss joint; 2. *(el)* junction; 3. *(ry)* [railway-] junction
Knotenpunktbeschlag *m (aero)* nodal fitting
Knotenpunktgleichung *f (el)* nodal equation
Knotenpunktverbindung *f (build)* nodal joint
Knotenregel *f (el)* law of networks
Knotenschaltung *f (el)* mesh connection
Knotenstrahl *m* node beam
Knotenverschiebung *f* nodal shifting *(waveguides)*
Knotenzwirn *m* knobbed (slub, spot) twist
Knoter *m (ag)* knotter
Knoterapparat *m (ag)* knotting mechanism
knotig knaggy *(wood)*
Knudsen-Effekt *m* Knudsen effect
Knudsen-Manometer *n* Knudsen gauge
Knudsen-Strömung *f* Knudsen flow, free molecule diffusion
knüllen to crumble; to crease
knüpfen to tie
Knoten k. to tie knots
Knüpfer *m (ag)* knotter
Knüpferschnabel *m (ag)* knotter beak
Knüpfmaschine *f (text)* tying-in machine
Knüpfteppich *m* hooked rug
Knüpftrikot *m(n) (text)* mesh pattern open work
Knüppel *m (met)* billet
Knüppelbohrer *m* billet driller
Knüppelholz *n* round billet wood
Knüppelschere *f* billet shear
Knüppelwalze *f* billet roll
Knüppelwalzwerk *n* billet mill
Koagulat *n* coagulate; coagulum
Koagulation *f* coagulation
koagulierbar coagulable
koagulieren to coagulate, to congeal
Koagulierungsmittel *n* coagulant
koaxial coaxial
Koaxialantenne *f* coaxial aerial
Koaxialkabel *n* coaxial cable
Koaxialleiter *m* coaxial conductor
Koaxialleitung *f* coaxial line
Koaxialstecker *m* coaxial plug
Koaxkabel *n s.* Koaxialkabel
Kobalt *n* cobalt
Kobaltbestrahlungsanlage *f* cobalt telecurie unit
Kobaltblau *n* cobalt blue
Kobaltblüte *f* cobalt bloom, erythrite
Kobaltbombe *f* cobalt bomb
Kobaltdotierung *f* cobalt impurities
Kobalteinheit *f* cobalt telecurie unit
Kobalterz *n* cobalt ore
Kobaltgitter *n* cobalt lattice
Kobaltglanz *m* cobalt glance
Kobaltglas *n* cobalt glass
kobalthaltig cobaltiferous
Kobaltkanone *f* cobalt telecurie unit
Kobaltkies *m (min)* linnaeite
Kobalt(II)-verbindung *f* cobaltous compound
Kobalt(III)-verbindung *f* cobaltic compound
Kochapparat *m s.* Kocher
kochbeständig resistant to boiling
Kochdauer *f* duration of boiling

kochecht fast to boiling
kochen 1. to boil; 2. *(met)* to flutter; 3. to steam *(wood)*; 4. to gas *(accumulator)*
Kochendwasserreaktor *m s.* Siedewasserreaktor
Kocher *m* 1. boiler, cooker; 2. digester *(papermaking)*
  elektrischer K. electric boiler
Kocherauskleidung *f* digester lining
Kocherei *f* digester house
Kochereischiff *n* floating whale-oil factory, whale-factory ship
Kochergrube *f* blow pit *(papermaking)*
Kocherhaus *n* digester house
Kochherd *m* range; kitchener
Kochkessel *m* 1. boiling vessel, boiler; 2. digester, pulp boiler *(papermaking)*
Kochkolben *m* boiling flask
Kochlauge *f* cooking liquor
Kochlaugenprüfung *f* digester test *(papermaking)*
Kochleoide *f (math)* cochleoid
Kochnische *f* kitchenette
Kochperiode *f* cooking cycle *(papermaking)*
Kochplatte *f* boiling plate
  elektrische K. electric boiling plate
Kochplattenumschalter *m* three-heat switch
Kochprobe *f s.* Kochversuch
Kochprozeß *m* 1. cooking process; 2. pulping process *(papermaking)*
Kochpunkt *m* boiling point
Kochsalz *n* common salt, sodium chloride
Kochsäure *f* cooking acid *(papermaking)*
Kochschema *n* cooking chart, set of cooking conditions *(papermaking)*
Kochung *f* pulping, digestion *(papermaking)*
  K. in Kochereinsätzen basket cooking
  K. ohne Umwälzsystem cooking without liquor circulation
Kochversuch *m* boiling test *(cement; corrosion)*
Kochwasserreaktor *m s.* Siedewasserreaktor
Kockpit *m* cockpit, well
Kode *m* code
  K. / ohne code void
  dicht gepackter K. close-packed code
  fehlerberichtigender K. error-correcting [telegraph] code
  fehlererkennender K. error-detecting [telegraph] code
  fehlerverbessernder K. error-correcting code
  selbstprüfender K. *(dat)* error-detecting code, self-checking code
Kodebake *f* code beacon
Kodeelement *n* code element
Kodein *n* codein
Kodeleser *m* code mark reader
Kodemuster *n* coded pattern
Ködermittel *n* bait
Kodescheibe *f* coded disk
Kodeschlüssel *m* key letter
Kodeumsetzer *m* code converter (translator)
Kodeumwandlung *f* code conversion
Kodewähler *m* code selector
Kodewandler *m* code converter (translator)
Kodezeichen *n* code letter
kodieren to [en]code
Kodierer *m* [en]coder
Kodierschablone *f* code disk
Kodierscheibe *f* code disk
Kodierung *f* coding
  numerische K. numeric coding
Kodierungsfolge *f* coding sequence
Kodleine *f* cod line *(of a net)*
Koeffizient *m* coefficient, factor
  K. der gegenseitigen Induktion mutual inductance
  K. der harmonischen Gesamtverzerrung coefficient of harmonic distortion
Koeffizienten *mpl* **/ unbestimmte** undetermined coefficients
Koeffizientenpotentiometer *n* coefficient potentiometer *(analogue computer)*
Koenzym *n* coenzyme
Koerzimeter *n* coercimeter
Koerzitivkraft *f* coercive force; coercivity; retentivity
Koffein *n* caffeine
Koffeinentzug *m* decaffeination
koffeinfrei without (free from) caffeine
Koffer *m* trunk, box
Kofferbeschlag *m* trunk fitting
Kofferboden *m* boot floor
Kofferbrücke *f* luggage grid
Kofferdamm *m (ship)* cofferdam
Kofferdammschott *n* cofferdam bulkhead
Kofferdeck *n* trunk deck
Kofferdecker *m s.* Kofferdeckschiff
Kofferdeckschiff *n* trunk[-decked] vessel
Kofferempfänger *m* portable receiver (set), self-contained radio set
Koffergerät *n* portable set
Koffergrammophon *n* portable phonograph
Koffergriff *m* trunk handle
Kofferkajüte *f* trunk cabin
Kofferkessel *m* wagon [head] boiler
Kofferleder *n* bag leather
Kofferprojektor *m* portable projector
Kofferraum *m* luggage compartment (boot) *(car)*, trunk; box room *(cabin)*
Kofferraumdeckel *m (aut)* trunk lid
Kofferschloß *n* trunk lock
Kofferschreibmaschine *f* portable typewriter
Kofferverstärker *m* portable amplifier
kogredient cogredient
Kogredienz *f* cogrediency
kohärent coherent, single-frequency
Kohärenz *f* coherence
Kohärenzeffekt *m* cohering effect

Kohäsion *f* cohesion
Kohäsionsdruck *m* intrinsic (cohesion) pressure
Kohäsionskraft *f* cohesion force
kohäsionslos cohesionless
kohäsiv cohesive
Kohle *f* 1. coal; 2. carbon *(e.g. electrode)*
  anthrazitartige K. anthracitic coal, semi-anthracite
  aufgelockerte anstehende K. pre-softened coal
  backfähige K. caking coal
  ballastreiche K. dirty coal
  bergehaltige K. bone coal
  bestmelierte K. [unscreened] ameliorated coal
  frische K. green coal
  gasarme K. low-volatile coal, non-gassing coal
  geringwertige K. low-grade coal
  gewaschene K. cleaned coal
  granulierte K. granulated coal
  hangende K. top coal
  hochflüchtige K. high-volatile coal
  hochwertige K. top-grade coal
  im Tiefbau gewonnene K. deep-mined coal
  kurzflammige K. furnace (short-flaming) coal
  langflammige K. long-flaming coal
  schiefrige K. *(Am)* bony coal
  schlackende K. clinkering coal
  schwefelfreie K. sweet coal
  schwefelkieshaltige K. drossy coal, brassil
  senkrecht zur Strebfront anstehende K. rib[side]
  tierische K. animal coal
  treibende K. swelling coal
  ungewaschene K. unwashed coal
  verkaufsfähige K. saleable coal
  verkokbare K. coking coal
  verkupferte K. *(el)* copper-plated carbon
  verwachsene K. bone coal
Kohleanode *f* carbon anode
kohlebeheizt coal-fired
Kohleboden *m* coal hopper *(drying)*
Kohlebogen *m* carbon arc
Kohlebogenlampe *f* carbon arc lamp
Kohlebürste *f* carbon brush
Kohledruckregler *m* carbon regulator
Kohledrucksäule *f* carbon pile
Kohledruckspannungsregler *m* carbon-pile voltage regulator
Kohleelektrode *f* carbon electrode
Kohlefaden *m* carbon filament
Kohlefadenlampe *f* carbon filament lamp
Kohlefeuerung *f* furnace for coal; coal firing
Kohlefilter *n* charcoal filter
kohlehaltig carbonaceous, carboniferous
Kohlehydrat *n* carbohydrate
Kohlehydratstoffwechsel *m* carbohydrate metabolism
Kohlehydrierung *f* hydrogenation of coal
Kohlelichtbogen *m* carbon arc
Kohlelichtbogen-Schweißverfahren *n* carbon arc welding process
Kohlemikrofon *n* carbon microphone
Kohlemikrofontonabnehmer *m* carbon contact pick-up
kohlen to soot; to coal
Kohlenart *f* class of coal, coal type
Kohlenaufbereitungsanlage *f* coal screening and grading plant
Kohlenausbruch *m* coal burst
  plötzlicher K. coal outburst
Kohlenbank *f* / **hangende** top coal
Kohlenbein *n* heel of coal
Kohlenbergwerk *n* colliery
Kohlenbrennerei *f* charcoal burning
Kohlenbrikett *n* coal briquette
Kohlenbunker *m (ship)* coal bunker, bung
Kohlenbunkerloch *n (ship)* bunker scuttle
Kohlendampfer *m* collier, coal steamer (carrier)
Kohlendioxid *n* carbon dioxide
Kohleneinsatz *m* coal consumed (consumption, input)
Kohleneisenstein *m* carbonaceous ironstone
Kohlenentgasung *f* coal carbonization
Kohlenfahrer *m* coal steamer (carrier)
Kohlenflöz *n* coal seam (vein)
Kohlenfördererboot *n* butty boat
Kohlengebirge *n* coal-bearing formation (rock)
kohlengeheizt coal-fuelled, coal-fired
Kohlengewinnung *f* coal mining
  schälende K. coal planing
Kohlengrube *f* colliery, coal pit
Kohlengrus *m* coal slack
Kohlenhalde *f* stockpile, stockyard
kohlenhaltig carboniferous *(seam)*, coal-bearing *(measure)*
Kohlenhauer *m* hewer, cutter
Kohlenhobel *m* coal planer (plough)
Kohlenkarre *f* coal barrow
Kohlenklein *n* duff, gum
Kohlenklumpen *m* knob
Kohlenkombine *f* cutter loader
Kohlenkörnermikrofon *n* carbon granule microphone
Kohlenkran *m* coaling crane
Kohlenladerampe *f* coaling stage
Kohlenlagerplatz *m* stockpile, stockyard
Kohlenleichter *m* coal lighter
Kohlenmeiler *m* charcoal pile
Kohlenmeßfahrt *f (ship)* coal consumption trial trip
Kohlenmonoxid *n* carbon monoxide
Kohlenoxid *n* carbon oxide
Kohlenoxidhämoglobin *n* carbon monoxide haemoglobin
Kohlenpflug *m* coal plow
Kohlenpier *m(f)* coal jetty

Kohlenprahm *m* coal lighter
Kohlensack *m* upper bosh line, belly *(blast furnace)*
Kohlensäure *f* carbonic acid; carbon dioxide
Kohlensäureausbruch *m* outburst of carbon dioxide
Kohlensäurediagramm *n* carbon dioxide diagram
kohlensäurehaltig carbonated
Kohlensäurelöscher *m* carbon dioxide fire extinguisher
Kohlenschaufel *f* coal shovel
Kohlenschiefer *m* black bat
Kohlenschiff *n* collier, coal carrier
Kohlensorte *f* class (commercial grade) of coal
Kohlenstaub *m* pulverized (powdered, dust) coal, coal dust
kohlenstaubbeheizt fired with pulverized coal
Kohlenstaubbunker *m* char-dust bin
Kohlenstaubexplosion *f* coal dust explosion
Kohlenstaubfeuerung *f* pulverized-coal firing (furnace)
Kohlenstaublunge *f* anthracosis
Kohlenstaubmotor *m* coal-dust engine
Kohlenstaubverbrennung *f* pulverized-coal combustion, powdered-coal combustion
Kohlenstoff *m* carbon
  freier K. uncombined carbon
  gebundener K. combined carbon
  graphitischer K. graphitic carbon
  ungebundener K. uncombined carbon
kohlenstoffabgebend carbonaceous
Kohlenstoffalter *n* radiocarbon age
kohlenstoffarm low-carbon
Kohlenstoffatom *n* carbon atom
  asymmetrisches K. asymmetric carbon atom
Kohlenstoffaufnahme *f* carbon pickup
Kohlenstoffelement *n* carbon-consuming cell
Kohlenstoffentziehung *f* decarbonization
Kohlenstoffgehalt *m* carbon content
kohlenstoffhaltig carbonaceous, carboniferous
Kohlenstoffkette *f* carbon chain
kohlenstoffreich rich in carbon
Kohlenstoffring *m* carbocycle, carboatomic ring
Kohlenstoffstahl *m* carbon steel
  unlegierter K. plain carbon steel
Kohlenstoff-Stickstoff-Kreislauf *m* Bethe (carbon-nitrogen) cycle
Kohlenstoffverbindung *f* carbon compound
Kohlenstoffzustellung *f* carbon lining
Kohlenstoß *m* / **schräggestellter** under-tipped face
  überkippter K. over-tipped face
Kohlentransporter *m* collier, coal carrier
Kohlentrimmer *m* trimmer, coal heaver
Kohlenübernahme *f* coaling
Kohlenverbrauch *m* coal consumption
Kohlenvorrat *m* coal supply, coaling
Kohlenwaage *f* coal weighing machine
Kohlenwäsche *f* coal washing plant, coal washery
Kohlenwasserstoff *m* hydrocarbon
  aromatischer K. aromatic hydrocarbon
  gesättigter K. saturated hydrocarbon
  halogensubstituierter K. halocarbon
  mehrfach halogenierter K. polyhalohydrocarbon
  perhalogenierter K. perhalo[hydro]carbon
kohlenwasserstoffhaltig hydrocarbonaceous
Kohlenwehr *n* auxiliary fire bridge
Kohlenwertstoff *m* by-product
Kohlenwertstoffbetrieb *m* by-product plant
Kohlenwinde *f* coal hoist (winch)
Kohlenzieher *m* trimmer, coal heaver
Kohlepapier *n* carbon paper
Kohleprobe *f* coal sample
Köhler *m* charcoal burner
Kohleschichtdrehwiderstand *m* carbon volume control
Kohleschichtwiderstand *m* carbon-film resistor, [film-type] carbon resistor; graphite resistance
  K. mit Wendel cracked-carbon resistor
Kohlestab *m* carbon rod (bar)
  K. für Bogenlampen *n* arc-lamp carbon
Kohletiegel *m* carbon crucible
Kohletrichter *m* coal hopper
Kohleveredlung *f* coal transformation
Kohleverflüssigung *f* hydrogenation of coal, coal liquefaction
Kohlevorkommen *n* coal deposit
Kohlewiderstand *m* carbon resistor; carbon pile resistance
Kohlezerkleinerer *m* coal crusher
Kohlezerkleinerung *f* coal crushing
kohlig carbonaceous
Kohlungsmittel *n* carburizing agent, carburizer
Koinzidenz *f* coincidence
  echte K. true coincidence
  verzögerte K. delayed coincidence
  zufällige K. *(nucl)* random (chance, accidental, spurious) coincidence
Koinzidenzanalysator *m* coincidence analyzer
Koinzidenzanordnung *f* coincidence array (arrangement, equipment)
Koinzidenzeinstellung *f* coincidence setting
Koinzidenzentfernungsmesser *m* coincidence range-finder
Koinzidenzkreis *m* interval-selector circuit
Koinzidenzlibelle *f* coincidence level
Koinzidenzmessung *f* coincidence measurement
Koinzidenzrate *f* coincidence rate
Koinzidenzschaltung *f* coincidence circuit
Koinzidenzspektrum *n* coincidence spectrum
Koinzidenzstrom *m* coincident current *(magnetic core store)*
Koinzidenzstromauslesen *n* coincident current read out *(magnetic core memory)*
Koinzidenzstufe *f* coincidence stage (circuit)

Koinzidenzszintillationsspektrometer *n* coincidence scintillation spectrometer
Koinzidenzzähler *m* coincidence counter
Koinzidenzzählrate *f* coincidence rate
Koinzidenzzählung *f* coincidence counting
koinzidieren to coincide
Koje *f* berth, bunk
 festeingebaute K. ready bunk
Kokain *n* cocaine
Kokerei *f* coking plant, cokery
Kokereiarbeiter *m* coke man
Kokereigas *n* coal (coke-oven) gas
Kokille *f* ingot (permanent metal) mould; chill
 aufklappbare K. split ingot mould
 gewellte K. corrugated mould
Kokillenanstrichmittel *n* ingot mould coating material
Kokillenbruch *m* ingot mould scrap
Kokillenfett *n* mould grease
Kokillenguß *m* chill (metal mould) casting
Kokillenschleuderguß *m* hot mould centrifugal casting
Kokillenschlichte *f* mould wash, ingot mould varnish (blackening)
Kokillenuntersatz *m* ingot mould stool
Kokon *m* cocoon
Kokosbutter *f* coconut butter (oil)
Kokosfaser *f* coconut fibre; coir
Kokosmatte *f* coir mat
Koks *m* coke
 glühender K. hot coke
 schaumiger K. spongy coke
 ungelöschter K. unquenched coke
Koksablagerung *f* coke formation (deposit)
Koksasche *f* breeze
Koksausbringen *n* coke yield
Koksausdrückmaschine *f* coke pusher (pushing machine)
Koksausdrückstange *f* coke ram
Koksbrand *m* coke charge
Koksbrecher *m* coke cutter
Koksfeinkohle *f* coking smalls
Koksfilter *n* coke filter
Koksführungsschild *m* coke guide
Koksführungswagen *m* coke guide car
koksgefeuert coke-fired
Koksgrus *m* fine coke, coke breeze
Kokskammer *f* coke (coking) chamber
Kokskohle *f* coking coal
Kokskohlenturm *m* coking coal bunker
Kokskorb *m* *(build)* brazier
Kokskuchen *m* coke cake
Kokskuchenführungswagen *m* coke guide car
Kokslöschen *n* quenching (damping down) of the coke
Kokslöschseite *f* coke-discharge side
Kokslöschturm *m* coke quenching tower
Kokslöschwagen *m* coke quenching car
Koksnaht *f* coke cleavage (parting)
Koksofen *m* 1. coke oven; 2. *(build)* brazier
Koksofenfüllwagen *m* coke oven charging car
Koksofengas *n* coke oven gas
Koksperle *f* coke grain
Koksplatz *m* coke bench
Koksrampe *f* coke bench (wharf)
Koksseite *f* coke bench side
Kokssieb *n* coke screen
Koksstaub *m* coke breeze (dust)
Kokstrommel *f* coke drum (trommel)
Kokswerk *n s.* Kokerei
Kokszieher *m* coke drawer
Kokung *f* coking
Kokungsvermögen *n* coking power
Kokungszeit *f* coking time
Kolben *m* 1. piston, plunger *(engine)*; 2. *(ch)* flask; 3. butt *(rifle)*; 4. *(el)* bulb, cone *(lamp, valve)* envelope *(vacuum tube)*; ram *(car steering)*; 6. iron, bit *(soldering)*; 7. spike *(maize)*
 K. mit Kreuzkopfführung crosshead piston
 doppelwandiger K. box piston
 einwandiger K. open piston
 flacher K. flat-topped piston
 gegenläufiger K. piston working in opposite direction
 geschlossener K. box piston
 getragener K. supported piston
 gleichläufiger K. piston working in the same direction
 hohler K. *s.* hohlgegossener K.
 hohlgegossener K. hollow-cast piston
 innenmattierter K. *(el)* satin-etched bulb
 offener K. open piston
 selbsttragender K. self-supporting piston, piston supported by the cylinder
 spitzenloser K. *(el)* tipless bulb
 verspiegelter K. *(el)* metal-coated bulb
 von der Zylinderwand getragener K. piston supported by the cylinder, self-supporting piston
 wechselläufiger K. piston working alternately
Kolbenabschwächer *m* piston attenuator *(waveguides)*
Kolbenaufgang *m* ascent of [the] piston
Kolbenauge *n* piston boss
Kolbenbeaufschlagung *f* piston load
kolbenbetätigt piston-operated, plunger-operated
Kolbenblasmaschine *f* glass-bulb forming machine
Kolbenblindteiler *m* piston attenuator *(waveguides)*
Kolbenboden *m* piston head (crown)
Kolbenbohrung *f* cylinder bore, bore of the piston
Kolbenbolzen *m* gudgeon (piston) pin

schwimmender K. floating gudgeon pin
Kolbenbolzenauge *n* gudgeon-pin boss
Kolbenbolzenauszieher *m* gudgeon-pin extractor
Kolbenbolzenbuchse *f* piston-pin bushing
Kolbenbolzenbüchse *f* gudgeon-pin bushing
Kolbenbolzenlager *n* **der Pleuelstange** small end bearing
Kolbenbüchse *f* piston-boss bushing
Kolbendampfmaschine *f* reciprocating (piston-type) steam engine
Kolbendecke *f s.* Kolbendeckel
Kolbendeckel *m* follower [plate], junk ring
Kolbendeckelschraube *f* junk ring pin, tap bolt
Kolbendiagramm *n* piston diagram
Kolbendichtung *f* piston packing
Kolbendruck *m* piston pressure
Kolbenfeder *f* piston spring, plunger spring
Kolbenfederteller *m* plunger spring plate
Kolbenfläche *f* piston area
Kolbenfressen *n* seizing of the piston
Kolbengasdurchlaß *m* piston-ring blow-by
Kolbengebläse *n* piston blower (compressor)
Kolbengeschwindigkeit *f* piston speed
mittlere K. average piston speed
Kolbenheizung *f* piston heating
Kolbenhemd *n* piston skirt
Kolbenhingang *m* forward piston stroke
Kolbenhöhe *f* length of the piston
Kolbenhub *m* piston stroke
Kolbenindikator *m* piston indicator
Kolbenklopfen *n* piston knock
Kolbenkörper *m* piston barrel
Kolbenkraftdiagramm *n* piston pressure-time diagram
Kolbenkrone *f* piston crown
Kolbenlader *m* displacement supercharger
Kolbenlauf *m* piston motion
Kolbenlaufbahn *f* working surface of [the] piston
Kolbenlaufbild *n* piston condition
Kolbenluftpumpe *f* piston air pump
Kolbenmanometer *n* piston-type pressure gauge
gewichtsbelastetes K. dead-weight piston gauge
Kolbenmantel *m* piston skirt
Kolbenmaschine *f* reciprocating engine
Kolbenmasse *f* piston mass
Kolbenmotor *m* reciprocating engine
Kolbennabe *f* piston boss
Kolbenniedergang *m* descent of [the] piston
Kolbennut *f* piston-ring groove
Kolbenpresse *f* ram-type press
Kolbenpumpe *f* piston pump
Kolbenring *m* piston (packing) ring; compression ring
K. mit überlapptem Stoß lap-ended piston ring
elastischer K. elastic (snap piston) ring, piston spring
Kolbenringfuge *f* piston-ring joint
Kolbenringnut *f* piston-ring groove
Kolbenringschleifmaschine *f* piston-ring grinding machine
Kolbenringschloß *n* piston-ring lock
Kolbenringsicherungsstift *m* piston-ring stop peg
Kolbenringspanner *m* piston-ring compressor (tightener)
Kolbenringspiel *n* piston-ring side clearance
Kolbenringstoß *m* end of [the] piston ring
Kolbenrückgang *m* return piston stroke
Kolbenschaft *m* piston skirt
Kolbenschieber *m* sleeve (piston-type of slide, sliding-spool type) valve, slide valve based on circular cylindrical surface
Kolbenschieberbüchse *f* / **eingesetzte** piston valve liner
Kolbenschiebereinsatz *m* piston valve liner
Kolbenschiebergehäuse *n* piston valve chest
Kolbenschiebersteuerung *f* piston valve gear
Kolbenschieberventil *n* spool (plunger slide-type) valve
Kolbenschlackern *n* piston slack
Kolbenschmierhahn *m* piston lubrication cock
Kolbenschwärzung *f (el)* bulb blackening
Kolbenseite *f* / **stangenfreie** head end of piston
Kolbensetzmaschine *f* piston jig
Kolbenspiel *n* piston play
Kolbenspielraum *m* piston clearance
Kolbenstange *f* piston rod; stem *(valve)*
durchgehende K. through piston rod
einseitige K. simple piston rod
wandernde K. travelling piston rod
Kolbenstangenbund *m* piston-rod collar (shoulder)
Kolbenstangenflansch *m* piston-rod flange
Kolbenstangenführung *f* piston-rod guide
eingleisige K. guide of the tail piston
hintere K. tail rod guide
Kolbenstangengewinde *n* piston-rod thread
Kolbenstangenmutter *f* piston-rod nut
Kolbenstangenschaft *m* rod of piston
Kolbenstangentraverse *f* crosshead beam
Kolbensteg *m* piston land
oberster K. piston top land
Kolbenstellantrieb *m* piston actuator
Kolbenstift *m* gudgeon pin
Kolbentakt *m* piston stroke
Kolbentotlage *f* piston dead centre [position]
Kolbentrimmer *m* piston trimmer
Kolbenüberlauf *m* over-run of the piston
Kolbenumsteuerung *f* reversal of the piston
Kolbenumsteuerventil *n* reversing piston valve, directional valve controlling the piston movement

Kolbenventil *n* piston-type valve, plunger-type valve
Kolbenverdrängung *f* piston displacement
Kolbenviskosimeter *n* piston-type viscosimeter
Kolbenwassermesser *m* piston (positive) water-meter
Kolbenweg *m* piston travel
Kolbenwegdiagramm *n* cylinder (volume, piston position-time) diagram
bogenförmiges K. curved piston-path diagram
Kolbenwegdruckdiagramm *n* piston position-time diagram
Kolbenweglinie *f* line indicating course of the piston
Kolbenzähler *m* piston-type flowmeter, reciprocating piston meter
Kolk *m (ship)* [deep] pool
Kolkbildung *f* cratering *(wear on the tool face)*
kolkfest crater-resistant
Kolktiefe *f* crater depth
Kolkung *f* 1. crater *(tool)*; 2. *(build)* underwashing, scour[ing]
Kolkverschleiß *m* crater wear
Kollagen *n* collagen, collogen
Kollektivmodell *n* **[des Atomkerns]** collective model of nucleus, collective nuclear model; model [of nucleus]
Kollektor *m* 1. *(el)* commutator, collector; 2. *(opt)* collecting lens
Kollektoranschluß *m* collector terminal (contact)
Kollektorbasisstrom *m* collector base current
Kollektorbürste *f* commutator brush
Kollektordiode *f* collector diode
Kollektorelektrode *f* collector electrode
Kollektorgeräusch *n* commutator noise
Kollektorglimmer *m* commutator mica
Kollektorgrenzschicht *f* collector transition region *(transistor)*
Kollektorhülse *f* commutator hub (sleeve)
Kollektorisolierhülse *f* commutator shell insulation
Kollektorkapazität *f* collector capacitance
Kollektorlamelle *f* collector segment, commutator section
Kollektormaschine *f* commutator machine
Kollektormotor *m* collector motor
K. mit Fremdbelüftung forced-cooled commutator motor
Kollektorpotential *n* collector potential
Kollektorschleife *f* collector loop
Kollektorschleifmaschine *f* commutator grinder
Kollektorspannung *f* commutator voltage
Kollektorspitzenspannung *f* maximum collector voltage, collector peak voltage *(semiconductor)*
Kollektorstrom *m* collector current
Kollektorteilung *f* lamellar division
Kollektorübergang *m* collector junction *(transistor)*
Kollektorwiderstand *m* collector resistance
Kollerbütte *f* breaker vat *(papermaking)*
Kollergang *m (ag, mine)* edge mill (runner); pugmill [mixer], muller mixer; pan grinder; putty chaser
Kollermühle *f* edge-runner mill
kollern to mull
Kollerstoff *m* broke
kollidieren to collide; to run into; to run down
Kollimator *m* collimator
kollinear collinear
Kollision *f* boarding, collision
Kollisionsmatte *f (ship)* collision mat
Kollisionsschott *n (ship)* collision bulkhead
Kollodium *n* collodion
Kollodiumverfahren *n* collodion process
Kollodiumwolle *f* collodion cotton, nitrocellulose, pyroxylin
Kolloid *n* colloid
lyophiles K. lyophilic colloid
lyophobes K. lyophobic colloid
kolloidal colloidal
Kolloidchemie *f* colloidal chemistry
Kolloidelektrolyt *m* colloidal electrolyte
Kolloidmühle *f* colloid mill
Kolonne *f* 1. *(ch)* column; 2. crew, gang, team
Kolonnenboden *m* column tray (plate)
Kolonnenionisation *f* columnar ionization
Kolonnenkopf *m* column cap (head)
Kolophonium *n* colophony, rosin
Kolophoniumlötdraht *m* rosin-cored solder
Kolorimeter *n* colorimeter
Kolorimetrie *f* colorimetry, colour densitometry (comparimetry)
kolorimetrisch colorimetric
Kolozinn *n* rosin-cored solder
Kolposkop *n* colposcope
Kolposkopie *f* colposcopy
Kolumne *f (print)* column
Kolumnenmaß *n* page-gauge
Kolumnenschiff *n* slip (column) galley
Kolumnenschnur *f* page cord
Kolumnentitel *m* heading
lebender K. running title
Kolur *m* colure *(great circle)*
Koma *f (opt)* coma
Komaabbildung *f* comatic image
Komabild *n* comatic image
Komarestfehler *m* residual comatic aberration
Kombinat *n* complex [of factory plants]
Kombination *f* combination
Kombinationsdrillmaschine *f* combine drill
Kombinationsdünger *m* compound fertilizer
Kombinationsfärbung *f* combination dyeing
Kombinationsleitwert *m (el)* combined conductivity

Kombinationspräparat *n* compound preparation
Kombinationsschwingung *f* combination band *(spectroscopy)*
Kombinationszange *f* combination pliers
Kombinatorik *f* combinatorial analysis
kombinatorisch combinatory
Kombine *f* combine (complete) harvester
Kombizange *f* combination pliers
Kombüse *f* galley, caboose
Kombüsenherd *m* galley range
Komet *m* comet
  künstlicher K. artificial comet
Kometenbahn *f* cometary orbit
Kometenkopf *m* comet head
Kometensucher *m* comet finder
Komma *n (math)* [decimal] point, radix point
  einstellbares K. adjustable point
  festes K. fixed point
  gleitendes K. floating point *(half-logarithmic)*
Kommaanzeige *f* decimal point indication,
Kommadarstellung *f* decimal point presentation
kommaförmig wedge-shaped *(chip)*
Kommanditgesellschaft *f* limited partnership
Kommanditist *m* limited partner
Kommandoanlage *f* control unit
Kommandoaufbauten *pl* navigating erections
Kommandobrücke *f* pilot (navigating, navigation) bridge
  achtere K. astern navigation bridge
Kommandoempfänger *m* cutoff signal receiver
Kommandogeber *m* control-signal generator
Kommandogerät *n* 1. fire-control director *(anti-aircraft)*; 2. *(aero)* sensory device
Kommandohilfsgerät *n* auxiliary fire director (predictor) *(anti-aircraft)*
Kommandolenkung *f (rock)* command guidance
Kommandorückmeldung *f (rock)* cutoff signal test
Kommandosender *m (rock)* cutoff signal transmitter
Kommandosenderendstufe *f* cutoff signal transmitter output stage
Kommaverschiebung *f (dat)* shift
kommen to come; to arrive; to approach *(train)*
  in die Leitung k. to enter a line
  zur Deckung k. to coincide
kommensurabel *(math)* commensurable
Kommensurabilität *f (math)* commensurability
Kommissionär *m* contractor [of the state]; government contractor
Kommissionsverkauf *m* sale by agent
Kommunalbetrieb *m* municipally-owned factory
kommunizierend communicating
Kommutation *f* commutation
kommutativ commutative
Kommutativgesetz *n* commutative law
Kommutator *m* collector, commutator
  elektronischer K. sampler
Kommutatoranker *m* commutator armature
Kommutatorbuchse *f* commutator sleeve
Kommutatorgleichrichter *m* commutator rectifier
Kommutatorhülse *f* commutator sleeve
Kommutatorisolierhülse *f* commutator shell insulation
Kommutatormaschine *f* commutating machine
  läufererregte K. machine with inherent self-excitation
Kommutatormotor *m* commutator (brush) motor
Kommutatorring *m* end ring
Kommutatorschalter *m* commutator-type switch
Kommutatorschritt *m* commutator pitch
Kommutatorsegment *n* commutator bar
kommutieren to commutate
Kommutierung *f* commutation
  erzwungene K. forced commutation
Kommutierungsachse *f* axis of commutation
Kommutierungsfeld *n* reversing (commutating, commutation) field
Kommutierungsfrequenz *f* commutator ripple frequency
Kommutierungskurve *f* commutation (magnetization) curve
Kommutierungswinkel *m* commutating angle, commutation margin
kompakt compact, self-contained
Kompaktbau *m* compact building; single-block factory plant
Kompaktbaustein *m* micromodule
Kompaktpuder *m* compact powder
Komparator *m* comparator
Komparatorprinzip *n* comparator principle
Kompaß *m* compass
  magnetelektrischer K. induction compass
Kompaßablenkung *f* compass deviation
Kompaßabweichung *f* compass declination (variation)
Kompaßaufhängung *f* compass suspension
Kompaßdiopter *m* azimuth mirror
Kompaßfehler *m* compass error
Kompaßhaus *n* binnacle
Kompaßkurs *m* compass heading
Kompaßmißweisung *f* compass declination
Kompaßmutter *f* master compass
Kompaßnadel *f* compass (magnetic variation) needle
Kompaßpeilung *f* compass bearing
Kompaßrose *f* compass card (rose, diagram)
Kompaßsäule *f* binnacle stand, compass pole
Kompaßstrich *m* bearing of the compass
Kompaßtochter *f* repeater compass
Kompatibilität *f* compatibility
Kompensation *f* compensation; balancing
  K. durch Integration integral equalization
  K. durch Vorhalt derivative equalization

Kompensationsantenne *f* balancing aerial
Kompensationsdiode *f* balancing diode
Kompensationsdrossel *f* compensating choke (reactor)
Kompensationsgerät *n* compensator
Kompensationsglied *n* equalizer
Kompensationshalbleiter *m* compensated semiconductor
Kompensationsinstrument *n* self-balancing instrument
Kompensationskammer *f* compensated [ionization] chamber
Kompensationskreis *m* bucking circuit
Kompensationsleitung *f* compensation lead
Kompensationslinienschreiber *m* recording potentiometer
Kompensationsmagnet *m* compensating (balancing) magnet
Kompensationsnetzwerk *n* equalizing network
Kompensationsokular *n* compensating eyepiece
Kompensationspendel *n* compensation (compensated) pendulum
Kompensationsrohr *n* expansion (compensating) pipe
Kompensationsröhrenvoltmeter *n* slide-back valve voltmeter
Kompensationsschaltung *f* compensating (equalizing) network, compensating (equalizing) circuit
Kompensationsschreiber *m* potentiometer (zero) recorder, recording potentiometer
Kompensationsspannung *f* balancing (bucking) voltage
Kompensationsspannungsmesser *m* compensating voltmeter
Kompensationsspule *f* compound (compensating) coil
Kompensationsstab *m s.* Trimmstab
Kompensationstransformator *m* neutralizing transformer
Kompensationsverfahren *n* compensation (balancing) method
Kompensationswicklung *f* compensation (balancing) winding
Kompensationswiderstand *m* compensating resistance
Kompensator *m* compensator, balance (differential) gear *(turbine)*; potentiometer
  komplexer K. complex compensator
  selbstabgleichender K. self-balancing potentiometer
  selbsttätiger K. automatic potentiometer
Kompensatornivellier *n* self-aligning level
kompensieren to compensate
  auf Null k. to back off
Kompensierung *f* balancing, neutralization
Kompensierungsbake *f* deviation bake
Kompensierungsboje *f* deviation buoy
Kompensierungsfahrt *f* trip for adjusting the compass, compass run
Kompensograf *m* potentiometer recorder, recording potentiometer
Kompilation *f* compilation *(programming method)*
komplanar *(math)* coplanar
Komplement *n* complement *(arithmetical and logical)*
komplementär complementary
Komplementärfarben *fpl* complementary colours
Komplementarität *f* complementarity [principle]
Komplementwinkel *m* complementary angle
komplettieren to complete, to finish
Komplettierung *f* completion, finish[ing]
komplex complex *(ratio)*
Komplex *m* aggregate; group of buildings; complex compound
Komplexbildner *m (ch)* complexing agent
Komplexbildung *f (ch)* complexing
Komplexion *n* complex ion
Komplexometrie *f (ch)* chelatometry
Komplexon *n (ch)* complexon
Komplexverbindung *f* complex compound
komplizieren to complicate
kompliziert complicated, complex, intricate
Komponente *f* component
  falschphasige K. *(el)* out-of-phase component
  gegenläufige K. *(el)* negative-sequence component
  gleichphasige K. *(el)* co-phase component, in-phase component
  harte K. hard (penetrating) component *(cosmic rays)*
  mitläufige K. *(el)* positive-sequence component, positive-direction component
  mitläufige symmetrische K. *(el)* positive-sequence symmetrical component
  primäre K. der kosmischen Strahlung primary cosmic-radiation component
  reguläre K. *(math)* regular component
  singuläre K. *(math)* singular component
  symmetrische K. *(el)* symmetrical component
  unsymmetrische K. *(el)* unbalanced component
Komponenten *fpl* **der Schnittkraft** components of the cutting force, cutting components
Kompositbau *m* composite building
Kompositbauweise *f* composite construction
Kompositschiff *n* composite[-built] vessel
Kompost *m* compost
Kompostierung *f* composting
Kompostschichter *m* compost stacker
Kompostwender *m* compost turner
Kompounddynamomaschine *f* compound-wound generator
Kompounderregung *f* compound excitation
kompoundieren to compound
Kompoundierung *f* compounding
Kompoundmotor *m* compound motor

Kompoundwicklung *f* compounding coil, [cumulative] compound winding
kompreß set solid
kompressibel compressible
Kompressibilität *f* compressibility
K. der Gase compressibility of gases
Kompressibilitätskoeffizient *m* coefficient of compressibility
Kompression *f* compression
Kompressionsarbeit *f* work done during compression
Kompressionsendspannung *f* final pressure of expansion, final compression pressure
Kompressionsmodul *m* modulus of volume elasticity, bulk modulus
Kompressionspunkt *m* point of compression
Kompressionsraum *m* 1. compression chamber; 2. volume of compression space
Kompressionsring *m* compression ring
Kompressionstakt *m* compression stroke
Kompressionsverhältnis *n* ratio of compression
Kompressionsverlust *m* compression loss *(turbine)*
Kompressionswelle *f* compression wave
hydromagnetische K. hydromagnetic compression wave
Kompressor *m* compressor
Kompressorkühlschrank *m* compressor (compression-type) refrigerator
Kompressormotor *m* supercharged (forced-induction) engine
komprimieren to compress
Komprimierung *f* compression
Konchoide *f (math)* conchoid
Konchoidenlenker *m* conchoidal motion
Kondensanz *f (el)* condensance, capacitance, capacitive (condensive) reactance
Kondensat *n* condensate, condensation product; water of condensation
Kondensatablauf *m* condensation run-off (discharge)
Kondensatbehälter *m* condensation tank
Kondensation *f* condensation
Kondensationsapparat *m* condensing apparatus
Kondensationsbetrieb *m* **der Dampfmaschine** steam engine working condensing
Kondensationsdampfmaschine *f* condensing steam engine
Kondensationsgefäß *n* condenser receiver
Kondensationsharz *n* condensation resin
Kondensationskalorimeter *n* steam calorimeter
Kondensationskeim *m* condensation centre, nucleus of condensation
Kondensationsmaschine *f* 1. condensing engine; 2. *(text)* curing machine
Kondensationspolymerisation *f* condensation polymerization
Kondensationsprodukt *n* condensation polymer (product)
Kondensationspumpe *f* condensing pump
Kondensationspunkt *m* condensation level
Kondensationsverlust *m* loss due to condensation
Kondensationswärme *f* heat of condensation
Kondensationswasser *n* condenser water, water of condensation
Kondensator *m* 1. capacitor, *(Am)* condenser; 2. condenser
K. mit Luftkühlung air capacitor
abgedichteter K. sealed capacitor
die Zeitkonstante bestimmender K. timing capacitor
durch Spannung veränderlicher K. voltage-variable capacitor
eingekapselter K. encapsulated capacitor
gedruckter K. printed capacitor
getrennter K. independent condenser
idealer K. perfect capacitor
in die Turbine eingebauter K. condenser combined with the turbine
induktionsfreier K. non-inductive capacitor
keramischer K. ceramic capacitor, ceramicon
logarithmischer K. mid-line capacitor
regelbarer K. adjustable capacitor
reihengeschalteter K. series capacitor
spannungsabhängiger K. varicap
trockener K. dry condenser
verlustbehafteter K. imperfect capacitor
verlustloser K. perfect (no-loss) capacitor
wellengerader K. straight-line wavelength capacitor
wellengleicher K. square-law capacitor, corrugated-plate capacitor
Kondensatorachse *f* capacitor spindle (shaft)
Kondensatorbatterie *f* bank of capacitors
Kondensatorbelag *m* capacitor coating, coating of a capacitor
Kondensatorbelegung *f* armature of a capacitor, capacitor armature
Kondensatordielektrikum *n* capacitor dielectric
Kondensatordruck *m* condenser pressure
Kondensatordurchführung *f* capacitor bushing; capacitor-type terminal
Kondensatorelektrometer *n* capacitor meter
Kondensatorelektroskop *n* capacitor electroscope
Kondensatorenbatterie *f* bank of capacitors
Kondensatorionisationskammer *f s.* Kondensatorkammer
Kondensatorkammer *f* condenser ionization chamber
Kondensatorladung *f* capacitor charge
Kondensatormikrofon *n* capacitor microphone
Kondensatormotor *m* capacitor motor
Kondensatorraum *m* condensation space (chamber)
Kondensatorrohr *n* condenser tube

Kondensatorrohrleitung *f* condenser piping
Kondensatorsaugstutzen *m* stand-pipe for condenser suction
Kondensatorsäule *f* capacitor stack
Kondensatorwiderstandskopplung *f* RC coupling
Kondensatorzündung *f* capacitor ignition
Kondensatpumpe *f* condensate (condensed-steam) pump
Kondensatsammelbehälter *m* condensate collector
Kondensatsammelgefäß *n* condensation trap, condensation-accumulator vessel
Kondenshahn *m* drain cock
kondensierbar condensable
kondensieren to condense
kondensiert condensed
Kondensmilch *f* condensed milk
Kondensor *m* condensing lens, condenser
 aplanatischer K. aplanatic condenser
 pankratischer K. pancratic condenser
Kondensorbewegung *f* substage control
Kondensorlinse *f* condenser lens
Kondenstopf *m* steam drier (separator, trap)
Kondenswasser *n* condensed (waste) water, water of condensation, condensate
Kondenswasserableiter *m* steam trap (separator)
Kondenswasserabscheider *m* steam drier (separator, trap)
Kondenswasseraustritt *m* condensed-water discharge
Kondenswasserbildung *f* condensate formation
Kondenswasserfilter *n* condensation water filter
Kondenswasserhahn *m* drain cock
Kondenswasserrückleiter *m* return-type steam trap
Kondenswasserrückspeiseanlage *f* apparatus for returning water of condensation
Konditionieranstalt *f* conditioning plant
Konditionierapparat *m* conditioning (moisture testing) oven
konditionieren to condition
Konditionierschein *m* conditioning certificate
Konditionierung *f* conditioning
Konditionierverlust *m* conditioning loss
Konduktanz *f (el)* conductance
Konduktometrie *f* conductometric analysis (titration), conductometry
Konfektion *f* garment industry
Konfektionierung *f* confectioning
Konfektionskleidung *f* ready-to-wear clothing
Konferenzgespräch *n (telec)* conference call
Konferenzschaltung *f* conference circuit
Konfiguration *f* configuration
 plane K. configuration in the plane
 räumliche K. configuration in the space
Konfitüre *f* jam
konfokal confocal
konform *(math)* conformal, angle-preserving, isogonal
Konglomerat *n (geol)* conglomerate
konglomerieren to conglomerate
Kongokopal *m* Congo copal
Kongorot *n* Congo red
kongruent congruent
Kongruenz *f* congruence
Koniferenöl *n* conifer oil
König *m (met)* regulus
Königsspeiche *f (ship)* king spoke *(steering wheel)*
Königsstange *f (mine)* king's hook, lifting rod
Königswasser *n* aqua regia
Königswelle *f* vertical (central) shaft
Königszapfen *m* trolley-pivot *(el. traction)*
konisch conical, taper[ed]
Konizität *f* conicity, degree of taper
konjugiert *(math)* conjugate
Konjunktion *f* conjunction *(logical operator)*
Konjunkturforschung *f* business research
konkav concave
konkavkonvex concavo-convex
Konkavkonvexlinse *f* concavo-convex lens
Konkavschleifen *n* grinding of concave surfaces
Konkavspiegel *m* concave mirror
Konkordanz *f* 1. concordance; 2. *(print)* 4-em quad
Konkordanzstück *n* 4-em piece
Konkurrent *m* competitor
Konkurrenz *f* competition
konkurrenzfähig competitive
Konnossement *n (ship)* bill of lading
Konode *f* conode
Konoid *n* conoid
Konoidschale *f* conoidal shell
konphas *s.* gleichphasig
Konserve *f* preserve, tin, *(Am)* can
Konservenfabrik *f* tinning factory, *(Am)* cannery
Konservenglas *n* preserve jar
konservieren to preserve; to tin, *(Am)* to can *(food)*, to bottle; *(leath)* to cure
Konservierung *f* conservation, preservation
Konservierungsmittel *n* preserving agent, preservative; *(leath)* curative agent
Konsistenz *f* consistency
Konsistenzzahl *f* consistency index
Konsolbalken *m* cantilever beam
Konsole *f* 1. *(mach)* knee *(milling)*; 2. *(build)* corbel, bracket, console, truss, cantilever
Konsolfräsmaschine *f* knee-and-column type milling machine, column-and-knee type milling machine
Konsolführung *f* knee slide
Konsolführungsbahn *f* knee-saddle way, *(Am)* knee-saddle bearing
Konsolklemmung *f* knee clamping
Konsollager *n* bearing supported by bracket

Konsolmaschine *f* column-and-knee type machine, knee-and-column type machine
Konsolsenkrechtfräsmaschine *f* knee-type vertical spindle milling machine
Konsolstützung *f* knee support *(milling machine)*
konstant constant, steady, stable, stabilized, fixed
Konstantan *n* constantan
Konstante *f* constant
K. der Bindungskraft bond-force constant
Avogadrosche K. Avogadro's constant
Bauersche K. Bauer's constant *(electric arc)*
beliebige K. arbitrary constant
Boltzmannsche K. Boltzmann constant
elastische K. elastic constant
Plancksche K. Planck['s] constant
Rydbergsche K. Rydberg constant
spannungsoptische K. stress-optical constant
willkürliche K. arbitrary constant
Konstantenmultiplikator *m* constant multiplier
Konstantenspeicher *m* constant memory (store)
Konstantförderpumpe *f* constant-displacement pump
Konstantgleichrichter *m* stabilized rectifier
Konstanthalter *m (el)* stabilizer, regulator
Konstantspannungsgenerator *m* constant-potential generator
Konstantstromtransformator *m* constant-current transformer
Konstanz *f* constancy, stability
Konstitutionsformel *f (ch)* constitutional (molecular, rational) formula
Konstitutionswasser *n* water of constitution
konstruierbar constructible
konstruieren to design *(machine)*; to construct *(geometry)*; to build
ein Dreieck k. to construct a triangle
konstruiert / für spezielle Arbeiten built around the job
für spezielle Teile k. built around the component
nach dem Baukastenprinzip k. unit-constructed
Konstrukteur *m* designer, design (designing, constructing) engineer
Konstruktion *f* design, construction; structure
K. nach dem Baukastensystem building-block construction
geschlossene K. packaged construction
patentierte K. patented design
robuste K. robust (rugged) construction
selbsttragende K. frameless (integral) construction *(car body)*
stabile K. rugged construction
tragende K. *(build)* supporting (load-bearing) structure
Konstruktionsabteilung *f* design department
Konstruktionsaufgabe *f* designing task
Konstruktionsbreite *f* breadth moulded, moulded breadth (beam)
Konstruktionsbüro *n* design office
Konstruktionselement *n* structural element (member, unit, component)
Konstruktionsfehler *m* construction fault
Konstruktionsgeschwindigkeit *f* speed of design, designed speed
Konstruktionsgrundsatz *m* design principle
Konstruktionslänge *f* length between perpendiculars
Konstruktionsleiter *m* chief designer
Konstruktionsmerkmal *n* design (constructional) feature, feature of construction
Konstruktionsmethode *f* design method
Konstruktionsprinzip *n* design principle
Konstruktionsproblem *n* design problem
Konstruktionsriß *m (ship)* lines, body plan
Konstruktionsspant *n (ship)* designed (body) section
Konstruktionsspantenriß *m (ship)* body plan
Konstruktionstiefe *f* moulded draught; designed draught
Konstruktionstiefgang *m* designed draught, draught designed; moulded draught
Konstruktionswasserlinie *f* designed (load) waterline
Konstruktionswasserverdrängung *f* designed displacement
Konstruktionswerkstoff *m* structural material
Konstruktionszeichner *m* design[er] draughtsman
konstruktiv structural
Konsumgüter *npl* consumer goods
Container *m (nucl)* container, coffin, casket
Kontakt *m* contact
K. mit Selbstreinigung self-cleaning contact
K. mit Spitzenüberzug tipped contact
beweglicher K. moving (movable) contact
enger K. intimate contact
federnder K. resilient contact
fester K. fixed contact
feststehender K. fixed contact
guter K. tight contact
inniger K. intimate contact
metallischer K. metal-to-metal contact
ohmscher K. ohmic contact
schlechter K. imperfect (poor) contact
ungewollter K. accidental contact
vergoldeter K. gold-plated contact
verschmorter K. scorched contact
versilberter K. silver-faced contact
Kontaktabgriff *m* contact-making pick-off, *(Am)* contact-making pick-up
Kontaktabstand *m* break distance, contact clearance (gap)
Kontaktabzug *m* contact print
Kontaktarm *m* wiper
Kontaktarmsatz *m* wiper set

Kontaktarmträger *m* wiper shaft
Kontaktbahn *f* contact path (bank)
Kontaktbehandlung *f* contacting
Kontaktbelastung *f* contact load
Kontaktberührung *f* contact touch
Kontaktbetätigung *f* contact action
Kontaktbestrahlung *f* contact irradiation
Kontaktbildung *f* contact formation
Kontaktbimetall *n* contact bimetal
Kontaktbock *m* contact jaw
Kontaktboden *m* contacting plate
Kontaktbügel *m* sliding bow
Kontaktbürste *f* [contact] brush, wiper
Kontaktdauer *f* duration of contact
Kontaktdauerstrom *m* current carrying capacity
Kontaktdraht *m* whisker, contact wire
Kontaktdruck *m* contact pressure
Kontaktdruckverstärkung *f* contact pressure increase
Kontaktelektrizität *f* contact (voltaic) electricity
Kontaktelektrode *f* contact electrode (conductor)
Kontakt-EMK *f* contact electromotive force
Kontaktfällung *f* contact precipitation
Kontaktfeder *f* contact spring (clip); brush spring
K. des Lochabstellers fall-out detector spring
Kontaktfedersätze *pl* contact assemblies
Kontaktfehler *m* contact fault
Kontaktfeld *n* contact panel
Kontaktfiltration *f (ch)* contact filtration
Kontaktfinger *m* contact finger, wiper
Kontaktfläche *f* contact surface
Kontaktgeber *m* contact maker, contactor
elektromechanischer K. contactor
Kontaktgeberzähler *m* meter relay
Kontaktgeschwindigkeit *f* contact velocity
Kontaktgestein *n* contact rock
Kontaktgift *n* contact poison (insecticide); catalyst poison
Kontaktgleichrichter *m* commutator (mechanical-contact) rectifier
Kontakthammer *m* trembler
Kontakthebel *m* contact lever; wiper
Kontaktheizfläche *f* contact heating surface
Kontaktherbizid *n* contact herbicide
Kontakthülse *f* contact bush
Kontaktinsektizid *n* contact insecticide
Kontaktinstrument *n* contact-making panel instrument
Kontaktkleber *m* close contact glue
Kontaktklotz *m* contact block
Kontaktknopf *m* contact stud
Kontaktkolben *m* contact piston
Kontaktkopiergerät *n* contact printing-down frame
Kontaktkrater *m* / **ausgefressener** pitted contact
Kontaktlagerstätte *f (geol)* contact deposit
Kontaktlamelle *f* contact bar (lamination)
Kontaktleiste *f* contact strip
kontaktlos non-contact[ing], contactless
Kontaktmesser *n* contact blade, knife-type contact
Kontaktmetallisieren *n* contact plating
kontaktmetamorph in contact metamorphosed
Kontaktmitnahme *f* contact follow
Kontaktniet *m* contact rivet
Kontaktofen *m* catalyst chamber (furnace), contact reactor
Kontaktöl *n* contact lubricant
Kontaktpinsel *m* cat's whisker
Kontaktplättchen *n* contact plate *(of lamp)*
Kontaktplatten *fpl* collector plates
Kontaktplattierung *f* contact plating
Kontaktpotential *n* contact potential
K. zwischen Flüssigkeiten liquid-junction potential
Kontaktpotentialwall *m* contact potential barrier *(semiconductors)*
Kontaktprellen *n* contact bounce *(of relays)*
Kontaktraster *m (print)* contact screen
Kontaktröntgentherapie *f* contact roentgen therapy, short focal-distance therapy
Kontaktsammelschiene *f* contact collecting bar
Kontaktsatz *m* bank of contacts; set of contacts
Kontaktschieber *m* adjusting slider
Kontaktschiene *f* contact bar
Kontaktschlitten *m* contact carriage
Kontaktschwefelsäure *f* contact acid
Kontaktspannung *f* contact potential
Kontaktspannungsabfall *m* contact drop
Kontaktspitze *f* contact point
Kontaktstecker *m* push-in plug
Kontaktstift *m* contact pin (stud), prong, tag
freier K. spare contact
Kontaktstoff *m* catalyst, catalyzer
Kontaktstück *n* contact maker (member)
Kontaktthermografie *f* contact thermography
Kontaktthermometer *n* electrical thermometer
Kontakttulpe *f* tulip-type contact
Kontaktverfahren *n* contact process *(H_2SO_4)*
Kontaktverschweißen *n* fusing of contacts
Kontaktweg *m* contact travel (path)
Kontaktweite *f* contact clearance
Kontaktwerkstoff *m* contact material
Kontaktwiderstand *m* contact resistance
Kontaktzünder *m (mil)* impact fuse
Kontaktzwillinge *mpl* juxtaposition twins *(crystallography)*
Kontamination *f* contamination
kontaminieren to contaminate
Kontenautomat *m* automatic account card feed device
Konteninhaber *m* account holder
Konterabdruck *m (print)* counterproof
Kontermarschstuhl *m (text)* positive lift loom

Kontermutter *f* lock nut, *(Am)* jam nut
Kontingent *n* quota, contingent
Kontinuebreitwaschmaschine *f* continuous open-width washer
Kontinuefärbeapparat *m* continuous dyeing machine
Kontinuekrempel *f* **mit Riemchenflorteiler** condenser card with tape divider
Kontinuemuster *n* continuous design
Kontinuespinnverfahren *n* continuous spinning system
kontinuierlich continuous
Kontinuität *f* continuity
Kontinuitätsgesetz *n* law of continuity
Kontinuitätsgleichung *f* continuity equation, equation (principle) of continuity
Kontinuum *n* continuous spectrum; continuum
Konto *n* account
Kontoauszug *m* statement of account
Kontoblatt *n* accounting form
Kontokarte *f* ledger card (sheet)
Kontonummernsucher *m* number detector
Kontor *n* office, bureau
kontragredient *(math)* contragredient
Kontragredienz *f* *(math)* contragredience
Kontraktion *f* contraction, shrinkage
Kontrapropeller *m* *(ship)* contra-propeller
Kontraruder *n* *(ship)* contra-guide rudder
Kontrast *m* contrast
kontrastarm poor in contrast
Kontrastblende *f* contrast stop
Kontrastdruck *m* *(text)* contrasting print
Kontrastgabe *f* accentuation of contrasts
Kontrastmittel *n* radioopaque medium, opaque matter
Kontrastregler *m* contrast control
kontrastreich high-contrast, rich in contrast, deeply contrasting
Kontraststeigerung *f* increase of contrast
Kontrastumfang *m* contrast range
Kontrastverhältnis *n* *(telev)* contrast ratio
Kontrastverminderung *f* contrast reduction
kontravariant *(math)* contravariant
Kontravarianz *f* *(math)* contravariance
Kontrollabschnitt *m* counterfoil, control coupon
Kontrollabteilung *f* inspection department
Kontrollampe *f* signal (pilot, indicator, warning, supervisory, tell-tale) light, signal (pilot, indicator, warning, supervisory, tell-tale) lamp
Kontrollanalyse *f* check analysis
Kontrollautsprecher *m* monitoring loudspeaker, monitor horn
Kontrollautstärke *f* side-tone level
Kontrollbezirk *m* *(telec)* control area
Kontrollbildgerät *n* monitor
Kontrollbit *n* *(dat)* check digit
Kontrollbohrung *f* inspection bore
Kontrolle *f* check-up, checking, monitoring; control; survey; verification
 100%ige K. screening, sorting *(statistics)*
Kontrolleinrichtung *f* **für den Arbeitsablauf** sequence-control equipment
Kontrollempfänger *m* monitor, monitoring radio receiver
Kontroller *m* controller; controlling magnet; camshaft controller
Kontrolleuchte *f* control lamp
Kontrollfolie *f* *(nucl)* monitoring foil
Kontrollfrequenz *f* check frequency
Kontrollgerät *n* monitor; control unit
Kontrollgrenze *f* control limit *(statistics)*
Kontrollicht *n* warning lamp
kontrollierbar controllable; checkable; verificable
kontrollieren to control; to check; to monitor
Kontrollingenieur *m* quality control engineer
Kontrollinie *f* control line
Kontrollinstrument *n s.* Kontrollgerät
Kontrollkarte *f* control chart
 K. für Standardabweichungen control chart for standard deviations
Kontrollkartentechnik *f* control-chart work
Kontrollkasse *f* cash register
Kontrollkassenrolle *f* cash register roll
Kontrollkassenrollenleser *m* cash register roll reader
Kontrollkommandoempfänger *m* *(rock)* cut-off test receiver
Kontrollkörner *m* *(mach)* centre punch; punch mark
Kontrollmarke *f* tally
Kontrollösung *f* check solution
Kontrollpult *n* monitoring console; mixer
Kontrollpunkt *m* check (testing) point
Kontrollpyrometer *n* control pyrometer
Kontrollraum *m* control centre (room)
Kontrollrechnung *f* tally
Kontrollrelais *n* control (supervisory) relay
Kontrollsender *m* tell-tale transmitter
Kontrollstand *m* control level
Kontrollstation *f* *(mach)* control station *(transfer line)*
Kontrollstreifen *m* control chart *(statistics)*, audit strip *(cash register)*
Kontrollstromkreis *m* pilot circuit
Kontrollsystem *n* control system; checking system
Kontrollübergabe *f* transfer of control
Kontrollverstärker *m* monitoring amplifier
Kontrollwecker *m* *(el)* pilot alarm, *(Am)* pilot bell
Kontrollzählrohr *n* *(nucl)* monitor counter
Kontrollziffer *f* check digit
Konturschrift *f* outline type
Konus *m* cone
Konusantenne *f* cone aerial

Konuseinsatz *m* taper adapter
Konuskupplung *f* cone (conical) clutch
Konuslautsprecher *m* cone loudspeaker
Konusschärmaschine *f* cone warping machine
Konusschatten *m* cone shadow
Konussinkscheider *m* conical dense-medium bath
Konustrichter *m* conical horn
Konvektion *f* convection
erzwungene K. forced convection
freie K. free convection
Konvektionsentladung *f* convective discharge
Konvektionsheizkörper *m* convection heater
Konvektionsofen *m* / **elektrischer** electric convector
Konvektionspotential *n* convection potential
Konvektionsschalter *m* convection circuit-breaker
Konvektionsstrom *m* convection current
konvektiv convective
Konvektor *m* convector
Konventionalstrafe *f* fine for breaking a contract
konvergent *(math)* convergent
absolut k. absolutely convergent
bedingt k. conditionally convergent
Konvergenz *f* convergence
unbedingte K. permanent convergence
Konvergenzbestrahlung *f* convergent-beam irradiation *(therapy)*
Konvergenzelektrode *f* convergence electrode *(colour picture tube)*
Konvergenzfläche *f* convergence plane *(colour picture tube)*
Konvergenzkreis *m* convergence circuit
Konvergenzkriterium *n* criterion (test) of convergence
Konvergenzradius *m* radius of convergence
Konvergenzsatz *m* convergence theorem
Konvergenzverhältnis *n* convergence ratio
Konvergenzwinkel *m* convergence angle, angle of convergence
Konvergenzzüchtung *f* convergence breeding
konvergieren to converge [toward]
Konversion *f* conversion *(e.g. of electrons, breeding material)*
innere K. internal conversion
Konversionselektron *n* [internal] conversion electron
Konversionskoeffizient *m* conversion ratio, [internal] conversion coefficient, conversion fraction *(fuel)*
Konversionslinie *f* internal conversion peak, conversion line
Konversionsreaktor *m* converter [plant], converter reactor, regenerative converter (reactor)
Konversionsverhältnis *n* conversion ratio
Konverter *m* 1. *(met)* converter, Bessemer apparatus; 2. *(nucl)* converter, converter plant (reactor), regenerative converter (reactor)
basischer K. *(met)* basic converter
liegender K. *(met)* barrel converter
saurer K. *(met)* acid converter
seitlich blasender K. *(met)* side-blower converter
stehender K. *(met)* upright converter
Konverterbauch *m (met)* converter belly
Konvertereinsatz *m (met)* converter charge
Konverterfutter *n (met)* converter lining
Konverterkupfer *n* converter copper
Konverterstahl *m* converter steel
Konverterwindkasten *m (met)* converter blast box
konvertieren to convert
Konvertierung *f* conversion
konvex convex
Konvexfräsen *n* **in Gesenken** cherrying
Konvexität *f* convexity
konvexkonkav convexo-concave
Konvexkonkavlinse *f* convexo-concave lens; negative meniscus
Konvexschleifen *n* grinding of convex surfaces
Konvexspiegel *m* convex mirror
Konvex- und Konkavfräsen *n* cherrying
Konzentrat *n* concentrate
Konzentration *f* 1. concentration *(solution)*; strength *(acid)*; 2. enrichment
maximal zulässige K. maximum permissible concentration, MPC
zulässige K. permissible concentration
konzentrationsbedingt *(ch)* colligative
Konzentrationselement *n (ch)* concentration cell
Konzentrationsgefälle *n* concentration gradient
Konzentrationspolarisation *f* concentration overvoltage
Konzentrationsverhältnis *n* concentration ratio
konzentrieren to concentrate
Konzentrierspule *f (el)* concentrating coil
konzentriert *(ch)* concentrated
konzentrisch concentric
k. bohren to drill round
Konzession *f* licence; *(mine)* grant
Konzessionsgebühr *f* licence fee *(telephony)*
Konzessionsinhaber *m* licensee
Koordinate *f* coordinate
Koordinaten *fpl* / **dimensionslose** dimensionless coordinates
ebene K. planar coordinates
kartesische K. Cartesian coordinates
rechtwinklige K. rectangular coordinates
schiefwinklige K. skew coordinates
Koordinatenachse *f* coordinate [axis], axis of coordinates
Koordinatenbewegung *f* coordinate movement
Koordinatenbohren *n* coordinate drilling
Koordinatenbohrmaschine *f* coordinate boring machine

Koordinatenbohrtisch *m* coordinate boring (drilling) table
Koordinatendrehung *f* coordinate rotation
Koordinatendreibein *n* coordinate trihedral
Koordinatenebene *f* coordinate plane
Koordinateneinstellung *f* *(mach)* coordinate positioning (setting)
Koordinatenmanipulator *m* *(nucl)* rectilinear manipulator
Koordinatenmeßgerät *n* coordinate measuring instrument
Koordinatenmeßtisch *m* coordinate [measuring] table, coordinate [measuring] stage
Koordinatenpapier *n* graph paper
Koordinatenraum *m* coordinate space
Koordinatenschalter *m* cross-bar selector
Koordinatenschreiber *m* coordinate plotter
Koordinatensystem *n* coordinate system
  krummliniges K. curvilinear coordinate system
  linkshändiges K. left-handed coordinate system
  rotierendes K. rotating coordinate system
Koordinatentisch *m* coordinate table
Koordinatenverschiebung *f* coordinate displacement
Koordinatenwandler *m* resolver
Koordinatenzahl *f* coordination number *(lattice)*
Koordination *f* coordination
Koordinationsbindung *f* semipolar (coordinate) bond
Koordinationsverbindung *f* coordination compound
Koordinationszahl *f* coordination number
Kopaivabalsam *m* copaiba balsam
Kopal *m* copal
Kopalharz *n* copal resin
Köper *m* *(text)* twill
  dreibindiger K. three-end twill
  gebrochener K. broken twill
  gleichseitiger K. balanced (even-sided) twill
  krummer K. curved twill
  steiler K. upright twill
  versetzter K. transposed twill
  vierbindiger K. four-end twill, four-leaf twill
Köperatlas *m* satin twill
Köperbindung *f* serge weave, twill [fabric] weave
Köpergrat *m* twill line
kopernikanisch Copernican
Köperschatten *m* shaded twill
Kopf *m* 1. head *(machine; rail)*; 2. top; upper end; 3. cheek *(in finished leather)*
  K. / mit pilzförmigem mushroom-headed
  pilzförmiger K. mushroom head
  verlorener K. dead (shrinkage, feeding, lost) head *(casting)*
Kopfanker *m* beam tie
Kopfansicht *f* head-on view
Kopfbahnhof *m* reversing (dead-end, terminus) station, *(Am)* sub-end station
Kopfbalken *m* capping beam
Kopfband *n* brace, knee (angle) brace
Kopfdrehmaschine *f* s. Plandrehmaschine
Kopfdüngung *f* top dressing
Köpfeinrichtung *f* topping unit *(beet harvester)*
Kopfende *n* *(build)* butt end; *(text)* head end *(fabric)*; top
Kopfflanke *f* face *(toothed gear)*
Kopffreiheit *f* clearance height, headroom
Kopfführung *f* *(mine)* front guiding
kopfgesteuert valve-in-head, overhead-valve
Kopfgoldschnitt *m* *(print)* gilt top edge
Kopfguß *m* top casting
Kopfhöhe *f* 1. addendum *(gear)*; 2. *(build)* clearance height, headroom *(furnace)*
Kopfholz *n* *(mine)* crusher block, lid
Kopfhörer *m* headphone, telephone head-gear receiver, earphone, headset
Kopfhöreranschluß *m* headphone connection
Kopfhörerbügel *m* head band, headphone bow
Kopfhörerlautstärke *f* headset level
Kopfkontakt *m* vertical off-normal contact *(telephony)*
Kopfkreis *m* outside (addendum, top) circle *(gear)*
Kopfkreishalbmesser *m* radius of [the] outside circle
Kopfkreuz *n* *(text)* head (end-and-end) lease
Kopfkürzung *f* tip relief *(gear)*
Kopflampe *f* *(mine)* cap lamp
Kopflast *f* / **kapazitive** *(el)* capacity top loading
kopflastig trimmed (down) by the head, bow-heavy; nosing down *(aeroplane)*
  stark k. nose dive
Kopflastigkeit *f* nose heaviness; trim by the head (bow)
Kopfleiste *f* *(print)* head-piece
Kopfleitwerk *n* forward tail group
Kopfmacher *m* snap, rivet set, punch-type dolly
Kopfmeißel *m* *(mach)* end-cutting tool; square-nosed tool
Köpfmesser *n* topping knife
Kopfplatte *f* top of blade *(turbine)*
  K. des Stativs tripod head plate
Kopfprodukte *npl* *(ch)* head products
Kopfrechnen *n* mental arithmetic
Köpfschlitten *m* sugar beet topping machine
Kopfschraube *f* cap-screw
Kopfschußbohrloch *n* *(mine)* back hole, header
Kopfschutz *m* head shield
Kopfseite *f* face
Kopfspiegel *m* head mirror *(magnetic tape mirror)*
Kopfspiel *n* clearance *(amount by which dedendum in a given gear exceeds the addendum in its mating gear)*

Kopfsteg *m (print)* head stick

Kopfstein *m* header, half bat *(brickwork)*; cobble [stone]

Kopfsteinpflaster *n* cobble stone paving

Kopfstrecke *f (mine)* head (tail, top) gate, top road

Kopfstück *n* 1. *(build)* headstone, header; 2. boring head; 3. centre bed *(knitting machine)*

Kopfstütze *f (aut)* head rest

Kopfteil *m* head

Kopfverschluß *m* pot head *(of cables)*

Kopfverstärker *m* head amplifier

Kopfwand *f* end wall

Kopfwelle *f* head wave

Kopfwipper *m* end tippler, kick-back dump

Kopfzünder *m* nose fuse

Kopie *f* copy; duplicate; print

negative K. negative print

positive K. positive print

schwer lesbare K. hard copy *(e.g. punched tape)*

Kopieband *n* copy tape

Kopierbewegung *f (mach)* copying motion

Kopierbezugsstück *n (mach)* copying master

Kopierdreheinrichtung *f* copy-turning attachment

kopierdrehen to copy-turn

Kopierdrehen *n* copy-turning

Kopierdrehmaschine *f* copying lathe

Kopiereffekt *m* print-through effect *(magnetic tape)*

Kopiereinrichtung *f* copying (tracing) attachment, copying (tracing, duplicating) device, duplicator

kopieren 1. to copy *(photography)*; 2. *(mach)* to duplicate, to copy, to contour, to trace

Kopieren *n* / **spiegelbildliches** *(mach)* reverse mirror copying

kopierfähig capable of being copied

Kopierfräsautomat *m* automatic tool room copy milling machine, automatic die sinking machine

hydraulischer K. automatic hydraulic copying miller

Kopierfräsen *n* copy milling

dreidimensionales K. die-sinking

Kopierfräsmaschine *f* copy milling machine

Kopiergerät *n* printing apparatus

Kopierhub *m* copy stroke

Kopierlampe *f* printing lamp

Kopierlineal *n (mach)* guide (master) plate

Kopiermaschine *f* 1. step-and-repeat machine; 2. *(mach)* copying (contouring) machine

Kopiernocken *m* master cam

Kopierpapier *n* duplicating (copying) paper

Kopierrahmen *m* / **pneumatischer** vacuum frame

Kopierrolle *f (mach)* roller [follower]

Kopierschablone *f (mach)* copying (master) template

Kopierschlitten *m (mach)* copying saddle (slide)

Kopierstift *m (mach)* guide pin

Kopier- und Gesenkfräser *m* copy and die milling cutter

Kopierverfahren *n* 1. printing process; 2. *(mach)* copying (duplicating) process

Kopiervorlage *f* original tracing *(technical drawing)*; *(mach)* master pattern

Kopierwaagerechtstoßmaschine *f* copying shaper

Kopierwerkzeugmaschine *f* copying (contouring) machine

koplanar coplanar

Kopolymer *n (ch)* copolymer

Koppel *f* 1. connecting rod *(kinematics)*; 2. *(ag)* enclosure

Koppelbesteck *n (ship)* [dead] reckoning

Koppelblende *f* coupling iris

Koppelbogen *m* bend coupling *(waveguides)*

Koppelfaktor *m (el)* coupling coefficient

Koppelkondensator *m* coupling capacitor

Koppelkurs *m (ship)* compound (combined) course

Koppelleitung *f* strapping *(of a multiple-cavity magnetron)*

koppeln 1. to couple; 2. *(ship)* to keep the dead-reckoning

Koppelnavigation *f* dead-reckoning navigation, navigation by dead reckoning, traverse sailing

Koppelnetzwerk *n* interstage network

Koppelort *m* dead-reckoning position *(navigation)*

Koppelquarz *m* piezoelectric coupler

Koppelraum *m* interaction space

Koppelschleife *f* coupling loop

Koppelschlitz *m* window *(waveguides)*

Koppelstange *f (mach)* connecting rod

Koppelstufe *f (el)* buffer stage

Koppelträger *m* cantilevered and suspended beam

Koppelweide *f* rotation pasture

Koppler *m* coupling capacitor

Kopplung *f (el)* coupling

K. zweier Hohlleiter durch Schlitze slit coupling of waveguides

K. zweier Schwingkreise durch eine Serienkapazität top capacity coupling

direkte K. direct coupling

elektrische K. electrostatic coupling

enge K. *s.* feste K.

feste K. tight (close) coupling

galvanische K. conductive (direct) coupling

gemischte K. combined coupling

induktive K. inductance (inductive) coupling

kapazitive K. capacitive coupling

kritische K. critical coupling

lose K. loose (weak) coupling

magnetische K. magnetic coupling
stabile K. stable coupling
starke K. strong coupling
überkritische K. *s.* feste K.
ungewollte K. stray coupling
unterkritische K. *s.* lose K.
wilde K. stray coupling
Kopplungsbeiwert *m* coupling coefficient
Kopplungselement *n* coupling element
Kopplungsenergie *f* coupling energy
Kopplungsfähigkeit *f* coupling ability
Kopplungsfaktor *m* coefficient of coupling, coupling factor (coefficient)
Kopplungsglied *n* coupling link
Kopplungsgrad *m* degree (coefficient) of coupling
Kopplungskapazität *f* coupling capacitance
Kopplungskoeffizient *m s.* Kopplungsfaktor
Kopplungskondensator *m* coupler capacitor
Kopplungskreis *m* coupling circuit
Kopplungsleitwert *m* transfer admittance
Kopplungsölkondensator *m* coupling oil capacitor
Kopplungsschaltung *f* interstage network
Kopplungsschleife *f* coupling loop
Kopplungsschwankung *f* coupling variation
Kopplungsschwingungen *fpl* oscillations in coupled circuits
Kopplungsspule *f* coupling coil; coupler
Kopplungsstück *n* connector *(waveguides)*
Kopplungstransformator *m* coupling (repeating) transformer, jigger
Kopplungsübertrager *m* coupling transformer
Kopplungswiderstand *m* coupling (interaction, transfer) impedance
Kops *m (text)* cop
Kopsspulmaschine *f* bobbin winder for cops
Kopulation *f* copulation, splice-grafting, whip graft
Kopuliermesser *n* copulating knife
Korallenriff *n* coral reef
Korb *m* basket; cage *(harvester)*; *(build)* bucket
umlaufender K. *(text)* revolving cage
Korbball *m* basket ball
Korbbodenspule *f (el)* basket coil
Korbbogen *m* elliptical arch
K. aus drei Leierpunkten three-centred arch
K. aus fünf Leierpunkten five-centred arch
Korbflasche *f* demijohn, carboy
Korbleiter *f (mine)* safety cage ladder
Korbrostfeuerung *f* furnace with basket grate
Korbscheider *m* basket-jig
Korbspule *f (el)* basket coil
Korbstich *m* basket stitch *(knitting)*
Korbwicklung *f (el)* lattice (imbricated, barrel, basket) winding
Kord *m* cord
Kordel *n* diamond-shaped knurling *(metal)*
Kordel *f* twine; cord, string
kordeln to knurl *(with double-wheel knurling tool)*
Kordeln *n* diamond (double-wheel) knurling
Kordelrädchen *n* knurl, knurling wheel *(having teeth which incline to the right on one knurl and to the left on the opposite knurl)*
Kordelteilung *f* pitch of the teeth of a knurl
Kordelwerkzeug *n* double-wheel knurling tool, knurling tool with two wheels
Kordgewebelage *f* cord ply
kordieren to knurl *(with double-wheel knurling tool)*
Kordierhalter *m* double-wheel knurl holder
Kordierspindel *f (text)* cabling spindle
kordiert diamond-knurled
Kordierwerkzeug *n* double-wheel knurling tool
Kordreifen *m* cord tyre
Kordsamt *m* corduroy
Kordzellstoff *m* cord pulp
Kork *m* cork
Korkeiche *f* cork oak
Korkformstück *n* cork tile
Korkmehl *n* cork powder
Korkplatte *f* cork slab (sheet), *(build)* cork-board
Korksäure *f* suberic acid
Korkschrot *n* cork dust
Korksohle *f* cork sole
Korkzieherregel *f* corkscrew rule
Korn *n* 1. grain *(material, abrasive, photographic layer)*; particle; 2. fore-sight *(rifle)*; 3. *(ag)* rye; corn
gebundenes K. bonded abrasive [grain]
geklemmtes K. incorrectly centred front sight *(rifle)*
gestrichenes K. medium sight *(rifle)*
links geklemmtes K. taking the tip of the foresight to the left *(of the centre of the backsight notch) (rifle)*
rechts geklemmtes K. taking the tip of the foresight to the right *(of the centre of the backsight notch) (rifle)*
Kornabstufung *f* particle (aggregate) gradation, particle (aggregate) grading
Kornbildung *f* crystallization, granulation *(crystallography)*
Kornboden *m (ag)* corn-loft
Korndichte *f* grain density *(crystallography)*
körnen to grain
Körner *m (mach)* punch mark, centre punch
Körneranschlag *m* centre-punch mark
Körnergebläse *n (ag)* grain blower
Körnermarke *f (mach)* [centre-]punch mark
Körnermikrofon *n* granular microphon
Körnersammler *m (ag)* grain saver
Körnersieb *n (ag)* second dresser riddle
Körnersilo *m* partitioned bulk ventilation grain drier

Körnerspitze *f (mach)* centre
 mitlaufende K. live centre
 nicht mitlaufende K. dead centre
 umlaufende K. live centre
Körnertrockner *m* hot-air drier
Körnertrocknungsanlage *f* grain drier
Kornflächenätzung *f* contrast etching
Kornform *f* grain (particle) shape
Korngefüge *n* grain structure
Korngrenze *f* grain boundary *(of crystals)*
Korngrenzenangriff *m* grain boundary attack
Korngrenzenätzung *f* etching bringing out grain boundaries in relief
Korngrenzenbruch *m* intercrystalline failure
Korngrenzenebene *f* grain boundary plane
Korngrenzenmaximum *n* grain boundary peak
Korngrenzenriß *m* grain boundary crack
Korngrenzensperrschichten *fpl* grain boundary barriers *(of semiconductors)*
Korngrenzentransistor *m* grain boundary transistor
Korngröße *f* grain (particle) size *(crystallography; emulsion)*
 mittlere K. average grain (particle) size
Korngrößenbereich *m* [grain] size range
Korngrößenverteilung *f* grain size distribution (classification)
körnig grained; granular, gritty
Körnigkeit *f* granularity, granulation
Kornklapper *f* combined winnowing and sieving machine
Kornklasse *f* size range; grade
Körnmaschine *f* graining machine, grainer
Kornprüfer *m (ag)* grain tester
Kornraster *m* granulated screen
Kornscheide *f* effective screen cut-point
Kornschott *n* grain bulkhead
Kornsilo *m* grain silo
Kornspeicher *m* granary
Kornstruktur *f* grain structure
Körnung *f* 1. grain, grain (grit) size; granulation size; *(opt)* graining *(print)*; *(build)* fraction; 2. punch mark
 K. des Mittelpunktes *(mach)* centre-punch mark
Körnungsband *n* size range
Körnungsbereich *m* size range
Körnungskennlinie *f* size-distribution curve
Körnungsnummer *f* grain size [number] *(grinding wheel)*
Körnungspunkt *m* [centre-]punch mark
Kornverfeinerung *f (met)* grain refinement
Kornvergröberung *f (met)* grain coarsening
Kornverteilungskurve *f* size-distribution curve
Kornwachstum *n* grain growth
Korona *f* corona *(s.a.* Koronaentladung)
Koronaanfangsspannung *f s.* Koronaeinsatzspannung
Koronaeinsatzspannung *f* corona threshold
Koronaentladung *f* corona discharge, corona [effect]
Koronaentladungszähler *m* corona counter
Koronaerscheinung *f* corona phenomenon (effect)
Koronastrahl *m* coronal ray (streamer)
Koronastrom *m* corona current
Koronaverluste *mpl* corona loss[es]
Koronograf *m (opt)* coronograph
Körper *m* 1. *(ph)* body; solid *(solid body)*; substance; 2. *(text)* body *(needle cylinder; hide)*; 3. *(ship)* hull; 4. *(math)* body *(geometry)*; field *(range)*
 K. des Satelliten body of the satellite, satellite body
 K. mit einer Spitze pointed body
 durchscheinender K. translucent body
 durchsichtiger K. transparent body
 fester K. *(ph)* solid
 flüssiger K. liquid body
 gasförmiger K. gaseous body
 grau streuender K. grey diffuser
 grauer K. *(ph)* grey body
 lichtstreuender K. diffuser
 lichtundurchlässiger K. opaque body
 neutral streuender K. grey diffuser
 nichtselektiv streuender K. non-selective diffuser
 rotationssymmetrischer K. body of revolution
 schwarzer K. *(ph)* black body, full radiator
 spitznasiger K. pointed body
 störender K. perturbing body
 streuender K. *(ph)* diffuser
 vollkommen streuender K. perfect diffuser
 zugemischter K. admixed substance *(turbine)*
Körperbelastung *f* body burden
Körperberechnung *f* mensuration of solids, cubature
Körperchen *n* corpuscle
Körperdiagonale *f* body diagonal
Körperdosis *f (nucl)* whole body dose
Körpererziehung *f* physical education
Körperhöhlenbestrahlung *f (nucl)* intracavitary irradiation
Körperinhalt *m* cubic content *(of a solid)*
Körperkante *f* edge, outline of part
Körperschall *m* impact noise (sound); sound in solids; solid-borne vibration
Körperschallmikrofon *n* surface (contact) microphone
Körperschichtaufnahme *f* body section roentgonography, stratigraphy, tomography
 K. mittels Planigraf planigraphy
Körperschluß *m (el)* body contact
Korpuskel *f* corpuscle
korpuskular corpuscular

Korpuskularstrahl *m* corpuscular ray
Korpuskularstrahlung *f* corpuscular (particle) radiation
 K. der Sonne corpuscular radiation of the sun, sun's corpuscular radiation
 sekundäre K. associated corpuscular emission
Korpuskulartheorie *f* corpuscular theory
Korrektion *f* correction
 optische K. optical correction
 relativistische K. relativistic correction
Korrektionsfaktor *m* / **quantentheoretischer** quantum mechanical correction factor
Korrektionsfilter *n* corrective filter
Korrektionsknopf *m* correcting knob
Korrektionskreis *m* corrector circuit
Korrektionsschaltung *f* shaping network
Korrektionssignal *n (el)* correcting signal, *(Am)* error signal
Korrektor *m* proof-reader
Korrektur *f* 1. *(mach)* readjustment, compensation; 2. *(opt)* correction; 3. *(print)* proof correction; proof sheet
 K. des Zahnprofils tooth-shape correction
 optische K. optical correction
Korrekturabzug *m* proof [of type], copy-proof, proof-sheet
Korrektureinrichtung *f (mach)* compensating device *(jig borer)*
Korrekturfahne *f* slip (galley) proof
Korrekturfaktor *m* 1. *(el)* correcting (correction) factor; 2. *(opt)* corrective factor
 K. für das Übersetzungsverhältnis *(el)* ratio correction factor
Korrekturfilter *n* equalizer *(frequency)*
Korrekturglied *n (dat)* compensating network; equalizing network
Korrekturkommando *n (rock)* correction signal
Korrekturkurve *f* correction curve
Korrekturlesen *n* proof-reading
Korrekturnetzwerk *n (dat)* compensating network; equalizing network
Korrekturschaltung *f (dat)* compensating network; equalizing network
Korrekturspule *f* **für hohe Frequenzen** peaking coil
Korrekturtaste *f* cancelling (reset) key
Korrekturzeichen *n* reader's mark, proof sign (mark)
Korrekturzeile *f* corrected line
Korrelation *f* correlation
Korrelationsfunktion *f* correlation function
Korrelationskoeffizient *m* correlation coefficient
Korrelationsrechner *m* correlation computer
Korrelationsrechnung *f* correlation calculation
korrelieren to correlate
Korrespondentreeder *m* ship's husband, managing owner
Korrespondenzprinzip *n (ph)* correspondence principle
Korrespondierende *pl (el)* corresponding wires
korrodierbar corrodible; corrosive; corroding
Korrodierbarkeit *f* corrodibility
korrodieren to corrode
korrodierend corrosive
Korrosion *f* corrosion
 anodische K. electro-dissolution
 elektrolytische K. local action
 interkristalline K. intercrystalline (intergranular) corrosion
 lokale K. localized corrosion
 örtliche K. localized corrosion
 selektive K. selective corrosion
Korrosionsanfälligkeit *f* susceptibility to corrosion
Korrosionsanfressung *f* pin holes
Korrosionsangriff *m* corrosion attack
korrosionsbeständig corrosion proof (resistant, resisting), non-corrosive, resistant to corrosion
Korrosionsbeständigkeit *f* corrosion resistance, resistance to corrosion
Korrosionselement *n* corrosion cell
korrosionsempfindlich susceptible to corrosion
Korrosionsempfindlichkeit *f* susceptibility to corrosion
Korrosionsermüdung *f* corrosion fatigue
Korrosionsfestigkeit *f* corrosion resistance
korrosionsfördernd stimulating (developing) corrosion
korrosionsfrei non-corrosive
Korrosionsgeschwindigkeit *f* corrosion rate
korrosionshemmend inhibiting corrosion
Korrosionshemmittel *n* corrosion inhibitor
korrosionshindernd corrosion-preventive, inhibiting corrosion, anticorrosive
Korrosionsinhibitor *m* corrosion inhibitor
Korrosionsmittel *n* corrosive
Korrosionsneigung *f* susceptibility to corrosion
Korrosionsprüfung *f* corrosion test
Korrosionsschutz *m* protection against corrosion
 galvanischer K. cathodic protection
Korrosionsschutzfarbe *f* anti-corrosive paint
Korrosionsschutzmittel *n* preventive against corrosion, anti-corrosive agent
Korrosionsverhalten *n* corrosion behaviour
korrosionsverhindernd anti-corrosive, inhibiting corrosion
Korrosionsvorgang *m* corrosive action, corrosion mechanism
Korrosionswiderstand *m* corrosion resistance
Kortdüse *f (ship)* Kort nozzle
Korund *m* corundum
Körung *f (ag)* inspection, selection and certification of sires
Koschenille *f* cochineal

Kosekans *m* cosecant

Kosinus *m* cosine

hyperbolischer K. hyperbolic cosine

kosinusförmig cosinusoidal

Kosinusgesetz *n* / **Lambertsches** Lambert's law

Kosinuslinie *f* cosinusoid

Kosinusregler *m* cosine governor

Kosinussatz *m* cosine formula (law), law of cosines

Kosmetika *npl* cosmetics

Kosmetikstift *m* stick

kosmetisch cosmetic

kosmisch cosmic

Kosmochemie *f* cosmochemistry

Kosmogonie *f* cosmogony

Kosmografie *f* cosmography

kosmografisch cosmographical

Kosmologie *f* cosmology

Kosmonaut *m* cosmonaut, spaceman

Kosmonautik *f* astronautics; cosmonautics; interplanetary (interstellar, space) aviation

Kosmonautin *f* woman (female) cosmonaut

Kosmotron *n* cosmotron

kosten 1. to be, to cost; 2. to taste

Kosten *pl* costs, expense[s], charge[s]

K. per Einheit unit costs

Kostenanschlag *m* estimate, cost esti ation, tender

Kostenanteil *m* / **abgeschriebener** expir costs

Kostenaufstellung *f* cost sheet (record)

Kostenaufwand *m* expenditure [of finan

K. für ungenützte Kapazität idle apacity cost (expenditure)

Kostenbeleg *m* cost records

Kostenberechnung *f* / **gesonderte** extr harge

kostengünstig low-cost

kostenlos free of charge

Kostenplaner *m* quantity surveyor, es ator

Kostenpreis *m* cost price

Kostenrechner *m* cost accountant

Kostenrechnung *f* cost accounting

Kostenrechnungsstelle *f* accounting department

Kostenrechnungssystem *n* **für auftragsweise Fertigung** job order costing

Kostensenkung *f* cost reduction (cutting)

Kostensenkungsprogramm *n* cost cutting programme

kostensparend cost-saving

Kostenüberschlag *m* estimate

Kostenvergleich *m* cost comparison, comparison of costs

Kostenvoranschlag *m* tender, quotation, estimate

Kostenvorschau *f* cost prediction

Kotangens *m* cotangent

hyperbolischer K. hyperbolical cotangent

Kotangenssatz *m* cotangent law

Kotblech *n (aut)* fender

Kotbrett *n (ag)* dropping board

Kote *f* level *(geometry, surveying)*

Kotflügel *m (aut)* mud guard, wing

Kotflügelleuchte *f* wing lamp

Kotflügelstütze *f* wing stay

Kotrinne *f (ag)* manure channel

Kotstoffe *mpl* faecal (fecal) matter, faeces, feces

Kötzer *m* cop

kovalent covalent

Kovalenz *f* covalence

Kovalenzbindung *f* covalent linkage (bond)

kovariant covariant

Kovariante *f* covariant

Kovarianz *f* covariance

Kovolumen *n* co-volume

Kozymase *f* cozymase

Krabben *n (text)* crabbing

Krabbmaschine *f (text)* crabbing machine

Krachen *n (text)* scroop

Krachfilter *n* scratch filter

Krachgriff *m (text)* scroopy feel

Krachtöter *m (el)* noise killer (limiter, silencer, suppressor), silencing device

Krachtöterschaltung *f* squelch circuit *(radio)*

Krachtötung *f (el)* quieting

Krackanlage *f (ch)* cracking plant

Krackbenzin *n* cracked petrol, *(Am)* cracked gasoline

Krackdestillation *f* cracking distillation

kracken *(ch)* to crack

Krackgas *n* cracked gas

Krackung *f (ch)* cracking

Krad *n* motorcycle

Kradmelder *m (mil)* motorcycle messenger

Kradschütze *m (mil)* motorcycle rifleman

Kraft *f* 1. *(ph)* force; power *(water, etc.)*; 2. strength *(physical)*; 3. thrust

abstoßende K. repulsive force, force of repulsion

aufgeprägte elektromotorische K. impressed electromotive force

äußere K. external force

bewegende K. moving power

Coriolissche K. Coriolis force

dissipative K. dissipative force

eingetragene K. applied force

elektromagnetische K. electromagnetic force

elektromotorische K. electromotive force, emf

gegenelektromotorische K. counter-electromotive force

generalisierte K. generalized force

innere K. internal force

konstante K. constant force

magnetische K. magnetic force

magnetomotorische K. magnetomotive force

mechanische K. mechanical force

mittige K. axial (concentric) force

nach oben gerichtete K. upthrust

ponderomotorische K. ponderomotive force
resultierende K. resultant force
thermoelektrische K. thermoelectromotive force
Kraftabgabe *f* supply of power, power output
Kraftabgabeseite *f* **des Motors** power end of the engine
Kraftangriff *m* application of [a] force
mittiger K. centric application of [a] force
Kraftanlage *f* power plant
Kraftantrieb *m* power drive
Kraftarm *m* force arm
Kraftbedarf *m* power consumption
kraftbetätigt power-operated
kraftbetrieben power-driven, power[-operated]
Kraftbeziehung *f* force (dynamic) relation
Kraftdreirad *n* motor tricycle
Krafteck *n* polygon of forces, force polygon
Kräftediagramm *n* force diagram
Kräftedreieck *n* triangle of forces
Kräftegleichgewicht *n* equilibrium of forces
Krafteinheit *f* unit of force
Kräftemaßstab *m* scale of power
Kraftentnahme *f* absorption of power
Kräftepaar *n* [force] couple
Kräfteparallelogramm *n* parallelogram of forces
Kräfteplan *m* force diagram
Kräftepolygon *n* polygon of forces, force polygon
Krafterzeugung *f* power generation
Kräftesystem *n* system of forces
ebenes K. system of coplanar forces
Kräftevieleck *n* polygon of forces
Kräftezerlegung *f* resolution of forces
Kräftezusammensetzung *f* composition of forces
Kraftfahrzeug *n* motor car (vehicle)
Kraftfahrzeugbestand *m* motorcar population
Kraftfahrzeuggasturbine *f* automotive gas turbine
Kraftfahrzeughalter *m* car (motor vehicle) owner
Kraftfahrzeughauptbaugruppe *f* major motor-vehicle component group
Kraftfahrzeugingenieur *m* automotive engineer
Kraftfahrzeugmotor *m* automotive (motorcar) engine
Kraftfahrzeugpark *m* motor-vehicle fleet
Kraftfahrzeugtechnik *f* auto mechanics; automotive engineering
Kraftfahrzeugtechniker *m* automotive engineer
Kraftfahrzeugwesen *n* automotive engineering; automobilism
Kraftfahrzeugzubehör *n* automotive supplies
Kraftfahrzeugzulassung *f* motor vehicle (car) licence
Kraftfeld *n* force field, field of force[s]
Kraftfluß *m (ph)* direction of force lines; *(el)* magnetic flux
Kraftfutter *n* concentrates
Kraftgas *n* fuel gas
Kraftheber *m* lift
hydraulischer K. hydraulic lift
kräftig 1. *(ch)* vigorous; 2. *(mach)* heavy *(cut)*; 3. powerful *(engine)*
Kraftkomponente *f* force component, component of a force
Kraftkonstante *f* force constant
Kraftleitung *f* power [transmission] line
Kraftlinie *f* line of force, field line
elektrische K. electric line of force, line of electric force
magnetische K. magnetic line of force, line of magnetic force
Kraftlinienbündel *n* tube of [lines of] force
Kraftliniendichte *f* / **magnetische** flux density
Kraftlinienfeld *n* field of force
Kraftlinienfluß *m* flux of lines of force
Kraftlinienrichtung *f* direction of lines of force
Kraftlinienstreuung *f* / **magnetische** flux leakage
Kraftlinienverlauf *m* path of lines of force
Kraftlinienverteilung *f* distribution of lines of force
Kraftlinienzahl *f* / **remanente** remanent flux density
Kraftmaschine *f* prime mover
Kraftmaschinenregler *m* engine governor; throttle governor
Kraftmeßdose *f* load cell
Kraftmesser *m* dynamometer
Kraftmeßschlüssel *m* torque wrench
Kraftpapier *n* kraft [paper]
gebleichtes K. bleached kraft paper
Kraftpolygon *n* force polygon
Kraftquelle *f* power source
Kraftrad *n* motorcycle
Kraftreserve *f* power reserve
Kraftrichtung *f* direction (sense) of force
Kraftröhre *f* power valve
Kraftschalter *m s.* Verstärker
Kraftschaufel *f* power shovel
kraftschlüssig non-positive; frictional; unclosed *(pair)*
Kraftschnitt *m (mach)* heavy cut
Kraftsinn *m* sense (direction) of force
Kraftspeicherantrieb *m* stored energy (power) drive
Kraftstation *f* power station
Kraftstellwerk *n (ry)* power locking frame
Kraftstoff *m* fuel; propellant
K. ohne Zusätze undoped fuel
flüssiger K. liquid (oil) fuel
gebleiter K. leaded fuel
Kraftstoffabschneider *m* fuel trap

Kraftstoffabsperrschieber *m* fuel cutoff valve
Kraftstoffanlage *f* fuel system
Kraftstoffbehälter *m* fuel reservoir (tank, container)
Kraftstoffdüse *f* fuel-metering nozzle; fuel jet *(petrol engine)*
Kraftstoffeinspritzer *m* fuel injector
Kraftstoffeinspritzpumpe *f* fuel-injection pump
Kraftstoffeinspritzung *f* fuel injection
Kraftstoffempfindlichkeit *f* **eines Motors** fuel sensitivity of an engine
Kraftstoffentlüfter *m* fuel deaerator
Kraftstoffhahn *m* fuel valve (shut-off cock)
Kraftstoffheizung *f* fuel heating
Kraftstofflager *n* fuel dump
Kraftstoffleitung *f* fuel pipe (line)
Kraftstoff-Luft-Gemisch *n* air fuel mixture
fettes K. rich air-fuel mixture
mageres K. lean fuel-air mixture
Kraftstoff-Luft-Verhältnis *n* fuel-air ratio
Kraftstoff-Masse-Verhältnis *n* fuel-weight ratio
Kraftstoffmesser *m* petrol gauge
Kraftstoffnadelventil *n* carburetter needle valve
Kraftstoffnebel *m* fuel mist (vapour)
Kraftstoffnormverbrauch *m* level road fuel consumption
Kraftstoff-Öl-Gemisch *n* lubricated petrol
Kraftstofförderpumpe *f* fuel [feed] pump
Kraftstoffpumpe *f* fuel [feed] pump
Kraftstoffpumpeneinstellung *f* fuel pump timing
Kraftstoffpumpenmembran *f* fuel pump diaphragm
Kraftstoffreiniger *m* fuel strainer
Kraftstoffringbehälter *m* *(rock)* annular fuel tank
Kraftstoffrücklaufbehälter *m* fuel-reclaiming tank
Kraftstoffsaugpumpe *f* fuel lift pump
Kraftstoffschnellablaß *m* fuel jettisoning gear
Kraftstoffstrahl *m* fuel jet
Kraftstoffuhr *f* fuel meter
Kraftstoffüllstutzen *m* fuel filler cap
Kraftstoffventil *n* fuel valve
Kraftstoffverbrauchsmesser *m* fuel-consumption gauge (indicator)
Kraftstoffverdampfer *m* fuel vaporizer
Kraftstoffvorratsanzeiger *m* fuel gauge
Kraftstoffzerstäubersystem *n* fuel-atomizer system
Kraftstoffzetanzahl *f* cetane number of a fuel
Kraftstrom *m* power current; household-tariff current
Kraftstromleitung *f* power circuit
Kraftstromschalttafel *f* power switchboard
Kraftübertragung *f* 1. *(el)* power transmission, transmission of power; 2. *(mech)* power train
elektrische K. transmission of power
Kraftübertragungsmanipulator *m* *(nucl)* force reflecting manipulator
Kraftübertragungsweg *m* *(mech)* power (gear) train
Kraftumlenkung *f* reversal of the force lines
Kraftverbrauch *m* power consumption
Kraftvergleichsregler *m* force-balance controller
Kraftverstärker *m* power amplifier (unit)
Kraftverstärkerröhre *f* power-amplifier valve
Kraftverteilung *f* *(el)* power distribution; force distribution
Kraftwaage *f* force-balance
Kraftwagen *m* motor vehicle (car), automobile
K. mit Kernenergieantrieb atomic automobile (car)
Kraftwagenantenne *f* car (automobile) aerial
Kraftwagenbatterie *f* car (automobile) battery
Kraftwagenempfänger *m* car (automobile) radio
Kraftwagenfähre *f* car ferry
Kraftwagenflak *f* mechanized antiaircraft gun (artillery)
Kraftwagenförderung *f* *(mine)* truck haulage
Kraftwagenindustrie *f* automobile industry
Kraftwagenpark *m* motor-vehicle fleet
Kraftwerk *n* power [generating] station, electric generating (power) station
grubennahes K. mine-mouth power station
vollautomatisches K. unattended power station
Kraftwerksleistung *f* station capacity
installierte K. generating station capacity
Kraftwerksreaktor *m* power reactor
Kraftwerkswirkungsgrad *m* generating station efficiency
Kraftwirkung *f* / **elektrische** electrical-force action
Kraftwirkungsfiguren *fpl* stretcher lines
Kraftwirtschaft *f* power supply
Kraftzellstoff *m* kraft pulp
Kraftzellstoffbleiche *f* kraft bleaching
Kraftzentrale *f* [electric] central station, generator room
Kraftzerlegung *f* resolution (decomposition) of a force
Kraftzurichtung *f* underlay
Kragarm *m* cantilever
Kragbalken *m* semi-beam, corbel (cantilever[ed]) beam
Kragdach *n* canopy
Kragenwendemaschine *f* collar turning machine
Kragspant *n* *(ship)* cantilever frame
Kragträger *m* cantilever[ed] (corbel) beam, semi-beam
Krähennest *n* *(ship)* crow's nest
Krählarbeit *f* rabbling operation
Krählarm *m* rabble arm
Krählhalter *m* rabble arm

Krählofen *m* rabble furnace, reverberatory calciner
Krallenverbinder *m* claw-type fastener *(belt)*
Krämer-Maschine *f (el)* Kraemer generator
Krammeisen *n* rake
Krampe *f* staple, clip, cramp
krampferzeugend spasmodic
Krampstock *m (met)* slag skimmer
Kran *m* crane
  K. mit Drehausleger swing-jib crane
  K. mit festem Ausleger fixed cantilever crane
  K. zum Masteinsetzen masting sheers
Kranarm *m* crane jib
Kranausleger *m* crane jib (boom)
Kranbahn *f* crane track
Kranbahnschiene *f* crane rail
Kranbahnträger *m* crane girder
Kranbalken *m* jib, cathead, cat-head
Kranbereich *m* crane area
Kranfahrerstand *m* [crane] driver's cab
Kranfahrmotor *m* long travel motor
Kranführer *m* crane driver (operator)
Krangebühr *f (ship)* cranage
Krangeld *n (ship)* cranage
krängen *(ship)* to heel [over], to incline
Krängung *f (ship)* heel[ing], list
Krängungsanzeiger *m* bank indicator
Krängungsfehler *m* heeling error *(compass)*
Krängungsmesser *m* inclinometer
Krängungstank *m (ship)* heeling tank
Krängungsversuch *m (ship)* inclining experiment
Krängungswinkel *m (ship)* angle of heel[ing], heeling angle
Kranhaken *m* crane hook
Krankenbahre *f* stretcher
Krankengeld *n* sick[ness] benefit
Krankenhauskleidung *f* hospital clothing
Krankenkasse *f* sick [insurance] fund
Krankenwagen *m* ambulance
Kranleine *f (ship)* boom lift
Kranpfanne *f (met)* crane ladle
Kranportal *n* gantry
Kransäge *f* pit saw
Kransägeangel *f* pit saw tiller
Kransägezahn *m* gullet tooth
Kranwagen *m* derrick wagon
Kranz *m* 1. gear rim *(gear, wheel)*; 2. row of blades *(turbine)*; 3. wreath
Kranzbohrloch *n (mine)* trimming hole
Kranzbrenner *m* annular-type burner
Kranzbruch *m* fracture of the rim *(flywheel)*
Kranzgesims *n* cornice
Kranzmasse *f* weight of the rim *(flywheel)*
Kranzschuß *m (mine)* easer
Kranzverbindung *f* connection of rim segments *(flywheel)*
Krapp *m* madder
Krapplack *m* madder lake
Krappwurzel *f* madder root
krarupisieren *(el)* to krarupize
Krarupisierung *f (el)* continuous loading
Krarup-Kabel *n* Krarup (continuously loaded) cable
Krarup-Wicklung *f* Krarup winding
Kratzbandklassierer *m* scraper-type classifier
Krätzblei *n* lead dross, slag lead
Krätze *f* dross, skimmings
Krätzebehälter *m* cinder pot
Kratzeisen *n* scraper
kratzen 1. to scratch, to mar, to scrape; 2. *(text)* to card
Kratzenbeschlag *m (text)* card clothing
  K. der Abnehmerwalze doffer filleting
Kratzendraht *m (text)* card wire
Kratzenleder *n* card[ing] leather
Kratzensetzmaschine *f (text)* card setting machine
Kratzenzahn *m (text)* card staple
Kratzer *m* scratch, mar
Krätzer *m* scraper *(shotfiring)*
Kratzerförderer *m* cham and flight conveyor
Kratzerkette *f* / **umlaufende** endless chain with slats *(mechanical gutter cleaner)*
Kratzerkettenförderer *m* scraper chain conveyor
Kratzfestigkeit *f* scratch resistance
Krätzfrischen *n (met)* refining of waste
Kratzgeräusch *n (el)* scratching (contact) noise, spluttering, frying, crackling
Kratzkettenförderer *m* scraper-chain conveyor
Krätzlöffel *m* skimming ladle, skimmer
Krätzverlust *m* dross loss
Kräuselgarn *n* crimped (stretch) yarn
Kräusellack *m* wrinkle finish, crinkle varnish
kräuseln to curl, to crisp; to crimp, to goffer
  sich k. to ripple *(water)*
Kräuseln *n* **und Rüschen** *n* fluting *(trimming)*
Kräuselspannung *f* ripple voltage, voltage ripple
Kräuselstrom *m* ripple current
Kräuselung *f* 1. rippling *(water)*; 2. *(text)* crimp
  flachbogige K. slightly curved crimp
  hochbogige K. high curved crimp
  regelmäßige K. even-running crimp
  schlichte K. faint crimp
Kräuselungsbogen *m* **der Faser** crimping arc of the fibre
Kräuselzwirn *m* waved twist
Krauseminzöl *n* curlymint oil
Kraut *n* [potato] haulm, tops *(beets)*; cabbage
Kräuteressig *m* aromatic vinegar
Krautschläger *m* potato haulm cutter
Krauttrennkette *f* haulm separating chain
Krawatte *f* cravat
Kraweelbau *m* carvel work (building)

Kraweelbeplankung *f (ship)* carvel planking
Kraweelgang *m (ship)* inside strake
Kreatin *n* creatine
Kreatinin *n* creatinine
krebsig affected with cancer *(wood)*
Kredit *m* credit
kreditieren to credit
Kreditorenauszug *m* accounts payable remittance
Kreide *f* chalk
Bologneser K. Bologna (French) chalk
Kreideabbau *m* chalk quarrying
Kreideboden *m* chalky soil
Kreidebruch *m* chalk quarry
Kreidemergel *m* chalk marl
Kreidereliefzurichtung *f* chalk relief makeready
Kreis *m* 1. *(math)* circle; 2. *(el)* circuit
abgestimmter K. tuned circuit
angekoppelter K. coupled circuit
angeschriebener K. excircle
Apollonischer K. Apollonius's circle
aufgeschnittener K. *(el)* open loop
betriebsbereiter K. armed circuit
dämpfungsloser K. all-pass network
eingeschriebener K. incircle
elektrischer K. [electric] circuit
Feuerbachscher K. nine-point circle
geerdeter K. earthed circuit
gekoppelter K. coupled circuit
geschlossener K. closed circuit
impulsformender K. pulse-shaping circuit
magnetischer K. magnetic (metallic) circuit
Mohrscher K. Mohr circle
nichtgeschlossener magnetischer K. imperfect magnetic circuit
offener K. open circuit
pupinisierter K. loaded circuit
selektiver K. selective circuit
stabiler K. stable circuit
stromführender K. live circuit
symmetrischer K. balanced circuit
Kreisablaufbahn *f (ship)* cambered standing ways
Kreisabschnitt *m* segment of [a] circle, circular segment
Kreisausschnitt *m* sector of [a] circle, circular sector
Kreisbahn *f* 1. circular path (orbit); circular trajectory; cam ring *(vane pump)*
Kreisbahngeschwindigkeit *f* orbital (circular orbit) speed, circular velocity
Kreisbahnhelling *f* cambered slipway
Kreisbeschleuniger *m* cyclic accelerator
Kreisbewegung *f* circular motion (movement), rotatory motion (movement), angular motion
Kreisblattschreiber *m* circular chart recorder
Kreisbogen *m* 1. *(math)* arc of circle, circular arc; 2. *(build)* bull's-eye arch *(masonry)*
kreisbogenverzahnt [teeth] curved on the arc of a circle
Kreisdiagramm *n* circle diagram
Kreisel *m* gyroscope
freier K. free gyroscope
Kreiselbewegung *f* gyroscopic motion
Kreiselbrecher *m* gyratory (rotary) crusher
Kreiselegge *f* rotary harrow
Kreiselektron *n* spinning electron
Kreiselement *n* circuit element
Kreiselgebläse *n* turbo-blower
Kreiselgerät *n* gyroscope
Kreiselhorizont *m* artificial horizon
Kreiselkompaß *m* directional gyro, gyrocompass, gyro-compass
Kreiselkondensator *m* centrifugal condenser
Kreisellot *n* gyro vertical
kreiselmagnetisch gyromagnetic
Kreiselmischer *m* impeller, gyro mixer
Kreiselmodell *n* **des Moleküls** gyroscopic model of the molecule
Kreiselmoment *n* gyroscopic moment
Kreiselmutter *f (ship)* master gyro (compass)
Kreiselpendel *n* gyro[scope] pendulum
Kreiselpumpe *f* centrifugal pump
Kreiselsextant *m* sextant with artificial horizon
Kreiselsichter *m* rotating-disk deduster
Kreiseltochter *f (ship)* repeater compass
Kreiselvertikale *f* gyro vertical
Kreiselwipper *m* rotary dump, rotating tippler
Kreiselwirkung *f* gyroscopic action, gyroscopic (gyrostatic) effect
Kreiselzetter *m (ag)* rotary tedder
kreisen to rotate, to circulate, to gyrate
in seiner Bahn k. to circle in its orbit
kreisend rotating, circulating; vertical *(flow)*
Kreisfläche *f* area of [a] circle
Kreisförderer *m* loop conveyor
kreisförmig circular
Kreisfrequenz *f* gyro-frequency, angular frequency, pulsatance
Kreisfunkbake *f* circular radio-beacon
Kreisfunkfeuer *n* circular radio-beacon
Kreisfunktion *f* circular function
Kreisgleichung *f* equation of circle
Kreisgüte *f (el)* factor of quality
Kreisgütemeßgerät *n* circuit magnification meter
Kreiskamm *m (text)* comb segment
Kreiskegel *m* circular cone
Kreiskolbengebläse *n* disk piston blower
Kreiskolbenmotor *m* rotary engine *(Wankel)*
Kreiskorn *n (mil)* circular fore sight
Kreiskornvisier *n* ring sight
Kreislauf *m* circuit *(hydraulic)*; cycle; circulation
K. der Kochung *s.* Kochperiode
erster K. *s.* Primärkreislauf

geschlossener K. closed cycle (system), closed circuit [system]
hydraulischer K. mit gleichbleibendem Umlaufvolumen constant-volume hydraulic circuit
offener K. open cycle, open circuit [system]
kreisläufig cycloidal
Kreislaufreaktor *m* circulating [fuel] reactor
Kreislaufüberwachung *f* circuit survey (control)
Kreismessung *f* cyclometry
Kreismusterung *f (text)* circular pattern
Kreisnut *f* circumferential groove
Kreisplatte *f* circular plate
Kreispolarimeter *n* circle polarimeter
Kreisprozeß *m* cycle *(Carnot, Rankine)*
Carnotscher K. Carnot cycle
Clausiusscher K. Clausius' cycle
geschlossener K. complete (closed) cycle
idealer K. ideal cycle
umkehrbarer K. reversible cycle
vollständiger K. closed (complete) cycle
Kreisquerschnitt *m* circular [cross] section
Kreisrauschen *n (el)* circuit noise
Kreisrechenschieber *m* circular slide-rule
Kreisring *m* annulus, circular ring
Kreisringplatte *f* circular ring slab
Kreisringspule *f* single ring coil
Kreisringträger *m* circular beam
Kreisrost *m* circular grate
kreisrund circular
Kreissäge *f* circular saw (sawing machine)
hinterlochte K. perforated circular saw
konische K. bevelled circular saw
Kreissägeblatt *n* circular saw blade
Kreissägenschärfautomat *m* automatic circular saw sharpener
Kreisschaltung *f* closed-loop circuit
Kreisscheibe *f* circular disk
Kreisschere *f* circle shear
Kreisschieberumsteuerventil *n* disk-type rotary selector
Kreisschieberventil *n* rotary disk valve, disk rotary valve, rotary valve based on flat surface
Kreisschnitt *m* 1. *(mach)* circular cut; 2. circular section
Kreisschwingsieb *n* circular vibratory screen
Kreissegment *n* circular segment
Kreisseiltrieb *m* continuous rope drive
Kreissektor *m* circular sector, sector of circle
Kreisskale *f* circular scale
Kreisspannung *f* voltage across the circuit
Kreisstapel *m (ship)* cambered standing ways
Kreisstrom *m* circular current; tank current
Kreisteilung *f* 1. *(mach)* circular division; 2. *(math)* cyclotomy
Kreisverkehr *m* roundabout
Kreisverstärkung *f (el)* loop gain
Kreiszweieck *n (math)* crescent
Kreiszylinder *m* circular cylinder
gerader K. right circular cylinder
Kreiszylinderschale *f* circular cylindrical shell
Krempel *f* carding engine (machine)
K. mit Sägezahndrahtbeschlag garnet clothing card
K. mit selbsttätiger Reinigung card with automatic stripper
K. mit Stichelhaarausscheidevorrichtung card with dog's hair extractor
K. mit Vortrommel card with breast roller
K. mit Vorwalze card with breast roller
Krempelausputz *m* card strips
Krempeldeckel *m* card flat
Krempelei *f* card room
Krempelflor *m* card web
Krempelflorteilung *f* division of the card web
Krempelkette *f* card chain
krempeln to card; to scribble; to garnet
Krempelreinigung *f* card stripping
Krempelsatz *m* set of cards
Krempelspeiser *m* card feeding apparatus
Krempelwalzenschleifmaschine *f* card roller grinding machine
Krempelwolf *m* carding willow
Krempler *m* carder
Kreosol *n* creosol
Kreosot *n* creosote
Kreosotöl *n* creosote oil
Krepp *m* crêpe, crape
geprägter K. embossed crêpe
Kreppapier *n* crêpe paper
Kreppgarn *n* crêpe yarn
Kreppkalander *m* calender for crêping
Kreppungswinkel *m* angle of filing
Kresol *n* cresol, cresylic acid
Kresolharz *n* cresol (cresylic) resin
Kresolmoschus *m* musk ambrette
Kresylat *n* cresylate
Kretonne *m(f)* cretonne
Kreuz *n* cross
Kreuzantenne *f* turnstile aerial
kreuzartig cruciform
Kreuzaufspanntisch *m* compound-type of workholding table
Kreuzband *n* [news] wrapper
Kreuzbohrmeißel *m* star bit
Kreuzbruch *m (print)* right angle fold
Kreuzbruchfalzer *m (print)* double folder
Kreuzdeckmuster *n (text)* cable stitch
Kreuzdipol *m* turnstile dipol
Kreuzdrillen *n* cross blocking
kreuzen 1. *(ag)* to cross, to hybridize; 2. to cross; to traverse *(wires)*
sich k. to intersect *(wires)*
Kreuzer *m* cruiser
Kreuzerheck *n* cruiser stern

Kreuzfach *n* cross shed
Kreuzfaltdipol *m* cross-type folded dipole
Kreuzfeldmultiplikator *m* crossed-fields electron-beam multiplier
Kreuzfeuer *n* cross fire
Kreuzfeuerbestrahlung *f* cross firing
Kreuzflügler *m* cross-winged guided missile
Kreuzfuchs *m* cross fox
Kreuzgaze *f* net gauze
Kreuzgelenk *n* swivel joint, universal joint (coupling)
Kreuzgelenkgabel *f* universal joint yoke
Kreuzgelenkkupplung *f* flexible coupling
Kreuzgewölbe *n* groined (cross) vault
Kreuzglied *n* lattice network
Kreuzgliedfilter *n* lattice-type filter
Kreuzgliedregler *m* lattice network regulator, lattice-type filter
Kreuzhacke *f* [flat] pick, pickaxe
Kreuzhaspeln *n* cross reeling
Kreuzhaspelung *f* cross reeling
Kreuzhiebfeile *f* double-cut file
Kreuzholz *n* quarter timber
Kreuzkarte *f* dot chart
Kreuzklampe *f (ship)* kevel
Kreuzknoten *m (ship)* carrick bend
Kreuzköper *m* cross twill
Kreuzkopf *m* crosshead
  gabelförmiger (gegabelter) K. forked crosshead
  viergleisiger K. crosshead for four slide bars, four bar type of crosshead
Kreuzkopfbolzen *m* crosshead pin, gudgeon
Kreuzkopfdeckel *m* crosshead cap (cover)
Kreuzkopfende *n* **der Kolbenstange** crosshead (small) end of the piston rod
Kreuzkopfführung *f* crosshead guide
Kreuzkopfkolben *m* crosshead piston
Kreuzkopflager *n* crosshead bearing
Kreuzkopfschleifer *m* slide (guide) block
Kreuzkopfzapfen *m* crosshead pin, gudgeon
Kreuzkopfzapfenlager *n* crosshead-pin bearing
Kreuzlenker *m* cross-shaped link
Kreuzlibelle *f* cross level
Kreuzlochschraube *f* capstan screw
Kreuzmeißel *m* cross-cut chisel
Kreuzmodelung *f s.* Kreuzmodulation
Kreuzmodulation *f* cross modulation
Kreuzmodulationsstörung *f* cross channel interference
Kreuznagel *m* lease peg (pin)
Kreuzneutralisation *f (el)* cross neutralization
Kreuzpeilung *f* cross-bearing
Kreuzplan *m (math)* cross-over design
Kreuzpoller *m (ship)* cross-bitt, bollard with cross-piece, cruciform bollard; kevel head *(wood)*
Kreuzprodukt *n (math)* cross (outer) product
Kreuzrahmen *m s.* Kreuzrahmenantenne
Kreuzrahmenantenne *f* crossed-coil aerial
Kreuzrahmenpeilverfahren *n* Bellini-Tosi system
Kreuzrändel *n (mach)* knurling having the ridges at right angles
Kreuzrohrstück *n* cross
Kreuzrute *f (text)* lease rod
Kreuzschalter *m* intermediate switch
Kreuzschaltung *f* cross (four-wire) connection
Kreuzscheibenkupplung *f* cross-keyed coupling
Kreuzschichtung *f (geol)* current-bedding
Kreuzschieber *m* compound rest *(lathe)*; saddle *(milling machine)*
Kreuzschiene *f* lease rod
Kreuzschienenverteiler *m (el)* cross-connecting board, cross-bar distributor
Kreuzschlagseil *n* cross-laid rope, cross-lay rope, ordinary lay rope
Kreuzschleife *f* Scotch yoke
Kreuzschlitten *m* compound slide *(lathe)*
Kreuzschlüssel *m* four-way rim wrench
Kreuzschnürung *f (text)* cross tie
Kreuzsee *f* beam sea
Kreuzsegel *n* jigger sail
Kreuzsprossenmaschine *f* window sash bar making machine
Kreuzspule *f* cross coil, cross-wound bobbin
Kreuzspulmaschine *f (text)* cone winder
  K. mit Schlitztrommel split drum winder for cheeses
Kreuzspulmeßinstrument *n* crossed-coil measuring instrument
Kreuzstab *m (text)* lease stick
Kreuzsteg *m (print)* gutter[-stick]
Kreuzstich *m* cross-stitch
Kreuzstoß *m* double-tee joint *(welding)*
Kreuzstrebe *f* diagonal cross brace
Kreuzstück *n* cross *(fitting)*
Kreuzsupport *m (mach)* compound rest
Kreuztisch *m (mach)* compound table; mechanical stage
Kreuztonne *f* junction (crossing) buoy
Kreuzung *f* 1. *(ag)* crossing, hybridization; 2. *(build)* crossing, intersection, junction; cross roads; 3. crossing, transposition *(overhead line)*
  niveaugleiche K. level (grade) crossing
Kreuzungsabschnitt *m* transposition section
Kreuzungsfolge *f* transposition sequence *(overhead line)*
kreuzungsfrei unintersected
Kreuzungsgestänge *n* transposition pole
Kreuzungsklemme *f* cross-over clamp *(overhead line)*
  K. für Luftkabeltragseile stay clamp
Kreuzungsmast *m* transposition tower (pole)
Kreuzungspunkt *m* junction

Kreuzungsweiche *f (ry)* slip points
  doppelte K. double-slip points
  einfache K. single-slip points
Kreuzungswinkel *m (mach)* angle of shafts *(transmission)*
Kreuzungszucht *f* cross-breeding
Kreuzverband *m* English cross bond *(brickwork)*
Kreuzverbinder *m* cross jointer
Kreuzverschwertung *f (build)* X-bracing, diagonal cross bracing
Kreuzverstrebung *f (build)* diagonal cross bracing, X-bracing
Kreuzverzahnung *f* staggered [arrangement of] teeth, alternately right-hand and left-hand angled teeth
Kreuzweife *f* cross reel
Kreuzwicklung *f* cross (honey-comb) winding
Kreuzzahnscheibenfräser *m* side and face cutter, staggered-tooth cutter, alternate angle side and face cutter
  geteilter K. staggered tooth type interlocking [side] mill, staggered tooth type interlocking [side] milling cutter
Kreuzzeiger *m* cross pointer
Kreuzzeigergerät *n* zero-reader
Kreuzzeigerinstrument *n* cross-pointer instrument
Kricketschläger *m* [cricket] bat
Kriechanlasser *m* inching (jogging) starter
Kriechdehnung *f* creep strain
kriechen 1. to creep *(material)*; 2. to leak, to creep *(current)*
Kriechen *n* 1. *(el)* creep[ing], [surface] leakage; 2. creep[ing], plastic flow *(material)*
Kriechendwert *m* ultimate creep value
Kriechentladung *f (el)* charge dissipation, creep leakage
Kriechfaktor *m* creep factor
Kriechfestigkeit *f* creep strength, resistance to creep
Kriechgang *m* creep feed (motion), creeping speed; inching
Kriechgeschwindigkeit *f* rate of creep, creep rate
Kriechgrenze *f* creep limit, limiting creep stress
Kriechstrecke *f* leaking (leakage) path, creeping (leakage) distance
Kriechstrom *m* creepage, leakage (creeping) current, current leakage
  K. gegen Erde earth leakage
  K. in Starkstromanlagen sneak current
Kriechstrombeständigkeit *f* non-tracking quality
kriechstromfest leakage-proof
Kriechstromverlust *m* creep leakage
Kriechverhalten *n* creep behaviour *(material)*
Kriechverlust *m s.* Kriechstromverlust
Kriechversuch *m* creep test *(material)*
Kriechweg *m* leaking path, creeping distance, leakage distance (path)
Krieg *m* war
  chemischer K. chemical war
Kriegsflagge *f* flag of war
Kriegshafen *m* military (naval) port
Kriegsmarine *f* navy
Kriegsschiff *n* man-of-war, warship, augmentation vessel
Kriegsschiffbau *m* naval construction
Kriegstechnik *f* military engineering
Krimpen *n* **des Windes** *(ship)* backing of the wind
Krippe *f (ag)* manger
Krispelholz *n* graining board, pommel
Krispelmaschine *f* graining machine
Kristall *m* crystal
  allochromatischer K. allochromatic crystal
  anisotroper K. anisotropic crystal
  doppelbrechender K. birefrigent crystal
  dotierter K. doped crystal
  einachsiger K. uniaxial crystal
  empfindlicher K. sensitive crystal
  fehlgeordneter K. imperfect crystal
  gedopter K. doped crystal
  gestörter K. imperfect crystal
  gezogener K. pulled crystal
  homöopolarer K. homopolar crystal
  idealer K. perfect crystal
  inhomogener K. inhomogeneous crystal
  komplexer K. complex crystal
  kubischer K. cubic crystal
  künstlicher K. synthetic crystal
  leitender K. conducting crystal
  nadelförmiger K. acicular crystal
  natürlicher K. natural crystal
  organischer K. organic crystal
  piezoelektrischer K. piezoelectric crystal
  reiner K. pure crystal
  synthetischer K. synthetic crystal
  ungestörter K. perfect crystal
  verformter K. deformed crystal
  verunreinigter K. doped crystal; contaminated crystal
Kristallachse *f* crystal axis
kristallähnlich crystalloid
Kristallanordnung *f* crystal texture
Kristallauflockerung *f (met)* loose place
Kristallautsprecher *m* piezoelectric loudspeaker
Kristallbau *m* crystal structure
Kristallbaufehler *m* crystal defect
Kristallbildgleichrichter *m* crystal video rectifier
Kristallbildung *f* crystallization, crystal formation
Kristalldetektor *m* crystal detector
Kristalldiode *f* crystal diode
Kristalle *mpl* / **kreuzartig verwachsene** cruciform twins

Kristallebene *f* crystal[lographic] plane
Kristallempfänger *m* crystal receiver
Kristallfehler *m* crystal imperfection (defect)
Kristallfehlordnung *f* crystal disorder
Kristallfilter *n* crystal filter
Kristallfilterneutralisation *f* crystal phasing
Kristallfläche *f* crystal face
Kristallform *f* crystal form; crystalline form
Kristallgitter *n* crystal lattice
  gestörtes K. imperfect crystal lattice
  ideales K. ideal crystal lattice
  kubisches K. cubic lattice
Kristallgitterfehler *m* crystal imperfection
Kristallglas *n* crystal glass
Kristallgleichrichter *m* crystal rectifier
Kristallhaufwerk *n* crystalline aggregation
Kristallichtverstärker *m* photistor
kristallin crystalline
Kristallinse *f* crystalline lens
Kristallisation *f* crystallization
  erzwungene K. forced crystallization
  fraktionierte K. fractional crystallization
Kristallisationsgeschwindigkeit *f* speed of crystallization
Kristallisationskeim *m* centre (nucleus) of crystallization, seed crystal
Kristallisationskern *m* centre of crystallization
Kristallisationsvermögen *n* crystalline force
Kristallisationswärme *f* latent heat of solidification
kristallisierbar crystallizable
kristallisieren to crystallize
kristallisierend / trigonal crystallizing in the trigonal system
  triklin k. crystallizing in the tricline system
Kristallisiergefäß *n* crystallizing pan
Kristallisierschale *f* crystallizing dish
Kristallit *m* crystallite
Kristallkante *f* edge of a crystal, crystal edge
Kristallkeim *m* crystal nucleus, seed (nucleus of) crystal
Kristallkern *m* crystal nucleus, nucleus (centre) of crystallization
Kristallkernbildung *f* crystallization, nucleation, germination
kristallklar *(plast)* crystal-clear
Kristallkorn *n* crystal grain
Kristallkunde *f* crystallography
Kristallmikrofon *n* crystal microphone
Kristallmodulator *m* frequency-changer crystal
Kristallografie *f* crystallography
kristallografisch crystallographic
Kristalloid *n* crystalloid
Kristallphosphor *m s.* Leuchtstoff
Kristallphysik *f* crystal physics
Kristallquarzfenster *n* crystal quartz window
Kristallrichtung *f* crystal orientation
Kristallrohling *m* crystal blank
Kristallschwinger *m* crystal resonator
Kristallschwingung *f* crystal vibration
Kristallseigerung *f* microsegregation
Kristallseitenkante *f* lateral edge of a crystal
Kristallsoda *f* washing soda
Kristallspektrograf *m* crystal spectrograph
Kristallspektrometer *n* crystal (Bragg) spectrometer
Kristallsteuerstufe *f* quartz-excited control stage
Kristallsteuerung *f* crystal control
Kristallstörung *f* crystal imperfection
Kristallstruktur *f* crystal structure
Kristallsymmetrie *f* crystal symmetry
Kristallsystem *n* crystal system
Kristalltriode *f* crystal triode
Kristallumineszenz *f* luminescence of crystallization, crystalloluminescence
Kristallwachstum *n* crystal growth
Kristallwasser *n* water of crystallization (constitution, hydration), combined water
kristallwasserfrei anhydrous
Kristallzähler *m* crystal [conduction] counter
Kristallziehen *n* crystal pulling
Kristallzüchtung *f* growing [of crystals], crystal growing (growth)
Kristallzucker *m* crystal sugar
Kristallzwilling *m* twin crystal
Kriterium *n* criterion
  K. der Regelgüte control criterion
  K. des mittleren quadratischen Fehlers rms error criterion *(statistics)*
kritisch critical
  k. werden to go critically *(reactor)*
Kritisch-Sein *n* critical state *(of reactor)*
Kritizität *f (nucl)* criticality
  prompte K. prompt criticality
Krokodilklemme *f (el)* crocodile (alligator) clip
Krokodilkontakt *m* ramp *(el. traction)*
Krokoit *m (min)* crocoi[si]te
Krokydolit *m (min)* blue asbestos, crocidolite
Krone *f* 1. *(build)* crown, crest *(wall)*, summit, top *(dam)*; 2. bit *(tool)*; 3. top, head *(tree)*
  abnehmbare K. detachable bit
  aufsteckbare K. detachable bit
Kronenarmatur *f* top fittings *(blast furnace)*
Kronenmutter *f* [hexagon] castle nut, [hexagon] castellated nut
Kronenmuttereinfräsung *f* castellation
Kronenmutterschlitz *m* castellated slot
Kronenschneide *f* six point bit
Kronglas *n* crown glass
Kronierung *f (mach)* castellation
Kronleuchter *m* electrolier
Kronrad *n* crown wheel
  K. mit Ritzel crown wheel and pinion
kröpfen to crank
Kröpfung *f (mach)* gap *(bed)*; offset

Krotonaldehyd *m* crotonaldehyde
Krotonöl *n* croton oil
Krotonsäure *f* crotonic acid
Krotylalkohol *m* crotonyl alcohol
Krücke *f* 1. iron rod; 2. rake, rabble; 3. larry; 4. boning rod *(surveying)*
Krückenisolator *m* spur (raker) insulator
Krümel *m* crumb *(of soil)*
Krümelbildung *f* crumb formation
krümeln to crumble
Krümelstruktur *f* crumb structure
Krümelwalzenegge *f* culti-pack roller
Krumenpacker *m* furrow press[er]
Krumenverdichtung *f* cap
krumm curved; bent; buckled; warped; crooked
krümmen to deflect; to buckle; to warp; to curve; to bend
Krümmer *m* elbow, bend, knee
Krümmerdurchflußmesser *m* elbow flowmeter
Krummfaser *f* spiral grain *(wood)*
krummfaserig wavy grained *(wood)*
Krummhaue *f* adz[e]
Krummholz *n* 1. curved (compass) timber, crook; 2. *(ship)* wale
Krümmling *m* wreath piece *(wooden stairs)*
krummlinig curvilinear
Krummschnitt *m* slanting cut
Krümmung *f* curvature; kurtosis; deflection; curve, bend
  abnormale K. *(math)* allokurtosis *(curve)*
  doppelte K. *(build)* double curvature
  mittlere K. *(math)* mesokurtosis *(curve)*
  normale K. *(math)* isokurtosis *(curve)*
Krümmungshalbmesser *m* radius of bend (curvature)
Krümmungskreis *m* circle of curvature
Krümmungsmittelpunktskurve *f* evolute
Krümmungsradius *m* radius of curvature (curve, bend)
Krümmungswinkel *m* angle of curvature
Krummzapfen *m* pin of bent crank
Krummziehen *n* **der Rohre** crooking of the tubes
krümpen to hot press
krumpfbeständig shrink resistant
krumpfecht ausrüsten to make unshrinkable
krumpfen to shrink
Krumpfen *n* shrinking process
krumpffrei non-shrinkable
Krumpfmaß *n* shrinkage
Krüppelwalmdach *n* false hip-roof
Kryochemie *f* cryochemistry
Kryolith *m (min)* cryolite
Kryosar *m* cryosar *(superconducting switching device with two electrodes)*
Kryoskopie *f* cryoscopy; cryoscopic method
kryoskopisch cryoscopic
Kryostat *m* cryostat
Kryotron *n* cryotron *(superconducting switching device)*
Kryotronik *f* cryotronics
Kryptoanalyse *f* cryptoanalysis
Kryptografie *f* cryptography
kryptokristallin cryptocrystalline
Krypton *n* krypton
Kryptoskop *n* cryptoscope
Kubatur *f* cubical contents, cubage, cubature
Kübel *m* bucket, bowl, skip, trough; pot, tub, vat
Kübelaufzug *m* bucket elevator, skip hoist
Kübelbegichtung *f* bucket charging *(blast furnace)*
Kübelkipper *m* dumper
Kubierung *f* cubage, cubature
Kubikinhalt *m* cubic (volumetric) capacity
Kubikmeter *n* cubic metre
  K. gewachsener Abraum *(mine)* bank cubic metres of overburden
Kubikwurzel *f* cubic root
Kubikzentimeter *n* cubic centimetre
Kubikzoll *m* cubic inch
kubisch cubic
  k. kristallisierend crystallizing in the cubic system
kubisch-flächenzentriert face-centred cubic
kubisch-raumzentriert body-centred cubic
kubisch-zentriert body-centred, cubically centred
Küche *f* kitchen; *(ship)* galley
Kuchen *m* pat *(cement)*
Küchenabfälle *mpl* garbage, kitchen waste
Kuchenerz *n* cake-ore
Küchenpersonal *n (ship)* catering staff
Kufe *f* 1. *(aero)* skid, sled; skid *(braking)*; runner; 2. dye vessel
Kufenflugzeug *n* sled plane
Küfer *m* cooper
Küferschlichthobeleisen *n* cooper's plane iron
Kugel *f* 1. *(math)* sphere; 2. *(mach)* ball; 3. shot *(sports)*; 4. pellet
  dielektrische K. dielectric sphere
  eingeschriebene K. *(math)* insphere
  polarisierte K. polarized sphere
  Ulbrichtsche K. photometric sphere
  umbeschriebene K. *(math)* circumsphere
Kugelabschnitt *m* spherical segment
Kugelantenne *f* isotropic aerial, *(Am)* unipole
Kugelausschnitt *m* spherical sector
Kugelbakentonne *f* globe buoy
Kugelbakterie *f* spherobacterium
Kugelblitz *m* ball lightning
Kugelboje *f* spherical buoy
Kugeldreheinrichtung *f* spherical turning attachment
Kugeldrehen *n* ball turning
Kugeldreieck *n* spherical triangle
Kugeldruckhärte *f (met)* ball thrust hardness

Kugeldruckhärteprüfung *f* indentation hardness test with ball indentor
Kugeleindruck *m* *(met)* indentation of the ball
Kugelendmaß *n* spherical-end measuring gauge
Kugelfallhärte *f* *(met)* scleroscope hardness
Kugelfallviskosimeter *n* falling (drop) ball viscosimeter
Kugelfender *m* *(ship)* spherical (globular) fender
Kugelfläche *f* spherical surface
Kugelform *f* spheroidal form
kugelförmig *(math)* spherical; ball shaped; globular
Kugelfotometer *n* globe photometer
Kugelfräsen *n* cherrying
Kugelfräser *m* cherry
Kugelfunkenstrecke *f* sphere (ball spark) gap
Kugelfunktion *f* *(math)* spherical function (harmonic)
Kugelfunktionsmethode *f* spherical harmonics method
Kugelgelenk *n* ball-and-socket joint
Kugelgelenkkrümmer *m* bend with ball and socket joint
Kugelgeometrie *f* spherical geometry
Kugelgestalt *f* sphericity
Kugelgraphit *m* spheroidal (nodular) graphite
Kugelgraphitguß *m* spheroidal (nodulized) graphite cast iron
Kugelgraphitgußeisen *n* / **ferritisches** ferritic ductile iron
Kugelgriff *m* ball handle
Kugelhaube *f* spherical surface
Kugelhaufenreaktor *m* pebble[-bed] reactor
Kugelintegrator *m* flyball integrator
Kugelionisationskammer *f* spherical ionization chamber
Kugelkalotte *f* spherical cup
Kugelkammer *f* spherical ionization chamber
Kugelkeil *m* spherical wedge
Kugelknopf *m* knob
Kugelkocher *m* spherical boiler (digester) *(papermaking)*
rotierender K. spherical rotary digester
Kugelkohle *f* coal apples, nigger heads
Kugelkondensator *m* spherical capacitor
Kugelkoordinatenwandler *m* spherical (ball) resolver
Kugelkopfabspannisolator *m* ball-headed strain insulator
Kugelkühler *m* ball (bulb, globe-shaped) condenser
Kugelkurbel *f* ball crank
Kugellager *n* ball bearing
zweireihiges K. double-row ball bearing
Kugelläppen *n* lapping of spherical surfaces
Kugelmühle *f* ball (pebble) mill
Kugelnickel *n* nickel pellets
Kugelpackung *f* sphere (spherical) packing *(lattice)*
Kugelpfanne *f* ball cup
Kugelpfannenlager *n* hemispherical bearing
Kugelpipette *f* bulb pipette
Kugelrückschlagventil *n* automatic ball steam pipe isolating valve
Kugelschale *f* *(mach)* ball socket, spherical shell *(mechanics)*
Kugelschaltung *f* ball-and-socket gear change
Kugelschaufler *m* revolving cutter head dredger
Kugelschlagbohren *n* pellet impact drilling
Kugelschlaghärteprüfer *m* ball-impact hardness tester
Kugelschlagprüfung *f* ball impact test
Kugelschott *n* *(ship)* round hatch
Kugelschreiber *m* ball-point pen
Kugelschwingmühle *f* vibrating ball mill
Kugelsegment *n* spherical segment
Kugelsektor *m* spherical sector
Kugelspiegel *m* spherical mirror
Kugelspiegelleuchte *f* ball reflector lamp
Kugelsteuerschieber *m* ball baffle
Kugelstirn *f* spherical end
Kugelstoßen *n* putting of the shot, shotput, shot putting
Kugelstrahlen *n* *(mach)* shot peening
Kugelstrahler *m* isotropic radiator
kugelsymmetrisch sphero-symmetric[al], spherically symmetric
Kugeltonne *f* *(ship)* spherical buoy
Kugelumlauflenkung *f* *(aut)* recirculating ball steering
Kugelvariometer *n* spherical coil-type variometer
Kugelventil *n* ball[-type] valve
Kugelviskosimeter *n* ball viscosimeter
Kugelwelle *f* spherical wave
Kugelzählrohr *n* spherical counter
Kugelzapfen *m* ball journal
Kugelzielfernrohr *n* hull gunner's periscope *(tank)*
Kugelzone *f* spherical zone
Kuhl *f* waist *(warship)*
Kühlanlage *f* refrigeration (cold storage) plant; cooling plant
Kühlapparat *m* cooling apparatus
Kühlbad *n* cooling bath
Kühlband *n* cooling baffle (conveyor, guide vane)
Kühlbassin *n* cooling pond (basin)
Kühlbecken *n* cooling basin
Kühlbehälter *m* cooltainer
Kühlbremsmittel *n* moderator coolant *(reactor)*
kühlen 1. to cool *(engines, machines)*; 2. to refrigerate *(food)*
Kühler *m* 1. *(aut)* radiator; 2. condenser *(distillation)*; 3. refrigerator

K. mit eingeschmolzenem Schlangenrohr condenser with spiral tube sealed to body
K. mit Schlangenrohr condenser with spiral condensing tube
ausziehbarer K. extractable (draw-out) radiator
einziehbarer K. retractable radiator
freifahrender K. externally mounted radiator
in Luftführung eingeschlossener K. tunnel-type radiator
Kühlerabdeckung *f* radiator shutter
Kühlerauslaufstutzen *m* radiator outlet connection
Kühlerblock *m* radiator core (block)
Kühlereinlaufstutzen *m* radiator inlet connection
Kühlerfuß *m* radiator mounting (support)
Kühlergehäuse *n* radiator frame
Kühlerhaube *f* radiator bonnet
Kühlerlamelle *f* radiator segment (lamination)
Kühlermantel *m* condenser jacket
Kühlerschutzbügel *m* radiator guard
Kühlerschutzgitter *n* radiator grille
Kühlerschutzring *m* radiator safety ring
Kühlerspritzblech *n* radiator baffle plate
Kühlerstrebe *f* radiator strut
Kühlerstütze *f* radiator support
Kühlerteilblock *m* radiator element
Kühlerverkleidung *f* radiator cowl (grille)
Kühlfalle *f* cooling trap
Kühlfläche *f* cooling surface
Kühlflügel *m* **der Elektrode** electrode radiator
Kühlflüssigkeit *f* cooling (heat-transfer) fluid, [liquid] coolant
Kühlflüssigkeitskanal *m* oil (coolant) channel
Kühlflüssigkeitspumpe *f* coolant pump
Kühlgas *n* cooling gas
Kühlgefäß *n* cooling tank
Kühlgraben *m* cooling conduit
Kühlgut *n* chilled cargo (goods)
Kühlhalle *f* cold store
Kühlhaube *f* cooling cap
Kühlhaus *n* cold storage (store)
Kühljackett *n* water jacket
Kühlkammer *f* cooling (refrigerating) chamber
Kühlkanal *m* cooling pipe; fuel channel; hot channel
Kühlkasten *m* cooling box, tuyère block (arch cooler)
Kühlkörper *m* heat sink *(semiconductor)*
Kühlkreislauf *m* heat-transfer circuit (cycle), thermal cycle
Kühlladeraum *m* insulated (refrigerated) hold
Kühlladung *f* refrigerated cargo
Kühllamelle *f* cooling fin (rib)
Kühlleitung *f* cooling pipe
Kühlluft *f* cooling air, air coolant
Kühlluftgebläse *n* air-cooling fan, cooling blower
Kühlluftleitblech *n* cooling baffle (guide vane)
Kühlluftmantel *m* cooling-air jacket
Kühlluftregelklappe *f* cowl flap
Kühlluftregler *m* radiator shutter (damper)
Kühlluftschlitz *m* cooling-air slot
Kühlmaische *f* cooler crystallizer
Kühlmansarde *f* cooling chamber
Kühlmantel *m* cooling (water) jacket
Kühlmantelriß *m* crack in the cooling jacket
Kühlmaschinenwagen *m* mechanically refrigerated wagon
Kühlmittel *n* coolant, cooling agent (medium); *(nucl)* heat-transfer agent; refrigerant; freezing mixture
K. des Primärkreislaufs primary coolant
K. für direkte Kühlung internal coolant
flüssiges K. liquid coolant, cooling fluid
Kühlmittelaktivität *f* coolant activity
Kühlmittelaufbereitung *f* coolant make-up
Kühlmittelbehälter *m (mach)* coolant container (tank, reservoir),suds tank
Kühlmitteldüse *f (mach)* coolant nozzle
Kühlmittelfilter *n (mach)* coolant filter
Kühlmittelfluß *m (mach)* coolant flow
Kühlmittelpumpe *f (mach)* coolant (suds) pump
Kühlmittelrückflußleitung *f (mach)* coolant return [line]
Kühlmittelumlauf *m (mach)* coolant circulation
Kühlmittelumlaufpumpe *f* circulating-coolant pump
Kühlmittelversorgung *f* coolant supply
Kühlmittelzuführung *f* coolant feed (supply)
Kühlnebel *m* coolant mist
Kühlnische *f* cooling plate box
Kühlofen *m* [annealing] lehr *(glass)*
Kühlöl *n* cooling oil
Kuhlorohr *n (el)* conduit wire
Kühlplatte *f* cooling disk
Kühlraum *m* refrigerator (cold storage) room; *(ship)* refrigerating hold
Kühlring *m* cooling ring
Kühlrippe *f* cooling fin (rib), radiator fin
Kühlriß *m* **durch Wasser** water crack
Kühlrohrleitung *f* circulating water mains
Kühlrührmaische *f* cooling cristallizer with stirring device
Kühlschelle *f* heat radiator *(semiconductor)*
Kühlschiff *n* refrigeration (refrigerated cargo) vessel; *(Am)* reefer [ship]
Kühlschlange *f* cooling (condensing) coil
Kühlschlitz *m* air (ventilating, cooling) duct
Kühlschrank *m* refrigerator
K. mit Kompressor compression-type refrigerator
elektrischer K. electric refrigerator
Kühlsieb *n* cooling sieve, hopper cooler
Kühlsole *f* brine; refrigerant
Kühlstock *m s.* Kühlstockwerk

Kühlstockwerk *n* cooling floor
Kühlstoffaustrittstemperatur *f* coolant exit temperature
Kühlstoffbehälter *m* coolant tank reservoir
Kühlstoffraum *m* coolant jacket
Kühlstofftemperaturmesser *m* coolant temperature gauge
Kühlsystem *n* cooling system (circuit)
Kühlteich *m* cooling pond
Kühlturm *m* cooling tower
Kühlung *f* 1. refrigeration; 2. cooling *(engine)*
K. durch Kompression compression refrigeration
K. durch Rückbelüftung reverse-flow cooling
K. durch Zerstäubung spray cooling
Kühlungsausfall *m* breakdown of the cooling system
Kühlungsnut *f* cooling slot
Kühlventilator *m* cooling fan
Kühlverlustleistung *f* cooling-power loss
Kühlvermögen *n* **der Kühlanlage** refrigerator capacity
Kühlversuch *m* cold test
Kühlvorrichtung *f* cooling (chilling) device
Kühlwagen *m* refrigerator wagon (car, truck), *(Am)* reefer
Kühlwalze *f* chill roll *(foils)*
Kühlwandung *f* cooling wall
Kühlwanne *f* cooler box, cooling tube
Kühlwasser *n* cooling water
Kühlwasserablauf *m* cooling water discharge
Kühlwasserableitung *f* cooling-water discharge pipe
Kühlwasseranzeiger *m* temperature gauge, calorimeter
Kühlwasserausfall *m* cooling-water failure
Kühlwasserauslaufstutzen *m* [cooling] water outlet connection
Kühlwasseraustritt *m* cooling-water outlet (discharge), circulating-water outlet (discharge)
Kühlwasserbedarf *m* quantity of cooling water required
Kühlwasserbehälter *m* cooling-water tank
Kühlwassereinlaufstutzen *m* water inlet connection
Kühlwassereintritt *m* entrance of cooling water
Kühlwasserleitung *f* cooling water pipe
Kühlwasserlieferung *f* cooling-water supply
Kühlwassermantel *m* water jacket
Kühlwasserpumpe *f* cooling water [circulating] pump
Kühlwasserumlauf *m* cooling-water circulation
Kühlwasserverteilung *f* breaking up of the cooling water
Kühlwasserzufuhrrohr *n* inlet pipe for circulating water
Kühlwasserzuleitung *f* supply of circulating water
Kühlwerk *n* cooling tower
Kühlwirkung *f* cooling effect (action)
Kühlzelle *f* cooling cell
Kühlzug *m* refrigerator train
Kühlzylinder *m* chilling roller, refrigerating cylinder
Kuhmist *m* cow dung
Kuhstall *m* cowhouse, cow-shed, byre
Küken *n* stopcock
Kükenaufzucht *f* chicken rearing
Kükenaufzuchtstall *m* floor brooder
kulieren *(text)* to form into loops, to link
Kulierexzenter *m* draw cam; stitch cam *(knitting machine)*
Kuliergewirke und -gestricke *npl* weft knitted fabrics
Kulierware *f* filling knit fabric, hosiery
Kulisse *f* 1. *(mach)* main driving link, trunnion-mounted link, slotted link (arm), rocker arm; 2. *(text)* slider; 3. sidescene *(theater)*
Kulissenantrieb *m* link drive
Kulissenkrümmung *f* curvature of swan neck
Kulissenmaler *m* scene painter
Kulissenrad *n* crank (main driving, bull) gear, crank (bull) wheel
schrägverzahntes K. single helical toothed bull gear
Kulissenschalter *m* door switch
Kulissenschaltung *f (aut)* gate change
Kulissenstange *f (mach)* link rod
Kulissenstein *m (mach)* sliding block
Kulissensteuerung *f* link motion
Kulmination *f (astr)* culmination
K. eines Fixsterns culmination of a star
Kulminationshöhe *f* height of culmination
Kulminationspunkt *m* culmination point
kulminieren to culminate
Kultivator *m* cultivator; grubber, scarifier
kultivieren to grow; to cultivate; to reclaim *(land)*
Kultur *f* crop
frostempfindliche K. frost sensitive (tender) crop
frostharte K. frost (winter) hardy crop
säureunverträgliche K. acid intolerant crop
säureverträgliche K. acid tolerant crop
Kulturmaßnahmen *fpl* cultivation measures
Kulturpflanzen *fpl* cultivated plants
Kulturphase *f* brush (nursery) stage *(of conifers)*
Kumarin *n* coumarin
Kumaron *n* coumarone
Kumaronharz *n* coumarone resin
Kumarsäure *f* coumaric acid
Kümmelöl *n* caraway oil
Kummet *n* [horse] collar
Kummetleder *n* collar leather
Kümo *n s.* Küstenmotorschiff

Kumol *n* cumene
Kumpel *m* 1. miner, mine worker; collier; 2. fellow-worker
Kümpelpresse *f* dishing (circular flanging) press
Kunde *m* customer
Kundendienst *m* [after-sales] service
kündigen to give notice
Kündigungsfrist *f* term of notice
Kunstantenne *f* artificial aerial
Kunstbuchbinder *m* art bookbinder
Kunstdarm *m* paper skin
Kunstdruckkarton *m* art board
Kunstdruckpapier *n* art (coated, enamelled) paper
Kunstdünger *m* [artificial] fertilizer
Kunsteis *n* manufactured ice
Kunsteisherstellung *f* ice making
Kunstfaser *f* artificial fibre
Kunstfaserzellstoff *m* dissolving pulp
Kunstfeinkorn *n* artificial fine grain, finely ground grain
Kunstflieger *m* acrobatic pilot
Kunstflug *m* acrobatic flight
Kunstfoliekondensator *m* polystyren (artificial-foil) capacitor
Kunstgewerbe *n* applied arts
Kunstgießerei *f* art foundry
Kunstguß *m* artistic casting
Kunstharz *n* synthetic resin
Kunstharzausrüstung *f* resin finish
Kunstharzbindemittel *n* resin binder
kunstharzgebunden [synthetic] resinoid-bonded *(grinding wheel)*
Kunstharzkleber *m* synthetic resin glue (adhesive, cement)
Kunstharzklebeverbindung *f* resin-bonded joint
Kunstharzlack *m* synthetic enamel
Kunstharzleim *m* synthetic resin glue
Kunstharzpappe *f* resin board
Kunstharzüberzug *m* resin coating
Kunstharzüberzugslack *m* synthetic [resin] finish
kunstharzverleimt resin-bonded
Kunsthonig *m* artificial honey
Kunsthorn *n* artificial horn
Kunstlauf *m* figure skating
Kunstlaufen *n* figure skating
Kunstlaufschlittschuh *m* figure skate
Kunstleder *n* imitation leather, leatherette, leatheroid; leather-cloth *(book)*
Kunstlederlack *m* artificial leather finish (coating)
Kunstleitung *f (el)* balancing network, artificial line
künstlich artificial, synthetic
Kunstlichtpapier *n* developing paper
Kunstpergament *n* artificial parchment paper
Kunstschaltung *f (el)* balancing network
Kunstschleiferstein *m* artificial pulpstone
Kunstschlittschuhlauf *m* figure skating
Kunstschlittschuhlaufen *n* figure skating
Kunstseide *f* rayon
Kunstseidenlackdraht *m* enamelled imitation-silk covered wire
Kunststein *m* artificial (cast) stone
Kunststoff *m* plastic, synthetic material
  elastomerer K. elastomer, elastomeric plastic
  glasartiger K. glazing material
  gummiartiger K. flexible plastic
Kunststoffbehälter *m* plastainer
Kunststoffkabel *n* plastic cable
Kunststoffmaterial *n* / **weiches** non-rigid plastic
Kunststoffolie *f* plastics sheet *(trimmed)*, plastics sheeting *(endless)*
Kunststoffplatte *f* plastic plate
Kunststoffplattenführung *f (mach)* plastic [guide] way
Kunststoffszintillator *m* plastic scintillator
Kunzit *m (min)* kunzite
Küpe *f* vat; boiler
Kupellieren *n (met)* cupelling
Küpenfärberei *f* vat dyeing
Küpenfarbstoff *m* vat dye[stuff]
Kupfer *n* copper
  feingranuliertes K. feathered shot copper
  gediegenes K. native copper
  kaltbrüchiges K. underpoled copper
Kupferabdichtung *f* packing with copper
Kupferabscheidung *f* copper deposition
Kupferanreicherung *f* copper enrichment
Kupferasbestdichtung *f* copper asbestos gasket
Kupferätzung *f* copper etching
Kupferazetat *n* / **basisches** cupric subacetate
Kupferazetylid *n* cuprous acetylide
Kupferband *n* copper tape (strip)
Kupferbanderder *m* copper strip earth conductor
Kupferbergwerk *n* copper mine
Kupferblech *n* sheet copper
Kupferbürste *f (el.m.)* copper brush
Kupfer-Chromnachbehandlung *f (text)* copper-chrome aftertreatment
Kupferdichtung *f* copper gasket
Kupferdichtungsscheibe *f* copper washer
Kupferdraht *m* copper wire
Kupfereinlage *f* copper packing
Kupfererz *n* copper ore
Kupferfolie *f* copper foil
Kupfergewinnung *f* extraction (winning) of copper
Kupferglanz *m* copper glance, chalcocite
Kupferglimmer *m* micaceous copper
kupferhaltig cupriferous
Kupferhütte *f* copper works
Kupferindigo *m* covellite, indigo copper
Kupferkalkbrühe *f* Bordeaux mixture

Kupferkarbonat *n* copper carbonate
basisches K. cupric subcarbonate
Kupferklischee *n* copper block
Kupferkohle *f (el.m.)* copper-plated carbon
Kupferlackdraht *m* enamelled copper wire
Kupferlasur *m* chessylite, azurite
Kupferleiter *m* copper conductor; copper track
Kupferlitze *f* copper strand, stranded copper wire
Kupfermanteldraht *m* copper-clad wire
Kupfernickel *n* niccolite, copper nickel, kupfernickel
Kupferoxid *n* copper oxide
Kupferoxidgleichrichter *m* copper-oxide rectifier
Kupferoxydulgleichrichter *m* copper-oxide rectifier
Kupferoxydulzelle *f* copper-oxide cell, cuprous oxide cell
Kupferrohr *n* copper pipe
Kupfersamterz *n* velvet copper ore, cyanotrichite
Kupfersau *f* matte of copper
Kupferschwamm *m* copper sponge
Kupferseide *f* cuprammonium rayon (filament)
Kupferspat *m* malachite
Kupferspule *f* copper coil
Kupferstecher *m* copperplate engraver
Kupferstein *m* copper matte
Kupferstich *m* copperplate engraving
Kupfertiefdruck *m* photogravure
Kupfertiefdruckpresse *f* intaglio printing press
Kupferunterlegscheibe *f* copper washer
Kupfer(I)-verbindung *f* cuprous compound, cuprocompound
Kupfer(II)-verbindung *f* cupric compound, cupricompound
Kupferverluste *mpl (el)* copper losses
Kupfervitriol *n* cupric sulphate, sulphate (vitriol) of copper, blue copperas
Kupferzahl *f* copper number *(cellulose)*
Kupfer-Zinkelement *n* blue-stone battery
Kupferzug *m (met)* coppering
kupieren to crop
Kupolofen *m* cupola furnace
K. mit erweitertem Herd compound cupola
Kupolofenausfütterung *f* cupola lining
Kupolofenbegichtung *f* cupola charging
Kupolofengattierung *f* cupola mixture
Kupolofenschacht *m* cupola shaft
Kupolofenvorherd *m* cupola receiver
Kuppe *f (build)* summit, crest
Kuppel *f (build)* cupola, dome
Kuppelachse *f* coupled axle
Kuppelachswelle *f* coupled axle-pin
Kuppeldach *n* domical (domed) roof
Kuppelfeld *n (el)* coupling panel (section)
Kuppellafette *f (mil)* cupola-type gunmount
Kuppelmagnet *m* magnetic clutch, coupling magnet
kuppeln 1. to connect, to engage; *(mach)* to couple, to clutch; 2. to couple *(diazo-compound)*
fest k. to couple rigidly
lose k. to couple loose
starr k. to couple rigidly
Kuppelradsatz *m* coupled wheel pair on axle
Kuppelschalter *m* coupling circuit-breaker
Kuppelstange *f* coupling (tie) rod
Kuppeltransformator *m* coupling transformer
Kuppelzelle *f (el)* coupling cubicle
Kuppenausrundung *f (build)* summit curve
Kupplung *f* 1. *(mach)* coupling *(shaft coupling; permanent coupling)*; clutch *(loose)*; 2. *(el)* flexible lead connector, coupling; connector *(electronics)*
aufgeschmiedete K. solid-forged coupling
aufgesetzte K. separate coupling
ausrückbare K. loose coupling, clutch
einstellbare K. flexible coupling
elastische K. flexible membrane clutch
elektrische K. electrical coupler *(motor vehicles)*
elektromagnetische K. [electro]magnetic clutch
formschlüssige K. positive clutch (coupling)
hydraulische K. hydraulic coupling
kraftschlüssige K. friction clutch
nicht schaltbare K. permanent coupling
schaltbare K. loose coupling, clutch
selbsteinrückende K. self-actuating clutch
selbsttätige K. automatic clutch
starre K. rigid coupling
Kupplungsauslösewelle *f* clutch withdrawal shaft
Kupplungsausrückgabel *f* clutch-throwout fork
Kupplungsausrücklager *n* throw-out bearing of the clutch
Kupplungsausrückmuffe *f* clutch release sleeve
Kupplungsausrückwelle *f* clutch withdrawal shaft
Kupplungsbelag *m* clutch facing
Kupplungsbolzen *m* coupling bolt
Kupplungsbremse *f* clutch brake
Kupplungsdaumen *m* clutch cam
Kupplungsdeckel *m* clutch cover
Kupplungsdose *f* coupler socket; microphone socket
Kupplungsdruckfeder *f* clutch thrust spring
Kupplungsdrucklager *n* clutch thrust bearing, clutch-throwout collar
Kupplungsdruckscheibe *f* clutch pressure (thrust) plate
Kupplungsexpansionsring *m* spreader of clutch
Kupplungsflächen *fpl* contact surfaces of clutch [halves]

Kupplungsflotte *f (text)* developing liquor
Kupplungsführungslager *n* clutch guide bearing
Kupplungsgehäuse *n* clutch housing
Kupplungsgelenk *n* clutch linkage
Kupplungsgestänge *n* clutch linkage
Kupplungshälfte *f* coupling half (member)
Kupplungskabel *n* clutch control-lever cable
Kupplungsklaue *f* clutch dog
Kupplungskopf *m* brake line coupling; coupler (drawgear) head
Kupplungslamelle *f* clutch segment, disk of a clutch
Kupplungsmuffe *f* clutch (coupling) sleeve, clutch (coupling) box
Kupplungsmuffenführung *f* clutch-sleeve guide
Kupplungspedal *n* clutch pedal
Kupplungsreaktion *f (ch)* coupling reaction
Kupplungsschalter *m* linked switch
Kupplungsscheibe *f* clutch (coupling) disk
Kupplungsschlupf *m* clutch slip
Kupplungsschwund *m* clutch fading
Kupplungsseite *f* **eines Generators** *(el)* back of a generator
Kupplungsspiel *n* clutch pedal clearance
Kupplungsstange *f* coupling rod
Kupplungssteckdose *f (el)* coupler socket
Kupplungsstecker *m (el)* coupler (adapter) plug
Kupplungssteuerung *f* clutch-and-brake steering *(of crawler-type tractors)*
Kupplungsstück *n (el)* coupling
Kupplungstreibscheibe *f* clutch drive plate
Kurbel *f (mach)* crank; lever
  K. zur Senkrechtverstellung der Konsole crank for vertical adjustment of knee
  außen aufgesteckte K. overhung crank
  dreifach gekröpfte K. three-throw crank
  überhängende K. overhung crank
Kurbelantrieb *m* 1. *(mach)* crank drive; 2. *(aut)* connecting-rod assembly
Kurbelarm *m* crank arm (web)
Kurbelbilge *f* crank race
Kurbelblatt *n* crank web
Kurbeldrehwinkel *m* crank[-rotation] angle, angle of rotation of the crank
Kurbelfenster *n* crank-operated window
Kurbelfolge *f* succession of cranks
Kurbelgehäuse *n* crankcase
Kurbelgehäuseentlüfter *m* crankcase breather (bleeder)
Kurbelgehäuseentlüftung *f* crankcase ventilation system
Kurbelgehäuseoberteil *n* crankcase top half
Kurbelgehäuseölverdünnung *f* crankcase dilution
Kurbelgehäusespülung *f* crankcase scavenging
Kurbelgehäuseunterteil *n* crankcase bottom half
Kurbelgehäuseverdichtung *f* crankcase compression
Kurbelgestänge *n* crank assembly
Kurbelgetriebe *n (mach)* crank [driving] mechanism
  durchschlagendes K. crossed crank mechanism, crossed link quadrilateral
  gleichschenkliges K. crank quadrilateral with isosceles members, isosceles crank mechanism
Kurbelgetriebeschwungradpumpe *f* crank and flywheel pump
Kurbelgriff *m* crank handle
Kurbelhandbohrmaschine *f* hand drill
Kurbelhub *m* crank throw (stroke)
Kurbelinduktor *m* magneto generator (inductor), hand generator
  K. für Widerstandsmessung magneto-ohmmeter
Kurbelkammer *f* crank chamber
Kurbelkasten *m* crankcase, crank chamber
Kurbelkastenentlüfter *m* crankcase breather
Kurbelkastenvorverdichter *m* crankcase supercharger
Kurbelkette *f* / **viergliedrige konische** crank quadrilateral with converging links, four bar motion with axes of rotation tending to a point
  zylindrische K. four-bar motion with axes of rotation parallel, linked quadrilateral
Kurbelkopf *m* crank end
Kurbelkreis *m* crank circle
Kurbelkröpfung *f* cranked portion of shaft
Kurbelkröpfungswange *f* crank cheek (web)
Kurbelkupplung *f* crank-connecting link, crank coupling
Kurbellager *n* crankshaft (main) bearing
Kurbellagerbock *m* crankshaft (main) bearing pedestal
Kurbelmast *m* extension (telescopic) mast
Kurbelmechanismus *m* crank mechanism
Kurbeln *fpl* / **gegenläufige** cranks moving in opposite directions, crossed parallelogram
  gleichläufige K. cranks rotating in the same direction
Kurbelnabe *f* crank boss
Kurbelpresse *f* crank-driven press
Kurbelradius *m* crank radius
Kurbelraum *m* crankcase
Kurbelschalter *m* radial arm switch
Kurbelscheibe *f* crank cam (disk, web), disk crank
  umlaufende K. rotating crank
Kurbelschenkel *m* crank cheek (web)
Kurbelschleife *f* Scotch yoke, crank-guide
  rotierende K. turning-block linkage
  schwingende K. oscillating lever (link), swinging arm; oscillating crank-gear

umlaufende K. turning-block linkage, Whitworth's variable speed gear
Kurbelschleifmechanismus *m* swinging-block mechanism (linkage)
Kurbelschwinge *f* crank (rocker) arm, crank and rocker, main driving link, slotted link (arm), trunnion-mounted link
Kurbelschwingsieb *n* crank shaking screen, screen with twin crank-shaft drive
Kurbelstellung *f* crankshaft position
Kurbelstuhl *m (text)* crank loom
Kurbeltrieb *m* crank assembly (mechanism), crankshaft drive
gleichschenkliger K. crank quadrilateral with isosceles members, isosceles crank mechanism
Kurbeltriebwerk *n* crank-drive assembly
Kurbelversetzung *f* angle between cranks, crank displacement
Kurbelviereck *n* / **konisches** crank quadrilateral with converging links, four bar motion with axes of rotation tending to a point
zylindrisches K. four-bar motion with axes of rotation parallel, linked quadrilateral
Kurbelwalke *f (leath)* crank (faller) stock
Kurbelwange *f* crank web
Kurbelwanne *f* lower part of crankcase, oil pan
Kurbelweg *m* crank path
Kurbelwelle *f* crankshaft
K. mit Gegengewicht counterweighted crankshaft
K. mit Kreuzkopf crankshaft with cross head
doppelt gekröpfte K. two-throw crank shaft, shaft with two cranks
dreifach gekröpfte K. shaft with three cranks, three-throw crank shaft
einfach gekröpfte K. shaft with one crank, one-throw crank
fünfmal gelagerte K. five-bearing crankshaft
gekröpfte K. cranked shaft
zusammengebaute K. built-up crankshaft
Kurbelwellenauge *n* crank eye
Kurbelwellenbund *m* crankshaft shoulder
Kurbelwellendrehmaschine *f* crankshaft lathe
Kurbelwellendurchbiegung *f* crankshaft deflection
Kurbelwellenfräsmaschine *f* crankshaft milling machine
Kurbelwellengegengewicht *n* crankshaft counterweight
Kurbelwellengrad *m* crankshaft degree *(valve timing)*
Kurbelwellenkröpfung *f* crankshaft throw
Kurbelwellenkupplung *f* shaft coupling
Kurbelwellenlager *n* crank (crankshaft, main) bearing
Kurbelwellenrad *n* crankshaft timing gear
Kurbelwellenschaft *m* crankshaft journal, body of the crankshaft
Kurbelwellenschwingungen *fpl* crankshaft vibrations
Kurbelwellenstumpf *m* tailshaft
Kurbelwellenverlängerung *f* crankshaft extension
Kurbelwellenzapfen *m* crank pin
Kurbelwiderstand *m* decade (lever-type) resistance box, [rotary] rheostat
Kurbelwinkel *m* crank angle
Kurbelzapfen *m* crank pin, wrist
Kurbelzapfenlager *n* crank-pin bearing
Kurbelzylinder *m* crank cylinder *(hydraulics)*
Kurierboot *n* duty signal boat
Kurkumapapier *n* curcuma paper
Kurre *f (ship)* beam trawl
Kurs *m* course; heading
anliegender K. heading [course]
beschickter K. real course
eingedrehter K. preselected heading
mißweisender K. magnetic course
rechtweisender K. true course
Kursabweichung *f* swerve, course deviation
Kursanzeige *f* / **automatische** automatic tracking
Kursanzeiger *m* course deviation
automatischer K. offset-course computer *(radar)*
Kursbeständigkeit *f (ship)* steering qualities
Kursbuch *n* time-table
Kürschner *m* furrier
Kursdreieck *n* set squares for chart work, course triangle
Kursfeuer *n* leading light
Kursführungsanlage *f* track guide
Kursgeber *m* course setting device
Kursivschrift *f* italics, oblique type
Kurskoppler *m* automatic navigator; dead reckoner
Kurslicht *n* range light
Kurslinie *f* course track
vorgewählte K. selected course track
Kursprojektion *f* track *(radar)*
Kursrose *f* directional gyro azimuth
Kursschreiber *m* course recorder, avigraph
Kurssteueranlage *f* / **automatische** autopilot, automatic pilot
Kurssteuerung *f* automatic flight control
astronomische K. celestial guidance
automatische K. automatic tracking
Kursverbesserung *f* azimuth correction
Kursversetzung *f* heading displacement
Kursvisier *n* course (drift) sight
Kurswagen *m* through carriage (coach), *(Am)* through car
Kurswähler *m* azimuth selector
Kurszeichen *n* course signal

Kurve *f* 1. curve, characteristic, graph, plot; trace; response; 2. *(build)* curve, bend; 3. *(aero)* turn
K. der flutbaren Längen *(ship)* floodable lengths curve
K. der spektralen Durchlässigkeit spectral transmission curve
K. des Muffenreglers sleeve-pressure curve
K. dritten Grades *(math)* cubic
K. gleicher Beleuchtungsstärke isophot curve
K. gleicher Lichtstärke isocandela curve
K. im Steigflug climbing turn
K. vierten Grades *(math)* quartic
adiabatische K. adiabatic curve
ausgezogene K. solid curve
ballistische K. ballistic curve
ebene K. plane curve
enge K. *(aero)* sharp turn
gestrichelte K. dotted-line curve
gezogene K. *(aero)* climbing turn
gut strakende K. *(ship)* kindly curve
ideale K. ideal curve
isodynamische K. isodynamic curve
isothermische K. isothermal curve
jungfräuliche K. initial magnetization curve
linksgewundene K. *(math)* sinistrorsum
logarithmische K. logarithmic curve
metazentrische K. *(ship)* curve of metacentric height
polytropische K. polytropic curve
rechtsgewundene K. *(math)* dextrorsum
Steinersche K. tricuspid
zyklische K. cycloidal curve
Kurven *fpl* / **Bonjeansche** *(ship)* Bonjean's curves
Kurvenabtaster *m* curve (graph) follower
Kurvenanalysator *m* curve analyzer
Kurvenausschlag *m* trace reflection
Kurvenbahnfräsen *n* track milling [of cams]
Kurvenband *n* curving conveyor
kurvenbetätigt *(mach)* cam-actuated
Kurvendarstellung *f* graphical representation
Kurvenelement *n* path element
Kurvenfestigkeit *f (aut)* cornering stability
Kurvenflug *m* / **steiler** steep-bank flight
Kurvenform *f* wave shape, waveform, curvature
K. des Reststromes residue curve
kurvenförmig 1. curved; 2. *(mach)* camlike
Kurvenfräseinrichtung *f* cam milling attachment
Kurvenfräsen *n* cam milling
Kurvenfräsmaschine *f* cam miller (cutting machine)
automatische K. automatic cam milling machine
Kurvenführung *f* curved guiding mechanism
kurvengängig curve-going *(conveyor)*
kurvengesteuert *(mach)* cam-controlled
Kurvengetriebe *n* crank
Kurvenhobeleinrichtung *f* cam planing attachment
Kurvenintegral *n* curvilinear integral
Kurvenknickungspunkt *m* point of discontinuity of a curve
Kurvenkrümmung *f* knee of the curve *(electron tubes)*
Kurvenlage *f* 1. *(aut)* cornering characteristics; 2. *(aero)* bank
Kurvenleser *m* graph (trace) reader
Kurvenlineal *n* irregular (French) curve
kurvenlos *(mach)* camless
Kurvenmesser *m* curvometer
Kurvenneigung *f* lateral sway
Kurvenrolle *f (mach)* cam roller
Kurvenschar *f* family of curves
Kurvenscheibe *f* cam plate (disk)
Kurvenschieber *m* sliding cam plate
Kurvenschiefe *f* skewness of curve
Kurvenschreiber *m* curve (graph) plotter, curve tracer, plotter
Kurvenschubgetriebe *n* mechanism for movement in a curve
Kurvenspitze *f* peak of the curve
Kurvensteilheit *f* peakedness
Kurvenstück *n* element of curve
Kurventrommel *f* drum (cylindrical, barrel) cam
kurvenverzahnt curved-tooth
Kurvenzug *m* combined curve
Kurvenzweig *m* element of [a] curve
kurz 1. short *(length, time)*; 2. accelerated *(test time)*; 3. abridged, concise *(book)*
Kurzanalyse *f* proximate analysis
Kurzanschrift *f* cable address
Kurzarbeit *f* short time working
Kurzbalken *m (ship)* half beam
Kurzbespulung *f* short-section loading
Kurzbewitterung *f* accelerated weathering
kurzbrennweitig of short focal distance, short-focus
kürzen *(math)* to cancel *(a number out of a fraction)*
Kürzen *n* 1. cancelling *(fraction)*, reduction to the simplest form; 2. shortening *(length)*
kurzfaserig short-fibre, short staple[d]
Kurzfaserzellstoff *m* short-fibred pulp
Kurzfassung *f* abridgement
Kurzformverschlüsselung *f* alphabetic coding
kurzgeschlossen *(el)* short-circuited
Kurzgewinde *n* short thread
Kurzgewindefräsen *n* milling of short threads, short-thread milling, multiple-cutter method of thread milling
Kurzgewindefräsmaschine *f* short-thread milling machine, multiple-cutter type of thread milling machine
Kurzhobel *m (mine)* light-duty plough
Kurzhobelmaschine *f* shaping machine

kurzhobeln to shape
Kurzhobeln *n* **mit Hartmetallmeißel** carbide shaping
Kurzhobler *m* shaping machine, shaper
Kurzholzstauung *f* short stowage
Kurzhubhonen *n* superfinishing
kurzhubig short-stroke
Kurzhubschaufellader *m* short-swing shovel-loader
kurzlebig short-lived
Kurzprüfung *f* accelerated test
kurzschließen *(el)* to short circuit
gegen Erde k. to short to earth
Kurzschließen *n* *(el)* short-circuiting, shorting
Kurzschließer *m* *(el)* shorting bar
Kurzschluß *m* short circuit
K. zwischen Phasen line-to-line fault
äußerer K. zero burden *(of instrument transformer)*
direkter K. dead short-circuit
ideeller K. virtual short-circuit
satter K. saturated (dead) short-circuit
ununterbrochener K. continuous short-circuit
vollständiger K. dead short-circuit
Kurzschlußanker *m* short-circuit[ed] armature, squirrel-cage armature
Kurzschlußart *f* nature of short-circuit
Kurzschlußbremsung *f* dynamic braking
Kurzschlußbrücke *f* shorting bridge (jumper)
Kurzschlußbügel *m* shorting plug; jumper wire, U-link *(telephony)*
Kurzschlußdauer *f* short-circuit time (period, duration)
Kurzschlußdrossel *f* current-limiting reactor (inductor)
Kurzschlußersatzschaltbild *n* constant current equivalent circuit
Kurzschlußersatzschaltung *f* constant current equivalent circuit
kurzschlußfest short-circuit proof
Kurzschlußfestigkeit *f* short-circuit strength
Kurzschlußfortschaltung *f* automatic short-circuit reclosing
automatische K. automatic reclosing
Kurzschlußfrequenz *f* short-circuit frequency
Kurzschlußgenerator *m* short-circuit current generator
Kurzschlußglied *n* short-circuit component
Kurzschlußimpedanz *f* closed-end impedance
Kurzschlußkäfig *m* squirrel-cage winding
Kurzschlußkennlinie *f* short-circuit characteristic
Kurzschlußkolben *m* piston, plunger *(waveguides)*
kapazitiver K. choke piston *(waveguides)*
Kurzschlußkraft *f* [dynamic] short-circuit force
Kurzschlußlasche *f* shorting link
Kurzschlußläufer *m* short-circuited rotor, squirrel-cage rotor
Kurzschlußläuferinduktionsmotor *m* squirrel-cage induction motor
Kurzschlußläufermotor *m* squirrel-cage motor
Kurzschlußleistung *f* short-circuiting power; short-circuiting capacity
Kurzschlußleitwert *m* short-circuit conductance
Kurzschlußlichtbogen *m* short-circuit arc
Kurzschlußring *m* shading ring
Kurzschlußschieber *m* short-circuit plunger (piston) *(waveguides)*
Kurzschlußschleife *f* short-circuited loop
Kurzschlußschutz *m* short-circuit protection
kurzschlußsicher short-circuit proof
Kurzschlußspannung *f* short-circuit voltage
Kurzschlußstecker *m* bridging (short-circuiting) plug
Kurzschlußstichleitung *f* closed stub *(aerial)*
Kurzschlußstrom *m* short-circuit current
K. der Batterie flash current
maximaler K. peak cathode fault current
thermischer K. *(Am)* thermal current rating
Kurzschlußstromimpuls *m* / **maximaler einmaliger** maximum surge current
Kurzschlußtaste *f* short-circuit key
Kurzschlußverhältnis *n* **im Leerlauf** short-circuit ratio
Kurzschlußversuch *m* short-circuit test
Kurzschlußwechselstrom *m* / **transitorischer** transient short-circuit current
Kurzschlußwicklung *f* short-circuited winding, squirrel-cage winding
Kurzschlußwiderstand *m* short-circuit resistance; closed-end impedance
Kurzschlußwindung *f* shorted turn
Kurzsignal *n* short-code signal *(radio)*
Kurzspan *m* broken (discontinuous, short) chip *(metal cutting)*
kurzspanend short-chip *(material)*
Kurzspinnverfahren *n* abbreviated spinning process
kurzstapelig short staple[d]
Kurzstreckenrakete *f* short-range rocket
Kurzstreckentransport *m* short-haul transport
Kurzstrohsieb *n* cavings riddle
Kurztagspflanze *f* winter (short day) flowering plant
Kurztagsweide *f* strip-grazed pasture where less than the half daily ration of pasturage is provided for the herd
Kurztauchverfahren *n* brief dipping process
Kurztemperung *f* *(met)* short annealing
Kürzung *f* 1. *(math)* cancellation; 2. shortening *(tensile test)*
Kurzunterbrechung *f* *(el)* automatic reclosing
Kurzwaren *fpl* small wares
Kurzwelle *f* short wave

Kurzwellenamateur *m* ham; short-wave amateur
Kurzwellenantenne *f* short-wave aerial
Kurzwellenband *n* high-frequency band
Kurzwellenbereich *m* short-wave range, high-frequency range
Kurzwellenempfang *m* short-wave reception
Kurzwellenempfänger *m* short-wave receiver
Kurzwellenradaralarmsystem *n* microwave early warning
Kurzwellensender *m* short-wave transmitter
Kurzwellenübertragung *f* short-wave transmission
Kurzwellenvorsatz *m* short-wave adapter (converter)
Kurzzeichen *n* abbreviation
Kurzzeitbestrahlung *f* acute exposure (irradiation), short irradiation
Kurzzeitbetrieb *m* short-time duty
Kurzzeitenenergieverbrauch *m* / **höchster** maximum demand
kurzzeitig short-time, short-term
Kurzzeitmesser *m* chronoscope
Kurzzeitpeilung *f* transient bearing
Kurzzeitstandfestigkeit *f* short-time creep strength
Kurzzeitstörung *f* **durch Sonneneinwirkung** impulsive solar noise
Kurzzeitstrom *m* short-time current
Kurzzeittechnik *f* short-time technique
Kurzzeitverhalten *n* short-time behaviour
Kurzzeitversuch *m* rapid (accelerated, short-time) test
Kurzzeitzünder *m* short-delay electric detonator
Küste *f* shore, coast
  felsige K. iron bound coast
Küstenbauten *mpl* coastal structures (works)
Küstenbauwerk *n* coastal structure
Küstenbefestigung *f* coastwise (coastal) fortification
Küstenbefeuerung *f* coast lighting
Küstenbrechung *f* coastal refraction *(radar)*
Küstendampfer *m* coast steamer
Küstendienst *m* inter-coastal service
Küstenecho *n* bench mark *(radar)*
Küsteneffekt *m* coastal deviation *(radar)*
küsteneinwärts inshore
Küstenfahrer *m (ship)* coaster
Küstenfahrt *f* coastwise trade; intercoastal navigation, coasting, cabotage
  große K. short-sea trade
  kleine K. small coasting trade
Küstenfeuer *npl* coastal lights
Küstenfischerei *f* inshore fisheries; coastal fishing
Küstenfrachter *m* short-sea coaster; coastal cargo-liner
Küstenfunkstelle *f* coastal radio station
Küstengewässer *n* coastal waters
Küstenkabel *n* shallow-water cable, shore-end cable
Küstenkreuzer *m* coastwise cruiser *(yacht)*
Küstenlastkahn *m* coastal barge
Küstenleuchtfeuer *n* coast light
Küstenlotsenboot *n* coast pilot boat
Küstenmeer *n* territorial waters
Küstenmotorschiff *n* motor-coaster
Küstennavigation *f* coast piloting
Küstennebel *m* sea fog
Küstenölbohrinsel *f* off-shore oil rig
  schwimmende K. off-shore drilling barge
Küstenpackeis *n* coastal pack
Küstenpeilstelle *f* coastal direction-finding station *(radar)*
Küstenradar *n* shore-based radar
Küstenrettungsboot *n* coastal lifeboat
Küstenrettungsdienst *m* lifeguard (life-boat) service
Küstensaum *m* back shore
Küstenschiff *n* coaster, coastal (coasting) vessel
Küstenschiffahrt *f* coastal navigation, coasting trade
Küstenschutz *m (build)* coastal engineering; coast protection
Küstenschutzbauten *mpl* coastal (coast protection) works
Küstenwache *f (Am)* coast guard
Kustode *m (print)* catchword
Kutter *m* cutter *(fishing boat)*
Kuttertakelung *f* cutter rig
Kuvert *n* envelope
Küvette *f* cuvette, [glass] cell
Küvettenraum *m* cell compartment
kyanisieren to kyanize *(wood)*
Kybernetik *f* cybernetics
  technische K. engineering cybernetics
Kymograf *m* kymograph

L

L-Antenne *f* L[-shaped] aerial
L-Dock *n* off-shore dock
L-Filter *n* L-section filter *(radio)*
L-Kettenglied *n (el)* mid-shunt termination
L-Kondensator *m* lacquer film capacitor
L-Profil *n* angle [section] *(steel)*
  gleichschenkliges L. equal [leg] angle
  ungleichschenkliges L. unequal [leg] angle
L-Schaltung *f* L-network
L-Schirm *m* L-scope *(radar)*
L-Winkelmesser *n (ag)* L (side) hoe
Labferment *n* rennin
labil instable, labile
Labilität *f* instability, lability
Labilitätszustand *m* state of instability (lability)
Labkasein *n* rennet casein

Labor *n* lab[oratory]
heißes L. *(nucl)* hot laboratory
radiochemisches L. radiochemical laboratory, tracerlab
warmes L. *(nucl)* warm (semi-hot) laboratory
Laborant *m* assistent chemist, laboratory worker
Laboratorium *n* laboratory (*s. a.* Labor)
Laboratoriumsmaßstab *m* laboratory size
Laboratoriumsreaktor *m* laboratory reactor
Laboratoriumstisch *m* bench
Laboratoriumsversuch *m* laboratory test
Laboraufbau *m (el)* setup
Laborinterferometer *n* laboratory interferometer
Laborkoordinatensystem *n s.* Laborsystem
Labormaßstab *m* laboratory scale
Laborprüfung *f* laboratory testing
Laborreaktor *m* laboratory reactor
Laborschnüre *fpl (el)* wander leads
Laborsystem *n (nucl)* laboratory system
Laborübung *f* laboratory exercise
Laborversuch *m* laboratory test
Labyrinthbüchse *f* labyrinth box (ring)
Labyrinthdichtung *f* labyrinth packing (seal)
Labyrinthkolben *m* labyrinth piston (packing)
Labyrinthring *m* labyrinth washer
Lachapellekessel *m* vertical cross tube boiler
Lachgas *n* laughing gas, nitrous oxide
Lachsfänger *m* salmon boat
Lachston *m* salmon shade
Lack *m* lacquer, varnish; lac; japan
hitzebeständiger L. heat-proof lacquer
lichtempfindlicher L. photosensitive lacquer
magerer L. low-grade varnish
säurefester L. acid-proof varnish
Lackanstrich *m* varnish coat
Lackband *n* varnished tape
Lackbaumwolldraht *m* cotton-covered enamelled wire
Lackbaumwollkabel *n* varnished cambric cable
Lackbenzin *n* white spirit, mineral spirits
Lackdraht *m* enamel[led] wire, enamel-insulated wire
lötfähiger L. self-fluxing enamelled wire
Lackdruck *m (text)* lacquer print
Lackentferner *m* varnish-remover
Lackfarbe *f* varnish colour
Lackfirnis *m* shellac varnish
Lackgewebe *n* varnished fabric
Lackharz *n* resin for paints and varnishes, coating (paint) resin
lackieren to lacquer, to varnish; to japan *(copal oil varnish, stoved)*
Lackierfabrik *f* japanning works
lackierfähig varnishable
Lackiermaschine *f* varnishing machine
Lackiertrommel *f* tumbler, tumbling barrel
Lackierung *f* 1. varnishing *(against current leakage)*; finishing; japanning *(copal-oil varnish, stoved)*; 2. coat of varnish (lacquer)
Lackkabel *n* enamelled-core cable
Lackleder *n* patent leather *(for bags and shoes)*, enamelled (japanned) leather
Lacklösemittel *n* lacquer solvent
Lacklösungsmittel *n s.* Lacklösemittel
Lackmus *n* litmus
Lackmuspapier *n* litmus paper
Lackpapier *n* varnished (impregnated) paper
Lackpapierdraht *m* varnished paper-insulated wire, paper-insulated enamelled wire
Lackpapierkabel *n* varnished-paper cable
Lackpflegemittel *n* car polish
Lackrohstoff *m* raw material for varnishes
Lackschicht *f* layer of varnish
Lackschlauch *m (el)* flexible varnished tubing, varnished sleeving (hose)
Lackvachetten *pl (leath)* enamelled hides
Lackverdünner *m* varnish thinner
Lade *f* 1. box, chest; drawer; 2. *(text)* batten, lay, lathe, loom slay
freifallende L. free falling lathe (slay)
Ladeaggregat *n* battery charger, charging set
Ladearbeit *f* 1. loading; 2. mucking *(stone)*
Ladeband *n* loading conveyor
Ladebaum *m* [cargo] derrick; *(Am)* cargo boom
Ladebaumhanger *m* derrick span
Ladebaumpfosten *m* Samson post
Ladebereitschaftsnotiz *f* notice of readiness
Ladebrücke *f (ship)* loading bridge (gangway)
Ladebuch *n (ship)* cargo-book
Ladebühne *f* loading dock (platform)
Ladedauer *f* time of charge *(batteries)*
Ladedienst *m* goods handling service
Ladedruck *m* supercharger (boost) pressure, boost
Ladedruckkurve *f* boost curve
Ladedruckregler *m* manifold-pressure control
Ladeeinrichtung *f* 1. *(mach)* loading device (mechanism); 2. *(el)* charging equipment
Ladeeinrichtungen *fpl (ship)* cargo-handling appliances
Ladefähigkeit *f* loading (net, carrying, cargo-carrying, deadweight loading) capacity, useful deadweight, burden
L. für Raumfracht *(ship)* measurement capacity
Ladefaktor *m (el)* charging factor
Ladefläche *f* loading area
Ladeflügel *m (mine)* flight
Ladeförderer *m* stage loader
Ladefrist *f* time for loading; loading days; period allowed for loading
Ladegebühr *f* loading charges (expenses)
Ladegenerator *m (el)* charging generator
Ladegerät *n* 1. loading-tackle; *(ag)* loader; 2. *(el)* [accumulator] charging unit, [accumulator] charging device, charger

L. **für Pufferbetrieb** *(el)* float charger
ladegerecht of manageable size *(debris)*
Ladegeschirr *n (ship)* cargo [handling] gear, discharging (loading) gear
L. für gekuppelte Ladebäume spar and boom gear
Ladegewicht *n* normal load
Ladegleichrichter *m* charging rectifier
Ladehafen *m* port of loading
Ladehemmung *f (mil)* jam, stoppage
Ladehub *m* intake stroke
Ladekai *m* wharf; dock
Ladekanal *m* charging tube *(reactor)*
Ladekapazität *f (el)* charging capacity
Ladekasten *m* 1. *(mine)* tub-charging hopper; 2. *(mach) (Am)* tote box
Ladekennlinie *f (el)* charging characteristic
Ladeklappe *f (aut)* tail board
Ladekoeffizient *m* charge coefficient, coefficient of charge
Ladekondensator *m (el)* charging capacitor; filter capacitor
L. für Zeitsteuerung timing capacitor
Ladekopf *m (ag)* pick-up head
Ladekran *m* cargo crane
Ladekühlraum *m* refrigerated cargo space, insulated hold
Ladeleitrad *n (aut)* diffuser
Ladelinie *f* load line, Plimsoll mark
Ladeluke *f* cargo hatch[way], loading hatch
große L. main hatchway
Lademarke *f (ship)* load-line disk, load line mark
Lademaschine *f* charging machine
Lademaß *n* loading gauge
Lademaßüberschreitung *f* out of [loading] gauge
Lademeister *m* loading foreman
Lademittel *n* loading-tackle
laden 1. to load *(machine, vehicle)*; 2. *(mine)* to load; to fill *(coal)*; to muck *(debris)*, to shovel; to charge *(bore hole)*; 3. *(el)* to charge; to activate; 4. *(nucl)* to charge *(reactor)*; 5. *(mil)* to load *(rifle)*
eine Batterie l. to charge a battery
Ladenanschlag *m (text)* beating-up *(lay)*
L. bei offenem Fach blow of the lay with open shed
Ladenantrieb *m* lathe driving
Ladenarm *m* lay sword
Ladenbaum *m* lay check
Ladenbewegung *f* lay movement
Ladendeckel *m* lay cap (top)
Ladenfuß *m* lathe (lay) sword
Ladenhub *m* stroke of lathe
Ladenklotz *m* lay check
Ladenrückgang *m* backward movement of the lay
Ladenschere *f* crank arm, lay
Ladenschlag *m* battening
Ladenschwingung *f* oscillation of the lay
Ladenweg *m* path of lay
Ladenzapfen *m* batten pin
Ladenzugstange *f* crank arm, lay
Ladeöl *n (ship)* oil in bulk, bulk (cargo) oil
Ladeperiode *f (aut)* charging period
Ladepforte *f (ship)* bow-port, raft-port, cargo port (door)
Ladepfosten *m (ship)* Samson (king) post; *(Am)* derrick post
Ladepfostenverbindungsbrücke *f* king bridge
Ladeplan *m (ship)* cargo (stowage, capacity) plan
Ladeplatte *f* stillage
Ladeplatz *m (ship)* loading berth; *(Am)* loading point; *(build)* loading bay
Ladepritsche *f* loading platform; *(ship)* board (platform) sling, loading scow
hydraulisch kippbare L. loading platform with hydraulic tipping control
Ladepumpe *f (aut)* charge pump
Ladepunkt *m (mine)* loading station
Lader *m* 1. loader *(machine, vehicle)*; 2. *(aut, aero)* supercharger
L. der Verdrängerbauart positive-displacement supercharger
obergriffiger L. *(mine)* loader attacking from above
seitengriffiger L. *(mine)* loader attacking from the side
vordergriffiger L. *(mine)* loader attacking from the front
Laderampe *f (ry)* loading platform
Laderaum *m* 1. [cargo] hold *(ship)*; 2. hopper *(dredger)*
L. mit unten schräggezogenen Schotten hoppered cargo hold
hinterer L. after hold
isolierter L. insulated hold
vorderer L. fore hold
Laderauminhalt *m* cargo (hold) capacity
L. für Ballen bale cubic capacity
Laderaumschott *n* hold bulkhead
Laderaumseitenschott *n* wing bulkhead
Laderäumte *f (ship)* cargo space
Laderaumwegerung *f* hold ceiling
Laderaupe *f* crawler-mounted loader
Laderohr *n* charging tube *(reactor)*
Laderutsche *f* loading-chute
Ladeschaffner *m (ry)* guard, member of train crew, *(Am)* conductor
Ladeschalter *m* battery (cell) switch
Ladeschalttafel *f* charger switchboard, battery charging panel
Ladeschaufel *f* bucket; front-ender loader, tractor shovel
Ladeschicht *f (mine)* filling shift
Ladeschlitten *m (ship)* skid
Ladeschurre *f* loading-chute

Ladeschütte *f (ship)* cargo chute
Ladeschütze *m (mil)* loader
Ladeseil *n (ship)* cargo whip
Ladestation *f (mach)* loading station *(transfer line)*
Ladestelle *f* loading station; *(Am)* loading point
Ladestock *m (mine)* beater, tamping bar (rod, stick)
Ladestraße *f* yard for loading and unloading
Ladestrecke *f (mine)* loader gate
Ladestreifen *m (mil)* strip, magazine (cartridge) clip
Ladestrom *m* charging current
Ladestromkontrolleuchte *f* charge control lamp
Ladestromkreis *m* charging circuit
Ladestromstärke *f* charge-rate, rate of charge
Ladestropp *m (ship)* cargo sling; *(Am)* strap
Ladetakelung *f (ship)* cargo rig
Ladetiefgang *m (ship)* load draught
Ladetonnage *f* deadweight tonnage
Ladeumformer *m (el)* charging converter
Lade- und Entladeeinrichtung *f (mach)* loading and unloading device
Lade- und Löschausrüstung *f (ship)* cargo-handling equipment
Lade- und Löscheinrichtung *f (ship)* cargo-handling appliances
Ladeverdrängung *f (ship)* [full] load displacement
Ladevermögen *n* cargo (deadweight carrying) capacity
Ladewasserlinie *f (ship)* load[-water]-line
Ladewiderstand *m (el)* charging resistance
Ladewinde *f* heavy-duty type hoist, loading [and unloading] winch; cargo winch
Ladezeitkonstante *f (el)* charging time constant
Ladezustandprüfer *m* charge tester
Ladung *f* 1. load *(contents of a wagon)*; 2. charge *(battery, reactor)*; 3. loading; 4. charging
  L. bei konstantem Strom constant-current charge
  L. bei konstanter Spannung constant-voltage charge
  L. eines Kondensators charge of a capacitor
  L. in Behältern containerized cargo
  L. in Säcken bagged cargo
  außergewöhnliche L. special loading
  bewegliche L. loose cargo
  effektive L. effective charge
  elektrische L. electric charge
  feuergefährliche L. inflammable cargo
  gebundene L. bound charge
  gemischte L. general cargo; *(Am)* mixed cargo
  geschichtete L. stratified charge
  gespeicherte L. stored charge
  gestaffelte L. *(mine)* non-continuous charge *(borehole)*
  induzierte elektrische L. induced electricity
  lebende L. live cargo (stock)
  leichte L. light cargo
  lose L. loose cargo
  nichtbrennbare L. incombustibles
  schwere L. heavy cargo
  spezifische L. specific charge
  spezifische L. des Elektrons electron-charge mass ratio
  verderbliche L. perishable cargo
  zurückbleibende L. *(el)* retained charge
Ladungen *fpl* / **gleichnamige** similar charges
  ungleichnamige L. unsimilar charges
Ladungsanschreiber *m (ship)* tallyman
Ladungsanzeige *f* charge indicator
Ladungsaustausch *m* charge exchange (transfer)
Ladungsaustauscherscheinung *f* charge-exchange phenomenon
Ladungsbehälter *m* [cargo] container
  flexibler L. flexible cargo container
Ladungsbild *n* charge image (pattern)
  elektrisches L. electrical image
Ladungsdeck *n* cargo deck
Ladungsdichte *f (el)* charge density
Ladungseinheit *f (el)* unit charge
  L. im elektrostatischen System unit quantity of electricity
  elektrostatische L. unit quantity of electricity
Ladungsfähigkeit *f* cargo capacity
Ladungshafen *m* port of loading (lading, shipment), lading port
Ladungskontrolleur *m (ship)* tallyman
Ladungsliste *f* loading-list
Ladungsnetz *n* cargo net
Ladungsspeicherröhre *f* charge-storage tube
Ladungsspin *m* charge spin
Ladungsträger *m* charge[d particle] carrier
Ladungsträgerdiffusion *f* charge carrier diffusion
Ladungsträgerinjektion *f* charge carrier injection
Ladungstragfähigkeit *f (ship)* paying deadweight
Ladungstransportband *n* charging belt *(of Van de Graaff generator)*
Ladungsübertragung *f* charge transfer
Ladungsverlust *m (ship)* spillage
Ladungsverteilung *f* distribution of charge
Ladungsverzeichnis *n* loading-list
Lafette *f* [gun-]carriage, gun mount[ing]
Lage *f* 1. position; location; 2. *(mine)* stratum, layer; 3. *(build)* course *(bricks)*; layer; ply
  L. des Hohlraumes location of cavity
  L. des Toleranzfeldes [zur Nullinie] disposition of the tolerance zone
  rechtwinklige L. squareness
  symmetrische L. symmetrical position
Lage-A-D-Umsetzer *m (dat)* spatial encoder
Lagebestimmung *f* localization
Lageeinstellung *f (mach)* positioning
  selbsttätige L. automatic positioning

Lageeinstellungsgenauigkeit *f* accuracy of positioning
Lageenergie *f* potential energy
Lagefehler *m (dat)* position error coefficient
Lagegenauigkeit *f* **der Bohrung** hole-location positioning accuracy, accuracy in hole positioning, accurate hole location
Lagegleichung *f* equation of position
Lagekreisel *m (cont)* displacement gyro
lagemäßig *(math)* topological
Lagemißweisung *f* site error
Lagen *fpl* sections
  L. / auf den *(mine)* end on, on end, on the face slips
  L. sitzend / auf den *(mine)* working on the face slip[s]
  L. / unter den *(mine)* on the back slips
  L. sitzend / unter den *(mine)* working on the back slip[s]
  aufgesteckte L. *(mine)* backward-hading slips
Lagenfalzung *f* quire folding
Lagenhöhe *f* elevation head
Lagensicherung *f* securing constant relative location; locating
Lagenwicklung *f (el)* layer winding
Lageplan *m* layout [plan], site plan
Lager *n* 1. store *(goods)*, warehouse, storehouse; stock; assortment; depot; dump; yard; 2. *(mach)* bearing; 3. *(mil)* camp; 4. *(geol)* deposit *(ore etc)*; seam, bed; vein
  L. / vom from stock, ex stock
  L. mit Belastung in Richtung der Längsachse thrust bearing
  L. mit Belastung senkrecht zur Längsachse journal bearing
  ausgelaufenes L. worn-out bearing
  doppelreihiges L. double-row bearing
  dreiteiliges L. three-part bearing
  geteiltes L. split bearing
  nach außen geneigtes L. outwardly inclined bearing
  nach innen geneigtes L. inwardly inclined bearing
  vierteiliges L. four-part bearing
  wanderndes L. mobile storage
  zweiteiliges L. two-piece bearing, two-part bearing
Lagerabdichtung *f* bearing sealing
Lagerarbeiter *m* storeman
Lageraufnahme *f* stock taking
Lagerausguß *m* babbitting; lining of the bearing, bearing lining
Lagerbalken *m* 1. *(mach, therm)* bearing frame; 2. *(text)* carrier bar *(loom)*
Lagerbalkenkopf *m* crank-shaft bearing-end of frame, front end of the frame
Lagerbehälter *m* storage bin
Lagerbelastung *f* bearing load
Lagerbelüftungsanlage *f* bulk ventilation grain drier
Lagerbestand *m* stock
Lagerbeständigkeit *f* storage stability
Lagerbier *n* lager beer
Lagerblock *m* 1. *(build)* bearing block; 2. *(ship)* plummer block
Lagerbock *m (mach)* pillow block
Lagerbohrung *f* bearing bore
Lagerbolzen *m* cap (bearing) bolt
Lagerbronze *f* copper bearing metal, gun metal
Lagerbüchse *f* bearing bush
Lagerbund *m* bearing collar
Lagerdeckel *m* bearing cap (cover)
Lagerdeckelplatte *f* flat bearing cap
Lagerdeckelschraube *f* bearing (cap) bolt
Lagerdruck *m* bearing pressure
Lageregel *f (math)* position rule
Lageregler *m* positioning controller
Lagerfähigkeit *f* storage stability (property); storage life *(battery)*
Lagerfähigkeitsprüfung *f* storage test
  beschleunigte L. accelerated storage test
Lagerfläche *f* 1. *(mach)* bearing surface; 2. *(build)* storage area, stacking ground; bed *(masonry)*
Lagerfrucht *f* laid (lodged) crop
Lagerfruchthaspel *f* pick-up reel
Lagerfuge *f* bed (horizontal) joint *(masonry)*
Lagerfuß *m* pedestal base, foot of bearing
Lagergang *m (geol)* bed (fissure) vein, sill
Lagergebühr *f* warehousing (storage) charges, warehouse rent (cost)
Lagergehäuse *n* bearing housing
Lagergeld *n s.* Lagergebühr
Lagergestell *n* stillage
Lagerhalle *f* warehouse
Lagerhals *m* **der Kurbelwelle** journal of the crank shaft
Lagerhaltung *f* stock keeping; inventory control
Lagerhaus *n* warehouse, store[house], storage shed
  schwimmendes L. *(Am)* wharfboat
Lagerhof *m* storage (stacking) yard
Lagerkäfig *m* bearing cage
Lagerkahn *m* store boat (ship)
Lagerkohle *f* stored coal
Lagerkörper *m* pedestal body
Lagerkosten *pl* storage expenses
Lagerlegierung *f* bearing (antifriction) alloy
Lagerluft *f* bearing clearance
Lagermeister *m* stores superintendent
Lagermetall *n* bearing (antifriction) metal
Lagermontage *f* bearing mounting
lagern *(ag)* to lodge; to store; *(mach)* to carry
Lagerplatz *m* storage area, stacking ground, storage (stacking) yard; dumping ground (place)
Lagerraum *m* depot, storage facility (space)

Lagerreibung *f* bearing friction
Lagerring *m* roller ring
Lagerrumpf *m* pedestal body
Lagerschale *f* bearing shell (bush)
  L. mit Schmiernut brasses with lubricating grooves
Lagerschalen *fpl* / **konzentrische** concentric bushes
Lagerschalenausguß *m* lining of the bearing
Lagerschalenschlüssel *m* hocked key for brasses
Lagerschild *m* end shield (bell), end-frame
Lagerschmieröl *n* bearing lubrication oil
Lagerschmierung *f* bearing lubrication
Lagerschuppen *m* storage shed
Lagersilo *m* storage silo
Lagerspiel *n* bearing clearance
Lagerstätte *f (geol)* deposit
  L. der magmatischen Folge deposit of primary magmatic origin, primary deposit
  L. der metamorphen Folge secondary deposit
  geschichtete L. seam
  magmatogene L. deposit of primary magmatic origin
  metasomatische L. replacement deposit
  schichtförmige L. bed
Lagerstättenerkundung *f* prospecting; examination of deposits; locating deposits
Lagerstättenteil *m* / **sehr reicher** bonanza
Lagerstein *m* jewel *(watch)*
Lagerstelle *f* point of support
Lagertank *m* storage tank
Lagerträger *m* support for [the] bearing
Lagerückführung *f* position feedback
Lagerung *f* 1. [holding in] storage, stocking, stock-piling, warehousing; 2. support; bearing; seating; bedding; mounting
  L. der Abfälle hold-up of the wastes
  L. der Drehspule moving-coil support
  L. des Hebels auf Schneiden knife-edge suspension of the lever
  L. für hin- und hergehende Bewegung *(mach)* flat (sliding) bearing
  L. in Wälzlagern anti-friction mounting
  L. mittels Tragkranz suspension by means of a segment of angle iron riveted on shell crown
  elastische L. elastic bedding (support)
Lagerungsschacht *m* storage well
Lagerungstest *m* storage test
Lagerunterteil *n* lower portion of [the] bearing
  geteiltes L. split lower portion of [the] bearing
  ungeteiltes L. solid lower portion of [the] bearing
Lagerverwalter *m* stock-keeper
Lagervorrat *m* stock [on hand], storage
Lagerweißmetall *n* babbit metal
Lagerwerkstoff *m* bearing material
Lageschwankung *f* / **senkrechte** *(el)* vertical hunting
Lagesteuerung *f (mach)* position control, positioning
Lageumsetzer *m s.* Lageverschlüßler
lageunabhängig position-independent
Lageverschlüßler *m* shaft position-to-digital converter; spatial encoder
lahmlegen to paralyze, to obstruct
Lahnfaden *m (el)* patchcord wire
Laibung *f* / **schräge** splayed jamp
Laichzeit *f* breeding season *(fish)*
Lakritze *f* licorice
Laktam *n* lactam
Laktat *n* lactate
Laktid *n* lactide
Lakton *n* lactone
Laktose *f* lactose
Lamawolle *f* lama wool
Lambda/4-Antenne *f* quarter-wave aerial, *(Am)* Marconi antenna
Lambda/4-Transformator *m* quarter-wave transformer
Lambda/2-Umwegleitung *f* half-wave balun *(aerials)*
Lambertfläche *f* uniform diffuser
Lamé *m* tinsel fabric
Laméfaden *m* metallized thread
lamellar lamellar
Lamelle *f* lamina
Lamellenbremse *f* multidisk brake
lamellenförmig laminated
Lamellenkühler *m* cellular radiator
Lamellenkupplung *f* multiple (friction) disk clutch
Lamellenmagnet *m* laminated (compound) magnet
Lamellenrohr *n* gilled (finned) tube, gilled (finned) pipe
Lamellensicherung *f* bridge (strip) fuse
Lamellenspannung *f (Am)* bar voltage
Lamellenstempel *m (mine)* lamellar prop
Lamellenteilung *f (el)* segment pitch
lamelliert laminated
laminar laminar
Laminarbewegung *f* laminar flow
Laminarströmung *f* laminar (smooth) flow
laminieren to laminate
  mit Schaumstoff l. to laminate with foam back
Laminieren *n* lamination
  L. durch Aufnähen lamination by quilting
  L. von Polyesterschaumstoff mit Maschenware durch Aufkleben adhesive lamination of polyester foam on knitted goods
  L. von Schaumstoff durch Aufkleben adhesive lamination of foam
Laminiermaschine *f* backing machine

L. für Waffelschaumstoff waffle foam backing machine
laminiert laminated
durch Wärmeschmelzverfahren l. heat-laminated
Lamm *n* lamb
Lammfell *n* / **geschorenes** shearling
Lammwolle *f* lamb's wool
La-Mont-Kessel *m* La Mont boiler
Lamor-Präzession *f* Lamor precession
Lampe *f* lamp; light bulb; lighting fixture (fitting), light fixture (fitting)
L. aus blendfreiem Glas anti-glare lamp
L. mit Edison-Sockel screw-type lamp
außenmattierte L. outside frosted lamp
durchgebrannte L. burnt-out bulb
freistrahlende L. unobscured lamp
gasgefüllte L. gas-filled bulb
innenmattierte L. inside frosted lamp
mattierte L. pearl (frosted) lamp, pearl (frosted) bulb
stoßfeste L. rough-service lamp
verspiegelte L. reflector (metallized) lamp, reflector (metallized) bulb
Lampenarm *m* lamp bracket
Lampenfassung *f* lampholder
Lampenfeld *n* *(telec)* signal-lamp panel
L. für Rufanzeige display panel
Lampenglocke *f* lamp globe
Lampenhaken *m* ceiling hook
Lampenhalter *m* lamp bracket
Lampenkolben *m* [lamp] bulb
Lampenlast *f* lighting load
Lampenmast *m* lamp standard
Lampennormal *n* secondary luminous standard, standard lamp
Lampenpaneel *n* lamp panel
Lampenputzraum *m* oil house, lamp room
Lampenruß *m* lamp-black
Lampenschaltung *f* light connection
Lampensignal *n* [electric] lamp signal
Lampensockel *m* lamp cap (base)
L. mit Edisongewinde standard screw socket
Lampenstock *m* stem
Lampenstromkreis *m* lamp circuit
Lampenstube *f* *(mine)* lamp room
Lampenwärter *m* lampman
Lampenwiderstand *m* lamp resistor; lamp resistance
Lancashire-Kessel *m* Lancashire (double flue) boiler
Land *n* 1. land; shore; 2. country [side] *(compared with town)*; 3. *(ag)* ground, soil, land; 4. country, territory, state
brachliegendes L. fallow land (soil)
fruchtbares L. fertile land (soil)
unfruchtbares L. infertile land (soil)
Landabsatz *m* *(mine)* landsale, local sale
Landanschluß *m* *(ship)* shore connection
Landarbeiter *m* agricultural worker, farmworker
Landaufnahme *f* land survey
tachymetrische L. tacheometric land survey
Landbau *m* farming, agriculture
Landbautechnik *f* agricultural technics
Landebahn *f* landing strip, runway
befeuerte L. flare path
Landebahnanflug *m* runway approach
Landebahnbefeuerung *f* contact lights
Landebahnfeuer *n* contact light[s]
Landebake *f* landing [radio] beacon
Landebremshaken *m* arresting hook
Landedeck *n* *(ship)* landing deck (space)
Landefackel *f* wing tip flare
Landé-Faktor *m* [Landé] g-factor, interval factor
Landeflugzustand *m* landing attitude
Landegeschwindigkeit *f* landing speed
Landehaken *m* *(mil)* arresting hook
Landehilfen *fpl* trailing-edge flaps
Landekurs *m* localizer course
Landekurssender *m* localizer
L. mit Phasenvergleich phase comparison localizer
Landeleitstrahl *m* / **dreidimensionaler** spatial landing path
Landelicht *n* approach (landing) light
Landemeldung *f* arrival message
landen 1. *(aero)* to land, to make a landing; 2. *(ship)* to land; to disembark
Landenge *f* *(geol)* neck; isthmus
Landescheinwerfer *m* board landing light
Landesimulator *m* landing simulator
Landeskoordinaten *fpl* cordinates of the state system
Landestreifen *m* landing strip
Landestreifenfeuer *n* strip light
Landeverfahren *n* landing procedure
Landezeitmeldung *f* report of landing time
Landfunkstelle *f* land station
Landgang *m* *(ship)* gangway, gangplank, brow
Landgangstreppe *f* gangway ladder
Landgewinnung *f* land reclamation, reclamation of land
Landkarte *f* map
Landkartenherstellung *f* mapping
Landkartenpapier *n* plan paper
Landkennungsfeuer *n* *(ship)* making (landfall) light
Landkessel *m* stationary boiler
Landmaschine *f* agricultural machine
Landmaschinen *fpl* agricultural (farm) machinery
Landnetzgruppe *f* *(telec)* rural district
Landpeilstation *f* land direction-finding station
Landpoller *m* *(ship)* checking (warping) bollard
Landrad *n* *(ag)* land wheel

Landrohr *n* shore delivery pipe *(reclamation dredger)*
Landröste *f* dew retting *(flax)*
Landschaftsarchitekt *m* *s.* Landschaftsgestalter
Landschaftsarchitektur *f* *s.* Landschaftsgestaltung
Landschaftsgestalter *m* landscape architect (planner)
Landschaftsgestaltung *f* landscape gardening (architecture, planning), landscaping
Landschaftsplaner *m* *s.* Landschaftsgestalter
Landstraße *f* high road, highway
Landtechnik *f* agricultural engineering (technology)
Landung *f* landing
L. mit großer Geschwindigkeit hot-speed landing
weiche L. soft landing
vom Boden kontrollierte L. ground control landing
Landungsbake *f* landing beam beacon (navigation)
Landungsboot *n* landing craft
Landungsbrücke *f* landing stage, [landing] pier
Landungsgebühren *fpl* landing dues (charges)
Landungshafen *m* port of landing (discharge, debarkation)
Landungsponton *m* landing pontoon
Landungsrakete *f* landing rocket
Landungsschiff *n* landing ship
Landungssteg *m* jetty, pier
Landungsunfall *m* accident whilst landing
Landvermesser *m* land surveyor
Landvermessung *f* land surveying
Landwind *m* off-shore wind, land breeze
Landwirt *m* farmer
Landwirtschaft *f* agriculture, farming
landwirtschaftlich agricultural
Landwirtschaftsbau *m* agricultural building, farm-building construction
Landwirtschaftswissenschaften *fpl* agricultural (agrarian) sciences
langbrennweitig of long focal distance, long-focus, long focal
Langdrahtantenne *f* long-wire aerial
Langdrahtlampe *f* squirrel-cage lamp
Langdrehautomat *m* sliding-head type automatic screw machine
Schweizer L. Swiss type automatic screw machine
Langdrehen *n* plain (cylindrical) turning, sliding
Langdrehmaschine *f* sliding lathe
Länge *f* 1. length; 2. longitude
L. des Arbeitshubs *(mach)* cutting stroke length
L. des Hohlraumes length of cavity
L. des Kolbenhubs amount of reciprocating motion of the piston, length of piston stroke
L. in der Konstruktionswasserlinie *(ship)* length of designed waterline
L. über alles length overall, overall length, extreme length over all
L. über Puffer *(ry)* length between (over) buffers
L. zwischen den Loten *(ship)* length between perpendiculars
Debyesche L. Debye shielding distance
geografische L. geographical longitude
größte L. length overall, overall length, extreme length over all
maßgenaue L. accurate length, dead length accuracy
verzahnte L. length of toothed portion *(saw)*
Länge-Durchschnittsverhältnis *n* L/D ratio *(ratio of length to diameter of a rocket motor combustion chamber)*
längen to lengthen, to stretch, to elongate, to extend
Längenänderung *f* change in length, length change
bleibende L. permanent linear deformation
Längenausdehnungskoeffizient *m* coefficient of linear expansion
Längenausgleich *m* length compensation
Längenausgleichsrinne *f* *(mine)* adjustment pan
Längenberechnung *f* calculation of length
Längenbestimmung *f* determination of longitude *(navigation)*
Längeneffekt *m* longitude effect
Längeneinheit *f* unit length
Längeneinstellung *f* length adjustment
Längengrad *m* degree of longitude
Längenkonstante *f* *(ship)* length constant
Längenkreis *m* circle of longitude
Längenmaßstab *m* scale of length
Längenmeßgerät *n* length (linear) measuring instrument
Längenmeßkomparator *m* length measuring comparator
Längenmessung *f* length (linear) measurement
Längenmeßvorrichtung *f* **für Yards** *(text)* yardage motion
längenmoduliert width-modulated *(pulses)*
Längennormal[e *f*] *n* length standard
Längennumerierung *f* *(text)* longitudinal count, number of fixed lengths per standard weight
Längenplan *m* *(ship)* longitudinal elevation
Längenprofil *n* longitudinal section
Längenschärfegrad *m* *(ship)* prismatical coefficient
längentreu length preserving
Längenverhältnis *n* *(ship)* beams (breadths) to length
Längenwachstum *n* upward growth
Längenzeichen *n* *(text)* cut mark

Längenzuwachs *m* lengthening, stretch, elongation, extension
Langettenrand *m* scalloped edge
langettieren to scallop
langfaserig long-fibre, having long fibres, long staple[d]
Langfräsmaschine *f* plano-milling machine
L. mit ausfahrbaren Ständern horizontal plano-miller with adjustable columns
langfristig long-term *(agreement)*
Langgewinde *n* long thread
Langgewindefräsen *n* single-cutter thread milling, long-thread milling
Langgewindefräsmaschine *f* single cutter type of thread milling machine, long-thread milling machine
Langgewindeschälmaschine *f* long-thread peeling machine
Langgut *n* lengthy materials
Langhobel *m* trying plane
Langholzmeßstab *m* log scaling rule
Langholzwagen *m* timber haulage vehicle
Langhubhonen *n* honing
langhubig long-stroke
Langhubräummaschine *f* long-stroke broaching machine
Langkesselüberhitzer *m* superheater for horizontal boiler
langkettig *(ch)* long-chain
Langlaufschi *m* racing (cross-country) ski
langlebig long-lived; long-half life; long-life
Langleine *f* long line
Langloch *n* 1. oblong hole, elongated slot, slotted perforation; 2. deep hole
Langlochbohren *n* deep-hole drilling
Langlochbohrer *m (mach)* deep-hole drill, D-bit, drill bit; gun drill
Langlochbohrhammer *m (mine)* deep-hole drill
Langlochbohrmaschine *f* longitudinal boring machine, slot mortizing (boring) machine
Langlochbrenner *m* long slot burner
Langlochfräser *m* [shank-type] slotting mill, [shank-type] slotting end mill, cotter (two-flute end) mill, [two-flute] slot drill
doppelseitiger L. two-flute double-end mill
Langlochfräsmaschine *f* slot miller (milling machine); groove milling machine
Langlochöffnung *f* oblong aperture
Langmachen *n* **der Kupplungen** loosening of couplings
Langmuireffekt *m* Langmuir effect
Langmuirsonde *f* Langmuir probe
Langnachformdrehen *n* plain contouring
Langnutzholz *n* long (stem) timber
Langoktav *n* oblong octavo
Langrips *m* weft rib
Langrost *m* long grate
längs longitudinal[ly]; along; lengthwise
l. der Küste *(ship)* along shore, coastwise
Längsaberration *f* longitudinal aberration
chromatische L. longitudinal chromatic aberration
sphärische L. longitudinal spherical aberration
Längsablauf *m (ship)* end launching
Längsabweichung *f s.* Längsaberration
Längsachse *f* longitudinal axis
magnetische L. direct axis
langsam slow *(speed; neutron)*
l. ansprechend slow operating
l. fahrend slow-moving
l. laufend slow-speed, slowly running
Langsambewegungsgetriebe *n* slow-motion link
Langsamfahrstelle *f (ry)* speed restriction on main track
Langsamfahrt *f (ry)* slow running
Langsamgang *m (mach)* inching travel
Langsamläufer *m (el)* low-speed motor
Langsamsteuerung *f* inching (jogging) control
Langsamstflug *m* stalling flight
Langsamunterbrecher *m* time-delay relay
Langsamvorschub *m (mach)* creep feed
Längsanker *m (mach)* longitudinal tie
Längsanschlag *m (mach)* length (end) stop
Längsansicht *f (ship)* inboard profile [plan], outboard profile
Längsaufriß *m* longitudinal view; *(ship)* longitudinal elevation
Längsbalken *m* stringer
Längsband *n (ship)* tie-plate *(deck)*
Längsbeblasung *f* longitudinal blowing *(circuit-breaker)*
Längsbelastung *f* axial thrust; thrust load
Längsbespülung *f* longitudinal [oil] scavenging *(circuit-breaker)*
Längsbewehrung *f* longitudinal reinforcement *(concrete)*
Langschaftdüse *f* long-stem nozzle
langschäftig long-boled
Langschermaschine *f (text)* longitudinal shearing machine
Langschienen *fpl* long welded rails
Langschnittsäge *f* pit saw
Längsdehnung *f* longitudinal strain
Längsdifferentialschutz *m (el)* biased differential protection
Längsdoma *n* brachydome *(crystallography)*
Längs-EMK *f* direct-axis component of the electromotive force
subtransitorische L. direct-axis subtransient electromotive force
transitorische L. direct-axis transient electromotive force
Längsentzerrer *m (el)* series equalizer
Längserregung *f (el)* longitudinal excitation
Längsfalz *m* longitudinal fold
Längsfaserspeisung *f* parallel fibre feed

Längsfeldachse *f (el)* direct axis
Längsfeld-EMK *f* direct-axis voltage
Längsfeldkomponente *f* direct-axis component *(of armature current)*
Längsfeldreaktanz *f* direct-axis synchronous reactance
Längsfeldspannung *f* direct-axis voltage
Längshobeln *n* longitudinal planing, planing lengthwise
Langsieb *n* flat screen, Fourdrinier wire *(papermaking)*
Langsiebmaschine *f* endless wire machine *(papermaking)*
  freitragende L. cantilever-type Fourdrinier machine
Langsiebpapiermaschine *f* Fourdrinier paper machine
Langsiebpartie *f* Fourdrinier section (part) *(papermaking)*
Langsiebwäscher *m* Fourdrinier-type washer *(papermaking)*
Längsimpedanz *f* series impedance
  subtransitorische L. direct-axis subtransient impedance
  synchrone L. direct-axis synchronous impedance
  transitorische L. direct-axis transient impedance
Längsinduktivität *f* series-arm inductance
Längskanal *m* longitudinal passage
Längskante *f* longitudinal edge
Längskeil *m (mach)* key
Längskeilverbindung *f* keyed joint
Längskompensation *f (el)* direct-axis compensation
Längskraft *f* longitudinal force (load), axial force (load)
  L. mit Biegung combined axial and bending force, direct (combined axial) stress and bending
Längskrümmung *f* longitudinal warping
Längslager *n* axial (thrust, journal) bearing
Längslenker *m* longitudinal control arm, trailing link
Längsmagnetisierung *f* solenoidal magnetization
Längsmagnetismus *m* solenoidal magnetism
Längsnaht *f* longitudinal seam *(riveting)*; longitudinal weld *(welding)*
Längsnut *f* longitudinal groove; *(mach)* spline [running lengthwise], keyway
Langspan *m* long continuous chip
langspanend stringy, long-chip *(material)*
Langspielplatte *f* long-playing record
Längsrauhen *n (text)* longitudinal raising
Längsregler *m (el)* longitudinal regulator
Längsrichtung *f* 1. longitudinal direction; 2. long grain *(paper)*
  L. / in lengthwise, longitudinal, in [a] longitudinal direction
Längsrips *m* weft rib
Längsriß *m* 1. *(ship)* sheer plan (draught); 2. longitudinal crack
längsschiffs along the ship
Längsschlagseil *n* Albert (Lang's) lay rope
Längsschliff *m* longitudinal grinding
Längsschlitten *m* saddle *(lathe)*
Längsschlitz *m* longitudinal (lengthwise) slot
Längsschneider *m* **und Umroller** *m* slitting and rewinding machine *(papermaking)*
Längsschnitt *m* longitudinal section *(techn. drawing)*; inboard profile [plan] *(of a vessel)*
Längsschott *n* longitudinal bulkhead
Längsschweißnaht *f* longitudinal weld
Längsschwingung *f* longitudinal (axial) vibration
längsseits side by side
Längsspalte *f (geol)* back
Längsspannung *f* 1. longitudinal stress; 2. *(el)* longitudinal (direct-axis component of the) voltage; 3. *(text)* warp tension
Längsspant *n (ship)* longitudinal frame
Längsspantenbauweise *f (ship)* longitudinal framing system
Längsspiel *n* axial (end) play, play in longitudinal direction
Längsstapellauf *m (ship)* end launching
Längsstrahler *m* [illegible]
Längs[illegible] *(aero)* jack stay
Längssüll *m (n) (ship)* side coaming
Langstabisolator *m* rod-type suspending (suspension) insulator
langstapelig long-staple[d]
Längsteilung *f* linear indexing
Längstisch *m (mach)* reciprocating table
Längsträger *m* longitudinal member, frame side bar; side member; stringer *(bridge)*
Längstragseil *n* longitudinal carrier cable
langstrahlig needle-like, elongated *(crystal)*
Langstrecken- long-distance; long-range
Langstreckenbomber *m* long-range bomber
Langstreckenjacht *f* long-range yacht
Langstreckenrakete *f* long-range rocket
Längstrikot *m (n)* longitudinal tricot
Langstrohschüttler *m* straw shaker
Längs- und Querschneidemaschine *f* duplex cutter *(papermaking)*
Längsverband *m (build)* longitudinal bracing
Längsvergrößerung *f* longitudinal magnification
Längsverluste *mpl (el)* line (transmission) losses
Längsversteifung *f* stringer
längsverstellbar longitudinally adjustable
Längsvorschub *m (mach)* longitudinal feed
Längswandbauweise *f* spine-wall construction
Längswelle *f* longitudinal wave
Längswiderstand *m (el)* series resistance

Längszug *m (mach)* power longitudinal feed, [power] longitudinal motion of carriage, [power] sliding motion of carriage
Langtagspflanze *f* summer (long day) flowering plant
Langtisch *m (mach)* rectangular (reciprocating) table
Langtischflächenschleifmaschine *f* surface grinding machine having a reciprocating work-table
Langträger *m* longitudinal girder, sole bar, side-sill of underframe
Längung *f* elongation
Langvorschub *m* sliding feed *(lathe)*, longitudinal feed
Langweggasküvette *f* long-path gas cell
Langwelle *f* long wave
Langwellenempfänger *m* long-wave receiver
Langzeitbestrahlung *f* protracted (long continued) irradiation
Langzeitbetrieb *m* continuous duty
Langzeitecho *n* echo of long duration, long-delay echo
Langzeitrelais *n* time-delay relay
Langzeitspeicher *m* long-time memory
Langzeitverhalten *n* long-time behaviour
Langzeitversuch *m* test of long duration, long-time test
Lanolin *n* lanolin
Lanthan *n* lanthanum
Lanthaniden *npl* lanthanides, lanthanide series, rare-earth elements
Lanzette *f* lancet
Lapislazuli *m* lapis lazuli
Laplace-Operator *m* Laplace operator
Laplace-Transformation *f* Laplace transformation
Lappen *m* 1. piece of cloth, rag; duster; 2. *(mach)* tang *(tool)*
läppen to lap
Läppen *n* lapping
  maschinelles L. machine lapping
Lappenkante *f* lap-edge
Lappetstickerei *f* lappet weaving
Lappetstuhl *m (text)* lappet loom
Läpphülse *f* / **zylindrische** lap for holes
Läppkäfig *m* spider-type work-holder for lapping
Läppkorn *n* abrasive grain for lapping
Läppmaschine *f* lapping machine
Läppmittel *n* lap material, lapping abrasive
Läpp-Pasta *f* lapping paste (compound), prepared lapping abrasive
Läppscheibe *f* / **ebene** flat lap, lap[ping] plate
Läppzugabe *f* lapping allowance
Lärche *f* larch
Lärchenholz *n* larch [wood]
Lärm *m* noise
Lärmbekämpfung *f* noise control; noise reduction
Larmor-Präzession *f* Larmor precession
Larsenschaltung *f (el)* Larsen's spark-suppression connection
Larvenvertilgungsmittel *n* larvicide
Laryngograf *m* laryngograph
Laryngoskop *n* laryngoscope
Lasche *f* butt (cover) strap; splice (flitch) plate, fish-plate; splice piece, tie block
  aufgearbeitete L. renovated fish-plate
Laschenanlagefläche *f* fishing surface
Laschenbolzen *m* fish-bolt, track-bolt
Laschenbolzenloch *n* fish-bolt hole
Laschenbruch *m* breakage of [the] fish-plate
Laschengelenk *n* shackle joint
Laschenkammer *f* fishing surface *(rail)*
Laschennietung *f* butt [riveted] joint, [riveted] butt joint with cover plate[s]
  doppelseitige L. double-strap [riveted] butt joint, riveted butt joint with cover plates each side
  einseitige L. riveted butt joint with single cover plate
  zweiseitige L. *s.* doppelseitige L.
Laschenstoß *m* fish[ed] joint, fishing
  zweischnittiger L. double-covered butt joint
Laschenverband *m* fish-plating
Laschenverbindung *f* fishing, fish[ed] joint, strap connection
Laser *m* laser *(light amplification by stimulated emission of radiation)*
  kontinuierlicher L. continuous wave laser
Laser-Anwendung *f* laser application
laserbetätigt laser-operated
Laser-Betrieb *m* laser operation
Laser-Forschung *f* laser research
Laser-Gerät *n* laser unit
Laser-Impuls *m* laser pulse
Laser-Leistung *f* laser performance
Laser-Material *n* lasing material
lasern to lase
Laser-Nachrichtensystem *n* laser communications system
Laser-Pumpen *n* laser pumping
Laser-Schweißgerät *n* laser welder
Laser-Sender *m* laser transmitter
Laser-Speicher *m* laser memory
Laser-Strahl *m* laser beam (ray)
Laser-Technik *f* laser technology
Laser-Verstärker *m* laser amplifier
Laser-Welle *f* laser mode
lasieren to glaze
Last *f* 1. load; weight; 2. *(ship)* burden; store room
  aufgebrachte L. applied load
  aufnehmbare L. load (load-carrying, bearing) capacity

bewegliche L. moving load
dynamische L. dynamic load
elektrische L. electric[al] load
gefederte L. suspended weight
gleichmäßig verteilte L. uniform[ly distributed] load; evenly shared load
induktive L. *(el)* reactive (inductance) load
kapazitive L. *(el)* capacitive (condenser) load
konzentrierte L. point load
kritische L. critical load[ing]
mittige L. axial (concentric) load
nacheilende L. *(el)* reactive load
rechnerische L. design load
ruhende L. static (dead) load
ständige L. permanent load (weight)
statische L. static load
verteilte L. distributed load
voreilende L. *(el)* condensive (capacitive) load
wandernde L. mobile (traffic, moving) load
zugeschaltete L. *(el)* added load
zulässige L. safe (allowable, working) load
zusammengesetzte L. combined load
zusätzliche L. additional load
Lastabhängigkeit *f* load sensitivity
Lastabschaltung *f* load interruption
Lastabwurf *m (el) (sudden)* disconnection of load
Laständerung *f* load change; load variation
Lastangriffspunkt *m* point of application of load
Lastannahme *f (build)* design load
Lastarm *m* weight arm
Lastaufnahme *f* / **maximale** maximum load-bearing capacity
Lastaufnahmefähigkeit *f* bearing value
Lastaufnahmemittel *n* load take-up *(conveying)*
Lastausschalter *m* load-break cut-out
Lastbegrenzungsrelais *n* load levelling (limiting) relay
Lastbild *n* load[ing] diagram, load[ing] scheme
Lastboot *n (ship)* hoy
Lastenaufzug *m* freight elevator, goods lift
Lastenheben *n* lifting loads
Lastensegler *m* troop-carrying glider, cargo-carrying glider
Lastenseilschwebebahn *f* cableway handling (conveying) materials
Lastenzug *m* bridge testing train, loading diagram
Lastfahrzeuganhänger-Transportschiff *n* trailer ship
Lastfaktor *m* load factor
Lastgüte *f (telec)* working Q, Q external *(with resonance systems)*
Lasthaken *m* lifting (load) hook
Lasthebedampfmaschine *f* hoisting steam engine
Lastkahn *m (ship)* barge, lighter
Lastkennlinie *f* load characteristic
Lastkraftwagen *m* lorry, *(Am)* truck
Lastkraftwagenreifen *m* lorry tyre
Lastkreis *m (el)* load circuit
Lastprüfung *f* load test
Lastregelung *f* load control
Lastrohr *n (ship)* tubular bulk cargo container
Lastschalter *m* [load-breaking] switch
Lastschiff *n* lighter, transport ship
Lastschriftzettel *m* debit note
Lastschwankung *f* load variation, variation in load
Lastschwerpunkt *m (el)* centre of distribution
Lastseil *n* fall rope; *(ship)* cargo runner (fall)
Lastseilsystem *n* / **gekoppeltes** married fall system, burton tackle system
Lastspannung *f* running voltage
Lastspiel *n* load (stress) cycle
Laststoß *m (el)* power impulse
Laststrom *m* running current
Laststufe *f* load increment
Lasttier *n* beast of burden
Lasttonne *f (ship)* ton burden
Lasttragtier *n* beast of burden
Lasttrennschalter *m* load isolator (disconnector, isolating switch)
Lastumschalter *m* on-load tap changer, on-load tap-changing equipment
Lastumschaltrelais *n* power transfer relay
Lastverluste *mpl (el)* load losses
Lastverstimmung *f* pulling by a load impedance *(electronic oscillators)*
Lastverstimmungsmaß *n* pulling factor (figure) *(oscillator)*
Lastverteiler *m (el)* load dispatcher
Lastverteilerzentrale *f* load dispatcher station
Lastverteilung *f* load distribution
Lastwechsel *m* stress (load) reversal
Lastwechselrhythmus *m* stress cycle
Lastwert *m* load rating
dynamischer L. dynamic load [rate]
wirklicher L. applied load [rate]
Lastzugbremsventil *n* tractor-trailer brake valve
Lasur *f* glaze, transparent coating
Lasurblau *n* ultramarine (lazulite) blue, azure
Lateibrett *n s.* Fensterbrett
latent latent
Latenzzeit *f* latent period *(radiology)*; *(dat)* latency
lateral lateral
Lateralschwerpunkt *m (ship)* centre of lateral resistance
Lateralwiderstand *m (ship)* lateral resistance
Laterna magica *f* magic lantern
Laterne *f* 1. hand lantern, storm lamp; 2. *(therm)* connecting (distance) piece, ring for the sealing liquid
Laternenfenster *n (therm)* opening in distance piece

seitliches L. side opening in distance piece
Laternenflansch *m (therm)* flange of distance piece
Laternenmast *m* lamp post (standard)
Laternenstütze *f* lamp bracket
Laternenventilkegel *m* developed V-port plug
Latex *m* latex
Latte *f* strip, batten, lath
klappbare L. folding-type of staff
Lattenabschnitt *m* staff intercept
Lattendeck *n* slatted deck
Lattengerüst *n* lattice work (frame)
Lattenschott *n (ship)* [batten and] space bulkhead, lattice (sparred) bulkhead
Lattentasche *f* batten pocket *(for sails)*
Lattentrommel *f (text)* lattice drum
Lattentür *f* batten (battened and braced) door
Latten- und Leistenschlitzmaschine *f* rebating machine
Lattenwand *f* batten wall
Lattenwegerung *f (ship)* spar (sparred, open) ceiling
Lattenzaun *m* trellis (pale) fence, paling
Lattenzielpunkt *m* staff target point
Laub *n* leaves
Laubenganghaus *n* gallery (balcony access) block
Laubholz *n* deciduous (hard) wood
Laubsägebogen *m* piercing saw frame
Laubsägeholz *n* fretsaw timber
Laue-Aufnahme *f* Laue pattern
Laue-Beugungsbild *n s.* Laue-Beugungsdiagramm
Laue-Beugungsdiagramm *n* Laue [diffraction] pattern, Laue photograph
Laue-Diagramm *n s.* Laue-Beugungsdiagramm
Laue-Fleck *m* Laue spot
Lauf *m* 1. *(mach)* run[ning]; course (river) movement; 2. *(build)* flight of stairs; 3. barrel *(e.g. rifle)*
beladener L. running under load
geräuscharmer L. silent (quiet) running
gezogener L. rifled barrel
Laufbahn *f* race[way], track *(bearing)*
Laufbild *n* animated picture
Laufbildwerfer *m s.* Projektor
Laufbohle *f* duckboard *(on roofs)*
Laufbrett *n* gang-board
Laufbrücke *f* flying (fore-and-aft) bridge, catwalk; gangway; fore-and-aft gangway *(tanker)*
Laufbuchse *f* bush; liner
nasse L. wet liner
trockene L. dry liner
Laufbügel *m* cradle *(Jacquard loom)*
Laufdecke *f* outer cover *(tyre)*
Laufeigenschaften *fpl* **eines Wagens** running qualities of a wagon
laufen *(mach)* to run
l. mit einer Geschwindigkeit (Drehzahl) von to run at a speed of
auf Grund l. to run (go) aground
aus dem Hafen l. to leave the harbour
eine Maschine l. lassen to run a machine
genau rund l. to run true
leer l. *(mach)* to run idle
nicht genau rund l. to run out of true (truth)
Schi l. to ski
schneller l. lassen to speed up
laufend current *(year)*; consecutive *(numbers)*
schlagfrei l. running true
Läufer *m* 1. *(build)* stretcher *(brick)*; 2. *(math)* runner, cursor *(slide rule)*; 3. *(el)* rotor; 4. *(mine)* runner *(support)*; 5. *(text)* porcupine; traveller; carpet runner
C-förmiger L. *(text)* C-shaped traveller
gewickelter L. *(el)* wound rotor
zurückbleibender L. *(text)* dragging traveller
Läuferanlasser *m* rotor[-resistance] starter, rotor rheostat
läufergespeist rotor-fed
Läuferkreis *m* rotor circuit
Läufermühle *f* edge mill
Läufernut *f* rotor slot
Läuferscheibe *f* **eines Zählers / stroboskopische** stroboscopic meter disk
Läuferschicht *f* stretcher course *(brickwork)*
Läuferspannung *f* rotor (inductor) voltage
Läuferstrom *m* rotor (inductor) current
Läuferstromkreis *m* rotor (inductor) circuit
Läufertrommel *f* rotor barrel
Läuferverband *m* stretcher bond *(brickwork)*
Läufervorkante *f (ship)* fore end of running ways
Läuferwelle *f* rotor shaft
Läuferwicklung *f* rotor (inductor) winding
Läuferzug *m (text)* traveller tension
Lauffeldröhre *f* **mit statischem Querfeld** transverse field travelling wave tube
Lauffläche *f* 1. *(mach)* bearing (sliding, running) surface; rubbing surface *(of a plain bearing)*; 2. tread *(tyre; wheel)*
Laufflächenprofil *n* tread pattern *(tyre)*
Laufgang *m* gangway, catwalk, walkway
Laufgenauigkeit *f* running accuracy
Laufgeschwindigkeit *f* 1. running speed; 2. floating speed (rate)
Laufgewicht *n* sliding weight
Laufgewichtswaage *f* Roman steelyard
Laufkarte *f* **für Werkstücke** routing card
Laufkatze *f* trolley, jenny, crab
L. für Hebezeuge hoist trolley
Laufkette *f* track
Laufknoten *m (ship)* running knot
Laufkran *m* [overhead] travelling crane
Lauflager *n* plummer block, line shaft bearing

Lauflänge *f* working length *(cylinder)*
Laufmasche *f* ladder
laufmaschenfrei run-resist
laufmaschensicher run-resist
Laufplanke *f* runner board, gang-board, brow; gangway
Laufrad *n* 1. turbine (rotor) wheel, runner *(turbine)*; 2. *(aero)* undercarriage wheel; *(aut)* road wheel; travelling wheel *(conveyor)*; 3. *(mach)* impeller
Laufradbremse *f* *(aero)* undercarriage wheel brake
Laufradscheibe *f* rotor disk
Laufradwelle *f* *(aut)* impeller shaft
Laufraum *m* drift space *(cavity resonator)*
Laufraumelektrode *f* drift tunnel
Laufrichtung *f* machine direction, direction of running; grain direction
Laufring *m* 1. race [track], track ring *(bearing)*; 2. *(print)* mackling ring
Laufrinne *f* *(met)* launder
Laufrolle *f* castor; [roller] idler
Laufruhe *f* *(mach)* smooth (quiet, silent) running; quietness in operation, quietness of running
Laufschaufel *f* turbine (moving, rotating) blade
Laufschaufelkanal *m* impeller channel
Laufschaufelkranz *m* blade ring
Laufschaufelsegment *n* bucket segment
Laufsohle *f* outsole *(shoe)*
Laufspindel *f* bearing spindle *(turbine)*
Laufspule *f* revolving bobbin
Laufstall *m* cow yard, loose housing stable
Laufstallhaltung *f* *(ag)* loose housing
Laufsteg *m* *(build)* catwalk, gangway; duckboard *(on roofs)*
  L. zwischen Wagen *(ry)* running board, footbridge, gangway
Laufstreifen *m* felt seam mark
Laufweg *m* *(nucl)* path[way], trajectory
Laufwelle *f* intermediate (line) shaft
Laufwerk *n* trolley *(conveying)*
Laufwerkpanzerung *f* *(mil)* track shield
Laufwiderstand *m* resistance to rolling
Laufwinkel *m* *(el)* transit [phase] angle
  effektiver L. effective bunching angle
Laufzeit *f* time of flight, flight (transit) time; delay *(electron valve)*; *(cont)* dead time, transportation (distance velocity) lag
  L. eines Schutzrechtes term of a protective right
Laufzeitanalysator *m* *(nucl)* time-of-flight analyzer
Laufzeitcharakteristik *f* *(el)* delay time characteristic
Laufzeiteffekt *m* sorting effect
Laufzeitentzerrer *m* delay equalizer
Laufzeitentzerrung *f* transit (rise) time correction, group delay correction
Laufzeitfehler *m* phase delay distortion
Laufzeitgenerator *m* electron oscillator
Laufzeitglied *n* delay line
Laufzeitkette *f* *(el)* retardation network, delay chain
Laufzeitkompensation *f* transit time compensation
Laufzeitmassenspektrometer *n* time-of-flight mass spectrometer
Laufzeitmodulation *f* velocity modulation
Laufzeitröhre *f* transit-time tube, travelling-wave tube
Laufzeitunterschied *m* *(el)* transient time difference
Laufzeitverzerrung *f* delay distortion
  L. in Datensystemen delay distortion in data systems
Laufzeitwinkel *m* *(el)* transit [phase] angle
Laufzettel *m* work-progress slip
Lauge *f* lye, leach
laugebeständig alkali-proof
Laugekessel *m* caustic pot
laugen to lixiviate, to leach
Laugenansetzkasten *m* lye preparation box
laugenecht fast to lye
laugenmerzerisiert mercerized in a caustic soda bath
Laugenwäsche *f* caustic scrubbing
Laugenzuführung *f* introduction of lye
Laugeprozeß *m* / **saurer** acid leach
Laugewaschanlage *f* caustic washing plant
Laugung *f* leaching
Laugungsmittel *n* leaching agent
Laumontit *m* *(min)* laumontite
Laurat *n* laurate
Laurinaldehyd *m* lauraldehyde
Laurinsäure *f* lauric acid
Laurit *m* *(min)* laurite
Lauritsen-Elektroskop *n* Lauritsen electroscope
laut loud; noisy
Laut *m* sound; tone
läuten to ring
Läuterelais *n* ringer relay
läutern to purify, to clarify; to defecate; to refine, to rectify
Läutewerk *n* alarm gear, [electric] bell
Lautheit *f* loudness
lautlos silent
Lautschrift *f* phonetic spelling (lettering)
Lautsprecher *m* loudspeaker, speaker
  dynamischer L. moving-coil loudspeaker, moving-conductor loudspeaker
  elektrodynamischer L. electrodynamic (coil-driven) loudspeaker
  elektromagnetischer L. electromagnetic loudspeaker
  elektrostatischer L. electrostatic (capacitor) loudspeaker

piezoelektrischer L. crystal loudspeaker
pneumatischer L. pneumatic loudspeaker
zweiter L. extension loudspeaker
Lautsprecherabschaltung *f* loudspeaker muting
Lautsprecheranlage *f* loudspeaker installation
Lautsprechergehäuse *n* loudspeaker housing
Lautsprecherkombination *f* multichannel loudspeaker
Lautsprecherkondensator *m* loudspeaker capacitor
Lautsprechermembrane *f* loudspeaker diaphragm
Lautsprecherschwingspule *f* loudspeaker voice coil
Lautsprecherseide *f* speaker cloth
Lautsprecherübertragungsanlage *f* public address system
Lautsprecherwagen *m* loudspeaker van
Lautstärke *f (ac)* loudness level; *(telec)* loudness, volume of sound, sound level, strength of reception
subjektive L. *(el)* equivalent loudness
Lautstärkebereich *m* volume range *(radio)*
Lautstärkeempfindung *f* perception of loudness
Lautstärkegrad *m* loudness level
Lautstärkemesser *m* sound level meter, phonometer
Lautstärkeregelung *f* [sound] volume control, volume adjustment
gehörrichtige L. acoustically compensated volume control
handbetätigte L. manual volume control
kompensierte L. compensated volume control
niederfrequente L. audio frequency volume control
physiologische L. compensated volume control
Lautstärkeregler *m* attenuator, gain controller, volume regulator
automatischer L. automatic volume range regulator
gehörrichtiger L. tone-compensated volume control
selbsttätiger L. automatic volume control
Lautstärkeschwankung *f* **im Empfang** variation in signal strength
Lautumschrift *f* phonetic lettering
Lautverständlichkeit *f* sound articulation
Läutwerk *n* / **elektrisches** electric (signal) bell
Lavaldüse *f* Laval nozzle
Lavendelöl *n* lavender oil
Lävulinsäure *f* levulinic acid
Lawine *f* avalanche *(electronics)*
Lawineneffekt *m* / **gesteuerter** controlled avalanche effect *(in semiconductor devices)*
Lazarettschiff *n* hospital ship; ambulance launch, clinic boat
Lazarettzug *m* ambulance (red-cross) train
Lazulith *m (min)* lazulite
LD-Verfahren *n* Linz-Donawitz process, LD process
Lebensdauer *f* life; cutting life *(tool)*; service (serviceable, useful, production, working) life, working service; *(met)* endurance; *(nucl)* lifetime
L. des Werkzeugs tool life
betriebliche L. service life
effektive L. effective lifetime
lange L. long life
mittlere L. mean (average) life, mean lifetime
scheinbare L. apparent lifetime
unbegrenzte L. unlimited life
Lebensdauerprüfung *f* life test; endurance test
Lebensdosis *f (nucl)* lifetime dose
Lebenserwartung *f (nucl)* life expectancy
Lebensgefahr *f* danger to life
Lebensmittel *npl* food[stuffs], victuals
Lebensmittelchemie *f* food[stuff] chemistry
Lebensmittelfarbe *f* foodstuff colour, nutriment colouring
Lebensmittelkonservierung *f* preservation of food
Lebensmittelrefraktometer *n* foodstuff refractometer
Lebensmitteluntersuchung *f* food investigation
Lebensrettung *f (ship)* life-saving
Lebensrettungseinrichtungen *fpl* life-saving appliances (equipment)
Lebensrettungsgeräte *npl* life-saving equipment (appliances)
Leberextraktionsanlage *f* liver-extraction plant *(on fishing vessels)*
Lebertran *m* cod liver oil
Lebertrankochanlage *f* liver[-boiling] plant *(on fishing vessels)*
Leberverwertungsanlage *f* liver plant *(on fishing vessels)*
lebhaft brisk *(reaction)*
Leblancverfahren *n* Leblanc process *(sodium carbonate)*
Lecherleitung *f* Lecher wires, Lecher circuit, parallel-wire line
Lechersystem *n* Lecher system
Leck *n* leak[age]
Leckage *f* leak[age]
leckdicht leakproof, leak-tight
lecken to leak *(fluid)*
Lecken *n* leakage *(fluid)*
Leckfaktor *m* non-leakage probability; marginal coefficient
Leckfinder *m* leak detector
Leckmatte *f (ship)* collision mat
Leckmenge *f* [amount of] leakage, leak rate
Leckölleitung *f* overflow oil line
Leckölleitungsanschluß *m* overflow oil line connection

Leckölmenge *f* leak rate
Leckprüfung *f* leak[age] test
Leckpumpe *f (ship)* flood pump
Leckrechnung *f* flooding calculation, calculation of the damaged ship
Leckschraube *f (ship)* drain plug, bleed cock
lecksicher leakproof *(fluid)*; leakage-proof
Leckstelle *f* leak[age], leaky area; point of leakage *(fluid)*
Leckstopfbeutel *m (ship)* damage-control splinter and stuffing box; jackass
Leckstopfmittel *npl (ship)* fothering means
Leckstrahlung *f* leakage radiation
Lecksucher *m* leak detector
Leckverlust *m* leakage loss
zulässiger L. leak tolerance
Leckwalze *f (print)* vibrator
Leckwasser *n* leak water; flooding water
Leclanché-Element *n (el)* Leclanché cell, sal ammoniac cell
Ledeburit *m (met)* ledeburite
Leder *n* leather
L. für Dichtungen hydraulic (valve) leather
L. für Einstechrahmen welting leather
gewalztes L. rolled leather
satiniertes L. satin leather
technisches L. mechanical leather
Lederbelag *m* facing of leather
L. / mit leather-faced
lederbezogen leather-faced, faced with leather, leather-covered
Lederdichtung *f* leather packing, hydraulic leather
Lederdicke *f* substance of the leather
Lederersatz *m* leather substitute
Lederfärberei *f* leather dyeing
Lederfaserpappe *f* leatherpulp board
Lederfett *n* dubbing
Ledergerbung *f* leather tanning, tanning of leather
Lederhaut *f (opt)* sclera
Lederherstellung *f* manufacture of leather
Lederkleber *m* leather bonding adhesive
Lederkohle *f* charred leather
Lederkoller *n* leather pea jacket
Lederkupplung *f* leather coupling
Lederlack *m* japan
Lederleinen *n* dowlas
Ledermanschette *f* leather packing
Ledermanschettendichtung *f* leather cup
Ledermeßmaschine *f* measuring machine *(shoe)*
Lederpappe *f* leather board
Lederriemen *m* leather belt *(strap)*
Ledersamt *m* duvetyne
Lederschmiere *f* stuffing
Lederschneidmaschine *f* rand cutting machine
Lederwichse *f* leather wax
Leduc-Righi-Effekt *m* Leduc effect

Lee *f* lee[-side]
L. / in alee
leegierig *(ship)* carrying lee helm, faint headed
leer empty; vacant *(lattice)*; blank *(paper)*
Leerbeförderung *f (ry)* empty run
Leerbelichtungssperre *f* empty-frame prevention
Leere *f (ph)* vacuum; emptiness
Torricellische L. Torricellian vacuum
leeren to empty
Leerfahrt *f* empty (flitting) run; light running
L. der Lokomotive idling (idle-run) of the locomotive
Leergang *m* idle gear
Leergangreibung *f* no-load friction
Leergewicht *n* weight empty; empty [running] weight; deadweight
Leergut *n* empties
Leerhub *m (mach)* non-cutting stroke *(metal cutting)*; idle [return] stroke; *(aut)* no-load stroke
Leerkarte *f* blank card
Leerklemme *f* vacant terminal
Leerkontakt *m* vacant (dead) contact, vacant terminal
Leerlauf *m* 1. *(mach)* idling, idle running; 2. *(aut)* idle gear; 3. *(el)* no-load operation; open-circuit operation
L. / im *(el)* in open circuit
L. eines Wählers *(telec)* idling of a selector
Leerlaufadmittanz *f* open circuit admittance
Leerlaufarbeit *f* no-load work
Leerlaufausschalter *m* no-load cut-out
Leerlaufbegrenzungsschraube *f (aut)* idle adjusting screw
Leerlaufbetriebszustand *m* ground idling conditions
Leerlaufcharakteristik *f* no-load characteristic
Leerlaufdrehzahl *f* idling (no-load) speed; *(aut)* idle running speed
Leerlaufdüse *f* slow-speed nozzle; slow-running jet; no-load nozzle *(turbine)*
Leerlaufeingangswiderstand *m (el)* self-impedance
Leerlaufeinstellschraube *f (aut)* slow-running screw
leerlaufen to idle; *(Am)* to loaf *(machine)*
leerlaufend 1. *(mach)* idle, idling; 2. *(el)* open-circuited
Leerlauferregung *f (el)* no-load excitation
Leerlaufersatzschaltbild *n* constant voltage equivalent circuit
Leerlaufgüte *f* Q unloaded *(electronics)*
Leerlaufgütefaktor *m* basic (non-loaded) Q
Leerlaufhub *m (aut)* no-load stroke
Leerlaufimpedanz *f* no-load impedance, open-circuit impedance
Leerlaufinduktivität *f* no-load inductivity

Leerlaufkennlinie *f* no-load characteristic (curve), open-circuit characteristic (curve)
Leerlaufleistung *f (el)* no-load [input] power
Leerlaufluftschraube *f (aut)* slow-running air screw
Leerlaufmagnetisierungsstrom *m* no-load magnetizing current
Leerlaufnummernschalter *m* telephone dial with pulse-spacing device
Leerlaufreaktor *m s.* Nulleistungsreaktor
Leerlaufschaltung *f* no-load connection
Leerlaufscheinleitwert *m* open-circuit admittance
Leerlaufspannung *f* no-load voltage, open-circuit voltage
Leerlaufstellung *f (aut)* slow-running position; neutral position *(of control lever)*
Leerlaufstrom *m* idle (no-load) current
Leerlaufverbrauch *m* idle-power consumption
Leerlaufverlust *m* open-circuit loss, no-load loss, stand-by loss
Leerlaufverstärkung *f* open loop gain *(semiconductors)*
Leerlaufversuch *m (el)* no-load test
Leerlaufwert *m (el)* quiescent value
Leerlaufwiderstand *m* open-circuit impedance *(of conductors)*; no-load resistance
Leerlaufzeit *f (el)* idle (unoccupied) time
Leerlaufzustand *m (el)* no-load condition
Leerrücklauf *m (mach)* idle return stroke
Leerschalter *m* disconnecting switch
Leerseite *f* blank page
Leerspaltensucher *m* blank column detector
Leerstation *f* idle station *(transfer line)*
Leerstecker *m (el)* feign plug
Leerstelle *f* vacant [lattice] site, [lattice] vacancy; defect; point imperfection
Leerstellendiffusion *f* vacancy diffusion
Leerstellenkonzentration *f* vacancy concentration
Leerstellenwanderung *f* vacancy migration
Leerstellung *f* feathering position
Leertiefgang *m (ship)* light draught
Leertrum *m* slack (free) side *(belt)*; return strand
Leerwagen *mpl* empties
Leerwagenverkehr *m* empty running
Leerwagenzug *m (mine)* empty journey
Leerweg *m (aut)* dead travel
Leerwert *m (opt)* blank value
Leerzeichen *n* ignore
Leerzug *m* 1. empty stroke; 2. *(mine)* empty journey
Leeseite *f* lee side
leewärts *(ship)* towards the weather side, alee
Leeweg *m (ship)* leeway
Legebatterie *f* hen laying battery
Legehenne *f* layer
Legemaschine *f* folding machine
legen to lay *(cable)*; to place; to put [down]; *(ag)* to plant *(potatoes)*, to sow; to lay down *(pavement, floor)*; to deposit; to put
  an Erde l. to short to earth
  auf Kiel l. to lay the keel
  auf Stapel l. to lay down *(on the stocks)*
  Embargo auf Schiffe l. to lay embargo on ships
  Schiff an die Pier l. to accommodate a vessel
Legende *f (print)* legend
Legeschiene *f* / **hintere** back guide bar
legierbar alloyable
Legierbarkeit *f* alloyability, alloying properties
legieren to alloy
Legierung *f* alloy
  L. mit niedrigem Schmelzpunkt low-smelting alloy
  binäre L. binary alloy
  halbmetallische L. semimetallic alloy
  Heuslersche L. Heusler alloy
  hochfeste L. high-strength alloy
  hochschmelzende L. high melting-point alloy
  kavitations- und erosionsbeständige L. anti-cavitation-erosion alloy
  kubisch flächenzentrierte L. face-centred cubic alloy
  leicht schmelzbare L. fusible alloy
  niedrigschmelzende L. low melting-point alloy
  pyrophore L. sparking allov
  quaternäre L. quaternary (four-component) alloy
  quasibinäre L. quasi-binary alloy
  ternäre L. ternary alloy
Legierungsanteil *m* alloying contribution
Legierungsaufbau *m* structure of alloys
Legierungsbestandteil *m* alloy constituent
legierungsdiffundiert alloy-diffused
Legierungsdiode *f* alloy diode
Legierungselement *n* alloying element
Legierungserz *n* alloy ore
Legierungsfolie *f* alloy film
Legierungsgehalt *m* alloy content
Legierungskomponente *f* alloying metal (element), constituent metal
Legierungsmittel *n* alloyer
Legierungspulver *n* alloy powder
Legierungstechnik *f* alloying technique
Legierungstemperatur *f* alloying temperature
Legierungstransistor *m* alloy[ed] transistor
Legierungszusammensetzung *f* alloy composition
Legierungszusatz *m* alloying addition
Legstahl *m (leath)* currier's laying steel
Leguminose *f* leguminose
Lehm *m* loam, clay
Lehmestrich *m* clay floor
Lehmform *f* loam mould *(foundry)*

Lehmformerei *f* 1. loam-moulding; 2. loam moulding shop
Lehmguß *m* loam casting
lehmhaltig argillaceous
lehmig loamy
Lehmkern *m* loam core *(foundry)*
Lehmmörtel *m* loam mortar
Lehmstein *m* adobe
Lehmziegel *m* adobe
Lehne *f* support; rest; back
lehnen gegen to lean against
Lehrabschlußprüfung *f* final examination of apprentices
Lehrausbilder *m* apprentice instructor
hauptberuflicher L. full-time apprentice instructor
Lehrbogen *m (build)* arch centring
Lehrbrief *m* instruction leaflet
Lehrbuch *n* school textbook
Lehrdorn *m* plug (male) gauge
Lehre *f* 1. gauge *(measuring)*; jig *(boring)*; template; strickle *(foundry)*; 2. apprenticeship; 3. teaching; tuition; 4. theory, science
L. absolvieren / eine to go through an apprenticeship
L. / eintreten in eine to enter an apprenticeship
L. mit anzeigenden Prüfmitteln indicating gauge
L. und Forschung *f* tuition (teaching) and research
einstellbare L. adjustable gauge
feste L. fixed gauge
lehren 1. to teach; 2. to gauge *(measuring)*
Lehrenbohrarbeit *f* jig boring operation
Lehrenbohren *n* jig boring
Lehrenbohrer *m* jig borer operator
Lehrenbohrmaschine *f* jig-boring machine
Lehrenbohr- und -fräsmaschine *f* jig boring and milling machine
Lehrenbohrwerk *n* jig-boring machine
L. in Einständerausführung openside jig boring machine
L. in Zweiständerausführung double column jig borer
Lehrenform *f* master form, templet
Lehrenfräsmaschine *f* jig mill[ing machine]
lehrenhaltig accurate to gauge, true
Lehrgang *m* course
Lehrgangsteilnehmer *m* course participant
Lehrgerüst *n (build)* staging, centring, falsework structure
Lehrgrenzdorn *m* plug limit gauge
Lehrhauer *m (mine)* apprentice hewer
Lehrkörper *m* teaching staff
Lehrling *m* apprentice
Lehrlingsausbildung *f* apprentice training
Lehrlingslohn *m* training allowance (award)
Lehrlingswohnheim *n* apprentice hostel
Lehrmittel *n* teaching aid
Lehrrevier *n (mine)* training district
Lehrring *m* ring gauge *(measuring)*
Lehrsatz *m* theorem; lemma
Lehrstreb *m (mine)* training face
Lehrvertrag *m* apprenticeship contract
Lehrwerkstatt *f* training (apprenticeship) workshop
Lehrzahnrad *n* precision master gear
Lehrzeit *f* apprenticeship [period], period of training, term of apprenticeship
Leibholz *n (ship)* margin (thwartship) plank
inneres L. inner waterway
Leibung *f* intrados; *(build)* flanning, reveal *(masonry)*; jamb *(door)*
Leibungsfläche *f* intrados surface
Leibwäschestoff *m* underwear fabric
Leibweitenrundstrickmaschine *f* **/ einfonturige** body size plain circular knitting machine
Leibweitenstrickmaschine *f* body fabric machine
Leiche *f (print)* missing word, out
Leichenbeförderung *f* transport of a corpse
Leichengift *n* cadaverine, ptomaine
leicht light in weight; easy
l. auswerfen to issue freely *(mould)*
l. schmelzbar easily fusible
l. zerspanbar sein to be readily machined
leichtangereichert *(nucl)* low-enriched, low-enrichment, slightly enriched
Leichtbau *m* light-weight building (construction)
Leichtbauplatte *f* building board
Leichtbauweise *f* light-weight construction
Leichtbenzin *n* light petrol; petroleum spirit, low-boiling naphtha
Leichtbeton *m* light-weight concrete
konstruktiver L. structural light-weight concrete
tragender L. structural light-weight concrete
Leichter *m* [towed] barge, lighter
L. für Eisenbahnwaggon rail-car barge
L. mit Eigenantrieb motorbarge
flexibler L. flexible barge
Leichterführer *m* lighterman
leichtern to lighten *(a vessel)*
Leichtern *n* lightening
Leichterrisiko *n* risk of craft (boat)
Leichterschiff *n* scow
Leichterverkehr *m* lighter (barge) traffic
leichtflüchtig volatile, easily volatilized
leichtflüssig flooding, mobile; easily fusible
Leichtgängigkeit *f* **der Steuerung** ease of steering
Leichtgut *n* light cargo (goods, freight)
Leichtigkeit *f* lightness *(weight)*; ease *(difficulty)*
leichtlöslich easily soluble
Leichtmatrose *f* ordinary seaman
Leichtmetall *n* light metal
Leichtmetallblech *n* light sheet metal

Leichtmetallegierung *f* light alloy
Leichtmetallgehäuse *n* light-alloy housing
Leichtmetallseil *n (el)* light-metal core
Leichtöl *n* light oil
Leichtölladung *f* white cargo
leichtschmelzbar low-melting point
leichtschmelzend rapidly melting
leichtsiedend low-boiling
leichtspaltbar easily split
Leichttrimmer *m* easy-trimmer *(vessel)*
leichtverderblich perishable
Leichtwasserlinie *f* light [load-]line, ballast waterline
Leichtwasserreaktor *m* light-water reactor
leichtzersetzlich easily decomposable
Leichtzuschlag *m* light-weight [concrete] aggregate
Leichtzuschlagbeton *m* light-weight aggregate concrete
Leichtzuschlagstoff *m* light-weight [concrete] aggregate
Leihbibliothek *f* lending library
Leim *m* glue; size, mucilage
Leimabbindezeit *f* open assembly time
Leimauftragen *n* spread coating; glue application
Leimauftragmaschine *f* glue applicator (spreading machine), spreader
Leimauftragwalze *f* gluing cylinder (roller)
Leimbütte *f* size vat
Leimdruck *m* leimtype
leimen to glue; to cement; to size *(paper)*
Leimfarbe *f* distemper; calcimine
Leimfarbenanstrich *m* distempering
Leimfuge *f* glue (plain) joint
Leimfugen-Scherfestigkeitsprüfung *f* glue-line shear test
Leimgrundierung *f* glue priming
Leimkocher *m* glue-pot, glue heating apparatus
Leimleder *n* glue stock
  gesalzenes L. green salted glue stock
Leimpresse *f* hot press
Leimsieder *m* glue boiler (heater)
Leimtopf *m* glue pot
Leimtrog *m* size vat
Leimung *f* glueing; sizing *(paper)*
Leimungsgrad *m* degree of sizing
Leimzwinge *f* glue press (clamp)
Lein *m* linseed plant
Leine *f* cord, line; leash *(dog)*; rope *(pulling)*
Leinen *n* linen
Leinengarnnumerierung *f* linen yarn numbering
Leinenpapier *n* linen-faced paper
Leinenprägung / mit linen-faced
Leinentsamungsmaschine *f* linseed thresher
Leinenwaren *fpl* lingerie
Leinenweberei *f* linen weaving mill
Leinenwurfrakete *f (ship)* line-throwing rocket
Leinkombine *f* complete flax harvester
Leinöl *n* linseed oil
Leinölfirnis *m* boiled linseed oil
Leinraufen *n* flax pulling
Leinraufmaschine *f* flax puller (pulling machine)
Leinsamen *m* linseed
Leinvollerntemaschine *f* complete flax harvester
Leinwand *f* linen, cloth; canvas
Leinwandbindung *f* calico (plain, linen) weave, tabby weaving
Leinwandeinlage *f* breaker *(tyre)*
Leinwandstuhl *m (text)* linen loom
Leiste *f* 1. *(build)* fillet, strip, lath, slat; batten; ledge, cleat; 2. *(mach)* gib [strip]; 3. *(text)* selvedge
leisten to perform; to do; to give; to carry out
  Bürgschaft l. to give security
  einen Dienst l. to render a service
  einen Eid l. to take an oath
  Ersatz l. to make restitution
  Hilfe l. to give assistance
Leisten *m* shoe (boot) tree
Leistenbauweise *f* strip-planking *(boat-building)*
Leistendämpfapparat *m (text)* selvedge steaming apparatus
Leistenelevator *m* ledge elevator
Leistenflachstrickmaschine *f* / **zweiköpfige** double-head flat border machine, flat double-head border machine
Leistenhobelmaschine *f* fillet moulding machine
Leistenneutralisiervorrichtung *f* selvedge neutralizing device
Leistenrundstrickmaschine *f* / **einfonturige** plain circular strong border knitting machine
Leistung *f* 1. *(mach)* power *(erg/sec)*, horsepower; performance; 2. output; yield; return; capacity; 3. efficiency; duty; *(el)* wattage
  L. / mit zu geringer underpowered
  L. am Radumfang power (output, rating) at wheel-rim
  L. am Zughaken output at the draw-bar *(electric train)*
  L. beim Streckenauffahren *(mine)* drivage (driving) rate
  L. der Feuerung rate of combustion
  L. des Reaktors reactor power [level]
  L. in PS horsepower rating
  L. in Tonnen tonnage
  L. in Watt wattage
  L. in Yard *(mine)* yardage
  abgegebene L. power output (given out)
  abgestrahlte L. radiated power *(of an aerial)*
  aufgenommene L. power input
  aufzuwendende L. input power
  augenblickliche L. momentary output
  ausgestrahlte L. radiated power
  effektive L. *(aut)* brake horsepower; effective output

einfallende L. incident power
elektrische L. electrical power
elektrische L. des Reaktors electric output (power) of the reactor
garantierte L. specified output
gleichbleibende L. consistent performance
indizierte L. indicated output
maximale L. maximum capacity
mechanische L. mechanical power
mittlere L. medium duty
natürliche L. inherent power
nutzbare L. brake horsepower
scheinbare L. apparent output
schwankende L. varying output
spezifische L. specific power
stabil übertragbare L. stably transmissible power
tatsächliche L. actual power
thermische L. thermal power *(reactor)*
übertragene L. transmitted (transferred) power
verfügbare L. available power
wirkliche L. real (actual) output
zugeführte L. power input
Leistungsabfall *m* decrease (drop) in power
Leistungsabgabe *f* power output (delivery)
Leistungsangabe *f (mach)* rating [of power]
Leistungsangaben *fpl (aut)* output data, engine rating
Leistungsanpassung *f (el)* power matching
Leistungsaufnahme *f* power input (consumption, absorption), absorption of power, [rate of horse]power input, power required; watts input
L. des Laders *(aut)* supercharger power absorption
Leistungsaufnahmemeßgerät *n s.* Wattmeter
Leistungsbedarf *m* power consumption (requirement)
Leistungsbedarfsspitze *f* maximum power demand
Leistungsbegrenzung *f (el)* load limitation
selbsttätige L. automatic load limitation
Leistungsbegrenzungsschutz *m* over-power protection, power-limiting protection
Leistungsberechnung *f* capacity rating
Leistungsbereich *m* range of capacity; power range
Leistungsbeschreibung *f* specification, quantity description
Leistungsbestimmung *f* power rating
Leistungsbeziehung *f* power relation
Leistungsbremse *f* absorption dynamometer
Leistungsbrutreaktor *m* power-breeder
Leistungsdichte *f* power density
Leistungsdichteüberhöhung *f (nucl)* peaking
Leistungseinheit *f* power unit, unit of power
Leistungsentnahme *f* power consumption
Leistungserhöhung *f* increase in power
leistungsfähig efficient, high-performance, high-capacity
Leistungsfähigkeit *f* performance, efficiency; capability, capacity; output
L. des Kessels boiler capacity
Leistungsfaktor *m (el)* [vector] power factor
L. von Eins unit[y] power factor
induktiver L. lagging (reactive) power factor
kapazitiver L. leading (capacitive) power factor
nacheilender L. reactive (lagging) power factor
voreilender L. leading (capacitive) power factor
Leistungsfaktormesser *m* power-factor meter (indicator), phase meter
Leistungsfaktorverbesserung *f* power-factor improvement
Leistungsfernmeßgerät *n* telewattmeter
Leistungsfütterung *f* rationing according to yield
Leistungsgarantien *fpl* guarantees of performance
Leistungsgewicht *n (aero)* power/weight ratio, power to weight ratio; weight per horsepower
ausgewogenes L. balanced power-to-weight ratio
Leistungsgewinn *m* power gain
Leistungsgleichung *f* power equation
Leistungsgrenze *f* power limit
Leistungsimpuls *m* power pulse
Leistungskennwerte *mpl* performance characteristics
Leistungskoeffizient *m* power coefficient
Leistungskondensator *m* power capacitor
Leistungskontrollabteilung *f* efficiency department *(factory)*
Leistungsmesser *m (el)* wattmeter, power meter; dynamometer
zählender L. integrating wattmeter
Leistungsmeßsender *m* high-power test generator
Leistungsmessung *f* power measurement; output measurement
Leistungsmodulation *f* power modulation
Leistungsnormal *n* power standard
Leistungsparameter *m* performance parameter
Leistungspegel *m* output (power) level
Leistungsprobe *f s.* Leistungsprüfung
Leistungsprüfung *f* efficiency test, performance trial
Leistungsrauschen *n (nucl)* power fluctuation
Leistungsreaktor *m* power reactor
Leistungsregelung *f* power control (regulation); governor gear
Leistungsregler *m* load governor, output regulator
Leistungsregulierung *f* power control (regulation)
Leistungsrelais *n* power relay

Leistungsreserve *f* power reserve
Leistungsrichtungsrelais *n* power direction relay
Leistungsrichtungsschutz *m* directional power protection
Leistungsröhre *f* power valve
Leistungsrückgang *m* decrease of output
Leistungsschalter *m* circuit-breaker, power switch (breaker)
 L. mit automatischer Wiedereinschaltung self-closing circuit breaker
 L. mit Freiauslösung trip-free circuit breaker
 L. mit Querbespülung cross-jet circuit breaker
 dreipoliger L. three-pole circuit breaker
 einpoliger L. single-pole circuit breaker
 ölarmer L. low oil-content circuit-breaker
 zweipoliger L. double-pole circuit-breaker
Leistungsschalterantrieb *m* circuit-breaker drive
Leistungsschaubild *n* performance diagram
Leistungsschild *n (mach)* rating plate
Leistungsschreiber *m* power [level] recorder
Leistungsschwingkreis *m* tank circuit
Leistungssender *m* power oscillator (transmitter)
Leistungssteigerung *f* increase of output
Leistungssteller *m (el)* power control, *(sl)* output pad
Leistungsstufe *f* power stage
Leistungsteil *m* power unit
Leistungsteiler *m* power divider *(waveguide)*
Leistungstransformator *m* power transformer
Leistungstransistor *m* power transistor
Leistungstrenner *m* load (power) isolator
Leistungstrennschalter *m* disconnecting circuit-breaker
Leistungsübertragung *f* power transmission (transfer); power transmitting capacity
 größte L. maximum power transfer
Leistungsumspanner *m* power transformer
leistungsunfähig inefficient in operation
Leistungsverbrauch *m* power input (absorption, consumption)
Leistungsverlust *m* loss of efficiency; power loss
Leistungsverminderung *f* degradation of performance
 L. durch Lagerung storage depreciation
Leistungsvermögen *n* output capacity
Leistungsverstärker *m* 1. booster relay *(pneumatics)*; 2. power amplifier *(electronics)*; 3. power unit *(hydraulics)*; *(telev)* image power amplifier
 pneumatischer L. air (booster) relay
Leistungsverstärkung *f* power (energy) amplification
Leistungsversuch *m* output test
Leistungsvorwählschalter *m (el)* demand setter
Leistungswandler *m (el)* power converter, transducer
Leistungswert *m* engine performance *(of an engine)*; hull performance *(of a ship)*
Leistungswicklung *f* power winding
Leistungszahl *f* figure of merit
Leistungszerhacker *m (el)* power inverter
Leistungsziffer *f* coefficient of performance
Leistungszulage *f* incentive bonus
Leitantenne *f* director
Leitapparat *m* 1. diffuser; 2. distributor *(turbine)*
Leitartikel *m* editorial
Leitbake *f* range beacon
Leitband *n* master tape *(with recorded data)*
Leitbaum *m (mine)* guide
Leitbaumverdickung *f* buffer beam
Leitblech *n* 1. *(mach)* chute; 2. *(ag)* guide plate *(seed cleaner)*; 3. *(aut)* deflector [guide]; 4. *(ship)* baffle [plate]
Leitdamm *m* jetty, pier
Leitebene *f* plane-of-flight reference
Leitelektron *n* conduction electron
Leitelement *n* tracer element
leiten 1. to conduct *(current)*; to pipe, to carry *(fluids)*; to drain; to pass (over, through); to convey; 2. to run [a factory]; to be in charge of; to guide; to direct; to manage; to control
Leiter *f* 1. ladder; 2. stay *(sail)*
Leiter *m* 1. manager; 2. *(el)* conductor
 L. 1. Klasse first class conductor
 L. mit schwebendem Potential ignored conductor
 äußerer L. external conductor
 biegsamer L. flexible conductor
 blanker L. plain (bare) conductor
 elektrischer L. conductor
 geerdeter L. earthed conductor
 geschichteter L. laminated conductor
 gestreckter L. straight conductor
 guter L. good conductor
 idealer L. perfect conductor
 isolierter L. insulated conductor
 massiver L. solid conductor
 mitschwingender L. resonant (equifrequent) conductor
 schlechter L. poor conductor
 spannungführender L. live conductor
 stromführender L. charged (current-carrying) conductor
 unter Putz verlegter L. hidden conductor
 verzinnter L. tinned conductor
 widerstandsbehafteter L. resistive conductor
Leiterabstand *m* conductor spacing
Leiteranordnung *f* conductor arrangement, system of conductors
Leiterausnutzungsfaktor *m* conductor utilization factor
Leiterband *n* strip conductor
Leiterbaum *m* ladder beam (side)
Leiterbock *m* ladder hoist gantry *(dredger)*

Leiterbruch *m* circuit (core, conductor) break
  L. im Kabel cable fault
  mechanischer L. break of line, line break
Leiterbruchausschalter *m* conductor break cut-out
Leiterbruchschutz *m* open-phase protection
Leiterbruchwächter *m* conductor break cut-out
Leiterbündel *n* bunch of conductors
Leiterdicke *f* conductor diameter (size); gauge
Leitererdspannung *f* voltage to neutral
Leitergang *m* *(geol)* ladder vein
Leitergerüst *n* ladder scaffolding
Leiterholm *m* ladder beam (side)
Leiterkessel *m* Climax boiler
Leiternetz *n* network
Leiterplatte *f* printed circuit board (panel)
Leiterquerschnitt *m* conductor cross-section
Leiterseele *f* core conductor
Leiterseil *n* stranded conductor
Leiterspannung *f* circuit (outer conductor, line-to-line, phase-to-phase) voltage, voltage between lines, voltage of a system, line voltage of a polyphase system
Leitersprosse *f* ladder round (rung); ladder step
Leiterstange *f* ladder pole
Leiter-Sternpunkt-Spannung *f* voltage to neutral
Leitertafel *f* *(math)* alignment chart, nomogram
Leiterunterbrechung *f* conductor disconnection (break), *(Am)* conductor dis-lead
Leiterwellenlänge *f* guide wavelength
Leitfaden *m* textbook
leitfähig conductive, conductible, capable of conducting
Leitfähigkeit *f* conductivity, conductibility, conducting capacity
  anisotrope L. anisotropic conductivity
  dielektrische L. permittivity
  elektrische L. electric conductivity (conductance)
  isotrope L. isotropic conductivity
  lichtelektrische L. photoelectric conductivity
  longitudinale L. longitudinal conductivity
  magnetische L. permeability, permeance
  molare L. molar conductivity (conductance)
  richtungsabhängige L. asymmetric[al] conductivity
  spezifische elektrische L. [electric] conductivity, conductibility
  spezifische L. des Erdbodens earth conductivity
  thermische L. thermal conductivity
  transversale L. transverse conductivity
  unendlich große L. infinite conductivity
Leitfähigkeitsdichtemesser *m* conductance-bridge hydrometer
Leitfähigkeitsfeuchtemesser *m* conductivity-type moisture meter
Leitfähigkeitsmeßfühler *m* conductivity sensor
Leitfähigkeitsmeßzelle *f* conductivity cell
Leitfähigkeitsmethode *f* conductance method
Leitfähigkeitsprüfung *f* conductivity test
Leitfähigkeitswasser *n* conductivity water
Leitfernrohr *n* guiding telescope
Leitfeuer *n* range (leading, guide) light
Leitflächenruder *n* *(ship)* contra-guide rudder
Leitflosse *f* *(ship)* guide fin
Leitfossil *n* *(geol)* guide (index) fossil
Leitfunkstelle *f* directing station
Leitgang *m* *(geol)* lead vein
Leitgerät *n* control (manual-automatic) station
Leitgewebe *n* conducting tissue *(wood)*
Leithorizont *m* datum horizon
Leitisotop *n* [isotopic] tracer, tracer isotope, indicator
  radioaktives L. radioactive tracer, radiotracer
Leitisotopenmethode *f* tracer technique (method), labelled atoms method
Leitkabel *n* guide (leader) cable; pilot cable *(navigation)*
Leitkarte *f* *(dat)* master card
Leitkartensortiervorrichtung *f* group sorting device
Leitklampe *f* *(ship)* cable chock
Leitkranz *m* rim of guide blade disk (diaphragm) *(turbine)*
Leitlack *m* *(el)* conductive lacquer
Leitlineal *n* *(mach)* guide (tangent, guiding) bar, former plate *(e.g. taper turning)*
Leitlinie *f* directrix
Leitlinienimpulsverfahren *n* multi-track pulse range
Leitlochstreifen *m* pilot tape
Leitmaschine *f* *(el)* guiding machine *(interconnected systems)*
Leitmauer *f* guide wall
Leitmineral *n* index mineral
Leitnocken *m* *(mach)* replica of the cam
Leitprogramm *n* executive routine, master programme
Leitrad *n* guide disk (blade) *(turbine)*
Leitradbüchse *f* bush (liner) of diaphragm *(turbine)*
Leitradnabe *f* boss of guide wheel (diaphragm) *(turbine)*
Leitradscheibe *f* diaphragm, guide blade disk *(turbine)*
Leitrohr *n* guiding telescope, guider
Leitrohrtour *f* conductor (surface) pipe, surface string
Leitrolle *f* guide (idle, jockey) pulley
Leitsalze *npl* conducting salts
Leitschaufel *f* guide blade (bucket) *(turbine)*
Leitschaufelkranz *m* vane ring
Leitschaufelsegment *n* guide blade segment *(turbine)*

Leitschaufelsektor *m s.* Leitschaufelsegment
Leitschaufelträger *m* inlet housing
Leitscheibe *f* diaphragm, guide blade disk *(turbine)*
Leitschicht *f* 1. *(el)* conductive layer; 2. *(geol)* key bed
Leitschiene *f (mach)* chute; guide bar; guard rail *(switch)*
Leitschienenzuführung *f* chute feeding, feed chute
Leitseil *n (build)* tagline
Leitsender *m* regional broadcasting station
Leitspindel *f (mach)* lead screw
Leitspindelmutter *f* lead screw nut
Leitstange *f s.* Lenkerstange
Leitstation *f* control station
Leitstelle *f* directing station
Leitstrahl *m* 1. *(el)* track beacon; 2. *(math)* position (radius) vector; 3. *(rock)* guide beam
Leitstrahlanflugverfahren *n* beam procedure
Leitstrahlaufschaltung *f (rock)* signal mixing for guide beam, guide-beam superposition
Leitstrahlbordgerät *n* guide-beam receiver
Leitstrahlempfänger *m* missile guide-beam receiver
Leitstrahlführung *f* beam control pilotage
Leitstrahlgerät *n* guide-beam unit
Leitstrahllandeverfahren *n* beam approach beacon system, BABS
Leitstrahllenkung *f (rock)* beam [follower] guidance, beam-rider guidance
Leitstrahllinie *f* equisignal line *(radar)*
Leitstrahlsender *m* directional radio beacon
Leitstrahlverfahren *n (rock)* localizer beam method
Leitstrom *m* conduction current
Leitton *m* course signal
Leit- und Zugspindeldrehmaschine *f* sliding and screw-cutting lathe
Leitung *f* 1. administration; management; 2. line, lead, circuit; conduit; mains
  L. für gerichteten Verkehr one-way circuit
  L. für wechselseitigen Verkehr bothway circuit, *(Am)* two-way circuit
  L. mit Induktionsschutz transposition line
  L. mit Stecker plug terminated cord
  abgeglichene L. balanced transmission line
  abgehende L. outgoing circuit
  abgeschirmte L. screened line
  am Ende kurzgeschlossene L. closed-end line
  angerufene L. *(telec)* called line
  ankommende L. *(telec)* incoming circuit
  belastete L. loaded line (circuit)
  besetzte L. *(telec)* busy line; engaged line
  bespulte L. coil-loaded line
  bewegliche L. flexible cord (cable, wire)
  biegsame L. flexible lead (cord, cable, wire)
  direkte L. direct line
  durchgehende L. transit line
  eindrähtige L. earth return circuit
  einseitig betriebene L. unidirectional circuit
  freie L. disengaged line
  geräuschfreie L. quiet circuit
  gestörte L. circuit out of order
  ideale L. lossless line
  inhomogene L. non-homogeneous line
  konzentrische L. coaxial line
  koronafreie L. corona-free line
  künstliche L. artificial line
  leerlaufende L. no-load line, line under no-load conditions
  lichtelektrische L. photoconduction
  mehradrige L. multi-core line
  mehrfache L. multiple line
  offene L. open-ended line
  positive L. positive wire
  provisorisch verlegte L. temporary wire
  symmetrische L. balanced circuit (transmission line)
  umhüllte L. sheathed wire
  unterirdische L. underground line
  verlustarme L. low-loss line
  verlustlose L. zero-loss circuit, lossless line
  vieladrige L. multi-core line
  wasserdichte L. watertight line
  wetterfeste L. weather-proof line
Leitungen *fpl (build)* services
Leitungsabschluß *m* line termination
Leitungsabschnitt *m* section of line
Leitungsabzweigung *f* branching of a circuit
Leitungsanschluß *m* line terminal
Leitungsaufseher *m* section lineman
Leitungsband *n* conduction band *(semiconductors)*
Leitungsbau *m* line construction (erection)
Leitungsbeheizung *f* **durch Kurzschluß** short-circuit line-heating
Leitungsbelag *m* propagation (linear electric) constant
Leitungsberührung *f* line-to-line fault
Leitungsblockierung *f* lock out *(telephony)*
Leitungsbruch *m* circuit-break
Leitungsbündel *n* bunch (group) of lines
Leitungsdämpfung *f* line attenuation; transmission loss
Leitungsdraht *m* conductor, conducting line
Leitungsdruck *m* 1. line pressure; 2. *(aut)* manifold pressure
Leitungsdurchhang *m* sag of an overhead line
Leitungseinführung *f* bush
Leitungselektron *n* conduction electron
Leitungsende *n* line terminal (termination)
Leitungsentzerrer *m* line equalizer
Leitungsentzerrung *f* line equalization
Leitungsersatzbild *n* equivalent circuit *(waveguides)*

Leitungserscheinungen *fpl* conductivity phenomena
Leitungserwärmung *f* conductor heating
Leitungsfehler *m (el)* line fault, fault on a line
Leitungsführung *f* running of wires, cable running
Leitungsglied *n* section
Leitungsgraben *m* utility (service) trench
Leitungshalterung *f* wire guide
Leitungsheizfläche *f* conduction heating surface
Leitungsinduktivität *f* line inductivity
Leitungsisolator *m* line insulator
Leitungskanal *m* service duct (channel), utility run
Leitungskapazität *f* line capacitance
Leitungsklemme *f* cable clamp (clip), line clamp *(overhead line)*
Leitungskoeffizient *m* coefficient of conductivity
Leitungskompensation *f* line compensation
Leitungskonstante *f* circuit (conductor) constant
Leitungskreuzung *f* crossing of lines
Leitungslänge *f (el)* line length
 elektrische L. electrical length of line
 scheinbare L. apparent length of line
 wirksame L. electrical length of [transmission] line
Leitungsmast *m* transmission-line pole (tower)
Leitungsmaterial *n* conductor material
Leitungsmethode *f* management technique
Leitungsnachbildung *f* line balance, balancing network, imitated line
Leitungsnetz *n* network
Leitungsparameter *m* circuit parameter, linear electrical constants
Leitungsplan *m* wiring diagram (scheme)
 L. für Rohrleitung piping (tubing, pumping) plan
Leitungsprobe *f* test call
Leitungsprüfer *m* continuity tester
Leitungsquerschnitt *m* line cross-section
Leitungsrauschen *n* line noise
Leitungsreaktanz *f* line reactance
Leitungsrelais *n* line relay
Leitungsrohr *n* conduit; pipe
Leitungsschleife *f* looped circuit, loop
Leitungsschnur *f* [flexible] cord, *(sl)* flex
Leitungsschutzschalter *m* automatic cut-out
Leitungsschwingkreis *m* resonant line
Leitungsstöpsel *m* plug
Leitungsstrom *m* conduction current; charge-carrier motion
Leitungsstück *n* / **gerades** straight-run
Leitungssystem *n* 1. piping *(fluid)*; 2. line system
 erdsymmetrisches L. balanced line system
Leitungstätigkeit *f* management
Leitungstheorie *f (el)* line theory; four-terminal network theory
Leitungsträger *m* wire carrier
Leitungsüberprüfung *f* examination of line
Leitungsüberschlag *m (el)* arc-over between phases
Leitungsübertrager *m* line transformer, repeating coil
Leitungsunterbrechung *f* [line] disconnection
Leitungsverbinder *m* cable connector, conductor joint
Leitungsverkürzung *f* [artificial] line shortening
Leitungsverlängerung *f* line extension
Leitungsverlegung *f* line installation, wiring
 L. auf Putz surface wiring
 offene L. exposed wiring
Leitungsverlust *m* line attenuation (loss)
Leitungsvermögen *n* conductivity
Leitungsverstärker *m* line amplifier
Leitungsverteiler *m* distribution frame
Leitungswähler *m* line (final) selector *(telephony)*
Leitungswasser *n* mains (tap) water
Leitungsweg *m* [forwarding] route *(goods)*; route of travel *(passengers)*
Leitungswiderstand *m* line resistance
Leitungswinkel *m* elbows *(waveguides)*
Leitungszahl *f* consignment number *(cables)*
Leitverkehr *m* guided system
Leitvermögen *n* conductivity
 spezifisches elektrisches L. electric conductivity
Leitwalze *f* guide roller
Leitwand *f (build)* guide wall
Leitweg *m* route, routing
Leitwegangabe *f* route indication
Leitwegkenngruppe *f* routing indicator
Leitwerk *n* control element (section, unit) *(computer)*; *(aero)* empennage, tail group
Leitwerkflosse *f* fixed tail surface
Leitwerksfläche *f* tail control surface
Leitwerksträger *m* tailboom
Leitwert *m* admittance, conductance
 L. bei kleinem Signal small-signal forward transadmittance
 L. einer Elektronenröhre anode conductance
 L. in der Stromrichtung forward conductance
 akustischer L. acoustic admittance
 elektrischer L. conductance
 komplexer L. [complex] admittance, vector admittance
 magnetischer L. permeance
 relativer L. normalized (reduced) admittance
Leitwertoperator *m* vector admittance
Leitwiderstand *m* conductive resistance
Leitzahl *f* 1. *(el)* routing code; 2. *(opt)* guide-number
Leitzelle *f (dat)* control word
Lektor *m* 1. publisher's reader; 2. tutor *(university)*

Lemma *n* lemma
Lemniskate *f* lemniscate
Lemongrasöl *n* lemon grass oil
Lenardfenster *n* Lenard window
Lenkachse *f (aut)* steering control shaft
lenkbar manœuvrable, steerable
Lenkbarkeit *f* manœuvrability, steerability
Lenkeinrichtung *f* / **radiogesteuerte automatische** radio autopilot coupler
Lenkerstange *f (mach)* radius rod *(radius planing)*
Lenkfinger *m* steering finger
Lenkgabel *f* steering fork
Lenkgehäuse *n* steering box (gear housing)
Lenkgestänge *n* steering linkage
Lenkgetriebe *n* steering gear
Lenkhebel *m* steering lever (knuckle); steering control arm
Lenkhebelanschlag *m* steering-arm stop
Lenkmutter *f* steering nut
Lenkrohr *n* steering tube
Lenkrohrlager *n* steering-column bearing
Lenkrohrstummel *m* steering tube extension
Lenkrolle *f* steering roller; steering idler
Lenksäule *f* steering pillar (column)
Lenksäulenhalter *m* steering-column bracket
Lenksäulenrohr *n* steering-column tube
Lenkschenkel *m* steering arm
Lenkschloß *n* steering-column lock
Lenkschneckenarm *m* pitman arm
Lenkschraube *f* steering cam
Lenkschubstange *f* steering drag link (rod)
Lenksegment *n* steering segment
Lenkspindel *f* steering column (screw)
Lenkspurhebel *m* drop arm, drag link
Lenkspurstange *f* steering track rod
Lenkstange *f* drag link; handle bar *(cycle)*
Lenkstock *m* steering column assembly
Lenkstockhebel *m* pitman (drop, steering-gear) arm
Lenkstockschaltung *f* steering-column gear change
Lenkstockspindelsegment *n* steering shaft sector
Lenkung *f* 1. *(aut)* steering; 2. *(rock)* guidance
  L. des Hangenden *(mine)* roof (strata) control
  L. mit Hilfe astronomischer Ortung *(rock)* star-tracking guidance
  L. mit hydraulischer Lenkhilfe *(aut)* hydraulically-assisted steering
  L. mit Kugelring *(aut)* ball-recirculating steering gear
  L. mit pneumatischer Lenkhilfe *(aut)* air-pressure-assisted steering
  selbstzentrierende L. *(aut)* self-centring steering
Lenkungsausschlag *m* steering lock
Lenkungsbock *m* steering-gear mounting
Lenkungseinschlag *m* steering lock
Lenkungskappe *f (therm)* guide-cap
Lenkungssystem *n (rock)* guidance system
Lenkungsventil *n (aut)* steering valve
Lenkverbindungsstange *f* connecting rod of steering assembly
Lenkzapfen *m (aut)* swivel (steering-knuckle) pin, steering pivot
Lenkzwischenhebel *m (aut)* idler arm
Lentz-Steuerung *f* Lentz valve gear
Lenzeinrichtung *f (ship)* freeing appliances
lenzen to scud; to free a ship, to pump out water ballast
Lenzhahn *m* bilge cock
Lenzleitung *f* bilge main (suction), bilge [pipe] line
Lenzpumpe *f* bilge pump
lenzpumpen to free *(a vessel)*
Lenzrohr *n* bilge pipe
Lenzsumpf *m* drain sump
Lenztank *m* drain tank
Lenz- und Ballasteinrichtungsplan *m* pumping plan
Lenz- und Ballastleitung *f* pumping line
Lenz- und Feuerlöschpumpe *f* general (water) service pump
Leonardsatz *m* Ward-Leonard set
Leonardschaltung *f* Ward-Leonard control
Lepidolith *m (min)* lepidolite, lithia mica
Lepolverfahren *n* Lepol process *(cement manufacture)*
Leporellofalzmaschine *f* interfolder
Leporellofalzung *f* accordion folding
Leporelloformular *n (dat)* continuous form
Leporelloformularführung *f* continuous form guide
Lepton *n* lepton
Lernen *n* / **maschinelles** machine learning
Lesbarkeit *f* readability
Leseband *n* picking belt
Leseberge *pl* hand-picked dirt
Lesegerät *n* reader, reading head
Lesegeschwindigkeit *f* reading speed
Leseglas *n* reader lens, reading glass
Lesekopf *m (dat)* reading head
Leseleuchte *f* reading lamp
Leselocher *m (dat)* punch reader
lesen 1. *(dat)* to read; to lecture *(university)*; 2. *(mine)* to hand-clean, to hand-pick; 3. *(ag)* to gather, to pick [up]
  Ähren l. to glean
  Zeichnungen l. to interprete (read) drawings
Lesen *n (dat)* reading
  destruktives L. destructive reading
  nichtlöschendes L. non-destructive reading
  zerstörendes L. destructive reading *(of magnetic core memories)*
  zerstörungsfreies L. non-destructive reading *(of magnetic memories)*

Leser *m* **für durchlaufende Rollen** continuous roll reader *(for cash registers)*
Lese-Schreibkopf *m* **/ kombinierter** *(dat)* read-record head
Leseverstärker *m* *(dat)* read amplifier
Lesezwischengut *n* intermediate product
letal lethal
Letaldosis *f* *(nucl)* lethal (fatal) dose
mittlere L. median lethal dose, MLD
Letalität *f* lethality
Letalmutation *f* lethal mutation
Letalzeit *f* **/ mittlere** median lethal time, MLT
Lethargie *f* [neutron] lethargy
Letten *m* clay
Lettenbesatz *m* clay stemming
Lettenbesteg *m* clay course (gouge), flucan, gouge, selvage, selvedge
lettenhaltig clayey
Letternmetall *n* type metal
lettig clayey
Leuchtbake *f* light beacon
Leuchtboje *f* lighted (light, gas) buoy
Leuchtdecke *f* luminous ceiling
Leuchtdichte *f* luminance; illuminating power per square metre, brightness
Leuchtdichtekontrast *m* contrast of brightness
Leuchtdraht *m* filament
doppelt gewendelter L. coiled-coil filament
gewendelter L. coiled filament
ringförmig angeordneter L. ring filament
Leuchte *f* light[ing] fitting, light[ing] fixture; lamp
indirekte L. ceiling reflecting lamp, indirect fitting
schlagwettergeschützte L. flameproof lighting fitting
Leuchtelektron *n* emitting (luminous) electron
leuchten to emit light; to shine *(star)*
leuchtend bright; luminescent *(dial)*
Leuchtenglocke *f* globe
Leuchtenschale *f* lamp basin, bowl
Leuchtenschirm *m* shade
Leuchtenwirkungsgrad *m* light output ratio of a fitting, efficiency of a lighting fixture
Leuchtfaden *m* filament
Leuchtfarbe *f* luminous (fluorescent) paint
Leuchtfeld *n* luminous field
Leuchtfeldblende *f* luminous-field diaphragm
Leuchtfeuer *n* [scintillating] light, beacon fire
L. mit wechselnden Farben combination light
unterbrochenes L. intermittent (occulting) light
Leuchtfeuerkennung *f* light's characteristic
Leuchtfeuersektor *m* light sector
Leuchtfeuerverzeichnis *n* list of lights, lights list, Admiralty List of Lights
Leuchtfleck *m* trace; light (luminous) spot *(cathode-ray valve)*
Leuchtfloß *n* light float
Leuchtgas *n* illuminating (lighting, town) gas
Leuchtgranate *f* star shell
Leuchtkorn *n* *(mil)* luminous fore (front) sight
Leuchtkörper *m* luminary
Leuchtkraft *f* 1. luminous power; 2. brilliancy *(shade)*
Leuchtnukleon *n* emitting nucleon
Leuchtöl *n* kerosene
Leuchtpatrone *f* Very light
Leuchtpistole *f* light (flare, signal, Very) pistol
Leuchtrakete *f* illuminating rocket
Leuchtröhre *f* tubular discharge lamp
Leuchtsatz *m* pyrotechnic composition
Leuchtschaltbild *n* illuminated *(mimic)* diagram
Leuchtschiff *n* lightship, lightvessel, floating light
Leuchtschirm *m* fluorescent (luminescent) screen, screen
L. einer Farbschriftröhre dark trace screen
Leuchtschirmbild *n* cathode ray trace (pattern)
Leuchtschirmfleck *m* **/ blinder** blind spot
eingebrannter L. blank spot
Leuchtschirmröhre *f* luminescent-screen tube
Leuchtskale *f* illuminated dial
Leuchtspitze *f* flare point *(torpedo)*
Leuchtspur *f* light trace
Leuchtspurgeschoß *n* tracer
Leuchtspurmunition *f* tracer ammunition
Leuchtstärke *f* luminous intensity
spezifische L. radiance
Leuchtstoff *m* luminescent material, luminous paint
Leuchtstofflampe *f* fluorescent (luminescent) lamp, gas discharge lamp
Leuchtstoffmesser *m* fluorescence meter
Leuchtstoffröhre *f* fluorescent lamp
Leuchtstoffröhrenlampe *f* tubular fluorescent (luminescent) lamp
Leuchtstrichlänge *f* length of fluorescent trace
Leuchttonne *f* lighted (light, gas) buoy
Leuchtturm *m* lighthouse
Leuchtturmtender *m* *(ship)* lighthouse tender *(vessel)*; *(Am)* Trinity-house tender
Leuchtturmwärter *m* lighthouse keeper
Leuchtuntersatz *m* translumination base
Leuchtwinkel *m* light sector *(navigation)*
Leuchtzeichen *n* signal lamp; *(mach)* telltale light
Leuchtzeiger *m* luminous hand *(clock)*; luminous pointer
Leuchtzifferblatt *n* luminous dial
Leukobase *f* leuco base
Leukofarbstoff *m* leuco dye[stuff]
Leukoverbindung *f* *(ch)* leuco compound
Leuzin *n* leucine
Leuzit *m* *(min)* leucite
Lezithin *n* lecithin
Lias *m* *(f)* *(geol)* lias

Liasformation *f* lias[ic] formation
Libelle *f* water-level, air (bubble) level
Libellenausschlag *m* deviation of the bubble
Libellenblase *f* bubble of the level
Libellenweg *m* run of the bubble
licht 1. clear *(e.g. span)*; 2. light *(red silver ore)*
Licht *n* light
anregendes L. exciting light
auffallendes L. incident light
ausgestrahltes L. emitted light
diffuses L. diffused light
einfallendes L. incident light
elektrisches L. electric light
farbiges L. coloured light
gebrochenes L. refracted light
gerichtetes L. directional lighting
linear polarisiertes L. plane polarized light
monochromatisches L. monochromatic light
polarisiertes L. polarized light
reflektiertes L. reflected light
ultraviolettes L. ultraviolet light
zerstreutes L. scattered light
Lichtabdeckschirm *m (telev)* flag
Lichtaberration *f* light aberration
Lichtabsorption *f* light absorption, absorption of light
Lichtabtastung *f* flying-spot scanning
Lichtanlage *f* lighting installation
Lichtanschlußschnur *f* lighting flex
Lichtäquivalent *n* / **mechanisches** mechanical equivalent of light
Lichtausbeute *f* scintillation response; light yield (output); luminous efficiency
L. einer Lichtquelle luminous efficiency, efficiency of a luminous source
Lichtausbreitung *f* propagation of light
Lichtausstrahlung *f* radiation of light
spezifische L. luminous emittance, light emissivity, radiance
Lichtband *n* lighting row
lichtbeständig fast to light, non-fading *(colour)*
Lichtbeständigkeit *f* light fastness (resistance)
Lichtbild *n* photograph
Lichtbildervortrag *m* illustrated lecture
Lichtblitz *m* flash of light; illuminating flash
Lichtblitzdauer *f* duration of flash
Lichtblitzstroboskop *n* Elverson oscilloscope
Lichtbogen *m* [electric] arc
L. mit Fremdheizkatode externally heated arc
abgeschirmter L. shielded arc
dynamischer L. dynamic arc
ruhiger L. silent arc
singender L. singing arc
stationärer L. steady-state arc
tönender L. singing arc
wandernder L. migrating arc
Lichtbogenabfall *m* arc drop *(of rectifier)*
Lichtbogenbespülung *f* scavenging of arc
lichtbogenbeständig arc-resistant
Lichtbogenbildung *f* arcing
Lichtbogenbohren *n* electric arc drilling
Lichtbogenbrennschneiden *n* arc cutting
Lichtbogendauer *f* arc duration
Lichtbogenentladung *f* arc discharge
Lichtbogenerdschluß *m* arcing earth
Lichtbogenfußpunkt *m* foot (footing point) of an arc
Lichtbogengleichrichter *m* arc rectifier
L. mit Funkensteuerung pilot spark arc rectifier
Lichtbogenheizung *f* [electric] arc heating
Lichtbogenhülle *f* arc shield
Lichtbogenkatode *f* arc cathode
Lichtbogenkern *m* arc centre (core)
Lichtbogenlampe *f* arc lamp
Lichtbogenlänge *f* arc length
Lichtbogenleistung *f* power consumption of arc
Lichtbogenlöscheinrichtung *f* quencher, arc break
Lichtbogenlöschkammer *f* arc quench chamber
Lichtbogenlöschmittel *n* [arc-]extinguishing medium, [arc-]quenching medium
Lichtbogenlöschung *f* arc quenching
Lichtbogenofen *m* [electric] arc furnace
direkter L. direct-arc furnace, electrode-hearth furnace
Lichtbogenpfad *m* arc path
Lichtbogenplasma *n* arc stream
Lichtbogenplasmabrenner *m* arc plasma torch
Lichtbogenschaden *m* arc deformation
Lichtbogenschaukelofen *m* rocking arc furnace
Lichtbogenschleife *f* arc loop
Lichtbogenschutzarmatur *f* arc-protective fitting, arcing device
Lichtbogenschutzkammer *f* arc deflector
Lichtbogenschutzring *m* arcing ring
Lichtbogenschweißen *n* arc welding
L. unter Schutzgas inert-gas arc welding
atomares L. atomic hydrogen welding
offenes L. unshielded arc welding
verdecktes L. shielded arc welding
Lichtbogenschweißtransformator *m s.* Lichtbogenschweißumspanner
Lichtbogenschweißumspanner *m* arc welding transformer
selbstregelnder L. self-regulating arc-welding transformer
Lichtbogenschweißverfahren *n* **mit Schweißelektrode** metal arc welding process
Lichtbogenschwingungen *fpl* arc oscillations
Lichtbogensender *m* arc transmitter
Lichtbogenspannung *f* arc voltage
Lichtbogenspritzen *n* arc spraying
lichtbogenstabilisierend arc-stabilizing
Lichtbogenüberschlag *m* arc[ing]-over

Lichtbogenventil *n* arc rectifier
Lichtbogenverlängerung *f* arc extension (lengthening)
Lichtbogenverluste *mpl* arc-drop losses
Lichtbogenwanderung *f* arc migration
Lichtbogenzeit *f* arcing time, arc duration
Lichtbogenzündung *f* arc ignition
lichtbrechend refractive, refracting, refringent
Lichtbrechung *f* refraction of light
Lichtbündel *n (ph)* light pencil, pencil of light
lichtdicht light-tight
Lichtdiffusor *m* diffuser
Lichtdrehschalter *m* light spindle switch
Lichtdruck *m* 1. *(ph)* light pressure, pressure of light; 2. *(print)* collotype [printing]; phototype
Lichtdruckfarbe *f* collotype ink
Lichtdurchlässigkeit *f* light transmittance (permeability, transmissivity), translucence, transparency
Lichtdurchlässigkeitsprüfer *m* transparency tester
lichtecht fast to light, non-fading
Lichtechtheit *f* fastness to light, light fastness, colour stability
Lichteinfall *m* light incidence
streifender L. grazing incidence
Lichteinheit *f* photometric unit
Lichteinwirkung *f* action of light
lichtelektrisch photoelectric
lichtempfindlich light-sensitive
l. machen to sensitize
Lichtempfindlichkeit *f* light-sensitivity; photosensitivity; speed of emulsion
statische L. static luminous sensitivity
Lichtempfindung *f* sensation of light
lichten 1. to clear *(forest)*; to thin out; 2. to weigh, to raise *(anchor)*
Anker l. to raise the anchor, to unmoor, to weigh anchor
Lichtenergie *f* luminous energy
Lichterführung *f* position of navigation lights, carrying of lights
Lichtfarbendruck *m* photochemical colour printing
Lichtfilter *n* light filter
Lichtfleck *m* light spot
Lichtfluß *m* light (luminous) flux
Lichtgeschwindigkeit *f* light velocity (speed), velocity of light
Lichtgeschwindigkeitsmessung *f* light velocity measurement
Lichthof *m* halo; halation *(electron ray)*
Lichthofbildung *f* halation
lichthoffrei anti-halo, nonhalating
Lichtintensität *f* light intensity
Lichtkabel *n* lighting cable
Lichtkasten *m* bank of light
Lichtkegel *m* illuminating pencil, cone of light, light (luminous) cone
Lichtkeim *m (ag)* chit, sprout
Lichtleiter *m* light pipe (guide, line)
Lichtleitung *f* electric lighting mains, lighting feeder (installation); *(aut)* light cable
Lichtmagnet *m* light magnet
Lichtmarke *f* light spot
Lichtmarkeninstrument *n* instrument with optical pointer
Lichtmaschine *f* dynamo
spannungsregelnde L. constant-voltage-control dynamo
stromregelnde L. third-brush dynamo, constant-current dynamo
Lichtmenge *f* quantity of light, luminous energy
Lichtmikroskop *n* light (optical) microscope
Lichtmikroskopie *f* light (optical) microscopy
Lichtmühle *f* / **Crookessche** Crookes's radiometer
Lichtnetz *n* lighting circuit (mains)
Lichtnetzantenne *f* mains aerial
Lichtpause *f* [blue]print, heliographic print, photostat
Lichtpausen *n* printing
Lichtpauspapier *n* sensitive paper for printing process; [blue]print paper; dyeline (black-line) paper
Lichtpausrahmen *m* printing frame
Lichtpausverfahren *n* printing process
Lichtpfadlänge *f* optical path length
Lichtpolarisation *f* polarization of light
Lichtpunkt *m* light spot
Lichtpunktabstand *m* light centre length
Lichtpunktabtaster *m* light-spot scanner, flying-spot scanner
Lichtpunktabtastung *f* light-spot method of scanning, flying-spot method of scanning
Lichtpunktlinienschreiber *m* light-spot recorder
Lichtpunktspeicher *m* flying-spot store
Lichtquant *n* photon, light quantum
Lichtquelle *f* light source
natürliche L. natural light-source
punktförmige L. punctiform light source
Lichtraum *m* clearance
Lichtraumprofil *n* **für Stromabnehmer** clearance gauge for pantographs
Lichtraumumgrenzung *f* structure gauge, maximum moving dimensions
Lichtreiz *m* light stimulus
Lichtrelais *n* photoelectric relay
Lichtrufanlage *f* lamp-signal call system
Lichtsatz *m* filmsetting; lumitype
Lichtschalter *m* installation (light, lighting) switch
Lichtschnittmeßgerät *n* light-sectioning tester
Lichtschnittverfahren *n* light-slit method
Lichtschranke *f* light barrier

Lichtschubschalter *m* light push switch
Lichtschutzbad *n* anti-fading bath
Lichtschutzpapier *n* / **schwarzes** black positive paper, black photo-paper
Lichtsetzmaschine *f* filmsetting machine, filmsetter, film composer
Lichtsignal *n* light signal, danger light
  elektrisches L. electric lamp signal
Lichtsirene *f* photoelectric siren
Lichtspalt *m* light slit (gap)
Lichtspektrum *n* light spectrum
Lichtspieltheater *n* cinema
Lichtspruch *m* light-signal message
Lichtspur *f* light track
lichtstark of high light intensity
Lichtstärke *f* luminous intensity; candle-power
  geometrische L. geometrical luminosity
  mittlere L. im Nutzkegel mean conical candle-power
  mittlere L. in der Nutzzone mean zonal candle-power
Lichtstärkeverteilungskurve *f* curve of intensity distribution
Lichtsteckdose *f* light socket, lamp receptacle
Lichtsteuerelektrode *f* control electrode
Lichtstrahl *m* light ray (beam), ray (beam) of light
  divergenter L. divergent light beam
  scharf gebündelter L. narrow light beam
Lichtstrahlabtastung *f* spot-light scanning
Lichtstrahloszillograf *m* moving-mirror oscillograph
Lichtstrahlung *f* light radiation
Lichtstreuung *f* light diffusion (scattering)
Lichtstrom *m* 1. luminous (light) flux; 2. mains current
Lichtstromdichte *f* luminous flux density
Lichtstromkreis *m* lighting circuit
Lichtstrommesser *m* integrating photometer
Lichtsumme *f* total emitted light, light sum
Lichttagessignal *n (ry)* day colour light signal
Lichttechnik *f* illumination engineering
Lichtteilchen *n* light corpuscle, photon
Lichttelefonie *f* optical telephony
Lichttelefoniegerät *n* photophone
Lichttheorie *f* theory of light
  elektromagnetische L. electromagnetic theory of light
Lichtton *m* optical sound
Lichttonverfahren *n* sound-on-film system
Lichttonverstärker *m* photocell amplifier
lichtundurchlässig opaque
Lichtung *f* clearing
Lichtventil *n* light valve (relay)
Lichtverlust *m* loss of light
Lichtverteilungskurve *f* luminosity (isophotic, light distribution) curve, isophote
  polare L. polar curve of light distribution
Lichtvisier *n* light beam localizer
Lichtweg *m* light path
Lichtwelle *f* light wave
Lichtwiderstand *m* light resistance
Lichtwirkung *f* luminous effect
Lichtwurflampe *f* projection lamp
Lichtzeiger *m* spot of light
  L. des Galvanometers galvanometer spot
Lichtzeigergalvanometer *n* light-beam galvanometer
Lichtzerfall *m* photodisintegration
Lichtzerhacker *m* light[-beam] chopper
Lichtzerlegung *f* decomposition of light
Liderring *m* piston (packing) ring
  schraubenförmiger L. coiled (helical) piston ring, coiled (helical) packing ring
  selbstspannender L. spring ring
Liderungsring *m s.* Liderring
Lieben-Röhre *f* Lieben-Reiss relay
Liebigkühler *m* Liebig condenser
Lieferant *m* supplier
lieferbar deliverable
Lieferbeton *m* ready-mix[ed] concrete
Lieferbetrieb *m* supplier, supplying factory
Lieferfrist *f* term of delivery
Liefergrad *m* / **volumetrischer** volumetric efficiency
Liefermenge *f* output; amount of delivery
liefern to supply; to deliver; to provide, to yield
Lieferschein *m* invoice, bill of delivery
Liefertermin *m* delivery date
Lieferumfang *m* schedule of parts supplied
Liefer- und Gütebestimmung *f* **[/ technische]** commercial standard
Lieferung *f* delivery, supply; yielding; consignment *(goods)*
  L. ab Zeche dispatches from mines
Lieferungsbedingung *f* condition (specification) of sale
Lieferungshafen *m* port of delivery
Lieferwagen *m* delivery van
Lieferwalze *f* feeding roller
Lieferzeit *f* delivery time
Lieferzylinder *m* feeding roller
Liegegebühr *f (ship)* demurrage
Liegegeld *n* demurrage
liegen to lie, to rest [on], to be [placed], to be situated (located)
  beigedreht l. to lie about *(vessel)*
  vor Anker l. to ride (lie) at anchor
Liegendes *n (mine)* bottom, floor, foot [wall], sill
Liegendkohle *f* bottom coal
Liegendschrämarm *m* undercutting jib
Liegeplatz *m* reclining berth, couchette; *(ship)* berth; lay-by
  fester L. appropriated berth
Liegeplatzkarte *f* sleeping-berth supplement

Liegesitz *m (ry)* reclining seat
Liegestall *m* resting stable
Liegewagen *m* coach with reclining berths (couchettes)
Liegewinkel *m (ship)* angle of repose
Liegezeit *f* lay days
Liek *n s.* Liektau
Liekgut *n s.* Liektau
Liektau *n (ship)* bolt rope
Liftslab-Verfahren *n (build)* lift-slab method, Youtz-Slick method
Ligatur *f* ligature, double letter
Lignin *n* lignin
Lignit *m* earthy brown coal, lignite
Ligroin *n* ligroin, petroleum (mineral) spirit
Limonen *n* limonene
Limonit *m* brown hematite (iron ore), limonite
Limousine *f* saloon car, limousine; *(Am)* sedan
Linalool *n* linalool
Linalylazetat *n* linalyl acetate
Linde *f* lime[-tree]
Lindeck-Rothe-Kompensator *m* deflection potentiometer
Lindenholz *n* lime wood
Lineal *n* straight edge, rule
L. zur Werkstückführung guide
linear linear
Linearbeschleuniger *m* linear [electron] accelerator
L. für Elektronen electron linear accelerator
Lineargleichungslöser *m* equation solver
linearisiert antiplane, linearized
Linearisierung *f* linearization, linearizing action
harmonische L. describing function; harmonic balance
Linearisierungswiderstand *m* linearizing (peaking) resistance
Linearität *f* linearity
Linearitätsregelung *f* frame linearity control, *(Am)* field linearity control
Linearplanung *f* linear programming
Linearverstärker *m* linear amplifier
Lineatur *f s.* Liniatur
Liner *m (ship)* liner
Liniatur *f* ruling, lining
Linie *f* line; [spectral] line; peak, maximum line; *(print)* rule
L. der elektrischen Feldstärke line of electric field strength
L. der magnetischen Feldstärke line of intensity of magnetic field
L. des Einfallens *(geol)* full dip
L. des geringsten Widerstandes *(mine)* line of least resistance
L. gleichen Druckes isobar, isobaric (constant-pressure) line
L. gleichen Mischungsverhältnisses *s.* L. gleicher Dampfmenge
L. gleichen Rauminhalts isopleric (constant-volume) line
L. gleicher Dampfmenge line of equal steam quantity
L. mit Gehrung mitred rule
atmosphärische L. atmospheric line
breite L. heavy line
doppelfeine L. double-fine rule
englische L. plain swelled rule
erlaubte L. *(nucl)* allowed line
feine L. plain (fine-faced) rule
fette L. full-faced rule
gemallte L. *(ship)* moulded line
gestrichelte L. dashed (broken) line
Lüdersche L. Lüders (Luders) line
Neumannsche L. Neumann band
punktierte L. dotted line
Linien *fpl* / **Fraunhofersche** Fraunhofer's lines
Linienabsorptionskoeffizient *m* line absorption coefficient
Linienabstand *m* line interval
Linienberührung *f* line contact *(gear)*
Linienbiege- und -schneidemaschine *f* rule bender and cutter
Linienblitz *m* forked lightning
Linienbreite *f* line width
Linienemission *f* line emission
Linienfahrer *m (ship)* liner
Linienfahrt *f* liner trade
Linienfahrtschiff *n* liner; mail boat
Linienflügel *m* wing of a line
Linienform *f* line shape
Linienfrachter *m* cargo liner
Linienführung *f (build)* route; route mapping
Linienhöhe *f* peak height
Linienintegral *n* line integral
Linienintensität *f* line intensity
Linienkern *m* core of a line
Linienkreuzung *f* crossing *(of lines)*
Linienladung *f (el)* line charge
Linienlast *f* line (linear) load
Linienmitte *f* line centre
Linienpassagierschiff *n* passenger liner
Linienquelle *f* line source
Linienraster *m* line screen
Linienreederei *f* liner company
Linienriß *m (ship)* lines (sheer draught) plan, lines [drawing]; mould-loft lines
Linienschiffahrt *f* ocean (merchant) shipping, liner trade
Linienschreiber *m* continuous recorder
Linienschwerpunkt *m* centre of gravity of a line
Linienserie *f* line series
Linienspektrum *n* line spectrum
Linienträgheitsmoment *n* linear moment of inertia
Linienverbreiterung *f* line broadening

Linienverkehr *m* line traffic
Linienverkettung *f* interlocking of lines
Linienverlauf *m* run of lines *(ship's form)*
Linienverschiebung *f* line shift
Linienverschmälerung *f* line narrowing
Linienverzweiger *m (el)* main cabinet
Linienwähler *m* peak selector
Linienziehgerät *n* ruling device
Linienzucht *f (ag)* line-breeding
linieren to rule
Liniermaschine *f* machine ruler, ruling machine
Linierung *f* lineation; ruling
Linke-Hand-Regel *f (el)* left-hand rule
links on the left; to the left; anti-clockwise *(rotation)*; left-hand
 nach l. to[wards] the left; leftwards
Linksabweichung *f* left-hand deviation
Linksanschlag *m* fully anti-clockwise position *(potentiometer)*
Linksappretiermaschine *f (text)* back-filling mangle
Linksarmmaschine *f* left-hand cylinder bed sewing machine *(shoe)*
Linksdrall *m* left-hand twist
linksdrehend 1. *(ph)* laevogyrate, laevorotatory, laevogyrous; 2. *(mach)* top-going *(mill)*; anti-clockwise rotating
Linksdrehung *f* 1. *(ph)* laevorotation; 2. *(mach)* anti-clockwise rotation; rotation to the left
Linkseinzug *m (text)* tie-up commencing at the left-hand side
linksgängig 1. *(math)* sinistrorse; 2. *(mach)* left-hand *(thread)*
Linksgewinde *n* left-hand thread
linksgewunden *(math)* sinistrorse
Linksimprägnierkalander *m (text)* back-filling mangle
Linkskrümmung *f* anti-clockwise curvature
Linkskurve *f* left-hand turn
Linkslauf *m* anti-clockwise rotation
Linkslenkung *f* left-hand steering
Links-Links-Flachstrickmaschine *f* flat purl knitting machine
Links-Links-Nadel *f* double-head needle
Linkspropeller *m* left-handed propeller
linksschneidend left-hand cutting
Linkssinn / im anti-clockwise
linksstricken to purl
Linkssystem *n* left-handed coordinate system
Links-Linksstrickmaschine *f* links-and-links machine
Linksverkehr *m* left-hand traffic
Link-Trainer *m* Link trainer
Linolensäure *f* linolenic acid
Linoleum *n* linoleum
Linolsäure *f* linoleic acid
Linolschnitt *m* lino cut
Linotype *f* linotype
Linotypemetall *n* linotype metal
Linotypezeile *f* machine (linotype) slug
Linoxin *n* linoxyn
Linse *f (opt)* lens
 achromatische L. achromatic lens
 asphärische L. non-spherical lens
 bikonkave L. biconcave (double-concave) lens
 bikonvexe L. biconvex (double-convex) lens
 elektrostatische L. dielectric (electrostatic) lens
 konkav-konvexe L. concavo-convex lens
 konvex-konkave L. convexo-concave lens
 magnetische L. electromagnetic lens
 plankonkave L. plane-concave lens
 plankonvexe L. plane-convex lens
 streuende L. diverging lens
Linsenachse *f* axis of a lens
Linsenfassung *f* lens mount
Linsenfehler *m* lens defect (error)
Linsenfernrohr *n* refracting telescope
Linsengang *m (geol)* lenticular vein
Linsenglied *n* lens component
Linsenkondensator *m* lens condenser
Linsenkranzabtaster *m (telev)* lens carrying scanning disk
Linsenlichtfleck *m (telev)* flare spot
Linsenobjektiv *n* lens objective
Linsenöffnung *f* aperture
Linsenscheitel *m* vertex of lens
Linsenspektrometer *n* lens spectrometer
Linsensystem *n* combination of lenses
Linsentrübung *f* clouding (opacity) of a lens
Lipase *f* lipase
Lipoid *n* lipoid
Lipolyse *f* lipolysis
Lipowitz-Metall *n* Lipowitz's alloy
Lippe *f* lip
Lippenapplikator *m* lip applicator
Lippendichtung *f* lip-type of packing
Lippenpfeife *f* flue (labial) pipe
Liquiduskurve *f* liquidus curve
Liquiduslinie *f* liquidus line
Lisene *f (build)* attached column
Lisseuse *f* lisseuse, back-washing machine
Liste *f* list; roll; register; catalogue
Liter *n* litre
Literaturangaben *fpl* bibliographical data
Literaturnachweis *m* references
Literleistung *f (aut, aero)* power per unit of displacement, specific power output
 L. des Motors engine output per 1000 cc. cylinder capacity
Litfaßsäule *f* advertising poster
Lithium *n* lithium
Lithiumbombe *f* lithium bomb
Lithiumboratglas *n* lithium borate glass
Lithiumbromid *n* lithium bromide

Lithiumfett *n* lithium base grease
Lithiumsilikatglas *n* lithium silicate glass
Lithiumgehalt *m* lithium content
Lithiumoxid *n* lithium monoxide, lithia
Lithograf *m* lithographer
Lithografenkalk *m* lithographic stone
Lithografenschiefer *m* lithographic stone
Lithografie *f* litho[graphy] *(process)*; lithograph *(print)*
Lithografiestein *m* litho stone
Lithopon[e *f***]** *n* lithopone
Litze *f* 1. *(text)* heald; *(Am)* [loom] heddle *(weaving)*; 2. *(el)* flexible wire, lace, strand, stranded wire (cable)
  doppeladrige L. twin flex[ible cord]
  feindrähtige L. multi-strand wire
  geflochtene L. braided wire
Litzenauge *n* heald eye, *(Am)* heddle eye *(weaving)*, harness (heald) eye, mail
Litzenberechnung *f* heald calculation
Litzendraht *m* flexible (stranded) wire
Litzenflechtmaschine *f* heald braiding machine
Litzenleiter *m (el)* stranded conductor
Litzenmaschine *f* plaiting machine
Litzenschiene *f (text)* ridge bar
Litzenschnur *f* flexible wire
Litzenseil *n* stranded rope
Livree *f* livery
Lizenz *f* licence
  eine L. erwerben to acquire a licence
Lizenzgebühr *f* licence fee
Lizenzinhaber *m* licensee, grantee of a licence
Lizenznehmer *m s.* Lizenzinhaber
Lizenzträger *m s.* Lizenzinhaber
Ljungström-Vorwärmer *m* Ljungstroem preheater
Löbbehobel *m (mine)* fast plough
Loch *n* hole; aperture, gap; pit; hollow; vacant electron site, defect electron
  schräg in den Stoß gebohrtes waagerechtes L. *(mine)* flanking hole
Lochabsteller *m (text)* fall-out detector
Lochband *n* punched tape
  ausgeschlossenes L. justified tape
  korrigiertes L. clean tape
lochbandgesteuert punch-tape controlled
Lochbandsteuerung *f* punched tape control
Lochblende *f* pin-hole stop, circular aperture
Lochdornstange *f* piercing rod
Locheisen *n* punch
Lochelektrode *f* apertured electrode disk
lochen to punch *(tape)*; to hole; to pierce
Locher *m* piercer; key punch machine; perforator, puncher
Löcherbesetzung *f* hole population
Löcherdichte *f* hole density
Löchereinfang *m* hole capture
Löcherfalle *f* hole trap
Locherkupplung *f* punch clutch
Löcherleitung *f* p-type conductivity, hole conduction
Lochermatrix *f* punch-die assembly
Löchertheorie *f* **/ Diracsche** Dirac hole theory
Löcherverteilung *f* hole distribution
Lochfeldsteuerung *f* field selection
Lochfraß *m* localized corrosion, pitting
Lochfraßkorrosion *f* pitting corrosion
Lochhobeleisen *n* cut plane iron
Lochkaliber *n* plug gauge
Lochkamera *f* pinhole camera
Lochkarte *f* punch[ed] card
Lochkarte-Magnetbandumsetzer *m* card-to-magnetic tape converter
Lochkartenablage *f* punched card file
Lochkartenablesegerät *n* card reading unit
Lochkartenabtaster *m* punched card reader
Lochkartenabteilung *f* punched card department
Lochkartenbett *n* duplicating rack
lochkartengesteuert punched-card controlled
Lochkartenmaschine *f* punched card machine
Lochkartenmuster *n* cardboard template
Lochkartenprogrammierung *f* punched card programming
Lochkartenschacht *m* hopper
Lochkartenspalte *f* card column
Lochkartensteuersystem *n* punched card control method
Lochkartensteuerung *f* punched card control
  L. / mit punched-card controlled
Lochkatode *f* perforated cathode
Lochkreis *m* hole (index, scribed) circle
Lochkreisdurchmesser *m* diameter of the pitch (hole) circle, diameter of the bolt hole circle
Lochleibung *f* intensity of pressure
Lochmaschenmuster *n* eyelet-stitch pattern *(knitting)*
Lochmaschine *f* punching machine
Lochmitte *f* centre point of hole
Lochmittenabstand *m* hole centre distance
Lochnadel *f* guide needle
Lochnadelbarre *f* guide bar
Lochpapierstreifen *m* punched paper tape
Lochplatte *f* 1. *(build)* core[d] panel, core[d] slab; 2. *(cont)* apertured plate
Lochprüfgerät *n* verifier
Lochprüfung *f* punching test
Lochraster *m* matrix of holes
Lochreihenabstand *m* distance between side rows of bolt-holes
Lochsäge *f* keyhole (sweep) saw
Lochscheibe *f (telev)* apertured disk
Lochschreiber *m* punching recorder; perforated-tape telegraph-recorder

Lochschriftübersetzer *m* alphabetical interpreter
Lochschweißung *f* plug and slot weld
Lochsirene *f* perforated disk siren
Lochspeicherung *f* hole storage
Lochstanze *f* perforator
Lochstein *m* 1. *(build)* perforated brick; 2. jewel hole, through-hole [jewel]
Lochstellung *f* punch position
Lochstempel *m* die; piercing die
Lochstreifen *m* punched (perforated) tape
Lochstreifenabtaster *m* paper-tape reader
Lochstreifenaufzeichnung *f* punched tape record
Lochstreifenempfänger *m* reperforator, receiving perforator
Lochstreifengeber *m* tape transmitter
lochstreifengesteuert punch-tape controlled
Lochstreifen-Lochkarten-Umsetzer *m* tape-to-card converter
Lochstreifenprüfungssystem *n* **durch quer verlaufendes Prüfen** cross-checking system
Lochstreifenredigierausrüstung *f* punched tape editing equipment
Lochstreifenschreiber *m* punched tape recorder
Lochstreifensender *m* automatic transmitter, autotransmitter
  vollautomatischer L. coupled reperforator and tape reader, *(Am)* fully automatic reperforator transmitter distributor
Lochstreifentechnik *f* punched tape technique
Lochstreifenübertragung *f* tape retransmission, perforated tape transmission, automatic routing
Lochtiefe *f* hole depth
Loch- und Öseneinsetzmaschine *f* hole-punching and eyeletting machine
Lochvisier *n* pin-hole sight
Lochwalze *f* distributing roll
Lochwalzwerk *n* piercing mill
Lochwandungsoberflächengüte *f* wall finish
Lochzange *f* [ticket] punch
Lochziegel *m* perforated (cellular) brick
locken to curl *(chip)*
Lockenspan *m* curled (continuous curly) chip
locker loose, slack
lockern to loosen, to slacken *(nut)*; to release; to break up, to hoe *(soil)*
Lockerstelle *f* loose place, flaw *(crystal defect)*
Lock-in-Verstärker *m (dat)* lock-in amplifier
Loctalröhre *f* loctal tube
Loden *m* unmilled woolen cloth
Löffel *m* 1. spoon; 2. dipper bucket *(dredge)*; bucket; shovel
  L. eines Hochlöffelbaggers shovel dipper
  L. eines Tieflöffelbaggers backshoe dipper
Löffelbagger *m* power shovel, navvy; spoon (dipper-bucket) dredger
  L. mit Löffelstiel push shovel
Löffelbohrer *m (wood)* spoon-auger
Löffelbug *m (ship)* spoon bow
Löffelheck *n (ship)* spoon (canoe) stern
Löffelinhalt *m* dipper capacity
Löffelprobe *f* spoon sample
Löffelschaber *m* half-round scraper
Löffelschaufelung *f* spoon-shaped blades *(turbine)*
Löffelschwimmbagger *m* dipper[-bucket] dredger
Löffelstiel *m* dipper boom (handle)
Löffelwalzmaschine *f* hunting spoon rolling mill
Löffler-Kessel *m* Loeffler boiler
Log *n* [speed] log
Logarithmenbasis *f* base of a logarithm
Logarithmierverstärker *m* logarithmic amplifier
logarithmisch logarithmic
Logarithmus *m* logarithm
  Briggsscher L. common (Briggsian) logarithm
  dekadischer L. common (Briggsian) logarithm
  gemeiner L. common logarithm
  gewöhnlicher L. common (Briggsian) logarithm
  hyperbolischer L. hyperbolic (Napierian, natural) logarithm
  natürlicher L. hyperbolic (Napierian, natural) logarithm
  Neperscher L. hyperbolic (Napierian, natural) logarithm
Logatom *n (ac)* logatom
Logbuch *n* [ship's] log, [ship's] log book
Logge *f* log
Logger *m* logger, gill-netter, drift boat, drifter *(fishing vessel)*
Logik *f* logic
  L. mit Tabellen[nachschlagen] table look-up logic
  anpassungsfähige L. adaptive logic *(capability of self repair)*
  digitale L. digital logic
  dreiwertige L. three-valued logic
  hydraulische L. hydraulic logic
  symbolische L. symbolic logic
  unterscheidende L. discriminating logic
Logikglied *n* logical element
Logikmodul *m* logic module
Logikschaltung *f* logical circuit
Logiktransistor *m* logic transistor
Logisraum *m (ship)* crew's quarter (space)
Logistik *f* logistics, mathematical logic
Logleine *f* log line
Logrolle *f* log reel
Lohbrechmaschine *f* bark breaking machine
Lohbrühe *f* tannery liquor

Lohe *f* tanning bark, tan
lohgar [vegetable-]tanned
Lohgerber *m* bark tanner
Lohgerberei *f* bark tannage
Lohgrube *f* tan pit
Lohhut *m* cover of the layers
Lohmühle *f* bark mill
Lohn *m* wages, hire, salary, payment
Lohnabzug *m* payroll deduction, pay cut
lohnanreizend incentive
Lohnarbeit *f* paid labour
Lohnarbeiter *m* labourer, paid workman
Lohnaufwand *m* cost of labour
Lohnempfänger *m* wage earner
lohnend remunerative, paying
Lohnerhöhung *f* increase in pay, rise in wages
Lohnetz *n* tan net
Lohngemeinkosten *pl* overheads
Lohnkosten *pl* cost of labour
Lohnliste *f* payroll
Lohntüte *f* wage packet
Lohnwalzen *n* job rolling
Lohpresse *f* spent tan press
Lok *f s.* Lokomotive
lokal local; selective *(heating)*
Lokalbahn *f (ry)* branch line, secondary railway
Lokalbestrahlung *f* local irradiation
Lokaldosis *f (nucl)* local dosage (dose)
Lokalelement *n* local cell
Lokalisator *m* [X-ray therapy] applicator, localizer, treatment cone
lokalisieren to localize, to locate
Lokalisierung *f* localization
Lokalkorrosion *f* local action, localized corrosion, pitting
Lokalmonitor *m* area monitor
Lokdienstleiter *m* chief of locomotive operations, locomotive running foreman
Lokomobile *f* portable steam engine
  feststehende L. semi-portable steam engine
  ortsfeste L. semi-portable steam engine
Lokomobilkessel *m* [semi-]portable boiler
  L. mit vorgehenden Heizröhren semi-portable Cornish multitubular boiler
  feststehender L. semi-portable boiler
  stationärer L. semi-portable boiler
Lokomobilschornstein *m* stack of portable engine
Lokomobilüberhitzer *m* superheater for semi-portable boiler
Lokomobilzylinder *m* cylinder of portable engine
Lokomotivantrieb *m* locomotive driver
Lokomotivdampfmaschine *f* locomotive steam engine
Lokomotive *f* locomotive
  dieselelektrische L. Diesel-electric locomotive
  kalte L. cold locomotive not under steam
Lokomotivfeuerkasten *m* locomotive fire box
Lokomotivführer *m* engine driver
Lokomotivheizer *m* locomotive fireman
Lokomotivkammer *f (mine)* loco garage
Lokomotivkessel *m* locomotive boiler
Lokomotivkran *m* locomotive crane
Lokomotivpark *m* locomotive stock
Lokomotivschuppen *m* locomotive (engine) shed
Lokomotivstehkessel *m* locomotive fire box
Lokomotivtender *m* locomotive tender
longitudinal longitudinal
Longitudinalschwingung *f* longitudinal oscillation (vibration)
Longitudinalwelle *f* longitudinal wave
Loran *n* loran *(long range navigation system)*
Lorbeeröl *n* bay-tree oil
Lorentz-Kraft *f* Lorentz force
Lorentz-Transformation *f* Lorentz transformation
Los *n* lot *(series)*
lösbar disconnectable *(pipe coupling)*; *(ch, math)* solvable, soluble
Lösbarkeit *f (mach)* disconnectability; *(ch, math)* solubility
losbekommen to get off (free) *(a stranded vessel)*
losbinden to untie
Losblatt *n (text)* loose reed
Losblatteinrichtung *f* loose reed mechanism
Losblattschützenwächter *m* loose reed warp protector
Losblattstuhl *m* loose reed loom
Losboden *m* removable bottom *(converter)*
Losbrechwiderstand *m* breakaway force *(el. train)*
Löschanlage *f* quenching plant
Löschbedingung *f (el)* [arc] quenching condition
Löschblattprobe *f* blotter (lubricating oil drop) test
Löschboden *m (met)* charcoal bottom
Löschdauer *f* arcing-time
Lösche *f* [coal] cinders, coal dust
löschen 1. to extinguish *(fire)*; *(el)* to quench, to extinguish *(arc)*; to clear; to erase, to reset *(stored data)*; to wash out *(tape)*; 2. to slake *(lime)*; 3. to unload, to unship, to lighten *(vessel)*
  Ballast l. to unballast
  Ladung l. to break bulk, to clear a ship
Löschen *n* **an der Luft** air slaking *(lime)*
Löscher *m* extinguisher; *(el)* quencher
Löschfunke *m (el)* quenched spark
Löschfunkenstrecke *f* quenched spark gap
Löschgas *n* quench[ing] gas
Löschgebühr *f (ship)* wharfage, landing charges, lighterage
Löschgeld *n s.* Löschgebühr
Löschgerät *n* erase unit, eraser *(magnetic tapes)*

Löschgeschirr *n (ship)* discharging gear
Löschhafen *m* port of discharge
Löschhebel *m (dat)* clearing lever
Löschhub *m* quenching distance *(arc)*
Löschimpuls *m* quench (reset) pulse
Löschkalk *m* slaked (hydrated) lime, quicklime
Löschkammer *f* quenching chamber, arc-control device, arc chute
Löschkammerkontakt *m* box contact
Löschkammerschalter *m* pot-type breaker
Löschkondensator *m* quenching capacitor; commutating (buffer) capacitor
Löschkontakt *m* reset contact
Löschkopf *m* erasing (erase) head
Löschkreis *m* quenching circuit
Löschleistung *f* quenching power
Löschmittel *n* quenching (arc-extinguishing) medium
Löschpapier *n* blotting paper
Löschplatz *m (ship)* place of discharge, discharging berth, landing place
Löschprinzip *n* [arc] extinguishing principle, quenching method
Löschrelais *n* cut-off relay, arc high-speed relay
Löschrohr *n* **der Sicherung** fuse (expulsion) tube
L. des Ableiters expulsion tube, *(Am)* muffle
Löschrohrableiter *m* expulsion-type arrester
Löschrohrsicherung *f* expulsion-fuse
Löschschaltung *f* quenching circuit
Löschseite *f* quencher side
Löschspannung *f* extinction (extinguishing) voltage, extinction (de-ionizing) potential
Löschsperre *f* erase cut-out key *(tape recorder)*
Löschspieß *m* poker
Löschspule *f* quenching coil
Löschtaste *f* cancel[ling] key, release key; erase button
Löschtorschaltung *f (dat)* eraser gate
Löschtransformator *m* neutral compensator, neutralizing transformer
Löschturm *m* quenching tower
Löschung *f* 1. *(el)* [arc] quenching; erasing *(record)*; 2. [zero] reset *(counter)*
Löschungsfrequenz *f* quenching frequency
Löschverfahren *n* [arc] quenching method
Löschverstärker *m (dat)* erase amplifier
Löschverteilerstutzen *m* discharge header *(on deck of a tanker)*
Löschwagen *m* quenching car
Löschwasser *n* water [supply] for fire protection
Löschwiderstand *m* quenching resistor
Löschwinkel *m* commutation margin, overlap angle *(of rectifiers)*
Löschwirkung *f (el)* [arc] quenching action
Löschzeit *f (el)* arcing-time; deionization time; erase time *(oscilloscope)*
Löschzusatz *m* quenching agent
Losdrehen *n* loosening
lose in bulk *(material)*; loose; movable; free
l. laufen [auf einer Spindel] to rotate freely [on a spindle]
Lose *f* backlash, lost motion
Lösearbeit *f* coal getting
Loseblatt- loose-leaf
Lösekessel *m* quencher
lösen to unscrew, to slacken *(nut)*; to loosen; to unclamp *(slide)*; to unlatch, to unlock; to release; to ease; to book *(tickets)*; *(ch)* to dissolve; *(ch, math)* to solve; *(mine)* to break down, to get; to part
die Bremse l. to release the brake
die Festklemmung l. to unlock
die Kupplung l. to declutch
Knoten l. to undo knots
sich l. 1. to loosen; 2. to dissolve
Löseneinbruch *m (mine)* drag cut
Lösewalzwerk *n* detaching mill
Losgröße *f* lot size (quantity)
kleine L. small lot (batch), short run
Loskiel *m (ship)* keel shoe, false (shoe of a) keel
losklopfen to rap *(mould)*
loslassen to release *(relay; push button)*
löslich soluble
l. machen to solubilize
Löslichkeit *f* solubility
Löslichkeitsänderung *f* change of solubility
Löslichkeitsgrenze *f* solubility limit
Löslichkeitskurve *f* solubility curve
Löslichkeitsprodukt *n* solubility product
Löslichkeitsveränderung *f* change of solubility
loslösen to detach; to liberate
loslöten to unsolder
losmachen to cast off, to unmoor *(vessel)*
losnieten to unrivet
Los-Qualitätssicherung *f* lot quality protection
losreißen 1. to tear off; to pull off; 2. *(ship)* to break free
Losreißmoment *n* break-away torque
Löß *m (geol)* loess
Lößablagerung *f* deposition of loess
Lößbildung *f* loess formation
Losscheibe *f (mach)* loose pulley
losschrauben to undo, to unbolt, to unscrew
Losstall *m* cow yard
Losumfang *m (mach)* lot size
Lösung *f* 1. *(ch)* solution; lotion; 2. *(math)* solution; 3. dissolution; 4. loosening
angenäherte L. asymptotic solution
angesäuerte L. acidified solution
echte L. true solution
eingeschleppte L. drag-in
elektrolytische L. electrolytic solution
explizite L. explicite solution
Fehlingsche L. Fehling's solution
feste L. solid solution

ganzzahlige L. integer solution
gesättigte L. saturated solution
herausgeschleppte L. drag-out
ideale L. ideal solution
isotonische L. isotonic solution
Kardanische L. Cardan's solution
kolloide L. colloidal solution, sol
normalisierte L. standardized solution
partikuläre L. particular solution
spezielle L. particular solution
triviale L. trivial solution
übersättigte L. supersaturated solution
verdünnte L. dilute[d] solution
wäßrige L. aqueous solution
zweinormale L. twice-normal solution
Lösungsanode *f* soluble anode
Lösungsansatz *m* trial solution
Lösungsdruck *m* solution pressure
elektrolytischer L. electrolytic solution pressure
Lösungselektrode *f* solvent electrode
Lösungsenergie *f* solvation energy
Lösungsglühen *n* solution heat-treatment
Lösungsmittel *n* [dis]solvent, solvency; carrier; *(plast)* vehicle
selektives L. preferential solvent
Lösungsmittelaufnahmefähigkeit *f* solvent tolerance
lösungsmittelecht fast to solvents
Lösungsmittelextraktion *f* solvent extraction
Lösungsmittelgemisch *n* solvent mixture
Lösungsmittelrückgewinnung *f* solvent recovery
Lösungsmittelverfahren *n* solvent method *(man-made fibre)*
Lösungspolymerisation *f* solution polymerization
Lösungspotential *n* solution potential
Lösungsreaktor *m* solution reactor; homogeneous reactor
Lösungsspannung *f* electrolytic solution voltage
Lösungstaste *f* release key
Lösungsverfahren *n* technique of solution
Lösungsvermittler *m* solubilizer
Lösungsvermögen *n* solvent power
Lösungswärme *f* heat of solution
loswerfen to cast off *(a vessel)*
Lot *n* 1. solder; hard solder, spelter; 2. lead, plumb, bob, plummet; 3. perpendicularity
L. / aus dem out of plumb
L. / außer out of plumb
akustisches L. echo sounder
optisches L. optical plummet
Lotablenkung *f* deflection of the plumb
Lotabweichung *f* deflection of the plumb (bob) line
Lotachse *f* vertical axis
Lötanschluß *m* soldering tag; soldering connection
Lötanschlußstück *n* soldered connection piece
Lötbad *n* solder bath, bath of molten solder
lotbar / nicht mehr *(ship)* off-sounding
lötbar solderable
Lötbarkeit *f* solderability
Lötbrunnen *m* cable jointing box
Lötdraht *m* solder[ing] wire
L. mit Kolophoniumseele resin-core solder
Loteinrichtung *f* sounding device
loten to sound, to cast; to plumb
nach Faden l. to fathom
löten to solder
Löten *n* soldering
Löter *m* solderer
Lötfähigkeitsprüfung *f* solder bath test
Lötfahne *f* soldering tag (lug, ear), pigtail
Lötfahnenanschluß *m* tag-end terminal
Lötfläche *f* soldering surface
Lötflußmittel *n* soldering (brazing) flux
lötfrei solderless
Lotfühler *m* pendulous erection device *(gyroscope)*
Lotgast *m (ship)* leadsman
Lotgerät *n* sounder
Lotion *f* lotion
Lötkabelschuh *m* sweating thimble
Lötklemme *f* soldering terminal
Lötkolben *m* copper bit, soldering iron
elektrischer L. electric soldering iron
Lötkolbenspitze *f* copper bit
Lötlampe *f* blow lamp for soldering, *(Am)* blow torch for soldering
Lötlasche *f* soldering tag
Lötlegierung *f* solder alloy
Lotleinenmarkierungen *fpl (ship)* marks and deeps
lötlos solderless
Lötmaschine *f* rotating soldering machine *(electron valve manufacture)*
Lötmittel *n* solder
Lötmuffe *f* soldering (splicing) sleeve
Lötöse *f* [soldering] lug, eyelet, soldering terminal, pierced solder tag, pigtail
Lötösenleiste *f* mounting (tag) strip, terminal board
Lötösenstreifen *m* soldering terminal strip
Lötpaste *f* soldering paste
lotrecht perpendicular
l. zu perpendicular to
Lötrohr *n* blowpipe
Lötrohrprobe *f* blowpipe test (analysis)
Lötrohrversuch *m* blowpipe test (analysis)
Lötschuh *m* sweating thimble
Lotse *m* pilot
Lötseilverbinder *m* splicing sleeve
lotsen to pilot
Lotsenamt *n* pilot office
Lotsenboot *n* pilot boat

Lotsenbrücke *f* pilot bridge
Lotsenfahrzeug *n* pilot boat (vessel)
Lotsenflagge *f* pilot flag (jack)
Lotsengebühr *f* pilotage
Lotsengeld *n* pilotage
Lotsenkunde *f* pilotage
Lotsenleiter *f* pilot's (jack) ladder
Lotsenstation *f* pilotage
Lotsentreppe *f* pilot's (jack) ladder
Lotsenversetzboot *n* pilot boat (vessel)
Lotsenwesen *n* pilotage, piloting
Lotsenzwang *m* compulsory pilotage
Lotsextant *m* sounding sextant
Lotstange *f* sounding pole
Lötstelle *f* surfaces to be joined by soldering (brazing); junction, soldered joint
  L. des Thermoelements thermocouple junction
  kalte L. dry junction, dry (cold) joint
  schlechte L. dry (cold) joint
  thermoelektrische L. thermojunction
Lötstift *m* soldering tag
Lotstock *m* sounding pole
Lötstützpunkt *m* soldering terminal
Löttemperatur *f* soldering temperature
Lotung *f* plumbing; sounding
Lotungslinie *f* fathom line
Lötverbinder *m* splicing sleeve
Lötverbindung *f* soldered (brazed) joint
Lötwasser *n* soldering fluid, zinc chloride flux
Lötzinn *n* tin-base solder, soft tin solder, soldering tin
Loxodrome *f (math)* rhumb
LP-Beton *m* air-entrained concrete
LP-Zusatz *m* air-entraining admixture, air-entrainment agent
Lücke *f* gap; interstice; gash, spacing *(tool)*; vacancy *(electronics)*
  L. / auf staggered
  oktaedrische L. octahedral site *(crystal)*
Lückenmessung *f* interstitial measurement
Lückensender *m* fill-in station
Lückenspiel *n* backlash *(play between mating teeth)*
Lückenwinkel *m* gash angle
Ludlow-Typograf *m* Ludlow typograph
Luft *f* 1. air; 2. *(mach)* play
  atmosphärische L. atmospherical air
  flüssige L. liquid air
  komprimierte L. compressed air
  staubhaltige L. dust-laden air
  umgebende L. ambient air
  verbrauchte L. vitiated air
  verdünnte L. rarefied air
  verflüssigte L. liquefied air
Luftabführung *f* air gate
luftabgeschreckt air-quenched *(steel)*
Luftabkühlung *f* air cooling, quenching in air
Luftablaß *m* de-aeration
Luftablaßventil *n* air escape valve
Luftabsaugeinrichtung *f* air-suction device
Luftabsauger *m* suction ventilator
Luftabscheider *m* air release (separator), deaerator
Luftabschluß *m* exclusion of air, hermetic seal
  L. / unter hermetically sealed, without access of air
Luftabschreckung *f* air quenching *(steel)*
Luftabsperrhahn *m* air shut-off valve
Luftabwehr *f* anti-aircraft defence
Luftabwehrgeschütz *n* anti-aircraft gun
Luftabwehrrakete *f* anti-aircraft rocket
Luftabzug *m* air vent
Luftabzugsrohr *n* air vent pipe
Luftaktivität *f* airborne radioactivity
Luftangriff *m* air raid
Luftansaugring *m* air-intake ring
Luftansaugrohr *n* air intake [pipe]
luftäquivalent *(nucl)* air-equivalent; air-wall
Luftaufbereitung *f* air conditioning
Luftaufnahme *f* aerial view (photograph); air photo
Luftausgleicher *m* air compensator, compressed-air equilibrator
Luftauslaß *m* air escape
Luftauslaßrohr *n* discharge header
Luftaustritt *m* air leak (outlet)
Luftaustrittsloch *n* air-exit hole
Luftbad *n (ch)* air bath; Babo flask holder
Luftballon *m* balloon
Luftbefeuchtung *f* air moistening (humidifying)
Luftbehälter *m* air bottle (reservoir)
Luftbehandlung *f* air conditioning
luftbereift pneumatic-tyred
Luftbereifung *f* pneumatic tyres
Luftbild *n* aerial photograph (view); air photo
Luftbildkamera *f* aerial survey camera
Luftbildmessung *f* aerial surveying
Luftbildumzeichner *m* sketchmaster
Luftblase *f* air bubble; blow hole; air void (pocket)
  eingeschlossene L. trapped air-bubble, entrapped bubble
Luftbremse *f* air brake
Luftbürste *f* air knife
Luftbürstenstreichmaschine *f* air brush coater
Luftbürstverfahren *n* micro-jet roll coating
luftdicht airtight
Luftdichte *f* atmospheric density
Luftdielektrikum *n* air dielectric
Luftdosis *f (nucl)* [free] air dose, in-air dose
Luftdrehkondensator *m* variable air capacitor
Luftdrossel *f* air-core choke
Luftdrosselspule *f* air-cored choke
Luftdruck *m* atmospheric (air, barometric) pressure
Luftdruckmesser *m* air-pressure gauge; *(aut)* tyre gauge

Luftdruckmessung *f* barometry
Luftdruckminderventil *n* air-pressure release valve
Luftdruckregler *m* air-reducing valve
Luftdruckschalter *m* air-pressure switch
Luftdruckschreiber *m* barograph
Luftdurchflußmenge *f* rate of air flow
Luftdurchlaß *m* air passage
luftdurchlässig penetrable by air
Luftdurchlässigkeit *f* permeability to air; venting property *(mould)*
Luftdüse *f* air [intake] nozzle
Lufteinblasemaschine *f* air-injection engine
Lufteinblasung *f* air injection
Lufteinlaß *m* air inlet (intake)
Lufteinlaßhahn *m* air-inlet cock
Lufteinlaßkanal *m* air-intake duct
Lufteinlaßklappe *f (aut)* air strangler
Lufteinschluß *m* air inclusion, entrapped (entrained) air, air void (pocket)
Luftelektrizität *f* atmospheric electricity
lüften to ease *(valve)*; to lift *(tool)*; to air, to ventilate *(room)*; to aerate *(sand)*
Luftentladung *f* air discharge
luftentzündlich pyrophoric, inflammable in contact with air
Lüfter *m* fan, blower, *(aut)* ventilator
  blasender L. blower
  mehrflügeliger L. multi-bladed fan
  saugender L. exhauster
Lüfterachse *f* fan-fixed shaft
Lüfterbock *m* fan bracket
Lüfterflügel *m* fan blade
Lufterhärtung *f* air curing *(concrete)*
Lüfterhaube *f* fan shroud (cowl)
Lufterhitzer *m* air heater
Lüfterkanal *m* fan drift
Lüfterkopf *m (ship)* ventilator head, cowl
Lüftermotor *m* fan motor
Lüfternabe *f* fan hub
Lüfterübersetzungsgetriebe *n* cooling-fan increasing gear
Lüfterwelle *f* fan drive shaft
Luftfach *n* air compartment *(jig)*
Luftfahrt *f* aeronautics
  L. mit gasgetragenen Luftfahrzeugen aerostation
Luftfahrtindustrie *f* aeronautical industry
Luftfahrttechnik *f* aeronautical engineering
Luftfahrtwerkstoff *m* aircraft material (metal)
Luftfahrzeug *n* aircraft
  gasgetragenes L. aerostat, lighter-than-air craft
Luftfänger *m* scoop-type air intake
Luftfeder *f* air spring
Luftfederung *f* air (pneumatic) cushioning
Luftfender *m (ship)* inflatable fender
Luftfeuchtigkeit *f* air humidity, atmospheric humidity (moisture)
  relative L. relative air humidity
Luftfeuchtigkeitsmesser *m* hygrometer, psychrometer
Luftfilter *n* air filter (strainer, cleaner)
Luftfilter-Ansauggeräuschdämpfer *m (aut)* air filter intake silencer
Luftfrischen *n (met)* air-refining
Luftführungsanlage *f* air ducting system
Luftführungskasten *m (aut)* air duct
Luftfüllung / mit pneumatic
Luftgang *m (ship)* air course
Luftgangentwicklung *f (text)* developing by air-passage
Luftgehalt *m* **des Wassers** air content of water
luftgehärtet air-hardened *(steel)*
luftgekühlt air-cooled; *(aut)* fan-cooled
luftgeschmiert air-lubricated
Luftgeschwindigkeit *f* air velocity (speed)
  gemessene L. calibrated air speed
luftgetrocknet air-dried, air-seasoned
lufthärtend air-hardening
Lufthärter *m* air-hardening steel
Lufthärtung *f* air hardening
Luftheizung *f* warm-air heating
Luftherd *m* dry-cleaning table
Luftionisationskammer *f* air[-filled] ionization chamber
  offene L. free-air ionization chamber, open-air ionization chamber
Luftionisierung *f* air ionization, ionization of air
Luftkabel *n* overhead (aerial) cable
Luftkabelleitung *f* aerial cable line
  auf Masten verlegte L. cable pole line
Luftkabeltragseil *n* messenger wire
Luftkalk *m* air-hardening lime
Luftkammer *f (mine)* aeration chamber; *(therm)* air-chamber, air-box
Luftkanal *m* air chute (duct, channel, vent)
Luftkasten *m* plastic-covered bow *(racing boat)*
  L. im Rettungsboot air chamber
Luftkastenschiff *n* captured air bubble over water vehicle, CAB vessel
Luftkissen *n* air cushion
Luftkissenboot *n s.* Luftkissenfahrzeug
Luftkissenfahrzeug *n (ship)* hydroskimmer, air-cushion vehicle, surface effect ship, ground effect machine, hovercraft, hovercoach
Luftklappe *f* ventilation (air) flap, choke
Luftklappenknopf *m* choke-control knob
Luftkompressor *m* air compressor
Luftkondensator *m* air capacitor
Luftkorridor *m (aero)* aerial corridor
Luftkraft *f* aerodynamic force
Luftkühler *m* air cooler
Luftkühlung *f* air cooling, cooling by air
Luftlage *f* tactical air situation
Luftlager *n* air bearing

Luftlagetisch *m* surveillance plotting board
Luftlandetruppen *fpl* airborne troops
luftleer evacuated, exhausted
l. gepumpt exhausted of air
l. machen to evacuate, to exhaust
Luftleitblech *n* cooling baffle (guide vane)
Luftleiter *m* aerial conductor
Luftleitung *f (el)* air main, overhead line
Luftlinie *f* bee-line
Luftloch *n* 1. *(aut)* air vent; 2. *(ship)* air course
Luftmangel *m* lack of air
Luftmantel *m* layer of air; air casing
Luftmatratze *f* inflatable (camp) mattress, air-bed
Luftmeßbild *n* survey photograph
Luftmesser *n* air knife
Luftmörtel *m* non-hydraulic mortar
Luftnachrichtentruppen *fpl* air force signal corps
Luftnotüberwachung *f* alerting service
Luftöffnung *f* / **kleine** bleed hole
Luftpendelverkehr *m* air shuttle service
Luftperspektive *f* aerial perspective
Luftpinselstreichmaschine *f* air coater
Luftpinselverfahren *n* **mit Walzenauftrag** air-knife coating
Luftplankton *n* aerosol
Luftplasma *n* air plasma
Luftpore *f* air void *(concrete)*
Luftporenbeton *m* air-entrained concrete
luftporenbildend air-entraining
Luftporenbildner *m* air-entraining admixture, air-entrainment agent
Luftporenbildung *f* air entraining (entrainment)
Luftporenzement *m* air-entraining cement
Luftpost *f* air mail
Luftpostpapier *n* air-mail paper
Luftpresser *m s.* Luftkompressor
Luftpuffer *m* air buffer
Luftpumpe *f* air pump, inflator
zweistufige L. double-stage air-pump
Luftpumpenbalancier *m* air-pump beam
Luftpumpendruckraum *m* feed-water reservoir [of the air-pump]
Luftpumpenkolben *m* air-pump piston
Luftpumpenmetallventil *n* metal air-pump valve
Luftpumpenzisterne *f* feed-water reservoir [of the air-pump]
Luftrakel *f* air knife
Luftrakelstreichmaschine *f* air knife coater
Luftrakelstreichverfahren *n* floating knife coating
Luftrakelverfahren *n* **mit Walzenauftragung** floating knife roll coating
Luftraum *m* 1. plenum *(ventilation)*; 2. *(mil)* sky, air; 3. air space
eingeschlossener L. air pocket *(in fabric)*
Luftraumkabel *n* air-space cable
Luftreibung *f* air friction; windage
Luftreiniger *m* air cleaner (purifier, filter), deodorizer
Luftreinigung *f* air cleaning (filtering, purifying), deodorizing
Luftröhrenkühler *m* air-tube radiator
Luftsack *m* air pocket
Luftsauerstoff *m* atmospheric oxygen
Luftsauerstoffbatterie *f* air-cell battery
Luftsauerstoffelement *n* air[-depolarized] cell
Luftsauerstoffzelle *f s.* Luftsauerstoffelement
Luftsauger *m* exhauster
Luftsaugerohr *n* air suction (intake) pipe, air suction (intake) hose
Luftsaugstutzen *m* stand-pipe for air-pump suction
Luftsäule *f* column of air
Luftschacht *m* air shaft
Luftschall *m* air-borne sound
Luftschalter *m* air[-break] circuit breaker
Luftschaltstrecke *f* air break
Luftschauer *m* air shower
ausgedehnter L. extensive (giant) air shower, Auger shower
großer L. *s.* ausgedehnter L.
Luftschicht *f* air layer, layer of air; atmospheric layer
Luftschiff *n* airship
halbstarres L. semirigid airship
lenkbares L. dirigible airship
Luftschiffahrt *f* aerial (air) navigation, aeronautics, aerostation
Luftschlauch *m* inner tube *(tyre)*
Luftschlauchventil *n* inflation valve
Luftschleier *m* air curtain
Luftschleiertür *f* air curtain door
Luftschleuse *f* air lock
Luftschlitz *m* vent (venting, ventilating) slot
Luftschraube *f* airscrew, aircraft propeller
L. mit automatischer Verstellung im Flug automatic-pitch airscrew
L. mit Verstellung beim Flug durch Pilot controllable-pitch airscrew
einstellbare L. adjustable-pitch airscrew
feste L. fixed-pitch airscrew
gegenläufige L. contra-rotating propeller
stehende L. dead stick
Luftschraubenblattspitze *f* blade tip
Luftschraubennabe *f* airscrew hub
Luftschraubenverstellung *f* change of pitch
Luftschütz *n* air-break contactor
Luftschutzbunker *m* air-raid shelter
luftseitig downstream
Luftsenkkasten *m* compressed-air caisson
Luftsetzmaschine *f* air (pneumatic) jig
Luftsortierung *f* air classification
Luftspalt *m (el)* air gap;
Luftspaltdrossel *f* air-gap choke
Luftspaltinduktion *f* air-gap flux density

Luftspaltmagnetometer *n* flux-gate magnetometer
Luftspaltspannung *f* air-gap voltage
Luftspaltstreuung *f* peripheral (main) leakage
Luftspaltwiderstand *m* **[/ magnetischer]** *(el)* gap reluctance
Luftspeicherbremszylinder *m* air-cell brake cylinder
Luftspiegelung *f* mirage, fata morgana
Luftspule *f* air-cored coil (solenoid)
Luftspülung *f* air flushing
Luftstechen *n (met)* venting *(core)*
Luftstickstoff *m* atmospheric nitrogen
Luftstörungen *fpl* atmospherics
Luftstrahl *m* jet of air
Luftstrahltriebwerk *n* air jet-propulsion engine
Luftstraße *f* airway
Luftstrecke *f* air gap; air path
Luftstrom *m* current of air, air blast (stream, flow)
Luftstrudel *m (aero)* wash
Lufttasche *f* air pocket
Lufthermometer *n* air thermometer
Lufttorpedo *m* aerial torpedo
Lufttransformator *m* oilless (air-cooled, air-core) transformer
L. mit Fremdbelüftung air-blast transformer
Lufttrennung *f* air-gap overlap span *(trolley wire)*
Lufttrichter *m (aut)* Venturi tube (pipe), choke tube
lufttrocken air-dry, air-dried; air-seasoned *(timber)*
Lufttrockenmaschine *f* air-drying machine
Lufttrockner *m* air dehumidifier
Lufttrocknung *f* air seasoning (drying) *(timber)*
lufttüchtig airworthy
Lufttüchtigkeit *f* airworthiness
Lufttüchtigkeitsschein *m* certificate of airworthiness
Luftüberhitzer *m* blast superheater
Luftüberschuß *m* excess of air
Luftüberschußmesser *m* excess air meter
luftübertragen air-borne *(sound)*
Luftüberwachung *f* air control (survey)
Luftüberwachungsgerät *n* air monitor (survey instrument)
Luftumwälzofen *m* air-circulation furnace
Luftumwälzung *f* air circulation
Lüftung *f* 1. ventilation *(room)*; 2. venting *(mould)*; 3. aeration *(water)*; 4. lift[ing] *(tool)*
L. des Speisewassers aeration of the feed water
Lüftungsbecken *n* aeration tank *(sewage treatment)*
Lüftungskanal *m* ventilation duct (conduit), air duct (chute)
Lüftungsloch *n s.* Lüftungsöffnung
Lüftungsöffnung *f* vent hole, air drain
Lüftungsrohr *n* air chimney (pipe), flue, [air] vent
Lüftungsschacht *m* ventilation trunk (duct)
Lüftungsschlitz *m* venting (air) slot, vent hole
Lüftungsventil *n* vent valve
Lüftungsziegel *m s.* Lochziegel
Luftventil *n* atmospheric valve
Luftverbesserer *m* deodorizer
Luftverbrauch *m* air consumption
Luftverdrängung *f* aerodynamic volume displacement
Luftverdünnung *f* rarefaction of the air
Luftverfilzer *m* air-felting unit
Luftverfilzungsverfahren *n* air-felting process
Luftverflüssigung *f* liquefaction of air
Luftverflüssigungsanlage *f* air liquefying plant
Luftverkehr *m* air service (traffic)
L. auf kurzen Strecken short-haul service
planmäßiger L. scheduled air service
Luftverkehrslinie *f* air line
Luftverkehrsnetz *n* air-line network
Luftverkehrssicherung *f* air-traffic control
Luftverkehrsvorschriften *fpl* air-traffic rules
Luftvermessung *f* aerial surveying (mapping)
Luftverseuchung *f* air[-borne] contamination
Luftverteiler *m* air diffuser
Luftverunreinigung *f* air contamination, air (atmospheric) pollution
Luftvorwärmer *m* air-preheater, air heater *(boiler)*, economizer
Luftwändekammer *f* air-wall ionization chamber
Luftwärmer *m* air heater
Luftwäscher *m* air washer
Luftwechsel *m* air change, ventilation
Luftwelle *f* air wave
Luftwiderstand *m* [aerodynamic] drag; air resistance
Luftwirbel *m* air vortex (eddy)
Luftziegel *m* air brick, adobe
Luftzuführung *f* air admission (supply)
Luftzuführungskanal *m* air duct (passage)
Luftzug *m* air draught (draft)
Luftzutritt *m* air admission, access of air
Lügendetektor *m* lie detector
Luke *f* hatch[way]
arbeitsfähige L. workable hatch
blinde L. blind hatch
kleine L. scuttle
öldichte L. oiltight hatch
Lukenabdeckung *f* hatchway covering
Lukenbaum *m* hatch boom
Lukenbezug *m* hatch covering
Lukenboot *n* hatch boat
Lukendeckel *m* hatch cover
glatt mit dem Deck abschließender L. flush-fitting hatch cover
Lukendeckelauflage *f* hatch rest
Lukenendbalken *m* hatch-end beam
Lukengeländerstütze *f* hatch stanchion
Lukengräting *f* grated hatch, hatchway grating

Lukenkeil *m* securing (hatch) wedge
Lukenlängsbalken *m* hatchway fore-and-after, bridle beam
Lukenlängssüll *m (n)* hatchway side-coaming
Lukenlängsträger *m* hatch-side girder
Lukenleiter *f* hatchway (pillar) ladder
Lukenprinzip *n* / **größtes** workable (working) hatch
Lukenquersüll *m (n)* hatchway end coaming; headledge *(wooden ship)*
Lukenrampe *f* brow plate
Lukenschacht *m* trunked hatchway, hatchway trunk
Lukenschalkkeil *m* hatch wedge
Lukenscherstock *m* hatch (king) beam
Lukenschiebebalken *m* hatchway (hatch, blind) beam
Lukensicherung *f* hatch securing, securing of hatch
Lukensicherungsriegel *m* hatch securing bar
Lukenstringer *m* lock stringer (strake), [hatchway] tie-plate
Lukenstütze *f* hatch stanchion
Lukensüll *m (n)* hatch[way] coaming, coaming
Lukensüllstütze *f* coaming stay
Lukenübermaß *n* excess of hatchway
Lukenüberschuß *m* excess of hatchway
Lukenverschlußbügel *m* tie beam
Lumen *n* lumen
Lumenmesser *m* lumen meter
Lumenstunde *f* lumen-hour
Luminal *n* phenobarbital
Lumineszenz *f* luminescence
Lumineszenzeigenschaft *f* luminescence property
Lumineszenzemission *f* luminescent emission
Lumineszenzlicht *n* luminescent light
Lumineszenzlöschung *f* luminescence quenching
Lumineszenzmikroskop *n* luminescence microscope
Lumineszenzschwelle *f* threshold of luminescence
Lumineszenzspektrum *n* luminescent spectrum
Lumineszenzstrahlung *f* luminescent radiation
Lumineszenzzentrum *n* luminescent centre
lumineszieren to luminesce
Luminophor *m* luminescent material, luminophor
Lumpen *pl* rags *(papermaking)*
Lumpenentstaubungstrommel *f* rag shaker
Lumpenfäule *f* rotting vat
Lumpenhalbstoff *m* rag pulp
Lumpenklauber *m* rag picker
Lumpenpapier *n* rag paper
Lumpenreinigungsmaschine *f* rag cleaning machine
Lumpenreißen *n* rag grinding
Lumpenreißer *m* shredder
Lumpenschmälzmaschine *f* rag oiler
Lumpenschneider *m* rag chopper (cutter, cutting machine)
Lumpenwolf *m* rag tearing machine
Lumpenwolle *f* reclaimed wool, shoddy
Lumpenzerreißer *m* rag breaker
Lünette *f* steady, back rest
Lunge *f* / **eiserne** iron lung
Lunker *m* shrinkage (pipe) cavity, shrink hole, void
  primärer L. primary pipe
Lunkerabdeckmasse *f* anti-pipe compound
Lunkerbildung *f* piping, shrinking, liquid shrinkage
lunkerfrei pipeless, sound
Lunkerfreiheit *f* soundness
Lunkern *n* piping, shrinking
Lunkerpulver *n* anti-piping compound (powder)
Lunkerstelle *f* shrinkage cavity
Lunkerung *f* piping [effect]
lunkerverhütend anti-piping
Lunkerverhütungsmittel *n* pipe eliminator
Lunte *f* 1. slow match wick *(explosive)*; 2. slubber, slubbing *(spinning)*
Luntenführer *m* traverse guide *(slubbing)*
Lupe *f* loupe, magnifier, magnifying glass
Lupenaufnahme *f* low-power photomicrograph
Lupenbrille *f* magnifier spectacles
Lupenpasser *m* dot-on-dot register
Lupenstativ *n* magnifier stand
Lupine *f* lupine
Luppe *f (met)* puddled ball, [ore] bloom, loop
Luppeneisen *n* ball iron
Luppenfeuer *n* fining forge, forge hearth, Catalan furnace, bloomery
Luppenfrischarbeit *f* fining forge, forge hearth
Luppenherd *m* bloomery forge
Luppenmühle *f* burden (coffee mill) squeezer *(puddling)*
Luppenquetsche *f* ball squeezer
Luppenstab *m* puddle bar
Luppenstahl *m* bloom steel
Lüsterklemme *f* lustre (block) terminal, coupler terminal strip
lüstrieren *(text)* to glaze, to lustre
Lüstriermaschine *f* lustring machine
Luteolin *n* luteolin
Lutetium *n* lutecium, cassiopeium
lutro *s.* lufttrocken
Lutte *f (mine)* air duct, ventilating pipe
  zusammenschiebbare L. concertina-type ducting
Luv *f (ship)* weather side, luff, windward
Luvanker *m* weather anchor
Luvballast *m* shifting ballast
Luvbaum *m* outrigger, spreader
luven / in den Wind to luff into the wind
luvgierig carrying weather helm, griping

Luvgierigkeit *f* ardency
Luvkante *f* luff *(sails)*
Luvreling *f* weather rail
Luvseite *f* weather side, luff, windward
luvwärts [to] windward
Luvwinkel *m* crab angle
Lux *n* lux, metre-candle
Luxmasse *f* iron hydroxide powder
Luxmeter *n* luxmeter
Luxuskajüte *f* stateroom
Luxuspapier *n* fancy paper
Luxusschiff *n* luxury liner
Lykopin *n* lycopene
Lykopodiumpulver *n* lycopodium powder
Lyman-Serie *f (nucl)* Lyman series
Lymphserum *n* lymphatic serum
lyophil lyophilic
lyophob lyophobic
lyotrop lyotropic
Lysol *n* lysol
Lyxose *f* lyxose

M

M-Bande *f* M-band
M-Darstellung *f* M-display *(radar)*
M-Schirm *m* M-scope *(radar)*
M-Tank *m* mixed acid tank
Mäanderspannung *f* square-wave voltage
Mäanderwelle *f* meander wave
Maat *m* petty officer *(marine)*
Machart *f* make, style, type, design
machen to make; to produce, to manufacture, to construct; to prepare; to do; to render; to effect, to cause; to shape, to form; to perform; to cost
  Besteck m. to prick *(navigation)*
  Bestellungen m. to place orders
  blank m. to blank
  bündig m. to flush
  ein Glied gestellfest m. to fix a link
  Format m. to make up the margin
  Probefahrt m. to undergo a speed trial
  Stand m. *(print)* to position
  wetterfest m. to weatherproof
Machmeter *n* machometer
mächtig large, big, wide; *(geol)* thick
Mächtigkeit *f* 1. *(geol)* thickness, richness, depth; 2. *(math)* cardinality; potency
  gebaute M. working thickness
Mächtigkeitsverringerung *f* convergence
Mach-Zahl *f* Mach number
MacIntoshit *m* macintoshite *(uranium-bearing mineral)*
Mac-Intyre-Doppelboden *m (ship)* Mac-Intyre tank
Madenschraube *f* grub screw
Madrasgardinen *fpl* madras curtains
Maerz-Boelens-Ofen *m* pork pie furnace
Magazin *n* 1. *(build)* store; 2. *(mach)* magazine *(feeding; rifle)*
Magazinbau *m (mine)* magazine mining, overhand stoping on ore, shrinkage stoping, *(Am)* back stoping, *(Am)* lay system
Magazingatter *n (text)* supply creel
Magazinieren *n* storing, storage [in a magazine]
Magazinladevorrichtung *f (text)* battery loading mechanism
Magazinschleifer *m* magazine grinder *(papermaking)*
Magazin- und Blockbruchbau *m* / **kombinierter** *(mine)* shrinkage stope and pillar-caving, *(Am)* pillar-caving
Magazinwechsel *m* magazine change
Magazinzuführung *f (mach)* magazine feed
mager 1. *(print)* light[-faced]; 2. lean, weak *(concrete, ore)* poor
Magerbeton *m* lean concrete (mix), weak concrete
Magererz *n* lean ore
Magergas *n* lean gas
Magerkalk *m* poor (lean) lime
Magerkohle *f* lean (non-baking, non-caking, semibituminous) coal, semi-anthracite
Magerkoks *m* lean coke
Magermilch *f* skim milk
Magma *n (geol)* magma
Magnesia *f* magnesia, bitter earth
  gebrannte M. calcined magnesia
  kalzinierte M. magnesium oxide
Magnesiabinder *m* magnesium oxichloride cement
Magnesiaeisenglimmer *m* biotite, iron mica
Magnesiamixtur *f* magnesia mixture
Magnesiamörtel *m* magnesia cement
Magnesiazement *m* magnesia cement
magnesiothermisch magnesiothermal
Magnesit *m* magnesite
Magnesitauskleidung *f* magnesite lining *(furnace)*
Magnesitbinder *m s.* Magnesiabinder
Magnesitstein *m* magnesite brick
Magnesitziegel *m s.* Magnesitstein
Magnesitzustellung *f* magnesite lining
Magnesium *n* magnesium
Magnesiumdampf *m* magnesium vapour
Magnesiumeisen *n* brown magnesite
Magnesiummetall *n* metallic magnesium
Magnesiummolybdat *n* magnesium molybdate
Magnesiumsulfit *n* magnesium sulphite
Magnet *m* 1. magnet, 2. *(aut)* magneto
  künstlicher M. artificial magnet
  natürlicher M. natural magnet
  permanenter M. permanent magnet
  ringförmiger M. ring-shaped magnet

supraleitender M. superconducting magnet
temporärer M. temporary magnet
Magnetabscheider *m* magnetic grate
Magnetanker *m* holding-on magnet *(relays)*
Magnetanlasser *m* solenoid starter, magneto booster
Magnetantrieb *m* magnet drive *(switchgear)*
Magnetausleser *m (ag)* magnetic [seed] separator
Magnetausschalter *m* field-break[ing] switch
Magnetband *n* magnetic tape
Magnetbandaufnahmegerät *n* magnetic tape recorder
Magnetbandaufzeichnung *f* magnetic recording
Magnetbandbefehl *m* magnetic tape order *(computer)*
Magnetbanddatenschützer *m* tape data preserver
Magnetbandeinheit *f* magnetic tape handler
Magnetbandgerät *n* tape unit
Magnetbandleser *m* magnetic tape reader
Magnetband-Lochkarte-Umsetzer tape-to-card converter
Magnetbandschreiber *m* **/ kartengesteuerter** card-to-magnetic tape converter
Magnetbandspeicher *m* magnetic tape store
Magnetbandsteuerung *f* magnetic tape control
Magnetbandstörungsschreiber *m* magnetic-tape fault recorder
Magnetbandtransport *m* magnetic recording tape transport
Magnetbelastung *f* magnet load
Magnetblasschalter *m* magnetic blow-out circuit-breaker
Magnetbremse *f* solenoid (magnetic) brake
Magnetdraht *m* magnetic wire *(for data storage)*
Magneteisen *n* magnetic iron
Magneteisenerz *n s.* Magneteisenstein
Magneteisenstein *m* magnetic iron ore, magnetite
Magneterzscheidung *f* magnetic ore concentration
Magnetfeld *n* magnetic field
M. der Erde earth's (terrestrial) magnetic field
äußeres M. external magnetic field
gleichförmiges M. uniform magnetic field
homogenes M. homogeneous magnetic field
konstantes M. constant magnetic field
kraftfreies M. force-free magnetic field
meridionales M. meridional magnetic field
periodisch umkehrendes M. periodically reversing magnetic field
permanentes M. permanent magnetic field
pulsierendes M. pulsating magnetic field
remanentes M. remanent magnetic field
schwaches M. low magnetic field
starkes M. high magnetic field
stellares M. stellar magnetic field
stetiges M. steady magnetic field
symmetrisches M. symmetric magnetic field
toroidales M. toroidal magnetic field
turbulentes M. turbulent magnetic field
ungleichförmiges M. non-uniform magnetic field
verdrilltes M. twisted magnetic field
wirbelfreies M. irrotational magnetic field
Magnetfeldabkühlung *f* cooling down in a magnetic field
Magnetfeldbereich *m* magnetic field range
Magnetfelddichte *f* magnetic field (flux) density
Magnetfelderzeugung *f* magnetic field generation
Magnetfeldindex *m* [magnetic-]field index
Magnetfeldmesser *m* magnetometer; flux indicator
Magnetfeldregler *m* field rheostat
Magnetfeldröhre *f* (linear) magnetron
Magnetfluß *m* magnetic flux
Magnetfilmspeichergerät *n* magnetic film storage unit
Magnetflußdichtemesser *m* fluxmeter
Magnetgestell *n* magnet frame
Magnetheber *m* electro-lifting magnet *(conveying)*
magnetisch magnetic
magnetisierbar capable of being magnetized, magnetizable
Magnetisierbarkeit *f* ability to be magnetized, magnetizability
magnetisieren to magnetize
Magnetisierung *f* magnetization
freie M. natural (free, unimpeded) magnetization
quergerichtete M. transverse magnetization
spontane M. spontaneous magnetization
zurückbleibende M. residual magnetization
zyklische M. cyclic magnetization
Magnetisierungsarbeit *f* magnetization work
Magnetisierungskurve *f* magnetization curve (characteristic), saturation curve
Magnetisierungsrichtung *f* direction of magnetization
Magnetisierungsstrom *m* magnetizing current
Magnetismus *m* magnetism
remanenter M. remanent (residual) magnetism, remanence
Magnetit *m* magnetic iron ore, magnetite
Magnetkarte *f* magnetic card
Magnetkartenspeicher *m* **mit beliebigem Zugriff** card random access memory
Magnetkern *m* magnetic core *(memory)*; core of a magnet, magnet core
lesegestörter M. read-disturbed magnetic core
Magnetkernspeicher *m* magnetic core memory
Magnetkernspeichermatrix *f* magnetic core memory matrix
Magnetkies *m* magnetic pyrite, pyrrhotite

Magnetkompaß *m* magnetic compass
Magnetkontokarte *f* magnetic ledger card
Magnetkopf *m* magnetic [replay] head, recording head
schwebender M. floating magnetic head
Magnetkopfspalt *m* head gap
Magnetkraftlinie *f* magnetic line of force
Magnetkran *m* magnet crane
Magnetkreis *m* / **verkoppelter** interlinked electromagnetic circuit
Magnetlautsprecher *m* [electro]magnetic loudspeaker
Magnetlegierung *f* magnetic alloy
Magnetlocher *m* electric punch
Magnetnadel *f* magnetic needle
Magnetochemie *f* magnetochemistry
magnetografisch magnetographic
Magnetohydrodynamik *f* magnetohydrodynamics
magnetomotorisch magnetomotive
Magneton *n* / **Bohrsches** [electronic] Bohr magneton
magnetooptisch magneto-optic[al]
Magnetopyrit *m* magnetic pyrite, pyrrhotite
Magnetorotation *f* magneto-optic rotation
magnetostatisch magnetostatic
Magnetostriktion *f* magnetostriction
Magnetostriktionsschwinger *m* magnetostriction oscillator
Magnetplattenspeicher *m* magnetic disk (plate) memory, magnetic disk (plate) store
Magnetpulver *n* ferromagnetic powder
Magnetpulverkupplung *f* magnetic particle coupling (clutch)
Magnetpulverprüfung *f* magnetic particle testing
Magnetregelung *f* magneto timing *(ignition)*
Magnetregler *m* field regulator
Magnetring *m* annular magnet
Magnetron *n* magnetron
Magnetrongenerator *m* magnetron oscillator
Magnetronverstärker *m* magnetron amplifier
Magnetschalter *m* contactor; solenoid (magnetically actuated) switch
Magnetscheiden *n* magnetic separation
Magnetscheider *m* magnetic [ore] separator, magnetic [ore] grader
Magnetschenkel *m* magnet limb (leg)
Magnetschicht *f* magnetic film
Magnetschienenbremse *f* electro-magnetic rail-brake
Magnetschreiblocher *m* electric printing punch
Magnetschriftdrucker *m* magnetic ink imprinter
Magnetschrifterkennung *f* magnetic ink character recognition
Magnetschriftfarbband *n* magnetic ink ribbon
Magnetschriftleser *m* magnetic ink reader
Magnetschriftsortiermaschine *f* magnetic character sorter
Magnetschütz *n* magnetic contactor (cut-out)
Magnetspannfutter *n (mach)* magnetic chuck
Magnetspeicher *m* magnetic memory
Magnetspeichermatrix *f* magnetic memory matrix
Magnetspektrometer *n* magnetic spectrometer
Magnetspule *f* magnet (magnetizing, field) coil
Magnetstabantenne *f* magnet-core aerial
Magnetstabspeicher *m* magnetic rod memory
Magnetstahl *m* magnetic steel
Magnetstarter *m* magnetic starter
Magnettonband *n* magnetic (recording) tape
Magnettonbandgerät *n* magnetic tape recorder
Magnettongerät *n* magnetic recorder
Magnettonwiedergabe *f* magnetic sound-playback
Magnettrommel *f* magnetic (storage) drum
Magnettrommelspeicher *m* magnetic drum memory
Magnetunterbrecher *m* magnetic interrupter
Magnetventil *n* solenoid valve
Magnetverstärker *m* transductor, magnetic amplifier
Magnetverstärkerkippschaltung *f* / **bistabile** magnetic amplifier flip-flop
Magnetwicklung *f* magnet winding; field winding
Magnetzünder *m* magneto
Magnetzündung *f* magneto ignition
Magnus-Effekt *m* Magnus effect
Mahagoni *n* mahogany
australisches M. bastard mahogany
Mähbalken *m* [mower] cutter bar
Mähbinder *m* [combined reaper and] binder
Mahd *f* mowing; hay crop
Mähdrescher *m* harvester-thrasher, harvester-thresher
mähen to mow; to reap
Mäh[feld]häcksler *m* cutter loader, forage harvester
Mähläder *m* cutter loader
Mahlanlage *f* crushing mill, pulverizing equipment
Mahlbarkeit *f* grindability, millability
mahlen to crush *(mineral)*; to pulverize, to mill; to grind
Mahlgang *m* milling cycle
Mahlgerät *n* beater *(papermaking)*
Mahlgrad *m* beating degree, freeness [index], slowness
Mahlgradprüfer *m* freeness (beaten stuff) tester
Mahlgut *n* grinding stock; beating material *(papermaking)*
Mahlsand *m s.* Treibsand
Mahlstein *m* mill stone

Mahltrockner *m* crusher-drier
Mahltrocknung *f* in-the-mill drying
Mahlung *f* beating, milling, grinding, disintegration
Mähmaschine *f* mowing machine, mower, reaper
Mähwerk *n* mower
Maierform *f (ship)* Maier form
Maierschiffsform *f* Maier form
Maihak-Saitendehnungsmesser *m* vibrating wire strain gauge
Mailleuse *f (text)* loop wheel
Mais *m* maize, *(Am)* corn
Maischapparat *m* masher; wort masher; scum mixer
Maischbottich *m* fore-masher
Maische *f* mash
Maischwürze *f* mash wort
Maisdreschmaschine *f* maize sheller
Maisentkörnungsmaschine *f* maize sheller
Maisentlieschmaschine *f* de-husker
Maiserntemaschine *f* maize harvester
Maishackgerät *n* maize hoe
Maiskolbenpflückmaschine *f* maize cob picker
Maiskombine *f* complete maize harvester
Maislegemaschine *f* maize planter
Maismehlkleber *m* zein
Maisreb[b]ler *m* maize sheller
Maissämaschine *f* maize planter
Maisstärke *f* maize starch
Maisvollerntemaschine *f* complete maize harvester
Majorana-Kraft *f* Majorana force
Majorana-Teilchen *n* Majorana particle
Majorante *f (math)* majorant
Majoritätsemitter *m* majority emitter
Majoritätsträger *m* majority carrier
Majuskeln *fpl* capital letters, caps
Makadam *m(n)* macadam
Makadamdecke *f* macadam surface (surfacing)
Makadamstraße *f* macadam road
Maklergebühr *f* brokerage
Makobaumwolle *f* maco cotton
Makogarn *n* maco yarn
Makoimitation *f* imitation maco
Makotuch *n* maco cloth
Makroachse *f* macroaxis *(crystals)*
Makroätzung *f* macroetch
Makrofotografie *f (opt)* photomacrography
Makrogefüge *n* macrostructure
Makrokode *m* macrocode
Makrokosmos *m* macrocosm
Makromolekül *n* macromolecule
Makrophysik *f* macrophysics
Makroschliff *m* macrosection
makroskopisch macroscopic
Makrostruktur *f* macrostructure
Makroteilchen *n* macroparticle
Makroturbulenz *f* macroturbulence
Makrozustand *m* macrostate
Makulatur *f* waste, misprints
Malachit *m* malachite, green copper ore
Malachitgrün *n* malachite green
Malat *n* malate
Maleat *n* maleate
Maleinsäure *f* maleic acid
Maleinsäureanhydrid *n* maleic anhydride
Maler *m* painter, decorator
Malerfarbe *f* paint
Malerhellegat *n (ship)* paint locker
Malerleinwand *f* canvas
Malerpinsel *m* painter's brush
Mall *n (ship)* mould
Mallbreite *f* **des Balkens** moulding of a beam
mallen *(ship)* to mould
Malonat *n* malonate
Malonsäure *f* malonic acid
Maltase *f* maltase
Malteserkreuz *n (mach)* Geneva cross, maltese cross
Malteserkreuzbewegung *f* Geneva motion
Maltose *f* maltose, malt-sugar
Malz *n* malt
Malzdarre *f* malt kiln
Mälzerei *f* malt house
Malzschrot *n(m)* malt grist
Malzzucker *m* malt sugar
Mammutbaum *m* sequoia
Mammutpumpe *f* air-lift [pump], mammoth pump
Mammutschiff *n* crack liner (steamer)
Mandarinenschalenöl *n* oil of mandarin[e] peels
Mandelkleie *f* almond powder
Mandelöl *n* almond oil
Mandelsäure *f* mandelic (phenylhydroxyacetic) acid
Mandrill *m* mandrel
Mangan *n* manganese
Manganat *n* manganate
Manganbronze *f* manganese bronze
Manganchlorid *n* manganese chloride
Mangandioxid *n* manganese dioxide, black oxide of manganese
Manganerz *n* manganese ore
manganhaltig manganesian, manganic, manganiferous, manganous
Manganhartstahl *m* high-manganese steel
Manganin *n* manganin
Manganit *m (min)* manganite
Mangankarbid *n* manganese carbide
Mangankupfer *n* manganese copper
Manganlegierung *f* manganese alloy
Manganosit *m (min)* manganosite
Mangansalz *n* manganese salt
Manganspat *m* manganese spar, rhodochrosite
Manganstahl *m* manganese steel

siliziumhaltiger M. silico-manganese steel
Mangan(II)-verbindung *f* manganous compound, mangano-compound
Mangan(III)-verbindung *f* manganic compound
Mangel *m* lack; deficiency; deficit; shortness
M. an lack of, shortage of
natürlicher M. inherent vice
Mangel *f (text)* mangle, ironer *(hot pressing)*
Mangelelektron *n* hole electron
mangelhaft defective, imperfect, faulty
Mangelhalbleiter *m* p-type semiconductor
Mangelleiter *m* defect (p-type) conductor
Mangelleitung *f* hole (p-type) conduction *(semiconductor)*
mangeln *(text)* to mangle
Manilahanf *m* abacá, manila [fibre], manila hemp
Manilakarton *m* manila cardboard
Manilapapier *n* manila paper
Maniokastärke *f* starch of manioc
Manipulation *f* manipulation, handling
Manipulator *m* manipulator, positioner
ferngesteuerter M. remote-control manipulator
hydraulischer M. hydraulic manipulator
Manipulieren *n* manipulating, manipulation
Mann über Bord! man overboard!
Mannigfaltigkeit *f (math)* manifold
geschlossene M. circuit
Mannit *m* mannitol
Mannloch *n* manhole
Mannlochbügel *m* manhole dog (cross-bar)
Mannluke *f* manway
Mannonsäure *f* mannonic acid
Mannose *f* mannose
Mannschaft *f* [ship's] crew, lower deck personnel; gang, team
Mannschaftslampe *f (mine)* driver's lamp
Mannschaftsmesse *f* crew's mess-room
Mannschaftsraum *m* crew's quarter (space)
Mannschaftsschlafraum *m* [crew's] bunks
Mannschaftstransportwagen *m* / **gepanzerter** armoured personnel carrier
Mannschaftsverzeichnis *n (ship)* station bill
Manometer *n* manometer, pressure gauge
M. mit Rostdeckel pressure gauge with protecting screen
M. mit selbsttätiger Druckaufzeichnung recording pressure gauge
M. mit Sicherheitsgitter pressure gauge with protecting screen
Manometerablesung *f* pressure-gauge reading
Manometeranschlußstück *n* nipple for pressure gauge
Manometerprobierpumpe *f* pressure-gauge test pump
Manometerprüfwaage *f* dead-weight pressure-gauge tester
Manostat *m* manostat
manövrierbar manoeuvrable
Manövrierbarkeit *f* manoeuvrability
manövrieren to work *(a vessel)*; to manoeuvre
manövrierfähig manageable *(vessel)*
nicht m. unmanageable
Manövrierfähigkeit *f* manoeuvrability, manoeuvring capacity, handiness
Manövrierprüfung *f* manoeuvring trial
Manövrierschalter *m* motor control switch
Manövrierstand *m* manoeuvring station
manövrierunfähig not under command (control), unmanageable
Manövrierunfähigkeitslaterne *f* not-under-command light
Manövrierversuch *m* manoeuvring test; manoeuvring experiment *(with a model)*
Mansarddach *n* mansard (curb) roof
Mansarde *f* mansard
Manschette *f* collar, sleeve; lip-type of packing
V-förmige M. vee-leather
Mantel *m* 1. *(text)* overcoat; 2. circumferential (cylindrical) surface *(cylinder)*; 3. outer cover *(tyre)*; 4. sheath *(cable)*; 5. jacket *(cooling water, boiler)*; 6. shell *(mould, blast furnace)*
zweireihiger M. double-breasted overcoat
Manteldruck *m (geol)* lateral pressure
Mantelelektrode *f* coated (covered) electrode
Mantelfläche *f* area of the cylindrical surface
Mantelheizung *f* jacket heating
Mantelhülse *f* thimble *(micrometer)*
Mantelkabel *n* sheathed cable
Mantelkern *m* shell-type core *(transformer)*
Mantelkondensat *n* jacket water *(steam engine)*
Mantelkühlung *f* jacket cooling
Mantelkurve *f (mach)* barrel (drum) cam
Mantellinie *f* **eines Kegels** element of a cone
Mantelmagnet *m* pot (shell-type) magnet, *(Am)* boxed magnet
Mantelofen *m* jacket furnace
Mantelreibung *f* skin friction *(piles)*
Mantelreibungspfahl *m* suspended pile
Mantelrohr *n (build)* casing; protecting tube; *(mach)* [outer] sleeve *(radial drill)*
Mantelschichten *fpl* casing plies *(of tyres)*
Mantelstoff *m* overcoating
Manteltalje *f (ship)* runner [and] tackle
Manteltransformator *m* shell-type transformer
Mantelverlust *m* [cable-]sheath loss
Mantisse *f* mantissa
Manuskript *n* matter, copy, manuscript
druckfertiges M. fair copy
Manuskripthalter *m* copy-holder
Margarine *f* margarine
Margarinsäure *f* margaric acid
Marginale *n* marginal (side) note
Marginalkosten *pl* marginal cost
Marine *f* marine *(merchant)*; navy

Marinearchitekt *m* naval architect
Marineartillerie *f* naval artillery
Marineflieger *m* naval flyer
Marineflugzeug *n* navy aircraft
Marinehafen *m* naval port
Marineingenieur *m* naval constructor; marine engineer
Marinekessel *m* marine boiler
Marineradar *n* marine radar
Marineseeflugzeug *n* naval seaplane
Marinesender *m* navy transmitter
Marinewerft *f* [marine] dockyard, naval yard, navy-yard
marinieren to pickle, to marinade
maritim maritime
Mark *n* marrow *(bone)*; pith *(plant)*
Markasit *m* marcasite, white iron pyrite
Marke *f* mark; trade-mark; brand, make, label, type; [postage] stamp
M. für den höchsten Druck maximum pressure index
Markengeber *m* / **elektronischer** electronic marker generator
Markenkamera *f* trade-marked camera
Markeur *m (ag)* marker
markieren to label, to tag; to mark
radioaktiv m. to label (tag) with a radioactive isotope
Markierer *m (telev)* selector position-marker
markiert labelled, tagged; marked
stabilisotop m. labelled with a stable isotope
Markierung *f* 1. labelling, tagging; marking; 2. label, mark[er], tag
M. durch unsauberen Schützen *(text)* shuttle mark
radioaktive M. radioactive tracing
Markierungsboje *f* watch (dan) buoy, mark-buoy, marking-buoy
Markierungsdifferenz *f* rate of change *(radar)*
Markierungselement *n* tracer element
Markierungsfunkfeuer *n* marker beacon
Markierungsgenerator *m* fixed range mark generator
Markierungsimpuls *m* marker pulse
Markierungsisotop *n* isotopic tracer, tracer isotope
Markierungsstrich *m* graduation mark
Markierungssynthese *f (nucl)* labelling synthesis
Markise *f* awning, sun-blind
Markisenstoff *m* awning cloth
Markow-Prozeß *m* Markov process
Markpapier *n* / **chinesisches** rice paper
Markscheide *f (mine)* boundary
Markscheidekunde *f* mine surveying
Markscheider *m* mine (underground surveyor
Markscheiderausrüstung *f* mine surveyor's equipment
Markscheiderei *f* mine surveying department
Markscheidergehilfe *m* chainman
Markscheidesicherheitspfeiler *m* boundary pillar
Markscheidezug *m* dialling, lining
Markstrahl *m* medullary ray *(timber)*
Markt *m* market
marktfähig marketable, saleable, fit for sale
Marktfähigkeit *f* marketability, saleability, fitness for sale
Marktforschung *f* market research
Markthalle *f* covered market, market hall
Marmor *m* marble
Marmorieren *n* marbling *(book)*
Marmorschnitt *m* marbled edge
Maroquineinband *m* morocco-binding
Mars *m (ship)* top, scuttle
Marschgeschwindigkeit *f (mine)* haulage (cutter) speed
Marschturbine *f* cruising turbine
Martensit *m* martensite
Martensitbildung *f* martensite formation
Martensitgefüge *n* martensitic structure
Martensitstahl *m* martensitic steel
Martinofenschlacke *f* open-hearth cinder
Masche *f (text)* loop, mesh, stitch; mesh *(wire)*
M. eines elektrischen Netzes mesh (loop) of an electric network
Maschenanode *f* meshed anode
Maschenbildung *f (text)* formation of stitches, loop formation (forming)
Maschendraht *m* wire cloth (fabric, mesh, netting), screen (mesh) wire
maschenfest ladder-proof *(stocking)*
Maschennetz *n (el)* meshed (interlinked) system, [fully intermeshed] network
Maschennetzschalter *m* reverse-current protective switch
Maschenreihe *f (text)* course
Maschenschaltung *f (el)* mesh connection
Maschenscheinwiderstand *m* mesh impedance
Maschenstäbchen *n (text)* wale
Maschenstrom *m* mesh current
Maschenweite *f* mesh [size]; mesh aperture (size)
lichte M. aperture width, clear mesh
Maschine *f* machine; engine; *(el)* motor
M. für Linksstoß *(mine)* left-handed machine
M. für Rechtsstoß *(mine)* right-handed machine
M. für ununterbrochenes Räumen continuous broaching machine
M. für zweiseitigen Druck duplex-printing machine
M. mit ausgeprägten Polen *(el)* motor (generator) with salient poles
M. mit hin- und hergehender Hauptbewegung reciprocating machine

M. mit mehreren Arbeitsstellen multiple-station machine
M. mit verstellbarer Spindel adjustable-spindle machine
M. nach Baukastensystem unitized machine tool
analytische M. analytical engine *(first programmed calculator designed by Babbage in 1833)*
lehrende M. teaching machine
lehrsatzbeweisende M. theorem-proving machine
rotorgespeiste M. inverted machine
schlagwettergeschützte M. flameproof (fire-damp-proof) machine
schwallwassergeschützte M. watertight machine
spielende M. game playing machine *(e.g. chess)*
statistische M. statistical machine

maschinell by machine; mechanical
Maschinen *fpl* **und Geräte** *npl* **/ landwirtschaftliche** farm machinery and tools
umlaufende elektrische M. rotating electric machinery
Maschinenabzug *m (print)* machine proof
Maschinenanlage *f (ship)* engine plant, propelling machinery, machinery layout
Maschinenanordnung *f (ship)* layout of engines
Maschinenantrieb *m* machine drive
Maschinenarbeit *f* machine work
Maschinenarbeiter *m* [machine] operator, [machine] attendant
Maschinenauftrag *m (text)* machine coating
Maschinenausfallzeit *f* machine down (idle) time
Maschinenausgangsseite *f* exit side of the machine
Maschinenausnutzungsgrad *m* **/ tatsächlicher** machine effective utilization index
Maschinenbau *m* machine building
Maschinenbauart *f* type of machine, machine type
Maschinenbauer *m* machine builder
Maschinenbaugruppe *f* machine (constructional) element
Maschinenbauteil *n* machine detail
Maschinenbedienungsmann *m* machine operator (attendant)
Maschinenbett *n* machine bed
Maschinenbügelsäge *f* power hacksaw
Maschinenbügelsägeblatt *n* power hacksaw blade
Maschinenbütte *f* machine chest *(papermaking)*
Maschinenbütten *n* mould-made paper
Maschinendruck *m* machine printing
Maschineneinrichter *m* machine setter
Maschinenelement *n* machine part (element, detail)
Maschinenfabrik *f* machine-building factory
Maschinenfahrstand *m (ship)* [engine] control platform
Maschinenfett *n* machine (engine) grease
Maschinenformerei *f* machine moulding
Maschinenfundament *n* machine foundation; engine seating (foundation)
Maschinenfuß *m* pedestal [of a machine], column base
Maschinengeländer *n* hand rail
Maschinengestell *n* [main] frame of [a] machine
maschinengestrichen machine-coated *(paper)*
Maschinengewehr *n* machine gun
starres M. fixed gun
Maschinengewehrdrehkranz *m* gun ring
Maschinengewehrkanzel *f* gunner's cockpit
Maschinengewehrsteuerung *f* synchronization (timing) of machine guns
Maschinengewindebohrer *m* machine tap
maschinenglatt machine-finished *(paper)*
Maschinengleichung *f* machine equation
Maschinengrundplatte *f* machine base plate
Maschinengrundreibahle *f* machine reamer of the rose type
Maschinenguß *m* machine casting
Maschinenhalle *f* machine shop
Maschinenhammer *m* power hammer
Maschinenhaus *n* engine house
Maschinenhobelmesser *n* machine plane iron
Maschineninstandhaltung *f* machine maintenance
Maschinenkapazität *f* **eines Betriebs** capacity of a shop's machinery
Maschinenkatalog *m* machine[ry] catalogue
Maschinenkegelreibahle *f* machine taper reamer
Maschinenkunde *f* theory of machines
Maschinenlagerung *f (ship)* engine seating (foundation)
Maschinenlaufzeit *f* machine running time
Maschinenleitstand *m* engine control platform
Maschinenmelken *n* machine milking
Maschinenmitte *f* centre of machine
Maschinenmutternbohrer *m* machine nut tap
Maschinennähen *n* machine sewing
Maschinennebenzeit *f* machine ancillary (idle) time
Maschinennieten *n* machine riveting
Maschinennietlochreibahle *f* machine bridge reamer
Maschinenoberlicht *n (ship)* engine skylight
Maschinenöl *n* machinery oil
Maschinenperiode *f (dat)* machine cycle
Maschinenpersonal *n (ship)* engine room staff
Maschinenprototyp *m* prototype machine
Maschinenrahmen *m* machine frame (bed); engine bed-plate
Maschinenraum *m* engine room, machinery space *(on vessels)*

Maschinenraumanordnung *f* engine room layout
Maschinenraumflurplatten *fpl* engine room floor plates
Maschinenraumhilfsmaschinen *fpl* engine room auxiliaries
Maschinenraumkran *m* engine room crane
Maschinenraumluke *f* engine room hatchway
Maschinenraumpersonal *n* engine room staff
Maschinenraumschott *n* engine room bulkhead
Maschinenreibahle *f* machine reamer
Maschinenrichtung *f* machine direction *(paper-making)*
Maschinensatz *m* machine (mechanical) composition
Maschinenschacht *m* *(ship)* machinery (engine) casing
Maschinenschaden *m* machine defect, breakdown of the machine
Maschinenschere *f* [power-driven] shear, shearing machine
Maschinenschliff *m* machine grinding
Maschinenschlitten *m* machine slide
Maschinenschraube *f* machine bolt
Maschinenschraubstock *m* machine vice
Maschinenschrift *f* type-writing, type script
in M. type-written
Maschinenseite *f* pusher side, ram bench (side)
Maschinensender *m* automatic (alternator) transmitter
Maschinenspindel *f* machine spindle
Maschinenspitze *f* 1. *(text)* machine-made lace; 2. *(mach)* centre
Maschinensprache *f* machine language; common language
Maschinenstall *m (mine)* stable [hole]
Maschinenständer *m* machine column
Maschinensteuergerät *n* machine control unit
Maschinensteuerung *f* machine control
Maschinenstillstandszeit *f* machine down time
Maschinenstrich *m* machine coating *(of paper)*
Maschinenstricken *n* machine knitting
Maschinenteil *n* machine part (detail)
geradlinig bewegtes M. rectilinear moving (reciprocating) mechanism
Maschinentelegraf *m* automatic telegraph, perforated strip-type transmitter; engine[-room] telegraph *(vessel)*, ship's telegraph
Maschinentelegrafenglocke *f* *(ship)* jingle [bell]
Maschinentisch *m* machine table
Maschinen- und Gerätepark *m* equipment fleet, *(Am)* spread; machinery and equipment *(factory)*
Maschinenunterbau *m* engine foundation
Maschinenwärter *m* machine (engine) attendant, operator
Maschinenweber *m* power-loom weaver
Maschinenwelle *f* machine (engine) shaft
Maschinenwerkhalle *f* machine shop
Maschinenwerkzeug *n* machine tool
Maschinenwirkzeit *f* machine-available time
Maschinenzahlengeber *m* pulse machine
Maschinenzeichnen *n* machine drawing
Maschinenzeit *f* machine-controlled time
verfügbare M. machine-available time
Maschinenzentrale *f* central power plant
Maschinist *m* engine operator
Maser *m* maser *(microwave amplification by stimulated emission of radiation)*
kontinuierlicher M. continuous wave maser
optischer M. optical maser, laser
maserig veined *(wood)*
Maserknollen *m* twig burr
Maser-Leistung *f* maser performance
masern to grain
Maserpapier *n* fancy paper
Maser-Strahl *m* maser beam
Maser-Verstärker *m* maser amplifier
Maser-Wirkung *f* maser action
Maserwuchs *m* veined growth *(wood)*
Maske *f* 1. mask; 2. shell *(moulding)*
Maskenmikrofon *n* mask microphone
Maskentechnik *f* masking technique
maskieren to mask
Maß *n* dimension *(technical drawing)*, size; measure; degree; gauge
M. / auf to size
M. / genau nach accurate[ly] to size
nichttoleriertes M. dimension without tolerance
toleriertes M. dimension with tolerance
Maßabweichung *f* variation in size, dimensional variation
zulässige M. amount of variation permitted, permissible amount of dimensional variation
Maßanalyse *f* titration, volumetric analysis
elektrometrische M. electrometric titration
Maßbeständigkeit *f* dimensional stability
Maßbuchstabe *m* dimension letter *(technical drawing)*
Masse *f* 1. *(ph)* mass; weight; 2. compound *(e. g. for sealing purposes)*; 3. *(el)* earth, ground *(for earthing or return purposes)*
M. bei Brennschluß *(rock)* burn-out weight
M. der letzten Raketenstufe final mass
akustische M. acoustic inertia (mass)
bewegte M. moving mass
inerte M. inert mass
kritische M. *(nucl)* critical mass; critical (chain-reacting) amount
passive M. passive mass
sich verzögernde M. retarding mass
spezifische M. specific mass
Masseäquivalent *n* mass equivalent
M. der Energie mass-energy equivalent
Masseeinsparung *f* weight-saving
Masseelektrode *f* earth electrode *(spark plug)*

Masse-Energie-Beziehung *f* mass-energy relation, principle of equivalence of mass and energy
Masse-Energie-Gleichung *f s.* Masse-Energie-Beziehung
massefrei *(el)* floating
massegefärbt coloured-all through
Maßeinheit *f* unit of measurement
M. für das Stricken knitting gauge
Maßeinteilung *f* graduation
M. / mit graduated
Massekabel *n* solid-type cable, lead cable with [impregnated] paper insulation, compound impregnated cable
Massekern *m* powder (powdered, mass, pressed-dust, dust, iron-dust) core
Massekernspule *f* powdered-core loading coil
Massel *f* pig
gekerbte M. notched pig
Masselbett *n* pig bed
Masselbrecher *m* pig breaker
Masselform *f* pig mould
Masselformmaschine *f* pig-moulding machine
Masselgießmaschine *f* pig-casting machine
Masselgraben *m* sow channel
masselos without mass, massless, zero-mass
Massenabsorptionskoeffizient *m* mass coefficient of absorption, mass-absorption coefficient
Massenanziehung *f* gravitation
Massenäquivalent *n* mass-energy equivalent
Massenauflösung *f* mass resolving power
Massenausgleich *m (ph)* balancing of masses; *(mach)* system of balancing; *(build)* balanced excavation (earthworks); balancing of quantities (masses)
Massenauszug *m* bill of quantities
Massenbestimmung *f* mass determination
Massenbeton *m* mass concrete
Massenbewegung *f* mass motion
Massenbremsvermögen *n* mass stopping power
Massendefekt *m* mass defect (effect), packing effect
Massendichte *f* mass density
Massendruck *m* pressure of the mass
Masseneinheit *f* unit [of] mass; atomic mass unit, amu
atomare M. *s.* atomphysikalische M.
atomphysikalische M. atomic mass unit, amu
kernphysikalische M. *s.* atomphysikalische M.
tausendstel M. millimass unit, mamu
Massenelektrode *f* electrode used for welding plain carbon steels
Massenerhaltung *f* conservation of mass
Massenerzeugung *f* mass production
Massenfertigung *f* mass (quantity) production
Massenfertigungsmethode *f* mass production technique
Massenfilter *n* mass filter
Massengut *n* bulk, bulk cargo (goods); heavy traffic conveyed in bulk
Massengutfrachter *m* bulk freighter (carrier)
Massenherstellung *f* mass (lot) production, large-scale manufacture
Massenkern *m* pressed dust core, iron-powder core
Massenkorrektur *f* mass correction
Massenleimung *f* resin (engine) sizing
Massenleitfähigkeit *f* bulk conductivity
Massenmittelpunkt *m* mass centre, centre of mass (inertia, gravity)
Massenmittelpunktsystem *n* centre-of-inertia system
Massenmoment *n* moment of the (a) mass
Massenproduktion *f* mass production
Massenpunkt *m* point particle
Massenschwächungskoeffizient *m* mass attenuation coefficient
Massenschwungrad *n* heavy flywheel
Massenspektrograf *m* mass spectrograph
Massenspektrometer *n* mass spectrometer
doppeltfokussierendes M. double-focusing mass spectrometer
Niersches M. Nier-type mass spectrometer
Massenspektrometrie *f* mass spectrometry
massenspektrometrisch mass-spectrometric
Massenspektroskopie *f* mass spectroscopy
Massenspektrum *n* mass spectrum
Massenstrahler *m* mass radiator
Massenteil *n* mass-produced part
Massenteilchen *n* particle of mass
diskretes M. discrete particle
Massenträgheit *f* inertia
Massenträgheitsdurchflußmesser *m* momentum flowmeter
Massenträgheitskoeffizient *m* inertia coefficient
Massenträgheitsmoment *n* mass moment of inertia, inertia moment of a body (solid)
Massentransport *m* mass handling
Massentrenner *m* mass separator
elektromagnetischer M. electromagnetic [mass] separator
Massenübergang *m* mass transfer
Massenüberschuß *m* mass excess
Massenverhältnis *n* mass ratio
Massenverlust *m* bulk loss
Massenverteilung *f* mass distribution
kontinuierliche M. continuous mass distribution
räumliche M. space distribution of mass
Massenverzeichnis *n* bill of quantities
Massenwert *m* atomic mass
Massenwiderstand *m* **des Dampfes** inertia of the steam *(turbine)*
Massenwirkung *f* action of the mass, mass action
Massenwirkungsgesetz *n* law of mass action

Massenwirkungskonstante *f* equilibrium (mass-action) constant
Massenzahl *f* mass number, rough atomic weight
Massenzunahme *f* mass increase
relativistische M. relativistic mass increase
Masseplatte *f* pasted plate *(of batteries)*
Masseprozente *npl* weight percentage
Masseschlußklemme *f* body contact terminal
Masseverbinder *m* bonder, bonding connector
Massewalze *f (print)* composition roller
Massewiderstand *m* composition[-type] resistor
Massezylinder *m (plast)* injection (heating) cylinder
maßgenau accurate in size, true to size
Maßgenauigkeit *f* dimensional accuracy, accuracy to size
maßgerecht of correct size (dimensions), true to size
maßgeschneidert tailor-made
maßgleich *(math)* isometric
Maßgleichheit *f (math)* isometry
Maßgrenze *f* [dimensional] limit
Maßgrenzen *fpl* / **enge** close limits
Maßgüter *npl* measurement goods (cargo)
Maßgüterladung *f* measurement goods (cargo)
Maßgutladung *f* measurement goods (cargo)
maßhaltig dimensionally stable, accurate to size
Maßhaltigkeit *f* dimensional stability, accuracy to size, holding of size, size consistency
Maßhaltigkeitsprüfung *f* test of accuracy
Maßhilfslinie *f* extension (witness, reference) line *(technical drawing)*
Massicot *m* massicot
massig bulky; *(geol)* massive
massiv 1. solid *(tool)*; 2. heavy; strong; 3. massive, solid, pure *(noble metal)*
Massivdraht *m* solid wire
Massivholz *n* solid timber
Massivholzbauweise *f* all-wood construction
Massivinduktor *m* one-piece inductor
Massivisolation *f* solid insulation
Massivkabel *n* solid cable
Massivsilo *m* brick (concrete) silo
Maßkahn *m* standard barge
Maßkolben *m* volumetric (measuring) flask
Maßkontrolle *f* dimensional inspection
Maßlehre *f* limit gauge
Maßlehren *n* limit gauging
Maßlinie *f* dimension line *(technical drawing)*
Maßlösung *f* volumetric (standard) solution
Maßnahme *f* / **gesetzwidrige** illegal measure
Maßordnung *f* modular coordination
Maßpfeil *m* arrowhead [of dimension line]
Maßstab *m* scale; rule
M. / in großem large-scale
M. / in vergrößertem scaled-up
M. / in verkleinertem scaled-down
logarithmischer M. log scale
verkleinerter M. reduced scale
Maßstabbeiwert *m* scale factor
Maßstabfaktor *m* scale factor
Maßstabfehler *m* imperfection of the scale
maßstäblich full scale
m. gezeichnet drawn [true] to scale
m. sein to be to scale
Maßstabseffekt *m* scale effect
Maßstabseinfluß *m* scale effect
Maßstabsfaktor *m* model correlation factor *(model testing)*
maßstabsgetreu to scale
Maßstabsmodell *n* scale model
Maßsteuerung *f* self-resetting
Maßsystem *n* system of units
Giorgisches M. Giorgi units
Maßtoleranz *f* dimensional tolerance
Maßtonne *f* freight ton
Maßverlust *m* size loss
Maßwalzwerk *n* sizing mill
Maßwerk *n (build)* tracery
Maßzahl *f* dimension [specification], dimension figure *(technical drawing)*
Mast *m* 1. *(el)* pole; tower; standard; pylon; mast *(aerial)*; 2. *(ship)* mast, stick
abgespannter M. guyed mast
abgestagter M. stayed mast
ausfahrbarer M. collapsible mast
einziehbarer M. lowering mast
freitragender M. self-supporting mast
klappbarer M. collapsible mast
nichtabgestagter M. non-stayed mast
umlegbarer M. lowering mast
Mastabschnitt *m* mast section
Mastabspannung *f* pole guy
Mastabstand *m* span, pole (tower) clearance
Mastanker *m* [pole] guy, anchor
Mastaufführung *f* post head *(cable)*
Mastausleger *m* cantilever
Mastbeleuchtung *f* tower lighting
Mastenbauweise *f* pole-type construction, lightweight precast framing system
Mastenkran *m* sheerlegs; derrick
Masterdungswiderstand *m* tower earthing resistance
Mastfalte *f (leath)* fat wrinkle
Mastfell *n (leath)* veal skin
Mastfuß *m* tower (mast) base, pole butt
Mastfußverstärkung *f* reinforcement of [the] pole butt
Mastgeschirr *n* mast-mooring gear
Mastgründung *f* tower (pole) foundation
Mastholz *n (ship)* inch mast; hand mast
Mastikation *f* mastication
Mastix *m* [gum] mastic
mastizieren to masticate
Mastknopf *m (ship)* truck
Mastkondensator *m* pole-type capacitor

Mastkopf *m* pole head
Mastkorb *m (ship)* crow's nest, top
Mastkran *m (ship)* masting sheers
Mastliek *n (ship)* luff, mast leech
Mastschalter *m* pole-mounted circuit-breaker, pole switch
Mastschuh *m* reinforcement of [the] pole butt
Mastsonde *f* pole tester
Mastspreizung *f* spread of tower
Maststütze *f (ship)* outrigger
Masttransformator *m* pole[-type] transformer
Masttrenner *m* pole-type isolator (isolating switch)
Mastumspanner *m* pole transformer
Mastwerk *n* masting *(on vessels)*
Masurium *n* masurium
Masut *n* mazut, masut, heavy fuel
Mater *f (print)* flong; matrix, mat
Material *n* material, stock; substance, matter
aktives M. *(nucl)* active material
ausgesuchtes M. selected material
eingeschweißtes M. weld material
gewebeäquivalentes M. *(nucl)* tissue-equivalent material
hitzehärtbares M. *(plast)* thermoset
pulveriges M. fines *(ore)*
radioaktives M. active material; contaminant, contaminating material
rollendes M. *(ry)* rolling stock
schallschluckendes M. sound-absorbent material
undurchsichtiges M. opaque material
Materialaufgabe *f* feed
Materialaufwand *m* expenditure of materials, material expenditure
Materialausnutzung *f* material economy
Materialbeanspruchung *f* internal stress, stress imposed on material
Materialdurchleuchtung *f* industrial fluoroscopy
Materialeinhängen *n (mine)* material lowering, supply hoisting
Materialeinsatz *m* material inventory
Materialeinsparung *f* saving in material
Materialfehler *m* defect (fault) in [the] material
Materialfestigkeit *f* strength of materials
Materialfeuchtemesser *m* moisture-content meter
Materialfluß *m* flow of material *(manufacture)*
Materialforschung *f* materials research
Materialknappheit *f* material shortage
Materialkonstante *f* matter constant
Materialliste *f* bill of materials
Materialprüflaboratorium *n* material[s] testing laboratory
Materialprüfreaktor *m* material[s] testing reactor, MTR
Materialprüfung *f* material[s] testing, testing of materials
Materialprüfungsreaktor *m* material[s] testing reactor, MTR
Materialstrecke *f (mine)* supply gate
Materialtransport *m* matter transport
Materialüberhauen *n (mine)* supply raise
Materialumschlag *m* material handling
Materialverbrauch *m* material consumption
Materialverschwendung *f* material wastage
Materialzugabe *f (mach)* stock allowance
Materie *f (ph)* matter; substance
interstellare M. interstellar matter
materiefrei free of matter
materiell *(ph)* material
Materiestrahl *m* matter ray
Materiewelle *f* matter wave
Maternpappe *f* matrix board
Materntrockner *m* matrix (flong) drier
Mathematik *f* mathematics
angewandte M. applied mathematics
reine M. pure mathematics
Mathematiker *m* mathematician
mathematisch mathematical
Matratze *f* mattress
Matratzendrell *m* mattress duck
Matrix *f (math)* matrix; *(geol)* ground-mass
alternierende M. alternating matrix
inverse M. reciprocal matrix
Matrixspeicher *m* matrix memory, memory matrix
Matrize *f* matrix; [lower] die, bottom die, female piece, swage, cavity [block]; casting mould (bed) *(concrete construction)*
positive M. positive matrix
Matrizenform *f* matrix form
Matrizenkarte *f* master card
Matrizenmechanik *f* matrix mechanics
Matrizenprägepresse *f* moulding (matrix embossing) press
Matrizenschaltung *f (telev)* matrix unit
Matrizenstahl *m* die steel
Matrizentrockner *m* matrix drier (drying drum)
Matrizenwalze *f* bottom roll
Matrose *m* sailor, seaman; mariner
matt dull *(surface)*; mat; dim
m. geschliffen frosted *(glass)*
m. werden to dull
Mattanstrich *m* dull finish
Mattblech *n* dull-finish sheet, terne plate
Mattbrenne *f* matte dip
mattbrennen to dead dip
Mattdekatur *f* dull decating
Matte *f* 1. gymnastic (ground, gym) mat; 2. *(build)* mat
harzbesprühte M. loaded mat (roving)
Mattenbewehrung *f* wire-mesh reinforcement, steel-fabric reinforcement
Mattenbindung *f (text)* mat weave
Mattenpreßverfahren *n* mat moulding

Mattenverkleidung *f (build)* mat revetment
Mattglas *n* frosted glass *(lamp)*, opal (depolished, obscured) glass
mattieren to dull; to delustre *(textiles)*; to frost
mattiert / durch Ätzung acid-etched
Mattierung *f* dull polish
Mattierungsmittel *n* delustrant, dulling (flatting) agent
Mattkohle *f* clarain, dull coal, durain
Mattlack *m* flat varnish
Mattpaste *f* non-gloss paste
Mattscheibe *f* focusing (diffusing) screen; ground-glass plate
Mauer *f* wall
M. ohne Öffnungen blank (blind) wall
Mauerabdeckung *f* coping
schräge M. feather-edged coping
Mauerabschluß *m* coping
Mauerabschnitt *m* wall section
Maueranker *m* wall tie, anchor
Mauerausbau *m (mine)* bricklining, masonry lining
Mauerbohrer *m* stone drill; wall drilling machine
Mauerbügel *m* wall bracket
Mauerdurchführung *f (el)* wall bushing
Mauerecke *f* quoin
Mauerflucht *f* wall line
Mauerfuß *m* walling crib
Mauerfutter *n* wall lining
Mauergewölbe *n* brick arching
Mauergründung *f* wall footing
Mauerkranz *m* walling crib
Mauerkreuzung *f* junction
Mauerlatte *f* wall plate
Mauermörtel *m* structural mortar
Mauerpfeiler *m* counterfort, buttress
Mauerschlitz *m* chase
Mauerseite *f* / **bündige** facework, fair face
Mauerstein *m* brick
feuerfester M. refractory brick
Mauerung *f* bricking; masonry
M. mit Luftkanälen bricking with air passages
Mauerverband *m* bond
Mauerwerk *n* masonry, brickwork
durchgehendes M. blank (blind) wall
scharriertes M. axed work
Mauerwerksanker *m* wall tie, anchor
Mauerwerksfuge *f* brick joint
bündige M. flat joint
Mauerwerksverankerung *f* tying of the brickwork
Mauerziegel *m* brick
Maul *n* jaw *(snap gauge)*
Maulbeerseidenspinner *m* silkworm fed on mulberry leaves, mulberry silk moth
Maulesel *m* hinny
Maulkorb *m* muzzle
Maulschlüssel *m* open-end wrench
Maulwurf *m s.* Maulwurfpflug
Maulwurfdrän *m* mole drain
Maulwurfdränung *f* mole drainage
Maulwurfpflug *m* mole plough (drainer)
Maulwurfrohrdränmaschine *f* mole drain machine
Maulwurfsfell *n* moleskin
Maurer *m (build)* mason; bricklayer
Maurerkelle *f* trowel
Maurerpolier *m* bricklayer's foreman
Maurerschnur *f* bricklayer's line
Maureske *f* Moresque
maurisch Moorish *(architecture)*
Mäusegift *n* ratsbane, rodenticide
Maxigraf *m* maxigraph
maximal maximum, maximal
Maximalausbeute *f* X-ray output at full ratings *(roentgen tubes)*
Maximalauslöser *m (el)* overload release
Maximalauslösung *f (el)* overload tripping, overcurrent circuit breaking
Maximalausschalter *m* ultimate limit switch, overload circuit breaker
Maximaldosis *f (nucl)* maximum dose
Maximaldruckventil *n* pressure-limiting valve, overload relief valve
Maximalfüllung *f* maximum admission
Maximalleistung *f* maximum output
Maximalrelais *n* maximum [current] relay, overcurrent (overload) relay
Maximalschalter *m* overcurrent circuit-breaker, maximum current circuit-breaker, overload switch
M. mit Nullspannungsauslösung overcurrent undervoltage circuit-breaker
Maximalsignalspannung *f* peak-signal level
Maximalspannungsausschalter *m* overvoltage circuit-breaker
Maximalsteigung *f* ruling grade (up-gradient)
Maximalstromrelais *n* overcurrent relay
Maximalwert *m* maximum value
Maximum *n* maximum, peak
M. an Schwarz *(telev)* peak black
M. an Weiß *(telev)* peak white
breites M. broad maximum
flaches M. flat maximum
spitzes M. sharp peak
zulässiges M. safety maximum
Maximumanzeige *f* maximum reading
Maximumanzeiger *m* **[eines Zählers]** demand attachment [for a meter]
M. mit aussetzender Registrierung restricted hour maximum-demand indicator
Maximum-Minimum-Thermometer *n* maximum and minimum thermometer
Maximumstromzeiger *m* maximum ammeter
Maximumthermometer *n* maximum thermometer

Maximumzähler *m* maximum-demand indicator, demand meter; peak hour-meter
Maximumzeiger *m s.* Maximumstromzeiger
Maximum-zu-Minimum-Verhältnis *n* peak-to-valley ratio
Maxwell *n* maxwell
Maxwell-Brücke *f* Maxwell bridge
Maxwell-Verteilung *f* Maxwell[ian] distribution
mazerieren to macerate
McLeod-Manometer *n* McLeod gauge
Mechanik *f* mechanics
  M. der festen Körper mechanics of solid bodies
  M. der Flüssigkeiten mechanics of liquids (fluids)
  M. der Gase mechanics of gases, pneumatics
  M. ruhender Körper statics
  klassische M. classical mechanics
  statistische M. statistical mechanics
Mechaniker *m* mechanic[ian]
Mechanikerdrehmaschine *f* bench lathe
  M. zum Feindrehen precision lathe
mechanisch mechanical
Mechanisierung *f* mechanization
  M. der Landwirtschaft farm mechanization
Mechanismus *m* mechanism *(kinematics)*
Mediankarte *f* median control chart
Medianwert *m* median
Mediäval *f* old face (style)
Medium *n* medium
  anisotropes M. anisotropic medium
  brechendes M. refractive medium
  dichtes M. dense medium
  dünnes M. rare medium
  festes M. solid medium
  fließendes M. fluid
  flüssiges M. liquid medium
  gasförmiges M. gas medium
  isotropes M. isotropic medium
  streuendes M. scattering medium
  turbulentes M. turbulent medium
Medizinalrinde *f* officinal bark
Medizinball *m* medicine ball
Medizinelektronik *f* medical electronics
Meer *n* sea, ocean
  M. / auf dem [out] at sea, on the sea
  offenes M. open sea, high seas *(outside territorial waters)*
  weites M. *s.* offenes M.
Meerbusen *m* bay, gulf
Meerenge *f* sound, straits, narrow channel
Meeresalgen *fpl* seaweed
Meeresgrund *m* sea bed, [sea] bottom
Meereskunde *f* oceanography, hydrography
Meeresspiegel *m* sea level
  M. / über dem above sea level
Meeresufer *n* sea-shore, beach
Meerwasser *n* ocean (sea) water
meerwasserecht *(text)* fast to sea water
Megaampere *n* megampere
Megabar *n* megabar
Megaelektronenvolt *n* mega-electron volt, million electron volt, MeV
Megafon *n* megaphone
Megahertz *n* megacycles per second, mcps
Megawatt *n* megawatt
Megawatt-Tag *m* megawatt-day, MWd
Megohm *n* megohm
Megohmmeter *n* megohmmeter, *(sl)* megger
Mehl *n* meal *(rock)*; flour, powder, dust
Mehlbleichmittel *n* flour improver
Mehlkern *m* flour core *(foundry)*
Mehlkleister *m* flour paste
Mehlsand *m (geol)* silt
Mehrachsenfahrzeug *n* multi-axle vehicle
Mehradressenbefehl *m (dat)* multi[ple]-address instruction
mehradrig multicore, multi-core *(cable)*
Mehranodenventil *n* multi-anode rectifier
Mehrarbeit *f* extra work
mehratomig polyatomic
Mehraufwand *m* extra expenditure
Mehrbandantenne *f* multi-band aerial
mehrbasisch polybasic
Mehrbefehlskode *m* multiple-address code
Mehrbelastung *f* surplus load
Mehrbereichmeßgerät *n* multi-range meter
Mehrbereichsöl *n* multigrade oil
Mehrdecker *m* multiplane
Mehrdecksiebmaschine *f* multi-deck screen
mehrdeutig *(math)* ambiguous
Mehrdeutigkeit *f* ambiguity, ambiguousness
mehrdimensional multidimensional, polydimensional
Mehrdrahtantenne *f* multiple-wire aerial
Mehrdrahtaufhängung *f* compound catenary suspension
Mehrdüsenvergaser *m* multiple-jet carburettor
Mehrebenenstufenschalter *m* rotary wafer (stud) switch
mehreckig polygonal
Mehreckprofil *n* polygonal shape
mehreindeutig *(math)* many-to-one
Mehrelektrodenröhre *f* multi-electrode valve
Mehrerzeugung *f* surplus production
Mehretagenofen *m* multiple-story furnace
mehrfach multiple
  m. abgesetzt stepped, multi-diameter
  m. gekröpft multiple-throw
  m. geladen multiply charged
  m. gelagert continuous, supported at more than two points
  m. gespiegelt (reflektiert) multiply reflected
Mehrfachabschirmung *f* multiple shielding
Mehrfachanruf *m* multiple call
Mehrfachantenne *f* multiple-tuned aerial

dielektrische M. polyrod (dielectric) aerial
Mehrfachaufspannung *f (mach)* multiple set-up
Mehrfachausnutzung *f (telec)* multiplexing, channel working
Mehrfachbahnentrockner *m* multipass drier
Mehrfachbelichtung *f* multiple exposure
Mehrfachbeschleuniger *m s.* Ringbeschleuniger
Mehrfachbohren *n* multi-drilling
Mehrfachbündel *n* multiple-bunch conductors
Mehrfachecho *n* multiple echo
Mehrfacheinzug *m (text)* corkscrew drawing-in draft
Mehrfachelektrode *f* polyelectrode
Mehrfachempfang *m* diversity reception
Mehrfachempfänger *m* **zum Fadingausgleich** diversity receiver
Mehrfacherdschluß *m* polyphase earth
Mehrfachexpansionsmaschine *f* multiple-expansion engine
Mehrfachfasermaterial *n* multiple-fibre material
Mehrfachform *f* multiple-cavity mould, multi-impression mould
Mehrfachfunkenstrecke *f* multigap, multiple spark gap
Mehrfachgewindebohren *n* multi-tapping
Mehrfachgleitung *f* multiple slip
Mehrfachhalbleitersystem *n* multiple-unit semiconductor device
Mehrfachimpulsgeber *m* multiple pulse generator
Mehrfachinterferenz *f* multiple interference
Mehrfachionisation *f* multiple ionization
Mehrfachkette *f (el)* multiple insulator-chain; multiple chain
Mehrfachkondensator *m* tapped (multiple-unit) capacitor
Mehrfachleitung *f* multiple line
Mehrfachmeißelhalter *m (mach)* multi-tool post (holder)
Mehrfachmodulation *f* multiple modulation
Mehrfachnebenwiderstand *m* universal shunt
Mehrfachnocken *m* multi-lobe cam
Mehrfachoszillografie *f* multiple oscilloscopic presentation
Mehrfachplan *m* sequential-sampling plan, sequential scheme *(quality control)*
Mehrfachpreßform *f* multiple-cavity mould *(die casting)*
Mehrfachprogramm *n* multiple programme
Mehrfachprozeß *m* multiple process
Mehrfachpunktschweißmaschine *f* multiple-spot welding machine
Mehrfachräumwerkzeugsatz *m* multiple-broach assembly
Mehrfachrautenantenne *f* multiple rhombic aerial
Mehrfachreflexion *f* multiple reflection
Mehrfachregelung *f* multivariable control system
Mehrfachröhre *f* multiple valve
Mehrfachrollenkette *f* multiple-strand roller chain
Mehrfachschalter *m* multiple contact switch
Mehrfachschaltung *f* multiple connection
Mehrfachschelle *f s.* Sammelschelle
Mehrfachschmetterlingsantenne *f* super-turnstile aerial
Mehrfachschweißmaschine *f* multiple-operator welding unit
Mehrfachserienbohren *n* multiple repetition drilling
Mehrfachspaltung *f s.* Spallation
Mehrfachspeisung *f* multiple feed
Mehrfachspiegelung *f* multiple reflection
Mehrfachspule *f* multiple coil
Mehrfachstecker *m* multiple (multi-contact) plug, socket outlet adapter
Mehrfachsteckverbindung *f* multi-way connector (plug and socket)
Mehrfachstichprobenverfahren *n* multiple sampling
Mehrfachstreuung *f* multiple (plural) scattering
Mehrfachstromrichter *m* multiple convertor
Mehrfachtarifzähler *m* multiple tariff meter
Mehrfachtransformation *f* cascade transformation *(voltage transformer)*
Mehrfachtransistor *m* multiple transistor
Mehrfachtypendruckverfahren *n* printing multiplex system
Mehrfachübertragung *f* multiplex radio transmission
Mehrfachumschalter *m* multiple-changeover switch, multi-way switch
Mehrfachunterbrechung *f* multiple interruption *(circuit breakers)*
Mehrfachuntersetzer *m (el)* multi-scaler
Mehrfachverdampfer *m* multiple-effect evaporator
Mehrfachvervielfacher *m* multistage multiplier
Mehrfachwerkzeug *n* multiple-cavity die *(die casting)*
Mehrfachzeichen *n* eccentricity signal *(radar)*
Mehrfachzug *m* multiple drawing
Mehrfachzwirn *m* cabled yarn
mehrfädig multi-filament
Mehrfarbendruck *m* colour printing
mehrfarbig multi-colour[ed]; polychromatic
Mehrfasenbohrer *m* multi-cut drill
Mehrfeldbrücke *f* continuous bridge
Mehrfeldplatte *f* continuous slab
Mehrfeldrahmen *m* continuous (multiple-span, multiple-bay) frame
mehrfeldrig multi-span, multi-bay
Mehrfeldträger *m* continuous beam (girder)
Mehrflächentransistor *m* multiple-area transistor

Mehrflözbau *m (mine)* simultaneous extraction of several seams
mehrförmig multi-tee *(slot)*
Mehrfrequenzzeichen *n* compound signal
Mehrfurchenpflug *m* multi-furrow plough
mehrgängig multi-start; multi[ple]-threaded
Mehrgangkondensator *m s.* Mehrfachkondensator
mehrgeschossig multi-storey
Mehrgewicht *n* excess weight, overweight, weight in excess
Mehrgitterröhre *f* multi-electrode valve, multigrid valve
Mehrgratköper *m (text)* combined (stitched) twill
Mehrgruppentheorie *f (nucl)* multi-group theory
Mehrheitsentscheidungselement *n (dat)* majority-decision element
Mehrheitslogik *f (dat)* majority logic
Mehrimpulspunktschweißen *n* pulsation welding
Mehrkammerklystron *n* multicavity klystron
Mehrkammermühle *f* compartment mill
Mehrkanalanalysator *m* multichannel analyzer
Mehrkanalgerät *n* multi-channel instrument
Mehrkanalübertragung *f* multi-channel transmission
mehrkantig multi-edged
Mehrkantprisma *n* polyhedral prism
Mehrkantprofil *n* polygon profile
Mehrkeilprofil *n* multi-spline profile
mehrkernig multinuclear; polynuclear
Mehrknickboot *n* multi-chine boat
Mehrkolbenpumpe *f* multi-piston pump
Mehrkopfflachkämmer *m (text)* rectilinear comber
Mehrkopfstrecke *f (text)* multiple-head draw frame
Mehrkörperkraft *f* multi-body force
Mehrkörperproblem *n* many-body problem
Mehrkosten *pl* extra cost[s], additional (extra) expenses, additional (extra) charges
mehrkreisig multiple-circuit *(radio)*
Mehrkreistriftröhre *f* multicavity velocity modulated valve
Mehrkristall *m* polycrystal
Mehrkurvenautomat *m (mach)* automatic screw machine having independent cam control for each tool; drum-cam operated screw machine
Mehrkurvensteuerung *f (mach)* independent cam control for each operation
Mehrlagenpappe *f* filled board
mehrlagig composed of several layers, multilayer; built-up
mehrläufig multiple-flight *(stairs)*
Mehrlaufregelung *f* multi-speed floating control
Mehrlaufwirkung *f* multi-speed action
Mehrleiterkabel *n* multiconductor (multi-core) cable
Mehrleiternetz *n* polyphase (multiphase) network
Mehrleitersystem *n* polyphase (multiphase) system
Mehrlochdüse *f* multi-hole nozzle
Mehrlocheinspritzdüse *f* multiple-orifice injector
Mehrlochferritkern *m* multi-hole ferrite core
Mehrlochkabelkanal *m* multiple-duct conduit, duct route
Mehrlochmagnetkern *m* multihole (multipath) magnetic core
Mehrmantelkabel *n* multiple-sheath cable
Mehrmaschinenbedienung *f* multiple-machine assignment (attendance)
mehrmehrdeutig *(math)* many-to-many
Mehrmeißelhalter *m* gang-type toolholder
Mehrmetallplatte *f* multi-metal plate
Mehrmotorenantrieb *m* multi-motor drive
mehrmotorig multi-engine[d]
Mehrnadelflachmaschine *f* multi-needle flat bed sewing machine *(shoe)*
Mehrnadelnähmaschine *f* multi-needle sewing machine
Mehrnährstoffdünger *m* compound fertilizer
mehrnutig multi-flute *(tool)*
Mehrphasengenerator *m* multiphase (polyphase) generator
Mehrphasengleichrichter *m* polyphase rectifier
Mehrphasenkreis *m* polyphase (multiphase) circuit
Mehrphasenmotor *m* polyphase motor
Mehrphasennebenschlußmotor *m* polyphase shunt commutator motor
 ständergespeister M. double-fed polyphase shunt commutator
Mehrphasenreihenschlußmotor *m* polyphase series commutator motor
 M. mit Begrenzung der Leerlaufdrehzahl polyphase compound commutator motor
Mehrphasenschaltung *f* polyphase (multiphase) circuit
Mehrphasenstrom *m* polyphase (multiphase) current
Mehrphasensystem *n* polyphase (multiphase) system
 symmetrisches M. balanced polyphase system
Mehrphasentransformator *m* polyphase transformer
Mehrphasenwechselstrom *m* polyphase alternating current
Mehrphasenzähler *m* polyphase meter
mehrphasig polyphase, multiphase
Mehrplattenkondensator *m* multi-vane capacitor
mehrpolig multipolar, multi-pole
Mehrpreis *m* extra charge (cost)
mehrprofilig multi-rib[bed], multi-edge[d] *(tool)*
Mehrpunktregelung *f* multi-step action control

Mehrpunktverhalten *n* multi-step action
mehrreihig genietet multiple-riveted
Mehrrillen[gewinde]fräser *m* multi-ribbed thread cutter, thread milling hob, multiple-type of thread milling cutter
Mehrscharpflug *m* multi-furrow plough
Mehrscheibenkupplung *f* multi[ple]-disk clutch
Mehrschichtenfarbfilm *m* multi-layer colour film
Mehrschichtenfilm *m* multi-emulsion film
Mehrschichtenplatte *f* *(build)* multiple sandwich slab, multi-layer board
mehrschichtig multi-layer, laminated, built-up
mehrschiffig multi-bay
Mehrschneckenpresse *f* multi-screw extruder
Mehrschneidenwerkzeug *n* multi[ple]-cutting edge tool, multipoint tool
mehrschneidig multi[ple]-edged, multi-toothed, multipoint
Mehrschraubenschiff *n* multiple-screw vessel
Mehrseilförderung *f* *(mine)* multi-rope hoisting (winding)
Mehrseilgreifer *m* multi-rope grab bucket
mehrseitig multilateral
mehrsitzig multi-seat
Mehrspeichersystem *n* multiple-capacity system
Mehrspindelanordnung *f* *(mach)* multi-spindle arrangement
Mehrspindelautomat *m* multi-spindle automatic
Mehrspindelbauart *f* multi-spindle type (model, design, construction)
Mehrspindelbohrkopf *m* multiple drill head
Mehrspindelbohrmaschine *f* multiple (multi-spindle) drilling machine
M. mit einstellbaren Spindeln drilling machine with multiple spindle head, adjustable spindles
M. mit fester Bohrspindelanordnung multi-spindle drilling machine with fixed spindles
Mehrspindelfutterautomat *m* multiple-spindle chucking automatic
mehrspindelig multiple-spindle
Mehrspindelmaschine *f* multi[ple]-spindle machine
Mehrspindelstangenautomat *m* multi[ple]-spindle bar automatic
mehrspindlig multi[ple]-spindle
Mehrspurmagnetonkopf *m* multitrack head
mehrstellig multi-digit, multi-place
mehrstöckig multi[ple]-storey
mehrstrahlig multi-jet
Mehrstrahloszillograf *m* multi-beam oscilloscope
Mehrstromgenerator *m* multiple-current generator
Mehrstückspannvorrichtung *f* *(mach)* multiple-type fixture
Mehrstufenbleiche *f* multistage bleaching
Mehrstufenboot *n* multistep hydroplane
Mehrstufeneindampfer *m* multiple-effect evaporator
Mehrstufenfräsen *n* multiple-step cutting
Mehrstufenpressen *n* multiple-shot moulding
Mehrstufenrakete *f* stage (staged, composite, multi-stage, step, multistaged) rocket
M. mit Hintereinanderanordnung der Stufen multistage tandem rocket
M. mit paralleler Stufenanordnung multistage cluster rocket
Mehrstufenschalter *m* multipoint switch
Mehrstufenverfahren *n* long tube vertical multiple effect distillation process *(drinking water generator)*
mehrstufig multi-stage, multistage; multi-step
Mehrtarifzähler *m* multi-rate meter
mehrteilig parted, built-up, composite, multipart; sectional *(broach)*
Mehrwalzenkalander *m* calender with several rollers
Mehrwebstuhlsystem *n* multi-loom system
Mehrwegebohrmaschine *f* multiway (way-type) drilling machine
Mehrwegeempfang *m* multipath reception
Mehrwegemaschine *f* multiway (way-type) machine
Mehrwegeschieber *m* multi-way valve
Mehrwegeventil *n* multi-way valve
Mehrwegrutsche *f* multiple-way chute
Mehrwegübertragung *f* multipath transmission
Mehrwert *m* surplus
mehrwertig multivalent, polyvalent, plurivalent; polyhydric *(alcohol)*
Mehrwertigkeit *f* multivalency, polyvalency
mehrzahnig multi-toothed *(tool)*
Mehrzonenreaktor *m* multi-region reactor
Mehrzweckfahrzeug *n* general-purpose vehicle
Mehrzweckflugzeug *n* utility (general-purpose, multi-purpose) aircraft, utility (general-purpose, multi-purpose) plane
Mehrzweckförderer *m* general-purpose elevator
Mehrzweckfrachter *m* general-purpose cargo vessel, general trader
Mehrzweckfräsmaschine *f* multi-purpose milling machine
Mehrzweckgeschütz *n* multi-purpose gun
Mehrzweckmaschine *f* multiple-purpose machine
Mehrzweckmotor *m* multi-purpose motor
Mehrzweckofen *m* all-purpose furnace
Mehrzweckreaktor *m* multi-purpose reactor
Mehrzweckschiff *n* all-round ship (vessel)
Mehrzwecktransportschiff *n* three-in-one transport ship
Mehrzweckwalzwerk *n* universal rolling mill
Mehrzylinderdampfmaschine *f* multi-cylinder steam engine
Meile *f* mile
gemessene M. measured course

Meilenfahrt *f* trials upon the measured mile, measured mile trip, ship propulsion trial
Meilenzähler *m* milage recorder
Meiler *m* 1. charcoal kiln; 2. *s.* Reaktor
Meilerholz *n* charcoal wood
Meilerofen *m* charcoal kiln
Meilerverkohlung *f* kiln-burning
Meilerwagen *m* trailer for guided missiles
Meißel *m* 1. *(build)* chisel; 2. *(mach)* tool *(lathe, planer)*
  M. mit abgerundeter Ecke nose-radius tool
  abgesetzter M. offset tool
  beiderseits abgesetzter M. offset tool having blade central
  gebogener linker M. left-bent tool
  gebogener rechter M. right-bent tool
  gekröpfter M. offset tool
  hartmetallbestückter M. carbide-tipped tool
  linksschneidender M. left-hand tool
  massiver M. one-piece tool, solid tool
  rechtsschneidender M. right-hand tool
Meißelabhebeeinrichtung *f* tool lifting gear
Meißelabheber *m* tool lifter
Meißelabhebung *f* relief [of the tool], tool relief (relieving, lifting)
  hydraulische M. hydraulic tool relief
Meißelbesetzung *f* / **gleichmäßige** regular pitch *(cutter chain)*
  ungleichmäßige M. irregular pitch *(cutter chain)*
Meißeleinsatzhalter *m (mach)* tool-bit holder
Meißelfanghaken *m* bit hook
Meißelhalter *m (mach)* 1. tool holder (post), 2. [pick] box *(cutter chain)*
  M. mit Abhebung relieving tool holder
Meißelhalterklappe *f* tool apron
Meißelklappe *f* clapper [block], clapper-type tool block, clapper[-type tool] holder, tool holder flap, relieving toolbox
Meißelklappenbolzen *m* hinge pin [of the clapper]
Meißelklappenträger *m* clapper (flapper) box
meißeln to chip
Meißelnachschliff *m (mach)* tool regrind[ing]
Meißelquerschnitt *m (mach)* tool section
Meißelreihe *f* pick line *(cutter chain)*
Meißelschlitten *m (mach)* tool slide; head slide *(shaper)*; ram head slide *(slotter)*
Meißelschneide *f (mach)* tool tip (cutting edge)
Meißelstandzeit *f (mach)* tool life, life of tool
Meißelzinken *m* chisel tine
Meißnerschaltung *f* Meissner oscillator (circuit)
Meister *m* general foreman; champion
Meistergang *m* gas gallery
Meisterprofilrolle *f (mach)* master [crusher] roll
Meisterquartier *n* [queen] closer *(brickwork)*
Meisterschalter *m* master controller
Meisterstück *n (mach)* master
Meisterwalze *f* master (barrel) controller, drum switch
Melamin *n* melamine
Melaminformaldehydharz *n* melamine-formaldehyde resin
Melaminharz *n* melamine resin
Melangegarn *n* blended yarn
Melangegewebe *n* blended fabric
Melangemuster *n* blended design
Melangeuse *f* mixing gill box
Melasse *f* molasses
Melassebehälter *m* molasses tank
Melasseschlempe *f* molasses slop
Melasseschnitzel *npl* molasses pulp
Meldeamt *n (telec)* record office
Meldeanlage *f* indicator system
Meldeboot *n* aviso, advice (dispatch) boat
Meldeeinrichtung *f* alarm
Meldelämpchen *n* lamp repeater
Meldelampe *f* signal (indicating) lamp; pilot signal (lamp)
Meldeleitung *f* record line (circuit)
melden to signal *(train)*
  sich m. to answer [a call] *(telephony)*
Meldeorgan *n* sensing element *(governor)*
Melder *m* alarm, annunciator
Melderakete *f* message-carrying rocket
Melderelais *n* signal relay; record (pilot) relay
Meldeschalter *m* indicating switch
Meldeschiff *n* aviso, advice (dispatch) boat
Meldetafel *f* indicator panel
  optische M. diagram panel
Meldezettel *m* advice note
Meldung *f* message
  dringende M. priority message *(teletype)*
Melierschützen *m (text)* shuttle with several bobbins
meliert mottled *(cast iron)*; mixed
Melioration *f* land improvement, amelioration
meliorieren to ameliorate
Melissenöl *n* balm oil
Melissylalkohol *m* mellissyl (myricyl) alcohol
Melkanlage *f* milking plant
Melkbecher *m* milk cup
Melkeimer *m* milk bucket (pail)
melken to milk
Melker *m* milker
Melkflur *m* milking pit
Melkgeschwindigkeit *f* milking rate
Melkkanne *f* milk bucket (pail); milk churn (can)
Melkmaschine *f* milking machine, milker
  elektrische M. electric milking machine
  transportable M. portable milking machine
Melkpumpe *f* milking pump
Melkrohrsystem *n* pipeline milkers
Melkschemel *m* milking stool
Melkstand *m* milking parlour

fahrbarer M. mobile milking parlour
Melkstandgebäude *n* milking parlour
Melkzeug *n* cluster
Mellitsäure *f* mellitic acid
Membran *f* diaphragm, membrane
halbdurchlässige M. *s.* semipermeable M.
semipermeable M. semipermeable membrane
Membranakkumulator *m* diaphragm accumulator
Membranantrieb *m* diaphragm motor
Membranausgleichvorrichtung *f* diaphragm expansion joint
Membrandose *f* pressure capsule
Membrane *f s.* Membran
Membrangefäß *n* diaphragm tank *(measuring transformer)*
Membrankammer *f* valve top
Membranmanometer *n* membrane (diaphragm) pressure gauge
Membranmengenmesser *m* flowmeter with diaphragm-type pressure gauge
Membranmotor *m* diaphragm motor
federbelasteter M. spring-loaded diaphragm motor
Membranpumpe *f* diaphragm (check-valve suction) pump
Membranringmodusfilter *n s.* Membranringwellentypfilter
Membranringwellentypfilter *n* diaphragm-ring mode filter *(waveguides)*
Membranschalter *m* pressure-sensitive switch, diaphragm-box level detector
Membranspannung *f* diaphragm (membrane, direct) stress
Membranspannungszustand *m* membrane state of stress
Membranstellmotor *m* diaphragm motor
Membranstift *m* vibrating pin
Membrantheorie *f* membrane theory
Memoryeffekt *m* memory effect
Mendelevium *n* mendelevium
Menge *f* 1. amount; quantity; 2. set, assemblage
kompakte M. compact set
überfallende M. discharge over a weir
mengen to blend; to mix
Mengenausbringen *n* recovery
Mengenfertigung *f* quantity (large-scale) manufacture
Mengenkontrolle *f* mass inspection
Mengenlehre *f* theory of sets, set theory
Mengenleistung *f* output, rate of production, production rate
Mengenmesser *m* flowmeter
volumetrischer M. displacement-type flowmeter, positive-displacement flowmeter
Mengenmessung *f* volume measurement (determination)
Mengenregelung *f* flow (volume) control; quantity governing *(turbine)*
kontinuierliche M. flow control
Mengenregelventil *n* regulator (flow-regulating) valve
Mengenregler *m* volume governor *(hydraulics)*
Mengenstrombild *n* weighted flow-sheet
Mengenverhältnis *n* quantity relative number
Mengenverstärker *m* air-relay, booster *(pneumatics)*
Mengenvoreinstellung *f* quantity selection *(positive displacement meter)*
Meniskus *m* meniscus *(capillarity; lens)*
Meniskuslinse *f* meniscus lens
Mennige *f* red lead
Mennigekitt *m* red-lead cement
Mensa *f* refectory; students' canteen
Mensur *f* graduated cylinder, measure
Menthan *n* menthane
Menthol *n* menthol
Mergel *m* marl
Mergelton *m* chalky clay
Meridian *m* meridian
Meridianebene *f* meridian plane
Meridiankreis *m* meridian circle
Meridiansucher *m* meridian finder
Meridionalschnitt *m* meridional section
Merinoschaf *n* merino sheep
Merkaptan *n* mercaptan
Merkaptid *n* mercaptide
merkbar noticeable
Merklampe *f* telltale lamp
Merkmal *n* feature; criterion
Merkmalsvariablen / mit zwei bivariate
Merkmalwert *m* variate
Merkurierung *f* mercuration
Meroedrie *f* merohedrism
meroedrisch merohedral
Merzerisieranlage *f* mercerizer
Merzerisierapparat *m* mercerizing frame
Merzerisierechtheit *f* fastness to mercerizing
merzerisieren to mercerize
Merzerisierfoulard *m* mercerizing mangle
Mesadiode *f* mesa diode
Mesatransistor *m* mesa transistor
Mesitylen *n* mesitylene
Mesomerie *f* mesomerism
Mesomolekül *n* mesic molecule
mesomorph mesomorphous
Meson *n* meson
neutrales M. neutral meson
schweres M. heavy meson
Mesonatom *n* mes[on]ic atom
Mesoneffekt *m* mesonic effect
Mesonenbahn *f* meson track
Mesoneneinfang *m* meson capture
Mesonenfeld *n* meson field
Mesonenquelle *f* meson source

Mesonenreaktion *f* meson reaction
Mesonentheorie *f* **der Kernkräfte** meson theory of nuclear forces
Mesonenzerfall *m* meson decay
Mesothorium *n* mesothorium
Mesoweinsäure *f* mesotartaric acid
Meßablesung *f* test reading
Meßader *f* test core (pair), second wire
Meßanordnung *f* measuring equipment
Meßapparat *m* / **hydraulischer** hydraulic testing unit
Meßausrüstung *f* / **optische** optical measuring equipment
meßbar measurable
Meßbarkeit *f (ph)* measurability, mensurability
Meßbatterie *f* testing battery
Meßbecher *m* graduated jug
Meßbereich *m* measuring range, scope of measurement
Meßblende *f* calibrated orifice; measuring shade
Meßbrief *m* rating certificate, bill of admeasurement, bill of tonnage *(of a vessel)*
Meßbrücke *f* measuring bridge [network], bridge [network] for measuring electrical resistances
Meßbrückenkapazitätsanalysator *m* bridge-type capacitance analyzer
Meßbrückenverstärker *m* bridge amplifier
Meßbuchse *f* test socket
Meßdorn *m* test mandrel
Meßdraht *m* taut [measuring] wire, slide (thread-measuring) wire
Meßdruck *m s.* Meßkraft
Messe *f* fair
Meßebene *f* useful plane *(illumination)*
Messegelände *n* fair grounds (site, area)
Messehalle *f* exhibition hall
Meßeinrichtung *f* measuring device (unit)
Meßeinrichtungen *fpl* metering equipment
Meßeinteilung *f (ph)* graduation
Messekatalog *m* fair catalogue
Meßelektrode *f* test electrode
Meßelement *n* gauge; sensor
messen to measure; to gauge; to check; to determine
  die Emission einer Röhre m. to check a tube
Messen *n* **nach dem Arbeitsgang** post-process gauging
  M. während des Arbeitsganges in-process gauging
  mittelbares M. comparative measurement
  unmittelbares M. direct measurement
Messer *m* gauge; meter, measuring instrument
Messer *n* knife; *(mach)* [cutting] blade
  M. aus einem Stück solid cutter in one piece
  M. der Häckselmaschine chaff knife
  aufgeschraubtes M. blade locked by screws *(milling cutter)*
  geriffeltes M. serrated blade *(mill)*
  wolframkarbidbestücktes M. tungsten-carbide tipped blade
Messerantrieb *m* blade activation *(coal plough)*
Messerbalken *m* cutter bar, knife beam
Messerbefestigung *f* **durch Keil** [blade] wedge clamping *(milling cutter)*
Messeregge *f* scarifier, scuffler
Messerentkörnungsmaschine *f (text)* Macarthy gin
Messerfeile *f* knife-edge file
Messerführung *f* blade guide *(contacts)*
Messerfurnier *n* knife (sliced, carved) verneer
Meßergebnis *n* measurement (test) result, measurement (test) reading
Messergriff *m s.* Messerheft
Messerheft *n* knife handle
Messerkasten *m* griff[e] box *(Jacquard loom)*
Messerklemme *f (mach)* cutter clamp
Messerklinge *f* knife blade
Messerkontakt *m* blade (V-type) contact
Messerkopf *m (mach)* cutter head, built-up type of face mill, face milling cutter with inserted blades
  M. mit gekrümmten Messern cutter block with curved knives
  M. mit hohl aufliegenden Messern cutter block with hollow faces
Messerkopfschleifmaschine *f* face-milling cutter grinder
Messerkopfzapfen *m* cutter shank
Messerleiste *f (el)* plug assembly, multiple connector plug
Messermaschine *f* knife-veneer cutting machine
messern to slice
Messerradhäcksler *m* flywheel (rotary) cutter
Messerschälmaschine *f* knife barking machine
Messerschalter *m* knife [blade] switch
  mehrpoliger M. tandem knife switch
Messerschleifmaschine *f* knife (cutter) grinder
Messerschneidemaschine *f* knife-cutting machine
Messersech *n* knife coulter
Messerspindel *f* cutter block spindle
Messerspitzbohrer *m (mach)* flat drill with inserted cutting edges
Messerstahl *m* cutlery steel
Messerstandzeit *f (mach)* cutting-blade life
Messerstangensenker *m* counterbore with removable cutters (blades)
Messerstecker *m* blade (knife) plug
Messerstütze *f* knife supports
Messerträger *m* knife holder
Messer- und Federleiste *f (el)* plug-and-socket connector
Messerwalze *f* beater roll
Messerwelle *f* cutter block spindle
Messerzeiger *m* knife-edge pointer
Messerzinken *m* spike-tooth tine

Messestand *m* fair stand
Meßfehler *m* measuring error, error in measurement
Meßfläche *f* gauging surface
Meßformel *f (ship)* tonnage formula
Meßfrequenz *f* test frequency
Meßfühler *m* sensing (detecting, primary measuring) element, sensor, *(Am)* transducer
Meßfühlhebel *m* indicating contactor
Meßfunkenstrecke *f* measuring (calibrated) spark-gap
Meßgenauigkeit *f* measuring (measurement) accuracy, accuracy of measurement
Meßgerät *n* measuring instrument (device), meter (*s. a.* Meßinstrument)
  M. für elektrische Größen electrical measuring instrument
  M. für zwei Meßbereiche double-range instrument
  M. mit Gleichrichter rectifier instrument
  anzeigendes M. indicating instrument
  elektrisches M. für nichtelektrische Größen electrically measuring instrument
  integrierendes M. integrating meter
  kernphysikalisches M. nuclear (nucleonic) instrument
  registrierendes M. recording instrument
  schreibendes M. registering meter
  summierendes M. summation instrument
  thermoelektrisches M. thermo-junction instrument
  zählendes M. registering meter
Meßgeräteausrüstung *f* instrumentation *(of satellites)*
Meßgerätewagen *m* trolley, *(Am)* testomobile
Meßgestell *n* [repeater] test rack
meßgesteuert with automatic size control and compensation, size-controlled
Meßgitter *n* graticule
  geeichtes M. ruled grating
Meßgitterrahmen *m* escutcheon
Meßgitterrasterscheibe *f* graticule
Meßgitterscheibe *f* calibrated scale
Meßgleichrichter *m* meter (instrument) rectifier
Meßglied *n* discriminating (metering) element; sensor, *(Am)* transducer
Meßgrenze *f* limit of measurement
Meßgröße *f* measured quantity
Meßhütchen *n* contact tip
  M. mit Kugelfläche ball-pointed contact tip
  M. mit Schneide knife-edge contact tip
Messing *n* brass
Messingabschirmung *f* brass shield
messingartig aerose, brassy
Messingblech *n* sheet (latten) brass
Messingdraht *m* brass wire
Messingfassung *f* brass holder
Messinghartlot *n* brazing spelter
Messingkatode *f* brass cathode
Messinglinien *fpl* brass rules
Messinglot *n* brazing (hard) solder
Messingrohling *m* **[zum Pressen]** brass slug
Messingüberzug *m* brass plating
Meßinstrument *n* measuring instrument; meter (*s. a.* Meßgerät)
  M. mit Fernanzeige remote measuring instrument, *(Am)* tallymeter
  M. mit Hängelager single-pivot instrument
  M. mit Nullpunkt in der Skalenmitte zero-centre instrument
  M. mit unterdrücktem Nullpunkt suppressed-zero instrument
  aperiodisches M. highly-damped instrument
  elektrisches M. electric instrument
  elektrostatisches M. electrostatic instrument
  schreibendes M. recording instrument
  summierendes M. summation instrument
  träges M. sluggish measuring instrument
  zählendes M. registering instrument
Meßkabel *n* measuring cable (lead)
Meßkanal *m* measuring channel; instrument hole; flume *(Venturi)*
Meßkern *m* measuring core *(measuring transformer)*
Meßkette *f* measuring (land) chain
Meßklinke *f* metering jack
Meßkoffer *m* portable measuring set
Meßkolben *m (ch)* measuring (graduated) flask
Meßkopf *m* measuring head
Meßkraft *f* measuring force
Meßkreis *m (el)* measuring circuit
Meßkreisel *m* impeller
Meßkunde *f* metrology
Meßlänge *f* gauge length
Meßlättchen *n* midget staff
Meßlatte *f* measuring rod; staff
Meßleitung *f* instrument lead[s]; measuring circuit; waveguide standard *(waveguides)*; C-wire *(telephony)*
  geschlitzte M. slotted measuring section *(waveguides)*
  koaxiale M. slotted line
Meßlupe *f* measuring magnifier
Meßmarke *f (el)* marker pulse, pip *(oscilloscopy)*; datum point, gauge mark
Meßmaschine *f* measuring machine
Meßmethode *f* method of measurement, measuring technique
Meßmikrofon *n* standard microphone
Meßmikroskop *n* measuring microscope
Meßmoment *n* deflecting torque
Meßort *m* location of the sensor in a control system
Meßpipette *f* graduated pipette
Meßplan *m (Am)* schedule of periodic tests
Meßplatte *f* surface plate

Meßplatz *m* measuring device, test rack
Meßprogramm *n* testing schedule
Meßpumpe *f* metering (control, controlled volume) pump
Meßpunkt *m* measuring (test) point
Meßreihe *f* series of measurements
Meßrohr *n* / **graduiertes** graduated measuring tube
Meßscheibe *f* zero disk *(jig borer table)*
Meßschieber *m* 1. vernier caliper; 2. metering valve
Meßschleife *f (el)* measuring (test) loop
Meßschnabel *m* measuring jaw
Meßschnur *f* instrument leads, measuring cord
Meßschraube *f* micrometer; micrometer screw
  M. für Außenmessungen micrometer for external measurements, external (outside) micrometer
  M. für Innenmessungen micrometer for internal measurements, internal (inside) micrometer
  M. für Tiefenmessungen depth micrometer
Meßsender *m* signal generator; standard-signal oscillator; test oscillator
Meßsonde *f* measuring (test) probe
Meßspannung *f* measuring (test) voltage
Meßspindel *f* adjustable (measuring) spindle; measuring screw
Meßspule *f* measuring coil
Meßstation *f* testing office
Meßstelle *f* test (measuring) point
Meßstellenumschalter *m* selector switch; scanner
Meßstellung *f* gauging (measuring) position
Meßsteuereinrichtung *f (mach)* sizing (size-control) device
Meßsteuerung *f (mach)* size control, sizing
Meßstift *m* test pin
Meßstrahl *m* measuring beam
Meßstrecke *f* gauge length
Meßstreifen *m* measuring tape; strain gauge; strip chart *(recorder)*
Meßsystem *n* 1. system of measurement; 2. test set
  verschleißfreies M. measuring system free from wear
Meßtasche *f (mine)* measuring pocket
Meßtaster *m* probe with magnet
Meßtechnik *f* measurement technology, measuring engineering; measuring techniques
Meßtechniker *m* metrologist, measuring technologist
Meßtrichter *m* scale hopper
Meßtrommel *f* **mit Meßstrichen** graduated drum
  M. mit Nonius graduated drum with vernier
Meßtrupp *m* testing crew
Meßuhr *f* dial gauge (indicator)
Meßuhrdickenmesser *m* dial thickness gauge
Meßuhrtiefenmaß *n* dial depth gauge
Meßumformer *m* transmitter
Meßumschalter *m* throw-over switch
Meß- und Aufwickelmaschine *f* measuring and rolling machine
Meß- und Legemaschine *f* measuring and folding machine
Meß- und Mischsystem *n* dosing and mixing system
Messung *f* measurement; gauging; determination
  M. durch Strahlverschiebung slide-back measurement
  kalorimetrische M. calorimetric measurement
  markscheiderische M. survey measurement
  statische M. steady-state measurement
Meßventil *n* metering (volume-control) valve
Meßverfahren *n* measuring (measurement) technique, measuring (measurement) method
  ionometrisches M. ionometric measurement method
Meßverstärker *m* measuring amplifier, phantom repeater
Meßvorrichtung *f* measuring apparatus
Meßwagen *m* dynamometer car
Meßwandler *m* instrument (pilot) transformer, transmitter
  M. in Sparschaltung instrument autotransformer
  kompensierter M. compensed instrument transformer
Meßwarte *f* control station (centre, room)
Meßwerk *n* indicator operating mechanism; movement
Meßwerkregler *m* electromechanical controller on the base of an indicating system
Meßwert *m* measured value; indicated value
Meßwertaufzeichnung *f* notification, registering
Meßwertgeber *m* detecting (primary measuring) element, *(Am)* transducer
Meßwertschreibeinrichtung *f (ship)* data logging equipment
Meßwertschreiber *m* data recorder
Meßwertsender *m* [air-to-ground] data transmitter
Meßwertspeicher *m* data logger
Meßwertübertrager *m* data transmitter
Meßwertübertragungseinrichtung *f* data transmission equipment
Meßwertübertragungssystem *n* data transmission system
Meßwertwandler *m* transducer, transmitter
Meßwicklung *f* search (measuring) winding; standard winding (coil)
Meßwiderstand *m* measuring rheostat
Meßzähnezahl *f* number of teeth to gauge (measure over), number of teeth included in chordal measurement

Meßzelle *f* detector
Meßzerhacker *m* chopper
Meßzylinder *m* graduated cylinder, measuring glass
Metabolismus *m* metabolism
Metaborsäure *f* metaboric acid
Metadyne *f* metadyne
Metadynmotor *m* metadyne motor
Metadynumformer *m* metadyne converter
Metakieselsäure *f* metasilicic acid
Metaldehyd *m* metaldehyde
Metall *n* metal
  M. erster Schmelzung primary metal
  M. zweiter Schmelzung secondary metal
  edles M. noble metal
  granuliertes M. metal in grains
  reines M. native metal
  Rosesches M. Rose['s] alloy
  überhitztes M. burnt metal
  unedles M. base metal
Metallabschirmung *f* metal screen
  geflochtene M. braided metal screen
Metallader *f* metallic lode (vein)
Metallaffinität *f* metal affinity
Metallampensockel *m* all-metal lamp base
Metallanfärbung *f* / **galvanische** metallochromy
Metall-Asbest-Gewebedichtung *f* metal-asbestos fabric-gasket
Metallauflage *f* metal coating
Metallauftrag *m* plating charge
Metallbad *n* molten metal bath
Metallbadtauchen *n* dipping in molten metal bath
Metallbandsteuerung *f* metal tape control
Metallbarometer *n* aneroid barometer
Metallbearbeitung *f* metal working (cutting and forming)
  elektro-erosive M. electro-erosion cutting
Metallbehälter *m* metal container
Metallbohren *n* metal drilling
Metalldach *n* metallic roof
Metalldampf *m* metal vapour
Metalldampflampe *f* metal-vapour lamp
Metalldampfplasma *n* metal-vapour plasm[a]
Metalldichtung *f* metal[lic] packing, solid metal gasket
Metalldrahtlampe *f* metal-filament lamp
Metalldrehspan *m (mach)* metal turning
Metalldrückbank *f* metal-spinning lathe
Metalle *npl* **der seltenen Erden** metals of rare earth
Metalleinlage *f* metal insert
Metalleinsatz *m* metal charge
Metallelektrode *f* metal electrode
Metallentfettung *f* metal degreasing
Metallfaden *m* metallic filament
Metallfadenlampe *f* filament lamp
Metallfaser *f* metallic fibre
Metallfiberarmierung *f* metal-fibre reinforcement
Metallfolie *f* metal foil (leaf)
Metallfolienabschirmung *f* / **gelochte** perforated metal screen
Metallfolienkondensator *m* tin-foil capacitor
Metallfügemethode *f* metal-jointing method
metallführend metalliferous
Metallgarn *n* metallic yarn
Metallgaze *f* metal gauze
metallgebunden metal-bonded
Metallgeflecht *n* metal gauze (braid)
Metallgefüge *n* metal structure
Metallgehalt *m* **[des Erzes]** grade of [the] ore
Metallgewebe *n* wire cloth (fabric, mesh, netting)
Metallgewinnung *f* / **elektrolytische** electrowinning [of metal]
Metallgicht *f* metal charge
Metallgießerei *f* non-ferrous foundry
Metallglanz *m* metallic lustre
Metallgleichrichter *m* metallic (tank) rectifier
Metallgranalien *fpl* metal in grains
Metallgravierung *f* metal engraving
Metallguß *m* non-ferrous casting; cast metal, non-ferrous castings
Metallgußstück *n* metal (non-ferrous) casting
metallhaltig metalliferous
Metallhaut *f* sheathing *(ship's bottom)*
Metallhinterlegung *f* metal backing
Metall-Holz-Laminierung *f* metal-to-wood laminating
Metalliderung *f* metallic packing
Metallinertgasschweißen *n* inert-gas metal-arc welding
metallisch metallic
metallisieren to metallize
  elektrolytisch m. to electroplate
Metallisierung *f* metallization
Metallkassette *f* metal dark-slide *(camera)*
Metallkeramik *f (nucl)* metal-ceramic, cer[a]met
Metall-Keramik-Magnet *m* metal-ceramic-magnet
Metallkies *m* metal grit
Metallklebemittel *n* plastic bonding agent for bonding metals
Metallkleben *n* adhesive bonding of metal [to metal], metal glu[e]ing
Metallklebeverbindung *f* plastic-bonded joint of metals
Metallkohlebürste *f* compound brush
Metallkomplexfarbstoff *m* metallized dyestuff
Metallkönig *m* metallic regulus
Metallkonstruktion *f* metal construction
Metallkrätze *f* waste metal
Metallkristall *m* metallic crystal
Metallkühlung *f* metal cooling
Metallkunde *f* science of metals
Metallmembran *f* metal diaphragm

Metallmikroskop *n* metallurgical microscope, metalloscope
Metallmikroskopie *f* microscopic metallography
Metallmischung *f* metallic mixture
Metallmodell *n* metal pattern
Metallnebel *m* metal fog
Metallnitschelwerk *n (text)* metal rubbers
Metallografie *f* metallography
metallografisch metallographic
Metalloid *n* metalloid
metallorganisch organo-metallic
Metallöten *n* metal soldering
Metallpackung *f* metallic packing
Metallpapier *n* metallized paper
Metallpapierkabel *n* metal-covered paper-insulated cable
Metallpapierpackung *f* / **galvanische** electrodeposited metallic paper packing
Metallpartikel *f* metal particle
Metallphysik *f* metal physics
Metallplatte *f* metal plate
Metallplattenventil *n* metal disk valve
Metallplattierschicht *f* metal cladding
metallplattiert metal-clad
Metallpulver *n* metal powder
Metallradiografie *f* radiometallography
Metallringscheibe *f* metal washer
Metallrohr *n* metal tubing
Metallröhre *f* metal (canned) valve
Metallrohrpresse *f* metal tube-making press
Metallsäge *f* metal saw
Metallsalznachbehandlung *f* aftertreatment with a metallic salt
Metallscheidung *f* / **elektrolytische** electrolytic parting
Metallschild *m* metallic shield
Metallschlauch *m* [flexible] metal tube, [flexible] metal hose
  gefalzter, biegsamer M. interlocked type of flexible metal hose
Metallschliff *m* metallographic specimen, ground [metal] sample
Metallschlitzsäge *f* metal slotting saw
  M. mit Kreuzverzahnung staggered-tooth metal slotting saw
Metallschmelze *f* molten (fused, liquid) metal, melt of a metal
Metallschmelzenreaktor *m* liquid-metal reactor
Metallschuh *m* metal shoe
Metallschutzhülse *f* metal protector
Metallschwamm *m* metallic sponge
Metallsockel *m* metal base *(lamp)*
Metallspan *m* metal chip
Metallspanen *n* metal cutting
Metallspiegel *m* metal reflector; *(telev)* metallized screen *(valve)*
Metallspritzen *n* metal spraying
Metallspritzpistole *f* metal spraying pistol
Metallspritzverfahren *n* metal spraying process
Metallstrangpresse *f* metal-bar extruding press
Metallsuchgerät *n* metal detector (locator)
Metallthermometer *n* pyrometer, metallic thermometer
Metalltrennverfahren *n* metal severing process
Metallüberzug *m* metal coat[ing]
Metallurg[e] *m* metallurgist
Metallurgie *f* metallurgy
metallurgisch metallurgical
Metallventil *n* metal valve
metallverarbeitend metal-fabricating, metalworking
Metallverarbeitung *f* metal working
Metallvolumen *n* / **zerspantes** volume (rate) of metal removal, volume (rate) of metal removed
Metallwaben *fpl* honeycomb core
Metallwändekammer *f* metal-walled [ionization] chamber
Metallwaren *fpl* hardware
Metall-Wasser-Abschirmung *f* metal-water (hydrogen) shield
Metall-Wasser-Schichten *fpl* metal-water mixture
Metallwellring *m* **mit Weichauflage** corrugated sheet-metal gasket with filling
Metallzerspanung *f* metal cutting
metamer metameric
Metaphosphorsäure *f* metaphosphoric acid
Metasomatose *f* replacement
metastabil metastable
metazentrisch metacentric
Metazentrum *n* metacentre
  M. über Verdrängungsschwerpunkt metacentric radius
  falsches M. prometacentre
Meteorecho *n* meteor echo *(radar)*
Meteoreisen *n* meteoric iron
Meteoreisenstein *m* meteoric iron
Meteorit *m* meteorite
Meteorograf *m* meteorograph
Meteorschwarm *m* knot within a meteor stream
Meteorstrom *m* meteor shower (stream)
Meter *n* metre
Meterampere *n* metre-ampere
Metergewicht *n* weight per metre run between (over) buffers
Meterkilogramm *n* kilogramme-metre
Meter-Kilogramm-Sekunde *f* metre-kilogram[me]-second
Meterprototyp *m* prototype of metre
Meterware *f (text)* cut goods
Meterwelle *f* metric wave
Methakrylat *n* methacrylate
Methakrylsäure *f* methacrylic acid
Methan *n* methane
Methanabsaugung *f (mine)* firedamp (methane) drainage

Methanausgasung *f (mine)* methane emission
Methanol *n* methanol, methyl alcohol
Methanspürgerät *n (mine)* methane detector
Methanzählrohr *n* methane counter
Methingruppe *f* methine (methilidene) group
Methode *f* method
M. der kleinsten Quadrate method of least squares
M. wiederholter Momentenverteilung moment distribution method, Hardy Cross method
densimetrische M. densimetric method
experimentelle M. experimental technique
Methoxygruppe *f* methoxy group
Methylamin *n* methylamine
Methylat *n* methylate
Methylchlorid *n* methyl chloride
Methylenblau *n* methylene blue
Methylengruppe *f* methylene
Methylgruppe *f* methyl group
methylieren to methylate
Methylierung *f* methylation
erschöpfende M. exhaustive methylation
Methylorange *n* methyl orange
Methylviolett *n* methyl violet
Metol *n* metol
metrisch metric
Metrologie *f* metrology
Mettage *f* make-up section
Mettagetisch *m* imposing table
Metteur *m* make-up hand, maker-up, clicker
Metteurtisch *m* imposing (make-up) table
Mezzanin *n* entresol, mezzanine
Michell-Lager *n* Michell bearing
Michelson-Interferometer *n* Michelson interferometer
Miedergewebe *n* corset (brassière) cloth
mietbar rentable; hirable; charterable
Miete *f* 1. *(ag)* clamp; pit; rick, stack *(hay)*; 2. rent, hire; *(ry)* hire charges *(wagon)*
mieten to charter; to rent *(flat)*; to hire
Mietentschädigung *f* lodging allowance
mietfrei rentfree
Mietshaus *n* apartment house (block), block of flats
Mietwagen *m* hired wagon
Mietwohnung *f* rental unit
Mignonsockel *m (el)* miniature Edison screw cap, midget (mignon) cap, midget (mignon) socket
Migration *f (geol, ph)* migration
Migrationslänge *f s.* Wanderlänge
migrieren to migrate
Mikanit *n* micanite
Mikroampere *n* microampere
Mikroamperemeter *n* microammeter
Mikroanalysator *m* microanalyzer
Mikroanalyse *f* microanalysis
Mikroanalysenwaage *f* microbalance
Mikroätzung *f* micro-etching
Mikroautoradiografie *f* microauto[radio]graphy, micro[auto]radiograph, microautoradiogram
Mikrobar *n* microbar
Mikrobohren *n* micro-drilling, drilling of microscopic holes
Mikrobohrung *f* microscopic hole
Mikrobürette *f* microburette
Mikrochemie *f* microchemistry
mikrochemisch microchemical
Mikrochirurgie *f* micro-surgery
Mikrocurie *n* microcurie, μCi, μc
Mikrodetektor *m* microdetector
Mikroelektrode *f* micro-electrode
Mikroelement *n* microelement
logisches M. micrologic element
Mikrofilm *m* microfilm
Mikrofließen *n* micro flow
Mikrofon *n* microphone, transmitting set
M. mit kleiner Ausgangsleistung low-power microphone
M. mit Störgeräuschunterdrückung noise-cancelling microphone
auf Schalldruck ansprechendes M. pressure-actuated microphone
auf Schallschnelle ansprechendes M. velocity microphone
dynamisches M. moving-contactor microphone
elektrodynamisches M. electrodynamic (moving-conductor) microphone
elektromagnetisches M. electromagnetic microphone
elektrostatisches M. electrostatic (capacitor) microphone
gerichtetes M. directional microphone
piezoelektrisches M. crystal microphone
ungerichtetes M. omnidirectional microphone
Mikrofonaufnahme *f* microphone recording
Mikrofonbatterie *f* microphone battery
Mikrofoneffekt *m* microphone (microphonic) effect, microphonics, microphonism
Mikrofoneinblendung *f* fade-in of microphones
Mikrofonfassung *f* microphone mount
Mikrofongalgen *m* microphone gallows
Mikrofongehäuse *n* microphone housing
Mikrofongeräusch *n* side tone, microphone noise
Mikrofonkabel *n* microphone cable
Mikrofonkapsel *f* microphone capsule
Mikrofonleitung *f* microphone cable
Mikrofonmembran *f* diaphragm of microphone
Mikrofonrauschen *n* microphone hiss (noise)
Mikrofonsender *m (Am)* speech input equipment
Mikrofonspeisung *f* transmitter (speaking current) supply
Mikrofonstativ *n* microphone stand
Mikrofontransformator *m* microphone transformer
Mikrofonvorverstärkerröhre *f* microphone preamplifier valve

Mikrofotografie *f* microphotograph[y]
Mikrohärte *f* micro-hardness
Mikrohärteprüfer *m* microhardness tester
Mikroklin *m* microcline
Mikrokomponente *f* microcomponent
mikrokristallin microcrystalline
Mikrolunker *m* micropipe
Mikromanipulator *m* micromanipulator
Mikromanometer *n* micro pressure-gauge, micromanometer
Mikrometer *n* 1. *s.* Meßschraube; 2. micron, *μ*
Mikrometereinstellung *f* micrometer setting
  M. / mit micrometer-adjustable
Mikrometerschraube *f* micrometer screw
Mikromikrocurie *n s.* Picocurie
Mikrominiaturisierung *f* microminiaturization
Mikromodul *m (dat)* micro-module
Mikron *n* micron
Mikronetz *n (el)* micro network *(simulated network)*
Mikroohm *n* microhm
Mikroorganismus *m* micro-organism, microorganism
Mikroorientierung *f* micro-orientation
Mikrophysik *f* microphysics
Mikroprogrammierung *f* microprogramming
Mikroradiografie *f* microradiography; microradiograph, microradiogram
Mikrorille *f* micro-groove, fine-groove
Mikroriß *m* micro-flaw
Mikroschalter *m* miniature (micro) switch
Mikroschalttechnik *f* solid logic technology, S.L.T.
Mikroschliff *m* microsection
Mikroskop *n* microscope
  umgekehrtes M. inverted microscope
  zusammengesetztes M. compound microscope
Mikroskopie *f* microscopy
Mikroskopierleuchte *f* microscope illuminant
mikroskopisch microscopic
Mikroskopokular *n* microscopic eyepiece
Mikroskoptubus *m* microscope tube
Mikrostruktur *f* microstructure
Mikrosystem *n* microsystem *(quantum mechanics)*
Mikroteilchen *n* microparticle
Mikrotelefon *n* hand set
Mikrotom *n* microtome
Mikrotron *n* microtron
Mikroturbulenz *f* microturbulence
Mikrowelle *f* microwave
Mikrowellenantenne *f* microwave aerial
Mikrowellenbereich *m* microwave region
Mikrowellenelektronik *f* microwave electronics
Mikrowellenentladung *f* microwave discharge
Mikrowellenfrequenz *f* microwave frequency
Mikrowelleninterferometrie *f* microwave interferometry
Mikrowellenmessung *f* microwave measurement
Mikrowellenoszillator *m* microwave oscillator
Mikrowellenröhre *f* microwave valve
Mikrowellenspektroskopie *f* microwave spectroscopy
Mikrowellentechnik *f* microwave technique; microwave engineering
Mikrozustand *m* microstate
Mikrurgie *f* micrurgy
Milanesekettenwirkmaschine *f* Milanese warp knitting machine
Milchbehandlungsraum *m* milk room
Milchbuch *n* dairy herd-recorder
Milchdurchflußmeßgerät *n* milk flowmeter
Milchentrahmer *m* cream separator
Milchfilter *n* milk filter
Milchglas *n* opal (obscured) glass
milchig milky; opaque
Milchkammer *f* milk room
Milchkanne *f* churn; milk pail
Milchkühleinrichtung *f* milk cooler
Milchkühler *m* milk cooler
Milchleitung *f* pipe-line *(conveying milk)*
Milchprüfgerät *n* milk testing equipment
Milchpulver *n* milk powder, evaporated milk
Milchpumpe *f* milk pump
Milchsäure *f* lactic (hydroxypropionic) acid
Milchsäurebakterien *fpl* lactic bacteria
Milchsäuregärung *f* lactic fermentation
Milchschlauch *m* milk tube
Milchseparator *m* cream separator
Milchstapelbehälter *m* bulk tank
Milchtank *m* milk bulk-type storage and collection tank
Milchtauchkühler *m* immersion milk cooler
Milchtransportbehälter *m* milk bulk-type storage and collection tank
Milchtransportkanne *f* milk churn
Milchvieh *n* dairy cattle
Milchwaage *f* milk recording scales
Milchwirtschaft *f* dairy farm[ing]
Milchzentrifuge *f* cream separator
Milchzucker *m* lactose, milk sugar
militärisch military
Militärperspektive *f* cavalier perspective
Militärtechnik *f* military engineering
Militärwissenschaft *f* military science
Millerintegrator *m* Miller integrator
Millerit *m* millerite
Millerkreis *m* Miller circuit
Miller-Zeitbasis *f* Miller time-base
Milliamperemeter *n* milliammeter
Milliarde *f* milliard, *(Am)* billion
Millibar *n* millibar
Millibarn *n* millibarn, mb
Millicurie *n* millicurie, mCi, mc
Milliliter *n* millilitre
Millimeter *n* millimetre, *(Am)* millimeter

Millimeterpapier *n* graph paper with millimetre squares, scale (square) paper
Millimeterteilung *f* millimetre pitch
Millimeterwelle *f* millimetric wavelength
Millimikron *n* millimicron
Milliröntgen *n* milliroentgen, mR
Millisekundenschießen *n (mine)* millisecond-delay shotfiring
Millisekundenzünder *m (mine)* split-second detonator
Mimese *f* mimesis
mimetisch mimetic
Minderbetrag *m* deficit, shortage
Mindergewicht *n* shortage *(weight)*
Mindermaß *n* short measure
Minderung *f* **der Übertragungsgüte** transmission impairment
M. der Übertragungsgüte durch Frequenzbandbegrenzung distortion transmission impairment
M. der Übertragungsgüte durch Leitungsgeräusche noise transmission impairment
minderwertig low-grade, low-quality; poor
Mindestabstand *m* minimum distance; *(el)* minimum [conductor] spacing
Mindestansprechwert *m* minimum operating value *(relay)*
Mindestbesatzung *f* nucleus crew *(on vessels)*
Mindestbödenzahl *f* / **theoretische** *(nucl)* minimum number of theoretical plates
Mindestdrehzahl *f (mach)* minimum speed
Mindestdurchbruch *m* forward break-over *(semiconductors)*
Mindestentfernung *f* minimum distance
Mindestfahrpreis *m* minimum (lowest) charge
Mindestfestigkeit *f* minimum strength
Mindestfracht *f* minimum (lowest) charge
Mindestgeschwindigkeit *f* minimum speed (velocity)
Mindestgewicht *n* minimum weight
Mindesthärte *f* minimum hardness
Mindestlatenz *f* minimum latency
Mindestleistung *f* minimum efficiency (performance)
Mindestrücklaufverhältnis *n (nucl)* minimum reflux ratio
Mindestsuchzeit *f* minimum latency (access-time)
Mindestüberschlagspannung *f* minimum flash-over voltage
Mindestzugriffszeit *f s.* Mindestsuchzeit
Mindestzustandmaschine *f* minimal-state machine
Mine *f* 1. *(mine)* metalliferous mine; 2. *(mil)* mine
Minenfeld *n (mil)* mine field
Minenkammer *f* / **kleine** *(mine)* gopher hole
Minenleger *m* mine layer
Minenräumboot *n* mine sweeper
Minenräumer *m* mine sweeper
Minenräumgerät *n* mine-sweeper gear
Minensperre *f* mine barrage
Minensuchboot *n* mine sweeper
Minensuchgerät *n* mine detector
Mineral *n* mineral
Mineralader *f* mineral vein
Mineralaggregat *n* aggregation of minerals
Mineralbildung *f* mineral formation
Mineraldünger *m* fertilizer
Mineraldüngerverregnung *f* spraying of fertilizers
Mineralfaserstoff *m* mineral fibre
Mineralgerbung *f* mineral tannage
Mineralgewinnung *f* mineral winning
Mineralien *npl* / **beigemengte** accessory minerals
Mineraloge *m* mineralogist
Mineralogenhammer *m* mineralogist's hammer
Mineralogie *f* mineralogy
Mineralöl *n* mineral oil
Mineralölmischung *f* mineral oil blend
Mineralquelle *f* mineral spring
Mineralsäure *f* mineral acid
Mineralsäurehärte *f* non-carbonate hardness *(water)*
Mineralschmieröl *n* mineral lubricating oil
Mineralstoffe *mpl* mineral matter[s]
Mineralwasser *n* mineral water
Mineralwolle *f* mineral wool
Minette *f (geol)* minette
Miniatur *f* miniature
Miniaturbauelement *n* miniature component *(electronics)*
Miniaturbild *n* miniature display
Miniaturisierung *f* miniaturization
Miniaturröhre *f* miniature valve
Miniaturschaltung *f* miniature circuiting
Minimalablenkung *f* minimum deviation
Minimalfläche *f* minimal surface
Minimalfrequenz *f* minimum frequency
Minimalgeschwindigkeit *f* minimum speed
minimalisieren to minimize
Minimalrelais *n* underload (undercurrent) relay
Minimalschalter *m* minimum (undercurrent, underload) circuit-breaker
Minimalspannungsrelais *n* no-voltage relay
Minimalsuchzeit *f (dat)* minimum latency
Minimalsuchzeitprogrammierung *f* minimum-access programming
Minimax *n* minimax, saddle point *(theory of games)*
minimieren to minimize
Minimum *n* minimum, valley *(of curves)*
Minimumanzeige *f* minimum reading
Minimumionisation *f* minimum ionization

Minimumthermometer *n* minimum thermometer
Minkowski-Welt *f (math)* Minkowski world
Minorante *f (math)* minorant
Minoritätsemitter *m* minority emitter
Minoritätsträger *m* minority carrier
Minuend *m (math)* minuend
minus 1. minus; 2. *(el)* minus, negative
Minusader *f (el)* minus wire
Minusbohrer *m* undersize (reamer) drill
Minusbürste *f (el)* negative brush
Minuskel *f* minuscule, lower-case letter
Minusleiter *m (el)* negative conductor
  M. des Netzes negative main
Minuspol *m* negative (minus) pole
Minuspolklemme *f* minus (negative) terminal
Minute *f* minute *(time, arc)*
Minutenskale *f* minute scale
Minutenzeiger *m* minute hand
Mischbarkeit *f* miscibility, mixability
Mischbatterie *f* mixer tap, mixing valve
Mischbestand *m* mixed crop *(forest)*
Mischbettaustausch *m* mixed-bed ion exchange
Mischbühne *f* mixing stage (platform)
Mischbütte *f* mixing box
Mischdämpfung *f (el)* conversion loss
Mischdauer *f* mixing time
Mischdiode *f* mixer diode
Mischdünger *m* compound fertilizer
Mischdüse *f* mixing nozzle
Mischelement *n* multi-isotopic element
mischen to mix; to blend; to mingle; *(dat)* to merge, to collate *(punched card machinery)*
  von Hand m. to spade *(concrete)*
Mischen *n* **mit Aussteuern** *(dat)* match-merge
Mischer *m* mixer; blender; *(ac)* fader; *(dat)* collator *(punched-card machinery)*
  kontinuierlich arbeitender M. continuous type mixer *(concrete)*
Mischerfahrer *m* mixer driver (operator)
Mischerfahrzeug *n* truck [concrete] mixer
Mischerfertiger *m* mixer-paver, paving mixer *(road)*
Mischerinhalt *m* mixed batch capacity
Mischfeuer *n* fixed and flashing-light, mixed light
Mischfrucht *f* dredge corn, mixed crop
Mischgarn *n* blended yarn
Mischgatter *n* OR-circuit *(logical switching)*
Mischgerbung *f* combination tannage
Mischgewebefärberei *f* union dyeing
Mischgitter *n (el)* injection grid
Mischgrubber *m* spring-tine cultivator
Mischgut *n* mix; dry batch *(concrete)*
Mischheptode *f* pentagrid converter
Mischhexode *f* mixing hexode
Mischhöhen *fpl (telev)* mixed highs
Mischkammer *f (aut)* mixing chamber
Mischkatalysator *m* mixed catalyst
Mischkoks *m* run-of-oven coke
Mischkollergang *m* mixing runner
Mischkondensation *f* 1. condensation by mixing; 2. *(ch)* co-condensation
Mischkondensator *m* mixing condenser
Mischkonfiguration *f* configuration mixture
Mischkopf *m (el)* mixer head
Mischkreis *m (el)* mixer circuit
Mischkristall *m* solid solution, mixed crystal
Mischladung *f* mixed cargo
Mischleistungsrelais *n* arbitrary phase-angle power relay
Mischlichtlampe *f* mixed light lamp *(discharge lamp)*
Mischluftrührer *m* air-lift agitator
Mischmaschine *f* 1. *(build, ag)* mixer, mixing machine; 2. *(text)* mixing picker; blending machine
Mischofen *m* holding furnace
Mischoszillator *m* mixer oscillator
Mischpfanne *f* reservoir ladle *(cupola)*
Mischpolymerisat *n* copolymer, interpolymer
Mischpolymerisation *f* copolymerization, heteropolymerization
Mischpotential *n* mixed polyelectrode potential
Mischpult *n* **/ zentrales** master control
Mischpumpe *f* mixing pump
Mischquarz *m* mixer crystal
Mischraum *m* 1. gas mixing chamber; 2. *(text)* blending room
Mischreißer *m (ag)* combined food mixer and shredder
Mischrohr *n (aut)* choke tube, carburetter choke
Mischröhre *f* mixer (frequency-changer) valve
Mischsäule *f* mixed column
Mischsäure *f* mixed acid
Mischschaltung *f* mixing circuit
Mischschmelzpunkt *m* mixed melting point
Mischschmierung *f* mixed-film lubrication
Mischsignal *n* mixed signal
Mischsplitt *m* coated chippings *(highway construction)*
Mischstation *f* mixing plant *(concrete)*
Mischsteilheit *f (el)* conversion transconductance
Mischstelle *f (dat)* mixing (summing) point
Mischstrahlung *f* complex (mixed) radiation
Mischstrecke *f (text)* blending intersecting frame
Mischstufe *f (el)* mixer [stage]
Mischsystem *n* combined system *(sewerage)*
Mischtrichter *m (aut)* mixing funnel
Mischtrommel *f* tumbling mixer (barrel); pug mill
Mischturm *m* mixing tower *(concrete)*
Mischung *f* mixture; mix *(concrete, fibres)*; blend *(oil, tobacco)*; mixing
  M. des Dampfes mixing of steam *(turbine)*
  additive M. additive mixing

erdfeuchte M. stiff (low-slump) mix *(concrete)*
multiplikative M. multiplicative mixing *(radio)*
Mischungsdruck *m* pressure of mixture *(steam)*
Mischungskomponente *f (text)* blending component
Mischungslücke *f (met)* miscibility gap
Mischungsrechnung *f (math)* alligation
Mischungsregel *f s.* Mischungsrechnung
Mischungstemperatur *f* temperature of mixture
Mischungsturbine *f* mixture turbine
Mischungsverhältnis *n* mix, proportion *(concrete)*; mixture ratio; *(text)* blending ratio
Mischungsvorwärmer *m* feed heater with exhaust in contact
Mischungsvorwärmung *f* heating with exhaust in contact
Mischungswärme *f* mixing heat
Mischungszuführtuch *n (text)* mixing feed table
Mischventil *n* mixing valve
Mischverlust *m (el)* conversion loss
Mischverstärkung *f (el)* conversion (mixer) gain
Mischwähler *m (el)* hunting switch
Mischwalze *f* mixing mill (rolls), roll mill
Mischwalzwerk *n* mixing mill
Mischwasserkanal *m* combined sewer
Mischwerk *n* agitator
Mischwolf *m* blending feeder (willow)
Mischzeit *f* mixing time *(concrete)*
Mischzone *f* twilight zone *(navigation)*
Mißweisung *f* declination of a compass, magnetic declination (variation), indication error
Mißweisungswinkel *m* angle of declination
Mist *m* manure, dung
Mistbeet *n* hotbed
Mistbreiten *n* spreading of dung (manure)
Mistbreiter *m* manure spreader
Mistgabel *f* manure fork, pitch-fork
Misthaufen *m* dung hill
Miststapler *m* manure lifter and loader
Mistzerreißer *m* manure shredder
Mitanmelder *m* co-applicant *(patent)*
Mitarbeiter *m* collaborator, assistant; contributor *(book)*
wissenschaftlicher M. scientific worker; scientist
Mitbenutzungsrecht *n* right of co-use *(of patents)*
Mitchell-Abbauverfahren *n (mine)* Mitchell slicing system
Miterfinder *m* joint (fellow) inventor
mitfahren mit to go along with, to be taken along by
Mitfällung *f* coprecipitation
mitführen to carry along
Mitführungskoeffizient *m* / **Fresnelscher** Fresnel's convection coefficient
mitgenommen werden / durch Reibung to be carried around by frictional contact
mitgerissen entrained *(oil)*
Mitglied *n* member
Mitgliedverwaltung *f* administration party to a union (association), member-administration
Mithörapparat *m* monitoring set
mithören to listen in, to enter a circuit
Mithörklinke *f* listening (branching) jack
Mithörlautstärke *f* side-tone level
Mithörschaltung *f* monitoring (side-tone) circuit
Mithörspule *f* monitoring coil
Mithörstellung *f* listening position
Mithörtaste *f* listening button, monitoring key
Mithörtelefon *n* side-tone telephone set
Mithörverstärker *m* monitoring amplifier
Mitkomponente *f (el)* positive-direction component
Mitkoordinate *f* **eines Dreiphasensystems** positive sequence co-ordinate of a three-phase system
Mitkopplung *f* regenerative (positive) feedback
äußere M. separate self-excitation
innere M. [auto] self-saturation
mitlaufend 1. *(dat)* on-line; 2. *(mach)* rotating *(centre)*
Mitlaufwähler *m* repeating selector *(telephony)*
Mitlaufwerk *n s.* Mitlaufwähler
Mitlesedrucker *m* control printer
Mitlesegerät *n* local record *(teletype)*
Mitlesestreifen *m* home record
Mitnahme *f* 1. *(mach)* [work] drive, driving of the work; 2. *(el)* pulling into tune
M. des Werkstücks driving [of the work]
M. durch Keil key drive
formschlüssige M. positive drive
kraftschlüssige M. non-positive drive
zwangsläufige M. positive drive
Mitnahmebereich *m (el)* lock-in range
Mitnahmeoszillator *m* locked oscillator
Mitnahmeschaltung *f* intertripping *(relay)*
Mitnahmespannung *f* locking voltage
Mitnahmesynchronisierung *f (telev)* pull-in-type synchronization
mitnehmen to drive [the work] *(e.g. lathe)*; to take (carry) along
Mitnehmer *m (mach)* work driver, dog, carrier, driving tongue; flight [of a chain] *(conveying)*
Mitnehmerbolzen *m* driving pin, pin of the driving plate *(lathe)*
Mitnehmerfinger *m* carrier-rod dog *(knitting machine)*
Mitnehmerkeil *m (mach)* driving key
Mitnehmerklaue *f (mach)* engaging dog
Mitnehmerlappen *m s.* Lappen 2.
Mitnehmernase *f (mach)* driving lug
Mitnehmernut *f (mach)* drive slot
Mitnehmerscheibe *f* driving (carrier, catch) plate
Mitnehmerspitze *f (mach)* drive centre

Mitnehmerstange *f* flight bar *(conveying)*; grief stem, kelly *(rotary drilling)*
Mitnehmerstift *m (mach)* driver (driving) pin
Mitnehmerzapfen *m (mach)* driving lug
Mitreaktanz *f* positive-sequence reactance
Mitreeder *m* joint (part) owner
mitreißen to entrain, to carry along
Mitrofanow-Methode *f* Mitrofanov group machining method, group technology
Mitschreiben *n (dat)* logging
Mitschreiber *m (dat)* logger
mitschwingen to resonate
Mitschwingen *n* resonance
Mitschwingung *f* covibration, sympathetic vibration, resonance
Mitsprechdämpfung *f* side-tone attenuation
Mitsprechen *n* side-to-phantom crosstalk
Mitstreuungszerlegung *f* covariance analysis
Mitstrom *m (ship)* [fractional] wake
Mitstromformel *f* friction formula
Mitsystem *n* positive phase-sequence system
  mehrphasiges M. positive-sequence polyphase system
Mittagsschicht *f* afternoon shift, day turn (shift)
Mitte *f* centre
  M. / auf central[ly], at centre, concentric
  M. / genau auf dead central
  M. der Gezeit half tide
  M. Fahrwassertonne fairway (mid-channel) buoy
  M. Schiff midship, centre (middle) line of the ship
  M. Strom mid-river
Mittel *n* 1. *(math)* mean, average; 2. medium, agent; means
  M. gegen Kesselstein boiler fluid (composition)
  M. von Mittelwerten grand average
  M. zur wasserbeständigen Ausrüstung *(text)* waterproofing agent
  abbauverhinderndes M. antidisintegrant
  aggressives M. corrodent, corroding agent
  arithmetisches M. arithmetic mean, arithmetical average
  bakterizides M. germicide
  gelbildendes M. gelling agent
  geometrisches M. geometric mean
  oberflächenaktives M. surface active agent
  reduzierendes M. reducing agent
  sauerstoffabgebendes M. oxygen developing agent
  schlackenbildendes M. slag-former
  schmutzlösendes M. degreasing agent
  schweißtreibendes M. sudorific
  systemisches M. systemic preparation
  wasserentziehendes M. desiccant
Mittelabgriff *m* centre (mid-point) tap
Mittelachse *f* central axis
Mittelanzapfung *f* centre (mid-point) tap, centre (mid-point) tap connection
Mittelauge *n* heddle eye *(heald)*
Mittelbank *f (text)* intermediate frame
mittelbar indirect
Mittelblech *n* light plate
Mitteldamm *m (mine)* intermediate pack
Mitteldeck *n* middle deck; waist *(on warships)*
Mitteldecker *m* midwing [mono]plane
Mitteldruck *m* medium (intermediate) pressure
Mitteldruckdampfkessel *m* medium-pressure boiler
Mitteldruckdiagramm *n* intermediate-pressure diagram
Mitteldruckkessel *m* medium-pressure boiler
Mitteldruckmarschturbine *f* intermediate-pressure cruising turbine
Mitteldruckpumpe *f* medium-pressure pump
Mitteldruckzylinder *m* intermediate-pressure cylinder
Mittelebene *f (math)* bisector; middle plane *(plate)*
Mittelelektrode *f* secondary (centre) electrode *(spark plug)*
Mittelenergiephysik *f* medium-energy nuclear physics
Mittelfahrwassertonne *f* mid-channel buoy
Mittelflyer *m* intermediate frame
Mittelfraktion *f* heart cut, middle fraction
Mittelfrequenz *f* medium frequency
Mittelfrequenzgenerator *m* high-frequency [rotary] generator
Mittelfrequenzmotor *m* high-frequency motor
Mittelfrequenztachometermaschine *f* alternating-current tachometer
Mittelfrequenzumformer *m* high-frequency converter
Mittelgang *m (ry)* central aisle (corridor)
Mittelgerüst *n* intermediate roll stand
Mittelgrat *m* middle (dividing) edge *(turbine)*
mittelgroß medium-size[d]
Mittelgut *n* middlings
Mittelheckspant *n (ship)* horn timber
Mittelkern *m (el)* centre core
Mittelkettenförderer *m* centre-strand conveyor
Mittelkiel *m (ship)* centre keel
Mittelkielplatte *f* middle-line keel-plate, centre girder
  durchlaufende M. centre through-plate [keel]
Mittelkielschott *n* middle-line keel bulkhead
Mittelkielschwein *n* centre (middle-line) keelson
  durchlaufendes M. through-plate centre keelson
Mittelkraft *f* resultant [force]
  radiale M. resultant radial force *(turbine)*
Mittelkrempel *f (text)* second breaker
Mittellage *f* 1. central position; mid-position; 2. core, centre ply
  M. aus Stäbchen batten core

M. des Exzenters middle-position of the eccentric
Mittellängsschott *n* *(ship)* middle-line bulkhead, centre-line bulkhead
nichtdurchlaufendes M. non-throughed centre-line bulkhead
Mittellängswand *f* spine wall
Mittelleiste *f* inside selvedge *(fabric)*
Mittelleiter *m* middle (neutral, zero) wire; centre bar *(electrolysis)*
geerdeter M. earthed neutral conductor
nichtgeerdeter M. floating neutral, insulated middle wire
Mittellinie *f* centre line, bisector, midline
Mittelmast *m* middle-mast
Mittelmastaufhängung *f* centre pole suspension
Mittelöffnung *f* main span *(bridge)*
Mittelöl *n* medium (middle) oil
Mittelplatte *f* / **interkostale** *(ship)* intercostal keelson-plate
Mittelpufferkupplung *f* central buffer coupling
Mittelpunkt *m* centre; mid-point; *(el)* centre point, neutral
optischer M. optical centre
Mittelpunktanzapfung *f* centre (mid-point) tap
Mittelpunktaufweitung *f* centre expansion, expanded centre plan display *(radar)*
Mittelpunktskegelschnitt *m* central conic
Mittelpunktsspeisung *f* centre feed
Mittelpunktstrahl *m* radial line
Mittelpunkttransformator *m* static balancer
Mittelriegel *m* *(text)* centre traverse beam
Mittelrohrrahmen *m* central-tube frame
Mittelrostfeld *n* middle section of grate
Mittelschicht *f s.* Mittagsschicht
Mittelschiff *n* middle body
paralleles M. square (parallel middle) body
Mittelschlitten *m* middle slide *(milling machine)*
Mittelschneider *m* plug tap *(thread)*
mittelschnell intermediate
Mittelschnittbalken *m* medium-level cutter bar
Mittelschußwächter *m* central weft stop motion
Mittelschwelle *f* *(ry)* intermediate sleeper
Mittelschwert *n* *(ship)* centre-board
Mittelsenkrechte *f* perpendicular erected at the midpoint, mid-perpendicular
Mittelsieb *n* centre screen
Mittelspannung *f* *(el)* medium voltage; mean stress
Mittelspannungsleitung *f* medium-voltage line *(transmission)*
Mittelspant *n* square (midship) frame
Mittelspindel *f* centre screw
Mittelstamm *m* average stem, mean sample-tree
Mittelstellung *f* centre (central) position, mid-position; centre-travel position, mid-travel position *(potentiometer)*
M. / auf at mid-travel *(potentiometer)*
Mittelstempel *m* centre post
Mittelstift *m* centre pin
Mittelstraße *f* medium plate mill *(rolling)*
Mittelstrecke *f* *(text)* intermediate drawing frame
Mittelstreckenrakete *f* medium-range missile, intermediate-range missile
Mittelstreifen *m* central reservation, median strip
Mittelstromschiene *f* centre conductor-rail
Mittelstück *n* centre (connecting, distance) piece
Mittelstückhöhe *f* height of centre piece *(cloth)*
Mittelteil *m* centre portion
Mittelteilzelle *f* *(rock)* hull of centre section
Mittelträger *m* 1. spine beam; 2. *(ship)* centre girder *(of the double bottom)*
Mittelwalze *f* centre roll
Mittelwand *f* midfeather *(beater)*
Mittelware *f* *(text)* second lengths
Mittelwelle *f* medium wave
Mittelwellenbereich *m* medium-frequency band *(radio)*
Mittelwellenempfänger *m* medium wave broadcast receiver
Mittelwellensender *m* medium-frequency transmitter
Mittelwert *m* [arithmetic] mean, average
M. der Grundgesamtheit population (process) average
M. der Spannweiten average range
M. der Stichprobe sample average
M. des spezifischen Leitwertes combined conductivity
angenommener M. guess mean
quadratischer M. [root-]mean-square value, root-mean-square, rms; mean effective value
zeitlicher M. mean value with respect to time
Mittelwertanzeiger *m* average value indicator
Mittelwertkarte *f s.* $\bar{x}$-Karte
Mittelwertmesser *m* [counting-]rate meter, integrator, integrating instrument
Mittelwertsatz *m* mean-value theorem
Mittelwertsteuerung *f* *(telev)* mean-value brightness control
Mittelwertzeiger *m* average-value indicator, mean-impulse indicator
Mittelzapfen *m* central pintle
Mittelzeitverhalten *n* *(nucl)* average behaviour
Mittelzerkleinerung *f* intermediate crushing
mitten to centre
Mitten *n* centring
mitten im Fahrwasser mid-channel
Mittenabstand *m* [centre-to-]centre distance
M. haben / geringen to be closely situated (spaced)
M. haben / gleichen to be equally spaced, to be spaced evenly

M. haben / ungleichen to be spaced unevenly
Mittenbohrung *f* centre (central) hole
Mitteneinregelung *f* centring control
Mittenfrequenz *f* centre (mid-band) frequency
Mittenkreis *m* central circle
Mittenverlagerung haben to be out of centre
mittig at centre, concentric, central[ly]
genau m. dead central
mittschiffs [a]midships
Mittschiffsbiegemoment *n* amidship bending moment
Mittschiffsdecksplanke *f* king plank
Mittschiffsebene *f* centre-line plane
Mittschiffslinie *f* keel (fore-and-aft) line, centre-line, middle-line
Mitverbunderregung *f* cumulative [differential] compound excitation
Mitzieheffekt *m (el)* frequency pulling
Mitziehen *n* **von Oszillatorkreisen** synchronization of oscillators
Mixtur *f* mixture
Mizelle *f (ch)* micelle
MKS-Einheit *f* metre-kilogram[me]-second unit
MKS-System *n* metre-kilogram[me]-second system
Möbel *n* furniture
Möbeldamast *m* furniture damask
Möbelkattun *m* / **bedruckter** cretonne
Möbelkord *m* furniture cord
Möbelpolitur *f* furniture polish
Möbelrips *m* ribbed upholstery
Möbelstoff *m* upholstery fabric
Möbeltischler *m* cabinet maker
Möbelwagen *m* furniture van
Mobilbagger *m* self-propelled excavator
Mobilkran *m* mobile crane
Model *m (text)* block
kupferverstärkter M. coppered block
Modeldruck *m* block printing
Modell *n* model; pattern *(foundry)*; form
M. mit Größenverhältnis 1:1 full form model
geteiltes M. split pattern *(foundry)*
maßstäbliches M. scale model
nacktes M. *(ship)* naked model *(model without appendages)*
Modellack *m* pattern varnish *(foundry)*
Modellanlage *f* pilot plant
Modellaufspannplatte *f* model holding-plate
Modellausschmelzverfahren *n* investment moulding
Modellbauer *m* model maker
Modelleinfaßmaschine *f* pattern-binding machine *(shoe)*
Modellgradiermaschine *f* pattern-grading machine *(shoe)*
Modellheber *m* pattern lifter *(foundry)*
Modellholz *n* pattern lumber *(foundry)*
Modellierung *f (dat)* modeling
Modellmacherei *f* patternmakers' shop
Modellmaßstab *m (ship)* scale ratio of a model
Modellmetall *n* pattern (lead-model) metal
Modellrakete *f* rocket model
Modellsand *m* facing sand *(foundry)*
Modellschiff *n* ship's model
Modellschleppversuch *m (ship)* model experiment
M. mit selbstangetriebenem Modell self-propelled experiment
Modellschleppversuchsanstalt *f* model tank establishment
Modellschleppversuchstank *m* model [testing] tank
Modellstatik *f* [structural] model analysis
Modelltank *m (ship)* trial (model experimental) tank, model basin
Modelltischlerei *f* patternmakers' shop
Modellversuch *m* model experiment
Modellversuchstank *m* **für Seegangsversuche** sea-keeping tank
Modenfunktion *f* mode function
Modenzahl *f* mode number
Moderator *m* moderator
organischer M. organic moderator
Moderator-Brennstoff-Gitter *n* moderator-fuel lattice
Moderatortrimmung *f* moderator control
moderieren to moderate, to slow down
Modeton *m* fashion shade
Modifikation *f* modification
allotrope M. allotropic modification
modifizieren to modify
Modifizierung *f* modification
Modul *m* 1. *(math)* modulus; 2. metric module, module millimetres *(ratio of the pitch diameter in millimetres divided by the number of teeth in the gear)*
M. einer komplexen Zahl modulus of a complex number
Modulation *f* modulation
M. der Anodenspannung plate modulation
M. der Elektronendichte electron-density modulation
M. der Gitterspannung grid modulation
M. durch Funkenstrecke spark-gap modulation
M. durch selbsterregte Impulse self-pulse modulation
M. geringer Tiefe shallow modulation
M. mit hohem Pegel high-power modulation
additive M. upward modulation
gegenseitige M. intermodulation
magnetische M. magnetic modulation
mehrstufige M. multiple modulation
negative M. negative modulation
nichtlineare M. non-linear modulation
positive M. positive modulation

subtraktive M. downward modulation
unvollendete M. undermodulation
verzerrungsfreie M. linear modulation
vollkommene M. perfect modulation
Modulationsanode *f* modulating anode
Modulationsanzeiger *m* modulation indicator
Modulationselektrode *f* modulator electrode
Modulationsfähigkeit *f* modulation capacity
Modulationsfaktor *m* modulation percentage (factor), index of modulation
Modulationsfrequenz *f* modulation frequency
Modulationsgrad *m* modulation depth (factor), degree (amount) of modulation
effektiver M. effective percentage modulation
prozentualer M. percentage modulation
Modulationshüllkurve *f* modulation envelope
Modulationsindex *m* modulation index
Modulationskennlinie *f* modulation characteristic
Modulationsklirrfaktor *m* envelope distortion
Modulationskreis *m* modulation circuit
Modulationskurve *f* modulation characteristic
Modulationsmesser *m* modulation meter
Modulationspause *f* quiet interval
Modulationsprodukt *n* intermodulation product
Modulationsrauschen *n* modulation noise
Modulationssatz *m* auxiliary modulation set
Modulationsspannung *f* modulating (modulator) voltage
Modulationsspiegel *m* modulation mirror
Modulationsspule *f* modulation coil
Modulationsstufe *f* modulator stage
Modulationstiefe *f* modulation depth
Modulationstransformator *m* modulation transformer
Modulationsverfahren *n* method of modulation
Modulationsverflachung *f* reduction of detection efficiency
Modulationsverzerrung *f* modulation distortion
Modulationsvorverstärker *m* speech amplifier
Modulator *m* modulator
elektromagnetischer M. chopper, vibrator
magnetischer M. magnetic modulator
Modulator-Demodulator *m* modulator-demodulator
Modulatorkette *f* modulator chain
Modulatorröhre *f* modulator valve
Modulatorschaltung *f* modulator circuit
Modulbasis *f (math)* modular basis
Modulbaustein *m* modul component
Modulbauweise *f* modular design
Modulierbarkeit *f* modulation capability
modulieren to modulate
Modulkehrwert *m* diametral pitch
Modulplatte *f* wafer
Modulraster *m* modular grid
Modultechnik *f* modular technique
Modulteilung *f* diametral pitch
Modus *m* mode
unterkritischer M. evanescent mode *(waveguide)*
Modusfilter *n* mode filter [in a waveguide circuit]
Modusumformer *m* wave converter *(waveguides)*
Moduswandler *m* mode changer *(waveguides)*
Moellon *n s.* Dégras
Mögel-Dellinger-Effekt *m* Dellinger effect, radio fade out, impulsive solar noise
möglich possible; feasible; practicable
Möglichkeit *f* possibility; feasibility; practicability
Mohair *m s.* Mohär
Mohär *m* mohair
Mohärfaser *f* mohair fibre
Mohärgarn *n* mohair yarn
Mohnöl *n* poppy seed oil
mohnrot ponceau
Möhrenerntemaschine *f* carrot harvester
Möhrensortiermaschine *f* carrot grader
Moiré *m(n)* moiré
Moiréappretur *f* moiré finish
Moirémuster *n* moiré pattern
Moiréstörung *f (telev)* interference pattern
Mokett *m* moquette
Mokettstuhl *m* moquette loom
Mol *n* mol[e], gram-molecule
Molalität *f* molality
molar molar
Molarität *f* molarity
Mole *f* mole, pier, jetty, quay, wharf
Molekel *f s.* Molekül
Molekül *n* molecule
Molekülaggregat *n* molecular cluster
Molekülanordnung *f* molecular configuration
molekular molecular
Molekularbewegung *f* molecular motion (movement)
Brownsche M. Brownian movement (motion)
Molekulardestillation *f* molecular distillation
Molekularelektronik *f* molecular electronics
Molekularformel *f* molecular formula
Molekulargeschwindigkeit *f* molecular velocity
Molekulargewicht *n* molecular weight
Molekularkraft *f* molecular force
Molekularleitfähigkeit *f* molecular conductivity
Molekularluftpumpe *f* molecular air pump
Molekularphysik *f* molecular physics
Molekularpolarisation *f* molar polarization
Molekularpumpe *f* molecular pump
Molekularschaltung *f* molecular circuit
Molekularsieb *n* molecular sieve
Molekularspeicherung *f* molecular storage *(computer)*
Molekularspektrum *n* molecular spectrum
Molekularstrahl *m* molecular ray (beam)
Molekularströme *mpl* / **Amperesche** molecular Ampere currents

Molekularströmung *f* / **thermische** thermal molecular flow
Molekularverstärker *m s.* Maser
optischer M. *s.* Laser
Molekularwärme *f* molecular (molar) heat
Molekularzustand *m* molecular state, molecularity
Molekülbande *f* molecular band
Molekülbindung *f* molecular bond
Moleküldurchmesser *m* diameter of a molecule
Molekülgeschwindigkeit *f* molecular velocity
Molekülgitter *n* molecule (molecular) lattice
Molekülhaufen *m* molecular cluster
Molekülkristall *m* molecular crystal
Molekülmasse *f* molecular mass
Molekülrückstoß *m* aggregate recoil
Molekülspektrum *n* molecular spectrum
Molekülstruktur *f* molecular structure
Molekülteilchen *n* molecular particle
Molekülzustand *m* molecular state
metastabiler M. metastable molecular state
Molenbruch *m* molar (mole) fraction
Molenfeuer *n* pierhead (port, harbour) light
Molenkopf *m* mole head, pierhead
Molenzahl *f* number of [gram] moles
Molenzunge *f (ship)* apron
Molgewicht *n* molar (gram molecular) weight
Molke *f* whey
Molkerei *f* dairy
Molkereiwesen *n* dairying
Möller *m* burden, batch, ore and flux
Möllerausbringen *n* burden yield
Möllerberechnung *f* burden calculation
Möllerbrecher *m* burden squeezer
Möllergewicht *n* burden ratio
Möllergicht *f* burden charge
möllern to burden, to mix ores and fluxes
Möllersonde *f* charge (stock) level indicator
Möllerung *f* mixing of ores and fluxes, [furnace] burdening
Möllervorbereitung *f* mixture preparation
Möllerwagen *m* larry car
Mollierdiagramm *n* Mollier diagram
Molnormalvolumen *n* standard molar volume
Molpolarization *f* molar polarization
Molrefraktion *f* molar refraction
Molsuszeptibilität *f* molar susceptibility
Molverhältnis *n* molar ratio
Molvolumen *n* molar volume
Molwärme *f* molecular (molar) heat
partielle M. partial molar heat
Molybdän *n* molybdenum
Molybdänglanz *m* molybdenite
molybdänhaltig molybdeniferous
Molybdänit *m s.* Molybdänglanz
Molybdänsäure *f* molybdic acid
Molybdänsäureanhydrid *n* molybdic anhydride
Molybdänschnellstahl *m* molybdenum high-speed steel
Molybdän-II-verbindung *f* molybdous compound
Molybdän-III-verbindung *f* molybdenic compound
Molybdän-VI-verbindung *f* molybdic compound
Molybdat *n* molybdate
Moment *n* moment; momentum
M. der Bewegungsgröße angular momentum, momentum of momentum about an axis
M. des Kräftepaares moment of the couple of forces
M. einer Doppelquelle moment of a doublet
M. gegen Kippen *(ship)* anti-tipping moment
aufnehmbares M. moment capacity
eingetragenes M. applied moment
elektrisches M. electrical moment
kernmagnetisches M. nuclear magnetic moment
magnetisches M. magnetic moment
magnetisches M. Null zero-magnetic moment
statisches M. static moment
synchronisierendes M. pull-in torque
momentan prompt *(neutrons, gamma-rays)*; instantaneous, momentary
Momentanachse *f* instantaneous axis
Momentanangriffspunkt *m* momentary point of application
Momentanbahn *f* instantaneous orbit
Momentanbeschleunigung *f* instantaneous acceleration
Momentanbewegung *f* instantaneous motion
Momentandrehung *f* instantaneous rotation
Momentanfrequenz *f* instantaneous frequency
Momentangeschwindigkeit *f* instantaneous velocity (speed)
Momentankraft *f* instantaneous force
Momentankreis *m* instantaneous orbit
Momentanleistung *f* instantaneous power
Momentanstrom *m* instantaneous current
Momentanwert *m* instantaneous value
Momentanzentrum *n* instantaneous centre
Momentaufnahme *f* snapshot
Momentenausgleich *m* moment distribution
Momentenausgleichsverfahren *n* moment distribution method, Hardy-Cross-method
Momentenempfänger *m* synchro receiver
Momentenfläche *f* area of moments, moment area
Momentenflächenverfahren *n* area-moment method
Momentengeber *m* torque motor
Momentengleichung *f* moment equation
Momentengradient *m* torque gradient
Momentenhüllkurve *f* moment limits curve
Momentenkurve *f* moment curve
M. des Achsenreglers moment curve of the shaft governor

Momentenmethode *f* moments method
Momentenmotor *m* torque motor
Momentennullpunkt *m* point of contraflexure
Momentenplan *m* moment diagram
Momentensatz *m* theorem of moments
Momentenverstärker *m* torque amplifier
Momentrelais *n* instantaneous relay
Momentschalter *m* quick[-make-and-]break switch, high-speed switch
Momentschaltung *f* instantaneous connection
Momentzünder *m* instantaneous fuse (electric detonator)
Monatsbilanz *f* monthly balance-sheet
Monatsdurchschnitt *m* monthly average
Monazit *m* *(min)* monazite
Mönch *m* convex tile *(roofing)*
Mönchsbogen *m* blank sheet
Mond *m* moon
Mondbasis *f* *(astr)* lunar base
Möndchen *n* crescent, lune
Mondfinsternis *f* lunar eclipse
Mondgas *n* Mond gas
Mondgebirge *n* lunar mountains
Mondkrater *m* lunar crater
Mondlandung *f* lunar landing
Mondnickel *n* Mond nickel
Mondoberfläche *f* lunar surface
Mondrakete *f* moon rocket
Mondsatellit *m* moon (lunar) satellite
Mondsonde *f* moon (lunar) probe
Mondumlaufbahn *f* lunar orbit *(of a probe in orbit about the moon)*
Mondverfahren *n* *(met)* Mond process
Monelmetall *n* Monel metal
Moniereisen *n s.* Bewehrungsstahl
Monitor *m* monitor[ing instrument]
Monitorempfänger *m* monitor receiver
monoatomar monoatomic
Monoblocbohrer *m* integral (monobloc) steel
Monobromnaphthalin *n* monobrome-naphthalene
Monochloressigsäure *f* monochloracetic acid
Monochromasie *f* monochromatism
Monochromatfilter *n* monochromatic filter
monochromatisch monochromatic
Monochromator *m* monochromator
monoenergetisch monoenergetic, monoergic, homogeneous
Monofilfaden *m* monofil[ament]
Monofilgarn *n* monofilament yarn
Monogrammstickerei *f* monogram embroidery
Monokarbonsäure *f* monocarboxylic acid
Monokellupe *f* monocle magnifier
monoklin monoclinic, monocline *(crystal system)*
monokular monocular
Monolithbau *m* monolithic concrete construction
Monolithbeton *m* monolithic concrete
monolithisch hergestellt cast-in-place, poured-in-place, cast-in-situ, cast-in-position
monomer monomeric
Monomeres *n* monomer
monomolekular monomolecular
Monosaccharose *f* monosaccharose
Monosilan *n* silicomethane
Monotektikum *n* *(met)* monotectic
monotektisch *(met)* monotectic
Monotropie *f* monotropy
Monotypemetall *n* monotype metal
Monotypesetzmaschine *f* monotype composing machine
Monotypetaster *m* monotype keyboard
monovariant univariant
Monozelle *f* single cell
monozyklisch monocyclic
Montage *f* mounting, erection, installation, fitting, construction; *(mach)* assemblage, assembling, assembly
stoßsichere M. shock mounting
Montageabteilung *f* assembly section
Montageanweisung *f* fitting (assembling, assembly) instruction
Montagearbeit *f* assembly operation
Montageband *n* assembly line (conveyor)
Montagebauweise *f* precast [concrete] construction, dry construction
Montagebrigade *f* *(build)* erection crew
Montagedurchhang *m* erection sag *(of cables)*
montagefertig ready-to-assemble
Montagehalle *f* assembling hall, assembly bay (shop)
Montagekosten *pl* cost of erection (assembling)
Montagekran *m* erection crane
Montageleim *m* assembly adhesive (glue), structural adhesive
Montageleimung *f* constructional (secondary) gluing
Montageniet *m* field rievet
Montageplan *m* erection schedule
Montageplatte *f* mounting-plate, base plate, chassis
Montageplatz *m* erection site
Montagespannung *f* assembly stress
Montagestoß *m* site connection
Montagestraße *f* assembly line
Montagetechnik *f* assembly technique
Montagewerkstatt *f* assembly shop
Montagezeichnung *f* assembly drawing for use in the erecting shop
Montagezeit *f* erecting (assembling) time
Montanindustrie *f* mining industry
Montansäure *f* montanic acid
Montanwachs *n* montan wax
Monte-Carlo-Methode *f* Monte-Carlo method
Monteur *m* *(mach)* assembler, assembly worker; *(build)* erector; fitter *(installation)*

montieren to mount; *(mach)* to assemble; *(build)* to erect; to fit, to install
Montierung *f (opt)* mount
  englische M. English mount
  parallaktische M. equatorial mount
Montierungsdruck *m* assembling pressure *(turbine)*
Montierwerkstatt *f* assembling (erecting) shop
Montmorillonit *m (min)* montmorillonite
Moor *n* bog
Moorbad *n* mud bath
Moorboden *m* peaty soil
Moorbruch *m* bog burst
Mooringkette *f (ship)* ground chain
Moorkohle *f* moorcoal, sub-bituminous coal
Moorsprengung *f* bog blasting
Moos *n* / **Isländisches** Iceland moss
Moped *n* moped, petrol-assisted bicycle
Moräne *f (geol)* moraine
Moränenschutt *m* moraine débris
Morast *m* morass
Morasterz *n* bog ore *(iron ore)*
Morphium *n* morphine
morsch rotten, decayed
Morsealphabet *n* Morse code
Morseapparat *m* Morse telegraph
Morsedoppelstromsystem *n* polar direct-current telegraph system
Morseeinstromsystem *n* neutral direct-current telegraph system
Morseempfänger *m* Morse receiver
Morsegeber *m* Morse sender
Morsegerät *n* Morse telegraph
Morsekegel *m (mach)* Morse taper
Morsekegellehre *f* Morse-taper gauge
Morsekegelschaft *m* Morse-taper shank, M.T. shank
Morseleuchtfeuer *n* code beacon
Morselocher *m* Morse tape-perforator
morsen to morse
Morsen *n* morsing
Morsepunkt *m* [Morse] dot
Mörser *m (ch, mil)* mortar
  ballistischer M. ballistic mortar
Mörserkeule *f* pestle
Morseschrift *f* Morse code
Morsesender *m* Morse transmitter
Morsestreifen *m* Morse tape
Morsestreifenlocher *m* code tape-recorder
Morsestrich *m* [Morse] dash
Morsetaste *f* [Morse] sending key
Morsetastenlocher *m* keyboard perforator
Morsetelegraf *m* Morse telegraph
Morsetelegrafie *f* Morse telegraphy
Morsezeichen *n* Morse signal (code), code signal
Mortalität *f (nucl)* mortality
Mörtel *m* mortar
  feiner M. grout
Mörtelaufbereitung *f* mortar fabrication
Mörteleinpressung *f* mortar intrusion
Mörtelfuge *f* mortar joint
Mörtelinjektion *f* mortar intrusion
Mörtelkasten *m* hod
Mörtelkelle *f* trowel
Mörtelmischer *m* mortar mixer (mixing machine)
Mörteltrog *m* hod
Mörtelverstrich *m* torching *(roof tilting)*
Mosaik *n* mosaic
mosaikartig mosaic-like
Mosaikbelag *m s.* Mosaikfußboden
Mosaikblock *m* mosaic block *(crystal)*
Mosaikelektrode *f* mosaic electrode
Mosaikfußboden *m* mosaic flooring
Mosaikpflaster *n* tessellated pavement, mosaic paving
Mosaikpflasterstein *m* tessela, tessera
Mosaikpflasterweg *m* tesselated pavement
Mosaikstein *m* tessela, tessera
Moschus *m* musk
Moschuskörneröl *n* amber-seed oil
Mößbauer-Effekt *m* Mössbauer effect
Mostrich *m* mustard
Motor *m (el)* motor; *(aut)* engine
  M. für pendelnde Drehbewegung rotary jack
  M. für Schwerstbetrieb arduous-duty motor
  M. für Traktorenkraftstoff kerosine engine
  M. großer Leistung high-powered motor
  M. mit Achsaufhängung axle-hung motor
  M. mit Ankersteuerung armature-controlled motor
  M. mit Berührungsschutz protected (semi-enclosed) motor
  M. mit Doppelschlußverhalten motor with compound characteristic
  M. mit Drehzahlregelung adjustable-speed motor
  M. mit Drosselanlasser reactor-start motor
  M. mit Eigenbelüftung naturally ventilated motor
  M. mit Eigen- und Fremdbelüftung motor with combined ventilation
  M. mit elektrischer Zündung spark-ignition engine
  M. mit Feldsteuerung field-controlled motor
  M. mit gleichbleibender Drehzahl constant-speed motor
  M. mit hängenden Ventilen inverted engine
  M. mit Hauptstromcharakteristik inverse-speed motor
  M. mit kapazitiver Hilfsphase capacitor-start motor
  M. mit Kompensationswicklung compensated motor
  M. mit Kondensatoranlauf capacitor-start motor

M. mit Mantelkühlung ventilated totally enclosed motor
M. mit mehreren Drehzahlstufen multiple-speed motor, change-speed motor
M. mit Nebenschlußverhalten motor with shunt characteristic
M. mit regelbarer Drehzahl adjustable-speed motor, change-speed motor
M. mit Reihenschlußverhalten motor with series characteristic
M. mit Schnellanlauf quick-starting motor
M. mit Selbstzündung diesel (compression-ignition) engine
M. mit stehenden Ventilen L-head engine
M. mit stehender Welle torque motor
M. mit Stehlager pedestal-type motor
M. mit stufenloser Regelbarkeit infinitely variable speed motor
M. mit Untersetzungsgetriebe back-geared motor
M. mit unveränderlichem Schluckvolumen constant-capacity motor
M. mit veränderlicher Geschwindigkeit variable-speed motor
M. mit Verdichtungszündung compression-ignition engine
M. mit Widerstandsanlasser resistance-start motor
M. ohne ausgeprägte Pole smooth-core motor
M. ohne Drehzahlbeeinflussung uncontrollable motor
M. zur Festklemmung des Querbalkens cross-slide clamping motor
M. zur Senkrechtverstellung des Querbalkens cross-slide raising and lowering motor
ankergesteuerter M. armature-controlled motor
aufgeladener M. supercharged engine
belüfteter M. ventilated motor
blockierter M. stalled motor
direkt umsteuerbarer M. direct reversing engine
doppelpoliger M. bipolar (double-pole) motor
drehzahlumschaltbarer M. multi-speed motor
durch Gebläse gekühlter M. blower-cooled engine
eintouriger M. constant-speed motor
feldgesteuerter M. field-controlled motor
fremdbelüfteter M. forced-ventilated motor
gekapselter M. cased (enclosed) motor
geschlossener M. totally enclosed motor
hängender M. inverted engine
hochkomprimierter M. high-compression engine
hochtouriger M. high-speed engine
hochverdichteter M. high-compression engine
kopfgesteuerter M. valve-in-head engine, overhead-valve engine
langsamlaufender M. slow-running engine
liegender M. horizontal engine
linksläufiger M. left-handed motor
nicht steuerbarer M. uncontrollable motor
obengesteuerter M. overhead-valve engine, valve-in-head engine
pneumatischer M. air motor
polumschaltbarer M. pole-changing motor, pole-change motor, change-pole motor
rechtsläufiger M. right-handed motor
schiebergesteuerter M. sleeve-valve engine
seitengesteuerter M. L-head engine, side-valve engine
starker M. high-powered motor
stehender M. vertical engine
sternförmiger M. radial engine
steuerbarer M. control motor
stufenlos regelbarer M. infinitely variable-speed motor
umsteuerbarer M. reversible motor
untengesteuerter M. engine with side-by-side valve arrangement
untersetzter M. back-geared motor
vorverdichteter M. supercharged engine
zwangsgekühlter M. blower-cooled engine
zweipoliger M. bipolar (double-pole) motor

Motoranker *m* motor armature
Motoranlasser *m* engine-starter
Motoranlaßschalter *m* motor starting-switch
Motorantrieb *m* motor (engine) drive
Motoraufhängung *f* engine mounting
Motorbarkasse *f* motor launch (tender)
Motorbaumspritze *f* orchard sprayer
motorbetrieben engine-driven, engine-powered, motor-driven
Motorbetriebsstunden *fpl* engine hours
Motorblock *m* engine block
Motorbock *m* engine bearer (bed, mount)
Motorboot *n* motor (power) boat, motorboat;
Motorbremsschaltung *f* motor braking circuit
Motorbrummen *n* engine humming
Motordrehmoment *n* motor torque
Motordrehzahl *f* motor (engine) speed
M. bei Stundenleistung speed at one-hour rating
Motoreinbau *m* engine mounting
Motorengehäuse *n* motor casing
Motorenöl *n* motor (engine) oil
unlegiertes M. straight (non-additive type) motor oil
Motorenraum *m* engine room (space); *(el)* motor room (space)
Motorenschlosser *m* engine fitter
Motorenwart *m* engine mechanic
Motorfahrgastschiff *n* passenger motor-ship
Motorfahrzeug *n* motor vehicle; motor craft
Motorfertigungsstraße *f* engine production line

Motorflachstrickmaschine *f* mechanical knitting machine
Motorflugbuch *n* engine logbook
Motorfundament *n* engine bed
Motorgehäuse *n* motor frame
Motorgenerator *m* motor generator
Motorgeneratoraggregat *n* motor generator set
Motorgüterboot *n* motor cargo-boat
Motorhaube *f* engine bonnet (cowling, hood)
motorisieren to motorize
Motorisolator *m* solid-core insulator
Motorjacht *f* motor yacht, motor cruiser
M. mit Hilfsbesegelung motor yacht with auxiliary sails
Motorjacquardflachstrickmaschine *f* power-driven flat Jacquard knitting machine
Motorkenndaten *pl* engine (motor) data
Motorklemme *f* connection
Motorkonsole *f* engine bracket
Motorkreuzer *m* motor (power) cruiser *(yacht)*
Motorkühlschiff *n* refrigerated cargo-motorship
Motorlafette *f* engine gun mounting
Motorlager *n* engine mount
Motorleistung *f* motor (engine) performance; engine power, horsepower
Motorlocher *m* motor drive punch *(punched card machinery)*
Motorlochprüfer *m* motor drive verifier
Motorlogger *m* motor drifter
Motorprüfschein *m* engine-test certificate
Motorpumpe *f* / **gekapselte** canned (leakproof) motor pump
Motorrad *n* motor cycle
Motorraum *m* engine room (space); *(aut)* engine compartment; *(el)* motor room (space)
Motorregler *m* motor governor
Motorrennboot *n* racing motor boat, motor racer
Motorriemenscheibe *f* **im Tonbandgerät** motor pulley
Motorroller *m* motor scooter
Motorrostschutzmittel *n* fly-away oil
Motorsäge *f* power saw
Motorschaden *m* engine trouble (failure)
Motorschiff *n* motor vessel (ship); motor craft
Motorschild *n* engine plate
Motorschürfkübel *m* tractor-scraper
Motorschürfkübelwagen *m* tractor-scraper
Motorschutzblech *n* engine guard plate
Motorschutzschalter *m* protective motor switch, thermal (open-phase) circuit-breaker
Motorsegler *m* motor sailer, sailing ship with auxiliary motor
Motorsockel *m* engine base
Motorspeicherantrieb *m* drive with auxiliary power stored by winding-up motor *(circuit breakers)*
Motorsport *m* motoring
Motorsteuerung *f* *(el)* motor control; engine timing
M. durch Feldschwächung motor-field control
Motorstörung *f* engine trouble
Motorstraßenhobel *m* motor grader
Motorträger *m* engine bearer
Motortragplatte *f* bedplate of an engine
Motorunterbrecher *m* motor-driven interrupter
Motorverkleidung *f* engine cowling
Motorverstäuber *m* motor duster
Motorwähler *m* motor uniselector *(telephony)*
Motorwanne *f* lower part of the crankcase
Motorwelle *f* motor (engine) shaft
Motorwirkungsgrad *m* motor (engine) efficiency
Motorzähler *m* motor meter
Motorzylinder *m* engine cylinder
mottenabwehrend moth repellent
mottenbeständig moth resistant
mottengeschützt mothproof
Mottenkugel *f* mothball
Mottenpulver *n* insect powder
Mottenschutzmittel *n* moth repellent (proofing agent)
Moulinégarn *n* mouliné yarn
moulinieren to throw *(silk)*
Mp-Leiter *m* *(el)* neutral [conductor]
Mückenschleiertüll *m* mosquito netting
Mudderprahm *m* sludge vessel
Muffe *f* sleeve, bell, bush, socket *(pipe)*, muff, boss; *(el)* sealing (connecting) box, straight joint, coupling
abgeschirmte M. shielded joint
Muffel *f* 1. muffle; 2. *(rock)* flame damper
Muffelbrennofen *m* muffle kiln
Muffelofen *m* muffle furnace
Muffenausgußmasse *f* box compound *(cables)*
Muffendruck *m* sleeve pressure, pressure on [the] sleeve
Muffendruckkurve *f* sleeve-pressure curve
Muffenhub *m* lift of the sleeve, sleeve lift
Muffenkupplung *f* sleeve (muff, box) coupling
Muffenregler *m* governor with sleeve
Muffenrohr *n* bell-and-spigot pipe; socket pipe
Muffenrohrverbindung *f* spigot joint
Muffenverbinder *m* screwed coupler
Muffenverbindung *f* [screwed] sleeve joint, bell-and-spigot joint
Muffenwiderstand *m* joint resistance
Mugaspinner *m* Indian (muga) moth
Mühle *f* flour mill; [saw] mill; refiner; crusher, pulverizer
Mühlrad *n* mill wheel
Mukonsäure *f* muconic acid
mukös mucinlike
Mulde *f* 1. hollow; trough; tray; 2. *(geol)* synclinal fold, syncline
Muldenballenbrecher *m* pedal bale-breaker

Muldenband *n* troughed (trough-type) belt
Muldenbandförderer *m* troughed belt conveyor
Muldenbeschickkran *m* charging crane
Muldenbildung *f (geol)* synclinal formation
Muldenfeuerung *f* furnace with ridge grate
Muldengurtförderer *m* troughed belt conveyor
Muldenhinterkipper *m* end-dump truck
Muldenkipper *m* dumper, trough (truck) tipper
Muldenpresse *f* roller press
Muldenrolle *f* troughing (trough-shaped) idler
Muldenrost *m* trough grate
Muldenrostfeuerung *f* furnace with trough grate
Muldenroststab *m* fire bar of the trough grate
Muldenschar *n* trough-shaped share
Muldentragrolle *f* troughing idler
Muldenvorkrempel *f* shell breaker card
Muldenwäscher *m* trough washer
Muldung *f* troughing
Mule[maschine] *f* mule [spinning machine]
Mulespinnerei *f* mule spinning
Mull *m* muslin
Müll *m* waste, dust, rubbish, refuse; garbage
Müllabfuhr *f* refuse collection
Müllabfuhrarbeiter *m* dust man
Müllabfuhrwagen *m* refuse-collecting vehicle
Müllabladeplatz *m* refuse tip
Mullbausch *m* tampon
Müllbeseitigung *f* refuse (waste) disposal
Müllbrenner *m* [refuse] incinerator
Mülleimer *m* dust bin
Mullenprüfer *m* mullen tester
Müllfeuerung *f* refuse destructor furnace
Mullit *m* mullite
Müllschlucker *m* waste-disposal unit, garbage (refuse, rubbish) chute
Mülltonne *f* refuse disposal drum
Müllverbrennung *f* [refuse] incineration
Müllverbrennungsanlage *f* [refuse] incinerator, refuse destructor
multialkalisch multi-alkali
Multibandschwingkreis *m* multiband tank-circuit
Multigradöl *n* multigrade oil
Multihobel *m* multi-plough
Multiklon *m* multiclone
Multiplett *n* multiplet
  normales M. normal multiplet
  regelrechtes M. normal multiplet
  verkehrtes M. inverse multiplet
  verkettetes M. interlocked multiplet
Multiplettaufspaltung *f* multiplet splitting
Multiplexbetrieb *m* multiplex telephony
Multiplexpappe *f* multi-ply board
Multiplexsender *m* multiplex radio transmitter
Multiplikand *m* multiplicand
Multiplikation *f* multiplication
  abgekürzte M. short-cut multiplication
Multiplikationsfaktor *m* multiplication constant, *k*-factor
Multiplikationskonstante *f* stadia lines ratio
Multiplikationsregel *f* multiplication theorem
Multiplikator *m* multiplier
  mechanischer M. mechanical multiplying device
Multiplikatorpotentiometer *n* multiplier potentiometer
multiplizieren to multiply
  über Kreuz m. to cross-multiply
Multipliziergetriebe *n* mechanical multiplying device
Multiplizierglied *n* multiplier
Multipliziertrieb *m* mechanical multiplying device
Multiplizität *f* multiplicity
Multipol *m* multipole
Multipolentwicklung *f* multipole expansion
Multipolordnung *f* multipole order
Multipolstrahlung *f* multipole radiation
Multi-Vari-Karte *f* multi-vari chart
Multivibrator *m* multivibrator
  astabiler M. free-running multivibrator
  bistabiler M. bistable multivibrator, flip-flop [oscillator]
  monostabiler M. monostable (astable, one-cycle, start-stop, one-shot) multivibrator
Multivibrator[kipp]schaltung *f* multivibrator circuit
Multizellularvoltmeter *n* multiple electrometer
Multizellularzählrohr *n* multicellular counter tube
Multizyklon *m s.* Multiklon
Mu-Metall *n* mu-metal
Mundharmonika *f* mouth organ, harmonica
Mundleuchte *f* oral lamp
Mundschnur *f* subsidiary line
Mundstück *n* nozzle; neck *(turbine)*
Mündung *f* aperture *(flow)*; mouth, estuary *(river)*; junction; muzzle *(gun)*
Mündungsbremse *f* muzzle brake
Mündungsdruck *m* pressure at the outlet opening *(turbine)*
  kritischer M. critical pressure at the outlet opening
Mündungsebene *f* plane of the outlet opening
Mündungsenergie *f* muzzle energy
Mündungsfeuer *n* muzzle flash
Mündungsfeuerdämpfer *m* flash hider
Mündungsform *f* shape of outlet *(of the nozzle)*
Mündungsgebiet *n* estuary *(river)*
Mündungshafen *m* estuary harbour
Mündungskappe *f (mil)* muzzle cover (protector)
Mündungsschoner *m s.* Mündungskappe
Mundwasser *n* liquid dentrifice, mouthwash
Munition *f* ammunition

Munitionsaufzug *m* ammunition hoist, shell crane
Munitionsfabrik *f* ammunition factory
Munitionsfahrzeug *n* ammunition carrier
Munitionslager *n* ammunition store (dump)
Munitionsstreifen *m* slip of ammunition
Muntzmetall *n* Muntz metal
Münzautomat *m* coin-operated machine, penny-in-the-slot automat
Münze *f* 1. coin; 2. mint
Münzeinwurf *m* coin slot
Münzerz *n* money-ore
Münzfernsehen *n* pay-television, coin television
Münzfernsprecher *m* prepayment (coin-box) telephone
Münzlegierung *f* coinage alloy
Münzmeister *m* mintmaster
Münzstätte *f* mint
Münzstättenangabe *f* mint mark
Münzzähler *m* prepayment (slot) meter
Murexidprüfung *f* murexide test
Muringboje *f* mooring buoy
Muringstein *m* mooring clump (block); killick
Muschelbruch *m* clam shell markings (marked fracture)
Muschellinie *f* conchoid
musen *(ag)* to pulp
Muser *m (ag)* pulper
Musik *f* music
  elektrische M. electrosonic music
Musikbespulung *f (el)* music loading *(cables)*
Musikinstrument *n* musical instrument
Musiknote *f* music note
Musikwiedergabetreue *f* music fidelity
Musivgold *n* mosaic gold
Muskatöl *n* mace oil
Muskovit *m (min)* muscovite
Musselin *m* muslin
Muster *n* 1. sample, specimen; model; prototype; standard; example; 2. *(mach)* master *(copying)*; 3. *(text)* pattern, design
  durchbrochenes M. open work pattern
  durchlaufendes M. continuous pattern (design)
  kariertes M. checkerwork
  unterbrochenes M. broken pattern
Musterausnehmen *n* weave analysis
Musterbau *m* prototype building
Musterbuch *n* sampling book
Musterfaden *m* pattern thread
Musterfadenspule *f* whip roll
Musterformkasten *m* master flask
Mustergeschirr *n* figuring shafts
mustergetreu *(text)* perfectly matching
Musterharnisch *m (text)* figuring harness
Musterkette *f (text)* pattern chain
Musterkurve *f (mach)* master cam
Mustermesse *f* sample fair
Musterrapport *m (text)* repeat of pattern
Musterrolle *f (ship)* station bill, muster roll (bill)
Musterrundstrickmaschine *f* **mit Einschließplatinen** design sinker top knitting machine
Musterschläger *m (text)* card cutter
Musterschutz *m* trade-mark protection
Musterstück *n* 1. *(mach)* master [component], master piece, model *(copying)*; 2. sample, specimen
Mustertarif *m* model tariff
Mustertrommel *f (text)* pattern drum
Musterweberei *f* pattern weaving
Musterwechsel *m* pattern changing *(loom)*
Musterwirtschaft *f* model farm
Musterzylinder *m* **aus Kupfer** copper roller
Mutante *f* mutant
Mutation *f* mutation
Mutmaßlichkeit *f* likelihood *(statistics)*
Mutter *f* nut
  gerändelte M. knurled nut
  gesicherte M. locked nut
Mutteratom *n* parent [atom]
Mutterbahnhof *m* rail-head
Mutterblech *n* starting sheet
Mutterblechbad *n* stripper tank
Mutterboden *m* top soil
Mutterbohrer *m* nut tap
Muttererde *f* top soil
Mutterfels *m* bed-rock
Mutterfräsmaschine *f* nut milling machine
Muttergewinde *n* thread in a nut, internal thread
Muttergewindebohreinrichtung *f* nut tapping attachment
Muttergewindebohren *n* tapping of nuts, nut tapping
Muttergewindebohrer *m* nut tap
Mutterkern *m* precursor, original (parent) nucleus
  M. verzögerter Neutronen delayed neutron emitter
Mutterkompaß *m* master compass (gyro)
Mutterlauge *f* mother liquor
Mutterscheibe *f* plain washer
Mutterschiff *n* mother (depot) ship; floating whale-oil factory
Mutterschloß *n* split (clam) nuts [on lead screw]
Muttersender *m* master transmitter
Muttersicherung *f* nut locking
Mutterstecker *m* female plug
Muttersubstanz *f* parent substance
Mutteruhr *f* master (primary) clock
Mutterwerkstoff *m* parent metal *(welding)*
Mutterwolle *f* ewe's wool
Mutung *f* prospecting
Mützenfutter *n* cap lining
Mützenlampe *f* cap lamp
Myon *n (ph)* muon
Myoneneinfang *m* muon capture
Myonenstreuung *f* muon scattering
Myonenzerfall *m* muon decay

Myosin *n* myosin
Myristinsäure *f* myristic acid
Myrizylalkohol *m* myricyl alcohol
Myrtenöl *n* myrtle oil
Myzel *n* mycelium

N

n-Adressenkode *m (dat)* n-address code
n-Adressenrechner *m* n-address computer
n-fach n-fold
n-Halbleiter *m* n-type semiconductor
n-leitend n-conducting
n-Leiter *m* n-type conductor
n-Leitung *f* n-type conduction
n-Lösung *f* normal solution
n-Tor *n* n-port *(electronics)*
n-Typ-Halbleiter *m* n-type semiconductor
n-Verfahren *n (build)* elastic design (method), m-method, modular ratio method
n-zählig n-fold
n-Zone *f* n-region *(electronics)*
Nabe *f* boss, hub *(wheel)*
  N. mit Kupplungsnocken hub with clutch cam
  feste N. fixed hub
  getrennte N. split hub
  ringförmige N. armature sleeve
  ungeteilte N. solid hub
Nabenbohrmaschine *f* nave boring machine
Nabendeckel *m* hub cover (cap)
Nabenhülse *f (el)* armature sleeve
Nabennut *f* keyway in hub, hub keyway
Nabenschraube *f* boss (hub) joint bolt
Nabensitz *m* wheel seat
Nabenstern *m* hub spider
Nabenverbindung *f* boss joint
Nabenverkleidung *f* hub fairing
Nabenwirbel *m* root vortex
Nabenwirbelkavitation *f* boss cavitation
Nabenwulst *m* **[des Kreuzkopfes]** bead of the boss [of the crosshead]
Nabla *n (math)* nabla, del
Nablaoperator *m* nabla
Nachabgleich *m* post-equalization
Nachablenkung *f* post-deflection
nachahmen to copy, to duplicate, to imitate; to simulate
  Holzmaserung n. to imitate the grain of wood
Nachappretur *f* afterfinish
Nacharbeit *f* rework[ing], second operation
nacharbeiten to rework, to correct; to refinish; to finish
Nacharbeiten *n* **des Bohreranschnitts** correction of the drilling
  N. von Bohrungen finishing holes
Nachauftragnehmer *m* sub-contractor
Nachbaramt *n (telec)* adjacent exchange
Nachbaratom *n* neighbouring atom
Nachbarbildträger *m (telev)* adjacent picture carrier
Nachbarbildträgerwelle *f* adjacent channel picture-carrier
Nachbarfeld *n (mine)* adjoining concession
Nachbarfrequenz *f* adjacent frequency
Nachbarkanal *m* adjacent channel
Nachbarkanalstörung *f* adjacent-channel interference
Nachbarortsverkehr *m* adjacent-city telephone-traffic
Nachbarschaft *f (build)* [residential] neighbourhood unit; *(el)* proximity
Nachbartonträger *m* adjacent sound-carrier
Nachbarwindungen *fpl* adjacent turns
Nachbarzone *f* adjacent zone
nachbauen 1. to reproduce; 2. *(mine)* to repair *(road)*
Nachbauen *n (mine)* back ripping, re-rip
Nachbearbeitung *f* further processing; rework[ing]
Nachbeeisung *f* re-icing
nachbehandeln 1. to aftertreat; 2. to cure *(concrete)*
Nachbehandlung *f* 1. additional treatment; retreatment, aftertreatment; 2. curing *(concrete)*
  N. durch Abdecken mit feuchtem Sand sand curing *(concrete)*
  N. mit Dampf steam curing *(concrete)*
  N. mit Wasser wet (water) curing *(concrete)*
Nachbehandlungsanlage *f* retreatment plant
Nachbelichtung *f* post-exposure
Nachbeschickung *f* recharging, reloading
Nachbeschleunigung *f (el)* post-acceleration, after-acceleration
Nachbeschleunigungsanode *f* post-accelerating anode
Nachbeschleunigungselektrode *f* intensifying electrode
nachbestellen to reorder
Nachbestellung *f* reorder
Nachbezahlung *f* additional payment
Nachbild *n (opt)* after-image
nachbilden 1. *(mach)* to copy; 2. *(el)* to simulate *(e.g. circuits)*; to reproduce; to imitate
  analog n. to simulate
Nachbildfehlerdämpfung *f* active balance return loss
Nachbildmesser *m (Am)* return measuring set
Nachbildner *m* simulator
Nachbildprüfer *m* balance tester
Nachbildung *f* modelling; simulation *(real-time operation)*; mock-up; balance, balancing (bal-

ance) network; replica *(surface)*; copy, reproduction; imitation

angenäherte N. compromise balance network

Nachbildungsgestell *n* balancing network rack

Nachbildungsgüte *f* quality of balance

Nachbildungskurve *f* growth curve *(reactor)*

Nachbildungsprüfer *m* balance tester

Nachbildungssucher *m* impedance unbalance finder

Nachbildungsverfahren *n* balancing method

nachblasen to afterblow

nachbleiben to lag, to stay behind

Nachbleiben *n* **der Steuerung** lag of control

Nachblütenspritzung *f* petal fall spray

zweite N. fruitlet spray

Nachbohren *n* reboring; finish-boring

Nachbrecher *m* recrusher

Nachbrennen *n (rock)* afterburning

Nachbrenner *m (rock)* afterburner, tail-pipe chamber

nachbunkern *(ship)* to refuel, to rebunker, to recoal

nachchlorieren to post-chlorinate

Nachchromieren *n* back-chroming

nachchromiert chrome-topped

Nachchromierungsfarbstoff *m* chrome top dyestuff

Nachdechseln *n* re-adzing *(sleepers)*

Nachdichten *n* **von Decksfugen** *(ship)* refilling of deck seams

Nachdrehen *n (text)* additional twist

Nachdrehschleuder *f* after-worker *(sugar industry)*

Nachdrehung *f (text)* additional twist

Nachdruck *m* reprint; reproduction

N. verboten all rights reserved

unberechtigter N. pirated edition

nachdunkeln to deepen *(printing ink)*

nacheichen to recalibrate

Nacheichung *f* recalibration

nacheilen *(ph)* to lag [behind]

Nacheilen *n* lag[ging]

Nacheilung *f (ph)* lag[ging], lag[ging] of phase, time-lag

Nacheil[ungs]winkel *m* angle of lag

nacheinander successively, progressively

Nachen *m* boat, bark

Nachentladung *f* after-count, after-discharge, re-ignition

Nachentzerrung *f* de-emphasis *(radio)*

Nacherhebung *f* collection of additional charges

nachernten to glean

Nachfalldach *n (mine)* draw slate

nachfallen to break back *(coal)*

Nachfallkohle *f* backs

Nachfallpacken *m (mine)* claggy, clod, draw slate

nachfärben to dye again

Nachfaulung *f* secondary digestion *(sewage treatment)*

Nachfließen *n* after-flow

Nachfolgegerät *n* back-up unit

nachfolgen to follow

nachfolgend following, subsequent

Nachforderung *f* collection of additional charges

Nachformarbeit *f* copy (duplicate) machining operation

fühlergesteuerte N. tracing work

Nachformbewegung *f (mach)* copying motion

Nachformdreharbeit *f* contouring (contour-turning) operation

Nachformdrehen *n* contour (duplicate, copy) turning, repetition turning of curved profiles

Nachformdrehmaschine *f* copying (contouring) lathe

Nachformeinrichtung *f* copying (contouring) attachment, duplicator, duplicating device (attachment)

hydraulische N. hydraulic tracing attachment, hydraulic copying device

nachformen *(mach)* to copy, to duplicate; to contour *(two-dimensional)*

Nachformen *n (mach)* copy (duplicate) machining; copying, duplicating; contouring *(two-dimensional)*

spiegelbildliches N. reverse [mirror] duplicating

zweidimensionales N. two-dimensional tracing (contouring)

Nachformfähigkeit *f* repeatability

Nachformfräsautomat *m* automatic die-sinking machine

Nachformfräsen *n* copy milling

N. mit Pantograf pantographing

dreidimensionales N. cavity milling

Nachformfräsmaschine *f* copy-milling machine

zweidimensionale N. profiling (profile-milling) machine

Nachformfrässchablone *f* copy-milling template

Nachformfühler *m* form tracer

Nachformgenauigkeit *f* accuracy of reproduction

Nachformhobeleinrichtung *f* contour-planing attachment

Nachformhub *m* copy stroke

Nachformkante *f* **der Schablone** tracing edge of the template

Nachformmaschine *f* [form-]duplicating machine

Nachformschleifen *n* copy [profile] grinding

Nachformschlitten *m* copy saddle

Nachformventil *n* tracer valve

Nachformvorhobeln *n* rough contour-planing

Nachformwaagerechtstoßmaschine *f* copying shaper

Nachformwalze *f* follower-roller
Nachforschungsgebühr *f* investigation charge
Nachfrage *f* demand
Nachfräser *m* finishing cutter
Nachfrucht *f* aftercrop, successor crop
Nachführautomatik *f* automatic guidance system
Nachführregelung *f* master-and-slave control
Nachführung *f* follow-up [device]
  automatische N. automatic follow-up device
nachfüllen to replenish, to refill; to top up
Nachfüllluke *f* grain (trimming) hatch
Nachgärung *f* secondary fermentation, post-fermentation
Nachgeben *n* **/ ruckweises** *(mine)* load-shedding *(support)*
nachgerben to retan
  mit Eigelb n. to re-egg
Nachgerbung *f* retannage
nachgiebig elastic *(foundation)*
nachgießen *(met)* to pour again, to cast on, to refill *(mould)*
nachgleichen 1. to rebalance; 2. to retune *(radio)*
nachglimmen to afterglow
nachglühen 1. to re-anneal; 2. *s.* nachglimmen
Nachhall *m* reverberation, reverberant sound
nachhallend reverberant
nachhallgesteuert reverberation-controlled
Nachhallkammer *f* reverberation chamber
Nachhallmessung *f* measuring [of] the reverberation, reverberation measurement
Nachhallraum *m* reverberation chamber
Nachhallzeit *f* time of reverberation, reverberation time
Nachhalter *m (cont)* inverse-derivative element
Nachhärtungsfrist *f (plast)* aging time, conditioning period
Nachhieb *m* secondary felling *(forest)*
Nachhinken *n* **der Garung** retarded coking time
Nachimpuls *m* after-pulse
Nachklärbecken *n* secondary sedimentation-basin, final settling tank (basin)
nachkochen *(met)* to reboil
Nachkondensator *m* secondary condenser
Nachkühler *m* additional cooler, aftercooler
Nachkupfern *n (text)* copper aftertreatment
nachlackieren to recellulose, to revarnish
nachladen to recharge *(secondary cell)*
Nachladung *f* recharging; additional charge
nachlassen 1. to slow down, to slacken in speed; to abate *(wind)*; 2. to ease *(a wedge)*; to slacken, to loosen; to relax
Nachlassen *n* **des Leuchtstoffes** dark burn *(cathode ray tube)*
  N. eines Permanentmagneten weakening of a permanent magnet
Nachlaßvorrichtung *f* temper screw *(rope boring)*
Nachlauf *m* 1. wake *(fluid flow)*; 2. *(aut)* trail *(front wheels)*; 3. *(ch)* last runnings, tailings, tail fraction (run), distillation tail
Nachlaufeinrichtung *f* servomechanism
Nachlaufgeschwindigkeit *f* overrun speed
Nachlaufinstrument *n* servo-operated instrument
Nachlaufmotor *m* follow-up motor
Nachlaufpotentiometer *n* follow-up potentiometer
Nachlaufregler *m* follower, follow-up mechanism, servomechanism
Nachlaufschaltung *f* tracking circuit *(radar)*
Nachlaufsteuerung *f* overrun control
Nachlauftank *m* slop tank
Nachlaufwiderstand *m* self-balancing potentiometer
Nachlaufwinkel *m* castor angle *(front wheels)*
Nachlaufwirbel *m* trailing vortex
Nachlenzpumpensystem *n (ship)* stripping pumping system
Nachleuchtcharakteristik *f* persistence (decay) characteristics
Nachleuchtdauer *f* fluorescent life-time; persistence
nachleuchten to afterglow, to persist
Nachleuchten *n* persistence, afterglow; phosphorescence
Nachleuchtschirm *m* long-persistence screen
Nachleuchtschleppe *f* tail
nachliefern to supply (deliver) subsequently
Nachlinks-Schweißen *n* leftward welding, *(Am)* forehand welding
Nachluft *f* supplementary (additional) air
Nachmahd *f* aftermath, second hay
nachmessen to re-measure, to check [the dimensions], to verify
Nachmessung *f* check[ing], re-measurement
Nachnahme *f* reimbursement, refund
  N. / gegen cash on delivery, c.o.d.
Nachnahmegebühr *f* reimbursement charge
Nachnahmesendung *f* cash on delivery consignment, C.O.D. consignment
Nachnahmezustellschein *m* advice of availability of a paid-on for value
nachnehmen *(mine)* to dint
nachpantoffeln *(leath)* to regrain
Nachpotential *n* after-potential
  negatives N. negative after-potential
  positives N. positive after-potential
Nachpressen *n* re-pressing
Nachproduktmaische *f* after product crystallizer
nachprüfen to verify, to check
nachrechnen to check, to compute (calculate, reckon) again

Nachrechts-Schweißen *n* rightward (backward) welding, *(Am)* backhand welding
Nachregelung *f* re-adjustment
N. der Verstärkung gain adjustment
Nachreife *f* ultimate ripening
Nachreifen *n* ag[e]ing
Nachreißberge *pl (mine)* ripping dirt
nachreißen *(mine)* to brush down
Nachreißen *n* **des Hangenden in Abbaustrecken** ripping
Nachreißhauer *m* [back] brusher, ripper
Nachreißmaschine *f (mine)* ripping machine
Nachreißstelle *f (mine)* canch, caunch, ripping [lip]
Nachricht *f* message
nachrichten to re-level, to realign
Nachrichtenansager *m* newscaster
Nachrichtenaufzeichnung *f* / **automatische** automatic telecommunication log
Nachrichtenbandbreite *f* intelligence bandwidth
Nachrichteneinheit *f* bit *(information theory)*
Nachrichtenelektronik *f* communications electronics
Nachrichtenempfänger *m* communications receiver
Nachrichtenentropie *f* negative entropy, negentropy
Nachrichtengerät *n* signal apparatus; communication equipment
Nachrichteningenieur *m* communication engineer
Nachrichtenkanal *m* communication channel
benachbarter N. adjacent communication channel
Nachrichtenmenge *f* decision content *(information theory)*
Nachrichtennetz *n* [tele]communication network, communication system
Nachrichtensatellit *m* communications satellite
Nachrichtentechnik *f* telecommunication engineering
elektrische N. electrical telecommunication engineering
Nachrichtentheorie *f* communication theory
Nachrichtentruppe *f* signal corps (troops)
Nachrichtenübermittlung *f* communications
Nachrichtenübertragung *f* communications
Nachrichtenverarbeitung *f* data (information) processing
Nachrichtenverbindung *f* communication
Nachrichtenverbindungssystem *n* communication system
Nachrichtenverkehr *m* communications
Nachrichtenwesen *n* communications
Nachrutschen *n* slipping
Nachsaturation *f* final saturation
Nachsatzkonzentration *f (text)* concentration of the feeding liquor
Nachsaugen *n* **einer Form** contraction of a mould
Nachschaben *n (mach)* re-scraping
nachschärfbar resharpenable
nachschärfen to resharpen
Nachschärfung *f* resharpening
Nachschäumzelle *f* cleaner
Nachscheidesaft *m* post-defecation juice
Nachschiebefahrt *f* banking *(electrical traction)*
Nachschlagewerk *n* book of reference, reference book
nachschleifbar regrindable
Nachschleifen *n* regrinding; resharpening *(tools)*; *(aut)* refacing
Nachschliff *m* regrind; resharpening *(tool)*
Nachschneider *m* reamer
Nachschneidezahn *m* follower (square finish) tooth *(saw blade)*
Nachschreibstörimpuls *m* post-write disturb pulse
Nachschubflotte *f* train *(vessels)*
Nachschubpanzer[kraftwagen] *m* armoured supply and maintenance vehicle
Nachschwaden *m* after-damp, choke-damp *(blasting)*
Nachschwindung *f* after-contraction
nachsetzen *(met)* to recharge
Nachsetzmaschine *f* rewashing machine
Nachsichten *n* re-sorting *(flax)*
nachsortieren 1. to re-sort *(flax)*; 2. *(mine)* to re-clean, to rewash
Nachsortierung *f* final separation
nachspannen to tension, to re-stretch; *(mach)* to reclamp *(work)*
Nachspannen *n* **des Werkstücks** reclamping the work
Nachspannvorrichtung *f* tensioning device *(overhead line)*
selbsttätige N. automatic tension-regulator
nachstellbar [re]adjustable
Nachstellbarkeit *f* [re]adjustability
Nachstelldauer *f* [re]adjusting time
Nachstelldruckschraube *f* set-screw for adjusting the bearing
Nachstelleinrichtung *f* **des Lagers** adjusting device for the bearing
Nachstelleiste *f (mach)* [adjustable] gib strip, adjusting gib (strip), take-up strip
nachstellen to readjust; to reset
Nachstellkeil *m* tightening wedge
Nachstellung *f* readjustment; resetting
Nachstellvorrichtung *f* readjusting device
N. für Lagerbolzen readjusting device for bearing bolts
Nachstellzeit *f* integral action time
nachsteuern to follow-up

Nachstimmbereich *m* frequency control limits
Nachstimmdiode *f* diode for frequency control
nachstimmen to retune
Nachstimmröhre *f* reactance-tube for frequency control
nachstollen to restake *(shoes)*
Nachstrecken *n (text)* supplementary drawing
nachstreichen *(leath)* to rescud
Nachstrom *m* 1. *(ship)* [frictional] wake, wake current; 2. *(el)* post-arc current
Nachstromkoeffizient *m* wake coefficient
Nachstrompropeller *m* wake-adapted propeller
Nachstromziffer *f* wake factor (coefficient, fraction)
Nacht *f* night
Nachtalarmschalter *m* night alarm switch
Nachtankkupplung *f (rock)* oxygen-replenisher coupling
Nachtankleitung *f (rock)* replenishing line
Nachtankventil *n (rock)* oxygen-replenisher valve
Nachtankvorrichtung *f (rock)* replenishing device
nachtblind night-blind
Nachtblindheit *f* night-blindness
Nachteffekt *m* night effect (error), polarization error *(radar)*
Nachteil *m* drawback, disadvantage
Nachtfischen *n* owling
Nachtfrost *m* night frost
Nachthaus *n (ship)* binnacle, nighthood *(compass)*
Nachtjagdwarngerät *n* night-fighter detect[illegible] apparatus *(radar)*
Nachtlicht *n* life buoy light, luminou[illegible] [illegible]ght life) buoy *(navigation)*
Nachtrabant *m (telev)* post-equalizing [illegible]
Nachtrag *m* **zum Tarif** tariff supple[illegible]
Nachtragearbeit *f* supplementary [illegible]
Nachtrettungsboje *f* luminous ([illegible]t life) buoy
Nachtrockenpartie *f* end drier s[illegible]ion
Nachtrocknung *f* re-drying, sec[illegible]d[ary] drying
Nachtschalter *m* night alarm key (switch)
Nachtschicht *f* night shift
Nachtsehen *n* scotopic vision
Nachttarif *m* night tariff
Nachverbrenner *m (rock)* afterburner, tail-pipe chamber
Nachverbrennung *f* exhaust reheat *(turbo propulsion)*
nachverdampfen to re-evaporate
Nachverdampfen *n* re-evaporation
Nachvergrößerung *f (opt)* subsequent enlargement
nachwalken to restock
Nachwärme *f* after-heat, afterheat
Nachwärmofen *m (met)* reheating furnace
Nachwäsche *f (mine)* re-cleaning
nachwaschen *(mine)* to re-clean, to rewash
Nachwaschen *n (text)* second washing
Nachwaschkohle *f* second product coal
Nachwaschsetzmaschine *f* re-wash jig, rewash box
Nachweide *f* autumnal pasture
Nachweis *m* proof; verification; establishment; *(ch)* detection, indication, identification
  chemischer N. chemical detection
  fotografischer N. photographic detection
  radioaktiver N. radioactive detection
Nachweisbarkeit *f* detectability
Nachweisempfindlichkeit *f* detection sensitivity
nachweisen to prove; to verify; to establish; *(ch)* do detect, to identify, to indicate
Nachweisgerät *n* detector
Nachweisgrenze *f* detection limit, limit of detection
Nachweisung *f* / **monatliche** monthly statement
nachwiegen to check the weight
Nachwiegung *f* verification, checking of weight
Nachwinterspritzung *f* winter oil spraying
Nachwirkung *f (elastic)* after-effect, aftereffect
  dielektrische N. dielectric fatigue (hysteresis)
  elastische N. elastic lag (hysteresis)
  magnetische N. magnetic after-effect
  plastische N. plastic after-flow
Nachwirkungsbeiwert *m* **[bei Eisenkernspulen]** re[illegible]idual-induction coefficient
N[illegible]chwirkungsstrom *m* absorption (reversible, transient-decay) current
Nachwirkzeit *f* decay (hangover, partial-restoring) time
nachwuchten to rebalance
nachzeichnen to redraw
nachzerkleinern to bulldoze, to re-crush
Nachzerkleinerung *f* re-crushing
  N. durch Schießarbeit secondary blasting
  N. mit dem Hammer sledging
nackt bare *(electrode, wire)*
Nadel *f* needle; stylus *(e.g. grammophone; etching)*; *(mach)* bar, tool, broach
  kegelförmig zulaufende N. cone-shaped needle
  verdichtende N. burnishing bar
Nadelabweichung *f* magnetic declination *(compass)*
Nadelanode *f* needle anode
Nadelast *m* pin-knot
Nadelausleser *m (ag)* needle cylinder
Nadelausschlag *m* needle throw *(instrument)*
Nadelaustrieb *m (text)* raise of the needle
Nadelbarre *f* / **vordere** *(text)* front needle bar
Nadelbaum *m* conifer
Nadelbett *n* **einer Flachstrickmaschine** flat-needle bed
  hinteres N. back needle bed
Nadelbewegung *f* stylus trajectory (recording)
Nadelblei *n* needle lead
Nadelboden *m* pin-hole plug *(converter)*
Nadelbrecher *m* pick breaker

Nadelbrett *n* needle board
Nadelbrettverschiebung *f* displacement of the needle board
Nadelbruchabsteller *m* needle protector
Nadeldraht *m* pin wire
Nadeldruck *m* needle pressure
Nadeldrucker *m* wire printer
Nadeldurchlauf *m (text)* needle travel
Nadeleisenerz *n* goethite
Nadelelektrode *f* needle electrode
Nadelfach *n (text)* shed for wire
Nadelfeder *f* needle spring
Nadelfeile *f* needle file
nadelförmig acicular; needle-shaped
Nadelfunkenstrecke *f* needle-point gap
Nadelgalvanometer *n* moving-needle galvanometer
Nadelgehäuse *n (text)* needle box
  schwingendes N. swing needle box
Nadelgeräusch *n* needle scratch, surface noise
Nadelgeräuschfilter *n* scratch filter
Nadelhaken *m* / **eckig gebogener** excelsior hook *(latch needle)*
  zur Nadelachse vorgebogener N. central position of hook *(latch needle)*
Nadelholz *n* coniferous (soft) wood
Nadelimpuls *m* needle pulse, spike *(pulse)*
Nadelkette *f* needle creel
Nadelkristall *m* spicule crystal, whisker
Nadelkupplung *f* needle coupling *(turbine)*
Nadellager *n* needle bearing
  zweireihiges N. double-row needle roller bearing
Nadellattentuch *n (text)* spiked lattice
Nadelleistenrost *m (text)* pin bar grid
Nadelpaar *n* pair of needles *(measuring)*
  astatisches N. astatic system *(galvanometer)*
Nadelquelle *f* needle-shaped source
Nadelrauschen *n* needle scratch *(record)*
Nadelregelung *f (aut)* needle control
Nadelreihe *f* row of needles *(comber)*
Nadelrollenlager *n* needle roller bearing
Nadelschaltung *f* symmetrical heterostatic circuit
Nadelsegment *n* needle segment *(comber)*
Nadelspitze *f* 1. *(text)* needle lace; 2. *(mach)* needle point
Nadelstab *m (text)* faller, gill bar
Nadelstababstand *m* ratch of the fallers
Nadelstabfeld *n* set of fallers
Nadelstabstrecke *f* gill box
Nadelstabstreckwerk *n* **mit Kanne** can gill box
Nadelstich *m* pitch of needles
  grober N. coarse pitch of needles
Nadelstrahl *m* pencil beam
Nadelstreifen *m (text)* hairline stripe
Nadelteppich *m* tufted (needle loom) carpet
Nadelton *m* stylus sound
Nadeltrieur *m* needle cylinder
Nadelübergabe *f (text)* transfer of needles
Nadelüberwachungsgerät *n (text)* needle detection installation
Nadelventil *n* needle [point] valve
Nadelverschnürung *f (text)* needle tying
Nadelwalze *f (text)* porcupine
Nadelwalzenstrecke *f* hedgehog drawing, porcupine drawing frame
Nadelwebstuhl *m* needle loom
Nadelweg *m (text)* path of the needle
Nadelwehr *n* needle weir
Nadelzähler *m s.* Spitzenzähler
Nadelzug *m* drawing of needles
Nadelzunge *f* needle latch
Nadir *m* nadir
Nadirabstand *m* nadir distance
Nadirwinkel *m* nadir angle
Nagel *m* nail; spike *(large nail)*; sprig *(small, headless)*; tack *(with large head)*
Nagelausziehwiderstandsprüfung *f* nail withdrawal test
nagelbar nailable
Nagelbarkeit *f* nailability
Nagelbinder *m* nailed [timber] truss
Nageldurchziehprüfung *f* nail head pull-through test
Nageleisen *n* wrecking bar *(with claw)*
nagelfest scratch-proof *(printing ink)*
Nagelhaltefestigkeit *f* nail holding [property]
Nagelhautentferner *m* cuticle remover
Nagelkaliber *n* nail pass *(rolling)*
Nagelklaue *f* claw wrench
Nagelkopf *m* nail head
Nagellack *m* nail lacquer
Nagellackentferner *m* enamel remover
nageln to nail; to tack
Nagelung *f* / **verdeckte** secret nailing
Nagelzieher *m* nail-lifter
Nagyagit *m (min)* nagyagite
Nähahle *f* sewing awl
Nahaufklärer *m* close-reconnaissance aircraft
Nahaufnahme *f* close-up, close-up photograph (view)
Nahbereich *m* close-up area, local (short) range *(radar)*
Nahbestrahlung *f* short-distance irradiation
Nahbestrahlungstherapie *f* contact [radiation] therapy
Nahecho *n* near echo (radar)
Nahechodämpfung *f* local echo attenuation
Nahechodämpfungsschaltung *f* suppressor
nähen to sew; to suture *(lips of a wound)*
Nahentfernungsmeßbake *f* locator beacon *(radar)*
Nahentzerrung *f* de-emphasis
nähern / sich *(math)* to approximate
  sich asymptotisch n. to shallow out *(curves)*

sich einander n. to approach each other
Näherung *f* exposure; *(math)* approximation
adiabatische N. adiabatic approximation
Bornsche N. Born approximation
harmonische N. harmonic approximation
quasiklassische N. quasi-classical approximation
schräge N. oblique exposure
stufenweise N. *(math)* successive approximation
Näherungsanalyse *f* approximation analysis
Näherungsdarstellung *f* approximate representation
Näherungseffekt *m (el)* proximity effect
Näherungsformel *f* approximation formula
Näherungsgleichung *f* approximation equation
Näherungslösung *f* approximate solution
Näherungsmethode *f* approximation (approximate) method
Näherungsverfahren *n* approximation method
Näherungswert *m* approximate value
N. von Blasius Blasius's approximation
Nahewirkungsgesetz *n* law of proximity theory
Nahewirkungskraft *f* short-range force
Nähfadenlochnadel *f* sewing thread eye-needle
Nahfahrt *f* inshore navigation, coasting trade
Nahfeld *n* near field (zone)
Nahfeuervisier *n* open sight
Nahgüterzug *m* slow (pick-up) goods train
Nahkampfpanzer *m* close-support tank
Nahkurzschluß *m s.* Abstandskurzschluß
Nähmaschine *f* sewing machine
Nähmaschinennadel *f* sewing machine needle
Nahnebensprechdämpfung *f* near-end crosstalk attenuation
Nahnebensprechen *n* near-end crosstalk
Nahpeilung *f* close-range direction finding
Nahpunkt *m* near point
Nährboden *m* culture medium
Nährflüssigkeit *f* nutrient, nutritive fluid
Nährgewebe *n* storage tissue
Nährhefe *f* yeast nutrient (extract)
Nährlösung *f* nutrient solution
Nährmittel *pl* cereals
Nährstoff *m* nutritive substance, nutrient
Nährstoffbedürfnis *n* nutrient requirement; fodder requirement
Nährsubstrat *n* nutrient base, nutritive substance
Nahrung *f* 1. food; 2. emulsion *(tanning)*
Nahrungsmittel *n* aliment, foodstuff, nutrition, nutriment
Nahrungsmittel *pl* food[stuff], eatables, victuals
Nahrungsmittelindustrie *f* food industry
Nährwert *m* nutritive value, nutritional quality
Nahschiffahrt *f* short-distance navigation
Nahschwund *m* interference-type fading
Nahschwundzone *f* fading area
Nähseide *f* sewing silk
Nahsichtprüfgerät *n* near-sight tester
Nähspitze *f* needle [point] lace
Nahsteuerung *f* short-distance control
Nahstörung *f (telec)* close-range disturbing effect
Nahstreuung *f (telec)* short-distance scatter
Naht *f* 1. *(text)* seam; suture; 2. *(mach)* weld; joint, juncture
N. mit zwei Fäden cobbler's suture
falsche (imitierte) N. mock seam
ringsumlaufende N. *(mach)* weld all round
Nahtausreibmaschine *f* seam rubbing machine *(shoe)*
Nahteil *m* **[im Zweistärkenglas]** reading portion
Nahthöhe *f* throat *(welding)*
Nahtleiste *f (ship)* [seam] batten, batting
nahtlos seamless
Nahtoberfläche *f* weld face
Nahtschweißen *n* seam welding
Nahtspant *n* seam batten, ribband
Nahtüberhöhung *f* weld reinforcement
Nahverkehr *m* short-distance traffic; short-haul traffic *(transport)*, local transit (traffic)
Nahverkehrsamt *n (telec)* toll exchange, zone centre
Nahverkehrsbereich *m (aero)* terminal control area
Nahverkehrseinrichtung *f* short distance trunk equipment *(telephony)*
Nahverkehrsleitung *f* toll line; short distance line
Nahvorhänger *m* front hanger lens
Nähwirkstelle *f* stitch-bonding element *(non-wovens)*
Nähwirktechnologie *f* chain-stitch technique *(non-wovens)*
Nähwirkverfahren *n* stitch-bonding method *(non-wovens)*
Nahzünder *m* proximity fuse
Nähzwirn *m* sewing cotton
NaK-Legierung *f s.* Natrium-Kalium-Legierung
Nakrit *m (min)* nacrite
Namengebertext *m* answer-back code
Namenschild *n* nameplate; escutcheon
Namensticker *m* name embroiderer
Nanoampere *n* nanoampere, millimicroampere
Nanofarad *n* nanofarad, millimicrofarad
Nanosekunde *f* nanosecond, millimicrosecond
Nanosekundenimpuls *m* millimicrosecond pulse
Nanosekundenimpulsgenerator *m* millimicrosecond pulse generator
Napf *m (mach)* cup
napfziehen to cup, to dish, to shallow form
Napfziehversuch *m* cupping test
Naphtha *n* naphtha
Naphthalin *n* naphthalene
Naphthalinsulfonsäure *f* naphthalene sulfonic acid

Naphthalinzähler *m* naphthalene counter
Naphthenat *n* naphthenate
Naphthene *pl* cyclanes, naphthenes
Naphthensäure *f* naphthenic acid
Naphthionsäure *f* naphthionic acid
Naphthoesäure *f* naphthoic (naphthalene carboxylic) acid
Naphthol *n* naphthol
Naphtholfarbstoff *m* naphthol dye
Naphthylamin *n* naphthylamine
narben to shagreen, to cicatrize
Narben *m (leath)* grain
  aufgepreßter N. pebble grain
  brüchiger N. brittle grain
  dichter N. close grain
  gebrochener N. broken grain
Narbenbrüchigkeit *f* crackiness
Narbenfehler *m* grain defect
Narbenplatte *f* embossing plate
Narbenpreßrolle *f* embossing roller
narbenrein clean-grained
Narbenrolle *f* grain roller
Narbenspalt *m* grain split
narbig *(met)* pitted
Narkose *f* / **elektrische** electrical anesthesia
Narkotikum *n* narcotic
Narkotin *n* narcotine
narrensicher fool-proof
Nase *f* nose; *(mach)* lug *(projection)*; beak; catch
  N. des Nockens lobe of [the] cam
  N. des Pleuelkopfes dipper, splasher
Nasenisolator *m* spur insulator
Nasenkeil *m* gib-head key
Nasenklammer *f* brace
Nasenreizstoff *m* sternutator
Nasenring *m* bull ring (leader)
Nasenspindel *f (text)* hocked (neb) peg
Nasenzahl *f (text)* number of teeth *(opener)*
naß wet *(process)*; humid; damp *(wall)*; moist
Naßabbauhammer *m* wet pick
Naßappretur *f* wet dressing (finishing)
Naßaufbereitung *f* floatation: wet cleaning, washing
Naßausschuß *m* wet broke *(papermaking)*
Naßbagger *m* [floating] dredger, wet dredger
naßbaggern to dredge
Naßbaggerung *f* [underwater] dredging, dredger work
Naßbeize *f* wet seed disinfectant
Naßbeizmittel *n* liquid chemical seed dresser
Naßbeizung *f* wet disinfection
Naßbestäubungseinrichtung *f* liquid spray equipment
Naßbleiche *f* wet bleaching
Naßbügelechtheit *f* fastness to wet pressing
Naßbügeln *n* wet hot pressing
Naßdampf *m* wet steam
Naßdehnung *f* wet elongation
Naßdekatur *f* wet decating (processing method)
Naßdienst *m* wet section *(briquetting)*
Naßdruck *m* wet-on-wet printing
Nässe *f* humidity, wetness
naßecht fast to wet treatment
Naßelement *n* hydroelectric (wet) cell
Naßentrindung *f* / **hydraulische** hydraulic barking
Naßentstauber *m* wet-type dust collector
Naßfäule *f* wet rot
Naßfestigkeit *f* wet strength *(fabric)*
Naßfilter *n* oil-bath air cleaner
Naßfilz *m* press (wet, pick-up) felt
Naßfuttermischmaschine *f* vertical food mixer
Naßgasreinigungsanlage *f* wet precipitator
Naßgespinst *n* wet-spun yarn
naßgesponnen wet-spun
Naßherd *m* wet-washing table
Naß-in-Naß-Druck *m* wet-on-wet printing
Naßklassierung *f* hydraulic classification, wet screening
Naßkopfdüngung *f* liquid top dressing
Naßkupplung *f* oil-type clutch
Naßlichtechtheit *f* wet fastness to light
Naßlöschung *f* wet quenching *(coke)*
Naßluftfilter *n* oil-wetted air cleaner
Naßmagnetscheider *m* wet magnetic cobber
Naßmetallurgie *f* hydrometallurgy
Naßpartie *f* wet end *(paper)*
Naßpresse *f* couching roll
Naßputztrommel *f* wet tumbler
Naßrauhen *n (text)* wet raising
Naßreinigung *f* washing, wet purification
Naßreißwolf *m* **für Abfälle** wet rag-tearing machine
Naßringspinnmaschine *f* wet ring-spinning frame
Naßsandformen *n* green sand moulding
Naßscheidung *f* wet purification
Naßschleifeinrichtung *f* wet grinding attachment
Naßschliff *m* wet grinding
Naßschnitzel *npl* wet pulp *(sugar industry)*
Naßschrämarm *m* whale jib
Naßschrämen *n* wet cutting
Naßsiebung *f* wet screening
Naßsortiermaschine *f* washer
Naßspinnen *n* wet spinning
Naßspinnmaschine *f* wet spinning frame
Naßstäubemittel *n* wettable powder
Naßtrommeln *n* wet tumbling
Naß- und Trockenbestäuber *m* anti-set-off fluid and powder unit
Naßveredlung *f* wet processing
Naßverfahren *n* wet process (method) *(cement manufacture)*; wet method *(man-made fibre)*
Naßverzinkung *f* hot-dip galvanizing
Naßvliesbildung *f* wet sheet formation *(papermaking)*

Naßvliesformmaschine *f* wet sheet forming machine *(papermaking)*
Naßwaschmaschine *f* laundry washing machine
Naßzug *m* wet drawing
Naßzwirnen *n* wet twisting
Nasturan *n (min)* nasturan, pitchblende, uranite
naszierend nascent
Nationalflagge *f* national flag
Nationalitätszeichen *n (aut)* nationality plate
Natrium *n* sodium
  geschmolzenes N. liquid sodium
Natriumamalgam *n* sodium amalgam
Natriumamid *n* sodamide
Natriumazetat *n* sodium acetate
Natriumazid *n* sodium azide
Natriumbromid *n* sodium bromide
Natriumchlorat *n* sodium chlorate
Natriumchlorid *n* sodium chloride
Natriumdampflampe *f* sodium-vapour lamp
natriumgekühlt sodium-cooled
Natrium-Graphit-Reaktor *m* sodium-graphite reactor
Natriumhydrogenkarbonat *n* sodium bicarbonate (hydrogen carbonate), bicarbonate of soda
Natriumhydrogensulfit *n* sodium bisulphite
Natriumhydroxid *n* sodium hydroxide, caustic soda
Natriumjodid *n* sodium iodide
Natrium-Kalium-Legierung *f* NaK alloy, sodium-potassium alloy
Natriumlampe *f* sodium lamp
Natriummetall *n* metallic sodium
Natriumnitrat *n* soda nitre, [Chile] saltpetre, sodium nitrate
Natriumpresse *f* sodium press
Natriumpumpe *f* sodium pump
Natriumspektralleuchte *f* sodium spectral-lamp
Natriumsulfat *n* sodium sulphate
Natriumsulfid *n* sodium sulphide
Natriumthiosulfat *n* sodium thiosulphate
Natriumzyanid *n* sodium cyanide
Natronablauge *f* black liquor
Natronfeldspat *m (min)* soda feldspar
Natronglas *n* soda glass
Natronglimmer *m (min)* soda mica
Natronkalk *m* soda lime
Natronkochlauge *f* caustic soda pulping liquor
Natronlauge *f* soda lye, solution of caustic soda
Natronpapier *n* sodium paper
Natronsalpeter *m* sodium nitrate
Natronseife *f* soda soap
Natronwasserglas *n* soda water glass
Natronzellstoff *m* soda pulp
Naturalausgleich *m s.* Naturalleistung
Naturalleistung *f* payment in kind
Naturasphalt *m* natural asphalt
Naturfarbstoff *m* natural dye
  pflanzlicher N. vegetable dye
Naturfaser *f* / **pflanzliche** vegetable fibre
Naturfeinkorn *n* natural fine grain
Naturgas *n* natural (rock) gas
Naturgesetz *n* natural law, law of nature
naturgetreu true-to-nature
Naturglimmer *m* [block] mica
Naturgummi *m* rubber
Naturhärte *f* natural hardness
Naturharz *n* natural (vegetable) resin
natürlich natural
  n. gealtert naturally aged
Naturschleifstein *m* grindstone
Naturschutzgebiet *n* nature reserve
Naturstein *m* quarry stone
Natursteinindustrie *f* natural stone industry
Natursteinmauerwerk *n* rubble masonry (work), ashlar walling (stonework)
Natursteinplatte *f* quarry tile, flagstone
Natursteinquader *m* ashlar
Natursteinverblendung *f* ashlar facing
Natursteinverkleidung *f* ashlar facing
Naturumlauf *m* natural circulation
Natururan *n* natural[ly occuring] uranium, unenriched uranium
Natururanreaktor *m* natural-uranium reactor
Naturversuch *m* weathering test
Naturwachs *n* natural wax
Naturwissenschaft *f* [natural] science
Nautik *f* nautics, art of navigation
Nautiker *m* navigator
nautisch nautical
naval naval
Navigation *f* navigation
  astronomische N. astronomical (celestial) navigation, celo-navigation
  fehlerhafte N. improper navigation
  terrestrische N. coastal (terrestrial) navigation
Navigationsausrüstung *f* navigational aids *(of a ship)*
Navigationsfunkfeuer *n* navigation radio-light
Navigationshilfe *f* navigational aid
Navigationsinstrument *n* navigation instrument
Navigationskennfeuer *n* code beacon
Navigationsradar *n* navigational radar
Navigationsraum *m* chart room
Navigationsrechner *m* navigation computer
Navigationssystem *n* / **bodenunabhängiges** *(aero)* self-contained navigation system
navigatorisch navigerous *(capable of floating vessels)*
navigieren to navigate
Neapelgelb *n* Naples yellow
Nebel *m* fog, mist; *(astr)* nebula
  diffuser N. irregular nebula
Nebelbildung *f* formation of fog
Nebelgerät *n* 1. *(ag)* fogger; 2. *(mil)* smoke-laying apparatus

tragbares N. *(ag)* knapsack fogger
Nebelgeschoß *n* smoke projectile (shell)
Nebelgranate *f* smoke shell
Nebelhorn *n* fog (reed) horn
Nebelkammer *f* cloud chamber
kontinuierliche N. *s.* Diffusionsnebelkammer
Wilsonsche N. Wilson's chamber
zählrohrgesteuerte N. automatic (counter-controlled) cloud chamber
Nebelkammermethode *f* cloud-chamber method
Nebelkammerspuren *fpl* cloud-chamber tracks
Nebelkavitation *f* mist cavitation
Nebellampe *f* fog lamp
Nebellinie *f* nebular line
nebeln 1. *(ag)* to fog; 2. *(mil)* to lay a smoke-screen
Nebeln *n* 1. *(ag)* fogging; 2. *(mil)* laying a smoke-screen
Nebelscheinwerfer *m (aut)* fog light (lamp)
Nebelsignal *n* fog signal
Nebelspur *f* cloud[-chamber] track, cloud[-chamber] trail
Nebelstoffe *mpl* / **giftige** blanketing clouds
harmlose N. screening smokes
Nebelteilchen *n* fine fog particle
Nebelverfahren *n (ag)* fogging
Nebelwäscher *m* fume scrubber
Nebelwerfer *m (mil)* smoke[-shell] mortar
neben side by side with, alongside; near, close to
Nebenachse *f* minor axis *(ellipse)*
Nebenamt *n (telec)* satellite (minor) exchange
Nebenanschluß *m* extension set (station), private branch
Nebenanschlußleitung *f* extension line (wire)
Nebenantrieb *m (mach)* secondary (auxiliary) drive; feed drive
Nebenantriebssperre *f* auxiliary drive lock
Nebenarbeiten *fpl* ancillary operations
Nebenausstrahlung *f* spurious radiation *(radio)*
Nebenbahn *f* branch line
Nebenbedingung *f (math)* auxiliary (supplementary) condition
Nebenbefehl *m (dat)* branch order
nebenberuflich [in] part-time, [in] part-time occupation
Nebenbestand *m* secondary crop *(forest)*
nebeneinander side by side
nebeneinanderlegen to lay side by side
nebeneinanderliegend adjacent, laying side by side
nebeneinanderschalten to connect in parallel
Nebeneinanderschaltung *f* parallel connection (circuit)
N. der Überhitzerrohre superheating tubes connected in parallel
Nebeneingang *m* side entrance
Nebenerscheinung *f* secondary phenomenon
Nebenfehler *m* class 3 defect, minor A defect
Nebenfluß *m* tributary, affluent
Nebenfrequenz *f* secondary frequency
Nebengang *m (geol)* dropper, feeder
Nebengebühren *fpl* accessory (additional) charges
Nebengelaß *n* closet
Nebengeräusch *n (telec)* undesired (additional) noise
nebengeschlossen bridge-connected
Nebengestein *n (mine)* country (enclosing, surrounding) rock, wall rocks
Nebengetriebe *n* power take-off
Nebengewinnungsanlage *f* by-recovery plant
Nebengleis *n* siding, side track
Nebengruppe *f (math)* coset
nebenhergehend side by side with, parallel with
Nebenhörer *m* bridging telephone
Nebenkanal *m* tributary canal
Nebenkeule *f* side (minimum, subsidiary) lobe *(aerial)*
Nebenkläger *m* co-plaintiff
Nebenklasse *f (math)* coset
Nebenklemme *f* subterminal
Nebenkosten *pl* incidental expenses, contingencies
Nebenlager *n* outboard (tail) bearing
Nebenlautsprecher *m* extension loudspeaker
Nebenleitung *f* side line
Nebenlinie *f* 1. *(ph)* subordinate line; 2. *(ry)* branch line
Nebenluft *f* admixed air
Nebenlufteinlaß *m* air-bleed opening
Nebenpleuel *m* auxiliary rod
Nebenprodukt *n* by-product
Nebenproduktofen *m* by-oven
Nebenquantenzahl *f* secondary (azimuthal) quantum number
Nebenreaktion *f* consequent (secondary, side) reaction
Nebenrechnung *f* subroutine *(computing)*
Nebenresonanz *f* spurious response
Nebenrippe *f* false rib
Nebenrohrleitung *f* secondary piping
Nebenschalter *m* auxiliary switch
Nebenschicht *f* back-shift
Nebenschluß *m (el)* shunt, by-pass
einfacher induktiver N. single inductive shunt
induktiver N. inductive shunt
kapazitiver N. capacitive shunt
magnetischer N. magnetic shunt
Nebenschlußbogenlampe *f* shunt arc lamp
Nebenschlußdämpfungswiderstand *m* shunt damping resistor
Nebenschlußerregung *f* shunt excitation
Nebenschlußfelderreger *m* shunt field rheostat
Nebenschlußgenerator *m* shunt-wound generator
Nebenschlußkollektormotor *m* shunt induction motor

Nebenschlußkommutatormotor *m* / **läufergespeister** shunt induction motor
Nebenschlußmotor *m* shunt-wound motor
kompoundierter N. stabilized shunt-wound motor
regelbarer N. brush-shifting motor
Nebenschlußregelung *f* shunt regulation
Nebenschlußregler *m* / **automatischer** automatic shunt regulating resistance
Nebenschlußrelais *n* shunt relay
Nebenschlußsteller *m* shunt field rheostat
Nebenschlußübergangsschaltung *f* shunt transition
Nebenschlußverhalten *n* shunt characteristics
Nebenschlußwicklung *f* shunt winding (coil)
Nebenschlußwiderstand *m* shunt (leak) resistance
Nebenschneide *f (mach)* trail (secondary cutting) edge
Nebensender *m* repeater transmitter; slave station *(decca)*
Nebenserie *f* / **erste** diffuse series *(spectrum)*
zweite N. sharp series *(spectrum)*
Nebenspannung *f* 1. *(build)* secondary stress; 2. spurious voltage
Nebensprechabstand *m* signal-to-crosstalk ratio
Nebensprechdämpfung *f* crosstalk attenuation
Nebensprechdämpfungsmesser *m* crosstalk meter
Nebensprechdämpfungsnetzwerk *n* anti-induction network
Nebensprechen *n* crosstalk
unverständliches N. unintelligible (inverted) crosstalk
verständliches N. intelligible (uninverted) crosstalk
Nebensprechkopplung *f* [transverse] crosstalk coupling
Nebenstelle *f* extension *(telephone)*
N. mit freier Amtswahl subscriber's extension station
amtsberechtigte N. subscriber's extension station
Nebenstellenanlage *f* private branch exchange
Nebenstellenschrank *m* private branch exchange switchboard
Nebenstellenzentrale *f* private branch exchange
Nebenstrahlung *f* spurious radiation
Nebenstraße *f* subsidiary (secondary, minor) road
Nebenstrecke *f (mine)* branch road
Nebenstromfilter *n* by-flow filter
Nebenstromkreis *m* auxiliary (branch) circuit, subcircuit
Nebenuhr *f* secondary (repeater, slave) clock
Nebenvalenz *f* auxiliary (secondary) valence
Nebenvalenzbindung *f* auxiliary (secondary) valence bond
Nebenvalenzkräfte *fpl* uncompensated attractions
Nebenwellenselektion *f* second channel rejection
Nebenwiderstand *m* parallel resistance, shunt
kombinierter N. universal shunt
Nebenwinkel *m* supplementary angle
Nebenwirkung *f* side effect
Nebenzeit *f (mach)* idle machine time, non-productive time, non-machining time, machine-handling time
N. beim Schleifen non-grinding time in the grinding cycle
Nebenzipfel *m* side (minor, secondary, parasitic) lobe *(aerial)*
Nebenzipfelfreiheit *f* freedom from secondary lobes
Nebler *m (ag)* fogger
Neck *m (geol)* neck
negativ negative
n. elektrisch electronegative
n. geladen negatively charged
Negativ *n* 1. negative *(photography)*; 2. *(el)* negative matrix
Negativbild *n* negative picture
Negativfilm *m* negative film stock
Negativklischee *n* reversed block
Negativlinse *f* negative lens
Negativmodulation *f* downward (negative) modulation
Negativretusche *f* negative retouching
Negaton *n s.* Negatron
Negator *m (dat)* inhibiting gate, negator, NOT-gate
Negatoreingang *m* negatory input; NOT input
Negatron *n* negatron
nehmen to take; to receive; to seize
in Betrieb n. to start up
in Fracht n. to hire a vessel
in Schlepp n. to tow, to take in tow
ins Schlepptau n. to take in tow
Kurs n. auf to steer for
neigbar tiltable, inclinable
neigen to rake, to incline, to slope; *(mach)* to tilt, to incline
n. zu to tend to
Neigung *f* inclination, incline, acclivity, slope, pitch; gradient; rake, batter; *(mach)* tilt, inclination; fall; *(mine)* dip; hade *(mineral vein)*
N. der Bahnebene inclination of orbit
N. der Strebfront *(mine)* gradient on face
Neigungsachse *f* axis of inclination
Neigungsgrad *m* degree of inclination
Neigungskompaß *m* inclination (dipping) compass
Neigungsmesser *m* [in]clinometer
Neigungsstabilität *f (aut)* cornering stability
Neigungswaage *f* inclination balance
Neigungswinkel *m* angle of inclination; tilt angle

(wobble plate); *(mach)* angle of inclination of the cutting edge, angle of the cutting edge with the line of travel of the work *(tool)*

N. der elastischen Linie angle of inclination of the elastic line *(turbine)*

N. der Regelscheibe angle of inclination of the control (regulating) wheel axis

N. des Geländes degree of terrain sloping

N. gegen die Senkrechte angle of inclination to the vertical

N. zweier Flächen interfacial angle

Neigungszeiger *m* bank indicator

Nelkenöl *n* cloves oil

NE-Metall *n* non-ferrous metal

NE-Metallblech *n* non-ferrous metal sheet

Nennableitvermögen *n* rated (maximum permissible) arrester current

Nennabmaß *n* variation [between basic dimension and limit], maximum variation

oberes N. maximum variation between basic dimension and maximum limit

unteres N. maximum variation between basic dimension and minimum limit

Nennausschaltleistung *f* rated interrupting (breaking) capacity

Nennausschaltstrom *m* rated breaking current

Nennausschaltvermögen *n* rated breaking capacity

Nennbelastung *f* **beim Bruch** nominal stress at fracture

Nennbetrieb *m* rated duty *(motors)*; rated operation *(switchgear)*

Nennbürde *f* rated burden *(instrument transformer)*

Nenndaten *pl* / **elektrische** ratings, rated data

Nenndauerleistung *f* continuous rating

Nenndauerstrom *m* rated continuous current

Nenndrehmoment *n* torque at rated load [of a motor]

Nenndrehzahl *f* rated speed

Nenndurchmesser *m* nominal diameter

N. des Gewindes nominal diameter of [the] thread

Nenneinschaltstrom *m* rated making-current, making current rating

Nenneinschaltvermögen *n* rated making capacity

Nenner *m* denominator

gemeinsamer N. common denominator

kleinster gemeinsamer N. lowest common multiple

Nennfrequenz *f* rated (nominal, design) frequency

Nenngrenzfrequenz *f* nominal cut-off frequency

Nenngrößen *fpl* ratings

Nennkurzschlußstrom *m* rated short-circuit current

Nennkurzzeitstrom *m* rated short-time current, momentary current rating, *(Am)* thermal current rating

Nennlast *f* 1. *(el)* rated load; rated power, power rating; 2. *(mine)* nominal load-bearing capacity, rated yield load *(support)*

Nennlastausschaltstrom *m* load-break current rating

Nennleistung *f* [nominal] rating; nominal output (capacity, horsepower), rated output (capacity, horsepower)

Nennmaß *n* basic size (dimension) *(tolerance)*, nominal size (dimension)

Nennmeßbereich *m* **eines Instruments** rating of an instrument

Nennquerschnitt *m* rated cross-section *(e.g. of conductors)*

Nennspannung *f* 1. *(el)* rated [circuit] voltage, nominal voltage rating; 2. nominal stress *(statics)*

primäre N. rated primary voltage

Nennstrom *m* rated (nominal) current

Nennstrombelastung *f* rated current load

Nennstromstärke *f* current rating

Nennweite *f* nominal diameter

N. des Rohrs nominal bore (pipe size)

Nennwert *m* nominal value

Neodym *n* neodymium

Neon *n* neon

Neonfüllung *f* neon charge

Neonlampe *f* neon lamp

Neonröhre *f* neon tube

Neopren *n* neoprene, duprene

Neper *n* neper

Nephelinbasalt *m* nepheline-basalt

Nephelometer *n* nephelometer

Nephelometrie *f* nephelometric (turbidimetric) analysis, nephelometry

nephelometrisch nephelometric, turbidimetric

Neptunium *n* neptunium

Neptunium[zerfalls]reihe *f* neptunium series

neritisch neritic

Nernst-Brenner *m* Nernst burner

Nernst-Lampe *f* Nernst lamp

Nernst-Stäbchen *n s.* Nernst-Stift

Nernst-Stift *m* Nernst filament (glower)

Nerol *n* nerol

Neroliöl *n* neroli oil

Nervenbetäubungsmittel *n* nerve narcotic

Nervenelektrizität *f* neuroelectricity

Nervengift *n* paralyzant

Nervenimpuls *m* nerve impulse

Nerzfell *n* mink

Nesselfaser *f* nettle fibre

Nest *n* 1. pocket *(concrete)*; 2. *(nucl)* cluster

Nestei *n* nest egg

Nettogewicht *n s.* Nettomasse

Nettomasse *f* net weight

Nettoproduktion *f* net production

Nettoquerschnitt *m* net section
Nettoraumgehalt *m* net tonnage *(register ton)*
Nettoregistertonne *f* net [register] ton
Nettoregistertonnengehalt *m* registered tonnage
Nettotonnage *f* net tonnage
Nettotonnengehalt *m* net tonnage
Nettotonnenkilometer *m* net ton-kilometre
Nettovermessung *f* net tonnage
Netz *n* 1. *(el)* network, [electric] mains, system, electric supply line, supply; 2. *(text)* netting; 3. net; grid; lattice
  N. mit freiem Sternpunkt system with floating neutral
  N. mit geerdetem Nullpunkt system with earthed neutral
  N. mit isoliertem Sternpunkt system with insulated neutral
  N. mit verdrillten Leitungen network with interposed phases
  geerdetes N. earthed system
  gelöschtes N. *s.* kompensiertes N.
  isoliertes N. earth-free network
  kompensiertes N. compensated network
  starres N. rigid network
  stehendes N. stationary (fixed) net *(fishery)*
  vermaschtes N. *s.* Maschennetz
  Wulffsches N. Wulff net
Netzabschaltung *f* mains disconnection
Netzanalysator *m* network analyzer
Netzanode *f* grid anode
Net[illegible]ains (main-circuit) connection, mains supply
Netzanschlußempfänger *m* mains receiver, main receiving set
Netzanschlußgerät *n* power supply unit (set), mains set, power pack, mains (socket) power-unit
Netzanschlußkabel *n* power supply cable
Netzanschlußschnur *f* line cord
Netzanschlußteil *n* power supply unit, power pack
Netzanschlußwerte *mpl* power supply requirements
Netzantenne *f* mains aerial
Netzausfall *m* mains (power) failure
Netzbad *n (text)* wetting bath
Netzberechnung *f* network study
netzbetrieben mains-operated
Netzbewehrung *f* wire-mesh reinforcement, steel fabric reinforcement *(concrete)*
Netzbrumm *m* power-line hum, alternating-current hum
Netzbrummen *n* power-line hum, alternating-current hum
Netzdose *f* wall socket
Netzdrossel *f* line (ripple-filter) choke
Netzebene *f* 1. network plane *(telephony)*; 2. lattice plane *(crystallography)*
Netzebenenabstand *m* space-lattice distance
Netzegge *f* chain harrow
Netzeinholsystem *n* hauling system *(fishery)*
Netzeinschub *m* electrical mains plug-in unit
Netzeinschwingungsspannung *f* system transient voltage
Netzelektrode *f* wire gauze-electrode, net-shaped electrode
netzen to wet
Netzerdung *f* network earthing
Netzersatzanlage *f* emergency power system (plant)
Netzfilter *n* hum eliminator
Netzfischer *m s.* Netzleger
Netzflimmern *n* power-line ripple
Netzfrequenz *f* line (power, industrial) frequency
Netzgaze *f* net gauze
Netzgebiet *n (math)* network region
netzgebunden mains-borne
Netzgewölbe *n* reticulated vault[ing]
Netzgliederung *f* network structure
Netzgrund *m (text)* lace ground
Netzgruppenknoten *m* network-junction exchange *(telephony)*
Netzgruppenmittelpunkt *m* network centre
Netzgruppenschalter *m* network-group switch
Netzhautbeleuchtung *f* retinal illumination
Netzhautgrube *f* fovea centralis
Netzhautkamera *f* retinal camera
Netzheber *m* net lifter *(fishery)*
Netzimpedanz *f* network impedance
Netzkennlinie *f* load-governing characteristic
  natürliche N. natural load governing characteristic
Netzknotenpunkt *m* network centre (junction-point)
Netzknüpfmaschine *f* netting machine
Netzkuppelschalter *m* mains-coupling circuit breaker
Netzkurzschlußstrom *m* prospective current of a circuit
Netzleger *m* netlayer, netter *(ship)*
Netzmantelelektrode *f* fusarc electrode
Netzmantelelektrodenschweißen *n* fusarc welding [process]
Netzmikrometer *n* grid micrometer
Netzmittel *n* wetting agent; *(ch)* interlacing agent
Netzmodell *n (dat)* network analogue computer, circuit analyzer; *(el)* simulated network
Netzobjektträger *m* disk of wire screen
Netzöl *n* wetting oil
Netzpapier *n* coordinate paper
Netzplan *m* network
Netzpräparat *n* wetting formulation
Netzpunkt *m (math)* net point
Netzraster *m* embossing *(colour film)*
Netzregler *m* line-voltage regulator
Netzreihe *f (text)* net course

Netzschalter *m* mains (power-supply) switch, line breaker
Netzschlinge *f* cargo net, net sling *(fishery)*
Netzschnur *f* mains input lead
lösbare N. detachable power cord
Netzschutzschalter *m* mains protective switch; network protector
Netzschwankung *f* mains[-voltage] fluctuation
Netzsicherung *f* mains fuse
Netzsignalform *f* mains line waveform
Netzspannung *f* mains (power line) voltage
Netzspannungskonstanthalter *m* line regulator
Netzspannungskorrektion *f* mains-voltage compensation
Netzspannungsmesser *m* line voltmeter
Netzspannungsregler *m* line-voltage regulator
Netzspannungsschwankung *f* line variation, fluctuation in the main voltage
Netzspannungsumschalter *m* mains (line-voltage) selector
Netzspannungswähler *m* mains tapping panel
Netzspannungswahlschalter *m s.* Netzspannungsumschalter
Netzspeiseleitung *f* network feeder
Netzspeisung *f* **mit vorgegebenen Daten** rated-power supply
Netzsperre *f* mains filter
Netzspitze *f (text)* net lace
Netzsteckdose *f* wall socket
Netzstecker *m* mains (power) plug
Netzsternpunkt *m* system neutral
Netzstert *m (ship)* cod-end
Netzstrom *m* mains (line, supply) current
Netzstropp *m (ship)* cargo net, net sling
Netzstruktur *f (met)* network structure
Netzsynchronisation *f* mains hold locking
netzsynchronisiert locked to mains-frequency
Netzsynchronisierung *f* **in der Fernsehtechnik** vertical synchronization employing power-line frequency
Netztafel *f (math)* net chart
Netzteil *m* power supply, supply unit
Netztransformator *m* mains (distribution) transformer
Netzübertragungsverluste *mpl* mains leakage
Netzumspanner *m s.* Netztransformator
Netzumspannunterwerk *n* distribution sub-station
Netzunterbrechung *f* mains failing
Netzunterspannung *f* low-line voltage
Netzuntersuchung *f* network investigation
Netzverluste *mpl* mains leakage
Netzverriegelung *f* filtering of power leads
Netzverteiler *m* main distributor
Netzwerk *n* 1. *(el, plast)* network; 2. *(text)* mesh, net
aktives N. active network
differenzierendes N. differentiating network
dreipoliges N. three-terminal network
duales N. dual network
ebenes N. planar network
integrierendes N. integrating network
passives N. passive network
symmetrisches N. ladder (series-shunt) network
Netzwerkanalogierechner *m* network analogue computer
Netzwerkanalysator *m* circuit analyzer
Netzwerkkuppel *f* network dome
Netzwerkstruktur *f* cancellated structure
Netzwiderstand *m* network impedance
Netzwinde *f* trawl winch *(fishery)*
Netzzusammenschaltung *f* network interconnection (coupling)
neu auflegen to re-edit
n. entdeckt newly discovered
n. konstruieren to re-design, to design anew
Neuanschaffung *f* replacement; acquisition
Neuaufgabe *f* reforwarding
Neuauflage *f* re-edition
Neuaufspannung *f* re-clamping *(work)*; resetting
Neuaufstellung *f (mach)* rearranging the layout
Neuausmauerung *f* re-bricking
Neuausrüstung *f* re-equipment
Neuausstattung *f s.* Neuausrüstung
Neubau *m* new building, newly built house
fertiggestellter N. completion *(of a vessel)*
Neudruck *m* reprint, rerun
Neueinstellung *f* readjustment
Neuentwicklung *f (mach)* newly-developed machine
Neuerer *m* innovator
Neuerung *f* innovation
Neugrad *m* gon, grade *(1/100 of a right angle)*
Neuheitsprüfung *f* **einer Patentanmeldung** examination of a patent application for novelty
Neuklotzen *n (text)* re-padding
Neulandgewinnung *f* land reclamation, reclamation of land
Neumessing *n* yellow metal
Neumungo *m (text)* mungo from new rags
Neuneck *n* nonagon
Neunerkomplement *n (dat)* nine's complement
Neunpunktekreis *m* nine-point circle
Neunzehneck *n* nonadecagon
Neuordnung *f* rearrangement *(crystals)*
Neuristor *m* neuristor
Neuron *n* neuron
künstliches N. artificial neuron, information processing cell
Neusand *m* new (fresh) sand *(foundry)*
Neusilber *n* alpaca, nickel brass (silver), argentan, German silver
neutral neutral
elektrisch n. electrically neutral

Neutralglas *n* neutral filter
Neutralisation *f (ch)* neutralization
  induktive N. shunt (inductive) neutralization
Neutralisationskapazität *f* balancing (neutralizing) capacity
Neutralisationskondensator *m* balancing (neutralizing) capacitor
Neutralisationswärme *f* heat of neutralization
Neutralisationszahl *f* neutralization number
Neutralisator *m* neutralizer
neutralisieren *(ch)* to neutralize; *(el)* to compensate
Neutralisierung *f* neutralization
Neutralöl *n* neutral oil
Neutralsalz *n* normal salt
Neutretto *n* neutretto, neutral meson
Neutrino *n* neutrino
Neutrodenschirm *m* neutrode shield
Neutrodynempfänger *m* neutrodyne receiver
Neutrodynschaltung *f* neutrodyne circuit
Neutron *n* neutron
  epithermisches N. epithermal neutron
  freies N. free neutron
  gebremstes N. moderated neutron
  intermediäres N. *s.* mittelschnelles N.
  kaltes N. cold neutron $(E_m < 0{,}0018\ eV)$
  langsames N. slow neutron; thermal neutron
  mittelschnelles N. intermediate neutron
  polarisiertes N. polarized neutron
  promptes N. prompt [fission] neutron
  schnelles N. fast neutron
  thermisches N. thermal neutron
  unterthermisches N. *s.* kaltes N.
  verzögertes N. delayed neutron
Neutronenabfangstab *m s.* Absorberstab
Neutronenabsorber *m* neutron absorber (sponge), neutron-absorbing material
Neutronenabsorption *f* neutron absorption
Neutronenaktivierungsanalyse *f* neutron activation analysis
Neutronenalter *n* neutron age
neutronenarm neutron-deficient
Neutronenausfluß *m* [neutron] leakage
Neutronenbeschuß *m* neutron bombardment (irradiation), bombardment with neutrons
Neutronenbestrahlung *f* neutron bombardment (irradiation), bombardment with neutrons
Neutronenbeugung *f* neutron diffraction
Neutronenbilanz *f* neutron balance
Neutronenbreite *f* neutron width
Neutronenbremsung *f* retardation (moderation) of neutrons; thermalization
Neutronendefizit *n* neutron deficiency
  N. / mit neutron-deficient
Neutronendetektor *m* neutron detector
Neutronendichte *f* neutron density
Neutronendiffraktometer *n* neutron diffractometer
Neutronendiffusion *f* neutron diffusion
Neutronendosimeter *n* neutron dosimeter
Neutronendosimetrie *f* neutron dosimetry
Neutronendosis *f* neutron dose
Neutroneneinfang *m* neutron capture
Neutroneneinfluß *m* [neutron] importance
neutronenempfindlich neutron-sensitive
Neutronenenergie *f* neutron energy
Neutronenfänger *m* neutron absorber
Neutronenfluß *m* neutron flux
  schneller N. fast [neutron] flux
  thermischer N. thermal [neutron] flux
Neutronengas *n* neutron gas
Neutronengeneration *f* generation of neutrons
Neutronengenerator *m* neutron generator
Neutronenkammer *f* neutron chamber
Neutronenkonverter *m* neutron converter
Neutronenmonochromator *m* neutron monochromator, neutron [velocity] selector, chopper
Neutronenökonomie *f* neutron economy
Neutronenoptik *f* neutron optics
Neutronenphysik *f* neutron physics
Neutronenquelle *f* neutron source
Neutronenquerschnitt *m* neutron cross-section
Neutronenreflektor *m* neutron reflector
neutronenreich neutron-rich
Neutronenresonanz *f* neutron resonance
Neutronenschale *f* neutron shell
Neutronenschild *m* neutron shield
Neutronenschutz *m* neutron shield
Neutronenspektrometer *n* neutron spectrometer
Neutronenspektrometrie *f* neutron spectrometry
Neutronenspektrum *n* neutron spectrum
Neutronenstrahl *m* neutron beam; neutron ray
Neutronenstrahler *m* neutron emitter
Neutronenstrahlung *f* neutron radiation
Neutronentemperatur *f* neutron temperature
Neutronentherapie *f* neutron therapy
Neutronentransport *m* neutron transport
Neutronenüberschuß *m* neutron excess [number], isotopic (difference) number
  N. / mit neutron-rich
Neutronenverdampfung *f* neutron evaporation
Neutronenvergiftung *f* neutron poisoning (poison effect)
Neutronenverlust *m* [neutron] leakage
Neutronenvermehrung *f* neutron multiplication
Neutronenverzögerung *f* retardation of neutrons
Neutronenwelle *f* neutron wave
Neutronenwirkungsquerschnitt *m* neutron cross-section
Neutronenzahl *f* **im Kern** neutron number
Neutronenzähler *m* neutron counter
Neutronenzählrohr *n* neutron counter
Neutronenzyklus *m* neutron cycle
neutrophil neutrophil

Neuzustellung *f* re-lining, fresh lining *(furnace)*
NE-Werkstoff *m* non-ferrous material
Newton *n* newton
NF *s.* Niederfrequenz
Nichols-Diagramm *n* Nichols' (log-magnitude-angle) diagram
Nichrothermalstahl *m* nichrotherm steel
nicht beide not both, Sheffer stroke *(logical operator)*
nichtabschmelzbar non-consumable *(electrode)*
nichtaktiv *(nucl)* inactive; cold; dry
nichtangesteuert undriven *(valve)*
nichtätzend non-corrosive
nichtausgehärtet *(plast)* undercure
Nichtausübung *f* **eines Patentes** non-working of a patent
Nichtbefolgung *f* non-observance
Nichtbezahlung *f* non-payment
nichtbindig cohesionless, non-cohesive, granular
nichtdiffundiert undiffused, non-diffused
nichtdruckend non-printing
nichtdurchgehärtet *(plast)* cokey
Nichtedelmetall *n* base metal
Nichteisenmetall *n* non-ferrous metal
Nichtelektrolyt *m* non-electrolyte
nichtentflammbar flame-resistant
nichtentzündbar non-inflammable
Nichterfüllung *f* non-compliance *(contract)*, non-performance; non-fulfilment *(plan)*
Nichterhaltung *f* **der Parität** *(nucl)* non-conservation of parity
nichterregt unenergized
nichteuklidisch non-Euclidian
Nichtfachmann *m* inexpert, non-expert
Nichtfestigkeitsschott *n* *(ship)* partition (non-structural) bulkhead
nichtfluchtend misaligned
nichtflüchtig non-volatile
nichtfokussiert unfocused
nichtgelenkig rigid *(joint)*
NICHT-Glied *n* NOT-gate
nichthärtbar non-hardening
nichthärtend non-hardening *(steel)*
nichtionisiert unionized
nichtisotop non-isotopic, heterotopic
Nichtkarbonathärte *f* non-carbonate hardness
nichtkorrodierend non-corrosive
nichtleitend non-conducting
Nichtleiter *m* non-conductor, dielectric
  absoluter N. ideal (perfect) dielectric
  vollständiger N. ideal (perfect) dielectric
nichtlinear non-linear
Nichtlinearität *f* non-linearity
nichtmagnetisierbar non magnetizing
Nichtmetall *n* non-metal
nichtmetallisch non-metallic
Nichtnullsummenspiel *n* *(dat)* non-zero-sum game

nichtoxydierbar inoxidizable
nichtparallel non-parallel
nichtpolarisiert unpolarized
nichtradioaktiv stable, inactive, non-radioactive; cold
Nichtraucherabteil *n* non-smoking compartment
nichtredundant *(dat)* irredundant
nichtrostend non-rusting, antirost; stainless *(steel)*
Nichtrückkehr *f* **zu Null** *(dat)* non-return to zero
nichtrund non-circular
nichtsaugend non-absorbent
nichtschaltbar permanent *(joint)*
NICHT-Schaltung *f* *(dat)* inhibiting gate, NOT-gate
nichtschmelzbar *(ch)* infusible
nichtschwarz non-black *(radiator)*
nichtspaltbar non-fissionable, non-fissile
nichtspaltfähig non-fissionable, non-fissile
nichtsplitternd unsplinterable
nichtstabil unstable
nichtstabilisiert unstabilized *(steel)*
nichtstationär non-steady, unsteady *(flow)*
nichtsteuerbar uncontrolled
Nichtstrickstellung *f* float position *(knitting machine)*
nichttragend *(build)* non-[load]bearing, non-structural
nichtumkehrbar irreversible, non-reversible
NICHT-UND NAND, NOT AND *(logical operator)*
NICHT-UND-Schaltung *f* NOT-AND circuit *(logistics)*
NICHT-UND-Verknüpfung *f* NOT-AND connection *(logistics)*
nichtvergasbar non-gasifiable
nichtvergilbend non-yellowing
Nichtvorhandensein *n* absence, lack
nichtwirbelnd non-turbulent
Nickel *n* nickel
Nickelantimonglanz *m* nickel antimony glance, ullmanite
Nickelarsenkies *m* nickel arsenic glance, gersdorffite
Nickelbad *n* nickel bath
Nickelblech *n* nickel sheet
Nickelblende *f* capillary pyrite
Nickelblüte *f* nickel bloom, annabergite
Nickel-Eisen-Akku *m* Ni-Fe accumulator, nickel-iron-alkaline accumulator
Nickel-Eisen-Zelle *f* nickel-iron-alkaline cell
Nickelerz *n* nickel ore
nickelhaltig nickeliferous
Nickel-Kadmium-Akkumulator *m* cadmium-nickel storage battery
Nickelmünze *f* nickel coin
nickelplattiert nickel-clad

Nickelröhre *f* nickel tube
Nickelspeise *f* nickel speiss
Nickelstahl *m* nickel steel
Nickeltetrakarbonyl *n* nickel carbonyl
Nickel(II)-verbindung *f* nickelous (niccolous) compound
Nickel(III)-verbindung *f* nickelic (niccolic) compound
Nicol-Analysator *m* Nicol analyzer
Niederbordschiff *n* low-built ship
niederbrechen to break down
niederbringen to sink *(well, shaft)*
Niederbringen *n* **eines Bohrlochs** sinking a borehole
  N. eines Schachtes shaft sinking
Niederdruck *m* low pressure
Niederdruckakkumulator *m* / **pneumatischer** low-pressure air accumulator
Niederdruckdampfkessel *m* low-pressure boiler
Niederdruckdampfmaschine *f* low-pressure steam engine
Niederdruckdeckel *m* low-pressure cover
Niederdruckdiagramm *n* low-pressure diagram
niederdrücken to force (press) down; to depress
Niederdruckkammer *f* low-pressure chamber
Niederdruckkessel *m* low-pressure boiler
Niederdruckkurbel *f* low-pressure crank
Niederdrucknebelkammer *f* low-pressure cloud chamber
Niederdruckpresse *f* low-density baler
Niederdruckpreßverfahren *n (plast)* contact [pressure] moulding
Niederdruckpumpe *f* low-pressure pump
Niederdruckrad *n* low-pressure disk *(turbine)*
Niederdruckraum *m* low-pressure region
Niederdrucksammelpresse *f* low-density pick-up baler
Niederdruckspritzverfahren *n (plast)* contact [pressure] spraying
Niederdruckstufe *f* low-pressure stage *(turbine)*
Niederdruckzylinder *m* low-pressure cylinder *(steam engine)*
Niederdurchforstung *f* thinning in the suppressed wood
Niederenergiephysik *f* low-energy physics
niederentern *(ship)* to come down from aloft, to go down
Niederflußreaktor *m* low-flux reactor
Niederfrequenz *f* low frequency; audio-frequency
Niederfrequenzausgang *m* audio output channel
Niederfrequenzband *n* audio-frequency band
Niederfrequenzbereich *m* audio-frequency range
Niederfrequenzempfang *m* reception in the low-frequency band
Niederfrequenzerregung *f* low-frequency; excitation
Niederfrequenzfernsprechen *n* audio (voice) frequency telephony
Niederfrequenzofen *m* low-frequency furnace
Niederfrequenzprüfung *f* baseband testing
Niederfrequenzrufsatz *m* low-frequency ringer
Niederfrequenzrufumsetzer *m* ringing repeater
Niederfrequenzsperre *f* low-frequency rejection
Niederfrequenzsperrkreis *m* high-pass selective circuit
Niederfrequenztransformator *m* low-frequency transformer
Niederfrequenzverstärker *m* audio (low-frequency, speech) amplifier
Niederfrequenzverstärkung *f* audio (low-frequency) amplification
Niederfrequenzwiderstand *m* low-frequency resistance
Niederführung *f* down lead *(aerial)*
  abgeschirmte N. screened down-lead
Niedergang *m* 1. *(ship)* companion-way; 2. *(mach)* downward stroke
Niedergangsluke *f* companion hatch, ladderway
Niedergangstreppe *f* companion ladder
niedergehen *(mach)* to descend, to move downward
Niederhalter *m* hold-down, blank holder *(metal forming)*; toe dog *(clamping)*
Niederhubwagen *m* pallet truck
Niederlassen *n (mach)* lowering
niederlegen to lower, to hinge down *(e.g. mast)*
Niederleistungskernkraftwerk *n* low-power atomic energy plant
Niederleistungsreaktor *m* low-power reactor
niederohmig low-resistance, low-resistive
Niederrahmen *m (aut)* drop frame
Niederrahmenfahrgestell *n* low-frame chassis, low-built chassis
Niederschachtofen *m* low-shaft furnace
Niederschlag *m* 1. *(ch)* precipitate; deposit; sediment; fallout; 2. rainfall
  angebrannter N. burnt deposit
  elektrolytischer N. electrolytic precipitation
  galvanischer N. electro-deposit
  käsiger N. curds, curdy precipitation
  kristallinischer N. crystalline deposit
  radioaktiver N. radioactive fallout; [radio]-active deposit
Niederschlagelektrode *f* precipitation electrode
niederschlagen *(ch)* to deposit, to precipitate
Niederschlaggefäß *n* condensing vessel
Niederschlagsarbeit *f* precipitation process *(lead)*
Niederschlagsgebiet *n* catchment (drainage) area, *(Am)* watershed
Niederschlagsmenge *f* amount of rainfall
Niederschlagsverteilungsverhältnis *n* metal distribution ratio
Niederschlagswasser *n* atmospheric water
Niederschlagverlust *m* loss due to condensation
Niederschlagwasser *n* water of condensation
Niederspannung *f* low voltage (tension)

Niederspannungsanlage *f* low-voltage installation

Niederspannungsisolator *m* low-voltage insulator

Niederspannungskabel *n* low-voltage cable

Niederspannungskraftanlage *f* low-voltage power station (plant)

Niederspannungsmaschennetz *n* low-voltage meshed system

Niederspannungsofen *m* low-voltage resistor furnace

Niederspannungsröntgentherapie *f* low-voltage roentgen therapy

Niederspannungsschalter *m* low-voltage switch

Niederspannungsschaltgerät *n* low-voltage switchgear

niederspannungsseitig secondary *(e.g. of transformers)*

Niederspannungsstromkreis *m* low-tension circuit

Niederspannungsverteilung *f* low-voltage distribution

Niederspannungsverteilungsnetz *n* secondary distribution system

Niederspannungs-Verteilungssammelschiene *f* secondary distribution bus-bar

Niederspannungszündmaschine *f* low-voltage magneto

Niederstamm *m* dwarf-standard

Niedertemperaturreaktor *m* low-temperature reactor

Niederung *f* low ground (land)

Niedervakuumröhre *f* low-vacuum tube

Niedervoltelektrolytkondensator *m* low-voltage electrolytic capacitor

Niedervoltlampe *f* low-voltage lamp *(up to 24 volts)*

Niederwald *m* low forest

niedriggekohlt low-carbon

niedriginkohlt low-rank

niedriglegiert low-alloy[ed]

niedrigschmelzend low-melting

niedrigsiedend low-boiling

Niedrigtemperaturdestillation *f* low-temperature rectification

Niedrigwasser *n* low water (tide)

Niedrigwasserstand *m* height of low water

Niedrigwasserstandrufer *m* low-water alarm

Niello *n* niello

Nierencharakteristik *f* cardioid diagram *(aerials)*

nierenförmig kidney-shaped, reniform

Nierenplattenkondensator *m* square-law capacitor

Niet *m* rivet

Nietabstand *m* rivet pitch

nieten to rivet

Nietendöpper *m* rivet snap

Nietenschlagen *n* rivet driving

Niethammer *m* rivet[ing] hammer

  pneumatischer N. air riveter (riveting hammer)

Nietkopfsetzer *m* rivet set

Nietloch *n* rivet hole

Nietlochreibahle *f* structural reamer

Nietmaschine *f* riveting machine

Nietstahl *m* rivet steel

Nietteilung *f* rivet pitch, pitch of rivets

Nietverbindung *f* riveted joint (connection), rivet fastening

  dreireihige N. riveted joint with three rows of rivets

  einreihige N. rivetet joint with single row of rivets

  vierreihige N. riveted joint with four rows of rivets

  zweireihige N. riveted joint with double row of rivets

Nife-Akkumulator *m* storage battery of iron-nickel type

Nigrosin *n* nigrosine

Nikotin *n* nicotine

Nikotinpräparat *n* nicotine preparation

Nikotinsäure *f* nicotinic acid

Nikotinsäureamid *n* nicotinic amide

Niob *n* columbium, niobium

Niobat *n* niobate

Niobit *m (mine)* columbite

Niobsäure *f* columbic (niobic) acid

Nipkow-Scheibe *f* Nipkow disk

Nippel *m* nipple

Nipphochwasser *n* high-water neap

Nipptide *f* neap-tide

Nipptidenhub *m* neap range

Nische *f* recess, niche, alcove

Nitramid *n* nitramide

Nitrat *n* nitrate

Nitratseide *f* nitro-cellulose rayon

Nitratsprengstoff *m* nitrate explosive

Nitrid *n* nitride

nitrierbar nitratable, nitrifiable

nitrieren to nitride *(steel)*; to nitrify

nitrierfähig nitriding *(steel)*

nitrierhärten to nitride, to nitrogen-harden

Nitrierhärten *n* nitriding, nitrogen case-hardening

Nitrierhärteverfahren *n* nitriding (nitrogen case-hardening) process

Nitrierofen *m* nitriding furnace

Nitriersäure *f* nitrating (mixed) acid

Nitrierstahl *m* nitrided (nitralloy) steel

Nitrierung *f* 1. *(ch)* nitrification; 2. nitriding *(case-hardening)*

Nitrifikationsbakterien *fpl* nitrifying bacteria

Nitril *n* nitrile

Nitrit *n* nitrite

Nitrobenzol *n* nitrobenzene
Nitroglyzerin *n* nitroglycerin[e]
Nitrokörper *m* nitro-compound
Nitrolsäure *f* nitrolic acid
Nitronsäure *f* nitronic acid
Nitrosogruppe *f* nitroso group
Nitrosoverbindung *f* nitroso compound
Nitrosylschwefelsäure *f* nitrosulphuric acid, chamber crystals
Nitrozellulose *f* nitrocellulose
Nitrozelluloselack *m* nitrocellulose dope, pyroxylin lacquer
Nitschelfläche *f* rubbing surface
Nitschelhose *f* leather hose, rubber (rubbing) apron, rubbing leather
  geschnittene N. rubbing leather with cut
  kreuzgerippte N. cross-ribbed rubbing leather
nitscheln to rub slubbing into a cylindrical formation
Nitschelwalze *f* rubber gear, rubbing condenser (roller), traversing condenser roller
Nitschelzeug *n* rubber gear
Nitschler *m* rubber gear
Niveau *n* level
  angeregtes N. excited level
  stabiles N. stable level
Niveauabstand *m* level spacing
Niveaubreite *f* level width
Niveaufläche *f* equipotential surface
Niveauflasche *f* levelling bottle
niveaugleich at the same level
Niveauschema *n* energy level diagram, term diagram (scheme)
Niveaustandsmesser *m* level gauge (transmitter)
Niveaustandszeiger *m* level indicator
Niveauunterschied *m* difference of level
Nivellement *n* levelling [survey], boning
  N. hoher Präzision high-precision levelling
  N. unter Tage underground-mine levelling
nivellieren to level, to bone
Nivellieren *n* **aus der Mitte** levelling midway between two points
  markscheiderisches N. levelling underground
Nivelliergerät *n* [surveyor's] level
Nivellierinstrument *n* levelling instrument
Nivellierlatte *f* levelling staff, level rod (pole)
Nivellierlibelle *f* telescope level
Nivellierstab *m* levelling (boning) rod
Nivellierung *f* levelling, boning-in
Nobelium *n* nobelium
Nobelpreis *m* Nobel prize
Nobelpreisträger *m* Nobel prizewinner (laureate)
Nocke *f* cam; boss; dog
Nocken *m* cam
  harmonischer N. harmonic cam
Nockenbahn *f* cam travel (contact surface)
nockenbetätigt cam-actuated, dog-actuated
Nockenbuckel *m* cam lobe
Nockenerhebungszeit *f* cam dwell, dwell of cam
Nockenfahrschalter *m* cam controller
Nockenflanke *f* cam contact surface
Nockenform *f* cam contour
Nockenfräsen *n* cam milling
Nockenfräsmaschine *f* cam-cutting machine
nockengesteuert cam-controlled, dog-controlled
Nockenhub *m* cam lift
Nockenkontakt *m* cam[-actuated] contact
Nockenlaufbahn *f* cam path (track)
Nockenmeisterschalter *m* cam-controlled master switch
Nockenprofil *n* cam profile
Nockenrelais *n* cam-operated relay
Nockenschalter *m* cam-operated switch, camshaft controller
Nockenschaltwerk *n* camshaft gear
Nockenschar *n* cam-share
Nockenscheibe *f* cam disk (wheel)
Nockenschleifen *n* cam grinding
Nockensteuerung *f* *(mach)* cam control; *(aut)* cam gear
  N. mit Hilfsmotor camshaft and servo-motor equipment
Nockenstößel *m* cam follower
Nockenverweilzeit *f* cam dwell
Nockenwelle *f* camshaft
  obenliegende N. high-level camshaft
Nockenwellenantriebsrad *n* camshaft timing gear
Nockenwellengehäuse *n* camshaft casing
Nockenwellengetriebe *n* camshaft gears
Nockenwellenkettenantrieb *m* camshaft chain drive
Nockenwellenkettenrad *n* camshaft sprocket
Nockenwellenmeßeinrichtung *f* camshaft testing element
Nockenwellenmeßgerät *n* camshaft tester
Nockenwellenrolle *f* camshaft roller
Nockenwellenschaltwerk *n* camshaft gear
Nockenwellenschleifmaschine *f* camshaft grinding machine
Nockhaus *n* *(ship)* wing shelter
Nöggerathmaschine *f* acyclic machine
Nomenklatur *f* nomenclature
Nominalfrequenz *f* nominal frequency
Nominalkonzentration *f* nominal concentration
nomografisch nomographic
Nomogramm *n* nomogram; alignment (nomographic) chart
Nonienteilung *f* vernier scale
Nonius *m* vernier
  N. mit Lupe optical vernier
Noniusnullpunkt *m* nonius zero
Noniusnullstrich *m* zero mark of the vernier scale, zero line of vernier, vernier zero

Nonne *f* concave tile *(roofing)*
Noppe *f (text)* burl; nap
Noppeisen *n* burling iron
noppen *(text)* to burl; to nap
Noppengarn *n* nap yarn
 falsches N. imitation nap yarn
Noppensetzverfahren *n* tufting *(carpet)*
Noppenstoff *m* nap fabric
Noppenstreuapparat *m* nap distribution apparatus
Noppmaschine *f* burling frame
Nopprahmen *m* burling frame
Noppzange *f* burling iron
Nord north
Norden *m* north
nördlich northern
Nordlicht *n* northern lights, aurora borealis
nordmagnetisch north-magnetic
Nordosten *m* northeast
Nordpol *m* north pole
Nordpunkt *m* north point
Nordwesten *m* northwest
NOR-Glied *n (dat)* NOR-gate
NOR-Schaltung *f* NOR-circuit
Norleuzin *n* norleucine
Norm *f* standard; quota
 handelsübliche N. commercial standard
 vorläufige N. tentative standard
normal normal
 n. gekrümmt isokurtic
Normal *n* comparison standard; prototype
 abgeleitetes N. secondary standard
Normalabweichung *f* standard deviation
Normalanker *m* stocked (standard) anchor
Normalatmosphäre *f* normal (standard) atmosphere
 N. für Prüfungen atmospheric standard conditions
Normalatombombe *f s.* nominelle Atombombe
Normalausrüstung *f (mach)* standard equipment
Normalbatterie *f* standard cell
Normalbedingung *f* normal temperature and pressure
Normalbelastung *f* standard loading
Normalbenzin *n* regular grade of petrol
Normalbeschleunigung *f* normal acceleration
Normalbogen *m (el)* standard bend [of conduit]
Normaldosis *f* normal dose
Normaldrehkondensator *m* variable capacitance standard
Normaldrehmoment *n* normal torque
Normaldrehung *f* standard twist
Normaldruck *m* normal pressure; pressure normal to blade surface *(turbine)*
Normaldruckzählrohr *n* atmospheric-pressure counter
Normale *f* normal, perpendicular
Normalelektrode *f* normal electrode
Normalelement *n* standard cell
Normalfassung *f (opt)* standard mount; standard lamp-holder
Normalfernhörer *m* standard receiver
Normalform *f* / **disjunktive** disjunctive normal form *(logistics)*
 konjunktive N. conjunctive normal form *(logistics)*
Normalformat *n* standard size
Normalfrequenz *f* standard frequency
Normalfrequenzgenerator *m* standard frequency generator
Normalfrequenzsender *m* standard frequency station
Normalfüllung *f* normal admission (cut-off) *(steam)*
Normalgattierung *f* standard burden
Normalgenerator *m* standard-level generator
Normalgeometrie *f* standard geometry
Normalgeschwindigkeit *f* normal speed; service speed
Normalglühen *n* normalizing
Normalinstrument *n* reference instrument
normalisieren to normalize
Normalität *f (ch)* normality, equivalent concentration
 N. einer Lösung normality of a solution
Normalkabel *n* standard cable
Normalkerze *f* primary luminous standard
normalkettig *(ch)* straight-chain
Normalkomponente *f* normal component
Normalkondensator *m* standard (reference) capacitor
Normalkraft *f* normal force
Normalkurve *f* normal (bell-shaped) curve
Normallage *f* normal position
Normallampe *f* standard lamp
Normallast *f* standard load
Normallastenzug *m* standard train loading *(bridge design)*
Normallautsprecher *m* standard loudspeaker
Normallautstärke *f* normal loudness
Normalleistung *f* normal power (output); standard capacity
Normalleiter *m* normal conductor
Normallichtquelle *f* luminous standard
Normallösung *f* normal solution
Normalmaß *n* standard
Normalmaßstab *m* standard scale
Normalmikrofon *n* standard microphone
Normalobjektiv *n* standard lens
Normal-Ohm *n* standard ohm
Normalpegel *m* standard (reference) level
Normalpotential *n* standard electrode potential
Normalprobestab *m* standard test bar
Normalprüfsender *m* standard signal generator

Normalquarzplatte *f* standard quartz plate
Normalrauschgenerator *m* standard noise generator
Normalschliff *m* base pulp *(papermaking)*
Normalschwingung *f* normal mode
normalsichtig emmetropic
Normalsichtigkeit *f* emmetropia
Normalspannung *f* normal (axial) stress *(bar)*
elektrische N. normal voltage
Normalspule *f* standard loading coil
Normalspur *f (ry)* standard gauge
Normalstand / über above normal
Normalstellung *f* normal position
Normalstrahlung *f* normal radiation
Normalteilung *f* normal pitch *(gear)*
Normaltemperatur *f* standard temperature
Normaltiefe *f* standard depth *(dyeing)*
Normalton *m* standard musical pitch
Normaltonfrequenz *f* standard audio frequency
Normalverteilung *f* normal (bell-shaped) distribution *(statistics)*
Gaußsche N. *s.* Normalverteilung
Normalverzahnung *f* standard teeth
Normalwald *m* model forest
Normalwellenband *n* broadcast band
Normalwellenmesser *m* standard wavemeter
Normalwiderstand *m* standard resistor
Normalwinkelkamera *f* standard-lens camera
Normalzählrohr *n* standard counter
Normalzeit *f* standard time
Normalzubehör *n (mach)* standard equipment
Normalzustand *m* normal state
Normand-Kessel *m* Normand boiler
Normatmosphäre *f* standard atmosphere
Normbaustein *m* standard modular unit
Normbegriff *m* standard term
Normblatt *n* loose leaf standard
Normblende *f* standard orifice
normen to standardize
Normenausschuß *m* standards committee
Normenentwurf *m* draft standard specification
Normenvorschlag *m* standards proposal
Normfestigkeit *f* standard strength
normieren to calibrate, to normalize
normiert non-dimensional
n. orthogonal orthonormal
Normierung *f* calibration, normalization
Normierungsfaktor *m* normalizing factor
Normklima *n* standard atmosphere
Normlastreihe *f* standard load progression
Normquerschnitt *m* standard section
Normsand *m* standard sand *(testing of cement)*
Normschraubenzieher *m* standard screw driver
Normschrift *f* lettering recommended by the standard
Normspannung *f* standard voltage
Normteil *n* standard component
Normung *f* 1. determination of quota *(material consumption; work; production time)*; 2. standardization
Normversuch *m* standard test
Normvorschrift *f* standard specification
Normwandler *m (telev)* television-standard converter
Normzeit *f* standard time
Norton-Getriebe *n* Norton-type gear
Norton-Vorschubgetriebe *n* feed box of Norton type
Notabschaltung *f* emergency shut-down *(reactor)*
Notaggregat *n* emergency (stand-by) plant
Notanker *m* spare (security, jury) anchor
Notausbesserung *f* temporary repair
Notausgang *m* emergency door (exit), fire exit, escape door
Notausrüstung *f* emergency outfit (equipment)
Notausschalter *m* emergency cutout; emergency stop *(lift)*
Notausschaltung *f* emergency tripping
Notausstieg *m* emergency door (exit); escape hatch
Notbatterie *f* emergency battery
Notbeleuchtung *f* emergency lighting
Notbeleuchtungsanlage *f* emergency lighting system
Notbetrieb *m* emergency operation
Notbremse *f* emergency train stop (brake), emergency stop (brake)
Notbremsschalter *m* emergency braking switch *(electrical traction)*
Notbrennschluß *m (rock)* emergency cutoff
Note *f* note
ganze N. semibreve
geschwänzte N. quaver
halbe N. minim
Notendruck *m* music printing
Notenlinien *fpl* lines of the staff
Notenpapier *n* music paper
Notenstechen *n* music engraving
Notfall *m* emergency
Notflagge *f (ship)* distress flag
Notfrequenz *f* distress frequency
Nothafen *m* harbour of refuge, port of distress
Nothalt *m (mach)* emergency stop
Notieren *n* logging
Notizblock *m* [memo] pad
Notizbuch *n* notebook
Notkühlung *f* emergency cooling *(reactor)*
Notlandung *f* forced landing
N. auf dem Wasser ditching
Notlaterne *f* distress lantern, not-under-command light
Notlaufeigenschaft *f* resistance to seizure (scoring), antifrictional (antiseizure) property *(of bearing metal)*
Notleitung *f* emergency line

Notlenzpumpe *f* submersible (emergency bilge) pump
Notluke *f* escape hatch[way]
Notmast *m (ship)* jury mast
Notmeldung *f* distress message
Notpeilsender *m* emergency transmitter beacon
Notreife *f* premature (incomplete) ripeness
Notruder *n* jury rudder
Notrudermaschine *f* emergency steering gear
Notruf *m* distress call; emergency call
Notruffrequenz *f* distress frequency
Notschacht *m (mine)* escape shaft
Notschalter *m* emergency [shut-down] switch, scram button (switch)
Notschenkel *m* journal-collar
Notschlachtung *f* emergency (forced) slaughter
Notschleuse *f* emergency (escape) lock
Notsender *m* emergency transmitter
Notsignal *n* distress signal *(SOS)*; emergency signal
Notsignalfeuer *n (ship)* distress light
Notstandsstrahlenbelastung *f* emergency exposure
Notsteuereinrichtung *f (ship)* emergency steering gear
Notstich *m (met)* auxiliary tap hole
Notstromaggregat *n* mains failure set, emergency generator set
Notstromanlage *f* emergency power plant
Notstrombatterie *f* emergency storage battery
Notstromgenerator *m* emergency generator
Nottakelage *f (ship)* jury rig
Nottreppe *f* fire escape stair
Notwasserung *f (aero)* ditching
notwendig und hinreichend *(math)* necessary and sufficient
Notwindform *f* auxiliary tuyere
Novokain *n* novocaine
Novolak *n* novolak
np-Karte *f* control chart for number of defectives
np-Kontakt *m* n-p junction
np-Übergang *m* n-p junction
npn-Flächentransistor *m* n-p-n junction transistor
Nuance *f (text)* hue, shade
Nuclid *n s.* Nuklid
nuklear nuclear
Nuklearmedizin *f* nuclear medicine
nuklearrein nuclear pure; reactor-grade
Nukleinsäure *f* nucleic acid
Nukleon *n* nucleon
  erregtes N. excited nucleon
Nukleonengas *n* nucleon gas
Nukleonenkomponente *f* nucleonic component
Nukleoprotein *n* nucleoprotein
Nukleotid *n* nucleotide
Nuklid *n* nuclide
  radioaktives N. radioactive nuclide, radionuclide
Null *f* zero *(scale)*; naught *(decimal fraction)*; nil *(sport)*; cipher
  N. / über above zero
  N. / unter below zero
  N. werden to vanish, to become zero
Nullabgleich *m* nullification; null balance (balancing), zero balance (balancing)
Nullabgleichgerät *n* null metering unit
Nulladreßbefehl *m* zero-address instruction
Nulladreßsystem *n* zero-address system
Nullage *f* zero position
Nullamplitude *f* **eines Trägers** zero carrier
Nullanode *f* freewheeling rectifier
Nullanstellung *f (aero)* zero incidence *(wing)*
Nullanzeige *f* zero indication
Nullausschalter *m (el)* zero-current circuit-breaker, no-load circuit-breaker
Nulldetektor *m* null detector
Nulldurchgang *m* zero passage
  N. der Spannung zero voltage
  N. des Stroms zero current, current zero
Nullebene *f (ph)* polar plane; neutral plane *(bending)*
Nulleffekt *m* background; natural background radiation
Nulleffektmessung *f* background determination
Nulleffektzählrate *f* background counting rate
Nulleinstellung *f* zero setting (adjustment)
Nulleistung *f* zero output
Nulleistungsreaktor *m* zero power reactor, ZPR
Nulleiter *m* neutral (zero) conductor
  nichtgeerdeter N. floating neutral
Nulleiterdraht *m* middle wire
nullen *(el)* to neutralize, to connect to neutral; *(dat)* to nullify, to rollback; to restore; to reset
Nullen *fpl* **/ führende** *(dat)* leading zeros
Nullenergiereaktor *m s.* Nulleistungsreaktor
Nullenkomplementdarstellung *f (dat)* noughts-complement representation
Nullenzirkel *m* bow instrument (spring compass)
  N. mit Einsatz für Blei bow pencil
  N. mit Einsatz für Tusche bow pen
Nullexperiment *n s.* Nulleistungsreaktor
Nullfehler *m* zero error
Nullfolge *f* null sequence
Nullförderung *f* zero delivery
Nullfrequenz *f* zero frequency
Nullfüllung *f* zero admission
Nullhubstellung *f* zero (no-delivery, no-stroke) position *(pump)*
Nullhyperbel *f* lane *(radar)*
Nullhypothese *f* null hypothesis
Nullimpedanz *f* zero-sequence field impedance; zero phase-sequence impedance
Nullinie *f* zero (null) line; neutral line (axis) *(statics)*

Nullinienfesthalteschaltung *f (telev)* clamping circuit
Nullinstrument *n* null instrument, (detector, indicator), balance indicator, balancing (central-zero) apparatus
Nullkomponente *f (el)* zero-sequence component, zero-frequency component
symmetrische N. zero-sequence symmetrical component
Nullkontakt *m* zero blocking-contact *(telephony)*
Nullkontrolle *f* 0 clear signal, zero balancing *(calculators)*
Nullkurve *f* zero curve
Nullmethode *f* null (zero, balance, zero-compensating) method
Nullode *f* electrodeless tube, nullode
Nullparzelle *f* untreated plot
Nullpegel *m* dead (zero) level
Nullphasenwinkel *m* zero phase-angle
Nullphasenwinkelmodulation *f* zero phase-modulation
Nullpotential *n* zero (earth) potential
Nullpunkt *m* 1. zero point *(scale)*; 2. *(el)* neutral (zero) point, earthed neutral; 3. *(aero)* node
absoluter N. absolute zero
künstlicher N. artificial neutral point
lebendiger N. life zero *(signal)*
mechanischer N. mechanical zero
unterdrückter N. suppressed zero
Nullpunktabweichung *f* zero variation *(instrument)*
Nullpunktanomalie *f* origin distortion
Nullpunktdrossel *f* absorption coil
Nullpunkteinstellung *f* zero adjustment (setting), setting of zero reference
Nullpunktempfindlichkeit *f* zero-level sensitivity
Nullpunktenergie *f* zero-point energy, energy of absolute zero
Nullpunkterdung *f* neutral earthing (grounding)
Nullpunktfehler *m* origin distortion, residual deflection *(electron valve)*
Nullpunktfesthalteschaltung *f* d.c. restorer
Nullpunktisolation *f* neutral insulation
Nullpunktkorrektur *f* zero-point correction
Nullpunktleistung *f* zero-point energy
Nullpunktsfeld *n* zero-point field
Nullpunktspannung *f* earth (neutral-to-earth, zero) voltage
Nullpunktunterdrückung *f* range suppression; zero offset
Nullpunktwanderung *f* zero drift
Nullrate *f s.* Nulleffektzählrate
Nullreaktanz *f* zero-sequence reactance
Nullreaktor *m s.* Nulleistungsreaktor
Nullresistanz *f* zero-sequence resistance
Nullriß *m* zero line
Nullschalter *m* no-load circuit-breaker
Nullschwebungsmethode *f* zero-beat method
Nullspannung *f* zero voltage
Nullspannungsauslöser *m* no-voltage release, zero cut-out
Nullspannungsauslösung *f* undervoltage tripping, no-volt release
Nullspannungsrelais *n* undervoltage relay
Nullspannungsschalter *m* undervoltage (no-voltage) circuit-breaker
Nullspant *n* balanced frame; midship section (frame)
Nullstab *m* unstrained member *(truss)*
Nullstelle *f* zero
Nullstellen-Pol-Verteilung *f* pole-zero configuration
Nullsteller *m* zero adjuster
Nullstellung *f* zero adjustment (setting), neutral position; point of rest
automatische N. *(dat)* clearing device
Nullstrom *m* zero current
Nullstromauslösung *f* / **direkte** series undercurrent tripping
Nullstromausschalter *m* / **automatischer** automatic zero cut-out
Nullstromschalter *m* no-load circuit-breaker
Nullstromselbstausschalter *m* automatic zero cut-out
Nullsummenspiel *n (dat)* zero-sum game
Nullsystem *n* zero phase-sequence system
Nullsystemdrosselspule *f* zero-sequence system reactor
Nullsystemschutz *m* zero phase-sequence protective system
Nullung *f (el)* connection to neutral, multiple protective earthing
Nullunterdrückung *f* zero suppression
Nullvektor *m* null vector
Nullverfahren *n* zero method *(Wheatstone bridge)*
nullwertig zero-valent
Nullwertigkeit *f* zero valency
Nullzacke *f* null signal *(radar)*
Nullzählung *f s.* Nulleffektmessung
Nullzone *f* silent space; neutral plane *(multiphase machine)*
Nullzustand *m* zero state
numerieren to number
durchlaufend n. to number continuously
Numeriermaschine *f* numbering machine
Numerierung *f* numbering
metrische N. metrical numbering system
Numerierwerk *n* numbering apparatus (box)
numerisch numerical
n. gesteuert *(mach)* numerically controlled
Numerus *m* antilogarithm
Nummer *f* number
falsche N. wrong number *(telephony)*
laufende N. serial number
Nummernanzeigeeinrichtung *f* call announcer system

Nummernanzeiger *m* call indicator
Nummerneisen *n* number punch
Nummernfolge *f* numerical sequence
Nummerngabeschlußzeichen *n* end of pulsing signal
Nummernscheibe *f* [calling] dial
Nummernscheibenkontakt *m* dial key
Nummernschild *n (aut)* number (licence) plate
Nummernschildbeleuchtung *f* number-plate light (illumination lamp)
hintere N. rear number-plate light
vordere N. front number-plate light
Nummernschlucker *m* digit-absorbing selector
Nummernumrechnung *f* conversion of counts
Nummernumsetzer *m* register translator
Nummernwahl *f* dialling
Nummernwähler *m* numerical selector
Nurflügelbauart *f (aero)* flying-wing design
Nurflügelflugzeug *n* all-wing aeroplane
Nuß *f* 1. *(print)* pinion; 2. nut *(anchor)*; 3. boss *(propeller)*
Nußbaumholz *n* walnut
Nußgrieß[kohle *f***]** *m* nutty slack
Nußisolator *m* egg insulator
Nußkern *m* nut kernel
Nußkohle *f* nut (chestnut, graded) coal, graded sizes, nuts
Nußkoks *m* nut coke
Nußöl *n* nut oil
Nußplatte *f (ship)* boss plate
Nut *f* slot, groove *(metal)*; rabbet *(wood, stone)*; mortise *(carpentry)*; flute, groove *(tool)*
N. des Kolbenkörpers piston-ring groove
N. des Kurvenschiebers cam slot *(belt shifting)*
N. und Feder *f* groove and tongue; key and slot, joggle
prismatische N. vee-groove
schraubenförmige N. helical (twisted) groove
unbewickelte N. *(el)* empty slot
Nutation *f* **der Kreiselachse** nutation of the gyro-axis
Nutationswinkel *m* nutation angle
Nutbreite *f* width of the keyway (groove)
nuten to groove
Nutenauslauf *m* cutter sweep *(tool)*
Nutenbreite *f (mach)* slot width *(driving slot)*
Nutenfräser *m* grooving cutter, slot drill (milling cutter), *(Am)* keyway cutter
N. für Gewindebohrer milling cutter for fluting taps
dreizahniger N. three-lip keyway cutter
gekuppelter N. interlocking slotting cutter
Nutenfräsmaschine *f* flute milling machine; groove milling machine
Nutenfüllfaktor *m (el.m.)* slot space factor *(armature)*
Nutenkeil *m (el.m.)* slot wedge *(armature)*; cotter
Nutenkeilnut *f* hub keyway
Nutenlänge *f* flute length *(tool)*
Nutenscheibe *f* feed driving disk *(shaper)*
Nutenschritt *m* slot (coil) pitch
Nutenstanzautomat *m* notching press
Nutensteg *m* piston land
Nutensteigung *f* [lead of] helix *(tool)*
Nutenstoßmaschine *f* push-type slotter
Nutenstück *n* **des Kardangelenks** universal joint spline
Nutentiefe *f* slot depth *(driving slot)*
Nutenverschlußkeil *m* armature key
Nutenwelle *f (mach)* splined shaft
Nutenwicklung *f (el.m.)* slot winding
Nutenzahl *f (mach)* number of flutes *(tool)*
gerade N. even number of flutes *(reamer)*
ungerade N. odd number of flutes *(reamer)*
Nutfrequenz *f (el.m.)* slot (tooth pulsation) frequency
Nutgrund *m* bottom of the keyseat
Nuthobel *m* grooving plane
Nuthobelmaschine *f* matching machine
Nuthülse *f* slot liner *(insulation)*
Nutisolation *f* slot insulator; slot insulation
Nutkörper *m* slotted piece
Nutmaschine *f* grooving machine
Nutringmanschette *f* groove-ring collar, square-base U-ring
Nutsättigung *f (el.m.)* tooth saturation
Nutsche *f* porcelain funnel, suction filter
Nutteilung *f (el)* slot pitch (spacing)
Nut- und Federhobel *m* hand grooving and tonguing plane
Nut- und Federholz *n* match board
Nut- und Spundfräser *m* slot and key cutter
hinterdrehter N. backed-off slot and key cutter
Nut- und Spundmesser *n* slot and key iron
Nutzanwendung *f* utilization, practical application
Nutzarbeit *f* effective work
Nutzband *n (el)* useful band
nutzbar utilizable, usable
Nutzbildfläche *f (telev)* useful screen area
Nutzbremsschaltung *f* regenerative control
Nutzbremsung *f* / **elektrische** regenerative braking *(electrical traction)*
nutzbringend profitable
Nutzdämpfung *f (el)* volume (effective transmission) equivalent, useful damping
N. der Fernleitung trunk loss
Nutzdauer *f* endurance *(material)*
Nutzeffekt *m* efficiency, useful effect
Nutzen *m* / **wirtschaftlicher** commercial efficiency
Nutzfahrzeug *n* commercial vehicle
Nutzfläche *f* / **landwirtschaftliche** land in cultivation

Nutzförderhöhe *f* total head pumped against, operating head *(pump)*
Nutzhöhe *f* effective depth *(reinforced concrete design)*
Nutzholz *n* timber, lumber
Nutzholzbaum *m* timber tree
Nutzinhalt *m* effective capacity
Nutzkraft *f* effective power
Nutzladefähigkeit *f* useful load, paying deadweight
Nutzladung *f* useful (paying) load
Nutzland *n* arable land
Nutzlast *f* payload *(vehicle)*; *(build)*. working (superimposed, live, service) load, surcharge
N. der letzten Raketenstufe final payload
Nutzlast-Aufbau-Kraftstoff-Verhältnis *n (rock)* payload-structure-fuel weight ratio
Nutzlastverhältnis *n (rock)* payload ratio
Nutzleistung *f* effective (useful) power, effective (useful) horsepower; useful power output, output energy; net efficiency; *(el)* service output *(battery)*
N. einer Pumpe pump efficiency; hydraulic horsepower
größte N. available power
Nutzmeßbereich *m* effective range
Nutzpferdestärke *f* effective horsepower
Nutzquerschnitt *m (el)* useful cross-section; net section
Nutzreaktor *m s.* Leistungsreaktor
Nutzscheitholz *n* industrial split wood
Nutzspalt *m* front gap *(of a magnetic head)*
Nutzspaltholz *n* cloven (riven) timber
Nutzspannung *f* useful voltage
Nutzstrahl *m* useful beam
Nutzstrahlenbündel *n* useful beam
Nutztiere *npl* domestic stock
Nutztonnenkilometer *m* actual ton-kilometre
Nutzung *f* **der Kernenergie / friedliche** peaceful use (application) of nuclear energy
thermische N. thermal utilization
Nutzungsdauer *f* working (service) life
Nutzungshauptzeit *f (mach)* operating (cutting) time
Nutzungszeit *f* 1. *s.* Nutzungshauptzeit; 2. *s.* Nutzungsdauer
Nutzvieh *n* domestic stock
Nutzwiderstand *m* useful resistance
Nylon *n* nylon
Nylontrosse *f* nylon hawser
Nyquist-Diagramm *n* Nyquist diagram

O

O-Bande *f* O-band
O-Bus *m* trolley bus
O-Stab *m* unstrained member *(truss)*
O-Wagen *m* open goods wagon with high sides, high-sided open wagon
OB-Apparat *m* local battery telephone set
OB-Betrieb *m* local battery working *(telephony)*
OB-Klappenschrank *m* magneto switchboard *(telephony)*
obengesteuert valve-in-head, overhead-valve *(engine)*
obenliegend overhead
Obenschmieröl *n* upper cylinder lubricant
Oberärmel *m* upper sleeve
Oberauslauf *m* external force feed
Oberbank *f (geol)* top coal
Oberbau *m* superstructure; *(ry)* permanent way
Oberbauleiter *m* chief site engineer, field superintendent
Oberbaumaterial *n (ry)* permanent way material
Oberbecken *n* upper reservoir
Oberbereich *m* upper range *(of transformation)*
Oberdeck *n* upper deck
Oberdrempel *m (ship)* upper port sill *(dock)*
Oberfach *n (text)* upper (top) shed
Oberfaden *m* upper thread
Oberfelge *f* upper rim
Oberfeuer *n* updraft fire
Oberfläche *f* surface
O. / mit rauher rough-surfaced
O. der Schweißnaht face of the weld
O. mit gleichmäßigem Gewebeflor evenly napped surface
abkühlende O. des Zylinderdeckels cooling surface of the cylinder cover
aktivierte O. activated surface
benetzte O. wetted (immersed) surface
ebene O. plane surface
freie O. free surface
gekrümmte O. curved surface
lichtempfindliche O. photosensitive surface
pigmentierte O. pigmented layer
reibende O. der Turbinendüse friction surface
rutschsichere O. *(aut)* non-skid surface
spiegelnde O. mirror finish, reflective surface
thermodynamische O. thermodynamic surface
wirksame O. surface area (exposed)
Oberflächenabdruck *m* surface replica
Oberflächenaffinität *f* surface affinity
oberflächenaktiv capillary-active, surface-active
Oberflächenaktivität *f* surface activity
Oberflächenangabe *f* machining instruction
Oberflächenappretur *f* face finish
Oberflächenart *f* kind (character) of surface
Oberflächenauflage *f* surface coating
Oberflächenaufrauhung *f* surface roughening (wrinkling)
Oberflächenausdehnungskoeffizient *m* coefficient of surface expansion

Oberflächenbarrieredetektor *m* surface-barrier detector
Oberflächenbearbeitung *f* surface machining (finishing); surface treatment (dressing)
Oberflächenbearbeitungsangabe *f* machining instruction
Oberflächenbehandlung *f* surface dressing (treatment, finish); surface coating; [metallic] finishing
Oberflächenbeladung *f* activation (charging) of a surface
Oberflächenbeschädigung *f* surface damage
Oberflächenbeschaffenheit *f* surface texture (condition); surface finish
Oberflächenbestrahlung *f* surface irradiation
Oberflächencharakter *m* / **geordnet-rilliger** definite direction of lay
Oberflächendichte *f* density of surface charge
Oberflächeneffekt *m* surface effect
lichtelektrischer O. surface photoelectric effect
Oberflächenentkohlung *f* surface decarburization
Oberflächenentladung *f* surface discharge
Oberflächenentwässerung *f* surface drainage
Oberflächenerdung *f* surface earthing
Oberflächenfehler *m* surface flaw (imperfection, irregularity)
Oberflächenfeldstärke *f* surface field strength
Oberflächenfestigkeit *f* surface bonding strength
Oberflächenfotoeffekt *m* surface (external) photoelectric effect, photoelectric emission, photoemissive (Hallwachs) effect
oberflächengefärbt calender-coloured *(paper)*
Oberflächengefüge *n* surface structure
Oberflächengegenstromkondensator *m* countercurrent surface condenser
oberflächengeleimt tub-sized, machine-sized
Oberflächenglanz *m* lustre [of the surface]
Oberflächengüte *f* surface finish (quality)
geringe O. poor finish
Oberflächenhärte *f* surface hardness
Oberflächenhärten *n* surface hardening
Oberflächenhärtung *f* **mit induktiver Erwärmung** induction surface hardening
Oberflächenkitt *m* surfacer
Oberflächenkohlung *f* surface carburizing
Oberflächenkondensation *f* surface condensation
Oberflächenkondensator *m* surface condenser
Oberflächenkontamination *f* surface contamination
Oberflächenkonzentration *f* surface concentration
Oberflächenkraft *f* surface force
Oberflächenladung *f* surface charge
Oberflächenleimung *f* tub sizing *(paper)*
Oberflächenleitung *f* surface conduction
Oberflächenmagnetisierung *f* surface magnetization
Oberflächenmeßgerät *n* surface meter
Oberflächenmessung *f* surface measurement
oberflächennah near the surface
Oberflächennormale *f* surface specimen block
Oberflächenorientierung *f* surface orientation
Oberflächenparallelstromkondensator *m* parallel current surface condenser
Oberflächenprüfgerät *n* surface-finish tester
Oberflächenrauheit *f* surface roughness
Oberflächenreibung *f* surface friction; *(ship)* skin friction
Oberflächenreinigung *f* surface cleaning
Oberflächenrieselkondensator *m* evaporative surface condenser
Oberflächenriß *m* surface crack
Oberflächenrüttler *m (build)* surface vibrator
Oberflächenschicht *f* surface layer
Oberflächenschutz *m* surface protection
Oberflächenspannung *f* surface (interfacial) tension
Oberflächenstrom *m* surface current
Oberflächenstromdichte *f* surface-current density
Oberflächentemperatur *f* surface temperature
Oberflächentherapie *f* superficial [surface] therapy, surface therapy
Oberflächenthermometer *n* surface (external) thermometer
Oberflächenüberzug *m* surface coat[ing]
Oberflächenungenauigkeit *f* surface irregularity
Oberflächenverdichter *m* surface compactor
Oberflächenveredlung *f* surface refinement
Oberflächenvergütung *f* surface refinement
Oberflächenverschleiß *m* surface wear
Oberflächenverseuchung *f* surface contamination
Oberflächenvorbehandlung *f* surface preparation (pretreatment)
Oberflächenvorwärmer *m* surface feed heater
Oberflächenvorwärmung *f* surface heating
Oberflächenwasser *n* free (surface) moisture *(coal)*
Oberflächenwasserzufluß *m* surface water inflow
Oberflächenwelle *f* surface wave
Oberflächenwiderstand *m* electric surface resistance
Oberflächenzeichen *n* drafting symbol for surface roughness
Oberflächenzeichnung *f* lay
Oberflächenzustand *m* condition (character) of surface; *(mach)* surface finish
oberflächlich superficial[ly]
Oberflasche *f* top block
Oberflottenjigger *m* surface jig
Oberform *f (met)* upper box, cope

Oberfräse *f* recessing and shaping machine, top spindle moulder
obergärig top-fermented
Obergärung *f* top fermentation
Obergeländer *n (ship)* gallant rail
Obergeschoß *n* upper storey
Obergesenk *n* upper (top) die, upper (top) swage
Obergewebe *n* face cloth
Oberglocke *f* small bell *(furnace)*
Obergurt *m (build)* top boom (chord), upper chord (boom)
Oberguß *m* top casting
Oberhaupt *n* inner (upper) gates *(lock)*
Oberheizer *m* chief stoker (fireman)
Oberhemdenmanschette *f* cuff of shirt
Oberhieb *m* upcut
Oberhitze *f* upper heat
Oberingenieur *m* executive engineer
oberirdisch aerial; overground; overhead *(line)*
Oberkammer *f* **der Schleuse** upper chamber of the collector *(turbine)*
Oberkante *f* upper (top) edge
O. Kiel *(ship)* top (upper edge) of keel
Oberkasten *m* top flask *(foundry)*
Oberkessel *m* upper cylinder (drum)
Oberkesselstütze *f* support of the upper drum
Oberkette *f (text)* face warp
Oberkiel *m (ship)* upper keel
Oberkielschwein *n* rider keelson
Oberkohle *f (geol)* top coal
Oberlänge *f (print)* ascender
oberlastig top heavy
Oberleder *n* upper leather
Oberlederstanze *f* upper press knives *(shoe)*
Oberleitung *f* overhead contact wire (system), bus (train) line
Oberleitungs[omni]bus *m* trolley bus
Oberlicht *n* overhead light *(stage lighting)*; *(build)* roof-light, skylight
Oberlichtbezug *m (ship)* skylight cover
Oberlichtfenster *n* skylight
Oberluft *f* air entering above the grate, over-grate blast
Obermaat *m* chief mate
Obermesser *n* top (upper) blade, upper knife
Oberputz *m* finish[ing] coat
Oberschanzkleid *n (ship)* topgallant bulwarks
Oberschicht *f (geol)* upper stratum
Oberschiff *n* upper (dead) works *(that part of a vessel, which is above water when it is laden)*
oberschlächtig *(mine)* overcast *(shearer drum)*; overshot *(mill)*
Oberschlag *m (text)* overpick
Oberschlagstuhl *m* overpick loom
Oberschlinge *f (text)* upper loop
Oberschlitten *m* upper (top, compound, toolpost) slide, tool rest *(lathe)*
Oberschmiermittel *n* upper-cylinder lubricant
Oberschuß *m (text)* face pick (weft)
Oberschußfaden *m* face weft thread
Oberschwingung *f* harmonic [oscillation]
harmonische O. higher harmonic [oscillation], harmonic of fundamental sound, overtone
Oberschwingungen *fpl* / **ungeradzahlige** odd harmonics
Oberschwingungsfrequenz *f* harmonic frequency
Oberseil *n* headrope
Oberseite *f* top side (surface); *(text)* face
Obersieb *n* 1. *(ag)* top screen [pre-cleaner]; 2. face wire *(paper)*
Oberspannung *f* primary (higher) voltage *(of transformer)*
Oberspannungsanzapfung *f* primary (higher-voltage) tap *(of transformer)*
Oberspannungswicklung *f* primary (higher-voltage) winding
Obersteiger *m* senior overman, inside foreman
Oberstempel *m* 1. *(met)* counter die; 2. *(mine)* upper prop
Oberstrich *m* maximum telegraphy output; *(telev)* black-level picture signal
Oberstrichleistung *f* peak power output
Oberstrichwert *m s.* Oberstrichleistung
Oberteil *n* top; top band *(shoe)*
ausgelochtes O. punched galosh *(shoe)*
Oberton *m* / **charakteristischer** significant overtone
Obertor *n* top gate
Obertrum *n* top run *(conveyor)*
Oberwagen *m* superstructure *(excavator)*
Oberwalze *f* upper (top) roll
Oberwange *f* upper beam *(press brake)*
oberwasserseitig upstream
Oberwelle *f* harmonic wave; harmonic component, upper harmonic
geradzahlige O. even[-order] harmonic
ungeradzahlige O. odd harmonic
Oberwellenanhebung *f* accentuation of harmonic
Oberwellenanteil *m* harmonic content
Oberwellenemission *f* harmonic emission
Oberwellenerzeugung *f* harmonic generation
Oberwellenfilter *n* harmonic trap (filter)
Oberwellenkompensation *f* harmonic component compensation
Oberwellenkomponente *f* harmonic component
Oberwellenkristall *m* overtone crystal
Oberwellenleistung *f* harmonic power
Oberwellenmeßgerät *n* distortion bridge
Oberwellenoszillator *m* harmonic oscillator
Oberwellenquarz *m* overtone crystal
oberwellenreich rich in harmonics
Oberwellenschwingquarz *m (el)* overtone crystal
Oberwellenschwingungsanteil *m* harmonic content
Oberwellensieb *n* harmonic filter (suppressor)

Oberwellensiebschaltung *f* harmonic stopper
Oberwellenspannung *f* ripple voltage
Oberwellenverzerrung *f* modulation-frequency harmonic distortion
Oberwerk *n (ship)* upper (dead) works
Oberwind *m* overgrate blast, air entering above the grate
Oberzug *m* top (overtop, upper) flue
Objekt *n* specimen, object
fliegendes O. *(astr)* flying object
unidentifiziertes fliegendes O. unidentified flying object
Objektabstand *m* working distance
Objektfrontlinse *f* front lens of the objective
Objektführer *m* object traverser
Objekthalter *m* specimen holder
Objektiv *n* objective *(of microscopes)*; lens *(of cameras)*
O. mit Präparateschutz spring-protected objective
achromatisches O. achromatic lens
anastigmatisches O. anastigmatic objective
korrigiertes O. corrected objective
lichtstarkes O. objective of great light transmitting capacity
schwaches O. low-power objective
starkes O. high-power objective
zweilinsiges O. doublet objective
Objektivfassung *f* objective (lens) mount
Objektivprisma *n* objective prism
Objektivrevolver *m* objective revolving nose piece
Objektmarkierer *m* object marker
Objektmikrometer *n* stage micrometer
Objektschleuse *f* object sluice (air-lock)
Objektträger *m* specimen holder
Obmann *m* chairman; foreman
Observatorium *n* observatory
Obstanlage *f* orchard
Obstbau *m* fruit growing, orcharding
Obstbaum *m* fruit tree
Obstbaumschnitt *m* pruning
Obsterntegerät *n* fruit harvester
Obstkeltermaschine *f* fruit press
Obstkonserve *f* fruit preserve
Obstkonserven *pl* preserved fruits
Obstmadenspritzung *f* fruitlet spray
Obstpflücker *m* hand fruit picker
Obstsaft *m* fruit juice
Obstschiff *n* fruit ship, fruit-carrier
Obstsortiermaschine *f* fruit grader
Obstwein *m* fruit wine
Obus *m* trolleybus
Ochse *m* ox
Ochsenauge *n* bull's eye [window]
Ochsengalle *f* ox-gall
Ocker *m* ochre, ocher
gelber O. yellow ochre
Oddesse-Dampfpumpe *f* Oddesse steam pump
ODER *n* / **ausschließendes** EITHER-OR, exclusive OR, OR-ELSE *(logical operator)*
ODER *n* / **einschließendes** disjunction, inclusive OR *(logical operator)*
ODER-Glied *n* OR-gate, OR element
exklusives O. OR-ELSE gate
negiertes O. NOR-gate
ODER-Schaltung *f* OR-gate, OR buffer, OR circuit
Ödland *n* waste ground
Odograf *m* odograph
Oerstedt *n* oerstedt
Ofen *m (met)* furnace; kiln *(lime, bricks, pottery)*; stove; oven
elektrischer O. electric furnace
halbgeteilter O. half-divided oven (furnace)
koksbeheizter O. coke-fired furnace
kopfbeheizter O. gas gun oven
ölgeheizter O. oil furnace
stationärer O. stationary furnace
tiegelloser O. non-crucible furnace
unterwindgefeuerter O. forced-draft furnace
Ofenanschlußwert *m* furnace wattage
Ofenbär *m (met)* [furnace] bear
Ofenbaustein *m* refractory brick; kiln brick
Ofenbetonsohle *f* oven pad
Ofenbetrieb *m* operating of ovens (furnaces, kilns)
Ofenblei *n* pig lead
Ofenblock *m* setting
Ofenbodenstein *m* flagstone of the furnace
Ofendecke *f* roof of the furnace
Ofenfassungsvermögen *n* oven (furnace) capacity
Ofenfüllung *f* oven (furnace) charging
Ofengalmei *m* cadmia
Ofengang *m* furnace campaign, operating (working) of the oven, operating (working) of the furnace
Ofengewölbe *n* [arched] furnace roof, arch of the furnace (oven)
Ofengruppe *f* setting
Ofenkachel *f* stove tile
Ofenkopf *m* oven end (quoin)
Ofenleistung *f* output per (of the) oven (furnace)
Ofenloch *n* kiln hole
Ofenlöten *n* furnace soldering
Ofenmann *m* / **zweiter** first helper
Ofenmassiv *n* body of a furnace
Ofenprofil *n* furnace lines
Ofenregler *m* oven control
Ofenreise *f* furnace campaign (life)
Ofenrohr *n* stove pipe
Ofensau *f* [furnace] sow, salamander
Ofenschacht *m* furnace shaft
Ofenschalter *m* furnace circuit-breaker
Ofenschirm *m* screen

Ofenschließer *m* lid
Ofenschließstein *m* lid
Ofenschwärze *f* powdered blacklead
Ofensetzer *m* stove fitter
Ofenstirnseite *f* elevation of the oven (furnace)
Ofensumpf *m* sump (bottom) of a furnace
Ofentragplatte *f* oven deck pad
Ofentrocknung *f* oven drying
Ofenumspanner *m* furnace transformer
Ofenwarte *f* furnace control station
Ofenwolf *m* [furnace] bear
Ofenzunder *m* furnace scale
Ofenzuschlag *m* furnace addition
offen open, non-enclosed *(electric motor etc.)*; uncovered, undecked *(vessel)*
Offenfachjacquardmaschine *f* open-shed Jacquard machine
Offenfachschaftmaschine *f* open-shed dobby
Offenstall *m* open-front shelter, partly-covered yard
öffentlich public
offerieren to offer
Offerte *f* offer, tender
Offertingenieur *m* sales engineer
offiziell official
Offizier *m* / **erster** *(ship)* chief, officer; chief mate
Offizin *f* printing house (office)
öffnen to open *(door etc.)*; to unlock; *(el)* to break, to open *(switch)*; to disconnect *(line)*
Öffner *m* 1. opener; 2. *(el)* b-contact, break (opening, closed-circuit, normally closed auxiliary) contact; 3. pulper *(papermaking)*
  Ö. für Schweißwolle opener for greasy wool
Öffnerabfall *m* opener waste
Öffnertrommel *f* opening cylinder
Öffnung *f* opening; orifice *(flow, optic)*; *(opt)* aperture; *(aut)* port
  Ö. an der Druckseite delivery outlet
  Ö. eines Schalters opening of a circuit-breaker
  Ö. in der Abschirmung *(nucl)* shield opening
  abgerundete Ö. rounded orifice
  scharfkantige Ö. sharp-edged orifice
Öffnungsblende *f* aperture stop
Öffnungsdauer *f* 1. *(aut)* injection period; 2. *(el)* break period
Öffnungsdruck *m (aut)* injection pressure
Öffnungsfehler *m (opt)* spherical aberration, apertural error (defect)
Öffnungsfunke *m* spark at break
Öffnungsimpuls *m* break [im]pulse, start signal
Öffnungskontakt *m* opening contact; break contact
Öffnungskraft *f* opening force
Öffnungsstrom *m* breaking current
Öffnungsverhältnis *n (opt)* relative aperture, aperture ratio
Öffnungsweite *f* **der Backen** jaw opening
Öffnungswinkel *m* groove angle *(weld)*; beam width *(aerials)*
Öffnungszeit *f (el)* [contact] clearing time, opening time
Offsetandruckpresse *f* offset proofing press
Offsetdruck *m* offset printing
Offsetdruckfarbe *f* litho ink
Offsetdruckmaschine *f* offset press
Offsetdruckplatte *f* **mit Wischbeschichtung** wipe-on [plate]
Offsetplatte *f* / **hochgeätzte** relief etch litho plate
  mikrogekörnte O. micrograined plate
  vorbeschichtete O. presensitized litho plate
Offsetraster *m* litho screen
Offsetverfahren *n* offset process
Ohm *n* ohm
  absolutes O. true ohm
  akustisches O. mechanical ohm
Ohmmeter *n* ohmmeter
Ohmskale *f* ohmmeter scale
Ohmwert *m* ohmic value, ohmage
Ohne-Rückkehr-zu-Null non-return-to-zero *(change of the magnetic direction)*
Ohr *n* / **künstliches** artificial ear
Öhr *n* ear, eye
Ohrluppe *f* aural magnifier
OK-Terrain *n (build)* ground (grade) level
Ökonom *m* economist
Ökonometer *n* econometer
Ökonometrie *f* operations research
ökonomisch 1. economic; 2. economical *(saving)*
Oktaeder *n* octahedron
oktaedral octahedral
oktaedrisch octahedral
oktagonal octagonal
Oktaldarstellung *f* octal notation
Oktalfassung *f* octal socket
Oktalsockel *m* octal base
Oktalstecker *m* octal plug
Oktalziffer *f* octal digit
Oktan *n* octane
Oktantfehler *m* octantal error *(radar)*
Oktanzahl *f* octane number (rating), anti-knock value
  O. / **mit hoher** high-octane *(fuel)*
Oktanzahlbestimmung *f* knock rating
Oktanzahlerhöhung *f* raising of the octane number
Oktanzahlverbesserer *m* anti-knock agent, octane booster
Oktett *n* octet
Oktode *f* octode, hexagrid valve
Okular *n* eyepiece, ocular
  gebrochenes O. diagonal eyepiece
  Huygenssches O. Huygens' eyepiece
  monozentrisches O. monocentric ocular
  orthoskopisches O. orthoscopic eyepiece

Ramsdensches O. Ramsden eyepiece
stellbares O. focusable eyepiece
Okularblende *f* eyepiece diaphragm
Okularmeßschraube *f* micrometer [drum of the] eyepiece
Okularmikrometer *n* eyepiece micrometer
Okularmuschel *f* eyepiece cup, eye-cup
Okularprisma *n* prismatic eyepiece
Okularrevolverkopf *m* rotary eyepiece selector
Okularschraubenmikrometer *n* eyepiece screw micrometer
Okularspektroskop *n* ocular spectroscope
Okulation *f (ag)* budding
Okuliermesser *n (ag)* grafting (budding) knife
Öl *n* oil
Ö. auf Rizinusölbasis castor-base oil
abgetriebenes Ö. distilled oil
ätherisches Ö. essential (volatile) oil
dickflüssiges Ö. viscous oil
dünnflüssiges Ö. light (low-viscosity) oil
gekracktes Ö. refractory oil
gemischtbasisches Ö. mixed base oil
kaltgeschlagenes Ö. cold-drawn oil
leichtflüssiges Ö. light oil
mineralisches Ö. mineral oil
mitgerissenes Ö. entrained oil
nichtflüchtiges Ö. fixed oil
nichttrocknendes Ö. non-drying oil
pflanzliches Ö. vegetable oil
trocknendes Ö. drying oil
wasserlösliches Ö. soluble oil
Ölabdichtung *f* oil seal
Ölabfluß *m* oil outlet
Ölablaß *m* oil drain
Ölablaßhahn *m* oil drain cock
Ölablaßschraube *f* oil drain plug
Ölablaßstopfen *m* oil drain plug
Ölablaufwanne *f* oil drain pan
Ölablenkblech *n* oil deflector
Ölabscheider *m* settling tank, oil separator (trap, baffle); grease filter *(steam)*
Ölabscheider-Leichter *m* oil-separating barge
Ölabschreckung *f* oil quenching *(steel)*
Ölabstreicher *m* oil wiper
Ölabstreifring *m* oil scraper (control) ring, scraper (oil) ring
geschlitzter Ö. slotted scraper ring
Ölalterung *f* oil ag[e]ing
Ölanlage *f* oiling system
Ölanlasser *m* oil starter
Ölanstrich *m* oil painting, coat of oil paint
ölarm with low oil-level, with low oil-content, oil-poor *(circuit-breaker)*
Ölaufbereitung *f* oil treatment *(insulating oil)*
Ölauffangschale *f* oil pan
Ölauffangtank *m (ship)* oil save tank
Ölausdehnungsgefäß *n* conservator *(transformer)*
Ölaustrittsgeschwindigkeit *f* rate of escape of oil

Ölbad *n* oil bath
Ölbadfilter *n* oil-bath filter
Ölbadluftfilter *n* oil-bath air cleaner (filter)
Ölbadschmierung *f* oil-bath lubrication
Ölbehälter *m* oil reservoir (collector); oil container; oil supply tank *(hydraulics)*
Ö. der Hydraulik hydraulic [oil] tank
ölbeheizt oil-fired
Ölbeheizung *f* oil firing
ölbeständig oil-resistant
Ölbohranlage *f* / **schwimmende** oil-drilling vessel, oil-rig vessel
Ölbohrausrüstung *f* oil-drilling rig
Ölbohrinsel *f* oil-drilling rig, off-shore drill rig, mobile drilling platform
küstenferne Ö. off-shore drilling barge
Ölbohrung *f* oil well
Ölbremse *f* oil brake (dashpot)
Ölbrenner *m* oil burner
Ölbuchse *f* oil cup
Ölbürste *f* oil brush
Öldämpfer *m* oil damper
Öldampfsperre *f* oil-vapour baffle
Öldämpfung *f* oil damping
öldicht oil-tight
Öldichtung *f* oil seal
Öldiffusionspumpe *f* oil-diffusion pump
Öldocht *m* lubricating wick
Öldruck *m* 1. *(print)* oleography, lithochrome; oleograph; 2. *(mach)* oil pressure
Öldruckapparat *m* mechanical plunger lubricator
Öldruckbremse *f* oil (hydraulic) brake
Öldruckerei *f* oleography
Öldruckfederung *f* oil shock absorption
Öldruckkabel *n* oil-pressure cable
Öldruckkontrolleuchte *f (aut)* oil-pressure indicator lamp
Öldrucklager *n* bearing for forced lubrication
Öldruckpolster *n* oil cushion
Öldruckpresse *f* hydraulic jack
Öldruckpumpe *f* oil-feed pump
Öldruckverstärkerpumpe *f* booster pump *(hydraulics)*
Öldurchführung *f (el)* oil-filled bushing
Öldurchschlag *m* dielectric fault of oil, oil-puncture
Öldurchschlagsfestigkeit *f* dielectric strength of oil
ölecht fast to oil
Olefin *n* olefine
olefinisch olefinic
Öleinfüllöffnung *f* oil filler inlet
Öleinfüllstöpsel *m* oil filler plug
Öleinfüllstutzen *m* oil funnel (filler neck)
Öleinfüllverschraubung *f* oil filler plug
ölen to oil, to lubricate
Ölen *n* oiling, lubrication

Oleomargarin *n* oleomargarine
Öler *m* oil cup; lubricator, oiler
Ölerwärmung *f* oil heating
Oleum *n* oleum, fuming sulphuric acid
Ölextraktion *f* oil extraction
Ölfang *m* oil splash-guard; *(aut)* oil collector
Ölfangbehälter *m* oil-drainage container
Ölfangblech *n* oil baffle
Ölfänger *m* oil interceptor (catcher)
Ölfangring *m* oil catch-ring *(seal)*
Ölfangschale *f* oil pan
Ölfarbe *f* oil paint
Ölfeder *f* lubricator spring
Ölfeuerung *f* oil firing
Ölfilm *m* oil film
Ölfilmzerreißfestigkeit *f* oil-film rupture strength
Ölfilter *n* oil strainer (filter)
Ö. mit auswechselbarem Einsatz cartridge oil filter
Ölfilterpatrone *f* oil-filter cartridge
Ölfiltertopf *m* oil-filter cartridge
Ölfirnis *m* oil varnish
Ölflammofen *m* oil-fired air-furnace
ölfrei oilfree
Ölfüllung *f* oil filling
Ölgas *n* oil gas
Ölgehalt *m* oil length *(varnish)*
ölgehärtet oil-hardened
ölgeheizt oil-fuelled, oil-fired, oil-heated
ölgekühlt oil-cooled
ölgeschmiert oil-lubricated
ölgetränkt oil-impregnated
Ölgrube *f* oil sump *(transformer)*
Ölhafen *m* fuel (petroleum) wharf
Ölhalterring *m (aut)* oil control ring
ölhaltig oleiferous
ölhärtend oil-hardening *(steel)*
Ölhärter *m* oil-hardening steel
Ölhärtung *f* 1. *(met)* oil hardening *(steel)*; 2. *(ch)* hydrogenation of oil
Ölharzlack *m* oleo-resinous varnish
ölhydraulisch oil-hydraulic
ölig oily
Oligase *f* oligase
Ölimmersion *f* oil immersion
homogene Ö. homogeneous oil immersion
Olive *f* 1. olive; 2. *(mach)* burnishing shell *(broach)*
Olivenbaum *m* olive tree
Olivenöl *n* olive oil
Olivin *m* olivine
Ölkabel *n* oil-filled cable
Ölkanal *m* oil passage *(pump)*; oil hole; *(aut)* oilway
Ölkännchen *n* oil squirt
Ölkatarakt *m* dashpot
Ölkeil *m* wedge-shaped oil film
Ölkern *m* oil core *(foundry)*
Ölkessel *m* oil tank
Ölkesselschalter *m* dead-tank oil circuit-breaker
Ölkesselsicherung *f* oil-tank fuse
Ölkitt *m* oil putty
Ölkleidung *f* oilskin
Ölkohle *f* oil carbon, carbon deposits
Ölkoks *m* oil coke
Ölkondensator *m* oil-filled capacitor, oil-immersed capacitor
Ölkreislauf *m* oil circuit *(hydraulics)*
Ölkruste *f* oil deposit (crust)
Ölkuchen *m* oil (hemp) cake
Ölkuchenmühle *f* oil-cake mill (breaker)
Ölkühlung *f* oil cooling
Öllack *m* oil varnish
fetter Ö. long-oil-type varnish
Öllacksiedekessel *m* oil-boiling pot
Ölladung *f* oil cargo
Ölleichter *m* oil barge
flexibler Ö. flexible oil barge
Öllein *m* linseed flax
Ölleitung *f* oil line
Öllieferung *f* **/ theoretische** theoretical swept volume *(pump)*
Ölloch *n* oil hole
öllöslich oil-soluble
Ölmangel *m* oil starvation
Ölmanometer *n* oil-pressure gauge
Ölmenge *f* amount of oil
Ölmeßstab *m (aut)* oil dip-stick
Ölnase *f* **[an der Pleuelstange]** oil slinger (dipper, splasher)
Ölnebel *m* oil mist
Ölnebelkühlanlage *f* mist (spray) coolant system
Ölnebelkühlung *f* mist (spray) cooling
Ölnebelschmiereinrichtung *f* oil mist lubricator
Ölnebelschmierung *f* oil mist lubrication
Ölnut *f* oil groove (run), oilway
Ölpapierkondensator *m* oiled-paper capacitor
Ölpaste *f* paste paint
Ölpest *f* oil pollution
Ölpfütze *f* pool of oil
Ölprahm *m* oil barge (lighter)
Ölprobe *f s.* Ölprüfung
Ölprobiermaschine *f s.* Ölprüfer
Ölprüfer *m* oil tester
Ölprüfung *f* oil test
Ölpumpe *f* oil pump
Ölpumpengehäuse *n* oil-pump carrier
Ölraffination *f* oil refining, degumming of oil
Ölraffinerie *f* oil refinery
Ölreiniger *m* oil filter
Ölreinigungseinrichtung *f* oil filter
Ölring *m* oil ring, *(aut)* ring oiler
Ölrückführung *f* oil return
Ölrückgewinnungsschute *f* oil recovery barge
Ölrücklauf *m* oil return

Ölrücklaufleitung *f* oil return pipe
Ölrückstand *m* oil deposit (residue), residual oil
Ölrückstandskühler *m* residual oil cooler
Ölsammeltank *m* oil save tank
Ölsäure *f* oleic acid
Ölschalter *m* oil[-break] switch, oil circuit-breaker
Ö. mit isoliertem Kessel dead-tank oil circuit-breaker
Ölschauglas *n* oil-gauge glass
Ölschiefer *m* kerosene (oil) shale
Ölschirm *m* oil splash guard
Ölschläger *m* oil presser
Ölschlamm *m* oil deposit (residue, mud), sludge *(of an engine)*
Ölschleuderring *m* oil-thrower ring, oil-flinger ring, oil-slinger ring
Ölschleuse *f* oil reservoir (collector)
Ölschmierung *f* oil lubrication
Ölschöpfer *m* oil scoop (catcher, dipper)
Ölschütz *n* oil-filled contactor
Ölsicherung *f* oil-break fuse, oil-quenched fuse
Ölsieb *n* oil strainer (screen)
Ölspiegel *m* oil level, level of the oil
Ölspritzblech *n* oil splasher (splash guard)
Ölspritze *f* oil gun
Ölstand *m* oil level
Ölstand[s]anzeiger *m* oil[-level] gauge, contents gauge of the oil tank
Ölstandsmelder *m* oil-level transmitter, [minimum] oil-level alarm
Ölstein *m* oilstone
Ölstoff *m* empire cloth
Ölstoßdämpfung *f* oil shock absorption
Ölstrahlschalter *m* orthojector circuit-breaker
Ölströmung *f* oil flow
Ölströmungsschalter *m* oil-blast [circuit-]breaker
Ölsumpf *m* oil sump, sump pan, seepage
Öltank *m* oil [fuel] tank
Öltanker *m* bulk-oil carrier (carrying vessel), oil-tanker, oiler
Öltankleitung *f* oil line
Öltankschiff *n* oil-tank vessel, bulk-oil carrier (carrying vessel), vessel carrying petroleum in bulk, oil tanker, oiler
Öltasche *f* oil hole
Öltransformator *m* oil-immersed transformer, oil-insulated transformer, oil-cooled transformer
Ö. mit Anblasung forced air-cooled oil-immersed transformer
Ö. mit Lüfterkühlung forced air-cooled oil-immersed transformer
Öltrog *m* oil splash basin
Öltropfapparat *m* drip-feed lubricator, drop-feed lubricator
Öltropfenfall *m* oil dripping
Ölüberdruckventil *n* oil-pressure relief valve
Ölübernahme *f* fuel loading, oil fuelling
Ölüberströmventil *n* oil relief valve
Ölumlauf *m* oil circulation
Ölumlaufschmierung *f* oil-circuit lubrication
Ölumlaufsystem *n* oil-circulation system
Ölumleitventil *n* oil by-pass valve
Ölvergütung *f* 1. *(met)* oil treating; 2. oil-tempering *(paper-making)*
Ölversorgungsschiff *n* oil-supply vessel, oiler
Ölwandler *m* oil-immersed instrument transformer
Ölwanne *f* oil pan (sump), oilsump, crankcase bottom
Ölwärmetauscher *m* oil temperature equalizer
Öl-Wasser-Separator *m (ship)* oily water separator
Ölwechsel *m* oil renewal
regelmäßiger Ö. periodic oil-renewal
saisonbedingter Ö. seasonal oil-renewal
Ölwechselempfehlung *f (aut)* crankcase drain recommendation
Ölwechselperiode *f* oil renewal period
Ölwechselzeiten *fpl (aut)* crankcase drainage periods
Ölwerfer *m* oil splasher (dipper)
Ölwirkstoff *m* oil additive
Ölwurst *f* oil sausage *(flexible oil barge)*
schwimmende Ö. flexible barge
Olympiajolle *f* Olympic dinghy
Ölzähigkeit *f* oil viscosity
Ölzerstäuber *m* oil atomizer
Ölzeug *n* oilskin
Ölzufuhr *f* oil supply
Ölzuführung *f* oil supply
ungenügende Ö. oil starvation
Ölzusatz *m* oil dope (additive)
Ölzylinder *m* oil cylinder
Omnibus *m* [omni]bus
Omnibusfahrer *m* bus driver
Omnibusfahrpreis *m* bus fare
Omnibushof *m* bus depot (terminal)
Omnibusschaffner *m* bus conductor
Önanthaldehyd *m* enanthaldehyde, heptaldehyde, oenanthic aldehyde
Önanthsäure *f* enanthic (oenanthic, oenanthylic) acid
Ondograf *m* ondograph
Onyx *m* onyx
oolithisch oolitic
OP-Prozeß *m* Oppenheimer-Philips process
opak opaque, black
Opaleszenz *f* opalescence
Opalglas *n* opal[ine] glass
Opalglasleuchte *f* opal lighting fitting
Opalisiermittel *n* opacifier
Opallampe *f* opal (opaque) lamp
Opazität *f* opacity

Operandenadresse *f (dat)* operand address
Operateur *m s.* Reaktorfahrer
Operation *f* operation; handling, manipulation
binäre O. binary operation
logische O. logical operation
Operationalverstärker *m* operational amplifier
Operationsablauf *m* / **simultaner** concurrent operation
Operationsbefehl *m* operation code
Operationscharakteristik *f* operating characteristic
Operationshandleuchte *f* surgery hand-lamp
Operationshilfsleuchte *f* supplementary surgery lamp
Operationsleuchte *f* surgery lamp
Operationsmikroskop *n* surgery microscope
Operations Research operations research
Operationssaal *m* operating room *(hospital)*; operating theatre *(university)*
Operationssymbol *n* operator
Operationsteil *m* operation (instruction) part
Operationstisch *m* operating table
Operator *m (math)* operator
adjungierter O. adjoint operator
komplexer O. complex operator
Operatorenrechnung *f* operational calculus
Operngias *n* opera glass
Opernhaus *n* opera house
Ophthalmologe *m* ophthalmologist
Ophthalmologie *f* ophthalmology
Ophthalmometer *n* ophthalmometer
Ophthalmoskop *n* ophthalmoscope
Opiumpräparat *n* opiate
Oppenheimer-Philips-Prozeß *m* Oppenheimer-Philips process
Optik *f* optics
anamorphotische O. anamorphote lens
geometrische O. geometrical optics
Optiker *m* optician
Optikkitt *m* cement for optical elements
optimal optimum
Optimalbedingungen *fpl* optimum conditions
Optimalfilter *n* optimum filter
Optimalitätskriterium *n* optimality criterion
Optimalwert *m* optimum value
Optimalwertregelung *f* optimizing control
O. mit Maximumsuche peak-holding optimizing control
optimieren to optimize
Optimierung *f* optimization
lineare O. linear optimization
Optimierungsmaschine *f* automatic optimizer
Optimisator *m* / **automatischer** automatic optimizer
optisch optical
o. aktiv optically active
opto-elektronisch opto-electronic
Optometer *n* optometer
Orangenblütenöl *n* neroli oil
Orangenschalenöl *n* sweet orange oil
Orchester *n* orchestra pit *(theatre)*
ordentlich ordinary *(ray)*
Ordinate *f* ordinate
Ordinatenachse *f* ordinate axis, y-axis
ordnen to arrange; *(dat)* to collate
Ordner *m* folder; *(dat)* collator
Ordnungsstrafe *f* disciplinary fine
Ordnungszahl *f (math)* ordinal number; *(ch)* atomic number
O. der Oberschwingung harmonic number
effektive O. effective atomic number
Organ *n* element, member, organ
aperiodisch gedämpftes O. aperiodic element
astatisches O. astatic element
bewegliches O. moving element
kritisches O. *(nucl)* critical organ
Organdosis *f (nucl)* organ dose
Organe *npl* / **blutbildende** blood-forming system
Organiker *m* organic chemist
organisch organic *(chemistry)*
organisch-chemisch organic-chemical
organoleptisch organoleptic
Organometallverbindung *f* organometallic compound
Organosol *n* organosol
Organsinseide *f* organzine
Orgelstempel *m (mine)* battery stulls, breaker prop
orientiert *(math)* sensed; *(ph)* oriented *(crystal)*
Orientierung *f* alignment, sense; orientation *(crystal)*
automatische O. im Labyrinth automatic path-finding in the maze
vorzugsweise O. preferred orientation *(crystal)*
Originalabtastung *f* direct pick-up
Originalaufnahme *f* direct pick-up
Originalersatzteil *n* genuine part
Originalfrachtbrief *m* original consignment-note
Originalgalvano *n* master plate
Originalsendung *f* life broadcast
Originalvorlage *f* master copy
Originalzeichnung *f* original drawing
Ornamentglasleuchte *f* figured-glass lamp, ornamented-glass lamp
Ornamentguß *m* ornamental castings
Ornithin *n* ornithine
Orsatapparat *m* Orsat apparatus
Ort *n (mine)* end
Ort *m* 1. place; spot; site; 2. verge *(roof)*; 3. *(math)* locus
O. / **am** *(build)* in-situ
geometrischer O. locus
Ortbeton *m* site (cast-in-situ, poured-in-place, cast-in-place, in-situ) concrete

Ortbetonpfahl *m* cast-in-situ pile, cast-in-place pile
orten to locate, to position
Örterpfeilerbau *m* pillar mining method
Orthoborsäure *f* boric acid
orthochromatisch orthochromatic
Orthodeuterium *n* ortho-deuterium
Orthodiagrafie *f* orthodiagraphy
Orthodoma *n* orthodome *(crystallography)*
orthogonal orthogonal
Orthohelium *n* orthohelium
Orthokieselsäure *f* orthosilicic acid
Orthoklas *m* orthoclase
Orthophosphorsäure *f* orthophosphoric acid
Orthoröntgenoskopie *f* orthoradioscopy
orthoskopisch orthoscopic
Orthowasserstoff *m* ortho-hydrogen
Orthozustand *m* ortho-state
örtlich local[ly]; topological
  ö. begrenzt localized
Ortrolle *f* barge (verge) course
ortsabhängig locus-dependent
Ortsabhängigkeit *f* locus-dependency
Ortsältester *m (mine)* charge hand, chargeman, *(Am)* face boss, foreman
Ortsamt *n* local exchange; *(Am)* local centre office
Ortsbatterie *f* local battery
Ortsberge *pl (mine)* roadhead (first ripping) dirt
ortsbesetzt locally busy *(telephony)*
ortsbeweglich portable; movable
Ortsbrett *n (mine)* breast board
Ortsbrust *f (mine)* drift face *(stone)*, heading face *(coal)*, breast, face
Ortsdosis *f (nucl)* local dosage (dose)
Ortsempfänger *m* receiver for local reception
ortsfest fixed, stationary
Ortsfunktion *f (math)* position function
Ortsgebührenzone *f* local service area
ortsgebunden *(mach)* stationary; localized *(vector)*
Ortsgespräch *n* local call
Ortshöhe *f* elevation head
Ortskabel *n* local (subscriber's) cable
Ortsknotenvermittlung *f* local junction exchange
Ortskurve *f* geometric (transfer) locus
  O. des Frequenzganges frequency-response, Nyquist diagram, transfer locus
Ortsleitungsnetz *n* local telephone network
Ortsleitungswähler *m* local line selector
Ortsmünzfernsprecher *m* public telephone-booth for local service
Ortsnamenverzeichnis *n* index of place names
Ortsnetz *n (el)* local [distributing] system, local mains; *(telec)* local exchange (telephone) network
Ortspfeiler *m (mine)* face pillar
Ortsplatz *m* local position *(telephony)*
Ortsquerschlag *m (mine)* sublevel crosscut
Ortsstoß *m (mine)* bank, breast, end, face
Ortsvektor *m* localized (position, radius) vector
ortsveränderlich portable, mobile
Ortsverkehr *m (telec)* local transit
Ortsverkehrsbereich *m (telec)* local area
Ortszeit *f* local time
Ortung *f* positioning, locating; position finding (fixing)
  O. der Störung trouble shooting
  O. des Fehlers trouble shooting
  astronomische O. celestial navigation
Ortungsfehler *m* position error
Ortungsfrequenz *f* frequency used in radiolocation, radar frequency
Ortungsfunkstelle *f* radiodetermination station
Ortungsgenauigkeit *f (mach)* positioning accuracy, accuracy of location
Ortungsgerät *n* bearing compass, position (direction) finder
Öse *f* eyelet, eye, ear; loop *(spring)*
Ösenbrett *n* eyeletted board
Ösenmaschine *f* eyeletting machine *(shoe)*
Oskulation *f (math)* osculation
Oskulationskreis *m* osculating circle
Osmat *n* osmate
Osmiridium *n* osmiridium
Osmium *n* osmium
Osmiumglühfaden *m* osmium filament
Osmiumlegierung *f* osmide
Osmiumsäure *f* osmic acid
Osmometer *n* osmometer
Osmondit *m* osmondite
Osmose *f* osmosis
Osmosebleiche *f* osmotic bleaching
Osmoseverfahren *n* diffusion process
osmotisch osmotic *(pressure)*
Ostasienfahrer *m* Indiaman *(vessel)*
Östron *n* oestrone
Ostwaldverfahren *n* Ostwald process
Oszillator *m* oscillator
  O. mit abgestimmten Gitter- und Anodenkreisen tuned-grid tuned-anode oscillator
  O. mit abgestimmten Resonanzleitungen resonant-line oscillator
  O. mit abstimmbarer Spannung voltage-tunable oscillator
  O. mit fester Phasenbeziehung coherent oscillator
  O. mit hoher Abstimmgenauigkeit high-discrimination oscillator
  O. mit koaxialen Topfkreisen coaxial line oscillator
  O. mit negativem Widerstand dynatron oscillator
  anharmonischer O. anharmonic oscillator

harmonischer O. harmonic oscillator
Hertzscher O. Hertzian oscillator
impulsgesteuerter O. pulsed oscillator
kohärenter O. coherent oscillator
langsamer O. low-speed oscillator
leistungsfähiger O. powerful oscillator
quarzgesteuerter O. quartz oscillator
schneller O. high-speed oscillator
selbsterregter O. free-running oscillator
spannungsgesteuerter O. voltage-sensitive oscillator
symmetrischer O. balanced oscillator
synchronisierter O. locked oscillator
Oszillatorabgleich *m* oscillator alignment
Oszillatoraggregat *n* oscillator section
Oszillatoransprechzeit *f* **bei Tastung** pulsed-oscillator starting time
Oszillatorfrequenz *f* oscillator frequency
Oszillatorklystron *n* oscillating klystron
Oszillatormodell *n* oscillator model
Oszillatorröhre *f* oscillator valve
Oszillatorserienkondensator *m* oscillator padder
Oszillatorstärke *f* oscillator strength
Oszillatorstufe *f* oscillator stage
oszillieren to oscillate
Oszillograf *m* oscillograph, recording oscillometer
O. in Einschubtechnik modular oscillograph
O. mit Kalziumwolframatleuchtschirm blue-glow oscillograph
O. mit langer Nachleuchtdauer long-persistence oscillograph
schreibender O. oscillograph
Oszillografenschirm *m* oscillograph screen
Oszillogramm *n* oscillogram, pattern
Oszilloskop *n* oscilloscope
Otterfell *n* otter
Ottokraftstoff *m* petrol
Ottomotor *m* Otto engine
Ovalkaliber *n* oval pass (groove)
Ovalkessel *m* oval boiler
Ovalradzähler *m* oval gear meter
Ovalstich *m* bellied pass
Ovizid *n* ovicide
Oxalat *n* oxalate
Oxalsäure *f* oxalic acid
Oxazinfarbstoff *m* oxazine dye
Oxid *n* oxide
Oxidausschlackung *f* oxide slagging
Oxideinschluß *m* oxide inclusion
Oxidhaut *f* oxide skin
Oxidhäutchen *n* oxide film
oxidisch oxidic, oxygenic
Oxidkatode *f* oxide-coated cathode
Oxidkeramik *f* oxide ceramic
Oxidschicht *f* layer of oxide, oxidation film; oxide layer (film)
Oxim *n* oxime
Oxoformel *f* planar sugar formula
Oxoniumsalz *n* oxonium salt
Oxychinolin *n* hydroxyquinoline
Oxydase *f* oxidase
Oxydation *f* oxidation
innere O. inner oxidation
Oxydationsapparat *m* oxidizing machine
oxydationsbegünstigend pro-oxidant
oxydationsbeständig oxidation-resistant
Oxydationsbeständigkeit *f* oxidation stability
Oxydationsflamme *f* oxidizing flame
Oxydationsgrad *m* degree of oxidation
oxydationshemmend anti-oxidant
oxydationshindernd anti-oxidant
Oxydationsmittel *n* oxidizing agent, oxidizer, oxidant
Oxydationsprodukt *n* oxidate
Oxydationsschutz *m* oxidation inhibition
Oxydationsstufe *f* degree of oxidation, oxidation stage
Oxydationsvorgang *m* oxidation process
Oxydator *m* *(rock)* oxidizing component con-stituent), oxygen c ier, oxidizer
Oxydatorbehälte . oxidant (oxidizer) tank
Oxydatorstrab oxidizer jet
Oxydatorverb uch *m* oxidizer flow (consumption)
oxydierbar oxidizable, oxygenizable, capable of being idized
nic o. inoxidable
Ox erbarkeit *f* oxidizability, capability of idation
ydieren to oxidize
Oxydierungsmittel *n* oxidizing agent
Oxygenase *f* oxygenase
Oxysäure *f* hydroxy acid
Oxyzellulose *f* oxycellulose
Ozalidlichtpause *f* Ozalid print
Ozalidlichtpausverfahren *n* Ozalid printing (white print) process
Ozeandampfer *m* ocean steamer
Ozeanografie *f* oceanography
Ozeanschiff *n* ocean vessel, ocean-going ship
schnelles O. ocean greyhound
Ozeanschlepper *m* ocean-going tug
Ozeanwetterschiff *n* ocean weather ship
Ozokerit *m* ozokerite, earth wax
Ozon *n* ozone
ozonbeständig ozone-proof
Ozonbildung *f* ozone formation
Ozonbleiche *f* ozone bleaching
ozonfest ozone-resistant
ozonhaltig ozoniferous
Ozonid *n* ozonide
Ozonisator *m* ozonizer
ozonisieren to ozonize
Ozonlampe *f* ozone lamp

P

P-Abweichung *f* load static error
P-Anteil *m* proportional control factor
P-Band *n* P-band
P-Bereich *m* proportional band, *(Am)* proportioning band
vorübergehender P. temporary droop
P-Folgeregler *m* type 0 servomechanism
P-Glied *n* pure gain element
p-Halbleiter *m* p-type semiconductor
p-leitend p-conducting, p-type
p-Leitfähigkeit *f* p-type conductivity
P-Leitung *f* p-type conduction
P-Regelung *f* proportional control, *(Am)* proportioning control
P-Regler *m* proportional [acting] controller
p-Typ-Halbleiter *m* p-type semiconductor
P-Verhalten *n* proportional action
P-Verstärkung *f* proportional sensitivity
P-Wirkung *f* proportional action
P-Zacken *mpl* P-waves *(electrocardiogram)*
pa-Großbehälter *m* container fitted for special conveyance, pa-container
Paar *n* pair *(shoes; gears; electrons; etc.)*; couple *(forces)*
Paarbildung *f (nucl)* pair production (formation, creation)
Paarbildungsenergie *f* pairing energy
Paarbildungskoeffizient *m* pair-creation coefficient
Paarbildungsquerschnitt *m* pair production cross section
Paarbildungsreaktion *f* pair production (formation, creation)
paaren to mate *(gears)*, to conjugate; to pair; to match *(e.g. transistors)*; to couple
Paarerzeugung *f (nucl)* pair production (creation)
paarig in pairs, paired
Paarigkeit *f (telev)* twinning
Paarigstehen *n* **der Zeilen** *(telev)* twinning
Paarspektrometer *n* pair spectrometer
Paarung *f* pairing; mating *(fit)*
P. der Zahnräder mating of [the] gears
formschlüssige P. closed pair, mating
kraftschlüssige P. unclosed pair
Paarungslehre *f* functional gauge *(measurement)*
Paarungslehren *n* functional gauging *(measurement)*
Paarvernichtung *f (nucl)* positron-electron annihilation; pair annihilation
Paarverseilung *f (el)* pairing, pair stranding (twisting)
paarweise [angeordnet] [arranged] in pairs
Pacht *f* lease
pachtfrei rent-free
Pachtgeld *n* rent
pachtweise on lease
Pack *m* pack; bundle; bale
Packbindfaden *m* twine, cord
Päckchen *n* [small] parcel; bundle, sheaf, wad
packen to bundle, to pack [up]
Packen *m* bundle; bale; pack
Packer *m* 1. packer; 2. *(ag)* furrow press
Packfilm *m* packfilm
Packlage *f (build)* bottoming, pitching, telford base, hard-core
Packleinwand *f* packcloth
Packmaterial *n* packing material
Packpapier *n* packing (brown, wrapping) paper
Packraum *m* packing-room
Packtisch *m* packing table
Packtrockenmaschine *f (text)* package drying machine
Packung *f* package, packing; gasket
hexagonal dichteste P. hexagonal-close packing
kubisch dichteste P. cubic-close packing
Packungsanteil *m (nucl)* packing fraction
Packungsdichte *f* packing density *(of components)*
Packungseffekt *m (nucl)* mass (packing) effect
Packungsmaterial *n* packing material *(sealing)*
Packungsraum *m* packing space *(sealing)*
Packungsringe *mpl* / **ineinandergreifende** intersetting packing
Packungsscheibe *f* packing shim
Packungsstopfbüchse *f* stuffing box, packing gland
Packungstreiber *m* packing stick
Packungszieher *m* packing worm
Packwagen *m* luggage van
Packwagenschaffner *m* guard, *(Am)* baggage man
Paddelboot *n* paddling (paddle) boat, canoe
Paddingmaschine *f* padder, padding machine
Pagina *f* folio, page
paginieren to page, to paginate
Paginierwerk *n* numbering apparatus (box)
Pak *f s.* Panzerabwehrkanone
Paket *n* 1. parcel; 2. *(nucl)* packet, pack[age], bunch
Paketbildung *f* / **unterkritische** underbunching *(electronics)*
Paketboot *n* packet, packet (mail) boat
paketieren *(met)* to pile, to fag[g]ot
Paketieren *n* **[bei Schweißstahlerzeugung]** fag[g]oting
Paketierpresse *f* baling (scrap piling) machine
Paketierung *f (nucl)* bunching
Paketschalter *m* multi-section-type rotary switch, gang-switch
Paketschiff *n* mail (packet) boat
Paketstapeln *n* packaging

Paläobotanik *f* fossil botany
Paläontologie *f* fossilism, palaeontology
Palette *f* palette; plywood pallet; [loading] pallet *(materials handling)*
Palettenfrachtschiff *n* palletized cargo ship
palettieren to palletize *(materials handling)*
Palettisieren *n* palletization *(materials handling)*
Palisanderholz *n* rosewood
Palladium *n* palladium
Palladiumasbest *m* palladinized asbestos
Palladiummohr *n* palladium black
Palladiumschwamm *m* palladium sponge
Palladium(II)-verbindung *f* palladous compound
Palloidspiralkegelrad *n* palloid spiral bevel gear
Palmette *f* fan-trained tree
Palmitinsäure *f* palmitic acid
Palmkernöl *n* palm kernel oil
panchromatisch panchromatic
Paneel *n* panel
abgesperrtes und furniertes P. cross-banded veneered panel
PAN-Faser *f* polyacrylonitrile fibre
Panfilm *m* pan film stock
Pankreassaft *m* pancreatic juice
Panne *f* breakdown; puncture; failure
Panoramaaufnahme *f* panoramic exposure
Panoramadarstellung *f* panoramic display
Panoramaeinrichtung *f* all-round locking device
Panoramaempfänger *m* panoramic receiver
Panoramagerät *n* panoramic unit, [air-borne] all-allround search apparatus, radar landscape scanner, plan position indicator
Pantoffelholz *n* cork-board, pommel of cork
Pantograf *m* pantograph [mechanism]
P. für Raumformen (dreidimensionales Fräsen) three-dimensional pantograph
Pantografengraviermaschine *f* pantographic engraving machine
Pantografnachformfräsmaschine *f* pantomiller, pantograver
Pantometer *n* pantometer
Pantothensäure *f* pantothenic acid
Pantry *f (ship)* pantry
Panzer *m (mil)* tank; armour
Panzerabwehrkanone *f* antitank gun
Panzeraderleitung *f* armoured (metal-cased) conductor
Panzerbrandgranate *f* armour-piercing incendiary shell
panzerbrechend armour-piercing
Panzerbüchse *f* antitank rifle
reaktive P. antitank rocket launcher
Panzerdeck *n (ship)* protective deck
panzerdurchschlagend armour-piercing
Panzererkennungsdienst *m* tank identification service
Panzerfahrer *m (mine)* conveyor operator
Panzerfalle *f* tank trap
Panzerförderer *m (mine)* armoured conveyor
Panzergalvanometer *n* shielded galvanometer
Panzergraben *m* [anti]tank ditch
Panzerhindernis *n* antitank obstacle
Panzerkabel *n* shielded (armoured) cable
Panzerkanone *f* tank gun
Panzerkasten *m* tank hull
panzern 1. *(nucl)* to shield, to screen; 2. to hardsurface *(welding)*
Panzerplatte *f* armour plate; *(el)* box[ed] plate, box-type plate *(of storage battery)*
Panzerplattenhobelmaschine *f* armour-plate planer (planing machine)
Panzerrohr *n (el)* steel conduit
Panzerschlauch *m* armoured hose
Panzerschnur *f (el)* armoured cord
Panzerschott *n* armour bulkhead
Panzerspähwagen *m* armoured car, armoured reconnaissance (scout) car
Panzersperre *f* antitank obstacle
Panzerstahl *m* armour steel
Panzertransportfahrzeug *n* tank carrier
Panzerturm *m* armoured turret
Panzerung *f* armour[ing]; *(mine)* anti-wear lining
Panzerwarndienst *m* antitank warning system (service)
Papaverin *n* papaverine
Papier *n* paper
P. bester Güteklasse high-grade paper
P. für Notizbücher tablet paper
aufgearbeitetes P. recovered waste paper
beschichtetes P. coated paper
doppeltlogarithmisches P. log-log paper
einfachlogarithmisches P. semilogarithmic paper
einseitig gestrichenes P. semicoated (one-sided) paper
einseitig glattes P. machine-glazed paper, M. G. paper
fettdichtes P. grease-proof (grease-resistant) paper
geleimtes P. sized paper
geripptes P. laid paper
gummiertes P. gummed paper
gußgestrichenes P. cast-coated paper
halblogarithmisches P. ratio (semilogarithmic) paper
handgeschöpftes P. hand-made paper
harzgeleimtes P. resin-sized paper
hochsatiniertes P. supercalendered paper
holzfreies P. wood-free paper
kaschiertes P. laminated paper
leinengeprägtes P. linen-grained paper

lichtempfindliches P. sensitized paper
liniiertes P. ruled paper
maschinengeleimtes P. machine-sized paper
maschinengetrocknetes P. cylinder-dried paper
maschinenglattes P. machine-finished paper, M. F. paper, machine-glazed paper, M. G. paper
metallisiertes P. metallized paper
naßfestes P. wet-strength paper
ölimprägniertes P. oil-impregnated insulating paper
parfümiertes P. scented paper
satiniertes P. calendered (glazed, imitation art) paper
saugfähiges P. absorbent paper
scharfsatiniertes P. surglace
sensibilisiertes P. sensitized paper
stark gefülltes P. highly filled paper
ungeglättetes P. unfinished paper
ungeleimtes P. unsized (waterleaf) paper
ungestrichenes P. uncoated paper
Papierabfall *m* waste paper
Papierabgang *m* waste paper
Papierabreißvorrichtung *f* paper tearing device
Papierabschnitt *m* offcut
Papieraufhängevorrichtung *f* paper hanging device
Papierbahn *f* paper web
Papierbahnbreite *f* web width
Papierbahnspannung *f* web tension
Papierbaumwollkabel *n* paper-cotton-covered cable
Papierbeschneidemaschine *f* paper trimmer
Papierbespannung *f* paper covering
Papierbindfaden *m* paper twine
Papierbleikabel *n* paper-insulated lead-covered cable
Papierbogen *m* sheet of paper
Papierbrei *m* paper pulp
Papierchromatografie *f* paper chromatography
Papierchromatogramm *n* paper chromatogram
Papierdicke *f* caliper of paper
Papiereinschuß *m* paper interleaf
Papierfabrik *f* paper-mill
Papierfaserstoff *m* half-stuff, pulp *(prior to stuff or stock preparation)*
P. aus Holz wood pulp
P. aus Laubholz hardwood pulp
P. aus Nadelholz softwood pulp
gebleichter P. bleached pulp
halbgebleichter P. partly bleached pulp
ungebleichter P. unbleached pulp
Papierfeuchte *f* wetting device (bench)
Papierfilter *n* paper filter
getränktes P. treated-paper filter
Papierformat *n* paper (sheet) size
Papierführung *f* paper guide
Papiergewebe *n* paper cloth (fabric)
Papierhadern *pl* rags
Papierhersteller *m* paper maker, papermaker
Papierherstellung *f* paper making, papermaking
Papierhohlraumkabel *n* air-space paper-core cable
Papierisolation *f* / **ölimprägnierte** *(el)* oil-impregnated paper insulation
Papierisolierrohr *n (el)* paper conduit
papierisoliert paper-covered, paper-insulated
Papierisolierung *f* paper insulation
Papierkabel *n* paper[-insulating] cable
Papierkondensator *m* paper[-dielectric] capacitor
rohrförmiger P. tubular paper capacitor
Papierkorb *m* [waste-]paper basket
Papierleim *m* paper size
Papierleimung *f* paper sizing
Papierlochband *n* punched paper tape
Papiermaché *n* papier-mâché
Papiermacherfilz *m* paper-maker's felt
Papiermaschine *f* paper [making] machine
Papiermaschinenbütte *f* head box *(papermaking)*
Papiermaschinenrückwasser *n* backwater from the papermaking machine
Papiermasse *f* [paper] pulp
Papiermassefänger *m* pulp catcher
Papiermater *f* paper flong
Papiermaulbeerbaum *m* paper mulberry
Papiermaulbeerbaumfaser *f* paper mulberry fibre
Papiernormalmaß *n* paper gauge
Papiernummer *f* paper number
Papiersack *m* / **mehrfachwandiger** multi-wall paper sack
Papierschere *f* paper-scissors, paper-shears
Papierschneider *m* [paper] cutting machine, paper trimmer
Papierschnitzel *npl* paper shavings
Papierschnur *f* paper twine (string)
Papierschrumpfung *f* paper shrinkage
Papierserviette *f* paper napkin
Papierspannung *f* web tension, paper stress
Papierspannungsregelung *f* web tension control
Papierspule *f* paper bobbin
Papierstaub *m* paper dust
Papierstoff *m* paper stock, stuff
Papierstreifen *m* paper strip (band, tape)
Papiertapete *f* wall paper
Papiertrommel *f* paper drum *(indication)*
Papierverarbeitungsindustrie *f* paper-converting industry
Papierverarbeitungsmaschine *f* paper-converting machine
Papiervorschub *m* paper feed
Papierwaage *f* paper scales
Papierwarenhersteller *m* paper goods manufacturer

Papierwolf *m* paper shredder (shredding machine)
Papierwolle *f* paper shred
Papierzerfaserer *m* paper shredder
Papierzeug *n* stuff
Papierzuschuß *m* additional paper
Papierzwischenlage *f* intermediate layer of paper
Pappband *m* pasteboard-binding
Pappdach *n* felt roof[ing]
  mehrlagiges P. built-up roof[ing]
Pappe *f* 1. board; cardboard; millboard; boxboard; carton; 2. felt *(roofing)*
  biegbare P. domino board
  durchgefärbte P. board coloured throughout
  gezogene P. moulded board
  mehrlagige P. multi-ply board
  voluminöse P. bulking board
  zusammengeklebte P. pasted board
Pappel *f* poplar
Pappelholz *n* poplar wood
Pappenbedruckmaschine *f* board printing machine
Pappenbiegemaschine *f* [card]board bending machine
Pappenfabrik *f* board mill
Pappenform *f* board mould
Pappenmaschine *f* board machine
Papphülse *f* pasteboard tube
Pappkarton *m* cardboard box
Pappkreisschere *f* circular cardboard cutter
Pappschachtel *f* carton, cardboard box
Pappschere *f* cardboard cutter (shears)
Papyrolin *n* cloth-faced paper
Parabansäure *f* parabanic acid
Parabel *f* parabola
Parabelabschnitt *m* segment of the parabola
Parabelfachwerk *n* parabolic truss
Parabelmethode *f* parabola method
Parabelnietung *f* chain-spaced rows of rivets
Parabelpotential *n* parabolic potential
Parabelspektrograf *m* parabola spectrograph
Parabelträger *m (build)* parabolic girder
parabolisch parabolic
Paraboloid *n* paraboloid
Paraboloidschale *f* / **hyperbolische** hyperbolic paraboloid shell
Parabolreflektor *m* parabolic reflector
Parabolspiegel *m* parabolic mirror (reflector)
Parachor *m(n)* parachor
Paradeuterium *n* para deuterium
Paradoxon *n* / **hydrodynamisches** hydrodynamic paradox
  hydrostatisches P. hydrostatic paradox
Paraffin *n* paraffin
Paraffinbad *n* paraffin bath
Paraffingatsch *m* paraffin sludge (slack wax)
paraffinieren to paraffin
paraffinisch paraffinaceous, paraffinic
Paraffinkohlenwasserstoffe *mpl* paraffinic hydrocarbons
Paraffinkrätze *f* [paraffin-]rash
Paraffinleim *m* wax size
Paraffinöl *n* kerosene, paraffin oil
Paraffinpapier *n* waxed paper
Paraffinphantom *n* paraffin phantom
Paraform[aldehyd *m***]** *n* paraform[aldehyde]
Paragraphschiff *n* paragraph-vessel
Parahelium *n s.* Parhelium
Paraldehyd *m* paraldehyde
parallaktisch parallactic
Parallaxe *f* parallax
Parallaxenausgleich *m* parallax compensation
Parallaxenfehler *m* parallax error
parallaxenfrei free from parallax
Parallaxenrechner *m* parallax computer
parallel parallel
  p. mit dem Feld parallel to the field
Parallelausdrucken *n (dat)* line-a-time printing
Parallelausgang *m* parallel output
Parallelbetrieb *m* parallel operation, operation in parallel
Parallelbewegung *f* parallel motion
Parallelbruch *m* parallel fold
Paralleldämpfung *f* parallel damping
Paralleldrahtleitung *f* parallel-wire line, twin feeder
Paralleldrossel *f* shunt inductor
Parallelendmaß *n* block (slip) gauge
Parallelepiped *n* parallelepiped
Parallelfachwerk *n* flat (parallel chord) truss, N-truss
Parallelfeld *n* parallel field of forces
Parallelflach *n* parallelepiped
Parallelform *f* / **halbpermanente** semi-permanent parallel form
Parallelführung *f* parallel mechanism *(drafting machine)*; *(el)* parallel line running *(overhead lines)*; *(therm)* parallel guide
Parallelfunkenstrecke *f* parallel discharge-gap
Parallelimpedanz *f* leak impedance
parallelisieren to make parallel, to cause to lie parallel to one another *(fibres)*
Parallelisierung *f* parallelization *(fibres)*
Parallelität *f* parallelism
Parallelkapazität *f (el)* shunt capacity
  P. des Quarzes inherent capacity
Parallelkondensator *m* parallel (shunt) capacitor
Parallelkreis *m* shunt circuit
Parallelkurbel *f* parallel-crank mechanism
Parallelkurbelgetriebe *n* parallel-crank four-bar linkage
Parallellauf *m (el)* parallel operation
Parallellaufkreise *mpl* tracking circuits

Parallelmanipulator *m* master-slave manipulator
Parallelogramm *n* parallelogram
  P. der Geschwindigkeiten parallelogram of velocities
  P. der Kräfte parallelogram of forces
Parallelogrammführung *f* parallel mechanism
Parallelperspektive *f* parallel perspective
Parallelplattenionisationskammer *f* parallel plate [ionization] chamber
Parallelprojektion *f* parallel projection
  schiefe P. auf lotrechte Bildebene oblique projection
Parallelreduktion *f (dat)* on-line data reduction
Parallelreißer *m* surface gauge, scribing block
Parallelresonanz *f* parallel resonance
Parallelresonanzkreis *m* parallel resonant circuit
Parallelröhrenmodulation *f* choke modulation
Parallelsaumkreissäge *f* circular saw for parallel sawing
parallelschalten *(el)* to parallel, to connect (put) in parallel, to shunt, to connect across
  die Turbinen p. to connect the turbines in parallel
Parallelschaltung *f* parallel (shunt) connection, shunting; paralleling
Parallelschnitt *m* parallel section
Parallelschraubzwinge *f* parallel clamp
Parallelschwingkreis *m* parallel-resonance circuit
Parallelschwingung *f* parallel vibration
Parallelserienwandler *m (dat)* dynamicizer
Parallelspeicher *m* parallel memory (storage)
Parallelspule *f* bridging coil
Parallelstoß *m* edge joint
Parallelstromkondensation *f (therm)* parallel-flow condensation, parallel-current condensation
Parallelstromkondensator *m* parallel-current condenser
Parallelstück *n* parallel strip
Paralleltaktschaltung *f* push-push circuit
Parallelton *m (telev)* separate sound
Parallelträger *m* parallel-flanged girder
Parallelübertrag *m (dat)* simultaneous carry
Parallelverschiebung *f* parallel displacement (translation, shift)
Parallelverzweiger *m* shunt T, T-junction *(waveguides)*
Parallelwicklung *f (el)* parallel winding
  eingängige P. simple parallel winding
  mehrgängige P. multiple parallel winding
Parallelwiderstand *m (el)* parallel resistance; bleeder (shunt) resistance
Paralyse *f* paralysis
Paramagnetikum *n* paramagnet[ic material]
paramagnetisch paramagnetic
Paramagnetismus *m* paramagnetism
  atomarer P. atomic paramagnetism
Parameter *m* parameter
  P. der Flugbahn parameter of the orbit
  P. der Grundgesamtheit population parameter
Parameter *mpl* **/ parasitäre** stray parameters
Parameterstreuung *f* parameter scattering
  zulässige P. permissible parameter scattering
parametrisch parametric
Parapiassava *f* para piassava
Pararot *n* para-red
Parasitärstrahlung *f* spurious radiation
Parawasserstoff *m* para-hydrogen
Paraxialraum *m* paraxial region
Parazustand *m* para-state
Pardune *f* guy[-rope]
Pardunenausleger *m (ship)* backstay stool (outrigger)
Pardunenisolator *m* guy insulator
Parenthese *f* parenthesis
Parfüm *n* perfume, scent
Parfümerie *f* perf[illegible]ery
parfümieren to perfume, to scent
Parfümzerstäuber *m* perfume atomizer
Parhelium *n* parhelium
Parität *f* parity
Paritätsänderung *f* parity change
Paritätskontrolle *f* parity check
Paritätsprüfung *f* parity check
Paritätsziffer *f* parity digit
parken *(aut)* to park
parkern *(met)* to parkerize
Parkett *n* orchestra *(theatre)*; parquet[ry], inlaid floor
Parkettbrett *n* parquet board
Parkettfügemaschine *f* machine for parquetry work
Parkettfußboden *m* wood-block flooring, parquetry; inlaid floor
Parkettstab *m* parquetry-fillet
Parkettstabhobelmaschine *f* planing machine for parquetry fillets
Parkleuchte *f (aut)* parking lamp
Parklicht *n (aut)* parking light
Parkplatz *m* car park, parking lot
Parkspur *f* parking lane, layby
Parkuhr *f* park meter
Parkverbot *n* no parking!
Parkweg *m* layby *(motor way)*
Parlamentärschiff *n* cartel [ship]
Parsek *n* parsec
Partenreeder *m* part (joint) owner
Parterre *n* ground-floor
Partialbreite *f (nucl)* partial [level] width
Partialbruch *m (math)* partial fraction
Partialbruchzerlegung *f* decomposition into partial fractions

Partialdruck *m* partial pressure
Partialdruckmeßgerät *n* partial pressure measuring device
Partialquerschnitt *m (nucl)* partial cross section
Partialschild *m (nucl)* shadow shield
Partialturbine *f* partial admission turbine
Partialwelle *f* partial wave
Partie *f* lot, parcel
Partieerfolgsmatrix *f* pay-off matrix
partiell *(math)* partial
Partienummer *f (text)* lot number
Partikel *f* particle
Partikelausbruch *m* ejection of particles
Partikelfeld *n* field of a particle
Parzelle *f* allotment, plot
Parzellendreschmaschine *f* parcel thresher
Parzellendrillmaschine *f* parcel drill
Paschen-Back-Effekt *m* Paschen-Back effect
Paschen-Serie *f (nucl)* Paschen series
paspeln to braid, to bord
Paspelnaht *f* welt seam
Passage *f* 1. passage; voyage; 2. *(ship)* passageway
Passagier *m* passenger
  blinder P. stowaway; *(Am)* deadhead
Passagierflugzeug *n* air-liner
Passagiergut *n* luggage
Passagierschiff *n* passenger ship (vessel); passage boat
Paßarbeit *f* fitting
Paßbild *n* passport photograph
Paßdifferenz *f* register difference
Paßdorn *m* setting plug
passen to suit; to fit
passend suitable; [well-]fitting; applicable
  dicht p. tight-fitting
Passer *m* / **genauer** hairline register
Passerkreuz *n* register mark
Paßfeder *f* feather [key]
Paßfläche *f* mating surface
paßgenau true to size
passierbar passable
passieren to pass; to strain *(screen)*; to traverse
passiv passive *(metal)*; *(ch)* inert, inactive
passivieren to inactivate, to passivate
Passivierung *f* passivation
  anodische P. anodic passivation
Passivität *f* passivity, inactivity, inertness
Paßkreuz *n* register mark
Paßkreuzscheibe *f* register mark templet
Paßlehre *f* gauge
Paßring *m* gauge ring
Paßrohr *n* adapting (template) pipe, making-up piece
Paßschaufel *f* stop blade *(turbine)*
Paßschraube *f* precision (reamed, fitting, fitted, template) bolt; fitting [ring] screw, gauge ring
Paßstift *m* locating (set, fitting, steady) pin, dowel [pin]
Paßstück *n* fitting piece, mating part *(fits)*, adapter
Paßteil *n* fitting piece, mating part
Passung *f* fit
Passungsauswahl *f* selection of fits
Passungseinheit *f* tolerance unit
Passungsfläche *f* mating surface
Passungskurzzeichen *n* symbol designating [a] fit
Passungstafel *f* [standard] fit table
Paste *f* paste
Pastekatode *f* pasted filament
Pastellfarbstift *m* pastel[-crayon]
pasteurisieren to pasteurize
Pastille *f* pastil, lozenge, tablet
Pastillenpresse *f* tablet press
pastös pasty, soft
Patent *n* patent
  angemeldetes P. patent pending, filed patent
Patentablauf *m* expiration of [the] patent
Patentamt *n* patent office
Patentanker *m* patent (close stowing) anchor
Patentanmelder *m* patentee, claimant
Patentanmeldung *f* patent application
Patentanspruch *m* patent claim, claim of a patent
Patentanwalt *m* patent attorney (agent)
Patentausübung *f* / **obligatorische** compulsory working of patents
Patentbeschreibung *f* patent specification
Patentbrief *m* letters patent
Patenterneuerung *f* patent renewal
Patenterteilung *f* patent allowance (grant)
patentfähig patentable
Patentgebühr *f* patent fee
Patentgemeinschaft *f* patent pool
Patentgenehmigung *f* patent granting
Patentgesetz *n* patent statute (law, code)
Patentgesuch *n* patent application
Patentglas *n* patent glass
patentieren *(ec, met)* to patent
Patentierofen *m* patenting furnace
patentiert im In- und Ausland patented at home and abroad
Patentinhaber *m* patentee, patent holder
Patentklage *f* patent suit (action)
Patentlaufzeit *f* patent life
Patentliste *f* patent list
Patentlizenz *f* patent licence
Patentlog *n (ship)* screw (patent, taffrail) log, speedometer
Patentrecht *n* patent law (right)
Patentregister *n* register of patents, patent rolls
Patentschäkel *m (ship)* swivel (patent) shackle
Patentschrift *f* patent [specification]
Patentträger *m* patentee

Patentuntersuchung *f* patent survey
Patenturkunde *f* patent grant
Patentverletzung *f* patent infringement, infringement of patent
Patentzeichnung *f* patent-office drawing
Paternoster *m* paternoster (rotary) lift
Paternosteraufzug *m* paternoster (rotary) lift
Patina *f* verdigris; aerugo, patina
patinieren to patinate
Patinierfarben *fpl* greening lacquers
Patrize *f* *(met)* punch; *(print)* counter-die, punch[eon]; negative matrix
Patrone *f* 1. *(mil)* cartridge; 2. *(el)* refill unit, cartridge *(fuse)*; 3. *(mine)* cartridge; shell; 4. *(text)* draft; 5. point paper design
  lange P. *(mine)* king-size cartridge
  technische P. technical point paper design
  vollständige P. *(text)* completed draft
Patronengurt *m* cartridge belt
Patronengurtzuführung *f* cartridge belt feed
Patronenhülse *f* cartridge case
Patronenkammer *f* magazine *(rifle)*
Patronenpapier *n* design (drafting) paper
Patronenrahmen *m* charger, clip
Patronensicherung *f* cartridge fuse[-link], nonrenewable cartridge fuse, screw-in-type cartridge fuse
  träge P. surge resisting cartridge
Patronentasche *f* cartridge pouch, *(Am)* cartridge belt
patronieren *(text)* to draft
Patronit *m (min)* patronite
Patrouillenboot *n* patrol boat
Patschuliöl *n* patchouli oil
Pattingson-Prozeß *m* Pattingson process
Pauli-Prinzip *n* Pauli exclusion principle
Pauschalbetrag *m* lump sum
Pauschalfracht *f* lump freight
Pauschalsumme *f s.* Pauschalbetrag
Pauschaltarif *m* bulk tariff
Pausche *f* pommel *(parallel bars)*
Pauschenpferd *n* pommel[ed] horse
Pause *f* 1. tracing *(technical drawing)*; 2. interval, break; *(mach)* dwell
pausen to trace, to print
Pausenkode *m* space code
pausenlos *(mach)* continuous, uninterrupted
Pausenstrom *m* spacing current *(telegraphy)*
Pausenwelle *f (telec)* spacer-wave, spacing wave
Pausenzeichen *n* interval signal
Pausfähigkeit *f* printability
pausieren to stop, to [make a] pause
Pauspapier *n* tracing paper
Pavillon *m* pavilion
PC-Faser *f* polyvinyl chloride fibre
PD-Glied *n* proportional-plus-derivative element
PD-Regelung *f* proportional-plus-derivative control
Peak *m (nucl)* peak, maximum
Pebblereaktor *m* pebble[-bed] reactor
Pech *n* pitch
pechartig pitchy
Pechblende *f* pitchblende; nasturan; uranite *(uranium mineral)*
Pechdraht *m* pitched thread, waxed twine
pechen to pitch
Pechharz *n* mastic pitch
Pechkohle *f* pitch (bituminous, jet) coal
Pechkoks *m* pitch coke
Pechtorf *m* bituminous (black fuel) peat
Pechvorlage *f* pitch main
Pedal *n* pedal
Pedalschalter *m* foot (pedal) switch
Pedaltaste *f* pedal key *(organ)*
Pedalwebstuhl *m* tread loom
Pedion *n* pedion *(crystallography)*
Pegel *m* gauge, level; water mark, depth gauge
  P. bei abgeschlossener Leitung terminated level
  P. des Störfeldes high-frequency noise level
  mittlerer P. mean level
  relativer P. relative level
Pegelanzeiger *m* *(el)* transmission level indicator
Pegelausgleich *m* *(el)* equalization (compensation) of levels, level equalization (compensation)
Pegeldiagramm *n* level diagram, hypsogram *(telephony)*
Pegeldiskriminator *m* level discriminator
Pegeldrift *f* level drift
Pegeleinstellung *f* level setting
Pegelempfänger *m* level receiver unit
Pegelgenauigkeit *f* level accuracy
pegelgesteuert level-controlled
Pegelmeßeinrichtung *f* transmission measuring set
Pegelmesser *m* transmission level meter, level indicator; float (tape level) gauge
Pegelmessung *f* level measurement
Pegelplan *m* **einer Übertragung** transmission lay-out
Pegelregelung *f* level control; volume control *(electroacoustics)*
Pegelregler *m* level controller
  selbsttätiger P. instantaneous automatic gain control *(radar)*
Pegelschnellschreiber *m* high-speed level recorder
Pegelschreiber *m* recording transmission measuring set, *(Am)* level recorder
Pegelsollwert *m* nominal level
Pegelstand *m* level of the water-mark post
Pegelzeiger *m* volume (level) indicator, hypsometer
Pegmatit *m (min)* pegmatite

Peigneur *m (text)* doffer
Peilanlage *f* direction-finding installation
Peilantenne *f* direction finder aerial
Peilanzeiger *m* bearing indicator
Peilbereich *m* detection range
Peilboje *f* making buoy
Peilboot *n* surveying (sounding) boat
Peildeck *n (ship)* compass flat (platform, bridge), wheelhouse top
peilen to take a bearing, to bear; to gauge *(level)*; to sound
Peiler *m* direction finder
  selbsttätiger P. automatic direction finder
Peilfehler *m* error in bearings, direction-finding error
Peilfunk *m* radio detection finding; directional radio
Peilfunksender *m* direction-finding transmitter
Peilgerät *n* direction finder
  automatisches P. automatic direction finder
Peilkompaß *m* bearing (radio) compass
Peilkompaßstand *m (ship)* monkey island
Peillinie *f* bearing line
Peilminimum *n* position of minimum signal
Peilrahmen *m* direction-finding loop
Peilrichtung *f* bearing direction
Peilrohr *n* sounding pipe
Peilrose *f* azimuth dial
Peilsextant *m* sounding sextant
Peilsinn *m* sense of a bearing
Peilstab *m (aut)* dipstick
Peilstabablesung *f* dipstick reading
Peilstableuchte *f (aut)* side-marker lamp
Peilstange *f (aut)* side marker
Peilstock *m* gauge-rod; sounding pole
Peilstrahl *m* beacon course
Peiltabelle *f* bearing chart
Peilung *f* direction finding, bearing; sounding
  akustische P. und Entfernungsmessung *f* sound finding and ranging
  gegenseitige P. mutual (reciprocal) bearing *(ship and land)*
  konstante wahre P. maintenance true bearing *(radar)*
  mißweisende P. magnetic bearing
  rechtweisende P. true [compass] bearing
Peilungsrichtung *f* bearing direction
Peilvorrichtung *f* radio bearing installation
  P. mit Phasenverschiebung phase localizer *(radar)*
Peilwertübertragung *f* transmission of bearing information
Peilwinkel *m* radio bearing
Peilziel *n* bearing object (target)
Peitschenantenne *f* whip aerial
Peitschenleuchtenmast *m* curved lamp standard
Peitschenschnur *f* whip-cord
Pektin *n* pectin[e]
Pektinase *f* pectin[e] enzyme
Pektinsäure *f* pectic acid
Pelargonsäure *f* nonanoic (nonoic, nonylic, pelargonic) acid
pelletisieren to pelletize
Peltier-Effekt *m* Peltier effect
Peltier-Zelle *f* Peltier cell
Peltonturbine *f* Pelton wheel (turbine)
Pelz *m* / **künstlicher** imitation fur
Pelzbesatz *m* fur trimming
Pelzfutter *n* fur-lining
Pelzkrempel *f* second breaker
Pelzstiefel *m* fur-boot
Pelzstoff *m* fur fabric
Pelztrommel *f* lap drum
Pelzwerk *n* fur
Pendel *n* pendulum
  aperiodisches P. aperiodic pendulum
  ballistisches P. ballistic pendulum
  Foucaultsches P. Foucault's pendulum
  mathematisches P. simple pendulum
  physikalisches P. compound (physical) pendulum
  zusammengesetztes P. *s.* physikalisches P.
Pendelabweichung *f* pendulum deviation
Pendelachse *f (aut)* swing axle, jointed cross-shaft
Pendelarm *m* / **geknickter** bent pendulum-arm
Pendelbeschleunigungsmesser *m* pendulous accelerometer
Pendelbestrahlung *f* pendulum irradiation
Pendelbetrieb *m* shuttle service
Pendelbewegung *f (mach)* floating movement *(tool)*; reciprocating movement
Pendeldrehpunkt *m* centre of oscillation
Pendelelektron *n* oscillating electron
Pendelfräsen *n* pendulum (reciprocal) milling
Pendelfrequenz *f* oscillating frequency; quench[ing] frequency
Pendelgleichrichter *m* pendulum (tuned-reed, vibrating) rectifier
Pendelgurtförderer *m* shuttle-belt conveyor
Pendelhalter *m (mach)* floating [type of] holder *(reamer)*
Pendelhammerschlagwerk *n* pendulum impact testing machine
Pendelhebel *m* shuttle bar
Pendelhülse *f* floating adapter holder *(tool)*
Pendelkontakt *m (el)* pendulum-actuated contact
Pendelkreissäge *f* pendulum circular saw
Pendelkugellager *n* self-aligning ball bearing
Pendellager *n* self-aligning bearing
Pendelleuchte *f* pendant [fitting, light]
Pendelmesserreibahle *f* two-bladed floating reamer
Pendelmodulation *f* phase-angle modulation
Pendelmotor *m* swivel bearing motor

Pendelmühle *f* centrifugal grinder
pendeln to shuttle *(to move to and fro)*; to swing; to oscillate; to rock; to float *(tool)*; to hunt *(motor)*; to reciprocate
Pendeln *n* **des Wagens** swinging movement of the coach
pendelnd pendulous, floating *(tool)*
  p. angeordnet free to float (align and centre itself) *(lapping spindle)*
Pendelpfeiler *m* hinged (rocking) pier *(bridges)*
Pendelregler *m* pendulum governor
Pendelreibahle *f* floating reamer
Pendelrückkopplung *f (el)* superregeneration
Pendelrückkopplungsaudion *n* [self-quenching] superregenerative detector
  P. mit selbsterzeugter Pendelfrequenz self-quenched detector
Pendelrückkopplungsempfang *m* superregenerative reception
Pendelrückkopplungsempfänger *m* periodic trigger-type receiver, squegger superregenerative receiver, superregenerator
Pendelrückstrahler *m* oscillating reflector
Pendelsäge *f* pendulum (swing) saw
Pendelschlaghammer *m* **/ Charpyscher** Charpy impact testing machine
Pendelschlagwerk *n* pendulum impact-testing machine
Pendelschleifmaschine *f* swing-frame grinder
Pendelschnur *f* pendant flex (cord)
Pendelschutz *m (el)* hunting (out-of-step) protection
Pendelschwingung *f* hunting; pendular oscillation
Pendelstütze *f* hinged column
Pendeltherapie *f* pendulum therapy
Pendeltransport *m* shuttle haulage
Pendeltür *f* swing door
Pendeluhr *f* pendulum clock
Pendelumformer *m (el)* pendulum-type vibrator, synchronous self-rectifying vibrator
Pendelung *f* squegging; pendulum motion, hunting
Pendelverkehr *m* shuttle service (traffic)
Pendelvervielfacher *m* dynamic multiplier
Pendelwaage *f* pendulum weighing system
Pendelwagen *m* shuttle car (truck) *(conveying)*; wagon on shuttle service
Pendelwalze *f* pendulum roller *(leather)*
Pendelwinker *m (aut)* semaphore (oscillating) direction indicator
Pendelzähler *m* pendulum (oscillating) meter
Pendelzug *m* shuttle service train
Penicillin *n* penicillin
Pentaerythrit *n* pentaerythritol, pentaerythrite
Pentagondodekaeder *n* pentagonal dodecahedron, pyritohedron
Pentan *n* pentane
Pentaprisma *n* pentagonal prism
Pentathionsäure *f* pentathionic acid
Pentit *m (ch)* pentitol
Pentlandit *m (min)* pentlandite
Pentode *f* pentode
Pentose *f* pentose
Penumbra *f* penumbra, partial shadow, half shade
Pepsin *n* pepsin
Peptid *n* peptide
Peptidase *f* peptidase
Peptisation *f* peptization
peptisieren to peptizate
peptonerzeugend peptogenous
Perborat *n* perborate
Perchlorat *n* perchlorate
Perchlorsäure *f* perchloric acid
Perchromat *n* perchromate
perfekt perfect; concluded *(contract)*
perforieren to perforate, to punch
Perforierkamm *m* perforating comb
Perforierleiste *f* perforating bar
Perforierlinie *f* perforating rule
Perforiermaschine *f* perforating (punching) machine
perforiert perforated, punched
Perforierung *f* perforation
Pergament *n* parchment
Pergamentersatz *m* imitation parchment, pergamyne
Pergamentfarbe *f* parchment ink
Pergamentgerbung *f* parchment leather dressing
pergamentieren to parchmentize
Pergamentmembran *f* parchment membrane
Pergamentpapier *n* vellum (parchment) paper
Pergola *f* pergola
Perhydrol *n* perhydrol
Peridot *m* peridot, olivine
Peridotit *m* peridotite
Perigäum *n* perigee
Perihel *n* perihelion
Perimeter *n* perimeter
Periode *f* 1. *(ph)* period; 2. *(el)* cycle; 3. *(math)* repetend
  P. des Reaktors period [of the reactor], reactor period
  ganze P. complete cycle
  halbe P. half cycle
  volle P. complete cycle
Periodenbereich *m* instrument (period) range
Periodendauer *f* cycle duration
Periodendauermessung *f* period measurement
Periodendruck *m* periodic weight *(rock mechanics)*
Periodengesetz *n* **[von Mendelejew]** [Mendeleev's] periodic law
Periodenmesser *m* [reactor-] period meter

Periodensystem *n* **[der Elemente]** periodic system [of elements]
Periodenumformer *m* frequency converter (changer)
Periodenverstärker *m* period amplifier
Periodenzahl *f* number of cycles
periodisch periodic, cyclic
 p. gesteuert periodically controlled
 p. wiederkehrend periodically recurring
Periodizität *f* periodicity
Periodogramm *n (math)* periodogram
peripher peripheral
Peripherie *f* periphery, circumference
Peripheriegeschwindigkeit *f* circumferential (peripheral) speed
Peripheriewinkel *m* inscribed angle
Periskop *n* periscope
Perizykloide *f* pericycloid
Perjodat *n* periodate
Perjodsäure *f* periodic acid
Perldrossel *f* differential pressure regulator for bubbler gauge
Perle *f* pearl; bead; bubble *(water)*
Perlerz *n* pearl-ore
Perlit *m* pe[a]rlite
 feinstreifiger P. fine pearlite
 körniger (kugelförmiger) P. globular pearlite
 lamellarer P. lamellar pearlite
 streifenförmiger P. *s.* lamellarer P.
perlitisch pe[a]rlitic
Perlkoks *m* nut coke
Perlkondensator *m* bead capacitor
Perlleim *m* glue in pearls
Perlmutt[er *f***]** *n* mother of pearl
Perlpolymerisation *f* bead polymerization
Perlreaktion *f* bead test *(analytical chemistry)*
Perlrohr *n* bead tube
Perlschnurblitz *m* bead lightning
Perlstich *m (text)* links-and-links stitch
Perlweiß *n* pearl white
Permalloy *n* permalloy, Permalloy *(trade name)*
Permalloystufenschichten *fpl* fluted permalloy films *(memory)*
permanent permanent
Permanentappretur *f* durable (permanent) finish
Permanenthärte *f* permanent hardness *(water)*
Permanentmagnet *m* permanent magnet
Permanentspeicher *m* permanent memory
Permanentweiß *n* permanent white
Permanenz *f* permanency
Permanganat *n* permanganate
Permangansäure *f* permanganic acid
Permeabilität *f* permeability
 P. des leeren Raumes space permeability
 absolute P. absolute permeability
 differentielle P. incremental permeability
 komplexe P. complex permeability
 magnetische P. magnetic permeability
 relative P. relative permeability
 reversible P. reversible permeability
Permeabilitätsabnahme *f* **durch Alterung** time decrease of permeability
Permeabilitätsabstimmung *f* permeability tuning
Permeameter *n* permeameter
Permeanz *f* permeance
Permutation *f* permutation
permutieren to permute
Peroxid *n* peroxide
Peroxidbleiche *f* peroxide bleaching
Peroxydase *f* peroxidase
Perpetuum mobile *n* perpetuum mobile
Perron *m* platform
Persalz *n* persalt
Persäure *f* peracid
Persenning *f* tarpaulin
Persistor *m* persistor *(electronics)*
Personalausweis *m* identity card
Personalbestand *m* staff
Personalmangel *m* staff shortage
Personalüberwachung *f (nucl)* personnel monitoring
Personenaufzug *m* passenger elevator (lift)
Personen-Autofähre *f* passenger/car ferry
Personenbeförderung *f* passenger traffic; *(mine)* man-riding, men-riding
Personendosis *f (nucl)* personal dose
Personen-Fahrzeugfähre *f* passenger/vehicle ferry
Personenkilometer *n* passenger-kilometre
Personenkraftwagen *m* passenger car (automobile)
Personenmonitor *m* individual monitor; personnel monitor
Personenseilschwebebahn *f* cableway transporting persons
Personenstunde *f* manpower hour
Personentarif *m* passenger tariff (fare)
Personenverkehr *m* passenger traffic
Personenwagen *m (ry)* coach, carriage, *(Am)* passenger car; *(mine)* mantrip car
Personenwagenkilometer *n* coach-kilometre
Personenzug *m* slow (stopping) train; passenger train
Perspektive *f* perspective
perspektivisch perspective
Perspektivplan *m* outline (master) plan
Perspektivplanung *f* outline planning
PERT-Verfahren *n* PERT, programme evaluation and review technique *(network planning)*
Pertinax *n* pertinax
Perubalsam *m* balsam of Peru, Peru balsam
Peschelrohr *n* [overlapped] slit conduit
Petalit *m (min)* petalite
Petersen-Spule *f* Petersen (arc-suppressing) coil

Petrischale *f* Petri dish
Petroläther *m* petroleum ether, light petroleum, ligroin
Petrolatum *n* mineral (petroleum) jelly, petrolatum
Petrolchemie *f* petrol chemistry
Petroleum *n* petroleum; kerosene *(lamp oil; vaporizing oil)*
Petroleumbohranlage *f* petroleum-well drilling plant
Petroleuminjektor *m* petroleum feeder
Petroleumkai *m* petroleum wharf
Petroleummotor *m* kerosene engine
Petroleumtank *m* paraffin (kerosene) tank
Petrolkoks *m* petroleum coke
Petrolpech *n* petroleum pitch
Petzit *m (min)* petzite
Pfahl *m* pile; pole; post
  P. für Tiefgründung foundation pile
  schräger P. batter (raker, raking) pile
  schwebender P. floating (suspended) pile
  senkrechter P. plumb pile
  stehender P. end-bearing pile
Pfahlbau *m* pile work
Pfahlewer *m* well boat (smack), single pole-masted barge
Pfahlgründung *f* pile[d] foundation, *(Am)* caisson foundation
  P. / mit pile-supported
Pfahlkopf *m* pile head
Pfahlmast *m* pole mast
Pfahlneigung *f* pile batter
Pfahlprobe *f* standing-pull trial, static test *(of a tug)*
Pfahlramme *f* pile driver (hammer), piling hammer, pile [driving] plant
Pfahlrost *m* 1. *(build)* pile foundation grill; 2. *(therm)* pile grating
Pfahlschuh *m* pile shoe
Pfahltrossenzug *m* [static] bollard pull
Pfahlzieher *m* pile extractor
Pfahlzug *m (ship)* bollard pull
Pfahlzugprobe *f* bollard trials
Pfanne *f* 1. *(build)* pan-tile *(roofing)*; 2. *(met)* ladle; 3. *(mach)* socket; pan
  P. mit Ausgußschnauze *f (met)* dip-pour ladle
  P. mit Bodenausguß *(met)* bottom pouring ladle
  P. mit Stopfenausguß *(met)* teeming ladle
  stellbare P. *(mach)* adjustable slide
Pfannenausguß *m (met)* ladle lip
Pfannenauskleidung *f* ladle lining
Pfannenbär *m (met)* ladle skull
Pfannenführung *f (met)* ladle guide
Pfannengabel *f (met)* ladle shank
Pfannenkruste *f (met)* ladle skull
Pfannenschnauze *f (met)* ladle spout
Pfannenstopfen *m* / **undichter** *(met)* running stopper
Pfannenwagen *m (met)* ladle car
Pfannenzusätze *f (met)* ladle additions
Pfauenauge *n* peacock's eye
Pfauenaugenbindung *f* bird's-eye weave
Pfefferminzöl *n* peppermint oil
Pfefferstrauch *m* / **schwarzer** pepper [plant] *(piper nigrum)*
Pfeifabstand *m (telec)* singing margin
Pfeife *f* 1. pipe *(organ; tobacco)*; whistle; 2. *(mine)* dead hole, socket *(blasting)*
pfeifen to whistle; to sing
Pfeifen *n* singing; whistling
  P. des Schiebers hissing of the valve
Pfeifentabak *m* smoking tobacco
pfeiffrei free of (from) singing
Pfeifgrenze *f (telec)* singing limit
Pfeifneigung *f (telec)* tendency to sing
Pfeifpunkt *m (telec)* singing point
Pfeifpunktabstand *m (telec)* singing margin
Pfeifsicherheit *f (telec)* antisinging stability
Pfeifton *m (telec)* whistling tone
Pfeil *m* arrow; *(build)* rise, camber
Pfeilen *n* **[der Tragfläche]** *(aero)* sweep-back
Pfeiler *m* 1. *(build)* pier *(bridge)*; column; pillar; 2. *(mine)* board, *(Am)* bord, breast; room, stall; cog, pillar
  P. aus Betonscheiben *(mine)* concrete pancake column
Pfeilerabbau *m (mine)* pillar drawing (robbing)
Pfeilerbau *m (mine)* bord (room) and pillar method, pillar [and chamber] working, pillar (post) and stall method, rooming, stoop and room method
Pfeilerbauverfahren *n s.* Pfeilerbau
Pfeilergründung *f (build)* pier foundation
Pfeilerquerdehnung *f (mine)* transverse pillar extension
Pfeilerrauben *n s.* Pfeilerrückgewinnung
Pfeilerrückgewinnung *f (mine)* pillar drawing (extraction, recovery, robbing)
Pfeilerstrecke *f (mine)* entry, gangway, [stall] road
Pfeilerverband *m (build)* pier bond
Pfeilflügel *m (aero)* swept-back wing
Pfeilhöhe *f* height of the camber, rise
Pfeilprofil *n* herringbone profile *(tyre)*
Pfeilrad *n* herringbone gear
Pfeilradgetriebe *n* herringbone gearing
Pfeilstirnrad *n* herringbone gear
Pfeilverzahnung *f* herringbone (double-helical) teeth
  echte P. continuous double helical teeth
  unechte P. conventional herringbone teeth, teeth having circular recess
Pfeilzahn *m* herringbone tooth
Pfeilzahnradfräser *m* herringbone gear cutter

Pferchen *n (ag)* folding
Pferd *n* 1. horse; 2. vaulting horse *(sport)*
Pferdefuhrwerk *n* horse-drawn vehicle
Pferdegeschirr *n* harness
Pferdegespann *n* team of horses
Pferdehacke *f* horse hoe
Pferdehaut *f* horse-hide
Pferdeschweif *m* horse-tail
Pferdestall *m* stable
Pferdestärke *f* [continental] horsepower, hp
  effektive P. effective horsepower
  gemessene P. indicated horsepower
  indizierte P. indicated horsepower
Pferdezucht *f* horse breeding
Pferdezug *m* horse traction
Pfette *f* purlin
Pflanze *f* plant
Pflanzeisen *n* planting peg, dibble
pflanzen to plant
Pflanzen *n* planting
Pflanzenabdruck *m* impression of a plant
Pflanzenbau *m* / **erdeloser** hydroponics, soilless culture
Pflanzenchemie *f* phytochemistry
Pflanzeneiweißstoff *m* vegetable albumen
Pflanzenfarbstoff *m* plant pigment; vegetable dye (dyeing matter)
Pflanzenfaser *f* vegetable fibre
Pflanzenfett *n* vegetable shortening (fat)
Pflanzengift *n* vegetable poison
Pflanzengummi *n* [vegetable] gum
Pflanzenleim *m* vegetable glue (adhesive)
Pflanzennährstoff *m* plant nutrient
Pflanzenöl *n* vegetable oil
Pflanzenpresse *f* botanist's press
Pflanzenschleim *m* mucilage
Pflanzenschutz *m* plant protection
Pflanzenschutzforschung *f* plant-protection research
Pflanzenschutzmaßnahme *f* plant-protective measure
Pflanzenschutzmittel *n* plant protective, plant-protecting agent
Pflanzensetzmaschine *f* [trans]planter
Pflanzenwuchs *m* vegetation; plant growth
Pflanzenwuchsstoff *m* plant hormone (growth substance)
Pflanzenzüchter *m* grower
Pflanzenzüchtung *f* plant breeding (cultivation)
Pflanzgarten *m* nursery
Pflanzhacke *f* planting hoe
Pflanzholz *n* planting peg, dibble
Pflanzlochmaschine *f* plant hole digger
Pflanzmaschine *f* [trans]planter, planting machine
Pflanzstock *m* planting peg, dibble
Pflanzung *f* plantation
Pflaster *n* 1. *(build)* pavement; 2. plaster, adhesive tape
Pflasterdecke *f* block (sett) pavement
Pflasterer *m* paver, paviour
Pflasterholz *n* paving wood
Pflasterklotz *m* paving block
pflastern to pave
Pflasterstein *m* paving stone (sett, block)
Pflasterung *f* pavement
Pflatschdruck *m* slop-pad printing
Pflatschen *n* padding
Pflaumenbaum *m* plum tree
Pflege *f* 1. *(ag)* cultivation; 2. *(mach)* attendance, maintenance; service
  P. des Hangenden *(mine)* roof (strata) control
Pflegearbeiten *fpl* cultivation operations
Pflegedienststand *m* lube bay
Pflegeleichtigkeit *f* ease of care (attendance)
Pflegemaschinen *fpl* **und -geräte** *npl* cultivating machinery
pflegen 1. to tend, to look after; 2. *(ag)* to cultivate; 3. *(mach)* to maintain, to service
Pflegepersonal *n* service personnel
Pflegetraktor *m* row-crop tractor
Pflicht *f* duty; obligation
Pflichtenheft *n (el)* rating, specification sheet, [equipment] specifications
  P. für Kabel cable specification
Pflichtexemplar *n* [library] presentation copy
Pflichtlandung *f* compulsory landing
Pflichtwiederholung *f* compulsory repetition
Pflock *m* stake, peg; plug
pflücken to gather, to pick; to pluck
  Baumwolle p. to pick cotton
Pflücker *m* picker
Pflug *m* plough, *(Am)* plow
Pflugbalken *m* [plough] beam
Pflugbaum *m* [plough] beam
Pflugbrust *f* frame
Pflugeisen *n* coulter
pflügen to plough
Pflugfurche *f* furrow
Pflugkarren *m* fore-carriage
Pflugmesser *n* coulter
Pflugschar *n* plough share
Pflugsech *n* coulter
Pflugsohle *f* shin; plough pan
Pflugsterz *m* plough handle
Pflugzusatzeinrichtung *f* plough attachment
Pfosten *m (build)* standard, post, upright, jamb, pillar; *(aero)* soldier
Pfostenramme *f* post driver
Pfostenzieher *m* post puller
Pfriem[en] *m* awl
Pfropfbastard *m* graft hybrid
pfropfen to graft
Pfropfen *n* grafting
Pfropfen *m* cork, plug, stopper; tampon

Pfropfenbildung *f (plast)* plug-up
Pfropfpolymerisation *f* graft polymerization
Pfropfreis *n* [graft] scion
Pfropfung *f* **/ stumpfe** butt joint graft
Pfusch *m* slipshod work
pfuschen to botch
pH-Durchflußmeßelektrode *f* flow-type pH electrode
pH-Messer *m* pH meter, pH-acidometer, ionometer
pH-Meß- und -Kontrollgerät *n* pH meter-controller
pH-Meßzelle *f* pH-measuring cell
pH-Meter / registrierendes recording pH meter
pH-Wert *m* pH value, hydrogen ion concentration
Phänomen *n* phenomenon
  Purkinjesches P. Purkinje effect (phenomenon)
Phantasiebindung *f (text)* fancy weave
Phantasiepapier *n* fancy paper
Phantastron *n (el)* phantastron
Phantom *n* phantom
Phantombildung *f* phantoming
Phantomkammer *f* phantom chamber
Phantomkreis *m* phantom circuit
Phantommessung *f* phantom measurement
Phantomschaltung *f* phantom circuit
Phantomspule *f* phantom coil
Phantomsteuerung *f* phantom control
Phantomsubstanz *f* phantom (tissue-equivalent) material
Pharmakologie *f* pharmacology
Pharmakopöe *f* pharmacopeia
Pharmazeut *m* pharmacist
Pharmazeutika *pl* pharmaceuticals
Pharmazie *f* pharmacy
Phase *f (ph, el)* phase
  P. / außer out of phase
  beständige P. stable phase
  fehlerfreie P. sound phase
  feste P. solid phase
  flüssige P. liquid phase
  intermetallische P. intermetallic phase
  nacheilende P. retarded phase
  unbeständige P. unstable phase
  voreilende P. leading phase
Phasenabgleich *m* phase adjustment, phasing
Phasenabgleichspule *f* retardation coil
phasenabhängig phase-sensitive
Phasenabstand *m* phase spacing
Phasenänderung *f* phase change
  P. durch Einfügung eines Gliedes insertion phase change
Phasenanschnittsteuerung *f* phase-shifting control *(of thyratrons)*
Phasenanzeiger *m* phase indicator
Phasenausfall *m* phase failure
Phasenausgleich *m* phase compensation
Phasenausgleichsglied *n* phase (delay) equalizer
Phasenaussortierung *f* phase selecting
Phasenbedingung *f* phase condition
Phasenbelag *m* phase-change coefficient
Phasenbereich *m* region of phase
Phasenbestimmung *f* phase determination
Phasenbrücke *f* phase-angle bridge, discriminator
Phasenbündelung *f* bunching
Phasendetektor *m* phase detector
Phasendiagramm *n* phase (equilibrium) diagram
Phasendifferenz *f* phase difference; space phasing *(aerials)*
  P. der interferierenden Strahlen phase difference of the interfering beams
Phasendiskriminator *m* phase discriminator (detector)
Phasendreher *m* phase shifter
Phasendrehtransformator *m* interphase transformer
Phasendrehung *f* phase rotation
Phasenebene *f* phase portrait (plane); molecular plane
Phaseneinstellung *f* phase timing, phasing
Phasenentzerrer *m* phase compensator (equalizer)
Phasenentzerrung *f* phase equalization (correction)
phasenfalsch misphased, phase-shifted
Phasenfehler *m* phase error
Phasenfeld *n* phase field
Phasenfläche *f* [equi]phase surface
Phasenfokussierung *f* phase focusing
Phasenfolge *f* phase sequence, succession of phases
Phasenfolgeanzeiger *m* phase-sequence indicator
Phasenfolgewinkel *m* characteristic angular difference of phases
Phasengang *m* phase characteristic (response)
Phasengeschwindigkeit *f* phase (group) velocity
  P. einer Welle velocity of a periodic wave
Phasengesetz *n* phase rule
  Gibbssches P. Gibbs' phase rule
phasengleich in-phase
Phasengleichheit *f* synchronism of phases, in-phase state, phase coincidence (balance)
Phasengrenze *f* interphase
Phasengrenzschicht *f* phase boundary
Phasenintegral *n* phase integral, action variable
Phasenkette *f* phasing chain (network)
Phasenklemme *f* line terminal
Phasenkompensation *f* power-factor correction (improvement), phase compensation
Phasenkondensor *m* phase condenser
Phasenkonstante *f* phase[-change] coefficient

Phasenkontrast *m* phase contrast
Phasenkontrastbild *n* phase-contrast image
Phasenkontrastmikroskop *n* phase-contrast microscope
Phasenkontrastmikroskopie *f* phase-contrast microscopy
Phasenkontrastverfahren *n* phase-contrast method (technique)
Phasenlage *f* phase relation (position); phasing
Phasenlaufzeit *f* phase delay[-time]
Phasenleiter *m* phase conductor
Phasenmaß *n* phase constant, phase-change coefficient
Phasenmesser *m* phase meter, power-factor indicator
Phasenminimumsystem *n* minimum-phase system
Phasenmodulation *f* phase modulation
Phasenmodulator *m* **mit Reaktanzröhre** valve-reactor phase modulator
Phasennacheilung *f* phase lag (retard), lagging of the phase
Phasennormal *n* phase standard
Phasenopposition *f* out-of-phase *(switching)*
Phasenplatte *f* phase plate
Phasenporträt *n* phase portrait (plane)
Phasenrand *m* phase margin
Phasenraum *m* phase space
Phasenraumelement *n* phase-space cell
Phasenrauschen *n* phase noise
Phasenregel *f* phase rule
Phasenregler *m* phase regulator (shifter)
Phasenrelais *n* phase-sensitive relay
Phasenresonanz *f* velocity (series) resonance
phasenrichtig in proper phase
Phasenring *m* annular phase plate
Phasenrückdrehung *f* phase lag (retard)
Phasenschieber *m* phase shifter (converter, advancer), interphase transformer, induction regulator, balancer, *(Am)* phase-turner, static balancer, balancer-transformer
  P. ohne Ständer expedor phase transformer (advancer)
  asynchroner P. asynchronous capacitor
  rotierender P. rotary phase advancer
  synchroner P. synchronous capacitor
Phasenschieberkondensator *m* power-factor capacitor, phase-shifting capacitor
Phasenschieberschaltung *f* phase shift circuit
Phasenschiebertransformator *m* phase-shifting transformer
Phasenschiebung *f* power-factor improvement
Phasenschwankung *f* phase variation
Phasenspannung *f* phase voltage of a winding, star (phase-to-neutral) voltage, Y-voltage
Phasenspielraum *m* phase margin
Phasensprung *m* phase jump (shift, change)
phasenstabil phase-stable
phasenstarr locked to phase, phase-locked
Phasensteuerung *f* phase control
  selbsttätige P. automatic phase control
Phasenstrom *m* phase current
Phasenteiler *m* phase splitter (divider)
Phasenteilung *f* phase splitting (division)
Phasentransformator *m* induction regulator (advancer), phase transformer
Phasentrenner *m* phase splitter
Phasenumformer *m* phase changer
Phasenumkehr *f* 1. *(el)* reversal of phase, phase reversal (inversion); 2. *(ph)* product reflux [in a separation column]
Phasenumkehrschaltung *f* phase inverter (inversion circuit)
Phasenumkehrstufe *f* phase inverter (reversing stage)
Phasenumkehrung *f* phase inverting (inversion)
Phasenumwandlung *f* phase transformation
  P. erster Ordnung *(therm)* first-order transition
  P. zweit[illegible]ung *(therm)* second-order t[illegible]tion
Phasenungleichheit *f* phase unbalance
Phasenunterschied *m* phase difference
Phasenverhalten *n* phase response
phasenverkehrt in phase opposition, in opposite phase
  p. wirkend acting in opposite phases
Phasenverschiebung *f* phase displacement (shift, unbalance), alteration of phase
Phasenverschiebungswinkel *m* angle of phase difference
phasenverschoben dephased, out of phase, offset in phase, *(Am)* outphased
  p. um 90° in[-phase] quadrature
  p. um 180° 180° out of phase
phasenvertauscht misphased
Phasenverzerrung *f* phase distortion
Phasenverzerrungsgrad *m* phase distortion index
Phasenverzögerung *f* phase lag (delay), retardation in phase
Phasenvordrehung *f* phase lead (advance)
Phasenvoreilen *n* phase lead[ing]
Phasenvoreilung *f* phase lead (leading, advance)
Phasenvorhalt *m* phase advance (lead)
Phasenvorrat *m* phase margin
Phasenwähler *m* phase selector
Phasenwinkel *m* phase angle, angle of phase difference (displacement)
  P. der Stromrichterschaltung circuit angle
Phasenwinkelvergleichsschutz *m* phase-comparison protection
Phasenzeiger *m* phase indicator
Phasotron *n* synchro-cyclotron, f-m cyclotron
Phenanthren *n* phenanthrene
Phenazetin *n* phenacetine

Phenol *n* phenol
Phenolabscheider *m* dephenolizer
Phenolat *n* phenate, phenylate
Phenoläther *m* phenolic ether
Phenolharz *n* phenolic resin
Phenolharzkleber *m* phenolic cement
Phenolharzleim *m* phenolic adhesive (resin glue)
Phenolharzpreßmischung *f* phenolic moulding compound
Phenolharzschichtstoff *m* phenolic laminated sheet
Phenolkunststoffleiste *f (mach)* phenolic lining
Phenolphthalein *n* phenolphthalein
Phenoplast *m* phenol-formaldehyde plastic
Phenyläthylalkohol *m* phenyl ethyl alcohol
Phenylessigsäure *f* phenylacetic acid
Phenylhydrazin *n* phenylhydrazine
Phenylierung *f* phenylation
Phenylrest *m* phenyl residue
Phenylsenföl *n* phenyl mustard oil
Phiole *f* phial, vial
Phon *n* phon
Phonokardiograf *m* phonocardiograph
Phonolith *m (geol)* phonolite
Phonon *n* phonon
Phononenerzeugung *f* phonon production
Phosgen *n* phosgene
Phosphat *n* phosphate
Phosphatase *f* phosphatase
Phosphatid *n* phosphatide
phosphatieren to phosphatize
Phosphatieren *n* phosphatizing, phosphate coating (treatment)
Phosphid *n* phosphide
Phosphin *n* phosphine
Phosphit *n* phosphite
Phosphoniumsalz *n* phosphonium salt
Phosphor *m* phosphorus *(element)*; phosphor *(fluorescent substance)*
Phosphorbronze *f* phosphor bronze
Phosphoreszenz *f* phosphorescence
phosphoreszieren *(ch)* to phosphorate, to phosphoresce
phosphoreszierend phosphorescent
phosphorhaltig phosphoric
Phosphorit *m (min)* phosphorite, rock-phosphate
Phosphormolybdänsäure *f* phosphomolybdic acid
Phosphoroskop *n* phosphoroscope
Phosphorpentachlorid *n* phosphorus pentachloride, phosphoric chloride
Phosphorpentoxid *n* phosphorus pentoxide, phosphoric anhydride
Phosphorsalz *n* microcosmic salt, sodium ammonium hydrogen phosphate
Phosphorsalzperle *f* microcosmic bead
Phosphorsäure *f* phosphoric acid
Phosphorsäureester *m* phosphoric ester
Phosphortrichlorid *n* phosphorus trichloride, phosphorous chloride
Phosphor(III)-verbindung *f* phosphorous compound
Phosphor(V)-verbindung *f* phosphoric compound
Phosphorwasserstoff *m* phosphine
Phosphorwolframsäure *f* phosphotungstic acid
Phot *n* phot
Photo . . ., photo . . . *s.* Foto . . ., foto . . .
Photon *n* photon, light quantum
  absorbiertes P. absorbed photon
  angeregtes P. stimulated photon
Photonenabsorption *f* photon absorption
Photonenenergie *f* photon energy
Photonenfluß *m* photon flux
Photonenrakete *f* photon rocket
Photonenstrahlung *f* photon radiation
Photonenstrom *m* photon flux
Photonenübergang *m* photon transition
Photonenvernichtung *f* photon annihilation
Photonenverstärkung *f* photon amplification (gain)
Photonenverteilung *f* photon distribution
Photonenzähler *m* photon counter
Photonik *f* photonics
Phthalamid *n* phthalic amide
Phthalat *n* phthalate
Phthalozyanin *n* phthalocyanine *(dye)*
Phthalsäure *f* phthalic acid
Physik *f* physics
  P. hoher Energien high-energy physics
  angewandte P. applied physics
  chemische P. physical chemistry
  klassische P. classical physics
  theoretische P. theoretical physics
physikalisch physical
physikalisch-chemisch physico-chemical
Physiker *m* physicist
Physikochemiker *m* chemist and physicist
physikochemisch physico-chemical
physiologisch physiological
phytochemisch phytochemical
Phytosterin *n* phytosterol
PI-Drehzahlregler *m* isochronous governor
PI-Glied *n* proportional-plus-integral element
PI-Regelung *f* proportional-plus-integral control
PI-Regler *m* PI-acting controller, PI-controller, isochronous governor *(speed control of engines)*
pi-Tiefpaß *m* capacitor-input filter
Piassava *f* piassava
Piassavefaser *f s.* Piassava
Pick-à-pick-Spulenwechsel *m (text)* pick-and-pick bobbin changing
Pickelbad *n (leath)* rinsing (falling) solution
pickeln to pickle *(leather)*
Pick-up-Presse *f (ag)* pick-up baler
Pick-up-Reaktion *f* pick-up [reaction]

Pictogramm *n* pictogram
PID-Glied *n* three-action element
PID-Regler *m* PID-acting controller, three-response controller, three-term controller
Piek *f (ship)* peak
Piekbalken *m* panting beam
Piekdeck *n* panting deck
Piekschott *n* peak bulkhead
Piekspant *n* peak frame
Piekverstärkungsbalken *m* panting beam
Piepsen *n (telec)* squeak[ing]
Pier *m (f)* pier, wharf
Piezoeffekt *m* piezoelectric effect
piezoelektrisch piezoelectric
Piezoelektrizität *f* piezoelectricity
Piezokristall *m* piezoelectric crystal
piezomagnetisch piezomagnetic
Piezometer *n* piezometer
Piezoquarz *m* piezoelectric quartz
Pigment *n* pigment
  verschnittenes P. reduced pigment
Pigmentdruck *m* carbon print; *(text)* pigment print
Pigmentfarbstoff *m* pigment dye
Pigmentfärbung *f* pigment dyeing
Pigmentfüllstoff *m* loading pigment
pigmentiert pigmented
Pigmentklotzverfahren *n (text)* pigment padding process
Pigmentpapier *n* pigment paper
Pikee *m* piqué
Pikeebarchent *m* swansdown
pikieren 1. *(ag)* to prick out; 2. *(text)* to stitch
Pikiernadel *f* pricker needle
Pikocurie *n* micro-micro-curie
Pikofarad *n* micro-micro-farad
Pikolin *n* picoline
Pikotagering *m (mine)* piling ring
Pikraminsäure *f* picramic acid
Pikrinsäure *f* picric (picronitric) acid
Pilaster *m* pilaster
Pile *m s.* Kernreaktor
Pilgerdorn *m (met)* pilger mandrel, piercer, broach
Pilgermandrill *m s.* Pilgerdorn
Pilgerschrittschweißung *f* step back welding
Pilgerschrittverfahren *n (met)* pilger (pilgrim, reciprocating rolling) process
Pilgerschrittwalzwerk *n* pilgrim step rolling mill
  P. für nahtlose Rohre pilger step-by-step-type seamless tube rolling mill
Pilgerwalze *f* pilger (pilgrim) roll
Pilgerwalzwerk *n* pilger (pilgrim) mill
Pilieranlage *f* soap-refining outfit
Piliermaschine *f* soap mill
Pille *f* pellet *(e.g. electron tubes)*, tablet
Pilot *m* pilot
  automatischer P. automatic direction finder, ADF, robot pilot
Pilotanlage *f* pilot plant
Pilotballon *m* pilot balloon
Pilotempfänger *m* pilot receiver
Pilotensitz *m* cockpit
Pilotfrequenz *f* pilot frequency
Pilotpegel *m* pilot level
Pilotschwingung *f* reference pilot
Pilz *m* 1. mushroom; fungus; 2. *(nucl)* mushroom cloud
Pilzanker *m (ship)* mushroom anchor
Pilzbefall *m* fungus attack
Pilzbekämpfungsmittel *n* fungicide
Pilzdecke *f* mushroom (two-way flat) slab
  P. ohne Stützenkopfverstärkung flat slab
Pilzgrundplatte *f* inverted flat-slab foundation
Pilzisolator *m* mushroom insulator
Pilzkopf *m (build)* capital, drop panel, splayed head
Pilz[kopf]lüfter *m* mushroom ventilator
Pilzmischer *m* mushroom mixer
Pilzschale *f (build)* umbrella shell
Pilzsicherung *f (aut)* mushroom-type retainer
Pilzstößel *m (aut)* mushroom tappet
Pilzventil *n* mushroom valve
Pilzvertilgungsmittel *n* fungicide
Pimelat *n* pimelate
Pimelinsäure *f* pimelic acid
Pimentöl *n* pimento oil
Pinakoid *n* pinacoid
Pinakolin *n* pinacoline
Pinakolinumlagerung *f* pinacolic rearrangement
Pinasse *f* pinnace
Pinchantrieb *m* pinch drive
Pincheffekt *m* pinch effect
Pinkenheck *n (ship)* pink (high) stern
Pinne *f (ship)* tiller, helm
Pinole *f (mach)* quill, sleeve
Pinolendurchmesser *m* sleeve (quill) diameter
Pinolenfestklemmung *f* sleeve-lock
Pinolenrückzug *m* withdrawal of the quill (sleeve)
Pinsel *m* brush
Pinselanstrich *m* brush painting
Pinselputz *m* brush plaster
Pinselschrift *f* brush lettering
Pinzette *f* pincette, [pair of] tweezers
Pion *n* pion, pi-meson
Pionenerzeugung *f* pion production
Pionierpanzerkampfwagen *m* engineer tank
Pioniertruppen *fpl* engineer troops
Piperazin *n* piperazine
Piperidin *n* piperidine
Piperin *n* piperine
Piperinsäure *f* piperic acid
Piperonal *n* piperonal, heliotropin
Pipette *f* pipette

Pipettengestell *n* pipette stand
pipettieren to pipette
Pirani-Manometer *n* Pirani gauge, hot wire [gauge]
Pisébau *m*, **Piseebau** *m* pisé de terre
Pistazit *m (min)* pistacite
Piste *f* runway
Pistenbefeuerung *f* runway lighting (lights)
Pistill *n* pestle
Pitot-Rohr *n* Pitot tube
P.I.V.-Antrieb *m* P.I.V. drive
Pivotfenster *n* opening side-light
Plakat *n* poster
Plakatdruck *m* poster printing
Plakatfarbe *f* lithographic colour
Plakette *f* badge
Plan *m* plan; schedule, project; diagram; layout *(engine room)*
  P. für regelmäßige Messungen maintenance testing schedule
plan flat, plane; transverse
Plananschlag *m (mach)* facing stop
Planarbeit *f (mach)* facing operation
Planbewegung *f (mach)* transverse traverse (movement)
Plandreharbeit *f (mach)* surfacing operation
plandrehen *(mach)* to [sur]face
Plandrehen *n (mach)* [sur]facing
Plandrehmaschine *f* face lathe
Plandrehschlitten *m (mach)* [sur]facing slide
Plandrehsupport *m (mach)* facing head (slide)
Plane *f* tarpaulin, wagon sheet
Planelektrode *f* plane electrode
planen 1. to plan, to project, to design, to schedule; 2. *(mach)* to face
Planen *n* 1. planning, projecting, designing, scheduling; 2. *(mach)* facing
  P. von Versuchen design of experiments
Planengestell *n (aut)* tarpaulin frame
Planer *m* planner
Planerfüllung *f* plan fulfilment
Planet *m* planet
  künstlicher P. artificial (man-made) planet
Planetarium *n* planetarium
Planetenbahn *f* planetary orbit
Planetengetriebe *n* planetary gears (gearing, transmission), epicyclic gearing (transmission)
Planetenmodell *n* **des Atoms** Rutherford atom model
Planetenrad *n* planet pinion, planet[ary] wheel
Planetenradträger *m* planet carrier
Planetenrührwerk *n* planetary paddle mixer
Planetenschleifspindel *f* planetary wheel spindle
Planetenspindel *f* planetary spindle
Planetenspindelfräsmaschine *f* planetary milling machine
Planetenspindelgewindefräsmaschine *f* planetary type of thread milling machine
Planetenspindelschleifmaschine *f* planetary-type [internal] grinding machine
Planfeuertafel *f* lead chart *(anti-aircraft)*
Planfilm *m* cut (flat, pack) film
Planfilmverfahren *n* flat film method
Planförderung *f (mine)* target production
Planfräsmaschine *f* fixed-bed type milling machine, solid-bed type milling machine, adjustable-spindle milling machine, *(Am)* manufacturing [bed-]type milling machine
  P. mit verstellbarem Spindelkopf adjustable-head horizontal milling machine
plangemäß according to the plan
Plangitter *n* plane grating
Planglas *n* optical flat
Planglasplatte *f* optical flat
Planherd *m* slime-concentrating table
Planhobelmaschine *f* surface planing machine
planieren to level, to plane; to planish
Planiergerät *n* grader, bulldozer; planer
Planierloch *n* leveller door
Planierlöffel[bagger] *m* skimmer
Planierraupe *f* [bull]dozer, grade-builder
Planierschild *m* dozer blade, pusher-blade
Planierstange *f* leveller, levelling rod
Planiertür *f* leveller (levelling) door
Planierung *f* levelling, grading
Planimeter *n* planimeter
Planimetrie *f* plane geometry
Planimetrierung *f* measurement by planimeter
Plankatode *f* plane cathode
Planke *f* plank, structural timber, board
Plankenende *n (ship)* hood end
Plankengang *m (ship)* main wale, [plank] strake
Plankenholz *n* plank timber
Plankensäge *f* plank saw
plankonkav plano-concave
Plankonkavlinse *f* plano-concave lens
plankonvex plano-convex
Plankonvexlinse *f* plano-convex lens
Plankurve *f (mach)* face cam
Plankurvenfutter *n (mach)* cam-type chuck
planmäßig planned, systematic, scheduled
Plannachformdrehen *n* contour facing
Planoauslage *f* flat sheet delivery
Planoformat *n* full size
Planopapier *n* flat paper
planparallel parallel-sided, plane-parallel
Planparallelschleifen *n* grinding work parallel
Planplattenmikrometer *n* rocking-plate micrometer
planpolarisiert plane-polarized
planpoliert flat-polished
Planquadrat *n* grid square
Planrad *n* crown wheel
Planrätter *m* gyratory screen
Planrost *m* flat (horizontal) grate

Planrostfeuerung *f* furnace with horizontal grate
Planrostinnenfeuerung *f* furnace with internal horizontal grate
Planroststab *m* bar of a horizontal grate
Planrostunterfeuerung *f* furnace with horizontal grate underneath the boiler
Planrostvorfeuerung *f* extended furnace with horizontal grate
Planschbecken *n* paddling (wading) pool
Planscheibe *f (mach)* faceplate
P. der Karusselldrehmaschine table of the vertical boring and turning mill
Planscheibendrehzahl *f* table speed *(vertical turning and boring mill)*; faceplate speed
Planscheibenklaue *f* faceplate jaw
Planscheibenspannschlitz *m* faceplate clamping slot
Planschleifeinrichtung *f* attachment for face grinding
Planschlitten *m* cross slide *(lathe)*
Plansenken *n (mach)* spot-facing
Plansenker *m* facing tool; spot facer
Plansichter *m* sifter; screening table
Plansieb *n* plane sifter, flat screen
Plansortierer *m* flat screen
Planspiegel *m* flat (plane) mirror
Planspiel *n* management game
Planspindel *f (mach)* cross-slide-screw
Planspirale *f* plane spiral
Planstoßherd *m* slime-concentrating table
Plantrockenschlichtmaschine *f* flat drying and sizing machine
Planübergang *m* level crossing
Planum *n (build)* formation, subgrade
Plan- und Ausdrehkopf *m (mach)* facing and boring head
Planung *f* planning; design
langfristige P. long-term planning
operative P. operational planning
technologische P. process planning
Planungsabteilung *f* planning department
Planungsforschung *f* operational (operations) research
Planungsleiter *m* planning manager
Planungstheorie *f* planning theory
Planvorschub *m (mach)* cross feed, surfacing feed *(lathe)*
Planwelle *f* plane wave
Planwerkzeugträger *m (mach)* facing head
Planwirtschaft *f* planned economy
Planzahl *f* planned figure
Planziel *n* plan target, target of the plan
Planziffer *f* target figure
Planzug *m (mach)* transverse motion (traverse), cross feed, power cross-feed, power surfacing-motion
Plasma *n* plasma
ionisiertes P. ionized plasma
Plasmaantrieb *m* plasma propulsion
Plasmabeschleuniger *m* plasma accelerator
Plasmabrenner *m* plasma (electronic) torch
Plasmadruck *m* plasma pressure
Plasmaeinschließung *f* plasma confinement (isolation)
Plasmafaden *m* plasma channel (cylinder)
Plasmafrequenz *f* plasma frequency
Plasmagleichgewicht *n* plasma balance
Plasmalichtbogenbrenner *m* plasma arc torch
Plasmaphysik *f* plasma physics
Plasmaschwingung *f* plasma oscillation
Plasmaspaltungsreaktor *m* fission plasma reactor
Plasmastrahl *m* plasma beam
Plasmaströmung *f* plasma flow
Plasmawelle *f* plasma wave
Plasmolyse *f* plasmolysis
Plast *m* plastic [material]
hitzehärtbarer P. thermosetting plastic
mit Textilien verstärkter P. textile-reinforced plastic
verstärkter P. reinforced plastic
plastbeschichtet plastic-coated
Plastfolie *f* thin-sheet plastic, sheet of [a] plastic material
Plastfolienzelt *n* plastic barn cloche
Plastformteil *n* moulded plastic, plastic moulding
Plastifikator *m* plasticizer, plastifying agent
plastifizieren to plasticize, to plastify, to soften
Plastifizierleistung *f (plast)* melting capacity
Plastifizierung *f* plasticization
Plastifizierungsmittel *n* plasticizer
Plastik *f* sculpture
Plastikszintillator *m s.* Plastszintillator
Plastilin *n* plasticine
plastisch plastic
Plastizität *f* plasticity
vollkommene P. perfect plasticity
Plastizitätsgrenze *f* plastic limit
Plastizitätstheorie *f* plastic theory
Plastograf *m* plastograph
Plastometer *n* plastometer
Plastpreßstück *n* compression-moulded plastic
Plastschaum *m* foamed plastic
Plastspritzflasche *f* squeeze bottle
Plastspritzgußstück *n* injection-moulded plastic
Plaststoff *m* plastic material
Plastszintillator *m* plastic scintillator
Plasttechnik *f* plastics engineering
Plastüberzug *m* plastic coat[ing]
Plastvermehrungszelt *n (ag) (Am)* plastic wigwam
Platane *f* plane-tree
Plateau *n (nucl)* plateau; voltage (Geiger) plateau

Plateauanstieg *m* plateau slope
Plateaubereich *m* plateau [region]
Plateaubreite *f s.* Plateaulänge
Plateaucharakteristik *f s.* Zählrohrcharakteristik
Plateaulänge *f* length of plateau, plateau length
Plateauneigung *f* plateau slope
Platforming *n (ch)* platforming
Platin *n* platinum
Platinanode *f* platinum anode
Platinasbest *m* platinized asbestos
Platinat *n* platinate
Platinblech *n* platinum foil (sheet)
Platindraht *m* platinum wire
Platine *f* 1. *(met)* sheet bar (billet); sheet metal blank *(metal forming)*; 2. *(text)* sinker; hook *(Jacquard loom)*
blinde P. *(text)* bluff sinker
Platinelektrode *f* platinum electrode
Platinenboden *m* bottom board *(Jacquard machine)*
Platinenexzenter *m* sinker (jack) cam *(knitting)*
Platinenexzenter *mpl* / **ineinandergreifende** interlocking sinker cams
Platinenexzenterring *m* cam-ring of sinker
Platinennase *f (text)* catch of the hook, lifter end (nose)
Platinenrad *n (text)* looping wheel
Platinenschnur *f (text)* neck cord
Platinenwalzwerk *n* sheet-bar rolling-mill
Platinenwärmofen *m* pair furnace
Platingefäß *n* platinum vessel
platinhaltig platiniferous, platinum-bearing
platinieren to platinize
Platiniridiumdraht *m* platinum-iridium wire
Platinkatode *f* platinum cathode
Platinkontakt *m* platinum contact
Platinlot *n* platinum solder
Platinmetall *n* platinum metal
Platinmohr *n* platinum black
Platinplattierung *f* platinum cladding
Platinrhodiumdraht *m* platinum-rhodium wire
Platinschiffchen *n (ch)* platinum boat
Platinschwamm *m* platinum sponge
Platinschwarz *n* platinum black
Platinthermometer *n* platinum resistance thermometer
Platintiegel *m* platinum crucible
Platin(II)-verbindung *f* platinous compound
Platin(IV)-verbindung *f* platinic compound
platonisch Platonic
platt flat; level, even, plane
Plattbodenboot *n* flat-bottomed vessel, flat [boat], kettle bottom
Plättbrett *n* ironing board
Plättchen *n* 1. lamina; 2. *(mach)* tool tip; 3. wafer, slab *(of crystals)*
hart aufgelötetes P. brazed-on tip
Platte *f* 1. plate *(mechanics, photography, battery)*; 2. *(build)* panel; slab *(concrete)*; tile; board; 3. *(print)* stereo plate; 4. bed plate *(beater)*; 5. *(plast)* sheet; 6. record *(gramophone)*
P. mit Siebabdruck screen-back board
P. zur Innenausstattung interior finish board
bituminöse P. bituminous board
einachsig (einfach) bewehrte P. one-way [reinforced] slab
einstufengeätzte P. powderless-etched plate
extraverdichtete P. special-densified hardboard
faserverfilzte P. fibrous-felted board
flachgepreßte P. flat-platen pressed board
gipsbelegte P. plaster-lath board
grundierte P. prime-coated board
halbharte P. semi-hard board
harte P. mit Metallüberzug metal-faced hardboard
kreuzweise bewehrte P. two-way [reinforced] slab, slab reinforced in two directions
maschinell gelochte harte P. perforated hardboard
ölgetränkte P. oil-treated board
ölvergütete P. oil-tempered board
orthotrope P. orthotropic plate
pastierte P. pasted plate *(battery)*
vergütete P. tempered board
vorgefertigte P. prefinished board
Plätte *f (ship)* kettle bottom, flat[boat]
plätten to iron, to smooth; to press
Plätten *n* pressing; ironing
Plattenableiter *m* plate-type lightning arrester
Plattenanlage *f* plate assembly *(Cottrell precipitator)*
P. in stehender Ausführung set of vertical plates
Plattenbalken *m* tee beam, T-beam, tee section, T-section, beam and slab
Plattenbalkendecke *f* slab and beam floor
Plattenband *n* slat *(conveying)*; steel-plate conveyor
Plattenbandförderer *m* slat (apron) conveyor
Plattenbankhobeleisen *n* skew plane iron
Plattenbau *m* [large] panel construction, [large] panel system; large-panelled structure
Plattenbauweise *f* [large] panel construction, [large] panel system
Plattenbearbeitung *f* plate finishing
Plattenbelag *m* flagging, tiling, tile flooring
Plattenbeschichtung *f* plate coating
Plattenbohrwerk *n* floor plate-type horizontal boring mill, floor-type horizontal boring and facing machine
Plattenbreite *f* / **mitwirkende** effective [flange, slab] width
Plattenbrücke *f* slab bridge
Plattenbug *m (ship)* plate (soft) bow, soft-nosed stem

abgerundeter (runder) P. barrel bow
Plattendekatur *f* plate decating
Plattenelektrode *f* plate electrode
Plattenelement *n* plate-type fuel element
Plattenentkörnungsmaschine *f (text)* Macarthy gin
Plattenfederindikator *m* metallic diaphragm indicator
Plattenfedermanometer *n* diaphragm (Schaeffer type) pressure gauge
Plattenfilzmaschine *f* plate felting machine
Plattenförderer *m* plate conveyor
Plattenformat *n* plate size
Plattenfundament *n (build)* raft (slab, mat) foundation; *(print)* [mounting] base, mount
Plattengang *m (ship)* [plate, knuckle] strake
Plattengründung *f (build)* raft (slab, mat) foundation
Plattengummi *m* sheet rubber
Plattenherstellung *f* platemaking
Plattenionisationskammer *f* parallel plate [ionization] chamber
Plattenkamera *f* plate camera
Plattenkammer *f (nucl)* parallel plate [ionization] chamber
Plattenklemme *f* plate clamp *(plate whirler)*
Plattenkondensator *m* plate capacitor
Plattenkorn *n* plate grain
Plattenkörnen *n* plate-graining
Plattenkultur *f* plate culture
Plattenkurzschluß *m* plate short-circuit *(storage battery)*
Plattenmast *m (ship)* plated mast
Plattenmaterial *n* sheet material
Plattenmontage *f* panel mounting
Plattenofen *m* multiple-hearth roaster
Plattenpaket *n* element, plate group *(storage battery)*
Plattenrüttler *m s.* Rüttelplatte
Plattenschleuder *f* plate whirler
Plattenschneider *m* / **elektrischer** electric phonograph recorder
Plattensenge *f (text)* plate singer
Plattenspieler *m* record player, phonograph
Plattenspielerzusatz *m* gramophone attachment
Plattensteven *m (ship)* soft-nosed stem, fashion plate stem, plate bow
Plattentechnik *f (nucl)* [nuclear] emulsion technique, photographic plate technique
Plattenteller *m* turntable *(gramophone)*
Plattentiefdruck *m* plate gravure printing
Plattenüberhitzer *m* Heizmann's plate superheater
Plattenventil *n* disk (plate) valve
Plattenverbinder *m* plate fastener *(belt)*
Plattenverkleidung *f* stone facing
Plattenvorsteven *m* fashion plate stem
Plattenvorwärmer *m* baffle (plate) feed heater
Plattenwechsler *m* record changer
Plattenzähler *m (nucl)* parallel plate counter
Plattenzylinder *m* plate cylinder
Plattform *f* platform
Plattformball *m* punching bag *(boxing)*
Plattformdeck *n* flat, platform deck
wasserdichtes P. watertight flat
Plattformhubwagen *m* elevating platform truck
Plattformstropp *m (ship)* platform (board) sling
Plattformträger *m (aut)* platform outrigger
Plattformtuch *n (ag)* bottom canvas
Plattformwagen *m* flat wagon; fixed platform truck
Plattformzwischendeck *n (ship)* watertight flat
Plattgatt *n (ship)* square stern
Plattgattboot *n* square-stern[ed] boat
Plattgattheck *n* square[-tuck] stern
plattieren 1. *(met)* to clad, to bond; 2. *(text)* to plate
Plattierfaden *m* plating yarn
Plattierfadenführer *m* plating thread carrier
Plattiergarn *n* plating yarn
Plattierschicht *f (met)* facing, cladding, clad plate
Plattierung *f* cladding, plating
Plattierungsmaterial *n (met)* cladding material
Plattstich *m* broad stitch
Plattstichstickerei *f* flat-stitch embroidery
Platz *m* 1. circus *(round)*; square; 2. site *(building)*; ground *(sports)*; 3. place; spot; point; 4. position; 5. space, room; seat
besetzter P. occupied position *(telephony)*
freier P. reserve position *(telephony)*
leerer P. vacant site *(in the crystal lattice)*
Platzanflughöhe *f* initial approach altitude
Platzanzeige *f* place indication
Platzarbeiter *m* yardman
Platzausnutzungsgrad *m* seat occupancy coefficient
Platzbedarf *m (mach)* [floor] space required, [floor] space requirement
Platzbefeuerung *f* air-field lighting
Platzeinflugzeichen *n* boundary (inner) marker
platzen to burst; to explode; to crack, to puncture, to blow out *(tyre)*
Platzen *n* **des Reifens** blowout
Platzersparnis *f* space saving
Platzfeuer *n* air-field lighting
Platzkartengebühr *f* reserved seat charge
Platzlampe *f* calling (pilot) lamp *(telephony)*
Platzmangel *m* shortage of seats; lack (limitation) of space
Platzpatrone *f* blank cartridge
platzsparend room-saving, space-saving
Platzwechsel *m* transposition crossing *(lines)*
Platzwettervorhersage *f* aerodrome meteorological forecast

Platzzähler *m* position meter *(telephony)*
Pleochroismus *m* pleochroism, dichroism
Pleuel *m* connecting rod
Pleuelauge *n* small end of conrod, small-end boss
Pleuelbuchse *f* small-end bushing
Pleuelkopf *m* connecting-rod head
Pleuellager *n* connecting-rod bearing, big-end bearing
Pleuelschaft *m* connecting-rod blade (shank)
Pleuelschraube *f* connecting-rod bolt
Pleuelstange *f* connecting rod, conrod; piston rod
Pleuelstangenausschlag *m* connecting-rod angle
Pleuelstangenbügel *m* connecting-rod strap
Pleuelstangenkopf *m* connecting-rod head
Pleuelstern *m* master connecting-rod assembly
Plicht *f (ship)* cuddy, cockpit
  selbstlenzende P. self-draining cockpit, automatic baler (bailer)
Pließten *n* polishing with leather-covered wooden disk
Pließtscheibe *f* leather-covered wood polishing wheel
Plinthe *f* plinth *(architecture)*
Plissee *n* pleating
Plisseeform *f* pleating pattern
Plisseerock *m* pleated skirt
Plissiermaschine *f* pleating machine
Plombe *f* 1. seal; 2. filling *(tooth)*
Plombendraht *m* safety (sealing) wire
plombieren to seal
Plombierzange *f* sealing tongs (pliers)
Plubberrohr *n* bubbler gauge *(level indicator)*
Plumbat *n* plumbate
Plumbit *n* plumbite
Plunger *m* plunger
Plungerkolben *m* plunger piston
Plusbürste *f (el)* positive brush
Plüsch *m* plush
  geschnittener P. cut plush
  gezogener P. uncut plush
plüschartig plushy
Plüschgewebe *n* plush fabric
Plüschleder *n* velvet leather
Plüschrundstrickmaschine *f*/ **einfonturige** pile fabric plain circular knitting machine
Plüschwalze *f* clearer roller *(comber)*
Plüschweberei *f* plush weaving
Plüschwebstuhl *m* plush loom
Plusklemme *f* positive terminal
Plusleiter *m* positive conductor (main, lead)
Pluspol *m* positive pole
Pluszeichen *n* plus sign
Plutonium *n* plutonium
Plutoniumbombe *f* plutonium bomb
Plutoniumerzeugung *f* plutonium production
Plutoniumerzeugungsreaktor *m* plutonium [-producing] reactor, production[-type] reactor
Plutoniumfabrik *f* plutonium factory
Plutoniumisotop *n* plutonium isotope
Plutoniumreaktor *m* plutonium reactor (pile)
Plutoniumrückführung[sprozeß *m***]** *f* plutonium recycle
Plutoniumrückgewinnung *f* plutonium regeneration
Plutoniumspaltung *f* plutonium fission
Pluviometer *n* pluviometer
pn-Flächentransistor *m* p-n junction transistor
pn-Fotoelement *n* photovoltaic cell
pn-Gleichrichter *m* p-n rectifier
pn-Kontakt *m* p-n junction
pn-Sperrschichtzelle *f* photovoltaic cell
pn-Übergang *m* p-n junction (transition) *(semiconductor)*
Pneumatik *f* pneumatics
pneumatisch pneumatic
pneumatisch-mechanisch pneumatic-mechanical
pneumohydraulisch hydro-pneumatic
Pneumomotor *m* air motor
Pneumonik *f* compressed air techniques
pnp-Transistor *m* p-n-p transistor
pnpn-Element *n* p-n-p-n device
Pochgestein *n* stamp rock
Pochtrübe *f* stamp pulp
Pochwerk *n* stamp[ing] mill
Pockholz *n* pock-wood, Guaiacic wood
Pockholzlager *n* pock-wood pillow, lignum-vitae pillow
Podest *n(m)* landing; platform
Podestbalken *m* bearer, landing slab
Podestträger *m* bearer
Podium *n* dais, podium
Podsol *m* podsol
Pointierungsokularkopf *m* guiding ocular head
Poise *n* poise
Pökellauge *f* pickle, brine
pökeln to salt, to corn, to pickle
Pol *m* 1. *(ph, el)* pole; terminal; instantaneous centre; 2. *(text)* pile *(closed pile loop)*
  ausgeprägter P. salient pole
  geerdeter P. earthed pole
  laminierter P. stamped pole
  magnetischer P. magnetic pole
  nacheilender P. lagging pole
  negativer P. minus (negative) pole
  positiver P. plus (positive) pole
  voreilender P. leading pole
Polabstand *m* pole distance
Polachse *f* direct axis
Polanker *m* armature with salient poles, pole armature
polar polar
Polarachse *f* polar axis
Polardiagramm *n* polar diagram

horizontales P. curve of horizontal distribution
vertikales P. curve of vertical distribution
Polare *f s.* Polardiagramm
Polarform *f* **einer komplexen Zahl** polar form of a complex number
Polarimeter *n* polarimeter
Polarisation *f* polarization
P. der Elektronenhülle electronic polarization
dielektrische P. dielectric polarization
elektrolytische P. electrolytic polarization
elliptische P. elliptic polarization
lineare P. plane polarization
molare P. molar polarization
partielle P. partial polarization
Polarisationsachse *f* polar axis
Polarisationsapparat *m* polariscope, polarizer
Polarisationsausbeute *f* polarization efficiency
Polarisationsbatterie *f* polarization battery
Polarisationsblindwiderstand *m* polarization reactance
Polarisationsbrille *f* polarizing spectacles
Polarisationsdreher *m* polarization rotator
Polarisationsebene *f* plane of polarization, polarization plane
Polarisationsellipse *f* polarization ellipse
Polarisationsfehler *m* polarization error
Polarisationsfilter *n* polarization filter
Polarisationsfotometer *n* polarizing photometer
Polarisationsgrad *m* degree of polarization, polarization coefficient
Polarisationskapazität *f* polarization capacitance
Polarisationskraft *f* polarization force
Polarisationsladung *f* polarization charge
Polarisationsmagnet *m* polarizing magnet
Polarisationsmikroskop *n* polarization microscope
Polarisationsprisma *n* polarizing prism
Polarisationsrichtung *f* polarization direction
Polarisationsschwund *m* polarization fading, fading by polarization
Polarisationsspannung *f* polarizing (polarization) potential
Polarisationsstrom *m* polarizing (polarization) current
Polarisationsvermögen *n* polarizing power
Polarisationswinkel *m* polarization (polarizing) angle
Polarisationswirkwiderstand *m* polarization resistance
Polarisationswolke *f* polarization cloud
Polarisator *m* polarizer
polarisierbar polarizable
Polarisierbarkeit *f* polarizability
polarisieren to polarize
polarisiert polarized
elliptisch p. elliptically polarized
geradlinig p. plane-polarized
linear p. plane-polarized
teilweise p. partially polarized
zirkular p. circularly polarized
Polarität *f* polarity
Polaritätsanzeige *f* sign indication
Polaritätsumschlag *m* **der Thermospannung** thermoelectric inversion
Polaritätsunterscheidung *f* polarity discrimination
Polarkegelschnitt *m* polar conic
Polarkoordinate *f* polar coordinate
Polarkoordinatenmodus *m* polar mode
Polarkoordinatograf *m* polar coordinatograph
Polarlicht *n* polar light *(aurora borealis or australis)*
Polarografie *f* polarography
Polarogramm *n* polarogram
Polarplanimeter *n* polar planimeter
Polarpotentiometer *n* bias potentiometer
Polarprojektion *f* polar projection
Polarstern *m* polar star
Polarwinkel *m* amplitude, argument
Polarzölostat *m* polar coelostat
Polbahn *f* locus of instantaneous centre
bewegliche P. movable path of instantaneous centre
feste (ruhende) P. fixed path of instantaneous centre
Polbogen *m* pole arc
Polbolzen *m* terminal pillar *(of battery)*
Polbrücke *f* cell connector
Polder *m* polder
Poldihammer *m* Poldi hardness tester
Pole *mpl* / **gleichnamige** like poles
ungleichnamige P. unlike (opposite) poles
versetzte P. staggered poles
Poleck *n* polar diagram
polen to pole *(molten metal)*
Polfaden *m (text)* pile thread
Polgewebe *n* / **gezogenes** *(text)* uncut pile fabric
Polianit *m (min)* polianite
Polier *m (build)* [general] foreman
Polierballen *m* polishing cup
Polierband *n* abrasive belt [for polishing]
Polierdorn *m* buffing cone
Polierdrücken *n* burnishing
polieren to polish; to burnish
farbig p. to stain
Polieren *n* / **elektrolytisches** electrolytic polishing, electro-polishing
Polierfeile *f* smoothing file
Polierfilz *m* polishing felt
Poliergerät *n* polisher
elektrolytisches P. electrolytic polisher
Poliergerüst *n* planishing stand

Polierläppen *n* buffing
Poliermaschine *f* polishing machine
Polierpulver *n* polishing powder
Polierrollen *n* burnishing
Polierrot *n* rouge, ferric oxide
Polierscheibe *f* polishing wheel (pad)
Polierschlagen *n* planishing
Polierschleifen *n* abrasive-belt polishing
Polierstahl *m* burnisher
Poliertrommel *f* polishing drum (barrel)
Polierwachs *n* polishing wax
Polierwalzengerüst *n* bull head
Polierwalzwerk *n* planishing mill
Politur *f* polish, finish, gloss, lustre
Polkante *f* pole horn (edge); terminal edge
  ablaufende P. trailing pole horn
  auflaufende P. leading pole horn
  hintere P. trailing pole horn
  vordere P. leading pole horn
Polkern *m* pole core
Polkette *f (text)* pile warp
Polklemme *f* pole terminal (binder), cell (electrode) terminal, terminal post
Polkopf *m* polar head
Polleiste *f* battery bus-bar, cell connector
Poller *m* 1. *(ship)* bollard[s], bitton, bitt[s], mooring bitts, timber head; 2. *(build)* bollard
Pollerkopf *m s.* Poller 1.
Pollux *m* pollucite, pollux
Polmagnetisierung *f* longitudinal magnetization
Polodiekegel *m* polhode cone
Polonium *n* polonium
Poloschläger *m* polo-stick
Polpaarzahl *f* number of pole pairs
Polplatine *f (text)* pile sinker
Polprüfer *m* polarity indicator, pole tester
Polprüfpapier *n* pole[-finding] paper
Polrad *n* magnet (cog) wheel, field (pole) spider
Polradspannung *f* magnet wheel voltage
Polreagenzpapier *n s.* Polprüfpapier
Polschalter *m* pole-changing switch
Polschenkel *m* pole piece (shank)
Polschermaschine *f (text)* pile-shearing machine
Polschuh *m* pole shoe (piece, tip)
Polschuhabstand *m* magnet gap
Polstärke *f* pole strength
  magnetische P. magnetic pole strength
Polstärkeeinheit *f* / **elektromagnetische** unit magnetic mass
Polster *n* cushion, pad
Polstermaterial *n* packing material, upholstery backing (material)
Polstermöbelstoff *m* tapestry
polstern to pad, to stuff; to upholster
Polstern *m (el)* field (pole) spider
Polsterschmierung *f* oil pad (waste) lubrication
Polstersitz *m (ry)* upholstered seat
Polsucher *m* polarity indicator
Polteilung *f* pole pitch
Polumkehr *f* pole reversal
polumschaltbar pole-change[able], pole-changing, change-pole
Polumschalter *m* pole reverser, [pole] reversing switch, pole-changing switch
Polumschaltung *f* reversal of poles, pole changing; pole-changing control
Polungsschutz *m* polarity guard
Polvieleck *n* polygon of instantaneous centre
Polware *f* / **doppelflächige** double-faced pile fabric
Polwechsel *m* change (alternation) of polarity, pole changing
Polwechselschalter *m* pole reverser, [pole] reversing switch, pole-changing switch
Polwechsler *m s.* Polwechselschalter
Polwender *m* commutator, pole changer
Polwendung *f* reversal of poles, commutation
Polwicklung *f* field coil (winding)
Polyaddition *f* polyaddition
Polyakrylnitril *n* polyacrylonitrile
Polyakrylnitrilfaser *f* polyacrylonitrile fibre
Polyakrylsäure *f* polyacrylic acid
Polyamid *n* polyamide
Polyamidfaser *f* polyamide fibre
Polyäthylen *n* polyethylene
Polybasit *m (min)* polybasite
Polychloropren *n* polychloroprene
polychromatisch polychromatic
polydispers polydisperse
Polyeder *n* polyhedron
polyedrisch polyhedral
polyenergetisch polyenergetic
Polyesterfaser *f* polyester fibre
Polyesterfolie *f* polyester foil
Polyesterharz *n* polyester resin
Polygon *n* polygon
polygonal polygonal
Polygonausbau *m* cockering
Polygonausrüstung *f* traversing equipment
polygonieren to traverse
Polygonierung *f* polygonization, traversing
Polygonisation *f s.* Polygonierung
Polygonrost *m (therm)* polygonal grate
Polygonschaltung *f* polygon[al] connection, simply mesh-connected circuit
Polygonseite *f* traverse leg
Polygonspannung *f (el)* polygonal voltage
Polygonzug *m (el)* polygonal connection; traverse
Polygonzugverfahren *n (math)* polygon method
Polygrafie *f* printing [industry]
Polyisobutylen *n* polyisobutylene
Polykarbonat *n* polycarbonate
Polykondensat *n* polycondensation product
Polykondensation *f* polycondensation

polykonisch polyconic
polykristallin polycrystalline
Polymer *n* polymer
polymer polymeric
Polymerbenzin *n (Am)* polymer gasoline
Polymerisat *n* polymer
Polymerisation *f* polymerization
strahlungsinduzierte P. radiation-initiated polymerization
zyklisierende P. cyclo-polymerization
Polymerisationsanlage *f* polymerization plant (reactor)
polymerisationsfähig polymerizable
Polymerisationsinhibitor *m* polymerization inhibitor
polymerisierbar polymerizable
Polymerisierbarkeit *f* polymerizability
polymerisieren to polymerize
Polymethakrylat *n* polymethacrylate
polymorph polymorphous, polymorphic
Polymorphie *f* polymorphism
Polynom *n* polynomial
polynom polynomial
Polynomrechner *m* polynomial computer; equation solver
Polypeptid *n* polypeptide
Polypgreifer *m (mine)* orange-peel bucket, cactus grab
Polyphenyl *n* polyphenyl
Polyphosphat *n* polyphosphate
Polyphosphornitrilchlorid *n* polyphosphonitrilic chloride
Polypropylenfaser *f* polypropylene fibre
Polysaccharid *n* polysaccharide
Polysiloxan *n* polysiloxane
Polystyrol *n* polystyrene
Polysulfid *n* polysulphide
polytechnisch polytechnic[al]
Polyterephthalsäureester *m* polyterephthalic acid ester
Polytetrafluoräthylen *n* polytetrafluoroethylene
Polythionsäure *f* polythionic acid
Polytrope *f* polytropic curve
polytropisch polytropic
Polyurethan[harz] *n* polyurethane [resin]
Polyurethanschaum *m* polyurethane foam
Polyvinylalkohol *m* polyvinyl alcohol
Polyvinylazetal *n* polyvinyl acetal
Polyvinylazetat *n* polyvinyl acetate
Polyvinylchlorid *n* polyvinyl chloride
nachchloriertes P. postchlorinated polyvinyl chloride
Polyvinylchloridfaser *f* polyvinyl fibre
Polyvinylchloridpaste *f* PVC paste
Polyvinylidenchlorid *n* polyvinylidene chloride
Polyvinylkarbazol *n* polyvinyl carbazole
Polyvinylpyrrolidon *n* polyvinylpyrrolidone
Polyvinylschaumstofflaminat *n* polyvinyl foam laminate
polyzyklisch polycyclic
Polzahl *f* number of poles
Pomeranzenöl *n* orange oil
Ponceau *n* ponceau *(azo dye)*
Pond *n* gram[me] weight, *(Am)* gram force
Ponton *m* pontoon
Pontonbrücke *f* pontoon (floating) bridge
Pontondock *n* pontoon dock
Pontonkran *m* pontoon (floating) crane
Pontonwagen *m* pontoon carrier
Pool-Palette *f* pool-pallet *(conveying)*
Poop *f (ship)* poop
Poopdeck *n* poop deck
Popeline *f* poplin
Porcupine-Öffner *m (text)* porcupine opener
Pore *f* pore; void; pin hole *(paint)*
Porenbeton *m* cellular concrete
Porenflüssigkeitsanelastizität *f* pore-fluid anelasticity
porenfrei non-porous
Porenfüller *m* pore filler, primer, sealer
Porenleitfähigkeit *f* pore conductivity
Porenraum *m* void spaces [between grains]
porenschließend pore-damming
Porenvolumen *n* **eines Papiers** porosity of paper
Porenwasser *n* pore water *(soil mechanics)*; *(mine)* void water; interstitial water
Porenziffer *f* void ratio *(soil mechanics)*
porös porous
Porosität *f* porosity; sponginess *(metal)*
Porphyr *m* porphyry
porphyrisch porphyritic
Portal *n (build)* portal; *(mach)* bridge
Portalbauweise *f* portal (gantry) design
Portalhubwagen *m* straddle carrier
Portalkran *m* gantry crane (hoist), portal (frame) crane
Portallehrenbohrwerk *n* jig borer with uprights and cross rail
Portalmast *m* portal mast, gantry tower
Portalmasten *mpl (ship)* goal-post masts; pair masts
Portalmastenbrücke *f (ship)* king bridge
Portalrahmen *m* portal frame, bent
Portefeuilleleder *n* pocket book leather
Portikus *m* portico
Portionsweide *f* strip-grazed pasture
Portlandzement *m* Portland cement
Porto *n* postage
portofrei post-free; carriage paid
portopflichtig liable to postage
Porzellan *n* porcelain; china[-ware]
Meißener P. Dresden china
unglasiertes P. biscuit porcelain

Porzellanabdampfschale *f* porcelain evaporating dish
porzellanartig porcelaneous, vitreous
Porzellanbrei *m* porcelain slip
Porzellandielektrikum *n* porcelain dielectric
Porzellandurchführung *f* porcelain bushing
Porzellaneinsatz *m* porcelain shelf *(desiccator)*
Porzellanemail *n* porcelain enamel
Porzellanerde *f* china clay, kaolin
Porzellanfarbe *f* porcelain colour
Porzellanfassung *f* porcelain lamp holder
Porzellangeschirr *n* china-ware
Porzellanisolator *m* porcelain insulator
Porzellankondensator *m* porcelain capacitor
Porzellanmanufaktur *f* porcelain factory
Porzellanschiffchen *n* porcelain boat
Porzellantiegel *m* porcelain crucible
Porzellantülle *f* porcelain elbow (outlet) *(conduits)*
Porzellanüberwurf *m (el)* porcelain cap
Posamenten *npl* trimmings
Posamentenwebstuhl *m* lacing loom
Position *f* position
Positionär *m (cont)* valve positioner
Positioniereinrichtung *f (mach)* positioning device
positionieren to position
Positionierung *f* positioning; position control
  P. des Tisches table positioning
Positionierungseinrichtung *f (mach)* positional control system, positioning device
Positionierungsgenauigkeit *f (mach)* accuracy of positioning *(table)*; locating precision *(hole)*
Positionsbestimmungsgerät *n* air position indicator
Positionsfadenmikrometer *n* position filar micrometer
Positionsgenauigkeit *f (mach)* positioning accuracy
Positionshalten *n (aero)* station-keeping
Positionslampe *f (aut)* side-lamp
Positionslaterne *f* running (regulation) light, navigating lantern
Positionslicht *n* 1. position light; 2. *s.* Positionslaterne
Positionsmanschette *f (el)* marker sleeve
Positionsmeldung *f* position message (report)
positiv positive
  p. geladen positively charged
  p. gerichtet positive-going
  p. werden to go positive
Positivdruck *m* positive printing
Positivfarbauszug *m* positive separation
Positivfarbfilm *m* colour-print film
Positivretusche *f* positive retouching
Positron *n* positron
Positronenstrahler *m* positron emitter (radiator)
Positronenvernichtung *f* positron annihilation
Positronenzerfall *m* positron decay, β^+-decay
Positronium *n* positronium
Postament *n* pedestal
Postangestellter *m* postal clerk
Postdampfer *m* mail steamer (boat, packet), liner
Posten *m* 1. place; post; station; 2. amount; item; *(mach)* lot *(series)*, batch; 3. *(math)* addend
  magnetisch gedruckter P. *(dat)* magnetically printed item
  mangelhafter P. bad lot
  offener P. *(dat)* delinquent account
  qualitätsgerechter P. good lot
Postenumdrucker *m (dat)* posting machine
Postenzähler *m (dat)* transaction counter; item counter
Postkarte *f* postcard
Postpapier *n* bond paper
Postschiff *n* mail boat (packet), packet [boat]
Postwagen *m (ry)* railway mail car, mail-van, *(Am)* mail-car
Potential *n* potential
  anziehendes P. attractive potential
  elektrisches P. electric potential
  elektrokinetisches P. electrokinetic potential
  elektrostatisches P. electrostatic potential
  konstantes P. constant potential
  magnetisches P. magnetic potential
  magnetostatisches P. magnetostatic potential
  retardiertes P. retarded potential
  skalares P. scalar potential
  thermodynamisches P. thermodynamic potential
  transversales P. transverse potential
  vektorielles P. vector potential
Potentialabfall *m* voltage drop, fall of potential
Potentialabnahme *f* decline in potential
Potentialänderungen *fpl* potential variations
Potentialanstieg *m* potential rise
Potentialausgleich *m* potential equalization
Potentialberg *m* potential barrier
Potentialbild *n* potential diagram *(of an electron-optical system)*
Potentialdifferenz *f* potential difference
  elektrische P. electric potential difference
Potentialentkoppler *m* isolating transformer
Potentialfeld *n* potential field
Potentialfläche *f* level surface
Potentialfunktion *f* potential function
Potentialgefälle *n* potential gradient
Potentialgleichung *f* potential (Laplace's) equation
Potentialgradient *m* potential gradient (slope)
Potentialkurve *f* potential curve
Potentialmulde *f* potential well
Potentialnapf *m* potential trough

Potentialschwelle *f* potential threshold (barrier)
Potentialsenke *f* potential trough
Potentialsprung *m* potential jump
Potentialsteuerung *f* potential guiding
Potentialstreuung *f* potential scattering
Potentialströmung *f* potential flow
Potentialtopf *m* potential pot, potential well [with rounded edges]
Potentialtopfmodell *n s.* Schalenmodell
Potentialtransformator *m* potential transformer
Potentialunterschied *m* potential difference
Potentialverlauf *m* potential course (distribution, curve)
Potentialverteilung *f* potential distribution, distribution of potential
Potentialwall *m* potential barrier
 Coulombscher P. Coulomb barrier
potentiell potential
Potentiometer *n* potentiometer
 induktives P. inductive potentiometer
 konusförmiges P. tapered potentiometer
Potentiometer *npl* / **achsgekoppelte** ganged pair of potentiometers
Potentiometerabschnitt *m* potentiometer segment
Potentiometerarm *m* moving arm of the potentiometer
Potentiometereinstellung *f* potentiometer setting
Potentiometerkörnigkeit *f* potentiometer granularity
Potentiometerschaltung *f* potentiometer connection
Potentiometerschreiber *m* potentiometer recorder
Potentiometerstreifenschreiber *m* potentiometer chart recorder
Potentiometerteilung *f* potentiometer graduation
Potentiometrie *f* potentiometric analysis
potentiometrisch potentiometric
Potenz *f (math)* power
Potenzgesetz *n* power law
potenzieren *(math)* to raise to a power
Potenzierung *f (math)* involution; raising to a power
Potenzreihe *f* exponential (power) series
Potier-EMK *f* Potier's electromotive force
Pottasche *f* potash, potassium carbonate
Pottermeter *n* pottermeter, turbine flowmeter
Pottingechtheit *f (text)* fastness to potting
Power-and-Free-System *n* power-and-free system
Prädikatenkalkül *m (dat)* predicate calculus
Prägedruck *m* embossed print[ing]
Prägefolie *f* embossed sheet
Prägeform *f* coining die; stamping mould
Prägegalvano *n* electroblock
Prägekalander *m* embossing (goffering) calender
prägen to coin, to strike; to hob; *(print)* to emboss, to stamp; to mould *(matrix)*; to mint *(coins)*
Prägepoliermaschine *f* burnishing lathe
Prägepresse *f* embossing press
Prägestempel *m* punch, die
Prägestock *m* matrix, die
Prägevergoldung *f* gold blocking
Prägewalze *f* engraved roller
Prägung *f* coinage, mintage *(coins)*; embossing
Prahm *m* barge, pram, praam, punt, scow; pontoon
 P. für Abortabwässer sewerage pram (barge)
Prahmfähre *f* ocean-carrier, sea-barge clipper (carrier) *(suitable for the carriage of commercial cargo over long sea passages in specially designed barges)*
Prahm-Mutterschiff *n s.* Prahmfähre
Praktikant *m* student (graduate) apprentice
Praktiker *m* practician, expert
Praktikum *n* / **chemisches** laboratory course
praktisch practical *(application)*; feasible, working; serviceable
Prallblech *n* baffle [plate], deflector
prallen [auf] to bounce [against]
Prallfläche *f* baffle wall (surface), deflecting wall (plate, surface)
Prallklappe *f* deflector
Prallmühle *f* impact mill, rebound crusher
Prallplatte *f* baffle sheet (plate, board), bounce plate; deflecting (deflection, deflector) plate; baffle, flapper *(pneumatics)*; target *(electronics)*
Pralluftschiff *n* non-rigid airship
Prämiensystem *n* bonus scheme
Präparat *n* preparation; drug; source; preserve; specimen
 galenisches P. galenical preparation
 gefärbtes P. stained specimen
 mikroskopisches P. microscopic slide
 radioaktives P. radioactive source (drug, preparation)
Präparateschutz *m* slide-protective spring mount
Präparatträger *m* source holder; drug carrier
Präparatwechsler *m s.* Probenwechsler
Präparierbesteck *n* dissecting utensils
präparieren to prepare, to treat, to cure, to preserve
Präparierlupe *f* dissecting magnifier
Präpariermikroskop *n* dissecting microscope
Präparierwechselkondensor *m* dissecting change-over condenser
Praseodym *n* praseodym[ium]
Präserve *f* preserve
prasseln to crackle, to decrepitate; *(telec)* to splutter
Präventivbestrahlung *f* preventive irradiation

Praxis *f* practice
Präzession *f* precession
  adiabatische P. adiabatic precession
  progressive P. progressive precession
  pseudoreguläre P. pseudo-regular precession
  regelmäßige P. regular precession *(gyro-axis)*
  retrograde P. retrograde precession
Präzessionsachse *f* axis of precession
Präzessionsbewegung *f* precessional motion
Präzessionsgeschwindigkeit *f* velocity of precession
Präzessionskegel *m* precession cone
Präzessionswinkel *m* angle of precession
Präzessionswinkelgeschwindigkeit *f* angular velocity of precession
präzipitieren to precipitate
präzis precise, exact
Präzision *f* precision
Präzisionsarbeit *f* precision work
Präzisionsfunkenstrecke *f* micrometer gap
Präzisionsgewindespindel *f* precision screw
Präzisionsguß *m* precision casting
Präzisionsinstrument *n* precision instrument
Präzisionsmeßgerät *n* high-accuracy measuring device
Präzisionsmessung *f* high-accuracy measurement
Präzisionsnivellement *n* precise levelling
Präzisionsschliff *m* precision grind[ing]
Präzisionsskale *f* precision measuring scale
Präzisionsverfahren *n* precision method
Präzisionswaage *f* precision balance
Präzisionswellenmesser *m* precision (high-accuracy) wavemeter
Präzisionswiderstand *m* precision (standard) resistor
Preemphasis *f* pre-emphasis
Preis *m* price, cost
Preisangabe *f* quotation
Preisgabe *f* abandonment *(insurance term)*
Preisspiegel *m* cost picture
Preisvektor *m* price vector *(operational research)*
Prellbalken *m* fender beam
Prellblock *m* bumping post, buffer stop *(train-ferry)*
Prellbock *m (ry)* buffer stop, *(Am)* bumper
prellen to bounce, to rebound; to chatter *(relay)*
Prellfilter *n* bounce filter
Prellpfahl *m* fender pile (pole), pole fender
Prellstein *m* bollard, curb
Prellwand *f* baffle
Prepaktbeton *m* Colcrete, prepacked (preplaced) aggregate concrete
Pre-slung-cargo-Methode *f (ship)* pre-slung cargo method
Preßautomat *m* [fully] automatic pressing machine *(glass)*
Preßbank *f* press-bench *(papermaking)*
Preßbengel *m* press jack (stick)
Preßbernstein *m* pressed amber
Preßdruck *m* press[ing] power
Preßdüse *f* extrusion die
Presse *f* press *(metal forming, printing)*; brake; jack
  hydraulische P. hydraulic press
  mechanische P. power press
Preßeinleger *m (ag)* press charger (feeder)
pressen *(met)* to press, to form by press-working; *(plast)* to [compression] mould; to squeeze; *(text)* to calender
  Narben p. to emboss the grain *(leather)*
Pressen *n* pressing; press-working *(metals)*; *(text)* calendering; *(plast)* compression moulding; squeezing
  P. von Verbundbauteilen sandwich moulding
Pressendienst *m s.* Pressenhaus
Pressengestell *n* press frame *(metal forming)*
  C-förmiges P. C-frame
Pressenhaus *n* press house
Pressenmaulentstaubung *f* stamp de-dusting
Pressenpartie *f* press end (section) *(papermaking)*
Pressentisch *m (mach)* press bed
Presseur *m* cylinder, pressure bowl
Preßfehler *m* moulding fault, pit, segregation
Preßfettschmierung *f* pressure grease lubrication
Preßfläche *f* / **voll ausgenutzte** press-filling size *(papermaking)*
Preßform *f* die block; *(plast)* compression mould
Preßfuß *m* pressed-glass base *(tubes)*
Preßgaskondensator *m* compressed-gas capacitor
Preßgasschalter *m* compressed-gas circuit breaker
Preßglanz *m* gloss, lustre
Preßglanzdekatiermaschine *f* lustre shrinking machine
preßglanzpoliert press-polished *(plastics)*
Preßglas *n* moulded glass
Preßglassockel *m* pressed glass base *(photomultiplier tube)*
Preßgrat *m* fin, flash
Preßguß *m* die casting
Preßharz *n* moulding resin
Preßholz *n* compreg, laminated wood
Preßkohle *f* briquette
Preßkopf *m* head of a press
Preßkuchen *m* filter (press) cake
Preßling *m* pressing; compact *(powder metallurgy)*; briquette, moulding
Preßluft *f* compressed air
Preßluftabbauhammer *m (mine)* pneumatic coal pick hammer
Preßluftanlage *f* compressed air plant
Preßluftantrieb *m* compressed-air drive
Preßluftarmatur *f* compressed-air fitting
Preßluftbohrer *m* air (pneumatic) drill

Preßluftfeuerung *f* furnace with forced draught
Preßluftfilter *n* pressure filter
Preßluftförderer *m* airflow conveyor
Preßluftfutter *n (mach)* air chuck
Preßlufthammer *m* pneumatic (compressed-air) hammer; *(mine)* pneumatic pick
Preßlufthaspel *f (mine)* tugger
Preßluftleuchte *f* compressed-air lamp
Preßluftlokomotive *f* compressed-air locomotive
Preßluftmotor *m* compressed-air engine, air (pneumatic) motor
Preßluftramme *f* air (pneumatic) rammer
Preßluftstampfer *m* pneumatic (air) tamper, pneumatic (air) rammer
Preßluftsteuerung *f* compressed-air control
Preßlufttechnik *f* compressed-air technique
Preßluftwerkzeug *n* pneumatic tool
Preßmantelelektrode *f* extruded coating electrode
Preßmaschine *f (text)* calender
Preßmasse *f (plast)* compression moulding material
Preßnaht *f* fin
Pressostat *m* manostat
Preßpappe *f* pressboard
Preßpartie *f* couch and press part
Preßpassung *f* interference fit; *(Am)* tight fit; force (press) fit
Preßplatte *f* pulpwood; press plate
Preßpulver *n* moulding powder
Preßschmierung *f* pressure lubrication
preßschweißbar pressure-weldable
Preßschweißen *n* pressure welding
  aluminothermisches P. pressure thermit welding
Preßsitz *m* interference (press) fit
Preßsitzmontage *f* press-fit mounting
Preßspan *m* pressboard
Preßspritzen *n (plast)* transfer-moulding, flow-moulding, pot-moulding
Preßspritzform *f (plast)* transfer mould (tool)
Preßspritzkolben *m (plast)* transfer [moulding] plunger
Preßspritzverfahren *n (plast)* transfer moulding process
Preßspritzzylinder *m (plast)* transfer [moulding] cylinder
Preßstahlrahmen *m* pressed-steel frame[work]
Preßstempel *m* plunger, heading tool, punch
Preßstoff *m (plast)* compression moulding material; moulded article, pressed material
Preßstoffgehäuse *n* moulded [bakelite] jacket
Preßstoffisolator *m* moulded insulator
Preßstoffisolierung *f* moulded insulation
Preßstück *n s.* Preßteil
Preßstumpfschweißung *f* upset butt welding
Preßtablette *f* moulding pellet
Preßtechnik *f* moulding technique
Preßteil *n* 1. *(met)* pressed (pressure-forged) part; compact *(powder metallurgy)*; 2. *(plast)* compression moulding
Preßtisch *m* platen
Preßtuch *n* press (filter) cloth
Pressung *f* pressure; compression
Pressungswinkel α *m* pressure angle *(gear)*
Preßverfahren *n* moulding, pressing method, pellet technique
Preßwalzenwaschmaschine *f (text)* roller press washing machine
Preßwasser *n* power water
Preßwerkzeug *n* pressing tool
Preßzeit *f* squeezing time *(spot welding)*
Preßziegel *m* press-brick
Preußischblau *n* Prussian blue
prim *(math)* prime
primär primary
Primärabschirmung *f* primary shield[ing]
Primäranker *m* primary (fixed) armature
Primäraustauscher *m* primary heat exchange
Primärbatterie *f* primary (galvanic) battery
Primäreffekt *m* primary effect
Primärelektron *n* primary (initiating) electron
  auftreffendes P. incident primary electron
Primärelement *n* primary cell
Primärfokus *m* prime focus
Primärimpuls *m* primary pulse
Primärinduktion *f* primary inductance
Primärionisation *f* primary ionization
Primärkornbildung *f* ingotism
Primärkreis *m* primary circuit
Primärkreislauf *m* primary circuit
Primärlagerstätte *f (geol)* primary deposit
Primärmetall *n* primary (virgin) metal
Primärreaktion *f* simple reaction
Primärreaktor *m* primary reactor
Primärregler *m* master controller
Primärrelais *n* primary (main-current) relay
Primärschild *m* primary shield[ing]
Primärseite *f (el)* input end
Primärspannung *f* primary voltage
Primärspule *f* primary coil
Primärstrahler *m* primary radiator
Primärstrahlung *f* primary radiation (emission); primary cosmic radiation
Primärstrom *m* primary current
Primärstromkreis *m* primary circuit
Primärteilchen *n* initiating (primary) particle, primary
Primärwicklung *f* primary coil (winding)
Primärwirkung *f* primary action
Primzahl *f* prime number
Primzahlcharakter *m* primality
Primzahleigenschaft *f s.* Primzahlcharakter
Prinzip *n* principle
  P. der Äquivalenz von Masse und Energie principle of mass-energy equivalence

P. der Entropiezunahme principle of increase of entropy

P. der Erhaltung der Energie principle of the conservation of energy

P. der kleinsten Wirkung least action principle, principle of least action

P. des kleinsten Zwanges principle of least resistance (constraint)

P. von Braun und Le Châtelier Le Chatelier-Braun principle

Archimedisches P. Archimedes' principle

Babinetsches P. Babinet's principle

d'Alembertsches P. d'Alembert's principle

Maupertuissches P. least action principle

zugrundeliegendes P. underlying principle

Prinzipschaltbild *n* general wiring-diagram, schematic diagram

Prise *f (ship)* prize, capture

Prisma *n* prism

Abbesches P. Abbe prism

achromatisches P. achromatic prism

bildumkehrendes P. image-inverting prism

Cornusches P. Cornu prism

doppelbrechendes P. double-refracting prism

Fresnelsches P. Fresnel prism

gerades P. right prism

geradsichtiges P. direct-vision prism

Nicolsches P. Nicol prism

stark fächerndes P. highly dispersive prism

totalreflektierendes P. total-reflecting prism

Prismahängelade *f (text)* cylinder in swing

Prismastehlade *f (text)* horizontal cylinder

prismatisch prismatic

Prismatoid *n* prismatoid

Prismenaufsatz *m* penta-prism view finder

Prismendioptrie *f* prism dioptre

Prismenfeldstecher *m* prism binoculars

Prismenfestigkeit *f* prism strength

Prismenfläche *f* prism plane *(crystal)*

Prismenfräser *m* equal-angle cutter, symmetrical-angle cutter

Prismenführung *f* prismatic guideway (way, bearing-surface), inverted vee-guide

Prismenglas *n* prismatic lens

Prismenmeißel *m* angle tool

Prismensatz *m* / **Porroscher** Porro prism set

Prismenspektrograf *m* prism spectrograph

Prismenspektrum *n* prismatic spectrum

Prismenstück *n (mach)* vee block

Prismensucher *m* prismatic viewfinder

Prismenumkehrsatz *m* prism erecting set

Prismenvorsatz *m* prism attachment

Prismenwirkung *f* prismatic effect

Pritschenwagen *m* lorry with platform body

Privatgespräch *n* private call

dringendes P. urgent private call

gewöhnliches P. ordinary private call

Privatgleisanschluß *m (ry)* private siding

Privatgüterwagen *m* privatly-owned wagon

Privatleitung *f* **mit begrenzter Mietdauer** part-time private wire circuit *(telephony)*

Privatnebenstellenanlage *f* [dial] private branch exchange *(telephony)*

Probe *f* 1. sample, specimen, coupon; 2. *s.* Prüfung

gesinterte P. sintered specimen

Probeabzug *m* proof sheet (print)

Probeanruf *m* test call

Probeaufnahme *f (telev)* test shot

Probeauftrag *m* trial order

Probebelastung *f* pile loading test *(foundation)*

Probebetrieb *m* trial (test) operation, trial (test) run

Probebild *n* test picture; *(telev)* definition chart

Probebohrung *f* trial boring

Probedruck *m* test pressure

Probeentnahme *f* sampling, taking (obtaining) of samples

Probefahrt *f* trial; trial (test bed) operation; trial run[ning]; trial trip (test); ship (sea) trials

P. mit Schubschlepper push-towing trial

Probefahrtgeschwindigkeit *f* speed at trial, trial speed

Probefahrtmeile *f* measured course, trial range

Probeflug *m* trial (check) flight

Probekörper *m* test specimen, sample

Probelast *f* test load

Probelauf *m* test (trial) run, running test

P. bei Vollast power test run

Probenahme *f* sampling, obtaining (taking) samples

Probenahmegerät *n* sampler, sampling device

Probenbehälter *m* sample container

Probenehmer *m* sampler

Probenentnahme *f* sampling, taking (obtaining) of samples

Probenhalter *m* sample holder

Probenhalterung *f* sample holder

Probenherstellungstechnik *f* sample preparation technique

Probennehmer *m* sampling apparatus

Probensatz *m (cont, dat)* sampling theorem

Probentafel *f* prepatch panel, problem board

Probenwechsler *m* sample changer

Proberammung *f* driving of test piles

Probestab *m* [sample] test bar

Probestück *n* specimen, sample

Probewürfel *m* test cube

Probezeit *f* trial period

Probezylinder *m* test cylinder

Probierbrille *f (opt)* trial frame

probieren to assay, to test, to try, to check

Probierglas *n (ch)* test tube (glass); *(opt)* trial lens

Probierhahn *m* test (gauge) cock

Probierkorn *n* assay button

Probierkunst *f (ch)* assaying
Probierofen *m* assay furnace
Probierstein *m* touchstone
Probiertiegel *m* test (assay) crucible
Probierverfahren *n (dat)* trial-and-error method
Probierwaage *f* assay balance
Probierzange *f* assayer's tongs
Problem *n* / **technisches** engineering problem
Problemstellung *f* problem
Prochlorit *m (min)* prochlorite
Produkt *n* product; produce
  äußeres P. outer product
  logisches P. logical product
  skalares P. scalar product
  vektorisches P. cross (outer, vector) product
Produktion *f* production; output, yield; manufacture
  P. je Kopf per capita output
  serienmäßige P. large-scale production
Produktionsabteilung *f* workshop
Produktionsarbeit *f* production work
Produktionsarbeiter *m* shop worker
Produktionsausstoß *m* production output
Produktionsbad *n* commercial tank *(plating)*
produktionsfördernd production-raising
Produktionsgang *m* manufacturing process
Produktionsgenossenschaft *f* co-operative association
  gärtnerische P. horticultural production co-operative, co-operative market garden
  handwerkliche P. artisan co-operative
  landwirtschaftliche P. agricultural production co-operative, co-operative farm
Produktionskosten *pl* production (manufacturing) cost
Produktionsleistung *f* productive capacity
  stündliche P. [production] rate per hour
Produktionsleiter *m* production manager (superintendent, supervisor)
Produktionsmaßstab *m* production scale
Produktionsmethode *f* manufacturing method
Produktionsmittel *npl* means of production
Produktionsplan *m* production plan
Produktionsplanung *f* production planning
Produktionsprozeß *m* production process
Produktionsreaktor *m* production[-type] reactor, plutonium[-producing] reactor
Produktionssonde *f* producing well
Produktionssteigerung *f* production increase, output raising
Produktionssteuerungsplan *m* production control scheme
Produktionsverfahren *n* manufacturing process
Produktionsvorausplanung *f* forward production planning
Produktionsziffer *f* production rate
produktiv productive
Produktivität *f* productivity
Produktivkraft *f* productive force
Produktkern *m* product nucleus
Produzent *m* producer
produzierbar producible
produzieren to produce, to manufacture; to yield
Professur *f* professorship
Profil *n* profile; contour; outline, configuration; section *(steel)*; shape
  P. einer Strecke line profile
Profilabrichteinrichtung *f* profile truing attachment *(grinding wheel)*
Profilansicht *f* outline profile *(of a vessel)*
Profilanzeigegerät *n* edgewise indication meter
Profilaufnahme *f* plotting of profiles
Profilaußendurchmesser *m* major diameter *(spline shaft)*
Profilbild *n* profile chart
Profildraht *m* profile (section, sectional) wire
Profile *npl* sections *(steel)*
  gespritzte P. *(plast)* extruded sections
Profileisen *n* sectional iron
Profilfahrdraht *m* grooved trolley wire
Profilform *f* structural shape *(steel)*
Profilfräsen *n* profile milling
Profilfräser *m* profile cutter
Profilgerüst *n (met)* section mill frame
profilgeschliffen profile-ground
Profilglasleuchte *f* figured glass lamp
profilhinterschliffen shaped-profile; profile-relieved
profilieren to profile, to shape, to contour
profiliert profiled; *(aut)* non-skid *(tyre)*; grooved *(roll)*
Profilinnendurchmesser *m* minor diameter *(spline shaft)*
Profilinstrument *n* edgewise meter
Profilkalander *m* embossing calender
Profilkaliber *n* shaping groove
Profillehre *f* profile gauge
Profillinie *f* **des Hochofens** construction lines of brickwork
Profilprojektion *f* outline (contour) projection
Profilprojektor *m* contour projector
Profilquerschnitt *m* shape (profile, contour) of the cross section
Profilradius *m* hydraulic radius
Profilrauhtiefe *f* / **größte** maximum peak-to-valley height
Profilregistrierung *f* profile graphing
Profilreibahle *f* form reamer
Profilrolle *f* crusher roll *(grinding wheel)*
  P. mit eingearbeitetem Profil formed crusher roll
Profilscheibe *f* formed wheel
Profilschiene *f (el)* sectional bus bar
Profilschleifen *n* profile grinding
Profilschleifmaschine *f* profile grinding machine

optische P. optical profile grinding machine
Profilschleifscheibe *f* formed wheel
Profilschnittverfahren *n* silhouetting section method
Profilschrämarm *m (mine)* curved jib
Profilschrieb *m* profile chart (record)
Profilschuß *m (mine)* trimmer
Profilstahl *m* sectional steel
Profilstrahldarstellung *f* shaped beam display
Profilstrahlröhre *f* mask tube
Profilstufensenker *m (mach)* step counterbore
Profilverzerrung *f* distortion of profile
Profilwalzblock *m* shaped ingot
Profilwalze *f* pass roller, shape roll
Profilwiderstand *m* profile drag
Progesteron *n* progesterone
Programm *n* programme, routine, instruction
P. zum Überwachen tracing routine
allgemeines P. general routine
auswechselbares P. prewired instruction
externes P. external programme
kodiertes P. coded programme
lineares P. linear programme
Programmablauf *m (mach)* sequence of operations
Programmband *n* programme tape
Programmbearbeiter *m s.* Programmierer
Programmbefehl *m* programme instruction
Programmbibliothek *f* library
Programmendwinkel *m (rock)* theoretical cut-off angle
Programmerprobung *f* programmed checking
Programmfertigung *f* programming, coding
Programmgeber *m* [control] timer
programmieren to programme
Programmieren *n* / **lineares** linear programming
relatives P. relative programming
Programmierer *m* programmer
Programmierung *f* programming; setup procedure *(analogue computer)*
mathematische P. mathematical programming
schnelle P. minimum-access programming
Programmierungsaufgabe *f* programming problem
Programmierungsfeld *n* patch bay
Programmierungsmethode *f* programming method
Programmierungssprache *f* programming language
Programmierungssystem *n* programming system
Programmierungstafel *f* patchboard, plugboard
Programmnumerierung *f* programme numbering
Programmothek *f* library
Programmplan *m* programme
Programmplaner *m* programmer
Programmregelung *f* programme control
Programmregler *m* process (programme, timed) controller
Programmschalter *m* controller, sequence (programme) switch
Programmschaltplatte *f* programming plugboard
Programmschleife *f* programme loop
Programmschleifen *fpl* / **iterative** iterative programme loops
Programmschritt *m* programme step
Programmspeicher *m* programme counter (register)
Programmsteuerung *f* programme [cycle] control
P. des Rechners computer programme control
P. mit Lochkarte punch card programming
elektronische P. electronic programme control
Programmüberblender *m* fade-in-and-out control
Programmüberprüfung *f* proofreading
Programmvorwahl *f (mach)* programme preselection
Programmwerkzeug *n* subland (multi-cut) drill, combination tool
Programmzeitschalter *m* timer switch
Projekt *n* project
projektieren to project, to plan, to design
Projektierung *f* projecting, planning, designing
Projektil *n* moving particle; projectile, missile, shell
Projektilsteuerung *f* missile control
Projektion *f* projection
mechanische P. mechanical projection
stereografische P. stereographic projection
Projektionsachse *f* axis of projection
Projektionsanzeige *f* projection display
einzeilige P. in-line projection display
Projektionsanzeiger *m* projection indicator
Projektionsapparat *m* projector
Projektionsbildlupe *f* enlarging focusing magnifier
Projektionsebene *f* plane of [the] projection
Projektionselektrometer *n* projection electrometer
Projektionsempfänger *m* projection receiver
Projektionsentfernung *f* projection distance
Projektionsfernsehempfänger *m* projection receiver
Projektionsfläche *f* [projection] screen *(pictures)*
Projektionsglühlampe *f* projector-type filament lamp
Projektionslampe *f* projection (projector) lamp
Projektionslinse *f* projector lens
Projektionsmikroskop *n* projection microscope
Projektionsperimeter *n* projection perimeter
Projektionsschirm *m* [projection] screen

Projektionsskaleninstrument *n* projected-scale instrument
projektiv projective *(geometry)*
Projektor *m* projection lantern, projector
Projektoroptik *f* projection optics
projizieren to project
Prolamin *n* prolamin[e]
Prolin *n* proline
Promenade / offene airing space *(on deck of a vessel)*
Promenadendeck *n* promenade deck
Promethium *n* promethium
Promoter *m* promoter
prompt prompt
promptkritisch prompt-critical
Prontosil *n* prontosil
Propan *n* propane
Propargylalkohol *m* propargyl alcohol
Propeller *m* propeller, screw *(of a vessel)*, impeller
P. mit abnehmbaren Flügeln detachable-blade propeller
fester P. non-feathering propeller
rechtsdrehender P. right-hand propeller
singender P. singing propeller
Propelleranemometer *n* propeller-type anemometer
Propellerantrieb *m* propeller drive
Propellerbelastungsgrad *m* propeller load coefficient
Propellerblatt *n* propeller blade
Propellerbock *m* propeller strut
Propellerbrunnen *m* propeller (screw) aperture, propeller well; *(Am)* wheel port
Propellerdämpfungskoeffizient *m* propeller damping coefficient
Propellerdrehmoment *n* propeller torque
Propellerdrehzahl *f* propeller speed
Propellerdruckseite *f* propeller face
Propellerdurchmesser *m* propeller diameter, diameter of the propeller
Propellerdüsenring *m* propeller nozzle ring
Propellerfläche *f* propeller area
Propellerflächenverhältnis *n* propeller area ratio
Propellerflügel *m* propeller (screw) blade
Propellerflügelhinterkante *f* trailing edge
Propellerflügelneigung *f* rake of [the] propeller blade
Propellerfreischlag *m* tip clearance of the propeller
Propellergebläse *n* propeller fan
Propellerkreisfläche *f* propeller disk
Propellerlog *n* impeller-type log
Propellermassenschwungmoment *n* propeller inertia
Propellernabe *f* [propeller] hub, [propeller] boss
Propellersaugfläche *f* drag surface of [the] propeller
Propellerschirm *m* propeller guard *(twin-screw vessel)*
Propellerschlag *m* turn of the propeller
Propellerschub *m* propeller thrust, thrust of the propeller
Propellerschwanzwelle *f* propeller tail shaft
Propellerschwungmoment *n* centrifugal moment of the propeller, second moment of area
Propellerspiel *n* tip clearance
Propellersteigung *f* propeller pitch
Propellersteven *m* propeller frame (post)
Propellerstrom *m* propeller (screw) wake, backwash
Propellerströmungsfelder *npl* propeller flows
Propellerträgheit *f* propeller inertia
Propellerumdrehungszahl *f* propeller speed
Propellerversuch *m* **mit freifahrendem Propeller** open-water propeller test
Propellerwelle *f* propeller (tail, screw) shaft
Propellerwellentunnel *m* propeller shaft tunnel
Propellerwellenüberzug *m* shaft liner
Propellerwind *m* slipstream
Propellerwirbelstrom *m* propeller eddy
Propellerwirkungsgrad *m* propeller efficiency
P. des freifahrenden Propellers open-water efficiency of the propeller
Prophylaktikum *n* prophylactic
prophylaktisch preventive, prophylactic
Propionat *n* propionate
Propionsäure *f* propionic (propanoic) acid
proportional proportional
direkt p. directly proportional
umgekehrt P. inversely proportional
Proportionalabweichung *f (cont)* load static error
Proportionalbereich *m* proportional region; proportional band, *(Am)* proportioning band
Proportional-Differential-Regelung *f* proportional-plus-derivative control
Proportionalfernmeßgerät *n* direct-relation telemeter
Proportional-Integral-Regelung *f* proportional-plus-integral control
Proportionalionisationskammer *f* proportional ionization chamber
Proportionalität *f* proportionality
begrenzte P. limited proportionality
direkte P. direct proportionality
strenge P. strict proportionality
umgekehrte P. inverse proportionality
Proportionalitätsannahme *f* assumption of proportionality
Proportionalitätsbeziehung *f* proportionality relationship
Proportionalitätsgrenze *f* limit of proportionality, proportional limit

Proportionalitätskonstante *f* proportionality constant
Proportionalkammer *f* proportional ionization chamber
Proportionalregelung *f* proportional control
Proportionalregler *m* proportional [acting] controller
Proportionalverstärkung *f* proportional control factor
Proportionalzähler *m* proportional counter [tube]
Proportionalzählrohr *n* proportional counter tube
Propulsionswiderstand *m* *(ship)* resistance to propulsion
Propulsionswirkungsgrad *m* *(ship)* propulsive efficiency, efficiency of propulsion
Propylalkohol *m* propanol, n-propyl alcohol
Propylen *n* propylene
Prospekt *m* 1. drop *(theatre)*; 2. prospectus, folder
Prospektieren *n* prospecting
Prospektion *f* *(nucl)* prospecting, search
Prospektionszähler *m* radioactive ore detector
Protaktinium *n* protactinium
Protease *f* protease
Protein *n* protein
Proteinfaser *f* protein [base] fibre
Proteinplast *m* protein plastic
Proteolyse *f* proteolysis
Protium *n* protium
Protokatechusäure *f* protocatechuic acid
Protokoll *n* record, minute
protokollarisch [festgehalten] recorded
Protokollführer *m* recorder
Protokollwerte *mpl* record data
Proton *n* proton
Protonenakzeptor *m* proton acceptor
Protonenbeschleuniger *m* proton accelerator
Protonenbeschuß *m* proton bombardment
Protonenbestrahlung *f* proton irradiation
Protonenbremsstrahlung *f* proton bremsstrahlung
Protonenbündel *n* proton beam
Protonendonator *m* proton donor
Protoneneinfang *m* proton capture
Protonenenergiespektrum *n* proton spectrum
Protonenmikroskop *n* proton microscope
Protonennachweis *m* proton detection
Protonenquelle *f* proton source
Protonenquerschnitt *m* proton cross section
Protonenreichweite *f* proton range
Protonenrückstoßzähler *m* proton recoil counter
Protonenspektrometer *n* proton spectrometer
Protonenspektrum *n* proton spectrum
Protonenspin *m* proton spin
Protonenstern *m* proton-induced star
Protonenstoß *m* proton collision
Protonenstrahl *m* proton beam
Protonenstrahler *m* proton emitter
Protonenstreuung *f* proton scattering
Protonensynchrotron *n* proton-synchrotron
Protonenübermikroskop *n* proton microscope
Protonenüberschuß *m* proton excess
Protonenwelle *f* proton wave
Protonenzahl *f* atomic (proton) number
Proton-Neutron-Modell *n* **des Atomkerns** proton-neutron model of the nucleus
Proton-Neutron-Reaktion *f* proton-neutron reaction
Protoplasma *n* protoplasma
Protoplast *m* protoplast
Prototyp *m* prototype
Protrahierung *f* [dose] protraction, protracted treatment
Protuberanz *f* prominence, protuberance
Protuberanzenbeobachtungsansatz *m* prominence observation attachment
Protuberanzenspektroskop *n* prominence spectroscope
Protzöse *f* trail eye *(gun)*
Proustit *m* proustite, ruby (light-red) silver ore
Provenienz *f* provenance
Provenienzkreuzungen *fpl* crossing products of strains
Proviantboot *n* bumboat
Proviantschiff *n* food (supply) ship, victualling vessel
provisorisch provisional, temporary
Provisorium *n* provisional (temporary) arrangement
Provitamin *n* provitamin, vitamin precursor
Prozedur *f* procedure
Prozent *n* percent, per cent
Prozentdifferentialrelais *n* percentage differential relay
Prozentsatz *m* percentage
prozentual per cent, percentage
Prozentvergleichsschutz *m* percentage differential protection
Prozeß *m* process, operation
  diskontinuierlicher P. batch (intermittent) process
  irreversibler P. irreversible process
  kontinuierlicher P. continuous process
  reversibler P. reversible process
  technologischer P. production process
  thermodynamischer P. thermodynamic transformation
Prozeßanalyse *f* activity analysis *(economical statistics)*
Prozeßdurchschnitt *m* process average *(statistics)*
Prozeßkosten *pl* legal costs
Prozeßrechner *m* process computer
Prozeßregler *m* process control system

Prüfader *f* test wire, C-wire *(telephony)*
Prüfanleitung *f* *(mach)* test chart (instructions)
Prüfanstalt *f* testing station
Prüfanweisung *f* instruction for the [inspection and] testing
Prüfatmosphäre *f* testing atmosphere
Prüfbatterie *f* testing battery
Prüfbelastung *f* test load
Prüfbericht *m* test report
Prüfbrennerzahl *f* test burner number
Prüfbuchse *f* test jack
Prüfbürste *f* *(el)* pilot brush
Prüfdauer *f* duration of [the] test
Prüfdorn *m* *(mach)* test mandrel
Prüfdraht *m* *(el)* holding (pilot) wire, C-wire
Prüfdruck *m* test pressure
Prüfechosender *m* performance-monitor transmitter *(radar)*
Prüfeinrichtung *f* checking device, test set (unit, equipment)
prüfen 1. to test; to check; to verify; to gauge *(dimension)*; to analyze, to inspect *(products)*; to assay; 2. to examine *(university)*
 p. auf to test (check) for
 auf Schlag p. to check for eccentricity *(spindle)*
 die Sicherungen p. to verify the fuses
 eine Leitung p. to examine a line
 eine Röhre p. to test (check) a valve
Prüfen *n* testing; checking
 P. von Größen testing of sizes, gauging
 maßliches P. measuring
 räumliches P. three-dimensional inspection
 zerstörungsfreies P. non-destructive testing
Prüfer *m* 1. inspector *(quality control)*; 2. *(dat)* verifier; 3. examiner
Prüfergebnis *n* test result
Prüffeld *n* 1. test hall (laboratory, room, shop); proving ground; 2. test panel (board)
 P. für elektrische Hochspannungstechnik high-voltage test laboratory
Prüffeldingenieur *m* test engineer
Prüffinger *m* [contact] test-finger
Prüffläche *f* measuring face
Prüffolge *f* test sequence
Prüffrequenz *f* test frequency
Prüfgas *n* test gas
Prüfgelände *n* proving ground
Prüfgenauigkeit *f* testing accuracy
Prüfgerät *n* testing apparatus (set, equipment, instrument), inspection instrument; analyzer
Prüfgestell *n* test frame *(switchgear)*
Prüfhörer *m* receiver watch
Prüfkabel *n* measuring cable
Prüfklemme *f* test (calibrating) terminal
Prüfklemmenblock *m* test terminal box
Prüfklinke *f* test jack
Prüfklinkenfeld *n* test-jack panel
Prüfkontaktarm *m* private wiper *(telephony)*
Prüfkopf *m* probe; gauge
Prüfkörper *m* test specimen, sample
Prüfkreis *m s.* Prüfstromkreis
Prüflampe *f* test lamp
Prüflast *f* test load
Prüfleiter *m* test conductor *(of a cable)*
Prüfleitung *f* test wire (line)
Prüfling *m* specimen, [test] sample *(material)*
Prüfmaschine *f* testing machine
Prüfmethode *f* testing technique, checking method
Prüfmittel *n* testing medium
Prüfmittel *npl* testing equipment
Prüfplatz *m* test position *(telephony)*
Prüfprogramm *n* test routine (schedule)
Prüfprotokoll *n* inspection record *(quality control)*
Prüfpunkt *m* check (test) point
Prüfraum *m* check (test) room
Prüfreaktor *m* testing reactor
Prüfrelais *n* test relay
Prüfsatz *m* test set
Prüfschalter *m* test key, feeler switch
Prüfschaltung *f* test connection; monitoring (test) circuit
 P. für Durchlaßrichtung forward test circuit
 P. für Sperrichtung reverse test circuit
 synthetische P. synthetic test connection
Prüfschein *m* test certificate
Prüfschneide *f* *(mach)* testing edge
Prüfschnur *f* *(el)* test cord
Prüfschrank *m* *(el)* test cabinet (board)
Prüfsender *m* [test] signal generator
Prüfsieb *n* testing sieve *(laboratory)*
Prüfsignal *n* test signal
Prüfsignalgeber *m* *(telev)* test-signal generator
Prüfsignalgemisch *n* composite test signal
Prüfsignalgerät *n* test waveform generator (generating equipment
Prüfsonde *f* test probe
Prüfspannung *f* test[ing] voltage
Prüfspitze *f* [test] prod
Prüfspule *f* test (search, exploring) coil, search winding
Prüfstand *m* test stand (rig)
Prüfstandversuch *m* bench test
Prüfstation *f* inspection station *(transfer machine)*
Prüfstecker *m* test plug
Prüfstelle *f* measuring (contact) point; test site; testing point
Prüfstift *m* [test] prod
Prüfstöpsel *m* test plug
Prüfstrom *m* test[ing] current
Prüfstromkreis *m* test circuit
Prüfstück *n* specimen, sample, test piece
Prüfstufe *f* inspection level
Prüftabelle *f* check list
Prüftaste *f* test key
Prüftaster *m* testing spike

Prüftechnik *f* test technique
Prüfton *m* test tone
Prüf- und Meßschrank *m* test desk (board), trunk test rack
Prüfung *f* 1. test[ing]; checking; verification; inspection *(material)*; 2. examination *(university)*
P. am Einsatzort field test
P. an Ort und Stelle in-situ test, on-the-spot check, on-site inspection
P. auf Betriebsfähigkeit clear test *(telephony)*
P. auf Durchgang *(el)* continuity test
P. auf Umgebungseinflüsse environmental testing
P. außerhalb der Schaltung out-of-circuit testing
P. bei aussetzender Entladung intermittent test *(batteries)*
P. der Erweichungstemperatur cut-through test *(of winding wire insulation)*
P. der mechanischen Festigkeit der Isolierung cut-through test
P. durch Hörvergleich mit Sprache voice-ear test
P. innerhalb der Schaltung in-situ testing
P. modulo n modulo n check
augenscheinliche P. visual test, inspection
mehrfache P. multiple check
mündliche P. oral examination
normale P. normal inspection *(quality control)*
programmierte P. programmed checking
redundante P. redundant check
reduzierte P. reduced inspection *(quality control)*
verschärfte P. tightened inspection *(quality control)*
zerstörungsfreie P. non-destructive test
Prüfungsanforderungen *fpl* examination requirements *(university)*
Prüfungsfach *n* subject of examination
Prüfungslösung *f (dat)* check solution
Prüfungsprotokoll *n* test record
Prüfverfahren *n (mach)* test[ing] procedure; test (checking) method; inspection technique *(quality control)*
P. der ununterbrochenen Folgen run test
Prüfvorgang *m* test cycle; operation of testing
Prüfvorrichtung *f* test jig, inspection device
Prüfvorschrift *f* test specification
Prüfwähler *m* test [final] selector *(telephony)*
Prüfwechselspannung *f* alternating current test voltage
Prüfwert *m* test value
Prüfwicklung *f* search winding (coil), test winding
Prüfzeichen *n* test mark
Prüfzeit *f* inspection time
Prüfzelle *f* pilot cell *(battery)*; test cell
Prüfziffer *f (dat)* check digit
Prüfzwischensockel *m* test adapter
Prüfzylinder *m* gauge (test) cylinder
Prussiat *n* prussiate
Pseudoadresse *f (dat)* floating address
Pseudobase *f* pseudo base
Pseudobefehl *m (dat)* pseudo instruction
pseudodielektrisch pseudodielectric
Pseudohalogen *n* pseudohalide
Pseudokatalyse *f* pseudocatalysis
Pseudokode *m* pseudo code *(not directly understood by machine)*
Pseudokumol *n* pseudocumene
pseudomorph pseudomorphous
Pseudomorphose *f* pseudomorphism, pseudomorphosis
Pseudostreuergebnis *n* pseudo-random result
Psi-Funktion *f* wave function
Psilomelan *m (min)* psilomelane
Psychrometer *n* psychrometer
ptolemäisch Ptolemaic
Puddelblock *m* boiled bar
Puddeleisen *n* puddle iron
Puddelherd *m* puddling hearth
Puddelhütte *f* puddle iron works
Puddelkrücke *f* rabble
Puddelluppe *f* mill (muck) bar, puddle ball
puddeln to puddle
Puddelofen *m* puddling furnace
Puddelschlacke *f* puddle cinder, tappings
Puddelstab *m* puddler bar
Puddelstahl *m* puddle[d] steel
Puddelstahlwalzwerk *n* muck mill
Puddelstraße *f* puddle train
Puder *m* powder
Puffer *m* buffer; pad; dashpot
Pufferbatterie *f* floated (buffer, balancing) battery
Pufferbetrieb *m* buffer-battery system; floating [operation], trickle operation
Pufferbohle *f* buffer beam, head stock
Pufferfeder *f* buffer spring
Puffergehäuse *n* buffer box
Pufferhülse *f* buffer guide (case, box)
Pufferkreis *m* buffer circuit
Pufferladegerät *n* trickle charger, *(Am)* battery booster
Pufferladung *f* floating (trickle, compensating) charge *(battery)*
Pufferlösung *f (ch)* buffer solution
puffern *(el)* to float, to cushion; to buffer *(battery)*; to poise; to trickle *(charge)*; *(ch)* to poise
Pufferschaltung *f* buffer circuit
Puffersilo *m* reserve storage-silo
Pufferspeicher *m* buffer memory (storage) *(input or output memory)*
Pufferstange *f* buffer rod
Pufferstufe *f (el)* buffer stage, separator

Pufferteller *m* buffer disk (head)
Pufferung *f* buffering, buffer action; cushioning; floating *(battery)*; built-in damping *(piston)*
Pufferverstärker *m* buffer amplifier
Pufferverzweiger *m* buffer cabinet
Pulfrich-Fotometer *n* Pulfrich photometer
Pulfrich-Refraktometer *n* Pulfrich refractometer
Pull *m (ship)* pull
pullen *(ship)* to pull, to row, to ply the oars
Pulpe *f* pulp
Puls *m* pulse [train]
Pulsamplitude *f* pulse amplitude
Pulsamplitudenmodulation *f* pulse-amplitude modulation, PAM
Pulsation *f* pulsation
Pulsationsdämpfer *m* / **pneumatischer** *(Am)* snubber
pulsationsfrei pulsation-free
Pulsator *m* pulsator; pulsating machine
Pulsatorsetzmaschine *f* pulsating jig
Pulsatron *n* pulsatron
Pulsbreitenmodulation *f* pulse-width modulation
Pulsdauermodulation *f* pulse-duration modulation
Pulserzeugung *f* pulse generation
Pulsformung *f* pulse shaping
Pulsfrequenz *f* pulse repetition frequency
Pulsfrequenzmodulation *f* pulse-frequency modulation
pulsieren to pulsate
Pulskodemodulation *f* pulse-code modulation
Pulslängenmodulation *f* pulse-length modulation
Pulsmodulation *f* pulse modulation
Pulsometer *n* pulsometer pump
Pulsphasenmodulation *f* pulse-phase modulation, PPM
Pulsradar *n* pulsed radar
Pulsrückbildung *f* pulse regeneration
Pulsschlauch *m* pulse tube
Pulsschreiber *m* pulse recorder
Pulssteigungsmodulation *f* pulse-slope modulation
Pulszahlkode *m (dat)* unit-counting code
Pulszahlmodulation *f* pulse-code modulation
Pult *n* [control] desk
Pultdach *n* shed (penthouse, lean-to) roof
Pultlampe *f* panel lamp
Pultsteuerung *f* desk control
Pulver *n* powder
Pulverflasche *f* reagent bottle
Pulverform *f* powder form
pulverig powdery
pulverisierbar pulverizable
pulverisieren to powder, to pulverize; to comminute
Pulverkern *m* mass core
Pulvermetallurgie *f* powder metallurgy
pulvermetallurgisch powder-metallurgical
Pulvermethode *f* powder method *(Debye)*
Pulversatz *m* powder train (ingredients)
Pulversintern *n* powder sintering
pumpbar pumpable
Pumpbereich *m* pump[ing] region
Pumpbeton *m* pumped concrete, pumpcrete
Pumpdiagramm *n* pumping scheme *(laser)*
Pumpe *f* pump
  P. für Handbetrieb hand-operated pump
  P. mit äußerer Beaufschlagung pump in which the supply and discharge take place from the outer ends of the cylinders
  P. mit beweglichem Widerlager rotary-abutment pump
  P. mit elektrischem Antrieb motor-driven pump
  P. mit gleichbleibender Fördermenge *s.* nichtverstellbare P.
  P. mit innerer Beaufschlagung pump in which supply and discharge take place by way of ports in a stationary central shaft
  P. mit Preßluftantrieb air-operated pump
  P. mit Turbinenantrieb turbine-driven pump
  P. mit umkehrbarer Förderrichtung reversible-flow type pump
  P. mit unbeweglichem sichelförmigem Widerlager crescent pump
  P. mit unterbrochener Leistung intermitter
  P. mit veränderlicher Fördermenge variable-delivery pump
  einstufige P. single-stage pump
  feste P. stationary (standing) pump
  gestörte P. foul pump
  nichtverstellbare P. constant-delivery pump, constant-displacement pump, constant-flow pump, constant-volume pump
  schlagende P. knocking pump
  sichtbar messende P. visible pump
  unklare P. choked (foul) pump
  verstopfte P. choked pump
  ziehende P. drawing pump
  zweistufige P. two-volume pump
pumpen to pump
Pumpen *n* pumping
  P. der Kolbenringe pump-action of piston rings
  P. eines Schalters pumping of a circuit-breaker
  magnetisches P. magnetic pumping
  optisches P. optical pumping
Pumpenaggregat *n* direct-connected motor-driven pump
Pumpenanlage *f* pumping plant (installation)
Pumpenansaugkorb *m* pump strainer
Pumpenansaugleitung *f* pump suction-line
Pumpenauslaßventil *n* pump outlet-valve

Pumpenbagger *m* suction (pump, hydraulic) dredger
Pumpenbeschlag *m* pump gear
Pumpenbock *m* pumping jack
Pumpendruck *m* pumping pressure
Pumpendüse *f* pump jet
Pumpeneinlaßventil *n* pump inlet-valve
Pumpenergie *f* pump[ing] energy
Pumpenfernsteuerung *f* pump telecontrol
Pumpenflügel *m* pump-impeller vane (blade)
Pumpenflügelrad *n* impeller of pump
Pumpenförderleistung *f* pump duty (delivery)
Pumpengehäuse *n* pump casing
Pumpengestänge *n* pump rods
Pumpengestängerechen *m* sucker-rod hanger (jack)
Pumpengestell *n* pump jack
Pumpenhaube *f* pump cover (bonnet)
Pumpenhebel *m* pump [piston] lever
Pumpenherz *n* lower pump box
Pumpenkammer *f* pump room (chamber)
Pumpenkasten *m* pump stock
Pumpenkolben *m* pump piston (plunger)
 P. mit Klappe pump bucket
Pumpenkolbenrohr *n* plunger working barrel
Pumpenkolbenventil *n* plunger valve
Pumpenkorb *m* pump strainer
Pumpenlamelle *f* vane of [vacuum] pump
Pumpenleistung *f* horsepower delivered by the pump
Pumpenliderung *f* pump packing
Pumpenpeilstock *m* gauge rod of a pump
Pumpenplatte *f* pump plate
Pumpenprobehahn *m* pet cock for pumps
Pumpenrad *n* pump impeller
 getriebenes P. pump idler-gear
 treibendes P. pump drive-gear
Pumpenrahmen *m* / **nicht schwenkbarer** fixed yoke *(axial pump)*
Pumpenraum *m* pump room (chamber)
Pumpenregler *m* pump governor
Pumpenrohr *n* tubing [of the pump]
Pumpenröhre *f s.* Pumpenrohr
Pumpenrohrhänger *m* tubing hanger (spider)
Pumpenrohrkrebs *m* tubing catcher
Pumpensatz *m* pump set, set of pumps
Pumpenschacht *m* pump shaft
Pumpenschwengel *m* pump handle
Pumpenselbstschalter *m* electropneumatic regulator
Pumpensod *m* pump well
Pumpenspritze *f* spray pump
Pumpenstange *f* sucker rod
Pumpenstempelantrieb *m* pump plunger-drive
Pumpenstiefel *m* barrel (body, chamber) of a pump
Pumpenstock *m* pump staff
Pumpensumpf *m* pump well
Pumpenträger *m* bracket of pump
Pumpenumlaufkühlung *f* pump circulated cooling
Pumpenventil *n* pump valve
Pumpenverstellung *f* displacement control on pumps
Pumpenwelle *f* pump shaft
Pumpenwerk *n* pumping gear
Pumpenwinde *f* pumping jack
Pumpenzylinder *m* pump barrel (cylinder)
Pumpfähigkeit *f* pumpability
Pumpgetriebe *n* pumping gear
Pumpgrenze *f* surge (compressor pulsation) limit
Pumpleistung *f* pump[ing] power, pumping capacity
Pumploch *n* exhaust port *(electron valves)*
Pumpniveau *n* pump level
Pumpquelle *f* / **konstante** steady pumping source *(laser)*
Pumprad *n* pump wheel
Pumprohr *n* exhaust tube; tubing
Pumpschaltung *f* pump circuit
Pumpspannung *f* pumping voltage
Pumpspeicher *m* pumped storage *(in a hydroelectric station)*
Pumpspeicherbecken *n* pumped storage reservoir
Pumpspeicherkraftwerk *n* pump-fed power station, pumped-storage power station
Pumpspill *n* windlass
Pumpspitze *f* tip *(electron valves)*
Pumpstand *m* evacuating plant
Pumpstation *f* pumping (drainage) station
Pumpsteiger *m* venting channel *(foundry)*
Pumpstengel *m* pumping stem, exhaust tube *(bulbs)*
Pumpstock *m* pricker *(foundry)*
Pumpstrom *m* pump current
Pumpübergang *m* pump[ing] transition
Pumpvorgelege *n* pump jack
Pumpvorrichtung *f* pump jack
Pumpwerk *n* pumping station (plant)
Pumpwerkdampfmaschine *f* pumping engine
Pumpwirkung *f* pumpage
Punkt *m* point; centre; *(print)* point *(unit)*; full stop; spot; dot; pivot
 P., an dem sich die Grenzschicht ablöst burble point
 P. des Phasenraumes point of phase space
 anallaktischer P. anallactic point
 auf einen Kreis bezogener inverser P. inverse of a point with respect to a circle
 isoelektrischer P. isoelectric point
 isolierter P. einer Kurve acnode
 kritischer P. critical point; break point
 singulärer P. singular point
 toter P. dead centre (point), neutral point
 trigonometrischer P. triangulation point, control point *(of known position and elevation)*

unendlich ferner P. *(math)* point at infinity
unfreier P. non-free point
Punktablenkung *f* spot deflection
Punktanstrahlung *f* spot lighting
Punktball *m* small punch ball *(boxing)*
Punktberührung *f* point contact
Punktbildung *f* dot formation
Punktdurchmesser *m* spot diameter
Punkte *mpl* / **konjugierte** conjugates
Punktfehlordnung *f* point defect (imperfection)
Punktfolgeverfahren *n* *(telev)* dot sequential system
punktförmig point[like]
Punktfrequenz *f* dot frequency
punktgeschweißt spot-welded
Punktgitter *n* point lattice
Punkthaufen *m* aggregate of particles
Punkthochhaus *n* point (tower) block
punktieren to point, to dot
Punktkatode *f* point cathode
Punktkontaktdiode *f* point-contact diode
Punktkontakttransistor *m* point-contact transistor
Punktkorrosion *f* localized attack
Punktladung *f* point charge
Punktlast *f* concentrated (point) load
Punktleimverfahren *n* spot glu[e]ing
Punktlichtquelle *f* point source of light
Punktnaht *f* spot weld
Punktpasser *m* dot-on-dot register
Punktpolmeßtaster *m* point-pole probe and magnet arrangement
Punktquelle *f* point source
Punktraster *m* dot screen
Punktscheinwerfer *m* spotlight lamp, spot *(stage lighting)*
Punktschreiber *m* printing-type recorder
punktschweißen to spot-weld
Punktschweißen *n* spot welding
Punktschweißmaschine *f* spot-welding machine
Punktschweißnaht *f* spot weld
Punktschweißung *f* spot weld[ing]
Punktschweißzange *f* portable spot-welding gun
Punktschweißzeuge *npl* spot-welding equipment
Punktsonde *f* point probe
Punktsprungverfahren *n* *(telev)* dot interlacing
Punktsteuerung *f* point-to-point control [system] *(machine tool)*; position control, positioning
Punktstrahllampe *f* spot lamp
Punktsucher *m* spot locator *(electronics)*
Punkttransformation *f* / **erweiterte** extended point transformation
punktuell abbildend point-focal
Punktur *f* points
Punktwohnhaus *n* tower (point) block
Punktwolke *f* scatter diagram
Punktzahl *f* *(dat)* score
Pünte *f* *(ship)* scow, punt
Punzentiefe *f* depth of set
Pupille *f* pupil
Pupillenerweiterung *f* pupillary dilatation
pupinisieren to pupinize, to coil-load
Pupinisierung *f* coil (series) loading, pupinization
besonders leichte P. extra-light loading
gleichmäßige P. continuous loading
mittelschwere P. medium loading
Pupinkabel *n* [coil-]loaded cable
Pupinleitung *f* coil-loaded circuit
Pupinspule *f* loading (Pupin) coil
Puppe *f* 1. doll *(toy)*; 2. puppet *(theatre)*; 3. *(plast)* billet; 4. *(ag)* stack [of sheaves]
Purin *n* purine
Purpur *m* purple
purpurrot purple, scarlet, crimson
Putreszin *n* putrescine
Putz *m* 1. *(build)* plaster[work], parget[ting], rendering, stucco; 2. *(text)* millinery
dreilagiger P. render, float and set; three-coat work on walls
schallschluckender P. acoustic plaster
zweilagiger P. render and set, two-coat work on walls
Putzarbeit *f* fettling *(foundry)*
Putzbeitel *m* planing chisel
Putzeinrichtung *f* second sifting cleaner *(beet harvester)*
Putzeisen *npl* weaver's nippers
putzen 1. *(build)* to plaster, to render, to parget; 2. *(met)* to scour, to clean *(castings)*; to trim, to burr, to deseam, to snag
Putzen *n* **von Gußstücken** cleaning of castings
Putzen *m* *(mine)* ore pocket; *(mach, therm)* boss
Putzer *m* *(build)* plasterer, pargeter
Putzerei *f* 1. *(text)* blowing (opening) room; 2. *(met)* dressing shop, cleaning (fettling) room
Putzereimaschine *f* *(text)* blow-room machine
Putzerkelle *f* plastering trowel
Putzhobel *m* fine plane
Putzlappen *m* cleaning rag
Putzloch *n* cleaning hole
Putzmacherin *f* milliner
Putzmesser *n* back iron, finishing (unhairing) knife
P. zum Entfernen der Grundhaare short hair knife
feststehendes P. fixed back iron
Putzmittel *n* cleaning material, polish
Putzmörtel *m* plastering mix, plaster mortar
Putzmühle *f* combined winnowing and sieving machine
Putzöffnung *f* cleaning hole
Putzschicht *f* [plaster] coat
Putzspan *m* finishing chip
Putzstern *m* rattler star *(castings)*

Putzträger *m* *(build)* lathwork, [counter]-lathing
Putztrommel *f* tumbling barrel
Putzwolle *f* [cleaning] cotton-waste
Puzzolan *n* pozz[u]olana
puzzolanartig pozz[u]olanic
Puzzolanzement *m* pozz[u]olanic cement
p-v-Diagramm *n* pressure-volume diagram, p-v diagram
Pyknometer *n* pycnometer
Pylone *f* pylon, tower
Pyramide *f* pyramid
  dreiseitige P. triangular pyramid
  regelmäßige P. regular pyramid
Pyramideneinbruch *m* *(mine)* diamond (pyramid) cut
Pyramidenrohr *n* taper *(waveguides)*
Pyramidenstumpf *m* frustum of a pyramid
Pyran *n* pyrane
Pyrargyrit *m* *(min)* pyrargyrite
Pyrazol *n* pyrazole
Pyrazolonfarbstoff *m* pyrazolone dye
Pyren *n* pyrene
Pyrheliometer *n* pyrheliometer
Pyridin *n* pyridine
Pyrimidin *n* pyrimidine
Pyrit *m* [iron] pyrite
Pyritabbrand *m* pyrite cinder
Pyriterz *n* pyrite ore
pyritisch pyritic *(smelting)*; pyritiferous
Pyritoeder *n* pyritohedron
Pyritofen *m* pyrites furnace
Pyrochemie *f* pyrochemistry
pyroelektrisch pyroelectric
Pyroelektrizität *f* pyroelectricity
Pyrogallol *n* pyrogallol
pyrogen pyrogenic
Pyrolusit *m* *(min)* pyrolusite
Pyrolyse *f* pyrolysis
Pyrometallurgie *f* pyrometallurgy
Pyrometer *n* pyrometer
  fotoelektrisches P. photoelectric pyrometer
  optisches P. optical pyrometer
Pyrometerkegel *m* pyrometer cone
Pyrometrie *f* pyrometry
Pyromorphit *m* *(min)* pyromorphite
pyrophor pyrophoric
Pyrophosphorsäure *f* pyrophosphoric acid
Pyroschwefelsäure *f* pyrosulphuric (disulphuric) acid
Pyrotechnik *f* pyrotechny, pyrotechnics
Pyroxen *m* *(min)* pyroxene
Pyrrhotin *m* *(min)* pyrrhotite
Pyrrol *n* pyrrole
Pyrrolidin *n* pyrrolidine
pythagoreisch Pythagorean
PZ-Regler *m* proportional acting controller with disturbance feed-forward

Q

Q-Band *n* Q-band
Q-Elektron *n* Q-electron
Q-Faktor *m* Q-factor, quality factor
Q-Schale *f* Q-shell
Q-Wert *m* Q-value, nuclear disintegration energy
Q-Zacken *mpl* Q-waves *(electrocardiogram)*
Quader *m* 1. *(math)* cuboid, rectangular parallelepiped; 2. *(build)* ashlar, square stone; 3. block *(electronics)*; 4. quadrangular prism
Quadermauerwerk *n* ashlar stonework (walling)
Quadrant *m* quadrant
Quadrantenelektrometer *n* quadrant electrometer
Quadrantenschaltung *f* heterostatic circuit
Quadrantinstrument *n* square-scale instrumer
Quadrantrudermaschine *f* *(ship)* quadr ing gear
Quadrat *n* square
  Q. / ins squared
  lateinisches Q. Latin square
  mittleres Q. root mean square
quadratisch 1. square; 2. quadratic[al], of the second degree
Quadratkaliber *n* *(met)* square groove
Quadratkilometer *n* square kilometre
Quadratliniennetz *n* square net
Quadratmeter *n* square metre
Quadratmillimeter *n* square millimetre
Quadratnetz *n* net of squares
Quadratpflanzverfahren *n* planting on the square
Quadratspundung *f* square grooving and tonguing
Quadratstahl *m* square [bar] *(reinforcement)*
Quadratur *f* *(math)* quadrature
Quadratwelle *f* square wave
Quadratwurzel *f* square root
Quadratwurzelrechner *m* square-root calculator
Quadratzentimeter *n* square centimetre
Quadratzoll *m* square inch
quadrieren to square
Quadrierglied *n* *(dat)* square-law function generator, squaring circuit
Quadrupelpunkt *m* quadruple point
Quadruplexsystem *(el)* quadruplex system
Quadrupol *m* quadrupole
Quadrupolaufspaltung *f* quadrupole splitting
Quadrupolkopplung *f* quadrupole coupling
Quadrupollinse *f* quadrupole lens
Quadrupolmoment *n* quadrupole moment
Quadrupolresonanzmethode *f* quadrupole resonance method
Quadrupolstrahlung *f* quadrupole radiation
Quadrupolwelle *f* quadrupole wave
Quai *m* *s.* Kai
Qualifikation *f* qualification[s]
Qualifizierung *f* qualification, training

Qualität *f* quality
handelsübliche Q. commercial grade
qualitativ qualitative
Qualitätsarbeit *f* quality work
Qualitätsbeanstandung *f* quality complaint
Qualitätsbewußtsein *n* quality-mindedness
Qualitätsblech *n* quality sheet
Qualitätseisen *n* high-grade iron
Qualitätserwartung *f* quality expectancy
Qualitätsfaktor *m* quality factor *(radiobiology)*
Qualitätsgerät *n* high-grade instrument
Qualitätsgrad *m* quality grade
Qualitätsindex *m* quality index
Qualitätskontrolle *f* quality control (checking)
statistische Q. statistical quality control, quality control by statistics
Qualitätslage *f* quality level
akzeptierbare Q. acceptable quality level
gelegentlich akzeptierte Q. tolerable quality level
Qualitätsmerkmal *n* quality characteristic (feature)
Qualitätsminderung *f* reduction in quality
Qualitätsstandard *m* quality standard
Qualitätsüberwachung *f* quality surveillance
Qualitätsveränderliche *f* / **stetige** continuous variable
Qualitätsverlust *m* degradation, loss of quality
Qualitätsverminderung *f* debasement; lowering of quality
Qualm *m* smoke; *(aut)* exhaust fumes
Quant *n* quantum
quanteln to quantize
Quantelung *f* quantization
Quantenabsorption *f* absorption of quanta
Quantenanregung *f* quantum excitation
Quantenausbeute *f* quantum yield (efficiency)
Quantenbahn *f* quantum orbit
Quantenbiologie *f* quantum biology
Quantenchemie *f* quantum chemistry
Quantenelektrodynamik *f* quantum electrodynamics
Quantenelektronik *f* quantum electronics
Quantenemission *f* quantum emission
Quantenenergie *f* quantum energy
Quantenertrag *m* quantum yield
Quantenfeldtheorie *f* quantum field theory
Quantenmechanik *f* quantum mechanics
quantenmechanisch quantum-mechanical
Quantenoptik *f* quantum optics
Quantenphysik *f* quantum physics
Quantensprung *m* quantum jump (leap)
Quantenstatistik *f* quantum statistics
Quantenstrahlung *f* quantum radiation
Quantenstreuung *f* quantum scattering
Quantentheorie *f* quantum theory
Quantenübergang *m* quantum transition (jump)
Quantenzahl *f* quantum number
äquatoriale Q. total angular-momentum quantum number
azimutale Q. orbital (azimuthal) quantum number
innere Q. inner quantum number
magnetische Q. magnetic quantum number
Quantenzustand *m* quantum state
Quantisierer *m (dat)* digitizer, quantizer
Quantisierung *f* quantization
Quantität *f* quantity
quantitativ quantitative
Quantitätsregelung *f* quantity governing *(pump)*
Quarantäne *f* quarantine
Quarantäneboot *n* quarantine boat
Quarantäneflagge *f* quarantine flag, yellow jack
Quarantänehafen *m* quarantine harbour, port of quarantine
Quark *m* curd
Quarkmühle *f* curd mill
Quartband *m* quarto volume
Quarterboot *n* quarter boat *(any boat hung on davits over the ship's quarter)*
Quarterdeck *n* quarter deck
erhöhtes Q. raised quarter deck
Quarterdecker *m* [raised] quarterdecker *(type of vessel)*
Quarterdeckschiff *n* [raised] quarter-deck ship
Quarterdeckschott *n* break bulkhead of raised quarter deck
Quarterschiff *n* quarter of a vessel
Quartett *n* quartet
Quartowalzgerüst *n* four-high stand
Quarz *m* quartz
geschmolzener Q. fused quartz
klarer Q. clear quartz
linksdrehender Q. left-handed quartz
synthetischer Q. synthetic quartz
Quarzblockfilter *n* crystal block filter
Quarzeicher *m* crystal calibrator
Quarzeichung *f* crystal check
Quarzfaden *m* quartz thread (filament, fibre)
Quarzfadendosimeter *n* quartz-fibre dosimeter
Quarzfadenelektrometer *n* quartz-fibre electrometer
Quarzfadenmanometer *n* quartz-fibre manometer
Quarzfassung *f* crystal holder
Quarzfenster *n* crystal window
Quarzfilter *n* crystal filter
quarzführend quartz-bearing
quarzgesteuert crystal-controlled
Quarzglas *n* quartz (flint) glass
Quarzhalter *m* crystal holder
quarzhaltig quartziferous
Quarzit *m* quartzite
Quarzkeil *m* quartz wedge
Quarzkolben *m* quartz envelope

Quarzkompensator *m* **nach Babinet** Babinet compensator
Quarzkristall *m* quartz crystal
Quarzlampe *f* quartz lamp
Quarzmonochromator *m* quartz monochromator
Quarznormal *n* crystal check facility
Quarzofen *m* crystal oven
Quarzoptik *f* quartz lens system
Quarzoszillator *m* crystal oscillator, quartz [crystal] oscillator
Quarzplanglas *n* quartz flat
Quarzporphyr *m* quartz-porphyry
Quarzprisma *n* quartz prism
Quarzrohr *n* quartz tube
Quarzröhrenabdichtung *f* quartz tube seal
Quarzsand *m* silica sand, arenaceous quartz
Quarzsender *m* quartz oscillator
Quarzspektralfotometer *n* quartz spectrophotometer
Quarzstab *m* quartz rod
Quarzsteuerung *f* quartz control, crystal control (drive)
Quarzsynchronisierung *f* crystal-lock [system]
Quarzuhr *f* quartz-crystal clock, crystal-controlled clock
Quarzverzögerungsleitung *f* quartz delay line
Quarzwandler *m* quartz transducer
quasiadiabatisch quasi-adiabatic
quasibinär quasi-binary
Quasiimpuls *m* quasi-momentum
quasimetallisch quasi-metallic
quasineutral quasi-neutral
Quasispannweite *f* quasi-range
quasistationär quasi-stationary
quasistatisch quasi-statical
Quassiaholz *n* quassia [wood]
quaternär quaternary
Quaternionen *fpl (math)* quaternions
Quaterphenyl *n* quaterphenyl
Quebracho *n* quebracho
Quebrachoextrakt *m* quebracho extract
Quebrachoholz *n* quebracho [wood]
Quecksilber *n* mercury
Quecksilberbadprüfung *f* mercury test *(insulation test for winding wires)*
Quecksilberbarometer *n* mercury barometer
Quecksilberbogen *m* mercury arc
Quecksilber(I)-chlorid *n* mercurous chloride, calomel, black precipitate
Quecksilber(II)-chlorid *n* mercuric chloride, chloride of mercury, corrosive sublimate
Quecksilberdampf *m* mercury vapour
Quecksilberdampfgleichrichter *m* mercury vapour rectifier
  Q. mit flüssiger Katode pool-type [mercury-vapour] rectifier
Quecksilberdampflampe *f* mercury-vapour lamp, mercury-arc lamp, Cooper-Hewitt lamp
Quecksilberdampfumrichter *m* mercury-arc inverter
Quecksilberdampfventil *n* mercury-arc rectifier
Quecksilberdampfwechselrichter *m* mercury-arc inverter
Quecksilberdiffusionspumpe *f* mercury diffusion pump
Quecksilberelektrode *f* mercury electrode
Quecksilberelement *n* mercury cell
Quecksilberfänger *m* mercury collector *(rectifier)*
quecksilberhaltig mercurial, mercuric
Quecksilberhochdrucklampe *f* high-pressure mercury vapour lamp
Quecksilberhöchstdrucklampe *f* extra high-pressure mercury vapour lamp
Quecksilberkatode *f* mercury[-pool] cathode
Quecksilberkontakt *m* mercury contact
Quecksilberkontaktthermometer *n* mercury contact thermometer
Quecksilberlichtbogen *m* mercury arc
Quecksilbermanometer *n* mercury gauge
  Q. mit verkürzter Skale mercury pressure-gauge with reduced scale
Quecksilberniederdrucklampe *f* low-pressure mercury vapour lamp
Quecksilberpendel *n* mercurial pendulum
Quecksilberpumpe *f* / **rotierende** mercury rotating pump
Quecksilbersalbe *f* mercurial ointment
Quecksilbersäule *f* column of mercury, mercury column
Quecksilberschalter *m* mercury switch
Quecksilberschaltröhre *f* mercury-contact tube, mercury switch valve
Quecksilberspektrallampe *f* mercury spectral lamp
Quecksilberstrahlgleichrichter *m* mercury-jet rectifier [equipment], jet wave rectifier
Quecksilbersulfid *n* mercuric sulphide
Quecksilberthermometer *n* **mit Stickstofffüllung** mercurial nitrogen thermometer
Quecksilbertrockenbatterie *f* dry mercury battery
Quecksilberultraschallspeicher *m* mercury memory
Quecksilbervakuummeter *n* mercury vacuum gauge
Quecksilber(I)-verbindung *f* mercurous compound
Quecksilber(II)-verbindung *f* mercuric compound
Quecksilberverdrängerrelais *n* mercury-plunger relay
Quecksilbervergiftung *f* hydrargyrism, mercurialism
Quecksilberverzögerungsleitung *f* mercury delay line
Quecksilberzähler *m* mercury hour-meter

Quelldichte *f* source density, density of source distribution; divergence *(vector field)*
Quelle *f* 1. spring, fountain; 2. *(ph)* source; injector
ausgedehnte Q. extended source
ebene Q. plane source; flat source
geschlossene Q. encapsulated (sealed) source
konstante Q. stable source
nadelförmige Q. needle-shaped source
offene Q. unsealed (free) source
punktförmige Q. point source
rohrförmige Q. tubular source
quellen 1. to swell; to imbibe *(saturate with moisture)*; 2. to steep *(soak in water)*; 3. *(mine)* to creep, to heave, to lift *(floor)*; 4. to expand *(concrete)*
Quellen *n* **der Sohle** *(mine)* floor lift, crowning
Quellenfeld *n (el)* source field
quellenfrei source-free; zero
Quellenfunktion *f* source function
Quellenleitwert *m* source admittance
optimaler Q. optimum source admittance
Quellenstärke *f* source strength, strength of source
Quellenverteilung *f* source distribution
Quellenwiderstand *m* source resistance
Quellfähigkeit *f* swelling capacity
Quellfestappretur *f* swell-proof finish
Quellfluß *m* source flux
Quellmittel *n* swelling agent
Quellneutron *n* source neutron
Quellsenkenverfahren *n (math)* source-sink method
Quellstärke *f* source strength, strength of source
Quell[ungs]vermögen *n (plast)* expansive capacity, swelling property
Quellwasser *n* spring water
Quellwert *m* swelling (water-imbibition) value
Quellwertbestimmung *f* determination of the swelling value
Quellzement *m* expansive (expanding) cement
quer transverse; cross
q. ab abeam off, abreast of, on the beam *(navigation)*
q. zu across
q. zur Faser across the grain *(timber)*; *(met)* cross-grained
Queraberration *f* transverse spherical aberration
Querablauf *m s.* Querstapellauf
Querabmessung *f* lateral dimension
Querachse *f* transverse axis
Queramperewindung *f* cross ampere-turn
Querast *m* wing-shaped knot
Queraussteifung *f (build)* cross-bracing; *(mach)* cross-ribbing
Queraxt *f* cross grain axe
Querbalken *m (build)* diaphragm beam *(bridge)*; cross-beam; *(mach)* cross-rail, cross-bar, cross-slide *(planer)*; trimmer
Q. der Einständerhobelmaschine cross arm
feststehender Q. *(mach)* top rail, *(Am)* top housing
Querbalkenfestklemmung *f* rail clamping (lock) *(planer)*
Querbalkenklemmeinrichtung *f (mach)* cross-rail clamping device
Querbau *m (mine)* cross-cut method
abwärtsgeführter Q. cross-cut method in descending order
aufwärtsgeführter Q. cross-cut method in ascending order
Querbeanspruchung *f* transverse strain
Querbeblasung *f* transversal blowing *(circuit-breaker)*
Querbelastung *f* lateral (transverse) load
Querbeplankung *f* cross planking
Querbeplattung *f* cross plating
Querbespantung *f* transverse framing
Querbespülung *f* transversal [oil] scavenging *(circuit-breaker)*
Querbett *n* cross bed
Querbewegung *f* cross (transverse) movement
Querbewehrung *f* distribution steel, transverse (secondary) reinforcement
Querbeziehung *f* cross-correlation
Querbiegefestigkeit *f* transverse strength
Querbiegeversuch *m* transverse test
Querbohrung *f* cross hole
Querbrechen *n (mine)* slice-drift, sub-level
Querbruchbau *m (mine)* scramming, slicing-and-caving, top-slicing [and cover-caving]
Querbruchfestigkeit *f* transverse strength
Querbunker *m (ship)* cross bunker
Querbürstmaschine *f* cross-brushing machine
Querdehnung *f* transverse (lateral) strain
Querdehnungsempfindlichkeit *f* transverse strain sensitivity
Querdiffusion *f* transverse diffusion
Querdraht *m* cross-span
Quer-EMK *f* quadrature-axis component of the electromotive force
subtransiente Q. quadrature-axis subtransient electromotive force
transiente Q. quadrature-axis transient electromotive force
Querentzerrer *m* shunt admittance equalizer
Quererregung *f (el)* transversal excitation
Querfalz *m* cross fold
Querfaserspeisung *f* transverse fibre-feed
Querfeder *f* transverse spring *(car)*
Querfeld *n (el)* cross (transverse) field
Querfeldamperewindungen *fpl* cross ampere-turns
Querfelddynamo *m* self-stabilizing dynamo

Querfeldgenerator *m* Rosenberg (cross-field) generator
Querfeldinstrument *n* transverse field device
Querfeldkomponente *f* quadrature-axis component
Querfeldmaschine *f* armature-reaction excited machine
Querfeldschaltsystem *n* cross-field switching system
Querfeldverstärkermaschine *f* amplidyne
Querfeldwechselwirkung *f* crossed-field interaction
Querfestigkeit *f* lateral (transversal) strength
Querförderer *m* transverse (cross) conveyor
Querformat *n* broadside format; landscape
Querführung *f* cross-rail *(shaper)*
Quergabelstapler *m* side loader
Quergatter *n* cross-cut frame saw
Quergefälle *n* transverse slope, cross fall, camber
quergelocht with transverse hole[s]
quergestellt transverse
Quergleitung *f* cross slide
Querglied *n* shunt component (arm, element)
Querhalter *m* cross-tie
Querhaupt *n* *(mach)* crosshead, top rail, cross-beam, *(Am)* top housing
Querheftung *f* side stitching, stabbing
Querhelling *f* *(ship)* sideslip[way]
Querhobeleinrichtung *f* cross-planing feature (attachment)
Querhobelmaschine *f* squaring-up machine
Querhobeln *n* cross (transverse) planing, planing crosswise
Querhobelsupport *m* transverse-cutting toolhead, cross-[rail] planing head, right-angle planing head
Querholz *n* *(ship)* cross piece
Querholzheck *n* transom stern
Querimpedanz *f* leak (shunt) impedance
  subtransiente Q. quadrature-axis subtransient impedance
  synchrone Q. quadrature-axis synchronous impedance
  transiente Q. quadrature-axis transient impedance
Querinduktivität *f* shunt-arm inductance
Querkeil *m* *(mach)* cotter
Querkeilverbindung *f* cotter joint
Querkompensation *f* transversal compensation
Querkomponente *f* shunt (quadrature-axis) component
Querkontraktion *f* lateral deformation
Querkopfziehform *f* cross head *(extrusion)*
Querkraft *f* lateral (shearing) force
Querlage *f* *(aut)* lateral sway; *(aero)* bank
  leichte Q. gentle bank
Querlager *n* thrust (radial) bearing
Quer-Längslager *n* bearing taking combined thrust and journal loading
Querlast *f* transverse load
querlastig *(aero)* lopsided
Querleiste *f* cleat
Querleitwert *m* transverse conductance
Querlenker *m* *(aut)* transverse link (control arm)
Querlibelle *f* transverse level
Querlochstromwandler *m* slot-type current transformer
Querluftspannrahmen *m* stenter with lateral ventilation
Querlüftung *f* through ventilation
quermagnetisch transverse (cross) magnetic
Quermagnetisierung *f* transverse (cross) magnetization
Querneigung *f* 1. *(build)* camber, transverse slope, cross fall; 2. *(aut)* lateral sway; 3. *(aero)* bank
Querneigungsmesser *m* *(aero)* bank indicator
Quernut *f* cross-slot
  Q. zur Mitnahme driving slot
Querort *n* *(mine)* cross-face, cross-head, cross heading
Querpolarisation *f* cross polarization
Querpropeller *m* lateral thrust units; transverse propeller
Querprüfungssystem *n* cross-checking system *(for punched tapes)*
Querrauhen *n* *(text)* cross raising
Querräumer *m* *(mine)* deflector plate
querrechnen to crossfoot
Querrestkontrolle *f* *(dat)* modulo-n check
Querrichtung *f* cross (transverse) direction
  Q. / in crosswise, transversely
Querriegel *m* *(build)* diaphragm beam; cross bracket *(pole)*
Querrippe *f* *(mach)* cross rib
Querrippen / mit cross-ribbed
Querrips *m* *(text)* warp rib
Querriß *m* transverse crack
Querruder *n* *(aero, ship)* aileron
  angesetztes Q. external aileron
Querrudertiefe *f* depth of aileron
Quersäge *f* trim (cross-cut, reciprocating) saw
Querschaltung *f* *(el)* cross-connection
Querschermaschine *f* *(text)* cross-cutting machine, cross-shearing machine
Querschieber *m* *(mach)* saddle
querschiffs [a]thwartship
Querschlag *m* *(mine)* cross-adit, cross-cut
Querschlepper *m* transfer *(rolling)*
Querschlitten *m* *(mach)* cross slide *(lathe)*; transverse carriage; saddle *(shaper)*
Querschneide *f* chisel edge (point) *(twist drill)*
Querschneidenwinkel ψ *m* complementary angle of the chisel edge angle
Querschneider *m* cross-cutter, sheeter

Querschnitt *m* 1. [cross] section; midship section; 2. cross-cut[ting]
kreisförmiger Q. circular cross-section
mehrteiliger Q. built-up section
schematischer Q. schematic section
verjüngter Q. tapered section
Querschnittsänderung *f* cross-sectional variation
Querschnittsdarstellung *f* outline drawing of the cross-sectional shape
Querschnittsfeldverteilung *f* cross-sectional field distribution
Querschnittsfläche *f* [cross-]sectional area
Querschnittsform *f* shape (form) of [the] section
Querschnittsphasenverteilung *f* cross-sectional phase distribution
Querschnittsschwächung *f* weakening of the cross section
Querschnittsübergang *m* section change, change in section
Querschnittsverengung *f* **durch Kontraktion** reduction of cross-section due to contraction *(turbine)*
Querschnittsverminderung *f* reduction of cross-sectional area, diminution of section; *(mine)* closure *(roadway)*
Q. durch Schmieden draught by forging
Q. durch Walzen draught by rolling
Querschott *n* [cross] bulkhead, transverse bulkhead
Querschwingung *f* transverse (transversal, lateral) oscillation, transverse (transversal, lateral) vibration
Quersiederkessel *m* flue-boiler with cross tubes
Querslip *m (ship)* sideslip
Querspannung *f* transverse (quadrature-axis component of the) voltage
Querspant *n (ship)* balanced (transverse, square) frame
Querspanten *npl* transverse framing
Quersplint *m* cotter pin
Querspritzkopf *m (plast)* angle extrusion head, head for side extrusion
Querspule *f* bridging coil
Querstabilität *f* lateral (transverse) stability; *(ship)* lateral statical stability
Querstange *f* cross bar
Querstapellauf *m* side launching
Quersteifigkeit *f* transverse rigidity
Querstift *m (mach)* cotter [pin]
Querstoß *m* transverse impact
Querstrahlanlagen *fpl (ship)* lateral thrust units
Querstrahlung *f* broadside radiation
Querstrahlwanderfeldröhre *f* transverse beam travelling-wave tube
Querstrebe *f* cross-tie
Querstrecke *f (mine)* cross heading, crosscut, butt entry
Querstrich *m* bar
Querstrom *m (el)* transverse (transversal, shunt) current; cross-flow
Querstromlöschkammer *f (Am)* turbulator
Querstromschalter *m* cross-jet circuit breaker
Quersüll *m (n) (ship)* transverse (end) coaming
Quersummenrestkontrolle *f (dat)* modulo-n check
Quersupport *m* rail head, cross-slide tool box, tool box of the cross slide *(planer)*; cross slide *(lathe)*
linker Q. LH-rail head
rechter Q. RH-rail head
Querträger *m (build)* cross girder (beam), cross-head, diaphragm (secondary) beam; *(mach)* grate bar support (bearer)
Q. mit U-Profil channel-section cross member
Quertrikot *m* transverse tricot
Querverband *m* transverse (cross) bracing
Querverbinder *m (el)* cross jointer
Querverbindung *f (el)* cross connection; tie line, *(Am)* tie trunk
Querverbindungsleitung *f* junction (cross-connection) line
Quervergleichsschutz *m* transverse differential protection
querverrippt *(mach)* cross-ribbed
Querverschiebung *f* cross-feed; transverse displacement
Querversetzung *f* screw dislocation *(crystals)*
Querversteifung *f* cross bracing
querverstellbar adjustable transversely
Querverstellung *f* transverse (cross) adjustment
Quervorschub *m* transverse (cross) feed
selbsttätiger Q. automatic cross-feed
Quervorschubspindel *f* cross-feed screw *(shaper)*
Querwalzen *n* transversal laminating
Querwand *f (build)* cross-wall, diaphragm wall; *(aut)* bulkhead
aussteifende Q. buttressing (tie) wall
Querwandbauweise *f* cross-wall construction
Querwelle *f* transverse wave
Querwiderstand *m* shunt (cross, leak) resistance
Querzahl *f* Poisson's ratio
Quetschdruck *m* squeeze pressure
Quetsche *f* squeezer; crusher
quetschen to squeeze, to squash; to crush, to bruise; to roll *(flax)*; to soften *(jute)*
Quetschflüssigkeit *f* trapped fluid *(pump)*
Quetschflüssigkeitsbildung *f* trapping of fluid *(pump)*
Quetschfuß *m* squash, pinch *(lamp-bulb)*; glass pinch, pinched base, squeeze *(electron tube)*
Quetschfußdurchführung *f* pinch-type lead-in
Quetschgrenze *f* compressive yield point
Quetschhahn *m* pinchcock
Quetschholz *n (mine)* crusher (squeeze) block, lid

Quetschklemme *f* [spring] clip
Quetschkondensator *m* compression[-type] capacitor
Quetschleitung *f* squeeze section *(waveguides)*
Quetschmühle *f (ag)* corn bruiser, crushing (bruising) mill; cake crusher
Quetschöl *n* trapped oil *(pump)*
Quetschölbildung *f* trapping of the oil between the teeth *(gear pump)*
Quetschpresse *f (text)* roller press
Quetschtechnik *f* crimp technique
Quetschwalze *f (met)* crushing roller
Quetschwalzwerk *n (met)* squeezer
Queue *n* cue
Quickmetall *n* amalgamated metal
quietschen to squeal, to squeak
Quietschen *n* squealing, squeak
quinär quinary
Quintett *n* quintet
quirlförmig whorled
quittieren to sign, to receipt, to acknowledge;
Quittung *f* receipt
  Q. / gegen against (on) receipt
Quittungskreis *m (el)* acknowledging circuit
Quittungsschalter *m* acknowledger, acknowledging (position-indicating, forestalling) switch
Quittungszeichen *n* acknowledgement signal
Quotenstichprobe *f* quota sample
Quotient *m* quotient
Quotientenmesser *m* ratiometer
Quotientenmeßwerk *n* cross-coil meter system
Quotientenrelais *n* quotient relay
Quotientenverstärker *m* ratio amplifier

R

R *s.* Rydberg-Konstante
r-Einheit *f s.* R-Einheit
R-Einheit *f* roentgen, r-unit, R
R-Gespräch *n* collect (transferred-charge) call
R-Karte *f s.* Spannweitenkarte
r-Meter *n* r-meter
R-Schirm *m* R-scope *(radar)*
R-Winkelmesser *n* A hoe
R-Zacken *mpl* R waves *(electrocardiogram)*
Rachen *m* jaw *(snap gauge)*
Rachenlehre *f* snap gauge
Rachenreizstoff *m* sternutator
rad-Einheit *f (nucl)* rad [unit]
Rad *n* wheel *(car)*; gear *(toothed)*
  R. aus einem Stück wheel in one piece
  R. mit Stumpfverzahnung stub-tooth gear
  abnehmbares R. detachable wheel
  außenverzahntes R. external gear
  Barlowsches R. *(el)* Barlow disk
  einfaches R. single wheel
  festes R. fixed wheel
  fliegendes R. overhung wheel *(turbine)*
  getriebenes R. driven gear, follower
  innenverzahntes R. internal gear, annular-toothed wheel
  phonisches R. phonic (tone) wheel
  schrägverzahntes R. helical gear
  treibendes R. impeller [gear]; driving gear
Radabstand *m* wheel-centre distance
Radabweiser *m* bollard, fender post
Radabzieher *m* wheel puller
Radantenne *f* cartwheel aerial
Radar *n* radar
  R. für absolute Bewegungsanzeige true-motion radar
Radarabtaster *m* scanner unit
Radarantenne *f* radar aerial
  schnellrotierende R. high-speed scanning aerial
Radaranzeige *f* radar display
Radarastronomie *f* radar astronomy
Radarausbildung *f* radar training
Radarauswertung *f* radar plotting
Radarbedeckung *f* radar coverage
Radarbereich *m* radar coverage
Radarbewegtziel *n* moving radar target
Radarbild *n* radar trace
Radarbildentzerrer *m* radar presentation restitutor
Radarbildregistriergerät *n* radarscope recorder
Radarecho *n* radar response
Radarerfassung *f* radar contact
Radarerkennungsgerät *n* radar recognition equipment
Radarfunkfeuer *n* radar beacon
Radargerät *n* radar equipment; radar set
  R. für Rollfeldverkehr airfield surface movement indicator
  R. zur Abtastung der Erdoberfläche surface search radar
  R. zur Frühwarnung early-warning radar
Radargeschwindigkeitsmesser *m* radar speedometer
radargesteuert radar-controlled
Radargleichung *f* radar equation
Radarhaube *f* radome
Radarhöhenmesser *m* radar altimeter
Radarhorizont *m* radar horizon
Radaridentifikation *f* radar identification
Radarinsel *f* radar island
Radarkenngerät *n* radar identification unit
Radarkette *f* radar chain
Radarkeule *f* radar lobe
Radarkuppel *f* radome
Radarlandegerät *n* approach control radar
Radarleitdienst *m* radar surveillance
Radarnavigation *f* radar navigation
Radarnetz *n* radar link

Radaroptik *f* radar optics
Radarortung *f* radar detection
Radarortungsgerät *n* radar detector
Radarreflektor *m* radar reflector
Radarschatten *m* radar shadow
Radarschaubild *n* radar plot
Radarschaubildrechner *m* radar plotting computer
Radarschirm *m* radar screen; radar panel
Radarschirmbild *n* radar display
Radarschirmbilddarstellung *f* radar presentation
Radarsender *m* radar transmitter
Radarsicherungsfahrzeug *n* radar picket-vessel
Radarstörsender *m* radar jammer (jamming station)
Radarstörung *f* radar interference
Radarstrahl *m* radar beam
Radarsuchgerät *n* radar search unit
Radarsystem *n* radar system
  R. zum Kursabsetzen radar plotter system
Radartarnüberzug *m* radar-swamping coverage
Radartäuschung *f* radar deception
Radartrugziel *n* phantom radar target
Radarüberwachung *f* radar watch
Radarwarngerät *n* early warning equipment
Radarzeichen *n* radar trace
Radarzeichnung *f* radar plotting
Radarziel *n* radar target
Radarzielflug *m* / **gelenkter** radar homing-guidance
Radausfluchtung *f* track alignment
Radball *m* cycle ball, [bi]cycle football
Radbefestigungsbolzen *m* wheel mounting bolt
Radbefestigungsmutter *f* wheel mounting nut
Raddampfer *m* paddle[-wheel] steamer, paddle wheeler, side-wheel steamer
Raddruck *m* wheel load
Raddurchmesser *m* wheel diameter
Radeffekt *m* spoking *(radar)*
Radeinstechmaschine *f* infeed gear cutting machine
Rädelerz *n* wheel ore, bournonite
Räder *npl* / **entgegengesetzt laufende** disks running in opposite directions *(turbine)*
  gegenläufige R. disks running in opposite directions *(turbine)*
  gepaarte R. mating gears
Räderabmessungen *fpl* gear size (dimensions)
Räderantrieb *m* geared drive
räderfrei gearless
Rädergetriebe *n* train of gears (pulleys); gear train
  R. mit einfacher Übersetzung simple gear train
Räderkasten *m* gear box
  sechsstufiger R. six-speed gear box
räderlos ungeared
Räderpaar *n* pair of wheels (gears)
  miteinander in Eingriff stehendes R. mating (meshing, engaging) gears
Räderraupenkampfwagen *m* hermaphrodite tank
Räderraupenpanzer *m* *s.* Räderraupenkampfwagen
Räderspindelkasten *m* *(mach)* geared head
Räderspindelstock *m* *(mach)* geared headstock
Rädertrieb *m* gear drive, drive by toothed gears (pulleys, sprockets), gear wheel transmission
Räderübersetzung *f* gear transmission
Räderverhältnis *n* gearing ratio
Rädervorgelege *n* back gears
radfahren to [ride on a] cycle
Radfahrer *m* cyclist
Radfahrweg *m* cycle track (path), bicycle track (path)
Radfelge *f* felloe
Radfelgenbiegemaschine *f* felloe bending machine
Radfelgenkranz *m* [wheel] rim
Radflattern *n* wheel shimmy
Radgehäuse *n* paddle-box *(paddle-wheel steamer)*
Radhacke *f* push hoe
radial radial
Radialbeanspruchung *f* radial stress (loading)
Radialbelastung *f* radial loading
Radialbeschleunigung *f* circumference (radial) acceleration
Radialblechkern *m* laminated radial[-shaped] core
Radialbohren *n* radial drilling
Radialbohrmaschine *f* radial drilling machine
  R. auf Fahrbett traversing-type radial drilling machine
  R. mit Höhenverstellung elevating-arm radial drilling machine
  R. ohne Höhenverstellung radial drilling machine with fixed height arm
  einfache R. plain radial drilling machine
Radialbürste *f* radial brush
Radialdavit *m* *(ship)* radial (luffing) davit
Radialdurchschlag *m* *(el)* radial breakdown (puncture)
Radialfärbeapparat *m* radial dyeing-apparatus
Radialfilter *n* radial grating *(waveguides)*
Radialfokussierung *f* radial focusing
Radialgebläse *n* radial compressor
Radialgeschwindigkeit *f* radial velocity
Radialhohlleiterantenne *f* radial waveguide-aerial
Radialkammer *f* *(nucl)* radial-motion chamber
Radialkolbenpumpe *f* radial piston pump
Radialkomponente *f* radial component
Radialkraft *f* radial force, thrust force *(tool)*
Radiallager *n* radial (thrust) bearing
Radialleitung *f* radial transmission feeder

Radialleitungsresonanz *f* radial line resonance
Radialluft *f (mach)* radial play
Radiallüfter *m* radial blower (ventilator)
Radialnut *f* radial groove
Radialpolarisation *f* radial polarization
Radialriß *m* boxheart checking
Radialschlag *m* radial runout
Radialschnitt *m* quarter sawing
Radialschwindung *f* radial shrinkage
Radialspannvorrichtung *f* radial stretching device
Radialspiel *n* radial play
Radialtriangulierung *f* radial triangulation
Radialturbine *f* radial flow turbine
Radialverfahren *n* **zur Schneckenradherstellung** worm-wheel hobbing by the radial-feed method
Radiant *m* 1. radiant; 2. radian *(circular measure)*
Radiationschemie *f s.* Strahlenchemie
Radiator *m* radiator
Radiaxlager *n* bearing taking combined radial and axial load
Radienabrichteinrichtung *f* radius dressing fixture, [grinding] wheel dresser for shaping radii
Radienabrichten *n* radius dressing *(grinding wheel)*
radieren to erase; to etch *(arts)*; to grind *(tyre)*
Radieren *n* **des Reifens** *(aut)* grinding of the tyre
Radierschablone *f* erasing shield
Radikal *n* 1. *(math)* radical sign; 2. *(ch)* radical
  freies R. *(ch)* free radical
Radikalfänger *m* scavenger
Radikalreaktion *f* radical reaction
Radikalwanderung *f* migration of radicals
Radikand *m* radicand
Radio *n* radio [set]
Radioaktinium *n* radioactinium
radioaktiv *(nucl)* [radio]active
  künstlich r. artificially radioactive
  natürlich r. naturally radioactive
Radioaktivität *f* [radio]activity
  künstliche R. artificial (man-made, induced) radioactivity
  natürliche R. natural radioactivity
Radioakustik *f* radioacoustics
Radioamateur *m* radio amateur, *(sl)* ham
Radioanlage *f* radio installation
Radioansager *m* radio announcer
Radioapparat *m* wireless (radio) set
Radioastronom *m* radio astronomer
Radioastronomie *f* radio astronomy
Radiobastler *m* radio constructor
Radiobiologie *f* radiobiology
Radiochemie *f* radiochemistry
  angewandte R. applied radiochemistry
Radiochemiker *m* radiochemist
radiochemisch radiochemical
Radiochromatografie *f* radiochromatography
Radioelement *n* radioactive element, radioelement
Radioempfang *m* radio (broadcast) reception
Radioempfänger *m* wireless [receiving] set
  R. mit Schallplattenwiedergabe radio-gramophone
radiogen radiogenic
Radiografie *f* 1. radiography; 2. radiogram, radiograph
Radiogramm *n* radio-autograph, radiogram, radiograph
Radiohörer *m* listener
Radioisotop *n* radioactive isotope, radioisotope
Radiojod *n* radioactive iodine; radioiodine
Radiokäfig *m* radio control cage
Radiokanal *m* radio channel
Radiokobalt *n* radioactive cobalt; radiocobalt
Radiokohlenstoff *m* radioactive carbon; radiocarbon
Radiokolloid *n* radio-colloid
Radiokompaß *m* radio compass, [automatic] direction finder
Radiologe *m* radiologist
Radiologie *f* radiology
  medizinische R. radiological medicine
  physikalische R. radiological physics
radiologisch radiologic[al]
Radiolumineszenz *f* radioluminescence
Radiolyse *f* radiolysis
Radiometallografie *f* radio metallography
Radiometer *n* radiometer
  R. nach Crookes Crookes' radiometer
Radiometrie *f* radiometry
radiometrisch radiometric
Radiomimetika *npl* radiomimetics
Radionavigation *f* radio navigation
Radionuklid *n* radioactive nuclide, radionuclide
  natürliches R. natural radioactive nuclide
Radionuklidproduktion *f* radionuclide production
Radiophosphor *m* radioactive phosphorus; radiophosphorus
Radioröhre *f* radio valve
Radioschwefel *m* radioactive sulphur; radiosulphur
Radiosendezeit *f* broadcasting time, time on air
Radiosendung *f* radio broadcast (transmission)
Radioskalenlampe *f* wireless panel lamp
Radiosonde *f* radiosonde system, radio sonde
Radiospektrometer *n* radio spectrometer
Radiostation *f* radio station
Radiostudio *n* radio studio
Radiotechnik *f* radio engineering
Radiotelefonie *f* wireless telephony
Radioteleskop *n* radio telescope

Radiothorium *n* radiothorium
Radiowelle *f* radio wave
Radiowellenausbreitung *f* radio-wave propagation
Radium *n* radium
Radiumalter *n* radium age
Radiumbestrahlung *f* radium irradiation
Radiumbombe *f* teleradium unit
Radiumemanation *f* radon, radium emanation
Radiumfamilie *f s.* Uranreihe
Radiumfernbestrahlung *f* radium teleirradiation
Radiumfernbestrahlungsgerät *n* teleradium unit
Radiumkanone *f* teleradium unit
Radiumkapillare *f* radium seed
Radiumkapsel *f* radiode, radium seed
Radiumkontaktbestrahlung *f* radium contact irradiation
Radiumnadel *f* radium needle
Radiumpräparat *n* radium source (drug)
Radiumreihe *f s.* Uranreihe
Radiumspickung *f* radium implantation
Radiumstandard *m* / **internationaler** international radium standard
Radiumtherapie *f* radium therapy
Radiumzerfallsreihe *f s.* Uranreihe
Radius *m* radius
  Bohrscher R. Bohr radius
  hydraulischer R. hydraulic radius
Radiusfräser *m* radius (corner-rounding) cutter
  R. für Ecken cutter for cornering
  einseitiger R. single-corner rounding cutter
Radiuslehre *f* radius (fillet) gauge
Radiusparameter *m* radius parameter
Radiusvektor *m* radius vector
Radixschreibweise *f (dat)* radix notation
radizieren to draw roots, to extract a root
Radkappe *f* hub cap
Radkasten *m* paddle-box, sponson *(paddle-wheeler)*
Radkastenbalken *m* / **äußerer** sponson-beam
Radkastendeck *n* paddle deck
Radkörper *m* wheel body (centre)
  unbearbeiteter R. gear blank
Radkranz *m* [wheel] rim; wheel tyre
Radkurve *f* cycloid
Radlast *f* wheel load
Radlinie *f* cycloid
Radmagnetron *n* cavity magnetron
Radmutternschlüssel *m* wheel-nut spanner
Radnabe *f* [wheel] hub, boss of wheel
Radnabenbohr- und -stemmaschine *f* nave boring and mortising machine
Radnabenbuchse *f* hub bushing
Radnabendeckel *m* hub cap
Radon *n* radon, emanation, emanon
Radonhohlnadel *f* radon seed
Radpaarung *f* pair of mating gears
Radpolo *n* [bi]cycle polo
Radpropeller *m* wheel propeller, paddle wheel
Radreibung *f* wheel friction
Radreifen *m* tyre
Radreifenbohrwerk *n* [wheel] tyre vertical boring and turning mill
Radreifensprengring *m* tyre-clip, tyre-clasp, spring-ring
Radsatz *m* wheel set, set of wheels
Radsatzdrehmaschine *f* wheel lathe
Radschalter *m* rotary switch
Radscheibe *f* **aus Blech** wheel disk stamped out of a plate
Radschlepper *m* paddle tug
Radsitz *m* wheel seat
Radspeiche *f* wheel spoke
Radspeichenhobelmaschine *f* wheel-spoke planing machine
Radspreizungswinkel *m* king-pin angle
Radspur *f* wheel track
Radspurlockerer *m (ag)* track loosener
Radspurmesser *m* track alignment-gauge
Radstand *m* wheel base [of a vehicle]
  großer R. long wheel base
Radstellungsanzeiger *m* wheel-position indicator
Radstern *m* [spoke] wheel centre
Radsturzwinkel *m* camber angle
Radtraktor *m* wheel tractor
Radtrommel *f* drum
Radverpackung *f* wheel package *(electronic gear in rockets)*
Radweg *m* bicycle track, cycle path (track)
Radzylinder *m* wheel cylinder
raffen to gather [up]
Raffiabast *m* raffia
Raffinade *f* refined sugar
Raffinat *n* raffinate, refined product
Raffinatblei *n* refined lead
Raffination *f* refining; purification; clarification
  elektrolytische R. electro-refining
Raffinationsanlage *f* refining (purification) plant
Raffinatkupfer *n* refined (selected) copper
Raffinerie *f* refinery; *(met)* refining plant
Raffineriemelasse *f* barrel sirup, refinery molasses
Raffineurstoffbütte *f* refiner stock-chest *(papermaking)*
raffinieren to refine
Raffinierofen *m* refining furnace
Raffinierung *f* / **partieweise** batch refining
Raffinose *f* raffinose
Ragger *m* ragger *(papermaking)*
Rah[e] *f (ship)* yard
Rähm *m* purlin, head runner (rail, plate, piece) *(carpentry)*
Rahmen *m (mach)* frame; *(aut)* chassis; *(build)* frame[work]; plough-beam; welt *(shoe)*; edge, border, rim; clip *(bullets)*
  R. mit biegesteifen Knoten rigid frame

bettförmiger R. frame with circular end
biegesteifer R. rigid frame
doppelt gekröpfter R. *(aut)* double-drop frame
eingespannter R. fixed frame
gekröpfter R. dropped frame
geschlossener R. closed frame
starrer R. stiff frame
Rahmenantenne *f* circular (coil, frame, loop) aerial
drehbare R. rotating frame aerial
Rahmenausbau *m (mine)* goal-post system, square-set timbering
Rahmenbalken *m* frame girder
Rahmenbrille *f* frame centre rest
Rahmenfenster *n* guide opening
Rahmenflansch *m* frame flange
Rahmenfuß *m* foot of frame
Rahmengabel *f* frame fork
Rahmengestell *n* panelled framework
Rahmenhöhe *f* ground clearance of the frame
Rahmenholm *m* frame side bar
Rahmenkante *f* seam *(shoe)*
Rahmenkonstruktion *f* framework
Rahmenkopf *m* front-end of the frame, crankshaft bearing-end of the frame
Rahmenkröpfung *f* frame drop
Rahmenlibelle *f* frame level
Rahmenmaschine *f* box frame assembling machine
Rahmenpeiler *m* frame direction finder
Rahmenpflug *m* frame plough
Rahmenplan *m* outline (master) plan
Rahmenplatte *f* frame plate *(storage battery)*
Rahmenpresse *f* frame press
Rahmenschott *n (ship)* partial bulkhead
Rahmenschrämarm *m* frame-jib
Rahmenschrämlader *m* frame-jib cutter-loader
Rahmenspaltsäge *f* jig saw
Rahmenspant *n (ship)* web frame
Rahmenstrahler *m* closed-loop radiator
Rahmensucher *m* frame view-finder
Rahmenunterzug *m* frame trussing
Rahmenverlängerung *f* frame extension
Rahmenversteifung *f* trussing of the frame
Rahmenweite *f* / **lichte** inside width of a frame
Rahmenwerk *n* frame[d] structure
Rahmenzimmerung *f (mine)* square-set timbering, goal-post system
Rahmerhitzer *m* cream heater
Rahmseparator *m* cream separator
Rahnock *n (f) (ship)* yard arm
Rahschoner *m (ship)* schooner
Rahtakel *n (ship)* jeer
Rain *m* ridge [of fields]; border
Rakel *f* doctor [blade], excess paste removal blade
Rakelappreturapparat *m* doctor finishing machine
Rakelfarbwerk *n* doctor-blade inking unit
Rakelführer *m* doctor ruler
Rakellineal *n* doctor ruler
Rakelmesser *n* doctor blade
Rakelschleifmaschine *f* doctor blade grinder; squeegee dresser *(screen printing)*
Rakelstreichmaschine *f* knife coater
Rakelstreichverfahren *n* knife coating
Rakelverfahren *n* **mit Walzenauftrag** doctor kiss coating
Rakelwalze *f* doctor roller
Rakete *f* rocket, *(mil)* missile
R. mit mittlerer Reichweite intermediate-range rocket
R. ohne Stabilisierungsflächen finless (unfinned) rocket
R. zur Erforschung der oberen Schichten der Atmosphäre exploratory (high-altitude research, upper-air research) rocket
ballistische R. ballistic rocket
bemannte R. manned (man-carrying) rocket
chemische R. chemical rocket
dreistufige R. three-step rocket, three-stage rocket
einstufige R. single-step rocket, single-stage rocket
flossengedämpfte R. fin-stabilized rocket
gelenkte R. controlled (guided) rocket
gesteuerte R. controlled (guided) rocket
ihr Ziel verfehlende R. rogue rocket
instrumentetragende R. instrumented (instrument-carrying) rocket
interkontinentale ballistische R. intercontinental ballistic missile
interkontinentale R. intercontinental missile
interplanetarische R. interplanetary rocket
kosmische R. astronautic (cosmic, space) rocket
mehrstufige R. multistaged (multi-stage, composite, step) rocket, stage[d] rocket
meteorologische R. meteorological (exploratory) rocket
unbemannte R. unmanned rocket
ungelenkte R. free-fly rocket, free[-flight] rocket
ungesteuerte R. free[-flight] rocket, free-fly rocket
vierstufige R. four-step rocket, four-stage rocket
vom Ziel abweichende R. rogue rocket
von einem Ballon aus gestartete R. rockoon (balloon-supported, balloon-borne, balloon-launched) rocket *(used for high-altitude cosmic ray studies)*
zweistufige R. two-step rocket, two-stage rocket
Raketenabschußbasis *f* missile base, rocket launching site
Raketenabwehr-Laser *m* missile destroy laser
Raketenabwehrsystem *n* antimissile system
Raketenachse *f* rocket centre line
Raketenantrieb *m* rocket propulsion (drive)
Raketenaufstieg *m* rocket sounding

Raketenbatterie *f* rocket battery
Raketenboden *m* rocket base
Raketenbrennkammer *f* rocket combustion chamber
Raketenbrennstoff *m s.* Raketentreibstoff
Raketendüse *f* rocket nozzle
Raketenelektronik *f* rocket electronics
Raketenfachmann *m* rocket expert, rocketeer
Raketenflug *m* rocket flight
Raketenflugzeug *n* rocket-driven aircraft
Raketengeschütz *n* **mit mehreren Läufen** cluster rocket launcher
Raketengrundgleichung *f* fundamental equation of reaction propulsion, fundamental equation of rocket motion
Raketeninneres *n* interior of a rocket
Raketeninstrumentierung *f* rocket instrumentation
Raketenkörper *m* rocket body (airframe)
Raketenlenkung *f* rocket guidance
Raketenschwerpunkt *m* rocket centre
Raketensondierung *f* rocket sounding
Raketenstabilisierung *f* rocket stabilization
Raketenstart *m* rocket launching
Raketenstartanlage *f* / **fahrbare** [rocket-]launching vehicle
Raketenstartbeschleuniger *m* auxiliary take-off rocket unit
Raketenstarteinrichtung *f* [rocket-]launching installation
Raketenstartplatz *m* rocket-launching site
Raketensteuerung *f* rocket control
Raketenstufe *f* rocket step (stage), stage of a rocket
  R. mit Triebwerk propulsion [rocket] stage
  dritte R. third stage
  erste R. first stage, mother missile
  vierte R. fourth stage
  zweite R. second-stage [rocket], daughter missile
Raketentechnik *f* rocket technique[s] (engineering), rocketry
Raketenträger *m* rocket-carrier, launching vehicle
Raketentreibsatz *m* rocket-propelling charge
Raketentreibstoff *m* rocket fuel; propellant
Raketentriebwerk *n* rocket propulsion (motor assembly, engine)
Raketenvernichtungssystem *n* antimissile kill system
Raketenversuchsgelände *n* proving establishment
Ramanspektrum *n* Raman spectrum
Ramie *f* China grass, ramie
Rammanlage *f* piling plant (hammer)
Rammbär *m* hammer, monkey, tup
Rammbug *m (ship)* ram (tumble-home) bow
Ramme *f* punner, pommel *(foundry)*; *(build)* ram[mer]
rammen to drive *(piles)*; to ram
Rammformel *f* [pile] driving formula
Rammgasse *f (mine)* ram track
Rammgerät *n (mine)* ram
Rammhaube *f* helmet, cushion block, driving cap
Rammkasten *m (mine)* ram box
Rammkette *f (mine)* ram chain
Rammkörper *m (mine)* ram body
Rammpfahl *m* ram-pile
Rammschrapper *m (mine)* ram scraper-box
Rammsteven *m (ship)* ram bow (stem)
Rammung *f* **von Pfählen** driving of piles
Rammzinken *m (mine)* ram horn
Rampe *f* ramp; platform
Rampenantwort *f (dat)* ramp response
Rampenfunktion *f (dat)* ramp function
Rampenlicht *n* proscenium light
Rand *m* edge; rim; border
  abgekettelter R. purl-edge *(knitting)*
Randabstand *m* **der Niete** distance [from centre line of the row of rivet holes] to the edge of plate
Randaufbruch *m (telev)* edge flare
Randauflösung *f* edge acuity *(of an image)*
Randbalken *m* edge (boundary, rim) beam
Randbeanspruchung *f* boundary stress
Randbedingung *f* boundary condition
Randbefeuerung *f* boundary lights
Randbemerkung *f* marginal note
Randblasen *fpl* subcutaneous blowholes *(casting)*
Randdichte *f* boundary density
Randdicke *f* thickness at point of tooth *(saw)*
Randdruckeinrichtung *f (text)* selvedge-printing device
Randeffekt *m* edge effect; boundary effect; fringe effect
Randeinfassung *f* edging
Rändel *n* knurling *(having straight parallel grooves)*
Randelektron *n* boundary electron
Rändelhalter *m* single-wheel knurl holder
Rändelknopf *m* knurled [control] knob
Rändelmutter *f* hand nut
rändeln to knurl straight-lined patterns
Rändeln *n* knurling straight-lined patterns
Rändelradantrieb *m* edge-operated control
Rändelrädchen *n* knurling roller, straight knurl
Rändelschraube *f* knurled screw
Rändelwerkzeug *n* single-wheel knurling tool, straight knurl
Randentkohlung *f (met)* edge (skin) decarburization
Randentladung *f* boundary (marginal) discharge
Ränderiermaschine *f* stitch wheeling machine *(shoemaking)*
rändern to border
Randfaser *f* extreme (outer) fibre

Randfehler *m (opt)* spherical aberration
Randfeld *n* fringing (stray) field
Randfeldstärke *f* boundary field strength
Randfläche *f* lateral face *(crystallography)*; boundary [surface]
Randformen *fpl* **der Schleifscheibe** [standard] wheel faces
Randgebiet *n* outskirts, fringe area, suburb; allied field
Randintegral *n* integral around a closed contour
Randkoeffizient *m* marginal coefficient
Randkonzentration *f* boundary concentration
Randkraft *f* marginal (boundary) force, end reaction
Randlinie *f* boundary line
Randlochkarte *f* edge-punched card, marginal-punched card
Randlochkartenleser *m* edge-punched card reader
Randmasche *f (text)* selvedge loop
Randmoment *n* end moment
Randplanke *f (ship)* margin (boundary) plank
Randpunkt *m* end point
Randrisse *mpl* edge-cracks
Randschicht *f* peripheral (boundary, barrier) layer, boundary surface, transition region
Randschichtdicke *f* barrier-layer thickness
Randschichteffekt *m* barrier-layer effect
Randschichtkapazität *f* barrier capacitance
Randschichtpotential *n* boundary (barrier) potential
Randschichttheorie *f* barrier layer theory
Randsomholz *n (ship)* fashion piece
Randspannung *f* edge (extreme-fibre, boundary) stress
radiale R. radial rim strain *(turbine)*
Randstein *m* kerb-stone, curbstone
Randstörung *f* edge perturbation (disturbance)
Randstrahl *m (opt)* peripheral (marginal) ray
Randstreifen *m* verge, edge (shoulder, marginal, margin) strip *(motorway)*
befestigter R. hard shoulder *(motorway)*
Randträger *m* boundary (edge, rim) beam
Randveränderung *f* **während spannungsoptischer Messungen** time-edge effect in photoelastic measurements
Randverdunkelung *f* limb darkening *(sun)*
Randverletzung *f* surface scratch (defect)
Randwert *m* marginal (boundary) value
Randwertaufgabe *f* boundary value problem
Randwertprüfung *f* marginal checking
Randwertschaltung *f* clamping circuit
Randwinkel *m* contact (rim) angle, angle of contact
Randwirbel *m* rim vortex
Randwulst *m (f)* beaded edge
Randzeche *f* marginal mine
Randzone *f* border zone; edge

Rang *m* 1. rate, class, order; degree, grade; 2. rank; 3. gallery, balcony *(theatre)*
R. einer Matrix rank of a matrix
Rangieranlage *f* marshalling yard
Rangierbahnhof *m* marshalling (sorting) yard, shunting yard goods station
Rangierbewegung *f* shunting movement
Rangierdienst *m* shunting operations, switching
Rangierdraht *m* jumper wire
rangieren to shunt, to switch
Rangieren *n* **am Ablaufberg** shunting (switching) by gravitation, hump (gravity) shunting
R. durch Abstoßen fly shunting (switching), pushing off the wagons
R. verboten shunting (switching) forbidden
Rangierer *m* shunter
Rangierfunk *m* radio in marshalling yard
Rangiergleis *n* shunting track, sorting line in marshalling yard, *(Am)* classification yard line
Rangierleiterstelle *f* **am Ablaufberg** hump cabin
Rangierlitze *f* flexible pigtail, *(Am)* jumper flex
Rangierlokomotive *f* shunting (switching) locomotive
Rangiermeister *m* foreman shunter, passenger (goods) yard foreman
Rangierstellwerk *n* shunting tower
Rangierstoß *m* shunting impact
Rangierzettel *m* cut card, switching list
Rangkorrelation *f* rank correlation
Rangkorrelationskoeffizient *m* rank correlation coefficient
Rankheit *f* crankness, tenderness *(of a vessel)*
ranzig rancid
Rapidstahl *m s.* Schnellarbeitsstahl
Rapport *m (text)* repeat; report
Rapportbreite *f* repeat of warp threads
Rapporteinstellung *f* repeat mechanism
Rapporthöhe *f* repeat of weft thread
rapportieren *(text)* to repeat according to pattern
Rapputz *m* rough cast [plaster]
Rapsöl *n* rape[seed] oil, colza oil
Raschelmaschine *f* raschel warp knitting machine
Raschigringe *mpl* Raschig rings
Rasen *m* lawn, turf, grass
Rasenbesen *m* sweep
Rasenbleiche *f* grass bleach[ing]
Raseneisenerz *n* marsh (lake, bog) iron ore, limonite
Raseneisenstein *m* meadow ore
Rasenhängebank *f (min)* pithead bank, pit-top
Rasenmähmaschine *f* lawn mower
Rasenschere *f* edging shears; lawn shears
Rasenschneider *m* turf cutter
Rasensode *f* sod
Rasensprenger *m* lawn sprinkler

Rasierklinge *f* razor blade
Rasierseife *f* shaving soap
Raspel *f* rasp
raspeln to rasp
rasseln to rattle, to clatter
Rasselwecker *m* buzzer
Rast *f* 1. bosh *(blast furnace)*; 2. stop, locking, arresting; rest
Rastankerring *m* bosh band *(blast furnace)*
Rastblende *f* click stop
Raste *f* notch, retainer
Rastenrad *n* notched wheel
Rastenschalter *m* drum (multi-position) switch
Raster *m* screen, mesh; *(telev)* raster; grid *(printed circuit)*, modular grid; *(el)* spill shield, louvre *(lamps)*
dreidimensionaler R. modular space grid
unverzerrter R. undistorted raster
Rasterablenkung *f* frame deflection
Rasterätzung *f* halftone block, autotype
Rasterbild *n (Am)* picture frame
Rasterblende *f* scanning diaphragm
Rasterentzerrung *f (el)* frame tilt
Rasterfarbauszug *m* screened colour separation
Rasterfeinheit *f* fineness of scanning, definition of image
Rasterfolgeverfahren *n (telev)* sequential colour system
Rasterfrequenz *f* vertical (frame) frequency
Rastergrundmaß *n* [structural] module
Rasterhöhenregler *m* vertical size control
Rasterimpuls *m* frame pulse; scanning pulse
Rasterklischee *n* halftone block
Rasterkorn *n* silver plug (globule) *(photography)*
Rasterleuchte *f* louvre luminaire, spill-shield lamp
Rasterlinie *f (build)* modular (grid) line
Rastermarkierung *f* graticule marking
Rastermaß *n (build)* structural module
Rastermethode *f* grid method
Rastermikroskop *n* raster (electron-scan, scanning electron) microscope
rastern to screen
Rasterplatte *f* grid board
Rasterpunkt *m* [halftone] dot
Rasterscheibe *f* scanning disk
Rasterscheibenrahmen *m* escutcheon
Rastertherapie *f* raster (grid) therapy
Rastertiefdruck *m* photogravure
Rasterverformungsentzerrung *f* frame tilt
Rasterwechselverfahren *n* frame-sequential system, sequential-colour system
Rasterwinklung *f* screen angle
Rasthöhe *f* height of bosh *(blast furnace)*
Rastkühlkasten *m* bosh cooling plate *(blast furnace)*
Rastlinie *f* arrest line
Rastmantel *m* shell of bosh *(blast furnace)*
Rastpanzer *m* bosh casing *(blast furnace)*
Rastpolkegel *m* herpolhode cone
Rastpunkte *mpl* lock-points
Rastral *n* music pen
Rastscheibe *f* locking disk
Raststätte *f* roadside restaurant
Raststellung *f* register; indexed position
Raststift *m* plunger pin; *(Am)* latch pin
R. der Teilscheibe locating plunger
rückwärtiger R. back pin, stop-pin at the rear *(index plate)*
Rastvorrichtung *f* arresting facility
Rastwinkel *m* bosh angle *(blast furnace)*
Rastzapfen *m* drop-in pin
Rat *m* **für gegenseitige Wirtschaftshilfe** Council for Mutual Economic Aid, CMEA, Comecon
Rate *f* 1. instalment [payment]; 2. rate, velocity
Ratemeter *n* ratemeter
Ratemeterschaltung *f* ratemeter circuit
ratenweise by instalment
ratifizieren to ratify
ratinieren *(text)* to frieze
Ratiniermaschine *f (text)* friezing machine
rational *(math)* rational
r. machen *(math)* to rationalize
rationalisieren to rationalize *(industry)*
Rationalisierung *f* rationalization *(of industry)*
Rationalitätsgesetz *n* law of rationality, rationality law
rationell rational, systematic; economical
rationieren to ration
Rationsweide *f* strip-grazed pasture
rationsweise in rations
Ratsche *f* ratchet [stop]
Rattenfalle *f* rat-trap
Rattenfraß *m (leath)* place gnawed by rats *(fault)*
Rattengift *n* rat poison, rodenticide
ratterfrei chatter-free, chatterless
Rattermarke *f (mach)* chatter mark
rattern *(mach)* to chatter; to rattle
Rattern *n (mach)* chatter[ing]
regeneratives R. regenerative chatter
Ratterschwingung *f* chatter vibration
Raubbau *m (mine)* robbing working
rauben *(mine)* to draw off, to rob, to withdraw
Rauben *n* **eines Pfeilers** *(mine)* abstraction of a pillar
Raubgerät *n* **für Streckenbögen** *(mine)* arch-saver
Raubschicht *f (mine)* prop-drawing shift
Raubvorrichtung *f (mine)* dog-and-chain
Raubwinde *f (mine)* withdrawal winch
Rauch *m* smoke; *(ch)* fume
Rauchabzug *m* chimney hood
Rauchabzugskanal *m* flue
Rauchbelästigung *f* smoke nuisance

rauchdicht fume-proof, gas-proof
Rauchdichte *f* smoke density
optische R. optical density of smoke
Rauchdichtemesser *m* smoke-density indicator
rauchend *(ch)* fuming
Rauchentwicklung *f* smoke development
Raucherabteil *n* *(ry)* smoker, smoking compartment
Räucherkerze *f* fumigating candle
Räuchermittel *n* fumigant
räuchern 1. to smoke *(tanning)*; 2. to smoke-dry *(meat)*; 3. to cure; 4. to fumigate *(room)*
Räucherpatrone *f* smoke cartridge
Räucherstreifen *m* pressed pasteboard fumigant
Räuchertablette *f* fumigant tablet
Räucherverfahren *n* fumigation
Raucherzeuger *m* smoke generator
Rauchfärbung *f* coloration of smoke
Rauchfleisch *n* smoked meat
rauchfrei smokeless
Rauchgas *n* smoke (flue, furnace) gas, gaseous products of combustion, burnt gases
Rauchgasanalysator *m* flue-gas analyzer
Rauchgasanzeiger *m* smoke detector
Rauchgasexplosion *f* flue-gas explosion
rauchgasfest fume-proof
Rauchgasgeschwindigkeit *f* velocity of the flue-gas
Rauchgaskanal *m* waste gas flue
Rauchgasmesser *m* combustion indicator
Rauchgasprüfer *m* flue-gas tester (detector)
Rauchgaspumpe *f* flue-gas pump
Rauchgerbung *f* smoke tanning
Rauchglas *n* smoked glass
Rauchgranate *f* smoke shell
Rauchkammer *f* smoke box *(locomotive)*
Rauchkammerüberhitzer *m* smoke-box superheater
Rauchkanal *m* flue
Rauchklappe *f* revolving (swivel) damper
Rauchmeldeanlage *f* smoke-detector fire alarm
Rauchpatrone *f* smoke cartridge
Rauchpilz *m* mushroom cloud
Rauchquarz *m* smoky quartz
Rauchrohr *n* flue; smoke tube
Rauchrohrkessel *m* smoke tube boiler
R. mit wenigen Rauchrohren boiler with a small number of fire (smoke) tubes
Rauchrohrüberhitzer *m* smoke-tube superheater
plattenförmiger R. Heizmann's plate superheater
Rauchschieber *m* damper regulator, slide (sliding-flue, smoke) damper
Rauchschieberöffnung *f* slide-damper opening
Rauchschleier *m* smoke screen
Rauchschott *n* *(ship)* fire bulkhead
Rauchsignalpatrone *f* smoke signal cartridge
Rauchskale *f* smoke scale
Rauchverzehrer *m* smoke consumer
Rauchwaren *fpl* [clipped and dyed] furs
Rauchwarenindustrie *f* fur industry
Raufe *f* rack
raufen to pull [from the ground] *(flax)*; to pluck
Raufmaschine *f* pulling machine *(flax)*
Raufwolle *f* plucked wool
rauh rough *(surface)*; coarse
Rauheit *f* roughness *(surface)*
rauhen *(text)* to raise; to tease; to nap
Rauhen *n* *(text)* raising; teaseling; napping
Rauhfutter *n* roughage
Rauhfutterautomat *m* roughage self-feeder
Rauhfutterraufe *f* hay rack
Rauhgriffigkeit *f* *(text)* coarse feel
Rauhkarde *f* *(text)* teasel
Rauhmaß *n* nominal measure
rauhnarbig harsh-grained *(leather)*
Rauhputz *m* *(build)* rough cast [plaster]
Rauhreif *m* hoarfrost
rauhspund rough-sawn *(timber)*
Rauhtiefe *f* peak-to-valley height
Rauhwalze *f* *(ag)* ribbed (toothed) roller; *(text)* teaseling roller
Rauhwert *m* peak-to-valley value
Raum *m* space; room *(buildings)*; zone; *(mach)* chamber; *(aut)* volume
abstrakter R. abstract space
affiner R. affine space
Berner R. *(ry)* Berne rectangle, free coupling space
beschränkter R. restricted space
bestrichener R. *(mil)* danger[ous] zone, danger[ous] space, beaten zone (space)
dreidimensionaler R. three-dimensional space
ebener R. homaloid
erdnaher R. earth's near space environment, space near the earth
erschütterungsfreier R. shock-free room
euklidischer R. *s.* ebener R.
extraterrestrischer R. outer space
feldfreier R. field-free space
freier R. clear space
freiflutender R. flooding space
Gaußscher R. paraxial region
intergalaktischer R. intergalactic space
interplanetarischer R. interplanetary space
interstellarer R. interstellar space
lichter R. *(build)* clearance
linearer R. linear space
luftleerer R. vacuum
luftverdünnter R. rarefied air space
mehrdimensionaler R. multidimensional space
schädlicher R. clearance *(cylinder)*
schalltoter R. soundproof chamber
schwerefreier R. gravity-free space
toter R. dead space

umbauter R. enclosed space
unbegrenzter R. unlimited space
Raumakustik *f* room acoustics
Raumänderung *f* volume change
Raumanzug *m* space suit
Räumarbeit *f* 1. *(build)* clearing job; 2. *(mach)* broaching operation
R. in zwei Durchgängen *(mach)* two-pass broaching operation
raumartig space-like
Raumauflösung *f* spatial resolution (resolving power)
Räumaufspannung *f (mach)* broaching setup
Raumausdehnungszahl *f* thermal coefficient of cubical expansion
Raumausnutzungsfaktor *m* space factor
Raumbedarf *m* space requirement; *(mach)* floor space occupied; *(ship)* stowage factor
Raumbeleuchtung *f* room lighting
Raumbeschränkung *f* space limitation
raumbeständig low-volume-change
r. sein to retain a constant volume
Raumbild *n* stereogram, stereograph, stereoscopic picture (image)
Raumbildentfernungsmesser *m* stereoscopic range-finder
Raumdehnung *f* cubical (volume) strain
Raumdiagramm *n* cylinder (piston position time, volume) diagram
Raumdichte *f* **der elektrischen Ladung** density of volume charge
Räumdorn *m (mach)* push broach
Raumdosis *f s.* Volumendosis
Raumdrehung *f* space rotation
Räumdurchgang *m (mach)* broaching pass
Räume *mpl* / **vermessene** *(ship)* included spaces
Raumeinheit *f* unit volume
räumen 1. *(build)* to clear; 2. *(mach)* to broach; 3. to deplete
Räumen *n* / **ununterbrochenes** *(mach)* continuous broaching
Räumer *m (mine)* clearing (scavenging) plough, reamer
statischer R. clearing (static) plough *(coal cutter)*
Raumfachwerk *n* space frame (truss, framework)
Raumfahrer *m* astronaut, space traveller
Raumfahrt *f* interstellar flight, space travel (flight, flying), interplanetary aviation
bemannte R. manned space flight
interstellare R. interstellar flight
Räumfahrt *f (mine)* cleaning-up run
Raumfahrtaufgaben *fpl* space missions
Raumfahrtelektronik *f* space electronics
Raumfahrtfernmessung *f* aerospace telemetry
Raumfahrtmedizin *f* space medicine
Raumfahrtphysiologie *f* space-flight physiology
Raumfahrttechnik *f* astronautics, space engineering
Raumfahrtzwecke *mpl* space applications
Raumfahrzeug *n* space ship (vehicle, craft)
raumfest space-fixed
Raumfeuchtigkeit *f* room humidity, RH
Raumflug *m* space flight (travel) (*s. a.* Raumfahrt)
bemannter R. manned space flight
Raumformfräsen *n* cavity milling
Raumforscher *m* space explorer
Raumforschung *f* space research (exploration)
Raumfracht *f* measurement cargo
Raumfrequenz *f* space frequency
Raumgehalt *m* capacity; tonnage
Räumgeschwindigkeit *f (mach)* broaching speed
Raumgestaltung *f* interior decoration
raumgetreu stereophonic *(sound)*
Raumgewicht *n* weight per unit volume, volumetric weight, [bulk] density
Raumgitter *n* space lattice *(crystal)*
hexagonales R. hexagonal space lattice
kubisch-flächenzentriertes R. face-centred cubic lattice
kubisch-raumzentriertes R. body-centred cubic lattice
rhomboedrisches R. rhombohedral space lattice
tetragonales R. tetragonal space lattice
raumgroß room-sized
Raumgruppe *f* space-group *(crystal)*
Raumheizung *f* space heating
Räumhub *m (mach)* broaching stroke
Rauminhalt *m* volume; cubic[al] contents; capacity
R. der Verdrängung *(ship)* volume of displacement
Raumintegral *n* volume integral
Raumkabine *f* space capsula
Raumkapsel *f* space capsula
bemannte R. manned space capsula
Räumklappe *f* non-return flap
Raumkoordinate *f* space (spatial) coordinate
Raumkraft *f* body force
Räumkraft *f* broaching thrust (capacity), capacity of broaching machine
R. in Tonnen tonnage of the broaching machine
Raumkurve *f* space curve
Raumladestrom *m* space charge current
Raumladung *f* 1. spatial (space) charge; 2. *(ship)* inboard cargo
Raumladungsabstoßung *f* space-charge repulsion
Raumladungsbegrenzung *f* space-charge limiting
Raumladungsdichte *f* space-charge density, density of volume charge
Raumladungsdiode *f* space-charge diode
Raumladungseffekt *m* space-charge effect
Raumladungsgebiet *n* space-charge region

Raumladungsgitter *n* space-charge grid
Raumladungsgrenzschicht *f* space-charge boundary
Raumladungsimpuls *m* cloud pulse
Raumladungskapazität *f* space-charge capacitance
Raumladungskennlinie *f* three-halves-power characteristic
Raumladungskonstante *f* perveance
Raumladungskraft *f* space-charge force
Raumladungsmodulation *f* space-charge modulation
Raumladungsneutralität *f* space-charge neutrality
Raumladungspolarisation *f* space-charge polarization
Raumladungsschicht *f* space-charge layer (region), depletion layer of barrier
Raumladungssperrschicht *f* space-charge barrier layer
Raumladungsstrom *m* space-charge current
Raumladungswelle *f* space-charge wave
Raumladungswirkung *f* space-charge effect
Raumladungszone *f* space-charge region, depletion [layer] region
Raumladungszustand *m* space-charge-limited-current state
Räumlänge *f (mach)* length of broaching cut
Raumleiter *f (ship)* pillar (hold) ladder
räumlich spatial; three-dimensional; triaxial
Raumlicht *n* ambient light
Raummangel *m* shortness of space (room), limitation (lack) of space
Raummangels / infolge on account of space limitation
Räummaschine *f* broaching machine
Raummaß *n* solid measure, measure (unit) of capacity, standard of volume
Raummeter *n* cubic meter *(of staked wood)*
Raummodell *n* space (spatial) model
Raummodul *m* bulk modulus of elasticity
raumnachformen to duplicate (reproduce) in three dimensions
Raumnachformen *n* three-dimensional reproduction (duplication)
Raumnachformfräseinrichtung *f* three-dimensional milling device
Raumnachformfräsen *n* die-sinking, three-dimensional copy-milling
Räumnadel *f (mach)* [internal] broach
  glättende R. burnishing broach
  verdichtende R. burnishing broach
Räumnadelhalter *m* broach holder
Räumnadelziehkopf *m* broach pull head
Räumpflug *m (mine)* pan, gathering plough
Räumpresse *f (mach)* broaching press
Raumquantelung *f* directional (space) quantization
Raumraster *m* [modular] space grid
Raumrecht *n* space (cosmic) law
Raumschiff *n* space vehicle (ship), spacecraft
Raumschiffahrt *f* space nautics
Räumschild *m* dozer blade
Räumschlitten *m (mach)* broach tool slide (ram)
Raumsonde *f* space probe
Raumspant *n (ship)* hold frame
raumsparend space-saving
Raumspiegelung *f* space reflection
Raumstation *f* space station
Räumsteg *m (mine)* flight
Räumstellung *f (mach)* broaching position
Räumstößel *m (mach)* broach ram
Raumstrahlung *f* radiation background
Raumstütze *f (ship)* hold stanchion (pillar)
Räumte *f* [hold] capacity *(of a vessel)*
Raumteile *mpl* quantities by volume
Raumtemperatur *f* room (inside) temperature
Raumtiefe *f (ship)* registered depth, depth of hold
Raumton *m* stereophonic sound
Raumtoneffekt *m* binaural effect
Raumtonne *f (ship)* ton measurement
Raumüberwachung *f* area monitoring
Raumüberwachungsgerät *n* area monitor
Räum- und Glättwerkzeug *n* combination broach-burnisher tool
Räum- und Sammelpresse *f* combined hay (crop) crusher and lifter
Räum- und Zentriermaschine *f* broach and centre machine
Räumung *f (build)* clearing, clearance; evacuation; removal
Räumungsarbeiten *fpl* clearing
Raumvektor *m* space vector
Räumverfahren *n (mach)* broaching process
Raumverteilung *f* space (spatial, directional) distribution
Räumvorrichtung *f (mach)* broaching fixture
Raumweger *m (ship)* hold ceiling
Raumwegerung *f (ship)* hold ceiling
Raumwelle *f* free [space] wave, indirect (sky) wave
Räumwerkzeug *n (mach)* broach[ing tool]
  R. mit kreisförmigem Querschnitt circular broach
  mehrteiliges (zusammengebautes) R. sectional broach
Räumwerkzeugschärfmaschine *f* broach sharpener
Räumwerkzeugträger *m* broach carrier
Räumwerkzeugzubringer *m* broach-handling mechanism
Räumwerkzeugzubringeschlitten *m* broach-handling slide
Raumwinkel *m* solid angle
Raumwinkeleinheit *f* unit solid angle

Räumzahn *m* raker tooth; broach cutting tooth
Raumzeigerdiagramm *n* space vector diagram
Raumzeit *f* space-time
Raum-Zeit-Vorstellung *f* space-time conception
Raumzelle *f* three-dimensional element (block), building box
Raumzellenbauweise *f* box construction
raumzentriert body-centred, space-centred *(crystals)*
Räumzeug *n* surface (external) broach
Räumzeugwerkstoff *m* broach material
Räumzeugzubringeeinrichtung *f* broach-retriever mechanism
Räumzugabe *f* broaching allowance
Raupe *f* caterpillar, crawler tread (track)
Raupenfahrgestell / auf crawler-mounted
Raupenfahrwerk *n* caterpillar truck
  R. / auf crawler-mounted
Raupenfahrzeug *n* tracked (track-type) vehicle
Raupenkette *f* crawler tread (track)
Raupenkran *m* caterpillar (crawler-type) crane
Raupenrad *n* track sprocket
Raupenschlepper *m* crawler (caterpillar) tractor
Raupenschlepprad *n* crawler-tractor truck wheel
Rauschabstand *m (el)* signal-[to-]noise ratio
Rauschabstimmung *f* noise tuning
Rauschanpassung *f* noise matching
Rauschanteil *m* noise component
Rauschanzeige *f* noise indication
Rauschbandbreite *f* noise bandwidth
Rauschbezugstemperatur *f* noise-standard temperature
Rauschbild *n* noise pattern
Rauschdiode *f* noise diode
Rauschdiodentechnik *f* noise diode technique
Rauscheigenschaft *f* noise property
Rauschen *n* [background] noise; random noise; boiling noise *(electron tube)*; hiss *(microphone)*
  R. bei kurzgeschlossenem Eingang short-circuit noise
  R. bei offenem Eingang open-circuit noise
  R. des ersten Schwingkreises first-circuit noise
  elektrisches R. electrical noise
  erratisches R. random noise
  galaktisches R. galactic noise
  Gaußsches R. Gaussian noise
  magnetisches R. magnetic noise
  statistisches R. fluctuation (random) noise
  thermisches R. thermal noise
  thermisches R. im elektrischen Kreis circuit noise
  unterbrochenes R. impulsive (intermittent) noise
  weißes R. white noise
  weißes R. mit Gaußscher Verteilung white (Gaussian) noise
rauschend noisy
Rauschersatzwiderstand *m* equivalent noise resistance
Rauschfaktor *m* noise factor (figure)
  effektiver R. spot noise figure
  maximaler R. maximum noise figure
Rauschfeldstärke *f* radio noise field-intensity
Rauschgelb *n s.* Auripigment
Rauschgift *n* narcotic, drug, *(sl)* dope
Rauschgiftdezernat *n* narcotics (dope) squad
Rauschgold *n* tinsel, Dutch metal
Rauschgüte *f* noise figure
Rauschleistung *f* noise power (output) *(aerials)*
Rauschleistungsspektrum *n* noise power spectrum
Rauschmechanismus *m* noise mechanism
Rauschminimum *n* noise minimum
rauschmoduliert noise-modulated
Rauschpegel *m* noise level
Rauschpegeleinregelung *f (telev)* shading
Rauschröhre *f* noise tube
Rauschschwelle *f* noise threshold
Rauschsender *m* noise generator
Rauschsignal *n* noise signal
Rauschspannung *f* noise voltage
Rauschstrom *m* noise current
Rauschsummer *m* standard noise generator
Rauschtemperatur *f* noise temperature
Rauschthermometer *n* noise thermometer
Rauschuntergrund *m* noise background
Rauschverhalten *n* noise behaviour
Rauschvierpol *m* noise twoport
Rauschwiderstand *m* [equivalent] noise-resistance
Rauschzahl *f* noise figure (factor)
Raute *f* lozenge, rhombus
Rautenantenne *f* rhombic aerial
Rayleigh-Destillation *f* Rayleigh distillation
Rayleigh-Scheibe *f* Rayleigh disk
Razemat *n* racemate
Razemie *f* racemism
razemisch racemic
razemisieren to racemize
Razemisierung *f* racem[iz]ation
RBW-Dosis *f* RBE dose
RC-Generator *m* resistance-capacitance oscillator, resistance-tuned oscillator
RC-Konstante *f* RC-constant
RC-Kopplung *f* RC coupling
RC-Netzwerk *n* RC network
RC-Verstärker *m* RC amplifier, resistance-capacitance coupled amplifier
Reagens *n* reagent, [chemical] agent
  Neßlers R. Nessler's reagent (solution)
Reagenzflasche *f* reagent bottle
Reagenzglas *n* test tube
Reagenzglasgestell *n* test tube stand
Reagenzglashalter *m* test tube holder
Reagenzpapier *n* test paper

reagieren *(ch)* to react; to respond *(instrument)*
reagierend reactive
Reaktanz *f* reactance
  induktive R. inductive reactance
  kapazitive R. capacitive reactance
Reaktanzglied *n* reactor
Reaktanzrelais *n* reactance relay
Reaktanzröhre *f* reactance tube
Reaktanzsatz *m* reactance theorem
Reaktanzschaltung *f* reactance circuit
Reaktanzschutz *m* reactance protection
Reaktanzspannung *f* reactance voltage
Reaktanzspule *f* [limiting] reactor
Reaktanztransformator *m* reactance transformer
Reaktion *f (ch; ph)* reaction; retroaction; response; *(math)* reverse action
  R. dritter Ordnung third-order reaction
  R. erster Ordnung reaction of the first order
  R. zweiter Ordnung second-order reaction
  basische R. alkaline (basic) reaction
  beendete R. spent reaction
  bimolekulare R. bimolecular reaction
  Cannizzarosche R. Cannizzaro reaction
  chemische R. chemical reaction
  endotherme R. endothermal reaction
  exotherme R. exothermic reaction
  fotonukleare R. photonuclear reaction, [nuclear] photodisintegration
  Friedel-Craftssche R. Friedel and Crafts' reaction
  gekoppelte R. linked reaction
  reversible R. balanced reaction
  saure R. acid reaction
  thermonukleare R. thermonuclear reaction, [nuclear] fusion, fusion reaction
  unvollständig verlaufende R. incomplete reaction
Reaktionsarbeit *f* work of reaction
Reaktionsausbeute *f* reaction yield
Reaktionsbedingung *f* reaction condition
Reaktionsbeschleuniger *m* catalyst, catalyzer, catalytic agent, accelerator
Reaktionsdampfer *m* reaction steamer
Reaktionsdruck *m* reaction pressure
Reaktionsenergie *f* reaction energy
reaktionsfähig reactive; responsive
Reaktionsfähigkeit *f* [re]activity; responsiveness, responsivity
Reaktionsgefäß *n* reactor, [reaction] apparatus
Reaktionsgemisch *n* reaction mixture
Reaktionsgenerator *m* reluctance generator
Reaktionsgeschwindigkeit *f* 1. *(ch)* reaction velocity (rate); 2. speed of response *(instrument)*
Reaktionsgleichung *f* reaction equation
Reaktionsgrad *m* degree of reaction
Reaktionskanal *m* reaction channel (species)
Reaktionskette *f* reaction chain
Reaktionskinetik *f* reaction kinetics
Reaktionskonstante *f* reaction constant
Reaktionskraft *f* reaction (reactive) force
Reaktionsmechanismus *m* reaction mechanism
  bimolekularer R. bimolecular law
Reaktionsmedium *n* reaction medium
Reaktionsordnung *f* order of reaction, reaction order
Reaktionspartner *m* co-reactant, reacting agent
Reaktionsprinzip *n* reaction principle *(turbine)*
Reaktionsprodukt *n* reaction product
Reaktionspropeller *m* hydraulic (reaction) propeller
Reaktionsquerschnitt *m* reaction cross section
Reaktionsrad *n* reaction wheel *(turbine)*
Reaktionsrichtung *f* reaction direction
Reaktionsschiff *n* reaction vessel
Reaktionsschmelzen *n* smelting
Reaktionsschwelle *f* threshold of the reaction, reaction threshold
Reaktionsstufe *f* reaction step
Reaktionstechnik *f* reaction technique
Reaktionsteilnehmer *m* reactant
reaktionsträge inert, less active, passive
Reaktionstriebwerk *n* reaction propulsion engine
Reaktionsturbine *f* reaction turbine
Reaktionsvermögen *n* reagency, reactive power; response, respondence
Reaktionswahrscheinlichkeit *f* reaction probability
Reaktionswärme *f* heat of reaction
Reaktionsweg *m s.* Reaktionskanal
Reaktionswirkungsquerschnitt *m s.* Reaktionsquerschnitt
Reaktionszeit *f* reaction time
Reaktivfarbstoff *m* reactive dye
reaktivieren to reactivate
Reaktivierung *f* reactivation
Reaktivität *f* reactivity
  verzögerte R. delayed reactivity
Reaktivitätskoeffizient *m* reactivity coefficient
Reaktivitätsmesser *m* reactivity meter
Reaktivitätsrechner *m* reactivity computer
Reaktivitätssteuerung *f* reactivity control
Reaktivitätsüberschuß *m* reactivity excess
Reaktivitätsvorrat *m* reactivity excess
Reaktor *m* reactor, pile; reaction apparatus
  R. mit Reflektor reflected reactor
  R. ohne Reflektor naked (bare) reactor, naked (bare) pile
  beweglicher R. mobile reactor (chain-reacting unit); transportable reactor
  druckwassergekühlter R. pressurized water reactor
  epithermischer R. epithermal reactor
  gasgekühlter R. gas-cooled reactor
  graphitmoderierter R. carbon (graphite) reactor, carbon (graphite) pile

heterogener R. heterogeneous reactor
homogener R. homogeneous reactor
intermediärer R. intermediate reactor
jungfräulicher R. cold reactor
kalter R. cold reactor
keramischer R. ceramic reactor
kritischer R. critical reactor, chain-reacting pile
langsamer R. slow (thermal) reactor
luftgekühlter R. air-cooled reactor
mittelschneller R. intermediate [spectrum] reactor
nackter R. *s.* unreflektierter R.
natriumgekühlter R. sodium-cooled reactor
nichtbrütender R. non-breeder [reactor]
nichtkritischer R. non-critical reactor (pile)
organisch moderierter R. organic moderated reactor, OMR
ortsbeweglicher R. transportable reactor
ortsfester R. stationary (non-mobile) reactor
prompt-kritischer R. prompt-critical reactor
quasihomogener R. quasi-homogeneous reactor
reflektierter R. reflected reactor
schneller R. fast[-neutron] reactor
selbstregelnder R. self-regulating reactor
stationärer R. stationary (non-mobile) reactor
thermischer R. thermal reactor
thermonuklearer R. thermonuclear reactor
transportabler R. [trans]portable reactor; mobile reactor (chain-reacting unit); package reactor
überkritischer R. supercritical reactor
unendlich ausgedehnter R. infinite reactor
unreflektierter R. naked (bare) reactor, naked (bare) pile
unterkritischer R. subcritical reactor
virtueller R. image reactor
wassergekühlter R. water-cooled reactor
wassermoderierter R. water-moderated reactor (pile)
wäßrig-homogener R. aqueous homogeneous reactor
zylindrischer R. [right circular] cylindrical reactor
Reaktorabschirmung *f* reactor shielding
Reaktoranlage *f* reactor plant
Reaktorantrieb *m* nuclear propulsion
Reaktorausladung *f* unloading of the reactor
Reaktorbau *m* reactor engineering, [nuclear] reactor art
Reaktorbehälter *m s.* Reaktorgefäß
Reaktorbeladung *f* reactor loading; reactor feed (charge), fuel charge
R. mit Betriebsunterbrechung *s.* Kaltbeladung
R. ohne Betriebsunterbrechung *s.* Heißbeladung
Reaktorbeschickung *f s.* Reaktorbeladung
Reaktorbetrieb *m* reactor process (operation)
Reaktordruckgefäß *n* reactor vessel (tank)
Reaktordurchgang *m* runaway of the reactor
Reaktorfahrer *m* operator [of reactor]
Reaktorgefäß *n* reactor vessel (tank)
Reaktorgeometrie *f* reactor (pile) geometry
Reaktorgift *n* reactor (fission) poison
Reaktorgitter *n* reactor lattice
Reaktorgleichung *f* reactor equation
Reaktorinstrumentierung *f* instrumentation of the nuclear reactor, reactor instrumentation
Reaktorkanal *m* reactor channel, hole in the pile
Reaktorkuppel *f* reactor sphere
Reaktormantel *m* reactor shell (envelope)
Reaktorneutron *n* reactor (pile) neutron
Reaktorneutronenspektrum *n* pile spectrum
Reaktoroszillator *m* pile (reactor) oscillator
Reaktorperiode *f* reactor period, period [of the reactor]
Reaktorphysik *f* reactor physics
Reaktorquerschnitt *m* pile-spectrum mean cross section
Reaktorregelung *f* reactor control
Reaktorreinheit *f* reactor purity
Reaktorschiff *n* nuclear-propelled ship
Reaktorschild *m* reactor shield
Reaktorsicherheitstest *m* reactor safety test
Reaktorsimulation *f* reactor simulation
Reaktorsimulator *m* reactor simulator
Reaktorsteuerpult *n* reactor control board
Reaktorsteuerung *f* reactor control
Reaktorstrahlung *f* reactor (pile-emitted, reactor-emitted) radiation
Reaktortechnik *f* reactor (pile) technology, reactor engineering
Reaktortechniker *m* reactor (pile) technologist
Reaktortheorie *f* reactor theory
Reaktorwerkstoff *m* reactor material
Realgar *m* realgar
Realkristall *m* real crystal
Realteil *m (math)* real part
Realzeit *f* real time
Realzeitrechner *m* real time computer
Reaumurskale *f* Réaumur scale
Rebbelmaschine *f* maize sheller
reb[b]eln to shell maize
Rebenspritze *f* vineyard sprayer
Receiverdampfmaschine *f* steam engine with receiver
Rechen *m (ag)* rake; cradle; *(build)* screen
Rechenanlage *f* computing machinery, computer
elektronische R. electronic computer
programmgesteuerte R. programme-controlled computer
speicherprogrammierte R. stored programme computer
Rechenauslage *f* flyer (shoe-fly) delivery

Rechenautomat *m* computer, computing machine
digitaler R. digital computer
elektronischer R. electronic computer
programmgesteuerter R. programme controlled computer
Rechenautomatenbetrieb *m* computer operation
Rechenbalken *m* electromechanical computing relay
Rechenblatt *n* nomogram, alignment chart
Rechenfehler *m* computing error
Rechengang *m* mathematical procedure
Rechengerät *n* computer
R. mit Kartenbilddarstellung pictorial computer
elektronisches R. electronic computer
Rechengeschwindigkeit *f* calculating speed
Rechenglied *n* computing relay, computer component
Rechengröße *f* operand
Rechengrößenadresse *f* operand address
Rechenheuwender *m (ag)* swath turner
Rechenklassierer *m* rake classifier
Rechenkode *m* computer code
Rechenkorb *m (ag)* basket
Rechenkunst *f* arithmetic
Rechenleistung *f* calculating power
Rechenlocher *m* calculating punch[er]
elektronischer R. electronic calculating punch[er]
Rechenmaschine *f* calculator, computing machine
druckende R. listing calculator
elektronische R. electronic computer
Rechennetzwerk *n* computing network
Rechenoperation *f* arithmetic operation
Rechenprobe *f* numerical check
Rechenradwender *m* combined side-delivery rake, swath turner and tedder
Rechenregel *f* algorithm
Rechenregler *m* computing controller
Rechenrelais *n* operational relay
Rechenschaltung *f* computing circuit (network)
Rechenschema *n* [flow] diagram
Rechenschieber *m* slide rule
Rechenschwadenwender *m* combination side-delivery rake and swath turner
Rechenspeicher *m* computer memory [store], computer store
Rechenstab *m* slide rule
Rechenstelle *f* computing centre
Rechentafel *f* computation table, alignment chart, nomogram
Rechentechnik *f* computer technique; computer engineering
maschinelle R. computation
Rechenverstärker *m* operational amplifier
Rechenvorschrift *f* algorithm
Rechenwerk *n* arithmetic element (unit) *(computer)*; calculating register
Rechenzentrum *n* computing (computation) centre
rechnen to reckon, to calculate, to compute; to estimate *(cost)*
das Besteck r. *(ship)* to calculate the position
elektronisch r. to compute
Rechnen *n* arithmetic; computing
binäres R. binary arithmetic
Rechner *m* calculator; computer, computing machine
R. zur Auflösung linearer Gleichungssysteme equation solver
bionischer R. bionic computer
elektronischer R. electronic computer
hydraulischer R. hydraulic computer
pneumatischer R. pneumatic computer
programmgesteuerter R. programme-controlled computer
Rechneranlage *f* computer system
Rechnerbaustein *m* computer assembly
Rechnerbetrieb *m* computer operation
Rechnerdiode *f* computer diode
Rechnergleichung *f* machine equation
rechnerisch computational, by calculation, by computation
Rechnerlogik *f* computer logic
Rechnermagnetbandgerät *n* computer tape unit
Rechnernetzgerät *n* computer power pack
Rechnerperiode *f* [computing] machine cycle
Rechnerschaltung *f* / **logische** computer logic circuit
Rechnersprache *f* machine language
Rechnervariable *f* computer (machine) variable
Rechnerwirkzeit *f* available-machine time
Rechnerwort *n* machine word
Rechnung *f* 1. computation, calculation; 2. bill, invoice
Rechnungsabschluß *m* balancing of accounts
Rechnungsart *f* operation; algorithm
Rechnungsausstellung *f* invoice writing
Rechnungsdienst *m* accounting department
Rechnungsgang *m* calculating procedure, method of calculation
Rechnungsjahr *n* financial year
Rechnungsprüfer *m* accountant
Rechnungsprüfungskarte *f* audit card
Rechnungsschreibung *f* billing
Rechnungsstelle *f* account section
Rechnungsverfahren *n* **mit fortschreitenden Näherungswerten** trial-and-error method
Rechnungsvordruck *m* invoice form
Rechnungsvorschub *m* bill-feed device
Rechnungswesen *n* / **betriebliches** cost accounting
Rechteck *n* rectangle
Rechteckbalken *m* rectangular beam

Rechteckdiagramm *n* bar chart
Rechteckfeder *f* rectangular plate spring
Rechteckferrit *m* rectangular ferrite
Rechteckflügel *m* rectangular wing
Rechteckform *f* rectangular shape
Rechteckformung *f* squaring
Rechteckgenerator *m* rectangular-wave generator, square-wave generator
Rechteckhohlleiter *m* rectangular waveguide
rechteckig rectangular
Rechteckigkeit *f* squareness *(pulse)*
Rechteckigkeitsverhältnis *n* squareness ratio *(magnetic core)*
Rechteckimpuls *m* square[-sided] pulse, rectangular (square-wave) pulse
Rechteckkorn *n* square sight *(rifle)*
Rechteckkreis *m* squaring circuit
Rechteckleiter *m* rectangular (square-section) conductor
Rechteckquerschnitt *m* rectangular section
Rechteckspannung *f* square-wave voltage
Rechteckstromimpuls *m* square-wave current pulse
Rechteckwelle *f* rectangular (square) wave
negativgehende R. negative going square wave
steil ansteigende R. sharp-attack square wave
Rechteckwellenform *f* rectangular (square) waveform
Rechteckwellenfrequenzgang *m* square-wave response characteristic
Rechteckwellenimpuls *m* rectangular (square-wave) pulse
Rechteckwellenmodulation *f* square-wave modulation
Rechteckwellenwechselrichter *m* square-wave inverter
Rechtehandregel *f* right hand rule, Fleming's second rule
Rechtflach *n* rectangular parallelepiped
rechts / nach to[wards] the right
Rechtsanschlag *m* maximum clockwise-rotation *(potentiometer)*; fully clockwise position
Rechtsanspruch *m* legal claim (title)
Rechtsanwaltgebühren *fpl* lawyer's fees
Rechtsbefugnis *f* competence
Rechtsbeistand *m* legal assistance
Rechtsberater *m* judicial adviser, lawyer
Rechtsberatung *f* legal advice
Rechtsdrall *m* right-hand helix twist, right-handed helix
R. / mit right-hand helical
rechtsdrehend *(mach)* clockwise; top-coming *(milling-cutter)*; dextrorotatory, dextrogyrated, dextrogyric
Rechtsdrehung *f* clockwise (right-hand) rotation, rotation to the right; dextrorotation
rechtsgängig *(mach)* right-handed; dextrorse
Rechtsgewinde *n* right-hand thread
rechtsgewunden dextrorse
Rechtsgratköper *m (text)* right-hand twill
rechtsgültig legal, valid
Rechtsgültigkeit *f* legality
Rechtsirrtum *m* mistake of law
rechtskräftig legally binding
Rechtskrümmung *f* clockwise curvature
Rechtslauf *m* clockwise rotation
Rechtsmasche *f* plain stitch *(knitting)*
Rechtsmittel *n* legal redress
Rechtsnachfolge *f* succession in title
Rechtspropeller *m* right-hand propeller
Rechtsschleife *f* dextrorotatory loop
rechtsschneidend *(mach)* right-cut, right-hand cut[ting]
Rechtsschutz *m* legal protection
Rechtsspirale *f* right-handed helix
Rechtssystem *n* right-hand vector system, right-handed system of coordinates
rechtsunwirksam legally inoperative, of no legal effect
Rechtsweg *m* legal procedure
rechtswendig dextrorse, right-handed
Rechtswicklung *f (el)* right-hand winding
rechtswirksam legally operative
rechtweisend true [heading]
rechtwinklig right [angular], right-angle[d], rectangular
r. beschneiden to square up
r. zueinander at right angles to each other, square with each other
Rechtwinkligkeit *f* squareness, perpendicularity, right-angularity, rectangularity
Reck *n* horizontal bar
Reckalterung *f* strain ag[e]ing, strain age hardening
reckalterungsbeständig non-strain-ag[e]ing, resistant to strain ag[e]ing
Reckbelastung *f* tensile stress (load)
recken to stretch *(wire, plastic)*; to lengthen, to extend, to elongate; to draw out *(forge)*
Recken *n* **des Drahtes** removing of kinks, stretching of the wire
Reckfähigkeit *f (plast)* stretchability
Reckspannung *f* tensile stress
Reckumformen *n* drawing
Reckwalzwerk *n* merchant roll
redigieren to edit
Redox-Brennstoffelement *n* redox fuel-cell
Redox-Potential *n* redox (oxidation-reduction) potential
Reduktase *f* reductase
Reduktion *f* reduction
Reduktionsätze *f* reduction discharge
Reduktionsätzmittel *n* reduction discharge agent
Reduktionsbedingung *f* reducing condition

Reduktionsentfernungsmesser *m* reducing telemeter
Reduktionsfähigkeit *f* reducibility
Reduktionsflamme *f* reduction flame
Reduktionsgetriebe *n* reduction (step-down) gear
Reduktionsmaßstab *m* reduction scale
Reduktionsmittel *n* reducing agent
Reduktionsofen *m* smelting furnace
Reduktionspunkt *m* reduction point
Reduktionsrolle *f* reducing pulley
Reduktionsschlacke *f* white (final) slag; first-run slag *(tin)*
Reduktionstachymeter *n* reducing tacheometer
Reduktionsvermögen *n* reducing power
Reduktionszone *f* reduction zone *(blast furnace)*
redundant redundant
Redundanz *f* redundancy
Redundanzprüfung *f* redundant (digit) check
Redundanztechnik *f* redundancy technique
reduzierbar reducible
Reduziereinsatz *m (mach)* collet
Reduziereinsatzhülse *f* **für Fräsmaschinen** milling machine collet
reduzieren to reduce, do diminish, to cut down
  auf ein Minimum r. to minimize
  das Potential r. *(el)* to drop the level
  stufenweise r. to reduce by steps
reduzierend reducing, reductive
Reduzierfassung *f* reducing (reduction) socket *(lamp)*
Reduziergerüst *n (met)* reducing roll stand
Reduzierkrümmer *m* reducing elbow *(pipe)*
Reduziermuffe *f (el)* reducing sleeve; reducing [pipe] joint, reducing [pipe] socket
Reduzierrohr *n* reducing (diminishing) pipe
Reduzierstation *f s.* Reduzierventil
Reduzierstück *n (mach)* adapter; reducer
Reduzierventil *n* [pressure] reducing valve
Reduzierwalzwerk *n* sinking mill
Redwood-Sekunde *f* Redwood second
Reede *f (ship)* road[stead]
  R. / auf in the roads
  geschlossene R. good road[stead], shut anchorage
  offene R. open road[stead], open anchorage
Reeder *m* [ship]owner
Reederei *f* shipping (navigation) company
Reedereiflagge *f* burgee, house flag
Reedereigewerbe *n* shipping trade
Reedrelais *n* reed relay
reell real *(numbers)*
Reemission *f* re-emission
Reetdach *n* thatched roof
reffen *(ship)* to reef
Refiner *m* disk mill *(papermaking)*
reflektieren to reflect
Reflektor *m* reflector; tamper; target *(radar)*
  gespeister R. fed reflector *(aerials)*
  passiver R. passive reflector
  strahlungsgekoppelter R. parasitic reflector
  ungespeister R. parasitic reflector
Reflektorelektrode *f* reflector electrode
Reflektoren *mpl* reflector fittings *(aerials)*
Reflektorersparnis *f* reflector savings
Reflektorgewinn *m* reflector savings
Reflektorlampe *f* reflector (reflecting) lamp
Reflektormaterial *n* reflecting material
Reflektorscheibe *f* reflecting disk
Reflektorspannung *f* reflector voltage
Reflektorstrahler *m* emission reflector
Reflektorwand *f* rear curtain
Reflektoskop *n* reflectoscope
Reflexblendung *f* reflected glare
Reflexempfang *m* reflex reception
Reflexempfänger *m* dual receiver
Reflexion *f* reflection
  R. am Erdboden ground reflection
  R. durch die Seeoberfläche sea returns *(radar)*
  diffuse R. diffuse reflection
  gerichtete R. regular reflection
  gestreute R. diffuse reflection
  regelmäßige R. direct (regular) reflection
  selektive R. selective reflection
  totale R. total reflection
  zerstreute R. diffuse reflection
Reflexionsansauggeräuschdämpfer *m (aut)* resonator intake silencer
Reflexionsdämpfung *f (el)* return loss
Reflexionsdämpfungsmesser *m (el)* return-loss measuring set, reflection measuring set
Reflexionsdickenmessung *f* thickness control (gauging) by backscattering
Reflexionseffekt *m* reflection effect
Reflexionselektrode *f* **im Klystron** repeller
Reflexionsfaktor *m* reflection (back-scattering, return-current) coefficient; albedo
Reflexionsfaktormesser *m* reflectometer
Reflexionsgesetz *n* law of reflection
Reflexionsgitter *n* reflection grating
Reflexionsglas *n* reflecting glass
Reflexionsgoniometer *n* reflecting goniometer
Reflexionsgrad *m* reflection coefficient (factor)
Reflexionskoeffizient *m* reflectance, reflection coefficient (factor); mismatch (transition) factor
Reflexionskörper *m* reflecting body
Reflexionspolariskop *n* reflection polariscope
Reflexionsschalldämpfer *m (aut)* resonant-absorption silencer
Reflexionsschicht *f* reflection layer
Reflexionstypfilter *n* reflection-mode filter *(waveguides)*
Reflexionsverlust *m* reflection loss
Reflexionsverlustkoeffizient *m* reflection (mismatch) factor

Reflexionsvermögen *n* reflectivity, reflectance, reflecting capacity (power), albedo
Reflexionswasserstandglas *n* gauge glass protector and reflector
Reflexionswinkel *m* angle of reflection, specular angle
reflexmindernd anti-reflection
Reflexschaltung *f* reflex circuit *(radio)*
Reflexverstärkung *f* dual amplification
Reflexvisier *n (mil)* reflex sight
Reformieren *n* reforming *(petrol)*
Reformingbenzin *n* reformed petrol (gasoline)
Refraktärstadium *n* refractory state
Refraktion *f* refraction *(radiation)*
Refraktometer *n* refractometer
Refraktometrie *f* refractometry
Refraktor *m* refractor, refracting telescope
Regal *n* rack
Regale *npl* shelving
Regalwagen *m* rack truck
Regatta *f* race, regatta
Regel *f* rule; principle
  Abeggsche R. Abegg's rule
  Avogadrosche R. Avogadro's hypothesis
  Cramersche R. Cramer's rule
  Dulong-Petitsche R. Dulong and Petit's law, law of Dulong and Petit
  Geiger-Nuttallsche R. Geiger-Nuttall relation (rule)
  Laportesche R. Laporte parity rule
  Lenzsche R. Lenz's law
Regelabweichung *f* [control] deviation; [output] error
  R. bei Sollwertänderung static error
  R. unter Störgrößeneinwirkung load static error
  bleibende R. steady-state error, offset, *(Am)* droop
  momentane R. deviation
  stationäre R. steady-state error
  vorübergehende R. transient error
Regelanlage *f* automatic control system; control assembly
Regelanlasser *m* starter (rotor) rheostat, rheostatic starter
Regelantrieb *m* drive control, controlled drive
Regelation *f (ph)* regelation
Regelband *n* control band
regelbar controllable
  in Stufen r. adjustable in steps
  stufenlos r. infinitely (steplessly) variable
Regelbarkeit *f* controllability
Regelbereich *m* control range (zone); range of adjustment
Regeldetri *f* rule of three
Regeldiode *f* regulator diode
Regeldrossel *f (el)* regulating (variable) inductor; variable-area nozzle
Regeldrosselklappe *f* control valve
Regeleinrichtung *f* automatic controller; control assembly (device)
Regelfaktor *m* error-ratio
  dynamischer R. deviation ratio
Regelfläche *f* control area
Regelgenauigkeit *f* control accuracy
Regelgerät *n* controller, control unit, governor
Regelgetriebe *n* variable-speed drive
Regelglied *n* attenuator, control element
Regelgröße *f* controlled condition (variable)
Regelkennlinie *f* control characteristic
Regelkondensator *m* variable capacitor
Regelkreis *m* regulating (regulation) circuit, automatic control loop (system)
  geschlossener R. closed control loop
  mehrschleifiger R. multiloop system
  vermaschter R. interconnected control system
Regelkreisanalysator *m* transfer function analyzer
regellos *(math)* random
regelmäßig regular; uniform
Regelmotor *m* control (change-speed) motor
regeln to control *(closed loop)*; to govern; t[o] adjust
Regelorgan *n* controller, control member
  R. für die Fördermenge delivery control member
Regelpotentiometer *n* control potentiometer
Regelpult *n* control panel
Regelpumpe *f* control (variable-delivery, controlled-volume) pump
Regelpunkt *m* control point
Regelquerschnitt *m (build)* typical cross-section
Regelrelais *n* regulating relay
Regelröhre *f* control (regulating) valve
  sechspolige R. fading hexode
Regelschalter *m* regulating switch, regulator
Regelschar *f (math)* regulus
Regelscheibe *f* regulating (control) wheel *(grinding)*
Regelscheibenabrichteinrichtung *f* regulating (control) wheel truing attachment
Regelschieber *m* control valve
Regelspannung *f* control voltage; bias *(voltage)*
  verzögerte R. delay bias
Regelspindel *f* regulating (control) wheel spindle *(grinding)*
Regelspindellagerung *f* regulating (control) wheel spindle bearing
Regelspur *f* standard gauge
Regelstab *m* control rod; regulating (fine control) rod *(reactor)*
Regelstange *f* control rack
Regelstrecke *f* controlled system (plant, process, load)
Regelstrom *m* controlling (correcting) current

Regelsystem *n* regulating (control) system
Regeltechnik *f s.* Regelungstechnik
Regeltransformator *m* adjustable (variable-ratio, regulating) transformer
Regelung *f* [closed-loop] automatic control; regulation; governing; control adjustment
R. der Bilddrehung picture-rotate control *(radar)*
R. der Kippfrequenz hold control
R. der Luftfeuchtigkeit humidity control
R. der Vertikalablenkung vertical hold control
R. des Zeilenkippgenerators horizontal hold control
R. des Zuges draught regulation
R. durch Änderung der Frequenz frequency control
R. durch Änderung der Spannung variable-voltage control
R. durch Dampfdrosselung governing by throttling the steam supply, throttle governing
R. durch Feldschwächung shunted-field control
R. durch Gruppierung der Motoren series-parallel control
R. durch Kaskadenschaltung cascade control
R. durch Polumschaltung pole-changing control
R. mit Ersatzregelgröße *(Am)* indirect control
R. mit Hilfsenergie indirect (power-operated) control
R. mit Hilfsregelgröße open-closed loop control
R. mit Störgrößenaufschaltung open-closed loop control
R. mit Zusatztransformator auxiliary-transformer regulation
R. ohne Hilfsenergie direct control
automatische R. automatic control
automatische R. des Empfanges automatic sensitivity control
differenzierend wirkende R. derivative control
digitale R. digital control
direkte R. direct control
elektronische R. electronic control
indirekte R. indirect (power-operated) control
kontinuierliche R. continuous adjustment
mechanische R. mechanical control
mittelbare R. pilot valve governing *(turbine)*
optimale R. optimal control
selbstanpassende R. self-adaptive control system, self-adjusting control system
selbsteinstellende R. self-adjusting control system, self-adaptive control system
selbstoptimierende R. self-adaptive control system, self-adjusting control system
selbsttätige R. automatic regulation (control)
stetige R. continuous control
stufenlose R. continuous control
stufenweise R. step-by-step regulation, stepped control
übergeordnete R. imposed (supplementary) regulation
unstetige R. discontinuous control
vermaschte R. correlated (interconnected) control system
Regelungsgeschwindigkeit *f* regulation (control) speed
Regelungsgesetz *n* control mode, mode of control
Regelungsingenieur *m* control engineer
Regelungssystem *n* [automatic] control system, closed-loop control system
Regelungstechnik *f* control technology (engineering); regulation (automatic control) technique
Regelventil *n* control valve
R. mit logarithmischer Kennlinie equal-percentage control valve
gleichprozentiges R. equal-percentage control valve
reversierbares R. reversible-globe control valve
thermostatisches R. expansion valve *(refrigeration system)*
Regelverstärker *m* variable-gain amplifier
Regelwiderstand *m* rheostat, variable (adjustable) resistor; regulating (adjustable) resistance
Regelwidrigkeit *f* abnormality
Regelzeit *f* acting time
Regelzelle *f* regulating cell
Regen *m* rain[fall]
Regenbekleidung *f* rain wear
Regenbogen *m* rainbow
regendicht rainproof, showerproof
Regendichtimprägnierung *f* rainproof (showerproof) impregnation
Regenecho *n* rainecho *(radar)*
Regenenttrübung *f* anticlutter rain *(radar)*
Regeneratfaser *f* regenerated fibre
Regeneratgummi *m* reclaim, reclaimed rubber
Regeneration *f* regeneration
R. der Ablauge recovery of the digestion liquor
Regenerationsofen *m* regenerative furnace
Regenerationsvermögen *n* recovery power
regenerativ regenerative
Regenerativgasfeuerung *f* regenerative gas firing
Regenerativkammer *f* regenerative chamber
Regenerativofen *m* regenerative furnace
Regenerator *m* regenerator
regenerieren to regenerate, to recover
Regenerierstrahl *m* holding beam *(memory)*
Regenerierung *f* regeneration, recovery

R. von Altöl filtering of used oil
Regenerierungsanlage *f* recovery plant
Regenerierungsgeschwindigkeit *f* speed of recovery
Regenerierungskessel *m* regeneration boiler
Regenfallrohr *n* rainwater pipe
Regenkappe *f* rain cap (shield) *(chimney)*
Regenleiste *f (ship)* watershed; eye brow; *(aut)* weatherstrip
Regenmesser *m* rain gauge, pluviometer, ombrometer, hyetometer
Regenprüfung *f* wet (artificial rain) test
Regenrinne *f (build)* rainwater gutter[ing], gutter channel; *(aut)* drip moulding
rechteckige R. box gutter
Regenschreiber *m* rainfall recorder, pluviograph
Regenüberschlag *m (el)* wet arcing (flashover)
Regenüberschlagsspannung *f* wet arc-over voltage
Regenwasser *n* rainwater
Regieassistent *m* assistant director
Regiegeräte *npl* control equipment
Regiepult *n* control desk
Regieraum *m* central control room
Regierungsfahrzeug *n* government vessel
Regiesignal *n (telev)* cue
Regietechnik *f* production technique
Regietisch *m* control desk
Regisseur *m* director
Register *n* 1. register *(for storing of machine words)*; index *(book)*; 2. director; 3. register, stop *(organ)*
Registerhafen *m* port of registry; *(Am)* home port
Registermaße *npl (ship)* tonnage measures
Registermontagetisch *m* line-up and register table
Registerpapier *n* ledger (recording) paper
Registerpartie *f* wire-cloth rollers
Registerschelle *f* conduit rack
Registerschneidemaschine *f* index cutting machine
Registertonne *f (ship)* register ton
Registertonnengehalt *m* register tonnage
Registrieradaptometer *n* recording adaptometer
Registrierapparat *m* recorder
Registrierballon *m* sounding (test, recording) balloon
Registrierbuchungsautomat *m* automatic listing and book-keeping machine
Registrierbürette *f* / **automatische** automatic recording burette
registrieren to record, to register; to log
Registrierfotometer *n* recording photometer
Registriergerät *n* recorder, recording instrument
Registriergeschwindigkeit *f* recording speed
Registrierhafen *m* port of registry; *(Am)* home port
Registrierinstrument *n* chart-recording instrument
Registrierkasse *f* / **elektrische** electric cash-register
Registriermanometer *n* recording pressure-gauge
Registrierpapier *n* recording chart
Registrierpyrometer *n* recording pyrometer
Registrierstreifen *m* recorder tape (strip, band), strip chart
Registriertechnik *f* recording technique
Registrierteil *n* recording element
Registriertrommel *f* recording drum
Registrierung *f* recording; plotting; registration; filing
R. von Prozeßdaten data logging
Registrierverfahren *n* filing method
Registriervoltmeter *n* recording voltmeter
Registrierwerk *n* recording clockwork (unit)
Registrierzähler *m* recording meter
Regler *m* [automatic] controller, controlling unit; regulator *(switch, dynamo)*; governor *(speed control of engines)*
R. mit breitem P-Bereich wide-band controller
R. mit gekreuzten Stangen governor with crossed arms
R. mit Kreuzschleife governor with fixed centre of oscillation
R. mit Meßinstrument indicating controller
R. mit rhombischer Aufhängung governor with rhombic suspension
R. mit schmalem P-Bereich narrow-band controller
R. mit Schubkurbelgetriebe governor with open links
R. mit Sprungschaltung snap-acting controller
R. mit umgekehrter Pendelaufhängung governor with inverted suspension
R. ohne Hilfsenergie self-actuated controller
anzeigender R. indicating controller
astatischer R. astatic governor
automatischer R. automatic controller
direktwirkender R. self-actuated controller
elektropneumatischer R. electro-pneumatic controller
fliegend gelagerter R. overhung governor
geführter R. submaster (secondary) controller *(cascade control system)*
halbautomatischer R. semi-automatic controller
in der Anlage montierter R. field controller
indirekter R. rheostatic regulator
integral wirkender R. integral action controller
labiler R. unstable governor
pseudoastatischer R. pseudo-astatic governor

schreibender R. recorder controller
selbsttätiger R. automatic controller
selbsttätiger R. des Rauchschiebers automatic damper regulator
stabiler R. stable governor
statischer R. static governor
stetiger R. continuous controller
unstetiger R. discontinuous controller
Reglerarm *m* regulating lever
Regleraufsatz *m* governor hood *(turbine)*
Reglerausschlag *m (therm)* governor deflection
Reglerbüchse *f* governor bush *(turbine)*
Reglerdeckel *m* governor cover *(turbine)*
Reglerdiode *f* regulator diode
Reglereinstellung *f* governor setting
Reglereinwirkung *f* governor influence
Reglerelement *n* control element
Reglerfeder *f* governor spring
Reglergehäuse *n* governor casing
Reglergestänge *n* plunger
Reglergewicht *n* governor balance-weight, weight of fly-balls
Reglerhaube *f* governor cover
Reglerhebel *m* speed lever
Reglerhülse *f* governor sleeve
Reglerkegel *m* governor cone
Reglerkorb *m* governor frame *(turbine)*
Reglerkugel *f* governor [fly-]ball
Reglermuffe *f* governor collar
Reglerrelais *n* control relay
Reglerschalter *m* regulator and cut-out
Reglerschaltung *f* control circuit
Reglerschieber *m* governor piston valve *(turbine)*
Reglerspindel *f* control (governor) spindle
Reglerstellung *f* governor setting (position)
Reglersystem *n* control system
Reglertanzen *n* governor dancing
Reglertechnik *f* regulator technique
Reglerventil *n* governor valve *(turbine)*
Reglerverstärker *m* servo-amplifier
Reglette *f* reglet, clump
Regner *m* sprinkler
Regreßanspruch *m* right of recovery
Regression *f* regression
lineare R. linear regression
Regressionsgerade *f* regression line
Regressionsgleichung *f* regression equation
Regressionskoeffizient *m* regression coefficient
Regressionslinie *f* regression line
Regreßpflicht *f* liability to recourse
Regreßrecht *n* right of recovery
regulär cubic *(crystal)*; regular *(precession)*
Regulex *m (cont)* regulex
regulierbar controllable, capable of being regulated (timed, adjusted)
Regulierdrehklappe *f* regulating flap valve
regulieren to regulate; to control; to time *(ignition)*; to adjust; to govern
Reguliergestänge *n (ry)* governor rodding
Regulierscheibe *f s.* Regelscheibe
Regulierschüttfeuerung *f* furnace with adjustable feeding device
Regulierung *f (therm)* governing; regulation; control; timing *(ignition)*; adjustment, governing
Regulierventil *n* regulator valve
Regulierwicklung *f* regulating winding
regulinisch reguline
Regulus *m* regulus
Regulusofen *m* regulus furnace
Reibahle *f* reamer
R. mit eingesetzten Messern inserted-blade reamer
R. mit Führungsschaft jig reamer
R. mit Messereinstellung adjustable blade reamer
R. mit Spreizkörpereinstellung expanding reamer
R. mit ungleicher Zahnteilung increment cut reamer, reamer with unequally spaced teeth
kreuzverzahnte R. duplex spiral-type reamer
Reibahlendurchmesser *m* reamer diameter
Reibahlennutfräser *m* reamer fluting cutter
Reibahlenpendelhülse *f* floating reamer-holder
Reibahlenschaft *m* reamer shank
Reibahlenschneidenteil *n* reamer body
Reibbacken *mpl* rubbing pads
Reibchemismus *m* fretting corrosion
Reibebrett *n* float *(plastering)*
reibecht fast to rubbing
Reibechtheit *f* fastness to rubbing
Reibechtheitsprüfer *m* rubbing-fastness tester
reiben to rub; *(mach)* to ream
auf Maß r. to ream to size
fertig r. to finish ream
halbfertig r. to semi-finish ream
Reiben *n* reaming
R. auf Fertigmaß reaming to the final size
R. mit Maschinenwerkzeug machine-ream[er]ing
Reibestange *f* reaming bar
Reibfläche *f* rubbing surface
Reibgetriebe *n* friction drive
Reibhärte *f* abrasive hardness
Reibhemmung *f* frictional resistance
Reibholz *n (ship)* spar fender
Reibkontakt *m* sliding contact
Reibkorrosion *f* fretting corrosion
Reibkraft *f* friction[al] force
Reibkupplung *f* friction clutch
Reiblöten *n* tinning
Reibmesser *n (mach)* reamer blade
Reibradantrieb *m (mach)* friction drive
Reibrädergetriebe *n* train of friction gears
Reibradgetriebe *n s.* Reibradtrieb 2.

Reibradintegrator *m* ball-and-disk integrator, wheel-and-disk integrator
Reibradtrieb *m* 1. *(mach)* friction-gear drive; 2. wheel-and-disk device
Reibsäge *f* friction saw
Reibsägen *n* friction sawing *(band and circular)*
Reibschleifen *n s.* Läppen
reibschlüssig by frictional resistance
Reibschlußverbindung *f* frictional resistance transmission
Reibtrennen *n* abrasive friction cutting
Reib- und Schnelltrennsäge *f* steel friction-disk metal-sawing machine
Reibung *f* friction
  R. mitgenommen werden / durch *(mach)* to be carried around by frictional contact
  R. beim Stapellauf launching friction
  R. der Bewegung kinetic friction
  R. der Ruhe static friction
  R. fester Körper solid friction
  bohrende R. pivoting friction
  Coulombsche R. Coulomb friction
  dynamische R. kinetic friction
  gleitende R. sliding friction
  innere R. internal friction
  laminare R. laminar friction
  rollende R. rolling friction
  trockene R. Coulomb friction
  turbulente R. turbulent friction
  wälzende R. rolling friction
Reibungsarbeit *f* work of (done by, spent in overcoming) friction
Reibungsbeiwert *m* friction coefficient, coefficient of friction
Reibungsbelag *m* friction facing
Reibungsdruck *m* compressive shear
Reibungselektrizität *f* triboelectricity, frictional electricity
Reibungserwärmung *f* friction heating
Reibungsfestigkeit *f* abrasion resistance
Reibungsfläche *f* frictional surface
Reibungsgesetz *n* friction law
Reibungsgewicht *n* adhesive (adhesion) weight *(locomotive)*
Reibungskegel *m* cone of friction, friction cone
Reibungskoeffizient *m* friction coefficient, coefficient of friction
Reibungskompensationsspule *f* starting coil *(measuring)*
Reibungskraft *f* force of friction, frictional force
Reibungskupplung *f* friction clutch
Reibungsmaterial *n* friction material *(brake)*
Reibungsmitstrom *m (ship)* friction[al] wake
Reibungsmoment *n* friction torque
Reibungsnachstrom *m (ship)* friction[al] wake
Reibungspfahl *m* friction pile
Reibungspferdestärke *f* frictional horsepower
Reibungsrelais *n* friction relay
Reibungsschweißen *n* friction welding
Reibungsspannung *f* friction stress
Reibungsstoßdämpfer *m* friction shock-absorber
Reibungsströmung *f* viscous flow
Reibungsturbine *f* friction turbine
Reibungs- und Zahnradlokomotive *f* friction and rack locomotive
Reibungsverhalten *n* frictional behaviour
Reibungsverlust *m* friction loss, loss due to friction
  R. durch Ölviskosität oil drag
Reibungsvorstrom *m (ship)* friction[al] wake
Reibungswärme *f* frictional heat
Reibungswiderstand *m* frictional resistance; *(ship)* skin resistance
  R. des Motors drag of the engine
Reibungswinkel *m* angle of friction (repose)
Reibungszahl *f* friction coefficient, coefficient of friction
Reibwalze *f (print)* distributor
Reibwolf *m* crushing machine
Reibzugabe *f (mach)* allowance for reaming, reaming allowance
reich an rich in *(harvest, mineral)*
Reichblei *n* rich lead
reichen bis to reach to, to extend to
Reicherz *n* rich ore
Reichgas *n* rich gas
Reichgasvorlage *f* rich gas collecting main
reichlich copious, abundant; ample
Reichweite *f* reach, range [of transmission], working distance
  R. / außer out of range
  R. / in within range
  R. / von großer of long range
  R. am Erdboden ground range
  R. bei elektromagnetischen Wellen maximum range of radio waves
  R. eines Empfängers coverage of a receiver
  eigentliche R. intrinsic range
  endliche R. finite range
  größte R. maximum transmission range
  horizontale R. ground range
  mittlere R. mean range
  radiooptische R. radio-optical range
  tatsächliche R. actual range
  wirksame R. effective range
Reichweite-Energie-Beziehung *f* range-energy relation
Reichweitemessung *f* range measurement
Reif *m* hoarfrost
Reifen *n* maturing, ripening; ageing
Reifen *m* 1. *(aut)* tyre, *(Am)* tire; 2. hoop
  hochelastischer R. cushion tyre
  rutschsicherer R. non-skid tyre
  schlauchloser R. tubeless tyre
  wulstloser R. straight-side tyre
Reifenabnutzung *f* tyre wear

Reifenaufdrückmaschine *f* hoop driving machine
Reifenauftreibmaschine *f* hoop driving machine
Reifenaufzieheisen *n* tyre iron
Reifenaufziehmaschine *f* claw trussing machine
Reifenaufziehpresse *f* tyre press
reifend / spät late maturing
Reifendruck *m* tyre pressure
Reifendruckmesser *m* tyre gauge
Reifeneinsenkung *f* tyre deflection
Reifenfülldruck *m* inflation pressure
Reifenfülldruckmesser *m* tyre gauge
Reifenfüller *m* tyre inflator
Reifenfüllflasche *f* tyre-inflator cylinder
Reifenfüllhahn *m* tyre inflating cock
Reifengewebegarn *n* tyre yarn
Reifenheber *m* tyre iron
Reifenhülle *f* tyre casing
Reifenkord *m* cord tyre fabric, tyre cord fabric
Reifenkordwebeanlage *f* tyre-fabric weaving-machine
Reifenlaufdecke *f* tyre casing
Reifenlaufmantel *m* tyre casing
Reifenlaufzeit *f* tyre mileage
Reifenluftdruck *m* inflation pressure
Reifenmantel *m* tyre cover
Reifenmontage *f* tyre fitting
Reifenpanne *f* puncture, blowout
Reifenprofil *n* tread pattern, pattern of [the] tyre tread
  rutschsicheres R. non-skid pattern
Reifenreparatur *f* tyre repair (fixing)
Reifenschaden *m* tyre failure
Reifenschulter *f* tyre shoulder
Reifenspur *f* tyre marks
Reifenunterbaubewertung *f* ply-rating
Reifenwächter *m* kerb feeler
Reifeprozeß *m* curing *(concrete)*
Reifung *f (ch)* digestion of emulsion
Reihe *f* 1. row *(rivets; houses)*; 2. short production run, series; batch *(manufacture)*; 3. *(math)* series; 4. *(mach)* line *(linked)*; gang *(boring machine)*; progression *(speeds)*; 5. sequence; string; succession; set; number
  R. /in *(el)* in series *(connected)*
  R. der Durchmesserbereiche series of diameter ranges
  absolut-konvergente R. absolutely convergent series
  arithmetische R. arithmetic series
  buchstabengekennzeichnete R. letter series *(drill)*
  buchstabengekennzeichnete R. der Durchmesserbereiche letter series [of diameter ranges]
  Fouriersche R. Fourier series
  geometrische R. *(math)* geometrical series; *(mach)* geometrical progression *(speeds)*
  homologe R. homologous series
  isoelektronische R. isoelectronic sequence
  logarithmische R. logarithmic series
  metrische R. der Durchmesserbereiche metric series [of diameter ranges]
  unendliche R. infinite series
  verkettete R. von Maschinen mit geradliniger Werkstückeingabe linked line of machines with straight in-line transfer of workpieces
reihen *(text)* to baste
Reihenabstand *m* **der Niete** distance between rows of rivets
Reihenanlage *f* serial intercommunication set
Reihenanmeldung *f* sequence calling *(telephony)*
Reihenanordnung *f* series arrangement
Reihenaufspannung *f (mach)* line setup
Reihenbau *m* series production
Reihenbebauung *f* terraced housing
Reihenbelüftung *f* series ventilation
Reihenbetrieb *m* tandem operation *(telephony)*
Reihenbohren *n* gang (in-line) drilling
Reihenbohrmaschine *f* in-line multiple spindle drilling machine, straight-line type of drilling machine, gang drill
Reihenentwicklung *f (math)* series expansion
Reihenersatzschaltbild *n* series representation
Reihenfertigung *f* series (multiple, small-lot, batch, duplicate) production
Reihenfolge *f* order [of sequence]; sequence
  R. / in fallender *(mine)* underhand
  R. / in steigender *(mine)* overhand
  umgekehrte R. reverse order
Reihengarage *f* multiple garages
Reihenhaus *n* terrace house
Reiheninduktivität *f* series inductance
Reihenklemme *f* coupler terminal strip, terminal block
Reihenkondensator *m* series capacitor
Reihenkultur *f* row crop
Reihenlampenbeleuchtung *f* closed-loop series street-lighting system
Reihenmotor *m* in-line engine
  R. mit mehreren Zylinderreihen in-line multibank engine
Reihenparallelanlasser *m* series-parallel starter
Reihenparallelschaltung *f* parallel-series connection, series-parallel connection, bridge transition
Reihenparallelumschaltung *f* **mit Unterbrechung** open-circuit transition
Reihenparallelwicklung *f* series-parallel winding, multiplex-wave winding
Reihenresonanz *f* series [phase] resonance
Reihenresonanzkreis *m* resonant circuit connected in series
Reihenrückkopplung *f* series feedback
Reihensaat *f* row crop
Reihensämaschine *f* ridge drill, sowing machine

Reihenschaltstufe *f* series notch
Reihenschaltung *f* series connection, cascade [connection]
Reihenschlußerregung *f* series excitation
Reihenschlußgenerator *m* series-wound generator
Reihenschlußkommutatormotor *m* series commutator motor
Reihenschlußmotor *m* series[-wound] motor
Reihenschlußspule *f* series coil
Reihenschlußverhalten *n* series characteristics
Reihenspannung *f* series [insulation] voltage
Reihenspektrum *n (ph)* series spectrum
Reihenspule *f* series coil
Reihenstelle *f (telec)* series-connected station
Reihenstempel *m (mine)* breaker prop
Reihensteuerung *f* cascade control
Reihenwicklung *f* series winding
Reihenwiderstand *m* series resistance
Reihenzieher *m* furrow opener
Reihgarn *n (text)* basting thread
Reihkamm *m (text)* raddle
Reihzug *m (text)* order of drawing in
rein pure *(metal)*; clean
  chemisch r. chemically pure
  spektral r. spectroscopically pure
  technisch r. commercially pure
Reinaluminium *n* pure aluminium
Reinbenzol *n* refined benzol
Reindarstellung *f* preparation in a pure state, purification
Reinelement *n* pure element
Reinhadernpapier *n* pure rag paper
Reinheit *f* purity *(metal)*; cleanliness
  radiochemische R. radiochemical purity
Reinheitsgrad *m* percentage purity
  R. der Baumwollfaser cleanliness of the cotton fibre
Reinheitsprüfung *f* purity test
reinigen to clean, to cleanse; to purify; to scour *(wool)*; to clear; to refine; to scavenge
Reinigung *f* cleaning, cleansing; purification; scouring *(wool)*
  anodische R. anodic (reverse-current) cleaning
  chemische R. dry cleaning
  elektrolytische R. electrolytic cleaning
  katodische R. cathode (cathodic) cleaning
Reinigungsarbeiten *fpl (mine)* cleanup work
Reinigungsbassin *n* cleaning basin, deposition tank
Reinigungsechtheit *f (text)* fastness to cleaning
Reinigungsemulsion *f* liquefying cleansing cream
Reinigungsexpansion *f* cleaning expansion
Reinigungsfällung *f* scavenging
Reinigungsfeld *n* clearing (sweeping) field
Reinigungskrem *f* cleansing cream
Reinigungsmittel *n* cleansing agent, purifier, detergent
  ablösendes R. detergent
  chemisches R. dry cleaning solvent
  oberflächenaktives R. detergent
Reinigungsöffnung *f* clean-out hole
Reinigungstür *f* cleaning door
Reinigungsvermögen *n* detergency
Reinkohle *f* pure (homogeneous) coal
Reinkohlenbogenlampe *f* pure carbon arc lamp
Reinquerschnitt *m* / **axialer** axial net cross-section *(turbine)*
Reinschrift *f* fair copy
reinseiden all-silk
Reinstaluminium *n* super-purity aluminium
Reinstmetall *n* pure metal
Reinststoff *m* [extra-]pure substance
Reinwasserabfluß *m* discharge of clear water
Reinwasserbassin *n* clear water tank
reinwollen all-wool
Reise *f* 1. travel, journey *(land)*; voyage, travel, trip *(sea)*; 2. campaign *(blast furnace)*
Reiseboot *n* touring boat
Reisebüro *n* travel bureau (agency), touring office
Reisebus *m* motor coach
Reisecharter *m (ship)* voyage charter; trip charter
Reisecharterindex *m* trip charter index
Reisedauerleistung *f* / **höchstzulässige** *(aero)* maximum economy cruising power
Reiseempfänger *m* portable [set]
Reiseflughöhe *f* **über der Wolkendecke** on-top cruising level
Reiseflugzeug *n* touring aircraft
Reisegepäck *n* luggage, *(Am)* baggage
Reisegeschwindigkeit *f* cruising (service) speed; commercial speed; schedule speed
Reisemikroskop *n* travelling microscope
Reisender *m* / **abgefertigter** passenger conveyed
Reiserholz *n* brushwood
Reiseweg *m* route of travel
Reisigauflader *m* pruning sweep
Reisiggradierwerk *n* brushwood cooler (cooling stack)
Reisigholzhäcksler *m* scrub flail cutter
Reispapier *n* rice paper
Reißahle *f* marking awl; *(ship)* scrieve hook, scriber *(mould-loft tool)*
Reißbrett *n* drawing board; copy board *(process camera)*
Reißbrettstift *m* drawing pin, *(Am)* thumbtack
reißen to rupture, to crack, to fracture; to tear; to check *(casting)*; to break *(filament)*
Reißer *m (ag)* shredder, lacerating device
Reißfeder *f* drawing (ruling) pen
Reißfestigkeit *f* breaking strength *(e. g. cotton)*; tenacity
Reißgeschwindigkeit *f* breaking speed
Reißhaken *m (ship)* scrieve hook, scriber *(mould-loft tool)*

Reißhakenhobel *m (mine)* drag-hook plough
Reißkraft *f* biting (crowding) force *(excavator)*
Reißkrempel *f* breaker card, first breaker
Reißlack *m* brittle lacquer (varnish)
Reißlänge *f* breaking length
Reißlast *f* maximum tensile-load
Reißmaß *n (cont)* height gauge
Reißmikrofon *n* carbon microphone
Reißnadel *f* scriber, scribing point
Reißöl *n* dust binding oil
Reißschiene *f* T-square, tee square
Reißschienenparallelführung *f* parallel attachment
Reißschnur *f* rip cord
Reißspan *m* tear (discontinuous, segmental) chip
Reisstärke *f* rice starch
Reißverschluß *m* zip[per], zip-fastener
Reißverschlußband *n* zipper-closed belt
Reißwalze *f* toothed roller
Reißwinkelkasten *m* box parallel (angle-plate)
Reißwolf *m* tearing machine, teazer; opener; shredder
  R. für Schweißwolle opener for greasy wool
Reißwolle *f* reclaimed wool, shoddy
Reißwollgarn *n* reclaimed wool yarn
Reißzeug *n* drawing set, set of drawing (drafting) instruments; *(mach)* marking-off tool
Reißzwecke *f* drawing pin, *(Am)* thumbtack
Reiter *m* rider *(weight)*
  spanischer R. *(mil)* knife rest
Reiterheizer *m* clamp-on heater *(pipe)*
Reiterlibelle *f* striding level
Reiterschaufelturbine *f* turbine with rimstraddling buckets
Reithose *f* riding breeches
Reitnagel *m (mach)* dead tailstock centre
Reitpferd *n* saddle horse
Reitstiefel *m* riding boot
Reitstock *m (mach)* tailstock
Reitstockkörper *m* tailstock body
Reitstockoberteil *n* tailstock barrel
Reitstockpinole *f* tailstock [spindle] sleeve
Reitstockspitze *f* tailstock centre
Reitstockunterteil *n* tailstock base
Reiz *m* / **einschleichender** accumulating stimulus
Reizgas *n* irritant gas
Reizgift *n* irritant poison
Reizmittel *n* irritant, stimulant
Reizschwelle *f* threshold of response (sensation)
Reizzeitstromkurve *f* strength-duration curve
Reklamepapier *n* weddings
reklamieren to raise a claim; to complain about
Rekombination *f* recombination
  R. von Ionen recombination of ions
  strahlende R. radiative recombination
Rekombinationsgeschwindigkeit *f* recombination rate
Rekombinationskinetik *f* recombination kinetics
Rekombinationskoeffizient *m* coefficient of recombination, recombination coefficient (rate)
Rekombinationsleuchten *n* recombination light
Rekombinationsprozeß *m* recombination process
Rekombinationsrate *f* recombination rate, rate of recombination
Rekombinationsreaktion *f* back reaction
Rekombinationsstrahler *m* recombination radiator
Rekombinationsstrahlung *f* recombination radiation
Rekombinationszentrum *n* centre of recombination, recombination centre
rekombinieren to recombine, to reunite
Rekonstruktion *f* re-equipment *(factory)*
Rekristallisation *f* recrystallization
Rekristallisationsschicht *f* fused junction *(semiconductors)*
Rekristallisationszwilling *m* recrystallization twin
rekristallisieren to recrystallize
Rektaszension *f* right ascension
Rektifikation *f (math, ch)* rectification
rektifizierbar rectifiable
rektifizieren *(math, ch)* to rectify; *(ch)* to redistil, to fractionate
rekultivieren to restore; *(mine)* to reclaim, to recultivate
Rekultivierung *f* restoration; *(mine)* reclamation, recultivation
Rekuperativofen *m* recuperative furnace
Rekuperator *m* recuperator
Rekursion *f (math)* recurrence, recursion
Rekursionsformel *f* recursion formula
rekursiv *(math)* recurrent
Relais *n* relay; contactor
  R. mit Ansprechverzögerung slow-operating relay
  R. mit magnetischem Nebenschluß shunt-field relay
  R. mit verzögerter Auslösung slow-release relay
  R. mit Vormagnetisierung biased relay
  R. mit Wechselkontakten relay with make-and-break contacts
  abfallverzögertes R. slow-release relay
  ankerloses R. relay without armature
  durch Stromstöße weiterschaltendes R. trigger relay
  einpoliges R. single-contact relay
  einseitig eingestelltes R. biased relay
  einspuliges R. single-coil relay
  elektrodynamisches R. electrodynamic relay
  elektronisches R. electronic relay
  empfindliches R. sensitive relay
  fotoelektrisches R. light relay

gepoltes R. polarized relay
induktionsfreies R. non-reactive relay
kapazitätsarmes R. low-capacitance relay
koaxiales R. coaxial relay
magnetisches R. magnetic relay
magneto-dynamisches R. magnetic-electric relay
neutrales R. non-polarized relay
polarisiertes R. polarized (neutral, centre-zero) relay
rückfallendes R. self-reset
schnellansprechendes R. high-speed relay
selbsthaltendes R. locking (lock-in) relay
steckbares R. plug-in relay
thermisches R. temperature (thermostat) relay
unpolarisiertes R. neutral (non-polarized) relay
unverzögertes R. instantaneous relay
verzögertes R. [time-]delay relay
Relaisabfallzeit *f* relay release time
Relaisanker *m* relay armature
Relaisanrufsucher *m* call-finder relay *(telephony)*
Relaisanzeige *f* relay indication
Relaisbetätigung *f* relay operation
Relaisbetrieb *m* relay operation
relaisgesteuert relay-operated
Relaiskern *m* relay core
Relaiskette *f* relay chain
relaislos relayless
Relaismagnet *m* relay magnet
Relaismesser *m* relay meter
Relaisrechenmaschine *f* relay computer
Relaisrechner *m* relay computer
Relaisröhre *f* trigger tube
Relaissatz *m* relay set
Relaisschaltung *f* relay circuit
Relaisschirm *m* relay screen
Relaissender *m* relay [broadcasting] station
Relaisspule *f* relay coil
Relaissteuerung *f* relay control
Relaisstufe *f* relay stage
Relaisuhr *f* relay clock
Relaisverstärker *m* relay amplifier
Relaisvorwähler *m* relay preselector
Relaiswähler *m* relay-type selector, relay group
Relaiswicklung *f* relay winding
Relaiszähler *m* relay counter
Relation *f* relation
relativ relative
Relativbeschleunigung *f* relative acceleration
Relativbewegung *f* relative movement
Relativempfindlichkeit *f* relative response
Relativgeschwindigkeit *f* relative velocity
relativistisch relativistic
Relativität *f* relativity
Relativitätselektrodynamik *f* relativistic electrodynamics
Relativitätsmechanik *f* relativistic (relativity) mechanics
Relativitätspostulat *n* relativity postulate
Relativitätsprinzip *n* relativity principle, principle of relativity
Relativitätstheorie *f* theory of relativity, relativity [theory]
allgemeine R. general theory of relativity
spezielle R. restricted (special) theory of relativity, special relativity
Relativitätsthermodynamik *f* relativistic (relativity) thermodynamics
Relativmessung *f* relative measurement
Relaxation *f* relaxation
magnetische R. magnetic viscosity
Relaxationsentfernung *f* relaxation distance
relaxationsfrei relaxation-free
Relaxationslänge *f* relaxation length
Relaxationspolarisation *f* relaxation polarization
Relaxationsstrecke *f* relaxation length
Relaxationsverfahren *n* relaxation method
Relaxationszeit *f* relaxation (equilibrium) time
relaxierend relaxing
Relief *n* relief; embossing
Reliefautotypie *f* relief-halftone
Reliefklischee *n* relief-block
reliefpolieren to relief-polish, to attack-polish
Reling *f (ship)* railing, breast rail
Relingstütze *f* rail stanchion, bulwark-stay
Reluktanz *f* reluctance
spezifische R. reluctivity
Reluktanzgenerator *m* reaction alternator
Reluktanzmotor *m* reluctance motor
rem *s.* rem-Einheit
rem-Einheit *f* roentgen equivalent [man unit]
remanent remanent, residual
Remanenz *f* remanence, remanent (residual) induction, remanent (residual) magnetization, remanent (residual) flux density
Remanenzfeld *n* residual (retentive) field
Remanenzfluß *m* remanent flux
Remanenzinduktion *f* residual induction, magnetic retentivity
Remanenzrelais *n* remanent relay
Remanenzspannung *f* residual voltage
Remission *f* reflectance
Remissionsgrad *m (opt)* luminance factor
Rendezvousmanöver *n* rendezvous manoeuvre *(astronautics)*
Renkfassung *f* bayonet holder
Renkverschluß *m* bayonet cap
Rennauto *n* racing car
Rennboot *n* racing boat; race boat, racer
Rennfeuer *n (met)* bloomery
Rennherd *m* bloomery hearth, Catalan furnace
Rennjacht *f* racing yacht, racer
Rennklassenboot *n* / **nationales** nationally organized racer
Rennkraftstoff *m* racing fuel
Rennrodel *m* racing-sled

Rennschlitten *m* bobsleigh, bobsled
Rennschlittschuh *m* speed (racing) skate
Rennschuh *m* field (spiked, running, track) shoe
Rennstahl *m* ore steel, bloomery iron
Rennverfahren *n* bloomery (direct-iron, native-forge) process
Rennzweier *m* **mit Steuermann** two-oared shell with coxswain
renovieren to face-lift; to renovate, to renew, to restore
Renovierung *f* face-lift; renovation, renewal, restoration
rentabel profitable
Rentabilität *f* profitability
Rentenalter *n* pensionable age
Rentner *m* old-age pensioner
rep *s.* physikalisches Röntgenäquivalent
rep-Einheit *f s.* physikalisches Röntgenäquivalent
Reparatur *f* repair; overhaul
 vorläufige R. temporary repair
Reparaturarbeit *f* repair work
reparaturbedürftig in need of repair
Reparaturhauer *m (mine)* repairer
Reparaturmechaniker *m* maintenance technician
Reparaturschiff *n* repair ship
Reparaturschweißung *f* repair welding
Reparaturtechniker *m* service engineer
Reparaturwerft *f* repair yard
Reparaturwerkstatt *f*, **Reparaturwerkstätte** *f* repair (maintenance) shop
reparieren to repair; to overhaul; to recondition; to mend
repetieren to repeat
Repetiergerät *n* reset generator
Repetierkamera *f* step-and-repeat camera
Repetitionstheodolit *m* repetition theodolite
Replikation *f (math)* replication
Reprise *f (text)* moisture regain
Reproduktion *f (mach)* duplication; reproduction; reprint
Reproduktionsfaktor *m s.* Vermehrungsfaktor
Reproduktionsfotografie *f* process photography
Reproduktionskamera *f* process camera
reproduzierbar repeatable; reproducable
Reproduzierbarkeit *f* repeatability, reproducibility
 R. der Anzeige consistency of reading *(of a meter)*
 R. der Regelgröße control precision
reproduzieren to reproduce; *(mach)* to duplicate; to repeat; to reprint
Repulsionsinduktion *f* repulsion induction
Repulsionskraft *f* repelling force (power)
Repulsionsmeßwerk *n* dynamometer-type meter movement [system]
Repulsionsmotor *m* repulsion (doubly-fed) motor
 R. mit Käfiganker repulsion induction motor
 R. mit kurzgeschlossenem Feld inverted repulsion motor
 R. mit Zusatzkäfig repulsion induction motor
 kompensierter R. single-phase commutator motor with self-excitation
Reserve *f* 1. reserve; 2. *(text)* resist
Reserveader *f (el)* reserve wire
Reserveanker *m (ship)* spare anchor
Reserveanlage *f* stand-by plant
Reservebatterie *f* spare battery
Reservebuganker *m* spare (best bower) anchor
Reservedampfmaschine *f* stand-by steam engine
Reservedruck *m (text)* resist print
Reservefilter *n* spare filter
Reservegenerator *m* stand-by generator
Reserveglas *n* spare gauge glass
Reservekabel *n* spare cable
Reservekessel *m* spare (stand-by) boiler
Reservekreis *m (Am)* fall back circuit
Reservepinne *f (ship)* emergency tiller
Reservepropeller *m* spare propeller
Reserverad *n (aut)* spare wheel
Reserveradhalter *m* spare-wheel carrier
Reserverohrleitung *f* spare piping
Reservesatz *m* stand-by unit *(rotating machinery)*; spare (stand-by) set
Reserveschwimmfähigkeit *f (ship)* reserve buoyancy
Reservestreb *m* spare face
Reserveverdrängung *f (ship)* reserve buoyancy
Reserveverstärker *m* spare repeater
Reservewelle *f* spare shaft
reservieren 1. to reserve; 2. *(text)* to resist
 bunt r. to resist in colours
Reservoir *n* reservoir, storage tank
Residuum *n* residue
 elektrisches R. electric residuum
Resinat *n* resinate
Resistenz *f* resistance
Resitol *n (plast)* resitol
Resnatronröhre *f* resnatron
Resol *n* resol
Resolharz *n* one-stage resin, one-step resin, single-stage resin, single-step resin
Resolvente *f (math)* resolvent
Resonanz *f* resonance
 R. / in at resonance
 R. mit der Eigenschwingung period resonance
 anharmonische R. anharmonic resonance
 einfache R. single resonance
 mechanische R. mechanical resonance
 optische R. optical resonance
 paramagnetische R. electron permagnetic resonance
 parasitäre R. parasitic resonance
 scharfe R. sharp resonance
 unerwünschte R. spurious resonance

Resonanzabsorber *m* resonance absorber
Resonanzabsorption *f* resonance absorption
Resonanzaktivierung *f* resonance activation
Resonanzanpassung *f* resonance matching
Resonanzanregung *f* resonance excitation
Resonanzanstieg *m* resonant rise
Resonanzbereich *m* resonance range
Resonanzbeschleuniger *m* resonance accelerator
Resonanzbodenholz *n* timber for sounding boards
Resonanzbreite *f* resonance width
Resonanzbrücke *f* resonance bridge
Resonanzbuckel *m* resonance hump
Resonanzdetektor *m* resonance detector
Resonanzdrossel *f* resonance choke
Resonanzdrosselspule *f* resonance choke
Resonanzeinfang *m* resonance capture [of neutrons]
Resonanzenergie *f* resonance energy
Resonanzentkommwahrscheinlichkeit *f* resonance escape probability
Resonanzerschütterung *f* resonant vibration
Resonanzflucht *f* resonance escape
Resonanzfluchtwahrscheinlichkeit *f s.* Resonanzentkommwahrscheinlichkeit
Resonanzfluoreszenz *f* resonance fluorescence (scattering)
resonanzfrei resonance-free
Resonanzfreischwingsieb *n* non-resonance vibrating screen
Resonanzfrequenz *f* resonant frequency, frequency of resonance
Resonanzgrundfrequenz *f* first resonating frequency
Resonanzintegral *n* resonance integral
Resonanzkreis *m* resonant circuit
Resonanzkreisfrequenz *f* angular resonance frequency
Resonanzkurve *f* resonance curve
  scharfe R. narrow resonance curve
  schmale R. narrow resonance curve
  spitze R. pointed resonance curve
Resonanzleitung *f* resonant line
Resonanzleitwert *m* resonance conductance
Resonanzlinie *f* resonance line
Resonanzmaschine *f* resonant machine *(fatigue testing)*
Resonanzmodus *m* resonant mode *(waveguides)*
Resonanznebenschluß *m* resonant shunt
Resonanzneutron *n* resonance neutron
Resonanzneutronenreaktor *m* resonance reactor
Resonanznichteinfangwahrscheinlichkeit *f s.* Resonanzentkommwahrscheinlichkeit
Resonanzpunkt *m* resonance point
Resonanzrelais *n* resonance (tuned) relay
Resonanz[ring]schalter *m* ring switch *(waveguides)*
resonanzscharf sharply-resonant
Resonanzschärfe *f* sharpness of resonance, magnification of the circuit *(oscillators)*
Resonanzschwingsieb *n* resonance vibro-screen
Resonanzschwingung *f* resonant mode
Resonanzserie *f* resonance series
Resonanzsieb *n* resonance screen
Resonanzspannung *f* resonance potential (voltage)
Resonanzspitze *f* resonance peak
Resonanzstrahlung *f* resonance radiation
Resonanzstreuung *f* resonance (resonant) scattering
Resonanzstruktur *f* resonating structure *(laser)*
Resonanzsystem *n* resonant system
Resonanztheorie *f* resonance theory
Resonanztransformator *m* resonance (resonant, tuned) transformer
Resonanztypfilter *n* resonant-mode filter *(waveguides)*
Resonanzüberhöhung *f* resonance step-up, overvoltage of resonance
Resonanzverlust *m* resonance loss
Resonanzverstärker *m* resonance (tuned, ringing) amplifier
Resonanzwiderstand *m* resonant impedance
Resonanzzustand *m* resonance state, condition of resonance
Resonator *m* resonator
  abstimmbarer R. tunable resonator
  akustischer R. acoustical resonator
  kugelförmiger R. spherical resonator
  schwach gedämpfter R. persistent resonator
Resonatorraum *m* resonant chamber
Resonatorschwingung *f* resonator mode
resorbieren to re[ab]sorb
Resorzin *n* resorcinol
Ressort *n* department
Rest *m* 1. *(math)* remainder; 2. *(ch)* residue; 3. remnant, remanent
Restablenkung *f* residual (remanent) deviation
Restaffinität *f* residual affinity
Restaktivität *f* residual activity
Restauftrieb *m* *(ship)* reserve buoyance
Restbestand *m* remaining articles *(workpieces)*; remainder
Restbetrag *m* remainder
Restblock *m* butt ingot
Restdämpfung *f* overall attenuation loss
Restentladung *f* residual discharge
Restfehler *m* residual error; residual aberration
Restfeld *n* residual [magnetic] field
Restfettgehalt *m* residual fat content
Restfeuchtigkeit *f* residual moisture
Restgas *n* residual gas *(electron tube)*
Restimpuls *m* residual pulse
Restinduktivität *f* saturation inductance
Restintensität *f* residual intensity

R. in der Linienmitte residual central intensity

Restitutionszyklus *m* / **elektrischer** [electrical] recovery cycle

Restkapazität *f* residual capacity

Restkern *m* residual nucleus, nuclear fragment

Restklasse *f (math)* coset, residual (residue) class *(carryless computing)*

Restklassenkode *m* residue code

Restklassenprobe *f* casting out

Restklassenzahlensystem *n* residue number system

Restkräuselung *f* residual stretch

Restkrumpfung *f* residual shrinkage

Restladung *f (el)* residual charge

Restleitfähigkeit *f* post-arc conductivity *(of an arcing distance)*

Restmagnetisierung *f* residual magnetization

Restpfeiler *m (mine)* remnant, residual pillar

Restpunktpolverfahren *n* point-pole method

Restradioaktivität *f* residual radioactivity

Restreaktanz *f* saturation reactance

Restschmelze *f* residual heat (liquid)

Restschrumpfung *f* residual shrinkage

Restseitenband *n* vestigial sideband

Restspannung *f* 1. *(el)* residual voltage; 2. *(mech)* residual stress

Reststrahlen *mpl* residual rays

Reststrahlung *f* residual [nuclear] radiation

Reststrom *m* residual current; leakage current

Reststromstoß *m* residual pulse

Reststück *n* remnant

Restvalenz *f* residual valence

Restwärme *f* after-heat, afterheat

Restwiderstand *m* 1. residual (end) resistance; 2. *(ship)* residuary resistance

resublimieren to resublime

Resultante *f* resultant, eliminant

Resultat *n* result

resultierend resultant

Resultierende *f* resultant

Resümee *n* summary

Retardierung *f* retardation

Retardierungseffekt *m* retardation effect

Retardierungsfaktor *m* retardation factor

Retention *f* retention

Retentionsfaktor *m* initial body retention; retention coefficient

Retentionszeit *f* retention time

Retigraf *m* retigraph

Retorte *f* retort, batch still

Retortenkohle *f* retort carbon

Retortenkoks *m* retort coke

Retortenofen *m* retort furnace

Retortenverkohlung *f* retort coking

retten to save; to rescue

Rettung *f* rescue; saving, salvage

Rettungsausrüstung *f* life-saving equipment

Rettungsboje *f* life (safety) buoy

Rettungsboot *n* life-boat *(on board a ship)*

selbstlenzendes R. self-bailing boat; *(Am)* self-baling boat

Rettungsbootstation *f* **an Land** *(Am)* coast guard station

Rettungsdienst *m* rescue service (team); life-guard (life-boat) service

Rettungseinrichtungen *fpl* life-saving appliances (equipment)

Rettungsfloß *n* liferaft, life float; life-saving float

aufblasbares R. inflatable liferaft

Rettungsgerät *n* life preserver; live-saving equipment

Rettungsgürtel *m* life jacket (belt, preserver)

Rettungskasten *m* first-aid box (chest)

Rettungsleine *f* hand line

Rettungsmannschaft *f* rescue team; *(mine)* helmet crew

Rettungsrakete *f* life (line-throwing) rocket

Rettungsring *m* life preserver; *(ship)* life belt (ring), safety (life) buoy

Rettungsschiff *n* rescue ship

tankähnliches R. life car

Rettungsschlepper *m* salvage (rescue) tug

Rettungsschwimmen *n* life-saving swimming

Rettungsstelle *f* rescue centre

Rettungswagen *m* breakdown vehicle

Retusche *f* retouching

retuschieren to retouch

Retuschierfarbe *f* retouching paint

Retuschierfeder *f* retouching pen

Retuschiergerät *n* retouching device

Retuschierpinsel *m* retouching brush

Retuschierpult *n* retouching desk

Reuse *f* 1. eel-basket; 2. *(mine)* safety fence *(borehole)*

Reusenantenne *f* cage (bow-net-type, drum-net-type, sausage, squirrel-cage) aerial

Reuter *m* tripod

Revers *n* revers

Revers *m* reverse *(of a coin)*

reversibel reversible

Reversierduostraße *f* two-high reversing mill

reversierend reverse-acting *(pneumatics)*

Reversiermotor *m* reversible motor

Reversierturbine *f* reversing turbine

Reversionslibelle *f* reversible level

Reversionspendel *n* reversible pendulum

revidieren to revise

Revier *n* 1. *(mine)* area, district, region; 2. estuary *(of a river)*

Revierfahrt *f* estuary trading

Revierlotse *m* branch pilot

Reviermarkscheider *m* district surveyor

Revierschlepper *m* estuary tug

Reviersteiger *m* deputy, *(Am)* section boss

Revision *f* inspection; auditing

Revisionsbogen *m* final proof, machine-revise
Revisor *m* auditor
Revolver *m* 1. revolver; 2. *(mach)* turret; capstan
automatisch schaltender R. capstan
Revolverdrehmaschine *f* turret lathe; capstan lathe
R. mit Revolverschlitten auf Support capstan lathe
Revolverkopf *m (mach)* turret head
offener R. hollow turret
Revolverkopfbohrmaschine *f* turret drilling machine
Revolverkopfschlitten *m* turret carriage
Revolverlade *f (text)* circular box lathe
Revolverschlitten *m (mach)* turret carriage (slide)
Revolverwechsel *m (text)* circular box motion
Revolverwerkzeugschlitten *m* turret carriage (slide)
Reyon *m (n) s.* Viskoseseide
Rezept *n* 1. *(ch)* recipe; 2. prescription *(medicine)*
Rezipient *m* receiver, container; *(met)* recipient; bell-jar
reziprok reciprocal
Reziprokwert *m* reciprocal [value]
Reziprozitätsgesetz *n* reciprocity law
Reziprozitätssatz *m* reciprocal theorem
Rhamnit *n* rhamnitol
Rhenium *n* rhenium
Rheniumlegierung *f* rhenium alloy
Rheoapparat *m* rheo-box
Rheologie *f* rheology
Rheopexie *f* rheopexy *(chemistry of colloids)*
Rheostat *m* rheostat
Rheostriktion *f* rheostriction, pinch effect
Rheotron *n* rheotron
Rheowäsche *f* rheolaveur washery
rhm-Einheit *f* roentgen per hour at one meter, rhm
Rhodamin *n* rhodamine
Rhodanat *n* sulphocyanate
Rhodanid *n* rhodanide
rhodanieren to rhodanize
Rhodanierung *f* rhodanizing
Rhodanwasserstoffsäure *f* rhodanic (thiocyanic) acid
Rhodanzahl *f* thiocyanogen value
Rhodium *n* rhodium
rhodiumplattiert rhodium-plated
Rhodochrosit *m (min)* rhodochrosite
Rhodonit *m (min)* rhodonite
Rhombendodekaeder *n* rhombic dodecahedron, rhombododecahedron
rhombisch [ortho]rhombic
Rhomboeder *n* rhombohedron
rhomboedrisch rhombohedral
Rhomboid *n* rhomboid
Rhombus *m* rhombus
Rhombusantenne *f* rhombic aerial
Rhönrad *n* gyro-wheel
Rhumbatron *n* rhumbatron
Ribonukleinsäure *f* ribonucleic acid
Ribose *f* ribose
Richtantenne *f* directional aerial
einseitig arbeitende R. unidirectional aerial
Richtantennenanlage *f* directive aerial system
Richtantennengruppe *f* aerial array
Richtantennennetz *n* aerial array
Richtbake *f* direction beacon
Richtbohren *n* controlled (directional) drilling
Richtcharakteristik *f* directional characteristic (diagram, pattern)
senkrechte R. vertical directive pattern
Richtdiagramm *n* directive pattern *(of aerials)*; pickup pattern *(microphony)*
Richtdipol *m* fishbone aerial
Richtefest *n (build)* topping-out ceremony
Richteffekt *m* directional effect
Richtempfang *m* directional receiving (reception)
Richtempfänger *m* directional receiver
richten 1. to direct *(radio)*; 2. *(mach)* to align; to straighten; 3. *(mil)* to aim, to point; to lay; 4. to erect, to put up
Richtfähigkeit *f* directivity
Richtfest *n (build)* topping-out ceremony
Richtfeuer *n* leading light; directional light
Richtfunkbake *f* radio[-range] beacon
Richtfunkendstelle *f* radio link terminal
Richtfunkrelaiskette *f* microwave-relay system
Richtfunkstation *f* beam station
Richtfunkstrecke *f* radio link (relay transmission system)
Richtfunkverbindung *f* radio link
Richtkeil *m (mine)* whipstock
Richtkennlinie *f* directional characteristic, beam pattern of an aerial array
Richtkennlinienfeld *n* characteristic-curve family *(rectifier)*
Richtkoppler *m* directive feed assembly, directional coupler
Richtkraft *f* directing (directive, directional) force
Richtkreisel *m* directional gyro[scope]
Richtlatte *f (build)* straight-edge; *(mil)* aiming post
Richtmagnet *m* control (controlling, reset) magnet
Richtmaschine *f (mil)* layer
Richtmaß *n* nominal measurement (size, dimension)
Richtmikrofon *n* directional microphone (receiver)
Richtmoment *n* restoring moment (torque); directing (driving) moment
Richtmomentzunahme *f* torque gradient

Richtpfahl *m (build)* guide pile
Richtphasenschieber *m* directional phase-shifter
Richtplatte *f* surface plate
Richtreihenansatz *m* test-strip attachment
Richtrohr *n* **für Paßmarken** register mark aligning tube
Richtschärfe *f (telec)* sharpness of directivity
Richtscheit *n (build)* straight-edge
Richtschiene *f (mach)* parallel
Richtsendung *f* directional transmission, beam emission
Richtspant *n (ship)* chief (principal, balanced, square) frame
Richtstand *m (rock)* sighting platform
Richtstange *f* scaffold pole
Richtstollen *m* pilot tunnel (drift)
Richtstrahl *m* radio (directional) beam
Richtstrahlantenne *f* beam aerial
  rotierende R. rotary-beam aerial
Richtstrahler *m* directional radiator, directional [transmitting] aerial
Richtstrahlsender *m* beam (directional) transmitter
Richtstrahlung *f* directive transmission
Richtstrecke *f* drift, *(Am)* lateral [road] *(ore mining)*
Richtstreifen *m* felt seam mark
Richtstrom *m* rectified current
Richtung *f* direction; sense; course; *(mil)* alignment
  abfallende R. decreasing direction
  entgegengesetzte R. opposite direction
  rechtweisende R. *(aero)* track bearing
Richtungen / nach beiden bidirectional[ly], in two directions
richtungsabhängig directional, anisotropic
Richtungsabhängigkeit *f* directionality, anisotropy, directional dependence
Richtungsabweichung *f (mach)* misalignment
Richtungsänderung *f* change in direction, directional change
Richtungsangabe *f* indication of direction
Richtungsanzeiger *m (aut)* direction indicator, trafficator
  mechanischer R. semaphore direction indicator
Richtungsausscheidung *f* automatic routing of toll (long-distance) calls
Richtungsbestimmung *f* direction finding
Richtungsblockierung *f (ry)* directional interlocking
Richtungsbündelung *f* directionality *(of laser-beams)*
Richtungsdiffusität *f* directional diffusion
Richtungsdoppeldeutigkeit *f* sense ambiguity
Richtungsdoppelfokussierung *f* two-directional focusing
Richtungseigenschaft *f* directional property
Richtungsfehler *m* directional error (distortion)
Richtungsfeuer *n* range (leading) light
Richtungsfokussierung *f* directional (space, angular) focusing
Richtungsgeber *m (aero)* turn control
Richtungsgenauigkeit *f* directional accuracy, trueness of direction
Richtungsgewinn *m* directive gain *(of aerials)*
Richtungsglied *n (el)* directional element
Richtungskoppler *m* directional coupler *(waveguides)*
Richtungskosinus *m* direction cosine
richtungslos not directional
Richtungsprüfung *f* directional measurement
Richtungsrelais *n* [current] directional relay
Richtungsschalter *m* direction switch
Richtungssinn *m* sense
  R. der Geschwindigkeitskomponente direction of the velocity component *(turbine)*
Richtungsstabilität *f (aero)* directional stability
Richtungsübertragung *f* transfer of direction
  mechanische R. mechanical transfer of direction
Richtungsumkehr *f* reversal of the direction
Richtungsverlauf *m* **der Rauheit** lay, direction of predominant surface pattern
Richtungsverteilung *f* directional distribution
Richtungswechsel *m* change of direction
richtungswechselnd changing the direction; reversing
Richtungswender *m* reversing switchgroup, reverser
Richtungswinkel *m* slope (elevation) angle
Richtverstärker *m* directional amplifier
Richtvorrichtung *f (mil)* laying gear
Richtwaage *f* level
Richtwalze *f* straightening roll, roll straightener
Richtwert *m (mach)* recommended value, recommendation; reference value (datum)
Richtwiderstand *m* [rectifier] load resistance, directional resistance
Richtwirkung *f* directionality, directivity *(radio)*; rectifying effect
  R. / ohne non-directional
  einseitige R. unidirectional action *(of aerials)*
  horizontale R. horizontal directivity
Richtzeichnung *f* assembly drawing for use in the erecting shop
Richtzylinder *m* alignment cylinder
Riebeckit *m (min)* riebeckite
riechend / stechend with pungent odour
Riechstoff *m* odorous constituent; odoriferous substance, scent, odorant, perfume
Ried *n* reed
Rieddach *n* thatched roof
Riedkamm *m (text)* dividing comb
Riefe *f* ridge
Riefelung *f* fluting

riefen to corrugate, to flute, to ridge, to groove
riefenfrei *(mach)* without [tooling] marks
Riegel *m* 1. *(build)* tie (cross, ledger) beam; bolt, latch; waler, waling; 2. *(aut)* interlock device; 3. piece, cake *(soap)*
Riegelbohrer *m* brad-awl
Riegelfeder *f* interlock spring
Riegelhebel *m (ry)* lock lever *(signal)*
Riegelkugel *f (aut)* interlock ball
Riegelleitung *f (ry)* lock line
Riegelmaschen *fpl (text)* coiled loops
Riegelschieber *m (ry)* tappet rod *(signal)*
Riegelstopfen *m (aut)* interlock plug
Riemchen *n* [queen] closer *(brickwork)*
Riemchenflorteiler *m (text)* tape condenser (divider)
Riemchensteuerung *f (text)* strap fork mechanism
Riemen *m* 1. *(mach)* belt; belting; 2. sweep, oar *(rowing)*; scull
 gedrehter R. twisted strap
 gekreuzter R. *(mach)* crossed belt
 offener R. *(mach)* open belt
Riemenantrieb *m* belt driving (drive)
Riemenauflegen *n* putting on a (the) belt
Riemenaufleger *m* belt-slipper
Riemenausrücker *m* belt-shifter
Riemenboot *n* rowing (pulling) boat; *(Am)* row boat
Riemenbreite *f* belt width
Riemenbremse *f* belt brake
 selbstregulierende R. self-regulating belt-brake
Riemendehnung *f* belt stretch
Riemendicke *f* belt thickness
Riemendolle *f (ship)* thole pin
Riemendrehdolle *f* swivel rowlock
Riemenförderer *m* belt conveyor
Riemengabel *f* belt guide (fork)
Riemengabelzapfen *m* fork pivot
Riemengetriebe *n* train of pulleys, cone-pulley transmission
Riemenkitt *m* belt cement
Riemenlänge *f* belt length
Riemenleder *n* belt leather, belting
Riemenrettungsboot *n* pulling lifeboat
Riemenrolle *f* pulley
Riemenschalter *m* belt shifting mechanism
Riemenscheibe *f* [belt] pulley
 ballige R. crowning pulley
 geteilte R. split[-type] pulley
Riemenscheibengetriebe *n* train of pulleys
Riemenscheibengewindebohrer *m* pulley tap
Riemenscheibenschwungrad *n* belt-pulley flywheel
Riemenschlupf *m* belt slip, slip of [the] belt
Riemenschmiere *f* belt lubricant
Riemenspannrolle *f (aut)* fan-belt tensioner; *(mach)* idler pulley
Riemenspannung *f* belt tension[ing]
Riemenspeisepumpe *f* belt-driven feed pump
Riementachometer *n* belt tachometer
Riementrieb *m* belt drive
 R. / mit belt-driven
 gekreuzter R. crossed belt-drive
 halbgeschränkter R. half-crossed belt drive
 offener R. open belt-drive
Riemenübertragung *f* belt transmission
Riemenumlegegabel *f* belt shifting-fork
Riemenverbinder *m* belt fastener
Riemenverbindung *f* belt joint
Riemenvorgelege *n* countershaft
Riemenzug *m* belt tension
Ries *n* ream *(paper)*
Rieselanlage *f* stacking *(cooling tower)*
Rieselfeld *n* sewage-irrigated field
Rieselfelderwirtschaft *f* sewage farming
Rieselgut *n (mine)* spillage
Rieselkondensator *m* open-surface condenser
Rieselkühler *m* spray (open-surface) cooler, dripping (trickling) cooling plant
rieseln to run, to trickle
Rieselrost *m* grate with sprinkling of the fire bars
Rieselrostfeuerung *f* furnace with grate-bars cooled by spraying water on them
Rieselturm *m* spray tower (column), wash column, washer, scrubber
Rieselverdampfer *m* open-surface evaporator
Rieselwasser *n* trickling water
Riesenmolekül *n* giant molecule
Riesenresonanz *f* giant resonance
Riesenschauer *m* extensive (Auger, giant air) shower
Riesenschiff *n* super-liner
Riet *n (text)* comb
Rietstab *m* split of the reed
Rietzahn *m* dent of the reed, reed wire
Riff *n* reef
Riffelbildung *f* corrugation *(roaring rails)*
Riffelblech *n* chequer[ed] plate
Riffelkamm *m* ripple *(flax)*, rippling comb
riffeln *(ag)* to ripple, to treat with ripple; to flute
Riffelsegment *n (text)* fluted segment
Riffelstahl *m* fluting tool
Riffelung *f* fluting
Rigolen *n (ag)* double-digging
Rille *f* groove; scratch; *(text)* eye *(needle)*
Rillenabstand *m* roughness width *(surface)*
Rillenherd *m* riffled table
Rillenisolator *m* corrugated insulator
Rillenkugellager *n* grooved ball bearing
Rillenscheibe *f* sheave, grooved pulley
Rillenschiene *f* grooved (channel) rail
Rillentiefenmessung *f* scratch depth checking
Rillenzieher *m (ag)* drill hoe

Rill- und Schneidemaschine *f* creasing and cutting machine
Rind *n* cattle
Rinde *f* bark, rind
Rindenflecke *mpl* bark specks *(in the paper)*
rindenfrei bark-free
Rindenkratzer *m* debarker, bark scraper
Rindenprozent *n* allowance for bark
Rindenreiniger *m* tree scraper
Rindenschäler *m* barker, barking iron
Rindenschälmaschine *f* bark peeling machine, barker
Rinderhaltung *f* cattle farming
Rinderoffenstall *m* open-type cattle shelter
Rinderstall *m* cowshed, byre, cattle shelter (shed)
Rinderzucht *f* cattle breeding; rearing cattle
Rindleder *n* / **chromgares** crome-side leather
 glattes R. smooth-hide leather
Rindsportleder *n* / **wasserdichtes** waterproof chrome side leather
Rindviehhaltung *f* cattle farming
Ring *m* ring; annulus; washer; *(ch)* cycle
 R. eines Kabels bight of a cable
 ebener R. plane annulus
 federnder R. elastic (snap-piston) ring, piston spring
 stopfbuchsenartig eingreifender R. gland-shaped ring
Ringanalyse *f* hydrocarbon group analysis, *(Am)* round-robin test
Ringanker *m (el)* ring armature; *(build)* ring beam
Ringausschnitt *m* opening of a (the) ring
Ringbalken *m* ring beam
Ringbank *f* ring rail *(spinning)*
Ringbankbewegung *f* movement of the ring rail
Ringbankhub *m* lift of the ring rail
Ringbankschubvorrichtung *f* ring-rail motion
Ringbeschleuniger *m (nucl)* cyclic accelerator
Ringbildner *m (ch)* annular producer
Ringblende *f* annular stop
Ringbohren *n (mach)* trepanning
Ringbrenner *m* collar burner
Ringbrennkammer *f (rock)* annular combustion chamber
Ringbüchse *f* sleeve (bush) packing
Ringdichtung *f* ring packing
Ringdüse *f* annular-clearance nozzle
Ringdynamo *m* ring dynamo
Ringe *mpl* flying rings *(sports)*
 Newtonsche R. Newton's rings
Ringelektrode *f* annular electrode
Ringelektromagnet *m* ring-shaped electromagnet
Ringeln *n (ag)* branch ringing
Ringelung *f (ag)* branch ringing
Ringelwalze *f (ag)* ring roller
Ringfassung *f* annular lampholder
Ringfeder *f* ring (annular) spring
Ringfeuerung *f* annular furnace

Ringfläche *f* torus
ringförmig 1. *(ch)* cyclic; 2. ring-shaped, annular, toroidal
Ringgarn *n* ring yarn
Ringhybride *f* hybrid ring *(waveguides)*
Ringkammer *f* [vacuum] doughnut, toroid
Ringkanal *m* guide channel
Ringkern *m* toroid[al] core
Ringkerntransformator *m* toroidal transformer
Ringkernwandler *m* toroidal current transformer
Ringklemme *f (mach)* lock nut
Ringkolbenzähler *m* oscillating-piston flowmeter
Ringkörper *m* torus
Ringkühler *m* annular radiator
Ringläufer *m* ring traveller *(ring spinning)*
Ringleiste *f* circular band
Ringleitung *f (el)* ring mains, closed-loop system; ring system
Ringlinse *f* annular lens
Ringmagnet *m* annular magnet
Ringmodulator *m (telec)* bridge modulator
Ringmodusfilter *n* ring mode filter
Ringnetz *n (el)* ring mains
Ringnut *f* annular (circular, ring) groove
Ringofen *m* ring (annular) kiln, circular furnace
Ringriß *m* ring (cup) shake *(timber)*
Ringröhre *f* [vacuum] doughnut, toroid
Ringschäle *f* cup (internal annual) shake, shakiness
ringschälig cup-shaken, cup-shaky
Ringschaltung *f* ring connection (circuit), closed-loop connection
Ringschieber *m* annular slide valve
Ringschluß *m (ch)* cyclization, ring closure
Ringschlüssel *m* 12 point opening socket wrench
Ringschmierlager *n* ring-oiled bearing, oil-ring bearing, self-lubricating bearing
Ringschmierung *f* ring oiling (lubrication)
Ringschmiervorrichtung *f* ring lubrication device
Ringsegment *n* annular segment *(overhead-line conductors)*
Ringsicherung *f* ring retainer (lock)
Ringspalt *m* annular clearance *(of a nozzle)*, annulus; radial clearance *(turbine)*
Ringspaltdüse *f* slotted rocket jet
Ringspaltung *f (ch)* ring cleavage
Ringspannung *f* circumferential (hoop) stress
Ringspiel *n* piston gap
Ringspindel *f* water spindle
Ringspinnen *n* ring spinning
Ringspinnerei *f* ring spinning
Ringspinnmaschine *f* ring [spinning] frame
 R. für Schußkötzer ring spinning frame for pin cops
 R. für Zettelkötzer ring spinning frame for warp cops
 R. mit Drehkopf ring spinning frame with twisting head

Ringspule *f (el)* toroid (toroidal, astatic) coil

Ringstauchprüfung *f* ring crush *(paper)*

Ringsteg *m* piston (ring) land

Ringstruktur *f* annular structure [of proton]; *(ch)* ring structure

Ringstück *n* ring-segment

ringsumlaufend all around *(weld)*

Ringsystem *n* / **kondensiertes** *(ch)* fused nucleus compound

Ringtisch *m* annular table *(machine tool)*

Ringtischmaschine *f* rotary table machine with the machine heads inside the table

Ringträger *m* ring beam (support, carrier)

Ringübertrager *m* toroidal repeating coil

Ringverbindung *f* cyclic compound

Ringverkleben *n (aut)* ring sticking

Ringverschmelzung *f* ring seal *(electron tubes)*

Ringvorspannung *f* circumferential prestressing *(prestressed concrete)*

Ringwaage *f* ring balance

Ringwalzenpresse *f* ring-roll press

Ringwelle *f* annular wave

Ringwicklung *f* toroidal (ring) winding

Ringzieher *m (aut)* ring extractor

Ringzonenreaktor *m* annular-zone reactor *(Argonaut-type)*

Ringzwirnmaschine *f* ring doubling (twisting) frame, ring twister

Rinmanns Grün *n (ch)* Rinmann's green

Rinne *f (build)* gully, channel, hollow; *(mach)* chute, channel *(conveying)*
  V-förmige R. *(build)* arris gutter

Rinnenkessel *m (build)* rainwater head

Rinnenschnauze *f (met)* spout nozzle

Rinnenschuß *m (mine)* conveyor pan

Rinnenstrang *m (mine)* line of pans

Rinnstein *m* curb[stone], kerb-stone

Rippe *f* 1. *(mach)* rib; web; 2. *(build)* stem *(beam)*; groin *(vaulting)*; 3. *(el)* rib *(porcelain insulator)*; 4. *(text)* wale *(defect)*

Rippenbildung *f* **am Bohrlochtiefsten** toe *(blasting)*

Rippendecke *f* ribbed (slab and joist) floor

Rippengußdeckel *m* ribbed cover

Rippenheizkörper *m* radiator

Rippennaht *f (aero)* rib stitching

Rippenplatte *f* ribbed slab

Rippenprofil *n* ribbed-type tread *(tyre)*

Rippenrohr *n* finned (gilled) tube, [externally] finned pipe, gilled pipe, ribbed tubing

Rippenröhrenvorwärmer *m* feed heater with ribbed tubes

Rippenrohrkessel *m* boiler with ribbed flue

Rippenrohrüberhitzer *m* ribbed (gilled) tube superheater

Rippenroststab *m* ribbed fire-bar

Rippensamt *m* rib velvet

Rippenstahl *m* ribbed steel [bar], grip bar

Rippenversatz *m (mine)* strip packing

Rippung *f* ribbing

Rips *m (text)* ribs
  figurierter R. figured ribs

Ripsbindung *f* rib weave

Risikofunktion *f* risk function

Riß *m* 1. *(met)* fissure, fracture, crack; shake, split *(timber)*; crevice; seam; 2. view *(technical drawing)*, map, sketch, design, plan, draft
  R. zwischen den Nietlöchern seam rip
  innerer R. *(met)* internal fissure
  intrakristalliner R. grain-boundary crack

Rißanfälligkeit *f* inclination to cracking

Rißausbreitung *f* crack propagation

Rißbildung *f* formation of cracks, crack formation, cracking; *(plast)* crazing; checking *(paint)*; fissuring

Risseprüfer *m* crack detector

rissig cracky
  r. werden to crack

Rißlinie *f* scribed (marking-out) line, *(Am)* layout line

Rißneigung *f* inclination to cracking

Rißprüfer *m* crack detector

Rißprüfung *f (met)* crack detection

Ritz *m* crevice, cleft, crack; scratch

Ritzel *n (mach)* pinion

Ritzelmesser *n* scoring knife

Ritzelwelle *f* pinion shaft

ritzen to scratch

Ritzhärte *f* scratch hardness; abrasive hardness

Ritzhärteprobe *f* surface scratching specimen

Ritzhärteprüfer *m* sclerometer

Ritzhärteprüfung *f* abrasion (surface scratching) test

Ritzmaschine *f* scoring machine

Ritzmesser *n* scoring knife

Ritzversuch *m* scratch test

Ritzzeichen *n* screeve mark

Rizinoleat *n* ricinoleate

Rizinolsäure *f* ricinoleic acid

Rizinusöl *n* castor oil

Robbenfänger *m* sealer, sealing vessel

Roboter *m* robot

Robotersteuerung *f* automatic direction finder

robust robust, sturdy, rugged, rigid

Rochelle-Salz *n* Rochelle salt

Rochonprisma *n* Rochon prism

Rockwellhärte *f* Rockwell hardness

Rockwellhärteprüfmaschine *f* Rockwell hardness tester

Rockwellhärteprüfung *f* Rockwell hardness test

Rodegeräte *npl* land clearing tools

Rodehacke *f* clearing (grub) hoe, mattock

Rodekörper *m* lifting body

Rödelbalken *m* ribband

Rödeldraht *m* tie (splicing, binding) wire

Rodelschlitten *m* sled[ge]

roden to uproot, to lift, to stock up
Rodentizid *n* rodenticide
Rodung *f* clearing
Roebelstab *m* Roebel rod
Rogen *m* spawn, eggs *(of fishes)*
Roggen *m* rye
Roggenmehl *n* rye flour
Roggenstroh *n* rye straw
roh crude *(metal, oil)*; blank *(workpiece)*; raw *(material)*; native *(lime, sulphur)*; rough *(surface)*; *(text)* greige
Rohabwasser *n* crude sewage
Rohbau *m* carcass, carcase; carcassing
rohbaufertig topped out
Rohbaumwolle *f* raw cotton
Rohbenzol *n* crude benzol
Rohblei *n* crude lead, lead bullion
Rohblock *m (met)* raw (rough-rolled) ingot, bloom
Rohbogenformat *n* sheet size
Rohbramme *f* slab ingot, ingot slab
Rohbraunkohle *f* raw brown coal, raw (crude) lignite
Rohedelmetall *n* bullion
Roheisen *n* pig iron
  feinkörniges R. close-grained pig
  graues R. grey (graphitic) pig iron
  heißerblasenes R. hot-blast pig iron
  hochsiliziertes R. high-silicon pig iron
  im Masselbett vergossenes R. sand-cast pig
  kaltgeblasenes R. cold-blast pig iron
  siliziumarmes R. dry (low-silicon pig) iron
  übergares R. black pig iron
Roheisenabstichrinne *f* iron runner
Roheiseneinsatz *m* pig iron charge
Roheisen-Erz-Verfahren *n* pig-and-ore process, pig iron-ore process
Roheisen[massel]gießmaschine *f* pig machine
Roheisenmischer *m* [pig iron] mixer, hot-metal mixer
Roheisenpfanne *f* hot-metal ladle
Roheisen-Schrott-Verfahren *n* pig-and-scrap process
Roheisensorte *f* grade of pig iron
Roheisenverfahren *n* pig process
Roherz *n* crude (raw) ore; run-of-mine ore
Rohfeinkohlenbunker *m* bunker for unwashed fines
Rohfilm *m* raw film stock
Rohförderkohle *f* run-of-mine coal
Rohförderung *f* bulk (gross, raw-coal) output, raised-and-weighed tonnage
Rohformat *n* untrimmed size
Rohgang *m* cold working *(blast furnace)*
Rohgas *n* crude gas
Rohgaswasser *n* crude ammonia liquor
Rohgewebe *n* greige cloth
Rohglas *n* raw glass
Rohglimmer *m* mica slab, block (natural) mica
Rohgold *n* crude gold, gold bullion
Rohgummi *m* raw rubber
Rohguß *m* undressed castings, rough-casting
Rohhaut *f* rawhide
Rohhäutelager *n* rawhide store
Rohhumus *m* raw humus
Rohkarton *m* base board
Rohkette *f (text)* grey warp
Rohknüppelbohrmaschine *f* billet drilling machine
Rohkohle *f* green (raw, uncleaned, run-of-mine, R.O.M.) coal, through-and-through [coal]
Rohkohlenturm *m* raw-coal bunker
Rohkupfer *n* crude (blister, black) copper
Rohleinen *n* unbleached linen
Rohling (mach) [work] blank, raw piece; *(plast)* blank, cold-pressed moulding
Rohluppe *f* puddle ball
Rohmaß *n* rough size
Rohmaterial *n* raw material (stock)
Rohmehl *n* raw meal *(cement manufacture)*
Rohmetall *n* crude metal
Rohniet *m* rivet [before driving]
Rohnietdurchmesser *m* diameter of rivet before driving
Rohnietlänge *f* length of rivet
Rohofen *m* ore smelting furnace
Rohöl *n* crude oil, petroleum
Rohölschlepper *m* internal-combustion tractor
Rohöltanker *m* petroleum tanker, vessel carrying petroleum in bulk
Rohpapier *n* base (body) paper
Rohpappe *f* crude board
Rohprodukt *n* raw material
Rohr *n* tube, pipe; duct; conduit
  R. mit aufgelötetem Flansch pipe with flange brazed on
  R. mit aufgenietetem Flansch pipe with flange riveted on
  R. mit aufgeschraubtem Flansch pipe with flange screwed on
  R. mit festem Flansch pipe with fixed flange
  R. mit losem Flansch pipe with loose flange
  ausziehbares R. telescopic tube
  biegsames R. flexible conduit
  erdverlegtes R. underground pipe
  geglättetes R. smoothing tube
  hoch geripptes R. high-fin tubing
  isoliertes R. pipe covered with non-conducting material
  kaltgepilgertes R. rocked tube
  Kundtsches R. Kundt's tube
  nacktes R. bare pipe
  nahtloses R. seamless tube
  niedrig geripptes R. low-fin tube
  Pitotsches R. Pitot tube
  rundes R. circular pipe
  stumpfgeschweißtes R. butt welded tube

wärmeisoliertes R. lagged pipe
Rohrabschneider *m* pipe cutter
Rohrabzieher *m* take-up machine *(extruding)*
Rohrabzweigung *f* pipe branching
Rohransatz *m* flanged socket (nozzle)
Rohraufhängung *f* pipe suspension
Rohraufsatz *m* cowl
Rohraufweitpresse *f* tube-expanding press
Rohrauskratzer *m* tube scraper
Rohrbefestigung *f* fixing of pipes
Rohrbekleidung *f* pipe lagging (covering)
Rohrbiegemaschine *f* pipe bender
Rohrbiegung *f* pipe bend
Rohrboden *m* tube plate
Rohrbogen *m* pipe bend
R. von 90° quarter bend
Rohrbohrung *f* pipe bore
Rohrbruch *m* pipe fracture
Rohrbruch- und -absperrventil *n* / **kombiniertes** combined steam-pipe isolating and stop valve
Rohrbruchventil *n* automatic steam-pipe isolating valve
Rohrbrücke *f* pipe bridge
Rohrbund *n* nest of tubes
Rohrbündel *n s.* Rohrbund
Röhrchen *n* tubule
Röhrchenplatte *f* tubes-type plate *(battery)*
Röhrchensicherung *f* glass-enclosed fuse
Röhrchenwebstuhl *m* tube loom
Rohrdach *n* thatched roof
Rohrdichtung *f* pipe packing
Rohrdipol *m* sleeve dipole
Rohrdraht *m* conduit (metal-sheathed) wire
Rohrdrehmaschine *f* tube lathe
doppelte R. double-ended tube lathe
Rohrdurchlaß *m (build)* pipe culvert
Röhre *f* valve, *(Am)* tube
R. großer Steilheit high-slope valve
R. mit Elektronenbündelung aligned-grid valve
R. mit Geschwindigkeitsmodelung (Geschwindigkeitssteuerung) velocity-modulated valve
R. mit hochgesetzter Katode self-biased valve
R. mit Hochspannungsheizfaden high-voltage heater valve
R. mit hohem Innenwiderstand high plate-resistance valve
R. mit Mikrofoneffekt microphonic valve
R. mit veränderbarem Durchgriff valve with variable slope, variable mu valve
abgeschmolzene R. sealed-off valve
anodenlose R. anodeless valve
antimikrofonische R. anti-microphonic valve
Braunsche R. oscillograph (cathode-ray, electron beam, Braun) tube
demontierbare R. assembled-parts valve
elektrodenlose R. electrodeless valve, nullode
gasgefüllte R. ionic (gas-filled, soft-vacuum) valve
geschwindigkeitsmodulierte R. velocity-modulated valve
indirekt geheizte R. indirectly heated valve; cathode-heater valve
kalte R. unlighted valve
metallisierte R. metallized valve
rückgekoppelte R. back-coupled valve
schlecht evakuierte R. gassy valve
sockellose R. footless valve
stark rauschende R. noisy valve
steile R. high-slope valve
wechselstromgeheizte R. mains valve
weiche R. soft valve
Röhrenabschirmung *f* valve shield
Röhrenalterung *f* valve ageing
Röhrenanheizzeit *f* valve heating time
Röhrenanlage *f* tube assembly *(Cottrel precipitator)*
Röhrenarbeitswiderstand *m* dynamic plate resistance
Röhrenausfall *m* valve failure
Röhrenauspumpstutzen *m* glass-tipped vacuum seal
Röhrenauszieher *m* valve extractor
röhrenbestückt valved
Röhrenbestückung *f* complement of valves
Röhrenbrumm *m* valve hum
Röhrenbussole *f* tubular compass
Röhrendetektor *m* [thermionic] valve detector
Röhrendiode *f* thermionic diode
Röhreneinbruch *m (mine)* burn (shatter) cut
Röhrenelektrode *f* valve electrode
Röhrenelektronik *f* valve electronics
Röhrenempfänger *m* valve receiver
Röhrenersatz *m* valve replacement
Röhrenfassung *f* valve base (socket, holder)
federnde R. anti-vibration valveholder, cushioned socket
Röhrenfedermanometer *n* spring-type pressure gauge, Bourdon-type pressure gauge, spring-tube manometer
röhrenförmig tubular
Röhrenfuß *m* stem assembly, foot
Röhrengenerator *m* thermionic (valve) generator, valve oscillator
Röhrengestell *n* electron-valve rack
Röhrengießerei *f* pipe foundry
Röhrengleichrichter *m* valve (thermionic) rectifier
Röhrenhalterung *f* valve clamp (retainer)
Röhrenheizfläche *f* fire-tube heating surface
Röhrenkabel *n* conduit (duct) cable
Röhrenkapazität *f* interelectrode capacitance
Röhrenkennlinie *f* valve characteristic
Röhrenkessel *m* tubular boiler

Röhrenklingen *n* microphony, microphonism
Röhrenkolben *m* [valve] envelope, [valve] bulb
Röhrenkopplung *f* intervalve coupling
Röhrenkühler *m (aut)* tubular radiator
Röhrenlager *n* tubular bearing
Röhrenlampe *f* tube (tubular) lamp
Röhrenlebensdauer *f* valve life
Röhrenleistung *f* valve output (power)
Röhrenlibelle *f* glass-tube, vial having cylindrical body ground internally to a barrel shape
Röhrenofen *m* tube furnace, pipe still
Röhrenprüfgerät *n* valve tester (checker)
Röhrenprüfung *f* valve rejection test; valve testing (checking)
Röhrenrauschen *n* valve noise (hiss)
Röhrenreinigungskette *f (therm)* chain tube-cleaner
Röhrenrückkopplungsschaltung *f* **nach Meißner / induktive** Meissner-circuit
Röhrensatz *m* set of valves
Röhrenschaltung *f* valve circuit
Röhrensender *m* [thermionic] valve transmitter
Röhrensockel *m* valve (plug) base
Röhrensockelstift *m* valve prong (pin)
Röhrenspannungsgleichhalter *m* valve stabilizer
Röhrensummer *m* valve oscillator
Röhrentrockner *m* tubular drier
Röhrenverdampfer *m* tubular vaporizer
Röhrenverstärker *m* valve amplifier
Röhrenvoltmesser *m* valve (thermionic) voltmeter
Röhrenvorwärmer *m* tubular-feed heater
  R. mit Heizung durch Abgase economizer, waste gas feed heater
Röhrenwärmetauscher *m* tubular recuperative air-heater
Röhrenwechsel *m* valve replacement
Röhrenzwischenstecker *m* valve adapter
Rohrerder *m* tubular earthing bus
Rohrfahrt *f* / **letzte** oil string
Rohrfeder *f* Bourdon (spring) tube
Rohrfernleitung *f* pipeline
Rohrflansch *m* pipe flange
Rohrformerei *f* pipe moulding
rohrförmig tubular
Rohrformstück *n* pipe fitting
Rohrgeflechtfender *m (ship)* cane fender
Rohrgewinde *n* pipe thread
Rohrguß *m* pipe casting
Rohrhaken *m* wall hook
Rohrhalter *m* pipe bracket
Rohrhaspelmaschine *f* tubes strip winding machine
Rohrhaube *f* cowl
Rohrisolierung *f* tubular insulation
Rohrkanal *m* service channel (duct); pipe trench (channel)
Rohrkolben *m* plunger
Rohrkondensator *m* tubular capacitor
Rohrkopfbenzin *n* casinghead (natural) petrol
Rohrkreisförderer *m* tubular conveyor
Rohrkrümmer *m* pipe bend, elbow, knee
Rohrkühlung *f* barrel cooling
Rohrlagerung *f* bedding of pipes
Rohrleger *m* pipe fitter, plumber
Rohrlegerhandwerk *n* plumbing trade (craft)
Rohrlegerschiff *n* pipe line vessel
Rohrleitung *f* pipe conduit (line), line, piping, tubing
  biegsame R. flexible pipe
  fallende R. descending piping
  oberirdische R. overhead pipe installation
  steigende R. ascending piping
  unterirdische R. piping under the floor, underground pipe installation
Rohrleitungsplan *m* general pumping arrangement, diagrammatic plan of pipes, pipe leads
Rohrleitungszubehör *n* conduit fittings
Rohrliderung *f* pipe packing
Rohrluppe *f* tube bloom
Rohrmast *m* tubular mast
Rohrmelkanlage *f* pipeline milking installation
Rohrmelksystem *n* pipeline milking system
Rohrmontierung *f* **mit offener Gitterkonstruktion** skeleton-type tube mount
Rohrmuffe *f* pipe bell
Rohrmühle *f* rod (tube) mill
Rohrmündung *f (mil)* muzzle
Rohrnetz *n* network of pipes
Rohrofen *m* tube furnace
Rohrpackung *f* pipe packing
Rohrplan *m* plan of pipe-installation, piping plan
Rohrplattenheizfläche *f* tube-plate heating surface
Rohrpost *f* pneumatic tube conveyor, shuttle, rabbit
  hydraulische R. hydraulic rabbit
Rohrpostbüchse *f* shuttle, rabbit
Rohrpostgebläse *n* pneumatic tube blower
Rohrpostkanal *m* rabbit tube
Rohrpostventil *n* pneumatic valve
Rohrpresse *f* metal-tube-making press, tube extruding press
Rohrquerschnitt *m* pipe section
Rohrquerträger *m* tubular cross member
Rohrrahmen *m (aut)* tubular frame
Rohrreduktionsstück *n* pipe reducer
Rohrreibung *f* pipe friction
Rohrreibungsverlust *m* pipe friction loss
Rohrreiniger *m* tube scraper
  verstellbarer R. adjustable tube-scraper
Rohrreinigungsbürste *f* tube brush
Rohrrinnen *n* dripping from a pipe joint
Rohrrippe *f* gill
Rohrsäule *f (mine)* pipe column
Rohrschaber *m* tube scraper

Rohrschacht *m* pipe trench (channel)
Rohrschaft *m* tubular shank
Rohrschelle *f* pipe clip (clamp), wall clamp
Rohrschlange *f* coil, coiled pipe
Rohrschlitzantenne *f* slot aerial
Rohrschlüssel *m* pipe wrench
Rohrschraubenschlüssel *m* pipe spanner
Rohrschraubstock *m* pipe vice
Rohrschweißwalzwerk *n* pipe-welding rolling mill
Rohrsicherung *f* tube fuse
Rohrstäbchenmatte *f* cane mat
Rohrsteckschlüssel *m* L-handled socket wrench
Rohrstempel *m (mine)* tubular prop
Rohrstopfer *m* tube stopper
Rohrstoßbank *f* tube push-bench
Rohrstrang *m* pipe line, line of pipes
Rohrstrangpresse *f* tube extruding press
Rohrstreifenwalzwerk *n* skelp rolling mill
Rohrstück *n* **/ gerades** straight [pipe] length
Rohrstütze *f* pipe support
Rohrstutzen *m* flanged socket (nozzle)
Rohrtour *f* pipe column
wasserabsperrende R. water string
Rohrtrum *m (mine)* pipe[way] compartment, pipeway
Rohrtunnel *m (ship)* pipe tunnel
Rohrumhüllung *f* pipe lagging (covering)
Rohrunterstützungsrolle *f* pipe-supporting roller
Rohrverbindung *f* tubing connection, pipe joint; jointing of pipes; union, coupling
lösbare R. disconnectable coupling
Rohrverbindungsflansch *m* pipe connection flange
Rohrverbindungsstück *n* pipe joint
Rohrverengung *f* baffled (constricted) tube
Rohrverlegung *f* pipe laying (installation), laying of pipes; *(el)* conduit wiring
Rohrverschluß *m* pipe closer (stopper)
Rohrverschraubung *f* screwed (bolted) pipe-joint
Rohrverteiler *m* manifold
Rohrverzweigung *f* pipe branching, junction [in a pipeline], manifold
Rohrwalzwerk *n* pipe (tube rolling) mill
Rohrwand *f* pipe wall, wall of pipe (tube)
Rohrwanddicke *f* thickness of pipe
Rohrwanddickenmesser *m* tube thickness gauge
Rohrwandung *f* pipe wall, wall of pipe (tube)
Rohrwasser *n* untreated water
Rohrwechsel *m* retubing
Rohrweite *f* internal diameter of pipe (tube)
Rohrwiderstand *m* tubular-type resistor
Rohrzange *f* pipe wrench (tongs)
Rohrzerstäuber *m* tubular sprayer (atomizer)
Rohrziehen *n* tube drawing, tubing
Rohrzucker *m* cane sugar
Rohrzuckersaft *m* cane [sugar] juice
Rohsaft *m* raw juice
Rohsalpeter *m* caliche, Chile saltpetre
Rohsalz *n* crude salt
Rohschlacke *f* mat-scoria, tap cinder
Rohschlamm *m* slurry *(cement manufacture)*
Rohschmieröl *n* black oil
Rohseide *f* raw silk
Rohseidenputzmaschine *f* raw-silk cleaning machine
Rohsilber *n* silver bullion
Rohsoda *f* black ash
Rohstahl *m* crude steel
Rohstahlausstoß *m* crude-steel output
Rohstahlblock *m* steel ingot
Rohstahlerzeugung *f* crude-steel manufacture
Rohstein *m* matte
Rohstoff *m* raw material
Rohstrecker *m* gill box minder *(spinning)*
Rohteer *m* crude tar
Rohteil *n s.* Rohling
Rohwollwäsche *f* raw-wool scouring
Rohzink *n* raw zinc, spelter
Rohzucker *m* muscovado, raw sugar
Rolladen *m* roller blind
Rolladenfabrik *f* wood blind factory
Rollapparat *m* rolling apparatus (set); reeling device
Rollbahn *f* taxiway
Rollbahnbefeuerung *f* taxiway lighting
Rollbewegung *f* rolling motion
Rollblock *m* paper (roller-type) capacitor
Rollbock *m* truck
Rollboden *m* moving apron
Rollbrücke *f* retractable roller bridge
Rolle *f* pulley *(tackle)*, idler, block *(pulley)*; roll; *(text)* reel *(yarn)*; mangle; coil *(wire)*; pacing roller *(pace-maker motorcycle)*; *(mine)* boxhole, chute
R. zum Gewinderollen thread roll
feste R. fixed (fast) pulley
gefederte R. cushion (featherbed) idler
gesteuerte langsame R. *(aero)* hesitation roll
glatte R. flat roll
lose R. movable (loose) pulley
Rollegge *f* rotary harrow
rollen to roll
sich r. to curl up, to coil
Rollenanklebevorrichtung *f* reel paster
automatische R. autopaster
Rollenbahn *f* roller conveyor (track)
R. ohne Antrieb gravity-type roller conveyor
angetriebene R. powered roller conveyor
Rollenbock *m* bogie
Rollenbreite *f* width of a reel *(yarn)*
rollend rolling *(friction)*
Rollendraht *m* coiled wire[s]
Rollendrehscheibe *f (ry)* turntable on roller bearings

Rollendrehstern *m* reel star
Rollendruckmaschine *f* web printing press
Rolleneinhebwagen *m* reel lifting truck
Rollenelektrode *f* circular electrode
Rollenförderband *n* roller conveyor
Rollenförderer *m* roller conveyor
Rollenförderstrecke *f* roller conveyor
Rollengabel *f (el)* trolley-shield
Rollengang *m* power-operated roller conveyor, roller rail
  R. vor einer Walze front mill table
Rollenhebevorrichtung *f* roller table
Rollenkette *f* roller chain
Rollenkettenrad *n* roller chain sprocket
Rollenkettentrieb *m* roller chain drive
Rollenlager *n* roller bearing
Rollenoffset *m* web offset
Rollenpackmaschine *f* end bander
Rollenquetscher *m* roller squeegee
Rollenschneid- und -wickelmaschine *f* slitting and rewinding machine
Rollenschweißgerät *n* roll spot welding machine
Rollenspriegel *m* roller bow
Rollenstern *m* reel star
Rollenstößel *m* roller follower (tappet)
Rollenstromabnehmer *m* trolley roller
Rollentiefdruck *m* rotogravure
Rollentisch *m* roller table
Rollenzählwerk *n* drum-type counting system, drum-type counter mechanism
Rollfeld *n (aero)* tarmac
Rollfeldringstraße *f* perimeter track
Rollfilm *m* roll film
Rollfuhrdienst *m* collection and delivery service, cartage department
Rollgang *m* roll table, roller gear bed, table roller
  R. hinter der Walze back mill table
Rollgeschwindigkeit *f* velocity of pure rolling
Rollgeschwindigkeitsmesser *m (aero)* roll rate indicator
Rollhaken *m* roller-type hook
rollig cohesionless, non-cohesive, granular, unconsolidated *(soil)*
Rollkalander *m* rolling (swissing) calender
Rollkanal *m* gravity track (conveyor)
Rollkolbenkompressor *m (aut)* lobe-type compressor
Rollkondensator *m* paper (roller-type) capacitor
Rollkörper *m* rolling member *(bearing)*
Rollkreis *m* generating (rolling) circle *(gear)*
Rollnaht *f* roll spot weld
Rolloch *n (mine)* boxhole, chute
  in Holz ausgebautes R. crib chute
Rollochbohrmaschine *f (mine)* boxhole (chute) borer
Rollochverschluß *m (mine)* chute-gate
Rollofen *m* roll-over type furnace
Roll-on-roll-off-Schiff *n* trailer (roll-on-roll-off) ship
Rollperiode *f (ship)* period of roll[ing]
Rollpunkt *m* instantaneous centre
Rollreibung *f* rolling friction
Rollsäumer *m (text)* foot for rolled hems
Rollschicht *f* barge (brick-on-edge) course *(brickwork)*
Rollschuh *m* roller skate
Rollschuhbahn *f* roller-skating rink
Rollschuhlauf[en *n***]** *m* roller-skating
Rollschuhläufer *m* skater
Rollsichter *m* fractionator, roll screen
Rollsitz *m* slide, sliding seat *(boat)*
Rollstab *m* roller rod *(hatchery)*
Rolltreppe *f* escalator, moving staircase
Rolltür *f* sliding door
Rollversuch *m* **[eines Flugzeuges]** taxiing trial
Rollvorhang *m* roller blind
Rollwagen *m* wheeled truck
Rollwiderstand *m (ry)* running resistance, resistance to rolling; *(aut)* drag road
Rollzeug *n* rolling set (apparatus)
Rollziegel *m* brick-on-edge
Romankalk *m* Roman (natural) cement
rommeln to rumble, to tumble
Rommeltrommel *f* rumble, tumble, tumbling barrel
Ronde *f* circular blank
Rondellennickel *n* nickel rondelles
röntgen to X-ray, to radiograph
Röntgen *n* roentgen, r-unit, R, r
Röntgenanlage *f* X-ray unit
Röntgenapparat *m* X-ray apparatus
Röntgenäquivalent *n* roentgen equivalent
  biologisches R. roentgen equivalent, man, roentgen equivalent man unit, rem
  physikalisches R. roentgen equivalent physical; physical roentgen equivalent, rep
Röntgenassistent[in *f***]** *m* X-ray assistant
Röntgenaufnahme *f* roentgenogram, radiograph, X-ray photograph
röntgenbestrahlt X-[ir]radiated
Röntgenbestrahlung *f* X-ray radiation, irradiation by X-rays
Röntgenbild *n* X-ray pattern
Röntgenblitz *m* X-ray flash
Röntgenblitzröhre *f* X-ray flash tube
Röntgendiaskop *n* X-ray diascope
Röntgendosis *f* X-ray dose, quantity of X-rays
Röntgendosismesser *m* X-ray dosemeter
Röntgendurchstrahlung *f* X-ray penetration [through]
Röntgenemission *f* X-ray emission
Röntgenfeinstruktur *f* X-ray structure
Röntgenfeinstrukturprüfung *f* X-ray crystal analysis
Röntgenfilm *m* X-ray film

Röntgenfluoreszenzstrahlung *f* X-ray fluorescence radiation
Röntgengerät *n* X-ray unit (apparatus), roentgen (radiological) apparatus
Röntgengrobstrukturuntersuchung *f* macroscopic X-ray analysis
Röntgenkammer *f* roentgen (X-ray) chamber
Röntgenkater *m* nervous irritation after X-ray irradiation
Röntgenkinematografie *f* roentgen cinematography, cineradiography
Röntgenkontakttherapie *f* X-ray contact therapy
Röntgenkunde *f* roentgenology
Röntgenlinie *f* X-ray line
Röntgenlumineszenz *f* roentgen (X-ray) luminescence
Röntgenmaterialprüfung *f* radio-materiology, radio-examination of materials
Röntgenmesser *m s.* Röntgenmeter
Röntgenmeßkunde *f* X-ray measurement
Röntgenmetallografie *f* radiometallography, X-ray metallography, metal radiography
Röntgenmeter *n* r-meter, roentgen meter, roentgenometer
Röntgenmikroskop *n* X-ray microscope
Röntgenografie *f* roentgenography, radiography
Röntgenogramm *n* radiograph, roentgenogram, X-ray picture
Röntgenoskopie *f* fluoroscopy, roentgenoscopy, radioscopy
Röntgenphysik *f* physics of X-rays
Röntgenprüfung *f* roentgenomateriology, X-ray materiology
Röntgenquant *n* X-ray quantum
Röntgenröhre *f* roentgen (X-ray) tube
  gasgefüllte R. gas X-ray tube
  weiche R. gas X-ray tube
Röntgenröhrenvakuum *n* X-ray vacuum
Röntgenschirm *m* fluoroscope
Röntgenschirmbildgerät *n* photofluorograph[ic unit]
Röntgenserie *f* X-ray series
Röntgenspektrograf *m* X-ray spectrograph
Röntgenspektrometer *n* X-ray spectrometer
Röntgenspektrum *n* roentgen (X-ray) spectrum
Röntgenstrahl *m* X-ray
Röntgenstrahlen *mpl* X-rays, X-ray radiation, X-radiation
  charakteristische R. characteristic X-rays
  harte R. hard X-rays
  weiche R. soft X-rays
Röntgenstrahlenabsorption *f* absorption of X-rays
Röntgenstrahlenbeugung *f* X-ray diffraction
Röntgenstrahlenbündel *n* X-ray beam
Röntgenstrahlendosis *f* X-ray dose
Röntgenstrahleneinheit *f* X-ray unit
Röntgenstrahlenfeinstruktur *f* X-ray fine structure
Röntgenstrahlenintensität *f* intensity of X-rays
Röntgenstrahlenquelle *f* X-ray source
Röntgenstrahlenschutz *m* X-ray protection
Röntgenstrahlung *f* X-radiation, X-ray radiation, roentgen-radiation, X-rays
  charakteristische R. characteristic X-radiation, characteristic X-ray radiation
  kontinuierliche R. continuous X-radiation, continuous X-ray radiation
Röntgenstreustrahlung *f* scattered X-rays
Röntgenstrukturuntersuchung *f* X-ray crystallography
Röntgentechnik *f* X-ray technique
Röntgentherapie *f* roentgen (X-ray) therapy
  intrakavitäre R. intracavitary roentgen therapy
Röntgentiefbestrahlung *f* deep X-ray irradiation
Röntgentiefenbehandlung *f* X-ray deep therapy
Röntgentiefentherapie *f* deep roentgen therapy, X-ray deep therapy
Röntgentransformator *m* X-ray transformer
Röntgenuntersuchung *f* X-ray examination (investigation)
Röntgenverbrennung *f* radiation (X-ray) burn
Röntgenwerkstoffprüfung *f* roentgenomateriology, X-ray materiology, X-ray materials testing
Rootsgebläse *n* Roots blower
Rootslader *m* Roots blower
Rosenberggenerator *m s.* Querfeldgenerator
Rosenholz *n* rose wood
Rosenöl *n* attar, rose oil
Rosenquarz *m* rose-quartz
Rosette *f* 1. rosette; 2. *(el)* base block *(surface switch)*; 3. *(ry)* ventilating grille; circular air grid; 4. draught regulating register
Rosettenrädchen *n* register (draught regulating, circular-air-grid) wheel
Rosettenschneide *f* six point bit
Rosmarin *m* rosemary
Rosmarinöl *n* rosemary oil
Rosolsäure *f* rosolic acid
Roßchevreaux *n* glazed horse
Roßhaar *n* horse hair
Roßleder *n* horse-hide leather
Rost *m* 1. rust; 2. grate, grating; grid, grill[e]; screen
  R. mit Wasserberieselung der Roststäbe grate with sprinkling of the fire bars
  beweglicher R. shaking grate, furnace with movable grate-bars
  fester R. stationary (fixed) grate
  geteilter R. sectional grate
  mechanisch bewegter R. reciprocating grate
Röstanlage *f* calcining plant
  R. für Erze ore roasting plant

Rostanstrengung *f* coal burnt per unit grate area
Rostausdehnung *f* expansion of the grate bars
Rostbelag *m* rust coating
Rostbelastung *f* coal burnt per unit grate area
Rostbeschickung *f* grate charging, charging of the grate *(boiler)*
  mechanische R. mechanical stoking (charging)
rostbeständig resistant to rusting, rust-resisting
Rostbeständigkeit *f* stainless (rust-resisting) property
Röstbett *n* roasting bed
Rostbildung *f* rust formation, rusting
Rostdock *n (ship)* gridiron
rosten to rust, to corrode
rösten to calcine; to roast *(ore)*; to ret *(flax)*
Rösten *n* roasting
  oxydierendes R. oxidation roasting
Rostentfernungsmittel *n* rust removing agent
Rösterz *n* calcined ore
rostfarben rubiginous
Rostfeld *n* grate section
rostfest anticorrosive, rustproof
Rostfläche *f* grate area *(boiler)*
Rostflächenbeanspruchung *f* coal burnt per unit grate area
rostfrei stainless *(steel)*, non-rusting, antirust
Rostfuge *f* grate opening
Rostgestell *n* grate bar frame
Röstherd *m* roasting hearth
rostig rusty
Rostkammer *f* grizzly chamber
Rostkette *f* grate-chain
Rostkettentrommel *f* grate-chain drum
Rostkitt *m* rust cement
Rostkühlung *f* grate cooling
Rostneigung *f* 1. inclination of grate; 2. tendency to rust
Rostoberfläche *f* grate area (surface)
Röstofen *m* roasting furnace (kiln); calciner
  mehrherdiger R. multihearth roasting furnace
Rostplatte *f* grate plate
  perforierte R. pin-hole grate
Rostrahmen *m* grate-bar frame
Röstreaktionsarbeit *f* roast-reaction process, air-reduction method of smelting
Röstreduktionsarbeit *f* roast-reduction process, blast-furnace method; extraction by fusion *(bismuth)*
Röstreduktionsverfahren *n* roasting reduction method
Röstreife *f* retting ripeness
Röstrückstand *m (met)* residue from roasting
Röstschachtofen *m* roasting blast-furnace
Rostschicht *f* rust coating
Röstschlacke *f* scoria of raw matt
Röstschmelzen *n* roasting-smelting
Rostschutz *m* rust prevention, rust-proofing
Rostschutzanstrich *m* anti-rust paint, rust-protecting paint
rostschützend rust-preventive
Rostschutzfarbe *f* rust inhibiting (preventive) paint, anti-rust paint
Rostschutzmittel *n* rust preventive, antirusting (anticorrosive) agent
Rostsohle *f (mine)* grizzly level
Roststab *m* grate (fire) bar; grizzly bar
  dreifacher R. triple grate bar
  fischbauchförmiger R. fish-bellied grate bar
  keilförmiger R. wedge-shaped grate bar
  prismatischer R. prismatic grate bar
Roststabkopf *m* head of the grate bar
Roststabträger *m* grate bar support (bearer)
Rostträger *m* grate bar frame
Rosttreppe *f* grate step
Rost- und Korrosionsschutzfett *n* rust proofing grease
Röstung *f* roasting
  magnetisierende R. magnetic roasting
  sulfatisierende R. sulphating roasting
rostverhütend rust-preventive
Rostwiderstand *m* stain-resistance
Rotamesser *m* rotameter, rotameter (variable-area) flowmeter
Rotangtauwerk *n (ship)* rattan rope
Rotarybohren *n* rotary drilling
Rotarytiefbohrverfahren *n* rotary [deep drilling] process
Rotarytisch *m* rotary table
Rotaryverfahren *n s.* Rotarytiefbohrverfahren
Rotation *f* rotation
  R. eines Vektors curl of a vector
Rotationsachse *f* rotation axis, axis of rotation
Rotationsanregung *f* rotational excitation
Rotationsbande *f* rotational band
Rotationsbestrahlung *f* rotation irradiation
Rotationsdampfmaschine *f* rotary steam engine
Rotationsdiffusion *f* rotational diffusion
Rotationsdispersion *f* rotary dispersion
Rotationsdruck *m* rotary printing
Rotationsellipsoid *n* spheroid
  R. um die große Achse prolate spheroid
  R. um die kleine Achse oblate spheroid
  abgeplattetes R. oblate spheroid
  gestrecktes R. prolate spheroid
Rotations-EMK *f* rotational electromotive force
Rotationsenergie *f* rotational energy
Rotationsfeinstruktur *f* rotational fine structure
Rotationsfilz *m* news felt
Rotationsfläche *f* surface of rotation (revolution)
Rotationsfreiheit *f* rotational freedom
Rotationsfrequenz *f* rotational frequency
Rotationshackgerät *n* rotary hoe
Rotationshyperboloid *n* hyperboloid of revolution

rotationsinvariant rotation-invariant
Rotationskegel *m* cone of revolution
Rotationskonstante *f* rotational constant
Rotationskörper *m* solid of revolution
Rotationslinie *f* rotational line
Rotationsmaschine *f* rotary press
Rotationsniveau *n* *(ph)* rotational level
Rotationspaar *n* couple of rotations
  momentanes R. couple of instantaneous rotations
Rotationsparaboloid *n* paraboloid of revolution
Rotationspolarisation *f* optical activity
Rotationspressen *n* *(plast)* rotation moulding
Rotationspumpe *f* rotary pump
Rotationsquantenzahl *f* rotational quantum number
Rotationsquerschneider *m* rotary cross cutter, *(Am)* rotary sheeter
Rotationsschablonendruckmaschine *f* rotary stencil printing machine
Rotationsschale *f* rotational shell, shell of revolution
Rotationsschwingungsbande *f* rotation-vibration band
Rotationsschwingungsenergie *f* rotation-vibration energy
Rotationsschwingungskonstante *f* rotation-vibration constant
Rotationsschwingungsspektrum *n* rotation-vibration spectrum
Rotationsspektrum *n* rotation[al] spectrum
Rotationssymmetrie *f* rotational symmetry
rotationssymmetrisch rotationally symmetrical
Rotationsterm *m* rotational level
Rotationstherapie *f* rotation therapy
Rotationstiefdruck *m* rotogravure
Rotationstrocknung *f* rotary drying
Rotationstrommeltrockner *m* rotating drum drier
  R. mit stationärem Gehäuse fixed-drum-type rotary drier
Rotationsverbreitung *f* rotational broadening *(spectrum)*
Rotationsviskosimeter *n* rotational-type viscosimeter, drag-torque viscosimeter
Rotationszentrum *n* centre of rotation (gyration)
Rotationszustand *m* rotational state
Rotationszwilling *m* rotation twin
Rotbande *f* red band
Rotbleierz *n* crocoi[si]te
Rotbruch *m* red shortness
rotbrüchig red-short
Rotbuche *f* common beech
Roteiche *f* / **Australische** bastard mahogany
Roteisenstein *m* haematite, blood stone, kidney ore
rotempfindlich red-sensitive
Rotfilter *n* red filter
Rotfuchs *m* European fox
Rotgerber *m* bark tanner
rotglühend red-hot
Rotglut *f* red heat
Rotgluthärte *f* red-hardness
Rotgültigerz *n* / **dunkles** pyrargyrite, dark-red silver ore
  lichtes R. proustite, light-red silver ore
Rotguß *m* red brass
Rotholz *n* redwood
rotkernig red-hearted *(wood)*
Rotkupfererz *n* red copper ore
Rotmessing *n* pinchbeck, red brass
Rotnickelkies *m* niccolite
Rotor *m* 1. *(el)* rotor; 2. *(math)* curl *(vector)*
  gewickelter R. wound rotor
Rotoranlasser *m* rotor starter
Rotoren *mpl* / **ineinanderkämmende** *(aero)* intermeshing rotors *(helicopter)*
Rotorflügel *m* rotor vane
Rotornutenfräsmaschine *f* rotor slot milling machine
Rotorschiff *n* rotor ship (vessel)
Rotorspannung *f* rotor voltage
Rotorspindel *f* rotor shaft
Rotorspule *f* rotor coil
Rotorstrom *m* rotor current
Rototrol *n* rototrol
Rotschlamm *m* red mud
Rotsilber *n* ruby silver ore
rotstichig red-tinged
Rotstift *m* red pencil
Rottanne *f* spruce
Rotte *f* crew, gang, team
Rottenarbeiter *m* platelayer, lengthman, *(Am)* trackman
Rottenmeister *m* foreman
Rottenwagen *m* gang car
Rotunde *f* rotunda
Rotverschiebung *f* red shift *(spectral lines)*
Rotwein *m* red wine, claret
Rotzinkerz *n* red oxide of zinc
Rouleau *n* sun-blind
Rouleauxdruck *m* roller printing
Rouleauxdruckmaschine *f* cylinder printing machine
Roulette *f* roulette *(engraving)*
Route *f* *(ship)* track, route
Routineanalyse *f* routine analysis
  qualitative R. qualitative routine analysis
Routinebetrieb *m* routine operation
Routineboot *n* trot boat
Routinegerät *n* routine instrument
routinemäßig routinely
Routinemessung *f* routine check
Routineoperation *f* *(dat)* *(Am)* red-tape operation
Routinestart *m* normal start-up
Rowland-Gitter *n* Rowland grate

Royalsegel *n* royal sail
Rubeanwasserstoff *m* rubeanic acid, rubean hydride
Rübenaufladegerät *n* beet elevator (loader)
Rübenausdünngerät *n* gapping (thinning) machine, [along-the-row] gapper, down-the-row thinner
Rübenblattableger *m* beet top depositor
Rübenblattaufladegerät *n* sugar-beet top collector
Rübenblätter *npl* beet tops
Rübendrückrolle *f* single-wheel roller
Rübenerntemaschine *f* beet harvesting machine
Rübenhebegerät *n* sugar beet lifter
Rübenhöhenförderer *m* beet elevator
Rübenkernauslesemaschine *f* inclined conveyor-type grader *(sugar beet seed)*
Rübenkombine *f* complete sugar-beet harvester
Rübenköpfer *m* [beet] topper
Rübenmusmühle *f* beet pulper
Rübenrodegabel *f* lifting (beet) fork
Rübenrodegerät *n* root lifter
Rübenrodekörper *m* beet lifting body
Rübenrodepflug *m* beet [lifting] plough
Rübensaft *m* beet juice
Rübensamenauslesemaschine *f* beet seed cleaner, inclined conveyor-type grader *(sugar-beet seed)*
Rübensammelroder *m* complete sugar-beet harvester
Rübenschaufler *m* beet scoop
Rübenschneider *m* root cutter
Rübenschnitzelmaschine *f* beet slicing machine
Rübenschrägförderer *m* beet elevator
Rübenspaten *m* lifting (beet) fork
Rübenstampfmaschine *f* beet pulper
Rüben- und Kartoffelgabel *f* lifting fork
Rübenverhackgerät *n* gapping (thinning) machine, [along-the-row] gapper, down-the-row thinner
Rübenverladegerät *n* root elevator
Rübenverziehgerät *n* root thinner
Rübenvollerntemaschine *f* complete sugar-beet harvester
Rübenvorreinigungs- und -aufbereitungsmaschine *f* root cleaner and cutter
Rübenwaschmaschine *f* beet washer
Rübenzucker *m* beet sugar
Rubidium *n* rubidium
Rubidiumfluorid *n* rubidium fluoride
Rubin *m* ruby
Rubin-Laser *m* ruby laser
Rubin-Maser *m* ruby maser
rubinrot rubinous, ruby-coloured
Rüböl *n* colza oil
Rubrik *f* box [head]
Rübsenöl *n s.* Rüböl
Ruck *m* jerk
Rückansicht *f* rear view
Rückantwort *f* reply
Rückarbeitsverfahren *n* back-to-back method *(e. g. of generators)*
ruckartig jerky
rückbar flittable *(conveyor)*; shiftable
Rückbau *m s.* Rückwärtsbau
rückbilden to re-form; to reshape
Rückblickspiegel *m* driving (rear-view) mirror
Rückbreite *f* **eines Förderers** advance per conveyor move
Rückbrennung *f* passing of the flames to the front of furnace
rückdiffundieren to diffuse back
Rückdiffusion *f* back diffusion
Rückdrehmoment *n* restoring moment (torque)
Rückdrehung *f* **der Spindeln** reversing of the spindles
Rückdruck *m* back pressure
  R. auf die Düse reaction upon the nozzle *(turbine)*
Rückdruckturbine *f* reaction turbine
Rückdruckventil *n* speed-control valve, back-pressure valve
rucken to jerk
rücken to shift; to jack; to shove; to throw *(lever)*; to move over, to push
Rücken *n (wood)* [re]moving
  R. des Förderers conveyor advancing
Rücken *m* 1. back *(book, knife)*; 2. *(el)* wave tail; 3. crest *(weir)*
  R. des Spiralbohrers body clearance of the twist drill
  gerader R. square back
rückenappretiert back-filled
Rückenbreite *f* **am Räumwerkzeug** tooth thickness of broach
Rückendurchmesser *m* body clearance diameter *(twist drill)*
Rückenfräsmaschine *f* back router
Rückenführung *f* back guide *(saw)*
Rückenhalbwertzeit *f* duration to half-value of wave-tail
Rückenkante *f* heel *(twist drill)*
Rückenlage *f* **bei Furnieren** back ply
Rückenlehne *f* / **klappbare** *(ry)* reversible back of seat
Rückenlehnenkarton *m* buggy board
Rückenpfeilung *f (aero)* positive sweepback
Rückenrundemaschine *f* back rounding machine
Rückenspritze *f* knapsack sprayer
Rückentitel *m* back title
Rückenwäsche *f* washing on sheep's back
Rückerstattung *f* reimbursement, refund
Rückextraktion *f* reextraction
Rückfahrgeschwindigkeit *f (mine)* flitting speed *(coal cutter)*
Rückfahrkarte *f* return ticket

Rückfahrscheinwerfer *m* reverse (back-up) lamp
Rückfahrt *f* return (homeward) journey; return trip; *(mine)* flitting run *(coal cutter)*
rückfedernd springy; springing back; elastic
Rückfederung *f* spring rebound, bounce back of springs; elastic recovery *(material)*
Rückflanke *f (aero)* trailing edge
Rückflugschein *m* round trip ticket
Rückfluß *m* return flow; reflux[ing] *(distillation)*
Rückflußdämpfung *f (el)* regularity (structural) return loss
Rückflußdämpfungsmesser *m* reflection (reflecting) measuring set
Rückflußkoeffizient *m* regularity return current coefficient
Rückflußkühler *m* reflux (back-flow) condenser
Rückflußleitung *f* return line; exhaust line *(hydraulic transmission)*
Rückförderpumpe *f (aero)* scavenger pump
Rückfracht *f* inbound (home) freight
Rückfragehäufigkeit *f* repetition rate *(telephony)*
ruckfrei jerk-free
Rückfrontsynchronisierung *f (telev)* back-flank synchronization, trailing-edge synchronization
Rückführbahn *f (mach)* return rail *(track)*
Rückführdifferenz *f* return difference *(signal flow graph)*
Rückführdrehmoment *n* restoring moment (torque)
Rückführgeber *m* feedback element (pick-off)
Rückführglied *n* feedback element; monitoring element *(servomechanism)*
Rückführgröße *f* feedback variable
unmittelbare R. primary feedback variable
Rückführkreis *m* feedback loop
Rückführpumpe *f* condensate pump; recirculating pump
Rückführung *f* recycling, recirculation; feedback
nachgebende R. temporary feedback action *(isochronous governor)*
rückgängig machen to cancel
Rückgangsverhältnis *n* re-setting ratio *(relay)*
Rückgangszeit *f* re-setting time *(relay)*
rückgewinnen to reclaim, to recover
Rückgewinnung *f* recovery; reclamation
Rückgewinnungsanlage *f* reclaiming installation (plant)
rückgeworfen reflected *(ray)*
Rückgitterstrom *m* reverse grid current
Rückgratkonformation *f (ch)* backbone conformation
Rückhaltebecken *n* retaining basin
Rückhalteträger *m (nucl)* hold-back carrier (agent)
Rückhaltezeit *f* retention time
Rückholfeder *f* return spring
Rückhörbezugsdämpfung *f* reference equivalent of side-tone, sidetone reference equivalent
Rückhördämpfung *f* sidetone reduction
Rückhöreffekt *m* sidetone
Rückhören *n* sidetone
Rückimpuls *m* back (reflected) impulse; return echo pulse *(radar)*
Rückkehr *f* return
R. nach Null return to zero *(magnetic zero level)*
Rückkehradresse *f (dat)* return address
Rückkehrkreis *m* turn-up circle
Rückkehrpunkt *m* cusp[idal point]
R. einer Kurve *f* stationary (turning) point of a curve
rückkohlen to recarburize
Rückkohlung *f* recarburization
rückkoppeln to feedback
Rückkopplung *f* feedback
akustische R. acoustic feedback
induktive R. inductive feedback
kapazitive R. capacitive feedback
magnetische und elektrische R. mixed retroaction
negative R. negative feedback (reaction, retroaction
optische R. optical feedback
positive R. positive feedback, regeneration
Rückkopplungsaudion *n* self-interference audion, regenerative [valve] detector
Rückkopplungsbegrenzer *m* feedback limiter
Rückkopplungsdrehkondensator *m* variable reaction capacitor
Rückkopplungsempfänger *m* regenerating (self-heterodyne) receiver
Rückkopplungsfaktor *m* coefficient of reaction
Rückkopplungsfilter *n* feedback filter
rückkopplungsfrei non-regenerative, without feedback
Rückkopplungsgenerator *m* feedback oscillator
Rückkopplungsgleichrichtung *f* regenerative detection
Rückkopplungsgrad *m* reaction coefficient
Rückkopplungskondensator *m* reaction capacitor
Rückkopplungskreis *m* feedback (regenerative) circuit
Rückkopplungsmagnetron *n* feedback magnetron oscillator
Rückkopplungsregelung *f* feedback control
Rückkopplungsschaltung *f* feedback circuit
Rückkopplungsschleife *f* feedback loop
Rückkopplungsschleifenverstärkung *f* loop gain
Rückkopplungssperre *f* anti-reaction device, feedback (reaction, singing) suppressor
sprachgesteuerte R. voice-operated antisinging device

Rückkopplungssperrvorrichtung *f* anti-singing arrangement
Rückkopplungsspule *f* tickler (feedback) coil
Rückkopplungsverstärker *m* feedback (reaction) amplifier
Rückkopplungsverstärkung *f* feedback (regenerative) amplification
Rückkopplungsverzerrung *f* distortion due to feedback
Rückkopplungsweg *m* feedback (singing) path
Rückkopplungswicklung *f* feedback winding
Rückkopplungswirkung *f* back-feed
Rückkraft *f* thrust force *(tool)*
Rückkühlanlage *f* recooling plant
rückkühlen to recool
Rückkühlturm *m* recooling tower
Rückkühlzone *f* cooling section (zone) *(hot-air grain drier)*
Rückladung *f* return cargo
Rücklauf *m* 1. *(mach)* reverse (return, non-cutting) stroke; return, reverse, withdrawal *(drill)*; 2. *(ch)* reflux; runback; 3. *(telec)* homing action; fly-back *(of an electron beam)*; run-back *(magnetic tape recorder)*
R. der Nummernscheibe return of the dial
R. zum Ausspänen *(mach)* swarf relieving stroke
Rücklaufachse *f* reverse idler shaft
Rücklaufaustastung *f* blanking of the retrace, retrace blanking
Rücklaufbehälter *m* reclaiming tank
Rücklaufboden *m* / **erster** *(ag)* top shoe
zweiter R. *(ag)* bottom shoe
Rücklaufbuchse *f* reverse idler gear bushing
Rücklaufdoppelrad *n* reverse twin gear
Rücklaufdunkeltastung *f* flyback blackout
Rücklaufgeschwindigkeit *f (mach)* return speed
R. des Räumwerkzeugs broach-return speed
Rücklaufhemmung *f* 1. escapement mechanism; 2. *(el)* escapement barrier *(of meters)*
rückläufig retrograde[d]
Rücklaufleitung *f* return pipe
Rücklauflinie *f* return line *(cathode ray tube)*
Rücklaufölbehälter *m* return basin
Rücklaufrad *n* reverse idler gear
Rücklaufrohr *n* return pipe
Rücklaufschrott *m* mill scrap
Rücklaufstrahlaustastung *f* retour trace blanking
Rücklauftank *m* return tank
Rücklauftaste *f* rewinding key
Rücklauftrum *m* return strand
Rücklaufturbine *f* astern (back) turbine
Rücklaufverdunkelung *f* return-trace blanking
Rücklaufverhältnis *n* reflux ratio
Rücklaufweg *m* return trace
Rücklaufzeit *f* re-setting time *(relay)*; *(telev)* return interval
Rückleistung *f* reverse power
Rückleistungsauslösung *f* reverse-power tripping
Rückleistungsrelais *n* reverse power relay
Rückleistungsschalter *m* reverse power circuit breaker
Rückleistungsschutz *m* reverse-power protection
rückleiten to reconduct
Rückleitung *f (el)* return line (wire, circuit)
isolierte R. insulated return wire
Rückleitungsdraht *m* return wire
Rückleitungskabel *n* return cable
Rückleitungsstutzen *m* return connection
Rücklicht *n* tail (rear) light
Rückmeldeanlage *f (telec)* revertive communication apparatus
rückmelden to feedback
Rückmelder *m (el)* position indicator; control transformer *(synchro)*
Rückmeldescheibe *f* answer-back drum
Rückmeldetelegraf *m* repeating telegraph
Rückmeldetrommel *f* answer-back drum
Rückmeldung *f* 1. *(cont)* feedback; 2. *(telec)* audible ringing signal; 3. *(ry)* repeating the signal
R. eines Zuges announcing the arrival of a train
Rückprall *m* rebound, recoil
Rückprallhärte *f* rebound hardness
Rückprojektor *m* background projector
Rückreaktion *f (ch)* back (counter, opposing) reaction
Rückrechnung *f* recharge
Rückrechnungskarte *f* recharge invoice
Rückreise *f* return trip
R. / auf der homeward bound
Rückruf *m (telec)* ring back
Rückrufbetrieb *m* delay working
Rückruftaste *f* ring-back key
Rückrufwähler *m* reverting call switch
Rucksack *m* knapsack
Rückschaltrelais *n* step-back relay
Rückschaltwerk *n* ratchet-and-pawl mechanism
Rückschlag *m* recoil, rebound; back-stroke, return shock; backfiring *(flame)*; *(el)* back-kick
Rückschlagimpuls *m* back-kick pulse
Rückschlagtopf *m (aut)* flame trap
Rückschlagventil *n* non-return valve
Rückschluß *m* / **magnetischer** *(el)* magnetic back-circuit
Rückschmelztransistor *m* meltback transistor
diffundierter R. diffused-meltback transistor
Rückschwefelung *f* resulphurization
Rückseite *f* rear [side]; back; backing *(fabric)*; reverse *(coin)*
Rückseitenbearbeitungsmaschine *f* back finishing machine
Rückseitenfräsmaschine *f* milling and plate back finishing machine

Rücksitz *m (aut)* rear seat
Rückspannung *f* inverse (back) voltage
Rücksprung *m (build)* offset, recess *(wall)*; rebound
Rücksprunghärte *f* scleroscope (rebound) hardness
rückspulen to rewind
Rückspülwasser *n* backwash water
Rückstand *m* residue; deposit; remainder; refuse; *(el)* residuum; tailings *(ore)*
  hochsiedender R. bottoms
  kohleartiger R. carbonaceous deposit
  paraffinhaltiger R. wax tailings
Rückstandsöl *n* residual oil
Rückstau *m* back pressure; reflux, back-surge *(turbine)*
Rückstauwasser *n* back-water
Rückstellanode *f* reset anode
Rückstelleinrichtung *f* zero adjuster
rückstellen to reset *(counter)*; to restore *(spring)*
Rückstellgrad *m* degree of reset
Rückstellimpuls *m* reset pulse; clear pulse
Rückstellklappe *f* self-restoring drop (indicator) *(telephony)*
Rückstellknopf *m* restoring key, reset button
Rückstellkraft *f* restoring force
Rückstellmagnet *m* reset magnet
Rückstellmoment *n* restoring moment (torque), self-aligning torque
Rückstellregelungskreis *m* reset control circuit
Rückstelltaste *f* restoring (reset) key; reset button
Rückstellung *f* resetting
Rückstellvorrichtung *f* resetting device
Rückstellzeit *f* reset time
Rückstich *m* 1. *(met)* return pass; 2. *(text)* backstitch
Rückstoß *m* rebound; recoil; rearward action
  radioaktiver R. radioactive recoil
Rückstoßantrieb *m* reaction propulsion
Rückstoßatom *n* recoil (knock-on) atom
Rückstoßeffekt *m* recoil effect
Rückstoßelektron *n* recoil electron
rückstoßfrei recoilless
Rückstoßprinzip *n* reaction principle
Rückstoßproton *n* recoil proton
Rückstoßprotonenzähler *m* proton recoil counter
Rückstoßspektrum *n* recoil spectrum
Rückstoßstrahlen *mpl* recoil rays
Rückstoßtrennung *f* isotope separation by recoil
Rückstoßzähler *m* recoil counter
Rückstrahl *m* return beam
rückstrahlen to reradiate; to reflect
Rückstrahler *m* 1. [rear] reflector *(vehicle)*; 2. reradiator
Rückstrahlkamera *f* back-reflection camera
Rückstrahlung *f* 1. back reflection; 2. backward radiation; reradiation
Rückstrahlvermögen *n* reflecting capacity (ability, power), albedo
Rückstreichkamm *m (text)* evener comb
Rückstreudiagramm *n* back-scatter pattern
Rückstreudickenmessung *f* thickness control by backscattering
rückstreuen to back-scatter
Rückstreufehler *m* backscattering error
Rückstreukoeffizient *m* backscattering coefficient
Rückstreuung *f* backscattering, back-scatter
Rückstreuungsverluste *mpl* spill losses
Rückstrom *m* 1. return (counter, reverse, back) current; 2. *(ship)* backwash [current]
Rückstromauslöser *m* reverse current release, directional tripping magnet
Rückstromauslösung *f* reverse-power tripping
Rückstromautomat *m* discriminating cut-out
Rückstromrelais *n* directional (power-reversal, reverse-current, reverse-power) relay
Rückstromschalter *m* discriminating (reverse-current) circuit-breaker
Rückstromschütz *n* reverse-current cut-out
Rückströmung *f* backwash *(vessel)*
Rücktaste *f* back spacer
Rücktitration *f* back titration
rücktitrieren to titrate back
Rücktransformation *f* re-transformation
Rücktrittbremse *f* back-pedalling brake
Rücktrocknung *f* **von Oberflächenfurnier** redrying of face veneer
Rücküberschlag *m (el)* backward arc-over
Rückübertrag *m* end-around-carry *(at complementing adding circuits)*
rückübertragen to transfer back, to retransfer
Rückübertragung *f* retransfer, retransmission
rückverformen / sich to recover
Rückvergrößerungsgerät *n* re-enlarger
rückverwandeln to reconvert
  sich r. in to change back to
Rückvorrichtung *f* shifting gear
Rückvorschub *m* intermittent feed
rückwägen to reweigh
Rückwand *f* back wall; rear panel
Rückwandfenster *n (aut)* rear window
Rückwandtür *f (aut)* rear door
rückwärts rearward, back[wards]; *(mach)* [in] reverse [direction]; *(ship)* astern
Rückwärtsbau *m (mine)* back work, back-coming, retreat [mining, working], retreating [working], working home
Rückwärtsbewegung *f* rearward motion (movement); stern-board *(of a vessel)*
Rückwärtsdiode *f* backward diode
Rückwärtsfahrt *f* 1. reverse run; 2. *(ship)* sternway
Rückwärtsfahrtprobe *f (ship)* astern trial
Rückwärtsgang *m (aut)* reverse gear
Rückwärtsganganschlag *m* reverse gear stop

Rückwärtsgleitbahn *f s.* Rückwärtsgleitfläche
Rückwärtsgleitfläche *f* backward stroke guide surface
Rückwärtsgleitschiene *f* back plate
Rückwärtskipper *m* end dump truck
Rückwärtskupplung *f* reverse clutch
rückwärtspullen *(ship)* to back the oars
Rückwärtsregelung *f* backward-acting control
rückwärtsrudern to back the oars
Rückwärtsscheinleitwert *m* feedback admittance
Rückwärtsscheinwerfer *m* back-up light
Rückwärtsschlagen *n* going astern, backing *(of a propeller)*
Rückwärtsschneiden *n* back cutting
Rückwärtsschutz *m* back-up protection
Rückwärtsständer *m* backward stroke standard (column)
Rückwärtsturbine *f* back (astern) turbine
Rückwärtsübersetzung *f (el)* inverse-current ratio, inverse-voltage ratio
Rückwärtswelle *f* backward wave
Rückwärtswirkung *f* rearward action
Rückwärtszähleinrichtung *f* count-down facility
Rückwäsche *f* back wash
Rückweg *m (el)* return path
ruckweise *f (mach)* intermittently; jerky
Rückweisekriterium *n* rejection criterion *(inspection)*
Rückweisewahrscheinlichkeit *f* probability of rejection
Rückweisezahl *f* rejection number
Rückweisung *f* rejection
Rückwicklung *f* rewind
Rückwinderegler *m (text)* backing-off chain tightening motion
rückwirken to react, to retroact, to transfer back
Rückwirkung *f (el)* reaction, retroaction, interaction, back effect
  R. der Steuerung reaction of the valve gear
rückwirkungsfrei *(cont)* non-reactive; unidirectional
Rückwirkungskapazität *f* reflected capacity
Rückwirkungsleitwert *m* feedback admittance
Rückwirkungsspannung *f* back voltage
Rückwirkungswiderstand *m* reflected resistance
rückzahlbar repayable
rückziehbar retractable *(tool)*
Rückzugfeder *f* release spring
rückzünden *(el)* to arc back
Rückzündspannung *f* restriking voltage
Rückzündung *f (el)* arc-back; reignition
Ruder *n* 1. *(ship, aero)* rudder; 2. *f* Steuerfläche
  R. / hart full rudder
  nicht ausbalanciertes R. unbalanced rudder
  nicht ausgeglichenes R. unbalanced rudder
  unausbalanciertes R. unbalanced rudder
Ruder *npl* **/ blockierte** *(aero)* jammed controls

Ruderabgleichkasten *m* vane alignment box
Ruderanlage *f* **/ dampfhydraulische** steam-hydraulic steering gear
Ruderantrieb *m* vane motor
Ruderanzeiger *m* holm indicator
Ruderautomat *m* electronic helmsman
Ruderbank *f* [rowing] thwart, rower's seat
Ruderbarkasse *f* launch
Ruderblatt *n* rudder blade, wash
Ruderboot *n* rowing (pulling) boat, yawl; *(Am)* moses boat
Ruderducht *f* [rowing] thwart
Ruderer *m* oarsman, rower
  vorderster R. im Boot bowman
Rudergänger *m* helmsman, man at the wheel; *(Am)* wheelman
Rudergast *m* helmsman
Rudergeschirr *n* steering gear
Rudergräting *f* wheel grating
Ruderhacke *f* rudder heel
Ruderhaus *n* wheel (pilot) house
Ruderjoch *n* rudder yoke
Ruderkasten *m* fixed tub
Ruderlagenzeiger *m* helm (rudder-angle) indicator
Rudermanöver *n* steering manoeuvre
Rudermanövrierprobe *f* turning trial
Rudermaschine *f* steering gear
  R. mit direktem Antrieb direct-connected steering gear
Rudermaschinenraum *m* steering flat, steering gear flat (compartment)
Rudermoment *n* restoring moment due to vanes
rudern to row, to play the oars; to pull
Ruderpfosten *m* rudder main piece, main piece of the rudder
Ruderpinne *f* tiller, rudder tiller (helm)
Ruderrennen *n* rowing match
Ruderrettungsboot *n* pulling lifeboat
Ruderrückstellmoment *n* restoring moment due to vanes
Ruderschaft *m* shaft loom
Ruderschlag *m* stroke
Rudersmann *m* helmsman, oarsman
Rudersollwinkel *m* required vane angle
Ruderstellung *f (aero)* position of [the] control surface; *(rock)* vane position
Rudersteuerung *f* **/ elektrische** remote-control steering gear
Rudersteven *m* sternpost, rudder (back) post
Ruderstock *m* main piece of the rudder, rudderstock
Ruderstütze *f* outrigger
Ruderträger *m* dumb chalder *(wooden ship)*
Ruderversuch *m* manoeuvring experiment *(model)*
Ruderzeiger *m* rudder-angle indicator
Ruf *m* call, ringing *(telephony)*

R. mit Handinduktor generator signalling
abgekürzter R. abridged call
abgestimmter R. tuned ringing
akustischer R. audible signal
direkter R. immediate ringing
drahtloser R. wireless call signal
halbselbsttätiger R. semi-automatic ringing, machine-key ringing
handbetätigter R. manual ringing
mangelhafter R. defective ringing
selbsttätiger R. automatic signalling
verzögerter R. delayed ringing
Rufadresse *f (dat)* call address
Rufanschalterelais *n* ringing[-trip] relay
Rufanzeiger *m* call indicator
Rufbatterie *f* call-signal battery
Rufeinrichtung *f* call device
rufen to call, to ring *(telephony)*
Ruffolge *f (dat)* calling sequence
Ruffrequenz *f* calling frequency
Rufgenerator *m* ringing generator
Rufinduktor *m* calling magneto, magneto ringer
Rufkode *m* ringing code
Ruflampe *f* calling lamp
Rufleitung *f* calling line
Rufmaschine *f* ringer, ringing machine (generator)
Rufnummer *f* call (subscriber's) number
Rufnummernanzeiger *m* call indicator
Rufnummernverteilung *f* numbering scheme
Rufphase *f* ringing periodicity
Rufprüfung *f* signalling test
Rufrelais *n* calling (ringing, signalling) relay
Rufsatz *m* ringing set
Rufschalter *m* ringing (calling) key
Rufspannung *f* signalling voltage
Rufstellung *f* calling position
Rufstörung *f* ringing failure, defective ringing, signalling fault (trouble)
Rufstrom *m* ringing current
niederfrequenter R. low-frequency signalling current
überlagerter R. superposed ringing current
Rufstromanzeiger *m* ringing current indicator
Rufstromdynamo *m* ringing dynamo
Rufstromfrequenz *f* signalling (ringing) frequency
Rufstromgeber *m* ringer, ringing generator
Rufstrommaschine *f* ringing dynamo (generator), ringer
Rufstromschaltung *f* ringing connection
Ruftaste *f* ringing key
Ruf- und Signalmaschine *f* ringing and signalling machine
Ruf- und Sprechschalter *m* speaking and ringing key, *(Am)* talking-ringing key
Rufwecker *m* call bell
Rufzeichen *n* call sign
abgekürztes R. abridged call sign[s]
Rufzeichenerzeuger *m* ringing interrupter
Rufzeit *f* calling time
Ruhe *f* 1. rest; dwell *(machine movement)*, non-operation *(instrument)*; 2. silence, quietness *(running of machine)*
Ruhebereich *m* region of non-operation *(relay)*
Ruhebühne *f (mine)* ladder sollar, landing
Ruhefach *n (text)* open shed
Ruhekontakt *m* rest (break, closed-circuit) contact, normally closed [auxiliary] contact
R. des Hebmagneten vertical interrupter contact *(telephony)*
Ruhelage *f* position of rest, rest position; equilibrium [position]
R. der Relaiskontakte normal position of relay contacts
Ruhemasse *f* rest mass
R. des Elektrons rest mass of the electron
ruhend *(mach)* stationary; at rest *(liquid)*; still *(air)*; dead *(load)*; static
Ruhenergie *f* rest energy
Ruheperiode *f* period of rest (dwell)
Ruhepotential *n* resting potential
Ruhereibung *f* static friction
Ruhesitz *m (mach)* permanent fit
Ruhespannung *f* normal-operation voltage, voltage in normal position *(switchgear)*; rest potential
Ruhestellung *f* off-position *(switch)*; position of rest; neutral (resting, unoperated) position
Ruhestrom *m* spacing (closed-circuit, rest, static) current
Ruhestromalarmgerät *n* closed-circuit alarm device
Ruhestromanlage *f* closed-circuit system
Ruhestromauslöser *m* no-volt[age] release
Ruhestromauslösung *f* no-voltage release
Ruhestrombatterie *f* closed-circuit battery
Ruhestrombetrieb *m* closed-circuit operation
Ruhestromkreis *m* closed circuit
Ruhestromprinzip *n* closed-circuit current principle
Ruhestromschaltung *f* closed-circuit connection, circuit-opening connection
Ruheträgerfrequenz *f* centre frequency
Ruhewert *m* steady value *(electron valve)*
Ruhezeit *f* rest period; dwell period *(machine movement)*
Ruhezustand *m* static condition, passive state; state of rest
R. / im at rest
R. des Multivibrators multivibrator-waiting condition
ruhig *(mach)* quiet *(running)*; calm *(weather)*; smooth *(action)*
Ruhmasse *f* rest mass
Rühreinrichtung *f* agitator; stirring device
rühren to agitate, to stir

Rührer *m* agitator, stirrer
Rührfahrzeug *n* *(build)* agitating (agitator) truck
Rührflügel *m* mixing arm, agitating vane, scoop
Rührfrischen *n s.* Puddeln
Rührhaken *m* rabble
Rührkrücke *f* rabble
Rührofen *m* rabble furnace
Rührschneckenförderer *m* paddle screw conveyor
Rührwelle *f* agitator
Rührwerk *n* agitator, stirrer
 gegenläufiges R. double-motion mixer
Rührwerksantrieb *m* agitator (stirrer) driving gear
Rührwerkszelle *f* agitator-froth machine
Rumbrennerei *f* rum distillery
Rumpf *m* body, trunk; *(ship)* hull; *(aero)* fuselage, body
 R. des Pfluges frame of the plough
Rumpfelektron *n* inner electron
Rumpfende *n* *(aero)* tail
Rumpfoberholm *m* *(aero)* upper longeron
Rumpfoberholz *n* *(aero)* top longeron
Rumpfstrebe *f* *(aero)* fuselage strut
Rumpfumströmung *f* *(aero)* flow around fuselage
Rumpfwiderstand *m* *(aero)* fuselage drag
Rumpfzweidecker *m* biplane fuselage model
rund round; circular; rounded; cylindrical
 genau r. true [round]
Rundbau *m* rotunda
Rundbiegevorrichtung *f* circular bending apparatus
Rundbleche *npl* circular laminations *(armature)*
Rundblickaufnahme *f* panoramic (wide-angle) view
Rundblock *m* round bloom
Rundbodenboot *n* round bottom boat
Rundbogen *m* Norman arch
Rundbrecher *m* gyratory crusher
Rundbrechmaschine *f* circular breaking machine *(flax)*
Rundbug *m* rounded (barrel) bow
Rundbürste *f* circular brush, brushing wheel
Runddach *n* compass roof
Runddämpfer *m* cottage steamer
Runddeck *n* whaleback, turtleback
Runddeckdampfer *m* whaleback steamer
Runddraht *m* round wire
Runddrehmaschine *f* **für gebogene Hölzer** bent wood turning machine
Runddurchschlag *m* round punch
Runddurchschläger *m s.* Runddurchschlag
Rundeisen *n* round bar iron
runden to round off, to truncate; to approximate *(calculus)*; *(mach)* to radius
runderneuern to rebuild, to retread *(tyre)*

43*

Rundfahrtboot *n* roundabout
Rundfeile *f* round file
Rundfenster *n* bull's eye [window]; compass window
Rundfeuer *n* commutator flashing
Rundfeuerschutz *m* flash barrier
Rundfilz *m* tubular felt
Rundfräsarbeit *f* circular milling operation
Rundfräseinrichtung *f* circular milling attachment, rotary attachment [in miller]
Rundfräsen *n* circular (rotary) milling
 pausenloses R. continuous circular milling
Rundfräsmaschine *f* circular milling machine
Rundfrästisch *m* circular milling table
Rundführung *f* cylindrical guide; clamp-bed *(milling machine)*
Rundführungsflansch *m* flange of the circular guide
Rundführungsrahmen *m* frame with circular guide
Rundfunk *m* radio; broadcasting; wireless
 R. für Verkaufsstätten storecasting
Rundfunkansager *m* announcer
Rundfunkapparat *m* **mit Stationswählertasten** push-button radio
Rundfunkapparatesicherung *f* radio fuse
Rundfunkband *n* broadcast band
Rundfunkeingang *m* **am Tonbandgerät** broadcast frequency input
Rundfunkempfänger *m* radio receiver
 schnurloser R. cordless-radio
Rundfunkentstörung *f* radio interference suppression
Rundfunkentstörungsdienst *m* radio interference suppression service
Rundfunkfrequenz *f* radio frequency
Rundfunkgerät *n* radio (wireless receiving) set
 R. mit Plattenspieler radiogram; radio record player
Rundfunkhörer *m* broadcast listener
Rundfunkleitung *f* programme line
Rundfunksender *m* broadcasting station; broadcasting transmitter
Rundfunksonderdienst *m* special broadcasting message
Rundfunktechnik *f* radio engineering
Rundfunkübertragung *f* broadcast [transmission]
Rundfunkübertragungsverbindung *f* programme circuit
Rundfunkvermittlung *f* programme distribution system, wire programme distribution
Rundfunkvermittlungsanlage *f* radio exchange installation, radio relay exchange
Rundfunkzubringerleitung *f* broadcast relay circuit
Rundfutter *n* *(mach)* circular [magnetic] chuck
rundgeschliffen ground circular, cylindrically ground

Rundgesprächsanlage *f* conference calling equipment
Rundgesprächseinrichtung *f s.* Rundgesprächsanlage
Rundgewinde *n* knuckle thread
Rundgliederkette *f* coil chain
  steglose R. plain coil chain
Rundheck *n (ship)* elliptical (round) stern
Rundheit *f* roundness
Rundherd *m* circular-deck table
Rundhobel *m* compass (hollowing) plane
Rundhobelapparat *m* planer centres
Rundhobeleinrichtung *f* radial (radius) planing device, circular planing (shaping) attachment, circular motion attachment *(shaper)*
Rundhobeln *n* radius planing
Rundholz *n* 1. round timber; pole; 2. *(ship)* spar
Rundholztabelle *f* round timber measurement
Rundholztrocknung *f* drying in the log
Rundkaliber *n (met)* round pass
Rundkämmaschine *f* circular comber (combing machine)
Rundkatode *f* cylindrical cathode
Rundkerbe *f* semi-circular notch
Rundkettelmaschine *f* circular linking machine
Rundkettenstuhl *m* circular warp loom
Rundkettenwirkmaschine *f* circular warp knitting machine
Rundkipper *m* all-round tipping wagon
Rundkolben *m* round[-bottom] flask
Rundkopfschraube *f* cheese-head screw
Rundkulierwirkmaschine *f* circular weft knitting machine
Rundläppen *n* cylindrical lapping
Rundlauf *m* true running
  genauer R. true running
rundlaufen / genau to rotate in truth, to run true
rundlaufend rotating; circular, true
Rundlauffräsmaschine *f* rotary-table milling machine
Rundlaufmeßgerät *n* concentricity tester
Rundlauftoleranz *f* tolerance given for running true
Rundleistenmaschine *f (text)* circular string border machine
Rundlitzenseil *n* round strand rope
Rundloch *n* round perforation (hole)
Rundlochbrenner *m* straight slot burner
Rundlochkarte *f* round-hole card
Rundmesser *n (text)* circular blade
Rundmutter *f* round nut
Rundnaht *f* circular seam *(riveting)*
Rundpassung *f* cylindrical fit
Rundplattengießmaschine *f* curved plate casting machine
Rundprobe *f (met)* round specimen
rundpumpen *(nucl)* to pump through the loop
Rundrahmenantenne *f* circular aerial
Rundrändermaschine *f (text)* round knitting frame
Rundraster *m* circular (round) screen
Rundreise *f* closed-circuit routing, circular journey (tour), round trip
Rundreisefahrschein *m* circular-tour ticket
Rundsäule *f* circular (round) column
Rundschacht *m* circular shaft
rundschalten *(mach)* to index around
Rundschaltmaschine *f* rotary-indexing machine
  R. der Ringtischbauart centre-column rotary indexing machine
Rundschalttisch *m* indexing rotary table, [circular] indexing table
Rundschalttischmaschine *f* rotary-indexing machine
Rundschleifen *n* cylindrical grinding, grinding cylindrical work
Rundschleifmaschine *f* cylindrical grinding machine
  spitzenlose R. centreless cylindrical grinding machine
Rundschliff *m* cylindrical (circular) grinding
  R. / mit cylindrically ground
Rundschliffase *f* cylindrical margin (land), land ground circular
Rundschmelzmaschine *f* glazing mill *(electron tube manufacture)*
Rundschnittverfahren *n* method of joints *(truss analysis)*
Rundschruppschleifen *n* cylindrical rough grinding
Rundschweißnaht *f* circular weld
Rundsech *n (ag)* disk coulter
Rundsichtanzeiger *m* plan-position indicator
Rundsichtgerät *n* plan-position indicator
  in Nordrichtung stabilisiertes R. north-stabilized plan-position indicator
Rundsichtverglasung *f* panorama windscreen, wrap-around windshield
Rundsieb *n* rotary screen *(cardboard)*
Rundsiebfilz *m* **/ grauer** grey cylinder felt
Rundsiebmaschine *f* board (cylinder vat, mould) machine *(cardboard)*
Rundsiebmaschinenpappe *f* cylinder board
Rundsiebpappenmaschine *f* board machine
Rundsiebzylinder *m* cylinder mould
Rundskala *f* [circular] dial, circular scale
Rundspantboot *n* round-bottom boat, round-framed boat
Rundspulamperemeter *n* repulsion ammeter
Rundstab *m (build)* round [bar], rod; round specimen *(testing)*
Rundstabhobel *m* hollowing (bead) plane
Rundstahl *m* round bar steel, round [bar]
Rundstahlfeder *f* round wire spring
Rundstereo *n* curved (round) stereo

rundstirnig round-ended
Rundstoßen *n (mach)* circular slotting
Rundstoßnachformen *n (mach)* rotary tracing (slotting)
Rundstrahlantenne *f* omni-directional aerial, non-directional aerial, non-directive aerial
Rundstrahlbrenner *m* fan-tailed burner
Rundstrahler *m (el)* equi-radial aerial, omni-aerial
Rundstrickmaschine *f* **für Oberkleiderstoffe** dress-fabric circular knitting machine
  R. mit Hakennadeln bearded-needle circular knitting machine
  einfonturige R. circular knitting machine with one set of needles
  einfonturige R. mit Hakennadeln plain circular spring-beard needle knitting machine
Rundstrickrippmaschine *f* circular rib knitting machine
Rundstrickstrumpfware *f* circular knit hosiery
Rundstuhl *m* circular loom; circular knitting machine
Rundsuchantenne *f* scanner, scanning aerial *(radar)*
Rundsuchradar *n* radar landscape scanner, surveillance radar
Rundtaktmaschine *f* rotary indexing automatic, rotary transfer machine
Rundteilen *n (mach)* circular spacing (division, indexing)
Rundteilung *f* **des Werkstücks** circular division of work
Rundtisch *m (mach, opt)* circular (rotary, round) table
  R. mit Teileinrichtung *(mach)* circular dividing table
  R. mit Teilkreis *(opt)* optical rotary table
Rundtischantrieb *m* rotary attachment feed gear
Rundtischflächenschleifmaschine *f* surface grinding machine having a rotating work-table, rotary table surface grinding machine
Rundtischfräsmaschine *f* rotary-table milling machine
Rundtischmaschine *f* rotary-table machine with the machine heads outside the table
Rundtischräummaschine *f* rotary broaching machine
Rundtischschaltmaschine *f* rotary-table transfer machine, *(Am)* rotary indexing unit
Rundtischvorschub *m* circular-table feed
Rundung *f (mach)* radius
Rundungsfehler *m (dat)* rounding (truncation) error
Rundvorschub *m (mach)* rotary (circular) feed
Rundwaschmaschine *f* circular washer
Rundwebstuhl *m* circular loom
Runge *f (ry)* stanchion
Rungenöse *f* stanchion socket
  selbstauslösende R. self-opening stake pocket
Rungenwagen *m* stanchion car, wagon with stanchion
runzelfrei wrinkle-free
Runzellack *m* wrinkle finish (paint)
Runzelröhre *f* corrugated guide
rupfen to pick, to peel; to pluck
Rupffestigkeit *f* picking resistance
Ruschelzone *f (geol)* fractured zone
Ruß *m* soot; carbon (lamp, hydrocarbon) black
Rußabscheidung *f* deposit of soot
Rußblasen *n* soot blowing
Rußbläser *m* soot blower
Russel-Saunders-Kopplung *f* Russel-Saunders coupling, LS-coupling
Rüsselvorrichtung *f (text)* rope piling device
Rußflocken *fpl* smuts
rußgeschwärzt smoked
Rußglas *n* smoked glass
Rußvorlage *f* soot receiver (collector, filter)
Rüstbaum *m (build)* gin (scaffolding) pole
Rüstbrett *n* scaffold board
rüsten 1. *(build)* to scaffold; 2. *(mach)* to set up, to tool; to prepare
Rüstleine *f (ship)* shank painter
Rüstloch *n* putlog hole
Rüstrohr *n* scaffold tube
Rüstung *f (build)* scaffold[ing]
Rüstzeit *f (mach)* tooling (set-up) time; *(mine)* make-ready time *(drilling)*
Rute *f (text)* pile wire
Rutenstuhl *m* pile-wire loom
Rutenteppichstuhl *m* carpet wire pick loom
Rutenzugwebstuhl *m* truvet loom
Ruthenium *n* ruthenium
Rutherford *n* rutherford, rd
Rutherford-Einheit *f* rutherford, rd
Ruths-Speicher *m* Ruths' accumulator
Rutil *m* rutile
Rutsche *f* chute; slide; shoot
rutschen to chute *(workpieces)*; to slip, to slide *(crystallography)*; *(aut)* to skid *(wheels)*; to slip *(clutch)*
Rutscherschützen *m (text)* sliding shuttle
rutschfest anti-skid, non-slip, non-skid *(floor, tyre)*
Rutschfestigkeit *f (aut)* anti-skid properties
Rutschfläche *f (geol)* slickenslides
Rutschgefahr *f* **[bei der Landung]** danger of skidding
Rutschkupplung *f* safety coupling (clutch), friction coupling [adjusted to slip at a predetermined torque]
Rutschschere *f* [drilling] jar
rutschsicher non-skid *(tyre)*
Rutschsicherheit *f (aut)* skid resistance

Rutschung *f* slipping, slippage
Rutschzuführung *f* chute feeding
Rüttelbeton *m* vibrated concrete
Rüttelbohle *f* vibrating beam (screed)
Rüttelbohlenfertiger *m* vibrating beam finisher
Rüttelflasche *f* poker *(concrete)*
Rüttelformmaschine *f* jolt (jar) moulding machine *(foundry)*
Rüttelgerät *n* vibrator *(concrete)*
Rüttelherd *m (met)* bump (sweep) table, percussion frame
Rüttelmaschine *f* vibrator *(concrete)*; jarring machine *(foundry)*
rütteln to vibrate *(concrete)*; to rock; to shake [up]; to jar, to jolt *(moulding)*
Rüttelplatte *f* plate vibrator
Rüttelpreßformmaschine *f* jolt squeeze machine
Rüttelpreßwendeplattenformmaschine *f* jarring pressure turn-over moulding machine
rüttelsicher safe against loosening under vibration
Rüttelsieb *n* shaking sieve, vibrating (vibratory) screen
Rüttelstampfer *m* vibrating tamper
Rütteltisch *m (build)* table vibrator, vibratory (vibrating) table; jolt (jarring) table *(foundry)*
Rüttelverdichter *m* vibratory (vibrating) compactor
Rüttelverdichtung *f* dynamic compaction, consolidation by vibration *(concrete)*
Rüttelverfahren *n* jar ramming method *(foundry)*
Rüttelwalze *f* vibrating (vibratory) roller
Rüttelzetter *m (ag)* kick-action tedder
Rüttler *m* vibrator *(concrete)*
Rydberg *n (nucl)* rydberg
Rydberg-Konstante *f* Rydberg constant

S

S-Bahn *f* elevated railway
S-Band *n (el)* S band
S-Drehung *f (text)* S-twist
S-Haken *m* S-link
S-Kurve *f* reverse curve
s-Leitfähigkeit *f* s-conductivity *(of semiconductors)*
S-Zacken *pl* S waves *(electrocardiogram)*
Saalgeräusch *n (ac)* room (hall) noise
Saalverdunkelung *f* illumination darkening
Saalverdunkelungseinrichtung *f* dimming device
Säapparat *m s.* Sämaschine
Saat *f* seeding, sowing; seeds
Saatbeet *n* seed bed
Saatbeetbereitung *f* seed bed preparation, preparation of seed beds
Saatbeize *f* seed dressing
Saatelemente *npl (nucl)* seed [elements]
Saatgut *n* seed
Saatgutaufbereitung *f* seed cleaning and grading
Saatgutaufbereitungsmaschine *f* seed cleaner and grader
Saatgutbeizmittel *n* seed dresser
Saatgutbereiter *m* seed dresser
Saatgutbereitung *f* seed dressing
Saatgutherrichtung *f* seed cleaning and grading
Saatgutmechanisierungskette *f* complete flow-line seed processing system
Saatgutnaßbeize *f* wet seed dresser
Saatgutprobe *f* [seed] sample
Saatgutreinigung *f* seed cleaning
Saatgutreinigungsanlage *f* seed cleaner
Saatgutsortiermaschine *f* seed grader
Saatgut- und Düngerdrillmaschine *f* seed and fertilizer drill
Saatkasten *m* seed hopper (box)
Saatkristall *m* seed crystal
Saatleitung *f* seed feed tube
Saatleitungsrohr *n* drill tube
Saatpuder *m* seed powder
Saattermin *m* seeding deadline
Saattiefe *f* depth of sowing
Saatwalze *f* drill land roller
Saatwechsel *m* crop rotation
Saatzucht *f* nursery gardening
Saatzuchtgut *n* seed-nursery
säbar sowable
Säbelast *m* sword-shaped knot
Saccharase *f* saccharase, invertase
Saccharimeter *n* saccharimeter
Saccharin *n* saccharin
Saccharomyces *f* saccharomyces *(yeast)*
Saccharose *f* saccharose, sucrose
sachdienlich relevant, pertinent
sachgemäß proper, pertinent, suitable
Sachkatalog *m* subject guide
Sachkonto *n* general account
Sachregister *n* [subject] index
Sachverständiger *m* expert
Sachwörterverzeichnis *n* [subject] index
Sack *m* sack, bag
Sackaufzug *m* sack hoist
Sackbahnhof *m* dead-end station
Sackbohrer *m* pocket drill; sack borer
Sackförderer *m* bag conveyor
Sackgasse *f* blind alley, cul-de-sac, dead-end street
Sackgut *n* bag cargo, bagged goods
Sackhalter *m* sack holder
Sackheber *m* sack lifter, sacklifter
Sackkarre *f* bag truck
  S. mit Fußtritthub bag truck with pedal lifting mechanism
Sackkarren *m* sack barrow
Sackladung *f* bag cargo, bagged goods
Sackleinen *n* hessian [cloth], burlap

Sackleinwand *f* sack cloth
Sackloch *n* blind hole
Sacknadel *f* packing needle
Sackspant *n (ship)* U-shaped frame
sackweise by the bag; in bags
Sackzement *m* sacked cement
säen to sow, to seed
Saffian *m* saffian
Saffianeinband *m* morocco-binding
Saffianleder *n* morocco
Safran *m* saffron
Safranin *n* safranine *(dyestuff)*
Saft *m* juice
 saturierter S. carbonated juice *(sugar industry)*
Saftfutterzerkleinerer *m* forage chopper (cutter-blower)
Saftgehalt *m* sappiness
Saftreinigungsmittel *n* defecant *(sugar industry)*
Saftverdrängungsverfahren *n* Boucherie process
Säge *f* saw
 S. mit geradem Rücken straight-back saw
 S. mit geschweiftem Rücken skew-back saw
 S. mit hin- und hergehender Schnittbewegung reciprocating saw
 S. mit stehendem Zahn pad tooth saw
 gefaßte S. framed hand saw with stretchers
Sägeangel *f* saw blade holder, tang for web blade
Sägeblatt *n* saw blade
 S. für Bügelsägemaschine power hacksaw blade
 S. mit eingesetzten Zähnen inserted-tooth blade
 S. mit gerader Zahnspitzenlinie straight-tooth blade
 S. mit Zahnsegmenten segmental-type blade
 endloses S. für Bandsäge band-saw blade, saw band
 gekrümmtes S. curved blade, blade with convex tooth edge, blade with round breast
 geschränktes S. straight-set blade
 gewelltes S. wave-set blade
Sägeblattanordnung *f* saw-tooth system *(support)*
Sägeblattführung *f* saw-blade guidance
Sägeblattschärfmaschine *f* saw sharpening machine
Sägeblattschleifmaschine *f s.* Sägeblattschärfmaschine
Sägebock *m* saw horse, sawing jack
Sägebügel *m* saw frame
Sägedach *n* saw-tooth roof
Sägeegreniermaschine *f* saw gin
Sägefeile *f* saw file
Sägefurnier *n* sawn veneer
Sägemaschine *f* sawing machine, machine (bench) saw
Sägemehl *n* sawdust
Sägemehllager *n* sawdust room
sägen to saw
Sägengewinde *n* buttress screw thread, saw-tooth thread
Sägenut *f* saw groove, kerf
Sägeschärfmaschine *f* saw gummer
Sägeschärfschleifscheibe *f* saw grinding wheel
Sägeschlitten *m* saw [blade] carriage
Sägeschnitt *m* saw kerf (cut)
Sägespäne *mpl* sawdust
Säge- und Feilmaschine *f* / **kombinierte** combined sawing and filing machine
Sägewerk *n* saw mill
 fliegendes S. portable saw mill
Sägewerker *m* sawyer
Sägezahn *m* [saw] tooth, saw-shaped tooth; *(telev)* sawtooth, ramp
Sägezahnantenne *f* zigzag aerial
Sägezahndrahtbeschlag *m* saw-tooth wire filleting
Sägezahnform *f* saw-tooth form
Sägezahngenerator *m* saw-tooth[ed] generator (oscillator)
 kapazitiver S. capacitive saw-tooth generator
Sägezahnimpuls *m* saw-tooth pulse
Sägezahnkurve *f* saw-tooth characteristic, saw-tooth[ed] waveform
Sägezahnmodulation *f* saw-tooth modulation
Sägezahnprofil *n* corrugated (serrated) profile *(turbine)*
Sägezahnsegment *n* saw segment
Sägezahnspannung *f* saw-tooth voltage
Sägezahnwellenform *f* saw-tooth waveform
sagittal sagittal
Saisonarbeit *f* seasonal work
Saite *f* chord, string; catgut
Saitenelektrometer *n s.* Fadenelektrometer
Saitengalvanometer *n* string galvanometer
Saiteninstrument *n* string[ed] instrument
Saitenoszillograf *m* string oscillograph
Saitenspannungsmesser *m* vibrating wire strain gauge
Säkasten *m* hopper
Säkulargleichung *f* secular equation
Salband *n (geol)* selvage, selvedge, rider
Salbe *f* ointment, salve
Salbeiöl *n* sage oil
Salbengrundlage *f* ointment base
Saldenaufnahme *f* / **automatische** automatic balance pick-up
Saldenauswahl *f* balance selection
Saldenvortrag *m* balance forward, brought-forward balance, carried-forward balance
Saldenwähler *m* balance selector
Saldiermaschine *f* adding machine
Saldierwerk *n* adding register
Saldoübertrag *m* transfer of balance

Saldovortrag *m* brought-forward balance, carried-forward balance
Saline *f* saline, salt mine
Salinenwasser *n* saline water
Salinität *f* salt content, salinity
Salizin *n* salicin
Salizylat *n* salicylate
Salizylsäure *f* salicylic acid
Salleiste *f (text)* list[ing]
Salmiak *m* ammonium chloride, sal-ammoniac, ammoniac salt
Salmiakgeist *m* aqueous (spirits of) ammonia
Salondeck *n* saloon deck
Salonwagen *m* saloon carriage (coach), parlour car
Salpeter *m* saltpetre, nitre
salpetersauer nitric
Salpetersäure *f* nitric acid
  konzentrierte S. concentrated nitric acid, aqua fortis
  rauchende S. fuming nitric acid
Salpetersäureanhydrid *n* nitric anhydride, nitrogen pentoxide
Salpetrigsäureanhydrid *n* nitrous anhydride
Salz *n* salt
  basisches S. basic salt, subsalt
  kesselsteinbildendes S. scale-producing salt
  kristallwasserhaltiges S. hydrous salt
  Mohrsches S. Mohr's salt
  saures S. acid (hydrogen) salt
  Schlippesches S. Schlippe's salt
  wasserfreies S. anhydrous salt
Salzablagerung *f* salt deposit
Salzader *f* salt vein
Salzausblühung *f* bloom, efflorescence
Salzbad *n* salt bath
Salzbadhärteofen *m* salt-bath furnace
Salzbadhärtung *f* salt-bath case hardening
Salzbadofen *m* salt-bath furnace
Salzbadpatentierung *f* salt-bath patenting
Salzbergwerk *n* salt mine
salzbildend salt-forming
Salzbildner *m* halogen, salt former
Salzbildung *f* salification, salt forming
salzen to salt
Salzgehalt *m* salinity, salt content
Salzglasur *f* salt glaze
Salzgrübchen *npl* salt pits *(fault of leather)*
salzhaltig saliferous, salinous, salty
salzhart hard-cured *(hide)*
Salzhorst *m* salt plug
Salzkorn *n* grain of salt
Salzkupfererz *n* atacamite
Salzlagerstätte *f* salt deposit
Salzlake *f* salt brine, pickle
Salzlast *f* salt pen *(on fishing vessels)*
Salzlaugung *f* brine leaching
Salzlecke *f* salt lick
Salzlösung *f* salt solution, brine
salzsauer hydrochloric
Salzsäure *f* hydrochloric acid
Salzschmelzenreaktor *m* molten-salt reactor
Salzsee *m* salt (saline) lake
Salzsole *f* brine
Salzspritzmaschine *f (mine)* salt gun
Salzstrichhaut *f* painted hide
salztrocken salt-dried *(hide)*
Salzung *f* **in Salzlake** brining
Samarium *n* samarium
Samariumvergiftung *f* samarium poisoning
Samarskit *m* samarskite *(uranium mineral)*
Sämaschine *f* drill, sowing machine
Samen *m* seed
Samenbaum *m* seed tree
Samenbaumwolle *f* seed cotton
Samenfaser *f* seed fibre
Samenmuster *n* [seed] sample
Samenprobe *f s.* Samenmuster
Samenreinigungs- und -sortiergeräte *npl* seed cleaning and grading machinery
Samenunkräuter *npl* ephemerals
sämischgerben to chamois
Sämischleder *n* chamois [leather], chammy
Sämling *m* seedling
Sammelanode *f* anode collector
Sammelanruf *m* traffic list
Sammelanschluß *m* collective numbers
Sammelanschlußteilnehmer *m* subscriber with several lines
Sammelband *n s.* Sammelförderband
Sammelbecken *n* collecting tank (bond), catch basin
Sammelbehälter *m* storage tank, reservoir
Sammelelektrode *f* collecting (collector) electrode
Sammelfahrschein *m* party (collective) ticket
Sammelfehler *m* accumulative (accumulated) error; composite error *(gear)*
Sammelfehlerprüfung *f* composite test *(gear)*, composite check of errors
Sammelförderband *n* trunk conveyor
Sammelfuchs *m* collecting flue
Sammelganggeschwindigkeit *f* accumulating (tab) speed *(punched card machinery)*
Sammelgebiet *n* drainage (catchment) area
Sammelgefäß *n* collecting vessel, reservoir, storage tank
Sammelgespräch *n* conference call
Sammelgesprächseinrichtung *f* conference calling equipment
Sammelgleis *n* train formation line (track)
Sammelgleisgruppe *f* set of formation sidings
Sammelgut *n* groupage
Sammelgüterzug *m* pick-up goods train, collective freight train, *(Am)* way freight train

Sammelgutspediteur *m* sender of groupage traffic *(forwarding agent)*
Sammelhefter *m* gatherer-stitcher
Sammelkanal *m* service duct; collector, interceptor, collecting main
Sammelkonnossement *n (ship)* omnibus bill of lading
Sammelkontakt *m* master number contact
Sammelkristallisation *f* accretive crystallization
Sammelladung *f* collective truck; *(Am)* collective shipment
Sammelleitung *f (el)* ring (collecting) main; *(aut)* manifold
Sammellinse *f* condensing (converging, collecting, focusing) lens
sammeln *(ag)* to gather, to pick, to harvest; to collect *(stamps)*; to select *(scientific work)*; to compile
Sammelplattform *f* picking platform
Sammelpresse *f* pick-up baler
Sammelroder *m* complete potato harvester
Sammelrohr *n* header; collecting pipe; *(aut)* manifold
Sammelrollgang *m (met)* collecting roller table
Sammelschalter *m* concentration switch
Sammelschaltung *f (telec)* conference connection
Sammelschelle *f (el)* multiple conduit clip, conduit rack
Sammelschiene *f* [main] bus-bar, collecting bar; trunk line
  S. mit Abgriffen contact collecting bar
  hochkant angeordnete S. edge bus-bar
Sammelschienendrosselspule *f* bus reactor
Sammelschienenfehler *m* bus-bar fault
Sammelschienenisolator *m* bus-bar support (insulator)
Sammelschienenkraftwerk *n* bus-bar-type power station
Sammelschienenkuppelfeld *n* bar coupling panel
Sammelschienenlängstrenner *m* bus-bar sectionalizer (sectionalizing switch)
Sammelschienenschutz *m* bus-bar protective device
Sammelschienenstütze *f* bus-bar support
Sammelstück *n (ag)* clawpiece
Sammler *m* 1. *(el)* storage (accumulator) battery, accumulator [storage battery], secondary cell (battery) (*s.a.* Batterie); 2. *(build)* interceptor, collector; 3. *(print)* assembler
  alkalischer S. alkaline accumulator
  tragbarer S. portable storage battery
Sammleranlage *f* accumulator plant
Sammlerbatterie *f* storage battery
Sammlergefäß *n* accumulator container (tank, box), containing cell
Sammlergestell *n* accumulator rack
Sammlerglas *n (el)* accumulator jar
Sammlersäure *f* accumulator acid
Sammlerschlitten *m (print)* assembler slide
Sammlerspannung *f* accumulator voltage
Sammlerzelle *f* storage cell
Sampling-Oszillograf *m* sampling oscilloscope
Sampling-Prozeß *m* sampling process
Sampling-Theorem *n* sampling theorem *(information theory)*
Samt *m* velvet
  gezogener S. uncut velvet
Samtblende *f (min)* goethite
Samtgewebe *n* pile fabric
Samtkette *f* nap warp
Samtnadel *f* pile wire
Samtrute *f* pile wire
Samtschneidemaschine *f* velvet cutting machine
Samtstuhl *m s.* Samtwebstuhl
Samtweberei *f* pile (velvet) weaving
Samtwebstuhl *m* velvet (plush) loom
Sand *m* sand
  gebrauchter S. used sand *(foundry)*
  scharfer S. sharp sand
Sandader *f* sand mark *(casting)*
sandartig gritty, sandy
Sandaufbereitung *f* sand preparing *(foundry)*
Sandauslaufrohr *n* sand nozzle
Sandbank *f* sand-bar, sandbank, sands
  wandernde S. shifting-sand
Sandbaum *m (text)* sand roller
Sandbett *n* sand bed *(foundry)*
Sandboden *m* sandy soil
Sandeinfüllstutzen *m* sand filling socket *(locomotive)*
Sandelholz *n* sandalwood
Sandelholzöl *n* sandalwood oil
Sanden *n* grit blasting
Sandfang *m* sand trap, riffle *(wood pulp)*
Sandfilter *n* sand filter
Sandform *f* sand mould *(foundry)*
Sandformerei *f* sand moulding
Sandfüllung *f* sand filling
Sandgrube *f* sand working (pit)
Sandguß *m* sand casting[s]
Sandhaken *m* gagger, lifter *(foundry)*
Sandherd *m* sand bottom
sandig gritty, sandy
Sandkasten *m* sand box, sander *(locomotive)*
Sandkern *m* sand core *(foundry)*
Sandkruste *f* sand skin *(foundry)*
Sandmischer *m* sand mixer
Sandpapier *n* abrasive paper, sandpaper
Sandpolster *n (build)* sand cushion
Sandpolstergründung *f* compacted earth fill foundation
Sandsack *m* 1. sand-bag, ballast bag *(boat)*; 2. *(mil)* sand-bag; 3. heavy punching bag *(boxing)*
Sandsackmodell *n s.* statistisches Kernmodell
Sandschiefer *m* sandy shale

Sandschleuderformmaschine *f* sand slinger
Sandschwimmverfahren *n* chance (sand floatation) process
Sandsieb *n (ag)* tailings riddle
Sandslinger *m s.* Sandschleuderformmaschine
Sandstein *m* sandstone
  bituminöser S. blae[s], blaze
  eisenhaltiger S. brownstone
  fester schiefriger S. hazle
Sandstrahlbehandlung *f* sand-blasting
sandstrahlen to sand-blast
Sandstrahlen *n* sand-blasting
Sandstrahlgebläse *n* sand-blast [machine]
Sandstrahlputzen *n* sand-blast cleaning
Sandstrak *n (ship)* garboard strake
Sandstrand *m* sandy beach
Sandstreuer *m* sand distributor
Sandstreugerät *n* gritter
Sandstreuvorrichtung *f (ry)* sanding device
Sanduhr *f* sand clock
Sandverdichtung *f* sand packing *(foundry)*
Sandwichbauweise *f* sandwich construction
Sandwichbestrahlung *f* sandwich irradiation
Sandwichplatte *f (build)* sandwich panel
Sandwichquelle *f* sandwich[ed] source
Sanforisieren *n (text)* sanforizing
Sanidin *m* sanidine
sanierungsbedürftig blighted
sanitär sanitary
Sanitäranlage *f* sanitary fitting (fixture)
Sanitärpumpe *f (ship)* sanitary pump
Sanitärtank *m* sanitary tank
Sanitärtechnik *f* sanitary (public health) engineering
Sanitäts[kraft]wagen *m* ambulance car
Sankt-Elms-Feuer *n* St. Elmo's fire
Sansibarkopal *m* Zanzibar copal
Säpflug *m* drill plough
Saphir *m* sapphire
Saponin *n* saponine
Sappanholz *n* sap[p]an-wood
Sappe *f (mil)* sap
Sapropel *n* sapropel
sapropelitisch sapropelic
Särad *n* seed-barrel
Sargbeschlag *m* coffin mounting
Sargdeckel *m* 1. *(mine)* bell, cauldron-bottom, pot hole; 2. coffin lid *(joinery)*
Sarkosin *n* sarcosine
Säschar *n* coulter
Satellit *m* satellite
  gesteuerter S. guided satellite
  künstlicher S. artificial (man-made) satellite
  meteorologischer S. meteorological satellite
  nichtgesteuerter S. non-guided satellite
Satellitenortung *f* satellite tracking
Satellitenstadt *f* satellite town
Satellitenstart *m* launching of a satellite
Satellitenverfolgung *f* satellite tracking
Satellitimpuls *m* satellite pulse
Satelloid *n* satelloid
Satinage *f* machine finishing
Satinageriß *m* calender broke
Satinbindung *f* satin weave
Satindrell *m* satin ticking
Satinglanz *m* satin glaze
satinieren to finish, to glaze, to satin[-finish], to calender *(paper)*; to lay the grain *(leather)*
Satinieren *n* **mit Friktion** friction glazing *(papermaking)*
Satinierkalander *m* supercalender
Satiniermaschine *f* glazing machine, pendulum jigger
satiniert calendered, [calender-]finished, glazed
satt saturated, deep *(colour)*
Sattdampf *m* dry (saturated) steam
Sattdampfbetrieb *m* working with saturated steam
Sattel *m* 1. saddle *(horse)*; 2. *(geol)* anticlinal fold, anticline
Sattelanhänger *m* semitrailer
Sattelbefestigung *f (aut)* cradle mounting
Satteldach *n* gabled (gable-ended, gable-type) roof
Sattelfläche *f* saddle surface
Sattelfüllkörper *m* saddle packing, saddles
Sattelgurt *m* saddle girth
Sattelisolator *m* shell insulator
satteln to saddle
Sattelpunkt *m* saddle point *(of strategic games)*, saddle *(phase plane analysis)*
Sattelpunktmethode *f* saddle-point method
Sattelschlepper *m* semitrailer, *(Am)* truck tractor
  S. mit aufgesatteltem Anhänger articulated vehicle
Sattelschleppzug *m* semitrailer unit
Satteltasche *f* saddle bag; wallet, case holding repairing tools for bicycle
Sattelwagen *m (ry)* semitrailer
sättigen to saturate, to satiate
  mit Kohlendioxid s. to carbonate
Sättigung *f* saturation; emission limitation *(electron tube)*
  magnetische S. magnetic saturation
Sättigungsaktivierung *f* saturation activation
Sättigungsaktivität *f* saturated activity
Sättigungsanteil *m* fraction of saturation
Sättigungsbedingung *f* saturation condition
Sättigungsbereich *m* saturation range
Sättigungsdichte *f* saturation density
Sättigungsdrossel *f* saturable reactor
Sättigungsdruck *m* vapour (saturation) pressure
Sättigungseffekt *m* saturation effect
Sättigungsgebiet *n* zone of saturation
Sättigungsgrad *m* degree of saturation
  S. der Luft degree of saturation of the air

natürlicher S. field moisture equivalent *(soil mechanics)*

Sättigungsgrenze *f* saturation limit

Sättigungsinduktion *f* saturation induction

Sättigungsinduktivität *f* saturation inductance

Sättigungskapazität *f* saturation capacity

Sättigungskennlinie *f* saturation curve (characteristic)

Sättigungsknie *n* saturation bend

Sättigungskonzentration *f* saturation concentration

Sättigungslinie *f* saturation line

Sättigungsmagnetisierung *f* saturation magnetization

Sättigungspunkt *m* saturation point

Sättigungsreaktanz *f* saturation reactance

Sättigungsschaltung *f* saturation connection

Sättigungsspannung *f* saturation voltage

Sättigungsspule *f* saturable inductor

Sättigungsstrom *m* saturation current; filament emission *(electron tubes)*

Sättigungstemperatur *f* saturation temperature

Sättigungswert *m* saturation value

Sättigungszustand *m* saturation state

Sattler *m* saddler

Sattstrahl *m* jet

Saturateur *m* saturator

S. für Kohlensäure carbonator *(sugar industry)*

saturieren to saturate, to satiate; to carbonate *(sugar industry)*

Satz *m* 1. set *(of tools, etc)*; 2. batch *(work pieces)*; 3. *(print)* composition, type-matter; 4. *(mine)* round *(shot firing)*; 5. law, lemma, theorem, thesis *(sciences)*; 6. assortment *(goods)*

S. des Apollonius Apollonius's theorem

S. des Ceva Ceva's theorem

S. vom Parallelogramm der Kräfte principle of the parallelogram of forces

S. von der Ersatzspannungsquelle Thevenin's theorem

S. von der Ersatzstromquelle Norton's theorem

S. von der Zunahme der Entropie principle of increase of entropy

bleiloser S. *(print)* cold composition

durchschossener S. *(print)* leaded matter

gequirlter S. *(print)* squabble

gesetzter S. *(print)* type-matter

gespaltener S. *(print)* column-matter, half measure

glatter S. *(print)* text matter

großer Fermatscher S. Fermat's last theorem

Helmholtzscher S. 1. Thevenin's theorem *(referring to voltage)*; 2. Norton's theorem *(referring to current)*

Heßscher S. Hess's law

kleiner Fermatscher S. Fermat's theorem

kompresser S. *(print)* solid matter

mathematischer S. *(print)* mathematical composition

Maxwellscher S. [Maxwell] reciprocity theorem, Maxwell's theorem

Poyntingscher S. Poynting's theorem

verfischter S. *(print)* pie

Satzbetrieb *m (mach)* batch operation

Satzbreite *f (print)* measure

Satzfräser *m* gang cutter, cutter gang

Satzgewindebohrer *m* serial tap

Satzherstellung *f (print)* type-setting

lochbandgesteuerte S. tape-controlled type-setting

Satzkoks *m* coke per charge *(cupola)*

Satzspiegel *m (print)* layout, type area

Satzspiegelhöhe *f (print)* depth of column

Satztrockner *m (ag)* batch-type tray drier

Satzverständlichkeit *f (telec)* intelligibility of phrases

Sau *f (met)* bear, pig, sow

Sauberkeit *f* cleanliness

Sauberkeitsschicht *f* sub-base *(highway construction)*

sauer *(ch)* acid; acidic; sour

Sauerfutter *n* [en]silage

säuerlich *(ch)* acidiferous; acidulous; acescent

säuern *(ch)* to acidulate, to acidify; to sour

Sauerstoff *m* oxygen

absorbierter S. absorbed oxygen

flüssiger S. liquid oxygen

Sauerstoffanalysator *m* oxygen analyzer

paramagnetischer S. paramagnetic-thermal oxygen analyzer

sauerstoffangereichert oxygen-enriched

Sauerstoffanreicherung *f* oxygenation

Sauerstoffanschluß *m* oxygen point *(in the aeroplane)*

sauerstoffarm poor in oxygen

Sauerstoffatmungsgerät *n* oxygen inhaling (breathing) apparatus

Sauerstoffaufblasverfahren *n (met)* LD process

Sauerstoffaufnahme *f* absorption of oxygen

Sauerstoffbeatmungsgerät *n* oxygen breathing apparatus

Sauerstoffbedarf *m* oxygen demand

biochemischer S. biochemical oxygen demand

biologischer S. biological oxygen demand

Sauerstoffbehälter *m* oxygen cylinder (flask)

Sauerstoffbleiche *f* oxygen bleaching

Sauerstoffbrücke *f* oxygen bridge

Sauerstoffflasche *f* oxygen flask (cylinder)

sauerstofffrei oxygen-free, containing no oxygen

sauerstoffhaltig oxygenous

Sauerstoffkaltbleichverfahren *n* oxygen cold bleaching process

Sauerstofflanze *f (met)* oxygen lance

Sauerstoffleerstelle *f* oxygen vacancy
Sauerstoffmangel *m* oxygen deficiency
Sauerstoffmesser *m* oximeter
Sauerstoffmessung *f* oximetry
Sauerstoffschneideelektrode *f* oxy-arc cutting electrode
Sauerstoffschneiden *n* / **elektrisches** oxy-arc cutting
Sauerstoffstelle *f* oxygen site *(crystal lattice)*
Sauerstoffstrahl *m* oxygen jet, jet of oxygen
Sauerstoffträger *m* oxidizing constituent (component), oxygen carrier, oxidizer, oxidant
Sauerstoffträgerbehälter *m* oxidizer (oxidant) tank
Sauerstoffträgerstrahl *m* oxidizer jet
Sauerstoffträgerverbrauch *m* oxidizer consumption (flow)
Sauerstoffverbindung *f* oxide, oxygen compound
Sauerstoffverbrauch *m* oxygen demand (consumption)
Sauerteig *m* leaven
Sauerwerden *n (ch)* acescency
Sauganode *f* first anode *(cathode-ray tubes)*
Saugbagger *m* suction (hydraulic, pump) dredger
S. mit drehendem Saugkopf rotary suction dredger
S. mit Schneidkopf cutter-suction dredger, [suction] cutter dredger
Saugball *m* suction ball
Saugbeton *m* vacuum concrete
Saugdrossel *f (el)* balance coil, interphase reactor (transformer), compensating choke
Saugdruck *m* suction pressure
saugen to suck; to absorb
luftleer s. to evacuate, to exhaust
Saugen *n* **einer Form** contraction of a mould
saugfähig absorbent
Saugfähigkeit *f* absorbability, absorbency, absorbing capacity
Saugfilter *n* suction (notching) filter
Saugfiltration *f* edge filtration
Saugflasche *f* aspirator (suction) bottle, suction flask
Sauggas *n* suction gas
Sauggebläse *n* exhauster
Sauggeschwindigkeit *f* rate of evacuation
Saugheber *m* siphon; pipette
Saughöhe *f* absorptive height, capillary rise; suction head, [suction] lift, height of suction *(pump)*
statische S. static suction lift
Saughopperbagger *m* suction hopper-dredger
Saughub *m* charging (intake, inlet) stroke
Saugkammer *f* inlet chamber, suction chamber (port), intervalve space
Saugkanal *m* induction port
Saugkasten *m* **der Gautsche** couch box *(papermaking)*
Saugkopf *m* **mit Schneidwerk** suction cutter *(dredger)*
Saugkopflüfter *m* uptake ventilator
Saugkorb *m* [suction] strainer, suction basket
Saugkraft *f* **des Kondensators** aspiration of the condenser
Saugkreis *m (el)* absorber (absorbing, resonant) circuit
Saugleistung *f* suction speed
Saugleitung *f* intake line *(pump)*; suction pipe line; *(mine)* suction gas main
Saugluft *f* intake air; *(aero)* indraft
Saugluftbehälter *m* vacuum reservoir
Saugluftbremse *f (ry)* vacuum brake; *(aut)* vacuum servo brake
Sauglüfter *m* exhaust fan, uptake (exhaust, suction) ventilator
Saugluftförderer *m* vacuum conveyor tube
Saugluftleitung *f* vacuum line
Saugluftpumpe *f* vacuum pump
Saugluftschaltung *f (aut)* vacuum-power change
Saugluftschaltzylinder *m* vacuum shift cylinder
Sauglufttrockner *m* suction air drier
Sauglüftung *f* exhaust ventilation
Saugluftverteiler *m* vacuum distributor
Saugmassel *f* shrink head *(casting)*
Saugöffner *m* / **kastenförmiger** pneumatic box opener
Saugpapier *n* blotting paper
Saugpostpapier *n* duplicating paper
Saugpumpe *f* suction pump
Saugraum *m* suction chamber
Saugrohr *n* suction pipe
S. zum Wasserheber *(ry)* elevator suction pipe, inlet (suction) pipe
Saugrohrhebegeschirr *n* pipe hoisting gear *(dredger)*
Saugrohrreibung *f* suction pipe friction
Saugrüssel *m* suction pipe discharge (line) *(grain unloading)*
Saugseite *f* intake side *(pump)*
Saugseitenkavitation *f* back cavitation
Saugsieb *n* [suction] strainer, rose
Saugspülbohrverfahren *n (mine)* reverse circulation drilling
Saugteller *m* vacuum cup
Saugtransformator *m* draining (interphase, sucking, negative boosting) transformer
Saugventil *n* suction valve
Saugventilator *m* exhauster
Saugventilbuchse *f* suction valve bushing
Saugventilfeder *f* suction valve spring
Saugventilkegel *m* suction valve cone
Saugventilschlitz *m* inlet valve port
Saugwalze *f* suction roller
Saugwirkung *f* suction effect, sucking action
Saugzug *m* suction (induced) draught; upward pull *(funnel)*

Saugzugsinterung *f* down-draft sintering
Saugzugventilator *m* induced draught fan
Säule *f* pillar, post, column, stanchion, upright
bügelbewehrte S. tied (hooped) column *(reinforced concrete)*
durchlaufende S. continuous column
eingebundene S. attached column
eingespannte S. fixed column
mehrteilige S. spaced column
spiralbewehrte S. spirally reinforced column *(reinforced concrete)*
thermische S. thermal column
umschnürte S. spirally reinforced column
Säulenbohrmaschine *f* 1. *(mach)* upright drilling machine [round column model]; round-column upright drilling machine; pillar drill; vertical round-column upright drilling machine, vertical drilling machine with column of round section; 2. *(mine)* bar-mounted drifter, column-mounted drill
Säulenchromatografie *f* column chromatography
Säulendiagramm *n* bar chart
Säulendrehkran *m* post-mounted swing jib crane
Säulenfüllung *f* column packing
Säulenfundament *n* column footing
Säulenfuß *m* base of column, plinth
Säulengang *m* colonnade
Säulenhalle *f* portico
Säulenionisation *f* columnar ionization
Säulenklemmung *f (mach)* column clamp
säulenlos astylar *(architecture)*
Säulenmaschine *f* post bed sewing machine *(shoe)*
Säulenplatte *f* plinth
Säulenpreßformmaschine *f* post squeezer
Säulenrüttelpreßformmaschine *f* post jar squeezer
Saum *m* hem, seam; border, margin
säumen to hem, to seam
Säumer *m* / **verstellbarer** *(text)* adjustable hemmer
Säumfüßchen *n (text)* hemmer foot
Saumleiste *f* fillet
Säumling *m* bark edge
Säummaschine *f* hemming machine
Saumstich *m* hemstitch
Säure *f* acid
arsenige S. arsenious acid
chlorige S. chlorous acid
diphosphorige S. pyrophosphorous acid
dischweflige S. pyrosulphurous acid
dithionige S. hyposulphurous acid
dreibasische S. tribasic acid
einbasische S. monoacid, monohydric acid
konzentrierte S. concentrated (strong) acid
phosphorige S. phosphorous acid
pyrophosphorige S. pyrophosphorous acid
pyroschweflige S. pyrosulphurous acid
salpetrige S. nitrous acid
schwache S. weak acid
schweflige S. sulphurous acid
selenige S. selen[i]ous acid
starke S. strong acid
tellurige S. tellurous acid
unterbromige S. hypobromous acid
unterchlorige S. hypochlorous acid
unterjodige S. hypoiodous acid
unterphosphorige S. hypophosphorous acid
untersalpetrige S. hyponitrous acid
unterschweflige S. sulphoxylic acid
verdünnte S. dilute acid
vierbasische S. tetrabasic acid
zweibasische S. dibasic acid
Säureabbau *m* degradation of acids
Säureamid *n* acid amide
Säureauslaugung *f* acid leach
Säurebad *n (text)* souring (bleaching) bath
Säureballon *m* carboy
Säurebehälter *m* acid reservoir
Säurebehandlung *f* acid treatment
säurebeständig acid-proof, acid-resisting, acid-resistant
Säurebestimmung *f* acidimetry
säurebildend acid-forming
Säurebildung *f* acid formation, acidification
Säurebottich *m* acid vat
Säurechlorid *n* acid chloride
Säureechtheit *f* fastness to acids
Säurefarbstoff *m* acid dye[stuff]
säurefest acid-proof, acid-resistant, acid-resisting
säurefrei acidless, non-acid, non-corrosive
Säuregehalt *m* acidity, acid content; sourness
Säuregrad *m* acidity [degree]; acid value; sourness
säurehaltig acidiferous, acidic
Säurehärter *m* acid catalyst
Säureharz *n* acid sludge
Säureheber *m* acid siphon
Säurekitt *m* acid-proof cement
Säurekochechtheit *f* fastness to acid boiling
Säureküpe *f* acid vat
Säuremesser *m* acidimeter
Säuremessung *f* acidimetry
Säureprüfer *m* acid tester
Säureradikal *n* acid radical
Säurerest *m* acid residue
Säurescheidung *f* acid parting
Säureschlamm *m* acid sludge
Säureschwemmverfahren *n* acid-flotation method (process)
Säurespaltung *f* acid cleavage
Säureteer *m* acid sludge

Säureturm *m* acid tower
Säureverfahren *n* acid process *(digestion of chemical pulp with a bisulphite liquor)*
säurewalkecht fast to acid milling
Säurezahl *f* acid (neutralization) number
Säurezentrifuge *f* acid hydroextractor
Sävorrichtung *f* seeding mechanism
Säwelle *f* pinion wheel
Scavenging *n (nucl)* scavenging
Schäbe *f (ag)* boon, shive
Schabebaum *m* fleshing beam *(leather)*
Schabeisen *n* fleshing knife *(leather)*; *(mach)* scraper
Schabemaschine *f* fleshing machine *(leather)*
schaben *(mach)* to scrape *(ways)*; to take a shaving cut
Schaben *n (mach)* scraping
  S. mit gekreuzten Achsen [rotary] crossed-axes shaving
  S. mit parallelen Achsen parallel-axes shaving
  S. mit umlaufenden Rädern rotary method of gear shaving
  S. mit Zahnstange rack method of gear shaving
  S. von Hand hand scraping
  S. von Zahnrädern shaving of gears
Schaber *m (mach)* scraper
Schabermesser *n* doctor blade
Schabeschnitt *m (mach)* shaving cut
Schabezahn *m (mach)* shear cutting tooth, straight tooth *(broach)*
Schabhobel *m* **mit Klinge / eiserner** iron spoke shave with blade *(wood)*
Schabkunst *f* mezzotint
Schablone *f* templet, template; model, pattern; *(print)* stencil; cam; gauge; strickle *(moulding)*
Schablonendrehmaschine *f* copying lathe
Schablonendruck *m* stencil printing
Schablonenfarbe *f* stencil paint
Schablonenformerei *f* template moulding
Schablonenfräsmaschine *f* profile milling machine
Schablonenlack *m* screen varnish
schablonenmäßig stereotyped, mechanical
Schablonennachformeinrichtung *f (mach)* template copying attachment
Schablonenrollenspindel *f (mach)* cam-roller spindle
Schablonenspule *f* preformed (form-wound) coil
Schablonenwicklung *f* preformed (form, contact) winding
Schabmaschine *f* shaving machine *(for gears)*
Schabotte *f* anvil block (bed)
Schabrad *n* rotary gear shave (shaving) cutter
schachbrettartig chequer-board [effect]
Schachbrettmuster *n (telev)* chequer-board pattern; chequerwork
Schacht *m* 1. *(mine)* shaft; 2. well *(elevator)*; 3. stack *(blast furnace)*; 4. *(ship)* casing, trunk, trunk[ed] casing
  S. für Materialförderung materials shaft
  S. mit rechteckiger Schachtscheibe rectangular shaft
  S. mit runder Schachtscheibe circular shaft
  einziehender S. intake shaft
  seigerer S. vertical shaft
  tonnlägiger S. incline, incline[d] shaft
Schachtabteufen *n* shaft sinking
Schachtansatzpunkt *m* shaft placing point
Schachtanschlag *m* shaft inset
Schachtausbau *m* **mit Schrotzimmerung** curbing
Schachtausmauerung *f* stack lining *(blast furnace)*
Schachtbagger *m* hopper dredge[r]
Schachtband *n* stack band *(blast furnace)*
Schachtbohrer *m (mine)* trepan
Schachtbohrverfahren *n* shaft boring (drilling technique)
Schachtdeckel *m* shaft cover
Schachtel *f* box
Schachtelkarton *m* box board
Schachtelklebmaschine *f* box pasting machine *(matchbox)*
schachteln to nest
Schachtelpappe *f* container board
Schachtelspan *m* box veneer *(match)*
Schachtelspanteilmaschine *f* box veneer cutting machine
Schachtelung *f (el)* lamination overlap *(of transformer cores)*, multiplexing *(signal transmission)*; nest
schachten to excavate, to sink *(a shaft)*
Schachtfalle *f (mine)* bearing-up stop, catch, kep
Schachtförderleistung *f* hoisting (winding) capacity
Schachtförderung *f* shaft hoisting
Schachtfutter *n* stack lining *(blast furnace)*
Schachtgatter *n (mine)* lifting guards
Schachtgebäude *n* head house, heapstead
Schachtgeviert *n* crib-ring, cribwork
Schachthalle *f* shaft hall
Schachthauer *m* shaft-repairer
Schachtkabelübergangskasten *m* **/ begehbarer** *(el)* manhole junction box
Schachtkapazität *f (mine)* hoisting (winding) capacity
Schachtklappe *f (mine)* trap door
Schachtluke *f (ship)* trunked hatchway
Schachtmauerwerk *n* stack brickwork *(blast furnace)*
Schachtmündung *f* [shaft] collar
Schachtofen *m* shaft furnace (kiln), blast furnace
Schachtofenprozeß *m* **/ reduzierender** reducing smelting process

Schachtöffnung *f* manhole *(cable)*
Schachtpanzer *m* stack casing *(blast furnace)*
Schachtpfeiler *m* shaft [safety] pillar
Schachtpumpenbagger *m* hopper suction dredger
Schachtquerschnitt *m* shaft cross-section
Schachtring *m (mine)* drum-curb
Schachtsäule *f* shaft structure
Schachtscheibe *f* shaft [cross-]section
Schachtschleuse *f* shaft air-lock
Schachtsicherheitspfeiler *m* shaft [safety] pillar
Schachtsignal *n* hoisting signal
Schachtsignalanlage *f* shaft signalling device
Schachtsignalschreiber *m* shaft signal recorder
Schachtstuhl *m (mine)* inset frame
Schachttor *n* shaft gate
Schachttragring *m* mantle ring *(blast furnace)*
Schachttrockner *m (text)* tower drier
Schachtumfahrungsstrecke *f* shaft by-pass
Schachtumtrieb *m* loop, pit bottom circuit, shaft by-pass
Schachtwand *f* manhole wall *(cable)*
Schaden *m* damage, breakdown, defect
Schadenersatz *m* consequential damage, compensation, idemnification
Schadenersatzanspruch *m* claim for compensation
schadenersatzpflichtig liable for damages
Schadensmeldung *f* advice of damage, damage note
schadhaft faulty, defective, damaged
schädigen to damage
Schädigungsrate *f* damage rate
schädigungssicher fail-save
schädlich detrimental, damaging
Schädling *m* pest
Schädlingsbekämpfung *f* pest control
Schädlingsbekämpfungsgeräte *npl* pest control equipment
Schädlingsbekämpfungsmittel *n* pesticide
Schadwagenzettel *m (ry)* repair label
Schafblöße *f* / **gepickelte** *(leath)* pickled sheep pelt
Schaffell *n* sheep pelt
  entwolltes S. dewool[l]ed sheepskin
Schaffner *m* guard, member of train crew, conductor
Schaffplatte *f* **[eines Ofens]** fore plate
Schafftür *f* charging door
Schaffußwalze *f (build)* sheep-foot roller
Schafgarbenöl *n* yarrow oil
Schafscherer *m* clipper
Schafschermaschine *f* sheep shearing machine
Schafschur *f* sheep shearing
Schafstall *m* shed, pen
Schaft *m* shank *(tool)*; shaft *(column)*; upper *(shoe)*; stock *(rifle)*; handle; stem *(insulator)*
  S. mit Anzugsgewinde tapped-end shank
  zylindrischer S. parallel shank
Schaftaufhänger *m (text)* harness wire
Schaftausbohrung *f* rod bore
Schaftdurchmesser *m* shank diameter, diameter of shank *(tool)*; stem diameter *(pin-type insulator)*
  reduzierter S. diameter of recess *(tool)*
Schäften *n* **von Furnieren** splicing of veneers
Schaftfräser *m* end mill cutter; shank cutter
  S. mit Anzugsgewinde tapped-shank end mill
  S. mit nockenverriegeltem Schaft cam-lock shank end mill
Schafthülse *f* sleeve, barrel *(micrometer)*
Schaftmaschine *f (text)* dobby, tie-up Jacquard
  S. für Hoch- und Tieffach centre-shed dobby, double-lift dobby
Schaftsohle *f* base of shaft *(chimney)*
Schaftstiefel *mpl* top boots
Schaftstuhl *m (text)* chain (dobby, harness) loom
Schafttrommel *f* shaft section
Schaftwebstuhl *m* **mit Hebel** negative lever dobby
Schafwolle *f* sheep wool
Schafzucht *f* sheep breeding
Schäkel *m* shackle [with screw eye pin], clevis
Schäkelbolzen *m* shackle bolt
Schäkelisolator *m* shackle insulator
Schälblock *m* peeler
Schalboden *m* soffit boards
Schalbrett *n* shuttering board *(concrete construction)*
Schäldrill *m* combined disk plough and seeder
Schale *f* 1. dish, basin, cup, bowl, pan; 2. scale, pan *(of scales)*; 3. shell *(of electrons)*; 4. *(math)* nappe; 5. leaf *(brickwork)*; 6. *(ag)* peel; skin *(of potatoes)*; 7. dome, gong; 8. bowl *(of lamps)*
  abgeschlossene S. *(nucl)* filled (completed, closed) shell
  besetzte S. filled shell
  doppelt gekrümmte S. doubly curved shell
  dünnwandige S. thin shell
  einfach gekrümmte S. singly curved shell
  vollbesetzte S. filled (completed, closed) shell
Schäleinleger *m (ag)* self feeder
Schäleisen *n* barking iron, bark scraper
schälen 1. to peel; to bark *(tree)*; 2. *(mine)* to plane, to plough *(coal)*; 3. *(met)* to peel, to scalp; to scale off
Schälen *n (ag)* shallow ploughing; stubble cleaning; barking *(tree)*
Schalenabschluß *m (nucl)* closure of the shell
Schalenätzung *f (print)* trough etching
Schalen[auf]bau *m* shell structure *(of nucleus)*
Schalenbauweise *f (aero)* stressed-skin construction, monocoque design
Schalendach *n* shell roof
Schalenelektrode *f* dished electrode
Schalenfeuerung *f* kettle (trough) furnace

Schalengewölbe *n* shell vault
Schalengreifer *m* clamshell grab
Schalenhartguß *m* chilled cast iron
Schalenkabel *n* shell cable
Schalenkonstruktion *f (build)* shell structure
Schalenkreuzanemometer *n* rotating-cup anemometer, revolving-cup anemometer
Schalenkupplung *f* box (muff, clamp, compression) coupling
Schalenlager *n* shell bearing
Schalenmodell *n* **des Atomkerns** [nuclear] shell model, shell model of nucleus
Schalenrumpf *m (aero)* monocoque fuselage
Schalenstruktur *f* shell structure *(of atomic nucleus)*
Schalentheorie *f (build)* theory of shells
Schälfurnier *n* moving (rotary) veneer
Schalgerüst *n* falsework *(concrete construction)*
Schälholz *n* peeled billet-wood
Schall *m* sound
  diffuser S. diffuse sound
  direkter S. direct sound
schallabsorbierend sound-absorbing
Schallabsorption *f* acoustic (sound) absorption
  molekulare S. molecular acoustic (sound) absorption
Schallabsorptionsgrad *m* degree of sound absorption
Schallabsorptionsstoff *m* material for sound absorption
Schallabsorptionsvermögen *n* acoustical absorptivity
Schallaufzeichnung *f* sound record[ing]
Schallausbreitung *f* sound propagation
Schallausbreitungsgeschwindigkeit *f* rate of travel of sound, rate of sound propagation
Schallbekämpfung *f* suppression of noise
Schallbeugung *f* sound diffraction
Schallbündelung *f* sound focusing
schalldämmend sound absorbing (insulating)
Schalldämmplatte *f* acoustic board
Schalldämmstoff *m* sound absorbant
Schalldämmung *f* sound damping (attenuation), acoustical (sound) insulation, sound-proofing, suppression of noise
Schalldämmzahl *f* coefficient of sound damping
schalldämpfend sound absorbing (absorbent, deadening), sound-insulated
Schalldämpfer *m (aut)* [exhaust] silencer; *(el)* sound absorber *(circuit-breaker)*
  S. mit Resonatoren resonant-absorption silencer
  S. mit schallschluckenden Materialien acoustic-absorbent silencer
  S. mit Wirbelung baffle-type silencer
Schalldämpfung *f* sound attenuation (absorption), sound-proofing
  S. durch Wirbelung *(aut)* baffle silencing
schalldicht sound-proof
Schalldichte *f* sound-energy density
Schalldruck *m* acoustic (sound, sonic) pressure
Schalldruckmesser *m* sound level meter
Schalldruckmikrofon *n* pressure microphone
Schalldruckpegel *m* sound pressure level
Schalleitfähigkeit *f* sound conductivity
schallen to sound
Schallenergie *f* sound energy
Schallenergiedichte *f* sound-energy density
Schallerregung *f* sound generation (excitation)
Schallerzeuger *m* generator of sound, acoustic (sound) generator
Schallerzeugung *f* sound generation
Schallfeld *n* sound field
Schallfluß *m* sound-energy flux
Schallfortpflanzung *f* sound propagation
Schallfortpflanzungsgeschwindigkeit *f* velocity of sound propagation
Schallgeschwindigkeit *f* sound velocity, velocity of sound; sonic speed
Schallhärte *f* acoustic inertia (stiffness), sound hardness
Schallimpedanz *f* acoustic impedance
Schallimpuls *m* sound pulse
Schallintensität *f* acoustic intensity per unit area, sound intensity, intensity of sound
  momentane S. instantaneous acoustic intensity per unit area
schallisoliert sound-insulated
Schallisolierung *f* sound insulation
Schallkammer *f* reverberation chamber
Schallmauer *f* sonic (sound) barrier
Schallmeßaufklärung *f* sound ranging
Schallmessung *f* sound [level] measurement
schallnahe transonic
Schallnut *f* sound groove
Schallöffnung *f* earpiece opening
Schallortung *f* sound ranging
Schallot *n* sonic depth finder
Schallotung *f* acoustic sounding
Schallpegel *m* sound level
Schallpegelmesser *m* sound level meter
Schallplatte *f* [gramophone] record, [gramophone] disk
Schallplattenaufnahme *f* disk recording
Schallplattenaufnahmegerät *n* disk recorder
Schallplattenaufzeichnung *f* disk recording
Schallplattenherstellung *f* gramophone-record manufacture
Schallplattenrille *f* sound groove
Schallplattentonfilm *m* photogramophone
Schallplattenverstärker *m* gramophone amplifier
Schallquelle *f* sound source
Schallradar *n* sound ranging
schallreflektierend sound-reflecting
Schallreflexion *f* acoustic (sound) reflection
Schallreiz *m* acoustic (sound) stimulus

Schallrückstrahlung *f* reverberation of sound
Schallschirm *m* baffle [board]
schallschluckend sound-absorbent, sound-absorbing, sound-deadening
Schallschutz *m* sound control, sound-proofing
Schallschwelle *f* sound (sonic) threshold
Schallsiebmaschine *f* sonic-type screening machine
Schallspektroskopie *f* sound spectroscopy
Schallspektrum *n* sound spectrum
Schallstärke *f* sound intensity, intensity of sound
Schallstrahl *m* acoustic (sound) ray
Schallstrahlungsdruck *m* acoustic (sound) radiation pressure
schalltot anechoic, acoustically inactive (dead)
Schalltrichter *m* sound horn, trumpet *(loudspeaker)*
Schallübertragung *f* sound transmission
Schallursprungsort *m* source of sound
Schallverzögerungsleitung *f* sonic delay line
Schallwand *f* baffle [board], loudspeaker baffle
Schallwelle *f* acoustic (sound) wave
Schallwellenecholot *n* sonic depth finder, sonic echo sounder
Schallwiedergabe *f* sound reproduction
Schälmaschine *f* rotary (scalping) lathe *(veneer)*
Schälmesser *n* drawing (shave) knife *(veneer)*
Schalöl *n* form (mould) oil *(concrete construction)*
Schalplan *m* shuttering plan *(concrete construction)*
Schälprügel *m* peeled billet *(wood)*
Schälqualität *f* rotary veneer quality
Schälreibahle *f* diemaker's reamer
  S. mit großem Drall steep-spiral reamer
Schälrestrollen *fpl* waste veneer core
Schälriß *m* cup shake *(wood)*
Schälrissigkeit *f* shakiness *(wood)*
Schälschrapper *m (mine)* scraper box
schälspiralgenutet steep-spiral *(reamer)*
Schalstück *n* [rigging] batten *(sailing ship)*
Schaltader *f* jumper wire
Schaltafel *f* shuttering (form) panel *(concrete construction)*
Schaltalgebra *f* Boolean algebra
Schaltanlage *f (el)* switch plant; distributing board; switchgear
Schaltanordnung *f* circuit arrangement; switching arrangement
Schaltarbeit *f* electric switching power
Schaltarm *m* wiper *(rotary switch)*
schaltbar controllable; switch selected; loose *(coupling)*; *(mach)* indexing *(turret)*
Schaltbarkeit *f* / **feinstufige** sensitivity
Schaltbewegung *f* control (indexing) movement
Schaltbild *n* circuit (wiring) diagram, diagram of connections
Schaltbock *m (aut)* gear-shift lug
Schaltbrett *n* switchboard, switch panel, dashboard
Schaltdeckel *m (aut)* gear-shift cover
Schaltdetektor *m* / **trägergesteuerter** recycling detector
Schaltdiode *f* switching diode
Schaltdraht *m* jumper (tracker, hook-up, distributing) wire
Schaltdrahtverbindung *f* jumper connection
Schaltdrossel *f* switching reactor
Schalteinheit *f* stepping unit
Schalteinrichtung *f* switching equipment (device, facility)
Schaltelement *n (el)* circuit (switching) element; *(mach)* operating element, control
  konzentriertes S. lumped element
  sinnfälliges S. *(mach)* directional control
schalten 1. *(el)* to switch *(on, off)*, to connect *(in series, parallel)*, to wire *(circuit)*; 2. *(mach)* to control, to operate, to index *(turret)*; *(aut)* to shift a gear, to put into gear; to clutch
  eine Leitung s. to line up
  gemischt s. to connect in series parallel
  in Reihe s. to connect in series
  nach Schaltbild s. to connect up as in diagram
Schalten *n* **des Revolverkopfes** *(mach)* indexing of the turret
  sinnfälliges S. *(mach)* directional control
Schalter *m* 1. *(el)* switch, circuit-breaker, interrupter; 2. ticket-window
  S. für die Meißelabhebung *(mach)* tool-lifting switch
  S. mit automatischer Wiedereinschaltung automatic-reclosing circuit-breaker
  S. mit Betätigungsstange rod-operated switch
  S. mit doppelter Unterbrechung double-break circuit-breaker
  S. mit Drehstützer rotating-insulator switch
  S. mit Druckluftbetätigung pneumatically operated switch
  S. mit einer Kontaktebene single-bank wafer switch
  S. mit Einfachunterbrechung single-break switch
  S. mit Kontakttulpe plunger switch
  S. mit Kraftantrieb mechanically operated switch
  S. mit Kurzschlußkontakten dead-end switch
  S. mit magnetischer Beblasung magnetic blow-out circuit-breaker
  S. mit senkrechter Schaltstrecke vertical break switch
  S. mit Sicherung fuse [disconnecting] switch, fused switch
  S. mit Sperre locking switch
  S. mit thermischer Auslösung thermal circuit-breaker

S. mit Überstrom- und Unterspannungsauslösung overcurrent undervoltage circuit-breaker
S. mit verriegeltem Deckel locked-cover switch
S. mit Verriegelung locking switch
S. mit verzögerter Auslösung time-delay circuit-breaker
S. mit Vielfachunterbrechung multibreak circuit-breaker
S. mit waagerechter Schaltstrecke horizontal break switch
S. unter Verschluß secret switch
automatischer S. mechanically-operated circuit-breaker
berührungssicherer S. shock-proof switch
doppelpoliger S. double-pole switch
dreietagiger S. three-storey switch
dreipoliger S. three-pole switch (circuit-breaker), triple-pole circuit-breaker
dreiwegiger S. three-way switch
druckempfindlicher S. pressure-sensitive switch
einpoliger S. single-pole switch (circuit-breaker), single-contact switch
erschütterungsfester S. shock-proof switch
halbversenkter S. semi-flush switch, semi-recessed switch
kapazitätsarmer S. anticapacitance switch
magnetisch betätigter S. solenoid switch
mehretagiger S. multi-storey switch
mehrpoliger S. multi-pole switch (circuit-breaker), multiple-pole switch (circuit-breaker), multiple-contact switch
mehrstufiger S. multiple-point switch
nichteinrastender S. non-locking key
ölarmer S. low oil-content circuit-breaker, oil-minimum circuit-breaker
ölloser S. oilless circuit-breaker
schlagwettergeschützter S. flameproof switch
stoßfester S. shock-proof switch
stoßsicherer S. shock-proof switch
verschlossener S. asylum (locked-cover) switch
versenkter S. sunk switch
vielstufiger S. multipoint switch
zweipoliger S. two-pole switch, double-pole switch

Schalter *mpl* / **gekuppelte** linked switches
Schalteramt *n (telec)* crossbar exchange
Schalteranschluß *m* switch connection
Schalterantrieb *m* circuit-breaker drive
Schalterbeanspruchung *f* circuit-breaker stress
Schalterbirne *f* pear [switch]
Schalterdeckel *m* switch cover
Schalterdiode *f* booster (series-efficiency) diode
Schalterdurchführung *f* circuit-breaker bushing
Schalterebene *f* switch wafer, contact plate
Schaltereinheit *f* switch unit
Schalterfassung *f* key holder
Schalterfeder *f* switch spring
Schaltergrundplatte *f* switch base
Schalterhalle *f* ticketing building
Schalterkappe *f* switch cover
Schalterknopf *m* switch knob
Schalterkontakt *m* switch contact
Schalterlampe *f* switch lamp
Schalternippel *m* bat handle
Schalterplatte *f* flush plate *(for recessed switch)*
Schalterpol *m* circuit-breaker pole
Schalterreihe *f (telec)* bank of keys
Schaltersatz *m (telec)* key unit
Schalterscheibe *f (telec)* wafer
Schaltersockel *m* switch base
Schalterstellung *f* switch position
Schalterstromkreis *m* switch circuit
Schalterstufe *f* switch position
Schaltfeder *f* switch spring
Schaltfeld *n* switch panel (section), switchboard section, jumper-field; *(cont)* patch bay
Schaltfolge *f* switching sequence
Schaltfrequenz *f* switching rate; *(telev)* sampling frequency
Schaltfunke *m* switching spark
Schaltfunktion *f* switching function; logic function
Schaltgabel *f* selector (gear-shift) fork
Schaltgenauigkeit *f (mach)* accuracy of index[ing] *(turret, table)*
Schaltgerät *n (el)* switchgear, switching device; *(mach)* control gear
S. für Hochspannungsanlage high-voltage switchgear
S. für Niederspannungsanlage low-voltage switchgear
S. mit Phasentrennung separated-phase switchgear
gußgekapseltes S. iron-clad switchgear
tropfwassergeschütztes S. drip-proof switchgear

Schaltgeräusch *n* commutator noise
Schaltgeschwindigkeit *f* switching speed
Schaltgestänge *n* switching hook (stick)
Schaltgetriebe *n (mach)* speed-changing mechanism; *(aut)* change-speed gear
Schaltglied *n* contact assembly; connecting device; logic element
Schaltgrenze *f (cont)* switching line
Schaltgriff *m (aut)* selector; *(mach)* control handle
Schaltgruppe *f (el)* connection group; vector group *(transformer)*
Schalthaken *m* switch hook
Schalthäufigkeit *f* switching frequency (rate)
Schalthebel *m (el)* switch lever, actuating arm; *(mach)* operating (control) lever; *(aut)* gear-shift lever

sinnfällig angeordneter S. *(mach)* directional control lever
Schaltimpuls *m* switching pulse
Schaltkammer *f (el)* arc [quenching] chamber, arc chute
Schaltkapazität *f* circuit (wiring) capacity, circuit stray
Schaltkasten *m (aut)* switch box; *(mach)* control box
Schaltklaue *f* shift dog
Schaltklinke *f (el)* jack; *(mach)* pawl
Schaltknebel *m* switch knob
Schaltknopf *m* control button
Schaltknopfstellung *f* control-button position
Schaltkreis *m* switching circuit
S. einer Selbststeueranlage steering pilot
kombinatorischer S. combinational switching circuit
logischer S. logical circuit
Schaltkreisanordnung *f* circuit layout
Schaltkreislogik *f* circuit logic
Schaltkupplung *f* clutch
Schaltkurve *f (cont)* switching line
Schaltleiste *f* connecting block
Schaltleistung *f* switching (rupturing, breaking) capacity
Schaltlichtbogenlöschung *f* arc extinguishing
Schaltlitze *f* flexible pigtail
Schaltmagnet *m* driving (switching) magnet
Schaltmaschine *f* barring engine
Schaltmuffe *f (aut)* gear-shift sleeve
Schaltnocken *m* trip cam
Schaltorgan *n (mach)* control element
Schaltort *m (el)* switch point
Schaltpaket *n* switch unit
Schaltpause *f* circuit blue-print
Schaltpendel *n* pendant switch
Schaltpilotfrequenz *f* switching [control] pilot
Schaltplan *m* wiring (connection, circuit) diagram, wiring (connection, circuit) scheme
S. der Fernleitungen trunk switching scheme
Schaltpult *n* control console (desk), control panel
Schaltrad *n (mach)* notched wheel, feed ratchet wheel *(shaper)*
Schaltraum *m* switch (control) room
Schaltreaktion *f (mach)* control response
Schaltrelais *n* switching relay
Schaltröhre *f* electronic switch, switching (contact) valve
Schaltsäule *f* switch pillar (column), control column, pillar-type switchgear
Schaltscheibe *f* switch wafer
Schaltschema *n s.* Schaltplan
Schaltschiene *f* gear-shift rail
Schaltschloß *n (el)* latch
Schaltschnecke *f (mach)* control worm
Schaltschrank *m (mach)* control cabinet; *(telec)* switch cabinet
Schaltschritt *m* switching step
Schaltschütz *n* contactor
automatisches S. automatic (self-acting) contactor
Schaltspiel *n* switch[ing] cycle
Schaltstange *f (el)* operating pole, actuating rod, stick; *(aut)* gear-shift bar
Schaltstecker *m* jack-plug
Schaltstelle *f* switch point
Schaltstellung *f* switch position
Schaltstellungsänderung *f* switch-position changing
Schaltstellungsanzeiger *m* [switch-]position indicator; three-position relay
Schaltstellungsmeldung *f* [switch-]position signalling, position acknowledgement
Schaltstörung *f* contact noise
Schaltstrecke *f* contact-break distance
Schaltstromkreis *m* switching circuit
Schaltstück *n* contact member (stud), contact [maker], disconnecting blade (contact, link)
bewegliches S. moving contact [member]
festes S. fixed contact [member]
stationäres S. stationary contact [member]
Schaltstückbewegung *f* contact movement (motion)
Schaltstückpaar *n* pair of contact members
Schaltsymbol *n* wiring (graphical) symbol
Schaltsystem *n* switching circuit (system)
Schalttafel *f (el)* switchboard, [distribution] panel; *(cont)* plug board, patchboard; *(aut)* instrument panel, dashboard; *(mach)* control panel
Schalttafelanschluß *m* switchboard connection
Schalttafelaufbau *m* panel mounting
Schalttafeleinbau *m* panel-frame mounting
Schalttafelinstrument *n* panel (switchboard) instrument, panel indicator (mounting meter)
Schalttafelregler *m* back-of-board controller, board-mounted controller
Schalttafelschalter *m (el)* panel switch
Schalttafelsicherungselement *n* switchboard fuse unit
Schalttafelsteuerung *f* panel control
Schalttakt *m* switching cycle
Schalttechnik *f* switching technique; circuitry
Schaltteilliste *f* list of circuit elements
Schalttor *n (cont)* gate
Schalttransistor *m* switching transistor, transistor switch
Schaltturm *m (aut)* gear-shift dome
Schaltüberspannung *f* switching surge, transient voltage, break impulse
Schaltuhr *f (el)* clock (time) switch; *(opt)* timer
Schaltung *f* 1. *(el, cont)* circuit; circuitry; circuit wiring, connection; switching [operation];

2. *(mach)* control; indexing *(automatics, turret lathe)*; 3. *(aut)* shifting [gears], gear-changing
S. mit konzentrierten Parametern lumped-constant network
S. mit mitlaufender Ladespannung bootstrap circuit
S. zur Versteilerung eines Auslöseimpulses trigger sharpener
aktive elektrische S. active electric network
bistabile S. bistable circuit
einmalig kippende S. trigger circuit
elektrische S. electrical circuit
elektronische S. electronic circuit
erprobte S. proven circuit
geätzte S. etched circuit
gedruckte S. printed circuit
gemischte S. series-parallel connection
geräuschdämpfende S. anti-sidetone device
getrennte S. *(aut)* separate gear change
heterostatische S. heterostatic circuit
idiostatische S. idiostatic method
integrierte S. integrated circuit
logische S. logical circuit
mehrphasige S. multi-phase circuit
mikroelektronische S. microelectronic circuit
mitsprechfreie S. sidetone reduction wiring
monostabile S. monostable circuit
nichtstabile S. astable circuit
parametrische S. parametric circuit
passive S. passive electric network
phasenempfindliche S. phase-sensitive circuit
plattierte S. plated circuit
symmetrische S. balanced circuit
vereinfacht dargestellte S. simplified circuit
vollintegrierte S. complete integrated circuit
volltransistorisierte S. all-transistor circuit
vorgesetzte S. *(aut)* dashboard gear change
Schaltungsanalysator *m (dat)* circuit analyzer
Schaltungsanordnung *f* circuit arrangement (layout); circuitry
logische S. logical arrangement of the circuit
Schaltungsbaustein *m* packaged circuit module
Schaltungsbegrenzung *f* circuit limitation
Schaltungselement *n* circuit element
Schaltungsentwurf *m* circuit design
Schaltungsfehler *m* circuit fault
Schaltungsfrequenz *f* circuit frequency
Schaltungskapazität *f* circuit wiring capacity, stray wiring capacitance
Schaltungskenngröße *f* circuit parameter
Schaltungslogik *f* circuit logic
Schaltungsmodul *m* circuit module
Schaltungsplatte *f* circuit board
Schaltungsschema *n* interconnection pattern
Schaltungstechnik *f* circuit technique; circuit engineering
Schaltungsweg *m* circuit path
Schaltventil *n* control (manipulating) valve
Schaltverhältnis *n* switching ratio
Schaltvermögen *n* switching capacity; breaking (making) capacity
Schaltverzug *m* time-lag, time-delay
Schaltvorgang *m (el)* switching operation; *(mach)* indexing movement *(automatic screw machine)*
Schaltwächter *m* controller
Schaltwagen *m* truck-type switchgear, removable truck
Schaltwarte *f* switch-gallery, switching station
Schaltweg *m* contact travel
Schaltweise *f* switching mode
Schaltwelle *f (el)* interrupter (selector) shaft; *(mach)* control shaft; *(aut)* gear-shift-lever shaft
Schaltwerk *n (el)* switch (operating) mechanism *(of a switchgear)*; *(therm)* barring gear
S. mit Hebel lever barring gear
S. mit Schnecke worm barring gear
doppeltwirkendes S. double-acting barring gear *(flywheel)*
einfachwirkendes S. single-acting barring gear
Schaltzeichen *n* wiring (graphical) symbol
Schaltzeit *f (el)* operating (switching) time; *(mach)* indexing time *(automatic screw machine)*
Schaltzeitkonstante *f* switching time constant
Schaltzelle *f (el)* [switchgear] cubicle; *(mach)* control cubicle
Schaltzustand *m (el)* connection state
Schalung *f* shuttering, form[work], mould *(concrete construction)*
verlorene S. permanent formwork (shuttering)
Schalungsauskleidung *f* form[work] lining, form liner
Schalungsform *f* mould, form
Schalungsgerüst *n* falsework
Schalungsöl *n* form (mould) oil
Schalungsrüttler *m* form (clamp-on) vibrator
Schalungstafel *f* shuttering (form) panel
Schaluppe *f* sloop, quarter boat, barge
Schälwunde *f (wood)* scar of peeling
schamfilen *(ship)* to chafe, to damage by friction
Schamfilungsleiste *f (ship)* chafing piece
Schamfilungsmatte *f* puddening, hanging mat; *(Am)* pudding
Schamotte *f* fireclay, refractory clay
Schamotteauskleidung *f* fireclay lining
Schamotteerde *f* fireclay, refractory clay
Schamottemörtel *m* fireclay mortar
Schamottestein *m* fire-brick, fireclay (refractory) brick
Schamotteziegel *m s.* Schamottestein
Schandeck *n (ship)* plank sheer, gunwale, covering-board
Schandeckel *m s.* Schandeck
Schandeckleiste *f (ship)* sheer rail
Schanzentisch *m* take-off, platform *(ski-jumping)*
Schanzkleid *n (ship)* bulwark; closed railing
hohes S. topgallant bulwark

Schanzkleidpforte *f* clearing (bulwark, freeing) port
Schanzkleidreling *f* bulwark rail
Schanzkleidstütze *f* bulwark stay (stanchion)
Schappeseide *f* schappe
Schappespinnerei *f* schappe [silk] spinning
Schar *n (ag)* share, coulter
Schar *f (math)* family
einparametrige S. pencil
Schäranlage *f (text)* warping plant
Schärbank *f* bobbin warping creel
Schärbaumgestell *n* warp beam creel
Schärbrief *m* card of warping particulars
schären *(text)* to warp
Schären *n (text)* warping
S. in Abschnitten sectional warping
S. von Hand hand warping
Schärenkreuzer *m* skerry cruiser
scharf sharp; defined; pointed; caustic; keen *(edge)*
Scharf *n* **des Vorschiffes** entrance *(bow)*
Scharfabstimmung *f* sharp (fine) tuning
selbsttätige S. automatic frequency control, automatic tuning [correction]
Schärfapparat *m* sharpening device
Schärfe *f* sharpness *(radio)*; keenness *(edge)*; *(opt)* definition
Scharfeinstellung *f (opt)* critical focusing
schärfen to sharpen *(tools)*; to scarf
Schärfentiefe *f (opt)* depth of focus
scharfkantig angular, feather-edged, sharp-edged, sharp-cornered, sharp-crested *(weir)*
Scharflaufen *n* **des Spurkranzes** sharpening of the flange
Schärfmaschine *f* sharpening machine, sharpener; skiving machine *(shoe)*
Scharfschleifen *n* resharpening, regrinding, tool and cutter grinding
Scharfschleifmaschine *f* sharpening (regrinding, tool and cutter grinding) machine
Scharfschütze *m* sniper
Schärgang *m (text)* group of threads
Schärgatter *n* beaming creel
Schärgestell *n* beaming creel
Schärholz *n* lease rod
Schärkanter *m* bobbin warping creel
Schärkegel *m* warping cone
Schärlänge *f* length of beaming
Schärlatte *f* bobbin frame
Schärmaschine *f* warping frame *(short chain)*, warping mill *(long chain)*
S. für Bänder warping mill for ribbons
Schärmuster *n* warp pattern
Scharnier *n* [piano] hinge
Scharnierband *n* hinge band, butt hinge
Scharnierbolzen *m* hinge bolt
Scharniersäge *f* joint saw
Scharnierschiff *n* pinjoint ship
Scharnierstift *m* hinge pin
Schärrahmen *m (text)* warping frame
Scharriereisen *n* bush chisel
scharrieren to boast, to bush, to char, to axe
Scharrierhammer *m* bush hammer
scharriert bush-hammered, axed, charred
Schärriet *n (text)* warping reed
Schärstock *m* bobbin warping creel
Schartrichter *m (ag)* coulter tube
Schärtrommel *f (text)* warp cylinder, warping drum
Schärzettel *m* card of warping particulars
Schatten *m* shade, shadow; *(astr)* umbra
Schattenbild *n* shadow image
Schattenbildverfahren *n* silhouetting method
Schattendeck *n* shade-deck
Schattendeckschiff *n* shade-deck vessel
Schattendruck *m (text)* shadow print
Schatteneffekt *m (nucl)* shadow (umbra) effect
Schattenfärbung *f* ombré dyeing
Schattengebiet *n* shadow area (region) *(radar)*
Schattengenerator *m (telev)* shading generator
Schattengrenze *f* shadow boundary, boundary of shadow
Schattenholz *n* tolerant tree
Schattenholzart *f* shady kind [of trees]
Schattenkompensationssignal *n (telev)* shading-compensation signal
Schattenlinien *fpl* shade lines
Schattenschild *m (nucl)* shadow shield
Schattenschutz *m (nucl)* shadow shield
Schattensektor *m* blind sector *(radar)*
Schattenstreuung *f (nucl)* shadow scattering
Schattenzeigerinstrument *n* shadow column instrument
schattieren to hachure *(map)*; to shade *(drawing)*
Schattierung *f* hachure *(map)*; shading *(drawing)*
schätzbar estimable
schätzen to estimate, to appreciate, to value, to rate
Patentwert s. auf to value a patent at
Schätzung *f* estimation, evaluation
Schätzwert *m* estimate[d value]
Schaubild *n* diagram, graphic illustration, graph, plot
Schauer *m* shower [of particles]
kosmischer S. cosmic-ray shower
Schauerleute *pl* deck gang (crew); holdsmen; stevedores
Schauermann *m* stevedore; docker; holdman
Schaufel *f* shovel; blade *(turbine)*, bucket, vane; paddle *(mixing machine)*; beater, scoop
S. mit Gabelzinken blade with pronks
aus Rohr gepreßte S. tubular blade
bis zur Austrittstelle gekrümmte S. blade curved throughout
flache S. plain back shovel
gekrümmte S. curved blade

kantige S. square point shovel
schwalbenschwanzförmig übergreifende S. dovetailed overlapping blade
seitlich nicht begrenzende S. open blade
zweiseitige S. double bucket
Schaufelanordnung *f* blading *(turbine)*
Schaufelarbeit *f (mine)* hand mucking (loading), shovelling
Schaufelbagger *m* dipper dredger
Schaufelbefestigung *f* **durch Nietung** fastening of the blade by riveting rings *(turbine)*
Schaufelblock *m* blade base
Schaufelbreite *f*/ **axiale** width (axial length) of blade
Schaufeldruck *m* blade pressure
Schaufelfläche *f* blade face
führende S. guiding face of blade
Schaufelform *f* shape of blade
Schaufelfuß *m* root of blade
schwalbenschwanzförmiger S. dovetailed root of blade
Schaufelgeschwindigkeit *f* blade velocity
Schaufelhöhe *f*/ **freitragende** free height of blade
radiale S. radial height of blade
Schaufelhöhlung *f* blade hollow
Schaufelkanal *m* passage between blades
verengter S. contracted passage between blades
Schaufelkerbe *f* blade groove
Schaufelkettenbagger *m* scoop-chain dredger, bucket-chain dredger
Schaufelkopf *m* blade head (tip)
Schaufelkranz *m* blade rim
Schaufelkrümmung *f* blade curvature
Schaufellader *m* tractor (loading) shovel, front-end loader, [rocker] shovel loader
Schaufellänge *f*/ **axiale** width (axial length) of blade
schaufeln to shovel
Schaufelpackung *f* measured shovel-packing
Schaufelprofil *n* blade section
Schaufelrad *n* bucket (blade, centrifugal) wheel *(turbine)*; paddle wheel
Schaufelradbagger *m* bucket wheel excavator
Schaufelradlader *m* bucket-wheel loader
Schaufelradschlepper *m* paddle tug
Schaufelreihe *f* row of blades
Schaufelrücken *m* back of blade
Schaufelsalat *m* blade failure (breakdown)
Schaufelsatz *m* set of blades
Schaufelspitze *f* blade head (tip)
Schaufelstellung *f*/ **schiefe** inclined blade position
Schaufelstiel *m* shovel handle
Schaufelstreifen *m* lashing strip *(turbine)*
Schaufeltasche *f* [blade] pocket
Schaufelteilung *f* blade pitch (spacing)
Schaufelträger *m* blade carrier
Schaufelverlust *m* blade loss
Schaufelwinkel *m* blade angle
Schaufelwurzel *f* root of blade
Schaufelzylinder *m* turbine casing
Schaufenster *n* show-window
Schaufensterdekorateur *m* window dresser
Schaufler *m (mine)* loading shovel
Schauglas *n* inspection (sight) glass, oil-flow indicator
S. für Druckölstand oil-sight gauge
S. für Ölstand oil-level gauge
Schaukasten *m* display box, show-case
Schaukel *f* swing
Schaukelbewegung *f* rocking motion
Schaukelförderer *m* swing tray conveyor
Schaukelofen *m* tilting (rocking-type) furnace
Schaukelrinne *f* rocking runner
Schaukelschwingung *f* rocking vibration *(spectroscopy)*
Schauloch *n* eye[sight], peep (sight) hole
Schaum *m* foam, froth, skimming *(melt)*, scum; lather; suds *(soap)*
Schaumabstreifer *m* foam (froth) skimmer
schaumartig foamy, frothy
Schaumaufbereitung *f* froth flotation
Schaumbeton *m* foamed (aerated) concrete, aero-concrete
Schaumbildner *m* foaming agent, foamer
Schaumbildung *f* frothing, formation of foam (froth, scum)
Schaumbildungsvermögen *n* capacity for forming foam
Schaumbrecher *m* froth breaker (killer)
Schaumdämpfer *m* antifoam
Schaumdämpfungsmittel *n* foam inhibitor
Schaumdämpfungsöl *n* antifroth oil
schäumen to foam, to froth; to lather *(soap)*; to effervesce *(bubble)*
Schäumen *n* foaming, frothing; effervescence; violent bubbling *(melt)*
S. der Emaille sparkling of the enamel
übermäßiges S. foam overblow
ungenügendes S. foam collapse
Schaumentwässerung *f* foam drainage
Schäumer *m* frother, frothing agent
Schaumerzeuger *m* foam generator
Schaumfähigkeit *f (plast)* expandibility
Schaumfang *m* foam trap
Schaumfeuerlöscher *m* foam extinguisher
Schaumflotation *f* froth flotation *(waste paper)*
Schaumgummi *m* sponge (froth, foam, cellular) rubber
schaumig foamy, frothy, scummy
Schaumkammer *f* frothing chamber
Schaumkleber *m* foam glue
Schaumkopf *m* scum riser *(casting)*
Schaumlamelle *f* foam film

Schaumlöschverfahren *n* fire fighting by foam systems
Schaummattentrocknung *f* foam-mat drying
Schaummittel *n* foam[ing] agent
Schaumschlacke *f* pumice (foamed) slag
Schaumschutzmittel *n* antifoaming agent
Schaumschwimmaufbereitung *f (met)* froth flotation *(ore)*
Schaumschwimmverfahren *n (met)* froth flotation *(ore)*
Schaumstoff *m* foamed (expanded, aerated) plastic; [foam] laminate, foam material
schwammförmiger S. sponge
Schaumstofflaminate *pl* fabric-foam laminates
Schaumstofflaminieren *n* fabric bonding with foam, foam lamination
Schaumstoffolie *f* sheet of foam
Schaumstoffrückseite *f* foamback *(textiles)*
Schaumstoffschicht *f* expanded plastic film
Schaumstrahl-Monitor *m (ship)* jet foam monitor *(fire extinguisher)*
Schaumtrichter *m* skimmer gate *(casting)*
Schaumverbesserer *m* lather booster
Schaumverhinderungsmittel *n* antifrothing agent
Schaumwein *m* sparkling wine, champagne
Schaumwert *m* lather value
Schaumzerstörungsdüse *f* froth-destroying nozzle
Schauöffnung *f* peep (sight) hole
Schauzeichen *n* indicator, annunciator
Scheckabrechnungsmaschine *f* bank proof machine
Scheckbearbeitung *f* / **automatische** automatic cheque handling
Scheckkontrollabschnitt *m* cheque counterfoil
Scheelit *m (min)* scheelite
Schegg *n (ship)* false stem
Scheibe *f* disk; *(build)* diaphragm, plate; pane *(window)*; *(aut)* windscreen; slice, cut; washer *(nut)*; *(mach)* pulley, sheave; wheel *(grinding, potter)*; *(el)* wafer *(switch)*; *(mil)* target; puck *(icehockey)*
S. mit Ausgleichlöchern disk with balancing holes *(turbine)*
unrunde S. cam
Scheibenabschwächer *m* disk attenuator *(waveguides)*
Scheibenanker *m* disk-type armature
Scheibenanode *f* disk-shaped anode
Scheibenantenne *f* disk aerial
Scheibenaufnahme *f* wheel mount *(grinding)*
Scheibenbau *m (mine)* slicing [method]
Scheibenbremse *f* disk brake
Scheibenbruchbau *m (mine)* cover-caving, scramming, slicing and caving [method], topslicing
Scheibendiode *f* disk seal diode
Scheibendrehzahl *f* [grinding] wheel speed
Scheibendrillschar *n* disk coulter
Scheibenegge *f* disk harrow
Scheibeneisen *n* disk pig
Scheibenelektrometer *n* disk-type electrometer
Scheibenexzentrizität *f* eccentricity of the disk *(turbine)*
Scheibenfeder *f (el)* disk (plate) spring
Scheibenfräser *m* cylindrical (side and face milling) cutter
kreuzverzahnter S. staggered-tooth side and face cutter, alternate-gash plain mill
Scheibenfunkenstrecke *f* disk discharger
Scheibenfutter *n* disk chuck
Scheibengelenk *n* disk joint
Scheibengenerator *m* disk-type electrostatic generator
Scheibenglas *n* sheet glass
Scheibengleichrichter *m* disk rectifier
Scheibenhantel *f* disk-type dumb-bell
Scheibenhobelmaschine *f* revolving disk planing machine
Scheibenhochhaus *n* slab block
Scheibenhülse *f (text)* double-flanged bobbin
Scheibenkolben *m* disk (solid) piston
Scheibenkollektor *m* disk (radial) commutator
Scheibenkondensator *m* disk-type capacitor
Scheibenkreiselmischer *m* disk impeller
Scheibenkupfer *n* rose copper
Scheibenkupplung *f* disk (single-plate) clutch
Scheibenkurve *f (mach)* disk (plate) cam
Scheibenmeißel *m* disk bit
Scheibenmühle *f* 1. disk crusher (mill); 2. disk attrition mill *(papermaking)*
Scheibenpflug *m* disk plough
Scheibenprofil *n* / **hyperboloidisches** hyperbolic disk profile *(turbine)*
Scheibenrad *n* disk *(turbine)*; disk wheel
Scheibenradhäcksler *m* flywheel (rotary grass) cutter
Scheibenrand *m* rim of disk *(turbine)*
hinterdrehter S. dovetailed disk rim
Scheibenrelais *n* cage relay
Scheibenröhre *f* disk-seal valve; disk tube
Scheibenrost *m* disk screen, rotary-disk grizzly
Scheibenrotor *m* built-up rotor
Scheibenschar *n (ag)* disk share
Scheibenschardrillmaschine *f* disk drill
Scheibenschießen *n* target practice (shooting)
Scheibenschlepper *m* target-towing aircraft
Scheibenschwerpunkt *m* centre of gravity of the disk *(turbine)*
Scheibensech *n (ag)* disk coulter
Scheibenspule *f* disk (frame, pancake, sectional) coil
Scheibenspulenwicklung *f* disk winding
Scheibenstange *f* disk bar *(weight lifting)*
Scheibentriode *f* planar (disk-seal) triode

Scheibentrocknung *f* disk drying *(wooden chips)*
Scheibenturbine *f* disk turbine
Scheibenunterbrecher *m* chopper disk
Scheibenventil *n* disk valve
Scheibenwalzstrecke *f (text)* disk plate drawing frame
Scheibenwicklung *f* disk (sandwich) winding, pile-wound coil
Scheibenwirkung *f* diaphragm action
Scheibenwischer *m* windscreen wiper
elektrischer S. electric windscreen wiper
Scheibenwischermotor *m* windshield wiper motor
Scheibenzähler *m* disk indicator
Scheideanstalt *f (met)* parting workshop (plant)
Scheidearbeit *f* parting work
Scheidebad *n* separating bath
Scheidebeil *n* dressing hammer
Scheideerz *n* spalling (bucking) ore
scheidefähig separable
Scheidegold *n* parting gold
Scheidehammer *m* spalling (cobbing) hammer
Scheidekapelle *f* cupel
Scheidekessel *m* parting kettle
Scheidekonus *m* separatory cone
Scheidemittel *n* separating (parting) agent
Scheidemünze *f* subsidiary unit, small coin
scheiden to refine; to separate; to sort; to defecate; *(met)* to part; to cob
Scheideofen *m* parting furnace
Scheidepfanne *f* defecator
Scheider *m* separator; *(ch)* liner
Scheidesilber *n* parting silver
Scheidetisch *m* picking table
Scheidetrichter *m* separatory (separating) funnel
Scheidewand *f* partition wall; separating wall *(crosshead)*; midfeather *(beater)*
Scheidewasser *n* aqua fortis, concentrated nitric acid; aqua regia *(gold)*
Scheidung *f* separation; parting *(gold)*; dividing; *(ch)* decomposition
Schein *m* shine *(light)*; brightness; appearance; note, bill *(money)*
Scheinaktivität *f* zero-time activity *(programming)*
Scheinangriff *m* feint attack
Scheinbefehl *m (dat)* dummy instruction
Scheinbeschleunigung *f* apparent acceleration
Scheinbewegung *f* apparent motion
Scheindämpfung *f* apparent attenuation
Scheinenergie *f* apparent energy
Scheinfenster *n* dead window
Scheinfrequenz *f* apparent frequency
Scheinfuge *f* dummy joint
Scheingewölbe *n (mine)* apparent arch
Scheinkomponente *f* apparent component
Scheinköper *m (text)* false twill
Scheinkraft *f* apparent (fictitious) force
Scheinleistung *f* apparent power (watts), volt-amperes
Scheinleistungsmesser *m* volt-ampere meter
Scheinleitung *f* apparent conduction
Scheinleitwert *m* admittance
S. einer offenen Leitung open-circuit admittance
komplexer S. complex admittance
Scheinnaht *f (text)* mock seam
Scheinpermeabilität *f* apparent permeability
Scheinstrom *m* apparent current
Scheinverbrauchszähler *m* apparent power meter, [kilo]voltampere-hour meter
Scheinversuch *m* mock-up test
Scheinwerfer *m* projector, searchlight; spot light; *(aut)* headlight, headlamp
eingelassener S. flush headlight
Scheinwerferbeleuchtung *f* spot lighting
Scheinwerferlampe *f* headlamp bulb
Scheinwerferstütze *f* headlamp socket
Scheinwiderstand *m* impedance
akustischer S. acoustic impedance
inverser S. inverse impedance
mechanischer S. mechanical impedance
negativer S. negative impedance
Scheinwiderstandsbrücke *f* impedance bridge
Scheinwiderstandsebene *f* impedance plane
Scheinwiderstandskettenleiter *m* impedance network
Scheinwiderstandsmeßbrücke *f* impedance bridge
Scheinwiderstandsmesser *m* impedance measuring unit
Scheinwirkungsgrad *m* apparent efficiency
Scheit *n* piece of wood, billet
Scheitel *m* apex; vertex *(angle)*; *(el)* crest, peak; mode; sagitta *(arc)*; *(build)* crown, crest
Scheitelfaktor *m* peak (crest) factor *(current)*
Scheitelkreis *m* **einer Ellipse** eccentric circle of an ellipse
S. eines Kegelschnittes auxiliary circle of a conic
Scheitellinie *f (el)* crest line
Scheitelplatte *f* vertex plate *(waveguides)*
Scheitelpunkt *m* vertex, apse, summit *(trajectory)*; crown *(arch)*
Scheitelspannung *f* peak (crest) voltage
Scheitelspannungsmesser *m* peak (crest) voltmeter
Scheitelstrom *m* peak current
Scheitelwert *m (el)* peak value *(current)*, crest value, maximum amplitude; mode
S. der positiven Sperrspannung peak blocked voltage
S. der Spannung crest voltage
S. der Stromwelle peak current
S. des Anodenstroms peak plate current
S. des Einschaltstroms peak making current
S. des Katodenstroms peak cathode current
Scheitelwertmesser *m* peak programme meter

Scheitelwinkel *m* vertical angle
Scheitholz *n* split fire-wood
Schellack *m* shellac; lac
Schellackbindung *f* shellac bond
Schellackfirnis *m* shellac varnish
schellackgebunden shellac-bonded
Schellacklösung *f* shellac solution
Schellackwachs *n* shellac wax
Schelle *f* 1. bell, gong; 2. clip, clamp, fitting, bracket *(pipe)*; holderbat *(discharge pipe)*; 3. *(el)* sliding tap
Schelleisen *n* snap
Schellhammer *m* rivet set, punch-type dolly
Schellharz *n* barras
Schema *n* block diagram, array; layout; pattern
Hornersches S. Horner's method
optisches S. optical layout
Schemadiagramm *n* block diagram
Schemaschaltbild *n* schematic diagram
schematisch schematic[al], diagrammatic[al]
Schemazeichnung *f* schematic drawing
Schemelwagen *m* timber truck, bogie car
Schenkel *m* arm; side; leg *(L-steel, angle)*; lateral side *(isosceles triangle)*; prong *(tuning fork)*
Schenkelpolgenerator *m* homopolar generator
Schenkelpolläufer *m* pole wheel
Scherbeanspruchung *f* shear[ing] stress
Scherbelastung *f* shear[ing] load
Scherben *m* body of the ware *(ceramics)*
Scherblatt *n* *(text)* wraith
Scherblatthalter *m* wraith holder
Scherbrett *n* *(ship)* otter board; trawl board
Scherbrettnetz *n* otter-trawl
Scherbruch *m* shear fracture
Scherdegen *m* *(leath)* German spring flesher
Scherdübel *m* shear connector
Schere *f* 1. [pair of] scissors; *(ag)* pruner *(tree)*; 2. shear *(scissors of large size)*; 3. *(mach)* change-gear bracket; swing plate (frame) *(lathe)*
Scherebene *f* shear plane, plane of shear
scheren to shear *(sheet metal, furs, textile)*; to clip; to sheer *(ship)*; *(leath)* to shave
Scherenbolzen *m* movable (intermediate) stud
scherend *(mach)* shearing *(cut)*
Scherenfernrohr *n* scissors telescope, stereo-telescope
Scherenkran *m* *(ship)* shears, shear (sheer) legs, shear-leg derrick, stiff-leg derrick
Scherenkranhulk *m* shear (sheer) hulk
Scherenmanipulator *m* *(nucl)* mechanical pantograph manipulator
Scherenpfanne *f* *(met)* bull ladle
Scherenschwingung *f* scissor vibration *(spectroscopy)*
Scherenstromabnehmer *m* pantograph
Scherer *m* *(text)* warper
Scherfestigkeit *f* shear[ing] strength
Scherfleisch *n* *(leath)* fleshings
Scherflocken *fpl* *(text)* shearings
Schergang *m* *(ship)* sheer strake
Scherhaspel *f* horizontal warping reel
Scheringbrücke *f* Schering bridge *(dielectrics)*
Scherkraft *f* shear[ing] force
Scherling *m* *(leath)* shearling
Schermaschine *f* *(met)* shearing machine; *(text)* shearing frame
elektrische S. *(ag)* electric clippers
Schermeißel *m* shear tool
Schermesser *n* *(text)* shearing blade
Schermodul *m* shear modulus, modulus of shearing
Schermoment *n* shearing moment
Scherschwingung *f* shear vibration
Scherspannung *f* shear[ing] stress
Scherstift *m* shear pin
Scherstock *m* *(ship)* hatchway beam
Scherung *f* shearing [action]
Scherungsviskosität *f* shear viscosity
Scherversuch *m* shearing test
Scherwinkel *m* shear angle
Scherwirkung *f* shearing action (effect)
scheuerfest *(text)* wear-resistant
Scheuerfestigkeit *f* wear (abrasion) resistance; *(print)* rub resistance
Scheuerkiel *m* *(ship)* ground[ing] keel
Scheuerleiste *f* skirting board, baseboard, mop-board, washboard; *(ship)* permanent fender, fender guard
scheuern to gall *(metals)*; to scrub, to scour
Scheuerpfahl *m* *(ship)* fender pile, dolphin
Scheuerplatte *f* *(ship)* bolster plate
Scheuerprahm *m* *(ship)* [side] punt
Scheuerpulver *n* scouring powder
Scheuerstelle *f* *(text)* chafe mark *(error)*
Scheuertrommel *f* scouring barrel *(wire)*
Scheune *f* barn
Schi *m* ski
Schibacken *mpl* toe irons
Schicht *f* 1. layer, ply, couch; 2. coat, film, deposit, lamina; coating *(paint)*; 3. *(build)* layer, lift *(concrete)*; couch *(bricks)*; 4. tier *(cable)*; 5. shift, turn *(working time)*
angrenzende S. adjacent layer
anodische S. anode layer
atmosphärische S. atmospheric layer
atomare S. atomic layer
aufgedampfte S. deposited film
auskragende S. oversailing course *(masonry)*
dielektrische S. dielectric layer
diffundierte S. diffused layer
dotierte S. doped junction *(semiconductors)*
dünne magnetische S. thin magnetic film *(for data storage)*
dünne S. [thin] film
fotoempfindliche S. photosensitive layer

fotografische S. photographic deposit (emulsion)
gedopte S. doped junction *(semiconductors)*
ionisierte S. ionized layer
katodische S. cathode layer
lichtempfindliche S. photosensitive layer
magnetische S. magnetic film
monomolekulare S. monolayer, monomolecular layer
reflexmindernde S. anti-reflection coating
sensibilisierte S. sensitized layer
tragfähige S. hardplan *(foundations)*
wasserdurchrieselte S. *(therm)* layer traversed by stream of water
Schichtarbeit *f* shift work
Schichtaufnahmeverfahren *n* tomography, laminography, planigraphy
Schichtband *n* coated tape
Schichtbildaufnahme *f* tomography, laminography, planigraphy
Schichtbildung *f* lamination
Schichtdicke *f* thickness of layer; stratum thickness
äquivalente S. equivalent thickness
Schichtemulsionstechnik *f* sandwich technique
schichten to laminate; to arrange in layers; to stack *(wood)*, to pile up
Schichtenaufwand *m (mine)* number of shifts worked
Schichtenbuch *n* accountancy book
Schichtendetektor *m* sandwich detector
Schichtenfolge *f (geol)* strata sequence (series)
Schichtengitter *n* layer (stratified) lattice
Schichtengruppe *f (geol)* formation
Schichtenmauerwerk *n* ashlar (coursed rubble) walling
hammerrechtes S. squared rubble in courses
regelmäßiges S. regular coursed rubble (ashlar)
Schichtenpaket *n (geol)* series of formations
Schichtenpapier *n* laminated paper
Schichtenquelle *f (nucl)* sandwich[ed] source
Schichtenschi *m* laminated ski
Schichtenschnitt *m (geol)* strata section
Schichtenstörung *f (geol)* disturbance
Schichtenströmung *f* laminated (laminar) flow
Schichtfläche *f (geol)* bedding plane, cleat
Schichtfolge *f (geol)* succesion of strata
Schichtfolie *f* laminating sheet
Schichtgefüge *n* laminated structure
Schichtglas *n* laminated (compound) glass
Schichthöhe *f* thickness of fire, depth of fuel layer
Schichtholz *n* laminated wood; stacked wood
verleimtes S. glued laminated wood
Schichtkavitation *f* sheet cavitation
S. an eintretender Kante leading edge sheet *(cavity)*
Schichtkomplex *m (geol)* group of strata
Schichtkonstruktion *f* / **verleimte** glued laminated construction
Schichtmaß *n* piled (stacked) measure
schichtmäßig laminar *(flow)*
Schichtnutzholz *n* corded (stacked) timber
Schichtpreßstoff *m* laminated plastic, multilayer moulded material
Schichtseite *f* 1. coated side; 2. active surface
S. des Films emulsion side of the film
Schichtspaltung *f* delamination
Schichtsteiger *m (mime) (Am)* shift boss
Schichtstoff *m (plast)* [synthetic resin bonded] laminate, laminated (reinforced) plastic; *(text)* laminated fabric
Schichtstreichen *n (geol)* bearing, direction of strata
Schichtstück *n (mach)* chip segment
Schichtträger *m* emulsion base (carrier), film base; substrate
Schichttransistor *m* junction transistor
Schichtung *f* lamination; *(geol)* stratification, bedding
S. in waagerechter Richtung horizontal lamination
vorgepreßte S. prepress lamination
Schichtungsdiskordanz *f (geol)* discordance
Schichtverband *m* interlaminar bonding
Schichtwechsel *m* change of shift
schichtweise laminated; by layers; in shifts *(work)*
Schichtwiderstand *m (el)* 1. composition (carbofilm, film, deposited-film) resistor; 2. film resistance
Schiebebalken *m (ship)* hatch (shifting) beam
Schiebebuchse *f* slip bushing
Schiebebühne *f* sliding platform; *(ry)* travelling platform, traverser, transfer table
Schiebedach *n (aut)* sliding roof; *(ship)* movable deck cover
Schiebedachverschluß *m* sliding-roof fastener
Schiebefahrt *f* propelling movement *(electric traction)*
Schiebefenster *n* sliding hatch; sliding sash; *(aut)* sliding window
Schiebefestappretur *f* non-slip finishing
Schiebegerüststapler *m* reach truck
Schiebegitterstreuung *f (ag)* reciprocating plate feed
Schiebehülse *f* sliding collar
Schiebekappe *f (mine)* slide bar
Schiebekasten *m (ship)* sliding dock gate
Schiebekeil *m* feather key, spline
Schiebekontakt *m* sliding contact
Schiebekupplung *f* sliding coupling
Schiebelehre *f* caliper rule, vernier caliper
Schiebelokomotive *f* bank[ing] engine, booster (pusher) locomotive

Schiebeluke *f* sliding hatch

schieben to move, to push, to slide; to shove; *(text)* to slip *(fabric)*; *(el)* to shift [phases] *(power-factor correction)*

Schieber *m* slide valve *(hydraulics, steam)*; sleeve; blade *(Enor pump)*; *(aut)* sliding vane; ram, head (vertical) slide *(planer)*; movable head, sliding member *(vernier caliper)*; bolt, bar; *(el)* sliding contact

S. des Hobelmaschinensupports planer (tool) head slide, tool-holder slide [of the planer]

S. mit kreisförmiger Schieberplatte circular slide valve

entlasteter S. balanced gate (parallel slide stop) valve, gate valve with by-pass

freifallender S. gravity-operated shutter

Schieberad *n (aut)* sliding mesh gear; *(mach)* sliding (shifting) gear

Schieberäderblock *m* cluster (sliding) gears

Schieberädergetriebe *n* sliding-mesh gear train

Schieberauslenkung *f* distance of the valve from the centre position

Schieberbewegungsrichtung *f* direction of the travel of the slide valve

Schieberdampfmaschine *f* steam engine with slide valve

Schieberdiagramm *n* slide valve diagram

ovales S. oval slide valve diagram

Reuleaux-Müllersches S. Reuleaux-Müller valve diagram

Schieberegister *n* shift register *(for data)*

Schieberegler *m* sliding control

Schieberellipse *f* slide valve ellipse

Schieberführung *f* damper guide

schiebergesteuert sleeve-valve [controlled] *(engine)*

Schiebergrube *f* damper pit

Schieberhub *m* stroke (travel) of the slide valve

Schieberkante *f* / **steuernde** distributing edge of the slide valve

Schieberkasten *m* slide valve chest (box), steam chest

Schieberkastendeckel *m* steam chest cover

Schieberkette *f* damper chain

Schieberkreis *m* slide-valve circle

Schieberluftpumpe *f* slide-valve pump

Schiebermittellage *f* mid-position of the valve

Schiebermotor *m* sleeve-valve engine

Schiebernadel *f (text)* slide needle

Schiebernadelblei *n* slide-needle lead

Schieberöffnung *f* sleeve port

Schieberost *m* sliding grate

Schieberoval *n* oval slide valve diagram

Schieberregulierung *f* slide damper regulation

Schieberschlitz *m* sleeve port

Schieberschubrichtung *f* direction of the travel of the slide valve

Schieberstange *f* valve rod

Schieberstangendichtung *f* valve rod packing

Schieberstangenliderung *f* valve rod packing

Schieberstangenpackung *f* valve rod packing

Schiebertotlage *f* position of the valve at the dead point

Schieberventil *n* sleeve valve; slide [gate] valve

Schiebeschalter *m* slide switch

Schiebespule *f* sliding inductance

Schiebestange *f (cont)* stem *(of a key of a keyboard)*

Schiebestück *n (aut)* forked sleeve

Schiebetisch *m (mach)* receding table

Schiebetor *n* sliding gate

Schiebetür *f* sliding door

Schiebevorrichtung *f* / **n-stellige** n-place shifter *(for data)*

Schiebewand *f* sliding panel

Schiebewelle *f* sliding shaft

Schiebewicklung *f* shift winding

Schiebewiderstand *m* slide (sliding) resistor, rheostat

Schiebezahnrad *n* sliding gear

Schieblehre *f* sliding vernier caliper, slide caliper (gauge)

Schiebung *f* translation, displacement; shear deformation

Schiebungsbruch *m* shear failure

Schiedsanalyse *f* arbitration analysis

Schiedsgericht *n* court of arbitration

schief skew, oblique, inclined, sloping

schiefachsig oblique-axial

Schiefe *f* obliquity

Schiefeinfall *m* oblique incidence

Schiefer *m* slate, shale, schist

schieferartig slaty, shaly, schistose

Schieferbergbau *m* slate quarrying

Schieferbruch *m* 1. slate quarry; 2. fibrous fracture *(material)*

Schieferdecker *m* slater

Schieferdeckung *f* slating *(roofing)*

Schiefergips *m* foliated gypsum

schieferhold *(min)* cleavable

Schieferkohle *f* banded coal

Schieferöl *n* shale oil

Schieferplatte *f* slate

Schieferton *m* slate coal, slate clay

Schieferung *f* [slaty] cleavage, foliation cleavage

Schieflage *f* subsidence slope, tilt

Schieflast *f* load unbalance, unbalanced (asymmetrical) load

Schiefscheibe *f* non-rotating crank *(pump)*

schiefsymmetrisch skew-symmetric

schiefwinkelig skew, oblique-angled

schielen to squint

Schiene *f* rail; beam *(vernier caliper)*; bar; shin guard (pad) *(hockey)*

S. mit breitem Fuß rail with wide base

Schienenbefestigung *f* rail fastenings

Schienenbohrmaschine *f* rail drill[ing machine]
Schienenbremse *f* / **elektromagnetische** electromagnetic shoe brake
Schienenbruch *m* rail break, breaking (rupture) of rail
Schienenfahrzeug *n* rail vehicle
Schienenfederklammer *f* spike anchor
Schienenführung *f (mach)* bar guide
Schienenfuß *m* rail base, base (foot) of the rail
schienengängig track-mounted
schienengebunden rail-borne, rail-mounted
Schienenkontakt *m* track contact; electrical depression bar
Schienenkopf *m* rail-head, head of rail
Schienenkran *m* rail-borne crane, rail-mounted crane
Schienenlasche *f* fishplate
Schienenlauffläche *f* rail surface, tread (running-surface) of the rail
schienenlos trackless *(vehicle)*
Schienennagel *m* rail (dog, crew) spike
Schienenoberkante *f* top of rail
Schienenräumer *m (ry)* rail guard
Schienenrückleitung *f* track return system
Schienenschalter *m* track switch
Schienenschläger *m (text)* beater scutcher
Schienenspurweite *f* track gauge
Schienenstahl *m* rail steel
Schienensteg *m* web of the rail
Schienenstoß *m* rail joint
  isolierter S. block joint
Schienenstrang *m* rail track, stretch of rails
Schienenstrom *m* rail current
Schienenstromabnehmer *m* track sliding contact
Schienenstromabnehmerkontaktschuh *m* shoe-gear
Schienenstrombremse *f* magnetic track braking
Schienenstromwandler *m* bus-bar transformer, bar-type current transformer, single-turn current transformer, open-primary current transformer
Schienenübergang *m* [railway] grade crossing, [railway] level crossing
Schienenüberhöhung *f* superelevation of outer rail
Schienenverbinder *m* [track] rail bond, bond
Schienenwagen *m* rail car (wagon)
Schienenwalzwerk *n* rail-rolling mill
Schienenwandler *m s.* Schienenstromwandler
Schienenweg *m* rail track
Schießarbeit *f (mine)* shot-firing, blasting
Schießausbildung *f* training in firing
Schießbaumwolle *f* gun cotton
Schießberechtigter *m (mine)* blaster, powder man, shot-firer
schießen 1. to shoot; 2. *(mine)* to blast; 3. *(ag)* to bolt, to run to seed
Schießen *n* 1. shooting; 2. *(mine)* shot-firing, blasting
  S. aus dem Vollen off-the-solid shot-firing
  S. mit Preßluft airbreaking, Armstrong blasting
  S. mit untergelegter Ladung snakeholing, undermining
Schießhauer *m* blaster, powder man, shot-firer
Schießleitung *f (mine)* leading wire, shot-firing cable
Schießmeister *m* shot-firer
Schießpulver *n* gunpowder
Schießscharte *f* loop-hole
Schießschwaden *mpl (mine)* fumes
Schießstand *m* rifle (target) range
Schießsteiger *m* shot-firing deputy
Schießübung *f* target practice
Schießvorschrift *f* gunnery (ballistics) manual
Schiff *n* 1. ship, vessel, boat; craft; 2. *(build)* nave, aisle; 3. *(text)* shuttle
  S. außer Betrieb inactive vessel
  S. für flüssige Ladung liquids carrier
  S. für Heilbuttfang halibuter
  S. für Schellfischfang haddocker
  S. in der Inlandfahrt home trade ship
  S. in der Linienfahrt liner
  S. in Eisnot icebound vessel
  S. in Fahrt running ship
  S. mit Atomantrieb atomic-powered ship
  S. mit dieselelektrischem Antrieb Diesel-electric ship
  S. mit Düsenantrieb jet-propelled vessel
  S. mit Eigenantrieb self-propelled vessel
  S. mit Freibord reduced draught vessel
  S. mit geraden Seitenwänden wall-sided vessel
  S. mit gewellter Außenhaut corrugated vessel
  S. mit glatten, gut strakenden Linien clean-lined ship
  S. mit großem Aktionsradius long-legged vessel
  S. mit großem Tiefgang long-legged vessel
  S. mit hohlen Wasserlinien hollow-ended vessel
  S. mit Kernenergieantrieb nuclear[-powered] ship
  S. mit nichtdurchlaufenden Aufbauten partial superstructure vessel
  S. mit Plattensteven soft-ended vessel
  S. mit Propellern vorn und achtern double-ended vessel *(e.g. ferry)*
  S. mit Schlagseite lopsided (lapsided) vessel, lopsider, lapsider
  S. mit sehr hohem Sprung vorn und achtern moon-sheered ship
  S. mit teilweisem Sturmdeck partial awning deck vessel

S. mit teilweisen Aufbauten partial superstructure vessel
S. mit U-Spantform U-shaped vessel
S. mit V-Spantform V-shaped vessel
S. mit völligen Enden full-ended ship
S. ohne Auftrag idle ship
S. ohne Besatzung manless ship
S. ohne Deck undecked ship
S. ohne Doppelboden single-bottom vessel
S. ohne Eigenantrieb non-propelled vessel
S. ohne Ladung idle ship
S. vor Anker liegend ship at anchor
abgetakeltes S. dismantled ship, hulk
achterlastiges S. ship down by the stern
allein fahrendes S. single-cruising vessel
auf Reede liegendes S. roadster
aufgelegtes S. laid-up vessel
außer Dienst gestelltes S. ship out of commission, laid-up ship
beladenes S. loaded (laden) ship
bewachsenes S. foul ship
eingeeistes S. icebound vessel
eingefrorenes S. icebound vessel
einmastiges S. single-masted vessel, single sticker
fabrikmäßig gebautes S. fabricated ship
fabrikmäßig hergestelltes S. fabricated ship
ferngelenktes S. remote (distance) controlled vessel
flachbodiges S. flat-bottomed vessel
flachgehendes S. shallow draught vessel; *(Am)* shoal-draft vessel
ganz elektrisch geschweißtes S. all-electric-welded ship
geschweißtes S. welded ship
gleichlastiges S. ship on an even keel
im Bau befindliches S. ship under construction
im Seegang ruhig liegendes S. easy [working] vessel
in Sektionen gebautes S. fabricated ship
kursstabiles S. stable ship
kursunstabiles S. unstable ship
neutrales S. neutral bottom (ship)
offenes S. undecked (open) ship, all-hatch ship
rankes S. tender ship
schmutziges S. dirty ship *(tanker for dirty oil)*
schwerbedienbares S. heavy-working ship
schwerbeladenes S. heavily laden ship
seeklares und eingesetztes S. active vessel
seetaugliches S. able vessel
seetüchtiges S. able vessel
sehr breites S. beamy ship
selbstentladendes S. self-unloader
selbstlöschendes S. self-unloading vessel, self-unloader
tiefgehendes S. deep-drawing vessel
überaltertes S. out-of-date ship
unbeladenes S. light ship
unbemanntes S. unattended ship
ungedecktes S. undecked (open) ship
unmagnetisches S. non-magnetic vessel
verlassenes S. derelict
völliges S. full-bodied ship
vorgefertigtes S. prefabricated ship
wendiges S. easy steerer
Schiffahrt *f* navigation, shipping
Schiffahrtbefeuerung *f* navigation lights, navigational lighting
Schiffahrtsbehörde *f* shipping authority
Schiffahrtsgesellschaft *f* shipping (navigation) company
Schiffahrtshindernis *n* navigational dangers, obstruction to navigation
Schiffahrtskunde *f* navigation, nautics
Schiffahrtslinie *f* shipping line
Schiffahrtsrinne *f* [navigational] channel, fairway
Schiffahrtsstraße *f* shipway, ocean (water) highway, navigable waterway, track
Schiffahrtsweg *m s.* Schiffahrtsstraße
schiffbar / nicht unnavigable
Schiffbau *m* ship building, naval architecture
S. in Serienproduktion multiple ship production
Schiffbauer *m* shipbuilder; shipfitter
Schiffbauholz *n* ship timber
Schiffbauingenieur *m* naval (marine) architect
Schiffbaukonstruktionsbüro *n* ship designing department
Schiffbaukunst *f* naval architecture
Schiffbauplatz *m* stocks for shipbuilding, shipbuilding yard
Schiffbaustahl *m* shipbuilding (ship, mild) steel
Schiffbauversuchsanstalt *f* model[-testing] tank establishment, shipbuilding research establishment
Schiffbauwesen *n* shipbuilding
Schiffbruch *m* shipwreck
schiffbrüchig shipwrecked, castaway
Schiffbrüchiger *m* castaway
Schiffbrücke *f* floating (pontoon) bridge
Schiffchen *n* shuttle *(sewing machine)*
Schiffchenstickmaschine *f* shuttle embroidery machine
Schiffe *npl* **/ angekommene** arrivals
schiffen to navigate
Schiffer *m* [ship-]master; skipper
Schifferausdruck *m* nautical term
Schifferknoten *m* sailor's (running) knot
Schifferpatent *n* master's certificate
Schiffsabort *m* head
Schiffsabwracker *m* ship-breaker
Schiffsanhänge *mpl* appendages *(shaft bossings)*

Schiffsanstrichfarbe *f* marine paint
Schiffsantrieb *m* [ship's] propulsion, marine propulsion
magnet-hydrodynamischer S. magnetohydrodynamic ship's propulsion
Schiffsantriebsanlage *f* marine propulsion machinery
Schiffsantriebsmaschine *f* propelling engine
Schiffsantriebsmotor *m* propelling motor *(electric motor)*
Schiffsantriebsreaktor *m* ship-propulsion reactor
Schiffsartillerie *f* naval ordnance
Schiffsartilleriebeobachtungsflugzeug *n* gunnery fleet spotter
Schiffsaufbau *m* superstructure
Schiffsaufzug *m* marine elevator
Schiffsaugen *npl* eyes of a ship
Schiffsausdruck *m* nautical term
Schiffsausmessung *f* gauging of ships
Schiffsausrüster *m* ship-supplier
Schiffsausrüstung *f* ship's equipment (outfit)
Schiffsbedarf *m* ship stores
Schiffsbefrachtung *f* affreightment
Schiffsbergung *f* salvage service
Schiffsbeschläge *mpl* ship fitments
Schiffsbesichtiger *m* surveyor
Schiffsbewegung *f* shipmotion, motion of a ship
Schiffsbewegungen *fpl* / **sanfte** kindle motions of a vessel
Schiffsboden *m* ship's bottom
bewachsener S. foul bottom
Schiffsbodenanstriche *mpl* anti-foulings
Schiffsbodenfarbe *f* antifouling paint, bottom composition
Schiffsbodenverlauf *m* run
Schiffsboot *n* ship's boat
Schiffsbreite *f* beam, breadth of beam (the ship)
Schiffsbug *m* bow, nose
Schiffsdampfmaschine *f* marine steam engine
Schiffsdienst *m* service afloat
Schiffsdieselmotor *m* marine Diesel engine
Schiffsdisponent *m* ship's husband
Schiffsdock *n* dock
Schiffsdurchfahrt *f* fairway span *(bridge)*
Schiffseichung *f* gauging of ships
Schiffseigner *m* [ship]owner
Schiffsende *n* / **stumpfes** blunt end
trogförmiges S. flume end
Schiffserprobungen *fpl* ship trials
Schiffsfarbanstrich *m* marine coating
Schiffsfarbe *f* marine paint
Schiffsform *f* ship's shape (form), hull form
S. unter Wasser run
Schiffsformgütegrad *m* hull efficiency
Schiffsfracht *f* [ship's] freight
Schiffsfrachtbrief *m* bill of lading
Schiffsführer *m* [ship]master; skipper
Schiffsführung *f* navigation, ship's management; conning
Schiffsgalgen *m* davit
Schiffsgerät *n* apparel
Schiffsgerippe *n* sceleton of a vessel; carcass *(of a stranded vessel)*
Schiffsgetriebe *n* naval propulsion gear
Schiffsgewicht *n* weight of ship
Schiffshebewerk *n* lift lock, ship [canal] lift
Schiffshilfsmaschine *f* hull auxiliaries
Schiffshygiene *f* vessel sanitation
Schiffsingenieur *m* marine engineer
Schiffsisolierung *f* marine insulation
Schiffsjournal *n* official logbook
Schiffskammer *f* berth
Schiffskessel *m* marine boiler
Schiffskesselüberhitzer *m* superheater for marine boiler
Schiffsklasse *f* class of a ship; ship's rating
Schiffskörper *m* hull, ship's hull (body)
Schiffskörperform *f* hull form
abwickelbare S. developable hull form
hydrokonische S. hydroconic hull form
Schiffskörpermodell *n* hull model *(model testing)*
Schiffskreisel *m* [gyro] stabilizer, gyrostatic stabilizer
Schiffsküche *f* galley, caboose
Schiffskurbelwelle *f* marine crank[shaft]
Schiffsladung *f* cargo, load, shipment
Schiffsladungsverpfändung *f* respondentia
Schiffslast *f* freight, burden; shipload
Schiffslazarett *n* ship's hospital; sick bay
Schiffsleitung *f* management of the ship
Schiffsliegeplatz *m* dock; berth *(ship's place at wharf, at anchor)*
Schiffslift *m* marine elevator
Schiffslinien *fpl* lines of a ship
Schiffsmakler *m* ship broker; loading broker
S. im Außendienst water clerk
Schiffsmaklerei *f* ship's agent
Schiffsmannschaft *f* ship's crew
Schiffsmaschine *f* marine engine
Schiffsmaschinenbau *m* marine engineering
Schiffsmaschinist *m* marine engineer
Schiffsmeßbrief *m* tonnage certificate (bill)
Schiffsmittellinie *f* centre line [of a ship]
Schiffsmittenlinie *f* keel line [of a ship]
Schiffsmodell *n* ship['s] model, model [of a ship]
aus mehreren Segmenten bestehendes S. multi-segmented ship's model
drahtlos ferngesteuertes S. radio-controlled ship's model
Schiffsmodell-Versuchstank *m* ship's model testing tank
Schiffsmotor *m* marine engine; marine motor *(electric motor)*
Schiffsnase *f* nose, bow

Schiffsort *m* ship's position
astronomischer S. position by astro-calculation
Schiffsortung *f* location of ships
Schiffsposition *f* ship's position
Schiffspropeller *m* [screw] propeller, screw; *(Am)* wheel
Schiffsprototyp *m* type ship
Schiffsradar *n* naval radar
Schiffsraum *m* [cargo] hold, shiproom, tonnage
Schiffsreaktor *m* ship-propulsion reactor
Schiffsreeder *m* shipowner
Schiffsregister *n* register book (of shipping), ship's register
Schiffsreise *f* voyage, trip
Schiffsreparatur *f* ship repair
Schiffsrumpf *m* [ship's] hull, body of a ship
Schiffsschornstein *m* ship's funnel
Schiffsschraube *f* [ship's] propeller, screw [propeller]
linksgängige S. left-handed propeller
Schiffsschwingung *f* ship oscillation
Schiffssicherheit *f* safety of a ship
S. durch wasserdichte Unterteilung watertight integrity
Schiffsstab *m* ship's staff *(personnel)*
Schiffsstandort *m* position of the ship
Schiffstagebuch *n* ship's log (journal), official log book
Schiffstau *n* marine rope
Schiffstaufe *f* naming (christening) of a ship
Schiffstelegraf *m* ship's telegraph
Schiffstonne *f* ton of shipping
Schiffstrümmer *pl* wreck[age]
Schiffsturbine *f* marine turbine
Schiffstyp *m* type of vessel
Schiffsuhr *f* ship's clock
Schiffsumlauf *m* turn-around *(in a port)*
Schiffsumschlagseinrichtung *f* marine cargo handling equipment
Schiffsunfall *m* marine disaster, [marine] casualty
Schiffsunkosten *pl* disbursements
Schiffsuntersetzungsgetriebe *n* marine reduction gear
Schiffsverlust *m* casualty
Schiffsvermessung *f* measurement of ships; tonnage measurement
Schiffsvermessungsamt *n* ship's measurement office
Schiffsvermessungsordnung *f* tonnage law
Schiffsvorderteil *n* prow *(bow)*
Schiffsvorräte *mpl* sea (ship) stores
Schiffswache *f* watch
Schiffswechselsprechanlage *f* marine correspondence system
Schiffswelle *f* marine shaft
Schiffswerft *f* shipyard, shipbuilding yard; shipwright's yard
Schiffswiderstand *m* ship resistance, resistance of ship
Schiffswrack *n* wreck *(of a vessel)*
Schiffswurm *m* shipworm, cappanus
Schiffszeit *f* ship's time
Schiffszertifikat *n* certificate of registry (identity)
Schiffszusammenstoß *m* collision [of ships]
Schifter *m* jack rafter
Schiftersparren *m* jack rafter
Schikjöring *n* ski joring
Schilauf[en *n*] *m* skiing
Schild *n* sign, [sign-]board
Schild *m* [protective] shield, protective wall; barrier; *(build)* blade
biologischer S. biological shield
thermischer S. thermal shield
Schildkrötendeck *n (ship)* turtle-deck, turtleback deck
Schildpatt *n* tortoise shell
Schildvortrieb *m (build)* shield tunnelling
Schilf *n* reed, rush
Schilfdach *n* thatched roof
Schilfmatte *f* cane mat
Schilfpapier *n* reed paper
Schilfrohr *n* reed
Schilfschneidemaschine *f* reed mower
Schimmel *m* mould, mildew, white film
schimmelbeständig mildew-resistant
schimmeln to mould
Schimmelpilz *m* mould [fungus]
schimmern to shimmer, to gleam, to glitter
Schindel *f* shingle
schinden / den Motor to labour the engine
Schinden *n* **des Motors** labouring of the engine
Schiplatte *f* foot plate *(ski)*
Schippe *f* [hand] shovel, digging shovel
Schirm *m* [protective] screen, shield
elektromagnetischer S. electromagnetic screen
elektrostatischer S. electrostatic screen
geerdeter S. earthed screen
kapazitiver S. electrostatic screen
magnetischer S. magnetic screen
metallhinterlegter S. metallized (metal-backed) screen
metallisierter S. metallized screen
Schirmanker *m (ship)* mushroom anchor
Schirmantenne *f* umbrella (cone) aerial
Schirmbild *n* screen image; image; pattern
Schirmbildaufnahme *f* fluorography, photofluorograph
Schirmbildfilm *m* fluorographic film
Schirmbildfotografie *f* screen photography
Schirmbildgerät *n* photofluorograph[ic unit], photo-roentgen unit, PR unit
Schirmbildübertragung *f* representation transmission *(radar)*

Schirmdurchmesser *m* petticoat (bell) diameter *(chain insulator)*
Schirmeinbrennen *n* screen burning
Schirmfaktor *m* shielding factor
S. eines Gitters screen-factor [of a grid], degree of screening
Schirmgenerator *m* umbrella-type generator, parasol-type generator
Schirmgewebespeicher *m* woven-screen memory *(for data)*
Schirmgitter *n* screen grid
Schirmgitterbelastung *f* screen-grid dissipation
Schirmgitterblock *m* screen by-pass capacitor
Schirmgitterdurchgriff *m* reciprocal of screen-grid amplification
Schirmgittermodulation *f* screen modulation
Schirmgitterpotential *n* screen-grid bias
Schirmgitterröhre *f* screened (screen-grid, screened-plate, shielded-grid) valve, screened pentode
Schirmgitterspannung *f* screen[-grid] voltage
feste S. constant screen-grid voltage
gleitende S. sliding screen-grid voltage
Schirmgitterspannungsquelle *f* screen-grid voltage supply
Schirmgitterstrom *m* screen-grid current
Schirmgitterverlustleistung *f* screen-grid dissipation
Schirmgittervorwiderstand *m* screen dropping resistor
Schirmgitterwiderstand *m* screen-grid dropping resistor
Schirmglucke *f (ag)* hover, brooder
Schirmisolator *m* umbrella-type insulator
Schirmkabel *n* screened cable
Schirmkrümmung *f* screen curvature
Schirmkultur *f (ag)* nurse crop
Schirmring *m (el)* guard ring
S. in Quecksilberdampfröhren splash ring
Schirmschott *n (ship)* screen bulkhead
Schirmstoff *m* umbrella cloth
Schirmträger *m* face-plate *(cathode-ray tube)*
Schirmverspiegelung *f (telev)* metal backing
Schirrkammer *f* harness room
Schirting *m* white calico
Schispannzeug *n* ski press
Schispringen *n* ski-jumping
Schitiefenzug *m* tension hook for down-run *(skiing)*
Schivorderstrammer *m* front tightener of ski
Schlabberkohle *f* spillage coal
Schlabberkohlenband *n* spillage conveyor
Schlachtdeck *n* flensing deck *(whale factory ship)*
hinteres S. after-plan
Schlachtflugzeug *n* [ground-]attack aircraft, [ground-]attack airplane
Schlachthaus *n* slaughter house
Schlachthaushaut *f* slaughtered hide
Schlachthof *m* abattoir
Schlachtwolle *f* skin wool
Schlacke *f* slag, cinder; scoria, scum, dross; clinker
aggressive S. scouring slag
angereicherte S. enriched slag
basische S. basic slag (clinker)
dünnflüssige S. wet slag
erstarrte S. frozen slag
glasige S. vitreous clinker
granulierte S. granulated slag
kristallinische S. crystalline clinker
nichtschmelzende S. dry slag
schlackebildend slag-forming
Schlackebildner *m* slag-former
Schlackenabscheider *m* slag skimmer
Schlackenabstich *m* slag spout
Schlackenabstichloch *n* slag notch (hole)
Schlackenabstichrinne *f* slag spout
Schlackenabzug *m* slag-off
Schlackenangriff *m* slag action
schlackenartig slaggy, scoriaceous
Schlackenaufbereitung *f* refining of waste
Schlackenauge *n* slag-opening *(blast furnace)*
Schlackenband *n* slag conveyor
Schlackenbehandlung *f* slag treatment
Schlackenbeton *m* slag concrete
Schlackenbildner *m* slag-forming constituent
Schlackenbildung *f* slag (ash) formation, scorification, slagging
Schlackenblech *n* dam plate
Schlackeneinschluß *m* slag entrapment (inclusion)
Schlackenentfernung *f* removal of the clinker from the grate, clinkering
Schlackenerz *n* glassy silver ore
Schlackenform *f* slag hole (tuyere), cinder notch *(blast furnace)*; chill mould
schlackenfrei slagless, free from slag; drossless *(ore)*; free from clinker *(coal)*
Schlackenfrischen *n* lime boiling
Schlackenfrischreaktion *f* lime boil
Schlackenführung *f* slag control (regulation)
basische S. basic slag practice, working with basic slag
saure S. acid slagging practice, working with acid slag
Schlackengehalt *m* percentage of clinker
Schlackengranulierung *f* granulation of slags
Schlackengrube *f (met)* scale (ladle) pit
Schlackenhaken *m* slag hook, poker iron
Schlackenhalde *f* cinder dump
schlackenhaltig slag-bearing
Schlackenkammer *f* slag pocket
Schlackenkarre *f* slag car
Schlackenkasten *m* slag catcher
Schlackenkopf *m* scum riser
Schlackenkuchen *m* slag cake; cake of clinker
Schlackenloch *n* pit hole

Schlackenmasse *f* cake of clinker
Schlackenmehl *n* ground slag
Schlackenofen *m* slag furnace
Schlackenpfanne *f* cinder pot; slag ladle
Schlackenpresse *f* slag press
schlackenreich rich in slag
Schlackenrinne *f* slag (cinder) spout, slag (cinder) channel
Schlackenrost *m* dumping (clinker) grate
Schlackenrösten *n* slag roasting
Schlackensammler *m* slag pocket
Schlackenstein *m* slag sett (brick), cinder concrete block
Schlackenstich *m* cinder notch
Schlackenstichloch *n* slag hole (notch)
Schlackenstreifen *m* streak of slag *(materials testing)*
Schlackentasche *f* slag pocket
Schlackentopf *m* slag pot
Schlackentrichter *m* slag riser
Schlackenüberlauf *m* slag overflow
Schlackenwagen *m* slag ladle car
Schlackenwechsel *m* change of slags
Schlackenwehr *n* sand stopping
Schlackenwolle *f* slag wool
Schlackenzeile *f* slag streak *(materials testing)*
Schlackenzement *m* slag cement
Schlackenziegel *m* slag brick
Schlackenzieher *m* poker iron
Schlackenzuschlag *m* slag addition
Schlackeverlust *m (met)* dross lost
schlaff slack *(rope)*; loose
Schlaffseilschalter *m* slack cable switch
Schlafkoje *f* berth, bunk
Schlafmittel *n* hypnotic, soporific; sedative
Schlafplatz *m* berth
Schlafstadt *f* dormitory town
Schlafwagen *m* sleeping car
Schlag *m* 1. blow, impact; 2. *(el)* shock; 3. *(mach)* eccentricity *(crankshaft etc)*; 4. lay *(cable)*; 5. *(ag)* field; 6. *(mine)* bell *(signal)*; 7. beat *(pulse)*; beating *(heart)*; 8. pull, stroke *(oar)*
  elektrischer S. [electric] shock
Schlagbeanspruchung *f* impact load
Schlagbettungsstrebe *f (ship)* dog (dagger) shore
Schlagbiegefestigkeit *f* impact bending strength
Schlagbiegeversuch *m* impact bending test
Schlagblech *n* batten plate *(knitting)*
Schlagbohren *n* percussive drilling; churn drilling
Schlagbohrer *m* percussion drill; churn drill
Schlagbohrmaschine *f* piston drill
Schlagbolzen *m* striker, firing pin
Schlagbolzenfeder *f* striker spring
Schlagbrecher *m* impact[or] crusher
Schlagbug *m (ship)* long (good) board
Schlagdrehversuch *m* torsion impact test

Schlageinteilung *f (ag)* assigning of fields within a crop rotation
Schlägel *m* mall[et]
schlagen 1. to blow, to strike; 2. to drive *(rivets)*; to beat *(forge)*; 3. *(mach)* to run out of truth; to run untrue, to be off centre *(spindle)*; 4. to wobble, to whip *(drill)*; 5. to flap *(rope, belt)*; 6. *(text)* to scutch *(flax)*; to beat
Schlagen *n* **der Lenkung** steering wobble
  S. des Bohrers wobbling of the drill
  S. des Riemens flapping of the belt
  seitliches S. des Kolbens *(aut)* piston slap
Schläger *m (text)* beater [opener]; scutcher *(flax)*
Schlägerabfall *m* scutcher waste
Schlägeranschlag *m* catch for picking arm
Schlägerdeckel *m* beater lid
Schlägerrost *m* beater grid bar
Schlägerschaufel *f* blade of hockey stick
Schlägerschiene *f* beater bar
Schlägersicherung *f* beater lock
Schlagexzenter *m (text)* picker cams
Schlagfeder *f* impact spring
Schlagfehlerprüfgerät *n* concentricity tester
Schlagfehlerprüfung *f* testing for true running, concentricity testing
schlagfest impact-proof
Schlagfestigkeit *f* impact strength; resistance to shock
Schlagfließpressen *n* impact extrusion
Schlagfräser *m* fly cutter
  S. mit einem Meißel single-blade fly cutter
  S. mit mehreren Meißeln multiple fly cutter
Schlaggold *n* gold leaf
Schlaghacke *f (ag)* hoe, weeder
Schlaghammer *m* strike hammer
Schlaghärte *f* impact hardness
Schlaghärteprüfer *m* impact hardness tester
Schlaghärteprüfung *f* impact hardness test[ing]
Schlaghaube *f* helmet, cushion block, driving cap *(piling)*
Schlagholz *n* [cricket] bat
Schlagknickversuch *m* impact buckling test
Schlagkorbmühle *f* disintegrator, peg mill
Schlagkreuzmühle *f* cross beater (hammer) mill, fixed-hammer mill
Schlaglänge *f* length of lay *(cable)*
Schlagleiste *f* baffle plate
Schlagloch *n* pot-hole
Schlaglot *n* brazing (spelter, hard) solder
Schlagmaschine *f (text)* beater; scutcher *(flax)*
  doppelte S. double scutcher
  einfache S. single scutcher
Schlagmesser *n* activated blade *(coal plough)*
Schlagmessing *n* beaten brass
Schlagmetall *n* gilding metal

Schlagmühle *f* hammer mill; beater [mill], centrifugal mill, crusher, disintegrator
Schlagpatrone *f* blasting cap, igniter, primer [cartridge]
Schlagplatte *f (ship)* wash (baffle) plate *(tank)*
Schlagpresse *f* matrix striking press
schlagpressen *(plast)* to impact mould
Schlagrapport *m (text)* picking repeat
Schlagregen *m* driving rain
Schlagrichtung *f* direction of winding
Schlagschott *n (ship)* [s]wash bulkhead
Schlagschweißen *n* percussion welding
Schlagseite *f* lopside, lapside, list
S. haben to list
Schlagstauchversuch *m* compression impact test
Schlagstockführung *f (text)* picking stick motion
Schlagversuch *m* impact test
Schlagwalze *f* scutching roller *(flax)*
Schlagwasserplatte *f (ship)* [s]wash plate
Schlagweite *f* pole clearance, striking distance; clearance *(to earth)*; spark gap *(between conducting parts)*
Schlagwerk *n* 1. striking work; impact testing machine; 2. striking mechanism *(clock)*
Schlagwetter *n* fire-damp
Schlagwetteranzeiger *m* indicating methane detector, fire-damp indicator
Schlagwetterexplosion *f* explosion of fire-damp
schlagwetterfrei non-fiery, non-gassy, clear of gas
schlagwetterführend fiery, gassy
schlagwettergeschützt flameproof
Schlagwettergrube *f* fiery (gaseous, gassy, safety-lamp) mine
Schlagwetterschutz *m* flame protection
schlagwettersicher flame-proof
Schlagwiderstand *m* shock-resisting property
Schlagwirkung *f* percussion effect
Schlagwort *n* key word *(for information retrieval)*
Schlagzähigkeit *f* impact strength; notched bar test toughness
Schlagzahn *m* fly cutter
Schlagzahndorn *m* fly cutter arbor
Schlagzahnfräsarbeit *f* fly cutter operation
Schlagzahnfräsen *n* fly cutting
Schlagzahnfräser *m* fly cutter
S. mit zwei Messern two-blade fly cutter
Schlagzahnfräsverfahren *n* fly cutter method of cutting
Schlagzeile *f* headline
Schlagzerreißbeanspruchung *f* tension impact stress
Schlagzerreißversuch *m* impact pulling (tearing) test
Schlagzugbeanspruchung *f* tension impact stress
Schlagzugversuch *m* tension impact test
Schlagzündung *f* percussion priming

Schlamm *m* 1. mud; sludge *(sewage treatment)*; slurry, dredge, suspension, residue, sediment, silt, slime; 2. slime *(anode)*; 3. scum *(sugar)*
aktivierter S. activated sludge
belebter S. activated sludge
Schlammablagerungen *fpl* sludge deposits
Schlammabsetzbecken *n* mud settling pond
Schlammabsüßen *n* scum washing *(sugar industry)*
Schlämmanalyse *f* elutriation analysis
Schlammansammlung *f* accumulation of mud
Schlammaufbereitung *f* slimes treatment
Schlammbad *n* mud bath
Schlammbelebung *f* sludge activation
Schlammbelüftung *f* sludge aeration
Schlammbildung *f* sludge formation
Schlammbohrer *m* mud auger
Schlämme *f (build)* slurry
Schlammeinbruch *m (mine)* mud broke (inflow, inrush), mudrush
schlämmen to levigate *(to reduce to powder)*; to elutriate, to wash
Schlammentwässerung *f* dewatering of sludge
Schlammfang *m* silt box *(gully)*
Schlammfaulen *n* sludge digestion
Schlammfaulgas *n* sewage gas
Schlammfaulraum *m* sludge digestion chamber (tank)
Schlammfaulung *f* sludge digestion
Schlammfilter *n* mud filter
Schlammgefüge *n* muddy structure
Schlämmgerät *n (ch)* elutriator
Schlammgewinnung *f* recovering of tailings
Schlammgrube *f* slime pit
Schlammhahn *m* mud cock
schlammig muddy, sludgy; silty, slimy
Schlämmkreide *f* whiting
Schlammlöffel *m* fluke, sludger
Schlammpresse *f* sludge press
Schlämmprobe *f* sedimentation test
Schlammpumpe *f* sludge pump
Schlämmrohr *n (mine)* calyx
Schlammsaft *m* carbonation (scum) juice *(sugar industry)*
Schlammscheider *m* slime separator
Schlammschicht *f* slime layer
Schlammschiff *n* sludge disposal vessel
Schlammsprudel *m* mud volcano
Schlämmstoffgehalt *m* clay substance
Schlammsumpf *m* sludge bottom, slime pit; sludge sump
Schlammteich *m* [mud] settling pond; sludge lagoon
Schlammventil *n* mud valve
Schlämmverfahren *n* washing (elutriation) process
Schlammvulkan *m* mud volcano
Schlammwasser *n* mud water

Schlammzersetzung *f* sludge digestion
Schlange *f* coil *(heating)*
Schlangenbohrer *m* centre bit; auger bit
Schlangengift *n* [snake] venom, snake poison
Schlangenkühler *m* serpentine cooler, submerged-coil condenser, spiral radiator (condenser)
Schlangenlinie *f* wavy rule
Schlangenrohr *n* coil-pipe, winding pipe; spiral tube
Schlangenrohrüberhitzer *m* zigzag superheater
Schlangenroststab *m* serpentine (transversely corrugated) grate bar
Schlangenüberhitzerrohr *n* zigzag superheating tube, superheater coil
Schlangenventil *n* coil valve
schlank slender, slight
schlankgenutet *(mach)* slow-helix *(tool)*
Schlankheit *f* fineness *(ship's form)*
Schlankheitsgrad *m* slenderness ratio; *(ship)* coefficient of fineness; length constant
Schlankheitsverhältnis *n* fineness ratio
Schlauch *m* hose, flexible tube (tubing)
S. der Bremsleitung *(aut)* flexible pipe
S. mit Gewebeeinlage fabric-reinforced [type of] hose
baumwollarmierter S. cotton-braid hose
Schlauchanschluß *m* hose connection
Schlaucharmatur *f* pipe assembly
Schlauchboot *n* rubber dinghy, rubber (inflatable rubber, bubble) boat
Schlauchdraht *m* hose cable
Schlauchfilter *n* bag filter
Schlauchgestrick *n* tubular knit[ted] fabric
Schlauchgewebe *n* tubular woven fabric
Schlauchhaspel *f* hose reel
Schlauchklemme *f* hose (spring) clip, tube clamp
Schlauchkopsautomat *m (text)* automatic tubular cop winder
Schlauchkopsspinnmaschine *f* tubular cop spinning frame
Schlauchkupplung *f* hose coupling
selbstabdichtende S. non-spill coupling
Schlauchkupplungsgewinde *n* hose-coupling thread
Schlauchleitung *f* hose line
Schlauchteil *m* **eines Hebers** circular cam *(knitting machine)*
Schlauchventilbrücke *f (aut)* inner-tube valve fitting
Schlauchventileinsatz *m (aut)* inner-tube valve insert
Schlauchverschraubung *f* hose coupling
Schlechte *f (geol)* cleat, cleavage
Schlechtgrenze *f* lot tolerance percent defective
Schlechtläufer *m* bad runner, slow running wagon
Schlechtlehre *f* not-go gauge

schlechtwüchsig badly shaped
Schlegelernter *m* flail harvester
Schlegeltrommel *f (ag)* flail rotor
Schleichdrehzahl *f* crawling speed
Schleichfahrtmotor *m* micro-speed motor
Schleichgang *m (mach)* inching, creep feed
Schleichganggeschwindigkeit *f* creep speed
Schleier *m* 1. *(text)* veil; 2. *(opt, ch)* fog; haze; mist; blur *(glass)*
Schleierbildung *f* **des Fadens** ballooning of the yarn
Schleierschutz *m* fog prevention
Schleierschwärzung *f* fog density
Schleifahle *f s.* Honahle
Schleifarbeit *f* grinding work (operation)
Schleifaufmaß *n* grinding allowance
Schleifautomat *m* automatic grinder
Schleifband *n* abrasive-coated belt
Schleifbock *m s.* Schleifspindelstock
Schleifbügel *m* sliding (contact) shoe *(of trolleys)*
Schleifdraht *m (el)* [differential] slide wire, potentiometer wire
Schleifdrahtkompensator *m* slide-wire potentiometer
Schleifdrahtmeßbrücke *f* metre (slide-wire) bridge
Schleife *f* 1. *(el, ph)* loop; 2. *(ag)* drag *(sledge)*; 3. *(text)* tie; kink, loop
hängende S. suspended loop *(galvanometer)*
stehende S. upright loop
Schleifeinrichtung *f* grinding attachment
S. für Spiralbohrer twist drill point grinding attachment
schleifen 1. *(mach)* to grind; 2. to drag; to draw; to trail; 3. *(mil)* to dismantle; 4. to buff *(leather)*; 5. to cut *(gem)*
eine Leitung s. to loop a line
Keilflanken s. to grind splines
planparallel s. to grind two parallel surfaces simultaneously
Schleifen *n* **des Krempelbeschlages** *(text)* card grinding
S. im Stücklohn piece-work grinding
S. im Zeitlohn day-work grinding
spitzenloses S. centreless grinding
Schleifendrahtbrücke *f* loop-test bridge, loop-wire bridge
Schleifengalvanometer *n* [single-]loop galvanometer
Schleifengarn *n* loop yarn
Schleifenkapazität *f (el)* wire-to-wire capacity
Schleifenoszillograf *m* loop (bifilar, moving-coil, moving-mirror, galvanometer) oscillograph
Schleifenprobe *f* echo check[ing]
Schleifenspeicher *m (dat)* tank
Schleifenstrom *m* loop current

Schleifensystem *n* loop dialling, two-wire automatic telephone system
Schleifentrockner *m* loop drier
Schleifenumkehrung *f* sweep reversion, loop reversal of flow *(turbine)*
Schleifenverstärkung *f* loop gain
Schleifenwahl *f* loop dialling
Schleifenwandler *m* multi-loop current transformer
Schleifenwicklung *f* lap (multi-circuit, parallel) winding
Schleifenwiderstand *m* loop resistance
Schleifer *m* 1. grinder, polisher; 2. wiper *(potentiometer)*
Schleiferei *f* grinding plant (shop), grindery
Schleifergehäuse *n* grinder case
Schleiferstein *m* pulp stone
Schleifertrog *m* grinder pit
Schleiffehler *m* grinding error, error in grinding
Schleifflüssigkeit *f* coolant for grinding
Schleiffunkenprüfung *f* spark testing
Schleifgrat *m* sharpening burr *(tool)*
Schleifkontakt *m* sliding (rubbing) contact
Schleifkopf *m* [grinding] wheel head
Schleifkorn *n* abrasive grain [for grinding]
Schleifkörper *m* abrasive tool, bonded abrasive
gerader S. straight wheel
Schleifkugeln *fpl* glass grinding marbles
Schleifkurbelgetriebe *n* / **schwingendes** oscillating-crank gear
Schleiflack *m* rubbing varnish
Schleiflänge *f* grinding length
Schleifleinen *n* abrasive cloth
Schleifleinenscheibe *f* cloth abrasive disk
Schleifleiste *f* 1. *(mach)* honing (bonded abrasive) stick; 2. *(el)* contact strip *(trolley)*
Schleifmarkierung *f* sanding machine marking
Schleifmaschine *f* 1. *(mach)* grinding machine; 2. *(leath)* buffing machine
einscheibige S. single-wheel grinding machine
Schleifmittel *n* abrasive
gepulvertes S. flour abrasive
Schleifmühle *f* attrition mill
Schleiföl *n* grinding oil
Schleifpapier *n* abrasive paper
Schleifring *m* 1. *(mach)* ring (cylinder) wheel; 2. *(el)* slip (collecting, collector) ring, collector
Schleifringanker *m* slip-ring armature
Schleifringanlasser *m* slip-ring starter
Schleifringhülse *f* slip-ring bush
Schleifringläufer *m* slip-ring rotor
Schleifringleistung *f* slip-ring power
Schleifringmotor *m* slip-ring [induction] motor, wound-rotor induction motor
Schleifrippe *f* rib of the [grinding] wheel
Schleifriß *m* grinding crack (check)
Schleifrohstoff *m* grinding [raw] material
Schleifscheibe *f* grinding (abrasive) wheel
dreiprofilige S. triple-edge [thread] grinding wheel, three-ribbed [thread] grinding wheel
einprofilige S. single-rib[bed] grinding wheel
kegelige S. taper wheel
mehrprofilige S. multi-rib grinding wheel
Schleifscheibenabrichteinrichtung *f* grinding-wheel truing attachment
Schleifscheibenabrichter *m* grinding wheel dresser
Schleifscheibenaufnahme *f* [grinding-]wheel mount
Schleifscheibenbefestigung *f* [grinding-]wheel mount
Schleifscheibenbindung *f* grinding-wheel bond
Schleifscheibendurchmesser *m* diameter of [grinding] wheel
Schleifscheibenflansch *m* grinding-wheel flange
Schleifscheibenfutter *n* [grinding-]wheel chuck
Schleifscheibenhärte *f* grade of the grinding wheel
Schleifscheibenschutzhaube *f* grinding-wheel guard
Schleifscheibensupport *m* [grinding-]wheel slide
Schleifscheibenumfläche *f* [cutting] periphery of the grinding wheel
Schleifscheibenzustellung *f* infeeding [of the grinding wheel], feeding of the wheel to work
Schleifschlamm *m* grinding sludge
Schleifschlitten *m* wheel slide
Schleifschmant *m* grinding swarf
Schleifschütze *f (text)* sliding shuttle
Schleifsohle *f (ag)* shin
Schleifspalt *m* grinding throat, gap between the grinding wheels
Schleifspindel *f* grinding-wheel spindle
in Kugellagern laufende S. ball-bearing spindle
Schleifspindelachse *f* [grinding-]wheel spindle axis
Schleifspindellagerung *f* [grinding-]wheel spindle bearing
Schleifspindelschlitten *m* [grinding-]wheel slide
Schleifspindelstock *m* grinding-wheel spindle head
Schleifspindelstockdrehteil *n* base of the wheel head
Schleifspule *f (text)* fixed bobbin, pirn
Schleifstein *m* grindstone
Schleifstelle *f (text)* chafe mark *(error)*
Schleifstift *m* pencil grinder
Schleifstück *n* shoe *(sliding contact)*
Schleifsupport *m* grinding head
Schleifteil *n* part to be ground
Schleiftiefe *f* depth of cut [in grinding]
Schleiftisch *m* grinding table, work table [in grinding]
Schleiftrommel *f* sand drum

Schleiftuch *n* abrasive cloth
Schleifversuch *m* abrasion test
Schleifwärme *f* grinding heat, heat generated in grinding
Schleifwirkung *f* abrasion, attrition
Schleifzeit *f* grinding time
Schleifzone *f* grinding area
Schleifzugabe *f* grinding allowance
Schleifzylinder *m (leath)* bluffing wheel
Schleim *m* slime
schleimbildend slime-forming
Schleimgärung *f* mucous (viscous) fermentation
Schleimharz *n* gum resin
schleimig muciform, mucous, slimy
Schleimsäure *f* mucic acid
Schleißblech *n* wearing plate
schleißen to strip *(hemp)*
Schleißplatte *f* anti-wear liner, wearing plate
Schlempekohle *f* charred slop
Schleppanhang *m (ship)* train of barges, tow
Schleppanker *m* kedge anchor, kedger; drag [anchor] *(launch)*
Schleppantenne *f (aero)* reel aerial, [low-drag] trailing aerial
Schleppbalken *m (ag)* drag beam
Schleppbarkasse *f* launch tug; towing launch
Schleppbeting *m (f) (ship)* towing bitt (bollard)
Schleppbogen *m* tow rail *(tug)*
Schleppboot *n* tow boat, tugboat
Schleppbügel *m (ship)* tow-hook traveller
Schleppdampfer *m* steam tug, towing steamer
Schleppdienst *m* tug service
Schleppe *f* tail; *(ag)* drag
Schleppeinrichtung *f (ship)* towing carriage *(model testing)*
schleppen 1. *(ship)* to tug, to tow, to track; 2. to tram *(coal)*
  achteraus s. to tow astern
Schleppen *n* towage, towing
Schlepper *m* 1. tractor; 2. *(ship)* tug[-boat], towing vessel, tow [boat], water tractor; 3. *(mine)* barrow (haulage) man, drawer, hauler, hurrier
  S. für Leichterbetrieb lighter marshalling tug
  S. mit Schleppzuganhang tug and tow
  S. und Tender *m* tug and tender
Schlepperfahrer *m* tractor driver
Schlepperförderung *f (mine)* hand tramming
Schleppergebühr *f* towage
Schlepperhackmaschine *f* tractor-mounted hoe
Schlepperheck *n* tugboat (cut-away) stern
Schlepperhilfe *f* towage service, assistance of tug towing
Schlepper-Leichter-Verband *m* tug-barge combination
Schlepperpflug *m* tractor plough
Schlepperschaufel *f* tractor shovel
Schleppfähre *f* trail-ferry
Schleppfischdampfer *m* [steam] trawler
Schleppgalgen *m (ship)* tow rail
Schleppgas *n* carrier gas *(gas chromatography)*
Schleppgebühr *f* towage
Schleppgerät *n* trawl gear *(fishing vessel)*
Schleppgeschirr *n (ship)* towing gear
Schleppglied *n* drag-link
Schlepphaken *m* 1. *(ship)* towing hook, tow-hook; 2. *(mine)* haulage clip
Schleppkabel *n (el)* trailing cable *(e.g. excavators)*
Schleppkahn *m* [towed] barge
Schleppkanal *m (ship)* model tank
Schleppkette *f* drag-chain
Schleppkettenförderer *m* drag-link conveyor, (drag-chain conveyor, drag-flight conveyor, U-link conveyor
Schleppklingenstreichmaschine *f* trailing blade coater
Schleppkopfsaugbagger *m* trailing suction dredger
Schleppkurve *f* tractrix [curve]
Schleppleistung *f (ship)* towing power
Schlepplöffel *m* dragline bucket
Schlepplöffelbagger *m* boom dragline, dragline excavator
Schlepplog *n (ship)* towing log
Schlepplohn *m* towage, tug charge
Schleppnetz *n* trawl, drag[-net], dredge
Schleppnetzfischer *m* dredge[r]
Schleppnetzfischerboot *n* dragger
Schleppnetzfischerei *f* trawling
Schleppnetzfischkutter *m* trawl smack
Schleppnetzgeschirr *n* trawl gear
Schleppoller *m (ship)* towing bitt (bollard)
Schlepprakel *f* trailing blade
Schlepprechen *m (ag)* horse hoe
Schlepprinne *f (ship)* model (trial) tank
Schlepprohrsaugbagger *m* trailing suction dredger
Schleppsaugkopfbagger *m* drag suction dredger
Schleppschaufel *f (build)* scraper (dragline) bucket; *(ag)* winch-operated scoop *(mechanical gutter cleaner)*
Schleppschaufelanlage *f* mechanical gutter cleaner with a winch-operated scoop
Schleppschaufelbagger *m* dragline excavator
Schleppscheibe *f (mil)* tow[ed] target
Schleppschiff *n* tug, towing vessel, lighter
Schleppschiffahrt *f* towing, towage
Schleppseil *n* elastic rope, tow-rope, shock cord *(gliding)*; *(mine)* tail rope
Schleppspannung *f* tractive stress
Schlepptank *m (ship)* towing tank *(model testing)*
Schlepptau *n* tow rope
Schlepptauschäkel *m* towing shackle
Schlepptrosse *f* tow-rope
Schlepptrossenzug *m* tow-rope pull

Schlepptrossenzugmessung *f* towing pull test
Schleppversuch *m* *(ship)* towing experiment, resistance test
S. mit Modellen model-tank testing
Schleppversuchsanstalt *f* *(ship)* model-tank establishment
Schleppversuchstank *m* model basin, trial (model experimental, ship test) tank
Schleppwagen *m* *(ship)* 1. cradle *(launching)*; 2. towing carriage *(model testing)*
Schleppwiderstand *m* tow-rope resistance
Schleppwinde *f* towing winch (engine)
Schleppwirkungsgrad *m* towing efficiency
Schleppzeiger *m* maximum pointer
Schleppzug *m* 1. tractor train; 2. *(ship)* [tug and] tow, train of barges, barge train, linked barges
Schleppzugeinheit *f* *(ship)* barge train unit
Schleuder *f* 1. centrifuge, hydroextractor; 2. spin-drier
Schleuderbeton *m* spun (centrifugally cast) concrete
Schleuderbetonrohr *n* spun [concrete] pipe
Schleuderdrehzahl *f* *(el.m.)* overspeed, excess speed
Schleuderfilter *n* centrifugal filter
Schleuderförderer *m* centrifugal conveyor
Schleuderformmaschine *f* sand slinger, slinger moulding machine
Schleudergebläse *n s.* Zentrifugalgebläse
Schleudergeschwindigkeit *f* ramming rate [of the sand slinger]
Schleudergießerei *f* centrifugal casting
Schleudergießmaschine *f* **für Rohre** pipe-spinning machine
Schleudergießverfahren *n* centrifugal casting method
Schleuderguß *m* centrifugal casting
Schleudergußstück *n* centrifugal casting
Schleuderkopf *m* ramming head [of the sand slinger]
Schleuderkorb *m* *(text)* hydroextracting cage
Schleuderlader *m* centrifugal supercharger
Schleuderlüfter *m* centrifugal fan
Schleuderluftfilter *n* centrifugal air cleaner
Schleudermühle *f* centrifugal (peg) mill, disintegrator
schleudern 1. to sling, to throw; 2. *(aut)* to bounce, to slide-slip, to skip; *(ry)* to slip *(wheels)*; 3. to centrifuge; *(build)* to cast centrifugally; to spin
Schleudern *n* **der Räder** *(ry)* slipping of the wheels
S. des Anhängers *(aut)* bouncing of the trailer
Schleuderprüfstand *m* whirling test stand
Schleuderprüfung *f* dynamic balance test
Schleuderpsychrometer *n* sling psychrometer
Schleuderpumpe *f* centrifugal pump
Schleuderrad *n* bladed (bucket, centrifugal) wheel; impeller
Schleuderradroder *m* spinner, rotary digger
Schleuderroder *m* potato spinner
Schleuderschmierung *f* centrifugal (splash) lubrication
Schleudersieb *n* centrifugal screen
Schleuderstern *m* *(ag)* auxiliary spinner
Schleuderverfahren *n* centrifugal casting, spinning *(concrete)*
Schleuderversatz *m* *(mine)* high-speed belt stowing, mechanical stowing with slingers
Schleuderversatzmaschine *f* *(mine)* centrifugal stower
Schleuderverteiler *m* *(ag)* spreading rotor
Schleuderzucker *m* centrifugal sugar
Schleuse *f* gate, lock, sluice *(river)*; drain *(street)*; [air-]lock; by-pass
Schleusenbecken *n* scouring basin, sluicing pond
Schleusendeckel *m* manhole cover
Schleusendock *n* wet basin, wet (closed) dock
Schleusenkammer *f* coffer; canal lock chamber; entrance lock
Schleusenkiel *m* sluice keel
Schleusenmeister *m* lock-keeper, lock-master
Schleusenponton *m* caisson
Schleusenschieber *m* sluice (slide gate) valve
Schleusenschiff *n* caisson
Schleusenschottschieber *m* sliding watertight gate
Schleusenschütz *n* slide gate valve
Schleusentor *n* gate, dock (flood, slide, sluice) gate; penstock; caisson, gate vessel
Schleusenwehr *n* lock-weir
Schlichtarbeit *f* *(mach)* finishing (finish machining) operation
Schlichtbaum *m* *(leath)* perch
Schlichtdrehmeißel *m* finish-turning tool, finishing lathe tool
Schlichte *f* 1. *(text)* sizing material, size; 2. slur *(moulding)*, cinder paste; 3. *(ag)* drag
schlichtecht *(text)* fast to sizing
Schlichtefleck *m* *(text)* size stain
Schlichtegemisch *n* *(text)* sizing mix
Schlichtekocher *m* *(text)* size mixing apparatus
Schlichtemasse *f* *(text)* sizing paste
Schlichtemittel *n* *(text)* size, sizing agent
schlichten 1. *(mach)* to finish-machine; 2. *(text)* to size; to dress, to smooth
Schlichten *n* 1. finishing *(workpiece)*; 2. *(text)* dressing, sizing
S. der Kettfäden sizing of the warp threads
Schlichtetrog *m* size box
Schlichtfräser *m* finishing [milling] cutter
Schlichtfrässpindel *f* finish milling spindle
Schlichthobeleisen *n* uncut plane iron
Schlichthobeln *n* finish planing

Schlichtmaschine *f (text)* dressing frame, sizing machine
S. für Riemchen tape sizer
S. mit Heißlufttrocknung hot-air sizing machine
Schlichtmeißel *m* finishing tool
Schlichtmittel *n* sizing agent
Schlichtmond *m* perching knife *(shoemaking)*
Schlichtnachformhobeln *n* finish contour planing
schlichträumen to finish broach
Schlichträumwerkzeug *n* finishing broach
Schlichtschneideneinsatz *m* finishing shell (insert)
Schlichtschnitt *m* finishing cut
Schlichtspan *m* finishing cut
Schlichtverfahren *n* finishing process
Schlichtzahn *m* finishing tooth *(broach)*
Schlichtzeichen *n* finish mark
Schlichtzugabe *f* finish[ing] allowance
Schlicker *m* slip *(ceramics)*; *(leath)* setting slicker
Schliere *f* schliere, streak, *(Am)* stria
Schlierenapparat *m* schlieren apparatus
Schlierenaufnahme *f* schlieren photography
Schlierenbild *n* schlieren picture
Schlierenbildung *f* striation
Schlierenblende *f* schlieren edge [and slit]
Schlierenmethode *f* schlieren method
Schlierentechnik *f* striation technique
Schließdrahtblei *n (text)* closing wire lead
schließen 1. to lock, to shut, to close; 2. to close down *(factory)*; 3. to finish; to fill, to seal; to fit *(without clearance)*; 4. *(el)* to make *(circuit)*; to complete
einen Stromkreis s. to complete (make) a circuit
Schließen *n* **der Abstichöffnung** filling of the tap-hole
S. der Bleche clamping together of the plates
S. des Ventils valve dropping
Schließer *m (el)* a-contact, closing (make, making) contact
Schließkontakt *m s.* Schließer
Schließkopf *m* driven head *(of a rivet)*
Schließplatte *f* imposing surface (stone)
Schließrahmen *m* chase
Schließrahmenmittelsteg *m* cross-bar
Schließrolle *f* pinion
Schließsäge *f* sash saw
Schließseil *n* closing rope *(grab bucket)*
Schließung *f* closure *(factory)*
Schließungsimpuls *m* make pulse
Schließungslichtbogen *m* closing arc
Schließungsstromstoß *m* make pulse
Schließzeit *f (el)* make (closing) time
Schließzeug *n* quoins
Schliff *m (mach)* grind; glass grind; polished section; pulp *(papermaking)*
mugeliger S. [en]cabochon
röscher S. free pulp *(papermaking)*
Schliffbild *n* micrograph
Schliffentwässerungsmaschine *f* pulp drying machine *(papermaking)*
Schliffgüte *f (mach)* grinding quality
Schliffhahn *m* ground glass stopcock
mit Quecksilber gedichteter S. mercury-seal stopcock
Schliffhülse *f* ground-joint female part
Schliffkern *m* ground-joint male part
Schliffstopfen *m* ground-glass stopper
Schlifftrocknung *f* pulp drying *(papermaking)*
Schlinge *f* 1. *(text)* loop *(rope)*; shoe braid; uncut pile; 2. *(el)* curl *(wire)*; 3. *(ship)* sling; 4. loop *(theory of graphs)*
Schlingenfaden *m* loop thread
Schlingenfestigkeit *f* loop tenacity
Schlingengarn *n* loop (snarl) yarn
Schlingenheber *m* loop lifter
Schlingenkanal *m* looper
Schlingenkurve *f (math)* lemniscate
Schlingenspanner *m* looper
Schlingerbewegungen *fpl* corkscrewing, corkscrew motion *(vessel)*
Schlingerdämpfer *m* stabilizer, anti-rolling stabilizer
Schlingerdämpfung *f* roll quenching
Schlingerdämpfungseinrichtungen *fpl* anti-rolling devices
Schlingerkiel *m* bilge keel, rolling (bilge) chock
Schlingerkoje *f* swing berth
Schlingerleiste *f* fiddle *(at a table)*
herunterklappbare S. disappearing fiddle
Schlingern *n* / **leichtes** easy roll
Schlingerperiode *f* period of roll
Schlingerplatte *f* wash plate
Schlingerschott *n* [s]wash bulkhead, shifting boards
Schlingerstag *n (ship)* preventer stay
Schlingertanks *mpl* anti-rolling tanks
Schlingertisch *m* suspended table
Schlingerverband *m (build)* sway bracing
Schlingerwulst *m (f)* anti-rolling bulge
Schlingfaden *m* catch thread
Schlinghelfe *f* link heald
Schlingkette *f* chain sling
Schlitten *m* 1. sled[ge], sleigh; 2. *(mach)* carriage, slide, saddle, slide head *(broaching machine)*; slide (guide) block; 3. *(ship)* sliding (running) ways *(launching)*
zweisystemiger S. double-lock carriage *(knitting machine)*
Schlittendrehlager *n* poppet *(launching)*
vorderes S. fore poppet *(launching)*
Schlittendruck *m* launching pressure

Schlittenführung *f* **für den Zylinderfuß** slide for the cylinder base
doppelseitige S. double guide
ebene S. flat guide
einseitige S. single guide
Schlittenkeil *m* launching wedge
Schlittenklemmung *f (mach)* saddle clamp
Schlittenkolben *m* piston with slipper
Schlittenkufe *f* [sledger-]runner
Schlittenläufer *m* runner, running ways *(launching)*
Schlittenständer *m (ship)* dog-shore
Schlittenstütze *f* [launching] poppet
vordere S. fore poppet *(launching)*
Schlittenvorkante *f* fore end of running ways *(launching)*
Schlittenvorschubgatter *n* log frame with rag carriage feed
Schlittenweg *m* sled-road
Schlittschuh *m* skate
Schlittschuhkufe *f* blade of skate
S. mit Hohlschliff hollow-ground blade
Schlittschuhlaufen *n* skating
Schlittschuhläufer *m* skater
Schlitz *m* slot, kerf, groove; slit; nick *(screw head)*; port *(engine)*; *(mine)* kerf; cleft, rift; window *(e.g. transformer)*
Schlitzanode *f* split anode
Schlitzanodenmagnetron *n* split-anode magnetron
Schlitzantenne *f* slot aerial
Schlitzbemusterung *f (geol)* slot sampling
Schlitzblende *f (opt)* slit-stop; slot (slit) diaphragm
Schlitzbrenner *m* slit (long-slot) burner
Schlitzbunker *m* slot bunker
Schlitzdüse *f* air jet
Schlitzelektrode *f* slotted electrode
schlitzen *(mach)* to slot[-mill]; to split; to slash; *(mine)* to cut, to kerf, to shear
Schlitzen *n* **von Kronenmuttern** *(Am)* castellating
Schlitzfräsen *n* slot milling
Schlitzfräser *m* slitting (keyseating) cutter
Schlitzkanalstromschiene *f* conduit conductor rail system
Schlitzkegel *m* skirted valve plug, V-port plug *(valve)*
Schlitzkeilanker *m (mine)* slot-and-wedge [roof-]bolt
Schlitzlochung *f* slot perforation
Schlitzmagnetron *n* slot-type magnetron
Schlitzmantelkolben *m* slotted piston
Schlitzmaschine *f* slot mortising machine
Schlitzmomentverschluß *m (opt)* roller-blind instantaneous shutter
Schlitzmutter *f* slotted nut
Schlitzprobe *f* channel sample
Schlitzring *m* slit ring *(turbine)*; *(aut)* grooved piston ring
Schlitzrohr *n (el)* slotted conduit
Schlitzsäge *f* sash (slitting) saw
Schlitzscheibe *f* grooving cutter block
Schlitzschraube *f* slotted (recessed-head) screw
Schlitzstrahler *m* slot radiator *(radar)*
unabgestimmter S. non-resonant slot radiator
Schlitzstrahlerkombination *f* slot array *(aerial)*
Schlitzstreuung *f (ag)* plate and flicker feed
Schlitzstromzuführung *f* slotted conduit
Schlitztrommel *f* slit drum
Schlitz- und Zapfeneisen *n* cutter for tenons and mortises
Schlitzverschluß *m (opt)* focal-plane shutter
Schlitzweite *f* **der Rostspalten** width of the grate openings
Schlitzwellenfilter *n* mode filter slot *(waveguides)*
Schlitzzerstäuber *m* slot sprayer (atomizer)
Schloß *n* 1. lock *(door, rifle)*; padlock *(detachable)*; snap; 2. cam *(knitting machine)*; 3. clasp *(e.g. book)*; 4. buckle *(belt)*
eingelassenes S. flush-encased lock
Schlößchen *n* cocking piece *(rifle)*
Schlosser *m* locksmith; fitter, mechanic
Schlosserei *f* fitter's shop
Schlosserhammer *m* locksmith's hammer
Schlosserhandwerk *n* locksmith's trade
Schlosserwerkstatt *f* locksmith's shop
Schloßführung *f* lock guide
Schloßkanal *m* cam groove (raceway) *(knitting machine)*
Schloßkappe *f* lock cap
Schloßkasten *m* 1. apron *(lathe)*; 2. cam box *(knitting machine)*
Schloßkeil *m* locking wedge
Schloßmantel *m* cam ring *(knitting machine)*
Schloßmutter *f* half (clasp) nut *(lathe)*
Schloßplatte *f* 1. *(mach)* [lathe] apron; 2. cam plate (section) *(knitting machine)*
Schloßriegel *m* bolt of a lock
Schloßring *m* locking ring *(turbine)*
Schloßschalter *m* interlocked switch
Schloßschraube *f* lock screw; carriage bolt
Schloßteil *n* cam *(knitting machine)*
S. für Schlaucharbeit cam for tubular knitting
im Schlitten montiertes S. carriage cam
Schloßteiltragering *m (text)* cam retaining ring
Schlot *m* chimney [stack], [smoke] stack, funnel, flue; *(geol)* canal, pipe
schlucken / Schall to absorb sound
Schluckgrad *m* coefficient of absorption
Schluckstoff *m* sound absorber (deadening material)

Schluckvermögen *n* 1. absorbing power; *(mine)* water absorption capacity *(well)*
Schluckvolumen *n* capacity *(hydraulic motor)*
Schluckzahl *f* sound absorption coefficient
Schluff *m* silt, watery clay
schluffartig silty
Schlupf *m* slip[page], slipping; percentage slip; *(ac)* drift
  S. der magnetischen Feldlinien slip of the magnetic field lines
Schlupfbrüter *m* hatching unit
schlupffrei non-slip
Schlupffrequenz *f* slip frequency
Schlupfkasten *m* *(ag)* hatching box, breeding apparatus
Schlupfmotor *m* cumulative compound motor
Schlupfregler *m* *(el)* slip regulator
  automatischer S. automatic slip regulator
Schlüpfrigkeit *f* **des Schmiermittels** lubricity of the lubricant
Schlupfvariable *f* slack variable *(linear programming)*
Schlupfverlust *m* slip loss, loss due to slippage
Schluß *m* **des Zuges** rear of the train
Schlüssel *m* 1. key *(lock)*; *(mach)* key *(chuck)*; spanner, wrench *(nut)*; 2. code; cipher
  verstellbarer S. adjustable open-end wrench
Schlüsseladresse *f* *(dat)* key address
Schlüsselbart *m* key bit
Schlüsselfeile *f* warding file
Schlüsselfräser *m* **für Maulweiten** wrench cutter
Schlüsselgruppe *f* *(dat)* key
Schlüsselloch *n* keyhole
Schlüssellochung *f* *(dat)* code punching
schlüsseln to code; to cipher
Schlüsselschalter *m* key-type switch, key-operated switch
Schlüsselschild *n* escutcheon
Schlüsselzahl *f* code number
Schlußgeschwindigkeit *f* *(therm)* closing speed
Schlußglied *n* *(ph)* end (final) product *(decay series)*
Schlußglühung *f* finish anneal[ing]
Schlußhärten *n* final hardening
Schlußkontakt *m* *(el)* terminal contact
Schlußkontrolle *f* final examination
Schlußlampe *f* *(telec)* clearing lamp
Schlüßler *m* coder
Schlußleuchte *f* *(aut)* rear lamp, tail light
Schlußlicht *n* *(ry)* rear (tail) lamp, rear (tail) light
Schlußrelais *n* clearing relay
Schlußstein *m* keystone, crow-stone, apex stone *(arches)*
Schlußstück *n* end filling piece, blade stopper
Schlußtaste *f* clearing key *(telephony)*
Schlußvignette *f* tail piece
Schlußzeichen *n* clearing (clear-back, sign-off, disconnect) signal; *(Am)* supervisory (on-hook) signal; clearing pulse
Schlußzeichenwecker *m* clearing signal bell *(telephony)*
Schlußzeile *f* **einer Seite** bottom line
schmal narrow; thin
Schmalband *n* narrow band; *(text)* narrow braid
Schmalbündelantenne *f* pencil-beam aerial
Schmalfilm *m* substandard film; ciné film
Schmalführung *f* *(mach)* narrow guide
Schmalgewebe *n* narrow fabric
Schmalsatz *m* *(print)* narrow measure
Schmalschnitt *m* pump edge *(shoe)*
Schmalspur *f* *(ry)* narrow gauge
Schmalspurbahn *f* light (narrow-gauge) railway
Schmalspurbahnmotor *m* narrow-gauge traction motor
Schmalz *n* lard
Schmälze *f* oiling agent; application of oil to fibres, lubricant, lubricating oil
schmälzen to oil *(yarn)*, to lubricate
Schmälzen *n* **der Wolle** wool oiling
Schmälzwolf *m* oiling devil (willow)
Schmant *m* sludge *(boring)*
Schmantlöffel *m* bailer
Schmantseil *n* sand line
Schmanttrommel *f* sand reel
Schmasche *f* unborn lambskin
Schmelz *m* enamel
Schmelzausbringung *f* production cast
Schmelzbad *n* *(met)* melting bath; pool *(welding)*
  ruhiges S. quiet melt
schmelzbar fusible, meltable; smeltable
  schwer s. hard to melt
Schmelzbarkeit *f* fusibility, meltability; smeltability
Schmelzbetrieb *m* melting practice
Schmelzdauer *f* [s]melting period
Schmelzdecke *f* *(met)* tile
Schmelzdraht *m* fuse (fusible) wire
Schmelzdrahtsicherung *f* wire fuse
Schmelze *f* melt, fused (molten) metal; melting bath; blow *(converter)*; heat *(quantity of metal)*; fused electrolyte
Schmelzeinsatz *m* fuse[-link], refill unit
schmelzen to fuse *(wire, fibres)*, to melt; to smelt, to liquefy; to found *(glass)*
Schmelzerei *f* melting department
Schmelzfluß *m* melt, fused metal; complete fusion; coalescence *(welding)*
Schmelzflußelektrolyse *f* electrolysis in the dry way
schmelzflüssig molten, fusible
Schmelzflußstreichmaschine *f* extrusion coater *(papermaking)*
schmelzgegossen fused
Schmelzgeschwindigkeit *f* melting rate

Schmelzgut *n* [melting] charge, melting stock
Schmelzherd *m* [s]melting hearth
S. mit Verschlußdeckel closed hearth
sauer zugestellter S. acid melting hearth
Schmelzhütte *f* smelting plant
Schmelzkegel *m* Seger (melting, fusible, pyrometric) cone
schmelzkleben to heat-seal
Schmelzklebstoff *m* thermoplastic adhesive
Schmelzkurve *f* fusion (melting-point) curve; solidification curve, ice line
Schmelzleistung *f* melting efficiency, melting-down power *(furnace)*
Schmelzleiter *m* fuse (fusible, current-responsive) element, fusing conductor
Schmelzlöser *m* smelt-dissolving tank *(papermaking)*
Schmelzmittel *n* flux
Schmelzofen *m* melting furnace
Schmelzperle *f* blowpipe bead; enamel pearl
Schmelzpfropfen *m* fusible plug
Schmelzprobe *f* assay, sample
Schmelzpunkt *m* melting (fusion) point
Schmelzpunkterniedrigung *f* lowering of the melting point
Schmelzpunktröhrchen *n* melting-point tube
Schmelzsalz *n* smelting salt
Schmelzsäule *f* scaffold *(blast furnace)*
Schmelzschnittbandsäge *f* band-sawing machine designed for friction cutting
schmelzschweißbar fusion weldable
schmelzschweißen to fuse, to fusion-weld
Schmelzschweißen *n* fusion welding
Schmelzschweißnaht *f* fusion weld
Schmelzschweißung *f* fusion weld; fusion welding
Schmelzschweißverfahren *n* fusion [welding] process
Schmelzsicherung *f* [safety] fuse, fusible cut-out
Schmelzspinnen *n (plast)* melt extrusion
Schmelzstahl *m* natural (furnace, German) steel
Schmelzstreichüberzugverfahren *n* hot-melt coating *(papermaking)*
Schmelzstreifen *m* fuse (fusible metal) strip
Schmelztemperatur *f* melting (fusing) temperature
Schmelztiegel *m* [melting] crucible, melting pot
Schmelzung *f* fusion; heat; liquefaction, eliquation
Schmelzverfahren *n (met)* melting process
basisches S. basic [melting] process
Schmelzverlaufkarte *f* cast history sheet
Schmelzverlust *m* melting loss
Schmelzwärme *f* heat of fusion, melting heat
Schmelzzement *m* aluminous cement
Schmelzzone *f* melting zone, zone of fusion
Schmelzzuschlag *m* flux, fusing addition
Schmergel *m s.* Schmirgel
Schmerzgrenze *f (ac)* threshold of feeling
schmerzlindernd lenitive
Schmetterlingsantenne *f* [super-]turnstile aerial
Schmetterlingsausgangskreis *m* butterfly output circuit
Schmetterlingsdach *n* butterfly (double lean-to) roof
Schmetterlingskreis *m* butterfly circuit
Schmetterlingsschale *f (build)* butterfly shell
schmiedbar forgeable, malleable
Schmiedbarkeit *f* forgeability, malleability
Schmiede *f* smithy, forge, blacksmith's shop
Schmiedeabfall *m* forge scrap
Schmiedeamboß *m* blacksmith's anvil
Schmiedeblock *m* forging ingot
Schmiedeeisen *n* forging-grade steel, wrought (malleable) iron
Schmiedefeuer *n* forge (smith's) fire
Schmiedegebläse *n* smith's bellows
Schmiedegesenk *n* drop-forging die
Schmiedehammer *m* forging hammer
Schmiedehandwerk *n* smith's trade
schmiedeheiß at the forging heat
Schmiedeherd *m* smith's hearth
Schmiedekohle *f* smith's coal; smithy coal
Schmiedekran *m* forging crane
Schmiedelegierung *f* forge (forging) alloy
Schmiedemaschine *f* forging machine
Schmiedemessing *n* forging brass
schmieden to forge, to hammer, to beat
von der Stange s. to forge from the bar
Schmieden *n* forging
S. im Gesenk drop forging
Schmiedenagel *m* spike
Schmiedeofen *m* forging (stock) furnace
Schmiedepresse *f* forging press
Schmiedeprobe *f* pit sample
Schmiedeprüfung *f* forging test
Schmiederiß *m* forging crack
Schmiedeschlacke *f* blacksmith's slag
Schmiedesinter *m* forge (forging) scale
Schmiedestahl *m* wrought iron
Schmiedestück *n* forging
Schmiedetechnik *f* forging practice
Schmiedeteil *n* forging
Schmiedetemperatur *f* forging temperature
Schmiedeversuch *m* forgeability test
Schmiedewalze *f* forging rolls
Schmiedewerkzeug *n* blacksmith's tool
Schmiedezunder *m* forging scale
Schmiege *f* bevel [rule]
Schmiegebene *f* osculating plane
Schmiegeegge *f* flexible harrow
Schmiegroststab *m* interlocking (zigzag) grate bar
schmiegsam flexible, pliant, supple
Schmiegungsebene *f* osculating plane
Schmiegungskreis *m* osculating circle

Schmiegungskugel *f* osculating sphere
Schmiegungszahl *f* osculation coefficient
Schmierapparat *m* lubricating device, lubricator
Schmierbohrung *f* oil (lubrication) hole
Schmierbüchse *f* oil cup
Schmierbügel *m* wiper
Schmierbürste *f* oil brush
Schmiere *f* grease *(vehicle)*
Schmiereigenschaften *fpl* lubricating qualities
schmieren to lubricate; to oil, to grease; *(leath)* to tallow
im Faß s. *(leath)* to drum-stuff
mit der Hand s. *(leath)* to hand-stuff
schmierfähig oily, lubricating
Schmierfähigkeit *f* lubricity, oiliness; lubricating quality
Schmierfänger *m* oil wiper
Schmierfeder *f* lubricator spring
Schmierfett *n* [lubricating] grease
Schmierfilm *m* lubricant film
Schmierflüssigkeit *f* lubricating oil, cutting solution
Schmiergefäß *n* lubricating cup
Schmiergerät *n* lubricating appliance
Schmierhäufigkeit *f* relubrication interval
schmierig smeary, greasy, oily
Schmierkamm *m* wiper
Schmierkanne *f* oil can
Schmierkeilfilm *m* wedge-shaped oil film
Schmierkeilfilmbildung *f* formation of a wedge-shaped oil film
Schmierkissen *n* lubricating (oil) pad
Schmierkopf *m* **für Fettschmierung** grease nipple
Schmierkraft *f* lubricity, oiliness
Schmierlager *n* grease bearing
Schmierleistung *f* lubricating efficiency
Schmierleitung *f* lubricant piping
Schmierloch *n* lubricating (oil) hole, oil run
Schmiermittel *n* lubricant
Schmiermittelaustritt *m* lubricant leakage
Schmiermittelkreis *m* lubricant circuit
Schmiermittelpumpe *f* lubricating pump
Schmiermittelreibung *f* friction between lubricated surfaces
Schmiermittelrückführung *f* lubricant return
Schmiermittelzufuhr *f* lubricant supply
Schmiernippel *m* grease (lubrication) nipple
Schmiernut *f* oil (lubricating) groove
längslaufende S. axial oil groove
umlaufende S. circular oil groove
Schmieröffnung *f* lubricating (oil) hole
Schmieröl *n* lubricating oil
hochviskoses S. bright stock
Schmierölbehälter *m* lubricating oil tank (container)
Schmierölpumpe *f* lubricating oil pump
Schmierplan *m* lubrication schedule (chart)
Schmierpolster *n* lubricating film; oil pad
Schmierpresse *f* grease gun, mechanical plunger lubricator
Schmierring *m* oiling ring *(bearing)*
Schmierrohr *n* oiling tube
Schmierseife *f* soft soap
Schmierstelle *f* lubrication (lubricating, oiling, greasing) point
Schmierstoff *m* lubricant
Schmierstoffbehälter *m* lubricant container
Schmierstoffentschäumer *m* jet oil-foam separator, air-oil separator
Schmierstoffversorgung *f* lubricant supply
Schmierstutzen *m* lubrication connection
Schmiersystem *n* oil system
Schmiertechnik *f* lubrication engineering
Schmierübersicht *f* lubricating schedule
Schmierung *f* lubrication, oiling
verlorene S. non-recovery lubrication, all-loss lubrication, total-loss lubrication
vollflüssige S. hydrodynamic lubrication
vollkommene S. perfect lubrication
zwangsläufige S. forced feed lubrication
Schmiervermögen *n* lubricity
Schmiervorrichtung *f* lubricator
automatische S. automatic (self-acting) lubricator
Schmiervorschrift *f* lubrication instructions
Schmierwalkfaß *n (leath)* stuffing drum
Schmierwert *m* lubricating value
Schmierwirkung *f* lubricating effect (action)
Schminke *f* make-up, paint
Schminkpuder *m* blemish concealers
Schmirgel *m* emery
Schmirgelleinen *n* emery (abrasive) fabric, emery (abrasive) cloth
Schmirgelleinwand *f s.* Schmirgelleinen
Schmirgel-Öl-Gemisch *n* emery cake (paste)
Schmirgelpapier *n* abrasive (coated, emery) paper, sandpaper, glass-paper
Schmirgelrohpapier *n* emery base-paper
Schmirgelscheibe *f* emery wheel
Schmirgelschleifmaschine *f* emery grinding machine
Schmitz *m (print)* slur
Schmitz-auf-Schmitz *(print)* bearer-to-bearer
schmitzen to mackle, to slur
Schmitzleisten *fpl* bed bearers, mackling rails
Schmitzliste *f* accountancy book
Schmitzring *m* [cylinder] bearer
Schmuckblattelegramm *n* de luxe telegram
Schmuckmetall *n* ornamental metal
Schmuggelschiff *n* smuggler; *(Am)* hovering vessel
Schmutz *m* dirt; *(leath)* scud
Schmutzablagerung *f* pollution (contamination, dirt) deposit
Schmutzabscheider *m* dirt separator *(wood)*

Schmutzbogen *m* set-off sheet
schmutzen *(text)* to get dirty
schmutzig dirty
Schmutzlösevermögen *n* soil removing capacity
Schmutzöltank *m* dirty oil tank *(tanker)*
Schmutzpunktzahl *f* dirt count
Schmutztitel *m* half (bastard) title
Schmutztragevermögen *n* soil suspending power
Schmutzwalke *f* grease milling
Schmutzwasser *n* waste (foul) water; sewage
Schmutzwolle *f* greasy wool
Schnalle *f* buckle, clasp
schnallen to buckle, to strap
Schnallenisolator *m* shackle insulator
Schnappbremse *f* snatch brake
schnappen to catch *(lock)*; to snap
Schnäpper *m (mach)* catcher, latch, pawl; trigger *(rifle)*
Schnapperkupplung *f (el)* pawl coupling
Schnappkontakt *m (el)* snap-action contact
Schnappring *m* snap ring
Schnappschalter *m* snap (quick-break) switch
Schnappschloß *n* spring catch (lock)
Schnappschuß *m* snapshot
Schnappverschluß *m* spring catch (lock)
Schnarchventil *n* snifting valve
Schnauze *f* spout *(ladle)*
Schnecke *f* 1. *(mach)* worm, endless screw; 2. *(build)* volute
  eingängige S. single-start worm
  mehrgängige S. multi-start worm
Schneckenantrieb *m* worm drive
Schneckenaufgabe *f* screw feeding
Schneckenaufgeber *m* screw feeder, feed screw
Schneckenbohren *n* auger drilling
Schneckenbohrer *m* twisted (worm) auger
Schneckenelevator *m* auger elevator
Schneckenförderer *m* screw (worm, spiral) conveyor
schneckenförmig helical, spiral
Schneckenfräsmaschine *f* worm miller (milling machine)
Schneckengetriebe *n* worm gearing, worm and worm wheel
Schneckengetriebekasten *m* worm box
Schneckenlenkung *f* worm-and-wheel steering gear, worm-and-sector steering gear
Schneckenmischer *m* screw mixer
Schneckenpresse *f* screw extruder (extrusion press)
Schneckenrad *n* worm wheel (gear)
Schneckenradantrieb *m* worm-wheel drive
Schneckenradfräsen *n* worm-gear cutting (milling)
Schneckenradsegment *n* [worm-wheel] quadrant
Schneckenradtrieb *m* worm-wheel drive
Schneckenradübersetzung *f* worm-wheel transmission
Schneckenradzahn *m* worm-wheel tooth
Schneckenreduziergetriebe *n* worm reduction gear
Schneckenreiniger *m* screw cleaner *(beet harvester)*
Schneckenspeiser *m* screw feeder, feed screw
Schneckentrieb *m* worm drive
Schneckentrieur *m (ag)* spiral separator
Schneckenwelle *f* worm shaft
Schnee *m* snow
Schneeabfall *m* [sudden] throwing-off of snow *(overhead lines)*
Schneebeseitigung *f* snow clearing
Schneebrille *f* snow goggles
Schneefanggitter *n* snow fence
Schneekette *f (aut)* non-skid chain; snow track
Schneelast *f* snow load
Schneepflug *m* snow-plough [with blades]
Schneeprofil *n* snow tread *(tyre)*
Schneeräumung *f* snow clearing
Schneesack *m* snow trap
Schneeschleudermaschine *f* rotary snow plough
Schneezaun *m* snow fence
Schneidarbeit *f (mach)* cutting action
Schneidbohrer *m* tap
schneidbrennen to flame-cut
Schneidbrenner *m* cutting torch (blowpipe), flame cutter, fusing burner
  autogener S. oxy-acetylene cutting torch
Schneiddose *f* mechanical recording head
Schneide *f* cutting edge *(tool)*; blade *(axe)*; bit; knife-edge *(balance)*
  S. mit Spanbrechernuten nicked cutting edge *(to break up chips)*
  S. mit Spanbrecherstufen stepped cutting edge
  bestückte S. tipped edge
  formrichtige S. proper cutting edge
  korrigierte S. modified lip *(twist drill)*
Schneideisen *n* die
  S. für Gewinde [button, threading] die
Schneidemaschine *f* cutter, guillotine
schneiden 1. to cut; to clip, to shear, to blank, to notch, to trim; 2. *(math)* to intersect; 3. *(ag)* to mow
  Gewinde s. to cut threads, to thread
  in Streifen s. to slice
  sich s. *(math)* to intersect
Schneiden *n* **aus dem Vollen** cutting from the solid
Schneidenabnutzung *f* cutting edge wear
Schneidenaufbau *m* built-up edge
Schneidenausbildung *f* point geometry, cutting surface shape
Schneidenbrust *f s.* Spanfläche
Schneidenecke *f* outer [edge] corner *(twist drill)*
Schneideneinsatz *m* insert, section *(broach)*, cutting edges of the shell design

S. im Räumwerkzeug shell section in the broaching tool
Schneidengeometrie *f* tool geometry, bit shape
Schneidenhalter *m* holder *(broach)*
Schneidenkopf *m* tool point, point [of the tool]; drill tip; drill bit, cutting head
Schneidenrücken *m* back of the cutting edge
drückender S. dragging back of the cutting edge
Schneidensatz *m* set of inserts
Schneidenschaft *m* body *(twist drill)*
Schneidenspektrograf *m* slitless (wedge) spectrograph
Schneidenstandzeit *f* life of the cutting edge
Schneidenteil *m* cutting part (portion, section) *(broach)*; body *(twist drill)*
Schneidenteilung *f* spacing of the cutting edges
ungleiche S. unequal spacing of the flutes (cutting edges)
Schneidenverschleiß *m* cutting-edge wear
Schneidenwerkstoff *m* tool tip material
Schneidenzähne *mpl* cutting teeth *(broach)*
Schneider *m* 1. cutter; 2. tailor
Schneiderbüste *f* tailor's dummy
Schneiderhandwerk *n* tailor's trade
Schneiderkreide *f* tailor's chalk
Schneidermeister *m* master tailor
Schneiderpuppe *f* dummy
Schneiderwerkstatt *f* tailor's workshop
Schneidetisch *m* editing table
Schneidewinkel *m* sharpness angle *(woodworking)*
Schneidfacette *f* cutting facet *(tool)*
Schneidfähigkeit *f* cutting quality; cutting property
Schneidfläche *f* cutting surface
Schneidflamme *f* cutting flame
Schneidflüssigkeit *f* cutting fluid
Schneidfuß *m* cutting toe
schneidhaltig capable of retaining the cutting edge
s. sein to hold (retain) a keen cutting edge
Schneidhaltigkeit *f* ability to maintain cutting power; ability to retain a cutting edge, cutting quality [of a tool]
Schneidhärte *f* cutting hardness
Schneidkamm *m* rack[-form] cutter *(gear generation)*
Schneidkante *f* cutting edge; lip *(twist drill)*
S. am Spiralbohrer lip of the twist drill
bändchenartige S. wear land
Schneidkantenversetzung *f* offset of cutting edges
Schneidkluppe *f* die-stock
Schneidkopf *m* rotary cutter *(suction dredger)*
Schneidkopfbagger *m* suction-cutter dredger
Schneidlegierung *f* alloy for cutting tools, cutting alloy
Schneidmetall *n* cutting metal
Schneidmittel *n* cutting compound
Schneidmühle *f* impeller breaker, rotary [knife] cutter
Schneidnadel *f* recording (cutting) stylus, cutter
Schneidöl *n* cutting oil
Schneidplatte *f* cutting [tool] tip, tool bit
Schneidplattenfreifläche *f* tip flank
Schneidplattenschneide *f* tip cutting edge
Schneidrad *n* **zum Wälzstoßen** cutter for gear-shaping, gear-shaper cutter, gear-generating shaper cutter
S. zum Zahnradschaben rotary gear shaving cutter
Schneidrolle *f* cutting roll *(shear)*
Schneidschaft *m* body *(twist drill)*
Schneidschraube *f* self-tapping screw
Schneidschuh *m* bottom ring *(drop shaft sinking)*, cutting edge
Schneidstichel *m* recording (cutting) stylus
Schneidstoff *m* cutting material
Schneidwerk *n (ag)* mower
Schneidwerkstoff *m* cutting material (medium)
Schneidwerkzeug *n* cutting tool
Schneidwerkzeugplättchen *n* cutting-tool tip
Schneidwirkung *f* cutting action
Schneidwurftrommel *f (ag)* cutting cylinder
Schneidzahn *m (mach)* cutting tooth *(tool)*; point *(dipper)*
Schneidzeug *n* cutting tool
S. für Gewinde thread cutting tool
schnell fast *(steady movement)*; quick; high-speed *(operation)*; speedy; rapid
schnellabbindend quick-setting
Schnellablaß *m (nucl)* dump
Schnellablaßventil *n* quick-release valve
Schnellabschalterelais *n* high-speed arc suppression relay
Schnellabschaltung *f (el)* rapid interruption, quick breaking, instantaneous disconnection, high-speed circuit protection; scram
Schnellabstimmung *f* fast tuning
Schnellabtastmotor *m* motor for rapid scanning, *(Am)* slewing motor *(radar)*
Schnellabtastung *f (telev)* rapid scan[ning]
Schnellade *f (text)* fly-shuttle lathe
Schnelladestrom *m* fast-charging current
Schnelladung *f (el)* quick (rapid) charge
Schnellalterungsversuch *m* accelerated aging test
Schnellamt *n* no-delay office
Schnellamtsverkehr *m* demand working
Schnellanalyse *f* quick (rapid) analysis, quick (rapid) determination
Schnellanlauf *m* rapid starting
schnellansprechend fast-response; rapid-acting, quick-operating
Schnellarbeitsstahl *m* high-speed steel
kobalthaltiger S. cobalt high-speed steel

Schnellaufbohrmaschine *f* high-speed [motor-driven] drilling machine
schnellaufend high-speed
Schnelläufer *m* high-speed [printing] press
Schnelläuferkettenstuhl *m* high-speed tricot machine
Schnelläuferstuhl *m* high-speed loom
Schnellaufspindel *f* high-speed attachment (auxiliary spindle)
Schnellauslöser *m* instantaneous release
Schnellauslösung *f* instantaneous tripping
Schnellbearbeitung *f (mach)* high-speed machining
Schnellbeschickung *f (mach)* rapid loading
Schnellbestimmung *f* / **betriebliche** rapid industrial determination
Schnellbinderzement *m* quick-setting cement
Schnellbohreinrichtung *f* high-speed drilling attachment
Schnellbohrgerät *n* high-speed boring head
Schnellbohrverfahren *n* high-speed drilling method (operation)
Schnellboot *n* speedy (high-speed, fast) boat, speedboat
S. mit Raketenantrieb rocket-powered speedboat
Schnellbremse *f* quick-acting brake
Schnelldampfer *m* express (high-speed cargo) liner, fast steamer
Schnelldämpfer *m* rapid ager
Schnelldienst *m* no-delay service *(telephony)*
Schnelldreharbeit *f* high-speed lathe work
Schnelldrehen *n* high-speed turning, high-velocity turning
Schnelldrehmaschine *f* high-speed lathe
Schnelldrehmeißel *m* high-speed turning tool
Schnelldrucker *m* automatic (high-speed) printer *(computer)*
xerografischer S. high-speed xerographic printer
Schnelldruckwerk *n s.* Schnelldrucker
schnellen to whip-lash
Schnellentladung *f* rapid discharge
Schneller *m (text)* hank
Schnellerregung *f (el)* high-speed excitation
Schnellessigverfahren *n* quick vinegar process
Schnellfading *n* rapid fading *(radio)*
Schnellfernmeldetechnik *f* high-speed telecommunication engineering
Schnellfeuergeschütz *n* quick-firing gun, Q.F. gun
Schnellfeuerwaffe *f* rapid-fire weapon
Schnellfotometer *n* rapid photometer
Schnellfräseinrichtung *f* high-speed milling attachment
Schnellfräser *m* high-speed cutter
Schnellfräskopf *m s.* Schnellfräseinrichtung
Schnellfrequenzmotor *m* high-frequency motor, increased-frequency motor
Schnellgang *m* rapid travel
Schnellgerbung *f* accelerated tannage
Schnellgespräch *n* no-delay call *(telephony)*
Schnellgetriebe *n* high-speed gear
schnellhärtend *(plast)* fast-setting, quick-curing
Schnellhefter *m* folder
Schnellheizröhre *f* instant-heat valve
Schnellhobel *m (mine)* fast (rapid) plough
Schnellhobelmaschine *f* high-speed planer
Schnellhöhe *f* sag *(overhead line)*
Schnellhorizontiergerät *n* rapid levelling instrument
Schnelligkeit *f* rapidity, speed, fastness
schnellklebend quick-sticking
Schnellmotorschiff *n* express liner
Schnellnachrichtentechnik *f* high-speed telecommunication engineering
Schnellocher *m* gang punch
Schnellot *n* quick solder
Schnellpresse *f* high-speed press; *(print)* cylinder machine (press)
Schnellpreßmasse *f (plast)* fast-curing moulding compound
Schnellquerabschneider *m* high-speed guillotine
Schnellrechner *m* high-speed computer (computing machine)
Schnellregler *m* [automatic] high-speed regulator, [automatic] quick-acting regulator
Schnellrelais *n* quick-acting relay, high-speed relay
Schnellrücklauf *m* fast rewind *(magnetic tape)*; *(mach)* quick return
Schnellschalter *m* fast-action circuit-breaker, high-speed circuit-breaker
Schnellschleifen *n* high-speed grinding
Schnellschluß *m (nucl)* scram
Schnellschlußregler *m* safety governor
Schnellschlußschalter *m* scram button (switch) *(reactor)*
Schnellschlußstab *m* scram (safety, shut-off, emergency shut-off) rod *(reactor)*
Schnellschlußventil *n* quick-acting valve
Schnellschneidemaschine *f* high-speed cutter (guillotine)
Schnellschraubenschlitzmaschine *f* high-speed screw cutter
Schnellschreiber *m* high-speed recorder
Schnellschrupp[schleif]scheibe *f* high-speed roughing (snagging) wheel
Schnellschützen *m (text)* flying shuttle
Schnellschwingsieb *n* high-speed jigging screen
Schnellsegler *m* clipper, fast sailer
Schnellsenkrechtfräseinrichtung *f* high-speed vertical milling attachment
Schnellspaltfaktor ε *m* fast fission (multiplication) factor

Schnellspaltung *f* fast [neutron] fission
Schnellspaltungsfaktor ε *m* fast fission (multiplication) factor
Schnellspannfutter *n* quick-action chuck
Schnellspeicher *m* rapid (high-speed) memory, fast-access storage
Schnellstahl *m* high-speed steel
Schnellstahlmeißel *m* high-speed steel tool, HSS tool
Schnellstahlreibahle *f* high-speed steel reamer, HSS reamer
Schnellstartlampe *f* rapid-start [discharge] lamp
Schnellsteuerverstärker *m* air (booster) relay, pilot *(pneumatics)*
Schnellstopptaste *f* temporary stop key
Schnellstreckenauffahren *n (mine)* speedy advancing (tunnelling)
Schnellstreckenvortrieb *m (mine)* rapid drivage, speedy advancing
Schnelltelegraf *m* automatic (high-speed) telegraph
Schnelltriebwagen *m* fast rail car
dieselelektrischer S. fast Diesel-electric rail car
schnelltrocknend fast-drying, quick-drying
Schnelltrockner *m* flash drier
Schnelltrocknung *f* quick drying
Schnellübertrager *m* air (booster) relay, pilot *(pneumatics)*
Schnellumformen *n* **mit hoher Energie** high-energy-rate forming
schnellumlaufend *(mach)* high-speed
Schnellunterbrecher *m* ticker
Schnellverkehr *m* 1. no-delay service, express telephone service; 2. express traffic *(road)*
Schnellverkehrsamt *n* no-delay exchange *(telephony)*
Schnellverkehrsflugzeug *n* high-speed commercial airplane
Schnellverkehrsleitung *f* no-delay circuit
Schnellverkehrsschrank *m* toll switchboard
Schnellverkehrsstraße *f* express road (way)
kreuzungsfreie S. freeway
Schnellverschluß *m* quick-seal coupling
Schnellverstärker *m* air (booster) relay, *(Am)* pilot *(pneumatics)*
Schnellverstellung *f (mach)* rapid adjustment
Schnellvorlauf *m* fast forward *(magnetic tape)*
Schnellvorlauftaste *f* key for accelerated forward speed
Schnellwechselfutter *n* quick-change chuck, quick-acting chuck
Schnellwechselgetriebe *n* quick-change-gear mechanism
Schnellwechselkegel *m* quick-release taper *(tool)*
Schnellwiedereinschaltung *f (el)* rapid reclosing
automatische S. automatic [rapid] reclosing
Schnellzeitzünder *m* short-delay electric detonator
Schnellzerspanen *n* high-speed machining, machining at high speeds, high-speed stock removal
Schnellzerspanung *f s.* Schnellzerspanen
Schnellzuführung *f* rapid approach *(tool)*; high-speed feed
S. bis Fräsbeginn fast run up to the cutting position
Schnellzug *m* express train
Schnellzugriff *m (dat)* immediate access
Schnellzugriffsspeicher *m* quick-access store
Schnellzugzuschlag *m* supplement (surcharge) for express train
Schnippmaschine *f (text)* jute (root hackling) machine, snipper
Schnitt *m* 1. cut, slit, incision, notch; cutting; 2. cutting die (punch), die set; 3. intersection *(two lines)*; 4. sectional drawing; section; 5. *(ac)* dubbing; 6. face *(letter)*; 7. *(ag)* harvesting, mowing; edge *(book)*
gebrochener S. broken section *(technical drawing)*
Goldener S. golden section
kräftiger S. heavy cut *(machining)*
Ritterscher S. Ritter method of sections
sauberer S. accurate cut
schälender S. *(mach)* shearing cut
scherender S. shearing cut
schräger S. angular cut
Schnittbalken *m (ag)* cutter bar
Schnittbau *m* die making
Schnittbedingungen *fpl* cutting conditions
Schnittbildmeßsucher *m* split-image range finder
Schnittbreite *f* width of cut
Schnittbrenner *m* slit (batwing) burner
Schnittdarstellung *f* sectioning; sectional view (drawing)
Schnittdruck *m* 1. cutting pressure; 2. gravity feed *(cutting pressure by weight of the saw and the frame)*
Schnitte *mpl* buttock lines *(of a vessel)*
S. und Stanzen *fpl* dies and punches
Schnittebene *f* cutting (section) plane
Schnitter *m* mower, reaper
Schnittergebnis *n* output in sawing
Schnittfläche *f* cut surface *(technical drawing)*; shoulder of the cut *(metal cutting)*; area of cut
Schnittflächenschraffur *f* section lining *(technical drawing)*
Schnittfrequenz *f* cross-over frequency
Schnittfuge *f* flat joint *(brickwork)*; [saw] kerf, notch
Schnittfugenbreite *f* width of kerf *(sawing)*
Schnittgeräte *npl (ag)* pruning tools
Schnittgeschwindigkeit *f* cutting speed

S. in Fuß/min cutting ft. per min.
Schnittgeschwindigkeitsanzeiger *m* cutting speed indicator
Schnittgeschwindigkeitstabelle *f* cutting speed chart
Schnittholz *n* saw[n] wood, saw[n] timber
S. in Bündeln packaged timber
gehobeltes S. planed sawn wood
unbehobeltes S. rough timber, *(Am)* rough lumber
Schnittholzhäcksler *m (ag)* brushwood flail cutter
Schnittiefe *f* cutting depth, depth of cut
S. je Zahn cut per tooth
schnittig streamlined
Schnittkraft *f (mach)* cutting force; internal force *(statics)*
spezifische S. cutting pressure
Schnittkraftkomponente *f* cutting component, component of the cutting force
tangentiale S. tangential cutting force component
Schnittlänge *f (mach)* length of cut
Schnittmeister[in *f*] *m* cutter
Schnittplatte *f* die plate
Schnittpunkt *m* intersection point, point of intersection; junction *(wires)*
Schnittreife *f* binder ripeness
Schnittrichtung *f (mach)* direction of cutting, hand of [the] cut
S. des Spiralbohrers rotation of cutting of the twist drill
S. einer Reibahle hand of a reamer cut
Schnitt-, Stanz- und Ziehwerkzeuge *npl* press tools
Schnittstelle *f* cutting site
Schnittstellung *f* cutting position
Schnittverfahren *n* **/ Rittersches** Ritter method, method of sections
Schnittvergoldung *f (print)* edge gilding
Schnittverhältnisse *npl* cutting conditions
Schnittverlust *m* waste *(sawing)*, *(Am)* stumpage
Schnittware *f* yardage goods *(textiles)*
Schnittweite *f* focal intercept
Schnittwelle *f* section wave
Schnittwerkzeug *n* die
Schnittwinkel *m* 1. cutting angle; 2. intersecting angle
Schnittzeichnung *f* sectional drawing
Schnittzeit *f* cutting time
Schnitzel *npl* chips, slices; pulp
Schnitzelfülldichte *f (wood)* chip packing
Schnitzelmaschine *f* slicer *(sugar industry)*
Schnitzelmasse *f* chopped-filled plastic
Schnitzelrutschrinne *f (wood)* chip chute
Schnitzelverfahren *n* chip process *(man-made fibre)*
Schnüffelventil *n* snift[ing] valve
Schnur *f* 1. twine, cord; string; 2. *(build)* line; 3. *(el)* [flexible] cord, *(sl)* flex, flexible lead, lace
einadrige S. single-conductor cord
zweiadrige S. double-conductor cord
Schnurbaum *m (ag)* cordon
senkrechter S. upright cordon
waagerechter S. horizontal cordon
Schnürboden *m* mould loft [floor], drawing loft *(of a shipyard)*
Schnürbodenarbeit *f* laying off, lofting
Schnürbodenaufmaße *npl* mould-loft offsets
Schnürbrettloch *n* comber board hole
Schnurführung *f* cord guide
Schnurkranz *m* cord rim
Schnürloch *n* lacing hole
Schnurlot *n* plumb line
Schnürnadel *f (text)* bodkin
Schnurpaar *n* pair of cords
Schnurpackung *f* cord packing
Schnurplatine *f* knot (neck) cord
Schnurschalter *m* pendant switch
S. mit Druckknopf pressel (pear) switch
Schnurschaltung *f* cord circuit
Schnürsenkel *m* shoe lace (string)
Schnurspanner *m* cord tightener
Schnürstiefel *mpl* lace boots
Schnürung *f (text)* harness mounting; tie-up
bandförmige S. repeating tie-up in bands
gerade S. straight tie-up
mehrfache S. compound tie-up
Schnurverbinder *m (el)* flexible lead connector
Schnurverstärker *m* cord circuit repeater
Schober *m* rick, stack
Schoberbau *m* ricking, stacking
Schobersetzvorrichtung *f* hay-stacker
Schollenbrecher *m (ag)* clod crusher
Schöndruck *m* first printing (forme print)
Schöndruckseite *f* recto side
schönen *(leath)* to flame
Schoner *m* schooner
Schonerbark *f* main topsail schooner, barkentine, barkentyne
Schonerbrigg *f* brigantine *(two-masted ship)*
Schonersegel *n* boom fore-sail of a schooner
Schonersegelbaum *m* fore-boom of a schooner
Schonganggetriebe *n* overdrive
Schöpfbagger *m* dipper dredger
Schöpfblech *n* baffle *(papermaking)*
Schöpfbütte *f* dipping vat
Schöpfeimer *m* bailer
Schöpffinger *m* oil catcher (dipper)
Schöpfkelle *f* scoop, ladle
Schöpfnetz *n* bailing dip-net *(fishery)*
Schöpfpapier *n* mould-made paper
Schöpfprobe *f (met)* ladle (pit) sample
Schöpfprobenuntersuchung *f* ladle analysis
Schöpfrad *n* cup-feed disk

Schöpfradschmierung *f* scoop-wheel lubrication
Schöpfrahmen *m* board mould *(papermaking)*
S. für Bütten hand mould, mould frame, deckle
Schore *f (mine)* ore pass
Schorf *m (met)* scab
Schörl *m* schorl *(black tourmaline)*
Schornstein *m* chimney [stack], stack, flue; funnel *(boat, engine)*
S. mit Etagenfutter chimney with stepped lining
S. mit Funkenfängersieb funnel with spark arrester
S. mit Schamottefutter chimney with fire clay lining
blinder S. *(ship)* dummy funnel
doppelwandiger S. cavity (double-walled) chimney
einziehbarer S. *(ship)* lowering funnel
freistehender S. independent chimney
niedriger S. short chimney
umlegbarer S. hinged funnel
Schornsteinabsatz *m* chimney step
Schornsteinabzeichen *n* funnel mark
Schornsteinaufsatz *m* chimney pot (cap, top), cowl
Schornsteinbrand *m* chimney fire
Schornsteinfeger *m* chimney sweeper
Schornsteinfuchs *m* chimney flue
Schornsteinfuß *m* chimney foot
Schornsteinfutter *n* chimney lining
Schornsteingruppe *f* chimney stack
Schornsteinhaube *f* chimney pot (cap, top), cowl
Schornsteinkappe *f* chimney pot (cap, top), cowl
Schornsteinkragen *m (ship)* umbrella
Schornsteinkrone *f* chimney pot (cap, top), cowl
Schornsteinleistung *f* chimney duty
Schornsteinluke *f* chimney (funnel) opening
Schornsteinmantel *m* chimney casing; *(ship)* outer funnel
Schornsteinmauerwerk *n* chimney brickwork
Schornsteinregenkragen *m (ship)* umbrella
Schornsteinsäule *f* chimney shaft
Schornsteinschacht *m* chimney (funnel) shaft
Schornsteinschuß *m* belt (ring) of an iron chimney
Schornsteinsockel *m* chimney base (socket)
Schornsteinsprengung *f* blowing up of the chimney
Schornsteinstag *n (ship)* funnel stay
Schornsteinüberhitzer *m* chimney superheater
Schornsteinumbau *m (ship)* fiddley
Schornsteinzug *m* chimney draught (draft); chimney flue
Schott *n (ship)* bulkhead; wainscot *(wooden)*
leichtes S. joiner bulkhead
versetztes S. stepped bulkhead
Schottdurchführung *f* bulkhead spigot
Schottendeck *n* bulkhead deck
Schotteneinteilung *f* subdivision *(of a vessel)*
Schottenladelinie *f* subdivision load line
Schottenmanöver *n* collision drill
Schottenmuster *n* plaid
Schottenrechnung *f* calculation of floodable length
Schottentiefgang *m* subdivision draught
Schotter *m* crushed rock, broken stone; *(ry)* ballast
Schotterbett *n (ry)* ballast bed
Schotterdecke *f* macadam (crushed rock) surfacing *(road)*
Schotterung *f (ry)* ballasting
Schotterwagen *m (ry)* ballasting wagon
Schottkurve *f* curve of admissible lengths, curve of bulkhead spacing, flooding (floodable lengths) curve
Schottky-Defekt *m* Schottky defect
Schottquerkurven *fpl (ship)* cross-curve diagram
Schottrechnung *f (ship)* flooding calculation
Schottschiebetür *f* sliding bulkhead door
Schottschleuse *f* sluice gate
Schottsüll *n* bulkhead coaming
Schottür *f* bulkhead door
Schottverschluß *m* bulkhead door
Schottversteifung *f* bulkhead stiffening; bulkhead stiffener
schraffieren to shade, to hachure, to hatch, to section *(technical drawing)*
kreuzweise s. to cross-hatch
Schraffierung *f* hachure, hatching, sectioning *(technical drawing)*
Schraffierungslinie *f* hachure line
Schraffur *f* hachure, hatching, shading, sectioning *(technical drawing)*
S. von Schnittflächen section lining
einfache S. plain sectioning *(technical drawing)*
Schraffurlinie *f* shade line; section line *(technical drawing)*
schräg oblique; inclined, slanting; sloping *(site)*; skew; tilted, canted *(body)*; raked *(mast, funnel)*; battered *(wall, terrace, bank)*
Schräganode *f* oblique anode
Schrägansicht *f* oblique view
Schrägantrieb *m* inclined drive
Schrägaufhauen *n (mine)* inclined raise
Schrägaufnahme *f* oblique photograph
Schrägaufzug *m* inclined elevator (lift), skip hoist (bridge) *(blast furnace)*
Schrägbandausleser *m (ag)* inclined conveyor-type grader
Schrägbau *m (mine)* obliquely stepped face
Schrägbeleuchtung *f* oblique illumination
schrägbohren *(mach)* to drill holes at an angle

Schrägbohren *n (mine)* controlled (directional) drilling
Schrägdach *n* pitched roof
Schräge *f* slope, obliquity; batter, cant, bevel, slant
Schrägeinstellung *f* angular setting
Schrage-Motor *m* Schrage motor *(shunt-characteristic polyphase commutator motor with double set of brushes)*
Schrägfach *n (text)* V-shed
Schrägförderer *m* inclined conveyor
Schrägförderung *f* inclined conveyance
Schräghaltung *f* angular positioning
Schräghobeln *n* angular planing
Schrägkammerofen *m* inclined-chamber oven
Schrägkante *f* chamfer; bevel
Schrägkippen *n* angular tilt
Schrägkugellager *n* angular contact ball bearing
Schräglage *f* inclination; angular position (location); *(aero)* bank
Schräglageanzeiger *m (aero)* bank indicator
schrägliegend inclined, angular, in angular position (location)
Schrägmaß *n* bevel rule
Schrägnut *f* helical armature slot, *(Am)* helix slot
Schrägpfahl *m* batter (raker, raking) pile
Schrägpolygonroststab *m* inclined polygonal grate bar
Schrägrad *n* **mit Schraubennuten** helical gear with helical splines
Schrägregelung *f* skew control
Schrägrevolverkopf *m* tilted turret
Schrägrohrmanometer *n* tilting (inclined-tube) manometer
Schrägrollengang *m (met)* skew roller table
Schrägrost *m* sloping (inclined) grate
Schrägrostfeuerung *f* furnace with inclined grate
Schrägrostinnenfeuerung *f* furnace with internal inclined grate
S. mit Rückbrennung (Gegenfeuer) furnace with inclined grate with back-draught arranged in the interior of the boiler, Tenbrink furnace
Schrägroststab *m* inclined grate bar
Schrägrostunterfeuerung *f* furnace with inclined grate underneath the boiler
Schrägrostvorfeuerung *f* furnace with external inclined grate
Schrägrostvorstellplatte *f* dead plate for inclined grate
Schrägschacht *m* incline[d] shaft
Schrägschleifen *n* grinding angular work; producing angular work by grinding
Schrägschneidemaschine *f* angle-cutting machine
Schrägschneiden *n* angular cutting; mitring
Schrägschnitt *m* bevel cut; oblique section *(drawing)*
Schrägschulterfelge *f* rim with taper bead-seats
Schrägschuß *m (text)* skewed filling
Schrägsieb *n* gravity (inclined) screen
Schrägspeisung *f* diagonal feed
Schrägspritzkopf *m (plast)* oblique extruder head, angle extrusion head
Schrägstab *m* inclined bar *(statics)*; *(build)* strut, diagonal brace (member)
schrägstellbar inclinable; tilting, tiltable; angularly adjustable
schrägstellen to set at an angle; to tilt; to adjust angularly
Schrägstellung *f* inclination; angular adjustment (tilt, positioning, setting, position); tilt
S. der Scheibe inclination of the disk *(turbine)*
Schrägstellungsbereich *m* angular range
Schrägstirnrad *n* helical gear
Schrägstrahler *m* angle reflector (lighting fitting)
Schrägtrommelmischer *m* tilted drum mixer
Schrägtubus *m* inclined tube
monokularer S. monocular inclined tube
Schrägungsfaktor *m (el)* skew factor *(winding)*
Schrägungswinkel β *m* angle of helix, helix angle *(acute angle between tooth and axis of gear)*
Schrägungswinkelmeßgerät *n* helix-angle tester
Schrägverband *m* raking bond *(brickwork)*
schrägverstellbar angularly adjustable, swivelling, tiltable
Schrägverstellbarkeit *f* angular adjustment (adjustability)
schrägverzahnt [single-]helical
Schrägverzahnung *f* helical (angular) teeth
Schrägwalzwerk *n* tube-piercing mill, skew-rolling mill
Schrägwicklung *f (el)* oblique (helical) winding
Schrägwinkel *m* oblique [rake] angle
Schrägzahn *m* helical tooth *(helical gear)*; skew tooth *(bevel gear)*
Schrägzahnkegelrad *n* skew bevel gear
Schrägzahnrad *n* helical gear
Schrägzahnstirnrad *n* parallel helical gear
Schrägzustellung *f* angular feed
Schram *m (mine)* cut
senkrechter S. shear cut
unterer S. bottom cut
Schrämarm *m* [cutting] jib
S. für das Naßschrämen whale jib
S. mit Schrämpilz mushroom jib
abgewinkelter S. curved jib
doppelt abgewinkelter S. cranked (swan-necked) jib
gebogener S. curved jib
gekröpfter S. cranked (swan-necked) jib
Schrämausleger *m* cutter bar, [cutting] jib
Schramdicke *f* cut height

schrämen *(mine)* to cut
zu Tal s. to cut downhill
Schrämen *n* **im Mittelteil des Flözes** middle-cutting
S. im oberen Teil des Flözes over-cutting
S. im unteren Teil des Flözes under-cutting
Schrämfahrt *f* cutting run
Schrämförderer *m* cutter conveyor
Schrämgasse *f* cutter track
Schrämhobel *m* cutter plough
Schrämkette *f* coal cutter chain
Schrämkettenantrieb *m* cutting chain gear
Schrämklein *n* cuttings, gummings
Schrämkleinräumblech *n* gummer bar
Schrämkleinräumer *m* gummer, gumthrower
Schrämkleinräumschnecke *f* scroll (spiral) gummer
Schrämkleinschleuder *f* gum flinger
Schrämkopf *m* cutting chain gear, gear head, jib turret, jib-head
Schrämladekette *f* loading cutter chain
Schrämlademaschine *f* cutter loader
Schrämlader *m* cutter loader
Schrämleitung *f* coal-cutter cable
Schrämmaschine *f* coal cutter, [coal] cutting machine
S. für kurzen Abbaustoß shortwall coal cutter
S. für Strebbau longwall coal cutter
S. für Streckenvortrieb arcwall coal cutter, arcwaller
S. mit dreifachem Schrämarm triplex jib coal cutter
Schrämmeißel *m* [cutter] pick
Schrämmeißelanordnung *f* **in der Kette** lacing pattern
symmetrische S. in der Schrämkette balanced lacing pattern
unsymmetrische S. in der Schrämkette unbalanced lacing pattern
Schrämpilz *m* mushroom
Schrämrahmen *m* frame jib
Schrämscheibe *f* cutting disk
Schrämschicht *f* cutting shift
Schrämschrapper *m* cutter-scraper
Schrämstange *f* cutter bar
Schrämtiefe *f* bearing-in, cutting depth, depth of cut
Schrämtiefstes *n* back of the cut
Schrämturm *m* cutting turret
Schrämwalze *f* cutting (shearer) drum
Schrämwalzenlader *f* shearer-loader
Schrank *m* cabinet, closet, locker, cupboard, wardrobe, dresser; *(telec)* board
schnurloser S. cordless board
Schranke *f* *(ry)* barrier, gates; *(math)* bound; barrier
obere S. *(math)* upper bound, majorant
untere S. *(math)* lower bound, minorant
Schränkeisen *n* saw-set
schränken to set *(saw teeth)*
abwechselnd rechts und links s. to set one tooth to the right and the next tooth to the left
Schranken *fpl* / **optische** optical limits
Schrankenwärter *m* [railway] crossing keeper, gate-keeper
Schränkmaschine *f* saw setting machine
Schranknagel *m* beamer's lease pin
Schränkschicht *f* diagonal course *(brickwork)*
Schrankstromkreis *m* cord circuit
Schränkung *f* set *(saw teeth)*
S. des Reglerkurbeltriebs horizontal displacement of the joints
Schränkungsmaß *n* horizontal displacement *(joints)*
Schränkverband *m* diagonal bond *(brickwork)*
Schrankwand *f* storage wall
Schränkzange *f* **für Sägezähne** saw set pliers
Schrapnell *n* shrapnel
Schrappbühne *f* *(mine)* slusher ramp (slide)
Schrapper *m* scraper, slusher
Schrapperbahn *f* scraper track
Schrapperbühne *f* slusher ramp (slide)
Schrapperfeld *n* scraper track
Schrapperhaspel *f* slusher hoist
Schrappkasten *m* scraper box *(coal)*, slusher bucket *(dirt)*
Schrapplader *m* scraper loader
Schrappversatz *m* mechanical stowing with scraper-buckets, slusher packing
Schraube *f* bolt *(with nut)*; screw *(wood etc)*; *(ship)* propeller, screw
abgesetzte S. necked-down bolt
doppelgängige S. double-thread[ed] screw
eingängige S. single-thread screw
eingelassene S. countersunk screw
eingepaßte S. fitted (reamed, template) bolt
flachgängige S. square-threaded screw
hochfest vorgespannte S. *s.* hochfeste S.
hochfeste S. high-strength [friction grip] bolt, high-tensile [prestressed] bolt, friction grip bolt
selbstschneidende S. self-tapping screw
versenkte S. countersunk screw
zweigängige S. double-threaded screw
schrauben to screw; to tighten the screw (nut)
Schrauben *fpl* / **gegenläufige** *(ship)* counter-turning screws
Schraubenachse *f* screw axis
Schraubenanschluß *m* *(el)* screw terminal
Schraubenantenne *f* helical aerial
Schraubenanziehmaschine *f* nut runner
schraubenartig helicoidal
Schraubenautomat *m* automatic screw machine
Schraubenbelastungsgrad *m* *(ship)* propeller load coefficient

Schraubenbewegung *f* helical (helicoidal, screw) motion
Schraubenblattfreigang *m (ship)* blade clearance
Schraubenbolzen *m* [screw] bolt
Schraubenbrunnen *m (ship)* propeller well, propeller (screw) aperture, aperture; *(Am)* wheel port
Schraubendampfer *m* screw steamer
Schraubendruckfeder *f* helical coil compression spring
Schraubendurchmesser *m (ship)* propeller diameter
Schraubenfeder *f* helical (wire coil) spring
zylindrische S. helical spring
Schraubenfläche *f* helicoid, helical (screw) surface; propeller area
Schraubenflächenverhältnis *n* propeller area ratio
Schraubenflügel *m* propeller (screw) blade
schraubenförmig helical, screw-shaped
Schraubenfutter *n* screw chuck
Schraubenganglenkung *f (aut)* cam-and-peg steering gear
Schraubengewinde *n* screw-thread
Schraubengewindelinie *f* helical line
Schraubenkapselwerk *n* **mit axialer Zuführung** axial pump
Schraubenkopf *m* bolt head; screw head
Schraubenkopffeile *f* screw-head file
Schraubenkopfschlitzmaschine *f* screw-head slotting machine
Schraubenkopfstauchen *n* bolt heading
Schraubenkurve *f* cochleoid
Schraubenlenkung *f (aut)* screw-and-nut steering gear
Schraubenlinie *f* helix, helical curve
Schraubenloch *n* bolt hole
Schraubenlochkreis *m* bolt[-hole] circle, pitch circle
Schraubenlüfter *m* axial fan
Schraubenmikrometer *n* screw micrometer
Schraubenmutter *f* nut; lock nut
Schrauben-, Mutter- und Rohrgewindeschneidmaschine *f* bolt, nut, and tube threading machine
Schraubennabe *f* propeller hub (boss)
Schraubennuß *f (ship)* stern boss
Schraubennut *f* helical groove
Schraubennutfräsen *n* spiral milling
Schraubenöffnung *f s.* Schraubenbrunnen
Schraubenpaar *n* screw pair *(kinematics)*
Schraubenpropeller *m* [screw] propeller
Schraubenpumpe *f* screw pump
S. mit zwei Spindeln twin-screw type of pump
Schraubenrad *n* helical gear
hyperbolisches S. hyperbolical gear
Schraubenräder *npl* crossed helical gears
Schraubenradgaszähler *m* rotary gas meter
Schraubenrahmen *m* propeller frame
Schraubenringspanner *m* coiled piston ring, helical packing ring
Schraubenrohling *m* blank bolt; screw blank
Schraubenrohrüberhitzer *m* helical tube superheater, spiral-tube superheater
Schraubenschiff *n* screw ship
Schraubenschlepper *m* screw tug
Schraubenschlitz *m* screw slot
Schraubenschlitzautomat *m* automatic screw-head slotter
Schraubenschlitzeinrichtung *f* screw-slotting attachment
Schraubenschlitzfräser *m* screw-slotting cutter, slotting saw
Schraubenschlitzmaschine *f* screw-slotting machine
Schraubenschlitzsäge *f* screw-slotting saw
Schraubenschlüssel *m* wrench; spanner; box spanner
verstellbarer S. universal (adjustable open-end) wrench
Schraubenschlüsselfräser *m* **zum Einfräsen der Maulweite** wrench cutter
Schraubenspindel *f* screw jack
Schraubenspindelsteuervorrichtung *f* screw steering gear
Schraubensteven *m* propeller frame (post)
Schraubenstrecke *f (text)* screw gill drawing frame
Schraubenstrom *m* propeller (screw) wake, backwash *(of a vessel)*
Schraubenverbindung *f* bolting, bolted joint; screwed connection; nipple *(pipe)*
gleitfeste S. high-tensile bolted structural joint; high-tensile bolting, friction-grip bolting
Schraubenversetzung *f* screw dislocation
Schraubenwelle *f* propeller (screw) shaft
Schraubenwellenbezug *m (ship)* screw-shaft liner
Schraubenwellentunnel *m (ship)* screw-shaft tunnel
Schraubenwendeleiste *f* spiral retarder
Schraubenwinde *f* [screw] jack, screw winch
Schraubenwindung *f* helix
Schraubenwirkungsgrad *m* propeller efficiency
Schraubenzieher *m* screw driver
Schraubfassung *f* screwed lamp-holder
Schraubfittings *npl (mpl)* screwed fittings
Schraubgetriebe *n* gears operating on crossed axes
Schraubkappe *f (el)* fuse holder, screw cap
Schraubkeilklemmung *f* screw release wedge clamping
Schraubkern *m (el)* threaded R. F. tuning-type dust core
Schraubklemme *f* screw terminal

Schraubkontakt *m* screwed contact
Schraubkopf *m* fuse-carrier
Schraublehre *f* micrometer
Schraubmuffe *f* screwed socket (coupler)
Schraubokular *n* screw-in eyepiece
Schraubräder *npl* gears on nonparallel, non-intersecting axes
Schraubsicherung *f (mach, aut)* screw retainer; *(el)* screw-plug cartridge fuse, screw-type fuse
Schraubsockel *m* screw base
Schraubspindel *f* screw
Schraubstecker *m* screwed plug, screw-plug box
Schraubstempel *m (mine)* screw-jack prop
Schraubstock *m* vice
  drehbarer S. swivel vice
  pneumatisch betätigter S. air vice
Schraubstockbacken *m* vice jaw
Schraubstockspannweite *f* width of vice jaws
Schraubstöpselsicherung *f* screw-plug cartridge fuse
Schraubstutzen *m (aut)* screw neck
Schraubtrieb *m* crossed helical gears
Schraubverbindung *f* 1. *(mach)* screw fastening (joint, connection); bolted connection (joint, union) *(pipe)*; 2. *(el)* screw connector
Schraubverschluß *m* screw plug
Schraubzwinge *f* C-clamp, screw clamp
Schreckplatte *f* chill *(foundry)*
Schreckschicht *f* chilled layer *(foundry)*
Schrecksekunde *f (aut)* reaction time; *(cont)* response latency
Schreibarm *m* straight arm *(ellipsograph)*
Schreibautomat *m* **/ lochstreifengesteuerter** tape-controlled typing machine
Schreibblock *m* pad
Schreibelektrode *f* recording electrode
Schreibempfänger *m* recording Morse receiver
schreiben to write; to type *(typewriter)*; to read in, to record; to plot *(graph)*
Schreiber *m* [pen] recorder; recording instrument; registering meter
  automatischer S. auto-plotter
Schreiberaufzeichnung *f* recorder presentation
Schreibfehler *m* typist's error *(typewriter)*; writing error *(handwriting)*
Schreibfeinzeiger *m* recording precision indicator
Schreibgerät *n* pen (graph) recorder, recorder, recording instrument
Schreibgeschwindigkeit *f* writing (recording) speed
Schreibimpuls *m* write pulse
Schreibkatode *f (Am)* writing gun
Schreibkopf *m* recording head
  ringförmiger S. ring-type recording head
Schreiblocher *m* typing [re]perforator
Schreibmaschine *f* typewriter
  elektrische S. electric typewriter
Schreibmaschinenband *n* typewriter ribbon
Schreibmaschinenlack *m* typewriter finish
Schreibmaschinenpapier *n* typewriter paper
Schreibmaschinenschrift *f* typewriter face
Schreibmaschinentastatur *f* typewriter keyboard
Schreibmaschinenteil *n* typewriter component
Schreibmaschinenwalze *f* typewriter roller
Schreibpapier *n* writing paper
Schreibprojektor *m* writing projector
Schreibsatz *m* typewriter composition
Schreibschrift *f* script
Schreibspitze *f* tip of the stylus
Schreibstange *f* writing bar (edge)
Schreibsteuerungsausgang *m* print selection common exit *(punched card machinery)*
Schreibsteuerungseingangsbuchse *f* print selection entry hub *(punched card machinery)*
Schreibstift *m* stylus
Schreibstiftbewegung *f* stylus trajectory
Schreibstiftführung *f* pencil guide
Schreibstrahl *m* sweeping beam
Schreibstrommesser *m* recording ammeter
Schreibtelegraf *m* recording telegraph
Schreibtrommel *f* recording drum
Schreibverstärker *m* write amplifier
Schreibwaren *fpl* stationery
Schreibweise *f* notation *(of data)*
Schreibwerk *n* recording mechanism (system) *(instrument)*
Schreiner *m* joiner
Schreinerei *f* joinery, joiner's shop
Schreiner-Kalander *m* Schreiner's calender
Schreitbagger *m* walking dragline (excavator)
Schreitkufenbagger *m s.* Schreitbagger
Schrenz *m* screens
Schrenzpackpapier *n* screens
Schrenzpappe *f* common board, plain grey cardboard
  kaschierte S. filled chip-board
Schrieb *m* plot; record
Schrift *f* [text] type
  abgequetschte S. worn types
  erhabene S. raised letters
  lichte S. open types
Schriftabnutzung *f* type wear
Schriftbild *n* face
Schriftfamilie *f* type family
Schriftgießen *n* type casting
Schriftgießerei *f* type foundry
Schriftgrad *m* type size
Schrifthobel *m* type planer
schrifthoch type-high
Schriftkasten *m* letter (type) case
Schriftkegel *m* [type] body, body (point) size
Schriftkunst *f* typography
schriftlich written; in writing
Schriftmetall *n* type (printer's) metal

Schriftmusterbuch *n* type [founder's] specimen book
Schriftprobe *f* line pattern *(character recognition)*
Schriftsatz *m* type matter
Schriftschablone *f* lettering guide
Schriftsetzer *m* type-setter, typographer
Schriftstückablage *f* document file
automatische S. automatic document file
Schrift- und Zahlzeichenzuordner *m* character translator *(character recognition)*
Schriftzeichen *n* letter, character
verschlüsseltes S. coded character
Schriftzeichenleser *m* / **automatischer** character reader
Schritt *m* 1. step; 2. pitch *(winding)*; *(telec)* elementary interval
Schrittfunktion *f* step function
Schrittgeschwindigkeit *f* telegraph speed
Schrittkode *m* equal-length code
Schrittlänge *f* duration of signal
Schrittmacher *m* pacer, pace-maker
Schrittmacherofen *m* rocker-bar type furnace
Schrittmotor *m (cont)* stepping motor
Schrittregelung *f (cont)* step-by-step control
Schrittregler *m (cont)* step-acting controller
Schrittschalter *m* stepping (step-by-step) switch
Schrittschaltoperation *f* stepping operation
Schrittschaltrad *n* step wheel
Schrittschaltrelais *n* stepping relay
Schrittschaltsystem *n* step-by-step system
Schrittschalttelegraf *m* step-by-step telegraph
Schrittschaltwähler *m* step-by-step switch
Schrittschaltwählsystem *n* step-by-step automatic system
Schrittschaltwerk *n (cont)* stepping relay; step-by-step switch
Schrittspannung *f* surface voltage gradient
Schrittumsetzer *m (telec)* code converter
Schrittwahl *f* step-by-step selection *(telephony)*
Schrittwähler *m* step-by-step switch *(telephony)*
Schrittwählersystem *n* step-by-step dial system *(telephony)*
schrittweise stepwise
Schrittweite *f* pitch
Schröckingerit *m* schroeckingerite *(uranium mineral)*
Schrödinger-Gleichung *f* Schroedinger's wave equation
Schrot *m* 1. metal (steel) shot; 2. shot
Schrotbohren *n* calyx-drilling, shot-boring, shot-drilling
Schrotbohrer *m* adamantine drill
Schrotbohrverfahren *n* shot-boring, shot-drilling
Schroteffekt *m* shot effect
schroten to shred, to kibble
Schrotflinte *f* shot-gun
S. für Vögel fowling piece
Schrothaken *m* heel tool
Schrotkrone *f* shot bit
Schrotkronenbohren *n* shot boring (drilling)
Schrotkugel *f* shot pellet
Schrotmühle *f* grinding (kibbling) mill
Schrotpressen *n (plast)* shot [bag] moulding
Schrotrauschen *n* shot noise
Schrotsäge *f* trim (pit, cross-cut) saw
Schrott *m* scrap
leichter S. fine scrap
paketierter S. faggot[s]
stückiger S. solid scrap
unsortierter S. unassorted scrap
Schrottbühne *f* scrapping floor
Schrottkohlungsverfahren *n* scrap-carbon process
Schrottlagerplatz *m* scrap stock yard
Schrottofen *m* bushelling furnace *(puddling)*
Schrottpaketieren *n* scrap faggoting
Schrottpaketierpresse *f* scrap baling (briquetting) press
Schrottplatz *m* scrap yard
Schrott-Roheisen-Verfahren *n* pig-and-scrap process
Schrottschere *f* scrap shear
Schrottschmelze *f* scrap heat
Schrottschmelzverfahren *n* allscrap process
Schrottverhüttung *f* scrap smelting
Schrottwert *m* residual (recovery) value
Schrottzuschlag *m* admixture of scraps
Schrotzimmerung *f* cribbing
Schrumpfband *n* shrunk ring
schrumpfecht shrink-resistant
schrumpfen to shrink; to contract
schrumpffest shrink-resistant
Schrumpfklammer *f* shrunk dowel
Schrumpflack *m* wrinkle finish
Schrumpflunker *m* shrinkage piping
Schrumpfmaß *n* 1. shrink rule; 2. shrinkage
Schrumpfmaßstab *m* shrink rule
Schrumpfpassung *f* expansion fitting
Schrumpfpreßpassung *f* shrink [press] fit
Schrumpfring *m* shrink ring
Schrumpfriß *m* casting (shrinkage, contraction, cooling) crack
Schrumpfspannung *f* stress due to shrinking, cooling stress
Schrumpfung *f* shrinking; shrinkage, contraction
schrumpfungsfrei shrink-resistant
Schrumpfungsprüfung *f* shrinking test
Schrumpfverbindung *f* shrink joint (fit)
Schrumpfzugabe *f* contracting allowance *(casting)*
Schrupparbeit *f (mach)* roughing (rough-machining) operation
Schruppdrehmaschine *f* roughing lathe
Schruppdrehmeißel *m* roughing lathe tool, rough-turning tool
Schruppdurchgang *m* roughing pass
schruppen to rough[-machine]

Schruppfräsen *n* rough-milling
Schruppfräser *m* roughing cutter
Schruppfrässpindel *f* rough-milling spindle
Schrupphobel *m* jack plane
Schrupphobeln *n* rough planing *(planer)*; rough shaping *(shaper)*
Schruppmeißel *m* roughing tool
Schruppräumwerkzeug *n* roughing broach
Schruppreibahle *f* roughing reamer
  S. für kegelige Bohrungen roughing taper reamer
schruppschleifen to rough-grind
Schruppschnitt *m* roughing cut
Schruppspan *m* roughing cut
Schruppspantiefe *f* roughing depth
Schruppteil *m* roughing end *(broach)*
Schruppwerkzeug *n* roughing tool
Schruppzahn *m* roughing tooth
Schub *m* thrust; shear
  S. des Raketentriebwerkes rocket power
  kreisförmiger S. curved cam surface effecting circular motion
  spezifischer S. specific thrust (impulse)
  statischer S. static thrust *(propeller)*
Schubabschaltung *f (rock)* thrust cut-off
Schubantrieb *m (cont)* linear actuator
Schubbeanspruchung *f* shear stress
Schubbeiwert *m* thrust [deduction] coefficient *(propeller)*
Schubbelastung *f* thrust load
Schubbeschleunigung *f* shear acceleration
Schubbetrieb *m* push-boating
Schubbewegung *f* sliding movement
Schubbewehrung *f* shear (web) reinforcement *(concrete construction)*
Schubboot *n* push-boat, pusher [tug], puffing tug
Schubbruch *m* shear fracture
Schubbugsierer *m* pusher tug
Schubdüse *f* thrust nozzle
Schubeinheit *f* barge train [unit], push tow
Schubfestigkeit *f* shear[ing] strength
Schubgrenze *f* displacing limit
Schubkahn *m* barge
Schubkarre *f* wheel-barrow
Schubkraft *f* thrust [force]; shear[ing] force; pushing force
Schubkugel *f (aut)* torque ball
Schubkugelgelenk *n* torque tube ball joint
Schubkurbel *f* sliding block linkage, crank
  umlaufende (rotierende) S. rotating crank
Schubkurbelgetriebe *n* crank mechanism
  exzentrisches S. eccentric crank mechanism
  geschränktes S. eccentric crank mechanism
Schubkurbelpresse *f* crank-mechanism extrusion press
Schublehre *f* vernier caliper
Schubleichter *m (ship)* barge
Schublinie *f* shearing stress line
Schubmessung *f* thrust measurement
Schubmittelpunkt *m (mach, therm)* centre of movement; shear centre
Schubmodul *m* shear modulus
Schubprahm *m (ship)* barge
Schubrad *n* fluted roller
Schubräumer *m (mine)* activated deflector plate
Schubraupe *f* push loader
Schubraupenbetrieb *m* push loading *(earthmoving)*
Schubschiffahrt *f* pusher navigation
Schubschlepper *m* tow boat, push towboat, pusher tug [unit], [push-type] towboat, puffing tug push-boat, push-tow-tug; sea mule *(in harbours)*
Schubschlepperverband *m* push tug unit
Schubschleppzug *m* push tow
Schubschwingungsart *f* shear mode of vibration
Schubspannung *f* shear[ing] stress
Schubstange *f* connecting rod; push bar, pusher
Schubstangenachse *f* centre line of the connecting rod
Schubstangenanlage *f* reciprocating rod and hinged slats
Schubstangenförderer *m* push-plate conveyor, push-bar conveyor, suspended-flight conveyor, roller flight conveyor
Schubstangengabelung *f* fork of the connecting rod
Schubstangenkopf *m* connecting-rod head
Schubstangenlänge *f*/ **unendliche** infinite length of the connecting rod
Schubstangenschaft *m* shank of the connecting rod
Schubtransformator *m* moving-coil transformer *(voltage control)*
Schubumformen *n* shearing
Schubverband *m* barge train, tow, push [tow]
Schubverkehr *m* push-boating
Schubverlust *m (ship)* thrust deduction
Schubverstärkung *f* thrust augmentation
Schubversuch *m* shearing test
Schubverzerrung *f* shearing strain
Schubwagenspeiser *m* reciprocating-plate feeder
Schubwiderstand *m* resistance to shear
Schuh *m* shoe; ferrule
Schuhabsatz *m* heel
Schuhahle *f* shoemaker's awl
Schuhband *n* shoe-lace, shoe-string
Schuhdraht *m* shoe thread
Schuhfabrik *f* shoe factory
Schuhkreme *f* shoe polish
Schuhleder *n* shoe-leather
Schuhleisten *m* shoemaker's last
Schuhoberteil *n* galosh, golosh
Schuhpappe *f* boot (shoe) board
Schuhprüfzählrohr *n* shoe counter
Schuhschnalle *f* shoe buckle

Schuhsohle *f* sole [of a shoe]
geklebte S. stuck-on sole
Schuko *m s.* Schutzkontakt
Schukosteckdose *f* socket (wall) outlet with earthing contact[s], socket with earthing contact[s]
Schukostecker *m* [safety] plug with earthing contact[s], plug with protective ground contact
Schülerzeitkarte *f* student's season ticket
Schulfernrohr *n* school telescope
Schülpe *f* scab *(casting)*
Schulschiff *n* training (drill) ship, schoolship, trainer
Schulter *f* shoulder, collar
Schulterdecker *m* high-wing monoplane
Schulternebelgerät *n* knapsack fogger
Schulterradschaben *n* shaving shoulder gears
Schulung *f* **des Personals** personnel training
Schulwesen *n* **/ technisches** engineering education (educational facilities)
Schulz-Kessel *m* Schulz boiler
Schummelloch *n* rabbling hole *(puddle furnace)*
schummern to hachure
Schummerung *f* hachure
Schuppe *f* scale; flake *(graphite)*
schuppenartig imbricated, scale-like
Schuppenbildung *f* scaling
Schur *f* clip, shearing
Schüreisen *n* poker, fire (coal) rake
schüren / das Feuer to clean the fire, to stir up; to agitate the fuel bed
Schürer *m s.* Schüreisen
Schürfbohrmaschine *f* prospecting drill
schürfen to costean, to prospect, to search, to trench; to dig, to cut *(coal)*
Schürfen *n* exploration, prospecting, search
Schürfgraben *m* costeaning ditch, trench
Schürfgrube *f* trial pit
Schürfkübel *m* bucket; scoop; scraper bowl; drag[line bucket]
Schürfkübelanhänger *m* scraper
Schürfkübelbagger *m* boom dragline, dragline excavator
Schürfkübelraupe *f* scraper-dozer
Schürfkübelwagen *m* motor scraper
Schürfloch *n* trial hole
Schürfmittel *n* digging element
Schürfschacht *m* test (bell) pit, prospect shaft
Schürfstelle *f* digging point
Schürfung *f* exploration, prospecting
geophysikalische S. geophysical prospecting
Schürfwagen *m* scraper
Schürhaken *m* fire iron (hook)
Schürkelle *f* rake
Schürplatte *f* coking (dead, choke) plate
Schürraum *m* stoke-hole
Schurre *f* chute; shoot
Schürspitze *f* slice [bar]
Schürstange *f s.* Schüreisen
Schurwolle *f* shorn wool
Schürze *f* apron; flashing *(sheet metal)*
Schuß *m* 1. *(mil)* shot, round *(ammunition)*, report *(sound)*; 2. *(text)* filling [pick], weft; 3. *(plast)* shot; 4. *(mine)* blasting; 5. pan, section *(conveyor)*; 6. lift, section *(columns etc.)*
fehlerhafter S. *(text)* mispick
Schußausrichter *m (text)* weft straightener
Schußdichte *f (text)* pickage
Schußdouble *m (text)* filling-backed fabric
gemusterter S. figured weft backed fabric
Schußeintragevorrichtung *f (text)* picking motion
Schüsselklassierer *m* bowl classifier
Schußentfernung *f (mil)* firing range
Schußfaden *m (text)* weft thread
einzelner S. pick
Schußfadenbruch *m* broken pick
Schußfadenlegeeinrichtung *f* layering mechanism for the weft threads
Schußfeinheit *f (text)* pick count
Schußflamme *f (mine)* primary flame
Schußflottung *f (text)* filling (weft) float
Schußfolgemuster *n (text)* filling pattern
Schußgabel *f (text)* filling fork
Schußgenauigkeit *f* accuracy of fire
Schußgewicht *n (plast)* shot (charge) weight
Schußkamera *f* gun-camera
Schußkötzer *m* pin cop
Schußleistung *f (plast)* shot capacity
Schußrapport *m (text)* repeat of the weft thread
Schußregler *m (text)* weft straightener
Schußrips *m (text)* weft rib
Schußsamt *m (text)* filling pick fabric
Schußschlag *m (text)* pick
Schußseide *f* tram
Schußspule *f* pirn
Schußspulmaschine *f* bobbin (weft) winder, quiller, quilling (weft winding) machine
spindellose S. spindleless pirn winder
Schußsuchen *n (text)* pick finding
Schußwächtergabel *f (text)* filling stop motion
Schußwechsel *m (text)* weft (filling) change
Schußwirkware *f* weft knit [fabrics]
Schußzahl *f* **je Zoll** *(text)* picks per inch
Schußzähler *m (text)* pick clock
Schusterahle *f* shoemaker's awl
Schute *f* lighter, scow, [open] barge
S. mit aufgezogenen Enden rake-ended barge
S. mit hochgezogenem Boden swim barge
selbstklappende S. self-dumping scow
Schutenführer *m* bargee
Schutt *m* wast[ag]e; rubbish, rubble; debris, ruins
Schuttabladeplatz *m* dump site, dumping ground, rubbish dump
Schüttbeton *m* poured (in-situ) concrete
Schüttdichte *f* bulk density
Schüttelaufgabetisch *m* shaker feeder

Schüttelbewegung *f* shaking motion
Schüttelherd *m* shaking (vibrating) table
Schüttelkultur *f* shake culture
Schüttelmaschine *f* shaking machine, mechanical shaker
schütteln to shake, to agitate, to vibrate
Schüttelrinne *f* shaking (shaker conveyor) trough; shaking chute
Schüttelrost *m* vibrating grate (grizzly)
Schüttelrostfeuerung *f* furnace with movable grate-bars, shaking grate
Schüttelrutsche *f* shaking chute, jigger (jigging, shaker) conveyor
Schüttelsieb *n* pulsating (vibrating, shaker) screen, riddle, shaking sieve, sifter; vibratory screen *(paper)*
Schüttelsortierer *m* shaking sliver screen *(paper)*
Schütteltrichter *m* separatory (shaking) funnel
schütten to pour *(liquid, bulk)*; to dump
  Beton s. to pour (place) concrete
Schüttgewicht *n* bulk weight, apparent (bulk, powder) density
  spezifisches S. bulk-specific gravity
Schüttgosse *f* feeding trough, chute *(grate)*
Schüttgut *n* bulk (loose) material; goods in bulk, bulk freight (cargo); bulk [goods]
Schüttgutbehälter *m* hopper, silo, bin
Schüttgutfrachter *m* bulk carrier
Schüttguthöhe *f* packed height
Schüttgutschaufel *f* bucket fork
Schüttgutumschlag *m* bulk handling
Schütthöhe *f* 1. dumping height; 2. thickness of fire, depth of fuel layer
Schüttkanal *m* charging chute (conduit)
Schüttkasten *m* [coal] hopper
Schüttladung *f* bulk, bulk (loose) cargo
Schüttmaterial *n* fill, fill material (earth)
Schüttofen *m* shelf furnace, self-feeding furnace
Schüttrinne *f* shoot, chute
Schüttrohr *n* elephant trunk, tremie
Schüttrost *m* chute (self-feeding, continuous-charging) grate
Schüttrostfeuerung *f* self-feeding furnace, continuous-charge furnace
Schüttrumpf *m* feed (receiving) hopper
Schütttrichter *m* [coal] hopper
Schüttung *f (build)* fill
Schüttvolumen *n* bulking value
Schüttweite *f* dumping reach
Schüttwichte *f* bulk density
Schüttwinkel *m* angle of repose
Schutz *m* 1. protection; 2. *(mach)* safeguard; cover; 3. shelter, refuge
  anodischer S. plate protection *(anti-corrosion)*
  harmonikaähnlicher S. concertina-type cover, bellows-type cover
  katodischer S. cathode protection *(anti-corrosion)*
  überlagerter S. *(el)* back-up protection
  ziehharmonikaähnlicher S. concertina-type cover, bellows-type cover
Schütz *n* 1. *(el)* contactor; 2. sluice
  S. mit Relais automatic tripping contactor
  elektromagnetisches S. magnetic contactor
Schutzabschirmung *f* guard shield
Schutzabstand *m (el)* protection ratio; protective distance
Schutzanstrich *m* protective coat[ing], protecting paint
Schutzanzug *m* [protective] frogsuit; overall
Schutzart *f* protective system
Schutzatmosphäre *f* protective atmosphere
Schutzbehandlung *f* preservation treatment
Schutzbelag *m* 1. protective layer (plating); 2. *(ship)* preventer deck
Schutzberechnung *f* shield calculation
Schutzbestimmungen *fpl* safety regulations
Schutzblatt *n* fly leaf
Schutzblech *n* mudguard *(cycle)*, wing; protective sheet *(furnace)*
Schutzblechmantel *m* protective iron lagging
Schutzbrille *f* safety glasses (goggles)
Schutzdach *n* canopy
Schutzdeck *n* shelter deck
Schutzdecke *f* screen cover
Schutzdecker *m* shelter-deck vessel
  geschlossener S. closed shelter decker
  offener S. open shelter decker
Schutzdeckschiff *n* shelter-deck vessel
Schutzdrossel *f (el)* protective reactor
Schutzeinrichtung *f* protective gear (equipment) *(power line)*; *(cont)* safety control
Schutzelektrode *f* guard electrode
schützen to protect; to guard; to screen, to shield; to shelter; to preserve *(wood)*
Schützen *m* shuttle *(loom)*
Schützenauswechselung *f* shuttle changing
  selbsttätige S. automatic shuttle changing
Schützenbahn *f* race board, shuttle race
Schützenblockierung *f* blocking device for shuttles
Schützenbremsung *f* shuttle braking
Schützenfangdraht *m* shuttle guard wire
Schützenfänger *m* shuttle catcher (guard)
Schützenkasten *m* shuttle box
Schützenkette *f (text)* throat of the shuttle; *(mil)* extended order
Schützenloch *n* foxhole
schützenlos *(text)* shuttleless
Schützenpanzerwagen *m* armoured personnel carrier
Schützenschalter *m* contactor
Schützenschlag *m (text)* pick[ing]
Schützenspule *f* quill

Schützensteuerung *f* contactor control (equipment)
Schützensteuerwalze *f* contactor controller
Schutzerde *f* protective (protection) earth
Schutzerdung *f* protective earthing
Schutzfenster *n* shielding window
Schutzfilter *n* protective filter
S. gegen einfallendes Nebenlicht *(telev)* ambient light filter
Schutzfrequenzband *n* guard band *(radio)*
Schutzfrucht *f* nurse (cover) crop
Schutzfunkenstrecke *f* protective [spark] gap, safety-gap
Schutzgas *n* inert gas *(welding)*
Schutzgasatmosphäre *f* protective atmosphere
Schutzgaslichtbogenschweißen *n* inert-gas arc welding
Schutzgaslötung *f* controlled-atmosphere furnace brazing
Schutzgasmantel *m* shielding gas envelope, shield of gas
Schutzgasofen *m* artificial-atmosphere furnace
Schutzgasrelais *n* reed relay
Schutzgasschweißen *n* inert-gas [shielded] arc welding
Schutzgastransformator *m* inert-gas transformer
Schutzgehäuse *n* protective casing (housing)
S. für die Mikrofonkapsel transmitter capsule protector
Schutzgeländer *n* safety fence; guard (hand) rail, crash barrier
Schutzgewebe *n* wire-gauge guard
Schutzgitter *n* 1. protective grating; 2. *(el)* screen grid
Schutzglas *n* 1. *(therm)* glass-tube protector; 2. *(aut)* safety glass
Schutzgrad *m s.* Schutzart
Schutzhafen *m* port of distress (refuge)
Schutzhandschuh *m* safety (protective) glove; *(nucl)* lead-rubber glove
Schutzhaube *f* *(mach)* protective hood (covering)
Schutzhülle *f* protecting envelope; protecting sheathing
Schutzhüllenverpackung *f* blister packaging
Schutzhülse *f* tube protector
Schutzkappe *f* *(el)* protecting cap *(shock protection)*
Schutzkegel *m* protective (protecting) cone
Schutzkennlinie *f* selectivity characteristic *(relay)*
Schutzkern *m* protective core, core for protection purposes *(current transformer)*
Schutzkiel *m* slab (false) keel
Schutzklasse *f* *(el)* degree of protection
Schutzkleid *n* *(ship)* dodger
Schutzkleidung *f* protective clothing (suit, outfit); *(nucl)* anticontamination clothing
Schutzklemme *f* *(el)* guard terminal
Schutzkolloid *n* protective colloid
Schutzkondensator *m* protective capacitor
Schutzkontakt *m* [protective] earthing contact, protective contact
Schutzkontaktstecker *m* plug with protective ground contact, safety plug
Schutzlack *m* protective lacquer
Schutzleiter *m* protective conductor
Schutzleitung *f* protective lead
Schutzmarke *f* trade-mark
Schutzmaske *f* protecting mask, face shield, respirator
Schutzmittel *n* preventive, preservative; shielding medium *(arc welding)*
Schutzmuffe *f* protective sleeve
Schutzmuster *n* protective design
Schutzniveau *n s.* Schutzpegel
Schutzpegel *m* protective [voltage] level
Schutzpfahl *m* *(ship)* pile fender
Schutzpolster *n* plastron *(fencing)*
Schutzraum *m* shelter
S. gegen radioaktiven Niederschlag fall-out shelter
Schutzrechte *npl* patent rights
Schutzrelais *n* protective (guard) relay
Schutzring *m* guard ring
Schutzrohr *n* 1. thermowell; 2. *(cont)* protecting tube
Schutzschalter *m* protective switch, thermal release, overcurrent cut-out, automatic circuit-breaker
Schutzscheibe *f* glass guard
Schutzschicht *f* *(opt)* protective layer; coating *(metal)*; protective film
Schutzschirm *m* **der Lampe** baffle
Schutzschlauch *m* **der Biegewelle** casing (covering) of the flexible shaft
Schutzschleuse *f* double safety (guard) gates
Schutzschott *n* screen bulkhead
Schutzschwelle *f* curb fender
Schutzsenkung *f* *(mach)* countersunk recess
Schutzspiere *f* *(ship)* defence boom
Schützsteuerwalze *f* contactor controller
Schutzstrecke *f* *(el)* gap section *(wire)*
Schutztransformator *m* isolation transformer
Schutztrenner *m* protective isolator
Schutzüberzug *m* protective coating
Schutzumschlag *m* [book] jacket, dust cover
Schutzvorhang *m* fire curtain *(theatre)*
Schutzvorrichtung *f* safety guard
Schutzwagen *m* guard (shock-absorbing, match, buffer, runner, protection) wagon
Schutzwald *m* reservation
Schutzwand *f* protective screen (shield, wall); *(el)* flash barrier
Schutzwandler *m* isolation transformer
Schutzwiderstand *m* 1. preventive (discharging) resistor, bleeder; 2. protective resistance

Schwabbelmaschine *f* buffing machine
schwabbeln to buff
Schwabbeln *n* buffing
Schwabbelscheibe *f* buffing wheel
schwachangereichert low-enriched, slightly enriched
schwachbeleuchtet dimly lit
Schwachbestrahlung *f* low-intensity irradiation
schwächen to weaken *(cross section)*; to lessen, to diminish; to extinct *(rays)*; *(text)* to tender
Schwachgas *n* lean (poor) gas
Schwachgasgenerator *m* lean-gas generator
Schwachgasheizung *f* lean-gas firing
Schwachlauge *f* weak liquor
schwachsauer faintly acid
Schwachstelle *f* weak point, trouble spot
Schwachstrom *m* weak current, low-voltage current
Schwachstromkabel *n* cable for communications circuit
Schwachstromtechnik *f* telecommunications [and electronics], light-current engineering, weak-current engineering; signal engineering
Schwächung *f* attenuation *(radiation)*; extinction, absorption; reduction; fading; weakening *(cross section)*
Schwächungsfaktor *m* shot-noise reduction factor
Schwächungsfilter *n* attenuation filter
Schwächungsfunktion *f* attenuation function
Schwächungskoeffizient *m* attenuation factor (coefficient)
Schwächungskurve *f* absorption (attenuation, transmission) curve
Schwächungsquerschnitt *m* attenuation cross section
Schwadblech *n* swath board
Schwadbrett *n* swath board
Schwaden *m* 1. *(ag)* swath; 2. *(mine)* vapour[s]
Schwadenaufnahmevorrichtung *f* swath pick-up attachment
Schwadenmäher *m* swath mower
Schwadenrechen *m* horse (hay) rake
Schwadenwender *m* swath turner
schwalben to dovetail
Schwalbennest *n s.* Senkkasten
Schwalbenschwanz *m* dovetail *(carpentry)*
  verdeckter S. lap dovetail
Schwalbenschwanzblatt *n* dovetail halving
Schwalbenschwanzeinschnitt *m* dovetail hole
schwalbenschwanzförmig dovetailed
Schwalbenschwanzführung *f* *(mach)* dovetail guide
Schwalbenschwanznut *f* dovetail groove
Schwalbenschwanzverbindung *f* dovetailing
Schwallblech *n* wash-plate
Schwallwasser *n* splash water
schwallwassergeschützt splash-proof
Schwamm *m* 1. sponge; fungus; 2. *(build)* dry rot
Schwammblei *n* sponge lead
Schwammeisen *n* sponge iron
Schwammfilter *n* sponge filter
Schwammgummi *m* sponge rubber
Schwammigkeit *f* sponginess
Schwammzirkonium *n* zirconium sponge
Schwanenhalsausleger *m* cranked (swan-necked) jib
Schwanenhalslüfter *m* swan-neck ventilator
schwanken to fluctuate, to vary
Schwankung *f* fluctuation *(e. g. of current)*; pulsation; variation
  S. der Netzspannung mains voltage variation
  absolute S. absolute fluctuation
  lokale S. local fluctuation
  relative S. relative fluctuation
  spontane S. spontaneous fluctuation
  zeitliche S. time fluctuation
Schwankungen *fpl* **der Flugbahnwerte** fluctuations in the parameters of the orbit
Schwankungsmaß *n* *(math)* measure of dispersion
Schwanz *m* tail
Schwanzbildung *f* tailing *(chromatography)*
Schwänzeln *n* *(aero)* shimmying
Schwanzende *n* spigot *(piping)*
Schwanzflosse *f* *(aero)* tail fin
Schwanzkufe *f* *(aero)* tail skid
schwanzlastig *(aero)* tail-heavy, nosing-up
Schwanzlastigkeit *f* tail-heaviness
Schwanzriemenleder *n* crupper leather
Schwanzsporn *m* *(aero)* tail skid
Schwanzwelle *f* propeller (tail, screw) shaft
Schwanzwolle *f* crutchings
Schwarm *m* shower *(of electrons)*; cluster; swarm
Schwarte *f* rind, slab
Schwartenholz *n* large slabs
schwarz black
schwarzästig black-knotty
Schwarzaustastimpuls *m* blanking-out pulse
Schwarzaustastung *f* *(telev)* blanking-out
Schwarzblech *n* black plate
Schwarzbrache *f* bare fallow
Schwarzbrot *n* brown bread
Schwarzbruch *m* *(met)* black shortness
Schwarzdecke *f* hydrocarbon (black-top, bituminous) pavement, asphalt
Schwarzdeckeneinbaumaschine *f s.* Schwarzdeckenfertiger
Schwarzdeckenfertiger *m* bituminous (black-top) paver-finisher, asphalt paver
Schwärze *f* 1. blackness; 2. *(met)* black[en]ing, black wash, mould facing
Schwärzegrad *m* degree of blackness
schwärzen to black[en]; to black-wash; to dark
Schwarzglühen *n* *(met)* black annealing

Schwarzguß *m* black-heart malleable cast iron, all-black malleable cast iron
Schwarzhörer *m* radio pirate, pirate listener
Schwarzkerntemperguß *m* black-heart malleable cast iron, all-black malleable cast iron
Schwarzkochung *f* burnt cook *(wood pulp)*
Schwarzkörper *m* black body
Schwarzkörperstrahlung *f* black-body radiation
Schwarzkupfer *n* black copper
Schwarzlauge *f* black liquor
Schwarzlücke *f (telev)* blanking interval
Schwarzmanganerz *n* hausmannite
Schwarzpegel *m* black level
Schwarzpulver *n* blackpowder, gunpowder
Schwarzsättigung *f (telev)* black saturation
Schwarzschmelze *f* black ash *(papermaking)*
Schwarzschulter *f (telev)* porch
  hintere S. back porch
  vordere S. front porch
Schwarzseher *m* television pirate
Schwarzseite *f* **der Kaue** *(mine)* dirty-clothes side of pitbath
Schwarzspitze *f (telev)* black peak, peak black
Schwarzsteuerdiode *f* direct current restorer (restoring) diode
Schwarzsteuerungskreis *m (telev)* clamping circuit
Schwarztastimpuls *m* blanking pulse
Schwarztreppe *f s.* Schwarzschulter
Schwärzung *f* blackening *(photographic deposit)*; [optical] density
Schwärzungsabstufung *f* density graduation *(photography)*
Schwärzungsbereich *m* density range *(photography)*
Schwärzungskurve *f* density curve *(photography)*
Schwärzungsmesser *m* [optical] densitometer
Schwärzungsmessung *f* [optical] densitometry
Schwärzungsring *m* halo
Schwärzungsschwelle *f* density threshold
Schwarzweißbild *n* black-and-white picture
Schwarzweißbildröhre *f* black-and-white picture tube
Schwarzweißfernsehbildgeberröhre *f* monotron
Schwarzweißfilm *m* black-and-white film
Schwarzweißkanal *m* monochrome channel
Schwarzwert *m (telev)* black level
Schwarzwertbezugspegel *m* reference black level
Schwarzwertrückgewinnung *f (telev)* clamping
Schwarzwertrückgewinnungsschaltung *f (telev)* clamping circuit
Schwebe *f (mine)* arch (level) pillar
Schwebebahn *f* cableway, suspended (elevated) railway
Schwebebalken *m* balance (balancing) beam
Schwebebühne *f (mine)* suspended platform
Schwebefähigkeit *f* drifting property
Schwebeflug *m* hovering flight *(helicopter)*
Schwebekolbendurchflußmesser *m* piston-type area flowmeter
Schwebekörper *m* float, plummet *(flowmeter)*
Schwebekörperdurchflußmesser *m* rotameter (variable-area) flowmeter
Schwebekörperviskosimeter *n* plummet viscosimeter
Schwebelage *f* suspended (hanging) position
Schwebemittel *npl (plast)* anti-settling agents
schweben to float, to be suspended; to hover
  frei im Raum s. to float freely in space
schwebend hanging, suspended; *(mine)* overhand
Schweberösten *n* flash roasting
Schweberuder *n (ship)* [balanced] underhung rudder, suspended rudder
Schwebeschlamm *m* float slime
Schwebestaub *m* airborne dust
Schwebestoffe *mpl* suspended matter
Schwebeteilchen *n (ch)* suspensoid, suspended particle
Schwebeträger *m* suspended beam
Schwebetrockner *m* flash drier
Schwebezustand *m* suspension state
Schwebstoff *m* aerosol
  radioaktiver S. radioactive aerosol
Schwebung *f* beat
  aperiodische S. dead beat
Schwebungsamplitude *f* beat amplitude, amplitude of beats
Schwebungsanzeige *f* beat indication
Schwebungsdauer *f* period of beat
Schwebungsempfang *m* beat (heterodyne) reception
Schwebungsempfänger *m* beat receiver
Schwebungsfrequenz *f* beat frequency
Schwebungslücke *f* zero beat
Schwebungsmethode *f* beat method
Schwebungsnull *f* zero-beat
Schwebungsoszillator *m* beat-frequency oscillator
Schwebungsperiode *f* beat cycle (period)
Schwebungsstrom *m* beating current
Schwebungssummer *m* beat frequency oscillator
Schwebungston *m* beat (combination, difference) tone
Schwebungstonhöhe *f* beat-note pitch
Schwebungsverstärker *m* heterodyne amplifier
Schwebungsvorgang *m* beating process
Schwefel *m* sulphur, *(Am)* sulfur
Schwefelabdruck *m (ch)* sulphur print
schwefelarm low-sulphur *(coke)*
schwefelartig sulphurous
Schwefeläther *m* sulphuric ether
Schwefelblei *n* lead sulphide
Schwefelblume *f s.* Schwefelblüte
Schwefelblüte *f* flowers of sulphur

Schwefeldampf *m* sulphur vapour
Schwefeldioxid *n* sulphur dioxide, sulphurous anhydride
schwefelecht fast to stoving
Schwefelentfernung *f* sulphur removal
Schwefelerz *n* sulphide ore
Schwefelfarbstoff *m* sulphur dye[stuff]
Schwefelfärbung *f* dyeing with sulphur dyes
Schwefelgehalt *m* sulphur content
Schwefelgeruch *m* sulphurous smell
Schwefelgrube *f* sulphur pit
schwefelhaltig sulphur-containing; sulphur[e]ous
Schwefelhexafluorid *n* sulphur hexafluoride
Schwefelkies *m* iron (sulphurous) pyrites
Schwefelkohlenstoff *m* carbon disulphide, disulphide of carbon
Schwefelleber *f* hepar sulphuris, liver of sulphur
Schwefelmilch *f* milk of sulphur
schwefeln 1. to sulphurate, to sulphurize; *(met)* to sulphate; 2. to fumigate
Schwefelnatriumäscher *m* sodium sulphite lime
Schwefelofen *m* sulphur burner
schwefelsauer sulphuric
Schwefelsäure *f* sulphuric (vitrolic) acid
  rauchende S. fuming sulphuric acid, oleum
  verdünnte S. dilute sulphuric acid
Schwefelsäureanhydrid *n* sulphuric anhydride, sulphur trioxide
Schwefelsilber *n* silver sulphur
Schwefelwasserstoff *m* hydrogen sulphide, sulphuretted hydrogen
Schwefelyperit *n* yperite, mustard gas
schweflig sulphur[e]ous
Schweifsäge *f* fret saw, sweep [cutting] saw
Schweifsägemaschine *f* fret (jig) saw machine
Schweigefrequenz *f* taboo frequency
Schweigegebiet *n* silent area
Schweigezone *f (el)* silent area
Schweineborste *f* hog's hair (bristle)
Schweinehof *m* piggery
Schweinestall *m* pigsty, piggery, *(Am)* hog house
Schweinewaage *f* pig weigher
Schweinsleder *n* pigskin
Schweiß *m* perspiration, sweat
  öliger S. oily yolk *(wool)*
Schweißaggregat *n* welding set (unit), welder
Schweißapparat *m* welding apparatus
Schweißarbeit *f* welding [job]
Schweißausrüstung *f* welding equipment
Schweißautomat *m* automatic welding machine
Schweißband *n* sweat band *(hat)*
schweißbar weldable
Schweißbarkeit *f* weldability
Schweißbeständigkeit *f (text)* perspiration resistance
Schweißbogen *m* welding arc
Schweißbrenner *m* welding torch
Schweißbrille *f* welding goggles
Schweißdraht *m* welding wire
Schweißdruck *m* welding pressure
Schweißdüse *f* welding tip
Schweiße *f* added metal
schweißecht fast to perspiration
Schweißeisen *n* wrought (puddle) iron
Schweißeisenblock *m* merchant bar
Schweißeisenluppe *f* charcoal bar
Schweißelektrode *f* welding (metal) electrode
Schweißempfindlichkeit *f* susceptibility to welding cracks
schweißen to weld
  autogen s. to weld autogenously
  elektrisch s. to electro-weld
  elektrolytisch s. to weld electrolytically
  stumpf s. to butt-weld
  überlappt s. to lap-weld
Schweißen *n* welding
  aluminothermisches S. aluminothermic welding
  maschinelles S. machine welding
  überlappendes S. lap welding
Schweißer[in *f*] *m* welding operator, welder
Schweißerschutzschild *m* [welder's] face shield
Schweißfehler *m* welding defect
Schweißflamme *f* welding flame
Schweißflußmittel *n* welding flux
Schweißfuge *f* weld[ing] groove
schweißhärten to weld-harden
Schweißhelm *m* welding helmet
Schweißherd *m* soaking hearth
Schweißhitze *f* welding heat
Schweißingenieur *m* welding engineer
Schweißkabel *n* welding cable
Schweißkappe *f* welding helmet
Schweißkonstruktion *f* welded construction
Schweißlage *f* bead
Schweißlatte *f (ship)* sweat board
Schweißleistung *f* welding capacity
Schweißlichtbogen *m* welding arc
Schweißmaschine *f* welder, welding machine
Schweißmittel *n* welding flux
Schweißnaht *f* weld[ing seam]
Schweißnahthöhe *f* weld throat
Schweißnahtoberfläche *f* weld face
Schweißnahtrissigkeit *f* cracking in welds
Schweißnahtüberhöhung *f* weld reinforcement
Schweißofen *m* welding furnace
Schweißpaste *f* welding paste
Schweißperle *f* welding bead
Schweißpistole *f* welding handle
Schweißprobe *f* welding specimen
Schweißpulver *n* granulated welding flux (composition), powdered (granular fusing) flux
Schweißpunkt *m* welding point (spot)
Schweißraupe *f* welding bead
Schweißrichtung *f* hand (direction) of welding

Schweißriß *m* welding crack
Schweißrißempfindlichkeit *f* susceptibility to welding fissures
Schweißrissigkeit *f* weld cracking, incomplete fusion *(of weld and parent metal)*
Schweißsand *m* welding sand
Schweißschlacke *f* welding cinder
Schweißschrumpfung *f* welding contraction
Schweißspannung *f* 1. *(el)* welding arc voltage; 2. welding stress
Schweißstab *m* welding rod (electrode)
ummantelter S. sheathed electrode
Schweißstahl *m* wrought steel (iron)
Schweißstoß *m* joint
Schweißstrom *m* welding current
Schweißtechnik *f* welding engineering; welding practice
Schweißteil *n* weldment
Schweißträger *m* welded girder
Schweißtransformator *m* welding transformer
schweißtreibend diaphoretic, sudorific
Schweißumformer *m* welding transformer
Schweißung *f* 1. welding; 2. weld
Schweißunterlage *f* backing strip
Schweißverbindung *f* weld[ed] joint, welded connection
dichte S. tight weld joint
Schweißverfahren *n* welding process
Schweißverzug *m* distortion by welding
Schweißvorrichtung *f* welder
Schweißwalzwerk *n* welding mill
Schweißwasserbildung *f* deposit of moisture
Schweißwasserverkleidung *f (ship)* sweat board
Schweißwolle *f* / **rohe** greasy wool
Schweizer *m (ag)* milker
Schweizerdegen *m* compositor and printer
Schwelanlage *f* low-temperature plant
Schwelbenzin *n* low-temperature coking gasoline
schwelen to smoulder, to carbonize at low temperature; to distil *(tar)*
Schwelen *n* low-temperature distillation (carbonization)
Schweler *m* low-temperature carbonizer
Schwelerei *f* low-temperature carbonization plant
Schwelgas *n* low-temperature [carbonization] gas; gaseous product of incomplete combustion
Schwelgenerator *m* low-temperature carbonizer
Schwelkohle *f* low-temperature carbonization coal, brown coal for low-temperature retort process
Schwelkoks *m* low-temperature coke
Schwellbeanspruchung *f* repeated loading *(in one direction)*
Schwelle *f* 1. *(build)* sole plate (runner), sill [plate] *(carpentry)*; 2. *(ry)* [cross-]tie, sleeper; 3. *(mine)* floor bar; 4. *(ph)* threshold [value]; barrier, limit, margin; 5. step *(pulse)*
schwellen to swell; to increase; to rise *(water)*
Schwellenabstand *m (ry)* distance between sleepers, spacing of sleepers
Schwellenbedingung *f* threshold condition
Schwellenbefeuerung *f* boundary lights
Schwellenbinden *n (ry)* hooping (binding) of sleepers
Schwellendosis *f* threshold dose
Schwellenenergie *f* threshold [energy]
S. der Reaktion threshold of the reaction, reaction threshold
Schwellenfeuer *f (aero)* threshold light
Schwellenfrequenz *f* threshold frequency
Schwellenholz *n* ground joist
Schwellenlogik *f* threshold logic
Schwellenniveau *n* threshold level
Schwellenreaktion *f* threshold reaction
Schwellenschraube *f (ry)* sleeper bolt (screw), sill screw
Schwellenspannung *f* threshold voltage
Schwellenstopfen *n (ry)* packing of sleepers, *(Am)* tamping ties
Schwellenstopfer *m (ry)* tamper
Schwellenwert *m* 1. threshold [value]; 2. underexposure region *(photography)*
S. der Empfindlichkeit threshold of sensitiveness
fotoelektrischer S. photoelectric threshold
Schwellenwertbegrenzer *m* peak chopper, clipper
Schwellenwertfunktion *f* majority decision function
Schwellenwertmessung *f* threshold measurement
Schwellenwertunterdrücker *m s.* Schwellenwertbegrenzer
Schwellfarbe *f (leath)* plumbing liquor
Schwellfestigkeit *f (met)* fatigue strength under pulsating stresses
Schwellspannung *f* 1. *s.* Schwellenspannung; 2. repeated stress [in one direction]
Schwellung *f (nucl)* swelling; expansion
Schwellwert *m* threshold level (value)
Schwellwertdetektor *m* threshold detector
Schwellzahl *f* coefficient of swelling *(soil mechanics)*
Schwelofen *m* low-temperature carbonizer (furnace, carbonization furnace)
Schwelteer *m* low-temperature tar
Schwelung *f* low-temperature carbonization
Schwelverhüttung *f* low-temperature smelting
Schwelzone *f* carbonization zone
Schwemmarbeit *f* **mittels Kanälen** *(mine)* ground sluicing
Schwemmbagger *m* reclamation dredger
schwemmen to rinse, to float, to wash
Schwengel *m* pendant lever; handle *(pump)*; clapper *(bell)*
S. mit Gegengewicht balance bob

Schwengelpumpe *f* beam pump
Schwenkachse *f* swivelling axis
Schwenkarm *m* swinging (pivoted) arm; pendant; swinging bar; slewing boom
Schwenkarmlader *m* gathering arm-loader
Schwenkbagger *m* full-revolving excavator
schwenkbar swivelling, arranged to swivel; rotatable; pivoting; slewable *(crane)*
Schwenkbewegung *f* swinging *(excavator)*; slewing *(crane)*; *(mach)* swivel motion
Schwenkbohrmaschine *f s.* Radialbohrmaschine
schwenken to swing [around]; to swivel; to rotate; to slew *(crane)*, to revolve, to reverse
die Keule s. to shift the lobe *(aerials)*
Schwenkgelenk *n* swivel hinge
Schwenkgetriebe *n* swivelling mechanism
Schwenkgreiflader *m* gathering arm-loader
Schwenkhahn *m* swivel tap
Schwenkhebel *m* rocking lever; pivoted lever
Schwenkkeil *m* toggle wedge
Schwenkkörper *m* tilting yoke *(axial piston pump)*
Schwenkkran *m* slewing (revolving, jib, swing) crane
Schwenkladebaum *m (ship)* slewing derrick
Schwenklader *m* swing loader
Schwenklager *n* swivel bearing
Schwenkpflug *m* reversible plough
Schwenkplatte *f* hinged plate
Schwenkrolle *f* castor
Schwenkschalter *m* lever switch
Schwenkschaufellader *m* swing loader
Schwenkschaufler *m* swing loader
Schwenkscheibe *f* tilting box *(pump)*
Schwenkschild *m s.* Schwenkschildplanierraupe
Schwenkschildplanierraupe *f* angling dozer, angledozer
Schwenktisch *m* swivel table
Schwenktrenner *m (el)* tumbler isolator
Schwenkwerk *n* slewing gear
Schwenkwinkel α *m* amount of angularity between the cylinder block and the drive shaft *(angle α, axial pump)*
schwer heavy *(load, duty)*; difficult, hard; weighty; gravitational
Schwerachse *f* gravity axis
Schweranlauf *m* heavy starting duty *(motor)*
Schwerbenzin *n* heavy petrol
Schwerbeton *m* heavy[-aggregate] concrete
Schwerbetonabschirmung *f* heavy-aggregate shield
Schwerdrehmaschine *f* large (heavy-duty) lathe
Schwere *f* gravity
künstliche S. artificial gravity
Schwerebeschleunigung *f* acceleration of (due to) gravity, gravity acceleration; gravitational constant
Schwerefeld *n* gravitational (gravity) field
schwerefrei gravity-free
Schwerelektronik *f* heavy electronics
schwerelos weightless
Schwerelosigkeit *f* zero-gravity; weightlessness
Schweremesser *m* gravimeter, gravity meter
statischer S. gravimeter
Schwereseigerung *f* gravity segregation
Schweretrennung *f* gravity separation
Schwerewelle *f* gravitational wave
Schwerflintglas *n* extra dense flint glass
schwerflüchtig difficult to volatilize
Schwerflüssigkeitssetzmaschine *f (mine)* dense-medium jig, *(Am)* heavy-media jig
Schwerflüssigkeitssortierung *f* dense-medium washing
Schwerflüssigkeitsverfahren *n* dense-medium process, *(Am)* heavy-media separation
Schwerflüssigkeitszyklon *m* dense-medium cyclone
Schwergewichtsmauer *f* gravity dam, gravity [retaining] wall
Schwergut *n* deadweight, deadweight (heavy) cargo, heavy lift (goods)
Schwergutbaum *m (ship)* heavy derric (boom)
Schwergutfahrt *f (ship)* heavy trade
Schwergutfrachter *m* heavy cargo (lift) ship
Schwergutgeschirr *n (ship)* heavy cargo-handling gear
Schwergutladebaum *m (ship)* heavy derrick (boom), jumbo boom
Schwergutladegeschirr *n (ship)* heavy cargo-handling gear, heavy-lift cargo gear
Schwergutmast *m* king post
Schwerindustrie *f* heavy industry
Schwerkraft *f* force of gravity, gravitational force
Schwerkraftakkumulator *m* gravity-loaded accumulator
Schwerkraftantrieb *m* gravity drive
Schwerkraftbeschleunigung *f* acceleration due to gravity
Schwerkraftdavit *m (n) (ship)* gravity davit
Schwerkraftförderer *m* gravity-type conveyor
Schwerkraftguß *m* gravity casting
Schwerkraftklassierer *m* gravity classifier
Schwerkraftseigerung *f* gravity segregation
Schwerkraftwirkung *f* effect of gravity
Schwerkraftzentrum *n* centre of gravity
Schwerlastanhänger *m* heavy-duty trailer
Schwerlastgut *n* heavy traffic conveyed in bulk
Schwerlastwagen *m* heavy-duty lorry, *(Am)* heavy-duty truck
Schwermaschinen *fpl* heavy machinery
Schwermaschinenbau *m* heavy engineering industry
Schwermetall *n* heavy (high-density) metal
Schwermetallabschirmung *f* heavy-metal shield
Schwermetallguß *m* heavy-metal casting

Schwermetallsalz *n* salt of heavy metals
Schwermetallschild *m* heavy-metal shield
Schwermineral *n* heavy mineral
Schweröl *n* heavy oil
Schwerölladung *f* black cargo
Schwerölmotor *m* heavy-oil engine
Schwerölvergaser *m* heavy-oil carburettor
Schwerpunkt *m* centre of gravity, centroid
S. des Treibungswiderstands centre of lateral resistance
Schwerpunktbestimmung *f* determination of the centre of gravity
Schwerpunktbewegung *f* centre-of-mass motion
Schwerpunktlage *f* position of the centre of gravity
Schwerpunktradius *m* radius of gyration (the centre of gravity)
Schwerpunktsabstand *m* centroidal distance
Schwerpunktsatz *m* theorem of centre of mass
Schwerpunktsdistanz *f* centroidal distance
Schwerpunktslage *f* position (location) of the centre of gravity
Schwerpunktswellenlänge *f* spectral centroid
Schwerpunktsystem *n* centre-of-mass system
Schwerpunktverlegung *f* displacement of the centre of gravity
schwerspaltbar difficult to split
Schwerspat *m* baryte, heavy spar
Schwerstange *f* drill collar (stem)
Schwerstbeton *m* heavy-weight concrete
Schwerstkronglas *n* extra-dense barium crown glass
Schwert *n* 1. sword; 2. *(ag)* knife-edged coulter; 3. *(ship)* centre-board
kurzes S. dagger board
Schwertakel *n* jeer *(sail)*
Schwertbrunnen *m* *(ship)* centre-board well (trunk)
Schwertfalzmaschine *f* knife-folder
Schwertfeile *f* slitting file
Schwertkasten *m* *(ship)* centre-board well (trunk)
Schwertrübe *f* dense (heavy) medium
Schwerwasser *n* heavy water
Schwerwasserdruckreaktor *m* pressurized heavy-water reactor
Schwerwassererzeugungsanlage *f* heavy-water plant
Schwerwasserreaktor *m* heavy-water reactor (pile); heavy-water moderated pile, deuterium-moderated pile
Schwerwassersiedereaktor *m* heavy-water boiling reactor
Schwesterschiff *n* sister ship
Schwesterschiffe *npl* twin ships
Schwimmaufbereitung *f* flotation *(ore)*
S. durch Öl oil flotation
Schwimmaufbereitungsverfahren *n* flotation (buoyancy) process
Schwimmbad *n* natatorium, swimming pool
Schwimmbadreaktor *m* swimming-pool[-type] reactor
Schwimmbagger *m* [floating] dredger
Schwimmbeckenreaktor *m* swimming-pool[-type] reactor
Schwimmblase *f* fish bladder
Schwimmboje *f* dan [buoy]
Schwimmbrücke *f* floating (pontoon) bridge
Schwimmdock *n* floating dock (dry-dock, pontoon)
Schwimmebene *f* plane of flotation
schwimmen 1. to swim; 2. to float *(object)*; 3. to drift
schwimmend / frei afloat
Schwimmer *m* *(aut)* float
Schwimmerdichtemesser *m* float-type densitometer
Schwimmerfüllstandsmesser *m* ball level gauge, buoyancy probe
Schwimmergehäuse *n* *(aut)* float chamber
Schwimmerkreisel *m* float-type gyroscope
Schwimmermanometer *n* float-type manometer (pressure gauge)
Schwimmermengenmesser *m* flowmeter with float-type manometer
Schwimmermesser *m* float gauge
Schwimmermethode *f* temperature-gradient method
Schwimmernadel *f* *(aut)* float needle
Schwimmernadelventil *n* float-needle valve
Schwimmerregel *f* / **Ampèresche** Ampère's rule
Schwimmerschalter *m* float switch
Schwimmerwagen *m* **für Seeflugzeuge** beaching gear
Schwimmfähigkeit *f* floating capacity, watertight integrity, buoyancy
Schwimmfender *m* *(ship)* fender spar
Schwimmflächenkurve *f* *(ship)* curve of flotation
Schwimmflächenskala *f* *(ship)* scale of waterplane areas
Schwimmflosse *f* frogman's flipper
Schwimmgleichgewicht *n* buoyant stability
Schwimmgürtel *m* lifebelt
Schwimmkasten *m* *(build)* floating box
Schwimmkörper *m* floating body
Schwimmkraft *f* buoyancy
Schwimmkran *m* floating (pontoon) crane
Schwimmlage *f* trim *(of a vessel)*
Schwimmpanzer[kraftwagen] *m* amphibious tank
Schwimmramme *f* floating pile-driver
Schwimmsand *m* quicksand, running sand
Schwimmseife *f* floating soap
Schwimmstoffe *mpl* floating matter
Schwimmtor *n* gate vessel, caisson *(lock)*

Schwimmweste *f* lifebelt, life-jacket, life-preserver
Schwindbewehrung *f* shrinkage reinforcement *(concrete)*
schwinden 1. to shrink, to contract; 2. to fade *(radio)*
Schwinden *n* 1. contraction, shrinkage; 2. fading *(radio)*
Schwindfähigkeit *f* shrinkability
Schwindgrenze *f* shrinkage limit
Schwindmaß *n* *(met)* shrinkage rule
Schwindmaßstab *m* patternmaker's rule
Schwindmaßzugabe *f* shrinkage allowance
Schwindriß *m* shrinkage crack
Schwindspannung *f* shrinkage stress
Schwindung *f* shrinkage, contraction
Schwindungshohlraum *m* shrinkage cavity
Schwindungslunker *m* shrinkage piping
Schwindungsriß *m* shrinkage crack
Schwindvermögen *n* shrinkability
Schwindzugabe *f* shrinkage allowance
Schwindzugspannung *f* tensile shrinkage stress
Schwingachse *f* *(aut)* flexible (floating) axle
Schwingamplitude *f* amplitude of oscillations *(crystallography)*
Schwinganker *m* oscillating (pendulum-type) armature *(electric clocks)*
Schwinganlage *f* swing feeder
Schwingarm *m* *(mach)* pawl arm; rocking arm *(shaper)*; *(aut)* suspension arm; swinging bar
Schwingarmgreifer *m* swinging-arm loader
Schwingaudion *n* autodyne (oscillating) detector, ultraaudion oscillator
Schwingaudionempfang *m* self-heterodyne reception
Schwingbeanspruchung *f* vibration loading
Schwingbeil *n* *(text)* scutch blade
Schwingbereich *m* oscillation range
Schwingbeschleunigung *f* vibration acceleration
Schwingbett *n* cot [bed]
Schwingbewegung *f* oscillatory motion
Schwingbühne *f* hinged loading ramp
Schwingdauer *f* vibration period
Schwingdrossel *f* *(el)* swinging choke
Schwingdurchmesser *m* *(mach)* swing [diameter]
Schwinge *f* 1. lever, beam *(kinematics)*; 2. *(mach)* rocker arm; 3. *(text)* scutch blade
Schwingebene *f* vibration plane
schwingen to oscillate, to vibrate; to rock; to hunt *(feedback system)*; *(text)* to swingle, to scutch *(flax)*
Schwingen *n* **der Dampfmaschine** oscillation of the steam engine
Schwinger *m* oscillator; transducer, vibratory unit *(ultrasonics)*
  eingliedriger S. simple oscillator
  gedämpfter harmonischer S. damped harmonic system
  magnetostriktiver S. magnetostriction oscillator
schwingfähig oscillable
Schwingfähigkeit *f* ability to vibrate
Schwingflachs *m* scutched flax
Schwingflügel *m* 1. *(build)* pivoted sash; 2. swinging vane *(pump)*
Schwingflügelfenster *n* pivot window
Schwingförderer *m* vibrating (shaker, jog) conveyor; *(mach)* vibratory feeder
Schwingförderrinne *f* shaker loader
Schwinggleichrichter *m* vibrating reed rectifier
Schwinggreifer *m* swing gripper
Schwinghebel *m* rocker arm, valve rocker; rocking lever
Schwinghebelblock *m* valve-rocker pedestal (bracket)
Schwinghebeldrehbolzen *m* valve-rocker fulcrum pin
Schwinghebelrolle *f* valve roller
Schwingherd *m* shaking (vibrating) table
Schwingkondensator *m* vibrating[-reed] capacitor
Schwingkondensatorelektrometer *n* vibrating-reed electrometer, vibrating-diaphragm electrometer
Schwingkontakt *m* oscillating (vibrating) contact
Schwingkreis *m* oscillating (resonant, oscillatory) circuit
  geschlossener S. closed oscillating circuit
  offener S. open oscillating circuit
Schwingkreiskopplung *f* tuned-anode coupling
Schwingkreisspule *f* oscillator coil
Schwingkreisstrom *m* circulating current
Schwingkreisumrichter *m* oscillating circuit inverter
Schwingkreiswechselrichter *m* oscillating circuit inverter
Schwingkristallmethode *f* oscillating-crystal method
Schwingleistung *f* oscillatory power
Schwingloch *n* oscillation stop
Schwingmesser *n* *(text)* scutch blade
Schwingmetall *n* bonded-rubber pad
Schwingmotor *m* motor with reciprocating movement
Schwingmühle *f* vibration (swing-sledge) mill
Schwingneigung *f* tendency to oscillation
Schwingpflug *m* turn-about plough
Schwingprüfmaschine *f* fatigue-test machine
Schwingquarz *m* quartz oscillator (resonator), crystal oscillator
Schwingsaitendehnungsmesser *m* vibrating-wire strain gauge
Schwingschiff *n* vibrating shuttle *(sewing machine)*
Schwingschleifen *n* *s.* Kurzhubhonen
Schwingsichter *m* **für Hackschnitzel** shaker-type chip screen

Schwingsieb *n* vibrating (oscillating, reciprocating, shaker, swinging) screen, swing sieve
Schwingspannung *f (mech)* vibration (cyclic) stress
Schwingspule *f* oscillator coil
S. des Lautsprechers voice coil
Schwingspulenzähler *m* oscillating meter
Schwingstrom *m* oscillating current
Schwingung *f* 1. oscillation, vibration; [single] cycle; 2. *(mine)* kick *(boring cable)*
abklingende S. dying oscillation (vibration)
akustische S. acoustical vibration
aperiodische S. aperiodic oscillation
aufgedrückte S. forced oscillation (vibration)
aufgezwungene S. forced oscillation (vibration)
eigenerregte S. self-oscillation
elastische S. elastic vibration
elektrische S. electrical oscillation
elektromagnetische S. electromagnetic oscillation
erzwungene S. forced oscillation (vibration)
freie S. free oscillation (vibration)
fremderregte S. constrained oscillation (vibration)
gedämpfte S. damped oscillation (vibration)
gekoppelte S. coupled oscillation (vibration)
harmonische S. harmonic [oscillation], sinusoidal oscillation (vibration)
induzierte S. induced oscillation
labile S. unstable oscillation (vibration)
mechanische S. mechanical oscillation (vibration)
modulierte S. modulated oscillation
parasitäre S. parasitic oscillation (vibration)
pendelartige S. pendulum-like vibration
reine S. pure oscillation
schwach gedämpfte S. weakly damped oscillation (vibration)
selbsterregte S. self-oscillation
stationäre S. steady-state oscillation (vibration)
stehende S. stationary oscillation
ungedämpfte S. undamped (sustained, continuous) oscillation, undamped (sustained, continuous) vibration
von außen her verursachte S. vibration from outside sources
wilde S. parasitic oscillation (vibration)
Schwingungen *fpl* **der positiven Ionen** positive-ion oscillations
S. im Flüssigkeitsstrom delivery pulsation *(hydraulics)*
S. je Sekunde cycles per second
pendelartige S. swaying, pendulum-like vibration
Schwingungsabtaster *m* vibration pick-up
Schwingungsachse *f* axis of oscillation (vibration)
Schwingungsamplitude *f* amplitude of vibration (oscillation)
Schwingungsanfälligkeit *f* vibration liability
Schwingungsanzeiger *m* oscillation detector
Schwingungsarbeit *f* **der Moleküle** molecular kinetic energy
Schwingungsart *f* mode of vibration
Schwingungsbauch *m* vibration antinode (loop), loop of oscillation
Schwingungsbetrag *m* amplitude
Schwingungsbewegung *f* vibratory movement
Schwingungsbogen *m* arc of oscillation
Schwingungsdämpfer *m* vibration absorber (damper, isolator), antivibrator
Schwingungsdämpfung *f* absorption (damping) of vibration
Schwingungsdauer *f* period (time) of vibration, period (time) of oscillation; time of swing
S. des Instrumentenzeigers period of an instrument
Schwingungsebene *f* plane of oscillation (vibration)
Schwingungseinsatz *m* starting (setting-in) of oscillations
Schwingungseinsatzpunkt *m* point of self-oscillation, singing point
wirksamer S. active singing point
Schwingungsenergie *f* oscillatory (oscillation) energy
Schwingungsenergieniveau *n* vibrational energy level
schwingungserregend vibromotive, exciting oscillations
Schwingungserregung *f* excitation of oscillations
schwingungserzeugend vibromotive, generating oscillations
Schwingungserzeuger *m* vibration generator
Schwingungsfestigkeit *f* dynamic strength; resistance to vibration
Schwingungsform *f* type (mode, form) of vibration, type (mode, form) of oscillation
schwingungsfrei vibration-free, non-vibrating, non-oscillating
Schwingungsfrequenz *f* frequency of vibration
Schwingungsgeber *m* vibration pick-up
Schwingungsgenerator *m* vibration generator
Schwingungsgleichung *f* equation of oscillation
Schwingungsgrenze *f* oscillation limit
Schwingungsgröße *f* magnitude of oscillation
Schwingungshöhe *f* amplitude of oscillation
Schwingungsknoten *m* oscillation (vibration) node, nodal point of vibration (oscillation)
Schwingungskraft *f* vibration force
Schwingungskreis *m s.* Schwingkreis
Schwingungslehre *f* theory of vibrations
Schwingungsmaximum *n* peak (crest) of oscillation
Schwingungsmesser *m* vibration meter

schreibender S. vibrograph
schreibender S. mit Verstärker amplifying vibrograph
Schwingungsmessung *f* vibration measurement
Schwingungsmittelpunkt *m* centre of oscillation
Schwingungsmodus *m* mode [of oscillation] *(waveguides)*
Schwingungsniveau *n* vibrational level
Schwingungsperiode *f* cycle of oscillation (vibration)
Schwingungspuffer *m* vibration isolator
Schwingungsquant *n* vibrational quantum
Schwingungsquantenzahl *f* vibrational quantum number
Schwingungsschreiber *m* recording oscillometer
Schwingungssignal *n* vibrational signal
Schwingungsspektrum *n* oscillation (vibrational) spectrum
Schwingungssynthese *f* synthesis of oscillations
Schwingungstechnik *f* vibration engineering
Schwingungsterm *m* vibration level
Schwingungstilger *m* vibration absorber
Schwingungstyp *m* mode [of oscillation] *(waveguides)*
akustischer S. acoustic mode
aperiodisch abklingender S. evanescent mode
räumlich gedämpfter S. evanescent mode
Schwingungsverhalten *n* quadratic response
Schwingungsviskosimeter *n* ultrasonic viscosimeter
Schwingungsweite *f* amplitude of oscillation (vibration)
doppelte S. total amplitude
Schwingungswelle *f* undulation
Schwingungswellenform *f* oscillatory (vibration) waveform
Schwingungswellentyp *m* vibrational mode
Schwingungszahl *f* number of oscillations, rate of vibration (oscillation)
Schwingungszug *m* cycle
Schwingungszustand *m* vibrational state
Schwingverdichter *m* vibrating (vibratory) compactor *(concrete)*
Schwingvoltmeter *n* generating voltmeter
Schwingweite *f* vibration amplitude
Schwingzahl *f* **je Minute** number of oscillations per minute
Schwirrgeschwindigkeit *f (aut)* whirling speed
Schwitzen *n* misting *(of the screen)*; getting damp (wet) *(wall)*
Schwitzwasser *n* sweat *(in a vessel)*
Schwitzwasserbildung *f* deposit of moisture, formation of condensed moisture
Schwödebrei *m* lime cream
schwöden to paint
Schwörer-Überhitzer *m* Schwörer superheater
Schwund *m* 1. shrinkage *(timber)*; 2. fading *(radio)*
selektiver S. selective (differential) fading
Schwundausgleich *m* [automatic] volume control
Schwundhäufigkeit *f* fading rate
Schwundmaß *n* fading rate
Schwundminderer *m* anti-fading device
Schwundmuster *n* fading pattern
Schwundperiode *f* fading period
Schwundregelung *f* automatic gain control
selbsttätige verzögerte S. instantaneous automatic gain control
unverzögerte S. simple automatic gain control
verzögerte S. suppressed automatic volume control
Schwundriß *m* shrinkage crack *(timber)*
Schwundstörung *f* fading trouble
Schwungenergie *f* energy stored up in a flywheel
Schwungkörper *m* fly-balls
Schwungkraft *f* centrifugal force; inertial power
Schwungkraftanlasser *m* inertia starter
Schwungkugel *f* governor [fly-]ball
Schwungmasse *f* gyrating (centrifugal) mass
Schwungmassenmittelpunkt *m* centre of gyration
Schwungmoment *n* flywheel effect
Schwungrad *n* flywheel
beiderseitig gestütztes S. flywheel arranged between two bearings
fliegendes S. overhung flywheel
gesprengtes S. split flywheel
geteiltes S. sectional flywheel
mehrteiliges S. sectional flywheel
vierteiliges S. flywheel in quarters
Schwungradabstimmung *f* flywheel-loaded tuning
Schwungradantrieb *m* flywheel drive
Schwungradberechnung *f* flywheel calculation
Schwungradeffekt *m* flywheel effect *(resonant circuit)*
Schwungradexplosion *f* flywheel explosion
Schwungradgenerator *m* flywheel generator
Schwungradgrube *f* flywheel pit
Schwungradkeil *m* flywheel key
Schwungradkranz *m* rim of the flywheel
Schwungradlichtanlaßzünder *m* flywheel starter-generator ignition unit
Schwungradmagnet *m s.* Schwungradmagnetzünder
Schwungradmagnetzünder *m* flywheel magneto
Schwungradmarke *f* flywheel timing mark
Schwungradregler *m* flywheel governor
Schwungradriemenscheibe *f* flywheel pulley
Schwungradschaltung *f* flywheel circuit
Schwungradspeisepumpe *f* feed-pump with flywheel
Schwungradsynchronisation *f (el)* flywheel synchronization
Schwungradumformer *m* flywheel converter
Schwungradverklammerung *f* flywheel dowelling

Schwungradverzahnung *f* flywheel toothing (cogging)
Schwungradwirkung *f* flywheel effect
Schwungradzahnkranz *m* toothed flywheel rim
Schwungscheibe *f* flywheel
Schwungwirkung *f* flywheel action
Scott-Schaltung *f* Scott connection *(transformers)*
Sebazat *n (ch)* sebacate
Sebazinsäure *f* sebacic acid
Sech *n (ag)* coulter
sechsatomig hexatomic
Sechseck *n* hexagon
sechseckig hexagonal
Sechseckschaltung *f* hexagon connection
sechsfach sixfold
Sechsfachkette *f* sextuple-strand chain
sechsgliedrig *(ch)* six-membered
Sechskant *n* hexagon
Sechskanteinsatz *m* hexagonal socket
Sechskantgegenmutter *f* hexagon lock nut
Sechskantkronenmutter *f* hexagon slotted nut
Sechskantmutter *f* hexagon[al] nut
Sechskantprofil *n* hexagon
Sechskanträumwerkzeug *n* hexagon broach
Sechskantrevolverkopf *m* hexagonal turret
Sechskantringschlüssel *m* hexagon ring spanner
Sechskantschlüssel *m* hexagonal spanner
Sechskantschraube *f* hexagon[al]-head bolt, hexagon[al]-head screw
Sechskantstahl *m* hexagon [bar], hexagon bar steel
Sechskantwelle *f* six-spline shaft
Sechsphasengleichrichter *m* six-phase rectifier
Sechsphasenspannung *f* hexagon voltage
sechsphasig hexaphase
sechspolig six-pole
Sechspolröhre *f* hexode
Sechsring *m* six-membered ring
Sechsringverbindung *f* hexacyclic compound
sechsseitig hexagonal
Sechsspindelautomat *m* six-spindle automatic
sechsspindelig six-spindle
Sechsspindelstangenautomat *m* six-spindle [horizontal] automatic bar machine
sechsstellig six-figure
Sechsstufenfilter *n* six-step filter
sechsstufig six-speed *(gear-box)*, six-step
sechswertig hexavalent, sexivalent
Sechswertigkeit *f* sexivalency
Sechszylindermotor *m* six-cylinder engine
Sechszylinderreihenmotor *m* six-cylinder in-line engine
Sechzehnerleitung *f* quadruple phantom circuit *(telephony)*
Sediment *n* sediment
Sedimentation *f* sedimentation, silting
Sedimentationsgeschwindigkeit *f* rate of sedimentation
Sedimentgestein *n* sedimentary rock
sedimentieren to sedimentate
Sedimentrohr *n* calyx
See *m* lake
See *f* sea; ocean
  S. / auf hoher at full sea
  S. von vorn head sea
  achtern auflaufende S. following seas
  auf Deck überkommende S. dollop
  aufkommende S. rising sea
  bewegte S. agitated sea
  durcheinanderlaufende S. confused sea
  gekreuzte S. beam sea
  hohe S. high seas
  kabbelige S. choppy sea
  leicht bewegte S. slight sea
  offene S. offing, high seas
  schwere S. high sea
  sehr schwere S. foaming sea
  spiegelglatte S. mirror (dead-smooth) sea
  überbrechende S. overfall
  wilde S. confused sea
Seeamt *n* Admiralty Court, Court of Admiralty, Court of Inquiry
Seeanker *m* sea (driving) anchor, drag
Seebäderschiff *n* sea-side excursion ship
Seebär *m* sea dog *(able seaman)*
Seebau *m* marine (maritime) construction
Seebauten *pl* marine structures (works)
Seebauwerk *n* marine structure
Seebeben *n* seaquake
Seebeck-Effekt *m* Seebeck (thermo-electric) effect
Seebeschädigungsbedingungen *fpl* sea-damaged terms
Seebetrieb *m* sea routine
Seeboot *n* sea boat
Seebulldog *m* sea mule *(harbour tug)*
Seed *n (nucl)* seed
Seedampfer *m* ocean (sea-going) steamer
seediensttauglich able-bodied
See-Echo *n* sea return *(radar)*
See-Eigenschaften *fpl* sea-keeping qualities
See-Eimerbagger *m* sea-going bucking dredger
See-Erz *n* lake iron ore
seefähig seaworthy
Seefähigkeit *f* weatherly qualities *(of a vessel)*
seefahrend seafaring
Seefahrer *m* mariner, seafarer, sailor
Seefahrt *f* sea (ocean) navigation
Seefahrtkreuzer *m* off-shore cruiser, sea-going cruiser *(yacht)*
Seefahrtsamt *n* Navigation Board
Seefahrtsschule *f* school of navigation, naval school
Seefahrzeit *f* sea turn *(of a vessel)*
Seefallreep *n* sea (rope, Jacob's) ladder
Seefernkabel *n* long-distance submarine cable
Seefernsprechkabel *n* submarine telephone cable

seefest seaworthy
Seefischerei *f* sea fishing (fisheries)
Seeflugzeug *n* seaplane
Seefracht *f* sea freight, *(Am)* ocean freight
Seefrachtbrief *m* bill of lading, shipping bill
Seefunk *m* marine radio communication
Seefunkdienst *m* marine radio service
Seefunkstelle *f* marine (naval) radio station
Seegang *m* sea[way], heave (motion) of the sea
Seegangechos *npl* sea returns *(radar)*
Seegangsbecken *n* sea-keeping basin *(model testing)*
Seegangschwundeffekt *m* roller fading *(radar)*
Seegangsreflex *m* sea (wave) clutter *(radar)*
Seegangstärke *f* force of sea
Seegatt *n* gut *(channel)*
Seegefahren *fpl* sea risk, risks at sea, maritime peril
seegehend sea-going, ocean-going
Seegericht *n* maritime court, Court of Admiralty
Seegrasfaser *f* sea-grass fibre
Seegüterverkehr *m* sea-borne goods traffic
Seehafen *m* sea-port, [trading] port
Seehafenbahnhof *m* harbour station
Seehafenspediteur *m* shipping agent
Seehahn *m* sea cock *(valve)*
Seehandbuch *n* sailing directions (directory book)
Seehandel *m* maritime commerce, sea-borne trade (commerce)
Seehorizont *m* visible (apparent) horizon, horizon
Seehundsfell *n* sealskin
Seekabel *n* submarine (submerged) cable
Seekabeltelegrafie *f* submarine telegraph system
Seekanal *m* ship (sea) canal
Seekarte *f* [sea] chart, nautical chart; hydrographical map
Seekennung *f* sea-mark; intelligence of the sea marks
Seekompaß *m* mariner's compass
Seekreuzer *m* ocean (off-shore, sea-going) cruiser *(yacht)*
Seeküste *f* seaboard, seashore
Seele *f* core *(cable)*; web
Seeleichter *m* coasting (sea-going) barge, coasting (sea-going) lighter
Seelenelektrode *f* cored (flux-core) electrode
Seelenfaden *m* core thread
Seelogbuch *n* sea log
Seemann *m* seaman
befahrener S. able seaman
Seemannsmesser *n* jackknife
Seemarke *f* sea-mark
Seemaschinist *m* marine engineer
Seemeile *f* sea (nautical) mile
Seemine *f* naval mine
Seemuli *n* sea mule *(harbour tug)*
Seendampfer *m* lake steamer
Seenot *f* distress at sea
Seenotboot *n* rescue craft, rescue (life-saving) boat
Seenotdienst *m* sea-rescue service
Seenotfahrzeug *n* rescue boat (craft)
Seenotflugzeug *n* sea-rescue aircraft
Seenotfrequenz *f* distress frequency
Seenotkreuzer *m* rescue cruiser
Seenotmeldedienst *m* reporting-service for distress at sea
Seenotruf *m* distress call
Seenotrufgerät *n* / **selbsttätiges** auto-alarm device
Seenotverkehr *m* *(telec)* distress traffic
Seenotwellenlänge *f* distress wavelength
Seeotter *m* sea otter
Seeotterfell *n* sea otter
Seepocken *fpl* barnacles
Seerecht *n* sea (maritime) law
Seereise *f* [ocean] voyage
kurze S. trip
Seeschiff *n* sea boat, seaboat, sea-going ship (vessel), ocean-going ship
Seeschiffahrt *f* maritime navigation, ocean (merchant) shipping, liner trade
Seeschiffahrtsstraßenordnung *f* Rules of the Road at Sea, rules of navigation
Seeschlepper *m* sea-going tug
Seeschleuse *f* sea gate
Seeschlitten *m* sea sled, inverted-V-bottom boat
Seestraße *f* sea-lane, water high-way
Seestraßenordnung *f* Rules of the Road at Sea, rules of navigation; International Regulations for Preventing Collisions at Sea *(London 1948)*
Seestreitkräfte *fpl* naval forces
leichte S. auxiliary fighting ships
Seetang *m* sea-weed
Seetörn *m* sea turn *(of a vessel)*
Seetransport *m* sea (marine) transport, sea conveyance
Seetransportversicherung *f* marine insurance
seetüchtig seaworthy, wholesome
Seetüchtigkeit *f* seaworthiness
Seetüchtigkeitsattest *n* seaworthiness certificate
Seeübernehmen *n* / **achterliches** pooping
Seeunfall *m* sea accident, maritime peril
Seeuntüchtigkeit *f* unseaworthiness
Seeventil *n* sea cock (valve)
Seeverhalten *n* **eines Schiffes** routing of a ship
Seeverhaltenversuche *mpl* sea-keeping trials
Seeverkehr *m* sea (marine, ocean) traffic, ocean (marine) transport
Seeverklarung *f* sea-protest
Seeversicherer *m* maritime insurer, underwriter
seewärts seaward[s]
seewasserbeständig sea-waterproof
Seewasserechtheit *f* fastness to sea water

Seewasserstraße *f* water (ocean) highway
Seewasserverdampfer *m* sea-water evaporator
Seeweg *m* seaway, track, sea route (lane)
Seewesen *n* maritime affairs
Seezeichen *n* sea-mark, balize, navigational aids
Seezeichenschiff *n* buoy layer, buoy-laying vessel
Segel *n* sail
Segelbaum *m* [sail] boom
Segelboot *n* sailing boat; *(Am)* sail boat
Segeldingi *n* sailing dinghy
Segelfähigkeit *f* sailing trim
Segelfläche *f* sail area
Segelfliegen *n* gliding
Segelflugzeug *n* glider
Segelführung *f* carrying of sails
Segelhandbuch *n* sailing directions
Segeljacht *f* **mit Hilfsmotor** auxiliary [yacht]
Segeljolle *f* yawl, sailing dinghy
Segellatte *f* batten
Segelliek *n* bolt rope
Segellogger *m* sailing drifter
Segelreffen *n* reefing of sails
Segelriß *m* sail plan
Segelschiff *n* sailing ship; *(Am)* sail ship
S. mit Hilfsmotor sailing ship with auxiliary motor, motor-sailer
Segelschiffahrt *f* sailing navigation
Segelschlitten *m* ice-boat, ice-yacht
Segelschulschiff *n* training sailing ship, sailing training ship
Segelschwerpunkt *m* centre of effort of sails
Segelsport *m* sailing, yachting
Segelstellung *f (aero)* feathered pitch
Segeltuch *n* canvas
Segelwerk *n* sails
Segerkegel *m* Seger [temperature] cone, pyrometric (melting) cone
Segler *m* sailing ship, sailer; *(Am)* sailship
schlechter S. heavy sailer
Segment *n* segment; section *(commutator)*
Segmentanker *m (el)* segment-core armature
Segmentbogen *m (build)* segmental arch, arched lintel
Segmentlager *n* segmental bearing
Segmentschleifkörper *m* segmental abrasive block
Segmentschleifscheibe *f* segmental grinding wheel
Segmentspannung *f* [commutator] segment voltage
Sehachse *f* axis of vision
Seheindruck *m* visual sensation
Sehen *n* vision
mesopisches S. mesopic vision
räumliches S. stereoscopic vision
Sehfehler *m* deficiency of sight
Sehfeld *n* field of view
Sehleistung *f* visual efficiency
Sehne *f* chord *(geometry)*
Sehnenlänge *f* chord length, length of the chord
Sehnenmaß *n* chordal dimension
Sehnenwicklung *f (el)* chord (fractional-pitch) winding
Sehnungsfaktor *m* pitch factor [of winding], winding (breadth) coefficient
Sehprobentafel *f* sight-testing chart
internationale S. international sight-testing chart
Sehpurpur *m* visual purple
Sehreiz *m* optical stimulus
Sehrohr *n* periscope
Sehschärfe *f* visual acuity
Sehstörung *f* perturbed vision
Sehwinkel *m* visual (optic, viewing) angle
Sehzeichenprojektor *m* test-figures projector
Seide *f* silk
geklöppelte S. braided silk
wilde S. wild silk
seidebesponnen silk-covered *(wire)*
Seidenabfall *m* silk waste
unentbasteter S. gum waste
seidenartig silky
Seidenbast *m* silk gum
Seidendraht *m (el)* silk-covered wire
Seidenfaden *m* silk thread (fibre)
Seidenfinisheffekt *m* silk finish effect
Seidenfinishkalander *m* calender for scrooping
Seidengewebe *n* silk fabric
Seidenglanz *m* silkiness, silky lustre
seidenglänzend silky [in lustre]
Seidenglanzkalander *m* Schreiner calender
Seidengriff *m* scroopy feel
Seidenhaspel *f* swift
Seidenherstellung *f* silk manufacture
Seidenisolation *f* silk insulation
Seidenkabel *n* silk-covered cable
Seidenlackdraht *m (el)* varnished silk-braided wire
Seidenleim *m* silk-gum, silk-glue, sericin
seidenmattiert satin-frostet
Seidenpapier *n* tissue paper
Seidenrasterdruck *m* silk-screen printing
Seidenraupe *f* silkworm
Seidenraupenzucht *f* 1. sericulture; 2. silkworm farm
Seidenraupenzüchter *m* sericulturist
Seidenschrei *m* scroop
Seidenspinner *m* silk moth
Seidenspulmaschine *f* silk-winding frame
Seidenstoff *m* silk fabric (cloth)
Seidenstuhl *m* silk loom
Seidenverarbeitung *f* silk manufacture
Seidenweberei *f* silk weaving
Seidenwebstuhl *m* silk loom
seideumsponnen silk-covered *(wire)*
seidig silky

seidigglänzend silken in lustre, silky
Seife *f* soap
ungefüllte S. genuine soap
seifenartig soapy, saponaceous
Seifenbergbau *m* placer mining
seifenbeständig soap-resistant
Seifenbildung *f* saponification
Seifenblase *f* soap bubble
Seifenerz *n* alluvial ore
Seifenflocken *fpl* soap flakes (chips), flake soap
Seifengold *n* placer (alluvial) gold
Seifenhaut *f* soap film
Seifenhautgleichnis *n* analogy of membrane, membrane analogy
Seifenherstellung *f* soap manufacture
seifenkochecht fast to boiling soap-suds
Seifenlauge *f* soap-suds
Seifenleim *m* soap paste
Seifenlösung *f* soap solution
Seifenpulver *n* soap powder
Seifenschaum *m* lather
Seifensiederei *f* soap factory
Seifenspiritus *m* soap tincture
Seifenstrangpresse *f* bar moulding machine
Seifenwalke *f* soap-milling
Seifenwasser *n* soap water
Seifenzinn *n* stream tin
Seifenzusatzstoffe *mpl* soap builders
Seifnachbehandlung *f* soaping aftertreatment
seiger vertical
Seigerarbeit *f* liquation process
Seigerherd *m* liquation hearth
seigern to segregate, to liquate
Seigern *n* segregation, liquation; sweating *(tin)*
Seigerofen *m* liquating furnace
Seigerpfanne *f* liquation pan
Seigerrückstand *m* liquation residue
Seigerschacht *m* vertical shaft
Seigerschlacke *f* liquation slag
Seigerstelle *f* segregated spot
Seigerung *f* segregation; liquation process
umgekehrte S. inverse segregation
Seigerungsstelle *f* segregated spot
Seigerungsstreifen *m* segregated band
Seigerungszelle *f* segregation streak
Seignettesalz *n* Seignette salt
Seignettesalzkristall *m* Seignette electric crystal
seihen to strain, to filter
Seihtuch *n* straining cloth
Seil *n* rope, cord, line, cable; string *(mechanics)*
aufgedrehtes S. unlaid rope
patentverschlossenes S. locked-coil rope
schlaffes S. slack rope
vierschäftiges S. four-stranded rope
Seilablenkung *f* rope fleeting
Seilablenkungswinkel *m* fleet angle
Seilauflegen *n* rope installation, roping up
Seilaufliegezeit *f* rope life
Seilausgleichsvorrichtung *f* rope-tension compensating equipment
Seilbagger *m* cable excavator
Seilbahn *f* cable railway [at ground level], funicular railway, ropeway, aerial chair-lift
Seilbandförderer *m* ropebelt conveyor
seilbetätigt cable-powered, cable-operated
Seilbiegungsfaktor *m* rope bending ratio
Seilbohren *n* cable-tool drilling
Seilbohrverfahren *n* cable-tool drilling
Seilbohrzeug *n* drilling string, string of tools
Seilbruchlast *f* rope breaking load
Seilbrücke *f* cable bridge
Seilbucht *f* rope loop
Seildehnung *f* / **bleibende** permanent rope stretch
Seildreher *m* rope twister
Seildruckluftbremse *f* air-operated cable brake
Seildurchhang *m* sag of rope
Seileck *n* funicular (link, string) polygon
Seileinband *m* rope capping
Seilende *n* / **aufgedrehtes** fag end
Seiler *m* ropemaker, cordmaker
Seilerware *f* cordage, rope-ware
Seilerwarenfaser *f* cordage fibre
Seilfähre *f* rope ferry
Seilfahrt *f* man-riding, men-riding
Seilfahrtboden *m* man-riding deck
Seilfahrtbühne *f* man-landing platform (level), man-riding platform (level)
Seilfahrtschacht *m* man shaft
Seilfanghaken *m* spear
Seilfaser *f* rope (cordage) fibre
Seilflasche *f* rope socket; rope block
Seilflaschenzug *m* wire rope hoist
Seilförderanlage *f* cable haulage machine
Seilförderband *n* cable-belt conveyor
Seilförderbandgerüst *n* rope framed conveyor structure
Seilförderung *f* rope haulage
Seilführung *f* 1. rope guide *(conveying)*; 2. *(text)* carriage band arrangement *(selfactor)*
Seilgeschirr *n* tackle *(overhead-line construction)*
Seilgeschwindigkeit *f* *(mine)* hoisting (rope, winding) speed
seilgesteuert cable-operated
Seilhängebahn *f* aerial ropeway
Seilhaspel *f* rope reel
Seilhülle *f* cable conduit
Seilklemme *f* rope clamp (clip, cramp); cable clamp; wire-rope grip
Seilkriechausgleichvorrichtung *f* rope creep-correction device
Seilkurve *f* funicular curve
Seillebensdauer *f* rope life
Seilleitrolle *f* rope sheave
Seillitze *f* rope strand
Seilpolygon *n* funicular (link, string) polygon

Seilreibung *f* rope friction, friction of the rope
Seilrille *f* rope groove
Seilrillenfutter *n* rope tread
Seilrolle *f* rope sheave
Seilrutsch *m* rope slip
Seilscheibe *f* rope sheave; *(mine)* head gear pulley, headsheave
lose S. hoisting block
Seilscheibengetriebe *n* train of rope sheaves
Seilscheibenschwungrad *n* rope-pulley flywheel
Seilschlag *m* rope lay
Seilschlagbohren *n* churn drilling (boring)
Seilschlagmaschine *f* rope-making machine
Seilschloß *n* rope clamp (clip, cramp)
Seilschmiermittel *n* rope lubricant
Seilschrämgerät *n (mine)* rope-suspended cutter
Seilschrapper *m* cable scraper
Seilschwebebahn *f* aerial cableway, overhead-cable railway
Seilspannklemme *f* draw-vice, draw-tongs *(overhead line)*
Seilspannung *f* rope tension *(overhead line)*
Seilspeer *m* spear
Seilsteuerung *f* rope (cable) control
S. / mit cable-controlled
Seilstrahl *m* ray of the polygon of forces
Seiltragrolle *f* top pulley
Seiltrennen *n* cutting-off of a rope
Seiltrieb *m* rope drive
Seiltrommel *f* hoisting drum; rope barrel (drum)
Seil- und Drehbohrgerät *n* / **kombiniertes** combination rig
seilverankert guyed
Seilwechsel *m* rope changing
Seilwinde *f* wire-rope hand-winch; cable winch
Seilzug *m* 1. *(el)* tow-line; 2. funicular (link, string) polygon *(mechanics)*; 3. *(aut)* cable control
Seilzugbetätigung / mit *(aut)* cable-operated
Seilzugbremse *f (aut)* cable brake
Seilzugdifferential *n (dat)* tape differential
Seilzugsteuerung *f (aut)* cable control
Seiner *m* seiner *(fishing boat)*
Seismik *f* seismology
Seismizität *f* seismicity
Seismograf *m* seismograph
Seismogramm *n* seismogram
Seismologie *f* seismology
Seismometrie *f* seismometry
Seismoskop *n* seismoscope
Seite *f* 1. side; page *(book)*; 2. *(math)* member *(equation)*, side *(geometry)*; 3. *(build)* wing
S. an Seite side by side *(of two vessels)*
gerade S. back (even, verso) page *(book)*
linke S. lefthand side
rechte S. righthand side
schmale S. *(wood)* planed measure
ungerade S. odd (recto) page *(book)*
Seitenaberration *f* / **sphärische** transverse spherical aberration
Seitenabfall *m* scrap side
Seitenabschnitt *m* scrap side
Seitenabweichung *f* lateral deviation
Seitenachse *f* lateral axis
Seitenanker *m (el)* side armature
Seitenankerrelais *n* side armature relay
Seitenanschlag *m* side stop
Seitenanschluß *m* side contact
Seitenansicht *f* side (end) view
linke S. left side view
rechte S. right side view
Seitenaufriß *m* side elevation
Seitenband *n* sideband
oberes S. upper sideband
übertragenes S. main sideband, *(Am)* transmitted sideband
unabhängiges S. independent sideband
unteres S. lower sideband
Seitenbandanalysator *m* sideband analyzer
Seitenbänder *npl* / **störende** spurious sidebands
Seitenbandfrequenz *f* sideband frequency
Seitenbandinterferenz *f* sideband interference
Seitenbandspektrum *n* sideband spectrum
Seitenbandstörung *f* adjacent-channel interference, infringement
Seitenbeplankung *f (ship)* side planking
Seitenbesäumer *m* side trimmer
Seitenbestimmer *m* sense finder *(radar)*
Seitenbestimmung *f* sense finding *(radar)*
Seitenbestimmungsantenne *f* sense aerial
Seitenbezirk *m* lateral area
Seitenboot *n* quarter (side) boat, davit craft
Seitenbühne *f* side stage, wing
Seitendavit *m (ship)* quarter davit
Seitendrehmeißel *m* side cutting turning tool
Seitendruck *m* lateral pressure
dynamischer S. dynamic lateral pressure
Seitendrucker *m* page printer
Seitendrucktelegraf *m* page-printing telegraph
Seitenecho *n* side-lobe echo *(radar)*
Seitenelektrode *f* side (lateral) electrode
Seitenentleerer *m (ry)* side-dump car, side-discharge wagon
Seitenfenster *n* side window; *(ship)* port light
Seitenfläche *f* side surface; facet
Seitenflosse *f (aero)* dorsal fin, vertical stabilizer
Seitenfrequenz *f* side frequency
Seitenführung *f* lateral guiding
Seitenfülling *f* air course board *(wooden ship)*
Seitengang *m* side corridor
seitengesteuert side-valve *(engine)*
Seitengewehr *n* side-arm, bayonet
Seitengrifflader *m (mine)* side-arm grab loader
Seitengummi *m* wall rubber *(tyre)*
Seitenguß *m* pouring at the joint line of the mould

Seitenhalter *m* registration arm *(trolley wire)*
Seitenhöhe *f* page depth *(book)*; *(ship)* moulded depth, depth moulded; framing depth *(frame)*
Seitenkasten *m (ship)* wing structure
Seitenkastendeck *n* wing deck *(floating dock)*
Seitenkette *f (ch)* side chain
Seitenkettenkonformation *f (ch)* side-chain conformation
Seitenkeule *f* side lope *(of aerials)*
Seitenkielschwein *n (ship)* sister keelson
Seitenkipper *m* side-dump truck
Seitenkipplader *m* side dump (tilting shovel) loader
Seitenklappen *fpl* side flaps
Seitenklüse *f (ship)* mooring pipe
Seitenkomponente *f* **des Bahndruckes** component at right angles to the plane of blade motion *(turbine)*
Seitenlage *f* lateral position
Seitenlampe *f* side-lamp
Seitenlänge *f* length of side, lateral length
S. von Schraubenmitte bis Schraubenmitte distance between end rows of bolt-holes
Seitenleiste *f* side plate
Seitenleitwerk *n (aero)* rudder unit; vertical tail surfaces
Seitenlichter *npl (ship)* side-lights
Seitenmeißel *m* shoulder tool
abgesetzter linker S. left-offset tool
abgesetzter rechter S. right-offset tool
Seitenneigung *f* lateral sway
Seitenpeilung *f* relative bearing
Seitenplatte *f* end shield
Seitenraddampfer *m* side paddle-wheel steamer, side wheeler
Seitenrampe *f (ry)* lateral loading platform
Seitenrand *m* margin *(book)*
äußerer S. fore-edge
innerer weißer S. back inner margin
oberer S. head
unterer S. tail
Seitenräumer *m* angledozer, angling dozer
Seitenräumerplanierraupe *f* angledozer, angling dozer
Seitenregistersteuerung *f* web guide
seitenrichtig true to side
Seitenrichtmaschine *f (mil)* traversing gear (mechanism)
Seitenring *m* side ring
Seitenriß *m* side elevation (view) *(technical drawing)*; sheer plan
Seitenrohrkessel *m* boiler with eccentric flue
Seitenrost *m* side section of grate
Seitenruder *n* quarter rudder *(boats)*
Seitenschalung *f* side form *(road building)*
Seitenschlichtmeißel *m* side cutting finishing tool
Seitenschneider *m* side cutting pliers
Seitenschott *n (ship)* side (wing) bulkhead
Seitenschruppmeißel *m* side cutting roughing tool
Seitenschwert *n* lee-board *(of small vessels)*
Seitenschwertboot *n* lee-boarder
Seitenstapler *m* side loader
Seitensteifigkeit *f* lateral stiffness
Seitenstraße *f* subsidiary (minor) road
Seitenstreifen *m* verge, shoulder *(highways)*
Seitenstringer *m (ship)* hog frame
Seitenstütze *f (ship)* side shore, quarter pillar
Seitensupport *m* side head (tool-box) *(planer)*
Seitentank *m (ship)* wing (side) tank
Seitenträger *m (ship)* side girder *(double bottom)*
Seitentrawler *m* side trawler
Seitentrum *m* dropper
Seitenumkehr *f (telev)* lateral inversion
Seitenverhältnis *n* **der Tragflügel** *(aero)* aspect ratio
seitenverkehrt left-and-right reversed; wrong reading
Seitenverkleidung *f* panelling
Seitenwand *f* side wall; *(aut)* side panel
Seitenwandrahmen *m* side-panel frame
Seitenwandschwebefahrzeug *n* side-wall hovercraft
Seitenwelle *f* axle (half) shaft
Seitenwiderstand *m* lateral resistance
Seitenwind *m (ship)* beam wind
Seitenwindfahrwerk *n* drift landing-gear
Seitenzug *m* side flue *(heating)*; *(mach)* side draw
seitlich sidewise, lateral
Sekante *f* secant
Sektion *f (ship)* section
vorgearbeitete S. preconstructed section
Sektionalkammerkessel *m* sectional header boiler
Sektionalkessel *m* sectional boiler
Sektionsbauweise *f* section building *(shipbuilding)*
Sektionsdock *n* sectional dock
Sektionspontondock *n* sectional pontoon dock
Sektor *m* sector
Sektorabtastung *f* sector scanning *(radar)*
Sektorbildung *f* sectoring *(radar)*
Sektorendiagramm *n* pie chart
Sektorinstrument *n (el)* sector pattern instrument
Sektorleiter *m (el)* shaped conductor
sekundär secondary
Sekundärabstimmung *f* secondary tuning
Sekundäranker *m* secondary armature
Sekundärauslösung *f* indirect tripping, shunt release
Sekundärbatterie *f* secondary battery
Sekundärecho *n* second-trace echo *(radar)*
Sekundärelektron *n* secondary electron

Sekundärelektronenausbeute *f* secondary yield, yield in secondary electrons, gain of secondary emission
Sekundärelektronenemission *f* secondary electron emission
Sekundärelektronenspektrum *n* secondary electron spectrum
Sekundärelektronenvervielfacher *m* secondary electron multiplier, electron multiplier [tube], photomultiplier [tube]
Sekundärelement *n* secondary cell
Sekundäremission *f* secondary emission
Sekundäremissionsausbeute *f* secondary-emission rate
Sekundäremissionselektrode *f* secondary-emission electrode
Sekundäremissionsfaktor *m* secondary-emission rate
Sekundärempfang *m* secondary reception
Sekundärionisation *f* secondary ionization
Sekundärklemme *f* secondary terminal
Sekundärkreis *m* 1. secondary circuit *(transformer)*; 2. secondary coolant circuit *(reactor)*
aperiodischer S. untuned secondary circuit
Sekundärkreislauf *m* secondary [coolant] circuit
Sekundärkühlkreis *m* secondary coolant circuit
Sekundärlagerstätte *f (geol)* secondary deposit
Sekundärmetall *n* secondary metal
Sekundärnormale *f* sub-standard
Sekundärreaktion *f* secondary reaction
Sekundärreaktor *m* secondary reactor
Sekundärregler *m* submaster (secondary) controller
Sekundärregulierung *f* secondary control
Sekundärschild *m* secondary shield[ing]
sekundärseitig secondary
Sekundärspannung *f* secondary voltage
Sekundärspule *f* secondary coil
Sekundärstoß *m* secondary collision
Sekundärstrahlenblende *f* anti-diffusion grid
Sekundärstrahlenschutzwand *f* secondary protective (protection) barrier
Sekundärstrahler *m* director; parasitic element
Sekundärstrahlung *f* secondary radiation
Sekundärstreuung *f* secondary leakage
Sekundärstrom *m* secondary current
Sekundärträger *m (telev)* subcarrier
Sekundäruhr *f s.* Nebenuhr
Sekundärwelle *f* secondary shaft
Sekundärwicklung *f* secondary winding
Sekundärwiderstand *m* secondary resistance
Sekunde *f* second
Sekundengeber *m* seconds contactor
Sekundenimpuls *m* seconds pulse
Sekundenzähler *m* seconds counter
Sekundenzeiger *m* sweep (seconds) hand
sekundlich per second
selbstabdichtend non-spill *(coupling)*; self-sealing
Selbstabgleich *m* automatic balancing; self-adjustment
selbstabgleichend self-balancing; self-adjusting
selbstablaufend automatic
Selbstabschirmung *f* self-shielding
Selbstabsorption *f* self-absorption
Selbstabsorptionskoeffizient *m* self-absorption coefficient, internal absorption coefficient
Selbstabstimmung *f* auto-tuning
selbstadjungiert self-adjoint
selbständig unaided, unassisted; independent; self-contained
Selbstanlasser *m* self-starter, automatic starter
Selbstanleger *m* automatic feeder
Selbstanpassung *f* automatic adjustment
selbstansaugend normally aspirated *(petrol engine)*; self-priming *(pump)*
Selbstanschlußamt *n* automatic exchange
Selbstaufheizkatode *f* ionic-heated cathode
Selbstaufheizung *f* internal heating
selbstaufrichtend self-righting *(boat)*; self-erecting *(crane)*
selbstaufzeichnend self-registering
Selbstausgleich *m* inherent regulation
Selbstausleger *m* automatic delivery
Selbstauslöschung *f* self-quenching
Selbstauslöseeinrichtung *f* automatic tripping device *(telephony)*
Selbstauslöser *m* automatic release
Selbstauslösung *f* automatic disengagement (release, trip)
Selbstberührungspunkt *m* tacnode *(curve)*
Selbstblock *m* automatic block
S. mit Achszählern automatic block with axle counters
S. mit Gleisstromkreis automatic block with track circuits
selbstbrennend self-baking *(electrode)*
selbstdichtend self-sealing
Selbstdiffusion *f* self-diffusion
selbstdockend self-docking
selbsteichend self-checking
Selbsteinleger *f* self feeder
Selbsteinschalter *m* self-closing circuit breaker
selbsteinstellend self-aligning, self-adjusting
Selbsteinstellung *f* self-alignment, self-adjustment, automatic adjustment
Selbstenergie *f* self-energy
Selbstentladeeinrichtung *f* self-discharging facility
selbstentladend self-unloading
Selbstentlader *m* automatic discharge wagon, automatic dump car; self-unloading vessel, self-unloader
Selbstentladeschiff *n* self-unloader, self-unloading vessel, self-discharging vessel
Selbstentladewagen *m* automatic-discharge hopper car

Selbstentladung *f* 1. self-discharge; 2. automatic unloading
selbstentzündlich spontaneously inflammable, pyrophoric, self-igniting
Selbstentzündung *f* autoignition, self-ignition, spontaneous ignition (combustion)
Selbstentzündungspunkt *m* autoignition point
Selbstentzündungstemperatur *f* autoignition temperature
selbsterregend self-exciting
selbsterregt self-excited
Selbsterregung *f* spontaneous excitation, self-excitation; squitter *(measuring)*
direkte S. auto self-excitation, self-saturation
getrennte S. separate self-excitation
kritische (volle) S. critical self-excitation
selbsterstarrend air-drying *(core)*
selbstfahrend self-propelled
Selbstfahrer *m* self-propelled vessel, motorbarge
schiebender S. pushing motorbarge
Selbstfahrlafette *f* self-propelled gun mounting, motorized cart
S. für Panzerbekämpfung *(Am)* tank destroyer
Selbstfernwahl *f* automatic toll dialling
Selbstfilterung *f* self-filtering; self-shielding
selbstfokussierend self-focusing
Selbstfokussierung *f* automatic focusing action
Selbstfütterer *m* self-feeder
Selbstgang *m (mach)* power traverse [motion]
selbstgebaut shop-built, self-made, home-made
Selbstgeber *m* self-feeder
selbstgefertigt shop-built, self-made, home-made
selbsthemmend self-retaining; self-locking
Selbsthemmung *f (mine)* self-locking *(prop)*
S. / mit retained by friction; self-retaining, self-holding
Selbstinduktion *f* self-induction
Selbstinduktionskoeffizient *m* coefficient of self-induction
Selbstinduktionsspule *f* inductance coil
selbstinduktiv self-inductive
Selbstinduktivität *f* self-inductance
Selbstionisation *f s.* Autoionisation
selbstkippend self-tipping
selbstklebend self-adhesive, self-sealing
Selbstkontrolle *f* automatic checking
Selbstkosten *pl* prime cost
Selbstkostenberechnungsabteilung *f* cost accounting department
Selbstkostenkalkulation *f* cost accounting
Selbstkostenpreis *m* cost price
Selbstkostenrechnung *f* cost accounting
Selbstkühlung *f* self-cooling *(transformer)*
S. / mit natural-cooled *(machine)*
Selbstladeschiff *n* self-loading ship
Selbstlaufdrehzahl *f* self-sustaining speed
Selbstlöschung *f* self-quenching
selbstöffnend self-opening
Selbstprogrammierer *m* compiler
Selbstprogrammierung *f* automatic programming
selbstprüfend automatically checking
Selbstprüfung *f* automatic checking *(computer)*
Selbstregelung *f* self-regulation; automatic control
Selbstreinigung *f* self-purification
Selbstsättigungsschaltung *f* self-saturation connection
selbstschaltend self-indexing *(turret)*; automatically engaging
Selbstschalter *m* automatic circuit-breaker
Selbstschalterkontakt *m* automatic switch contact
Selbstschaltgetriebe *n* self-shifting gear
selbstschärfend self-sharpening
Selbstschärfung *f* self-sharpening action *(grinding wheel)*
Selbstschlußapparat *m* automatically closing water gauge
S. mit Verschlußklappe automatic closing device with flap valve
Selbstschlußklappenventil *n* automatic flap steam pipe isolating valve
Selbstschlußkugelventil *n* automatic ball steam pipe isolating valve
Selbstschlußventil *n* automatic stop valve
selbstschmierend self-lubricating
selbstschreibend automatically recording, autographic
selbstsichernd self-locking
Selbstspalten *n* blistering *(hide)*
selbstspannend self-clamping
Selbstspanner *m* spring ring
selbstsperrend self-locking
Selbststabilisierung *f* self-stabilization
Selbststeuerapparat *m* gyro pilot control
Selbststeuereinrichtung *f* automatic pilot
Selbststeuerung *f (el)* automatic control; *(aero)* automatic direction finding; gyro pilot control
Selbststreuung *f* self-scattering
selbstsynchronisierend self-locking, self-synchronizing
selbsttätig automatic[al]; self-acting; unassisted, unaided; independent; self-contained
selbsttragend self-supporting
Selbsttränker *m (ag)* drinker
Selbsttrimmerschiff *n* self-trimming vessel
Selbstüberlagerer *m* autodyne, self-heterodyne
selbstüberlagernd autodyne, self-heterodyning
Selbstumkehrung *f* **von Spektrallinien** inversion (self-reversal) of spectral lines
Selbstunterbrecher *m* automatic interrupter, buzzer *(telephony)*
Selbstunterbrechung *f* automatic interruption
Selbstverstellung *f* automatic advance

Selbstvorschub *m* automatic gauge *(cutter)*
Selbstwählbetrieb *m* dial (automatic telephone) service
Selbstwählferndienst *m* trunk dialling service
Selbstwählfernverkehr *m* subscriber's trunk dialling, automatic long-distance service, toll line dialling
selbstzentrierend self-centring
Selbstzündung *f* autoignition, self-ignition, automatic (spontaneous) ignition
S. durch Verdichtung compression ignition
Selbstzündungstemperatur *f* self-ignition temperature
selektieren to pick out
Selektion *f* selection; selectivity *(radio)*
Selektionsdraht *m (dat)* selection wire
Selektionsgrad *m* degree of selectivity *(dat)*
Selektionskreis *m* gating circuit *(electronics)*
Selektionskurve *f* selectivity characteristic
selektiv selective
Selektivabsorption *f* selective absorption
Selektivfilter *n* selective filter
Selektivität *f* selectivity
Selektivschutz *m* selective protection
Selektivschutzsystem *n* discriminating protective system
Selektivstrahler *m* selective radiator
Selektronröhre *f (dat)* selectron
Selen *n* selenium
Selenat *n* selenate
Selenbeimengung *f* addition of selenium
Selendampf *m* selenium vapour
Selenelement *n* selenium photocell
Selenfotoelement *n* selenium photocell (photovoltaic cell)
Selengleichrichter *m* selenium rectifier
Selenid *n* selenide
selenig selenious
Selenit *n* selenite
Selensäure *f* selenic acid
Selensperrschichtfotoelement *n* selenium barrier cell
Selenüberzug *m* selenium coating
Selenwasserstoff *m* hydrogen selenide, selenium hydride
Selenwiderstand *m* selenium bridge
Selenwiderstandszelle *f* selenium conductive cell
Selenzelle *f* selenium cell
Selfaktor *m* mule [spinning machine], self-acting mule, self-actor
Selfaktorspindel *f* mule spindle
Selfaktorwagen *m* mule carriage
Selsyn *n* selsyn, synchro
selten rare *(metal)*
semantisch semantic
Semaphor *n* semaphore
Semikarbazid *n* semicarbazide
semipermeabel semipermeable
semipolar semipolar
semizyklisch semicyclic
Sendeanlage *f* transmitting station (plant)
Sendeantenne *f* transmitting (emitting) aerial
Sendeaufforderungszeichen *n* proceed-to-transmit signal
Sendeband *n* transmission band
Sendebereich *m* coverage *(of a transmitter)*
Sendeeinrichtung *f* transmitting apparatus
Sende-Empfangs-Funkgerät *n* wireless transmitting and receiving apparatus
Sende-Empfangs-Kontakt *m* send-receive contact
Sende-Empfangs-Relais *n* disabling relay
Sende-Empfangs-Schalter *m* transmit-receive switch
Sende-Empfangs-Umschalter *m* sending and receiving switch
Sende-Empfangs-Umschaltung *f* break-in system
Sendefreigabesignal *n* proceed-to-send signal
Sendefrequenz *f* transmitter frequency
Sendegebäude *n* transmitter house
Sendegerät *n* transmitting set
Sendekanal *m* transmission channel
Sendekondensator *m* signalling (transmitting) capacitor *(telegraphy)*
Sendeleistung *f* transmitting power
senden to transmit, to radio; to emit; to send; to forward
Strom s. to feed current
Senden *n* **mit unterdrückter Trägerwelle** suppressed carrier transmission
Sendepause *f* silent period, interval in transmission
Sendepegel *m* transmitting level
Sender *m* 1. transmitter; emitter; broadcasting station; 2. transmitting probe *(testing)*
S. einer Relaisstrecke link transmitter
S. mit Simultankanälen channellized transmitter
S. mit Steuerstufe controlled transmitter
S. mit unabhängiger Seitenbandübertragung independent sideband transmitter
amplitudenmodulierter S. amplitude-modulated transmitter
eigenerregter S. self-excited transmitter
extern gesteuerter S. driven transmitter
fahrbarer S. mobile transmitter
induktiv gekoppelter S. inductive transmitter
kristallgesteuerter S. crystal-controlled transmitter
ortsfester S. fixed transmitter
quarzgesteuerter S. crystal-controlled transmitter
quarzstabilisierter S. crystal-stabilized transmitter

rückgekoppelter S. feedback transmitter
stationärer S. fixed transmitter
stoßerregter S. shock-excitation-type transmitter
tragbarer S. portable transmitter
widerstandsgekoppelter S. rheodyne transmitting circuit
Senderabstimmung *f* transmitter tuning
Senderantenne *f* transmitting aerial
Senderaum *m* studio
Senderausfall *m* transmitter failure
Sendereichweite *f* transmission range
Sender-Empfänger *m* transceiver
Senderenergie *f* transmitter energy
Senderfrequenz *f* transmitter frequency
Senderleistung *f* transmitter power; capacity of a transmitter
Senderöhre *f* transmitting valve
zerlegbare S. demountable valve
Sendersystem *n* **mit versetzten Trägerwellen** offset-carrier system
Senderüberwachungsgerät *n* station monitor
Sendervorheizung *f* transmitter preheating
Senderwelle *f* outgoing wave
Sendesaal *m* studio
Sendeseite *f* transmitting (sending) generator end
Sendespeicher *m* transmitter storage
Sendesperrzelle *f* transmitter-blocker cell
Sendestärke *f* transmitting power
Sendestation *f* [transmitting] station
Sendestreifen *m* transmitting tape
Sendestrom *m* transmitting current
Sendesystem *n* transmitting system
Sendetaste *f* sending key
Sendeturm *m* radio tower (mast)
Sende- und Empfangs-Gerät *n* two-way radio
Sendeverfahren *n* / **verschlüsseltes** privacy system
Sendeverstärker *m* transmitter amplifier
Sendevolumenregler *m* / **automatischer** automatic volume range regulator
Sendezähler *m* meter-relay
Sendezeit *f* time of transmission
Sendezentrale *f* transmitter control station
Sendung *f* 1. transmission; broadcast *(radio)*; 2. consignment, shipment
fehlgeleitete S. diverted consignment
verschleppte S. diverted (wrongly routed) consignment
Senf *m* mustard
Senfgas *n* mustard gas
Senföl *n* mustard oil
Sengen *n* singeing; gassing, genapping *(yarn)*; flaming *(leather)*
Sengschaden *m* damage due to singeing
Senkbewegung *f* lowering motion
Senkblei *n* [plumb] bob, plummet; lead
Senkbohrung *f* / **kegelige** countersink
zylindrische S. counterbore
Senkbremssystem *n* lowering braking system *(cranes)*
Senkbrunnengründung *f* sunk well foundation
Senke *f* depression, hollow; sink
Senkel *m* lace
senken to lower; to depress; *(mach)* to counterbore *(cylindrically)*; to countersink *(conically)*
die Ausschußquote s. to lower the reject rate
die Fertigungskosten s. to cut the production costs
Senker *m* countersink *(conical)*; counterbore *(cylindrical)*
S. für Körnerloch mit Schutzsenkung bell-type countersink
S. für Körnerloch ohne Schutzsenkung regular countersink
S. zum Aussenken kegeliger Vertiefungen countersink[ing cutter]
S. zum Aussenken zylindrischer Vertiefungen counterbore
Senkgrube *f* catch pit, sink [hole]; cesspool
Senkkasten *m* caisson
Senkkastengründung *f* caisson foundation
Senkkopf *m* countersunk head *(rivet)*
Senkloch *n* sink hole
Senklot *n* plummet, bob; lead
Senkniet *m* countersunk-head rivet
senkrecht vertical; perpendicular; normal (*s. a.* vertikal-)
s. aufeinander mutually perpendicular
s. zu den Lagen *(mine)* on face
s. zueinander mutually perpendicular
s. zum Feld perpendicular to the field
Senkrechtabhebung *f* vertical relief movement *(tool)*
Senkrechtaufnahme *f* vertical photograph
Senkrechtbauart *f (mach)* vertical construction
Senkrechtbeförderung *f* vertical lift
Senkrechtbohren *n* vertical drilling
Senkrechtbohrmaschine *f* vertical drilling machine
Senkrechtbohrwerk *n* vertical boring mill
Senkrechtdrehwerk *n* vertical boring and turning mill
Senkrechte *f* vertical, perpendicular; normal
Senkrechtflachschleifmaschine *f* vertical surface grinding machine
Senkrechtförderer *m* elevator
Senkrechtfräseinrichtung *f* vertical head *(for a horizontal milling machine)*, vertical milling attachment
Senkrechtfräsen *n* vertical milling
Senkrechtfräsmaschine *f* vertical milling machine
Senkrechtfräs- und Lehrenbohrmaschine *f* vertical miller and jig borer
Senkrechthobeln *n* planing of vertical surfaces

Senkrechthonmaschine *f* vertical honing machine
Senkrechtkraft *f* vertical force
Senkrechtluftbild *n* vertical air photograph
Senkrechtmehrspindelbohrmaschine *f* **mit einstellbaren Spindeln** vertical drilling machine with multiple-spindle head
Senkrechträummaschine *f* vertical broaching machine
Senkrechtschnitt *m* vertical section
Senkrechtschrämarm *m* shearing jib
Senkrechtspindel *f* vertical screw
Senkrechtspindelmaschine *f* vertical spindle machine
Senkrechtstart *m (aero)* vertical take-off
Senkrechtstoßeinrichtung *f* slotting attachment
Senkrechtstoßen *n* slotting
Senkrechtstoßmaschine *f* slotting machine
 S. mit Kurbelschleifenantrieb crank slotting machine
 S. zum Bearbeiten von Eisenbahnrädern curvilinear slotting machine
Senkrechtstoßräumen *n* vertical push broaching
Senkrechtverstellung *f (mach)* vertical adjustment, elevation *(radial drill)*
Senkrechtvorschub *m* / **schräger** angular downfeed
Senkrechtvorschubschlitten *m* vertical feed slide
Senkrechtwebstuhl *m* vertical loom
Senkschacht *m* drop shaft; well
Senkschachtverfahren *n* **mit Überdruckkammer** pneumatic sinking method
Senkschieber *m* vertical (sliding) damper
Senkschraube *f* countersunk screw
Senkschuh *m (mine)* cutting edge *(drop shaft)*
Senkung *f* lowering, reduction; *(build)* sinking; settlement, setting; *(mine)* subsidence; *(mach)* counterbore *(cylindrical)*; countersink *(conical)*
Senkungsgeschwindigkeit *f (mine)* rate of subsidence
Senkungstrog *m* subsidence trough
Senkungszone *f* area of subsidence
Senkwerkzeug *n* counterboring tool *(cylindrical)*; countersink *(conical)*
 mehrstufiges S. step counterbore *(cylindrical)*
 zweischneidiges S. two-lip countersink *(conical)*
Sense *f* scythe
Sensibilisator *m* sensitizer, sensibilizer
sensibilisieren to sensitize, to sensibilize
Sensibilisierung *f* sensitization
 lichtelektrische S. photoelectric sensitization
Sensibilisierungsmittel *n* sensitizing agent
Sensitometer *n* sensitometer
Sensitometrie *f* sensitometry
Sensor *m* sensor
Sentinelpyrometer *n* sentinel pyrometer
Separator *m* separator *(for batteries)*
Sepialichtpause *f* sepia print
Sepialichtpausverfahren *n* sepia printing process
Sepiolith *m* sepiolite, meerschaum
septisch septic
Sequentialstichprobenverfahren *n* sequential sampling
Sequenzanalyse *f* sequential analysis
Serie *f* 1. series; 2. [production] run
 S. des Spektrums spectral series
 S. mit großer Stückzahl *(mach)* long-run repetition work
 diffuse S. diffuse series *(spectrum)*
 kleine S. *(mach)* short production run
 scharfe S. sharp series *(spectrum)*
Serienabgleichkondensator *m* padding capacitor
Serienarbeit *f (mach)* repetition (series-production) job
Serienaufnahme *f* serial radiography
Serienbau *m* serial (quantity) production, production work
Serienbetrieb *m (dat)* serial operation
Serienbohren *n* duplicate boring
Serienboot *n* stock boat
Serienfertigung *f* series (duplicate, repetition) production
 S. kleiner Stückzahlen series production of small batches
Seriengegenkopplung *f* series feedback
Serienherstellung *f* series (multiple) production
Serienkapazität *f* series capacitance
Serienklemme *f* series terminal
Serienkompensation *f* cascade compensation
Serienkondensator *m* series capacitor
serienmäßig in quantity
 s. gefertigt manufactured in series, volume-built, series-manufactured
Serienmotor *m* 1. *(el)* series[-wound] motor; 2. *(aut)* engine currently in production
Seriennummer *f* serial number
Serienparallelwandler *m* staticizor
Serienparallelwicklung *f* multiplex-wave winding
Serienproduktion *f* quantity (batch) production
Serienrechner *m* serial computing machine
Serienresonanz *f* series resonance
Serienresonanzkreis *m* series-resonant circuit, series-tuned circuit
Serienschalter *m* multicircuit (variable-control) switch
Serienschaltung *f* series connection
Serienspeicherung *f* serial storage [of data]
Serienspektrum *n* series spectrum
Serienstück *n* series-produced part, duplicate (repetitive) part
Serienteil *n s.* Serienstück
Serientransport *m (dat)* serial transfer

Serienübertrag *m (dat)* successive carry, serial transfer
Serienverzweiger *m* series T *(waveguides)*
Serife *f* serif
Serin *n* serine
Serizin *n* sericin[e], silk gum (glue)
Serizit *m (min)* sericite
Serpentin *m (min)* serpentine
Serpentinasbest *m (min)* chrysotyle, Canadian asbestos
Serum *n* serum
Serumalbumin *n* serum albumin
Servicemeßsender *m* radio-frequency signal generator
Serviceoszillator *m* signal generator
Servobremse *f* servo brake
Servoeinrichtung / mit servo-assisted
Servointegrator *m* servo integrator
Servolenkgetriebe *n* power-steering mechanism
Servolenkung *f* power-steering mechanism
  hydraulische S. hydraulically assisted steering
  pneumatische S. air-pressure-assisted steering
Servomotor *m (el)* servo-motor, pilot motor; booster *(hydraulic)*
  S. zur Bohrtischlageeinstellung table-positioning servo-motor
Servomultiplikator *m* servo multiplier
Servostempel *m (mine)* servo prop
Sesamöl *n* sesame oil
Sesquiterpen *n* sesquiterpene
Sesselbahn *f* chair lift; ski-lift, man-rider
Sesselform *f* chair form *(stereochemistry)*
Set *n* set *(monotype)*
Setkeil *m* set quoin *(monotype)*
Settrommel *f* justifying scale
Setzarbeit *f (mine)* jigging
Setzboden *m* charging level *(cupola)*
Setzdruck *m* / **erster** *(mine)* first weight *(support)*
setzen 1. to set [up]; to put, to place; 2. *(print)* to compose; 3. to erect *(prop)*; 4. to hoist *(flag)*; 5. to plant *(garden)*
  auf Strand s. *(ship)* to strand, to beach
  außer Betrieb s. to put out of operation; to take out of service; to switch off; to stop; to close [down] *(factory)*
  einen Mast s. to sink a pole
  in Betrieb s. to put into operation *(factory)*; to set to work; to start; to switch on; to set going *(engine)*
  in Klammern s. to bracket, to enclose within brackets
  in Wechselbeziehung s. to correlate
  Münzen in Kurs s. to issue coins
  Segel s. to make sail
  sich s. to settle *(ground)*
Setzer *m* type-setter, compositor
Setzerei *f* composing room
Setzholz *n* planting peg, dibble
Setzkasten *m (print)* [letter] case
Setzklaue *f (mine)* setting clamp
Setzkopf *m* die (set) head *(rivet)*
Setzkübel *m* drop-bottom bucket
Setzlast *f* setting load *(support)*
Setzlibelle *f* block level
Setzmaschine *f* 1. *(print)* typograph, composing (type-setting) machine; 2. *(mine)* jig, wash-box
  S. mit Fernsteuerung durch Lochband teletype-setter, TTS
  elektrische S. electrotypograph
  kolbenlose S. baum jig
  luftgesteuerte S. baum jig
Setzmaschinenzeile *f* [machine] slug
Setzöltank *m* settling tank
Setzpacklage *f (build)* telford foundation (base), hand-packed bottoming, hand-set pitching
Setzpult *n* composing frame
Setzregal *n* composing (case) rack
Setzriß *m (mine)* main roof break
  S. in den Dachschichten *(mine)* secondary roof break
Setzschiff *n* galley
Setzstock *m (mach)* steady [rest]
  feststehender S. fixed steady, steady rest
  mitlaufender S. moving (travelling) steady, follower rest
Setzstocklager *n* stay (back-rest) bearing
Setzstufe *f* riser *(stairs)*
Setztank *m* settling tank
Setzung *f* settlement *(soil)*
  bleibende S. permanent settlement
  gleichmäßige S. equal settlement
  ungleichmäßige S. irregular (differential) settlement
Setzungsspannung *f* settlement stress
Setzvorrichtung *f (mine)* setting device
Sexagesimalteilung *f* sexagesimal division
Sextant *m* sextant
Sextantenfernrohr *n* sextant telescope
Seziermesser *n* scalpel
Sgraffito *n* [s]graffito
Sheddach *n* saw-tooth roof, north-light roof
Shedschale *f* north-light shell
Shelterdecker *m* shelter-deck vessel
Shelterdeckschiff *n* shelter-deck vessel
sherardisieren to sherardize
Shim *m (nucl)* shim
Shim-Stück *n (nucl)* shim
Shore-Härte *f* Shore scleroscope hardness
Shore-Härteprüfung *f* Shore hardness test[ing]
shorterisieren to shorterize
Shunt *m (el)* shunt
  induktiver S. inductive shunt
  ohmscher S. shunting resistance
shunten to fit with a shunt
Sichel *f (ag)* reaping hook, sickle
Sichelschrämmaschine *f* sickle cutter

sicher safe; proof, protected
Sicherheit *f* safety; *(math)* confidence
inhärente S. inherent safety
statistische S. level of confidence
Sicherheitsabschaltung *f* safety shut-down *(reactor)*
Sicherheitsabstand *m* safety distance; clearance
Sicherheitsanschlagnocke *f* safety stop dog
Sicherheitsbeauftragter *m* safety officer
Sicherheitsbehälter *m* *(nucl)* containment vessel, safety container
Sicherheitsbereich *m* *(math)* confidence interval
Sicherheitsbühne *f* *(mine)* timber pentice
Sicherheitsdeck *n* safety deck *(floating dock)*
Sicherheitsdeckung *f* safety lap
Sicherheitsdruckregler *m* pressure-regulating valve
Sicherheitseinrichtung *f* safety device; safety controls
Sicherheitsfahrschaltung *f* dead-man's handle
Sicherheitsfaktor *m* safety factor, factor of safety
Sicherheitsfilm *m* safety film
Sicherheitsfunkenstrecke *f* safety gap
Sicherheitsgeländer *n* crash barrier, guard rail
Sicherheitsglas *n* safety (non-shattering, shatter-proof) glass
Sicherheitsgrad *m* degree of safety
Sicherheitsgurt *m* safety belt
Sicherheitshaken *m* safety hook
Sicherheitshandkurbel *f* safety crank
Sicherheitshub *m* / **astatischer** astatic lift
Sicherheitsinjektor *m* restarting injector
Sicherheitsinspektor *m* safety inspector
Sicherheitsintervall *n* *(math)* confidence interval
Sicherheitskiel *m* *(ship)* safety keel
Sicherheitskoeffizient *m* safety factor, factor of safety
Sicherheitskupplung *f* safety clutch
Sicherheitslampe *f* safety lamp
Sicherheitsleiter *f* *(mine)* safety cage ladder
Sicherheitsluke *f* *(ship)* escape hatch
Sicherheitsmantel *m s.* Sicherheitsbehälter
Sicherheitsmaßnahme *f* safety measure, precaution
Sicherheitsmindestflughöhe *f* minimum safe height
Sicherheitsnadel *f* safety pin
Sicherheitspfeiler *m* *(mine)* safety pillar
Sicherheitspunkt *m* *(math)* significance point
Sicherheitsregler *m* safety governor *(turbine)*
Sicherheitsschalter *m* safety switch
Sicherheitsschloß *n* safety lock
Sicherheitsschwelle *f* *(math)* confidence limit
Sicherheitsspanne *f* margin of safety
Sicherheitssprengstoff *m* safety (coal-mining) explosive
Sicherheitsstab *m* safety (scram) rod, [emergency] shut-off rod *(reactor)*
Sicherheitsstreichholz *n* safety match
Sicherheitsstreifen *m* *(aero)* shoulder *(runway)*
Sicherheitsüberwachung *f* safety control
Sicherheitsventil *n* safety valve
S. der Speisepumpe feed-relief valve
S. mit Dampfabblasestutzen enclosed-discharge safety valve
S. mit Federbelastung spring-loaded safety valve
S. mit Federwaage spring-loaded lever safety valve
S. mit Gewichtbelastung safety valve loaded with weight
S. mit Hebelbelastung lever safety valve
S. mit Hebelgewichtbelastung safety valve loaded with lever and weight
S. mit unmittelbarer Federbelastung safety valve with direct spring loading
S. mit unmittelbarer Gewichtbelastung Cowburn (dead weight safety) valve
S. ohne Dampfabblasestutzen open-discharge safety valve
doppeltes S. double-safety valve
geschlossenes S. enclosed-discharge safety valve
offenes S. open-discharge safety valve
Sicherheitsverblockung *f s.* Sicherheitsverriegelung
Sicherheitsverriegelung *f* safety interlock
Sicherheitsverriegelungseinrichtung *f* safety-interlock device
Sicherheitsvorrichtung *f* safety appliance
Sicherheitsvorschrift *f* safety rule (regulation)
Sicherheitswasserstand *m* safety [water] level
Sicherheitswinkel *m* margin of commutation *(rectifier)*
Sicherheitszahl *f* safety factor, factor of safety
Sicherheitszeichen *n* safety signal *(radio)*
sichern so secure; to lock; to guard; to fasten
die Kolbendeckelschraube s. to lock the bolt
sicherstellen to ensure
Sichertrog *m* *(mine)* horn, pan
Sicherung *f* 1. [safe]guard, locking; 2. *(el)* [safety] fuse; 3. *(mil)* safety catch
S. des Hangenden mit Ankern *(mine)* roof bolting
S. mit Auslösemechanismus tripping fuse
S. mit Ausschaltung drop-out fuse
S. mit Energiekennzeichen indicating fuse
S. mit Meldeeinrichtung indicating fuse
S. mit rückwärtigem Anschluß back-connected fuse
S. mit selbsttätiger Wiedereinschaltung reclosing fuse
S. mit Unterbrechungsmelder indicating fuse

S. unter Öl oil-quenched cut-out
abfallende S. drop-out fuse
automatische S. automatic cut-out
durchgebrannte S. blown fuse
einpolige S. single-pole cut-out
elektrische S. fuse
flinke S. instantaneous (quick-break, non-delayed) fuse
geschlossene S. enclosed fuse, fuse with enclosed fuse element
halbgeschlossene S. semi-enclosed fuse
offene S. plain (open-wire) fuse, renewable fuse unit
rücklötbare S. solder (head-coil-type) fuse
träge S. time-delay fuse, slow-burn fuse
unverzögerte S. instantaneous (non-delayed) fuse
Sicherungsanlage *f* protective device
Sicherungsautomat *m (el)* automatic cut-out
Sicherungsblasrohr *n* fuse muffler (expulsion tube)
Sicherungsblech *n* lock plate
Sicherungsbolzen *m* safety bolt
Sicherungsdose *f* fuse box
Sicherungsdraht *m* fuse wire
Sicherungseinsatz *m* fuse link
Sicherungsflügel *m (mil)* safety handle
Sicherungsfüllstoff *m* fuse filler
Sicherungsgriff *m* fuse holder (carrier)
Sicherungshalter *m* fuse holder
Sicherungsisolator *m* fused insulator
Sicherungskästchen *n* fuse box
Sicherungskasten *m* cut-out box
Sicherungsklinke *f* safety (locking) pawl
Sicherungsmutter *f* safety (lock) nut
Sicherungspatrone *f* cartridge fuse-link
Sicherungspilz *m* **für Kolbenbolzen** piston button
Sicherungsplatte *f* locking plate
Sicherungsring *m* snap ring
Sicherungsschalter *m* fuse switch
Sicherungsscheibe *f* locking plate
Sicherungsschloßteil *n* guard cam *(knitting machine)*
Sicherungsstöpsel *m* fuse plug
Sicherungsstreifen *m* fuse strip
Sicherungstafel *f* fuse board
Sicherungstrenner *m* fuse disconnecting switch, fused isolator
Sicherungsverteilungstafel *f* section fuse board
Sicherungszange *f* fuse tongs
Sicht *f* visibility
optische S. optical range
schlechte S. poor visibility
ungehinderte S. unobstructed view [of]
Sichtalarm *m* visual alarm
Sichtanflug *m* visual approach
Sichtanzeige *f* visual indication (display)
elektronische S. electronic display
Sichtauswertung *f* visual interpretation
sichtbar visible
Sichtbarkeitskurve *f* visibility curve
Sichtbarkeitsschwelle *f* threshold of visibility
Sichtbarmachung *f* visualization; display
Sichtbegrenzung *f* optical limitation
Sichtbehinderung *f* sight disturbance
Sichtbereich *m* visibility; visual range
Sichtbeton *m* decorative (architectural, exposed, facing) concrete
S. mit freigelegten Zuschlagstoffen exposed-aggregate concrete
sichten 1. to assort, to classify, to grade, to screen, to separate, to sift, to sort; 2. *(opt)* to sight
Sichterstaub *m* aspirated dust (fines)
Sichtetisch *m (text)* sorting table
Sichtfläche *f (build)* face, facing; *(mach)* non-working [visual] surface
Sichtgerät *n* display unit, indicator *(radar)*
Sichtkante *f* visual edge
Sichtlinie *f* line of viewing; optical path
Sichtmesser *m* visibility meter
Sichtpeiler *m* bearing indicator
Sichtpeilgerät *n* bearing indicator
Sichtprüfung *f* visual inspection (examination)
Sichtung *f* separating; [pneumatic] sizing; sorting
Sichtungsanlage *f* separating plant
Sichtverbindung *f* line-of-sight communication
Sichtvergleich *m* visual comparison
Sichtvergleichsnormale *f* visual comparison block
Sichtweite *f* luminous range *(beacon fire)*; range of vision
Sickenmaschine *f* flanging machine
Sickenwalze *f* flanging roller
Sickerbecken *n* oozing basin
Sickerdruck *m* seepage pressure *(soil mechanics)*
Sickerfeuerung *f* trickle-furnace
Sickergrube *f* soakaway
Sickerkanal *m* catch-water drain
Sickerlaugung *f* percolating leach
Sickerloch *n* weep hole, soakaway
sickern to seep, to percolate, to ooze, to trickle
Sickerrohr *n* filter drain
Sickerstrahlung *f* leakage [radiation], radiation leakage
Sickerung *f* percolation, seepage
siderisch sidereal
Siderit *m (min)* siderite
Sideroskop *n* sideroscope
Siderostat *m* siderostat
Sieb *n* screen, sieve; riddle; sifter; strainer; filter
S. mit Oberschwingungen screen with harmonics
Siebabstreicher *m* water deflector

Siebanalyse *f* sieve (screen) analysis; sizing test
Siebanlage *f* screening plant
Siebband *n* endless-belt screen
Siebbelag *m* screen cloth
Siebbestrahlung *f* grid (sieve-plate) irradiation
Siebblech *n* screen plate
Siebboden *m* screen deck (underframe); *(text)* false bottom; sieve plate, perforated bottom
Siebbodenkolonne *f* sieve-plate column, sieve-tray column, perforated-plate column, grid-tray column
Siebdrossel *f* filter reactor, smoothing (swinging) choke
Siebdruck *m* screen [process] printing, silk-screen printing, serigraphy
Siebdruckrakel *f* squeegee
Siebdurchfall *m* screenings
Siebdurchgang *m* passing; screen underflow, siftings, throughs, undersize
Siebdurchlauf *m* screen underflow
Siebeinsatz *m* mesh bottom, strainer
sieben to screen, to sieve, to riddle, to sift, to size, to strain; *(el)* to filter
siebenatomig heptatomic
Siebeneck *n* heptagon
siebeneckig heptagonal
siebenfach sevenfold; septiform
siebenfarbig septicoloured
siebenflächig heptahedral
Siebenflächner *m* heptahedron
siebengliedrig *(ch)* seven-membered
Siebenpolröhre *f* heptode
Siebenring *m (ch)* seven-membered ring
Siebenringverbindung *f* heptacyclic compound
siebenseitig seven-sided, septilateral
siebenwertig septivalent
Siebenwertigkeit *f* septivalence
Siebfege *f* gravity screen
Siebfeines *n* screen undersize, siftings, throughs
Siebfeld *n* screening panel
Siebfilter *n* gauze filter
Siebfläche *f* screen surface
  offene S. open screening area
Siebfraktionierung *f* screen classification
Siebgewebe *n* screen cloth
Siebglied *n* filter section
Siebgrobes *n* coarse of screen, screen oversize
Siebgütegrad *m* screening efficiency
Siebhammermühle *f* screen hammer mill
Siebkapazität *f* filter capacity
Siebkasten *m (ag)* screening chamber *(precleaner)*; screen frame; straining box
Siebkern *m* strainer grate core *(foundry)*; filter core
Siebkette *f* chain (sifting) web
  vordere S. first sifting chain *(beet harvester)*
Siebkettenförderer *m* screen-chain conveyor
Siebkettenroder *m* elevator digger with chain sifter
Siebkettenvorratsroder *m* elevator potato digger
Siebkies *m* screened gravel
Siebklassierer *m* sizer
Siebklassierung *f* screen classifying (grading, separation, sizing), screening
Siebkondensator *m* filter capacitor
Siebkreis *m* filter circuit
Siebkugelmühle *f* ball mill with screen walls
Siebkurve *f (build)* grading curve
Siebleistung *f* screening efficiency
Sieblinie *f (build)* grading curve
Siebmarkierung *f* wire mark *(papermaking)*
Siebmaschine *f* screening machine
Siebnummer *f* drainage area *(papermaking)*
Siebpartie *f* wire frame (part, end), wet end, screen; Fourdrinier part *(papermaking)*
Siebradroder *m* potato digger with reel
Siebränderbiegemaschine *f* frame bending machine
Siebreiniger *m* sieve cleaner
Siebrost *m* sieve grate, grizzly
Siebroststab *m* grizzly bar
Siebrückstand *m* oversize (rejected) material; retainings; coarse of screen, screen overflow, screens
Siebsatz *m* set of sieves, sieve set
Siebschablonendruck *m* screen stencil printing
Siebschaltung *f* filter [circuit, network], selective network
Siebschleuder *f (ch)* extractor-type centrifuge; *(mine)* whizz out with perforated basket
Siebseite *f* wire side
Siebsortierer *m* riddling equipment, riddle
Siebsortierung *f* screen classification
  S. nach Dicke grading by size with screens
Siebspule *f (el)* filter coil (reactor)
Siebstab *m* grizzly bar
Siebtrennungsgrad *m* screening efficiency
Siebtrommel *f* rotary (trommel) screen, sizing (sieve, straining) drum
  obere S. *(text)* top cage
  untere S. *(text)* bottom cage
Siebtrommelfilter *n (text)* dust cage filter
Siebtrommelroder *m* potato digger with sifting cage
Siebtuch *n* bolting cloth, screen
Siebüberlauf *m* coarse of screen, screen overflow
Siebung *f* screening, screen separation; straining; *(telec)* filtering
  S. der Anodenspannung filtering of the anode voltage
Siebverstopfung *f* blinding of screen
Siebversuch *m* sieve test
Siebwasser *n* white water
Siebzylinder *m* perforated cylinder

Siedeanalyse *f* analysis by boiling, fractional analysis
Siedebeginn *m* initial boiling point
Siedebereich *m* boiling range; distillation range *(petrol)*
Siedediagramm *n* boiling curve, liquid-vapour equilibrium diagram
Siedeende *n* end-boiling point, final-boiling point
Siedegrenzenbenzin *n* special boiling-point petrol
Siedehitze *f* boiling heat
Siedekessel *m* French (elephant) boiler
Siedekühlröhre *f* vapour-cooled tube
Siedekurve *f* boiling curve
sieden to boil
Sieden *n* boiling
  S. des Blutes blood boiling
  lokales S. local (surface, subcooled) boiling
  örtliches S. *s.* lokales S.
Siedepunkt *m* boiling point
Siedepunktserhöhung *f* boiling point elevation
  molare S. molar boiling point elevation
  molekulare S. molecular boiling point elevation
Siedereaktor *m s.* Siedewasserreaktor
Siederohrkessel *m* water tube boiler
Siedestein *m* boiling chip
Siedethermometer *n* barometric thermometer
Siedeverhalten *n* distillation (boiling) characteristics
Siedeverlauf *m* distillation characteristics; boiling progress (curve)
Siedeverzug *m* retardation of boiling (ebullition), delay in boiling
Siedewasserreaktor *m* boiling [water] reactor
  direkter S. direct-cooling boiling [water] reactor
Siedlung *f* housing estate
Siedlungsplanung *f* community planning
Siegel *n* seal
Siegellack *m* sealing wax
Siegelmarkendruck- und -prägepresse *f* seal printing and embossing machine
siegeln to seal
Siemens *n* mho, reciprocal ohm
Siemens-Drehspulfernschreiber *m* Siemens moving coil-type telewriter
Siemens-Martin-Ofen *m* open-hearth furnace, Siemens-Martin furnace
  basischer S. basic open-hearth furnace
  feststehender S. fixed open-hearth furnace
  saurer S. acid open-hearth furnace
Siemens-Martin-Ofenkopf *m* open-hearth furnace port end
Siemens-Martin-Stahl *m* open-hearth steel
  basischer S. basic open-hearth steel
Siemens-Martin-Verfahren *n* open-hearth process
  basisches S. basic open-hearth process
  saures S. acid open-hearth process
Siemens-Schnellschreiber *m* Siemens-multiplex printing telegraph
Siemensit *m (min)* siemensite
Siemob *n* screen with harmonics
Sienaerde *f* sienna
Signal *n* signal
  akustisches S. sound signal
  analoges S. analogue signal
  ankommendes S. incoming signal
  ausgeglichenes S. balanced signal
  ausgehendes S. outgoing signal
  echtes S. genuine signal
  entschlüsseltes S. decoded signal
  externes S. external signal
  gemischtes S. mixed signal
  großes S. large signal
  kleines S. small (low-level) signal
  nichtstetiges S. non-steady signal
  numerisches S. digital signal
  optisches S. visual signal
  pausenfreies S. sequence signal
  rechteckiges S. rectangular signal
  reflektiertes S. returned signal
  stetiges S. steady signal
  symmetrisches S. balanced signal
  umlaufendes S. circulating signal
  ungewobbeltes S. non-swept signal
  unsymmetrisches S. unbalanced signal
  zusammengesetztes S. multi-component signal
Signalabgabe *f* signal output
Signalabschwächung *f* signal attenuation
Signalabstand *m* signal-to-noise ratio
Signalamplitude *f* signal amplitude
Signalanlage *f* signal installation
Signalanode *f* signal plate
Signalantrieb *m* signal drive; signal machine *(electric traction)*
Signalauflösung *f* signal resolution
Signalauswertung *f* strobing, gating *(radar)*
Signalbake *f* marker
Signalball *m* black ball
Signalbuch *n* code book
  Internationales S. International Code of Signals
Signalbuchwimpel *m* answering pennant
Signaldarstellung *f* signal representation
Signaldraht *m (mine)* knocker line
Signaleingabe *f* signal input
Signaleinspeisung *f* signal input
Signalfeuer *n* warning-light
Signalfläche *f* signal area
Signalflanke *f* / **steile** steeply-falling side, steep slope
Signalflußdiagramm *n* signal-flow graph (diagram)

Signalflußgraf *m* signal-flow graph
Signalflußplan *m s.* Signalflußdiagramm
Signalflußschema *n s.* Signalflußdiagramm
Signalfrequenz *f* signal frequency
Signalfrequenzbereich *m* baseband *(of frequency)*
Signalgabe *f* signalling
Signalgeber *m* alarm
Signalgebung *f* signalling
Signalgemisch *n* composite [video] signal
Signalgenerator *m* signal generator
Signalgerät *n* alarm
Signalgeschwindigkeit *f* signal velocity
Signalglocke *f* signal[ling] bell
Signalgrubenlampe *f* signalling lamp
Signalhaltmelder *m* on-signal proving
Signalhammer *m (mine)* knocker
Signalhöhe *f* signal level
Signalhupe *f* hooter; [electric] horn
Signalisator *m* signal indicator
signalisieren to signal
Signalkelle *f (aero)* signal bat
Signalkode *m* signal code
Signallampe *f* signal lamp, indicator light
Signalleistung *f* signal power
Signalleitung *f* signalling line
  pneumatische S. pneumatic branch line
Signallicht *n* signal light
Signallichtüberwachung *f* signal-light proving
Signalmanometer *n* electric alarm pressure gauge
Signal-Nebensprech-Verhältnis *n (telec)* signal-to-crosstalk-ratio
Signalpatrone *f* Very light
Signalpegelblockierung *f* clamping of signal level
Signalpfeife *f* alarm whistle
Signalplatte *f* signal (pick-up) plate
Signal-Rausch-Quotient *m* signal-to-noise ratio
Signal-Rausch-Verhältnis *n s.* Signal-Rausch-Quotient
Signalrückführung *f* signal return
Signalrücklicht *n* signal backlight
Signalrückmelder *m* signal indicator
Signalschwanz *m* tail of signal *(pulse engineering)*
Signalsummierpunkt *m* signal-summation point
Signalsystem *n* signalling system
Signalthermometer *n* alarm thermometer, thermostat
Signalüberdeckung *f* blanketing, blanketing (masking) of signals
Signalüberwachung *f* signal proving
Signal- und Steuerleitungen *fpl* small wiring
Signalunterscheidung *f* signal discrimination
Signalverarbeitung *f* information processing
Signalverfolger *m* signal tracer
Signalverfolgung *f* signal tracing
Signalverknüpfung *f* signal processing
Signalverteilungskasten *m* signal mixer-unit *(radar)*
Signalverzerrung *f* signal distortion
Signalverzögerung *f* signal delay
Signalvorrichtung *f* signal device
Signalwandler *m* signal converter
Signalwelle *f* signal wave
Signalzuführung *f* signal injection
Signierfarbe *f* marking ink
Signierkreide *f* marking chalk
Signierstift *m* coloured wax pencil
Signifikanztest *m* test of significance
Sikkativ *n* siccative
Silage *f* [en]silage
Silan *n* silane, silicon hydride
Silbentrennung *f* syllabic division
Silbenverständlichkeit *f* syllabic articulation
Silber *n* silver
  massives S. solid silver
  ungeprägtes S. silver bullion
Silberamalgam *n* silver amalgam
silberartig silvery, argentine
Silberbarren *m* silver ingot
Silberbergwerk *n* silver mine
Silberbeschlag *m* silver lining
Silberblech *n* silver lamina
Silberbrennen *n* silver refining
Silberbromid *n* silver bromide
Silberbronze *f* silver bronze
Silberchlorid *n* argentic chloride
Silberdraht *m* silver wire
Silberelektrode *f* silver electrode
Silberfahlerz *n* argentiferous tetrahedrite
Silberfolie *f* silver foil
Silberfuchs *m* silver (black) fox
Silbergeld *n* silver money
Silbergewinnung *f* silver extraction
Silberglanz *m* 1. *(min)* silver glance, argentite; 2. silvery lustre
Silbergold *n (min)* argentiferous gold
Silbergrube *f* silver mine
Silberhalogenid *n* silver halide
silberhaltig argentiferous; silver-bearing
Silberhartlot *n* silver brazing alloy, hard silver solder
Silberhornerz *n (min)* hornsilver, corneous silver
Silberhütte *f* silver foundry
Silberkeim *m* silver nucleus *(photography)*
Silberlaugerei *f* silver leaching plant
Silberlegierung *f* silver alloy
Silberlot *n* silver solder
Silbermetall *n* silver metal
Silbermineral *n* silver mineral
Silbermünze *f* silver coin
silbern silvery, argent[ine]
Silberniederschlag *m* silver precipitate
Silbernitrat *n* argentic nitrate, nitrate of silver
Siberoxid *n* argentous oxide
Silberpackpapier *n* antitarnish paper

silberplattiert silver-clad
Silberplattierung *f* silver plating
Silberpunkt *m* melting point of silver
silberreich rich in silver
silberrein pure as silver
Silbersalz *n* silver salt
Silberscheideanstalt *f* silver refinery
Silberscheidung *f* silver refining
Silberschlaglot *n* silver solder
Silberschmelzsicherung *f* silver wire fuse
Silberspiegel *m* silver-coated mirror
Silberspießglanz *m (min)* antimonial silver
Silberspritzkabine *f* silver spraying booth
Silberstahl *m* silver steel
Silberstift *m* silver pointed style
Silbersulfid *n* silver sulphide
Silberwaren *fpl* silver goods
Silberweichlot *n* soft silver solder
Silberweiß *n* silver white
silieren to [en]silage
Silierung *f* [en]silaging, ensilage
Silifizierung *f (geol)* silification
Silikagel *n* silica gel
silikahaltig siliceous
Silikaspiegel *m* silica-spiegel
Silikastein *m* silica (Dinas) brick
Silikat *n* silicate
Silikatbindung *f* silicate bond *(grinding wheel)*
Silikatemaille *f* silicate enamel
Silikatglas *n* silica glass
Silikatschlacke *f* silicate clinker
Silikon *n* silicone
Silikondichtung *f* silicone packing
Silikonfett *n* silicone grease
Silikongummi *m* silicone rubber
Silikonharz *n* silicone resin
Silikonlack *m* silicone varnish
Silikonöl *n* silicone oil
Silikose *f* silicosis
Silitwiderstand *m* silit resistor
Silizid *n* silicide
silizieren to siliconize
Silizium *n* silicon
Siliziumblättchen *n* silicon chip
Siliziumblock *m* silicon ingot
Siliziumbronze *f* silicon bronze
Siliziumdetektor *m* silicon detector
Siliziumdiode *f* silicon diode
Siliziumdioxid *n* silicon dioxide, silica
Siliziumeinkristall *m* silicon block
Siliziumflächentransistor *m* silicon junction transistor
Siliziumfotoelement *n* silicon photocell
siliziumfrei non-silicious
Siliziumgelfett *n* silica gel grease
Siliziumgleichrichter *m* silicon rectifier
  gesteuerter S. silicon-controlled rectifier
siliziumhaltig silici[fer]ous, siliceous
Siliziumkarbid *n* silicon carbide, carbon silicide
Siliziumkristallgleichrichter *m* silicon crystal rectifier
Siliziummangan *n* silicon manganese
Siliziummetall *n* metallic silicide
siliziumorganisch organosilicon
Siliziumplättchen *n* silicon wafer
Siliziumsonnenzelle *f* silicon solar cell
Siliziumsperrschichtfotoelement *n* silicon photovoltaic cell
Siliziumstab *m* silicon rod
Siliziumstahl *m* silicon steel
Siliziumtransistor *m* silicon transistor
  legierter S. alloyed silicon transistor
Siliziumverbindung *f* silicon compound
Siliziumwasserstoff *m* silicomethane
Siliziumwiderstand *m* silicon resistor
Siliziumzelle *f* silicon cell
Sillimanit *m (min)* sillimanite, fibrolite
Silo *m* silo, bin, hopper
Silofutter *n* silo fodder, ensilage
Silohäcksler *m* silage cutter and blower, silage cutter-blower
Silorüttler *m* bin vibrator
Siloverschluß *m* bin door
Siloxen *n* siloxen
Silozangengreifer *m* silage grab
Silozelle *f* silo compartment
Silumin *n* Silumin, Alpax
Silurformation *f (geol)* silurian formation
Silurzeit *f (geol)* silurian period
Silvester *m (mine)* dog-and-chain
silylieren *(ch)* to silylate
Simmerring *m* oil-seal ring; shaft-sealing ring, oil seal
Simplexbetrieb *m (telec)* simplex working, single operation
Simplexbremse *f* simplex brake
Simplexkriterium *n* simplex criterion *(linear programming)*
Simplexpumpe *f* simplex steam pump
Simplexverfahren *n* **des linearen Programmierens** simplex algorithm
Sims *m (n)* cornice, ledge, moulding
Simsbrett *n* fascia board
Simshobel *m* rabbet plane
Simulation *f* simulation
Simulator *m* simulator
Simultanantenne *f* duplexer
Simultanbetrieb *m* composite working
Simultanempfang *m* simultaneous reception
Simultanfarbfernsehen *n* simultaneous colour television
Simultankontrast *m* simultaneous contrast
Simultanleitung *f* composite (simplex, superposed) circuit
Simultanrechner *m* time-shared computer

Simultanschaltung *f (telec)* composited arrangement
Simultantelegraf *m* superposed telegraph
Simultanverbindung *f* composited circuit
Simultanverfahren *n* **beim Farbfernsehen** frame simultaneous system
singulär singular *(point)*
Singularität *f* singularity
Singulett *n* singlet
Singulettzustand *m* singlet state
sinken to drop; to fall; to settle *(soil)*; to sink *(vessel)*
Sinken *n* drop[ping]; fall[ing]; foundering *(of a vessel)*
Sinkgeschwindigkeit *f* 1. *(build)* settling speed (velocity); 2. *(aero)* sinking speed, descent velocity
Sinkgeschwindigkeitsmesser *m (aero)* rate-of-change indicator
Sinkkasten *m* mud-trap; caisson
Sinkmittel *n* depressant *(flotation)*
Sinkprodukt *n* sink product
Sinkscheider *m* dense-medium washer, float-and-sink apparatus, *(Am)* heavy-media separator
Sinksicherheit *f (ship)* watertight integrity
Sinkstoff *m* sediment, settlings
Sinkstoffablagerung *f* sediment deposition
Sinn *m* sense, direction
Sinnbild *n* symbol
Sinnenprüfung *f* organoleptic test
Sinter *m* [rolling] scale
Sinteranlage *f* agglomerating plant; sintering plant
Sintereisen *n* sintered iron
Sintergrube *f* scale pit
Sinterhartmetall *n* cemented carbide
Sinterhartstoff *m* sintered (cemented) carbide
Sinterherd *m* fritted (fused) hearth bottom *(open-hearth furnace)*
Sinterhitze *f* sintering heat
Sinterkarbid *n* sintered (cemented) carbide
Sinterkarbidschneidwerkstoff *m* [cemented] carbide-cutting material
Sinterkohle *f* cherry coal
Sinterkuchen *m* agglomerated (sintered) cake
Sintermagnet *m* metal-ceramic magnet
Sintermetall *n* sintered metal
sintern *(met)* to sinter, to cement; to clinker *(cement)*; to fuse; to agglomerate; to frit, to slag, to vitrify
Sinterofen *m* sintering furnace
Sinterpresse *f* sintering press
Sinterpunkt *m* sintering point
Sinterquelle *f* sintered source
Sinterröstung *f* blast roasting
Sinterschlacke *f* clinker, scale
Sinterstahl *m* sintered steel
Sinterstück *n* sintering
Sinterteil *n* blank *(powder metallurgy)*
Sintertemperatur *f* sintering heat; clinkering temperature *(cement manufacture)*
Sintertränklegierung *f* impregnated sinter-alloy
Sinterwerkstoff *m* sintered material
Sinterzone *f* clinkering zone *(cement manufacture)*
Sinus *m* sin[e]
hyperbolischer S. hyperbolic sine
Sinusbedingung *f* sine condition
Sinusbussole *f* sine galvanometer
Sinuselektrometer *n* sine electrometer
sinusförmig sinusoidal
Sinusgeber *m* sine-wave generator
Sinusglied *n* sine term
Sinusgröße *f* sinusoidal quantity
gedämpfte S. damped sinusoidal quantity
Sinuskurve *f* sine curve
Sinuslineal *n (mach)* sine bar
Sinuslinie *f* sinusoid; sine curve
Sinusoide *f* sinusoide
Sinusoszillator *m* harmonic oscillator
Sinusreihe *f (math)* sine series
Sinussatz *m* law of sines, sine law
Sinusschwingung *f* sinusoidal oscillation, harmonic motion
gedämpfte S. damped sinusoidal motion
Sinussignal *n* sinusoidal (sine-wave) signal
Sinusspannung *f* sinusoidal (sine-wave) voltage
Sinusstrom *m* sine (harmonic) current
Sinustisch *m* sine table
Sinuswelle *f* sine (sinusoidal) wave
ebene S. plane sinusoidal wave
Sinuswellenanregung *f* sine-wave excitation
Sinuswellenempfindlichkeit *f* sine-wave response
Sinuswellenerzeuger *m* sine-wave generator
Siphon *m* siphon; trap
Sirene *f* siren [generator] *(ultrasonics)*
elektrische S. electric siren
Sirup *m* syrup
Sitz *m* 1. seat[ing] *(valve)*, socket; bearing area; 2. fit
konischer S. conical seating
Sitzangebot *n* available seats, seating capacity
Sitzbezug *m* seat cover
sitzen in *(mach)* to be seated (fitted) into
Sitzfeder *f* seat spring
Sitzfläche *f* bearing surface; seat[ing], seating face
S. des Ventiltellers valve seat
Sitzgestell *n* seat frame
Sitzkissen *n* seat cushion
Sitzlehne *f* seat back
Sitzpolster *n* seat upholstery
Sitzringbiegemaschine *f* seat frame bending machine
Sitzstange *f* roost

Sitzungszimmer *n* board-room
Sitzventil *n* seated valve
automatisches S. automatic seated valve
Sitzverstellung *f* seat adjuster
Skala *f s.* Skale
skalar scalar
Skalar *m* scalar; scalar quantity
Skalarfeld *n* scalar field
Skalarfunktion *f* scalar function
Skalarmatrix *f* scalar matrix
Skalarprodukt *n* scalar product
Skale *f* scale, dial
S. für Einstellung der Tiefenauslösung feed trip dial
S. mit Gradeinteilung graduated scale
S. mit unterdrücktem Nullpunkt suppressed zero scale, set-up scale
S. ohne Nullpunkt suppressed zero scale, set-up scale
gedehnte S. expanded scale
gedrängte S. restricted scale
gedruckte S. printed scale
kontinuierliche S. continuous scale
Skalenablesung *f* scale (dial) reading
Skalenantrieb *m* dial drive
Skalenauflösung *f* scale discrimination
Skalenausschlag *m* swing on the dial
Skalenbeleuchtung *f* dial illumination
Skalenbeleuchtungslampe *f* instrument lamp
Skalenbereich *m* measuring range
Skalenbezifferung *f* scale numbering
Skaleneichung *f* scale calibration
Skaleneinstellung *f* dial setting
Skaleneinteilung *f* graduation
Skalenendwert *m* full-scale value
Skalenfaktor *m* scale factor
Skalengalvanometer *n* scale galvanometer
Skalenlampe *f* dial (instrument) lamp
Skalenlänge *f* scale length
Skalenlinearisierung *f* scale linearization
Skalenmarke *f* scale mark
Skalenmeßgerät *n* direct-reading instrument
Skalenmitte *f* midscale
Skalenrad *n* dial-cord pulley
Skalenring *m* collar, dial, graduated collar (ring)
direkt ablesbarer S. direct-reading dial
Skalenscheibe *f* dial [plate], graduated disk
Skalenschwungrad *n* dial-cord pulley
Skalenseil *n* driving string, pulley (drive) cord
Skalenstrich *m* scale mark
Skalenteil *n* scale division
Skalenteilgröße *f* width of scale division
Skalenteilung *f* scale division
Skalentrommel *f* collar
Skalenübereinstimmung *f* dial tracking
Skalenverzerrung *f* scale distortion
Skalenwert *m* scale interval; reading
Skalenzeiger *m* dial pointer (needle)
Skalpell *n* scalpel
Skandium *n* scandium
Skapolith *m* scapolite, wernerite
Skatol *n* skatole
Skelett *n (build)* frame[work], skeleton framing, building framework
Skelettbau[weise *f*] *m* frame[d] (skeleton) construction
Skelettplattenbauweise *f* panel frame construction
Skelettsubstanzen *fpl (ch)* builders
Skiaskop *n* skiascope
S. mit elektrischer Beleuchtung luminous skiascope
Skiaskopierscheibe *f* **nach Lenz** Lenz skiascopy disk
Skiatron *n* skiatron
Skineffekt *m* skin (Heaviside) effect
Skiograf *m* skiograph
Skip *m (mine)* gunboat *(in inclined shafts)*; skip
S. mit Bodenentleerung drop-bottom skip, bottom-discharge skip
kippbarer S. Kimberley (overturning-discharge, self-dumping) skip
Skizze *f* sketch
Skizzenblech *n* sketch plate
Skizzieren *n* sketching
Sklera *f* sclera
Skleraleuchte *f* scleral lamp
Sklerometer *n* sclerometer
Skleroskop *n* scleroscope
Shoresches S. Shore scleroscope
Skleroskophärte *f* scleroscope hardness
Skopolamin *n* scopolamine
skorbutverhütend antiscorbutic
Skotom *n* scotoma
Skrubber *m* scrubber, column washer, wash column
Skull *n* scull; oar
Skuller *m* sculler *(sporting boat)*
Slawjanow-Lichtbogenschweißverfahren *n* Slavianoff arc welding process
Slip *m (ship)* ways, [hauling] slip; railway slip (dry dock) *(launching)*
Slipanlage *f* railway dry-dock
Slipdock *n* slip dock
Sliphaken *m* tripping link
Slipstopper *m* slip stopper
Slipvorrichtung *f* tumbler *(anchor)*
Slipwagen *m* railway cradle
SM-Ofen *m* open-hearth furnace
SM-Stahl *m* open-hearth steel
SM-Verfahren *n* open-hearth process
Smalte *f* smalt
Smaragd *m* emerald
smaragdgrün emerald-green
Smaragdit *m* smaragdite
Smithsonit *m* smithsonite

Sockel *m* 1. *(build)* pedestal, plinth, footing stall, base; 2. *(el)* cap, *(Am)* base *(electric lamp)*, holder, socket; 3. understructure *(blast furnace)*
fünfpoliger S. five-prong base
Sockeladapter *m* valve adapter
Sockelanschluß *m* socket terminal
Sockelbuchse *f* cap sleeve
Sockelkitt *m* basing cement
Sockelkleber *m* lamp-capping cement
Sockelleiste *f* base-board, skirting [board]
Sockelmauer *f* plinth wall
sockeln to cap electron tubes
Sockelpunkt *m* pedestal control point
Sockelschalter *m* socket switch
Sockelschaltung *f* pin connection; base diagram *(radio)*
Sockelstift *m* contact pin; base pin
Soda *f* soda, carbonate of sodium, natron
kalzinierte S. soda ash, sodium carbonate
kaustische S. sodium hydroxide, caustic soda
kristallisierte S. soda crystals
unreine S. black ash
Sodaauszug *m* soda extraction *(analytical chemistry)*
Sodainspirator *m* soda aspirator
Sodakochechtheit *f* fastness to boiling soda
Sodalith *m* sodalite
Sodarückstände *mpl* soda residues *(Leblanc process)*
Sodazellstoff *m* soda pulp
Söderberg-Elektrode *f* Soderberg electrode
Sodraum *m (ship)* bilge
Soffittenfassung *f* tubular lamp holder
Soffittenlampe *f* festoon (tube, strip-light) lamp, festoon (tube, strip-light) bulb
Soffittenleitung *f* tubular lamp wire
Sofortruf *m (telec)* immediate ringing
Sofortverkehr *m (telec)* demand working, no-delay operation (service)
Sog *m (ship)* wake, run, dead water, suction; augmentation of resistance
Sogeinwirkung *f* vacuum effect
Sogwirkung *f* suction effect
Sogziffer *f* wake fraction, thrust deduction
Sohlbank *f* window sill
Sohldruck *m* bottom (base, soil) pressure
Sohle *f* 1. *(build)* floor, bottom; level; foot; base; footwall; 2. *(mine)* invert; horizon, lift; 3. sole *(shoe)*
Sohlenabstand *m (mine)* level interval
Sohlenauflegemaschine *f* sole-laying machine
Sohlenbelag *m (mine)* floor covering
Sohlenbergbau *m* horizon mining
Sohlenbogen *m (mine)* invert
Sohlengewölbe *n (mine)* floor arch
Sohlenhebung *f (mine)* floor lift
Sohlenheftmaschine *f* sole-tacking machine
Sohlenholz *n (mine)* floor bar, foot block, sill
Sohlenleder *n s.* Sohlleder
Sohlenpacken *m* floor coal
Sohlenschaufel *f (ag)* ditching shovel
Sohlenschneidemaschine *f* sole-cutting machine
Sohlenschußbohrloch *n* bottom hole
Sohlenschüsse *mpl (mine)* floor trimmers
Sohlenschutz *m* tip (sole) protector
Sohlenstrecke *f (mine)* footwall drift
Sohlenvorreißer *m* precutting floor blade *(coal plough)*
söhlig aclinal
Sohlleder *n* sole leather
altgegerbtes S. white German sole leather
lohgares S. vegetable tanned sole leather
säuregeschwelltes S. acid swollen sole leather
Sohllöseneinbruch *m (mine)* bottom-draw cut, toe-cut
Sohlplatte *f (build)* invert, base (bed) plate
Sohlpressung *f* base (bottom, soil) pressure
Sohn *m* stamper *(gramophone record)*
Sojaöl *n* soya bean oil
Sol *n* sol
Solarbatterie *f* solar battery
Solarelement *n* solar cell
Solarisation *f* solarization
Solarkonstante *f* solar constant
Solarkonverter *m* solar energy converter
Solarmaschine *f* solar engine
Solaröl *n* solar oil
Solarometer *n* solarometer *(navigation)*
Solbildung *f* solation
Sole *f* brine
Solekühler *m* brine cooler
Solenoid *n* solenoid
luftgekühltes S. air-cooled solenoid
Solenoidspektrometer *n* solenoidal spectrometer
Solenoidspule *f* solenoid
Solexgerät *n* Solex pneumatic micrometer
Soliduskurve *f* solidus curve (line)
Soliduslinie *f* solidus line (curve)
Solion *n* solion
Soliontetrode *f* solion tetrode
Sollabmessung *f* nominal (desired) dimension
Sollbahn *f* stable orbit
Sollbruchstelle *f (aero)* preset (predetermined) breaking point
Solldicke *f* desired thickness
Solleistung *f* required (nominal) output; design power; demand power
Sollförderung *f* planned output, target production
Sollfrequenz *f* required frequency
Sollgeschwindigkeit *f* cut-off velocity; required speed
Sollkreis *m* stable orbit
Sollkreisdurchmesser *m* equilibrium diameter
Sollmaß *n* required measure
Sollstellung *f* required (desired) position

Sollwert *m* rated (nominal) value; desired value; reference input, *(Am)* command
Sollwerteinsteller *m* reference element
Sollwerteinstellung *f* control point adjustment, desired value adjustment
Sollwertgeber *m* reference element
Sollwertumformer *m* input means *(controller)*
Solutionssalz *n* dissolving salt
Solva[ta]tion *f* solvation
solvatisiert solvated
Solventextraktion *f* solvent extraction
Solventnaphtha *n (f)* solvent (coal-tar) naphtha
Solvolyse *f* solvolysis
Sommerfahrplan *m* summer service
Sommerfreibord *m* summer freeboard
Sommerfreibordtiefgang *m* summer draught *(of a vessel)*
Sommerfrüchte *fpl* spring-sown crops
Sommergerste *f* spring barley
Sommergetreide *n* spring cereals
Sommersaat *f* spring-sown crops
Sommerschnitt *m* summer pruning
Sommertiefladelinie *f* summer load line *(of a vessel)*
Sommerung *f* spring-sown crops
Sonar *n* sonar *(sound navigation and ranging)*
Sonarempfänger *m* sonar receiver
Sonarimpuls *m* ping *(radar)*
Sonde *f* probe
  abgeschirmte S. screened probe
  akustische S. sonic probe
Sondenabtastung *f* probe scanning
Sondengerät *n* probe unit
Sondenkennlinie *f* probe characteristic
Sondenmessung *f* probe measurement
Sondenmethode *f* foil method; probe method
Sondenmikrofon *n* probe microphone
Sondenspitze *f* probe tip
Sondentechnik *f* probe technique
Sonderabdruck *m* separate print
Sonderausbildung *f* specialized training
Sonderausführung *f* special design
Sonderausrüstung *f* special equipment
  wahlweise S. optional extras
Sonderausstattung *f* auxiliary equipment
Sonderbauart *f* special design
Sonderbauweise *f* special type (design)
Sonderbewetterung *f* auxiliary ventilation
Sonderbohrer *m* **mit verstärktem Kern** special thick-web drill
Sonderbohrmaschine *f* special drilling machine
Sonderdrehmaschine *f* special[-purpose] lathe
Sondereinrichtung *f* special attachment
Sondereinsatz *m* special duty
Sonderfahrzeug *n* special-purpose vehicle
Sonderfall *m* special (particular) case
Sonderform *f* special form (shape)
Sondergewinde *n* special thread
Sondergüte *f* special quality
Sonderhobelmaschine *f* special-type planer
Sonderkonstruktion *f* special design
Sonderlänge *f* extra-long length *(drill)*
Sonderlüfter *m* auxiliary (booster) fan
Sondermessing *n* **für Kondensatorrohre** admiralty metal
Sonderpassung *f* special fit
Sonderprofil *n* special section
Sonderreibahle *f* special-purpose reamer
  S. in Vollhartmetall solid-carbide special-purpose reamer
Sonderroheisen *n* special pig iron
Sonderschicht *f* extra shift
Sonderschiff *n* special ship
Sonderschleifmaschine *f* special-purpose grinding machine
Sonderspannvorrichtung *f s.* Sonderspannzeug
Sonderspannzeug *n* special work-holding fixture, special work-fixture
Sonderspiralbohrer *m* special twist drill
Sonderstahl *m* special (alloy) steel
Sondersupport *m* special tool head
Sonderverfahren *n* special process
Sondervorrichtung *f (mach)* special fixture
Sonderwerkstoff *m* specific material
Sonderwerkzeug *n* special-purpose tool
Sonderzubehör *n* extra equipment, special attachments; extras
Sonderzug *m* extra (special) train
Sonderzweck *m* special (specific) purpose
Sonderzweckfräsmaschine *f* special-purpose milling machine
Sondierbohrung *f* exploratory boring
sondieren *n* to probe; to sound
Sone *n* sone
Soneskale *f* sone scale
Sonnenaktivität *f* solar activity
Sonnenatmosphäre *f* solar atmosphere
Sonnenausbruch *m s.* Sonneneruption
Sonnenbatterie *f* solar cell (battery)
Sonnenbeobachtung *f* solar observation
sonnenbeständig sun-resistant, sunfast
Sonnenbestrahlung *f* solar irradiation, insolation
Sonnenbleiche *f* sun bleaching
Sonnenblende *f (build)* brise-soleil; *(aut)* sun visor; *(opt)* camera hood
Sonnenblumenöl *n* sunflower oil
Sonnendach *n* sun-blind; *(ship)* awning; *(aut)* sliding roof
Sonnendeck *n* sun deck
Sonnenenergie *f* solar (sun) energy
Sonnenenergieumwandlung *f* solar energy conversion
Sonnenenergiewandler *m* solar energy converter
Sonneneruption *f* solar eruption
Sonnenfilter *n* solar filter

Sonnenfinsternis *f* solar eclipse
Sonnenfleck *m* sunspot
Sonnenfleckentätigkeit *f* sunspot activity
Sonnenheizung *f* solar heating
Sonnenhelligkeitsmesser *m* sunshine recorder
Sonnenkomponente *f* **der kosmischen Strahlung** solar component of cosmic rays
Sonnenkorona *f* solar corona
Sonnenkraftmaschine *f* solar engine
Sonnenkraftwerk *n* solar power station
sonnenlichtecht sunfast
Sonnenmagnetismus *m* solar magnetism
Sonnenmasse *f* sun's mass
Sonnenparallaxe *f* solar parallax
Sonnenpeilung *f* bearing by the sun
Sonnenprisma *n* sun prism
Sonnenrad *n* sun gear
Sonnenrand *m* solar limb
Sonnensegel *n* awning
Sonnenspektrograf *m* solar spectrograph
Sonnenspektrometer *n* / **automatisches** solar spectrum analyzer
Sonnenspektrum *n* solar spectrum
Sonnenspule *f (text)* flat conical cheese
Sonnenstrahlung *f* solar (sun's) radiation
Sonnensucher *m* sunseeker
Sonnensystem *n* solar system
Sonnentätigkeit *f* solar activity
Sonnenteleskop *n* solar telescope
Sonnentemperaturmesser *m* heliothermometer
Sonnenüberwachung *f* control of solar activity
Sonnenuhr *f* sundial
Sonnenwärme *f* solar heat
Sonnenzelle *f* solar cell (battery)
Sonntagsrückfahrkarte *f* sunday ticket
Sonometer *n* sonometer
Sorbens *n* adsorbent
Sorbinsäure *f* sorbic acid
Sorbit *m (met)* quenching troostite *(structure)*; *(ch)* sorbitol
Sorelmörtel *m* Sorel's cement
Soret-Phänomen *n* Soret effect, thermal diffusion
Sorption *f* sorption
Sorptiv *n* adsorbate
Sorten *fpl* / **geringwertige** inferior grades
Sortieranlage *f* picking (separating, sorting) plant; screen room *(papermaking)*
Sortierbarkeit *f* washability
sortieren to classify, to clean, to grade, to separate, to sort; to size
  der Größe nach s. to grade by size
Sortieren *n* **nach Größe** size-grading
Sortierer *m* sorter; centrifugal screen
Sortiergerät *n* sorter; sorting arrangement
Sortierleser *m* sorter reader
Sortiermaschine *f* grader; separator, sizing (sorting) machine; sorter
  fotoelektrische S. photo-electric sorter
Sortierraum *m* sorting room
Sortiersieb *n* separating screen
Sortierstift *m* sizing pin
Sortiertisch *m* sorting table
Sortiertrommel *f* sizing drum (trommel), revolving screen
  S. für Hackspäne chip screen
Sortierung *f* cleaning, grading, separation, sizing
Sortierzyklon *m* cyclone separator (washer)
Sortierzylinder *m* rotary screen
Sortiment *n* assortment
soßen to sauce *(tobacco)*
souplieren to degum *(silk)*
Souterrain *n* semi-basement
Soxhlet-Apparat *m* Soxhlet apparatus
Sozialgebäude *n* recreation (welfare) building
Sozialleistungen *fpl* social services
Soziusfahrer *m* pillion rider (passenger) *(cycle)*
Soziussitz *m* pillion
Spachtel *f (m)* 1. float, smoothing (finishing, flat) towel; putty knife; 2. *s.* Spachtelmasse
Spachtelboden *m (plast)* fleximer flooring
Spachtelkitt *m* filling putty
Spachtelmasse *f* putty, knotting, primer, undercoat, knifing glaze, stopping lac
spachteln to putty, to fill, to prime, to trowel
Spähschiff *n* scout
Spalier *n* espalier *(lattice-work)*
Spalierbaum *m* espalier tree
Spallation *f* spallation
Spallationsprodukt *n* spallation product; spallation fragment
Spallationsquerschnitt *m* spallation cross section
Spallationsreaktion *f* spallation reaction
Spalt *m* gap; fissure, cleavage; *(opt)* slit, aperture; clearance *(turbine)*; duct *(shield)*
  S. mit Versetzung stepped duct
  gekrümmter S. bent duct
  gerader S. straight slit
  schraubenförmiger S. helical duct
Spaltabsaugung *f (aero)* slot-syphoning
Spaltanlage *f* cracking plant
Spaltausbeute *f s.* Spaltproduktausbeute
Spaltausfüllung *f (geol)* fissure occupation
Spaltausleuchtung *f* slit illumination
spaltbar *(nucl)* fissile, *(Am)* fissionable; cleavable *(crystal)*
Spaltbarkeit *f (nucl)* fissility; cleavage, cleavability; delamination
  S. von Kristallen cleavage of crystals
  ebene S. cleavage
Spaltbenzin *n* cracked petrol (*Am* gasoline)
Spaltbild *n* slit image
Spaltblättchen *n* cleavage sheet
Spaltblende *f* slit-stop, slit diaphragm
Spaltbombe *f* fission bomb; atom[ic] bomb, A-bomb

Spaltbreite *f* width of clearance space *(turbine)*; gap clearance (length); gap width *(magnetic head)*
effektive S. effective gap length
Spaltbrüchigkeit *f* cleavage brittleness
Spaltbruchstück *n* fission fragment
Spaltdruck *m* clearance pressure *(turbine)*
Spaltdurchmesser *m* clearance diameter *(turbine)*
Spalte *f* 1. *(mine)* fissure; *(geol)* joint; 2. [card] column *(punched card)*
schwebende S. flat joint
Spaltebene *f* cleavage face (plane) *(crystallography)*
Spalteinfang *m (nucl)* fission capture
spalten to split; to cleave *(crystal)*; to crack; *(ch)* to decompose, to break; to knap *(stone)*; to delaminate
eine chemische Bindung s. to break a linkage
Nacken s. *(leath)* to neck
Narben s. *(leath)* to friz[z] off the grain
Spaltenabziehmaschine *f (print)* galley-proof press
Spaltenabziehpresse *f (print)* galley-proof press
Spaltenanzeiger *m* column indicating device *(punched card)*
Spaltenbreite *f* measure
Spaltenergie *f (nucl)* fission energy
Spaltenschiff *n* slip (column) galley
spaltfähig *(nucl)* fissile, *(Am)* fissionable; cleavable *(crystal)*
Spaltfeld *n* gap field *(magnetic head)*
Spaltfestigkeit *f* interlaminar strength; resistance to splitting, splitting resistance
Spaltfilter *n* gap-type filter; edge-type filter, plate-type oil filter; space filter *(hydraulics)*
hydraulisches S. hydraulic space filter
Spaltfläche *f* cleavage plane (surface) *(crystal)*
Spaltfuge *f* split
Spaltgatter *n* gang saw
Spaltgeschirr *n* splitting tool
Spaltgift *n* fission poison
Spaltglimmer *m* sheet mica
Spaltholz *n* cleft (split) timber
Spaltionisationskammer *f s.* Spaltungskammer
Spaltkammer *f s.* Spaltungskammer
Spaltkeil *m* splitting wedge
Spaltklappe *f (aero)* slotted flap
Spaltkristall *m* cleavage crystal
Spaltlampe *f* slit lamp
Spaltlänge *f* circumferential length of clearance space *(turbine)*; slit height
Spaltleiterschutz *m* divided-conductor protection
Spaltleitwert *m* gap admittance *(discharge)*
Spaltleuchtengerät *n* slit-lamp apparatus
Spaltmagnetron *n* slot-type magnetron
Spaltmaschine *f* splitting machine *(leather)*
Spaltmaterial *n* fissile (fissionable) material
Spaltmeißel *m* cleaving chisel
Spaltmesserschleifmaschine *f* bandknife grinding machine *(leather)*
Spaltneutron *n* fission neutron
Spaltneutronenspektrum *n* fission spectrum
Spaltpol *m* split pole
Spaltpolinstrument *n* split-pole instrument
Spaltpolmotor *m* split-pole motor
Spaltprodukt *n* fission product
Spaltproduktausbeute *f* fission[-product] yield
Spaltprüfung *f* split test; spalling test
Spaltquerschnitt *m* 1. *(nucl)* fission cross section; 2. roll gap *(rolling)*
Spaltrate *f* fission rate, rate of fission
Spaltring *m* split ring
Spaltriß *m* cleavage crack
Spaltrohstoff *m s.* Brutstoff
Spaltruder *n* split rudder
Spaltscheinleitwert *m* primary transit-angle gap loading
Spaltschwelle *f* fission threshold
Spaltspektrograf *m* slit spectrograph
Spaltspektrum *n* fission spectrum
Spaltspitzenspannung *f* peak alternating gap voltage
Spaltstoff *m* fissionable (fissile) material; nuclear fuel *(reactor)*
erschöpfter S. depleted fuel
Spaltstoffanordnung *f* fuel assembly
Spaltstoffanreicherung *f* fuel enrichment
Spaltstoffelement *n* fuel element
Spaltstoffregenerierung *f* nuclear-fuel regeneration
Spaltstoffrückgewinnung *f* nuclear-fuel regeneration
Spaltstoffverdopplung *f* fuel doubling time
Spalttiefe *f* gap depth *(magnetic head)*
Spaltüberdruck *m* absolute clearance pressure *(turbine)*
Spaltundichtigkeit *f* clearance leakage *(turbine)*
Spaltung *f* 1. [nuclear] fission; fission event; 2. cleavage *(mineral)*; cracking, decomposition, splitting
schnelle S. fast [neutron] fission
spontane S. spontaneous fission
symmetrische S. symmetrical fission
ternäre S. ternary fission, tripartition
thermische S. thermal [neutron] fission
ungesteuerte S. explosive (non-controlled) fission
unsymmetrische S. asymmetric fission
Spaltungsausbeute *f s.* Spaltproduktausbeute
Spaltungsbreite *f* fission width
Spaltungsebene *f* cleavage plane
Spaltungsenergie *f* fission energy, energy of fission
Spaltungskammer *f* fission chamber
spaltungslos non-fission *(capture)*
Spaltungsmethode *f* cleavage technique

Spaltungsneutron *n s.* Spaltneutron
Spaltungsquerschnitt *m s.* Spaltquerschnitt 1.
Spaltungsreaktion *f* fission reaction
Spaltungsreaktor *m* fission reactor
Spaltungstechnik *f* cleavage technique
Spaltungstonne *f (ship)* bifurcation buoy
Spaltungswärme *f* fission heat
Spaltungswirkungsquerschnitt *m s.* Spaltquerschnitt 1.
Spaltverkleidung *f* slot covering
Spaltverlust *m* clearance loss *(turbine)*
Spaltversuch *m* cleavage test
Spaltware *f* cloven timber
Spaltweite *f* width of port
Spaltzählrate *f* fission counting rate
Spaltzone *f* [reactor] core, active lattice *(reactor)*
Spaltzünder *m* high-tension electric detonator
Spaltzylinderprobe *f s.* Spaltzylinderprüfung
Spaltzylinderprüfung *f* split cylinder test *(concrete)*
Span *m* chip; cut *(metal)*; shaving, cutting *(wood)*
  aufgerollter S. curled chip
Spanabfall *m* dust content *(wood)*
Spanabfluß *m* chip flow, flow of [the] chips
Spanabflußrichtung *f* direction of chip flow
Spanabfuhr *f* chip disposal (clearance, removal), swarf disposal (clearance, removal)
Spanabführung *f* chip removal (clearance, disposal)
  innere S. central chip disposal *(deephole drilling)*
spanabhebend *s.* spanend
Spanabhebung *f s.* Spanung
Spanabnahme *f* chip removal; amount of metal removed
Spanabsaugeinrichtung *f* chip-removal suction device
Spanaufbereitung *f* particle preparation
Spanaufschüttung *f* chip distribution
spanbar / schwer hard-to-machine
Spanbildung *f* chip formation
Spanbildungsvorgang *m* mechanics of chip formation
Spanbrecher *m* chip breaker
Spanbrechernut *f* chip-breaking nick (groove)
Spanbreite *f* width of cut
Spandicke *f* chip thickness
  S. am Anfang des Schnittes commencement chip thickness
Späneabscheider *m* shavings separator *(wood)*
Späneauswurf *m* swarf clearance
Spänefang *m* chip chute
Spänefänger *m* shavings separator *(wood)*
Spänefangkasten *m* shavings bin *(wood)*; chip tray
Spänefluß *m* / **ungehinderter** unobstructed chip flow
Späneförderer *m* swarf (chip) conveyor
Spänekasten *m* swarf bin
Spänekeller *m* cellar for shavings *(wood)*
Spanelement *n* chip segment
spanen to cut, to machine
Spanen *n* cutting, stock removal
spanend [metal] cutting, metal removing; machining; chip-forming
Spänerutsche *f* chip chute
Spänesilo *m* chip charger
Spänestauung *f* chip clogging (congestion)
Spänetransporteinrichtung *f* chip conveyor
Spänetrog *m* [chip] trough
Spänevolumen *n* volume of chips [removed]
Spanfangschale *f* chip (swarf) tray, chip (swarf) pan
Spanfläche *f* cutting (tool, rake) face; *(Am)* chip-bearing surface; top face *(turning tool)*
  S. eines Fräserzahns tooth face
Spanfluß *m* flow of chips
Spanform *f* chip shape (form)
Spanformer *m* chip curler
Spanformstufe *f* chip-breaking shoulder
Spanleistung *f* cutting (metal removing) capacity, metal removal rate, chip removal
Spanleitblech *n* chip chute
Spanleitrille *f* land and groove type of chip breaker, ground-in groove type chip breaker
Spanleitstufe *f* ground-in step type chip breaker, step ground in the face of the tool, *(Am)* deflection shoulder
Spanloch *n* plane hole *(wood)*
spanlos chipless
Spanlücke *f* gash
Spanmasse *f* weight (mass) of metal removed
Spanmenge *f* amount of metal removed; rate of metal removal
  abgenommene S. amount of metal removed
  spezifische S. metal removal rate
Spannarm *m* bending arm
Spannauge *n (opt)* fixing lug
Spannbacke *f* [chuck, gripping] jaw; pulling jaw *(broach)*
Spannbahn *f* prestressing lane (bed) *(concrete)*
Spannbahnfertigung *f* long-line method (system) *(concrete)*
Spannbalken *m* tie-beam
Spannband *n* pressing ring
  S. für Kolbenringeinbau piston-ring clamp
Spannbandbefestigung *f* cradle mounting
Spannbank *f* saw sharpener's bench
Spannbeton *m* prestressed concrete
  S. mit nachträglichem Verbund post-tensioned concrete
  S. mit sofortigem Verbund pretensioned concrete
Spannbetonbau *m* prestressed-concrete construction
Spannbett *n* prestressing bed *(concrete)*

Spannbewehrung *f* prestressing steel *(concrete)*
endverankerte S. end-anchored reinforcement
Spannbock *m* poppet
Spannbolzen *m (mach)* clamp[ing] bolt; *(build)* pulling (stressing) bolt *(concrete)*
Spannbreite *f (text)* width of fabric
Spannbuchse *f* bushing [for holding work]
Spanndorn *m* mandrel, arbor
Spanndraht *m* 1. *(build)* prestressing wire *(concrete)*; 2. *(el)* suspension (guy, bracing, span) wire; pole guy; bridle
Spanndrahtaufhängung *f (el)* span-wire suspension, cross-span suspension
Spanndrahtgalvanometer *n* span-wire galvanometer
Spanndrahtschwelle *f (el)* anchor log, dead man
Spanndruck *m* clamping pressure
Spanneisen *n* holding strap, clamp *(workpiece)*; cross bar
spannen 1. *(mech)* to stress, to load, to tension; 2. *(mach)* to hold, to grip; to chuck, to set up the workpiece, to mount *(between centres)*; 3. to load *(spring)*; 4. to take up slack *(overhead line)*; 5. to tension *(concrete)*; 6. to strain *(hide)*; 7. to stretch; to tighten; to cock *(lever)*
Bohrer s. to hold drills
eine Schnur s. to stretch a line
Spannen *n* **der Werkstücke** setting up work; mounting of the workpiece *(centres)*; chucking; loading
S. des Sägeblattes tensioning (setting the tension) of the blade
S. im Futter chucking
S. langer zylindrischer Werkstücke supporting long thin work
S. mit Hilfe von Gewichten load-tensioning, weight-stretching *(concrete)*
S. mittels Schraubspindeln screw-tensioning *(concrete)*
S. zwischen Spitzen mounting on centres
thermisches S. heat-tensioning *(concrete)*
Spanner *m (text)* counter faller *(self-actor)*
Spannfinger *m* toe dog
Spannfutter *n* chuck
S. mit Dauermagnet permanent magnet chuck
elektromagnetisches S. electromagnetic holding device
Spanngeschirr *n* tackle *(overhead-line construction)*
Spanngitter *n* frame grid, frame-lock grid structure *(electron tubes)*
Spannglied *n* / **ausgepreßtes** bonded tendon
gekrümmtes S. depressed strand, draped tendon
Spanngliedkanal *m* cable (prestressing) duct, post-tensioning conduit *(concrete)*
Spannhebel *m* cocking lever
Spannkabel *n* [pre]stressing cable *(concrete)*
Spannkabelführung *f* cable profile *(concrete)*
Spannkabelrohr *n* cable housing (duct); conduit
Spannkanal *m* post-tensioning conduit, cable (prestressing) duct *(concrete)*
Spannkettenrad *n* idler sprocket
Spannkloben *m* clamp dog
Spannkopf *m* pulling (stressing) head *(concrete)*; pulling jaws *(broach)*; specimen-holding jaws *(testing)*
Spannkraft *f* prestressing force *(concrete)*
Spannlack *m* stiffening varnish
Spannmittel *n* holding (clamping) tool
Spannmutter *f* clamp nut
Spannplatte *f* / **fest fundamentierte** floor plate
Spannpresse *f* [pre]stressing jack *(concrete)*
Spannrahmen *m* stenter frame, tenter
Spannreck *n* portable horizontal bars
Spannring *m* pressing ring; locking ring; stretching ring *(capacitor microphone)*
hinterlegter S. internal (inner) spring ring, bottom ring
Spannrolle *f* guiding (idler) pulley *(belt)*; tension drum (pulley)
Spannsäge *f* frame (web) saw
Spannsägeblatt *n* **mit festen Angeln** span web blade with solid tangs
S. mit Nietangeln span web blade with riveted tangs
Spannschlitz *m* clamping slot
Spannschloß *n* screw shackle, turn-buckle swivel
Spannschraube *f* lock[ing] screw; drawback bolt, drawbolt *(milling spindle)*
S. mit Nutenstein poppet
Spannstab *m* stressing bar
Spannstahl *m* prestressing steel *(concrete)*
Spannstation *f* loading station; anchor station *(conveyor)*; tensioning (take-up) device
Spannstelle *f s.* Spannstation
Spannstift *m* brace
Spanntraverse *f* cross-head
Spanntreppe *f* step block
Spann- und Entspannstation *f* loading and unloading station
Spannung *f* 1. *(mech)* stress *(force per unit area)*, load; tension *(surface)*; 2. *(el)* voltage, potential [difference], tension
unter S. *(el)* alive
S. am Leitungsanfang sending-end voltage
S. am Leitungsende received voltage
S. aus Längskraft mit Biegung combined direct and bending stress
S. bei Belastung *(el)* closed circuit voltage
S. gegen Erde voltage to earth
S. im Millivoltbereich millivoltage
S. infolge Wärmerauschens thermal-agitation voltage, thermal-effect voltage
S. unter Last [on-]load voltage

S. zwischen den Außenleitern outer-conductor voltage
angelegte S. applied (impressed) voltage
äquivalente S. equivalent constant potential
aufgedrückte S. impressed voltage
aufgenommene S. absorbed voltage
aufgeprägte S. imposed (impressed) voltage
äußere S. 1. external stress; 2. external voltage
beruhigte S. smoothed voltage
bewertete S. weighted value of a voltage
dreieckige S. triangular voltage
effektive S. root-mean-square voltage
einachsige S. simple (uniaxial) stress
eingeprägte S. impressed voltage
elastische S. stress
elektrische S. electric voltage
erdsymmetrische S. balanced voltage
gleichmäßige S. steady voltage
gleichphasige S. in-phase voltage
größte verkettete S. diametral voltage *(of a polyphase system)*
hochfrequente S. radio-frequency voltage
höchste erreichbare S. ceiling voltage
höchstzulässige S. maximum potential difference; maximum allowable stress
induzierte S. induced voltage
innere S. 1. internal stress; 2. internal voltage
konstante S. stabilized voltage
kritische S. critical stress
magnetische S. magnetic potential difference
maximal erreichbare S. ceiling voltage
niedrigste S. minimum voltage (stress)
periodische S. periodic potential difference
pulsierende S. pulsating voltage
rechnerische S. working (design) stress
reduzierte S. 1. reduced stress; 2. reduced (abated) voltage
rein sinusförmige S. simple sinusoidal potential difference
resultierende S. resultant voltage (stress)
sinusförmige S. sinusoidal voltage
stabilisierte S. stabilized voltage
thermische S. thermal stress
untere S. bottom voltage
verbrauchte S. absorbed voltage
verkettete S. line voltage of a polyphase system, voltage between lines, voltage of a system, delta (resultant, interconnected, mesh) voltage
vorgegebene S. predetermined voltage
vorgeschriebene S. rated voltage
wiederkehrende S. recovery voltage
zugeführte S. injected (applied) voltage
zulässige S. 1. safe (allowable, permissible) stress; 2. permissible (admissible, allowable) voltage
zusammengesetzte S. composite stress
spannungführend active, [a]live
Spannungsabbau *m* loss of prestress, stress relaxation
Spannungsabfall *m* 1. loss of prestress, stress relaxation; 2. voltage drop, fall in voltage
S. auf der Leitung *(el)* line drop
S. in der Röhre *(el)* tube drop
S. über der Elektrode electrode drop
induktiver S. inductive voltage drop
innerer S. internal voltage drop
ohmscher S. ohmic drop in voltage
relativer S. relative voltage drop
spannungsabgestimmt voltage-tunable
Spannungsabgriff *m* voltage tap
Spannungsableiter *m s.* Überspannungsableiter
Spannungsabnahme *f* voltage drop; voltage pick-off
Spannungsänderung *f* voltage change (variation), change in voltage
Spannungsänderungsgeschwindigkeit *f* rate of voltage change
Spannungsanregung *f* voltage excitation
Spannungsanstieg *m* voltage rise, increase of voltage
induktiver S. bei kapazitiver Last inductive rise
linearer S. linear voltage rise
relativer S. relative voltage rise
Spannungsanteil *m* voltage component
Spannungsantrieb *m* voltage drive
Spannungsanzeige *f* voltage metering
Spannungsanzeiger *m* 1. stress indicator; 2. voltage indicator
Spannungsaufteilung *f* voltage division
Spannungsausgleich *m* 1. stress equalizing; 2. voltage equalizing (compensation), compensation of voltage
Spannungsausgleichschaltung *f* equalizing (equipotential) connection
Spannungsausgleichsverbindung *f* equipotential connection
Spannungsbauch *m* voltage maximum (loop)
Spannungsberechnung *f* stress analysis *(design)*, calculation of stresses
Spannungsbereich *m* voltage range
Spannungsbild *n* stress diagram
Spannungscharakteristik *f* voltage characteristic
Spannungsdämpfung *f* voltage attenuation
Spannungs-Dehnungs-Diagramm *n* stress-strain diagram
Spannungs-Dehnungs-Kurve *f* stress-strain curve
Spannungs-Dehnungs-Schaubild *n* stress-strain diagram
Spannungsdiagramm *n* pressure-volume diagram, p-v-diagram
Spannungsdoppelbrechung *f* strain birefringence

Spannungsdreieck *n* voltage triangle
Spannungsdrift *f* voltage drift
Spannungsdurchschlag *m (el)* puncture
Spannungseinbruch *m* drop-out *(magnetic television recording)*
Spannungseinstellung *f* voltage control
Spannungsempfindlichkeit *f* voltage sensitivity
spannungsentlastet stress-relieved *(material)*
spannungserhöhend 1. stress-raising; 2. voltage-increasing
Spannungserhöher *m (el)* positive booster; stress raiser
Spannungserhöhung *f* voltage increase; stress increase
Spannungsermittlung *f* stress determination (analysis) *(design)*
spannungserniedrigend step-down; stepping down of the voltage, voltage reducing; stress lowering
Spannungsfall *m* voltage drop
Spannungsfeld *n* stress field
Spannungsfernmeßgerät *n* televoltmeter
Spannungsfestigkeit *f* [di]electric strength, puncture strength
spannungsfrei 1. stress-free, unstressed *(material)*; 2. *(el)* dead
spannungsfreiglühen to stress-relieve [by annealing]
Spannungsfreiglühen *n* stress-free annealing
spannungsführend live
Spannungsfunktion *f* 1. stress function; 2. voltage response
Spannungsgefälle *n* voltage drop (gradient)
Spannungsgegenkopplung *f* voltage feedback
Spannungsgesetz *n* pressure law
spannungsgespeist voltage-fed
Spannungsgleichhalter *m* voltage stabilizer (regulator)
Spannungsgleichhaltung *f* voltage regulation
Spannungsgleichheit *f* voltage balance
Spannungsgrenze *f* stress limit *(material)*
Spannungshaltung *f* voltage stabilization
Spannungshochregelung *f* recovery of voltage
Spannungsknoten *m* voltage node (minimum)
Spannungskomponente *f* 1. component of stress; 2. component voltage
Spannungskonstanthalter *m* voltage stabilizer (regulator)
Spannungskorrosion *f* stress corrosion
Spannungskurve *f* 1. stress curve; 2. voltage curve
spannungslos 1. free from tension (stress); 2. *(el)* dead
Spannungsmeßbereichumschalter *m* voltage range switch
Spannungsmesser *m* voltmeter, voltage indicating meter
Spannungsnachweis *m* stress analysis *(design)*
Spannungsnulldurchgang *m* zero voltage
Spannungsoptik *f* photoelasticity
spannungsoptisch photoelastic
Spannungspegel *m* voltage level
  absoluter S. voltage level referred to 0.775 volt, true level
Spannungspfad *m* voltage (potential) circuit, voltage path
Spannungspolarität *f* sense of the voltage
Spannungsprüfer *m* 1. stress tester; 2. voltage detector (indicator); charge indicator
  optischer S. polarizing stress tester
Spannungsprüfung *f* 1. stress determination; 2. voltage (dielectric) test
  optische S. photoelasticity test
Spannungsquelle *f* power source
  stabilisierte S. regulated power source
Spannungsregelung *f* voltage control (regulation)
  S. durch Bürstenverschiebung brush-shifting control
Spannungsregelungsanlage *f* voltage-regulation system
Spannungsregler *m* [constant-]voltage regulator; magnestat
  selbsttätiger S. automatic voltage regulator
Spannungsreihe *f (el)* contact series
  elektrochemische S. electromotive series
  galvanische S. electromotive series
  thermoelektrische S. thermoelectric series
Spannungsrelais *n* voltage-control relay
Spannungsresonanz *f* voltage resonance
Spannungsresonanzkreis *m* parallel resonant circuit
Spannungsriß *m* stress crack
Spannungsrückwirkung *f* voltage reaction
Spannungsschnellregler *m* automatic voltage regulator
Spannungsschreiber *m* voltage recorder
Spannungsschutz *m* voltage protection
Spannungsschwankung *f* voltage fluctuation (variation, ripple)
Spannungssicherung *f* potential fuse
Spannungsspitze *f* 1. peak stress, stress concentration *(material)*; 2. voltage peak
Spannungssprung *m* voltage jump, step of voltage
Spannungsspule *f* voltage coil
Spannungsstabilisierung *f* voltage stabilization
Spannungssteuerung *f* voltage control
Spannungsstoß *m* voltage impulse (surge), surge
Spannungsstufe *f* step of voltage, voltage increment
Spannungsstufenregler *m* step voltage regulator
Spannungsteiler *m* voltage (potential) divider; potentiometer
  dekadischer S. decade voltage divider
  kapazitiver S. capacitive voltage divider

ohmscher S. resistive voltage divider
Spannungsteilerfaktor *m* voltage-divider factor
Spannungsteilung *f* voltage division
Spannungstensor *m* stress tensor
Spannungstrajektorien *fpl* trajectories of principal stresses
Spannungstransformator *m* voltage transformer
Spannungstrennstrecke *f* voltage [clearing] distance
Spannungsüberlagerung *f* voltage superposition
Spannungsübersetzung *f* voltage transfer
Spannungsübertragungsfaktor *m* response to voltage
Spannungsumlagerung *f* stress redistribution *(material)*
Spannungsumschalter *m* voltage tapping switch
Spannungsunterschied *m* difference of potential
Spannungsuntersuchung *f* stress investigation *(design)*
Spannungsvektor *m* voltage vector
Spannungsverdoppler *m* voltage doubler
Spannungsverdopplerschaltung *f* voltage doubler [circuit]
Spannungsverhältnis *n* voltage ratio
Spannungsverlauf *m* 1. flow of stress; stress distribution; 2. voltage curve
Spannungsverlust *m* 1. loss of prestress, prestress loss *(prestressed concrete)*; 2. voltage loss
Spannungsverstärker *m* voltage amplifier
Spannungsverstärkung *f* voltage amplification
Spannungsverstärkungsfaktor *m* voltage ratio (amplification factor)
Spannungsverteilung *f* 1. stress distribution *(material)*; 2. voltage (potential) distribution
Spannungsvervielfacher *m* voltage multiplier
Spannungswaage *f* absolute electrometer
Spannungswahlschalter *m* voltage selector
Spannungswahlstecker *m* voltage selector panel
Spannungswandler *m* voltage (potential) transformer
kapazitiver S. capacitor transformer
kapazitiver S. in Resonanzschaltung resonance capacitor transformer
Spannungswandlung *f* voltage conversion
Spannungswechsel *m* 1. stress reversal; 2. change of voltage
Spannungswelle *f* voltage surge (wave)
Spannungswelligkeit *f* voltage ripple
Spannungswicklung *f* voltage winding
Spannungszeiger *m* volt[age] vector
Spannungszone *f* region of stress
Spannungszuführungspunkt *m* point of voltage application
Spannungszunahme *f* increase of voltage; increase of stress
Spannungszustand *m* state of stress
dreiachsiger S. threedimensional stress, general case of stresses
ebener S. plane (biaxial, two-dimensional) stress
einachsiger S. uniaxial stress
räumlicher S. triaxial (three-dimensional) stress, general case of stresses
zweiachsiger S. biaxial (two-dimensional, plane) stress
Spannut *f* chip groove, flute
S. am Fräserkörper chip pocket *(milling cutter)*
gerade S. straight flute
mehrförmige S. multi-tee slot
Spannutfräser *m* fluting cutter
Spannvorrichtung *f* *(mach)* holding (clamping) fixture, chucking device; workholding equipment; tensioning device; tensioning equipment *(overhead-line construction)*
Spannwagen *m* *(mine)* balance car
Spannweite *f* span; span length; range *(testing)*; *(aero)* wing span; opening capacity *(e. g. vice jaws)*; span width, opening, bridge (clear) span
S. der Stichprobe sample range
S. des Schraubstocks opening capacity (width) of vice jaws
Spannweitenkarte *f* R-chart, range chart, control chart for ranges
Spannweitenmitte *f* mid-range
Spannwerkvollgatter *n* log frame with tension apparatus
Spannzange *f* collet [chuck]
Spannzeug *n* *(mach)* clamping (holding) tool; *(el)* tackle *(overhead-line construction)*
Spannzeuge *npl* clamping accessories, holding devices, devices for holding work
Spanplatte *f* chipboard, particle board
furnierbeplankte S. composite board including veneer-faced particle
hartplattenartige S. hard-board-type particle board
leichte S. low density particle board
mittelschwere S. medium density particle board
stranggepreßte S. extruded particle board
Spanplattentäfelungsmaterial *n* particle-board panel stock
Spanplattenvlies *n* particle board mat
Spanquerschnitt *m* sectional area of chip, chip cross-section
Spanraum *m* chip room (space, clearance)
Spanroller *m* chip curler
Spanschirm *m* chip guard
Spanstauchung *f* chip upset
Spanstauung *f* clogging (choking, crowding) of the chips, swarf congestion

S. im Zahngrund trapping [of] the chip in the tooth gullet
Spanstufe *f* chip-breaking shoulder
Spant *n* frame; timber *(wooden ships)*; rib *(boats)*
  ausfallendes S. sloping outward frame
  eingebogenes S. bent frame; bent-in timber, steam-bent timber
  eingeschnittenes S. built-in timber
  festes S. fixed timber
  gebautes S. compound (build-up) frame
  gewachsenes S. fixed (built-in, grown) timber
  in Dampf gebogenes S. steam-bent timber
Spantabstand *m* frame spacing
Spanten *npl* framework
Spantenplan *m* scri[e]ve board
Spantenriß *m* body plan, frame lines (stations)
Spantenskale *f* area of section
Spantensystem *n* / **kombiniertes** combination [system of] framing
Spantfläche *f* sectional area
Spantflächenkurve *f* area of sections, curve of areas of cross-sections
Spantfüllstreifen *m* frame liner
Spantglühofen *m* bending furnace
Spantholz *n* frame timber
Spantiefe *f* cutting depth, [depth of] cut, rate of cut
  S. / bei geringer on a light cut
  S. / bei großer on a heavy cut
  S. je Zahn chip [load] per tooth *(milling)*
  geringe S. light cut
  große S. heavy cut
Spantinhaltskurven *fpl (ship)* Bonjean's curves
Spantrahmen *m* frame ring
Spantumfang *m* girth of a vessel (frame, section)
Spantwerk *n* [ship's] framing
Spantwinkel *m* frame angle
Spanumfangswinkel *m* angle of approach *(miller)*
Spanung *f* cutting
  unterbrochene S. intermittent cutting
Spanungskennwert *m* machining characteristic
Spanungsleistung *f* efficiency of metal removal
Spanungsrichtwerte *mpl* machining data (recommendations)
Spanungsverfahren *n* cutting process
Spanvolumen *n* volume of metal removed, metal removal rate (volume)
  abgenommenes S. volume of metal removed
Spanwinkel *m* [tool] rake angle; cutting rake *(milling cutter, twist drill)*
Sparbeize *f* pickling inhibitor
Spardeck *n (ship)* spardeck
Spardeckschiff *n* spardeck ship
Spardiode *f* efficiency diode
Spardüse *f* economizer jet
Sparflamme *f* pilot flame
Sparflug *m (aero)* weak-mixture cruising
Sparfluggemisch *n* economical mixture
Sparganggetriebe *n* overdrive
Spargemisch *n (aero)* lean mixture
Sparglied *n* bluff link *(knitting machine)*
Sparkatode *f* low-consumption cathode
Sparrad *n (text)* bluff wheel
Sparren *m* rafter
  gekrümmter S. compass rafter
Sparrendach *n* couple roof
Sparrengebinde *n* couple
Sparrenpaar *n* couple
Sparrost *m* economical grate
sparsam economical
Sparsamkeit *f* economy
Sparschaltung *f* restrictive circuit
Spartränkverfahren *n* empty-cell process
Spartransduktor *m* auto-transductor
Spartransformator *m* auto-transformer
Sparumformer *m* booster converter, converter with differential winding
Sparwiderstand *m* empty resistor
spasmodisch spasmodic
Spat *m* spar
Spateisenstein *m* spathic iron, chalybite, siderite
Spatel *m* spatula
Spaten *m* spade
Spatenegge *f* diamond harrow
Spatenruder *n* spade rudder
Spätfolgen *fpl* **der Bestrahlung** delayed (late) radiation effect
Spätheu *n* aftermath, second hay
Spätholz *n* late (summer) wood
spationieren to space
Spationieren *n* letter-spacing
Spatium *n* space
Spätschaden *m* late radiation effect, latent injury
Spätschicht *f* late shift
Spättod *m* late radiation death
Spätwirkung *f* **der Strahlung** delayed (late) radiation effect
Spätzündung *f* retarded ignition
Speckkalk *m* lime putty
Speckschmierung *f* lard lubrication
Speckstein *m* soapstone, steatite
Specularith *m (min)* specular hematite
Speditionsgesellschaft *f* forwarding agency; shipping agency
Speer *m* javelin
Speiche *f* spoke
Speichenrad *n* spoke[d] wheel
Speicher *m* 1. loft, store, granary; storehouse, warehouse; 2. memory, storage *(computer)*; register, store *(data)*, 3. accumulator
  S. mit beliebigem Zugriff random access memory (storage)
  akustischer S. acoustic (sonic) delay line, acoustic (sonic) delay memory

assoziativer S. associative memory
äußerer S. external store
dynamischer S. dynamic memory *(data)*
eingebauter S. internal memory
elektrostatischer S. electrostatic store
fotochemischer S. photochemical memory
hydropneumatischer S. compressed-gas accumulator
langsamer S. slow store
leistungsabhängiger S. volatile memory
leistungsloser S. permanent memory
löschbarer S. erasable memory
magnetischer S. magnetic memory
nichtpermanenter S. volatile memory (store)
periodischer S. cyclic memory (store)
permanenter S. permanent (non-volatile) memory, permanent (non-volatile) store
supraleitfähiger S. supraconducting memory
zugriffzeitfreier S. zero-access memory (store)
Speicherbau *m (mine)* overhead stoping on ore, shrinkage stoping, *(Am)* back stoping, *(Am)* lay system
Speicherbecken *n* reservoir, storage basin
Speicherbinärstelle *f* binary cell
Speicherdauer *f* maximum retention (storage) time *(electronics)*
Speicherelektrode *f* energy-storage electrode, energy-accumulation electrode
Speicherelement *n* storage cell, memory element
Speicherenergieschweißen *n* stored-energy welding
Speicherfähigkeit *f* [storage] capacity
Speicherfehler *m* blemish *(storage elements)*
Speicherförderer *m* bunker conveyor
Speichergasanlage *f* motor-fuel gas storage
Speicherinhalt *m* store contents
Speicherintegrator *m* storage integrator
Speicherkapazität *f* memory (storage) capacity
Speicherkatode *f* storage cathode
Speicherkern *m* memory core
Speicherkondensator *m* reservoir capacitor
Speicherkraftwerk *n* storage power station
Speicherkreis *m* storage circuit
Speichermagazin *n (mach)* storage magazine
Speichermatrix *f* storage matrix
Speichermedium *n* storage medium
Speichermittel *n* storage medium
speichern to store; to record
Speichern *n* storage; recording
magnetisches S. auf dünnen Schichten thin-film magnetic storage
Speicherofen *m* / **elektrischer** storage space heater
Speicheroszillograf *m* storage oscilloscope
Speicherpaket *n* store (storage) stack
Speicherplatte *f* storage plate (target)
Speicherplattenelement *n* target element
Speicherplattenoberfläche *f* target surface
Speicherplatz *m (telec)* register; storage location *(computer)*
angesprochener S. addressed register
Speicherregister *n* memory register
Speicherröhre *f* memory (storage) tube *(data)*
Speicherschaltung *f* memory circuit
Speicherschiff *n* hulk *(vessel)*
Speicherschleife *f* storage loop
Speichersender *m* storage transmitter
Speicherstelle *f* location *(computer)*
Speichertastatur *f* storage keyboard
Speicherung *f* storage
S. auf Pufferbahnen buffer storage
elektrostatische S. electrostatic storage
leistungsabhängige S. volatile storage
löschbare S. erasable storage
unlöschbare S. non-erasable storage
zugriffzeitfreie S. zero-access storage
Speicherungszeit *f* storage time
Speicherverfahren *n* storage technique
Speichervermögen *n* storage capacity
Speicherwerk *n (dat)* memory element, storage section; accumulating register *(computer)*
Speicherzeit *f* hold time *(distillation column)*
Speicherzelle *f* storage location (cell) *(computer)*
Speigatt *n (ship)* deck drain, scupper [hole]
speilen to frame *(shoemaking)*
Speise *f (met)* speiss
Speiseabschnitt *m (el)* supply section
Speiseabteil *n (ry)* dining compartment
Speisebrücke *f* feeding (supply) bridge *(telephony)*
Speisedrossel *f* feed coil
Speisedruckrohr *n* feed delivery pipe
Speiseeis *n* icecream
Speisefeld *n (el)* feeding panel
Speisefett *n* nutrient (edible) fat
Speiseflasche *f* gravity feed tank, feed tank for feeding without a pump
Speisefrequenz *f* feed frequency
Speisegefäß *n* **zum Speisen ohne Pumpe** *s.* Speiseflasche
Speisehahn *m* feed cock
Speisekabel *n* feeder [cable]
Speisekanal *m* supply channel
Speisekasten *m* feeding hopper
Speisekopf *m* combined check and stop valve
Speiseleitung *f* feeder *(aerial)*; supply line, feed[ing] line
S. im Radialnetz radial distribution feeder
abgehende S. outgoing feeder
ankommende S. incoming feeder
direkte S. independent feeder
konzentrische S. coaxial (concentric pipe) transmission line
symmetrische S. balanced feeder
Speiseleitungsschutz *m* feeder protection
Speiseleitungsverteiler *m* feeder distribution centre

speisen to feed; to charge, to load
einen Verstärker s. to drive an amplifier
im Mittelpunkt s. *(el)* to centre-feed
Speiseöl *n* salad oil
Speisepumpe *f* feed[-water] pump
S. mit Exzenterantrieb eccentric-driven feed pump
S. mit Räderantrieb geared feed pump
S. mit Riemenantrieb belt-driven feed pump
Speisepunkt *m (el)* feed[ing] point, feeder junction point
Speiseregelung *f* **/ selbsttätige** automatic hopper feed
Speiseregler *m* feed regulator
selbsttätiger S. automatic hopper feeder
Speisereguliervventil *n* feed-regulating valve
Speiserohr *n* feed pipe, feeding line
Speiserohrverteiler *m* perforated feed pipe
Speiserufer *m* low-water alarm [with float]
Speiseschalter *m* feeder switch
Speisestrom *m* feed current
Speisestromkreis *m* feeder circuit
Speisestutzen *m* feed block (nozzle, pipe)
Speiseteil *m* power unit, *(Am)* power pack
Speisetrichter *m* feed hopper
Speiseventil *n* delivery valve; feed[er] valve; feed check-valve *(boiler)*
Speisevorrichtungen *fpl* **/ voneinander unabhängige** independent boiler feeds
Speisewagen *m* dining car
Speisewalze *f* feed roller
S. für Farbe colour furnisher
geriffelte S. fluted feed roller
Speisewasser *n* feed water
Speisewasserabsperrung *f* feed check *(boiler)*
Speisewasseraufbereitung *f* feed-water conditioning
Speisewasseraufbereitungsanlage *f* feed-water preparation (purifying) plant
Speisewasserbehälter *m* feed-water tank
Speisewasserbetätigungsvorrichtung *f* feed controller
Speisewasserenthärtung *f* feed-water softening
Speisewasserentlüfter *m* feed-water de-aerator
Speisewasserentlüftung *f* feed-water de-aerating
Speisewassererzeuger *m* feed-water generator
Speisewasserfilter *n* feed-water filter
Speisewasseroberflächenvorwärmer *m* feed-water surface heater
Speisewasserprüfer *m* electrolytic water-tester
Speisewasserpumpe *f* feed-water pump
Speisewasserregelung *f* feed-water control
Speisewasserregler *m* feed regulator
selbsttätiger S. automatic feed regulator
Speisewasserreiniger *m* feed-water purifier
Speisewasserreinigung *f* feed-water purification
Speisewassertank *m* feed[-water] tank
Speisewasseruntersuchung *f* feed-water analysis
Speisewasserventil *n* feed[-water] valve
Speisewasservorwärmer *m* feed-water [pre]heater, economizer
Speisewasservorwärmung *f* feed-water [pre]heating
Speiskobalt *m* smaltite
Speisung *f* feed[ing]; feed, supply; batch charge, input
S. über Druckerhöhungspumpe boosted feed
doppelpolige S. insulated return system
periodische S. periodical feeding
ununterbrochene S. continuous feeding
spektral spectral
Spektralanalysator *m* spectrum analyzer
Spektralanalyse *f* spectral (spectrum, spectroscopic) analysis
Spektralapparat *m* instrument for the examination of spectra
Spektralatlas *m* spectrum map
Spektralaufzeichnung *f* spectrum recording
Spektralbande *f* spectral band
Spektralbereich *m* spectral range (region)
weiter S. wide spectral range
Spektralempfindlichkeit *f* spectral sensitivity (response)
Spektralfarbe *f* spectral colour
Spektralfilter *n* spectral filter
Spektralfotometer *n* spectrophotometer
Spektralfotometrie *f* spectrophotometry
Spektralfrequenz *f* spectral frequency
Spektralgebiet *n* spectral region
Spektralkurve *f* spectral curve
Spektrallinie *f* spectral line
Spektrallinienverschiebung *f* displacement (shift) of spectral lines, spectral-line displacement
Spektralrauschzahl *f* spot noise factor
Spektralschwerpunkt *m* spectral centroid
Spektralserie *f* spectral series
Spektralterm *m* spectral term
Spektralverschiebung *f* spectral shift
Spektralverteilung *f* spectral energy distribution, spectral distribution [of energy]
Spektren *npl* **einheitlicher Richtung** unidirectional spectra
Spektrenprojektor *m* spectrum projector
spektrochemisch spectrochemical
Spektrograf *m* spectrograph
Spektrogramm *n* spectrogram
Spektroheliograf *m* spectroheliograph
Spektrohelioskop *n* spectrohelioscope
Spektrometer *n* spectrometer
doppeltfokussierendes S. double-focusing spectrometer
eisenloses S. iron-free spectrometer
magnetisches S. magnetic spectrometer
Spektrometrie *f* spectrometry
spektrometrisch spectrometric

Spektroskop *n* spectroscope
geradsichtiges S. direct-vision spectroscope
Spektrum *n* spectrum
S. der Eigenwerte eigenvalue spectrum
diskretes S. discrete spectrum
ebenes S. flat spectrum
energiegleiches S. equi-energy spectrum
geebnetes S. flattened spectrum
kanneliertes S. fluted spectrum
kontinuierliches S. continuous spectrum
linienreiches S. complex spectrum
prismatisches S. prismatic spectrum
sekundäres S. secondary spectrum
thermisches S. thermal spectrum
tertiäres S. tertiary spectrum
Spengler *m* pipe fitter, plumber
Spermazetöl *n* sperm (bottlenose) oil
Sperrad *n* ratchet (pawl) wheel
Sperraste *f* lock pin
Sperrballon *m* barrage balloon
Sperrbereich *m* filter attenuation band; stop band
Sperrboot *n* barrage vessel
Sperrdampfkondensator *m* gland condenser
Sperrdämpfung *f* stop band attenuation *(of a filter)*
Sperrdiode *f* blocking diode
Sperrdrossel *f* quenching coil
Sperre *f* lock; prevention device; interlock[ing]; *(mine)* stopping
S. für wilde Schwingungen parasitic stopper
elektrische S. electrical interlocking
mittelbare S. indirect interlocking
Sperreinrichtung *f (mach)* locking device
Sperrelektrode *f* blocking electrode
sperren 1. *(mach)* to lock, to arrest, to guard; 2. *(build)* to insulate; to waterproof, to damp-proof; 3. to cut off, to place (take) out of service *(telephone)*; 4. *(print)* to space
eine Frequenz s. to reject (restrict) a frequency
eine Röhre s. to bias off an electron tube
mit Drossel s. to choke
Sperrfaktor *m* blocking coefficient
Sperrfeder *f* retaining (lock) spring
Sperrfeuer *n* barrage
Sperrfilter *n* gate (suppression) filter; *(opt)* barrier filter
Sperrflüssigkeit *f* sealing (packing) liquid, sealing (packing) fluid, liquid seal *(turbine)*
Sperrgatter *n* inhibiting circuit *(logical circuit)*
Sperrgebiet *n* cut-off region
Sperrgetriebe *n* ratchet-and-pawl mechanism
Sperrgitter *n* barrier grid
Sperrgitterröhre *f* barrier-grid tube
Sperrgleichrichter *m* blocking rectifier
Sperrglied *n* blocking element
mechanisches S. pawl (ratchet) wheel
Sperrgut *n* bulky load (goods), awkward load
Sperrhaken *m (mach)* catch
Sperrhebel *m* 1. safety catch; 2. ratchet lever *(telephony)*
Sperrholz *n* plywood
S. mit gleich dicken Lagen plywood with equal plies
gebogenes S. bent plywood
mit harter Platte beplanktes S. hardboard-faced plywood
Sperrholzleim *m* laminating glue
Sperrholzplatte *f* / **dreifache** plywood three-ply
Sperrholzschäftung *f* / **kontinuierliche** continuous plywood scarf jointing
Sperrichtung *f* inverse direction *(rectifiers)*; reverse direction *(semiconductors)*
Sperriegel *m* plunger block
sperrig awkward, unwieldy; bulky
Sperring *m* retainer ring *(teleprinter)*
Sperrklinke *f (mach)* pawl, catch; dog; holding (stop) pawl *(telephony)*; latch
Sperrkondensator *m* blocking capacitor
Sperrkreis *m* rejector (anti-resonance, anti-resonant, block, parallel-resonant) circuit, wave trap
Sperrmagnet *m* electric lock
S. mit mechanischer Abstützung forced drop lock *(relay)*
Sperrmauer *f* dam
Sperrmechanismus *m* barrier mechanism
Sperröhre *f* cut-off valve (tube)
Sperrpappe *f* building paper
Sperrplatte *f* check plate
Sperrschalter *m* holding key
Sperrschaltung *f* block circuit
Sperrschicht *f* 1. *(el)* barrier (blocking) layer, rectifying film; junction *(semiconductors)*; 2. *(build)* waterproofing (damp-proofing) course
magnetische S. magnetic barrier layer
Sperrschichtbreite *f* junction width
Sperrschichtdicke *f* junction thickness
Sperrschichtfotoeffekt *m* photovoltaic (boundary) effect, p-n junction photovoltaic effect
Sperrschichtfotozelle *f* photovoltaic junction (barrier-layer) cell, junction photocell, barrier-layer rectifier
Sperrschichtgleichrichter *m* [p-n] junction rectifier
Sperrschichtkurve *f* junction curve
Sperrschichtmaterial *n* barrier material
Sperrschichttemperatur *f* junction temperature
Sperrschichttransistor *m* depletion-layer transistor
Sperrschichtzelle *f* photovoltaic (barrier-layer) cell, junction photocell
Sperrschleuse *f* double guard (safety) gates
Sperrschritt *m* stop signal

Sperrschwinger *m* blocking (self-quenching) oscillator, blocking (self-quenching) generator, squegger
monostabiler S. single-shot blocking oscillator
Sperrspannung *f* inverse (gate, back, paralysing, anode) voltage, biasing potential
Sperrspannungsscheitelwert *m* peak inverse voltage *(static converter)*
Sperrspitzenspannung *f* inverse peak voltage
Sperrstoff *m* insulant
Sperrstrom *m* backward (inverse, reverse) current
Sperrtaste *f* locking key
Sperrtopf *m* choke joint *(waveguides)*
Sperrträgheit *f* recovery time *(of a diode)*
Sperrung *f* 1. *(mach)* locking, interlock; 2. *(build)* damp-proofing; waterproofing; insulation; 3. *(print)* spacing
Sperrventil *n* check (by-pass) valve
Sperrvorrichtung *f* locking (plugging-up) device *(telephony)*
Sperrwasserdichtung *f* water-sealed gland
Sperrwerk *n* ratchet-and-pawl mechanism, pawl-and-ratchet mechanism
Sperrwiderstand *m* back (reverse) resistance
Sperrylith *m* sperrylite
Sperrzahnkranz *m* ratchet drum *(telephony)*
Sperrzeit *f (telec)* blanking time, idle period, insensitive interval, off-period; inverse period *(rectifier)*; dead time
S. der Torschaltung gate-width
positive S. blocking period
Spesen *pl* expenses
Spessartin *m* spessartine, spessartite
Spezialausführung *f* special version (construction, design)
Spezialbeton *m* special concrete
Spezialdrehmeißel *m* special lathe tool
Spezialfassung *f* special holder
Spezialgarn *n* specialty yarn
spezialisieren auf / sich to specialize on
Spezialisierung *f* specialization
Spezialnutmotor *m* double-squirrel-cage motor
Spezialrechner *m* special-purpose computer
speziell special
Spezies *f* species
spezifisch specific
spezifizieren to specify
Sphalerit *m (min)* sphalerite
sphärisch spherical
Sphäroguß *m* spheroidal-graphite cast iron
sphärolithisch spheroidal, nodular
Sphärometer *n* spherometer
Sphenoid *n* sphenoid *(crystal form)*
spicken 1. *(text)* to lubricate; 2. *(nucl)* to spike
Spickmethode *f* spiking
Spickmittel *n (text)* lubricant
Spicköl *n (text)* lubricating oil
Spickungspräparat *n* implant *(radiology)*
Spiegel *m* 1. *(opt)* mirror, reflector; 2. level *(liquid)*; 3. *(build)* web *(in coffered panels)*; 4. *(leath)* shell; 5. *(wood)* splash, silver grain; 6. transom [board]; escutcheon *(of a vessel)*
asphärischer S. aspherical mirror
ebener S. plane mirror
halbdurchlässiger S. semi-permeable mirror
hyperbolischer S. hyperbolic mirror
magnetischer S. magnetic mirror
parabolischer S. parabolic mirror
sphärischer S. spherical mirror
Spiegelablesung *f* mirror reading
Spiegelabstimmpunkt *m* repeat point *(radio)*
Spiegelantenne *f* reflector aerial
Spiegelbelag *m* mirror (reflector) coating
Spiegelbelegung *f* silvering
Spiegelbild *n* mirror image
Spiegelbildkopiereinrichtung *f (mach)* reverse image attachment
spiegelbildkopieren *(mach)* to produce mirror-image parts
spiegelbildlich mirror- *(copying)*, mirror-inverted
spiegelblank mirror-bright; highly-polished
Spiegelebene *f* mirror (reflection) plane
Spiegeleffekt *m* image effect *(of aerials)*
Spiegeleisen *n* spiegel iron
Spiegelexophthalmometer *n* reflecting exophthalmometer
Spiegelfernrohr *n* reflecting telescope
Spiegelfrequenz *f* image (second-channel) frequency
Spiegelfrequenzempfindlichkeit *f* image response
Spiegelfrequenzselektion *f* image rejection
Spiegelfrequenzstörung *f* second-channel interference
Spiegelfrequenzunterdrückung *f* image suppression
Spiegelfrequenzverwerfungsgrad *m* image-frequency rejection ratio
Spiegelgalvanometer *n* mirror (reflecting) galvanometer
Spiegelglas *n* plate glass
spiegelgleich mirror-symmetric[al]
Spiegelheck *n* transom (square) stern, flat counter
Spiegelhochschnitt *m* edge grain *(wood)*
Spiegelkerne *mpl* mirror nuclei
Spiegelkluft *f* clean split *(wood)*
Spiegelkondensor *m* mirror (catoptric) condenser
Spiegellinsenfernrohr *n* catadioptric telescope
Spiegelmaß *n* planed measure *(wood)*
Spiegelmikroskop *n* reflecting microscope
Spiegelmonochromator *m* mirror monochromator
Spiegelnuklide *npl* mirror nuclides
Spiegeloberfläche *f* mirror surface

Spiegelobjektiv *n* catoptric objective
Spiegelöffnung *f* aperture of the reflector *(aerials)*
Spiegeloptik *f* catoptrics
Spiegeloszillograf *m* mirror oscillograph
Spiegelrad *n (telev)* mirror scanning wheel
Spiegelreflexkamera *f* reflex camera
  einäugige S. single-lens reflex camera
  zweiäugige S. twin-lens reflex camera
Spiegelrinde *f* silver bark *(wood)*
Spiegelschnitt *m* back cutting; vertical grain *(wood)*
Spiegelsymmetrie *f* mirror (specular) symmetry
Spiegelteleskop *n* reflector (reflecting) telescope
Spiegelübergang *m (nucl)* mirror transition
Spiegelung *f* [regular] reflection, specular effect, direct reflection; glare
  diffuse S. diffuse reflection
  regelmäßige S. regular reflection
  selektive S. selective reflection
Spiegelungsfläche *f* reflecting surface
spiegelverkehrt mirror-inverted
Spiegelwelle *f* reflected wave
Spiegelwirkung *f* mirror (reflecting, cut-off) effect
Spiel *n* 1. *(math, cont)* game; 2. *(mach)* play, backlash, clearance; allowance for fit (clearance) *(fits)*; free travel *(clutch)*; cycle *(loading, stressing)*
  S. mit n Personen n-person game
  endliches S. finite game
  unendliches S. infinite game
  zulässiges S. permissible clearance
Spielausgleich *m* backlash (play) compensation; backlash eliminator (eliminating device)
Spielausgleicheinrichtung *f* blacklash eliminator (eliminating device), *(Am)* anti-backlash device
Spielfilm *m* feature film
  abendfüllender S. full-length feature film
spielfrei no-backlash, backlash-free; free from play
Spielmatrix *f* pay-off matrix
Spielpassung *f* clearance (loose) fit
Spielraum *m* clearance; backlash, play; allowance
Spielsitz *m* clearance (loose) fit
Spieltheorie *f* theory of games
Spieluhr *f* musical clock
Spielwarenindustrie *f* toy industry
Spielwert *m* value of game
Spielzahl *f* cycle index
Spielzahlanzeiger *m* cycle index
Spielzeugfabrik *f* toy factory
Spielzeuglack *m* toy lacquer
Spiere *f (ship)* boom, spar
Spierentonne *f* spar buoy
Spieß *m (print)* rising space
Spießglanzblende *f* kermesite
Spießkantkaliber *n* diamond-pass of rolls
Spill *n* capstan; windlass
Spillkopf *m* capstan (trundle) head
Spillkörper *m* capstan barrel
Spillspake *f* capstan bar
Spillwelle *f (mine)* cat shaft *(boring)*
Spin *m* spin
  halbzähliger S. half-integral spin
Spinabsättigung *f* spin saturation
Spin-Bahn *f* spin orbit
Spin-Bahn-Kopplung *f* spin-orbit coupling
Spind *n(m)* locker
Spindel *f* 1. *(ph)* areometer, hydrometer; 2. *(mach)* spindle *(main movement)*; screw; scroll *(screw pump)*
Spindelantrieb *m* spindle drive
Spindelantriebsmotor *m* spindle drive motor
Spindelaräometer *n* hand hydrometer
Spindelbank *f* bobbin (speed) frame
Spindelbohrer *m* single-lipped rifle drill; gun drill
Spindelbohrung *f* spindle bore (hole)
Spindelbremse *f (mach)* spindle brake; *(ry)* screw brake
Spindelbund *m* spindle collar
Spindeldrehrichtung *f* direction of spindle rotation
Spindeldrehrichtungsumkehr *f* spindle reversal
Spindeldrehzahl *f* spindle speed
Spindeldrehzahländerung *f* spindle speed change, variation in spindle speed
Spindelführung *f* spindle guiding
Spindelfußlager *n (text)* spindle footstep
Spindelhalslager *n (text)* spindle collar
Spindelhülse *f* spindle sleeve (quill)
Spindelkegel *m* spindle taper; nose
Spindelkopf *m* **mit Gewinde** screwed spindle nose
  S. mit Spindelflansch flanged spindle nose
Spindelkopfspitze *f* head centre
Spindellagerplatte *f* cluster plate
Spindellenkgetriebe *n* cam-and-peg steering gear, screw-and-nut steering gear
Spindelöl *n* spindle oil
Spindelpinole *f* spindle quill (sleeve)
Spindelpresse *f* screw press
  mechanische S. power screw press
Spindelrücklauf *m* return of the spindle
Spindelrundlaufgenauigkeit *f* true running spindle accuracy
Spindelschaft *m* centre shaft
Spindelschaftdichtungsring *m* centre shaft washer
Spindelschleifkörper *m* pencil grinder
Spindelschnur *f* spindle band
Spindelspitze *f* spindle point (tip)
Spindelstillsetzung *f* spindle stop
Spindelstock *m* headstock *(lathe)*, spindle head

beweglicher S. travelling head
Spindelstockgetriebe *n* headstock gearing
Spindelteilung *f (text)* pitch of spindles
Spindeltrommel *f* spindle carrier drum
Spindelventil *n* poppet valve
Spindelvorgelege *n* back gear
Spindelwirtel *m* spindle whirl
Spindrehimpuls *m* spin angular momentum
Spinechospeicher *m* spin echo memory *(for data)*
Spinell *m* spinel
Spinellschicht *f* spinel layer
spinlos spinless
Spinmatrix *f* spin matrix
Spinmodus *m* spin mode
Spinmoment *n* spin moment; spin
Spinmultiplett *n* spin multiplet
Spinnbadstrecke *f* effective length of spinning bath
spinnbar spinnable
Spinnbarkeit *f* spinnability
Spinndüse *f* spinneret, spinning jet
Spinndüsenöffnung *f* aperture of the spinning nozzle
Spinneigenschaft *f* spinning quality
spinnen to spin
Spinnen *n* **des Kokons** cocoon spinning
S. mit Nachdraht spinning with twisting at the head
Spinnerei *f* spinning mill
Spinnereiabfall *m* spinning waste
Spinnereivorwerk *n* card room
Spinnfäden *mpl* spider threads
Spinnfärbung *f (Am)* dope dyeing
Spinnflüssigkeit *f* spinning dope
spinngefärbt spin-dyed, solution-dyed
Spinngut *n* spinning material
Spinnkabel *n* tow
Spinnkanne *f* [sliver] can
Spinnkuchen *m* cake
Spinnöl *n* lubricating oil
Spinnrad *n* spinning wheel
zweispuliges S. double spinning wheel
Spinnverfahren *n* / **amerikanisches** American system of spinning
deutsches S. German system of spinning
Spinnzentrifuge *f* centrifugal box
Spinoperator *m* spin operator
Spinor *m* spinor [particle]
Spinquantenzahl *f* spin quantum number
Spinraum *m* spin (isobaric, isotopic) space
Spinrichtung *f* direction of spin, spin direction
Spinterm *m* spin term
Spinthariskop *n* spinthariscope
Spinumkehr *f s.* Umklappen des Spins
Spinwelle *f* spin wave
Spinwellenfunktion *f* spin wave function
Spinzustand *m* spin state
Spiralantenne *f* helical aerial
Spiralaufgabevorrichtung *f* spiral feeder
Spiralbahn *f* spiral orbit
Spiralbahnspektrometer *n* spiral orbit spectrometer
Spiralbandrohr *n* spiral pipe
Spiralbewehrung *f* spiral (helical) reinforcement *(reinforced concrete)*
Spiralbohrer *m* twist drill
S. mit äußerer Ölzuführung oil-groove drill, twist drill with exterior oil channel
S. mit innerer Ölzuführung oil-hole drill, twist drill with interior oil channel
S. mit schraubenförmiger Verschiebung der Freifläche spiral point drill
S. mit veränderlichem Drall drill with increase twist
S. mit Zylinderschaft jobber's length drill *(jobber's length, shank of the same diameter as the drill body)*
kurzer S. short-length twist drill
langer S. long-length twist drill
Spiralbohrerausspitzmaschine *f* drill point thinner
Spiralbohrerbenennungen *fpl* twist drill nomenclature
Spiralbohrerhinterschleifmaschine *f* twist drill relief grinding machine
Spiralbohrerschleifeinrichtung *f* twist drill grinding attachment
Spiralbohrerschleifmaschine *f* twist drill grinding (sharpening) machine
Spiralbohrerspitzenschleifmaschine *f* twist drill point grinding (sharpening) machine
Spiraldruckfeder *f* helical compression spring
Spirale *f* spiral
S. / mit steiler steep-spiral *(reamer)*
archimedische S. Archimedes' spiral
hyperbolische S. hyperbolic spiral
logarithmische S. logarithmic (equiangular) spiral
Spiralen *fpl* / **Airysche** Airy's spirals
Spiralendiagramm *n* spiral diagram
Spiralfeder *f* spiral [coiled] spring; whisker *(crystal detector)*
Spiralfederdetektor *m* cantilever detector
Spiralformfräseinrichtung *f* spiral-form milling attachment
spiralförmig spiral-shaped
Spiralfräseinrichtung *f* spiral milling attachment
Spiralfräsen *n* spiral milling
Spiralfräser *m* / **grobzahniger** cutter of the coarse-tooth spiral type
Spiralfräskopf *m* spiral milling head
Spiralfutter *n* spiral chuck
Spiralgehäuse *n* spiral housing
spiralgenutet with helical flutes (cutting edges), spiral-fluted

spiralgeschweißt spiral welded
Spiralheftung *f* spiral binding
Spiralkegelrad *n* spiral bevel gear
Spirallänge *f* flute length *(twist drill)*
Spirallinie *f* spiral curve, spiral (helical) line
Spiralmikroskop *n* spiral microscope
Spiralnebel *m* spiral nebula
Spiralnut *f* spiral flute (groove)
Spiralnutenfräsen *n* spiral flute milling
Spiralpotentiometer *n* helical potentiometer
Spiralpumpe *f* volute pump
Spiralrohr *n* pipe coil
Spiralrohrüberhitzer *m* spiral superheater
Spiralsaatleitung *f* spiral seed feed tube
Spiralscheider *m* spiral cleaner
Spiralschlauch *m* spiral hose
Spiralsenker *m* counterbore with spiral flutes
Spiralspan *m* spiral chip
Spiralspule *f* spiral coil
Spiralsteigung *f* [lead of] helix
Spiralstrecke *f (text)* spiral drawing frame
spiralverzahnt spiral-toothed; spiral-fluted
Spiralverzahnung *f* spiral teeth
Spiralwendeleiste *f* spiral retarder
Spiralwinkel *m* helix angle
Spiralzahnkegelrad *n* spiral bevel gear
Spirituosen *pl* spirits
Spiritus *m* spirit, spirits of wine
Spirituslack *m* spirit varnish (lacquer)
Spirule spirule *(root locus)*
spitz pointed; acute *(angle)*
Spitzboden *m (build)* cock loft
Spitzbogen *m* pointed (lancet) arch
Spitzbohrer *m* spade (flat) drill
Spitzboje *f* conical (nun) buoy
Spitze *f* 1. point; tip; top; peak *(cone; load)*; *(math)* apex; vertex; spinode; *(mach)* centre *(holding work)*; crest *(thread)*; 2. *(text)* lace
  S. / mit scharfer sharp-pointed
  S. des Anodenstromes anode peak current
  S. des Kegels apex of the cone
  S. eines Gewindes crest of a thread
  abgeflachte S. cut-away centre
  durch chemische Verfahren hergestellte S. artificial lace
  federnde S. spring centre
  feste S. dead centre
  feststehende S. dead centre
  irische S. Cork lace
  mitlaufende S. live centre
spitzen to sharpen *(pencil)*
Spitzenabstand *m* point separation; *(mach)* centre distance
Spitzenanhebungskreis *m* peaking circuit
Spitzenanodenstrom *m* peak anode current
  maximaler S. maximum peak anode current
Spitzenband *n* edge lace braid
Spitzenbelastung *f* peak load
Spitzenbelastungszeit *f* maximum power-demand time
Spitzende *n* spigot *(piping)*
Spitzendiode *f* point-contact diode
Spitzendrehmaschine *f* centre lathe
Spitzendrehmoment *n* peak torque
Spitzendruckpfahl *m* point-bearing pile
Spitzeneffekt *m* point effect
Spitzenelektrode *f* point electrode
Spitzenentladung *f* point discharge
Spitzenfunkenstrecke *f* point gap
Spitzengeschwindigkeit *f* peak velocity
Spitzengleichrichter *m* point-contact rectifier
Spitzenhöhe *f* height of centres, centre height
Spitzenimpulsleistung *f* peak pulse power
Spitzenkamm *m (nucl)* spray points
Spitzenkapazität *f* top capacity
Spitzenkatode *f* point cathode
Spitzenklöppelei *f* pillow-type lace making
Spitzenkollektor *m* point collector
Spitzenkontakt *m* point contact
Spitzenlast *f* peak load
Spitzenlastkraftwerk *n* peak load station
Spitzenlastpunkt *m (nucl)* hot spot
Spitzenlattentuch *n* spiked lattice
Spitzenleistung *f* peak efficiency (capacity, performance); peak load; maximum output
Spitzenlinderung *f (text)* bobbin finings
spitzenlos centreless
Spitzennadel *f* spring [beard] needle
Spitzenprüfgerät *n* instrument-compartment test set
Spitzenqualität *f* top quality
Spitzenrundung[shalbmesser *m***]** *f* nose radius *(tool)*
Spitzenschnürung *f (text)* pointed tie up
Spitzenschreiber *m* maximum recording attachment, maxigraph
Spitzenspannung *f* 1. peak (crest, needle-point) voltage; 2. peak stress *(material)*
Spitzenspannungsmesser *m* peak[-to-peak] voltmeter
Spitzenspanwinkel *m* top (back) rake, top (back) rake angle
Spitzensperrspannung *f* peak inverse voltage
Spitzenspiel *n* crest (bottom, root) clearance
Spitzenstoßstrom *m* peak surge current *(semiconductor)*
Spitzenstrom *m* peak current
Spitzenstromwert *m* peak current value
Spitzentragfähigkeit *f* end bearing capacity *(piles)*
Spitzentransformator *m* series peaking transformer
Spitzentransistor *m* point-contact transistor *(semiconductor)*
Spitzenüberlastung *f* peak overload
Spitzenweite *f* distance between centres

Spitzenwert *m* peak value
S. der Anodenspannung in Durchlaßrichtung peak forward anode voltage
S. der Anodenspannung in Sperrichtung peak inverse anode voltage
maximal zulässiger S. der Sperrspannung blocking voltage rating, permissible peak inverse voltage
Spitzenwertbildung *f* peaking
Spitzenwertzeiger *m* peak indicator
Spitzenwinkel *m* nose angle
S. beim Spiralbohrer point angle
S. des Kegels angle of taper
Spitzenwirbel *m* tip vortex *(cavitation)*
Spitzenwirbelkavitation *f* tip cavitation
Spitzenzähler *m (el)* excess (maximum-demand, peak-load) indicator, excess (maximum-demand, peak-load, peak) meter; *(nucl)* point counter [tube]
S. mit Zeitverzögerung lagged-demand meter
Geigerscher S. Geiger point counter
Spitze-zu-Null peak-to-zero
Spitze-zu-Spitze peak-to-peak
Spitze-zu-Spitze-Amplitude *f* peak-to-peak amplitude, double-amplitude peak
Spitzfeile *f* taper file
Spitzgatt *n (ship)* sharp end (stern)
Spitzgattboot *n* double-ended boat, double ender
Spitzgeschoß *n* pointed bullet (projectile)
Spitzgewinde *n* vee (V-shaped, sharp) thread
Spitzhacke *f* pick axe
Spitzhammer *m* pole pick
Spitzkasten *m* spitzkasten, box classifier
Spitzkelle *f* pointing trowel
Spitzkerbe *f* V-notch
Spitzkolben *m* flask with tapering neck
Spitzlicht *n* high light
Spitzlötkolben *m* pointed copper bit
Spitzpunkt *m* cusp
Spitzsäule *f* obelisk
Spitzschar *n (ag)* pointed share
Spitzschnürung *f (text)* diamond tie-up
Spitzschraube *f* pointed-top screw
Spitzsenker *m* countersink[ing cutter]
S. mit Führungszapfen pilot form of countersink
Spitzstampfer *m* peg[ging] rammer
Spitztonne *f* conical (nun) buoy
Spitztrichter *m* conical hopper
spitzwinklig acute-angled
Spitzwinkligkeit *f* acuteness
Spitzzange *f* taper-nose pliers
Spitzzirkel *m* divider
spleißen to splice
Spleißstelle *f* cable splice
Spleißung *f* splice, splicing
Spleißverbindung *f* spliced connection
Splint *m* split pin; cotter pin; sap-wood
Splintanker *m* foundation bolt secured with a cotter
Splintbläue *f* sap-stain
Splintfleck *m* sap-stain
Splintholz *n* sap-wood, alburnum
Splintkohle *f* splint coal
Splintloch *n* split-pin hole
Splintverbindung *f* cotter joint
Splintverschluß *m* cotter lock
Splitt *m* chippings, chips, grit, crushed rock
Splitter *m* splinter, strand, shiver
splitterbar splinterable
Splitterbombe *f* fragmentation (shrapnel) bomb
Splitterfang *m s.* Splitterfänger
Splitterfänger *m* bull screen, knot catcher, knotter, strainer
splitterfrei non shatterable, shatterproof
Splitterholz *n* wood chips
Splitterkohle *f* splint coal
splittern to splinter, to split
Splittern *n (met)* scaling
splitternd splinterable
Splitterring *m (mil)* fragmentation ring *(grenade)*
Spodumen *m* spodumene, triphane
spontan spontaneous
Spontanemission *f* spontaneous emission
Spore *f* spore
Sporn *m (aero)* tailspar, tailskid
Spornfederung *f* tailskid springing
Sportbekleidung *f* sports clothing
Sportboot *n* sporting boat, pleasure launch
Sportflugzeug *n* sporting plane
Sportruderboot *n* sport rowing boat
Sportwagen *m* sport[s] car, sporting car
Sportzweisitzer *m* roadster
Sprachanalysator *m* speech analyzer
Sprachanalyse *f* linguistic analysis *(machine translation)*
Sprachaufnahme *f* voice recording
Sprachdehner *m (telec)* expander
Sprachdeutlichkeit *f* intelligibility
Sprache *f* language; speech
chiffrierte S. coded language
gedehnte S. expanded speech
maschineneigene S. hardware representation
offene S. plain speech
verschlüsselte S. inverted (scrambled) speech
verschwommene S. blurred voice
Spracherkenner *m (dat)* word recognizer
Spracherkennung *f* **/ phonetische** speech recognition
Sprachfehler *m* speech defect
Sprachfrequenz *f* speech (voice) frequency
Sprachfrequenzband *n* speech band
sprachgesteuert voice-controlled, voice-operated
Sprachgüte *f* quality of speech
Sprachmodulation *f* speech (voice) modulation
Sprachsignal *n* speech signal

Sprachübertragung *f* voice communication
Sprachverständlichkeit *f* intelligibility
Sprachverzerrer *m* speech scrambler equipment (apparatus), scrambler
Sprachwiedergabe *f* voice reproduction
Spratzprüfung *f* crackle test *(insulating oils)*
Sprechader *f (telec)* speaking wire
Sprechaufnahme *f* speech recording
Sprechbatterie *f* speaking (microphone) battery
Sprechbereich *m* telephone area
sprechbereit ready to speak
Sprecher *m* announcer *(broadcasting)*
Sprecherecho *n* talker echo
Sprechfrequenz *f* voice (speech) frequency
Sprechfrequenzbereich *m* speech frequency range
Sprechfunk *m* radiotelephony
Sprechgarnitur *f* operator's set
Sprechgeschwindigkeit *f* elocution velocity *(telephony)*
Sprechkanal *m* speech channel
Sprechkapsel *f* microphone capsule, transmitter unit (inset)
Sprechkopf *m* reproducing head
Sprechkreis *m* speaking circuit
Sprechleistung *f* speech power
Sprechprobe *f* voice test[ing]
Sprechschalter *m* speaking key
Sprechstrom *m* speech (voice) current
Sprechstromkreis *m* telephone circuit
Sprechtaste *f* press-to-talk button (switch); speaking key
Sprechtrichter *m* mouthpiece of microphone (transmitter)
Sprech- und Mithörschalter *m* combined listening and speaking set
Sprechzeug *n* operator's telephone set
Spreizdorn *m* expansion arbor
  fliegend aufgespannter S. stub expansion arbor
Spreize *f (mine)* nog *(cut)*, stull
spreizen to spread, to expand
Spreizenstapler *m* straddle truck
Spreizfassung *f* expanding-block-type cutter mounting *(boring bar)*
Spreizhobel *m (mine)* scissor-frame plough
Spreizhülsenanker *m (mine)* expansion-sleeve bolt, expansion-shell bolt
Spreizkegel *m* expanding cone; taper expanding plug, tapered plug
Spreizklappe *f (aero)* split flap *(wing)*
Spreizkonus *m s.* Spreizkegel
Spreizlafette *f* split-trail carriage
Spreizmechanismus *m* expander
Spreizreibahle *f* expanding (expansion) reamer
Spreizringkupplung *f* radially expanding clutch
Spreizstempel *m (mine)* sprag
Sprengapparat *m (ag)* sprinkler
Sprengbohrloch *n* blast hole
  geladenes S. charged shothole
Sprengbombe *f* demolition bomb
  überschwere S. blockbuster [bomb]
sprengen 1. to blast, to shotfire *(with explosive)*; 2. to sprinkle *(liquid)*
Sprengen *n* 1. blasting, shooting, shot-firing; 2. sprinkling
  S. mit einem kleinen ausgekesselten Bohrloch gopher-hole blasting, gophering
Sprenggelatine *f* blasting (explosive) gelatine
Sprenggranate *f* high-explosive shell
Sprenghöhe *f* **einer Blattfeder** camber of a leaf spring
Sprengkammer *f* demolition chamber *(bridge)*
Sprengkapsel *f* blasting cap
  einfache S. plain detonator, squib
Sprengkopf *m* warhead
Sprengkraft *f* explosive power
Sprengladung *f* explosive (blasting) charge, charge of explosive
Sprengloch *n* shot-hole, blast hole
Sprengmeister *m* shot-firer, blaster
Sprengniet *m* explosive rivet
Sprengring *m* snap (retaining) ring; lockwasher
  S. der Bandage *(ry)* shoulder of tyre, tyre-clip, tyre-clasp, spring-ring
Sprengsalpeter *m* blasting powder
Sprengstoff *m* explosive
  brisanter S. high explosive
  nichtummantelter S. unsheathed explosive
  schiebend wirkender S. low (weak) explosive
  ummantelter S. sheathed explosive
Sprengstoffbehälter *m* explosives canister
Sprengstoffkammer *f* explosives magazine
Sprengstofflager *n* explosives store
Sprengstoffmagazin *n* explosives magazine
Sprengtrupp *m (mil)* demolition squad
Sprengung *f* blast[ing]; bursting
Sprengvorrichtung *f* sprayer, sprinkler *(liquid)*
Sprengwagen *m* water[ing] cart, motor sprinkler, street (road) flusher, street sprinkler (washer)
Sprengwerk *n (build)* strut frame
Sprengzünder *m* blasting fuse
Sprenkler *m (ag)* sprinkler
Spreugebläse *n* chaff blower (fan)
Spriegel *m (aut)* tarpaulin frame
Sprietbaum *m* sail sprit
Springblende *f* / **automatische** *(opt)* all-automatic diaphragm
Springduowalzwerk *n* jump roughing mill
springen 1. to crack, to fissure; to burst; 2. to jump *(function)*
Springflut *f* spring tide
Springhochwasser *n* high-water spring
Springlein *m* rattle flax
Springluke *f* [quick-acting] scuttle
Springniedrigwasser *n* low water springs

Springschreiber *m* start-stop apparatus *(telegraphy)*
Springschreibertelegrafie *f* start-stop telegraphy
Springstagkompaß *m* jumper stay compass
Springtide *f* spring tide
Springtidenhub *m* spring range
Springwagen *m* shuttle carriage
Sprinkleranlage *f* automatic sprinkler equipment, fire sprinkler, sprinkler installation *(fire extinguishing system)*
Sprit *m* spirit
Spritlack *m* alcohol-diluted lacquer
Spritzbeton *m* gunned (cement-gun, air-placed) concrete, jetcrete, shotcrete, gunite
Spritzblech *n* splash-guard; deflector plate
Spritzbrühe *f* spraying solution, spray wash
Spritzdecke *f* spray apron *(folding canoe)*
Spritzdraht *m* spraying wire
Spritzdruck *m (plast)* splutter print
Spritzdüse *f* spray nozzle
Spritze *f* sprayer, injector, syringe; injection
spritzen to spray, to apply by spray *(paint)*; to syringe; *(plast)* to injection-mould; to extrude *(fibre)*
Spritzen *n* extrusion *(man-made fibre)*; ink fly, ink-misting *(printing ink)*
elektrostatisches S. electrostatic spraying
Spritzenboot *n* fire float (boat)
Spritzentladung *f* spraying discharge
Spritzer *m* splash
Spritzfähigkeit *f* sprayability
Spritzfehler *m (plast)* injection moulding defect
Spritzfett *n* spraying grease
Spritzflasche *f* wash bottle
Spritzgerät *n* [high-volume] sprayer
Spritzgießen *n (met)* die-casting; *(plast)* moulding by injection, injection moulding
Spritzguß *m (met)* die-casting
Spritzgußform *f (met)* die-casting mould (die); *(plast)* injection mould
Spritzgußlegierung *f* die-casting alloy
Spritzgußmaschine *f (met)* die-casting machine; *(plast)* injection machine
Spritzgußmasse *f (plast)* injection moulding compound
Spritzgußstück *n (plast)* [injection] moulded plastic
Spritzgußtechnik *f (met)* die-casting practice
Spritzgußteil *n (met)* die-casting
Spritzkabine *f* spray booth *(paint)*
Spritzkegel *m* spray-cone *(condenser)*
Spritzkolben *m (plast)* injection plunger
Spritzkonzentrat *n* spraying concentrate
Spritzkopf *m* extruder head, separate pot mould
Spritzlack *m* spraying lacquer
Spritzleder *n* splash-board leather
Spritzling *m (plast)* injection moulding
Spritzloch *n* spray-hole
Spritzmaschine *f* injection machine; spraying machine
Spritzmetallisierung *f* sprayed metal coating; metal coating by spraying
Spritzmittel *npl (ag)* sprays
Spritznaht *f (plast)* partition line
Spritzpistole *f* spray gun
Spritzpresse *f* injection press
Spritzpulver *n* spraying powder
Spritzputz *m* sprayed mortar
Spritzring *m* oil-thrower ring
Spritzrohr *n* lance
Spritzschmierung *f* splash (oil-bath) lubrication
Spritzschutz *m* splash-guard
Spritzteller *m* splash cup
Spritzung *f (plast)* injection shot
Spritzversteller *m* injection timing device (mechanism)
Spritzverstellernabe *f* injection control hub
Spritzverstellmuffe *f* injection timing collar
Spritzwascher *m* spray scrubber
spritzwassergeschützt splash-proof
Spritzwasserschutz *m* splash guards
Spritzzündung *f* jet ignition *(rectifier)*
Sprödbruch *m* brittle fracture
spröde brittle; short
Sprödglanzerz *n* brittle silver ore, stephanite
Sprödigkeit *f* brittleness; shortness; rottenness *(steel)*
Sprödigkeitspunkt *m (plast)* freezing temperature
Sprosse *f* rung *(ladder)*
Sprossenförderer *m* cross-bar conveyor
Sprossenschrift *f* variable density track (recording)
Sprossenwand *f* wall bars
sprudeln to bubble
Sprudelrohr *n* bubbler gauge
Sprudeltabletten *fpl* bubble bath
Sprühdüse *f* spray nozzle
Sprühelektrode *f* emitting (ionizing, discharge) electrode
sprühen 1. to spray, to drizzle, to sprinkle; 2. *(telec)* to fizz; to sputter *(gas discharge)*
Sprühen *n* 1. spraying; 2. sputtering
Sprühentladung *f (ph)* brush (corona) discharge
Sprühgerät *n* sprayer
Sprühkäfig *m* spray-protection device
Sprühkamm *m s.* Spitzenkamm
Sprühkühldüse *f* mist cooling nozzle
Sprühkühlung *f* mist cooling
Sprühmittel *n* spray
Sprühöl *n* spray oil
Sprühschutz *m* anti-brushing fittings
Sprühversuch *m* spray test
Sprühwäscher *m* column washer
Sprung *m* 1. crack, fissure *(material)*; shake *(wood)*; 2. *(geol)* [normal] fault; 3. increment *(values)*; jump *(voltage)*; transient, step function

(*electronics*); transition (*electrons*); (*dat*) jump, transfer; 4. (*ship*) sheer [of a deck]
S. ins Hangende (*geol*) upthrown fault
S. ins Liegende (*geol*) downthrown fault
bedingter S. conditional jump (transfer of control) (*programming*)
gewöhnlicher S. (*geol*) hade-slip fault
isothermer S. isothermal jump
nichtbedingter S. unconditional jump (transfer) (*programming*)
unbedingter S. *s.* nichtbedingter S.
Sprungantwort *f* step-function response
Sprunganweisung *f* (*dat*) go to-statement
Sprungbalken *m* taking-off mark
Sprungbefehl *m* (*dat*) jump instruction (order), transfer order
Sprungbildung *f* 1. crack formation; 2. (*geol*) faultage, faulting; 3. jog formation (*crystallography*)
Sprungbock *m* vaulting buck
Sprungbrett *n* diving board (*swimming*); springboard, beating board, beat-board
Sprungcharakteristik *f* transient (response) characteristics
Sprungdeckel *m* snap-off cover
Sprungeingang *m* step input
Sprungentfernung *f* skip distance (*radio*)
Sprungfunktion *f* transitional function, Heaviside unit-function; step function
Sprunggrube *f* pit
Sprunghöhe *f* (*geol*) fault throw
Sprungkasten *m* vaulting box
Sprungkennlinie *f* surge characteristic, unit function response
Sprungkontakt *m* snap-action contact
Sprunglatte *f* cross-bar
Sprunglauf *m* ski-jumping
Sprungleitung *f* (*el*) jumper connection
Sprungpferd *n* vaulting horse
Sprungschalter *m* quick-break switch
Sprungschanze *f* jumping hill, ski[ing] jump
Sprungschi *m* jumping ski
Sprungschicht *f* (*nucl*) boundary (bounding) surface
Sprungschienen *fpl* skip bars (*punched card machinery*)
Sprungseil *n* skipping rope
Sprungspannung *f* initial inverse voltage (*electronics*)
Sprungstab *m* pole
Sprungständer *m* jumping stand, post
Sprungtemperatur *f* transition temperature
Sprungtischfräsen *n* intermittent feed milling
Sprungtischvorschub *m* jump feed
Sprungturm *m* diving stages (platform, tower), tower for high diving
Sprungvorschub *m* intermittent [table] feed (*milling*)
S. für Leerwege im Eilgang skip-feeding for rapid traverse between gaps in the work
Sprungweite *f* 1. (*math*) amount of jump; 2. (*geol*) fault heave
Sprungwelle *f* crack wave; surge
Sprungwinkel *m* (*geol*) angle of slip
Sprungzeit *f* transition time (*electrons*)
Spucken *n* stalling (*of the engine*); slopping (*of the converter*)
Spülabort *m* lavatory
Spülanlage *f* sluicing system (*ash*)
Spulart *f* (*text*) type of winding
Spülbagger *m* dredger
Spülbecken *n* sink; scouring basin (*harbour, lock*)
spülbohren to jet
Spülbohren *n* wash boring, jetting
Spülcharge *f* (*met*) wash-out heat
Spule *f* 1. (*el*) coil; 2. (*text*) reel, spool, bobbin; 3. hub (*for magnetic tape*)
S. für ein Wellenband band coil
S. mit Gleitkontakt slider coil
S. mit Unterbrecher trembler coil
S. mit Wabenwicklung honeycomb coil
S. mit Widerstand resistive coil (inductor)
S. zur Brummunterdrückung hum-bucking coil
angezapfte S. tapped coil
astatische S. astatic coil
blinde S. idle coil
drehbare S. revolving (rotable) coil
einlagige S. single-layer coil
erregte S. energized coil
feste S. fixed coil
feststehende S. fixed coil
kernlose S. air-core coil
körperlose S. coreless coil
kreuzgewickelte S. cross-wound coil, lattice[-wound] coil
luftgekühlte S. air-cooled solenoid
mehrlagige S. multi-layer coil
primäre S. primary coil
quadratische S. square coil
rechteckige S. rectangular coil
sekundäre S. secondary coil
verlustarme S. low-loss coil
voreilende S. (*text*) leading bobbin
widerstandsbehaftete S. resistive coil (inductor)
spulen to reel, to spool; to wind
spülen to cleanse; to rinse; to flush, to wash [out]; to scavenge, to jet
Spulen *fpl* / **übereinandergreifende** crossed coils
Spulenabgriff *m* coil tap
Spulenabstand *m* coil (load) spacing
Spulenachse *f* axis of coil
Spulenantenne *f* coil (helical) aerial

Spulenanzapfung *f* coil tap[ping]
Spulenaufstecker *m* bobbin feeder
Spulenbankgetriebe *n (text)* gearing of the fly frame
Spulenbehälter *m* coil container
S. mit Abspuleinrichtung pay-out container
spulenbelastet coil-loaded
Spulenbelastung *f* coil loading
Spulenblindwiderstand *m* coil reactance
Spulenbremse *f* bobbin brake
Spulenbrett *n* bobbin board
Spulendraht *m* winding (coil, magnet) wire
Spulenfeld *n* coil section
Spulenfeldergänzung *f* building-out section
Spulenfluß *m* flux linking a coil
Spulenform *f* 1. profile of a bobbin; 2. coil configuration
Spulengabel *f* bobbin holder (carrier frame)
Spulengalvanometer *n* coil galvanometer
Spulengatter *n* bobbin creel, spool rack
Spulengestell *n* bobbin stand (table)
Spulengüte *f* coil quality
Spulenhalter *m* coil holder (frame, base)
Spulenhülse *f* bobbin case *(sewing machine)*
Spulenkabel *n* coil-loaded cable
Spulenkern *m* 1. coil core; 2. bobbin
Spulenkette *f (el)* low-pass filter
Spulenkontakt *m* coil contact
Spulenkopf *m (el)* end winding
Spulenkörper *m* bobbin
Spulenlack *m (el)* coil varnish
Spulenprofil *n* profile of a bobbin
Spulenrahmen *m* coil former
Spulenreaktanz *f* reactance of coil, coil reactance
Spulensatz *m* coil pack (assembly), set of coils
Spulenschalter *m* coil switch
Spulenschlitten *m* **mit leerer Spule** carriage with spent bobbin
Spulenschlittenscheibe *f (text)* carriage disk
Spulenseitenteilung *f* unit interval [in a winding]
Spulenständer *m (text)* creel
Spulenständerstift *m (text)* creel peg
Spulenteilung *f* unit interval [in a winding]
Spulenteller *m (text)* cone plate
Spulentisch *m* bobbin rail, creel table
Spulenträgerarm *m* bobbin holder (shaft)
Spulenvorwiderstand *m* swamping resistance
Spulenwalze *f* bobbin cylinder
Spulenwechsel *m* bobbin changing
Spulenwechselautomat *m* automatic bobbin changer
Spulenwechseleinrichtung *f* bobbin changer
Spulenweite *f* coil span (width)
S. in Nutteilungen coil pitch
Spulenwickelgerät *n* coil winding set
Spulenwickelmaschine *f* coil winding machine
Spulenwicklung *f* coil winding; armature winding
konzentrische S. concentric winding
Spulenwiderstand *m* coil resistance
Spulenzündung *f* coil ignition
Spuler *m (text)* bobbin winder
Spülgas *n* circulation (scavenging, flush, sweep) gas
Spülgasschwelung *f* carbonization involving gas recirculation
Spülicht *n* rinsings
Spülkasten *m* cistern, flush-box, rinsing box
Spülkippe *f* sluicing dump
Spülklosett *n* flush (wash-down) toilet, water closet
Spülklosettbecken *n* water-closet pan
Spülkopf *m* swivel
Spülluft *f* scavenging air
Spulmaschine *f (text)* spooler; winding frame
Spülmaschine *f* rinsing (washing) machine
S. mit Treiber rinsing machine with propeller
Spülmittel *n* rinsing agent
Spülöl *n* flushing oil
Spülprobe *f* sludge sample
Spülpumpe *f* sluice pump
Spülrinne *f* water sluice *(ash)*
Spülsäule *f* mud column
Spülschiff *n* reclamation vessel
Spülschlitz *m* scavenge port
Spülschmelze *f* wash-out heat
Spültakt *m* scavenge stroke
Spülteer *m* circulation tar
Spültest *m* rinsing test
Spültrübe *f* fluid mud *(for flushing the cuttings)*, mud fluid
Spülung *f* rinsing, scouring; washing; cleansing; flushing; scavenging
Spülversatz *m (mine)* flushing, hydraulic filling (stowing), silting
Spülwasser *n* flushing (wash) water; *(mine)* drilling water
Spülwassertank *m* sanitary tank
Spülwirkung *f* scavenging action
Spund *m* [groove and] tongue
Spundbohle *f* sheet pile, pile plank
Spundbrett *n* match[ed] board
Spunddrehmaschine *f* bung (chive) turning machine
spunden to match *(boards)*
Spunden *n* **von Brettern** grooving and tonguing of boards
Spundfräser *m* tonguing cutter
Spundhobel *m* match plane
Spundloch *n* bunghole
Spundlochbohrmaschine *f* bunghole boring machine
Spundmaschine *f* matcher

doppelte S. double tonguing and grooving machine
Spundpresse *f* / **selbsttätige** automatic chive pressing machine
Spundung *f* match (tongue and groove) joint
Spundwand *f* sheet-pile wall, sheet piling; steel sheet piling
Spundwandverfahren *n* sheet piling *(shaft sinking)*
Spur *f* trace; *(nucl)* trace[r] amount; track *(tape)*; traffic lane *(roads and highways)*; gauge *(railway)*
einfache S. single trace
frequenzmodulierte S. frequency modulated track
verriegelte S. locked trace
vorrückende S. precessing track *(magnetic drum or magnetic tape)*
Spuramplitude *f* amplitude of the trace
Spurbildung *f* track condensation *(cloud chamber)*
Spurbreite *f* diffusion breadth; gauge
spuren to keep the track, to follow the rut
Spuren *n* track keeping *(of wheels)*
spüren to feel; to perceive; to track; to search
Spurenanalyse *f* detection of trace elements, tracer analysis
Spurenchemie *f* microchemistry, *(Am)* trace chemistry
Spurenelement *n* trace[r] element
Spurenfinder *m* tracer detector
Spurenkonzentration *f (nucl)* trace concentration
Spurenlänge *f* track length
Spurenmetall *n* rare metal
Spurennachweis *m* trace detection
Spurenräumer *m (ag)* divider
Spurenverseuchung *f* trace contamination
Spurenverunreinigung *f* trace impurity
Spurenzahl *f* track population
Spurerweiterung *f* gauge widening (clearance)
Spürgerät *n* probe; detector
Spurhaltigkeit *f* track keeping *(of wheels)*
Spurhebel *m (aut)* track arm
Spürisotop *n s.* Leitisotop
Spurkette *f* non-skid chain
Spurkranz *m* flange [of wheel], wheel flange
abgenutzter S. worn flange
Spurlager *n* thrust (journal, footstep) bearing
Spurlatte *f (mine)* guide rail
Spurlockerer *m (ag)* track loosener
Spurmaß *n* gauge measure
Spurofen *m* gutter furnace
zweiherdiger S. spectacle furnace
Spurplatte *f* breast plate
verstellbare S. adjustable footstep plate *(turbine)*
Spurplattenträger *m* footstep pillow *(turbine)*
Spurreißer *m (ag)* marker
Spurstange *f* tie (track) rod
Spurstangengabel *f* tie-rod yoke
Spurstangenhebel *m* track control arm
Spurstangenkopf *m* track-rod end
Spurweite *f (aut)* wheel base, track gauge; [railway] gauge
Sputnik *m* artificial (man-made) satellite
staatlich state; government-owned
Staatsanwaltschaft *f* prosecuting authority
Staatsdienstschiff *n* public vessel
Staatsexamen *n* government examination
Staatsfahrzeug *n* public vessel
Staatsgespräch *n* government [telephone] call
Staatsgewalt *f* executive power
Staatsrecht *n* public law
Staatstelegramm *n* government telegram
Stab *m* bar, rod; column *(buckling)*; member *(truss)*
aufgebogener S. bent-up bar (rod), inclined bar
Stababstand *m* pitch between rods
Stabanker *m* bar-type armature, bar-wound armature
Stabantenne *f* rod aerial
Stabantrieb *m (nucl)* rod drive
Stabausdehnungsthermometer *n* rod-and-tube thermometer, solid-rod thermostat (expansion thermometer)
Stabaustausch *m (build)* bar substitution
Stabbatterie *f* cylindrical (tube) cell
Stabbewehrung *f* bar (rod) reinforcement *(concrete)*
Stabbrandbombe *f* stick-type incendiary bomb
Stabbrettdecke *f* beaded boarding
Stäbchen *n* pin, little rod
stäbchenförmig rod-shaped
Stäbchenmittellage *f* laminated board (core)
Stabdiagramm *n* bar chart
Stabeisen *n* bar (rod) iron
Stabelektrode *f* bar (pencil) electrode
Stabelement *n (nucl)* rod-type fuel element, fuel rod; *(build)* bar, member
stäben to mould timber
Staberder *m* earthing rod
stabförmig rod-shaped
Stabfräser *m* astragal cutter
Stabfunkenstrecke *f* [protective] rod gap
Stabfußboden *m* wood-block flooring
Stabhobel *m* round plane
Stabhobeleisen *n* round moulding iron
Stabhochsprung *m* pole jump (vault)
stabil stable *(equilibrium)*; rugged; rigid, robust
Stabilisation *f* stabilization
Stabilisationsnetzwerk *n* stabilizing (equalizing) network
Stabilisator *m* stabilizer, balancer; constant-voltage regulator; antivibrator; *(ch)* stabilizer; gum inhibitor

hydraulischer S. hydraulic stabilizer
Stabilisatorröhre *f* stabilizer
stabilisieren to stabilize; to regulate
Stabilisierung *f* stabilization; equalization
Stabilisierungsfaktor *m* regulation characteristic
Stabilisierungsflosse *f (ship)* stabilizing fin, stabilizer
Stabilisierungsglied *n* equalizer, stabilizer
Stabilisierungsnetz *n* stabilizing (equalizing) network
Stabilisierungsplatte *f* fin plate
Stabilisierungstransformator *m* stabilizing transformer
Stabilisierungswulste *mpl (ship)* stabilizing bulges
Stabilität *f* stability *(equilibrium)*; *(mech)* ruggedness, rigidity; balance, equilibrium
S. des Gleichgewichts stability of equilibrium
S. im Mittel stability in the mean
S. im Seegang sea stability
S. in Wahrscheinlichkeit stability in probability
dynamische S. mechanic (transient) stability
Stabilitätsbedingung *f* stability condition
Stabilitätsgrad *m* **des Reglers** degree of irregularity of the governor
Stabilitätsgrenze *f* stability limit, limit (range) of stability
Stabilitätskriterium *n* stability criterion
Stabilitätsnachweis *m* stability test
Stabilitätsregel *f* stability rule
Stabilitron *n* stabilitron
Stabkern *m* rod core
Stabkörper *m* body of the slide rule
Stablot *n* plumb rod
Stabmagnet *m* bar (rod) magnet
Stabmaterial *n* bar stock
Stabmühle *f* rod mill
Stabparkettfußboden *m* parquetry, parquet floor
Stabquerschnitt *m* cross section of the bar
Stabreflektor[wand *f***]** *m* rod mirror
Stabrost *m* grizzly; bar screen
Stabrostband *n* bar belt
Stabstahl *m* steel bar; steel rod
gezogener S. drawn steel bar
glatter S. smooth (plain) bar
Stabstahlwalzwerk *n* light-section rolling mill
Stabstandsanzeiger *m* position indicator
Stabstraße *f* rod (bar) mill
Stabstromwandler *m* bar-type current transformer, single-turn current transformer, open-primary current transformer
Stabwandler *m s.* Stabstromwandler
Stabwerk *n* framework, framed structure
Stabwicklung *f* bar winding
Stabzug *m (build)* bar chain
Stacheldraht *m* barbed wire
Stachelkettenbreithalter *m (text)* spiked chain temple
Stachelradführung *f (cont)* pin-feed platen device
Stachelscheibenbreithalter *m (text)* ring temple
Stachelwalze *f (text)* temple roller; toothed roller
Stachelwalzenbrecher *m* toothed roll crusher
Stachelwalzenführung *f (cont)* pin-feed platen device
Stachelwalzwerk *n* toothed roll crusher
Stadion *n* stadium
Stadium *n* stage
Stadtautobahn *f* urban expressway
Stadtbahn *f* elevated railway *(in Berlin)*; metropolitan railway
Stadtbild *n* townscape, cityscape
Städtebau *m* municipal (town, urban) planning; municipal engineering
Städtebauer *m* town (urban, municipal) planner
Stadtentwässerung *f* sewerage
Städteplaner *m* urban (municipal, town) planner
Städteplanung *f* municipal (town, urban) planning; municipal engineering
Stadtgang *m (aut)* low range of gears
Stadtgas *n* town gas
Stadtkanalisierung *f* town drainage
Stadtkern *m* town centre
Stadtlandschaft *f* townscape, cityscape
Stadtplaner *m s.* Städteplaner
Stadtplanung *f s.* Städteplanung
Stadtrandzone *f* outskirts, fringe area, suburbs
Stadtstraße *f* street
Staffelanleger *m (print)* stream feeder
Staffelbild *n* histogram
Staffelei *f* easel
Staffellauf *m* relay race
staffeln to graduate; to stagger, to grade *(telephony)*
Staffelrost *m* multi-stage grate
Staffelrostfeuerung *f* furnace with multi-stage grate
Staffelstab *m* [relay] baton
Staffelsummen *fpl (dat)* progressive totals
Staffelsystem *n* echelon telegraphy, extended telegraph circuit
Staffelung *f (build)* echelon; *(telec)* grading
unsymmetrische S. unsymmetrical grading
Staffelungsgruppe *f (telec)* grading group
Staffelwalze *f (met)* stepped roll, roll in steps
Stag *n (ship)* stay
Stagfockbaum *m* jumbo boom
Stagkante *f* luff
Stagliek *n* luff
Stagnation *f* stagnation
Stahl *m* 1. steel; 2. *s.* Meißel
S. mit hartem Kern hard-centred steel
S. mit weichem Kern mild-centred steel
alitierter S. calorized steel

angelassener S. tempered steel
antimagnetischer S. non-magnetic steel
ausgegarter S. dead steel
basischer S. basic steel
beruhigter S. killed (dead) steel
blasenfreier S. sound steel
blasiger S. unsound steel
eutektischer S. eutectic steel
ferritischer S. ferrite steel
gehärteter S. hardened steel
halbberuhigter S. balanced steel
hitzebeständiger S. heat-resisting steel
hochfester S. high-strength steel, high-tensile steel
hochgekohlter S. high-carbon steel
hochlegierter S. high-alloy steel
hochzugfester S. high-tensile steel, high-strength steel
kaltgewalzter S. cold-rolled steel
kaltgezogener S. cold-drawn steel
kohlenstoffarmer S. low-carbon steel, mild-carbon steel
kohlenstoffarmer weicher S. soft carbon steel
korrosionsbeständiger S. corrosion-resisting steel
laugenbeständiger S. alkali-proof steel
legierter S. alloy (special) steel
luftgehärteter S. air-quenched steel
lufthärtender S. air-hardening steel
nichtrostender S. stainless steel
nickellegierter S. nickel steel
ölhärtender S. oil-hardening steel
rostfreier S. stainless steel
säurebeständiger S. acid-resisting steel
selbsthärtender S. natural steel
steigender S. unkilled (effervescent) steel
übergarer S. overblown steel
unberuhigter S. unkilled (effervescent) steel
unlegierter S. unalloyed (plain) steel
unmagnetischer S. non-magnetic steel
vergüteter S. heat-treated steel, hardened and tempered steel
verzugsfreier S. non-warping steel
wassergehärteter S. water-hardening steel
weichmagnetischer S. soft-magnetic steel
zunderbeständiger S. scale-resisting steel
Stahlabdeckung *f (ry)* steel flooring
Stahlakkumulator *m* alkaline cell
Stahl-Alu-Leitung *f* steel-cored aluminium conductor
Stahlaluminiumseil *n* aluminium-steel cable, steel-cored aluminium conductor
stahlartig steely, steel-like
Stahlausbau *m (mine)* steel casing
Stahlausstoß *m* steel output
Stahlbacken *m* **zum Gewindewalzen** flat[-faced] thread rolling die
Stahlband *n* steel band
S. für Rohre skelp
Stahlbandage *f* steel tyre
Stahlbandantrieb *m* steel-belt drive
Stahlbandbewehrung *f* steel-band armouring
Stahlbandförderer *m* steel-band conveyor
Stahlbandregelung *f* steel-band governing *(turbine)*
Stahlbau *m* structural steel erection; steel construction, steelwork
Stahlbauer *m* erector, fabricator
Stahlbaumonteur *m* erector
Stahlbeton *m* reinforced concrete
Stahlbetonbalken *m* reinforced concrete beam
Stahlbetonbau *m* reinforced concrete construction (engineering)
Stahlbetonbestimmungen *fpl* reinforced concrete specifications
Stahlbetonbrücke *f* reinforced concrete bridge
Stahlbetondecke *f* reinforced concrete floor
Stahlbetondeckenplatte *f* reinforced concrete floor slab
Stahlbetonfertigteil *n* precast [reinforced] concrete unit, precast [reinforced] concrete product
Stahlbetonkuppel *f* reinforced concrete dome
Stahlbetonmast *m* ferroconcrete mast
Stahlbetonpfahl *m* concrete pile
Stahlbetonplatte *f* reinforced concrete slab
einachsig (einfach) bewehrte S. one-way reinforced concrete slab
kreuzweise bewehrte S. two-way reinforced concrete slab
Stahlbetonrahmen *m* reinforced concrete frame
Stahlbetonrippendecke *f* ribbed concrete floor, slab and joist floor
Stahlbetonschiff *n* reinforced concrete vessel
Stahlbetonschwelle *f* reinforced concrete sleeper
Stahlbetonskelett *n* reinforced concrete frame
Stahlbetonskelettmontagebauweise *f* precast concrete skeleton construction
stahlblau steel blue
Stahlblech *n* steel sheet (plate)
S. mit blaugrauer Walzhaut blue steel plate
kadmiertes S. cadmium-plated sheet steel
verbleites S. lead-coated steel sheet
Stahlblechmantel *m* sheet-metal shell
Stahlblechschablone *f* sheet-steel template
Stahlblechscheibenrad *n* sheet-steel disk wheel
Stahlblechverkleidung *f* sheet steel panelling
Stahlblechzellenkasten *m* steel container *(battery)*
Stahlbolzen *m* steel pin
Stahlboot *n* / **nahtloses** seamless boat
Stahldickenmesser *m* steel thickness gauge
Stahldraht *m* steel wire
gespannter S. taut steel wire
gezogener S. drawn steel wire
vernickelter S. nickel-plated steel wire
Stahldrahtbürste *f* steel-wire brush
Stahldrahtgewebe *n* steel-wire cloth

Stahldrahtlitze *f* cast steel wire heald
Stahldrahtseil *n* steel cable
Stahleisen *n* cast iron for steel production; open-hearth pig iron; oligiste
Stahlelektrode *f* steel electrode
Stahlempfangsröhre *f* metal receiving valve
stählern steely
Stahlfach *n* safe
Stahlfeder *f* 1. steel spring; 2. steel (metallic) pen
Stahlfeinblech *n* steel sheet, sheet steel
Stahlfläche *f s.* Stahlquerschnitt
Stahlflasche *f* steel cylinder
Stahlformguß *m* cast steel; steel casting
Stahlformgußstück *n* steel casting
Stahlformsand *m* steel moulding sand
Stahlfräserrohling *m* steel cutter blank
Stahlfrischherd *m* steel finery
Stahlgewinnung *f* steel making (manufacture)
Stahlgewinnungsverfahren *n* steel-making process
Stahlgießerei *f* steel[-casting] foundry
Stahlgießpfanne *f* steel ladle
Stahlgittermast *m* steel tower (lattice mast)
Stahlglanzblech *n* blue steel plate
Stahlgliederband *n* apron (steel-plate) conveyor
Stahlgliedersiebkette *f* sifting chain with steel links
Stahlgliedertransportband *n* steel plate conveyor
Stahlgräting *f (ship)* grill flooring
stahlgrau steel grey
Stahlgravierung *f* steel engraving
Stahlguß *m* cast steel, steel castings; steel casting
Stahlgußform *f* steel mould, mould for casting steel
Stahlgußscheibenrad *n* cast-steel disk wheel
Stahlgußstück *n* **mit eingefrästen Düsen** steel casting with milled nozzles *(turbine)*
Stahlgußventil *n* cast-steel valve
Stahlgüte *f* grade of steel
Stahlhärte *f* steeliness
Stahlherstellung *f* steel making, manufacture of steel
Stahlhochbau *m* steel erection
Stahlkabel *n* steel cable
Stahlkern *m* steel core
Stahlkerngeschoß *n* steel-core bullet
Stahlkies *m* steel grit (shot)
Stahlkokille *f* steel mould (die)
Stahlkonstruktion *f* steel structure
Stahlkugel *f* steel ball
Stahlkupferleiter *m* copper-clad steel conductor
Stahlkupferseil *n* steel copper-rope
Stahllamelle *f* fixed plane iron
Stahllamellenkupplung *f* steel lamination coupling
Stahllineal *n* steel rule, straight-edge of steel
Stahllukendeckel *m* steel hatch cover
Stahlluppe *f* steel ball
Stahlmast *m* steel tower
Stahlmuldenkipper *m* steel-body tipping wagon
Stahlnabe *f* steel hub
Stahlofen *m* steel furnace
Stahlpanzer *m* steel armour
Stahlpanzerrohr *n (el)* [rigid] steel conduit, screwed conduit
Stahlpfahl *m* steel pile
Stahlpfanne *f* steel ladle
Stahlplatte *f* steel plate
Stahlplattenpanzer *m* steel-plate liner
Stahlplattenpanzerung *f* steel-plate lining
Stahlplattenstartbahn *f (aero)* iron-matted runway
Stahlpunze *f* steel punch
Stahlquecksilberthermometer *n* steel-tube mercury thermometer
Stahlquerschnitt *m* area of steel (reinforcement) *(reinforced concrete)*
Stahlregal *n* steel shelf
Stahlreißnadel *f* steel scriber
Stahlringausbau *m (mine)* tubbing
Stahlrohr *n* tubular steel, steel pipe (tube, tubing)
Stahlröhre *f* metal valve *(electronics)*
Stahlröhrensockel *m* all-metal tube base
Stahlrohrgerüst *n* tubular [steel] scaffolding
Stahlrohrrumpf *m (aero)* fuselage of steel-tubing framework
Stahlsaite *f* steel string
Stahlsaitenbeton *m* prestressed wire concrete
Stahlsandblasen *n* shot-peening
Stahlschalung *f* steel formwork (shuttering)
Stahlschiene *f* steel rail
Stahlschiff *n* steel ship
Stahlschiffbau *m* steel shipbuilding
Stahlschmelzofen *m* steel melting furnace
Stahlschrott *m* steel (iron, ferrous) scrap
Stahlschutzblech *n* sheet iron cover
Stahlschwelle *f (ry)* steel sleeper
Stahlseil *n* steel rope
Stahlseilgurt *m* steel rope belt, steel-cord belt conveyor
Stahlseilreifen *m* studded tyre
Stahlsiebkasten *m* all-steel cleaning shoe
Stahlskelett *n* steel frame (framing)
Stahlsorte *f* steel grade
Stahlspant *n* [steel] frame
Stahlspeichenrad *n* steel-spoked wheel
Stahlstege *mpl (print)* steel furniture
Stahlsteindecke *f* Kleine (hollow-block) floor
Stahlstich *m* steel engraving; die-stamping
Stahlstichimitation *f* thermography, imitation die stamping
Stahlstichimitationsdruck *m* thermography
Stahlstützenschalenlager *n* steel-backed bearing

Stahlträger *m* steel beam (girder)
Stahltresor *m* [steel] safe
Stahltürstock *m (mine)* steel set
Stahlveredler *m* stabilizer [for steel]
Stahlwerk *n* steel mill (works)
Stahlwerker *m* steel smelter (metallurgist)
Stahlwerksfrischerz *n* steel refining ore
Stahlwinkel *m* steel square
Stahlwolle *f* metal (steel) wool
Stahlzellendecke *f* cellular flooring
Staketenzaun *m* pale fence (fencing), paling
Stakudraht *m* copper-clad steel wire
Stalagmit *m (geol)* stalagmite
Stalaktit *m (geol)* stalactite
Stall *m* animal shelter, stable
  oberer S. *(mine)* top stable
  unterer S. *(mine)* bottom stable
Stalldung *m* farmyard manure
Stalldungstapler *m* manure lifter and loader
Stalldungstreuer *m* manure spreader
Stallschrämmaschine *f* stable-hole machine
Stallwagen *m (ry)* horse box car
Staluseil *n* steel-cored aluminium conductor
Stamm *m* trunk *(tree)*
Stammabschnitt *m* log end
Stammbaum *m* flow-sheet
Stammbesatzung *f* nucleus crew *(of a vessel)*
Stammblatt *n* saw-blade body, disk *(saw)*
Stammdurchgang *m* inside width of a frame
Stammfarbe *f* primary colour
Stammholz *n* stem timber
Stammholzboden *m* log deck
Stammleitung *f* side (real) circuit *(telephony)*
Stammlösung *f* stock (original) solution
Stammschiene *f (ry)* stock rail, main (through) line
Stammspule *f* side-circuit loading coil *(telephony)*
Stammsubstanz *f* parent substance
Stammwinde *f* log hoist
Stammwürze *f* original wort
Stammzeichnung *f* original drawing
Stampfappretur *f (text)* beetled finish
Stampfasphalt *m* compressed asphalt
Stampfauskleidung *f* rammed lining
Stampfbeton *m* rammed concrete
Stampfbetrieb *m* stamping
Stampfbewegung *f* pitching [motion] *(of a vessel)*
Stampfboden *m* tamped bottom *(furnace)*
Stampfbohle *f* tamping board (beam), compacting beam
Stampfe *f* stamper, rammer; trip hammer
stampfen to tamp, to ram, to pun; to bank *(hearth)*; *(text)* to beetle; *(ship)* to heave and set, to pitch
  vor Anker s. to pitch at anchor
Stampfen *n* tamping, punning, puddling, rodding, ramming; *(ship)* pitching
Stampfer *m* tamper, rammer, punner
Stampffertiger *m* road tamping machine
Stampfherd *m* rammed bottom lining
Stampfkalander *m (text)* beetler
Stampfkohle *f* compressed coal
Stampfkuchen *m* compressed coal cake
Stampflehm *m* rammed clay
Stampfmasse *f* tamped clay; ramming mixture *(foundry)*
Stampfmühle *f* stamp
Stampfplatte *f* tamping plate
Stampfschwingungen *fpl (ship)* pitching
Stampfsteven *m* straight stem
Stampfstöße *mpl* ship slamming, slamming loads
Stampfwerkzeug *n* ramming tool
Stampfwinkel *m (ship)* angle of pitch, pitch angle
Stampfwirkung *f* ramming effect
Stand *m* 1. state; status; 2. stand *(fair)*, stall; box *(stable)*; 3. *(ry)* platform, foot-plate *(locomotive)*; 4. position; 5. level *(liquid)*
  S. der Entwicklung state of the development
  S. der Technik state of technology
Standanzeige *f* level indication
Standanzeiger *m* level indicator
Standard *m* standard
Standardabweichung *f* standard deviation
  S. einer Grundgesamtheit population standard deviation
  S. einer Stichprobe sample standard deviation
Standardausführung *f* standard version (model)
Standardausgangssignal *n* **des Bildkanalverstärkers** standard picture-line amplifier output signal
Standardausrüstung *f* standard equipment; standard component
Standardbauelement *n* standard component
Standardbaugruppe *f* standard assembly
Standardbaustein *m* standard building brick; standard module
Standardbefehl *m (dat)* stereotyped command
Standardbildsendereingangssignal *n* standard visual-transmitter input signal
Standardfehler *m* standard error
Standardfurniere *npl* standard sheet stock
Standardgehäuse *n* standard casing
Standardgeometrie *f* standard geometry
Standardgröße *f* standard size
Standardionisationskammer *f s.* offene Luftionisationskammer
standardisieren to standardize
Standardisierung *f* standardization
Standardlampe *f (opt)* secondary standard
Standardlösung *f* standard solution
Standardmensch *m (nucl)* standard man

Standardneutronenquelle *f* neutron standard, standard neutron source
Standardobjektiv *n* standard lens
Standardpräparat *n (nucl)* standard [source]
radioaktives S. radioactive (radioactivity) standard, reference source
Standardquelle *f (nucl)* standard [source]
Standardschiff *n* standard ship (vessel)
Standardsignal *n* / **amerikanisches** American television standards
Standardstichprobenplan *m* lot plot plan
Standardtonsendereingangssignal *n* standard aural-transmitter input signal
Standardvorschrift *f* standard specification
Standdruck *m* static (stagnation) pressure
Ständer *m* puncheon, upright, pillar, stanchion, post; *(mach)* column, housing, upright, standard; *(build)* vertical member; *(aut)* prop; *(el)* stator *(of rotating machinery)*
geschlossener S. closed housing *(rolling mill)*
waagerecht verschiebbarer S. travelling column
Ständeranlasser *m* stator[-resistance] starter, reduced-voltage starter
Ständerbau *m* post-and-beam construction
Ständerbein *n* leg of standard
Ständerblech *n (el)* stator lamination
Ständerbohrmaschine *f* upright (column-type) drilling machine
Ständereisenverlust *m* field iron loss
Ständerführungsbahn *f* **der Konsolfräsmaschine** knee-to-column way, *(Am)* knee-to-column bearing
Ständergehäuse *n* stator frame
Ständerlampe *f s.* Ständerleuchte
Ständerleuchte *f* standard lamp
Ständermitte *f* centre of column
Ständernut *f (el)* stator slot
Ständerspule *f* stator coil
Ständerstab *m* stator conductor
Ständerstrom *m* stator current
Ständerwicklung *f* stator winding
standfest stable
Standfestigkeit *f* stability
Standgefäß *n (ch)* fixed vessel, storage flask, thickwalled beaker, jar
Standgeld *n* demurrage, siding rent
Standguß *m* static casting
ständig permanent; continuous; regular; dead *(load)*
Standkolben *m (ch)* flat bottomed flask
Standlängenkurve *f* tool life curve *(Taylor)*
Standlicht *n* parking lamp
Standmast *m* fixed (gin) pole, mast
Standmenge *f* output per sharpening, parts between regrinds, amount of work per grind
S. des Bohrers drill life *(yards of hole)*
Standmoment *n* restoring force; righting moment
Standmotor *m (aero)* upright-type engine
Standort *m* location, position, place, stand, station; station point
Standortanzeiger *m* position indicator
Standortbestimmung *f* 1. position finding; 2. *(aero)* position fixing; 3. siting, position plotting *(e. g. industrial plant)*
S. mittels Richtungstelegrafie position fixing by directional wireless telegraphy
Standortpeilung *f* position fixing
Standortplanung *f* planning of the siting of industry
Standortwahl *f* siting
Standpfahl *m* pile bearing at the point
Standpfeiler *m (mine)* permanent chock
Standprobe *f (ship)* dock trial
Standrohr *n* stand pipe; conductor (surface) pipe, surface string
Standschraublehre *f* bench micrometer
Standschub *m* static thrust *(of a propeller)*
Standseilbahn *f* underground (passenger) cable railway, funicular railway
standsicher stable
Standsicherheit *f* [static] stability
Standspur *f (aut)* hard shoulder, parking lane
Standvermögen *n* staying power
Standversuch *m* creep test[ing]
Standweide *f* free-range pasture
Standweite *f* in[ter]-row spacing
Standzeit *f* tool life *(between regrinds)*, between-[re]grind life; residence time; *(nucl)* down time
S. eines Ofens furnace life
lange S. und Lebensdauer *f (mach)* long life and ability to be resharpened many times *(tool)*
Standzeitkurve *f* tool-life curve
Standzeitprüfung *f* tool-life test
Standzeituntersuchungen *fpl* tool-life studies
Standzylinder *m* cylinder
Stange *f* stick; bar, rod; *(el)* pole
S. mit Verspannung trussed pole, *(Am)* truss-guyed pole
angeschuhte S. shoed (stub-reinforced) pole
getränkte S. treated pole
rohe S. untreated pole
verstrebte S. strutted pole
Stangenarbeit *f (mach)* bar work
Stangenautomat *m* automatic bar machine
Stangenbiegemaschine *f* wooden bar bending machine
Stangenform *f* rod (bar) form
Stangenhalbautomat *m* bar-type automatic lathe
Stangenholz *n* pole timber
Stangenkupfer *n* bar (ingot) copper
Stangenmagazin *n (mach)* bar magazine
Stangenmaterial *n* bar stock
Stangenrost *m* grizzly

Stangensalpeter *m* nitre in bars
Stangenschalter *m* rod-operated switch
Stangenschere *f* rod shears; long-handled pruner
Stangenschleifen *n* bar grinding
Stangenschuh *m* pole shoe
Stangenschwefel *m* roll sulphur
Stangensilber *n* bar (ingot) silver
Stangenspat *m* barrel spar
Stangensteckschlüssel *m* socket wrench
Stangenstromabnehmer *m* rod collector, trolley
Stangenvorschub *m* bar feed
Stangenwähler *m* *(telec)* panel selector; panel switch
Stangenwählersystem *n* panel system
Stangenwerkstoff *m* bar stock
Stangenziehbank *f* bar-drawing bench
Stangenzinn *n* bar tin
Stangenzirkel *m* trammel point
Stangenzylinder *m* straight cylinder *(hydraulics)*
Stannat *n* stannate
Stanniol *n* tin foil
Stanniolband *n* tin-foil tape
Stanniolbelag *m* layer of tin foil, tin-foil coating
Stanniolelektrode *f* tin-foil electrode
Stanniolkondensator *m* tin-foil capacitor
Stanzautomat *m* automatic puncher (punching machine)
Stanzblech *n* punching sheet
Stanze *f* puncher
stanzen to punch, to stamp, to blank
Stanzer *m* perforator, puncher
Stanzerei *f* press room
Stanzform *f* matrix, mould
Stanzklötze *mpl* cutting blocks
Stanzlesegerät *n* card read punch *(punched card machinery)*
Stanzloch *n* *(sl)* punch
Stanzmesser *npl* caster knives
Stanzpappe *f* friction board
Stanzpresse *f* punching press
Stanzstempel *m* punch, die
Stanztiegel *m* punching platen
Stapel *m* 1. pile, stack; *(text)* staple; 2. *(ship)* [keel]block, stocks for shipbuilding, ways; 3. *(mine)* blind shaft
  geschnittener S. cut staple
Stapelablage *f* pile delivery
Stapelanlage *f* layout of pile
Stapelanleger *m* pile feeder
stapelbar stackable
Stapelfaser *f* staple fibre
Stapelfüllholz *n* *(ship)* packing, filling pieces, making up
Stapelgestell *n* skid, stillage
Stapelhöhe *f* stacking height
Stapelklotz *m* *(ship)* wedge block
Stapelkran *m* stacker crane
Stapellänge *f* *(text)* staple length
Stapellauf *m* launch[ing]
  S. auf zwei Schlitten double-way launching, two-way launching
  S. mit Mittelschlitten single-way launching
Stapellaufbahn *f* standing (fixed) ways, ways
Stapellaufbahnneigung *f* declivity of standing ways
Stapellaufgewicht *n* launching weight
Stapellaufkeil *m* launching wedge, sliver
Stapellaufschleppketten *fpl* launching drags
Stapellaufschlitten *m* runner, sliding ways
Stapellaufschmiere *f* launching grease
stapeln to stack, to tier, to pile
Stapeln *n* **in Blockform** piling in log form
Stapelofen *m* batch-type furnace
Stapelplatz *m* stacking ground (yard); *(ship)* yard, stocks
Stapelprüfung *f* *(text)* stapling test
Stapelregler *m* stacked-diaphragm controller *(pneumatics)*
Stapelschacht *m* staple shaft
Stapelschlitten *m* *(ship)* cradle
Stapelsorte *f* *(text)* staple class
Stapelung *f* *(ship)* slipway, blocking, cribbing
Stapelziehen *n* *(text)* drawing of the staple
Starglas *n* cataract lens
  asphärisches S. aspherical cataract lens
Stärke *f* 1. starch, amylum; 2. strength; intensity
  S. der Strahlungsquelle source strength, strength of source
  S. des magnetischen Blattes strength of the magnetic shell
Stärkeabbau *m* peptonization
Stärkebildner *m* amyloplast
Stark-Effekt *m* Stark effect
Stärkegummi *n* starch gum, dextrin
stärkehaltig starchy, amylaceous
Stärkekleister *m* starch paste
Stärkekorn *n* starch granule
Stärkeleim *m* starch glue
Stärkemehl *n* amylum, starch (arrowroot) flour
stärken 1. to starch; 2. to strengthen
Starkgas *n* rich gas
Starkholz *n* large (heavy) timber
Starkpappe *f* solid fibreboard
Starkstrom *m* medium-voltage current *(voltage exceeding 24 volts)*; power (high-voltage, heavy) current
Starkstromanlage *f* power plant
Starkstromentladung *f* high-current discharge
Starkstromfreileitung *f* overhead power line
Starkstromgeräusch *n* power-induction noise
Starkstromingenieur *m* electrical power engineer, heavy-current electrical engineer
Starkstromkabel *n* power [supply] cable
Starkstromkontakt *m* power contact

Starkstromleitung *f* high-power transmission line, power circuit (line)
Starkstromnetz *n* power mains (supply system)
Starkstromsammelschiene *f* power bus bar
Starkstromschalter *m* heavy-current switch
Starkstromsicherung *f* high-voltage fuse
Starkstromtechnik *f* electrical power engineering, heavy current engineering
Stärkungsmittel *n* restorative, tonic
Starkwasser *n* strong liquor
starr rigid
Starregge *f* rigid tine harrow
Starrheit *f* rigidity
Starrkörperverschiebung *f* rigid-body displacement
Starrluftschiff *n* rigid airship
Starrluftschraube *f* fixed-pitch propeller, fixed-blade propeller
starrplastisch rigid-plastic
Starrschmiere *f* semi-solid lubricant
Start *m* take-off, start; *(rock)* launch[ing]
S. mit Düsenstarthilfe jet-assisted take-off
S. mit Seitenwind cross-wind take-off
S. mit Starthilfe assisted take-off
Startanlage *f* launching installation
Startanode *f s.* Starteranode
Startaufbau *m* launching (firing) platform, launching emplacement
Startbahn *f* runway
Startbeschleuniger *m* assisted take-off unit; booster
Startdauer *f* take-off time
Starteigenschaft *f* starting ability
starten to start, to set going; *(rock)* to launch
Starter *m* 1. starter [anode], auxiliary anode; 2. car starter *(road vehicles)*; cranking motor
Starteranode *f* starter anode
Starteranodensteuerspannung *f* starter anode trigger voltage
Starteranodenvorspannung *f* starter anode bias voltage
Starterbatterie *f* starter battery
Starterdrosselklappe *f* starting butterfly valve
Starterelektrode *f* primer electrode
Starterlaubnis *f* permission to take off
Startfläche *f* take-off surface
Startfreigabe *f* take-off clearance
Startgestell *n* launching rail (track), rocket track, rail rocket launcher
Starthilfe *f* ignition aid
Startimpuls *m* release (initiating) pulse; pilot pulse *(radar)*
Startkatapult *m(n)* launching catapult
startklar ready for take-off
Startmasse *f* launching (take-off, initial) weight
Startperiode *f* start-up period
Startplattform *f* firing (launching) platform, launching emplacement
Startplatz *m* aerodrome of departure; launching site (base), firing base (site)
Startrakete *f* assisted launching (take-off) rocket, booster (launcher) rocket, rocket-assisted take-off gear, RATO-rocket
Startrampe *f* launching ramp
Startreagens *n* trigger *(chain reaction)*
Startreaktion *f* initiating reaction
Startschiene *f* launching track (rail), rocket track
Startschub *m* initial thrust
Startschwierigkeit *f* launching difficulty
Startspiel *n* starting cycle
Start-Stop-Multivibrator *m* start-stop multivibrator *m*, single-shot multivibrator
Start-Stop-Telegrafie *f* start-stop telegraphy
Startstromstoß *m* start [im]pulse
Starttisch *m* launching cradle (pad, table)
Startturm *m* servicing tower
Start- und Landebahn *f* runway
Startvergaser *m* starting carburetter
Startvermögen *n* starting ability
Startvorrichtung *f* launching installation
Startwagen *m* launching carriage
Startwinkel *m* launching angle
Startzeit *f* start-up time
Startzwischenfall *m* start-up accident
Staßfurtit *m (min)* stassfurtite
Statex-Relais *n* mercury-plunger relay
Statik *f* 1. statics; 2. *s.* Proportionalbereich
S. plastischer Massen plasticostatics
angewandte S. applied statics
Station *f* station; ward *(hospital)*
halbautomatische S. partial-automatic station
interplanetare S. extraterrestrial base
ortsfeste S. fixed station
stationär fixed, stationary; steady *(flow)*
Stationsansage *f* station identification
Stationswecker *m* station ringer
statisch static[al]
Statistik *f* statistics
Statistiker *m* statistician
statistisch statistic[al]
Stativ *n* stand; tripod; support
zusammenschiebbares S. telescopic (sliding) tripod
Stativaufsatz *m* stockhead
Stativbein *n* tripod leg
Stativkopf *m* tripod head
Stativschraube *f* tripod screw
Stator *m* stator
Stator-EMK *f* stator e. m. f.
Statorfeld *n* stator field
Statorspule *f* stationary (stator, field, magnetizing) coil
Statorstrom *m* stator current
Statoskop *n* statoscope
Statuenbronze *f* statuary bronze

Statuenmarmor *m* statuary marble
Stau *m* blanked-up water level, damming up, impounding *(of water)*; congestion *(traffic)*
Stauanlage *f* barrage
Staub *m* dust; meal *(stone)*
äolischer S. windblown dust
kosmischer S. cosmic dust
lungengängiger S. respirable dust
radioaktiver S. radioactive dust
Staubablagerung *f* dust accumulation, pollution
Staubabsaugeeinrichtung *f* dust extraction (collecting) equipment, dust exhaustor (exhaust system)
staubabsaugend dust-extracting
Staubabsauger *m* suction dust remover
Staubabsaugung *f* dust exhaust (aspiration)
Staubabscheider *m* dust collector (separator, precleaner), deduster, precipitator; cyclone, dust (settling) chamber
Staubabscheidung *f* / **elektrische** electrical precipitation
Staubbekämpfung *f (mine)* dust suppression
staubbeladen *s.* staubhaltig
Staubbelästigung *f* dust nuisance
Staubbelastung *f (mine)* dust exposure
staubbeständig dust-resistant
Staubbildung *f* dust formation
Staubbindeöl *n* dust laying oil
staubdicht dust-tight, dust-proof, sealed against dust
Staubdichte *f (mine)* dustiness
Staubdichtung *f* dust-proof seal
Staubecken *n* reservoir
Staubeckenmauer *f* reservoir dam
Stäubegeräte *npl* dusters
Stäubemittel *n* dust, powder
stauben to fluff *(paper)*
stäuben to dust, to powder; to spray *(fluid)*
Staubentferner *m* dust remover
Staubentwicklung *f* formation of dust
Staubexplosion *f* dust explosion
Staubfall *m* dust fall
Staubfang *m* dust catcher
Staubfänger *m* dust collector (catcher)
Staubfeuerung *f* pulverized coal firing, coal dust firing
Staubfilter *n* aerosol (aerial dust) filter; fabric collector
staubförmig *(ch)* finely divided, powdered, dusty
staubfrei dust-free
staubgeschützt dust-proof
staubhaltig dusty
Staubkammer *f* dust-settling chamber
Staubkappe *f* dust cap
Staubkohle *f* powdered (dust, fine, pulverized) coal
Staublunge *f* pneumo[no]koniosis
Staubmaske *f* dust respirator
Staubniederschlagung *f* dust precipitation (suppression)
Staubpartikel *f* dust particle, speck of dust
Staubpilz *m* mushroom cloud
Staubsack *m* dust catcher *(blast furnace)*
Staubsammler *m* dust collector (remover, trap)
Staubsauger *m* vacuum cleaner
Staubschott *n* screen bulkhead
Staubschutz *m* dust guards
ziehharmonikaähnlicher S. bellows-type dust guards
Staubschutzhülle *f* dust-protective cover
Staubteilchen *n* dust particle, speck of dust
staubtrocken set to touch *(lacquer)*
Staubwolke *f* dust cloud
Staubzucker *m* sugar powder
stauchbar malleable, compressible
Stauchbarkeit *f* malleability, compressibility
stauchen *(mech)* to compress; *(mach)* to upset; to head *(bolts)*, to machine forge; to crush *(paper test)*; to stook *(hemp)*
Stauchgerüst *n* edging mill stand
Stauchkaliber *n* edging (upset) pass, upsetting die
Stauchmaschine *f* upsetting machine (press)
Stauchmatrize *f* upsetting die
Stauchpresse *f* upsetting press (machine)
Stauchschweißen *n* butt welding with pressure
Stauchsetzmaschine *f* movable-sieve jig
Stauchstich *m* edging (upset) pass
Stauchung *f* linear compression (contraction)
Stauchversuch *m* upsetting (bulging) test; crushing test *(tube)*
Stauchwalze *f* edging roll
Stauchweg *m* upset *(welding)*
Stauchwerkzeug *n* bulging die
Stauchwert *m (mach)* chip compression
Staudamm *m* retaining (embarkment type) dam
Staudensetzer *m* shrub planter
Staudensetzmaschine *f* shrub planter
Staudruck *m* dynamic (stagnation) pressure
Staudruckdurchflußmesser *m* Pitot-tube flowmeter
Staudruckfahrtmesser *m (aero)* velocity-head speed indicator, hydrodynamical speedometer
Staudrucklog *n* hydrodynamical log
stauen 1. to crowd, to stagnate, to choke; to be jammed; to dam up *(water)*; 2. to stow [away] *(goods)*
sich s. to clog *(chips)*; to rise *(water)*
Stauen *n (ship)* stowage
S. der untersten Ladung flooring off
Stauer *m* stevedore
Stauereiarbeiter *m* holdsman
Stauerlohn *m* stowing (stevedoring) charges
Staufferbüchse *f* grease cup, Stauffer screw-down greaser
Staufferfett *n* cup grease

Stauhöhe *f* height of swell; height of back water; overflow level
Stauholz *n* dunnage [wood]
Staukante *f* edge of rise *(condenser)*
Staukoeffizient *m* stowage factor
Staukurve *f* back-water curve
Staulinie *f* back-water curve
Staulücke *f* broken-stowage
Staumauer *f* [retaining] dam, barrage wall
aufgelöste S. buttress[ed] dam
Stauplan *m s.* Stauungsplan
Staupunkt *m* stagnation point; loading point *(packed column)*
Stauraum *m* storage capacity; stowage [space]
Staurohr *n* Pitot (Prandtl) tube, [impact] tube
Staurolith *m (min)* staurolite
Stauscheibe *f* measuring orifice; baffle plate
Stauscheibenförderer *m* retarding disk conveyor
Stausee *m* reservoir
Staustoff *m (ch)* retarder
Staustrahltriebwerk *n* ram jet
Staustufe *f* barrage
Staustufenkraftwerk *n* barrage power station
Stauung *f* damming [up], stemming *(water)*; swell, eddy *(flow)*; stowage *(goods)*
S. mit Staulücken broken stowage
Stauungsplan *m* stowage (capacity, cargo) plan
Stauwasser *n* back-water, still (slack) water
Stauwehr *n* barrage, spillway dam, weir
Stauweite *f* amplitude of swell
Stauwelle *f* back-wave
Stearat *n* stearate
Stearin *n* stearin
Stearinkerze *f* stearic candle
Stearinsäure *f* stearic acid
Stearylalkohol *m* stearylic alcohol
Steatit *m (min)* steatite, soapstone
Stechbeitel *m* ripping chisel
S. mit Dülle socket firmer chisel
Stechdrehmeißel *m* end-cut turning tool
stechen to pierce, to hole; to engrave, to cut, to carve; to punch *(time clock)*
Stechheber *m* pipette, siphon
Stechkarte *f* clocking-in card
Stechmeißel *m* parting[-off] tool, cut-off tool
linker S. left-hand parting tool
rechter S. right-hand parting tool
Stechpaddel *n* single paddle
Stechsäge *f* tree saw
Stechthermometer *n* mercury-in-steel thermometer
Stechuhr *f* time clock
Stechzirkel *m* divider
Steckachse *f* fully floating axle
Steckanschluß *m* plug-in connection
Steckbaustein *m* plug-in unit (module)
Steckbrett *n* patchboard, patch panel
Steckbuchse *f (el)* receptacle
Steckdose *f* socket, connector (plug-in, wall) socket, socket outlet, plug connector, receptacle [outlet], mounting (convenience) receptacle
dreipolige S. three-pole socket [outlet], three-pin socket [outlet]; *(telec)* three-pole jack
mehrpolige S. *(telec)* multi-contact jack
vielpolige S. *s.* mehrpolige S.
Steckdosenanschluß *m* socket outlet
Steckdosenstromkreis *m* cord circuit *(for domestic appliances)*
Steckeinheit *f* plug-in unit (device)
Steckel-Walzwerk *n* Steckel mill
steckenbleiben to stick, to get stuck, to jam
im Werkstück s. to wedge in the work *(drill)*
Stecker *m (el)* plug, connector; attachment plug
S. für Wandsteckdose wall plug
berührungssicherer S. shock-proof plug
dreipoliger S. three-pin plug, three-contact plug
fünfpoliger S. five-way plug
konzentrischer S. coaxial connector
mehrpoliger S. multiple-pin plug
n-poliger S. n-pin plug
passender S. mating plug
umschaltbarer S. switch-over plug
vielpoliger S. multi-contact plug
zweipoliger S. two-pin plug
Steckeranschluß *m* plug connection
S. / mit plug-connected
Steckeranschlußstelle *f* receptacle outlet
Steckerbuchse *f* plug socket, bush
Steckerfeld *n* plugboard, peg board
Steckerhülse *f* pin bushing
Steckerschalter *m* plug switch
Steckerschnur *f* cord and plug; patch cord
Steckersicherung *f* plug fuse
Steckerstift *m* plug pin, male plug, prong
Steckfassung *f* socket; plug-socket holder
Steckfeld *n* peg board, plugboard
Steckkontakt *m* plug
Steckkupplung *f* adapter jack, coupler, plug connection
Steckmast *m* plug-in mast; dagger (withdrawable) mast
Steckokular *n* plug-in eyepiece
Steckruder *n* spade (balanced underhung) rudder
Steckschlüssel *m* socket wrench, box spanner
rohrähnlicher S. mit einsteckbarem Griff pin-handle socket wrench
Steckschwert *n (ship)* drop (sliding) keel
Stecksockelrelais *n* plug-in relay
Steckspindel *f* [loose] inserted peg
S. mit Federhaltung inserted peg with spring-catch
Steckspule *f* plug-in coil, adjustable-plug coil
Steckstift *m* pin plug
Steckverbindung *f* plug [and socket] connection, connector

S. von Kettengliedern stab connection of links
dreipolige S. three-way connector
konzentrische S. concentric plug and socket
mehrpolige S. multi-terminal connector
stirnseitige S. end connector
Steckverbindungsfeld *n* patchbay, patchcord bay
Steckvorrichtung *f* [plug and socket] connector; plug-in device, coupler
Steert *m s.* Stert
Steg *m* 1. web *(beam)*; stem; bridge, stud *(chain)*; 2. *(el)* cell connector *(battery)*; 3. *(print)* stick; 4. *(ship)* gang-board; 5. path
S. einer Schiene web of a rail
eingesetzter S. inserted bridge piece *(turbine)*
Stegbewehrung *f* web reinforcement
Stegblech *n* web plate
Stegbreite *f* width of land *(tap)*
Stegfach *n (print)* furniture case
Steghohlleiter *m* ridge waveguide
Stegkette *f* coil chain with a stud (bridge) across the coil
Stegkettenförderer *m* / **leichter** light-duty scraper conveyor
schwerer gepanzerter S. heavy-duty armoured conveyor
Steglasche *f* bracket clip
Stegleitung *f* ribbon cable
Stegrippe *f (aero)* web rib
Stegrundung *f* radius of web
Stehbild *n* still [picture]
Stehbildprojektor *m* still projector
Stehblech *n* web plate
Stehbolzen *m* stay bolt, stud
Stehbolzenteilung *f* pitch of stay bolts, distance between stay bolts
Stehbolzenzieher *m* stay bolt driver
stehen to stand [upright]; to be at rest
fest s. to be stationary
stehenbleiben to run down, to stall, to stop
Stehenbleiben *n* **des Werkzeugs im Schnitt** stalling of the tool when cutting
stehend standing, stationary; vertical, upright
Steher *m* long-distance rider *(cycling)*
Stehfach *n (text)* open shed
Stehfeldbestrahlung *f* irradiation in a stationary field
Stehlampe *f* standard lamp, floor (portable) standard
Stehprobe *f* dock trial
Stehsatz *m (print)* standing (live) matter
Stehschaft *m* stationary shaft
Stehspannung *f* withstand voltage
Stehstoßspannung *f* impulse withstand voltage
Stehwechselspannung *f* power-frequency withstand voltage
S. im beregneten Zustand power-frequency wet withstand voltage
S. im betauten Zustand power-frequency dew withstand voltage
S. im trockenen Zustand power-frequency dry withstand voltage
Stehwelle *f* standing wave
Stehwellenmeßgerät *n* standing wave indicator
Stehwellenverhältnis *n* standing-wave ratio
steif stiff, rigid; dry *(concrete)*
Steife *f* 1. *(build)* puncheon, shore, stay, prop, brace; rib, stiffener; 2. *(ch)* stiffening agent
Steifezahl *f* coefficient of rigidity; coefficient of compressibility *(soil mechanics)*
Steifheit *f* stiffness, rigidity; inflexibility
steifheitsbedingt dependent on rigidity
Steifheitszahl *f* coefficient of rigidity
Steifigkeit *f* stiffness, rigidity; inflexibility
magnetische S. magnetic rigidity
Steifigkeitsmodul *m* stiffness module
Steifkappe *f* cased toe *(shoe)*
Steifleinen *n* calico
Steifsäge *f* bucking saw
Steigeisen *n* stirrup [iron]
Steigeisen *npl* climbing irons, [pole] climbers
S. an der Außenhaut sea ladder
Steigeleitung *f* rising main
steigen to climb; to mount, to ascend; to rise; to increase
an die Wasseroberfläche s. to be floated, to rise to the surface
Steiger *m* 1. *(met)* rising gate, riser outgate *(mould)*; 2. *(mine)* deputy, foreman, *(Am)* shift boss
Steigerabteilung *f* deputy's district
steigern to increase
Steigerohr *n* ascending pipe
Steigerrevier *n (mine)* deputy's district
Steigerung *f* increase
Steigestütze *f* pole climber
Steigfähigkeit *f* hill-climbing ability
Steigflug *m* climb
Steigflugleistung *f* climbing power
Steiggeschwindigkeit *f* rate of climb, climbing speed
S. in Bodennähe initial rate of climb
Steiggeschwindigkeitsmesser *m* rate-of-climb indicator
Steiggitter *n (text)* upright lattice
Steighöhe *f* ceiling; height of ascent (rise)
S. in Kapillaren capillary rise, elevation
Steigkasten *m (text)* drop (rising) box
Steigkastenwechsel *m* drop-box motion
Steigkastenwechselwebstuhl *m* drop box loom
Steiglattentuch *n (text)* upright lattice
Steigleitung *f* rising main, riser
Steigpresse *f* reverse press *(papermaking)*
Steigrohr *n* ascending pipe
Steigsichter *m (ag)* aspiration duct
Steigstromvergaser *m* up-draught carburetter

Steigung *f* gradient, incline, inclination, slope, rise; *(mach)* pitch *(thread)*, lead *(single-start thread, helix)*
S. eines Kegels taper [per unit length]
befahrbare S. climbing gradient
konstante S. uniform pitch *(propeller)*
zunehmende S. increasing pitch *(propeller)*
Steigungsabweichung *f* slope deviation *(radar)*
Steigungsanzeiger *m* pitch indicator
Steigungsfehler *m* lead (pitch) error
Steigungsgenauigkeit *f* accuracy in lead (pitch)
Steigungssinn *m* direction (hand) of helix *(gear)*
S. entgegengesetzt teeth of opposite hand
S. gleichgerichtet teeth of the same hand
entgegengesetzter S. opposite hand
Steigungsverhältnis *n* pitch ratio *(propeller)*
Steigungsverstellbereich *m (aero)* pitch range
Steigungswinkel *m* angle of ascent (slope, inclination); lead angle *(thread)*; helix angle
Steigvermögen *n* hill-climbing ability
Steigwind *m* ascending wind
Steigwindfege *f (ag)* combined aspiratory and sieving machine
Steigwinkel *m (aero)* flight path angle, angle of climb
Steigzeit *f* duration of ascent, time of climb
steil steep; abrupt
Steilband *n* high-angle belt conveyor
Steilböschung *f* scarp
Steildach *n* pitched roof
Steilfeuer *n* high-trajectory fire
Steilförderer *m* steeply inclined conveyor; *(mine)* steep seam box transporter
steilgängig steep-spiral, coarse-pitch *(thread)*
Steilgewinde *n* quick thread
Steilheit *f* steepness; *(math)* kurtosis, slope; *(el)* mutual (grid-plate) conductance
S. der Charakteristik slope of the emission characteristic *(electron tube)*
Steilkegel *m (mach)* steep [angle] taper, short taper
Steilkurve *f* sharp turn
Steilsichtprisma *n* steep-sight prism
Steilsieb *n* fixed grizzly
Steilstrecke *f* steep course
Stein *m* stone; brick; *(met)* matte
armer S. low-grade matte
blechummantelter S. ferroclad brick
feuerfester S. refractory brick
glasierter S. vitrified brick
halber S. half bat *(brick)*
reicher S. high-grade matte
Steinalaun *m* stone-alum
steinartig ston[e]y
Steinbank *f (geol)* stone bed
Steinbearbeitungsmaschine *f* stone-working machine
Steinblock *m* block of stone
Steinboden *m* stony ground; stone floor
Steinbohrer *m* jumper, auger, rock (wall, stone) drill
Steinbruch *m* quarry; quarrying
Steinbrucharbeiter *m* quarrer, quarrier
Steinbruchbagger *m* quarry (rock) shovel
Steinbruchbetrieb *m* quarrying
Steindamm *m* mole
Steindruck *m* lithography
Steinfall *m (mine)* fall of ground (roof)
Steinfang *m* stone trap
Steinformat *n* brick size
Steinfräsmaschine *f* natural stone milling machine
Steinfundament *n* stone foundation
Steingut *n* stoneware, earthenware, crockery
Steinharz *n* dammar
Steinholz *n* artificial flooring cement, magnesite composition, xylolite, magnesium oxychloride
Steinholzestrich *m* composition (magnesite, magnesium-oxychloride) flooring
Steinholzfußboden *m s.* Steinholzestrich
Steinindustrie *f* stone industry
Steinkeil *m* stone wedge
Steinkiefer *f* stone pine
Steinkohle *f* hard (mineral, pit) coal
Steinkohlenader *f* coal seam
Steinkohlenaufbereitung *f* coal dressing [process]; coal dressing plant
Steinkohlenbergbau *m* [hard] coal mining
Steinkohlenbrikett *n* hard-coal briquet[te]
Steinkohlendestillation *f* carbonization of [bituminous] coal
Steinkohlenentgasung *f* [bituminous-]coal carbonization
Steinkohlenfeuerung *f* [bituminous] coal firing
Steinkohlenflöz *n* [bituminous] coal seam
Steinkohlengas *n* [bituminous-]coal gas
Steinkohlenhochtemperaturteer *m* high-temperature coal tar
Steinkohlenschwelteer *m* low-temperature coal tar
Steinkohlenschwelung *f* coal carbonization
Steinkohlenteer *m* coal tar
Steinkohlenteeröl *n* coal-tar oil
Steinkohlenteerpech *n* coal-tar pitch
Steinkohlenverkokung *f* coal carbonization coking
Steinmauerwerk *n* [stone] masonry, walling
Steinmeißel *m* stone chisel
Steinmeteorit *m* stone meteorite
Steinmetz *m* [stone] mason
Steinmüller-Kessel *m* Steinmüller boiler
Steinobst *n* stone fruit
Steinöl *n* mineral (rock) oil
Steinpackung *f* rip-rap[ping]
Steinpflaster *n* stone pavement
Steinplatte *f* stone slab
Steinpresse *f* brick-making press

Steinsäge *f* stone[-cutting] saw
Steinsalz *n* rock (common) salt; halite
Steinsalzprisma *n* rock-salt prism
Steinsalzstruktur *f* rock-salt structure
Steinschaufel *f* stone shovel
Steinscheibe *f* flange *(grinder)*
Steinschlag *m* broken stone (rock)
Steinschmirgel *m* stone-emery
Steinschotter *m* ballast stone
Steinschraube *f* rag-bolt, stone bolt, wall screw
Steinschutt *m* ballast, broken stones; *(geol)* debris
Steinschüttdamm *m* rock-fill dam
Steinschüttung *f* rip-rap[ping]
Steinsetzer *m* paver, paviour
Steinsplitt *m* chippings, chips
Steinwolf *m* lewis
Steinwolle *f* rock wool
Steinzeichnung *f* construction lines of brickwork *(blast furnace)*
Steinzeug *n* earthenware *(insulating material)*; stoneware, pottery
Stellage *f* rack; *(ship)* staging
Stellantrieb *m* [servo, valve] actuator, servo drive
S. mit Drehbewegung rotary actuator
S. mit Linearbewegung linear actuator
S. mit Winkelbewegung rotary actuator
stellbar adjustable
Stellbereich *m* adjusting (controlling) range; rangeability *(valve)*
Stellbogen *m* quadrant
Stelldetektor *m* adjustable detector
Stelldruck *m* actuating pressure
Stelle *f* point; site; place, spot; *(math)* digit, place, figure; column
benannte S. *(dat)* denominational order
flache S. *s.* Untiefe
harte S. *(met)* hard spot
höchste S. most significant digit
singuläre S. singularity
Stelleinrichtung *f* regulating unit
stellen 1. to put, to set, to place; to put (set) upright; 2. to adjust; to control, to regulate
s. auf to locate to; to set to
auf Mitte s. to centralize
auf Null s. to reset; to set on zero
außer Dienst s. to put out of commission
fest s. to lock
in Dienst s. to [put into] commission
in Rechnung s. to charge [items]
Stellendauer *f* digit period
Stellenschreibweise *f* *(dat)* positional notation
Stellenverschiebung *f* arithmetic shift
Stellenwert *m* place value *(number)*, column
Stellenwertverschiebung *f* arithmetic shift *(from digit place to digit place)*
Stellenzahl *f* / **maximale** capacity
Stellenziffer *f* place digit
Steller *m* regulating unit
Stellfeder *f* adjusting spring
Stellgeschwindigkeit *f* regulating speed
Stellgetriebe *n* actuating mechanism, valve-actuating gear
Stellglied *n* final control element, correction device
Stellgröße *f* control medium, *(Am)* manipulated variable
Stellhebel *m* control (adjustable) lever, actuating arm
Stellit *m(n)* stellite
Stellkeil *m* adjusting key, tightening wedge
Stellklaue *f* adjusting dog
Stellknopf *m* adjusting knob
Stellkolben *m* power (servo) piston
Stellkondensator *m* adjustable capacitor
Stellmacher *m* 1. wheelwright, cartwright; 2. *(cont)* [valve] positioner
Stellmacherei *f* wheelwright's shop
Stellmacherholz *n* timber used by the cartwright
Stellmagnet *m* operating magnet
Stellmotor *m* servomotor, servo (control) motor; change-speed motor
hydraulischer S. hydraulic servomotor
pneumatischer S. pneumatic servomotor
Stellmutter *f* adjusting nut
Stellnetz *n* stationary (fixed, set) net *(fishery)*
Stellort *m* regulating point *(location of the final control element)*
Stellpumpe *f* variable capacity (delivery, displacement) pump
Stellring *m* guide (locating) ring; collar; cursor; adjusting ring
Stellsäge *f* frame saw
Stellscheibe *f* setting dial
Stellschiene *f* slide rail
Stellschraube *f* adjusting (regulating, set) screw
Stellspindel *f* adjusting screw
Stellstift *m* set (adjusting) pin
Stellstiftwagen *m* pin carriage *(office machine)*
Stellstrom *m* controlling current
Stellstromkreis *m* control circuit
Stellstückwagen *m* pin carriage *(office machine)*
Stelltransformator *m* adjustable (variable-ratio) transformer
Stellung *f* position; setting
S. des Kommas position of the decimal point
S. des Leuchtfleckes spot position *(cathode ray tube)*
freie S. unfilled vacancy *(job)*
gestreckte S. *(leath)* stretched pattern
oberste S. top (uppermost) position
Stellungnahme *f* comment
Stellungsgeber *m* [remote] position indicator
Stellungslichter *npl* aircraft navigation lights

Stellungsmacher *m* [valve] positioner
Stellungsmelder *m* position indicator
Stellungszeiger *m* vane-type relay
Stellverhältnis *n* rangeability *(valve)*
Stellwerk *n* 1. *(cont)* servo actuator; [valve] positioner; *(therm)* barring gear; 2. *(ry)* signal cabin; interlocking cabin, *(Am)* interlocker; signal box
elektrisches S. electrical signal box
elektromechanisches S. electro-mechanical signal box
Stellwerksanlage *f* interlocking plant
Stellwerksturm *m* switch tower
Stellwiderstand *m* variable (sliding) resistor, rheostat
Stellzeit *f* regulating time *(travel time of the final control element)*
Stellzeug *n* adjusting gear *(governor)*
Stellzeugwiderstand *m* resistance of the adjusting gear
Stelze *f* stilt *(governor)*
Stemmbrett *n* footpiece, stretcher *(boat)*
Stemmen *n* cutting *(woodworking)*
Stemmfuge *f* caulking joint
Stemmtor *n* dock gate
Stempel *m* 1. stamp; 2. punch; die *(press)*; 3. *(build)* puncheon, prop, shore; 4. *(mine)* leg, post, prop; strut, piston
S. mit Keilschloß wedge prop
frühtragender S. early bearing prop
soforttragender S. immediate-bearing prop
spättragender S. late bearing prop
starrer S. rigid prop
verstellbar starrer S. rigid-extensible prop
Stempelabstand *m* prop spacing
Stempeldichte *f* prop density
Stempelfarbe *f* stamp pad ink
stempelfrei prop-free
Stempelglanz / mit uncirculated *(coin)*
Stempelholz *n* pit props
Stempelkissen *n* stamp pad
Stempelkopf *m* **für Schiebekappen** *(mine)* slide-bar head
stempeln to stamp, to affix a stamp to; to mark
Stempelpresse *f* stamping press
Stempelrauber *m (mine)* prop drawer
Stempelsetzer *m (mine)* timber setter
S. hinter einem Schrämlader back-end man
Stempelsetzvorrichtung *f* prop-setting jack
Stempeluhr *f (Am)* time clock
Stempel- und Werkzeugprofilhobelmaschine *f* punch-and-tool profile shaping machine
Stengel *m* stem *(of a key of a keyboard)*
Stengelansatzmaschine *f* tubulating machine *(electron tubes)*
Stengelfaser *f* stem (stalk) fibre
Stengelgefüge *n* columnar structure
Stengelkristall *m* columnar crystal
Stenograf *m* shorthand writer, stenographer
Stenografie *f* shorthand, stenography
stenografieren to write shorthand
Stenografin *f* shorthand writer, stenographer
Stenogramm *n* shorthand writing (report), stenograph
Stenotypist[in *f***]** *m* shorthand-typist
Stephanit *m* stephanite, brittle silver ore
Steppdecke *f* quilt
Steppdeckeneinlage *f* quilting interlining
steppen to quilt *(bed-coverlet)*
Stepperei *f* closing room *(shoe)*
Steppstich *m* quilt [stitch], running stitch
Sterad *m (math)* steradian, spheradian
Steradiant *m (math)* steradian, spheradian
Sterbling *m (leath)* fallen hide
Sterblingswolle *f (text)* dead wool
Stereo *n* stereo
Stereoaufnahme *f s.* Stereobild
Stereobild *n* stereogram, stereo[photo]graph
Stereochemie *f* stereochemistry
stereochemisch stereochemical
Stereodurchleuchtung *f* stereofluoroscopy, stereoradioscopy
Stereoeffekt *m* stereo effect
Stereoentfernungsmesser *m* stereoscopic range finder
Stereofernsehen *n* stereoscopic television
Stereofilm *m* stereo[scopic] film
stereofon stereophonic[al]
Stereofotografie *f* stereoscopic photography
Stereogerät *n* stereo apparatus
Stereografie *f* stereo[radio]graphy
stereografisch stereographic
Stereohöhenmesser *m* stereoscopic height finder
Stereoisomer *n* stereoisomer
Stereoisomerie *f* stereoisomerism
Stereokamera *f* stereoscopic camera
Stereokartierung *f* stereoscopic mapping
Stereokomparator *m* stereocomparator
Stereomeßkamera *f* photogrammetric stereocamera
Stereomeßkammer *f* photogrammetric stereocamera
Stereometall *n* stereotype metal
Stereometrie *f* geometry of solids, solid geometry, stereometry
Stereomikroskop *n* stereomicroscope
Stereoplatte *f* stereo
Stereoprojektor *m* stereoprojector
Stereoschallplatte *f* stereophonic disk
Stereoskop *n* stereoscope
stereoskopisch stereoscopic
Stereotelemeter *n* stereoscopic range finder
Stereoton *m* stereo sound
Stereotypdruck *m* stereotyping
Stereotypeur *m* foundry-man

Stereotypie *f* stereotyping, stereotype process; foundry *(department)*
Stereoverstärker *m* stereo amplifier
Stereovorsatz *m* stereo attachment
steril sterile
Sterilisation *f* sterilization
S. in der Trockenkammer kiln sterilization
Sterilisator *m* sterilizer, sterilizing apparatus
sterilisieren to sterilize
Sterilisierungsdosis *f* sterilization dose
Sterin *n* sterol
sterisch steric
Stern *m* star
künstlicher S. artificial star
veränderlicher S. variable star
Sternanker *m (el)* multipole armature
Sternanordnung *f* radial arrangement *(cylinders)*
Sternantenne *f* star aerial
Sternbild *n* constellation
Sternbildprojektor *m* constellation-figures projector
Sternbildung *f* star production
Sternbuchse *f* stern gland (bush)
Sternbuchsenschott *n* after peak bulkhead
Sterndämpfer *m* star steamer
Sterndreieck *n (el)* star-delta
Sterndreieckanlasser *m* star-delta starter
Sterndreieckanlauf *m* star-delta starting
Sterndreieckschalter *m* star-delta switch (starter)
Sterndreieckschaltung *f* star-delta connection
Sterndreieckumschalter *m* star-delta throw-over switch
Sternerzeugung *f* star production
sternförmig radial *(engine)*
Stern-Gerlach-Versuch *m* Stern-Gerlach experiment
sterngeschaltet star-connected, Y-connected
Sternhaspel *f* star reel
Sternjahr *n* sidereal year
Sternkette *f* star chain *(radar)*
Sternkunde *f* astronomy
Sternkurve *f* astroid
Sternmotor *m* radial engine
Sternplatte *f* tripod headplate
Sternplattenfotometer *n* star-plate photometer
Sternplattenkomparator *m* star-plate comparator
Sternpunkt *m* neutral (centre, zero) point, centre
geerdeter S. earthed neutral point
isolierter S. insulated neutral point
künstlicher S. artificial neutral point
unmittelbar geerdeter S. directly earthed neutral point
wirksam geerdeter S. effectively earthed neutral point
Sternpunkterdung *f* neutral earthing
Sternpunktklemme *f* neutral terminal *(of a polyphase machine)*
Sternpunktleiter *m* neutral [conductor]
Sternpunktsanode *f* zero-point anode
Sternpunktspannung *f* neutral voltage
Sternpunktstrom *m* neutral current
Sternrad *n* star wheel
Sternradgetriebe *n* star-wheel mechanism
Sternradwender *m (ag)* revolution side delivery rake
Sternrevolverdrehmaschine *f* horizontal-turret lathe, vertical-axis-turret lathe
Sternrevolverkopf *m* horizontal-turret head
Sternschaltung *f* star-connection, Y-connection, star-connected circuit
Sternschauzeichen *n* [white] star indicator *(telephony)*
Sternschnuppe *f* falling (shooting) star
Sternschnuppenschwarm *m* meteoric shower
Sternspannung *f* star voltage, Y-voltage, voltage to neutral
Sternspektrum *n* stellar (star) spectrum
Sternsternschaltung *f* star-star-connection
Sternverseilung *f* star-twisting *(cable)*, spiral quad formation
Sternviererkabel *n* star quad cable
Sternwalze *f (ag)* toothed roller
Sternwarte *f* observatory
Sternwartenkuppel *f* observatory dome
Sternzeit *f* sidereal time
örtliche S. local sidereal time
Sternzeituhr *f* clock for sidereal time
Sternzickzackschaltung *f* star-zig-zag connection, interconnected star connection
Steroid *n* steroid
Stert *m* tail *(of a vessel)*
Sterz *m* plough tail
Stethoskop *n* stethoscope
stetig *(math)* continuous, steady *(function)*
Stetigbahnsteuerung *f* continuous-path control [system] *(machine tool)*
Stetigförderer *m* continuous conveyor
Stetigfördern *n* conveying
Stetigholländer *m* continuous beater
Stetigkeit *f (math)* continuity *(function)*
S. der Bewegung continuity of the motion
S. der Kavitation steadiness of cavitation
gleichmäßige S. einer Funktion uniform continuity of a function
Stetigmischer *m* continuous mixer
Stetigschleifer *m* continuous grinder *(pulp)*
Steuer *n* wheel; rudder *(rowing)*
Steueralgorithmus *m* control algorithm
Steuerautomat *m* aircraft director, autopilot
Steuerautomatik *f* automatic flight control
Steuerband *n* control tape
steuerbar controllable; manoeuvrable
Steuerbarkeit *f* controllability; manoeuvrability
Steuerbefehl *m* control command
Steuerbefehlsregister *n* control register

Steuerbefehlsspeicher *m* control program[me] register, control register, program[me] counter
Steuerbelag *m (el)* [potential] control coating
Steuerbewegung *f* distributing valve motion *(steam)*
Steuerbord *n* starboard
  hart S. hard-a-starboard
Steuerbordruder *n* right rudder
Steuerbrücke *f* control bars *(of accounting machine)*
Steuerdraht *m (telev)* grid wire; *(telec)* pilot wire
Steuerdrossel *f* magnetic modulator, modulation reactor
Steuerdruck *m* operating (control) pressure
Steuereinlage *f (el)* [potential] control insertion
Steuereinrichtung *f* control assembly *(open-loop control)*
Steuerelektrode *f* control (modulator) electrode
Steuerelement *n* control element (device), positioning actuator
Steuerempfänger *m* control receiver; control transformer *(synchro)*
Steuerempfindlichkeit *f* control sensitivity
steuerfähig werden / wieder to regain control
Steuerfähigkeit *f* 1. steerage, steering qualities; 2. controllability
Steuerfähigkeitsprobe *f* manoeuvring trial
Steuerfaktor *m* **bei Gasentladungsröhren** control ratio
Steuerfeld *n* control panel
Steuerfläche *f* control surface
Steuerfluß *m* control flux
Steuerfrequenz *f* pilot (driving) frequency
Steuerfrequenzempfänger *m* pilot receiver
Steuerfrequenzgeber *m* control transmitter
Steuergehäuse *n (aut)* timing housing
Steuergeneratorensatz *m* motor-generator set
Steuergerät *n* controller; operating mechanism; control gear (unit)
  elektronisches S. electronic control unit
Steuergestänge *n* control linkage
Steuergetriebe *n* control gear
Steuergitter *n* control (signal) grid
Steuergitterbelastung *f* control-grid dissipation
Steuergittereinsatzspannung *f* [control] grid cut-off voltage
Steuergitterfotozelle *f* three-electrode cell
Steuergittergleichstrom *m* d-c control grid current
Steuergittergoldschicht *f* gold coating *(electron tubes)*
Steuergittermodulation *f* control-grid modulation
Steuergitterspannung *f* control grid voltage
Steuergriff *m* master handle *(manipulator)*
Steuerhaus *n* wheel house
Steuerhebel *m* operating (control) lever
Steuerhilfsventil *n* pilot valve
Steuerhysterese *f* control hysteresis *(radio)*
Steuerimpuls *m* command (pilot, driving) pulse; main bang *(radar)*
Steuerkabel *n* control cable
Steuerkabine *f* cabin *(crane)*
Steuerkanal *m* pilot channel
Steuerkennlinie *f* transfer characteristic; control characteristic
Steuerkette *f* [open-loop] control system; *(aut)* timing chain
Steuerknüppel *m* control column (stick)
Steuerkolben *m* actuating piston, spool
Steuerkolbentauchspule *f* control piston solenoid
Steuerkompaß *m* steering compass
Steuerkontakt *m* control contact
Steuerkörper *m* / **schwenkbarer** tilting yoke *(axial piston pump)*
Steuerkreis *m* control circuit
Steuerkreisumspanner *m* control-circuit transformer
Steuerkurve *f (mach)* controlling cam; cam path
steuerlastig down by the stern
Steuerlastigkeit *f* trim by the stern
  konstruktive S. drag
Steuerleine *f* rudder line, yoke ropes
Steuerleistung *f* control (grid-driving) power
Steuerleitung *f* pilot[-wire] circuit; cab cable
Steuerleitungen *fpl* control wiring
Steuermagnet *m* control[ling] magnet
Steuermann *m* coxswain; helmsman *(boating)*
  erster S. chief mate *(sailing ships)*
Steuermannsmaat *m* second mate
Steuermechanismus *m* control (timing) mechanism
Steuermotor *m* servomotor
steuern 1. to control; to drive, to monitor; to regulate; to time; 2. to steer, to pilot, to navigate
  eine Röhre s. to drive a valve
  einen Verstärker s. to drive an amplifier
Steuernocken *m* manoeuvring (sequence-switch) cam
Steuerorgan *n* control element; controller
Steueroszillator *m* master oscillator
  quarzgesteuerter S. crystal oscillator
Steuerprogramm *n* control programme; *(dat)* executive routine
Steuerpult *n* control desk (console)
Steuerquarz *m* control (oscillator) crystal
Steuerrad *n* steering wheel
Steuerraum *m* input resonator *(waveguides)*; buncher space *(electronic tubes)*; control room *(radio)*
Steuerrelais *n* control (pilot) relay
Steuerröhre *f* control (modulating) valve
Steuerruder *n* rudder
Steuersäule *f* control column

Steuerschalter *m* controller, control switch; camshaft controller
handbetätigter S. manual controller
indirekt angetriebener S. pilot controller
Steuerschalttafel *f* control panel
Steuerschaltung *f* control circuit
S. mit Servomotor servo-motor-operated controller
Steuerschaltwalze *f* pilot controller
Steuerschieber *m* control slide valve; shifter; switch blade
S. mit freiem Durchfluß by-pass valve
Steuersender *m* control transmitter, pilot oscillator
Steuersignal *n* driving (drive) signal
Steuersitz *m* coxswain's seat *(boat)*
Steuerspannung *f* control (driving) voltage
S. eines elektronischen Tores gate voltage
Steuerspeicher *m* programme counter (register)
Steuerstab *m* control rod *(reactor)*
Steuerstabkennlinie *f* control-rod efficiency function, control characteristic
Steuerstand *m* steering stand, pilot station; manoeuvring station *(of a vessel)*
Steuerstange *f (aero)* push and pull rod
Steuerstrecke *f* controlled system *(open loop control)*
Steuerstrich *f* lubber's mark (line) *(compass)*
Steuerstrom *m* control current
kritischer S. critical controlling current *(of a cryotron)*
Steuerstromkreis *m* control (pilot) circuit
Steuerstromstoß *m* directing pulse
Steuerstromunterbrecher *m* control cut-out switch
Steuerstufe *f* control (exciter) stage, driver; master oscillator *(radio)*
Steuersystem *n* control (regulating) system
Steuertafel *f* control panel
Steuertransformator *m* adapting transformer
Steuerung *f* 1. control; drive *(amplifier)*; timing; 2. control gear; timing gear; 3. steering
S. der Trockenkammer kiln control
S. des Prozesses processing adjustment
S. durch Lochstreifen tape control
automatische S. automatic control, automatic control (steering) apparatus
bedingte S. conditional control
drahtlose S. telecontrol, wireless control
elektrische S. electric control
elektromagnetische S. electromagnetic control
elektromechanische S. electromechanical drive
elektronische S. electronic control
fotoelektrische S. photoelectric control
kontaktlose S. static control (switching system)
lichtelektrische S. photoelectric control
lineare S. linear control
logische S. logical control
numerische S. numerical control
ruhende S. static switching system
selbstanpassende S. autoadapting control
zentrale S. central control
Steuerungsdiagramm *n* valve (distribution) diagram
Steuerungsgerät *n* control gear
Steuerungssystem *n* control system
S. zur Lageeinstellung *(mach)* positional control system
Steuerungsventilkasten *m* distributing valve chest
Steuerungsvorgang *m* process of distribution
Steuerungswahl *f* / **koinzidente** coincident drive selection *(magnetic core memory)*
Steuerventil *n* control (pilot, check, manipulating) valve; distributor (triple) valve
S. mit Überlappung in der Ruhelage closed-centre valve
Steuervermögen *n* steerage
Steuerverstärker *m* control amplifier
Steuerwagen *m* driving trailer, vehicle with driving cab
Steuerwalze *f* master (barrel) controller
Steuerwelle *f* 1. camshaft, distribution shaft; 2. pilot wave
Steuerwellengehäuse *n* camshaft casing
Steuerwerk *n* steering mechanism; controls; control section (unit, circuits) *(computer)*
Steuerwicklung *f* drive winding, control coil
Steuerwiderstand *m* control resistor
Steuerwirkung *f* control action; action of the distributing-valve motion *(steam)*
Steuerzeichen *n* pilot signal
Steuerzylinder *m* 1. piston valve cylinder; 2. control (modulator) electrode *(electron valve)*
Steven *m* stern post (frame); [stem] post
ausfallender S. flare out bow
stark auflaufender S. cut-away forefoot
überhängender S. raked-stem
Stevenbeschlag *m* dovetail plate
Stevenfuß *m* stem foot
Stevenlager *n* stern bearing
Stevenlauf *m* / **runder** cut-away forefoot
Stevenlog *n* hydrodynamical log
Stevenrohrbuchse *f* stern gland (bush)
Stevenrohrlager *n* stern bush, stern tube bushing (bearing)
Stevenrohrlängsschiene *f* retaining strip
Stevenschwalbe *f* dovetail plate
Stibitz-Kode *m* excess-three code
Stich *m* 1. *(text)* stitch, pitch *(non-wovens)*; tack; 2. engraving, cut; 3. *(build)* camber, rise; 4. *(met)* pass *(rolling)*; 5. hue *(colour)*
verlorener S. lost pass

Stichabnahme *f* reduction per pass
Stichausreißfestigkeit *f (leath)* stitch resistance
Stichaxt *f* mortise axe
Stichbogen *m* segmental arch, arched lintel
Stichdrehzahlmesser *m* chronometric tachometer, shaft counter
Stichel *m* graver
Stichelhaus *n* tool post
Stichflamme *f* darting (pointed-jet) flame, burst of flame; flareback *(furnace)*
Stichflammenbildung *f* formation of a blowpipe flame
Stichhöhe *f (build)* rise, camber
Stichleitung *f (el)* tie line; open feeder, stub
Stichleitungsantenne *f* stub aerial
Stichloch *n* 1. tapping hole *(furnace)*; iron notch *(blast furnace)*; 2. needle hole *(sewing machine)*
Stichlochstopfmaschine *f* clay (tap-hole, blast-furnace) gun
Stichmaß *n* inside micrometer
Stichofen *m* tap furnace
Stichplatte *f* needle plate *(sewing machine)*
Stichprobe *f* sample [taken at random]
  repräsentative S. representative sample
  systematische S. systematic sample
Stichprobendatensystem *n* **/ digitales** digital sampled-data system
Stichprobenentnahme *f* sampling
Stichprobenmittel *n* sample mean
Stichprobenplan *m* sampling plan
Stichprobenprüfung *f* sampling [inspection]
Stichprobentabelle *f* sampling table
Stichprobenumfang *m* sample size
  idealer S. rational subgroup
  kumulativer S. combined (cumulative) sample size
Stichprobenverfahren *n* sampling [procedure]
  S. für messende Prüfung sampling by variables
Stichprobenversuch *m* spot test
Stichsäge *f* keyhole saw
Stichseite *f* **eines Ofens** vent-side of a furnace
Stichsteller *m* stitch regulator
Stichwort *n (print)* catchword, key word; cue *(theater)*
Stichwortverzeichnis *n* subject index
Stickarbeit *f* embroidery
sticken to embroider
Stickerei *f* embroidery
Stickereiwebstuhl *m* lappet loom
Stickkette *f* figuring warp
Stickoxid *n* nitric (nitrogen) oxide *(blasting)*
Stickrahmen *m* tabaret
Stickstoff *m* nitrogen
stickstoffabgebend nitrogenous
Stickstoffanregung *f* nitrogen excitation
Stickstoffbakterien *fpl* nitrifying bacteria
Stickstoffbestimmung *f* nitrogen estimation
  S. nach Kjeldahl Kjeldahl nitrogen method
Stickstoffbrücke *f* nitrogen bridge
Stickstoffdioxid *n* nitrogen dioxide
Stickstoffdünger *m* nitrogen fertilizer
stickstofffrei nitrogen-free, non-nitrogenous
Stickstoffgehalt *m* nitrogen content
Stickstoffgewinnung *f* nitrogen fixation
stickstoffhaltig nitric, nitrogenous; nitrogenated *(compounds)*
Stickstoffkatastrophe *f (nucl)* nitrogen catastrophe
Stickstoffkreislauf *m* nitrogen cycle
stickstoffreich highly nitrogenous
Stickstoffsammler *m* nitrogen storing plant
Stickstoff(III)-verbindung *f* nitrous compound
Stickstoff(V)-verbindung *f* nitric compound
Stickstoffwasserstoffsäure *f* hydrazoic acid
Stiefel *mpl* boots
  geschraubte S. screwed boots
  holzgenagelte S. pegged boots
Stiefelsohle *f* boot sole
Stiel *m* 1. handle; helve *(axe)*; stick; 2. *(build)* vertical member; column, leg *(frame)*; 3. strut *(aeroplane)*
  eingespannter S. fixed column (leg) *(frame)*
Stielfeilkloben *m* small hand vice
Stielhandgranate *f* stick grenade
Stielstich *m* backstitch
Stielstichstickerei *f* backstitch embroidery
Stielstrahlung *f* stem radiation
Stift *m* pin; stick; tack, spike, stud; pencil
  S. des Klappenhalters clapper pin
  desodorierender S. stick deodorant
  elektrisch leitender S. conductive pencil
Stiftabgriff *m (telec)* pin reader
Stiftbolzen *m* stud bolt
Stiftelektrode *f* pin electrode
Stiftendreschmaschine *f* pin-type thresher
Stiftführung *f* pin guide
Stiftgeradführung *f* parallel-guiding of the pencil
Stiftlagerung *f* pin support (suspension)
Stiftloch *n* pin hole
Stiftlochreibahle *f* taper-pin reamer
Stiftmühle *f* pinned (iron) disk mill
Stiftnietung *f* pin riveting
Stiftöler *m* pin-type lubricator
Stiftschraube *f* stud bolt
  geschlitzte S. slotted set screw
Stiftsockel *m* pin base
Stiftstecker *m* pin-plug
Stiftwandler *m (el)* probe transformer
Stiftzahn *m* post-crown
stigmatisch stigmatic
Stigmatismus *m* stigmatism
Stigmator *m* stigmator
Stilben *n* stilbene
stillegen to put out of service; to close, to shut down; to abandon *(permanently)*

Stillegung *f* shut-down, closure *(mine, factory)*
stillsetzen to bring to rest, to stop
Stillsetzen *n* stopping
Stillsetzung *f* stoppage
Stillstand *m* rest; stop[page]; dwell; stalling; standstill
  S. / im at rest
Stillstandsdauer *f (mach)* dwell period, period of dwell
Stillstandzeit *f* idle (down) time; dwell time *(at reverse)*; stoppage time
stillstehen to be at rest; to stall
stillstehend motionless; stationary
Stillwasser *n* slack water
stimmen to tune [up]
Stimmgabel *f* tuning fork, pitchfork
  schreibende S. tuning-fork chronoscope
Stimmgabelgenerator *m* tuning-fork oscillator
Stimmgabelstabilisierung *f* tuning-fork stabilization
Stimmgabelsteuerung *f* tuning-form control
Stimmschlüssel *m* tuning key
Stimmung *f* musical pitch
  internationale S. international musical pitch
  physikalische S. physical standard-pitch
  temperierte S. tempered music-pitch
Stimulusimpuls *m* stimulus pulse
Stippen *fpl* flakes
Stippenbildung *f* stippling
Stirn *f* front, face, end
  S. / mit halbrunder radius-end, radiused
Stirnabschreckversuch *m* end quench test
Stirnansicht *f* **des Ofens** front of the oven
Stirnblech *n* face plate
Stirnen *n* face milling
Stirnfenster *n* end window
Stirnfläche *f* front (top) surface, [end] face
  S. / mit runder round-ended
Stirnflachnaht *f* square-edge weld
Stirnfräser *m* facing (face-milling) cutter
Stirnfugennaht *f* edge-joint single-V groove weld
Stirnführung *f* front guiding
Stirnkühler *m* front radiator
Stirnkurbel *f* overhung crank
Stirnleuchte *f* cap (head-band) lamp
Stirnmauer *f* head wall
Stirnnaht *f* edge weld
Stirnplatte *f* cluster plate *(drilling machine)*
Stirnrad *n* spur gear
  S. mit Doppelpfeilverzahnung triple-helical gear
  S. mit Pfeilverzahnung herringbone (double-helical) gear
  geradverzahntes S. straight spur gear
Stirnraddifferential *n* [cylindrical] spur differential
Stirnradgetriebe *n* [spur] gears on parallel axes
Stirnradpumpe *f* spur-gear pump
  geradverzahnte S. straight spur-gear pump
Stirnradschraubgetriebe *n* spiral gears, *(Am)* pair of crossed helical gears
Stirnradverzahnung *f* [straight] spur-gear teeth
Stirnradwälzfräser *m* spur-gear hob
Stirnreibahle *f* bottoming reamer
Stirnschleifen *n* face grinding
Stirnschneide *f* face cutting edge *(face mill)*, end-cutting edge
Stirnseite *f* front end (face)
Stirnsenker *m* facing tool; spotfacer
Stirnspreize *f (build)* face waling (piece)
Stirnstreuung *f* end (brow) leakage
Stirnstreuverlust *m* face-ring stray loss *(armature)*
Stirnteilung *f* transverse pitch
Stirnverbindung *f* [armature] end connection, face-ring connection, end winding
Stirnwand *f (aut)* dashboard; end wall, drop end *(wagon)*; front cylinder cover, top cover *(cylinder)*
  abklappbare S. drop end *(wagon)*
Stirnwandstütze *f* cowl support
Stirnzahn *m* end (radial) tooth *(milling cutter)*; front (end) cutting edge
Stirnzapfen *m* end journal
Stirnzeit *f* rise time *(wave)*
stochastisch stochastic, random
stochern to stoke, to stir, to poke; to rod, to puddle *(concrete)*
Stocherstange *f* tamping rod
Stöchiometrie *f* stoichiometry
stöchiometrisch stoichiometric
Stock *m* 1. stick; staff *(flag)*; 2. *(mine)* boss, stock; 3. *(build)* floor, storey
Stockanker *m* stock (stocked, common) anchor
stocken 1. to stop, to end, to cease; to jam *(engine)*; 2. to congeal; to get (turn) mouldy
Stockfäule *f* rotting of the trunk
Stockholz *n* butt wood; root-wood
Stocklack *m* lac
Stockpresse *f* screw press
Stockpunkt *m* point of congelation, pour (setting) point *(lubricating oil)*
Stockpunktbestimmung *f* pour test
Stockpunkterniedriger *m* pour-point depressant
Stockpunktverbesserer *m* pour-point improver
Stockung *f* stop[page], stagnation
Stockwerk *n* storey, floor; *(mine)* stockwork
Stockwerkrahmen *m* multi-storey frame
Stockwerksanzeiger *m* [lift] landing signal lamp, lift car annunciator
Stockwerkschaltung *f* **/ automatische** automatic floor-stop operation *(lifts)*
Stockwerksruf *m* waiting-passenger indicator
Stoff *m* 1. *(ch)* substance, material; matter; 2. *(text)* fabric, cloth; 3. subject; 4. stock, stuff, pulp *(papermaking)*

adsorbierter S. adsorbate
amorpher S. amorphous substance
beschichteter S. coated fabric
eingedickter S. slush stock
ferromagnetischer S. ferromagnetic material
fester S. solid
gelöster S. solute
gewirkter S. knitted cloth
haariger S. nap fabric
leicht bleichbarer S. soft pulp
moirierter S. cloudy fabric
nichtleitender S. non-conducting substance
oberflächenaktiver S. surface-active agent
paramagnetischer S. paramagnetic substance
radioaktiver S. radioactive material
röscher S. *(Am)* fast stuff
schattierter S. ombré fabric
schmieriger S. slow draining stuff
schwer bleichbarer S. hard pulp

Stoffänger *m* pulp saver, save-all *(papermaking)*
Stoffansatz *m* stock preparation *(papermaking)*
Stoffaufbereitung *f s.* Stoffansatz
Stoffauflauf *m* [stock] inlet; flow (breast, stuff) box *(machine)*
dreifach ausgebildeter S. three-compartment headbox
Stoffauflöser *m* pulper *(papermaking)*
Stoffaufrollvorrichtung *f* cloth roll-up
Stoffausschläger *m* refiner *(papermaking)*
Stoffaustauschzahl *f (nucl)* mass-transfer coefficient
Stoffballen *m* bale of cloth
Stoffbereitung *f* stock preparation *(papermaking)*
Stoffbespannung *f* fabric covering *(glider)*
Stoffbreithalterarm *m* cloth-spreader arm
Stoffbütte *f* stuff chest *(papermaking)*
Stoffdichteregler *m* consistence regulator *(papermaking)*
Stoffdichteüberwachung *f* consistence control *(papermaking)*
Stoffdruck *m* textile (fabric) printing
Stoffeintrag *m* pulp furnish *(papermaking)*
Stofffluß *m* flow of the pulp *(papermaking)*
Stoffgrube *f* dump (pulp) pit *(papermaking)*
Stoffherstellung *f* stock preparation *(papermaking)*
Stoffkasten *m* drainer *(papermaking)*
Stoffkette *f* material chain *(chemical kinetics)*
Stoffleimung *f* engine (resin) sizing *(papermaking)*
stofflich material
Stofflichkeit *f* material nature *(particles)*
Stofflöser *m* dissolver
Stoffpumpe *f* pulp pump
Stoffschauer *m* examiner
Stoffübergang *m* mass transfer
Stoffverteilung *f* flow of distribution
Stoffwechsel *m* metabolism
Stoffwechselprodukt *n* metabolic product, product of assimilation
Stoffwechselvorgang *m* metabolic process
Stokes *n* stoke *(unit of kinematic viscosity)*
Stollbock *m* draw beam *(shoe making)*
Stolleisen *n* arm stake *(shoe making)*
stollen to stake; to put calks to horseshoes
Stollen *m* 1. *(mine)* adit, day-drift, tunnel; *(build)* drift, gallery *(cable)*, heading; 2. calkin *(horseshoe)*
Stollenbergbau *m* drift (adit-cut) mining
Stollenbetrieb *m* drift mine
Stolleneingang *m* adit
Stollengrube *f* drift mine
Stollensohle *f (mine)* level
Stollkrücke *f* crutch stake
Stollmaschine *f* staking machine
Stolln *m s.* Stollen 1.
Stollpfahl *m* draw beam
Stolperdraht *m* trip wire, turbulence rod *(model testing)*
Stolzit *m (min)* stolzite
Stop-Bad *n* acid stop bath
Stopfbeutel *m* hawse-bag *(hawse hole)*
Stopfbuchsdichtung *f* stuffing box packing
Stopfbuchse *f* stuffing (packing) box
S. aus Einzelringen stuffing box with separate rings
S. aus Messingblechen brass-foil packing
Stopfbuchsenbehälter *m* / **angeschraubter** loose (inserted) stuffing box
Stopfbuchsenbrille *f* gland
Stopfbuchseneinsatz *m* loose (inserted) stuffing box
Stopfbuchsenmuffe *f* gland, packing
Stopfbuchsenschott *n* stern tube bulkhead, after peak bulkhead
Stopfbuchsensperrdampf *m* gland-sealing service steam
Stopfbuchsensperrluft *f* gland-sealing service air
stopfen 1. to tamp *(railway track)*; 2. *(text)* to darn; 3. *(met)* to stop *(tapping hole)*; to plug
Stopfen *m* plug; screw plug; stopper, cork; bolt
Stopfenmaschine *f* notch gun *(blast furnace)*
Stopfenpfanne *f (met)* teeming (bottom-tap) ladle
Stopfenwalzwerk *n* plug (automatic) mill
Stopfenzug *m* bar (mandrel) drawing
Stopfgarn *n* darning yarn
Stopfhacke *f* tamping pick, packer
Stopfmatte *f (ship)* collision mat
Stopfmeister *m (text)* darner
Stopfnadel *f* darning needle
Stopfschaufel *f* packing shovel, tamping tool
Stoppelauslesemaschine *f* inclined conveyor-type grader
Stoppelfrucht *f* catch crop
Stoppelpflug *m* Kentish broadshare

Stoppelräumung *f* stubble clearing
Stoppelsaat *f* catch crop
Stoppelsturz *m* shallow ploughing; stubble cleaning
Stopper *m* dagger shore *(launching)*
Stopphebel *m* launching trigger
Stopplicht *n (aut)* stop light
Stoppstraße *f* stop arteria
Stoppstromstoß *m* stop impulse *(telephony)*
Stoppuhr *f* stop-watch
Stoppzylindermaschine *f* stop-cylinder press
Stöpsel *m* stopple, stopper; plug, peg *(telephony)*
geschlitzter S. split plug
gesteckter S. plug in position
gezogener S. plug out of box
Stöpseleinrichtung *f* patching facility *(telephony)*
Stöpselflasche *f* stoppered bottle
Stöpselhahn *m* stopcock
Stöpselhülse *f* plug sleeve
Stöpselkasten *m* resistance box *(telephony)*
Stöpselkörper *m* plug body
Stöpselloch *n* plug hole
stöpseln to plug [in]
Stöpselring *m* plug ring
Stöpselringanschluß *m* sleeve wire *(telephony)*
Stöpselschalter *m* plug switch (cut-out)
Stöpselsicherung *f* plug fuse
Stöpselsitz *m* plug seat
Stöpselspule *f* [adjustable] plug coil
Stöpselverbindung *f* plug connection
Störabstand *m* signal-to-noise ratio
Störanfälligkeit *f* incidence of faults, probability of failure, susceptibility to trouble; degree of ruggedness *(instrument)*
Störatom *n* impurity atom
Störbefreiung *f* elimination of interference, static elimination *(radio)*
Störbefreiungskondensator *m* anti-interference capacitor *(radio)*
Störbefreiungsschaltung *f* static eliminator *(radio)*
Störbegrenzer *m* noise limiter
Störbegrenzung *f* noise limitation, limitation of interference
Störbereich *m* disturbance range
Störbeseitiger *m* noise trap *(radio)*
Störbeseitigung *f* debugging *(computer)*
Storchschnabel *m* pantograph [mechanism]
S. für Raumformen three-dimensional pantograph
Storchschnabelgraviermaschine *f* pantographic engraving machine
Storchschnabelmanipulator *m* mechanical pantograph manipulator
Störeffekt *m* interference (parasitic, spurious) effect
Störempfindlichkeit *f* susceptibility to interference
stören to disturb; to interfere with; to jam
absichtlich s. to jam
Störfaktor *m* interference (influence) factor; signal-to-noise ratio
Störfeld *n* interference field; fringing (stray) field
Störfeldstärke *f* noise field intensity
Störfilter *n* noise filter
Störflugzeug *n* nuisance raider
störfrei trouble-free, undisturbed
Störfreiheit *f* freedom from interference
Störfrequenz *f* interfering (spurious) frequency
Störfunkstelle *f* radio jamming station
Störgebiet *n* interference (mush) area *(radio)*
Störgenerator *m* interference generator
Störgeräusch *n* disturbing (static, undesired, additional) noise, parasitics; ripple
atmosphärisches S. atmospheric noise
Störgröße *f* disturbance (load) variable
Störgrößenaufschaltung *f* feedforward control
Störhalbleiter *m* extrinsic (impurity) semiconductor
Störimpuls *m* interference (disturb) pulse
Störkraft *f* disturbance force *(oscillation)*
Störleiter *m s.* Störhalbleiter
Störleitfähigkeit *f* intrinsic (impurity) conductivity
Störleitung *f* defect (impurity) conduction
Störmodulation *f* spurious modulation
stornieren to cancel
Storniertaste *f* cancellation key
Störniveau *n* noise level; impurity (defect) level
Störpegel *m* noise level; background level
Störpegelabstand *m* signal-to-noise ratio
Störpegelmesser *m* interference meter
Störpegelzählung *f* background count
Störquelle *f* noise source
Störrauschen *n* spurious noise
Störreflektor *m* parasitic reflector
Störschutz *m* radio shielding, noise suppression; mains filter, interference eliminator
Störschutzband *n* interference guard band
Störschutzdiode *f (telev)* black spotter
Störschutzfilter *n* radio interference filter
Störschutzkondensator *m* interference (spark, anti-interference, anti-noise) capacitor
Störschwingung *f* spurious (parasitic) oscillation; undesired oscillation
Störsender *m* unwanted (interfering) station, unwanted (interfering) transmitter; jamming station, jammer
Störsignal *n* spurious (parasitic) signal
S. bei Bildaufnahmeröhren shading signal
Störspannung *f* disturbing (noise) voltage
gleichförmig auftretende S. continuous random noise
induktive S. inductive interference voltage
vorübergehende S. sporadic impulse noise
Störspannungsanzeiger *m* noise level meter

Störspannungsmessung *f* noise-level measurement
Störsperre *f* interference suppressor
Störsperrschaltung *f* interference rejection circuit
Störspiegel *m* noise level
Störspitze *f* pulse spike
Störstelle *f* lattice distortion (imperfection, defect), defect, point imperfection *(crystal)*
  kristalline S. crystalline imperfection
  Schottkysche S. Schottky defect
Störstellenabstand *m* extrinsic distance *(semiconductors)*
Störstellenatom *n* impurity atom
Störstellendiffusion *f* impurity diffusion
Störstellenhalbleiter *m* extrinsic semiconductor
Störstellenladung *f* impurity charge
Störstellenleitung *f* extrinsic (defect) conduction
Störstellenverteilung *f* impurity distribution
Störstrahlung *f* stray radiation; perturbing radiation; spurious emission (radiation)
Störstrom *m* disturbing (noise) current
Störstromsuchgerät *n* interference locator
Störton *m* interfering tone
Störübergangsfunktion *f* **der Regelstrecke** recovery curve
Störung *f* 1. *(mach)* breakdown, disturbance, trouble, failure, fault; 2. perturbation *(fluid flow)*; 3. interference *(broadcast)*; 4. *(dat)* malfunction; 5. disarrangement, defect, [lattice] imperfection, [lattice] distortion
  S. der Ionosphäre ionosphere disturbance
  S. des Kristallgitterbaus lattice dislocation
  S. durch den Nachbarkanal adjacent-channel interference
  S. durch Drahtbruch disconnection fault
  S. durch Schwund propagation disturbance
  S. durch Sender im gleichen Kanal co-channel interference
  absichtliche S. jamming
  atmosphärische S. atmospherics, statics
  gegenseitige S. mutual interference
  induktive S. inductive disturbance
  ionosphärische S. ionospheric disturbance
  kristalline S. crystalline imperfection
  lichthofartige S. halo disturbance
  lokalisierte S. localized perturbation
  magnetische S. magnetic disturbance (perturbation)
  meteorologische S. meteorological disturbance
  selektive S. selective interference
  zeitweise S. bug *(computer)*; intermittent interference *(radio)*
Störungen *fpl* / **überlagerte** ripple
Störungsanzeige *f* failure indication
störungsarm anti-interference
Störungsbericht *m* fault report
Störungsbeseitigung *f* fault clearance, failure recovery
Störungsgrad *m* degree of disturbance (interference)
Störungslinie *f (opt)* interference line
Störungsmeldung *f* fault indication (signal); fault report
Störungsrate *f* rate of failure
Störungsrechnung *f* perturbation method
Störungsstelle *f* trouble spot
Störungssuche *f* trouble shooting (tracking, locating); interference location
Störungssucher *m* faultsman, trouble man
Störungstheorie *f* disturbance (perturbation) theory
Störungsüberwachungsplatz *m* exchange testing position *(telephony)*
Störungsunterdrücker *m* parasitic suppressor
Störungsunterdrückung *f* noise suppression
Störungszeichen *n* trouble tone; clutter *(radar)*
Störunterdrückung *f* noise rejection
Störursache *f* cause of trouble, source of variation *(statistical control)*
  nachweisbare S. assignable cause
  nicht nachweisbare S. unassignable cause
  zufällige S. unassignable cause
Störverhalten *n* disturbance response
  S. bei Laststörungen load sensitivity
Störverhältnis *n* signal-noise ratio
Störwelle *f* interference (disturbing, parasitic) wave
Störzipfel *m* parasitic lobe
Stoß *m* 1. impact *(mechanics)*; percussion; shock; collision, encounter; jerk; pulse; 2. joint *(rail)*; junction *(boiler)*; splice; 3. pile, heap; pack *(cards)*, deck; bundle *(paper)*; 4. *(mine)* face; side, wall
  S. erster Art collision of the first kind
  S. zweiter Art collision of the second kind
  elastischer S. elastic impact (collision)
  elektrischer S. impulse, surge; shock
  exzentrischer S. eccentric impact
  fokussierender S. focusing collision
  gerader S. direct impact
  Hoffmannscher S. *(nucl)* burst
  normal zur Schlechtenrichtung verlaufender S. *(mine)* butt
  oberer S. *(mine)* heel of stope *(stoping)*; toe of stope *(bench working)*
  schiefer S. oblique impact
  schräger S. oblique butt joint
  schwebender S. suspended joint
  stumpfer S. butt joint
  superelastischer S. superelastic collision
  unelastischer S. inelastic impact (collision)
  unterer S. *(mine)* heel of stope *(bench working)*; toe of stope *(stoping)*
  unterster S. *(mine)* cutting-out stope

zentrischer S. central impact (collision)
Stoßabscheider *m* baffle separator
Stoßachse *f* axis of thrust
Stoßamplitude *f* surge amplitude
Stoßanregung *f* impact (collision) excitation, excitation by collision
Stoßantwort *f* impulse response
stoßartig discontinuous; intermittent *(drive)*
Stoßaufnahme *f* shock absorption
Stoßausbreitung *f* shock propagation
Stoßbalken *m* putting board *(putting of the shot)*
Stoßbank *f* jointing plane, jointer; push bench
Stoßbankluppe *f* push bench bloom
Stoßbau *m (mine)* working by stopes
stoßbeansprucht subjected to shock loads
Stoßbeanspruchung *f* shock load; impact stress
Stoßbegrenzung *f* shock limitation
Stoßbeiwert *m* coefficient of impact
Stoßbelastung *f s.* Stoßbeanspruchung
Stoßbewegung *f* percussive motion
Stoßblech *n* splice plate
Stoßbohren *n* percussive (churn) drilling
Stoßbohrer *m* percussion (churn) drill; jumper
Stoßbohrmaschine *f* piston drill[ing machine], drifter [machine]
Stoßbohrschuh *m* spudding shoe
Stoßbohrvorrichtung *f* spudder
Stoßboot *n* push-boat, [pusher] tug, puffing tug, tow boat
stoßdämpfend shock-absorbing
Stoßdämpfer *m* shock absorber; dashpot; shock mount
doppeltwirkender S. double-acting shock absorber
hinterer S. rear shock absorber
hydraulischer S. hydraulic (dashpot-type) shock absorber
ölgefüllter S. oil dashpot
vorderer S. front shock absorber
Stoßdämpferbock *m* shock-absorber bracket
Stoßdämpferhalter *m* shock-absorber mounting
Stoßdämpfung *f* shock absorption, collision damping; *(el)* reflection loss
Stoßdaumen *m* actuating cam
Stoßdichte *f* collision density
Stoßdruck *m* 1. shock (impact) pressure; 2. side pressure
Stoßeinrichtung *f (mach)* slotting attachment
Stößel *m (mach)* ram *(shaper)*; *(aut)* cam follower; tappet *(valve)*; *(aero)* valve lifter
Stößelantrieb *m* ram drive
Stößelausgleichsgewicht *n* ram balance weight
Stoßelektron *n* impact electron
Stößelfeder *f* injection spring
Stößelfestklemmung *f* ram lock
Stößelfläche *f* tappet face
Stößelführung *f (mach)* ram way (slide, guide); *(aut)* cam-follower guide; *(aero)* tappet guide *(engine)*
schrägstellbare S. tilting body *(slotter)*
Stößelführungsschraube *f* tappet guiding screw
Stößelhublage *f* ram position
Stößelhublänge *f* ram-stroke length
Stößelhubverlagerung *f* ram positioning
Stößelhubverstellung *f* ram stroke control
Stößelhubzahl *f* **je min** ram reciprocations per min
Stößelklemmung *f* ram clamp
Stößelkopf *m* ram face, tool head, head of the ram
Stößelrolle *f* tappet roller
Stößelrücklauf *m* ram return stroke
Stößelschlitten *m* saddle
Stößelschutz *m* ram guard *(shaper)*
Stößelspiel *n* tappet clearance
Stößelstange *f* [valve] push-rod
Stößelstelleiste *f* ram gib
Stößelteller *m* tappet head
Stoßempfindlichkeit *f* susceptibility to shocks
stoßen to push, to strike, to hit; *(mach)* to slot
waagerecht s. to shape
Stoßentladung *f* impulse discharge
Stoßerregung *f* shock excitation; impact excitation
Stoßfaktor *m* impulse (surge) factor, coefficient of impact
Stoßfänger *m* fender, bumper
stoßfest shock-resistant, shock-proof
Stoßfestigkeit *f* shock resistance; impact (shock) strength
Stoßfläche *f* shock surface *(turbine)*; abutting end
stoßfrei shockless, shock-free, without shock; pulsation-free *(hydraulics)*
Stoßfuge *f* vertical joint *(brickwork)*; joint gap *(clearance)*; *(ry)* expansion joint
Stoßfühler *m* push finger *(loom)*
Stoßfunktion *f* collision function; transient function
Stoßgalvanometer *n* ballistic galvanometer
Stoßgenerator *m* pulse (impulse) generator
Stoßhaltespannung *f s.* Stehstoßspannung
Stoßhäufigkeit *f* collision frequency *(in the ionosphere)*
Stoßherd *m (met)* bump[ing] table, percussion table
Stoßhöhe *f (mine)* bench height
Stoßionisation *f* collision (impact) ionization, ionization by collision
stoßionisiert impact-ionized
Stoßionisierung *f s.* Stoßionisation
Stoßkalander *m* beetling mill
Stoßkissen *n* pad piece *(on life boats)*
Stoßkissenfender *m* swifter
Stoßkomponente *f* shock component

Stoßkraft *f* impact force
Stoßkurzschlußreaktanz *f* transient reactance
Stoßkurzschlußstrom *m* asymmetric short-circuit current
Stoßlade *f* shooting board *(planing)*
Stoßlasche *f* fish (splice) plate, splice piece, butt strap
Stoßleuchten *n* collision light
Stoßlinie *f* line of impact
Stoßlöseneinbruch *m* *(mine)* side-draw cut *(blasting)*
Stoßmagnetisierung *f* flash magnetization
Stoßmatrix *f* collision matrix
Stoßmeißel *m* *(mach)* slotting tool; *(mine)* intermediate knife *(fast plough)*
Stoßmittelpunkt *m* centre of percussion, impact centre
Stoßmühle *f* impact pulverizer
Stoßnaht *f* butt joint
Stoßofen *m* continuous-type furnace, pusher-type furnace
Stoßoszillator *m* blocking oscillator
Stoßparameter *m* impact (collision) parameter
Stoßpegel *m* impulse (basic impulse insulation) level
Stoßplatte *f* baffle, butt strap
Stoßpolarisation *f* impact polarization
Stoßprüfung *f* bumping test
Stoßquerschnitt *m* collision cross section
stoßräumen to push broach
Stoßräumen *n* push broaching
Stoßräumpresse *f* press-type broaching machine, *(Am)* broaching press
Stoßreaktanz *f* transient reactance
Stoßreinigung *f* baffle separation
Stoßrelais *n* rate-of-change relay
Stoßring *m* thrust collar
Stoßsäge *f* frame (span) saw
Stoßschaufellader *m* *(mine)* duckbill
Stoßschlepper *m* pusher tug
Stoßschweißung *f* percussive welding
Stoßsegment *n* *(mine)* side section
Stoßseiten *fpl* *(leath)* rolled bellies
stoßsicher shockproof
Stoßsieb *n* impact screen
Stoßspannung *f* impulse (transient, surge) voltage
Stoßspannungscharakteristik *f* impulse flash-over volt-time characteristic
Stoßspannungsfestigkeit *f* impulse strength
Stoßspannungspegel *m* impulse level
Stoßstange *f* slotting bar *(slotting machine)*; *(aut)* bumper
Stoßstelle *f* joint; abutting surface; intersection; junction; interface
S. im Magnetkreis magnetic joint
Stoßstellen *fpl* / **versetzt liegende** staggered joints
Stoßstrahlung *f* impact radiation
Stoßstrom *m* impulse (surge, transient) current
Stoßtränkrohr *n* *(mine)* infusion tube
Stoßtränkung *f* *(mine)* water infusion
Stoßüberschlag *m* impulse flash-over
Stoßüberschlagsprüfung *f* impulse spark-over test
Stoßüberspannung *f* voltage surge
Stoß- und Zugvorrichtung *f* buffer and draw gear
Stoßverbindung *f* butt joint
Stoßverhalten *n* impulse response
Stoßverlust *m* *(nucl)* collision loss; loss due to shocks *(turbine)*
Stoßversuch *m* impact test
Stoßvorgang *m* surge event
Stoßwahrscheinlichkeit *f* probability of collision
Stoßweglänge *f* collision mean free path
stoßweise intermittent, pulsating; discontinuous
Stoßwelle *f* shock (impulse) wave; impulse, surge
hydromagnetische S. hydromagnetic shock wave
Stoßwellenrohr *n* shock tube
Stoßwinkel *m* angle of shock *(turbine)*
Stoßwulst *m* puddening
Stoßzahl *f* impact coefficient; shock (collision) frequency; collision number
Stoßzug *m* *(ship)* tow
stottern to sputter, to galop *(engine)*
Strabismus *m* *(opt)* strabismus
straff tight, tensioned; taut *(rope, wire)*
Straffer *m* top-sheet, drawsheet
Strafgesetz *n* Criminal Law
Strafgesetzbuch *n* Penal Code
Strafprozeß *m* criminal case
Strafverfahren *n* criminal procedure
Strafvollstreckung *f* execution of judgment
Strahl *m* 1. ray, beam (*s. a.* Strahlen); 2. jet *(fluid flow)*; 3. *(math)* half-line; radius
S. nullter Ordnung zero-order ray
abgelenkter S. deflected ray
achsnaher S. paraxial ray
auftreffender S. impinging (incident) ray
ausfallender S. emergent ray
ausfließender S. outflowing jet
außeraxialer S. abaxial (extraaxial) ray
außerordentlicher S. extraordinary ray
austretender S. emergent ray
direkter S. straight-through ray *(aerial)*; fundamental wave
einfallender S. incident ray
energiereicher S. high-energy beam
feingebündelter S. finely focused beam
gebrochener S. refracted ray
gebündelter S. focused beam
gerichteter S. directed beam
kegelförmiger S. conical beam

kontrahierender S. contracting jet
konvergenter S. convergent beam
ordentlicher S. ordinary ray
projizierender S. *(math)* projector
scharf begrenzter S. sharply defined beam
schiefer S. oblique (slant) ray
schmaler S. narrow beam
Strahlablenkung *f* beam (ray) deflection; deflection of the steam jet *(turbine)*
doppelte S. divided (double) deflection of the steam jet
einseitige S. one-sided deflection of the steam jet
zweiseitige S. *s.* doppelte S.
Strahlablenkungspropeller *m* jet deflection propeller
Strahlabtastung *f* beam scanning
Strahlantrieb *m* jet propulsion; hydro-jet propulsion
Strahlauftastung *f* beam trigger
Strahlaustritt *m* jet exit *(turbo propulsion)*
Strahlaustrittsdruck *m* pressure at the jet exit
Strahlaustrittsgeschwindigkeit *f* jet velocity
Strahlbegrenzung *f* beam limiting; radiation stop
Strahlbehandlung *f* **mit Stahlkies** cloud-burst treatment
Strahlbelastung *f* beam load
Strahlblende *f (opt)* beam stop
Strahlbreite *f* 1. beam width; 2. spread *(aerial)*
Strahldichte *f* radiant intensity per unit area, radiance, radiant (radiation) density, energy density of radiation
Strahldicke *f* diameter of [the] steam jet *(turbine)*
Strahldivergenz *f* beam divergence
Strahldüse *f* [supply-pressure] jet
Strahleinfallsrichtung *f* direction of beam incidence
Strahleinspritzung *f* solid injection
Strahleinstellung *f (telev)* beam positioning
elektrische S. beam control
strahlen to radiate, to emit [rays]
Strahlen *mpl* rays (*s. a.* Strahl)
harte S. hard rays
konvergierende S. converging rays
kosmische S. cosmic rays
positive S. positive (canal) rays
ultrarote S. infrared rays
ultraviolette S. ultraviolet rays
Strahlen... *s. a.* Strahlungs...
Strahlenanalysator *m* radiation analyzer
Strahlenabschirmung *f* radiation barrier (protection, shielding)
Strahlenauffänger *m* beam catcher
Strahlenausbeute *f* X-ray output
Strahlenbahn *f* ray path
Strahlenbegrenzung *f* beam limiting, radiation stop
Strahlenbehandlung *f* radiation treatment (processing)
Strahlenbelastung *f* radiation dose (dosage); exposure
chronische S. chronic exposure
natürliche S. dose of natural background radiation
zulässige S. radiation tolerance
Strahlenbiologie *f* radiobiology
strahlenbiologisch radiobiologic[al]
Strahlenbrechung *f* refraction
Strahlenbündel *n* beam; pencil of rays
paralleles S. beam of parallel rays
Strahlenbüschel *n* pencil of rays
Strahlenchemie *f* radiation chemistry, actinochemistry
strahlenchemisch radiation-chemical, actinochemical
strahlend radiating, radiant, beaming
Strahlendiagnostik *f* diagnostic radiology
Strahlendosis *f* radiation dosage (dose)
Strahlendurchgang *m* passage of rays
doppelter S. double passage of light (rays)
strahlendurchlässig transparent
Strahlendurchlässigkeit *f* transmissibility for radiation; transparency
Strahleneinfall *m* incidence of rays
streifender S. glancing incidence of rays
Strahlenempfindlichkeit *f* radiosensitivity
Strahlenexponierung *f* / **nichtberufsbedingte** non-occupational exposure
Strahlenfänger *m* beam catcher
Strahlenfeld *n* radiation field
strahlenfest radioresistant
Strahlenfilter *n* ray (light) filter
strahlenförmig radial
Strahlengang *m* path of rays, beam (optical, light) path
telezentrischer S. telecentric path of rays
Strahlengefahr *f* radiation hazard
Strahlengefährdung *f* radiation hazard
Strahlengemisch *n* mixed radiation
Strahlengenetik *f* radiation genetics
strahlengenetisch radiation-genetic
Strahlenhärte *f* penetrating power of radiation, radiation hardness
Strahlenhygiene *f* radiation hygiene
strahleninduziert radiation-induced
Strahlenintensität *f* radiation intensity
Strahlenkanal *m* beam hole
S. für biologische Versuche biological hole
Strahlenkater *m* radiation sickness
Strahlenkegel *m* cone (pencil) of rays
Strahlenkrankheit *f* radiation sickness
akute S. acute radiation sickness
Strahlenkunde *f* radiology
Strahlenlaboratorium *n* radiation laboratory
Strahlenletalität *f* radiation lethality

Strahlenmaske *f* mask *(coloured television)*

Strahlenmesser *m* actinometer

Strahlenmeßtechnik *f* technology of radiation measurement

Strahlenmessung *f* radiation measurement; radiometry; dosimetry

Strahlennachweis *m* radiation detection

Strahlennachweisgerät *n* radiation detector (indicator), radiation detection instrument

Strahlenoptik *f* geometric[al] optics

Strahlenphysik *f* radiation physics

Strahlenreichweite *f* range of rays

strahlenresistent radioresistant

Strahlenresistenz *f* radioresistance

Strahlenring *m* halo

Strahlenriß *m* star shake *(wood)*

Strahlenschaden *m* radiation damage; radiation injury

Strahlenschild *m* radiation shield

Strahlenschleuse *f* radiation maze (trap)

Strahlenschranke *f* radioactive relay; radiation barrier

Strahlenschutz *m* radiological protection; radiation barrier (protection); health physics (precaution); nuclear shielding
  chemischer S. chemical protection of radiation; radiation blocker
  physikalischer S. health physics (precaution)

Strahlenschutzarzt *m* radiologist

Strahlenschutzbeauftragter *m* health physicist

Strahlenschutzinstrument *n* survey instrument

Strahlenschutzkleidung *f* protective clothing

Strahlenschutzphysik *f* health physics

Strahlenschutzplakette *f* film badge

Strahlenschutztechnik *f* [radiological] protection technique

Strahlenschutzüberwachung *f* radiation (protection) survey, radiation (protection) control

Strahlenschwächung *f (nucl)* radiation attenuation; *(opt)* beam variation

Strahlenspätschaden *m* late radiation effect

Strahlensterblichkeit *f* radiation mortality

Strahlensticheinrichtung *f (opt)* micro-beam research equipment

Strahlensyndrom *n* radiation syndrome

Strahlenteilung *f (opt)* beam-splitting

Strahlentherapie *f* radiation therapy, radiotherapy

Strahlentod *m* radiation death

Strahlentransformator *m s.* Betatron

Strahlenüberwachung *f* [radiation] monitoring, radiation survey; health physics (precaution)

Strahlenüberwachungsgerät *n* radiation monitor (survey meter), radiation monitor[ing] instrument

strahlenundurchlässig radio[o]paque

Strahlenundurchlässigkeit *f* radioopacity

Strahlenvereinigung *f* fusion of rays

Strahlenverlauf *m* path of rays

Strahlenwirkung *f* radiation effect
  genetische S. genetic effect of radiation

Strahler *m* 1. emitter, radiator, radioactive source; radiation source, source of radiation; irradiator, irradiation source; 2. aerial
  grauer S. *(opt)* grey body
  isotroper S. isotropic radiator
  magnetischer S. magnetic radiator
  nichtselektiver S. non-selective radiator
  schwarzer S. black body, full radiator

Strähler *m (mach)* chaser *(thread)*

Strahlerabstand *m* element spacing *(aerial)*

Strahlerelement *n* elementary aerial

Strahlergruppe *f* multiple aerial

Strahlerkasten *m* aerial unit

Strahlerreihe *f* set of elements *(aerial)*

Strahlerwand *f* multi-element array *(aerial)*

Strahlerweiterung *f (telev)* spread of beam

Strahlerzeile *f* row of radiators *(aerial)*

Strahlfokussierung *f* beam focusing

Strahlführung *f* jet guiding *(turbine)*
  divergente S. divergent jet guiding
  parallele S. parallel jet guiding

Strahlgeometrie *f* beam geometry

Strahlintensität *f* beam intensity

Strahlkies *m* white iron pyrite

Strahlkondensation *f* jet condensation

Strahlkondensator *m* ejector condenser

Strahlkontraktion *f* contraction of jet *(turbine)*

Strahlkonvergenz *f* beam convergence

Strahlkonzentrierung *f* focusing
  magnetische S. magnetic focusing

Strahlläppen *n (mach)* vapour blasting

Strahlmischer *m* jet agitator

Strahlmittel *n (mach)* shotblasting medium, abrasive for blasting

Strahlöffnung *f* scanning aperture

Strahlpentode *f* beam pentode

Strahlprofil *n* beam profile

Strahlpumpe *f* jet pump

Strahlquerschnitt *m* beam aperture (width)

Strahlreinigung *f* blast cleaning, shot blasting

Strahlrohr *n* jet pipe

Strahlrohrdurchlaufblankglühofen *m* radiation-tube continuous bright-annealing furnace

Strahlrohrrahmen *m* spray boom

Strahlrohrregler *m* jet-pipe regulator

Strahlrohrverstärker *m* jet-pipe valve, jet relay

Strahlrücklauf *m* beam return

Strahlruder *n* exhaust (jet, gas) vane

Strahlsaugapparat *m* ejector *(turbine)*

Strahlschutz *m* heat-shield *(of a cathode)*

Strahlspannung *f* beam voltage

Strahlspektrometer *n* beam spectrometer

Strahlsperrspannung *f* blanking voltage

Strahlstärke *f* intensity of radiation, radiant intensity

Strahlsteuerung *f* jet control
Strahlstreuung *f* beam divergence
Strahlstrom *m* beam current
Strahltetrode *f* beam tetrode
Strahltrieb *m* jet propulsion
Strahltriebwerk *n* jet power plant, jet (reaction) engine
Strahlung *f* radiation; emission
S. aus dem Reaktor reactor [emitted] radiation
aktinische S. actinic radiation
anregende S. exciting radiation
antistokessche S. antistokes radiation
außerfokale S. stem radiation
charakteristische S. characteristic radiation
durchdringende S. penetrating radiation
elektromagnetische S. electromagnetic radiation
energiereiche S. energetic (hard) radiation
gerichtete S. directional radiation
gestreute S. scattered radiation
graue S. grey [body] radiation
harte S. hard radiation
heterogene S. heterogeneous radiation
homogene S. homogeneous radiation
infrarote S. infrared radiation
inkohärente S. incoherent (spontaneous) radiation
ionisierende S. ionizing radiation
kohärente S. in-step radiation
kontinuierliche S. steady (continuous) radiation
konvergente S. convergent radiation
kosmische S. cosmic rays (radiation)
kurzwellige S. short wavelength radiation
langwellige S. long wavelength radiation
monochromatische S. monochromatic radiation
polarisierte S. polarized radiation
radioaktive S. radioactive radiation
reflektierte S. reflected radiation
schwarze S. black-body radiation
sekundäre kosmische S. secondary cosmic radiation
sichtbare S. visible radiation
thermische S. thermal radiation
ultrarote S. infrared radiation
ultraviolette S. ultraviolet radiation
unsichtbare S. invisible radiation
zusammengesetzte S. complex radiation
Strahlungs... *s. a.* Strahlen...
Strahlungsabsorption *f* radiation absorption; photoelectric absorption
Strahlungsanalyse *f* radiation analysis
Strahlungsanregung *f* radiation excitation
Strahlungsäquivalent *n* / **fotometrisches** luminous efficiency
Strahlungsart *f* type of radiation
Strahlungsausbeute *f* yield of radiation; G-value
Strahlungsbedingung *f* radiation condition
Strahlungsbereich *m* range of radiation
Strahlungsbreite *f* radiation width
Strahlungscharakteristik *f* radiation characteristic; field pattern, lines of forces
Strahlungsdämpfung *f* radiation damping
Strahlungsdetektor *m* radiation detector (detection instrument)
Strahlungsdiagramm *n* radiation pattern; field pattern
Strahlungsdichte *f* radiant intensity per unit area, radiance, radiant (radiation) density, energy density of radiation
Strahlungsdiffusion *f* imprisoned radiation
Strahlungsdosis *f* radiation dose (dosage); absorbed dose
berufsbedingte S. occupational radiation dosage
Strahlungsdruck *m* radiation pressure
Strahlungsdurchgang *m* radiation transmission
Strahlungseinfang *m* radiative capture
Strahlungseinheit *f* radiation (cascade) unit
Strahlungselement *n* radiating element *(aerial)*
Strahlungsempfänger *m* radiation detector (detection instrument)
Strahlungsempfindlichkeit *f* radiosensitivity
Strahlungsenergie *f* radiant (radiation) energy
Strahlungsenergiedichte *f* energy density of radiation
Strahlungserscheinung *f* radiation phenomenon
strahlungsexponiert radiation-exposed, irradiated
Strahlungsfeld *n* radiation (radiated) field
Strahlungsfläche *f* radiating surface; emitting area
Strahlungsfluß *m* radiant flux (power), radiation flux
Strahlungsflußdichte *f* radiant flux density
Strahlungsfunktion *f* relative spectral energy distribution
Strahlungsgefährdung *f* radiation hazard
Strahlungsgesetz *n* radiation law, law of radiation
Strahlungsgleichgewicht *n* radiative (radiation) equilibrium
Strahlungsgröße *f* radiometric quantity
Strahlungsgürtel *m* radiation zone, trapped-radiation belt
äußerer S. outer radiation zone
innerer S. inner radiation zone
Strahlungsheizfläche *f* radiation heating surface
Strahlungsheizung *f* radiant (radiation) heating
eingebaute S. built-in radiation heating
Strahlungsindikator *m* radiation indicator
strahlungsinduziert radiation-induced
Strahlungsintensität *f* *(nucl)* radiation intensity, intensity of radiation; radiance of a surface, luminous flux per unit area

Strahlungsionisation *f* radiation ionization
Strahlungskeule *f* radiation lobe *(aerial)*
Strahlungskonstante *f* radiation constant
Strahlungsleistung *f* emissive power, radiant intensity; radiated energy (power)
strahlungslos non-radiative *(Auger effect)*
Strahlungsmagnetisierung *f* radiation magnetization
Strahlungsmanometer *n* radiometric gauge
Strahlungsmaximum *n* radiation peak
Strahlungsmenge *f* quantity of radiation
Strahlungsmeßgerät *n* radiation meter (measuring instrument)
Strahlungsmeßtechnik *f* technology of radiation measurement
Strahlungsmessung *f* radiation measurement; dosimetry; radiometry
Strahlungsniveau *n* radiation level
Strahlungsnormal *n* radiation standard
Strahlungsofen *m* radiation (indirect-arc) furnace
  elektrischer S. electric radiator
Strahlungspegel *m* radiation level
Strahlungspyrometer *n* radiation pyrometer
Strahlungsquant *n* photon
Strahlungsquelle *f* radiation (radioactive) source, source of radiation; radiographic source; radiant
  geschlossene S. encapsulated (sealed) source
  offene S. unsealed (free) source
  punktförmige S. point source
  umschlossene S. sealed source
Strahlungsrekombination *f* radiation recombination
Strahlungsschaden *m* irradiation damage
Strahlungsschutz *m* nuclear shielding; radiological safety
Strahlungsstärke *f* intensity of radiation, radiant intensity
Strahlungsstoß *m* radiation burst
Strahlungsstrom *m* radiant flux
Strahlungstechnik *f* radiation technology
Strahlungstheorie *f* theory of radiation
Strahlungsthermografie *f* projection thermography
Strahlungstod *m* radiation death
Strahlungstransport *m* radiation streaming (transport)
Strahlungsuntergrund *m* radiation background
Strahlungsverlust *m* radiative (radiation) loss
Strahlungsvermögen *n* intrinsic radiance; radiating capacity, radiating (radiative) power
Strahlungswachstum *n* radiation-induced growth
Strahlungswärme *f* radiating (radiant) heat
Strahlungswarngerät *n* radiation monitor[ing instrument], radiation survey meter, monitor
Strahlungswiderstand *m* radiation resistance *(aerial)*
Strahlungswinkel *m* angle of radiation
Strahlungswirkungsgrad *m* radiation efficiency *(aerial)*
Strahlungszentrum *n* radiative centre
Strahlungszipfel *m* lobe *(aerial)*
Strahlungszusammensetzung *f* composition of radiance
Strahlverbreiterung *f* spreading, widening *(of the beam)*
Strahlverlauf *m* beam path (trajectory)
Strahlverschiebung *f* beam positioning
Strahlvortrieb *m* jet propulsion
strahlwassergeschützt hose-proof
Strahlwinkel *m* gas angle *(gas turbine)*
Strahlwirkung *f* jet action *(turbine)*
Strahlwobblung *f* spot wobble
Strahlzerlegungsspiegel *m* beam-splitting mirror
Strahlzerstäubung *f* jet spraying
Strähne *f* hank
  abgebundene S. skeined hank
Straight-run-Benzin *n* straight-run petrol (*Am* gasoline)
Strak *n (ship)* sheer
Straken *n* fairness
Straklinie *f* fairline *(of the hull)*
Strand *m* shore, beach, strand
stranden to run ashore
Strandgut *n* wreck
  angeschwemmtes S. jetsam
Strandhafen *m* dry harbour
Strandräuber *m* wrecker
Strandwade *f* beach seine *(net)*
Strang *m* 1. rope, cord; 2. *(text)* skein; 3. *(met)* billet; 4. *(el)* phase-belt; 5. track *(rails)*; 6. line *(piping)*
  S. / im in rope-form
  S. ohne Ende endless rope
Strangausbreiter *m* rope expander
Strangbleiche *f* rope bleaching
Strangbleicherei *f* bleaching in rope form
Strangfärben *n* dyeing in the hank, skein dyeing
Strangform *f* rope form
Stranggarn *n* hank yarn
stranggefärbt dyed in the yarn, skein-dyed
Strangguß *m* continuous casting
Stranggußverfahren *n* continuous casting method
Strangklemme *f (el)* line terminal
Strangpresse *f* extrusion press, extruder
  hydraulische S. *(plast)* stuffer
strangpressen to extrude
Strangpressen *n* extrusion
Strangpreßmasse *f (plast)* extrusion-moulding material
Strangpreßprofil *n* extruded shape, extrusion
Strangpreßverfahren *n (met)* extrusion process; *(plast)* extrusion moulding

Strangschlichtmaschine *f (text)* hank sizing machine
Strangspannung *f* phase voltage [of a winding]
Strangtrockenapparat *m (text)* skein drier; rope drier
Strangwalze *f (met)* strand roll
strapazierfähig for hard wear
Straß *m* strass
Straße *f* 1. street; road, highway; route; 2. *(mach)* line *(linked)*; train, mill *(rolling)*
Straßenabziehmaschine *f* road scraping machine
Straßenanordnung *f (met)* arrangement of the mill train
Straßenanschnitt *m* road shelf
Straßenaufbruchhammer *m* road drill (breaker)
Straßenaufreißer *m* road ripper
Straßenausbesserung *f* road repair
Straßenausrüstung *f* street (road) furniture (equipment)
Straßenbahn *f* tram[way], *(Am)* street car
  elektrische S. electric tramway
Straßenbahndepot *n* tram depot
Straßenbahngleise *npl* tram[way] tracks
Straßenbahnhof *m* tram depot
Straßenbahnoberleitung *f* tram mains
Straßenbahnoberleitungsdraht *m* overhead tramway wire
Straßenbahnschiene *f* tramway rail
Straßenbahnwagen *m* tram-car
Straßenbau *m* road (highway, street) building, road (highway, street) construction, road-making
Straßenbauingenieur *m* highway engineer
Straßenbauschlacke *f* road building slag
Straßenbaustelle *f* road site
Straßenbaustoff *m* paving material
Straßenbefestigung *f* pavement, paving
Straßenbelag *m* pavement, paving; surface, surfacing
Straßenbeleuchtung *f* [overhead] street lighting
Straßenbepflanzung *f* roadside planting
Straßenbeschotterung *f* road gravelling
Straßenbettung *f* road bed
Straßenbrücke *f* highway (road) bridge
Straßendamm *m* causeway
Straßendecke *f* pavement, paving; surface, surfacing
Straßendeckenmischung *f* road mix
Straßeneinlauf *m* gully
Straßenfahrzeug *n* road vehicle
Straßenfertiger *m* road finisher (finishing machine), spreader-finisher
Straßengabelung *f* Y-junction
Straßengang *m (aut)* high range
straßengängig roadholding
Straßengraben *m* road ditch
Straßenhobel *m* [blade, road] grader
Straßenkampf *m (mil)* block-by-block fighting
Straßenkehrmaschine *f* street (motor) sweeper
Straßenklinker *m* paving brick, paviour
Straßenknoten *m* intersection
Straßenkocher *m* road kettle
Straßenkreuzung *f* cross roads; junction, intersection
  S. in getrennten Ebenen road interchange
Straßenkrone *f* road crown
Straßenlage *f (aut)* road-holding, roadability
Straßenmarkierung *f* markings
Straßenmarkierungsfarbe *f* traffic paint
Straßenmarkierungsgerät *n* line marker
Straßenmeister *m* road overseer
Straßenmeisterei *f* highway depot
Straßennetz *n* road net[work]
Straßenpflaster *n* pavement, paving
Straßenplanierer *m* [road, blade] grader
Straßenquerschnitt *m* [road] cross section
Straßenrad *n* road wheel
Straßenreinigung *f* road cleansing
Straßenreparatur *f* road repair
Straßenrinne *f* gutter [channel], road channel
Straßenscheitel *m* vertex of a road
Straßenschleifmaschine *f* road grinder
Straßenschotter *m* road (crushed) stone
Straßensprengwagen *m* street washer (flusher, sprinkler), road flusher
Straßenteer *m* road tar
Straßentransport *m* road (highway) transport
Straßentunnel *m* vehicular (road) tunnel
Straßenüberführung *f* [overpass] bridge
Straßenunterbau *m* road foundation (bed)
Straßenverhalten *n (aut)* roadability, roadholding
Straßenverkehr *m* road traffic
Straßenverkehrsordnung *f* highway code
Straßenverkehrssicherheit *f* highway safety
Straßenverkehrssignal *n* **an schienengleichen Wegübergängen** level-crossing road signal
Straßenverkehrstechnik *f* traffic engineering
Straßenverkehrsunfall *m* road traffic accident
Straßenverkehrszeichen *n* road (traffic) sign
Straßenwalze *f* road roller
Straßenwärter *m* lengthsman
Strategie *f* strategy
  gemischte S. mixed strategy *(theory of games)*
  optimale S. optimal strategy *(theory of games)*
  originäre S. pure strategy *(theory of games)*
Stratigrafie *f* stratigraphy
Stratonaut *m* stratonaut
Stratosphäre *f* stratosphere
Stratuswolke *f* stratus cloud
Strauchegge *f* bush-harrow
Strauchpflanzgerät *n* bush planter
Streb *m (mine)* longwall face, panel
  S. im Rückbau retreating face
  S. im Vorwärtsbau advancing face
  einflügeliger S. single-unit face

fallender S. dip face
gestundeter S. stationary face
in Betrieb befindlicher S. working face
mit Schrämlader ausgerüsteter S. power-loaded face
schwebender S. rise face
streichender S. strike face
streichender S. mit zusätzlichem Einfallen zum Alten Mann strike face dipping towards the goaf
streichender S. mit zusätzlichem Einfallen zum Kohlenstoß strike face dipping towards the coal-face
zweiflügeliger S. double-unit panel
Strebausbau *m* face support
Strebbau *m* longwall working, *(Am)* longwalling
S. mit breitem Blick broadwall method
Strebbruch *m* face collapse, heavy (roof) fall *(face)*
Strebe *f* brace, strut, angle (diagonal) brace; shore; *(el)* tie *(mast)*, stay; *(mine)* cockermeg, [pit] prop, stull
Strebebogen *m* arch[ed] buttress
Strebenfachwerk *n* strut frame (bracing)
Strebepfeiler *m* [arc, arched, flying] buttress, counterfort
Strebestempel *m* anchor post
Strebfortschritt *m* face advance
Strebfront *f* / **söhlige** face level
Strebstellung *f* face positioning
Strebvorfeld *n* zone ahead the face
Strebzuschnitt *m* face layout
Streckband *n (text)* drawing (drawn) sliver
Streckbank *f* wire stretching machine
streckbar ductile, [ex]tensible
Streckbarkeit *f* ductility, extensibility
Streckbug *m* good board
Streckdraht *m* stretched wire
Strecke *f* 1. distance; segment of a line, line segment; 2. *(mine)* drift, gallery, gate, heading, road, tunnel, *(Am)* entry; 3. *(text)* draw[ing] frame *(spinning)*; 4. route, *(ry)* line; 5. *(cont)* plant, process
S. für Materialtransporte *(mine)* supplies road
S. im Alten Mann *(mine)* brush, gob-road
S. im Liegendgestein *(mine)* footwall drift
S. mit Begleitort *(mine)* double entry
ausgesetzte S. *(mine)* spur *(of road)*
durch den Versatz getriebene S. *(mine)* scouring
durchfahrene S. day's run
einfallende S. *(mine)* dip, incline
eingleisige S. *(ry)* single track
flache S. *(mine)* incline
schwebende S. *(mine)* jinny
steil ansteigende S. *(mine)* brow
Streckeisen *n (leath)* breaking iron
strecken 1. to stretch, to yield, to lengthen, to elongate; to draw out *(forging)*, to rough down; 2. *(ch)* to admix, to dilute; 3. to break *(hide)*
Strecken *n* 1. drawing out *(forging)*; 2. drafting, drawing *(spinning)*; 3. plastic flow *(material)*
Streckenabschnitt *m (ry)* section of line; *(aero)* route segment
Streckenabstellung *f (text)* stopping of the drawing frame
Streckenabzweigung *f (mine)* [roadway] junction, [roadway] bifurcation
Streckenarbeiter *m* track-layer
Streckenauffahren *n (mine)* drifting, tunnel driving, tunnelling
S. mit strossenartiger Arbeitsweise head-and-bench tunnelling
Streckenauffahrmaschine *f* heading (tunnelling) machine
Streckenausbau *m (mine)* roadway (tunnel) support
Streckenband *n (text)* drawing (drawn) sliver
Streckenbefeuerung *f* airway lighting
Streckenbelastung *f* volume (density) of traffic, load on section
Streckenblock *m* section blocking
selbsttätiger S. automatic block
Streckenbogen *m (mine)* [roadway] arch
S. mit abnehmbarem Fußteil loose-footed roadway arch
breitbeiniger S. splay-legged roadway arch
geradbeiniger S. straight-legged roadway arch
Streckenbruch *m (mine)* heavy fall *(roadway)*
Streckendamm *m (mine)* gateside (roadside) pack
Streckendämpfung *f* path attenuation
Streckendämpfungsmesser *m* transmission measuring set
Streckendämpfungsmessung *f* transmission efficiency test
Streckenentblockung *f* section clearing
Streckenfernmeldeverkehr *m* point-to-point communication
Streckenfernsprecher *m* portable telephone set
Streckenfeuer *n* course light
Streckenförderband *n* gate conveyor
Streckenförderung *f* haulage
Streckenfunkfeuer *n* course beacon
Streckengetriebe *n (mine)* side spiling
Streckengewölbe *n* roadway arch *(rock mechanics)*
Streckenlast *f* uniform[ly distributed] load
Streckenmessung *f* distance measurement
Streckennachrichtenverkehr *m* point-to-point communication
Streckenort *n* forehead, roadhead
Streckenpfeiler *m (mine)* entry stump
Streckenraum *m* roadway excavation

Streckenschalter *m* section circuit-breaker, sectionalizing (track, overhead-line) switch; gate-end box (switch unit)
S. mit Hornkontakten horn-break switch
Streckenschutz *m* pilot protection
S. mit Hilfsleitung pilot-wire protection
S. mit Trägerfrequenzverbindung carrier-current protection
Streckensenkmaschine *f* dinting machine
Streckenstellwerk *n* centralized control point
Streckensteuerung *f* straight-line control *(numerical)*
Streckentrenner *m (el)* section isolator, air section break
Streckentriebwerk *n (text)* draw frame gearing
Streckenunterhaltung *f* roadway maintenance
Streckenverblockung *f (ry)* section blocking
Streckenvortrieb *m* drifting, tunnel driving, tunnelling
Streckenvortriebsmaschine *f* heading (roadway ripping, tunnelling) machine
Streckenwärter *m* lengthsman
Streckenzimmerhauer *m (mine)* backbye deputy
S. beim Pfeilerbau bordroom-man
Streckenzimmerling *m (mine)* backbye deputy
Strecker *m* header, bondstone *(brickwork)*
Streckerschicht *f* heading (bonding) course *(brickwork)*
Streckfestigkeit *f* resistance to stretching
Streckformen *n (plast)* stretch forming
Streckformverfahren *n (plast)* drape (slip) forming
Streckkaliber *n* drawing pass
Streckmaschine *f* 1. stretching machine; 2. *(text)* draw frame; 3. *(leath)* breaking machine
Streckmetall *n* expanded metal
Streckmittel *n* diluent; extender; filler
Streckpassage *f (text)* passage of drawing
Strecksäge *f* framed hand saw
Streckspannung *f* yield stress
Streckspinnverfahren *n* stretch-spinning process
Streckstahl *m* expanded steel
Streckung *f* axial elongation; tension *(yarn)*
Streckwalze *f* 1. *(met)* cogging-down roll; 2. *(text)* drawing roller
Streckwalzenpaar *n (text)* pair of drawing rollers
Streckwerk *n (text)* drawer, draw[ing] frame
S. der Spulenbank fly-frame drawing rollers
geneigtes S. inclined drawing rollers
Streckzwirnmaschine *f* draw-twister
Streckzylinder *m (text)* drawing roller
streichbar paintable, brushable
Streichbaum *m (text)* back rail; *(Am)* back rest [of a loom]; scudding beam *(leather)*
Streichblech *n* mould-board, breast; trowel *(moulding)*
Streichbrett *n* mould-board, breast
Streicheisen *n* scudding knife *(leather)*
streichen to brush, to paint; to coat *(paper)*
Streichen *n* spread coating *(leather)*; *(mine)* strike
allgemeines S. general trend
Streichfähigkeit *f* brushing property
streichfertig ready mixed *(paint)*
Streichgarn *n* [carded] wool yarn, woollen thread (yarn)
Streichgarngewebe *n* carded woollen goods
Streichholz *n* match
Streichkalander *m* spreading calender
Streichlack *m* brushing lacquer
Streichlinie *f (geol)* line of strike
Streichmaschine *f* coater, coating machine *(paper)*; spread coater, rubber spreader
Streichpfahl *m* fender pile (pole)
Streichpresse *f* coating press *(paper)*
Streichschiene *f* tailpiece
Streichung *f* deletion
Streifband *m* news (postal) wrapper
Streifbandumlegemaschine *f* newspaper wrapping machine
Streifblech *n* wiper
streifen to stripe; to graze, to touch lightly
Streifen *m* strip [chart]; [paper] tape; stripe; fringe; streak, band; barré, barry mark *(fault)*
endloser S. endless strip
magnetischer S. magnetic strip *(preprinted on documents)*
vertikaler schwarzer S. *(telev)* vertical black band
vertikaler weißer S. *(telev)* vertical white band
Streifenabschwächer *m* strip attenuator; flap (vane) attenuator *(waveguide)*
Streifenabsorptionsteiler *m* vane attenuator *(waveguide)*
Streifenabtastung *f* rectilinear scanning
Streifenaufwicklung *f* paper take-up
Streifenbild *n* fringe pattern
Streifenbildung *f* striation
Streifenblatt *n* strip chart
Streifenblech *n* strip steel, steel band
Streifenboot *n* coastal defence craft
Streifendiagramm *n* strip chart
Streifendichtung *f* strip sealing
Streifendruck *m* tape print
Streifendrucker *m* tape printer
Streifenentladung *f* striated (stratified) discharge
Streifenfundament *n* continuous (wall) footing, strip foundation (footing)
Streifengeber *m* [perforated] tape transmitter
Streifengeschwindigkeit *f* chart speed
Streifenhobelmesser *n* stripe iron
Streifenkavitation *f* streak cavitation *(fluid flow)*
Streifenkode *m* bar code *(of automatic character recognition)*
Streifenkohle *f* banded [bituminous] coal

Streifenleitung *f* strip line, microstrip *(waveguides)*
Streifenlesekopf *m* tape reader (reading head)
Streifenlocher *m* tape punch, perforator
kartengesteuerter S. card-to-tape converter
Streifenschreiber *m* strip-chart recorder
Streifensicherung *f* [metal-]strip fuse
Streifenstauchprüfung *f* flat crush *(paper)*
Streifenvorschub *m* tape (paper) feed
Streifenwiderstand *m* strip resistor
Streifenzerlegung *f* zone television
Streifenzieher *m* tape puller
streifig striated; *(text)* barry *(fault)*; *(geol)* interstratified
strengflüssig sluggish, viscous
Streu *f (ag)* litter, bedding
Streu... *s. a.* Streuungs...
Streuamplitude *f* scattering amplitude
magnetische S. magnetic scattering amplitude
relativistische S. relativistic scattering amplitude
Streuband *n* [scattering] range; spread band
Streubereich *m* margin of error *(statistics)*
Streubrechung *f* scattered (dispersed) refraction
Streudüse *f* spray nozzle
Streudüsenzerstäuber *m* nozzle sprayer (atomizer)
Streuecho *n* scatter echo
Streueffekt *m* scattering effect
Streuelektron *n* scatter (stray) electron; scattered electron
Streuemission *f* side emission
streuen to spread, to strew; to scatter; *(el)* to leak, to stray; to disperse *(e. g. electron rays)*; to broadcast
Streufaktor *m* leakage factor; fringing factor
Streufeld *n* leakage (stray) field; fringing field
elektrostatisches S. stray electrostatic field
magnetisches S. stray magnetic field
Streufläche *f* scattering surface
Streufluß *m* leakage (stray) flux
magnetischer S. magnetic leakage
Streufrequenz *f* scattering frequency
Streugrenze *f* limit of scattering
Streuimpedanz *f* leakage impedance
Streuindikatrix *f* indicatrix of diffusion
Streuinduktivität *f* leakage inductance
Streuintensität *f* scattering intensity
Streukapazität *f* stray capacitance
Streukegel *m* cone of dispersion
Streukoeffizient *m* scattering (dispersion, leakage) coefficient
Streukopplung *f* stray (spurious) coupling
Streukörper *m* diffuser *(illumination engineering)*
Streukreis *m (opt)* circle of confusion
Streulicht *n* adventitious (scattered, diffused) light
Streulichtmethode *f* scattered-light method
Streumaterial *n* scatterer, scattering medium
Streumatrix *f* scattering matrix
Streumittel *npl* scattering formulation
Streuneutron *n* scattered neutron; stray neutron
Streupfad *m* / **magnetischer** magnetic leakage path
Streuprozeß *m* scattering process
Streuquerschnitt *m* scattering cross section
Streureaktanz *f* leakage reactance; [quadrature-axis] subtransient reactance
Streureflexion *f* scattered (dispersed) reflection
Streusand *m* blinding sand
Streuscheibe *f (el)* diffusing screen
Streuschirm *m* diffusing screen
Streuschwund *m* scatter fading
Streuspalt *m* leakage air-gap *(magnet)*
Streuspannung *f* reactance (leakage) voltage
Streustrahlen *mpl* stray (scattered) rays
Streustrahlenblende *f* anti-diffusing screen
bewegliche S. moving (reciprocating) grid
feststehende S. stationary grid
Streustrahlübertragung *f* scattering transmission
Streustrahlung *f (ph)* scattered (stray) radiation
Streustrahlungsdosis *f* scattered radiation dose
Streustrahlverbindung *f* / **troposphärische** tropospheric forward scatter
Streustrahlverbindungsstrecke *f* troposcatter hop
Streustrom *m* dispersion (stray, leakage) current, current leakage
Streutheorie *f* theory of scattering
Streutransformator *m* constant-current transformer
Streuüberlagerung *f* stray interference
Streuung *f (nucl)* scattering; scattering event; straggling, spread; fluctuation; *(math)* scatter, dispersion, variance, variation; leakage
S. des Dampfstrahls fanning of the steam jet *(turbine)*
S. des Lichtes light scattering
S. von Funkwellen scattering of radio waves
akustische S. acoustic scattering
Braggsche S. Bragg scattering
doppelt verkettete S. double-linkage leakage
einfache S. single scattering
elastische S. elastic scattering
elektromagnetische S. electromagnetic leakage
inkohärente S. incoherent scattering
innere S. internal scattering
instrumentell bedingte S. instrumental straggling
kohärente S. coherent scattering
magnetische S. magnetic leakage (dispersion)
mehrfache S. multiple scattering
mittlere quadratische S. *(math)* mean square deviation
nichtkohärente S. incoherent scattering

statistische S. statistical straggling
unelastische S. inelastic scattering
Streuungs . . . *s. a.* Streu . . .
Streuungsbereich *m* zone of dispersion
Streuungsbild *n* dispersion diagram
Streuungskegel *m* dispersing cone, cone of dispersion
Streuungskurve *f* scatter curve
Streuungslinse *f* diverging lens
Streuungsmaß *n* measure of dispersion
Streuungsquerschnitt *m* scattering cross-section
Streuungstheorie *f* theory of scattering
Streuungsverlust *m* dispersion loss *(turbine)*; *(el)* leakage
Streuungszerlegung *f* *(math)* variance analysis
Streuverbindung *f* scatter system
Streuverlust *m* *(el)* leakage, scattering loss
Streuvermögen *n* diffusion factor *(illumination engineering)*; throwing power; scattering power
Streuvorgang *m* scattering event
Streuweg *m* leakage path
magnetischer S. magnetic leakage path
Streuweglänge *f* scattering mean free path
Streuwelle *f* scattered wave
Streuwinkel *m* scattering angle, angle of scattering; angle of divergence, divergence angle
Streuwirkung *f* scattering effect
Streuzahl *f* leakage coefficient; dispersion coefficient
Streuzentrum *n* scattering centre
Strich *m* 1. line *(drawing)*; 2. *(math)* accent; 3. stroke *(fraction)*; 4. dash *(Morse code)*; 5. *(min)* streak; 6. coating
Shefferscher S. NOT BOTH, Sheffer stroke *(logical operator)*
Strichabstand *m* distance between lines
Strichappretieren *n* napping
Strichappretur *f* nap finish
Strichauflösung *f* streak resolution
Strichdicke *f* width of line *(technical drawing)*
stricheln to dash; to shade *(hatching)*
Strichendmaß *n* hair-line gauge block
Strichfokus *m* line focus
Strichfokusröhre *f* line-focus tube
Strichklischee *n* line block
Strichkreuz *n* cross-hair[s]
Strichlänge *f* dash length *(telegraphy)*
Strichlehre *f* length gauge
Strichlinie *f* dash[ed] line
Strichliste *f* frequency tabulation
Strichlistenauswertung *f* frequency table
Strichmarke *f* division (locating) mark; hairline
Strichmaß *n* line standard
Strichperforierung *f* slot perforation
Strichplatte *f* graticule
strichpunktiert dash-dotted
Strichpunktlinie *f* dash-and-dot line
Strichraster *m* 1. *(print)* line screen; 2. *(telev)* bar pattern
Strichzeichnung *f* line drawing (illustration, diagram)
Strick *m* line, rope
stricken to knit
Strickfuß *m* knitting butt
Strickgarn *n* knitting yarn
Strickheber *m* knitting cam
Strickhöhe *f* knitting level
Strickkurve *f* knitting curve
Strickleiter *f* rope ladder
Strickmaschine *f* knitting machine
Stricknadel *f* knitting needle
Strickschloß *n* cam assembly (system) *(knitting machine)*
hinteres S. back lock
vorderes S. front lock
Strickstellung *f* latch-clearing position *(needle)*
Strickware *f* knit goods
Strickwolle *f* knitting wool
Striegel *m* *(ag)* curry-comb; flexible harrow
Strieme *f* 1. *(text)* wale *(fault)*; 2. striation *(material)*
Striktion *f* striction
Stringer *m* stringer
strippen to strip
Stripper *m* stripper
Stripperkran *m* stripper (ingot stripping) crane
Stripperwerk *n* stripper mechanism
Stripping *n* *(nucl)* stripping
Strippingfilm *m* stripping emulsion (film)
Strippingquerschnitt *m* stripping cross section
Strippingreaktion *f* *(nucl)* stripping reaction
Stroboskop *n* stroboscope
stroboskopisch stroboscopic[al]
Stroboskopröhre *f* stroboscopic tube
Strobotron *n* strobotron
Stroh *n* straw
Strohballenpresse *f* straw baler
Strohbinder *m* straw trusser
Strohdach *n* thatched roof
Stroheinleger *m* straw charger
Strohfaserplatte *f* straw-board
Strohfaserstoff *m* straw pulp
Strohmischpapier *n* mixed straw paper
Strohmischpappe *f* mixed straw-board
Strohpapier *n* straw paper
Strohpappe *f* straw-board
gelbe S. semichemical straw-board
Strohpresse *f* straw baler
Strohquetscher *m* straw crusher
Strohreißer *m* straw cutter
Strohschüttler *m* straw shaker
Strohseil *n* straw rope
Strohstoff *m* [semichemical] straw pulp
Strohstoffpappe *f* semichemical straw-board
Strohzellulose *f* straw pulp

Strom *m* 1. *(el)* current; 2. stream *(fluid flow)*
S. / gegen den against the tide
S. / unter active, live
S. gleichbleibender Richtung unidirectional current
S. in Durchlaßrichtung forward current
S. in Sperrichtung reverse current
abgegebener S. available current
abgehender S. outgoing current
abklingender S. decaying current
akustischer S. acoustical current
ankommender S. incoming current
ausgeglichener S. smoothed current
dielektrischer S. dielectric current
dynamischer S. mechanical current
eingeprägter S. impressed current
eingeschwungener S. steady-state current
eingespeister S. applied current
einwelliger S. sinusoidal (harmonic) current
elastischer S. elastic flow
elektrischer S. [electric] current
Faradischer S. faradic current
fotoelektrischer S. photoelectric current, photocurrent
galvanischer S. voltaic (galvanic) current
geglätteter S. smoothed current
gleichgerichteter S. rectified current
gleichphasiger S. in-phase current
induzierter S. induced current
intermittierender S. intermittent current
kommutierter S. commutated current
konstanter S. constant (stable) current
kritischer S. critical current
lichtelektrischer S. photoelectric current, photocurrent
maximal zulässiger S. maximum permissible current
mehrwelliger S. harmonic current
nacheilender S. lagging current
nachfolgender S. follow-on current
nichtstationärer S. non-stationary current
niedergespannter S. low-voltage current
oszillierender S. oscillating current
parasitärer S. parasitic current
periodischer S. periodic current
phasenverschobener S. out-of-phase current
positiver S. forward current
primärseitiger S. primary current
pulsierender S. pulsating current
pyroelektrischer S. pyroelectric current
quasistationärer S. quasistationary current
raumladungsbegrenzter S. space current
schwacher S. low current
sekundärseitiger S. secondary current
sinusförmiger S. harmonic (sinusoidal, sine) current
stationärer S. steady[-state] current
stetiger S. continuous current
tetanisierender S. tetanizing current
theoretisch zu erwartender S. prospective current
thermoelektrischer S. thermocurrent
überlagerter S. super[im]posed current
vagabundierender S. vagabond (stray) current
verketteter S. interlinked current
voreilender S. leading current
wahrnehmbarer S. appreciable current
wattloser S. wattless (reactive) current
welliger S. ripple current
stromab downstream *(fluid flow)*
Stromabfall *m* current drop, decrease (decay) of current
Stromabgabe *f* current delivery (output)
Stromabgleich *m* current balance
stromabhängig current-dependent; current-controlled, current-operated
Stromabnahme *f* decrease of current; current collection
Stromabnehmer *m* sliding contact, trolley, current collector; user of electric power
Stromabnehmerbock *m* trolley base
Stromabnehmerbügel *m* sliding (collector) bow
Stromabnehmerbürste *f* brush *(of the collector)*
Stromabnehmerfänger *m* pole retriever
Stromabnehmerkopf *m* trolley head
Stromabnehmerlöffel *m* contact slipper
Stromabnehmerrolle *f* trolley wheel
Stromabnehmerstange *f* trolley pole, *(sl)* boom
Stromabschaltung *f* power cut, disconnection of the power supply
stromabwärts downstream, down-stream, down river
Stromaggregat *n* generating set
Stromamplitude *f* current amplitude
Stromänderung *f* current change, change in current
Stromänderungsgeschwindigkeit *f* rate of current change
Stromanregung *f* current response
Stromanstieg *m* rise (increase, build-up) of current
Stromanstiegsrelais *n* surge relay
Stromanzeiger *m* current indicator, galvanometer
Stromart *f* kind (type, nature) of current
stromauf upstream
Stromaufnahme *f* current consumption, taking of current
Stromaufnahmewert *m* charging rate *(battery)*
stromaufwärts upstream
Stromausbeute *f* electrical (current) efficiency
anodische S. anode efficiency
katodische S. cathode efficiency
Stromausbreitung *f* current propagation
Stromausgleich *m* current balance; compensation of current

Stromausgleichsbatterie *f* floating trickle
Strombahn *f* current path
Strombauch *m* current antinode (loop)
Strombedarf *m* current demand; current requirement
Strombegrenzer *m* current (demand) limiter
Strombegrenzung *f* current limitation
Strombegrenzungsdrossel *f* current-limiting coil (reactor, inductor)
Strombegrenzungssicherung *f* current-limiting fuse
Strombegrenzungswiderstand *m* current-limiting resistor, load-limiting resistor
Strombelastbarkeit *f* current[-carrying] capacity, carrying (electrical-load) capacity
Strombelastung *f* electrical (power, current) loading, electrical (power, current) load, load-carrying capacity
  zulässige S. safe carrying capacity
Strombereich *m* current range
strombetätigt current-operated
Stromdamm *m s.* Uferdamm
Stromdämpfung *f* current attenuation
Stromdichte *f* current density
Stromdichteverlauf *m* current density distribution
Stromdieb *m* pirate power-consumer *(person or equipment)*
Stromdifferentialschutz *m* current phase-balance protection
stromdurchflossen current-carrying, live
Stromdurchfluß *m* current flow (passage)
Stromdurchgang *m* current passage
Stromdurchlässigkeit *f* permeability to current
Ströme *mpl* **entgegengesetzter Richtung** unlike currents
  entgegengesetzte S. opposing currents
  gleichgerichtete S. rectified (like) currents
  symmetrische S. balanced currents
  vagabundierende S. stray (leakage, circulating) currents
  verzweigte S. branched currents
Stromeinheit *f* unit of current
Stromeintrittszone *f* cathodic (negative) area
Stromeisen *n* current magnet
Stromempfindlichkeit *f* current sensitivity
strömen to flow, to stream, to pass
Stromentnahme *f* current drain
Stromerhöhung *f* rise of current
Stromerzeuger *m* current generator
Stromerzeugung *f* electric power generation, generation of current
Stromerzeugungsstation *f* electric generating station
Stromfaden *m* 1. *(el)* current path (propagation); 2. stream line *(turbine)*; liquid filament, thread (filament) of stream
Stromfernmeßgerät *n* teleammeter
Stromfluß *m* current flow
Stromflußwinkel *m* [cut-off] angle of current flow
stromführend [a]live, active; current-carrying; conducting
Stromfunktion *f* 1. current function; 2. stream function
Stromgegenkopplung *f* current feedback
Stromgeschwindigkeit *f* current velocity (speed, rate)
stromgespeist current-fed
stromgesteuert current-controlled
Stromgleichrichter *m* [electric] current rectifier
Stromhafen *m* estuary (river) harbour
Stromimpuls *m* pulse of current, current impulse
  abklingender S. decaying current impulse
  nicht abklingender S. steady current impulse
Stromintensität *f* current intensity, intensity (strength) of current
Stromkern *m* series stator
Stromklemme *f* feeder clamp
Stromknoten *m* current node
Stromkomponente *f* current component
Stromkonstante *f* galvanometer constant
Stromkorrektur *f* current correction
Stromkorrosion *f* anodic corrosion
Stromkreis *m* [electric] circuit
  S. der Hilfsbetriebe auxiliary circuit
  abgestimmter S. tuned circuit
  angeschlossener S. branch circuit
  aperiodischer S. aperiodic circuit
  äquivalenter S. equivalent circuit
  äußerer S. external circuit
  dreiphasiger S. three-phase circuit
  geschlossener S. closed (complete) circuit
  gestörter S. faulty circuit
  induktionsfreier S. non-inductive circuit
  induktiver S. inductive circuit
  induktivitätsfreier S. non-inductive circuit
  induzierender S. inducing circuit
  innerer S. internal circuit
  magnetischer S. magnetic circuit
  offener S. incomplete circuit
  überlagerter S. superposed circuit
  ungeerdeter S. insulated circuit
Stromkreise *mpl* / **in Reihe (Serie) geschaltete** series circuits
  parallelgeschaltete S. shunt circuits
Stromkreistheorie *f* theory of circuits
Stromlauf *m* current flow
Stromlaufplan *m* [schematic] wiring diagram, [schematic] circuit diagram
Stromleiter *m* conductor
Stromleitung *f* current conduction
Stromlieferung *f* current supply, supply of current
Stromlinie *f* stream line; line of flow *(turbine)*
Stromlinienform *f* streamline form

stromlinienförmig stream-lined
Stromlinienruder *n* hydrofoil rudder
stromlos dead, currentless
Stromlosigkeit *f* absence of current
Strommenge *f* amount of current, quantity of electricity
Strommeßbereich *m* current (amperage) range
Strommesser *m* 1. *(el)* ammeter, amperemeter, current measuring instrument; 2. current meter *(hydraulics)*
aperiodischer S. aperiodic ammeter
Strommessung *f* current measurement
Stromnennwert *m* current rating
Stromnetz *n* electric mains (supply line)
Stromnulldurchgang *m* zero current
Strompfad *m* current circuit; current path
Strompfeiler *m* water (river) pier *(bridge)*
Stromphase *f* current phase
Stromquadratmeßgerät *n* current square meter
Stromquelle *f* current (power) source, source of supply
Stromrauschen *n* current noise
Stromrauschpegel *m* current noise level
Stromregelung *f* current regulation
Stromregler *m* current regulator
Stromrelais *n* current relay
Stromresonanz *f* current resonance
Stromresonanzkreis *m* parallel resonant circuit
Stromresonanzpunkt *m* anti-resonance peak
Stromrichter *m* [static] converter; inverter; rectifier
Stromrichtertransformator *m* converter (rectifier) transformer
Stromrichtung *f* sense (sign) of current, direction of [current] flow, current direction
Stromrichtungsumkehr *f* current reversal
Stromrückgang *m* decrease of current
Stromrückkopplung *f* current feedback
Stromrücklauf *m* return of current
Stromrückleitungskabel *n* return cable
Stromsammelschiene *f* current collecting busbar
Stromschicht *f* diagonal course *(brickwork)*
Stromschiene *f* power supply bar, contact (conductor) rail
seitliche S. side conductor rail
Stromschienenfestpunkt *m* conductor-rail anchor
Stromschienenlichtraumprofil *n* contact-rail gauge
Stromschienensystem *n* conductor-rail system
Stromschienentrennung *f* contact-rail gap
Stromschienenverankerung *f* conductor-rail anchor
Stromschiffahrt *f* river sailing (traffic)
Stromschluß *m* circuit closing
Stromschreiber *m* current recorder
Stromschritt *m* signal element *(telephony)*
Stromschrittfolge *f* current combination
Stromschutz *m* current protection
gerichteter S. directional current protection
ungerichteter S. non-directional current protection
Stromschwankung *f* current (power) fluctuation
Strom-Spannungs-Beziehung *f* current-voltage relation[ship]
Strom-Spannungs-Charakteristik *f* current-voltage characteristic
Strom-Spannungs-Kennlinie *f* current-voltage characteristic
Strom-Spannungs-Kurve *f* current-voltage characteristic
Stromspeisung *f* current feed *(aerial)*
Stromsperre *f* power cut, disconnection of the power supply
Stromspitze *f* maximum (peak) current
Stromsprung *m* step of current
Stromspule *f* current coil
Stromstabilisator *m* current stabilizer
Stromstabilisierung *f* current stabilizing
Stromstärke *f* current intensity (strength), amperage
Stromsteuerung *f* current drive
Stromstoß *m* current impulse (surge, rush)
Stromstoßdauer *f* burst duration; impulse period *(automatic telephony)*
Stromstoßrelais *n* impulse relay
Stromstrecke *f* reach *(of a navigable river)*
Stromteiler *m* current divider, transition coil
Stromtor *n* thyratron
Stromträger *m* current (electrical) carrier
Stromtragfähigkeit *f* current-carrying capacity
Stromtransformator *m* current transformer
Stromübergang *m* leakage *(current)*
Stromüberlagerung *f* current superposition
Stromübersetzung *f* current transfer
Stromübertragungsfaktor *m* response to current
Stromübertragungsglied *n* conductor
Stromübertragungskonstante *f* current transmission coefficient
Stromumkehr *f* reversal of current
Strömung *f* [fluid] flow, motion of a fluid; flux, stream
S. in einer Richtung one-way flow
S. in offenen Gerinnen open-channel flow
S. in zwei Richtungen two-direction flow
S. mit Kavitation cavity flow
divergierende S. *(Am)* broomy flow
drallfreie S. irrotational flow
laminare S. laminar (streamline, viscous) flow
molekulare S. molecular flow
nichtstationäre S. non-stationary flow (motion), unsteady flow (motion)
rücklaufende S. backwash
stationäre S. stationary (steady) flow, stationary (steady) motion

turbulente S. turbulent (sinuous) flow
verlustlose S. unresisted flow, flow without loss
wirbellose S. irrotational flow
Strömungsbild *n* flow pattern (diagram)
Strömungsdoppelbrechung *f* birefringence of flow, flow birefringence
Strömungsdüse *f* flow nozzle
Strömungsenergie *f* energy of [out]flow
Strömungserscheinungen *fpl* flow phenomena
strömungsfähig fluid
Strömungsfigur *f* flow pattern
Strömungsgebilde *n* flow formation
Strömungsgeschwindigkeit *f* velocity (rate) of flow, [fluid] flow velocity, stream velocity
Strömungsgetriebe *n* hydraulic (fluid) transmission, hydraulic (fluid) drive
Strömungskraftübertragung *f* fluid transmission of power
Strömungskupplung *f* hydraulic (fluid) coupling
Strömungslehre *f* fluid dynamics, fluid flow
Strömungslinie *f* stream (flow) line
Strömungsmaschine *f* flow (fluid-kinetic) machine
Strömungsmenge *f* flow rate
S. je Zeiteinheit rate of volume flow per unit time, volume rate of flow
Strömungsmesser *m* flow (current) meter *(hydraulics)*
Strömungspotential *n* streaming potential
Strömungsregelventil *n* flow regulating valve
Strömungsrichtung *f* direction of flow
Strömungsschalter *m* jet-type circuit breaker; flow switch
Strömungsteilung *f* flow split
Strömungsumwandler *m* torque converter
Strömungsverlust *m* flow loss
Strömungswächter *m* airfoil (vane) flowmeter, flow indicator
Strömungswiderstand *m* drag
Strömungsziffer *f* liquid-flow coefficient
Stromunterbrecher *m* current interrupter
Stromunterbrechung *f* interruption of current
Stromvektor *m* current vector
Stromverband *m* diagonal bond *(brickwork)*
Stromverbrauch *m* power (current) consumption
Stromverbraucher *m* current consumer
Stromverbrauchszähler *m* energy (electric-supply) meter
Stromverdrängung *f* current displacement (loss); skin effect
Stromverdrängungsfaktor *m* current displacement factor
Stromverhältnis *n* current ratio
Stromverlauf *m* current path (flow)
Stromversorgung *f* current (electrical, power) supply
konstante S. stabilized power supply
Stromversorgungsanlage *f* generating plant
Stromversorgungsgerät *n* power supply unit
Stromversorgungskabel *n* power cable
Stromversorgungsnetz *n* public (supply) mains
Stromversorgungsschwankung *f* power supply variation
Stromverstärkung *f* current amplification; current gain *(transistor)*
Stromverstärkungsfaktor *m* current ratio (amplification factor) *(transformer)*
Stromverstimmung *f* pushing
Stromverstimmungsmaß *n* pushing figure
Stromverteilung *f* current (power) distribution
Stromverteilungsgesetz *n* / **Belowsches** Below's law of current distribution
Stromverteilungsrauschen *n* partition noise
Stromverteilungstafel *f* distribution board
Stromvervielfachung *f* current multiplication
Stromverzweigung *f* electric (current) network, branching network of wires, current branching
Stromwaage *f* current (electrodynamic) balance; current weigher
Stromwächter *m* automatic controller
Stromwandler *m* current transformer
S. mit stromproportionaler Zusatzmagnetisierung auto-compounded current transformer
S. mit Zusatzmagnetisierung compound-wound current transformer
Stromwärme *f* electric heat
Stromwechsel *m (telec)* current reversal
Stromweg *m* current path
Stromwelle *f* current wave
Stromwellenform *f* current wave form
Stromwender *m* commutator, current-reverser
Stromwendermaschine *f* commutating machine
Stromwenderschritt *m* commutator pitch
Stromwendespannung *f* commutating voltage
Stromwendung *f* commutation, current reversal
Stromwinkel *m* angle of flow
Stromzähler *m* current (electric-supply) meter
Strom-Zeit-Kennlinie *f* time-current characteristic *(e. g. of fuses)*
Stromzuführung *f* current (electrical) supply; leading-in wire, power (supply) lead
Stromzunahme *f* rise of current
Stromzweig *m* branch circuit
Strontianit *m (min)* strontianite
Strontium *n* strontium
Strontiumalter *n* strontium age
Strontiumchlorid *n* strontium chloride
Strontiumhydroxid *n* strontium hydroxide
Strontiumkarbonat *n* strontium carbonate
Strontiumnitrat *n* strontium nitrate
Strontiumoxid *n* strontia, strontium oxide
Strontiumsulphat *n* strontium sulphate
Strophantin *n* strophantin
Stropp *m* sling, garland; strap
Strosse *f (mine)* bench, lift *(open cut)*; face

Strossenband *n* bench belt conveyor
Strossenbau *m* bottom-stoping, underhand stoping
  gelöster S. Cornish method of underhand stoping
Strossenböschungswinkel *m* slope of bench
Strossenfuß *m* bank toe
Strossenhöhe *f* bank height
Strossenkante *f* bank crest (edge)
Strudel *m* eddy, whirl, vortex
Struktur *f (met, ch)* structure; *(text, geol)* texture
  adrige S. nerve structure
  amorphe S. amorphous structure
  blättrige S. lamellar (foliated) texture
  dendritische S. dendritic pattern
  faserige S. fibrous structure; texture
  fehlgeordnete S. imperfect structure
  flächenzentrierte S. face-centred structure
  geordnete S. ordered structure
  gestörte S. imperfect structure
  lamellare S. lamellar structure
  linienhafte S. line structure
  raumzentrierte S. body-centred structure
  zellenartige S. cellular (honeycomb) structure
Strukturanalyse *f* structure analysis
Strukturänderung *f* structural change
Strukturchemie *f* structural chemistry
strukturell structural
Strukturermüdung *f* structural fatigue
Strukturfaktor *m* structure factor *(X-rays)*; atom[ic] form factor
Strukturfehler *m* structural fault (defect)
Strukturformel *f* structural (constitutional) formula
strukturiert textured
Strukturisomerie *f* structural isomerism
strukturlos structureless, amorphous
Strukturviskosität *f (plast)* intrinsic viscosity
Strumpflänge *f* leg length
Strumpf- und Trikotwaren *fpl* hosiery
Strumpfwirkerei *f* hosiery knitting
Strunkdurchmesser *m s.* Schaftdurchmesser
Strusen *pl* silk waste
Strychnin *n* strychnine
Stuck *m* stucco
Stück *n* piece; workpiece, part; lump; component
  kurzes S. short length
Stuckdecke *f* stucco (floating) ceiling
stückeln to cut into pieces
Stückelstanze *f* small clicking press
Stückerz *n* knockings, lump ore, lumps
Stückfärben *n* piece dyeing
Stückfertigungsgrundzeit *f* floor-to-floor time
Stuckgips *m* plaster of Paris, flower of gypsum
Stückgut *n* general (package, parcelled) cargo, parcels, packages, part loads, piece goods, individual parts (articles), *(Am)* package freight, *(Am)* less carloads (than carload freight)
Stückgutfrachter *m* general trader
Stückgutladung *f* general cargo
Stückgutscheider *m* scalper
Stückgutsendung *f* part-load consignment
stückig lumpy
Stückkalk *m* lump lime
Stückkohle *f* coarse (large, lump, round) coal
Stückkoks *m* large coke
Stückkosten *pl* piece-production cost, cost per component
Stückliste *f* list of parts, part list; bill of materials
Stücklohn *m* piece wages, wage for piece work
Stückmattierung *f (text)* delustring on the piece
Stückpfortenklappe *f* / **halbe** *(ship)* half port
Stuckplastik *f* stucco moulding
Stückseife *f* bar soap
  synthetische S. synthetic bar
Stückware *f (text)* piece goods
stückweise in pieces
Stückzahl *f* number of pieces, batch-quantity
  S. je Stunde pieces (rate) per hour
Stückzeit *f* piece time, total production time for one part; machining time per component
Stückzeitermittlung *f* piece-rate setting
Studentverteilung *f* Student distribution
Studiengebühr *f* fee
Studienjahr *n* academic year
Studio *n* studio
  nachhallfreies S. dead studio
  nachhallreiches S. live studio
  schalltotes S. dead studio
Studioausrüstung *f* studio equipment
Studiokulisse *f* movable scene
Studiosendung *f* studio broadcast
Studioverstärker *m* programme amplifier
Stufe *f* 1. *(build)* stair, step; 2. stage, digit; degree; grade; 3. rank *(tensor)*; 4. level *(telephony)*
  erste S. *(rock)* first (primary) stage, mother missile
  letzte S. *(rock)* final (last) stage, final (last) step
  oberste S. *(rock)* last (final, upper) stage, last (final) step
  polarografische S. polarographic wave
  untere S. *(rock)* bottom step, lower (earlier, primary) stage
  zweite S. *(rock)* second stage, daughter missile
Stufenanker *m* double-winding armature
Stufenantrieb *m* cone pulley drive
stufenartig graduated, graded, stepped
Stufenbarren *m* uneven parallel bars
Stufenbatterie *f* cascade battery
Stufenblock *m* guide block
Stufenbohrer *m* stepped drill
  S. für dreifach abgesetzte Bohrung three-stepped drill

Stufenboot *n* hydroplane, step-bottomed boat, saw-formed boat
Stufenbreite *f (build)* going
Stufendrehschalter *m* rotary stud switch
Stufeneichleitung *f* step attenuator
Stufenfilter *n* step filter
stufenförmig stepped
Stufenfotometer *n* step photometer
Stufenfräsen *n* step milling
Stufenfundament *n* benched foundation
Stufengetriebe *n* change-speed drive
stufengewickelt *(el)* bank-wound
Stufengitter *n* echelon grating
Stufengruppe *f* group of stages *(turbine)*
Stufenhärtung *f* interrupted quenching
Stufenhobel *m* stepped-knife plough
Stufenisolation *f* stepped insulation *(transformers)*
Stufenkegel *m* step[ped] wedge
Stufenkeil *m* step[ped] wedge
Stufenkolonne *f (nucl)* plate column
Stufenkopplung *f* intervalve coupling
Stufenleiter *f* step ladder
Stufenlinse *f* echelon lens
stufenlos stepless
Stufenmultiplikator *m* step multiplier
Stufenplan *m* diagram of stages *(turbine)*
Stufenpresse *f* transfer press
Stufenprozeß *m* step process
Stufenrädergetriebe *n* train of gears; change-speed geared drive
Stufenrakete *f* stage (staged, composite, multi-stage) rocket, step-rocket
Stufenregelung *f* stepped control
Stufenreichweite *f* reach *(e. g. of a relay)*
Stufenrelais *n* two-step relay
Stufenrohrkessel *m* boiler with stepped flue
Stufenrost *m* step[ped] grate
Stufenschalter *m* on-load tap changer, multi-point (step, stepping, polystage, multiple-contact, multi-contact, multiple-throw) switch
Stufenschaltung *f* step switching
Stufenscheibe *f* cone pulley, stepped [cone] pulley, step cone
Stufenscheibenantrieb *m* cone [pulley] drive
  S. / mit step-cone pulley-driven
  S. mit Zahnrädervorgelege cone pulley drive with back gear
Stufenschott *n* recessed (stepped) bulkhead
Stufenschütz *n* tapping contactor
Stufensektor *m* stepped sector
Stufensenker *m* step counterbore
Stufensprung *m* progressive ratio; increment
Stufenspule *f* stepped coil
Stufensteller *m (el)* step-control[ler]
Stufentransformator *m* regulating (step-down) transformer
Stufentrennung *f (rock)* stage separation
Stufentrennungshöhe *f* altitude of stage separation, staging altitude
Stufentrocknung *f* gradual drying
Stufentrommel *f* stepped drum *(turbine)*
Stufenturbine *f* stage turbine
Stufenverschlüßler *m* binary weighing AD-converter
Stufenversetzung *f* edge dislocation
Stufenverstärker *m* cascade amplifier
Stufenverstärkung *f* gain of the stage
Stufenwalze *f (met)* stepped roll
Stufenwäsche *f* fractional washing *(papermaking)*
Stufenwäscher *m* scrubber with stages
stufenweise stepwise, step by step, successively, gradually
Stufenwicklung *f* bank[ed] winding
Stufenwiderstand *m* decade resistor
Stufigkeit *f (cont)* granularity
Stufung *f* increments *(values)*; grading *(grain size)*
  S. der Längenzunahme increments of increase of length
Stuhl *m* 1. chair, seat; 2. *(text)* loom
Stuhlabfall *m* loom waste
Stuhlbaum *m* loom beam
stuhlfertig in the loom state
Stuhlflug *m* loom fly
Stuhlindustrie *f* chair-making industry
Stuhllehne *f* back of a chair
Stuhllehnenbiegemaschine *f* chair back bending machine
stuhlroh loom-finished
Stuhlsitzdreh- und -falzmaschine *f* seat turning and folding machine
Stuhlware *f* loom-state fabric
Stuhlzettel *m (text)* tie-up
Stukkateur *m* plasterer
stumm silent, quiet; mute
Stummabstimmung *f* quiet (silent) tuning, squelching
Stummelblock *m* butt ingot
Stummfilm *m* silent film
Stummfilmprojektor *m* silent projector
stumpf dull *(tool edge)*; blunt; obtuse *(angle)*
Stumpf *m (math)* frustum
  schiefer S. *(math)* ungula
stumpfgeschweißt butt welded
Stumpfheit *f* dullness
Stumpfkegel *m* truncated cone
Stumpfnaht *f* butt weld *(welding)*
stumpfschweißen to butt-weld
Stumpfschweißen *n* butt welding
Stumpfschweißung *f* butt weld[ing]
Stumpfstoß *m* butt (butted, square, abutting) joint
  schräger S. scarf joint *(soldering)*
Stumpfung *f* chamfer *(corner of cutter tooth)*
Stumpfverzahnung *f* stub teeth

Stumpfwerden *n* dulling *(tool)*
stumpfwinklig obtuse-angled
Stunde *f* hour
stunden to abandon *(temporarily)*, to stop
Stundenachse *f* hour axis
Stundenantrieb *m* hour-axis drive
Stundenbetrieb *m* one-hour duty
Stundenkilometer *n* kilometre per hour
Stundenkreis *m* hour circle
Stundenleistung *f* production rate per hour, [machine's] hourly output, capacity *(pieces per hour)*; power consumption per hour
Stundenlohn *m* hourly earning
Stundenplan *m* duty chart
Stundenschmelzleistung *f* melting rate [in tons per hour]
Stundenzähler *m* **[für Traktoren]** hour recorder
Stundenzeiger *m* hour hand
Stundenzugkraft *f (ry)* tractive effort at hourly rating
Stundungsfrist *f* period of extension
Sturm *m* gale, storm
 geomagnetischer S. geomagnetic storm
 magnetischer S. magnetic storm
Sturmball *m* black ball
Sturmboot *n* assault boat
Sturmdeckschiff *n* awning-decked vessel, awning deck vessel; *(Am)* hurricane deck vessel
Sturmgeschütz *n* assault gun
Sturmhaken *n* window fastener (stay), casement stay
Sturmlaterne *f* hurricane lamp, wind lantern
Sturmpanzer *m* assault tank
Sturmpfahl *m* pile fender
sturmsicher storm-proof
Sturmsignal *n* storm signal
Sturmwarnung *f* gale (storm) warning
Sturmwarnungsdienst *m* gale (storm) warning service
Sturz *m* 1. fall; 2. *(build)* lintel, header; 3. *(mine)* batter *(prop)*
 gewölbter S. arched lintel
stürzen *(math)* to transpose
Stürzen *n* **der Gicht** fall of the scaffold *(blast furnace)*
Sturzflug *m* [nose] dive
Sturzflugautomatik *f* automatic dive control
Sturzflugbremse *f* dive brake
Sturzhelm *m* crash helmet
Sturzkampfflugzeug *n* dive bomber
Sturzkocher *m* tilting digester
Sturzmaschine *f* folding machine *(shoe)*
Sturzsee *f* heavy (green, toppling) sea, breaker[s]
 grundliche S. green sea
Sturzspirale *f (aero)* spinning nose-dive
Sturzträger *m (build)* lintel beam, breastsummer
Sturzversatz *m (mine)* gravity stowing
Sturzwellen *fpl* breakers
Stützarm *m* bracket
Stützbalken *m* shore; brace, stringer
Stützblech *n* gusset
Stützblock *m* backing block
Stützbock *m* [planer] jack
 S. der Kolbenstange pedestal of the piston rod guide
Stütze *f (build)* pillar, post, column, stanchion, shore; *(mach)* support, rest; brace, bracket
 betonummantelte S. *(build)* cased column
 eingespannte S. *(build)* fixed column
 mehrteilige S. *(build)* spaced column
stutzen to prune
stützen to support; to prop *(vertically)*; to back *(horizontally)*; to retain; to shore *(wall)*
Stutzen *n* pruning
Stutzen *m* nozzle; connection *(pipe)*, feed pipe, connecting piece, socket, union; stem *(electron valve)*
 S. auf der Druckseite delivery connection
Stutzenbruch *m* fracture of a stand pipe (nozzle)
Stützenfundament *n* column footing
Stützenisolator *m s.* Stützisolator
Stützenleiter *f (ship)* pillar ladder
Stützensenkung *f (build)* displacement (lowering) of support, support movement
Stützer *m s.* Stützisolator
Stützfläche *f* supporting surface
Stützfuß *m* **/ ausfahrbarer** outrigger *(truck crane)*
Stützhülse *f* carrying bracket
Stützisolator *m* supporting (spreader-type, pin-type, pin, base) insulator
Stützkonsole *f* supporting bracket
Stützkraft *f* supporting force (reaction), support reaction
Stützlager *n* step bearing
Stützlänge *f* span [length]
Stützlinie *f* resistance (thrust) line
Stützmauer *f* retaining wall
Stützmoment *n* moment at support
Stützmotorsteuerung *f (aero)* power-assisted control
Stützpfeiler *m* pillar, bearer
Stützplatte *f* bracket plate; backing plate *(battery)*
Stützplattenbodenwrange *f* [deep] bracket floor *(double bottom)*
Stützpunkt *m* 1. point of support, bearing surface; pivot *(rotation)*; 2. *(aero)* air base; 3. *(mil)* base, stronghold
Stützpunktschiff *n* depot vessel
Stützring *m* back-up ring (washer)
Stützrohr *n* support tube
Stützrolle *f* idler pulley
Stützsäule *f* strut
Stützschale *f* backing shell *(bearing)*
Stützscheibe *f* back-up ring (washer)

Stützsieb *n* backing wire
Stützstoff *m* stabilizer
Stützung *f* / **elastische** elastic supports
Stützwalze *f (met)* back[ing]-up roll
Stützwand *f* retaining wall
Stützweite *f* span
Stützwinkel *m* support bracket
Stützzapfenlager *n* pivot bearing
Styphninsäure *f* styphnic acid
Styptikum *n* styptic
Styrol *n* styrene
Styrolharz *n* styrene resin
Styrolkunststoff *m* styrene plastic
Styrolplast *m s.* Styrolkunststoff
subatomar subatomic
Suberat *n* suberate
Suberinsäure *f* suberic acid
Subharmonische *f* subharmonic
Subkorn *n* subgrain
Subkorngrenze *f* sub-boundary
subkutan subcutaneous
Subkutanpräparat *n* hypodermic product
Sublimat *n* sublimate
Sublimation *f* sublimation
Sublimationskurve *f* sublimation curve, hoar frost line
Sublimationslinie *f s.* Sublimationskurve
Sublimationspunkt *m* sublimation point (temperature)
Sublimationswärme *f* [latent] heat of sublimation
sublimierbar sublimable
Sublimierbarkeit *f* sublimability
sublimieren to sublime
submikroskopisch submicroscopic
Subminiaturbauweise *f* subminiature construction *(electronics)*
Subminiaturgerät *n* subminiature unit
Subminiaturröhre *f* subminiature valve
Subnormale *f* subnormal
Substandard *m* substandard
substantiell material
substantiv substantive *(dye)*
Substantivfarbe *f* substantive dye
Substanz *f* substance, material, matter; essence
  fadenbildende S. filament-forming substance
  körnige S. granulated substance
  oberflächenaktive S. surface-active material
  waschaktive S. active detergent
Substituent *m* substituent
substituieren to substitute, to replace; *(dat)* to extract
substituiert / dreifach triple-substituted
Substitution *f* substitution, replacement; extraction
Substitutionsatom *n* substitutional atom
Substitutionsleitung *f* substitution conduction
Substitutionsmischkristall *m* substitutional solid solution
Substrat *n* substratum; substrate
Substratosphäre *f* substratosphere
Subtangente *f* subtangent
Subtrahend *m* subtrahend
subtrahieren to subtract
Subtraktion *f* subtraction
Subtraktionszähler *m* balance counter
subtraktiv subtractive *(colour)*
Subtraktivverfahren *n* subtractive process
Subtransientreaktanz *f* **des Längsfeldes** direct-axis subtransient reactance
Subtransientzeitkonstante *f* **des Längsfeldes bei Kurzschluß** direct-axis subtransient short-circuit time constant
  S. des Längsfeldes bei Leerlauf direct-axis subtransient open-circuit time constant
Subunternehmer *m (build)* sub-contractor
Succinat *n s.* Sukzinat
Succinimid *n* succinimide
Suchanker *m* drag anchor
Suchantenne *f* sensing aerial
Suche *f* search; *(mine)* prospecting; *(opt)* view finding; tracing, locating *(fault)*
Sucheinrichtung *f* view finder
Suchempfänger *m* search receiver *(radio)*
suchen to look for; to search; to hunt for; to trace *(e. g. a fault)*; to find [out] *(e. g. a trouble)*
Sucher *m* locator; detector; *(aut)* spot lamp; selector; *(opt)* view-finder
  S. für unbewegte Konten detector of inactive accounts
  elektronischer S. electronic view finder
  optischer S. optical view finder
Suchereinrichtung *f* view finder *(photography)*
Sucherfernrohr *n* spotting scope, finding (finder) telescope
Sucherkontakt *m* cat's whisker
Sucherlampe *f* spot-lamp bulb
Sucherobjektiv *n* finder lens
Sucherschacht *m (opt)* waist-level finder
Sucherschuh *m (opt)* accessory shoe
Suchgerät *n* locator, detector, searcher
Suchmethode *f (dat)* searching method
Suchradargerät *n* search radar
Suchscheinwerfer *m* searchlight, spot light
Suchschiff *n* searcher *(vessel)*
Suchspannung *f* search voltage
Suchspule *f* exploring (pick-up, search) coil
Suchverfahren *n* search method *(computer)*
Suchzähler *m* probe counter
Suchzeichen *n* inquiry signal
Suchzeit *f* searching time; access time *(computer)*
Sud *m* decoction
Süd south
Sudkessel *m* brew kettle
südmagnetisch south magnetic
Südost south-east
Südpol *m* south pole

Südwest south-west
Suezkanalvermessung *f* Suez Kanal tonnage
Sukzinat *n* succinate
Sulfaminsäure *f* sulphanic (thionamic) acid
Sulfanilsäure *f* sulphanilic acid
Sulfat *n* sulphate
sulfatieren to sulphate, to sulphatize
Sulfatisierofen *m* sulphating roasting furnace
Sulfatzellstoff *m* sulphate [wood] pulp
  S. aus Nadelholz softwood sulphate pulp
Sulfid *n* sulphide
Sulfideinschluß *m* sulphide inclusion
Sulfiderz *n* sulphide ore
sulfidieren to sulphidize
Sulfidiertrommel *f* xanthating churn
sulfidisch sulphidic
Sulfidschwefel *m* sulphide sulphur
Sulfinsäure *f* sulphinic acid
Sulfit *n* sulphite
Sulfitablauge *f* sulphite (waste) liquor
Sulfitverfahren *n* sulphite process
Sulfitzellstoff *m* sulphite wood pulp
Sulfochlorid *n* sulphonic chloride
Sulfonamid *n* sulphonic amide
Sulfonat *n* sulphonate
sulfonieren to sulphonate
Sulfonieren *n* sulphonation
Sulfo[n]säure *f* sulphonic acid, sulphoacid
Sulfoxylat *n* sulphoxylate
Sulfoxylsäure *f* hydrosulphurous (sulphoxylic) acid
sulfurieren to sulphurize, to sulphonate
Sulfurylchlorid *n* sulphuryl chloride
Süll *m (n)* coaming, sill [of a dock]
Süllfenster *n* coaming level window
Süllrand *m* wash-board
Süllschott *n* coaming bulkhead
Süllversteifungsprofil *n* coaming stiffener
Sulze *f* salt-lick
Sülze *f* jelly
Sulzer-Kessel *m* Sulzer boiler
sumachgar sumach tanned *(leather)*
Summand *m* addend[um]
  erster S. augend[um]
Summation *f* summation
Summationsformel *f* summation formula
Summationsindex *m* summation index
Summationsverfahren *n* summation method
Summationsverstärker *m* summator amplifier
Summator *m* summer, summing amplifier
Summe *f* sum
  S. der Quadrate sum of squares
  algebraische S. algebraical sum
  runde S. lump sum
  vektorielle S. vector[ial] sum
summen to hum
Summen *n* low-frequency drone, humming
Summendrossel *f* summation reactor
Summendrucktaste *f* key for printing totals
Summenfehler *m* cumulative error
Summenformel *f (ch)* empirical (molecular) formula
Summenfrequenz *f* sum frequency
Summengang *m* total cycle
Summengleichung *f* summation equation
Summenhäufigkeit *f* cumulative frequency
Summenkarte *f (dat)* summary card
Summenkontrolle *f* summation check
Summenkurve *f* cumulative frequency curve
Summenlocher *m* [gang] summary punch
Summenprobe *f* summation check
Summenprüfung *f (dat)* parallel balance
Summenregel *f* sum rule
Summenschreiber *m (dat)* summary recorder
Summenschreibung *f (dat)* total printing
Summenstanzer *m s.* Summenlocher
Summenstromwandler *m* summation transformer
Summenteilfehler *m* accumulative indexing error; accumulated tooth spacing error
Summenübertragung *f (dat)* total transfer
  S. im Kartengang card cycle total transfer *(punched card machinery)*
Summenvektor *m* vector sum, resultant vector
Summenverteilung *f* cumulative distribution
Summenzählwerk *n* totalizator, totalizer
Summer *m* buzzer; howler, hummer
Summeranlaßzündung *f* buzz-type starting ignition
Summeranruf *m* buzzer call
Summerelement *n* buzzer cell
Summerempfang *m* buzzer reception, reception by buzzer
Summergerät *n* buzzer set
Summerrelais *n* buzzer relay
Summerröhre *f* radiotron
Summerschaltung *f* audiooscillator circuit
Summerton *m* buzzer sound
Summerunterbrecher *m* buzzer interrupter
Summerzusatz *m* buzzer set, tone oscillator
summieren to sum, to add up
Summierglied *n* summator
Summierintegrator *m* summing integrator
Summierstelle *f* summing point
Summiertrieb *m* mechanical adding device *(computer)*
Summierung *f* summation; addition
  zeitliche S. summation in time
Summierverbindungsstelle *f (dat)* summing junction
Summierverstärker *m* summer, summing amplifier *(computer)*
Summierwerk *n* summing circuit
Summton *m* buzzing sound
Sumpf *m* 1. swamp, marsh, bog; 2. *(mach)* oil sump; 3. *(mine)* sump; 4. waveguide termination

Sumpfeisenstein *m* bog iron ore
sümpfen 1. to bail *(water)*; 2. *(mine)* to re-open
Sümpfen *n* 1. bailing *(water)*; 2. *(mine)* re-opening, re-sinking
Sumpferz *n* bog iron ore
Sumpfgas *n* marsh gas; sewer gas
Sumpfphasenhydrierung *f* liquid-phase hydrogenation
Sumpfsprengung *f* bog blasting
Sumpfstrecke *f* *(mine)* drainage road, drainway
Sumpfstrecken *fpl* *(mine)* water standages
Sümpfung *f* dewatering; drainage
Sund *m* sound
Super *m* superhet[erodyne receiver]
Superakzeptor *m* superacceptor
Superaerodynamik *f* superaerodynamics
Superballonreifen *m* super-balloon tyre
Superdonator *m* superdonor
Superfinish-Verfahren *n* superfinishing
Superhet *m* superhet[erodyne receiver]
Superheterodynempfang *m* superheterodyne reception
Superhet[erodyn]empfänger *m* superhet-[erodyne receiver]
Superikonoskop *n* *(telev)* supericonoscope, super-emitron
Superkavitation *f* supercavitation
Superorthikon *n* image orthicon
Superoxid *n* peroxide
Superphosphat *n* superphosphate
Superposition *f* superposition
Superpositionsgesetz *n* principle of superposition
Superpositionsprinzip *n* principle of superposition
Supertanker *m* supertanker
Supplement *n* supplement
supplementär supplementary
Supplementwinkel *m* supplementary angle
Support *m* [lathe] carriage; slide rest; tool box (head) *(planer)*
Supraelektron *n* superelectron
Supraflüssigkeit *f* superfluidity
Supraleiter *m* superconductor
supraleitfähig superconductive
Supraleitfähigkeit *f* superconductivity
Supraleitung *f* superconduction
Suprawärmeleitfähigkeit *f* thermal superconductivity
suspendieren *(ch)* to suspend
Suspension *f* suspension
  wäßrige S. water suspension (slurry)
Suspensionskolloid *n* suspensoid
Suspensionspolymerisation *f* suspension polymerization
Suspensionsreaktor *m* suspension (slurry) reactor
Suspensionsspritzmittel *n* suspension spray
Suspensionsspritzpulver *n* suspension spray powder
Suspensionstrockner *m* suspension-type drier
Suspensoid *n* suspensoid
Süßen *n* sweetening process
Süßerde *f* beryllium oxide
Süßholzsirup *m* liquorue, licorice
Süßmost *m* pasteurized fruit juice
Süßwaren *fpl* sweet meats, sweets
Süßwasser *n* fresh water
Süßwasserbehälter *m* fresh-water storage tank
Süßwasserboot *n* fresh-water boat
Süßwein *m* dessert (sweet) wine
Suszeptanz *f* susceptance
Suszeptibilität *f* [magnetic] susceptibility, intrinsic permeability
  magnetische S. magnetic susceptibility
  molare S. molar susceptibility
Swanfassung *f* bayonet holder
Swansockel *m* bayonet cap (base)
Syenit *m* syenite
Syllogismus *m* syllogism
Sylvanit *m* sylvanite
Sylvin *m* sylvine
Sylvinit *m* sylvinite
Symbol *n* symbol; character
  S. der Kristallfläche symbol of the crystal face
  logisches S. logical (functional) symbol
Symbollogik *f* mathematical logic
Symmetrie *f* symmetry; balance *(network)*
  kubische S. cubic symmetry
Symmetrieachse *f* axis of symmetry, symmetry axis
Symmetrieebene *f* plane of symmetry
Symmetrieeigenschaft *f* symmetry property
Symmetrieelement *n* symmetry element
Symmetriegesetz *n* law of symmetry
Symmetriegruppe *f* symmetry group
Symmetrieoperation *f* symmetry operation
Symmetriepunkt *m* point of symmetry (balancing) *(network)*
Symmetriereinrichtung *f* *(el. m.)* balancer
Symmetrierglied *n* line-balance converter
Symmetrierstufe *f* balancing network, balancer
Symmetriertopf *m* line-balance converter *(aerials)*
Symmetriertransformator *m* balancing transformer
Symmetrierung *f* symmetrization, balancing
Symmetrierungsschleife *f* balancing loop
Symmetrierwiderstand *m* balancing resistor
Symmetrieverstärker *m* paraphase amplifier
Symmetriezentrum *n* centre of symmetry
symmetrisch symmetrical; *(el)* balanced
  s. gegen Erde balanced to earth
  s. zur Nullinie symmetrical about the zero axis
Synchro *n* synchro
Synchrodetektor *m* synchronous detector
Synchrodynempfang *m* homodyne reception

synchron synchronous; in step
Synchronabtastung *f* synchronous scanning
Synchronantrieb *m* synchronous drive
Synchrondrehzahl *f* synchronous speed
Synchroneinheit *f* synchromesh unit
Synchronfeder *f* synchronizing spring
Synchrongenerator *m* synchronous generator, alternator
Synchrongetriebe *n* *(aut)* synchromesh gear; *(aero)* synchromesh transmission
Synchronimpulsgeber *m* synchronizer
Synchronisation *f* synchronization
Synchronisationsabtrennung *f* synchronizing clamp
Synchronisationsbereich *m* retaining zone, synchronization range
Synchronisationsfehler *m* *(telev)* horizontal hunting
Synchronisationsimpuls *m* **/ gezahnter vertikaler** serrated vertical pulse
Synchronisierabtrenner *m* *(telev)* amplitude separator
Synchronisierbereich *m* lock-in range
Synchronisiereinrichtung *f* synchronizing mechanism
synchronisieren to synchronize
Synchronisierfrequenz *f* synchronizing frequency
Synchronisierimpuls *m* sync (synchronizing, timing) pulse
Synchronisiersignal *n* sync[hronization] signal
  zusammengesetztes S. composite sync[hronization] signal
Synchronisierspannungsgenerator *m* selsyn generator
Synchronisierstufe *f* synchronizer
Synchronisierung *f* synchronization; locking-in
  örtliche S. independent time control
Synchronisierungsdrossel *f* synchronizing reactor
Synchronisierungsimpuls *m* synchronizing pulse
Synchronisierungsschalter *m* paralleling switch
Synchronisierungsschiene *f* synchronizing busbar
Synchronisiervorrichtung *f* synchronizer
Synchronisierwelle *f* timing shaft
Synchronismus *m* synchronism
  S. / außer out of synchronism (step)
Synchronkegel *m* synchronizing cone
Synchronkontakt *m* synchronizing contact
Synchronkörper *m* detent
Synchronkugel *f* synchronizing ball
Synchronkupplung *f* synchromesh unit
Synchronmaschine *f* synchronous machine
Synchronmotor *m* synchronous motor
  S. mit Selbstanlauf *s.* selbstanlaufender S.
  selbstanlaufender S. self-starting synchronous motor, auto-synchronous motor

Synchronoskop *n* synchronoscope, rotary synchronizer
Synchronphasenschieber *m* synchronous (rotary, dynamic) capacitor, synchronous phase advancer
Synchronreaktanz *f* direct-axis synchronous reactance
Synchronrechenautomat *m* synchronous computer
Synchronrechner *m* synchronous computer
Synchronriegel *m* synchronizing block
Synchronscheibe *f* synchronizing disk
Synchronschiebehülse *f* synchronizing slide collar
Synchronsteuerung *f* synchronization control
Synchrontor *n* synchronous gate
Synchronuhr *f* synchronous clock (timer)
Synchrophasotron *n* synchro-phasotron
Synchrotron *n* synchrotron
Synchrotronstrahlung *f* synchrotron (acceleration) radiation
Synchrozyklotron *n* synchro-cyclotron, f-m cyclotron
Syndiazotat *n* syn-diazo compound
Synerese *f* syneresis
Syn-Form *f* syn-isomer
Synthese *f* synthesis
  S. optisch aktiver Verbindungen asymmetric synthesis
Synthesefaser *f* synthetic fibre
Synthesegas *n* synthesis gas
synthetisch synthetic
synthetisieren to synthesize, to synthetize
System *n* system
  S. der absoluten elektrischen Einheiten absolute system of electrical units
  S. Einheitsbohrung basic-hole system
  S. Einheitswelle basic-shaft system
  S. erster Ordnung first-order system
  S. minimaler Phase minimum-phase system
  S. mit einem Freiheitsgrad one-degree-of-freedom system
  S. mit geerdetem Mittel- oder Sternpunkt earthed neutral system
  S. mit mehreren Freiheitsgraden many-degrees-of-freedom system
  S. mit überkritischer Dämpfung overdamped system
  S. mit unterkritischer Dämpfung underdamped system
  S. zweiter Ordnung second-order system
  abbildendes S. image-forming system
  abgestimmtes S. tuned system
  absolutes S. absolute system
  allpaßfreies S. minimum-phase system
  aperiodisches S. overdamped system
  benachbartes S. neighbouring system
  berührungsloses S. non-contact system

bewegliches S. moving system
entzerrtes S. equalized system
gebundenes mechanisches S. constrained system
geschlossenes S. für Luftpolsterabgrenzung captive air bubble concept, CAB-vessel
gestörtes S. faulty system
in Ruhe befindliches S. rest system
klassisches S. classical (non-quantized) system
konstantes S. zufälliger Ursachen constant system of chance causes
kubisches S. isometric system
lineares S. linear system
mehrschleifiges S. multiloop system
metrisches S. metric system
ortsfestes S. static system
pankratisches S. pancratic system
periodisches S. periodic chart (table, system of elements)
quantisiertes S. quantized system
reguläres S. *(min)* isometric system
ruhendes S. rest system
selbsteinstellendes S. self-adjusting system
selbstorganisierendes S. self-organizing system
selbstsynchronisierendes S. self-aligning system
statisch ausgeglichenes S. statically balanced system
ungeerdetes S. insulated system
zielstrebiges S. goal-seeking system

Systemanordnung *f* system configuration
Systematik *f* systematics
systematisch systematic
Systemfehler *m* system error
Systemingenieur *m* system engineer
Systemkonstrukteur *m* system designer
Systemlast *f* system loading
Systemlinie *f* grid line
Systemprogrammierung *f* system programming
Systemtechnik *f* system engineering
Systemverstärkung *f* system gain
Szilard-Chalmers-Effekt *m* Szilard-Chalmers effect
Szintigrafie *f* scintigraphy
Szintigramm *n* scintigram
Szintillation *f* scintillation, flash, twinkling
Szintillations-Gamma-Spektrometer *n* gamma-ray scintillation spectrometer, scintillation gamma-ray spectrometer
Szintillationskopf *m* scintillation head
Szintillationskristall *m s.* Szintillatorkristall
Szintillationsmeßkopf *m* scintillation head
Szintillationssonde *f* scintillation probe
Szintillationsspektrometer *n* scintillation spectrometer
Szintillationszähler *m* scintillation (photomultiplier) counter
Szintillator *m* scintillator, scintillating material
flüssiger S. liquid scintillator
flüssiger organischer S. organic scintillating solution
organischer S. organic scintillator
plastischer S. plastic scintillator

Szintillatorkristall *m* scintillation crystal
szintillieren to scintillate
Szintillometer *n* scintillometer, portable scintillation counter
Szintilloskop *n* scintilloscope
Szintiscanner *m* scintiscanner

T

T-Antenne *f* T[-shaped] aerial
T-Dampfrohr *n* tee[-piece]
T-Dose *f* **[für I-Rohr]** tee-conduit box
T-förmig tee-shaped, T-shaped
T-Glied *n* T-network
symmetrisches T. H-network
überbrücktes T. bridged-T-network

T-Kettenglied *n* mid-series termination
T-Klemme *f (el)* tee-joint
T-Leitung *f* telegraph line
T-Nut *f* T-slot, tee-slot
T-Nutenfräser *m* T-slot cutter, tee-slot cutter
kreuzverzahnter T. alternate angle staggered tooth-type tee-slot cutter

T-Profil *n* tee [section], T-section
T-Querschnitt *m* tee [section], T-section
T-Regler *m* T-section attenuator
T-Schaltung *f* T-network
T-S-Diagramm *n* entropy diagram
T-Schweißung *f* T welding
T-Stahl *m* tee-steel, T-steel
T-Stoß *m* tee-joint, T-joint
T-Streb *m (mine)* double unit face
T-Stück *n* tee-piece, T-piece
T. von Schutzrohren *(el)* conduit-tee
koaxiales T. coaxial T-connector

T-Träger *m* T-beam
T-Verbindung *f* T-joint
T-Verzweigung *f* T-junction *(waveguides)*
T-Zacken *m* T wave *(electrocardiogram)*
Tabaksbrühe *f* tobacco juice
Tabakzurichter *m* tobacco manipulator
tabellarisch tabular
tabellarisieren to tabulate
Tabelle *f* table, chart
Tabellenform / in in tabular form
Tabellenkopf *m* box head[ing]
Tabellensatz *m* tabular matter
Tabellenspalte *f* table column
Tabellenwert *m* tabular value
Tabelliermaschine *f* tabulating (electric accounting) machine, tabulator

alphabetschreibende T. alphabetical accounting machine
druckende T. *s.* schreibende T.
schreibende T. printing tabulator
Tableau *n* panel; indicator board
Tablett *n* tray *(conveying)*
Tablette *f* tablet, tabloid; pellet
tablettieren to tablet; to pellet
Tablettierfähigkeit *f* pelleting property
Tablettiermaschine *f (plast)* pelleter, tabletting (pilling) machine
Tabulator *m* tabulator
Tabulatortaste *f* tabulator key
Tabuliermaschine *f* tabulator
Tabulierung *f* tabulation
Tabulierungsbefehl *m* tabulation instruction
Tachogenerator *m* tacho-generator
Tachograf *m* tachograph, recording tachometer
Tachometer *n* tachometer, speedometer
registrierendes T. tachograph, recording tachometer
Tachometerantrieb *m* speedometer drive
Tachometerantriebsdeckel *m* speedometer drive cover
Tachometerantriebsgehäuse *n* speedometer drive housing
Tachometerantriebsrad *n* speedometer drive gear
Tachometerantriebsritzel *n* speedometer drive pinion
Tachometertester *m* tachometer tester
Tachoskop *n* tachoscope
Tachymeter *n* tachymeter
Tachymeterkeil *m* / **logarithmischer** logarithmic tachymetric wedge
Tachymetertheodolit *m* tachymeter theodolite
Tachymetrie *f* tachymetry
tachymetrisch tachymetric
Tacitron *n* tacitron
Täckheftmaschine *f* tacking machine
Tafel *f* 1. *(build)* slab, panel; sheet, plate; 2. *(math)* chart, table, array; 3. *(el)* panel, board
Tafelbauweise *f* [large] panel construction
Tafelblech *n* sheet steel
Tafelblei *n* sheet lead
Tafeldifferenz *f* proportional part *(logarithms)*
Tafeleinbau *m* flush mounting
tafelförmig tabular
Tafelglas *n* plate glass
Tafelleim *m* cake-glue, sheet-glue, bone-glue in block[s]
täfeln to panel *(wall)*; to inlay *(floor)*
Tafelsignalausrüstung *f* target-signal equipment
Täfelung *f* panelling, wainscot[ting]
Tafelwaage *f* platform balance
Täfler *m (text)* cuttler, folder
taftbindig taffeta-weave
Taftbindung *f* plain (taffeta) weave
Tag *m* 1. day; 2. *(mine)* surface
Tagaufklärung *f (mil)* daylight reconnaissance
tagblind day-blind, hemeralopic
Tagblindheit *f* day-blindness, hemeralopia
Tage / über *(mine)* on the surface, above ground
T. / unter *(mine)* below ground, underground
Tagebau *m* daylight (strip) mine, open cast (cut, pit)
Tagebaugewinnung *f* strip (open-cast, open-cut) mining, open pitting, stripping
Tagebaukante *f* open pit rim
Tagebaukohle *f* open-cast coal
Tagebuch *n* job diary (record), log
Tagesbelastung *f* day load *(power plant)*
Tagesetmal *n (ship)* 24-hour run, day's run (work)
Tagesgang *m* day-night effect, diurnal (daily) variation
Tagesgezeit *f* diurnal tide
Tagesleistung *f* output per day
Tagesleuchtfarbe *f* daylight fluorescent ink
Tageslicht *n* daylight
künstliches T. artificial daylight
Tageslichtaufnahme *f* daylight photo
tageslichtempfindlich sensible to daylight
Tageslichtentwicklung *f* daylight development
Tageslichtentwicklungsdose *f* daylight developing-box
Tageslichtfilm *m* daylight film
Tageslichtkassette *f* daylight loading system
Tageslichtlampe *f* daylight lamp
Tageslichtquotient *m* daylight factor
Tageslichtspule *f* daylight loading spool
Tagesoberfläche *f (mine)* day, surface
Tagesschacht *m* air-shaft
tonnlägiger T. surface drift
Tagessehen *n* photopic vision
Tagesverbrauch *m* daily rate of consumption
Tageszeit *f* daytime
täglich daily; diurnal
Tagschicht *f* day-turn, day shift
Tag- und Nachtgleiche *f* equinox
Tag- und Nachtpunkt *m* equinoctial point
Taifun *m* typhoon
Taillenabnäher *mpl* waistline darts
Taillenschraube *f* necked-down bolt
Takel *n* tackle; purchase
Takelage *f* rig[ging], tackling
Takelriß *m* rigging plan
Takelung *f* rig[ging]
T. und Ausrüstung *f* furniture *(ship)*
Takelwerk *n* rig[ging]
Takt *m* stroke; cycle; *(mach)* timed sequence; bar *(notes)*; beat *(music)*
T. / im in time (beat); *(mach)* in timed sequence
ganzer T. whole bar
Taktablauf *m* cycling
Taktfertigung *f* cyclic work
Taktfeuer *n* periodical (rhythmic) light

Taktförderer *m* indexing conveyor
Taktfrequenz *f* clock frequency (rate)
Taktgeber *m* master clock, timing element (pulse generator); time tapper; clock generator
Taktgeberbetrieb *m* fixed-cycle operation
Taktgeberfrequenz *f* clock frequency
Taktgebermaß *n* clock rate
Taktgeberspur *f* clock track
Taktgerät *n* timing unit
Taktimpuls *m* clock (timing) pulse
Taktstabilität *f* timing stability
Taktstraße *f* automated transfer (machine) line
Taktverfahren *n* cyclic work
Taktwerk *n* clock unit
Taktzeichen *n* cadence signal
Taktzeit *f (mach)* cycle time; clock time *(computer)*
Tal *n* 1. valley; 2. trough *(wave)*
Talbrücke *f* viaduct
Talfahrt *f* down-stream run (navigation), run downstreams
T. der Schrämmaschine flitting
Talfahrtbremse *f (ry)* down-hill brake
Talg *m* tallow
talgig tallowy
Talje *f* tackle
Talk *m* talc
Talkschnur *f* talc coiling *(sealing)*
Talkum *n* talc[um]
Tallöl *n* tall oil
Tallölharz *n* sulphate wood resin
Talsperre *f* [hydro]dam; reservoir
talwärts down-stream, down the river
Tamm-Niveau *n* Tamm level
Tampon *m* tampon
Tandemaufstellung *f* tandem arrangement
Tandembetrieb *m* tandem operation
Tandemdampfmaschine *f* tandem [steam] engine
Tandemförderung *f (mine)* tandem hoisting
Tandemgenerator *m* tandem-type generator
Tandemmaschine *f* tandem engine
Tandemmotor *m* tandem motor
Tandempotentiometer *n* tandem potentiometer
Tandemschichtdrehwiderstand *m* tandem [ganged] control
Tandemstand *m* tandem parlour
Tandemverbunddampfmaschine *f* tandem compound [steam] engine
Tandemwalze *f* tandem roller
Tandemwalzwerk *n* tandem mill
Tang *m* seaweed
Tangens *m* tangent
T. hyperbolicus hyperbolic tangent
Tangente *f* tangent
Tangentenbild *n* tangent image
Tangentenbogen *m* tangent arc
Tangentenbussole *f* tangent galvanometer
Tangentensatz *m* tangential law
Tangentenschnittpunkt *m* point of intersection of tangents
Tangentenvektor *m* tangential vector
Tangentenvieleck *n* polygon circumscribed about a circle
tangential tangential
Tangentialbeanspruchung *f* tangential stress
Tangentialbeschleunigung *f* tangential acceleration
Tangentialdruck *m* tangential pressure
Tangentialdruckdiagramm *n* tangential pressure diagram, crank effort diagram
Tangentialdruckfläche *f* tangential pressure area
Tangentialebene *f* tangent plane
Tangentialgeschwindigkeit *f* tangential velocity (speed)
Tangentialguß *m* tangential casting
Tangentialkeil *m* tangential key
Tangentialkomponente *f* **der Kraft** tangential [component of] force
Tangentialkraft *f* tangential force
Tangentialschnitt *m* tangential section; *(wood)* slash-grain
Tangentialspannung *f* tangential stress
Tangentialvorschub *m* tangential feed
Tangentialwiderstand *m* tangential resistance
Tangentkeil *m* tangential key
tangieren to be tangent to
tangierend tangent
Tangsäure *f* algic acid
Tank *m* cistern, basin, reservoir, tank
Tankabschlußdeck *n* crown of the tank
Tankanlage *f* berth for fuel-oil bunkering
Tankaufnahmekapazität *f* tankage
Tankbelüftung *f* tank pressurizing
Tankbelüftungsleitung *f* tank air charging (pressurizing) line
Tankbelüftungsrohr *n* tank standpipe
Tankdecke *f* tank top *(double bottom)*
Tankdruckregelleitung *f* oxygen tank pressure control line
tanken to fuel
Tanken *n* fuel[l]ing
Tanker *m* tank vessel, tanker
Tankfassungsvermögen *n* tank capacity, tankage
Tankfüllstandsmesser *m* tank gauge
T. mit Fernanzeige remote-reading tank gauge
Tankfüllstandszeiger *m* tank level indicator
Tankgebühr *f* tankage
Tankgelegenheit *f* servicing (fueling, fuelling) facilities
Tankgenerator *m* tank-type generator
Tankhof *m* fuel[l]ing point
Tanklager *n* fuel depot, oil storage tank
Tankleichter *m* tank barge
Tankofen *m* tank furnace

Tankreaktor *m* tank-type reactor
Tankreinigungsprahm *m* hopper-barge
Tankschiff *n* tankship, tanker, tank vessel
Tankstandsmesser *m* tank gauge (level transmitter)
Tankstelle *f* [roadside] filling station, petrol station, *(Am)* gasoline station
Tankverschluß *m* tank screw-cap
Tankvordruck *m* initial tank pressure
Tankwagen *m* *(aut)* bulk lorry; cistern (tank) wagon
Tankwart *m* fuel[l]ing attendant, refuel[l]er
Tanne *f* fir
Tannenbaumantenne *f* fishbone (pine-tree array type) aerial
Tannenbaumkristall *m* pine (fir-tree) crystal, dendrite
Tannenbaumkrone *f* Christmas-tree bit
Tannenbaumprofil *n* fir-tree profile (form)
Tannenbaumstruktur *f* dendritic structure
Tannenholz *n* fir wood
Tannennadel *f* fir needle
Tannenzapfen *m* fir cone
Tannin *n* tannin, tannic acid
Tanninbuntätzdruck *m* coloured tannin discharge print
Tantal *n* tantalum
Tantalanode *f* tantalum anode
Tantalat *n* tantalate
Tantalerz *n* tantalum ore
Tantalfaden *m* tantalum filament
Tantalgleichrichter *m* tantalum rectifier
Tantalit *m* tantalite
Tantalkarbid *n* tantalum carbide
Tantalkondensator *m* tantalum capacitor
Tantallampe *f* tantalum[-filament] lamp
Tantaloxid *n* tantalum oxide
tantalplattiert tantalum-clad
Tantalschicht *f* tantalum film
Tanzeffekt *m* *(telev)* vertical hunting
Tanzen *n* hunting *(governor)*; skipping; jerking *(steam engine)*
T. der Dampfmaschine jerking of the steam engine
T. des Bildes trembling
Tapete *f* wallpaper
abwaschbare T. washable (sanitary) wallpaper
Tapetendruckmaschine *f* wallpaper machine
Tapetenmuster *n* wallpaper pattern
Tapetenpapier *n* long elephants
Tapetentür *f* secret (jib) door
tapezieren to hang wallpaper, to paper
Tapezieren *n* paperhanging
Tapezierer *m* paperhanger
Tapezierung *f* paperhanging
Tapisseriegewebe *n* tapestries
Tara *f* tare [weight]
Target *n* target
Targetelement *n* target element
Targetkern *m* target nucleus
Targetmaterial *n* target material
Targetstrom *m* target current
tarieren to tare
Tarierwaage *f* tare balance
Tarif *m* tariff, rate[s]
ermäßigter T. cheap fare (rate, tariff), reduced fare (rate, tariff)
Tarifabschnitt *m* section of a tariff
Tarifbahnhof *m* tariff station
Tarifbestimmungen *fpl* tariff conditions (regulations)
Tarifbildungsweg *m* chargeable route, route on which charges are based
Tarifentfernung *f* chargeable distance, distance charged for distance conveyed
Tariferhöhung *f* tariff increase
Tarifermäßigung *f* tariff reduction
Tarifieren *n* tariffing, fixing the rate (fare)
Tarifierungsgrundlage *f* basis on tariffs, *(Am)* rate base
tarifmäßig according to tariff
Tarifmaßnahmen *fpl* tariff arrangements
Tarifnachtrag *m* tariff supplement
Tarifrecht *n* law relating to tariffs
Tarifschnittpunkt *m* tariff connecting point
Tarifspanne *f* tariff range
Tarifstaffelung *f* scale gradations, *(Am)* rate scale
Tariftonnenkilometer *n* ton-kilometre charged
Tarifverband *m* tariff union (agreement)
Tarifvereinheitlichung *f* standardization of tariffs
Tarifvertrag *m* tariff regulation
Tarnanstrich *m* dazzle painting, camouflage paint[ing]
tarnen to camouflage
Tarnmittel *n* camouflage material
Tarnnetz *n* netting
Tarnscheinwerfer *m* masked (screened, hooded, blackout) head lamp
Tarnung *f* camouflage
Tartrat *n* tartrate
Tasche *f* 1. pocket; [hand]bag; case *(map)*; 2. pouch *(cartridge)*; wallet; satchel; 3. compartment *(silo)*
Taschenbandelevator *m* apron conveyor
Taschenbuch *n* pocket-book
Taschendosimeter *n* pocket [dose]meter
Taschenformat *n* pocket size
Taschenionisationskammer *f* ionization pocket chamber
Taschenkompaß *m* pocket compass
Taschenlampe *f* torch lamp, flashlight
Taschenlampenbatterie *f* flashlight battery, torch cell
Taschenmesser *n* pocket (clasp) knife

Taschenmeßinstrument *n* pocket meter
Taschenplatte *f* pockets-type plate *(battery)*
Taschenradioempfänger *m* pocket receiver (radio set)
Taschensender *m* pocket transmitter
Taschenuhr *f* watch
Taschenumdrehungszähler *m* pocket counter
Taschenzähler *m* pocket counter
Täschnerleder *n* pocket-book leather
Tasse *f* cup; *(ag)* bay [of a barn]
Tastatur *f* keyboard
  T. mit Verriegelung shift-lock keyboard
Tastaturlocher *m* / **druckender** printing keyboard perforator
Tastbetrieb *m s.* Tippbetrieb
Tastbügelregler *m* chopper-bar controller
Taste *f* [push-button] key, key (press) button; signalling key *(telegraphy)*
  T. mit fester Stellung locking (push-button) key
  T. mit selbsttätiger Auslösung press-button locking with automatic release
  nichtrastende T. non-locking button
  vollautomatische T. electronic key
Tastelektrode *f* electrode probe
tasten 1. *(mach)* to feel, to trace; 2. to scan *(e.g. radar)*; to key *(e.g. telegram)*
Tastenfeld *n* keyboard
Tastengeber *m* keyboard transmitter
Tastenhebel *m* key lever
Tastenlocher *m* key punch
Tastenreihe *f* bank of keys
Tastensatz *m* key set
Tastenschalter *m* push-button switch
Tastenstengel *m* key stem
Tastensteuerung *f* press-button control, push-button control
Tastenstreifen *m* strip of keys *(telephony)*
Tastenwahl *f* push-button dialling *(telephony)*
Taster *m* 1. *(mach)* tracer, feeler *(copying)*; 2. *(el)* push-button switch; 3. caliper[s] *(measuring)*; 4. probe
Tasterdraht *m* feeler wire
Tasterspitze *f (cont)* probe tip
Tasterventil *n* stylus valve
Tasterzirkel *m* caliper
Tastfrequenz *f* repetition (keying) frequency
Tastfunk *m* wireless telegraphy
Tastgeräusch *n (telec)* key thump (click)
Tasthub *m* frequency shift *(telegraphy)*
Tastimpuls *m* sampling (gating) pulse
Tastkopf *m* search (scanning) unit, pick-up coil, probe
Tastkopfverstärker *m* probe amplifier
Tastnase *f* tracer, follower
Tastpause *f* time interval between pulses
Tastrad *n* feeler wheel
Tastrolle *f* follower roll
Tastspitze *f* probe tip
Tastspule *f* [surface] probe coil
Tastspulgerät *n* probe-coil instrument; eddy-current instrument *(sigmatest)*
Taststift *m* tracer (feeler, tracing) pin, tracer (feeler, tracing) finger, tracer (feeler, tracing) point *(copying)*
  T. der Meßuhr feeler pin of the dial gauge
Tastteiler *m* series probe *(oscilloscope)*
Tastung *f* keying; gating *(radar)*; scanning
  T. durch Frequenzwechsel two-source frequency keying
  T. im Anodenkreis anode keying
Tastverhältnis *n* pulse duty factor, duty ratio; mark-to-space relation (ratio)
Tastverstärker *m* probe amplifier
Tastwelle *f* marking wave
  negative T. spacing (back) wave
Tastzeit *f* keying period *(radar)*; sampling time
Tatbestandsaufnahme *f* factual report
tätigen to carry out, to effect *(sale, purchase)*
Tätowierzange *f (ag)* tattooing forceps
Tatzlagerantrieb *m* nose suspension
Tatzlagermotor *m* nose[-and-axle]-suspended motor, axle-hung motor
Tau *m* dew
Tau *n* rope, hawser; cable
  geteertes T. black rope
taub 1. *(ac)* deaf; 2. *(min)* dead, barren, valueless *(gangue)*; 3. *(ag)* empty
Taubenmist *m* pigeons' dung
Taubfeld *n (mine)* barren track
Taubildung *f* dew-formation, formation of dew *(conductor)*
Taubkohle *f* blind coal
Tauchalitierung *f (met)* alumetizing by dipping
Tauchankerrelais *n* plunger-type relay
Tauchbad *n* splashing, dipping pan, oil bath
Tauchbadschmierung *f* splash (oil-bath, pan dipping) lubrication
Tauchband *n* dredge conveyor
Tauchbatterie *f* plunging battery
Tauchbehälter *m* dip tank
Tauchbeize *f* dip dressing
Tauchbohrhammer *m* down-the-hole drill
Tauchboot *n* submersible [boat], diving boat
Tauchbrett *n* baffle-board
Tauchbütte *f* dipping vat
Tauchelektrode *f* coated (dipped) electrode
Tauchelement *n* bichromate cell
tauchen to submerge; to dive *(e. g. submarine)*; to dip, to immerse, to plunge
Taucher *m* diver
  Cartesischer T. Cartesian diver
Taucheranzug *m* diving suit
Taucheraufnahme *f* submarine photograph
Taucherboot *n* dive logger, diving boat (vessel)
Taucherglocke *f* diving bell

Taucherglockenprahm *m* diving-bell barge
Taucherprahm *m* diving barge (lighter)
tauchfähig submersible
Tauchfärbverfahren *n* dip dyeing
Tauchförderer *m* dipping conveyor
Tauchfräsen *n* plunge milling
Tauchfräsmaschine *f* plunge-milling machine
Tauchglocke *f* floating bulb; bell gauge, pressure bell
Tauchglockendurchflußmesser *m* bell flowmeter
Tauchglockenmanometer *n* bell-pressure gauge
Tauchgrenze *f* margin line
Tauchhärten *n (met)* liquid hardening, hardening by immersion
Tauchkern *m* plunger
Tauchkernrelais *n* plunger relay
Tauchkernspule *f* plunger (sucking) coil
Tauchkolben *m* plunger [piston]
Tauchkolbenmotor *m* trunk-type engine
Tauchkolbenpumpe *f* plunger pump
Tauchkondensator *m* plunger-type capacitor
Tauchkontaktrelais *n* dipper relay
Tauchkorrosionsversuch *m* immersion test
Tauchlack *m* dipping varnish
tauchlackieren to dip-coat
Tauchlötbad *n* dip-soldering bath
tauchlöten to dip-solder, to dip-braze
Tauchlöten *n* dip soldering (brazing)
Tauchlötverfahren *n* dip-soldering method
Tauchmetallisieren *n* immersion (dip) plating, immersion (dip) coating; metal replacement
Tauchpresse *f* steeping press *(man-made textiles)*
Tauchprozeß *m (met)* dipping process; immersion method *(testing)*
Tauchpumpe *f* submersible pump
Tauchrelais *n s.* Tauchkontaktrelais
Tauchrüttler *m* spud (internal, immersion, needle, poker) vibrator *(concrete)*
Tauchschmierung *f* splash (oil-bath) lubrication
Tauchschwingungen *fpl* heaving *(of a vessel)*
Tauchschwingungsamplitude *f* scend *(of a vessel)*
Tauchsieder *m* immersion heater
Tauchspule *f* plunger-type coil
Tauchspulenlautsprecher *m* moving-coil loudspeaker
Tauchspulenmikrofon *n* moving-coil microphone
Tauchspulkupplung *f* electrodynamic loudspeaker clutch
Tauchspulmagnet *m* plunger electromagnet
Tauchstreichmaschine *f* dip coater
Tauchtank *m* diving tank *(submarine)*
Tauchteiler *m* guillotine attenuator *(waveguide)*
Tauchthermoelement *n* immersion thermocouple
Tauchtiefe *f* depth of immersion; underwater (undersea) level *(submarine or diver)*; moulded (designed) draught *(vessel, surface craft)*
Tauchtrommel *f* dipping drum
Tauch- und Stampfbewegung *f* heaving and pitching motion *(of a vessel)*
Tauchung *f* draught
Tauchungsänderung *f* alteration (change) of trim
Tauchverfahren *n* hot dipping *(coats)*; immersion method *(testing)*
  T. / im by immersion
Tauchverschluß *m* liquid seal
Tauchversuch *m* trial dive *(submarine)*
tauchverzinken to hot-dip galvanize
Tauchwaage *f* hydrometer, areometer
Tauchwalze *f* dipping (immersion) roller
Tauchwiderstand *m* immersed rheostat
Tauchzähler *m* dipping counter
Tauchzelle *f* dip cell
Tauchzündung *f* dipper-anode ignition *(rectifier)*
Tauenpapier *n* Manilla (cartridge) paper
tauglich suited, suitable; usable, serviceable; worthy
Tauglichkeit *f* suitability; ability, fitness; qualification; usefulness; worthiness
Tauklüse *f* bow-chock mooring pipe *(of a vessel)*
Taukragen *m (ship)* garland
Taumelfehler *m* drunkenness *(thread)*
taumeln to run untrue *(flywheel)*, to nutate
Taumelscheibe *f* wobbling disk; wobble plate
  fest mit Antriebswelle verbundene T. Z-crank
Taumelscheibenmotor *m (aero)* wobble (swash-plate) engine
Taumelscheibenpumpe *f* wobble-plate pump, rotary swash-plate pump
Taumelscheibenzähler *m* nutating-piston meter
Taumelsieb *n* tumbler screen
Taupunkt *m* dew point
Taupunktanzeiger *m* dew-point indicator
Taupunkthygrometer *n* dew-point hygrometer
Taupunktmesser *m* dew-point meter
Taurin *n* taurine
Taurocholsäure *f* taurocholic acid
Tauröste *f* dew retting *(flax)*
Tauschlinge *f* sling
Täuschung *f (telec)* deception
  optische T. optical illusion
Täuschungsanstrich *m s.* Tarnanstrich
Tausendkorngewicht *n* weight-per-thousand
Taustopper *m (ship)* deck (lanyard) stopper
Taustropp *m (ship)* rope sling (strap), baker
tautomer *(ch)* tautomeric
Tautomerie *f (ch)* tautomerism, dynamic isomerism
tautomerisieren to tautomerize
Tauwerk *n* ropes, cordage
Taxameter *n(m)* taximeter
Taxgewicht *n* chargeable weight, *(Am)* weight for assessment of freight charge
Taxichauffeur *m* taxi driver
taxieren to taxe, to assess, to rate

Taylor-Reihenentwicklung *f* *(math)* Taylor series expansion
Taylor-Schaltung *f* Taylor connection
TE-Welle *f* transverse electric wave
Teakholz *n* teak [wood]
Technetium *n* technetium
Technicolorverfahren *n* Technicolor
Technik *f* technology *(science)*, engineering *(practical application)*; technique[s] *(method, process)*; art, skill; workmanship
  experimentelle T. experimental technique
Techniker *m* engineer; technologist
Technikum *n* technological (engineering) school, technological (engineering) college
technisch technological, engineering; technical; industrial
Technologe *m* process-planning engineer, production engineer (technologist); *(Am)* methods engineer
Technologie *f* production engineering (technology); planning technology; manufacturing research; production technique
  chemische T. chemical engineering
technologisch production-technological, production-engineering
Teckel *m* *(mine)* bogie
Teekannenpfanne *f* teapot ladle *(foundry)*
Teer *m* tar
  destillierter T. distilled tar
  entwässerter T. dehydrated tar
Teerabscheider *m* tar extractor (collecting tank, settler), detarrer
  elektrischer T. electrostatic detarrer
Teerabscheidung *f* tar extraction; detarring
Teeranstrich *m* tar covering
Teerbeton *m* tar concrete
Teerbetondecke *f* tar concrete pavement
Teerbildung *f* tar formation
Teerbitumengemisch *n* tar-asphaltic bitumen mixture, *(Am)* tar-asphalt bitumen mixture
Teerblase *f* tar still
Teerbrenner *m* tar burner
Teerdampf *m* tar vapour
Teerdestillat *n* tar distillate
Teerdruckscheidung *f* tar decantation under pressure
Teereimer *m* tar pail
teeren to tar; to pitch *(a vessel)*
Teerfarbe *f* coal-tar dye *(aniline dye)*
Teerfarbstoff *m* coal-tar dye[stuff]
Teerfaß *n* tar barrel
Teerfleck *m* tar stain
Teergewinnung *f* tar extraction
Teergrube *f* tar pit, tar-storage tank
teerig tarry
Teerkessel *m* tar boiler
Teerkondensat *n* tar condensate
Teerkühler *m* tar cooler-box
Teerleitung *f* tar main
Teermakadam *m (n)* tarmacadam
Teernebel *m* tar mist
Teeröl *n* [coal-]tar oil
Teerpappe *f* tar-saturated felt, tarred board
  mit Zwirn verstärkte T. threaded felt board
Teerpech *n* artificial (coal-tar) asphaltum, tar-pitch
Teersammler *m* tar collector
Teersäure *f* tar acid
Teerscheider *m* tar extractor (separator, collecting tank)
Teerscheidung *f* tar separation
Teerschlauch *m* tar hose
Teerschotter *m* tarred stone
Teerschwelerei *f* tar distillery
Teersplitt *m* tarcoated chipping[s]
Teerspülung *f* tar circulation
Teerung *f* tarring
Teervorlage *f* hydraulic (tar-collecting, tar) main
Teerwagen *m* tar tank car
Teerwäsche *f* tar extraction plant
Teflon *n* teflon, polytetrafluor ethylene
Teich *m* pond, pool
Teig *m* dough, paste
  warmer T. warm doughy mass *(foil)*
teigartig doughy
teigig pasty
Teigmulde *f* kneading trough
Teigpreßverfahren *n* *(plast)* dough moulding
Teigreife *f* pasty condition *(of maize)*
Teigwaren *fpl* noodles
Teil *m* part; section, portion, share; element
  gefederter T. suspended (hanging) portion
  kurzwelliger T. short wavelength part
  regulärer T. regular component
  singulärer T. singular component
  spannungführender T. live part
  stromführender T. current-carrying part
  ungefederter T. non-suspended portion, non-hanging portion
  untrennbarer T. integral part
  verjüngter T. necked-down portion
Teil *n* component, part, workpiece; element
  ausführendes T. slave end *(manipulator)*
  bewegliches T. moving part
  einwandfreies T. non-defective part
Teilamt *n* subexchange, satellite exchange *(telephony)*
Teilansicht *f* partial view
Teilapparat *m* dividing apparatus
Teilaufbauten *mpl* partial superstructures
Teilausschnitt *m* section
Teilauswerfeinrichtung *f* part-ejection device
Teilband *n* sub-band
Teilbandförderer *m* sectional-belt conveyor
Teilbandstrecke *f* *(text)* divided sliver delivery
teilbar divisible

durch einen gemeinsamen Teiler t. commensurable
Teilbarkeit *f* divisibility, divisibleness
T. durch einen gemeinsamen Teiler commensurability
Teilbaum *m (text)* sectional beam
Teilbelastung *f* partial load
Teilbereich *m* subdomain
Teilbetrag *m* partial amount; instalment *(to pay)*
Teilbewegung *f* indexing movement
T. / mit indexing
Teilbild *n (opt)* component picture; sample *(cathode-ray tube)*; *(telev) (Am)* field
Teilbildablenkung *f* frame deflection
Teilbildaustastung *f (telev)* frame blanking
Teilbildfrequenz *f* vertical (frame) frequency
Teilbildfrequenzteiler *m (telev)* frame divider
Teilbildhöhenregler *m* vertical size control
Teilbildunterdrückung *f (telev)* frame blanking
Teilbildverzerrung *f* frame distortion
Teilblock *m* sub-block *(computer)*
Teilblockkühler *m* block radiator
Teilbrache *f* bastard fallow
Teilbremsprobe *f* partial trial of brake
Teilchen *n* particle; corpuscle
adsorbiertes T. adsorbed particle
atomares T. atomic particle
beschleunigtes T. accelerated particle
festes T. solid particle
fremdes T. strange particle
geladenes T. charged particle
heißes T. hot particle
kräftefreies T. free particle
masseloses T. massless particle
neutrales T. neutral (uncharged) particle
nichtrelativistisches T. non-relativistic particle
relativistisches T. relativistic particle
seltsames T. strange particle
stabiles T. stable particle
Teilchen *npl* / **energiegleiche** equal-energy particles
getrennte T. discrete particles
ungleiche T. inequivalent particles
Teilchenbahn *f* particle path (trajectory)
Teilchenbeschleuniger *m* [particle] accelerator
Teilchenbewegung *f* particle motion
Teilchendichte *f* particle density
Teilchenfeld *n* particle field
Teilchenfluenz *f* [particle] fluence
Teilchenflußdichte *f* particle flux density
Teilchengeschwindigkeit *f* particle velocity
Teilchengröße *f* particle size
Teilchennachweis *m* particle detection
Teilchenstoß *m* particle impact
Teilchenstrahlung *f* corpuscular (particle) radiation
Teilchenstrom *m* corpuscular stream
Teilchenverlust *m* particle loss
Teilchenverschiebung *f* particle displacement (shift)
Teilchenzustand *m* particle state
Teildock *n* sectional pontoon dock
Teildruck *m* partial pressure
Teildruckgefälle *n* partial drop of pressure
Teildurchbruch *m* partial breakdown
Teildurchführung *f* crippled leap-frog test
teildurchlässig semipermeable, semitransparent
Teile *npl* / **fügende** mating parts
typengleiche T. same-type parts
Teileheber *m* elevator *(for elevating work)*
Teileinrichtung *f* dividing (indexing) fixture, dividing (indexing) attachment
Teileliste *f* parts list *(assembly)*
teilen *(math)* to divide; *(mach)* to index, to space, to divide; to split *(line)*; to separate, to graduate *(material)*; to section[ize]; to share *(work)*
Teilen *n* **eines Kreisumfangs** circumferential indexing, circular spacing
T. nach Winkelmaß angular indexing
T. unter Einhaltung enger Toleranzen close-limit spacing
einfaches T. plain (simple) indexing
ungleiches T. unequal spacing
unmittelbares T. direct (rapid) indexing
Teiler *m* divider; *(math)* divisor
ganzzahliger T. submultiple
größter gemeinsamer T. highest (greatest) common divisor
teilerfremd prime, aliquant
Teilerkette *f* divisor chain
Teilerschaltung *f* dividing (divider, divide-down) circuit
Teilerstufe *f* dividing stage
teilevakuiert partially evacuated
Teilfarbennegativ *n* separation negative
Teilfehler *m* indexing (dividing) error *(dividing head)*, pitch error *(gear)*
Teilfläche *f* 1. sub-area; 2. sub-critical area of extraction
Teilflugstrecke *f* leg
Teilfolge *f* subsequence, partial sequence
Teilfrequenz *f* component frequency
Teilfuge *f* joint
schräge T. oblique split, inclined joint
Teilgenauigkeit *f* indexing accuracy, accuracy of indexing (spacing)
Teilgesamtheit *f* sub-population *(statistics)*
Teilhärtung *f* selective hardening
Teilionisation *f* partial ionization
Teilkammerkessel *m* sectional header boiler
Teilkapazität *f* partial capacitance
T. zwischen zwei Elektroden direct interelectrode capacitance
Teilkegel *m* pitch cone *(gear)*
Teilkegelwinkel *m* pitch angle *(gear)*

Teilkerngeschoß *n* sectional bullet
Teilkonstrukteur *m* detailer
Teilkonvergenz *f* incomplete convergence
Teilkopf *m (mach)* dividing head
T. für weite Bereiche wide-range dividing head
optischer T. optical dividing head
Teilkopfspindel *f* indexing spindle
Teilkopfzeigerpaar *n* dividing-head sector
Teilkörperbestrahlung *f* partial body irradiation
Teilkraft *f* component of [a] force
Teilkreis *m* pitch circle *(gear)*; angular scale, graduated circle *(measuring)*
Teilkreisbogen *m* arc of the pitch circle *(gear)*
Teilkreisdurchmesser *m* pitch [circle] diameter *(gear)*
Teilkreisloch *n* indexing hole
Teilkurbel *f* index crank
Teilkurbelgriff *m* indexing handle
Teillast *f* partial load, sub-load
Teillastnadel *f* part-load needle
Teillastventil *n* partial-load valve
Teilleiter *m* conductor element
Teillichtbogen *m* partial arc
Teillineal *n* dividing ruler
Teilmarke *f* index
Teilmenge *f (math)* subset
Teilmodul *m* sub-module
Teilmontage *f* sub-assembly; sub-assembling, assembling of sub-assemblies
Teilnachwirkzeit *f* partial restoring time
teilnehmen to participate, to take part; to be subscribed to, to be a customer
Teilnehmer *m* subscriber *(telephony)*
angerufener T. called party (subscriber)
anrufender T. calling subscriber, caller
Teilnehmeranschluß *m* subscriber's station
Teilnehmeranschlußleitung *f* subscriber's line
Teilnehmerantenne *f* viewers' aerial
Teilnehmerapparat *m* subscriber's set
Teilnehmerendverstärker *m* loudspeaker telephone set
Teilnehmerfernschreibdienst *m* telex [service]
Teilnehmerfernschreibsystem *n* telex system
Teilnehmerfernwahl *f* subscriber's trunk dialling
Teilnehmerkabel *n* subscriber's cable
Teilnehmerklinke *f* subscriber's (local, answering) jack
Teilnehmerleitung *f* subscriber's line
Teilnehmernummer *f* subscriber's number
Teilnehmersprechstelle *f* subscriber's set
Teilnehmerzentrale *f* private branch exchange
Teilplatte *f* separator
Teilproblem *n* subproblem
Teilprodukt *n* subproduct
statisches T. static subroutine
Teilprogramm *n (dat)* subprogramme, subroutine
dynamisches T. dynamic subroutine
geschlossenes T. closed subroutine
offenes T. open subroutine
Teilpunkt *m (math)* point of division
Teilraum *m* subspace
Teilreaktion *f* partial reaction
Teilresonanz *f* sub-multiple resonance
Teilsättigung *f* partial saturation
Teilschaltung *f* subcircuit
Teilscheibe *f* index (graduated, dividing) plate
Teilscheibenraste *f* indexing notch
Teilscheibenraststift *m* index plunger
Teilschnecke *f* indexing worm
Teilschnitt *m* drawing shown partly in section, part section
Teilschott *n* partial (partition) bulkhead
Teilschritt *m* partial pitch *(winding)*
Teilschwingung *f* partial vibration (oscillation)
Teilschwingungen *fpl* / **harmonische** harmonic components
Teilsenkung *f* partial subsidence
Teilsohle *f* intermediate level, sub-level, *(Am)* sub
Teilsohlenbruchbau *m* sub-drift caving, sub-level caving (slicing); *(Am)* subbing, subslicing
Teilsohlenkammerbau *m* sub-level stoping, sub-stoping
Teilsohlenquerschlag *m* sub-level crosscut
Teilsohlenweitungsbau *m* sub-level stoping, sub-stoping
Teilspannung *f* component voltage
Teilsperrzeit *f* partial restoring time
Teilstichprobe *f* / **feste** quota sample
Teilstrahlungspyrometer *n* partial-radiation pyrometer
Teilstrecke *f (ry)* track section
Teilstrich *m* scale division, graduation [mark], graduation (scale) line
Teilstrichwaage *f* graduated scale
Teilstrom *m* component current
Teilstromkreis *m* divided circuit
Teilstück *n* segment
Teiltabelle *f* index[ing] table
Teilübertrag *m (dat)* partial carry
Teilumrißnachformen *n* segment copying
Teilung *f* division; graduation; partition; separation; splitting; spacing *(workpiece, tool)*; pitch *(gear)*
T. / mit gleicher uniformly spaced
T. / mit großer wide-pitched *(gear)*
T. nach Winkelmaß angular division
grobe T. *(text)* coarse pitch of needles
ungleiche T. irregular spacing; *(mach)* increment cut
Teilungsebene *f* parting plane *(foundry)*; plane of division (joint)

Teilungsfaktor *m* division factor
Teilungsfehler *m* error of the division; pitch error; spacing error; dividing (indexing) error
Teilungsgenauigkeit *f* spacing accuracy, accuracy of spacing (division, pitch)
Teilungslinie *f* dividing line
Teilungspunkt *m* point of division
Teilungswürfel *m* beam-splitting cube
Teilvakuum *n* partial vacuum
Teilvektor *m* component vector
Teilverhältnis *n* indexing ratio
Teilversatz *m* partial stowing
Teilversetzung *f* partial dislocation *(crystals)*
Teilverzug *m (text)* partial draft
Teilvielfachfeld *n (telec)* partial multiple
Teilwalze *f (text)* dividing roller
Teilwelle *f* component (partial) wave
Teilzahlung *f* instal[l]ment (partial) payment
Teilzeichnung *f* detail drawing
Teilzirkel *m* bow compass, dividers
Teilzuführungseinrichtung *f* hopper; [part] feeding device
Teilzylinder *m* pitch cylinder *(worm)*
Tektonik *f* tectonics
tektonisch tectonic
Telecurietherapie *f s.* Tiefentherapie
Telefon *n* telephone [set], phone
  batterieloses T. sound-powered telephone
  dynamisches T. sound-powered telephone
  lautsprechendes T. loudspeaker telephone set
Telefonanlage *f* telephone installation
Telefonanschluß *m* telephone connection
Telefonapparat *m* telephone set
Telefonbetrieb *m* telephone working
Telefonbuch *n* directory, *(Am)* telephone directory
Telefonbuchse *f* telephone jack
Telefondienst *m* telephone service
Telefondraht *m* telephone wire
Telefongebühr *f* telephone charges
Telefongespräch *n* telephone call (conversation)
  dringendes T. urgent call
Telefonhaken *m* telephone switch hook
Telefonhörer *m* telephone receiver
Telefonie *f* telephony
  drahtlose T. radiotelephony
Telefonieempfang *m* telephone reception
telefonieren to telephone
Telefoniesender *m* radiotelephone transmitter
Telefonieträgerstrom *m* telephone carrier current
telefonisch telephonic[al]
Telefonist[in *f*] *m* telephone operator
Telefonkabel *n* telephone cable
Telefonkabine *f s.* Telefonzelle
Telefonleitung *f* telephone circuit
Telefonnummer *f* subscriber's number
Telefonordnung *f* telephone regulations
Telefonschnur *f* telephone flex (cord)
Telefonverbindung *f* telephone line (circuit)
Telefonverkehr *m* telephone traffic
Telefonwählerscheibe *f* telephone dial
Telefonzelle *f* telephone box (booth, kiosk)
Telegraf *m* telegraph
  optischer T. optical telegraph
Telegrafenalphabet *n* telegraph alphabet (code)
Telegrafenamt *n* telegraph office (station)
Telegrafenanlage *f* telegraph plant
Telegrafenbautrupp *m* telegraph construction gang
Telegrafendienst *m* telegraph service
Telegrafendraht *m* telegraph wire
Telegrafenisolator *m* telegraph insulator
Telegrafenkabel *n* telegraph cable
Telegrafenkode *m* telegraph code
Telegrafenleitung *f* telegraph line
Telegrafenmast *m* telegraph pole
Telegrafennetz *n* telegraph network
Telegrafenrelais *n* telegraph relay
Telegrafenstange *f* telegraph pole
Telegrafenstelle *f* public telegraph office
Telegrafentaste *f* telegraph key
Telegrafentechnik *f* telegraph engineering
Telegrafie *f* telegraphy
  T. mit Aufzeichnung von kodierten Zeichen signal-recording telegraphy
  abgestimmte drahtlose T. syntonic wireless telegraphy
  automatische T. automatic telegraphy
  drahtlose T. wireless telegraphy, radio-telegraphy
  gerichtete drahtlose T. directional radio-telegraphy
  kommerzielle T. commercial radio-telegraphy
  optische T. visual telegraphy
Telegrafieempfang *m* telegraphy reception
Telegrafiefunk *m* radiotelegraphy
Telegrafiefunker *m* radiotelegraph operator
Telegrafiegeschwindigkeit *f* telegraphic speed
Telegrafiemodulation *f* modulation keying
telegrafieren to telegraph, to send a telegram, to wire; to cable *(by ocean cable)*
Telegrafierfehler *m* operator's error
Telegrafierfrequenz *f* signalling frequency
Telegrafiergeräusch *n* telegraph noise, *(Am)* thump
Telegrafiergeschwindigkeit *f* telegraphic speed
Telegrafiernebensprechen *n* telegraph cross-talk
Telegrafierzeichen *n* code character, telegraph signal
Telegrafierzeichenelement *n* telegraph signal element
Telegrafierzeichenschritt *m* telegraph signal element

Telegrafieübertrager *m* / **entzerrender** telegraph regenerative repeater
Telegrafiezeichen *n* code signal, telegraph character
Telegrafiezeichenunschärfe *f* telegraph-pulse distortion
telegrafisch telegraphic[al]
Telegramm *n* telegram
dringendes T. urgent telegram
gewöhnliches T. ordinary telegram
verstümmeltes T. mutilated telegram
zugesprochenes T. phonogram
Telegrammadresse *f* telegraphic[al] address
Telegrammformular *n* message (telegram) blank
Telegrammschalter *m* telegram counter
Telegrammschlüssel *m* telegraph code
Telegrammwort *n* code word
Telegrammzustellung *f* delivery of telegrams
Telemechanik *f* telemechanics
Telemeter *n* telemeter
Telemetrie *f* telemetry
Teleobjektiv *n* telephoto lens
Teleran *n* TELERAN *(Television and Radar Navigation)*
Teleskop *n* telescope
teleskopisch telescopic
Teleskopmast *m* telescopic mast
Teleskoprohr *n* telescopic tube
Teleskopschornstein *m* telescopic funnel
Teleskopspindel *f* telescopic screw
Teleskopstoßdämpfer *m* telescopic shock absorber
doppeltwirkender T. dual-action telescopic shock absorber
Teleskopwelle *f* telescopic shaft
Teleskopzylinder *m* telescopic cylinder
Teletherapie *f* teletherapy
Teletherapiegerät *n* teleradium unit
Teletypesetter *m* teletype-setter
Telexleitung *f* telex line
Telexverkehr *m* Telex
telezentrisch telecentric
Telleranode *f* disk (plate) anode *(electron tubes)*
Tellerbohrer *m* disk auger
Tellerdüngerstreuer *m* disk manure distributor
Telleregge *f* disk harrow
Tellerfeder *f* Belleville spring
tellerförmig plate-shaped
Tellerfräser *m* plate-shaped cutter
Tellerisolator *m* plate insulator
Tellermine *f* flat anti-tank mine
Tellermühle *f* dish crusher
Tellerrad *n* crown wheel; ring (bevel drive) gear
Tellerschleifscheibe *f* dish wheel
Tellerstanze *f* open-ended clicking press
Tellerstreuer *m* disk distributor
Tellertrockner *m* disk drier
Tellerventil *n* poppet (disk, quick-opening) valve
Tellur *n* tellurium
Tellurat *n* tellurate
Tellurerz *n* telluric ore
tellurhaltig telluriferous
Tellurid *n* telluride
tellurig tellurous
Tellurit *n* tellurite
Tellursäure *f* telluric acid
Tellurwasserstoff *m* hydrogen telluride
Telomer *n* telomer
Telomerisation *f* telomerization
TEM-Typ *m* transverse electric and magnetic mode *(waveguides)*
Temperafarbe *f* distemper
T. auf Ölgrundlage oil-bound distemper
Temperatur *f* temperature
T. des anstehenden Gebirges virgin rock temperature
T. unter Null temperature below zero
absolute T. absolute temperature
betriebszulässige T. safe operating temperature
erhöhte T. elevated temperature
gleichbleibende T. steady temperature
gleichmäßige T. uniform temperature
kritische T. critical temperature
schwarze T. black-body temperature
tiefe T. low temperature
Temperaturabfall *m* drop in temperature, fall of temperature, temperature fall (drop)
temperaturabhängig temperature-dependent
Temperaturabhängigkeit *f* temperature dependence
Temperaturänderung *f* temperature variation, variation of temperature
Temperaturanstieg *m* temperature rise (increase), rise (increase) in temperature
Temperaturanstiegsrate *f* thermal response
Temperaturanzeiger *m* temperature indicator
Temperaturäquivalent *n* temperature equivalent
Temperaturausgleich *m* temperature compensation
Temperaturbeiwert *m* temperature coefficient
Temperaturbereich *m* range of temperature, temperature range
Temperaturbeständigkeit *f* ability to resist high temperatures; temperature (thermal) stability
Temperaturbestimmung *f* temperature determination
kalorimetrische T. calorimetric determination of temperature
Temperaturdifferenz *f* temperature differential
Temperatureffekt *m* temperature effect
Temperatureinfluß *m* influence of temperature
temperaturempfindlich temperature-sensitive

Temperaturempfindlichkeit *f* temperature sensitivity
Temperaturerhöhung *f* rise of temperature, temperature rise
Temperaturfaktor *m* temperature factor
Temperaturfeld *n* temperature field
Temperaturfühler *m* temperature sensor, temperature-sensing element, thermometer probe
T. mit Schutzrohr thermometer tube
Temperaturfühlglied *n* thermo-sensitive element
Temperaturgang *m* function of temperature; temperature variation; temperature tracking
Temperaturgeber *m* temperature transmitter
Temperaturgefälle *n* temperature differential (drop)
Temperaturgradient *m* temperature gradient
Temperaturgrenze *f* temperature limitation
Temperaturintervall *n* temperature interval
Temperaturkennlinie *f* temperature characteristic
Temperaturkoeffizient *m* temperature coefficient
T. der Durchbruchsspannung temperature coefficient of Zener voltage *(semiconductor)*
T. der Reaktivität temperature coefficient of reactivity
Temperaturkompensation *f* thermal compensation
temperaturkonstant thermally stable
Temperaturkonstanz *f* temperature stability
Temperaturkorrektion *f* temperature correction
Temperaturkurve *f* temperature curve
Temperaturlabilität *f* temperature instability
Temperaturleitfähigkeit *f* temperature (thermal) conductivity
Temperaturleitzahl *f* thermal conductivity
Temperaturmesser *m* temperature measuring device
Temperaturmeßgerät *n* temperature measuring device, thermometer
elektrisches T. electric thermometer
Temperaturmessung *f* thermometry
Temperaturprofil *n* temperature distribution
Temperaturrauschen *n* temperature noise
Temperaturregelung *f* temperature control
Temperaturregler *m* thermostat; thermoregulator; temperature controller (control unit)
T. mit Federthermometer remote-bulb thermostat
T. ohne Hilfsenergie thermostatic valve
direktwirkender T. thermostatic valve
Temperaturrückgang *m* drop in temperature
Temperaturschaden *m* temperature damage
Temperaturschreiber *m* recording thermometer, temperature recorder
Temperaturschwankung *f* temperature variation (change)
Temperaturskale *f* temperature (thermometric) scale, scale of temperature
absolute thermodynamische T. [Kelvin's] absolute scale of temperature
Temperaturspannung *f* 1. temperature equivalent; 2. *s.* Wärmespannung
Temperatursprung *m* jump in temperature
temperaturstabil temperature-stable
Temperaturstrahlung *f* temperature radiation
Temperatursturz *m* thermal shock
temperaturunabhängig temperature-independent
Temperaturunabhängigkeit *f* temperature independence
Temperaturunterschied *m* temperature difference
Temperaturverteilung *f* temperature distribution
Temperaturwächter *m* thermostat
Temperaturwechselbeanspruchung *f* thermal cycling
Temperaturwechselbeständigkeit *f* resistance to thermal shocks
Temperaturwelle *f* temperature wave
Temperaturzunahme *f* increase (rise) of temperature
Temperbedingung *f* annealing condition
Tempererz *n* malleablizing (annealing) ore
Tempergießerei *f* malleable-iron foundry
Temperglühofen *m* malleable-annealing furnace
Temperguß *m* malleable cast iron
amerikanischer T. American malleable cast iron
deutscher T. whiteheart malleable cast iron
schwarzer T. blackheart malleable cast iron
weißer T. *s.* deutscher T.
Temperiereinrichtung *f* temperature-regulating device
Temperkohle *f* temper carbon
Tempermittel *n* packing material
tempern *(met)* to malleablize; to anneal *(photoelastic materials)*
Temperofen *m* malleablizing (malleable-annealing) furnace
Temperroheisen *n* malleable pig iron
Temperrohguß *m* unannealed malleable iron, malleable hard iron
Tempertopf *m* annealing pot
Temperverfahren *n* malleablizing (annealing) process
Tenakel *n* copyholder
Tenbrinkfeuerung *f* Tenbrink furnace *(furnace with inclined grate with back-draught arranged in the interior of the boiler)*
Tendenz *f* tendency; trend
fallende T. downward tendency
steigende T. upward tendency
Tender *m* tender, cockboat

Tenderlokomotive *f* tank locomotive
Tennantit *m (min)* tennantite
Tenne *f* threshing floor
Tennis *n* lawn tennis
Tennisplatz *m* tennis court
Tennisschläger *m* [tennis] racket
Tenside *pl* surface active agents
Tensidstruktur *f* constitution of surface agents
Tensimeter *n* tensimeter
Tensiometer *n* tensiometer
Tensionsthermometer *n* vapour-actuated thermometer
Tensor *m* tensor
Tensoralgebra *f* tensor algebra
Tensorbedingung *f* tensor condition
tensoriell tensorial
Tensorkraft *f* tensor force
Tensorrechnung *f* tensor calculus
Teppich *m* carpet, rug *(small)*
  T. mit Schaumstoffrückseite foam-back carpet
  hochfloriger T. deep-pile carpet
  maschinell gewebter T. power-loomed carpet
  nicht zerdrückbarer T. non-crush carpet
Teppichbänder *npl* carpet binding
Teppichbelag *m* road mat (carpet), carpeting *(road)*
Teppichförderer *m (mine)* bottom loading belt, shuttle-belt conveyor
Teppichschützen *m* carpet shuttle
Teppichstuhl *m* carpet loom
Teppichweberei *f* carpet weaving
Teppichwebstuhl *m* carpet loom
  indischer T. Indian type loom
Teppichwolle *f* carpet wool
Terbium *n* terbium
Terephthalsäure *f* terephthalic acid
Term *m* term, [energy] level, energy state
  Diracscher T. Dirac term
  langsamer T. slow state
  optischer T. optical level
  schneller T. fast state
Termabstand *m* level distance
Termanalyse *f* energy-level analysis
Termaufspaltung *f* term splitting
Termdichte *f* level density
Termin *m* fixed term, [target] date, deadline
Terminkalender *m* appointments diary
Terminologie *f* terminology
Terminplan *m* [time] schedule
Terminplanung *f* scheduling
Termkonzentration *f* density of states
Termschema *n* energy level diagram, term scheme (diagram)
  T. eines Festkörpers energy spectrum of a solid
Termsystem *n* level system
  T. der Kerne nuclear system of terms
Termverschiebung *f* level shift
Termzuordnung *f* term assignment
ternär ternary
Terpen *n* terpene
Terpenalkohol *m* terpene alcohol
Terpentin *n* turpentine
Terpentinersatz *m* white spirit
Terpentinöl *n* turpentine oil, oil of turpentine
Terpentinölersatz *m* adulterated turpentine
Terphenyl *n* terphenyl
Terpinen *n* terpinene
Terpineol *n* terpineol
Terpinolen *n* terpinolene
Terrain *n* grade, ground
Terrainhöhe *f* grade (ground) level
Terrakotta *f* terra-cotta, terracotta
Terrakottaton *m* terracotta clay
Terrasse *f* terrace
Terrassenanbau *m* storey cropping (farming), terraced cropping (farming)
Terrassendach *n* terrace (platform) roof
Terrassenhaus *n* terrace house
Terrassenschnitt *m* cascade cut
terrassieren to terrace, to step
Terrazzo *m* terrazzo
terrestrisch terrestrial
Territorialgewässer *npl* territorial waters
Tertiär[formation *f*] *n* tertiary system
Tertiärgebirge *n* tertiary overburden
Tertiärwicklung *f* tertiary (stabilizing) winding
Terylene *n* terylene
Tesla-Lampe *f* Tesla lamp
Tesla-Röhre *f* Tesla tube
Tesla-Spule *f* Tesla coil
Tesla-Transformator *m* Tesla transformer
Test *m* test; check
Testband *n* test tape
Testbedingungen *fpl* test requirements
Testbenzin *n* painter's naphtha, white spirit
Testbild *n (telev)* test pattern
Testbildröhre *f* monoscope
testen to test
Testfilm *m* test film
Testprogramm *n (dat)* diagnostic routine
Testreiz *m* electrical test stimulus
Teströhre *f* test electron tube
Tetartoeder *n* tetartohedron
tetartoedral tetartohedral
Tetraborat *n* tetraborate
Tetraborsäure *f* tetraboric acid
Tetrachloräthan *n* tetrachloroethane
Tetrachlorkohlenstoff *m* carbon tetrachloride, *(Am)* carbon tet
Tetrachlorogoldsäure *f* gold hydrochloric acid
Tetrade *f* tetrad
Tetraeder *n* tetrahedron
tetraedral tetrahedral
tetraedrisch tetrahedral

Tetraedrit *m* tetrahedrite
Tetrafluoräthylen *n* tetrafluoroethylene
Tetrafluorkohlenstoff *m* carbon tetrafluoride
tetragonal tetragonal
Tetrahalogenid *n* tetrahalide
Tetrahydrofuran *n* tetrahydrofurane
Tetrakisazofarbstoff *m* tetrakisazo-dyestuff
Tetralin *n* tetralin[e]
Tetrathionsäure *f* tetrathionic acid
Tetrazin *n* tetrazin
Tetrode *f* tetrode
Tetrose *f* tetrose
Tetryl *n* tetryl
Teufe *f (mine)* depth
teufen to sink, to deepen
Teufenanzeiger *m* depth indicator
Tex-Numerierung *f* tex-system of yarn counts
Textildruck *m* fabric printing
Textilien *pl* textiles
Textilindustrie *f* textile industry
Textilöl *n* textile oil
Textilseife *f* textile soap
Textilveredlung *f* textile finishing, finishing of textiles
Textur *f* texture
Texturblech *n* grain-orientated lamination *(transformer core)*
Texturgarn *n* textured yarn
texturiert textured
Thallium *n* thallium
thalliumaktiviert thallium activated
Thalliumchlorid *n* thallium chloride
thalliumhaltig thalliferous
Thallium(I)-verbindung *f* thallous compound
Thallium(III)-verbindung *f* thallic compound
Thematik *f* subject matter
Thenardsblau *n* Thénard's blue
Theobromin *n* theobromine
Theodolit *m* theodolite
Theophyllin *n* theophylline
Theorem *n* theorem
  T. der übereinstimmenden Zustände theorem of corresponding states
Theoretiker *m* theorist
Theorie *f* theory
  T. der freien Elektronen free electron theory
  T. der plastischen Stabilität *(ship)* plastic stability theory
  T. der Schaltkreise network theory
  T. der spezifischen Wärme theory of specific heat
  T. der Spiele game theory, theory of games
  T. kleiner Signale small signal theory
  Debye-Hückelsche T. Debye-Hückel theory
  kinetische T. der Wärme kinetic theory of heat
  molekularkinetische T. molecular-kinetic theory
Therapeutikum *n* therapeutic
Therapiedosis *f* therapeutic (therapy) dose
Thermalhärtung *f* time quench hardening
Thermalisierung *f* thermalization
Thermalpräzipitator *m* thermal precipitator
Thermalsäule *f* thermal column
Thermion *n* thermion
Thermionenentladung *f* thermionic discharge
Thermionenquelle *f* thermionic ion source
thermionisch thermionic
thermisch thermal
Thermistor *m* thermistor
Thermistorkalorimeter *n* thermistor calorimeter
Thermistorsonde *f* thermistor head
Thermistorwiderstand *m* thermistor resistance
Thermit *n* thermit[e]
Thermitpreßschweißen *n* pressure thermit welding
Thermitschmelzschweißen *n* thermit fusion welding
Thermitschweißen *n* thermit welding, aluminothermics
Thermitverbundschweißen *n* thermit combined welding
Thermitverfahren *n* thermit (aluminothermic) process
Thermoanalyse *f* thermal analysis
Thermoaräometer *n* thermohydrometer
Thermobarograf *m* thermobarograph
Thermobarometer *n* thermobarometer
Thermobohrer *m* jet piercing lance
Thermochemie *f* thermochemistry
thermochemisch thermochemical
Thermochromfarbe *f* thermocolour
Thermodiffusion *f* thermal diffusion, Soret phenomenon
Thermodiffusionskoeffizient *m* thermal diffusion coefficient
Thermodiffusionskolonne *f* thermal diffusion column
Thermodiffusionspotential *n* potential of thermodiffusion
Thermodynamik *f* thermodynamics
  irreversible T. irreversible thermodynamics
  statistische T. statistical thermodynamics
Thermodynamiker *m* thermodynamicist
thermodynamisch thermodynamic
Thermoelastizität *f* thermoelasticity
thermoelektrisch electrothermal, thermo-electric, thermoelectric
Thermoelektrizität *f* thermoelectricity
Thermoelement *n* thermocouple, thermoelectric couple
  edles T. rare-metal couple
  unedles T. base-metal couple
Thermoelementkontaktstelle *f* thermoelectric junction, thermojunction

Thermo-EMK *f* thermal e. m. f.
Thermofixierung *f* thermosetting
Thermogalvanometer *n* thermo-galvanometer
Thermograf *m* thermograph
Thermografie *f* thermography
Thermoionisation *f* thermal ionization
Thermokraft *f* thermoelectric force (power), thermal force (power)
  absolute T. absolute thermoelectric force
Thermokreuz *n* thermocouple, thermal converter, thermo-junction
Thermoleistungsmesser *m* thermocouple wattmeter
Thermolumineszenz *f* thermoluminescence
Thermolumineszenzdosimeter *n* thermoluminescence dosimeter
Thermolyse *f* thermolysis
thermomagnetisch thermomagnetic
Thermomeßinstrument *n* thermal instrument
Thermometer *n* thermometer
  elektrisches T. electric thermometer
  manometrisches T. pressure thermometer
  nasses T. wet-bulb thermometer
  thermoelektrisches T. thermoelectric thermometer
  trockenes T. dry-bulb thermometer
Thermometerfaden *m* thermometric column
Thermometerhülse *f* thermometer well
Thermometerkugel *f* thermometer bulb
Thermometerregler *m* thermometer controller
Thermometersonde *f* thermometer probe
Thermometrie *f* thermometry
thermonuklear thermonuclear
Thermonuklearbombe *f* fusion bomb
Thermopaar *n* thermocouple, thermo-junction
thermophil thermophilic
Thermoplast *m* thermoplastic [material]
Thermoplastbandaufzeichner *m* thermoplastic tape recorder
Thermoplastfolie *f* thermoplastic sheet
thermoplastisch thermoplastic
Thermopren *n* thermoprene
Thermoregler *m* thermoregulator
Thermorelais *n* thermoelectric (thermal) relay
Thermosäule *f* thermoelectric pile, thermopile
Thermoschalter *m* temperature switch
Thermosflasche *f* thermos flask
Thermospannung *f* thermoelectric voltage (tension, potential)
Thermospannungsmesser *m* thermocouple voltmeter
thermostabil thermostable
Thermostat *m* thermostat
thermostatisieren to thermostate
Thermostatregler *m* thermostatic controller
Thermostatrelais *n* thermostat relay
Thermostatsteuerung *f* thermostatic control
Thermostrom *m* thermoelectric current
Thermoumformer *m* thermocouple, thermal converter
Thiamin *n* thiamine, aneurin, vitamin B_1
Thioalkohol *m* mercaptane, thioalcohol
Thioäther *m* thioether, sulphur ether, sulphoether
Thioharnstoff *m* thiourea, sulphourea, thiocarbamide
Thioindigo *n* vat red
Thiokol *n* thiokol
Thionylchlorid *n* thionyl chloride
Thiophen *n* thiophene
Thiophenol *n* thiophenol
Thioplast *m* thioplast
Thiosäure *f* thioacid, thioic acid
Thioschwefelsäure *f* thiosulphuric acid
Thiosulfat *n* thiosulphate
thixotrop thixotropic
Thixotropie *f* thixotropy
Thomasbirne *f* basic Bessemer converter, Bessemer converter with a basic refractory lining
Thomas-Fermi-Modell *n (nucl)* Thomas-Fermi model
Thomasmehl *n* Thomas meal, ground basic slag
Thomasroheisen *n* Thomas (basic Bessemer) pig iron
Thomasschlacke *f* basic slag
Thomasstahl *m* Thomas-Gilchrist steel, basic converter (Bessemer) steel
  kohlenstoffarmer T. Thomas low carbon steel
Thomasverfahren *n* Thomas-Gilchrist process, basic Bessemer process
Thomsonbrücke *f* double bridge
Thomsonstreuung *f* Thomson (classical) scattering
Thorerde *f* thoria, thorium dioxide
Thorianit *m* thorianite
thorieren to thoriate
Thorit *m* thorite
Thorium *n* thorium
Thoriumemanation *f* thorium emanation
Thoriumerz *n* thorium ore
Thoriumfamilie *f s.* Thoriumreihe
thoriumhaltig thorium-containing
Thoriumkreislauf *m* thorium cycle
Thoriumreaktor *m* thorium reactor
Thoriumreihe *f* thorium series
Thoriumspaltung *f* thorium fission
Thoriumzerfallsreihe *f* thorium series
Thornycroft-Kessel *m* speedy-type Thornycroft boiler
Thoron *n* thoron
Thortveitit *m (min)* thortveitite
Threonin *n* threonine
Thulium *n* thulium
Thuliumoxid *n* thulia
Thymianöl *n* thyme oil

Thymol *n* thyme camphor
Thymolblau *n* thymol blue
Thymonukleinsäure *f* thymonucleic acid
Thyratron *n* thyratron, gas triode, gas-filled three-electrode valve, gas-tube switch
Thyristor *m* thyristor
Thyroxin *n* thyroxine
Tidebecken *f* tidal basin
Tidehafen *m* tidal harbour
Tidenatlas *m* tide tables, tidal atlas
Tidengewässer *n* tidal waters
Tidenhub *m* tide range, range of tide
  täglicher T. diurnal range
Tidenmesser *m* tide-gauge
Tidenstillstand *m* stand of the tide
Tidentiefstand *m* set of tide
tief deep; low
Tief *n* depression, low [pressure area]
tiefätzen to deep-etch
Tiefätzung *f* final (deep) etching
Tiefaufreißer *m (build)* rooter
Tiefbaggerung *f* deep cut digging
Tiefbau *m* 1. civil engineering; 2. *(mine)* deep level working, underground (drift) mining
  städtischer T. municipal engineering
Tiefbaubetrieb *m (mine)* underground stoping
Tiefbaugrube *f* underground mine
Tiefbaustelle *f* below-grade site
Tiefbehälter *m* ground tank
Tiefbestrahlung *f* deep irradiation
Tiefbettanhänger *m* low-loading trailer
Tiefbettfelge *f* drop-base rim, drop-centre rim, well-base rim
Tiefbohranlage *f* deep well drilling plant
Tiefbohrarbeit *f* deep-boring
tiefbohren to deep bore
Tiefbohren *n* deep well drilling
  drehendes T. rotary [deep] drilling
  schlagendes T. percussion [deep] boring
  stoßendes T. percussion [deep] boring
Tiefbohringenieur *m* engineer for artesian well-drilling
Tiefbohrtechnik *f* deep-hole boring
Tiefbohrung *f* deep boring
Tiefbrunnen *m* deep well
Tiefbrunnenpumpe *f* deep-well pump
Tiefbunker *m* ground storage bin
Tiefdecker *m* low-wing aircraft
Tiefdruck *m* 1. low pressure; 2. photogravure, photogravure (intaglio) printing
Tiefdruckautomat *m* automatic gravure press
Tiefdruckfarbe *f* gravure ink
Tiefdruckgebiet *n* low-pressure area
Tiefdruckmaschine *f* gravure [printing] press
Tiefdrucknäpfchen *n* gravure cell
Tiefdruckpapier *n* gravure paper
Tiefdruckraster *m* gravure screen
Tiefdruckzentrum *n* centre of low pressure
Tiefe *f* depth
  optische T. optical depth
Tiefebene *f* low-level flat, lowland
Tiefeinbrandelektrode *f* deeply penetrating electrode, deep-penetration electrode
Tiefeinbrandschweißen *n* deep-penetration welding
Tiefenanhebung *f* bass boost *(radio)*
Tiefenanschlag *m* depth [control] stop *(boring)*
Tiefenanzeiger *m* 1. depth indicator; 2. bass indicator *(radio)*
Tiefenauslösung *f (mach)* depth (feed) trip
Tiefenbegrenzer *m* depth limiter
Tiefenbestrahlung *f* deep irradiation
Tiefendosis *f (nucl)* depth dose
  prozentuale T. percentage depth dose
Tiefendüngung *f* subsoil manuring (fertilization)
Tiefeneinstellung *f* depth control (adjustment)
Tiefenfeuereinrichtung *f* searching-fire device
Tiefenlehre *f* depth gauge
Tiefenlinie *f* fathom line, line of equal depths *(nautical chart)*
Tiefenmaß *n* depth [gauge] micrometer
Tiefenmelder *m* submarine sentry
Tiefenmesser *m* depth gauge; *(ship)* fathometer
Tiefenmeßgerät *n* depth gauge; *(ship)* depthometer; bathymeter; sounder
Tiefenmessung *f* depth measurement; *(ship)* bathymetry
Tiefenraum *m (opt)* space depth
Tiefenregler *m* tone control for bass *(radio)*
Tiefenschärfe *f (opt)* depth of focus (definition)
Tiefenskale *f* depth scale
Tiefensondierung *f* deep sounding
Tiefenstaffelung / mit arranged for depth cutting *(broach)*
Tiefensteuermaschine *f* depth-regulating gear
Tiefenstufe *f* gradient
  geothermische T. geothermal (rock-temperature) gradient
Tiefentherapie *f* deep (telecurie) therapy
Tiefentherapieanlage *f* telecurie [therapy] unit, teletherapy unit
Tiefenunschärfe *f (opt)* insufficient focus
Tiefenvergrößerung *f (opt)* depth magnification
Tiefenvorschub *m* vertical feed
  schräger T. angular downfeed
Tiefenwachstum *n* depth growth, growth in depth *(crystals)*
Tiefenwirkung *f* depth effect; penetrative effect *(plant protection)*
Tiefenzustellung *f* downfeed
Tieffach *n (text)* bottom (lower) shed *(Jacquard)*
tieffliegen to hedge-hop
Tiefflöz *n* deep seam
Tieffußnadel *f* low butt needle
Tiefgang *m* draught
  T. achtern draught aft

beschränkter T. reduced draught
geringer T. easy draught
kritischer T. beim Docken critical docking draught
zulässiger T. permissible draught
Tiefgangsänderung *f* change (alteration) of trim
Tiefgangsanzeiger *m* draught indicator (gauge)
Tiefgangsmarken *fpl* draught-marks
Tiefgangsskale *f* draught marks (scale), *(Am)* immersion scale
Tiefgarage *f* underground (deep-level) garage
Tiefgrubber *m* rigid tine cultivator, sub-drill cultivator
Tiefgründung *f* deep foundation
Tiefkühlanlage *f* deep freezer, cooling plant for low temperature
tiefkühlen to deep-freeze
Tiefkühlen *n* deep freezing
Tiefkühlfach *n* freezer [bin]
Tiefkühlhecktrawler *m* freezer stern trawler
Tiefkühlschrank *m* deep-freezing cupboard
Tiefkühltrawler *m* deep-freeze trawler
Tiefkühltruhe *f* freezer chest, chest freezer
Tiefkultur *f* deep ploughing
Tiefkulturpflug *m* grub breaker
Tiefladeanhänger *m s.* Tieflader
Tiefladelinie *f* load line, load (deep) water line
gesetzliche T. statutory load line
Tieflademarke *f* freeboard (load-line, Plimsoll) mark
Tieflader *m* low loader, low-bed trailer
Tiefladewagen *m (ry)* well wagon
tieflegen *(print)* to deep-etch
Tieflochaufbohren *n* deep-hole boring
Tieflochbohrarbeit *f* depth-drilling operation (job), deep-hole operation (job)
Tieflochbohren *n* deep-hole drilling
Tieflochbohrer *m* deep-hole drill; D-bit, drill bit; gun drill
Tieflochbohrkopf *m* gun-boring cutter head, cutting head for boring long bores
Tieflochbohrmaschine *f* deep-hole drilling (boring) machine
Tieflochreibahle *f* gun reamer
Tieflochspiralbohrer *m* crankshaft drill
Tieflockerer *m* depth tormentor, subsoiler
Tieflöffel *m* drag shovel; [back]hoe dipper
Tieflöffelbagger *m* backhoe, backacter, ditch digger, ditcher; drag shovel; backdigger, pull-scoop, trench hoe
Tieflot *n* deep-sea lead
Tieflotungsapparat *m* flying sounder
Tieflotungsgerät *n* flying sounder
Tiefofen *m* pit-type furnace, soaking pit [furnace]
beheizter T. live soaking pit
Tiefofenkran *m* soaking pit crane
Tiefpaß *m* low-pass filter *(radio)*
T. in Pi-Schaltung shunt capacitor input filter
T. in T-Schaltung series inductance input filter
Tiefpaßfilter *n* low-pass filter
Tiefpflug *m* grub breaker
Tiefpflügen *n* deep ploughing
Tiefpunkt *m* trough
Tiefschaft *m (text)* lowering shaft
Tiefschleifen *n* abrasive machining
Tiefschnittbagger *m* down-dredge
Tiefschnittmähbalken *m* low-level cutter-bar
Tiefschweißung *f* deep penetrating welding
Tiefschwund *m* blackout fading *(radio)*
Tiefsee *f* deep sea
Tiefseefischerei *f* off-shore fisheries
Tiefseefischfang *m* pelagic fishing
Tiefseelotung *f* deep-sea sounding, bathymetry
Tiefseemesser *m* bathymeter
Tiefseetauchboot *n* deep diving submarine, bathyscaph
Tiefseetauchschiff *n* bathyscaph
Tiefseewellen *fpl* deep-sea waves
tiefsiedend low-boiling
Tiefsilo *m* pit silo
Tiefstall *m* deep litter cowshed
Tiefstallmistwender *m* deep litter turner
Tiefstrahler *m* narrow angle lighting fitting, concentrating reflector; ring lamp *(boxing)*
Tiefstreu *f* deep litter
Tiefstreulegestall *m* deep litter hen house
Tieftank *m* deep tank
Tieftemperaturbehandlung *f* sub-zero treatment
Tieftemperaturbereich *m* low-temperature region
Tieftemperaturchemie *f* cryochemistry
Tieftemperaturdestillation *f* low-temperature distillation
Tieftemperaturkautschuk *m* cold rubber
Tieftemperaturphysik *f* low-temperature physics
Tieftemperaturteer *m* carbocoal (coalite) tar
Tieftemperaturvergasung *f* low-temperature distillation
Tieftemperaturverkokung *f* low-temperature carbonization
Tieftonabschwächung *f* under-colour removal
Tieftonblende *f* bass control
Tieftonlautsprecher *m* boomer, low-frequency speaker, woofer
Tieftonleistungsverstärker *m* subsonic power amplifier
Tieftonverstärker *m* subsonic amplifier
Tiefung *f* depth of the cup
Tiefungsmesser *m* deep-drawing tester
Tiefungsprüfgerät *n* cup (deep-drawing) testing device
Tiefungsversuch *m* cupping test

Tiefungswert *m* depth of the cup
Tiefziehblech *n* deep-drawing [quality] sheet
Tiefzieheignung *f (met)* deep-drawing capacity; *(plast)* swaging capacity
tiefziehen *(met)* to deep-draw; *(plast)* to swage
Tiefziehen *n (met)* deep drawing; *(plast)* swaging
Tiefziehverfahren *n (met)* cupping (deep-drawing) process
Tiefzug *m* deep drawing
Tiegel *m* crucible; melting pot
 feuerfester T. refractory crucible
 leitender T. conducting crucible
Tiegelautomat *m* automatic platen
Tiegeldruckpresse *f* platen press
Tiegelgußstahl *m* crucible (pot) steel
Tiegelofen *m* crucible furnace
Tiegelprobe *f* teeming hole
Tiegelschmelzofen *m* crucible (pot) melting furnacc
Tiegelstahl *m* crucible (pot) steel
Tiegelverfahren *n* crucible process
Tiegelzange *f* crucible tongs
Tiegelziehzange *f* crucible-lifting tongs
Tiekholz *n s.* Teakholz
Tierarzt *m* veterinary surgeon, *(sl)* vet
Tierbehälter *m* animal container
tierisch animal
Tierkohle *f* animal charcoal
Tierleim *m* animal size (glue)
Tieröl *n* bone oil
 Dippelsches T. Dippel['s] oil
Tierpfleger *m (ag)* stockman
Tierzucht *f* animal (live-stock) breeding; live-stock farming
tilgen to quench; to extinguish; to absorb *(vibration)*; to amortize; *(dat)* to delete; to erase
Tilger *m* absorber *(vibration)*
Tilgung *f* quenching; extinction; absorption *(vibration)*; amortization; *(dat)* deletion; erasion
Tilt-up-Bauweise *f* tilt-up construction (method)
Time-division-Multiplikator *m (dat)* time division multiplier
Tinkal *m (min)* native borax, tincal
Tinktur *f* tincture
Tinte *f* [writing] ink
 leitfähige T. electrographic ink
 lichtelektrische T. photoconductive ink
 magnetische T. magnetic ink
 unauslöschliche T. indelible ink
 Tintenfüllpatrone *f* ink cartridge
Tintenschreiber *m* siphon (ink) recorder, indicating pen, pen unit
Tintenstrich *m* / **ausgelaufener** feathering
Tintometer *n* tintometer
Tippbetrieb *m* inching, jogging *(operation)*
Tipper *m (aut)* carburetter tickler (primer)
Tippschalter *m* finger-tip switch
Tippschaltung *f* finger-tip control

Tisch *m (mach)* platen; table; bed *(press)*
 heizbarer T. heatable stage *(microscope)*
 hobelmaschinenähnlicher T. planer-type table
Tischanalogrechner *m* analogue desk computer
Tischanschlag *m* table stop
Tischantrieb *m* table drive
Tischapparat *m* desk set *(telephone)*
Tischaufspannfläche *f* table working surface
Tischausführung *f* table-mounted version *(telephone)*, bench[-mounting] type
Tischbohrmaschine *f* bench[-type] drilling machine
 feinfühlige T. sensitive drill [press]
Tischbohrwerk *n* table-type horizontal boring and facing machine
Tischbreite *f* table width
Tischdecke *f* table cloth
Tischdrehmaschine *f* bench lathe
Tischeilgangsknopf *m* table rapid-traverse button
Tischeinstellung *f* table positioning
Tischfarbwerk *n* slab inking unit
Tischfernsprecher *m* table telephone set
Tischfläche *f* table area, area of the table
 nutzbare T. working area of the table
Tischförderer *m* conveyor table
Tischführung *f* table track, platen sliding surface
Tischgehäuse *n* bench-mounting cabinet
Tischgerät *n* desk (bench-type) instrument
Tischgeschwindigkeit *f* table speed
Tischgröße *f* table size, size of the table
Tischhaltesteuerung *f* table dwell control
Tischhubspindel *f* table screw
Tischkasten *m* table drawer
Tischknagge *f* table trip
Tischkochplatte *f* / **elektrische** electric table cooker
Tischkreissäge *f* circular saw bench
Tischlampe *f* table lamp (standard), portable standard
Tischlänge *f* table length
Tischlängsgang *m* longitudinal table movement
Tischleiste *f* table strip
Tischleitspindel *f* table leadscrew
Tischler *m* joiner
Tischlerei *f* joiner's shop, joinery
Tischlerholz *n* benchwood
Tischlerkitt *m* plastic wood
Tischlerleim *m* joiner's glue
Tischlerplatte *f* batten [plywood] board
 T. mit Stäbchenmittellage laminated plywood board
Tischlerpolitur *f* wood finish
Tischlersteifsäge *f* cabinet saw
Tischlötgerät *n* bench soldering machine
Tischoberfläche *f* table top surface, top of the table

Tischputzmaschine *f* table cleaning machine
Tischrechenmaschine *f* top-of-desk calculator
Tischrechner *m* desk computer
Tischrücklaufgeschwindigkeit *f* table return speed
Tischrüttelformmaschine *f* plain jolt-moulding machine *(foundry)*
Tischrüttler *m* vibrating table *(concrete)*
Tischsäge *f* **mit Handvorschub** hand fed push bench
Tischschaltung *f* **für selbsttätigen Arbeitsablauf** automatic-cycle table-control
Tischseitenfläche *f* table side surface
Tischstillstanddauer *f* table dwell
Tischstütze *f* supporting bracket, table support *(shaper)*
Tischstützenklemmung *f* table support clamp
Tischtennis *n* table tennis, ping-pong
Tischtennisschläger *m* bat
Tischträger *m* cross-rail *(shaper)*; table support arm *(drilling machine)*
Tischträgerklemmung *f* rail clamp *(shaper)*
Tischumdrehung *f* revolution of table, table revolution
Tischumdrehungen *fpl* **je min** table r. p. m.
Tischumkehr *f* table reversal
Tischumsteuerhebel *m* table reverse lever
Tischumsteuerknagge *f* table reversal (reversing) dog
Tischumsteuerkupplung *f* table reverse clutch
Tischumsteuerung *f* table reversal, reversal of the table
Tischumsteuerungsgenauigkeit *f* table-reversal accuracy
Tischumsteuerventil *n* table-reversal valve
Tischvorschub *m* table feed
Tischvorschubantrieb *m* table-feed drive
Tischvorschubspindel *f* table-feed screw
Tischweg *m* table travel (traverse)
Tischzahnstange *f* table rack
Titan *n* titanium
Titanat *n* titanate
Titanblech *n* titanium sheet
Titandioxid *n* titanium dioxide, titania
Titaneisen[erz] *n* titaniferous iron ore, ilmenite
Titanerz *n* titanite
titanhaltig titaniferous, titanium-bearing
Titanit *m* titanite
Titankarbid *n* titanium carbide
Titansäure *f* titanic acid
Titanschwamm *m* titanium sponge
Titan(III)-verbindung *f* titanous compound
Titan(IV)-verbindung *f* titanic compound
Titanweiß *n* titania
Titel *m* heading
Titelbild *n* frontispiece
Titelblatt *n* frontispiece
Titelschrift *f* titling type
Titelzusatzgerät *n* movie titler
Titer *m* titre
Titereinstellung *f* establishment of titre, standarization [of titre]
Titration *f* titration
  coulometrische T. coulometric titration
  potentiometrische T. potentiometric titration
  radiometrische T. radiometric titration
Titrationslösung *f* titrant, volumetric (standard) solution
Titrierbecher *m* titrating beaker
titrieren to titrate
Titriermesser *m* titrimeter
titrimetrisch titrimetric, volumetric
TM-Welle *f* transverse magnetic wave
Tochteraktivität *f (nucl)* daughter activity
Tochteranzeigegerät *n* slave indicator
Tochteratom *n* daughter atom
Tochterelement *n* daughter element
Tochterfraktion *f* daughter fraction
Tochtergesellschaft *f* affiliated (subsidiary) company
Tochterkern *m* daughter nucleus
Tochterkompaß *m* auxiliary (repeater) compass
Tochterprodukt *n s.* Tochtersubstanz
Tochtersender *m* repeater (satellite) transmitter
Tochterstoff *m s.* Tochtersubstanz
Tochtersubstanz *f* daughter, decay product
Tochteruhr *f* slave clock
Tochterzerfall *m* daughter decay
Todesstrahlen *mpl* death rays
Toilette *f* lavatory, toilet, water-closet
Toilettenbecken *n* pan, toilet bowl
Toilettenseidenpapier *n* toilet tissue paper
Toilettenseife *f* toilet soap
Tokopherol *n* tocopherol, vitamin E
Toleranz *f* allowance, tolerance, clearance, limits
  enge T. close tolerance
  natürliche T. natural tolerance
  vorgewählte T. pre-set tolerance
  weite T. wide tolerance
Toleranzbereich *m* tolerance band
Toleranzfeld *n* tolerance zone
Toleranzgrenze *f* tolerance limit
toleranzhaltig holding to close limits (tolerances)
Toleranzmeßbrücke *f* limit bridge
Toleranzreihe *f* series of tolerances
Toleranzskale *f* limit dial
Toleranzstufe *f* grade of tolerance (accuracy)
Tolerieren *n* determination of suitable tolerances
Tolubalsam *m* Tolu balsam, balsam of Tolu
Toluol *n* toluene
Toluolmoschus *m* toluene musk
Toluolsulfochlorid *n* toluene sulfonyl chloride
Toluylsäure *f* toluic acid
Tolylierung *f* tolylation
Tombak *m* tombac, pinchbeck, red brass
Tombakschlauch *m* tombac sylphon

Tomografie *f* tomography, laminography, planigraphy
Ton *m* 1. clay; 2. *(ac)* note, tone, sound; 3. shade, tone *(colour)*
T. im Bild *(telev)* sound on vision
einfacher T. pure tone
fetter T. unctuous clay
feuerfester T. fire (refractory) clay
ganzer T. whole step (tone)
gebrannter T. burned clay
gewöhnlicher T. low-grade clay
hoher T. high-pitched note
magerer T. lean clay
plastischer T. plastic clay
reiner T. pure sound
stetiger T. steady sound
tiefer T. low-pitched note
unreiner T. ragged note
unstetiger T. rapidly fluctuating sound
Tonabnehmer *m* [sound] pick-up
dynamischer T. dynamic pick-up
elektrodynamischer T. moving-coil pick-up
elektrostatischer T. capacitor pick-up
kapazitiver T. capacitor pick-up
magnetischer T. magnetic pick-up
magnetodynamischer T. variable-reluctance pick-up
mechanischer T. acoustic pick-up
piezoelektrischer T. crystal pick-up
Tonabnehmereinsatz *m* pick-up cartridge
Tonabnehmerzusatz *m* gramophone attachment
Tonabstimmung *f* tone tuning
Tonabstufung *f* grading of tones
Tonalität *f* tonality
Tonarm *m* tone (pick-up) arm
Tonarmlager *n* tone arm bearing
Tonarmverriegelung *f* tone arm locking device
tonartig argillaceous, clayey
Tonaufbereitung *f* clay preparation
Tonaufnahme *f* record[ing]
Tonaufnahmeraum *m* sound-recording studio
Tonaufzeichnung *f* sound recording; canning
Tonaufzeichnungsgerät *n* sound recorder
Tonausfall *m (telev)* loss of sound
Tonbad *n* toning bath *(photography)*
Tonbagger *m* clay excavator
Tonband *n* 1. tape; 2. *(geol)* clay band
Tonbandaufnahme *f* record on a tape
Tonbandgerät *n* tape recorder
Tonbank *f* clay bank
Tonbeton *m* mechanical [soil] stabilization
Tonblende *f* tone control
Tonboden *m* clay soil
Tondichtungskern *m* clay core
Tondreieck *n* pipeclay triangle
Tonempfänger *m* sound receiver
Tonemulsion *f* clay emulsion
tonen 1. to tone *(photography)*; 2. to tint
Tonen *n* 1. toning *(photography)*; 2. tinting
tönen to sound
Tonerde *f* 1. alumina, alum (argillaceous) earth; 2. clay soil
essigsaure T. acetate of alumina
Tonerdebeize *f* red liquor
Tonerdegehalt *m* alumina content
tonerdehaltig aluminiferous
Tonerdeschamotte *f* clay-bond fire-clay
Tonerdeschmelzzement *m* aluminous cement
Tonerdesilikat *n* aluminium silicate
Tonerdezement *m* aluminous (high-alumina, alumina) cement
Tonfall *m* cadence
Tonfenster *n* sound gate
Tonfilm *m* sound film
Tonfilmgeschwindigkeit *f* sound-film speed
Tonfilmprojektor *m* sound picture projector
Tonfilter *n* tone filter
Tonfixierbad *n* [fixing and] toning bath *(photography)*
Tonfräse *f* clay shredder
Tonfrequenz *f* audio (voice, sonic) frequency
Tonfrequenzamplitudenmodulation *f* audio-tone amplitude modulation
Tonfrequenzband *n* audio-frequency band
Tonfrequenzfernwahl *f* voice frequency dialling
Tonfrequenzkanal *m* audio channel
Tonfrequenzoszillator *m* audio oscillator
Tonfrequenzrelais *n* voice-frequency relay
Tonfrequenzruf *m* voice frequency ringing
abgestimmter T. harmonic selective ringing
Tonfrequenzrufumsetzer *m* voice-frequency signalling relay set
Tonfrequenzschwebung *f* audio frequency beat
Tonfrequenzspektrometer *n* audio-frequency spectrometer
Tonfrequenzspektrum *n* audible spectrum
Tonfrequenzteil *m* audio-frequency section
Tonfrequenztelegrafie *f* audio-frequency telegraphy
Tonfrequenzübertrager *m* audio-frequency transformer
Tonfrequenzverstärker *m* audio amplifier
zweistufiger T. double-note amplifier
Tonfrequenzzeichengebung *f* tonic train signalling
Tonfrequenzzerhacker *m* audio-frequency chopper
Tonfüllkörper *m* clay tile filler
Tonfunk *m* sound radio
Tongefäß *n* earthenware container
Tongehalt *m* clay content
Tongemisch *n* complex sound
Tongenerator *m* tone (audio) generator
tongesteuert tone-operated
Tongewinnung *f* getting of clay
Tongips *m* argillaceous gypsum

Tongrube *f* clay pit
Tongut *n* porous earthenware (ceramic)
tonhaltig aluminous, argillaceous, clayey
Tonhobel *m* clay planer
Tonhöhe *f (ac)* pitch
  T. des Kammertons standard pitch
  T. des Zeichens pitch of the signal note
  niedrige T. low pitch
Tonhöhenabstimmung *f* note tuning
Tonhöhenschwankung *f* pitch variation
tonig argillaceous, clayey
Tonikum *n* tonic
Toningenieur *m* audio engineer
Toninjektion *f* pressure grouting with clay
Tonintervall *n* tone interval
Tonkalk *m* argillaceous limestone, argillocalcite
Tonkamera *f* sound camera
Tonkanal *m* sound channel
Tonkardiogramm *n* phonocardiogram
Tonkern *m* clay core
Tonkomponente *f* clay fraction
Tonkonstanz *f* constancy of pitch
Tonkopf *m* sound head
Tonlagerstätte *f* clay deposit
Tonlampe *f* exciter lamp
Tonlehm *m* clay loam
Tonleiter *f* [musical] scale
  chromatische T. chromatic scale
  wohltemperierte T. just scale
Tonmeister *m* monitor man; sound engineer
Tonmergel *m* clayey marl, loam
Tonmischeinrichtung *f* audio mixer
Tonmischpult *n* sound mixer
Tonmischung *f* sound mixing
tonmoduliert sound-modulated, modulated at audible frequencies
Tonmontagegerät *n* editing recorder
Tonmotor *m* capstan motor *(tape recording)*
Tonnage *f* tonnage, shiproom
Tonnagevermessung *f* tonnage measurement
Tonne *f* 1. barrel, cask; 2. buoy *(navigation)*; 3. metric ton, tonne *(1000 kg)*
  T. Tragfähigkeit ton burden (deadweight)
  T. zur Bezeichnung einer Baggerschüttstelle spoilground buoy
  amerikanische T. short ton *(2000 lbs. = 907 kg)*
  befeuerte T. lighted buoy
  englische T. long ton *(2240 lbs. = 1016 kg)*
  metrische T. metric ton *(1000 kg)*
  stumpfe T. cylindrical buoy
Tonnendach *n* barrel (compass) roof
tonnenförmig barrel-shaped
Tonnengehalt *m (ship)* equipment tonnage
Tonnengewölbe *n* circular (annular, barrel) vault
Tonnenlager *n* barrel-type roller bearing
Tonnenleger *m* buoy layer, buoy-laying vessel; lighthouse tender
Tonnenretortenofen *m* drum-type retort furnace
Tonnenschale *f* barrel (cylindrical) shell
Tonnenverzeichnung *f* positive (barrel) distortion
tonnenweise by tons
tonnlägig *(mine)* inclined
Tonometer *n* tonometer
Tonpfeiler *m (geol)* clay pillar
Tonplatte *f* clay tile
Tonqualität *f* tonal quality
Tonquelle *f* sound source
Tonrad *n* phonic (tone) wheel
Tonregler *m* sound corrector
Tonreinheit *f* purity of tone (sound)
Tonreißer *m* clay shredder
Tonrohr *n* clay pipe
Tonrohrdränung *f* tile drainage
Tonrolle *f* capstan
Tonsand *m* argillaceous sand
Tonsandstein *m* argillaceous sandstone
Tonscherben *m* clay body
Tonschiefer *m* clay schist (slate), shale, argillaceous rock
  aufgeblähter T. expanded shale (slate)
tonschieferartig argillitic
Tonschlämme *f* clay wash *(moulding)*
Tonschneider *m* clay cutter; pug mill
Tonschreiber *m* mechanical recording head
Tonschrift *f* tone print
Tonschürze *f* clay blanket
Tonseife *f* aluminous soap; soap earth
Tonsender *m* sound (aural) transmitter
Tonsieb *n* audio (tone) filter, tone control
Tonsignal *n* sound signal
Tonsilo *m* clay storage bin
Tonspektrum *n* sound spectrum
Tonsperrkreis *m* sound trap
Tonspülung *f* clay base mud
Tonspur *f* sound track
Tonstärke *f* intensity of tone, sound-energy flux density
Tonstreifen *m* sound track
Tonstudio *n* recorder room
Tonsuspension *f* clay suspension
Tontaube *f s.* Wurftaube
Tontechnik *f* acoustic (audio) engineering
Tonteil *m (telev)* sound section
Tonteller *m* porous plate
Tontiegel *m* fire-clay crucible
Tonträger *m* sound carrier
Tontrockner *m* clay drier
Tonüberlagerung *f* modulation at audio frequency
Tonumfang *m* sound volume, tonal range
Tönung *f s.* Tonen
Tönungskraft *f* tinting strength
Tonunterdrückung *f* sound rejection, rejection of the accompanying sound

Tonverstärker *m* sound (note) amplifier
Tonverstärkung *f* sound (audio-frequency) amplification
Tonvorkommen *n* clay deposit
Tonware *f* clay product
Tonwert *m* tonal value
Tonwertwiedergabe *f* tone reproduction
T. von Farben colour value rendition
Tonwiedergabe *f* sound reproduction, reproduction of sound
räumliche T. stereophonic reproduction
Tonwirkung *f* sound effect
Tonzelle *f* 1. porous pot (cell); 2. sound box
Tonzeug *n* impermeable earthenware
Tonziegel *m* clay brick (tile)
Tonzuschlag *m* aluminous flux
Topas *m* topaz
Topf *m* cask, jar; pot; body *(hand-grenade)*
Topfanode *f* can anode *(electron tube)*
Töpfer *m* potter
Töpferei *f* pottery
Töpferglasur *f* potter's lead
Töpferscheibe *f* potter's wheel
Töpferton *m* potter's clay
Töpferwaren *fpl* pottery[-ware], earthenware
Topffärben *n* potting
topfförmig cup-shaped *(tools)*
Topffräser *m* cup-shaped cutter
Topfglühofen *m* pot (pan-type) annealing furnace
Topfkern *m (el)* pot core
Topfkernspule *f* pot-type coil
Topfkreis *m* closed resonator, pot (shell) circuit
Topfmagnet *m* cylindrical (pot) magnet, shielded electromagnet
Topfmanschette *f* cup seal
Topfpflanze *f* pot-plant, potted plant
Topfschleifscheibe *f* cup wheel
gerade T. straight cup wheel
kegelige T. flaring cup wheel
Topfsockel *m* shaped base
Topfspindel *f* can spindle
Topfspinnen *n* can spinning
Topfspinnmaschine *f* can spinning frame
Topfstrecke *f (text)* can gill box
Topfwagen *m* carboy (jar) wagon
Topfwandler *m* potted (potted instrument, insulator-type) transformer
Topochemie *f* topochemistry
topochemisch topochemical
Topograf *m* topographer
Topografie *f* topography
topografisch topographic[al]
Topologie *f* topology
topologisch topological
Topp *m* top [end]; [mast-]head
Toppbenzin *n* topped gasoline
topplastig top-heavy
Topplaterne *f* steaming light, mast-head lantern, top lamp
Topplicht *n* mast-head lantern, steaming light, top lamp
Topprückstand *m* long residue, topped crude
Toppseitenspant *n* cantilever frame
Toppseitentank *m* topside (cantilever, gunwale, corner) tank
Tor *n* 1. *(build)* gate[way]; 2. gate *(pulse engineering)*; 3. goal
zweiteiliges T. folding gate
Toranschluß *m* gate connection; gate terminal
Torbernit *m (min)* torbernite
tordieren to twist
Tordurchfahrt *f* gateway
Toreinfahrt *f* doorway
Torelektrode *f* gate electrode
Torf *m* peat
Torfbagger *m* peat excavator
Torfboden *m* peat soil
Torfbruch *m* peat swamp
Torfgas *n* peat gas
Torfgrube *f* peat pit
Torfkohle *f* peat charcoal
Torfkoks *m* peat coke
Torfmoor *n* peat bog
Torfmull *m* peat dust, ground peat
Torfpflug *m* peat cutter
Torfpresse *f* peat cake press
Torfschneider *m* peat cutters
Torfstaub *m* peat dust
Torfstechen *n* peat digging
Torfstecher *m* peat digger
Torfverkohlung *f* peat charring
Torimpedanz *f* gate impedance
Torimpuls *m* gating impulse
torisch toric
Torkran *m* frame (portal) crane
Torkreis *m* gate circuit
Torkret *m s.* Torkretbeton
Torkretbeton *m* gunite, shotcrete, gunned (air-placed) concrete, jetcrete
Torkretgerät *n s.* Torkretiergerät
torkretieren to gunite
Torkretieren *n* guniting
Torkretiergerät *n* cement (concrete, shotcreting) gun
Torkretkanone *f s.* Torkretiergerät
Torkretputz *m* gunite, shotcrete, jetcrete, gunned (air-placed) concrete
Torkretverfahren *n s.* Torkretieren
Torlatte *f* crossbar
Tornisterempfänger *m* pack (kit-bag) receiver
Tornisterfunkgerät *n* knapsack transmitter, pack-type radio-set
Tornistergerät *n* portable set
Tornisterkessel *m* knapsack boiler
Toröffnungszeit *f* gate time

Toroid *n (math)* toroid
Toroidröhre *f* [vacuum] doughnut, toroid
Toroidspule *f* toroid[al coil], single-ring coil
torpedieren to torpedo
Torpedo *m* 1. *(mil)* torpedo; 2. *(mine)* safety plug *(in a borehole)*
Torpedo[blasen]bahn *f* torpedo wake
Torpedoboot *n* topedo-boat
Torpedobootsheck *n* torpedo-boat stern
Torpedoflugzeug *n* torpedo aircraft
Torpedokopf *m* torpedo warhead
Torpedolaufbahn *f* torpedo track
Torpedorohr *n* torpedo tube
Torpfosten *m* goal post
Torr *n* torr
Torröhre *f* gate (gating) valve; dual-control heptode
Torschaltung *f* gate (gating) circuit *(switching circuit)*
Torschaltungslogik *f* gate logic
Torschiff *n* gate vessel, caisson
Torsignal *n* gate signal
Torsiograf *m* torsiograph
Torsion *f* torsion, twist[ing]
Torsionsbandfeinzeiger *m* torsion balance
Torsionsbeanspruchung *f* torsional load (stress)
Torsionsdynamometer *n* torsion dynamometer
Torsionselastizität *f* torsion elasticity
Torsionselektrometer *n* torsion electrometer
Torsionsfaden *m* torsion thread
Torsionsfeder *f* torsion spring
Torsionsfestigkeit *f* torsional strength
Torsionsgalvanometer *n* torsion galvanometer
Torsionsindikator *m* torsion indicator
Torsionskraft *f* torsional (twisting) force
Torsionsmoment *n* torsional (external twisting) moment
Torsionspendel *n* torsional pendulum
Torsionsprüfmaschine *f* torsion-testing machine
Torsionsschwingung *f* torsional oscillation (vibration)
Torsionsspannung *f* torsional stress
Torsionsstabfederung *f* torsion-bar springing
torsionssteif stiff against torsion
Torsionssteifigkeit *f* torsional rigidity
Torsionsverband *m* torsion frame
Torsionsversuch *m* torsional test
Torsionswaage *f* torsion balance
Torsionswelle *f* torsional wave
  hydromagnetische T. torsional hydromagnetic wave
Torsionswiderstand *m* torsional resistance
Torsionswinkel *m* angle of torsion (twist)
Torsteuerung *f* gating
Torus *m* torus
Torweg *m* gateway
Tosbecken *n* stilling basin
Toskammer *f s.* Tosbecken
Tosylierung *f* tosylation
total total; complete
Totalbestrahlung *f* whole-body irradiation, total-body irradiation
Totalbetrag *m* total amount (sum)
Totalbruchbau *m (mine)* full caving, total caving [method]
Totaldurchgang *m* total transmission
Totalkontrolle *f* total inspection
Totalnutzeffekt *m* overall efficiency
Totalreflexion *f* total reflection
  innere T. inner total reflection
Totalschwund *m* fade-out *(waves)*
Totalverlust *m* total loss
totbrennen to deadburn, to overburn *(lime)*
Totgang *m* lost motion, backlash
Totholz *n (ship)* dead (rising) wood, skeg
Totlage *f* dead-centre position
Totlagenstellung *f* dead-centre position
Totlast *f* dead weight
Totmannbremse *f* dead-man brake
Totmanneinrichtung *f* dead man's handle
Totmannkurbel *f* dead-man's handle
Totmannsknopf *m* dead man's button
Totpunkt *m* dead centre; slack point
  oberer T. top dead centre
  unterer T. bottom dead centre
totpunktfrei without dead centre
Totpunktlage *f* dead-centre position
Totpunktsmarke *f* timing mark
Totpunktzündung *f* dead-centre ignition
Totraum *m* wake space
Totreife *f* dead ripeness, blank ripe
Totrösten *n* dead roasting *(ores)*
Totspritzen *n (ag)* haulm destruction
Tötung *f* / **elektrische** electrocution
Totwasser *n* still (slack, dead) water; wake
Totwassergeschwindigkeit *f* overground speed, speed in still water
Totzeit *f* dead time, distance velocity lag, *(Am)* transportation lag; *(mach)* down (non-cutting) time
  innere T. recovery time
  reine T. pure time delay
Totzeitkorrektion *f* coincidence (dead-time) correction
Touren *pl* revolutions
Tourendampfer *m* [regular] trader, liner
Tourenfahrrad *n* roadster
Tourenregler *m* speed control
Tourenstellzeug *n* **des Reglers** adjusting gear of the governor
Tourenversteller *m* speed changer
Tourenzahl *f* number of revolutions [per unit time], engine revolutions, shaft speed
Tourenzähler *m* revolution counter
Touristendampfer *m* excursion (pleasure) steamer
Touristenkarte *f* tourist ticket

Touristenklasse *f* tourist (economic) class
Touristik *f* touring, travel for pleasure
Tournaiteppich *m* cut pile carpet
Tournaivelourteppich *m* Wilton carpet
Townsend-Entladung *f* Townsend discharge
primäre T. first Townsend discharge
toxikologisch toxicological
Toxikum *n* toxic agent
Toxin *n* toxin, toxicant
toxisch toxic
Toxizität *f* toxicity, toxity
Trabant *m* 1. tail *(impulse)*; 2. satellite
künstlicher T. artificial satellite
Trabanten *mpl (telev)* equalizings
Trabantenstadt *f* satellite town
Tracer *m* tracer, indicator
Tracerchemie *f* tracer chemistry
Tracerlaboratorium *n* tracer lab[oratory]
Tracermenge *f* tracer amount, trace [amount]
Tracerverfahren *n* tracer technique (method), labelled atoms method, radioactive tracer method
Tracht *f* 1. sea gauge *(draught of a laden ship)*; 2. *(ag)* crop
Trächtigkeit *f (ship)* burden [tonnage]
Trachyt *m (geol)* trachyte
Trachytlava *f* trachytic lava
Track *m* sea route, track
Trackbeginn *m* track initiation
Trackbezeichnung *f* track designation
Trackdarstellung *f* track display
tracken to track
Trafo *m s.* Transformator
Traganteil *m* percentage of contact area, ratio of bearing contact area to total area *(surface)*
Tragarm *m* cantilever, support[ing] arm
Tragbahre *f* stretcher, litter
Tragbalken *m* column, standard, girder
Tragband *n* carryable belt conveyor
tragbar 1. *(mach)* portable; 2. *(text)* wearable
Tragbild *n* surface appearance *(piston etc)*
Tragbock *m* boiler bearer (cradle, saddle)
stopfbuchsenartiger T. pedestal forming a stuffing box
Tragdraht *m* suspension (supporting, span) wire
träge *(ch)* inert; *(el)* sluggish *(fuse)*, slow-acting
Trageeigenschaft *f (text)* wearability
tragen 1. to support, to bear; to carry; 2. *(text)* to wear
Tragen *n* **der Zähne** bearing of the teeth
tragend supporting, load-bearing *(wall)*, structural
Träger *m* 1. support; *(build)* beam, girder, joist; carrier *(radiochemistry)*; supporting material (substance, base) *(chromatography)*
T. auf zwei Stützen simple (simply supported) beam
T. mit Auflagerschrägen haunched beam
aufgefrischter T. reconditioned carrier
betonummantelter T. cased beam
durchlaufender T. continuous beam
eingeschwungener T. steady carrier
eingespannter T. fixed-end girder, fixed beam
frei aufliegender T. simple beam
hochfrequenter T. radio-frequency carrier
impulsmodulierter T. pulse-modulated carrier
isotoper T. isotopic carrier
schräger T. raker beam
sprachgesteuerter T. voice-controlled carrier
starker T. strong carrier
unmodulierter T. unmodulated carrier, continuous waves
vollwandiger T. web girder
wandartiger T. deep girder (beam), diaphragm beam
Trägerachse *f* axis of the beam
Trägeramplitude *f* carrier amplitude
Trägeramplitudenabweichung *f* carrier shift
Trägeramplitudenregelung *f* carrier amplitude regulation
Trägeranalyse *f* carrier analysis
Trägerausgangsleistung *f* carrier power output rating
Trägerbeweglichkeit *f* carrier (Hall) mobility
Trägerdichte *f* carrier density
Trägerdiffusion *f* carrier diffusion
Trägerdrift *f* excursion of the carrier
Trägereinfang *m* carrier trapping
Trägerelement *n* carrier
Trägererzeugung *f* carrier production
Trägerfalle *f* carrier trap
Trägerfällung *f* carrier precipitation
Trägerfluß *m* carrier flow
Trägerflüssigkeit *f* vehicle, carrying agent, liquid carrier
trägerfrei carrier-free *(radiochemistry)*
Trägerfrequenz *f* carrier frequency
feste T. resting carrier frequency
Trägerfrequenzabweichung *f* carrier shift
Trägerfrequenzbereich *m* carrier frequency range
Trägerfrequenzdrahtfunk *m* carrier-frequency wire-broadcasting
Trägerfrequenzendeinrichtung *f* carrier terminal set
Trägerfrequenzfernsprechen *n* carrier telephony
Trägerfrequenzkonstanz *f* carrier-frequency stability
Trägerfrequenzleitung *f* carrier circuit
Trägerfrequenzsperre *f* line trap
einwellige T. single-frequency line trap
Trägerfrequenztechnik *f* carrier art
Trägerfrequenztelefonie *f* carrier[-current] telephony
Trägerfrequenztelegrafie *f* carrier telegraphy
Trägerfrequenzüberlagerer *m* homodyne
Trägerfrequenzumsetzer *m* carrier converter

Trägerfrequenzunterdrückung *f* carrier suppression
Trägerfrequenzverstärker *m* modulated-carrier amplifier
Trägergas *n* carrier gas
trägergesteuert carrier-controlled
Trägerhöhe *f* depth of beam (girder)
Trägerimpuls *m* pulse carrier
Trägerinjektion *f* carrier injection
Trägerkielschwein *n* *(ship)* detached keelson
Trägerkonzentration *f* carrier density
Trägerlage *f* position of the carrier
Trägerlaufzeit *f* carrier transit time
Trägerlawine *f* electron avalanche
Trägerlebensdauer *f* carrier lifetime
Trägerleistung *f* carrier power
Trägerleitung *f* carrier frequency line
trägerlos *(plast)* unsupported
Trägermaterial *n* base material; substrate; support, carrier (back-up) material
Trägerplatte *f* backing plate; base
Trägerrakete *f* carrier (mother) rocket, satellite vehicle
Träger-Rausch-Abstand *m* carrier-to-noise ratio
Trägerrauschen *n* carrier noise
Träger-Rausch-Verhältnis *n* carrier-to-noise ratio
Trägerreichweite *f* carrier range
Trägerrost *m* girder (beam) grillage *(foundation)*
Trägerschwingung *f* carrier wave
  impulsmodulierte T. carrier-frequency pulse
Trägerspannung *f* carrier-frequency voltage
Trägerstab *m* supporting rod *(electron tubes)*
Trägersteuerung *f* carrier control
Trägerstrom *m* 1. carrier current; 2. carrier stream
Trägerstromfernsprechen *n* carrier [current] telephony
Trägerstromtelefonieeinrichtung *f* carrier telephone system
Trägerstromtelegrafie *f* carrier [frequency] telegraphy
Trägerstromverstärker *m* carrier repeater
Trägersubstanz *f* base (back-up) material
Trägertelefonieverstärker *m* carrier telephony repeater
Trägerverbindung *f* carrier compound
Trägerwalzwerk *n* beam (girder) rolling mill
Trägerwelle *f* carrier [wave]
  angehobene T. reconditioned carrier
  modulierte T. modulated continuous wave
  unterdrückte T. suppressed carrier wave
Trägerwellenamplitude *f* **für Weißmaximum** *(telev) (Am)* carrier-reference white level
trägerwellengesteuert carrier-operated
Trägerwellenschrift *f* carrier script *(for data processing)*
Trägerwerkstoff *m* base material
Trägerzusatz *m* carrier addition
Trägerzusatzeinrichtung *f* *(telec)* speech restorer
Tragestange *f* **des Spulentellers** *(text)* cone plate bar
  T. für den Stoffbreithalter cloth spreader support shaft
  mittlere T. des Spulengestells centre shaft
Tragewelle *f* **des Spulengestells** *(text)* bobbin support shaft
Tragezapfen *m* trunnion
Tragfähigkeit *f* load-carrying capacity (ability), bearing capacity (power, resistance, strength), supporting capacity (power, resistance, strength); lifting (hoisting) capacity *(crane)*; *(ship)* deadweight all told; net (deadweight loading, deadweight carrying, carrying) capacity, shipload
  T. an reiner Ladung useful deadweight, net capacity, shipload
  T. der Reifendecke load capacity of the tyre cover
  T. des Baugrundes bearing power of the soil, load-bearing capacity of the soil
  T. des Versatzes supporting strength of filling
Tragfähigkeitsfaktor *m* deadweight cargo factor
Tragfähigkeitskoeffizient *m* deadweight/displacement coefficient
Tragfähigkeitstonne *f* deadweight ton
Tragfeder *f* suspension spring
Tragfestigkeit *f* breaking (ultimate) strength
Tragfilz *m* carrier felt *(paper)*
Tragfingerling *m* bearing pintle *(rudder)*
Tragfläche *f* bearing (supporting) surface; *(aero)* airfoil, wing
  durchgehende T. *(aero)* one-piece wing
Tragflächenantenne *f* *(aero)* notch aerial
Tragflächenboot *n* hydrofoil [boat]; *(Am)* hydrofoil seacraft
Tragflächenende *n* *(aero)* wing tip
Tragflächenhautfräsmaschine *f* skin milling machine
Tragflächenhinterkante *f* *(aero)* trailing edge
Tragflächenkanone *f* wing cannon
Tragflächenmaschinengewehr *n* wing gun
Tragflächenspitze *f* *(aero)* tip
Tragflächenstrebe *f* *(aero)* wing strut
Tragflächenüberhang *m* *(aero)* wing tip rake
Tragflächenzuspitzung *f* *(aero)* wing taper
Tragflügel *m* *(aero)* airfoil
Tragflügelboot *n* hydrofoil boat
Tragflügelprofil *n* *(aero)* airfoil profile; *(ship)* hydrofoil section
Tragflügelpropeller *m* hydrofoil propeller
Traggas *n* lifting gas
Traggerüst *n* support rack
Traggeviert *n* *(mine)* bearer, bearing set
Traggewölbe *n* arch support
Traggriff *m* lifting handle

Traggurt *m* shoulder strap
Trägheit *f (mech)* inertia; *(ch)* inertness, inactivity
  T. eines Empfängers receiver response time *(radio)*
Trägheitsachse *f* axis of inertia
  mittlere T. mean axis of inertia
Trägheitsbeanspruchung *f* inertia stress
Trägheitsbewegung *f* inertial motion
Trägheitsellipse *f* ellipse of inertia
Trägheitsellipsoid *n* inertia ellipsoid
Trägheitsflug *m* inertial flight
trägheitsfrei inertialess
Trägheitsgesetz *n* law of inertia
Trägheitshalbmesser *m* radius of gyration (inertia)
Trägheitskoeffizient *m* inertia coefficient
Trägheitskraft *f* force of inertia, inertial (mass) force
Trägheitskreis *m* area moment circle
Trägheitslenkung *f* inertial guidance
trägheitslos inertialess
Trägheitsmittelpunkt *m* centre of inertia
Trägheitsmoment *n* moment of inertia
  axiales T. axial moment of inertia
  polares T. polar moment of inertia
Trägheitsnavigation *f* inertial navigation
Trägheitsortung *f s.* Trägheitsnavigation
Trägheitsradius *m* radius of gyration (inertia)
Trägheitsvermögen *n* inertia
Trägheitswiderstand *m* inertia resistance
  T. des ausströmenden Dampfes inertia of the outflowing steam *(engine)*
Tragkette *f* suspension chain
Tragklotz *m* shoe, segment *(segment bearing)*
Tragkonstruktion *f* supporting structure
Tragkraft *f* load (load-carrying, supporting, carrying) capacity; lifting power (capacity); burden *(cargo-carrying capacity)*
Tragkranz *m* lintel girder *(blast furnace)*; *(therm)* suspension segment; *(mine)* wedging crib
Tragkranzsäule *f* mantle pillar *(blast furnace)*
Traglager *n* journal bearing; bearing support; *(mach)* arbor support (yoke) *(milling)*; *(ship)* plummer block
Traglast *f* ultimate (collapse) load
Traglastverfahren *n* limit (ultimate load, ultimate strength) design, ultimate load (strength) method, load-factor method
Tragmast *m* suspension tower (pole); supporting pole (mast)
Tragmasthängekette *f* suspension-tower insulator chain
Tragring *m* thrust collar *(shaft)*; lintel girder *(blast furnace)*; ring support (carrier)
Tragrohr *n* suspension tube
Tragrolle *f* carrying (idler) roller
Tragsäule *f* strut
Tragschicht *f* base [course] *(highway construction)*
Tragschlaufe *f* sling
Tragschrauber *m* autogiro
Tragseil *n* standing rope; carrying (messenger, catenary, supporting) cable, bearer-cable, bracer
Tragseilgurtbandförderer *m* cable-belt conveyor
Tragseite *f* carrying side *(belt)*
Tragstück *n* **des Kolbens** piston shoe
Tragstuhl *m* boiler support
Tragstütze *f* supporting member (bracket)
Tragvermögen *n* supporting (load-carrying, carrying) capacity
Tragweite *f* reach, range; geographic range
Tragwerk *n* framework, supporting structure; *(aero)* wing assembly
  räumliches T. space structure (frame, framework)
Tragzapfen *m* pivot, trunnion
Trailerschiff *n* trailer ship
Trajekt *m (n)* ferry bridge
Trajektorie *f* [phase] trajectory
Trajektschiff *n* car ferry, sea train
Trakt *m* tract
Traktor *m* tractor
  landwirtschaftlicher T. farm tractor
Traktorenanhängevorrichtung *f* tractor hitch
Traktorenkraftstoff *m* tractor fuel (vaporizing oil); power kerosene
Traktorenmotor *m* vaporizing oil engine, tractor engine
Traktorist *m* tractor driver
Traktormähbinder *m* tractor-drawn binder
Traktorpflug *m* tractor plough
Traktorrechen *m* tractor-mounted hay rake
Traktorzug *m* tractor haulage
Traktrix *f* tractrix [curve]
Tramper *m s.* Trampschiff
Trampoline *f* [elastic] springboard
Trampschiff *n* tramp [ship], tramper, all-round tramp
Trampschiffahrt *f* tramping
Tramseide *f* tram
Tran *m* train-oil, liver (fish) oil; blubber
Tränengas *n* tear-gas
tränenreizend lachrymatory
Tränenschwamm *m* dry rot fungus
tränken to impregnate, to soak, to steep; to saturate; *(mine)* to infuse *(face)*
  mit Harz t. *(plast)* to resin[ate]
  mit Öl t. to impregnate with oil
  mit Teeröl t. to creosote
  mit Zinkchlorid t. to burnettize
Tränkflüssigkeit *f* impregnation liquid
Tränkharz *n (plast)* resin varnish
Tränkkessel *f* impregnating boiler
Tränkkopf *m (mine)* infusion-tube head
Tränklack *m* impregnating varnish
Tränklegierung *f* alloy for impregnation

Tränklösung *f* impregnating solution
Tränkmasse *f* impregnating compound (matter)
Trankochanlage *f* liver[-boiling] plant
Tränkrohr *n (mine)* infusion tube
Tränkschlauch *m (mine)* infusion hose
Tränkstelle *f* watering place; watering halt (stop) *(animals)*
Tränkung *f* impregnation, soaking, steeping; *(ag)* watering
T. der Schwellen impregnation of sleepers
Tränkungsdauer *f* impregnation time
Tränkungsschießen *n (mine)* pulsed infusion shotfiring
Tränkwanne *f* resin tank *(coating)*; coating pan
Transadmittanz *f* transadmittance
Transaktor *m* transactor
Transatlantikdampfer *m* transatlantic steamer
Transcuriumnuklid *n* transcurium nuclide
Transduktor *m* transductor, amplistat, magnetic amplifier
T. in Parallelschaltung parallel transductor
T. in Reihenschaltung series transductor
Transduktorregler *m* transductor regulator
Transduktorverstärker *m* transductor (magnetic) amplifier
Transfermaschine *f* transfer machine
Transferstraße *f* transfer line
Transfiguration *f* transfiguration
Transfluxor *m* transfluxor *(multi-apertured magnetic core)*
Transformation *f* transformation
additive T. additive transformation
inverse T. inverse transformation
Transformationsbereich *m* transformation [temperature] range
Transformations-EMK *f* transformer electromotive force
Transformationsgleichung *f (text)* transformation equation
Transformationsglied *n* matching section
Transformationshülse *f* slug *(waveguides)*
Transformationsmatrix *f* matrix of the transformation
Transformationsrechnung *f* transform calculus
Transformationstheorie *f* transformation theory
Transformationsverhältnis *n* transformation ratio
Transformator *m* transformer
T. mit beweglicher Sekundärwicklung moving-coil transformer
T. mit geschlossenem Kern closed-core transformer
T. mit geteilter Wicklung split transformer
T. mit Gleichrichter transrector
T. mit konstanter Primärspannung constant-potential transformer
T. mit Luftkühlung air-cooled transformer, dry-type transformer
T. mit Luftstromkühlung air-blast transformer
T. mit unterteiltem Eisenkern air-gap transformer
T. mit Wasserkühlung water-cooled transformer
abgeschirmter T. shielded (screened) transformer
abgestimmter T. tuned transformer
eisenloser T. air-core transformer
luftgekühlter T. ventilated transformer
ölgekapselter T. oil-immersed transformer
ölloser T. *s.* Trockentransformator
ruhender T. static transformer
schlagwettergeschützter T. flame-proof transformer
spannungserniedrigender T. step-down transformer
umschaltbarer T. switch-over transformer
Transformatoranlasser *m* compensator starter
Transformatoranpassung *f* adaption of the transformer
Transformatoranzapfung *f* transformer tap
Transformatorenbank *f* bank of transformers, transformer bank
Transformatorenbau *m* construction of transformers
Transformatorenblech *n* transformer plate (sheet)
Transformatorenhaus *n* transformer house
Transformatorenkopplung *f* transformer coupling
Transformatorenschalter *m* transformer switch
Transformatorenstation *f* transformer station
Transformatorenzelle *f* transformer vault
Transformatorersatzschaltung *f* equivalent circuit of a transformer
Transformatorkern *m* transformer core
Transformatorleistung *f* transformer capacity
Transformatormantel *m* transformer shell
Transformatoröl *n* transformer oil
Transformatorölausdehnungsgefäß *n* transformer conservator
Transformatorsatz *m* transformer theorem
Transformatorschaltgruppe *f* connection-group of a transformer
Transformatorspule *f* transformer coil
Transformatorwicklung *f* transformer winding
primäre T. transformer primary
sekundäre T. transformer secondary
Transformatorwirkungsgrad *m* transformer efficiency
transformierbar transformable
Transformierbarkeit *f* transformability
transformieren to transform
Transformierte *f (math)* transform
Transformierung *f* transformation; transforming

Transiederei *f* blubber (train-oil, fish-oil) boilery
Transientreaktanz *f* transient reactance
T. des Längsfeldes direct-axis transient reactance
Transientzeitkonstante *f* **des Längsfeldes bei Kurzschluß** direct-axis transient short-circuit time constant
Transisomeres *n (ch)* trans-isomer
Transistor *m* transistor
T. hoher Leistung high-powered transistor
T. mittlerer Leistung medium-power transistor
analoger T. analog transistor
diffundierender T. diffused transistor
dreischichtiger T. three-layer transistor
gesteuerter T. controlled transistor
legiert diffundierter T. post-alloy diffused transistor
legierter T. alloy transistor
lichtempfindlicher T. phototransistor
optischer T. optical transistor
rauschender T. noisy transistor
Transistoralterung *f* transistor aging
Transistoraufbau *m* transistor structure
Transistorausfall *m* transistor failure
transistorbestückt transistorized
Transistorbestückung *f* transistorization
teilweise T. part transistorization
Transistor-Dioden-Logik *f* transistor diode logic
Transistorelektronik *f* transistor electronics
Transistorempfänger *m* transistor radio [receiver]
Transistorenfertigung *f* transistor manufacture
Transistorersatzschaltung *f* transistor equivalent circuit
transistorgetrieben transistor-driven
Transistorgröße *f* transistor size
Transistorgrundschaltung *f* basic transistor circuit
transistorisiert transistorized
Transistorisierung *f* transistorization
Transistorkühler *m* transistor cooler
Transistorlebensdauer *f* transistor life
Transistorleistung *f* transistor performance
Transistorprüfgerät *n* transistor tester (testing instrument, checker)
Transistorradio *n s.* Transistorrundfunkgerät
Transistorrundfunkgerät *n* transistor radio [receiver]
Transistorschaltelement *n* transistor switching element
Transistorschalter *m* transistor switch
Transistorsender *m* transistor transmitter
Transistorstabilisierung *f* transistor stabilization
Transistortechnik *f* transistor engineering; transistor technology; transistor technique
Transistortheorie *f* transistor theory
Transistortriode *f* transistor triode
Transistorverfahrenstechnik *f* transistor process technology
Transistorverlustleistung *f* transistor dissipation
Transistorverstärker *m* transistor amplifier
Transistorverstärkung *f* transistor gain
Transistorwirkung *f* transistor action
Transistorzerhacker *m* transistor chopper
Transit *m* transit
Transitgüter *npl* transit goods, goods in transit
Transithafen *m* transit port, port of transit
Transitladung *f* transit cargo
Transitlagerschuppen *m* transit shed
Transitleitung *f* via circuit *(telephony)*
Transitronoszillator *m* negative-resistance oscillator
Transitverkehr *m* through traffic
Transitverwaltung *f* intermediate (transit) administration
transkristallin transcrystalline
Transkristallisation *f* transcrystallization
Translation *f* translation
Translationsbewegung *f* translation movement (motion), translational movement (motion)
Translationsfläche *f* translation surface
Translationsgeschwindigkeit *f* velocity of translation
Translationsgitter *n* [Bravais] lattice, translation lattice
Translationsschale *f* translational shell
translatorisch translatory
Transmission *f* 1. transmission *(radiation)*; 2. *(mach)* transmission, [line] shafting
Transmissionsdampfmaschine *f* steam engine driving shafting
Transmissionsdynamometer *n* transmission (belt) dynamometer
Transmissionsexperiment *n (nucl)* transmission experiment
Transmissionsgitter *n* transmission grating
Transmissionsgrad *m* transmission factor (coefficient), transmittance
Transmissionspumpe *f* pump driven by power
Transmissionsriemen *m* driving belt
Transmutation *f (nucl)* transformation, transmutation
Transomheck *n (ship)* transom stern
Transomplatte *f (ship)* transom floor
Transomwrange *f (ship)* transom floor
transozeanisch transoceanic
transparent transparent, translucent
Transparentfolie *f (plast)* transparent film (foil)
Transparentpapier *n* transparent paper
Transparentzeichenpapier *n* [glazed] tracing paper
Transparenz *f* transparence, transparency, transmittancy, transmissivity, diaphaneity
transponieren *(math)* to transpose

Transponierung *f (math)* transposition
Transponierungsempfänger *m* frequency-changing receiver
Transport *m* 1. transport *(by road, rail, or shipping)*; transportation; haulage; shipment; 2. handling *(materials)*; transfer *(workpiece)*; conveyance; carriage; 3. *(mach)* drive; moving
innerbetrieblicher T. inter-bay transport, inter-factory transport; material (mechanical) handling
schrittweiser T. *(mach)* intermittent drive
transportabel portable
Transportanlage *f* conveying (handling) plant
Transportanlagen *fpl* conveying machinery
Transportband *n* conveying belt, belt conveyor
Transportbefehl *m (dat)* transfer instruction
Transportbehälter *m* coffin, casket; [transfer] container; shipping container
Transportbeton *m* truck-mixed concrete, transit-mix[ed] concrete
Transportbetonmischer *m* agitator (agitating, transit-mix) truck, transit (transit-truck, truck) mixer
Transportbetonwerk *n* ready-mix[ed concrete] plant
Transportbügel *m* lifting (handling) hook *(precast concrete)*
Transporte *mpl* / **außergewöhnliche** special consignments
Transporteinrichtung *f* transfer unit, conveying device, handling equipment *(factory)*
Transportentfernung *f* haul[age] distance
mittlere T. average haul distance
Transporter *m* transporter, carrier *(vessel)*
Transporteur *m* plain protractor
transportfähig transportable
Transportfahrzeug *n* transporting vehicle, hauling unit, hauler
Transportflugzeug *n* cargo transport airplane
Transportgang *m* aisle, gangway *(factory)*
Transportgeräte *npl* handling gear *(factory)*
Transportgeschwindigkeit *f (mach)* feed rate
Transportgleichung *f* transport equation
Boltzmannsche T. Boltzmann transport equation
Transporthaken *m* handling (lifting) hook *(precast concrete)*
transportieren 1. to transport, to convey, to haul, to ship; 2. to handle, to transfer *(workpiece)*; 3. *(mach)* to drive, to move
Transportkasten *m* tote box; carrying case
Transportkette *f* chain-type carrier
Transportkontainer *m* transfer container
Transportkontrolle *f (dat)* transfer check
Transportkübel *m* skip *(mixing machines)*
Transportleistung *f* traffic performance
Transportlochung *f* sprocket holes
Transportmischer *m* agitator (agitating, transit-mix) truck, transit [truck] mixer, truck mixer
Transportnäherung *f* transport approximation
Transportöse *f* lifting (handling) hook *(precast concrete)*
Transportpalette *f* pallet
Transportpfanne *f* transfer ladle
Transportquerschnitt *m (nucl)* transport cross section
Transportrolle *f* sprocket *(film)*
Transportschaden *m* transportation damage
Transportschiff *n* transport vessel (ship)
Transportschlaufe *f* lifting loop (hook), handling hook *(precast concrete)*
Transportschnecke *f* worm (screw) conveyor
Transportstange *f* transfer bar
Transportsystem *n* material-handling system *(factory)*
Transporttheorie *f (nucl)* transport theory
Transportwagen *m* transporter, haulage vehicle, truck
Transportwasser *n (geol)* flush water
Transportweglänge *f (nucl)* transport mean free path
Transportwesen *n* transport industry
transsonisch transsonic
Transuran *n* transuranium (transuranic) element
transuranisch transuranic
transversal transverse, transversal
Transversalbeschleunigung *f* transverse acceleration
Transversalbewegung *f* transverse motion
Transversaldrehung *f (telev)* rolling
Transversale *f (math)* transversal
Transversalfeld *n* transverse field
Transversalschwingung *f* transverse vibration
Transversalspannung *f* transverse voltage
Transversalstrom *m* transverse current
Transversalwelle *f* transverse wave
transzendent transcendental
Trap *m* trap *(semiconductor)*
Trapez *n (math)* trapezium, *(Am)* trapezoid
Trapezbinder *m (build)* hip (pitched) truss
Trapezeffekt *m (telev)* trapezium effect
Trapezentzerrung *f (telev)* keystone correction
Trapezfachwerk *n* pitched (hip) truss
Trapezfehler *m* trapezium (keystone) distortion *(in a cathode-ray tube)*
Trapezflügel *m (aero)* tapered wing
trapezförmig trapeziform, *(Am)* trapezoidal
Trapezgelenkeck *n* trapezoidal linkage
Trapezgewinde *n* acme thread
Trapezoeder *n* trapezohedron
Trapezraster *m* trapezoidal raster
Trapezschaltung *f* tetragon connection
Trapezverzerrung *f (telev)* trapezium (keystone) distortion
Trapezwelle *f* trapezoidal wave

Traß *m* trass
Trasse *f* route; line; *(ry)* right of way
Trassenvermessung *f* route surveying
trassieren to lay out *(road)*; to plot *(cable run)*
Trassierung *f* route mapping; alignment survey; plotting *(cable run)*
Traßzement *m* trass cement
Traubensaft *m* grape juice
Traubenzucker *m* glucose, grape sugar
Trauerkleidung *f* mourning
Traufbohle *f* eave (eaves, fascia, gutter) board
Traufbrett *n* eave (eaves, fascia, gutter) board
Traufe *f* eave
Träufelrinne *f (mine)* water garland
Träufelspule *f* mush-wound coil
Träufelwicklung *f* drop-in winding
Traverse *f* bridge (top) beam; *(mach)* top rail, cross-beam, cross girth (head); cross bar, suspension bracket
Travertin *m* travertine
Trawler *m* trawler
Treber *pl* [distiller's] grains, spent grain (residue)
Treffer *m* hit
Trefferanzeiger *m* / **automatischer** *(mil)* auto-spotter, automatic spotter
Treffertheorie *f* hit (target) theory
Treffgenauigkeit *f* accuracy of fire
Treffplatte *f* target
Treibachse *f* driving axle
Treibachswelle *f* driving axle-pin
Treibanker *m* driving (sea) anchor, drag
Treibbake *f* floating beacon
Treibbeet *n* hotbed, forcing bed
Treibbeetfenster *n* frame, Dutch light
Treibdruck *m* swelling pressure
Treibeis *n* drift (floating) ice
treiben 1. *(mach)* to drive; to propel; 2. *(met)* to cupel; to hammer; 3. to hoist, to raise *(shaft winding)*; *(mine)* to wind; to swell; to drive, to tunnel *(road)*; 4. to float, to drift; 5. *(ag)* to force *(plants)*
treibend adrift *(vessel)*
Treiber *m* driver, driving gear; jigger sail
Treiberei *f (ag)* forcing
Treiberimpuls *m* driving pulse
Treiberröhre *f* driver valve
Treiberscheinung *f* driving phenomenon *(chain insulator)*
Treibersignal *n* driving signal *(electronics)*
Treiberstufe *f* driver, driving (driver) stage *(electronics)*
Treiberstufenleistung *f* driving power
Treibflasche *f* drift bottle
Treibgas *n* fuel gas
Treibgaserzeuger *m* gasifier *(gas turbine)*
Treibhaus *n* hot-house
Treibkeil *m (mach)* drive-fitted key
Treibkette *f* driving (power-transmission) chain
Treibmine *f* floating mine
Treibmittel *n* 1. *(mach)* hydraulic fluid; propellant; 2. *(plast)* blowing (expanding, foaming, gas-developing) agent
Treibmittelkreis *m* hydraulic circuit
Treibmittelpumpe *f* fuel diffusion pump
Treibnetz *n* drift net
Treibnetzfischereifahrzeug *n* drifter
Treibnetzfischereischiff *n* gill-netter
Treibofen *m* cupel[lation furnace]
Treiböl *n* oil; [Diesel] fuel; hydraulic oil
Treiböltank *m* oil fuel tank
Treibölübernahmestelle *f* fuel-loading position
Treibölzusatzmittel *n* fuel oil additive
Treibpflanze *f* forced plant
Treibrad *n* driving wheel
Treibriemen *m* driving belt
Treibriemenleder *n* belting leather
Treibsand *m* shifting-sand; quicksand
Treibscheibe *f* traction sheave, friction (Koepe) wheel
Treibscheibenaufzug *m* traction-sheave elevator
Treibscheibenfördermaschine *f* friction-type hoist
Treibsegel *n* drift-sail
Treibstoff *m* fuel, propellant
  T. mit Bleitetraäthylzusatz ethylized fuel
  fester T. solid propellant
  flüssiger T. liquid propellant
  gasförmiger T. gaseous propellant
  klopffester T. anti-knock fuel, anti-detonating fuel
  nuklearer T. nuclear propellant
Treibstoffaufnahme *f* [re]fuelling
Treibstoffbehälter *m* fuel tank (reservoir)
Treibstofflager *n* fuel dump
Treibstoffschnellentleerung *f* fuel jettisoning
Treibstofftank *m* oil fuel tank
Treibstoffübernahme *f* fuel loading, [re]fuelling
Treibstoffverbrauch *m* fuel (propellant flow) rate, propellant consumption
Treibstrom *m* hydraulic fluid stream
Treidellokomotive *f* towing tractor *(electric locomotive)*
treideln to tow [along], to track, to warp, to haul
Treidelpfad *m* horse path
Trennanlage *f* separation plant, separator
Trennapparatur *f* separator
Trennarbeit *f (mach)* cutting-off operation, parting-off operation
Trennband *n* separator
trennbar separable; *(el)* disconnectable; disengageable
Trennbruch *m* separation fracture
Trenndiode *f* buffer diode
Trenndüse *f* separation nozzle
Trenneinsatz *m* / **sichelförmiger** crescent-shaped spacer *(internal gear pump)*

trennen to separate, to divide; *(ch)* to grade, to classify; *(el)* to disconnect; to interrupt, to clear; to break *(connection)*; *(mach)* to disengage; to cut off, to part *(in lathe)*; to resaw *(wood)*
eine Verbindung t. *(el)* to break a connection; to disconnect a call *(telephone)*
Trennendverschluß *m (el)* terminal separating box
Trenner *m* separator; isolator
Trennfaden *m* lacing thread; draw thread *(knitting)*
Trennfaktor *m* separation factor
elementarer T. simple process factor
theoretischer T. ideal separation factor
Trennfestigkeit *f* rupture strength
Trennfilter *n* separating (channel) filter
Trennfläche *f* surface of discontinuity *(fluid flow)*
Trennflüssigkeit *f* liquid seal; stationary phase *(chromatography)*
Trenngatter *n* horizontal band saw, resaw
Trennholm *m* bolting (connecting) longeron
Trennimpuls *m* spacing pulse
Trennisolator *m* test insulator
Trennklinke *f (telec)* break jack
Trennkolonne *f* separating column (tower)
mehrstufige T. multi-stage separating column
Trennkondensator *m* isolating capacitor
Trennkontakt *m* break (spacing) contact
Trennkraft *f* separative power
Trennkreissäge *f* ripping circular saw
Trennlänge *f* cut-off length
Trennlasche *f* disconnecting link
Trennmaschine *f* cutting-off machine
Trennmauer *f* cut-off wall
Trennmesser *n* slitting knife; cutting (ripping) knife; *(el)* disconnecting blade *(switch)*
Trennmittel *n* parting compound, [mould] release agent
Trennmuffe *f* section box
Trennpappe *f* layer board
Trennrelais *n* cut-off relay, ringing-trip relay *(telephony)*
Trennrohr *n* diffusion separator, separation tube, thermal diffusion pipe
Clusiussches T. Clusius column
Trennröhre *f* cut-off valve; buffer valve
Trennrohrverfahren *n* separation of isotopes by thermal diffusion
Trennsäge *f* cut-off saw
Trennsäule *f* separation column (tower)
gefüllte T. packed column *(gas chromatography)*
Trennschalter *m* disconnecting (isolating) switch, cut-out
trennscharf selective
nicht t. unselective
Trennschärfe *f* 1. [channel] selectivity; discrimination *(filter)*; 2. *(mine)* accuracy of separation
Trennschärfefaktor *m* selectivity factor
Trennschärferegelung *f* selectivity (bandwidth) control
selbsttätige T. automatic selectivity control
Trennscheibe *f (mach)* abrasive cut-off wheel *(metal)*; *(build)* abrasive blade *(concrete saw)*
Trennschicht *f* separating layer
Trennschleifen *n* abrasive cutting[-off]
Trennschleifmaschine *f* abrasive-wheel cutting-off machine
Trennschleifscheibe *f* cutting-off wheel, abrasive saw
Trennschnitt *m (mach)* parting cut
Trennschott *n* partition, partition (intervening, non-structural) bulkhead
Trennschutzschalter *m* isolating switch
Trennsicherung *f* disconnecting (bridge) fuse, fuse disconnecting switch, combined fuse and cut-out, disconnecting link
Trennstelle *f* testing point *(telephony)*
Trennstrecke *f (el)* [air] break
Trennstreifen *m* dividing strip *(road)*
Trennstrom *m* spacing current
Trennstufe *f* isolator (buffer) stage; separator
Trennstufenhöhe *f (nucl)* height equivalent to a theoretical plate (stage), HETP, HETS
Trennsystem *n* separate (two-pipe) system *(sewerage)*
Trenntransformator *m* isolating transformer
Trennung *f* 1. separation, division; 2. *(ch)* grading, classification, precipitation; 3. *(el)* disconnection, isolation; clearing *(telephony)*, release; 4. *(mach)* cutting-off, parting *(lathe)*; 5. fall-away *(rocket)*
T. der Variablen separation of variables
T. des Satelliten von der Trägerrakete satellite separation
chemische T. chemical separation
chromatografische T. chromatographic separation [method]
radiochemische T. radiochemical separation
sofortige T. quick release
verzögerte T. *(el)* slow release
vorzeitige T. premature disconnection
Trennungsblech *n* baffle-plate
Trennungsebene *f* plane of division (joint) *(turbine)*; division (junction) plane
Trennungsfläche *f* separation surface; surface of contact *(liquids)*; parting (cleavage) surface, parting (cleavage) plane; natural joint *(crystal)*
Trennungsgang *m* / **chemischer** analytical procedure
Trennungslinie *f* separating line; line of demarcation
Trennungsmauer *f* **der Längszüge** mid-feather wall
Trennungsschnitt *m* ripping cut *(slaughtering)*

Trennungswinkel *m* angle of take-off *(between web and felt)*
Trennungszeichen *n* break sign *(telephony)*; double dash *(Morse code)*
Trennungszunge *f* spacing (separating) lug, distance piece
Trennverfahren *n* cutting-off process; separation process; discriminatory analysis *(inspection)*
Trennvermögen *n* selectivity; separative power
Trennverstärker *m* trap (buffer) amplifier
Trennvorlage *f* pressure transmitting seal, *(Am)* chemical attachment
Trennwagen *m* disconnecting carriage
Trennwand *f* diaphragm, membrane; *(build)* partition [wall]; *(aut)* partition panel
  aussteifende T. tie (buttressing) wall
  bewegliche T. movable partition
  leichte T. partition [wall]
  zweischalige T. double partition
Trennwanddiffusion *f* barrier diffusion
Trennwandfenster *n* partition panel window
Trennwandverkleidung *f* partition panel lining
Trennwerkzeug *n* parting tool
Trennwiderstand *m* cohesive resistance
Trennwirkung *f (ch)* separation effect, selectivity
Trennzeit *f (telec)* splitting time
Trense *f* snaffle
Treppe *f* staircase, stairway, stairs; ladder *(vessel)*
Treppenabsatz *m* [stair] landing
Treppenarm *m* flight [of stairs]
Treppenauge *n* open well
Treppenautomat *m* automatic staircase lighting switch
Treppenbeleuchtung *f* stair lighting
Treppendurchlaß *m* cascade culvert
treppenförmig stepped
Treppenfunktion *f* step function
Treppenfunktionsgenerator *m (Am)* box-car generator
Treppengeländer *n* hand-rail [of stairs]
Treppenhaus *n* stair well, staircase
Treppenhausautomat *m* automatic staircase lighting switch
Treppenimpuls *m* staircase pulse
Treppenkurve *f (math)* staircase curve
Treppenlauf *m* stairway, flight [of stairs]
  gerader T. straight stair
Treppenöffnung *f* **im Deck** ladderway *(vessel)*
Treppenpfosten *m* newel
Treppenpodest *n (m)* [stair] landing
Treppenpolygon *n (math)* staircase curve
Treppenrost *m* step[ped] grate
Treppenrostfeuerung *f* step grate furnace, furnace with stepped grate
Treppenrostunterfeuerung *f* furnace with stepped grate arranged underneath the boiler
Treppenrostvorfeuerung *f* external furnace with stepped grate
Treppenschalter *m* landing switch
Treppenspannung *f* staircase voltage
Treppenspannungsgenerator *m* staircase generator
Treppenstufe *f* step
Treppenwange *f* string [board]; notch board *(timber)*; step grate side bearers
Treppenwicklung *f* stepped winding
Tresor *m* safe; *(nucl)* bunker
Tresse *f* **/ zweiflechtige** double-strand braid
Trester *pl* pomace, grape-marc
Tretanlasser *m* kick-starter
Tretdynamo *m* pedal dynamo
treten to tread; to pedal, to treadle
  über die Ufer t. to overflow
Tretkurbel *f* kick-starter
Tretschalter *m* foot switch
Tretschemel *m* foot treadle
Tretspinnrad *n* treadle wheel
Tretwebstuhl *m* tread[le] loom
Tretweise *f (text)* order of treadling
Treue *f* **der Wollfaser** evenness of the wool fibre
Triangel *m* triangle
Triangulation *f* triangulation
Triangulierung *f* triangulation
Triäthanolamin *n* triethanolamine
Triäthylamin *n* triethylamine
Triäthylenglykol *n* triethylene glycol
Triazetat *n* triacetate
Tribüne *f* grand stand
Tribünenbau *m* grand stand
Tributylphosphat *n* tributhylphosphate, TBP
Trichloräthylen *n* trichlorethylene
Trichloressigsäure *f* trichloroacetic acid
Trichlorsilan *n* silicochloroform
Trichter *m* funnel; hopper; *(ac)* horn; *(mine)* glory-hole; mill[-hole]
  dampfbeheizter T. steam funnel
Trichterbau *m (mine)* glory-hole work, milling; glory-hole method, underground milling
trichterförmig funnel-shaped
Trichterhorn *n* taper *(waveguides)*
Trichterlautsprecher *m* horn[-type] loudspeaker
Trichtermühle *f* hopper mill
Trichteröffnung *f* horn mouth
Trichterspinnverfahren *n* funnel spinning method
Trichterspulmaschine *f* cup winding machine
Trichterwagen *m (ry)* hopper car (wagon), travelling loading hopper
Trickaufnahmevorrichtung *f* trick switching arrangement
Trickfilm *m* cartoon film
Trickfilmkamera *f* cartoon camera
Tricktaste *f* trick (blend-in) button *(magnetic recording)*
Tricktisch *m* cartoon-camera table

Tricküberblendung *f (telev)* wipe
Tridymit *m* tridymite
Trieb *m* 1. gear *(small pinion)*; 2. *(mach)* drive; transmission
Triebachse *f* drive axle
Triebfahrzeug *n* tractive (motor-power) unit
Triebfeder *f* main [watch] spring; moving spring
Triebhebel *m* cam lever *(for the traverse rail)*
Triebkasten *m* motion box
Triebknopf *m* pinion knob
Triebkraft *f* motive power; propelling force, propulsion power
Triebkranz *m* scroll gear
Triebmotor *m* traction motor
Triebrad *n* driving wheel
Triebschraube *f* drive screw
Triebstange *f* driving bar
Triebstock *m* leaf, pin, stave, round
Triebstockkranz *m* lantern pinion
Triebstockzahnrad *n* lantern wheel *(pinion)*, trundle wheel
Triebwagen *m* motor coach
Triebwageneinheit *f (ry)* indivisible train unit (set)
Triebwasserleitung *f* penstock
Triebwerk *n* 1. *(aero, rock)* engine; propulsion unit, power plant; 2. *(mach)* driving mechanism (gear), power transmission equipment
T. der ersten Raketenstufe first-stage engine
T. der zweiten Raketenstufe second-stage engine
Triebwerkmechaniker *m* jet engine fitter
Triebwerkschlosser *m s.* Triebwerkmechaniker
Triebwerkspult *n* propulsion-control desk
Triebwerksregler *m* propulsion-unit governor
Triebwerksteile *npl* driving parts, power-transmission elements
Triebwerkswelle *f* power-transmission shaft
Trieur *m (ag)* indented cylinder
Trieurauslese *f* **nach Länge** grading by shape with indented cylinder
Trieurzellenweite *f* cell width of [the] indented cylinder
triften to drift
Triftgang *m (ag)* walk-through
Triftgeschwindigkeit *f* drift velocity
Triftholz *n* drift wood
Triftkanal *m* floating channel
Triftpfad *m* drift (side) path
Triftröhre *f* drift tube, klystron
Trigger *m* trigger
binärer T. binary trigger
Triggerartenschalter *m* trigger source switch
Triggerimpuls *m* trigger [im]pulse
Triggerimpulsversteilerungskreis *m* trigger sharpener
triggern to trigger
Triggerpegel *m* trigger level
Triggerschalter *m* trigger switch
Triggerschaltkreis *m* trigger circuit
Triggerschaltung *f* trigger circuit
monostabile T. single-shot trigger circuit, single-trip trigger circuit
Triggersperre *f* hold-off circuit, retriggering hold-off
Trigistor *m* trigistor
trigonal trigonal
Trigonometrie *f* trigonometry
sphärische T. spherical trigonometry
trigonometrisch trigonometric[al]
triklin triclinic
Trikot *m(n)* tights
Trikotagen *fpl* tricot underwear
Trikotrundstrickmaschine *f* / **einfonturige** jersey circular knitting machine
Trikresylphosphat *n* tricresyl phosphate
trimer trimeric
Trimethylamin *n* trimethylamine
Trimm *m* trim
Trimmabschaltstab *m* shim safety rod *(reactor)*
Trimmänderung *f (ship)* change (alteration) of trim
trimmen to trim
Trimmen *n* trimming; *(nucl)* shimming
Trimmer *m (ship)* trimmer; *(el)* trimmer, trimming (adjustable) capacitor
Trimmklappe *f (aero)* trimming tab
Trimmlage *f* trim
Trimmluke *f* trimming hatch, escape hole
Trimmoment *n* trimming moment
Trimmotor *m* trim tab motor
Trimmstab *m* shim (coarse control) rod *(reactor)*
Trimmtank *m* trimming tank
trimolekular trimolecular, termolecular
Trinitrobenzol *n* trinitro-benzene
Trinitroglyzerin *n* glycerol trinitrate
Trinitrotoluol *n* trinitrotoluene, trotyl, TNT
Trinitrotoluoläquivalent *n* trinitrotoluene(TNT) equivalent
Trinkbrunnen *m* drinking fountain
Trinkwasser *n* potable (drinking) water
kein T. non-drinkable water
Trinkwasseraufbereiter *m* water purification unit
Trinkwasseraufbereitungsanlage *f* drinking water generator
Trinkwasserbehälter *m* drinking (potable) water tank
Trinkwasserfluorierung *f* drinking water fluoridation
Trinkwasserlast *f* drinking water tank
Trinkwasserprüfung *f* analysis of [drinking] water
Trio *n s.* Triowalzwerk
Trioblechstraße *f* three-high plate mill train
Trioblockstraße *f* three-high cogging mill train

Triode *f* triode
T. hoher Verstärkung high-gain triode
T. mit hohem Verstärkungsfaktor high-mu triode
Triode-Hexode-Mischröhre *f* triode hexode frequency changer
Triodenröhre *f* triode valve
Triodenschaltung *f* triode connection
Triodenverstärker *m* triode amplifier
Triogerüst *n* three-high [rolling] stand
Trioroststab *m* triple grate bar
Triostraße *f* three-high [rolling mill] train
Triowalzgerüst *n* three-high [rolling] stand
Triowalzwerk *n* three-high [rolling] mill
Trioxid *n* trioxide, teroxide
Tripel *m (min)* Tripoli powder, tripolite
Tripelbindung *f* triple bond
Tripelpunkt *m* triple point
Tripelspiegel *m* triple mirror
Tripeptid *n* tripeptide
Triphenylkarbinol *n* trityl alcohol
Triphenylmethanfarbstoff *m* triphenylmethane dye
Triphenylmethylradikal *n* trityl radical
Triplet *n* triplet lens
Triplett *n* triplet
Triplettlinie *f* triplet line
Triplettspektrum *n* triplet spectrum
Triplettzustand *m* triplet state
Triplexkarton *m* **mit Stoffmittellage** cloth-centered board
gegautschter T. double white patent-coated board
Triplexpappe *f* triplex board
Trippleine *f* tripping line
Trisektion *f* trisection *(angle)*
Trisilan *n* silicopropane
tristabil tristable
Trithiokohlensäure *f* [tri]thiocarbonic acid
Trithionsäure *f* trithionic acid
Tritid *n* tritide
tritieren to tritiate
tritiieren *s.* tritieren
Tritium *n* tritium
Tritiumkern *m* triton
Triton *n* triton
Tritt *m* step; pedal, foot treadle
außer T. sein to be out of step
Trittbrett *n (ry)* footboard, tail board, step; *(aut)* running board
Trittbrettantenne *f (Am)* running-board antenna
Trittbretthalter *m* running-board support
Trittfolge *f* order of treadling
Trittleiter *f* stepladder
Trittmaschine *f (text)* tie-up Jacquard
Trittschall *m* impact noise (sound)
Trittstufe *f* tread *(stairs)*
Trittvorrichtung *f* treadle motion
Trittwebstuhl *m* tappet (treadle) loom
Trochoidallaufzeitspektrometer *n* trochoidal time-of-flight mass spectrometer
Trochoidenmassenspektrometer *n* trochoidal mass analyzer
Trochotron *n* trochotron
trocken dry *(friction, steam)*; arid; barren *(land)*; non-fluid *(friction)*
absolut t. bone-dry
Trockenanalyse *f* dry analysis
Trockenanlage *f* drying plant (equipment)
Trockenapparat *m* drier, drying apparatus
Trockenappretur *f* dry finishing
Trockenaufbereitung *f* dry treatment
Trockenausschuß *m* dry broke
Trockenbagger *m* excavator
trockenbaggern to excavate
Trockenbarometer *n* aneroid (dry) barometer
Trockenbatterie *f* dry battery
Trockenbauweise *f* dry construction
Trockenbeizautomat *m* automatic chemical powder dresser
Trockenbeizmaschine *f* seed dusting machine
Trockenbeizung *f* dry disinfection
Trockenbestäubungseinrichtung *f* powder-spray equipment
Trockenboden *m (ag)* drying loft
Trockenbohren *n* dry drilling
Trockenbügeln *n* dry pressing
Trockendampf *m* dry steam
Trockendarre *f* drying tunnel (kiln)
trockendekatieren to decatize with dry steam
Trockendekatur *f* dry decating
Trockendestillation *f* destructive distillation
Trockendienst *m* drying section
Trockendock *n* dry dock, graving [dry] dock
Trockeneis *n* dry-ice, solid carbon dioxide, carbon dioxide snow
Trockenelement *n* dry cell
Trockenentwicklung *f* dry-development
Trockenfähigkeit *f* drying properties
Trockenfärben *n (leath)* brush-staining
Trockenfäule *f* dry rot
Trockenfestigkeit *f* dry strength
Trockenfeuerlöscher *m* dry-chemical fire extinguisher, powder-type fire extinguisher
Trockenfilzspanner *m* drier felt tightener
Trockenfirnis *m* siccative varnish
Trockenförderer *m* drying conveyor
Trockenformen *n* boarding *(hosiery)*
Trockenfrachter *m* dry cargo vessel (freighter), dry-stores carrier
Trockenfräsen *n* dry milling
Trockenfutterapparat *m* dry hopper
Trockenfuttermischmaschine *f* horizontal food mixer
Trockengelenk *n* dry-disk joint
Trockengemüse *n* dried vegetables

Trockengerüst *n* drying rack (stand)
Trockengießform *f* dry mould
Trockengleichrichter *m* metal (dry-plate, bimetallic) rectifier
Trockengräting *f (ship)* gridiron
Trockenherd *m* drying hearth
Trockenkammer *f* drying stove *(foundry)*; drying kiln
  T. mit abschnittweiser Beladung compartment kiln
  T. mit Außenventilator external fan kiln
  T. mit Innenventilator internal fan kiln
  T. mit künstlichem Zug forced draught kiln
  T. mit natürlichem Zug natural draught kiln
  kontinuierlich arbeitende T. progressive kiln
Trockenkammergerät *n* kiln apparatus
Trockenkanal *m* drying tunnel
Trockenklebrigkeit *f (plast)* dry (aggressive) tack
Trockenkupplung *f* dry coupling; dry-disk clutch
Trockenladeraum *m* dry[-stores] cargo room
trockenlegen to dewater; to drain
Trockenlegung *f* dewatering
Trockenlöscher *m* dry-chemical fire extinguisher, powder-type fire extinguisher, dry quencher
Trockenlöschung *f* dry quenching
Trockenluft *f* drying air
Trockenmagnetscheider *m* dry magnetic separator
Trockenmahlen *n* dry crushing
Trockenmauer *f* dry [rubble] wall, dry walling
Trockenmauerwerk *n* dry [rubble] wall, dry walling (masonry)
Trockenmilch *f* evaporated (desiccated) milk
Trockenmischung *f* dry batch *(concrete)*
Trockenmittel *n* drier, drying agent, [de]siccative, desiccant, dehumidifier
Trockenofen *m* drying furnace (stove), seasoning kiln, baking oven; foundry stove
Trockenoffset *m* dry offset, letterset
Trockenpartie *f* dry[ing] end, dry[ing] section
Trockenpatrone *f* desiccator cartridge
Trockenplatte *f* dry plate
Trockenpresse *f* drying press
Trockenrasierer *m* electric razor
Trockenrauhen *n (text)* dry raising
Trockenraum *m* drying chamber (space), kiln
  T. mit Dampfniederschlagvorrichtung drying chamber with steam condensing apparatus
  T. mit mechanischer Luftbewegung blower kiln
  T. mit natürlicher Luftbewegung ventilated kiln
Trockenraumtrocknung *f* kiln drying of lumber
Trockenreiniger *m* dry cleaner
Trockenreinigung *f* dry cleaning
Trockenreinigungsmittel *n* dry cleaning solvent
Trockenriß *m* shrinkage shake, season check *(timber)*
Trockenrückstand *m* dry residue
Trockensandformen *n* dry sand moulding
Trockenscheibenkupplung *f* dry-plate clutch
Trockenschlamm *m* dry sludge *(sewerage)*
Trockenschleifen *n* dry grinding
Trockenschleuder *f* centrifugal drier
Trockenschliff *m* dry grinding
Trockenschnitzel *npl* dried chips (slices)
Trockenschrank *m* hot cabinet, cabinet (shelf, tray) drier, drying oven (cabinet)
Trockensortierung *f* dry cleaning
Trockenspinnen *n* dry spinning
Trockenspinnmaschine *f* dry spinning frame
Trockenspinnverfahren *n* dry spinning method
Trockenständer *m* drying rack
Trockenstoff *m* drier, siccative
Trockensubstanz *f* dry matter
Trockensubstanzgehalt *m* percentage of solids
Trockensumpfschmierung *f* dry-sump lubrication, gravity-feed lubrication
Trockensuspensionsreaktor *m* dry-suspension reactor
Trockenteil *n (text)* drying section
Trockentemperatur *f* dry-bulb temperature
Trockenthermometer *n* dry-bulb thermometer
Trockentransformator *m* oilless (air-cooled) transformer
Trockentrommel *f* rotary drier, drying drum, rotating kiln; steam-heated cylinder
Trockenüberschlag *m* dry arc-over
Trockenüberschlagsfestigkeit *f* dry arc-over strength
Trockenverfahren *n* dry method *(cement manufacture)*
Trockenvorrichtung *f* **mit Luftumwälzung** forced air oven
Trockenwandler *m* oilless instrument transformer
Trockenzug *m* dry drawing *(wire)*
Trockenzungenrelais *n* reed relay
Trockenzwirnen *n* dry twisting
Trockenzylinder *m* drying (steam-heated) cylinder
trocknen to dry; to dehumidify, to desiccate; to season *(timber)*
  auf der Stange t. to pole-dry
Trocknen *n* **an der Luft** natural drying
  T. von Stangen pole seasoning
Trockner *m* drier
  T. mit hoher Temperatur high flash point drier
  T. mit Rührwerk paddle-trough type drier
  fahrbarer T. mobile grain drier
  pneumatischer T. pneumatic [tower and drum] drier
Trocknerzapfen *m* drier journal

Trocknung *f* drying; desiccation; seasoning *(timber)*
T. im Rahmen drying on frames
T. im Rührwerk *(wood)* paddle-trough drying
T. im Trockenhaus loft drying
T. von außen nach innen drying from the surface
chemische T. chemical (salt) seasoning
elektrische T. electro-desiccation
natürliche T. air-seasoning
technische T. kiln drying of lumber
Trocknungsanlage *f* crop drier
Trocknungskapazität *f* drying capacity
Trog *m* trough; vat; pan; tub; tank; tray
elektrolytischer T. electrolytic tank (trough)
Trogbandförderer *m* troughed belt conveyor
Trogbatterie *f* trough battery
Trogkettenförderer *m* chain-and-flight conveyor, drag-link conveyor, troughed-chain conveyor
Trogmischer *m (ch)* trough (pan) mixer
Trogtränkung *f* open tank method
Trolleybus *m* trolleybus
Trolleyleitung *f* twin-contact wires
Trommel *f* drum; rotor; barrel; cylinder; reel *(winding)*
kegelförmige T. conical drum *(turbine)*
umlaufende T. revolving drum
zylindrische T. cylindrical drum *(turbine)*
Trommelabsatz *m* chimney step
Trommelabscheider *m* drum separator
Trommelabtaster *m* drum scanner
Trommelanker *m (el)* drum armature
Trommelaufzug *m* winding-drum elevator
Trommeldrehfilter *n* rotary drum filter
Trommelfilter *n* drum filter
Trommelfördermaschine *f* drum-type hoist
Trommelfräsmaschine *f* drum-type milling machine, rotary-drum fixture milling machine
Trommelgalvanisierung *f* barrel electro-plating
Trommelgießpfanne *f* pouring drum
Trommelhäcksler *m* cylindrical cutter
Trommelinduktor *m* non-salient pole rotor
Trommelkompaß *m* barrel (drum) compass
Trommelkonverter *m* barrel converter
Trommelkurve *f* drum cam
Trommellager *n* drum bearing
Trommelläufer *m (el)* cylindrical (non-salient pole) rotor
trommellos without drum[s] *(boiler)*
Trommelmagazin *n (mach)* drum-type magazine; *(mil)* ammunition drum
Trommelmagnetscheider *m* drum-type magnetic separator
Trommelmaschine *f* 1. drum-type transfer machine; 2. tumbling barrel
Trommelmischer *m* drum mixer
Trommelmühle *f* tumbling barrel, drum (cylinder) mill
trommeln to rumble, to tumble *(cleaning)*
Trommelnase *f* cylinder tooth
Trommelofen *m* rotary (barrel-shaped) furnace
Trommelöffner *m* cylinder opener
T. mit großer Trommel large-cylinder opener
T. mit mehreren kleinen Trommeln multiple-cylinder opener
T. mit Schläger und Wickelvorrichtung cylinder opener combined with beater and lap machine
Trommelprobe *f* rattler test
Trommelputzwalze *f (text)* card fancy
Trommelräummaschine *f* rotary-drum broaching machine
Trommelrechenwender *m (ag)* combination side-delivery rake and swath turner
Trommelrevolver *m* 1. revolver *(handgun)*; 2. *(mach)* horizontal-axis turret, vertical turret
Trommelrevolverdrehmaschine *f* vertical turret lathe
Trommelrevolverkopf *m* vertical turret head
Trommelrundsiebsortierer *m* rotary potato sorter
Trommelschalter *m* turret switch
Trommelschlichtmaschine *f* cylinder sizing machine
Trommelschreiber *m* drum-chart recorder
Trommelsieb *n* rotary (revolving, drum) screen, cylindrical sieve
Trommelsiebbrecher *m* bradford breaker
Trommelsinkscheider *m* dense-medium washing drum
Trommelsinterofen *m* rotary sintering kiln
Trommelskale *f* cylinder dial
Trommelspeicher *m (dat)* drum memory (store)
Trommelstirnfläche *f* face of drum *(turbine)*
Trommeltrockenschlichtmaschine *f* slasher
Trommeltrockner *m (ag)* rotary-drum drier; *(mine)* drying drum
Trommelturbine *f* drum turbine
Trommelversuch *m* rattler test *(steel balls)*; drum test
Trommelwassermesser *m* drum water-meter
Trommelwehr *n* drum weir
Trommelwender *m* swath turner
Trommelwicklung *f* drum winding
Trommelzähler *m* drum-type meter
Trommelzetter *m* rotary tedder
Trompete *f* trumpet intersection *(motorway)*
Trompetenrohr *n* circular syphon
Troostit *m* quenching troostite
Tropasäure *f* tropic acid
Tropenausrüstung *f* tropical kit
Tropenbedingungen *fpl* tropical conditions
Tropenfarbe *f* tropical paint
tropenfest tropicalized, tropic-proof
tropengeschützt tropicalized
Tröpfchen *n* droplet

Tröpfchenmodell *n* liquid drop-model *(of the nucleus)*
Tropfelektrode *f* dropping electrode
tröpfeln to drop, to drip; to trickle
tropfen to drop, to drip; to trickle, to percolate; to leak
Tropfen *m* drop
Tropfen *n* **der Flanschverbindung** leaking of the flange joint
Tropfenanzeiger *m* drop lubricator
Tropfenfallmethode *f* *(nucl)* falling drop [specific densimeter] method
Tropfenprobe *f* lubricating-oil drop test
Tropfenschlagkavitation *f* impingement attack
tropfenweise dropwise, drop by drop
Tropfenzähler *m* dropper, stactometer, stalagmometer
Tropffeuerung *f* drip-furnace
Tropfflasche *f* dropping bottle
Tropfkante *f* dripping rim *(insulator)*
Tropfkörper *m* percolation (biological) filter, bacteria (contact, filter) bed
Tropfkörperbrocken *m* filter block
Tropfloch *n* weep hole
Tropfnase *f* water (weather) drip; dripstone, label
Tropföl *n* dripping oil
Tropföler *m* drip-feed lubricator, drip oiler
Tropfpunkt *m* drop[ping] point *(grease)*
Tropfschmiergefäß *n* drip-feed oiler (lubricator), sight-feed oiler (lubricator)
Tropfschmierung *f* drip-feed lubrication
Tropfsteinhöhle *f* stalactite cave
Tropftrichter *m* dropping funnel
tropftrocken drip-dry
Tropfwasser *n* dripping water
tropfwassergeschützt drip-proof
Tropfwasserschutz *m* protection from dripping water
Troß *m* baggage
Trosse *f* hawser
Trossenkneifer *m* cable compressor
Trossenrolle *f* hawser reel
Trossenruder *n* hawser rudder
Trossenwinde *f* hawser reel (winch), towing engine (winch)
Trossenzug *m* line (tow-rope) pull
  T. am Pfahl static bollard pull
Troßschiff *n* fleet supply ship
Trotyl *n* trinitrotoluene
trüb cloudy, opaque, turbid
Trübe *f* [liquid] ore pulp
Trübglasprisma *n* turbid-glass prism
Trübpunkt *m* cloud point
Trübung *f* cloudiness, opaqueness, turbidity; *(plast)* haze
Trübungsmesser *m* nephelometer, opacimeter, turbidimeter
Trübungsmessung *f* turbidimetry, nephelometry, opacimetry
Trübungsmittel *n* opacifier *(paint)*
Trübungspunkt *m* cloud point
Trudelbewegung *f* *(aero)* tailspin
Trudeln *n* *(aero)* [abnormal] spin
trudelsicher *(aero)* non-spinning
Trum *m* 1. strand *(conveyor)*; 2. compartment *(shaft)*; 3. side *(belt)*
Trümmer *pl* rubble, wreckage; debris, fragments, fission products
Trümmergestein *n* conglomerate
Trümmermasse *f* *(geol)* detritus
Trunkdeck *n* trunk deck
Trunkdeckschiff *n* trunk deck vessel
Truppentransporter *m* troop carrier (ship), trooper
Trypaflavin *n* trypaflavine
Trypsin *n* trypsin
Tryptophan *n* tryptophan
Tschebyscheff-Charakteristik *f* Tschebyscheff (Chebishev) shape
Tschebyscheff-Funktion *f* Tschebyscheff (Chebishev) function
Tübbing *m* *(mine)* tubbing
  deutscher T. inside-flange tubbing ring
  englischer T. outside-flange tubbing ring
Tübbingausbau *m* tubbing
Tube *f* [collapsible] tube
Tubulus *m* tubule
Tubus *m* tube, tubus
  bildaufrichtender T. image-erecting tube
  binokularer T. binocular tube
  gerader T. straight tube
  monokularer T. monocular tube
Tubusaufsatz *m* tube attachment
Tubusträger *m* tube support
Tubusverlängerung *f* tube extension
Tuch *n* *(text)* cloth
Tuchbürstmaschine *f* brushing machine
Tuchdekateur *m* cloth shrinker
Tuchfilter *n* filter cloth, cloth filter
Tuchschermaschine *f* cropping (cloth shearing) machine
Tuchweberei *f* cloth (wool) weaving
Tuchwebstuhl *m* cloth-weaving loom
Tuff *m* tuff
Tufferde *f* tufaceous earth
Tuffstein *m* tufa
Tüll *m* net[ting], tulle, bobbinet
Tülle *f* nozzle
Tüllmaschine *f* bobbinet machine
Tüllwebstuhl *m* net machine
Tulpennaht *f* single-U groove weld
Tumblerschalter *m* tumbler (trigger) switch
Tumordosis *f* *(nucl)* tumor dose
Tünche *f* whitewash, lime-wash
tünchen to whitewash, to whiten, to lime-wash

Tuner *m (el)* tuner
Tungöl *n* tung oil
Tunnel *m* tunnel, gallery
Tunnelbau *m* tunnelling, tunnel construction
Tunnelbauer *m* tunneller
Tunnelbeleuchtung *f* tunnel lighting [system]
Tunnelbrunnen *m* tunnel well
Tunneldiode *f* tunnel diode
Tunneldiodenkenngröße *f* tunnel diode parameter
Tunneldiodenkennlinie *f* tunnel diode characteristic
Tunneldiodenkippschwingungsgenerator *m* tunnel diode relaxation oscillator
Tunneldiodenschaltung *f* tunnel diode circuit
Tunneldiodenverstärker *m* tunnel diode amplifier
Tunneleffekt *m* tunnel effect
Tunneleingang *m* tunnel portal
Tunnelgerüst *n* / **fahrbares** jumbo
Tunnelheck *n* tunnel stern
Tunnelkiel *m* duct keel
Tunnelofen *m* tunnel furnace (kiln)
Tunnelpropeller *m (ship)* tunnel screw
Tunnelschlepper *m* tunnel stern tug
Tunnelsohle *f* tunnel invert
Tunnelstrom *m* tunnel[ling] current
Tunnelübergang *m (el)* tunnel junction
Tunnelwellenlager *n* plummer block, tunnel shaft bearing
Tüpfelanalyse *f* drop[-test] analysis
  T. auf Filterpapier filter paper analysis
Tüpfelindikator *m* external indicator
tüpfeln to speckle, to spot, to test by the spot method, to stipple
Tüpfelplatte *f* drop plate
Tüpfelreaktion *f* spot (drop-test) reaction
tupfen to stipple
Tür *f* door
  T. mit Bleieinlage *(nucl)* lead door
  T. mit glattem Blatt flush door
  einflügelige T. single-wing door
  verglaste T. glazed door
  vorgetäuschte T. blank door
  zweiflügelige T. double-wing door
Türabhebemaschine *f* door lifting machine
Türabheber *m* door extractor (lifter)
Türabhebevorrichtung *f* door extracting device
Türangel *f* hinge pivot
Türanschlag *m (aut)* door catch
Türaufschlag *m* door swing
Türaußenblech *n* outside door panel
Türaußengriff *m (aut)* outside door handle
Türbeschlag *m* cabin hook
Türbeschläge *mpl* door furniture
Türbetätigungszylinder *m (aut)* door-operating cylinder
Turbine *f* turbine
  T. für Rückwärtsgang back (astern) turbine
  T. für Vorwärtsgang ahead turbine
  T. mit äußerer Beaufschlagung inward-flow turbine
  T. mit gemischter Beaufschlagung mixed-flow turbine
  T. mit innerer Beaufschlagung outward-flow turbine
  T. mit mehrfacher Beaufschlagung re-entry turbine
  T. mit partieller Beaufschlagung fractional-supply turbine
  T. mit radialer Dampfströmung radial-flow turbine
  T. mit voller Beaufschlagung full-supply turbine
  eingehäusige T. single-cylinder turbine
  einstufige T. single-stage turbine
  gegenläufige T. turbine with nozzles and blades rotating in opposite directions
  gemischt beaufschlagte T. mixed-admission turbine
  mehrstufige T. multi-stage turbine
  reibungslose T. frictionless turbine
  umsteuerbare T. reversible turbine
  viergehäusige T. four-cylinder turbine
  zweigehäusige T. two-cylinder turbine
Turbinenanlage *f* turbine plant
  T. mit Freikolbengaserzeuger free-piston gasifier turbine unit
Turbinenbohren *n* turbine drilling
Turbinenbohrer *m* turbo drill
Turbinendampfer *m* turbine steamer
Turbinendoppelrad *n* turbine double-wheel
Turbinendurchflußzähler *m* turbine-flowmeter
Turbinendüse *f* turbine nozzle
Turbinendüsenschaufel *f* turbine nozzle blade
Turbinenfundament *n* turbine foundation
Turbinengebläse *n* turbine blower
Turbinengehäuse *n* turbine casing (case, cylinder)
Turbinenhaus *n* turbine house
Turbinenkreuzer *m* turbine cruiser
Turbinenlaufrad *n* turbine rotor
Turbinenlaufradschaufel *f* movable blade
Turbinenlokomotive *f* turbine locomotive
Turbinenluftstrahltriebwerk *n* turbojet [engine]
Turbinenmantel *m* turbine casing (case)
Turbinenmotor *m* turbine engine
Turbinenrad *n* turbine wheel
Turbinenregelung *f* turbine regulation
Turbinenregler *m* turbine governor
Turbinenrohrreiniger *m* turbine tube cleaner
Turbinenrotor *m* turbine rotor
Turbinensatz *m* turbine set (unit)
Turbinenschaufel *f* turbine blade (vane)
Turbinenschiff *n* turbine ship
Turbinenstrahlmotor *m* turbojet engine

Turbinenstrahltriebwerk *n* turbojet engine
Turbinentorpedoboot *n* turbine torpedo boat
Turbinentrockner *m* turbine drier
Turbinenverlust *m* turbine loss
Turbinenverschalung *f* turbine casing
Turbinenwelle *f* turbine shaft (spindle)
Turbinenzähler *m* turbine-flowmeter
Turboalternator *m* turbo alternator
turboaufgeladen turbo-charged
Turboaufladung / mit turbo-charged
Turbobrenner *m* turbo-burner
Turbodynamo *m* turbo-dynamo
turboelektrisch turbo-electric
Turbofeile *f* rotary file
Turbogebläse *n* turbo-blower
Turbogenerator *m* turbo-generator, turbine-driven generator
Turbogeneratorsatz *m* turbo-generator set
Turbokompressor *m* turbo-supercharger, turbo-compressor
Turbolader *m* turbo-supercharger
Turbolöser *m* pulp disintegrator
Türbolzen *m* door bolt
Turbomischer *m* turbine (centrifugal-impeller) mixer, impact [wheel] mixer, radial-flow impeller
  T. mit Gasabsorption turbo gas absorber
Turbomotor *m* turbine[-type] engine
Turboproptriebwerk *n* turbo-prop engine
Turbostrahltriebwerk *n* turboprop [engine]
turbulent turbulent *(flow)*
Turbulenz *f* sinuosity, turbulence
Turbulenzbewegung *f* turbulent flow (motion)
Turbulenzspektrum *n* spectrum of turbulence
Türdichtungsstreifen *m* door weather strip
Türflügel *m* door leaf
Türführung *f* door guide
Türfüllung *f* panel
Türfüllungsplattenmaschine *f* panel raising machine
Türfutter *n* jamb liner (lining)
Türgriff *m* door-handle
Türinnenblech *n* inside door panel
Türkis *m* turquoise
Türkischrotöl *n* Turkey-red oil
Türklinke *f* latch
Türkontakt *m* car-door electric contact
Türkontaktschalter *m (aut)* door switch (contact interrupter)
Türleibung *f* reveal
Turm *m* tower
Turmalin *m* tourmaline
  blauer T. indicolite
  farbloser T. achroite
  roter T. rubellite
  schwarzer T. schorl
Turmantenne *f* tower aerial
Turmdach *n* steeple (spire) roof
Turmdeck *n* turret deck
Turmdecker *m* turret-deck vessel
Turmdeckschiff *n* turret-deck vessel
Turmdestillation *f* flash distillation
Turmdrehkran *m* [revolving] tower crane, turret (mono-tower, tower type) crane
Turmfördermaschine *f* tower-mounted winder
Turmfrequenz *f* tower frequency
Turmgeschütz *n* turret gun
Turmhelm *m* spire
Turmluk *n* conning tower hatch *(submarine)*
Turmluke *f* turret hatch
Turmrolle *f (mine)* crown block *(boring)*
Turmsäure *f* tower acid
Turmschiff *n* turret ship, monitor
Turmschrämmaschine *f* turret overcutter (coal-cutter)
Turmspeicher *m* stacking pillar
Turmspitze *f* spire
Turmuhr *f* tower clock
Turmwäscher *m* column scrubber
Turnanzug *m* gym suit
Türnase *f* door catch
Turnhalle *f* gymnasium
Turnschuh *m* gym shoe
Türoberlicht *n* fan-light
Türöffner *m* **/ elektrischer** electric door operator
Türrahmen *m* door frame[work]
Türriegel *m* latch
Türsäule *f* door post (pillar)
Türschale *f* door skin
Türschalter *m* door switch
Türscharnier *n* door hinge
Türscharniersäule *f* door hinge pillar
Türschließanlage *f* door closing device
Türschließer *m* [automatic] door-closer
Türschließzylinder *m* door lock cylinder
Türschloßsäule *f* door lock pillar
Türschwelle *f* door sill
Türstock *m (mine)* drift (timber) set
  dreiteiliger T. three-piece timber set
  halber T. half set
  normaler T. three-quarter set
Türsturz *m* lintel
Türverkleidung *f* door panel
Türverriegelung *f* door latch
Türwinde *f* door winch
Türzarge *f* door frame[work]
Türziehgriff *m* door pull handle
Tusche *f* ink
Tuscheoriginal *n* original ink drawing
Tuschezeichnung *f* ink drawing
Tuschfarbe *f* water colour
Tuschierplatte *f* surface plate
Tussahseide *f* tussah [silk]
Tussahspinner *m* tussah silkworm
Tüte *f* [paper]bag, sack *(cement)*
Tütenpapier *n* bag paper

Twistor *m* twistor *(data store)*
Tympanon *n* tympanum *(architecture)*
Tyndalleffekt *m* Tyndall effect
Tyndallkegel *m* Tyndall cone
Typ *m* standard [building] design, type, design
Type *f* type *(letter)*
Typenbezeichnung *f* type designation
Typendruck *m* type printing
Typendruckempfänger *m (telec)* type printing receiver
typengleich of the same type
Typenprojekt *n* standard [building] design
Typenschild *n* identification plate
Typenstangen *fpl* / **alphanumerische** *(dat)* alphanumerical type bars
typisieren to standardize
Typisierung *f* standardization
Typograf *m* typographer
Typografie *f* typography
Typometer *n* line gauge
Tyramin *n* tyramine
Tyrosin *n* tyrosin

U

U-Bahn *f* underground railway, subway
U-Boot *n* sub[marine]
U-Bootabwehr *f* anti-submarine warfare
U-Bootfalle *f* decoy vessel, Q-ship
U-Bootjäger *m* submarine hunter, *(Am)* submarine chaser
U-Bootmutterschiff *n* submarine depot ship, *(Am)* submarine tender
U-Dock *n* U-dock, box dock
U-Eisen *n* channel [iron]
U-förmig U-shaped
U-Leitung *f (el)* unloaded line
U-Naht *f* single U-butt weld
U-Profil *n* [rolled steel] channel
U-Querschnitt *m* channel
U-Rohr *n* U-tube; return (double) bend, syphon
U-Rohrmanometer *n* fluid manometer, liquid-column gauge
U-Schienenführung *f* U-rail guide
U-Spant *n (ship)* U-shaped frame
U-Strahl *m* channel, U-section
übelriechend ill-smelling
üben to exercise; to practise; to train
Überalterung *f* / **technische** obsolescence
Überanpassung *f* over-match *(radio)*
Überballonreifen *m* giant tyre
Überbandmagnetscheider *m* top-mounted magnetic separator
Überbau *m* superstructure
überbauen *(mine)* to over-work
Überbeanspruchung *f* overload; overtension
überbelasten to superload
Überbelastung *f* excessive load, overload; surplus load
überbelichten to overexpose
Überbelichtung *f* overexposure
Überbestrahlung *f* overexposure, overirradiation, excessive irradiation
überblasen *(met)* to overblow
überblatten to halve *(carpentry)*; to overlap
Überbleichen *n (text)* overbleaching
überblenden to fade out and in, to fade over; to dissolve
Überblender *m* fader
Überblendung *f* change-over, fade-over, wipe, fading
  weiche Ü. *(telev)* fade-over, fade-in, fade out
Überblendungseinrichtung *f (telev)* fader, fading control
Überbordwerfen *n* jettison
überbrücken *(build)* to span, to straddle; *(el)* to bridge; to shunt; to connect across; to by-pass
Überbrückung *f (el)* bridging
  Ü. der Zündkerzenelektroden point bridging
Überbrückungskabel *n* bridging cable; jumper cable
Überbrückungskondensator *m* shunt (by-pass, bridging) capacitor
Überbrückungskontakt *m* bridging contact
Überbrückungsschalter *m* by-pass switch, field-breaking switch
Überbrückungsschaltung *f* bridging circuit, bridge connection
Überbrückungsschwingkreis *m* resonant shunt
Überbrückungsspule *f* leak coil
Überbrückungsvorrichtung *f* **für Zeitschalter** omitting device
Überbrückungswiderstand *m* field-breaking resistor
Überchlorsäure *f* perchloric acid
überdachen to roof [over]
Überdachung *f* roofing; canopy
überdecken 1. to [over]lap; 2. to mask; to cover; 3. to sweep
Überdecken *n* **der Schaufelenden** overlapping of the blade ends *(turbine)*
Überdeckung *f* 1. lap[ping], overlapping; 2. blanketing *(signal)*; 3. contact *(gear)*
  äußere Ü. steam (outside) lap
  innere Ü. inside (exhaust) lap
Überdeckungsgrad *m* contact ratio *(gear)*
Überdeckungsstoß *m* lap joint, lapped splice
überdehnt werden to stretch; to be overstressed
überdestillieren to distill over, to overdistill
überdimensionieren to oversize
Überdosierung *f* overdosage, overdose; overfeeding
überdrehen *(mach)* to turn cylindrically along the outside *(in lathe)*; to strip *(thread)*; to overwind *(spring)*

überdreht / stark hard-twisted *(yarn)*
Überdrehzahl *f* overspeed
Überdruck *m* 1. excessive (positive) pressure; overpressure; 2. overprint
Überdruckbogen *mpl* overs
Überdruckbuntreserve *f* coloured overprint resist
Überdruckdiagramm *n* overpressure diagram
überdrücken to over-compress *(briquette)*
Überdruckgrad *m (therm)* degree of reaction
Überdruckionisationskammer *f* overpressure ionization chamber
Überdruckkabine *f (aero)* sealed cabin
Überdruckregler *m* excess-pressure regulator
Überdruckschalter *m* maximum-pressure switch
Überdruckturbine *f* reaction turbine
  Ü. mit Druckabstufung reaction turbine with pressure stages, multistage reaction turbine
  mehrstufige Ü. reaction turbine with pressure stages, multistage reaction turbine
Überdruckventil *n* [pressure-]relief valve
Überdruckwächter *m (el)* maximum-pressure governor
Überdruckwindkanal *m* compressed-air tunnel
Übereinandergreifen *n* overlapping
  Ü. der Schüsse overlapping of the belts
übereinandergreifend overlapping; imbricated *(gutter-tile)*
übereinstimmen to agree, to be in agreement, to coincide; to match together
  ü. mit to conform to, to comply with *(standards)*
Übereinstimmung *f* agreement, conformity, consistency; correspondence; coincidence
  genaue Ü. dead match
überentwickeln to overdevelop *(photography)*
Überentwicklung *f* overdevelopment *(photography)*
übererlaubt *(nucl)* superallowed
Übererregung *f (el)* overexcitation, extra-excitation
übereutektisch hypereutectic
übereutektoid hypereutectoid
Überexponierung *f* overirradiation, overexposure
überfahren to run over; to cross; to pass
Überfahrt *f* passage, crossing, voyage
Überfall *m (build)* overflow
Überfallanruf *m* emergency call
Überfalldamm *m* overflow dam
Überfalldurchflußmesser *m* weir flowmeter
überfällig overdue
Überfallkrone *f* spillway crest
Überfangglas *n* flashed glass
Überfärbeechtheit *f (text)* fastness to topping (cross-dyeing)
Überfärben *n (text)* cross-dyeing
  saures Ü. cross-dyeing in acid liquor
überfärbt top-dyed, cross-dyed
Überfärbung *f* cross-dyeing
Überfettungsmittel *n (ch)* superfatting agent
Überfläche *f* supercritical area of extraction
überfliegen to fly across (over)
Überflurförderer *m* overhead conveyor
Überfluß *m* abundance
überflüssig machen to eliminate
Überflüssigkeit *f* redundancy
überfluten to inundate, to flood; to overflow
Überflutung *f* inundation, flood[ing]
Überflutungspunkt *m (ch)* flooding point
Überform *f* mantle, coat *(foundry)*
überführen to lead across; to carry over; to transfer; to convert; to transform
  die Papierbahn ü. to blow over the paper
Überführung *f* transfer[ence]; transmission; conversion; overpass [bridge], overbridge; fly-over *(road)*
Überführungsdose *f* terminal socket
Überführungsendverschluß *m* test (terminal) box
Überführungsflug *m* ferrying flight
Überführungskasten *m (telec) (Am)* cross-connecting terminal
Überführungsschnur *f (telec)* patchcord
Überführungstrommel *f* transfer (delivery) drum
Überführungswärme *f* heat of transfer
Überführungszahl *f (ch)* transport (transference) number
  Hittorfsche Ü. Hittorf number
Übergabe *f* delivery; transfer; handing-over
  schonende Ü. anti-breakage transfer
Übergabebahnhof *m* interchange point (station)
Übergabefuß *m* transfer butt
Übergabeklemme *f* consumer's terminal
Übergabeleistung *f (el)* delivery output
Übergabestation *f* transfer point
Übergabestelle *f* transfer point
Übergabewalzwerk *n* pass-over hot mill, pull-over hot mill
Übergabezeit *f* change-over time
Übergabezylinder *m* delivery cylinder
Übergang *m* 1. passage cross, crossing; change-over; 2. junction; transition; transfer; 3. jump; 4. *(geol)* graduation
  Ü. zu neuer Serie *(mach)* job change-over
  abgestufter Ü. graded junction
  allmählicher Ü. graded junction
  angeregter Ü. excitation transition *(photoconduction)*
  direkter Ü. direct transition
  elektromagnetischer Ü. electromagnetic transition
  elektronischer Ü. electron transition
  erlaubter Ü. allowable (allowed) transition
  fotovoltaischer Ü. photovoltaic transition
  gemischter Ü. mixed transition
  induzierter Ü. induced transition

intramolekularer Ü. intermolecular transition
isomerer Ü. isomeric transition
legierter Ü. alloy junction
optischer Ü. optical transition
radioaktiver Ü. radioactive transformation
schienengleicher Ü. level [railway] crossing
strahlender Ü. radiative transition
strahlungsloser Ü. radiationless transition
übererlaubter Ü. superallowed transition
verbotener Ü. forbidden transition
zulässiger Ü. allowable transition
Übergangsbahnhof *m* transit (junction, exchange) station
Übergangsbearbeitung *f (dat)* transfer operation
Übergangsbereich *m* transition region
Übergangsbetrieb *m* transient behaviour *(reactor)*
Übergangsbogen *m* easement (junction, transition) curve
parabolischer Ü. parabolic transition curve
Übergangsdüse *f (aut)* change-over jet
Übergangseffekt *m* transition effect
Übergangseisen *n* off-grade iron
Übergangselement *n (ch)* transition element
übergangsfähig suitable for transit
Übergangsfaktor *m* contamination factor
Übergangsfassung *f* diminishing socket
Übergangsfläche *f* interface
Übergangsform *f* junction shape
Übergangsfrequenz *f (telec)* transition (cross-over) frequency
Übergangsfunktion *f* transfer (transition) function; *(cont)* step function (response); transient response
Ü. der Regelstrecke reaction curve
Übergangsgebiet *n* 1. transition zone *(fluid flow)*; 2. [junction] transition region *(semiconductor)*
Übergangsgleichgewicht *n* transient (dynamic) equilibrium
radioaktives Ü. transient radioactive equilibrium
Übergangshülse *f (mach)* sleeve, adapter
Übergangskapazität *f* transition capacitance
Übergangskennlinie *f* transfer characteristic
Übergangskurve *f* easement (transition) curve *(road building)*
Übergangsmantel *m* autumn (spring) coat
Übergangsmatrix *f* transition matrix
Übergangsmetall *n* transition metal
Übergangsmuffe *f (el)* reducing sleeve (socket, pipe-joint)
Übergangspassung *f* neutral (transition) fit
Übergangspunkt *m* transition point; break point
Übergangsroheisen *n* off-iron
Übergangsschaltung *f* transition
Übergangssehen *n* mesopic vision
Übergangssitz *m* neutral (transition) fit
Übergangsstecker *m* plug adapter
Übergangsstelle *f* junction; transitory point
Übergangsstrom *m* transient current
Übergangsströmung *f* transition flow
Übergangstemperatur *f* transition temperature
Übergangsverhalten *n* transient behaviour *(reactor)*; transient response
Übergangsverlust *m* contact (transition) loss
Übergangsvorgang *m* transient
Übergangswahrscheinlichkeit *f* transition probability
Übergangswiderstand *m* contact (transition) resistance
Übergangswinkel *m (Am)* secondary chamfer angle *(reamer)*
Übergangszone *f* transition zone
Übergangszustand *m* transient [state] *(of reactor)*; compound (intermediate, nonsteady) state *(of nucleus)*
übergar overrefined; dry *(metal)*; black *(pig iron)*
übergehen / auf eine andere Welle to shift to another wave
Übergehen *n* **der Ladung** *(ship)* shifting of cargo
Übergeschwindigkeitsbegrenzer *m* overspeed limiter
Übergreifen *n (telec)* slipped grading
Übergröße *f* outsize; oversize
Übergruppenfrequenzband *n* supergroup band
Übergruppenmodulationseinrichtung *f* supergroup translating equipment
Überhang *m* overhang; foldover *(of a curve)*
überhängen to overhang, to hang over
Überhauen *n (mine)* raise, rise
Ü. zur Vorrichtung einer neuen Kammer development raise
Überhauenbohrloch *n* staple hole
Überhebetisch *m* relifting table *(rolling)*
Überhebevorrichtung *f* lifting blocks *(rolling)*
Überhebewalzwerk *n* pass-over mill
überheizen to overheat
überhitzen to superheat, to overheat
Überhitzer *m* superheater, overheater
Ü. in aufgesetzter Kammer superheater placed in a superposed chamber
Ü. mit eigener Feuerung independently fired superheater
Ü. mit Luftkühlung air-cooled superheater
Ü. mit Teilung des Dampfstromes superheater with division of the steam current
Ü. mit versenkbarem Schieber superheater with sliding damper
ausschaltbarer Ü. superheater with by-pass flue
ausziehbarer Ü. withdrawable superheater
durch Strahlung geheizter Ü. superheater heated by radiation

Heizmannscher Ü. Heizmann's plate superheater
hinter dem Kessel eingebauter Ü. downtake[-type of] superheater
kammerloser Ü. superheater not placed in a chamber
Überhitzeranlage *f* superheating plant
Überhitzerdampfkammer *f* superheater steam collector (header)
Überhitzerdoppelrohr *n* **mit gerader Trennungswand / glattes** plain superheating tube with flat diaphragm
glattes Ü. mit verdrehter Trennungswand plain superheating tube with twisted diaphragm
Überhitzerentwässerung *f* drainage of the superheater
Überhitzerfeuerkiste *f* superheater chamber in smoke box
Überhitzerfläche *f* superheating surface
Überhitzerflachrohr *n* flat superheater coil
Überhitzerheizfläche *f* superheating surface
Überhitzerheizschlange *f* zig-zag superheating tube, superheater coil
Überhitzerheizung *f* heating of superheater
Überhitzerkammer *f* superheater chamber
aufgesetzte Ü. mit Regulierdrehklappe superposed superheater chamber with swivel damper
Überhitzerregelung *f* superheater regulator
Überhitzerrippenrohr *n* ribbed (grilled) superheater tube
Überhitzerrohr *n* **/ schlangenförmiges** zig-zag superheating tube, superheater coil
Überhitzersammelkammer *f* superheater steam collector (header)
Überhitzerschraubenrohr *n* helical (spiral) superheating tube
Überhitzerspiralrohr *n* flat spiral superheating tube, superheater spiral
überhitzt superheated *(steam)*, overheated
Überhitzung *f* superheating *(steam)*, overheating
örtliche Ü. local overheating
überhöhen *(build)* to apply superelevation; to step *(tool teeth)*; to raise, to elevate
überhöht banked, superelevated *(curves in highways and railways)*; *(math)* leptokurtic; stepped *(teeth)*
Überhöhung *f* *(build)* superelevation, banking, cant *(curves)*; camber *(cross section)*; *(math)* leptokurtosis; degree of incline (slope); stepping, size of steps *(teeth)*; *(nucl)* peaking
Ü. der Schweißnaht reinforcement of the weld
Ü. je Zahn step (rise) per tooth, tooth increment *(broach)*
Ü. von Zahn zu Zahn increment rise per tooth
überholen 1. to overhaul, to recondition; 2. to pass, to overtake
Überholklauenschaltung *f* *(aut)* override clutch gear change
Überholsignalgerät *n* *(aut)* passing-signal indicator
Überholspur *f* acceleration lane
Überholung *f* repairing, repairs, overhauling
Überholungsbahnhof *m* overtaking station
Überholungsgleis *n* siding, passing track
Überkaro *n* *(text)* overcheck
Überkavitation *f* supercavitation
Überkehrsieb *n* *(ag)* upper (capes, chob) riddle
überkippen to tip over; to overturn
überkochen to boil over
Überkohlung *f* supercarburization
Überkompensation *f* overcompensation
Überkompoundierung *f* overcompounding
überkomprimiert overcompressed, supercompressed
Überkopfauswurf *m* overhead delivery outlet *(forage cutter-blower)*
Überkopfbohrer *m* *(build)* stoper
Überkopflader *m* overhead (overshot) loader; *(mine)* shovel (front-end) loader, rocker shovel [loader]
Überkorn *n* oversize [pieces], coarse-grained parts
Überkorrektion *f* overcorrection
überkorrigiert overcorrected
überkritisch supercritical, hypercritical
überladen 1. to overload; to overburden; 2. to overcharge *(battery)*
überlagerbar superposable
Überlagerer *m* beat oscillator; local oscillator
überlagern to heterodyne, to super[im]pose *(frequency)*
sich ü. to superheterodyne *(radio)*
Überlagerung *f* super[im]position; beat
Ü. von Absorptionslinien overlap of absorption lines
lineare Ü. linear superposition
Überlagerungsdruck *m* *(mine)* roof (normal rock) pressure
Überlagerungseffekt *m* beat effect
Überlagerungsempfang *m* heterodyne (double-detector, supradyne, superheterodyne) reception
Überlagerungsempfänger *m* [super]heterodyne receiver
Überlagerungsfernschreiber *m* ultra-acoustic teleprinter
Überlagerungsfrequenz *f* heterodyne frequency; beat frequency
Überlagerungsfrequenzmesser *m* heterodyne frequency meter
Überlagerungskode *m* superimposed code
Überlagerungskomponente *f* beat component
Überlagerungskreis *m* superposed circuit

Überlagerungsoszillator *m* beat frequency oscillator
Überlagerungspfeifen *n* heterodyne interference, self-whistle
Überlagerungsprinzip *n* [super]heterodyne principle, principle of superposition
Überlagerungsschaltung *f* superimposed (superheterodyne) connection
Überlagerungssender *m* heterodyne (local) oscillator
Überlagerungssteilheit *f* conversion transconductance
Überlagerungsstufe *f* mixer stage
Überlagerungssummer *m* multi-frequency heterodyne generator
Überlagerungstechnik *f* heterodyne technique
Überlagerungstelegrafie *f* superimposed (super-audio, super-acoustic) telegraphy
Überlagerungstelegrafieempfang *m* heterodyne code reception
Überlagerungston *m* beat note, note of pitch
Überlagerungsvorsatzgerät *n* superheterodyne converter
Überlagerungswellenmesser *m* heterodyne (oscillator) wavemeter
Überlagerungswirkung *f* beat effect
Überlandfreileitung *f* overhead transmission line
Überlandfreileitungsgittermast *m* overhead transmission line tower
Überlandfreileitungsholzmast *m* overhead transmission line [wooden] pole
Überlandleitung *f* *(el)* [overhead] transmission line
Überlandleitungsmast *m* transmission line tower (pole), electric power pylon
Überlandverkehr *m* long-distance road haulage
überlang overlong
überlappen to [over]lap
Überlappen *n* overlap[ping]
  Ü. der Absorptionslinien overlap of absorption lines
überlappend overlapping; imbricated *(gutter-tile)*
Überlappnaht *f* lap weld
Überlappnahtschweißung *f* lap seam weld
Überlappschweißung *f* lap weld
Überlappstaffelung / mit overlapping *(broach)*
Überlappstoß *m* lap joint
Überlappnietung *f* lap riveted joint
Überlappung *f* lap[ping], overlapping; lap joint
Überlappungsgrad *m* degree of overlap
Überlappungsmodell *n* overlap model
Überlappungsnietung *f* lap [riveted] joint
Überlappungsperiode *f* *(el)* overlapping period
Überlappungsstoß *m* lap joint
Überlappungswinkel *m* *(el)* overlap angle
Überlast *f* excessive load, overload
Überlastbarkeit *f* overload capacity
  elektrische Ü. electrical overload capacity
überlasten to overload, to superload
Überlastfaktor *m* *(nucl)* hot-channel factor
Überlastrelais *n* overload relay
  thermisches Ü. thermal overload relay
Überlastschalter *m* overload circuit-breaker
Überlastschutz *m* overload protection, safeguard against overload
Überlaststrom *m* overload current
  Ü. bei Elektrodenschluß fault electrode current
Überlastung *f* overload[ing]
  Ü. / empfindlich gegen vulnerable (susceptible) to overload
  Ü. durch Einschaltspitze inrush load
  betriebsmäßige Ü. operating overload
  kurzzeitige Ü. momentary overload
Überlastungsanzeige *f* overload indication
Überlastungsanzeiger *m* overload indicator
Überlastungsauslösung *f* overload release
Überlastungskupplung *f* overload (safety) clutch
Überlastungspunkt *m* overload point
Überlastungsschutz *m* overload protection
Überlastungsschutzeinrichtung *f* overload protection device
Überlastungssicherung *f* overrange protection
Überlastungsspitze *f* peak overload
Überlastungsventil *n* overload (by-pass) valve
Überlastungszähler *m* congestion meter *(telephony)*
Überlauf *m* overflow; overflow pipe; down-flow; waste weir; *(Am)* spill-over; *(dat)* overflow; *(mach)* overtravel, overrunning *(slide)*
Überlaufdamm *m* overflow (spillway) dam
überlaufen to overflow *(liquid)*; *(mach)* to overrun *(slide)*
Überlaufen *n* **der Kolbenringe** overrunning of the piston rings
Überlaufmelder *m* overflow indicator
Überlaufrinne *f* overflow gutter
Überlaufrohr *n* overflow pipe, down-flow
Überlaufstutzen *m* branch overflow (nozzle)
Überleben *n* survival
Überlebenskurve *f* survival curve
Überlebenszeit *f* survival time
überlegen / sich to heel over *(vessel)*
überleiten to lead over; to conduct over
überleitfähig superconducting
Überleitfähigkeit *f* superconductivity
Überleitrohr *n* by-pass pipe *(turbine)*
Überleitungsamt *n* *(telec)* transfer exchange
überliegen *(ship)* to list
Überliegetage *mpl* *(ship)* demurrage, overlaying days
Übermangansäure *f* permanganic acid
Übermaß *n* interference, allowance *(fit)*, allowance for fit (interference); oversize

Ü. an Luken excess of hatchway
Ü. bei Langholz excess length
übermäßig excessive; *(math)* redundant
Übermikroskop *n* ultra-microscope; electron microscope
Übermikroskopie *f* ultra-microscopy; electron microscopy
übermikroskopisch electron-microscopical
übermitteln *(telec)* to transmit
eine Gesprächsanmeldung ü. to pass a booking
Übermittlung *f* transmission
Übermittlungsgeschwindigkeit *f* *(telec)* transmission speed
Übermöllerung *f* overburdening
Übernachtungsraum *m* dormitory
übernähen to stitch-bond
Übernahme *f* transfer; taking over; acceptance
Übernahmebedingung *f* condition of acceptance
Übernahmekennlinie *f* transfer characteristic
Übernahmepumpe *f* fueling (fuel loading) pump
Übernahmezeit *f* changeover time
übernehmen to accept, to receive; to take charge
Elektronen ü. to collect electrons
überpolen to overpole *(copper)*
Überproduktion *f* surplus production
überprüfen to verify; to check *(dimensions)*; to overhaul *(repair)*; to examine, to inspect
ü. auf to check for
Überprüfung *f* 1. [control] check, checkout; 2. overhaul *(repair)*
Überputz *m* finishing coat
überqueren to traverse, to cross
überregeln to overtravel
Überreiber *m* pyramid (distributing) roller, distributor
Überreichweite *f* super range
Überreichweiteausbreitung *f* transhorizon propagation
Überreißen *n* **des Wassers** priming (entrainment) of the water
Überrest *m* residue, remainder, rest
übersättigen to supersaturate, to oversaturate
übersättigt supersaturated
Übersättigung *f* oversaturation, supersaturation
Ü. mit Kohlendioxid overcarbonation
Übersäuerung *f* superacidulation
Überschall *m* ultrasound
Überschallbearbeitung *f* ultrasonic machining
Überschallflugzeug *n* supersonic aircraft
Überschallfrequenz *f* supersonic frequency
Überschallgeschwindigkeit *f* supersonic velocity (speed)
Überschallprüfung *f* ultrasonic (supersonic) testing
Überschallsonde *f* ultrasonic probe
Überschallströmung *f* flow at supersonic velocity, supersonic flow
Überschallwelle *f* supersonic (ultrasonic) wave
Überschaltdrossel *f* transition coil
überschalten to override
Überschaltwiderstand *m* transition (preventive) resistance
überschätzen to overrate, to overestimate
überschäumen to bubble over
Überschicht *f* overtime shift
überschichten to add a layer
Überschiebung *f* *(geol)* reversed fault, thrust
Überschießen *n* *(text)* picking above
überschlächtig overshot
Überschlag *m* 1. *(el)* arcing, arc-over, flash-over, spark-over, striking an arc (a spark), spark jump; 2. estimate *(cost)*; 3. overturn
roher Ü. rough estimate *(cost)*
rückwärtiger Ü. *(el)* backward arc-over
überschlagen 1. *(el)* to flash (arc) over, to break over, to spark over; 2. to estimate *(cost)*; 3. *(aero)* to nose over, to overturn
Überschlaggefahr *f* *(aero)* danger of nosing-over
Überschlagsberechnung *f* approximate computation
Überschlagsdistanz *f* *s.* Überschlagsstrecke
Überschlagsfunke *m* jumping (flash-over) spark
überschlagsicher arc-over proof
Überschlagspannung *f* sparking (flash-over, spark-over) voltage
Ü. zwischen Spitzen needle-point voltage
Überschlagsprüfung *f* *(el)* flash-over test, spark-over test
Überschlagsrechnung *f* estimate calculation
Überschlagsspannung *f* *s.* Überschlagspannung
Überschlagsstrecke *f* *(el)* spark-over distance, arc-over distance, flash-over distance
Überschlagsverzögerung *f* impulse inertia
Überschlagswert *m* coarse estimate
Überschleifen *n* **des Kanals** passing of the slide beyond the port
überschneiden to overlap
Überschneidungsfrequenz *f* cross-over frequency
Überschneidungsgebiet *n* **von Funksignalen** equiphase zone
überschreiten to exceed; to overshoot; to overtravel
Überschreiten *n* transgression
Ü. der zulässigen Geschwindigkeit speeding
Ü. des Grenzwertes *(el)* over-ranging *(measurement)*
Überschreitung *f* overshoot; exceeding
Überschreitungswahrscheinlichkeit *f* risk of passing off-standard production *(quality control)*
Überschrift *f* heading
Überschuß *m* [positive] excess; surplus
Ü. an Arbeit excess of work
Überschußalkali *n* excess alkali

Überschuß-Drei-Kode *m* excess-three code
Überschußelektron *n* excess electron
Überschußfaktor *m* excess multiplication factor
Überschußflüssigkeit *f* surplus liquid
Überschußgas *n* surplus gas
Überschußhalbleiter *m* excess (n-type) semiconductor
überschüssig surplus
Überschußleitung *f* n-type conductivity
Überschußmetall *n* excess of metal
Überschußneutron *n* surplus neutron
Überschußreaktivität *f* excess (built-in) reactivity
Überschußstrom *m* excess current
Überschußtechnik *f* redundancy technique
Überschußventil *n (nucl)* dump valve
überschwemmen to inundate, to flood
Überschwemmung *f* inundation, overflow, flood[ing]
überschwingen 1. to overtravel; to overshoot; 2. to ring *(pulse)*
Überschwingen *n* overshoot, overswing; ballistic factor
Überschwingfaktor *m* overshooting factor
überschwingfrei aperiodic, *(Am)* dead-beat
Überschwingung *f* overshoot, overswing
Überschwingweite *f* / **maximale** total overshoot
Überseefunk *m* transoceanic radio communication
Überseekabel *n* transoceanic (transatlantic) cable
Überseeschiff *n* oversea vessel
Überseeverkehr *m (telec)* transoceanic communication
 drahtloser Ü. transoceanic radio service
übersetzen 1. to ferry *(river, canal)*; 2. *(met)* to overcharge the furnace; 3. to translate
Übersetzen *n* **ins Langsame** gearing down, speed reduction (reducing)
 maschinelles Ü. machine translation
Übersetzer *m* coder; transcriber *(computer)*; translater *(telegraphy)*; converter
Übersetzfähre *f* transfer ferry
übersetzt / niedrig low-geared
Übersetzung *f* 1. *(mach)* transmission ratio; 2. *(el)* transformation ratio; 3. conversion; 4. translation
 Ü. des Achsantriebs axle-drive ratio
 Ü. einer Radpaarung speed ratio [of a pair of gears]
 Ü. ins Langsame speed reduction, reduction of speed, reduction [gear] ratio
 Ü. ins Schnelle speed increase
 hohe Ü. high range [of gears]
 niedrige Ü. low range [of gears]
 schnelle Ü. *(aut)* high range [of gears]
Übersetzungsfehler *m (el)* ratio error
Übersetzungsgetriebe *n* speed increaser, speed-increasing drive; transmission gear; step-up gear
 Ü. ins Langsame reduction gearing
Übersetzungsmaschine *f* machine translator, electronic interpreter, mechanical dictionary
Übersetzungsstufe *f* speed ratio
Übersetzungsverhältnis *n* speed (velocity) ratio, ratio of gears (transmission); *(el)* transformation ratio, ratio of transformation; step-up ratio; increasing ratio
 Ü. bei Zahnrädern ratio of driving and driven wheels
 Ü. ins Langsame reduction [gear] ratio
übersichtlich angeordnet clearly arranged
Übersichtskarte *f* sketch map
Übersichtsplan *m (build)* key plan; *(ship)* general plan
Übersichtszeichnung *f* general drawing
überspannen *(build)* to span; to overstress
Überspannung *f (el)* overvoltage, excess[ive] voltage; voltage surge
 atmosphärische Ü. overvoltage of atmospheric origin
 konzentrationsbedingte Ü. concentration overvoltage
 momentane Ü. voltage transient
Überspannungsableiter *m* overvoltage (surge) arrester, surge diverter (dissipator), arrester
 Ü. mit Wasserwiderstand water-column arrester
 elektrolytischer Ü. electrolytic lightning arrester
Überspannungsdurchschlag *m* breakdown due to overvoltage
Überspannungsfaktor *m* overvoltage coefficient
Überspannungsrelais *n* maximum-voltage relay
Überspannungsschalter *m* overvoltage (maximum-voltage) circuit-breaker
Überspannungsschutz *m* overvoltage preventer (protective device); overvoltage (surge) protection
Überspannungssicherung *f* overvoltage circuit-breaker, excess voltage cut-out
Überspannungsstoß *m* overvoltage transient
Überspannungsverhältnis *n* magnification of the circuit *(oscillators)*
überspielen to transmit; *(ac)* to re-record
Überspielen *n* **von Bändern** re-recording of tapes
Übersprechdämpfung *f* crosstalk attenuation
Übersprechen *n* cross-talk, cross induction
Übersprechkopplung *f* cross (transverse cross-talk) coupling
Übersprechstörung *f* cross-talk trouble (interference)
überspringen 1. to skip; 2. to flash (spark) over

Überstabilität *f* overstability
Überstau *m* flood irrigation
überstehen to project, to overhang
überstehend projecting, overhanging
übersteigen to exceed
übersteuern to override, to overdrive; to overmodulate
Übersteuerung *f* overmodulation, overamplification; overload[ing]
Übersteuerungsanzeiger *m* overload indicator
Übersteuerungsschutz *m* overrange protection
Überstrahlung *f* *(telev)* bloom; *(telec)* overshoot distortion
überstrecken to overstretch
überstreichen to sweep [over] *(mechanics)*; to scan; to traverse
  einen Bereich ü. *(el)* to cover a range
Überstreichung *f* sweep; scanning; coverage
Überstrom *m* overcurrent, excess current, surge *(semiconductor)*
Überstromauslöser *m* overcurrent (overload) release, maximum cut-out, overload trip
Überstromauslösung *f* overcurrent (overload) tripping
überströmen to overflow, to flow over, to flood, to spill
Überstromgebiet *n* overcurrent range *(current transformer)*
Überstromgenauigkeit *f* overcurrent accuracy *(current transformer)*
Überströmkanal *m* transfer port, passage way
Überstromkontakt *m* overcurrent contact
Überstromrelais *n* overcurrent (maximum-current, overload) relay
Überströmrohr *n* by-pass pipe
Überstromschalter *m* overcurrent circuit-breaker, overload switch
Überströmschlitz *m* spill port
Überstromschutz *m* overload (overcurrent) protection
überstromsicher excess current-proof
Überstromspule *f* overload coil
Überströmventil *n* relief valve
Überstromziffer *f* overcurrent factor
Überstruktur *f* superlattice structure
überstumpf *(math)* reflex
Überstunden *fpl* overtime
Überstundenzuschlag *m* overtime rate (premium)
übersynchron hypersynchronous
übertönen *(ac)* to drown
Übertrag *m* carry *(from digit place to digit place)*; amount brought forward *(bookkeeping)*
  autonomer Ü. self-instructed carry
  selbsttätiger Ü. self-instructed carry
Übertragausbreitung *f* carry propagation
Übertragbarkeit *f* transmissibility
übertragen to transmit, to convey; to carry over (forward); to transfer; to telemeter
  eine Information ü. *(telec)* to convey an intelligence
  eine Last ü. auf to bear on
Übertrager *m* [coupling] transformer, transducer; repeating coil
  angepaßter Ü. matched transformer
  entzerrender Ü. regenerative repeater
  symmetrischer Ü. symmetrical transformer
Übertragerbrücke *f* transformer bridge
Übertragersatz *m* gain booster
Übertragerspule *f* repeating coil
Übertragerverlust *m* transformer loss
Übertragerverzerrung *f* transformer distortion
Übertragerwiderstand *m* transfer resistance
Übertragheber *m* *(text)* transfer cam
Übertragschaltung *f* carry circuit (gate)
Übertragung *f* transmission, transfer; translation
  Ü. unter schrägem Einfall *(telec)* oblique incidence transmission
  Ü. von Fernschreibzeichen über Funk telex on multiplex
  drahtlose Ü. radio transmission
  gerichtete Ü. directive transmission
  indirekte Ü. indirect transmission
  kapazitive Ü. capacity transfer
  verlustarme Ü. low loss transmission
Übertragungsabschnitt *m* *(telec)* transmission link
Übertragungsamt *n* repeating office
Übertragungsanlage *f* / **stereofone** stereophonic sound system
Übertragungsband *n* transmission band
Übertragungsbereich *m* transmission range (region)
Übertragungscharakteristik *f* transfer characteristic
Übertragungsdämpfung *f* transmission loss
Übertragungsdauer *f* picture transmission period
Übertragungsfähigkeit *f* transmission capacity
Übertragungsfaktor *m* proportional gain constant; transfer factor
  Ü. eines Vierpols quadripole propagation factor
  axialer Ü. axial response [factor]
Übertragungsfehler *m* *(telev)* transmission error
Übertragungsform *f* transmission mode
Übertragungsformel *f* radio transmission formula
Übertragungsfunktion *f* transfer function
  Ü. des gesamten Systems overall transfer function
Übertragungsgeschwindigkeit *f* transmission speed; transfer rate
Übertragungsgleichung *f* transfer equation
Übertragungsglied *n* transfer element
Übertragungsgröße *f* transmission performance
Übertragungsimpuls *m* transmitting pulse

Übertragungskanal *m* [communication] channel; transmission channel
Übertragungskennlinie *f* transfer characteristic; transient response
Übertragungskonstante *f* propagation constant
Übertragungsleistung *f* power-handling capacity
Übertragungsleitung *f* transmission line; transfer line
Ü. zwischen zwei Sendestationen programme loop
symmetrische Ü. balanced transmission line
Übertragungsleitwert *m* forward transadmittance; transfer admittance
Übertragungsmaß *n* propagation constant [per section]
konjugiert-komplexes Ü. conjugate transfer coefficient (constant)
Übertragungsmechanismus *m* mechanism of transfer
Übertragungsmeßgerät *n* transmission measuring set
Übertragungsmittel *n* communication medium
Übertragungsnetz *n (el)* [overhead] transmission system
Übertragungsnorm *f* transmission standard
Übertragungsnormal *n* transfer standard
Übertragungspegel *m* transmission level
Übertragungsquerschnitt *m* transmission cross-section
Übertragungsrelais *n* repeating relay
Übertragungsschalter *m* transfer switch
Übertragungsschaltung *f* / **elektrische** electric transducer
Übertragungssystem *n* communication system
Übertragungsverbindung *f* transfer connection
Übertragungsverhalten *n* performance; transmission characteristic; transient response
Übertragungsverhältnis *n* transmission ratio; transfer ratio
Übertragungsverlust *m* transmission loss, loss in transmission
Übertragungsverzögerung *f* transmission lag
Übertragungsweg *m* transmission channel (path); transmission chain
Übertragungswiderstand *m* transmission (transfer) resistance
Übertragvorschub *m* overflow ejection *(punched card machinery)*
Übertreffen *n* exceeding
übertreiben 1. to overdraw; 2. *(mine)* to overwind *(shaft)*
Übertreiben *n* overwinding *(shaft)*
Übertreibschutzvorrichtung *f (mine)* cage arresting device
Übertritt *m* transgression
Überverbunderregung *f* overcompounding, overcompound excitation *(generator)*
überverdichtet supercompressed
überwachen to control, to monitor; to inspect; to supervise
überwachsen overgrown
Überwachung *f* monitoring, control; survey; checking; production control, inspection
Ü. der Feuerung stoke-hole control
akustische Ü. audible monitoring
elektronische Ü. electronic supervision
laufende Ü. routine test
zwangsläufig wirkende Ü. *(ry)* compulsory checking
Überwachungsanlage *f* monitoring system, monitor equipment
Überwachungsbezirk *m* observation area
Überwachungseinrichtung *f* monitor facility, supervising (checking, supervisory) equipment, monitoring system
Überwachungsempfänger *m* monitoring radio receiver
Überwachungsgerät *n* monitor[ing instrument]; survey meter (instrument); controlling instrument
Überwachungsgestell *n (telec)* supervisory rack
Überwachungsinstrument *n s.* Überwachungsgerät
Überwachungsklinke *f (telec)* monitor jack
Überwachungskontakt *m* controlling (detector) contact
Überwachungskreis *m* supervising circuit; guard circuit *(reception)*
Überwachungslampe *f (telec)* pilot (telltale) lamp, pilot (telltale) signal
Überwachungsmeßgerät *n* survey meter (instrument)
Überwachungsmessung *f* maintenance test
Überwachungsprogramm *n (dat)* tracing routine
Überwachungspult *n* monitor (control) desk
Überwachungspunkt *m* monitoring point
Überwachungsraum *m* control room
Überwachungsrelais *n* supervisory relay
Überwachungsschalter *m* monitoring switch
Überwachungssystem *n* protection system; *(ry)* vigilance control system
Überwachungstafel *f* control panel
Überwachungszeichen *n* pilot signal
Überwärmung *f* overheating
Überwasserfahrzeug *n* surface craft
Überwasserform *f* above-water form, shape above
Überwassergeschwindigkeit *f* surface speed *(submarine)*
Überwasserschiff *n* surface vessel, upper work
Überwasserstabilität *f (ship)* surface stability
Überwasserstreitkräfte *fpl* surface forces
Überwasserteile *npl* topsides *(of a vessel)*
Überweg *m* overpath
überweisen to transfer
Überweisung *f* transfer

Überweisungsamt *n* transfer exchange *(telephony)*
Überweisungsfernamt *n* transfer trunk exchange
Überweisungsleitung *f* transfer line, toll switching trunk
Überweisungsmeldeleitung *f* recording trunk
Überweisungstaste *f* assignment key
Überweitwinkelobjektiv *n* super-wide-angle objective
Überweitwinkelprojektor *m* super-wide-angle projector
Überwendlichnähen *n* overcasting
Überwendlichnähmaschine *f* overcasting machine
überwinden to overcome, to get over; to surmount
überwölben to [over]arch
Überwucht *f* amount of overbalance (counterweight) *(turbine)*
Überwuchtmasse *f s.* Überwucht
Überwurf *m* [porcelain] cap
Überwurfmutter *f* sleeve (cap, box, union, coupling, spigot) nut, screwed cap
überzählig redundant *(statics)*
Überzähligkeit *f (math)* redundancy
Überzähligkeitsmeldung *f* advice of excess parcels
überziehen to coat, to laminate, to line, to plate
mit Metall ü. to metallize
Überziehen *n* coating, cladding, plating, jacketing, canning; metallizing
Überziehpresse *f* coating press
überzogen coated
Überzug *m* coating, clad[ding material]; film; skin; plating; cover[ing]; *(build)* upstand beam
abstreifbarer Ü. strip coating
anodischer Ü. anodic coating
anorganischer Ü. inorganic coating
galvanischer Ü. electrodeposit
gelartiger Ü. gel-coat finish
metallischer Ü. metallic coating
nichtmetallischer Ü. non-metallic coating
organischer Ü. organic coating
Überzugsfehler *m* coating defect
Überzugsgewebe *n* cover cloth
Überzugslack *m* protective lacquer, coating (finishing) varnish, top coat
Überzugswerkstoff *m* coating material
Übungsboot *n* drill boat
Übungsflug *m* practice flight
Übungsgerät *n* training device
Übungsrakete *f* practice rocket
Übungsschießen *n* target (artillery) practice
Ufer *n* bank *(river)*; shore, strand
Uferbefestigung *f* embankment
Uferdamm *m* levee, quay
Ufererosion *f* bank erosion
Ufermauer *f* quay wall
Uferschutz *m* revetment; bank protection
ug-Kern *m* odd-even nucleus
UHF-Band *n* ultra-high frequency band, u.h.f. band
UHF-Mischer *m* ultra-high frequency mixer
UHF-Triode *f* / **leuchtturmförmige** lighthouse tube
Uhr *f* clock; watch
U. mit elektrischem Aufzug electrically wound clock
U. mit Stromstoßfortsteuerung impulse-driven clock
astronomische U. astronomical clock
elektrisch geregelte U. electrically controlled clock
elektrische U. electric clock
elektrische U. mit Gangreserve electric clock with reserve power
impulsgesteuerte U. impulse-driven clock
stimmgabelgesteuerte U. tuning-fork clock
stromstoßgeschaltete U. impulse clock
Uhrenfeder *f* watch spring
Uhrengehäuse *n* watch case
Uhrenindustrie *f* watch and clock industry
uhrgesteuert clock-controlled
Uhrglas *n* watch-glass; *(ch)* beaker cover
Uhrmacherschraubenzieher *m* watchmaker's screwdriver
Uhrstein *m* jewel
Uhrwerk *n* movement of a watch; clockwork, clock drive; chart drive *(recorder)*
uhrwerkbetrieben clock-work driven
Uhrwerktriebfeder *f* main watch spring
Uhrwerkzünder *m* clockwork fuse
Uhrzeiger *m* clock hand
Uhrzeigersinn / gegen den anti-clockwise, counter-clockwise
U. / im clockwise
UKW VHF, v.h.f., very high frequency, ultra-short wave
UKW-Autofunk *m* VHF vehicle calling
UKW-Band *n* VHF band
UKW-Baustein *m* VHF tuner unit
UKW-Bedeckung *f* VHF coverage
UKW-Bereich *m* VHF region
UKW-Drehfunkfeuer *n* VHF omnirange
UKW-Empfang *m* VHF reception
UKW-Frequenzmesser *m* VHF frequency meter
UKW-Peiler *m* VHF direction finder
UKW-Schwingungen *fpl* / **wilde** VHF parasitics
UKW-Sender *m* VHF transmitter
UKW-Träger *m* VHF carrier
UKW-Verkehr *m* / **ziviler** citizen VHF radio broadcasting
Ulbricht-Kugel *f* Ulbricht globe
Ulme *f* elm
Ulmenholz *n* elm [wood]
ultraakustisch superaudible

Ultradynempfang *m* ultradyne reception
Ultradynempfänger *m* ultradyne receiver
Ultrafilter *n* colloidal filter
Ultrafiltration *f* ultrafiltration
ultrahart ultrahard
Ultrahochfrequenzentladung *f* ultra-high frequency breakdown
Ultrahochvakuumanlage *f* ultra-high vacuum installation
Ultraionisation *f* ultraionization
Ultrakronglas *n* extra heavy crown glass
Ultrakurzwelle *f s.* UKW
Ultramarin *n* ultramarine
Ultramikroanalyse *f* ultramicroanalysis
Ultramikrochemie *f* ultramicrochemistry
Ultramikroskop *n* ultra-microscope
Ultramikroverfahren *n* ultramicro method
Ultraschall *m* ultrasonic sound beam, ultrasonics
Ultraschallbaustein *m* ultrasonic device
Ultraschallbereich *m* ultrasonic (supersonic) frequency range
Ultraschallbestrahlung *f* exposure to ultrasonic waves
Ultraschallbohrmaschine *f* ultrasonic drilling machine
Ultraschalldämpfung *f* ultrasonic attenuation
Ultraschalldefektoskop *n* ultrasonic flaw detector
Ultraschalldickenmeßgerät *n* ultrasonic thickness testing instrument
Ultraschalldurchflußmesser *m* sonic (acoustic) flowmeter
Ultraschallecholot *n* supersonic echo-sounder
Ultraschalleistung *f* ultrasonic power
Ultraschallempfänger *m* ultrasonic receiver
Ultraschallerzeuger *m* ultrasonic source
Ultraschallfehlersuche *f* ultrasonic crack detection
Ultraschallfrequenz *f* ultrasonic (supra-acoustic) frequency
Ultraschallgeber *m* ultrasonic transmitter
Ultraschallgenerator *m* ultrasonic power generator
Ultraschallintensitätsmesser *m* ultrasonic intensity meter
Ultraschallot *n* ultrasonic depth gauge
Ultraschallotung *f* ultrasonic sounding
Ultraschallprüfgerät *n* ultrasonic inspection equipment, ultrasonic flow detector, ultrasonic testing device (apparatus)
Ultraschallprüfung *f* ultrasonic (supersonic) testing
Ultraschallreinigungsanlage *f* ultrasonic cleaning equipment
Ultraschallrelais *n* ultrasonic relay
Ultraschallsender *m* ultrasonic transmitter
Ultraschallsignal *n* ultrasonic signal
Ultraschallspeicher *m* acoustic memory
Ultraschalltiefenmesser *m* ultrasonic depth gauge
Ultraschalltopf *m (opt)* ultrasonic vessel
Ultraschallumformer *m* ultrasonic transducer
Ultraschallverbindung *f* supersonic communication
Ultraschallverstärkung *f* ultrasonic amplification
Ultraschallverzögerungsleitung *f* sonic delay line
Ultraschallwelle *f* ultrasonic (supersonic) wave
Ultraschwarzgebiet *n (telev)* blacker-than-black region
Ultraschwarzpegel *m (telev)* blacker-than-black level
Ultraschwärzung *f* infrablack condition
ultraviolett ultraviolet
Ultraviolettbandenspektrum *n* ultraviolet band spectrum
ultraviolettdurchlässig ultraviolet transmitting
Ultraviolettechnik *f* ultraviolet engineering
Ultraviolettkatastrophe *f* ultraviolet catastrophe
Ultraviolettmikroskopie *f* ultraviolet microscopy
ultraweich ultrasoft
Ultrazentrifuge *f* ultra-centrifuge, high-speed centrifuge
Ultrazentrifugierung *f* ultra-centrifugation
umarbeiten to change, to convert, to modify, to rework
  u. auf to form to
Umarbeitung *f (mach)* reworking
  U. der Probe sample treatment
Umbau *m* rebuilding; alteration; reconstruction; conversion
  U. des Gleises repair of the track
umbauen to reconstruct; to rebuild; to alter
Umbauhobel *m (mine)* conversion plough
Umbehandlung *f* re-registration, re-invoicing
Umbehandlungsbahnhof *m* re-invoicing station
Umber *m* umber
umbiegen to bend over
Umbiegung *f (geol)* drag
umbilden to re-form, to transform
Umbildung *f* **eines Zuges** re-forming of a train
Umbördeln *n* flanging
Umbra *f* umber
umbrechen *(ag)* to plough up, to turn; *(print)* to make up
Umbruch *m (ag)* ploughed up land; *(print)* make-up
  geklebter U. *(print)* paste-up
umbuchen to repost
Umbuchung *f* bookkeeping transfer
umdecken to retile *(roof)*
umdrehen to rotate, to revolve, to turn [round]
Umdrehung *f* revolution, turn
  volle U. complete rotation (turn)

Umdrehungen *fpl* **je Minute** revolutions per minute, r.p.m.
U. je Zeiteinheit revolution frequency
Umdrehungsanzeiger *m* speed (turns) indicator, tachometer
Umdrehungsellipsoid *n* ellipsoid of revolution
Umdrehungsfläche *f* surface of revolution
Umdrehungsgeschwindigkeit *f* speed of revolution
Umdrehungskörper *m* solid of revolution
Umdrehungszahl *f* number of revolutions, speed
Umdrehungszähler *m* chronometric tachometer, motometer, spin (revolution) counter
Umdrehungszählwerk *n* cyclometer
umdrucken to transfer
Umdruckfarbe *f* transfer ink
Umdruckpapier *n* [chalk] transfer paper
Umdruckpresse *f* transfer press
Umeichung *f* gauge transformation
umeinandergezwirnt *(text)* doubled on one another
Umesterung *f* ester interchange, interesterification, alcoholysis
Umfällung *f (ch)* reprecipitation
Umfang *m* 1. circumference; perimeter, periphery; 2. size; volume; 3. range; extend
U. des eingetauchten Hauptspants submerged girth of midship-section
benetzter U. wetted perimeter
Umfangdruck *m* peripheral (circumferential) pressure
Umfangkomponente *f* **der Dampfgeschwindigkeit** peripheral component of the steam velocity *(turbine)*
Umfangkraft *f* peripheral (circumferential) pressure
Umfanglänge *f* circumferential length
Umfangsabrichten *n* **der Schleifscheibe** peripheral dressing of the grinding wheel
Umfangsdichtung *f* peripheral seal
Umfangsfläche *f* circumferential surface
Umfangsgeschwindigkeit *f (mach)* peripheral speed (velocity)
Umfangskanal *m* peripheral passage
Umfangsrichtung *f* circumferential direction
Umfangsschneide *f* peripheral cutting edge *(milling cutter)*
Umfangsteilung *f* circular pitch
Umfangszahn *m* flute edge *(reamer)*, tooth on the periphery
umfärben to redye
Umfärbeverfahren *n* redyeing process
umfassen to comprise, to comprehend, to include, to cover
einen Bereich u. to cover a range
umfassend comprehensive; full-scale
Umfassungswinkel *m* angle of contact
Umfläche *f (Am)* land *(twist drill)*
Umflechtmaschine *f* braiding machine, braider
Umflechtung *f* braid
Umformarbeit *f* deformation work
umformbar *(met)* workable, formable
Umformbarkeit *f (met)* formability, workability
umformen 1. to transform; 2. *(el)* to convert; 3. *(met)* to form, to work
Umformen *n* forming, working *(metal)*
Umformer *m (el)* transformer; transducer; converter, dynamotor, motor-generator, current changer; reshaper
Umformergruppe *f* motor-generator set
Umformerlokomotive *f* mutator locomotive
Umformmaschine *f* metal forming machine
Umformtechnik *f* metal forming (working)
Umformung *f* 1. metal forming (working); 2. *(el)* conversion
Umfriedung *f* enclosure
Umführung *f* repeater, looping mill, turn over *(rolling)*
Umführungskanal *m* peripheral passage, guide channel
Umführungsventil *n* by-pass valve
umfüllen to decant
Benzin u. to decant petrol
Umgang *m* 1. handling, manipulation; 2. *(build)* ambulatory
Umgangsventil *n* by-pass valve
umgebend ambient, surrounding
Umgebung *f* environment, neighbourhood, vicinity
Umgebungsänderung *f* ambient change
Umgebungsdruck *m* ambient pressure
Umgebungsfeuchtigkeit *f* ambient humidity
Umgebungsstabilisation *f* environment stabilization
Umgebungsstrahlung *f* background radiation
Umgebungstemperatur *f* ambient (environmental) temperature
umgehen to by-pass
Umgehung *f* by-pass
Umgehungsfilter *n* by-pass filter
Umgehungsleitung *f* by-pass line
Umgehungsstraße *f* by-pass highway (road), by-pass
Umgehungsventil *n* by-pass valve
umgekehrt inverse; reverse; reciprocal
u. proportional inversely proportional
umgestalten to modify; to transform
umgießen to decant, to transfuse, to refound, to refuse; *(met)* to repour
Umgleiser *m* shunting frame, traverser
Umgleisschiebebühne *f s.* Umgleiser
Umgrenzung *f* boundary
U. der festen Teile load limit gauge
U. des lichten Raumes loading (clearance, train) gauge

Umgrenzungslinie *f* / **allgemeine** universal gauge
umgruppieren to rearrange, to regroup
Umgruppierung *f* rearrangement
umhaspeln to rereel
Umhaspeln *n* double reeling, rereeling *(silk)*
umhüllen to can, to coat, to cover, to mantle, to encase; to wrap; to sheathe; to jacket
Umhüllende *f* envelope
Umhüllung *f (el)* wrapping, sheath[ing], covering *(of cable)*; coating; encasement; encapsulation; enclosure; jacket
Umhüllungskurve *f* envelope
  Mohrsche U. intrinsic curve *(rock mechanics)*
Umhüllungsmaterial *n* sheathing material
Umkapselung *f* **der Spule** coil jacket
Umkehr *f* 1. reversal; 2. turning
  U. der Bewegungsrichtung reversal of travel direction
  U. der Förderrichtung delivery reversal
  U. der Spindeldrehrichtung spindle reversal
  automatische U. automatic reversal
Umkehranlasser *m* reversing starter
umkehrbar reversible *(direction)*; *(math)* invertible
  eindeutig u. *(math)* one-to-one
  nicht u. irreversible
Umkehrbarkeit *f* reversibility
Umkehrbetrieb *m (el)* reversible (reversing) operation
Umkehrdampfmaschine *f* reversing steam engine
umkehren to reverse; to return
Umkehrentwicklung *f* reversal development
Umkehrfilm *m* reversible film
Umkehrfrequenz *f* inverse frequency
Umkehrfunktion *f* inverse function
Umkehrkanal *m* reversing nozzle
Umkehrkrümmer *m* return bend
Umkehrlinse *f* lenticular erecting system
Umkehrmotor *m* reversing (reversible) electric motor
Umkehrmotorantrieb *m* reversing motor drive
Umkehrofen *m* U-type furnace
Umkehrpflug *m* reversible plough
Umkehrprisma *n* [image-]erecting prism
Umkehrprobe *f (math)* reversal test
Umkehrpropeller *m* reversible (feathering screw) propeller
Umkehrprozeß *m* inverse process
Umkehrpunkt *m* inversion point; arrest point; dead centre
Umkehrreaktion *f (ch)* counter reaction
Umkehrrichtung *f* reverse direction
Umkehrsatz *m (opt)* image-erecting set
Umkehrschalter *m* reversing switch (commutator)
Umkehrschaufel *f* reversing bucket *(turbine)*
  schraubenförmige U. helical reversing bucket
Umkehrspektrum *n* inversion spectrum
Umkehrspiegel *m* reversing (inverting) mirror
Umkehrspülung *f (aut)* loop scavenging
Umkehrstraße *f* reverse rolling mill
Umkehrstromrichter *m* two-way inverter
Umkehrsystem *n (opt)* reversing system
Umkehrtemperatur *f* reversal temperature
Umkehrtrommel *f (mine)* return (tail) drum; *(build)* reverse drum *(mixing machines)*
Umkehrturbine *f* reversing turbine
Umkehrung *f* inversion; conversion; reversal; reversing
  U. der Exzenterbewegung reversal of the eccentric motion
  U. der Polarität reversal of polarity
  U. des Kreisprozesses inversion of the cycle
  Waldensche U. Walden inversion
Umkehrwalzenbeschichtung *f* reverse roll coating
Umkehrwalzwerk *n* reverse rolling mill
umkippen 1. to upset, to overturn, to tilt over; 2. *(el)* to run away *(circuit)*
Umklappen *n* **des Spins** flip-over of spin, spin flip
Umklappprozeß *m (nucl)* umklapp (flip-over) process
umkleiden to sheathe, to jacket, to coat, to cover, to encase
Umkleideraum *m* dressing room
Umkleidung *f* jacket[ing]; casing; sheathing *(cable)*; coating
Umklöppelmaschine *f* braider, braiding machine
umklöppeln to braid *(cable)*
Umklöppelung *f* braid[ing] *(cable)*
Umkodierer *m* converter
Umkodierung *f* code conversion
umkonstruieren to redesign
Umkreis *m* perimeter, circumcircle
umkreisen to orbit, to circle; to spin; to revolve
  die Erde u. to circle the earth
Umkreismittelpunkt *m* circumcentre
Umkristallisation *f* recrystallization, fractional crystallization
umkristallisieren to recrystallize
Umladebahnhof *m* tran[s]shipment station
Umladebehälter *m* cask, flask
Umladebühne *f* tran[s]shipment platform, *(Am)* transfer platform
Umladeeinrichtung *f* loading and tran[s]shipment installation
Umladegebühr *f* tran[s]shipment charges
Umladegleis *n* tran[s]shipment line (track, siding), *(Am)* transfer line
Umladegut *n* goods to be tran[s]shipped
Umladehafen *m* port of tran[s]shipment
Umlademaschine *f (nucl)* reloading (loading, recharging, charging) machine

umladen to tran[s]ship, to shift *(cargo)*, *(Am)* to transfer
Umladeprinzip *n* recharging principle
Umladung *f* 1. charge exchange (transfer); reloading *(of reactor)*; 2. overside delivery *(from ship to barge)*; tran[s]shipment *(cargo)*
Umladungsprozeß *m* recharging process
Umladungsquerschnitt *m* charge exchange cross section
Umlagerung *f* rearrangement, transformation [and rearrangement]; *(nucl)* discomposition, knocking-out; displacement
U. von Kristallen rearrangement of crystals
Beckmannsche U. Beckmann molecular transformation
intramolekulare U. intramolecular rearrangement
Umlagerungsverlust *m* compression and eddy losses *(turbine)*
Umlauf *m* circulation, revolution; recirculation; turnover; *(dat)* cycle; *(mach)* pass round *(gear shaping)*
U. des Schmiermittels circulation of the lubricant
U. des Wagens turn round of wagons
U. im Uhrzeigersinn clockwise rotation
U. um die Erde revolution round the earth
Umlaufbahn *f* orbit, trajectory
Umlaufbewegung *f* rotation; orbital motion
U. des Werkstücks work rotation
Umlaufdauer *f (nucl)* circulation period
Umlaufdruckschmierung *f* circulating lubrication [system] using force feed
Umlaufdurchmesser *m* swing [diameter]
Umläufe *mpl* / **notwendige** passes round taken *(gear shaping)*
umlaufen to circulate *(fluid)*; to rotate *(work, machine part)*; *(dat)* to cycle
u. lassen *(mach)* to give rotary motion; to circulate *(liquid)*
u. um *(ph)* to orbit around
wiederholt u. *(dat)* to recycle
umlaufend live *(centre)*; revolving; rotating; circulating
Umlauffeile *f* rotary file
Umlauffrequenz *f* rotational frequency *(of electrons)*
Umlaufgenerator *m* rotary generator
Umlaufgeschwindigkeit *f* 1. rotation[al] speed, number of revolutions per unit time; 2. cycle rate
Umlaufgetriebe *n* epicyclic gear[ing]
Umlaufkessel *m* circulation boiler
Umlaufkühlung *f (nucl)* closed-circuit cooling; *(aut)* forced-circulation cooling
Umlaufmenge *f* amount of fluid discharged
Umlaufmotor *m* rotary engine
Umlaufperiode *f* orbital period
Umlaufpotentiometer *n* infinite-turn potentiometer
Umlaufpumpe *f* recirculating (circulating, rotary, drum) pump
Umlaufrädergetriebe *n* hypocyclic drive
Umlaufregler *m* speed regulator (governor)
Umlaufrichtung *f* direction of rotation
Umlaufschmierung *f* circulating lubrication
Umlaufschrott *m* home scrap
Umlaufsinn *m* direction (sense) of rotation
Umlaufspannung *f* circulation voltage
Umlaufspeicher *m* [re]circulating memory, [re]circulating register
Umlaufsystem *n* recirculation system
Umlauftrockner *m* furnace gas circulation drier
Umlauftrocknung *f* drying by circulating air
Umlaufübertrag *m (dat)* end-around carry
Umlaufvakuumpumpe *f* rotary vacuum pump
Umlaufventil *n* rotating (by-pass) valve
Umlaufvorrichtung *f* recycling facility
Umlaufzahl *f (dat)* cycle criterion
Umlaufzeit *f (nucl)* revolution period, rotation time; turn-round time; orbital period; time of revolution (rotation); *(rock)* orbital period
umlegen to pull down; to tilt; to throw, to shift *(belt, lever)*; *(mine)* to snake over, to turn over
Umlegeriemen *m* shifting belt
Umlegeschicht *f (mine)* turning shift
Umleghebel *m (ry)* point lever
Umlegungszeichen *n (el)* transfer signal
Umleimer *m* edge-strip, veneer tape, banding core
Umleimung *f* veneer taping
umleiten to divert *(traffic)*; to by-pass
Umleitung *f* by-pass; detour *(traffic)*; diversion
Umlenkblech *n* baffle [plate]
Umlenkdüse *f* reversing nozzle
Umlenker *m* circulator
Umlenkgeschwindigkeit *f* angular program[me] velocity
Umlenkprisma *n* deviating prism
Umlenkprogramm *n* climbing program[me]
Umlenkrolle *f* tail (angle, corner, deflection) pulley, tail (angle, corner, deflection) sheave *(conveyor)*; idler
Umlenkschaufel *f* corner vane
Umlenkstelle *f* drape point *(prestressed concrete)*
Umlenktrommel *f* return (tail) drum
Umlenkung *f* turn[ing]
U. der Dampfstromrichtung alteration of the direction of the flow
umlernen to train for a new job
Umluftheizer *m* recirculatory heater
Umluftheizung *f* heating by circulating air
Umlufttrockner *m* through-circulation drier
Umluftverfahren *n* air-return system *(ventilation)*
Ummagnetisierung *f* remagnetization, reversal of magnetization, magnetic reversal

Ummagnetisierungsverlust *m* magnetic hysteresis loss

ummanteln to jacket, to cover, to coat; to encase, to surround; to sheathe *(cable)*

ummantelt jacketed; *(build)* [en]cased

Ummantelung *f* casing, encasement, jacket[ing]; sheathing *(cable)*, covering

umordnen to rearrange; to redistribute

Umordnung *f* rearrangement; permutation

Umorientierung *f* reorientation

umpflanzen to replant, to transplant

umpflügen to plough up, to turn

Umpflügen *n* ploughing up

umpfropfen to top-graft

Umpfropfen *n* top-grafting

umplanen to re-schedule, to re-plan

Umplanen *n* re-scheduling, re-planning

Umpolarisierung *f* pole (polarity) reversal, reversal of polarity

Umpoleffekt *m (telev)* turnover

umpolen to reverse the polarity

Umpolen *n* pole-changing, reversal of polarity

Umpoler *m* reversing commutator

Umpolschalter *m* reversing (commutating) switch

Umpolung *f* pole reversal, reversal of poles (polarity); *(telev)* turnover

Umpolungsspannung *f* turnover voltage

Umpolungssystem *n* reversing system

umprägen to recoin

umpumpen to recirculate, to recycle

Umradlinie *f s.* Perizykloide

umrahmen to frame

umranden to border

Umrandung *f* border

umrechnen to convert

Umrechner *m* translator *(telephony)*; coder

Umrechnung *f* translation *(telephony)*; conversion

Umrechnungsfaktor *m* conversion factor; *(ch)* reduction (gravimetric) factor

Umrechnungstabelle *f* conversion table

umreißen 1. to outline, to sketch; 2. to pull down

umrichten 1. *(mach)* to reset; 2. *(el)* to convert

Umrichten *n (mach)* resetting *(cutting tools)*

Umrichter *m* frequency converter (changer); inverter; transverter

Umrichtezeit *f (mach)* resetting (set-up) time

Umriß *m* outline, contour

Umrißbohren *n* contour boring

Umrißdiagramm *n* contour diagram

Umrißfräsen *n* contour milling

Umrißkopieren *n* contour machining

Umrißnachformen *n* contour machining

Umrißnachformhobeln *n* contour planing

Umrißnachformsteuerung *f* contour controls

Umrißschnitteinrichtung *f* contour sawing attachment

Umrißversteilerung *f (telev)* crispening

umrollen to swage *(metal)*; to rewind

Umroller *m* rewinder

Umrollformmaschine *f* rollover moulding machine

Umrollmaschine *f* rewinding machine

Umrollrüttelformmaschine *f (met)* jar ramming rollover moulding machine

umrühren to agitate, to stir

umrüsten *(mach)* to retool, to reset, to change the tool set-up

Umrüsten *n (mach)* resetting, [tooling] change-over

U. für eine andere Serie change-over and resetting

Umrüsthäufigkeit *f* change-over frequency

Umrüstzeit *f* change-over time

Umsatz *m* turnover; sale

Umsatzkarte *f* accounting detail card *(punched card machinery)*

Umsatzstatistik *f* sales analysis

umsäumen to hem

umschaltbar switch-selected, switch-over

Umschalteinheit *f* switchboard unit

umschalten to switch over; to change over; to commutate

Umschalten *n* change-over switching

Umschalter *m* polystage (change-over, throw-over, two-way, selector, double-throw) switch

U. für Lastregelung on-load tap changer

U. für Spannungsregelung unter Last on-load tap-changing equipment

einpoliger U. single-pole double-throw switch

elektronischer U. electronic transfer switch

rotierender U. revolving commutator

unterbrechungsfreier U. transfer switch

zweipoliger U. [double-pole] double-throw switch, double-pole change-over switch

Umschalterelais *n* switching relay

Umschaltgerät *n* transfer unit

Umschaltklinke *f* transfer jack

Umschaltkontakt *m* switching-over contact, make-and-break contact, double-throw contact, throw-over contact, change-over contact, two-way contact

U. mit Unterbrechung two-way break-before-make contact *(relay)*

unterbrechungsloser U. two-way make-before-break contact *(relay)*

Umschaltröhre *f* contact tube

Umschalttor *n* double-throw gate

Umschaltung *f* switching; *(el)* changing-over; commutation

U. mit Unterbrechung open-circuit control

automatische U. automatical switching [off] *(tape recorder)*

elektronische U. electronic switching

stoßfreie U. bumpless transfer

Umschaltungsschnur *f* patchcord

umschichtig alternate; in layers
umschiften to shift the cargo
Umschlag *m* 1. cover *(book)*; 2. envelope; 3. *(text)* turn-up *(trousers)*; 4. handling, tran[s]shipment; turnover, turn-around
Umschlaganlage *f* handling installation
Umschlaganlagen *fpl* tran[s]shipment (transient) facilities *(of a port)*
Umschlageinrichtungen *fpl* handling appliances *(of a port)*
Umschlaggebühren *fpl* tran[s]shipment fees, wharfage
Umschlaghafen *m* tran[s]shipment (final) port
Umschlaghalle *f* transfer shed
Umschlagrelais *n* throw-over relay
Umschlagspunkt *m (ch)* final point, point of change
Umschlagverfahren *n* swing-round method *(testing)*
Umschlagverkehr *m* tran[s]shipment traffic
Umschlagzeit *f* transit time *(relay)*; change-over time
umschließen to enclose, to surround
Umschließung *f* enclosure
Umschlingung *f* contact [of belt and pulley]
Umschlingungswinkel *m* angle of contact (wrap) *(belt drive)*, arc of contact
Umschmelze *f* remelt heat
umschmelzen to remelt
Umschmelzlegierung *f* remelted alloy
Umschmelzmetall *n* remelted metal
Umschmelznickel *n* remelted nickel
Umschmelzofen *m* remelting furnace
Umschmelzverfahren *n* remelting process
umschmieden to reforge
Umschnitt *m* re-recording
Umschnürung *f* hooping *(reinforced concrete columns)*; reinforcement
umschotten to provide with bulkheads
Umschottung *f* bulkheading; boundaries, boundary bulkheads
umschreiben 1. to circumscribe *(geometry)*; 2. to transcribe *(computer)*
Umschulung *f* retraining
umsetzbar transformable, interconvertible
Umsetzbarkeit *f* transformability, interconvertibility
umsetzen 1. to convert; to transform; to encode; to translate; 2. to reset *(support)*
  Dampfdruck in Geschwindigkeit u. to convert the steam pressure into velocity
Umsetzer *m (dat)* converter, [en]coder; translator
  U. nach dem Subtraktionsprinzip binary weighing encoder
Umsetzgeschwindigkeit *f* speed of decomposition reaction *(explosive)*
Umsetzung *f* conversion; transposition; translation; rearrangement; transformation, change; reaction; decomposition *(explosive)*
Umsetzungsmöglichkeit *f* **in die Praxis** practicability
Umsetzungswärme *f* conversion heat
Umspannanlage *f (el)* transformer plant (station)
umspannen 1. *(el)* to transform; 2. *(mach)* to reset *(workpiece, tool)*, to change the set-up, to rechuck, to relocate *(work)*, to reclamp; 3. to span
Umspanner *m (el)* transformer
Umspannergruppe *f* transformer bank
Umspannerzelle *f* transformer vault (cubicle)
Umspannstation *f* transforming station
Umspannung *f* 1. *(el)* transformation; change of voltage; 2. *(mach)* resetting, repositioning, changing the set-up, reclamping, rechucking
  U. / ohne without changing the position of the workpiece (work, part)
Umspannunterwerk *n* transformer substation
Umspannwerk *n* transformer (transforming) station
umspeichern to exchange *(computers)*
Umspeicherung *f* exchange *(computer)*
umspielen to re-record
umspinnen to braid *(wire, conductor)*
Umspinnung *f* braiding, wrapping, whipping, screen
umsponnen braided; covered
  mit Baumwolle einfach u. single-cotton covered
  mit Seide einfach u. single-silk covered
Umspuleinrichtung *f* film-rewinding outfit
umspulen to rewind, to re-reel
Umstauung *f* restowing
umstechen *(ag)* to turn *(grain)*
umstecken to unplug; to change
  Anschlüsse u. to reverse connections
Umsteckwalzwerk *n* looping-type rolling mill
umsteigen to change *(trains)*
umstellen to shift [from one position to another], to throw *(lever etc)*
  die Weiche u. to reverse (throw over) the points
Umsteller *m* tap changer
Umstellhebel *m* shifter lever
Umstellklappe *f* flop gate *(bin)*
Umstellung *f* inversion; repositioning, rearrangement
umsteuerbar *(mach)* reversing, reversible
Umsteuerbarkeit *f (mach)* reversibility
Umsteuerdrehkolbenventil *n* rotary piston-type of selector valve
Umsteuerdrehschieber *m* rotary selector, rotary piston-type of selector valve
Umsteuerhebel *m (mach)* reversing lever
Umsteuerknagge *f* reverse (reversal) dog

Umsteuerkolbenlängsschieber *m* selector of the slide type
Umsteuerkolbenschieber *m* slide selector
Umsteuerkolbenventil *n* piston-type reverse valve
Umsteuerkupplung *f* reversing clutch
Umsteuerlängsschieber *m* reversing valve of the sliding piston type
umsteuern to reverse; *(dat)* to modify
Umsteuern *n* **der Bewegung** *(mach)* reversal in the direction
Umsteuerpropeller *m s.* Verstellpropeller
Umsteuerpumpe *f* reversible flow-type pump
Umsteuerschalter *m* reversing switch
Umsteuerschieber *m* reverse (selector) valve, port-type selector, directional valve controlling the piston movement
Umsteuerung *f* reversal
 U. der Drehrichtung reversal of the direction of rotation
Umsteuerungsgetriebe *n* reversing gear
Umsteuerungsturbine *f* reversing turbine
Umsteuerventil *n* reversing (reverse) valve
Umsteuervierwegeschieber *m* directional-control four-way valve
Umsteuervorrichtung *f (el)* discriminator, reversing device
umströmen to flow about (around) an obstacle
Umsublimation *f* resublimation
umtopfen to re-pot
Umtriebsweide *f* rotation pasture
umveredeln to top-graft
Umveredlung *f* top-graft[ing]
umwalzen to re-roll
umwälzen to circulate *(liquid)*
Umwälzen *n* piping, recirculation
Umwalzer *m* looper, looping roller
Umwälzkocher *m* circulating digester *(papermaking)*
Umwälzkochung *f* circulation cook *(papermaking)*
Umwälzleitung *f* fuel-return line, by-pass line
Umwälzpumpe *f* [re]circulating pump
Umwälzverfahren *n* forced-air system
umwandelbar *(ch)* convertible, changeable; *(el)* transformable
Umwandelbarkeit *f* convertibility, changeability; transformability
umwandeln to convert, to change; *(dat)* to translate; *(el)* to transform
Umwandler *m* converter; transducer
Umwandlung *f* conversion, change; metabolism; decay, disintegration; transformation, transition, transmutation
 U. von Elementen transmutation of elements
 U. von mechanischer Arbeit transformation of mechanical work
 allotrope U. allotropic transformation
 innere U. internal conversion
 isobare U. isobaric transformation
 kristallografische U. crystallographic transformation
 radioaktive U. radioactive transformation (transmutation)
 thermonukleare U. thermonuclear transmutation
Umwandlungsbereich *m* transformation range
Umwandlungsempfindlichkeit *f* conversion sensitivity
Umwandlungsfaktor *m (nucl)* conversion ratio
Umwandlungsgeschwindigkeit *f* velocity of transformation
Umwandlungshärtung *f* isothermal quenching
Umwandlungsprodukt *n* transmutation (conversion) product
Umwandlungspunkt *m* critical (transformation) point
 magnetischer U. magnetic transition point
Umwandlungsreihe *f s.* Zerfallsreihe
Umwandlungstemperatur *f* transformation (transition) temperature
Umwandlungswahrscheinlichkeit *f* probability of the [nuclear] transformation
Umwandlungswärme *f* heat of transformation, transformation (latent) heat
Umwandlungszeit *f* conversion time; transformation period
Umwegschalter *m* by-pass switch
Umweiselung *f* requeening *(bee-keeping)*
Umweltbedingung *f* environmental condition
Umwelteinfluß *m* environmental factor
umwenden to turn over; to turn round; to invert
umwickeln to wrap; to lap; to re-reel
 mit Isolierband u. to tape
umzäunen to fence in
Umzeichnen *n s.* Nachzeichnen
unabgeschirmt unshielded
unabgestimmt untuned
unabhängig independent
Unabhängigkeit *f* independence; invariance
Unabhängigkeitsprinzip *n* independence theorem *(Newton)*
unabzählbar indenumerable
unähnlich dissimilar *(mechanics)*
unaktiviert unactivated
unangelassen untempered *(steel)*
unangemessen inadequate; undue; unsuitable
unauflöslich indissoluble; *(math, ch)* insoluble
unausführbar impracticable
Unausführbarkeit *f* impracticability
unausgebaut *(mine)* unlined
unausgeglichen unbalanced, non-equilibrium
Unausgeglichenheit *f* unbalance
unausgewuchtet unbalanced
unbauwürdig *(mine)* inexploitable
unbeeinflußt unaffected
unbefahrbar unnavigable *(for ships)*

unbegrenzt infinite; unlimited, limitless; indefinite
unbehandelt untreated
unbehauen unhewn
unbeherrscht uncontrolled
unbehindert unimpeded; clear of
unbekannt unknown
Unbekannte *f* unknown
unbeladen unloaded, without cargo (load)
unbelastet unloaded, no-load
unbelichtet unexposed, non-exposed
unbemannt unmanned; crewless; pilotless; unattended, unwatched *(beacon, lightship)*
unbenutzt unused; disused
unberechnet free of charge
unberuhigt *(met)* unkilled, rimmed, open-poured, effervescing
unbesäumt round-edged
unbeschichtet uncoated
unbeschränkt unlimited, unrestricted
unbeschwert *(plast)* unpigmented
unbesetzt vacant *(shell)*; unfilled, unpopulated, empty *(state)*; unoccupied
unbeständig unstable, unsteady
Unbeständigkeit *f* instability
Unbeständigkeitskonstante *f* *(nucl)* instability constant
Unbestellbarkeitsmeldung *f* advice of traffic on hand
unbestimmt indefinite, indeterminate, undetermined *(statics)*; uncertain
  hochgradig statisch u. statically indeterminate to a large degree
  hochgradig u. indeterminate to a large degree *(statics)*
  statisch u. statically indeterminate
Unbestimmtheit *f* indefiniteness; indeterminacy *(statics)*; uncertainty
  statische U. statical indeterminacy
Unbestimmtheitsprinzip *n* uncertainty principle
Unbestimmtheitsrelation *f* uncertainty relation
  Heisenbergsche U. Heisenberg principle of uncertainty
unbestrahlt unirradiated, unexposed
unbestückt solid, one-piece *(tool)*
unbewacht unattended, unwatched *(beacon, lightship)*
unbewaffnet unarmed; unaided
unbeweglich immovable; stationary
unbewegt stagnant *(liquid)*
unbewehrt non-reinforced, plain *(concrete)*; non-armoured *(cable)*
unbewickelt not surrounded by a winding
unbrauchbar werden to become useless, to fail
Unbrauchbarwerden *n* failure
unbrennbar non-combustible, non-inflammable
UND AND *(logical operator)*
UND-Gatter *n* AND-gate
UND-Glied *n* AND-element
UND-Schaltkreis *m* AND-circuit *(logical switching)*
UND-Schaltung *f* AND-circuit
undefinierbar indefinable
undehnbar inextensible
Undekan *n* undecane
Undekansäure *f* undecanoic acid
undicht leaky, leaking; porous
Undichte *f* leakage
Undichtheit *f* leak[ing], leakage
Undichtheitsverlust *m* loss due to leakage
Undichtigkeit *f* leakage; porosity
Undichtigkeitskoeffizient *m* coefficient of leakage
Undichtigkeitsverlust *m* leakage loss, loss due to leakage
undissoziiert undissociated
undotiert undoped
Undulationstheorie *f* theory of wave motion, undulatory theory
Undurchdringbarkeit *f* impermeability
undurchdringlich impermeable, impenetrable
undurchlässig impermeable, opaque; tight
Undurchlässigkeit *f* impermeability, opacity
undurchschossen *(print)* solid
undurchsichtig opaque, non-transparent
Undurchsichtigkeit *f* opacity
uneben uneven *(surface)*, undulated, irregular
Unebenheit *f* unevenness
unecht imitated, false; spurious; *(math)* improper *(fraction)*
unedel base, ignoble *(metal)*
uneigentlich improper
unelastisch inelastic
Unelastizität *f* inelasticity
unempfindlich *(ch)* inert; sluggish; insensitive *(instrument)*; immune
Unempfindlichkeit *f* 1. inertness; sluggishness; insensitivity; 2. robustness
Unempfindlichkeitsbereich *m* dead (neutral) zone, *(Am)* differential
Unempfindlichkeitsgrad *m* degree (coefficient) of insensitiveness
Unempfindlichkeitszeit *f s.* Totzeit
unendlich [groß] infinite
  u. klein infinitesimal
Unendlichkeit *f* infinity
unentwickelt undeveloped
unerforscht unexplored
unergiebig unproductive; poor
unermeßlich immeasurable, immense
Unermeßlichkeit *f* immeasurability
unerreichbar unobtainable
unerschöpflich inexhaustible
unfachgemäß inexpert
Unfall *m* accident; *(ship)* casualty
  U. beim Anfahren des Reaktors start-up accident

tödlicher U. fatal accident
Unfallanruf *m* emergency call
Unfallanzeige *f* accident report
Unfallboot *n* accident (emergency, sea) boat
Unfallgefahr *f* accident hazard
Unfallhäufigkeit *f* accident incidence
Unfallmeldung *f* accident report
Unfallschutzmaßnahme *f* safety precaution, protective measure
unfallsicher accident-proof
Unfallstation *f* ambulance station
Unfallstelle *f* scene of an (the) accident
Unfallstrahlenbelastung *f* accidental exposure
Unfallverhütung *f* accident prevention, prevention of accidents
Unfallversicherung *f* accident insurance
Unfallwagen *m* ambulance car
unfrei carriage forward, charges to pay
unfruchtbar infertile, sterile; *(ag)* poor
Unfruchtbarkeit *f* infertility, sterility
ungeätzt unetched
ungebleicht unbleached, raw
ungebunden 1. *(ch)* free, unbound; 2. loose *(abrasive)*
ungedämpft undamped, sustained
ungedeckt undecked, open *(vessel)*
ungedopt undoped
ungeeicht uncalibrated
ungeerdet unearthed
ungefalzt flat
ungefärbt undyed
ungefedert springless
ungefiltert unfiltered
ungeglüht *(met)* unannealed
ungehärtet *(met)* unhardened
ungehindert unrestricted
ungekerbt unnotched
ungekühlt uncooled
ungekürzt unabridged *(book)*
ungeladen uncharged
ungelernt unskilled
ungelöscht unslaked *(lime)*
ungelöst *(ch)* undissolved
ungemustert plain *(cloth)*
ungenau inaccurate, incorrect
Ungenauigkeit *f* inaccuracy
Ungenauigkeitsrelation *f s.* Unbestimmtheitsrelation
ungenießbar inedible, uneatable
ungenügend insufficient
ungenutet unfluted, ungrooved
ungeordnet random; disorder
ungepolt non-polarized
ungepuffert unbuffered
ungerade out of truth; odd *(number)*
Ungerade-gerade-Kern *m* odd-even nucleus
Ungerade-ungerade-Kern *m* odd-odd nucleus
Ungeradheit *f* oddness *(number)*; out-of-straightness
ungeradzahlig odd-numbered
ungeregelt unregulated
ungereinigt unpurified
ungerichtet undirected *(ray)*; random
ungeröstet unprocessed, green *(coffee)*
ungesättigt unsaturated
ungeschickt heavy-handed
ungeschliffen unground
ungeschmiert unlubricated
ungeschult untrained
ungeschützt unprotected
ungesperrt *(print)* set solid
ungesteuert uncontrolled
ungestört perfect, undistorted *(crystal lattice)*
ungestreut unscattered
ungestrichen uncoated, unpainted
ungestützt unsupported
ungewobbelt non-swept
Ungeziefer *n* vermin, pest
ungiftig nonpoisonous
unglasiert unglazed
ungleich *(math)* unlike *(signs)*, unequal
u. geteilt unequally (unevenly) spaced
ungleicharmig unequal-armed *(lever)*
ungleichartig inhomogeneous, heterogeneous
Ungleichartigkeit *f* inhomogeneity, heterogeneity
ungleichförmig non-uniform; inhomogeneous; irregular, discontinuous
Ungleichförmigkeit *f* non-uniformity; inhomogeneity; irregularity; discontinuity; *(el)* notching ratio *(of control)*
Ungleichförmigkeitsfaktor *m* non-uniformity factor
Ungleichförmigkeitsgrad *m* coefficient of cyclic variation, degree of uniformity
U. des Reglers degree of irregularity of the governor
Ungleichförmigkeitsverhältnis *n (mach, therm)* coefficient of cyclic variation, degree of uniformity
Ungleichheit *f* inequality, inequation
ungleichmäßig non-uniform, irregular; uneven; unequal; inconstant
Ungleichmäßigkeit *f* unevenness, non-uniformity; inconstancy; discontinuity
lineare U. central tendency
Ungleichmäßigkeitsfaktor *m* non-uniformity factor
ungleichnamig *(math)* dissimilar, unlike
ungleichschenklig with unequal legs
ungleichseitig scalene, inequilateral
Ungleichung *f* inequality, inequation
Tschebyscheffsche U. Tschebyscheff's (Chebishev's) inequality
ungleichwertig inequivalent
Unglücksfall *m* accident

ungrundiert non-primed
ungültig invalid, void
unheilbar irreparable; incurable
unhörbar inaudible
Uniformstoff *m* army cloth
unimolekular unimolecular
unipolar unipolar
Unipolargenerator *m* acyclic (non-polar) generator
Unipolarinduktion *f* unipolar induction
Unipolarmaschine *f* homopolar (non-polar) generator, acyclic machine
Unipolarsystem *n* unipolar [electrode] system
Unipolartransistor *m* unipolar transistor
unisoliert uninsulated
unitär *(math)* unitary
univariant univariant
universal universal
Universalauslegerbohrmaschine *f* universal radial drilling machine
Universalbagger *m* multi-purpose excavator, fully versatile excavator
Universaldiode *f* general-purpose diode
Universaldreibackenfutter *n* three-jaw universal chuck
Universalentwickler *m* allround developer
Universalförderer *m* general-purpose elevator
Universalfräseinrichtung *f* universal milling attachment
Universalfräsmaschine *f* universal milling machine
  U. zum Nuten von Spiralbohrern universal twist drill fluting machine
Universalfutter *n* universal (concentric, self-centring, gear scroll) chuck
Universalgelenk *n* universal (ball-and-socket) joint
Universalindikator *m (ch)* universal indicator
Universalinjektor *m* **mit zwei Düsensätzen** double-tube injector, injector with two sets of cones (nozzles)
Universalkondensator *m* universal condenser
Universalmanipulator *m* general-purpose manipulator
Universalmeßbrücke *f* universal bridge
Universalmeßinstrument *n* multi-meter
Universalprüfmaschine *f* universal testing machine
Universalradar *n* general-purpose radar
Universalrechner *m* general-purpose computer
Universalrevolverdrehmaschine *f* universal turret lathe
Universalröhre *f* universal valve
Universalrundschleifmaschine *f* universal cylindrical grinding machine
Universalschalter *m* general-use switch
Universalschleifmaschine *f* universal grinding machine
Universalschlüssel *m* monkey wrench
Universalschmelzeinsatz *m (el)* universal fuse link
Universalspindelkopf *m* swivelling attachment, swivel head
Universalspiralfräskopf *m* universal spiral index head
Universalteilkopf *m* universal dividing head
Universaluntersetzer *m* multiscaler
Universalwalzwerk *n* universal rolling mill
Universalwaschmittel *n* all-purpose washing agent
Universalwerkzeugschleifmaschine *f* universal tool and cutter grinder
Universalwerkzeug- und -gesenkfräsmaschine *f* universal tool and die milling machine
Universalwinkelmesser *m* universal bevel protractor
Universalzündholzmaschine *f* universal match machine
Universum *n* universe
Univibrator *m* univibrator
unkenterbar non-capsizable
unkompensiert unbalanced
unkompliziert straightforward
Unkosten *pl* expenses, cost[s]
  feststehende U. assured cost[s]
  laufende U. running cost[s]
Unkrautbekämpfung *f* weed control
Unkrautbekämpfungsmittel *n* weed killer, herbicide
Unkrautegge *f* weeder, chain harrow
Unkrautsamen *m* weed seed
Unkrautsieb *n* tailings riddle
Unkrautstriegel *m* weeder
Unkrautvernichtung *f* weed killing
Unkrautvertilgungsmittel *n* weedicide, weed killer, herbicide
unkristallisierbar uncrystallizable
unkritisch acritical
unlegiert unalloyed
unlenkbar *(aero)* ingovernable
unlesbar illegible
unliniert unruled, plain *(paper)*
unlösbar permanent *(joint)*; *(math)* insolvable
unlöschbar inextinguishable
unlöslich *(ch)* insoluble
Unlösliches *n (ch)* insoluble matter
Unlöslichkeit *f* insolubility
unmagnetisch non-magnetic
unmeßbar unmeasurable, immeasurable
Unmeßbarkeit *f* unmeasurability, immeasurability
unmischbar immiscible
unmittelbar direct; intermediate
unmittig off-centre
Unmittigkeit *f* off-centre position

unmoduliert unmodulated
Unordnung *f* disarrangement, disorder
unoxydierbar inoxidizable
unpaar[ig] unpaired, odd *(electron)*
unparallel out-of-parallel
unpassierbar impassable *(road)*
unperforiert unperforated
unplattiert unclad
unpolar non-polar
unpolarisiert unpolarized
unpraktisch unpractical; impracticable
unproduktiv unproductive; non-productive
unpünktlich unpunctual
Unrat *m* dirt, rubbish
unregelmäßig irregular
Unregelmäßigkeit *f* irregularity
Unregelmäßigkeitsmeldung *f* advice (notice) of irregularity
unreif unripe; dead *(cotton)*; immature
unrein impure; unclean, dirty
unrentabel not profitable
Unrichtigkeit *f* incorrectness
Unruh *f* balance *(watch)*
Unruhe *f* / **atmosphärische** atmospheric agitation
unrund untrue, out of truth (round), non-circular
Unrundheit *f* lack of roundness, out-of-roundness
  U. eines Radreifens ovalization of a wheel tyre
Unrundschleifen *n* contour grinding
Unrundwerden *n* **des Zylinders** wearing out-of-truth of the cylinder
unsatiniert unglazed
unscharf *(opt)* blurred *(image)*; unsharp, out-of-focus
Unschärfe *f* *(opt)* confusion, lack of definition; unsharpness; diffuseness
Unschärfeprinzip *n s.* Unbestimmtheitsrelation
unschiffbar innavigable, unnavigable
Unschiffbarkeit *f* innavigability, unnavigability
unschmelzbar infusible
unschraffiert unshaded
unschweißbar unweldable
unselektiv non-selective
Unselektivität *f* non-selectivity
Unsicherheit *f* uncertainty *(statistics)*
Unsicherheiten *fpl* unknown factors
Unsicherheitsfaktor *m* factor of uncertainty
Unsicherheitsprinzip *n s.* Unbestimmtheitsrelation
unsichtbar invisible
unsinkbar unsinkable, insubmersible
Unsinkbarkeit *f* insinkability, unsinkability
unstabil unstable
Unstabilität *f* instability, unstability
unstetig discontinuous; unsteady
Unstetigkeit *f* discontinuity; unsteadiness *(movement)*
  behebbare U. removable discontinuity
  nicht behebbare U. non-removable discontinuity
Unstetigkeitsfläche *f* surface of discontinuity, discontinuity surface
Unstetigkeitsstelle *f* discontinuity
Unstetigkeitswelle *f* discontinuity wave
Unsymmetrie *f* dissymmetry, unsymmetry; *(el)* unbalance, asymmetry *(breaking current)*
Unsymmetriestrom *m* out-of-balance current
unsymmetrisch asymmetric, unsymmetric[al]; *(el)* unbalanced
untauglich unserviceable, unfit, useless
unteilbar indivisible
untengesteuert with side-by-side valve arrangement
unter under; bottom, lower; below *(zero)*
Unterabteilung *f* subdivision
Unteramboß *m* anvil block
Unteramt *n* branch (dependent) exchange *(telephony)*
Unteransicht *f* bottom view
unteratmosphärisch sub-atmospheric
Unterauslauf *m* internal force feed
Unterbank *f* *(mine)* bottom coal *(seam)*
Unterbau *m* *(build)* substructure; base, foundation; sub-base, base (foundation) course; *(aut)* basic structure, underframe, subframe; *(ry)* road bed, subgrade
unterbauen *(mine)* to undercut, to undermine, to underwork; *(build)* to underpin; *(mach)* to pack, to shim, to jack *(work)*
Unterbaugruppe *f* ancillary unit
Unterbauung *f* *(mach)* packing *(work)*
Unterbeanspruchung *f* understressing
Unterbelastung *f* *(el)* fractional load
unterbelichten to underexpose
unterbelichtet underexposed
Unterbelichtung *f* underexposure
unterbemannt short-handed
Unterbereich *m* *(math)* sub-range; subdomain
unterbesetzt short-handed
unterbinden *(ch)* to inhibit
  das Austreten u. to retain
Unterboden *m* *(ag)* subsoil
Unterbodenlockerung *f* subsoiling
unterbrechen to interrupt; *(el)* to disconnect; to break, to open; to cut off; to make inoperative
  eine Leitung u. to break a line
Unterbrecher *m* *(el, aut)* interrupter; [contact] breaker; chopper; vibrator
  umlaufender U. *(aut)* rotary interrupter *(ignition)*
Unterbrecherabstand *m* breaker point gap
Unterbrecherfunke *m* break spark

Unterbrecherfunkenstrecke *f* breaking spark gap
Unterbrecherhammer *m* rocker (breaker) arm, breaker lever
Unterbrecherhebel *m* rocking (breaker) lever, contact arm
Unterbrecherkontakt *m* contact breaker; interrupting contact
Unterbrecherkontaktabstand *m* contact-breaker gap
Unterbrechernocken *m* ignition cam; interrupter cam
Unterbrecherrelais *n* vibrating relay
Unterbrecherspule *f* trip coil
Unterbrechung *f* *(el)* break; interruption; disconnection; abruption
U. einer Leitung [mains] disconnection
augenblickliche U. fleeting break
verzögerte U. slow release
Unterbrechungsbad *n* [short-]stop bath
Unterbrechungskontakt *m* break contact
Unterbrechungsstöpsel *m* infinity plug
Unterbrechungstaste *f* interruption key, break-key
Unterbrechungszeit *f* interruption time
Unterbrennerofen *m* underfired (underjet) oven
unterbringen to place; to store; to accommodate; to house
an Bord u. to accommodate *(passengers etc)*
Unterbringung *f* placing; storage; accommodation; housing
Unterbringungsmöglichkeit *f* accommodation
unterbrochen interrupted; disconnected; *(text)* a-jour threaded
Unterbündelung *f* underbunching *(electronics)*
Unterdachtrocknung *f* *(ag)* barn hay drying
Unterdamm *m* *(mine)* low-side pack
Unterdeck *n* *(ship)* lower deck
Unterdeckraum *m* underdeck space
Unterdeckraumgehalt *m* underdeck tonnage
Unterdecktonnage *f* underdeck tonnage
Unterdeterminante *f* minor, subdeterminant
Unterdrehpflug *m* turnabout plough
Unterdruck *m* 1. *(ph)* low pressure, underpressure, pressure below atmospheric, subpressure, [partial] vacuum, suction; 2. *(print)* ground tint; underprinting
Unterdruckbeton *m* vacuum concrete
Unterdruckbremsanlage *f* *(aut)* vacuum brake system
unterdrücken to suppress *(amplitude modulation)*; to reject; to damp out *(vibration)*
Unterdruckförderer *m* vacuum pump element
Unterdruckförderpumpe *f* vacuum fuel pump
Unterdruckmesser *m* vacuum gauge
Unterdruckregler *m* vacuum governor
Unterdruckschalter *m* minimum-pressure switch
Unterdrucktragfläche *f* suction aerofoil
Unterdrückung *f* suppression, rejection; cutting-off
U. der Bildträgerfrequenz image rejection
U. der Konvektion *(ph)* inhibition of convection
U. der Trägerwelle suppression of carrier
U. gleichphasiger Signale common-mode rejection
gleichphasige U. in-phase rejection
Unterdrückungsschwingung *f* quench oscillation
Unterdruckwächter *m* minimum-pressure governor
Unterdruckzone *f* *(aero)* zone of negative pressure
Unterdruckzündversteller *m* *(aut)* vacuum advance mechanism
Unterdruckzündverstellung *f* *(aut)* vacuum timing control; vacuum ignition adjustment (advance)
Untereinheitenanordnung *f* sub-unit array *(protein)*
unterentwickelt under-developed
Untererregbarkeit *f* [electrical] subnormality
untererregen to under-excite
untereutektisch hypoeutectic
Unterfach *n* *(text)* bottom (lower) shed
Unterfaden *m* *(text)* lower thread
unterfahren *(mine)* to undercut, to undermine, to underwork
Unterfamilie *f* *(nucl)* branch
unterfangen to underpin
Unterfangung *f* underpinning
Unterfeuerung *f* under-grate firing, furnace under the boiler
Unterflasche *f* bottom (fall) block
Unterflottenhaspel *f* *(text)* winch below the surface of the liquor
Unterflurbewässerung *f* sub-irrigation
Unterflurförderer *m* underground (under-bed) conveyor
Unterflurkühler *m* underground cooler
Unterflurofen *m* underground (melting-hole) furnace
Unterfrequenz *f* subharmonic
Unterführung *f* underbridge, underpass, subway [leading to platforms], underground passage, tunnel
Unterfunktion *f* sub-function
Unterfutterung *f* firring
untergärig bottom fermenting
untergebracht in housed in
untergehen to sink, to founder *(vessel)*
untergeordnet subordinate
Untergesenk *n* lower (bottom) die, bottom swage
Untergestell *n* [under]frame, base [frame], understructure
untergetaucht submerged

untergießen / die Dampfmaschine to grout the steam engine
Unterglaskultur *f* greenhouse crop
Unterglocke *f* lower (big, main) bell *(blast furnace)*
untergossen embedded in a foundation *(concrete)*
untergraben *(build)* to undermine
Untergriff *m (mach)* undercut
Untergröße *f* undersize
Untergrund *m* subsoil; bottom; background; *(geol)* C-horizon
  lichtdurchlässiger U. transparent background
  tragfähiger U. hardpan *(foundations)*
Untergrundaktivität *f* background activity
  künstliche U. man-created environmental activity
Untergrundbahn *f* subway, underground railway
Untergrundbewässerung *f* sub-irrigation
Untergrundfarbe *f (text)* bottom shade
Untergrundlockerung *f* subsoiling
Untergrundmonitor *m* background monitor
Untergrundmuster *n* background pattern, mechanical tint
Untergrundpacker *m* furrow press
Untergrundpflug *m* subsoiler
Untergrundspeicherung *f* down storage
Untergrundstrahlung *f* background radiation
  kontinuierliche U. steady background
  kosmische U. cosmic-ray background
  natürliche U. natural background radiation
Untergrundüberwachung *f* background monitoring
Untergrundzählrate *f* background counting rate
Untergruppe *f* subgroup
Untergurt *m* bottom (lower) chord, lower boom
Untergurtförderer *m* bottom belt conveyor, bottom loading belt
Unterguß *m* backing; *(met)* bottom (uphill) casting
unterhalten to maintain
Unterhaltung *f* maintenance
  U. einer Strecke maintenance of a line
  regelmäßige U. routine maintenance
Unterhaltungsarbeiten *fpl* maintenance work
Unterhaltungshalbleiter *m* entertainment semiconductor
Unterhaltungskosten *pl* maintenance cost[s], cost of upkeep
Unterhaupt *n* lower (outer) gates *(lock)*
Unterhebestab *m (text)* lower twilling bar
Unterhieb *m* overcut
unterhöhlen to undermine
unterirdisch underground
Unterkammer *f* **der Schleuse** lower chamber of the collector *(turbine)*
Unterkante *f* bottom (lower) edge
  U. Kiel bottom of keel
Unterkasten *m* base; drag, nowel *(moulding)*
unterkellert with basement
  nicht u. slab-on-grade; without basement
Unterkessel *m* lower cylinder (drum)
Unterkette *f (text)* ground warp
Unterkettenbahn *f (mine)* chain creeper
Unterkiel *m* lower keel *(wooden ship)*
unterklemmen *(el)* to secure under terminals
Unterkorn *n* undersize; sub-grain
Unterkorrektion *f* undercorrection
unterkorrigiert *(opt)* undercorrected
unterkritisch subcritical
unterkühlen to supercool, to undercool, to subcool
Unterkühlen *n* supercooling, sub-zero cooling
unterkühlt supercooled; subcooled
Unterkühlung *f* supercooling; subcooling; undercooling
Unterkultur *f* undercrop
Unterkunft *f* lodging quarter, accommodation
Unterlage *f* backing, support; base; substrate; *(ag)* stock; *(build)* bedding
  elastische U. elastic foundation (bedding)
Unterlagefehler *m (nucl)* backscattering error
Unterlagen *fpl* data; information; documentation
Unterlager *n* bottom bearing *(meter)*
Unterlagerungsfernwahl *f* low-frequency dialling
Unterlagerungsmodelung *f* subcarrier modulation
Unterlagerungstelegrafie *f* infra-acoustic telegraphy, sub-audio telegraphy
Unterlagestück *n (mach)* shim
Unterlagsplatte *f (ry)* bearing (sole, base, supporting) plate, tie-plate; *(build)* bearing (bed) plate
  U. mit geneigter Oberfläche inclined sole plate, inclined tie-plate
Unterlänge *f* descender
Unterleder *n* bottom leather
Unterlegeblock *m (mach)* packing
Unterlegeklotz *m (mach)* [packing] block
unterlegen to bolster; to line; to underlay
Unterlegeplatte *f (mach)* shim
Unterlegestück *n (mach)* packing piece
Unterlegscheibe *f* washer; lock washer; plain washer; packing shim *(workpiece)*
  U. für die Mutter bolt washer
  aufgebogene U. safety washer, square washer with corner bent-up to secure the bolt
  federnde U. wrinkle washer
Unterlegstreifen *m* liner
Unterleitung *f* conduit conductor rail system
Untermaß *n* undersize
  U. haben to be undersize
  U. für das Aufreiben undersize for reaming
  U. zum Reiben reaming allowance

Untermaßbohrer *m* undersize (reamer) drill
Untermatrix *f* sub-matrix
Untermenge *f (math)* subset
Untermoderierung *f (nucl)* undermoderation
Untermöllerung *f* underburdening
Unternehmensforschung *f* operational (operations) research
Unternehmungsentscheidung *f* executive decision
Unterniveau *n* sublevel
unternormal substandard
Unternormal *n* secondary standard
unteroxydiert underoxidized
unterpolt underpoled *(copper)*
Unterprogramm *n (dat)* subroutine
  dynamisches U. dynamic subroutine
  statisches U. static subroutine
Unterpulverschweißen *n* submerged arc welding
Unterputz *m* rough (scratch) coat, undercoat, coarse stuff
Unterputzanlage *f* buried installation, concealed wiring
Unterputzleitungsverlegung *f* buried wiring
Unterputzschalter *m* flush (recessed, sunk) switch
Unterputzsteckdose *f* flush socket
Unterreibmaß *n* undersize for reaming
Unterrostung *f* sub-surface corrosion
Untersaat *f* undersown crop
Untersatz *m* pedestal; console [base]; socket; plinth
Unterschale *f (nucl)* subshell
Unterschallflugzeug *n* subsonic aircraft
Unterschallgeschwindigkeit *f* subsonic (infrasonic, subacoustic) speed, subsonic (infrasonic, subacoustic) velocity
Unterschallströmung *f* flow at subsonic velocity
Unterschallwelle *f* infrasonic (subaudio) wave
unterschätzen to underrate, to underestimate
Unterschaufeln *n* / **maßgenaues** measured shovel-packing
unterscheiden to distinguish, to discriminate
Unterscheidung *f* differentiation, discrimination
Unterscheidungsmerkmal *n* distinguishing feature *(statistics)*
Unterscheidungsschwelle *f* discrimination threshold
Unterschicht *f* sublayer *(fluid flow)*
Unterschieber *m (mach)* cross slide
Unterschied *m* difference
  statistisch gesicherter U. significant difference
Unterschiedsmessen *n* comparative measurement, comparing (transferring) dimensions
Unterschiedsmessung *f* differential measurement
Unterschienenschweißen *n* firecracker welding *(Hafergut method using bar of copper)*
Unterschießen *n (text)* picking below
Unterschiff *n* quick work *(submerged part of a floating vessel)*
unterschlächtig undershot
Unterschlagrevolverwebstuhl *m* circular box underpick loom
Unterschlagstuhl *m* underpick loom
Unterschlitten *m (mach)* lower slide
unterschneiden to undercut
Unterschneidung *f* undercut
Unterschnitt *m* undercut
unterschnitten kerned *(letter)*; undercut *(metal)*
unterschnittfrei without undercut
Unterschnittwinkel *m* undercut angle *(broach)*
unterschrämen *(mine)* to undercut
Unterschrämen *n (mine)* hewing
unterschreiten to fall [below]
Unterschubfeuerung *f* underfeed furnace
Unterschuß *m* 1. deficiency; deficit; 2. *(text)* back pick (filling)
Unterschwingen *n (el)* undershoot
Unterseeboot *n* submarine
Unterseebootreaktor *m* submarine reactor
Unterseefrachtboot *n* submarine cargo-carrier
Unterseekabel *n* submarine cable
Untersee-Laser-System *n* undersea laser system
Unterseil *n* balance rope, tailrope
Unterseite *f* bottom side (surface), underside; *(text)* back[ing]; *(build)* intrados *(arches and vaults)*; soffit
Untersetzer *m* scaler, scaling circuit
  U. in Ringschaltung ring scaler
  U. mit angekoppeltem Schreibgerät scale[r]-printer
  binärer U. binary counter (scaler), scale-of-two scaler
Untersetzerschaltung *f* scaler, scaling circuit *(for pulses)*
  U. 1 zu n scale-of-n-circuit
  rückgekoppelte U. ring scaler
Untersetzung *f* 1. *(nucl)* scale, scaling; 2. step-down ratio; 3. *s.* Übersetzung ins Langsame
Untersetzungsfaktor *m (nucl)* scaling factor (ratio)
Untersetzungsgetriebe *n* reduction (step-down) gear
Untersetzungsverhältnis *n* reduction (step-down) ratio
Untersicht *f* soffit; intrados *(arches, vaults)*; underside view; underside elevation
Untersieb *n* bottom screen; backing wire
Unterspannung *f* undervoltage *(below rated voltage)*; secondary voltage *(of transformers)*
Unterspannungsauslösung *f* undervoltage release (tripping)
Unterspannungsrelais *n* undervoltage (minimum-voltage) relay
Unterspannungsschalter *m* undervoltage circuit-breaker

Unterspannungsspule *f* no-volt coil
Unterspannungswicklung *f* low tension winding
unterspülen to underwash
Unterspülung *f* underwashing, scour[ing]
Unterstempel *m (mine)* lower prop
unterstopfen to tamp *(railway tracks)*
Unterstopfen *n* tamping *(railway tracks)*
Unterstromauslösung *f* undercurrent release
Unterstromschalter *m* [automatic] minimum circuit-breaker, underload circuit-breaker
Unterstromwächter *m* underload (automatic minimum, undercurrent) circuit-breaker
unterstützen to support; to carry; to prop up
Unterstützung *f* support
untersuchen to examine, to investigate; to analyze, to test, to trace, to study, to check
quantitativ u. to investigate quantitatively
Untersuchung *f* investigation, examination, study, test[ing], check[ing], analysis, exploration
experimentelle U. experimental study
mikroskopische U. microscopic examination
röntgenografische U. radiographical investigation
statische U. statical analysis
Untersuchungsbohrung *f (mine)* exploratory boring
Untersuchungsergebnis *n* result of investigation
Untersuchungshäuschen *n* test hut *(telephony)*
untersynchron subsynchronous, hyposynchronous
Untertagevergasung *f* gasifying in place, underground gasification
untertägig underground
Unterteil *n* lower part; base
unterteilen to subdivide; to calibrate; to sectionalize; to break up *(chips)*
Unterteilung *f* subdivision; calibration; partition
U. des Druckgefälles subdivision of the drop of pressure
wasserdichte U. *(ship)* watertight subdivision
unterthermisch *(nucl)* cold
Untertisch *m* lower work table
Untertitel *m* subtitle, subheading
Untertonbereich *m* sub-audio frequency range
Unterträger *m* sub-carrier
Untertrikotagen *fpl* / **regulär gewirkte** full-fashioned underwear
Untertrum *m* bottom strand *(conveyor)*
Unterverbunderregung *f* under-compound excitation *(generator)*
Unterverzweiger *m* sub-cabinet
Unterwagen *m* bogie, undercarriage
Unterwalze *f* bottom (lower) roll
Unterwange *f* lower beam *(press brake)*
Unterwanten *pl (ship)* lower shrouds
Unterwasser *n (mine)* back (under-screen) water *(jigger)*
Unterwasseranstrich *m* underwater painting (coat)
Unterwasserantenne *f* underwater (submerged) aerial
Unterwasserauspuff *m* underwater exhaust
Unterwasserbeton *m* underwater concrete
Unterwassererzfrachter *m* submarine ore-carrier
Unterwasserexplosion *f* underwater explosion (burst), deep-water explosion
Unterwasserfahrzeug *n* underwater vehicle
Unterwasserfernmeßtechnik *f* underwater telemetry
Unterwasserfernsprechkabel *n* submarine telephone cable
Unterwasserfläche *f* wetted surface *(of a vessel)*
Unterwasserform *f* underwater form, shape of hull under water, shape of submerged part of hull
Unterwasserfotometer *n* underwater photometer
Unterwassergeschwindigkeit *f* submerged speed *(submarine)*
Unterwasserhelling *f (ship)* lower launch slip
Unterwasserhorchgerät *n* hydrophone
Unterwasserkabel *n* submerged cable
Unterwasserkondensator *m* submerged condenser
Unterwasserlautsprecher *m* subaqueous loudspeaker
Unterwassermanipulator *m (nucl)* underwater manipulator
Unterwassermotor *m* submersible motor
Unterwassermotorpumpe *f* submersible pump
Unterwasserreaktor *m* underwater reactor
Unterwasserschallsender *m* submarine (impact) oscillator
Unterwasserschiff *n* underwater body, submerged part of hull, submarine vessel
Unterwasserschutz *m* underwater protection
Unterwasserschutzanstrich *m* anti-fouling paint
Unterwasserschweißen *n* underwater welding
Unterwasserströmung *f* submarine (underwater) current
Unterwasserteil *n* submerged (underwater) parts, underbody *(of a vessel)*
Unterwassertiefe *f (ship)* underwater level
Unterwassertunnel *m* underwater tunnel
Unterwasserwaffe *f* underwater weapon
Unterwasserzünder *m* submarine electric detonator
Unterwegsbahnhof *m* intermediate station
unterwerfen to subject to
Unterwerk *n (el)* substation
bemanntes U. staffed substation
unbesetztes U. unattended (automatic) substation
Unterwerksbau *m (mine)* dip working
Unterwind *m* undergrate blast

Unterwindfeuerung *f* forced draught furnace, closed-ashpit furnace
Unterwindgebläse *n* undergrate blower
Unterzangenbacke *f* bottom nipper plate *(comber)*
Unterzentrale *f* sub-exchange, branch exchange *(telephony)*
Unterzug *m (build)* main beam, girder, bearer; *(mine)* girt, scab, stringer; bottom flue; *(ship)* [deck] girder, runner
  durchlaufender U. continuous girder
Untiefe *f* shoal, shallow, flat, sandbank
Untiefentonne *f* sandbank buoy, buoy for a shelf, shoal buoy (marker)
untrennbar inseparable
Untreue *f* **der Wollfaser** unevenness of the wool fibre
unüberbrückt *(el)* unshunted
ununterbrochen uninterrupted; continuous
unveränderlich *(math)* anallagmatic; invariable, stable, constant; steady-state
unverändert unaltered, unchanged
unverarbeitet raw; untreated, unprocessed
unverbrannt unburned
unverbrennbar incombustible, non-combustible, fire-proof
unverdaulich indigestible
unverdünnt undiluted
unvereinbar incompatible
Unvereinbarkeit *f* NAND, NOT AND *(logical operator)*
unverestert unesterified
unverfälscht *(ch)* unadulterated; true
unverfärbt uncoloured
unverfestigt unconsolidated; not work-hardened *(after machining)*
unverformt undeformed, unstrained
unvergärbar unfermentable
unverhüttbar unsmeltable
unvermengt neat
unvermischt unmixed; pure; neat
unverritzt *(mine)* unworked, virgin
unverschieblich fixed, rigid
unverschleißbar unwearable
unverschlüsselt *(dat)* absolute
unverschnitten *(ch)* unadulterated, unextended
unverseifbar unsaponifiable
unverstärkt unamplified; not reinforced
unverstellbar nonadjustable
unverwechselbar non-interchangeable *(e.g. plug)*
Unverwüstlichkeit *f* robustness
unverzehrbar non-consumable *(electrode)*
unverzerrt undistorted
unverzinkt ungalvanized
unverzögert instantaneous; non-delayed, quick-action *(e.g. fuse)*
unverzweigt *(ch)* unbranched
unvollendet unfinished, uncompleted
unvollständig incomplete
Unvollständigkeit *f* incompleteness
unvorsichtig imprudent; incautious
unwägbar unweighable
unwirksam *(ch)* ineffective; inefficient; inert
Unwirksamkeit *f* inefficiency, ineffectiveness
unwirtschaftlich uneconomical
Unwucht *f* unbalance, out-of-balance, lack of balance
unwuchtfrei in good balance; true
unzählbar uncountable, innumerable
unzählig innumerable
unzerbrechlich unbreakable
unzerlegbar indecomposable
unzerreißbar untearable
unzersetzt undecomposed
unzerspanbar non-machinable, unmachinable
unzerstörbar indestructible
Unziale *f (print)* uncial
unzugänglich inaccessible
unzugerichtet rough-tanned
unzulässig undue
unzusammendrückbar incompressible
Unzusammendrückbarkeit *f* incompressibility
unzweckmäßig unsuitable
  u. konstruiert ill-designed
unzweideutig unequivocal
unzyklisch acyclic
UOP-Faktor *m (ch)* characterization factor
UP-Schweißen *n* submerged arc welding
Urämie *f* uremia
Uran *n* uranium
  abgebranntes U. depleted uranium
  abgereichertes U. depleted uranium
  angereichertes U. enriched uranium
  natürliches U. natural[ly occuring] uranium
  nichtangereichertes U. unenriched uranium, natural[ly occuring] uranium
Uran-235 *n* uranium-235, actino-uranium
Uran-Aktinium *n s.* Uran-235
Uranat *n* uranate
Uranbombe *f* uranium bomb
Uranbrennelement *n* uranium fuel element
Urandioxid *n* uranous oxide, uranium dioxide
Uranchlorierung *f* uranium chlorination
Uranerz *n* uranium ore
Uranfamilie *f s.* Uranreihe
Uranfluorierung *f* uranium fluorination
Urangelb *n* yellow uranium oxide
Urangitter *n* uranium lattice
Uranglas *n* uranium glass
Uranglimmer *m* uranite, uranglimmer
Uran-Graphit-Reaktor *m s.* Graphit-Uran-Reaktor
uranhaltig containing uranium, uraniferous, uranium-bearing
Uranhexafluorid *n* uranium hexafluoride, uranic fluoride
Uranhydrid *n* uranium hydride

Uraninit *m* uraninite *(uranium mineral group)*
Uranit *m* uranite[-lime] *(uranium mineral)*
Urankarbid *n* uranium carbide
Urankern *m* uranium nucleus
Urankonzentrat *n* uranium concentrate, yellow cake
Uranmaschine *f s.* Uranreaktor
Uranmeiler *m s.* Uranreaktor
Uranoxid *n* uranium oxide
Uranpechblende *f* pitchblende, nasturan, uranite *(uranium mineral)*
Uranpecherz *n s.* Uranpechblende
Uran-Radium-Familie *f s.* Uranreihe
Uran-Radium-Reihe *f s.* Uranreihe
Uran-Radium-Zerfallsreihe *f s.* Uranreihe
Uranreaktor *m* uranium reactor (pile)
Uranreihe *f* uranium series
Uransalz *n* uranium salt
Uransäure *f* uranic acid
Uran-Schwerwasser-Reaktor *m* uranium and heavy water pile
Uranspaltprodukt *n* uranium fission product
Uranspaltung *f* uranium fission
Urantetrafluorid *n* uranium tetrafluoride, green salt, uranous fluoride
Uran(IV)-verbindung *f* uranous compound
Uran(VI)-verbindung *f* uranic compound
Uranylazetat *n* uranyl acetate
Uranylnitrat *n* uranyl nitrate
Uranylsulfat *n* uranyl sulfate
Uranzerfall *m* uranium decay
Uranzerfallsreihe *f* uranium series
Urbarmachen *n* cultivation
Urease *f* urease
Ureat *n* urate
Ureichkreis *m* master telephone transmission reference system
Ureid *n* ureide
Urenastengelfaser *f* urena fibre
Urethan *n* urethane, ethyl carbamate
Urheber *m* author
Urheberrecht *n* copyright
Urheberrechtsverletzung *f* infringement of copyright
Urkilogramm *n* prototype kilogramme
Urkunde *f* document
Urlaub *m* / **bezahlter** holiday with pay
Urlaubsschicht *f* leave shift
Urlehre *f* master gauge
Urliste *f* ungrouped frequency table
Urmeter *n* international prototype metre
Urmodell *n* master form *(galvanics)*
Urnuklid *n* **verzögerter Neutronen** precursor
Urotropin *n* urotropine, hexamethylenetetramine
Urplasma *n* ylem
Urraster *m* **für gedruckte Schaltung** printed circuit master grid
Urspannung *f* impressed electromotive force, primary e.m.f.
Ursprung *m* origin; source
ursprünglich original; initial *(strength of materials)*
Ursprungslänge *f* original dimension
Ursprungspunkt *m* point of origin
Ursprungsverzerrung *f* origin distortion
Urstoff *m* primary (parent) matter, primordial substance
Urteer *m* low temperature (carbonization) tar
Urtiter *m* original titer, titrimetric standard
Urtitersubstanz *f* standard titrimetric substance, titrimetric standard
Urtyp *m* prototype
Urwaldbestand *m* original (virgin, primeval) stand, first growth
Urzeichnung *f* original drawing
Urzelle *f* primordial cell
US-Schweißen *n* firecracker welding *(Hafergut method using bar of copper)*
uu-Kern *m* odd-odd nucleus
UV . . . *s.* Ultraviolett . . .

V

V-Antenne *f* V-aerial
V-Bahn *f (mach)* V-way, vee
V-Band *n* V band
V-Bodenboot *n* V-bottom boat, vee-bottom boat
V-Bodenform *f (ship)* V-bottom form, vee-bottom form
V-Form *f (aero)* dihedral
V-Formboden *m* V-bottom, vee-bottom, vee-shaped bottom *(of a vessel)*
V-förmig ausgebildet vee-type, V-shaped, *(Am)* V-
V-Gespräch *n* personal call
V-Heck *n (ship)* V-stern
V-Karte *f* control chart for coefficients of variation
V-Motor *m* vee (V-type) engine
V-Schweißnaht *f* V-weld
V-Spant *n (ship)* V-shaped frame
V-Spantbodenboot *n* V-bottom boat
V-Stab *m* vertical [member] *(truss)*
V-Stellung *f (aero)* dihedral
V-Stumpfnaht *f* single-V-butt weld
Vakat *n* blank page
Vakuum *n* vacuum, evacuated space
  V. / im in vacuo
  echtes V. real vacuum
  erreichbares V. practicable (attainable) vacuum
Vakuumanlage *f* vacuum equipment (system)
Vakuumaufdampfung *f* vacuum deposition
Vakuumaufnahmeplatte *f* vacuum pick-up plate

Vakuumbedampfung *f* vacuum deposition (evaporation)
Vakuumbeton *m* vacuum concrete
Vakuumblitzableiter *m* vacuum lightning arrester
Vakuumbremse *f* vacuum brake
Vakuumdämpfeinrichtung *f* vacuum steaming equipment
Vakuumdestillation *f* vacuum distillation (topping), distillation under reduced pressure
vakuumdicht hermetic, vacuum-tight
Vakuumdichtung *f* vacuum seal
Vakuumeindampfer *m* vacuum drier (evaporator)
Vakuumexsikkator *m* Scheibler's desiccator
Vakuumfaktor *m* vacuum factor; gas ratio *(electron tube)*
Vakuumfett *n* vacuum grease
Vakuumfilmhalter *m* vacuum back *(process camera)*
Vakuumfotozelle *f* vacuum photocell (tube)
vakuumgetempert thermally treated in vacuum
Vakuumgießen *n* suction (vacuum) casting
Vakuumgießpfanne *f* vacuum pan
Vakuumgreifer *m* vacuum grab
Vakuumguß *m* vacuum casting
Vakuumherstellung *f* vacuum generation
Vakuumimprägnierung *f* vacuum impregnation
Vakuumkammer *f* vacuum chamber
V. des Betatrons [vacuum] doughnut, toroid
ringförmige V. [vacuum] doughnut, toroid
Vakuumkolben *m* vacuum envelope
Vakuumkondensator *m* vacuum capacitor
Vakuumlampe *f* vacuum lamp
Vakuumleistungsschalter *m* vacuum circuit-breaker
Vakuumlichtgeschwindigkeit *f* velocity of light in vacuum
Vakuumlinie *f* vacuum curve (line)
Vakuummantelgefäß *n* vacuum flask
Vakuummesser *m* vacuum gauge
Vakuummetallisieren *n* vacuum metallizing
Vakuummetallurgie *f* vacuum metallurgy
Vakuummeter *n* vacuummeter, vacuum gauge
Vakuummeteranzeige *f* indication of vacuum
Vakuumofen *m* vacuum furnace
Vakuumpolarisation *f* vacuum polarization
Vakuumpräparationsverfahren *n* vacuum preparative process
Vakuumpumpanlage *f* evacuating machine *(electron tubes)*
Vakuumpumpe *f* vacuum pump
Vakuumreglerventil *n* vacuum breaker
Vakuumröhre *f* vacuum valve (tube)
Vakuumröhrenelektronik *f* vacuum tube electronics
Vakuumröhrengleichstrom *m* direct vacuum tube current
Vakuumröhrenverstärker *m* vacuum tube amplifier
Vakuumrundsiebmaschine *f* vacuum cylinder former *(papermaking)*
Vakuumschalter *m* vacuum switch; vacuum interrupter
Vakuumschicht *f* vacuum layer
Vakuumschlauch *m* pressure tubing
Vakuumschleuse *f* vacuum lock
Vakuumschmelzanlage *f* vacuum melting plant
Vakuumschwankung *f* fluctuation of vacuum
Vakuumspektrograf *m* vacuum spectrograph
Vakuumsystem *n* evacuated system
Vakuumtechnik *f* vacuum engineering
Vakuumtiefziehverfahren *n* *(plast)* straight-vacuum forming
Vakuumtrockenschrank *m* vacuum drying oven, vacuum shelf drier
Vakuumverdampfer *m* vacuum evaporator
Vakuumverformung *f* *(plast)* vacuum forming
Vakuumvermehrer *m* vacuum augmenter
Vakuumzelle *f* vacuum tube (photocell)
Vakzin *n* vaccine
Valenciennespitze *f* Valenciennes [bobbin] lace
Valentinit *m* *(min)* valentinite
Valenz *f* valency, valence; atomicity
Valenzbahn *f* valence orbital
Valenzband *n* valence band
Valenzbandstruktur *f* valence band structure
Valenzbandverzerrung *f* valence band distortion
Valenzbandzustand *m* valence band state
Valenzbindung *f* valence bond (link, linkage)
Valenzelektron *n* valence (bonding, peripheral, outer-shell) electron
Valenzelektronenkonzentration *f* valence electron concentration
valenzgesättigt valency-saturated
Valenzgitter *n* valence lattice
Valenzkräfte *fpl* valence forces
Valenzschwingung *f* valence vibration
Valenzwinkel *m* valence angle, angle of valence
Valenzzustand *m* valence orbital, atomic orbit
Valeraldehyd *m* valeric aldehyde, valeraldehyde
Valeriansäure *f* valeric (valerianic) acid
Valin *n* *(ch)* valine
Van-Allen-Strahlung *f* Van Allen radiation
Van-de-Graaff *m* *s.* Van-de-Graaff-Generator
Van-de-Graaff-Generator *m* Van de Graaff generator (accelerator)
van der Waals[sche] Kräfte *fpl* van der Waals' forces
Vanadat *n* vanadate
vanadinhaltig vanadium bearing
Vanadinit *m* *(min)* vanadinite
Vanadinkarbid *n* vanadium carbide
Vanadinsäure *f* vanadic acid
Vanadinstahl *m* vanadium steel

Vanadium *n* vanadium
Vanillearoma *n* vanilla flavour
Vanillezucker *m* vanillin sugar
Vanillin *n* vanillin
variabel variable
Variable *f* variable
  abhängige V. dependent variable
  induktive V. inductive variable
  ohmsche V. resistive variable
  unabhängige V. argument, independent variable
Variablenmerkmal *n* variable
Variablenprüfung *f* inspection by variables
Variante *f* variant
Varianz *f* variance
Varianzanalyse *f* analysis of variance, variance analysis
Variation *f* **der Konstanten** variation of constants
Variationsgleichung *f* variational equation
Variationskoeffizient *m* coefficient of variation (variability)
Variationsprinzip *n* variational principle
Variationsproblem *n* variational problem
Variationsrechnung *f* calculus of variations, variational method (procedure)
Variationswellenfunktion *f* variational wave function
Varicap *(cont)* varicap
variieren to vary
Variometer *n* [mutual] inductometer, radiometer
Varistor *m* varistor
Varmeter *n (el)* varmeter
Vaseline *f* vaseline, petrolatum, petroleum (mineral) jelly
Vaselinöl *n* vaseline (petrolatum, white) oil
vegetabilisch vegetable
vegetarisch vegetarian
Vegetation *f* vegetation
Vegetationsperiode *f* growing season
Vegetationszeit *f* growing season
Veilchenöl *n* violet oil
Vektor *m* vector
  axialer V. axial vector
  freier V. free vector
  kontragradienter V. contragradient vector
  kontravarianter V. contravariant vector
  magnetischer V. magnetic vector
  ortsgebundener V. localized vector
  polarer V. polar vector
  Poyntingscher V. Poynting's vector
Vektoraddition *f* vector addition
Vektoralgebra *f* vector algebra
Vektoranalysis *f* vector analysis
Vektorbild *n* vector diagram
Vektordarstellung *f* vector[ial] representation
Vektordiagramm *n* vector diagram
Vektorelektrokardiogramm *n* vector electrocardiogram
Vektorfeld *n* vector[ial] field
Vektorgerüst *n* vector model
Vektorgröße *f* vector quantity
vektoriell vectorial
Vektorkomponente *f* vector component
Vektormodell *n s.* Vektorgerüst
Vektorpotential *n* vector potential
Vektorprodukt *n* vector (cross, outer) product
Vektorraum *m* vector space
Vektorrechnung *f* vector calculus
Vektorrichtung *f* direction of vector
Vektorsumme *f* vector sum
Vektorzerleger *m (dat)* resolver
Velin *n* wove paper
Velodyne *(dat)* velodyne integrator
Velourieren *n (text)* emerizing
Velourpapier *n* velvet paper
veloutieren *(leath)* to velour-finish
Veloxkessel *m* velox boiler
Venn-Diagramm *n* Venn diagram
Ventil *n* valve
  V. für Stillstandspause time-delay valve
  V. mit Alternativsteuerung two-way pilot valve *(pneumatics)*
  V. mit auswechselbarem Ventilsitz valve with renewable valve seat
  V. mit flüssiger Katode pool rectifier
  V. mit freiem Durchlaß bei Mittelstellung by-pass valve
  V. mit kombinierter Einlaß-Auslaß-Düsensteuerung three-way pilot valve
  V. mit Rückkehr auf Ausschaltstellung dead-man valve
  V. mit Vakuumhaltung pumped rectifier
  V. zum Einschalten des automatischen Arbeitsganges cycle-starting valve
  ausbalanciertes V. balanced valve
  druckempfindliches V. pressure-sensitive valve
  druckluftbetätigtes V. air-operated valve
  elektrisches V. rectifier, electric valve
  elektrochemisches V. electrochemical valve
  elektropneumatisches V. electro-pneumatic valve
  gesichertes V. dead-man valve
  gesteuertes V. mechanically operated valve
  hängendes V. overhead (drop, inverted) valve
  kegeliges V. cone-plug cock
  liegendes V. horizontal valve
  magnetisches V. magnetic valve
  mit steigendem Steuerdruck öffnendes V. normally closed valve, reverse-acting valve
  mit steigendem Steuerdruck schließendes V. normally opened valve, direct-acting valve
  obengesteuertes V. overhead valve
  oberflächenvergütetes V. hard-faced valve
  pilzförmiges V. poppet valve

pumpenloses V. *(el)* sealed rectifier
seitliches V. side valve
stationär eingebautes V. valve of the seated type
stehendes V. vertical valve
Ventilableiter *m* *(el)* non-linear-resistance arrester
Ventilabstandsmesser *m* valve clearance gauge
Ventilanhubstange *f* valve push rod
Ventilanordnung *f* valve-gear arrangement, valve location
Ventilantrieb *m* valve gear (operating mechanism, train)
Ventilation *f* ventilation
Ventilationsarbeit *f* ventilation (windage) work *(turbine)*
Ventilationsklappe *f* ventilation flap
Ventilationsrad *n* runner, fan propeller
Ventilationssystem *n* ventilation system
Ventilationsverlust *m* ventilation (windage) loss *(turbine)*
Ventilationswiderstand *m* ventilation (windage) resistance
Ventilator *m* ventilator; fan; blower; aspirator
V. für Unterwindfeuerung forced-draught blower
V. mit Zahnradantrieb gear-driven fan
blasender V. pressure fan
Ventilatorflügel *m* fan vane (blade)
Ventilatorgehäuse *n* fan casing
Ventilatorgradierwerk *n* cooling stack with fan
Ventilatorkanal *m* fan drift
Ventilatormotor *m* fan motor
Ventilatorpressung *f* pressure produced by the fan
Ventilatorrad *n* impeller
Ventilatorriemenspannrolle *f* fan-belt idler
Ventilatorwind *m* fan blast
Ventilaufsatz *m* valve cap (cover)
Ventilauslaß *m* valve outlet
Ventilbelastung *f* load on the valve
Ventilbetätigung *f* actuation of valves
Ventilbetätigungsmechanismus *m* valve-operating mechanism
Ventildampfmaschine *f* steam engine with drop valve gear
Ventildeckel *m* valve cap
Ventildichtungsfläche *f* valve-sealing surface
Ventildichtungsscheibe *f* valve packing disk
Ventildrehvorrichtung *f* valve rotator
Ventileinsatz *m* valve-seat insert
Ventileinsatzring *m* valve-seat insert
Ventileinschleifen *n* seating valves
Ventileinschleifpaste *f* abrasive for seating valves
Ventileinstellschlüssel *m* tappet wrench
Ventileinstellschraube *f* valve-adjusting stud
Ventileinstellung *f* valve timing
Ventilentlastung *f* balancing of the valve
Ventilentwässerung *f* valve drainage
Ventilerhebungszeit *f* injection period
Ventilerhöhung *f* valve lift
Ventilfeder *f* valve spring
Ventilfedergehäuse *n* valve-spring cap
Ventilfederhaube *f* valve-spring cap
Ventilfederkeil *m* valve-spring seat retainer
Ventilfedermitnehmer *m* valve-spring lifter
Ventilfederplatte *f* valve-spring retainer
Ventilfederschwingung *f* valve-spring surge
Ventilfederteller *m* valve-spring cap (retainer, washer)
Ventilführung *f* valve [stem] guide
Ventilgehäuse *n* valve casing (body, chamber, cover, housing)
Ventilgehäusedeckel *m* valve cover plate
Ventilgehäuseverschluß *m* valve plug
Ventilgestänge *n* valve gear
Ventilgetriebe *n* valve control motor
Ventilhandsteuerungseinrichtung *f* manual valve positioner
Ventilhaube *f* valve hood
Ventilhebel *m* valve arm (lever)
Ventilhebezange *f* valve remover
Ventilhub *m* valve lift (travel, stroke)
Ventilhubbegrenzung *f* valve-lift stop
Ventilkäfig *m* valve cage
Ventilkammer *f* valve chamber
Ventilkanal *m* valve port
Ventilkappe *f* *(aut)* valve hood (cap)
Ventilkasten *m* valve box (chest)
Ventilkegel *m* valve cone (plug, poppet)
konischer V. tapered plug
Ventilklappe *f* valve flap
Ventilkolben *m* valve piston; bucket
Ventilkolbenstange *f* valve stem
Ventilkorb *m* valve cage
Ventilkörper *m* valve body
Ventillage *f* valve location
Ventilmechanismus *m* valve train (gear)
Ventilmitnehmer *m* valve lifter
Ventilnachschleifmaschine *f* valve refacer
Ventilnadel *f* [valve] needle plug
Ventilöffnung *f* valve port
Ventilöffnungsdauer *f* valve-lift period, valve-opening duration
Ventilöffnungsdruck *m* valve-opening pressure
Ventilöffnungsgeschwindigkeit *f* valve-lifting velocity
Ventilöffnungszeit *f* valve-lift period
Ventilplatte *f* valve plate
Ventilregelstange *f* valve control rod
Ventilring *m* valve ring
Ventilschaft *m* valve stem (shaft)
Ventilschalter *m* valve switch
Ventilschaltung *f* gate *(electronics)*
Ventilschieber *m* valve actuator

Ventilschiebestange *f* valve rod
Ventilschluß *m* valve-seat contact
Ventilschubstange *f* valve pushrod
Ventilschutzhaube *f* valve petticoat
Ventilsitz *m* valve seat
Ventilsitzbearbeitung *f* valve seat machining
Ventilsitzfläche *f* valve face
Ventilsitzfräseinrichtung *f* [valve] seat dresser
Ventilsitzfräser *m* valve reseater
Ventilsitzfräsmaschine *f* valve-seating machine
Ventilsitznachschleifmaschine *f* valve reseater
Ventilsitzring *m* valve-seat ring
Ventilspiel *n* valve (tappet) clearance
Ventilspindel *f* valve rod (stem)
Ventilstellgetriebe *n* valve-actuating gear
Ventilstellschraube *f* valve-adjusting stud
Ventilstellungsmelder *m* valve-position indicator
Ventilsteuergestänge *n* valve train, valve-control linkage
Ventilsteuerhebel *m* valve rocker
Ventilsteuerhebelstütze *f* valve-rocker bracket
Ventilsteuerhebelzapfen *m* valve-rocker fulcrum pin
Ventilsteuermechanismus *m* valve gear
Ventilsteuerung *f* valve gear; valve timing; poppet valve distribution (gear)
Ventilsteuerungsorgan *n* valve gear
Ventilsteuerungsvorrichtung *f* valve-operating mechanism, valve-timing gear
Ventilsteuerzeit *f* valve timing
Ventilstößel *m* valve lifter (rocker, tappet); cam follower
  V. mit automatischer Spieleinstellung self-adjusting hydraulic tappet
Ventilstößelführung *f* valve-tappet guide
Ventilstößelrollenbolzen *m* valve-tappet roller pin
Ventilstößelstange *f* valve push rod
Ventilstoßstange *f* valve push rod
Ventilstoßstangenkugel *f* valve push rod ball
Ventilteller *m* valve disk; bevel plug
Ventilträger *m* valve support
Ventiltrieb *m* valve gear
Ventilverkleidung *f* valve-gear housing
Ventilverlängerung *f* valve extension *(tyre)*
Ventilverschluß *m* valve shutter
Ventilverzögerung *f* valve lag
Ventilvoröffnung *f* valve lead
Ventilwirkung *f* valve action
Ventilzylinder *m* cylinder for drop (poppet) valves
Venturidüse *f* Venturi meter
Venturikanal *m* flume
Venturirohr *n* Venturi tube
Verachsung *f* [shaft] misalignment
verachtfachen [/ sich] to octuple
verallgemeinern to generalize
Verallgemeinerung *f* generalization
veraltet obsolete, antiquated
Veraluminieren *n* aluminizing
Veranda *f* veranda[h], *(Am)* porch
Verandadeck *n (ship)* veranda[h] deck
veränderlich varying, variable; changeable
  stufenlos v. infinitely (steplessly) variable, variable without steps
Veränderliche *f* variable
  abhängige V. dependent variable
  unabhängige V. independent variable
Veränderlichkeit *f* variability
verändern to change, to alter, to vary, to modify
  die Signalform v. to modify the waveform
verändert werden / stufenlos to be varied steplessly
Veränderung *f* change, variation, alteration; modification
verankern to anchor, to guy *(pole)*
  die Dampfmaschine v. to secure the steam engine to the foundation
  rückwärtig v. *(build)* to tie back *(e. g. sheet-piling)*
verankert anchored, guyed
  in sich v. self-anchored
Verankerung *f* anchoring, anchorage, guying, stiffening, bracing; moor[ing]
  V. der Firste *(mine)* roof bolting
  V. der Sohle *(mine)* floor bolting
  V. des Hangenden *(mine)* roof bolting
  V. des Liegenden *(mine)* floor bolting
Verankerungsbuchse *f* anchor-bush
Verankerungsisolator *m* strain insulator
Verankerungspfeiler *m* anchor[age] pier *(bridges)*
Verankerungsplatte *f* anchor plate *(prestressed concrete)*
Verankerungszone *f* anchor[age] zone *(prestressed concrete)*
veranlassen to cause
veranschaulichen to illustrate
veranschlagen to rate *(power)*; to estimate *(cost)*
verarbeitbar workable, processible
Verarbeitbarkeit *f* workability, processibility
verarbeiten to work [up], to process, to convert, to treat
  einen Frequenzbereich v. to handle a frequency range
Verarbeitung *f* manufacturing; processing; working; machining
  automatische V. automatic processing
  elektronische V. electronic processing
  logische V. logical processing
Verarbeitungsanlage *f* processing system
Verarbeitungsfähigkeit *f* workability, processibility
Verarbeitungsgeschwindigkeit *f* processing speed
Verarbeitungshilfe *f* handling aid

Verarbeitungsstufe *f* stage of manufacture
verarmen to deplete
Verarmung *f* depletion; impoverishment; exhaustion
Verarmungszone *f* depletion layer; zone of reduced density
veraschen to ash, to incinerate
Veraschung *f* ashing, incineration
verästelt tree-like *(crystal)*
Verästelung *f* ramification
veräthern to etherify, to alkoxylate
Veratrumsäure *f* veratric acid
Verätzung *f* causticization
Verband *m* joint; bond *(masonry)*; bracing *(bridge)*; dressing, bandage; assembly; union
  Märkischer V. monk bond
Verbandholz *n* framing timber
Verbandmull *m* aseptic gauze
Verbandseigenschaften *fpl (mine)* structural properties
Verbandsschott *n (ship)* structural bulkhead
Verbandsstart *m (aero)* formation take-off
Verbandstarif *m* joint tariff
Verbandwatte *f* surgical wool, absorbent cotton
Verbau *m* lining *(trenches)*
verbessern to improve
Verbesserung *f* improvement
Verbesserungsvorschlag *m* improvement suggestion, suggestion [for improvement]
Verbeulung *f* buckling
  V. des Kessels bulging of the boiler
verbiegbar bendable
verbiegen to bend; to buckle; to deflect; to distort
Verbiegung *f* bending; deflection; distortion; buckling
verbinden to tie, to bind, to unite; *(telec)* to connect; *(ch)* to combine; to join *(welding, soldering, etc)*; to couple together
  durch Gießen v. to join by casting
Verbindung *f (ch)* compound; binding, joining; *(el, telec)* connection; interconnection; communication; coupling; junction; link[age]
  V. durch Vermittlung switched connection
  V. von Hohlleitern junction of waveguides
  abgehende V. outgoing junction
  abgelehnte V. refused call
  aliphatische V. aliphatic compound
  ankommende V. incoming junction
  anorganische V. inorganic compound
  aromatische V. aromatic compound
  binäre V. binary compound
  chemische V. chemical compound
  drahtlose V. radio communication
  einfache V. simplex communication
  fluorsubstituierte aliphatische V. fluoroaliphatic compound
  heteropolare V. ionic compound
  hochfest verschraubbare V. high-tensile bolted structural joint; friction-grip bolting, high-tensile bolting
  homöopolare V. covalent compound
  leitende V. conductive connection
  lösbare V. fastening joint
  lose V. slack joint
  lötlose V. solderless joint
  markierte V. labelled (tagged) compound
  metallorganische V. organometallic compound
  nicht zustande gekommene V. ineffective call
  nichtstöchiometrische V. interstitial compound
  organische V. organic compound
  perizyklische V. peri-cyclo compound
  polare V. polar compound
  semipolare V. coordination compound
  siliziumorganische V. silicone
  unlösbare V. permanent joint
  zustande gekommene V. effective call
  zyklische V. cyclic (cycle, ring) compound
Verbindungsbau *m* linking block
Verbindungsbolzen *m* tie bolt
Verbindungsbrücke *f* flying (fore and aft, connecting) bridge *(tanker)*
Verbindungsdose *f (el)* joint box, access fitting
Verbindungsdraht *m* interconnecting wire
Verbindungselement *n* constituent element; fastening [device]; connector
  lösbares V. fastening [device], fastener
  unlösbares V. joint
Verbindungsfläche *f* bonding surface *(soldering)*
Verbindungsfräser *m* assemblage cutter
Verbindungsgang *m* binding strake *(wooden vessel)*
Verbindungsgestell *n* connecting rack
Verbindungsglied *n* connecting joint; link; latch
Verbindungshalbleiter *m* compound semiconductor
Verbindungskabel *n* junction (connecting) cable
Verbindungsklemme *f* connector, connecting terminal, bonding clip
Verbindungsleine *f* securing line
Verbindungsleitung *f* connection line, transfer circuit, connecting wire
Verbindungslinie *f* connecting line
Verbindungsmuffe *f* connection sleeve, union socket; *(el)* coupling (cable-jointing, joint) sleeve, connecting (cable-junction) box; *(build)* coupler *(tubular scaffold)*
Verbindungsöse *f* splicing ear *(cable)*
Verbindungsplatte *f* joint piece, junction plate; butt plate
Verbindungsschäkel *m (ship)* bending shackle; joining shackle
Verbindungsschalter *m* position coupling (grouping) key

Verbindungsschiene *f* connector bar
Verbindungsschlauch *m (aut)* flexible connection
Verbindungsschnur *f (el)* connecting (flexible, patch) cord, cord
doppeladrige V. twin flex
Verbindungsschweißen *n* joining [metals, plastics] by welding, welding
Verbindungsstange *f* connecting (tie) rod; link
untere V. bottom spacer shaft *(knitting machine)*
Verbindungsstecker *m* connecting plug, connector
Verbindungsstelle *f* joint; junction
Verbindungsstift *m* tag; connecting pin
Verbindungsstrecke *f (mine)* connecting passage
V. zwischen Überhauen *(Am)* scram drift
Verbindungsstrich *m* **über einem Ausdruck** *(math)* vinculum
Verbindungsstück *n (el)* coupling, jointing fitting, tie; fitting
Verbindungstechnik *f (text)* structural technique
Verbindungsteil *n* connector
Verbindungstrakt *m (build)* linking block
Verbindungsverfahren *n* joining process
Verbindungsverlust *m (telec)* junction loss
Verbindungsweg *m* connection path
Verbindungswinkel *m* **des Vorstevens mit dem Kiel** bearding angle
Verbindungszug *m* fantail *(open-hearth furnace)*
verblasbar blowable
verblasen to blow
Verblasen *n* **einer Stahlschmelze** blowing of a steel melt
V. in flüssigem Zustand converting
Verblaserost *m* sintering grate
Verblaseröstung *f* blast (pot) roasting
Verblaseverfahren *n* blasting method
verblassen to fade
verblatten to halve *(carpentry)*
Verbleibwahrscheinlichkeit *f (nucl)* non-leakage probability
verbleien to lead-coat
Kraftstoff v. to lead fuel
Verbleien *n* leading, lead coating
verblenden *(build)* to face
Verblender *m* facing brick
Verblendmauerwerk *n* facework, facing brickwork
verblocken to interlock, to interconnect
verblockt interconnected
miteinander v. interlocked with each other
Verblockung *f (el, mach)* [safety] interlock, [safety] interlocking; *(mach)* fastening, latching
verbodmen to head a cask
verbogen bent; distorted
verbolzen to bolt
Verbolzung *f* bolting; *(mine)* strutting
Verbot *n* forbiddenness *(of a transition)*
V. der Kernwaffen prohibition of nuclear weapons
verboten forbidden, prohibited
Verbrauch *m* consumption, expenditure, expense; use; dissipation
durchschnittlicher V. average consumption
spezifischer V. specific consumption
verbrauchen to consume; to dissipate
Leistung v. to dissipate power; to absorb power
Strom v. to consume current
Verbraucher *m* consumer; user
Verbraucher[anschluß]klemme *f* consumer's terminal
Verbraucherkreis *m* load circuit
Verbrauchernetz *n (el)* public mains
Verbraucherort *m* power consuming point
Verbraucherseite *f* receiving end
Verbrauchsbehälter *m* ready-use tank
Verbrauchsgüter *npl* consumer goods
Verbrauchsziffer *f* consumption figure
verbraucht worn-out, used up; finished; spent; stale *(air)*
verbreitern to broaden; to widen
Verbreiterung *f* broadening; widening
V. der Basis widening of the base
V. einer Bande broadening of a band
V. einer Spektrallinie broadening of a spectral line
V. von Interferenzlinien broadening of interference lines
Verbreitung *f* occurence; spreading, distribution
verbrennbar combustible
Verbrennbarkeit *f* combustibility
verbrennen to burn; to incinerate; to combust; to cremate *(corpses)*; to scorch
Verbrennung *f* combustion, burning; deflagration, incineration
V. durch Wärmestrahlung flash (thermal) burn
Verbrennungsanalyse *f* combustion analysis
Verbrennungsgas *n* flue (combustion) gas; burning gas
Verbrennungsgeschwindigkeit *f* speed (rate) of combustion, burning rate
Verbrennungshub *m (aut)* ignition cycle (stroke)
Verbrennungskapillare *f* capillary combustion tube
Verbrennungskraftmaschine *f* internal combustion engine
Verbrennungsluft *f* combustion air, air for combustion
Verbrennungsmarkierung *f (plast)* burned spot
Verbrennungsmotor *m* internal combustion engine
Verbrennungsofen *m* combustion tube furnace; incinerator
Verbrennungsperiode *f* period of combustion

Verbrennungsprodukt *n* product of combustion, combustion (end) product
Verbrennungsprozeß *m* process of combustion
Verbrennungsraum *m (aut)* combustion chamber
V. mit flachem Boden flat-topped combustion chamber
Verbrennungsrohr *n* combustion tube
Verbrennungsrückstand *m* combustion residue, residue from combustion
Verbrennungsschiffchen *n (ch)* combustion boat
Verbrennungstemperatur *f* combustion temperature
Verbrennungswärme *f* heat of combustion, combustion heat
Verbrennungswirkungsgrad *m* efficiency of the combustion, combustion efficiency
Verbrennungszone *f* tuyere zone *(cupola)*
Verbrennungszustand *m* state of combustion
verbrettern to sheathe, to plank, to clad with boards
Verbretterung *f* sheathing, boarding, planking, furring, cladding
Verbruch *m (mine)* caved area (waste), goaf, gob
verbrühen to scald
Verbrühen *n* scalding
V. durch ausströmenden Dampf scalding by escaping steam
verbuchen to book, to register; to enter
Verbund *m* bond *(concrete)*
Verbundamt *n (telec)* minor exchange
Verbundbau *m s.* Verbundbauweise
Verbundbauweise *f* composite [building] construction
Verbundbergwerk *n* combined mine
Verbundbetrieb *m* interconnection, grid operation *(power-systems)*
Verbunddampfmaschine *f* compound steam engine
V. mit doppelter Expansion double-expansion compound steam engine
dreizylindrige V. mit dreifacher Expansion triple-expansion steam engine
mehrzylindrige V. multi-cylinder compound steam engine
verbunden interconnected
drehbar v. pivoted [on a pin]
gelenkig v. pin-connected, pin-jointed, hinged
mit Bolzen v. pin-connected, pin-jointed
Verbunderregung *f* compound excitation
V. für gleichbleibende Spannung level-compound excitation *(generator)*
Verbundfenster *n* double-glazed casement (window)
Verbundfolie *f* composite film
Verbundgenerator *m* compound generator
Verbundgießen *n* composite casting
Verbundglas *n* compound (laminated, safety) glass
Verbundglasröhre *f* compound-glass tube
Verbundguß *m* composite (compound) casting
Verbundkarte *f* dual card *(punched card)*
Verbundkern *m* compound (intermediate) nucleus
Verbundkessel *m* combination boiler
Verbundkettenfahrleitung *f* compound catenary suspension line
Verbundkoksofen *m* compound coke oven
Verbundkonstruktion *f* composite structure
Verbundkreiselpumpe *f* two-stage centrifugal pump
Verbundlager *n* composite bearing
Verbundleuchte *f (mine)* lamp with firedamp detector
Verbundlokomobile *f* compound portable steam engine
Verbundlokomotive *f* compound locomotive
Verbundmaschine *f* compound motor
Verbundmetall *n* composite metal
Verbundmotor *m* compound motor
Verbundnetz *n* grid [system], [inter]national grid
Verbundplatte *f* composite board (panel), sandwich-type panel, sandwich board, blockboard
Verbundpulver *n* compound powder
Verbundröhre *f* compound valve, multiple[-unit] valve
Verbundsäule *f* composite (combination) column
Verbundschaltung *f* compound connection
Verbundschmälze *f (text)* combined greasing agent
Verbundspule *f* compound coil
Verbundstahl *m* compound (composite) steel
Verbundstoff *m (plast)* combined plastic, compact material, composite
Verbundteilen *n (mach)* compound indexing
Verbundträger *m* built-up girder, composite girder (beam)
Verbundtransformator *m* compound transformer
Verbundtransistor *m* compound transistor
Verbundturbine *f* compound turbine
Verbundwicklung *f (el)* compound winding
Verbundwirkung *f* composite (bond) action
Verbundzeche *f* compound mine
Verbundzentrifugalpumpe *f* two-stage centrifugal pump
Verbundzustand *m (nucl)* compound (intermediate) state
verchromen to chromize, to chromium-plate
verchromt chromium-plated
Verchromung *f* chromizing, chromium-plating
verdämmen to tamp; to dam
verdampfbar vaporizable, volatilizable
Verdampfbarkeit *f* vaporability, volatility
V. des Schmiermittels vaporability of the lubricant

verdampfen to evaporate, to vaporize, to volatilize
Verdampfer *m* evaporator, [e]vaporizer
Verdampferhalter *m* vaporizer holder
Verdampferreaktor *m s.* Siedewasserreaktor
Verdampferteil *m* evaporation section
Verdampfung *f* vaporizing; vaporization, evaporation
V. durch Sonnenenergie solar evaporation
V. fester Körper evaporation of solids
Verdampfungsapparat *m* evaporator
Verdampfungsenergie *f* evaporation energy
Verdampfungsfähigkeit *f* evaporating capacity
Verdampfungsgeschwindigkeit *f* rate (rapidity) of vaporization, evaporation rate
Verdampfungskolonne *f* flash column (tower)
Verdampfungskühlung *f* cooling by evaporation, evaporative cooling
Verdampfungsnukleon *n* evaporation nucleon
Verdampfungspunkt *m* vaporization (steam) point
Verdampfungsreaktor *m s.* Siedewasserreaktor
Verdampfungsröstung *f* volatilization roasting
Verdampfungsrückstand *m* residue on evaporation
Verdampfungssubstanz *f* evaporant
Verdampfungsvergaser *m* *(aut)* vaporizing carburetter
Verdampfungsverlust *m* evaporation loss, loss by evaporation
Verdampfungsvermögen *n* evaporating capacity
Verdampfungsversuch *m* evaporation test
Verdampfungswärme *f* [latent] heat of evaporation
Verdampfungsziffer *f* steam raising coefficient
verdauen to digest
Verdeck *n* [folding] top, hood
Verdeckbezug *m* hood covering
verdecken to cover; to mask; to hide
Verdeckgestell *n* folding-top structure
Verdeckhülle *f* folding-top cover
Verdecklager *n* folding-top base
Verdeckrahmen *m* folding-top frame
Verdeckspriegel *m* folding-top bow
verdeckt hidden; concealed; covered; interior; enclosed
verdecktliegend covered
Verdeckung *f* cover; hood; *(telec)* masking
Verdeckverschluß *m* folding-top clamp
verderben to spoil; to rot
verderblich perishable *(goods)*
verdichtbar condensable; compactible; compressible
Verdichtbarkeit *f* compactibility; compressibility
verdichten to condense, to condensate; to pack; to densify, to compact; to compress; *(mach)* to burnish

Verdichter *m* compressor; condenser; *(build)* compactor
Verdichterschaufel *f* compressor blade
Verdichterturbine *f* compressor turbine
Verdichterventil *n* compressor valve
Verdichterzylinder *m* compressor cylinder
Verdichtung *f* condensation; pack[ing]; compaction, densification; compression
Verdichtungsanlage *f* densener *(mould)*
Verdichtungsdruck *m* compression pressure
Verdichtungsenergie *f* compacting energy
Verdichtungsgerät *n* compactor
Verdichtungsgrad *m* degree of condensation, bulk factor
Verdichtungshub *m* compression stroke (cycle)
Verdichtungsraum *m* compression volume
Verdichtungsring *m* compression ring
Verdichtungsstoß *m* surging shock, rebound *(turbine)*
Verdichtungstakt *m* compression cycle
Verdichtungsverflüssigung *f* compression liquefying
Verdichtungsverhältnis *n* ratio of compression
Verdichtungswalze *f* compaction roller
Verdichtungswärme *f* heat of compression
Verdichtungswelle *f* shock wave
verdicken to concentrate, to thicken, to body, to inspissate *(fluid)*; to increase the thickness
Verdickung *f* concentration, thickening, bodying, inspissation *(fluid)*; increasing of the thickness
V. der Schaufelplatte thickening of the blade root
V. des Schaftes increase of the cross-sectional area
mit Stärke angesetzte V. starch-thickened gum
Verdickungsmittel *n* thickener
Verdienst *m* earnings *(worker)*; profit *(business)*
Verdienstspanne *f* profit margin
Verdieselung *f* dieselization
Verdolpapier *n* endless paper for Jacquard machines
verdoppeln to double
Verdoppler *m* doubler; frequency doubler; voltage doubler
Verdopplerabstimmkästchen *n* tuning box for frequency doubler
Verdopplung *f* doubling, replication
Verdopplungszeit *f* *(nucl)* [breeding] doubling time
verdrahten to wire [up], to interconnect
Verdrahtung *f* connecting-up, [circuit] wiring
feste V. permanent wiring
gedruckte V. printed wiring
kapazitätsarme V. low-capacity wiring
Verdrahtungsfehler *m* faulty wiring
Verdrahtungsmaschine *f* wiring machine
Verdrahtungsplan *m* wiring diagram

Verdrahtungsschablone *f* wiring jig
verdrallen to twist
verdrängen to deplete; to displace *(material)*
Verdränger *m* displacer
Verdrängerhub *m* pressure stroke
Verdrängung *f* displacement; depletion; volume of displacement
  V. auf Spanten *(ship)* moulded displacement
  V. voll beladen *(ship)* [full] load displacement
Verdrängungsboot *n* displacement boat
Verdrängungskurve *f (ship)* curve of areas of cross-sections, displacement curve
Verdrängungsschiff *n* displacement ship
Verdrängungsschwerpunkt *m (ship)* centre of buoyancy
  V. der Länge nach longitudinal centre of buoyancy
Verdrängungsstrom *m* displacement current
Verdrängungstonne *f (ship)* displacement ton
Verdrängungsvölligkeitsgrad *m (ship)* block coefficient
Verdrängungswiderstand *m* driving resistance
Verdrängungszähler *m* positive-displacement meter
verdrehbeansprucht stressed in torsion, torsionally stressed, subjected to torsion
Verdrehbelastung *f* torque loading, torsion load
verdrehen to twist; to distort
Verdrehfestigkeit *f* torsional strength
Verdrehflankenspiel *n* backlash in the direction of the pitch circle
Verdrehmoment *n* torsional moment
Verdrehspannung *f* twisting stress
Verdrehsteifigkeit *f* torsional rigidity
verdreht twisted
Verdrehung *f* twist[ing], torsion; distortion
  spiralförmige V. winding *(of wood)*
Verdrehungsbeanspruchung *f* torsional load; torsional stress
Verdrehungsbelastung *f* torsion load
Verdrehungsbruch *m* torsion failure (fracture)
Verdrehungselastizität *f* torsion elasticity
Verdrehungsfestigkeit *f* torsional strength
Verdrehungskraft *f* torsional (twisting) force
Verdrehungsprüfung *f* torsional test
Verdrehungsschwingung *f* torsional oscillation (vibration)
Verdrehungsstab *m* torsion rod
verdrehungssteif torsionproof, stiff against torsion
Verdrehungssteifigkeit *f* torsional stiffness
Verdrehungsversuch *m* torsional test
Verdrehungswelle *f* torsional wave
Verdrehungswiderstand *m* twisting (torsional) resistance
Verdrehungswinkel *m* distortion angle, angle of twist (torsion)
verdreifachen to treble, to triple
verdrillen to twist; to distort; to transpose
Verdrillprüfung *f* twist test *(of paired wires)*
Verdrillung *f* torsion, twist[ing]; *(el)* transposition *(of overhead lines)*
Verdrillungsisolator *m* transposition insulator *(overhead line)*
Verdrillungsmast *m* transposition pole
verdrücken *(mine)* to squeeze
Verdrückung *f (mine)* nip (pinch) out
verdübeln to peg, to dowel
Verdübelung *f* pegging, dowelling
verdunkeln to black-out; to darken
verdunkelt daylight-proof
Verdunkelung *f* black-out; *(Am)* dim-out
Verdunkelungsdrossel *f* reactance dimmer
Verdunkelungspapier *n* black-out paper
Verdunkelungsschalter *m* dimming (dimmer) switch
Verdunkelungstransformator *m* dimmer transformer
Verdunkelungswiderstand *m* dimming resistor, [reactance, rheostat] dimmer
verdünnbar rarefiable *(gas)*
Verdünnbarkeit *f* rarefiability *(gas)*
verdünnen to thin; to dilute *(liquid)*; to rarefy *(gas)*
Verdünner *m* thinner *(paint etc)*
verdünnt diluted *(liquid)*; rarefied *(gas)*; aqueous *(acid)*
Verdünnung *f* dilution *(gas)*; rarefaction *(gas)*
Verdünnungsanalyse *f* dilution analysis
Verdünnungsfaktor *m* diluting ratio *(varnish)*
Verdünnungsgesetz *n* **[/Ostwaldsches]** [Ostwald's] dilution law
Verdünnungsmittel *n (ch)* diluent, attenuant, thinner *(paint)*; *(print)* ink reducer
  nichtlösendes V. non-solvent
Verdünnungswärme *f* dilution heat, heat of dilution
Verdünnungswelle *f* wave of rarefaction, rarefaction (expansion) wave
verdunsten to evaporate, to vaporize, to volatilize
Verdunstung *f* evaporation, vaporization, volatilization
Verdunstungsdruck *m* vapour pressure (tension)
Verdunstungskühlung *f* evaporative cooling
Verdunstungsmesser *m* evaporimeter
Verdunstungsverlust *m* evaporation loss, loss due to evaporation
veredeln *(ag)* to graft, to bud; to improve; *(text)* to finish; *(met)* to refine
Veredlung *f (ag)* graft; improvement; *(text)* finishing; refining *(pulp)*; up-grading
  alkalische V. alkaline refining
  einjährige V. *(ag)* maiden
Veredlungsanlage *f* refining unit (assembly)
Veredlungsindustrie *f* processing industry

Veredlungsverfahren *n* refining procedure (process)
Vereinbarung *f* declaration *(programming)*
vereinfachen to simplify
Vereinfachung *f* simplification
vereinheitlichen to unify; to standardize
Vereinheitlichung *f* unification; standardization
vereinigen to unite, to combine, to connect
Vereinigung *f* union; junction; combination; joining; association
Vereinigungsklasse *f (math)* class sum
Vereinigungsmenge *f (math)* set union
Vereinigungsreaktion *f s.* thermonukleare Reaktion
Vereinzelmaschine *f (ag)* gapper, down-the-row thinner
vereinzeln to single; to isolate
vereinzelt isolated; discrete; scattered
vereisen to frost, to glaciate, to freeze, to ice
verenge[r]n to narrow, to contract, to restrict, to reduce
  sich v. to converge
Verengung *f* narrowing, restriction, contraction *(pipe)*, constriction, reduction
  V. des Kanals narrowing of the steam passage
veresterbar esterifiable
verestern to esterify
Veresterung *f* esterification
Verfahren *n* process; procedure; method, technique; treatment
  V. bei Funkverkehrsausfall radio communication failure procedure
  analytisches V. analytical method
  diskontinuierliches V. batch system *(papermaking)*
  experimentelles V. experimental procedure
  fehlerkorrigierendes V. error-control procedure
  gesetzlich geschütztes V. patented procedure
  methodisches V. orderly procedure
  n-freies V. limit (ultimate load, ultimate strength) design, load-factor method, ultimate strength (load) method
  spanabhebendes V. *s.* spanendes V.
  spanendes V. machining process (technique), cutting process (technique)
  spanloses V. chipless process (forming technique)
Verfahrensfehler *m* truncation error; error of approximation
Verfahrensforschung *f* process research; operational research
Verfahrenskammer *f* processing chamber
Verfahrenskurs *m* **über Grund** *(aero)* procedure track
Verfahrensregelung *f* process control
  industrielle V. industrial process control
Verfahrensregler *m* process controller
Verfahrensstammbaum *m* flow-sheet
Verfahrenstechnik *f* chemical engineering
Verfahrenstechniker *m* chemical engineer
Verfahrensüberwachung *f* process control
Verfahrenszeit *f* process time
Verfall *m* decay, ruin; dilapidation
verfallen to decay, to dilapidate
verfälschen *(ch)* to adulterate; to vitiate
Verfälschung *f* adulteration *(addition of foreign substances having less value)*
Verfälschungsmittel *n* adulterant
verfärben to discolour, to change [the] colour; to stain
Verfärbung *f* colouration; discolouration; stain[ing]
  elektrolytische V. electrolytic colouration
  homogene V. uniform colouration
  sichtbare V. visible colouration
Verfärbungsgrad *m* degree of colouration
Verfärbungsprozeß *m* colouring process
Verfärbungstiefe *f* depth of colour
verfeinern to refine, to improve
  das Faserband v. to attenuate the sliver
verfestigen to grout; *(build)* to stiffen, to consolidate; *(ch)* to solidify; *(met)* to work-harden
Verfestigung *f (build)* consolidation; *(met)* work hardening
verfestigungsfähig *(met)* work-hardenable
Verfilzen *n* felting
Verfilzer *m* felter
Verfilzungsmaschine *f* felter
verfischt *(print)* wrong *(fount)*
verfitzen to entangle
Verflacherschaltung *f (el)* graduator
verflechten to interlace
verflüchtigen to volatilize
Verflüchtigung *f* volatilization
verflüssigen to liquefy; to condensate
Verflüssiger *m (build)* plasticizer
Verflüssigung *f* liquefaction; condensation
Verflüssigungsleistung *f (plast)* plasticizing capacity
Verflüssigungsmittel *n* liquefacient
Verflüssigungspunkt *m* liquefaction (flooding) point
verfolgen to track *(radar)*
Verfolgen *n* tracking *(radar)*
Verfolgung *f* tracking *(radar)*
  automatische V. automatic tracking
Verfolgungsauswertegerät *n* tracking plotter
Verfolgungskurve *f* curve of pursuit, pursuit line
Verfolgungsradar *n* tracking radar
Verfolgungsscheinwerfer *m* illuminating pencil
Verfolgungssystem *n* tracking system
verformbar deformable; *(Am)* workable, formable *(metal)*
  kalt v. formable (workable) in cold state

warm v. thermoplastic; hot workable, warm formable
Verformbarkeit *f* deformability; workability, formability *(metal)*
verformen to deform, to shape
sich v. to distort, to deform
Verformung *f* forming, working *(metal)*; deformation; distortion
bleibende V. permanent deformation (set)
elastische V. elastic deformation
spanlose V. chipless (non-chipping) forming technique
thermale V. thermal strain *(semiconductor)*
Verformungsarbeit *f* work of deformation
Verformungsbruch *m* plastic fracture
Verformungsgeschwindigkeit *f* strain rate (velocity)
Verformungsschwingung *f* deformation vibration
Verformungswiderstand *m* resistance to deformation
Verformungszustand *m* state of deformation
verfrachten to load; to ship; to let to freight; to charter
Verfrachter *m* carrier, ship owner
verfügbar available
verfugen to point *(brickwork)*
Verfugung *f* pointing *(brickwork)*
Verfüllboden *m (build)* backfill[ing], fill
verfüllen to fill [in], to back-fill *(soil)*
Verfüllmasse *f* backfill[ing], fill
Verfüllung *f s.* Verfüllmasse
vergällen *(ch)* to denature
Vergällungsmittel *n* denaturant
vergärbar fermentable
Vergärbarkeit *f* fermentability
vergären to ferment
vergasbar gasifiable
vergasen to gasify, to vaporize, to carburet
Vergaser *m (aut)* carburet[t]or, carburetter; gasifier
V. mit Auspuffheizmantel exhaust-jacketed carburetter
V. mit ölgespeistem Heizmantel oil-jacketed carburetter
V. mit Unterdrucksteuerung vacuum-controlled carburetter
Vergaseranschlußstutzen *m* carburetter flange
Vergaserbrand *m* carburetter fire
Vergaserbrennstoffnadelventil *n* carburetter needle valve
Vergaserdüse *f* spray nozzle
Vergasereinstellung *f* carburetter tuning (adjustment)
Vergaserfrostschutzvorrichtung *f* carburetter anti-icer
Vergasergehäuse *n* carburetter main body
Vergasergestänge *n* carburetter control linkage
Vergaserkraftstoff *m* petrol, *(Am)* gasoline
Vergaserlufteinlaß *m* carburetter induction system
Vergaserluftfänger *m* carburetter-air scoop
Vergaserlufttrichter *m* choke
Vergaserluftvorwärmer *m* carburetter-air heater
Vergasermischrohr *n* Venturi tube
Vergasermotor *m* petrol (spark-ignition) engine, *(Am)* gasoline engine
Vergaserwarmluftklappe *f* carburetter hot-air door
Vergaserzwischenstück *n* carburetter adapter
Vergasung *f* gasification
Vergasungsberge *pl* carbonaceous refuse
vergeben to let *(work)*; to place *(orders)*
Arbeit an Zulieferbetrieb v. to sublet
vergießbar *(plast)* castable, pourable
Vergießbarkeit *f (plast)* castability, pourability
vergießen *(plast, met)* to cast, to teem, to pour; to seal; *(Am)* to pot
matt v. to cold cast
vergiften to poison
Vergiftung *f* poisoning; contamination; intoxication
V. des Reaktors durch Neutronenabsorber neutron poisoning (poison effect)
V. mit Spaltprodukten fission-product poisoning
Vergiftungsanlage *f (nucl)* poison-injecting equipment
Vergiftungsfaktor *m (nucl)* danger coefficient
vergilben to yellow
Vergilben *n* brightness reversion *(pulp)*
vergittert trellized, latticed, laced
Vergitterung *f* lacing, lattice-work
verglasen *(build)* to glaze; *(ch)* to vitrify
Verglasen *n* **der Schleifscheibe** glazing of the grinding wheel
verglast glass-enclosed; glazed *(grinding wheel)*
Verglasung *f* glazing; vitrification
doppelte V. *(build)* double glazing
Vergleich *m* 1. comparison; 2. error detection *(controller)*
logischer V. logical comparison
Vergleicher *m (cont)* 1. comparator; 2. error measuring device
Vergleicherkreis *m (el)* comparator
Vergleichsabsorber *m* reference absorber
Vergleichsapparat *m* reference instrument
Vergleichsdiagramm *n* reference pattern
Vergleichseinrichtung *f* comparator
Vergleichselektrode *f* reference electrode
Vergleichsfläche *f* plane of comparison
Vergleichsfrequenz *f* standard (reference) frequency
Vergleichsglied *n* 1. comparison element; 2. error measuring device
Vergleichsinstrument *n* reference (comparison) instrument

Vergleichskörper *m* reference body
Vergleichslampe *f* comparison lamp
Vergleichslänge *f* gauge length
Vergleichsmessung *f* comparison measurement
Vergleichsmethode *f* comparison method
Vergleichsmikroskop *n* comparison microscope
Vergleichsnormale *f* comparison standard
Vergleichsokular *n* comparison eyepiece
Vergleichsoszillator *m* reference oscillator
Vergleichspegel *m* reference level
Vergleichsprobe *f* reference specimen
Vergleichsprüfung *f* comparison test
Vergleichsquellle *f* reference source; standard neutron source
Vergleichssatz *m (math)* comparison theorem
Vergleichsschaltung *f* differential connection
Vergleichsspektrum *n* comparison spectrum
Vergleichsstelle *f* 1. mixing point *(controller)*; 2. reference junction *(thermocouple)*
Vergleichston *m* reference tone
Vergleichsverfahren *n* comparison process
Vergleichsweißpegel *m (telev)* reference white level
Vergleichswert *f* comparative value
Vergleichswiderstand *m* reference (comparative) resistance
Vergleichszahl *f* comparative figure
Vergleichszähler *m* substandard (reference) meter
verglichen mit [as] compared with
Verglühung *f (met)* faulty annealing
Vergnügungsboot *n* pleasure (excursion) boat
Vergnügungsdampfer *m* pleasure (excursion) steamer
Vergnügungsreise *f (ship)* pleasure trip, cruise
Vergnügungsschiff *n* excursion vessel
vergolden to gild
  galvanisch v. to gold-plate; to electro-gild
Vergoldepresse *f* blocking and embossing press
Vergoldung *f* gold-plating, gilding
  galvanische V. electrogilding
vergossen / ruhig killed *(steel)*
vergraben to bury *(radioactive waste)*
Vergrämungsmittel *n (ag)* repellant
vergrößern to magnify *(optics)*; to enlarge *(photograph)*; to increase *(amount, number)*
  maßstäblich v. to scale up
vergrößernd / stark highly magnifying
Vergrößerung *f* 1. *(opt)* magnifying; magnification; 2. enlarging; enlargement *(photography)*; increase *(amount, number)*
  förderliche V. useful magnification; magnifying power
  lineare V. linear magnification
Vergrößerungsapparat *m* enlarger, enlarging machine (apparatus)
Vergrößerungsdifferenz *f* **/ chromatische** chromatic difference of magnification
Vergrößerungsfaktor *m* magnification factor
Vergrößerungslinse *f* magnifying lens
Vergrößerungsmikroskop *n* magnification microscope
Verguß *m* seal[ing]
Vergußharz *n* casting resin
Vergußkammer *f* sealing chamber
Vergußmasse *f* sealing (insulating) compound *(cable)*
Vergußtechnik *f* potting technique
Vergütbarkeit *f* heat-treating property
vergüten *(met)* to harden and temper, to temper after hardening
Vergüteofen *m* quenching and tempering furnace
Vergütungsstahl *m* quenched and subsequently tempered steel
Verhackgerät *n (ag)* gapper, down-the-row thinner
verhalten / sich to behave; to perform; to response, to react
Verhalten *n* behaviour *(material)*; performance *(instrument, machine)*; response, reaction; attitude
  V. auf See behaviour at sea *(vessel)*
  anodisches V. anodic behaviour
  asymptotisches V. asymptotic behaviour
  entartetes V. degenerate behaviour
  stationäres V. steady state behaviour
  zeitliches V. time behaviour
Verhältnis *n* relation; ratio; proportion
  V. / im quadratischen in proportion to the square
  V. / in geradem in direct proportion (ratio)
  V. Abraum : Kohle *(mine)* overburden-to-coal ratio
  V. der stehenden Wellen *(el)* standing wave ratio
  V. des Helligkeitssignals zum Synchronisationssignal picture to synchronizing ratio
  elektro-akustisches V. eines Empfangssystems electric-acoustic index
  gyromagnetisches V. gyromagnetic ratio
  stöchiometrisches V. stoichiometric ratio
  umgekehrtes V. inverse ratio
Verhältnisdetektor *m* ratio detector
Verhältnisregelung *f* ratio control
Verhältniszahl *f* ratio; proportionality factor
verhärten to harden; to set *(cement)*
verharzen to gum up; to resinify
Verharzung *f* gumming-up; gummy deposit; gum formation; resinification; asphaltization
  V. des Schmieröls resinification of the lubricating oil
verhauen to get, to mine, to win, to work
verheuern to freight; to charter *(a vessel)*
Verhieb *m (mine)* extraction, getting, working
  fallender V. working to the dip
Verhiebsart *f (mine)* method of working
Verhiebsbreite *f (mine)* increment of face advance

Verhiebsfortschritt *m (mine)* face advance
Verhiebsrichtung *f (mine)* direction of working
verhindern to prevent; to inhibit; to obstruct
das Peilen v. to impede the taking of bearings
den Schmiermittelaustritt v. to retain the lubricant
Verhindern *n* **des Beschlagens** demisting *(screen)*
verhindernd / Harzbildung anti-gumming *(hydraulics)*
Schaumbildung v. anti-foam
Verholanker *m (ship)* kedge [anchor]
Verholboje *f* kedge (hauling off, warping) buoy
verholen to warp, to move *(vessel)*
Verholklampe *f (ship)* cable chock, fair-lead
Verholklüse *f (ship)* mooring (warping, bulwark-mooring) pipe, warping chock, hawser port
Verholleine *f (ship)* warp [rope], guess warp
Verholpoller *m (ship)* warping guide, checking bollard
Verholschlepper *m* docking (ship handling) tug
Verholspill *n (ship)* warping winch (capstan), gipsy capstan
Verholtonne *f (ship)* kedge (warping) buoy
Verholwinde *f (ship)* warping winch
verholzen to lignify
Verholzung *f* lignification
verhüten to prevent
verhütten to smelt
Verhüttung *f* smelting; metallurgical treatment
verhüttungsfähig smeltable
Verjährungsfrist *f* term of prescription (limitation)
verjüngen *(ag)* to rejuvenate
sich v. to taper, to narrow
verjüngt taper[ed]
Verjüngung *f* 1. *(ag)* rejuvenation; 2. constriction; taper; degree of taper; draft *(mould)*; decrease of cross-sectional area
V. eines Modells draught of a pattern
kegelige V. taper[ing]
Verjüngungshieb *m* regeneration (reproduction) felling
Verjüngungsmaßstab *m* reduction scale
verkabeln to cable
Verkabelung *f* cable-laying, cabling; wiring
verkadmen / galvanisch to cadmium-plate
Verkadmen *n* cadmizing
galvanisches V. cadmium plating
verkalken to calcify, to calcine
verkalkulieren to miscalculate
Verkalkung *f* calcification, calcination
verkämmen to cog *(carpentry)*
Verkämmung *f* cogging *(carpentry)*
verkanten to cant
verkantet / nach links [back] sight tilted to the left *(rifle)*
nach rechts v. [back] sight tilted to the right *(rifle)*
Verkapselung *f* encapsulation
Verkattanker *m (ship)* back anchor
verkatten *(ship)* to back
Verkauf *m* sale
Verkaufen *n* **/ automatisches** automatic vending
Verkäufer *m* sales executive *(industry)*
verkäuflich saleable
Verkaufsabteilung *f* sales department
Verkaufsautomat *m* slot machine
Verkaufsbedingungen *fpl* terms of sale
Verkaufsleiter *m* sales manager
Verkaufspreis *m* selling price
Verkaufsrecht *n* right to sell
Verkaufsvertrag *m* sales agreement
Verkehr *m* traffic; transport, carriage, conveyance; *(telec)* communication, traffic; service *(railway)*
abgehender V. *(telec)* originating (outgoing) traffic
ankommender V. *(telec)* incoming traffic
einspuriger V. single-lane traffic
innerbetrieblicher V. internal traffic
internationaler V. international traffic
verkehren *(telec)* to communicate; *(ry)* to run *(train)*
Züge v. lassen to run trains
Verkehrsabwanderung *f* loss of traffic
Verkehrsabwicklung *f* traffic handling, handling of traffic
Verkehrsader *f* traffic artery
Verkehrsampel *f* traffic light (signal)
Verkehrsandrang *m* rush of traffic
Verkehrsanforderung *f* traffic requirements
Verkehrsaufkommen *n* volume of traffic
Verkehrsbau *m* transport engineering
Verkehrsbauten *pl* traffic structures
Verkehrsbelastung *f* traffic load
Verkehrsbeobachtung *f* traffic observation
Verkehrsboot *n* passenger launch, commuter, runabout
schnelles V. high-speed runabout
Verkehrsdichte *f* density of traffic; motorcar population
Verkehrsempfänger *m* communications receiver
Verkehrsfahrzeug *n* **/ öffentliches** public-service vehicle
Verkehrsfläche *f* circulation area
Verkehrsflugzeug *n* commercial airplane; airliner
Verkehrsfrequenz *f* working frequency
Verkehrsgebiet *n* trading area
Verkehrsgesetzgebung *f* transport law
Verkehrsgüte *f (telec)* quality (grade) of service
Verkehrsinsel *f* street refuge, traffic island
Verkehrsknotenpunkt *m* junction point
Verkehrskontrolle *f* audit office
Verkehrskoordinierung *f* coordination of transports

Verkehrslast *f* live (traffic, superimposed) load
Verkehrsleistung *f* traffic performance
V. eines Leitungsbündels traffic capacity
Verkehrsmittel *npl* transport facilities, means of transport (conveyance)
Verkehrsordnung *f* traffic regulations
Verkehrsplanung *f* traffic planning
Verkehrspolizist *m* [traffic] patrolman
Verkehrsregelung *f* traffic control
Verkehrssicherheitsaktiv *n* Traffic Safety Committee
Verkehrssperrung *f* block
Verkehrsstockung *f* traffic congestion (jam)
Verkehrsstreife *f* traffic patrol
Verkehrsstrom *m* traffic stream
Verkehrssünder *m* traffic offender
verkehrstechnisch traffic-engineering
Verkehrsträger *m* [common] carrier
Verkehrstüchtigkeit *f* roadworthiness
Verkehrsunfall *m* road (traffic) accident
Verkehrsunfallzahl *f* accident toll
Verkehrswelle *f* traffic wave
Verkehrszähler *m* traffic meter
Verkehrszeichen *n* traffic (road) sign
verkehrt wrong; upside down; inverted
verkeilen to key, to wedge; to cotter
Verkeilung *f* keying, cottering
verkettbar *(mach)* linkable; *(el)* interlinkable
Verkettbarkeit *f* *(mach)* linkability; *(el)* interlinkability
verketten *(mach)* to link; *(el)* to interlink; *(math)* to correlate
mit einem Stromkreis v. to thread through a circuit
verkettet *(mach)* linked; *(el)* interconnected, meshed *(system)*
Verkettung *f* *(mach)* linking [together], link-up; *(el)* interlinking; *(math)* correlation
V. der Linien interlocking of lines
elektromagnetische V. electromagnetic linkage
Verkettungseinrichtung *f* *(mach)* transfer link, line linking element
verkettungsfähig *(mach)* linkable
Verkettungsfähigkeit *f* *(mach)* linkability
Verkettungsgesetz *n* / **Amperesches** circuital form of Ampere's law
Verkettungskreis *m* link circuit
verkieseln to silicify
verkieselt silicated
Verkieselung *f* silification
verkippen to spoil
verkitten to putty up, to cement *(lens)*, to lute, to paste together
Verkittung *f* puttying-up, luting; cementing *(lens)*
verklagen auf Schadenersatz to sue for damages
verklammern to cramp
Verklammerungen *fpl* **für Lagerausgüsse** grooves for anchoring white metal to liner bush
verkleben to bond; to adhere *(fibres)*; to [ag]glutinate, to cement, to glue, to seal, to stick *(piston ring)*
verkleiden to line, to coat, to face; to sheathe, to plank, to clad with boards
mit Wärmeschutzmasse v. to cover with non-conducting material
Verkleidung *f* lining; coating; *(aut)* jacket; panelling; cladding, sheathing; fairing; lagging *(boiler)*
in öldichter V. enclosed in an oil-tight case
Verkleidungsbrett *n* facing board
Verkleidungskappe *f* protective covering for cylinder cover
Verkleidungsplatte *f* facing stone (slab), cladding panel
Verkleidungsring *m* lagging (cleating) ring
verkleinern to reduce; to decrease, to diminish
maßstäblich v. to scale down
Verkleinerung *f* reduction *(scale)*, scaling down
Verkleinerungsmaßstab *m* scale of reduction
Verkleinerungsverhältnis *n* reduction ratio
verklemmen to jam; to stick, to seize *(valve)*
sich v. to become jammed, to jam; to stick, to seize *(valve)*
Verklemmung *f* jamming; sticking *(valve)*
verkneten to malaxate, to triturate
verknoten to tie; to knot; to kink; to entangle
Verknüpfung *f* *(dat)* gating *(of logical signals)*; combination; connection, configuration
logische V. logical circuit; logical operation
Verknüpfungsglied *n* *(dat)* [logical] decision element; gate
Verknüpfungsoperation *f* *(dat)* connective operation
Verknüpfungsschaltung *f* *(dat)* combinational switching circuit
Verkobaltung *f* cobalt plating
verkochen *(ch, plast)* to boil down, to concentrate; to kettle
verkohlen to carbonize, to char, to coke
Verkohlung *f* carbonization, charring, coking
Verkohlungsanlage *f* charring plant
verkoken to coke, to carbonize coal
Verkokung *f* high-temperature carbonization, carbonizing, coking
Verkokungsanlage *f* coking plant
Verkokungseigenschaft *f* coking property
Verkokungsfähigkeit *f* coking capacity
Verkokungsgeschwindigkeit *f* rate of coking
Verkokungsindustrie *f* carbonizing industry
Verkokungskammer *f* carbonizing (coking) chamber
Verkokungsnaht *f* carbonizing split

Verkokungsneigung *f* tendency to form carbon, carbonizing qualities, carbon-forming properties
Verkokungsrückstand *m* carbon residue (deposit)
Verkokungszeit *f* carbonizing period, carbonizing (coking) time
verkommen to decay, to rot
verkoppeln to couple [together]
verkorken to cork
verkrümmen to warp; to bend
Verkrümmung *f* warpage; bending
Verkrustung *f* formation of a cap; incrustation
verküpen to vat
verkupfern to copper *(surface)*
  galvanisch v. to copper-plate, to coat with electrolytically deposited copper
Verkupferung *f* coppering; copper plating *(electrolytically)*; oxidizing *(electrotyping)*
Verküpung *f* vatting
verkürzen to shorten; to cut *(machining time)*; to reduce; *(text)* to abridge; *(el)* to clip *(e. g. amplitudes)*
Verkürzungskapazität *f* shortening capacity
Verkürzungskondensator *m* shortening capacitor
Verladebrücke *f* loading platform, bridge crane, charging bridge
Verladebunker *m* loading bin
Verladeeinrichtung *f* loading plant
Verladehafen *m* l[o]ading port, port of loading (lading, shipment)
Verladekran *m* loading crane
verladen to load, to entrain; to ship
Verladen *n* loading
Verlader *m* loader; shipper
Verladerampe *f* loading ramp
Verladerutsche *f* loading chute (gate)
Verladeschurre *f* loading chute (gate)
Verladung *f* loading; entraining; shipment
  V. als Massengut *(ship)* loading in bulk
  mangelhafte V. faulty loading
Verladungsgebühr *f* loading charge
Verlag *m* publishing house; publisher[s]
verlagern to shift, to displace; to misalign
Verlagerung *f* shift, displacement
  V. der Wellen shaft misalignment
Verlagszeichen *n* signet
verlängern to lengthen; to prolong *(line; time)*; to extend
Verlängerung *f* elongation; prolongation; extension
  V. der Achsen extension of the axes
Verlängerungsachse *f* lengthening spindle
Verlängerungsblech *n* make-up pan *(conveyor)*
Verlängerungskabel *n* extension cord
Verlängerungsleitung *f* pad, extension line (circuit)
Verlängerungsschnur *f* *(el)* extension cord, *(sl)* flex
Verlängerungsspule *f* lengthening (extension) coil
Verlängerungsstück *n* appendage, lengthening bar (piece), extension piece
verlangsamen to slow [down], to slacken, to decelerate
Verlangsamung *f* slowing down, stopping; retardation, delay; deceleration
verlaschen to fish *(plate)*
Verlaschen *n* fish
Verlaschung *f* fishing, fish-plating; lashing
Verlauf *m* behaviour *(curve)*; course, progress; development
  sinusförmiger V. sinusoidal (sine-law) variation
  zickzackförmiger V. zigzag path
verlaufen 1. to spread *(paint)*; 2. *(text)* to blend into each other; 3. *(mach)* to enter the work off centre, to wander from the centre, to deviate from the true course, to run out of centre, to walk off centre *(drill)*
Verlaufen *n* off centre *(drill)*; untrue *(hole)*
verlaufen untrue *(hole)*
Verlauffilter *n* gradient density filter
Verlaufmittel *n* levelling agent
Verlaufskurve *f* subsidence development curve
verlegen 1. to transfer, to shift, to move; 2. to edit *(book)*; 3. to lay *(cable)*, to lay out; to bury *(underground)*; to install *(wires, pipes)*; to pipe; to place *(reinforcement)*
  auf Putz v. to wire on the surface
  Drähte oberirdisch v. to run wires overhead
  eine Leitung v. to run a cable
  eine Rohrleitung v. to assemble a pipe line
  in die Erde v. to lay underground, to bury
  Kabel v. to lay out a cable
  Leitungen v. to wire
  Platten v. to tile, to flag
Verleger *m* publisher
Verlegung *f* 1. transfer; shifting; 2. installation *(pipes)*; laying *(cable)*; burying *(underground)*
  V. auf Putz exposed wiring
  V. in Rohre conduit wiring
Verleihung *f* *(mine)* grant
verleimen to glue, to cement
Verleimung *f* / **geschlossene** closed assembly
  offene V. open assembly
Verleimzeit *f* assembly time
Verlesemaschine *f* seed (potato) cleaner
Verletzungspotential *n* *(el)* injury potential
verlitzen *(el)* to strand
verloren lost *(foundry)*
Verlust *m* loss; leakage; escape; expenditure, expense; dissipation; *(met)* roasting residue; casualty *(of a vessel)*
  V. an Fördermenge delivery loss
  V. an Seidenleim loss of gum in scouring
  V. durch Fehlanpassung *(el)* standing wave loss factor

V. eines Rechtes forfeiture of a right
hydraulischer V. hydraulic loss
mechanischer V. mechanical loss
ohmscher V. ohmic loss
Verlustanzeige *f* advice of loss
verlustbehaftet imperfect
Verluste *mpl* **im Zähler** meter losses
elektrische V. dielectric losses
ohmsche V. ohmic (resistance) losses
stromabhängige V. load (impedance) losses
verlustfrei no-loss, loss-free; perfect *(dielectric)*
Verlusthöhe *f* loss of head, head loss, head lost [in friction]
Verlustkoeffizient *m* coefficient of losses
Verlustkonstante *f* **eines Kabels** damping constant of a cable
Verlustleistung *f* power lost; dissipation
verlustlos ideal, perfect *(capacitor)*, lossless, no-loss
Verlustmenge *f* **je Zeiteinheit** waste quantity per unit time
Verlustmessung *f* wattmeter method *(machine)*
Verlustschmierung *f* non-circulating lubrication
Verluststrom *m* leakage current
Verlustwiderstand *m (el)* equivalent resistance
Verlustwinkel *m* [dielectric] loss angle, dielectric phase difference
Verlustzähler *m* loss meter
Verlustzeche *f* losing mine
Verlustzeit *f* idle time
Verlustziffer *f* loss factor
vermahlen to grind up, to pulverize, to comminute
Vermahlung *f* grinding, comminution
Vermahlungsgrad *m* **eines Anstrichstoffes** grind fineness of a paint
vermascht *(el)* interconnected, meshed
Vermaschung *f* interconnection *(of networks)*
vermauern to wall-up
im Verband v. to bond
Vermauern *n* walling-off
vermehrbar multipliable
vermehren *(nucl)* to breed *(fuel)*; to multiply *(neutrons)*; *(ag)* to propagate, to spread
Vermehrung *f (nucl)* breeding; multiplication; *(ag)* propagation
vegetative V. vegetative propagation
Vermehrungsfaktor *m (nucl)* multiplication constant, *k*-factor
effektiver V. effective multiplication factor k_{eff}
vermeidbar avoidable
vermeiden to avoid *(e. g. accidents)*
vermengen to mix, to mingle, to blend
vermessen to measure, to survey
Vermessingen *n* brassing, covering with brass
Vermessung *f* 1. survey[ing]; 2. tonnage [measurement] *(of vessels)*
V. nach Höhe levelling
Vermessungsboot *n* surveying boat
Vermessungsbreite *f* registered (tonnage) breadth
Vermessungsdeck *n* tonnage deck
Vermessungsfahrzeug *n* survey[ing] vessel
Vermessungsflugzeug *n* survey aircraft
Vermessungsformel *f (ship)* tonnage formula
Vermessungsfrachttarif *m (ship)* tonnage scale
Vermessungsingenieur *m* [land] surveyor
Vermessungskunde *f* surveying
Vermessungslänge *f (ship)* registered (tonnage, measurement) length
Vermessungsluke *f (ship)* tonnage hatch[way]
Vermessungsschiff *n* survey[ing] vessel
V. für hydrografische Messungen hydrographic survey ship
Vermessungsstelle *f* evaluation receiver
Vermessungstiefe *f* registered (tonnage) depth, depth of hold
Vermessungstonne *f* register ton
Vermiculit *m (min)* vermiculite
vermieten to let, to hire out
vermindern to decrease, to diminish; to lessen; to cut *(reject rate)*; to reduce; to lower
die abgegebene Leistung v. to reduce the output
die Helligkeit des Elektronenstrahls v. to dim the trace
Verminderung *f* decreasing; diminution; lessening, lowering; reduction
V. der Geschwindigkeit deceleration; reduction of speed
verminen to mine
vermischbar miscible
vermischen to blend, to mix, to mingle
vermischungsfähig miscible
Vermischungsfähigkeit *f* miscibility
vermitteln to plug through, to connect *(a call)*; to relay
Vermittlung *f* exchange *(telephony)*, retransmission
Vermittlungsamt *n* [central] exchange
Vermittlungsbereich *m* central office area
Vermittlungsgebühr *f* commission
Vermittlungsklinke *f* connecting jack
Vermittlungsplatz *m* operator's position
Vermittlungsschnur *f* patch (patching, switchboard) cord *(telephony)*
Vermittlungsschrank *m* [telephone] switchboard
Vermittlungsstelle *f* exchange; switching centre
vermodern to moulder, to rot
Vermögen *n* power, ability; fortune *(money)*
vermörteln to grout
Vermörtelung *f* grouting
vermuffen to joint
Vermuffung *f* [sleeve] jointing
vermusen to pulp
vernachlässigbar negligible
vernachlässigen to neglect

Vernachlässigung *f* neglect
vernebeln to atomize, to disperse finely
Vernebelung *f* atomization; fine dispersion
Vernebler *m* atomizer; [thermal] fogger
Verneinung *f* negation
Verneinungsschaltkreis *m* NOT-circuit *(logical switching)*
vernetzen *(plast)* to cross-link, to interlace
vernetzt branched; *(plast)* interlaced
Vernetzung *f* cross-linkage, interlacing, network
intermolekulare V. intermolecular cross linking
vernichten to annihilate; to dissipate *(power)*; to destroy
Vernichtung *f* annihilation; destruction
Vernichtungsoperator *m* *(dat)* annihilation operator
Vernichtungsspektrum *n* annihilation spectrum
Vernichtungsstrahlung *f* annihilation radiation; pair annihilation; positron-electron annihilation
vernickeln to nickelize; to nickel-plate
Vernickeln *n* nickel-plating
vernickelt nickel clad, nickel-plated
Vernickelung *f* nickel-plating
vernieten to rivet [together]
Vernieten *n* riveting; rivet closing
Vernunftschluß *(dat)* logical conclusion
Verödung *f* / **elektrische** electrodesiccation
Veröffentlichung *f* 1. publication; 2. published work
verpachten to lease
verpacken to pack, to wrap up
Verpackung *f* packaging
aufteilbare V. fractional package
mangelhafte V. faulty packing
verlorene V. one way packing
Verpackungsart *f* method of packing
Verpackungsbehälter *m* package
Verpackungsgewicht *n* tare
Verpackungsindustrie *f* packaging industry
Verpackungsmaschine *f* packer, packing machine, machine for packaging
Verpackungsmaterial *n* packing material
Verpackungsmittelindustrie *f* industry for packing materials
Verpflegungspersonal *n* catering staff *(on vessels)*
verpichen to pitch
Verpolung *f* reversal of the channels *(radio)*
Verpreßbarkeit *f* mouldability
verpressen to press; *(build)* to pressure-grout, to inject
verpuffen to deflagrate
Verpuffung *f* deflagration
verputzen *(build)* to plaster, to render; *(met)* to condition *(semifinished steel)*
verquellen to warp by moisture
verrechnen / sich to miscalculate
Verrechnungsabkommen *n* clearing agreement *(with foreign country)*
verreiben to distribute; to triturate
Verreibung *f* trituration
verriegeln to bolt, to bar; *(mach)* to block; to clamp, to lock *(turret)*; to lock [in position]
gegeneinander v. to interlock
verriegelnd / gegenseitig interlocking [with each other]
Verriegelung *f* *(mach)* lock; interlock[ing]; locking mechanism (device)
V. bei Abrissen break interlock *(papermaking)*
elektrische V. electrical interlocking
logische V. logical interlock
magnetische V. magnetic latching
mittelbare V. indirect interlocking
Verriegelungseinrichtung *f* interlock[ing] device
Verriegelungsschalter *m* [inter]locking switch
Verriegelungsspannung *f* gate voltage
Verriegelungsstift *m* locking pin
Verriegelungsstromkreis *m* interlock circuit
verringern to minimize, to reduce, to diminish, to decrease
verrippen *(mach)* to rib
verrippt / stark heavily ribbed
Verrippung *f* *(mach)* ribbing; finning
Verrohrung *f* piping; *(mine)* casing, lining; tubing
abgesetzte V. landed casing
festgewordene V. frozen casing
verrosten to rust, to corrode
Verrosten *n* rusting, rust formation
verrostet rusty
verrotten to rot, to decay
verrottungsbeständig resistant to rotting
verrücken to displace, to shift; to jack
Verrückung *f* displacement; jacking
virtuelle V. virtual displacement
verrühren to mix
versagen to fail, to break down; to malfunction; *(mine)* to miss *(shotfiring)*
Versagen *n* failure, breakdown; malfunction
zeitweises V. bug *(of computers)*
Versager *m* *(mine)* mis-fire
programmbedingter V. *(dat)* program-sensitive error
Versalien *mpl* caps, titlings
Versalschrift *f* titling type
Versammlungshalle *f* auditorium, assembly hall
Versand *m* shipment, shipping
Versandabteilung *f* shipping department
Versandbahn *f* forwarding railway
Versandbahnhof *m* forwarding station
Versandbehälter *m* **aus Wellpappe** corrugated shipping container
versanden to sand
versandfertig machen to prepare for shipping
Versandkarton *m* packing case

Versandkatalog *m* mail order catalogue
Versandpackung *f* packing case
Versandrechnung *f* forwarded abstract
Versandverwaltung *f* origination (departure, forwarding) administration
Versatz *m* 1. misalignment; 2. *(mine)* filling, gobbing, packing, stowing; stowing material; *(leath)* layer
mechanischer V. *(mine)* power stowing
nach dem Abbau eingebrachter V. *(mine)* delayed filling
Versatzarbeit *f (mine)* packing *(by hand)*; stowing *(mechanical)*
Versatzarbeiter *m (mine)* cogger, packer, stower
Versatzberge *pl (mine)* stowing material
Versatzdraht *m (mine)* stowing screen
Versatzdruckdose *f (mine)* pack dynamometer
Versatzfeld *n (mine)* stowed goaf, waste filled area
Versatzgeflecht *n (mine)* dry-packing wire netting
Versatzgewebe *n (mine)* stowing screen
Versatzgut *n (mine)* stowing material
Versatzkante *f (mine)* pack line
Versatzmaschine *f (mine)* [goaf] stower
Versatzmatte *f (mine)* dry-packing wire netting
wandernde V. mobile stowing screen
Versatzrippe *f (mine)* strip pack
Versatzrolle *f (mine)* fill chute (raise), waste chute
Versatzrolloch *n* fill chute (raise), waste chute
Versatzschicht *f (mine)* stowing shift
Versatzschleuder *f (mine)* centrifugal stower
Versatzseite *f (mine)* goaf side
versatzseitig *(mine)* goaf-side
Versatzstrecke *f (mine)* brush
Versatzüberhauen *n (mine)* filling (waste) raise
verschalen *(build)* to sheathe; to plank, to board; *(mine)* to lace up
verschalken *(ship)* to secure, to batten down *(a hatch)*
Verschalung *f (build)* sheathing, planking, boarding, cladding; *(mine)* lacing, lagging
Verschanzung *f* bulwark *(of vessels)*
verschiebbar sliding [relative to other member], slidable; traversing *(slide)*
Verschiebbarkeit *f* travel *(machine slide)*; slidability
Verschiebeanker *m* sliding rotor
Verschiebebahnhof *m* marshalling (sorting) yard
V. mit Ablaufbetrieb gravity marshalling (classification) yard
V. mit Gefälle gravitation marshalling yard
V. ohne Gefälle flat marshalling yard
Verschiebebefehl *m (dat)* shift order
Verschiebekran *m* shunting crane
Verschiebelokomotive *f* shunting engine (locomotive), shunter, *(Am)* switch[ing] locomotive
verschieben to shift; to displace; to slide, to traverse
parallel v. *(math)* to translate
sich v. to travel
Verschiebeoperation *f (dat)* shifting operation
Verschieberegister *n (dat)* shift register
Verschiebespindelschleifeinrichtung *f* traverse-spindle grinding attachment, slide-spindle grinding attachment
Verschiebung *f* shift[ing], displacement, translation; drift; lag; slipping
V. der Informationsspeicherstelle location shifting
V. der Spektrallinien displacement (shift) of spectral lines, spectral-line displacement
arithmetische V. arithmetic shift *(from digit place to digit place)*
dielektrische V. dielectric displacement
elektrische V. electric displacement
gegenseitige V. relative displacement
magnetische V. magnetic displacement
spektrale V. spectral shift
virtuelle V. virtual displacement
zyklische V. cyclic shift
Verschiebungsbruch *m* shear fracture, sliding rupture
Verschiebungsdiagramm *n* shift diagram
Verschiebungsdifferenz *f* displacement difference
Verschiebungsfaktor *m* displacement factor
Verschiebungsfluß *m* displacement (electric) flux
Verschiebungsgesetz *n* displacement law
radioaktives V. radioactive (Soddy-Fajans) displacement law
Wiensches V. Wien['s] displacement law
Verschiebungsplan *m* displacement diagram
Verschiebungsregler *m* shift control
Verschiebungssatz *m* **[von Soddy und Fajans]** [Soddy-Fajans] displacement law, radioactive displacement law
Verschiebungsstrom *m* displacement current
Verschiebungstransformator *m* phase advancer
Verschiebungswinkel *m* shift (displacement) angle
verschieden different, unequal, various
v. groß different-sized
verschiedenartig heterogeneous, various
Verschiedenartigkeit *f* heterogeneity
verschiedenfarbig heterochromatic
Verschießen *n (text)* fading
V. durch Lufteinflüsse atmospheric fading
V. in Abgasatmosphäre fume (gas) fading
verschiffen to send by water
Verschiffung *f* shipment, shipping
Verschiffungshafen *m* port of shipment
verschimmeln to mould
verschlacken to slag, to scorify
Verschlackung *f* slagging, scorification
Verschlackungsfähigkeit *f* slaggability

Verschlackungsvermögen *n* slaggability
Verschlag *m* pen
verschlammen to silt up
verschlämmen to fill with mud
Verschlammung *f* mudsilting, siltation
  V. der Bettung choking of the ballast
Verschlechterung *f* deterioration
  V. der Stabilität *(cont)* destabilization
Verschlechterungsfaktor *m* deterioration factor
Verschleierung *f* *(opt)* veiling glare, fogged condition
Verschleifen *n* rubbing down
Verschleifung *f* rounding *(of pulse)*
Verschleiß *m* wear, attrition, abrasion
  V. der Kaliber wear of the grooves
  V. durch Auskolken crater wear
Verschleißausgleich *m* adjustment for wear
Verschleißbarkeit *f* wearability
verschleißbeständig wear-resistant, hard-wearing
Verschleißbeständigkeit *f* wear resistance, anti-wear properties; *(text)* abrasion resistance
Verschleißbremse *f (aut)* anti-wear agent
verschleißen to wear [out]
Verschleißfase *f* wear land
  V. an der Freifläche flank wear land
  V. an der Schneide cutting-edge wear land
Verschleißfasenbreite *f* wear land width, width of wear land
verschleißfest wear-resistant, resistant to wear, hard-wearing
Verschleißfestigkeit *f* wear resistance, resistance to wear, anti-wear properties
Verschleißfortschritt *m* rate of wear
verschleißfrei free from wear, wear-free, unaffected by wear, non-wearing
Verschleißmarke *f* wear mark
Verschleißmarkenbreite *f* land wear
Verschleißmaß *n* rate of wear and tear
Verschleißmessung *f* wear determination
verschleißmindernd wear reducing
Verschleißprüfung *f* wear[ing] test
Verschleißschicht *f* wearing course (surface)
Verschleißschutzmittel *n* *(aut)* anti-wear additive
Verschleißschutzstoff *m (aut)* anti-wear agent
verschleppen to divert
verschließen to lock up; to seal, to protect
Verschlucken *n* ingestion
Verschluß *m* fastening; gate; trap; seal[ing]; closure; stopper, lock; *(opt)* shutter; *(mil)* breech mechanism, obturator *(gun)*
  V. der Weichenzungen locking of the switch blades
  V. des Wagens fastening of the wagon
  V. für Röntgenröhren tube shutter
  V. mit Gewinde cap
  freifallender V. gravity-operated shutter
Verschlußauslöser *m (opt)* shutter release
Verschlußdeckel *m* lid, cover [lid]
verschlüsseln *(dat)* to [en]code
Verschlüsselung *f* coding
  numerische V. numeric coding
Verschlüsselungsvorrichtung *f* coder
Verschlußkappe *f* screw-cap
Verschlußkupplung *f* locking coupling
Verschlüßler *m* [en]coder
  V. nach dem Subtraktionsprinzip binary weighing encoder
Verschlußmasse *f* sealing compound
Verschlußpfropfen *m* screwed (core-hole) plug
Verschlußplatte *f* closing plate
Verschlußschalter *m* secret (locking) switch
Verschlußschicht *f* seal coat
Verschlußschraube *f* indicator (screw) plug
Verschlußspriegel *m* shutter bow
Verschlußstopfen *m* vent plug
Verschlußstück *n* indicator plug
Verschlußzeiten *fpl (opt)* shutter speeds
verschmelzen to melt, to fuse; to blend, to amalgamate
Verschmelzung *f* [nuclear] fusion; coalescence
Verschmelzungsfrequenz *f* *(opt)* fusion frequency
Verschmelzungsreaktion *f* *(nucl)* fusion [reaction]
Verschmieren *n* **der Schleifscheibe** loading of the grinding wheel
verschmoren to scorch *(contacts)*
verschmutzen to contaminate
Verschmutzung *f* contamination, pollution
verschneiden to blend, to adulterate, to cut
Verschneiden *n* pruning *(tree)*
Verschnitt *m* cut-back
Verschnittbenzin *n* blended petrol (*Am* gasoline); petrol used for blending
Verschnittbitumen *n* cut-back bitumen
Verschnittfähigkeit *f* diluting ratio *(solvent)*
Verschnittmittel *n* blending agent; diluent, adulterant
Verschnürmaschine *f* tying machine
verschoben / in der Phase out of phase
verschollen missing; not to be heard of *(vessel)*
Verschorfung *f* / **elektrische** electrodesiccation
Verschränken *n (telec)* crossed grading
verschrauben to bolt, to screw
verschraubt bolted
Verschraubung *f* bolting [up]; screwing *(pipe)*
verschrotten to scrap
verschulen to transplant *(in a nursery)*
Verschulen *n* transplanting *(in a nursery)*
verschweißbar weldable
Verschweißbarkeit *f* weldability
verschweißen to weld
  unter Einwirkung von Wärme v. *(plast)* to thermoweld

verschwelen to carbonize [under vacuum] at low temperature
Verschwelung *f* low temperature carbonization, distillation under vacuum
verschwenkbar tiltable
verschwenden / Strom to waste watts
verschwerten *(build)* to brace
Verschwertung *f* *(build)* [X-]bracing, diagonal [cross] bracing
verschwimmen to blend into each other
verschwinden to disappear; to vanish
versehen to equip, to furnish, to provide
mit Drosselspulen v. to equip with chokes
mit Einkerbungen v. to serrate, to indent; to notch
mit Gegengewicht v. to counterbalance
verseifbar saponifiable
Verseifbares *n* saponifiable matter
verseifen to saponify
Verseifung *f* saponification
Verseifungszahl *f* saponification number (value)
verseilen to strand, to twist *(cable)*
Verseilmaschine *f* stranding (quadding) machine; closing machine
Verseilung *f* stranding, twisting *(cables)*
versenden to dispatch, to forward, to send
auf dem Wasserwege v. to send by water
Versenden *n* **von Importgütern** forwarding import goods
versenken 1. *(mach)* to counterbore; to countersink; 2. to sink, to let down *(a vessel)*; 3. *(leath)* to lay away
Versenken *n* 1. *(mach)* counterboring; countersinking; 2. sinking [down]
Versenkgrube *f* *(leath)* lay away pit
Versenkmaschine *f* top spindle moulder
versenkt flush [with the surface]
versetzen 1. *(mine)* to bash, to fill, to pack, to stow; 2. *(leath)* to lay away; 3. to offset; to stagger; 4. *(math)* to transpose; 5. to displace *(crystal)*; 6. *(ch)* to add, to mix
gegeneinander v. to stagger
im Verband v. *(build)* to bond
in Schwingung v. to throw into vibrations
Versetzen *n* 1. *(mine)* packing *(by hand)*; stowing *(mechanical)*; 2. addition, adding
versetzt offset, off-centre; filled up *(felt)*
v. angeordnet staggered
gegeneinander v. staggered, alternate
zueinander v. staggered
Versetzung *f* *(ph)* discontinuity; *(met)* dislocation *(crystal)*; offset; *(math)* permutation; *(ch)* adding
beschleunigte V. accelerated dislocation
spiralförmige V. spiral dislocation
stabile V. stable dislocation
wandernde V. moving dislocation
Versetzungsabstand *m* dislocation space
Versetzungsbewegung *f* dislocation motion
Versetzungsdichte *f* dislocation density
Versetzungsebene *f* dislocation plane
Versetzungsenergie *f* energy of dislocation
Versetzungsgeschwindigkeit *f* dislocation velocity
Versetzungslinie *f* dislocation line
Versetzungspotential *n* dislocation potential
Versetzungsspirale *f* dislocation spiral
Versetzungswanderung *f* dislocation migration
Verseuchung *f* *(ch)* contamination, pollution
V. der Luft air-borne contamination
radioaktive V. radioactive contamination
Verseuchungsstoff *m* contaminant, contaminating material
versicherbar insurable
Versicherung *f* insurance
Versicherungsbeitragsmarke *f* insurance stamp
Versicherungsmathematik *f* actuarial theory
versicherungsstatistisch actuarial
versickern to seep
Versickerung *f* percolation, seepage, infiltration
versiegeln to seal
Versiegelungsüberzug *m* sealer coat
versilbern to silver[-coat], to silver-plate
Versilberung *f* silvering, silver-plating, silver-coating
versinken to sink
versorgen to provide, to supply
ein Gebiet v. to serve a district *(radio)*
Versorgung *f* supply
V. aus dem Netz main supply
Versorgungsgebiet *n* supply area *(power station)*
V. eines Senders primary coverage area
Versorgungskabel *n* service-cable
Versorgungskanal *m* service channel
Versorgungsleitung *f* supply pipe (line), utility line, service run
Versorgungsleitungen *fpl* services
Versorgungsnetz *n* [power] supply system
Versorgungsrakete *f* supply rocket
Versorgungsschiff *n* tender, store-carrier, supply-ship, store ship
Versorgungsüberhauen *n* *(mine)* supply raise
versotten to soot
verspachteln to seal [with]
verspannen to deform; to guy *(pole)*
Verspannung *f* rigging; deformation
Verspannungsschaubild *n* deformation diagram
Verspätung *f* *(ry)* train delay
Verspätungsgebühr *f* penalty for delay, deduction for time lost
versperren to block; to obstruct
verspiegeln to silver, to mirror
Verspiegelung *f* mirror coating
Verspiekerung *f* **der Beplankung** *(ship)* nailing of the planks
versponnen spun [to]

verspritzen to spatter, to splash; *(plast)* to injection mould
verspröden to embrittle, to become embrittled
Versprödung *f* embrittlement
versprühen to break up into a fine mist
Verstählung *f* steel plating; steel-facing
Verstandenzeichen *n (telec)* understood sign
Verständigung *f (telec)* readability; audibility
 mangelhafte V. defective conversation
 schlechte V. poor audibility
Verständigungsgüte *f* audibility
verständlich understandable, clear; audible
Verständlichkeit *f* readability; audibility, articulation
 ideale V. ideal articulation
Verständlichkeitsäquivalent *n* articulation equivalent
Verständlichkeitsgrad *m* / **momentaner** immediate appreciation percentage
verstärken 1. to strengthen, to reinforce; 2. *(el)* to amplify, to gain, to boost; 3. *(ch)* to concentrate *(solution)*; 4. to intensify *(photography)*; 5. to augment *(pressure)*; 6. to back *(foil)*; 7. *(mine)* to double up *(support)*
 elektrolytisch v. to increase by electrodeposition
Verstärker *m* amplifier; intensifier; repeater, follower; booster *(pneumatics)*
 V. mit automatischer Lautstärkebegrenzung gain-adjusting amplifier, peak-limiting amplifier
 V. mit Dynamikregler compression amplifier
 V. mit Katodenausgang bootstrap amplifier
 V. mit Phasenumkehrung paraphase amplifier
 V. mit versetzter Abstimmung stagger-tuned amplifier
 abblasender pneumatischer V. bleed-type pneumatic relay
 abgestimmter V. tuned amplifier
 aperiodischer V. aperiodic (wideband) amplifier
 ausgeglichener V. balanced amplifier
 dielektrischer V. dielectric amplifier
 direktgekoppelter V. direct-coupled amplifier
 dynamikbegrenzender V. volume-limiting amplifier
 einstufiger V. single-stage amplifier
 entzerrender V. equalizing (correcting) amplifier
 fotoelektrischer V. photoelectric amplifier
 gegengekoppelter V. feedback amplifier
 lautstärkebegrenzender V. volume-limiting amplifier
 linearisierter V. linear amplifier
 logarithmischer V. logarithmic amplifier
 magnetischer V. transductor, magnetic amplifier, *(Am)* amplistat
 mehrstufiger V. cascade (multi-stage) amplifier
 nichtabblasender pneumatischer V. non-bleed pneumatic relay
 paramagnetischer quantenmechanischer V. paramagnetic quantum-mechanical amplifier
 parametrischer V. parametric amplifier
 platzsparender V. reduced-size amplifier
 pneumatischer V. booster (pneumatic) relay
 pneumatischer V. mit dauerndem Luftverbrauch bleed-type pneumatic relay
 pneumatischer V. ohne dauernden Luftverbrauch non-bleed pneumatic relay
 rauscharmer V. low-noise amplifier
 rotierender V. rotating amplifier
 rückgekoppelter V. feedback amplifier
 schnellansprechender magnetischer V. high-speed magnetic amplifier
 selektiver V. selective amplifier
 symmetrischer V. balanced amplifier
 widerstandsgekoppelter V. resistance coupled amplifier
 zweistufiger V. two-stage amplifier
Verstärkerabschnitt *m (telec)* repeater section
Verstärkerabstand *m (telec)* repeater spacing
Verstärkeramt *n* repeater (amplifying) station
Verstärkeranlage *f* amplifying apparatus
Verstärkerausgang *m* amplifier output
Verstärkerbatterie *f* booster battery
Verstärkerfeld *n* repeater section
Verstärkerfolie *f (nucl)* intensifying screen
Verstärkergestell *n* repeater rack
Verstärkermaschine *f (el)* amplifying motor
Verstärkerpumpe *f* booster pump *(pipeline)*
Verstärkerrauschen *n* amplifier noise
Verstärkerröhre *f* amplifying valve
Verstärkersäule *f* rectifying section *(distillation)*
Verstärkerstation *f* booster station *(pipeline)*
Verstärkerstufe *f* amplification (amplifier) stage
Verstärkertransistor *m* amplifying transistor
Verstärkertriode *f* triode amplifier
Verstärkerwicklung *f* amplifying winding
verstärkt amplified; intensified; reinforced
Verstärkung *f* amplification; intensification; gain; boost[ing]; strengthening, reinforcement
 differentielle V. differential gain
 geradlinige V. flat gain
 hohe V. high-gain amplification
 veränderliche V. variable gain
Verstärkungsabfall *m* fall-off in gain
Verstärkungsbad *n* intensifying bath
Verstärkungsbandbreiteprodukt *n* gain-bandwidth product
Verstärkungsbatterie *f* booster battery
Verstärkungsbereich *m* amplification range
Verstärkungsblech *n* reinforcing plate
Verstärkungsfaktor *m* amplification factor; intensifying (speed) factor; multiplier gain; gain constant

Verstärkungsfeinregler *m* fine gain control
Verstärkungsgestell *n (el)* amplifier rack
Verstärkungskoeffizient *m* amplification (gain) coefficient
Verstärkungskonstanz *f* gain stability
Verstärkungsmessung *f* gain measurement
Verstärkungspfeiler *m* buttress[ing pier], counterfort *(masonry)*
Verstärkungsplatte *f* stiffening plate
Verstärkungspunkt *m* gain point
Verstärkungsrand *m* amplitude (gain) margin
Verstärkungsregelung *f* gain (volume) control
  automatische V. automatic gain control
  sprachgesteuerte V. voice-operated gain adjusting device
  zeitabhängige V. fast time gain control
Verstärkungsregler *m* gain control
  automatischer V. automatic gain control
Verstärkungsrippe *f* strengthening rib
Verstärkungsscheitelfaktor *m* apex factor
Verstärkungsstufe *f* amplification stage
Verstärkungsvorrat *m* amplitude (gain) margin
Verstärkungsziffer *f* amplification factor
Verstäubbarkeit *f* dustability
Verstäuber *m* duster
verstauen to secure; to stow [away] *(cargo)*
versteifen to strengthen, to brace, to stiffen, to reinforce
versteift / durch Rippen ribbed
Versteifung *f* stiffening, bracing, strut
  V. der Schaufel reinforcement (lashing) of the blades *(turbine)*
Versteifungsblech *n* bracing plate
Versteifungsdraht *m* lashing wire *(turbine)*
Versteifungskiel *m (ship)* stiffening keel
Versteifungsplatte *f* stiffening plate
Versteifungsring *m* stiffening ring
Versteifungsrippe *f* stiffening rib
versteilern to steepen
Versteilerung *f* squaring *(pulse)*
Versteinerung *f* mineralization, petrification
Verstellanzeiger *m* pitch-change indicator *(airscrew)*
verstellbar adjustable; variable
Verstellbarkeit *f* adjustability; variability
  V. der Dampfverteilung variability of the steam distribution
Verstellbereich *m* range of adjustment; angle of rotation *(of aerials)*
  V. der Reibahle amount of [reamer] expansion possible
  V. der Zündung *(aut)* timing range
Verstellelement *n* control element
verstellen to adjust; to shift *(lever)*
Verstellen *n* **der Füllung** *(therm)* varying the point of cut-off
  V. der Reitstockspitze setting over the tailstock centre

Verstellgetriebe *n* variable speed gear
Verstellhebel *m* shift lever
  V. mit verstellbarem Vollastanschlag *(aut)* control lever with adjustable full-load stop
Verstellmotor *m* servo-motor, servomotor, brush-shifting motor
Verstellpropeller *m* controllable (variable) pitch-propeller, reversible (feathering screw) propeller
Verstellschieber *m* feed regulator
Verstellspindel *f* adjusting screw
  V. für den Tisch table elevating screw
Verstelltragfläche *f (aero)* variable-incidence wing
Verstelluftschraube *f* controllable-pitch airscrew
  V. mit automatischer Verstellung im Flug automatic-pitch airscrew
Verstellung *f* [re]adjustment, shifting
Verstellweg *m* traverse; path of positioning movement
verstemmen to caulk
versticken to nitride
verstiften to pin
Verstiftung *f* pinned fitting
verstimmen / sich to go off tune
Verstimmen *n* detuning, untuning
verstimmt off-tune
Verstimmung *f* misalignment, mismatch; frequency increment
Verstimmungsschutz *m* off-resonance trip *(telephony)*
verstopfen to blind, to clog, to become clogged, to plug; to choke
Verstopfen *n* plugging, stopping; stoppage; choking [up]; obstructing; blinding
  V. der Roste clinkering (choking up) of the fire bars
  V. mit Spänen choking (jamming) with swarf
Verstopfung *f* obstruction, blockade (*s. a.* Verstopfen)
Verstopfungsanzeiger *m* plugging indicator
Verstoß *m* infringement *(patent)*; offence, breach
verstreben to brace; to strut
Verstrebung *f* bracing; strutting
Verstreckbarkeit *f* stretchability
verstrecken to stretch
Verstrecken *n* **des Kammzugs** *(text)* worsted drawing
Verstreichbürste *f (text)* grinding brush
verstreichen to point; to coal
verstreuen to scatter, to disperse
verstümmeln *(telec)* to mutilate
Verstümmelung *f (telec)* mutilation
Verstümmelungshäufigkeit *f* mutilation rate
verstürzen / Abraum *(mine)* to spoil

Versuch *m* trial, test; attempt; *(ch)* assay, experiment; run
V. im Stand *(ry)* stationary (standing) brake test
V. in Blöcken mit zufälliger Anordnung randomized complete block experiment
V. in unvollständigen Blöcken incomplete block design
V. mit freifahrendem Propeller *(ship)* propeller test in open water
praktischer V. trial
versuchen to test, to try, to attempt
Versuchsanlage *f* pilot plant
Versuchsanleitung *f* experimental direction
Versuchsanordnung *f* experimental arrangement, procedure
Versuchsaufbau *m* 1. setup; 2. *(dat)* breadboard assembly
Versuchsbau *m* experimental building; *(mine)* trial works
versuchsbedingt experimental; necessary for the conditions of the experiment
Versuchsbedingungen *fpl* test conditions
Versuchsbohrung *f* trial boring
Versuchsdauer *f* time of experimentation (test)
Versuchseinheit *f* experimental unit
Versuchsergebnis *n* test result, experimental datum, trial solution (result)
Versuchsfahrt *f* trial run
Versuchsfahrzeug *n* experimental vehicle; experimental vessel
Versuchsfehler *m* experimental error
Versuchsfeld *n* test hall, proving ground
Versuchsfunkstelle *f* experimental radio station
Versuchsgelände *n* test site, experimental area, proving (trial) ground
Versuchsgut *n* experimental farm
Versuchshaus *n* experimental house
Versuchskanal *m (nucl)* experimental hole
Versuchskörper *m* test specimen
V. in Form eines Würfels [cubic] test block
Versuchslast *f* test (proof) load
versuchsmäßig experimental
Versuchsmaßstab *m* experimental scale
Versuchsnorm *f* tentative standard
Versuchsofen *m* assay furnace
Versuchsrakete *f* research rocket
Versuchsreaktor *m* experimental (pilot) reactor
Versuchsreihe *f* series of tests, test series
Versuchssatellit *m* test satellite
Versuchsschiff *n* experimental vessel
Versuchsschleife *f* experimental loop
Versuchsstart *m (rock)* trial (test) launching
Versuchsstreb *m (mine)* experimental face
Versuchsstrecke *f (mine)* experimental (testing) gallery
Versuchstank *m (ship)* model tank (basin), trial tank
Versuchstier *n* experimental animal *(astronautics)*
Versuchswagen *m (ry)* test wagon
versumpft marshy, water-logged
Versumpfung *f* formation of marsh (bog)
versus *(math)* versed
Vertäuanker *m (ship)* mooring anchor
Vertäuboje *f (ship)* cable (mooring, trunk) buoy
Vertäuboot *n* mooring lighter, chain boat
Vertaubung *f (geol)* getting barren
vertäuen to moor
Vertäuklüse *f (ship)* mooring pipe, hawser port
Vertäuklüsenrohr *n s.* Vertäuklüse
Vertäuleine *f (ship)* guess warp, mooring line
Vertäupfahl *m (ship)* mooring post
Vertäupoller *m (ship)* mooring bitts
Vertäuschäkel *m (ship)* mooring (joggle) shackle
vertauschbar *(math)* permutable; interchangeable
Vertauschbarkeit *f (math)* permutability; interchangeability
vertauschen *(math)* to permute, to transpose; to exchange
Leitungen v. to transpose
Vertauschung *f (math)* permutation; commutation; *(el)* transposition
Vertauschungsrelation *f* exchange (commutation) relation *(quantum mechanics)*
Vertäuung *f* **über Bug und Heck** fore-and-aft mooring
Verteerungszahl *f* tar value
verteilen to distribute
Verteiler *m* spreader, distributor; spreading (distributing) machine *(road building)*; distributing frame
kombinierter V. *(telec)* combined distribution frame
Verteilerbohle *f* spreading (spreader) beam
Verteilerdose *f (el)* junction (distributing, conduit, distribution) box
Verteilerfinger *m* distributor rotor arm
Verteilergestell *n (telec)* distributing (distribution) frame
Verteilergetriebe *n (aut)* power take-off gear; auxiliary gearbox
Verteilerkabel *n* distribution cable
Verteilerkanal *m* distribution duct
Verteilerkasten *m (el)* distributing (distribution) box
Verteilerkopf *m (aut)* distributor head
Verteilernetz *n (build)* distribution system
Verteilerpunkt *m (el)* distributing point
Verteilerrohr *n* manifold
Verteilersammelschiene *f (el)* distribution busbar, *(Am)* submain
Verteilersäule *f (el)* distributing pillar
Verteilerschalter *m* distributor
Verteilerschiene *f (el)* distributing bus-bar
Verteilerschloßteil *n* dividing cam *(knitting machine)*

Verteilerschnecke *f* spreading (spreader) screw
Verteilersicherungstafel *f* distribution fuse board
Verteilerstäbe *mpl* distribution (secondary) steel, secondary reinforcement *(concrete)*
Verteilertafel *f (el)* branch switchboard, cross-connecting board
Verteilfernamt *n (telec)* zone (via) centre
verteilt distributed
gleichmäßig v. evenly distributed
Verteilung *f (el)* distribution; distribution panel, branch switchboard; distribution *(statistics)*; *(ch)* dispersion, spread, dispersal, distribution
V. der Belastung spreading the load
V. der Gesamtleistung auf die einzelnen Räder division of the total output amongst the separate wheels *(turbine)*
V. der Grundgesamtheit parent distribution *(statistics)*
asymmetrische V. skew distribution
diskrete V. discontinuous distribution *(statistics)*
räumliche V. spatial (geometric) distribution
regenförmige V. rain-like distribution
schiefe V. skewed (lop-sided) distribution *(statistics)*
schleierförmige V. film distribution
spektrale V. spectral distribution
statistische V. statistical distribution
stetige V. continuous distribution
strahlenförmige V. jet-like distribution
zeitliche V. time distribution
Verteilungsausschalter *m* distribution cutout
Verteilungscharakteristik *f* / **spektrale** spectral-vesponse characteristic
Verteilungschromatografie *f* partition chromatography
V. mit Phasenumkehr reversed-phase partition chromatography
Verteilungsfaktor *m* distribution factor, DF *(radiobiology)*
Verteilungsform *f* form of distribution *(statistics)*
Verteilungsfunktion *f* partition function
Verteilungskabel *n* distribution cable
Verteilungskoeffizient *m* distribution ratio, distribution (partition) coefficient
Verteilungskurve *f* distribution curve
eingipflige V. unimodal curve
zweigipflige V. bimodal curve
Verteilungsleitung *f (el)* distribution feeder
Verteilungsmaß *n* measure of dispersion *(statistics)*
Verteilungspunkt *m (el)* distribution centre
Verteilungsrauschen *n (el)* distribution fluctuations
Verteilungsrohr *n* distributing pipe
Verteilungs[schalt]tafel *f (el)* distribution switchboard (panel)
Verteilungszustand *m (ch)* dispersed state
vertiefen to depress; to hollow out, to recess; to sink *(well)*
Vertiefung *f* depression; impression; hollow, cavity, hole; *(mach)* recess
schwalbenschwanzförmige V. dove-tailing
vertikal vertical, perpendicular
Vertikalablenkplatte *f* vertical deflecting electrode
Vertikalablenkung *f* vertical deflection
Vertikalamplitude *f (telev)* vertical amplitude
Vertikalantenne *f* vertical aerial
V. mit Anzapfspeisung shunt-fed vertical aerial
Vertikalantrieb *m* vertical drive
Vertikalauflösung *f* vertical definition (resolution)
Vertikalaufstieg *m (rock)* vertical ascent
Vertikalbalkengenerator *m (telev)* vertical-bar oscillator
Vertikalbombenmagazin *n* vertical bomb container
Vertikalcoelostat *m* vertical coelostat
Vertikaldiagramm *n* vertical radiation pattern *(of aerials)*
Vertikaldrahtantenne *f* vertical wire aerial
Vertikale *f* vertical, perpendicular line
Vertikalfräseinrichtung *f* vertical head for a horizontal milling machine, vertical milling attachment
Vertikalfrequenz *f (telev)* vertical frequency
Vertikalgatter *n* vertical log frame
Vertikalkamera *f* vertical camera
Vertikalkessel *m* vertical boiler
Vertikallinearität *f* vertical linearity
Vertikalprojektion *f* orthographic projection
Vertikalregler *m* vertical hold
Vertikalruder *n* vertical rudder
Vertikalschieber *m* tool (down-feed) slide *(shaper)*, ram-head slide *(slotter)*
Vertikalstab *m* vertical [member] *(truss)*
Vertikalstart *m* vertical launching
Vertikalturbine *f* turbine with vertical shaft, vertical turbine
Vertikalwinkel *m* vertical angle
Vertikalzweiraumreproduktionskamera *f* two-room vertical process camera
Vertikalzylinderbohrmaschine *f* vertical cylinder boring mill (machine)
Vertrag *m* contract; agreement
vertraglich [stipulated] by contract
verträglich *(ph, math)* compatible; consistent
Verträglichkeit *f (ph, math)* compatibility; consistency
V. von Gleichungen consistency of equations
Verträglichkeitsbedingung *f* compatibility condition
Vertragsbruch *m* breach of contract, default

vertragsbrüchig defaultive, defaulting, breaking the contract
Vertragsforschung *f* industry sponsored research
Vertragshafen *m* treaty port
Vertragsklausel *f* stipulation of a contract
vertragswidrig contrary to the contract
Vertrauen *n* confidence *(statistics)*
Vertrauensbereich *m* confidence interval
Vertrauensgrenze *f* confidence limit
vertraulich confidential *(letter)*
Vertreter *m* 1. representative, agent; 2. substitute
Vertretung *f* representation, agency
Vertrieb *m* sale, distribution, marketing; retail trade
Vertriebskosten *pl* cost of marketing, sales expenses
vertrimmt out of trim
Vertrimmung *f* alteration of trim
vertrocknen to dry up; *(ag)* to wither
verunreinigen to pollute *(water)*; to contaminate *(by contact or mixture)*; to adulterate *(by base admixture)*
Verunreinigung *f* contamination; impurity; pollution *(water)*
  V. der Gewässer durch Öl oil pollution
verursachen to cause
vervielfachen to multiply
Vervielfacher *m* multiplier
  dynamischer V. dynamic multiplier
  Farnsworthscher V. Farnsworth multiplier tube
  fotoelektrischer V. photoelectric multiplier
Vervielfacherfotozelle *f* photo-multiplier cell
Vervielfacherstufe *f* multiplier stage
Vervielfachung *f* multiplication
Vervielfachungsfaktor *m* multiplication factor *(of multiplier-type valve)*
vervielfältigen to duplicate; to copy, to reproduce
Vervielfältigung *f* duplication; copying, reproducing
Vervielfältigungsapparat *m* copying machine
Vervielfältigungspapier *n* duplicating paper
vervierfachen [/ sich] to quadruple
vervollkommnen to improve, to perfect
vervollständigen to complete
verwachsen to intergrow, to grow together; *(min)* to aggregate
Verwachsung *f* intergrowth, overgrowth; *(mine)* aggregation
Verwachsungsfläche *f* composition surface *(of crystals)*
verwackeln to blur *(photography)*
Verwackelung *f* **/ zeitliche** *(el)* time jitter
Verwaltung *f* administration, management; authority; trusteeship
  übergebende V. transferor administration
  übernehmende V. transferee administration
Verwaltungsarbeit *f* executive work
Verwaltungsbarkasse *f* administration launch
Verwaltungsbau *m* administrative building
Verwaltungsbüro *n* administrative office
Verwaltungsgebäude *n* administrative building
Verwaltungskosten *pl* cost of management
Verwaltungspersonal *n* management personnel (staff)
Verwaltungswissenschaft *f* management science
Verwaltungszentrum *n* **/ kommunales** civic centre
verwandelbar transformable; transmutable; convertible
verwandeln to transform, to convert *(e. g. pressure into velocity)*; to change
Verwandlung *f* transformation; conversion
Verwandtschaft *f (ch)* affinity, relation
Verwaschung *f (el)* blur[ring]
Verwaschungsgebiet *n* confusion region *(radar)*
Verweben *n* **von Kett- und Schußfäden** interlacing of warp and weft threads
Verwehung *f* drifting
verweigert refused *(supplies)*
Verweilzeit *f* sojourn (detention, holding, residence, retention) time; hold-up time; lingering period; *(mach)* dwell period; *(aut)* dwell of cam
verwendbar usable, applicable
Verwendbarkeit *f* usability, applicability
  vielseitige V. versatility
verwenden to use; to utilize; to apply
Verwendung *f* use, application, utilization
  praktische V. utilization
Verwendungsmethode *f* method of application
Verwendungsmöglichkeit *f* [possible] use; application; possibility, applicability
Verwendungszweck *m* purpose, application, use
verwerfen to warp; to deform; to distort; to disturb
Verwerfung *f* distortion; *(geol)* [normal] fault; *(ry)* track distortion, warping of the track
  inverse V. *(geol)* thrust fault
  rechtfallende V. *(geol)* hading against the dip fault
Verwerfungsebene *f (geol)* fault plane
Verwertbarkeit *f* utilization
Verwertung *f* **[/ praktische]** utilization, use
Verwertungsanlage *f* utilizing plant
Verwesung *f* putrefaction, decay, rotting
verwickelt intricate
Verwindeprüfung *f* twisting test
Verwindung *f* torsion
verwindungsfrei anti-torsion
Verwindungssteifheit *f* torsional rigidity
verwirbelt turbulent *(fluid flow)*
Verwirbelung *f* turbulence *(fluid flow)*
verwirklichen to realize

Verwirrung *f* **von magnetischen Feldlinien** tangling of magnetic field lines
Verwirrungsgebiet *n (el)* nuisance (mush) area, interference area *(radio)*
verwischen *(telev)* to blur
verwittern to weather
Verwitterung *f* weathering
chemische V. chemical weathering
physikalische V. physical weathering
Verwitterungsprodukt *n* decomposition product
verwölben to warp
Verwölbung *f* warping, distortion
verworfen warped; disturbed; faulted
Verwurfswinkel *m (geol)* [angle of] hade
Verwurfzone *f (geol)* fault space
verzahnen *(build)* to key, to joggle; to cut gears
Verzahner *m* gear cutter
verzahnt *(build)* joggled, keyed; toothed *(gear)*
Verzahnung *f (build)* key[ing]; denticulation; toothing *(gear)*
Verzahnungsfehler *m* gear error
Verzahnungsgesetz *n* law of gearing
Verzahnwerkzeug *n* gear cutting (generating, forming) tool
verzehrbar consumable *(electrode)*
verzehrend / Wärme heat consuming, endothermic, endothermal
verzeichnen *(opt)* to distort
Verzeichnung *f* distortion, faulty delineation
geometrische V. *(telev)* geometric distortion
kissenförmige V. *(telev)* trapezium distortion
perspektivische V. *(opt)* perspective distortion
tonnenförmige V. *(telev)* barrel distortion
verzeichnungsfrei distortionless, free from distortion
Verzeichnungswert *m* distortion value
verzerren to distort
Verzerrer[schaltung *f***]** *m (el)* distortion circuit
Verzerrung *f* distortion
V. durch Ein- und Ausschwingen transient distortion
V. durch Frequenzabweichung deviation distortion
V. von Harmonischen modulation-frequency harmonic distortion
einseitige V. bias (asymmetrical) distortion
harmonische V. harmonic distortion
individuelle V. individual distortion
kubische V. cubic distortion
lineare V. linear distortion
nichtlineare V. harmonic distortion; nonlinear distortion
quadratische V. quadratic distortion
starke V. severe distortion
unregelmäßige V. *(el)* fortuitous distortion
Verzerrungsbereich *m* distortion range
Verzerrungsfaktor *m* distortion factor
verzerrungsfrei non-distorting, distortion-free
Verzerrungsfreiheit *f* absence of distortion
Verzerrungsgrad *m* degree of distortion
zulässiger V. conventional degree of distortion
Verzerrungsmeßgerät *n* distortion analyser
Verzerrungsregelung *f* distortion control
Verzerrungstensor *m* **/ räumlicher** strain tensor
Verzerrungszustand *m* state of strain
verziehen 1. *(ag)* to single; 2. *(text)* to draft; 3. *(mach)* to distort, to warp, to buckle *(work)*
das Band v. to draw the sliver
sich v. to distort, to warp
Verziehen *n* 1. *(ag)* singling; 2. *(text)* drafting; 3. distortion, warpage *(work)*
Verziehung *f (mach)* distortion, warpage
Verzifferung *f* coding
verzimmern *(mine)* to crib
verzinken to galvanize, to coat with zinc
galvanisch v. to electro-galvanize
Verzinken *n* galvanizing, zinc coating, coating with zinc
Verzinkung *f* 1. *(build)* dovetailing; 2. *s.* Verzinken
verzinnen to tin[-plate]
Verzinnen *n* tinning, tin-coating, coating with tin
verzogen warped, distorted *(work)*
Verzögerer *m* restrainer, retarder
verzögern to delay, to retard, to decelerate; to lag; *(ch)* to inhibit, to retard; to quench *(volume control)*
ein Signal v. to delay
verzögert / abhängig inverse time-lag *(relay)*
Verzögerung *f* deceleration, retardation; lag; [time] delay; transfer lag
V. bei der Signalübertragung transmission lag
V. bei Tischumkehr *(mach)* table reversal dwell
V. der Messung measuring lag
V. der Stellgröße output transmission lag
V. des Meßglieds measuring lag
V. zweiter Ordnung *(cont)* quadratic factor
stromunabhängige V. *(el)* constant time lag
thermische V. thermal lag
Verzögerungsbremse *f (ry)* checking brake
Verzögerungsdauer *f (mach)* time of dwell
Verzögerungsdruck *m* retarding (retardation) pressure
Verzögerungsfläche *f* retardation area
Verzögerungsglied *n* delay line, lag element
V. 1. Ordnung *(cont)* first-order system
elektrisches V. electric delay line
magnetostriktives V. magnetostrictive delay line
Verzögerungskabel *n* delay line
Verzögerungskoinzidenzschaltung *f* delay coincidence circuit

Verzögerungskondensator *m* timing capacitor
Verzögerungskreis *m* time-delay circuit
Verzögerungsleitung *f* delay line
V. in einer Lauffeldröhre travelling-wave tube interaction circuit
V. mit Phasenkorrektur equalized delay line
akustische V. sonic delay line
elektrische V. electric delay line
magnetostriktive V. magnetostriction delay line
Verzögerungsmittel *n* retarder; delayer; *(ch)* inhibitor
Verzögerungsnetzwerk *n* lag network
Verzögerungsrelais *n* timing (time-delay, slow-acting) relay
Verzögerungsschalter *m* time-delay switch
Verzögerungsschaltung *f* lag (delay) circuit, lag (delay) network
Verzögerungsschnappschalter *m* snap-action delay switch
Verzögerungsspannung *f* delay voltage
Verzögerungsspule *f* time-delay coil, retardation coil
Verzögerungsventil *n* time-delay valve
Verzögerungswiderstand *m* timing resistance
Verzögerungswinkel *m* angle of lag, lag (retardation) angle
Verzögerungszeit *f* time delay (lag), delay time *(relay)*
Verzögerungszünder *m* delay-action fuse
verzollbar dutiable, liable to duty
verzollen to pay duty on
Verzuckerung *f* saccharification
Verzug *m* 1. hang-over, delay; 2. *(met)* warpage; 3. *(text)* draught; 4. *(mine)* lacing, lagging
Verzugsbrett *n (mine)* cover board
verzugsfrei *(met)* non-warping, warp-proof, non-distorting
Verzugsfreiheit *f (met)* non-distorting properties
Verzugsgröße *f (text)* degree of draft
Verzugshöhe *f s.* Verzugsgröße
Verzugsholz *n (mine)* backing deal
Verzugsplatte *f (mine)* cover board
Verzugswalze *f (text)* drawing roller
Verzugszeit *f* delay time
verzundern to scale
Verzunderung *f* scaling
verzweigen to branch
Verzweiger *m* junction box *(of cables)*
verzweigt branched
Verzweigung *f (el)* network; branch, junction; *(nucl)* multiple disintegration (decay), branching; *(math)* ramification
reflexionsfreie V. matched junction
Verzweigungsanteil *m (nucl)* branching fraction
Verzweigungsbereich *m* cabinet (distribution) district
Verzweigungseinrichtung *f* distribution cabinet
Verzweigungsgesetz *n (el)* law of networks
Verzweigungsglied *n* / **angepaßtes** matched junction *(waveguide)*
Verzweigungskabel *n* distribution (secondary) cable
Verzweigungslinie *f* branch line
Verzweigungsmuffe *f* dividing box
Verzweigungspunkt *m (dat)* break point; *(math)* branch point
Verzweigungssatz *m (el)* law of networks
Verzweigungsstelle *f* branch (takeoff) point *(block diagram)*
Verzweigungsverhältnis *n (nucl)* branching ratio
verzwillingen to twin
Vestibül *n* vestibule, lobby
Vesuvian *m (min)* vesuvianite
Vetiveröl *n* vetiver oil
Viadukt *m* viaduct
Vibration *f* vibration
Vibrations . . . *s. a.* Schwingungs . . .
Vibrationsanhängewalze *f* towed vibratory roller
Vibrationsbeanspruchung *f* vibration stress
Vibrationsdämpfer *m* vibration absorber
Vibrationseinwirkung *f* vibration effect
Vibrationserzeuger *m* vibrating unit
Vibrationsförderer *m* vibrating feeder
Vibrationsfreiheit *f* absence of vibration
Vibrationsgalvanometer *n* vibration galvanometer
Vibrationsglättbohle *f* vibrating screed (beam)
Vibrationsrelais *n* vibration relay
Vibrationssieb *n* screen riddle, impact (vibration) screen
Vibrationsstampfer *m* vibrating (vibratory) tamper
Vibrationstachometer *n* tuned-reed indicator
Vibrationstest *m* vibration test
Vibrationsverdichtung *f* dynamic compaction
Vibrationsversuch *m* vibration test
Vibrationswalze *f* vibrating (vibratory) roller
Vibrator *m* vibrator
elektromagnetischer V. electromagnetic vibrator
Vibratorrinne *f* vibrator conveyor
Vibratorsiebmaschine *f* high frequency screen
Vibrierbohle *f* vibrating beam (screed)
Vibrierbohlenfertiger *m* vibrating beam finisher
vibrieren to vibrate
Vibriertisch *m* vibratory (vibrating) table
Vibrobohren *n* rotary-percussive drilling
Vibromassagegerät *n* vibro-massage machine
Vibrosichter *m* vibrating-screen type deduster, vibro-deduster
Vicatapparat *m* Vicat needle (apparatus) *(testing)*
Vicatnadel *f s.* Vicatapparat
Vicatnadelgerät *n s.* Vicatapparat

vicinal vicinal
Vickershärte *f* Vickers [diamond pyramid] hardness
Vickershärteprüfung *f* Vickers method of hardness testing
Victory-Schiff *n (Am)* victory ship *(cargo-type vessel)*
Videoband *n* video band
Videodetektor *m* video detector
Videofrequenz *f* video frequency
Videogleichrichtung *f* video detection
Videokarte *f* video map *(radar)*
Videokartendarstellungsgerät *n* video mapping unit *(radar)*
Videosignal *n* video signal
Videoverstärker *m* video amplifier
Videovorverstärker *m* **in der Fernsehkamera** head amplifier
Vidikon *n* vidicon *(camera tube)*
Vidikonaufnahmeröhre *f (telev)* vidicon pick-up tube
Vidikonröhre *f* vidicon tube
Viehbestand *m* live stock
Viehdampfer *m* cattle steamer
Viehfutter *n* forage, fodder, stock food
Viehhaltung *f* live-stock farming
Viehhof *m* abattoir, slaughter-house
Viehladung *f (ship)* live-stock
Viehsalz *n* cattle lick
Viehtransportschiff *n* cattle ship (vessel)
Viehwaage *f* weigh-bridge
Viehwagen *m* cattle wagon, *(Am)* stock car
Viehweide *f* pasture
Viehzucht *f* animal breeding, live-stock breeding (farming)
Viehzüchter *m* stock raiser
vieladrig multicore, multiwire
vielatomig polyatomic
vieldeutig ambiguous
Vieldeutigkeit *f* ambiguity
Vieleck *n* polygon
Vieleckausbau *m (mine)* cockering
vieleckig polygonal
Vieletagenheißpresse *f* multi-platen hot press *(papermaking)*
vielfach multiple
Vielfachabstimmvorrichtung *f* multiple tuner
Vielfachabtastung *f* multiple scanning
Vielfachantenne *f* multiple aerial
Vielfachbeschleuniger *m (nucl)* multiple accelerator
Vielfachbeschleunigung *f* multiple acceleration
Vielfachbohrarbeit *f* multi-drilling operation
Vielfachdiagramm *n* multiple diagram
Vielfachecho *n* multiple echo
Vielfachempfang *m* multiple reception
Vielfaches *n* multiple
  ganzes V. integral multiple
  kleinstes gemeinsames V. least common multiple
Vielfachfunkenstrecke *f* multigap discharger (arrester)
Vielfachinstrument *n* multirange instrument
Vielfachisolatorenkette *f* multiple insulator chain
Vielfachkabel *n* multiple cable
Vielfachkette *f s.* Vielfachisolatorenkette
Vielfachklinke *f (telec)* multiple (multi-contact) jack
Vielfachkoinzidenz *f* multiple coincidence
Vielfachleiter *m* multiconductor
Vielfachmagnetron *n* cavity magnetron
Vielfachmeißelanordnung *f* multi-tooling
Vielfachmeßbrücke *f* multifunction bridge
Vielfachprüfgerät *n* multitester
Vielfachpunktschweißmaschine *f* multiple-spot welding machine
Vielfachschaltung *f* multiple connection
  gerade V. straight multiple
  gestaffelte V. partial multiple
Vielfachspaltung *f (nucl)* spallation
Vielfachsteuerung *f* coupled running
Vielfachstreuung *f* multiple scattering
Vielfachtonspur *f* multiple sound track
Vielfachübertragung *f* multiple transmission
Vielfachumschalter *m* multiple switch
Vielfachunterbrechungsklinke *f (telec)* series-multiple jack
Vielfachuntersetzer *m (nucl)* multiscaler
Vielfachzug *m (met)* multiple drawing
Vielflach *n* polyhedron
vielflächig polyhedral
vielgestaltig multiform, polymorph
Vielgestaltigkeit *f* multiformity, polymorphism
Vielkanalanalysator *m* multichannel analyzer
Vielkanalimpulshöhenanalysator *m* multichannel pulse height analyzer
Vielkanalkoinzidenzschaltung *f* multichannel coincidence system
Vielkant *m* polyhedral
vielkantig polygonal
Vielkeilprofil *n* splines, multiple keys
Vielkeilverzahnung *f* multiple splining
Vielkeilwelle *f* multiple splined shaft
Vielkeilwellenräumen *n* multi-spline broaching
vielkernig polynuclear, polycyclic
Vielkörperproblem *n* many-body problem
Vielkristall *m* polycrystal
Viellinienspektrum *n* many-line spectrum, discrete-band spectrum
Vielmeißeldrehen *n* multi-tool turning
Vielmeißeldrehmaschine *f* multi-tool lathe
Vielmeißelhalter *m* multi-tool post (holder)
Vielnadelmaschine *f (text)* multi-needle machine
Vielnutnabe *f* multiple spline
vielpaarig multipair[ed]

vielphasig multiphase, polyphase
vielpolig multipolar
Vielröhrenkessel *m* multitubular boiler
stehender V. Cochran patent vertical multitubular boiler
Vielschlitzmagnetron *n* multi-segment magnetron
Vielschnittdrehmaschine *f* multi-cut lathe, multiple-tool lathe
vielseitig versatile; *(math)* multilateral
Vielseitigkeit *f (mach)* versatility
vielspindlig multi-spindle
Vielstoffgemisch *n* multi-component mixture
Vielstofflegierung *f* complex alloy
Vielstufenbleiche *f* multistage bleaching
Vielstufenwärmeaustauscher *m* multiphase heat exchanger
vielteilig multipartite
Vielwalzengerüst *n* cluster mill (roll stand)
Vielzweckmaschine *f* multi-purpose machine, general-purpose machine
Vielzweckschiff *n* all-round vessel
vierachsig four-axial
Vieradressenbefehl *m (dat)* four-address instruction
Vieradressenbetrieb *m (dat)* four-address operation
Vierartenrechner *m* four-rules calculator
vieratomig tetratomic
Vierbackenfutter *n* four-jaw chuck
vierbasisch quadribasic, tetrabasic, tetrahydric *(alcohol)*
Vierbogenmesserfalzmaschine *f* quadruple knife folding machine
Vierdecker *m* quadruplane
Vierdeckschiff *n* four-decked vessel
Vierdrahtgabel *f* four-wire terminating set
Vierdrahtleitung *f* / **symmetrische** four-wire balanced line
Vierdrahtverstärker *m* four-wire repeater
Viereck *n* quadrilateral, quadrangle
viereckig quadrilateral, quadrangular
Vierendeelfachwerk *n* Vierendeel truss
Vierendeelstütze *f* Vierendeel column
Vierendeelträger *m* Vierendeel girder
Vierer *m* 1. *(el)* four-wire unit, four-core cable, phantom circuit; 2. four[-oared shell] *(sporting boat)*
Viereranordnung *f* duplexing
Viererbelastung *f* phantom loading
Viererbündel *n* quad bunch conductor *(overhead lines)*
Viererbündelleitung *f* quad line
Viererkabel *n* quad (quadded, multiple-twin) cable
Viererkapazität *f* side-to-side capacity
Viererkreis *m* phantom circuit
V. mit Erdrückführung earth-phantom circuit
Viererleitung *f* phantom circuit
Viererpupinisierung *f* phantom loading
Viererschaltung *f* phantom circuit
Viererspule *f* superimposed circuit coil
Vierervektor *m* four-vector
vierfach fourfold, quadruple
v. halogeniert tetrahalogenated
v. substituiert tetrasubstituted
Vierfachbindung *f (ch)* quadruple bond
Vierfachkette *f* quadruple-strand chain
Vierfaktorformel *f* four-factor formula
Vierfarbendruck *m* four-colour printing
Vierfederspindel *f* **mit Ansatz** four spring peg with shoulder
Vierflächner *m* tetrahedron
Vierganggetriebe *n* four-speed drive
viergängig quadruple-threaded
Viergelenkgetriebe *n* four-bar linkage
Viergespann *n (mine)* four-piece timber set
viergleisig quadruple tracked
viergliedrig *(ch)* four-membered
vierhübig four-throw *(crankshaft)*
Vierkammerzähler *m* liquid-sealed drum-type gas meter
Vierkant *m* square
Vierkantdurchschlag *m* square punch
Vierkantdurchschläger *m s.* Vierkantdurchschlag
Vierkantfeile *f* square file
Vierkantkaliber *n (met)* square groove
Vierkantkopf *m* square head
Vierkantkopfschraube *f* square-head bolt
Vierkantlänge *f* length of square
Vierkantmeißel *m* square insert tool
Vierkantmesser *n* square blade *(milling cutter)*
Vierkantmutter *f* square nut
Vierkantrevolverkopf *m* square tool post
Vierkantschaft *m* squared shank, shank with a square cross section
verjüngter V. squared taper shank, taper shank with a square cross section
Vierkantschraube *f* square-head bolt
Vierkantstab *m* square-section specimen
Vierkantstahl *m* square [bar]
Vierkantstahlfeder *f* rectangular wire spring
Vierkantwalzkaliber *n* diamond-pass of rolls
Vierkursfunkfeuerempfänger *m (aero)* radio range receiver
Vierlingsflak *f* four-barreled antiaircraft gun
Vierlingsturm *m (ship)* quadruple turret
Viermastbark *f (ship)* four-mast[ed] bark
Viermastschiff *n* four-mast[ed] ship
Viermomentengleichung *f* four-moment theorem, four moments theorem
viernutig four-fluted
Vierparameterverfahren *n* four-parameter method *(determination of recovery voltage)*
Vier-Pi-Zähler *m (nucl)* 4π-counter, four pi counter

Vierplattenstoß *m* junction of four plates
Vierpol *m* four-pole network, four-terminal network, quadripole
  passiver V. passive four-terminal network
  symmetrischer V. balanced quadripole
Vierpoldämpfung *f* image attenuation coefficient
Vierpolgleichung *f* four-terminal network equation
vierpolig quadripolar
Vierpolkreuzglied *n* bridge network
Vierpolphasenfaktor *m* image phase factor
Vierpolröhre *f* tetrode
Vierpolübertragungsmaß *n* image transfer coefficient
Vierpolwinkelmaß *n* image phase[-change] coefficient
Vierpunktaufhängung *f* four-point suspension
Vierquadrantmultiplizierer *m* four-quadrant multiplier
Vierradantrieb *m* four-wheel drive
Vierradbremse *f* four-wheel brake
vierrädrig four-wheel[ed]
Vierring *m (ch)* four-membered ring
Vierringverbindung *f* tetracyclic compound
Vierschneidenbohrer *m* four-flute drill
Vierschneider *m* four-flute end mill; *(Am)* four-lip end mill
Vierschornsteinschiff *n* four-stacker
Vierschraubenschiff *n* quadruple screw ship
vierseitig quadrilateral; square-based *(pyramid)*
Vierspeziesrechner *m* four-rules calculator
Vierspindelautomat *m* four-spindle automatic
Vierspindelbohrmaschine *f* four-spindle drilling machine
Vierspindeldrehhalbautomat *m* four-spindle semi-automatic
Vierspindelfutterautomat *m* four-spindle automatic chucking machine
Vierspindelstangenautomat *m* four-spindle bar automatic
vierspindlig four-spindle
vierstellig four-place, four-figure
Vierstofflegierung *f* quarternary alloy
Vierstufenrakete *f* four-stage rocket, four-step rocket
vierstufig four-step
Viertakter *m* four-stroke engine
Viertaktlauf *m* four-stroke cycle
Viertaktmotor *m* four-stroke engine
Viertaktverfahren *n* four-stroke cycle
vierteilig quadripartite; four-component
Vierteladdierer *m* quarter adder
Vierteldrehung *f* quarter-turn
Viertelfeder *f* quarter-elliptic spring
Viertelflächner *m* tetartohedral crystal
Viertelkreis *m* quarter of a circle
Viertelkreisbogen *m* quadrant
Viertelkreisfräser *m* / **nach innen gewölbter** single-corner rounding cutter
Viertelquadratmultiplizierer *m* quarter-square multiplier
Viertelspunkt *m* quarter point
Viertelwelle *f* quarter-wave
Viertelwellenantenne *f* quarter-wave aerial
Viertelwellenlängenplättchen *n s.* Viertelwellenplatte
Viertelwellenlängenplatte *f s.* Viertelwellenplatte
Viertelwellenplättchen *n s.* Viertelwellenplatte
Viertelwellenplatte *f* quarter-wave[length] plate
vierundzwanzigflächig icositetrahedral
Vierundzwanzigflächner *m* twenty-four-sided solid, icositetrahedron
Vierwalzengerüst *n* four-high rolling stand
Vierwegebohrmaschine *f* four-way drilling machine
Vierwegehahn *m* four-way tap (cock)
Vierwegeschieber *m* four-way valve
Vierwegestück *n* four-way pipe
vierwertig quadrivalent, tetravalent
Vierwertigkeit *f* quadrivalence, tetravalence
vierzählig four-fold
Vierzehneck *n* tetradecagon
Vierzehnstiftsockel *m* diheptal base
Vierzylindermotor *m* four-cylinder engine
Vignette *f* stock block, vignette, floret
Vignettierung *f* vignetting *(photography)*
Vigognegarn *n* vigogne
Vigoureuxdruck *m* Vigoureux (Vigoreaux) printing
Villardschaltung *f* Villard rectifying circuit
Vinylalkohol *m* vinyl alcohol
Vinylazetat *n* vinyl acetate
Vinylchlorid *n* vinyl chloride
Vinyldispersion *f* vinyl dispersion
Vinylgruppe *f* vinyl group
Vinylharz *n* vinyl resin
Vinylharzkunststoff *m* vinyl plastic
Vinylidenchloridharz *n* vinylidene chloride resin
Vinylidenverbindung *f* vinylidene compound
Vinylierung *f* vinylation
Vinylkunststoff *m* vinyl plastic
Vinylmassen *fpl* / **weichgestellte** vinyl elastomeric compounds
Vinylverbindung *f* vinyl compound
Violettverschiebung *f* violet shift *(spectral lines)*
Virialkoeffizient *m* virial coefficient
Virialsatz *m* virial law
virtuell virtual *(image)*
Visiereinrichtung *f* sighting arrangement (device), sights *(rifle)*
Visierfernrohr *n* sighting telescope
Visierlinie *f* line of sight
Visierschieber *m* backsight (rear-sight) slide
Visierwinkel *m* angle of elevation (sight)

viskoelastisch visco-elastic
Viskogramm *n* viscosity (blending) chart
Viskometer *n* viscometer
viskoplastisch visco-plastic
viskos viscous
Viskose *f* viscose
Viskosefaser *f* viscose staple fibre
Viskosefaserstoff *m* viscose rayon
Viskoseschwamm *m* viscose sponge
Viskoseseide *f* viscose rayon
 mattierte V. dull rayon
 pigmentierte V. pigment rayon
Viskoseverfahren *n* viscose process
Viskosezellstoff *m* dissolving pulp
Viskosimeter *n* visco[si]meter
Viskosität *f* viscosity
Viskositätsgrad *m* degree of viscosity
Viskositätsindex *m* viscosity index
Viskositätsmesser *m* viscosimeter
Vitalfärbung *f* vital staining
Vitalfluorochromierung *f* intra-vitam fluorescent dye staining
Vitamin *n* vitamine
Vitaminmangel *m* hypovitaminosis
Vitriol *n* vitriol, copperas, iron sulphate
Vitriolöl *n* oil of vitriol
Vitrit *m (min)* vitrain
vitro / in in vitro
Vivianit *m (min)* vivianite
vivo / in in vivo
Vlies *n (text)* fleece
Vlieskrempel *f* second breaker
Vliesstoff *m* fleece, non-woven material
Vliesstoffindustrie *f* non-woven fabric industry
Vliestechnik *f* fleece technique
Vliestrommel *f* lap drum
Vlieswolle *f* fleece wool
Vogelzungenfeile *f* crossing file
Voigt-Effekt *m* Voigt effect
Voith-Schneider-Propeller *m* Voith-Schneider propeller
Volant *m* steering wheel
Volkseigener Betrieb *m* nationally-owned factory (undertaking), state-owned factory (undertaking)
Volksgut *n* nationally-owned farm
voll full, filled; solid *(material)*; compact *(hide)*
Volladdierer *m (dat)* full adder *(binary)*
Vollager *n* solid bearing
Vollanode *f* heavy (solid) anode
Vollast *f* full load
Vollasterregung *f* full-load excitation
Vollastfördermenge *f (aut)* fuel supply at full load
Vollastnadel *f (aut)* full-load needle
Vollastspannung *f* running voltage
Vollausschlag *m* maximum (full-scale) deflection
Vollausschlagbereich *m* full-scale range
Vollautomat *m* full automatic, fully-automatic machine
vollautomatisch fully automatic
Vollautomatisierung *f* full-scale automation
Vollbalken *m* solid beam
Vollbelastung *f* full load
Vollbelegung *f (telec)* occupancy
Vollbestrahlung *f* total body irradiation
Vollbohrrollenmeißel *m* tricone bit
Vollbordschiff *n* vessel of full scantlings
Vollbremsung *f (ry)* full brake application
Vollbruchlast *f* full rupturing load
Volldampf *m* full steam
Volldecke *f* solid ceiling
Volldeckschiff *n* full scantling vessel
Volldisjunktion *f (cont)* maxterm
Volldraht *m* solid wire
Volldruckdampfmaschine *f* steam engine working without expansion
Volldrucklinie *f* steam admission line
Volldruckperiode *f* period of admission
Volldünger *m* compound fertilizer
Volleinschlag *m (aut)* steering lock
Volleistung *f* full load
Volleistungshöhe *f (aero)* full throttle altitude
vollelektrisch all-electric
vollelektronisch all-electronic, full electronic
Vollen / aus dem from the solid
vollenden to complete
Vollerntemaschine *f* complete harvester
vollfett *(print)* full face
Vollfläche *f* critical area of extraction
Vollflächner *m* holohedral crystal, holohedron
Vollform *f* male form
Vollgas *n* full throttle (load)
Vollgatter *n* multiple blade sawframe (frame saw)
Vollgummibereifung *f* solid rubber tyres
Vollgummireifen *m* solid rubber tyre
Vollhartguß *m* white cast iron
Vollhartmetallbohrer *m* solid carbide drill
Vollhartmetallschneidenkopf *m* solid carbide tip
Vollhartmetallspiralbohrer *m* solid carbide twist drill
Vollhartmetallwerkzeug *n* solid carbide tool
Vollhobelmeißel *m* one-piece planer tool
Vollholz *n* solid wood
Vollholzscheibe *f* solid-wood pulley
Vollhub *m* full lift
vollhydraulisch all-hydraulic
völlig full *(vessel)*
Völligkeit *f* fineness, fullness, stumpiness *(of a vessel)*
 V. des Diagramms efficiency of cycle *(steam)*
Völligkeitsgrad *m (ship)* coefficient of fineness
 V. der Konstruktionswasserlinie waterplane coefficient

V. der Spantflächenskala longitudinal (prismatical) coefficient
V. der Verdrängung block coefficient
V. des Hauptspantes midship [section] coefficient, coefficient of fineness of midship section
V. in der Wasserlinie waterplane (waterline) coefficient
Vollinie *f* full (continuous, unbroken) line
dicke V. heavy (thick) full line
dünne V. light (thin) full line
Vollkaskoversicherung *f* full comprehensive insurance
Vollkegel *m* external taper
Vollkernisolator *m* solid insulator
Vollkernstützer *m* solid supporting insulator
vollkommen complete
Vollkonjunktion *f (cont)* minterm
Vollkorn *n (mil)* full sight
Vollkornbrot *n* whole-meal rye bread
Vollkreislicht *n* all-round light *(navigation)*
Vollmantelzentrifuge *f* bowl centrifuge
Vollmaß *n* full size
Vollmaßbohrer *m* size drill
Vollmatrose *m* able [bodied] seaman
Vollmilch *f* unskimmed milk
vollmotorisiert fully mechanized
Vollnickelgalvano *n* all nickel electric shell
Vollniet *m* solid rivet
Volloperation *f (dat)* complete operation
Vollpipette *f* transfer (volumetric) pipette
Vollprägen *n* full coining
Vollreife *f* full ripeness
Vollsäge *f* gang saw
Vollschmierung *f* hydrodynamic lubrication
Vollschnitt *m* full section *(technical drawing)*
Vollschrotausbau *m (mine)* close cribbing
Vollschrotzimmerung *f (mine)* close cribbing
Vollschweberuder *n* spade rudder
Vollschwingaufhängung *f (aut)* fully-floating suspension
Vollseil *n* solid wire-rope *(overhead lines)*
vollselbsttätig fully automatic
Vollsenkung *f* complete subsidence
Vollsichtkanzel *f* glas-walled cab
Vollsichtskale *f* full-vision scale
Vollsichtverglasung *f (aut)* wrap-around windshield
Vollspant *n* solid frame
vollständig complete
Vollstein *m* solid brick
Vollträn kverfahren *n* full-cell process
volltransistorisiert completely transistorized, all-transistor
Volltreffer *m* direct hit
Volltrommelmotor *m* smooth-core motor
Vollturbine *f* full admission turbine
Vollumrißfräseinrichtung *f* 360° profiling mechanism
Vollumrißfräsen *n* 360° profile milling
Vollumrißkopieren *n* 360° profiling, 360° peripheral copying
Vollvermittlungsstelle *f* main exchange *(telephony)*
Vollversatz *m (mine)* solid stowing
Vollwagen *m (mine)* full (loaded) tub
Vollwagenzug *m (mine)* full journey
Vollwählbetrieb *m* full-dial service
Vollwandbinder *m* plate (solid web) girder
Vollwandbogen *m* solid-rib arch
Vollwandbrücke *f* box-section bridge
vollwandig solid-webbed
Vollwandträger *m* plate (solid web) girder
Vollweggleichrichter *m* full-wave rectifier
Vollwegkristalldetektor *m* full-wave crystal detector
Vollweide *f* day pasture
Vollwelle *f* solid shaft
Vollwellengleichrichtung *f* full-wave rectification
Vollwinkel *m* perigon
Vollziegel *m* solid brick
Vollzug *m (mine)* full journey
Volt *n* volt
Voltaelement *n* two-fluid cell
voltaisch voltaic
Voltameter *n* voltameter, coulo[mb]meter
Voltampere *n* volt-ampere
Voltamperemeter *n* voltammeter
Voltamperesekunde *f* volt-ampere-second
Voltamperestunde *f* volt-ampere-hour
Voltaspannung *f* voltaic electricity
Voltmeter *n* voltmeter
hochohmiges V. high-impedance voltmeter
Voltmeteranzeige *f* voltmeter indication
Voltmeterumschalter *m* voltmeter switch
Voltolgleitöl *n* voltol sliding oil
Volumausdehnung *f* volume expansion
Volumdehnung *f* volume strain
Volumeinheit *f* volume unit
Volumelastizität *f* volume elasticity
Volumen *n* volume
V. der Masseeinheit volume of the unit of weight
V. der verdrängten Flüssigkeit volume of the displaced liquid
V. des zerspanten Werkstoffes volume of material removed
berechnetes V. calculated volume
empfindliches V. sensitive volume
schädliches V. dead space
strahlenempfindliches V. sensitive volume (region) *(radiobiology)*
wirkliches V. real volume
zerspantes V. [metal] removal rate
Volumenabnahme *f* decrease of volume

Volumenänderung *f* variation (change) of volume; dilatational (volume) strain
Volumendiagramm *n* cylinder (piston position time, volume) diagram
Volumendiffusion *f* volume diffusion
Volumendosierung *f* volumetric batching
Volumendosis *f* volume dose
Volumeneffekt *m* volume effect
Volumeneinheit *f* unit volume
Volumenelastizitätsmodul *m* bulk modulus of elasticity
Volumenenergie *f* volume energy, energy of volume
Volumenfläche *f* characteristic surface
Volumenfotoeffekt *m* volume photoeffect (photoelectric effect)
Volumenionisierung *f* volume ionization
Volumenkontraktion *f* contraction in volume
Volumenladung *f* volume charge
Volumenleitfähigkeit *f* bulk conductivity
Volumenleitung *f* volume conduction
Volumenmesser *m* electrical speech level meter *(telephony)*
Volumenprozent *n* percentage by volume
Volumenrekombination *f* volume recombination
Volumenverringerung *f* volumetric reduction
Volumenverschiebung *f* volume displacement
Volumenverunreinigung *f* volume impurity
Volumenvoltameter *n* volume voltameter
Volumenzähler *m* volumetric displacement flowmeter
Volumenzunahme *f* increase in (of) volume
Volumetrie *f* volumetric analysis
volumetrisch volumetric
voluminös voluminous, bulk
Volumkraft *f* volumetric force
Volumverlust *m* loss in volume
Vorabscheider *m* prefractionator, presettling tank
Vorabsenkung *f* initial convergence
Vorabsiebung *f* scalping
Voralarm *m* fore-warning
voraltern to pre-age
Voralterung *f* pre-aging
Voramt *n (telec)* control station
vorankommen to make progress
Voranschlag *m* estimate
Voranstrich *m* primary (prime, priming) coat, priming
Vorarbeit *f* preliminary work
Vorarbeiter *m* foreman
Vorarbeiterin *f* fore-woman
Vorätzung *f* first etching
Vorauflaufspritzung *f (ag)* pre-emergence spray
Vorausanzeige *f* ahead (heading) indication *(radar)*
vorausbestimmt predetermined
vorausplanen to plan ahead
Vorausplanung *f* advance planning
Voraussetzung *f* assumption, supposition, condition
  notwendige V. prerequisite
Vorausströmung *f* exhaust lead, inside lap
Vorauszahlung *f* payment in advance; prepayment
Vorbad *n* preliminary bath *(film)*
vorbauen / frei to cantilever *(bridge construction)*
Vorbaureck *n (mine)* track extension unit
vorbearbeiten / spanend to rough[-machine]
Vorbearbeitung *f* prior machining operation[s]
  spanende V. roughing
Vorbedingung *f* preliminary condition, precondition; prerequisite
vorbehandeln to pretreat
Vorbehandlung *f* pretreatment, preconditioning
vorbeifahren to pass
vorbeifliegen to fly past
vorbeifließen to flow past
vorbeiströmen to by-pass, to flow past
Vorbeize *f* bottom mordant
Vorbeizen *n (Am)* bottoming
vorbelasten to pre-load
Vorbelastungswiderstand *m* bleeder
Vorbemessung *f s.* Vorberechnung
Vorberechnung *f* preliminary computation
vorbereiten to prepare
  einen elektrischen Kreis v. to arm a circuit
Vorbereitung *f* preparation
  V. des Kabelendes dressing of the cable
Vorbereitungsarbeit *f* preliminary operation
Vorbereitungszeichen *n* prefix signal
Vorbereitungszeit *f* set-up time, start-up time
Vorbeschleunigung *f (nucl)* preacceleration
Vorbestellung *f* advance order, prior booking *(hotel)*
Vorbestrahlung *f* pre-irradiation
Vorbestrahlungsanalyse *f* pre-bombardment analysis
Vorbestrahlungswert *m* pre-irradiation value
vorbeugend prophylactic; preventive
Vorbeugungsmittel *n* preservative; preventive
Vorbildung *f* qualification, [level of] education
Vorblasen *n* fore-blow
Vorblech *n* sheet bar
Vorblock *m* cogged ingot
vorblocken to cog, to bloom
Vorblütenspritzung *f* pre-blossom spray, winter washing
  erste V. green-cluster spray, bud-burst spray, mouse-ear spray
  zweite V. pink-bud spray
vorbohren to predrill
Vorbohren *n* preliminary drilling, predrilling
  stufenweises V. step drilling
Vorbohrloch *n (mine)* advance borehole, pilot hole

Vorbohrung *f* predrilled (starting, previously drilled) hole
V. zur Führung des Werkzeugs pilot hole
Vorbramme *f* roughed slab
Vorbrecher *m* preliminary breaker, primary crusher
Vorbrennkammer *f* precombustion chamber
vorbringen *(mach)* to advance
Vordach *n* canopy, portico
Vordeck *n (ship)* foredeck
Vorderachse *f* front axle
Vorderachsgabel *f* front-axle fork
Vorderachskörper *m* front-axle beam
Vorderachsschenkel *m* stub axle
Vorderachsstrebe *f* front-axle bracing rod
Vorderachswelle *f* front-axle shaft
Vorderachszapfen *m* king pin, steering-knuckle pivot
Vorderansicht *f* front view (elevation)
Vorderblatt *n* vamp *(shoe)*
Vorderdeck *n (ship)* forecastle, foredeck
Vorderfeder *f (aut)* front spring
Vorderfederbock *m* front spring hanger
Vorderfederstütze *f* front spring support
Vorderfläche *f* frontal area, [front] face; drag surface *(of a propeller)*
Vorderflanke *f* forward (leading) edge *(of a pulse)*
Vorderfront *f* facade
Vordergeschirr *n (text)* pressure harness
Vorderglied *n* antecedent
Vordergrund *m* foreground
Vorderholm *m (aero)* front spar
Vorderkammerlinse *f* anterior chamber lens
Vorderkante *f* front edge; leading edge *(wing)*
Vorderkappe *f* **/ spitze** pointed toe *(shoe)*
Vorderkipper *m (aut)* front tipper; *(ry)* end tipping wagon
Vorderlader *m* muzzle-loader
vorderlastig *s.* kopflastig
Vorderliek *n (ship)* luff
Vorderplicht *f (ship)* forward cockpit
Vorderrad *n* front wheel
Vorderradantrieb *m* front-wheel drive, front-axle drive
Vorderradbremse *f* front-wheel brake
Vorderradlenkzapfen *m* steering-knuckle pin
Vorderradsatz *m (ry)* front wheel set
Vorderradspreizung *f (aut)* king-pin inclination
Vorderseil *n* pull rope *(slusher)*
Vorderseite *f* 1. front, face; 2. *(text)* face *(fabric)*; 3. obverse *(coin)*
Vordersitz *m* front seat
Vorderspant *n (ship)* fore-frame
Vordersteven *m (ship)* stem
Vorderteil *n (ship)* prow, bow, head, nose
Vordertreppe *f* front stairs
Vorderwand *f* front panel
Vorderwandrahmen *m* front-panel frame
Vorderwerk *n (text)* pressure harness
Vordock *n (ship)* basin, entrance dock
vordringen to propagate *(crack)*
Vordrossel *f (el)* series inductance; fixed nozzle *(pneumatics)*
Vordruck *m* form; *(text)* bottom print
Vordruckbuntreserve *f (text)* coloured pre-print resist
vordrücken to pre-compress
Vordruckreserve *f (text)* pre-printed resist
Vordruckwalze *f* dandy roll
Vordüse *f (aut)* pilot jet
voreilen to advance; to lead *(phase)*
Voreilen *n* advance; lead[ing]
V. beim Walzprozeß forward slipping
V. der Spule *(text)* bobbin lead
V. für den Austritt linear exhaust lead
V. für den Eintritt linear admission lead
äußeres lineares V. linear admission lead
inneres lineares V. linear exhaust lead
lineares V. linear lead
Voreilung *f* advance; [phase] lead
zeitliche V. time lead
Voreil[ungs]winkel *m* angle of lead (advance), lead angle
Voreinflugzeichen *n* outer marker
Voreinspritzdüse *f* pilot jet
voreinstellen to preset
Voreinstellung *f* presetting
V. des Arbeitsablaufs cycle setting
Voreinströmung *f* admission lead, outside lap
Voreintritt *m* admission lead, outside lap
Voreinweisung *f* preadjustment of homing device
Vorentschwefelung *f* primary desulphurization
Vorentwurf *m* preliminary design
Vorentzerrung *f (el)* de-emphasis
Vorentzerrungsschaltung *f* de-emphasis network
Vorerhitzer *m* preheater
Vorerhitzung *f* preheating
Voreröffnen *n* linear lead
Vorerzeugnis *n* semi-finished product
vorevakuieren to pre-evacuate
Vorfabrikation *f* prefabrication
Vorfahrt *f* right of way
Vorfahrtstraße *f* through street, major road
Vorfeinpolieren *n* fine polishing
Vorfeld *n* apron *(airport)*
vorfertigen / Bewehrungskörbe to pre-tie reinforcement cages
vorfertigschleifen to semi-finish grind
Vorfertigung *f* fabrication *(structural steelwork)*; preforming *(reinforcement)*; precasting *(concrete)*; prefabrication, pre-finishing
V. im Betonwerk factory precasting
Vorfertigungsfläche *f* [pre]casting area
Vorfertigungsplatz *m* [pre]casting yard

Vorfeuerung *f* extension furnace, furnace in front of the boiler
Vorflügel *m (aero)* slat
Vorflugrecht *n* right of way
 V. bei Landung landing priority
Vorfluter *m* outfall (drainage) ditch, feeder
Vorflyer *m (text)* slubbing frame
vorformen to preform
Vorformen *n (met)* preparatory shaping
Vorformfräser *m* roughing cutter
 V. für Zahnräder roughing cutter for gear teeth, roughing gear milling cutter
Vorformling *m (plast)* preform
Vorfräsen *n* rough-milling
Vorfräser *m* roughing cutter
vorfrischen to prefine
Vorfrischen *n* primary purification
Vorfrischmischer *m* active hot metal mixer
Vorfrischofen *m* primary furnace
vorfristig ahead of schedule, pre-schedule
Vorfrucht *f* preliminary (proceeding) crop *(in a crop rotation)*
Vorführapparat *m* demonstration apparatus
vorführen to demonstrate
 auf einem Schirm v. to render visible on a screen
Vorführer *m* projectionist *(cinema)*; demonstrator
Vorführgerät *n* demonstration set
Vorführung *f* **von Maschinen** demonstration of machines
Vorführungsbus *m* demonstration bus
Vorführungsschiff *n* demonstration vessel
Vorfüllung *f* full admission
Vorfunkenstrecke *f* auxiliary spark-gap
 äußere V. external series gap
Vorgabe *f (mine)* burden *(blasting)*; intended depth of cut *(coal plough)*; offset angle *(drilling)*
Vorgabewinkel *m (mine)* offset angle *(drilling)*
Vorgabezeit *f (ec)* allowed time
Vorgalvanisierbad *n* strike bath
Vorgalvanisierung *f* strike deposit
Vorgang *m* 1. process, reaction, operation, phenomenon; action; event; 2. mechanism *(mechanical)*
 einmaliger V. non-recurrent phenomenon, single transient
 irreversibler V. irreversible process
 isobarer V. isobaric process
 isochorer V. isochoric process
 isothermer V. isothermal process
 quasistatischer V. quasi-static process
 reversibler V. reversible process
 schwingungsfreier V. aperiodic phenomenon
Vorgänger *m* progenitor; *(math)* predecessor
Vorgarn *n* roving
 feines V. fine roving
 grobes V. coarse roving, slubbing
 mittleres V. middle roving
Vorgarnzähler *m* roving indicator
Vorgautschwalze *f* squeeze roll *(papermaking)*
vorgebaut / freitragend cantilevered
vorgebohrt [pre]drilled, previously drilled
vorgedrückt pre-compressed
vorgefertigt preformed *(reinforcement)*; fabricated *(structural steelwork)*; precast *(concrete)*; prefabricated, pre-finished
vorgeformt preformed
vorgekragt cantilevered, corbelled
Vorgelege *n* back gear
Vorgelegeachse *f* idler shaft
Vorgelegehebel *m* [speed change] back gear lever
Vorgelegerad *n* layshaft gear
Vorgelegetriebrad *n* countershaft drive gear
Vorgelegewelle *f (mach)* countershaft
Vorgelegezahnradblock *m* layshaft gear cluster
vorgepfändet *(mine)* cantilevered
Vorgerüst *n* cogging stand, rougher
vorgeschmiedet rough-forged
vorgeschrieben specified
vorgeschrumpft pre-shrunk
Vorgesenk *n* preliminary forging die
vorgespannt preloaded *(bearing)*; prestressed *(concrete)*
 beschränkt v. partially prestressed
 mit nachträglichem Verbund v. post-tensioned
 mit sofortigem Verbund v. pre-tensioned
 mittels Spannpresse v. jack-tensioned
 teilweise v. partially prestressed
Vorgespinst *n (text)* slubbing
 feines V. roving
vorgesteuert / hydraulisch pilot-operated
Vorgewende *n (ag)* headland
Vorglühen *n* preliminary annealing; *(aut)* preheating
Vorgriff *m (mine)* buttock
Vorhaben *n* project
Vorhafen *m* outport, outer port (harbour)
Vorhalle *f* lobby, vestibule, hall
Vorhalt *m* allowance *(aiming)*; *(cont)* derivative action, phase lead
Vorhaltdrossel *f* derivative restriction *(pneumatics)*
Vorhaltewinkel *m* dropping angle *(bombing)*; angle of lead *(shelling)*
Vorhaltnetzwerk *n (el)* lead network
Vorhaltschaltung *f s.* Vorhaltnetzwerk
Vorhaltstabilisierung *f* derivative equalization (stabilization)
Vorhalttransformator *m* stabilizing transformer
Vorhaltung *f* **der Aus- und Vorrichtung** *(mine)* lead of development work over winning
Vorhaltwirkung *f* derivative action
Vorhaltzeit *f* derivative action time
Vorhammer *m* sledge

Vorhang *m* curtain
Vorhangchintz *m* blind chintz
Vorhängeschloß *n* padlock
Vorhangwand *f* curtain wall[ing]
Vorheizen *n* pre-heating
Vorhelling *f (ship)* launch of the slipway, lower launch slip, ways-end
Vorherd *m* forehearth; casting shoe
Vorhersage *f* forecast *(weather)*; prediction
Vorhersagekarte *f* prognostic chart
Vorhobeln *n* rough planing
Vorholer *m* advancing chain creeper; recuperator
Vorholfeder *f* recuperator spring
Vorimprägnierverfahren *n* prepreg process
Vorimpuls *m* prepulse, prefix signal
vorionisiert pre-ionized
Vorisolation *f* pre-insulation
Vorkaliber *n (met)* preceding (shaping) pass
Vorkaliberwalzgerüst *n* strand mill
Vorkammer *f (auto)* pre-combustion chamber
Vorkammerdieselmotor *m* pre-ignition Diesel engine
Vorkammermotor *m* Diesel engine with pre-combustion chamber
Vorkarde *f (text)* breaker card
vorkeimen to chit, to sprout
Vorkeimlampe *f* potato chitting lamp
Vorkeimung *f* vernalization
Vorklärbecher *m* preliminary clarification tank
Vorklärbecken *n* detritus pit
Vorklassierung *f* pre-screening
Vorkommando *n* preliminary cutoff signal
Vorkommandogeschwindigkeit *f* velocity at preliminary cutoff signal
Vorkommen *n* occurrence; deposit
gediegenes V. native occurrence
Vorkondensator *m* 1. preliminary condenser; 2. *(el)* series capacitor
vorkonditionieren to pre-condition *(test preparation)*
Vorkontaktschalter *m* carbon (auxiliary-contact) switch
Vorkonzentrat *n* pre-concentrate, rough concentrate
Vorkorrektur *f* first proof
Vorkorrosion *f* pre-corrosion
vorkragen to corbel, to overhang, to cantilever
Vorkrempel *f (text)* breaker card, first breaker
Vorkühler *m* precooler
Vorlage *f* receiver *(distillation)*; gas main
Vorlagedruck *m* back pressure
Vorlagemuster *n* outturn sheet
Vorlagenhalter *m* copy-holder
Vorland *n* foreshore
vorlastig down (trimmed) by the head, bow-heavy
Vorlastigkeit *f* trim by the head
Vorlauf *m* 1. *(mach)* forward movement (stroke), advance; 2. *(ship)* pre-launch[ing]; 3. preliminary data collection *(statistics)*; 4. *(ch)* fore-runnings, first runnings, prerun; 5. *(aut)* lead *(front wheels)*; 6. pouring basin *(casting)*
V. der Spindel advance of the spindle
Vorlaufbehälter *m* heads tank *(distillation)*
Vorlaufbehandlung *f* pre-emergence treatment
Vorläufer *m* precursor *(series)*; previous model *(instrument)*
Vorlaufimpuls *m* transmitted (incident) pulse
Vorlegeschloß *n* padlock
Vorlegierung *f* preliminary (master) alloy
Vorlesungsreihe *f* course of lectures
vorlicher als dwars *(ship)* forward of the beam
vorlochen to pre-punch; to pre-pierce
Vorluftbehälter *m (aut)* preliminary air tank
vormagnetisieren to pre-magnetize
Vormagnetisierung *f* superposed (bias) magnetization, magnetic bias[ing]
Vormagnetisierungsfeld *n* polarizing field
Vormagnetisierungsfeldstärke *f* polarizing magnetizing force
Vormagnetisierungsinduktion *f* polarized induction
Vormagnetisierungswicklung *f* bias winding
Vormahlen *n* primary grinding
Vormaterial *n (met)* feed stock
Vormauerziegel *m* facing brick
Vormeldeeinrichtung *f* / **selbsttätige** automatic advice
Vormelden *n* **der Züge** train announcing
Vormeldestromkreis *m* warning circuit
Vormeldung *f* warning; *(ry)* train announcement
Vormelken *n* taking foremilk, foremilking
Vormetall *n* blown metal; premelted metal
vormischen to pre-mix
Vormontage *f* sub-assembly
vormontieren to pre-assemble
Vornorm *f* tentative standard
Voröffner *m (text)* preliminary opener, pre-opener
V. mit Nasenwalze porcupine opener
Voröffnungsstrecke *f (therm)* linear lead
Voröffnungswinkel *m (therm)* lead angle
Vorort *m* suburb
Vorortbahn *f* suburban train (line)
Vorortverkehr *m* suburban traffic
Vorortzug *m* suburban train
Vorpanzerung *f* additional plating in spaced armour
vorpfänden *(mine)* to block
Vorpfänden *n (mine)* horse-heading, fore poling
Vorpfändkappe *f (mine)* cantilevered bar
Vorpfeilung *f (aero)* negative sweep
Vorpiekbalken *m (ship)* panting beam
Vorplanung *f* preplanning, outline (preliminary) planning
Vorplastifizieren *n* preplastifying

Vorplatz *m* forward square *(passenger vessel)*
Vorpostenboot *n* outpost patrol boat, coastal defence craft
Vorpressen *n (met)* rough pressing
Vorprodukt *n* initial (crude, primary) product; intermediate product
vorprüfen to pre-test
Vorprüfung *f* preliminary (quick) test
Vorpumpe *f* backing pump; auxiliary pump
Vorpumpenventil *n* backing-pump valve
magnetisch-automatisches V. magnetic backing-pump valve with retarding relay
Vorraffination *f* improving *(lead)*
V. im Schmelzfluß preliminary fire refining
Vorraffinierofen *m* improving furnace
Vorrangmatrix *f (dat)* precedence matrix
Vorrat *m* stock, provision, supply
vorrätig in stock
Vorratsader *f* spare wire *(telephony)*
Vorratsbehälter *m* stock bin, storage tank; magazine, storage hopper
Vorratsbildung *f* storage *(conveyor)*
V. im Zwischenlager verketteter Maschinen in-process work storage
Vorratsbütte *f* stock chest
Vorratsflasche *f* stock bottle
Vorratsgabelschaltung *f* spare termination set
Vorratsgefäß *n* supply vessel, reservoir
Vorratshalde *f* stock-pile
Vorratskatode *f* dispenser cathode
Vorratslenkung *f* inventory control
Vorratslösung *f* stock solution
Vorratsraum *m* store [room], storage room
Vorratsroder *m* digger elevator, potato digger
Vorratsschiff *n* store boat, depot ship
Vorratssilo *m* storage bin (silo)
Vorraum *m* ante-room, ante-chamber; coking (distillation) chamber
Vorräumen *n* rough broaching
Vorreibahle *f* semi-finishing reamer
vorreiben to rough-ream
Vorreiben *n* rough reaming
Vorreiber *m* casement fastener (turn)
Vorreiniger *m* pre-cleaner
Vorreinigungsmaschine *f* precleaner, winnowing machine *(grain)*
Vorreinigungsverfahren *n* precleaning process
Vorreiz *m* / **elektrischer** conditioning stimulus
Vorrichtblock *m* gauge block
vorrichten *(mine)* to develop, to open cut *(seam)*
Vorrichtung *f (mach)* jig *(tool)*; fixture *(work)*; development work *(in coal)*
V. mit drei Spannstellen *(mach)* three-position fixture
selbstspannende V. *(mach)* self-clamping fixture
Vorrichtungsarbeiten *fpl* preparatory work
Vorrichtungsausleihstation *f* fixture leasing station
Vorrichtungsbau *m* toolroom [work]
Vorrichtungskonstrukteur *m* jig and tool designer
Vorrichtungskonstruktion *f* jig and fixture design
Vorrichtungsstrecke *f (mine)* development road, heading
seitliche V. *(Am)* lateral [road] *(ore mining)*
Vorröhre *f* preceding valve
Vorröhrenmodulation *f* series modulation
vorrösten to preroast
Vor-Rück-Verhältnis *n* rear-to-front ratio *(aerial)*
Vor-Rückwärtszähler *m* bidirectional (two-way) counter
vorsättigen to pre-saturate
Vorsättigung *f* pre-saturation
Vorsatz *m* adapter
Vorsatz *n (print)* end-paper
äußeres V. outer end-paper
fliegendes V. inner end-paper
Vorsatzbeton *m* face mix (concrete), facing concrete
Vorsatzblatt *n* fly leaf
Vorsatzfernrohr *n* attachment telescope
Vorsatzgerät *n* adapter
Vorsatzklebemaschine *f* end-papering machine
Vorsatzlinse *f* telephoto attachment
Vorsatzobjektiv *n* supplementary lens
Vorsatzpapier *n* end-paper
Vorschacht *m (mine)* cellar, pilot shaft
Vorschäler *m (ag)* skim coulter
Vorschaltdiode *f* series diode
Vorschaltdrossel *f* series reactor, ballast
Vorschalteinheit *f* add-on unit
vorschalten to connect before (in series)
Vorschaltgerät *n* ballast *(luminescent lamp)*
Vorschaltkondensator *m* series capacitor
Vorschaltlampe *f* loading lamp
Vorschaltlinse *f* attachment lens
Vorschaltrad *n* additional turbine wheel
Vorschaltturbine *f* auxiliary turbine
Vorschaltung *f* 1. superposition; 2. *(el)* connection in series
Vorschaltwiderstand *m* compensating (voltage dropping) resistor; range multiplier
Vorschäumer *m* rougher
Vorscheidung *f* predefecation, preliming *(sugar industry)*
vorschieben *(mach)* to advance; to push forward
Vorschiff *n* forward quarter, forebody, stem; eyes
ausfallendes V. flare out bow
Vorschiffsauftrieb *m* bearing of a vessel
Vorschiffslinien *fpl* bow lines
Vorschiffsluke *f* fore hatch

Vorschiffsschärfe *f* entrance
Vorschiffsspant *n* forward frame (rib)
Vorschiffsspantenriß *m* forward body-plan
Vorschläger *m* preliminary opener *(cotton)*
Vorschlaghammer *m* aboutsledge
Vorschliff *m* rough grinding; preliminary surfacing
Vorschmelze *f* wash out heat
Vorschmelzeisen *n* premelted (first-smelting pig) iron
Vorschmelzofen *m* premelting furnace
Vorschmiedegesenk *n* blocker, blanker, rougher
vorschmieden to cog down
Vorschneider *m* *(mach)* rough cutter *(tool)*; entering tap *(threads)*; end cutting nipper *(tongs)*
Vorschneidezahn *m* *(mach)* leader (bevelled roughing, pre-cutter) tooth *(sawblade)*
Vorschneidmesser *n* roughing knife
Vorschoten *fpl* *(ship)* head sheets
vorschrämen to precut
Vorschrämmaschine *f* undercutting machine
vorschreiben to specify
Vorschrift *f* regulation; instruction; direction; specification
Vorschub *m* *(mach)* feed; advance
  V. / mit hydraulischem fluid-feed; with hydraulic feed
  V. des Löffels crowding *(power shovel)*
  V. je Zahn degree of taper *(broach)*
  automatischer V. *(mine)* power feed
  maschineller V. power-feeding [mechanism]
Vorschubantrieb *m* feed drive
Vorschubauslöseeinrichtung *f* feed-cut-off device
Vorschubauslösung *f* feed disengagement
  selbsttätige V. automatic feed disengagement
Vorschubausrückung *f* feed trip
Vorschubbereich *m* range of feeds
Vorschubbewegung *f* feed motion, [tool] feed movement
Vorschubeinrichtung *f* feeder; *(mach)* feeding mechanism
Vorschubeinrückhebel *m* feed engaging lever
Vorschubeinstellscheibe *f* feed dial
Vorschubeinstellung *f* feed adjustment (setting)
Vorschubgeschwindigkeit *f* rate of feed
Vorschubgetriebe *n* feeding mechanism (device, gear), feed train (gear, box)
Vorschubgetriebekasten *m* feed [gear]box
Vorschubgetriebemotor *m* feed motor
Vorschubgewindespindel *f* feed screw
Vorschubgröße *f* rate (amount) of feed
Vorschubhandhebel *m* feed handle
Vorschubhandrad *n* feed hand wheel
Vorschubklinkenrad *n* feed-screw sprocket wheel, feed ratchet wheel
Vorschubkraft *f* feed force; longitudinal force *(turning)*; *(mine)* thrust *(drilling)*
Vorschubkupplung *f* feed clutch
Vorschubmotor *m* feed motor
Vorschubräderkasten *m* feed gear box
Vorschubrichtung *f* direction of feed
Vorschubrichtwert *m* basic (reommended) feed
Vorschubriefe *f* feed mark
Vorschubschalter *m* feed switch
Vorschubschaltgerät *n* feed control
Vorschubscheibe *f s.* Regelscheibe
Vorschubschlitten *m* *(mine)* feed carriage *(drilling)*
Vorschubskale *f* feed [indicator] dial
Vorschubsperrgetriebe *n* feed-pawl mechanism
Vorschubsteuerung *f* *(dat)* carriage control
Vorschubstufe *f* rate of feed
Vorschubstütze *f* *(mine)* jack-leg
Vorschubventil *n* feed valve
Vorschubwählerknopf *m* feed selector knob
Vorschubwahlschalter *m* feed selector
Vorschubwelle *f* feed shaft
Vorschußzahlung *f* payment in advance
Vorsegel *n* jib
Vorseil *n* fish rope
Vorsetzmaschine *f* primary jig, rougher
Vorsetzwinkel *m* *(mine)* angle of lead of an overtipped face
Vorsichtsmaßregel *f* precaution
Vorsieb *n* fore-screen; scalping screen
Vorsieben *n* scalping
Vorsignal *n* preliminary (first marking) signal; *(ry)* distance signal
vorsintern to presinter
Vorsintern *n* presintering
vorsortieren to presort
Vorspannband *n* leader tape *(magnetic tape)*
Vorspannbeton *m s.* Spannbeton
vorspannen 1. to prestress; to pre-load *(concrete)*; 2. *(el)* to bias
  mit nachträglichem Verbund v. to post-tension
  mit sofortigem Verbund v. to pre-tension
Vorspannen *n* 1. tensioning, prestressing *(concrete)*; 2. *(el)* biasing
  V. mit nachträglichem Verbund [bonded] post-tensioning
  V. mit sofortigem Verbund pre-tensioning
  V. mittels Spannpresse jack tensioning
  V. ohne Verbund unbonded prestressing (tensioning)
  elektrothermisches V. electro-thermic tensioning
  thermisches V. heat-tensioning
Vorspannkraft *f* prestressing force
Vorspannturbine *f* auxiliary turbine
Vorspannung *f* 1. pretension, prestress *(concrete)*; *(mech)* preliminary stress; 2. *(el)* bias voltage

V. mit nachträglichem Verbund [bonded] post-tensioning
V. mit sofortigem Verbund pre-tensioning
V. ohne Verbund unbonded tensioning (prestressing)
magnetische V. magnetic bias
negative V. negative bias
verzögerte V. delay bias
zusätzliche V. additional bias
Vorspannungsbedingung *f* bias condition
Vorspannungseffekt *m* bias effect
Vorspannungsfeld *n* biasing field
Vorspannungsregelung *f* bias control
Vorspannungsrichtung *f* direction of bias
Vorspannungsstrom *m* bias current
Vorspannungsverlust *m* loss of prestress
Vorspannungszustand *m* bias condition
vorspeichern *(dat)* to prestore
Vorspeicherung *f (dat)* prestorage
Vorspektrum *n (ph)* preliminary spectrum
Vorsperröhre *f* pre-TR tube
Vorspinnkrempel *f* carding machine for slubbing
Vorspinnmaschine *f* billy, slubbing machine, roving frame
vorspringen to project, to overhang
Vorsprung *m* projection, lug; *(el)* shoulder *(voltage)*
Vorspur *f* toe-in *(motor car)*
Vorstadt *f* suburb
Vorstandard *m* tentative standard
Vorstecher *m* awl
Vorstechkamm *m (text)* punching comb
kreisförmiger V. circular punching comb
Vorsteckbolzen *m* cotter bolt
Vorsteckstift *m* stop pin
vorstehen to project
Vorstellplatte *f* dead (choke) plate
Vorsteuerelement *n* servomechanism
Vorsteuerpumpe *f* hydraulic governor
Vorsteuerschieber *m* pilot valve (selector)
Vorsteuerung *f* anticipatory control
V. / mit servo-controlled
hydraulische V. hydraulic detent
Vorsteven *m* stem [post]
ausfallender V. raked (raking) stem
gerader V. straight bow (stem)
senkrechter V. straight bow
überhängender V. raking stem
Vorstevenplatte *f* stem plate
Vorstevenrückenstück *n* independent piece
Vorstevenschuh *m* stem foot
Vorstevenunterteil *n* gripe of the stem
Vorstich *m (met)* roughing (cogging) pass
Vorstraße *f (met)* roughing (cogging) train
Vorstrecke *f* 1. *(met)* roughing train, roughing (cogging) stand of rolls; 2. *(text)* preparatory drawing frame
vorstrecken *(met)* to rough down
Vorstrecken *n* 1. *(met)* cogging, *(Am)* blooming; 2. *(text)* preparative drawing
V. des Kammzugs *(text)* topmaking
Vorstreckgerüst *n* pony roughing stand of rolls, pony-rougher
Vorstreckkaliber *n* cogging (preceding) pass
Vorstreckwalze *f* roughing (cogging) roll
vorstreichen to precoat
Vorstrom *m (ship)* wake
Vorstufe *f* prestage; preliminary step; preselector stage *(receiver)*
Vorstufenmodulation *f* low-level modulation
Vorteil *m* advantage
Vortopp *m (ship)* fore-mast
Vortrabanten *pl (telev)* preequalize pulses
Vortrag *m (dat)* amount brought forward
vortragen / den Saldo *(dat)* to carry forward the balance
Vortragskontrolle *f (dat)* check of balances carried forward
Vortragssaal *m* assembly hall, auditorium
Vortrieb *m* 1. *(aero)* propulsion, [forward] thrust; 2. *(mine)* advance rate
Vortriebsgeschwindigkeit *f (mine)* drift advance, drivage (driving) rate
Vortriebskraft *f* thrust
Vortriebsmaschine *f (mine)* header, heading (tunnelling) machine
Vortriebsmittel *n* propeller
Vortriebswiderstand *m* resistance to propulsion
Vortrockenpartie *f* pre-drier section *(papermaking)*
vortrocknen to pre-dry
Vorturbine *f* auxiliary turbine
vorübergehend temporary
Vorüberhitzer *m* pre-superheater
Vorübertrager *m (el)* input repeating coil
Vor- und Rückbau *m* **/ kombinierter** *(mine)* advancing-and-retreating working
Vor- und Rücklaufhobeln *n* two-way planing
Voruntersuchung *f* preliminary investigation
Vorvakuum *n* partial vacuum, pre-vacuum
Vorvakuumbehälter *m* interstage reservoir
Vorvakuumpumpe *f* forepump, pre-vacuum pump
vorverdichten 1. *(build)* to pre-consolidate, to pre-compact; 2. *(plast)* to preform; 3. to supercharge *(engine)*
Vorverdichter *m* 1. *(build)* precompactor; 2. supercharger *(engine)*
vorverdichtet 1. *(build)* pre-consolidated, pre-compacted; 2. supercharged *(engine)*
Vorverdichtung *f* 1. *(build)* pre-consolidation, pre-compaction; 2. crankcase compression, supercharging *(engine)*
vorverdicken to prebody *(lacquer)*
Vorvergasung *f* pre-distillation
Vorverstärker *m (el)* pre-amplifier; head amplifier

Vorverstärkerstufe *f* pre-amplifier stage
Vorverstärkung *f* pre-amplification
Vorversuch *m* preliminary test
Vorverzerrung *f* pre-emphasis
vorverzinnen to pre-tin
Vorverzug *m (text)* preliminary draft
Vorvulkanisation *f* prevulcanization
Vorwahl *f* preselection
  V. / mit preoptive; preselective
vorwählbar preselective, preoptive
Vorwähleinrichtung *f* preselector, preselection device
  V. / mit preoptive
vorwählen to pre-select, to select in advance
Vorwähler *m* preselector
Vorwählerscheibe *f* preselector dial
Vorwahlring *m* presetting ring
Vorwahlschalter *m* preselector [control]
Vorwahlschaltung *f* preselection, preselective switching
Vorwählschaltung *f (aut)* preselection change
Vorwählung *f* preselection
Vorwalze *f* roughing (cogging-down) roll
vorwalzen to rough down
Vorwalzgerüst *n* roughing (cogging-down) stand
Vorwalzstraße *f* roughing (cogging) train
Vorwalzwerk *n* cogging mill
Vorwärmeklappe *f (aut)* pre-heating valve
vorwärmen to preheat
  den Zylinder v. to warm-up the cylinder
Vorwärmen *n* preheating; warming-up
  V. des Arbeitszylinders warming-up of the cylinder
  V. für das Walzen soaking
Vorwärmer *m* preheater; economizer
  V. mit Doppelröhren feed heater with double tubes
  V. mit gebogenen Röhren feed heater with U-tubes
  mit Abdampf geheizter V. exhaust feed heater
  offener V. open feed heater
Vorwärmeranlage *f* feed heating plant
Vorwärmerfilter *n* feed water filter
Vorwärmerheizfläche *f* heating surface of the feed heater
Vorwärmermantel *m* shell of the feed heater
Vorwärmofen *m* preheating furnace, preheater
Vorwärmtisch *m* preheating bench *(electron tubes)*
Vorwärmung *f* **durch Abdampf** preheating by means of exhaust steam
vorwarnen to forewarn
Vorwarnfunkmeßgerät *n* long-range radar device
Vorwarnsystem *n* early-warning system
vorwärts forward; forth
Vorwärtsbewegung *f* forward movement
Vorwärtsgang *m (aut)* forward gear (speed)
Vorwärtsgleitbahn *f* forward stroke guide surface
Vorwärtshub *m* forward stroke
Vorwärtsimpedanz *f* forward impedance
Vorwärtskupplung *f* forward clutch
Vorwärtsregelung *f* forward-acting regulation, forward control
Vorwärtsrichtung *f* forward direction
Vorwärts-Rückwärtszähler *m* forward-backward counter
Vorwärtsruf *m* ring-forward signal *(telephony)*
Vorwärtsständer *m* forward-stroke standard (column)
Vorwärtsstreuung *f* forward scattering, forescattering
vorwärtstreiben to propel
Vorwärtsturbine *f* ahead turbine
Vorwäsche *f* bottoming *(bleaching)*
Vorwiderstand *m* series (dropping) resistor; series resistance; voltage multiplier *(of a multimeter)*
  äußerer V. instrument multiplier
Vorwiderstandslampe *f* ballast bulb
Vorwindfrischer *m* active (primary-refining) mixer
Vorzeichen *n (math)* sign; *(dat)* sign digit
  entgegengesetztes V. opposite sign
Vorzeichenbit *n (dat)* sign digit
Vorzeicheninverter *m (dat)* sign inverter
Vorzeichenregel *f* sign convention
Vorzeichenumkehrer *m (dat)* [phase] inverter, inverting circuit
Vorzeichenumkehrung *f (dat)* sign reversal
Vorzeichenumkehrverstärker *m* inverting amplifier, phase inverter
Vorzeichenziffer *f (dat)* sign digit
Vorzerkleinerung *f* pre-crushing
Vorzerklüftung *f* induced cleavage
Vorzerleger *m* order sorter
Vorziehkettenbahn *f (mine)* chain creeper
Vorziehvorrichtung *f (mine)* tub pusher
Vorzimmer *n* ante-room, ante-chamber
Vorzug *m* advantage; first draw *(deep drawing)*
vorzüglich extremely fine *(coin)*
vorzugsgerichtet with preferred orientation
Vorzugsmaß *n* preferred dimension
Vorzugsorientierung *f* texture, preferred orientation
Vorzugsreihe *f* preferred series
Vorzugsröhre *f* preferred type valve
Vorzugszölle *mpl* preferential tariff
Vorzünddistanz *f* pre-ignition distance
Vorzündung *f* early (advanced) ignition, spark advance
Vorzurichtepresse *f* pre-makeready press
Vorzurichtung *f* pre-makeready
Voute *f* [tapered] haunch *(reinforced concrete)*
Voutenbalken *m* haunched beam

Voutenträger *m* haunched beam
Vulkan *m* volcano
Vulkanfiber *f* vulcanized fibre (paper)
Vulkanfiberimitation *f* semivulcanized board
Vulkanfiberventil *n* vulcanized fibre valve
Vulkanisation *f* vulcanization, vulcanizing
Vulkanisationsbeschleuniger *m* vulcanizing accelerator
Vulkanisationsverzögerer *m* vulcanizing retarder
Vulkanisieranstalt *f* vulcanization works
vulkanisierbar vulcanizable
vulkanisieren to vulcanize, to cure
Vulkanisieren *n* vulcanizing, vulcanization
Vulkanisierkessel *m* vulcanizing pan
Vulkanisierung *f* vulcanization

W

Waage *f* balance, scale
  W. / in level
  aerodynamische W. wind-tunnel balance
  Coulombsche W. Coulomb's balance
  hydrostatische W. hydrostatic balance
  Langmuirsche W. Langmuir balance
  magnetische W. magnetic balance
Waagebalken *m* balance (scale) beam
Waagebalkenrelais *n* balanced-beam relay
waagerecht horizontal
Waagerechtaußenräummaschine *f* horizontal surface broaching machine
Waagerechtbohren *n* horizontal drilling (boring)
Waagerechtbohrfräswerk *n* horizontal boring and milling machine
Waagerechtbohrmaschine *f* **mit Plandrehsupport** horizontal boring and facing machine
Waagerechtbohr- und -fräswerk *n* **mit fester Aufspannplatte und waagerecht verschiebbarem Ständer** floor-type horizontal boring and milling machine with travelling column
  W. mit hobelmaschinenähnlichem Tisch planer-type horizontal boring mill
  W. mit [quer]beweglichem Ständer travelling column horizontal boring machine *(cross-bed machine)*
Waagerechtbohrwerk *n* horizontal boring mill
Waagerechte *f* horizontal
Waagerechtflachschleifmaschine *f* horizontal surface grinding machine
Waagerechtfräsmaschine *f* horizontal milling machine
Waagerechtgeschwindigkeit *f* horizontal velocity
Waagerechtinnenräummaschine *f* horizontal internal broaching machine
Waagerechträummaschine *f* horizontal broaching machine
Waagerechtrundtischschleifmaschine *f* horizontal [spindle] rotary table surface grinder, horizontal [spindle] rotary surface grinder
Waagerechtschleifmaschine *f* horizontal-spindle surface grinding machine
Waagerechtschnitt *m* sectional plan, plain section
Waagerechtstoßmaschine *f* shaping machine, shaper
  W. mit Kurbelschleifenantrieb crank shaper
  traversierende W. traverse shaper
Waageschneide *f* balance blade
Waagschale *f* scale pan
Wabe *f* honeycomb
wabenartig honeycomb
Wabenbauelement *n* honeycomb structure
Wabenbauweise *f* cellular construction
wabenförmig honeycomb
Wabenfundament *n* honeycomb furniture (mount)
Wabenkern *m* honeycomb core
Wabenkonstruktion *f* honeycomb structure
Wabenkühler *m* honeycomb radiator
Wabenspule *f (el)* honeycomb (duo-lateral) coil
Wabenstruktur *f* honeycomb structure
Wabenwicklung *f (el)* honeycomb (duo-lateral) winding
Wachboot *n* watch-boat
Wache *f* watch
Wachfahrzeug *n* patrol vessel (boat), guard ship
Wachfrequenz *f* / **regionale** regional guard frequency
Wacholderbeeröl *n* juniper [berry] oil
Wacholderharz *n* juniper gum
Wachs *n* wax
  rutschfestes W. no-rub wax
Wachsabdruck *m* wax impression
wachsartig wax-like, waxy, cereous
Wachsausschmelzverfahren *n* lost wax process, cire perdue process
Wachsbaumwolldraht *m* waxed cotton-covered wire
Wachsdraht *m* paraffined (wax-coated) wire
wachsen 1. to grow; to increase; 2. to wax
Wachsen *n* 1. growing; 2. waxing
Wachsfaden *m* waxed thread
Wachsform *f* wax form *(foundry)*
wachsfrei dewaxed
Wachsgehalt *m* wax content
Wachskarton *m* / **abgeschreckter** cold-waxed board
Wachsleim *m* wax size
Wachsmatrize *f* [duplicating] stencil
Wachsmodell *n* wax pattern
Wachspapier *n* waxed paper
Wachsphantom *n* wax phantom
Wachsprägung *f* wax mould
Wachsreife *f* pasty condition *(maize)*

Wachsreserve *f* batik resist
Wachsrohpapier *n* twisting paper
Wachstuch *n* oilcloth [for table covers], wax-cloth
Wachstum *n* growth
definiertes W. defined growth
dendritisches W. dendritic growth
spiralförmiges W. spiral growth
wachstumfördernd growth-accessory
Wachstumsfaktor *m* growth factor
Wachstumsgeschwindigkeit *f* growth rate, rate of growth
Wachstumskurve *f* growth curve
Wachstumsmesser *m* auxanometer
Wachstumsrate *f* rate of growth
W. der Industrie percentage growth of industry
Wachstumsrichtung *f* growth direction *(crystals)*
Wachstumsring *m* growth ring
Wachstumsschicht *f* growth layer
Wachstumsspirale *f* growth spiral *(crystals)*
Wachstumstempo *n* rate of growth
Wächter *m* automatic controller, governor *(overloads)*
Wächterkontrollanlage *f* watchman's control system
Wächterkontrolluhr *f* watchman's detector
Wackelkontakt *m* intermittent (defective, tottering, imperfect) contact, slack (imperfect contact) joint, loose connection
Wade *f* seine *(net)*
Wadenfischer *m* seiner, seine boat *(fishing vessel)*
Waffe *f* arm, weapon
thermonukleare W. thermonuclear weapon
Waffelbindung *f* honeycomb (waffle) weave
Waffelblech *n* goffered plate
Waffelboden *m* waffle floor
Waffeleisen *n* waffle iron
Waffenmeister *m* armourer
Waffenmeisterwagen *m* weapons repair truck
Waffenschein *m* fire-arms certificate, arms (gun) licence
wägbar weighable
Wägefläschchen *n* gravity bottle
wägen to balance, to weigh
Wagen *m* car, vehicle; wagon, coach; carriage; *(ag)* cart; *(mine)* bogie
W. mit Eigenantrieb self-propelled car
geschlossener W. box-car
zwangsläufig geführter W. *(text)* controlled carriage
Wagenachse *f* axle-tree
Wagenaufbau *m* coachwork, truck body
Wagenaufseher *m* carriage and wagon examiner, *(Am)* car inspector
Wagenausbesserungswerk *n* vehicle maintenance and repair shop
Wagenausfahrt *f (text)* outward run of the carriage
Wagenaustausch *m* exchange of rolling stock
Wagenbestellung *f* request for wagons
Wagenbezugleder *n* coach leather
Wagenbremsberg *m* car-gravity road
Wagendach *n* roof[ing]
Wagendeck *n* car-deck *(of a vessel)*
Wagendecke *f* tarpaulin, wagon sheet
Wageneinfahrt *f (text)* run-in of the carriage
Wagenfähre *f* car (vehicular) ferry
Wagenfänger *m* car catching device
Wagenfeder *f* carriage spring
Wagenfeststellvorrichtung *f (mine)* car stop *(cage)*
Wagenfett *n s.* Wagenschmiere
Wagenführer *m (mine)* truck driver
Wagengestell *n* car frame
Wagengießpfanne *f (met)* truck ladle
Wagengruppe *f* set (raft, rake) of wagons, set (raft, rake) of coaches, train set
Wagenguß *m* bogie casting
Wagenheber *m* hydraulic [automobile] jack, lifting (motor-car) jack, car lift
hydraulischer W. hydraulic [automobile] jack
Wagenherdofen *m* car-hearth furnace, car-bottom furnace
Wagenkasten *m* body of car (coach), car body
Wagenkipper *m* car dumper (tipper)
Wagenkuppler *m (mine)* coupler-on
Wagenlader *m* car loader
Wagenladung *f* complete (full) wagon load, *(Am)* truck-load
Wagenlauf *m* run of a wagon
Wagenmeister *m* carriage and wagon examiner, *(Am)* car inspector
Wagenmiete *f* hire charges for wagons
Wagenpark *m* rolling stock
Wagenreinigung *f* car cleaning; *(mine)* tub cleaning
Wagenrücklauf *m* carriage return
Wagenschmiere *f* axle (carriage) grease
Wagenschuppen *m* cart house, wagon shed, coach-house
Wagensperre *f (mine)* car (tub) stop
Wagenspiel *n (text)* play of the carriage
Wagenspinner *m (text)* [self-acting] mule, self-actor
Wagentag *m* wagon-day
Wagenübergang *m (ry)* bridge, gangway
Wagenumlauf *m (ry)* turn round of wagons; *(mine)* tub circulation
Wagenverzug *m (text)* carriage drag
Wagenvorschub *m* / **automatischer** automatic carriage
Wagenvorziehvorrichtung *f (mine)* car spotter
Wagenwäsche *f* car washing plant

Wagenwascher *m* carriage cleaner
Wagenwechsel *m (mine)* car changing (switching)
Wagenzähler *m* axle counter
Wagenzettel *m* routing list
Wagenzug *m (mine)* journey
Wagenzulassung *f* car registration
Wägestück *n* weight
Wägetechnik *f* weighing technique
Wägevorrichtung *f* weighing device
Waggon *m* railway (goods) wagon, goods van; *(Am)* freight car
  W. mit Bodenentleerung hopper bottom rail car
Waggondrehscheibe *f* car turntable
Waggonentlademaschine *f* waggon (car) unloader
Waggonfähre *f* railway ferry; *(Am)* railroad ferry
Waggonhebeanlage *f* railway car hoist
Waggonkipper *m* car dumper
Waggonwaage *f* railway-track scale
Wahl *f* selection; dialling
  direkte W. direct dialling
  erzwungene W. numerical selection
  freie W. hunting operation; hunting action
  unvollständige W. mutilated dialling
Wählamt *n* automatic exchange
Wählapparat *m* automatic telephone set
wählbar selectable
Wählbarkeit *f* selection ability
Wählbetrieb *m* automatic telephone system
Wähleinrichtung *f* dial system equipment
wählen to select; to dial
Wählen *n* selection; dialling
Wähler *m* selector
  erster W. first selector
  freier W. disengaged selector
  mehrstufiger W. multi-point selector
  schnellaufender W. high-speed selector
Wähleramt *n* automatic exchange, auto-exchange
Wähleranlage *f* selector installation
Wählerarm *m* wiper
Wählergeräusch *n* dialling effect
Wählergestell *n* selector rack
Wählerkontaktfeld *n* selector bank
Wähleröl *n* selector lubricating oil
Wählerrahmen *m* selector shelf
Wählerrelais *n* discriminating relay
Wählersaal *m* switchroom
Wählerschiene *f* permutation bar
Wählerstufe *f* rank of switches
Wählimpuls *m* selective (dialling) impulse
Wählkamm *m* selecting comb
wahllos random
Wählorgan *n* selecting mechanism
Wahlschalter *m* selector switch
Wahlscheibe *f* dial
Wahlscheibeneinstellung *f* dial-set feeding
Wählsternschalter *m* automatic line connector
Wählvermittlung *f* automatic exchange, auto-exchange
wahlweise optional
Wählwerk *n* selecting mechanism
Wählzentrale *f* automatic exchange
Wahnkante *f* natural level, defective angles
wahnkantig partly square
Wahrheitstafel *f* truth table *(of logical functions)*
wahrnehmbar perceptible
wahrscheinlich probable
Wahrscheinlichkeit *f* probability
  W. Null zero probability
  empirische W. empirical (a posteriori) probability
  mathematische W. mathematical (a priori) probability
  statistische W. statistical probability
  thermodynamische W. thermodynamic probability
Wahrscheinlichkeitsamplitude *f* probability amplitude
Wahrscheinlichkeitsdichte *f* density of probability, probability density
Wahrscheinlichkeitsfaktor *m* probability factor
Wahrscheinlichkeitsfunktion *f* probability function
Wahrscheinlichkeitsgesetz *n* probability law
Wahrscheinlichkeitskurve *f* probability curve
Wahrscheinlichkeitslehre *f* theory of probability
Wahrscheinlichkeitsnetz *n* probability paper
Wahrscheinlichkeitsrechnung *f* calculus (theory) of probability, probability calculus
Wahrscheinlichkeitsstrom *m* probability current
Wahrscheinlichkeitstheorie *f* probability theory
Wahrscheinlichkeitsverteilung *f* probability distribution
  bedingte W. conditional probability distribution
Wahrscheinlichkeitswelle *f* probability wave
Wahrscheinlichkeitswert *m* probable value
Waldarbeiter *m* forestry worker
Waldbestand *m* area of woodland
Waldbrand *m* forest fire
waldfrisch fresh from the forest
Waldkante *f* natural bevel
waldkantig partly square
Waldriß *m* heart shake
Waldsäge *f* cross-cut saw
Waldsteifsäge *f* one-man cross-cut saw
Walfangboot *n s.* Walfangschiff
Walfänger *m s.* Walfangschiff
Walfangmutterschiff *n* whale-factory ship, floating whale-oil factory
Walfangschiff *n* whaling vessel, whaler, whale catcher

Walkbrett *n* fulling board
Walkbrühe *f* milling liquor
Walke *f* milling machine
Walkechtheit *f* fastness to milling
walken to mill, to tumble, to full
Walken *n* **im Schmutz** milling in the grease
  letztes W. milling off
Walker *m* felter
Walkerde *f* fuller's earth
Walkfähigkeit *f* milling property
Walkfaß *n* drum tumbler
Walkflotte *f* milling liquor
Walkhilfsmittel *n* milling auxiliary
Walkholz *n* felting roller
Walkmaschine *f* milling machine, felting machine
Walkmühle *f* fulling mill
Walkprobe *f s.* Walkversuch
Walkverlust *m* milling loss
Walkvermögen *n* milling property
Walkversuch *m* milling test
Walkware *f* milled goods
Walkwürfel *m* fulling cube
Wall *m* bank, dam, dike *(earth)*; rampart *(stone)*; *(nucl)* barrier
wallen to bubble, to simmer
Wallstein *m* damstone *(blast furnace)*
Walm *m* hip [gable]
Walmdach *n* hip[ped] roof
Walmschifter *m* hip jack rafter
Walnußholz *n* walnut
Walratöl *n* sperm oil
Walrückendeck *n* turtleback deck, turtle-deck
Waltran *m* whale oil
Walverarbeitungsschiff *n* whale-oil factory vessel
Walzader *f* roll line
Walzaluminium *n* rolled aluminium
Walzasphalt *m* rolled asphalt
Walzaustritt *m* delivery side of rolls
Walzbahn *f* pass line
walzbar rollable
Walzbarkeit *f* rollability
Walzbarren *m* [rolling] slab
Wälzbewegung *f* generating motion *(gear manufacture)*
Walzblech *n* rolled sheet (plate)
Walzblei *n* rolled lead
Walzblock *m* rolling ingot, billet
Wälzbogen *m* rotating arc
Walzdorn *m* roll mandrel
Walzdraht *m* wire rod, rod wire
Walzdruck *m* roll pressure
Walze *f* roll[er], mill; drum; cylinder; barrel
  W. mit gehärteter Oberfläche part chill roll
  aufgerauhte W. ragged roll
  gegossene W. cast iron roll
  geriffelte W. ragged roll
  geschmiedete W. forged steel roll
  glatte W. plain roll
  halbkalibrierte W. half roll
  ungehärtete W. grain roll
walzen to roll; to mill
  Gewinde w. to roll threads
  Sohlleder w. to roll sole leather
Walzen *n* rolling; milling
Walzenabschleifmaschine *f* cylinder grinding machine
Walzenabstand *m* distance between rolls
Walzenangriff *m* bite of rolls
Walzenanordnung *f* arrangement of rolls
Walzenanstellung *f* lining of rolls
Walzenauftragmaschine *f* doctor roll, roller feeder
Walzenaustritt *m* delivery side of rolls
Walzenbackenbrecher *m* roll-jaw crusher
Walzenballen *m* roll surface (body)
Walzenballenlänge *f* surface length of the roll
Walzenbandage *f* shell of roll
Walzenbeschlag *m* cylinder covering
Walzenbezug *m* roller coating
Walzenbombierung *f* crowning of rolls
Walzenbrecher *m* roll crusher, set of crushing rolls
Walzenbreithalter *m* **ohne Nadelspitzen** cut temple roller
Walzenbruch *m* breakage of a roll
Walzenchronograf *m* drum-type chronograph
Walzendekatur *f* calender finish
Walzendrehmaschine *f* roll turning lathe
Walzendruck *m* roller printing
Walzenegreniermaschine *f* roller gin
Walzeneinstellung *f* lining of rolls
Walzenfahrschalter *m* cylindrical controller
Walzenfender *m (ship)* spar (wood) fender
walzenförmig cylindrical
Walzenfräser *m* cylindrical (plain milling) cutter
  kreuzverzahnter W. alternate gash plain mill
Walzenfreilauf *m* platen release mechanism
Walzengatter *n* [vertical] log frame with roller feed
Walzengerüst *n* stand of rolls
Walzengießerei *f* roll foundry
Walzengravur *f* roller engraving
Walzenhobelmaschine *f* planing machine with independent cutter head, planing machine with cylindrical cutters
Walzenkern *m* roll core
Walzenkessel *m* cylindrical (barrel) boiler
  W. mit Sieder elephant (French) boiler
  W. mit Vorwärmer cylindrical boiler with feed water heater
  einfacher W. single cylindrical boiler
  liegender W. horizontal cylindrical boiler
  mehrfacher W. combined (multiple cylindrical) boiler
Walzenkleeblattzapfen *m* roll wobbler (wabbler)

Walzenkörper *m* roll body
Walzenkrempel *f* roller card
W. mit Bandablieferung roller card with sliver delivery
Walzenlager *n* roller bearing
Walzenmangel *f* mangle
Walzenmantel *m* roll shell
Walzenmasse *f* roller composition
Walzenmühle *f* beater, roller mill, roll crusher
Walzennachformdrehmaschine *f* roll contouring lathe
Walzenpaar *n* pair (set) of rolls
Walzenpoliermaschine *f* roll grinder
Walzenpresse *f* roll[er] press
Walzenprofil *n* profile of a roll
Walzenquetschwerk *n* roller press
Walzenreinigungsrakel *f* cleaning doctor
Walzenrost *m* cylindrical (roller) grate
Walzenrostfeuerung *f* furnace with cylinder grate
Walzenschablone *f* pass templet
Walzenschalter *m* drum controller (switch), barrel switch
Walzenschleifmaschine *f* roll grinding machine
Walzenschloß *n* roller lock
Walzenschlupf *m* rolling slide, roll slippage
Walzenschulter *f* shoulder of rolls
Walzenspeiser *m* drum (roller) feeder
Walzenspindel *f* roller core, roll shaft
Walzenständer *m* stand of rolls
Walzenstirnfräsen *n* face (side) milling
Walzenstirnfräser *m* end face mill, shell end mill
Walzenstraße *f s.* Walzstraße
Walzenstreuer *m* roller manure distributor
Walzensystem *n* cylinder system
Walzentrockner *m* drum drier
kontinuierlicher W. continuous roller drier
kontinuierlicher W. mit Mehretagenanordnung continuous roller drier with multi-deck arrangement
Walzenumfangsgeschwindigkeit *f* roll surface velocity
Walzenvergüteofen *m* roll heat-treating furnace
Walzenverschleiß *m* wear of rolls
Walzenvorschub *m* roll feed [attachment]
Walzenwaschanlage *f* roller washing device
Walzenzapfen *m* roll neck (pin)
Walzenzugdampfmaschine *f* rolling mill steam engine
Walzerzeugnis *n* rolled product
Walzfaser *f* lamination
Walzfehler *m* rolling crack (fault)
Wälzfehler *m* total composite error
Walzfell *n* rolled (rolling) sheet
wälzfräsen to hob
Wälzfräsen *n* hobbing; peripheral milling
Wälzfräser *m* hob[bing cutter]
W. für Schneckenradherstellung worm gear hob
Wälzfräserdrehzahl *f* hob speed
Wälzfräserprüfgerät *n* hob-cutter testing machine
Wälzfräserscharfschleifmaschine *f* hob sharpening (grinding) machine
Wälzfräsmaschine *f* hobbing machine
Wälzfrästeil *n* hobbed part, part to be hobbed
Wälzfräsverfahren *n* hobbing process
Wälzgerade *f* rolling line
Walzgerüst *n* roll stand, stand of rolls
Walzgeschwindigkeit *f* rolling speed
Wälzgetriebe *n* gears on parallel and intersecting axes
Walzgrat *m* cold lap, burr, fin
Walzgrobeisen *n* merchant iron
Walzgut *n* material being rolled, rolling stock
Walzhaut *f* rolling skin; rolling (mill) scale
Walzhitze *f* rolling heat
Wälzhobelmaschine *f* **[für Zahnräder]** generating gear planer
Walzkaliber *n* roll pass; shaping pass, section groove
geschlossenes W. closed pass
Walzkante *f* rolled edge
Wälzkolbenzähler *m* oval gear meter
Wälzkontakt *m* rolling (rubbing) contact
Wälzkörper *m* rolling element (member)
Wälzkreis *m* [generating] rolling circle *(gear)*
Wälzkreisdurchmesser *m* generating pitch diameter; rolling circle diameter
Wälzkreisteilung *f* circular pitch
Wälzlager *n* antifriction (rolling-contact) bearing
Wälzlagerfett *n* anti-friction grease
Wälzlagerringdrehmaschine *f* lathe for bearing races
Walzplatte *f* sheet ingot; cake *(copper)*
Wälzplatte *f* roller plate
Walzplattieren *n* cladding by rolling
Walzprofil *n* rolled section
Wälzprüfgerät *n* rolling tester *(gear)*
Wälzpunkt *m* pitch point *(gear)*
Wälzräder *npl* gears on parallel or intersecting axes
Wälzregler *m* roller-type control
Wälzreibung *f* rolling friction
Walzrichtung *f* rolling direction, direction of rolling
Walzsand *m* blinding sand
Walzschlacke *f* mill cinder, roll slag
Walzschweißen *n* roll welding
Walzschweißplattierung *f* metal cladding
Walzsinter *m* mill scale
Walzspalt *m* roll gap
Wälzsprung *m* tooth to tooth spalt error
Walzstahl *m* rolled shape (steel section)

Walzstahlkonservierung *f* rolled steel preservation
Wälzstoßen *n* generating (gear) shaping
Wälzstoßmaschine *f* **für Zahnräder** generating gear shaper, gear shaping machine
Wälzstoßverfahren *n* gear-shaping process
Walzstraße *f* roll[ing] train, roll[ing] line, train of mill
  halbkontinuierliche W. semi-continuous [rolling] mill
  kontinuierliche W. continuous [rolling] mill
Walzstruktur *f* fibre structure
Walzteil *n* rolled part
Walztextur *f* rolling texture
Walztoleranz *f* rolling tolerance, mill limits
Walzträger *m* rolled steel joist
Wälzverfahren *n* generating process
Walzwerk *n* rolling mill
  W. mit Schrägwalzen pilger rolling mill
  eingerüstiges W. single rolling mill
  gemischtes W. mixing mill
  halbkontinuierliches W. repeating rolling mill
  mehrgerüstiges W. multiple-stand rolling mill
Walzwerkerzeugnis *n* rolled product
Walzwerkhilfsmaschine *f* mill auxiliary [machine]
Walzwerksaustrittsgeschwindigkeit *f* delivery speed
Walzwerksbau *m* rolling mill construction
Walzwerkserzeugnis *n* rolling mill product
Walzwerkshalle *f* rolling mill shop
Walzwerksingenieur *m* rolling mill engineer
Walzwerksprogramm *n* rolling mill schedule
Walzwerksrollgang *m* rolling mill table
Walzwerksschere *f* rolling mill shears
Walzwerkswesen *n* rolling mill matters
Wälzwiderstand *m* rolling friction[al resistance]
Wälzwinkel *m* angle rolled through
Walzzapfen *m* neck of a roll
Walzzink *n* sheet zinc
Walzzinn *n* laminated tin
Walzzunder *m* mill scale
Wand *f* wall; face *(earthwork, quarrying)*
  halbsteindicke W. half-brick wall
  schallschluckende W. dead (absorbing) wall
  vorgehängte W. curtain wall[ing]
Wandabsaugung *f* *(rock)* wall-siphoning
Wandabsorption *f* wall absorption
Wandanschlußdose *f* wall socket
Wandapparat *m* wall-type telephone, wall telephone set
Wandbefestigung *f* wall mounting
Wandbehang *m* wall rug
Wandbelag *m* wall covering
Wandbohrmaschine *f* wall drill[ing machine]
Wanddampfmaschine *f* wall steam engine
Wanddicke *f* wall thickness
Wanddickenmeßgerät *n* wall thickness gauge
Wanddickenmessung *f* wall thickness gauging (measurement)
Wanddrehkran *m* wall-mounted crane
Wanddurchführung *f* wall entrance
Wandeffekt *m* wall effect
Wandel *m* alteration
Wandelgang *m* ambulatory
Wanderausbau *m* *(mine)* self-advancing [roof-]support, self-moving roof-support
Wanderboot *n* touring boat
Wanderdeckelkrempel *f* *(text)* revolving flat card
Wanderfahrt *f* *(ship)* cruise
Wanderfeld *n* *(el)* travelling field
Wanderfeldröhre *f* travelling-wave tube
Wanderfeldverstärker *m* travelling-wave amplifier
Wanderfeuerung *f* furnace with travelling grate
Wanderjolle *f* centre-board cruiser, cruising centre-boarder
Wanderkasten *m* *(mine)* temporary chock
Wanderklemme *f* anti-creep device
Wanderlänge *f* *(nucl)* migration length
Wanderlast *f* moving load
Wandermatte *f* *(mine)* continuously advancing stowing screen *(steep seams)*
wandern *(ch)* to diffuse; to migrate, to shift; *(el)* to travel; to migrate *(ions)*; to move *(wave)*
Wanderpfeiler *m* *(mine)* temporary chock
Wanderrost *m* travelling grate (bar-grizzly)
Wanderrostfeuerung *f* travelling grate firing
Wanderschicht *f* fluidized bed; moving bed
Wandertisch *m* conveyor table
Wandertransformator *m* travelling track-bound transformer
Wanderung *f* migration; drift
  W. von Ionen migration of ions
  W. von Leerstellen migration of vacancies
  W. von Materie migration of matter
Wanderungsenergie *f* migration energy
Wanderungsgeschwindigkeit *f* rate of migration, drift (migration) velocity
Wanderwelle *f* moving (progressive, travelling) wave
  gegenläufige W. backward travelling wave
Wanderwellenbeschleuniger *m* travelling-wave accelerator
Wanderwellenlinearbeschleuniger *m* travelling-wave linear accelerator
Wanderwellenschutz *m* transient protective device, transient (surge) protection
Wanderzweier *m* double-touring canoe
Wandfaserplatte *f* wallboard
Wandfassung *f* batten lampholder

Wandfernsprecher *m* wall telephone set
Wandgroßplatte *f* room-size[d] wall panel
Wandhalter *m* wall holder
Wandisolator *m* wall insulator
Wandkran *m* wall crane
Wandladung *f* wall charge
Wandlampe *f* wall lamp
Wandleitwert *m* wall admittance
Wandler *m (el)* instrument transformer; transducer; converter; transmitter
 aktiver W. active transformer (transducer)
 asymmetrischer W. dissymmetrical transducer
 bilateraler W. bilateral transducer
 elektroakustischer W. electro-acoustic transducer
 elektromechanischer W. electromechanical transducer
 elektropneumatischer W. electro-pneumatic relay
 induktiver W. inductance transformer
 kapazitiver W. capacitance transducer
 lichtelektrischer W. photocell
 linear-logarithmischer W. linear to logarithmic converter *(computer)*
 magnetostriktiver W. magnetostrictive transducer
 passiver W. passive transducer
 reziproker W. reciprocal transducer
 symmetrischer W. symmetrical transducer
 unilateraler W. unilateral transducer
Wandlerauslösung *f* transformer tripping
Wandlerbaustein *m* transducer assembly
Wandlerempfindlichkeit *f* transducer sensitivity
Wandlerinstrument *n* transformer-type instrument
Wandlerleistung *f* converter performance
Wandlerregler *m* converter-controller
Wandlersicherung *f* transformer fuse
Wandlersignal *n* transduced signal
Wandlerzähler *m* transformer-connected meter
Wandleuchte *f* wall fitting
Wandlung *f* transformation, conversion
Wandpfeiler *m* pilaster
Wandplatte *f* wall panel; wall tile
 schallschluckende W. acoustic tile
Wandradialbohrmaschine *f* radial wall drilling machine
Wandrauheit *f* roughness of the wall surfaces
Wandreaktion *f* wall reaction
Wandrekombination *f (nucl)* wall recombination
Wandschalter *m* surface switch
Wandschalttafel *f* wall-pattern switchboard
Wandschalung *f* wall form
Wandstation *f* wall set *(telephony)*
Wandsteckdose *f* wall socket
Wandstreuung *f* room scattering
Wandtafel *f* wall panel
Wandtemperatur *f* wall temperature
Wandträger *m* bracket
Wandung *f* wall
Wandverkleidung *f* wall covering (facing)
Wandverlust *m* wall loss
Wandverschalung *f* panelling
Wandwange *f* wall string *(stairs)*
Wandwinde *f* wall[-mounted] hoist
Wange *f* cheek *(lathe)*; *(build)* string *(stairs)*; string board *(timber stairs)*
 abgeschrägte W. bevelled cheek
Wangenhobeleisen *n* cartwright's rabbet plane iron
Wanne *f* tub; tank; trough; pan; oil sump; tank *(foundation)*
 pneumatische W. pneumatic trough
Wannenform *f* boat form *(stereochemistry)*
wannenförmig pan-shaped
Wannengründung *f (build)* tanking
Wannenofen *m* tank furnace
Wannerpyrometer *n* Wanner optical pyrometer
Wanten *pl* rigging, shrouds
Wantklotje *f* truck
Wantstropp *m* selvage strop
Ward-Leonard-Schaltung *f* Ward-Leonard control
Ware *f* goods, ware, merchandise; *(text)* fabric
 gewirkte W. knitted material
 weiche W. software *(electronics)*
Warenabzug *m* **/ schwingender** *(text)* interwinding floating take-up
Warenabzugbaum *m* sand (take-up) roll
Warenaustausch *m* exchange of goods
Warenbaum *m* cloth beam
Warenbeschauer *m (text)* percher
Warenbreite *f* width of goods
Warendichte *f* gauge *(cloth)*
Warendoppler *m (text)* piece doubler
Wareneingang *m* receipt of goods
Warenlänge *f (text)* length of the piece
Warenstapel *m (text)* piled fabric
Warenzeichen *n* trade mark
Warenzylinder *m* cloth roller
warm warm, hot; low-activity, low-level
Warmaushärtung *f* aging with increased temperature, artificial aging
Warmbadhärten *n* martempering, hot quenching
Warmbandstraße *f* hot strip mill
Warmbearbeitbarkeit *f* hot workability
Warmbearbeitung *f* hot working
Warmbehandlung *f* heat curing *(concrete)*
Warmbeladung *f* hot[-reactor] loading
Warmbiegen *n* hot bending
Warmbildsamkeit *f* hot ductility
warmbrüchig hot short
Warmbrüchigkeit *f* hot shortness
Wärme *f* heat
 abgehende W. waste heat

durch Verdunstung entzogene W. heat absorbed by evaporation
eigenentwickelte W. self-generated heat
freie W. uncombined heat
fühlbare W. sensible (sensitive) heat
gebundene W. latent heat
Joulesche W. Joule heat
latente W. latent heat
spezifische W. specific heat
spezifische W. bei konstantem Volumen specific heat at constant volume
strahlende W. radiating (radiant) heat
zugeführte W. added (applied) heat, heat received
Wärmeabfall *m* heat drop
Wärmeabfluß *m* heat transfer
Wärmeabfuhr *f* heat dissipation (removal, removing)
Wärmeabführung *f* heat removal (removing)
Wärmeabgabe *f* heat output, emission of heat; dissipation of heat, thermal dissipation; heat loss
Wärmeabgabefläche *f* heat-emitting surface
wärmeabgebend exothermic
Wärmeableiter *m* heat dissipator; heat radiator
Wärmeableitfähigkeit *f* heat-abstracting characteristic
Wärmeableitung *f* heat dissipation (conduction, sink), dissipation of heat
Wärmeableitungsfehler *m* heat conduction error
Wärmeabnahme *f* decrease in temperature, temperature drop
Wärmeabschirmung *f* thermal shield
Wärmeabsorption *f* heat (thermal) absorption
Wärmeakkumulator *m* heat accumulator
Wärmeänderung *f* change of temperature
Wärmeansammlung *f* accumulation of heat
Wärmeäquivalent *n* heat equivalent
elektrisches W. electrical equivalent of heat
mechanisches W. mechanical equivalent of heat
Wärmearbeitswert *m* mechanical equivalent of heat
Wärmeaufnahme *f* heat absorption, gain of heat
Wärmeaufnahmefähigkeit *f* heat-absorption capacity
Wärmeaufnahmefläche *f* heat-absorbing surface
Wärmeaufnahmevermögen *n* heat absorptivity
wärmeaufnehmend heat-absorbing; endothermic, endothermal
Wärmeausbeute *f* heat yield
Wärmeausbreitung *f* thermal diffusion
Wärmeausbreitungsvermögen *n* thermal diffusivity
Wärmeausdehnung *f* thermal expansion, expansion due to heat
lineare W. linear thermal expansion
Wärmeausdehnungskoeffizient *m* coefficient of thermal expansion
Wärmeausdehnungsvermögen *n* thermal expansivity
Wärmeausdehnungszahl *f* coefficient of thermal expansion
Wärmeausgleich *m* heat equalization (compensation)
Wärmeausnutzung *f* heat (thermal) efficiency, heat utilization, utilization of heat
Wärmeausstrahlung *f* heat radiation, radiation of heat
Wärmeaustausch *m* heat exchange, interchange of heat
Wärmeaustauschapparat *m* heat exchanger (interchanger)
Wärmeaustauscher *m* heat exchanger
Wärmebeanspruchung *f* heat (thermal) stress
Wärmebedarf *m* heat (calorific) requirement
wärmebehandelt heat-treated
Wärmebehandlung *f* heat treatment
Wärmebehandlungsbad *n* heat-treatment bath
wärmebehandlungsfähig heat-treatable
Wärmebelastung *f* thermal stress
Wärmeberechnung *f* temperature-rise computation
wärmebeständig heat proof, heat-resisting; thermostable
Wärmebeständigkeit *f* heat resistance, thermal resistivity; *(plast)* thermal stability
Wärmebeständigkeitsklasse *f* temperature class
Wärmebewegung *f* thermal agitation (motion)
Wärmebilanz *f* heat balance
Wärmebild *n* temperature entropy diagram
Wärmebildner *m* heat producer
Wärmebindung *f* heat absorption
Wärmeblitz *m* heat flash
Wärmebrücke *f* thermal bridge
wärmedämmend heat-insulating
Wärmedämmplatte *f* heat-insulating board
Wärmedämmung *f* thermal (heat) insulation
Wärmedämmvermögen *n* thermal resistance
Wärmedehnung *f* thermal expansion
Wärmedehnungsmesser *m* heat dilatometer
Wärmediagramm *n* temperature entropy diagram
Wärmediffusion *f* thermodiffusion, thermal effusion
Wärmedissipation *f* heat dissipation
Wärmedrift *f* thermal drift
Wärmedruckprüfung *f* *(plast)* hot penetration test
Wärmedurchdringung *f* penetration of heat
Wärmedurchgang *m* passage of heat

Wärmedurchgangswiderstand *m* thermal resistance
Wärmedurchgangszahl *f* coefficient of heat transition, overall coefficient of heat transfer, overall heat-transfer coefficient, thermal transmittance
wärmedurchlässig diathermic, transparent to heat
Wärmedurchlässigkeit *f* diathermancy
Wärmedurchsatz *m* rate of heat transfer
Wärmedurchschlag *m* thermal breakdown
Wärmedynamik *f* thermodynamics
Wärmedynamometer *n* heat dynamometer
Wärmeeinheit *f* caloric (heat) unit
britische W. British thermal unit, B.T.U., Btu, BTU, B.t.u.
Wärmeeinsaugung *f* heat absorption
Wärmeeinschließung *f* thermal isolation (confinement)
Wärmeeinstrahlung *f* heat absorption
wärmeelastisch *(plast)* thermoelastic
wärmeelektrisch thermoelectric
Wärmeelektrizität *f* thermoelectricity
wärmeempfindlich heat-sensitive
Wärmeenergie *f* thermal (heat) energy
Wärmeentbindung *f* release of heat
wärmeentwickelnd heat-generating, exothermic
Wärmeentwicklung *f* generation (development) of heat
Wärmeentzug *m* heat extraction
Wärmeerhärtung *f* heat curing
Wärmeersparnis *f* saving of heat
Wärmeerzeugung *f* heat generation, generation (production) of heat
Wärmeexplosion *f* thermal explosion
Wärmefluß *m* thermal (heat) flow
Wärmegefälle *n* heat drop, temperature gradient
Wärmegeräusch *n* thermal noise
Wärmegleichgewicht *n* thermal (temperature) equilibrium
Wärmegleichung *f* heat equation
Wärmegrube *f (met)* heating pit
wärmehärtbar *(plast)* thermosetting
Wärmeimpulsschweißen *n (plast)* thermal impulse heat sealing
Wärmeinhalt *m* caloric capacity, heat content, enthalpy
Wärmeionisation *f* thermal ionization
Wärmeisolator *m* heat insulator
wärmeisoliert thermally insulated
Wärmeisolierung *f* heat insulation
W. mit Schaumstoff foam insulation
Wärmekapazität *f* calorific (thermal) capacity, capacity of heat
Wärmekippen *n* thermal instability *(of organic insulating materials)*
Wärmekontakt *m* thermal contact
Wärmekonvektion *f* thermal convection, convection of heat
Wärmekraft *f* thermal power
Wärmekraftmaschine *f* heat (thermal) engine
Wärmekraftwerk *n* thermal power station, *(Am)* thermal power plant
Wärmekreislauf *m* thermal cycle
Wärmelehre *f* thermodynamics
Wärmeleistung *f* heat output; thermal power *(of reactor)*
wärmeleitend heat conducting
Wärmeleiter *m* heat conductor
Wärmeleitfähigkeit *f* thermal (heat) conductivity
Wärmeleitfähigkeitsgasanalysator *m* hot-wire gas analyzer
Wärmeleitung *f* heat (thermal) conduction, conduction of heat
Wärmeleitungskoeffizient *m* coefficient of thermal conductivity
Wärmeleitungsmanometer *n* heat-conductivity pressure gauge, hot-wire pressure gauge
Wärmeleitungsvakuummeter *n* Pirani (heat-conductivity pressure) gauge
Wärmeleitungsvermögen *n* heat (thermal) conductivity
Wärmeleitungswiderstand *m* thermal resistance
spezifischer W. thermal resistivity
Wärmeleitvermögen *n* heat (thermal) conductivity
Wärmeleitzahl *f* coefficient of thermal conductivity
Wärmemangel *m* shortage of heat
Wärmemauer *f* heat barrier
Wärmemenge *f* heat quantity, quantity (amount) of heat
durch eine Scheidewand hindurchgehende W. quantity of heat passed through a partition
Wärmemengenzähler *m* heat meter
Wärmemessung *f* pyrometry
wärmen to warm, to heat
warmentgraten to hot trim
Wärmeplatte *f* hot plate
Wärmepotential *n* heat potential
Wärmepumpe *f* heat pump
Wärmequelle *f* heat source
Wärmerauschen *n* thermal[-agitation] noise
Wärmerauschgenerator *m* thermal noise generator
Wärmereaktor *m* [industrial] heat reactor, heat-producing reactor
Wärmerelais *n* thermally-operated relay
Wärmerückgewinnung *f* heat recovery
Wärmerückstrahlung *f* heat reflection
Wärmesatz *m* heat theorem
Wärmeschild *m* thermal shield

Wärmeschmelzverfahren *n* heat fusion process *(foambacks)*
Wärmeschrumpfung *f* thermal contraction
Wärmeschutz *m* heat (thermal) insulation; heat protection
Wärmeschutzband *n* tape covering
Wärmeschutzfilter *n* heat-absorbing filter
Wärmeschutzhülle *f* covering of non-conducting material
Wärmeschutzisolierung *f* heat insulation
Wärmeschutzkleidung *f* heat protective clothing
Wärmeschutzmasse *f* non-conducting material, lagging
Wärmeschutzmaßnahme *f* heat protection
Wärmeschutzmittel *n* non-conducting material, lagging
Wärmeschutzpanzer *m* heat-insulating casing
Wärmeschutzschirm *m* heat screen
Wärmeschutzschnur *f* cord covering
Wärmeschutzstoff *m* heat-insulating material
Wärmeschutzverkleidung *f* lagging
Wärmeschutzwagen *m* insulated (isothermic) wagon, *(Am)* insulated car
Wärmeschwingung *f* heat vibration
Wärmespannung *f* thermal stress
Wärmespeicher *m* heat regenerator
elektrischer W. electric heat accumulator
Wärmespeicherung *f* heat storage, accumulation of heat
Wärmespektrum *n* heat spectrum
Wärmestoß *m* thermal shock
Wärmestrahl *m* heat (thermal) ray
Wärmestrahler *m* heat (thermal) radiator
Wärmestrahlung *f* heat [transfer by] radiation, radiant heat
Wärmestreuung *f* dispersion of heat
Wärmestrom *m* heat flow, flow of heat; thermal (heat) current
Wärmestromdichte *f* heat flow density
Wärmeströmung *f* heat flow
Wärmetauscher *m* heat exchanger
Wärmetechnik *f* heat engineering
Wärmetheorie *f* theory of heat
Wärmetisch *m* / **elektrischer** electric hot table
Wärmetod *m* heat death
Wärmetönung *f* heating effect (tone, change)
Wärmetönungssauerstoffanalysator *m* catalytic combustion oxygen analyzer
Wärmeträger *m* heat-transfer agent (medium), cooling medium, coolant
Wärmeübergang *m* heat transmission (transfer)
W. durch Konvektion transfer of heat by convection
W. durch Leitung transfer of heat by conduction
W. durch Strahlung transfer of heat by radiation
Wärmeübergangszahl *f* coefficient of heat transmission, heat transfer factor (coefficient)
Wärmeüberschuß *m* excess of heat
Wärmeübertrager *m s.* Wärmeträger
Wärmeübertragung *f s.* Wärmeübergang
Wärmeübertragungsmittel *n s.* Wärmeträger
Wärmeübertragungszahl *f s.* Wärmeübergangszahl
Wärmeumformer *m* thermo-converter
Wärmeumlaufkühlung *f* thermo-siphon cooling
Wärmeumwälzung *f* heat convection
wärmeundurchlässig adiathermic
Wärmeverbrauch *m* heat consumption
wärmeverbrauchend heat-consuming, endothermal, endothermic
Wärmeverlust *m* heat loss, loss of heat
W. durch Luftüberschuß heat loss due to excess of air
W. durch Rußbildung heat loss due to formation of soot
W. durch unverbrannte Gase heat loss due to unburnt gases
W. durch unvollkommene Verbrennung heat loss due to incomplete combustion
Wärmewelle *f* heat (thermal) wave
Wärmewert *m* calorific (heat, thermal) value
W. der mechanischen Arbeit thermic equivalent
Wärmewiderstand *m* thermal resistance
Wärmewirkung *f* thermal effect
Wärmewirkungsgrad *m* thermal efficiency
Wärmezerstreuung *f* heat dissipation
Wärmezufuhr *f* heat supply
Wärmezustand *m* thermal state
Warmfederindikator *m* indicator with internal spring
warmfest high-temperature resistant
Warmfestigkeit *f* high-temperature strength
Warmfließgrenze *f* yield-point at elevated temperatures
Warmformen *n* hot forming (working)
Warmformgebung *f* hot forming (working)
warmgelaufen warmed up
warmgeschlagen hot formed *(rivet)*
Warmgesenkpresse *f* hot stamping press
warmgewalzt hot-rolled
warmgewickelt hot-wound, hot-coiled *(spring)*
Warmhalteofen *m* holding furnace
Warmhärte *f* hot (red) hardness
Warmkaliber *n* hot groove
Warmkreissäge *f* hot circular saw
warmlaufen to run hot
Warmlaufen *n* **des Lagers** overheating of the bearing
Warmluftfront *f* warm front
Warmluftheizung *f* warm-air heating
Warmluftkörnertrockner *m* hot-air grain drier
Warmluftofen *m* warm-air furnace

Warmnebelgerät *n (ag)* thermal [aerosol] fogger
Warmnebelmittel *n (ag)* thermal [aerosol] fog
warmnieten to hot-rivet
Warmnieten *n* hot riveting
Wärmofen *m (met)* soaking furnace
warmpressen to hot-press
Warmpressen *n* hot pressing
Warmrekristallisation *f* recrystallization in hot-worked material
Warmriß *m* hot crack
Warmsäge *f* hot saw[ing machine]
Warmschleifen *n* high-temperature grinding
Warmschliff *m* hot-ground pulp
Warmsprödigkeit *f* hot (red) shortness
Warmstartlampe *f* hot-start lamp
Warmstauchen *n* hot upsetting
Warmstraße *f* hot rolling train
Warmstreckgrenze *f* tensile yield point at elevated temperature
warmumformbar hot-workable
Warmumformbarkeit *f* hot workability
Warmumformung *f* hot forming (working)
Warmversprödung *f* hot embrittlement
Warmversuch *m* elevated-temperature test
warmwalzen to hot-roll
Warmwalzen *n* hot rolling
Warmwalzwerk *n* hot rolling mill
Warmwasser *n* hot water
Warmwasserbereiter *m* [electric] water heater
Warmwasserdurchlauferhitzer *m* geyser
Warmwasserheizung *f* hot-water heating
Warmwasserheizungsanlage *f* hot-water [domestic] heating system
Warmwasserinjektor *m* hot-water injector
Warmwasserspender *m* instantaneous water-heater
Wärmzeit *f* heating time
Warmzerreißprüfung *f* tensile test at elevated temperature, hot-tensile test
warmziehen to hot-draw
Warnanlage *f* warning device
Warngerät *n (nucl)* monitor
Warnglocke *f* warning (alarm) bell
Warngrenze *f* warning limit
Warnhupe *f* alarm horn
Warnkreuz *n (ry)* warning cross (signal)
Warnlampe *f* warning lamp; warning light, pilot lamp
Warnmeldung *f* warning
Warnpfeife *f* alarm whistle
Warnsignal *n* warning (alarm) signal
Warnung *f* warning, monitoring
Warnungsboje *f* danger (mark) buoy
Warnungstafel *f* warning board (notice)
Warnzeichen *n* warning sign[al]
  optisches W. warning light
Warnzone *f* caution zone
Warpanker *m* kedge [anchor], warp anchor
Warpklampe *f (ship)* cavil, cavel
Warpleine *f (ship)* [guess] warp
Warte *f* control panel; switchboard gallery; observation post; control room
Wartefunkfeuer *n* holding beacon
Wartehalle *f* shelter
warten to [at]tend; to service, to maintain
Wartesaal *m* waiting-room
Warteschlangentheorie *f (dat)* queueing theory
Warteschleife *f (aero)* holding pattern
  normale W. standard holding pattern
Warteverfahren *n (aero)* holding procedure
Wartezeit *f (telec)* delay[s], setting-up time, speed of answer, *(Am)* answering interval; waiting time
Wartezeitbetrieb *m* delay working *(telephony)*
Wartezeittheorie *f (dat)* queueing theory
Wartung *f* servicing, maintenance; human attendance, operator attention
  leichte W. ease of maintenance
  vorbeugende W. preventive maintenance
Wartungsaufwand *m* maintenance requirements
wartungsfrei without operator attention, without attention by the operator; maintenance-free, requiring no maintenance
Wartungshinweise *mpl* servicing information
Wartungskosten *pl* maintenance cost, cost of attendance
Wartungsmethode *f* maintenance method
Wartungsmöglichkeit *f (Am)* serviceability
Wartungspersonal *n* service personnel
Wartungsplan *m* maintenance chart
Wartungsprogramm *n (dat)* service routine
Wartungsvorschrift *f* servicing schedule
Warze *f* lug, boss; projection, embossment *(projection welding)*
Warzenblech *n* nipple plate
Warzenschweißen *n* projection welding
was zu beweisen war *(math)* quod erat demonstrandum, q.e.d., which was to be proved
waschaktiv surface active
Waschanlage *f* scrub section *(gas)*; washing (cleaning) plant
Waschautomat *m* automatic washer
waschbar washable
Waschbarkeit *f* washability
Waschbecken *n* wash basin
Waschbehandlung *f* washing treatment
Waschbenzin *n* [pure] benzine
Waschberge *pl* washery dirt (refuse)
Waschbeständigkeit *f* fastness to washing
Waschbord *m (ship)* washboard, wash rail; *(Am)* manger
Waschbrett *n* washboard
Waschbütte *f* purification tub
Wäsche *f* cleaning plant, washery

W. mit geschlossenem Wasserkreislauf closed washery
alkalische W. alkali extraction *(pulp)*
waschecht fast to wash[ing]
Waschechtheit *f* fastness to washing
Waschechtheitsprüfung *f* washing test
Wäschedesinfektionsanlage *f* linen-disinfection plant
Wascheigenschaften *fpl* detergent characteristics
Wascheinrichtung *f* washing plant
waschen to wash, to clean, to purify; to scrub *(gas)*
Wäscher *m* 1. *(ch)* washer, wash column, scrubber; 2. *(text)* laundrymen
Wäscherei *f* laundry
Wäschereieinrichtung *f* / **mechanische** *(text)* power laundry equipment
Wascherz *n* wash ore
Wäschesteiger *m (mine)* washery superintendent
Wäschestoff *m* linen fabric
Wäschetinte *f* ink for marking linen
Wäschezimmer *n* linen room
Waschfalte *f* crease formed in the washer
Waschfaß *n* dolly
waschfest *s.* waschecht
Waschfläche *f* scrubbing surface
Waschflasche *f (ch)* washing bottle
Waschflotte *f (text)* detergent solution
Waschgold *n* alluvial (placer) gold
Waschgrube *f (wood)* blow pit
Waschhaus *n* laundry
Waschhilfsmittel *npl* builders
Waschkaue *f (mine)* pit[head] bath
Waschkraft *f* detergent efficiency (power)
Waschlauge *f* washing liquor
Waschleder *n* wash leather
Waschlösung *f* scrub, strip; cleaning solution
Waschmaschine *f* washing machine
elektrische W. electric washing machine
Waschmittel *n* washing agent, detergent
geformtes W. shaped washing agent
synthetisches W. synthetic detergent
Waschöl *n* wash oil
abgetriebenes W. stripped wash oil
Waschplatte *f (ship)* wash plate
Waschpulver *n* washing powder
Waschschott *n (ship)* weather board; *(Am)* manger
Wasch- und Trageausrüstung *f (text)* wash and wear finish
Waschwasser *n* wash[ing] water
Waschwirkung *f* detergency, washing power
Waschzusatzmittel *n* scouring addition
Waschzyklon *m* cyclone separator (washer)
Washitastein *m* Washita stone
Wasser *n* water
angesäuertes W. acidulated water
beschränktes W. *(ship)* shallow water
destilliertes W. distilled water
eisenhaltiges W. ferruginous water
entlüftetes W. de-aired water
festes W. bound water
flaches W. *(ship)* shallow, shallow (shoal) water
fließendes W. running water
fossiles W. connate water
gewöhnliches W. ordinary (natural, light) water
hartes W. hard water
herabrieselndes W. trickling water
hygroskopisches W. hygroscopic water
kalkhaltiges W. carbonate [of lime] water
leichtes W. light water
mitgerissenes W. water entrained with the steam, priming water
natürliches W. ordinary (natural, light) water
offenes W. open water *(free from ice)*
salinisches W. saline water
schlammhaltiges W. muddy (sedimentary) water
schweres W. heavy water
stehendes W. dead water
stillstehendes W. stagnant water
übergerissenes W. priming water, water entrained with the steam
unbegrenztes W. open water
ungereinigtes W. untreated water
Wasserabflußhahn *m* drain cock
Wasserablaßloch *n* drain hole
Wasserablauf *m* drip cap; gully
Wasserabscheider *m* water separator
Wasserabschirmung *f* water shield[ing]
Wasserabspaltung *f (ch)* elimination of water, dehydration
Wasserabstoßen *n* bleeding, sweating *(concrete)*
wasserabstoßend *(ph)* water-repellent, hydrofuge, hydrophobic
wasserabweisend water-repellent, water-repelling
Wasserabweisung *f* water repellency
wasseranziehend hydrophilic, hygroscopic
Wasseraufbereitung *f* water treatment (purification)
Wasseraufbereitungsanlage *f* water-treating equipment
Wasseraufnahme *f* water absorption
Wasseraufnahmefähigkeit *f* water absorbing property
Wasseraufnahmevermögen *n* absorptive capacity for water *(soil)*
Wasserbad *n* water bath
Wasserball *m* water polo
Wasserballast *m* water ballast
W. hinten after-tank
W. vorn fore-tank
Wasserballastkasten *m* / **hoher** side tank

Wasserballasttank *m* [water] ballast tank
Wasserballasttankdecke *f* top (crown) of tank
Wasserbarometerhöhe *f* height of water column
Wasserbassin *n* water basin
Wasserbau *m* hydraulic engineering (construction)
 landwirtschaftlicher W. agricultural hydraulics
Wasserbauingenieur *m* hydraulic engineer
Wasserbauten *pl* hydraulic structures
Wasserbauwerk *n* hydraulic structure
Wasserbauzement *m* hydraulic cement
Wasserbecken *n* [water] reservoir, water basin
Wasserbedarf *m* water requirements, quantity of water required
Wasserbedarfsschwankung *f* variation of quantity of water required
Wasserbehälter *m* water reservoir (tank)
Wasserbelastungsverfahren *n* water-load method *(measuring technique)*
Wasserbenetzung *f* water wetting
Wasserberieselung *f* irrigation, water spraying
Wasserbesatz *m (mine)* water stemming
wasserbeständig water-resistant, waterproof
Wasserbeständigkeit *f* waterproof property
Wasserbombe *f* depth charge (bomb)
Wasserbreite *f* breadth of water *(of a river)*
Wasserbremse *f* water brake
Wasserchemie *f* water chemistry
Wasserdamm *m* dam, bank for water, watertight stopping
Wasserdampf *m* steam, water vapour
Wasserdampfabsorption *f* absorption due to water vapour
Wasserdampfdestillation *f* steam distillation
Wasserdampfdestille *f* steam still
Wasserdampfdruck *m* aqueous tension
Wasserdampfröste *f (text)* steam retting
wasserdicht watertight, waterproof; *(mach, el)* submersible
Wasserdichtigkeit *f* watertightness
Wasserdichtung *f* waterproof seal
Wasserdispersionsfarbe *f* water paint
Wasserdruck *m* hydraulic pressure; hydrostatic pressure
Wasserdruckkolben *m* hydraulic ram
Wasserdruckkraft *f* hydraulic power
Wasserdruckmanometer *n* hydraulic pressure gauge
Wasserdruckrelais *n* water-pressure relay
Wasserdruckversuch *m* hydraulic test
Wasserdurchflußmesser *m* water flowmeter
Wasserdurchlaß *m (build)* culvert
wasserdurchlässig pervious, permeable
Wasserdurchlässigkeit *f* water permeability
Wassereimer *m* water bucket
Wassereinbruch *m (mine)* water inrush (inflow)
Wassereindringung *f* intrusion of water
Wassereinspritzung *f* water injection
Wassereintritt *m* water inlet
Wasserenthärtung *f* demineralization of water, water softening
Wasserenthärtungsmittel *n* water softener
Wasserentkrautung *f* water weed control
Wasserentlüftungsanlage *f* air separator for water plants
Wasserentöler *m* oil separator [for water], grease filter [for water]
Wasserentölung *f* separation of water from oil
Wasserentsalzung *f* water desalination
wasserentziehend dehydrating
Wasserentziehung *f* dehydration, removal of water
Wasserfahrzeug *n* water craft, ship
 funkgesteuertes W. drone
Wasserfarbe *f* water paint
Wasserfaß *n* [water] breaker
Wasserfilter *n* water filter
Wasserfläche *f* surface of water
Wasserflugzeug *n* hydro[-air]plane, hydroaeroplane, seaplane, waterplane
Wasserflut *f* water flood
Wasserförderung *f* raising of water
Wasserfracht *f* water carriage
wasserfrei anhydrous; absolute *(alcohol)*
wasserführend water-bearing
Wassergang *m* / **innerer** *(ship)* inner waterway
Wassergas *n* water gas
Wassergasgenerator *m* water gas producer (generator)
Wassergasschweißen *n* water gas welding
Wassergeflügel *n* water fowl
Wassergehalt *m* moisture content; water content
Wassergehaltsspanne *f* difference in the moisture contents
wassergekühlt water-cooled
wassergesättigt waterlogged
Wasserglas *n* water glass
Wassergraben *m* water ditch
Wasserhahn *m* [bib-]cock, tap, faucet
wasserhaltig aqueous, hydrated, hydrous
 chemisch w. hydrated
Wasserhaltung *f* ground-water lowering, dewatering, draining, drainage, pumping
 W. mit Filterbrunnen well point pumping
Wasserhaltungsdampfmaschine *f* mine pumping steam engine
Wasserhaltungsstollen *m* drainage adit
Wasserhärte *f* water hardness
 bleibende W. permanent water hardness
 vorübergehende W. temporary water hardness
Wasserhärten *n (met)* water hardening
wasserhärtend water-hardening
Wasserhärter *m* water hardening material
Wasserhärteskale *f* scale of water hardness

Wasserhaushalt *m* water balance
Wasserhebung *f* drainage, water hoisting
Wasserhochbehälter *m* high level water tank
Wasserhöhe *f* water level
Wasserhorizont *m* water table
Wasserjoch *n (mine)* water-ring
Wasserkalk *m* hydraulic lime
Wasserkalorimeter *n* water calorimeter
Wasserkammerkessel *m* water tube boiler with header
Wasserkanal *m* water channel
Wasserkanone *f (mine)* hydraulic giant (gun), monitor
Wasserkante *f* waterfront, water's edge, seaboard
Wasserkasten *m* water cistern; *(auto)* radiator tank
Wasserkastenwalze *f* damping fountain roller
Wasserkesselreaktor *m s.* Wasserkocherreaktor
wasserklar waterwhite
Wasserklosett *n* water closet
Wasserkocherreaktor *m* water boiler [reactor]
Wasserkraft *f* water power
Wasserkraftanlage *f s.* Wasserkraftwerk
Wasserkraftgenerator *m* hydro-power generator
Wasserkraftmaschine *f* hydraulic motor
Wasserkraftwerk *n* hydroelectric (water) power station, *(Am)* hydroelectric (water) power plant
  oberliegendes W. above-lying hydroelectrical power station
Wasserkran *m* water crane (column), hydrant
Wasserkreislauf *m* automatic return of water
Wasserkühler *m* water-cooled radiator
Wasserkühlung *f* water cooling
Wasserlast *f (ship)* water hold (storage tank)
Wasserlauf *m (ship)* water course; course [of water]; channel
Wasserleitung[srohr *n***]** *f* water pipe
Wasserlinie *f (ship)* water line
  W. Null base line
Wasserlinien *fpl* laid lines, wire marks *(paper)*
Wasserlinienebene *f (ship)* water plane
Wasserlinienflächenkurve *f* curve of floatations
Wasserlinienlänge *f (ship)* waterline length
Wasserlinienriß *m (ship)* waterline (half-breadth) plan
Wasserlinienvölligkeitsgrad *m (ship)* waterplane (load-water-line) coefficient
Wasserlinienzylinder *m (print)* laid dandy roll
wasserlöslich water-soluble, soluble in water
Wasserlösung *f (mine)* drainage
Wasserlösungsreaktor *m* aqueous homogeneous reactor
Wassermanometer *n* water draught gauge
Wassermantel *m* water jacket
Wassermantelofen *m* water-jacketed furnace
Wassermasse *f* body of water
Wassermessung *f* water metering, hydrometry
Wassermörtel *m* hydraulic mortar
Wassermotor *m* hydraulic (water) motor
Wassermühle *f* water mill
wassern to land on water, to alight upon water *(seaplane)*
wässern to wash; to hydrate, to rinse, to purify
Wassernase *f* water (weather) drip
Wasserpapier *n* waterleaf paper
Wasserpaß *m* boot-top line, level of the water
Wasserpaßriß *m* half-breadth plan
Wasserpegelschreiber *m* water level recorder
Wasserpforte *f* bulwark (clearing, freeing) port, waterport
Wasserpumpe *f* water pump
Wasserpumpengehäuse *n* water pump body
Wasserpumpenkühlung *f* forced-circulation cooling
Wasserquellbarkeit *f* water soaking properties
Wasserquerschnitt *m* water cross-section *(of a river or channel)*
Wasserrad *n* water wheel
Wasserradiolyse *f* radiolysis of water, water decomposition
Wasserreaktor *m* light-water reactor
Wasserreinigung *f* water purification
  chemische W. chemical purification of water
  mechanische W. mechanical water purification
Wasserreinigungsanlage *f* water purification (softening) plant
Wasserreserve *f* water reserve
Wasserring *m (mine)* [water] garland, water ring
Wasserrinne *f s.* Wasserring
Wasserriß *m* water crack
Wasserrohr *n* water pipe (main casing)
Wasserrohrdampfkessel *m* water-tube boiler
Wasserrohr-Heizrohr-Kessel *m* combined water tube and fire tube boiler
Wasserrohrkessel *m* water tube boiler
  W. mit einer Wasserkammer water tube boiler with one header
  W. mit gebogenen Rohren water tube boiler with curved (bent) tubes
  W. mit geneigten Rohren water tube boiler with inclined tubes
  W. mit geraden Rohren water tube boiler with straight tubes
  W. mit geteilten Wasserkammern water tube boiler with divided headers
  W. mit Kapselverbindungen an den Rohrenden water tube boiler with junction boxes at the tube ends
  W. mit senkrechten Rohren water tube boiler with vertical tubes
  W. mit ungeteilten Wasserkammern water tube boiler with single header
  W. mit Wasserkammern water tube boiler with headers

W. mit zwei Wasserkammern water tube boiler with two headers
engrohriger W. water tube boiler with small tubes, express type of water tube boiler
stehender W. von Schuchoff Schuchoff's vertical water tube boiler
weitrohriger W. water tube boiler with large tubes
Wasserrösche *f* *(mine)* drainage ditch, flume, gutter
Wasserröste *f* water ret
Wasserrückhaltevermögen *n* water retention property
Wasserrückkühlung *f* water recooling
Wassersammler *m* steam drier (separator, trap)
Wassersäule *f* water head (column)
Wassersäulenzugmesser *m* water draught gauge
Wasserschaden *m* damage caused by water
Wasserscheide *f* watershed
Wasserschenkel *m* dripstone, label; water (weather) drip
Wasserschieber *m* water sluice valve
Wasserschild *m* water shield
Wasserschlag *m* water hammer
Wasserschlauch *m* water hose (tube)
Wasserschleier *m* water film
Wasserschutzpolizei *f* waterguard, *(Am)* coast guard
Wassersohle *f* drainage water
Wasserspeicher *m* cistern; reservoir
Wasserspeier *m* gargoyle, gurgoyle *(architecture)*; water-spout
Wassersperre *f* water barrier
Wassersperrung *f* *(build)* waterproofing
Wasserspiegel *m* water level (table)
Wassersport *m* aquatics, water (aquatic) sport
Wassersprengwagen *m* water spraying cart
Wassersprung *m* hydraulic jump
Wasserspülbohren *n* boring with water pressure
Wasserspülkasten *m* closet cistern
Wasserspülrohr *n* flush pipe
Wasserstag *n* *(ship)* bobstay
Wasserstand *m* water level; water table *(ground)*; height of water; sea level
mittlerer W. mean sea level
Wasserstandsanzeiger *m* water gauge (level indicator)
W. mit Schwimmer float gauge (water level indicator)
doppelter W. double water gauge
einfacher W. single water gauge
Wasserstandsgehäuse *n* arm of the water gauge
Wasserstandsglas *n* water gauge, gauge glass
W. mit Probierhahnreihe water gauge with test (gauge) cocks
Wasserstandsglasrohr *n* gauge glass
Wasserstandshahn *m* test (gauge) cock
Wasserstandshahnkopf *m* water gauge arm with cock
Wasserstandsmarke *f* water mark (level pointer)
Wasserstandsventilkopf *m* water gauge arm with valves
Wasserstau *m* **nach oben** rise of water
Wasserstaubecken *n* water reservoir
Wasserstoff *m* hydrogen
atomarer W. atomic hydrogen
gebundener W. bound hydrogen
leichter W. light hydrogen, protium
naszierender W. nascent hydrogen
schwerer W. heavy hydrogen, deuterium
überschwerer W. tritium
wasserstoffähnlich hydrogenlike
Wasserstoffanlagerung *f* hydrogenation
Wasserstoffatmosphäre *f* atmosphere of hydrogen
Wasserstoffatom *n* hydrogen atom
Wasserstoffaustauschreaktion *f* hydrogen exchange reaction
Wasserstoffbindung *f* hydrogen bond
Wasserstoffbombe *f* hydrogen bomb, H-bomb
Wasserstoffbombenreaktion *f* hydrogen thermonuclear reaction
Wasserstoffbrüchigkeit *f* hydrogen embrittlement
Wasserstoffbrücke *f* hydrogen bridge
Wasserstoffbrückenbindung *f* hydrogen bridge linkage
Wasserstoffeffekt *m* hydrogenium phenomenon *(circuit-breakers)*
Wasserstoffelektrode *f* hydrogen electrode
Wasserstoffentladungslampe *f* hydrogen discharge lamp
Wasserstoffentwicklung *f* development of hydrogen
Wasserstofferzeuger *m* hydrogen generator
Wasserstoffflasche *f* hydrogen cylinder
Wasserstoffgas *n* hydrogen gas
Wasserstoffgehalt *m* hydrogen content
wasserstoffgesättigt hydrogenated
Wasserstoffgewinnungsanlage *f* hydrogen producing plant
wasserstoffhaltig hydrogenous
Wasserstoffion *n* hydrogen ion
Wasserstoffionenkonzentration *f* hydrogen ion concentration, pH value
Wasserstoffkernwaffe *f* hydrogen weapon, H-weapon
Wasserstoffkorrosion *f* hydrogen attack
Wasserstoffkrankheit *f* *(met)* hydrogen unsoundness *(of copper)*
Wasserstoffperoxid *n* hydrogen peroxide
Wasserstoffrückstoßzähler *m* *s.* Protonenrückstoßzähler
Wasserstoffsprödigkeit *f* hydrogen embrittlement

Wasserstoffsuperoxid *n* hydrogen peroxide
Wasserstoffverbindung *f* hydrogen compound
Wasserstoffwertigkeit *f* hydrogen valence
Wasserstoffzahl *f* hydrogen ion concentration
Wasserstoffzyklus *m* proton-proton chain
Wasserstoll[e]n *m (mine)* drainage adit (tunnel); *(build)* water tunnel
Wasserstoß *m* impact of water
Wasserstrahl *m* water jet
Wasserstrahlantrieb *m* hydro-jet propulsion
Wasserstrahlelektrodenkessel *m* water-jet heater
Wasserstrahlgebläse *n* water-jet blower
Wasserstrahlpropeller *m* [water-]jet propeller
Wasserstrahlpumpe *f* aspirator (water-jet, filter) pump
Wasserstrahlputzen *n* high-pressure water cleaning of castings
Wasserstraße *f* water highway (route), waterway
Wasserstraßennetz *n* waterway system
Wasserstrecke *f (mine)* water level; draining level
Wasserströmungswächter *m* water flow relay
Wassersumpf *m (mine)* water standage
Wassertank *m* water reservoir (tank)
Wassertaxi *n* runabout, water-taxi
Wassertiefe *f* depth of water
W. über Pallen draught over keel blocks
Wassertragflächenboot *n* hydrofoil boat
Wassertransport *m* water carriage (transport)
Wassertransportboot *n* water carrier
Wassertrecker *m* water tractor *(vessel)*
Wassertropfen *m* drop of water
Wasserturbine *f* hydraulic (water) turbine
Wasserturm *m* water tower, elevated tank
Wasserüberwachung *f (nucl)* water monitoring (survey)
Wasserüberwachungsgerät *n* water [radioactivity] monitor
Wasseruhr *f* water gauge (meter)
Wasserumlaufbütte *f* hydrocycle tank *(papermaking)*
Wasserumlaufkühlung *f* cooling by means of circulating water
Wasserumlaufpumpe *f* water circulating pump, circulating-water pump
wasser- und aschefrei dry ash-free
wasser- und mineralstofffrei dry mineral-matter-free
wasserundurchlässig impervious, impermeable
Wässerung *f* wash[ing]
Wässerungsbad *n* washing bath
wasserunlöslich insoluble in water
Wasser-Uran-Gitter *n* uranium-water lattice
Wasserverdrängung *f* displacement
wasserverdünnt water-thinned
Wasserverschluß *m* water seal
Wasserversorgung *f* water supply
Wasserverteilungswand *f* baffle plate *(condenser)*
Wasserverunreinigung *f* water pollution
Wasservorrat *m* water reserve
Wasserwaage *f* water-level
Wasserwagen *m* water tank wagon
Wasser-Wasser-Reaktor *m* water-water reactor, pool-type reactor
Wasserweg *m* water route, waterway
Wasserwerk *n* waterworks
Wasserwert *m* water equivalent *(calorimetry)*; water factor *(viscosimetry)*
W. des Kalorimeters water equivalent of the calorimeter
W. des Kondensatorkörpers water-equivalent of the condenser body
kalorimetrischer W. calorimetric water equivalent, thermal capacity
Wasserwiderstand *m* water resistance
Wasserwirbelbremse *f* fluid brake
Wasserzeichen *n* water mark
Wasserzeichengerät *n* water mark apparatus
Wasserzeichenwalze *f* dandy roll
Wasser-Zement-Faktor *m* water-cement ratio (factor)
Wasser-Zement-Verhältnis *n s.* Wasser-Zement-Faktor
Wasser-Zement-Wert *m s.* Wasser-Zement-Faktor
Wasserzufluß *m* water feeding (supply); *(mine)* water influx
wäßrig aqueous, hydrous
Waten *n* **[im Öl]** churning
Watergarn *n* ring-spun yarn
Watfähigkeit *f (aut)* fording capacity
Watson-Effekt *m* Watson effect *(radar)*
Watt *n* 1. *(el)* watt; 2. bank of sand (clay), sands, flats
Watte *f* cotton wadding
Wattebausch *m* tampon
Wattkomponente *f (el)* active component
Wattleistung *f (el)* real power
wattlos *(el)* wattless *(e.g. component)*
Wattmeter *n (el)* wattmeter, power meter
Wattsekunde *f (el)* watt second
Wattstunde *f (el)* watt hour
Wattstundenzähler *m* active-energy meter, watt-hour meter
Wattverbrauch *m (el)* watt consumption, wattage
Wattverlust *m (el)* loss of watt
Wattzahl *f (el)* wattage
Wau *m* yellowroot
Webart *f* mode of weaving
Webblatt *n* reed
Webblattzahn *m* dent of the reed, reed wire
Webbreite *f* cloth width
Webekante *f s.* Webkante
Webelitze *f* heald
weben to weave

Weber *m* weaver
Weber *n* weber *(unit of magnetic flux)*
Weberei *f* mill
Weberglas *n* pick glass
Weberknoten *m* weaver's knot
Weberschiffchen *n* shuttle
Weberspule *f* bobbin
Webervogel *m (text)* picker
Weberzange *f* weaver's nippers
Webfach *n* loom shed
Webfehler *m* weaving fault, defect (fault) in weaving
Webfilz *m* felt-lined woven fabric
Webgeschirr *n* loom harness
Webkante *f* list[ing], selvedge, selvage
Webketten *fpl* / **gezettelte** warped ends
Webleiste *f s.* Webkante
Webmuster *n* weaving pattern
Webnest *n* skip
Webschiffchen *n* shuttle
Webschützen *m* shuttle
Webspitzenmaschine *f* lace weaving machine
Webstuhl *m* loom
  W. mit beweglichem Kamm loose-reed loom
  W. mit feststehendem Blatt fast-reed loom
  W. mit fliegendem Blatt loose-reed loom
  W. mit Revolverwechsel circular box loom
  W. mit Schützenwechsel box loom
  halbmechanischer W. dandy loom
  mechanischer W. power loom
  oberbauloser W. low-built loom
Webstuhlsaal *m* loomery
webtechnisch referring to the technology of weaving
Webware *f* woven material
Wechsel *m* 1. change[-over]; alternation, inversion; 2. *(build)* trimmer; trimming *(carpentry)*; 3. runway, haunt *(deer)*;
Wechselanteil *m (el)* oscillating component
Wechselbalken *m* trimmer [beam, joist] *(carpentry)*
Wechselbeanspruchung *f* alternating stress, cycling
Wechselbewegung *f* reciprocating motion
Wechselbeziehung *f* correlation; interrelation
Wechselbiegefestigkeit *f* resistance to alternating lateral bending
Wechselbiegeprüfung *f* alternating bending test
Wechselblende *f* alternating shutter
Wechselbolzen *m* stud
Wechseldruck *m* alternating (pulsating load) pressure
Wechsel-EMK *f* oscillating e.m.f.
Wechselenergie *f* alternating current energy
Wechselfeld *n* alternating field
  magnetisches W. alternating magnetic field
Wechselfeldhysterese *f* static hysteresis
Wechselfeldkreis *m* / **magnetischer** periodic magnetic circuit
Wechselfestigkeit *f* reversed fatigue strength
Wechselfeuer *n (ship)* combination (alternation, mixed) light
Wechselfluß *m* alternating flux (flow)
Wechselfranse *f (text)* intermittent fringe stitch
Wechselgetriebe *n* change-speed gear, multi-speed gear, variable-speed gear
Wechselglied *n* pulsating (oscillating) component
Wechselgröße *f* alternating (periodic) quantity
  sinusförmige W. sinusoidal quantity
Wechselgrünland *n* ley
Wechselkomponente *f* pulsating (oscillating) component
Wechselkontakt *m* alternating contact
Wechselkontaktschaltvorrichtung *f* make-and-break switching device
Wechsellade *f (text)* change-box slay, multiple-box slay
Wechsellage *f* alternating position *(crank)*
Wechsellager *n* double-thrust bearing
Wechsellast *f* alternating load
wechselläufig working alternately *(piston)*
Wechsellicht *n* alternating light
Wechsellichtempfindlichkeit *f* dynamic [luminous] sensitivity
Wechselmagnetisierung *f* alternating magnetization
wechseln to change; to alternate; to exchange
  den Ankerplatz w. to shift a berth
  Vorzeichen w. to change sign
Wechselpflug *m* turnabout plough
Wechselpolfeldmagnet *m* heteropolar field magnet
Wechselpolgenerator *m* heteropolar generator
wechselpolig heteropolar
Wechselpolmaschine *f* heteropolar machine
Wechselpotential *n* alternating potential
Wechselpunkt *m* point of alternation *(crank)*; change point
Wechselrad *n* interchangeable gear
Wechselräder *npl* change gears
Wechselräderberechnung *f* change-gear calculation
Wechselrädergetriebe *n* change-gear train
Wechselrädergetriebekasten *m* change-gear box
Wechselrädersatz *m* set of change gears
Wechselräderschere *f* quadrant
Wechselräderübersetzung *f* / **doppelte** compound train
  einfache W. simple gear train
Wechselräderverhältnis *n* change-gear ratio
Wechselreiber *m (print)* distributing roller
Wechselrelais *n* centre-zero relay
Wechselrichter *m* inverter
  einphasiger W. single-phase inverter

statischer W. static inverter
Wechselrichtung *f* a.c. conversion
Wechselschalter *m* change-over switch, double-throw switch
Wechselschaltung *f* two-way wiring (switching circuit)
Wechselspannung *f* 1. alternating stress; 2. *(el)* alternating [current] voltage, a.c. voltage
abgegebene W. output a.c. voltage
impulsförmige W. pulsed alternating voltage
impulsmodulierte W. pulse-modulated a.c. voltage
Wechselspannungsamplitude *f* alternating-voltage amplitude
Wechselspannungsanalogrechner *m* a.c. analogue computer
Wechselspannungsmesser *m* a.c. voltmeter
Wechselspannungsmessung *f* alternating voltage measurement
Wechselspannungsnetz *n* a.c. network
Wechselspannungsnormal *n* a.c. standard
Wechselsperre *f* reciprocal interlocking *(of relay)*
Wechselsprechanlage *f* correspondence (intercommunicating) system
Wechselsprechverkehr *m* two-way speech communication
Wechselstrom *m* alternating current, a.c.
einphasiger W. single-phase alternating current
gleichgerichteter W. rectified alternating current
mehrphasiger W. polyphase alternating current
wechselstrombetätigt a.c. operated
Wechselstrombetrieb *m* a.c. operation
Wechselstrombrücke *f* a.c. bridge
Wechselstromdrosselspule *f* / **gesättigte** a.c. saturated reactor
wechselstromgekoppelt a.c. coupled
Wechselstromgenerator *m* alternator
W. mit konzentrierter Wicklung monotooth alternator
W. mit umlaufender Kommutatorwicklung alternating current generator with revolving armature
kompoundierter W. compound alternator
Wechselstromgerät *n* a.c. equipment
wechselstromgespeist a.c.-powered
Wechselstromgleichrichter *m* a.c. rectifier
Wechselstrom-Gleichstrom-Umformer *m* a.c. to d.c. converter
Wechselstromglied *n* alternating (ripple, varying) component
Wechselstromimpuls *m* a.c. pulse
Wechselstromkomponente *f* a.c. component
reine W. balanced periodic component
Wechselstromkreis *m* a.c. circuit
Wechselstromleistung *f* a.c. output power
Wechselstromleitfähigkeit *f* a.c. conductivity
Wechselstromlichtbogen *m* a.c. arc
Wechselstrommagnetisierung *f* a.c.-magnetizing
Wechselstrommaschine *f* alternator, a.c. machine
Wechselstrommeßbrücke *f* a.c. [impedance] bridge
Wechselstrommesser *m* a.c. ammeter
Wechselstromnebenschlußmotor *m* a.c. shunt motor
Wechselstromnetz *n* a.c. mains (supply)
Wechselstromschalter *m* a.c. switch
Wechselstromschweißgenerator *m* welding alternator
Wechselstromschweißgeräte *npl* a.c. welding equipment
Wechselstromspannung *f* alternating voltage
Wechselstromspeisungsapparat *m* a.c.-supply apparatus
Wechselstromsprung *m* sinusoidal transient
Wechselstromstörung *f* a.c. interference
Wechselstromstoß *m* burst of alternating current
Wechselstromtachodynamo *m* induction (a.c.) tachometer
Wechselstromtachometermaschine *f* induction (a.c.) tachometer
Wechselstromtechnik *f* a.c. engineering; a.c. technique
Wechselstromtelegrafie *f* voice-frequency carrier telegraphy
Wechselstromverhalten *n* a.c. behaviour
Wechselstromversorgung *f* a.c. power supply
Wechselstromwahl *f* a.c. dialling *(telephony)*
Wechselstromwecker *m* magneto bell, polarized bell (ringer)
Wechselstromwiderstand *m* impedance, variational (a.c.) resistance
spezifischer W. effective resistivity
Wechselstromwirkleitwert *m* effective conductance
Wechselstromwirkwiderstand *m* effective a.c. resistance
Wechselstromzähler *m* a.c. meter
Wechselstuhl *m* check loom
Wechseltaste *f* shift key *(telephony)*
Wechseltubus *m (opt)* change-over tube
mikrofotografischer W. photomicrographic change-over tube
Wechselumrichter *m* frequency changer
Wechselventil *n* shuttle (two-way) valve
Wechselwebstuhl *m* multiple box (shuttle) loom
Wechselweide *f* ley pasture
Wechselwinkel *mpl* alternate angles
äußere W. alternate-exterior angles
innere W. alternate-interior angles

Wechselwirkung *f* interaction
atomare W. atomic interaction
Coulombsche W. Coulomb interaction
Wechselwirkungsanteil *m* interaction component
Wechselwirkungsenergie *f* interaction energy
Wechselwirkungsfaktor *m* interaction factor *(transformers)*
Wechselwirkungsgesetz *n* interaction law
Wechselwirkungsquerschnitt *m* interaction cross section
Wechselwirkungsraum *m* interaction space
Wechselwirkungsverlust *m* interaction loss *(transformers)*
Wechselwirkungszeit *f* interaction time
Wechselwirtschaft *f* ley farming
Wechselzeichen *n* shift signal *(telephony)*
Weckamin *n* analeptic amine
Wecker *m* bell, ringer *(telephony)*
Weckerausschalter *m* bell stop *(telephony)*
Weckerfallklappe *f* bell indicator drop *(telephony)*
WEDER-NOCH *n* NOR, NOT OR *(logical operator)*
WEDER-NOCH-Glied *n* NOR-gate
WEDER-NOCH-Schaltkreis *m* [NEITHER-] NOR circuit
Weg *m* 1. trip, route, way, road; 2. *(ph)* path, distance; displacement *(rectilinear translation)*; 3. braking (stopping) distance; 4. technique; approach *(e.g. of solution)*
drahtloser W. radio path
freier W. free path
kritischer W. critical path *(network planning)*
zurückgelegter W. distance covered
Wegamplitude *f* path amplitude
Wegänderung *f* displacement *(vibration meter)*
wegätzen to etch away
Wegdifferenz *f* path difference
wegdrehbar swing-away
Wegebau *m* road construction, road-making
Wegerung *f (ship)* ceiling
feste W. close ceiling
lose W. loose ceiling
offene W. batten and space ceiling
Wegerungslatte *f (ship)* cargo batten
Wegerungsluke *f (ship)* ceiling hatch
Wegevorschrift *f (ry)* routing (forwarding) instructions
Wegfüllarbeit *f (mine)* mucking
wegfüllen *(mine)* to clear, to load, to muck [out]
Wegfüllen *n* **von Hand** *(mine)* hand mucking
weggießen to pour away
wegkürzen *(math)* to cancel out
wegladen *(mine)* to clear, to load, to muck [out]
Weglänge *f* path [length]
freie W. free path
mittlere freie W. mean free path
optische W. length of optical path
wegleiten to carry off
wegnehmen to remove
das Gas w. to decelerate
wegräumen to clear [away]
wegschwemmen to flood off
wegspitzen *(mine)* to cut away
wegspülen to flush away, to wash away
Wegstrecke *f* / **durchlaufene** distance moved, space passed over *(mechanics)*
wegstreichen to cross out; to delete
Wegübergang *m* road-crossing, level-crossing, *(Am)* grade-crossing
unbewachter schienengleicher W. unprotected level-crossing
Wegübergangssignal *n (ry)* level crossing signal
Wegvergleich *m* motion balance
Wegvergleichsregler *m* motion-balance controller, displacement-type controller
Wegwerfmesser *n (mach)* throw-away blade
Wegwerftextilien *pl* one-way articles
Wegzähler *m* distance gauge
Weg-Zeit-Kurve *f* space-time curve
Weg-Zeit-Schaubild *n* displacement-time diagram
Wehnelt-Blende *f* modulator electrode
Wehnelt-Elektrode *f* modulator electrode
Wehnelt-Katode *f* Wehnelt cathode
Wehnelt-Unterbrecher *m* Wehnelt interrupter
Wehnelt-Zylinder *m* Wehnelt cylinder, control electrode
Wehr *n* weir, dam, barrage
bewegliches W. movable weir
festes W. fixed weir
scharfkantiges W. sharp-crested weir
Wehrfeuerung *f* furnace with auxiliary fire bridge
Wehrkrone *f* crest of weir
Wehrpfeiler *m* weir pier
Wehrrücken *m* crest of weir
Wehrschwelle *f* weir sill
Wehrüberlauf *m* dam lip *(grinder)*
Wehrverschluß *m* sluice gate
Weibel-Schweißverfahren *n* Weibel [electrical fusion welding] process
weich soft, unhardened; *(nucl)* non-penetrating *(X-rays)*
Weichbeton *m* mushy (quaking) concrete
Weichblei *n* refined (soft) lead
Weichbraunkohle *f* earthy brown coal
Weiche *f* 1. [railway] switch, points and crossings; sorting gate; 2. *(el)* switch; cell; branching network; filter; 3. *(leath)* soaking
auffahrbare W. *(ry)* trailable points
elektrische W. separating filter
faule W. *(leath)* putrid soak
Weicheisen *n* soft iron
Weicheisenamperemeter *n* soft-iron ammeter

Weicheisenanker *m* soft-iron armature
Weicheiseninstrument *n* soft-iron instrument, moving-iron instrument
Weicheisenkern *m* soft-iron core
Weicheisenmeßwerk *n* soft-iron instrument, moving-iron instrument
Weicheisennadel *f* induced needle
Weicheisenvoltmeter *n* moving-iron voltmeter
weichen to soak; to soften, to make soft
Weichenantrieb *m (ry)* point operating gear
Weichenherzstück *n* switch frog
Weichenhobelmaschine *f* frog and switch planer
Weichenstellbock *m (ry)* switch stand
Weichensteller *m* signalman, pointsman, *(Am)* switching-man
Weichenstellwerk *n* point lock
Weichenzunge *f* point, switch blade (rail)
Weichenzungenhobelmaschine *f* rail switch-blade planer
Weicherwerden *n* softening *(X-ray tube)*
weichfaul soft spongy *(wood)*
Weichfuttertrog *m* mash hopper
weichgeglüht soft-annealed
weichgestellt non-rigid
Weichglühung *f* soft annealing
Weichgummi *m* soft rubber
Weichharz *n* soft resin
Weichhaspel *f* soaking wheel
Weichheitsgrad *m* degree of smearing
Weichholz *n* soft (coniferous) wood
Weichkasten *m (leath)* soak pit
Weichlot *n* soft solder
Weichlöten *n* [soft] soldering
weichmachen to soften, to plasticize
Weichmacher *m* softener, plasticizer; larry *(spade)*
weichmacherfrei unplasticized
Weichmacherspaten *m* larry
Weichmacherwanderung *f* migrating (migration) of plasticizer
weichmagnetisch magnetically soft
Weichpackung *f* soft packing
armierte W. fibrous and metallic packing
Weichparaffin *n* soft paraffin
weichplastisch soft-plastic
Weichpreßziegel *m* soft-mud brick
Weichstrahlröhre *f* soft X-ray tube
Weichstrahlung *f* soft radiation
Weichzeichner *m* soft-focus lens
Weide *f* 1. willow *(wood)*; 2. pasture
Weidemelkanlage *f* milking bail
Weidewirtschaft *f* long-ley system (farming)
Weidezaun *m* / **elektrischer** electric fence
Weife *f (text)* reel
Wein *m* wine
Weinbrand *m* brandy, cognac
Weinessig *m* vinegar of wine
Weingärung *f* vinous fermentation
Weingeist *m* spirit of wine, alcohol
Weinhefe *f* distiller's yeast
Weinkeller *m* wine cellar
Weinsäure *f* tartaric acid
Weinstein *m* [cream of] tartar
Weißblech *n* tin plate
Weißblechbüchse *f* tin; *(Am)* can
Weißbleierz *n* cerussite
weißen to lime-wash, to whitewash, to whiten
Weißgehalt *m* brightness *(of paper)*
weißgerben to taw
Weißgerber *m* alum tanner
Weißgerbermesser *n* whitener
Weißgerbstoff *m* tawing matter
Weißgerbung *f* alum tanning, tawing
weißgeschält barked and basted
Weißglühen *n* white annealing
weißglühend incandescent, white-hot
Weißglühhitze *f* incandescent heat
Weißglut *f* incandescence, white heat
Weißkalk *m* fat (high-calcium) lime
Weißlauge *f* white liquor *(papermaking)*
Weißmetall *n* 1. white (antifriction) metal, babbit; 2. white metal *(intermediate; copper)*
Weißmetallager *n* babbit[ted] bearing
Weißmetallausfütterung *f* white-metal lining
Weißmetallausguß *m* white metal lining; babbitting
Weißmetallstreifen *m* white-metal strip
Weißnickelkies *m* white nickel, c[h]loanthite
Weißpegel *m (telev)* picture white
Weißpegelsättigung *f* white saturation
Weiß-Schwarz-Amplitudenbereich *m (telev)* white-to-black amplitude range
Weißsilbererz *n* antimonial silver ore
Weißspießglanz *m* valentinite
Weißware *f* bleached goods
weißweben to weave white goods
Weißwein *m* white wine
Weißwert *m (telev)* white level
Weitabtrennschärfe *f* far-off selectivity
Weitbereichsausbreitung *f* long-distance propagation
Weitbereichsdosimeter *n* wide-range radiation dosimeter
Weitbereichsnachrichtenverbindung *f* long-distance communication
Weite *f* width; distance; breadth
W. des Druckgewölbes *(mine)* span of the pressure arch
lichte W. inside diameter; bore of pipe; effective diameter; clear span (width, distance)
weiten to widen; to expand; to flare *(tube)*
Weiterbearbeitung *f* further treatment
weiterbefördern to transfer
Weiterbehandlung *f* further treatment
Weiterbildungseinrichtung *f* further education establishment

Weiterführung *f* extension; continuation
Weitergabe *f (telec)* retransmission
weitergeben to transfer, to transmit, to convey; *(telec)* to retransmit; to repeat *(relay)*
weiterlaufen *(mach)* to coast; to overrun
Weiterreißfestigkeit *f (leath)* stitch tear resistance
weiterschalten *(mach)* to index to the next position *(turret)*
Weiterschaltung *f (el)* step-by-step action; *(mach)* indexing to the next position
weiterteufen *(mine)* to deepen
Weiterteufen *n (mine)* deepening
weitgespannt long-span *(beam)*
Weitgrifflader *m* rake loader
weitmaschig coarse-mesh
weitreichend long-range
Weitstrahler *m (aut)* long-distance beam
Weitungsbau *m (mine)* random stoping, *(Am)* coyoting, *(Am)* gophering
Weitverkehrsleitung *f* long-distance circuit *(telephony)*
Weitverkehrstelegrafie *f* long-distance telegraphy
Weitverkehrsverbindung *f* long-distance call (connection) *(telephony)*
Weitwinkelbeobachtung *f* wide-angle view
Weitwinkeldüse *f* wide-angle spray jet
Weitwinkelobjektiv *n* wide-angle lens
Weitwinkelprojektor *m* wide-angle projector
Weitwinkelstreuung *f* wide-angle scattering
Weizenmehl *n* wheat flour
Welkgras *n* wilted grass
Wellasbesttafel *f* corrugated asbestos sheet
Wellblech *n* corrugated iron (metal)
Wellblechdach *n* corrugated-iron roof
Welldeck *n (ship)* well deck
Welldeckschiff *n* well-deck vessel, well-decker
Welle *f* 1. wave; 2. *(mach)* shaft
  W. für Hublängenverstellung stroke adjusting (setting) shaft
  W. für Verlegung des Hubs stroke positioning shaft
  ankommende W. incoming wave
  aperiodische W. aperiodic wave
  ausgebohrte W. hollow shaft
  ausgestrahlte W. radiated wave
  austretende W. outgoing wave
  biegsame W. flexible shaft
  durchgelassene W. transmitted wave
  durchlaufende W. passing shaft *(e.g. propeller-shaft)*
  ebene W. plane wave
  einfallende W. incident wave
  einlaufende W. incoming wave
  elektrische W. electric wave
  elektromagnetische W. electromagnetic wave
  elektrotonische W. electrotonic wave
  empfangene W. received wave
  fortschreitende W. travelling (advancing, progressive) wave
  fortschreitende ebene W. travelling plane wave
  frequenzmodulierte W. frequency modulated wave, F.M. wave
  gebrochene W. refracted wave
  gedämpfte W. aperiodic (damped, discontinuous) wave
  gekröpfte W. cranked shaft
  gerichtete W. guided wave
  gestreute W. scattered wave
  getastete W. keyed wave
  getastete ungedämpfte W. key-controlled continuous wave, telegraph modulated wave, type A1 wave
  getriebene W. driven shaft
  kontinuierliche W. continuous wave
  konvergierende W. converging wave
  kreispolarisierte W. circularly polarized wave
  kurze W. short wave
  lange W. long wave
  längsveränderliche W. telescopic shaft
  liegende W. horizontal (lying) shaft
  linear polarisierte W. plane polarized wave
  longitudinale W. longitudinal wave
  magnetohydrodynamische W. magnetohydrodynamic wave
  mittlere W. medium wave
  modulierte W. modulated wave
  orthogonalisierte ebene W. orthogonalized plane wave
  periodisch zerhackte W. interrupted wave
  polarisierte W. polarized wave
  polarografische W. polarographic wave
  pulsierende W. pulsating wave
  querlaufende W. transverse wave
  rechtslaufende W. dextropropagating wave
  reflektierte W. reflected (indirect) wave
  rückwirkende W. reactive wave
  senkrechte W. vertical shaft
  sinusförmige W. sine wave
  sprachmodulierte W. speech-modulated wave
  stehende W. 1. standing (stationary) wave; 2. vertical shaft
  stehende elektrische W. vertical electric wave
  transversale W. transverse wave
  treibende W. driving shaft
  ungedämpfte W. undamped (continuous, sustained) wave
  unsymmetrische W. asymmetrical wave
  unterbrochene W. subdivided shaft
  wandernde W. travelling (advancing, progressive) wave
  zylindrische W. cylindrical wave
wellen to corrugate

Wellenabsatz *m* chamfer of shaft
Wellenabsorption *f* wave absorption, absorption of waves
Wellenabsorptionssaugkreis *m* absorption wave trap
Wellenabstrahlung *f* radiation of waves
wellenähnlich wavelike
Wellenamplitude *f* height of waves, wave amplitude
Wellenamplitudenmaximum *n* wave-amplitude peak
Wellenanalysator *m* wave analyzer
Wellenanzeiger *m* wave detector
Wellenart *f* type of wave
wellenartig wavelike
Wellenausbreitung *f* wave propagation, propagation of waves
  W. über dem Stadtgebiet wave propagation over built-up areas
  W. über freiem Gelände wave propagation over open country
Wellenausrichtung *f* shaft alignment
Wellenauswahl *f* mode selection
Wellenbahn *f* wave path
Wellenband *n* wave band
Wellenbanddehnung *f* band spread
Wellenbauch *m* antinode [of wave], wave loop
Wellenbereich *m* wave band (range)
Wellenbereichsschalter *m* wave changing switch
Wellenbereichumschaltung *f* band switching
Wellenberg *m* wave crest, peak (top) of wave
Wellenberuhiger *m* *(ship)* wave subduer
Wellenberuhigungsöl *n* wave-subduing oil, wave-calming oil
Wellenbeugung *f* wave diffraction
Wellenbewegung *f* wave (undulatory) motion
Wellenbild *n* wave image (picture, pattern)
Wellenbinder *m* wave binder *(motorboat)*
Wellenbinderform *f* V-bottom form
Wellenbock *m* shaft bracket (strut), propeller-shaft-stay, spectacle frame, [propeller] strut
  zweiarmiger W. two-leg propeller strut
Wellenbockarm *m* shaft-strut, shaft bracket, propeller strut
Wellenbrecher *m* breakwater, jetty
Wellenbund *m* shaft collar (shoulder)
Wellenbündel *n* wave packet (beam)
Wellendetektor *m* wave detector
Wellendichtung *f* shaft packing (seal)
Wellendurchbiegung *f* shaft deflection
Wellenecho *n* wave echo
Wellenentartung *f* mode-degeneracy
Wellenerreger *m* oscillator, wave generator
Wellenerzeugung *f* wave generation
Wellenfalle *f* [absorption] wave trap; ditch *(of a waveguide)*; line trap
Wellenfallenrand *m* skirt
Wellenfeld *n* wave field
Wellenfilter *n* wave filter
  ringförmiges W. ring mode filter *(waveguides)*
Wellenfläche *f* wave surface
Wellenflansch *m* shaft flange
Wellenfolge *f* wave train (sequence)
Wellenform *f* waveform, wave shape; mode
  rechteckige W. rectangular waveform
  verzerrte W. distorted waveform
Wellenformgeber *m* waveform generator
wellenförmig undulatory
Wellenformverzerrung *f* waveform distortion
Wellenformwandler *m* mode changer, waveform converter
Wellenformverzerrung *f* waveform distortion
Wellenfortpflanzung *f* propagation (transmission) of waves
Wellenfortschrittsgeschwindigkeit *f* speed of advance of waves
Wellenfrequenz *f* wave frequency
Wellenfront *f* wave front
  ebene W. plane wave front
  geneigte W. tilt wave front
  steile W. steep wave front
Wellenfrontschreiber *m* wave-front plotter
Wellenfrontwinkel *m* wave tilt
Wellenfunktion *f* wave function
Wellengenerator *m* shaft[-driven] generator
Wellengeschwindigkeit *f* wave velocity
Wellengleichung *f* wave equation
Wellengruppe *f* wave group
Wellenhosen *fpl* [shell] bossings, shaft bossings
Wellenhosenplatte *f* boss plate
Wellenhosenspanten *npl* boss frame, bossed framing
Wellenkamm *m* crest (ridge) of a wave, wave top
Wellenkeilnut *f* shaft keyway
Wellenknoten *m* wave node, node of a wave
Wellenkontakt *m* shaft contact *(automatic telephony)*
Wellenköper *m* *(text)* waved twill
Wellenkupplung *f* shaft coupling
Wellenkupplungsflansch *m* shaft coupling flange
Wellenlager *n* steady bearing, [line] shaft bearing
Wellenlänge *f* wavelength
  gewogene mittlere W. weighted mean wavelength
  maximale W. peak wavelength
  verschwindende W. zero wavelength
Wellenlängenbereich *m* wavelength coverage (range)
Wellenlängenempfindlichkeit *f* wavelength sensitivity
Wellenlängengrenze *f* wavelength threshold (cut-off, limit)
Wellenlängenskale *f* wavelength scale

Wellenlehre *f* wave theory
Wellenleistung *f* shaft horsepower
Wellenleiter *m* waveguide
  biegsamer W. flexible waveguide
  dielektrischer W. dielectric waveguide
  geschlitzter W. slotted waveguide
  keilförmiger W. taper *(waveguide)*
Wellenleiterfolie *f* waveguide shim
Wellenleiterknie *n* waveguide elbow
Wellenleiteröffnung *f* spout, nozzle
Wellenleitung *f* [line] shafting
Wellenlinie *f* wave trace (line, curve)
Wellenlinienschreiber *m* ondograph
Wellenmechanik *f* wave mechanics
wellenmechanisch wave-mechanical
Wellenmesser *m* wavemeter, ondometer
Wellenmodus *m* mode
Wellennatur *f* **des Lichtes** wave nature of light
Wellennuß *f (ship)* boss, shaft bossing
Wellennut *f* shaft keyway, keyslot
Wellenölbehälter *m* storm oil tank
Wellenoptik *f* physical optics
Wellenpaket *n* wave packet
Wellenparameter *m* wave parameter
Wellenpferdestärken *fpl* shaft horsepower, s.h.p.
Wellenpunktauslöser *m* point-on-wave trigger
Wellenquantisierung *f* wave quantization
Wellenreflexion *f* sea (wave) clutter *(radar)*
Wellenreiten *n* surf-riding *(sporting)*
Wellenroststab *m* longitudinally corrugated grate bar
Wellenschalter *m* 1. band switch; 2. shaft clutch
Wellenschaltung *f* wave connection
Wellenschatten *m* blind spot
Wellenscheibe *f* shaft ring *(thrust bearing)*
Wellenschlucker *m* 1. *(el)* surge absorber, line trap; 2. *(ship)* wave swallower
Wellenschreiber *m* ondograph
Wellenschwingung *f* undulation
Wellenschwundloch *n* fading hollow
Wellenseele *f (mach)* shaft core
Wellenspannung *f* ripple (surge) voltage
Wellenstirn *f* wave (surge) front
  steile W. sharp wave front
Wellenstopfbuchse *f* shaft stuffing box
Wellenstoß *m* wave impulse
Wellenstrahlung *f* wave radiation
Wellenstrang *m (mach)* length (line) of shafting
Wellenstrecker *m (el)* line lengthener
Wellenstrom *m* undulatory (ripple, surge) current
Wellenstumpf *m (mach)* shaft butt (extension)
Wellental *n* wave trough
Wellentheorie *f* wave (undulatory) theory
  W. des Lichtes wave (undulatory) theory of light
Wellenträger *m (ship)* propeller-shaft stay
Wellentunnel *m (ship)* tunnel, shaft alley
Wellentyp *m* [wave] mode
Wellentypfilter *n* mode filter *(in a waveguide circuit)*
Wellentypringfilter *n* ring mode filter
Wellentyptransformator *m* mode changer
Wellentypwandler *m* mode changer
Wellenumschalter *m* wafered (wave band, coil, band) switch
Wellenverteilung *f (el)* frequency (channel) allocation
Wellenwicklung *f (el)* wave winding
Wellenwiderstand *m* 1. *(el)* iterative (surge, characteristic) impedance; 2. *(ship)* wave-making resistance
Wellenzahl *f* wave number
Wellenzapfen *m* [shaft] journal
Wellenzentrum *n* origin of wave
Wellenzone *f* wave zone
Wellenzug *m* wave train, train of waves
  harmonischer W. train of harmonic waves
Wellenzuteilung *f* frequency allocation
Welle-Teilchen-Dualismus *m* wave-particle duality (parallelism)
Wellfaserplatte *f* corrugated fibre board
wellig wavy, undulated
Welligkeit *f* waviness; ripple [factor]
Welligkeitsentzerrung *f* elimination of standing waves
welligkeitsfrei free of ripples
Welligkeitsfrequenz *f* ripple frequency
Welligkeitsgrad *m* ripple percentage
Welligkeitskurve *f* standing-wave pattern
Welligkeitsmesser *m* standing-wave meter
Wellpappe *f* corrugated paper (board)
  doppelgedeckte W. double-faced corrugated board
Wellpappendeckenlage *f* kraft liner
Wellpappenmittellage *f* corrugating medium
Wellplatte *f* corrugated sheet
Wellrohr *n* bellow, corrugated pipe
Wellrohrbogen *m* corrugated pipe bend
Wellrohrkessel *m* boiler with corrugated flue tube
Weltachse *f* celestial axis
Weltall *n* universe
Weltfunknetz *n* universal (world) radio network
Weltmeer *n* ocean, [main] sea
Weltmeere *npl* Seven Seas
Weltniveau *n* world technological standards, highest world quality
Weltraum *m* universal (cosmic, outer) space
Weltraumfahrer *m* astronaut, space traveller
Weltraumfahrt *f* space (interplanetary, interstellar) flight
Weltraumforschung *f* exploration of space, space research

Weltraumnachrichtenverkehr *m* space telecommunication[s]
Weltraumrakete *f* space rocket
Weltraumschiff *n* space ship
Weltraumschiffahrt *f* interstellar (interplanetary, space) navigation
Weltraumstrahlung *f* cosmic rays
Weltstand *m* world technological standards (*s. a.* Weltniveau)
Weltzeit *f* universal time, U.T.
Wende *f* turn, bend; reversal
Wendeanlasser *m (el)* reversing starter
Wendebahnhof *m* reversing (dead-end, stub-end) station
Wendebecken *n* turning (manoeuvring) basin
Wendeeinrichtung *f (mach)* manipulator, turn-over unit *(transfer line)*
Wendefeld *n (el.m.)* commutation (commutating) field
Wendefeldwicklung *f* commutating winding
Wendegetriebe *n* reversing gear (gears, mechanism), spindle reverse [gear]
W. zum Gewindebohren tapping attachment
Wendeherz *n* quadrant
Wendel *f* helix, coil, spiral, coiled filament *(lamp)*
Wendelabtastung *f* spiral (helical) scanning
Wendelantenne *f* helical aerial
Wendelausleser *m (ag)* spiral separator
Wendelbohrer *m* twist drill
Wendeldraht *m* coiled filament *(lamp)*
wendelförmig helical, in the form of a helix
Wendelgangpotentiometer *n* multi-turn potentiometer
wendelgenutet helically grooved (fluted)
Wendelgurtförderer *m* inverted-belt conveyor, twisted belt
Wendelhohlleiter *m* helix waveguide
Wendelnut *f* helical flute
Wendelrutsche *f* spiral chute
Wendelscheider *m* spiral cleaner
Wendelspan *m* spiral (continuous curly) chip
Wendeltreppe *f* corkscrew (helical, spiral) staircase, helical (spiral) stairs
Wendelwicklung *f* multiple-lap winding *(transformer)*
Wendemaschine *f* turning machine
Wendemotor *m (el)* reversible motor
Wendepol *m* 1. centre of inflections; 2. *(el)* commutating (auxiliary, reversing) pole, interpole
Wendepolnebenschluß *m* auxiliary pole shunt
Wendepolshunt *m* auxiliary pole shunt
Wendepolwicklung *f* commutation winding
Wendepresse *f* reverse press
Wendepropeller *m s.* Verstellpropeller
Wendepunkt *m (math)* point of inflection, inflection point *(curve)*; reversal point *(movement)*
W. der Kurve point of inflection of the curve
Wender *m* 1. reverser *(battery)*; 2. *(text)* clearer, stripping roller
W. der Karde card stripper
Wendesaugpresse *f* reverse suction-press *(papermaking)*
Wendeschalter *m* current reverser, reversing switch
Wendeschaltung *f* reversing circuit
Wendeschaufel *f* reciprocating blade
Wendeschütz *n* reversing contactor
Wendestange *f* turner (turning) bar
Wendestation *f* turn[ing]-over station, roll-over station *(transfer line)*
Wendetangente *f* tangent drawn at the point of inflection
Wendevorrichtung *f* roll-over fixture, turn-over device *(transfer line)*
Wendewalze *f (text)* clearer, stripper, stripping roller
Wendezeiger *m* rate-of-turn gyroscope
wendig manoeuvrable *(vehicle)*
Wendigkeit *f* handiness, manoeuvrability, easy steerability
taktische W. *(mil)* battlefield mobility
WENN-UND-NUR-WENN IF-AND-ONLY-IF *(logical operator)*
Wentzel-Kramers-Brillouin-Methode *f (nucl)* WKB method
Werbeabteilung *f* advertising department
Werbeaufsteller *m* poster
Werbefachmann *m* publicity man
Werbefernsehen *n* commercial television
Werbefilm *m* advertising film
Werbekosten *pl* advertising expense
werben to advertise
werfen to throw, to cast; to pull *(burden)*
Anker w. to cast anchor
Schützen w. to throw the shuttle
über Bord w. to cast overboard, to jettison
Werft *f* [ship]yard, shipbuilding yard, dockyard
Werftarbeiter *m* [dock]yard labourer
Werfteinrichtungen *fpl* shipyard facilities
Werftgelände *n* shipyard site
Werftleichter *m* yard-service lighter
Werftleiter *m* yard manager
Werftliegezeit *f* period in shipyard
Werftprobefahrt *f* builder's (yard) trial trip
Werfung *f* warping *(torsion)*
Werg *n* oakum, tow
Werggarn *n* tow yarn
Wergkrempel *f* tow card
Wergspinnmaschine *f* tow spinning frame
Wergvorspinnmaschine *f* tow roving frame
Werk *n* works, mill, factory; element *(computer)*; works *(watch)*
lebendes W. live work *(submerged parts of a vessel)*

totes W. dead work *(topsides of a vessel)*
Werkbank *f* work bench
Werkblei *n* crude (raw) lead
Werkdruck *m (print)* jobbing
Werkhalle *f* workshop, factory hall
Werkholz *n* industrial wood
Werkküche *f* factory canteen
Werkleiter *m* works manager
Werknummer *f* factory serial number
Werkrad *n* work gear, gear blank (being machined)
Werksatz *m (print)* bookwork
Werksgelände *n* works area (site)
Werkshafen *m* works harbour
Werkskizze *f* working sketch
Werksnorm *f* works standard [specification]
werksseitig eingestellt auf factory-set at
Werkstatt *f* [work]shop
mechanische W. machine shop
Werkstattarbeit *f* workmanship
Werkstattausdruck *m* shop term
Werkstattkniffe *mpl* shop kinks (practices)
Werkstattleiter *m* master mechanic
Werkstattmeßmikroskop *n* shop microscope
Werkstattmontage *f* shop assembly, workshop assembly (erection)
Werkstattniet *m* shop rivet
Werkstattschiff *n* repair ship
Werkstattschweißung *f* shop weld[ing]
Werkstattwagen *m* truck-mounted workshop, repair wagon
Werkstattzeichnung *f* working drawing
Werkstein *m* ashlar, quarry stone
Werksteinmauerwerk *n* ashlar stonework (walling), masonry
Werksteinverblendung *f* ashlar facing
Werksteinverkleidung *f* ashlar facing
Werkstoff *m* material
kurzspanender W. non-continuous chip material, short-chip material
langspanender W. continuous chip material
magnetischer W. magnetic material
plastischer W. plastic material
synthetischer W. synthetic material
technischer W. engineering material
thermoplastischer W. thermoplastic
werkstoffabhängig material-dependent
Werkstoffabnahme *f* stock removal
Werkstoffdämpfung *f* material damping
Werkstoffdurchlaß *m* **durch die Zange** collet capacity
Werkstoffeinsparung *f* saving in material
Werkstoffermüdung *f* fatigue of materials
Werkstoffestigkeit *f* strength of materials
Werkstofffluß *m* flow of materials
Werkstoffhandhabung *f* materials handling
Werkstoffkenngröße *f* characteristic of the material
werkstoffkennzeichnend indicating specific material[s]; indicating characteristics of the material
Werkstoffknappheit *f* material shortage
Werkstoffkonstante *f* material constant, constant of the material
Werkstoffkunde *f* materials science
Werkstoffmuster *n* specimen of the material
Werkstoffördereinrichtung *f* material handling device
Werkstoffpaarung *f* mating of materials *(bearing)*
Werkstoffprobe *f* sample of the material
Werkstoffprüflaboratorium *n* material[s] testing laboratory
Werkstoffprüfmaschine *f* material[s] testing machine
Werkstoffprüfung *f* testing of materials, material[s] testing
W. mittels radioaktiver Strahlung radiomateriology
zerstörungsfreie W. non-destructive testing
zerstörungsfreie W. mit Gamma-Strahlen gamma-ray materiology (material testing)
zerstörungsfreie W. mittels radioaktiver Strahlung radiomateriology
Werkstoffprüfungsreaktor *m* material[s] testing reactor
Werkstofftechnik *f* materials engineering
werkstoffunabhängig material-independent
Werkstoffverschwendung *f* material wastage
Werkstoffvolumen *n* / **abgenommenes** amount of stock removed
Werkstoffzugabe *f* material (stock) allowance
Werkstück *n* work[piece]; component, part
rundes W. circular work
unbearbeitetes W. work blank
Werkstückabmessung *f* component dimension
Werkstückachse *f* work axis, axis of the work
Werkstückanlage *f* work seat
Werkstückauflage *f* work seat (support)
Werkstückauflageschiene *f* **[für spitzenloses Schleifen]** work [supporting] blade, work rest blade
Werkstückaufspanndorn *m* work mounting mandrel
Werkstückaufspannfläche *f* work clamping area
Werkstückaufspannplatte *f* work holding plate
Werkstückaufspannung *f* work clamping (set-up), setting (set-up) of the work, workpiece mounting (set-up)
Werkstückaufspannvorrichtung *f* fixture
Werkstückauswerfeinrichtung *f* part-ejection device, work ejector mechanism
Werkstückauswerfer *m* work ejector
Werkstückbeschickungsanlage *f* / **hydraulisch betätigte** hydraulic loading mechanism
Werkstückbock *m* work rest (support)
Werkstückfluß *m* work (parts, components) flow

Werkstückfördereinrichtungen *fpl* work handling equipment
Werkstückförderung *f* work handling (conveying)
Werkstückgütekontrolle *f* parts inspection
Werkstückhandhabung *f* work handling
Werkstückhöhe *f* work height
Werkstücknocken *m* workpiece cam contour
Werkstückoberflächengüte *f* work surface finish
Werkstückobertisch *m* upper work table
Werkstückprüfeinheit *f* work-testing unit
Werkstückrohling *m* blank
Werkstückspannfutter *n* work-holding chuck
Werkstückspannvorrichtung *f* work-holding fixture
Werkstückspindelstock *m* work head[stock]
Werkstücktisch *m* work table
Werkstückträger *m* work-holding fixture, pallet *(transfer line)*
Werkstückträgerrücklauf *m* return of pallet[s] to loading station *(transfer line)*
Werkstücktransport *m* work transfer (handling)
Werkstücktransporteinrichtungen *fpl* work-handling equipment, workpiece transfer equipment
Werkstückumspannung *f* work relocation (repositioning), relocating of the workpiece
Werkstückuntertisch *m* lower work table
Werkstückvorschub *m* work feed
Werkstückwerkstoff *m* work material
Werkstückzuführeinrichtung *f* work loading device, feeder
Werkstückzuführung *f* loading, feeding
Werksvorfertigung *f* factory prefabrication; factory precasting *(concrete)*
Werkvertriebskosten *pl* factory selling expenses
Werkzeichnung *f* working drawing
Werkzeug *n* tool; instrument; *(plast)* mould
  W. mit Hartmetallschneide cemented carbide-tipped tool
  einschneidiges W. single-point tool, single-edged tool
  hartmetallbestücktes W. carbide-tipped tool
  mehrschneidiges W. multipoint (multi-edged) tool
  spanendes W. cutting tool
Werkzeugabhebung *f* tool relief (relieving)
  selbsttätige W. automatic tool relief
Werkzeuganordnung *f* tooling arrangement (diagram), tool set-up
Werkzeugaufbewahrung *f* tool storage
Werkzeugauflage *f* tool support
Werkzeugaufspannzeit *f* tool set-up time
Werkzeugausgabe *f* tool crib
Werkzeugauswerfer *m* tool ejector
Werkzeugbringer *m* **/ automatischer** automatic tool handling mechanism *(broaching)*
Werkzeugdurchgang *m* passage of the tool
Werkzeugeinstellehre *f* tool setting gauge
Werkzeughalter *m* **des Zugorgans der Räummaschine** broach pull head
Werkzeughöheneinstellehre *f* tool setting gauge
Werkzeughöhlung *f* mould cavity
Werkzeuginstandhaltung *f* tool maintenance
Werkzeugkasten *m* tool box; service kit
Werkzeugkegel *m* machine taper
Werkzeugkombination *f s.* Programmwerkzeug
Werkzeugmacher *m* toolmaker
Werkzeugmacherdrehmaschine *f* toolroom (toolmaker's) lathe
Werkzeugmacherei *f* toolmaking department (shop)
Werkzeugmacherei *f* **und Vorrichtungsbau** *m* toolroom
Werkzeugmaschine *f* machine tool
  W. nach Baukastenbauweise unitized machine tool
Werkzeugmaschinenantrieb *m* machine tool drive
Werkzeugmaschinenausfuhr *f* machine tool exports
Werkzeugmaschinenausstellung *f* machine tool exhibition
Werkzeugmaschinenbau *m* machine tool building
Werkzeugmaschinenbaugruppe *f* machine tool element
Werkzeugmaschineneinrichter *m* machine tool setter
Werkzeugmaschinenfachmann *m* machine tool expert
Werkzeugmaschinenforschungsinstitut *n* machine tool research institute
Werkzeugmaschinenführung *f* machine tool slideway
Werkzeugmaschinenhersteller *m* machine tool builder
Werkzeugmaschinenhydraulik *f* machine tool [hydraulic] circuit
Werkzeugmaschinenkatalog *m* machine tool catalogue
Werkzeugmaschinenkonstrukteur *m* machine tool designer
Werkzeugmaschinenkonstruktion *f* machine tool design
Werkzeugmaschinentisch *m* machine tool table
Werkzeugmaschinenverkaufsabteilung *f* machine tool sales department
Werkzeugmikroskop *n* toolmaker's microscope
Werkzeugnachschliff *m* tool grind[ing]
Werkzeugrad *n* rotary gear shave cutter
Werkzeugsatz *m* gang tool
Werkzeugschaft *m* tool shank
Werkzeugscharfschleifmaschine *f* tool grinding (sharpening) machine

Werkzeugschieber *m* tool slide *(vertical turning and boring machine)*
Werkzeugschleifmaschine *f* tool grinding machine
Werkzeugschlitten *m* carriage *(lathe)*; slide rest
W. mit Schloßplatte carriage and apron unit
Werkzeugschneide *f* tool cutting edge, cutting tool tip
Werkzeugschonerschaltknopf *m* tool-saver control button
Werkzeugspannung *f* tool clamping
Werkzeugspeicher *m* tool-storage unit
Werkzeugstahl *m* tool steel
unlegierter W. carbon tool steel
Werkzeugstandzeit *f* tool life, life of tool
Werkzeugträger *m* tool box *(planer)*, tool carrier
Werkzeug- und Fräserschleifmaschine *f* tool and cutter grinding machine
Werkzeug- und Gesenkfräsmaschine *f* tool and die miller
Werkzeug- und Gesenkmacherei *f* tool and die [making] shop
Werkzeugwechsel *m* tool changing (interchange)
Werkzeugwechseleinrichtung *f* automatic tool changing mechanism
Werkzeugwiege *f* tool cradle
Werkzeugwinkel *m* tool angle
Werkzeugzubringung *f* tool handling *(broaching)*
Wermutöl *n* wormwood (artemisia absinthium) oil
Wert *m* value; quantity; magnitude; result
W. der Übertragungsgüte *(telec)* transmission performance rating
W. des Arbeitspunktes operating-point value
absoluter W. absolute value
angenommener W. assumed value
effektiver W. der Spannung root mean square voltage
experimenteller W. experimental value
geschätzter W. estimated value
günstigster W. optimum value
häufigster W. mode *(statistics)*
konstanter W. plateau value
kritischer W. critical value
numerischer W. numerical value
registrierter W. recorded value
reproduzierbarer W. reproducible value
reziproker W. reciprocal [value]
stationärer W. steady[-state] value
wahrer W. true value
zulässiger W. permissible (safe) value
Wertangabe *f* rating
Werte *mpl* data; values
Wertepaar *n* pair of variates *(statistics)*
Werterhaltung *f* maintenance
Werterhaltungsarbeit *f* maintenance work
Wertevorrat *m (math)* range of values *(function)*
wertig valent
Wertigkeit *f* valency
wertmäßig in value
Wertminderung *f* depreciation
Werttarif *m* ad valorem tariff
Wertzuwachsmessung *f* incremental measurement
Weston-Element *n* Weston cell
Weston-Normalelement *n* Weston normal standard [electromotive force] cell
Wettbewerb *m* competition
wettbewerbsfähig competitive
Wettbewerbsteilnehmer *m* competitor
Wetter *n* weather
Wetter *pl (mine)* ventilation, air; damp
einziehende W. intake air
frische W. fresh air
matte W. dead air, blackdamp
nicht atembare W. irrespirable air
schlagende W. firedamp
schlechte W. foul (vitiated) air
stickige W. choke damp
stinkende W. stink damp
Wetterabteilung *f (mine)* ventilating district
Wetterabzug *m (mine)* air flue (escape)
Wetteranzeiger *m (mine)* air indicator
Wetterbeobachtung *f* meteorological (weather) observation
Wetterbeobachtungsschiff *n* ocean weather ship
Wetterbohrloch *n (mine)* bore-hole for ventilation
Wetterdach *n* awning, eaves
Wetterdeck *n* weather deck
freiliegendes W. exposed weather deck
wetterdicht weathertight, weatherproof
Wetterdurchhieb *m (mine)* air slit
Wetterfahne *f* wind indicator
wetterfest weather resistant, weatherproof, weathertight
Wetterfestigkeit *f* weather resistance
Wetterführung *f (mine)* ventilation
abfallende W. descensional ventilation
ansteigende W. ascensional ventilation
Wetterfunk *m* meteorological broadcast
Wetterfunkstelle *f* radio weather broadcast station
Wettergehilfe *m (mine)* bailiff
Wettergeschwindigkeit *f* air velocity
Wetterkanal *m (mine)* air course
Wetterkarte *f* weather map, synoptic [weather] chart, meteorological chart
Wetterkreuz *n (mine)* air crossing, overcast
Wetterkunde *f* meteorology
Wetterlampe *f (mine)* flame (Davy) safety lamp
Wetterlutte *f (mine)* air conduit (duct, pipe), ventilating pipe
Wettermann *m (mine)* bailiff
Wettermauer *f (mine)* ventilation rib

Wetterprobe *f* mine air sample
Wetterrakete *f* meteorological rocket
Wettersatellit *m* meteorological satellite
Wetterschacht *m* air (ventilation) shaft, wind road (way)
ausziehender W. upcast shaft
einziehender W. downcast (intake) shaft
Wetterscheider *m (mine)* abatis, air split, [air-]-brattice
Wetterschiff *n* weather ship
Wetterschleuse *f (mine)* air-lock
Wettersprengstoff *m* permissible (coal mining, safety, short-flame) explosive
Wettersteiger *m (mine)* ventilation deputy
Wetterstrecke *f (mine)* air course, airway, *(Am)* monkey [way]
ausziehende W. return gate-road
einziehende W. intake gate-road
Wetterströmung *f (mine)* airflow
Wettertrum *m (mine)* air compartment
Wettertuch *n (mine)* curtain, drop-sheat, brattice cloth
wettertüchtig weatherly *(vessel)*
Wettertür *f (mine)* air gate
Wetterüberhauen *n (mine)* air raise
Wetterumkehr *f* airflow-reversal
Wetterverschlechterung *f* deterioration of weather
Wettervorhersage *f* weather forecast[ing]
Wetterweg *m (mine)* airway
Wettfahrt *f (ship)* race
wetzen to sharpen edge tools, to hone
Wetzstein *m* grindstone
Wetzwinkel *m* primary clearance angle
Weymouthskiefer *f* Weymouth pine, [American] white pine
Wheatstone-Brücke *f* Wheatstone bridge
Wheeler-Prahm *m (ship)* Wheeler hopper barge *(for tank-cleaning)*
Wheeler-Tankreinigungssystem *n* Wheeler tank-cleaning system, Wheeler system of tank cleaning
Whiting-Verfahren *n* Whiting system *(shaft winding)*
Whitworth-Gewinde *n* Whitworth [screw] thread
Whitworth-Rohrgewinde *n* Whitworth pipe thread
wichsen to wax
Wichte *f* specific weight
Wichteanalyse *f* float-and-sink analysis
Wickel *m* batting; lap *(spinning)*
Wickelblock[kondensator] *m* paper capacitor
Wickelbock *m* winding stand
Wickelbreite *f* layer width *(transformer)*
Wickeldorn *m* winding mandrel, mandrel for spring winding
Wickeldraht *m* winding wire
Wickeldrahtverbindung *f* wire-wrapping
Wickelei *f* winding shop
Wickelgerät *n* merry-go-round *(prestressed concrete)*
Wickelgeschwindigkeit *f* winding speed
Wickelhaspel *f* cable (reeling) winch
Wickelkondensator *m* paper capacitor
Wickelkopf *m (el)* coil winding head
Wickelkopfkappe *f (el)* end bell
Wickelkörper *m (el)* winding form
Wickelmaschine *f* merry-go-round *(prestressed concrete)*; *(el)* winding machine
Wickelmotor *m* supply (take-up, torque) motor *(tape recording)*
wickeln to wind, to coil *(spring)*; to reel
Wickelplatte *f* wrap-around plate
Wickelprüfung *f* wrapping test
Wickelraum *m* winding space *(transformer)*
Wickelschritt *m (el)* [winding] pitch
Wickelspule *f (el)* spool
Wickelstrecke *f (text)* lap drawing frame, ribbon lap machine
Wickelteller *m* clutch spindle *(tape recorder)*
linker W. left-hand spindle
rechter W. right-hand spindle
Wickelwalze *f (text)* lap roller
Wicklung *f (el)* winding
W. mit verkürztem Schritt short-pitch winding, fractional-pitch winding
W. mit verlängertem Schritt long-pitch winding
äußere W. outer winding
bifilare W. bifilar (non-inductive, two-wire) winding, double-spiral coil
dreietagige W. three-range winding
eingängige W. simplex winding
einlagige W. single-layer winding
einschichtige W. single-layer winding
gefädelte W. threaded winding
hochohmige W. high-impedance winding
in sich geschlossene W. endless (ring, closed-coil) winding
induktionsfreie W. non-inductive winding
innere W. inner winding
konzentrierte W. concentrated winding
kurzgeschlossene W. short-circuited winding
linksgängige W. left-handed winding
mehrgängige W. reentrant winding
mehrphasige W. polyphase winding
mitkompoundierende W. cumulative compound winding
offene W. open-circuit winding
primäre W. primary winding
rechtsläufige W. right-handed winding
stillstehende W. stationary winding
übersehnte W. long-chord winding
unifilare W. unifilar winding
wiedereintretende W. reentrant winding

zweietagige W. two-range winding, two-plane winding, two-tier winding
zweifach geschlossene W. duplex winding
zweilagige W. double-layer winding
Wicklungsanzapfung *f* winding tap
Wicklungsdraht *m* winding (coil) wire
Wicklungsfaktor *m* winding (slot) factor, breadth coefficient
Wicklungsinduktivität *f* winding inductivity
Wicklungsisolierung *f* winding insulation
Wicklungskapazität *f* winding capacitance
Wicklungsnut *f* winding slot
Wicklungsquerschnitt *m* cross-sectional area of winding
Wicklungsschema *n* winding diagram
Wicklungsschritt *m* [winding] pitch, pitch of a winding
kollektorseitiger W. front pitch
Wicklungssinn *m* sense of winding
Wicklungsteilung *f* winding pitch
Widder *m* ram
hydraulischer W. hydraulic ram
Widerdruck *m* thrust; second printing, verso print
Widerdruckzylinder *m* perfecting cylinder
Widerhall *m* reverberation, echo
widerhallen to reverberate, to resound, to echo
Widerhallzeit *f* reverberation time
Widerlager *n* abutment *(arch)*
aufgelöstes W. hollow abutment
geschlossenes W. closed abutment
verlorenes W. dead abutment
vorspringendes W. projecting abutment
Widerlagerdruck *m* abutment pressure
Widerlagerpfeiler *m* abutment pier
Widerlagerstein *m* abutment stone, bearing pad
Wideröe-Bedingung *f* *(nucl)* Wideroe condition, 1:2 condition
Widersee *f* head sea; back-wash, surf
Widerspiegelung *f* reflection, mirror image
Widersprüchlichkeit *f* contradictoriness
widerspruchsfrei free from contradiction
Widerspruchsfreiheit *f* consistency
W. von Gleichungen consistency of equations
Widerstand *m* 1. *(el)* resistor; rheostat; 2. *(el)* resistance; resistivity; 3. *(aero)* drag
W. beim Anlassen starting resistance
W. der Zuleitungen lead resistance
W. des Reglers resistance of the governor
aerodynamischer W. [aerodynamic] drag
akustischer W. acoustic resistance
äquivalenter W. equivalent resistance
auf den Kurbelkreis bezogener W. resistance reduced to the crank circle
auf Keramikplatte gewickelter W. card resistor
differentieller W. differential (incremental) resistance
drahtgewickelter W. wire-wound resistor
effektiver W. effective resistance
eingebauter W. integrated resistor
elektrischer W. electrical resistance
flüssiger W. liquid resistance
gedruckter W. printed resistor
hochohmiger W. high-value resistor
induktionsfreier W. non-inductive resistor
induktiver W. inductive resistance (reactance)
innerer W. anode (plate) resistance
kapazitiver W. capacitive reactance
kombinierter W. joint resistance
komplexer W. complex impedance
kritischer W. für die Erregung critical build-up resistance
lichtelektrischer W. photoresistance; photoelectric resistance cell
magnetischer W. [magnetic] reluctance
negativer W. negative resistance
ohmscher W. ohmic resistance
parallelgeschalteter W. shunt resistor
phasenreiner W. *s.* ohmscher W.
regelbarer W. adjustable (variable) resistor, variable rheostat
reiner W. non-reactive resistance
resultierender W. resultant resistance
schädlicher W. *(aero)* parasite drag
scheinbarer W. apparent resistance
spannungsabhängiger W. voltage-dependent resistor, varistor
spezifischer W. specific resistance, resistivity
stetig regelbarer W. continuously adjustable resistor (rheostat)
tatsächlicher W. actual resistance
temperaturabhängiger W. thermistor
transformierter W. eines Steuerkreises reflected control circuit resistance
veränderlicher W. variable resistor
vorgeschalteter W. series resistor
Widerstandsabbrennschweißung *f* resistance flash-butt welding
Widerstandsabgriff *m* rheostat slider
Widerstandsabnahme *f* resistance drop
Widerstandsabschmelzschweißung *f* resistance flash welding
Widerstandsänderung *f* resistance change
Widerstandsanpassung *f* impedance match[ing]
Widerstandsarbeit *f* work of resistance
Widerstandsarm *m* **der Brücke** resistive arm of the bridge
Widerstandsausgleich *m* resistance balance
Widerstandsbeiwert *m* *(aero)* coefficient of [aerodynamic] drag; resistance coefficient, friction factor
Widerstandsbremsschalter *m* rheostatic braking controller
Widerstandsbremsung *f* resistance (rheostatic) braking

Widerstandsbrücke *f* resistance (Wheatstone) bridge
Widerstandsbuckelschweißung *f* resistance projection welding
Widerstandsdämpfung *f* resistance loss (attenuation)
Widerstandsdehnungsmeßstreifen *m* resistance strain gauge
Widerstandsdraht *m* resistance wire
Widerstandselement *n* resistance (resistive) element
Widerstandserdung *f* resistance earthing
Widerstandserwärmung *f* resistance heating
widerstandsfähig resistant, resistive, refractory, robust, proof
w. gegen Staubeinwirkung dust-resistant
w. sein gegen to resist
Widerstandsfähigkeit *f* resistivity; ruggedness, stability, strength
Widerstandsfolie *f* resistive film
Widerstandsgeber *m* **/ selbstabgleichender** potentiometric recorder
widerstandsgekoppelt resistance-coupled
Widerstandsgerade *f* load line
Widerstandsgewebe *n* resistance mat
Widerstandsheizung *f* resistance heating
Widerstandshöhe *f* friction head
Widerstandskomponente *f* resistance component
W. der Elektrodenimpedanz electrode differential resistance
Widerstandskopplung *f* resistance (resistive) coupling
Widerstandskörper *m* 1. resistor core; 2. obstacle *(fluid flow)*
Widerstandskraft *f* resisting force
Widerstandslampe *f* resistance (ballast) lamp
Widerstandslinie *f* line of resistance
Widerstandsmanometer *n* resistance pressure gauge
Widerstandsmatrix *f* impedance matrix
Widerstandsmeßbrücke *f* resistance (Wheatstone) bridge
Widerstandsmesser *m* ohmmeter
Widerstandsmeßgerät *n* **für kleine Widerstände** ducter
Widerstandsmessung *f* resistance measurement
Widerstandsmoment *n* moment of resistance; section modulus
polares W. section modulus of torsion
Widerstandsnetz *n* resistance net; resistor net
Widerstandsnormal *n* standard resistance
Widerstandsofen *m* resistance furnace
Widerstandsoperator *m* vector impedance
Widerstandspunktschweißung *f* resistance spot welding
Widerstandsregelung *f* resistance (rheostatic) control
Widerstandsregler *m* rheostat
widerstandsreziprok impedance-reciprocal, reciprocal with respect to impedance
Widerstandsröhre *f* ballast tube
Widerstandssäule *f* carbon pile resistance
Widerstandsschicht *f* resistive film
Widerstandsschleifdraht *m* slide-wire; resistance wire tapped off by a slider
Widerstandsschweißelektrode *f* resistance welding electrode
Widerstandsschweißen *n* resistance welding
Widerstandsspule *f* resistive coil
Widerstandsstumpfschweißung *f* resistance butt welding
Widerstandsteiler *m* resistive divider
Widerstandsthermometer *n* resistance thermometer, thermometer resistor
Widerstandstransformation *f* impedance transformation
Widerstandsventil *n* resistance valve
Widerstandsverdunkler *m* rheostat dimmer
Widerstandsverlust *m* rheostatic loss
Widerstandswerkstoff *m* resistor material
Widerstandswert *m* resistance value
Widerstandszelle *f* photoconductive cell
Widerstandszunahme *f* increase of resistance
Widerstandszünder *m* discharger
Widerstandszweig *m* resistive branch
widerstehen to resist; to withstand
Widerstrom *m* counter current
Wiedemann-Effekt *m* Wiedemann effect
Wiederanreicherung *f (nucl)* re-enrichment
Wiederansaugen *n* **/ selbsttätiges** automatic restarting
wiederaufarbeiten to reprocess, to rework, to regenerate
Wiederaufbau *m* reconstruction, rebuilding, restoration
wiederaufbauen to rebuild, to reconstruct, to restore
Wiederaufbereitung *f* reprocessing
Wiederauffinden *n* retrieval *(computer)*
W. digitaler Information retrieval of digital information
Wiederaufkohlung *f* carbon restoration
wiederaufladen *(el)* to recharge; to reload
Wiederaufladen *n (el)* recharging; reloading
Wiederaufladung *f* recharge
Wiederaufnahme *f* **in den Kreislauf** recirculation
Wiederaufwickeln *n* rewinding
Wiedereinführung *f* reintroduction, reinsertion
W. des Trägers reinsertion of the carrier
wiedereinschalten to reclose, to reconnect
Wiedereinschaltrelais *n* reclosing relay
Wiedereinschaltung *f* reclosing, reconnection
Wiedereinschaltzeit *f* reclosing time
wiedereinschmelzen to remelt
wiedereinsetzen to replace; *(math)* to resubstitute

Wiedereinstellgenauigkeit *f* repeatable accuracy
Wiedereinstellung *f* resetting
wiedereintreten to reenter
Wiedereintrittsgeschwindigkeit *f* reentry velocity
Wiedererkennen *n* **gesprochener Worte / maschinelles** *(dat)* machine recognition of spoken words
wiedererwärmen to reheat
Wiedergabe *f* 1. reproduction *(recording)*; display, playback; 2. *(print)* replica
dreidimensionale W. three-dimensional display
getreue W. faithful reproduction
telegrafische W. telegraphic restitution
verrauschte W. speckled background *(radar)*
Wiedergabecharakteristik *f* playback characteristic
Wiedergabeentzerrung *f* reproducing equalization *(records)*
Wiedergabefrequenz *f* reproducing frequency
Wiedergabegerät *n* reproducing equipment, reproducer, replay unit
Wiedergabegeschwindigkeit *f* replay speed
Wiedergabegüte *f* quality (fidelity) of reproduction
hohe W. high fidelity
Wiedergabekopf *m* replay (reproducing, playback) head
Wiedergabekurve *f* fidelity curve
Wiedergabenatürlichkeit *f* realism of reproduction
Wiedergabepegel *m* playback level
Wiedergabeteil *m* reproduce section *(tape recorder)*
Wiedergabetreue *f* fidelity
W. eines Verstärkers fidelity of an amplifier
hohe W. high fidelity
Wiedergabeverstärker *m* reproducing (replay) amplifier
Wiedergabeverzögerung *f* restitution delay *(telegraphy)*
wiedergeben to reproduce; to display
wiedergewinnen to recover; to reclaim; to recuperate; to regain, to regenerate
Wiedergewinnung *f* recovery; reclamation; salvage; regain *(moisture)*
wiederherstellen 1. to reestablish; to restore; 2. to recondition, to repair
Wiederherstellung *f* 1. reestablishment; restoration; 2. reconstruction; repair
vollständige W. total recovery
Wiederherstellungszeit *f (nucl)* recovery time
wiederholbar repeatable, repetitive
Wiederholbarkeit *f* repeatability, reproducibility
wiederholen to repeat, to rerun
periodisch w. to cycle
Wiederholerbake *f* racon, responder beacon, transponder *(radar)*
Wiederholgenauigkeit *f* repetitive accuracy
Wiederholprogramm *n (dat)* rerun programme; return routine
Wiederholpunkt *m (dat)* rerun (return) point
Wiederholtaste *f* repeat-key
Wiederholteil *n* repetition part
Wiederholung *f* repetition; iteration; recurrence
W. des Arbeitsganges recycling
regelmäßige W. periodicity
Wiederholungsbefehl *m (dat)* repetitive (iterative) instruction
Wiederholungsfehler *m* repetitive error
Wiederholungsfrequenz *f* repetition (recurrence) frequency, [sequence] repetition rate
Wiederholungsgeschwindigkeit *f* repetition rate
Wiederholungskonstanz *f* repetitive stability
Wiederholungslampe *f* ancillary lamp *(telephony)*
Wiederholungsprüfung *f* replication *(statistics)*
Wiederholungssender *m* repeater transmitter
wiederinstandsetzen to recondition, to reservice; to repair
Wiederinstandsetzung *f* reconditioning, reservicing; repair
Wiederkehr *f* recurrence, return
Wiederkehrspannung *f* recovery voltage
Wiederverbindung *f* recombination
wiedervereinigen / sich to recombine
Wiedervereinigung *f* recombination *(of charged particles)*
W. von Ionen recombination (reunion) of ions
Wiedervereinigungsprozeß *m* recombination process
Wiederverkäufer *m* retailer
wiederverwendbar recoverable *(material)*, reusable
wiederverwenden to reuse, to re-employ
Wiederverwendung *f* reuse
Wiederzündspannung *f* reignition (restriking) voltage
Wiederzündung *f* reignition, after-count, afterdischarge
wiegbar weighable
Wiege *f (mach)* cradle *(generating)*
Wiegebalken *m* weigh beam
Wiegebehälter *m* weighing bucket (hopper), weigh box
Wiegehäuschen *n* weigh-bridge
wiegen to weigh, to scale
Wiegenfeder *f (ry)* bolster (elliptic) spring
Wiegenkipper *m (ry)* tipping wagon
Wiegenlafette *f (mil)* cradle mounting
Wien-Brücke *f (el)* Wien bridge
Wiese *f* meadow
Wiesenegge *f* grass harrow
Wiesenerz *n* meadow ore

Wiesenpflug *m* grassland plough
Wiesenwalze *f* grassland roller
Wig-Schweißen *n s.* Wolfram-Inertgas-Schweißen
Wigner-Effekt *m* discomposition (Wigner, knocking-out) effect
Wigner-Energie *f* Wigner energy
Wigner-Kern *m* Wigner kernel
Wigner-Modell *n* uniform (statistical, Wigner) model of nucleus
Wikingerheck *n* canoe stern
Wildfell *n* deerskin
Wildhaut *f* exotic (overseas) hide
Wildleder *n* deerskin leather
Willemit *m* willemite
Williot-Verschiebungsplan *m* Williot diagram
willkürlich arbitrary; random
Willkürvorgang *m* random process
Wilsonelektroskop *n* Wilson electroscope
Wilsonkammer *f* Wilson cloud chamber
Wimpel *m* pennon, pennant, streamer
Wind *m* 1. wind; 2. blast *(furnace)*
  ablandiger W. off-shore wind
  böiger W. gusty wind
  elektrischer W. electric wind (breeze)
  elektrostatischer W. convective discharge
  heißer W. hot blast
  stürmischer W. gale
  umspringender W. baffling wind
Windablenkung *f* deflection of the wind
Windangriffsfläche *f* wind catching area
Windbelastung *f (build)* wind load[ing]
Windbruch *m* wind fall *(forest)*
winddicht wind-proof
Winddruck *m* 1. wind pressure (load); windage *(outdoor switches)*; 2. blast pressure *(cupola)*
Winddruckverteilung *f (build)* wind pressure distribution
Winde *f* 1. *(text)* reel *(spinning)*; 2. winch, capstan, windlass; jack; hoist
  W. mit Umsteuergetriebe reversible winch
  handbetriebene W. hand[-operated] winch
  hydraulische W. hydrostatic winch
  selbstaufspulende W. self-rendering winch
Windeisen *n (mach)* tap wrench
Winderhitzer *m* hot-blast stove
Windevorrichtung *f (text)* winding motion
Windfahne *f* wind vane
Windfahnenknopf *m* acorn
Windfang *m (build)* [draught] lobby, porch
Windfege *f (ag)* winnower
Windfeuerbrücke *f* split bridge
Windform *f* tuyere, blast inlet
Windformkühlkasten *m* tuyere cooling plate
Windfrischen *n* air blowing
Windfrischverfahren *n* Bessemer process
Windkanal *m* wind tunnel (channel); air duct (passage)
  offener W. non-return-flow wind tunnel
Windkasten *m (therm)* air chamber (box); wind belt; wind box *(furnace)*, blast box *(Bessemer)*
Windkessel *m* dashpot; air vessel; pneumatic accumulator (receiving tank)
Windkomponente *f* wind component
Windkraft *f* wind power (force)
Windkraftmaschine *f* wind-driven generator; wind motor
Windkraftwerk *n* boreal power station
Windlast *f* wind force (load, pressure)
Windleitblech *n* smoke deflector [plate] *(locomotive)*
Windmantel *m* air chamber (belt) *(furnace)*
Windmesser *m* anemometer; anemograph
Windmotor *m* wind motor
Windmühle *f* windmill
Windpfeife *f* vent pipe *(moulding)*
Windrad *n* wind wheel
Windrichtung *f* wind direction
Windrichtungsanzeiger *m* wind direction indicator
Windring *m* air chamber (belt) *(furnace)*
Windringleitung *f* bustle-pipe *(blast furnace)*
Windrose *f* compass card
Windschatten *m* lee
windschief twisted
windschnittig streamlined
Windschreiber *m* anemograph
Windschutzplane *f* windbreak
Windschutzscheibe *f* windscreen, windshield
Windschutzscheibenaussteller *m* windscreen opener
Windschutzscheibenkopfleiste *f* windshield header
Windschutzscheibenrahmen *m* windscreen frame
Windschutzstreifen *m* windbreak
Windseite *f* 1. weather side; 2. blast side *(furnace)*
Windsichter *m* air separator (sifter), air-classifier, air-elutriator, pneumatic deduster
Windsichtung *f* air separation (classification); pneumatic sizing
  W. nach Schwere grading by weight with aspirator (winnower)
Windspitzengeschwindigkeit *f* maximum gust speed
Windstärke *f* wind velocity (speed), weight of wind
Windstärkemesser *m* wind speed indicator
Windstein *m* stone rest
Windstille *f* calm
Windstoß *m* flurry; blast of wind
Windstrebe *f* wind brace
Windtunnel *m* wind tunnel (channel) *(model testing)*
Windung *f* 1. coil *(spring)*; turn *(of a winding)*; 2. twist

federnde W. active (effective) coil
tote W. *s.* unbelegte W.
unbelegte W. idle turn
wirksame W. active coil
Windungen *fpl* / **benachbarte** adjacent turns
gegenmagnetisierende W. demagnetizing turns
Windungsbelag *m* closeness of winding
Windungsfluß *m* flux linking a turn
Windungsrichtung *f* hand (direction) of coils *(spring)*
Windungsschlußschutz *m* protection for inter-turn short-circuits
Windungsverhältnis *n* turns ratio
W. Eins unity turn ratio
Windungszahl *f* number of coils *(spring)*; number of turns
Windverband *m* transverse (lateral) bracing *(bridge)*; wind bracing
Windversetzung *f* set due to wind
Windverteilungskasten *m* draught distributing box
Windvorwärmung *f* air-preheating *(furnace)*
Windwerk *n* double-drum winch; haulage gear *(coal cutter)*
Windwerksseil *n* hoist rope *(excavator)*
Windzapfen *m* hollow trunnion *(Bessemer converter)*
Windzuleitung *f* blast pipe *(furnace)*
Winkel *m* 1. *(math)* angle; 2. square, triangle *(engineering drawing)*
W. an der Schneide tool angle
W. an der Spitze apex (face) angle
W. der inneren Reibung angle of internal friction
brechender W. refracting angle
brechender W. des Prismas refracting angle of the prism
einspringender W. reentrant angle
Eulerscher W. Eulerian angle
gestreckter W. straight angle
kritischer W. critical angle
räumlicher W. solid angle
rechter W. right angle
schiefer W. oblique angle
spitzer W. acute angle
stumpfer W. obtuse angle
überstumpfer W. reflex angle
vorspringender W. salient angle
Winkelabhängigkeit *f* angle dependence
Winkelablenkung *f* angular deflection; *(opt)* deviating power
Winkelabrichteinrichtung *f* *(mach)* angle dressing fixture, [grinding] wheel dresser capable of being used for shaping angles
Winkelabrichten *n* dressing at an angle *(grinding wheel)*
Winkelabstand *m* angular distance
Winkelabweichung *f* angular displacement
Winkelabzweigdose *f* angle conduit box
Winkeländerung *f* angular deformation *(mechanics)*
Winkelantrieb *m* *(mach)* angular (right-angle) drive; bell crank drive
Winkelarm *m* angled arm
Winkelaufblendung *f* angle fade-in
Winkelauflösung *f* angular resolution; angular discrimination *(radar)*
Winkelbereich *m* angular range
Winkelbeschleunigung *f* angular acceleration
Winkelbeschleunigungsmesser *m* angular accelerometer
Winkelbetrag *m* angular amount
Winkelbewegung *f* angular motion
Winkelbeziehung *f* angular relationship
Winkelbreite *f* angular width
Winkeldispersion *f* angular dispersion, dispersive power
Winkeldrehmoment *n* angular momentum
Winkeldrehpflug *m* one-way plough
Winkeleinfluß *m* angular effect
Winkeleinstellung *f* angular setting
Winkeleisen *n* angle [bar], angle iron (plate)
Winkelfehler *m* *(build)* angular error; *(el)* phase-angle error, squint *(aerials)*
Winkelfernrohr *n* elbow telescope
Winkelfläche *f* angular surface
Winkelflansch *m* angle flange
Winkelfotometer *n* goniophotometer
Winkelfräser *m* angular [half side] mill
W. mit Zylinderschaft angle end mill
Winkelfrequenz *f* angular frequency
Winkelfunktion *f* function of angles, trigonometric function
Winkelgelenk *n* angle joint
Winkelgenauigkeit *f* angular accuracy
Winkelgeschwindigkeit *f* angular velocity (speed)
momentane W. instantaneous angular velocity
Winkelgleichung *f* angle equation
Winkelgrad *m* angular degree
Winkelgröße *f* **im Bogenmaß** arc length in radians
Winkelhaken *m* [composing] stick, joint hook; *(Am)* setting stick
Winkelhalbierende *f* bisectrix, bisector
Winkelhebel *m* bent lever
Winkelhebelregler *m* bell-crank governor
winkelig angular
Winkelkonstante *f* phase constant
Winkelkorrelation *f* angular (angle, directional) correlation
Winkellage *f* angular position
Winkellehre *f* angle gauge
Winkelmaß *n* 1. angular dimension, angle measure; 2. *(el)* wavelength constant; image phase change coefficient; 3. *(math)* square rule

W. der Schrägstellung amount of angularity
Winkelmast *m* angle pole (tower)
Winkelmessen *n* / **repetitionsweises** repetition angular measurement
Winkelmesser *m* protractor; goniometer
Winkelmesser *n* angled knife
Winkelmeßgerät *n* instrument for angular measurement
Winkelmeßokular *n* goniometer ocular
Winkelmessung *f* measurement of angles, angular measurement; goniometry
Winkelminute *f* angular minute
Winkelminutennonius *m* angular vernier
Winkelmodulation *f* angle modulation
Winkelmoment *n* angular momentum
Winkelmuffe *f* split elbow *(conduit wiring)*
Winkelprisma *n* prismatic square
Winkelprofil *n* angle [section]
Winkelprüfkopf *m* angle scanning unit
Winkelreflektor *m* corner reflector
Winkelrichtvermögen *n* angular directivity
Winkelring *m* junk ring, follower [plate]
Winkelrinne *f* angular trough
Winkelrostplatte *f* angular grate
Winkelschenkel *m* angle leg
Winkelschiene *f* set square
Winkelschraubenzieher *m* offset screw driver
Winkelspiegel *m (opt)* corner reflector
Winkelstahl *m* angle [steel]
gleichschenkliger W. equal[-leg] angle
ungleichschenkliger W. unequal[-leg] angle
Winkelstation *f* angle station *(conveyor)*
Winkelstellung *f* angular position
Winkelstirnfräser *m* single-angle cutter, angular half side mill
Winkelstrahler *m* angle reflector
Winkelstück *n* elbow, corner *(waveguide)*
Winkelstützmauer *f* buttress (cantilevered retaining, counterfort) wall
Winkelteilung *f* angular spacing
Winkelteilungsprüfgerät *n* angle-division tester
Winkelthermometer *n* angle (bent) thermometer
Winkeltisch *m* knee *(milling)*
winkeltreu angle-preserving, isogonal
Winkeltreue *f* isogonality
Winkeltrieb *m* bell-crank drive
Winkeltrisektion *f* trisection of angles
Winkelumsetzer *m* angle encoder, shaft position-to-digital converter
Winkelventil *n* angle valve *(tyre)*
Winkelvergrößerung *f (opt)* angular magnification
Winkelverhältnis *n (opt)* angular magnification, convergence ratio
Winkelverschiebung *f* angular displacement
Winkelverschlüßler *m* angle encoder, shaft position-to-digital converter
Winkelverschraubung *f* elbow
Winkelverteilung *f* angular (directional) distribution
Winkelverzahnung *f* double-helical gearing
Winkelvoreilung *f (el)* angular advance
Winkelzahn *m* herringbone tooth
Winker *m (aut)* direction indicator arm, semaphore direction indicator, trafficator
Winkerkontrolleuchte *f* direction-indicator control lamp
Winkflagge *f (ship)* hand flag
Winsch *f (ship)* winch
winterfest wintertight
Winterfestigkeit *f* frost hardiness
Winterfreibord *m* winter freeboard
Winterfrüchte *fpl* autumn sown crops
Wintergetreide *n* autumn [sown] cereals
Winterhafen *m* wintering harbour
Winterladelinie *f* winter load-line
Wintersaat *f* autumn sown crops
Winterschutzhafen *m* wintering harbour
Winterspritzmittel *n* winter sprayer
Winterspritzung *f* green-bud spray, mouse-ear spray, green-cluster spray, winter washing
Winterung *f* autumn sown crops
Wippbrett *n* hinged driver's platform
Wippdavit *m(n) (ship)* luffing davit
Wippdrehkran *m* luffing and slewing crane
Wipperboden *m* tippler floor
Wippkran *m* [level-]luffing crane
Wirbel *m* eddy, whirl, vortex *(fluid flow)*
gebundener W. bound vortex
trichterförmiger W. Rankine's combined vortex
Wirbelachse *f* axis of eddy (vortex)
Wirbelbereich *m* vorticity zone
Wirbelbeschleuniger *m* electrostatic accelerator
Wirbelbett *n* fluidized bed
Wirbelbewegung *f* vortex (swirling, eddy) motion
Wirbelbildung *f* eddy (vortex) formation, formation of eddies (vortices)
Wirbelbrenner *m* turbulent burner
Wirbeldiffusion *f* decay of vorticity
Wirbelerhaltung *f* conservation of vorticity
Wirbelfaden *m* vortex filament, thread of the vortex
Wirbelfeld *n* rotational field, field of vorticity *(fluid flow)*
Wirbelfläche *f* vortex (vorticity) surface
Wirbelfluß *m* flux of vorticity
wirbelfrei free from eddies, non-vortical, irrotational
Wirbelgebiet *n* vortex region
Wirbelhaken *m* swivel-hook
wirbelig vortical
Wirbelkammer *f* swirl chamber
Wirbelkammermotor *m* swirl-chamber engine
Wirbelkern *m* core of vortex, vortex core
Wirbellinie *f* vortex line

wirbellos non-eddying
Wirbelmischer *m* impeller
wirbeln to eddy *(fluid flow)*; *(met)* to swirl
wirbelnd turbulent
Wirbelpunkt *m* fluidizing point
Wirbelraum *m* eddy space *(turbine)*
Wirbelreihe *f* vortex row
Wirbelring *m* vortex ring
Wirbelröhre *f* vortex (whirl) tube
Wirbelschäkel *m (ship)* swivel shackle (piece)
Wirbelschicht *f* fluidized bed
brodelnde W. bubbling fluidized bed
durchbrochene W. channelling fluidized bed
homogene W. particulate fluidized bed
inhomogene W. aggregative fluidized bed
periodisch arbeitende W. batch fluidized bed
stetig arbeitende W. continuous fluidized bed
stoßende W. slugging fluidized bed
turbulente W. turbulent fluidized bed
Wirbelschichtreaktor *m* fluidized bed reactor
Wirbelschichtverfahren *n* fluid catalyst process, fluidization
Wirbelschleuder *f* centrifugal dirt catcher
Wirbelsichter *m* curlator
Wirbelsintern *n* dip coating in powder, fluidized bed coating
Wirbelstärke *f* vorticity, strength of a vortex
Wirbelstraße *f* vortex street
Wirbelstrom *m* eddy current
Wirbelstrombremse *f* eddy-current brake
Wirbelstrombremsung *f* eddy-current braking
Wirbelstromdämpfung *f* eddy-current damping
Wirbelstromgleisbremse *f* eddy-current rail brake
Wirbelstromglühverfahren *n* eddy-current heating method
Wirbelstromkupplung *f* eddy-current coupling
Wirbelstromtachometer *n* drag-torque tachometer
Wirbelströmung *f* turbulent (vortical) flow
Wirbelstromverlust *m* eddy-current loss
Wirbelung *f* swirl; *(met)* whirling motion
W. durch Ansaugluft induction air swirl
Wirbelverlust *m* eddy loss *(turbine)*
Wirbelwiderstand *m (ship)* wake (eddy) resistance
Wirbelwind *m* eddy wind, whirlwind
Wirbelzentrum *n* vortex centre
Wirbelzone *f* eddy zone
Wirbligkeit *f* vorticity
Wirkdruck *m* differential pressure; pressure drop *(flowmeter)*
wirken 1. to act, to operate; 2. *(text)* to knit
Wirken *n (text)* knitting
wirkend / einfach single-acting, single-action
einseitig w. unidirectional, acting in one direction
Wirkenergie *f* active energy
Wirkfläche *f* effective area *(aerial)*
Wirkkomponente *f (el)* effective (real, active, resistive, power, energy, co-phase, in-phase) component
Wirklast *f* active (actual) load
Wirkleistung *f (el)* active (actual, effective, real, non-reactive, wattful) power
W. der Gegenkomponente negative-sequence active power
W. des mitläufigen Systems positive-sequence active power
aufgenommene W. active input
Wirkleistungsmesser *m* active power meter
Wirkleistungsrelais *n* active power relay
Wirkleistungsübertragung *f* active power transmission
Wirkleitwert *m* conductance, active admittance
Wirkmaschine *f* knitting machine
Wirkquerschnitt *m* useful cross-section *(conductor)*
wirksam effective, active, efficient
Wirksamkeit *f* efficiency, effectiveness, efficacy effectivity; activity
biologische W. biological effectiveness (efficiency)
relative biologische W. relative biological effectiveness *(of radiation)*, RBE
Wirkschaltbild *n* functional diagram
Wirkspannung *f* active (in-phase) voltage
Wirkstoff *m* 1. *(ag)* active principle, agent; *(ch)* dope, additive; 2. *(text)* knitted cloth
Wirkstrom *m* active (in-phase, useful, wattful, true) current
Wirkstuhl *m* knitting frame
Wirkung *f* action, activity, effect
W. / mit gerichteter unidirectional
W. und Gegenwirkung *f* action and reaction
biologische W. biological effect *(of radiation)*
dämpfende W. dashpot action
dirigierende W. eines Substituenten *(ch)* directing effect of a substituent
dynamische W. dynamic action
fotochemische W. photochemical activity
progressive W. progressive action
proportionale W. proportional action
schiebende W. propellent force *(explosive)*
schwundmindernde W. anti-fading effect *(aerials)*
treibende W. der Massenkräfte driving effect of inertia
weichmachende W. *(plast)* plasticizing efficiency
zerspanende W. cutting action
zertrümmernde W. shattering force *(explosive)*
Wirkungsbereich *m* range of effectiveness
Wirkungsdauer *f* duration of effect
Wirkungsebene *f* plane of action

Wirkungsgrad *m* efficiency
W. bei der Zerspanung cutting efficiency
W. der Anode anode efficiency
W. der Kraftübertragung transmission efficiency *(e.g. of a thermoelectric vehicle)*
W. des freifahrenden Propellers efficiency in open water
W. des hinter dem Schiff fahrenden Propellers *(ship)* efficiency behind
W. einer Antenne radiation efficiency of an aerial
W. in Amperestunden ampere-hour efficiency
adiabatischer W. adiabatic efficiency
energetischer W. energy efficiency
hydraulischer W. hydraulic efficiency
hydraulischer W. der Pumpe hydraulic efficiency of the pump
idealer W. ideal efficiency
indizierter W. indicated efficiency
mechanischer W. mechanical efficiency
relativer W. relative efficiency
scheinbarer W. apparent efficiency
thermischer W. thermal efficiency
thermischer W. des Kessels thermal efficiency of the boiler
volumetrischer W. volumetric efficiency
Wirkungsgradbestimmung *f* efficiency test
Wirkungsgröße *f* action quantity
W. der Kolonne height of a transfer unit, HTU
Wirkungskraft *f* effective force
Wirkungslinie *f* line of action
Wirkungsquantum *n* quantum of action
Plancksches W. Planck's constant, quantum of action
Wirkungsquerschnitt *m* cross section
absoluter W. absolute cross section
atomarer W. atomic cross section
differentieller W. differential cross section
effektiver W. effective cross section
fotoelektrischer W. photoelectric cross section
kernphysikalischer W. nuclear cross section
makroskopischer W. macroscopic cross section
mikroskopischer W. microscopic cross section
mittlerer W. average cross section
partieller W. partial cross section
totaler W. total cross section
unelastischer W. inelastic cross section
Wirkungsrichtung *f* direction of action (effect)
Wirkungssphäre *f* sphere of action
Wirkungsvariable *f* action variable
Wirkungsweise *f* [mode of] action; operating mode; mechanism
Wirkverbrauchsmesser *m* active power meter
Wirkverbrauchszähler *m* active-energy meter, watt-hour meter
Wirkverstärkung *f* transducer gain
Wirkware *f* knitted material
reguläre W. full-fashioned fabric, fully fashioned fabric
Wirkwiderstand *m* active (effective, ohmic) resistance
akustischer W. acoustical resistance
ohmscher W. [ohmic] resistance
Wirkzeit *f (dat)* machine-available time
Wirrspan *m* / **zäher** stringy chip
Wirt *m* host
Wirtatom *n* host atom
Wirtkristall *m* host crystal
wirtschaftlich economical; commercial
Wirtschaftlichkeit *f* economy
Wirtschaftlichkeitsgrenze *f (mine)* break-even point
Wirtschaftsflug *m (ag)* aerial spraying (dusting, seeding, fertilizing), aircraft spraying (dusting, seeding, fertilizing)
Wirtschaftsgebäude *npl* outhouse, farm buildings
Wirtschaftsgeografie *f* economic geography
Wirtschaftsplanung *f* economic planning
Wirtschaftswissenschaften *fpl* economics
Wischer *m* 1. *(aut)* wiper; 2. *s.* Wischkontakt
Wischerarm *m* wiper arm
Wischerblatt *n* wiper blade
wischfest wiping-proof
Wischkontakt *m* wipe (passing) contact *(switch)*
Wischpolieren *n* lap polishing
Wischtest *m* rubbing test
Wischwalze *f* damping roller
Wischwasser *n* fountain (damping) solution
Wischwasserkasten *m* fountain
Wismut *n* bismuth
Wismutat *n* bismuthate
Wismutblende *f* bismuth blende, eulytite
Wismutdraht *m* bismuth wire
Wismutglanz *m* bismuth glance, bismuthinite
Wismutglätte *f* bismuth litharge
Wismutlegierung *f* bismuth alloy
Wismutlot *n* bismuth solder
Wismutnitrat *n* bismuth nitrate
Wismutocker *m* bismite, bismuth ochre
Wismutspirale *f* bismuth spiral
Wismut(III)-verbindung *f* bismuthous compound
Wismut(V)-verbindung *f* bismuthic compound
Wissenschaft *f* science
angewandte W. applied science
reine W. pure science
Wissenschaftler *m* scientist, scientific worker
wissenschaftlich scientific[al], learned
Witherit *m (min)* witherite
Witterung *f* weather
Witterungsbedingungen *fpl* atmospheric conditions

witterungsbeständig weather-proof
Witterungsbeständigkeit *f* weather resistance, resistance to atmospheric conditions
witterungsecht fast to weather
Witterungseinfluß *m* atmospheric influence
Witterungsschutz *m* weather protection
WKB-Methode *f s.* Wentzel-Kramers-Brillouin-Methode
Wobbelbereich *m* sweep width
Wobbelbetrieb *m* sweep operation
Wobbelfrequenz *f* wobbling (sweep) frequency
Wobbelgenerator *m* wobbler, wobbulator, sweep generator
Wobbelgeschwindigkeit *f* sweep rate
Wobbelmeßsender *m* sweep generator
wobbeln to wobble, to wobbulate
Wobbeln *n* frequency sweeping, wobbling, wobble modulation
Wobbeloszillator *m* sweep oscillator (generator)
Wobbelsender *m* wobbulator; sweep signal generator; sweep frequency generator
Wobbelton *m* warble tone
Wobbler *m* wobbulator; sweep generator; warbler
Wobblerfrequenz *f* sweep frequency; warble frequency
Wochendosis *f (nucl)* weekly dose
Wochenkarte *f* weekly season ticket
Wochenlohn *m* weekly earning
wogen to wave
Wöhlerkurve *f* Wöhler (stress-number, fatigue) curve
wohlriechend aromatic, aromatous
Wohnblock *m* apartment block, block of flats
Wohndichte *f* housing density
Wohneinheit *f* dwelling (residential) unit
Wohnfläche *f* floor area
Wohngebiet *n* residential area (district)
Wohngebietszentrum *n* community centre
Wohnhaus *n* residential house
Wohnhochhaus *n* tower block
Wohnkomplex *m* housing estate, neighbourhood (community) unit
Wohnküche *f* multi-purpose kitchen, all-purpose kitchen
Wohnlager *n* **für Arbeiter** labour camp
Wohnraum *m (ship)* [living] accomodation, living quarter
Wohnraumdeck *n (ship)* berth-deck
Wohnraumschott *n (ship)* accomodation bulkhead
Wohnsammelstraße *f* collector road
Wohnschiff *n* houseboat, barracks ship
Wohnschlafzimmer *n* bed-sitting room
Wohnsiedlung *f* housing estate
Wohnstraße *f* residential street
Wohnung *f* flat, apartment, dwelling
Wohnungsanschluß *m* residence telephone
Wohnungsbau *m* housing [construction], residential construction, domestic building
  W. in Plattenbauweise panellized housing
  industrieller W. industrialized housing
Wohnungsbauvorhaben *n* housing project (scheme)
Wohnungseinheit *f* dwelling (residential) unit
Wohnungseinrichtung *f* furniture
Wohnungstrennwand *f* party wall
Wohnunterkunft *f* living quarter
Wohnwagenanhänger *m* trailer coach, caravan
Wohnzimmer *n* living (sitting) room
wölben 1. *(math)* to curve; 2. *(build)* to vault; to turn *(arches)*; 3. to bow *(sheet metal)*, to camber
Wölbkehlnaht *f* convex fillet weld
Wölbnaht *f* convex weld
Wölbspiegel *m* convex mirror
Wölbstrecke *f (mine)* arched roadway
Wölbstück *n (ship)* bridge piece *(stern frame)*
Wölbung *f* 1. *(math)* curvature, kurtosis; 2. *(build)* vault; camber, crowning *(pulley)*
Wölbungsverhältnis *n* camber ratio *(e.g. of a propeller)*
Wolf *m* lewis *(stone)*
Wolfram *n* tungsten, wolfram
Wolframanode *f* tungsten anode
Wolframat *n* tungstate
Wolframbogenlampe *f* tungsten arc lamp
Wolframdraht *m* tungsten filament
Wolframdrahtgeflecht *n* tungsten mesh
Wolframdrahtlampe *f* tungsten [filament] lamp
Wolframelektrode *f* tungsten electrode
Wolframfaden *m* tungsten filament
  thorierter W. thoriated tungsten filament
Wolframgleichrichter *m* tungsten rectifier
Wolframheizdraht *m* tungsten filament
Wolframid *n* tungstide
Wolfram-Inertgas-Schweißen *n* inert gas-shielded arc welding [using tungsten electrode]
Wolframit *m* wolframite
Wolframkarbid *n* tungsten carbide
  gesintertes W. cemented tungsten carbide
wolframkarbidbestückt tungsten-carbide tipped
Wolframkarbideinsatz *m* tungsten-carbide insert
Wolframkarbidkugel *f* tungsten carbide ball
Wolframkarbidmesserkopf *m* tungsten-carbide face milling cutter
Wolframkarbidplättchen *n* tungsten-carbide tool tip
Wolframkarbidschneide *f* tungsten-carbide tool tip
Wolframkatode *f* tungsten cathode
Wolframkontakt *m* tungsten contact
Wolframlampe *f* tungsten filament lamp
  gasgefüllte W. gas-filled tungsten lamp *(Langmuir)*

Wolframsäure *f* tungstic acid
Wolframscheibe *f* tungsten disk
Wolframschnellstahl *m* tungsten high speed steel
Wolframsonde *f* tungsten probe
Wolframstahl *m* tungsten steel
Wolfram-Titan-Karbid *n* tungsten-titanium carbide
Wolfstift *m (text)* willow pin
Wolfszahn *m* gullet tooth
Wolke *f* cloud
  pilzförmige W. mushroom cloud
Wolken / über den *(aero)* on top
Wolkenbank *f* cloud bank
Wolkenbruch *m* cloud-burst
Wolkenentladung *f* cloud discharge
Wolkenhöhe *f* cloud ceiling
Wolkenhöhenmesser *m* cloud ceilometer
Wolkenkavitation *f* cloud cavitation
Wolkenpilz *m (nucl)* mushroom cloud
Wolkenschicht *f* cloud stratum
Wolkensegeln *n* cloud soaring
Wolkental *n* cloud trough
Wolkenwand *f* cloud bank
Wollastondraht *m* Wollaston wire
Wollastonprisma *n* Wollaston prism
Wollbett *n (text)* built-up stack of wool
Wolldichtung *f* wool packing
Wolldickenmesser *m* eriometer
Wolle *f* wool
  entschweißte W. degreased wool
  gechlorte W. chlorinated wool
  minderwertige W. doggy wool
  verklettete W. burred wool
  wiederverarbeitete W. reprocessed wool
Wollentschweißung *f* scouring of wool
Wollfarbstoff *m* wool dye
Wollfaser *f* wool fibre
  gekräuselte W. crimply (curly) wool fibre
  wellenförmige W. wavy wool fibre
  wellentreue W. evenly waved wool fibre
  wellenuntreue W. irregulary waved wool fibre
Wollfett *n* wool fat (grease)
Wollfettsäure *f* degras acid
Wollfettschlamm *m* grease scum
Wollkammzug *m* wool top
Wollklassierung *f* wool grading
Wollschafzucht *f* breeding (farming) of sheep for wool
Wollschuppenschicht *f* wool scale
Wollsortierung *f* wool sorting
Wollstrumpfwirkstuhl *m* worsted stocking loom
Wolltuch *n* / **gehlztes** baize
Wollwäsche *f* wool scouring
Wollwaschmaschine *f* wool washing machine
Wollweberei *f* wool weaving
Woltman-Zähler *m* turbine flowmeter
Woodmetall *n* Wood's alloy
Worpbalken *m (ship)* transom beam
Wort *n (dat)* word
Wortabfrage *f* word selection
Wortauswahl *f* word selection
Wortdarstellung *f* word representation
Wortelement *n* bit
Wörterblock *m* block *(computer)*
Wörterbuch *n* / **mechanisches** mechanical dictionary
Wortgeber *m* word generator *(computer)*
Wortgenerator *m* word generator
  achtstelliger W. eight digit word generator
Worthingtondampfpumpe *f* Worthington steam pump
Wortlaufzeit *f* word time
Wortmarke *f* word mark
Wortverstümmelung *f* clipping of words *(telegraphy)*
Wortwahl *f* / **externe** external word selection
Wortzähler *m* word counter
Wortzählung *f* word counting
Wrack *n* wreck, wrecked vessel
  gesunkenes W. sunk wreck
Wrackboje *f* wreck-marking buoy
Wrackgut *n* wrecked goods, wreckage
Wrackschiff *n* wreck-marking vessel
Wrackteile *npl* wreckage
Wracktonne *f* wreck buoy
Wrange *f (ship)* transom
Wrasen *m* water vapour
Wrasenabzug *m* air chimney (flue, vent), ventilation flue
Wrasenhaube *f* vapour hood
Wrasenrohr *n* flue pipe
Wringmaschine *f* wringing machine
wuchsfördernd growth promoting
wuchshindernd growth inhibiting
Wuchsrichtung *f* direction of growth
Wuchsstoff *m* growth substance (promoter)
Wuchsstoffherbizid *n* growth-promoting herbicide, hormone-type weed-killer
Wuchsstoffmittel *npl* growth[-promoting] substances
Wucht *f* kinetic energy
Wuchtfehler *m* unbalance
Wuchtzustand *m* **der Schleifscheibe** grinding wheel balance
Wühlgrubber *m* tine cultivator
  W. ohne Gegenfeder curved-tine cultivator, C-spring tine cultivator
Wulfenit *m* wulfenite
Wulst *f* 1. slub *(yarn)*; 2. bulge; 3. astragal, bulb; collar; bead *(tyre)*
  W. der Ankerklüse blister of the hawse hole
wulstartig collar-shaped
Wulstband *n* clincher band
Wulstbug *m* bulb[ous] bow, club-foot

Wulstfelge *f* clincher rim
Wulstferse *f* heel of bead
Wulstform *f* bulbous bowform
Wulstfußauflage *f* bead seat
Wulstheber *m* bead breaker
Wulstkiel *m* bulb keel
Wulstkieler *m* bulb keeler
Wulstkieljacht *f* bulb-keel yacht
Wulstreifen *m* clincher tyre
Wulstrohrkondensator *m* beaded rim tubular capacitor
Wulstschutzstreifen *m* chafer
Wulstspitze *f* toe of bead
Wulstwinkel *m* angle bulb, bulb angle
Wünschelrute *f* divining rod (stick)
Wünschelrutengänger *m* dowser, waterfinder
Wurf *m* throw
  lotrechter W. vertical throw
  schiefer W. inclined throw
Wurfanker *m* stream anchor, kedge
Wurfbahn *f* trajectory
Würfel *m* cube
Würfelbindung *f* basket (hopsack) weave
Würfeldruckfestigkeit *f* cube strength
würfelförmig cube-shaped, cubical
Würfelnickel *n* nickel cube
Würfelsymmetrie *f* cubic symmetry
Würfelzucker *m* lump sugar
Wurffeuerung *f* shovel stoker
Wurfförderrinne *f* reciprocating trough conveyor
Wurfgranate *f* mortar shell
Wurfleine *f* heaving (hauling) line
Wurfmaschine *f* trap *(trap shooting)*
Wurfrad *n* *(ag)* rotating head; *(therm)* fuel-throwing wheel
Wurfradgehäuse *n* casing of the fuel-throwing wheel
Wurfradhäcksler *m* chaff cutter with flywheel cutter head and impeller blades
Wurfschaufel *f* shovel
  schwingende W. swinging shovel
  umlaufende W. fuel-throwing wheel
Wurfschaufellader *m* rocker shovel, shovel loader
Wurftaube *f* clay pigeon
Wurftaubenschießen *n* trap shooting, shooting at clay pigeons
Wurftrichter *m* funnel-shaped crater *(blasting)*
Wurftrommelhäcksler *m* forage harvester with cylinder cutter head
Wurfweite *f* length of throw
Wurfwidder *m* swinging shovel
Wurfwinkel *m* *(mine)* angle of projection (throw)
Würgemaschine *f* wiring jig *(e.g. manufacture of cables)*
Würgeprüfung *f* torsional test *(lacquer)*
Würgewalze *f* top roller
Würgezange *f* *(mine)* crimper
Würgverbindung twist joint
wurmartig vermicular, worm-shaped
Wurtzit *m* *(min)* wurtzite
Würze *f* spice, flavouring, seasoning; wort *(brewery)*
Wurzel *f* root
  W. aus root of
  dritte W. cubic root
  zweite W. square root
Wurzelausdruck *m* radical
Wurzelbereich *m* root *(weld)*
Wurzelexponent *m* order of a root
Wurzelfehler *m* root crack *(weld)*
Wurzelfrüchte *fpl* root crops
Wurzelort *m* *(dat)* root locus
Wurzelortskurve *f* root locus
Wurzelzeichen *n* radical sign
Wurzelziehen *n* extraction of roots (a root)
würzen to spice, to season, to flavour
Würzmenge *f* length of wort *(brewery)*
Würzmittel *n* condiment
WZ-Faktor *m* water-cement ratio (factor)

X

X-Ablenkung *f* horizontal (x-axis) deflection *(X-ray tube)*
x-Achse *f* abscissa, x-axis
X-Amplitude *f* X-amplitude
X-Band *n* X-band
X-Dehnung *f* X expansion
X-Einheit *f* X-unit *(X-ray tube)*
$\bar{x}$-Karte *f* $\bar{x}$ chart, control chart for averages
X-Komponente *f* X-wave
X-Kontakt *m* X-contact
x-Koordinate *f* abscissa, x-coordinate
X-Naht *f* double V butt weld
X-Schneide *f* X-bit
X-Schnitt *m* X-cut *(of crystals)*
X-Stoß *m* double V joint
X-Strahlen *mpl* X-rays
X-Verschiebung *f* X shift
X-Verstärker *m* X-amplifier, horizontal amplifier
X-Y-Darstellung *f* X-Y presentation
X-Y-Schreiber *m* graph (X-Y, two-axis) plotter
Xanthin *n* xanthine, urinous acid
Xanthogenat *n* xanthate, xantho[ge]nate
Xanthogenoxid *n* xanthic oxide
Xanthogensäure *f* xanthic (xanthogenic, xanthonic) acid
Xanthon *n* xanthone
Xanthophyll *n* xanthophyll
Xanthoproteinreaktion *f* xantho-protein reaction
Xanthorrhöaharz *n* acaroid resin
xenomorph xenomorphic
Xenon *n* xenon
Xenonblitzlichtlampe *f* xenon flash lamp

Xenonbogenlampe *f* xenon arc lamp, xenon [discharge] lamp
Xenonszintillationszähler *m* xenon scintillation counter
Xenonvergiftung *f* xenon poisoning
Xenotim *m (min)* xenotime
Xerografie *f* xerography
Xeroradiografie *f* xeroradiography
xeroradiografisch xeroradiographic
Xylan *n* xylane
Xylenolharz *n* xylenol resin
Xylit *m* xylitol
Xylol *n* xylene
Xylolmoschus *m* musk xylene, xylene musk
Xylose *f* xylose, wood sugar

Y

Y-Ablenkung *f* vertical (y-axis) deflection *(X-ray tube)*
y-Achse *f* ordinate, axis of ordinates, y-axis
Y-Ebene *f* Y-plane
Y-Kontakt *m* Y-contact
y-Koordinate *f* ordinate
Y-Kopplung *f* wye junction *(waveguide)*
Y-Legierung *f* Y-alloy
Y-Schale *f (build)* butterfly shell
Y-Schnitt *m* Y-cut *(of crystals)*
Y-Stück *n* y-tube, y-pipe
Y-Verschiebung *f* Y shift
Y-Verstärker *m* Y-amplifier, vertical amplifier
Yacht *f* yacht
Yagi-Antenne *f* Yagi aerial
Yard *n* yard *(0,9144 m)*
Yarrow-Kessel *m* Yarrow boiler
Ylang-Ylang-Öl *n* ylang-ylang [oil]
Ylem *n (nucl)* ylem
Yperit *n* yperite, mustard gas
Ytterbium *n* ytterbium
Yttererde *f* yttrium oxide, yttria
Yttererden *fpl* yttrium sub-group of rare earth metals
Ytterspat *m (min)* xenotime
Yttrium *n* yttrium
Yukawa-Feld *n (nucl)* Yukawa field
Yukawa-Quant *n s.* Yukawa-Teilchen
Yukawa-Teilchen *n (nucl)* Yukawa particle, yukon
Yukon *n s.* Yukawa-Teilchen

Z

z-Achse *f* z-axis
Z-Antrieb *m* z-propulsion *(of a vessel)*
Z-Bake *f* Z marker
Z-Drehung *f* Z-twist
Z-Profil *n (build)* zee, zed
Z-Schneide *f* Z-bit
Zacke *f* 1. jag; 2. notch; 3. blip, pip *(radar)*; 4. serration
zackenförmig jagged
Zackenschnitt *m* serrated cut
Zackenschrift *f* variable area track *(film)*
zäh 1. tough; 2. *s.* zähflüssig
zähflüssig viscous, viscid, heavy-bodied
zähflüssig werden to become viscous
Zähflüssigkeit *f* viscosity
Zähflüssigkeitsgrad *m* degree of viscosity
Zähgehalt *m* carbon content *(of steel)*
Zähigkeit *f* 1. toughness *(solid)*; 2. viscosity *(liquid)*
  absolute Z. absolute viscosity
  absolute dynamische Z. absolute dynamic viscosity
  absolute kinematische Z. absolute kinematic viscosity
  kinematische Z. kinematic viscosity
Zähigkeitsbeiwert *m* coefficient of viscosity
Zähigkeitsmaß *n* coefficient of viscosity
Zähigkeitsmesser *m* viscosimeter
  Englerscher Z. Engler viscometer
Zahl *f* number; figure; *(dat)* quantity
  Z. der Freiheitsgrade number of degrees of freedom
  absolute Z. absolute number
  achtstellige Z. eight-digit number
  Avogadrosche Z. Loschmidt number (constant)
  beliebige Z. random number
  benannte Z. denominate number
  bevorzugte Z. *s.* magische Z.
  einstellige Z. one-place number
  endliche Z. finite number
  Faradaysche Z. Faraday number
  Froudesche Z. Froude number
  ganze Z. integer [number]
  geltende Z. significant number
  gerade Z. even number
  halbganze Z. half-integer
  imaginäre Z. imaginary number
  in Faktoren zerlegbare Z. composite number
  irrationale Z. irrational number, surd [number]
  komplexe Z. complex number
  Loschmidtsche Z. Avogadro's number (constant)
  magische Z. *(nucl)* magic number
  mehrstellige Z. multi-digit number
  n-stellige Z. n-place number
  natürliche Z. natural number
  Reynoldssche Z. Reynolds number
  unbenannte Z. abstract number
  ungerade Z. odd number
Zähllader *f* marked core *(cable)*
Zählapparat *m* counter

Zählausbeute *f (nucl)* counter (counting) efficiency, efficiency of counter (counting)
zahlbar payable
Zählbereich *m* counter range
Zählblatt *n* census card
Zählblock *m* automatic block with axle counters
Zählcharakteristik *f (nucl)* plateau (counting rate-voltage) characteristic, counting-response
Zähldekade *f* counting decade
Zähldraht *m* counter wire
Zahlen *fpl* / **konjugiert komplexe** conjugate imaginaries (complex numbers)
zählen to count, to register
Zählen *n* counting
  Z. mit Zeitvorwahl preset-time counting
Zahlenanzeige *f* numerical display
Zahlenbeispiel *n* numerical example
Zahlenbruch *m* numerical fraction
Zahlendarstellung *f* number representation
  halblogarithmische Z. floating-point representation
Zahlenebene *f* number plane
Zahlenfolge *f* sequence of numbers
Zahlengeber *m* number generator; key sender *(telephony)*
Zahlengleichung *f* numerical equation
Zahlengröße *f* numerical quantity
Zahlengruppe *f* word *(computer)*
Zahlenkapazität *f (dat)* capacity
Zahlenkomplex *m* complex of figures
Zahlenkontinuum *n* continuum of numbers
Zahlenkörper *m* number field
zahlenmäßig numerical
Zahlenmenge *f* assemblage of numbers
Zahlenmodul *m* number module
Zahlenrechnung *f* numerical computation
Zahlenreihe *f* number sequence (series)
Zahlenscheibe *f* figure wheel
Zahlensystem *n (dat)* number system
  binäres Z. binary number system
  dezimales Z. decimal number system
Zahlentafel *f* numerical table, table of figures
Zahlentheorie *f* theory of numbers, number theory
  elementare Z. elementary theory of numbers
Zahlenverhältnis *n* numerical ratio
Zahlenwert *m* numerical value
Zahlenwerte *mpl* numerical values; data
Zähler *m* 1. counter; counter [tube]; 2. *(el)* [electric] meter; 3. *(math)* numerator *(fraction)*; 4. scaler, scaling circuit; 5. integrating (hour) meter
  Z. für die Besetzt- und Verlustfälle congestion meter *(telephony)*
  Z. mit Maximumzeiger meter with demand indicator
  Z. zum Addieren add counter
  Z. zum Subtrahieren subtract counter
  astatischer Z. astatic meter
  binärer Z. scale-of-two circuit, binary scaler
  elektrischer Z. electric meter
  elektrodynamischer Z. Thomson meter
  elektromechanischer Z. electromechanical counter
  elektronischer Z. electronic counter
  mechanischer Z. mechanical register
  oszillierender Z. oscillating meter
  richtungsabhängiger Z. directional counter
  richtungsunabhängiger Z. non-directional counter; Milligoat counter
  summierender Z. totalizing counter
  voreingestellter Z. preset counter
Zähleranker *m* meter armature
Zähleranzeige *f* count
Zählerauflösungszeit *f* counter resolving time
Zählerausgabe *f* counter output reading
Zählerausgang *m* **für Posten** counter list exit
Zählerbatterie *f* meter battery
Zählerbeschleuniger *m* counter accelerator
Zählereichung *f* meter calibration
  stroboskopische Z. stroboscopic calibrating of a meter
Zählergehäuse *n* meter case
Zählergrundplatte *f* meter base
Zählerkapazität *f* counter capacity
Zählerkonstante *f* constant of a meter
Zählerleerlauf *m* meter creeping
Zählerrelais *n* metering relay
Zählerskale *f* / **runde** meter dial
Zählerstand *m* result of count
Zählerstellen *fpl* counter positions
  unmittelbar saldierende Z. net balance counter positions
Zählerüberprüfung *f* service test of a meter
Zählerwicklung *f* meter winding
Zählfaktor *m* scaling factor
Zählgas *n (nucl)* counter gas
Zählgeometrie *f* counting geometry
Zählgerät *n* counter; scaler; scaling circuit
  elektromechanisches Z. electromechanical counter
Zählgeschwindigkeit *f* counting rate (speed)
Zählglied *n* metering element
Zählimpuls *m* counting pulse
Zählkammer *f (nucl)* counting chamber; pulse-type chamber
Zählkanal *m (nucl)* counting channel
Zählkarte *f (math)* census card
Zählkette *f (dat)* counting chain
Zahlknopf *m* pay button *(telephone)*
Zählkreis *m* scaling (counting) circuit
Zählmagnet *m* magnetic counter
Zählmechanismus *m* counting mechanism
Zählperiode *f* count[ing] period
Zählplateau *n s.* Plateau
Zählrate *f* counting rate

Zählrelais *n* magnetic counter; metering relay

Zählrohr *n* counter (counting) tube, counter

  abgeschirmtes Z. shielded counter

  dampfgefülltes Z. steam-filled counter

  richtungsunabhängiges Z. non-directional counter; Milligoat counter

  selbstlöschendes Z. self-quenching counter tube

  xyloldampfgefülltes Z. xylene-vapour filled counting tube

Zählrohranzeige *f* counting-tube read-out

Zählrohrcharakteristik *f* plateau (counting rate-voltage) characteristic, counting response

Zählröhre *f* [decade] counter tube; electron counter; digital tube

  dekadische Z. decade counter tube

Zählrohrkatode *f* counter body

Zählrohrplateau *n s.* Plateau

Zählrohrspannung *f* counter voltage

Zählrohrteleskop *n* [counter] telescope

Zählrohrtotzeit *f* counter dead time

Zählschaltkreis *m* counting circuit

Zählschaltung *f* counting circuit; scaling circuit

Zählspannung *f s.* Zählrohrspannung

Zählstoß *m* count

Zählstufe *f* counting stage; *(Am)* binary

  binäre Z. binary counter [stage]

Zähltaste *f* meter key

Zähltechnik *f* counting technique

Zählüberwachungslampe *f* register pilot lamp

Zählung *f* counting; metering; registering; census

  Z. nach der Zeitdauer *(Am)* timing

  falsche Z. miscount

Zählunterdrückung *f* non-metering *(telephony)*

Zählverhinderungsrelais *n* non-metering relay

Zählverlust *m* counting loss

Zählvorgang *m* counting period; counting cycle

Zählvorrichtung *f* **am Lochkartenmischer** collator counting device

Zählwerk *n* counter, counting mechanism, register; scaling unit; integrating device

  Z. eines Zählers register of a meter

  elektrisches Z. electric counting mechanism

  elektromechanisches Z. electromechanical counter

  mechanisches Z. mechanical register (recorder)

Zählwerklogik *f* counter logic

Zahlzeichenlesemaschine *f* numeral reading machine

Zählzeit *f* counting period (time)

Zählzeitverzögerung *f* counter time lag

Zahmhaut *f* domestic hide

Zahn *m* 1. *(mach)* tooth *(gear)*; 2. tine *(harrow)*; 3. split *(weaving)*

  Z. des Webblatts split of the reed

  Z. mit Hartmetallschneide carbide-tipped tooth

  Z. und Trieb *m* rack and pinion

  gefräster Z. fluted tooth

Zahnabstände *mpl* tooth spacing

Zahnabstufung *f* tooth increment *(broach)*

Zahnbogen *m* tooth sector

Zahnbreite *f* tooth width, width of the tooth

Zahnbruch *m* tooth breakage

Zahndicke *f* thickness of tooth; circular thickness *(length of arc)*

  Z. am Teilkreis tooth thickness along the pitch circle

Zahndickensehnenmaß *n* chordal thickness of the pitch circle

Zahndurchmesser *m* diameter of tooth

Zähne *mpl* **mit Tiefenstaffelung** teeth arranged for depth cutting *(broach)*

Zähneanzahl *f s.* Zähnezahl

Zähnezahl *f* number of teeth *(gear)*; number of flutes *(reamer)*

Zähnezahlverhältnis *n* gear ratio *(number of teeth in gear / number of teeth in pinion)*, ratio of the number of teeth

Zahnflankenrichtungsfehler *m* profile error (deviation)

Zahnflankenspiel *n* backlash

Zahnformfräser *m* form cutter for gear teeth

Zahnformkorrektur *f* tooth-form correction

Zahnformvorfräser *m* gear stocking cutter, stocking gear milling cutter

Zahnfräsmaschine *f* gear tooth milling machine

Zahnfreiwinkel *m* tooth clearance angle

Zahnfrequenz *f* tooth pulsation frequency

Zahnfüllung *f* dental filling

Zahnfuß *m* dedendum

Zahnfußflanke *f* flank

Zahnfußhöhe *f* dedendum

Zahngold *n* dental gold

Zahngrund *m* tooth gullet, root *(broach)*

Zahngrundrundung *f* gullet [radius], fillet of the gullet, root radius, radius at the root

Zahnhöhe *f* [whole] depth of tooth, tooth height

Zahnholz *n (leath)* tooth-roller

Zahninduktion *f* tooth induction

Zahnkantenabrundmaschine *f* gear tooth rounding and chamfering machine

Zahnkantennachformfräsmaschine *f* tooth edge copy milling machine

Zahnkette *f* silent (inverted tooth) chain

  geräuscharme Z. silent chain

Zahnkettenkupplung *f* chain coupling

Zahnkettentrieb *m* silent-chain drive

Zahnkolbenstange *f* piston rod formed with rack teeth

Zahnkopfflanke *f* face

Zahnkopfhöhe *f* addendum

Zahnkopfspiel *n* clearance

Zahnkranz *m* rim of the gear; ring (internal) gear

Zahnkupplung *f* / **schaltbare** tooth[ed] clutch

Zahnkurve *f* tooth curve
Zahnlücke *f* tooth space, gash, gullet
Zahnlückenfläche *f* gullet area
Zahnmarke *f* tooth mark
Zahnoberteil *n* crest of the tooth
Zahnpasta *f* tooth-paste
Zahnpflegemittel *n* dentifrice
Zahnplatte *f s.* Zahnstange
Zahnprofil *n* gear-tooth profile (outline)
Zahnrad *n* [toothed] gear
balligtragendes Z. gear with localized tooth bearing, crowned gear
schrägverzahntes Z. helical gear
Zahnradbahn *f* rack railway
Zahnradblock *m* gear cluster
Zahnraddauerprüfmaschine *f* gear fatigue-testing machine
Zahnraddynamometer *n* toothed wheel dynamometer
Zahnrädergetriebe *n* train of toothed gears
Zahnrädervorgelege *n* back gear
Zahnradfehler *m* gear error
Zahnradfeinbearbeitung *f* gear finishing
Zahnradflüssigkeitsmotor *m* hydraulic gear motor
Zahnradfräseinrichtung *f* gear cutting attachment
Zahnradfräsen *n* gear milling
Z. im Wälzverfahren gear hobbing
Zahnradfräser *m* gear cutter
Zahnradfräsmaschine *f* gear milling machine
Zahnradhonen *n* gear honing
Zahnradhonmaschine *f* gear honing machine
Zahnradhydromotor *m* rotary motor of the gear type
Zahnradkörper *m* body of the gear
Zahnradkranz *m* gear rim
Zahnradläppmaschine *f* gear-lapping machine
Zahnradlehre *f* gear gauge
Zahnradlokomotive *f* rack locomotive
Zahnradnabe *f* gear hub
Zahnradpaar *n* pair of toothed gears
Zahnradprüfmaschine *f* gear-testing machine
Zahnradpumpe *f* gear pump
Zahnradräummaschine *f* gear broaching machine
Zahnradrohling *m* gear blank
Zahnradschaben *n* gear shaving
Z. mit gekreuzten Achsen [rotary] crossed-axes gear shaving
Z. mit parallelen Achsen parallel-axes shaving
Z. mit umlaufenden Rädern rotary method of gear shaving
Z. mit Zahnstange rack method of gear shaving
Zahnradschabmaschine *f* gear-shaving machine
Zahnradschiene *f* rack rail
Zahnradschleifmaschine *f* gear-grinding machine
Zahnradschneiden *n* gear cutting
Zahnradschneidrad *n* gear-generating shaper cutter, gear-shaping cutter
Zahnradsegment *n* segment gear
Zahnradtrieb *m* toothed-gear drive
Zahnradübersetzung *f* gearing ratio; gear transmission
Z. ins Langsame gear reduction
Zahnradübersetzungsverhältnis *n* gear transmission ratio
Zahnradvorgelege *n* back gearing
Zahnradwälzfräsen *n* gear hobbing
Zahnradwälzfräser *m* gear hob
Zahnradwälzfräsmaschine *f* gear hobbing machine
Zahnradwälzfräsverfahren *n* gear hobbing technique
Zahnradwälzhobeln *n* gear planing, generation of gears by rack
Zahnradwerkzeug *n* master gear [tool] (*s. a.* Schneidrad)
Zahnradzahn *m* gear tooth
Zahnriefe *f* tooth mark
Zahnrücken *m* tooth back, back of tooth
Zahnrückenbreite *f* width of land
Zahnrückendicke *f* size of land *(tool)*
Zahnsättigung *f (el.m.)* tooth saturation
Zahnschabrad *n* rotary gear shave cutter
Zahnscheibe *f* 1. *s.* Planrad; 2. toothed lock-washer
Zahnsegment *n* segment gear, tooth[ed] segment
Zahnspiel *n* clearance *(addendum)*; backlash *(face)*
Zahnspitze *f* crest of a tooth
gestauchte Z. swaged tooth point
Zahnspitzenlinie *f* top edge
Zahnstange *f (mach)* [gear] rack; pusher beam
Z. mit Geradverzahnung spur rack
Z. mit Pfeilverzahnung herringbone rack
Z. mit Schrägverzahnung helical rack
Zahnstangenfräseinrichtung *f* rack cutting (milling) attachment
Zahnstangenfräser *m* rack cutter
Zahnstangenfräsmaschine *f* rack cutting machine
Zahnstangenheber *m* rack-and-pinion jack
Zahnstangenkolben *m* piston formed with rack teeth
Zahnstangenlenkgetriebe *n* rack-and-pinion steering gear
Zahnstangenlenkung *f* rack-and-pinion steering gear
Zahnstangenritzel *n* rack pinion
Zahnstangenteileinrichtung *f* rack-indexing attachment, attachment for rack dividing
Zahnstangentrieb *m* drive by rack and pinion

Zahnstangenverzahnung *f* rack teeth
Zahnstangenwerkzeug *n* rack-form cutter (*s. a.* Schneidkamm)
Zahnstaucher *m* swaging die
Zahnstollenbreite *f* width of land *(tap)*
Zahnstreifen *m* gear mark
Zahnstütze *f* tooth rest
Zahnteilung *f* tooth pitch, spacing of teeth; spacing of flutes *(reamer)*
  Z. / mit ungleichmäßiger variable pitch teeth *(broach)*
  Z. bei Sägen saw pitch
  gleiche Z. uniform pitch of teeth
  gleichmäßige Z. regular spacing of teeth
  ungleiche Z. unequally spaced teeth, differential tooth spacing *(broach)*, irregular spacing of teeth
Zahntiefe *f* tooth depth, depth of teeth
Zahntragbild *n* tooth bearing
Zahntragen *n* tooth bearing
  balliges Z. localized tooth bearing
Zahntragfläche *f* tooth bearing area
Zahntrommel *f* porcupine cylinder
Zahnungseinsatz *m* **aus Hartmetall** carbide insert
Zahnungseinsätze *mpl* built-up teeth
Zahnwalze *f* toothed roller
Zahnweitenmaß *n* chordal dimension over two teeth
zähpolen to toughen by poling
zähreich high-carbon
zähschleimig mucous
Zange *f* [pair of] pliers, tongs; tie beam *(carpentry)*; chop *(wood vice)*; *(text)* nippers
Zangenbremse *f* shoe (clasp) brake
Zangenmanipulator *m (nucl)* expansion tong
Zangenmaul *n (text)* nose end of the nippers
Zangenschnabel *m (text)* nose end of the nippers
Zangenspannfutter *n* collet chuck
Zangenspannung *f* collet chucking
Zangenstrommesser *m* prong-type ammeter
Zangenwandler *m* split-core type transformer
Zäpfchen *npl* **und Stäbchen** *npl* cones and rods
Zapfen *m (mach)* journal *(bearing)*; *(build)* tenon *(wood)*; gudgeon *(metal pivot)*; dowling pin
Zapfenaufhängung *f* pivot suspension
Zapfenbohrer *m* pin drill
Zapfendruck *m* journal pressure
Zapfendüse *f* pintle nozzle *(Diesel)*
Zapfengelenkklappe *f* pivot-jointed bar
Zapfenloch *n* mortise
Zapfenreibung *f* journal friction
Zapfenreibungsmoment *n* moment of journal friction
Zapfenreibungszahl *f* coefficient of journal friction
Zapfensäge *f* dovetail saw
Zapfenschneidgerät *n* tenoning apparatus
Zapfensenker *m* counterboring tool with pilot
  Z. mit auswechselbaren Messern und auswechselbaren Führungszapfen counterbore with interchangeable blades (cutters) and guides
  Z. mit eingeschraubtem Führungszapfen screwed peg-type counterboring cutter
Zapfenverbindung *f* mortise and tenon joint
Zapfenzahn *m s.* Triebstock
Zapfenzahnrad *n s.* Triebstockzahnrad
Zapfgetriebe *n* power take-off [gear], p.t.o.
Zapfhahn *m* discharge nozzle, nozzle trigger
  nichttropfender Z. anti-drip nozzle
Zapfkolophonium *n* gum rosin
Zapfloch *n* bunghole, taphole; mortise
Zapfsäule *f* petrol pump, *(Am)* gasoline pump
Zapfstelle *f* service (roadside filling) station, refuelling point
Zapfwelle *f* power take-off, p.t.o. shaft
Zapfwellenmähbinder *m* power take-off driven tractor binder
Zaponlack *m* zapon lacquer
Zarge *f* border, edge, rim, frame, section
Zargenbeleimmaschine *f* edge gumming machine
Zargenriffelmaschine *f* border crimping machine
Zäsium *n* caesium, *(Am)* cesium
Zäsiumfotozelle *f* caesium photoelectric cell
Zäsiumresonator *m* caesium resonator
Zäsiumschicht *f* caesium film
Zäsiumzusatz *m* caesium addition
Zaum *m* bridle
  Pronyscher Z. Prony brake
Zaumschnalle *f* bridle slide
Zaumzeug *n* bridle gear
Zaun *m* fence
  elektrischer Z. electric fence
Zaunlatte *f* pale
Zaunpfahl *m* fence (boundary) post
ZB-Apparat *m* common battery telephone set
Zeche *f* [coal] mine; colliery
Zechenbahn *f* mine railway
Zechenbahnhof *m* mine station
Zechenkohle *f* mine coal
Zechenkoks *m* by-product coke
Zechenplatz *m* pit yard
Zechenselbstverbrauch *m* internal colliery consumption
Zechstein *m (geol)* zechstein
Zeckenschaden *m (leath)* tick mark *(fault)*
Zedernholz *n* cedar
Zedernholzöl *n* cedar wood oil
Zeeman Effekt *m* Zeeman effect
Zehneck *n* decagon
Zehnerdekade *f* tens decade
Zehnerkanadier *m* canoe C-10
Zehnerkomplementform *f* tens complement form

Zehnerlogarithmus *m* common logarithm
Zehnerpotenz *f* decimal power, power of ten
Zehnerstelle *f* tens digit (place)
Zehnerstufe *f* tens digit *(automatic telephony)*
Zehnerübertragimpuls *m* tens carry pulse
Zehnerübertragkontakt *m* carry contact *(punched card machinery)*
Zehnerübertragmechanismus *m* tens-carrying mechanism
Zehnfachuntersetzer *m* decade scaler, decascaler, scale of ten circuit
Zehnpointsmaschine *f s.* Zehnpunktmaschine
Zehnpunktmaschine *f* bobbin frame with ten needles per inch
Zehntelgrad *m* tenth of a degree
zehntelnormal decinormal
Zehntelskale *f* tenths' scale
Zehntelwertschicht *f (nucl)* tenth-value layer, TVL
Zehntelwertsdicke *f* tenth-value thickness
Zeichen *n* 1. sign; signal; mark; symbol; 2. blip *(radar)*; 3. character *(information theory)*
  Z. mit Wiederholung bis zur Bestätigung *(telec)* repeated-until-acknowledged signal
  hörbares Z. audible signal
  schwaches Z. weak signal
  sichtbares Z. visual signal
  unmoduliertes Z. unmodulated signal
  zusammengesetztes Z. compound signal
Zeichen *npl* / **aufeinanderfolgende** consecutive symbols
Zeichenabstand *m* figure space; spacing interval
Zeichenblock *m* pad
Zeichenbrett *n* drawing board
Zeichenbüro *n* drawing office
Zeichendreieck *n* set square
Zeichendrucker *m* symbol printer
Zeichenebene *f* plane of the paper
Zeicheneinrichtung *f* plotter
Zeichenerkennung *f* character (pattern) recognition *(computer)*
  Z. mit magnetischer Tinte magnetic ink character recognition
  automatische Z. automatic character (pattern) recognition
Zeichenerklärung *f* legend; key
Zeichenerzeugung *f* character generation
Zeichenfeder *f* ruling pen
Zeichenfolge *f* signal-train
Zeichenfrequenz *f* signal frequency
Zeichengeber *m* character generator
Zeichengebung *f* signalling
  selbsttätige Z. automatic signalling
Zeichengenerator *m* character generator
Zeichengerät *n* drawing (drafting) instrument
Zeichenintensität *f* signal intensity
Zeichenkarton *m* drawing cardboard
Zeichenkode *m* signal code
Zeichenkomponente *f* signal component
Zeichenlesung *f* mark sensing *(punched card machinery)*
Zeichenlochkarte *f* mark-sensing punched card
Zeichenlochung *f s.* Zeichenlesung
Zeichenmaschine *f* drawing (drafting) machine
Zeichenmaßstab *m* ruling scale
Zeichenpapier *n* drawing paper
Zeichenregel *f* rule of signs *(Descartes)*
Zeichenreihe *f* string *(propositional calculus)*
Zeichensaal *m* drafting room
Zeichenschärfe *f* definition
Zeichenspiegel *m* drawing mirror
Zeichenstab straight edge
Zeichenstärke *f* signal strength (intensity)
Zeichenstereometer *n* tracing stereometer
Zeichenstift *m* drawing pencil
Zeichenstrom *m* marking current *(telegraphy)*
Zeichenstromkontakt *m* marking contact *(telegraphy)*
Zeichentisch *m* drawing table
Zeichentrickfilm *m* animated drawings
Zeichentusche *f* drawing ink
Zeichenübertragung *f* letter transmission; character transfer
Zeichenübertragungsrate *f* character transfer rate
Zeichenumsetzer *m* translator *(telegraphy)*
Zeichenumsetzrelais *n* translating relay *(telegraphy)*
Zeichenverzerrung *f* [telegraph] signal distortion
Zeichenvorrat *m* character set
Zeichenwelle *f* signal wave; keying (marking) wave *(telegraphy)*
Zeichenzerlegungsverfahren *n* mosaic telegraphy
Zeichenzuordner *m* character translator *(character recognition)*
zeichnen to draw; to draft; to trace *(curve)*
  eine Bindung z. to plot a weave
  maßstäblich z. to draw to scale
Zeichnen *n* / **technisches** engineering drawing
Zeichner *m* draughtsman, draftsman, drawer
zeichnerisch diagrammatic[al], graphical
Zeichnung *f* drawing; graph; plan
  perspektivische Z. perspective drawing
  technische Z. engineering drawing
Zeichnungsbeschriftung *f* lettering of drawings
Zeichnungsformat *n* size of [the, a] drawing
Zeichnungsschrank *m* drawing (plan) chest
Zeichnungsschriftfeld *n* title of a drawing
Zeichnungstoleranz *f* specification (blueprint) limit
zeigen 1. to show; to point out; to exhibit; 2. to prove, to betray; 3. to indicate
  auf Null z. to register zero
Zeiger *m* pointer, needle, indicator *(instrument)*; hand *(clock)*; *(mach)* sector *(indexing)*; vector

geknickter Z. bent pointer

großer Z. minute hand

kleiner Z. hour hand

Zeigeranschlag *m* stop

Zeigerausschlag *m* pointer (needle, meter) deflection, needle deviation (throw)

Zeigerdiagramm *n* vector diagram

Zeigerdosimeter *n* indicating dosimeter

Zeigerdrehknopf *m* pointer knob

Zeigerfrequenzmesser *m* pointer (direct reading) frequency meter

Zeigergalvanometer *n* pointer galvanometer

Zeigerinstrument *n* indicating (pointer) instrument

Zeigerknopf *m* pointer knob

Zeigerokular *n* pointer eyepiece

Zeigerpaar *n (mach)* sector (pair of) arms *(indexing)*

Z. für Teilscheibe adjustable index sector

Zeigerschreibinstrument *n* direct-acting recording instrument

Zeigerstellung *f* pointer position

Zeigertelegraf *m* pointer (dial) telegraph

Zeigerthermometer *n* pointer-type thermometer

Zeile *f (print)* [type] line; *(math)* row; *(telev)* scanning line

Z. mit Bildinhalt *(telev)* active line

blinde Z. *(print)* white (blank) line

gegossene Z. *(print)* slug

neue Z. *(print)* new paragraph

schmale Z. *(print)* short line

signalisierte Z. signalled line

Zeilenablenkfrequenz *f* horizontal sweep frequency

Zeilenablenkgerät *n* sweep unit

Zeilenablenkung *f* horizontal (line) sweep

Zeilenabstand *m* 1. *(print)* leading; 2. line spacing

Zeilenabtastdauer *f (telev)* trace interval

Zeilenabtasten *n* line[-by-line] scanning

Zeilenausgangstransformator *m* line output transformer

Zeilenausschluß *m* justification *(line of type)*

Zeilenausstoßhebel *m* ejector lever

Zeilenaustastimpuls *m* horizontal blanking pulse

Zeilenaustastlücke *f* horizontal blanking interval

Zeilenaustastübertrager *m* blocking line transformer

Zeilenautomat *m (dat)* automatic carriage

Zeilenbebauung *f* terraced housing

Zeilenbreite *f* line width, width of line; spot diameter

Zeilenbreitenregelung *f* horizontal size control

Zeilenbreitenregler *m* line amplitude control

Zeilendrucker *m* line printer, line[-at]-a-time printer

Zeileneinstellung *f* **beim Buchen / feste** fixed-line posting

Zeilenfinden *n* **/ automatisches** *(dat)* automatic line find

Zeilenflimmern *n* interline flicker

Zeilenfolgeabtastung *f* sequential scanning

Zeilenfolgesystem *n* line sequential system *(colour)*

Zeilenfräsen *n* line-by-line milling, parallel stroke milling

Zeilenfrequenz *f* line frequency

Zeilenfrequenzteiler *m* line divider

Zeilenfüllvorrichtung *f* quadding device *(monotype)*

Zeilengenerator *m* **für veränderliche Abtastgeschwindigkeit** variable velocity scanning generator

Zeilenhacker *m* lead and slug cutter

Zeilenhaus *n* terrace house

Zeilenhonorar *n* linage

Zeilenimpuls *m* line (horizontal synchronization) pulse

Zeilenkanal *m* assembler channel *(monotype)*

Zeilenkippgenerator *m* horizontal time-base generator

Zeilenkippschaltung *f* line scan circuit

Zeilenkontrollröhre *f* line monitoring tube

Zeilenlänge *f* measure

Zeilenlinearität *f* horizontal linearity

Zeilenlöschimpuls *m* blanking pulse

Zeilenmittel *n (math)* row average

Zeilennennlänge *f* nominal line width

Zeilenraster *m* line-scanning pattern

Zeilenregelung *f* hold control

Zeilenregler *m* horizontal hold

Zeilenrücklauf *m* horizontal fly-back

Zeilenrücklaufimpuls *m* line fly-back pulse

Zeilensäge *f (print)* slug saw

Zeilenschalter *m* booster diode

Zeilenschreiber *m* line tracer

Zeilensetzmaschine *f* slug composing machine, linesetter

Zeilensprung *m* line interlacing

Zeilensprungabtastung *f* interlaced scanning

Zeilensprungverfahren *n* interlaced (line-jump) scanning

Zeilensprungzerlegung *f* line jump scanning

Zeilensynchronisation *f* line (horizontal) synchronization

Zeilensynchronisationsimpuls *m* horizontal synchronization pulse, line synchronizing signal

Zeilentransformator *m* line transformer

Zeilentransistor *m* line scanning transistor

Zeilenübertrager *m* line transformer

Zeilenvorschub *m* line feed *(typewriter)*

zeilenweise line by line

Zeilenzahl *f* linage

Zeilenzähler *m* line counter

Zeilenzeitbasis *f* line output
Zeilenzittern *n* bounce
Zeit *f* time, period
Z. des Gesprächsbeginns starting time *(of the conversation)*
angezeigte Z. displayed time
erforderliche Z. für Werkzeugwechsel downtime required to change tools
Greenwicher Z. Greenwich time
längere Z. extended time
regelbare Z. controllable time
tote Z. dead time *(radar)*
unempfindliche Z. *s.* Totzeit
verkehrsschwache Z. slack period
verkehrsstarke Z. rush hours
vorgewählte Z. preset time
zeitabhängig time-dependent; age-dependent
Zeitabhängigkeit *f* time dependence
Zeitablauf *m* time lapse, lapse of time
Zeitablenkbreite *f* sweep width
Zeitablenkdauer *f* sweep duration
Zeitablenkgenerator *m* time-base generator, scanning generator
Zeitablenkgerät *n* time base, sweep unit *(oscilloscope)*
Zeitablenkgeschwindigkeit *f* sweep speed (rate, velocity) *(oscilloscope)*
Zeitablenkschaltung *f* sweep circuit *(cathode ray oscillograph)*
Zeitablenkung *f* sweep *(oscilloscope)*; time-base deflection *(oscillogram)*
einmalige Z. single-shot sweep
elektronische Z. electronic time base
lineare Z. linear sweep
sinusförmige Z. sinusoidal sweep
Zeitablenkungsbereich *m* time-base range
Zeitablenkungselektrode *f* timing electrode
Zeitabschnitt *m* period, time interval
Zeitabstand *m* time spacing (interval)
diskreter Z. discrete time interval
Zeitachse *f* time axis
Zeitachsendrehung *f* trace rotation system *(radar)*
Zeitanalysator *m* time analyzer
Zeitansagedienst *m* speaking-clock service *(telephony)*
Zeitauflösung *f* time resolution (resolving power)
Zeitauflösungsvermögen *n* time resolution (resolving power)
Zeitaufwand *m* time used
Zeitauslöseeinrichtung *f* time-limit attachment
Zeitball *m* time ball
Zeitbasis *f* time base
gesteuerte Z. driven time base
Zeitbasis-A-D-Umsetzer *m (dat)* time-base encoder
Zeitbasisbreite *f* scan width
Zeitbasisdauer *f* sweep speed
Zeitbasisdehnung *f* time-base extension
Zeitbasisschalter *m* sweep selector
Zeitbasisverschlüßler *m (dat)* time-base encoder
Zeitbasisverzögerung *f* sweep delay
Zeitbereich *m* time domain; time range
Zeitbestimmung *f* determination of time, timing
Zeitcharterindex *m (ship)* time charter index
Zeitdauer *f* time duration (period); time lapse
Z. für den Energieaustausch energy-exchange time, time of energy exchange
Zeitdehnung *f* time magnification *(oscilloscope)*; *(opt)* retarded action
Zeitdiagramm *n* piston-pressure time diagram
Zeitdifferenz *f* time difference
Zeitdiskriminator *m* time discriminator
Zeitdiskriminatorkreis *m* interval-selector circuit, time-selector transducer
Zeitdrucker *m* time printer
Zeiteinheit *f* unit [of] time
Zeitersparnis *f* time saving, saving in time
Zeitfestigkeit *f* endurance limit
Zeitfolge *f* time sequence
feste Z. fixed time sequence
Zeitfolgeverfahren *n (telev)* field sequential system
Zeitfolgezählung *f* elapsed time counting
Zeitfront *f* time front
Zeitfunktion *f* function of time
Zeitgeber *m* timing element (pulse generator); master clock
selbsttätiger Z. autotimer
Zeitgeberbetrieb *m* fixed-cycle operation
Zeitgeberfrequenz *f* clock frequency
Zeitgebermaß *n* clock rate
Zeitgeberspur *f* clock track
Zeitgleichung *f* equation of time
Zeitglied *n* time (timing) element; computing relay *(with a special time response)*
Zeitimpuls *m* time pulse
Zeitintegral *n* time integral
Zeitintervall *n* time interval
Zeitintervallmesser *m* timer, time interval meter
Zeitkapazität *f* time store
Zeitkarte *f* season-ticket, *(Am)* commutation ticket
Zeitkenngröße *f* time characteristic [value]
Zeitkomponente *f* time component
Zeitkonstante *f* time constant (response)
Z. der Auferregung exciter build-up time constant
Z. des Gleichstromglieds time-constant of the aperiodic component
Zeitkonstanz *f* time stability
Zeitkontakt *m* time closing contact
Zeitkontakteinrichtung *f* timing interrupter
Zeitkontrolle *f* time check

Zeitkontrollimpuls *m* trigger timing pulse, time control pulse *(radar)*
zeitlich time, temporal
Zeitlinie *f* time base; timing line; sweep
Zeitlupe *f* slow-motion *(camera)*
Zeitlupenkamera *f* high-speed camera
Zeitmarke *f* time mark
Zeitmarken *fpl* / **eingetastete** superimposed time markers
Zeitmaß *n* measure of time
Zeitmaßfaktor *m* time scale factor
Zeitmaßstab *m* **des Oszillografen** time base of the oscilloscope
Zeitmaßstabsfaktor *m* time scale factor
Zeitmesser *m* elapsed time meter, timer
Zeitmessung *f* time measurement, timing
Zeitmittelwert *m* mean time value
Zeitmodulation *f* time modulation
Zeitmultiplex *m* *(telev)* time division multiplex
Zeitmultiplexsystem *n* *(telev)* time-division multiplex system
Zeitmultiplextelegrafie *f* time-division telegraph system
Zeitmultiplexverstärker *m* time-shared amplifier
Zeitnahme *f* timing *(work study)*
Zeitplangeber *m* [control] timer
Zeitplanregelung *f* programme control
Zeitplanregler *m* programme controller
Zeitplansteuerung *f* programme control *(open loop)*
Zeitplantemperaturregler *m* clock thermostat
Zeitpunkteinstellung *f* timing
Zeitraffer *m* time lapse motion, quick-motion *(camera)*
Zeitrafferfilm *m* time-lapse film
Zeitraffertechnik *f* time-compression technique
Zeitraffung *f* quick (fast) motion effect
Zeitrasterung *f* time quantization
zeitraubend time-consuming
Zeitraum *m* period, term
Zeitrelais *n* deferring (time-lag, timing) relay
  abhängiges Z. dependent time-lag relay
  begrenzt abhängiges Z. inverse time-lag relay with definite minimum
  reziprok abhängiges Z. inverse time-lag relay
  unabhängiges Z. independent time-lag relay
Zeitschachtelung *f* *(dat)* time multiplexing, time-sharing
Zeitschalter *m* time[-limit] switch, timer
  fotoelektrischer Z. phototimer
Zeitschaltung *f* timing circuit
Zeitschreiber *m* chronograph
Zeitschriftenliteratur *f* periodical literature
Zeitschwelle *f* *(ph)* threshold of time
Zeitschwingfestigkeit *f* fatigue strength at a stated number of cycles

Zeitselektor *m* *(dat)* time gate
Zeitsicherung *f* *(el)* time delay fuse, time-limit circuit breaker
Zeitsignal *n* time signal (tone)
Zeitspanne *f* time interval, period of time
zeitsparend time-saving
Zeitstandprüfmaschine *f* creep testing machine
Zeitstandversuch *m* creep-rupture test
Zeittafel *f* / **geologische** geologic clock
Zeitteilung *f* *(dat)* time multiplexing, time-sharing
Zeit-Temperatur-Umwandlungs-Schaubild *n* time-temperature-transformation diagram
zeitunabhängig time-independent
Zeitung *f* newspaper
  täglich erscheinende Z. daily [paper]
  wöchentlich erscheinende Z. weekly [paper]
Zeitungsbündelpresse *f* newspaper baling press
Zeitungsdruck *m* newspaper printing
Zeitungsdruckfarbe *f* news ink
Zeitungsdruckfilz *m* news felt
Zeitungsdruckpapier *n* newsprint
Zeitungsexpedition *f* newspaper mailroom
Zeitungskopf *m* newspaper heading
Zeitungspapier *n* newsprint
Zeitungsrotation[smaschine] *f* newspaper rotary press
Zeitunterschied *m* time difference
zeitvariabel time-variant
Zeitvektor *m* time vector
Zeitvergleich *m* time comparison
Zeitverhalten *n* time response
  Z. des Reaktors time behaviour of the reactor
  Z. 2. Ordnung *(cont)* quadratic response
Zeitverlust *m* time loss
Zeitverschiebung *f* time shift
Zeitverzögerung *f* time (transfer) lag
  unabhängige Z. fixed time lag
Zeitverzögerungsgerät *n* time delay unit
Zeitvorwahl *f* preset time
Zeit-Weg-Kurve *f* time-path curve; curve showing the time as a function of the course
Zeitzähler *m* hour (time) meter
Zeitzählung *f* timing, time metering
Zeitzeichen *n* time signal *(radio)*
Zeitzeichenanlage *f* time signalling system
Zeitzeichendienst *m* time signal service
Zeitzone *f* time zone
Zeitzonenzähler *m* time and zone meter *(telephony)*
Zeitzonenzählung *f* time and zone metering *(telephony)*
Zeitzünder *m* *(mine)* delay action exploder, delay electric detonator; *(mil)* time (delay) fuse
Zeitzyklus *m* timing cycle
zellähnlich cellular, celloid
Zelldrahtglas *n* chicken-wire glass

Zelle *f* 1. cell *(accumulator)*; 2. cubicle *(switchgear)*; 3. location *(computer)*; 4. compartment *(silo)*; 5. *(text)* mesh; 6. booth; cabinet; 7. *(aero)* airframe; 8. *(nucl)* cave
heiße Z. hot cave (cell)
lichtelektrische Z. photocell, photoelectric cell, electric eye
lichtempfindliche Z. light-reactive cell
poröse Z. *(ch)* porous pot
schalldichte Z. camera booth
zellenartig cellular
Zellenaufbau *m* cell construction
Zellenausleser *m (ag)* indented cylinder
Zellenbauweise *f* cellular construction
Zellenbeton *m* cellular concrete
Zellenbrücke *f* intercell connector *(battery)*
Zellendeckel *m* battery cell lid
Zellendoppelboden *m (ship)* cellular double bottom
Zellenfang[e]damm *m* cellular cofferdam
Zellenfilter *n* cellular (plate, revolving) filter
Zellenfiltereinsatz *m* filter cell unit
Zellenfüllung *f (text)* filling-in of the meshes
Zellengefäß *n* jar *(storage battery)*
Zellenkühler *m* honeycomb radiator
Zellenprüfer *m* cell-tester *(storage battery)*
Zellenrad *n* star feeder; *(ag)* cell-wheel
Zellenradblasmaschine *f (mine)* paddle-type stower
Zellensaugwalze *f* suction couch roll *(papermaking)*
Zellenschalter *m* cell (battery, accumulator, sectionalizing, regulating, discharge) switch
Zellenstopfen *m* battery cell plug
Zellentiefofen *m (met)* soaking pit furnace
Zellenverdichter *m (aut)* sliding-vane compressor
Zellflüssigkeit *f* protoplasm
zellförmig cellular
Zellgewebe *n* cellular tissue
Zellglas *n* cellophane
Zellglasfolie *f* acetate foil
Zellgummi *m* froth rubber
Zellhorn *n* celluloid
Zellkern *m* nucleus
Zellophan *n* cellophane
Zellstoff *m* chemical wood pulp
eingedickter Z. slush stock
gebleichter Z. bleached pulp
ungebleichter Z. brown pulp
weichgekochter Z. high-boiled pulp
Zellstoffaufbereitungsanlage *f* virgin pulp system
Zellstoffbehälter *m* pulp cake
Zellstoffflocken *fpl* loose pulp
Zellstoffkocher *m* digester
Zellstoffkochung *f* pulp cooking
Zellstoffpappe *f* millboard, pulp board
Zellstofftrockenmaschine *f* pulp drying machine
Zellstofftrocknung *f* pulp drying
Zellstoffwatte *f* artificial cotton, cellucotton
Zellstruktur *f* cellular structure
Zelluloid *n* celluloid
Zelluloidschablone *f* celluloid templet
Zellulose *f* cellulose
reine Z. chemical cotton
Zelluloseäther *m* cellulose ether
Zelluloseazetat *n* cellulose acetate
Zelluloseazetobutyrat *n* cellulose acetate butyrate
Zellulosederivat *n* cellulose derivative, cellulosic
Zelluloseester *m* cellulose ester
Zellulosefaser *f* cellulose fibre
Zelluloselack *m* cellulose lacquer
Zellulosenitrat *n* cellulose nitrate
Zelluloseplast *m* cellulosic
Zelluloseregeneratfaser *f* regenerated cellulose fibre
Zellulosexanthogenat *n* cellulose xanthogenate
Zellwolle *f* viscose staple fibre
Zeltplatz *m* camping site
Zement *m* cement
frühfester Z. rapid hardening cement
frühhochfester Z. high-early strength cement
langsamabbindender Z. slow-setting cement
loser Z. bulk cement
schnellabbindender Z. quick-setting cement
schnellerhärtender Z. rapid hardening cement
sulfatresistenter Z. sulphate-resistant cement
Zementanhänger *m* cement trailer unit
Zementationsmittel *n (met)* carburizing agent
Zementbedarf *m* cement requirement
Zementbrennofen *m* cement kiln
Zementdosierung *f* cement batching
Zementdrehofen *m* rotary cement kiln
Zementeinpressung *f* grouting with liquid cement, injection of liquid cement, cement injection
Zementestrich *m* cement finish (screed)
Zementfabrik *f* cement mill
Zementfüllstück *n (ship)* cement chock
Zementfundament *n* cement foundation
zementieren 1. *(met)* to cement, to carburize; 2. *(build)* to grout
Zementinjektion *f* cement injection
Zementit *m* cementite
kugliger Z. spheroidal cementite
Zementklinker *m* cement clinker
Zementkuchen *m* cement pat
Zementkupfer *n* cement copper
Zementmilch *f* laitance, cement slurry
Zementmörtel *m* cement mortar; cement grout, neat cement
reiner Z. neat cement
verlängerter Z. cement-lime mortar

Zementofen *m* cement kiln
Zementpaste *f* neat cement, laitance
Zementprüfung *f* cement testing
Zementputz *m* cement finish (rendering, facing)
Zementschachtofen *m* vertical cement kiln
Zementschlämme *f* cement slurry, neat cement, laitance
Zementschlempe *f* laitance; cement slurry, neat cement
Zementsilber *n* cement silver
Zementsilo *m* bulk cement silo
Zementstahl *m* blister (cemented) steel
Zementstein *m* hydrated cement
Zementumschlaganlage *f* cement handling facility
Zementvermörtelung *f* cement grouting
Zement-Wasser-Verhältnis *n* cement water ratio
Zenerdiode *f* Zener diode
Zenereffekt *m* Zener effect
Zenerkennlinie *f* Zener characteristic
Zenerwiderstand *m* Zener impedance
Zenit *m* zenith
Zenitabstand *m* zenith distance
Zenitokular *n* diagonal ocular
Zenitspiegel *m* zenith mirror
Zenitteleskop *n* zenith telescope
zentesimal centesimal
Zentibar *n* centibar
Zentimeter *n* centimetre, *(Am)* centimeter
Zentimeter-Gramm-Sekunde *f* centimetre gram[me] second
Zentimeterwelle *f* microwave, centimetre wave
Zentimeterwellenerzeuger *m* centimetre-wave oscillator
Zentimeterwellentechnik *f* centimetre-wave technique
Zentralachse *f* central axis
Zentralamt *n (telec)* central office
Zentralbatterie *f* central (common) battery
Zentralbewegung *f* revolution
Zentralbild *n* central image
Zentrale *f* central station; centre line *(crank position)*
Zentralebene *f* **des Kräftesystems** plane of the force system
Zentralgenerator *m* central producer
Zentralheizung *f* central heating [system]
Zentralintensität *f (ph)* central intensity
Zentralisation *f* centralization
Zentralkartei *f* master file
Zentralkondensationsanlage *f* central condensing plant
Zentralkraft *f* central force
Zentrallinie *f* centre line, line of centres
Zentralrohrsilo *m* centrally ventilated silc
Zentralschaltung *f* centralized control
Zentralschmierapparat *m* central lubricant distributing device (manifold)
Zentralschmierstößel *m* central lubrication plunger
Zentralschmierung *f (mach)* centralized [forced-feed] lubrication; *(aut)* central chassis lubrication
Zentralschmiervorrichtung *f* oil distribution box
Zentralspeicher *m* central storage unit
Zentralsteuerung *f* centralized control
Zentralstoß *m* central (head-on) collision
Zentralstrahl *m* central ray
Zentraluhr *f* master clock
Zentraluhrenanlage *f* electrical time-distribution system
Zentralverankerung *f (ship)* centre-point mooring
Zentralverschluß *m* between-the-lens shutter
Zentralwert *m* central value
Zentrierachse *f* centring spindle
Zentrierbohren *n* centre drilling
Zentrierbohrer *m* **mit beiderseitig abgesetztem Schneidenteil** combination centre (centring) drill
 Z. mit Schutzsenkung bell-type combination centre (centring) drill
 Z. ohne Schutzsenkung plain-type combination centre (centring) drill
Zentrierbohrkopf *m* centring head
Zentrierbohrkopfgehäuse *n* centre drilling head casing
Zentrierbohrung *f* centre hole (bore)
Zentrierdrehmaschine *f* centring lathe (machine)
zentrieren to centralize, to centre, to true
 z. mit to adjust central with
 sich z. to align itself
 sich z. auf to position itself on
Zentrierfassung *f* pre-centred base
Zentrierfeder *f* centring spring
Zentriergenauigkeit *f* accuracy of positioning
Zentrierglocke *f* bell (self-centring) punch
Zentrierleiste *f* centring shoulder
Zentriermaschine *f* centring machine
Zentriermikroskop *n* centring microscope
Zentrierring *m* locating (centring) ring
zentriersenken to centre-drill
Zentriersenken *n* centre drilling
Zentriersenker *m* combined drill and countersink
 Z. für Zentrierbohrung mit Schutzsenkung bell-type countersink
Zentrierspindel *f* centring spindle
Zentrierspitze *f* centre point *(drill)*
Zentrierung *f* centring, centration *(lenses)*
 mechanische Z. mechanical centring
Zentrierwinkel *m* centre square
Zentrierwulst *m* shoulder *(shell)*

zentrifugal centrifugal
Zentrifugalabscheider *m* centrifugal scrubber (separator)
Zentrifugalanlasser *m* centrifugal starter
Zentrifugalbeschleunigung *f* centrifugal acceleration
Zentrifugalbewegung *f* centrifugal motion
Zentrifugalgebläse *n* centrifugal blower
Zentrifugalguß *m* centrifugal casting
Zentrifugalkraft *f* centrifugal force
zusammengesetzte Z. compound centrifugal force
Zentrifugalkraftregler *m* flyweight governor
Zentrifugallüfter *m* centrifugal fan
Zentrifugalmoment *n* centrifugal moment
Zentrifugalmühle *f* centrifiner *(papermaking)*
Zentrifugalpumpe *f* centrifugal pump
Zentrifugalregler *m* centrifugal governor (regulator)
Zentrifugalreiniger *m* centricleaner *(papermaking)*
Zentrifugalschmiervorrichtung *f* centrifugal lubricator
Zentrifugalsortierer *m* centriscreen, centrifugal (rotary) screen *(papermaking)*
Zentrifugalspinnmaschine *f* pot spinning frame
Zentrifugaltrockenmaschine *f* centrifugal drying machine
Zentrifugalzerstäuber *m* centrifugal sprayer (atomizer)
Zentrifuge *f* centrifuge; hydro-extractor
zentrifugieren to separate, to centrifuge
Zentrifugierung *f* centrifugation, centrifuging operation
Zentriklon *n* centriclone
Zentripetalbeschleunigung *f* centripetal acceleration
Zentripetalbewegung *f* centripetal motion
Zentripetalkraft *f* centripetal force
zentrisch central *(impact)*
Zentriskop *n* centriscope
Zentriwinkel *m* centre-angle
Zentrum *n* centre
lokalisiertes Z. localized centre
Zeolith *m* zeolite
Zer *n* cerium
Zer-Mischmetall *n* misch metal
Zer(III)-verbindung *f* cerous compound
Zer(IV)-verbindung *f* ceric compound
zerbrechen to break
zerbrechlich breakable; frail
zerbröckeln to crumble
zerdrücken to crush; to squash
Zerenerverfahren *n* Zerener process
Zeresin *n* ceresine wax
Zerfall *m* decay, disintegration; decay event; decomposition, dissociation; destruction, breakup, breakage
dualer Z. dual decay; multiple disintegration (decay), branching
radioaktiver Z. radioactive decay (disintegration)
spontaner Z. spontaneous decay (disintegration)
verzögerter Z. delayed disintegration
zerfallen to decompose, to decay, to degrade, to disintegrate, to split
Zerfallselektron *n* disintegration (decay) electron
Zerfallsenergie *f* disintegration (decay) energy
Zerfallsfamilie *f s.* Zerfallsreihe
Zerfallsgeschwindigkeit *f* disintegration rate
Zerfallsgesetz *n* radioactive decay law, law of radioactive disintegration
Zerfallsgrenze *f* limit of dissociation
Zerfallskonstante *f* [radioactive] decay constant, disintegration constant
Zerfallskurve *f* decay (disintegration) curve
Zerfallsprodukt *n* disintegration (decay) product, daughter
radioaktives Z. [radio]active product
Zerfallsreihe *f* radioactive (decay) series, radioactive (decay, disintegration, transformation) chain
radioaktive Z. radioactive [disintegration] series
Zerfallsschema *n* mode of decay, decay scheme
Zerfallschlacke *f* slaking slag
Zerfallsstern *m* disintegration star
Zerfallswahrscheinlichkeit *f* decay probability
Zerfallswärme *f* heat of radioactivity
Zerfallszeit *f* decay time
Zerfaserer *m* defibrator, pulper, shredder, unravelling machine; stuff grinder
zerfasern to shred, to unravel, to defibre, to fiberize
Zerfasern *n* disintegration
Zerfaserung *f* shredding, defibring
Zerfließen *n* diffusing *(of a packet)*; dissolving; smearing-out
Zerfluorid *n* cerium fluoride
zerfressen to corrode, to pit *(metal)*; to cauterize
zerhacken to chop
Zerhacker *m (el)* [synchronous] vibrator, chopper, cutter, vibratory converter
fotoelektrischer Z. photoelectric chopper
mechanischer Z. mechanical chopper
Zerhackerfeder *f* synchronous reed
Zerhackerpatrone *f* vibrator unit
Zerhackerverstärker *m* chopper amplifier
zerhackt chopped up *(telephony)*
Zerhackungsfrequenz *f* chopping frequency
Zerit *m* cerite
Zeriterde *f* cerite earth
Zerkleinerer *m* crusher, disintegrator, pulverizer

zerkleinern to crush; to pulverize, to reduce in size, to comminute; to grind, to subdivide
schonend z. to crush gently
Zerkleinern *n* mechanical subdivision of solids; crushing
Zerkleinerung *f (mine)* comminution, fragmentation *(shotfiring)*, size reduction, mechanical subdivision
mechanische Z. comminution
Zerkleinerungsmühle *f* crusher
zerkratzen to mar, to score
zerlegbar decomposable; *(mach)* built-up; detachable
in Faktoren z. factorable
Zerlegbarkeit *f* decomposability
Z. in Faktoren factorability
zerlegen *(ch)* to analyze, to decompose, to resolve, to separate; *(telev)* to break down [into lines]; to disassemble
z. in *(math)* to break up into
in Faktoren z. to factorize
Zerlegung *f (ch)* decomposition; *(telev)* breaking-down in lines; disassembling; *(el)* scanning, dispersion
Z. in Bestandteile *(ch)* disaggregation
Z. in Faktoren factoring, factorization
Z. von Geschwindigkeiten resolution of velocities
Z. von Kräften resolution of forces
zermahlen to crush, to grind, to pulverize
Zerotinsäure *f* ceroti[ni]c acid
zerquetschen to squash, to crush, to squeeze; to mash
zerreiben to powder, to pulverize, to rub to powder, to triturate
Zerreißdiagramm *n* load-extension diagram
Zerreißdruck *m* bursting pressure
zerreißen to break *(material, thread)*; to tear
Zerreißer *m (ag)* shredder, lacerating device
Zerreißfestigkeit *f* breaking (ultimate tensile) strength
Zerreißgrenze *f* breaking (rupture) limit
Zerreißmaschine *f* tensile testing machine
Zerreißprobestab *m* tension specimen, tensile test bar
Zerreißprüfung *f* tensile test[ing]
Z. mit ruhender Belastung static tensile test
Zerreißschaubild *n* stress-strain curve
Zerrüttung *f* fatigue cracking
zersetzbar decomposable
zersetzen *(ch)* to decompose; to disintegrate; to digest *(sludge)*
Zersetzung *f* decomposition, decay; disintegration
chemische Z. chemolysis
fotochemische Z. photo-decomposition
galvanische Z. electro-dissolution
strahlenchemische Z. radiation damage
Zersetzungsdestillation *f* destructive (dry) distillation
Zersetzungsprodukt *n* decomposition product
Zersetzungspunkt *m* decomposition point
Zersetzungsspannung *f* decomposition voltage
Zersetzungszelle *f (el)* denuder
zerspanbar machinable
Zerspanbarkeit *f* machinability, cutability
Zerspanbarkeitsindex *m* machinability index
Zerspanbarkeitsprüfung *f* machinability test[ing]
Zerspanbarkeitswert *m* machinability datum (characteristic)
zerspanen to cut, to reduce to chips (chip form)
Zerspanung *f* cutting, reduction to chip form
Zerspanungsarbeit *f* cutting operation
Zerspanungseigenschaft *f* cutting quality (property)
Zerspanungsforschung *f* metal removal research
Zerspanungslabor *n* metal-cutting laboratory
Zerspanungslehre *f* metal cutting science
Zerspanungsleistung *f* cutting capacity (efficiency); rate of metal removal, metal removal performance (rate)
Zerspanungsmethode *f* cutting technique
Zerspanungsmittel *n* cutting medium
Zerspanungsprüfung *f* test-machining, test cutting
Zerspanungsrichtwerte *mpl* machining data
Zerspanungsverfahren *n* cutting process
Zerspanungsvorgang *m* cutting operation (mechanism)
Zersplitterung *f s.* Spallation
zersprengen to burst
zerspringen to burst, to explode; to splinter
zerstampfen to crush, to pound; to pulverize
zerstäubbar atomizable
zerstäuben to atomize; to pulverize; to spray; to break up into a fine mist
Zerstäuber *m* atomizer; pulverizer; sprayer; vaporizer; *(ag)* duster
Zerstäuberdüse *f* atomization jet, spray diffuser
zerstäubt reduced to dust *(chip)*
Zerstäubung *f* atomization, spraying
Zerstäubungsdüse *f* atomization nozzle
Zerstäubungsfeuerung *f* spraying furnace
Zerstäubungsmittel *n* atomizing agent, spray
Zerstäubungstrockner *m* atomizing (spray) drier
Zerstäubungsvergaser *m* jet carburetter
zerstörbar destructible
Zerstörbarkeit *f* destructibility
zerstören to destroy
Zerstörer *m* destroyer
Zerstörerstrahlung *f (ph)* annihilation radiation
Zerstörung *f* destruction; decay
zerstörungsfrei non-destructive
zerstoßen to crush, to powder, to pulverize, to triturate
Zerstrahlung *f* annihilation [radiation]

zerstreuen to disperse, to dissipate; to scatter
Zerstreuung *f* dissipation; diffusion
 Z. der Energie energy dissipation
 anomale Z. anomalous scattering
Zerstreuungskoeffizient *m* coefficient of dispersion
Zerstreuungskreis *m* blur circle, circle of diffusion
Zerstreuungslinse *f* divergent (dispersion, dispersing, dispersive) lens
Zerstreuungsspiegel *m* convex mirror
zerteilen to disperse, to separate; to part
Zerteilung *f* division; parting
zertrümmern to shatter
Zertrümmerung *f* destruction, fragmentation, fission
Zertrümmerungsstern *m* [nuclear] star; cosmic-ray star
Zertrümmerungszone *f (geol)* shear zone
Zerussit *m* cerussite
Zetanzahl *f* cetane number (value)
Zettel *m (text)* warp
Zettelfahrkarte *f* leaflet ticket
Zettelmaschine *f (text)* warper
zetteln *(text)* to warp
Zettelspule *f (text)* warp bobbin
Zettelwalze *f (text)* warping roller
Zetylalkohol *m* cetol
Zeug *n* material, matter, stuff; *(text)* cloth
Zeugdruck *m* cloth (textile, fabric) printing, application of designs to fabric
Zeugeberg *m (geol)* island mount
Zeughaus *n* ordnance
Zeugkasten *m (print)* lead case (box)
Zibet *n* zibet, civet
Zickzackduo *n* staggered mill
Zickzackegge *f* zigzag harrow
Zickzackfalzung *f* concertina (zigzag) folding
Zickzackkurs *m* zigzag course
Zickzackmagazin *n* zigzag magazine
Zickzacknietung *f* staggered rows of rivets
Zickzackschaltung *f (el)* zigzag connection, interconnected Y-connection
Zickzackschott *n (ship)* zigzag bulkhead
Zickzacktrio *n* cross-country mill
Ziegel *m* brick
 feuerfester Z. firebrick, refractory brick
 luftgetrockneter Z. air brick, adobe
Ziegelausfachung *f* brick nogging
Ziegelausfütterung *f* brick lining
Ziegelauskleidung *f* brick lining
Ziegelausmauerung *f* brick lining
Ziegelbau *m* brick construction, brickwork
Ziegelbauweise *f* brick construction, brickwork
Ziegelbrecher *m* brick crusher
Ziegelbrennen *n* firing bricks; brick burning
Ziegelbrennofen *m* brick kiln
Ziegelbruch *m* spall
Ziegelei *f* brick yard, brickwork[s]
Ziegeleindeckung *f* roof tiling
Ziegelgut *n* brick clay
Ziegellehm *m* pug *(clay)*
Ziegelmauerwerk *n* brickwork
Ziegelmehl *n* brick dust
Ziegelofen *m* brick kiln
Ziegelpflaster *n* brick paving
ziegelrot brick-red
Ziegelsäge *f* brick saw
Ziegelschicht *f* / **auskragende** belt (oversailing) course
Ziegelsplitt *m* crushed brick
Ziegelsplittbeton *m* crushed brick aggregate concrete
Ziegelstapler *m* brick stacker
Ziegelstein *m* brick
 spaltbarer Z. cutter, rubber
Ziegelton *m* brick earth (clay)
Ziegelverband *m* bond
Ziegelverblendung *f* brick facing
Ziegelverkleidung *f* brick facing
Ziegelwerkfundament *n* brick foundation
Ziegenhaut *f* goatskin
Ziegenleder *n* goatskin leather
Ziehbank *f* drawbench
ziehbar drawable
Ziehbarkeit *f* drawability
Ziehbereich *m* tuning range, wavelength coverage
Ziehbrett *n* strickle board *(foundry)*
Ziehdorn *m* mandrel
Ziehdüse *f* drawing die, die hole, die [plate]
Zieheinrichtung *f* broach puller, pulling mechanism *(broach)*
Zieheisen *n* wiredrawing die, die plate
ziehen 1. to pull; 2. to draw *(metal)*; 3. to tow, to haul *(a vessel)*
 eine Wurzel z. to extract a root
 Fahnen z. *(telev)* to blur
 Flachs z. to pull the flax
 Gräben z. to dig ditches
 in Stufen z. to reduce by steps
Ziehende *n* **eines Ofens** discharge end of a furnace
Zieherscheinung *f (el)* frequency pulling, drawing out phenomenon
ziehfähig drawable
Ziehfähigkeit *f* drawability
Ziehfeder *f* ruling pen
Ziehfeld *n (nucl)* sweeping (clearing) field
Ziehfett *n* drawing grease
Ziehgrenze *f* formability limit
Ziehgüte *f* drawing grade
Ziehhacke *f (ag)* [bow-holder] weeder
Ziehherd *m* discharge end of the furnace *(rolling mill)*

Ziehkeil *m* sliding key
Ziehkeilgetriebe *n* driving-key transmission
Ziehkopf *m* draw (pulling) head *(broach)*
  Z. mit Backenspannung jaw-type pull head *(broaching)*
  Z. mit Steckkeilbefestigung key-type pull head
  selbstkuppelnder Z. automatic pull head
Ziehlehre *f* strickle *(foundry)*
Ziehlochplatte *f* draw plate
Ziehmatrize *f* draw-die
Ziehmittel *n* drawing compound
Ziehnadel *f (text)* drawing needle
Ziehpappe *f* moulded board, board for pressing
Ziehpresse *f* drawing press
Ziehräumen *n* pull broaching
Ziehräummaschine *f* **mit nach unten gehendem Arbeitsgang** *(Am)* broach pull down machine
Ziehräumnadel *f* pull broach
Ziehriefen *fpl* stretcher strains
Ziehring *m* die
Ziehriß *m* draw (pull) crack
Ziehschleifahle *f* honing tool
ziehschleifen to hone
Ziehschleifmaschine *f* honing machine
Ziehschleifstein *m* honing stick (stone)
Ziehschleifvorrichtung *f* honing fixture
Ziehstahl *m* draw plate
Ziehstein *m* die; wire drawing die
Ziehstempel *m* punch
Ziehtrichter *m* die, [forming] bell *(tubing)*
Ziehtrommel *f* draw block
Ziehvermögen *n (text)* affinity to the fibre
Ziehwagen *m* dog
Ziehwerk *n* broach puller, pulling mechanism
Ziehzange *f* dogs, nippers
Ziel *n* aim; target
Zielauffindung *f* homing *(radar)*
Zielbahnhof *m* terminal (receiving, destination) station
Zielbohrung *f* directional drilling
Zieleinrichtung *f* sights *(rifle)*
Zielfernrohr *n* sighting telescope
Zielflugbake *f* homing beacon
Zielflugempfänger *m* homing receiver
Zielfluggerät *n* homing device (apparatus)
Zielfolgeradar *n* automatic tracking radar
Zielführung *f* / **selbsttätige** automatic tracking *(radar)*
Zielfunktion *f (dat)* objective function
Zielhafen *m* terminal [port]
Zielkern *m s.* Targetkern
Ziellinie *f* line of sight
Zielplatte *f s.* Target
Zielpunkt *m* apex
Zielrechenmaschine *f* predictor
Zielscheibe *f* target
Zielsuche *f (aero)* target-homing, homing guidance
  aktive Z. active homing
Zielsuchen *n* homing
Zieltafel *f (opt)* target signal
Zielverfolgung *f* tracking
Zielverfolgungsradar *n* tracking radar; zone position indicator
Zielverfolgungsverfahren *n* tracking system
Zielweg *m* track
Zielwegdarstellung *f* track display
Zielweite *f* sighting distance
Zielwinkel *m* angle of elevation (sight)
Zielwinkelentfernungsmesser *m* coincidence range finder
Zielzeichen *n* target blip *(radar)*
Zierband *n* band moulding (fillet) *(architecture)*
Zierdeckel *m* ornamental hub cap
Zierfaden *m* fancy thread
Ziergarn *n* fancy yarn
Zierhaube *f* ornamental hood, false cover
Zierleiste *f* ornamental border (ring, band); [band] moulding, band fillet *(architecture)*; *(aut)* moulding
Zierscheibe *f* ornamental disk
Zierschrift *f* fancy type
Zierstahl *m* ornamental steel
Zierstich *m* blanket stitch
Zierstück *n* ornament, vignette, flourish
Zierverband *m* decorative bond *(brickwork)*
Zierzwirn *m* effect (fancy) twist
Ziffer *f* figure, numeral; *(dat)* character; digit
  arabische Z. Arabic number (figure)
  bedeutsamste Z. most significant digit
  geltende Z. significant digit
  römische Z. Roman figure
  Zahlenwert besitzende Z. significant digit
Zifferblatt *n* dial
Ziffernanzeige *f* numerical indication (display, read-out)
Ziffernauswahl *f* digit selection
Zifferndarstellung *f* numerical display
  Z. mit gemischter Basis mixed base notation
Ziffernkasten *m* figure-case
Ziffernlesemaschine *f* numeral reading machine
Ziffernlochung *f* digit punching
ziffernmäßig digital
Ziffernraddrucker *m* counter-wheel printer
Ziffernrechenautomat *m s.* Ziffernrechner
Ziffernrechenmaschine *f s.* Ziffernrechner
Ziffernrechner *m* digital computer
Ziffernstelle *f* digit place (position)
Zifferntrommel *f* digit figure drum
Ziffernwerk *n* numbering machine
Ziffernzeichen *n* numerical digit
Zigarettenpapier *n* cigarette paper
Zigarrenanzünder *m* cigar lighter
Zigarrendeckblatt *n* wrapper

Zille *f* flatboat, swim barge
Zimmer *n* room
Zimmerantenne *f* inside (indoor) aerial
Zimmerer *m* carpenter
Zimmererhandwerk *n* carpentry
zimmergroß room-size[d]
Zimmerhauer *m (mine)* jerry man, timberer, timberman
Zimmerleitungsdraht *m* wire for indoor installation
Zimmerling *m (mine)* jerry man, timberer, timberman
Zimmermann *m* carpenter
Zimmermannsdach *n* [timber] framed roof
Zimmermannsnagel *m* spike
zimmern *(mine)* to timber
Zimmerplatz *m* timber yard
Zimmerspäne *mpl* parings
Zimmerung *f* timbering
Zimtaldehyd *m* cinnamic aldehyde, cinnamaldehyde
Zimtalkohol *m* cinnamic (cinnamyl) alcohol
Zimtöl *n* cinnamon (cassia) oil
Zimtrinde *f* cinnamon bark
Zimtsäure *f* cinnamic acid
Zimtsäureamid *n* cinnamamide
Zimtsäurenitril *n* cinnamo nitrile
Zineol *n* cineole, eucalyptole
Zink *n* zinc
  schwefelsaures Z. sulphate of zinc
Zinkasche *f* zinc ashes
Zinkat *n* zincate
Zinkätzung *f* zinc engraving (etching)
Zinkbad *n* zinc bath, bath of molten zinc
Zinkbatterie *f* zinc battery
Zinkblech *n* sheet zinc, zinc plate
Zink-Blei-Sammler *m* zinc-lead accumulator
Zinkblende *f* zinc blende, sphalerite
Zinkblume *f* flowers of zinc, spangle
Zinkblüte *f* hydrozincite
Zinkdampf *m* zinc vapour
Zinkdestillierofen *m* zinc distilling furnace
Zinkdiäthyl *n* zinc ethide
Zinkdiffusion *f* zinc diffusion
Zinke *f* 1. dovetail *(woodwork)*; 2. *s.* Zinken
  gerade Z. corner locking
  schräge Z. diagonal corner locking
Zinkeinkristall *m* zinc single crystal
zinken to dovetail
Zinken *m* tine, point; prong, spike
  halbstarrer Z. curved tine
  halbstarrer Z. mit Gegenfeder spring-loaded tine, spring-mounted tine
  starrer Z. rigit (straight) tine
Zinkenegge *f* spike harrow
Zinkenfräse *f* dovetailing machine
Zinkenfräsmaschine *f* corner locking machine, spindle moulder for dovetailing
Zinkenkette *f* second sifting tine chain *(beet harvester)*
Zinkenlänge *f* depth of pin
Zinkensäge *f* dovetailing machine with circular saw
Zinkenschneidmaschine *f* dovetailing machine
Zinkenteilung *f* pitch of dovetails
Zinkentsilberung *f* desilvering of zinc
Zinkentsilberungsanlage *f* zinc desilverization plant
Zinkerz *n* zinc ore, calamine
Zinkerzrösthütte *f* zinc ore roasting plant
Zinkgelb *n* zinc yellow
Zinkgewinnung *f* zinc extraction
Zinkgranalien *fpl* mossy zinc
zinkhaltig zinciferous
Zinkit *m* zincite
Zinklegierung *f* zinc alloy
Zinkmuffelofen *m* zinc distilling furnace
Zinkmuffelverfahren *n* zinc distilling process
Zinkografie *f* zincography
Zinkplatte *f (print)* zinco[graph]
Zinkplattenstreckapparat *m* zinc-plate stretching apparatus
Zinkpulver *n* zinc powder
Zinkraffinerie *f* zinc refinery
Zinkschaum *m* zinc scum
Zinkschicht *f* zinc film
Zinkschlicker *m* zinc dross
Zinkschmelze *f* molten zinc, zinc melt
Zinkschwamm *m* cadmia, tutty
Zinkspat *m* zinc spar, calamine, hydrozincite
Zinkstaub *m* zinc powder (dust)
Zinkstaubätze *f (text)* zinc dust discharge
Zinkstaubdestillation *f* zinc dust distillation
Zinkstaubkalkküpe *f* zinc-lime vat
Zinksulfidweiß *n* lithopone
Zinküberzug *m* zinc coating
Zinkung *f* dovetailing *(woodwork)*
Zinkvitriol *n* white vitriol
Zinkweiß *n* zinc white
Zinn *n* tin
Zinnasche *f* tin ashes
Zinnaustrag *m* tin yield
Zinnbelag *m* tin foil
Zinnblech *n* tin sheet
Zinnbronze *f* tin bronze
Zinnbutter *f* butter of tin
Zinn(IV)-chlorid *n* tin tetrachloride, stannic chloride
Zinndioxid *n* stannic oxide
Zinndisulfid *n* stannic sulphide
Zinnerschwerung *f* tin weighting *(silk)*
Zinnerz *n* tin ore (roughs), tinstone, cassiterite
Zinnflammofen *m* reverberatory tin furnace
Zinnfolie *f* tin foil
Zinngehalt *m* tin content
Zinngeschrei *n* tin cry, crackling sound of tin

Zinngießerei *f* pewtery
zinnhaltig stanniferous
Zinnkies *m* tin pyrite, stannite
Zinnkrätze *f* tin dross (waste)
Zinnlagermetall *n* white metal bearing alloy
Zinnober *m* cinnabar
  roter Z. vermilion
Zinnoxid *n* tin oxide
Zinnpest *f* tin pest (disease)
Zinnpochwerk *n* craze mill
Zinnpulver *n* grain-tin
zinnreich tin-rich
Zinnsalz *n* tin (stannic) salt
Zinnsäure *f* stannic acid
Zinnschrei *m* tin cry, creaking of tin
Zinnseife *f* stream tin
Zinnstein *m* tin-stone, cassiterite
Zinn(II)-verbindung *f* stannous compound
Zinn(IV)-verbindung *f* stannic compound
Zinnwaldit *m (min)* zinnwaldite
Zinsenberechnung *f* interest calculation
zinsfrei interest-free
Zipfel *m* 1. ear *(deep drawing)*; 2. lobe *(aerial)*
Zipfelbildung *f* earing *(deep drawing)*
Zipfelzugprüfung *f* wedge-draw test
Zirkel *m* pair of compasses, compass; calipers, callipers; dividers; bow instrument
  Z. mit Einsatz für Blei pencil compass, compass with pencil point
  Z. mit Einsatz für Tusche pen compass, compass with pen point
Zirkon *m (min)* zircon
Zirkon *n* zirconium *(element)*
Zirkonat *n* zirconate
Zirkondioxid *n* zirconia, zirconium oxide
Zirkonium *n* zirconium *(element)*
Zirkoniumdioxid *n* zirconia, zirconium oxide
Zirkoniumerde *f* baddeleyite
Zirkoniumnitrat *n* zirconium nitrate
Zirkoniumsilikat *n* zirconium silicate
Zirkonlampe *f* zirconium lamp
Zirkonylchlorid *n* zirconyl chloride
Zirkularbeschleuniger *m s.* Ringbeschleuniger
Zirkularpolarisation *f* circular polarization
Zirkulation *f* circulation, circulatory flow *(fluid flow)*
Zirkulationskessel *m* circulation boiler
Zirkulationsvorwärmer *m* circulation feed water heater
zirkulatorisch circulatory *(fluid flow)*
Zirruswolke *f* cirrus cloud
Zischeffekt *m (el)* hiss effect
zischen *(el)* to hiss, to fizz
  leise z. to hissel
Zischtondämpfer *m* hiss silencer
Zissoide *f* cissoid
Zisterne *f* cistern
Zitral *n* citral
Zitrat *n* citrate
Zitronellal *n* citronellal
Zitronellol *n* citronellol
Zintronengrasöl *n* lemon grass oil
Zitronenöl *n* lemon oil
Zitronensäure *f* citric acid, acid of lemon
Zitronensäuredehydrase *f* citrico dehydrogenase
zitronensäurelöslich citric soluble
Zitronensäurezyklus *m* tricarboxylic acid cycle
Zitronenschale *f* rind of a lemon, lemon peel
Zitrullin *n* citrulline
Zitrusfrucht *f* citrus fruit
Zitterbewegung *f* oscillatory motion
Zitterelektrode *f* vibratory (vibrating) electrode
zitterfrei *(telev)* jitterfree
Zitterkontakt *m* oscillating contact
Zittern *n* vibrating; jitter
  Z. der Turbine vibrating of the turbine
Zitterpappel *f* asp
Zobel *m* sable
Zobelpelz *m* sable [fur]
Zoisit *m* zoisite
Zölestin *m* celestine
Zoll *m* 1. inch *(25.4 mm)*; 2. customs; duty, tariff
Zollabfertigung *f* clearance [of goods]
Zollabgabe *f* duty
Zollager *n (ship)* bonded warehouse
Zollanmeldung *f* customs declaration
Zollbegleitschein *m* permit, transire, customs declaration for traffic in transit
Zolleinteilung *f* graduation in inches
Zollfahrzeug *n* revenue vessel
Zollfreischein *m* shipping bill
Zollgewinde *n* inch-measure screw thread
Zollgut *n* goods under customs bond (seal)
Zollhafen *m* port in the customs territory, port of entry
Zollhaus *n* custom house
Zollkreuzer *m* customs cruiser
Zollkutter *m* customs (revenue) cutter
Zollpapiere *npl* customs documents, clearance papers
zollpflichtig liable to duty
Zollschiff *n* customs cutter
Zollschuppen *m* bonded warehouse, customs shed
Zollstab *m* yard stick
Zollstock *m* folding rule
Zonarbau *m* zoning *(crystals)*
Zone *f* zone, area
  Z. des Schweigens zone of silence *(sound)*
  aktive Z. *s.* Spaltzone
  empfangslose Z. silent (skip) zone
  gemäßigte Z. temperate zone
  heiße Z. 1. torrid zone; 2. *(nucl)* hot area
  kalte Z. frigid zone
  primäre stille Z. *(telec)* primary skip zone

tote Z. blind spot *(radar)*; *(telec)* skip (silent) zone; *(cont)* differential gap, *(Am)* neutral zone; threshold
verbotene Z. forbidden gap (region, zone)
Zonenabkühlung *f (nucl)* directional cooling
Zonenachse *f* zone axis
Zonenbake *f* **mit vertikalem Leitstrahl** zone marker *(radar)*
Zonenfaktor *m (el)* spread factor *(of winding)*
Zonengesetz *n* zone law *(crystallography)*
Zonenhärtung *f* selective hardening
Zonenlochung *f (dat)* zone punching
Zonenplatte *f* zone plate
Zonenreinigungsverfahren *n* zone purification *(semiconductors)*
Zonenschmelzen *n* zonal melting
Zonenschmelzofen *m* zone melt furnace
Zonenschmelzverfahren *n* zone melting (refining) process; zonal melting
Zonenschrägbunker *m* inclined zonal bunker
Zonenschraube *f* duct adjusting screw
Zonentarif *m* zone tariff, zonal rate
Zonenteil *m* zone *(of a punched card)*
Zonenzählung *f* zone metering (registration) *(telephony)*
Zonung *f* zoning *(of punched cards)*
Zoosterin *n* zoosterine, zoosterol
Zopf *m* top *(wood)*
Zopfende *n* top end *(wood)*
Zopfmaß *n* top measure *(wood)*
Zopfmuster *n (text)* plaited pattern
Zopfstich *m (text)* braid stitch
Zopfware *f* top logs *(wood)*
Zopfwinde *f* ragger *(preparation of waste paper)*
ZTU-Schaubild *n* TTT-diagram, time-temperature-transformation diagram
Zubehör *n* accessories; attachments; armature; furniture, fittings
Zubehörteile *npl* accessories, fittings
Zuber *m* tub
zubereiten to prepare, to adjust, to fit
Zubringeeinrichtung *f (mach)* feeding device; broach handling unit, retriever unit *(broaching)*
Zubringeförderer *m* infeeding (feeder, loading) conveyor
zubringen to handle *(broach)*; to feed *(work)*, to load
Zubringer *m* [mechanical] feeder, loader; [magazine] follower *(rifle)*
Zubringerband *n* stage-loader
Zubringerdienst *m* feeder air service
Zubringerfeder *f* magazine spring *(rifle)*
Zubringerkopf *m (mach)* retriever head
Zubringerleitung *f (telec)* offering trunk
Zubringerluftverkehr *m* feeder air service
Zubringerrutsche *f* feed chute
Zubringerschiff *n* tender
Zubringerstraße *f* feeder (collector, approach) road *(traffic)*
Zubringung *f* **des Räumwerkzeugs** broach handling
Zubruchgehen *n (mine)* cover caving, collapsing
Zuchtbuch *n* herd-book
züchten 1. to grow; 2. to rear
Kristalle z. to grow crystals
Zuchtkeim *m* seed
Zuchtstoff *m s.* Brutstoff
Züchtung *f* 1. growing *(of crystals)*; 2. *(ag)* rearing; breeding, cultivation
zucken to flash
Zucker *m* sugar
Zucker *pl* carbohydrates
zuckerartig saccharine, saccharoid, sugary
Zuckerbildung *f* saccharification
Zuckerdicksaft *m* molasses
zuckerhaltig sacchariferous, saccherated, saccharine, sugary
Zuckerherstellung *f* sugar manufacture
Zuckerhut *m* cone of sugar
Zuckerindustrie *f* sugar industry, industry of sugar
Zuckermühle *f* sugar crushing mill
Zuckerraffinade *f* refined sugar
Zuckerrohr *n* sugar cane
Zuckerrohrquetsche *f* sugar-cane crusher
Zuckerrohrrückstände *mpl* bagasse, cane trash
Zuckerrübe *f* sugar beet
Zuckersäure *f* saccharic acid
Zuckerschleuder *f* sugar centrifugal machine
Zuckersirup *m* molasses, sugar syrup
Zuckerskale *f* sugar scale
Zuckertrockner *m* sugar drier, granulator
Zuckerwaren *fpl* sweets
zudecken to cover; to mask
Zudrücken *n* **eines Falzes** dressing [edges] down flat
Zuerkennungshafen *m* port of adjudication
Zufahrt *f* driveway, access way; entrance *(of a port)*
Zufahrtsbrücke *f* approach viaduct
Zufahrtskanal *m* access canal
Zufahrtsrampe *f* approach ramp (embankment)
Zufahrtsstraße *f* approach road
Zufall *m* accident; chance; random
zufällig accidental; random, stochastic
Zufälligkeit *f* randomness
Zufalls / unter Wahrung des at random
Zufallsauswahl *f* random sampling
Zufallsbelegung *f (telec)* random occupancy
Zufallsbrett *n (math)* Galton's board
Zufallsentscheidung *f* random sampling
Zufallsfolge *f* random sequence *(information theory)*; random train *(of pulses)*
Zufallsgröße *f* random number, stochastic variable

Zufallskoinzidenz *f* random (chance, accidental) coincidence
Zufallsmaschine *f* randomizer *(information theory)*
Zufallsprogrammierung *f (dat)* random excess programming
Zufallsschwankung *f* random variation, chance fluctuation
Zufallsstichprobe *f* random sample
Zufallsstreuung *f* empirical (sampling) dispersion
Zufallsvariable *f* random (stochastic) variable
Zufallsveränderliche *f* random variable
Zufallszahl *f (dat)* random number
zufließen to flow in
Zufluchtshafen *m* port of refuge
Zufluß *m* afflux, inflow, influx, supply
Zuflußmesser *m* head meter
Zuflußregler *m* flow regulator
Zuflußrohr *n* admission tube, feed (inlet) pipe
zufügen to add; to affix *(an index)*
Zufuhr *f* addition, admission, supply
zuführen to feed; to load; to supply
 durch Schwerkraft z. to feed by gravity *(workpieces)*
 Öl z. to feed oil
 Spannung z. to inject a voltage
Zufuhrrollgang *m* approach roller *(rolling mill)*
Zuführrutsche *f (mach)* loading (inlet) chute, chute loader
Zuführschrägrinne *f* inclined trough, feeding chute
Zuführtuch *n (text)* feeding apron
 Z. des Warenzylinders cloth roller apron
Zuführung *f (mach)* feeding, loading *(work)*; supply; feed line
 Z. von Werkstücken über Leitblech chute feed
 beschleunigte Z. high-speed feed
Zuführungsband *n (text)* feed ribbon
Zuführungseinrichtung *f* feeding (loading) device
 trichterförmige Z. feed hopper
Zuführungsleitung *f* supply line (mains); supply lead
Zuführungsmagazin *n (mach)* feed magazine; magazine feed
Zuführungsrolle *f* feed reel
Zuführungsspule *f* feed reel
Zuführwalze *f (met)* feeding roller
Zug *m* 1. *(mech)* tension; 2. draught *(air)*, cross-ventilation; 3. traction; 4. *(mach)* pull; draw, draft, drawing *(metal forming process)*; 5. flue *(heating)*; 6. *(ry)* train; 7. *(mil)* rifling *(in barrel e. g. of rifle)*
 Z. beanspruchen / auf to subject to tension
 Z. mit Postdienst train with mail service
 absteigender Z. flue with downward draught
 flauer Z. weak (poor) draught
 ganzer Z. *s.* geschlossener Z.
 geschlossener Z. *(ry)* complete train load
 künstlicher Z. artificial draught
 natürlicher Z. natural draught (draft)
Zugabe *f* 1. addition; 2. *(mach)* allowance
 Z. der Füllstoffe in den Holländer beater loading
Zugabfertiger *m* checker
Zugabsperrklappe *f* **für Flammrohre** flue damper (shutter), draught check plate, split fire-bridge
Zugang *m* access; accessibility; entrance; inlet
Zugangkanal *m* access duct
zugänglich accessible
Zugänglichkeit *f* accessibility
 Z. des Kessels von außen accessibility of the exterior of the boiler
Zugangspflockluke *f (ship)* access plug
Zugangsstollen *m* adit, access tunnel
Zuganker *m* [tie] bar, tie rod; tension tie
Zugauslastung *f (ry)* full use of a train load
Zugbalken *m* tractor toolbar
Zugband *n* tie [bar], tie rod; tension tie
Zug-Bandbetrieb *m* / **kombinierter** *(mine)* combined truck conveyor system
zugbeansprucht tensioned, subjected to tension
Zugbeanspruchung *f* tension [loading], tensile loading (stress)
 Z. / unter in tension
Zugbegleitpersonal *n (ry)* train personnel (staff, crew)
Zugbelastung *f* tensile load[ing]
Zugbereich *m* tensile range
Zugbetrieb *m* locomotive haulage
Zugbewehrung *f* tensile reinforcement *(concrete)*
Zugbildung *f (ry)* forming (making-up) trains
Zugbildungsbahnhof *m* formation yard
Zugbildungsgleis *n* train formation line (track)
Zugbildungsplan *m* train formation diagram
Zugbrücke *f* drawbridge
Zugbrückensignal *n (ry)* drawbridge signal
Zugdeckungssignal *n (ry)* train covering signal
Zugdehnung *f* tensional strain
Zugdichte *f* density of trains
Zugdifferenz *f* difference in draught
Zug-Druck-Dauerfestigkeit *f* tension-compression fatigue strength
Zug-Druck-Schwingungsprüfmaschine *f* alternating direct stress testing machine
Zug-Druck-Verhältnis *n* ratio of tension to thrust
Zug-Druck-Versuch *m* tension and compression test
Zug-Druck-Wechselfestigkeit *f* fatigue limit under reversed tension-compression stresses
Züge *mpl* **im Gewehr** rifling grooves
zugeben to add, to charge
Zügel *m* bridle[-rein]; ribbon

zugelassen / behördlich permitted by the authorities, authorized
Zugelastizität *f* elasticity of extension
Zügelleder *n* line leather
Zugende *n (ry)* rear of a train
zugeordnet conjugated; associated
  einander eindeutig z. *(math)* isomorphic
zugerichtet wrought; dressed
Zugerzeugung *f* production of draught
  mechanische Z. mechanical draught production
zugespitzt pointed
zugestellt / sauer acid-melting
Zugfeder *f* tension (helical extension) spring
zugfest strong in tension
Zugfestigkeit *f* tensile strength
Zugfolgestelle *f (ry)* block post, intermediate train distancing point
Zugförderung *f* locomotive haulage
  elektrische Z. electric traction
  thermoelektrische Z. thermo-electric traction
Zugführer *m (ry)* guard, conductor
Zugfunk *m* railway communication
Zuggeschirr *n* lowering harness
Zuggestänge *n* tractor toolbar
Zugglied *n (build)* tension tie (member), tie
Zuggurt *m (build)* tension boom (chord) *(truss)*
Zughacke *f* [bow-holder] weeder
Zughackenlader *m (mine)* rake loader
Zughaken *m* tow hook
Zughakenleistung *f* drawbar power
Zugholzzelle *f* gelatinous fibre
zügig steady *(load)*; tacky
Zügigkeit *f* tack
Zugkettengurtförderer *m* chain belt *(conveyor)*
Zugklappe *f* fan damper
Zugkontakt *m* pull contact
Zugkraft *f* 1. tensile force (load), pull; 2. *(ry)* tractive force; traction; 3. *(mach)* pulling capacity *(broach)*; 4. lifting power *(magnet)*
  Z. am Radumfang *(ry)* tractive effort
  Z. bei Dauerleistung tractive effort at continuous rating
Zugkraftdose *f* tensile dynamometer
Zugkraftglied *n* driving medium
Zugkraftmesser *m* die load meter *(metal forming)*
Zuglaufschreiber *m (ry)* train running indicator
Zuglaufüberwachung *f (ry)* train running section
Zugluft *f* draught
Zugmaschine *f* industrial tractor; towing vehicle
Zugmeldestelle *f (ry)* train announcing (recording, reporting) point
Zugmeldung *f (ry)* train announcing
Zugmesser *m* draught gauge (indicator)
  registrierender Z. recording draught gauge
Zugnummer *f (ry)* train number
Zugorgan *n* drive medium *(elevator)*
  Z. der Räummaschine broach pulling mechanism
Zugpause *f (ry)* headway, interval between two trains, train succession
Zugpendel *n* fall (counterweight) pendant
Zugpferd *n* draught horse
Zugpflaster *n* vesicatory
Zugprobe *f* tensile test specimen
Zugprüfmaschine *f* tensile testing machine
Zugprüfung *f* tensile testing
Zugregler *m* draught regulator
Zugrichtung *f* direction of draw *(metal forming)*
Zugriff *m* access *(computer)*
  beliebiger Z. random access
  wahlweiser Z. random access
  willkürlicher Z. random access
Zugriffsgeschwindigkeit *f* access speed
Zugriffsperiode *f* access cycle
Zugriffszahl *f* read around number (ratio)
Zugriffszeit *f* access time
Zugsäge *f* two-men saw
Zugschalter *m* ceiling (pull) switch
Zugschaufel *f* dragline bucket
Zugschaufelbagger *m* dragline excavator, boom dragline
Zugschiene *f* toolbar
Zugschluß *m (ry)* rear of a train
Zugschlußlicht *n* tail lamps of train
Zugschraube *f* 1. draw-in bolt; 2. *(aero)* puller (tractor) airscrew, puller (tractor) propeller
Zugschwellbereich *m* range of fatigue under pulsating tensile stresses
Zugschwellfestigkeit *f* fatigue strength under pulsating tensile stresses
Zugseil *n* draw-in wire *(cable)*; pulling rope; pull rope *(slusher)*
Zugseilbremse *f* hauling rope brake
Zugseilfördergurt *m* cable belt *(conveyor)*
zugsicher draught-proof
Zugsicherung *f* protection of train running
Zugspannung *f* tensile stress, tension
Zugspindel *f* feed screw; feed rod (shaft) *(lathe)*
Zugstab *m* 1. *(build)* tension tie (member), tie; 2. tension bar *(materials testing)*
Zugstange *f* 1. draw rod; tie rod; drag bar; 2. drawbar *(milling spindle)*; 3. *(mine)* pitman *(boring)*
Zugstärke *f* strength (intensity) of draught
Zugsteuerung *f (ry)* train control
Zugstrebe *f (build)* diagonal tie
Zugtier *n* beast of draught, draught animal
Zugtrennung *f (ry)* braking loose of a train
Zugtrum *m* tight (driving) side *(belt)*; traction strand
Zugtrumspannung *f* tight side tension [of belt]
Zug- und Leitspindeldrehmaschine *f* engine lathe [having a lead screw], screw-cutting lathe

Zug- und Stoßvorrichtung *f (ry)* buffer and draw gear
Zugverformung *f* tensile strain; drawing
Zugverholeinrichtung *f (mine)* car spotter
Zugverholung *f (mine)* car spotting
Zugversuch *m* tensile test; elongation test *(wire)*
Zugverzögerer *m* draught retarder
Zugvieh *n* draught animals
Zugvormeldekreis *m* approach circuit *(electric traction)*
Zugwagenbremsventil *n* tractor-brake valve
Zugwiderstand *m (ry)* traction resistance
Zugwirkung *f* **des Kamins** draught effect of the cooling tower
Zugzusammensetzung *f (ry)* train formation
Zugzusammenstoß *m* train crash (collision)
Zuhörerraum *m* auditorium
zukorken to stop, to cork up
Zulage *f* allowance *(wages)*
Zulageeisen *n* continuity rod *(reinforced concrete)*
zulassen to allow; to admit; to license
zulässig allowed, allowable, permissible, admissible
Zulassung *f* licence, registration *(car)*; permission
  Z. für Luftfahrtpersonal personnel licensing
Zulassungsschein *m* licence
Zulauf *m* intake, entrance, feed; gate *(casting)*
Zulaufleitung *f* feed line
Zulegestoff *m* solder [metal]
zulegieren to alloy
zuleimen to glue up
Zuleitung *f* 1. [electric] supply line, [electric] supply mains, lead-in, feeder; 2. feed (supply) pipe, feed (supply) line
  abgeschirmte Z. *(el)* screened lead
Zuleitungskabel *n* power supply cable; conducting (service) cable
Zuleitungsrohr *n* feed pipe
Zuleitungswiderstand *m* lead (branch, wiring) resistance
Zulieferarbeit *f* supplying (sub-contract) work
Zulieferbetrieb *m* supplying (sub-contracting) factory, supplier, outside contractor firm, contractor
Zulieferindustrie *f* supplying industry
zuliefern / geordnet to deliver orderly
Zuluft *f* intake (fresh) air
Zumachebrett *n (mine)* breast board
zumachen to close, to shut
zumessen to meter; to batch, to gauge, to proportion *(concrete)*
Zumessung *f* proportioning, batching *(concrete)*
zumischen to admix, to add
zunageln to nail up
Zunahme *f* increase; regain *(moisture)*; increment *(steps)*; growth, rise
Zunahmefinger *m (text)* back (double) picker
Zunahmekurve *f* growth curve *(statistics)*
Zündanlaßschalter *m* ignition switch
Zündanode *f* igniting (starting) anode
Zündapparat *m* ignition apparatus, magneto
Zündbedingung *f* ignition condition
Zündbereich *m* ignition range
Zündbeschleuniger *m* ignition accelerator
Zünddrehmoment *n* firing torque
Zünddrehzahl *f* firing speed
Zünddynamo *m* ignition dynamo
Zündeigenschaft *f* ignition quality
Zündeinsatzsteuerung *f (el)* phase control *(rectifier)*
Zündeinstellung *f* spark setting
Zündelektrode *f* ignition (starting) electrode
zünden to ignite, to fire; to initiate, to light
  ein Gasentladungsrohr z. to trip a gas tube
Zunder *m* scale; tinder
Zünder *m* igniter; *(mine)* detonator; fuze, fuse
  Z. für Raketentriebwerk rocket igniter
  Z. mit Zündkerze squib-initiated igniter
  elektrischer Z. spark igniter; electric blasting cap
zunderbeständig scale-resisting, non-scaling
Zunderbeständigkeit *f* resistance to scale, non-scaling property
Zunderbildung *f* scale formation, formation of scale
Zünderdraht *m* leading wire
zunderfest non-scaling, scale-resisting
Zunderfestigkeit *f* scaling resistance
Zunderflecken *m* **auf Feinblech** flashing
zunderfrei scale-free
zundern to oxidize, to scale
Zunderschicht *f* scale (oxide) layer
Zünderstellmaschine *f* automatic fuse cutter (setter), fuse-setting machine, fuse setter
Zünderstellung *f* fuse-setting
Zündfackel *f* fuse lighter
zündfähig ignitable
Zündfolge *f* 1. *(aut)* ignition timing; 2. *(mine)* firing order, delay pattern, order of firing
zündfreudig ignitable
Zündfunke *m* 1. igniting (trigger) spark; 2. *(mine)* incentive spark
  satter Z. fat ignition spark
Zündfunkengleichrichter *m* pilot spark arc rectifier
Zündfunkenstrecke *f* starter gap
Zündgang *m (mine)* round *(of shots)*
Zündgenauigkeit *f* ignition accuracy
Zündgeschwindigkeit *f* ignition velocity, rate of flame propagation
Zündgrenzen *fpl* ignition range
Zündhilfsmittel *n* ignition aid
Zündholz *n* match
Zündholzmaschine *f* match making machine
Zündhütchen *n* blasting cap; percussion cap

Zündkabel *n* ignition cable (wire), sparking plug wire
Zündkapsel *f* primer (priming) cap
Zündkennlinie *f (el)* control characteristic
Zündkerze *f* spark[ing] plug
Z. mit hohem Wärmewert cold plug
Z. mit niedrigem Wärmewert hot plug
verölte Z. oiled-up sparking plug
Zündkerzenabbrand *m* spark erosion
Zündkerzenanordnung *f* spark-plug position
Zündkerzenelektrode *f* spark-plug electrode
Zündkerzenstecker *m* spark[ing]-plug terminal
Zündkerzenstörung *f (telev)* spark-plug interference
Zündkerzenverschmutzung *f* sparking-plug fouling
Zündkirsche *f* ignition pellet
Zündleitung *f* ignition lead
Zündloch *n* vent
Zündmagnet *m* magneto
Zündmaschine *f* blasting machine, exploder
Zündmittel *n* igniting mixture, detonating agent; *(mine)* blasting accessories
Zündmomenteinstellung *f* spark timing
Zündprüfung *f* ignition test
Zündpunkt *m* ignition temperature, ignition (flash) point
Zündpunkteinstellung *f* magneto timing
Zündrakete *f (mine)* load spitter
Zündsatz *m (mine)* fusehead, fuse (detonator) composition
Zündschloß *n* ignition lock
Zündschlüssel *m* car ignition key
Zündschnur *f* blasting (safety, Bickford) fuse, firing tape
detonierende Z. detonating fuse
Zündspannung *f* ignition (starting) voltage; breakdown voltage
Zündspule *f* ignition (spark) coil
Zündstäbchen *n* fuse lighter
Zündstein *m* lighter-flint
Zündstift *m* ignitor; fuse lighter
Zündstiftsteuerung *f* ignitor control *(rectifier)*
Zündstörung *f* ignition interference
Zündstrecke *f* **bei Gasentladungsröhren** starter gap
Zündstrom *m* ignition current
Zündstromkreis *m* ignition circuit
Zündstufe *f (mine)* delay stage
Zündtemperatur *f* ignition temperature; *(ch)* inflammation temperature
zündträge having a poor ignition quality
Zündträgheit *f* poor ignition quality
Zündtransformator *m* ignition transformer
Zündung *f* ignition, firing; striking *(arc)*
abgeschirmte Z. screened ignition
elektrische Z. electric ignition
unregelmäßige Z. imperfect ignition
Zündverhalten *n* ignition characteristics
Zündversager *m* misfire
Zündverstellbereich *m (aut)* ignition-timing range, spark-timing range
Zündversteller *m (aut)* timing control
Zündverstellung *f* spark adjustment, ignition-timing; advance mechanism
automatische Z. automatic advance
drehzahlabhängige Z. varying the spark in relation to the engine speed, sparking-time control in relation to the engine speed
Zündverstellvorrichtung *f*/ **automatische** automatic advance mechanism
Zündverteiler *m* ignition distributor
Zündverteilernocken *m* distributor cam
Zündverteilerwelle *f* ignition-distributor shaft
Zündverzögerung *f* delayed ignition, ignition delay (lag)
Zündverzögerungswinkel *m* delay angle, angle of delay
Zündverzug *m* delayed ignition, ignition delay (lag)
Zündverzugwinkel *m* delay angle; angle of delay
Zündvoreilung *f* leading ignition *(rectifier)*
Zündvorrichtung *f* ignition device, primer
zündwillig having a good ignition quality
Zündwilligkeit *f* ignition properties (performance, quality), ability of fuel to ignite; inflammability
Zündwinkel *m* ignition angle; blocking period
Zündzeitpunkt *m* firing (ignition) point; spark timing
Zündzeitpunkteinstellung *f* ignition timing
Zündzeitpunktverstellung *f* spark-timing control, ignition timing adjustment
Zündzeitpunktverstellvorrichtung *f* spark-advance and retard mechanism
Zündzeitverstellung *f* timing, ignition control
zunehmen to increase; to grow
Zunge *f* 1. slide, tongue *(slide rule)*; 2. *(ry)* switch rail; 3. toe wing *(shoe)*; 4. latch *(knitting machine)*
Zungenfrequenz *f* reed frequency
Zungenfrequenzmesser *m* reed-type frequency meter, vibrating reed instrument
Frahmscher Z. Frahm frequency meter
Zungenfrequenzrelais *n* tuned-reed relay
Zungenkontakt *m* wiping spring contact
Zungenlautsprecher *m* reed loudspeaker
Zungennadel *f* latch needle
Zungenpfeife *f* reed pipe
Zungenrelais *n* reed relay
Zungenresonanzkreis *m* tuned-reed relay
Zungenschiff *n* slice galley
Zungenschrägroststab *m* tongue-shaped inclined bar
Zungenunterbrecher *m* vibrating-reed break
Zungenweiche *f (ry)* point switch

Zungenwurzel *f* heel of points (blade)
zuordnen to coordinate; to associate; to translate, to collate
einander z. *(math)* to map
Zuordner *m (dat)* interpreter, translator, collator
Zuordnerkode *m* interpreter code
Zuordnerprogramm *n* interpretative routine
Zuordnung *f* coordination; correspondence
Zuordnungsprogramm *n (dat)* interpretive routine
Zupfinstrument *n* plucked instrument
zupfropfen to plug
Zurechtfinden *n* **im Labyrinth / automatisches** *(dat)* automatic path-finding in the maze
Zurichtebogen *m* overlay (register) sheet
Zurichtefolie *f* makeready foil
zurichten 1. to make ready; 2. *(text)* to finish; 3. to dress *(timber)*
glatt z. to plain finish
Zurichterei *f* finishing shop
Zurichtung *f* 1. *(print)* makeready; 2. leather dressing
zurren to seize, to lash
seefest z. to lash
Zurrung *f* lashing
zurückbewegen to back out *(tap)*
zurückbiegen to bend back
zurückdrehen to turn down (backward)
zurückerstatten to refund
zurückfahren to flit *(coal cutter)*; to back [in], to move backwards, to return; to put back *(vessel)*
zurückfallen to fall back; to reset, to jump back *(voltage)*
zurückfedern to spring back
zurückfließen to flow back
zurückfließend refluent, flowing back
zurückführen *(ch)* to recycle, to recirculate; *(dat)* to restore
zurückgestreut backscattered
zurückgewinnen to reclaim, to recover
Zurückhaltekraft *f* retentivity
zurückhalten to retain
zurückkehren to return, to revert
in die Ruhelage z. to return to normal
zur Anfangsstellung z. to revert to the initial state *(multivibrator)*; to return to the initial position
zurückklappen to fold back; to flap back *(cutter bit)*
zurücklaufen to return
im Eilgang z. to fast-return
zurückleiten to reconduct
zurückliefern to give back *(energy)*
zurückmelden to report back
Zurücknahme *f* withdrawal
zurückprallen to rebound, to recoil
zurückrufen to call back *(telephony)*
zurückschalten to switch back
zurückschnellen to spring back
zurückschrauben to screw back
zurücksetzen *(dat)* to reset *(to zero)*, to re-zero
zurückspringen to rebound
zurückspulen to rewind
zurückstellen to reset
zurücktitrieren to retitrate
zurückwägen to reweigh
zurückweisen to reject
zurückwerfen to reflect
zurückziehbar retractable
zurückziehen to withdraw, to retract, to back out
Zurückziehen *n* withdrawal, retraction
Zusammenarbeit *f* cooperation, working partnership
wirtschaftliche Z. economic cooperation
zusammenarbeiten to cooperate
zusammenbacken to agglomerate
Zusammenbacken *n* agglomeration
zusammenballen to conglomerate, to agglomerate, to aggregate, to cake; to ball up, to spheroidize *(graphite)*
Zusammenballung *f* agglomeration, accumulation, clustering; balling-up, spheroidization *(graphite)*
Zusammenbau *m* assembly, assembling
nacharbeitsfreier Z. assembly (assembling) without reworking
zusammenbauen to assemble
zusammenblatten to scarf
zusammenbrechen to break down, to collapse; *(mine)* to jack-knife
zusammendrückbar compressible
Zusammendrückbarkeit *f* compressibility
zusammendrücken to compress
Zusammendrücken *n* compression
zusammenfallen 1. to collapse; 2. to coincide
Zusammenfallen *n* 1. collapse; 2. coincidence
zusammenfalten to fold [up]
zusammenfassen to combine, to unite; to summarize
Zusammenfassung *f* summary
zusammenfritten to frit together
Zusammenfrittung *f* fritting together
zusammenfügen to join; to fix together
zusammengebaut built-up, sectional *(broach)*
zusammengedrückt compressed
zusammengefaltet folded [up]
zusammengekuppelt coupled together
zusammengerechnet aggregate
zusammengeschraubt bolted together
zusammengesetzt assembled, built-up; composite; complex, combined *(force)*; composed
aus Baueinheiten z. unit-constructed
zusammengezählt aggregate
zusammenhalten to hold together

Zusammenhang *m* connection; correlation; context
zusammenhanglos incoherent
Zusammenhangskraft *f* adhesiveness
zusammenkeilen to wedge together
zusammenkitten to cement together
zusammenklappbar collapsible
zusammenklappen to fold up
zusammenkleben to cement (glue) together; to stick together *(contacts)*
Zusammenkleben *n* **von Kernteilen** pasting of core parts
zusammenkuppeln to couple together
zusammenlaschen to scarf [together]
zusammenlaufen to converge; to run together; to curdle
Zusammenlaufen *n* convergence; running together; curdling
zusammenlegbar folding, collapsible
zusammenlegen to fold up; to combine
zusammennieten to rivet together
Zusammenprall *m* collision
zusammenprallen to collide
zusammenpreßbar compressible
zusammenpressen to compress
zusammenrechnen to add [up]
zusammenrollen to roll up
zusammenschalten to interconnect; to join; to couple
  kapazitiv z. to join capacitively
Zusammenschaltung *f* interconnection
zusammenschnüren to tie together; to lash
zusammenschrumpfen to shrink
zusammenschweißen to weld up (together)
zusammensetzen to assemble; to compose
  sich z. aus to consist of
  wieder z. to re-assemble
Zusammensetzung *f* composition; structure
  Z. von Kräften composition of forces
  isotope Z. isotopic composition
  natürliche isotope Z. natural isotopic composition
zusammenstauchen to butt together
zusammenstellen to compile *(statistics)*
  in Tabellenform z. to tabulate
Zusammenstellungszeichnung *f* [general] assembly drawing *(showing machine in outline merely to illustrate its general arrangement)*
Zusammenstellzeichnung *f s.* Zusammenstellungszeichnung
Zusammenstoß *m (ph)* encounter, collision; occlusion *(of warm and cold fronts)*; *(ry)* [end-on] collision, head-on collision, buffer impact
  enger Z. close collision
  entfernter Z. distant encounter
zusammenstoßen to collide
  stumpf z. to butt-joint
Zusammenstürzen *n* collapse *(fluid flow)*
zusammentragen to collate, to gather
Zusammentragmaschine *f* gathering (collating) machine
Zusammenwirken *n* interaction; interworking
zusammenzählen to add up
Zusammenzählung *f* addition, summation
zusammenziehbar contractible
zusammenziehen to draw together; to constrict
  sich z. to shrink, to contract
Zusatz *m* additive, addition, admixture
  Z. zu Beizbädern pickling compound
  chemischer Z. additive
Zusatzanode *f* extra anode
Zusatzauftrieb *m* additional lift
Zusatzausrüstung *f* auxiliary equipment
Zusatzbatterie *f* booster battery
Zusatzdampf *m* additional steam *(turbine)*
Zusatzdraht *m* welding wire, filler rod *(welding)*
Zusatzdüse *f* auxiliary jet
Zusatzdynamo *m* booster dynamo
Zusatzeinrichtung *f* additional (auxiliary, special) attachment
Zusatzfilter *n* added filter
Zusatzgerät *n* accessory unit, auxiliary apparatus, complementary instrument; attachment
Zusatzgetriebe *n* auxiliary transmission
Zusatzkorrektur *f (rock)* correction by addition
Zusatzlast *f* additional load
Zusatzlautsprecher *m* extension loudspeaker
zusätzlich added, additional
Zusatzluftbehälter *m* auxiliary air reservoir
Zusatzmaschine *f* [positive] booster
  Z. für Zu- und Gegenschaltung reversible booster
  Z. in Gegenschaltung negative booster
  Z. mit Differentialerregung differential booster
Zusatzmetall *n* added (filler) metal *(welding)*
Zusatzmittel *n* admixture *(mortar, concrete)*, dope
  Z. zur Erhöhung der Oxydationsbeständigkeit anti-oxidation dope
Zusatzpatent *n* additional patent
Zusatzpumpe *f* ancillary (stand-by) pump
Zusatzschalter *m* auxiliary switch; booster switch
Zusatzscheinwerfer *m* auxiliary head lamp
Zusatzspannung *f* 1. boosting voltage; 2. secondary stress
Zusatzstoff *m* dope; admixture
Zusatztransformator *m* booster transformer
Zusatzventil *n* by-pass valve
Zusatzverlust *m (el)* supplementary loss
Zusatzverstärker *m* booster amplifier
Zusatzwasser *n* make-up water
Zusatzwerkstoff *m* filler metal (rod), welding rod, adding (joining) metal *(welding)*
Zusatzwiderstand *m* additional resistance

zuschalten to switch on; to connect, to engage; to charge; to add *(load)*
Zuschärfungslänge *f (ship)* length of entrance *(waterline)*
Zuschauerraum *m* auditorium
Zuschauertribüne *f* grandstand
zuschaufeln to shovel up
Zuschlag *m* 1. [concrete] aggregate; addition; flux; 2. additional fee
Zuschlaghammer *m* aboutsledge
Zuschlagporto *n* excess postage
Zuschlagstoff *m* [concrete] aggregate
ungesiebter Z. all-in aggregate
zuschmelzen to seal
zuschnallen to strap up *(belt)*; to buckle
Zuschneidebrett *n* clicking board *(shoe)*
Zuschneidemaschine *f* cutting[-out] machine
zuschneiden to cut up (out)
auf Maß z. to trim to size
Zuschneider *m* [male] cutter
Zuschneiderei *f* clicking room *(shoe)*
Zuschneiderin *f* [female] cutter
Zuschnitt *m* 1. *(mine)* layout; 2. sheet blank *(metal forming)*
zuschnüren to lace up; to cord up
Zuschuß *m* 1. extra pay; 2. *(print)* overs
Zuschußbetrieb *m* factory requiring a subsidy
Zuschußbogen *mpl* overs
zuschütten to fill in
zusetzen to add, to adulterate, to mix; to alloy *(metal)*
Vitamine z. to vitaminize
Zusetzen *n* **des Rasters** *(print)* filling-in
Zustand *m* state, condition; operating condition *(of reactor)*
Z. / im eingebauten *(el)* in situ
Z. / in kräftefreiem in the absence of forces
Z. der Schwerelosigkeit state of weightlessness
analoger Z. analogue state
angeregter Z. excited state
astabiler Z. astable state
atomarer Z. atomic state
betriebsfähiger Z. working condition
dampfförmiger Z. vapour state
eingeschwungener Z. steady state [condition]
einwandfreier Z. perfect condition
entarteter Z. degenerate state
erlaubter Z. allowed state
erregter Z. energized state
fehlerfreier Z. perfect condition
fehlerhafter Z. fault condition
fester Z. solid state
flüssiger Z. liquid state
gallertartiger Z. gelatiniform state
gasförmiger Z. gaseous state
gelöschter Z. clear condition
gerader Z. even state
gesättigter Z. saturated state
gleichbleibender Z. steady state [condition]
inaktiver Z. inactive state
isomerer Z. isomeric state
kolloider Z. colloidal state
kristalliner Z. crystalline state
kritischer Z. critical state
lecker Z. damaged condition
leerer Z. empty state
magnetischer Z. magnetic state
metastabiler Z. metastable state
nichtangeregter Z. unexcited state
nichtentarteter Z. non-degenerated state
optisch angeregter Z. optically excited state
physikalischer Z. physical state
quasistabiler Z. quasi-stable state
quasistationärer Z. quasi-stationary state; virtual state
reduzierter Z. reduced state
reiner Z. pure state
stabiler Z. stable state
stationärer Z. steady (stationary, bound) state
supraleitender Z. superconductive state
thermodynamischer Z. thermodynamical state
unerregter Z. de-energized state
ungebundener Z. unbound state
ungeordneter Z. disordered state
ungerader Z. odd-state
ungestörter Z. unperturbed state; undisturbed condition
unmodulierter Z. zero-modulation
virtueller Z. virtual state
Zustandsänderung *f (ph)* state change, change of state, process; constitutional change
adiabatische Z. adiabatic change of conditions
isenthalpische Z. isenthalpic process
isodynamische Z. isodynamic process
isoenergetische Z. isoenergetic process
isopotentielle Z. isopotential process
isotherme Z. isothermal expansion
polytropische Z. polytropic process
umkehrbare Z. reversibility of state
Zustandsdiagramm *n* equilibrium (temperature-concentration, constitutional) diagram
Zustandsfunktion *f* state function
Zustandsgleichung *f* equation of state
barotrope Z. barotropic equation of state
Clapeyronsche Z. Clapeyron's equation
piezotrope Z. piezotropic equation of state
Zustandsgröße *f* parameter of state, state variable
Zustandskurve *f* phase trajectory
Zustandsparameter *m* characteristic parameter
Zustandsschaubild *n s.* Zustandsdiagramm
Zustandssumme *f (nucl)* partition function
Zustellbetrag *m (mach)* feeding rate
Zustellbewegung *f (mach)* in-feed movement

zustellen 1. to line with refractories *(furnace)*; 2. *(mach)* to feed; 3. to deliver
Zusteller *m (telec)* messenger
Zustellgebühr *f (telec)* portage
Zustellhandrad *n* feed wheel
Zustellspindel *f* downfeed screw
Zustellung *f* 1. lining *(a furnace)*; 2. *(mach)* in-feed; 3. delivery
Z. der Honsteine radial forcing out of the honing sticks, stone expansion, feed-out of the honing tool, outfeed
Z. der Schleifscheibe wheel spindle infeed
Z. eines Ofens lining (brickwork) of a furnace
Z. zwischen zwei Zähnen step per tooth *(broach)*
saure Z. acid lining
Zustellungskosten *pl* fettling cost *(furnace)*
Zustellungsmasse *f* lining mass *(furnace)*
Zustopfung *f (telec)* wipe-out
Zustreicher *m (ag)* furrow coverer
Zustrom *m* influx, afflux
zuströmen to flow in
Zuteilschalter *m* assignment (transfer) key
Zuteiltaste *f* assignment key
Zutritt *m* access, admission
zuverlässig dependable; reliable
Zuverlässigkeit *f* dependability, reliability
Zuverlässigkeitsgrad *m* dependability
Zuverlässigkeitsprüfung *f* reliability testing
Zuverlässigkeitssteigerung *f* reliability increase
Zuwachs *m* increment, accretion; growth, gain, increase
jährlicher Z. annual increment
Zuwachsfaktor *m* build-up factor
Zuwachsinduktor *m* incremental inductor
Zuwasserfieren *n* launching, lowering *(of boats)*
zwängen to force
Zwanglauf *m* constrained movement, restraint
zwangläufig *s.* zwangsläufig
Zwangsberieselung *f* interlocked spray-water supply
Zwangsbremsung *f* automatic train stop
Zwangsdurchlauf *m* forced circulation
Zwangsentlüftung *f* exhaust ventilation
zwangsgekühlt blower-cooled
Zwangskommutierung *f* forced commutation
Zwangskraft *f (ph)* reactive force
zwangsläufig positive, solid, forced; non-slip
Zwangsläufigkeit *f* constrained movement
Zwangslüftung *f* forced-air system, forced draught
Zwangsmischer *m* paddle (pan-type, positive, pug-mill) mixer
Z. mit Planetrührwerk planetary mixer
Zwangsschmierung *f* forced-feed lubrication
Zwangsschwingung *f* forced oscillation
Zwangsumlaufkessel *m* forced circulation boiler
Zwangszentrierung *f* controlled centring
Zwangsumlauf *m* forced circulation
Zwanzigeck *n* eiconagon
zwanzigflächig icosahedral
Zwanzigflächner *m* icosahedron
Zweck *m* / **industrieller** industrial application
Zweiabteilungsschiff *n* two-compartment vessel
zweiachsig biaxial, two-dimensional
Zwei-Adressen-Befehl *m (dat)* two-address instruction
Zwei-Adressen-System *n (dat)* two-address system
zweiadrig two-stranded, two-conductor, two-core, double-wire
Zweiarmflansch *m* two-armed flange
zweiarmig two-armed
Zweiarmnabe *f (aut)* twin-sector clutch hub
zweiatomig biatomic, diatomic
Zweibackenfutter *n* two-jaw chuck
Zweibadfärbung *f* double-bath dyeing
Zweibadgerbung *f* double-bath process
Zweibahnablauf *m (ship)* double-way launching, two-way launching
Zweibahnenbett *n (mach)* double-track bed
Zweibändermodell *n* two-band model
zweibändig two-volume
Zweibandkabel *n* twin-band telephone cable
zweibasisch dibasic
Zweibeinlafette *f* bipod mounting
Zweibeinmast *m (ship)* bipod (sheer) mast
Zweibereichschalter *m* two-position range selector
Zweibettsetzmaschine *f* two-compartment jig
Zweibildkartiergerät *n* stereoscopic mapping instrument
Zweidecker *m* two-decker
Zweideckschiff *n* two-decked vessel
zweideutig ambiguous
Zweideutigkeit *f* ambiguity
Zweideutigkeitsaufhebung *f* sense finding (research) *(radar)*
zweidimensional two-dimensional; biaxial
Zweidornverfahren *n* **[der Fluchtungsprüfung]** two-mandrel method [of alignment testing], double-mandrel method [of alignment testing]
Zweidrahtbetrieb *m (telec)* two-wire operation
Zweidrahtleitung *f (telec)* two-wire circuit
Zweidrahtschaltung *f* two-wire connection
Zweidrahtübertragungsweg *m* two-wire circuit
Zweidrahtverstärker *m* two-wire repeater
Zweidruckreaktor *m* dual cycle reactor
Zweieck *n* lune
ebenes Z. plane lune
sphärisches Z. spherical lune
Zweielektronenrekombination *f* dielectronic recombination
Zweier *m* **ohne Steuermann** pair oar
Zweierbob *m* boblet

Zweierbündel *n* double-bunch conductor *(overhead line)*
Zweierstern *m (nucl)* two-pronged star
Zweierstoß *m (ph)* binary collision
zweifach twofold
Zweifachabstimmkondensator *m* double-gang tuning capacitor
Zweifachexpansionsdampfmaschine *f* double-expansion compound steam engine
Zweifachgarn *n* twofold (two-ply) yarn
Zweifachkabel *n* twin cable
Zweifachkette *f* double-strand chain
Zweifachkoinzidenz *f* double coincidence
Zweifachkondensator *m* two-gang capacitor
Zweifachkonusantenne *f* biconical aerial
Zweifachrollenkette *f* double (two-strand) roller chain
Zweifachschieber *m* non-by-pass valve
Zweifachschreibung *f* dual printing
Zweifachstecker *m* two-pin plug
Zweifachuntersetzer *m* scale of two circuit, binary scaler
Zweifachuntersetzung *f* scale-of-two
zweifachwirkend double-action, double-acting *(press)*
Zweifachzellenausleser *m (ag)* dual indented cylinder
Zweifarbenschnellpresse *f* high-speed two-colour press
zweifarbig two-colour[ed]
Zweifeldrahmen *m* two-bay frame
zweifeldrig two-bay, two-span
Zweiflach *n* dihedron
zweiflächig dihedral
Zweiflächner *m* dihedron
Zweiflammrohrkessel *m* Lancashire (double-flue) boiler
Zweiflanschnabe *f* double-flange hub
Zweifurchenpflug *m* two-furrow plough
Zweig *m* 1. branch; 2. twig
  gestörter Z. *(el)* faulty branch
  kapazitiver Z. *(el)* capacitance branch
  ohmscher Z. *(el)* resistive branch
zweigängig two-start *(thread)*
Zweigbahn *f (ry)* branch line
Zweigbetrieb *m* dispersal plant
Zweigelenkbogen *m* two-hinged arch, two-pinned arch
Zweigelenkrahmen *m* two-hinged frame
Zweigespanntrawl *n* bull trawl
zweigeteilt bipartite
zweigipflig bimodal *(curve)*
Zweigkabel *n* distribution cable
zweigleisig double-tracked
Zweigleitung *f* branch line
zweigliedrig *(math)* binomial
Zweigprodukt *n (ph)* branch product
Zweigrohrleitung *f* branch piping
Zweigruppenmodell *n (nucl)* two-group model
Zweigschalter *m* section switch
Zweigstrom *m* branch (shunt) current
Zweigströme *mpl* auxiliary flows *(paper sorting)*
Zweigstromkreis *m* branch circuit
Zweigutscheidung *f* two-product separation
zweihöckerig double-humped *(band filter)*
Zweihöckerkurve *f* double-peaked response
Zweihüllenboot *n* double-hulled boat
Zweihüllenschiff *n* double-hulled vessel
Zweikammerbremszylinder *m (aut)* two-chamber brake cylinder
Zweikammerkessel *m* water tube boiler with two headers
  Z. mit Umlaufvorrichtung double-header boiler with circulating device
  Z. mit zylindrischen Wasserkammern double-header boiler with cylindrical headers
Zweikammerklystron *n* two-cavity klystron
Zweikanalgerät *n* two-channel unit
Zweikanalverstärker *m* two-channel amplifier
Zweikettenkratzerförderer *m* double-chain conveyor
Zweikolbenspritzpreßverfahren *n (plast)* plunger moulding
Zweikomponentenklebstoff *m (plast)* separate application adhesive
Zweikörperproblem *n* two-body problem
Zweikraftstoffverfahren *n* bi-fuel system
Zweikreisabstimmung *f* two-circuit tuning
Zweikreisempfänger *m* double-circuit receiver
Zweikreissiedewasserreaktor *m* two-circuit boiling water reactor
Zweikrempelsatz *m (text)* two-card set
Zweikurbelpumpe *f* double-throw pump
zweilagig two-coat *(plaster)*
Zweilaufregelung *f* single-speed floating control, constant-speed floating control
Zweilaufwirkung *f (cont)* two-speed action
Zweileiterkabel *n* twin (two-core, duplex) cable
Zweileitersystem *n* two-wire system
Zweimantelisolator *m* double-petticoat insulator, double-shell insulator
Zweimaster *m* two-master
Zweimastschiff *n* two-masted ship
zweimolekular bimolecular
Zweimotorenantrieb *m* double-motor drive
zweimotorig twin-engined
Zweinormenempfänger *m* dual-standard television receiver
zweinutig two-flute
Zweiphasendosimeter *n* two-phase dosimeter
Zweiphasen-Dreileiterschaltung *f* two-phase three-wire connection
Zweiphasengenerator *m* two-phase alternator
Zweiphasengleichrichter *m* two-phase rectifier
Zweiphasenreaktor *m* two-phase reactor
Zweiphasenschaltung *f* two-phase connection

Zweiphasensystem *n* two-phase system
unverkettetes Z. two-phase four-wire system
verkettetes Z. interconnected two-phase system
Zweiphasenzähler *m* two-phase meter
zweiphasig two-phase
Zweiplattenkammer *f (nucl)* two-plate chamber
Zweiplattenruder *n* side (double-) plate rudder
Zweipol *m* two-terminal network
Zweipoladmittanz *f* driving-point admittance
Zweipolfunktion *f* driving-point function
zweipolig double-pole, bipolar, two-terminal
Zweipolimpedanz *f* driving-point impedance
Zweipolröhre *f* diode
Zweipolscheinleitwert *m* driving-point admittance
Zweipolscheinwiderstand *m* driving-point impedance
Zweipoltheorie *f* Thevenin's theorem
Zweiprismenjacquardmaschine *f* double-cylinder Jacquard
Zweiprismenschaftmaschine *f* double-cylinder dobby loom
Zweipunktabstimmung *f* double-spot tuning *(radio)*
Zweipunktniveauregelung *f* high-low level control
Zweipunktregelung *f* two-step action
Z. mit Hysterese two-step action with overlap
Zweipunktregler *m* two-step controller
Zweipunktverhalten *n* two-step action
Zweipunktwirkung *f s.* Zweipunktverhalten
Zweiquadrantenmultiplizierer *m (dat)* two-quadrant multiplier
Zweiräderblock *m (mach)* sliding double gear
Zweiraumschiff *n* two-compartment ship
zweireihig double-row *(bearing)*
Zweirichtungsbetrieb *m (dat)* push-pull action
Zweirichtungsimpuls *m* bidirectional pulse
Zweirumpfboot *n* catamaran *(vessel)*
Zweisammlerkrempel *f (text)* two-doffer card
zweisäurig diacid[ic]
Zweischalengreifbagger *m* clamshell grab
Zweischeibenkupplung *f (aut)* double-plate clutch
Zweischeibenläppmaschine *f* double-lap machine
Zweischeibenschar *n* double-disk coulter
Zweischichtenfilm *m* double-coated film
Zweischichtenpflug *m* subsoiler, double-depth plough
Zweischichtwicklung *f (el)* double-layer winding
zweischiffig two-bay, two-span
Zweischlitzmagnetron *n* split-anode magnetron
Zweischneider *m (mach)* two-flute cutter, *(Am)* two-lip end mill
zweischneidig *(mach)* two-edged, two-lipped, with two edges
zweischnittig double-shear *(shearing test)*
Zweischottentanker *m* twin-bulkhead tanker
Zweischraubenfahrgastschiff *n* twin-screw passenger ship
Zweischraubenschiff *n* twin-screw vessel
Zweischurwolle *f* double-clip wool
Zweiseitenband *n (el)* double sideband
Zweiseitenbandübertragung *f* double sideband transmission
Zweiseitenbetrieb *m (cont)* push-pull action
Zweiseitenempfindlichkeit *f* bilateral sense
Zweiseitenkippanhänger *m* two-way tipping trailer
zweiseitig bilateral; double-end[ed]
zweispaltig double-column, half-measure
Zweispännertrawl *n (ship)* bull trawl
Zweispindelautomat *m* two-spindle automatic
Zweispindelschraubenpumpe *f* twin-screw type of pump
zweispitzig bicuspid
Zweispitzniet *m* bifurcated rivet
zweispurig dual-lane, two-lane
Zweiständerexzenterpresse *f* straight-side eccentric press, double-sided eccentric press
Zweiständerkarusselldrehmaschine *f* double-column vertical turning and boring machine
Z. mit fester Planscheibe vertical boring and turning mill of the fixed table type
Z. mit verschiebbarer Planscheibe vertical boring and turning mill of the sliding table type
Zweiständersenkrechträummaschine *f* two-column vertical broaching machine
Zweistärkenglas *n* bifocal lens
Zweistärkenstarglas *n* **/ asphärisches** Franklin-type aspherical cataract bifocal lens
zweistellig two-figure *(number)*
zweistielig two-legged
Zweistiftsockel *m* bi-post cap
zweistöckig two-storey *(house)*
Zweistockwagen *m (ry)* double-deck car
Zweistofflegierung *f* binary alloy
Zweistoffsystem *n* binary system
Zweistöpselschnur *f* double-ended cord circuit
Zweistrahlbetrieb *m* dual-trace operation *(oscilloscope)*
Zweistrahloszillograf *m* dual-trace oscillograph, dual-channel oscillograph, two-trace oscillograph
Zweistückspannvorrichtung *f (mach)* two-station fixture
Zweistufenionisierung *f* ionization in two stages
Zweistufenprozeß *m* two-stage process
Zweistufenrakete *f* two-stage rocket, two-step rocket, *(Am)* bumper missile
Zweistufensulfataufschluß *m* two-stage sulphate digestion
Zweistufenverstärker *m* two-stage amplifier

zweistufig two-stage; two-step; two-volume *(pump)*
Zweitakter *m* *(aut)* two-stroke engine, two-stroker
Zweitaktgemisch *n* lubricated petrol, petrol-oil mixture
Zweitaktlauf *m* two-stroke cycle
Zweitaktmischung *f s.* Zweitaktgemisch
Zweitaktmotor *m* two-stroke engine
Zweitaktschmierung *f* petrol lubrication
Zweitaktverfahren *n* two-stroke cycle
Zweitarifzähler *m* *(el)* two-rate meter, double-range meter
Zweiteilchenzerfall *m* two-body decay
zweiteilig two-piece; split *(mould)*
Zweitemperaturverfahren *n* *(nucl)* dual-temperature process
Zweitfruchtfutterbau *m* fodder catch cropping
Zweitourendruckmaschine *f* two-revolution [printing] press
Zweitourenmaschine *f s.* Zweitourendruckmaschine
Zweitquantelung *f* second quantization
Zweiträgerkran *m* double-girder overhead-crane
Zweitschnitt *m* *(ag)* aftermath, second hay
Zweitschriftrechnung *f* *(dat)* follow-up invoice
Zweitwicklung *f* *(el)* secondary [winding]
Zweiunddreißigerleitung *f* octuple-phantom circuit *(telephony)*
Zweiwalzenbrecher *m* double-roll crusher
Zweiwalzenschrämlader *m* double-drum shearer
Zweiwegautomat *m* *(dat)* two-way automaton
Zweiwegebohren *n* two-way boring (drilling)
Zweiwegebohrmaschine *f* two-way boring (drilling) machine
Zweiwegeförderer *m* two-way conveyor
Zweiwegehahn *m* plug (two-way) cock
Zweiwegeschieber *m* two-way valve
Zweiweggleichrichter *m* full-wave rectifier
Zweiweggleichrichtung *f* full-wave rectification, two-wave rectification
Zweiwegschaltung *f* double-way connection
Zweiwelligkeit *f* two-wave property *(crystal)*
Zweiwellrohrkessel *m* Lancashire boiler with corrugated flues
zweiwertig *(ch)* divalent, bivalent; dihydric *(alcohol)*
Zweiwertigkeit *f* bivalence
zweizählig binary; two-fold
Zweizimmerwohnung *f* two-room[ed] flat
Zweizonenofen *m* double-fired furnace
Zweizonenreaktor *m* two-region reactor
Zweizugware *f* *(text)* double-stretch articles
Zweizweckreaktor *m* dual-purpose reactor
Zweizylinderdampfmaschine *f* two-cylinder steam engine
Zweizylindergarn *n* condensed yarn
Zweizylinderspinnerei *f* condenser spinning
Zwergempfänger *m* midget set
Zwerggiebel *m* *(build)* gablet
Zwergglimmlampe *f* miniature glow-discharge tube
Zwergkessel *m* dwarf boiler
Zwerglampe *f* miniature bulb
Zwergröhre *f* miniature (midget) valve
Zwergsignal *n* dwarf (ground) signal
Zwergstabbatterie *f* penlite cell
Zwergtriode *f* miniature triode
Zwickel *m* 1. *(build)* spandrel; 2. clock *(stocking)*; 3. gore *(sail)*
Zwickerei *f* lasting room *(shoe)*
Zwiebelfisch *m* *(print)* pie
Zwiebelrodegerät *n* onion harvester
Zwielicht *n* twilight
Zwilling *m s.* Zwillingskristall
Zwillingsachse *f* twinning axis *(crystal)*
Zwillingsausleger *m* duplex jib
Zwillingsaußenräummaschine *f* / **senkrecht arbeitende** vertical duplex-type of surface broaching machine
Zwillingsbereifung *f* twin tyre
Zwillingsbildung *f* twinning, formation of twin crystals
Zwillingsbrikettpresse *f* twin briquetting press
Zwillingsdampfmaschine *f* twin steam engine
Zwillingsebene *f* twinning plane
Zwillingsenergie *f* twinning energy
Zwillingsfahrleitung *f* double-contact wire system
Zwillingsfläche *f* twin plane *(crystal)*
Zwillingsflak *f* twin-barreled antiaircraft gun
Zwillingskabel *n* twin cable
Zwillingskessel *m* twin Cornish boiler
Zwillingskondensator *m* twin capacitor
Zwillingskontakt *m* twin contacts
Zwillingskristall *m* twinned crystal, twin [crystal]
Zwillingslinse *f* binary lens
Zwillingsmaschinengewehr *n* twin-barreled machine gun
Zwillingsprüfung *f* *(dat)* duplication (twin) check
Zwillingspumpensatz *m* twin-pump set
Zwillingspumpensystem *n* twin-pump system
Zwillingsrad *n* twin wheel
Zwillingsräummaschine *f* duplex-type broaching machine
Zwillingsreifen *m* dual tyre
Zwillingsriemenpumpe *f* twin-pump with belt drive, duplex-pump with belt drive
Zwillingsschrämarm *m* duplex jib
Zwillingsschrämpilz *m* twin mushroom
Zwillingssimplexpumpe *f* twin simplex pump
Zwillingsstecker *m* biplug, pair of plugs
Zwillingsstreifen *m* twin band
Zwillingstandemdampfmaschine *f* twin tandem engine

Zwillingstrommelmischer *m* twin-cylinder mixer
Zwillingswalzen *fpl (met)* duo rolls
Zwillingswalzwerk *n (met)* duo mill
Zwillingszugofen *m* twin-flued oven
Zwinge *f* ferrule; clamp
Zwirn *m* folded (plied) yarn, twist[ed yarn]
zwirnen to twist
Zwirnen *n* throwing, twisting [silk]; forming of a strand *(rope)*
Zwirnflügel *m (text)* twisting flyer
Zwirnkette *f (text)* doubled warp
Zwirnmaschine *f* twisting machine, twister
Zwirnring *m (text)* twister ring
Zwirnselfaktor *m (text)* twiner mule
Zwirnspindel *f (text)* twister (twisting) spindle
Zwirnspule *f (text)* twisting bobbin
zwischen den Arbeitsgängen inter-operational
Zwischenabbildung *f (opt)* intermediate depiction
Zwischenablesung *f* intermediate reading
Zwischenabschrägung *f* taper lead, *(Am)* starting taper *(machine reamer)*
Zwischenachsanbaugerät *n (ag)* mid-mounted implement
zwischenatomar interatomic
Zwischenaufenthalt *m* stop off
Zwischenbahn *f* intermediate orbit
Zwischenband *n* intermediate band
Zwischenbaustück *n* carrier piece *(rolling mill)*
Zwischenbeschleuniger *m (nucl)* interaccelerator
Zwischenbestand *m* secondary crop *(forest)*
Zwischenbild *n* intermediate image
Zwischenbildebene *f* intermediate image plane
Zwischenbildspektrometer *n* intermediate image spectrometer
Zwischenblock *m (mine)* extension block *(plough)*
Zwischendeck *n (ship)* between ('tween) deck, berth-deck
Zwischendecke *f* false floor, inserted ceiling
Zwischendecklukendeckel *m (ship)* 'tweendeck hatchcover
Zwischendeckvermessung *f (ship)* 'tweendeck tonnage
Zwischendose *f (el)* pull box
Zwischendruck *m* clearance pressure
Zwischenelektrode *f* dynode
Zwischenergebnisspeicher *m* working store
Zwischenergebnisspeicherung *f* working storage
Zwischenfall *m* accident
Zwischenfilmverfahren *n (telev)* intermediate-film method
Zwischenfläche *f* interface
Zwischenflansch *m* intermediate flange
Zwischenflug *m* intermediate approach
Zwischenförderer *m (mine)* intermediate conveyor
Zwischenfrequenz *f* intermediate frequency, i.f.
Zwischenfrequenzabgleich *m* i.f. alignment
Zwischenfrequenzröhre *f* i.f. valve
Zwischenfrequenzsperre *f* i.f. rejector
Zwischenfrequenzübertrager *m* i.f. transformer
Zwischenfrequenzverstärker *m* i.f. amplifier
Zwischenfrucht *f (ag)* intermediate crop
Zwischenfutter *n (text)* interlining
Zwischengerüst *n (met)* strand rolls
Zwischengeschoß *n (build)* mezzanine, entresol
Zwischengitter *n* interlattice
Zwischengitteratom *n* displaced (interstitial) atom, interstitial
Zwischengitterlage *f* interstitial site
Zwischengitterplatz *m* interstitial site
Zwischengleis *n (ry)* intermediate track
Zwischenglied *n* interconnection; intermediate medium
Zwischenglühen *n* intermediate annealing
Zwischengröße *f* intermediate size
Zwischenhafen *m* intermediate port, port of call
Zwischenhilfsträger *m (telev)* intermediate subcarrier
Zwischenhülse *f* adapter [sleeve]
Zwischenkabel *n* intermediate cable
Zwischenkern *m* compound (intermediate) nucleus
Zwischenkettenrad *n* idler sprocket
Zwischenkreis *m* intermediate circuit (loop)
Zwischenkreislauf *m* intermediate circuit (loop)
Zwischenkreisspule *f (el)* link coil
Zwischenkühler *m* intercooler
Zwischenkühlung *f* intercooling
Zwischenküstenfahrt *f (ship)* intercoastal trade
Zwischenlage *f (mach)* spacer; washer; *(text)* sandwich arrangement
  elastische Z. washer of compressible material
Zwischenlagekarton *m* slip-sheet board
Zwischenlager *n* intermediate (line shaft) bearing
Zwischenlagerung *f (geol)* interbedding, interstratification
Zwischenlandung *f* intermediate landing
Zwischenlegeblech *n* shim
Zwischenleitschaufel *f* intermediate guide blade *(turbine)*
Zwischenmagazin *n (mach)* buffer store (stock) *(transfer line)*
Zwischenmessung *f (mach)* in-process gauging
Zwischenmodulation *f* intermodulation
zwischenmolekular intermolecular
Zwischenphase *f* interphase, intermediate phase
Zwischenplatte *f* baffle plate; intermediate plate
Zwischenplattenunterzug *m (ship)* intercostal girder
Zwischenpodest *n(m)* half-landing
Zwischenpol *m* interpole
Zwischenprodukt *n* intermediate [product]
Zwischenproduktsynthese *f* intermediate preparation

Zwischenprüfung *f* intermediate test
Zwischenpuffer *m* storage bank *(transfer line)*
Zwischenpunktabtastung *f (telev)* dot interlacing
Zwischenpunktflimmern *n (telev)* interdot flicker
Zwischenpunktverfahren *n* dot interlacing
Zwischenrad *n (mach)* intermediate gear (wheel), idler gear; intermediate timing gear
Zwischenraum *m* clearance; interspace; gap of the joint; *(dat)* blank
  atomarer Z. atomic interspace
  falscher Z. split *(telegraphy)*
Zwischenraumbildung *f* / **selbsttätige** unshift-on-space *(telegraphy)*
Zwischenraumzeichen *n* space signal *(telephony)*
Zwischenrelais *n* intermediate (supplementary) relay
Zwischenring *m* spacing collar *(milling machine)*
Zwischenschacht *m* false lining *(blast furnace)*
Zwischenschale *f* sub-shell *(atom)*
zwischenschalten to insert; to interconnect
Zwischenschalter *m* intermediate switch
Zwischenschaltung *f* interposition; interconnection
  Z. von Leiträdern insertion of guide disks (blades) *(turbine)*
Zwischenschicht *f* intermediate layer
Zwischensender *m* intermediate (fill-in) transmitter, retransmitter
Zwischensockel *m* adapter
Zwischensohle *f (mine)* intermediate level, sub-level
Zwischenspant *n (ship)* intermediate frame
Zwischenspeicher *m (dat)* temporary storage
Zwischenspeicherung *f (dat)* temporary (intermediate) storage
Zwischenspindel *f* jack shaft *(rolling mill)*
Zwischenstecker *m* adapter [plug]
Zwischenstopp *m (dat)* break point *(during the run of a programme)*
Zwischenstück *n* adapter; connecting piece *(switch)*; intermediate piece; sleeve; binding post adapter
  nockenverriegeltes Z. cam-lock adapter
Zwischenstufe *f* intermediate stage
Zwischenstufengefüge *n* transformation stage structure
Zwischenstufenvergütung *f* austempering
Zwischenstütze *f* intermediate support
Zwischensupport *m* intermediate saddle *(turret lathe)*
Zwischenträger *m (build)* secondary beam; *(telev)* subcarrier
Zwischenträgerverfahren *n (telev)* intercarrier sound system
Zwischentraglager *n (mach)* intermediate bearing support
Zwischentransformator *m* interstage (adapter, matching) transformer
Zwischenüberhitzer *m* intermediate superheater
Zwischenüberhitzung *f* intermediate superheating, superheating between expansion stages
Zwischenumkehre *f (mine)* intermediate return *(attachment plough)*
Zwischenverbindung *f (ch)* intermediate compound
Zwischenverstärker *m* intermediate repeater
Zwischenverteiler *m (telec)* intermediate distribution frame
Zwischenwand *f (build)* partition wall; baffle *(liquid)*
Zwischenwelle *f* intermediate (line) shaft
Zwischenzeilenabtastung *f (telev)* interlaced scanning
Zwischenzeilenverfahren *n* / **vierfaches** *(telev)* quadrupled scanning
Zwischenzerkleinerung *f* intermediate crushing
Zwitterion *n* zwitterion, ampholite ion
Zwölfeck *n* duodecagon
Zwölfflächner *m* / **regelmäßiger** dodecahedron
Zyan *n* cyanogen
Zyanamid *n* cyanamide
Zyanat *n* cyanate
Zyanbadhärtung *f* cyaniding
Zyaneinsatzhärtung *f s.* Zyanbadhärtung
Zyanessigsäure *f* cyanoacetic acid
Zyanhalogenid *n* cyanogen halide
Zyanhydrin *n* cyanhydrin
Zyanid *n* cyanide
Zyanidlaugerei *f* cyaniding, leaching with cyanide solution
Zyanin *n* cyanine
Zyaninfarbstoff *m* cyanine dye
Zyankali *n* potassium cyanide
Zyanose *f* cyanosis
Zyanotypie *f* cyanotype, blue printing
Zyansäure *f* cyanic acid
Zyanursäure *f* cyanuric acid
Zyanwasserstoff *m* hydrogen cyanide
Zyanwasserstoffsäure *f* hydrocyanic (prussic) acid
zyklisch cyclic
zyklisiert cyclic
Zyklisierung *f* cyclization
Zyklobutan *n* cyclobutane
Zyklogramm *n* cyclogram
Zyklohexan *n* cyclohexane, hexahydrobenzene
Zyklohexanol *n* cyclohexanol
Zyklohexanon *n* cyclohexanone
zykloid cycloidal
Zykloide *f* cycloid[al curve]
Zykloidenmassenspektrometer *n (nucl)* trochoidal mass analyzer
Zykloidenpendel *n* cycloidal pendulum
Zykloidenprofil *n* cycloidal tooth profile
Zykloidenverzahnung *f* cycloidal teeth
Zyklometrie *f* cyclometry

zyklometrisch arc-trigonometric
Zyklon *m* cyclone separator (dust extractor)
Zyklonfeuerung *f* cyclone firing
Zyklonnachwäsche *f* cyclone re-washing plant
Zyklonverdampfer *m* cyclone-type evaporator
Zykloparaffin *n* cycloparaffin
Zyklopenmauerwerk *n* cyclopean masonry, polygonal rubble
Zyklopentan *n* cyclopentane, pentamethylene
Zyklopropan *n* cyclopropane, trimethylene
Zyklotron *n* cyclotron
  frequenzmoduliertes Z. frequency-modulated cyclotron
  kosmisches Z. cosmic cyclotron
Zyklotronfrequenz *f* cyclotron frequency
Zyklotronresonanz *f* cyclotron resonance
Zyklotronwelle *f* cyclotron wave
Zyklus *m* cycle
Zykluszeit *f* cycle time
Zylinder *m* cylinder
  Z. mit Dampfmantel cylinder with steam-jacket
  Z. mit Pufferung cushioned cylinder, jack (cylinder) with built-in damper
  doppeltwirkender Z. double-acting cylinder
  einfachwirkender Z. single-acting cylinder
  einseitig beaufschlagter Z. single-acting cylinder
  freihängender Z. overhung cylinder
  freitragender Z. overhung cylinder
  gerader Z. right cylinder
  koaxialer Z. coaxial cylinder
  schief abgeschnittener Z. oblique frustum of cylinder, obliquely cut cylinder, truncated right circular cylinder
  schwebender Z. overhung cylinder
Zylinderabstellung *f* cylinder throw-off
Zylinderabziehpresse *f* cylinder press
Zylinderanbohrung *f* opening into the cylinder
Zylinderanordnung *f* cylinder arrangement
  Z. in V-Form V-cylinder block, cylinders arranged in a vee
Zylinderantenne *f* cylindrical aerial
Zylinderarmatur *f* cylinder fittings
Zylinderauflagerung *f* cylinder support
Zylinderaufzug *m (print)* cylinder dressing
Zylinderauskleidung *f* cylinder liner (lining)
Zylinderausrüstung *f* cylinder fittings
Zylinderbefestigungsschraube *f* cylinder holding-down bolt
Zylinderblock *m* cylinder block
  Z. aus einem Gußstück monobloc
Zylinderblockfertigungsstraße *f* cylinder block production line
Zylinderboden *m*/ **abnehmbarer** cylinder cover, cover of casing
Zylinderbodendeckel *m* bottom (back cylinder) cover
Zylinderbohrmaschine *f* cylinder boring machine
Zylinderbohrung *f* cylinder bore
Zylinderbuchse *f*/ **eingestemmte** caulked-in liner
Zylinderdeckel *m (aut)* cylinder head; cylinder cover
  Z. ohne Durchbruch blind end [of cylinder]
Zylinderdeckelhaube *f* protective covering for the cylinder cover
Zylinderdichtungsring *m* cylinder head gasket
Zylinderdruckmaschine *f* cylinder printing press
Zylinderdurchmesser *m* cylinder diameter
  innerer Z. cylinder bore
Zylindereinsatz *m* cylinder liner
Zylindereinspritzung *f* direct injection
Zylinderfalz *m* cylinder fold
Zylinderfläche *f* circumferential (cylindrical) surface
Zylinderflansch *m* cylinder base
Zylinderfunktion *f* cylindrical function
Zylinderfuß *m* cylinder foot
Zylinderfußgestell *n* cylinder foot
Zylinderfutter *n* cylinder liner
Zylindergarnitur *f* cylinder fittings
Zylindergrube *f (print)* cylinder gap
Zylinderhandreibahle *f* hand reamer for parallel holes, parallel hand reamer, *(Am)* straight hand reamer
Zylinderheizraum *m* steam (cylinder) jacket
Zylinderhuf *m* portion of cylinder having smaller base area than base circle
Zylinderinhalt *m* cylinder displacement (capacity), swept volume
Zylinderkanal *m* 1. *(print)* cylinder gap; 2. *(ship)* cylinder passage
Zylinderkessel *m* cylindrical (Scotch) boiler
Zylinderkoeffizient *m (ship)* longitudinal (prismatical) coefficient
Zylinderkoordinaten *fpl* cylindrical coordinates
Zylinderkopf *m* cylinder head
Zylinderkopfdeckel *m* cylinder-head cover
Zylinderkopfdichtung *f* cylinder-head gasket
Zylinderkörper *m* cylinder block; plunger-carrying body *(plunger pump)*
Zylinderkristall *m* cylindrical crystal
Zylinderkühlrippe *f* cylinder cooling rib (fin)
Zylinderlaufbild *n (aut)* liner condition, cylinder bore wear pattern, wear condition of liner
Zylinderlaufbuchse *f* cylinder liner (barrel)
Zylinderlauffläche *f* cylinder bore (working surface)
Zylinderleistung *f* cylinder output
Zylinderlinse *f* cylindrical lens
Zylindermantel *m* cylinder barrel (jacket)
Zylindermaschine *f* side-wheel cylinder sewing machine *(shoe)*
Zylindernachbohrmaschine *f* cylinder reboring machine

Zylinderöl *n* cylinder oil
Zylinderpfahl *m* foundation cylinder
Zylinderreaktor *m* [right circular] cylindrical reactor
Zylinderreflektor *m* cylindrical reflector *(aerial)*
Zylinderreibahle *f* parallel (straight) reamer
Zylinderreihe *f* bank of cylinders
Zylinderrollenlager *n* cylindrical roller bearing
Zylinderschaft *m* parallel shank, *(Am)* straight shank
Zylinderschale *f* cylindrical (barrel) shell
Zylinderschlitz *m* cylinder port
Zylinderschmieröl *n* cylinder oil
Zylinderschnecke *f* cylindrical (parallel) worm
Zylinderschraube *f* fillister-head screw
Z. mit Innensechskant hexagonal socket-head bolt; internal wrenching bolt
Zylinderschutzmantel *m* cylinder lagging (cleating)
Zylindersenge *f (text)* cylinder singeing machine
Zylindersenkung *f* counterbore
Zylindersicherheitsventil *n* cylinder safety valve
Zylindersieb *n* cylinder screen (sifter)
Zylinderspule *f* cylindrical coil
einlagige Z. single-layer winding
Zylinderstift *m* plain (cylindrical) pin
Zylinderstirnwand *f* front (top) cylinder cover
Zylinderstopfbuchse *f* cylinder stuffing box
Zylinderstütze *f* cylinder foot
Zylinderstützung *f* cylinder support
Zylinderteil *n* cylindrical part (portion) *(of the reamer)*
Zylindertragbild *n* wear condition of liner, bore wear pattern
Zylindertragfuß *m* cylinder foot
Zylindertrockner *m* can drier
Zylindertrommel *f* cylinder barrel
Zylinderverkleidung *f* cylinder lagging (cleating)
Zylinderverschalung *f* cylinder lagging (cleating)
Zylindervolumen *n* cylinder capacity
Zylindervolumenverhältnis *n* ratio of cylinder volumes
Zylinderwand[ung] *f* cylinder wall
Zylinderweite *f* cylinder diameter
Zylinderwicklung *f* concentric (cylindrical) winding *(transformer)*
zylindrisch cylindrical; parallel, *(Am)* straight *(bore, shank)*
Zylindrizität *f* parallelity, straightness
Zymase *f* zymase
Zymol *n* cymene
Zypresse *f* cypress
Zystein *n* cysteine
Zysteinschutz *m (nucl)* cysteine protection
Zystin *n* cystine
Zystoskop *n* cystoscope
Zytolyse *f* cytolysis
Zytosin *n* cytosine

VERZEICHNIS WICHTIGER DEUTSCHER ABKÜRZUNGEN

A

A Additionsreaktion; Angriffspunkt; Arbeit; Atomgewicht; Anion; Aktivierungsenergie; Auftriebskraft; Ampere; Amperemeter; Anode; Auslösung; Amplitude; Austastsignal; Abfrage; Alphabetfaktor; Amplitudenmodulation; Aussteuerungsfaktor; alterungsbeständig; angelassen *(Stahl)*; Atlas; Äthanol; Alkohol; Auftrieb; Anker; Abteilung; Armee; Auslaß

Ä Äther

Å Ångström[einheit]

a Anker; Dämpfungsmaß; Anode; Arbeitskontakt *(Relais)*; Jahr; Jahre; Ausladung; Antriebs-; Arbeits-; Abdräng-; äußere[r]; Ar *(1 a = 1 dam² = 100 m²)*; Atto *(1 a = 10^{-18} Einheiten)*; Temperaturleitzahl

a. achtern

AA Arbeitsanweisung; Arbeitsausschuß; doppelt asphaltierte Kabelumhüllung

aa Arbeit-Arbeitskontakt *(Relais)*; zu gleichen Teilen *(ana partes aequales)*

AAD Auftastimpuls-Auslösediode

a.a.O. am angeführten Ort

AB Anfangsbestand; aussetzender Betrieb; Abflachschaltung; Amplitudenbegrenzer; Anodenbatterie

A.B. Autobahn

a.B. auf Befehl

ABA Allgemeine Bedingungen für Anschlußbahnen

ABB Ausschuß für Blitzableiterbau

Abb. Abbildung

ABF Arbeiter-und-Bauern-Fakultät

Abh. Abhandlung

Abk. Abkürzung

A.B.L. Armeebetriebsstofflager

ABPV Allgemeine Bergpolizeiverordnung

AbrPA Abrechnungspostamt

ABS Akrylnitril-Butadien-Styrol-Kopolymer[es]

Abs. Absender; Absatz

abs. absolut

Abschn. Abschnitt

Abst. Abstand

abs.Temp. absolute Temperatur

Abst.Zd. Abstandszünder

Abt. Abteilung

ABUS Ausrüstung von Bergbau und Schwerindustrie

ABW Außenbogenweiche

Abw. Abwehr

Ac Aktinium; Acyl

ac Azetat

a.c. anni currentis; à conto

Acc. Akzept

ACH Abteilungschef

ACHEMA, Achema Ausstellung chemischer Apparate

Acid. acidum

A/cm Ampere/Zentimeter

ACTH adrenokortikotropes Hormon

AD Artilleriedivision

ADMV Allgemeiner Deutscher Motorsportverband

ADN Allgemeiner Deutscher Nachrichtendienst

Adr. Adresse

ADS Allgemeine Deutsche Seeversicherungsbedingungen

ADSp Allgemeine Deutsche Spediteurbedingungen

ADU Analog-Digital-Umsetzer

AE Anrufeinheit; Ausfuhrerklärung; Ascheelementaranalyse; astronomische Einheit; Ångströmeinheit

Ae Äther

AeDTA, ÄDTA Dinatriumäthylendiamintetraazetat

AeDTE, ÄDTE Äthylendiamintetraessigsäure

AEF Ausschuß für Einheiten und Formelgrößen

aeq äquivalent

AEÜ Archiv für elektrische Übertragung

AF Auslandsfernamt; Ablenkfaktor; Ackerfläche

af aschefrei

AFB amtliches Fernsprechbuch; Abraumförderbrücke

AFF Ansteuerungsfunkfeuer; Amt für Fernnetze

AFN automatische Frequenznachstimmung

AFP Arbeitsflugplatz

AG Aschegehalt; Arbeitsgruppe

Ag Atomgewicht; Silber; Alfagras

a.gl.O. am gleichen Ort

AGM gummiertes Außenkabel mit Bleimantel

AGP Armeegerätepark

A.Gr. Armeegruppe

AGW Auslandsgruppenwähler

AH Ahorn; Außenhandel

Ah, ah Amperestunde

AHA Abhörapparat

AH-Salz Hexamethylendiammoniumsalz der Adipinsäure

AK Arbeitskraft; Absorptionskoeffizient; Außenkante; Armeekorps

A-K Anode-Katode

Ak Angora[kanin]wolle

AKA Absorptionskälteapparat

a.K.B. außer Kriegsbereitschaft

AKL Anrufkontrollampe

A-Kohle Aktivkohle

A.K.P. Armeekraftwagenpark

Akt. Aktivität

akt. aktiv

AKV Anzeigekatodenverstärker

AK voraus äußerste Kraft voraus

AK zurück äußerste Kraft zurück

AL. Alginatchemieseide; Alkalilöslichkeit

Al Anschlußleitung; Anruflampe; Aluminium; Alginat; Alpakawolle
Ala Alanin
AlBz Aluminiumbronze
Alk Alkyl
alkal. alkalisch
alkoh. alkoholisch
Al-Leg. Aluminiumlegierung
allg. allgemein
AlMBz Mehrstoffaluminiumbronze
ALR automatische Lautstärkeregelung
AlR Reinstaluminium
AM Amplitudenmodulation
AM 1 Sb Einseitenband-Amplitudenmodulation
AM 2 Sb Zweiseitenband-Amplitudenmodulation
Am Amerizium
A/m Ampere/Meter
A/m² Ampere/Quadratmeter
am. amorph
AMK Arbeitsmittelkarte
AMP Adenosinmonophosphat
Amp. Ampere
AMS Anlaßstufenschalter
Am-Sup Ammoniumsuperphosphat
amu atomare Masseeinheit *(1 amu = 1,659750 · 10^{-27} kg)*
AMW Auslandsmischwähler
AN Anpassungsnetzwerk
An Aktinon; Akon; Anpassungsnetzwerk
analyt. analytisch
ang. angular
Anh. Anhang
anh., anhydr. anhydrisch
Anl. Anlage
anl Anlauf
Anm. Anmerkung
anorg. anorganisch
AnT Anlaßtaste
Ant. Antenne
Anw Anweisung; Postanweisung
Anw. Anweisung; Anwendung
AnwAbk. Postanweisungsabkommen
Anz. Anzahl
AO Artillerieoffizier
AOK Armeeoberkommando
AP Ammoniumperchlorat; Anilinpunkt; Asbestpappe
A.P. Amerikanisches Patent
Ap Alpakawolle
apl. außerplanmäßig
App Apparat, Apparatur
APr Asbestpapier
Aq., aq, aq. Aqua
Äq Äquivalent
äq äquivalent
Äqu.-Gew. Äquivalentgewicht
äquiv. äquivalent
AR Anzeigeröhre; Rohrableiter; automatische Regelung; Rektaszension; Ardein
Ar Argon; Aryl; Arcatomschweißen; Abraum
ar Arbeit-Ruhekontakt *(Relais)*
Arak Atomrakete
Arb.Anw. Arbeitsanweisung
arccos, arc cos Arkuskosinus
arccot, arc cot Arkuskotangens
Arch.Pharm. Archiv der Pharmazie
arcosh, ar cosh Areakosinus
arcoth, ar coth Areakotangens
arcsin, arc sin Arkussinus
arctan, arc tan Arkustangens
ARFF Allrichtungsfunkfeuer
Arg Arginin
ARS Anrufrelaissatz
arsinh, ar sinh Areasinus
Art. Artikel
artanh, ar tanh Areatangens
AS Anrufsucher; Außenstation; Absperrschieber
A-S Altostratus
As Amperesekunde; Schwenkabsetzer; Arsen; Azetylenschweißen; Asbest
as. asymmetrisch
ASAO Arbeitsschutzanordnung
ASB Aussetzschaltbetrieb
AS-Boot Artillerieschnellboot
As/cm² Amperesekunde/Quadratzentimeter
ASK Arbeitsschutzkommission; Armeesportklub
ASKL Anruf- und Schlußkontrollampe
ASL Anruf- und Schlußlampe
As/m² Amperesekunde/Quadratmeter
ASN automatische Sternnavigation
ASO Arbeitsschutzobmann
Asp Asparaginsäure
asph. asphaltiert
ASprK Atomsprengkörper
ASt Abfragestöpsel; Amtsstelle; alterungsbeständiger Stahl
AsT Systemauslösetaste
AstM Aussteuerungsmesser
astr. Einh. astronomische Einheit
ASV Armeesportverein
ASVO Arbeitsschutzverordnung
asymm. asymmetrisch
AT Achtertelegrafie; Auslösetaste; aluminothermisch; Abfragetaste; Aufschaltetaste; Abgasturbine
At Astatin
äT äußerer Totpunkt
at technische Atmosphäre *(1 kp/cm²)*; Atom
at a technische Atmosphäre; absoluter Druck
ATAG Aerodynamische Versuchsanstalt in Göttingen
ATG Auftastgenerator
AtG Atomgewicht
At.-Gew. Atomgewicht

äther. ätherisch
ATL Abgasturbolader
atm physikalische Atmosphäre
(1 atm = 101325 N/m²)
Atomgew. Atomgewicht
ATP Adenosintriphosphat
atro. absolut trocken
AU Anpassungsübertrager, Ausgangsübertrager
Au Audion; Gold
au Arbeitsumschaltekontakt *(Relais)*
aubo. außenbord
Aufl. Auflage
AufT Aufschaltetaste
AuftrAbk. Postauftragsabkommen
Ausb. Ausbeute
Ausg. Ausgabe
AuslPAnw Auslands-Postanweisung
AusnT Ausnahmeaufschaltetaste
äuß. äußerlich
ausschl. ausschließlich
AusT Platzauslösetaste
AV Auslaßventil; Ventilableiter
A. V. Ausbildungsvorschrift
A/V Ampere/Volt
AVR automatische Verstärkungsregelung
AW Anpassungsnetzwerk; Amperewindung; Antiklinalwände
Aw Amperewindung
A-Waffen Atomwaffen
Aw/cm Amperewindungen/cm
AWE automatische Wiedereinschaltung
AWF Ausschuß für Wirtschaftliche Fertigung
AWP Adria-Wien-Pipeline
AZ Alkalitätszahl; Alkoholzahl; Abstandszünder
AZKW Amt für Zoll und Kontrolle des Warenverkehrs
AZO Allgemeine Zollordnung
AZV automatische Zielverfolgung

B

B Breite; Brennstoff; Bor; Berge; Kraftstoffverbrauch; Barometerstand; Werkstückbreite; Bessemerstahl; basisch; beste Bearbeitbarkeit; Faltversuch; Bandeisenbewehrung; Basis; Basisanschluß; Blindleitwert; Baumwolle *(Kabel)*; Bandbreite; Bel, bel *(Dämpfungsmaß)*; Schiffsbreite, größte Breite auf Spanten; Raum- oder Zwischendeckbalkenlage; Bodenbesichtigung im Dock oder auf dem Slip; Buche
b Beschleunigung; Walzgurtbreite; Bandbreite; Basisanschluß; Kolbenbeschleunigung
BA Bergakademie; Bergamt; Betriebsabteilung; Betriebsamt; Bildaustastung
Ba Barium; Back
Bals. Balsamum
Balta ballistische Tageseinflüsse
BAM Bundesanstalt für Materialprüfung
BAS Bildaustastsynchron[signal]
bas. basisch
BaT Betriebssicherungstaste
BB Bergbehörde; Braunkohlenbrikett; Butan-Butylen-Gemisch; Block-Basis-Schaltung; Breitband; Buchungsbeleg; Baukastenbohreinheit; Birnbaum
B-B- Boden-Boden-
Bb., B.b. Backbord
Bb.-Maschine Backbordmaschine
BBo Bugsierboot
B_{CWL} *(Schiff)* Breite in der Konstruktionswasserlinie
BD Bergungsdampfer
Bd Frequenzband
B. d. A. Befehlshaber der Aufklärungsschiffe
B-Dienst Beobachtungsdienst
BDTS Bescheid-Teilnehmerschaltung
B. d. U. Befehlshaber der Unterseeboote
BdW Bootsmann der Wache
Bd. Z[dr]. Bodenzünder
BE Betriebserde; Bohreinheit
Be Beryllium
Bef Pz Befehlspanzer
Beil. Beilage
Bel. Beleg
Bem. Bemerkung
benz. benzoicum
beob. beobachtet
Ber. Berichte
ber. berechnet
Bergd Bergungsdampfer
Bergs-M Bergungsmotorschiff
Beschl. Beschluß
BesL Besetztlampe
Best. Bestand; Bestimmung; Bestellung; Bestandteil
Betr. Betrag; Betrieb
betr. betreffend
Betr. Spg. Betriebsspannung
BEV Böschungsentfernungsverhältnis
BeV, bev Billion Elektronenvolt
Bev. Bevollmächtigter
Bew. Bewaffnung
bez. bezogen
BF Bandfilter; Brikettfabrik; Feinbohreinheit
Bf Bandfilter
Bfbtl Briefbeutel
BfE Büro für Erfindungs- und Vorschlagswesen
Bfk Brikettfabrik
BfN Büro für Neuererwesen
bfn, bfn., b. f. n. brutto für netto
BG Bandgerät; Baukastengewindebohreinheit
Bg Blechschraubengewinde
Bg. Bogen
bg Bremsgitter

bgft *(Brg)* bergfeucht
Bg Pz Bergepanzer
BHC Benzolhexachlorid
Bhrg Bohrung
BHS Batteriehauptschalter
BHT- Braunkohlenhochtemperatur-
BI Birke
Bi Biot *(1 Bi = 10 Ampere)*; Wismut
bidest. bidestilliert
Bin. Binnenschiff
BK Braunkohle; biologische Kampfmittel; Kastenständerbohrmaschine; zugblank hart *(Rohr)*; Betriebskontrolle
Bk Bogheadkohle; [Elektro-]Ballistokardiograf; Berkelium; Bark
bk blank
BKE Betriebskontrolleinrichtung
BKG Elektroballistokardiografie
Bkg Ballistokardiogramm
BKH Bikarbonathärte
BKK Braunkohlenkombinat
BKL Belegungskontrollampe
BKr Bogenkreuzung
BkVA Blindkilovoltampere
BKW Braunkohlenwerk; zugblank weich *(Rohr)*
B. Kw. Beobachtungskraftwagen
BkW Blindkilowatt
BL Belegtlampe; Betriebsstofflager
B-L- Boden-Luft-
Bl Blech
Bl. Blatt *(Norm)*
bl blank
bl. blau
BLA Boden-Luft-Abwehr
Blätt. Blättchen
BLK. Blinkfeuer
BLM Blindleistungsmaschine
Bls Blaugasschmelzschweißen
BM Mehrspindelbohrkopf; Bildmischeinrichtung
Bns Benzolschweißen
Bo Benzol
BOA Bau- und Betriebsordnung für Anschlußbahnen
BOB Bau- und Betriebsordnung für Werkbahnen im Braunkohlenbergbau über Tage
bodstg. bodenständig
boric. boricum
BOZ Bled-Manifold-Oktanzahl
BP Bandpaß
BPA Bahnpostamt; Beginn der Paraffinausscheidung
BPF Bezirksdirektion für Post- und Fernmeldewesen
BPS Bremspferdestärke
Bpw Bahnpostwagen
BR Belastungsregler; Radialbohrmaschine
Br Brom; Brennschneiden; *(Schiff)* Brückenhaus
br. breit; braun; brutto
Br + Ba Brückenhaus und Back verbunden
BRD Bergungs- und Rettungsdienst
BRF Betriebslaboratorium für Rundfunk und Fernsehen
BrFl Breitflachprofil
Brk. *(Schiff)* Brücke
Brlg. Brennlänge
bromat. bromatum
brosch. broschiert
BRT Bruttoregistertonne *(1 BRT = 2,832 m^3)*
Brtkm Bruttotonnenkilometer
Brü Pz Brückenlegepanzer
BrV Brennstoffventil
BS Beobachtungsschalter; Brennschluß; Säulenbohrmaschine; Befehlsstand
B. S. Bremsstrecke
Bs Buckelschweißen
BSAO Brandschutzanordnung
BSB biochemischer Sauerstoffbedarf
Bsch Brandschiefer
BSG Elektroballistospinografie; Betriebssportgemeinschaft
BSO Blindschachtordnung
BSv Betriebsrichtungsschalter
BT Belegungstaste; Bildtelegrafie; Bildträger; Tischbohrmaschine
B/T *(Schiff)* Breiten-Tiefengrad
BTÄ Bleitetraäthyl
BTB Branchentelefonbuch
BTG Brennstofftechnische Gesellschaft
bto. brutto
BTr Bautrupp
BTT Braunkohlentieftemperatur[-Koks]
btto. brutto
BTW Bild/Ton-Weiche
BU Bandumsetzer; Buche
BÜ Befehlsübermittlung
bü bündig
Büa, B. ü. a. Breite über alles
Buck Bl Buckelblech
BUS Batterieumschalter
BV Besteckversetzung; Bandverstärker; Bauvorschrift
BVG Berliner Verkehrsgesellschaft
BvT Betriebsrichtungstaste
BW Baumwollwachsdraht; Batteriestromkreiswiderstand; Bogenweiche
Bw Baumwolle
BWVO Binnenwasserstraßenverkehrsordnung
BZ Besetztzeichen; Bedienungszeit; Brennzünder; Beilstein-Zitat; Blechschraube mit Zapfen
Bz Bronze
B. z. B. Bord zu Bord
B. z. E. Bord zu Erde
BzK Bezirkskabel
Bzl. Benzol
Bzn. Benzin

C

C Coulomb; Kohlenstoff; Kollektoranschluß; Kapazität; Curie; Celsius; Chemisches Zentralblatt; Kompression
°C Grad Celsius
c Chromatschutzschicht; Konzentration; Zenti-; Absolutgeschwindigkeit; Kolbengeschwindigkeit
CA Zelluloseazetat
Ca Kalzium
CAB Zelluloseazetobutyrat
cal Kalorie *(1 cal = 4,1868 Joule)*
calc. kalziniert
CAP Zelluloseazetopropionat
carb. carbonicum
Cat-A Katalysator-Aktivität[stest]
CB Kollektor-Basis-Schaltung *(Transistor)*
cbar Zentibar
CBL Zementbondlog
CC3P zyklisch-katalytisches 3-Phasen-Verfahren
Cd Kadmium
cd Candela
cd/cm² Candela/Quadratzentimeter
cdyn Zentidyn
Ce Zer
Cell. Zellulose
Celltriac. Zellulosetriazetat
Cf Kalifornium
CFF kritische Fusionsfrequenz des Auges
cg Zentigramm *(1 cg = 10^{-2} g)*
CGSE, cgse Zentimeter-Gramm-Sekunde elektrostatisches System
CGSM, cgsm Zentimeter-Gramm-Sekunde elektromagnetisches System
chem. chemisch
Chem. Ber. Chemische Berichte
Chemie-Ing.-Techn. Chemie-Ingenieur-Technik
Chem. Techn. Chemische Technik
Chem. Zbl. Chemisches Zentralblatt
Chlf. Chloroform
chlorat. chloratum
chloric. chloricum
Ci Curie; Zirrus
Ci-Cu Zirrokumulus
Ci-S Zirrostratus
citric. citricum
Cl Mattkohle, Clarit; Chlor; Clausius
cl claritisch; Zentiliter
Cm Curium; Coulombmeter *(1 Cm = 1 m As)*
cm Zentimeter
CMC Karboxylmethylzellulose
cmgs Zentimeter-Gramm-Sekunde
cmkg Zentimeter-Kilogramm
cm/s Zentimeter/Sekunde
cm WS Zentimeter Wassersäule
CN Zellulosenitrat; Nitratchemieseide
Co Kobalt; Kompanie
CoA Koferment A
cos Kosinus
cosh Hyperbelkosinus
cot Kotangens
coth Hyperbelkotangens
Cp Kassiopeium
cP Zentipoise *(1 cP = 10^{-2} p = 10^{-3} Ns/m²)*
cp konstantes Potential
cp. zusammengesetzt
C-Platz Speicherfernplatz
Cr Chrom
crud. crudum
Cs Zäsium
cs Zentisekunde
cSt Zentistokes *(1 cSt = 10^{-2} St = 10^{-2} cm²/s)*
Cu Kupfer; Curie
CuB Kupferdraht mit einfacher Baumwollumflechtung
CuBB Kupferdraht mit doppelter Baumwollumflechtung
CuL Kupferlackdraht
CuS Kupferdraht mit einfacher Seidenumspinnung
CuSS Kupferdraht mit doppelter Seidenumspinnung
CW Kondensator/Widerstand
CWL Konstruktionswasserlinie
cycl. zyklisch
Cys Zystein

D

D Drehstrom, Dreiphasenstrom; Dichte; Druckluftantrieb, Druckmittelantrieb; Doppelschluß; Deka-; Diode; Drahtfunk; Durchgriff; Direktor *(Antenne)*; Deuterium; Masseneffekt; Diffusionskoeffizient; Diffusionsbeiwert; Deckgebirge; Doppeltorbagger; Durit; Dekkungsbahn; Deklination; Dauerhub; Durchlauf; Drehen; Drehmaschine; Durchmesser; Durchgangsbohrung; Dampfschiff; Deck; Propellerdurchmesser; Schiffsgewicht; Deplacement in Tonnen; Deplacement auf Außenhaut in Süßwasser; Deplacement auf Außenhaut in Seewasser; Deplacement auf Spanten [in m³]; Dieselmotor; Dampfleistung; Dampfdurchsatz; Druckseite; Rohraußendurchmesser
D Division
D- rechts [konfiguriert]
d Diodenanode; Durchmesser; Dicke *(Schicht)*; Deuteron; Tag[e]; Differential; Dezi-; dynamisch; Rohrinnendurchmesser; spezifischer Dampfdurchsatz; duritisch; dünn umhüllt *(Elektrode)*; doppeltwirkend
d- rechts[drehend]
DA Dienstanruf
D.A. Dienstanweisung

Da Außendurchmesser, äußerer Durchmesser
da Deka-
DAB Dauerbetrieb mit aussetzender Belastung; Deutsches Arzneibuch
DaB Dienst an Bord
DAFuSTöMD Dienstanweisung für den Funkstörungsmeßdienst
dag Dekagramm
DAGK Deutsche Arbeitsgemeinschaft Kybernetik
DAKf Dienstanweisung für das Postkraftfahrwesen
DAL, DaL Dienstanruflampe
DAl Drahtfunkanschlußleitung
dal Dekaliter *(1 dal = 10 l)*
DAlsp Drahtfunktanschlußspeiseleitung
dam Dekameter *(1 dam = 10 m)*
DAMW Deutsches Amt für Meßwesen und Warenprüfung
D. Anw. Dienstanweisung
DAP Diallylphthalat; Deutsches Ausschließungspatent
DAPkt Dienstanweisung für den Paketdienst
Darst. Darstellung
DASch Dienstanweisung für die Postscheckämter
DAT Dienstanschlußtaste
DaT Dienstabfragetaste
DAW Deutsche Akademie der Wissenschaften
DB Dauerbetrieb; Durchlaßbereich
Db Doppelboden
dB Dezibel
DBKrW doppelte Bogenkreuzungsweiche
DBO Deutsche Bauordnung
DBR Deutsche Binnenreederei
dBW Dezibel bezogen auf ein Watt
DC Dünnschichtchromatografie
DCTL direktgekoppelte Transistorlogik
DD Dynamikdehner
DDa Drahtfunkanschaltdose
DDT Dichlordiphenyltrichloräthan
DDv Drahtfunkverzweigungsdose
DE Dielektrizitätskonstante
De Demodulator; Detektor
Deb. Debit, Debitor
dek Sonderqualität für anodische Oxydierbarkeit mit dekorativer Wirkung
dest. destilliert
DET Detektor
DF Durchgangsfernamt
Df Drahtfunk
DFF Drehfunkfeuer; Deutscher Fernsehfunk
dfg. dienstfähig
DFL Dampffischlogger
DfTln Drahtfunkteilnehmer
dg Dezigramm
DGMA Deutsche Gesellschaft für Meßtechnik und Automatisierung
DGMK Deutsche Gesellschaft für Mineralölwissenschaft und Kohlechemie
DGRS Deutsche Gesellschaft zur Rettung Schiffbrüchiger
DGW Dienstgruppenwähler
DH Diensthabender; Doppelhub
Dh Deckshaus
dH deutsche Härte
°dH deutscher Härtegrad
DHC Druck-Hydrogenium-Kracken
DHD- Druck-Wasserstoff-Dehydrierungs-
DHfK Deutsche Hochschule für Körperkultur
DHP Drahtfunkhochpaß
DHR Druck-Hydrogenium-Raffination
Di Innendurchmesser, innerer Durchmesser
DIA Deutscher Innen- und Außenhandel
Dia Diapositiv
Diam. Diamantgitter
DIC Dreieckimpulscharakteristik
dil. dilutus
DIN Deutsche Industrienorm
Diss. Dissertation
diss. dissoziiert
Dissoz. Dissoziation
Div. Division
DK Dielektrizitätskonstante; Dezimalklassifikation; Dieselkraftstoff, Dieseltreibstoff
D. K. Düsenkanone
DKB Dauerbetrieb mit kurzzeitiger Belastung
DKl Drahtfunk-Knoten[amts]leitung
DKM Drehkolbenmotor
Dkn Durchladeknopf
DKrW doppelte Kreuzungsweiche
DKS Deutsches Kontor für Seefrachten
DKv Drahtfunk-Verteilknoten
DL Dampflogger; Doppellafette
Dl Dienstleitung
dl Deziliter
DLCO dekadischer LC-Oszillator
DLH Deutsche Lufthansa
DLW Dienstleitungswähler
DM Dieselmotor; Dynamotor
Dm Drehmaschine
dm Dezimeter
DMF Dimethylformamid
Dmr Durchmesser
DMSO Dimethylsulfoxid
DMT Dienst- und Mithörleitungstaste
DMW Dezimeterwelle
dmW Dezimeterwelle
DN Deutsche Notenbank
DNB Deutsche Notenbank
DNP Dinitrophenyl
DNS Desoxyribonukleinsäure
DOP Dioktylphthalat
dopol doppelseitig poliert
Dopp. T. Doppelturm
Dos. Dosierung
dos. let. Dosis letalis

DOT. Dotierung
DP Deckpeilung; Deutsche Post; Dynamikpresser
dpt Dioptrie
DR Deutsche Reichsbahn
DRT Doppelstrom-Ruhestrom-Telegrafie
DS Doppelsternschaltung
D. S. Dampfschiff
Ds Drehstrom; Doppeltorschwenkbagger
DSB Durchlaufschaltbetrieb
DSend Drahtfunksender
DSendA Drahtfunksendeamt
DSM Deutsche Schiffsmaklerei; Doppelsternmotor
DSR Deutsche Seereederei
DSRK Deutsche Schiffsrevision und -klassifikation
DST Dünnschichttransistor
DSZ Deutsche Sommerzeit
DT Dampfturbine; Decktank; Dienstleitungstaste
Dt Schnelltriebwagen
dt Dezitonne
DTA Differentialthermoanalyse
DTG thermogravimetrisch[e Grundkurve]
DTL Dioden-Transistor-Logik
DTLZ Dioden-Transistor-Logik mit Zenerdioden
DTP Drahtfunktiefpaß
Dtz[d], Dtzd. Dutzend
DÜ Differenzübertrager
Dub. Dubiosa, Dubiosen
DÜh Drahtfunkhauptübertrager
Durchm. Durchmesser
DÜs Drahtfunksonderübertrager
DÜz Drahtfunkzapfübertrager
DV Dienstvorschrift; dielektrischer Verlustfaktor; Datenverarbeitung
DVL Drahtfunkverbindungsleitung
DW doppelte Weiche; Drahtwiderstand; Dienstwähler; Drehwähler
DWa Drahtfunkamtsweiche
DWD Deutscher Wetterdienst
DWP Deutsches Wirtschaftspatent
DWR Druckwasserreaktor
DWT Wechselstromtelegrafie mit Doppeltonmodulation
DWt Drahtfunkteilnehmerweiche
DWt/f Drahtfunkteilnehmerweiche für Freileitung
DWt/k Drahtfunkteilnehmerweiche für Kabelleitung
DWu Drahtfunkumgehungsweiche
DX Duplex
Dy Dysprosium
DZ doppeltwirkender Zweitaktmotor
DZ, D. Z. Doppelzünder
DZl Drahtfunkzuführungsleitung

E

E Energie, Einphasenstrom; elektromotorische Kraft; Elektrizität; Extinktion; Erdungsgerät mit einseitiger Erdungseinrichtung; Edison-Gewinde; elektrische Feldstärke; Erstarrungspunkt; Empfindlichkeit; Elektrolyt; Emitter, Emitteranschluß; Entzerrer; Einlauf-; Eingriffspunkt; Einzelhub; Lichtbogenschweißen; Elektrostahl; einsatzgehärtet; erhaben; Eilzug; Einlaß; Elastizitätsmodul; East, Ost; Schiff mit verstärktem Bug für Fahrt durch Eis; Eiche
E- Elektrolyt-
°E Englergrad
e Elementarentladung; Emission; Emitteranschluß; Einbauhöhe; effektiv; End-; einfachwirkend
° e englischer Härtegrad
EA elektronischer Analogrechner; Elektroaffinität; Endamt; elektrische Anlage
EA- Einanker-
EAM Einseitenband-Amplitudenmodulation
EAU Einankerumformer
EAW Elektroapparatewerke
EB Einfuhrbewilligung; Endbestand; Emitter-Basis-Schaltung
eB Edelbohrung
Ebf Einschreibbrief
EBKrW einfache Bogenkreuzungsweiche
EC Äthylzellulose
ECG Elektrokortikografie
ED [relative] Einschaltdauer; Einzeldosis
Ed. Edition
EDG Elektrodermatografie
Edg Elektrodermatogramm
EDU Energiedirektumwandlung
EDV elektronische Datenverarbeitung
EDVA elektronische Datenverarbeitungsanlage
EE Erdungsgerät mit zweiseitiger Erdungseinrichtung
Ee Elektroenzephalograf
EEF Erdnußeiweißfaser
EEG Elektroenzephalografie
Eeg Elektroenzephalogramm
EF Elektrofilter; Endfernamt
E. F. Eisenbahnfähre
eF Edelfestsitz
eff. effektiv, Effektivwert
EFL Endfernleitung
EFM Einzelkanalfrequenzmodulation
Eg Edelgas; Entgiftung
eG Edelgleitsitz
EGG Elektrogastrografie
eGmbH, e. G. m. b. H. eingetragene Genossenschaft mit beschränkter Haftpflicht
EGR Elektrogasreiniger; elektrische Gasreinigung

EGW Endamtsgruppenwähler
EH Elin-Hafergut[-Schweißen]
eH Edelhaftsitz
EHA Elektroden-Haut-Abstand
EHV Elin-Hafergut-Verfahren
EI Eiche
EIG Elektrointestinografie
Eig. Eigenschaft
Eing. Eingang
Einh Einheit
einpol. einpolig
Eins F Erste Flottille
Einw. Einwirkung
Ei. Z. Einschießziel
EK Kegelelektrode; Entnahmekondensation[sturbine]
E. K. Eisenbahnkanone
Ek Elektrokardiograf
EKB Elektrochemisches Kombinat Bitterfeld
EKG Elektrokardiografie; Elektrokardiogramm
Ekg Elektrokardiogramm
EKrW einfache Kreuzungsweiche
Eky Elektrokymogramm
EL Erkennungslicht; Endamtsleitung; enger Laufsitz; extra leicht
EL. Einstecklauf
el elektrisch
Ela Elektroakustik
ela elektroakustisch
El. A. Z. elektrischer Aufschlagzünder
Elira Elektro-Linde-Rapid[schweißen]
Elko Elektrolytkondensator
E-Lok elektrische Lokomotive
ELT Eltinor-Methode
Elt Elektrolokomotive
EM Einseitenbandmodulation; Elektromagnet
Em Emanation
EMA Eßkohlen-, Magerkohlen- und Anthrazit-[-Zechen]
E-Maat Elektrikermaat
EMD Edelmetallmotordrehwähler
EME, E. M. E., emE elektromagnetische Einheit
E-Messer Entfernungsmesser
EMG Elektromyografie
EMK elektromotorische Kraft
EMKe elektromotorische Kräfte
Empf. Empfänger; Empfang
EMT Elektromeßtechnik
emw elektromagnetische Welle
en Äthylendiamin
END Elektroneuraldiagnostik
Ends Endsatz
ENG Elektronystagmografie
Entf. Entfernung
enth. enthält
entw. entwickelt
EO Einheitsoktaeder
EOG Elektrookulografie
EOS elektrooptisches Streckenmeßgerät
EP Erweichungspunkt; Hochdruck[schmiermittel]; Endpentode; Epoxyd; effektive Pferdestärke
E. P. Englisches Patent
Ep Elektropermeagraf
ep Erstarrungspunkt
EPD elektrisch-pneumatisches Durchladeventil
EPG Elektropermeagrafie
Epg Elektropermeagramm
EP-Öle Öle für extreme Lagerdrücke
EPS effektive Pferdestärke; effektive Schlepppferdestärke
EPS_F Restpferdestärke
EPS_R Reibungspferdestärke
EPZ Eisenportlandzement
ER Empfangsrelais; Erle
Er Erbium
Erf. Erfinder
ERG Elektroretinografie
Erg Elektroretinogramm
erg Erg *(1 erg = 10^{-7} Joule)*
Erl. Erläuterung
ERS Ersatz
ES Elektroschlackeschweißen; Esche; extra schwer; Empfangssieb; Schraubenelektrode
Es Einsteinium; erzsauer, erzsaurer Typ *(Elektrode)*
eS Edelschiebesitz
ESB Einseitenband
ESE Ostsüdost
E. S. E., e. s. E. elektrostatische Einheit
ESi Einzelsicherung
ESp Echosperre
ESR Elektronenspinresonanz
ESS elektrisches Schnellschweißen
Ess. Essigsäure
EStBV Vorschriften für die technische Sicherheit und den Arbeitsschutz im Erzbergbau
E-Stelle Erprobungsstelle
ET Einheitstetraeder; Energotechnologie; Einschaltetaste am Rundschreibplatz; Eintontelegrafie; Endtriode; Einzeltiter
E-T Einzeltiter
eT Edeltreibsitz
ETB einheitliche technische Baubestimmungen
ETH Eidgenössische Technische Hochschule Zürich
ETL Emitterfolger-Transistor-Logik
ETM Einheitstrimmoment
ETT Eintontelegrafie
ETt Endtetrode
ETZ Elektrotechnische Zeitschrift
EU Einankerumformer
Eu Europium
EUG Elektrouterografie
EV Empfangsverstärker; Emaniervermögen; Energieversorgung; Endverschluß

eV Elektronenvolt
ev Evolventenfunktion
EVG Elektrovaginografie
EVT Ersatz- und Verschleißteiltypung
EVU Elektrizitätsversorgungsunternehmen
EW Einweggleichrichter, Einweggleichrichtung; Eisenwasserstoffwiderstand; einfache gerade Weiche
E. W. Entschärferwerk
eW Edelwelle
EWZ Ersatzteile, Werkzeuge, Zubehör
Ex explosionsgeschützt
Ex. P. Exerzierpatrone
exp Exponentialfunktion; Schiff versuchsweise klassifiziert
exp. experimentell
expl. explodiert
EZ Esterzahl
E-Zeichnung Entwurfszeichnung

F

F Fernmeldewesen, Fernmelde-, Fernsprechen, Fernsprech-; Fern-; Frequenzmodulation; Frequenz; Farad; Faradaykonstante; Freiluftausführung; Fremderregung, fremderregt; Filter; Faksimile; Fluor; freie Energie; Flanken-; Federung; Festsitz; Festigkeitswert; Flammofen; Fang *(Spinnerei)*; Überwasserseitenfläche; Fischereifahrzeug; Formschwerpunkt, Verdrängungsschwerpunkt, Freibord; Froudesche Zahl; Frischwasserfreibord, Frischwassergebiet; Flottille; Föhre
°F Grad Fahrenheit
F 1 Feinkornbrikett mit der Körnung 0 bis 4 mm; Frequenzumtastung
F 2 Feinkornbrikett mit der Körnung 0 bis 3 mm; Telegrafie, tönend
F 3 Feinkornbrikett mit der Körnung 0 bis 2 mm; Fernsprechen
F 4 Feinstkornbrikett mit der Körnung 0 bis 1 mm; Bildfunk
F 5 Fernsehen
F 9 gemischte Übertragungen
f Frequenz; Impuls[folge]frequenz; Heizfaden; Brennweite; Fräser-; Fasen-; fest; fein; senkrecht fallend *(Schweißnaht)*
°f französischer Härtegrad
FA Fahrabteilung; Fachausschuß; Fernmeldeamt
Fa. Firma
faa Folge-Arbeit-Arbeitskontakt *(Relais)*
FAB Fernmeldeanlagenbaubetrieb
f. a. B., f. a. b. frei an Bord, frei Schiff
Fabr. Fabrik
FaChemFa Fachgruppe Chemische Herstellung von Fasern
FAD Fernsprechauftragsdienst
FAG Fernmeldeanlagengesetz
FAKRA Fachnormenausschuß Kraftfahrzeugindustrie
FALR Fla-Abfanglenkrakete
FAM Frequenz- und Amplitudenmodulation gleichzeitig
FAR Fernrufrelaissatz
far Folge-Arbeit-Ruhekontakt *(Relais)*
Farbst. Farbstoff
FAS Fußabblendschalter; Forschungsanstalt für Schiffahrt, Wasser- und Grundbau
FAX Faksimile-Gerät
FAZ Fernanrufzeichen
FB Fernmeldebau
Fb Flockenbast
Fb. Freibord
fb. fahrbereit
FBA Farbbildaustastung; Fernmeldebauamt
FBAS Farbbildaustast-Synchrongemisch
FBD Fernmeldebaudienst
FBG Fernmeldebaugerät
fbl. farblos
FBO Fernmeldebauordnung
F-Boot Geleitboot
F.-Boot Fernlenkboot
FC Fußballklub
FCH Flottillenchef
FD Fernmeldedienst, Fernsprechdienst; Fern-D-Zug; Fischdampfer; Frischdampf; Froschdosis
Fd Durchgangsferngespräch
Fd. Faden *(= 1,642 m)*
FDGW Ferndienstgruppenwähler
FDM Fehlerdämpfungsmesser
FDNB Fluordinitrobenzol
FDSt Fernmeldedienststelle
FDt Fernschnelltriebwagen
FE Filtereinsatz
F/E Forschung und Entwicklung
Fe Eisen
Fe-Elektrode hocheisenpulverhaltige Elektrode, Eisenpulverelektrode
Ferak Feststoffrakete
FES Fachnormenausschuß Stahl und Eisen
FF Formfläche
Ff Flockenflachs
FG Fernschaltgerät; Fernsprechgerät; Feingestalt; Gewichtmomenthebel; Farbträgergenerator
Fg abgehendes Ferngespräch
FGS Fischereifahrzeug- und -gerätestation
FGt Fernschaltgerät
FGV Fernsprechgebührenvorschriften
FGW Gruppenwähler für Fernverkehr
Fh Flockenhanf
FHD Fokus-Haut-Distanz
FHZ Freihandelszone

FI Fehlerstrom; Fichte
FIA Fluoreszenz-Indikator-Adsorptionsverfahren; Forschungsinstitut für Aufbereitung
FID Flammenionisationsdetektor
Fig. Figur
Fischd Fischdampfer
Fischereif Fischereifahrzeug
Fj Fallschirmjäger
FK Feuerkohle; Fliegerkorps; ferngelenkter Körper; Formschwerpunkt über Kiel; Verdrängungsschwerpunkt über Oberkante Kiel
F. K. Feldkanone
Fk ankommendes Ferngespräch; Feldklappenschrank; Fernkabel; Faserkohle
Fk-EV Fernkabelendverschluß
Fki Feinkies
FKL Füllsender kleiner Leistung
Fkl. Funklenk-
FkO Ortsfernkabel
FKR Flugkörperrakete
FkTF Trägerfernkabel
FKtr Fischkutter
FkVL Fernkabelverlängerungsleitung
FL Fernlenk-; Fischlogger
F. L. fernlenkbar; Fernlenk-
Fl Flöz; flüchtige Bestandteile; Fernlenkung; Fernleitung; Flachs; Flachprofil
Fl. Flüssigkeit
fl flüssig
Fla Flugabwehr, Flugzeugabwehr
Fla. A. O. Flugabwehrartillerieoffizier
Flabrakete Flugabwehrrakete
FlaggRG Flaggenrechtsgesetz
Fla-Mg Fliegerabwehrmaschinengewehr
Fla Pz Fliegerabwehrpanzer
Fla-R Flugabwehr-Rakete
Fla-R-Waffe Flugabwehrraketenwaffe
Flb. Flugbahn
flb flacher Boden, platter Boden
Fle Endfernleitung
FLgr mD Fischereilogger mit Dampfmaschine
Flh Hauptfernleitung
FlHrd Flachhalbrundprofil
FlK Fernleitungskabel; Flözkohle
Flm Pz Flammenwerferpanzer
Flor. Flores
Flot. Flotation
Flp. Flammpunkt
FlS Flüssigkeitssäule
Fls Flüssiggasschmelzschweißen
FLT Freileitungstelegrafie
FLÜ Fernleitungsübertrager
Fluk Flugkörper
Flürak Flüssigkeitsrakete
FLW Leitungswähler für Fernverkehr
Fl. W., Flw. Flammenwerfer
FlZ Zubringerfernleitung
FM Frequenzmodulation; Fachnormenausschuß Maschinenbau; Fischereimotorschiff
Fm Fermium
fm Festmeter
FMA Fernmeldeamt
FMG Funkmeßgerät
FmL Füllsender mittlerer Leistung
FMQ frequenzmodulierte Quarzschaltung
FMS Funkmeßstation
FMW Fernmeldewerk
FN Fachnormenausschuß; Fernnetz
F. N. Funknavigation
FNA Fachnormenausschuß
FNA FuO Fachnormenausschuß Feinmechanik und Optik
FN BAU Fachnormenausschuß Bauwesen
FNE Fachnormenausschuß für Elektrotechnik
FNF Fachnormenausschuß Farbe
FNK Fachnormenausschuß Kunststoffe
FNM Fachnormenausschuß Materialprüfung
FNNE Fachnormenausschuß Nichteisenmetalle
FNS Fachnormenausschuß Schiffbau
FNS/E Arbeitsausschuß Elektrotechnik im Fachnormenausschuß Schiffbau
FO Fernsprechordnung; Funkortung
Fo Formungseigenschaften *(Legierung)*
Fol Folie
Fol. Folia
FP Formprofil
F. P. Festpunkt; Fusionspunkt; Französisches Patent
Fp Funkpeilgerät
Fp, Fp. Schmelzpunkt
FPG Fischereiproduktionsgenossenschaft
FPS Forschungsinstitut für Physik der Strahlantriebe
FR Fernrakete, Flüssigkeitsraketenstrahltriebwerk; Folgeregler
Fr Frankium
Fr. Flammenwerfer
fr franko, frei
fra Folge-Ruhe-Arbeitskontakt *(Relais)*
frfr. frachtfrei
FRG Fernräumgerät
Fr. Kpt. Fregattenkapitän
fro. franko, frei
FRSt Fernmelderechnungsstelle
Frt. Fracht
Fruct. Fructus
Fr. W. Frischwasser
FS Fernschreiber; Fernsehen; Fußanlaßschalter; Feinsand; Flüssigkeitssäule; Formstoff; Flachsenkform; Fährschiff
FSG Fernsehgerät
FSK Frequenzumtastung
FsOl Fernsehortsleitung
FSt Feuerstellung
F. St. Funkstelle

FSU Fernsehumsetzer
F-Sync. Farbsynchronisierung
FT Farbträger
FT, F. T. Funktelegrafie
Ft Faserkohle; Fusinit; Triazetatfolie
FTA Fernteilnehmeranschluß
FTO Funktechnischer Offizier, Funkoffizier
FTr Flugzeugträger
FT-Raum Funktelegrafieraum, Funkraum
FTZ Fernmeldetechnisches Zentralamt; Fernmeldetechnische Zeitschrift
FU Frequenzumsetzer, Frequenzumformer
FÜ, Fü Funküberwachung[sstelle]
FuA Funkamt
FuD Funkdienst
FuG Funkgerät; Futterrohrgewinde
Füllst. Füllstoff
FuM Funkmeß-; Funkmeßtechnik
FU M. B. Funkmeß- und -beobachtungsgerät
FuMG Funkmeßgerät
FU M. O. Funkmeß- und -ortungsgerät
Fu MSt Funkmeßstation
Fu Pz Funkpanzer
FuSt Funkstelle
Fusta. Funkstation
FuStöMD Funkstörungsmeßdienst
FuVO Funkvollzugsordnung
fUW fahrbares Unterwerk
FV Fachverband
FVD Fernsprechvermittlungsdienst
FVE Fernverkehrseinheit
FVF Flimmerverschmelzungsfrequenz
FVSt Fernvermittlungsstelle
FW Fesa-Weibel[-Schweißen]; Funkwerk
Fw Fehlweisung
FWA Fernwählamt
F-Wert Festigkeitswert
FWl Fernwahlleitung
FZ Freizeichen; Frequenzzeiger
FZA Fernmeldezentralamt; Fernmeldezeug[amt]; Funktechnisches Zentralamt
FZG Forschungsstelle für Zahnräder- und Getriebebau
FZZA Fernmeldezentralzeugamt

G

G Giga *(10^9)*; Guß; Gleichstrom; Gabel[schaltung]; Generator; Gewinn; Gitter; Gummiisolierung; Gewichtsdurchsatz; Gravitationskonstante; [Zahn-]Grund; Gleitsitz; Grundbohrung; Sandguß; weichgeglüht; gegossen *(Magnet)*; mit Gleitlagereinsatz; Gasschweißen; giftig; Geschwader; Gewichtsschwerpunkt, Systemschwerpunkt des Schiffes
G- gegossen
g Gramm; Gravitationskonstante; Gewindeschneiden; Grobsitz; abgehend *(Leiter)*; Steuergitter, Gitter; glatt; grob; gasförmig
g. gelb
GA Gemeinschaftsanschluß; Außengewindeschneidmaschine mit Schneidkopf; Gefechtsabschnitt
G. A. Geradlaufapparat
Ga Gabel[schaltung]; Gallium
Gal Galeasse
G-AlBz Gußaluminiumbronze
Gall. Gallone
galv galvanisch
GAMM Gesellschaft für angewandte Mathematik und Mechanik
GAN Generalauftragnehmer
g-Äqui. Grammäquivalent
GAS grenzflächenaktive Stoffe
gasf. gasförmig
GAsVr Grundstückanschlußverstärker
g-Atom Grammatom
gau Getrennt-Arbeit-Umschaltekontakt *(Relais)*
GAV Gasabgabevermögen
GB Gitterbasis[schaltung]; Gitterbatterie
gB Grobbohrung
gb gebeizt
GBa Grabenbagger
GB-AB Gitterbasis-Anodenbasis-Schaltung
GBFA Gebührenbuch für den Fernsprechauslandsdienst
GBK blankgeglüht
GBl Gesetzblatt
gbl gebläut
G. B. M. mechanische Geräuschboje
GBR gelenkte ballistische Rakete
GBT Gebührenbuch für Telegramme
G. B. T. Geräuschbojenturbine
Gbz Gebührenzettel
G-Bz Gußbronze
GC Gaschromatografie
GCH Gruppenchef
GCu Kupferguß
GD Druckguß; Gegendruck[turbine]; Geodätischer Dienst
Gd Gadolinium
GDBH Gesellschaft Deutscher Berg- und Hüttenleute
GDCh Gesellschaft Deutscher Chemiker
GDMB Gesellschaft Deutscher Metallhütten- und Bergleute
GD-Ms Druckgußmessing
GE Gewichtseinheit
Ge Germanium; Germaniumdiode; Gußeisen
ge gelb
Geb Gebühr
geb. gebunden
gef. gefällt
Gef. P. Gefrierpunkt
gegl. geglüht
Geh Gehäuse
geh. geheftet

gel. geländegängig
Gel Bl gelochtes Blech
Gel.-Fl. Geleitflottille
Gem. Gemisch
gem. geminal; gemischt
GEMK, gEMK gegenelektromotorische Kraft
Gen Generator
GEP Gegentaktendpentode
gep. gepanzert
Gepr. geprüft
gepr. gepreßt
Ges. Gesellschaft
ges gesamt, Gesamt- *(Index)*
ges. gesättigt
gesch. geschüttet; geschützt
geschl. geschlossen
Geschw. Geschwindigkeit
ges. gesch. gesetzlich geschützt
Gest Gestell
GET Gegentaktendtriode
GeV Gigaelektronenvolt
GEW Geländeentgiftungswagen
Gew. Gewicht; Gewehr
gew. gewachsen
Gew. Gran. Gewehrgranate
Gew.-HL-Gr. Gewehrhohlladungsgranate
GewO Gewerbeordnung
Gew.-Proz., Gew% Gewichtsprozent
Gew. Pzgr. Gewehrpanzergranate
Gew. T. Gewichtsteil
GF Große Fahrt
gf. gasförmig
GFK Kurzgewindefräsmaschine; glasfaserverstärkter Kunststoff
GFL Langgewindefräsmaschine
GFLS Langgewindeschlagzahnfräsmaschine
GFP glasfaserverstärkter Polyester
GFZ weichgeglüht und entzundert
GG Grauguß; Grobgestalt
GGG Gußeisen mit Kugelgraphit
ggT größter gemeinsamer Teiler
Ggw. Gegenwart; Gutgewicht
GH Hartguß; Gesamthärte
GHz Gigahertz
GI Innengewindeschneidmaschine; Gegeninduktivität
Gie Gießbarkeit
Gießh. Gießharz
GINA Fachnormenausschuß Gießereiwesen
GK Gummikabel; Kokillenguß; Gewichtsschwerpunkt über Oberkante Kiel
°GK Grad Grubenklima
gk abgehend und ankommend
Gki Grobkies und Schotter
GK-Ms Kokillengußmessing
GL Glasfaser; Gestellampe; Gleichrichter
GL, G. L. Germanischer Lloyd
Gl Geschiebelehm; Gleichrichter
Gl. Gleichung
gl glänzend
g/l Gramm/Liter
GLC Gas-Flüssig-Chromatografie
Gleichgew. Gleichgewicht
GLF Glasfaser
GliF Glimmerfeingewebeband
Glk Glanzkohle
GLS Glaskunstseide; Gasverteilungschromatografie
Glu Glutaminsäure
GLW Großsammelleitungswähler; gleichwertiger Wasserstand
Gly Glysin
Glyz. Glyzerin
glz glänzend
glzd. glänzend
GM Gebrauchsmuster
GmbH, G. m. b. H. Gesellschaft mit beschränkter Haftung
Gme Geschiebemergel
g/ml Gramm/Milliliter
G-Ms Gußmessing
GO idealgeometrische Oberfläche; Goniometer, Peiler; Gewerbeordnung
GÖV Gas-Öl-Verhältnis
GP idealgeometrisches Profil
Gp Gefrierpunkt
gp. SFL gepanzerte Selbstfahrlafette
GPV Großpumpversuch
G. Pz. Gr. Gewehrpanzergranate
Gr. Grain
gr. grün
grad Gradient
gran. granuliert
graph graphitiert
grd Grad
grenz Grenzwert *(Index)*
grL besonders große Luken
Gr. m. P. Granate mit Panzerkopf
GRR *(Text)* gekreuztes Rechts/Rechts
gru Getrennt-Ruhe-Umschaltekontakt *(Relais)*
Gr. W. Granatwerfer; Grenzwelle
GS Stahlguß; Gesamtschwefel; Grobsand; mit Gleichstrom in Schwefelsäure; Gefechtsstand; Gefechtsstation; Geleitschiff
Gs Gleichspannung; Gleichstrom; Grenzschicht; Glasseide, lackiert; Gasschmelzschweißen
gs gut schweißbar
GSA Außengewindeschleifmaschine
GSC Gasadsorptionschromatografie
GSI Innengewindeschleifmaschine
Gsk Glanzstreifenkohle
Gs. Motor Gasmotor
G-SnBz Gußzinnbronze
G-SoMs Gußsondermessing
GSt Gewindestrehlmaschine
G. St. Gefechtsstand

GSU Universalgewindeschleifmaschine
Gsu Glasseide, [lackiert] 1 × umflochten
GT Temperguß; Gewichtsteil; Gasturbine
GTA Gasturbinenanlage
GTMS Gasturbinenmotorschiff
GTS schwarzer Temperguß
gtt. Tropfen
GTW weißer Temperguß
GU Gruppenumsetzer, Gruppenumsetzung
GÜ Gebührenüberwachung
Gu Guanakowolle; Gummi
GUm Umschalter für Gemeinschaftsanschlüsse
GÜSt Gebührenüberwachungsstelle
GV Gruppenverstärker
Gv Garnverarbeitung
GVW Gemeinschaftsvorwähler
GW Gleichstrom/Wechselstrom; Gleichspannung Wechselspannung; Gruppenwähler; Grundwasser; Gleichwelle
GWh Gruppenweiche; Gigawattstunde
Gwr. Gewinderohr
Gwsp. Grundwasserspiegel
GX mit Gleichstrom in Oxalsäure
GZ Gesamtzeit; Gesprächszeit
GZM Gesprächszeitmesser

H

H Wasserstoff; Henry; Handantrieb; Hauptschluß; Heptode; Hexode; Höhenregler *(Radio)*; Hochdruck; hart; gehärtet; Härte; halbberuhigt [vergossen] *(Stahl)*; Heizwert; Hub; Haftsitz; Haupt-; hohl; horizontal; Haufwerksprobe; Seitenhöhe des Schiffes, gemallte Höhe; Förderhöhe der Pumpe
h Hekto (10^2); Stunde; Hyperbelbahn; hin- und hergehend; Höhe
h. heiß
HA Hartasphalt; ausgehärtet; Hauptpostamt
H. A. Hilfsantenne
Ha Hanf
ha Hektar; aus hartem Zustand ausgehärtet
Haf. Hafen
Haka Hafenkapitän
Hako Hafenkommandant
H. A. L. Hamburg-Amerika-Linie
HAN Hauptauftragnehmer
HAPAG Hamburg-Amerika-Linie
HAPUG Modulationsschaltung nach Harbich, Pungs und Gerth
Hauwewa. Hauptwetterwarte
HB Brinellhärte; Brinellhärtezahl; Hainbuche
H. B. Hafenbecken
Hbf. Hauptbahnhof
HBH Hauptbuchhalter
HBL Harnstoff-Bisulfit-Löslichkeit
Hbmr. Halbmesser
HCH Hexachlorzyklohexan
H-Chef Halbflottillenchef
HD Hochdruck; Duroskophärte; Heißdampf; Hauptdeck
H/D Steigungsverhältnis
Hd. Hochdruck
Hdb. Handbuch
HDL höchste Dauerleistung
HD-Öle Hochleistungsöle
HDT Hochdruckturbine
HE Hydraulikeinheit
He Helium; Henequen
HEC Hydroxyäthylzellulose
Helv. chim. Acta Helvetica chimica Acta
Herb. Herba
Herkon hermetisch abgeschlossener Kontakt
Herst. Herstellung
hex. hexagonal
HEZ Haupteinflugszeichen
HF Hochfrequenz *(3 bis 30 MHz)*; Fallhärte; flammengehärtet; Hüllfläche; Halbe Fahrt
H. F. Hauptfeuerstellung
Hf Hafnium
HF/DF Hochfrequenzpeilung
HFG Hochfrequenzgesetz *(Gesetz über den Betrieb von Hochfrequenzanlagen)*
H. F. G. Hohlstabfernräumgerät
HFL Hauptfernleitung
Hfl. Halbflottille
HFO Halbleiterwerk Frankfurt/Oder
HFR homogener Flüssigkeitsreaktor
Hfrzbd. Halbfranzband
Hfs Hyperfeinstruktur
HF-TR HF-Telefonrundspruch
HG Hauptgetriebe
H. G. Handgranate
Hg Quecksilber; Quecksilbersäule; Herstellungsgenauigkeit
hg Hektogramm
Hgi Hartgummi
hgl[z] hochglänzend
H. Gr. Heeresgruppe
HGS Hauptgefechtsstand
HGW Hauptamtsgruppenwähler
Hgw Hartgewebe
HH Hochspannungs-Hochleistungs-Sicherung
HHF Höchstfrequenz
HHI Heinrich-Hertz-Institut
HI oberflächeninduktionsgehärtet
H. I. Halbinsel
Hi-Fi hohe Wiedergabetreue
His Histidin
HK Halbedelkorund; Hinterkante; Hilfskessel; Hefnerkerze
H. K. Hilfskessel; Hilfskreuzer
Hk Hefnerkerze
HKE Hörkopfersatzschaltung
HKL Hauptkampflinie
H. K. S. Hilfskriegsschiff

HKW Heizkraftwerk
HL Hartlegierung; Halbleiter; Hauptleitung; Hilfsschlußlampe; hinteres Lot
Hl Hauptamtsleitung
hl Hektoliter; holoedrisch, vollflächig
Hlm Hefner-Lumen
Hlw. Halbleinwand, Halbleinen
Hlx Hefner-Lux
HM Hartmetall
hm Hektometer
H. M.-Boot Hilfsminensuchboot
hMe Halbmetall
H-Mg Hüttenmagnesium
HN Hauptnenner; Normalhöhe *(NN + 14 cm)*
HNA Handelsschiffbaunormenausschuß
Ho Holmium; Hilfsoszillator
Hochsp. Hochspannung
HOM Homöopathisches Arzneibuch
Homa Holzmeßanweisung
hopolf hochglanzpolierfähig
HOZ Hochofenzement
HP Hochpaß; Höhenfestpunkt; Hüllprofil; Hermann-Preßschweißen
Hp Hartpapier
HQu Hauptquerschlag
HR Rockwellhärte; Handregelung; Hohlraumresonator
HRB, HRb Rockwellhärte B
HRC, HRc Rockwellhärte C
HRD Hertzsprung-Russell-Diagramm
HRStr Hauptrichtstrecke
HS Hochspannung
Hs Hasenhaar; Hochspannung
H. S. F. Hafensicherungsfahrzeug
HSh Rückprallhärte *(Shore)*
HSi Hauptsicherung
HSl Schnellamtsleitung
Hsp Hochspantenbauart
HSS Hochleistungsschnellstahl
HST hohe Seitentanks
HSt Hilfsstelle
Hst Hintersteven
HT Hochtemperatur; Hochtiefbaggerung
HTA Haupttelegrafenamt
HTH Hochtemperaturhydrierung
HTP Hochtemperaturpyrolyse
HU hochunempfindlich
H. üb. NN Höhe über Normalnull
H. ü. M. Höhe über Meeresspiegel
HV Vickershärte, Pyramidenhärte; Handvermittlung
HVFe Hauptverwaltung für Fernmeldewesen
HVFu Hauptverwaltung für Funkwesen
HVP Hauptverwaltung für das Postwesen
HVPF Hauptverwaltung für das Post- und Fernmeldewesen
HVZL Hauptverwaltung der Zivilen Luftfahrt
HW Halbwelle; Hochwasser
hW Hektowatt
HWD Halbwertdosis; Halbwertdicke
HWF Halbwertflächengewicht
HWR Heißwasserrakete
HWS Halbwertschicht
HWZ Halbwertzeit
Hy Henry
hydrat. hydratisiert
hydrochl. hydrochloricum
hydrolyt. hydrolytisch
hygr. hygroskopisch
HZ Zeithärte, Herbert-Pendelhärte
Hz Hertz

I

I Ionisationsenergie; Inertinit; Innenraumausführung; Binnenfahrt; Trägheitsmoment; Stromstärke; Inklination
IAEO Internationale Atomenergieorganisation
IAF Internationale Astronautische Föderation
IAM Impuls-Amplitudenmodulation
IBG Internationales Büro für Gebirgsmechanik
IBK Internationale Beleuchtungskommission
IBW Innenbogenweiche
ICF Interkostalfelder
i. D. im Dampf
IE Internationale Einheit
IF Impulsformer
IFF Impulsfolgefrequenz
IFV Internationaler Fernmeldevertrag
IGJ Internationales Geophysikalisches Jahr
IKW Industriekraftwerk
IKZ Impulskennzeichen
IL intermittierendes Luftstrahltriebwerk
i. L. im Lichten
Ile Isolenzin
IM ideale Modulation; Intermodulation
i. M. im Mittel
IMU Internationale Mathematische Union
IMUK Internationale Mathematische Unterrichtskommission
In Indium
i. N. im Normalzustand
inakt. inaktiv
Ind. Indikatorstoff; Industrie
ind indiziert
Ind. Ök. Industrieökonomie
Indusi induktive Zugsicherung
INFLEX induktives Frequenzreflexionssystem
inn. innerlich
Ins. Insel
int. international
IN-Verfahren Indanthrennormalfärbeverfahren
Invest Investition
IO Istoberfläche
Io Ionium
IPBM Papierbaumwollkabel mit Bleimantel

IPS indizierte Pferdestärke
IR Infrarot
Ir Iridium
I. S. internationales Segelbuch
ISA Internationale Standardatmosphäre
ISO Internationale Standardorganisation
Is. V. Isotopieverschiebung
IT Impulstelegrafie; Irrungstaste; Integration
iT innerer Totpunkt
i. T. in der Trockenmasse
i. Tr. in der Trockenmasse
ITU Internationale Vereinigung für Fernmeldewesen
IU Impulsunterdrücker; Induktorumschalter
IÜG Internationales Übereinkommen über den Güterverkehr
i. V. innere Verbindung
IVC Industrievereinigung Chemiefaser
IWS Internationales Wollsekretariat
IZ Ingenieurtechnische Zentralstelle; Internationale Zusammenarbeit

J

J Joule; Jod
JFK Jagdfliegerkräfte
jodat. jodatum
J. prakt. Chem. Journal für praktische Chemie
Ju Jute
JZ Jodzahl

K

K Kalium; Kalorie; Katode; Klirrfaktor; Kopplungsfaktor; Kopfkontakt an Hebdrehwählern; Kreuzmodulationsfaktor; Kabel [mit Bleimantel]; Kaltelektrode; Kurzwelle; Konstante; Kenngröße; Kegelkuppe; Kegelbohrung, keglige Bohrung; kaltverformt; Kurbelwelle; Kerbzähigkeit; Kohle; Kutikula; Köper; Kaliber; Kanone; Kapitän; Kommandant; Kolben; Kompressor; Kondensat
°K Grad Kelvin
k Katode; Konstante
k. kalt
KA Knotenamt; Kasein
K-A Katode-Anode
Ka Abfrageklinke, Amtsklinke
kA Kiloampere
KAF Kaseinfaser
K. A. K. Küstenabschnittskommando
KAL Kanalanruflampe
Kal. Kalium; Kaliber
Kan. Kanone
Kap Kapazität
KAR Kleinanalogrechner
kart. kartoniert
KAS Kaseinkunstseide
kat. katalytisch
KB Katodenbasisschaltung; Katodenbasisstufe; kurzzeitiger Betrieb, Kurzzeitbetrieb; Kernbohrprobe; Kraftstoffbehälter; Konstruktionsbüro; Kamerabediengerät
Kb Klinke für besondere Leitungen, kalkbasischer Typ *(Elektrode)*
KB-AB Katodenbasis-Anodenbasis-Schaltung
KBAV Kurzschließ- und Bürstenabhebevorrichtung
KB-GB Katodenbasis-Gitterbasis-Schaltung
Kblg. Kabellänge *(1 Kblg. = 185,2 m)*
Kbt. Kanonenboot
KBV Vorschriften für die technische Sicherheit und den Arbeitsschutz im Kali- und Erzbergbau
Kc Kupferkunstseide
kcal Kilokalorie
(1 kcal = 10^3 cal = 4186,84 Joule)
kCi Kilocurie
kCu Kilocurie
Kd Dienstleitungsklinke
Kda Dienstabfrageklinke
KdT Kammer der Technik
Kdt. Kommandant
kdyn Kilodyn
(1 kdyn = 10^3 dyn = 10^{-2} Newton)
KE Kohlenelementaranalyse; Kühlmitteleinheit
Ke Kohleelektrode
KEG Kabelendgestell
keg kegelig
KER Keramik
Kess. Kessel
KEV Kabelendverteiler
keV Kiloelektronenvolt
KF Karl-Fischer[-Lösung]; Korrosionsfestigkeit; Kleine Fahrt
K. F. Kommutationsfaktor
Kf Fernleitungsklinke; Kuff
KFK Kunstfeinkorn
K. F. K. Kriegsfischkutter
KFL Karl-Fischer-Lösung
Kfl. Kühlfläche
KFLR Kurzstrecken-Fla-Lenkrakete
KfR Kommission für Raumfahrttechnik
K. F. R. G. Kabelfernräumgerät
Kfrg. Kesselfeuerung
KFS Kugelfunkenstrecke
Kf-Tlk Kuff-Tjalk
Kfw. Kampfwagen
Kfz. Kraftfahrzeug
KG Kaltgas; Gewichtsschwerpunkt über Kiel
Kg Kühlgas
kg Kilogramm
kgcm Kilogrammzentimeter
Kgg Gegenkurs
kg/kWh Kilogramm je Kilowattstunde
Kgl Konglomerat

kgV kleinstes gemeinsames Vielfaches
KGW Knotenamtsgruppenwähler
KH Kunstharz[preßstoff]; Karbonathärte
Kh Hilfsklinke
KHA Azetatkunststichelhaar
K. H. S. Kennely-Heaviside-Schicht
KHV Viskosekunststichelhaar
kHz Kilohertz
KI Kiefer
KIB Konstruktions- und Ingenieurbüro
kJ Kilojoule
KK Kleinkaliber; Kraftzug-Kanone; Kompaßkurs; Kamerakopf
K. K. Kraftwagenkolonne
Kk Kännelkohle
KKM Kreiskolbenmaschine
Kkw. Krankenkraftwagen
KL Kondensatorleitung; Kontrollampe
Kl Leitungsklinke; Klinke, Abfrageklinke; Klirrfaktor; Knotenamtsleitung
kl Kiloliter
klm Kilolumen
KM Kontrollmonitor; Kriegsmarine
K. M. Kriegsmarine
Km Kolbenmotor; Mithörklinke; Kamelhaar[wolle]
km Kilometer
km/h Kilometer/Stunde
KML Kurz-, Mittel- und Langwellenbereich
KN Kerbnagel; Keilfeder-Nut; Kartennull
Kn Kaninhaar; Kraftstoffnormverbrauch
Kn. Kanone
kN Kilonewton
kn Knoten, Seemeilen in der Stunde *(1 kn = 1 sm/h = 1852 m/h)*
Ko Kokos[faser]; Korrosionsbeständigkeit gegen Seewasser; Ortsklinke; Kompression
Koeff. Koeffizient
KOM, Kom Kraftomnibus
Komm. Kommission
Kon. Adm. Konteradmiral
Konst[.] Konstante
konst konstant
Konz. Konzentration
konz. konzentriert
Kopo. Kopolymerisat
Kor. Kpt. Korvettenkapitän
korr. korrigiert
KP Kompaßpeilung
Kp Kristallisationspunkt; Kochpunkt; bespultes Kabel; Prüfklinke; Kapok; Kolbenpumpe
Kp. Kochpunkt, Siedepunkt
kp Kilopond
Kpfw Kampfwagen
kpm Kilopondmeter
KpN Kompaßnord
KPV Kurzprüfverfahren; Koppers-Pottascheverfahren; Kurzpumpversuch
KR Korrekturregler
K. R. Kesselraum
K.-R. Korrespondentreeder
Kr Krarupkabel; Kreuzung; Krypton
KRA Azetatkunstroßhaar
Krh Kreuzerheck
Krist. Kristalle; Kristallisation
krist. kristallin; kristallisiert
kristsch kristallinisch
KrK Kreiselkompaßkurs
KrP Kreiselkompaßpeilung
KRV Viskosekunstroßhaar
kryoskop. kryoskopisch
Krz. Kreuzer
KS keramische Schneidstoffe; Kerbstift; kosmische Strahlung; [Erweichungspunkt nach] Kraemer-Sarnow; Kühlstärke; Kuppelstelle; Kampfstoff; Küstenschnellboot; Kranschiene
Ks Schnellklinke
KSA Azetatkunstseide
K. S.-Boot Küstenschnellboot
KSK Kupferkunstseide
KSN Nitratkunstseide
KSP Kegelschmelzpunkt
KSS Küstenschutzschiff
Kst Störungsklinke
Kst. Küste
Kstb. Kohlenstaub
Kstfrg. Kohlenstaubfeuerung
Kstr Kohlestreifen
KSV Viskosekunstseide
KT Kreiseltheodolit; Kamerakabeltrommel
Kt Teilnehmerklinke
kt Kilotonne
ktex Kilotex *(1 ktex = 1 kg je 1000 m)*
Ktr Kutter
KU Kanalumsetzer; Kaskadenumformer
KÜ Überwachungsklinke
Kü Überweisungsleitungsklinke
kub. kubisch
Kümo Küstenmotorschiff
künstl. künstlich
KV Kabelverzweiger, Kanalverstärker
Kv Verbindungsklinke
kV Kilovolt
kVA Kilovoltampere
kvar Kilovar
KVD Kurzhub-Viertakt-Dieselmotor
KVO Kraftverkehrsordnung
KW Kohlenwasserstoff; Kraftwerk; Kurzwelle; Kurbelwelle; Kurbelwinkel
°KW Grad Kurbelwinkel
K. W. Kartenwinkelmesser
kW Kilowatt
kWB Kilowatt Bildleistung
kWh Kilowattstunde
Kw. K. Kampfwagenkanone
KWL Konstruktionswasserlinie

kWT Kilowatt Tonleistung
KZ Kennziffer; Koordinationszahl; künstlicher Zug
Kz Zuschalteklinke; Kerze
KZD Kurzhub-Zweitakt-Dieselmotor

L

L Lager; Laufsitz; Leichtmetall; Lot; Linsenkuppe; Länge; Lambert; Lautstärkeregler; Leuchtschirm; Langwelle; lackisoliert; laugenrißbeständig *(Stahl)*; Löslichkeit; Ligand; Lehm; Librationszentrum; Leinwandbindung; Ladelinienlänge; Arbeit; Luftmenge
L. Lieferung; Ladung
l Lumen; löslich; Leuchtschirm; Liter; Leitspindel; Längung [beim Drahtziehen]; spezifische Arbeit; spezifischer Luftverbrauch
l- links[drehend]
LA Logablese
La Lampe; Anruflampe; Überwachungslampe; Lanthan
Lä Längstwelle
lab. labil
lact. lactis
Lad. R. Laderampe; Laderaum
LAg Silberlot
LAH. Lithiumaluminiumhydrid
LAl Aluminiumlot
LAS Abblendschalter
LAT Leitungsaufschalttaste
LAW Leitungsaufschaltewähler
L/B Schlankheitsgrad der Kielwasserlinie
L-B- Luft-Boden-
L:B Längenverhältnis eines Schiffes
Lb Logberichtigung; Luftpostbriefsendung
LBF Lichtbogenfestigkeit
LBS Blinkschalter
LD Liniendrossel; Letaldosis; Leerlaufdüse
L. D. Lotsendampfer
Ld. Land
l. D. lichter Durchmesser
Ldg. Ladung
L. d. H. *(Schiff)* Länge des Hinterraumes
L. d. V. *(Schiff)* Länge des Vorraumes
LE Leitungsentzerrer; Ladungseinheit; Elektrostahl
Leg. Legierung
Leu Leuzin
LF Langsame Fahrt; Leistungsschalter
Lfb. Laufbahn
lfd. m laufendes Meter
lfd. Nr. laufende Nummer
LFF Landefunkfeuer
Lfg. Lieferung
LFK Lenkflugkörper
L. FL. Lehrflottille
LFLR Langstrecken-Fla-Lenkrakete
Lfv. Leuchtfeuerverzeichnis
LG leicht geglättet
Lg Lager; Länge
Lg. Ligroin
lgcpl Logarithmus complementi
Lgeh luftgehärtet
LgPbSb Lagerhartblei
Lgr Logger
Lgs Leuchtgasschweißen
LGW Leitungsgruppenwähler
L/H Längen-Höhengrad
l. H. lichte Höhe
l/h Liter je Stunde
LI Leitender Ingenieur; Linde
Li Lithium
Lign. Lignum
lim Limes, Grenzwert
linksdr. linksdrehend
linksg. linksgängig
Liqu. Liquor
LK Luftkühlung
LKD Leichtkassettendecke
Lkl. Leistungsklasse
LKV Lochkartenverfahren
LKW, Lkw Lastkraftwagen
LL leichter Laufsitz; Links/Links; Länge zwischen den Loten
L-L- Luft-Luft-
Ll Länge zwischen den Loten
ll., l. l. leicht löslich
L. L. C. Ladelinienkomitee
LLS Linienleitergleisschleife
Lm Lamawolle
lm Lumen
lMg leichtes Maschinengewehr
LMK Lang-, Mittel- und Kurzwellen; leichtes Marinekabel
LMKU Lang-, Mittel-, Kurz- und Ultrakurzwellenbereich
LMs Messinglot, Schlaglot
LMsAg Silber-Messing-Lot
lm/W Lumen/Watt
Lö Löß
Lok Lokomotive
Löl Lößlehm
Lös. Lösung
lösl. löslich
Lösungsm. Lösungsmittel
LP Lackpapierdraht; Linearprogrammierung
LPA Leitpostamt
LPb Bleilot
LPF Lauge-, Fäll- und Flotationsprozeß
LPV Langpumpversuch
LR Lurgi-Ruhrgas[-Verfahren]
LR, L. R. Lloyd's Register of Shipping
LS Lautsprecher; Leitungsschutzsicherung; Luftschutz
L. S. Luftschutz; Lastensegler

Ls Lichtbogenschweißen; Lichtbogen, stabilisierender Typ
LSD Lysergsäurediäthylamid
Lsg. Lösung
Lsgsm. Lösungsmittel
LSK Luftstreitkräfte
LSn Zinnlot
L-Stand Leitstand
Lstd. Leitstand
LT Brieftelegramm des Inlanddienstes; Leerlaufprüftaste
lt. At. m. U. laut Arzneitaxe mit Umsatzsteuer
LtBz Leitbronze
Ltg Leitung
LTK Kanallampentaste
Ltkm Leistungstonnenkilometer
LtPV Leistungspumpversuch
LTS-Boot Leichttorpedoschnellboot
LTÜ Übergabelampentaste
LU Langsamunterbrecher
Lu Lutetium
Lüa, L. ü. a. Länge über alles
Luvo Luftvorwärmer
LV Linienverzweiger; Leistungsverstärker; Luftvorwärmer
LW Langwelle *(30 bis 300 kHz)*, Langwellen-; Leitungswähler
Lw Lawrencium
l. W. lichte Weite
Lwk Leitwerk
LWL Lade-Wasserlinie, Länge in der Wasserlinie
LX Luxustelegramm
lx Lux
lxs Luxsekunde
Ly Lysin
LzL Länge zwischen den Loten
LZnAl Aluminium-Zink-Lot

M

M Mega *(10^6)*; Maxwell; Metazentrum; Masse; Massenzahl; Machsche Zahl; Modulator; Mikroton; Molekulargewicht; Molmasse; Metall; Mithören; Maß [der Bandbreite]; höchste Modulationsfrequenz; Maschine; Maschinenbau; metrisches Gewinde; Motor; Motorantrieb; Moment, Drehmoment; Modell; magnetisiert; Siemens-Martin; Mugelkohle; Mikrinit; Mast; Motorschiff
M. Marine; Meile; Meereshöhe
M-% Masseprozent
m Milli *(10^{-3})*; mittlerer; mittel; mitteldick umhüllt *(Elektrode)*; matt; Meter; Meile; Masse; Massendurchsatz; molar; magnetische Quantenzahl; äußerer Abschirmmantel; Modulation[sgrad]; Modul
m- meta-
Ma Machzahl; Masurium
mA Milliampere
maf im Gleichgewichtswassergehalt und aschefrei
Magn. Magnesium
MAH Mahagoniholz
MAK maximale Arbeitsplatzkonzentration [gesundheitsschädlicher Stoffe]
Ma. N magnetisch Nord
Mar. Marine
marit. maritim, zur See gehörig
MAS Materialausgabeschein; Maschinenausleihstation
MaS Makrospore
mAs Milliamperesekunde
Maßst. Maßstab
max, max. maximal; Maximum
MAZ magnetische Aufzeichnung; Meldeanrufzeichen
M_B Breitenmetazentrum
mb[ar] Millibar
MBF Meßempfänger mit Bandfiltern
Mbl Meßtischblatt
M-Boot Minensuchboot
Mcal Megakalorie
mcd Millicandela
MD Maximal-Einzel-Dosis; mechanisierte Division; Mitteldruck
Md Mitteldruck; mittlere Korngröße; Mendelevium
ME Masseeinheit[en]; Masseneinheit, Mengeneinheit; Mache-Einheit *($1\ ME = 3{,}64 \cdot 10^{-10}\ Cu/l$)*
Me Metall[e]; Mergel; Methyl; Metallelektrode
me metallisiert
Mech. Mechanik
mech. mechanisch; mechanisiert
MEK Mehrfacheinzelkanal; Methyläthylketon; Meßempfänger mit Einzelkreisen
Mer. Meridian
M. E. S. magnetischer Eigenschutz
Met Methionin
met. metallisch
Meth. Methode
MeV Millionen Elektronenvolt, Megaelektronenvolt
MEZ mitteleuropäische Zeit
MF Mittelfrequenz; Melaminformaldehyd; Metazenterhöhe über Verdrängungsschwerpunkt
M. F. P. Marinefährprahm
MFS Motorfährschiff
MfS Mitteilungen für Seefunkstellen
MG Molekulargewicht; Massenwirkungsgesetz; Mehrfachgruppen-; Maschinengewehr; metazentrische Höhe, Metazenterhöhe
M. G. Molekulargewicht
Mg Magnesium; Maschinengewehr
mg Milligramm; mittelgrob

mGal Milligal
MGG Mehrfachgrundgruppen-
MgH Magnesiahärte
mgh Milligrammstunde
MGL Maschinen- und Geräteliste
MGZ Maschinengewehrzieleinrichtung
M. G. Z. mittlere Greenwichzeit
MH Mikrohärte; Mohs-Härte
Mh Maschine im Hinterschiff
mH Millihenry
MHD magneto-hydrodynamisch
M. H. W. Mittelhochwasser
MHw mittleres Hochwasser
MHz Megahertz
mHz Millihertz
Mi Mikrofon
MIG Metall-Inertgas-Schweißen
MIK maximale Immissionskonzentration
mikrochem. mikrochemisch
Min. Minute
min, min. Minute; Minimum; minimal
mind. mindestens
Mio. Million
Mipo. Mischpolymerisat
mischb. mischbar
Mivometer Millivoltmeter
Mixt. mixtura
MK Morsekegel; Maschinenkanone; Magnetisierungskennlinie; mittlere Korngröße; Marinekabel, Schiffskabel; Metazenterhöhe über Oberkante Kiel
Mk Mattkohle
MKG mittlere Korngröße
Mki Mittelkies
mkl. monoklin
MKO Marinekabel ohne Bewehrung, Schiffskabel ohne Bewehrung
mkp Meterkilopond
MKS Meter-Kilogramm-Sekunde
ML Motorluftstrahltriebwerk
M_L Längenmetazentrum
mL Millilambert
ml Milliliter
MLP Minenlegepanzer
mm Millimeter
Mmin Mannminute
MMK magnetomotorische Kraft
mMol Millimol
mm/U Millimeter je Umdrehung
MMW Meldemischwähler
mm Hg Millimeter Quecksilbersäule
mm WS Millimeter Wassersäule
Mn Mangan
mNp Millineper
MNw mittleres Niedrigwasser
MO Megaohm
Mo Molybdän; Mohair; Metalloid
Mod-Verst. Modulationsverstärker
MoGe Motorgenerator
Mol Molekül
mol molekular
mol. molekular, molar
Mol.-G[ew]. Molekulargewicht
Mol.-Refr. Molekularrefraktion
Mom. Moment
mon. monoklin
Mot, mot Motor
mot. motorisiert
MOZ Motoroktanzahl
MP Mischpolymerisat; Mineralölprodukt; Metallpapier
Mp Mittelpunkt, Sternpunkt; Megapond
m. P. mit Panzerkopf
Mpc Megaparsec
mph Milliphot
MPI Max-Planck-Institut
MPi Maschinenpistole
MPko Metallpapierkondensator
MR Multiplikatorregister; Morsekode
M. R. Maschinenraum
mR Milliröntgen
Mrd. Milliarde
MRM magnetischer Ringmodulator
MRS Minenräumschiff
MS Mann und Schicht, Mannschicht; Mittelsand; Mikrosekunde
MS, M. S. Mitte Schiff; Motorschiff
M. S. Monate nach Sicht
Ms Messing; Mitsprechen; Maulbeerseide, echte Seide
ms Millisekunde
ms- meso
m/s Meter je Sekunde
msb Millistilb
msec Millisekunde
Msk Mattstreifenkohle
MSR Messen-Steuern-Regeln
MSt Meßstöpsel
M_{St} Stabilitätsmoment
mSt Millistokes
MsTh Mesothorium
MT Metall; Metalldraht; Magnetton; Mithörtaste; Mittelfrequenztelegrafie; Mahltrocknungsanlage
Mt. Maat
mt matt
MTB Mitte der Telegrafenbatterie
MTD. maximale Tagesdosis
MTF Metalldrahtfaser
MTG Magnettongerät
Mtr. Matrose
mtr. metrisch
MTS Maschinen-Traktoren-Station
MTW Mannschaftstransportwagen
Mu Mutter
mul mit dem Uhrzeiger laufend
Mun. Munition

MV Megavolt; Minutenvolumen; Multivibrator; verpreßte Masse; Mischungsverhältnis; mit magnetischer Vorzugsrichtung; Magnetverstärker
mV Millivolt
MVA Megavoltampere
mval Milligrammäquivalent
MVAr, MVar Blindmegavoltampere
Mvar Megavar
MVB Multivibrator
MVg Magnetventil für Gas
MVN Materialverbrauchsnorm
MW Megawatt; Mittelwelle; Mischwähler
M. W. Mittelwasserstand; Minenwerfer
M. W., Mw Mittelwasser
Mw Mißweisung
mW Milliwatt
M. W. G. Massenwirkungsgesetz
MWh Megawattstunde
mwK, mw Kurs mißweisender Kurs
mwN mißweisend Nord
mwP mißweisende Peilung
MZ Maschinenzeit; Massenzahl
MZG Mehrzweckgleisunterhaltungsgerät
MZK maximal zulässige Konzentration

N

N Neper; Newton; Nebenschluß; Netzwerk; Norm; Stickstoff; Normalität; Neutronenzahl; Nachnahme; Nenn-; Normal-; Nutz-; normalgeglüht; mit Ringnut; Nimbus; Umdrehungen je Minute
N, N. Nord, Norden
n Nano *(10^{-9})*; Neutron; normal; Pegel; Umdrehungen je Minute; neu, erneuert; Ausnutzungsfaktor
n. nördlich; netto
NA Nebelanlage
Na Natrium; Nachbildung; Nadellager
NABE Aluminiumnulleiter mit Bleimantel für Erdverlegung
Nachw. Nachweis
NAE Aluminiumnulleiter für Erdverlegung
NAN Nachauftragnehmer
Natr. Natrium
naut. nautisch
nav. naval
Navig. Navigation
Nawi nicht abwickelbare Membran
NB Nußbaum
Nb Nebelgeschoß; Niob
NBE Nulleiterdraht mit Bleimantel für Erdverlegung
NBK normalisierend blankgeglüht
n. Br. nördliche Breite
NBÜ Nachbrückenübertrager
NC- Nitrozellulose-
Nchbr. Nachbrenner
nCi Nanocurie
nCu Nanocurie
ND Niederdruck, Nenndruck
ND- Niederdruck-
Nd Neodym; Niederdruck
Nd. Niederschlag
NDL Norddeutscher Lloyd
ND-Turbine Niederdruckdampfturbine
NE Nichteisen; Nebensprecheinheit; Nulleiterdraht für Erdverlegung; Nachentzerrung; Nachrichteneinheit
Ne Neon
NEF Normalgewinde extra fein
neg. negativ
nep Neper
N. E. S. Nachterkennungssignal
neutr. neutral
NF, Nf Niederfrequenz
nF Nanofarad
nf naturfarben
NF-F Niederfrequenzfernsprechen
NFK Naturfeinkorn
N. f. S. Nachrichten für Seefahrer
NFZ normalisierend geglüht und entzundert
ng Nanogramm
NGl Netzgruppenleitung
NGlH Netzgruppenhauptleitung
NGlQ Netzgruppenquerleitung
NGM Netzgruppenmittelpunkt
NH Normalhöhenpunkt
Nh Naturharz
N. H. G. Navigationshorchgerät
Ni Nivellier; Nickel
Nife Nickeleisen
nitric. nitricum
nitros. nitrosum
N. J.-B. Nautisches Jahrbuch
NK Normalkorn[brikett]; Normalkorund
NKGG Nationalkomitee für Geodäsie und Geophysik
NKH Nichtkarbonathärte
NL Nachruflampe; Nulleiterdraht [nicht für Erdverlegung]
NL- Nachlinks- *(Schweißen)*
Nl Netzwerk für Leitungsabschluß
nl., n. l. nichtlöslich
NLA Aluminiumnulleiter [nicht für Erdverlegung]
NLC Kupfernulleiter [nicht für Erdverlegung]
Nm Newtonmeter *(1 Nm = 1 J = 1 Ws)*
nm Nanometer
N. M. f. S. nautische Mitteilungen für Seefahrer
NMP N-Methylpyrrolidon
NN Normalnull
NNO Nordnordost
NNW Nordnordwest
NO Nordost

NÖ Neutralöl
No Nobelium
Norw. P. Norwegisches Patent
n. o. T. nach dem oberen Totpunkt
Not. F. Notfeuer
NOW Niederschachtofenwerk
N. P. Nullpunkt
Np Neper; Neptunium
N-P-N, n-p-n negativ-positiv-negativ *(Halbleiter)*
NPT Netzplantechnik
NR- Nachrechts- *(Schweißen)*
nr. nichtrostend
NRT Nettoregistertonne
NS Nummernschalter; Normalschliff; Schiff mit Kernenergieantrieb
Ns Neusilber; [Rollen-]Nahtschweißen; Newtonsekunde
ns Nanosekunde; nichtschweißbar
NSA Nachtsignalapparat
nsa Nummernschalterarbeitskontakt
nsi Nummernschalterimpulskontakt
nsl Nummernschalterleerlaufkontakt
Ns Pz Nachschubpanzer
nsr Nummernschalterruhekontakt
NT nitriert; Neue Technik; Nadelton
N. T. nautische Tafel
Nt Netzwerk zum Leitungsabschluß beim Abtrennen
NTE Nitroloessigsäure
NTF Nachrichtentechnische Fachberichte
NTG Nachrichtentechnische Gesellschaft
NTK negativer Temperaturkoeffizient
nto netto
ntto netto
NTU Nevada-Texas-Utah[-Vergasungsverfahren]
NTZ Nachrichtentechnische Zeitschrift
Ntzl. Nutzlast
NÜ Nachübertrager
Nu Nummernschalter; Nusselt-Zahl
n. u. T. nach dem unteren Totpunkt
NV Niederfrequenzverstärker
NVD Normalhub-Viertakt-Dieselmotor
NW Nennweite; Nordwest; niedrigster Wasserstand
N. W. Niedrigwasser
NWO Nord-West-Ölleitung
nx Nox
NZ Neutralisationszahl; Normalzeit; Normzahl
NZD Normalhub-Zweitakt-Dieselmotor

O

O Sauerstoff; Oszillator; Oberkasten; Ost
Ö öffentliche Sprechstelle; ölbeständig
o- ortho-
OA Ortsamt; Organisationsanweisung
OB Ortsbatterie, Oberste Bergbaubehörde
OBA Oberbergamt
OBB Oberste Bergbehörde
OBd oberes Band
ÖBF Örtliche Bestimmungen für den Fahrbetrieb
Obv-Br Oberbauvorschrift für Werkbahnen im Braunkohlenbergbau über Tage
öd. öldicht
ODGW Ortsdienstgruppenwähler
OE Öleinheiten
OEG obere Explosionsgrenze
OEL Ortsempfangsleitung *(Radio)*
OEl örtliche Endamtsleitung
OEZ osteuropäische Zeit
OFLW Orts- und Fernleitungswähler
OFN Ortsfernsprechnetz
ÖGAAnl öffentliche Gemeinschaftsantennenanlage
Ögeh. ölgehärtet
O. G. G. Ottergeleitgerät
OGW Ortsamtsgruppenwähler
OK Vergaserkraftstoff, Ottokraftstoff; Oberkante
Ok Ortskabel
OKK Oberkante Kiel
OL Ortsverbindungsleitung
Ol. Oleum
ö. L. östliche Länge
OLÜ Ortsleitungsübertrager
OLW Leitungswähler für Ortsverkehr
ON Ortsnetz
ONO Ostnordost
ONW Ortsnetz mit Wählbetrieb
OPD Oberpostdirektion; Operativdienst
Opr Operationsregister
opt.-akt. optisch aktiv
opt.-inakt. optisch inaktiv
Op. Z. Operationszentrale
or. orange
ORG, O. R. G. Otterräumgerät
org. organisch
Org. Anw. Organisationsanweisung
Os Osmium
osgl optisch-spiegelglänzend
OSL Ortssendeleitung *(Radio)*; örtliche Schnellverkehrsleitung
OSL-m örtliche Schnellverkehrsmeldeleitung
OSL-v örtliche Schnellverkehrsvermittlungsleitung
OSO Ostsüdost
Osz., osz Oszillator
OT, oT, o. T. oberer Totpunkt
OTR operativ-taktische Rakete
OÜl örtliche Überweisungsleitung
OÜl-m örtliche Meldeleitung
OÜl-v örtliche Fernvermittlungsleitung

OVL Ortsverlängerungsleitung
OVSt Ortsvermittlungsstelle
OW Öl-in-Wasser *(Emulsion)*
oW ohne Antriebswelle
Ox oxydischer Typ *(Elektrode)*; anodische Oxydierbarkeit mit dekorativer Wirkung
ox Oxalat
Oxyd. Oxydation
oxyd. oxydiert; oxydatum
Oxydationsm. Oxydationsmittel
OZ Oktanzahl; Ordnungszahl; Oximzahl; Ölzahl; Ortszeit

P

P Phosphor; Poise; Preßsitz; Preßluft; poliert; preßschweißbar; Permanenterregung, permanenterregt; Pegel; Pegelbuchse; Pentode; Punktprobe; Panama; Parkplatz; *(Schiff)* Hütte, Poop
p Pond; Pico *(10^{-12})*; pyritisch; Proton; paarig; Pegel; Polpaarzahl
p- para-
PA Polyamid; Pappel; Postamt
Pa Protaktinium
p. a. pro analysi; per annum
PAB p-Aminobenzoesäure
PAF Polyamidfaser
Pag Papier, getränkt
Pak Panzerabwehrkanone
PAL Belegtlampe für den Platzanschaltwähler
PAM Pulsamplitudenmodulation
PAN Polyakrylnitril
PAnw Postanweisung
PAS Polyamidkunstseide; p-Aminosalizylsäure
PAT Platzanschalttaste
Pat. Patent
pat. patentiert
PatG Patentgesetz
Pau Papier, ungetränkt
PAW Platzanschaltwähler
PB Polybutadien
Pb Blei
PBS Platzbeobachtung
PBT Platzbezeichnungstaste
PBW Platzbezeichnungswähler
PC Polykarbonat; Papierchromatografie
PC- Polyvinylchlorid-
pc Parsec, parsec
PCM Pulskodemodulation
pcm Pondzentimeter
PCU Polyvinylchlorid, unchloriert
PD Potentialdifferenz
P. D. Panzerdurchschlagbombe
Pd Palladium
PEF Polyesterfaser
Pe. G. Peilgerät
perm-dyn permanent-dynamisch
Per/Sek Perioden/Sekunde
PES Polyesterkunstseide
PETP Polyäthylenterephthalat
PF Phenolformaldehyd; Perlfang; Panzerförderer
P. F. Personenfähre
pF Picofarad
PFA Postfuhramt
Pfd, Pfd. Pfund
Pfl. Pflanze
Pg Stahlpanzerrohrgewinde
Pgt Postgut
Pgtk Postgutkarte
Ph Phenyl
pH Picohenry; Wasserstoffexponent
ph Phot; Phosphatschutzschicht
pharm. pharmazeutisch; pharmazeutischen Anforderungen genügende Reagenzien
pharmak. pharmakologisch
Pharmaz. Pharmazie
pharmaz. pharmazeutisch
Phas Phasenschieber
Phe Phenylalanin
Phenolh. Phenolharz
phosphor. phosphoricum
phr phosphatrostgeschützt
phys.-chem. physikochemisch
PI proportional-integral wirkend
PKB Projektierungs- und Konstruktionsbüro
PKM Projektierungs-, Konstruktions- und Montagebüro
Pko Papierkondensator
Pkt Paket
Pktk Paketkarte
PKW, Pkw Personenkraftwagen
PL Prüflampe; Platzlampe
Pl Platz, Vermittlungsplatz
pl plattiert
plast. plastisch
PlB Platzbeobachtung
plötz abgeplötzt, in Wasser abgeschreckt
Plst. Pleuelstange
PLT Platztastenlampe
PLU Platzumschalter
PM Paraffinmasse; Permanentmagnet; Phasenmodulation; Pulsmodulation
Pm Promethium
pm Picometer; Pondmeter
PMMA Polymethylmethakrylat
PN Paßfeder-Nut
P-N positiv-negativ *(Halbleiter)*
Pn Päckchen
p-n positiv-negativ *(Halbleiter)*
P-N-P, p-n-p positiv-negativ-positiv *(Halbleiter)*
PO Polyolefin
Po Polonium
Pol Polierbarkeit
polym. polymerisiert
Pont. Ponton

POS, pos. positiv
PP Polypropylen
PPM Pulsphasenmodulation
PR Primärradar; Pulverraketenstrahltriebwerk
Pr Praseodym
Präp. Präparat
präp. präparativ
Preßh. Preßharz
prim. primär
Pro Prolin
Prod. Produkt
PröGA private öffentliche Gemeinschaftsantennenanlage
Prop. Propeller
proz. prozentig
PS Pferdestärke; Pegelsender; Polystyrol
Ps Punktschweißen
p/s Perioden je Sekunde
PSA Postsparkassenamt
PSch Postscheck
PSchA Postscheckamt
PSchG Postscheckgesetz
PSe effektive Pferdestärke
P. S. Gr. Panzerstahlgranate
PSh Pferdestärkenstunde
PSi indizierte Pferdestärke
PSK Präzisionsstereokomparator
PSp Postsparkasse
Psp Preßspan
PSpA Postsparkassenamt
PT Prüftaste
Pt Piektank; Platin
PTC positiver Temperaturkoeffizient
PTL Propeller-Turbinen-Luftstrahltriebwerk
PTS Prüfdienststelle für Technische Schiffsausrüstung
PTZ Physikalisch-Technisches Zentralinstitut
Pu Plutonium
PUF Polyurethanfaser
PUG Punktübertragungsgerät
Pulv. Pulver
PUR Polyurethan
PUS Polyurethankunstseide
PV Porenvolumen
Pv. Pulver
pv. pulverisiert
PVA Polyvinylalkohol; Polyvinylazetat
PVAC Polyvinylazetat
PVAL Polyvinylalkohol
PVB Polyvinylbutyrat
PVC Polyvinylchlorid
PVD Polyvinylidenchlorid
PVF Polyvinylfaser
PVG Pegelvergleichsgerät
PVI planmäßige vorbeugende Instandhaltung
PVP Polyvinylpropionat
PVS Polyvinylkunstseide
pW Picowatt
Py., py Pyridin
PZ Portlandzement
Pz Panzer
PZA Postzeitungsamt
Pz. B. Panzerbüchse
Pz Flak Panzerflak
Pzgr., Pz. Gr. Panzergranate
PzK Panzerkanone
Pzkpfw. Panzerkampfwagen
Pzm Panzermantel
PzSpW Panzerspähwagen
Pztr. Panzertruppen
Pz.-Zg. Panzerzug

Q

Q erhöhtes Deck, Quarterdeck; quetschbar, kaltstauchbar *(Stahl)*; Quer-
q Quernaht
QHM Quarzfaden-H-Magnetometer
q. l. beliebig *(quantum libet)*
QM Quantenmechanik
QS Quecksilbersäule
q. s. soviel wie nötig *(quantum satis)*
QT Quittungstaste
Qu Quarzit; Querschlag
qual. qualitativ
quant. quantitativ
quart. quartär
Quellungsm. Quellungsmittel
Qu.-F. Quadratfuß
QZ quantisiertes Zeichen

R

R Röntgen[-Einheit]; Réaumur; Rauhtiefe; Reibung; Rolle; Rück-, Rücken-; Resultierende; ruhig vergossen *(Stahl)*; Rips; Rakete; Regiment; Radikal; Rest; Refraktion; Register; Runddrahtbewehrung; Rundfunk, Radio; Retinit
r Röntgen[-Einheit]; Reflexionsfaktor; Ruhekontakt; Reflektor; Rekompensationskoeffizient; Radius; resultierend; rauh; Drehkreisradius
r. rechtsdrehend; rot
R. A. Restauftrieb; Ruderausschlag
Ra Radium; radioaktiv; Ramie
ra Ruhe-Arbeitskontakt
ra. radioaktiv
rac. razemisch
rad Radiant; Rad, Maßeinheit der Strahlungsdosis
Radd Raddampfer
RaEm Radiumemanation
Rak. Rakete; Raketen-

r. a. m. radioaktiv markiert
rauch. rauchend
RB Rohbraunkohle
Rb Rubidium
R_{BB} Interbasis-Widerstand
RBE relative biologische Effektivität
RBl Riffelblech
Rbl. F Rundblickfernrohr
R-Boot Minenräumboot
RBW relative biologische Wirksamkeit
R. D. rückwärtige Dienste
Rd Rundgewinde
rd Rad, Maßeinheit der Strahlungsdosis; Rutherford-Einheit
Rdbfr. Rundblickfernrohr
Rdm. Rudermaschine
RDO Rhein-Donau-Ölleitung
RE Reparatureinheiten
Re Reynoldssche Zahl; Rhenium
rechtsg. rechtsgängig
Red. Reduktion
red. reduziert; reduzierend
redest. redestilliert
Redukt. Reduktion
reg. regulär
Reg.-T. Registertonne
Rekrist. Rekristallisation
rekrist. rekristallisiert
resubl. resublimiert
Rev. Revier
RF Rundblickfernrohr; Rundfunk
Rf Rundfunk
RFF Richtfunkfeuer
Rfl Rundfunkleitung
RFZ Rundfunk- und Fernmeldetechnisches Zentralamt
RG Raumladegitterspannung; Reaktionsgeschwindigkeit
Rg Rotguß
Rg. Räumgerät
RGK Gegenkopplungswiderstand in der Katodenleitung
Rgt. Regiment
RGT-Regel Reaktionsgeschwindigkeitstemperaturregel
RGW Rat für Gegenseitige Wirtschaftshilfe
Rh Rhodium; Roßhaar
rh. rhombisch
Rhiz. Rhizoma
rhomb. rhombisch
Ri Rille
RIAG Arbeitsgemeinschaft für radioaktive Isotope
RIC Rechteckimpulscharakteristik
riech. riechend
RiT Richtungstaste
Rk. Reaktion
rk Richtungskontakt
RKL Rufkontrollampe
Rkr. Richtkreis
RKS radioaktiver Kampfstoff
RKZ Regelkennzeichen
RL Rechts/Links; Längsrips
RLd Rechts/Links durchbrochen
RLDF Rechts/Links-Doppelfang
RLF Rechts/Links-Fang
RLPF Rechts/Links Perlfang
RLS Reizleitungssystem
RM Rudermaschine
Rm Rotmetall
rm Raummeter
Rn Radon
RNS Ribonukleinsäure
rot Rotation; rotierend
ROZ Research-Oktanzahl
R. P. Richtpunkt
RPb Rohrblei
rph Radphot
RQ Querrips
RR besonders beruhigt *(Stahl)*; doppelte Runddrahtbewehrung; Rechts/Rechts
rr Ruhe-Ruhekontakt
RRB Rosin-Rammler-Bennett
RRP Rotterdam-Rhein-Pipeline
RRS Rufrelaissatz; Rosin-Rammler-Sperling
RS Resonanzrelais; Rufschalter; Rufsatz
Rs Raupenschwenkbagger
rs Resonator; rosa
RSM Ruf- und Signalmaschine
Rsp Rahmenspantenbauart
RStr Richtstrecke
RT Rückstelltaste; Ruftaste; Raumteile; Raumtiefe; Registertonne; Regietisch
rt rot
RTG Raketentauchgeschoß
RTL Resistor-Transistor-Logik
RTW relative Tiefenwirkung
RU Relaisunterbrecher; Richtungsumsetzung; Rufumsetzer; Rüster, Ulme
RÜ Rufübertragung
Ru Ruthenium
ru Ruhe-Umschaltekontakt
rü mit Rücksprung
RV Reduktionsvermögen; Regelventil; Rückschlagventil
RVM Röhrenvoltmeter
RW Regelwiderstand
R. W. Raketenwerfer; Reparaturwerkstatt
rw rechtweisend
RWB Raketenwasserbombe
r. w. K., rw. Kurs rechtweisender Kurs
rwN rechtweisend Nord
rwP rechtweisende Peilung
R. Z. A. Rohrzielapparat
rza Ruhe-Zwillingsarbeitskontakt

S

S Siemens; Sprungantrieb; Sechsphasenstrom; Steilheit; Scheinleistung; Sender; Sicherung; Siebfaktor; Strahlungsdichte; Streumatrix; Stromdichte; Schirm; Senden, Sendung; Schwefel; Sauerstoffträger; säurefest; Schwärzung; Substitution; Steigungs-; Streckgrenze; Schiebesitz; Serie; Schraube; spannungsfrei geglüht; schmelzschweißbar; Sicherheitsgrad; senkrecht; schwer; Schwerpunkt; Seite; Stratus; Süd, Süden
S. Segelschiff; Süd, Süden
s Sekunde; senkrecht; sehr dick umhüllt *(Elektrode)*; Scheitelwert; Schlupf; Schwungmasse; Schneidrad-, Schabrad-, Schleifkörper; Spantabstand
s. Sekunde; südlich
SA schwere Artillerie
S. A. Schiffsarzt; Selbstanlasser
Sacch. Saccharum
SAG Sozialistische Arbeitsgemeinschaft; Staatliche Aktiengesellschaft
SAH Hauptschalter
Sanka Sanitätskraftwagen
SB Seidenbaumwolldraht; Signalbatterie; Schappenbohrprobe; Siedebeginn
Sb Antimon
sB Schlichtbohrung
sb Stilb
SBK Seidenbaumwollkabel; Styrol-Butadien-Kautschuk
Sbp. Sublimationspunkt
s. Br. südliche Breite
SC Siliziumkarbid
Sc Skandium
Sch Schweißbarkeit; Schieferton; schlagwettergeschützt; Schlitzprobe; Schleppschiff
sch schlagwettergeschützt
scheu gescheuert
Schl. Schlepper
Schl. Bt. Schlauchboot
Schld Schleppdampfer
Schleppd Schleppdampfer
schliff geschliffen
Schl-M Motorschlepper
Schl Pz Schlepp-Panzer
Schmp. Schmelzpunkt
Scho Schoner; wasserdichtes Schott
Scho bis 2. D wasserdichtes Schott bis 2. Deck reichend
Schp. Schwerpunkt
Sch-Pz Schwimmpanzer
Schr Schrauben, Propeller
SchRs schwenkbarer Schaufelradbagger auf Raupen
Schr v. u. h. Schrauben (Propeller) vorn und hinten
Sch. W. Scheinwerfer
Schw Schweißeisen
schw. schwach; schwarz
SCu Kupferschweißdraht
S-Cu Stratokumulus
SD Spieldauer; Splitterbombe
S. D. Sichtwechsel
Sd. Siedepunkt
sd. siedend
S. D. G. Scherdrachengerät
Sdg Sendung
Sdp. Siedepunkt
SDW Sonderdrahtwiderstand
SE Sicherheitserde; Sekundärelektron; Sekundäremission; Selbsterregung; Spannungseinheit
Se Selen
sec Sekans; Sekunde
sec. sekundär
Seekdt. Seekadett
Seetr. Seetransport
SEF Sojabohnen[eiweiß]faser
Sek. Sekunde
sek. sekundär; Sekunde
SEL Stahl-Eisen-Liste
Ser Serin
SEV Sekundärelektronenvervielfacher; Sendereingangsverstärker
SF Semifusinit
Sf Plattenführungsschnitt
SFK Signaltechnik – Fernmeldetechnik – Kybernetik
Sfl Selbstfahrlafette
SFS Schnellfrequenzbohrmaschine für Sprenglöcher
Sfs Säulenführungsschnitt
Sfsg Gesamtschnitt mit Säulenführung
SFV Staatliche Flughafenverwaltung
Sfv Folgeschnitt, Schnitt mit Vorlocher
Sfzg Gesamtschnitt mit Zylinderführung
SG Summierungsgerät
Sg Spritzguß[legierung]
sG Schlichtgleitsitz
SG(CO_2) Schutzgaslichtbogenschweißen mit CO_2
SGE Sammelgesprächseinrichtung
SG(H_2) Schutzgaslichtbogenschweißen mit H_2
sgl[z] spiegelglänzend
SgSn Zinnspritzgußlegierung
SGW Schnellamtsgruppenwähler
SH Starthilfe
Shb. Seehandbuch
SHD Seehydrografischer Dienst
S. H. D. Sicherheits- und Hilfsdienst
SHF superhohe Frequenz *(3000 bis 30000 MHz)*; Staatssekretariat für das Hoch- und Fachschulwesen
SI Sicherheitsinspektion; Selbstinduktivität
Si Silizium; Sicherung; Siemens
Sial Silizium-Aluminium

sicc. wasserfrei
Sifa Sicherheitsfahrschaltung
SIK Selbstinduktionskoeffizient
Sima Silizium-Magnesium
sin Sinus
sinh Hyperbelsinus
SINT gesintert
SK Schaltkasten; Steinkohle; Sicherheitskoeffizient; Segerkegel; scharfe Kanten *(Stahl)*; Synchrokomparator
S. K. Schnellfeuerkanone, Schnelladekanone; schwerer Kreuzer
Sk Sendekontakte am Fernschreiber
sk Skot; Sammelkontakt
SKA Sorptionskälteanlage
SKE Steinkohleneinheit
SKF Seekriegsflotte
SKH Horndruckring
SKL Schlußkontrollampe
SKP Kippschalter
Skt[.] Skalenteilung, Skalenteil[e]
SKW schwere Kohlenwasserstoffe; Schwerlastkraftwagen
Skw Skalenwert
SKZ Siedekennziffer
SL Schlußlampe
Sl Schnellverkehrsleitung; Schlackenfasern
sL sandiger Lehm; Schlichtlaufsitz
sl., s. l. schwer löslich
sll., s. l. l. sehr leicht löslich
SLW Sammelanschlußleitungswähler
Sm Samarium; Messerschnitt
sm Seemeile *(1 sm = 1852 m)*
S. m. D. Segelschiff mit Dampfmaschine
Sm-FAZ Sammelfernanrufzeichen
SMG schweres Maschinengewehr
SMg Magnesiumgußschweißdraht
SmH Schiff mit Hilfsantrieb
sm/h Seemeilen in der Stunde
SmM Segelschiff mit Motor
Sm-MAZ Sammelmeldeanrufzeichen
Smp. Schmelzpunkt
SMW Speichermischwähler
Sn Zinn
S. N. D. Seenotdienst
SNi Nickelschweißdraht
SNS Neusilberschweißdraht
SO Solloberfläche; Schienenoberkante; Südost
SÖ Schweröl
So Sondertyp
So-GMs Sondergußmessing
SoMs Sondermessing
SOS Internationales Seenotzeichen
SOZ Straßenoktanzahl
SP Schmelz-Preß *(Schweißen)*; Schaltposten; Schußpunkt; Schützenpanzer; Schiffspeilung
Sp Span
spez. spezifisch
spez. G[ew]. spezifisches Gewicht, Wichte
spez. W. spezifische Wärme
SpG Spritzguß
Spg, spg Spannung
sp. G. spezifisches Gewicht
SpHw Springhochwasser
Spir. Spiritus
SpNw Springniedrigwasser
Spr Sprechzeug
Spritzg. Spritzguß
Spritzm. Spritzmasse
SPS Schubpferdestärke
SpT Sperrtaste
Spt. Spant
SPÜG Synchronimpulsüberwachungsgerät
SPV Spezialpumpversuch
sp. V. spezifisches Volumen
SPW Schützenpanzerwagen
Spw. Speisewasser
sp. W. spezifische Wärme
SpZ Sperrzeichen
Squ Squalan
SR Ruhestromschalter; Sekundärradar; Senderrelais; Sollwertregler
Sr Strontium
sr Steradiant
Sre. Säure
SRg Rotgußschweißdraht
SRs schwenkbarer Schaufelradbagger auf Raupen
SS Sendesieb, Störschutz; Schweißstahl; Schnellarbeitsstahl; Dampfschiff
S. S. Schraubenschiff
Ss Stumpfschweißen
ss schwer schweißbar
S. Sch. S. O. Seeschiffahrtsstraßenordnung
S. Sch. S. V. O. Seeschiffahrtsstraßenverkehrsordnung
SSF Siemens-Selenflachgleichrichter
S. S. G. Schleppspulengerät
SSK Seestreitkräfte
ssl., s. s. l. sehr schwer löslich
SSM Signalsendemodulator
SSO Südsüdost
SSS Segelschulschiff
SSt schweißbarer Stahl
S-St Schnittstanze
S. St. A. Schiffsstammabteilung
SSW Sonderschichtwiderstand; Südsüdwest
St Schlußtaste; Störträger; Stokes *($1\,St = 1\,cm^2/s$)*; Station; Stern, Sternverseilung; Stöpsel; Steine; Gesteinsfasern
St. Stahl; Stunde
sT sandiger Ton
st Stecker; Starter; Steuersteg
stab. stabil
Stb Biegestanze; Steinbank
Stb. Steuerbord
Stba Biegestanze mit Auswerfer

Stbk Biegestanze mit Keiltrieb
Stb.-Maschine Steuerbordmaschine
Stbua Biegestanze mit beweglichem Unterteil und Auswerfer
StBV Vorschriften für die technische Sicherheit und den Arbeitsschutz im Steinkohlenbergbau
StBz Stahlbronze
Std Stunde
Stdr. Stahldrahttau
steril. sterilisiert
Steu Steuerschalter
StF Gesteinsfaser
STG Schiffbautechnische Gesellschaft
STOL Kurzstartflugzeug
Stpl Planierstanze, Flachstanze
Stpr Prägestanze
Str Rollstanze
Stra Rollstanze mit Auswerfer
Strk Streifenkohle; Rollstanze mit Keiltrieb
Str. Kr. Straßenkreuzung
StrR Stromlinienruder
StSchAnw Starkstromschutzanweisung
StSt Startstellung
StÜe Stromstoßübertragung
Stu. G. Sturmgeschütz
StVG Straßenverkehrsgesetz
StVO Straßenverkehrsordnung
Su Schluff
subl. sublimiert
Subst. Substanz
sulfuric. sulfuricum
SV Sendeverstärker; Silbenverständlichkeit; spezifisches Volumen; Sicherheitsventil
S. V. Seilfahrtsverordnung; spezifisches Volumen
s. V. spezifisches Volumen
SVA Glühanlaßschalter
SVK Streckenvortriebskombine
SVO Seilverordnung
s. Vol. spezifisches Volumen
SW Schichtwiderstand; Suchwähler; Systemtrennweiche; spezifische Wärme; Südwest
S. W. Scheinwerfer
Sw Schraubenwelle
s. W. spezifische Wärme
sw schwarz
SWFV Selbstwählfernverkehr
Swissair Schweizerische Luftverkehr A.-G.
Swkw Schwimmkraftwagen
sWL weiter Schlichtlaufsitz
swl. sehr wenig löslich
SWO Seewasserstraßenordnung
Sw. R. G. schweres Räumgerät
Sws Schwimmsand
Sy Synchronisierung
Sygas Ammoniaksynthesegas
symm. symmetrisch
synth. synthetisch
SZ Säurezahl, Neutralisationszahl; Synchronisierungszeichen; Schauzeichen
S-Z Schnitt-Zug[-Verbundwerkzeug]
S-Z-S Schnitt-Zug-Schnitt
S-Z-St Schnitt-Zug-Stanze

T

T Tara; Tritium; Temperatur [°K]; Torr; Trägheit; Tangentialkraft; Teilchen; Telinit; Telefon; Telefonie; Telegraf; Telegrafie; Transistor; Triode; Tiefenregler *(Radio)*; Träger; Trägerschwingung; Tonband; Trimmer; Tisch; Drehzahl; Treibsitz; Thomas *(Stahl)*; Tank; Tauchtiefe; Tiefgang, Konstruktionstiefgang, Rechnungstiefgang; Tropenfreibord, Tropengebiet
T° Wärmegrad der absoluten Skale
2T Zweitakt
4T Viertakt
T. Tankschiff
t Temperatur in °C; Dauer, Zeit; Tangential-; tonig; Tiefgang ohne Kiel; Tonne *(1000 kg)*
$t_{1/2}$ Halbwertzeit
TA Telegrafenamt; Temperaturmeßapparat; Tanne; Tiefenablese am Echolot
T. A. Tiefenapparat
Ta Tantal; Tara
Tab. Tabelle
TAL Transalpine Erdölleitung
TAN technisch begründete Arbeitsnorm
tan Tangens
tanh Hyperbeltangens
TAS Tastenschalter
Tato Tagestonne
TB Telegrafenbatterie
Tb Terbium
TBA Telegrafenbauamt; tertiärer Butylalkohol
TBBI Technische Bezirksbergbauinspektion
TBG Telegrafenbaugerät
TBI Technische Bergbauinspektion
TBK Trockenbraunkohle
Tc Technetium; kritische Temperatur
T/cm^3 Teilchen je Kubikzentimeter
Tct. Tinctura
TD Tankdampfer; Drehtrenner; Telegrafendienst
TDSt Telegrafenstelle
tdw Tonnen Tragfähigkeit, Tonnen Deadweight *(1 tdw = 1000 kg)*
TE Trägerfrequenzerzeuger; Tischeinheit
T. E. Teileinheit; Truppeneinheit
Te Tetrode; Tellur
techn. technisch
Teilkr. Teilkreis
TEL Telefon; Telegramm; Bleitetraäthyl
Tel drahtlose Telegrafie
Tel. Telefon; Telegramm
tel telefonisch; telegrafisch

tel. Telefon; Telegramm
Temp Temperatur
tert. tertiär
tetr. tetragonal
TF Trägerfrequenz; Transformator
T.F. Tidenfall
Tf Torf; Trägerfrequenz; Tiefeinbrandelektrode
TF-F Trägerfrequenzfernsprechen
TFFK Trägerfrequenzfernkabel
TFK Transportfliegerkräfte
Tfl Tafel
TFS Turbinenfährschiff
TG Taktgeber; Tonfrequenzgenerator; Transduktorgerät
Tg Telegrafie; Telegrafen-
Tg, Tg. Tiefgang
Tgb Tagebau
Tgb. Tagebuch
TGeb Telefongebühr
Tgh Tiefgang hinten
TgHV Telegrafenhauptvermittlung
TGL Technische Gütevorschriften und Lieferbedingungen
tgl. täglich
Tgm Tiefgang mittschiffs
Tgv Tiefgang vorn, Tiefgang am vorderen Lot
TGZ Teilchengrößenzählgerät
TH, T.H. Technische Hochschule
Th Thorium
Theo Theodolit
theor. theoretisch
therapeut. therapeutisch
therm. thermisch
THF Tetrahydrofuran
Thr Threonin
THz Terahertz *(1 THz = 10^{12} Hz)*
TI Tiegelstahl
Ti Titan; Titanoxid-Typ *(Elektrode)*
TK Temperaturkoeffizient; Temperaturkompensation; Klapptrenner; Theodolitkreisel
T.K. Turmkanone
TKL Teilnehmerklinke
tkm, tkm. Tonnenkilometer
TKO Technische Kontrollorganisation
TKP Trikresylphosphat
TL Technische Lieferbedingungen; Turbinenluftstrahltriebwerk; Turbolader
Tl Thallium; Gewichtsteil, Teil
tL toniger Lehm
Tle. Gewichtsteile, Teile
TLF Tanklöschfahrzeug
Tlk Tjalk
Tln Teilnehmer
Tln. Teilen
tlw. teilweise
TM Tankmotorschiff
Tm Thulium
Tm.-Bk. Turmbake
TME tausendstel Masseeinheit; technische Masseeinheit
TMF Tonfrequenz-Multiplex-Fernsteuerung
T-Mi Tellermine
TMiZ Tellerminenzünder
tmot teilweise motorisiert
Tn Thoron
TNA, T.N.A. Trinitroanilin
TNT Trinitrotoluol
TO Telegrafenordnung
To. Torpedo
TOM technisch-organisatorische Maßnahmen
Tonn Bl Tonnenblech
tox. toxisch
TÖZ technisch-ökonomische Zielstellung
TP trigonometrischer Punkt; Tropfpunkt; Triangulationspunkt; Pantografen- oder Scherentrenner; Tiefpaß
Tp. Transport
TPS-Stahl T-Stahl mit parallelen Flansch- und Stegseiten
TR Rundtischeinheit; Telefonrundspruch; taktische Rakete; Temperaturregler
T.R. Torpedorohr
Tr Trapezgewinde; Trennschärfe; Trimm
Tr. Tropfen
tr Transformator
Trafo Transformator
Tragf. Tragfähigkeit
Tragk. Tragkraft
Transp. Transport
Trbst. Treibstoff
Trgf. Tragfähigkeit
Tri Trichloräthylen
tri. triklin
trig. trigonal
TrK Trockenkohle
Trp. Transport
Trsp. Transport
Try Tryptophan
TS Schalttischeinheit; Transportschiff; Teilnehmerschaltung; Trennschalter; Schubtrenner; Trennschleuder; Turbinenschiff
T.S. Tankschiff
Ts Thermitschweißen; Tussahseide, wilde Seide
ts tonig-sandig
TSB technische Sicherheit im Bergbau
TSC Turn- und Sportklub
TSch Tankschleppschiff
TsZ Tonerdeschmelzzement
TT Tonträger; Trenntaste; Tieftemperatur; Tischtennis
Tt Taschentanks
TTA Theonyltrifluorazeton
TTH Tieftemperaturhydrierung
TTK Turmtropfkörper
TTL Transistor-Transistor-Logik
TTT Tieftemperaturteer

TU, T.U. Technische Universität
TÜ Technische Überwachung
Tu Thulium; Tuch
TUS Tastenumschalter
TÜSt Telegrafieübertragungsstelle
TVB technologische Betriebsmittelvorbereitung
TVF technologische Fertigungsvorbereitung
tvF Tonne verwertbare Förderung
TVÜ Tagesverkehrsübersicht
TW Terawatt *(1 TW = 10^{12} W)*; Teilnehmerwähltelegrafie; Teilnehmerwählfernschreiben
Tw Triebwerk
TWB Temperaturwechselbeständigkeit
TWh Terawattstunde *(1 TWh = 10^{12} Wh)*
TWK technisch-wirtschaftliche Kennziffer
TWÜe Tonwahlübertragung
TwwF technisch-wissenschaftliche und wirtschaftliche Förderung
TX Telex
Tyr Tyrosin
TZ Tourenzahl
Tz Tibethaar, Haar der Tibetziege
T.Z.F. Turmzielfernrohr

U

U Uran; Umformer; Umschalter; Umschaltkontakt; Umfangs-, Umfang; Umdrehung; unmittelbar; Unterwasser[zünder]; Unterkasten; unruhig vergossen; unbehandelt *(Stahl)*; mit Unterlegscheibe
Ü Überlastwert; Übertrager
u vereinheitlichte atomare Masseeinheit *(1 u = 1,660277 · 10^{-27} kg)*
ü Übersetzungsverhältnis; Überkopf *(Schweißen)*
UB Universalbagger
UBd unteres Band
übersätt. übersättigt
Übertr. Übertrag
U.D. Ultradezimeterwelle
Ud dünn umhüllt *(Elektrode)*
u.d.M. unter dem Meeresspiegel
ü.d.M. über dem Meeresspiegel
Üds Überführungsdose
UDT Unterdecktonne
UE Untergestelleinheit
UED Ultrarotemissionsdiagnostik
UEG untere Explosionsgrenze
UF Harnstofformaldehyd
ÜF Überweisungsfernamt
ÜFESt Überseefunkempfangsstelle
UFK ungelenkter Flugkörper
UFO unbekanntes Flugobjekt
UGE Untergrundeffekt
ÜGO Übergruppenumsetzer
U/h Umdrehungen/Stunde
UHF Ultrahochfrequenz, ultrahohe Frequenz *(300 bis 3000 MHz)*
UHV Ultrahochvakuum
UK Unterkante
Ukg Ultraschallkardiogramm
UKK Unterkante Kiel
UKML Ultrakurz-, Kurz-, Mittel- und Langwellenbereich
UKW Ultrakurzwelle
ÜKZ Übergangskennzeichen
ÜL Überwachungslampe
Ül Überweisungsleitung
Ülm Meldeleitung
Ül-v Fernvermittlungsleitung
UM Unzustellbarkeitsmeldung
ÜM Übertragungshülse aus Messing
Um mittelstark umhüllt *(Elektrode)*
U/m Umdrehungen/Minute
u.M. unter dem Meeresspiegel
ü.M. über dem Meeresspiegel
Umf. Umfang
umgeb. umgebaut
UMI Universalmeßinstrument
U/min Umdrehungen/Minute
Ums. Umsetzer
ungegl. ungeglüht
unges. ungesättigt
Ungt. Unguentum
unl. unlöslich
unsymm. unsymmetrisch
Unters. Untersuchung
unzers. unzersetzlich
UP Unterpulver[schweißen]; ungesättigte Polyester
UpM, U.p.M., Upm Umdrehungen/Minute
UR Ultrarot
US Ultraschall; Unterspannung; Unterschienen[schweißen]
ÜS Übergabeschalter
Us stark ummantelt
ÜSL Übergabeschlußlampe
UT Umlegetaste; Unterlagerungstelegrafie
ÜT Überlagerungstelegrafie
u.T. unmodulierte Telegrafie; unter Tage; unterer Totpunkt
ü.T. über Tage
UTF Unterlagerungstelegrafie auf Freileitungen
UTV Untertagevergasung
UV ultraviolett, Ultraviolett-
UVV, U.V.V. Unfallverhütungsvorschriften
UW Umspannwerk; Unterwerk; Umsteuerwähler
UWM Unterwassermotorpumpe
UZn Umschmelzzink
UZO Unterseebootzieloptik

V

V Volt; Voltmeter; Volumen; Variation; Vakuum; Verstärkung; Verstimmung; Vierer *(Kabel)*; Vanadium; Glanzkohle, Vitrit; Ver-

größerung; Vorschub; Verlust; vergütet *(Stahl)*; Verdrängung, Verdrängungsvolumen, Wasserverdrängung in m^3; maximale Schiffsgeschwindigkeit [kn]
v Verzögerungsleitung; Spannungsverstärkung; Volumen; Geschwindigkeit; Voreil-; versetzt; mit Vorsprung; vitritisch
VA Voltampere
Val Valin
val Grammäquivalent
VAZ voraussichtliche Ankunftszeit
VB vorgeschobener Beobachter
vBa versenkte Back
Vbd Vielband
V-Bkg Vektorballistokardiogramm
vBr versenktes Brückenhaus
vc. kritisches Volumen
V/cm Volt/Zentimeter
VD Vordrossel
VDE Verband Deutscher Elektrotechniker
VDEh Verein Deutscher Eisenhüttenleute
VDI Verein Deutscher Ingenieure
Vdr Vierdraht
VE Verkehrseinheit; Verrechnungseinheit
v.E. vom Endwert
VEM Volkseigener Elektromaschinenbau
Verb. Verbindung
Verbr. Verbrauch; Verbrennung
Verd. Verdünnung
verd. verdünnt
Verdr. Verdrängung
Verf. Verfahren
verf. verfahren *(Schichten)*
Verg. Vergaser
Verg. Mot. Vergasermotor
Verh. Verhalten
Verk. Verkehr
VerkMin. Verkehrsministerium
verlg. verlängert
Vers. Versuch
Verst. Verstärkung; Verstärker
VerstA Verstärkeramt
VerstB Breitbandverstärker
VerstBe Breitbandendverstärker
VerstBst Breitbandsteuerverstärker
VerstK Kanalverstärker
VerstZ Zwischenverstärker
Verw. Verwendung
verz. verzinkt
VEZ Voreinflugzeichen
VF Videofrequenz
Vf Vulkanfiber
vF verwertbare Förderung
VG Vorschubgetriebe; Verzögerungsglied
Vg Gruppenverteiler
VH hydraulische Vorschubantriebseinheit
Vh Hauptverteiler
VhF Fernamtsverteiler
VI Viskositätsindex
Vi Vikunjawolle
vi violett
VizeAdm. Vizeadmiral
VK Vergaserkraftstoff; Vollkernisolator; Verkehrskreis; Vorkante
Vk Vektorelektrokardiograf; Vorkammer
VKA Verbrennungskraftanlage
VKG Vektorelektrokardiografie
VKm Vergaserkraftstoffmotor
VKMin Verkehrsministerium
VL Vorzugslehrenlänge; Verlängerungsleitung; vorderes Lot
VM Voltmeter; mechanische Vorschubantriebseinheit; Volksmarine
V/m Volt/Meter
VME Verstärkungsmeßeinrichtung
VO Verwaltungsoffizier; Vollzugsordnung
V.O. Verkehrsordnung
Vol. Volumen
Vol.-Tl. Volumenteil
Vork. Vorkommen
v.o.T. vor dem oberen Totpunkt
VP Viskositätspolhöhe
vP versenkte Aufbauten; versenkte Poop
Vp.-Boot Vorpostenboot
Vp.-Fl. Vorpostenflottille
VPM Vierpolmeßgerät
VPV vereinfachter Pumpversuch
VR Verbrennungsregler; Verhältnisregler
VRL Verstärker für Rundfunkleitungen
VRLh Hauptverstärker für Rundfunkleitungen
VRLhi Hilfsverstärker für Rundfunkleitungen
VRLz Zusatzverstärker für Rundfunkleitungen
V-RV Vorwärts/Rückwärts-Verhältnis *(Antenne)*
VS Verbindungssatz; Versuchsschnellboot
Vs Voltsekunde *(1 Vs = 1 Weber)*
VSA Voith-Schneider-Antrieb
Vs/A Voltsekunde je Ampere
VSP Voith-Schneider-Propeller
VSt Verbindungsstöpsel; Vermittlungsstelle
Vst Vordersteven
VStHand Vermittlungsstelle mit Handbetrieb
VStTW Telegrafiewählvermittlungsstelle
VStW Vermittlungsstelle mit Wählbetrieb
VStWoF Vermittlungsstelle mit Wählbetrieb in Ortsnetzen ohne Fernamt
V.T. Volumenteil
v.T. vom Tausend
VU Videoumschalter
VÜ Vorübertrager
vulk. vulkanisiert
v.u.T. vor dem unteren Totpunkt
VUW vereinfachtes Umspannwerk
VV Videoverstärker
V.V. Verkehrsvorschrift

VW Vorwähler
VZ Verseifungszahl; Verzoner
Vz Zwischenverteiler
vz verzinkt

W

W Wolfram; Werkstück; Werkzeugbau; Windfrischstahl; mit Wälzlagereinsatz; Wirk-; Widerstand; Watt; Wattmeter; Wechselstrom; Wasser; West, Westen; Wattfahrt
W. Wasser; Widerstand
w Windungszahl; Wendel; Werkstück; Wälz-; waagerecht, in Wannenlage; weich; Relativgeschwindigkeit
WA Wasseraufnahmevermögen [von Isolierstoffen]; Wärmeaustauscher; Wasserabscheider
wa aus weichem Zustand ausgehärtet
Wabo Wasserbombe
waf wasser- und aschefrei
Waff Bl Waffelblech
Warz Bl Warzenblech
WASA Wasserstraßenselbstwählanschluß
WB Wählbatterie
W. B. Wasserballast
Wb Weisbach; Weber
wd. wasserdicht
WdSS Widerstandsdraht mit Seide isoliert
WE Wärmeeinheit; Weide
W. E. Wärmeeinheit
Weichm. Weichmacher
Well Bl Wellblech
Werkst. Werkstatt
WESZ westeuropäische Sommerzeit
WEZ westeuropäische Zeit
W. F. Wagenfähre
Wf Wartefeld
wf wasserfrei
WFA Wählerfernamt
WFR Wanderfeldröhre
wfr. wasserfrei
Wfs Wasserstoffschweißen
WG Wassergehalt
Wg Wegerung
Wg. Wagen
Wgeh wassergehärtet
Wgl. Wagenladung
Wh Wattstunde
wh walzhart
WhS Weichensucher
WI Wachingenieur
WIG Wolfram-Inertgas-Schweißen
WK Wetterkühler; Wurfkörper; Wasserrohrkessel; Wasserkraft
WL Wartelampe; Wellenlänge; weiter Laufsitz; Wachleiter; Wasserlinie
w. L. westliche Länge
wl. wenig löslich
WLZ Wärmeleitfähigkeitsmeßzelle
W. L. Z. Werftliegezeit
WM Weißmetall; Widerstandsmaterial; Wachmaschinist
W. M. Wurfmine
wmf wasser- und mineralstofffrei
WNr Werkstoffnummer
WNW Westnordwest
WO Wobbler; Wasser-in-Öl *(Emulsion)*; Wachoffizier
Wo Wolle
WÖV Wasser-Öl-Verhältnis
Wpm Wörter pro Minute
WPS Wellenpferdestärke
WPV Weltpostverein
WS Wechselstrom; Wechselspannung; Wassersäule; Weltraumstation; Warteschaltungstaste; Werkzeugstahl; Wärmeschutz
Ws Wattsekunde; Wechselspannung; Wechselstrom; Wassergasschweißen
ws weiß
Wsp Wasserspiegel
Wss Widerstandsschweißen
wss. wäßrig
W. Str. Wasserstraße
WSW Westsüdwest
WT, W. T. Wechselstromtelegrafie
W. T. Wassertiefe
w. t. wasserdicht, wasserundurchlässig
W.-Tr. Wassertransport
WV Wählvermittlung; Wasserversorgung
WW Wartewähler
WWO Weltwetterorganisation
WZ Wählzeichen
W-Z Wasser-Zement[-Faktor]
Wzg. Werkzeug

X

XE Siegbahnsche Einheit, X-Einheit *($1\ XE = 10^{-13}\ m$)*
Xe Xenon

Y

Y Yttrium
Yb Ytterbium

Z

Z Zeit; Zahl; Ordnungszahl; Protonenzahl; Stoßzahl; Zweiphasenstrom; Zerstörungswert; Zündpunkt; Zustandsfunktion; zugeführt; Ziehstößel; Zug, Züge; Schleuderguß; ziehbar *(Stahl)*

Z. Zähigkeitszahl, Zähigkeitskoeffizient; Zersetzung; Zeit; Zoll
z Zündelektrode; Zählrate; Zahn; Zusatz-
z. zersetzlich
ZA Zahlungsabkommen; Zinsabkommen; Zentralamt; Zollamt
Za Züge mit Auswerfer
za Zwillingsarbeitskontakt *(Relais)*
ZAM Zweiseitenband-Amplitudenmodulation
ZB Zentralbatterie; Zwischenbasisschaltung
Zchg. Zeichnung
ZD Zwangsdurchlauf
Zdr Zweidraht
Ze Zellulose-Typ *(Elektrode)*
ZEF Maiseiweißfaser *(Zein)*
Zellst. Zellstoff
zem. zementiert
Zer Zerspanbarkeit
zerfließl. zerfließlich
Zerl. Zerleger
Zers. Zersetzung
zers. zersetzlich
Zers. p., Zers.-P. Zersetzungspunkt
ZF Zwischenfrequenz
Z. F. Zielfernrohr
Zf Zwischenfrequenz
ZFl Zubringerfernleitung
Z. Fl. Zielflug
Z-Flottille Zerstörerflottille
ZG Zahlengeber
Zgkw Zugkraftwagen
Zgm. Zugmaschine
ZgT Zahlengebertaste
ZGV Viskosezellgarn
ZGW Zentralamtsgruppenwähler
zh ziehhart
ZIG Zählimpulsgeber
Zi.-Temp. Zimmertemperatur
ZK Zwischenkühler
ZL Zahlengeberlampe
Zl Zentralamtsleitung
zll., z. l. l. ziemlich leicht löslich
ZLM Zündlichtschalter
ZM Zwischenmodulation
zmo zweimotorig
Zn Zink
ZPV Vor- und Zwischenpumpversuch
Zr Zirkon
ZRA Zeiß-Rechenautomat; zentrale Reparaturabteilung
ZSi Zwischensicherung
ZsL Zweigleitung
Z-St Zugstanze
ZT Zähltaste; Zwangstrennungstaste
ZTDSt Zentraltelegrafendienststelle
ZTG Zeittaktgeber
ZTL Zweistrom-Turbinen-Luftstrahltriebwerk
Ztr. Zentner
zul. zulässig
zul. Abw. zulässige Abweichung
Zus. Zusammensetzung; Zusatz
ZuT Zuschaltetaste
ZW Zellwolle; Zentralwerkstatt; Zweiweg-
ZWA Azetatzellwolle
zw. d. L. zwischen den Loten
ZWK Kupferzellwolle
Zw. P. Zwischenpunkt
ZwPA Zweigpostamt
ZWT Triazetatzellwolle
ZWV Viskosezellwolle
zykl. zyklisch
Zyl. Zylinder
Z-Z zickzack *(Nieten)*
ZZZ Zeitzonenzähler

RUDOLF WALTHER

POLYTECHNICAL DICTIONARY

POLYTECHNICAL DICTIONARY

ENGLISH-GERMAN

Edited by

Ing. RUDOLF WALTHER

with the cooperation of Dr. Martin Strauss

Containing about 100,000 terms

PERGAMON PRESS

Oxford · London · Edinburgh · New York
Toronto · Sydney · Paris · Braunschweig

VEB VERLAG TECHNIK BERLIN

Pergamon Press Ltd.
Headington Hill Hall, Oxford
4 & 5 Fitzroy Square, London W. 1

Pergamon Press (Scotland) Ltd.
2 & 3 Teviot Place, Edinburgh 1

Pergamon Press Inc.
44-01 21st Street, Long Island City, New York 11101

Pergamon of Canada, Ltd.
207 Queen's Quay West, Toronto, Ontario

Pergamon Press (Aust.) Pty. Ltd.
Rushcutters Bay, Sydney, N. S. W.

Pergamon Press S.A.R.L.
24 rue des Écoles, Paris 5e

Vieweg & Sohn G.m.b.H.
Burgplatz 1, Braunschweig

Library of Congress Catalog Card No. 67-25817

Printed in Germany

08 003312 1

PREFACE

This polytechnical dictionary is planned to bridge the gap between the general dictionary, i.e. that used for translations of general, non-technical language, and the special dictionaries covering the highly specialized terminologies of the individual fields of technology.
The use of the word "polytechnical" in the title suggests not merely that the dictionary contains terms used in different fields of technology and science, but also that it presents a selection of words with specific meanings in different fields. The selection incorporates the basic vocabulary common to several fields, supplemented by terminology peculiar to those technological sciences which are fundamental to more than one, if not all, fields. Some relevant general terminology has been included. On the other hand, it was found necessary to incorporate a significant number of special terms in important fields in order to cater for the growing complexity and comprehensiveness of technological sciences. There is a continuing tendency for individual branches to overlap each other, with the result that some terms are now common to several technologies.
The dictionary was compiled from material supplied by authors having considerable practical experience in the individual fields. The editor's task consisted in checking the vocabulary supplied and using the working principles which he has championed and used in all his publications to date. This has entailed using primary sources for verification, and endeavouring to ensure a high degree of accuracy in the definition of the concepts that the terms symbolize (which involves problems of standardization in English and in German speaking countries), bearing in mind the character of the systems of scientific terminology.
The methods used should ensure that this dictionary will be found very useful in practical work. The almost limitless abundance of material which had to be utilized, indicates that some errors of interpretation may exist, or that there may be some omissions. All criticisms and suggestions either about revisions or the inclusion of new terms are, therefore, very welcome and will be greatly appreciated.
The editor expresses his gratitude to all those who contributed to the vocabulary from their personal experience. Special thanks must be given to the staff of the publishing house, whose co-operation and assistance were of great value in the successful completion of this immense task.

Rudolf Walther

LIST OF COLLABORATORS

Dr. *Hans-Peter Angelé*, Leipzig; *Ernst-Georg Berends*, Berlin; Ing. *Horst Bauschke*, Berlin; Dr.-Ing. *Georg Brack*, Merseburg; *Helmut Buder*, Schulzendorf; Dr.-Ing. *Erich Bürger*, Karl-Marx-Stadt; *Horst Dupke*, Karl-Marx-Stadt; *Jutta Ernst*, Berlin; Dr.-Ing. *Kurt Fiedler*, Leipzig; Dr. *Anton Gabert*, Leipzig; Dipl.-Ing. *Georg Gerhardt*, Ilmenau; Dipl.phil. *Horst Görner*, Berlin; *Arno Gräf*, Berlin; Dipl.-Math. *Werner Heidrich*, Berlin; Obering. *F. Hodam*, Berlin; Ing. *Horst Hummelke*, Berlin; *Karl H. Jenisch*, Berlin; *Ruth Körner*, Dresden; *Helmut Kretschmar*, Strausberg; Dr. *W. Krug*, Berlin; Dipl.-Ing. *Ulrich Lagowitz*, Berlin; *Herbert Lasch*, Berlin; Dipl.-Ök. *Herbert Liebscher*, Leipzig; Ing. *Kurt Meyer*, Karl-Marx-Stadt; *Fritz Mühler*, Karl-Marx-Stadt; *Wolfgang Müller*, Leipzig; *Herbert Petzschner*, Freiberg; Dipl.-Ing. *Manfred Röhnisch*, Berlin; Dipl.-Gwl. *Helmut Schmidt*, Berlin; *Hans-Joachim Schuch*, Jena; Dr. habil. *G. Schulz*, Berlin; Dipl.-Math. *Ralf Sube*, Berlin; Dr. *Edmund Thews*, Berlin; Dipl.-Ing. *G. Weinzierl*, Berlin; *Peter Wennrich*, Berlin; Dipl.-Dolm. *Anneliese Werm*, Strausberg; Dipl.-Ing. oec. *Günter Wohlfahrt*, Berlin; Dipl.-Chem. *Gottfried Zimmermann*, Leipzig.

DIRECTIONS FOR USE · BENUTZUNGSHINWEISE

1. EXAMPLES OF THE ALPHABETICAL ARRANGEMENT

BEISPIELE FÜR DIE ALPHABETISCHE ORDNUNG

abutment
  a. pier
  a. pressure
  a. stone
  a. zone
abutting
  a. end
  a. joint
  a. surface
abvolt
abyss
abyssal realm
Abyssinian gold
Ac-point
acacia
  a. gum
academic year

time/to
time
time
  t.-base amplifier
  t.-base encoder
  t.-chart
timed control
  t. spark gap
timekeeper
timer
  t. switch
timing
tin/to
  t.-coat
  t.-plate
tin

2. SIGNS USED IN THE DICTIONARY

BEDEUTUNG DER ZEICHEN

/ screen/to = to screen

( ) alloying constituent (element) = alloying constituent *or* alloying element
unabhängige Variable (Veränderliche) = unabhängige Variable *oder* unabhängige Veränderliche

[ ] satellite period [of revolution] = satellite period *or* satellite periode of revolution
sheet pack[ed] = sheet pack *or* sheet packet
Selbst[aus]löschung = Selbstlöschung *oder* Selbstauslöschung
[ab]trennen = trennen *oder* abtrennen

( ) These brackets are used to include explanations
Diese Klammern enthalten Erklärungen

ABBREVIATIONS · ABKÜRZUNGEN

Ak Akustik / acoustics

Am Amerikanismus / Americanism

Astr Astronomie / astronomy

BMSR Regelungstechnik / control engineering

Brg Bergbau, Bergbautechnik / mining, mine engineering

Büro Bürotechnik / office machines

Bw Bauwesen / building and surveying

Ch Chemie, chemische Technik / chemistry, chemical engineering

Dat Datenverarbeitung / data processing

Eis Eisenbahnwesen / railway

El Elektrotechnik, Elektronik / electrical engineering, electronics

Elm Elektromaschinenbau / electric machines, transformers

f Femininum / feminine

Flg Flugwesen / aeronautics

Fmt Fernmeldetechnik / telecommunications

Fs Fernsehtechnik / television

Geol Geologie / geology

Holz Holz- und Papierindustrie / wood and paper industry

Kält Kältetechnik / refrigerating engineering

Kern Kernphysik, Kerntechnik / nuclear physics, nuclear engineering

Kfz Kraftfahrzeugtechnik / automotive engineering

Led Leder / leather

Lw Landtechnik / agricultural engineering

m Maskulinum / masculine

Masch Maschinenbau / machine building

Math Mathematik / mathematics

Mech Mechanik / mechanics

Met Metallurgie, Werkstoffprüfung, Gießereiwesen / metallurgy, materials testing, casting practice

Mil Militärtechnik / military engineering

Min Mineralogie / mineralogy

n Neutrum / neuter

Ök Ökonomie / economics

Opt Optik, optische Geräte / optics

Ph Physik / physics

pl Plural / plural

Plast Plast / plastics

Rak Raketentechnik / rockets

s. siehe / see

s. a. siehe auch / see also

Schiff Schiffbautechnik, Schiffahrt / ship construction

Strm Strömungslehre / fluid flow

Text Textiltechnik / textile industry

Typ Typografie / typography

Wärm Wärmelehre / thermodynamics

A

A-battery Heizbatterie *f*
A-bomb Atombombe *f*, A-Bombe *f*
A-frame *(Bw)* A-Rahmen *m*
A-gas *s.* aviation gasoline
A-horizon *(Geol)* A-Horizont *m*, Auslaugungshorizont *m*
a-jour threaded *(Text)* unterbrochen
A_1 point Haltepunkt A_1 *m*
A-power *s.* atomic power
A-stage A-Zustand *m (Gummi)*
A-sub *s.* atomic submarine
A-tower A-Mast *m*
A-type pole Bockgestänge *n*
A-waste *s.* atomic waste
a-weather *(Schiff)* luvwärts
AB class amplifier AB-Verstärker *m*
AB method AB-Betrieb *m (von Verstärkern)*
AB power back Stromversorgungsteil *m* für Batteriegerät
abacá Manilahanf *m*
abaft *(Schiff)* nach achtern zu, achterlich, achtern, hinten
abampere Abampere *n (elektromagnetische CGS-Einheit der Stromstärke)*
abamurus Stützmauer *f*; Strebepfeiler *m*
abandon / to aufgeben; abwerfen; stillegen
  a. a call *(Fmt)* abhängen; auflegen; einhängen
  a. the aeroplane *(Flg)* abspringen; aussteigen
abandonment Abandonnement *n*, Abandonnierung *f*, Aufgebung *f*, Preisgabe *f*, Abtretung *f (Versicherungswesen)*
abate / to nachlassen, abflauen *(Wind)*
abated voltage reduzierte Spannung *f*
abatement 1. Verminderung *f*, Nachlaß *m*, Aufhebung *f*; 2. Holzabfall *m*, Holzabfälle *mpl*, Holzverschnitt *m*, abgenommene Späne *mpl*
abatis *(Brg)* Wetterscheider *m*
abatjour Lichtöffnung *f*; Oberlicht *n*
abattoir Schlachthaus *n*, Schlachthof *m*, Viehhof *m*
abaxial achsenentfernt
  a. ray außeraxialer Strahl *m*
Abbe number Abbesche Zahl *f*
Abbe prism Abbe-Prisma *n*
Abbe spectrometer Abbe-Spektrometer *n*
Abbe substage Abbescher Beleuchtungsapparat *m*
Abbe vertical metroscope Abbe-Längenmeßgerät *n*
abbreviate / to [ab]kürzen; *(Math)* heben
abbreviated spinning process Kurzspinnverfahren *n*
abbreviation Kurzzeichen *n*; Abkürzung *f*
abbreviations *pl* **in radiocommunications** Abkürzungen *fpl* für den Funkdienst
A.B.C. telegraph Buchstabentelegraf *m*
abcoulomb Abcoulomb *n (elektromagnetische CGS-Einheit der Ladung)*
abduct / to ableiten *(z.B. Wärme)*
abeam [off] *(Schiff)* quer ab, dwars ab
Abelian integral Abelsches Integral *n*
aberration 1. Abweichung *f*; 2. *(Opt)* Abbildungsfehler *m*, Aberration *f*
  a. function Fehlerfunktion *f*
aberrational bearing mißweisende Peilung *f* *(Radar)*
aberrationless *(Opt)* fehlerfrei
abfarad Abfarad *n (elektromagnetische CGS-Einheit der Kapazität)*
abhenry Abhenry *n (elektromagnetische CGS-Einheit der Induktivität)*
abhesion Abhäsion *f (Verlust des Adhäsionsvermögens)*
abietic acid Abietinsäure *f*
ability Vermögen *n*, Fähigkeit *f*
  a. of fuel to ignite Zündwilligkeit *f*
  a. test Eignungsprüfung *f*
  a. to be deformed Formänderungsfähigkeit *f*
  a. to be magnetized Magnetisierbarkeit *f*
  a. to insulate Isolierfähigkeit *f*
  a. to maintain cutting power Schneidhaltigkeit *f*
  a. to retain a cutting edge Schneidhaltigkeit *f*
  a. to vibrate Schwingfähigkeit *f*
  a. to withstand overstressing by harmonics *(Fmt)* Oberwellenbelastbarkeit *f*
ablation *(Ch)* Abwaschen *n*; *(Geol)* Abtragen *n*, Ablation *f*
able-bodied *(Schiff)* befahren, voll ausgebildet, seediensttauglich
  a. seaman Vollmatrose *m*, befahrener Seemann *m*
  a. vessel seetaugliches (seetüchtiges) Schiff *n*
ablution Auswaschen *n*
abmho Abmho *n (elektromagnetische CGS-Einheit der Leitfähigkeit)*
abnormal anormal
  a. glow discharge abnormale Glimmentladung *f*
  a. radiation *(Fmt)* Funkstreuung *f*; ungewollte Strahlung *f*
  a. spinning *(Flg)* Trudeln *n*
aboard *(Schiff)* an Bord
abohm Abohm *n (elektromagnetische CGS-Einheit des Widerstandes)*
abordage *(Am) (Schiff)* Zusammenstoß *m*, Kollision *f*
aborted take-off *(Rak)* Fehlstart *m*
about 1. über; 2. ungefähr

a. an axis um eine Achse
a.-ready *(Schiff)* klar zum Wenden
aboutsledge Zuschlaghammer *m*, Vorschlaghammer *m*
above über
a. ground *(Brg)* über Tage
a. ground level über Erde
a.-lying hydroelectrical power plant obenliegendes Wasserkraftwerk *n*
a.-water form *(Schiff)* Überwasserform *f*
a. zero über Null
abradant *s.* abrasive
abrade / to abschleifen, abreiben, abscheuern, abschaben
abrading medium Schleifmittel *n*
abrasion 1. Abnutzung *f* *(durch Verschleiß)*; Abreibung *f*, Abrieb *m*; Abschliff *m*; Abtragung *f*; Verschleiß *m*; 2. spanende Bearbeitung *f* mit geometrisch unbestimmbarer Schneide; Schleifverfahren *n*
a. hardness *s.* a. resistance
a. measurement Verschleißmessung *f*
a. resistance 1. Abriebfestigkeit *f*; Abschleiffestigkeit; 2. *(Text)* Scheuerwiderstand[svermögen *n*] *m*, Verschleißfestigkeit *f*
a.-resistant verschleißfest
a.-resisting *s.* a.-resistant
a. strength Abriebfestigkeit *f*; Verschleißfestigkeit *f*
a. test Abschleifversuch *m*, Schleifversuch *m* *(Verschleißfestigkeitsprüfung)*; Ritzhärteprüfung *f*
a. tester Abrieb[prüf]gerät *n*
a. testing machine Abriebprüfmaschine *f*, Scheuerprüfmaschine *f*
a. wear Abriebsabnutzung *f*, Schleifabnutzung *f*
abrasionproof abriebfest, verschleißfest
abrasive Schleifmittel *n*, Schmirgel *m*; Poliermittel *n*, Putzsand *m*
a. action *(Masch)* Schleifwirkung *f*, Schleißwirkung *f*; Scheuerwirkung *f*
a. band Schleifband *n*
a. belt 1. Schleifband *n*; 2. Polierband *n*
a.-belt machine Bandschleifmaschine *f*
a.-belt polishing Polierschleifen *n*
a. blade Trennscheibe *f* *(Betonsäge)*
a. cloth Schleifleinen *n*, Schleiftuch *n*
a.-coated belt Schleifband *n*
a.-coated cloth Schleifleinen *n*, Schmirgelleinen *n*
a.-coated paper Schleifpapier *n*, Schmirgelpapier *n*
a. cut-off machine Trennschleifmaschine *f*
a. cut-off saw Trenn[schleif]scheibe *f*
a. cut-off wheel Trenn[schleif]scheibe *f*
a. cutting[-off] Trennschleifen *n*, Durchschleifen *n* mittels Trennscheibe
a. cutting machine Trennschleifmaschine *f*
a. cutting-off wheel Trennschleifscheibe *f*
a. disk Schleifscheibe *f*
a. dust Schleifstaub *m*
a. engineering practice Poliertechnik *f*, Schleiftechnik *f*
a. fabric Schmirgelleinen *n*
a. for blasting Strahlmittel *n*
a. for seating valves Ventileinschleifpaste *f*
a. friction cutting Trennschleifen *n*; Reibtrennen *n*
a. grain [for blasting] Strahlmittelkorn *n*
a. grain [for grinding] Schleifkorn *n*
a. grain [for lapping] Läppkorn *n*
a. hardness Ritzhärte *f*, Schmirgelhärte *f*; Abnutzungswiderstand *m*
a.-jet drilling Flüssigkeitsstrahlbohren *n*
a. material Schleifmittel *n*, Poliermittel *n*
a. paper Sandpapier *n*, Schleifpapier *n*, Schmirgelpapier *n*
a. pencil Retuschierstift *m*
a. powder Schmirgel *m*, Schleifpulver *n*
a. saw Trennschleifscheibe *f*
a.-stick dresser Abrichtstift *m*
a. stone for honing Honstein *m*
a. tool Schleifkörper *m*
a. wear Abnutzung *f*; Abrieb *m*
a. wheel Schleifscheibe *f*, Schmirgelscheibe *f*
a. wheel cutting-off machine Trennschleifmaschine *f*
a. wheel dresser Abrichter *m*, Abziehwerkzeug *n*
abraum salt Abraumsalz *n*
abreast *(Schiff)* dwars ab, quer ab
abreuvoir Mörtelfuge *f* zwischen Steinen *(einer Mauer oder eines Bogens)*
abridge / to *(Math)* [ab]kürzen *(Multiplikation, Division)*
abridged abgekürzt
abridgment Kurzfassung *f*, gekürzte Fassung *f*, Abriß *m* *(eines Buches usw.)*; Kürzung *f*
abrupt roll *(Flg)* gerissene Rolle *f*
a. voltage change Spannungssprung *m*
abscissa Abszisse *f*, Abszissenachse *f*, x-Achse *f*, x-Koordinate *f*; *(Masch)* Auftragslinie *f*
a. of point of cutoff *(Rak)* Brennschlußweite *f*
absence Abwesenheit *f*
a. of current Stromlosigkeit *f*
a. of distortion Verzerrungsfreiheit *f*
a. of play *(Masch)* Spielfreiheit *f*
absent subscriber service Fernsprechkundendienst *m*, Fernsprechauftragsdienst *m*
absentee service *(Fmt)* Auftragsdienst *m*
absenteeism Fehlschicht *f*, Fehlen *n*, Arbeitsversäumnis *f*; Arbeitsbummeln *n*
absolute 1. absolut, unumschränkt; 2. *(BMSR)* unverschlüsselt

a. acceleration Absolutbeschleunigung *f*
a. accuracy absoluter Fehler *m*
a. activity *(Kern)* absolute Aktivität *f*
a. address absolute Adresse *f (Ziffernrechner)*
a. age *(Kern)* absolutes Alter *n*
a. altitude absolute Höhe *f*
a. atomic weight Masseneinheit *f*, absolutes Atomgewicht *n*
a. attenuation cross section absoluter Schwächungsquerschnitt *m*
a. brightness absolute Helligkeit *f*
a. ceiling *(Flg)* Gipfelhöhe *f*
a. clearance pressure Spaltüberdruck *m* *(Turbine)*
a. colorimeter Absolutkolorimeter *n*
a. counter *(Kern)* Absolutzähler *m*
a. counting *(Kern)* Absolutzählung *f*
a. cross section absoluter Wirkungsquerschnitt *m*
a. disintegration rate absolute Zerfallsrate *f*
a. efficiency absoluter Nutzeffekt *m*
a. electrode potential absolutes Elektrodenpotential *n*
a. electrometer Spannungswaage *f*
a. entrance velocity absolute Eintrittgeschwindigkeit *f (Turbine)*
a. error absoluter Fehler *m*
a. humidity absolute Luftfeuchtigkeit *f*
a. index of refraction absolute Brech[ungs]-zahl *f*
a. inlet velocity absolute Eintrittgeschwindigkeit *f (Turbine)*
a. level *(El)* Meßpegel *m*, absoluter Pegel *m*
a. measurement Absolutmessung *f*
a. number absolute Zahl *f*, unbenannte Zahl
a. outlet velocity absolute Austrittgeschwindigkeit (Abflußgeschwindigkeit) *f (Turbine)*
a. permeability [absolute] Permeabilität *f*
a. permittivity absolute Dielektrizitätskonstante *f*
a. pressure absoluter Druck *m*, Absolutdruck *m*
a.-rest precipitation tank periodisch arbeitendes Absetzbecken *n (Sanitärtechnik)*
a. system of electrical units System *n* der absoluten elektrischen Einheiten
a. system of measurement absolutes Maßsystem *n*
a. temperature absolute Temperatur *f*, Kelvin-Temperatur *f*
a. terminal velocity absolute Austrittgeschwindigkeit (Abflußgeschwindigkeit) *f* *(Turbine)*
a. value 1. Absolutwert *m*, absoluter Wert *m*; 2. [absoluter] Betrag *m*
a.-value element Absolutwertelement *n* *(Analogrechner)*
a. velocity Absolutgeschwindigkeit *f*
a. zero absoluter Nullpunkt *m (der Temperatur)*
absolutely refractory state absolutes elektrisches Refraktärstadium *n*
absorb / to 1. absorbieren; aufsaugen; annehmen; neutralisieren; 2. dämpfen *(Stoß)*; 3. tilgen *(Schwingung)*; 4. aufnehmen *(Kraft, Wärme)*; 5. schlucken *(Schall)*; 6. aufnehmen, verbrauchen *(Leistung)*
a. energy [from] Energie entziehen
a. heat Hitze aufnehmen
a. shocks abfedern
absorbability Absorbierbarkeit *f*, Saugfähigkeit *f*
absorbable absorbierbar
absorbed dose *(Kern)* absorbierte Dosis *f*
a. particle *(Kern)* absorbiertes (eingefangenes, eingestrahltes) Teilchen *n*
a. voltage aufgenommene (verbrauchte) Spannung *f*
absorbency Absorptionsvermögen *n*, Absorptionsfähigkeit *f*; Aufnahmefähigkeit *f*; Saugvermögen *n*
absorbent aufsaugend, saugfähig
absorbent Absorbens *n*, Absorptionsmittel *n*
a. carbon Aktivkohle *f*
a. cotton Verbandwatte *f*
a. filter Saugfilter *n*, Absorptionsfilter *n*
a. finish *(Text)* Absorptionsveredlung *f*, Ausrüstung *f* durch Einsaugen des Ausrüstungsmittels
a. paper saugfähiges Papier *n*
absorber 1. Absorber *m*, Absorptionsapparat *m*, Schutzmaterial *n*; 2. Absorptionsmittel *n*; 3. Tilger *m (Schwingung)*
a. circuit Saugkreis *m*
a. tube *s.* a. valve
a. valve Ballaströhre *f*, Laströhre *f*
absorbing Absorptions-; Saug-
a. agent Absorbens *n*, Absorptionsmittel *n*
a. capacity Absorptionsvermögen *n*, Absorptionsfähigkeit *f*; Aufnahmevermögen *n*; Saugfähigkeit *f*
a. circuit Absorptions[strom]kreis *m*
a. isotope absorbierendes Isotop *n*, Neutronenfänger *m*
a. power Absorptionsvermögen *n*, Aufnahmevermögen *n*, Saugfähigkeit *f (Flüssigkeiten)*; Schluckvermögen *n (Schall, Wärme)*
a. screen Strahlenfilter *n*
a. spring dämpfende Feder *f*
a. well *(Bw)* Sickerschacht *m*
absorptance Absorptionsvermögen *n*; Absorptionsgrad *m*; Absorptionskoeffizient *m*
absorption 1. Absorption *f*, Aufsaugen *n*; Annahme *f*; Entzug *m*; 2. Schwächung *f (Strahlen)*; 3. Einfang *m ([n, γ]-Prozeß)*; 4. Schluk-

kung *f*, Schlucken *n (Schall)*; 5. Aufnahme *f (Wärme, Kraft, Leistung)*; 6. Dämpfung *f (Stoß)*; 7. Tilgung *f (Schwingung)*
a. band Absorptionsbande *f*
a. bottle Absorptionsgefäß *n*, Absorptionsflasche *f*
a. by materialization *(Kern)* Absorption *f* durch Paarbildung
a. capacity Absorptionsvermögen *n*, Aufnahmevermögen *n*
a. cell Absorptionsküvette *f*
a. coefficient Absorptionskoeffizient *m*, Schluckzahl *f (Schall)*
a. coil *(El)* Nullpunktdrossel *f*, Löschspule *f*
a. column Absorptionssäule *f*
a. control *(Kern)* Absorberregelung *f*, Reaktorregelung *f* durch Neutronenabsorption
a. cross section Wirkungsquerschnitt *m*, Absorptionsquerschnitt *m*
a. current Absorptionsstrom *m*; *(El)* Nachwirk[ungs]strom *m*
a. discontinuity Absorptionssprung *m*; Absorptionskante *f (Spektren)*
a. due to water vapour Wasserdampfabsorption *f (Strahlen)*
a. econometer Absorptionsökonometer *n*
a. edge Absorptionskante *f (Spektren)*
a. factor Absorptionsgrad *m*; Absorptionskoeffizient *m*
a. limit Absorptionskante *f (Spektren)*
a. limiting frequency untere Frequenzgrenze *f (durch ionosphärische Absorption)*
a. line Absorptionslinie *f*
a. mean free path mittlere freie Absorptionsweglänge *f*
a. of heat Wärmetönung *f*, Wärmefällung *f*
a. of power Leistungsaufnahme *f*
a. of vibration Schwingungsdämpfung *f*; Absorptionsdämpfung *f*, Stoßdämpfung *f*
a. of work Arbeitsaufnahme *f*
a. of X-rays Röntgenstrahlenabsorption *f*
a. power Absorptionsvermögen *n*
a. probability Absorptionswahrscheinlichkeit *f*
a. spectrum Absorptionsspektrum *n*
a. tube Absorptionsrohr *n*, Eudiometer *n*; Absorptionsgefäß *n*
a.-type frequency meter Absorptionsfrequenzmesser *m*, Frequenzmesser *m* im Absorptionsverfahren; Absorptionswellenmesser *m*
a.-type refrigeration machine Absorptionskälteanlage *f*, Absorptionskältemaschine *f*
a.-type refrigerator Absorptionskühlschrank *m*, Absorber[kühlschrank] *m*
a. vessel of variable depth of fluid Absorptionsgefäß *n* mit veränderlicher Schichtdicke
a. wave trap Wellenabsorptionssaugkreis *m*, Saugkreis *m*, Wellenfalle *f*
absorptive absorptiv, Absorptions-; lichtschluckend
a. attenuator Abschwächer *m*, Dämpfungsglied *n (Hohlleitertechnik)*
a. capacity Absorptionsvermögen *n*, Aufnahmevermögen *n*; Saugfähigkeit *f*
a. power Absorptionsvermögen *n*, Aufnahmevermögen *n*; Absorptionsgrad *m*
absorptivity Absorptionsfähigkeit *f*, Aufnahmevermögen *n*; Aufsaugfähigkeit *f*
abstergent Reinigungsmittel *n*
abstract / to entziehen; destillieren; ausziehen; abstrahieren
a. the heat Wärme abführen
abstract Auszug *m*
a. number abstrakte Zahl *f (Algebra)*
abstracting service Dokumentationsdienst *m*
abstraction 1. Abstraktion *f*; Absonderung *f*; 2. *(Ch)* Entziehung *f*
a. of heat Ableitung *f* der Wärme; Wärmeentziehung *f*
a. of pillar *(Brg)* Rauben *n* eines Pfeilers
abstractor Dokumentalist *m*
abstracts Auszüge *mpl*, Referate *npl (aus wissenschaftlichen und technischen Artikeln)*
abundance Überfluß *m*; Reichtum *m*; Häufigkeit *f (z. B. Isotope)*
a. ratio *(Kern)* relative Häufigkeit *f*
abuse / to falsch behandeln (bedienen, betätigen)
abuse Bedienungsfehler *m*; falsche Behandlung (Betätigung) *f*
abut / to 1. angrenzen, anstoßen; stumpf stoßen; *(mit Stirnfläche)* aneinanderstoßen; 2. bestoßen *(Holzbearbeitung)*; an Stoßflächen fügen
abutilon fibre Abutilonfaser *f*
abutment *(Bw)* Widerlager *n*, Auflager *n*, Endauflager *n*
a. pier *(Bw)* Widerlagerpfeiler *m*
a. pressure *(Masch)* Widerlagerdruck *m*; *(Bw)* Kämpferdruck *m*
a. stone Widerlagerstein *m*
a. zone Kämpferdruckzone *f (Tiefbau)*
abutting angrenzend, anstoßend, benachbart; Stoß-
a. end *(Masch)* Stoßfläche *f*
a. joint Stumpfstoß *m (Schweißen)*
a. surface Stoßfläche *f*
abvolt Abvolt *n (elektromagnetische CGS-Einheit der Spannung)*
abyss Abgrund *m*
abyssal realm Abyssalregion *f*, Tiefsee *f*
Abyssinian gold einseitig goldplattiertes Kupfer-Zink-Blech *n*
Abyssinian pump Abessinierpumpe *f*

Abyssinian well Abessinierbrunnen *m*
A.C., a.c. *s.* alternating current
Ac-point Haltepunkt *m (Schmelzkurve)*
acacia Akazie *f*
a. gum Akaziengummi *m*, Gummiarabikum *n*
academic year Studienjahr *n*
academically trained mit Hochschulbildung
acanthite *s.* argentite
acaroid resin Akaroidharz *n*, Erdschellack *m*, Xanthorrhöaharz *n*
accelerate / to 1. beschleunigen; 2. *(Kfz)* Gas geben
a. jerkily *(Kfz)* ruckartig (stoßweise) Gas geben
accelerated beschleunigt; kurzzeitig; Kurz-, Schnell-
a. aging test *(Met)* Schnellalterungsversuch *m*, Schnellalterungsprüfung *f*
a. stock Vorratsmischung *f* mit Beschleuniger *(Kunstharzherstellung)*
a. tannage Schnellgerbung *f*
a. test Kurz[zeit]versuch *m*, kurzzeitiger Versuch *m*
a. weathering Kurzbewitterung *f*
accelerating ability Beschleunigungsvermögen *n*
a. admixture *(Bw)* Abbindebeschleuniger *m*
a. agent *(Bw)* Abbindebeschleuniger *m*
a. anode Beschleunigungsanode *f*
a. area *(Kern)* Beschleunigungsbereich *m*
a. chamber *(Kern)* Beschleunigungskammer *f*
a. contactor *(El)* Anlaßschütz *n*
a. electrode Beschleunigungselektrode *f*
a. electronic lens Beschleunigungslinse *f (Elektronenoptik)*
a. field *(Ph)* Beschleunigungsfeld *n*; beschleunigendes Feld *n*
a. force Beschleunigungskraft *f*
a. gap *(Kern)* Beschleunigungsstrecke *f*, Beschleunigungsspalt *m*
a. grid Beschleunigungsgitter *n*
a. losses Anfahrverluste *mpl*, Beschleunigungsverluste *mpl*
a. pump *(Masch)* Beschleunigerpumpe *f*
a. region *(Kern)* Beschleunigungsbereich *m*
a. relay Beschleunigungsrelais *n*
a. slit *s.* a. gap
a. tube *(Kern)* Beschleunigungsrohr *n*, Beschleunigungsröhre *f*
a. voltage Beschleunigungsspannung *f*
acceleration *(Mech)* 1. Beschleunigung *f*; 2. *(Masch)* Überholung *f (Gewindeschneiden)*
a. along the path *(Kern)* Bahnbeschleunigung *f*
a. area Beschleunigungsfläche *f*
a. constant *(Elm)* Anlauf[zeit]konstante *f*
a. due to gravity Schwerebeschleunigung *f*, Schwerkraftbeschleunigung *f*, Fallbeschleunigung *f*
a. error *s.* a. misalignment
a. lag *s.* a. misalignment
a. misalignment *(BMSR)* Beschleunigungsfehler *m*
a. misalignment coefficient Beschleunigungskonstante *f (Folgeregler)*
a. nozzle Beschleunigungsdüse *f (Turbine)*
a. of gravity *s.* a. due to gravity
a. of translation Führungsgeschwindigkeit *f*
a. potential Beschleunigungspotential *n*
a. pump Beschleunigerpumpe *f (Vergaser)*
a. space Beschleunigungsraum *m*
accelerative force Beschleunigungskraft *f*
accelerator 1. Beschleuniger *m*, Akzelerator *m*, Beschleunigungsvorrichtung *f*; 2. *(Ch)* Beschleuniger; Härtebeschleuniger *m*, Beschleunigungsbad *n*; 3. *(Bw)* Abbindebeschleuniger *m*; 4. Warmwasserpumpe *f (Heizungstechnik)*; 5. *(Kfz)* Gaspedal *n*
a. anode Beschleunigungsanode *f*
a. pedal *(Kfz)* Fußgashebel *m*, Gaspedal *n*
a. plate Beschleunigerplatte *f*
accelerometer Beschleunigungsmesser *m*
accendibility Entflammbarkeit *f*
accent *(Math)* Strich *m*
accentuate contrasts / to die Kontrastwirkung erhöhen (verstärken)
accentuation of contrasts Kontrasterhöhung *f*; Kontrastgabe *f*
a. of harmonics *(Fmt)* Oberwellenanhebung *f*, Oberwellenverstärkung *f*
accept / to 1. *(Fmt)* annehmen *(Gespräch, Telegramm)*; 2. *(Masch)* abnehmen, aufnehmen
a. a call *(Fmt)* abfragen
acceptable quality level Gutlage *f*, akzeptierbare Qualitätslage *f*
acceptance *(Masch)* Abnahme *f*
a. band *(Fmt)* Durchlaßbereich *m*, Durchlaßbreite *f*
a. number Annahmezahl *f*
a. sampling Annahmekontrolle *f* durch Stichproben
a. test Abnahmeprüfung *f*
a. tolerance Abnahmetoleranz *f*
a. trial Abnahmeprüfung *f*
accepted standard Gebrauchsnorm *f*; Vollnorm *f*, anerkannter Standard *m*
a. stock Fertigstoff *m*, Gutstoff *m (Papierherstellung)*
acceptor Elektronenfänger *m*; Akzeptor *m (Transistortechnik)*
a. circuit Bandfilter *n*; Durchlaßkreis *m*, durchlässiger Kreis *m*; Serienresonanzkreis *m*
a. impurity Akzeptorverunreinigung *f*
a. level Akzeptorniveau *n*

accepts Fertigstoff *m*, Gutstoff *m* *(Papierherstellung)*
access Zutritt *m*; Zugang *m*; Zugänglichkeit *f*; Vorplatz *m*
a. board Laufplanke *f*
a. cover Schachtdeckel *m* *(Kabelschächte)*
a. door Kontrollöffnung *f*; Reinigungsöffnung *f*
a. duct *(Wärm)* Zugangkanal *m*
a. eye Reinigungsöffnung *f* *(Rohrleitungen)*
a. fitting *(El)* Verbindungsdose *f*, Abzweigdose *f*
a. plug *(Schiff)* Zugangspflockluke *f*
a. ramp Auffahrt[srampe] *f* *(Autobahn, Brücke usw.)*
a. road Zufahrt[sstraße] *f*, Zufahrtsweg *m*
a. time 1. *(Fmt)* Ablesezeit *f*, Abtastzeit *f*; 2. *(BMSR)* Suchzeit *f*, Zugriffszeit *f* *(Überführungszeit in Ziffernrechnern, z. B. bei Lese-/Schreibbefehl zwischen Speicher und Rechenwerk)*
accessibility 1. *(Brg)* Befahrbarkeit *f*; 2. Zugänglichkeit *f*
a. of the exterior of the boiler Zugänglichkeit *f* des Kessels von außen
a. of the interior and exterior of the boiler Befahrbarkeit *f* des Kessels
accessible zugänglich; erreichbar
accessories Zubehör *n*
accessory Zusatzgerät *n*
accessory Extra-; Ersatz-; Hilfs-; Zubehör-
a. agent *(Ch)* Hilfsstoff *m*
a. box Zubehörkasten *m*
a. drive Hilfsantrieb *m*
a. minerals beigemengte Mineralien *npl*
a. part Zubehörteil *n*
a. rack Geräteträger *m*
a. shoe Sucherschuh *m*, Aufsteckhalter *m* *(an Kameras)*
a. switch arrangement Abhängigkeitsschaltung *f*
accident Unfall *m*, Panne *f*; Störung *f*, Betriebsstörung *f*
a. boat *s.* emergency boat
a. hazard Unfallgefahr *f*
a. insurance Unfallversicherung *f*
a. preventer Unfallverhütungsvorrichtung *f*
a. prevention Unfallverhütung *f*
a.-proof unfallsicher
a. report Unfallmeldung *f*
a.-sensitive störanfällig
a. whilst landing Landungsunfall *m*
accidental coincidence zufällige Koinzidenz *f*, Zufallskoinzidenz *f*
a. contact zufällige Berührung *f*; ungewollter Kontakt *m*
a. earth *(El)* Erdschluß *m* *(als Störung)*
a. error zufälliger Fehler *m*
a. exposure zufällige Bestrahlung *f*
a. ground *s.* a. earth
a. jamming zufällige Störung *f*
a. radiation injury *(Kern)* zufälliger Strahlenschaden *m*
acclivity Böschung *f*; Gefalle *n*, Neigung *f*
accommodate / to 1. unterbringen *(Personen, z. B. auf Schiffen)*; 2. *(Opt)* akkommodieren *(Auge)*; 3. *(Masch)* aufnehmen
a. a vessel Schiff an die Pier legen
accommodation 1. Unterbringung *f* *(von Personen)*; Unterbringungsmöglichkeit *f*, Platz *m*; 2. *(Opt)* Akkommodation *f* *(Auge)*; 3. *(Masch)* Aufnahme *f*; 4. Anpassung *f*
a. bulkhead *(Schiff)* Einrichtungsschott *n*, Wohnraumschott *n*
a. ladder *(Schiff)* Fallreeptreppe *f*
accompanying metal Begleitmetall *n*
a. tank *(Mil)* Begleitpanzer[wagen] *m*
accomplish / to ausführen, zustande bringen
accordion 1. Akkordeon *n*, Ziehharmonika *f*, Handharmonika *f*, Harmonika *f*; 2. Falttür *f*, Harmonikatür *f*, Schiebetür *f*, Schiebewand *f*
a. door Falttür *f*, Harmonikatür *f*
a. folding *(Typ)* Leporellofalzung *f*
accost / to anlegen *(Schiff an die Pier)*
account / to berechnen, kalkulieren
account Konto *n*; Berechnung *f*; Rechnung *f*
a. holder Konteninhaber *m*
a. section Rechnungsstelle *f*
accountant Rechnungsführer *m*, Rechnungsprüfer *m*
accounting Buchführung *f*, Buchhaltung *f*
a. department Kostenstelle *f*, Rechnungsdienst *m*
a. detail card Umsatzkarte *f* *(Lochkartenmaschinen)*
a. form Buchungsformblatt *n*, Kontoblatt *n*, Kontokarte *f*
a. machine Buchungsmaschine *f*; Bürorechenmaschine *f*, Fakturiermaschine *f*
accounts payable remittance Kreditorenauszug *m*
a. receivable Debitorenauszug *m*, Debitorenbuchhaltung *f*
accoutrement Ausrüstung *f*
accretion of crystals Wachsen (Ansetzen) *n* von Kristallen
accretions of a shaft furnace *(Met)* Ansätze *mpl* eines Schachtofens
accumulate / to akkumulieren, anhäufen, speichern, aufspeichern, anreichern, sammeln
accumulated dose *(Kern)* Gesamtdosis *f*, akkumulierte Dosis *f*
a. error Gesamtfehler *m*, Sammelfehler *m*; resultierender Fehler *m*
a. pitch error Gesamtteilungsfehler *m*, Sammelteilungsfehler *m* *(Verzahnung, optisches Gitter usw.)*

a. tooth spacing error Summenteilfehler *m (Verzahnung)*
accumulating register Speicherwerk *n (z. B. von Buchungsmaschinen)*
a. speed Sammelganggeschwindigkeit *f (Lochkartenmaschinen)*
a. stimulus *(El)* einschleichender Reiz *m*
accumulation Ansammlung *f*; Häufung *f*; Speicherung *f*
a. electrode Speicherelektrode *f*
a. energy Energieaufspeicherung *f*
a. of fly Ansammlung *f* von Faserflug *(Spinnerei)*
a. of heat Wärmespeicherung *f*
a. point *(Math)* Häufungspunkt *m*
accumulative dosimetry *(Kern)* speichernde Dosimetrie *f*
a. error Sammelfehler *m*
a. indexing error Summenteilfehler *m*
a. process Anreicherungsprozeß *m*, Anreicherungsvorgang *m*
accumulator 1. *(El)* Sammler *m*, Akku[mulator] *m*; 2. *(BMSR)* Akkumulator, Zwischenspeicher *m (Register am Ende der Addier-/Subtrahierschaltung eines Ziffernrechners)*; 3. Flüssigkeitsspeicher *m*
a. acid Akkumulator[en]säure *f*, Füllsäure *f*
a. battery Sammlerbatterie *f*, Sammler *m*, Akkumulator *m*
a. box Akkumulator[en]kasten *m*; Batteriekasten *m*
a. car Akkumulator[en]fahrzeug *n*; Batteriefahrzeug *n*
a. cell Akkumulator[en]zelle *f*
a. charging apparatus (set, unit) Batterieladegerät *n*
a. container Batteriekasten *m*; Akkumulatorgefäß *n*; Sammlergefäß *n*
a. drive Batterieantrieb *m*
a. jar Akkumulatorglas *n*, Sammlerglas *n*
a. metal Akkumulatorhartblei *n*
a. plate Akkumulatorplatte *f*
a. storage battery Akkumulator *m*, Sammler *m*
a. switch Zellenschalter *m (Batterie)*
a. with liquid electrolyte Akkumulator *m* mit flüssigem Elektrolyt
accuracy 1. Genauigkeit *f*, Präzision *f*; 2. *(BMSR)* Fehlerfreiheit *f (in Ziffernrechnern)*; Genauigkeit *(in Analogrechnern)*; 3. Ungenauigkeit *f (Meßtechnik)*
a. in dividing Teilgenauigkeit *f*
a. in hole positioning Lagegenauigkeit *f* der Bohrung
a. in lead Steigungsgenauigkeit *f*
a. in measurement Meßgenauigkeit *f*
a. in reading Ablesegenauigkeit *f*
a. in size Maßhaltigkeit *f*, Maßgenauigkeit *f*
a. of alignment Fluchtgenauigkeit *f*
a. of bearings Peilgenauigkeit *f (z. B. Funkortung)*
a. of duplication Kopiergenauigkeit *f*
a. of fire *(Mil)* Treffgenauigkeit *f*, Schußgenauigkeit *f*
a. of fit Paßgenauigkeit *f*
a. of hole location Lagegenauigkeit *f* der Bohrung
a. of index[ing] Schaltgenauigkeit *f*; Teilgenauigkeit *f*
a. of location Ortungsgenauigkeit *f*; Lagegenauigkeit *f*
a. of measurement Meßgenauigkeit *f*
a. of pitch Steigungsgenauigkeit *f (Gewinde)*
a. of positioning [Lage-]Einstellgenauigkeit *f* [des Bohrtisches]; Einortungsgenauigkeit *f*; Positionierungsgenauigkeit *f*; Zentriergenauigkeit *f*
a. of reading Ablesegenauigkeit *f*
a. of reproduction Kopiergenauigkeit *f*
a. of separation *(Brg)* Trennschärfe *f*
a. of shape Formgenauigkeit *f*
a. of spacing *(Masch)* Teilungsgenauigkeit *f*
a. of work Arbeitsgenauigkeit *f*; Fertigungsgenauigkeit *f*
a. to gauge Maßhaltigkeit *f*, Lehrengenauigkeit *f*
a. to shape Formgenauigkeit *f*
a. to size Maßhaltigkeit *f*, Maßgenauigkeit *f*
accurate genau, maßgenau; präzis, fehlerfrei
a. cut *(Holz)* sauberer Schnitt *m*
a. hole location Lagegenauigkeit *f* der Bohrung
a. in size maßgenau
a. length maßgenaue Länge *f*
a. levelling *(Brg)* Feinhorizontierung *f*
a. position finder Zielsuchgerät *n (Radar)*
a. scanning Feinortung *f*
a. to geometry formgenau
a. to size genau nach Maß, maßhaltig
a. to three decimal places auf drei Dezimalstellen genau
accurately located (positioned) lagegenau
acerb herb
aceric acid Ahornsäure *f*
acescency Sauerwerden *n*
acescent sauer werdend; säuerlich
acetate Azetat *n*; Azetatfaserstoff *m*
a. fibre Azetatfaser *f*
a. film Azetatfilm *m*
a. foil Zellglasfolie *f*, Azetatfolie *f*
a. of copper Kupferazetat *n (Grünspan)*
a. rayon Azetatseide *f*
a. staple fibre Azetatfaser *f*
acetic acid Essigsäure *f*
a. anhydride Essigsäureanhydrid *n*
a. ether Äthylazetat *n*, Essigäther *m*

acetimeter Essigprüfer *m*
acetoacetic acid Azetessigsäure *f*
a. ester Azetessigester *m*
acetometer Azetometer *n*, Essigprüfer *m*
acetone Azeton *n*
a. lacquer Azetonlack *m*
a. resin Azetonharz *n*
acetylate / to azetylieren
acetylating agent Azetylierungsmittel *n*
acetylation Azetylierung *f*; Azetylieren *n*
a. number Azetylzahl *f*
acetylene Azetylen *n*
a. burner Azetylenbrenner *m*
a. cutter Brennschneider *m*
a. cylinder Azetylenflasche *f*
a. flame Azetylenflamme *f*
a. gas Azetylen *n*
a. generator Azetylenentwickler *m*
a. welding Azetylenschweißen *n*
Acheson graphite *(Am)* Elektrographit *m*, Acheson-Graphit *m*
achilleic acid Akonitsäure *f*
achromat Achromat *m (n)*
achromatic achromatisch
a. lens Achromat *m (n)*
achromaticity Achromasie *f*
achromatism Achromatismus *m*, Achromasie *f*
a. of magnification Achromasie *f* (Achromatismus *m*) der Bildgröße
a. of position Achromasie *f* des Bildortes, Achromatismus *m* der Bildlage
achromatization curve Achromatisierungskurve *f*
achromatize / to achromatisieren
acicular nadelartig, nadelförmig, nadelig
a. crystals nadelförmige Kristalle *mpl*
a. fracture nadelartiger Bruch *m*
acid sauer, gesäuert
acid Säure *f*
a. acceptor Säureakzeptor *m*
a. bath Säurebad *n*
a. bessemer process Bessemerverfahren *n*
a. brittleness *(Met)* Beizbrüchigkeit *f*, Beizsprödigkeit *f*
a. carboy Säureglasballon *m*
a. converter Bessemerbirne *f*
a. converter process Bessemerverfahren *n*
a. converter steel Bessemerstahl *m*
a. cure *(Plast)* Kaltvulkanisation *f*
a. density Säuredichte *f*
a. determination Säurebestimmung *f*
a. dye[stuff] Säurefarbstoff *m*
a. egg Druckbehälter *m*, Druckbirne *f*
a. etched durch Ätzung mattiert
a.-fast säurefest
a. fermentation saure Gärung *f*
a. fixer saures Fixiersalz *n*
a. flotation method (process) Säureschwemmverfahren *n*
a. for storage cells Akkumulatorsäure *f*
a.-forming säurebildend
a. fume scrubber Säurenebelwäscher *m*
a.-hardening *(Plast)* Säurehärten *n*
a. hydroextractor Säurezentrifuge *f*
a. leach Säureauslaugung *f*; *(Met)* saurer Laugeprozeß *m*
a. lined *(Met)* sauer zugestellt; mit saurer Ausmauerung
a. lining 1. säurefester Belag *m*; Säureverkleidung *f*; 2. *(Met)* saures Futter *n*, saure Zustellung *f*
a. melting hearth sauer zugestellter Schmelzherd *m*
a. number Säurezahl *f*
a. open hearth furnace saurer Siemens-Martin-Ofen *m*
a. parting Säurescheidung *f*
a. process Säureverfahren *n*
a.-proof säurebeständig, säurefest
a.-proof cement Säurekitt *m*
a.-proof lining säurefeste Auskleidung *f*
a. radical Säureradikal *n*
a. reaction saure Reaktion *f*
a.-resistant säurebeständig, säurefest
a.-resisting säurebeständig, säurefest
a.-resisting clothing Säureschutzkleidung *f*
a.-resisting paint säurefester Anstrich *m*
a.-resisting steel säurebeständiger Stahl *m*
a.-resisting varnish säurefester Lack *m*
a. salt saures Salz *n*
a. seal paint säurefeste Auskleidung *f*
a. siphon Säureheber *m*
a. slagging practice *(Met)* saure Schlackenführung *f*
a. sludge Säureabfall *m*; Säureschlamm *m*; Säureteer *m*
a. stop bath Unterbrechungsbad *n (Fotografie)*
a. swollen sole leather säuregeschwelltes Sohlleder *n*
a. syphon *s.* a. siphon
a. tank wagon Säurewagen *m*
a. tester Säureprüfer *m*, Säuremesser *m*
a. tower Säureturm *m*
a. treatment Säurebehandlung *f*
a. value Säurewert *m*, Säurezahl *f*
a. vat Säurebottich *m*
a. wash Absäuern *n*, Absäuerung *f*
acidic säurehaltig, sauer
acidiferous säuerlich, säurehaltig
acidification Säurebildung *f*; Ansäuern *n*, Sauermachen *n*
acidify / to 1. [an]säuern, azidifizieren; 2. sauer werden; 3. *(Text)* absäuern
acidimeter Säuremesser *m*

acidimetry Azidimetrie *f*, Säurebestimmung *f*, Säuremessung *f*
acidity Azidität *f*, Säuregehalt *m*, Säuregrad *m*
a. degree Aziditätsgrad *m*, Säuregrad *m*
acidizing of wells Erschließung *f* mit Säure *(Erdöllagerstätten)*
acidulate / to ansäuern, sauer machen, säuern
acidulated water wasserverdünnte Säure *f*
acidulous säuerlich
acierage *(Met)* Verstählen *n*; Aufkohlen *n*
acknowledger *(El)* Quittungsschalter *m*
acknowledging circuit *(El)* Quittungskreis *m*
acknowledgment of a signal Bestätigung *f* eines Signals *(Sendetechnik)*
a. signal Empfangsbestätigungssignal *n* *(Sendetechnik)*
aclinal *(Geol)* söhlig
acme [standard screw] thread *(Am)* Trapezgewinde *n*
a. thread tap Trapezgewindebohrer *m*
acnode isolierter Punkt *m* einer Kurve
acorn *(Schiff)* Windfahnenknopf *m*
a. nut Hutmutter *f*
a. tube *s.* a. valve
a. valve Kleinströhre *f*, Knopfröhre *f*, Acornröhre *f*, Eichelröhre *f*
acoustic akustisch
a.-absorbent silencer Absorptionsschalldämpfer *m*, Schalldämpfer *m* mit schallschluckenden Materialien
a.-absorbing material schallschluckendes Material *n*
a. absorption Schallabsorption *f*
a. absorption coefficient Schluckzahl *f*
a. absorptivity Schluckgrad *m*
a. admittance akustische Admittanz *f*, akustischer Leitwert *m*
a. board Akustikplatte *f*, Schalldämmplatte *f*
a. bridge akustische Brücke *f*
a. buoy *(Schiff)* Geräuschboje *f*
a. chamber Schallzelle *f*
a. clicks *(El)* Knackgeräusche *npl*
a. compliance *(El)* akustische Federung *f*, induktiver akustischer Blindwiderstand *m*
a. conductivity akustisches Leitvermögen *n*
a. delay line *(BMSR)* akustisches Laufzeitglied *n*
a. detector akustisches Ortungsgerät *n*
a. feedback akustische Rückkopplung *f*
a. figures Chladnische Klangfiguren *fpl*
a. flowmeter Ultraschall-Durchflußmesser *m*
a. generator Schallerzeuger *m*
a. grating akustisches Gitter *n*
a. impedance akustischer Scheinwiderstand *m*
a. inertance *(El)* kapazitiver akustischer Blindwiderstand *m*, akustische Trägheit *f*
a. inertia *(El)* 1. akustische Masse *f*; 2. Schallhärte *f (Verhältnis von Schalldruck zu Schallausschlag)*
a. irradiation Beschallung *f*
a. mass akustische Masse *f*
a. memory *(BMSR)* Ultraschallspeicher *m*, akustischer Speicher *m*
a. pattern Chladnische Klangfiguren *fpl*
a. power Schalleistung *f*
a. pressure Schalldruck *m*
a. properties of a room Raumakustik *f*
a. ray Schallstrahl *m*
a. reflection Schallreflexion *f*
a. resistance akustischer Widerstand *m*, Schall[wellen]widerstand *m*
a.-resistance-type dynamic microphone Rohrmikrofon *n (Typ eines Tauchspulenmikrofons)*
a. sounder *(Schiff)* Echolot *n*
a. sounding akustische Höhen- oder Tiefenmessung *f*
a. source Schallquelle *f*
a. stiffness Schallhärte *f*
a. storage *(BMSR)* akustischer Speicher *m*
a. tile schallschluckende Wandplatte *f*; Schluckplatte *f*
a. transmittivity Schalldurchlaßgrad *m*
a. velocity Schallgeschwindigkeit *f*, Schallschnelle *f*
a. wave Schallwelle *f*
acoustical absorption Schallabsorption *f*
a. absorption factor akustischer Absorptionsfaktor *m*
a. absorptivity Schallabsorptionskoeffizient *m*
a. board Akustikplatte *f*, Schalldämmplatte *f*
a. current akustischer Strom *m*, Schallstrom *m*
a. delay system akustisches Verzögerungssystem *n*
a. insulation Schalldämmung *f*, Schallisolierung *f*
a. ohm akustisches Ohm *n*
a. plate *(Holz)* Schalldämmplatte *f*
a. reactance akustischer Blindwiderstand (Wirkwiderstand) *m*
a. speech Sprechlautstärke *f*
a. switch akustischer Schalter *m*
acoustically compensated volume control gehörrichtige Lautstärkeregelung *f*
a. dead schalltot
a. inactive schalltot
a. live room Hallraum *m*
acoustics Akustik *f*, Schallehre *f*
acquire a craft / to einen Beruf erlernen
a. static electricity elektrisch werden
acquisition 1. Aufnahme *f (von Elektronen)*; 2. *(BMSR)* Datenerfassung *f*
a. radar Erfassungsradar *n*

acridine dye Akridinfarbstoff *m*
acrid[in]ic acid Akridinsäure *f*
acroleic acid Akroleinsäure *f*
across-flats dimensions Schlüsselweite *f (Werkzeuge)*
a. the grain quer zur Faser *(z. B. Holz)*
a.-the-line starter *s.* direct switching starter
acrylate Akrylat *n*, Akrylsäureester *m*
acrylic Akryl-
a. acid Akrylsäure *f*
a. fibre Akrylfaser *f*
a. plastic Akrylharzkunststoff *m*
a. resin Akrylharz *n*
acrylonitrile Akrylnitril *n*
act / to wirken, wirksam sein; einwirken; sich verhalten
a. in opposite phases *(El)* phasenverkehrt wirken
a. on wirken auf, angreifen an
a. on needles *(Text)* auf Nadeln einwirken
a. upon wirken auf, angreifen an
actinic radiation aktinische (chemisch wirksame) Strahlung *f*
a. screen Leuchtschirm *m*
actinium Aktinium *n*
a. emanation *s.* actinon
a. radioactive series Aktiniumzerfallsreihe *f*
a. series Aktiniumzerfallsreihe *f*
actinogram Aktinogramm *n*
actinograph Aktinograf *m*
actinometer Aktinometer *n*
actinon Aktiniumemanation *f*
actinotherapy UV-Therapie *f*
actinouranium Aktinouran *n*, Uran-235 *n*
action Gang *m*, Funktion *f*, Lauf *m (Maschine, Triebwerk)*
a. and reaction Wirkung *f* und Gegenwirkung *f*
a. current Aktionsstrom *m*
a. function Wirkungsfunktion *f*
a. of light Lichteinwirkung *f*
a. of the distributing-valve motion *(Wärm)* Steuerwirkung *f*
a. of the mass Massenwirkung *f*
a. period Abtastzeit *f (elektronische Speichertechnik)*
a. phase *s.* a. period
a. pitch tester Eingriffsteilung-Prüfgerät *n (Verzahnung)*
a. principle Wirkungsprinzip *n*
a. quantity Wirkungsgröße *f*
a. spot Abtastpunkt *m*
a. turbine Druckturbine *f*, Gleichdruckturbine *f*, Aktionsturbine *f*
a. turbine with pressure stages mehrstufige Druckturbine *f*, Druckturbine mit Druckabstufung
a. variable Wirkungsvariable *f*

activate / to 1. aktivieren, aktiv machen; anregen; 2. *(El)* laden; an Spannung legen; unter Strom setzen; 3. *(Brg)* beleben *(Flotation)*
activate Aktivierungsprodukt *n*, aktivierte Substanz *f*
activated aktiviert, belebt
a. carbon Aktivkohle *f*
a. charcoal Aktivkohle *f*, Adsorptionskohle *f*
a. earth aktivierte Bleicherde *f*
a. heater aktiver Heizfaden *m (von Elektronenröhren)*
a. molecule angeregtes Molekül *n*
a. sludge aktivierter (belebter) Schlamm *m*, Belebtschlamm *m*, Impfschlamm *m (Sanitärtechnik)*
a. sludge chamber Belebtschlammbecken *n*
a. sludge process (treatment) Belebtschlammverfahren *n (Sanitärtechnik)*
a. surface aktivierte Oberfläche *f*
activating agent Aktivierungsmittel *n*, Aktivator *m*
a. impulse *(Fs)* Betätigungsimpuls *m*
activation 1. *(Bw)* Aktivierung *f*, Schlammbelebung *f*, Schlammbelüftung *f (Sanitärtechnik)*; 2. *(Ch)* Aktivierung *(Reaktionskinetik)*
a. analysis *(Kern)* Aktivierungsanalyse *f*
a. cross section *(Kern)* Aktivierungsquerschnitt *m*
a. detector Aktivierungsdetektor *m*
a. energy *(Ch)* Aktivierungsenergie *f*
a. equation Aktivierungsgleichung *f*
a. of a surface Oberflächenaktivierung *f*, Oberflächenbeladung *f*
a. overvoltage Aktivierungsspannung *f (Galvanotechnik)*
a. yield *(Kern)* Aktivierungsausbeute *f*
activator Aktivator *m*, Aktivierungsmittel *n*, Beschleuniger *m*
active 1. *(Kern)* [radio]aktiv; 2. stromführend, spannungführend, unter Strom (Spannung); 3. in Betrieb befindlich; belastet, unter Last; 4. tätig
a. admittance *(El)* Wirkleitwert *m*
a. aerial gespeiste Antenne *f*
a. aerosol radioaktiver Schwebstoff *m*
a. air defence Luftabwehr *f*
a. balance return loss *(Fmt)* Nachbildfehlerdämpfung *f*
a. bleaching earth *s.* a. earth
a. coil wirksame (federnde) Windung *f*
a. component *(El)* Wirkkomponente *f*, Wattkomponente *f*
a. current *(El)* Wirkstrom *m*
a. cutting edge *(Masch)* Hauptschneide *f*
a. deposit *(Kern)* [radio]aktiver Niederschlag *m*
a. detergent *(Ch)* waschaktive Substanz *f*

a. earth Bleicherde *f*
a. earth pressure *(Bw)* aktiver Erddruck *m*
a. electrode aktive (differente) Elektrode *f*
a. emanation *(Kern)* [aktive] Emanation *f*
a.-energy meter *s.* watt-hour meter
a. force Belastungskraft *f*, angreifende Kraft *f*, auf einen Körper wirkende Kraft
a. fuse Betriebssicherung *f*
a. homing aktive Zielsuche (Zielansteuerung) *f (Flugkörper)*
a. hot metal mixer *(Met)* Vorfrischmischer *m*
a. input *(El)* aufgenommene Wirkleistung *f*
a. jamming absichtliche Störung *f*
a. lattice aktive Zone *f*, Spaltzone *f (eines Reaktors)*
a. line *(Fs)* wirksame Zeile *f*, Abtastlinie *f*, Zeile mit Bildinhalt
a. load Wirklast *f*
a. mass reagierende Masse *f*
a. material [radio]aktives Material *n*
a. mixer *(Met)* Vorwindfrischer *m*
a. oil seeps tätige Erdölaustrittsstellen *fpl*
a. oxygen aktiver Sauerstoff *m*
a. portion of a broach Schneidenteil *n* eines Räumwerkzeuges
a. power *(El)* Wirkleistung *f*
a. power meter Wirkleistungsmesser *m*, Wirkverbrauchsmesser *m*, Wirkstrommesser *m*
a. power relay Wirkleistungsrelais *n*
a. principle of pesticides Wirkstoff *m* der Schädlingsbekämpfungsmittel
a. product radioaktives Zerfallsprodukt *n*
a. profile of teeth tragende Zahnflanke *f (Verzahnung)*
a. resistance *(El)* Wirkwiderstand *m*, Ohmscher Widerstand *m*
a. return Echodämpfung *f*
a. return loss *(Am)* Echodämpfung *f*, Rückflußdämpfung *f*
a. rudder *(Schiff)* Aktivruder *n*
a. singing point *(Fmt)* wirksamer Schwingungseinsatzpunkt *m*
a. surface 1. *(Ch)* [katalytisch] wirksame Oberfläche *f*; 2. *(BMSR)* Schichtseite *f (von Aufzeichnungsträgern)*
a. thrust *(Bw)* aktiver Erddruck *m*
a. transducer 1. Energiewandler *m (mit Energiequelle)*; 2. aktiver Wandler *m*, aktives Übertragungssystem *n*
a. valence Normalvalenz *f*, aktive Wertigkeit *f*
a. vessel seeklares und eingesetztes Schiff *n*
a. voltage Wirkspannung *f*, Arbeitsspannung *f*
a. waste *(Kern)* radioaktiver Abfall *m*
a. zone *(Kern)* aktive Zone *f*, Spaltzone *f*
activity Aktivität *f*, Radioaktivität *f*; Wirksamkeit *f*
a. analysis Aktivitätsanalyse *f*, Prozeßanalyse *f (wirtschaftliche Statistik)*
a. curve Aktivitätskurve *f*
a. decay Aktivitätsabfall *m*
a. distribution Aktivitätsverteilung *f*
actual effektiv, Ist-, tatsächlich, vorhanden
a. address wirkliche Adresse *f (Ziffernrechner)*
a. cathode reelle Katode *f*
a. cost Gestehungskosten *pl*
a. deviation Ist-Abmaß *n*
a. dimension Istgröße *f*, Istmaß *n*, tatsächliche Größe *f*
a. drilling time reine Bohrzeit *f (Erdöl)*
a. efficiency *s.* efficiency
a. feed rate Ist-Vorschub *m*
a. frequency Arbeitsfrequenz *f*
a. grey-value grauwertrichtig *(Fotochemie)*
a. load Wirklast *f*
a. position Ist-Stellung *f*
a. power Wirkleistung *f*, tatsächliche Leistung *f*
a. resistance tatsächlicher Widerstand *m*
a. response curve *(Fmt)* wirksamer Bereich *m* einer Durchlaßkurve
a. size Istmaß *n*; natürliche Größe *f*
a. size of the reamer Nenndurchmesser *m* der Reibahle
a. steam consumption *(Wärm)* erreichter Dampfverbrauch *m*
a. value Istwert *m*, Effektivwert *m*, tatsächlicher Wert *m*
a. volumetric delivery tatsächliches Fördervolumen *n*
actuarial versicherungsmathematisch; versicherungsstatistisch
actuate / to betätigen; antreiben; auslösen; erregen
a. needles Nadeln betätigen *(Strickerei)*
a. the stop motion die Abstellvorrichtung auslösen
actuating 1. Betätigung *f (eines Schalters)*; 2. Ansprechen *n (eines Relais)*
a. appliance Betätigungsorgan *n*
a. arm Schalthebel *m*, Stellhebel *m*, Antriebshebel *m*, Betätigungshebel *m*, Auslösehebel *m*
a. cam *(Kfz)* Betätigungsnocken *m*, Stoßdaumen *m*
a. current 1. Betätigungsstrom *m*, Auslösestrom *m (Schaltgeräte)*; 2. Ansprechstrom *m (Relais)*
a. lever Betätigungshebel *m*
a. mechanism *(BMSR)* Stellgetriebe *n*, Stellantrieb *m*
a. piston Steuerkolben *m (Hydraulik)*
a. pressure *(BMSR)* Stelldruck *m*, Arbeitsdruck *m*

a. quantity erregende Größe *f (Relais)*
a. rod Schaltstange *f*
a. signal *(Am) (BMSR)* Regelabweichung *f*
a.-signal ratio *(Am) (BMSR) Übertragungsfunktion, die Regelabweichung und Führungsgröße [im geschlossenen Kreis] verknüpft*
a. switch Betätigungsschalter *m*
a. transfer function Betätigungsübertragungsfunktion *f*
actuation Betätigung *f*; Auslösung *f*
a. by foot Fußbetätigung *f*
a. of valves Ventilbetätigung *f*
actuator *(BMSR)* Stellantrieb *m*, Kraftschalter *m*
acuity Schärfe *f*, Auflösung *f (Fotografie)*
acutance Kantenschärfe *f (Fotografie)*
acute spitz; scharf
a. angle spitzer Winkel *m*
a. angled spitzwink[e]lig
a. exposure *(Kern)* kurzfristige Strahlenexponierung *f*, kurzfristige Bestrahlung *f*, Kurzzeitbestrahlung *f*
a. poisoning akute Vergiftung *f*
a. radiation injury *(Kern)* akuter Strahlenschaden *m*
a. triangle spitzwinkliges Dreieck *n*
acuteness Spitzwinkligkeit *f*; Schärfe *f*
acyclic azyklisch, unzyklisch
a. generator *(Elm)* Unipolargenerator *m*
a. irrotational motion *(Strm)* Potentialströmung *f* ohne Zirkulation
a. machine *(Elm)* Unipolarmaschine *f*, Nöggerath-Maschine *f*
acyclical azyklisch, unzyklisch
adamantine diamanthart, diamantartig
a. drill *(Brg)* Schrotbohrer *m*
a. luster Diamantglanz *m*
a. spar *s.* corundum
adapt / to 1. einrichten; 2. anpassen *(Unterprogramm in Ziffernrechnern)*
a. the laboratory information to everyday use Forschungsergebnisse für die Praxis anwendbar machen
adaptability Anpassungsfähigkeit *f*, Anpassungsvermögen *n*; Verwendungsbereich *m*
adaptation Adaptation *f*, Eignung *f*
a. brightness Adaptationsleuchtdichte *f*
a. phenomenon Anpassungserscheinung *f*
adapted for eingerichtet für, geeignet für
a. ring transformer *(Fmt)* Ringübertrager *m*
adapter 1. *(Masch)* Zwischenstück *n*; Flansch *m*, Paßstück *n*, Verbindungsstück *n*, Führungsbuchse *f*; Umspanndorn *m (Drehmaschinenspindel)*; Zwischenhülse *f (Schaftfräser)*; auswechselbares Einsatzstück *n (Halter einer Bohrerhülse)*; Futterscheibe *f (Klemmfutter)*; Schieber *m (Nietangelgattersäge)*; 2. *(El)* Zwischenstecker *m*, Adapter *m*, Zwischensockel *m*
a. amplifier Anpassungsverstärker *m*
a. bearing *(Masch)* Spannhülsenlager *n*
a. bushing *(Masch)* Anpaßbüchse *f*
a. jack *(El)* Steckkupplung *f*
a. plug *(El)* Anpaßstecker *m*, Übergangsstecker *m*, Kupplungsstecker *m*
a. ring Zwischenring *m*
a. sleeve Spannhülse *f (Lager)*
a. transformer Zwischentransformator *m*, Vorstecktransformator *m*
adapting Anpassung *f*
a. bottom tang untere Angel *f* zum Aufschieben *(Gattersäge)*
a. piece *s.* a. pipe
a. pipe Paßrohr *n*, Fassonrohr *n*, Formstück *n*
a. top tang obere Angel *f* zum Aufschieben *(Gattersäge)*
a. transformer Steuertransformator *m*, Anpaßtransformator *m*
adaption brightness Adaptionsleuchtdichte *f*
a. of the transformer Transformatoranpassung *f*
adaptive control system *s.* self-adaptive control system
a. logic anpassungsfähige Logik *f (Fähigkeit zu automatischer Störstellenbeseitigung, zuweilen mit Hilfe einer Fotowiderstandssteuermatrix)*
a. metallurgy Metallkunde *f*
adaptometer *(Opt)* Adaptometer *n*
adaptor *s.* adapter
adatom Adsorptionsatom *n*
Adcock direction finder Adcock-Peiler *m*
add / to 1. anfügen, anhängen, anlagern, beimengen, beimischen, hinzugeben; 2. *(Math)* addieren, zusammenzählen; 3. *(El)* zuschalten *(Last)*; 4. *(Met)* zugeben, zuschlagen, zulegieren *(zur Schmelze)*
a. a new joint to the drill pipe eine neue Stange an das Bohrgestänge anschrauben, nachsetzen
a. weight to the steam den Dampf beschweren *(Turbine)*
add-and-subtract counting chain Addier- und Subtrahierzählkette *f (automatische Rechner)*
a. gate *s.* adder
a. lister *s.* adding-listing machine
added heat zugeführte Wärme *f*
a. load *(El)* zugeschaltete Last (Normallast) *f*
a. metal Zusatzmetall *n*
addend Addend *m*, Summand *m*
addendum *(Masch)* Zahnkopfhöhe *f*, Kopfhöhe *f*
a. angle Zahnkopfwinkel *m (Kegelrad)*
a. circle Außenkreis *m*, Kopfkreis *m (Zahnrad)*

a.-corrected gearing *(Masch)* Stumpfverzahnung *f*, kopfkorrigierte Verzahnung *f*
a. cylinder Kopfkreiszylinder *m (Verzahnung)*
a. line Kopfkreis *m (Zahnrad)*
a. modification coefficient Profilverschiebungsfaktor *m (Verzahnung)*
adder 1. *(Fs)* Mischer *m*, Mischstufe *f*, Additionsstufe *f*; 2. *(BMSR)* Addierer *m*, Addierwerk *n*
a. accumulator Addierwerkakkumulator *m*, Addierwerkzwischenspeicher *m (in Ziffernrechenautomaten)*
adding-listing machine druckende (schreibende) Addiermaschine *f*
a. machine Additionsmaschine *f*
a. material Zusatzwerkstoff *m (beim Schweißen)*
a. of lime Kalkzuschlag *m*
a. register *(Büro)* Saldierwerk *n*
addition 1. Beimengung *f*, Beimischung *f*, Zugabe *f*, Zusatz *m*, Zutat *f*; 2. *(Math)* Addition *f*, Summierung *f*; 3. *(Bw)* Anbau *m*
a. agent Beizzusatz *m (Galvanotechnik)*; Dotierungsmittel *n*
a. compound *(Ch)* Anlagerungsverbindung *f*
a. polymerization Additionspolymerisation *f*
a. reaction *(Kern)* Anlagerungsprozeß *m*
additional zusätzlich, Zusatz-, Mehr-
a. air *(Wärm)* Nachluft *f*
a. amplifier Zusatzverstärker *m*
a. attachment Zusatzeinrichtung *f*
a. capacitor Zusatzkondensator *m*
a. charge Nachladung *f*, Zuladung *f*
a. equipment Zusatzgeräte *npl*
a. gun anode Nachbeschleunigungselektrode *f*
a. lift *(Flg)* Zusatzauftrieb *m*
a. load zusätzliche Last *f*, Zusatzlast *f*
a. noise *(Fmt)* Nebengeräusch *n*, Störgeräusch *n*
a. paper *(Typ)* Papierzuschuß *m (zum Einrichten)*
a. patent Zusatzpatent *n*
a. path *(Rak)* Umwegleitung *f*
a. plating in spaced armour Vorpanzerung *f*
a. steam Zusatzdampf *m (Turbine)*
a. treatment Nachbehandlung *f*
a. turbine wheel Vorschaltrad *n (Turbine)*
a. twist *(Text)* Nachdrehen *n*
a. twisting zusätzlicher Drall *m*
a. voltage *(El)* Zusatzspannung *f*
additions *(Met)* Zuschläge *mpl*, Zusätze *mpl*
additive *(Ch, Bw)* Zusatz *m*, Zusatzmittel *n*; Wirkstoff *m*
a. colour mixture additive Farbmischung *f*
a. colour process additives Dreifarbenverfahren *n* der Farbfotografie
a. colour system additives Farbensystem *n*
a. colour television system additive Farbfernsehmethode *f*
a. primary colour additive Grundfarbe *f*
a. process of colour photography additives Dreifarbenverfahren *n* der Farbfotografie
additivity Additivität *f*, additive Eigenschaft *f*
address *(Dat)* Adresse *f*
a. code Adressenkode *m*, Adressenteil *m (Teil eines Befehlswortes)*
a. modification Adressenumsteuerung *f (in Ziffernrechenautomaten)*
a. register Adressenregister *n (in Ziffernrechenautomaten)*
addressed memory *(BMSR)* volladressierter Speicher *m*
a. register *(BMSR)* angesprochener Speicherplatz *m*
adenosine triphosphoric acid Adenosintriphosphorsäure *f*
adhere / to [an]haften, kleben
a. by melting anschmelzen, zusammenschmelzen
adherence Anhaften *n*, Kleben *n*, Haften *n*; Haftfähigkeit *f*, Haftvermögen *n*; Adhäsion *f*
adherend *(Plast)* zu verklebender Körper *m*, Klebfläche *f*
adherent anhaftend, klebend
adhering Haften *n*, Kleben *n*
a. forces Haftkräfte *fpl*
adhesion 1. Adhäsion *f*, Haftfähigkeit *f*, Haftvermögen *n*; 2. *(Plast)* Klebfähigkeit *f*; Verklebung *f*
a. by molecular force *(Masch)* Ansprengen *n (von Endmaßen)*
adhesive anhaftend, klebrig, adhäsiv; gummiert
adhesive Klebemittel *n*, Kleber *m*, Klebstoff *m*
a. backing *(Text)* Klebebeschichtung *f*
a. bandage Heftpflasterverband *m*
a. binding *(Typ)* Klebebindung *f*
a. bonded foam *(Text)* adhäsiv verbundener Schaum *m*
a. bonded joint Klebeverbindung *f*
a. bonding Kleben *n*
a. capacity Adhäsionskraft *f*; Haftvermögen *n*
a. coat Klebeschicht *f*
a. dispersion *(Plast)* Klebdispersion *f*
a. film *(Plast)* Klebefolie *f*
a. foil Klebefolie *f*, selbstklebende Folie *f*
a. force Adhäsionskraft *f*, Bindekraft *f*; Haftvermögen *n*
a. laminating method *(Text)* Klebelaminierverfahren *n*
a. lamination of foam *(Text)* Laminieren *n* von Schaumstoff durch Aufkleben

a. lamination of polyester foam on knitted goods *(Text)* Laminieren *n* von Polyesterschaumstoff mit Maschenware durch Aufkleben
a. materials Kleb[e]stoffe *mpl*
a. mine Haftmine *f*
a. paste Kleister *m*
a. pitch Kabelwachs *n*
a. power Adhäsionskraft *f*; Adhäsionsfähigkeit *f*, Haftfähigkeit *f*, Haftvermögen *n*; Klebekraft *f*
a. tape 1. Klebeband *n*, Klebstreifen *m*; 2. Isolierband *n*
a. tension Haftspannung *f*
a. value *(Eis)* Haftwert *m* *(Schienen)*
a. weight Reibungsmasse *f* *(Zug, Lokomotive)*
adhesiveness Adhäsionsvermögen *n*, Haftvermögen *n*; Klebrigkeit *f*
adhesives *(Plast)* Klebstoffkitte *mpl*
adiabatic Adiabate *f*
adiabatic adiabatisch, ohne Wärmeausgleich verlaufend, wärmeundurchlässig
a. change [of conditions] adiabatische Zustandsänderung *f*
a. compression adiabatische Verdichtung *f*
a. condition adiabatische Bedingung *f*
a. curve Adiabate *f*, adiabatische Kurve *f*
a. demagnetization adiabatische Entmagnetisierung *f*
a. expansion adiabatische Ausdehnung (Expansion) *f*
a. graph *s.* a. curve
a. heat drop adiabatisches Wärmegefälle *n*
a. lapse rate *(Flg)* dynamische Erwärmung *f* und Abkühlung *f*
a. line Adiabate *f*, adiabatische Linie *f*
a. principle Adiabatenprinzip *n*
a. process adiabatische Zustandsänderung *f*
a. theorem Adiabatensatz *m*
adion *(Kern)* Adion *n*
adipamide Adipinsäureamid *n*
adipate Adipinsäureester *m*
adipocere *(Ch)* Fettwachs *n*, Leichenfett *n*
adit 1. *(Geol, Brg)* Stollen *m*, Stolln *m*; 2. *(Bw)* Zugangsstollen *m*, Stolleneingang *m*; 3. *(Am)* Lackpappe *f* *(Handelsname)*
a.-cut mining Stollenbergbau *m*
a. fan Seitenstollengebläse *n*
adjacent angrenzend, benachbart, Nachbar-; nebeneinanderliegend
a. angle Nebenwinkel *m*, Supplementwinkel *m*
a. channel *(Fmt)* benachbarter Nachrichtenkanal *m*, Nachbarkanal *m*
a. channel attenuation Trennschärfe *f* gegen Nachbarkanal
a. channel interference Seitenbandstörung *f*, Störung *f* durch den Nachbarkanal, Nachbarkanalstörung *f*
a. channel picture carrier Nachbarbildträgerwelle *f*
a. channel sensitivity Trennschärfe *f*
a. channel sound carrier Nachbartonträgerwelle *f*
a.-city telephone-traffic *(Fmt)* Nachbarortsverkehr *m*
a. double bonds *(Ch)* kumulierte Doppelbindungen *fpl*
a. frequency Nachbarfrequenz *f*, Seitenfrequenz *f*
a. picture carrier *(Fs)* Nachbarbildträger *m*
a. sound carrier *(Fs)* Nachbartonträger *m*
a. turns benachbarte Windungen *fpl*, Nachbarwindungen *fpl*
a. vision carrier *(Fs)* Nachbarbildträger *m*
adjective dye Beizenfarbstoff *m*, adjektiver Farbstoff *m*
adjoin / to adjungieren; *(Math)* anfügen; angrenzen [an]
adjoining concession *(Brg)* Nachbarfeld *n*
adjoint adjungiert, zugeordnet
a. function adjungierte (zugeordnete) Funktion *f*
adjust / to adjustieren, einstellen, justieren, regulieren; verstellen, ausrichten; einpassen, feinstellen, feststellen; nachstellen; richten; stellen; zusammenpassen, zurichten
a. a gauge for wear eine Lehre nachstellen
a. the governor den Regler einstellen
adjustability Einstellbarkeit *f*; Nachstellbarkeit *f*, Regelbarkeit *f*, Regulierbarkeit *f*; Verstellbarkeit *f*
a. of the slipper *(Masch)* Nachstellbarkeit *f* der Gleitschuhe
adjustable einstellbar, justierbar, nachstellbar; regelbar, regulierbar; verstellbar
a. bed press *(Masch)* Presse *f* mit verstellbarem Tisch
a. blade Verstellschaufel *f* *(Turbine)*
a. blade reamer Reibahle *f* mit Messereinstellung
a. clamp Schraubenzwinge *f*, Zwinge *f*, Klemmboden *m*, Klammer *f*
a. condenser *(El)* Stellkondensator *m*, Trimmer *m*
a. coupled axle *(Eis)* Lindner-Achse *f*
a. disk condenser Drehkondensator *m*
a. fence Anschlaglineal *n* mit verstellbaren Führungen *(Holzbearbeitung)*
a. footstep plate verstellbare Spurplatte *f* *(Turbine)*
a. head Schieber *m*, Schlitten *m* *(Schieblehre)*
a. head horizontal milling machine Planfräsmaschine *f* mit verstellbarem Spindelkopf
a.-height mit Höhenverstellung
a. hemmer *(Text)* verstellbarer Säumer *m*

a. **inductor** *(El)* regelbare Drossel *f*, Drossel[spule] *f* mit variabler Induktivität
a. **inside micrometer** verstellbare Lochlehre *f*, einstellbarer Schraublehrdorn *m*
a. **lever** Stellhebel *m*
a. **limit gap gauge** einstellbare Grenzrachenlehre *f*
a. **nut wrench** Rollgabelschlüssel *m*
a. **pipe tongs** Blitzrohrschlüssel *m*
a.**-pitch airscrew** Einstelluftschraube *f*
a. **plug coil** *(Fmt)* Steckspule *f*, Stöpselspule *f*
a. **point** einstellbares Komma *n* *(Ziffernrechner)*
a. **point gauge** Innentaster *m*
a. **rail milling machine** Fräsmaschine *f* mit verstellbarer Traverse
a. **reamer** nachstellbare Reibahle *f*
a. **resistance** *(El)* Regelwiderstand *m*
a. **resistor** *(El)* Abgleichwiderstand *m*, Einstellwiderstand *m*, Regelwiderstand *m*
a. **set square** Stellwinkel *m*
a. **slide** *(Wärm)* stellbare Pfanne *f*
a. **spanner** Universalschraubenschlüssel *m*
a.**-speed** drehzahleinstellbar
a.**-speed motor** Motor *m* mit Drehzahlregelung, Regelmotor *m*
a. **spindle** Meßspindel *f*
a.**-spindle drilling machine** Gelenkbohrmaschine *f* mit einstellbaren Spindeln
a. **stop** 1. *(Opt)* einstellbare Blende *f*; 2. *(Text)* Einstellanschlag *m*
a. **swivel rowlock** verstellbare Bügeldolle *f* *(Ruderboot)*
a. **swivel vice** *(Masch)* waagerecht und senkrecht verstellbarer Parallelschraubstock *m*
a.**-tension** mit einstellbarer Spannung *(Zugfeder)*
a. **thread guide** Kastenfadenführer *m* *(Strickmaschine)*
a. **transformer** Drehtransformator *m*; Regeltransformator *m*; Stelltransformator *m*
a. **tube scraper** verstellbarer Rohrreiniger *m*
a. **varying-voltage control** Drehzahlregelung *f* mit Kompoundcharakteristik
a. **voltage control** Drehzahlregelung *f* mit Nebenschlußcharakteristik
a. **wrench** verstellbarer Schraubenschlüssel *m*
adjuster Justiereinrichtung *f*, Einstelleinrichtung *f*, Einsteller *m*
adjusting apparatus Stellvorrichtung *f*
a. **clip** *(Masch)* Spurplättchen *n*
a. **collar** Einstellring *m*, Klemmring *m*, Stellring *m*, Verstellring *m*
a. **condenser** *(El)* Abgleicher *m*, Trimmer
a. **device** Justiereinrichtung *f*, Nachstellvorrichtung *f*, Regulierungsvorrichtung *f*, Stellvorrichtung *f*, Verstellvorrichtung *f*
a. **dog** *(Masch)* Stellklaue *f*
a. **equipment** *s.* a. device
a. **gauge** Einstellehre *f*
a. **gear** Stellzeug *n*
a. **gib** Nachstelleiste *f*, Einstelleiste *f*
a. **key** Stellkeil *m*, Stellmutter *f*, Stellschlüssel *m*
a. **knob** Stellknopf *m*, Drehknopf *m*
a. **ledge** Verstelleiste *f*
a. **linkage** Verstellgestänge *n*
a. **nut** Stellmutter *f*, Justiermutter *f*, Reguliermutter *f*
a. **piece** Paßstück *n*
a. **pin** Paßstift *m*, Justierstift *m*
a. **pivot** Einstellzapfen *m*
a. **point** Einstellspitze *f*
a. **potentiometer** Abgleichpotentiometer *n*
a. **ring** Einstellring *m*, Stellring *m*
a. **screw** 1. Einstellschraube *f*, Justierschraube *f*, Nachstellschraube *f*, Stellschraube *f*; Druckschraube *f* *(Walzwerk)*; 2. Stellspindel *f*, Verstellspindel *f*
a. **sleeve** *(Kfz)* Einstellhülse *f*
a. **slider** *(El)* Kontaktschieber *m*
a. **spring** Paßfeder *f*
a. **table** Justiertisch *m*
a. **valve** Regelventil *n*, regelbares Ventil *n*
a. **voltage** Einregulierspannung *f*
a. **wedge** Anzugskeil *m*, Stellkeil *m*, Verstellkeil *m*
a. **wrench** Einstellschlüssel *m*, Mutterstellschlüssel *m*
adjustment Adjustage *f*, Ausrichtung *f*, Berichtigung *f*, Einstellung *f*, Justage *f*, Justierung *f*, Nachstellung *f*, Regelung *f*, Regulierung *f*, Verstellung *f*
a. **bolt** Einstellbolzen *m*, Verstellbolzen *m*
a. **by hand wheel** Handradeinstellung *f*
a. **controls** *(BMSR)* Justierelemente *npl*, Einstellelemente *npl*
a. **cover** Einstelldeckel *m*
a. **mark** Einstellmarke *f*, Justiermarke *f*
a. **needle** Einstellnadel *f*; Einstellzeiger *m*
a. **of engine mixture** Gemischeinstellung *f* *(Verbrennungsmotor)*
a. **of errors** Fehlerausgleichung *f*
a. **of height** Höhenverstellung *f*
a. **pin** Justierstift *m*
a. **screw** Justierschraube *f*
adjutage Wasserabflußrohr *n*, Düse *f*
admeasure Aufmaß *n*, Maß *n*
administration Hauptverwaltung *f*, Leitung *f*
administrative office Verwaltungsbüro *n*
admiralty coefficient (constant) *(Schiff)* Admiralitätskonstante *f*
a. **court** Seeamt *n*, Admiralitätsgericht *n*
a. **knot** *(Schiff)* Admiralitätsknoten *m* *(Geschwindigkeitsmaß, 1 knot = 1,85318 km/h)*

a. list of lights *(Schiff)* Leuchtfeuerverzeichnis *n*, Feuerbuch *n*
a. metal Admiralitätslegierung *f (Sondermessing für Kondensatorrohre)*
a. metal brass Admiralitätsmessing *n*
a. unit *(El)* $^1/_{900}$ Mikrofarad *n*
admissible zulässig
a. length curve *(Schiff)* Schottenkurve *f*
admission Zulassung *f*, Einlassen *n*, Füllung *f*; Beaufschlagung *f (Turbine)*
a. branch Eintrittsstutzen *m (Turbine)*
a. chamber Eintrittskammer *f (Turbine)*
a. channel Eintrittskanal *m*
a. diameter Beaufschlagungsdurchmesser *m (Turbine)*
a. edge *(Wärm)* Einlaßkante *f*
a. lead *(Wärm)* Voreinströmung *f*, Voreintritt *m*
a. of steam Dampfeintritt *m*
a. pipe Einlaßrohr *n*, Eintrittsrohr *n*
a. port Eintrittskanal *m*, Dampfeinlaß *m*, Einlaßöffnung *f*, Einlaßschlitz *m*
a. pressure 1. *(Wärm)* Eintrittsdampfspannung *f*, Admissionsdruck *m*; 2. Einlaßdruck *m*
a. pressure curve *(Wärm)* Admissionsdruckkurve *f*
a. section *(Wärm)* Eintrittsquerschnitt *m*, Einlaßquerschnitt *m*
a. space *(Wärm)* Füllungsraum *m*
a. steam cock Dampfeinlaßhahn *m*
a. steam port Dampfeintrittskanal *m*
a. steam valve Dampfeintrittsventil *n*, Dampfeinlaßventil *n*
a. tube Einlaßrohr *n*, Einströmrohr *n*, Eintrittsrohr *n*, Zuflußrohr *n*
a. valve Einlaßventil *n*; Ansaugventil *n*
a. velocity Eintrittsgeschwindigkeit *f*; Einströmgeschwindigkeit *f*
admit / to 1. zulassen, einlassen, aufnehmen; 2. zuführen; 3. beaufschlagen
admittance *(El)* Scheinleitwert *m*, Admittanz *f*
admixed air *(Wärm)* Beiluft *f*, Nebenluft *f*
a. substance zugemischter Stoff *m*, Zusatzstoff *m*
admixture Beimengung *f*, Beimischung *f*; Begleitstoff *m*; Fremdbestandteil *m*, Verunreinigung *f*; Zusatz *m*, Zusatzmittel *n (Mörtel, Beton)*
a. of scraps Schrottzusatz *m*, Schrottzuschlag *m*
adobe *(Bw)* ungebrannter luftgetrockneter Ziegel *m*, Lehmziegel *m*, Luftziegel *m*
a. block *(Bw)* luftgetrockneter Lehmquader *m*
a. brick Lehmziegel *m*, Luftziegel *m*
a. clay lößartiger Lehm *m*, Tonerde *f*
adonitol *(Ch)* Adonit *n*
adornment *(Bw)* Schmuck *m*, Verzierung *f*
adrift treibend *(nicht verankert)*
adsorb / to adsorbieren, ansaugen, aufnehmen; anlagern; anreichern
adsorbability Adsorbierbarkeit *f*
adsorbable adsorbierbar
adsorbate Adsorbat *n*, adsorbierte Substanz *f*, aufgenommener Stoff *m*
adsorbed atom Adsorptionsatom *n*
a. layers adsorbierte Schichten *fpl*
adsorbent Adsorbens *n*, Adsorptionsmittel *n*, adsorbierender Stoff *m*, Sorbens *n*
a. bed Adsorptionsschicht *f*
adsorption Adsorption *f*; Anlagerung *f (eines Gases)*
a. capacity Adsorptionsfähigkeit *f*
a. chromatography Adsorptionschromatografie *f*
a. filtration Adsorptionsfiltration *f*, Filtration *f* durch eine adsorbierende Schicht
a. isotherm Adsorptionsisotherme *f*
a. power Adsorptionsvermögen *n*
a. thermostat Adsorptionsthermostat *m*
adsorptive precipitation adsorptive Fällung *f*
adulterant *(Ch)* Verfälschungsmittel *n*; Verschnittmittel *n*; Streckmittel *n*
adulterate / to *(Ch)* mischen, verschneiden, zusetzen, verfälschen, verunreinigen
adulterated turpentine Terpentinölersatz *m*
adulteration Verfälschen *n*, Verschnitt *m*
adumbration *(Opt)* Abschattung *f*
advance the timing of the ignition / to Frühzündung einstellen
advance 1. *(Masch)* Vorlauf *m*, Vorschub *m*, Beistellung *f*, Heranführen *n*; Zustellung *f (eines Werkzeuges)*; 2. Voreilung *f*
a. ball Polierkugel *f*
a. borehole *(Brg)* Vorbohrloch *n (Tiefbautechnik)*
a. copy *(Typ)* Signalexemplar *n*
a. element *(Fmt, Fs)* Differenzglied *n*; Vorhalteglied *n (Regelung)*
a. freight *(Schiff)* Frachtvorschuß *m*
a. heading *(Brg)* vorgesetzte Abbaustrecke *f*
a. ignition *(Kfz)* Vorzündung *f*
a. mechanism 1. *(Masch)* Vorschubmechanismus *m*; 2. *(Kfz)* Zündverstellung *f*
a. of the spindle Vorlauf *m* der Spindel
a. per conveyor move *(Brg)* Rückbreite *f* eines Förderers
a. per round *(Brg)* Abschlagstiefe *f*, Abschlagslänge *f*
a. planning Vorausplanung *f*
a. rate *(Brg)* Abbaufortschritt *m*, Abbaugeschwindigkeit *f*
a. sheet *(Typ)* Aushängebogen *m*, Aushänger *m*
a. winding *(El)* Steuerwicklung *f (eines Magneten)*

advanced heading *(Brg)* vorgehaltene Abbaustrecke *f*
a. ignition *(Kfz)* Frühzündung *f*
a. landing ground Feldflugplatz *m*
advancing Vorschieben *n*; Zustellung *f*
a. chain creeper *(Brg)* Vorholer *m*
a. face *(Brg)* Streb *m* im Vorwärtsbau
a. wave fortschreitende Welle *f*
adventitious light Streulicht *n*
advertising expense Werbekosten *pl*
a. film Werbefilm *m*
advice-boat *(Schiff)* Aviso *m*, Meldeboot *n*
advisory service Beratungsdienst *m*, Kundendienst *m*, technischer Dienst *m*
adz *s.*adze
a.-eye hammer Zimmererhammer *m*
adze / to behauen, beilen, dechseln, einblatten
adze 1. gerades Dachsbeil *n*, Dechsel *f*, Holzhaue *f*; 2. *(Brg)* Haue *f*
aerate / to 1. belüften; 2. mit Kohlensäure versetzen; 3. auflockern *(Formsand)*
aerated durchlüftet
a. concrete Schaumbeton *m*
a. plastic Schaumstoff *m*
a. sludge process Belebtschlammverfahren *n* *(Sanitärtechnik)*
aeration 1. Belüftung *f*, Bewetterung *f*, Lüftung *f*; 2. Versetzung *f* mit Kohlensäure; 3. Auflockerung *f* *(Formsand)*; Durchlüftung *f* *(Boden)*
a. chamber 1. *(Brg)* Luftkammer *f*; 2. *(Bw)* Lüftungsbecken *n* *(Sanitärtechnik)*
a. tank *(Bw)* Belüftungsbecken *n*, Lüftungsbecken *n* *(Sanitärtechnik)*
aerator Belüftungsanlage *f*, Belüfter *m*
a. basin Belüftungsbecken *n*
a. pipe Belüftungsrohr *n*
aerial 1. atmosphärisch; 2. oberirdisch, in der Luft; Luft-; Ober-
aerial Antenne *f*; Strahler *m*
a. amplifier insert Antennenverstärkereinschub *m*
a. array Richtantenne *f*, Richtstrahler *m*; Richtantennenanordnung *f*, Richtantennennetz *n*
a. attenuator Antennenabschwächer *m*
a. base Antennenfußpunkt *m*
a. booster Antennenverstärker *m*
a. cable Antennenkabel *n*; Freileitungskabel *n*; Luftkabel[leitung *f*] *n*
a. cableway Kabelkran *m*
a. car Seilbahnwagen *m*, Gondel *f* einer Schwebebahn
a. chair-lift Seilbahn *f*, Sessellift *m*
a. changeover switch Antennenumschalter *m*
a. choke Antennendrossel *f*
a. circuit Antennenkreis *m*
a. circuit keying Antennenkreistastung *f*
a. coil Antennen[kopplungs]spule *f*
a. conductor Luftleiter *m*
a. contact line Fahrdrahtoberleitung *f*
a. corridor Luftkorridor *m*, Flugschneise *f*
a. coupling feeder Antennenkopplungsleitung *f*
a. current Antennenstrom *m*
a. decrement Antennendämpfung *f*
a. dip Antennendurchhang *m*
a. directivity pattern Antennenrichtdiagramm *n*
a. drive Antennendrehvorrichtung *f*
a. effect Antenneneffekt *m*, Antennenwirkung *f* *(des Rahmens)*
a. extension Antennengebilde *n*
a. feed impedance Antenneneingangswiderstand *m*, Eingangsimpedanz *f* der Antenne
a. feeder Antennenspeiseleitung *f*, Antennenenergieleitung *f*
a. field gain Antennengüte *f*
a. gain Antennengewinn *m*, Antennenverstärkung *f* *(durch Richtwirkung)*
a. inductance Antennenabstimmung *f*; Antennenabstimmspule *f*
a. jack Antennenbuchse *f*
a. lead-in Antenneneinführung *f*, Antennenzuleitung *f*
a. loading Antennenbelastung *f*
a. loading coil Antennenverlängerungsspule *f*
a. loading inductance Antennenverlängerungsspule *f*
a. loss Antennenverlust *m*
a. loss damping Antennenverlustdämpfung *f*
a. mast Antennenmast *m*
a. matching and compensating means Antennenabgleichmittel *npl*
a. matching circuit Antennenanpassungskreis *m*
a. navigation Funknavigation *f*
a. noise Antennenrauschen *n*
a. overload protection Antennenüberspannungsschutz *m*
a. photograph Luftbild *n*
a. photography Luftbildtechnik *f*
a. power Antennenleistung *f*
a. power distributor Antennenleistungsverteiler *m*
a. reel Antennenhaspel *f*, Antennenrolle *f*
a. reflector Antennenspiegel *m*
a. ropeway Seilhängebahn *f*, Seilschwebebahn *f*
a. sailing Segelfliegen *n*
a. series condenser (capacitor) Antennenverkürzungskondensator *m*
a. shortening coefficient Antennenverkürzungsfaktor *m*
a. socket Antennenbuchse *f*

a. strain Antennenzug *m*
a. strand Antennenlitze *f*
a. support Abspannseil *n*, Abspannung *f*, Antennenstützpunkt *m*
a. survey Überwachung *f* durch Flugzeuge
a. survey camera Luftbildkamera *f*
a. surveying Luft[bild]vermesssung *f*, Fotogrammetrie *f*
a. suspension Antennenaufhängung *f*
a. telecommunication cable Fernmeldeluftkabel *n*
a. terminal Antennenanschluß *m*
a. torpedo Lufttorpedo *m*
a. tower Antennenmast *m*; Funkturm *m*
a. tramway *s.* a. ropeway
a. triangulation Aerotriangulation *f*
a. tuning capacitor (condensor) Antennenabstimmkondensator *m*
a. tuning coil Antennenabstimmspule *f*
a. tuning equipment Antennenabstimmittel *n*
a. tuning inductance Antennenabstimmspule *f*, Antennenverlängerungsspule *f*
a. unit Strahlerkasten *m*
a. view Luftaufnahme *f*, Luftbild *n*
a. wire change-over switch Antennenschalter *m*
aerials *(sl) (El)* Freileitung *f*
aeriform luftförmig, gasförmig
aerify / to *(Ch)* mit Luft vermischen
aero-engine Flugmotor *m*
a.-engine industry Flugmotorenindustrie *f*
aerobatic pilot Kunstflieger *m*
aerobatics Kunstflug *m*
Aerocrete *(Bw)* Gasbeton *m (Handelsname)*
aerodrome Flughafen *m*
a. beacon Flughafenleuchtfeuer *n*, Flughafenbake *f*
a. control radar Flughafenfunkmeßgerät *n*, Flughafenkontrollradar *n*
a. forecast Flughafenwettervorhersage *f*
a. of departure Abgangsflughafen *m*, Start[flug]platz *m*
a. of destination Bestimmungsflughafen *m*
aerodynamic aerodynamisch
a. centre of a wing section aerodynamischer Mittelpunkt *m* einer Tragflächeneinheit
a. drag Luftwiderstand *m*
a. heating aerodynamische Erwärmung *f*
aerodynamics Aerodynamik *f*
aerodyne Luftfahrzeug *n*, das schwerer als Luft ist *(Gegensatz: aerostat)*
aerofoil blade Tragflügelschaufel *f (Turbine)*
a. propeller Tragflügelpropeller *m*
a.-section blade Tragflügelschaufel *f (Turbine)*
aerolite *(Am)* Aluminiumlegierung *f*
aeromechanics Aeromechanik *f*
aeronautic station Bodenfunkstelle *f*
aeronautical beacon Luftfahrtleuchtfeuer *n*
a. engineering Luftfahrttechnik *f*
a. fixed service fester Flugmeldedienst *m*
a. fixed station feste Bodenfunkstation *f*
a. fixed telecommunication *s.* a. fixed service
a. industry Luftfahrtindustrie *f*
a. instrument Fluggerät *n*
a. lights Luftfahrtbefeuerung *f*
a. map Flugkarte *f*
a. radio Flugfunk *m*
a. radio service Flugfunkdienst *m*
a. station Luftfahrtfunkstelle *f*
aeronautics Luftfahrt *f*, Luftfahrtwesen *n*; Fluglehre *f*
aeroplane Flugzeug *n*
a. effect *(Ph)* Flugzeugeffekt *m*
a. flutter *(Fs)* Flugzeugflattern *n*
a. hangar Flugzeughalle *f*
aerose messingartig
aerosiderite Meteoreisenstein *m*
aerosol generator Gebläsezerstäuber *m*
aerostat Luftfahrzeug *n*, das leichter als Luft ist; gasgetragenes Luftfahrzeug
aerostatic aerostatisch
aerugo Rost *m*; Grünspan *m*, Patina *f*; Kupferoxid *n*
affect / to [ein]wirken
affected overburden Einwirkungsbereich *m (Gebirgsmechanik)*
a. with cancer krebsig *(Holz)*
affiliated company Tochtergesellschaft *f*
affine affin
a. connection *(Math)* affiner Zusammenhang *m*, lineare Übertragung *f*
a. transformation *(El)* affine Verzerrung *f*
affinity Affinität *f*
a. constant Affinitätskonstante *f*
a. to the fibre *(Text)* Anfärbevermögen *n*, Ziehvermögen *n*, Faseraffinität *f*
affix / to anhängen
a. an index einen Index anhängen, indizieren
affluent Zufluß *m*
afforest / to aufforsten, bewalden
afforestation Aufforstung *f*; aufgeforstetes Land *n*
affreightment Befrachtung *f*, Schiffsbefrachtung *f*
afloat *(Schiff)* flott, frei schwimmend, im Wasser befindlich
aft achtern, im Hinterschiff
after accelerator Nachbeschleunigungselektrode *f*
a.-bake *(Plast)* Nachhärten *n*
a. blow / to *(Met)* nachblasen
a. body Hinterschiff *n*

a. braces *(Schiff)* Achterbrassen *fpl*
a.-burning Nachbrennen *n*
a.-charging *(Met)* Nachsetzen *n*
a.-contraction *(Met)* Nachschwindung *f*
a. count Nachentladung *f (Zählrohr)*
a. cuddy *(Schiff)* Hinterplicht *f*
a. deck Achterdeck *n*
a.-effect Nachwirkung *f*
a.-end arrangement Achterschiff *n (alle Schiffsaufbauten achtern)*
a.-finish *(Text)* Nachappretur *f*
a. frames *(Schiff)* Achterspanten *npl*, Hinterschiffsspanten *npl*
a.-heat Restwärme *f (Reaktor)*
a. hold *(Schiff)* Achterraum *m*, hinterer Laderaum *m*
a. leech *(Schiff)* Achterliek *n*, Außenliek *n*
a.-machining Nach[be]arbeiten *n*
a. peak *(Schiff)* Achterpiek *f*, Hinterpiek *f*
a. peak bulkhead *(Schiff)* Stopfbuchsenschott *n*, Sternbuchsenschott *n*
a.-plan hinteres Schlachtdeck *n (eines Fischfabrikschiffes)*
a.-portion of the fuselage *(Flg)* Hinterteil *n* des Rumpfes, Heck *n*
a.-potential *(El)* Nachpotential *n*
a. pulse *(Kern)* unechter Impuls *m*
a. quarter Achterschiff *n*, Hinterschiff *n*
a. rib Heckspant *n*, Hinterschiffsspant *n*
a. run Schiffsformverlauf *m* achtern
a. sails *(Schiff)* Achtersegel *n*, Hintergeschirr *n*
a.-sales service Kundendienst *m*; Garantieleistung *f*; technischer Dienst *m*; Service *m*
a.-tank Wasserballast *m* hinten
a. worker Nachdrehschleuder *f (Zuckerfabrikation)*
afterage / to nachaltern
afterbody plan Hinterschiffsspantenriß *m*
afterburning *(Rak)* Nachverbrennung *f*
aftercooler Endkühler *m*, Nachkühler *m*, Rückkühler *m*
afterdamp Nachschwaden *m (Explosion)*
afterglow Nachleuchten *n*
a. duration Nachleuchtdauer *f*
afterheat Nachwärme *f*
aftermath Grummet *n*, Spätheu *n*, Zweitschnitt *m*
afternoon shift Mittelschicht *f*, Mittagsschicht *f*
afterpulse Nachimpuls *m*, unechter Impuls *m*
aftershave lotion Rasierwasser *n (nach der Rasur)*
aftertreat / to avivieren, nachbehandeln
aftertreatment Avivage *f*, Nachbehandlung *f*
a. with a metallic salt *(Text)* Metallsalznachbehandlung *f*
against-the-light picture Gegenlichtaufnahme *f*
a. the tide *(Schiff)* gegen den Strom

agalite Faserkalk *m*
agar-agar Agar-Agar *m (n)*
agate Achat *m*
a. mortar Achatmörser *m*
agatiferous achathaltig
agatine achatartig
agave fibre Sisalfaser *f*, Sisalhanf *m*
age / to 1. ablagern; 2. *(Met)* altern; aushärten; 3. reifen; 4. *(Text)* dämpfen, verhängen
a. artificially warmaushärten
a. harden altern *(Leichtmetalle)*; aushärten *(Stahl)*
age Alter *n*; Formation *f*, geologisches Alter; Zeitraum *m*
a.-class Altersklasse *f*; Bestandsalter *n (Holz)*
a.-coating Beschlagen *n (von Glühlampen)*
a. dependent zeitabhängig
a. determination Altersbestimmung *f*
a.-development *(Text)* Dampfentwicklung *f*
a. equation *(Kern)* Altersgleichung *f*, Fermi-Alter-Gleichung *f*
a.-hardenable alterungsfähig, aushärtbar
a. hardening 1. *(Met)* Aus[scheidungs]-härtung *f*; 2. *(Bw)* Zunahme *f* der Festigkeit mit dem Alter *(z. B. Beton)*
a. of stand *(Holz)* Bestandsalter *n*
a. of the beds Alter *n* der Schichten, Schichtenalter *n (Erdöl, Erdgas)*
a.-protecting agent Alterungsschutzmittel *n*
aged abgelagert, gealtert; gereift
a. black Hängeschwarz *n*, Oxydationsschwarz *n*
ag[e]ing 1. Altern *n*, Alterung *f*; Aushärten *n*; Anlagern *n*; 2. *(Text)* Dämpfen *n*, Dampfbehandlung *f*; Nachreifen *n (Chemiefaserherstellung)*
a. at room temperature Kaltaushärtung *f*
a. crack Alterungsriß *m*
a. hopper *(Ch)* Vorreifekasten *m*
a. in the air *(Text)* Oxydieren *n* an der Luft
a. process Alterungsverfahren *n*; Lagerungsprozeß *m*; Aushärtung *f*
a. resistance Alterungsbeständigkeit *f*
a. test Alterungsprobe *f*, Alterungsprüfung *f*, Alterungsversuch *m*
a. time *(Plast)* Nachhärtungsfrist *f*
a. treatment Alterungsbehandlung *f*, Auslagerungsbehandlung *f*
a. with increased temperature Warmaushärtung *f*
agent *(Ch)* Agens *n*, Faktor *m*, Medium *n*, Mittel *n*, Reaktionsmittel *n*
ager *(Text)* Dämpfer *m*
agglomerant Bindemittel *n*
agglomerate / to agglomerieren, zusammenbacken, zusammenballen, [zusammen]sintern
agglomerate *(Ch)* Agglomerat *n*, Anhäufung *f*, Sintererzeugnis *n*, Sinterkuchen *m*

agglomerated cake Agglomerat[kuchen *m*] *n*, Sinterkuchen *m*
agglomerating backend
a. plant Agglomerieranlage *f*, Sinteranlage *f*
agglomeration Agglomeration *f*, Anhäufung *f*, Ballung *f*, Sinterung *f*, Zusammenbacken *n*, Zusammenballen *n*, Zusammenballung *f*
agglutinant Klebemittel *n*, Bindemittel *n*
agglutinate / to ankleben, leimen, verkleben, verkleistern, zusammenleimen; binden; haften machen
agglutinating value *(Brg)* Backfähigkeit *f*
agglutinative verklebend
aggregate 1. Aggregat *n*; 2. Zuschlag[stoff] *m* *(Betontechnologie)*
a. grading Kornabstufung *f* *(Betontechnologie)*
a. grading curve Sieblinie *f* *(Betontechnologie)*
a. graduation Kornabstufung *f* *(Betontechnologie)*
a. of particles Punkthaufen *m*
a. recoil *(Ph)* Molekülrückstoß *m*
a. thickness *(Geol)* Gesamtflözmächtigkeit *f* *(einschl. Mittel)*
aggregation 1. Zusammenballung *f*, Aggregation *f*; 2. *(Min)* Verwachsung *f*
a. of minerals Mineralaggregat *n*
aggregative Gesamt-
a. polarization *(El)* Aggregatpolarisation *f*
aggressive tack *(Plast)* Trockenklebrigkeit *f*
aging *s.* ageing
agitate / to 1. [auf]rühren, aufrütteln, auflockern, bewegen, durchrühren, durchschütteln, erregen, in wirbelnde Bewegung versetzen, mischen, schütteln, umrühren; 2. schüren
agitated sea bewegte See *f*
agitating truck Transport[beton]mischer *m*, Rührfahrzeug *n*
a. vane Rührflügel *m*
agitation 1. Rühren *n*; Bewegung *f*; Schütteln *n*; 2. Schüren *n*
a.-froth machine *(Brg)* Rührwerkszelle *f* *(ohne Zusatzluft)*
agitator 1. Rührwerk *n*, Mischwerk *n*, Quirl *m*, Rührapparat *m*, Rührer *m*, Rühreinrichtung *f*, Schütteleinrichtung *f*; 2. Tonschneider *m* *(Elektroakustik)*
a. driving gear Rührwerkantrieb *m*
a. tank *(Masch)* Rührtank *m*
a. truck Rührfahrzeug *n*, Transport[beton]mischer *m*
agonic line Agone *f*
agricultural chemistry Agrikulturchemie *f*
a. fair Landwirtschaftsausstellung *f*
a. hydraulics landwirtschaftlicher Wasserbau *m*
a. implements Ackergeräte *npl*, landwirtschaftliche Geräte *npl*
a. machinery Landmaschinen *fpl*
a. soil Ackerboden *m*, Ackerland *n*
a. tools Ackergeräte *npl*, landwirtschaftliche Geräte *npl*
agriculture Ackerbau *m*, Landwirtschaft *f*
aground *(Schiff)* aufgelaufen
ahead indication Vorausanzeige *f* *(Radar)*
a. of schedule vorfristig
a. turbine Vorwärtsturbine *f*, Turbine *f* für Vorwärtsgang
aided tracking system 1. Radarzielverfahren *n*; 2. Flugsicherungsdienst *m*
aids to navigation radio control Navigationsfunkhilfe *f*
aiguille *(Bw)* Steinbohrer *m*
ailanthus silk *(Text)* Fagaraseide *f*
a. silkworm Ailanthusspinner *m*
aileron *(Flg)* Querruder *n*
a. balance *(Flg)* Ausgleich *m* des Querruders
aiming controller *(Mil)* Ziellöffel *m* *(Schießen)*
air / to auslüften, belüften, durchlüften, [ent]lüften, ablüften
air Luft *f*, Atmosphäre *f*
a. absorption Luftabsorption *f*, Luftaufnahme *f*
a. access Luftzutritt *m*
a. accumulator Druckluftspeicher *m*, pneumatischer Akkumulator *m*
a.-acetylene flame Brenngas-Luft-Flamme *f*
a.-actuated druckluftbetätigt
a. admission Luftzutritt *m*
a. atomization Luftzerstäubung *f*
a. baffle Luftleitblech *n*
a. bag Heizschlauch *m*
a. bag under pressure Drucksack *m*
a. base Luftstützpunkt *m*
a. bearing Luftlager *n*
a. bell Luftblase *f*; Glasblase *f*, Glasfehler *m*
a. blanket Luftschleier *m*, Luftmantel *m*
a. blast Gebläse[luft *f*] *n*, Gebläsewind *m*; Wind *m*; Druckluftstrom *m*
a.-blast breaker Druckluftschalter *m*, Druckgasschalter *m*
a.-blast circuit breaker Druckluft[leistungs]-schalter *m*, Expansionsschalter *m*
a.-blast transformer Transformator *m* mit Luftkühlung (Fremdbelüftung), Trockentransformator *m* mit Luftkühlung
a.-blast velocity Windgeschwindigkeit *f*
a. blasting 1. Druckluftgußputzen *n*; 2. Blasen *n*
a. bleed Belüftungsöffnung *f*; Entlüftungsöffnung *f*
a. bleed duct Entlüftungskanal *m*
a. bleed opening Nebenlufteinlaß *m*

a. bleed valve Ventil *n* zur Rückführung von Luft
a. bleeder *(Kfz)* Bremsluftbohrung *f*
a. blowing *(Met)* Windfrischen *n*
a.-borne *s.* airborne
a. bottle Druckluftflasche *f*, Druckluftbehälter *m*
a. box Unterdruckkammer *f*
a. brake 1. Druckluftbremse *f*, pneumatische Bremse *f*; 2. *(Flg)* Landeklappe *f*
a.-brattice *(Brg)* Wetterscheider *m*
a. break *(El)* Trennstrecke *f (des Kontaktes in Luft)*, Luftschaltstrecke *f*
a.-break circuit breaker *(El)* Luftschalter *m*
a.-break contactor *(El)* Luftschütz *n*
a. brick 1. Lochstein *m*, Lochziegel *m*, Lüftungsziegel *m*; 2. luftgetrockneter Ziegel *m*, Luftziegel *m*
a. brush coater *(Holz)* Luftbürstenstreichmaschine *f*
a. bubble Lufteinschluß *m*; Libellenblase *f*, Luftblase *(Wasserwaage)*
a. buffer [abgepuffertes] Luftpolster *n*, Luftpuffer *m*
a.-bulked luftgebauscht
a. buoyancy Luftauftrieb *m*
a. burst Explosion *f* in der Atmosphäre
a. caisson Druckluftsenkkasten *m*, Druckluftkammer *f*
a. capacitor Luftkondensator *m*
a.-car Autoflugzeug *n*
a. casing Kühlluftmantel *m*
a. cavity Lufteinschluß *m*
a. cell 1. *(El)* Luftsauerstoffelement *n*, Luftsauerstoffzelle *f*; 2. Luftspeicher *m (Dieselmotor)*
a. cell battery *(El)* Luftsauerstoffbatterie *f*
a. cell brake cylinder Luftspeicher-Bremszylinder *m*
a. cell combustion chamber Luftspeicherbrennraum *m (Dieselmotor)*
a. chamber 1. Luftbehälter *m*, Luftkammer *f*, Windkasten *m*; 2. *(Met)* Windmantel *m*, Windring *m (des Ofens)*; 3. Luftkasten *m (Rettungsboot)*
a. change *(Bw)* Luftumwälzung *f*, Luftwechsel *m*
a. channel *(Bw)* Luftkanal *m*, Lüftungskanal *m*
a. charged with moisture mit Feuchtigkeit durchsetzte Luft *f*
a. chuck *(Masch)* Preßluft[spann]futter *n*
a. chucking *(Masch)* Druckluftspannung *f*
a. chute *(Bw)* Luftkanal *m*, Lüftungskanal *m*
a. circuit breaker *s.* a.-break circuit breaker
a.-circulating furnace *(Met)* Luftumwälzofen *m*
a. circulation Luftumlauf *m*, Luftzirkulation *f*, Luftumschichtung *f*
a. circulation furnace Luftumwälzofen *m*
a. classification 1. Windsichtung *f*; 2. *(Holz)* Luftsortierung *f*
a. classifier Windsichter *m*
a. cleaner Luftfilter *n*, Luftreiniger *m*
a. cleaner bracket Luftansaugfilterstutzen *m*
a. cleaning Luftfilterung *f*, Luftreinigung *f*
a. coater *(Holz)* Luftpinselstreichmaschine *f*
a. cock *(Masch)* Lufthahn *m*
a. column Luftsäule *f*
a. compartment *(Brg)* 1. Luftfach *n (Setzmaschine; Aufbereitung)*; 2. Wettertrum *n (Tiefbautechnik)*
a. compression Luftverdichtung *f*
a. compressor Drucklufterzeuger *m*, Kompressor *m*, Luftverdichter *m*
a. condenser *(El)* Luftkondensator *m*; Kondensator *m* mit Luftkühlung
a. conditioner Klimaanlage *f*, Klimatisierungsanlage *f*
a.-conditioning Klimatechnik *f*; Klimatisierung *f (Heizung, Kühlung, Reinigung, Feuchtigkeitsregulierung der Luft)*
a.-conditioning of fabrics *(Text)* Klimatisierung *f* von Geweben
a.-conditioning plant (system) Klimaanlage *f*, Klimatisierungsanlage *f*
a. conduit *(Brg)* Wetterlutte *f*
a. contactor *s.* a.-break contactor
a. control 1. Luftregulierung *f*; 2. Preßluftsteuerung *f*
a. control valve *(Met)* Luftdrucksteuerventil *n*
a. coolant Kühlluft *f*
a.-cooled luftgekühlt
a.-cooled commutator motor Kollektormotor *m* mit Fremdbelüftung
a.-cooled transformer Lufttransformator *m*, Transformator *m* mit Luftkühlung
a.-cooled triode luftgekühlte Triode *f*
a.-cooled tube luftgekühlte Röhre *f*
a. cooler Luftkühler *m*
a.-cooling Luftkühlung *f*
a.-cooling fan Kühlluftgebläse *n*
a.-core choke *(El)* Luftdrossel *f*, Drossel *f* ohne Eisenkern
a.-core transformer Lufttransformator *m*, eisenloser Transformator *m*
a.-cored choke *(El)* Luftdrosselspule *f*, Drossel *f* ohne Eisenkern
a.-cored coil *(El)* Luftspule *f*, kernlose Spule *f*
a.-cored solenoid *(El)* Luftspule *f*, eisenlose Drossel[spule] *f*
a. correction jet *(Kfz)* Ausgleichluftdüse *f*, Luftkorrekturdüse *f*

a. course 1. Luftströmung *f*, Luftstrom *m*; 2. *(Brg)* Wetterkanal *m*, Wetterstrecke *f*; 3. Luftgang *m*, Luftloch *n (in Schiffsbauteilen)*
a. course board Fülling *f*, Seitenfülling *f (Holzschiff)*
a. crossing *(Brg)* Wetterkreuz *n*
a. curing Abbinden *n* und Erhärten *n* an der Luft, Lufterhärtung *f (Betontechnologie)*
a. curtain Luftschleier *m*
a. curtain door *(Bw)* Luftschleiertür *f*
a. cushion Luftkissen *n*, Luftziehkissen *n*, pneumatisches Ziehkissen *n*
a.-cushion vehicle Luftkissenfahrzeug *n*
a. cushioning Luftfederung *f*
a. cylinder Luftflasche *f (Anlasser)*
a. damper Luftpuffer *m*, Luftdämpfer *m*
a. data computer Bordrechner *m*, Flugzeugrechner *m*
a. defence Luftverteidigung *f*
a. deficiency Luftmangel *m*
a. density Luftdichte *f*
a. dielectric Luftdielektrikum *n*
a.-dielectric variable capacitor Luftdrehkondensator *m*
a. dilution Luftverdünnung *f*
a. discharge Luftentladung *f*
a. dope Bespannungslack *m*, Flug[zeug]lack *m*
a. draft Luftzug *m*, Zugluft *f*
a. drag Luftwiderstand *m*
a. drain *(Bw)* Luftkanal *m*, Luftloch *n (Lüftungsöffnung in einer Außenwand)*
a. drain petcock *(Masch)* Entlüftungshahn *m*
a. draught Luftzug *m*
a.-dried luftgetrocknet; lufttrocken
a.-dried hide lufttrockene Haut *f*
a.-dried timber lufttrockenes Holz *n*; luftgetrocknetes Holz
a. drill Preßluftbohrer *m*, Druckluftbohrer *m*
a. drilled well mit Luft[zusatz zur Spülung] heruntergebrachte Bohrung *f (Erdöl)*
a. drilling Spülbohren *n* unter Luftzusatz *(Erdöl)*
a. drive Ölförderung *f* mittels Druckluft
a. dry lufttrocken, lutro
a. drying Lufttrocknung *f*
a. drying machine *(Text)* Lufttrockenmaschine *f*
a. duct 1. Luftleitung *f*; 2. *(Masch)* Luft[zuführungs]kanal *m*, Windkanal *m*; 3. *(Bw)* Lüftungskanal *m*; 4. *(Brg)* Wetterlutte *f*, Lutte *f*; 5. *(Elm)* Kühlkanal *m*, Kühlschlitz *m*
a. ducting Luftführung *f*
a. ducting system Luftführungsanlage *f*
a. ejector Dampfstrahlluftpumpe *f*, Dampfstrahler *m*
a.-elutriator *(Brg)* Windsichter *m*
a. entering above the grate *(Wärm)* Oberluft *f*
a.-entrained concrete LP-Beton *m*, Luftporenbeton *m*
a. entraining *(Bw)* luftporenbildend; Luftporenbildung *f (Betontechnologie)*
a.-entraining admixture (agent) LP-Zusatz *m*, Luftporenbildner *m (Betontechnologie)*
a.-entraining cement Luftporenzement *m (Zement mit LP-Zusatz)*
a.-entrainment Luftporenbildung *f (Betontechnologie)*
a.-entrainment agent LP-Zusatz *m*, Luftporenbildner *m (Betontechnologie)*
a. escape Entlüftung *f*; Luftauslaß *m*; Luftabzug *m*; *(Brg)* Wetterabzug *m*
a. escape valve Entlüftungsventil *n*; Luftabzugsventil *n*
a. exhaust Luftabzug *m*, Luftauslaß *m*
a. exhauster Entlüfter *m*
a.-exit hole Luftaustrittsöffnung *f*
a.-felted fibreboard luftverfilzte Faserplatte *f*
a.-felting process *(Holz)* Luftverfilzungsverfahren *n*
a.-felting unit *(Holz)* Luftverfilzer *m*
a.-filled luftgefüllt, mit Luftfüllung; offen
a.-filled ionization chamber *(Ph)* Ionisationskammer *f* mit Luftfüllung
a. filter Luftfilter *n*, Luftreiniger *m*
a. filter intake silencer *(Kfz)* Luftfilter-ansauggeräuschdämpfer *m*
a. flap *(Kfz)* Luftklappe *f*
a. flooding Ölförderung *f* mittels Druckluft
a. flue Entlüftungsrohr *n*, Lüftungsrohr *n*; Abluftkanal *m*, Luftkanal *m*; Wetterabzug *m*, Wrasenabzug *m*, Wrasenrohr *n*
a. flushing Luftspülung *f (Tiefbautechnik)*
a. foam rubber Schaumgummi *m*
a. fog Luftschleier *m*
a. for combustion Verbrennungsluft *f*
a. force signal corps Luftnachrichtentruppen *fpl*
a. friction Luftreibung *f*
a. fuel mixture Kraftstoff-Luft-Gemisch *n*
a. funnel Schnorchel *m*
a. furnace Flammofen *m*, Wetterofen *m*, Windofen *m*
a. gap Luftspalt *m*; Polschuhspaltbreite *f (zwischen Magnetpolen)*; Luftstrecke *f (zwischen Elektroden)*; Luftzwischenraum *m*
a.-gap choke *(El)* Luftspaltdrossel *f*
a.-gap flux density *(El)* [magnetische] Luftspaltinduktion *f*
a.-gap lightning arrestor Luftblitzableiter *m*
a.-gap overlap span *(El)* Lufttrennung *f (Fahrdraht)*

a.-gap protector Funkenstreckenblitzableiter *m*; Schutzfunkenstrecke *f*
a.-gap transformer *(Fmt)* Übertrager *m* mit Eisenkern und Luftspalt
a. gate *(Brg)* Wettertür *f*; Luftabführung *f*
a. gun 1. Luftgewehr *n*; 2. Spritzpistole *f*
a. hammer Preßlufthammer *m*, Drucklufthammer *m*
a. hardenability Lufthärtbarkeit *f*
a. hardening an der Luft erhärtend, lufthärtend
a. hardening Lufthärtung *f*
a.-hardening lime *(Bw)* Luftkalk *m*
a.-hardening steel lufthärtender Stahl *m*, Lufthärter *m*
a. heater Heißlufterzeuger *m*, Lufterhitzer *m*; Luft[vor]wärmer *m*; Winderhitzer *m*
a. hoist Preßlufthebezeug *n*
a. hole Gasblase *f*
a. horn Luftstutzen *m (Verbrennungsmotor)*
a. humidifying Luftbefeuchtung *f (Klimatechnik)*
a.-hydraulic accumulator Druckluftspeicher *m*, pneumatischer Akkumulator *m*
a. inclusion Lufteinschluß *m*
a. induction coil Induktionsspule *f* ohne Eisenkern
a.-inflation indicator *(Kfz)* Füll-Luftmesser
a. injection Drucklufteinspritzung *f*; Lufteinblasung *f*
a.-injection engine Lufteinblasemaschine *f*
a. inlet *(Kfz)* Ansaugkrümmer *m*, Einlaßkrümmer *m*, Lufteinlaß *m*
a.-inlet cock Lufteinlaßhahn *m*
a.-inlet valve *(Opt)* Flutventil *n*
a. intake Luftansaugung *f*; Lufteinlaß *m*, Luftansaugrohr *n*
a. intake casing *(Rak)* Einlaßstutzen *m*, Eintrittsstutzen *m*
a. intake duct *(Kfz)* Lufteinlaßkanal *m (Lüftung)*
a. intake hose *(Kfz)* Luftsaugschlauch *m*
a. intake jet *(Kfz)* Luftdüse *f*
a. intake pipe *(Kfz)* Luftsaugrohr *n*
a. intake ring *(Wärm)* Luftansaugring *m*
a. intake silencer Geräuschdämpfer *m*
a. ionization Ionisierung *f* der Luft
a. jack *(Bw)* pneumatische Presse *f*
a. jet 1. Luftstrahl *m*, Luftstrom *m*; 2. *(Plast)* Schlitzdüse *f*
a.-jet loom [Luft-]Düsenwebmaschine *f*
a. jig *(Brg)* Luftsetzmaschine *f*
a. knife *(Holz)* Luftbürste *f*, Luftrakel *f*
a. knife coater 1. *(Plast)* Schlitzdüsenauftragmaschine *f*; 2. *(Holz)* Luftrakelstreichmaschine *f*
a. knife coating *(Led)* Luftpinselverfahren *n* mit Walzenauftrag *(Beschichtung)*
a. leg *(Brg)* Bohrknecht *m*, Bohrhammerstütze *f*
a. level *(Masch)* Libelle *f*
a. lift 1. Förderung *f* von Flüssigkeiten mit Hilfe von Druckluft; 2. *(Bw)* Druckluftheber *m*, Druckluftpumpe *f*, Preßluftförderer *m*; 3. *(Brg)* Mammutpumpe *f*
a.-lift agitator *(Ch)* Mischluftrührer *m*
a.-lift pump *(Brg)* Mammutpumpe *f*
a. line Luftlinie *f*; Fluglinie *f*
a. line *(Am) (El)* Freileitung *f*
a. liquefaction Luftverflüssigung *f*
a. liquefying plant Luftverflüssigungsanlage *f*
a. lock *(Bw)* 1. Lufteinschluß *m (in einem Rohrsystem für Flüssigkeiten)*; 2. Luftschleuse *f*, Wetterschleuse *f*
a. lubricated luftgeschmiert
a. lubrication Luftschmierung *f*
a. moistening Luftbefeuchtung *f (Klimatechnik)*
a. moisture Luftfeuchtigkeit *f*
a. monitor *(Kern)* Luftüberwachungsanlage *f*, Luftüberwachungsgerät *n*
a. motor Druckluftmotor *m*, Preßluftmotor *m*, pneumatischer Motor *m*, Pneumotor *m*
a. motor of the reciprocating piston type Hubkolbenmotor *m*
a. navigation Luftnavigation *f*
a. nozzle Luftdüse *f*
a. of combustion Verbrennungsluft *f (Verbrennungsmotor)*
a.-oil separator Schmierstoffentschäumer *m*
a.-operated pneumatisch, druckluftbetätigt, Druckluft-
a.-operated cable brake *(Kfz)* Seil-Druckluftbremse *f*
a.-operated chuck Preßluftfutter *n*, Spannfutter *n* für Preßluftspannung
a.-operated hydraulic brake *(Kfz)* druckluftbetätigte hydraulische Bremse *f*
a.-operated linkage brake *(Kfz)* Gestänge-Druckluftbremse *f*
a.-operated pump Pumpe *f* mit Preßluftantrieb
a. passage Luftdurchlaß *m*, Lüftungskanal *m*
a. penetration Lufteinbruch *m*
a.-photograph Luftbild *n*
a. photography Luftbildtechnik *f*
a. pipe 1. Entlüftungsrohr *n*, Lüftungsrohr *n*; *(Brg)* Wetterlutte *f*; 2. Gebläse *n (Hochofen)*
a. pit kleine Gasblase *f (in Gußstücken)*
a.-placed concrete Spritzbeton *m*, Torkretbeton *m*
a. pocket Luftblase *f*, Lufteinschluß *m*, Luftsack *m*, Lufttasche *f*, eingeschlossener Luftraum *m*

a. pollution Luftverunreinigung *f*
a. position indicator Positionsbestimmungsgerät *n*, Positionsanzeiger *m*
a. position radar Luftlageradar *n*
a.-powered mit pneumatischer Hilfsenergie; pneumatisch [betätigt]
a. preheater Luftvorwärmer *m*
a. pressure Luftdruck *m*
a.-pressure-assisted steering *(Kfz)* Lenkung *f* mit pneumatischer Lenkhilfe, pneumatische Servolenkung *f*
a.-pressure brake Druckluftbremse *f*, Luftdruckbremse *f*
a.-pressure circuit breaker *(El)* Druckluftschalter *m*
a.-pressure compensation Druckluftausgleich *m*
a.-pressure gauge 1. Luftdruckmesser *m*, Manometer *n*; 2. Luftdruckprüfer *m (für Reifen)*
a.-pressure release valve Luftdruckverminderungsventil *n*
a.-pressure switch *(El)* Luftdruckschalter *m*
a.-pressure test Druckluftprüfung *f*, Luftdruckprüfung *f*, Dichtigkeitsprüfung *f* mit Druckluft
a.-proof luftdicht
a. propeller Luftschraube *f*
a. protector Funkenstreckenblitzableiter *m*; Schutzfunkenstrecke *f*
a. pump Luftpumpe *f*; Vakuumpumpe *f*
a. pump beam *(Wärm)* Luftpumpenbalancier *m*
a. purifier Luftfilter *n*, Luftreiniger *m*
a. quenched steel luftgehärteter Stahl *m*
a. quenching Luftabschreckung *f (Stahl)*
a.-raid shelter Luftschutzbunker *m*, Luftschutzraum *m*
a. raise *(Brg)* Wetterüberhauen *n*
a. rammer *(Bw)* Preßluftramme *f*, Druckluftramme *f*; Preßluftstampfer *m*
a.-reducing valve Luftdruckregler *m*
a.-reduction method of smelting Röstreaktionsarbeit *f*
a.-refine *(Met)* Luftfrischen *n*
a.-relay *(Wärm)* pneumatischer Leistungsverstärker *m*, Schnell[steuer]verstärker *m*
a. release Luftaustritt *m*
a. release valve Entlüftungsventil *n*
a. reservoir Luftbehälter *m*; Windkessel *m*, Windsammler *m*
a. resistance Luftwiderstand *m*
a. return system Umluftverfahren *n (Lüftungstechnik)*
a. riveter pneumatischer Niethammer *m*
a. riveting hammer pneumatischer Niethammer *m*
a. route Fluglinie *f*
a. sampler *(Kern)* Filtergerät *n* zur Luftüberwachung
a. scoop Belüftungshaube *f*
a.-seasoned luftgetrocknet; lufttrocken
a.-seasoned timber lufttrockenes Holz *n*; luftgetrocknetes Holz
a. seasoning Lufttrocknung *f*, natürliche Trocknung *f*
a. section break *(El)* Streckentrenner *m (mit Luftisolation)*
a. separation Windsichtung *f*
a. separator Entlüfter *m*, Luftabscheider *m*, Windseparator *m*, Windsichter *m*
a. separator for water plants Wasserentlüftungsanlage *f*
a. shaft *(Brg, Bw)* Wetterschacht *m*, Luftschacht *m*
a. shower Luftschauer *m*
a. shut-off valve Luftabsperrhahn *m*
a. shuttle service Luftpendelverkehr *m*
a. sifter Windseparator *m*, Windsichter *m*
a. sifting *(Lw)* Windsichtung *f*
a.-sifting machine *(Lw)* Auslesemaschine *f (Getreide)*
a. silencer Luftberuhiger *m*; Ansauggeräuschdämpfer *m*
a.-slaked lime luftgelöschter Kalk *m*
a. slaking Löschen *n* von Branntkalk an der Luft
a. slit *(Brg)* Wetterdurchhieb *m*
a.-space blasting *(Brg)* Hohlraumschießen *n*
a.-space cable Hochfrequenzkabel *n*; Kabel *n* mit Luftraumisolierung
a.-space paper-core cable Papierhohlraumkabel *n*
a. spaced cable Kabel *n* mit Luftisolation; [konzentrischer] Hohlleiter *m*
a. speed *(Flg, Rak)* Eigengeschwindigkeit *f*
a.-speed indicator *(Flg)* Fahrtmesser *m*, Fluggeschwindigkeitsanzeiger *m*
a. split *(Brg)* Wetterscheider *m*
a. spring Luftfeder *f*
a. starting bottle (cylinder) Anlaßluftflasche *f*
a. starting equipment Druckluft-Anlaßvorrichtung *f*
a. station Fliegerhorst *m*
a. strainer Luftfilter *n*
a. strangler *(Kfz)* Lufteinlaßklappe *f*
a.-stream cleaner with by-pass after beater *(Text)* Luftstromreiniger *m* mit Umleitung hinter (nach) dem Schläger
a.-suction device Luftabsaugeinrichtung *f*
a.-supply line *(Kfz)* Fülleitung *f*
a.-supported structure *(Text)* Tragluftstruktur *f*
a. survey Luftüberwachung *f*

a. surveying Luftbildvermessung *f*, Luftvermessung *f*, Fotogrammetrie *f*
a. tank Luftbehälter *m*
a. temperature Lufttemperatur *f*
a. tempering *(Met)* Anlassen *n* in Luft
a. termination Auffangstange *f* mit Auffangleitung
a.-tight luftdicht
a.-tight cabin *(Rak)* hermetische Kabine *f*, Druckkabine *f*
a.-tight seal 1. luftdichte Abschmelzung *f* *(Lampenkolben usw.)*; *(Masch)* luftdichter Verschluß *m*
a.-to-air communication *(Flg)* Bord/Bord-Verkehr *m*
a.-to-air guided missile Luft/Luft-Rakete *f*
a.-to-air missile Luft/Luft-Rakete *f*
a.-to-air rocket Luft/Luft-Rakete *f*, Luftkampfrakete *f*
a.-to-ground data transmitter *(Rak)* Meßwertsender *m*
a.-to-ground missile Luft/Boden-Rakete *f*
a.-to-surface guided missile Flugzeugrakete *f* für Bodenziele, Luft/Boden-Rakete *f*
a.-to-vessel radar Schiffssuchsradar *n*
a. track Flugstrecke *f*; Luftziel *n (auf dem Bildschirm)*; Kurs *m (des Zieles)*
a. traffic control Luftraumüberwachung *f*
a. transformer Trockentransformator *m*, Lufttransformator *m*
a. trap Geruchverschluß *m (Rohrleitungen)*
a.-trap level meter manometrischer Füllstandsmesser *m*
a.-tube radiator Luftröhrenkühler *m*
a. up-take *(Met)* Luftkanal *m*, Luftzug *m*
a. valve Luftklappe *f*, Luftschieber *m*, Luftventil *n*, Windklappe *f*
a. vent 1. *(Masch)* Luftablaßventil *n*, Entlüftungsstutzen *m*, Luftauslaß *m*; 2. *(Kfz)* Entgasungsöffnung *f*; Luftloch *n*
a.-vent cock Entlüftungsventil *n*, Luftauslaßventil *n*
a.-vent pipe Entlüftungsrohr *n*
a.-vent screw Entlüftungsschraube *f* *(Batterie)*
a.-vessel Druckwindkessel *m*, Luftkessel *m*
a.-vessel end *(Wärm)* Kesselboden *m*
a. vice pneumatisch betätigter Schraubstock *m*
a. view Luftbild *n*; Vogelperspektive *f*
a. void Luftblase *f*, Lufteinschluß *m*, Luftpore *f*
a.-wall ionization chamber *(Kern)* luftäquivalente Ionisationskammer *f*
airborne luftübertragen
a. all-around search apparatus Panoramagerät *n*
a. contamination *(Kern)* [radioaktive] Luftverseuchung *f*
a. dust Flugstaub *m*, Schwebstaub *m*
a. early warning [radar] Frühwarngerät *n* *(Bordradargerät)*
a. instrument Bordgerät *n*
a. moving target indicator Bordradargerät *n* zur Anzeige beweglicher Ziele
a. radar Bordradar *n (Flugzeug)*
a. radio installation Flugbordfunkanlage *f*
a. radioactivity Radioaktivität *f* der Luft
a. rocket Bordrakete *f*, Flugzeugrakete *f*
a. search radar Flugzeugrundsuchradar *n*
a. sound Luftschall *m*
a. system *(Flg)* Bordanlage *f*
a. troops Luftlandetruppen *fpl*
airbreaking *(Min)* Schießen *n* mit Preßluft
aircraft Flugzeug *n*
a. armament Bordwaffen *fpl (Kampfflugzeuge)*
a. auxiliary Flugzeugsicherungsschiff *n*
a. battery Bordbatterie *f*
a. carrier Flugzeugträger *m*
a. compass Flugzeugkompaß *m*
a. construction Flugzeugbau *m*
a. direction-finder system Bordpeilanlage *f*
a. director Autopilot *m*, Steuerautomat *m*
a. flutter Störung *f* des Fernsehbildes durch Flugzeuge
a. interception Anpeilen *n* von Flugzeugen
a. intercommunication system Bordsprechanlage *f*
a. material Luftfahrtwerkstoff *m*
a. missile Flugzeugrakete *f*
a. navigation light Stellungslichter *npl (Flugsicherheitsbeleuchtung)*
a. radio Flugfunk *m*
a. radio equipment Flugzeugfunkeinrichtung
a. radio service Flugfunkdienst *m*
a. radio transmitter Flugzeugsender *m*
a. reporting service Flugmeldedienst *m*
a. rocket Flugzeugrakete *f*
a. skin Flugzeugaußenhaut *f*
a. twinboom Doppelrumpfflugzeug *n*
a. tyre Flugzeugreifen *m*
a. wireless service Flugfunkdienst *m*
airdry / to an der Luft trocknen
airdrying Lufttrocknung *f*, Windtrocknung *f*
airfield Flugplatz *m*, Landeplatz *m*
a. surface movement indicator Radargerät *n* für Rollfeldverkehr
airflow *(Brg)* Wetterströmung *f*
a.-reversal *(Brg)* Wetterumkehr *f*
airfoil *(Flg)* Tragflügel *m*, Tragfläche *f*; Flügelprofil *n*
a. flowmeter *(BMSR)* Klappendurchflußmesser *m*; Strömungswächter *m*
airing Belüftung *f*; Entlüftung *f*
a. space *(Schiff)* Decksfläche *f (als Promenade)*, offene Promenade *f*

a. tube Bewetterungsleitung *f*, Lutte *f*
airless life belt luftloser Rettungsring *m*
airplane Flugzeug *n*
a. aerial Flugzeugantenne *f*
a. base *(Mil)* Flugzeugstützpunkt *m*
a. control Flugzeugsteuerung *f*
a. engine Flugzeugmotor *m*
a.-engine supercharger Flugzeugmotorenlader *m*
a. photograph Flugzeugaufnahme *f*
a. position finding Flugzeugortung *f*
a. prospecting Flugzeugaufnahme *f*
a. pusher type *(Am)* Flugzeug *n* mit Druckschraube
a. receiving set Bordempfangsgerät *n*
a. tyre Flugzeugreifen *m*
a. wing Tragflügel *m*, Tragfläche *f*
airport Flughafen *m*
a. and surveillance radar Flughafen- und Luftwegüberwachungsradar *n*
a. beacon Flughafenleuchtfeuer *n*
a. lighting Flughafenbefeuerung *f*
a. localizer Landekurssender *m*
a. radar Flughafenradar *n*
a. surface detection radar Flugfeldüberwachungsradar *n*
a. surveillance radar Flughafenrundsichtradar *n*
airscrew Luftschraube *f*
a. blade Luftschraubenflügel *m*
a. pitch indicator Luftschraubenstellungsanzeiger *m*
airship Luftschiff *n*
airway 1. Fluglinie *f*; 2. *(Brg)* Wetterweg *m*, Wetterstrecke *f*
airworthy flugfähig; zum Flug zugelassen
Airy's spirals Airysche Spiralen *fpl*
aisle *(Bw)* Seitenschiff *n*; Gang *m*
a. roof *(Bw)* Pultdach *n*
aislometer *(Am)* *(El)* Hochspannungselektrometer *n*
ajutage *s.* adjutage
alabaster Alabaster[gips] *m*
a. board *(Typ)* Alabasterkarton *m*
alabastrite Alabastergips *m*
alarm Alarmapparat *m*; Warnapparat *m*, Meldeeinrichtung *f*, Melder *m*, Signalgeber *m*, Signalgerät *n*, Signalvorrichtung *f*
a. apparatus Alarmvorrichtung *f*; Warnvorrichtung *f*
a. bell Alarmglocke *f*; Signalglocke *f*
a. device Alarmvorrichtung *f*
a. gear Läutewerk *n*
a. lamp Signallampe *f*
a. locking Alarmverriegelung *f*
a. relay of grid potential Gitteralarmrelais *n*
a. thermometer *(Wärm)* Signalthermometer
albedo *(Kern)* Albedo *f*, Reflexionsfaktor *m*, Faktor *m* der diffusen Reflexion
Albert lay rope *(Brg)* Gleichschlagseil *n*, Längsschlagseil *n*
albite *(Min)* Albit *m*
albumen Eiweiß *n*
albumin glue Albuminleim *m*
alburnum *(Bw)* Splintholz *n*
alcatron Alkatron *n* *(Feldeffekt-Transistor)*
alclad mit Reinaluminium plattiertes Dural *n*, Alclad *m*
alcohol blended fuel Alkoholgemisch *n*, Alkoholtreibstoff *m*
a. burner Spiritusbrenner *m*; Spirituskocher *m*
a. diluted lacquers Spritlacke *mpl*
a. slug process Flutverfahren *n* mit einer Alkoholbank *(Tiefbohren)*
a. thermometer Alkoholthermometer *n*
alcoholate / to alkoholisieren, mit Alkohol umsetzen
alcoholate of plants Alkoholauszug *m* aus Pflanzen
alcoholic alkoholisch
alcoholization Sättigung *f* mit Alkohol
alcoholometer Alkoholometer *n*
alcoholometry Alkoholbestimmung *f*
alcoholysis Alkoholyse *f*, Umesterung *f*
alcove *(Bw)* Alkoven *m*, Bettnische *f*, Nische *f*
aldehyde resin Aldehydharz *n*
a. tanning Aldehydgerbung *f*
alder Erle[nholz *n*] *f*
Aldis lamp Handmorselampe *f*, Handscheinwerfer *m*
aldrey Aldrey *n* *(Al-Legierung)*
alee *(Schiff)* in Lee, leewärts, unter dem Winde
alembic *(Ch)* Destillierblase *f*
Alenčon lace Alenčon-Klöppelspitze *f*
Alexanderson alternator Alexanderson-Maschine *f* *(Wechselstrommaschine)*
algae control Algenverhütung *f*; Algenbekämpfung *f*, Algenvernichtung *f*
algaecide Algenvernichtungsmittel *n*, Algenvertilger *m*
algebra Algebra *f*, Buchstabenrechnung *f*
a. of residual classes Restklassenalgebra *f*
algebraic algebraisch
a. equation algebraische Gleichung *f*
algicide Algenvernichtungsmittel *n*, Algenvertilger *m*
alginic acid Alginsäure *f*
a. fibre *(Plast)* Alginatfaser *f*
algorithm Algorithmus *m*, Rechenregel *f*, Rechenvorschrift *f*; Rechnungsart *f*
alicyclic alizyklisch
alidade Alhidade *f*
alight upon water / to wassern *(Seeflugzeug)*
alighting Landung *f*

a. **gear** *(Flg)* Fahrwerk *n*
align / to [aus]richten, [ab]fluchten, [achsgerade] einstellen
a. **holes / to** Lochmittenunterschiede ausgleichen
align Axialität *f*, Fluchtlinie *f*
aligned-grid tube Röhre *f* mit Elektronenbündelung
aligning Ausrichten *n*, Einfluchtung *f*
a. **washer** *(Masch)* Einstellscheibe *f*
alignment Ausrichtung *f*, Einfluchtung *f*; Flucht *f*, Aneinanderreihung *f*; Axialität *f*; Bauflucht *f*, Absteckung *f*
a. **chart** *(Math)* Leitertafel *f*, Nomogramm *n*, Rechentafel *f*
a. **coil** Abgleichspule *f*
a. **cylinder** Richtzylinder *m*
a. **generator** Abgleichsender *m*; Empfangsprüfgenerator *m*
a. **of tuned circuits** *(Fmt)* Gleichlaufabstimmung *f*
a. **oscillator** Abgleichsender *m*; Empfangsprüfgenerator *m*
a. **pin** Paßstift *m*; Fixierstift *m*
a. **section** Führungsstück *n* *(Räumwerkzeug)*
a. **survey** Trassierung *f*
a. **testing** Fluchtungsprüfung *f*
a. **tool** Einstellschlüssel *m*; Trimmschlüssel *m*
a. **tools** *pl* Abgleichwerkzeug *n*; Abgleichbesteck *n*
alimentary substances *pl* Nährstoffe *mpl*
aline *(Am) s.* align
alinement *(Am) s.* alignment
aliphatic *(Ch)* aliphatisch
a. **compound** *(Ch)* aliphatische Verbindung *f*
a. **series** *(Ch)* aliphatische Reihe *f*
aliquant *(Math)* nicht aufgehend *(teilerfremd)*
aliquot *(Math)* ohne Rest aufgehend
alisonite *(Min)* Kupferbleiglanz *m*
alive *(El)* stromführend, unter Spannung befindlich; eingestellt
a. **wire** *(El)* Draht *m* unter Strom
alizarin dye[stuff] Alizarinfarbstoff *m*
alkalescence *(Ch)* Alkalität *f*, Alkaleszenz *f*
alkali Alkali *n*; Laugensalz *n*
a. **cellulose** Alkalizellulose *f*
a. **extraction stage** alkalische Wäsche *f* *(Zellstoffbleiche)*
a.-**proof** alkalibeständig, alkalifest, laugenbeständig
a.-**proof steel** laugenbeständiger Stahl *m*
alkaliferous alkalihaltig
alkalimetry Alkalimetrie *f*, Alkalibestimmung *f*
alkaline alkalisch; laugenartig
a. **accumulator** *(El)* alkalischer Akkumulator (Sammler) *m*
a. **counter-cell** *(El)* alkalische Gegenzelle *f*
a. **earth** Erdalkali *n*
a. **earth metal** Erdalkalimetall *n*
a. **process** Alkaliverfahren *n*
a. **pulping process** alkalischer Holzaufschluß *m*
a. **reaction** basische Reaktion *f*
a. **refining** *(Holz)* alkalische Veredlung *f*
a. **storage battery** alkalischer Akkumulator *m*
alkalinity Alkalität *f*
alkalize / to alkalisieren; mit Alkali umsetzen
alkaloid alkaloidal
alkaloid Alkaloid *n*, Pflanzenbase *f*
alkane Alkan *n* *(gesättigter Kohlenwasserstoff)*
alkene Alken *n* *(ungesättigter Kohlenwasserstoff mit C = C-Doppelbindung)*
alkine Alkin *n* *(ungesättigter Kohlenwasserstoff mit C ≡ C-Dreifachbindung)*
alkoxilate / to veräthern
alkyd resin Alkydharz *n*
all-aft arrangement ship Achterschiff *n* *(alle Decksaufbauten achtern)*
a.-**aft tanker** Achtertanker *m* *(alle Decksaufbauten betreffend)*
a.-**angle milling head** auf jeden Winkel verstellbarer Fräskopf *m*
a.-**automatic diaphragm** *(Opt)* automatische Springblende *f*
a.-**black malleable cast iron** Schwarzguß *m*
a.-**brick** *(Bw)* Ganzziegel-
a.-**channel decoder** Mehrkanaldekodierer *m*
a.-**clear position of clearing cam** Strickstellung *f* eines Strickhebers *(Strickmaschine)*
a.-**clear signal light** *(Rak)* Klarmeldelampe *f*
a.-**concrete** *(Bw)* Ganzbeton-
a.-**cotton cloth** reines Baumwollgewebe *n*
a.-**dry portable** *(Am)* tragbares Funkgerät *n* mit Trockenbatteriebetrieb
a.-**electric** vollelektrisch
a.-**electric set** Gerät *n* mit Vollnetzanschluß
a.-**electric-welded ship** vollständig geschweißtes Schiff *n*
a.-**glass construction** Allglasausführung *f*
a.-**hatch ship** „Offenes Schiff" *n*
a.-**hydraulic** vollhydraulisch
a.-**in aggregate** ungesiebte Zuschlagstoffe *mpl (Beton; Straßenbau)*
a.-**in gravel** ungesiebter Kies *m*
a.-**loss lubrication** *(Kfz)* Durchlaufschmierung *f*, verlorene Schmierung *f*
a.-**mains receiver (set)** Allstromempfänger *m*, Allstromgerät *n*
a.-**metal** Ganzmetall *n*
a.-**metal construction** Ganzmetallkonstruktion *f*, Ganzmetallbauweise *f*
a.-**metal lamp base** Metall-Lampensockel *m*
a.-**metal tube base** *(El)* Stahlröhrensockel *m*
a.-**or-nothing relay** Hilfsrelais *n* *(für Schutzzwecke)*

a.-pass network *(El)* Allpaß *m*; dämpfungsloser Kreis *m*
a.-plain sail area Amwind-Segelfläche *f*
a.-purpose adhesive Alleskleber *m*
a.-purpose computer Allzweckrechenanlage *f*
a.-purpose furnace *(Met)* Mehrzweckofen *m*
a.-purpose kitchen Wohnküche *f*
a.-purpose machine Allzweckmaschine *f*
a.-purpose tanker *(Schiff)* Allzwecktanker *m*
a.-purpose tool Allzweckgerät *n*, Universalwerkzeug *n*
a.-purpose washing agent *(Ch)* Universalwaschmittel *n*
a.-ready signal light *(Rak)* Klarmeldelampe *f*
a.-round box tipping wagon Muldenrundkipper *m*
a.-round developer Universalentwickler *m*
a.-round light *(Schiff)* Vollkreislicht *n*
a.-round looking device Panoramaeinrichtung *f*
a.-round ship Mehrzweckschiff *n*
a.-round tipping wagon Rundkipper *m*
a.-round tramp Trampschiff *n*
a.-silk reinseiden
a.-steel Ganzstahl-
a.-steel body Ganzstahlkarosserie *f*
a.-steel design Ganzstahlausführung *f*
a. told *(Schiff)* alles in allem
a.-watt motor phasenkompensierter Asynchronmotor *m*
a.-wave receiver Allwellenempfänger *m*
a.-weather outfit Allwetter[über]bekleidung *f*
a.-welded construction ganzgeschweißte Konstruktion *f*
a.-wheel air brake Allrad-Druckluftbremse *f*
a.-wheel drive Allradantrieb *m*
a.-wheel drive tractor Allrad-Schlepper *m*
a.-wing aeroplane Nurflügelflugzeug *n*
a.-wood construction Ganzholzbauweise *f*, Massivholzbauweise *f*
a.-wool reinwollen
Allan cell *(El)* Allan-Element *n*
allemontite *(Min)* Allemontit *m*, Arsenantimon *n*
Allen wrench *(Masch)* Schlüssel *m* für Innensechskantschrauben
allette Flügel *m (eines Gebäudes)*; Pfeiler *m*
alleyway *(Schiff)* Gang *m*, Laufgang *m*, Verkehrsgang *m (bei Wohnräumen)*
alligation *(Ch)* Mischungsregel *f*, Mischungsrechnung *f*; Versetzung *f*
alligator clip *(El)* Krokodilklemme *f*
a. connector *(Bw)* Alligator[zahnring]dübel *m (Holzbau)*
a. jaw *(Masch)* Zahnfederbacke *f*
a. pipe and nut wrench *(Masch)* Alligator-Rohr- und -Schraubenzange *f*
a. shear[s] *(Masch)* Alligatorschere *f*, Hebelschere *f*
a. wrench *(Masch)* Alligatorzange *f*; Rohrschlüssel *m*
allocate / to 1. aufnehmen, aufstellen; 2. *(BMSR)* zuweisen *(z. B. Datenspeicherplätze)*
allocation 1. Zuteilung *f*, Anordnung *f*, Aufstellung *f*; 2. Speicherverteilung *f (z. B. von Datenspeicherplätzen)*
a. of oil production within oil pools Förderzuteilung *f* innerhalb von Erdöllagerstätten
allokurtic *(Math)* abnormal gekrümmt *(bei der Häufigkeitskurve)*
allokurtosis *(Math)* abnormale Krümmung *f (der Häufigkeitskurve)*
allotment Zuteilung *f*; Parzelle *f*
a. of ore Erzfeld *n*
allotriomorphic allotriomorph
allotrope Modifikation *f (eines Stoffes)*
allotropic *(Ch)* allotrop[isch]
a. change allotrope Modifikation *f*
a. modification allotrope Modifikation *f*
a. transformation allotrope Umwandlung *f (eines Stoffes)*
allotter *(Fmt)* Anrufverteiler *m*, Anrufordner *m*
allotters *pl* **for calls** *(Fmt)* Anrufverteiler *m*
allotting circuit *(Fmt)* Zubringerleitung *f*
allow to deposit / to *(Ch)* absetzen lassen
a. to spark out [with the same setting] ausfunken lassen [ohne Zustellung]
allowable zulässig
a. deviation zulässige Abweichung *f*
a. error zulässiger Fehler *m*
a. load zulässige Belastung *f*
a. percent defective zulässiger Ausschußprozentsatz *m*, Ausschußquote *f* der Gutlage
a. stress zulässige Spannung *f*
a. transition *(Ph)* erlaubter (zulässiger) Übergang *m*
a. variation Toleranz *f*, zulässige Abweichung *f*
allowance 1. *(Masch)* Abmaß *n (Mindestspiel zwischen ineinandergreifenden Teilen)*; Kleinstspiel *n*, Spiel *n*, Toleranz *f*; Bearbeitungszugabe *f*, Übermaß *n*; 2. Entschädigung *f*, Gebührennachlaß *m*, Zulage *f (Lohn)*; 3. Erteilung *f (Patent)*
a. for Zugabe *f* für
a. for bark *(Holz)* Rindenprozent *n*
a. for fit (clearance) Spiel *n (Passung)*
a. for fit (interference) Übermaß *n (Passung)*
a. for reaming Reibzugabe *f*
allowed zulässig
a. orbit *(Kern)* zulässige Bahn *f*
a. time *(Ök)* Vorgabezeit *f*
a. transition *(Kern)* erlaubter Übergang *m*
alloy / to legieren, mischen, verschmelzen, versetzen, zusetzen, karatieren

alloy Legierung *f*
a. addition Legierungszusatz *m*
a. cast iron legierter Grauguß *m*
a. cladding Legierungsplattierschicht *f*
a. component Legierungsbestandteil *m*, Legierungskomponente *f*
a. content Legierungsgehalt *m*
a. for cutting tools Schneidlegierung *f*
a. steel legierter Stahl *m*, Sonderstahl *m*
a. that does not deteriorate with age alterungsfreie Legierung *f*
a. tool steel legierter Werkzeugstahl *m*
a. transistor legierter Transistor *m*
alloyable legierbar
alloyed state Legierungszustand *m*
a. steel legierter Stahl *m*
a. transistor Legierungstransistor *m*
alloying addition Legierungszusatz *m*
a. base Legierungsbasis *f*, Legierungsgrundlage *f*
a. constituent (element) Legierungskomponente *f*, Legierungsbestandteil *m*
a. metal einzulegierendes Metall *n*
a. property Legierbarkeit *f*
allscrap process Schrottschmelzverfahren *n*
alluvial deposit angeschwemmte Lagerstätte *f*, Ablagerung *f*
a. gold Waschgold *n*, in Anschwemmungen vorkommendes Gold *n*
a. ore Seifenerz *n*
a. soil angeschwemmter Boden *m*
a. stone Schwemmstein *m*
alluviation Ablagerung *f*
allyl mustard oil Allylsenföl *n*
allylate / to *(Ch)* allylieren
allylene oxide Dipropargyläther *m*
allylic alcohol Allylalkohol *m*
a. rearrangement *(Ch)* Allylumlagerung *f*
allylisothiocyanate Allylsenföl *n*
almandine *(Min)* Almandin *m*, Karfunkelstein *m*, roter Granat *m*
almandit *s.* almandine
almond Mandel *f*
a. furnace Schmelzofen *m*
alnico Alnico *n* *(Al-Ni-Co-Legierung für Dauermagneten)*
aloft *(Schiff)* über Deck
along shore längs der Küste
a. the ship längsschiffs
alongshoreman *(Am)* Hafenarbeiter *m*
alow *(Schiff)* unter Deck
alpaca Alpaka *n*, Neusilber *n*; Alfenid *n*
Alpax *(Am)* Silumin *n*, siliziumhaltiges Aluminium *n*
alpha bombardment Beschießen *n* mit Alpha-Strahlen
a. bronze *(Met)* Alpha-Bronze *f*
a. chamber *(Kern)* Alpha-Kammer *f*
a. counter *(Kern)* Alpha-Zähler *m*
a. counter tube *(Kern)* Alpha-Zählrohr *n*, Alpha-Kammer *f*
a. decay Alpha-Zerfall *m*, Alpha-Umwandlung *f*
a. disintegration *s.* a. decay
a. disintegration energy Alpha-Zerfallsenergie *f*
a. emitter Alpha-Strahler *m*
a. fibre Edelzellstoff *m*
a. iron Alpha-Eisen *n*, Ferrit *m*
a. meter *s.* a. counter
a. particle Alpha-Teilchen *n*
a.-particle bombardment Alpha[-Teilchen]-beschuß *m*
a.-particle emission Alpha-Strahlung *f*
a.-particle model of nucleus Alpha-Teilchenkernmodell *n*
a. pulp Alpha-Zellstoff *m*, Zellstoff *m* mit Alpha-Gehalt; Edelzellstoff *m*
a. radiation Alpha-Strahlung *f*
a. radioactivity Alpha-Aktivität *f*
a.-ray emitter Alpha-Strahler *m*
a.-ray spectrometer Alpha-Spektrometer *n*
a. sensitive alphaempfindlich
alphabetic code *(BMSR)* Kurzformkode *m*, Kurzformschlüssel *m*
a. coding *(BMSR)* Kurzformverschlüsselung *f*
alphabetical accounting machine *(Büro)* alphabetschreibende Tabelliermaschine *f* *(eine Art Lochkartenmaschine)*
a. collating device *(Büro)* Alphabetmischeinrichtung *f*, Alphabetmischvorrichtung *f*
a. duplicating punch *(Büro)* Doppler *m*, Duplizierlocher *m*, Wiederholungslocher *m*
a. interpreter *(Büro)* Alphabetlochschriftübersetzer *m*, Lochkartenbeschrifter *m*, Lochschriftübersetzer *m*
a. punch *(Büro)* Alphabetlocher *m*
a. verifier *(Büro)* Alphabetlochprüfer *m*
alphanumerical type bars *pl* *(Büro)* alphanumerische Typenstangen *fpl*
alphatron *(Kern)* Alphatron *n*
alphazurine Alphazurin *n* *(Triphenylmethanfarbstoff)*
alpinist's equipment Bergsteigerausrüstung *f*
altazimuth Altazimut *m (n)*
alter course / to Kurs ändern nach (auf)
a. the direction of the flow steam den Dampf ablenken *(Turbine)*
alteration Abänderung *f*, Änderung *f*, Veränderung *f*, Abwandlung *f*
a. of frequencies Frequenzänderung *f*
a. of phase Phasenverschiebung *f*
a. of the direction of the flow Umlenkung *f* der Dampfströmrichtung *(Turbine)*
a. of trim *(Schiff)* Vertrimmung *f*, Trimmänderung *f*, Tauchungsänderung *f*

altered rock zersetztes Gestein *n*
alternate abwechselnd, gegeneinander versetzt
a. airfield (airport) Ausweichflugplatz *m*, Landeflugplatz *m*
a. angles Wechselwinkel *mpl*
a. base Ausweichflugplatz *m*, Landeflugplatz *m*
a. bending strength Dauerbiegefestigkeit *f*
a. bending vibrations Dauerbiegeschwingungen *fpl*
a.-exterior angles äußere Wechselwinkel *mpl*
a. firing *(Wärm)* abwechselnde Beschickung *f*
a. gash plain mill kreuzverzahnter Walzenfräser (Scheibenfräser) *m*
a.-interior angles innere Wechselwinkel *mpl*
a. nicking gegeneinander versetzte Kerbung *f*, auf Lücke stehende Spanbrechernut *f*
a. plough Wechselpflug *m*
a. program Austauschprogramm *n (Ziffernrechner und Analogrechner)*
a. strength Wechselfestigkeit *f*, Schwingungsfestigkeit *f*
a. strength testing machine Dauerfestigkeitsprüfmaschine *f*
a. stress Dauerbeanspruchung *f*; Wechselspannung *f*
a. stress testing Wechselfestigkeitsprüfung *f*
a. tooth [milling] cutter Fräser *m* mit Kreuzverzahnung
a. tooth slitting saw Metallkreissäge *f* mit Kreuzverzahnung
a. torsional strength Drehwechselfestigkeit *f*
alternating schwingend, wechselnd
a. and direct current Allstrom *m*
a. beds *pl* wechsellagernde Schichten *fpl (Geol)*
a. bending test Wechselbiegeprüfung *f (Festigkeit)*
a. circuit Wechselstromkreis *m*
a. component Wechselstromanteil *m*, Wechselstromkomponente *f*
a. current, A.c., a.c. Wechselstrom *m*
a.c. ammeter Wechselstrommesser *m*
a.c. amplifier Wechselstromverstärker *m*
a.c. analogue computer Wechselspannungsanalogrechner *m*
a.c. anode voltage Anodenwechselspannung *f*
a.c. balancer Ausgleichstransformator *m*
a.c. bridge Wechselstrommeßbrücke *f*
a.c. component Wechselstromkomponente *f*
a.c. converter Wechselstromgleichrichter *m*
a.c. dialling Wechselstromfernwahl *f*
a.c. discharge tube Wechselstromentladungsröhre *f*
a.c. dynamo Wechselstromdynamo *m*
a.c. grid voltage Gitterwechselspannung *f*
a.c. hum Netzbrummen *n*; Brummen *n*
a.c. impedance bridge Wechselstrommeßbrücke *f*
a.c. line Wechselstromleitung *f*
a.c. mains Wechselstromnetz *n*
a.c. meter Wechselstromzähler *m*
a.c. multiple telegraphy Wechselstromtelegrafie *f*
a.c. operated mit Wechselstrom betrieben
a.c. pickup Wechselspannungseinstreuung *f*
a.c. plate resistance innerer Widerstand *m (Elektronenröhre)*
a.c. potential Wechselspannung *f*
a.c. power Wechselstromleistung *f*
a.c. resistance Wechselstromwiderstand *m*
a.c. supply Wechselstromnetz *n*; Wechselstromquelle *f*
a.c. supply apparatus Wechselstromspeisungsapparat *m*
a.c. tachometer Mittelfrequenztachometermaschine *f*, Wechselstromtachometer *n*
a.c. vacuum tube voltmeter Wechselspannungsröhrenvoltmeter *n*
a.c. voltage Wechselspannung *f*
a. direct stress testing machine Zug-Druck-Schwingungsprüfmaschine *f*
a. discharge Wechselentladung *f*
a. draw cut stroke Gegenzug *m (beim Sägen)*
a. field Wechselfeld *n*
a. gradient focusing *(Kern)* AG-Fokussierung *f*, Courant-Fokussierung *f*, starke Fokussierung *f*
a. grid voltage Gitterwechselspannung *f*
a. impact machine *(Met)* Dauerwechselschlagwerk *n*
a. light *(Schiff)* Wechselfeuer *n (Leuchtfeuer)*
a. load Wechsellast *f*, wechselnde Beanspruchung *f*
a. load deformation Wechselverformung *f*
a. magnetic field magnetisches Wechselfeld *n*
a. motion hin- und hergehende Bewegung *f*
a. part of a continuous current Wechselstromanteil *m* eines Gleichstroms
a. part of a direct current Wechselstromkomponente *f* eines Gleichstroms, Welligkeit *f*
a. position *(Wärm)* Wechsellage *f (Kurbelstellung)*
a. potential Wechselpotential *n*
a. quantity alternierende Größe *f*, Wechselgröße *f*
a. straining *(shop term)* Wechselbeanspruchung *f*
a. stress Dauerschwing[ungs]beanspruchung *f*, schwellende Belastung *f*, Wechselbelastung *f*
a. torsion fatigue test Dauerwechseldrehversuch *m*

a. voltage Wechselspannung *f*
alternation 1. Stromwechsel *m*; Polwechsel *m*; 2. Lastwechsel *m*; 3. Halbwelle *f*; 4. Disjunktion *f (logischer Operator)*
a. period *(El)* Halbperiodendauer *f*
alternative connection relay *(El)* Fortschaltrelais *n*
a. storage system Wechselstromspeichersystem *n*
alternator Wechselstromgenerator *m* (*s. a.* synchronous generator)
a. transmitter *(Fmt)* Maschinensender *m*
altimeter [barometrischer] Höhenmesser *m*
a. control equipment Höhenmeßkontrollgerät *n*
altitude 1. Höhe *f* [über Normalnull]; 2. Höhenlinie *f (z. B. im Dreieck)*
a. gauge Höhenmesser *m*
a. index level Höhenindexlibelle *f*
a. indicator Höhensichtgerät *n (Radar)*
a. of stage separation *(Rak)* Stufentrennungshöhe *f*
a. recorder Höhenschreiber *m*
a. supercharging Höhenaufladung *f*
a. valve *(BMSR)* Flüssigkeitsstandregler *m* ohne Hilfsenergie
aludur Aludur *n (Al-Mg-Legierung)*
alum Alaun *m*
a. leather Alaunleder *n*
a. shale Alaunschiefer *m*
a. tanner *(Led)* Weißgerber *m*
a. tanning *(Led)* Alaungerbung *f*, Weißgerbung *f*
alumel Alumel *n (Ni-Legierung)*
alumen Al-Zn-Cu-Legierung *f*
alumetizing Alumetierung *f*
alumina Tonerde *f*, Aluminiumoxid *n*
a. cement Tonerdezement *m*
a. hydrate Bauxit *m*
aluminiferous aluminiumhaltig, tonerdehaltig
aluminite process *(Am)* Eloxalverfahren *n*, anodische Oberflächenbehandlung *f* von Aluminiumlegierungen *(unter Verwendung von Schwefelsäure als Elektrolyt)*
aluminium Aluminium *n*
a. alloy Al-Legierung *f*
a. bronze Aluminiumbronze *f*
a. cable Aluminiumkabel *n*
a. cable, steel reinforced Aluminium-Stahl-Kabel *n*
a. casting Aluminiumguß *m*
a. clad aluminiumplattiert
a. coated aluminiert
a.-coating by the spraying process Alumetieren *n*
a. coil Aluminiumspule *f*
a. die casting Aluminiumspritzguß *m*
a. foil Aluminiumfolie *f*
a. paint Aluminiumbronze *f*; Aluminiumüberzug *m*
a. plated aluminiert
a. rectifier Elektrolytventil *n*, Elektrolytgleichrichter *m*, Aluminiumgleichrichter *m*
a.-rolling mill Aluminiumwalzwerk *n*; Aluminiumwalzstraße *f*
a. sheet Aluminiumblech *n*
a. silicate Tonerdesilikat *n*
a.-steel cable Stahlaluminiumseil *n*, Stalu-Seil *n*
aluminize / to aluminieren
aluminized [picture-tube] screen *(Fs)* aluminisierter Bildschirm *m*, Bildschirm mit aufgedampfter Aluminiumhaut
aluminizing Aluminieren *n*, Alitieren *n*
aluminothermic process Aluminothermie *f*
a. welding aluminothermisches Schweißen *n*
aluminothermics Aluminothermie *f*, Thermitverfahren *n*
aluminous cement Tonerde[schmelz]zement *m*
a. flux Tonzuschlag *m*
a. limestone Alaunstein *m*
a. soap Tonseife *f*
aluminum *(Am) s.* aluminium
alumstone Alaunstein *m*
alunite Alunit *m*, Alaunstein *m*
amalgam Amalgam *n*
amalgamate / to amalgamieren, [sich] vereinigen, [sich] vermischen, verquicken, verschmelzen
amalgamated metal Quickmetall *n*
amalgamating solution Amalgambad *n*
amalgamation Amalgamation *f*
a. pan Amalgamierherd *m*
a. process Amalgamationsverfahren *n*
amateur Amateur *m*, Funkamateur *m*
a. astro-camera Amateurastrokamera *f*
a. bands *pl (Fmt)* Amateurfrequenzbänder *npl*
a. radio link Amateurfunkverbindung *f*
a. reflecting telescope Amateurspiegelteleskop *n*
a. station Amateurfunkstelle *f*
a. telescope Amateurfernrohr *n*
a. wireless operator Funkamateur *m*
amber bernsteingelb
amber Bernstein *m*
a. glow lamp gelbe Signallampe *f*
a.-seed oil Moschuskörneröl *n*
ambergris Ambra *f*
amberoid Preßbernstein *m*
ambient umgebend, Umgebungs-
a. light *(Fs)* Raumbeleuchtung *f*
a. light filter *(Fs)* Schutzfilter *n* gegen einfallendes Nebenlicht
a. noise Nebengeräusch *n (Telefonie)*
a. temperature Umgebungstemperatur *f*, Außentemperatur *f*

ambiguity *(BMSR)* Mehrdeutigkeit *f (logisch)*
ambiguous mehrdeutig
ambiguousness Mehrdeutigkeit *f*
ambit Umkreis *m*; Kontur *f*; Kreislauf *m*; Außengrenze *f*
ambulance car Krankenwagen *m*, Sanitätskraftwagen *m*; Unfallwagen *m*
a. station Unfallstation *f*
ameliorate / to meliorieren
amelioration Bodenverbesserung *f*, Melioration *f*
American cloth Wachstuch *n*
A. malleable cast iron schwarzer Temperguß *m*
A. National Coarse *(Masch)* amerikanisches Grobgewinde *n*
A. National Fine *(Masch)* amerikanisches Feingewinde *n*
A. representation amerikanische Darstellungsweise *f (technisches Zeichnen)*
A. standard thread *(Masch)* amerikanisches Normalgewinde *n*, Sellersgewinde *n*
A. system of spinning amerikanisches Spinnverfahren *n*
A. type of boiler *(Wärm)* amerikanischer Kessel *m*
americium Amerizium *n (chem. Element)*
ametropia *(Opt)* Fehlsichtigkeit *f*
ametropic *(Opt)* fehlsichtig
amianthine asbestartig
amianthus Faserasbest *m*
Amici prism *(Opt)* Amici-Prisma *n*
amidship bending moment *(Schiff)* Mittschiffsbiegemoment *n*
a. section *(Schiff)* Hauptspant *n*
amidships mittschiffs
α-amino-acid α-Aminosäure *f*
ammeter Am[pere]meter *n*, Strommesser *m*
a. shunt Amperemeter-Nebenschlußwiderstand *m*
ammonia Ammoniak *n*
a. fume Ammoniakdampf *m*
a. scrubber Ammoniakwäscher *m*
a. soda process Ammoniak-Soda-Verfahren *n (Solvay)*
a. still Ammoniakabtreiber *m*
a. synthesis gas Ammoniaksynthesegas *n*
a. washer Ammoniakwäscher *m*
ammoniac[al] ammoniakhaltig, ammoniakalisch
ammonium Ammonium *n*
a.-base cooking liquor *(Holz)* Kochsäure *f* aus Ammoniumbisulfit
a.-base neutral sulfite semi-chemical process *(Holz)* Ammoniak-Bisulfit-Verfahren *n* zur Herstellung von Halbzellstoff
a. chloride Ammoniumchlorid *n*, Salmiak[salz *n*] *m*
a. nitrate-fuel oil explosive [selbsthergestellter] Sprengstoff *m* aus Ammonsalpeter und Heizöl
ammunition *(Mil)* Munition *f*
a. carrier Munitionsfahrzeug *n*
a. clip Ladestreifen *m*
a. drum Trommelmagazin *n*
a. hoist Munitionsaufzug *m*
amorphous amorph, gestaltlos, unkristallin
amortisseur winding *(Elm)* Dämpferwicklung *f*
amount Menge *f*, Betrag *m*, Größe *f*; Höhe *f*; Wert *m*; Maß *n*; Länge *f*; Volumen *n*
a. brought forward Vortrag *m*, Übertrag *m (in Buchungsvorrichtungen)*
a. field Betragsfeld *n (auf Lochkarten u. a.)*
a. of axial freedom Axialspiel *n*
a. of correction Berichtigungsgröße *f*
a. of eccentricity Exzentrizität[sgrad *m*] *f*
a. of energy Energiebetrag *m*
a. of expansion possible Verstellbereich *m* der Reibahle
a. of feed Beistellbetrag *m*, Vorschubgröße *f*
a. of fluid discharged Umlaufmenge *f*
a. of heat Wärmemenge *f*
a. of heat required Wärmebedarf *m*
a. of looseness *(Masch)* Spiel *n (Paßteile)*
a. of mudulation *(Fmt)* Modulationsgrad *m*
a. of off-centre Achsenabweichung *f*
a. of overbalance *(Masch)* Überwucht[masse] *f*
a. of resistance Widerstandswert *m*
a. of taper Ausmaß *n* (Grad *m*) der Verjüngung
a. of variation permitted zulässige Maßabweichung *f*
a. of work per grind Standmenge *f*
a. of work required Arbeitsbedarf *m*
amperage Stromstärke *f* [in A], Amperezahl *f*
a. range Strommeßbereich *m*
ampere Ampere *n*
a. balance *s.* current balance
a. bars per cm Strombelag *m*, AW/cm
a. conductors *pl* Ampere-Stäbe *mpl*
a.-hour Amperestunde *f*
a.-hour capacity Aufnahmefähigkeit *f*, Kapazität *f*, Leistung *f (Sammler, Batterie)*
a.-hour efficiency Wirkungsgrad *m* in Amperestunden
a.-hour meter Amperestundenzähler *m*, Ah-Zähler *m*
a. meter Am[pere]meter *n*
a.-second Amperesekunde *f*
a. turns Amperewindungen *fpl*, Amperewindungszahl *f*
Ampere's rule *(El)* Amperesche Schwimmerregel *f*, Daumenregel *f*
Amperian float law *(El)* Amperesche Schwimmerregel *f*

amperometry Amperometrie *f (elektrochemisches Analysenverfahren)*
amphibian Wasser-Land-Flugzeug *n*; Wasser-Land-Fahrzeug *n*, Amphibienfahrzeug *n*
a. tank Schwimmpanzer[kampfwagen] *m*, Amphibienpanzer *m*
amphibole Hornblendeasbest *m*
amphoteric amphoter
a. ion Zwitterion *n*
a. semiconduction amphotere Halbleitung *f*
amplidyne *(BMSR)* Amplidyne *f*, Querfeldverstärkermaschine *f*
a. generator Amplidyne *f*
amplification *(El)* Verstärkung *f*, Anfachung *f*; Intensivierung *f*
a. constant Verstärkungskonstante *f*
a. factor Verstärkungsfaktor *m*
a. gain *(Fmt)* Verstärkung *f*
a. stage *(Fmt)* Verstärkungsstufe *f*
amplifier Verstärker *m*
a. detector Audionverstärker *m*
a. for microwaves Dezimeterwellenverstärker *m*
a. noise Verstärkerrauschen *n*, Verstärkergeräusch *n*
a. rack Verstärkergestell *n*
amplify / to verstärken *(Fernsehen)*
amplifying apparatus Verstärkeranlage *f*
a. detector Audionverstärker *m*
a. pilot *(BMSR)* pneumatischer Leistungsverstärker (Schnellverstärker) *m*
a. station Verstärkeramt *n*
a. tube Verstärkerröhre *f*
a. valve Verstärkerröhre *f*
a. vibrograph schreibender Schwingungsmesser *m* mit Verstärker
a. winding *(El)* Verstärkerwicklung *f*
ampliscaler *(Fmt, Fs)* 100-fach-Zähler *m*, 100-fach-Untersetzer *m*
amplistat *(BMSR)* magnetischer Verstärker *m*, Magnetverstärker *m*, Transduktor *m*
amplitude 1. *(Math)* Amplitude *f*, Argument *n*, Azimut *m (n)*, Polarwinkel *m (bei Polarkoordinaten)*; 2. *(Ph)* Amplitude, Ausschlag[weite *f*] *m*, Schwingungsweite *f*; Scheitelwert *m (elektrischer Größen)*
a. and phase grating *(Opt)* Amplituden- und Phasengitter *n*
a.-change signalling Telegrafie-Amplitudenmodulation *f*
a. characteristic Amplitudencharakteristik *f*, Frequenzgang *m* [der Amplitude], Frequenzkurve *f*; Frequenzkennlinie *f*
a. control *(El)* Amplitudensteuerung *f*, Amplitudenregelung *f*
a. density Amplitudendichte *f*
a. discriminator Amplitudensieb *n (Fernsehempfangstechnik)*
a. distortion *(Fmt)* Amplitudenverzerrung *f*; Klirrverzerrung *f*; nichtlineare Verzerrung *f*
a. factor *(El)* Scheitelfaktor *m*
a. fading Amplitudenschwund *m*
a.-frequency plot *(El)* Aufzeichnung *f* des Amplitudenfrequenzganges
a.-frequency response characteristic *(El)* Amplituden-Frequenz-Kurve *f*
a. limiter *(Fmt, Fs)* Amplitudenbegrenzer *m*
a. margin *(BMSR)* Amplitudenvorrat *m*, Amplitudenrand *m*; Verstärkungsrand *m*, Verstärkungsvorrat *m*
a.-modulated transmitter amplitudenmodulierter Sender *m*
a. modulation Amplitudenmodulation *f*
a. of a symmetrical alternating quantity Amplitude *f* einer symmetrischen Wechselgröße, Scheitelwert *m*
a. of accommodation *(Opt)* Akkommodationsbreite *f*
a. of beats Schwebungsamplitude *f*
a. of frequency characteristic *(El)* Amplitudengang *m*
a. of swell *(Schiff)* Stauweite *f*
a. of the oscillations (vibrations) Schwingungsamplitude *f*
a. range *(Fmt)* Amplitudenaussteuerung *f*
a. relationship *(Fmt)* Amplitudenbedingung *f (für die Eigenerregung)*
a. response *(BMSR)* Amplitudengang *m*, Amplitudenverhalten *n*; *(Fmt)* Aussteuerungsabhängigkeit *f*
a. selection Amplitudenwahl *f (Analogrechner)*
a. separation *(Fmt)* Amplitudensiebung *f*
a. separator *(Fmt, Fs)* Amplitudenselektor *m*, Amplitudensieb *n*; Synchronisierabtrennung *f*
a. specimen *(Opt)* Amplitudenobjekt *n*
a. swing Amplitudensieb *n*; Amplitudenschwankung *f*
ampoule Ampulle *f*
ampulla *s.* ampoule
amyl Amyl *n*
amylaceous stärkehaltig
amylum Stärke[mehl *n*] *f*
AN-FO explosive *s.* ammonium nitrate-fuel oil explosive
anaerobic anaerob
anaglyphic method Anaglyphenverfahren *n*
anallactic point *(Opt)* anallaktischer Punkt *m*
anallagmatic anallagmatisch, unveränderlich
analog *s.* analogue
analog *s.* analogous
analogous analog
analogue Analogon *n*, Analogie *f*
a. computer Analogierechenmaschine *f*, Analogierechner *m*

a.-to-digital converter Analogdigitalumsetzer *m*, AD-Umsetzer *m*, Größenwerter *m*
a. transistor Analogtransistor *m*
analogy Analogie *f*
analysis 1. *(Ch)* Analyse *f*; 2. *(Math)* Analysis *f*
a. by boiling *(Ch)* Siedeanalyse *f*
a. funnel *(Ch)* Analysentrichter *m*
a. of variance Varianzanalyse *f*
analyst 1. analytischer Chemiker *m*; 2. Auswerter *m*; Laborant *m*
analytic geometry analytische Geometrie *f*, Meßgeometrie *f*
analytical analytisch
a. balance *(Ch)* Analysenwaage *f*
a. engine Bezeichnung *f* der von Babbage entworfenen ersten programmgesteuerten Rechenmaschine (1833)
a. geometry analytische Geometrie *f*
a.-grade *(Ch)* analysenrein
a. simulation analytisches Nachbilden *n* *(mit Hilfe von Rechenautomaten)*
analyze / to analysieren, untersuchen; zergliedern, zerlegen; auswerten
analyzer Analysator *m*, Analysenmeßgerät *n*
analyzing of picture *(Fs)* Bildzerlegung *f*
anamorphoser *(Opt)* Anamorphot *m*
anamorphote lens Anamorphot *m*; anamorphotische Optik *f*
anamorphotic *(Opt)* anamorphotisch
anastigmatic *(Opt)* anastigmatisch
a. folding magnifier *(Opt)* anastigmatische Einschlaglupe *f*
anastigmatism *(Opt)* Anastigmatismus *m*, Anastigmasie *f*
anastomose of the bark Baumschlag *m* *(Beschädigung durch fallende Bäume)*
anchor / to verankern; verklammern
anchor 1. *(El)* Anker *m*, Mastanker *m*; 2. Anker *m*, Schiffsanker *m*; 3. *(Bw)* Mauer[werks]-anker *m*; 4. Verklammerung *f*
a. agitator *(Ch)* Ankermischer *m*, Ankerrührer *m*, U-Rührer *m*
a. bolt *(Masch)* Ankerbolzen *m*, Fundamentanker *m*, Fundamentschraube *f*, Verankerungsbolzen *m*, Verankerungsschraube *f*
a. bolt tube *(Masch)* Ankerschraubenrohr *n*
a. chain *(Schiff)* Ankerkette *f*
a. chain bitter ends Ankerkettenendglied *n*
a. clamp *(El)* Abspannklemme *f*
a. hawse *(Schiff)* Ankerklüse *f*
a. hook Ankerhaken *m*, Klammerhaken *m*
a. log *(El)* Spanndrahtschelle *f*, Abspannpflock *m*
a. pier Verankerungspfeiler *m* *(Brückenbau)*
a. plate Ankerplatte *f*, Verankerungsplatte *f*
a. pocket *(Schiff)* Ankertasche *f*
a. pole *(El)* Abspannmast *m* *(Holzmast)*
a. post *(Brg)* Strebestempel *m*
a. rode *(Schiff)* Ankerleine *f*, Ankertau *n*
a. rope Ankertrosse *f*
a. screw Sicherungsschraube *f*
a. shank *(Schiff)* Ankerschaft *m*
a. tie Verankerung *f*
a. tower 1. Abspann[gitter]mast *m*; 2. Stützmast *m*
a. zone Verankerungszone *f* *(Spannbeton)*
anchorage 1. Verankerung; 2. Ankergrund *m*, Ankerplatz *m*
a. of lining Verklammerung *f* des Ausgusses *(Lager)*
a. pier Verankerungspfeiler *m* *(Brückenbau)*
a. zone Verankerungszone *f* *(Spannbeton)*
anchored befestigt, verklammert, verankert
anchoring 1. Verankern *n*; 2. Befestigen *n*; 3. Verklammern *n* *(Lagerausgüsse)*
a. bolt Ankerbolzen *m*
a. manœuvre Ankermanöver *n*
a. of the flock *(Text)* Verankerung *f* der Flockfaser
a. rod Zuganker *m*
a. rope Abspannseil *n*
a. wire Spanndraht *m*, Verankerungsdraht *m*
anchorize / to verankern
ancillary circuit *(El)* Hilfsstromkreis *m*
a. device Zusatzeinrichtung *f*
a. jack *(Fmt)* Aushilfsklinke *f*
a. lamp *(El)* Wiederholungslampe *f*, Rückmeldelampe *f*, Stellungsanzeigelampe *f*
a. position *(Fmt)* Hilfsplatz *m*
a. pump Hilfspumpe *f*, Zusatzpumpe *f*
ancon Tragstein *m*, Unterlage *f*
ancube *(Text)* Ancubeteppich *m* aus Belgien
AND *(BMSR)* UND *(logischer Operator)*
AND-circuit *(BMSR)* UND-Schaltung *f*, Torschaltung *f* *(logische Schaltung)*
AND-gate *(BMSR)* UND-Gatter *n*, UND-Tor *n*, UND-Glied *n*, UND-Schaltung *f*
AND-NOT-gate *(BMSR)* JEDOCH-NICHT-Schaltung *f*, JEDOCH-NICHT-Tor *n*, UND-NICHT-Schaltung *f*, UND-NICHT-Tor *n*
AND-operator *(BMSR)* logisches UND *n*
andalusite Andalusit *m*
Anderson bridge *(El)* Anderson-Brücke *f*, Anderson-Meßbrücke *f*
Anderson tube *(El)* Anderson-Röhre *f*
andesine *(Min)* Andesin *m*
andesite *(Min)* Andesit *m*
anechoic room echototer (schalltoter) Raum *m*
anelastic unelastisch
anelectric unelektrisch, nicht elektrisch
anemograph Anemograf *m*
anemology Anemologie *f*
anemometer Anemometer *n*, Gasströmungsmesser *m*, Windmesser *m*

aneroid altimeter barometrischer Höhenmesser *m*
a. barometer Aneroid[barometer] *n*
a. flowmeter Balgendurchflußmesser *m*
anesthetic Betäubungsmittel *n*, Narkosemittel *n*, Anästhetikum *n*
anesthetic anästhetisch, betäubend
angle Winkel *m*; Winkelstahl *m*, Winkeleisen *n*, Winkelprofil *n*; L-Profil *n*
a. advance Winkelvoreilung *f*
a. bar Eckschiene *f*, Winkeleisen *n*, Winkelstahl *m*, Winkelprofil *n*; L-Profil *n*
a. bead *(Bw)* Kantenschutz[leiste *f*] *m*, Kantenschutzschiene *f (Putz)*
a. bender Biegemaschine *f*, Biegetisch *m (Stahlbeton, Bewehrungsbau)*
a. between cranks *(Kfz)* Kurbelversetzung *f*
a. bisector Winkelhalbierende *f*
a. block Winkelendmaß *n*
a.-body valve Eckventil *n*
a. brace 1. *(Masch)* Winkelband *n*; 2. *(Bw)* Diagonalstab *m*, Kopfband *n*, Strebe *f*; 3. *(Holz)* Bug *m*, Büge *mpl (Zimmerung)*
a. bracket Befestigungswinkel *m*, Winkelkonsole *f*, Winkelauflage *f*, Winkelstütze *f*, Winkelverstärkung *f*
a. bulb *(Schiff)* Wulstwinkel *m*, Bulbwinkel *m*
a. conduit box *(El)* Winkel[abzweig]dose *f*
a. correlation Winkelbeziehung *f*
a. cutter *(Masch)* Winkelfräser *m*
a.-cutting machine *(Holz)* Diagonalschneider *m*, Schrägschneidemaschine *f*
a. dependence Winkelabhängigkeit *f*
a. diameter *(Am)* Flankendurchmesser *m (Gewinde)*
a.-division tester Winkelteilungsprüfgerät *n*
a. dressing fixture *(Masch)* Winkelabrichteinrichtung *f*
a. encoder *(BMSR)* Winkelumsetzer *m*, Winkelverschlüßler *m*
a. equation Winkelgleichung *f*
a. extrusion head Querspritzkopf *m*, Schrägspritzkopf *m*
a. fillet *(Bw)* Dreikantleiste *f*
a.-forming machine Eckenbiegemaschine *f*
a. gauge 1. Winkellehre *f*; 2. Winkelendmaß *n*
a. head plain grinding machine Schrägeinstechschleifmaschine *f*
a. hole *(Geol)* geneigtes Bohrloch *n*
a. iron 1. Winkel *m*; 2. Eckschutzschiene *f*, Eckeisen *n*, Winkeleisen *n*; L-Stahl *m*, Winkelprofil *n*
a. joint Winkelgelenk *n*, Winkelstoß *m*
a. leg Winkelschenkel *m*
a. lever Winkelhebel *m*
a. lighting fitting *(El)* Schrägstrahler *m*
a. measure Winkelmaß *n*
a. milling cutter Winkelfräser *m*
a. modulation *(Fmt)* Winkelmodulation *f*; Phasenmodulation *f*
a. of action Eingriffswinkel *m (Verzahnung)*
a. of advance *(Wärm)* Aufkeilwinkel *m (s. a.* advance)
a. of approach 1. *(Flg)* Zielkurs *m*, Annäherungswinkel *m*; 2. Einlauf[wälz]winkel *m (Zahnrad)*; 3. Eingriffswinkel *m*
a. of ascent *(Bw)* Neigungswinkel *m*, Steigungswinkel *m*
a. of aspect (attack) *(Flg)* Zielkurs *m*, Annäherungswinkel *m*
a. of attack Anstellwinkel *m*
a. of bank Neigungswinkel *m*, Kurvenwinkel *m*; Krängungswinkel *m*
a. of beam Strahlungswinkel *m*
a. of bedding *(Geol)* Fallwinkel *m* von Schichten
a. of bend Biegewinkel *m*, Abkantwinkel *m*
a. of break Bruchwinkel *m*
a. of chamfer Anschnittwinkel *m (Spiralbohrer)*
a. of circumference Polygonwinkel *m*, Umfangswinkel *m*
a. of clearance *(Masch)* Freiwinkel *m*
a. of climb *(Rak)* Steigwinkel *m*
a. of cone of the nozzle Kegelwinkel *m* der Düse *(Turbine)*
a. of connecting rod *(Kfz)* Pleuelstangenausschlag *m*
a. of contact 1. Berührungswinkel *m*; 2. *(El)* Umfassungswinkel *m*, Umschlingungswinkel *m*; 3. Randwinkel *m*
a. of convergence Konvergenzwinkel *m*
a. of correction *(Flg)* Korrekturwinkel *m*
a. of countersink Spitzenwinkel *m*, Versenkwinkel *m (Spitzsenker)*
a. of crank with the horizontal *(Masch, Wärm)* Erhebung *f* der Kurbel
a. of crossing Herzstückneigung *f (Weiche)*
a. of current flow *(El)* Stromflußwinkel *m*
a. of curvature 1. *(Bw)* Krümmungswinkel *m*; 2. *(Wärm)* Knickungswinkel *m (Regler)*
a. of cut Schnittwinkel *m*; Hiebwinkel *m (Feile)*
a. of deflection Ablenkungswinkel *m*, Ausschlagwinkel *m*
a. of delay *(Kfz)* Zündverzugwinkel *m*
a. of departure *(Rak)* Brennschlußwinkel *m*, Abgangswinkel *m*
a. of deviation Ablenk[ungs]winkel *m*; Abweichungswinkel *m*
a. of dip Inklinationswinkel *m (Radar)*
a. of divergence Austrittswinkel *m* eines divergenten Strahlenbündels; Erweiterungswinkel *m*, Divergenzwinkel *m*

a. of draw *(Brg)* Grenzwinkel *m*
a. of elevation Erhebungswinkel *m*; *(Mil)* Visierwinkel *m*, Höhenwinkel *m*
a. of emergence Austrittswinkel *m*, Ausfallswinkel *m*
a. of fall *(Mil)* Fallwinkel *m*; Einfallwinkel *m*
a. of filling *(Holz)* Kreppungswinkel *m (Maschine für Trockenkreppapier)*
a. of flow Stromwinkel *m*
a. of friction Reibungswinkel *m*
a. of gash Brustwinkel *m*, Spanwinkel *m (Fräserzahn)*
a. of glare Blendungswinkel *m*
a. of hade *(Geol)* Verwurfswinkel *m*
a. of heel[ing] *(Schiff)* Krängungswinkel *m*
a. of helix Spiralwinkel *m*; Schraubenlinienwinkel *m*, Schrägungswinkel *m (Schraubenrad)*
a. of impact Auftreffwinkel *m*
a. of incidence 1. Einfallwinkel *m (von Wellen)*; 2. *(Rak)* Anstellwinkel *m*
a. of inclination Neigungswinkel *m*
a. of inclination of the control (regulating) wheel axis Neigungswinkel *m* der Regelscheibe
a. of inclination of the elastic line Neigungswinkel *m* der elastischen Linie *(Turbine)*
a. of inclination to the vertical Neigungswinkel *m* gegen die Senkrechte
a. of inlet Eintrittwinkel *m (Turbine)*
a. of internal friction Winkel *m* der inneren Reibung
a. of intersection Durchdringungswinkel *m*
a. of lag Nacheil[ungs]winkel *m*, Verzögerungswinkel *m*
a. of lapping Ablaufwinkel *m (Seiltrieb)*
a. of lead Voreil[ungs]winkel *m*, Steigungswinkel *m*
a. of lead in diagram *(Wärm)* Diagrammvoreilwinkel *m*
a. of lip clearance with workpiece face Freiwinkel *α m* [beim Spiralbohrer]
a. of lock *(Kfz)* Ausschlagwinkel *m (Lenkung)*; Einschlagwinkel *m (Räder)*
a. of obliquity Eingriffswinkel *m (Zahnrad)*
a. of oscillation Ausschlagwinkel *m (z. B. von Schubstangen)*; Schwingungsweite *f (im Winkelmaß gemessen)*
a. of outlet Austrittwinkel *m*
a. of phase difference Phasen[verschiebungs]winkel *m*
a. of phase displacement Winkel *m* der Phasenverschiebung, Phasenwinkel *m*
a. of pitch 1. Neigungswinkel *m*; 2. Steigungswinkel *m*, Anstellwinkel *m (Luftschraube)*; Längsneigungswinkel *m (Flugzeug)*
a. of point Spitzenwinkel *m*, Querschneidenwinkel *m (Spiralbohrer)*
a. of polarization Polarisierungswinkel *m*
a. of precession *(Ph, Astr)* Präzessionswinkel *m*
a. of pressure Eingriffswinkel *m*, Flankenwinkel *m (Verzahnung)*
a. of rake Spanwinkel *m*
a. of rebound Aufprallwinkel *m*
a. of recess Auslauf[wälz]winkel *m (Zahnrad)*
a. of reflection Reflexionswinkel *m*, Ausfallwinkel *m*
a. of refraction Brechungswinkel *m*
a. of relief *(Masch)* Freiwinkel *m* [an der Fase hinter der Schneide]
a. of repose 1. natürlicher Böschungswinkel *m*; Schüttwinkel *m*; 2. Reibungswinkel *m*, Gleitwinkel *m*
a. of rotation Dreh[ungs]winkel *m*
a. of rotation of the crank Kurbeldrehwinkel *m*
a. of scattering *(Ph)* Streuwinkel *m*
a. of shafts Kreuzungswinkel *m (Getriebe)*
a. of shear Scherwinkel *m*
a. of shock Stoßwinkel *m (Turbine)*
a. of sight *(Mil)* Aufsatzwinkel *m*, Zielwinkel *m*
a. of site *(Am)* Zielwinkel *m*
a. of skid Schiebewinkel *m*
a. of slew *(Bw)* Drehbereich *m*
a. of slide Schiebungswinkel *m*
a. of slip *(Geol)* Sprungwinkel *m*
a. of slope Neigungswinkel *m*; Böschungswinkel *m*
a. of spiral Spiralwinkel *m*, Spiralsteigung *f*; Drallwinkel *m*
a. of take off *(Holz)* Abzugswinkel *m*, Trennungswinkel *m (der Papierbahn vom Filz)*
a. of taper Spitzenwinkel *m* [des Kegels]; Anzugwinkel *m (Keil)*; Seitenwinkel *m (Senker)*
a. of the tooth helix Drallwinkel *m (Verzahnung)*
a. of thread Flankenwinkel *m*
a. of throw Wurfwinkel *m*
a. of tilt 1. Neigungswinkel *m (z. B. Aufspannplatte)*, Winkelstellung *f*; 2. Nadirdistanz *f (Fotogrammetrie)*
a. of tooth *(Masch)* Zahnschräge *f*
a. of torque Verdrehungswinkel *m*
a. of torsion at break Bruchverdrehungswinkel *m*
a. of twist 1. Torsionswinkel *m*, Verdrehungswinkel *m*, Drehwinkel *m*; 2. Drallsteigungswinkel *m (Spiralverzahnung)*
a. of twist in the yarn *(Text)* Drehungswinkel *m* des Garns
a. of undercut *(Masch)* Unterschnittwinkel *m*

a. **of valence** *(Ch)* Valenzwinkel *m*
a. **of view** Blickwinkel *m*
a. **of wedge** *(Masch)* Keilwinkel *m*, Meißelwinkel *m*
a. **of yaw** Gierwinkel *m*
a. **plate** *(Masch)* Aufspannwinkel *m*, Spannwinkel *m*
a. **pole** Winkelmast *m*; Winkelstange *f (Holzmast)*
a. **post** *(Bw)* Eckpfosten *m*, Eckständer *m*, Eckstiel *m*
a.**-preserving** *(Math)* winkeltreu, konform *(Abbildungen)*
a. **rafter** *(Bw)* Gratsparren *m*
a. **reamer** *(Masch)* Winkelreibahle *f*
a. **reflector** 1. *(El)* Winkelstrahler *m*, Schrägstrahler *m*; 2. Winkelreflektor *m (Radar)*
a. **ridge** *(Bw)* Gratsparren *m*
a. **ring** *(Masch)* Winkelring *m*
a. **rolled through** *(Masch)* Wälzwinkel *m*
a. **scale** Winkelteilung *f*
a. **section** *(Masch)* Winkelprofil *n*
a. **setting** Winkel[ein]stellung *f*, Schrägstellung *f*
a. **shot** *(Fs)* verkantete Aufnahme *f*
a. **steel** Winkel[stahl] *m*
a. **stone** *(Bw)* Eckstein *m*
a. **support** Abspannmast *m*
a.**-to-digit converter** *(BMSR)* Winkelumsetzer *m*
a. **tool** Prismenmeißel *m*
a. **tower** Winkelmast *m (Gittermast)*
a. **valve** Eckventil *n*; Winkelventil *n (Luftreifen)*
a. **with equal legs** gleichschenkliger Winkel *m*
a. **with unequal legs** ungleichschenkliger Winkel *m*
angledozer *(Bw)* Schwenkschildplanierraupe *f*, Seitenräumerplanierraupe *f*
anglesite Anglesit *m*, Bleiglas *n*, Bleisulfat *n*, Bleivitriol *n*, Vitriolbleierz *n*
angling dozer *s.* angledozer
a. **drill hole** Bohrloch *n* mit schiefer Richtung *(Erdöl)*
a.**-type dozer** Schwenkschildplanierraupe *f*, Seitenräumerplanierraupe *f*
Angström unit Angström-Einheit *f* $(= 10^{-8}$ *cm)*
angular winkelig; kantig, schräg[liegend], spitzwinkelig; scharfgängig, scharfkantig
a. **acceleration** Winkelbeschleunigung *f*, Radialbeschleunigung *f*
a. **accelerometer** Winkelbeschleunigungsmesser *m*
a. **accuracy** Winkelgenauigkeit *f*
a. **adjustability** Winkelverstellbarkeit *f*, Schrägverstellbarkeit *f*
a. **adjustable** schrägstellbar, schwenkbar, winkelverstellbar
a. **adjustment** Winkel[ein]stellung *f*, Schrägverstellung *f*, Schwenkbarkeit *f*, Winkelverstellung *f*
a. **advance** Winkelvoreilung *f*
a. **alignment** Fluchten *n (Verzahnung)*
a. **aperture** Winkelöffnung *f*
a. **bearing** *(Masch)* Spurlager *n*, Traglager *n*
a. **belt drive** Riementrieb *m* mit sich kreuzenden Wellen
a. **bevel gears** Kegelräder *npl* mit von 90° abweichendem Achsenwinkel
a. **bit stock** Winkelbohrwinde *f*, winklig verstellbarer Werkzeughalter *m* für Bohrwinden
a. **butt strap** Winkellasche *f*
a. **clearance** *(Masch)* Winkelspiel *n*; Freiwinkel *m (Schnittplatte)*
a.**-contact bearing** *(Masch)* Schräglager *n*
a. **correlation** Winkelkorrelation *f*
a. **cut** Schrägschnitt *m*
a. **cutter** Winkel[stirn]fräser *m*
a. **deflection** Winkelausschlag *m (Meßinstrument)*
a. **degree** Winkelgrad *m*
a. **dependence** Winkelabhängigkeit *f*
a. **deviation** Winkelabweichung *f*
a. **deviation loss** Winkelabweichungsverlust *m*
a. **dimension** Winkelmaß *n*
a. **directivity** Winkelrichtvermögen *n*
a. **displacement** Winkelverschiebung *f*, Phasenverschiebung *f*
a. **distribution** Winkelverteilung *f*
a. **division** Teilung *f* nach Winkelmaß
a. **downfeed** Schrägzustellung *f*
a. **drive** Winkelantrieb *m*
a. **error** Winkelfehler *m*, Verdrehfehler *m*
a. **facing** *(Masch)* winkliges Plandrehen *n*
a. **feed** Schrägzustellung *f*
a. **focusing** Richtungsfokussierung *f*
a. **frequency** Kreisfrequenz *f*, Winkelgeschwindigkeit *f*
a. **function** *(Math)* Winkelfunktion *f*
a. **gear** Winkelzahnrad *n*
a. **grate** *(Wärm)* Winkelrostplatte *f*
a. **helical gears** zylindrisches Wälzschraubgetriebe *n*
a. **impulse** Flächenträgheitsmoment *n*
a. **indexing** *(Masch)* Teilen *n* nach Winkelmaß
a. **kinetic energy** Rotationsenergie *f*
a. **location** Schräglage *f*, Schrägstellung *f*
a. **magnification** *(Opt)* Winkelvergrößerung *f*
a. **measurement** Winkelmessung *f*
a. **meshing** Schrägeingriff *m (Verzahnung)*
a. **minute** *(Math)* Winkelminute *f*
a. **mirror** Winkelspiegel *m*
a. **momentum** Drall *m*, Drehimpuls *m*, Impulsmoment *n*

a. momentum flowmeter Drehimpulsdurchflußmesser *m*
a. momentum operator Drehimpulsoperator *m*
a. motion Winkelbewegung *f*
a. movement *s.* a. motion
a. offset of sparks *(El)* Funkenversetzung *f*
a. phase *(El)* Phasenwinkel *m*
a. planing Schräghobeln *n*
a. point Scheitelpunkt *m*
a. position Schrägstellung *f*; Winkelstellung *f*
a. program velocity *(Rak)* Umlenkgeschwindigkeit *f*
a. range Winkelbereich *m*
a. ratio Winkelverhältnis *n*
a. resolution Winkelauflösung *f*, Sehschärfe[frequenz]winkel *m*
a. resolving power Sehschärfegrenzwinkel *m*
a. resonance frequency Resonanz[kreis]frequenz *f*
a. scale Teilkreis *m*
a. setting Winkeleinstellung *f*
a. speed Winkelgeschwindigkeit *f*
a. teeth Schrägverzahnung *f*
a. thread *s.* vee thread
a. trough *(Brg)* Winkelrinne *f*
a. turning *(Masch)* Drehen *n* von Kegeln (nichtzylindrischen Formen)
a. variation 1. *(Masch)* Winkelabweichung *f*; 2. *(El)* Phasenschwankung *f*
a. velocity Winkelgeschwindigkeit *f*
a. wheel Winkelrad *n*
a. width Winkelbreite *f*
angularity Winkligkeit *f*, Winkel[ein]stellung *f*; Schrägstellung *f*
angularly adjustable schrägstellbar
anharmonic nicht-harmonisch, anharmonisch
anhydric wasserfrei
anhydrite *(Min)* Anhydrit *m*, wasserfreier Gips *m*
anhydro ring *(Ch)* Ring *m*, gebildet durch inneres Anhydrid
anhydrous *(Ch)* wasserfrei, anhydrisch, entwässert, kristallwasserfrei, nicht wäßrig
a. lime gebrannter Kalk *m*, Branntkalk *m*
a. sodium carbonate kalzinierte Soda *f*
aniline base fuel *(Rak)* Brennstoff *m* auf Anilinbasis
a. black *(Text)* Anilinschwarz *n*, Hängeschwarz *n*
a. dye Anilinfarbstoff *m*, Anilinfarbe *f*
a. [-formaldehyde] resin Anilin[formaldeyd]harz *n*
a. [rubber plate] printing *(Typ)* Anilingummidruck *m*
animal breeding Tierzucht *f*
a. charcoal Knochenkohle *f*
a. epidemic Tierseuche *f*
a. fibre tierische Faser *f*
a. shelter Stall[ung *f*] *m*
a. size *(Ch)* tierischer Leim *m*
a. traction Gespannzug *m*, Zugtierbetrieb *m*
animalize / to animalisieren *(Faser)*
animalized cotton *(Text)* animalisierte Baumwolle *f*
animated drawings Zeichentrickfilm *m*
a. picture Laufbild *n*
animating electrode *(El)* Ionisierungsgitter *n*
anion Anion *n*
a. exchange Anionenaustausch *m*
a. vacancy *(Ph)* Anionfehlstelle *f*
anionics anionenaktive Substanzen *fpl*
anise *(Ch)* Anis *m*
aniseed Anis[samen] *m*
anisic acid Anissäure *f*
anisometric *(Math)* anisometrisch *(Abbildung)*
anisotropic *(Ph)* anisotrop
a. components Anisotropieanteile *mpl*
a. distortion anisotropische Verzeichnung *f*
a. emission anisotrope Strahlung *f*
anisotropy Anisotropie *f*
a. due to crystalline lattice structure Anisotropie *f* im Gitterbau
ankerite *(Min)* Ankerit *m*
annabergite *(Min)* Annabergit *m*, Nickelblüte *f*
anneal / to [aus]glühen; kühlen *(Glas)*
anneal tempering Anlaßhärten *n*
annealed geglüht; angelassen *(Stahl)*
a. tensile strength *(Met)* Glühfestigkeit *f*
annealing *(Met)* Glühen *n*, Ausglühen *n*
a. arch *(Met)* Brennofen *m*
a. bath Glühbad *n*
a. bell Glühhaube *f*
a. box Glühkiste *f*, Glühkasten *m*
a. colour Glühfarbe *f*
a. cycle Glühzeit *f*
a. expansion Glühausdehnung *f*
a. furnace Glühofen *m*
a. installation Glüherei *f*
a. muffle Glühmuffel *f*
a. oil Härteöl *n*
a. period Glühzeit *f*
a. plant Glüherei *f*
a. plastics Spannungsfreimachen *n* von Kunststoffen
a. pot Glühtopf *m*, Tempertopf *m*
a. process Glühverfahren *n*
annex / to anhängen
annex[e] *(Bw)* Anbau *m*
annihilation *(Ph)* Zerstrahlung *f*, Vernichtung *f*, Annihilation *f*; Auflösung *f* *(von Versetzungen)*; Paarvernichtung *f*
a. of matter *(Ph)* Zerstrahlung *f* *(Übergang der Materie in Strahlung)*
a. radiation *(Ph)* Annihilationsstrahlung *f*; Vernichtungsstrahlung *f*

annual capacity Jahresleistung *f*
  a. **cut** *(Holz)* Hiebsatz *m*
  a. **fluctuation** Jahresschwankung *f*
  a. **growth ring** *(Holz)* Jahresring *m*
  a. **load diagram** *(El)* Jahresbelastungsdiagramm *n*
  a. **output** Jahresproduktion *f*
  a. **precipitation** jährliche Niederschlagsmenge *f*
  a. **ring** *(Holz)* Jahresring *m*
  a. **subscription** Jahresabonnement *n*, Jahresgebühr *f*
  a. **yield** Jahresausbeute *f*
annul each other / to *(Math)* sich gegenseitig aufheben
annular ringförmig, Ring-, Kreisring-
  a. **ball bearing** Axialkugellager *n*
  a. **borer** Kernbohrer *m*
  a. **clearance** *(Kfz)* Ringspalt *m (einer Düse)*
  a.**-clearance nozzle** *(Kfz)* Ringdüse *f*
  a. **combustion chamber** *(Rak)* Ringbrennkammer *f*
  a. **cone sound detector** *(Mil)* Ringtrichterrichtungshörer *m*
  a. **electrode** Ringelektrode *f*
  a. **exhaust** Abgasringkanal *m*
  a. **fuel tank** Kraftstoffringbehälter *m*
  a. **furnace** Ringfeuerung *f*, Ringofen *m*
  a. **fused-quartz spiral** Schmelzraupe *f*
  a. **gear** Zahnrad *n* mit Innenverzahnung
  a. **groove** Ringnut *f*
  a. **kiln** Ringofen *m*
  a. **magnet** Ringmagnet *m*
  a. **milling cutter** Ringfräser *m*
  a. **outflow** ringförmige Ausströmung *f*
  a. **producer** *(Ch)* Ringbildner *m*
  a. **radiator** Ringkühler *m*
  a. **ridges** *(Masch)* Riefen *fpl* in Form eines feinen Gewindes *(Bohrungswandung)*
  a. **ring** 1. *(Masch)* Ringmutter *f*; 2. Abstellring *m (Rundstrickmaschine)*
  a. **seat** *(Masch)* Ringsitz *m*
  a. **slide valve** Ringschieber *m*
  a. **sliding contact** Schleifringkontakt *m*
  a. **spring** Ringfeder *f*
  a. **stop** *(Opt)* Ringblende *f*
  a. **toothed wheel** Innenzahnrad *n*, Zahnrad *n* mit innerer Verzahnung
  a.**-type burner** *(Ch)* Kranzbrenner *m*
  a. **vault** *(Bw)* Tonnengewölbe *n*
  a. **wheel** Innenzahnrad *n*
  a. **wire loop** Drahtring *m*
annulated column *(Bw)* Bündelpfeiler *m*
annulus Kreisring *m*; Ringspalt *m*
  a. **of metal** ringförmiger Metallquerschnitt *m (Hohlbohren)*
annunciator *(BMSR)* Melder *m*, Schauzeichen *n*
  a. **drop** Anrufklappe *f*

anodal light *(El)* Anodenlicht *n*
anode Anode *f*
  a. **accumulator** Anodenakkumulator *m*
  a. **angle** Anodenwinkel *m*
  a. **arm** Anodenarm *m*
  a. **balancing coil** Anodenausgleichsdrossel *f*
  a. **battery** Anodenbatterie *f*
  a. **bend** Anodenknick *m (Anodenkennlinie)*
  a. **bend detection** Anodengleichrichtung *f*
  a. **bend detector** Anodengleichrichter *m*
  a. **bend rectification** Anodengleichrichtung *f*, Demodulation *f*
  a. **breakdown voltage** Anodenzündspannung *f*, Durchschlagspannung *f*
  a. **bridge** Anodenbrücke *f*
  a. **brighten / to** *(Met)* glanzeloxieren
  a. **butt** Anodenstumpf *m*
  a. **cage** Anodenkorb *m*
  a. **cap** Anodenkappe *f*
  a. **cathode gap** Anodenkatodenstrecke *f*
  a. **cathode voltage drop** innerer Spannungsabfall *m*
  a. **circuit** Anodenkreis *m*
  a. **circuit keying** Anodenkreistastung *f*
  a. **clamp** Anodenbügel *m*
  a. **converter** Anodenspannungsumformer *m*
  a. **current** Anodenstrom *m*
  a. **current characteristic** Anodenstromkennlinie *f*
  a. **current economy circuit** Anodenstromsparschaltung *f*
  a. **dark space** Anodendunkelraum *m*
  a. **density** Anodendichte *f*
  a. **detection** Anodengleichrichtung *f*
  a. **detector** Anodengleichrichter *m*
  a. **diaphragm** Anodenblende *f*
  a. **differential resistance** Innenwiderstand *m*
  a. **dissipation** Anodenverlustleistung *f*
  a. **drop** Anodenfall *m*
  a. **efficiency** anodische Stromausbeute *f*; Wirkungsgrad *m* der Anode
  a. **fall** Anodenfall *m*
  a. **family** Anodenkennlinienschar *f*
  a. **feed resistance** Anodenschutzwiderstand *m*
  a. **feedback** Anodenrückkopplung *f*
  a. **fin** Anodenflügel *m*
  a. **firing** Anodenzündung *f*
  a. **follower** Katodenbasisverstärker *m*
  a. **glow** Anoden[glimm]licht *n*
  a. **grid-plate capacitance** Anodengitterkapazität *f*
  a. **guard-net** Anodenschutznetz *n*
  a. **hum** Anodenrauschen *n*
  a. **keying** Anoden[spannungs]tastung *f*; Tastung *f* im Anodenkreis
  a. **layer** anodische Schicht *f*
  a. **lead** Anodenanschluß *m*

a. leading-in wire Anodendurchführung *f*
a. life Lebensdauer *f* der Anode
a. loop *s.* a. circuit
a. metal Anodenmetall *n*
a. mud Anodenschlamm *m*
a. peak current Anodenspitzenstrom *m*
a. plate Anodenplatte *f (Akkumulatoren)*
a. plate family Anodenstromkennlinie *f*
a. potential Anodenspannung *f*
a. rating zugelassene Anodenbelastung *f*
a. rays Anodenstrahlen *mpl*, Kanalstrahlen *mpl*, Ionenstrahlen *mpl*
a. reaction Anodenrückwirkung *f*
a. reactor Anodendrossel *f*
a. rejector circuit Anodensperrkreis *m*
a. residue Anodenrest *m*
a. resistance Innenwiderstand *m (Röhre)*
a. rest current Anodenruhestrom *m*
a. return Anodenrückleitung *f*
a. scrap Anodenabfall *m*, Anodenschlamm *m*
a. sheet metals Anodenbleche *npl*
a. shield Anodenschutz[rohr *n*] *m*, Anodenhülse *f*, Anodenschirm *m*
a. slime Anodenschlamm *m*
a. sludge Anodenschlamm *m (Kupferelektrolyse)*
a. sponge Anodenschwamm *m (Bleiherstellung)*
a. spot Anodenfleck *m*, Fokus *m*
a. sputtering Anodenzerstäubung *f*
a. starting Anodenzündung *f* durch eine Zündelektrode
a. straightener *(Am)* Anodenrichter *m*
a. supply Anodenspeisung *f*
a. tapping point Anodenspulenabgriff *m*
a. voltage Anodenspannung *f*
a. voltage characteristic Anodenkennlinie *f*
a. voltage degeneration Anodenspannungsgegenkopplung *f*
a. voltage drop Anodenspannungsabfall *m*
a. voltage supply Anodenspannungsquelle *f*
anodic anodisch
a. behaviour anodisches Verhalten *n*
a. cleaning anodische Reinigung *f*
a. corrosion Stromkorrosion *f*
a. current density Anodenstromdichte *f*
a. density Anodendichte *f*
a. etching elektrolytisches Anätzen *n*
a. oxidation Eloxieren *n*, Eloxierung *f*
a. passivation anodische Passivierung *f*
a. polarization Anodenpolarisation *f*
a. reaction Anodenreaktion *f*
a. treatment anodische Behandlung *f*, Eloxieren *n*
anodically oxidized eloxiert
anodige *(Am) (BMSR)* Analog-digital-Umsetzer *m*
anodization Eloxierung *f*
anodize / to eloxieren, anodisch oxydieren *(Aluminium)*
anodized eloxiert
a. coating *s.* a. oxide layer
a. oxide layer Eloxalschicht *f*
a. plate *(Typ)* eloxierte (anodisch oxydierte) Aluminium-Offsetplatte *f*
anodizing Eloxieren *n (Aluminium)*
anolyte *(El)* Anolyt *m*
anomalous anomal
a. abundance anomale Häufigkeit *f*
a. behaviour abweichendes Verhalten *n*
a. propagation unregelmäßige Ausbreitung *f*
a. scattering anomale Streuung *f*
anorthic triklin
anorthite *(Min)* Kalkfeldspat *m*, Anorthit *m*
anorthoclase *(Min)* Anorthoklas *m*
answer / to *(Fmt)* abfragen; sich melden; ansprechen *(Meßgeräte)*
a. on the circuit *(Fmt)* in die Leitung eintreten
a. the helm *(Schiff)* dem Ruder gehorchen
answer Antwort *f*; Ansprechen *n (Meßgeräte)*
a.-back code Kennziffer *f (Fernschreibtechnik)*
a.-back drum *(Fmt)* Rückmeldescheibe *f*, Rückmeldetrommel *f*
answering button Abfragetaste *f (Fernsprechtechnik)*
a. interval *(Am)* Wartezeit *f (Fernsprechtechnik)*
a. jack *(Fmt)* Teilnehmerklinke *f*, Abfrageklinke *f*
a. key *(Fmt)* Abfrageschalter *m*
a. pennant Antwortwimpel *m*, Signalbuchwimpel *m*
a. plug *(Fmt)* Abfragestöpsel *m*
a. time *(Am) (El)* Ansprechzeit *f (z. B. des Relais)*
antacid *(Ch)* säuretilgend, säurewidrig
ante-chamber Vorzimmer *n*, Vestibül *n*
a.-room Vorzimmer *n*, Vestibül *n*
a.-solarium Sonnenbalkon *m*
antecedent Vorderglied *n*, vorhergehendes Glied *n*; Vorgänger *m*; Voraussetzung *f*
antenna *(Am)* Antenne *f (s. a.* aerial)
antennafier *(Am)* Kombination *f* Empfangsantenne und Festkörperschaltung zur Verstärkung
antennaverter *(Am)* Kombination *f* Empfangsantenne und Festkörperschaltung zur Mischung
anterior chamber lens *(Opt)* Vorderkammerlinse *f*
anthra dyes Anthrafarbstoffe *mpl (Anthrachinon-Küpenfarbstoffe)*
anthracene Anthrazen *n*
a. dyestuff Anthrazenfarbstoff *m*
a. oil Anthrazenöl *n*
anthracite [coal] Anthrazit *m*, Glanzkohle *f*, Steinkohle *f* mit weniger als 8% flüchtigen Bestandteilen

anthracitization Anthrazitbildung *f*
anthracosis Kohlenstaublunge *f*
anthranilic acid Anthranilsäure *f*
anthraquinol Anthrahydrochinon *n*
anthraquinone Anthrachinon *n*
a. dyestuff Anthrachinonfarbstoff *m*
a. vat dye[stuff] Anthrachinon[küpen]farbstoff *m*
anti-aeroplane defence Luftverteidigung *f*
a. ager Alterungsschutzmittel *n*
a.-airborne [assault] defence Luftlandeabwehr *f*
a.-airborne protection Luftlandeabwehr *f*
a.-aircraft automatic Fla-MG *n*
a.-aircraft balloon Sperrballon *m*
a.-aircraft cannon *s.* a.-aircraft gun
a.-aircraft detection radar Ortungsstation *f*, Ortungsgerät *n*
a.-aircraft fire direction radar Geschützrichtstation *f*
a.-aircraft fire director Kommandogerät *n* *(Flak)*
a.-aircraft gun Fliegerabwehrkanone *f*, Flak *f*, Fliegerabwehrgeschütz *n*
a.-aircraft machine gun Fliegerabwehrmaschinengewehr *n*, Fla-MG *n*
a.-aircraft missile Flugabwehrrakete *f*
a.-aircraft rocket Luftabwehrrakete *f*, Fla-Rakete *f*
a.-aircraft searchlight Flakscheinwerfer *m*
a.-aircraft set Geschützrichtstation *f*
a.-aircraft situation Luftlage *f*
a.-aircraft tank Fla-Panzer *m*
a.-aircraft weapon Fla-Waffe *f*
a.-aviation defence Luftverteidigung *f*
a.-backlash mit Spielausgleich, Spielausgleich-
a.-bacterial finish *(Text)* Bakterizidausrüstung *f*
a.-blocking agents *(Plast)* Stoffe *mpl*, die unerwünschtes Kleben oder Backen verhindern
a.-breakage transfer schonende Übergabe *f*
a.-brushing fittings *(Elm)* Sprühschutz *m*
a.-chatter armature reed *(El)* Ankerreibfeder *f*
a.-clockwise gegen den Uhrzeigersinn, linksdrehend
a.-clockwise rotation Linksdrehung *f*, Linkslauf *m*, Drehung *f* entgegengesetzt dem Uhrzeigersinn
a.-clutter *(El)* Störunterdrücker *m*
a.-clutter gain control *(Fmt)* Verstärkung *f* mit gleichzeitiger Unterdrückung unerwünschter Echos *(Radartechnik)*
a.-clutter rain *(Fmt)* Regenenttrübung *f*
a.-clutter sea *(Fmt)* Nahechodämpfung *f*
a.-corrosive paint Korrosionsschutzfarbe *f*
a.-damping Anfachung *f*
a.-dazzle Blendschutz-
a.-detonant Antiklopfmittel *n* *(Verbrennungsmotor)*
a.-detonating fuel klopffester Treibstoff *m*
a.-detonation agent *(Kfz)* Klopfbremse *f*, Oktanzahlverbesserer *m*
a.-diffusion grid Streustrahlblende *f* *(Röntgen)*
a.-drip nozzle *(Kfz)* nichttropfender Zapfhahn *m*, nicht nachtropfende Düse *f*
a.-flooding agent Ausschwimmverhütungsmittel *n*
a.-fog agent *(Ch)* Antischleiermittel *n*
a.-fouling composition *(Schiff)* anwuchsverhindernde Farbe *f*
a.-fouling paint *(Schiff)* anwuchsverhindernde Farbe *f*
a.-halo lichthoffrei
a.-ICBM defence Raketenabwehr *f*
a.-ice measures *(Schiff)* Eissicherung *f* *(Schutz gegen Vereisung)*
a.-induction network *(Fmt)* Nebensprechdämpfungsnetzwerk *n*
a.-inductive arrangement Induktionsschutz *m*
a.-interference Entstörung *f*
a.-interference aerial störungsarme Antenne *f*
a.-interference aerial system störungsarmes Antennensystem *n*
a.-interference condenser Entstörungskondensator *m*
a.-jamming *(Fmt)* Gegenstörung *f*
a.-knock quality Klopffestigkeit *f*
a.-livering agent Eindickungsverhinderungsmittel *(Lackindustrie)*
a.-matter *(Ph)* Antimaterie *f*
a.-meson *(Ph)* Antimeson *n*
a.-microphonic valve antimikrofonische Röhre *f*
a.-missile defence Raketenabwehr *f*
a.-missile missile Gegenrakete *f*, Antirakete *f*, Antiraketen-Rakete *f*
a.-personnel bomb *(Mil)* Splitterbombe *f*
a.-pipe compound *(Met)* Lunkerabdeckmasse *f*
a.-piping lunkerverhütend
a.-reaction device *(El)* Rückkopplungssperre *f*
a.-reflection coating *(Opt)* reflexmindernde Schicht *f*
a.-rocket [missile] Gegenrakete *f*, Antirakete *f*
a.-sag agent Läuferverhinderungsmittel *n* *(Lackindustrie)*
a.-set-off roller *(Typ)* Abschmutzwalze *f*
a.-settling agents *(Plast)* Absetzverhinderungsmittel *npl*, Schwebemittel *npl*
a.-settling property Schwebefähigkeit *f*

a.-sidetone circuit *(El)* Dämpfungsschaltung *f*
a.-sidetone device *(El)* geräuschdämpfende Schaltung *f*
a.-stripping agent Haftmittel *n (Erdöl, Erdgas)*
a. submarine depth indicating control Ultraschallecholot *n*, Wasserschallortungsgerät *n*
a.-tank ditch Panzergraben *m*
a.-tank obstacle Panzerhindernis *n*, Panzersperre *f*
a.-tank rifle Panzerbüchse *f*
a.-tank rocket launcher reaktive Panzerbüchse *f*
a.-tank warning system Panzerwarndienst *m*
a.-tipping moment *(Schiff)* Moment *n* gegen Kippen *(beim Ablauf)*
a.-torsion verwindungsfrei
a.-trade winds *(Schiff)* Gegenpassat *m*
a.-transmit-receive-tube Sendersperröhre *f*
a.-vibration erschütterungsarm, schwingungsarm; erschütterungsfrei, schwingungsfrei
a.-vibration cap *(Fmt, El)* Antivibrationssockel *m*
a.-vibration lamp-holder Fassung *f* mit Vibrationsschutz, gefederte (schwingungsfreie) Fassung
a.-vibration mica Dämpfungsglimmer *m (Elektronenröhrenherstellung)*
a.-vibration mounting erschütterungsfreie Aufstellung *f*
a.-vibration valve holder federnde Röhrenfassung *f*
a.-wear additive Verschleißschutzmittel *n*
a.-wear agent Verschleißbremse *f*, Verschleißschutzstoff *m*
a.-wear liner *(Brg)* Schleißplatte *f*
a.-wear lining *(Brg)* Panzerung *f*
antiaircraft *s.* anti-aircraft
antibody Antikörper *m*, Gegenstoff *m*
anticapacitance kapazitätsarm
a. switch kapazitätsarmer Schalter *m*
anticatalyst *(Ch)* Paralysator *m*, Antikatalysator *m*
anticathode Antikatode *f*, Gegenkatode *f*
antichlorine Antichlor *n (Natriumthiosulfat)*
anticipatory control *s.* derivative action
anticlastic *(Math)* antiklastisch
anticlinal fold *(Geol)* Sattel *m*
anticline *(Geol)* Sattel *m*
anticoincidence *(Ph)* Antikoinzidenz *f*
a. analyzer *(Kern)* Antikoinzidenzanalysator *m*
a. circuit *(Kern)* Antikoinzidenzschaltung *f*
a. counter *(Kern)* Antikoinzidenzzähler *m*
anticontamination cloth Schutzkleidung *f*
anticorrosive korrosionsverhindernd, rostfest, rosthindernd, Korrosionsschutz-, Rostschutz-
a. agent Korrosionsschutzmittel *n*, Rostschutzmittel *n*
anticreaming agent Mittel *n* zur Verhinderung von Hautbildung *(bei Lacken)*
anticrease finish *(Text)* Knitterarmausrüstung *f*, knitterarme Ausrüstung *f*
anticyclone Antizyklone *f*
antidazzle cap *(Kfz)* Abblendhaube *f*
a. device Abblendeinrichtung *f*
a. glass Blendschutzglas *n*, Blendschutzscheibe *f*
a. lamp Blendschutzlampe *f*, Biluxlampe *f*
a. regulations Abblendvorschriften *fpl*
a. switch Abblendumschalter *m*
antidazzling screen Blendschutzscheibe *f*
antiderivative *(Math)* unbestimmtes Integral *n*
antidisintegrant *(Ch)* abbauverhinderndes Mittel *n*
antidistortion device *(Fmt, Fs, El)* Entzerrer *m*
antidribble / to abdichten
antidribble tropfsicher
antidyne circuit gegengekoppelter Sendekreis *m*
antielectron Positron *n*, Antielektron *n*
antifading aerial schwundmindernde Antenne *f*
a. bath *(Ch)* Lichtschutzbad *n*
a. device Schwundausgleicher *m*
a. effect schwundmindernde Wirkung *f (Antenne)*
antifatigue dauerschwingfest
antifermentative gärungswidrig
antiferroelectric antiferroelektrisch
antiferroelectrics Antiferroelektrika *npl*
antiferromagnetic antiferromagnetisch
antiferromagnetism Antiferromagnetismus *m*
antiflooding agent Ausschwimmverhütungsmittel *n (Lackindustrie)*
antifoaming agent Antischaummittel *n*, Entschäumer *m*, Schaumhemmungsmittel *n*, Schaumverhütungsmittel *n*
antifouling Anti-fouling-Anstrich *m (Schiffsbodenanstrich)*
a. paint anwuchsverhindernder Anstrich *m*, Unterwasser-Schutzanstrich *m*
a. temperature Selbstreinigungstemperatur *f (Zündkerze)*
antifreeze [mixture] Frostschutz[mittel *n*] *m*, Gefrierschutz[mittel *n*] *m*
antifreezing kältebeständig
a. quality Kältebeständigkeit *f*
antifriction alloy Lagerlegierung *f*
a. bearing Wälzlager *n*
a. bonderizing bath *(Met)* Gleitbonderflüssigkeit *f*
a. grease Wälzlagerfett *n*
a. metal Antifriktionsmetall *n*; Lagermetall *n*, Weißmetall *n*
a. properties Gleiteigenschaften *fpl*

antifrictional properties *pl* Notlaufeigenschaften *fpl*
a. quality Gleitfähigkeit *f (Lager)*
antifrost screen Frostschutzscheibe *f*
antifrostbite ointment Frostschutzsalbe *f*
antifroth oil Schaumdämpfungsöl *n*
antifrothing agent Schaumverhinderungsmittel *n*
antiglare device Blendschutz *m*
a. goggles Blendschutzbrille *f*
a. lamp Lampe *f* aus blendfreiem Glas, blendfreie Glühlampe *f*
a. screen Blendschutzscheibe *f*
antigorite *(Min)* Antigorit *m*
antihalation backing Antilichthofschicht *f*, Antireflexschutzschicht *f*
antihalo lichthoffrei
antihum capacitor Brummkondensator *m*
antihunt *(BMSR)* stabilisierend
a. action Stabilisierungswirkung *f*, Dämpfungswirkung *f*
a. transformer Beruhigungsdrossel *f*
antihunting circuit *(Fmt)* Beruhigungskreis *m*; Schaltung *f* zur Unterdrückung der Eigenschwingung
antiknock Antiklopfmittel *n (Verbrennungsmotor)*
antiknock klopffest, klopffrei *(Treibstoff)*
a. agent Klopfbremse *f*, Oktanzahlverbesserer *m*
a. blend Klopfbremse *f*
a. fuel klopffester Treibstoff *m*
a. fuel dope Antiklopfzusatzmittel *n (zum Treibstoff)*
a. value Klopf[festigkeits]wert *m*, Klopffestigkeitsgrad *m*, Oktanzahl *f (des Treibstoffs)*
antiknocking property Klopffestigkeit *f (Treibstoff)*
a. rating Klopfwertbestimmung *f*, Klopfwertangabe *f*
antileak cement Dichtungsmittel *n* gegen Flüssigkeitsdurchtritt
antilogarithm Antilogarithmus *m*, Numerus *m*
antilogous poles ungleichnamige Pole *mpl*
antimonate Antimonat *n*, Salz *n* der Antimonsäure
antimonial antimonartig, antimonhaltig, antimonisch; spießglanzartig
a. lead Hartblei *n*
a. silver ore Weißsilbererz *n*
antimonic antimonartig, antimonhaltig, antimonisch
a. acid Antimonsäure *f*
a. acid anhydride Antimonsäureanhydrid *n*, Antimonpentoxid *n*
a. compound Verbindung *f* des fünfwertigen Antimons
antimonite Antimonit *m*
antimonous anhydride Anhydrid *n* der antimonigen Säure, Antimontrioxid *n*
a. compound Verbindung *f* des dreiwertigen Antimons
antimony Antimon *n*
a. bloom Antimonblüte *f*, Antimontrioxid *n*; Valentinit *m*
a. crocus Legierung *f* aus antimoniger Säure und Antimontrisulfid
a. glass glasartiges Antimontrioxid *n*
a. golden sulfide Antimonpentasulfid *n*, Goldschwefel *m*
a. ore Antimonerz *n*
antimonyl potassium tartrate Antimonylkaliumtartrat *n*, Brechweinstein *m*
antineutrino Antineutrino *n*
antineutron Antineutron *n*
antinodal point Bauch *m (einer Schwingung)*
antinode Antinode *f*, Schwingungsbauch *m*, Wellenbauch *m*
antinous release Drahtauslöser *m*
antioxidant Antioxydationsmittel *n*, Oxydationsschutzmittel *n*
antiparallel coupling *(El)* Gegenschaltung *f*, Antiparallelschaltung *f*
antiparticle Antiteilchen *n*
antiperspirants *(Ch)* Antitranspirationsmittel *npl*
antiphase Gegenphase *f*
antiplane linearisiert
antipolar *(Math)* Gegenpolare *f*
antique lace Handklöppelspitze *f* schwerer Art
a. printing paper Antikdruckpapier *n*
antiresonance *(El)* Stromresonanz *f*
a. circuit Sperrkreis *m*; Stromresonanzkreis *m*
a. frequency Eigenfrequenz *f* des Sperrkreises
a. peak *(Fmt)* Stromresonanzpunkt *m (bei Filtern, absolute Sperre)*
antiroll bar *(Kfz)* Querstabilisator *m*, Drehstabstabilisator *m*
antirolling bulges *(Schiff)* Schlingerwülste *mpl*
a. device *(Schiff)* Schlingerdämpfungseinrichtung *f*
a. stabilizer *(Schiff)* Schlingerdämpfer *m*
a. tank *(Schiff)* Schlingertank *m*
antirot fäulnishindernd, fäulniswidrig
antirust rostfrei, nichtrostend, rostbeständig
a. paint Rostschutzanstrich *m*; Rostschutzfarbe *f*
antirusting agent Rostschutzmittel *n*
antiscale composition *(Wärm)* Kesselstein[löse]mittel *n*
antiseizing property Schmierfähigkeit *f*; Fähigkeit *f* eines Trenn- oder Gleitmittels, das Ankleben von Kunststoff an der Form zu verhindern

antiseizure properties Notlaufeigenschaften *fpl*
antiside tone Rückhördämpfung *f*
antisinging stability *(Fmt)* Pfeifsicherheit *f*
antiskid rutschfest
a. chain Gleitschutzkette *f*
a. pattern Gleitschutzprofil *n (Reifen)*
a. properties Rutschfestigkeit *f*
antiskinning agent *(Plast)* Hautbildungsverhinderungsmittel *n*
antislip chain Gleitschutzkette *f*
antisoaking properties Quellfestigkeit *f*
antispasmodic Antispasmodikum *n*, krampflösendes Mittel *n*
antistatic *(Plast)* Antistatikmittel *n*, Mittel *n* zur Verhinderung elektrostatischer Aufladungen
a. device Entelektrisiereinrichtung *f (für Papier)*
a. finish *(Text)* antistatische Ausrüstung *f*
antistokes radiation *(Ph)* antistokesche Strahlung *f*
antitarnish paper Silberpackpapier *n*
antithesis Antithese *f*
antivibrator *(BMSR)* Schwingungsdämpfer *m*, Dämpfer *m*, Stabilisator *m*
antizymotic gärungshemmend
anvil Amboß *m*
a. back Amboß *m (Klebschraubstock)*
a. base Amboßuntersatz *m*
a. beak Amboßhorn *n*
a. block Amboßstock *m*, Schabotte *f*
a. cap Schabotte-Einsatz *m*, Gesenkhalter *m*
a. face Amboßbahn *f*
a. heel Stauchamboß *m*
a. horn Amboßhorn *n*
a. punch Amboßhörnchen *n*, Spitzstöckel *m*
a. tool Amboßgesenk *n*, Gesenk *n*
a. with horn Amboß *m* mit eingesetztem Horn
apart mit Abstand von, auseinander
apartment Etagenwohnung *f*
a. block (building, house) Mietshaus *n*; Wohnblock *m*
a. house aerial Gemeinschaftsantenne *f*
a. unit Wohnungseinheit *f*; Wohnzelle *f*
apatite Apatit *m*, phosphorsaurer Kalk *m*
aperiodic aperiodisch
a. aerial Breitbandantenne *f*, aperiodische Antenne *f*
a. circuit aperiodischer Stromkreis *m*
a. component of a short-circuit current Gleichstromglied *n* des Kurzschlußstromes
a. discharge aperiodische Entladung *f*
a. element aperiodisch gedämpftes Organ *n*
a. instrument aperiodisch gedämpftes Instrument *n*
a. phenomenon schwingungsfreier (aperiodischer) Vorgang *m*
a. wave gedämpfte Welle *f*
apertometer *(Opt)* Apertometer *n*
apertural error *(Opt)* Öffnungsfehler *m*
aperture 1. *(Opt)* Apertur *f*; Öffnung[swinkel *m*] *f*; 2. *(Bw)* Aussparung *f*, Mauerwerksöffnung *f*; 3. *(Schiff)* Schraubenbrunnen *m (im Hintersteven)*
a. aberration Öffnungsfehler *m*
a. angle Öffnungswinkel *m*
a. compensation Öffnungsausgleich *m (Fernsehtechnik)*
a. correction Aperturkorrektur *f*
a. corrector with delay line Kosinusentzerrer *m (Fernsehaufnahmeröhren)*
a. diaphragm *(Opt)* Aperturblende *f*
a. lens Lochscheibenlinse *f*; Elektronenlinse *f*
a. limitation *(Opt)* Aperturbegrenzung *f*
a. of the beam Elektronenstrahlquerschnitt *m (Bildaufnahmeröhre)*
a. of the reflector Spiegelöffnung *f*
a. plate Aperturblende *f*
a. ratio *(Opt)* Öffnungsverhältnis *n (z. B. einer Linse)*
a. stop *(Opt)* Aperturblende *f*
a. width *(Brg)* [lichte] Maschenweite *f*
apertured disk Lochscheibe *f (Nipkow-Scheibe in der Fernsehtechnik)*
apertures Lochung *f (Siebe)*; Schlitze *mpl*
a. of the spinning nozzle Spinndüsenöffnungen *fpl*
apex 1. Scheitel[punkt] *m*; Spitze *f*; 2. Apex *m*, Zielpunkt *m*; 3. *(Brg)* obere Anschlagbühne *f*, Ausbiß *m*
a. angle Winkel *m* an der Spitze
a. factor *(Fmt)* Verstärkungsscheitelfaktor *m*
a. of trajectory *(Mil)* Gipfelpunkt *m* der Flugbahn
a. stone *(Bw)* Schlußstein *m*
aphakic eye aphakisches Auge *n*
aphonic schalltot *(Raum)*
apiarist Bienenhalter *m*, Bienenzüchter *m*; Imker *m*
apical an der Spitze, im Scheitelpunkt gelegen
apices *s.* apex
apiculture Bienenzucht *f*
aplanatic aplanatisch
a. condenser *(Opt)* aplanatischer Kondensor *m*
a. folding magnifier *(Opt)* aplanatische Einschlaglupe *f*
aplite Aplit *m*
apochromatic [lens] Apochromat *m (n)*
apogeal velocity Geschwindigkeit *f* im Apogäum, Geschwindigkeit in größter Erdferne
apogee Erdferne *f*, Apogeum *n*, Apogäum *n*
a. altitude Höhe *f* des Apogäums, Höhe der größten Erdferne
Apolonius's circle Apolonischer Kreis *m*
aport *(Schiff)* an Backbord, zur Backbordseite
apparatus Gerät *n*, Vorrichtung *f*, Apparat *m*

a. for analyzing flue gases Rauchgasanalysator *m*
a. for illumination Beleuchtungsapparat *m*
a. for returning water of condensation Kondenswasserrückspeiseanlage *f*
a. for testing the extensibility Dehnungsprüfapparat *m*
a. of resistance *(El)* Widerstandsmeßbrücke *f*
a. plug *(El)* Gerätestecker *m*
a. rack Rahmen *m*, Gestell *n*
a. room Wählersaal *m (Telefonie)*
apparel 1. *(Text)* Bekleidung *f*; 2. Schiffsgerät *n*, Gerätschaften *pl*
a. designer *(Text)* Couturier *m*
a. wool Wolle *f* für Kleidungszwecke
apparent acceleration scheinbare Beschleunigung *f*
a. arch Scheingewölbe *n*
a. attenuation Scheindämpfung *f*
a. conduction Scheinleitung *f*
a. current Scheinstrom *m*
a. density Schüttgewicht *n* (g/cm^3)
a. efficiency Scheinwirkungsgrad *m*
a.-energy meter *s.* volt-ampere-hour meter
a. force Scheinkraft *f*
a. half-life *(Kern)* scheinbare Halbwertzeit *f*
a. horizon *(Schiff)* sichtbarer Horizont *m*, Seehorizont *m*, Kimm *f*
a. life-time *(Kern)* scheinbare Lebensdauer *f*
a. motion scheinbare Bewegung *f*
a. ohmic resistance Bürde *f (eines Transformators)*
a. output scheinbare Leistung *f*
a. powder density Schüttgewicht *n (g/ml)*
a. power Scheinleistung *f*
a. power meter *(El)* Scheinverbrauchszähler *m*
a. resistance Impedanz *f*, Scheinwiderstand *m*
a. solubility Scheinlöslichkeit *f*
a. specific gravity *(Text)* Rohgewicht *n*
a. structure Scheinstruktur *f*
a. viscosity Scheinviskosität *f*
a. watts Scheinleistung *f*
appearence of fatigue Ermüdungserscheinung *f*
a. of fracture Bruchbild *n*, Bruchaussehen *n*
appendage Ansatz *m*; Ansetzstück *n*; Verlängerungsstück *n*
appendages Schiffsanhänge *mpl (Wellenhosen, Schlingerkiele usw.)*; Anhänger *m*
apple-shaped diagram Nierencharakteristik *f (Mikrofon)*
Appleton layer F-Schicht *f (Ionosphäre)*
appliance Einrichtung *f*, Vorrichtung *f*; Apparat *m*
a. plug Gerätestecker *m*
appliances elektrische Haushaltsgeräte *npl*
applicability Anwendbarkeit *f*, Anwendungsmöglichkeit *f*, Verwendungsmöglichkeit *f*
applicable anwendbar
application 1. Anwendung *f*, Ausführung *f* in Praxis, Einsatz *m*; 2. Aufbringen *n*, Auftragen *n*; 3. Antrag *m*
a. for grant of patent Patenterteilungsantrag *m*
a. for utility patent Gebrauchsmusteranmeldung *f*
a. of a force Angriff *m* einer Kraft; Belastung *f*
a. of adhesive Klebstoffauftrag *m*
a. of prestress Eintragung *f* der Vorspannung *(Spannbeton)*
a. printing *(Text)* Direktdruck *m*
applicator Elektrode *f*
a. coil Härtekopf *m (Induktionshärten)*
a. roll *(Holz)* Auftragswalze *f*
applied arts Kunstgewerbe *n*; angewandte Kunst
a. ignition *(Kfz)* Fremdzündung *f*
a. load 1. aufgebrachte Last *f*, Auflast *f*; 2. wirklicher Lastwert *m*; 3. *(Masch)* ideelle konstante Last *f (Lebensdauer von Wälzlagern)*
a. mathematics angewandte Mathematik *f*
a. mechanics angewandte Mechanik *f*
a. moment *(Bw)* eingetragenes Moment *n*
a. science angewandte Wissenschaft *f*
a. voltage angelegte Spannung *f*
appliqué *(Text)* Applikation *f*
a. pattern *(Text)* Aufnähmuster *n*
apply / to *(Bw)* anwenden; auftragen *(Farbe usw.)*; eintragen *(Kräfte usw.)*, aufbringen; anwendbar sein
a. a spanner einen Schraubenschlüssel ansetzen
a. a tension *(El)* eine Spannung anlegen
a. by brush aufstreichen
a. for a patent ein Patent anmelden
a. the brake gradually *(Kfz)* gleichmäßig bremsen
appointment call *(Fmt)* verabredeter Anruf *m*; Festzeitgespräch *n*
appointments diary Terminkalender *m*
appraisal Abschätzung *f*, Einschätzung *f*; Würdigung *f*
appraise / to abschätzen
apprentice Lehrling *m*
a. hostel Lehrlingswohnheim *n*
a. instructor Lehrausbilder *m*
a. workshop Lehrwerkstatt *f*
apprenticed ausgelernt; freigesprochen *(Geselle)*
apprenticeship Lehre *f*
a. contract Lehrvertrag *m*
a. period Lehrzeit *f*
a. workshop Lehrwerkstatt *f*
approach / to sich nähern; anfahren *(Seilfahrt)*
approach 1. *(Masch)* Anstellung *f*, Beistellung *f*, Zustellung *f (Werkzeug)*; 2. Heranfahren *n (Schlitten)*, Annähern *n*; 3. *(Flg)* Zielanflug *m*, Annäherung *f*

a. **aid** Anflughilfe *f*
a. **angle** Zielkurs *m*, Kurswinkel *m* des Zieles
a. **aspect** Zielkurs *m*
a. **beacon** Einflugbake *f*
a. **buoy** *(Schiff)* Ansteuerungstonne *f*
a. **circuit** Zug[vor]meldekreis *m*
a. **control** Anflugkontrolle *f*
a. **control radar** Anflugradargerät *n*, Radarlandegerät *n*
a. **corridor** Anflugschneise *f*
a. **light[s]** Anflug[richtungs]feuer *n* *(Flugsicherungsbeleuchtung)*
a. **lighting system** Anflug-Befeuerungsanlage *f*
a. **locking** *(El)* Annäherungsverschluß *m*
a. **marker beacon** Einflugzeichenbake *f*
a. **ramp** Auffahrt[srampe] *f*
a. **road** Auffahrt *f*, Zufahrtstraße *f*; Brückenauffahrt *f*
a. **roller** Zufuhrrollgang *m* *(Walzwerk)*
a. **side** Anschnittseite *f*
a. **signal lighting** *(El)* Annäherungsbeleuchtung *f*
a. **speed** Annäherungsgeschwindigkeit *f*
a. **viaduct** Zufahrtsbrücke *f*
approaching passage *(Plast)* Zugangskanal *m* *(zum Extruderkopf)*
a. **side** *(Masch)* Auflaufseite *f* *(Treibriemen)*
appropriated berth *(Schiff)* fester Liegeplatz *m*
approximate / to annähern; abrunden *(Zahlen)*
approximate method Näherungsverfahren *n*
a. **solution** Näherungslösung *f*
a. **value** Näherungswert *m*
approximately ungefähr
a. **elliptical motion** *(Wärm)* angenäherter Ellipsenlenker *m*
approximation *(Math)* Annäherung *f*
a. **calculus** Näherungsrechnung *f*
a. **method** *(Math)* Näherungsmethode *f*, Näherungsverfahren *n*
apron 1. Beforderungstuch *n*; Förderband *n*, Laufband *n*; 2. Räderplatte *f*, Schloßkasten *m*, Schloßplatte *f* *(Drehmaschine)*; 3. Querschlitten *m* *(Waagerechtstoßmaschine)*; 4. *(Schiff)* Binnenvordersteven *m*; Docksohle *f*; Molenzunge *f*; 5. Vorfeld *n* *(Flugplatz)*
a. **conveyor** Bandförderer *m*; Gliederband[förderer *m*] *n*, Plattenband[förderer *m*] *n*, Plattentransportband *n*, Taschenbandelevator *m*
a.**-conveyor dryer** Bandtrockner *m*
a. **feeder** *(Brg)* Aufgabeplattenband *n*
a. **lining** *(Holz)* Beschalung *f* der Treppenbalken
apse Extrempunkt *m*, Scheitelpunkt *m* *(einer Kurve in Polarkoordinaten)*
a. **line** Apsidenlinie *f* *(Astronomie)*
apsis *(Math)* Apsis *f*
aptitude Eignung *f*, Tauglichkeit *f*
a. **for flying** *(Flg)* Fliegertauglichkeit *f*
a. **for rolling** Walzbarkeit *f*
a. **test** Eignungsprüfung *f*
apyrous feuerfest
aqua fortis Ätzwasser *n*; *(Ch)* Scheidewasser *n*
a. **regia** *(Ch)* Königswasser *n*
aquamarine Aquamarin *m*
aquatics Wassersport *m*
aquation *(Ch)* Hydratation *f*
aqueduct Aquädukt *n*
aqueous wasserartig, wasserhaltig, wäßrig, verdünnt
a. **ammonia** Salmiakgeist *m*
a. **homogeneous reactor** *(Kern)* homogener Wasserreaktor *m*
a. **reactor** *(Kern)* Reaktor *m* mit flüssigem Kern
a. **solution** wäßrige Lösung *f*
a. **tension** Wasserdampfdruck *m*
a. **varnish** Wasserlack *m*
A_r point Haltepunkt A_r *m*
arabic gum Gummiarabikum *n*
Arabic number arabische Ziffer *f*
arabitol *(Ch)* Arabit *m*
arable kultivierbar, pflügbar; Acker-; Ackerbau-
a. **land** Ackerland *n*
a. **soil** Ackerboden *m*
arabonic acid Arabonsäure *f*
arachidic acid Arachidinsäure *f*
arachis oil Erdnußöl *n*
arachnoid netzartig, spinnwebenartig
aragonite *(Min)* Aragonit *m*
arbitrary willkürlich, beliebig
a. **constant** beliebige (willkürliche) Konstante *f*
a. **phase-angle power relay** *(El)* Mischleistungsrelais *n*
arbitration analysis *(Brg)* Schiedsanalyse *f*
arbor 1. Dorn *m*, Spanndorn *m* *(Fräsmaschine)*; 2. Halter *m* *(Reibahle)*; 3. Achse *f*, Welle *f*, Spindel *f*
a. **bearing sleeve** Laufbuchse *f* *(Fräserschaft)*
a. **brace** Gegenhalterschere *f* *(Fräsmaschine)*
a. **bracket** Dorntraglager *n*
a. **clamp** *(Masch)* Dornschraubzwinge *f*
a. **cutter** Aufsteckfräser *m*
a. **nut** *(Masch)* Fräsdornmutter *f*, Dornmutter *f*
a. **press** *(Masch)* Dornpresse *f*
a. **shank** *(Masch)* Dornschaft *m*
a. **spindle** *(Masch)* Fräswelle *f*
a. **support** Führungslager *n*, Traglager *n* *(Fräsdorn)*
a. **supporting bracket** Fräsdornstützlager *n*
a. **thickness** Dicke *f* der Schulter des Aufspanndornes

a. yoke Führungslager *n*, Traglager *n (Fräsmaschine)*
arboraceous baumartig, bewaldet
arborescent dendritisch, baumähnlich *(Kristallografie)*
a. crystal Dendrit *m*
arboriculture Baumzucht *f*
arbour Gartenlaube *f*, Laube *f*; Schattensitz *m* unter einem Baum
arc / to 1. funken, Lichtbogen bilden; 2. feuern *(bei elektrischen Maschinen)*
a. back rückzünden *(Stromrichter)*
arc 1. *(Math)* Bogen *m*; 2. *(El)* Lichtbogen *m*
a. back Rückzündung *f (Stromrichter)*
a. baffle Ablenkplatte *f*
a. blow Blaswirkung *f* des Lichtbogens
a. break *(El)* Funkenlöscher *m*
a. cathode Lichtbogenkatode *f*
a.-chute *(El)* Funkenkammer *f*; Lichtbogenlöschkammer *f (Löschrohrableiter)*
a. control Kohlenachschub *m*
a.-control device *(El)* Löschkammer *f (Überspannungsschutz)*
a. cosine Arkuskosinus *m*
a. cotangent Arkuskotangens *m*
a. cutting Lichtbogen[brenn]schneiden *n*
a. deflector *(El)* Blasmagnet *m*, Lichtbogenschutzkammer *f*
a. discharge Lichtbogenentladung *f*, Bogenentladung *f*
a. drop Lichtbogenabfall *m*; Brennspannung *f (bei Gleichrichtern)*
a.-drop loss Lichtbogenverlust *m*
a.-duration Lichtbogendauer *f (eines Schalters)*
a.-extinguishing medium Lichtbogenlöschmittel *n (Sicherung, Schalter usw.)*
a. function Arkusfunktion *f*
a. furnace Lichtbogenofen *m*
a. heating Lichtbogenheizung *f*
a. high-speed relay *(El)* Löschrelais *n*
a. ignition Bogenzündung *f*
a. lamp Lichtbogenlampe *f*, Bogenlampe *f*
a. lamp carbon Lichtbogenkohle *f*, Kohlestab *m*
a. length *(Math)* Bogenlänge *f*
a. length in radians Winkelgröße *f* im Bogenmaß
a. minute *(Math)* Bogenminute *f*
a. of action *(Masch)* Eingriffsbogen *m*, Walzbogen *m (Teilzylinder, Verzahnung)*
a. of approach Eingriffsbogen *m* vor dem Wälzpunkt *(Verzahnung)*
a. of circle Kreisbogen *m*
a. of contact 1. Umschlingungsbogen *m*; Umschlingungswinkel *m (Riementrieb)*; 2. Berührungsfläche *f (Schleifscheibe)*
a. of recess Eingriffsbogen *m* hinter dem Wälzpunkt *(Verzahnung)*
a. of the pitch circle Teilkreisbogen *m*
a.-over Überschlag *m (des Lichtbogens)*
a. projection Lichtbogenprojektion *f*
a.-proof lichtbogenbeständig
a. quenching medium Lichtbogenlöschmittel *n*
a. rectifier Lichtbogenventil *n*, Lichtbogengleichrichter *m*, Quecksilberdampfgleichrichter *m*, Quecksilberdampfventil *n*
a. resistance Lichtbogenfestigkeit *f*
a. resistance furnace Widerstandsofen *m*, Lichtbogenwiderstandsofen *m*
a.-resistant lichtbogenbeständig
a. shears Bogenschere *f*
a. shield *(El)* Lichtbogenhülle *f*
a. sine Arkussinus *m*
a. spark Bogenfunke *m*
a. spectrum Bogenspektrum *n*
a. stabilizing lichtbogenstabilisierend
a. stream *(El)* Lichtbogenplasma *n*
a.-stream burner Plasmabrenner *m*
a.-stream voltage Plasmaspannung *f*
a. striking Ziehen *n* des Lichtbogens *(Schweißen)*
a. suppression Gittersperrung *f*, Gitterabschaltung *f (eines Stromrichters)*; Lichtbogenunterdrückung *f*
a.-suppression coil Petersenspule *f*, Erdschluß[lösch]spule *f*, Blasspule *f*
a. tangent Arkustangens *m*
a. thickness Zahndicke *f* im Wälzkreis *(Verzahnung)*
a.-through *(El)* Durchzündung *f*
a. transmitter Lichtbogensender *m*
a.-trigonometric *(Math)* zyklometrisch
a. welding Lichtbogenschweißen *n*
a. welding converter Lichtbogenschweißumformer *m*
a. welding machine Lichtbogenschweißmaschine *f*
a. welding transformer Lichtbogenschweißumspanner *m*
arcade Arkade *f*, Bogengang *m*, Bogenreihe *f*
arch / to *(Masch)* überwölben, sich wölben
arch 1. *(Bw)* Bogen *m*; 2. *(Brg)* Streckenbogen *m (Ausbau)*
a. action *(Bw)* Gewölbebildung *f*; Gewölbewirkung *f*
a. barrel *(Am)* schiefliegendes Gewölbe *n*
a. bearing Bogenlager *n*
a. brick Formstein *m*, Gewölbestein *m*, Keilziegel *m*
a. bridge Bogenbrücke *f*
a. built-in at both ends beiderseits eingespannter Bogen *m*
a. buttress Strebebogen *m*, Strebepfeiler *m*
a. casing Bogenverschalung *f*

a. **cent[e]ring** Lehrbogen *m*, Bogenlehrgerüst *n*
a.-**core** *(Brg)* Gewölbekern *m*
a. **culvert** *(Bw)* Bogendurchlaß *m*
a. **dam** Bogensperrmauer *f*, Bogenstaumauer *f*, Gewölbesperrmauer *f*, Gewölbestaumauer *f*
a. **falsework** Bogenschalung *f*
a. **form** Bogenschalung *f*
a. **girder set** *(Brg)* Bogenausbau *m*
a. **hinged only at the abutments** Zweigelenkbogen *m*
a. **of the oven** *(Brg)* Ofengewölbe *n*
a. **panel** Bogenfeld *n*
a. **pillar** *(Brg)* Schwebe *f (zwischen Abbauraum und Streckensohle)*
a. **rib** Bogenrippe *f*
a. **rings** Gewölbescharen *fpl*
a. **springing** *(Bw)* Bogenkämpfer *m*
a. **stiffening** Bogenaussteifung *f*
a. **stone** Keilstein *m*, Gewölbestein *m*
a. **truss** *(Bw)* Bogenfachwerk *n*
arched bogenförmig, gewölbt
a. **beam** Bogenträger *m*
a. **bridge** Bogenbrücke *f*
a. **buttress** Strebebogen *m*, Strebepfeiler *m*
a. **construction** Gewölbe[ein]bau *m*
a. **crown** Gewölbe *n (des Ofens)*
a. **culvert** Bogendurchlaß *m*
a. **dam** Bogensperrmauer *f*, Bogenstaumauer *f*, Gewölbesperrmauer *f*, Gewölbestaumauer *f*
a. **dome** Kuppelgewölbe *n*
a. **furnace roof** Ofengewölbe *n*
a. **girder** Bogenbinder *m*
a. **gravity dam** Bogengewichtsmauer *f*, Bogengewichtssperre *f*
a. **lintel** *(Bw)* gewölbter Sturz *m*, Segmentbogen *m*
a. **roadway** Wölbstrecke *f*
a. **roof of hearth** *(Met)* Herdgewölbe *n*
a. **truss** Bogenfachwerk *n*
archer Bogenschütze *m*
archery Bogenschießen *n*
archetype Grundtyp *m*, Prototyp *m*; Modell *n*
Archimedean drill Drehbohrer *m*, Drillbohrer *m*
Archimedean screw Archimedes-Schnecke *f*
Archimedes' spiral archimedische Spirale *f*
arching Einwölben *n*, Wölben *n*; Aufwölbung *f*, Brückenbildung *f (Geologie)*
a. **action** *(Bw)* Gewölbebildung *f*
a. **effect** *(Bw)* Gewölbewirkung *f*
architectural ceiling untergehängte Decke *f*
a. **concrete** Sichtbeton *m*, Vorsatzbeton *m*
a. **design** architektonische Gestaltung *f*
a. **survey** Architekturvermessung *f*
archsaver *(Brg)* Raubgerät *n* für Streckenbögen
archway Bogengang *m*
arcilla ungereinigter Weinstein *m*
arcing *(El)* Lichtbogenbildung *f*, Überschlag *m (des Lichtbogens)*
a. **contact** *(El)* Lichtbogenkontakt *m*; Abbrennkontakt *m*; Abreißkontakt *m*; Kurzschlußkontakt *m*; Hilfskontakt *m*
a.-**over** *(El)* Bogenüberschlag *m*
a.-**time** *(El)* Löschdauer *f*, Löschzeit *f (von Schaltgeräten)*
a. **tips** *(El)* Abbrennkontakte *mpl*
arcotron Kaltkatodenröhre *f* mit Eigenschaften eines Tyratrons
arcuate bogenartig, bogenförmig, gebogen, gekrümmt
arcwall coal cutter *s.* arcwaller
arcwaller Schrämmaschine *f* für Streckenvortrieb
ardency *(Schiff)* Luvgierigkeit *f*
ardometer Ardometer *n*, Gesamtstrahlungspyrometer *n*
arduous duty motor Motor *m* für Schwerstbetrieb, Schwerstbetriebsmotor *m*
are Ar *n (100 m^2)*
area Fläche[ninhalt *m*] *f*; Gebiet *n*, Gegend *f*, Geländeabschnitt *m*
a.-**border ratio** *(Bw)* hydraulischer Radius *m*
a. **capacity** Flächenleistung *f (in der Landwirtschaft)*
a. **characteristic** Öffnungskennlinie *f (Ventil)*
a. **contact** Flächenberührung *f (Zahnrad)*
a. **control radar** Gebietskontrollradar *m*
a. **covered by the fire** *(Wärm)* Feuerbahn *f*
a. **flowmeter** Schwebekörperdurchflußmesser *m (Oberbegriff für Rotamesser und Schwebekolbenmesser)*
a. **load** Flächenlast *f*
a. **moment** Flächenmoment *n*
a.-**moment method** Momentenflächenverfahren *n (Statik)*
a. **monitor** *(Kern)* Raumüberwachungsgerät *n*
a. **monitoring** *(Kern)* Raumüberwachung *f*
a. **of audibility** Hörbarkeitsgebiet *n*
a. **of ball imprint** Kalottenfläche *f (Kugelhärteprüfung)*
a. **of base** Grundfläche *f*
a. **of circle** Kreisfläche *f*
a. **of contact** Berührungsfläche *f*; Eingriffsfeld *n (Verzahnung)*
a. **of cross section** Querschnittsfläche *f*
a. **of cut** Schnittfläche *f*; Kalottenoberfläche *f (Kugelhärteprüfung)*
a. **of drilling** Bohrrechteck *n*
a. **of extraction** *(Brg)* Einwirkungsfläche *f*
a. **of fracture** Bruchfläche *f*
a. **of indentation** Eindruckfläche *f (Kugelhärteprüfung)*
a. **of moments** Momentenfläche *f*
a. **of pressure** Druckfläche *f*

a. of reinforcement Stahlfläche *f*, Stahlquerschnitt *m* *(Stahlbeton)*
a. of section *(Schiff)* Spantflächenkurve *f*, Spantenskala *f*
a. of steel Stahlfläche *f*, Stahlquerschnitt *m* *(Stahlbeton)*
a. of subsidence *(Brg)* Senkungszone *f*
a. of table *(Masch)* Tischfläche *f*
a. of the diagram Diagramminhalt *m*, Diagrammfläche *f*
a. of woodland Waldbestand *m*
a.-preserving flächentreu *(Abbildung, Transformation)*
a. to be cut *(Holz)* Hiebfläche *f*, Schlagfläche *f*
areal density Flächendichte *f*
a. moment Flächenmoment *n*
a. velocity Flächengeschwindigkeit *f*
arenaceous sandig, brüchig
a. limestone sandiger Kalkstein *m*
a. quartz Quarzsand *m*
areometer Aräometer *n*, Senkwaage *f*, hydrostatische Waage *f*
Argand burner Argandbrenner *m*
argent silbern, silbrig
argentan Neusilber *n*, Alpaka *n*
argentiferous silberführend, silberhaltig
a. gold Silbergold *n*
a. tetrahedrite Silberfahlerz *n*, Weißgültigerz *n*
argentine *(Min)* Aphrit *m*
argentine silbern, silbrig
a. mica Silberglimmer *m*
argentite Argentit *m*, Glanzerz *n*, Glanzsilber *n*, Glaserz *n*, Schwefelsilber *n*, Silberglanz *m*, Silberglas *n*, Silberschwärze *f*
argentous oxide Silberoxid *n*
argil Ton[erde *f*] *m*
argilla Ton *m*, natürliches Aluminiumsilikat *n*
argillaceous lehmhaltig, tonartig, tonhaltig, tonig
a. earth Tonerde *f*
a. limestone Tonkalk *m*
a. marl Mergelton *m*
a. ore toniges Erz *n*
a. sandstone Tonsandstein *m*
a. shale Schieferton *m*
a. slate Tonschiefer *m*
argilliferous tonhaltig
argillite Argillit *m*, Tonschiefer *m*
argillization Kaolinisierung *f*
argillocalcite Tonkalk *m*
argon Argon *n*
a.-arc welding Argon-arc-Schweißen *n*
argument *(Math)* Argument *n*, unabhängige Variable (Veränderliche) *f*; Azimut *m*, Polarwinkel *m* *(bei Polarkoordinaten)*
a. place Leerstelle *f*
aridity Dürre *f*; Unfruchtbarkeit *f*
aridness Dürre *f*
arithmetic Arithmetik *f*; Rechenkunst *f*, Rechnen *n*
a. average arithmetisches Mittel *n*
a. mean arithmetisches Mittel *n*
a. operation Rechenoperation *f*
a. shift *(BMSR)* arithmetische Verschiebung *f*, Stellen[wert]verschiebung *f*
a. unit *(BMSR)* Rechenwerk *n*
arithmetical progression arithmetische Folge *f*
a. series arithmetische Reihe *f*
Arkansas file Arkansasfeile *f*
Arkansas round edge slip Arkansashohlmeißelstein *m*
Arkansas stone Arkansasstein *m*
arkose Kaolinsandstein *m*
arm 1. Schenkel *m* *(eines Winkels)*; 2. *(Masch)* Arm *m* *(Sägegestell)*; 3. Ausleger *m*, Hebel *m*, Tragbalken *m* *(Kran)*; 4. Ausleger *m*, Arm, Stößel *m* *(Werkzeugmaschine)*; 5. *(El)* Schiebekontakt *m*; 6. *(Mil)* Waffe[ngattung] *f*
a. brace *(Masch)* Gegenhalterschere *f*
a. of a force Hebelarm *m* *(einer Kraft)*
a. of eccentricity *(Bw)* Ausschlag *m* *(Statik)*; Außermittigkeit *f*
a. of the water gauge Wasserstandkopf *m*, Wasserstandgehäuse *n*
a. rest Armlehne *f*
a. signal Armsignal *n*, Flügelsignal *n*
a. stake *(Brg)* Stolleisen *n*
armament *(Mil)* Bewaffnung *f*, Ausrüstung *f*; Gefechtsteil *m*
armature 1. Anker *m* *(elektrische Maschine)*; 2. *(Masch)* Metallbeschlag *m*; 3. Armatur *f*, Zubehör *n*
a. band Ankerbandage *f*
a. bearing Ankerlager *n*
a. body Ankerkörper *m*
a. chatter Ankerprellung *f*
a. coil Ankerwicklung *f*, Ankerspule *f*
a. conductor Ankerstab *m*, Ankerleiter *m*
a. control Hauptstromregelung *f*
a. controlled motor ankergesteuerter Motor *m*, Motor mit Ankersteuerung
a. core Ankerkern *m*
a. core disk Ankerblech *n*
a. duct Ankerkühlkanal *m*, Ankerkühlschlitz *m*
a. end connections Stirnverbindungen *fpl* *(des Ankers)*
a. end plate Ankerdruckplatte *f*
a. field Ankerfeld *n*
a. hub Ankerbuchse *f*, Ankerbüchse *f*, Ankernabe *f*
a. key Nutenverschlußkeil *m* *(des Ankers)*
a. lamination Ankerblech *n*, Ankerpaket *n*
a. of a condenser Kondensatorbelegung *f*
a. punchings Ankerblechschnitte *mpl*

a. quill Ankernabe *f*
a. reactance Ankerreaktanz *f*
a. reaction Ankerrückwirkung *f*
a. reaction excited machine Querfeldmaschine *f*
a.-reed support Ankerfederlagerung *f*
a. shaft Ankerwelle *f*
a. sleeve Nabenhülse *f (des Ankers)*
a. slot Ankernut *f*
a. spider Ankernabenstern *m*
a. stroke Ankerhub *m*, Ankerspiel *n*
a.-voltage rectifier Ankerspannungsgleichrichter *m*
a. winding Ankerwicklung *f*
a. with salient poles Polanker *m*, Anker *m* mit ausgeprägten Polen
Armco iron Armco-Eisen *n*, Reineisen *n*, Weicheisen *n*
armor *(Am)* *s.* armour
armored *s.* armoured
armoring *s.* armouring
armour / to panzern, mit Panzerplatte versehen; bewehren *(Kabel)*
armour 1. Bewehrung *f*, Armierung *f*; Ausrüstung *f*; 2. *(Mil)* Panzerung *f*; Panzer *m*; 3. Taucheranzug *m*
a. of a blast furnace Hochofenausrüstung *f*
a.-piercing *(Mil)* panzerbrechend, panzerdurchschlagend
a.-piercing rifle grenade Gewehrpanzergranate *f*
a. plate Panzerplatte *f*
a.-plate planer (planing machine) Panzerplattenhobelmaschine *f*
a. plate rolling mill Panzerplattenwalzwerk *n*
armoured gepanzert; bewehrt, armiert
a. cable bewehrtes (armiertes) Kabel *n*, Kabel mit Bandage, Panzerkabel *n*
a. car Panzerspähwagen *m*, gepanzerter Kraftwagen *m*
a. combat car Panzerspähwagen *m*
a. concrete Stahlbeton *m*
a. conveyor *(Brg)* Panzerförderer *m*
a. cord *(El) (flexible)* Panzerschnur *f*
a. cruiser Panzerkreuzer *m*
a. fighter Panzerjäger *m*
a. glass Drahtglas *n*
a. patrol car Panzerspähwagen *m*
a. personnel carrier Schützenpanzerwagen *m*, gepanzerter Mannschaftstransportwagen *m*
a. supply and maintenance vehicle Nachschub-Panzerkampfwagen *m*, Nachschubpanzer *m*
armourer Büchsenmacher *m*
armouring 1. Armierung *f*, Bewehrung *f (von Kabeln)*; 2. Panzern *n*, Panzerung *f*
a. machine Bewehrungsmaschine *f*
arms Waffen *fpl*
Armstrong blasting *(Brg)* Druckluftschießen *n*, Schießen *n* mit Preßluft
army aircraft Infanterieflugzeug *n*
a. airport Feldflugplatz *m*
a. cloth Militärtuch *n*
Arnott valve *(Bw)* Wrasenklappe *f*
aromatic aromatisch, wohlriechend, würzig
a. fuel Kraftstoff *m* aus aromatischen Kohlenwasserstoffen
a. hydrocarbon aromatischer Kohlenwasserstoff *m*, Ringkohlenwasserstoff *m*
a. principle Duftstoff *m*
a. vinegar Toilettenessig *m*
aromatics *(Ch)* 1. Aromaten *pl*; 2. Riechstoffe *mpl*
aromatization Aromatisierung *f*
aromatous wohlriechend; würzig
Aron measuring circuit *(El)* Aron-Schaltung *f*
Aron meter *(El)* Amperestundenzähler *m*, Aronzähler *m*, Leistungsmesser *m* mit Aronschaltung
arranged to swivel schwenkbar
arrangement Anordnung *f*; Einbau *m*, System *n (Lager)*; Aufbau *m*
a. for arrest Arretiervorrichtung *f*
a. of atoms Atomanordnung *f*, Anordnung *f* der Atome
a. of cable conductors Adernlage *f (im Kabel)*
a. of furnace Feuerungseinrichtung *f*
a. of the crystals in the lattice Gitteranordnung *f*
array Anordnung *f*, Schema *n*, Tafel *f*
a. with variable polar diagram Antenne *f* mit schwenkbarer Charakteristik
arrest / to *(Masch)* arretieren, sperren; festhalten
a. carbon drop *(Ch)* Entkohlung aufhalten
arrest *(Masch)* Arretierung *f*, Festhaltevorrichtung *f*, Sperrvorrichtung *f*
a. line *(Met)* Rastlinie *f*
a. point Haltepunkt *m*, Umkehrpunkt *m*
arrestation point *(Ch)* Umwandlungspunkt *m*
arrester *(El)* Ableiter *m*; Überspannungsableiter *m (oft fälschlich: Blitzableiter)*
a. gap *(El)* Ableit[er]funkenstrecke *f*
arresting device Anschlagvorrichtung *f*, Sperrvorrichtung *f*
a. hook *(Mil)* Lande[brems]haken *m*
a. lever Sperrhebel *m*
arrhenic acid Methylarsinsäure *f*
arris Kante *f*, ausspringende Ecke *f*; First *m*, Grat *m (Dach)*
a. gutter *(Bw)* V-förmige Rinne *f*
a. hip tile Gratziegel *m (Dach)*
a. tile *s.* a. hip tile
arrival message *(Flg)* Landemeldung *f*
arrivals angekommene Schiffe *fpl*
arrowhead Maßpfeil *m*
arrowroot flour Stärkemehl *n*
arsanilic acid Arsanilsäure *f*

arsenate Arsenat *n*, Salz *n* der Arsensäure
arsenic Arsen *n*
a. acid Arsensäure *f*
a. apparatus Apparatur *f* zur Arsenbestimmung nach Marsh
a. bloom Arsenblüte *f*, Arsentrioxid *n*
a. cured mit Arsen behandelt (desinfiziert)
a. glass glasiges Arsenigsäureanhydrid (Arsentrioxid) *n*
a. hydride Arsenwasserstoff *m*, Arsin *n*
a. lime *(Led)* Arsenäscher *m*
arsenical arsen[ik]haltig, arsenig
arsenide Arsenid *n*, Arsen-Metall-Verbindung *f*
arseniferous arsenführend, arsenig
arsenious arsenig
a. acid arsenige Säure *f*
arsenite Arsenit *n*, Salz *n* der arsenigen Säure
arseniuretted hydrogen Arsenwasserstoff *m*, Arsin *n*
arsenolite *(Min)* Arsenolith *m*, Arsenit *m*; Arsentrioxid *n*, Arsenkies *m (verwittert)*; Giftmehl *n*
arsenopyrite Arsenkies *m*
arsenotungstic acid Arsenwolframsäure *f*
arsinate Arsinat *n*, Salz *n* der Arsinsäure
art Kunst *f*; Technik *f*, Fertigkeit *f*; Fach *n*
a. [card] board Kunstdruckkarton *m*
a. foundry Kunstgießerei *f*
a. of navigation Nautik *f*
a. paper Kunstdruckpapier *n*
a. skates Kunstlaufschlittschuhe *mpl*
a. skating Eiskunstlauf *m*
arterial drainage Verästelungssystem *n (Entwässerung)*
a. highway Fernverkehrsstraße *f*
a. road Ausfallstraße *f*, Fernverkehrsstraße *f*, Hauptverkehrsader *f*
artesian pressure artesischer Druck *m*
a. well artesischer Brunnen *m*
articulated gegliedert; gelenkartig, gelenkig; Glieder-
a. arch Gelenkbogen *m*
a. arm Gelenkarm *m*, Gelenkausleger *m*
a. bar *(Brg)* Gelenkkappe *f*
a. buckle Gelenkangel *f (Gattersäge)*
a. connecting rod Nebenpleuelstange *f*
a. loading shovel *(Brg)* Knickgelenkschaufellader *m*
a. locomotive zweiteilige Lok *f*
a. pipe Gelenkrohr *n*
a. rod Anlenkpleuel *m*
a. shaft Gelenkwelle *f*, Gliederwelle *f*
a. six-wheeler Sattelwagenzug *m*
a. vehicle *(Kfz)* Sattelschlepper *m* mit aufgesatteltem Anhänger
a.-yielding arch *(Brg)* nachgiebiger Gelenkbogen *m*
articulation Verständlichkeit *f (Fernsprechtechnik)*
a. equivalent *(Fmt)* Verständlichkeitsäquivalent *n*
a. reference equivalent *(Fmt)* Bezugsdämpfung *f* für die Verständlichkeit
artificial künstlich, synthetisch
a. aerial Ersatzantenne *f*, künstliche Antenne *f*
a. ag[e]ing künstliches Altern *n*
a. almond oil Benzaldehyd *n*, künstliches Bittermandelöl *n*
a. asphaltum Teerpech *n*
a. atmosphere furnace Schutzgasofen *m*
a. balancing cable Kabelnachbildung *f*
a. balancing line Leitungsnachbildung *f*
a. black signal Testsignal *n* für Schwarz *(Fernsehtechnik)*
a. blocking-layer künstliche Sperrschicht *f*
a. cement stone Zementkunststein *m*
a. cementation Baugrundverbesserung *f*, Baugrundverfestigung *f*
a. consolidation künstliche Bodenverdichtung *f*
a. cotton Zellstoffwatte *f*; Kunstbaumwolle *f*
a. dielectric künstliches Dielektrikum *n*, Isolierstoff *m*
a. drainage Drainung *f*, Drainage *f*
a. ear *(El)* künstliches Ohr *n*
a. earth satellite künstlicher Erdsatellit (Erdtrabant) *m*
a. extension line *(El)* Verlängerungsleitung *f*; Spulenfeldergänzung *f*
a. fertilizer Kunstdünger *m*, künstlicher Dünger *m*
a. fibres Kunstfasern *fpl (Oberbegriff für synthetische und halbsynthetische Fasern)*
a. fine grain *(Brg)* Kunstfeinkorn *n*
a. flooring cement Steinholz *n*
a. fog künstlicher Nebel *m*
a. foundation *(Bw)* künstliche Gründung *f*
a. gravity künstliche Schwere *f*
a. gum Dextrin *n*
a. harbour [künstlicher] Hafen *m*
a. horizon Kreiselhorizont *m*
a. horn Kunsthorn *n*
a. isotope künstlich erzeugtes Isotop *n*
a. lace durch chemische Verfahren hergestellte Spitze *f*
a. leather Kunstleder *n*, Ledererersatz *m*
a. leather coating *s.* a. leather finish
a. leather finish Kunstlederlack *m*
a. light exposure Kunstlichtaufnahme *f*
a. line *(Fmt)* Kunstleitung *f*, Leitungsnachbildung *f*
a. manure Kunstdünger *m*
a. mouth *(El)* künstlicher Mund *m*
a. mustard oil Allylsenföl *n*
a. neuron *(BMSR)* künstliche Nervenzelle *f (Information verarbeitende künstliche Zelle)*

a. neutral point *(Elm)* künstlicher Sternpunkt *m*; künstlicher Nullpunkt *m*
a. nuclear reaction künstliche Kernreaktion *f*
a. nuclear transformation künstliche Kernumwandlung *f*
a. parchment paper Kunstpergament *n*
a. radioactivity künstliche (indizierte) Radioaktivität *f*
a. resin Kunstharz *n*
a. satellite künstlicher Satellit *m*, Sputnik *m*
a. soil compaction künstliche Bodenverdichtung *f*
a. solidification *s.* a. cementation
a. star künstlicher Stern *m*
a. stone Beton[werk]stein *m*, Kunststein *m*
a. suntan künstliche Sonnenbräune *f*
a. transmutation *(Kern)* künstliche Kernumwandlung *f*
a. white signal Testsignal *n* für Weiß *(Fernsehtechnik)*
a. wood Kunstholz *n*
artificially *s.* artificial
artillery Artillerie *f*
a. caterpillar tractor Artillerieraupenschlepper *m*
a. prime mover Artillerieschlepper *m*, Artilleriezugmaschine *f*
a. range Artillerieschießstand *m*
a. survey Artillerievermessung *f*
artistic casting Kunstguß *m*
Artron *(BMSR)* Artron *(künstliche Entscheidungseinheit, die nach dem Prinzip „Lohn und Strafe" arbeitet)*
arts and crafts Kunstgewerbe *n*; Kunsthandwerk *n*
aryl residue Arylgruppe *f*, Arylrest *m*
arylate / to arylieren, Arylgruppe einführen
arylate Arylierungsprodukt *n*
arylating agent Arylierungsmittel *n*
arylation Arylierung *f*, Einführung *f* einer Arylgruppe
arylide arylsubstituierte Verbindung *f*, Metallverbindung *f* eines aromatischen Kohlenwasserstoffs
as a function of *(Math)* in Abhängigkeit von
as cast im gegossenen Zustand
as rolled im Walzzustand (gewalzten Zustand)
asbestic asbestartig
asbestoid asbestartig
asbestos Asbest *m*
a. apron Asbestschürze *f*
a. board Asbestkarton *m*, Asbestpappe *f*
a. canvas Asbestleinwand *f*
a. carpet Asbestteppich *m*
a. cement Asbestzement *m (n)*
a. cement board Asbestzementpappe *f*
a. cement sheeting Asbestzementplatte *f*
a. clothing Asbestkleidung *f*
a. cord Asbestschnur *f*, Asbestseil *n*
a. covered asbestumsponnen
a. covered wire Asbestdraht *m*
a. electrical insulation paper Asbestisolier[ungs]papier *n*
a. felt Asbestpappe *f*
a. fibre Asbestfaser *f*
a. fire-proof suit Asbestfeuerschutzanzug *m*
a. gasket *(Wärm)* Asbestzopf *m*, Asbestdichtung *f*
a. gloves Asbesthandschuhe *mpl*
a.-graphite gasket Asbest-Graphit-Dichtungsring *m*
a. joint Asbestdichtung *f*
a. layer Asbestzwischenlage *f*
a. lumber Asbestzementpappe *f*
a. milk Asbestaufschlämmung *f*
a. millboard Asbestpappe *f*
a. packing Asbestdichtung *f*, Asbestpackung *f*
a. paper Asbestpapier *n*, Asbestpappe *f*
a. screen Asbestschirm *m*
a. sheet Asbestplatte *f*
a. shingle Asbestzementschiefer *m*
a. slate Asbestschiefer *m*
a. textile Asbestgewebe *n*
a. wallboard Asbestbaupappe *f*, Asbestplatte *f*
a. washer Asbestdichtung *f*
a. wire gauze Asbestdrahtnetz *n*
a. wool Asbestwolle *f*
a. yarn Asbestgarn *n*
asbolane *(Min)* Asbolan *m*
asbolite *(Min)* Erdkobalt *m*, Kobaltmanganerz *n* (*s. a.* asbolane)
ascend / to [an]steigen, aufsteigen
ascending ansteigend; aufsteigend
a. main *(El)* Steigeleitung *f*
a. node aufsteigender Knoten *m*
a. pipe Steigrohr *n*
a. power steigende Potenz *f*
a. technique aufsteigende Technik *f (in der Chromatografie)*
ascensional cast steigender Guß *m*
a. ventilation *(Brg)* ansteigende Wetterführung *f*, steigende Bewetterung *f*
ascent 1. *(Rak)* Aufstieg *m*; 2. *(Brg)* Ausfahrt *f*; 3. *(Flg)* Steigflug *m*, Höhengewinn *m*; Auftrieb
a. of piston Kolbenaufgang *m*
a. trajectory Aufstiegsbahn *f*
ascertain a bearing / to *(Fmt)* peilen
aseptic gauze Verbandmull *m*
ash / to *(Plast)* polieren
ash 1. Asche *f*; 2. Esche[nholz *n*] *f*
a. box Aschenfall[blech *n*] *m*
a. cave Aschenfallraum *m (Ofen)*
a. cellar Aschenkeller *m*
a. chute Aschenabfuhrkanal *m*
a. constituent Aschebildner *m*, Aschebestandteil *m*

a. content Aschegehalt *m*
a. conveyor Aschenförderband *n*, Aschenförderanlage *f*
a. deposition Ascheablagerung *f*
a. drawer Aschenzieher *m*
a. ejector Aschenauswerfer *m*, Aschenejektor *m*
a. formation Aschenbildung *f*; Schlackenbildung *f*
a. fouling Ascheverschmutzung *f*
a.-free [and] dry asche- und wasserfrei
a. fusion temperature Ascheschmelztemperatur *f*
a.-handling plant Entaschungsanlage *f*
a. hoist Aschenwinde *f*
a.-lighter *(Schiff)* Aschprahm *m*
a. pan Aschenkasten *m*
a. pit Aschenraum *m*; Aschenfall *m*
a. pit damper Aschenfallklappe *f*
a. pocket Aschensack *m*
a. removal Entaschung *f*
a. shale Aschenschieferton *m*
a.-shoot *(Schiff)* Aschenschütte *f*
a. silo Aschenbunker *m*
a. stop Aschenschieber *m*
a. tray Aschenbecher *m*
a. wood Eschenholz *n*
ashing Äscherung *f*, Polieren *n*; Veraschung *f*
ashlar Natursteinquader *m*, Werkstein *m*, Quader *m*; Natursteinmauerwerk *n*, Bruchsteinmauerwerk *n*, Quadermauerwerk *n*, Werksteinmauerwerk *n*
a. facing Natursteinverkleidung *f*, Werksteinverkleidung *f*
a. masonry Werksteinmauerwerk *n*
a. stonework Natursteinmauerwerk *n*, Bruchsteinmauerwerk *n*, Quadermauerwerk *n*, Werksteinmauerwerk *n*
a. walling *s.* a. stonework
ashlering *(Holz)* innere Dachschalung *f*, Dachausschalung *f*
ashy component Ascheträger *m*
a. constituent Aschebildner *m*
asp Espe *f*, Zitterpappel *f*
asparacemic acid racemische Asparaginsäure *f*
asparagic acid Asparaginsäure *f*, α-Aminobernsteinsäure *f*
asparamide Asparagin[säureamid] *n*
aspartate Salz *n* der Asparaginsäure
aspartic acid Asparaginsäure *f*
aspect angle *(Flg)* Zielkurs *m*
a. of approach *(Flg)* Zielkurs *m*; Orientierung *f* [des Zieles] im Raum
a. ratio 1. Schlankheitsverhältnis *n*, Schlankheitsgrad *m*, Längenverhältnis *n (Knickfestigkeit)*; 2. *(Flg)* Reziprokwert *m* des Seitenverhältnisses; 3. *(Fs)* Bildseitenverhältnis *n*
asphalt / to asphaltieren
asphalt Asphalt *m*, Erdölbitumen *n*, Erdpech *n*, Mineralpech *n*
a.-base crude petroleum asphaltbasisches Rohöl *n*
a. block Asphaltpflasterstein *m*, Asphaltplatte *f*; Asphaltbrot *n*
a. built-up roof Asphaltdach *n*
a. cement Asphaltkitt *m*, Asphaltmastix *m*
a. concrete Asphaltbeton *m (Straßenbau)*
a. concrete paver *s.* a. finisher
a. finisher Schwarzdeckeneinbaumaschine *f*, Schwarzdeckenfertiger *m (Straßenbau)*
a. from petroleum oil Erdölasphalt *m*
a.-lac Asphaltlack *m*
a. paver *s.* a. finisher
a. rock Pechgang *m*
a. tar Asphaltteer *m*
asphalted felt Asphalt[bitumen]pappe *f*
asphaltic asphalthaltig, asphaltisch, erdpechhaltig
a. cement Asphaltkitt *m*, Asphaltmastix *m*
a. concrete Asphaltbeton *m (Straßenbau)*
asphalting Asphaltbelag *m*, Bitumenbelag *m*; Asphaltierung *f (Straßenbau)*
asphaltization Harzbildung *f (in Erdölprodukten)*, Verharzung *f*
aspherical asphärisch
a. cataract lens asphärisches Starglas *n*
a. mirror asphärischer Spiegel *m*
asphyxiant gas Lungengift *n*; chemischer Kampfstoff *m (früher: Grünkreuz)*
aspirated dust *(Brg)* Sichterstaub *m*
a. fines *(Brg)* Sichterstaub *m*
aspiration Ansaugen *n*
a. of the condenser *(Wärm)* Saugkraft *f* des Kondensators
aspirator 1. Entlüfter *m*, Sauggebläse *n*, Ventilator *m*; 2. Aspirator *m*; Luftsauger *m*; Druckflasche
a. bottle Saugflasche *f*
a. pump Wasserstrahlpumpe *f*
assault boat *(Mil)* Sturmboot *n*
a. gun *(Mil)* Sturmgeschütz *n*
assay / to Erzanalyse durchführen, analysieren
assay *(Met)* Erzanalyse *f*, Analyse[nprobe] *f*
a. apparatus Probenbesteck *n*, Probengerätschaft *f*
a. balance Probierwaage *f*
a. button Probierkorn *n*
a. crucible Probiertiegel *m*
a. furnace Kapellenofen *m*, Probierofen *m*, Versuchsofen *m*
a. of buddled ore Waschprobe *f*
a. of silver Herdprobe *f*
a. office Labor[atorium] *n*
assayer's tongs Probierzange *f*
assaying Probierkunst *f*
assemblage 1. Anhäufung *f*, Masse *f*; Montage *f*, Zusammenbau *m*; 2. *(Math)* Zahlenmenge *f*, Menge *f*

a. cutter *(Holz)* Verbindungsfräser *m*
a. of numbers Zahlenmenge *f*
assemble / to 1. montieren, zusammenbauen, zusammensetzen, anbringen, einbauen; 2. einfügen *(Programmierung von Ziffernrechnern)*
a. a pipe line eine Rohrleitung verlegen
a. by fitting reversing ends *(Masch)* stürzen *(Verzahnung)*
assembled parts tube *(El, Fmt)* demontierbare Röhre *f*
assembler 1. Monteur *m*; 2. *(Typ)* Sammler *m* *(z. B. an der Setzmaschine)*
assembling Zusammenbau *m*, Montage *f*
a. crane Montagekran *m*
a. pressure Montierungsdruck *m (Turbine)*
a. tool Montagewerkzeug *n*
a. without reworking nacharbeitsfreier Zusammenbau *m*
assembly 1. Anbringung *f*, Aufbau *m*, Aufstellung *f*, Einbau *m*, Montage *f*, Zusammenbau *m*, Zusammenstellung *f*; 2. Aggregat *n*, Baugruppe *f*, Montagegruppe *f*, Satz *m*; Baueinheit *f*, Einheit *f*
a. adhesive Montagekleber *m*, Montageleim *m*
a. bay Montagehalle *f*
a. bench Montagetisch *m*, Montagestand *m*
a. conveyor Montageband *n*
a. dimension Anschlußmaß *n*
a. drawing for use in the erecting shop Montagezeichnung *f*, Richtzeichnung *f*
a. fitter Monteur *m*
a. glue Montageleim *m*
a. glu[e]ing Montageleimen *n*
a. hall Versammlungshalle *f*, Vortragssaal *m*, Aula *f*
a. jig Montagevorrichtung *f*
a. line Fließband *n*; Montageband *n*; Montagestraße *f*
a.-line production Fließbandfertigung *f*, Fließverfahren *n*
a. machine Montagegerät *n*
a. of mating parts Paarung *f*
a. operation Montagearbeit *f*
a. screw Befestigungsschraube *f*, Verbindungsschraube *f*
a. section Montageabteilung *f*
a. shop Montagehalle *f*, Montagewerkstatt *f*
a. stand Montagebock *m*
a. time *(Plast)* Abbindezeit *f*, Verleimzeit *f*; Montagezeit *f*
a. wiring *(El)* Gesamtverdrahtung *f*
a. without reworking nacharbeitsfreier Zusammenbau *m*
a. worker Monteur *m*
asset *(Ök)* Aktivwert *m*; Aktivposten *m*; Guthaben *n*
a. item *(Ök)* Aktivposten *m*
assets *(Ök)* Aktiva *pl*, Aktivposten *mpl*
assign a trunk / to *(Am) (Fmt)* Nummer der Verbindungsleitung ansagen
assigned frequency *(Fmt)* Kennfrequenz *f*
assignee of a protective right Erwerber *m* eines Schutzrechtes
assignment key *(Fmt)* Überweisungstaste *f*; Zuteiltaste *f*
a. of freeboard *(Schiff)* Freibordbestimmung *f*
a. problem *(BMSR)* Ernennungsproblem *n*
assistance of tug towing *(Schiff)* Schlepperhilfe *f*
assistant Assistent *m*, Gehilfe *m*
a. cameraman Kameraassistent *m*
a. driver Beifahrer *m*
a. engineer Hilfsmaschinist *m*
a. holding bar relay *(Fmt)* Brückenhilfsrelais
a. supervisor *(Bw)* Bauführer *m*
assisted take-off *(Flg)* Start *m* mit Starthilfe *(z. B. Katapult)*
a. take-off rocket Startrakete *f*
a. take-off unit Startbeschleuniger *m*
assize Trommel *f (Säulenteilstück)*
Association cable Normenkabel *n*
associative interrogation register *(BMSR)* assoziatives Abfrageregister *n (Rückgewinnung gespeicherter Daten)*
a. memory *(BMSR)* assoziativer Speicher *m* *(Datenspeicherung, nach Art eines Warenhauskatalogs angelegt)*
associativity law *(Math)* Assoziativitätsgesetz *n*
assort / to klassifizieren, sortieren
assorting screen Sortiersieb *n*
assortment Sortiment *n*
assume / to annehmen
assumption Annahme *f*
assured costs feststehende Unkosten *pl*
astable circuit *(El)* instabiler Kreis *m*
a. multivibrator *(Fs)* astabiler (monostabiler) Multivibrator *m*
astatic astatisch
a. coil astatische Spule *f*, Ringspule *f*
a. controller *s.* integral controller
a. governor astatischer Regler *m*
a. lift *(Wärm)* astatischer Sicherheitshub *m*
a. needles astatisches Nadelpaar *n*, astatische Doppelnadel *f*
astaticity Astasie *f*
astatine *(Ch)* Astatin *n*
astern *(Schiff)* achteraus, rückwärts
a. running Rückwärtslauf *m*
a. trial *(Schiff)* Rückwärtsfahrtprobe *f*
a. turbine Rückwärtsturbine *f*, Turbine *f* für Rückwärtsgang, Rücklaufturbine *f*
astigmatic astigmatisch
astigmatism Astigmatismus *m*
a. of oblique pencils Astigmatismus *m* schiefer Bündel

astigmeter Astigmeter *n*
astillen *(Geol)* Salband *n*
Aston dark space *(El)* Astonscher Dunkelraum *m*
Astracon tube Astracon-Röhre *f* *(Sekundäremissionsvervielfacherröhre mit großer Sekundärelektronenausbeute)*
astragal Rohrschelle *f*; Rundstab *m*, Viertelstab *m*; Wulst *m*
  a. cutter Stabfräser *m (Holzbearbeitung)*
astrakhan *(Text)* Webpelz *m*, **Astrachan** *m (n)*
astrionics Weltraumforschung *f*
astro-hatch Luke *f* im Flugzeug für astronomische Ortung
astrograph Astrograf *m*
astrographic camera Astrokamera *f*
astroid *(Math)* Ast[e]roide *f*, Sternkurve *f*, vierspitzige Hyperzykloide *f*
astrometric astrometrisch
astrometry Astrometrie *f*
astronaut Astronaut *m*, Raumfahrer *m*, Weltraumfahrer *m*
astronautic rocket kosmische Rakete *f*
astronautics Astronautik *f*
astronomical clock astronomische Uhr *f*
  a. measurement by conjunction astronomische Anschlußmessung *f*
  a. telescope astronomisches Fernrohr *n*
  a. unit astronomische Einheit *f*, AE
astronomy Astronomie *f*, Sternkunde *f*
astrophysics Astrophysik *f*
astylar *(Bw)* säulenlos
astyllen *(Brg)* Wasserdamm *m* im Stollen
asylum switch *(El)* verschlossener Schalter *m* *(für Unbefugte unzugänglich)*
asymmetric asymmetrisch
  a. bars Stufenbarren
  a. carbon *s.* a. carbon atom
  a. carbon atom asymmetrisches Kohlenstoffatom *n*
  a. cell *(El)* Gleichrichterzelle *f*
  a. conductivity richtungsabhängige Leitfähigkeit *f*; Gleichrichterwirkung *f*
  a. deflection asymmetrische Ablenkung *f*
  a. fission *(Kern)* unsymmetrische Spaltung *f*
  a. heterostatic circuit *(El)* asymmetrisch-heterostatische Schaltung *f*
  a. resistance gerichteter (richtungsabhängiger) Widerstand *m*
  a. short-circuit current Stoßkurzschlußstrom *m*
  a. two wire system asymmetrische Doppelleitung *f*
  a. wave unsymmetrische Welle *f*
asymmetrical *s.* asymmetric
asymmetry Asymmetrie *f*, Unsymmetrie *f*, Abgleichfehler *m*
  a. potential Asymmetriepotential *n*

asymptote Asymptote *f*
asymptotic asymptotisch
  a. cone Asymptotenkegel *m*
  a. expression asymptotische Gleichung *f*
  a. focal length Asymptotenbrennweite *f*
  a. voltage Dauerdurchschlagsspannung *f*
asynchronism Ungleichlauf *m*
asynchronous asynchron
  a. admittance Asynchronadmittanz *f*
  a. alternator Asynchrongenerator *m (Wechselstrommaschine)*
  a. computer Asynchronrechner *m (Rechner, in dem eine Operation nur auf das Signal hin beginnt, daß die vorhergehende Operation beendet ist)*
  a. condenser *(Elm)* Blindleistungsmaschine *f*, [asynchroner] Phasenschieber *m*
  a. motor *s.* non-synchronous motor
  a. phase compensator *(El)* Asynchronphasenschieber *m*
  a. reactance Asynchronreaktanz *f*
  a. starting asynchrones Anlassen *n*
at anchor in the roads *(Schiff)* auf Reede zu Anker
at centre mittig
at earth potential *(El)* stromlos
at full sea *(Schiff)* auf hoher See
at full speed *(Schiff)* in voller Fahrt
at infinity unendlich fern
at normal temperature and pressure unter normalen Bedingungen
at one setting mit einer Aufspannung *(Werkzeugmaschinen)*
at resonance bei Resonanz; in Resonanz
at rest im Stillstand (Ruhezustand)
at right angles to each other rechtwinklig zueinander
at right angles to flight path *(Rak)* bahnsenkrecht
at the forging heat schmiedeheiß
at will willkürlich *(z. B. Auslösung)*
at zero pressure drucklos
atacamite *(Min)* Atakamit *m*, Salzkupfererz *n*
athermanous atherman, wärmeundurchlässig
athwartship *(Schiff)* querschiffs, dwars
atlas silk Fagaraseide *f*
atmospheric atmosphärisch, Luft-
  a. absorption Luftabsorption *f*
  a. agitation atmosphärische Turbulenz (Unruhe) *f*
  a. attenuation atmosphärische Dämpfung *f*
  a. conditions Witterungsbedingungen *fpl*, Witterungsverhältnisse *npl*
  a. corrosion Luftkorrosion *f*
  a. density Luftdichte *f*
  a. discharge atmosphärische Entladung *f*
  a. electricity Luftelektrizität *f*, atmosphärische Elektrizität *f*

a. excess pressure Atmosphärenüberdruck *m*
a. fading *(Text)* Verschießen *n* durch Lufteinflüsse
a. humidity Luftfeuchtigkeit *f*
a. influence Witterungseinfluß *m*
a. oxygen Luftsauerstoff *m*
a. perturbation atmosphärische Störung *f*
a. pollution Verunreinigung *f* der Luft
a. pressure Luftdruck *m*
a. suppressor *(Fmt)* Störsperre *f*
a. valve Luftventil *n*
a. water Niederschlagswasser *n*
a. weapon attack Luftüberfall *m*, Luftangriff *m*
atmospherical *s.* atmospheric
atmospherics *(El)* atmosphärische Störungen *fpl*
atom Atom *n*
a. bomb Atombombe *f*, Kernspaltungsbombe *f*
a.-bound electrons Atomelektronen *npl*
a. distribution Atomverteilung *f*
a. excitation Atomanregung *f*
a. form factor Atomformfaktor *m*, Strukturfaktor *m*
a. model Atommodell *n*
a.-powered mit Kernenergieantrieb, durch Kernspaltung, mit Atomantrieb
a.-powered icebreaker Atomeisbrecher *m*
a.-powered plane Atomflugzeug *n*
a.-powered ship Schiff *n* mit Kernenergieantrieb, Atomschiff *n*
a. transformation *s.* nuclear transformation
a. transmutation *s.* nuclear transmutation
atomic atomar, atomig
a. absorption atomare Absorption *f*
a. absorption coefficient atomarer Absorptionskoeffizient *m*
a. accelerator Atombeschleuniger *m*
a. age Atomzeitalter *n*
a. arc welding Arcatomschweißen *n*
a. attenuation coefficient atomarer Schwächungskoeffizient *m*, Atomschwächungskoeffizient *m*
a. blast Kernwaffenexplosion *f*
a. bomb Atombombe *f*, A-Bombe *f*; Kernbombe *f*
a. bomb burst Kernbombenexplosion *f*
a. bomb shelter Atombunker *m*, kernbombensicherer Bunker *m*
a. bombardment 1. Atombeschuß *m*; 2. Kernbombenabwurf *m*
a. bombardment particle atomares Geschoßteilchen *n*
a. bond *(Ch)* Atombindung *f*, atomare (homöopolare) Bindung *f*
a. charge Atomladung *f*
a. chart Atomgewichtstafel *f*
a. constant Atomkonstante *f*, atomare Konstante *f*
a. cross section atomarer Wirkungsquerschnitt *m*
a. diameter Atomdurchmesser *m*
a. distance in a molecule Atomabstand *m* im Molekül
a. electron shell Elektronenhülle *f* des Atoms, Atomhülle *f*; Atomschale *f*
a. excitation Atomanregung *f*
a. explosion trial Kernwaffenversuchsexplosion *f*
a. fission Kernspaltung *f*
a. form factor Atomformfaktor *m*, Strukturformfaktor *m*
a. formula Strukturformel *f*
a. heat Atomwärme *f*
a. heat of fission Kernfusionswärme *f*
a. hull Atomrumpf *m*
a. hydrogen atomarer Wasserstoff *m*
a. hydrogen welding Arcatom-Schweißung *f*, atomare Wasserstoffschweißung *f*
a. interspace atomarer Zwischenraum *m*
a. lattice Atomgitter *n*
a. level Atomniveau *n*
a. linkage Atomverkettung *f*
a. magnetic moment magnetisches Atommoment *n* (Moment *n* des Atoms)
a. mass Atommasse *f*, Isotopenmasse *f*, Massenwert *m*
a. mass unit atomare Masseneinheit *f (physikalische Skala)*
a. model Atommodell *n*
a. nucleus Atomkern *m*
a. number Kernladungszahl *f*, Ordnungszahl *f*, Atomnummer *f*
a. orbit Valenzzustand *m*
a. order atomare Größenordnung *f*
a. paramagnetism atomarer Paramagnetismus *m*
a. philosophy Atomistik *f*
a. physicist Atomphysiker *m*
a. physics Atomphysik *f*
a. plane Gitterebene *f*, Netzebene *f*
a. polyhedron Atompolyeder *n*
a. power Atomkraft *f*, Atomenergie *f*, Kernenergie *f*
a. power plant Atomkraftwerk *n*
a. power station Atomkraftwerk *n*
a.-powered *s.* atom-powered
a. projectile Atomgeschoß *n*
a. propulsion Atomantrieb *m*, Kernenergieantrieb *m*
a. radiation Atomstrahlung *f*
a. rays Atomstrahlen *mpl*
a. rearrangement atomare Umordnung *f*
a. research Atomforschung *f*; Kernforschung
a. retaliation *(Mil)* atomarer Gegenschlag *m*

a. **rocket** Atomrakete *f*
a. **scattering factor** Atomformfaktor *m*, Strukturfaktor *m*
a. **scientist** Atomforscher *m*, Atomwissenschaftler *m*; Kernforscher *m*
a. **shell** Atomhülle *f*, Elektronenhülle *f* des Atoms
a. **shelter** *s.* a. bomb shelter
a. **ship** Atomschiff *n*
a. **space lattice** Atomgitter *n*
a. **state** Energiezustand *m* des Atoms
a. **stopping power** atomares Bremsvermögen *n*
a. **structure** atomare Struktur *f*; Atom[auf]-bau *m*
a. **structure of diamant** Diamantgerüst *n*, Diamantgitter *n*
a. **submarine** Atom-U-Boot *n*
a. **susceptibility** atomare Suszeptibilität *f*, Atomsuszeptibilität *f*
a. **symbol** Atomsymbol *n*
a. **theory** Atomtheorie *f*
a. **timing device** Atomuhr *f*
a. **torso** Atomrumpf *m*
a. **trunk** Atomrumpf *m*
a. **union** Atomverband *m*, Atomverbindung *f*
a. **valence** Atomwertigkeit *f*
a. **vibration** Atomschwingung *f*
a. **volume** Atomvolumen *n*
a. **warhead** Atomsprengkopf *m*, atomarer Sprengkopf *m*
a. **waste** radioaktiver Abfall *m*, Atommüll *m*
a. **weapon** Atomwaffe *f*, Kernwaffe *f*, atomare Waffe *f*
a. **weight** Atomgewicht *n*
a. **weight unit** Einheit *f* des Atomgewichts, Atomgewichtseinheit *f*
atomical *s.* atomic
atomically dispersed atomdispers
atomicity *s.* valency
atomizable zerstäubbar *(Flüssigkeiten)*
atomization Zerstäubung *f*; Vernebelung *f*
a. **jet** Zerstäuberdüse *f*
a. **nozzle** Zerstäubungsdüse *f*
atomize / to zerstäuben, vernebeln, fein verteilen
atomizer Zerstäuber *m*, Zerstäubungsapparat *m*; Nebelspritze *f*; Einspritzventil *n*
a. **cone** Zerstäuberkegel *m*, Spritzkegel *m*
a. **plate** Düsenplatte *f (Zerstäuber)*
a. **valve** Einspritzventil *n*
atomizing agent Zerstäubungsmittel *n*
a. **drier** Zerstäubungstrockner *m*
a. **nozzle** Einspritzdüse *f*
atoxylic acid Arsanilsäure *f*
ATR-tube *s.* anti-transmit-receive-tube
atroglyceric acid α-Phenylglyzerinsäure *f*
atropic acid Atropsäure *f*, Phenylakrylsäure *f*

attach / to aufsetzen, befestigen, anbringen, anbauen, anhängen, vorsetzen
a. **a thread** anspinnen *(Faden)*
attached column *(Bw)* eingebundene Säule *f*, Halbsäule *f*, Lisene *f*
a. **implement** Anbaugerät *n*
attaching device Anhängevorrichtung *f*
attachment 1. Anbaugerät *n*, Zusatzgerät *n*; Zusatzeinrichtung *f*; Ansatz *m*, Aufsatz *m*, Einsatz *m*; Zubehörteil *n*; 2. Befestigung *f*, Anbringung *f*
a. **camera** Ansetzkamera *f*
a. **coefficient** Anlagerungskoeffizient *m*
a. **collar** Befestigungslasche *f*
a. **for face grinding** Planschleifeinrichtung *f*
a. **head piece** Anbaukopfstück *n*
a. **hook** Aufhängehaken *m*
a. **lens** Aufsteckglas *n*, Vorschaltlinse *f*
a. **link** Befestigungsglied *n*
a. **plug** *(El)* Anschlußstecker *m*
a. **rail** Befestigungsschiene *f*
a. **screw** Druckschraube *f*, Klemmschraube *f*
a. **support for implements** Anbaublock *m* für Geräte
a. **surface** Anlagefläche *f*
a. **telescope** Vorsatzfernrohr *n*
a. **telescope with dioptre scale** Dioptrienfernrohr *n*
attack / to *(Ch, Mil)* angreifen
a.**-polish** ätzpolieren, reliefpolieren
attack 1. *(Ch, Mil)* Angriff *m*; 2. *(Ch, Met)* Korrosionsangriff *m*
a. **aircraft** taktisches Jagdflugzeug *n*
a. **by chemical action** chemischer Angriff *m*, chemische Einwirkung *f*
a. **by vermin** Schädlingsbefall *m*, Schadfraß *m*
a. **curve** *(Flg)* Verfolgungskurve *f*, „Hundekurve" *f*
a. **fighter** taktisches Jagdflugzeug *n*
a. **sight** *(Flg)* Funkmeßvisier *n*
a. **time** Einregelzeit *f*; Einschwingzeit *f*
a. **vector** *(Flg)* Kampfkurs *m*
attal *s.* attle
attar Rosenöl *n*
attend / to warten, bedienen
attendance Wartung *f*; Bedienung *f*
a. **time** *(Ök)* bezahlte Anwesenheitszeit *f*
attendant Bedienungsmann *m*, Wärter *m*
attended station bemannte Station *f*
attention Wartung *f*, Bedienung *f*
a. **getter** *(Am)* Blickfang *m*
attenuant *(Ch)* Verdünnungsmittel *n*
attenuate / to dämpfen, schwächen
a. **the sliver** *(Text)* das Faserband verfeinern
a. **the voltage** *(El)* die Spannung herabsetzen

attenuation Schwächung *f*, Abschwächung *f*; [räumliche] Dämpfung *f*
a. characteristic *(El)* Dämpfungsverlauf *m*; Frequenzgang *m*
a. coefficient Schwächungskoeffizient *m*
a. compensator *(El, Fmt)* Dämpfungsentzerrer *m*
a. constant 1. *(Ph)* Absorptionskoeffizient *m*; 2. *(Fmt)* Dämpfungsmaß *n*; Kettendämpfung *f*
a. correction *(Fmt)* Dämpfungsausgleich *m*
a. cross section *(Kern)* Schwächungsquerschnitt *m*
a. curve *(Fmt)* Dämpfungsverlauf *m*; Frequenzgang *m*
a. distance Dämpfungslänge *f*
a. distortion *(El)* Dämpfungsverzerrung *f*; Amplitudenverzerrung *f*; Frequenzverzerrung *f*
a. equalizer *(Fmt)* Dämpfungsentzerrer *m*, Entzerrer *m*
a. equivalent *(Fmt)* Dämpfungsmaß *n*
a. factor *(Ak)* Dämpfungsmodul *m*
a. filter *(Opt)* Schwächungsfilter *n*
a. frequency distortion Dämpfungsverzerrung *f* eines elektroakustischen Apparates
a. in suppressed band *(Fmt)* Sperrdämpfung *f*
a. in the pass-band *(Fmt)* Durchlaßdämpfung *f*
a. in the space period *(Fmt)* Ruhedämpfung *f* *(Fernschreiber)*
a. increase *(El)* Dämpfungsversteilerung *f*
a. network *(Fmt)* Dämpfungsschaltung *f*
a. of radiation Schwächung *f* der Strahlung
a. of repeater section *(Fmt)* Verstärkerfelddämpfung *f*
a. of sound Schalldämpfung *f*
a. ratio *(Fmt)* Dämpfungsgrad *m*
a. response *(El)* Dämpfungsgang *m*
a. test *(El)* Dämpfungsmessung *f*
a. time *(Ph)* Dämpfungszeit *f*
attenuator *(BMSR)* Dämpfungsglied *n*, Abschwächungsglied *n*; Lautstärkeregler *m*, Regelglied *n*
a. pad Dämpfungsglied *n*
attic Boden *m*, Dachgeschoß *n*; Bodenkammer *f*, Dachstube *f*
attle *(Brg)* Gangmasse *f*, taubes Gestein *n*
attorney's fees Anwaltsgebühren *fpl*
attract / to anziehen
attraction Anziehung *f*
a. potential Anziehungspotential *n*
a. power Anziehungskraft *f*
a. type ammeter Flachspulamperemeter *n*
attractive force Anziehungskraft *f*
a. nuclear field anziehendes Kernfeld *n*
a. power Anziehungskraft *f*
attrited abgerieben, verschlissen
attrition Reibung *f*; Abreibung *f*, Verschleiß *m*, Abrieb *m*; Abbau *m*; Abtragung *f*; Zerschleifung *f*
a. mill *(Holz)* Schleifmühle *f*
a. resistant abriebfest
a. test 1. Verschleißprüfung *f*; 2. Abriebprobe *f* *(Bestimmung der Verschleißfestigkeit von Straßenpflasterstoffen)*
aubepine Anisaldehyd *n*, Aubepin *n*
audibility Hörbarkeit *f*; Verständigungsgüte *f* *(Funktechnik)*
a. acuity Hörschärfe *f*
a. factor Hörbarkeitsfaktor *m*
a. limit Hörbarkeitsgrenze *f*
a. threshold Hörschwelle *f*
audible hörbar; vernehmbar
a. enhancer akustische Anzeige *f* *(beim Funkortungsgerät)*
a. noise voltage Geräuschspannung *f*
a. range akustischer Bereich *m*, Hörbereich *m*
a. ringing signal *(Fmt)* Freizeichen *n*
a. spectrum Tonfrequenzspektrum *n*
audio amplification Niederfrequenzverstärkung *f*
a. amplifier Niederfrequenzverstärker *m*, Tonfrequenzverstärker *m*
a. control engineer Toningenieur *m*
a. degeneration Niederfrequenzgegenkopplung *f*
a. detector zweiter Detektor *m*; Zwischenfrequenzgleichrichter *m*
a. filter *(Fmt, Fs)* Tonsieb *n*
a. frequency Tonfrequenz *f*, Hörfrequenz *f*, Niederfrequenz *f*
a. frequency amplification Tonverstärkung *f*
a. frequency amplifier Niederfrequenzverstärker *m*, Tonfrequenzverstärker *m*
a. frequency chopper Tonfrequenzzerhacker *m*
a. frequency distribution niederfrequenter Drahtfunk *m*
a. frequency range hörbarer Frequenzbereich *m*
a. frequency rediffusion Leitungsrundfunk *m*
a. frequency section Tonfrequenzteil *m*, Niederfrequenzteil *m*
a. frequency telegraphy Tonfrequenztelegrafie *f*
a. frequency telephony Niederfrequenzfernsprechen *n*
a. frequency transformer Tonfrequenzübertrager *m*, Niederfrequenztransformator *m*, NF-Übertrager *m*
a. frequency volume control niederfrequente Lautstärkeregelung *f*
a. frequency wire broadcasting niederfrequenter Drahtfunk *m*

a. mixer Tonmischeinrichtung *f*, Mischtafel *f*, Mischpult *n*
a. oscillator circuit *(Fmt)* Summerschaltung *f*
a. output channel Niederfrequenzausgang *m*
a. range Hörfrequenzbereich *m*
a. reactor Tonfrequenzdrossel *f*
a.-tone amplitude modulation Tonfrequenzamplitudenmodulation *f*
a. transmitter Schallsender *m*
a. trap Tonsperrkreis *m*
a.-visual Fernsehvorführung *f*
audiometer Hörschwellenmeßgerät *n*
audion Audion[röhre *f*] *n*
a. receiver Audionempfänger *m*
a. rectifier Audiongleichrichter *m*; Gittergleichrichter *m*
audit card *(Büro)* Rechnungsprüfungskarte *f*
a. strip Kontrollstreifen *m* *(Registrierkassen)*
auditing Revision *f*
auditor Hauptbuchhalter *m*; Rechnungsbeamter *m*, Revisor *m*
auditorium Versammlungshalle *f*; Hörsaal *m*
auditory direction finder Horchgerät *n*, akustisches Ortungsgerät *n*
Auer metal Auermetall *n*
augend[um] *(BMSR)* der erste Summand *m*
auger Bohrer *m*, Erdbohrer *m*, Holzbohrer *m*, Löffelbohrer *m*, Nabenbohrer *m*, Schneckenbohrer *m*, Vorbohrer *m*
a. bit Schneckenbohrer *m*
a. conveyor Schneckenförderer *m*
a. drilling Schneckenbohren *n*
a. elevator Schneckenhöhenförderer *m*
a. feeder Schneckenförderer *m*
a. feeding device Schneckenförderer *m*, Schneckenzuleitung *f*
a. hole Handschürfbohrloch *n*
a. machine *(Brg)* horizontales Großlochbohrgerät *n*
a. mining *(Brg)* Abbau *m* mit horizontalem Großlochbohrgerät
a. type grain conveyor Kornförderschnecke *f*
a. type unloader Schneckenentlader *m*
Auger coefficient *(Ph)* Auger-Koeffizient *m*
Auger electron *(Ph)* Auger-Elektron *n*
Auger-transition *(Ph)* Auger-Übergang *m*
Auger yield *(Ph)* Auger-Elektronenausbeute *f*
augering *s.* auger mining
augite Augit *m*
augment / to vergrößern, verstärken, vermehren
augmentation Vergrößerung *f*, Verlängerung *f*
a. of lift *(Flg)* Auftriebserhöhung *f*
aural carrier Tonträger *m* *(Fernsehtechnik)*
a. dazzling akustische Blende *f*
a. detector Horchgerät *n*, akustisches Ortungsgerät *n*
a. magnifier *(Opt)* Ohrlupe *f*
a. radio range akustischer Leitstrahlsender *m*
a. reception Hörempfang *m*
a. transmitter *(Fs)* Tonsender *m*
aurate Aurat *n*, Salz *n* der Goldsäure
aureole 1. *(Astr)* Aureole *f*, Kranzerscheinung *f*, Hof *m*; 2. *(Brg)* Aureole *(Sicherheitslampe)*
auric acid Goldsäure *f*
a. compound Auriverbindung *f*, Gold(III)-Verbindung *f*
auriferous goldführend, goldhaltig
auro compound Auroverbindung *f*, Gold(I)-Verbindung *f*
aurora australis südliches Polarlicht *n*, Südlicht *n*
a. borealis nördliches Polarlicht *n*, Nordlicht *n*
auroral line Aurora[spektral]linie *f*
aurous compound Verbindung *f* des einwertigen Goldes
austempering *(Met)* Warmbadhärten *n*, Thermalhärten *n*, stufenweises Härten *n*, Zwischenstufenhärten *n*
austenite Austenit *m*
a. dissociation *(Met)* Austenitzerfall *m*
austenitic manganese steel Manganhartstahl *m*
Austrian cinnabar Chromzinnober *m*
author's correction *(Typ)* Autorkorrektur *f*
a. proof *s.* author's correction
auto abstracting [technique] *(Büro)* automatisches Ausziehen *n* *(z. B. von Büchern und technischen Artikeln)*
a.-abstracts *(Büro)* Autoabstracts *pl*, automatisch hergestellte Auszüge *mpl* (Zusammenfassungen *fpl*) *(z. B. eines Artikels)*
a.-alarm device Autoalarmgerät *n*, [selbsttätiges] Seenotrufgerät *n*
a.-compounded current transformer Stromwandler *m* mit stromproportionaler Zusatzmagnetisierung
a.-converter Gleichstromumrichter *m*
a.-cycle mit automatischem Arbeitsablauf
a. cycle Moped *n*
a.-dial *(Fmt)* Druckknopfwähler *m*
a.-exchange *(Fmt)* Wähleramt *n*, Wählvermittlung *f*
a.-feed selbsttätiger Vorschub *m*; automatische Beschickung *f*
a.-ferry Autofähre *f*
a.-follower Folgesystem *n*, Nachlaufsystem *n*
a. gyration Eigendrehung *f*
a. headlight Autoscheinwerfer *m*, Kraftfahrzeugscheinwerfer *m*
a.-ignition Selbstentzündung *f*
a.-manual exchange *(Fmt)* halbautomatische Vermittlung *f*; handbedientes Fernamt *n*
a. mechanics Kraftfahrzeugtechnik *f*
a.-rotation Autorotation *f*

a. self-excitation *(El)* direkte Selbsterregung *f*, innere Mitkopplung *f*
a.-sizing *(Masch)* selbsttätiges Bearbeiten *n* auf Sollgröße *(spanend)*
a.-synchronous motor Synchronmotor *m* mit Selbstanlauf, selbstanlaufender Synchronmotor
a.-tool changing *(Masch)* mit selbsttätigem Werkzeugwechsel
a.-transductor *(Elm)* Spartransduktor *m*, Autotransduktor *m*
a.-transformer *(El)* Spartransformator *m*, Autotransformator *m*
a.-transformer starting *(El)* Anlassen *n* mit Spartransformator
a. trip Selbstauslösung *f*
a.-truck crane Auto[mobil]kran *m*
a.-tuning *(Fmt, Fs)* Selbstabstimmung *f*
autoadapting control *(BMSR)* selbstanpassende Steuerung *f*
autobahn Autobahn *f*
autocatalysis Autokatalyse *f*, Selbstkatalyse *f*
autochrome printing *(Typ)* Autochromdruck *m*
autoclave Autoklav *m*
a. curing *s.* autoclaving
autoclaving Autoklavbehandlung *f*, Hochdruckdampfbehandlung *f (Betontechnologie)*
autocollimation *(Opt)* Autokollimation *f*
a. camera Autokollimationskamera *f*
autocollimator Autokollimationsfernrohr *n*
autodyne *(Fmt)* selbstüberlagernd
a. circuit *(El)* Schwingaudionschaltung *f*
a. connection *(El)* Dreipunktschaltung *f*
a. detector Schwingaudion *n*
a. reception Autodynempfang *m*, Eigenüberlagerungsempfang *m*
autoelectric emission Kaltemission *f*
autofunction Eigenfunktion *f*
autogenous autogen
a. cutting Autogenschneiden *n*, autogenes Schneiden *n*
a. cutting and welding apparatus autogenes Schneid- und Schweißgerät *n*
a. cutting torch Autogenschneidbrenner *m*
a. fusing burner Autogenschneidbrenner *m*
a. stack cutting Stapel-Brennschneiden *n*
a. welding Autogenschweißung *f*, Gasschmelzschweißung *f*
autogiro Tragschrauber *m*
autographic selbstschreibend, selbsttätig aufzeichnend
autogyro *s.* autogiro
autoignition Selbstzündung *f*
automate / to automatisieren
automated [machine] line Taktstraße *f*
automatic Automat *m*
automatic selbsttätig, automatisch
a. acceleration *(Fmt)* automatische Anfahrsteuerung *f*; automatische Beschleunigung *f*
a. account card feed device *(Büro)* Kontenautomat *m*
a. accounting machine Buchungsautomat *m*
a. adjustment Selbsteinstellung *f*, Selbstanpassung *f*
a. advance *(Kfz)* Selbstverstellung *f*, automatische Zündverstellung *f*
a. advance mechanism *(Kfz)* automatische Zündverstellvorrichtung *f*
a. advice *(El)* selbsttätige Vormeldeeinrichtung *f*
a. aiming selbsttätige Zielung *f*
a. air escape valve selbsttätiges Luftablaßventil *n*
a. amplitude control selbsttätige Amplitudenbegrenzung *f (Lautstärkeregelung)*
a. arc-welding machine Lichtbogenschweißautomat *m*
a. background [control] selbsttätige Helligkeitsregelung *f*
a. balance pick-up *(Büro)* automatische Saldenaufnahme *f*
a. balancing *(BMSR)* Selbstabgleich *m*
a. baler *(Schiff)* selbstlenzende Plicht *f*
a. ball closing selbsttätiger Kugelverschluß *m*
a. ball steam pipe isolating valve Selbstschlußkugelventil *n*, Kugelrückschlagventil *n*
a. bar machine Stangenautomat *m*
a. bass compensation automatischer Baßausgleich *m*, automatische gehörrichtige Lautstärkeregelung *f*; automatische Gitterspannungsregelung *f*; automatische Helligkeitsregelung *f (Fernsehen)*
a. bias selbsttätige Gittervorspannung *f (durch Katodenwiderstand)*
a. bias control automatische Gitterspannungsregelung *f*
a. binary computer mit dem Zweiersystem arbeitender automatischer Rechner *m*
a. black-level control *(Fs)* automatische Schwarzwerthaltung *f*
a. block *(El)* selbsttätiger Streckenblock *m*, Selbstblock *m*
a. block with axle counters *(El)* Selbstblock *m* mit Achszählern, Zählblock *m*
a. block with track circuits *(El)* Selbstblock *m* mit Gleisstromkreis, Gleisblock *m*
a. bobbin changer Spulenwechselautomat *m*
a. boring machine *(Holz)* Bohrautomat *m*, Holzbohrautomat *m*
a. break selbsttätige Unterbrechung *f*
a. brightness control *(Fs)* automatische Helligkeitsregelung *f*
a. buttonhole machine Knopflochautomat *m*
a. calculating machine Rechenautomat *m*

a. can changer *(Text)* automatischer Kannenwechsel *m*
a. carriage automatischer Wagenvorschub *m*, Zeilenautomat *m (Buchungs- und Fakturiermaschinen usw.)*
a. centre punch Federdruckkörner *m*
a. checking *(BMSR)* selbstprüfend
a. checking *(BMSR)* Selbstkontrolle *f*
a. cheque handling automatische Scheckbearbeitung *f*
a. chip disposal unit automatische Späneabführeinrichtung *f*
a. chive pressing machine *(Holz)* selbsttätige Spundpresse *f*
a. chroma control *(Fs)* automatische Farbregulierung (Farbeinstellung) *f (Fernsehen)*
a. chucking machine Futterautomat *m*
a. circuit-breaker *(El)* Schutzschalter *m*, Selbstschalter *m*, Sicherungsautomat *m*
a. circular saw sharpener Kreissägenschärfautomat *m*
a. clearing *(Fmt)* selbsttätiges Schlußzeichen *n*
a. closing device with flag valve *(Wärm)* Selbstschlußapparat *m* mit Verschlußklappe
a. clutch selbsttätige Kupplung *f*
a. computing equipment elektronische Rechenmaschine *f*
a. control automatische Steuerung *f*; Selbststeuerung *f*
a. control system Regelanlage *f*, Regeleinrichtung *f*; Regelkreis *m*, [automatisches] Regelungssystem *n*
a. controller [selbsttätiger] Regler *m*; Regeleinrichtung *f*, Steuerwerk *n*; Stromwächter *m*
a. curve follower *(BMSR)* automatischer Kurvenabtaster *m*
a. cut-out *(El)* 1. Leitungsschutzschalter *m*, Sicherungsautomat *m*; Selbstausschalter *m*, automatischer Ausschalter *m*, Schalter *m* mit Schütz; Ausschalter mit Sicherung, abgesicherter Ausschalter *(ohne Nenneinschaltleistung)*; 2. selbsttätige Auslösung *f*
a. cycle automatischer Arbeitstakt *m*, selbsttätiger Arbeitsablauf *m*
a. damper regulator *(Wärm)* selbsttätiger Regler *m* des Rauchschiebers
a. data processing automatische Datenverarbeitung *f*
a. deliverer *(Typ)* Selbstausleger *m (Schnellpresse)*
a. dial service *(Fmt)* Selbstwählbetrieb *m*
a. dicing machine Trennschleifmaschine *f* für Halbleiterplättchen
a. dieing press Schneidautomat *m*
a. direction finder selbsttätiger Peiler *m*, automatisches Peilgerät *n*, Radiokompaß *m*; automatischer Pilot *m*
a. discharge 1. *(El)* Selbstentladung *f*; 2. *(Masch)* Selbstentleerung *f*
a. discharge hopper car Selbstentladewagen
a. discharge wagon Selbstentlader *m (z. B. Sattelwagen)*
a. disengagement Selbstauslösung *f*
a. displacement control automatische Regelung *f* der Fördermenge *(Pumpe)*
a. dive control *(Mil)* Sturzflugautomatik *f*
a. document file automatische Belegablage (Schriftstückablage) *f*
a. door-closer [selbsttätiger] Türschließer *m*
a. double-cylinder hosiery knitting machine Doppelzylinderstrumpfautomat *m*
a. dump car Selbstentlader *m (z. B. Sattelwagen)*
a. equipment *(Fmt)* Wählereinrichtung *f*
a. error correction automatische Fehlerberichtigung (Fehlerbeseitigung) *f (Rechenautomaten)*
a. error detection automatische Fehlererkennung (Fehlerfeststellung) *f (Rechenautomaten)*
a. exchange *(Fmt)* Selbstanschlußamt *n*; Wählzentrale *f*, Vermittlungsstelle *f* mit Wählbetrieb
a. exposure timer *(Foto)* Belichtungsautomat
a. failure recovery *(BMSR)* automatische Fehlerberichtigung (Fehlerbeseitigung, Fehlerkorrektur) *f*
a. feed arc lamp sich selbst regulierende Bogenlampe *f*
a. feed drill *(Brg)* Bohrhammer *m* mit automatischem Vorschub
a. feed grate Wanderrost *m (Kesselfeuerung)*
a. feed of lathe Selbstgang *m (Drehmaschine)*
a. feed oil cup Selbstöler *m*
a. feed press Presse *f* mit automatischer Zuführungseinrichtung
a. feed regulator *(Wärm)* selbsttätiger Speisewasserregler *m*
a. feeder 1. *(Typ)* Selbstanleger *m*, Anleger *m (Schnellpresse)*; 2. *(Text)* Baumwollspeiser *m*
a. field weakening *(Elm)* selbsttätige Feldschwächung *f*
a. film processor *(Typ)* automatische Filmentwicklungsmaschine *f*
a. fire alarm selbsttätiger Feuermelder *m*
a. flap steam pipe isolating valve *(Wärm)* Selbstschlußklappenventil *n*
a. flight control *(Flg)* Kurssteuerung *f*
a. floor-stop operation automatische Stockwerksschaltung *f (bei Aufzügen)*
a. follow-up device *(Opt)* automatische Nachführung *f*
a. forming machine Fassonautomat *m*
a. frequency control *(El)* selbsttätige Scharfabstimmung *f*, automatische Frequenzregelung *f*, Frequenznachlauf *m*

a. fuse setter *(Mil)* Zünderstellmaschine *f*
a. gain control automatische Schwundregelung (Lautstärkeregelung) *f*, selbsttätige Verstärkungsregelung *f*
a. gain control characteristic *(El)* Regelkennlinie *f*
a. gain stabilization automatische Verstärkungsstabilisierung *f*
a. gauge *(Typ)* Selbstvorschub *m (Schneidemaschine)*
a. gravure press *(Typ)* Tiefdruckautomat *m*
a. grid bias selbsttätige Gittervorspannung *f (durch Katodenwiderstand)*
a. grid tension *(Ph)* automatische Gitterspannung *f*
a. grinder Schleifautomat *m*
a. ground control approach automatische Bodenanzeige *f (Radar)*
a. guidance system *(Flg)* Nachführautomatik *f*
a. hacksawing machine Bügelsägeautomat *m*
a. handling equipment automatische Werkstückbeschickungseinrichtung *f*
a. helmsman *(Schiff)* Selbststeuerer *m*
a. high speed voltage regulator Schnellregler *m (Sendetechnik)*
a. hobbing machine Wälzfräsautomat *m*
a. hold automatisches Halten *n (Analogrechner)*
a. holding device selbsttätige Halteeinrichtung *f*
a. homing device Zielsuchsystem *n*
a. homing head *(Rak)* Zielsuchkopf *m*
a. hopper feed *(Text)* selbsttätige Speiseregelung *f*
a. hopper feeder *(Text)* selbsttätiger Kastenspeiser (Speiseregler) *m*
a. hunting *(Fmt)* freie Wahl *f*
a. hypoid generator Hypoidwälzfräsautomat *m*
a. ignition Selbstzündung *f*
a. information pick-up automatische Informationsaufnahme *f (Dokumentation)*
a. ink density control *(Typ)* automatische Farbdichtemessung *f*
a. interlock Selbstsperrung *f*
a. interruption *(El)* Selbstunterbrechung *f*
a. key *(Fmt)* Maschinengeber *m*
a. lacing machine *(Text)* automatische Kardenbindemaschine *f*
a. lathe Halbdrehautomat *m*, Drehhalbautomat *m*
a. line find *(Büro)* automatisches Zeilenfinden *n*
a. listing and book-keeping machine Registrierbuchungsautomat *m*
a. load limitation selbsttätige Leistungsbegrenzung *f*
a. locking circuit automatische (automatisierte) Synchronisierschaltung *f*
a. long-di tance service *(Fmt)* Selbstwählfernverkehr *m*
a. loom Automatenwebstuhl *m*
a. lubricator automatische Schmiereinrichtung *f*, Selbstöler *m*
a. machine Automat *m*
a. message accounting *(Fmt)* automatische Gebührenverrechnung *f*
a. miller Fräsautomat *m*
a. minimum circuit-breaker Unterstromwächter *m*, Unterstromautomat *m*, Unterstromselbstschalter *m*
a. navigation computer system automatischer Navigationsrechner *m*
a. navigator *(Flg)* Kurskoppler *m*
a. offset correction *s.* integral action
a. optimizer automatischer Optimierer *m*, Optimisator *m*
a. path-findi g in the labyrinth (maze) *(BMSR)* automatische Orientierung *f* im Labyrinth
a. pattern recognition automatische Zeichenerkennung *f*, automatisches Mustererkennen *n* *(z. B. zum Verarbeiten von Bankbelegen)*
a. phase control selbsttätige Phasensteuerung *f*
a. picture control automatische Bildregelung *f*
a. pilot Autopilot *m*, automatische Kurssteueranlage *f*; Selbststeuereinrichtung *f*, Selbststeuergerät *n*
a. platen *(Typ)* Tiegel[druck]automat *m*
a. positioning selbsttätige Lageeinstellung *f*, Einfahren *n* in festgelegte Position
a. positioning boring machine Bohrmaschine *f* mit selbsttätiger Positionierung
a. potentiometer *(BMSR)* selbsttätiger Kompensator *m*
a. pressure diaphragm *(Opt)* automatische Druckblende *f*
a. printing press *(Typ)* Druckautomat *m*
a. private branch exchange *(Fmt)* Wählnebenstellenanlage *f*
a. processing machine *(Typ)* automatische Filmentwicklungsmaschine *f*
a. programming *(Dat)* Selbstprogrammierung *f*
a. pull head selbstkuppelnder Ziehkopf *m* *(Räummaschine)*
a. puncher Stanzautomat *m*
a. punching machine *s.* a. puncher
a. radio alarm selbsttätiger Funkruf *m*
a. range finder automatischer Entfernungsmesser *m*
a. range tracking automatischer Entfernungsmesser *m*
a. range unit automatischer Abstandsgeber *m*

a. reagent feeder *(Brg)* Dosiervorrichtung *f* *(Flotation)*
a. re-cycling automatische Fortsetzung (Wiederholung) *f* des Arbeitsganges
a. reclosing *(El)* automat. Wiedereinschaltung *f*; automat. Kurzschlußfortschaltung *f*
a. recording burette *(Opt)* automatische Registrierbürette *f*
a. recovery *(Rak)* automatische Bergung *f*
a. regulation selbsttätige Regelung *f*
a. regulator *s.* automatic controller
a. release 1. *(Foto)* Selbstauslöser *m*; 2. Selbstauslösung *f*
a. reset *s.* integral action
a. return of water *(Wärm)* Wasserkreislauf *m*; Dampfschleife *f*
a. return switch for direction indicator *(Kfz)* Winkerrückstellschalter *m*
a. rheostat selbsttätiger Feldregler *m*
a. rotating veneer clamp jointer automatische Furnierfügemaschine *f* mit rotierender Spannvorrichtung
a. route release *(El)* selbsttätige Auflösung *f* der Blockstrecke
a. routing *(Fmt)* [direkte] Lochstreifenübertragung *f*
a. routing of long-distance calls Richtungsausscheidung *f* *(Fernsprechtechnik)*
a. routing of toll calls Richtungsausscheidung *f* *(Fernsprechtechnik)*
a. safety switch cut-off *(Text)* automatischer Sicherheitsschalter *m* *(zum Ausschalten der Maschine bei Entnahme der Stoffrolle)*
a. sample changer automatischer Probenwechsler *m*
a. screw machine [Einspindel-]Drehautomat *m*, Schraubenautomat *m*, Automat *m*
a. screw steel Automatenstahl *m*
a. selectivity control automatische Bandbreiteneinstellung (Selektivitätsregelung) *f*, selbsttätige Trennschärferegelung *f*
a. sensitivity control *(El)* automatische Empfindlichkeitsregelung *f*, automatische Regelung *f* des Empfanges
a. service *s.* a. telephone service
a. shim rod follow-up *(Kern)* Trimmnachfolgesteuerung *f*
a. short-circuit reclosing Kurzschlußfortschaltung *f*
a. shunt regulating resistance automatischer Nebenschlußregler *m*
a. shutter automatischer Verschluß *m*
a. shuttle changing *(Text)* selbsttätige Schützenauswechselung *f*, selbsttätiger Schützenwechsel *m*
a. shuttle loading press automatische Presse *f* mit pendelnden Beschickungstischen *(Holzvergütung)*
a. signalling selbsttätiger Ruf *m*; selbsttätige Zeichengebung *f*
a. sizing grinder Kaliberschleifautomat *m*, selbsttätige Kaliberschleifmaschine *f*
a. slicing machine Genauigkeitstrennschleifautomat *m*
a. slip regulator *(Elm)* automatischer Schlupfregler *m*
a. spotter *(Mil)* automatischer Trefferanzeiger *m*
a. sprinkler equipment Sprinkleranlage *f*
a. staircase lighting switch Treppen[haus]-automat *m*
a. starter Selbstanlasser *m*
a. steam pipe isolating valve Rohrbruchventil *n*
a. steam-pressure reducer selbsttätiger Dampfdruckverminderer *m*
a. steel Automatenstahl *m*
a. steering apparatus automatische Steuerung[seinrichtung] *f*
a. stitch control lever *(Text)* automatisch arbeitender Festigkeitsveränderungshebel *m*
a. stop 1. automatische Ausschaltung *f*; 2. selbstauslösender Anschlag *m*
a. stop device Fahrsperre *f*
a. stop valve selbsttätig schließendes Absperrventil *n*; Selbstschlußventil *n*
a. stopping device selbsttätige Absperrvorrichtung *f*
a. stud driver *(Bw)* Bolzenschießgerät *n*
a. substation *(El)* unbesetztes (automatisch arbeitendes) Unterwerk *n*
a. switch contact Selbstschalterkontakt *m*
a. system *(Fmt)* Wählsystem *n*
a. table traverse Tischselbstgang *m*
a. telecommunication log automatische Nachrichtenaufzeichnung *f*
a. telegraph Maschinentelegraf *m*
a. telephone exchange Selbstanschluß[fernsprech]amt *n*, Wählamt *n*
a. telephone service Selbstwählbetrieb *m*
a. telephony *(Fmt)* Selbstanschlußbetrieb *m*
a.-tension drive selbstspannender Trieb *m*, Spannrollentrieb *m*
a. tension regulator selbsttätige Nachspannvorrichtung *f* *(Freileitung)*
a. thread grinding machine Gewindeschleifautomat *m*
a. threading *(Text)* selbsttätige Einfädelung *f*
a. three-knife trimmer *(Typ)* Dreischneider *m*, Dreimesserautomat *m*
a. timing control *(Kfz)* automatischer Zündversteller *m*
a. tipper Selbstentlader *m* *(z. B. Sattelwagen)*
a. tipple feeder *(Holz)* automatische Furnierzuführung *f*

a. toll dialling Selbstfernwahl *f*
a. tone compensation automatische Klangkorrektion *f*
a. tool changing mechanism automatische Werkzeugwechseleinrichtung *f*
a. tool handling mechanism automatischer Werkzeugzubringer *m*
a. tool relief automatische Werkzeugabhebung *f*
a. tracer duplicator control *(Masch)* selbsttätige Fühlernachformsteuerung *f*
a. tracking automatische Kursanzeige *f*; automatische Kurssteuerung *f*, selbsttätige Zielführung *f (Radartechnik)*
a. tracking radar Flugfolgeradar *n*; Zielfolgeradar *n*
a. trailer coupling selbsttätige Anhängerkupplung *f*
a. transfer line *(Am)* Taktstraße *f*
a. transmitter *(El)* Maschinengeber *m*, Maschinensender *m*
a. tripping selbsttätige Auslösung *f (Relais)*
a. tripping contactor Schütz *n* mit Relais
a. tripping device Selbstauslöseeinrichtung *f*
a. tubular cop winder *(Text)* Schlauchkopsautomat *m*
a. tuck bar automatisch einstellbarer Fangteil *m (Strickmaschine)*
a. tuning [correction] selbsttätige Scharfabstimmung *f*
a. tuning means Abstimmautomatik *f*
a. turning lathe Drehhalbautomat *m*
a. uncoupling Selbstauslösung *f*
a. undervoltage circuit-breaker Unterspannungswächter *m*, Unterspannungsautomat *m*, Unterspannungsselbstschalter *m*
a. unshift automatisches Zurückbewegen *n* auf Buchstaben *(Fernschreiber)*
a. vending automatisches Verkaufen *n*; Verkaufsautomatik *f*
a. voltage regulator Spannungs[schnell]-regler *m*
a. volume-contractor *(El)* Dynamikpresser *m*
a. volume control selbsttätige Lautstärkeregelung (Schwundregelung) *f*
a. volume control characteristic *(El)* Kennlinie *f* der selbsttätigen Lautstärkeregelung, Regelkennlinie *f*
a. volume control circuit *(El)* selbsttätiger Lautstärkeregler (Schwundregler) *m*
a. volume expander *(El)* Dynamikdehner *m*
a. volume expansion *(Fmt)* automatische Dynamikdehnung *f*
a. volume range regulator automatischer Lautstärkeregler *m*
a. watering trough Selbsttränkrinne *f*
a. welding Maschinenschweißung *f*
a. zero cut-out *(El)* Nullstromselbstausschalter *m*, automatischer Nullstromausschalter *m*
automatical switching off automatische Umschaltung *f (Magnettontechnik)*
automatically checking selbstprüfend
a. closing water gauge Wasserstandapparat *m* mit Selbstschluß
a. recording selbstaufzeichnend
automaticity Automatik *f*
automation Automatisierung *f*
automaton Automat *m*
automechanics Kraftfahrzeuginstandhaltung *f* und -reparatur *f*
automobile Automobil *n*, Kraftwagen *m*
a. battery Kraftwagenbatterie *f*
a. engine Kraftwagenmotor *m*
a. frame Autorahmen *m*, Kraftfahrzeugrahmen *m*
a. gear mechanism Schieberädergetriebe *n*
a. gearing Kraftwagengetriebe *n*
a. jack Wagenheber *m*, Automobilheber *m*
a. radio Autoradio *n*
a. tire cloth Automobilreifenkord *m*
a. tire cord Automobilreifenkord *m*
a. transmission Kraftwagengetriebe *n*
a. wrench verstellbarer Schraubenschlüssel *m*, „Franzose" *m*, „Engländer" *m*; Automobilgabelschlüssel *m*
automobilism Kraftfahrwesen *n*
automorphic *(Math)* automorph
automorphism *(Math)* Automorphismus *m*
automotive brake Kraftfahrzeugbremse *f*
a. brightwork Chromteile *npl* am Auto
a. engine Kraftfahrzeugmotor *m*
a. engineer Kraftfahrzeugtechniker *m*
a. engineering Kraftfahrzeugtechnik *f*, Kraftfahrzeugwesen *n*
a. free piston-turbine engine Freikolbenturbinentriebwerk *n* für Kraftfahrzeuge
a. gas turbine Kraftfahrzeuggasturbine *f*
a. industry Autoindustrie *f*, Kraftfahrzeugindustrie *f*
a. part Kraftfahrzeugteil *n*
a. socket wrench set Steckschlüssel *m* mit Einsätzen *(für Kraftfahrzeuge, Motorräder usw.)*
a. supplies *pl* Kraftfahrzeugzubehör *n*
a. textile Textilien *pl* für Kfz-Industrie
autonavigator *s.* automatic navigation computer system
autopaster *(Typ)* automatische Rollenanklebevorrichtung *f*, Autopaster *m*
autopilot[-navigator] Autopilot *m*, Flugregler *m*, Selbststeuergerät *n*, Steuerautomat *m*
autopneumatic circuit-breaker Druckluftleistungsschalter *m* mit Selbstbeblasung
a. lap end autopneumatische Wickelausstoßvorrichtung *f*

autopulse magnetic fuel pump elektrisch betätigte (magnetfeldbetätigte) Brennstoffpumpe *f*
autoscrutineer *(El)* Gleichlaufselbststeuerung *f*
autospotter *(Mil)* automatischer Trefferanzeiger *m*
autostoper *(Brg)* leichter Bohrhammer *m* mit automatischer Bohrstütze
autosyn Autosyn *(Firmenbezeichnung für eine Drehmeldertype)*
autotimer selbsttätiger Zeitgeber *m*
autotransmitter Lochstreifensender *m*
autotype Autotypie *f*, Rasterätzung *f*
autovac Unterdruckförderer *m*
autoxidation Autoxydation *f*
autunite *(Min)* Autunit *m*
auxiliary 1. Hilfsmittel *n*, Zubehör *n*; 2. *(Schiff)* Hilfsschiff *n*; Segeljacht *f* mit Hilfsmotor
auxiliary Hilfs-, Zusatz-, Neben-; zusätzlich
a. accessories akzessorische Bestandteile *mpl*
a. aerial Hilfsantenne *f*, Ersatzantenne *f*
a. air Zusatzluft *f*
a. air reservoir Zusatzluftbehälter *m*
a. air valve *(Kfz)* Zusatzluftregler *m*
a. attachment Hilfseinrichtung *f*, Zusatzeinrichtung *f*
a. boiler Hilfskessel *m*
a. bond *(Ch)* Nebenvalenzbindung *f*
a. bus-bars *(El)* Hilfssammelschienen *fpl*
a. calling relay Anrufhilfsrelais *n*
a. cam Hilfssenker *m (Strickmaschine)*
a. carrier cable *(El)* Hilfstragseil *n*, Hilfstragdraht *m*
a. circle of a conic Scheitelkreis *m* eines Kegelschnittes *(bei Ellipsen Hauptscheitelkreis)*
a. circuit 1. Stromkreis *m* der Hilfsbetriebe; 2. Hilfsstromkreis *m*, Hilfsschaltung *f (eines Gerätes)*
a. compass Tochterkompaß *m*
a. condenser Hilfskondensator *m*
a. condition Nebenbedingung *f*
a. controll system Reglerhilfssystem *n*
a. cruiser *(Schiff)* Hilfskreuzer *m*
a. device Hilfseinrichtung *f*
a. drive *(Masch)* Nebenantrieb *m*
a. drive lock *(Kfz)* Nebenantriebssperre *f*
a. electrode Hilfselektrode *f*
a. engine Hilfsmotor *m*
a. equipment Zusatzausrüstung *f*; Sonderausstattung *f*
a. fan Hilfslüfter *m*, Sonderlüfter *m*
a. fighting ships leichte Seestreitkräfte *fpl*
a. fire bridge *(Wärm)* Kohlenwehr *n*
a. fire director Kommandohilfsgerät *n (Flak)*
a. flows *(Holz)* Zweigströme *mpl*
a. frame Hilfsrahmen *m*
a. function Hilfsfunktion *f*
a. gearbox Verteilergetriebe *n*
a. generator set *(El)* Umformer *m* für Hilfsbetriebe
a. grid Hilfsgitter *n (Elektronenröhren)*
a. grid voltage Hilfsgitterspannung *f*
a. group relay *(Fmt)* Hilfsgruppenrelais *n*
a. head lamp Zusatzscheinwerfer *m*
a. housing Hilfsständer *m (Einständerhobelmaschine)*
a. idler shaft Hilfsachse *f*
a. jet Zusatzdüse *f*
a. machine Zusatzmaschine *f*
a. machinery *(Schiff)* Hilfsmaschinen[anlage *f*] *fpl*
a. memory äußerer Speicher *m*, Hilfsspeicher *m (Datenspeicherung)*
a. modulation set *(Fmt, Fs)* Modulationssatz *m*
a. motor Hilfsmotor *m*
a. piping Hilfsrohrleitung *f*
a. pole *(Elm)* Hilfspol *m*; Wendepol *m*
a. pole shunting Wendepolnebenschluß *m*, Wendepolshunt *m*
a.-powered sailing ship Segelschiff *n* mit Hilfsmotor, Motorsegler *m*
a. pump Vorpumpe *f*
a. radiator Hilfsstrahler *m*
a. sailplane Segelflugzeug *n* [ausgerüstet] mit Hilfsmotor
a. scale Hilfsmaßstab *m*, Nonius *m*
a. scale of graduation Hilfsteilung *f*
a. shaft Hilfswelle *f*
a. signal *(Am)* Schlußzeichenwecker *m*
a. spark-gap Hilfsfunkenstrecke *f*, Vorfunkenstrecke *f*
a. spray hole Hilfsdüse *f*
a. stage *(El)* Vorstufe *f*
a. steam engine Hilfsdampfmaschine *f*
a. suction lead *(Schiff)* Hilfslenzleitung *f*
a. switch Hilfsschalter *m*, Nebenschalter *m*; Schalter *m* für Hilfsbetriebe
a. system circuit Hilfsstromkreis *m*
a. take-off rocket unit Raketenstartbeschleuniger *m*
a. taphole *(Met)* Notstich *m*
a. track *(Masch)* Hilfslaufbahn *f*
a. track vehicle Hilfskettenfahrzeug *n*
a. transformer regulation Regelung *f* mit Zusatztransformator
a. transmission *(Kfz)* Zusatzgetriebe *n*
a. tube Nebenrohr[leitung *f*] *n*
a. turbine Vor[spann]turbine *f*, Vorschaltturbine *f*
a. tuyere *(Met)* Notform *f*
a. valence Nebenvalenz *f*
a. ventilation *(Brg)* Sonderbewetterung *f*
a. voltage Vorspannung *f*

auxochromes Auxochrome *npl*
available verfügbar, erhältlich
a. base *(Flg)* Ausweichflugplatz *m*, Landeflugplatz *m*
a. energy verfügbare Arbeit *f (Turbine)*
a. gain Verstärkung *f*
a. machine time *(BMSR)* nutzbare Rechenmaschinenzeit *f*, Rechnerwirkzeit *f*
a. power größte Nutzleistung *f*, verfügbare Leistung *f*
a. seats Sitzangebot *n*, Sitzgelegenheit[en *pl*] *f*
a. time *(BMSR)* Rechnerwirkzeit *f*
a. work verfügbare Arbeit *f (Dampfdiagramm)*
avalanche Lawine *f*
a. debris Lawinenschutt *m*
a. ionisation lawinenartige Ionisierung *f*, Ionenlawine *f*
a. of electrons Elektronenlawine *f*
a. of ions Ionenlawine *f*
aventurine quartz *(Min)* Aventurin *m*
average / to durchschnittlich betragen, sich durchschnittlich belaufen auf, das Mittel bilden
average 1. Durchschnitt *m*; Mittel[wert *m*] *n*; 2. *(Schiff)* Havarie *f*
average durchschnittlich, Mittel-, mittlerer
a. amount Durchschnittsbetrag *m*
a. brightness *(Fs)* mittlere Helligkeit *f*
a. calculating operation durchschnittliche (mittlere) Rechenoperationszeit *f*
a. capacity Durchschnittsleistung *f*
a. consumption Durchschnittsverbrauch *m*
a. cross section *(Kern)* mittlerer Wirkungsquerschnitt *m*
a. current mittlerer Wechselstromwert *m*
a. haul distance mittlere Transportentfernung *f*
a. hourly earning Durchschnittsstundenlohn *m*
a. life mittlere Lebensdauer *f*
a. mining rate vertically Abbaufortschritt *m* in die Teufe
a. noise factor *(Fmt)* Bandrauschzahl *f*
a. outgoing quality Durchschlupf *m*
a. outgoing quality limit Durchschnittsfehlerquote *f* im verkaufsfähigen Produkt
a. particle size mittlere Korngröße *f*; mittlere Teilchengröße *f*
a. performance Durchschnittsleistung *f*
a.-position control *(BMSR)* „Pendelregelung" *f (Spezialfall einer Zweipunktregelung)*
a. sample repräsentative Probe *f*
a. speed Durchschnittsgeschwindigkeit *f*
a. speed in rpm Durchschnittsdrehzahl *f*
a. stem *(Holz)* Durchschnittsstamm *m*, Mittelstamm *m*
a. value Mittel[wert *m*] *n*, Durchschnittswert *m*
a. value indicator Mittelwertanzeiger *m*
a. weight Durchschnittsmasse *f*
averaging Mittelwertbildung *f*
Avery test *(Met)* Tiefziehversuch *m*, Erichsen-Tiefungsversuch *m*, Tiefungsversuch *m* nach Erichsen
aviation Luftfahrt *f*, Flugwesen *n*
a. fuel Flugbenzin *n*
a. gasoline *(Am)* Flugbenzin *n*
a. lubricants Flugmotorenöl *n*
a. wear Fliegerbekleidung *f*
aviator Flugzeugführer *m*, Flieger *m*
avigraph Kursschreiber *m*
aviso Meldeschiff *n*; Meldeboot *n*
avivage Avivage *f*
Avogadro constant Avogadrosche Konstante *f*
Avogadro number Avogadrosche Zahl *f*
Avogadro's hypothesis Avogadrosches Gesetz *n*
avoid accidents / to Unfälle vermeiden (verhüten)
avoirdupois weight Handelsmassemaß *n (System der Masse- und Gewichtsbestimmung auf der Grundlage des pound, gültig für alle Waren außer für Edelmetalle, Edelsteine und in der Medizin)*
avometer Universalmeßgerät *n* für Spannung, Strom, Widerstand
award to the inventor Erfindervergütung *f*
awkward sperrig
awkwardly-shaped kompliziert (verwickelt) gestaltet
awl Ahle *f*, Pfriem *m*
awn-separator Entgranner *m (Dreschmaschine)*
awning 1. *(Bw)* Markise *f*, Sonnendach *n*, Wetterdach *n*; 2. *(Schiff)* Sonnensegel *n*
a. cloth Markisenstoff *m*
a. deck Awningdeck *n*, Sturmdeck *n*
a. deck vessel Sturmdeckschiff *n*, Awningdeckschiff *n*
a.-decked vessel *s.* a. deck vessel
axe Axt *f*, Haue *f*, Kaukamm *m*; Steinhammer *m*
axed arch Mauerwerksbogen *m* mit keilförmig zugehauenen Ziegelsteinen
a. work scharriertes Mauerwerk *n*
axes *s.* axis
axial achsrecht, axial
a. admission axiale Beaufschlagung *f (Turbine)*
a. alignment Fluchtung *f*, Fluchten *n*
a. angle Achsenwinkel *m*
a.-armature relay Achslagerrelais *n*; Flachrelais *n*
a. base pitch *(Masch)* Eingriffsteilung *f* im Axialschnitt
a. component Axialkomponente *f*

a. compressor Axialgebläse *n*, Axialverdichter *m*
a. contact ratio *(Masch)* Profilüberdeckungsgrad *m* im Axialschnitt
a. cylinder type hydraulic motor Axialkolbenhydraulikmotor *m*
a. cylinder type of piston (plunger) pump Axialkolbenpumpe *f*
a. deviation Abweichung *f* in Achsrichtung
a. distance Achs[en]abstand *m*
a. fan Schraubenlüfter *m*, Axiallüfter *m*
a. feed method Axialverfahren *n (Zahnradherstellung)*
a. flow axialsymmetrische Strömung *f*
a.-flow blower Axialgebläse *n*
a.-flow compressor Axialverdichter *m*
a.-flow supercharger Axialauflader *m*
a. flow turbine Axialturbine *f*
a. force Axialkraft *f*, Längskraft *f*, mittige Kraft *f*
a. length of blade axiale Schaufellänge *f*; axiale Schaufelbreite *f (Turbine)*
a. load Axiallast *f*, Längslast *f*, mittige Last *f*; Axialbelastung *f*, Beanspruchung *f* in Achsrichtung
a. magnification Tiefenvergrößerung *f*
a. moment of inertia axiales Trägheitsmoment *n*
a. movement Axialbewegung *f*
a. net cross-section axialer Reinquerschnitt *m (Turbine)*
a. oil groove längslaufende Schmiernut *f*
a. piston pump Axialkolbenpumpe *f*
a. pitch 1. Steigung *f (Gewinde)*; 2. Achsteilung *f (Zahnrad)*
a. plane 1. Achsebene *f*, Ebene *f* durch die Achsen von im Eingriff befindlichen Zahnrädern; 2. Axialschnitt *m (technisches Zeichnen)*
a. play Längsspiel *n*; Axialspiel *n*
a. porosity *(Met)* Fadenlunker *m*
a. pressure Axialdruck *m*
a. pressure angle Eingriffswinkel *m* im Axialschnitt
a. ratio Achsenverhältnis *n*
a. response axiale Empfindlichkeit *f*; axialer Übertragungsfaktor *m*
a. seal ring axiale Gleitringdichtung *f*
a. section Axialschnitt *m (technisches Zeichnen)*
a. section method *(Opt)* Achsenschnittverfahren *n*
a. sponginess *(Met)* sekundärer Lunker *m (im Gußblock)*
a. stress and bending Längskraft *f* mit Biegung
a. stretching device Axialspannvorrichtung *f*
a. thickness *(Masch)* Zahndicke *f* im Axialschnitt
a. thrust Axialschub *m*, Axialdruck *m*, Längskraft *f*; Axialbelastung *f*; Längsschub *m*
a. travel Axialbewegung *f*
a. vibration Axialschwingung *f*
axially loaded axial (mittig) belastet, axial (mittig) beansprucht
axinite *(Min)* Axinit *m*
axiom Axiom *n*
axiomatic majority-decision logic *(BMSR)* unanzweifelbare Majoritätsentscheidungslogik *f*
axis *(Math)* Achse *f*
a. communication Hauptnachrichtenweg *m*
a. line of pressure Drucklinie *f*, Linie *f* der Mittelkraft
a. of a crystal Kristallachse *f*
a. of a drill *s.* a. of drill
a. of a waveguide Hohlleiterachse *f (Wellenleiter)*
a. of abscisses X-Achse *f*
a. of commutation *(Elm)* Kommutierungsachse *f*
a. of drill Bohrerachse *f*
a. of drill rotation Bohrerdrehachse *f*
a. of guide Führungsachse *f*
a. of inertia Trägheitsachse *f*
a. of ordinates Y-Achse *f*
a. of oscillation Schwingungsachse *f*
a. of reference Bezugsachse *f*
a. of rotation Drehachse *f*, Rotationsachse *f*
a. of spin Drehimpulsachse *f*
a. of suspension Pendelachse *f*
a. of symmetry Symmetrieachse *f*
a. of the beam Trägerachse *f*
a. of the drill *s.* a. of drill
a. of the hole Bohrungsachse *f*
a. of twist Drillachse *f*, Verdrehachse *f*
a. of zero Nullinie *f*, Neutrale *f (Festigkeit)*
a. section Axialschnitt *m*
axisymmetric achsensymmetrisch
axle *(Masch)* Achse *f*
a. arrangement Achsfolge *f*
a. base *(Kfz)* Achs[ab]stand *m*
a. beam *(Kfz)* Starrachse *f*
a. bearing Achslager[ung *f*] *n*
a. bearing brass Achslagerschale *f*
a. bolt Achsbolzen *m*
a. box Achsgehäuse *n*, Achsenlager *n*, Lagerbüchse *f*, Lagerschale *f*, Achsbüchse *f*
a. box brass Achslagerschale *f*
a. box guide Achslagerführung *f*, Achslagersteg *m*
a. box guide plate Achslagergleitplatte *f*
a. box keep Achslagerunterkasten *m*
a. box key Achslagerstellkeil *m*
a. box lubricating cover (lid) Achslagerschmierdeckel *m*
a. box oil cap cover Achslagerschmierdeckel *m*

a. **box slide** Achslagerführung *f*
a. **box stirrup** Achslagerbügel *m*
a. **box yoke** Achsbüchsjoch *n*
a. **bracket** Achsenbock *m*, Achsenträger *m*; Achsstütze *f*
a. **camber** Achssturz *m*
a. **collar** Achsbund *m*
a. **counter** *(Eis)* Achszähleinrichtung *f*
a. **dip** Achssturz *m*
a. **drive** Achsantrieb *m*
a.**-drive bevel gear** großes Differentialantriebskegelrad *n*
a.**-drive bevel pinion** kleines Differentialantriebskegelrad *n*
a.**-drive ratio** Übersetzung *f* des Achsantriebs
a.**-driven generator** Achsdynamo *m*
a.**-driving shaft gasket** Achswellendichtung *f*
a. **finishing lathe** Achsenfertigdrehmaschine *f*
a. **grease** Achsenfett *n*, Wagenfett *n*, Wagenschmiere *f*
a. **guard** Achshalter *m*
a. **guide** Führungsgabel *f (Achse)*
a. **guide stay** Achsgabelsteg *m*
a.**-hung motor** Tatzlagermotor *m*, Motor *m* mit Achsaufhängung
a. **journal** Achshals *m*, Achsschenkel *m*, Achsstummel *m*, Achszapfen *m*, Lagerhals *m*
a.**-journal grinding machine** Achsschenkelschleifmaschine *f*
a. **journal with collar** Achsschenkel *m* mit Bund
a. **lathe** Achsendrehmaschine *f*
a. **load** Achsdruck *m*, Achslast *f*
a. **neck** Achsenhals *m*, Achsschenkel *m*, Achszapfen *m*
a. **oils** Achsenöle *npl*
a. **peeling lathe** Achsenschälmaschine *f*
a. **relief factor** Achsentlastungsfaktor *m*
a. **roughing lathe** Achsenschruppdrehmaschine *f*
a. **sequence** Achsfolge *f*
a. **shaft** Achswelle *f*, Seitenwelle *f*; Differentialseitenwelle *f*; Antriebswelle *f (der Hinter- oder Vorderräder)*
a. **spring** Achsfeder *f*
a. **stub** Achsschenkel *m*, Achsstummel *m*
a. **suspended motor** Achslagermotor *m*, Tatzlagermotor *m*
a.**-testing gauge** Achsprüflehre *f*
a.**-tree** Radachse *f*, Wagenachse *f*, Wellbaum *m*, Welle *f*
a.**-tree arm** Achszapfen *m*, Achsschenkel *m (Wagen)*
a.**-tree hoop** Nabenring *m (Wagen)*
a. **tube** Achsrohr *n*
Axminster Axminsterteppich *m (Wollgewebeteppich mit aufgeschnittenem Flor)*
axonometric drawing axonometrische Zeichnung *f*
azelaic acid Azelainsäure *f*
azeotropic azeotrop[isch]
a. **mixture** azeotropisches Gemisch *n*
azimuth Azimut *m*
a. **dial** *(Flg)* Peilrose *f*
a. **gyro** Azimutkreisel *m*
a. **mirror** Kompaßdiopter *m*
azimuthal azimutal
a. **quantum number** azimutale Quantenzahl *f*, Nebenquantenzahl *f*
azines Azine *npl*
azo dyes Azofarbstoffe *mpl*
azoic dye Azofarbstoff *m*; Eisfarbe *f*
azotic acid Salpetersäure *f*
azotize / to azotieren
azotometer Azotometer *n*
azurite Bergblau *n*; Berglasur *f*, Kupferlasur *f*

B

B-battery Anodenbatterie *f*
B-eliminator Ersatzgerät *n* für Anodenbatterie; Netzanschlußgerät *n*; Netzanode *f*
B-emission *(Fmt)* Aussendung *f* gedämpfter Wellen
B-horizon *(Geol)* B-Horizont *m*, Anreicherungshorizont *m*
b-line *(BMSR)* Index[register *n*] *m*
B-modulation *(Fmt)* B-Modulation *f*; Gegentakt-B-Modulation *f*
babassu [kernel] oil Palmkernöl *n*
Babbit's metal *s.* babbitt
babbitt / to ausgießen, ausfüttern *(ein Lager mit Weißmetall)*
babbitt Weißmetall *n*, Lagermetall *n*, Lagerfutter *n*
b. **bearing lining** Weißmetallausguß *m*
b. **file** Feile *f* zur Weißmetallbearbeitung
b.**-lined bearing** Weißmetallfutterlager *n*
b.**-lined bushing** Laufbuchse *f* mit Weißmetallfutter
b. **lining** Weißmetallfutter *n (eines Lagers)*
b. **metal** Lagerweißmetall *n*, Weißmetall *n*
babbitted bearing Weißmetallager *n*
babbitting Lagerausguß *m*
Babo flask holder *(Ch)* Babotrichter *m*; Luftbad *n*
baby Bessemer converter *(Met)* Kleinkonverter *m*
b. **Bessemer steel plant** *(Met)* Kleinbessemerei *f*
b. **bulldozer** Kleinplanierraupe *f*
b. **screwdriver** Taschenschraubenzieher *m*
back / to 1. stützen; Untergrund (Auflage, Anlage) bilden; 2. *(Schiff)* backholen, verkat-

ten; 3. *(Text)* kaschieren; 4. *(Typ)* widerdrucken, die Rückseite bedrucken
b. off 1. *(Masch)* hinterarbeiten, hinterdrehen, hinterfräsen, hinterschleifen; 2. *(El)* Gegenspannung anlegen an; [auf Null] kompensieren
b. out zurückbewegen, zurückziehen
b.-scatter rückstreuen
b. the oars *(Schiff)* rückwärtsrudern, rückwärtspullen
b. up 1. *(Bw)* abstützen, tragen; hintermauern; 2. *(Masch)* vorhalten, gegenhalten *(Nietung)*
b. water *(Schiff)* die Riemen streichen

back 1. *(Masch)* Rückseite *f*; Zahnrücken *m* *(Fräserzahn)*; Blattrücken *m* *(Säge)*; Schneidenrücken *m*; Deckel *m*; Aufspannbolzen *m*; 2. *(Typ)* Rücken *m* *(des Buches)*; Träger *m*, Grundmaterial *n* *(Kaschierung)*; Bund[steg] *m*; 3. *(Brg)* Hangendes *n*; Dach *n*, Firste *f*; das zwischen zwei Sohlen anstehende Erz *n*; Längsspalte *f*, Hauptschlechte *f*; 4. *(Text)* Unterseite *f*; Bottich *m*
b. abutment pressure *(Brg)* nacheilender Kämpferdruck *m*
b. ampere turns Gegenamperewindungen *fpl*
b. anchor *(Schiff)* Kattanker *m*
b. and front pitch *(El)* Teil[wicklungs]schritt *m*
b. angle Rückenwinkel *m* *(Verzahnung)*
b.-axle ratio Hinterachsenübersetzungsverhältnis *n*
b. balance *(Brg)* eintrümiger Bremsberg *m* mit Gegengewicht
b. beam Hinterbaum *m*, Streichbaum *m* *(Webstuhl)*
b. bias circuit Schwundregelung *f* nach rückwärts; Rückwärtsregelkreis *m*
b. bias control *(Elm)* Rückwärtsregelung *f*
b. bridge wall *(Met)* Fuchsbrücke *f* *(des Ofens)*
b. brusher *(Brg)* Nachreißhauer *m*
b. centre *(Masch)* Reitstockspitze *f*
b. chroming *(Am)* Nachchromieren *n*
b. clearance *(Masch)* Rückenspiel *n*
b. cloth *(Text)* Mitläufer *m*, Unterware *f*, Grundware *f*; Drucktuch *n*
b. coat Deckstrich *m* *(Lackierung usw.)*; Grundschicht *f*; Unterputz *m*
b.-coming *(Brg)* Rückbau *m*
b. cone *(Masch)* Rückenkegel *m*
b.-connected fuse Sicherung *f* mit rückseitigem (rückwärtigem) Anschluß
b.-connected switch Schalter *m* mit rückseitigem (rückwärtigem) Anschluß
b.-coupled valve rückgekoppelte Röhre *f*
b.-coupling Rückkopplung *f*
b. current *(El)* Rückstrom *m*, Gegenstrom *m*
b. cutting *(Holz)* Rückwärtsschneiden *n*; Spiegelschnitt *m* *(Rückenfurnier)*
b. cylinder cover *(Wärm)* Zylinderbodendeckel *m*
b. diffusion Rückdiffusion *f*
b.-diffusion loss Rückdiffusionsverlust *m*
b. digger *(Bw)* Grabenbagger *m*, Tieflöffelbagger *m*
b. drilling attachment Hinterbohreinrichtung *f*
b. echo Hinterbündelecho *n* *(Radar)*
b. effect Rückwirkung *f*
b. electromotive force gegenelektromotorische Kraft *f*, Gegen-EMK *f*
b. e. m. f. *s.* back electromotive force
b. emission Gegenemission *f*
b. end centre *(Masch)* Reitstockspitze *f*
b. end man *(Brg)* Stempelsetzer *m* hinter einem Schrämlader
b. entry *(Brg)* Begleitort *n*, Begleitstrecke *f*
b. face Rückenfläche *f*
b.-face reflection Rückwandecho *n*
b.-feed *(El)* Rückkopplung[swirkung] *f*
b.-flank synchronization Rückfrontsynchronisierung *f* *(Fernsehtechnik)*
b. flap hintere Klappe *f* *(eines Schutzumschlages)*
b. flow condenser Rückflußkondensator *m*, Rückflußkühler *m*
b. fork Hinterradgabel *f*
b. gear[s] Rädervorgelege *n*, Zahnrädervorgelege *n*, Spindelvorgelege *n*
b. gear lever Vorgelegehebel *m*
b. gear ratio Übersetzungsverhältnis *n*
b. gear shaft Rädervorgelegewelle *f*
b.-geared mit Vorgelege, mit Übersetzung *(ins Langsame)*
b.-gearing Vorgelege *n*, Zahnradvorgelege *n*
b. grey cloth *(Text)* Mitläufer *m* aus Rohgewebe
b. ground noise Grundgeräusch *n*, Störpegel *m*
b. guide Rückenführung *f* *(Sägemaschine)*
b. guide bar hintere Legeschiene *f* *(Wirkmaschine)*
b. hole *(Brg)* Firstbohrloch *n*
b. iron *(Holz)* Abziehmesser *n*, Putzmesser *n*
b. light Rücklicht *n*
b. lighting Hintergrundbeleuchtung *f*
b. lobe Hinterlappen *m* *(Antenne)*
b. lock *(Text)* hinteres Strickschloß *n*
b. margin *(Typ)* Bund *m*
b. mill table *(Met)* Rollgang *m* *(hinter dem Walzgerüst)*
b. needle bed *(Text)* hinteres Nadelbett *n*
b. of a generator Antriebsseite (Kupplungsseite) *f* eines Generators
b. of a motor Antriebsseite *f* eines Motors

b. of blade Schaufelrücken *m (Turbine)*
b.-of-board controller *(BMSR)* Schalttafelregler *m*
b.-off angle Freiwinkel *m (Räumwerkzeug)*
b.-off cam *(Masch)* Abdrückkurve *f*, Hinterdrehkurve *f*
b.-off clearance Hinterschliff *m*
b.-off lathe Hinterdrehmaschine *f*
b. panel *(Kfz)* Heck *n* der Karosserie
b. pick *(Text)* Unterschuß *m*
b. picker *(Text)* Zunahmefinger *m*
b. pin *(Masch)* rückwärtiger Raststift *m (Teileinrichtung)*
b. plate 1. Rückwand *f*; 2. *(Masch)* Futterscheibe *f (Spannfutter)*; 3. *(Wärm)* Rückwärtsgleitschiene *f*, übergreifende Gleitschiene *f*; 4. Gegenelektrode *f (Kondensatormikrofon)*
b. plate of a firebox Feuerbuchsrückwand *f*
b. ply Rückenlage *f* bei Furnieren
b. post *(Schiff)* Hintersteven *m*, Rudersteven *m*
b. potential *(El)* Gegenspannung *f*
b. pressure Gegendruck *m*, Vorlagedruck *m (Turbine)*
b. pressure engine Gegendruckmaschine *f*
b. pressure regulating valve Gegendruckregulierventil *n*
b. pressure-relief port Rückdruck-Entlastungsöffnung *f (Extruder)*
b. pressure roller *(Typ)* Andruckwalze *f*, Gegendruckwalze *f*
b. pressure turbine Gegendruckturbine *f*
b. pressure valve Rückdruckventil *n*, Gegendruckventil *n*, Halteventil *n*
b. pull Gegenzug *m (Drahtziehen)*
b. putty Fensterkitt *m*
b. rail Streichbaum *m (Webstuhl)*
b. rake [angle] *(Am) (Masch)* Spitzenspanwinkel *m*
b. reflection Rückstrahlung *f*
b. resistance Sperrwiderstand *m*
b. rest 1. *(Masch)* Stützlager *n*; Lünette *f*; Gegenhalter *m*; 2. Streichbaum *m (Webstuhl)*
b.-rest bearing Setzstocklager *n*
b.-ring *(Fmt)* Rückruf *m*
b. ripping *(Brg)* Nachbauen *n*, Durchbauen *n*
b. roller *(Text)* Hinterwalze *f (Streichwerk)*
b. rounding machine *(Typ)* Rückenrundemaschine *f*
b. router *(Typ)* Rückenfräsmaschine *f (Klebebindung)*
b. saw Rückensäge *f*, Fuchsschwanz *m*, Bogensäge *f*
b.-scatter factor Rückstreufaktor *m*
b.-scatter pattern Rückstreudiagramm *n*
b.-scattered [zu]rückgestreut
b.-scatterer *(Kern)* rückstreuendes Material *n*, Rückstreumaterial *n*
b.-scattering Rückstreuung *f*
b.-scattering error Rückstreufehler *m*
b.-scattering factor Rückstreufaktor *m*
b.-seat gunner *(Mil)* Heckschütze *m (des Bombenflugzeuges)*
b. shank *(Text)* Teil *m* des Schaftes hinter dem Fuß *(Zungennadel)*
b. shift *(Brg)* Nachtschicht *f*
b. shore 1. *(Schiff)* Aufklotzung *f*; 2. Küstensaum *m (der bei großem Sturm ausnahmsweise mit Wasser bedeckte Küstenteil)*
b.-sided timber wahnkantiges Holz *n*
b.-sight Kimme *f (Gewehr)*
b.-sized *(Text)* rückenappretiert, mit linksseitiger Erschwerung
b.-space key *(El)* Rückstelltaste *f*
b.-spacer *(El)* Rücktaste *f*
b. square *(Masch)* Anschlagwinkel *m*
b. squaring attachment *(Masch)* Rückenabrichtvorrichtung *f*
b. stern-post Achtersteven *m (Boot)*
b. stitch *(Text)* Hinterstich *m*
b.-stitch embroidery Stielstichstickerei *f*
b. stope *(Brg)* Firstenstoß *m*, alle über dem untersten Stoß liegenden Stöße *mpl (Firstenstoßbau)*
b. stoping *(Brg)* 1. Firstenbau *m (allgemein)*; Firstenbau mit abgesetzten Stößen; 2. Magazinbau *m*, Speicherbau *m*
b. strap *(Led)* Hinterbesatz *m*, Hinterriemen *m*
b. stripper *(Brg)* Hauer *m*, der zu großes Haufwerk zerkleinert
b.-surface echo Rückwandecho *n*, Bodenecho *n*
b.-surge Rückstau *m (Turbine)*
b. sweep *(El, Fs)* Rücklauf *m*
b. taper Durchmesserverjüngung *f*
b. title *(Typ)* Rückentitel *m*
b. titration Rücktitration *f*
b.-to-back method Rückarbeitsverfahren *n (Generatoren)*
b. turbine Rückwärtsturbine *f*, Rücklaufturbine *f*, Turbine *f* für Rückwärtsgang
b.-up brick Hintermauerziegel *m*, Hintermauerstein *m*
b.-up lamp *(Kfz)* Rückfahrscheinwerfer *m*, Rückwärtsscheinwerfer *m*
b.-up light *s.* b.-up lamp
b.-up material Füllstoff *m*, Unterschicht *f*
b.-up protection *(El)* Reserveschutz *m*, überlagerter Schutz *m*
b.-up ring *(Masch)* Stützring *m*
b.-up roll Stützwalze *f*
b.-up safety device Hilfssicherheitsvorrichtung *f*

b.-up washer *(Masch)* Stützscheibe *f*
b. view Rückansicht *f*
b. voltage Rück[wirkungs]spannung *f*, Sperrspannung *f*
b. wall Rückwand *f*
b.-wall barrier (blocking) layer cell *(El)* Hinterwandsperrschichtzelle *f*
b. warp *(Text)* Grundkette *f*
b. water 1. Abwasser *n*, Rückwasser *n*; 2. Stauwasser *n*, totes Wasser *n*; 3. Unterwasser *n (Setzmaschine)*
b.-water curve *(Bw)* Staukurve *f*, Staulinie *f*
b.-wave 1. *(Schiff)* Stauwelle *f*; 2. negative Tastwelle *f (Fernmeldetechnik)*
b. weft *(Text)* Grundschuß *m*
b. winding *(Text)* Wiederaufwickeln *n*; Rückspulen *n*; Aufwickeln *n*
b. work *(Brg)* Rückbau *m*, Heimwärtsbau *m (Abbaurichtung)*; Hilfsarbeit *f (Vorrichtung)*
b.-wound coil *(El)* verstürzte (verschachtelte) Spule *f*, Stufenspule *f*
b. yarn *(Text)* Grundfaden *m*, Grundgarn *n*
backacter Grabenbagger *m*, Tieflöffelbagger *m*
backacting excavator *s.* backacter
backbone conformation Rückgratkonformation *f (Proteinstruktur)*
backbreak *(Brg)* Böschungsausbruch *m*, Ausbruch *m (Tagebaustoß)*
backbye deputy *(Brg)* Streckenzimmerhauer *m*, Streckenzimmerling *m*
backed cloth *(Text)* Gewebe *n* mit Abseite
b. cloth weave *(Text)* Bindung *f* für Doubles
b. fabric *(Text)* Double *n*
b. foil kaschierte Folie *f*
b.-off hinterdreht; hinterschliffen; mit Freiwinkel
b.-off slot and key cutter hinterdrehter Nut- und Spundfräser *m*
backer *(Schiff)* Taustropp *m*
b.-up geführtes Flugzeug *n*
backfall *(Holz)* Kropf *m*, Gefälle *n*, Sattel *m*
backfill / to *(Brg)* hinterfüllen, versetzen
backfill temper *(Masch)* Formkastenklopfer *m*
backfilled *(Text)* stark gesteift
backfiller *(Bw)* Grabenverfüllgerät *n*
backfilling 1. Hinterfüllung *f*, Verfüllung *f*; Verfüllboden *m*, Verfüllmasse *f*, Versatz *m*; 2. *(Text)* Unterschuß *m*, Füllschuß *m*
b. finish *(Text)* einseitige Appretierung *f*, Füllappretur *f*
b. mangle *(Text)* Linksappretiermaschine *f*, Linksimprägnierkalander *m*
backfire 1. *(El)* Rückzündung *f*; 2. *(Kfz)* Frühzündung *f*
b. aerial Backfire-Antenne *f (Spezialtyp einer Yagi-Antenne)*
b. arrester Rückzündungsschutz *m*
backfiring 1. Flammenrückschlag *m*; Fehlzündung *f*; 2. Gleichrichterrückzündung *f*
backfocal length Anlagemaß *n (Fotografie)*
background 1. Hintergrund *m*, Untergrund *m*; 2. *(Kern)* Nulleffekt *m*, Strahlenuntergrund *m*; Untergrund[strahlung *f*]; 3. *(Fs)* Grundhelligkeit *f*; 4. *(Text)* Fond *m*
b. brightness *(Fs)* Grundhelligkeit *f*
b. control *(Fs)* Helligkeitsregelung *f*
b. count 1. *(El)* Störpegelzählung *f*; 2. *(Kern)* Nulleffekt *m*
b. counting rate *(Kern)* Nulleffektzählrate *f*, Untergrundzählrate *f*
b. level Grundniveau *n (Fernsehtechnik)*
b. noise Grundgeräusch *n*, Grundrauschen *n*, Nebengeräusch *n*, Störgeräusch *n*, Störpegel *m*; 2. *(El)* Eigenrauschen *n*, Rauschen *n*; Rauschton *m*
b. noise level Grundgeräuschpegel *m*
b. projector *(Opt)* Hintergrundprojektor *m*, Rückprojektor *m*
b. radiation Strahlenuntergrund *m*; Untergrundstrahlung *f*
b. scattering Untergrundstreuung *f*
backhand welding *(Am)* Nachrechtsschweißen *n*
backhoe *(Bw)* Grabenbagger *m*, Tieflöffelbagger *m*
b. dipper Löffel *m* eines Tieflöffelbaggers
backing 1. Unterstützen *n*, Stärken *n*, Verstärken *n*; 2. *(Bw)* Hintermauerung *f*; 3. *(Typ)* Unterguß *m*; 4. *(Led)* Verstärken von Kunststoff-Folie durch Gewebeunterlage; 5. *(Text, Led)* Abseite *f*, Rückseite *f*, Unterseite *f*; 6. Grundschicht *f*, Schichtträger *m*
b. anchor *(Schiff)* Kattanker *m*
b. block Stützblock *m*
b. brick Hintermauerungsstein *m*, Hintermauerungsziegel *m*
b. cloth Grundgewebe *n*
b. deal *(Brg)* Verzugsholz *n*
b. effect *(Brg) Bewegung einer Methanschicht in der Firste einer steigenden Strecke gegen den fallenden Wetterstrom)*
b. lamp Rückfahrscheinwerfer *m*
b. material Grundkörper *m*; Schichtträger *m*
b. of a propeller *(Schiff)* Rückwärtsschlagen *n* eines Propellers
b. of the wind *(Schiff)* Krimpen *n* des Windes
b. off 1. *(Masch)* Hinterdrehen *n*, Hinterfräsen *n*, Hinterschleifen *n*; Hinterschliff *m*; 2. *(Text)* Abwinden *n*; Abschlagen *n*
b.-off attachment Hinterschleifeinrichtung *f*; Hinterdreheinrichtung *f*
b.-off chain tightening motion *(Text)* Rückwinderegler *m*
b.-off device *(Masch)* Hinterdreheinrichtung *f*; Hinterschleifeinrichtung *f*

b.-off lathe Hinterdrehmaschine *f*
b.-off motion *(Text)* Rückwärtsbewegung *f*
b. pump Vorpumpe *f*
b.-pump valve Vorpumpenventil *n*
b. shell Stützschale *f*
b. store *(Am)* Ergänzungsspeicher *m (für Daten)*
b. support Unterlage *f*, Stütze *f*, Anlage *f*
b. thread *(Text)* Futterfaden *m*
b.-up *(Bw)* 1. Stützung *f*; 2. Hintermauerung *f*; 3. Unterlage *f*, Grundschicht *f*, Träger *m*
b.-up roll Stützwalze *f (Walzwerk)*
b. weft *(Text)* Futterschuß *m*
b. wire *(Holz)* Stützsieb *n*, Untersieb *n*
b. yarn *(Text)* Grundgarn *n (Florgewebe)*, Bindefaden *m*
backlash 1. *(Masch)* Spiel *n*, toter Gang *m*; 2. Flankenspiel *n (Zahnrad)*; 3. *(El)* Gleichrichterrückzündung *f*, positiver Gitterstrom *m*, unvollständige Gleichrichtung *f*
b. compensation Spielausgleich *m*
b. determined in the direction of the line of action Eingriffsflankenspiel *n (Zahnrad)*
b. elimination Ausgleich *m* des Totganges, Spielausgleich *m*
b. eliminator (eliminating device) Spielausgleicheinrichtung *f*, Gleichlauffräseinrichtung *f*
b.-free spielfrei; flankenspielfrei *(Zahnrad)*; ohne toten Gang *(Fräsmaschine)*
b. in the direction of the pitch circle Verdrehflankenspiel *n (Zahnrad)*
b. of thread Flankenspielraum *m (Gewinde)*
b. voltage Gegenspannung *f*; Sperrspannung *f*
backrope *(Schiff)* St[e]ert *m (beim Treibnetz)*
backscatter *s.* back-scatter ...
backsight Kimme *f (Visier)*
b. slide Visierschieber *m*
backsloper *(Bw)* Böschungshobel *m*
backstay 1. *(Masch)* Lünette *f*; 2. *(El)* Abspanndraht *m (vom Mast zum Anker)*; 3. *(Brg)* Fanghaken *m*
b. stool *(Schiff)* Pardunenausleger *m*
backstitch *(Text)* Hinterstich *m*, Rückstich *m*, Stielstich *m*
backstop Basketballständer *m*
backswept schräg nach hinten gestellt, gepfeilt *(Tragfläche)*
backward rückwärts, rückläufig
b.-acting regulation *(Fs)* Rückwärtsregelung *f*
b. current Rückstrom *m (Gleichrichter)*, Sperrstrom *m*
b. diode Rückwärtsdiode *f*, Sperrdiode *f*
b. direction *(El)* Sperrichtung *f*
b. movement of the slag *(Text)* Ladenrückgang *m*
b. propagation Rückwärtsausbreitung *f*
b. radiation Rückstrahlung *f*
b. recall signal Rückwärtsruf *m (Funktechnik)*
b. resistance *(El)* Sperrwiderstand *m*
b. stop Rücklaufanschlag *m*
b. stroke guide surface *(Wärm)* Rückwärtsgleitbahn *f*, Rückwärtsgleitfläche *f*
b. stroke standard *(Wärm)* Rückwärtsständer *m*
b. wave Rückwärtswelle *f (Wanderwellenröhren)*
b. wave oscillator Rückwärtswellenoszillator *m*
b. welding Nachrechtsschweißen *n*
backwash 1. *(Schiff)* Propellerstrom *m*, Schraubenstrom *m*, Rückströmung *f*; rücklaufende Strömung *f*; 2. *(Bw)* Rückwäsche *f*
b. water *(Holz)* Rückspülwasser *n*
backwashing *(Text)* Entfetten *n (Wolle)*
b. machine *(Text)* Entschweißmaschine *f*
backwater from the papermaking machine Papiermaschinen-Rückwasser *n*
bacteria bed Tropfkörper *m (Gesundheitstechnik)*
bacterial bate *(Led)* Bakterienbeize *f*
bacteriostatic finish *(Text)* bakterienwachstumshemmende (bakteriostatische) Ausrüstung *f*
bad-bearing sector Ungenauigkeitswinkel *m*, Gebiet *n* unsicherer Peilung *(Radar)*
b. bevel *(Holz)* Baumkante *f*
b. formation Unschärfe *f (Telegrafiezeichen)*
b. holding ground *(Schiff)* schlechter Ankergrund *m*
b. picture *(Foto)* wertlose Aufnahme *f*
b. reception schlechter Empfang *m (Radio, Fernsehen)*
b. salt stains *(Led)* Salzfraß *m*
baddeleyite *(Min)* Baddeleyit *m*
badly shaped schlechtwüchsig *(Bäume)*
badminton Federball[spiel *n*] *m*
baffle 1. Trennwand *f (vgl.* baffle plate); 2. Schutzschirm *m (Lampe)*; 3. *(Holz)* Schöpfblech *n*
b. board 1. *(El)* Schallschirm *m*; Schallwand *f*; 2. *(Brg)* Prallplatte *f (Zerkleinerung)*; Tauchbrett *n (Naßsetzen)*
b. breaker Prallblech *n*
b. chamber Blendengehäuse *n*, Staubkammer *f (mit Umlenkblechen)*
b. column Kolonne *f* mit Umlenkblechen *(Prallplatten)*
b.-column mixer Mischsäule *f* mit Prallplatten
b. feed heater *(Wärm)* Plattenvorwärmer *m*
b. grinding Prallzerkleinerung *f*
b. head Schwallwand *f (Tankwagen)*
b. plate 1. *(Masch)* Ablenkplatte *f*, Leitblech *n*, Leitwand *f*, Lenkwand *f*, Umlenkblech *n*; Verteilungsplatte *f*; Dämpfungsbrett *n*, Prallblech *n*, Prallwand *f*, Prallplatte *f*

(pneumatischer Verstärker); Stoßplatte *f (Dampfturbine)*; Sperre *f*, Stauscheibe *f*; Trennungsblech *n*, Trennungsscheibe *f*, Zwischenplatte *f*, Zwischenwand *f*; 2. *(Brg)* Schlagleiste *f*
b. ring Ringleitblech *n*
b. separation Stoßreinigung *f*, Stoßtrennung *f*
b. sheet Prallplatte *f*; Leitblech *n*, Trennblech *n*
b. silencing *(Kfz)* Schalldämpfung *f* durch Wirbelung
b. tightness enges Anliegen *n* der Kühlluftleitbleche
b.-type silencer Schalldämpfer *m* mit Wirbelung
b. wall *(Wärm)* Prallfläche *f*; Trennwand *f*
baffled mit Leitblechen
b. tube verengtes Rohr *n*; Rohrverengung *f*
baffling Richtungsänderung *f*, Umleitung *f* *(Gase oder Flüssigkeiten)*
b. wind *(Schiff)* umspringender Wind *m*
bag / to 1. einsacken, absacken; 2. vom Kurs abweichen; 3. bauschen *(Faden)*
bag Sack *m*; Tasche *f*; Tüte *f*; Beutel *m*
b. barrow Sackkarre *f*
b. binder Sackbindemaschine *f*
b. binding machine Sackbindemaschine *f*
b. cargo Sackladung *f*, Sackgut *n*
b. elevator Sackaufzug *m*
b.-filling scale Dosierwaage *f*
b. filter *(Brg)* Schlauchfilter *n*
b. filtering *(Brg)* Sichten *n*
b. hoist Sackaufzug *m*
b. leather Aktentaschenleder *n*; Kofferleder *n*
b. lifter Sackheber *m*
b. lifting device Sackheber *m*
b. maker Tütenhersteller *m*, Beutelhersteller *m*
b. moulding *(Plast)* Pressen *n* mit Gummisack
b. paper Tütenpapier *n*
b. sealing machine Beutelverschlußmaschine *f*, Tütenverschlußmaschine *f*
bagasse Bagasse *f*, Zuckerrohrrückstände *mpl*, Zuckerrohrtreber *pl*
b. furnace Bagassefeuerung *f*
baggage hold *(Schiff)* Gepäckraum *m*
bagged cargo *(Schiff)* Ladung *f* in Säcken
bagging aggregate Absackeinrichtung *f*
b. conveyor Einsackförderer *m*
b. device Absackeinrichtung *f*
b. machine Absackmaschine *f*
b. needle Packnadel *f*, Sacknadel *f*
baggy fabric *(Text)* bauschige Stelle *f* im Doppelgewebe
bagpipe Dudelsack *m*, Sackpfeife *f*
bail / to *(Min, Brg)* sümpfen, Wasser ziehen; Wasser ausschöpfen
bailer 1. *(Schiff)* Schöpfeimer *m*; 2. Schmantlöffel *m (Seilbohren)*
Bailey bridge Bailey-Behelfsbrücke *f (besteht aus Stahlfachwerksektionen, die zu beliebigen Spannweiten verbunden werden können)*
bailiff *(Brg)* Wettermann *m*, Wettergehilfe *m*
bailing Löffeln *n*, Schlämmen *n*, Schöpfen *n* *(Erdöl)*
bainite *(Met)* Bainit *m*, Zwischenstufengefüge *n*
baize *(Text)* gefilztes Wolltuch *n*
b. curtain Friesvorhang *m*
bakability Backfähigkeit *f*; Trockenfähigkeit *f*
bake / to 1. einbrennen, [nachträglich] mit Hitze behandeln; 2. hart werden; 3. trocknen *(durch Wärmeeinwirkung)*
bake-out period Härtezeit *f*
baked mould getrocknete Gußform *f*
bakelite Bakelit *n*, Formaldehyd-Kunstharz *n*
bakelized paper Hartpapier *n*
bakery oven Backofen *m*
baking *(Plast)* Ausbacken *n*, Festbacken *n*, Hitzehärtung *f*, Sintern *n*, Sinterung *f*; *(Holz)* Hitzebehandlung *f*; Trocknung *f* durch Wärmeeinwirkung
b. capacity Backvermögen *n*
b. enamel Einbrennemaille *f*, Einbrennlack *m*
b. finish Einbrennfarbe *f*, Einbrennlack *m*
b. oven 1. Trockenkammer *f*; Backofen *m*; 2. Trockenofen *m (Gießerei)*
b. powder Backpulver *n*
b. property Backvermögen *n*
b. temperature Einbrenntemperatur *f*
b. varnish *(Am)* Einbrennlack *m*
balance / to auslasten, abgleichen, ausbalancieren, auswuchten, sich aufheben
b. dynamically dynamisch auswuchten
b. the rotating masses die rotierenden Massen auswuchten
balance 1. Abgleich[ung *f*] *m*, Ausgleich *m*; Gleichgewicht *n*, Kompensation *f*, Symmetrie *f*, Druckausgleich *m*; 2. Waage *f*; 3. *(Masch)* Auswuchtung *f*
b. arm Waagebalken *m*
b. attenuation Fehlerdämpfung *f*
b. beam Waagebalken *m*
b. bob Ausgleichsgewicht *n*; Schwengel *m* mit Gegengewicht
b. bridge Klappbrücke *f*
b. brow *(Brg)* Gestellbremsberg *m*
b. car *(Brg)* Spannwagen *m (Förderung mit geschlossenem Seil)*
b. card Bestandskarte *f (Lochkartenmaschinen)*
b. chamber *(Schiff)* Auftriebskammer *f*
b. check *(BMSR)* Abgleichprüfung *f*, Ausgleichprüfung *f*
b. coil *(El)* Ausgleichspule *f*, Saugdrossel *f*

b. condenser *(El)* Abgleichkondensator *m*, Ausgleichkondensator *m*, Neutralisationskondensator *m*
b. converter Anpaßübertrager *m (Antennentechnik)*
b. counter *(BMSR, Büro)* Subtraktionszähler *m*
b. current *(El)* Ausgleichstrom *m*; Einschwingstrom *m*
b. equation Bilanzgleichung *f*
b. for end thrust *(Wärm)* Axialdruckausgleich *m*
b. forward *(Büro)* Saldenvortrag *m*
b. gear Kompensator *m (Turbine)*
b. handle *(Masch)* Schwungkurbel *f*
b. indicator 1. *(El)* Nullinstrument *n*; 2. Gleichgewichtsanzeiger *m*
b. mass *(Masch, Wärm)* Ausgleichmasse *f*
b. method Abgleichverfahren *n*
b. of momentum Impulsbilanz *f*
b. pan Waagschale *f*
b. payments in clearing *(Ök)* Zahlungsausgleich *m*
b. point *(El)* Abgleichpunkt *m*
b. profit *(Ök)* Gewinnsaldo *m*
b. ratio *(Schiff)* Balancierverhältnis *n (beim Balanceruder)*
b. rope *(Brg)* Unterseil *n*
b. sash Kipp[fenster]flügel *m*
b. selection *(Büro)* Saldenauswahl *f*
b. selector *(Büro)* Saldensteuerungsgerät *n*, Saldenwähler *m*
b. sheet *(Büro)* Bilanzbogen *m*
b. spring Unruhfeder *f (Uhr)*
b. transformer *(El)* Ausgleichtransformator *m*, Ausgleichübertrager *m*
b. weight *(Masch, Wärm)* Gegengewicht *n*, Zusatzgewicht *n*, Ausgleichgewicht *n*
b. wheel Schwungrad *n*
b. wheel of a watch Unruh *f*
balanced 1. im Gleichgewicht; entlastet; druckentlastet; ausgewogen; 2. *(Masch)* ausgewuchtet; 3. *(El)* abgeglichen, ausgeglichen; stromlos *(Meßbrücke)*
b. amplifier *(El)* Gegentaktverstärker *m*, symmetrischer Verstärker *m*
b.-beam relay *(El)* Waagebalkenrelais *n*
b. bridge transition *(El)* Ausgleichbrückenschaltung *f*
b. buffer gear Ausgleichpuffer *m*
b. circuit *(El)* symmetrischer Stromkreis *m*; symmetrische Schaltung *f*
b. condition Gleichgewichtszustand *m*
b. current relay *(El)* Differentialrelais *n*
b. currents erdsymmetrische Ströme *mpl*
b. differential transformer Ausgleichstransformator *m*; Ausgleichsübertrager *m*; Sprechspule *f*
b. earthworks Massenausgleich *m (Erdbau, Tiefbau)*
b. excavation Massenausgleich *m (Erdbau, Tiefbau)*
b. feeder *(El)* symmetrische Speiseleitung *f*
b.-flow vessel method Wägemethode *f (Dichtemessung)*
b. frame *(Schiff)* Richtspant *n*, Nullspant *n*, Querspant *n*
b. hoisting *(Brg)* Förderung *f* mit Seilgewichtsausgleich
b. lacing pattern *(Brg)* symmetrische Anordnung *f* der Schrämmeißel in der Schrämkette
b. line erdsymmetrische Leitung *f*
b. line system *(El)* erdsymmetrisches Leitungssystem *n*
b. load symmetrische Belastung *f*
b. mixer Brückenmischer *m (Wellenleiter)*
b. modulator *(Fmt)* Gegentaktmodulator *m* [zur Unterdrückung der Trägerwelle]
b. oscillator symmetrischer Oszillator *m*
b. periodic component reine Wechsel[strom]komponente *f*, reines Wechsel[strom]glied *n*
b. polyphase system symmetrisches Mehrphasensystem *n*
b. power-to-weight ratio *(Kfz)* ausgewogenes Leistungsgewicht *n*
b. pressure method Druckausgleichsverfahren *n*
b. pressure torch Gleichdruckbrenner *m*
b. reaction *(Ch)* reversible Reaktion *f*
b. roll Anstellwalze *f*
b. rudder *(Schiff)* Balanceruder *n*, Ausgleichsruder *n*
b. steel halbberuhigt vergossener Stahl *m*, halbberuhigter Stahl
b. to earth *(El)* erdsymmetrisch, symmetrisch geerdet (gegen Erde)
b. transmission line *(El)* symmetrische Übertragungsleitung *f*
b. twill *(Text)* gleichseitiger Köper *m*
b. twin *(El)* symmetrische Doppelleitung *f*
b. underhung rudder *(Schiff)* Balanceschweberuder *n*, Steckruder *n*, Schweberuder *n*
b. valve *(Wärm)* ausbalanciertes (entlastetes) Ventil *n*, Ventil (Schieber *m*) mit Druckausgleich
b. vane pump Flügelpumpe *f* mit Druckausgleich
b. voltage [erd]symmetrische Spannung *f*
balancer *(Elm)* Symmetrierstufe *f*, Symmetriereinrichtung *f*
b.-booster *(Elm)* Ausgleichregler *m*
b. set *(Elm)* Ausgleichmaschinensatz *m*
b.-transformer *(Elm)* Ausgleichtransformator *m*; Phasenschieber *m*

balancing 1. *(Masch)* Ausgleich *m* der Unwuchten, Auswuchtung *f*; 2. *(El)* Abgleich[ung *f*] *m*, Ausgleich[ung *f*] *m*; Nullabgleich *m*
b. apparatus *(Ph)* Nullinstrument *n*
b. battery *(El)* Ausgleichbatterie *f*; Pufferbatterie *f*
b. beam Schwebebalken *m*
b. capacitance Ausgleichkapazität *f*; Gegengewicht *n* *(Antennentechnik)*
b. capacitor *(El)* Ausgleichkondensator *m*, Neutralisationskondensator *m*
b. capacity *(El)* Ausgleichkapazität *f*, Neutralisierungskapazität *f*
b. circuit Ausgleichleitung *f*
b. device Ausgleicheinrichtung *f*
b. force Kompensationskraft *f*
b. loop *(El, Fmt)* Symmetrierungsschleife *f*
b. machine *(Masch)* Auswuchtmaschine *f*
b. mandrel *(Masch)* Auswuchtdorn *m*
b. mass Auswuchtmasse *f*
b. method *(Ph)* Abgleichverfahren *n*, Kompensationsmethode *f*
b. moment Ausgleichmoment *n* *(Statik)*
b. motor Abgleichmotor *m*, Ausgleichmotor *m*; Phasenschiebermotor *m*
b. network *(El)* Abgleichnetz *n*; Kunstschaltung *f*, Ausgleichleitung *f*, Leitungsnachbildung *f*
b. of accounts *(Ök)* Rechnungsabschluß *m*
b. of an operational amplifier *(BMSR)* Abgleichen *n* eines Rechenverstärkers *(eine Art Nullabgleich in Analogrechnern)*
b. of masses Massenausgleich *m* *(Erdbau, Tiefbau)*
b. of moments Momentenausgleich *m* *(Statik)*
b. of quantities *s.* b. of masses
b. of the valve *(Wärm)* Ventilentlastung *f*
b. relais Differentialrelais *n*
b. reservoir *(Bw)* Ausgleichbecken *n*
b. resistance *(El)* Ausgleichwiderstand *m*
b. resistor *(El)* Abgleichwiderstand *m*; Symmetrierwiderstand *m*
b. speed *(El)* Beharrungsgeschwindigkeit *f*
b. spring Ausgleichfeder *f*
b. stand Auswuchtbock *m*; Auswuchtwaage *f*
b. transformer Ausgleichtransformator *m*
b. voltage Kompensationsspannung *f*
b. washer Auswuchtscheibe *f* *(für Räder)*
b. weight Auswuchtstein *m*; Ausgleichgewicht *n*, Gegengewicht *n*
balata Balata *f*
balcony Balkon *m*, Altan *m*; Galerie *f*; Rang *m* *(Theaterbau)*
b. access block *(Bw)* Außenganghaus *n*, Laubenganghaus *n*
baldacchino *s.* baldachin
baldachin *(Bw)* Baldachin *m*
baldaquin *s.* baldachin
bale / to zu Ballen pressen
bale 1. Ballen *m*; 2. *(Schiff)* Bügel *m* *(am Mast)*
b. breaker *(Text)* Ballenbrecher *m*
b. breaker with travelling lattices *(Text)* Ballenbrecher *m* mit Lattentüchern
b. capacity *(Schiff)* Ballenladeraum *m*
b. cargo Ballenladung *f*
b. chute Ballenschurre *f*
b. cubic capacity *(Schiff)* Laderauminhalt *m* für Ballen
b. dyeing *(Text)* Färben *n* von Gewebe ohne vorheriges Waschen
b. handler Ballenförderer *m*
b. handling system Ballen[be]förderungssystem *n*
b. loader Ballenverlader *m*
b. sledge Ballenschlitten *m*
b. tara *(Text)* Baumwolltara *f*
b. transporter Ballenförderer *m*
b. truck *(Text)* Ballenkarre *f*
baled dry pulp getrockneter Zellstoff *m* in Ballen
b. fibres in Ballen verpackte Fasern *fpl*, zu Ballen gepreßte Fasern
baling capacity Ballenleistung *f* *(Sammelpresse)*
b. material Preßgut *n* *(Sammelpresse)*
b. press 1. *(Lw)* Ballenpresse *f*, Räumpresse *f*, Aufsammelpresse *f*; 2. *(Met)* Paketierpresse *f*
balize *(Schiff)* Seezeichen *n*
balk *(Bw)* 1. Kantholz *n* *(größer als 15 × 15 cm)*; 2. zwischen zwei Ausschachtungen stehengebliebener Boden *m*
b. ring Sperring *m*
ball 1. Kugel *f*; 2. Ball *m*; 3. *(Met)* Luppe *f*
b.-and-disk integrator *(BMSR)* Reibradintegrator *m*, Gonella-Trieb *m*
b. and socket bearing Kugelgelenklager *n*
b. and socket gear change *(Kfz)* Kugelschaltung *f*
b. and socket joint Kugelgelenk *n*, Kugelkupplung *f*, Kugelscharnier *n*, Universalgelenk *n*
b. and socket joint wrench Kugelgelenkschlüssel *m*
b. and socket type gear shifting *(Kfz)* Kugelschaltung *f*
b. baffle *(BMSR)* Kugelsteuerschieber *m*
b. bearing Kugellager[ung *f*] *n*
b.-bearing cage Kugellagerkäfig *m*
b.-bearing cup Kugellagergehäuse *n*
b.-bearing grease Kugellagerfett *n*
b.-bearing housing Kugellagergehäuse *n*
b.-bearing inner race Kugellagerinnenring *m*
b.-bearing outer race Kugellageraußenring
b.-bearing pillow block Kugelstehlager *n*, Stehkugellager *n*

b.-bearing race Kugellager[lauf]ring *m*
b.-bearing retainer Kugellagerringkäfig *m*
b.-bearing seat Kugellagersitz *m*
b.-bearing spacer Kugellagerringkäfig *m*
b.-bearing spindle in Kugellagern laufende Schleifspindel *f*
b. cage Kugelkäfig *m*
b. check valve Kugelrückschlagventil *n*
b. clay Töpferton *m*
b. condenser Kugelkühler *m*
b. crank Kugelkurbel *f*, Handkurbel *f*
b. crusher Kugelmühle *f*
b. cup *(Masch)* Kugelpfanne *f*, Kugelschale *f*
b. diameter Kugeldurchmesser *m*
b.-end cutter *(Masch)* Fräser *m* mit runder Stirn
b.-end end mill *(Masch)* Schaftfräser *m* mit runder Stirn
b.-ended connecting rod Kolbenstange *f* mit Kugelkopf, Druckstelze *f* mit Kugelkopf *(Presse)*
b. face Hammerbahn *f*
b. faced hammer Kugelhammer *m*
b. gauge Kugellehre *f*
b. groove *(Masch)* Kugellaufrille *f*
b. handle Ballengriff *m*; Kugelgriff *m*
b.-handle crank Handkurbel *f*, Kugelkurbel *f*
b.-hardness test Kugeldruckprüfung *f*
b. head for breaking glass Glasbrecher *m* *(Glasschneider)*
b.-headed strain insulator Kugelkopfabspannisolator *m*
b. impact hardness tester Kugelschlaghärteprüfer *m*
b. impact test Kugelschlaghärteprüfung *f*
b. impression Kugeleindruck *m*, Eindruckkalotte *f* *(Kugelhärteprüfung)*
b.-impression hardness Brinellhärte *f*, Kugeldruckhärte *f*
b. inclinometer *(Masch)* Kugelneigungsmesser *m*
b.-indentation test Kugeldruckhärteuntersuchung *f*, Kugeldruckhärteprüfung *f*, Brinellhärteprüfung *f*
b.-indentation testing apparatus Kugeldruckhärteprüfer *m*
b. iron Luppeneisen *n*
b. joint Kugelgelenk *n*
b. journal Kugelzapfen *m*
b. journal [bearing] Radialkugellager *n*
b. level gauge Schwimmerfüllstandsmesser *m*
b. lever Kugelhebel *m*
b. lightning Kugelblitz *m*
b. lock Kugelverblockung *f*, Kugelverschluß *m*
b. mill Kugelmühle *f*
b. mill with screen walls Siebkugelmühle *f*
b. pane gewölbte Hammerbahn *f*
b. penetrator Brinellkugel *f*
b. pin Kugelzapfen *m*, Kugelstück *n*
b. point Kugelkuppe *f*, kugelförmiger Taststift *m*; Kugelspitze *f* *(Kugelschreiber)*
b.-point pen Kugelschreiber *m*
b.-pointed contact tip *(Opt)* Meßhütchen *n* mit Kugelfläche
b. pump Kugelventilpumpe *f*
b. race *(Masch)* Kugellagerring *m*, Laufring *m*, Kugellaufkäfig *m*, Kugellaufsystem *n*
b. reception *(Fmt, Fs)* Ballempfang *m*
b.-recirculating steering gear *(Kfz)* Lenkung *f* mit Kugelring
b. reflector lamp Kugelspiegelleuchte *f*
b. resolver *(BMSR)* Kugelkoordinatenwandler *m*
b. retainer *(Masch)* Kugelkäfig *m*
b.-retainer ring *(Masch)* Ringkäfig *m*
b.-retainer valve *(Masch)* Kugelrückschlagventil *n*
b. screw Gewindespindel *f* mit Kugelmutter
b. seat *(Masch)* Kugelpfanne *f*, Kugelsitz *m*
b.-shaped kugelförmig, kugelig
b. socket *(Masch)* Kugelpfanne *f*, Kugelschale *f*
b. spacer Ringkäfig *m*
b. spark gap Kugelfunkenstrecke *f*
b. squeezer *(Met)* Luppenquetsche *f*
b. tap wrench Kugelwindeisen *n*
b. test Kugeldruckprüfung *f*
b. thrust bearing Druckkugellager *n*, Kugellängslager *n*, Längskugellager *n*
b. thrust hardness Kugeldruckhärte *f*
b. turning Kugeldrehen *n*
b. turning attachment Runddrehapparat *m*; Kugeldrehapparat *m*
b. turning lathe Kugeldrehmaschine *f*
b. turning rest Kugel[dreh]support *m*
b. valve Kugelverschluß *m*; Kugelventil *n*
b. viscosimeter Kugelviskosimeter *n*
b. warp beaming machine *(Text)* Bäummaschine *f* für gewickelte Ketten
ballast / to *(Bw)* [be]schottern
ballast 1. *(Bw)* Bettung *f*; Schotter *m*, Kiesschotter *m*, Steinschotter *m* *(Gleisbau, Straßenbau)*; 2. *(Bw)* Grobzuschlagstoffe *mpl* *(Beton)*; 3. Drossel *(Leuchtstofflampen)*; 4. Ballast
b. bed Schotterbett *n*
b. bulb *(El)* Vorwiderstand[slampe *f*] *m*
b. cleaner Bettungsreinigungsmaschine *f*
b. concrete Schotterbeton *m*
b. devices *(Schiff)* Ballasteinrichtungen *fpl*
b. donkey pump *(Schiff)* Hilfsballastpumpe *f*
b. fork Steinschlaggabel *f*
b. lamp Widerstandslampe *f*
b. leakage *(El)* Bettungsableitung *f*

b. resistance *(El)* Ballastwiderstand *m*, Vorwiderstand *m*, Belastungswiderstand *m (elektrische Größe)*
b. resistor Lastwiderstand *m*
b. tube *(Am) (El)* Ballaströhre *f*; Widerstandsröhre *f*
b. unit *(El)* Vorschaltgerät *n (z. B. bei Entladungslampen)*
b. voyage *(Schiff)* Ballastfahrt *f*
b. water Ballastwasser *n*
b. water-line *(Schiff)* Ballastwasserlinie *f*, Leichtwasserlinie *f*
ballasting Beschottern *n*, Schotterung *f*
b. wagon Schotterwagen *m*
balling 1. *(Met)* Luppen[machen] *n*, Zusammenballung *f*; 2. *(Text)* Knäuelwickeln *n*
b. furnace *(Met)* Schweißofen *m (für Rohschienenpakete)*
b. machine *(Text)* Knäuelwickelmaschine *f*
ballistic ballistisch
b. deflection error Beschleunigungsfehler *m (Kreiselkompaß)*
b. factor Überschwingung *f (eines Meßinstrumentes)*
b. galvanometer ballistisches Galvanometer *n*, Stoßgalvanometer *n*
b. pendulum ballistisches Pendel *n*
b. rocket ballistische Rakete *f*
ballistics *(Ph)* Ballistik *f*
b. manual *(Mil)* Schießvorschrift *f*
ballonnet Ballonet *n*
balloon Ballon *m*
b. ascent Ballonaufstieg *m*
b. barrage *(Mil)* Ballonsperre *f*
b.-borne rocket von einem Ballon aus gestartete Rakete *f*
b. control ring *(Text)* Ballonkontrollring *m*
b.-launched rocket von einem Ballon aus gestartete Rakete *f*
b. launching *(Rak)* Ballonstart *m*
b. sail Ballonklüver *m*, Ballonflieger *m*, Ballonsegel *n (Jacht)*
b.-supported rocket von einem Ballon aus gestartete Rakete *f*
b. theodolite Ballontheodolit *m*
b. tyre Ballonreifen *m*
b. without crew unbemannter Ballon *m*
ballooning of the yarn *(Text)* Ballonbildung *f*, Schleierbildung *f* des Fadens
balm oil Melissenöl *n*
Balmer discontinuity Balmersprung *m (H-Spektrum)*
balsam fir Balsamtanne *f*
b. of fir Kanadabalsam *m*
b. of Peru Perubalsam *m*
b. of Tolu Tolubalsam *m*
balsawood Balsa[holz] *n*, westindisches Korkholz *n*, Floßbaumholz *n*
Baltimore truss Fachwerkbinder *m* für weitgespannte Flachdächer
baluster Geländerstab *m*; Geländersprosse *f*, Geländerpfosten *m*
balustrade Geländer *n*; Brüstung *f*
bamboo [cane] Bambus[rohr *n*] *m*
banana carrier Bananenschiff *n*
b. jack Bananenbuchse *f*, Bananenstecker *m*
b. pin Bananenstecker *m*
b. plug Bananenstecker *m*
band 1. *(Masch)* Zugband *n*, Band *n*; Blatt *n (einer Säge)*; 2. *(Bw)* Böschung *f*; Damm *m*, Erddamm *m*, Wall *m*; Ufer *n*; 3. *(Ph)* Streifen *m*, Band, Zeile *f*; 4. *(Geol)* Band *(dünne Lage)*; 5. *(Text)* Baumwollband *n*, Band, Bund *m*
b.-armoured cable *(El)* bewehrtes Kabel *n*, Kabel mit Stahlbandbewehrung
b. brake *(Masch, Wärm)* Bandbremse *f*
b. coil Spule *f* für ein Wellenband
b. conveyor Bandförderer *m*, Förderband *n*, Gurtförderer *m*
b. elimination *(El)* Bandunterdrückung *f*
b. elimination filter *(El)* Bandsperre *f*
b. fillet Zierleiste *f*
b. filter *(El)* Bandfilter *n*, Bandpaß *m*
b. galvanometer Bändchengalvanometer *n*
b. head Bandenkopf *m (Spektrum)*
b. losses *(El)* Bandagenverluste *mpl*, Verluste *mpl* durch Bewehrung
b. mill train Bandeisenstraße *f*
b. moulding Zierleiste *f*
b. of crystal Kristallstreifen *m*
b. of frequencies Frequenzbereich *m*, Wellenband *n*
b. pass *(El, Fmt)* Durchlaßbreite *f*
b.-pass characteristic *(El, Fmt)* Durchlaßkurve *f*
b.-pass condenser Bandfilterkondensator *m*
b.-pass filter *(Fmt)* Bandfilter *n*
b.-pass response *(El, Fmt)* Bandfilterresonanzkurve *f*
b.-pass width of a filter *(El, Fmt)* Durchlaßbereich *m* eines Bandfilters
b. permeability Bandendurchlässigkeit *f*
b. progression Bandenfolge *f (Spektrum)*
b. rejection filter *(Fmt)* Bandsperre *f*
b. rope Flachseil *n*, Bandseil *n*
b. saw Bandsäge *f*
b. saw blade Bandsägeblatt *n*
b. saw file Bandsägenfeile *f*
b. saw for logs *(Holz)* Blockbandsäge *f*
b. saw mill Blockbandsäge *f*
b.-sawing machine designed for friction cutting Schmelzschnittbandsäge *f*
b. selector *(Fmt, Fs)* Bereichwähler *m*; Wellenschalter *m*
b. series Bandenserie *f (Spektrum)*

b. spectrum Bandenspektrum *n*
b. spread *(Fmt)* Wellenbanddehnung *f*
b. spreading Banddehnung *f*
b. steel Bandstahl *m*
b.-stop filter Bandsperrfilter *n*
b. stretcher Spannschnur *f*, Spannstrick *m*
b. switch *(Fmt)* Bereichs[um]schalter *m*; Wellen[um]schalter *m*
b. system *(Ph)* Bandensystem *n*, Bandenspektrum *n*
b. turning *(Masch)* Ringdrehen *n*
b. wheel Antriebswelle *f*, Hauptwelle *f* *(Seilbohren)*
bandage process Bandagenverfahren *n* *(Holzkonservierung)*
banded *(Geol)* gebändert
b. bituminous coal Streifenkohle *f*
b. coal Schieferkohle *f*, Streifenkohle *f*
b. filter *(Ch)* Streifenfilter *n*
b. structure *(Min)* Lagentextur *f*; Zeilenstruktur *f*
banding core *(Holz)* Umleimer *m* *(Furnier)*
bandknife grinding machine *(Led)* Bandmesserschleifmaschine *f*
b. splitting machine *(Led)* Bandmesserspaltmaschine *f*
bandless beater roll *(Holz)* bänderlose Holländerwalze *f*
bandwidth *(El, Fmt)* Bandbreite *f*; Durchlaßbereich *m*; Frequenzdurchlaßbereich *m* *(Filter)*; Frequenzgebiet *n*
b. characteristic *(El)* Frequenzcharakteristik *f* *(Bandfilter)*
banister *(Bw)* Geländersprosse *f*, Geländerstab *m*; Geländerpfosten *m*
banjo axle *(Kfz)* Banjoachse *f*
b. fitting Hohlschraubenverbindung *f*
bank / to *(Met)* dämpfen *(Hochofen)*; *(Met)* stampfen *(Herdsohle)*; in Schräglage fliegen (fahren); überhöhen
bank 1. Damm *m*, Fahrdamm *m* *(Straße)*; 2. *(Flg)* Kurvenlage *f*, Querlage *f*; Schräglage *f* *(des Flugzeuges in der Kurve)*; 3. Überhöhung *f* *(Kurve)*; 4. Abbaustoß *m*, Ortsstoß *m* *(Kohlenbergbau)*; 5. Arbeitstisch *m*; 6. Ufer[böschung *f*] *n*; 7. Flußufer *n*; 8. Reihe *f*
b. bill Kassenanweisung *f*, Kassenschein *m*
b. cable *(El)* Vielfachkabel *n*, Bündelkabel *n*
b. commission Bankprovision *f*
b. crest Strossenkante *f* *(Tagebau)*
b. cubicmetres of overburden Kubikmeter *npl* gewachsener Abraum
b. edge Strossenkante *f* *(Tagebau)*
b. engine Schiebelokomotive *f*
b. head *(Brg)* obere Anschlagbühne *f* *(Bremsberg)*
b. height *(Brg)* Strossenhöhe *f*
b. indicator 1. Neigungszeiger *m*; 2. *(Flg)* Querneigungsmesser *m*, Schräglageanzeiger *m*
b. line Uferlinie *f*
b. of a river Flußufer *n*
b. of capacitors Kondensatorenbank *f*, Kondensatorenbatterie *f*
b. of clay Sandbank *f*, Watt *n*, Bank *f*
b. of contacts *(El)* Kontaktkranz *m*
b. of cylinders Zylinderreihe *f*
b. of ditch Grabenböschung *f*
b. of keys Schalterreihe *f*; Tastenreihe *f*
b. of lights Lampenaggregat *n*
b. of sand Sandbank *f*, Watt *n*, Bank *f*
b. of transformers Gruppentransformator *m*, Transformatorenbank *f*; Transformatorenbatterie *f*, Transformatorengruppe *f*
b. proof machine *(Am)* *(Büro)* Scheckabrechnungsmaschine *f*
b. protection Uferschutz *m*
b. stabilization Uferbefestigung *f*
b. toe *(Brg)* Strossenfuß *m*
b. winding *(El, Fmt)* Stufenwicklung *f*
b.-wound *(El)* kapazitätsarm; stufengewickelt
banked blast furnace gedämpfter Hochofen *m*
b. winding *(Elm)* kapazitätsarme Wicklung *f*
banking 1. Überhöhung *f* *(Straßenbau)*; 2. *(Brg)* Abziehen *n* von Förderwagen auf der Hängebank; 3. Nachschiebefahrt *f*
b. industry Bankgewerbe *n*
b. out *(Brg)* Abziehen *n*
bankman *(Brg)* Hängebankanschläger *m*
banksman *(Brg)* Anschläger *m*
banquet Fensterbank *f*, Fenstersitz *m*
banquette Absatz *m*, Bankett *n*, Gehweg *m* *(Straße)*
bantam tube *(Am)* *s.* miniature valve
bar / to [ver]schließen, sperren, blockieren
b. down bereißen
bar 1. Bar *n* *(physikalische Einheit)*; Glied *n* *(Kinematik)*; 2. Stange *f*, Stab *m*, Traverse *f*, Querbalken *m*; Sprosse *f*; Führungsschiene *f*, Lineal *n*; Maßstab *m*, Maßschiene *f*; 3. Barren *m*, Balken *m*, Schiene *f*, Block *m*; 4. *(Bw)* Fenstersprosse *f*; 5. *(Brg)* Schrämausleger *m*, Schrämarm *m*; 6. Holländermesser *n* *(Papierherstellung)*; 7. Sandbank *f*; 8. [Barren-]Holm *m* *(Sport)*; 9. *(Math)* Strich *m*, Querstrich *m*
b.-and-post transformer *s.* cross-bar transformer
b. automatic Stangendrehautomat *m*
b. belt *(Brg)* Stabrostband *n*
b. bender Biegemaschine *f*, Biegetisch *m* *(Stahlbeton, Bewehrungsbau)*
b. bending machine Biegemaschine *f* *(Stahlbeton, Bewehrungsbau)*
b. buoy *(Schiff)* Barretonne *f*
b. capacity Stangendurchlaß *m* durch Spindel *(Drehautomat)*

b. chain *(Bw)* Stabzug *m*
b. chart Stabdiagramm *n*; Säulendiagramm *n*; Rechteckdiagramm *n*; Balkendiagramm *n*
b. code Streifenkode *m*, Strichkode *m (automatische Zeichenerkennung)*
b. coupling panel *(El)* Sammelschienenkuppelfeld *n*
b. cropping machine Knüppelkaltschere *f*
b.-cutting machine Blockschere *f*
b. draught Barretiefe *f (Niedrigwasserstand über einer Barre)*
b. drawing Stopfenzug *m (zur Herstellung nahtloser Rohre)*
b.-drawing bench *(Met)* Stangenziehbank *f*
b. electrode Stabelektrode *f*
b. fastener Klemmstück *n (Bohrstahlhalter)*
b. feed Stangenvorschub *m*
b. floor *(Bw)* Rohdecke *f*
b. folding and creasing Falzen *n* und Abkanten *n (von Folien)*
b. form Stangenform *f*
b. frame *(Masch)* Barrenrahmen *m*, Stabeisenrahmen *m*
b. generator *(Fs)* Balkengenerator *m*
b. grinding Stangenschleifen *n*
b. guide *(Masch)* Schienenführung *f*
b. joist *(Bw)* Gitterträger *m*
b. keel *(Schiff)* Balkenkiel *m*
b. magnet Stabmagnet *m*
b. mat Bewehrungsmatte *f (Stahlbetonbau)*
b. mill Stabeisenwalzwerk *n*; Stabeisenstraße *f*
b. moulding machine Seifenstrangpresse *f*
b.-mounted drifter *(Brg)* Säulenbohrmaschine *f*
b. of a crossing Schlagbaum *m (Übergang)*
b. of a horizontal grate *(Wärm)* Planroststab *m*
b. of an evener Wagenscheit *n* einer Wagenbracke *(Stellmacherei)*
b. pattern *(Fs)* Balkenmuster *n*
b. pilot *(Schiff)* Barrelotse *m*
b. port *(Schiff)* Barrehafen *m*
b. rack *(Masch)* Stangenlagergestell *n*
b. reinforcement Stabbewehrung *f (Stahlbeton)*
b. sealing *(Plast)* Hochfrequenzschweißen *n*
b.-shaped battery *(El)* Stabbatterie *f*
b. shears Knüppelschere *f*
b. shoe *(Brg)* Kappschuh *m*
b. silver Barrensilber *n*
b. skimming *(Masch)* Schnelldrehen *n* von rohem Stangenmaterial, Stangenschnelldrehen *n*
b. soap Stückseife *f*
b. spacer Abstandhalter *m (Stahlbeton)*
b. steadier Stangenwiderlager *n (Rohrwalzwerk)*
b. steel Stabstahl *m*
b. stem *(Schiff)* Balkenvorsteven *m*
b. stock *(Masch)* Stabmaterial *n*, Stangenwerkstoff *m*, Werkstoffstangen *fpl*
b. timbering *(Brg)* Kappenausbau *m*
b. tin Stangenzinn *n*
b. to be tested Probestab *m*
b. turning tool Vierkant[schrupp]drehmeißel *m*
b.-type armature *(Elm)* Stabanker *m*
b.-type automatic lathe Stangenhalbautomat *m*
b.-type [current] transformer Schienenstromwandler *m*, Stabstromwandler *m*; Durchsteckstromwandler *m*
b. voltage *(Am)* Segmentspannung *f*, Lamellenspannung *f*
b. winding Stabwicklung *f*
b. wound armature *(Elm)* Anker *m* mit Stabwicklung, Stabanker *m (gewickelt)*
b. zinc Barrenzink *n*
barb Widerhaken *m*; Grat *m*, Bart *m*
barbed hook Widerhaken *m*
b. wire Stacheldraht *m*
b. wire cutter Stacheldrahtschneidezange *f*
barbituric acid Barbitursäure *f*
barbotage Gasspülung *f*
bare / to *(El)* abisolieren, die Isolierung entfernen, abmanteln
bare blank *(Draht usw.)*
b. boat charter Charter *m* eines Schiffes *(ohne Mannschaft)*
b. cable blankes Kabel *n*
b. electrode nackte (blanke, mantellose) Elektrode *f*
b. hull charter Charter *m* eines Schiffes *(ohne Mannschaft)*
b. hyperon *(Kern)* nacktes Hyperon *n*
b. metallic welding wire Blankdraht *m*
b. nucleon *(Kern)* Nukleon *m*
b. pile *s.* b. reactor
b. pipe nacktes Rohr *n*
b. pole charter Charter *m* eines Schiffes *(ohne Mannschaft)*
b. reactor *(Kern)* unreflektierter (nackter) Reaktor *m*
b. wire Blankdraht *m*, blanker Leiter *m*
b.-wire welding *s.* unshielded welding
b. wiring *(El)* Blankverdrahtung *f*
barefooted well Bohrung *f* mit unverrohrtem Ölträger, Bohrung ohne Filterrohre *(Erdöl)*
barette file Barettfeile *f*
barge Schleppkahn *m*; Lastkahn *m*, Schute *f*, Prahm *m*, Frachtkahn *m*, Schaluppe *f*, Leichter *m*, Schubkahn *m*, Schubleichter *m*, Schubprahm *m*
b. course *(Bw)* Rollschicht *f (Mauerwerk)*; Ortrolle *f (Dach)*

b. lift Schiffshebeanlage *f*
b. operation *(Schiff)* Schleppbetrieb *m*
b. train *(Schiff)* Schleppzug *m*; Schubverband *m*, Schubeinheit *f*
bargee *(Schiff)* Ewerführer *m*, Schutenführer *m*
baric 1. barisch *(Luftdruck)*; 2. bariumhaltig
barite *(Am)* *s.* baryte
barium Barium *n*
b. carbonate Bariumkarbonat *n*
b. concrete Bariumbeton *m*
b. getter Bariumgettermetall *n*
bark / to 1. *(Holz)* abborken, abrinden, entborken, schälen; 2. *(Led)* gerben
b. tan *(Led)* mit Lohe gerben
bark 1. *(Holz)* Borke *f*, Rinde *f*; 2. *(Led)* Lohe *f*, Rinde; 3. Barke *f*, Barkschiff *n*, Bark *f* *(Dreimastschiff: Fock-, Groß- und Besanmast)*
b.-bared *(Holz)* abgeschält, entrindet
b.-blazer *(Holz)* Reißhaken *m*
b. breaker *(Led)* Lohbrechmaschine *f*
b. breaking machine *s.* b. breaker
b. crêpe *(Text)* Borkenkrepp *m*
b. edge Säumling *m*
b.-free *(Holz)* rindenfrei
b. mill Lohmühle *f*
b. paper Bastpapier *n*
b. peeling machine *(Holz)* Rindenschälmaschine *f*
b. scorching *(Holz)* Rindenbrand *m*
b. scraper Schäleisen *n*
b. specks Rindenflecke *mpl (Papier)*
b. tannage Lohgerbung *f*
b. tanned side leather lohgares Rindoberleder *n*
b. tanner *(Led)* Lohgerber *m*, Rotgerber *m*
barked and basted *(Holz)* entrindet und entbastet, weißgeschält
b. fracture *(Met)* splitterige Bruchfläche *f*
barkentine, barkentyne *(Schiff)* Barkschoner *m*, Schonerbark *f*, Barkantine *f*, Barkentine *f*
barker *(Holz)* Rindenschäler *m*; Rindenschälmaschine *f*
Barkhausen effect Barkhausen-Effekt *m*
Barkhausen jump Barkhausen-Sprung *m (Magnetisierung)*
Barkhausen-Kurz circuit *(El)* Barkhausen-Kurz-Schaltung *f*, Bremsfeldschaltung *f*
Barkhausen-Kurz oscillations Barkhausen-Kurz-Schwingungen *fpl*; Bremsfeldschwingungen *fpl*
Barkhausen-Kurz valve *(El)* Bremsfeldröhre *f*
barking *(Holz)* Entrinden *n*, Schälen *n*
b. drum *(Holz)* Entrindungstrommel *f*
b. iron *(Holz)* Rindenschäler *m*
barley 1. Gerste *f*; 2. gesiebter Anthrazitgrus *m*
b. coal gesiebter Anthrazitgrus *m*
b. malt Gerstenmalz *n*
b. shag Gerstenstroh *n*
barleycorn weave *(Text)* Gerstenkornbindung *f*
Barlow's wheel Barlow-Rad *n*
barn 1. Scheune *f*, Scheuer *f*; 2. Barn *n* *(physikalische Einheit)*
b. floor Scheunentenne *f*
barnacles Bewuchs *m (am Schiffskörper)*, Seepocken *fpl*, Barnackels *pl*, Anwuchs *m*
baroclinic baroklin
b. equilibrium *(Ph)* baroklines Gleichgewicht *n*
b. field *(Ph)* baroklines Feld *n*
b. fluid *(Ph)* barokline Flüssigkeit *f*
barograph Barograf *m*, Luftdruckschreiber *m*
barometer Barometer *n*
b. reading Barometerstand *m*
barometric barometrisch
b. column Quecksilbersäule *f*
b. discharge pipe *(Wärm)* barometrisches Abfallrohr *n*
b. fluctuation Luftdruckschwankung *f*
b. height barometrische Höhe *f*
b. levelling barometrische Höhenmessung *f*
b. measurement of altitude barometrische Höhenmessung *f*
b. pressure Luftdruck *m*, Normaldruck *m*
b. reading Barometerstand *m*
b. thermometer Siedethermometer *n*
barometry Luftdruckmessung *f*
baroscope Baroskop *n*
barotropic barotrop
b. equation *(Ph)* barotrope Zustandsgleichung *f*
b. equilibrium barotropes Gleichgewicht *n*
b. field Barotropiefeld *n*, barotropes Feld *n*
b. flow barotrope Strömung *f*
b. fluid barotrope Flüssigkeit *f*
barque *s.* bark
barquentine *s.* barkentine
barracks ship Wohnschiff *n*
barrage 1. *(Bw)* Sperrmauer *f*, Staudamm *m*, Staumauer *f*, Talsperre *f*; Stauwehr *n*; 2. *(Mil)* Sperrfeuer *n*
b. balloon *(Mil)* Sperrballon *m*
b. dam Sperrmauer *f*, Staudamm *m*, Staumauer *f*, Talsperre *f*
b.-fixe Steinstaumauer *f*
b. jamming Sperrstörung *f (Radar)*
b.-mobile Staudamm *m* mit Schleusen
barras *(Holz)* Barras *m*, Gal[l]ipot *m*, Scharrharz *n*
barre *(Text)* 1. Streifen *m (Fehler)*; 2. Gewebe *n* aus einfarbiger Kette und mehrfarbigem Schuß
barred speed range *(Masch)* gesperrter Drehzahlbereich *m*
barrel 1. Faß *n*; Tonne *f*; 2. *(Masch)* Walze *f*, Zylinder *m*; Trommel *f*; Hülse *f*, Schafthülse *f*; 3. *(Mil)* Lauf *m (einer Waffe)*; 4. *(Einheit, Zement)* = 376 lbs

b. bow *(Schiff)* Rundbug *m*, runder (abgerundeter) Plattenbug *m*
b. burst *(Mil)* Rohrkrepierer *m*
b. cam Mantelkurve *f*
b. cleaner *(Mil)* Rohrreiniger *m*
b. cleaning and reconditioning plant Faßreinigungs- und -aufbereitungsanlage *f*
b. controller *(El)* Steuerwalze *f*, Meisterwalze *f (Schalter)*
b. converter *(Met)* Trommelkonverter *m*, liegender Konverter *m*
b. cooling Rohrkühlung *f*
b. crozing machine *(Holz)* Faßkrösemaschine *f*
b. distortion *(Opt, Fs)* tonnenförmige Verzeichnung *f*
b. electro-plating Trommelgalvanisierung *f*
b. head Faßboden *m*
b. hoist Faßaufzug *m*
b. hoop Faßreifen *m*
b. lining Zylinderauskleidung *f*
b. mixer Trommelmischer *m*
b. of a pump Pumpenstiefel *m*
b. organ Drehorgel *f*
b. pitching machine Faßpichmaschine *f*
b. pump Faßpumpe *f*
b. roller bearing *(Masch)* Tonnenlager *n*
b. roof Tonnendach *n*
b. saw Faßsäge *f*
b.-shaped tonnenförmig, bauchig
b.-shaped furnace Trommelofen *m*
b.-shaped roller bearing *(Masch)* Tonnenlager *n*
b. shell *(Bw)* Tonnenschale *f*
b. sirup Raffineriemelasse *f*
b. sleeve Zylinderauskleidung *f*
b. stave Faßdaube *f*, Binderholz *n*
b. switch Walzenschalter *m*
b.-type roller bearing *(Masch)* Tonnenlager
b. vault *(Bw)* Tonnengewölbe *n*
b.-vault shell *(Bw)* Tonnenschale *f*
b. wheel Walzenrad *n*
b. winding *(Elm)* Trommelwicklung *f*, Korbwicklung *f*, Faßwicklung *f*
barren unfruchtbar
b. ground *(Geol)* taubes Gestein *n*
b. rock Abraum *m*
barretter *(El)* Ballastwiderstand *m*, Barretter *m*
barrier 1. *(Ph)* Barriere *f*, Potentialwall *m*, Wall *m*, Berg *m*; 2. Schranke *f*, Schlagbaum *m (Übergang)*; 3. *(El)* Isolation *f (z. B. der Wicklung)*, Isolierstoffschirm *m*
b. containing fixed space charges *(El)* Raumladungsrandschicht *f*
b. diffusion Diffusion *f* durch poröse Wände, Trennwanddiffusion *f*
b. factor *(Ph)* Durchlässigkeit[sfaktor *m*] *f* des Potentialwalles
b. filter *(Opt)* Sperrfilter *n*
b. grid tube Sperrgitterröhre *f*
b. height *(Ph)* Höhe *f* des Potentialwalls
b. layer Halbleiterschicht *f*, Sperrschicht *f*
b.-layer capacitance Kapazität *f* einer Sperrschicht
b.-layer cell Halbleiterfotozelle *f*, Fotoelement *n*; Sperrschichtfotozelle *f*
b.-layer photocell Sperrschichtfotozelle *f*
b.-layer photoelectric cell *s.* b.-layer photocell
b. of potential *(Ph)* Potentialschwelle *f*
b. penetration *(Ph)* Durchdringung (Durchlässigkeit) *f* des Potentialwalles
b. permeability *s.* b. penetration
b. pillar *(Brg)* Bergfeste *f*
b. plane photocell Halbleiterfotozelle *f*, Sperrschichtfotozelle *f*
barring *(Brg)* Einstrich *m*, Joch *n*
b. down *(Brg)* Bereißen *n*
b. engine *(Wärm)* Drehmaschine *f*, Schaltmaschine *f*
b. gear *(Wärm)* Drehvorrichtung *f*, Stellwerk *n*, Schaltwerk *n (Schwungrad)*
b. motor Durchdrehmotor *m*
barrow charging *(Met)* Karrenbegichtung *f*, Begichtung *f* von Hand
b. man *(Brg)* Schlepper *m*, Fördermann *m*
barry *(Text)* streifig *(Fehler)*
b. mark *(Text)* Streifen *m (Fehler)*
barvoys washer *(Brg)* Sophia-Jakobs-Verfahren *n*
barwood Sandelholz *n*
barycentre *(Ph)* Schwerpunkt *m*
barycentric baryzentrisch
barye *(Ph)* Mikrobar *n*, Dyn/cm^2
baryon *(Ph)* Baryon *n*
b. number *(Ph)* Baryonzahl *f*
baryt coating Barytage *f (Papier)*
baryta Baryt[erde *f*] *m*, Bariumoxid *n*, Schwererde *f*
b. water Barytwasser *n*
baryte Baryt *m*, Schwerspat *m*
b. concrete *(Bw)* Barytbeton *m (Strahlenschutzbeton)*
basal plane Grundebene *f*, Basisebene *f (Kristall)*
basalt Basalt *m*
b. nozzle inserted three-wing chisel *(Brg)* Dreiflügelmeißel *m* mit Basaltdüseneinlagen
basaltic Basalt-, basaltisch
b. iron ore Basalteisenerz *n*
b. lava Basaltlava *f*
bascule Basküle[verschluß *m*] *f*
b. bridge Klappbrücke *f*, Baskülbrücke *f*
base 1. *(Masch)* Boden *m*, Fußpunkt *m*, Grundfläche *f*, Grundlage *f*, Postament *n*, Sockel *m*, Unterbau *m*, Unterlage *f*, Untersatz *m*; 2. *(Masch)* Bett *n*, Unterteil *n*, Unter-

kasten *m*, Untergestell *n*, Grundplatte *f*; 3. *(Bw)* Sockel *m (Architektur)*; Tragschicht *f*, Unterbau *(Straßenbau)*; Fundament *n*; 4. *(Math)* Grundlinie *f (Geometrie)*; Basis *f (Zahlsysteme)*; 5. *(El)* Sockel; 6. *(Ch)* Base *f*; 7. Haftgrund *m (Farbanstrich)*

b. angle 1. Basiswinkel *m*; 2. *(Schiff)* unterer Gurtwinkel *m*, Grundwinkel *m (eines Schotts)*

b. block Rosette *f*, Fixrosette *f (Schalter usw.)*

b. circle Fußkreis *m (Zahnrad)*

b. component Grundkomponente *f*

b. connection *(El)* Röhrensockelschaltung *f*

b. contact *(El)* Basisanschluß *m*

b. course Straßenbefestigung *f*, Tragschicht *f*, Unterbau *m (Straßenbau)*

b. cylinder Fußkreiszylinder *m*, Grundzylinder *m*

b. diagram *(El)* Sockelschaltung *f*

b. diameter Fußkreisdurchmesser *m*

b. edges Basiskanten *fpl (von Kristallen)*

b. energizing Fußpunktspeisung *f (Antennenmast)*

b. exchange equipment Basenaustauscheranlage *f*

b. exchange softening *(Ch)* Basenaustauschverfahren *n (zur Speisewasserenthärtung)*

b. frame *(Masch)* Untergestell *n*

b. fuse *(Mil)* Bodenzünder *m*

b. helix angle Grundschrägungswinkel *m (Verzahnung)*

b. impedance Fußpunktwiderstand *m (von Antennen)*

b. insulator Stützisolator *m*, Fußpunktisolator *m (von Antennen)*

b. line Grundlinie *f*; Basislinie *f (Antennentechnik)*; Basis *f (Vermessung)*

b. material *(Bw)* Tragschichtmaterial *n*

b. measuring Basismessung *f*

b. metal 1. Grundmetall *n*; 2. unedles Metall *n*

b.-metal couple unedles Thermoelement *n*

b. molecule Grundmolekül *n*

b. of rail Schienenfuß *m*

b. of shaft Schaftsohle *f (Schornstein)*

b. of the wheel head Schleifspindelstockdrehteil *n*

b. of wheels *(Kfz)* Radstand *m*

b. ore armes Erz *n*

b. paper Rohpapier *n*

b. pay Grundlohn *m*, Grundgehalt *n*

b. period Basiszeitraum *m*

b. pin Sockelstift *m (Röhre)*

b. pitch Eingriffsteilung *f*, Grundkreisteilung *f (Zahnrad)*

b.-pitch tester Eingriffsteilungsmeßgerät *n (Zahnradprüfung)*

b. plate Fußplatte *f*, Grundplatte *f*, Sohlplatte *f*, Fundamentplatte *f*

b. plate of rail Unterlagsplatte *f (Schienen)*

b. post Ständer *m*

b. pressure *(Bw)* Sohldruck *m*, Sohlpressung *f*

b. pulp *(Holz)* Normalschliff *m*

b. register Indexregister *n*

b. resistance Basiswiderstand *m*

b. resistor *(El)* äußerer Basiswiderstand *m*

b. rim *(Kfz)* Grundfelge *f*

b. road *(Brg)* Grundstrecke *f*

b. shoe Ankerplatte *f (Mauerwerk)*

b. speed Nenndrehzahl *f*, Grunddrehzahl *f*

b. spreading resistance Basisbahnwiderstand *m*, Basisserienwiderstand *m (Halbleitertechnik)*

b. support 1. Trägerplatte *f (Lumineszenzschirm)*; 2. *(Rak)* Standholm *m*

b.-supported take-up *(Text)* im Maschinengestell gelagerter Abzug *m*

b. surface Grundfläche *f*; Basis *f*; Deckfläche *f*

b. surge *(Kern)* Basiswolke *f*

b. terminal *(El)* Basisanschluß *m*, Basiskontakt *m*

b.-three number Zahl *f* mit der Zahlenbasis Drei

b. with external contacts Außenkontaktsockel *m*

b. wood Blindholz *n*

baseband *(El)* Grundfrequenzbereich *m*, Signalfrequenzbereich *m*

baseboard 1. *(Bw)* Fußleiste *f*, Scheuerleiste *f*, Sockelleiste *f*; 2. *(Typ)* Rohkarton *m*

b.-layout *(El)* Brettschaltung *f*

basement [floor] Kellergeschoß *n*, Souterrain *n*

bash / to *(Brg)* versetzen *(Tiefbau)*

basic 1. Grund-; 2. *(Ch)* basisch

b. Bessemer (bessemer) pig iron Thomasroheisen *n*

b. Bessemer process Thomasverfahren *n*

b. Bessemer steel Thomasstahl *m*

b. bore *(Masch)* Einheitsbohrung *f*

b. cellulose Ausgangszellulose *f (Papierherstellung)*

b. circuit *(El)* Grundschaltung *f*

b. circuit diagram *(El)* Grundschaltbild *n*

b. clinker *(Wärm)* basische Schlacke *f*

b. colour Grundfarbe *f*

b. component Grundbestandteil *m*, Hauptbestandteil *m*

b. converter Thomasbirne *f*; basischer Konverter *m*

b. converter process Thomasverfahren *n*

b. converter steel Thomasstahl *m*

b. design Grundform *f*, typische Ausführung *f*

b. diameter Nenndurchmesser *m*

b. dimension *(Masch)* Nennmaß *n*, Normalmaß *n*
b. dye basischer Farbstoff *m*
b. engineering education technische Grundausbildung *f*
b. equation Grundgleichung *f*
b. feed Vorschubrichtwert *m*
b. form Grundform *f*
b. four-bar linkage Grundgelenkviereck *n*
b. frequency Grundfrequenz *f*
b. hole *(Masch)* Einheitsbohrung *f*
b.-hole system System *n* Einheitsbohrung
b. impulse insulation level *(El)* Grundstoßpegel *m*, Stoßpegel *m (Isolation)*
b. ingredient Grundbestandteil *m*
b. interference suppression *(El)* Grundentstörung *f*
b. ion *(Ch)* Kation *n*
b. law Grundgesetz *n (Naturwissenschaft)*
b. lead chromate basisches Bleichromat *n*; Chromrot *n*
b.-lined converter Thomasbirne *f*
b. lining *(Met)* basische Ausmauerung *f*
b. load Grundlast *f*
b. load rating Nennbelastung *f*, Nenn[grund]last *f*; Tragzahl *f (Wälzlager)*
b. machine type Grundmaschinentyp *m*
b. mat Grundbahn *f (Papier)*
b. mechanic Grundvorgang *m*
b. network *(El)* Grundschaltung *f*
b. noise *(El)* Eigenrauschen *n*
b. open-hearth process basisches [Siemens-] Martin-Verfahren *n*
b. open-hearth steel basischer Siemens-Martin-Stahl *m*
b. patent Grundpatent *n*
b. pig iron Thomasroheisen *n*; Phosphoreisen *n*
b. principle Grundprinzip *n*
b. profile Bezugsprofil *n*
b. Q *(El)* Leerlaufgütefaktor *m (einer elektrischen Impedanz)*
b. research Grundlagenforschung *f*, (reine) Erkundungsforschung *f*
b. rock basisches Gestein *n*
b. rule Grundregel *f*
b. shaft *(Masch)* Einheitswelle *f*
b.-shaft system System *n* Einheitswelle
b. shovel Grundbagger *m*
b. size Ausgangsmaß *n*, Nennmaß *n*, Normalmaß *n*; Grundmaß *n*
b. size gauge *(Masch)* Normallehre *f*, Urmaß *n*
b. skill Grundfertigkeit *f*
b. slag basische Schlacke *f*, Thomasschlacke *f*
b. slag practice basische Schlackenführung *f*
b. steel basischer Stahl *m*
b. structure *(Kfz)* Fahrwerk *n (ohne Motor)*, Unterbau *m*
b. substance Ausgangsstoff *m*
b. sulphate white lead Bleiweiß *n*; Bleisulfat *n*
b. technical knowledge technisches Grundwissen *n*
b. time Grundzeit *f*
b. training Grundausbildung *f*
b. unit Baueinheit *f*; Grundeinheit *f*
b. wage rate Grundlohn *m*
b. wiring *(El)* Grundschaltung *f*
b. workshop training Grundausbildung *f* in der Produktion
basicity *(Ch)* Basizität *f*
basifier *(Ch)* Basenbildner *m*
basifying *(Ch)* basenbildend
basils Basils *pl (Schafleder)*
basin 1. Becken *n*, Schüssel *f*; Bassin *n*; 2. Hafenbecken *n*; Binnenhafen *m*; Flußbecken *n*
basining *(Geol)* Beckenbildung *f*
basis of calculation Berechnungsgrundlage *f*
b. weight Flächengewicht *n* in lbs./Ries *(Brit.: zu 480 oder 500 Blatt, USA: zu 1000 Blatt)*
basket Korb *m*
b. centrifuge Korbzentrifuge *f*, Schälzentrifuge *f*
b. coil *(El)* Korb[boden]spule *f*
b. cook *(Holz)* Kochung *f* in Kochereinsätzen
b.-jig *(Brg)* Korbscheider *m*
b. stitch Korbstich *m (Stickerei)*
b. weave *(Text)* Würfelbindung *f*
b. winding *(Elm)* Faßwicklung *f*, Kettenwicklung *f*, Korbwicklung *f*
b. work Flechtwerk *n*
basketball basket (goal) Basketballkorb *m*
basketmakers' plier Korbmacherzange *f*
bass boost *(El)* Baßanhebung *f*
b. compensator *(El)* Baßentzerrer *m*
b. control *(El)* Tieftonblende *f*
b.-cut filter *(El)* Baßfilter *n*
b. drum große Trommel *f*
b. response Tieftonempfindlichkeit *f*
b.-treble control *(El)* Tonblende *f*
b. tuning *(El)* Baßabstimmung *f*
b. wood [amerikanische] Linde *f*, Lindenholz *n*
basset horn Bassethorn *n*
bassoon Fagott *n*
bast Bast *m*
b. coal Bastkohle *f*
b. fibre Bastfaser *f*
bastard file Bastardfeile *f*, Vorfeile *f*
b. mahogany Eukalyptus *m*, Jarrah *n*, australisches Mahagoni *n*, australische Roteiche *f*,
b. title Schmutztitel *m (eines Buches)*
baste / to *(Text)* reihen

baster Hefter *m*
basting Heftstich *m*
b. cotton *(Text)* Heftgarn *n*
b. thread *(Text)* Reihgarn *n*
bat 1. Schläger *m (Kricketschläger, Baseballschläger, Tischtennisschläger)*; 2. *(Bw)* Teil *m* eines Ziegelsteins *(z. B. Quartier, Kopf, Dreiquartier)*, Halbziegel *m*, Dreiviertelziegel *m*
batch / to 1. dosieren, zumessen *(Betontechnologie)*; 2. *(Text)* batschen
batch 1. Los *n*, Menge *f*, kleine Stückzahl *f*, Partie *f*, Posten *m*, Satz *m*; 2. *(Met)* Gicht *f*, Begichtung *f*, Möller *m*; 3. Mischung *f*, Mischgut *n (Betontechnologie)*
b. bin Chargierbunker *m*
b. boring Bohren *n* kleiner Lose
b. box Abmeßkasten *m*, Dosierkasten *m (Betontechnologie)*
b. counter 1. Gruppenzähler *m*; 2. Chargenzähler *m*
b. distillation Blasendestillation *f (Erdöl)*
b. furnace Ofen *m* für satzweise Beschickung; Stapelofen *m*, periodisch arbeitender Ofen
b. mixer Chargenmischer *m*
b. plant *(Bw)* Dosieranlage *f*; Mischstation *f (Betonherstellung)*
b. process Chargenbetrieb *m*, diskontinuierlicher Betrieb *m*; Chargenprozeß *m*, diskontinuierlicher Prozeß *m*
b. production Serienfertigung *f*; satzweise Fertigung *f*; Fertigung *f* kleiner Reihen
b.-quantity Stückzahl *f*, kleine Reihe *f*
b. recorder Chargenregistriergerät *n*
b. still *(Ch)* Retorte *f*; Destillierblase *f*
b. system diskontinuierliches Verfahren *n*
b. type furnace periodisch arbeitender Ofen *m*
batcher Dosieranlage *f*, Dosiervorrichtung *f*, Dosierwaage *f*
batching Dosierung *f*, Zumessung *f (z. B. Betonherstellung)*
b. bin *(Text)* Batschfach *n*
b. counter Dosierzähler *m*
b. in bins *(Text)* Batschen *n* im Fach
b. machine *(Text)* Batschmaschine *f*
b. medium *(Text)* Batschemulsion *f*
b. oil *(Text)* Batschöl *n*
b. plant Dosieranlage *f*; Mischstation *f (Betonherstellung)*
b. ripeness *(Text)* Batschreife *f*
batchwork Kleinreihenfertigung *f*
bate / to *(Led)* beizen
bate Beize *f*, Beizmittel *n*
b. shaving *(Led)* Falzen *n* nach der Beize
bath Bad *n*
b. current Badstrom *m (Galvanotechnik)*
b.-dyeing *(Text)* Flottenfärbung *f*
b. furnace *(Met)* Badofen *m*
b. lubrication Tauchschmierung *f*, Ölbadschmierung *f*
b. of lead Bleibad *n*
b. of molten salt Salzbad *n*
b. of molten solder Lötbad *n*
b. preparations *(Ch)* Badzusätze *mpl*
b. travel *(Text)* Badstrecke *f (Garn)*
b. voltage *(El)* Badspannung *f*
bathing clothes Badekleidung *f*
b. fashion Bademode *f*
b. mat Badevorleger *m*, Badeteppich *m*
bathometer Bathometer *n*, Bathymeter *n*, Gerät *n* zur Tiefenbestimmung
baths with several layers of liquid *(Text)* mehrschichtige Fällbäder *npl*
bathymeter *s.* bathometer
bathymetry Tiefenmessung *f*, Tiefseemessung *f*, Tiefseelotung *f*
bathyscaphe Tiefseetauchschiff *n*; Bathyscaph *m (n)*
batik *(Text)* Batik *f*, gebatikter Stoff *m*
b.-printed *(Text)* gebatikt
b. resist *(Text)* Batikreserve *f*, Wachsreserve *f*
bating *(Led)* Beizen *n*
batiste Batist *m*
baton 1. Stab *m (Staffellauf)*; 2. Knüppel *m*
bats-wing burner Fischschwanzbrenner *m*, Schnittbrenner *m*
batten / to verschalen *(mit Leisten)*
b. down *(Schiff)* verschalken *(Luke)*
batten 1. Bohle *f*, Latte *f*, Leiste *f*, *Richtscheit n*; Brett *n (Breite 4 bis 9", Dicke über 1½")*; 2. *(Text)* Lade *f*, Weblade *f*; Schlag *m (Webstuhl)*; 3. *(Schiff)* Latte *f*, Segellatte *f*; Schalstück *n*; Nahtleiste *f*
b. and space bulkhead *(Schiff)* Gitterschott *n*, Lattenschott *n*
b. and space ceiling *(Schiff)* Lattenwegerung *f*, offene Wegerung *f*
b. board Tischlerplatte *f*
b. core *(Holz)* Mittellage *f* aus Stäbchen
b. door Brettertür *f*; Lattentür *f*
b. lampholder Wandfassung *f*, Wandarmatur *f*
b. pin *(Text)* Ladenzapfen *m*
b. plate Schlagblech *n (Weberei)*
b. plywood board Tischlerplatte *f*
b. pocket *(Schiff)* Lattentasche *f (für Segel)*
b. wall Bretterwand *f*; Lattenwand *f*
battened and braced door Lattentür *f*
battening *(Text)* Ladenschlag *m*
batter / to nach oben zu in der Neigung abnehmen, sich verjüngen
batter *(Bw, Brg)* 1. Verjüngung *f*; 2. Anlauf *m*, Hang *m*, Neigung *f*, Schräge *f*, Sturz *m*
b. pile *(Bw)* Schrägpfahl *m*
battered abgeschrägt, geneigt, schräg, verjüngt

b. round *(Mil)* verbeulte Patrone *f*
b. wall Schrägwand *f*
battery 1. Batterie *f*, Sammler *m*; 2. *(Text)* Füllmagazin *n (Webstuhl)*
b. acid Akkumulatorensäure *f*
b. boiler *(Wärm)* Batteriekessel *m*
b. booster *(Am)* Pufferladegerät *n (für Batterie)*
b. box Batteriebehälter *m*, Elementbehälter *m*; Batteriegefäß *n*, Elementgefäß *n*
b. cell lid *(Kfz)* Zellendeckel *m*
b. cell plug *(Kfz)* Zellenstopfen *m*
b. charge regulations Batterieladevorschriften *fpl*
b. charger Batterieladevorrichtung *f*, Ladegleichrichter *m*
b. charging set Batterieladegerät *n*
b. clip *(El)* Batterieklemme *f*
b.-coil ignition *(Kfz)* Batteriezündung *f*
b. connecting terminal *(El)* Batterieanschlußklemme *f*
b. container *(Kfz)* Batteriekasten *m*
b. drive *(El)* Akkumulatorenantrieb *m*
b. electric traction Akkumulatorenzugförderung *f*
b. eliminator Batterieersatzgerät *n*; Netzanode *f*
b. heating Batterieheizung *f*
b. ignition *(Kfz)* Batteriezündung *f*
b. lamp Taschenlampe *f*
b. loading mechanism *(Text)* Magazinladevorrichtung *f*
b. mould Batterieform *f (Betonbau, Fertigteilherstellung)*
b. mounting *(Kfz)* Batterieträger *m*
b. mud Elementschlamm *m*
b. of boilers *(Wärm)* Kesselbatterie *f*, Kesselgruppe *f*, zusammenarbeitende Kesselreihe *f*
b. of secondary cells Akkumulatorenbatterie *f*, Sammlerbatterie *f*
b. operated receiver Batterieempfänger *m*
b. operation Batteriebetrieb *m (z. B. Radio)*
b. output Batterieausgangsleistung *f*
b.-powered batteriegespeist, batteriebetrieben
b. receiver Batterieempfänger *m*
b. resistance Batteriewiderstand *m*, innerer Widerstand *m* der Batterie
b. reversing switch Batterieumschalter *m*
b. sediment Elementschlamm *m*
b. stulls *(Brg)* Orgelstempel *mpl*
b. supply circuit noise Stromversorgungsgeräusch *n*, Stromversorgungsrauschen *n*
b. supply relay *(El)* Speiserelais *n*
b. switch *(El)* Zellenschalter *m*
b. syringe [Batterie-]Säureheber *m*
b. terminal Batteriekabelschuh *m*, Batterieklemme *f*
b. tester Elementprüfer *m*
b. top *(Brg)* Batteriedecke *f*
b. vent valve *(El)* Batterieentlüftungsventil *n*
b. voltage Batteriespannung *f*
batting *(Text)* Wickel *m*
battle anti-aircraft defence Truppenluftabwehr *f*
b. area air defence Truppenluftabwehr *f*
b. practice *(Mil)* Gefechtsschießen *n*
battlefield air defence (detection) Truppenluftabwehr *f*
b. mobility *(Mil)* taktische Beweglichkeit (Wendigkeit) *f*
baud *(Fmt)* Baud *n (Einheit der Telegrafiergeschwindigkeit, 1 Stromschritt je Sekunde)*
baulk *(Bw)* 1. zwischen zwei Ausschachtungen stehengebliebener Boden *m*; 2. Kantholz *n* größer als 15 × 15 cm, schwerer Holzbalken *m*
b. ring *(Am)* Sperring *m*
baum jig *(Brg)* kolbenlose (luftgesteuerte) Setzmaschine *f*
Baumann-Steinrück impact hardness tester Baumann-Hammer *m*
bauxite Bauxit *m*
bay 1. *(Schiff)* Bai *f*, Bucht *f*, Haff *n*, Meerbusen *m*; 2. *(Bw)* Erker *m*; Fensteröffnung *f*; Platte *f*, Feld *n*; Fläche *f*, Platz *m*; 3. *(Masch)* Abteilung *f* in einer Werkhalle; 4. *(Fmt)* Bucht *(Fernmeldetechnik)*
b. of barn Banse *f* der Scheune
b.-tree oil Lorbeeröl *n*
b. window *(Bw)* Erkerfenster *n*
bayonet *(Mil)* Bajonett *n*, Seitengewehr *n*
b. base *(El)* Bajonettsockel *m*, Swan-Sockel *m* *(Lampe)*
b. cap *(El)* Bajonettsockel *m*, Swan-Sockel *m* *(Lampe)*; *(Kfz)* Renkverschluß *m*
b. catch *(Masch)* Bajonettverschluß *m*, Bajonettkupplung *f*
b. fixing Bajonettverschluß *m*
b. frame *(Wärm)* Bajonettrahmen *m*
b. holder *(El)* Bajonettfassung *f*, Renkfassung *f*
b. joint Bajonettverbindung *f*, Bajonettverschluß *m*
b. lock Bajonettverschluß *m*
b. mount Bajonettfassung *f*
b. socket Bajonettfassung *f*
bazooka *(Fmt)* Symmetriertopf *m (Antennentechnik)*
BCD-code *s.* binary-coded decimal code
be a good fit in / to gut eingepackt sein in, gut sitzen in
be carried around by frictional contact durch Reibschluß mitgenommen werden
be closely situated (spaced) geringen Mittenabstand haben
be cut into chips zerspant werden

be employed on two shifts in zwei Schichten arbeiten
be equally spaced gleichen Mittenabstand haben
be fully supported volle Auflage haben
be in mesh kämmen, in Eingriff stehen *(Zahnräder)*
be in progress im Gange sein
be in soundings *(Schiff)* Grund haben
be light in weight leicht sein, wenig wiegen
be of opposite sign entgegengesetzte Vorzeichen haben
be of the same sign dieselben Vorzeichen haben
be off centre schlagen *(Spindel)*
be on the air im Funkbetrieb sein
be opposite in sign entgegengesetzte Vorzeichen haben
be out of centre Mittenverlagerung haben
be overstressed überdehnt werden
be rapid traversed im Eilgang bewegt (verstellt) werden
be rotated die Umlaufbewegung haben (ausführen)
be self-piloting sich selbst führen
be spaced evenly gleichen Mittenabstand haben
be spaced unevenly ungleichen Mittenabstand haben
be stationary fest stehen
be tangent to tangieren an
be the same size gleichgroß sein
be the wear *(sl) (Text)* Mode sein
be to scale maßstäblich sein
be true with (for) gelten (gültig sein) für
be undersize Untermaß haben, zu klein sein
be well supported gute Auflage haben

beach Strand *m*, Meeresufer *n*
b. robe Bademantel *m*
b. rock Muschelsandstein *m*
b. sand Küstensand *m*
b. seine Strandwade *f (Zugnetz)*

beaching Aufstrandholen *n (eines Bootes)*
b. gear Schwimmerwagen *m* für Seeflugzeuge

beacon / to befeuern

beacon Bake *f*, Ankerboje *f*; Leuchtfeuer *n*; Funkbake *f*, Feuer *n*, Funkfeuer *n*
b. aerial Bakenantenne *f*
b. airborne S-band Flugnavigationsleitverfahren *n* im cm-Band *(S-Band)*
b. airborne X-band Flugnavigationsleitverfahren *n* im X-Band
b. buoy Bakentonne *f*, Bakenboje *f*
b. course *(Fmt)* Peilstrahl *m*, Bakenkurs *m*
b. ground S-band Navigationsradar *n* im S-Band
b. light Flugstreckenfeuer *n*
b. transmitter *(Fmt)* Richtstrahlbake *f*

beaconing *(Schiff)* Abbakung *f*, Bebakung *f*

beacons and blind landing Blindlandungsfunkbaken *fpl*

bead / to auf Schnur ziehen; bördeln, kannelieren; sicken, sieken *(Faß)*; falzen, umlegen *(Blechkante)*

bead Kugel *f*, Pastille *f*, Perle *f*; Siedesteinchen *n*; Bördelrand *m*; Randwulst *m*; Sicke *f (beim Faß)*, Schweißraupe *f*; Schweißlage *f*
b. bend test Aufschweißbiegeprobe *f*
b. breaker *(Kfz)* Wulstheber *m*
b. capacitor *(El)* Perlkondensator *m*
b. chain Kugelkette *f (Antenne)*
b. lightning *(El)* Perlschnurblitz *m*
b. of the boss *(Wärm)* Nabenwulst *m (des Kreuzkopfes)*
b. plane Rundstabhobel *m*
b. polymerization *(Plast)* Perlpolymerisation *f*
b. seat *(Kfz)* Wulstfußauflage *f*
b. test *(Ch)* Perlenprobe *f (Analytik)*
b. transistor Perltransistor *m*
b. tube *(Ch)* Perlrohr *n*
b. wires *(Kfz)* Drahtkern *m (Reifen)*

beaded 1. perlschnurförmig; 2. *(Holz)* gestäbt *(Tischlerei)*
b. anvil Spitzamboß *m*, Bankhorn *n*, zweihörniger Amboß *m*
b. bar *(Masch)* Wulststab *m*
b. boarding *(Holz)* Stabbrettdecke *f*
b. edge *(Masch)* Randwulst *m*, Wulstrand *m*
b. rim tubular capacitor *(El)* Wulstrohrkondensator *m*
b. twist *(Text)* Perlschlung *m*

beading 1. *s.* bead; 2. *(El)* Durchführungsperle *f*, Einschmelzperle *f*; 3. *(Text)* mit Perlen besetztes Gewebe *n*; 4. Holzleiste *f*; 5. Sicken *n*
b. machine 1. Abkantmaschine *f*; Bördelmaschine *f*; Sickenmaschine *f*; 2. *(Text)* Spitzenmaschine *f*
b. plane Rund[stab]hobel *m*
b. plane iron Stabhobeleisen *n*

beak 1. Horn *n (Amboß)*; 2. Galion *n (Schiffsschnabel)*
b.-iron Amboßhorn *n*

beaked anvil Spitzamboß *m*
b. flute Schnabelflöte *f*

beaker *(Ch)* Becher[glas *n*] *m*; Kochbecher *m*
b. cover *(Ch)* Uhrglas *n*

beam 1. Strahl[enbündel *n*] *m*; 2. *(Bw, Statik)* Balken *m*, Träger *m*; Baum *m*; Unterzug *m*; Riegel *m (Rahmen)*; 3. Schiene *f*, Führungsschiene *f (Meßgerät)*; 4. *(Led)* Gerberbaum; *(Text)* Kettbaum *m*; 5. *(Schiff)* Decksbreite *f*, Schiffsbreite *f*; größte Breite *f*; 6. Schwinge *f (Getriebelehre)*
b. aerial Richt[strahl]antenne *f*

b. alignment *(El, Fs)* Strahljustierung *f (Elektronenoptik)*
b. analyzer *(Ph)* Strahlenanalysator *m*
b. and slab Plattenbalken *m (Stahlbeton)*
b. angle Strahlwinkel *m*
b. aperture *(Ph)* Bündelöffnung *f*; Strahlquerschnitt *m*; Öffnungswinkel *m* von Strahlen
b. bending and straightening machine Balkenbiege- und -richtmaschine *f*
b. bridge Balkenbrücke *f*
b. catcher *(Ph)* Strahlen[auf]fänger *m*
b. clamp *(Schiff)* Balk[en]weger *m*
b. compass *(Math)* Stangenzirkel *m*
b. control [elektrische oder magnetische] Strahleinstellung *f (Elektronenoptik)*
b. control pilotage *(Flg)* Leitstrahlführung *f*
b. coupling *(Ph)* Elektronenkopplung *f*
b. current *(Ph)* Strahlstrom *m*
b. deflection *(El, Opt)* Strahlablenkung *f*; Strahlungsablenkung *f*
b. deflection tube Elektronenstrahlröhre *f*
b.-deflector switch *(Kfz)* Abblendschalter *m*
b. dyeing *(Text)* Kettbaumfärben *n*
b. emission *(Fmt)* Richtsendung *f*
b. extractor *(Kern)* Ausschleusvorrichtung *f*, Vorrichtung *f* zur Bündelauslenkung
b. face *(Led)* Falzplatte *f*
b. follower guidance *(Rak)* Leitstrahllenkung *f*
b.-forming plate Bündelungselektrode *f*
b. gallows *(Schiff)* Baumgalgen *m*
b. grillage *(Bw)* Trägerrost *m*
b. guidance *(Rak)* Leitstrahllenkung *f*
b. hole *(Kern)* Bestrahlungskanal *m*; Strahlenkanal *m (für besonders schnelle Neutronen)*
b. limiting *(Ph)* Strahlenbegrenzung *f*
b.-limiting device *(Kern)* Strahlbegrenzer *m*
b. line *(Schiff)* Decksstrak *n*
b. moulded *(Schiff)* Breite *f* auf Spanten, gemallte Breite
b. movement Strahlbewegung *f*
b. of charged particles *(Ph)* Strahl *m* geladener Teilchen
b. of electrons Elektronenbündel *n*, Elektronenstrahl *m*
b. of fruit press Kelterbaum *m*
b. of parallel rays paralleles Strahlenbündel *n*
b. of press brake Wange *f* [der Abkantpresse]
b. of rays Strahlenbündel *n*
b. over all *(Schiff)* Breite *f* über alles
b. over fenders *(Schiff)* Breite *f* über Scheuerleisten
b. path Strahlengang *m*
b. pattern of an aerial array *(Fmt)* Richtkennlinie *f*
b. power Bündelleistung *f*
b. power amplifier tube [Elektronenstrahl-] Hochleistungsverstärkerröhre *f*
b. power tetrode Endtetrode *f* mit Elektronenbündelung
b. power tube *(Am)* Beam-Tetrode *f*
b. power valve Endröhre *f* mit Elektronenbündelung, Bündelendröhre *f*
b. procedure *(Flg)* Leitstrahlanflugverfahren *n*
b. property Biegefähigkeit *f*
b. pump *(Brg)* Schwengelpumpe *f*, Balkenpumpe *f*
b. reception *(Fmt)* Richtempfang *m*
b. reinforcement Balkenbewehrung *f*
b. relaxor Sägezahngenerator *m (Impulstechnik)*
b. return *(Fs)* Strahlrücklauf *m*
b.-rider guidance *(Rak)* Leitstrahllenkung *f*
b. rolling mill Trägerwalzwerk *n*
b. sea gekreuzte See *f*, Kreuzsee *f*, Dwarssee *f*
b.-splitting *(Opt)* Strahlenteilung *f*
b.-splitting cube *(Opt)* Teilungswürfel *m*
b.-splitting mirror Strahlzerlegungsspiegel *m*
b. station *(Fmt)* Richtfunkstation *f*
b. steam engine Balancierdampfmaschine *f*
b. stop Strahlblende *f*
b. stringer *(Schiff)* Balkenstringer *m*
b.-switching tube Elektronenstrahl-Schaltröhre *f*
b. system Beam-System *n (Richtstrahlantennenanlage)*
b. test Biegeversuch *m*
b. tetrode *(Fmt)* Bündeltetrode *f*, Strahltetrode *f*, Beam-Tetrode *f*
b. tie *(Bw)* Balkenanker *m*, Giebelanker *m*, Kopfanker *m*
b. to length *(Schiff)* Längenverhältnis *n* L : B *(Länge zu Breite)*
b. trammel Stangenzirkel *m*
b. transmitter *(Fmt)* Richtstrahlsender *m*
b. trap *(El)* Strahlenfalle *f*
b. trawl *(Schiff)* Kurre *f*; Baumnetz *n*, Baumgrundschleppnetz *n*
b. trigger *(Fs)* Helligkeitssteuerung *f*, Strahlauftastung *f*
b. tube Röhre *f* mit Elektronenbündelung
b.-type scale Balkenwaage *f*
b. variation *(Opt)* Strahlenschwankung *f*; Strahlenschwächung *f*
b. warper *(Text)* Schärmaschine *f*
b. warping *(Text)* Kettbaumschären *n*, Baumschären *n*
b. weakener Kompensationsblende *f (Fotometrie)*
b. width *(El, Opt)* Bündelweite *f*; Öffnungswinkel *m (eines Strahlenbündels)*; Strahldurchmesser *m (Antennentechnik)*, Strahlquerschnitt *m*
b. wind *(Schiff)* Dwarswind *m*; Seitenwind *m*
beamer *(Text)* Bäummaschine *f*

beamhouse *(Led)* Wasserwerkstatt *f*
b. weight *(Led)* Blößengewicht *n*
beaming 1. Wellenbündelung *f* *(Antenne)*; 2. *(Text)* Aufbäumen *n*
b. creel *(Text)* Schärgatter *n*, Schärgestell *n*
b. device *(El)* Bündelungsanlage *f*
b. frame *(Text)* Bäumstuhl *m*
b. machine *(Text)* Bäummaschine *f*, Aufbäummaschine *f*
beams Balkenlage *f* *(Zimmerei)*
beamy ship *(Schiff) sehr breites Schiff*
bean *(Brg)* Nußkohle *f*
b.-sheller Bohnenenthülser *m*
b. stack Bohnenschober *m*
beans harvester Bohnenerntemaschine *f*
bear / to 1. tragen, [unter]stützen; 2. [an]peilen *(Radar)*
b. on *(Bw)* auflagern auf, Last übertragen auf
bear *(Met)* Eisensau *f*, Ofenbär *m*, Bodensau *f*, Sau *f*, Ofenwolf *m*
bearance fulcrum Stützpunkt *m*, Hebelpunkt *m*
beard *(Text)* oberer elastischer Teil *m* des Hakens
bearded needle *(Text)* Hakennadel *f*
b. needle circular knitting machine Rundstrickmaschine *f* mit Hakennadeln
b. oats Fahnenhafer *m*
bearding angle *(Schiff)* Verbindungswinkel *m* des Vorstevens mit dem Kiel
bearer 1. *(Bw)* Podestbalken *m*, Podestträger *m*; Unterzug *m*; 2. *(Brg)* Traggeviert *n* *(Schachtausbau)*
b. cable *(El)* Tragseil *n*
b.-to-bearer *(Typ)* Schmitz-auf-Schmitz, aufeinanderlaufende Schmitzringe *mpl* *(an Offsetmaschinen)*
bearing 1. Lager *n* *(Maschinenelement)*; 2. *(Masch)* Lagerung *f*, Anlage *f*, Auflage *f*; 3. *(Am)* Führung[sbahn] *f*; Lagerstelle *f*; 4. *(Fmt)* Funkpeilung *f*, Peilung *f*, Ortung *f*; Peilwinkel *m*; Richtung *f*; 5. *(Fmt)* Gangstreichen *n*, Schichtstreichen *n*
b. against mit Anlage an; gedrückt gegen
b. alloy Lagerlegierung *f*, Lagermetall *n*
b. area Auflagefläche *f*, Tragfläche *f*, Sitzfläche *f*, Anlagefläche *f*
b. arrangement *(Masch)* Lagereinbau *m*, Lagerung *f*
b. axle Tragachse *f*
b. ball Lagerkugel *f*
b. beam Lagerbalken *m*
b. block 1. *(Bw)* Knagge *f* *(Holzbau)*; Lagerbock *m*, Lagerstuhl *m* *(Brückenbau)*; 2. Druckplatte *f* *(Prüfmaschine)*
b. bolt Lagerdeckelschraube *f*, Lagerbolzen *m*
b. bore Lagerbohrung *f*
b. box Lagergehäuse *n*, Lagerkasten *m*, Lagerkörper *m*, Lagerschale *f*
b.-box lining Lagerschalenfutter *n*
b. bracket Lagerarm *m*, Lagerbock *m*, Lagerstütze *f*
b. bronze Lagerbronze *f*
b. bush[ing] Lagerbuchse *f*, Lagerschale *f*
b. by stars Gestirnpeilung *f*
b. cage Lagerkäfig *m*
b. cap Lagerdeckel *m*
b. capacity aufnehmbare Last *f*, Tragfähigkeit *f*, Tragkraft *f*, Tragvermögen *n*
b. clearance Lagerspiel *n*, Lagerluft *f*; Führungsbahnenluft *f*
b. collar Lagerbund *m*, Lagerkragen *m*
b. compass Ortungsgerät *n*, Peilgerät *n*, Peilkompaß *m*
b. correction Peilungskorrektur *f*
b. cover Lagerdeckel *m*
b. crib *(Brg)* Traggeviert *n*, Tragkranz *m* *(Schachtausbau)*
b. curb *s.* b. crib
b. deviation indicator Mißweisungsanzeiger *m*
b. direction Peil[ungs]richtung *f*
b. distance *(Bw)* Stützweite *f*, lichte Weite *f*
b. error Peilfehler *m*, Funkfehlweisung *f*
b. fluctuations Peilschwankungen *fpl*
b. for forced lubrication *(Masch)* Öldrucklager *n*
b. frame Lagerbalken *m*
b. friction Lagerreibung *f*
b. hinge *(Masch)* Auflagergelenk *n*
b. housing Lagergehäuse *n*
b.-in *(Brg)* Schrämtiefe *f*
b. indicator 1. *(Fmt)* Sichtpeiler *m*; 2. *(Flg)* Azimuthanzeiger *m*, Azimuthanzeige[gerät *n*] *f*
b. length tragende Länge *f*
b. line Peillinie *f*
b. lining Lagerausguß *m*, Lagerfutter *n*
b. load Lagerbelastung *f*
b. luboil (lubricating oil) Lagerschmieröl *n*
b. lubrication Lagerschmierung *f*
b. material Lagerwerkstoff *m*
b. metal Lagermetall *n*
b. mounting Lagermontage *f*
b. neck *(Masch)* Lagerhals *m*
b. object Ortungsziel *n*, Peilziel *n*
b. of aircraft Azimut *m* *(n)* des Flugzeuges; Flugzeugpeilung *f*
b. of the compass Kompaßstrich *m*
b. of the teeth Tragen *n* der Zähne, Zahntragen *n*
b. out *(Bw)* vorkragend
b. pad 1. Auflagerblock *m*, Widerlagerstein *m* *(Brückenbau)*; 2. Führungsleiste *f* *(Tieflochbohrer)*
b. pedestal Lagerbock *m*, Lagerstuhl *m*

b. pile Druckpfahl *m (Grundbau)*
b. pile foundation stehende Pfahlgründung *f*
b. pintle *(Schiff)* Tragfingerling *m (des Ruders)*
b. plate *(Bw)* Auflagerplatte *f*, Unterlagsplatte *f*
b. play Lagerspiel *n*, Lagerluft *f*
b. power Tragfähigkeit *f*
b. power of soils Tragfähigkeit *f* des Baugrundes
b. pressure Anpreßdruck *m (Lagerung)*; Auflagerdruck *m*; Bodenpressung *f*; Lochleibung *f*
b. pulley Tragrolle *f*
b. race Laufring *m* des Lagers
b. rail Tragschiene *f*
b. roller Lagerrolle *f*
b. scraper Lagerschaber *m*, Dreikantlöffelschaber *m*
b. set *(Brg)* Traggeviert *n (Schachtausbau)*
b. shell Lagerschale *f*
b. slackness Lagerluft *f*
b. socket of turntable Drehpfanne *f* der Drehscheibe
b. spacer Distanzhülse *f*, Abstandhülse *f*
b. spindle Laufspindel *f (Turbine)*
b. spring Tragfeder *f*, Stützfeder *f*
b. strength Tragfähigkeit *f*
b. support Traglager *n (Fräsdorn)*
b. supported by hanger Hängelager *n*
b. surface 1. Auflagefläche *f*, Sitzfläche *f*, Tragfläche *f*; Anlage *f*; 2. Lauffläche *f*; Gleitfläche *f*; Führungsoberfläche *f*
b.-surface area Druckfläche *f*, Auflagerfläche *f*
b. surface of the foundation *(Masch, Bw)* tragende Fläche *f* des Fundaments
b. test Belastungsprobe *f*, Belastungsversuch *m*
b.-up stop *(Brg)* Aufsetzvorrichtung *f*, Schachtfalle *f*
b. value Tragfähigkeit *f*; Lastaufnahmefähigkeit *f*; zulässige Pressung *f*
b. way Gleitbahn *f*
b. wear Lagerabnutzung *f*, Lagerverschleiß *m*
b. with plastics lining in outer sleeve of metal kunststoffausgekleidetes Verbundlager *n*
beast of burden Tragtier *n*
b. of draught Zugtier *n*
beat / to 1. schlagen, klopfen; hämmern; 2. schwingen *(Flachs)*; zerfasern
b. out ausschlagen *(Gold)*
beat 1. Schlag *m (nicht rundlaufendes Werkzeug)*; 2. *(Ph)* Schwebung *f (Überlagerung von Wellen)*; 3. Ausschlag *m (eines Zeigers)*
b. amplitude *(Ph)* Schwebungsamplitude *f*
b. buzzer *(Fmt)* Schwebungssummer *m*
b. cycle Schwebungsperiode *f*
b. frequency Schwebungsfrequenz *f*
b. frequency oscillator Schwebungssummer *m*
b. indicator Schwebungsanzeiger *m*
b. method Schwebungsmethode *f*, Schwebungsverfahren *n*; Überlagerungsverfahren *n*
b. note Schwebungston *m*, Überlagerungston *m*
b. note pitch Schwebungstonhöhe *f*
b. note zero *(Fmt)* Schwebungslücke *f*, Schwebungsnull *f*
b. of the blast pipe *(Wärm)* Dampfschlag *m*
b. of the reed *(Text)* Blattanschlag *m*
b. oscillator Schwebungsoszillator *m*
b. period Schwebungsperiode *f*, Schwebungsdauer *f*
b. receiver *(Fmt)* Schwebungsempfänger *m*, Überlagerungsempfänger *m*
b. reception *(Fmt)* Schwebungsempfang *m*, Überlagerungsempfang *m*
b. tone Schwebungston *m*, Überlagerungston *m*
b.-up *(Text)* Noppenzahl *f* je Zoll
b. vibration *(Ph)* Schwebung *f*
beaten gold Blattgold *n*
b. zone *(Mil)* bestrichener Raum *m (Schießen)*
beater 1. Feinzeugholländer *m*, Flügel *m (eines Rührwerkes)*, Ganzzeugholländer *m*, Holländer *m*, Mahlholländer *m*; 2. *(Text)* Schläger *m*, Schlagmaschine *f*, Stampfer *m*, Stoffmühle *f*, Walkhammer *m*, Walzenmühle *f*, Schlagmühle *f (Asbest)*; Lade *f*; 3. *(Brg)* Ladestock *m*
b. additive Holländerzusatz *m*
b. bar *(Text)* Schlägerschiene *f*
b. bed-plate Holländergrundwerk *n*
b. blade Holländermesser *n*
b. coloured in der Masse gefärbt *(Papier)*
b. engine Holländer *m*
b. filling Holländereinsatz *m*
b. glue Holländerleim *m*
b. grid bar *(Text)* Schlägerrost *m*
b. hood Überwurfhaube *f* des Holländers
b. house Holländersaal *m*
b. knives Holländermesser *npl*
b. lid *(Text)* Schlägerdeckel *m*
b. loading Holländereintrag *m*, Zugabe *f* der Füllstoffe in den Holländer
b. lock *(Text)* Schlägersicherung *f*
b. mill Schlag[kreuz]mühle *f*
b. plate Holländergrundwerk *n*
b. roll Messerwalze *f*, Holländerwalze *f*
b. scutcher *(Text)* Schienenschläger *m*
b. sizing Leimen *n* im Stampfgeschirr *(Papierindustrie)*
b. trough Holländertrog *m*
b. tub Holländergeschirr *n*, Holländertrog *m*, Holländerbütte *f*

beating 1. *(Text)* Schlagen *n* *(Kokon mit Bürsten)*; 2. Knüpfende *n*; 3. Brechen *n* *(Flachs)*; 4. *(Ph)* Schwebung *f*
b. board Sprungbrett *n*
b. current Schwebungsstrom *m*
b. degree *(Holz)* Mahlgrad *m*
b. effect *(Fmt, El)* Interferenzeffekt *m*, Schwebungsvorgang *m*
b. engine Holländer *m*
b. mill Stampfkalander *m*, Stoßkalander *m*
b. of pulp *(Holz)* Ausmahlung *f*
b. spring board Federsprungbrett *n*
b. strip Schlagleiste *f* *(Dreschmaschine)*
b. tester *(Holz)* Mahlgradprüfer *m*
b.-up *(Text)* Anschlagen *n*, Ladenanschlag *m*
beats method *(Ak)* Schwebungsmethode *f*
Beaufort wind scale Beaufortskala *f* *(Windstärke)*
beaumontage Füllkitt *m*; Füllstoff *m*
beauty mask Gesichtsmaske *f*, Gesichtspackung *f*
beauxite *s.* bauxite
beaver board Holzfaserplatte *f*, Leichtbauplatte *f*
b. fur Biberpelz *m*
beck Bottich *m*
Beck arc Beck-Lichtbogen *m*
beck-iron *s.* beak iron
becket 1. Schlinge *f*; 2. *(Schiff)* Knebelstropp *m*
b.-rope *(Schiff)* Haltetau *n*, Festhaltetau *n*
becking mill Aufweit[e]walzwerk *n* *(Rohre)*
Beckmann molecular transformation Beckmannsche Umlagerung *f*
Beckmann thermometer Beckmann-Thermometer *n*
become clogged / to [sich] verstopfen
b. darker nachdunkeln
b. embrittled verspröden
b. jammed sich verklemmen
b. red-hot sich bis zur Rotglut erhitzen
b. stuck festfahren, hängen bleiben
b. twisted windschief werden *(Baum)*
b. warped sich krümmen, sich verwölben
Becquerel cell *(Ph)* Becquerel-Zelle *f*
Becquerel effect *(Ph)* Sperrschichteffekt *m*, Becquerel-Effekt *m*
bed / to einlegen, auflegen, betten, lagern; aufliegen, anliegen
b. squarely rechtwinklig anliegen, satt aufliegen
bed 1. *(Masch)* Maschinenbett *n*, Bett *n*; Grundplatte *f*; Untergestell *n*; 2. Tisch *m* *(Presse)*; 3. *(Brg)* schichtförmige Lagerstätte *f*, Flöz *n*, Lager *n*; 4. *(Bw)* Lagerfläche *f* *(Ziegelstein)*; Unterlage *f*; Lagerung *f*; Unterbau *m*; 5. Sohle *f* *(Herd)*; 6. Beet *n*
b. bearer *(Typ)* Schmitzleiste *f*
b. casting *(Masch)* Bettkörper *m*
b. joint Lagerfuge *f* *(Mauerwerk)*
b. knives *(Holz)* Holländergrundwerk *n*, Bodenplatten *fpl*
b. of a river Flußbett *n*
b. of straw Streu *f*
b. plate 1. *(Bw, Masch)* Auflagerplatte *f*, Unterlagsplatte *f*; Bodenplatte *f*, Fundament *n*, Grundplatte *f*, Sohlplatte *f*, Fußplatte *f*; 2. *(Text)* Tragering *m* für die Strickelemente
b.-plate box *(Holz)* Grundwerkfassung *f*
b.-plate junction *(Masch, Wärm)* Grundplattenanschluß *m*
b. plate of the engine *(Kfz)* Motortragplatte *f*
b. plate with condenser *(Wärm)* Grundplatte *f* mit Kondensationseinrichtung
b. plough Beetpflug *m*
b. rack Bettzahnstange *f*
b.-rock anstehender Fels *m*, Felsboden *m*, Felsuntergrund *m*, Mutterfels *m*, Muttergestein *n*
b. separation *(Brg)* Aufblätterung (Ablösung) *f* der Schichten
b.-separation cavities Webersche Hohlräume *mpl* *(Gebirgsmechanik)*
b.-sitting room Wohnschlafzimmer *n*
b. slides *(Masch)* Bettführungsbahnen *fpl*
b. ticking *(Text)* Bettbarchent *m*
b.-type milling machine *(Am)* Planfräsmaschine *f*
b. vein *(Geol)* Lagergang *m*
b. way Bettführungsbahn *f*
bedding 1. *(Bw)* Bettung *f*, Einbettung *f*, Lagerung *f*; Bett *n*, Unterlage *f*, Anlage *f*; 2. *(Geol)* Schichtung *f*; 3. Verlegung *f*
b. in *(Masch)* Einbau *m*, Einpassen *n*
b. of pipes Rohrlagerung *f*
b. plane *(Geol)* Schichtfläche *f*
b. putty Fensterkitt *m*, Glaserkitt *m*
b. without play ohne Spiel anliegend
bedrock anstehendes Gestein *n*, Anstehendes *n*
bedspread Bettüberwurf *m*, Bett[über]decke *f*
bee 1. *(Schiff)* Bugsprietklampen *fpl*; 2. Biene *f*
b.-keeper Bienenhalter *m*, Bienenzüchter *m*, Imker *m*
beech Buche[nholz *n*] *f*
b. charcoal Buchen[holz]kohle *f*
b. tar Buchen[holz]teer *m*
b. veneered mit Buche furniert
beechnut oil Bucheckernöl *n*
beef suet Rindertalg *m*
b.-up operation *(Plast)* Verstärkung *f* eines Sandwich-Schichtstoffes
beehive Bienenkorb *m*, Bienenstock *m*
b. oven *(Met)* Bienenkorbofen *m*, Wannenofen *m*
beer 1. Bier *n*; 2. *(Text)* 40 Fäden *mpl* *(Kettfäden)* 3. *(Led)* 36 Fäden

b. barrel Bierfaß *n*
b. bottle Bierflasche *f*
b. brewing Bierbrauen *n*
b. cellar Bierkeller *m*
b. cooler Bierkühlapparat *m*
b. fining Bierkläre *f*
b. mat Bieruntersetzer *m*
b. pump Bierpumpe *f*
beeswax Bienenwachs *n*
beet Runkelrübe *f*
b. elevator Rübenschrägförderer *m*, Rübenhöhenförderer *m*
b. grubber Rübenroder *m*
b. lifter Rübenroder *m*
b.-root Runkelrübe *f*
b.-root chips Rübenschnitzel *npl*
b.-root slicer Rübenschnitzelmaschine *f*, Rübenzerkleinerer *m*
b.-root slices Rübenschnitzel *npl*
b.-root washer Rübenwaschmaschine *f*
b.-root washing drum Rübenwaschtrommel *f*
b. scoop Rübenschaufler *m*
b. sugar Rübenzucker *m*
b. topper Rübenköpfer *m*
beetle / to *(Text)* beeteln, stampfen
beetle 1. Überhang *m*; 2. großer Holzhammer *m*; 3. Stampfer *m*
b. calender Beetlekalander *m*
b.-head Rammbär *m*
Beetle plastic *Handelsbezeichnung für Aminoplastgeschirr*
beetled finish *(Text)* Stampfappretur *f*
beetler *(Text)* Stampfkalander *m*
before-and-after test *(Kern)* Vor- und Nachbestrahlungsmessung *f*
beginner Anfangsbuchstabe *m (Signierschablone)*
beginning of stroke *(Wärm)* Hubanfang *m*, Hubbeginn *m*
b. of the exhaust *(Wärm)* Beginn *m* des Dampfaustritts
begohm *(El)* 10^9 Ohm
behaved wood imprägniertes Holz *n*
behaviour Verhalten *n (Werkstoff)*; Verlauf *m (Kurve)*
b. at sea Verhalten *n* auf See *(eines Schiffes)*
b. of an engine Betriebsverhalten *n* eines Motors
bel Bel *n* (*Zeichen* B)
belaying pin *(Schiff)* Belegnagel *m*
belfry *(Schiff)* Glockengalgen *m*
bell 1. Glocke *f*, Klingel *f*, Schelle *f*; 2. *(Fmt)* Wecker *m*; 3. Trichter *m*, Ziehtrichter *m (Rohrherstellung)*; 4. Muffe *f*, Muffenende *n (Rohre)*; 5. *(Brg)* Schlag *m (Signal bei der Schachtförderung)*; 6. *(Brg)* Sargdeckel *m (im Hangenden)*; 7. *(Met)* Gichtglocke *f*; 8. Schiffsglocke *f*; Glasen *pl (eight bells = 8 Glasen)*
b. and hopper *(Met)* Gichtverschluß *m*
b.-and-hopper arrangement Gichtverschluß *m*, Parrytrichter *m*
b.-and-spigot joint Muffenverbindung *f (Rohrleitungsbau)*
b.-and-spigot pipe Muffenrohr *n*
b. buoy *(Schiff)* Glockentonne *f*, Glockenboje *f*
b. button Klingelknopf *m*
b. clapper Glockenhammer *m*, Glockenklöppel *m*, Glockenschwengel *m*
b. counter *(Kern)* Glockenzählrohr *n*
b.-crank 1. *(Masch)* Kniehebel *m*, Winkelhebel *m*; 2. Glockenarm *m*, Schwengel *m*
b.-crank drive Winkelantrieb *m*
b.-crank governor *(Masch, Wärm)* Winkelhebelregler *m*
b.-crank lever Winkelhebel *m*, Kniehebel *m*
b. face flachrunde Hammerbahn *f*
b. flowmeter Tauchglockendurchflußmesser *m*
b. founder Glockengießer *m*
b. gauge Tauchglocke[nmanometer *n*] *f*
b. gear Glockenwerk *n*, Läutewerk *n*
b. gong Glockenschale *f*
b. hammer Glockenklöppel *m*
b. indicator drop *(Fmt)* Weckerfallklappe *f*
b. insulator Glockenisolator *m*
b. item counter Glockenzählwerk *n (Addiermaschine)*
b.-jar Rezipient *m*
b. key Läutetaste *f*
b. kiln Glockenofen *m*
b. loft Glockenstuhl *m*
b. metal Glockenbronze *f*
b. mouth Trichter *m*
b.-mouthed aufgetrichtert, mit trichterförmiger Öffnung; glockenförmig aufgeweitet
b. mouthing Aufweitung *f (Bohrung)*
b. operating gear Glockenwinde *f*, Gicht[glocken]winde *f (Hochofen)*
b. pit *(Brg)* Schurfschacht *m*
b. pressure gauge Tauchglockenmanometer *n*
b. pull Glockenzug *m (Kirchenglocken usw.)*
b. punch *(Masch)* Zentrierglocke *f*
b. push Klingelknopf *m*
b.-ringing apparatus Glockenläutemaschine *f*
b.-ringing transformer *s.* b. transformer
b. rope Glockenschnur *f*, Glockenseil *n*, Glockenstrang *m*
b. seal Glockenverschluß *m*
b.-shaped curve Glockenkurve *f*
b.-shaped ferrule Glockenzwinge *f*
b.-shaped insulator Glockenisolator *m*
b.-shaped magnet Glockenmagnet *m*
b.-shaped pulse *(El)* glockenförmiger Impuls *m*, Glockenimpuls *m*
b. signal Glockensignal *n*

b. stop Glockenzug *m*, Glockenregister *n* *(Orgel)*
b. striker Glockenklöppel *m*
b. transformer *(Elm)* Klingeltransformator *m*
b. trap Glockenverschluß *m*
b.-type annealing furnace *(Met)* Haubenofen *m*
b.-type combination centre (centring) drill Zentrierbohrer *m* mit Schutzsenkung
b.-type countersink Zentriersenker *m* für Zentrierbohrung *(Körnerloch)* mit Schutzsenkung
b.-type distributing system Glocke *f* *(Hochofen)*
b. wire Klingelleitung *f*
belled caisson *(Bw)* Kegelfußpfahl *m*
Belleville spring Tellerfeder *f*
bellied pass Ovalstich *m*
b. tow saw blade Bügelsägeblatt *n* mit Bauch
belling Wulsten *n*
Bellini-Tosi direction finder Goniometerpeiler *m*
Bellini-Tosi system *(Fmt)* Kreuzrahmen-Peilverfahren *n*
bellman *(Brg)* Anschläger *m*
bellmouth kegelige Erweiterung *f* einer Bohrung
bellow Ausdehnungsstück *n*; Faltenbalg *m*; Wellrohr *n*
b. flowmeter Balgendurchflußmesser *m*
b.-type expansion piece Wellrohrausdehnungsstück *n*
bellows Balg *m*, Blasebalg *m* *(Orgel; Schmiede)*
b. camera *(Opt)* Balgenkamera *f*
b. differential flowmeter Differentialbalgendurchflußmesser *m*
b. for close-ups *(Opt)* Balgennaheinstellgerät
b.-type cover [zieh]harmonikaähnlicher Schutz *m*
b.-type dust guard [zieh]harmonikaähnlicher Staubschutz *m*
b.-type folding camera Balgenkamera *f*
belly Kohlensack *m* *(Hochofen)*; *(sl)* *(El)* Schwingungsbauch *m*
b. blister *(Mil)* Bodenlafette *f*
b. of violin Geigendecke *f*
b.-shave / to *(Led)* Bauch und Klauen falzen
b. stay *(Schiff)* Bauchstag *n* *(Stag vom halben Mast)*
b. turret *(Mil)* Bodenwanne *f*
below deck *(Schiff)* unter Deck, im Laderaum
b. zero unter Null
Below region *(El)* Belowsches Gebiet *n*
Below's law of current distribution *(El)* Belowsches Stromverteilungsgesetz *n*
belt 1. Band *n*, Gurt *m*, Riemen *m*; 2. *(Masch)* Antriebsriemen *m*; Schleifband *n* *(Bandschleifmaschine)*; 3. Gürtel *m*; 4. *(Bw)* Gurtgesims *n*
b. adjuster Riemenspanner *m*, Spannrolle *f*
b. brake Riemenbremse *f*
b. cement Riemenkitt *m*
b. conveyor Bandförderer *m*, Förderband *n*, Gurtförderer *m*
b. conveyor bridge Bandbrücke *f*
b. conveyor structure Bandtraggerüst *n*
b. course *(Bw)* auskragende Ziegelschicht *f*
b. cover Riemenbelag *m*
b. drier Bandtrockner *m*
b. drive Riemen[an]trieb *m*
b.-driven pump Riemenspeisepumpe *f*, Speisepumpe *f* mit Riemenantrieb
b. driving Riemen[an]trieb *m*
b. drum 1. *(Mil)* Gurttrommel *f*; 2. Bandwalze *f*
b. drying Bandtrocknung *f*
b. fastener 1. Riemenband *n*, Riemenverbinder *m*, Riemenverschluß *m*; 2. *(Brg)* Bandhaken *m*
b. feeder Bandbeschicker *m*, Förderbandspeiser *m*
b. friction Riemenreibung *f*, Seilreibung *f*
b. generator Bandgenerator *m*
b. grease Treibriemenfett *n*
b. grinding machine Bandschleifmaschine *f*
b. guide Riemengabel *f*, Gabel *f* zur Riemenführung
b. joint Riemenverbindung *f*; Bandverbinder *m*
b. lace *(Led)* Binderiemen *m*
b. lacer Riemenverbinder *m*, Riemenverbindeapparat *m*
b. lacing machine Riemenverbindemaschine *f*
b. length Riemenlänge *f*
b. life Bandaufliegezeit *f*, Bandlebensdauer *f*
b.-lift drop hammer Riemenfallhammer *m*
b. loader Bandförderer *m*
b. loom *s.* Indian type loom
b. lubricant Riemenschmiere *f*
b. of an iron chimney Schornsteinanschluß *m*
b. pulley Gurtscheibe *f*, Riemenrad *n*, Riemenscheibe *f*
b.-pulley flywheel Riemen[scheiben]schwungrad *n*
b. punch Riemenlocher *m*
b. sanding machine Bandschleifmaschine *f*
b. separator Bandscheider *m*, Bandseparator *m*
b. shifter Riemenausrücker *m*, Riemenumrücker *m*
b.-shifter fingers Riemenausrückgabel *f*
b. shifting fork Riemenumlegegabel *f*
b. shifting mechanism Riemenschalter *m*
b. slip Riemenschlupf *m*
b.-slipper Riemenaufleger *m*
b. stretch Riemendehnung *f*

b. stretcher Riemenspanner *m*
b. tachometer Riementachometer *n*
b. tension Riemenspannung *f*
b. thickness Riemendicke *f*
b. tightener *(Masch)* Riemenspannrolle *f*, Spannrolle *f*
b.-tightener attachment Riemenspanneinrichtung *f*
b.-tightening slide rail Riemenspannschiene *f*
b.-transmission Riemenübertragung *f*
b.-type centrifugal discharge bucket *(Holz)* Gürtelelevator *m* mit Schleuderentleerung
b.-type generator Bandgenerator *m*
b. weaving Gurtweberei *f*, Schmalweberei *f*
b. weigher Bandwaage *f*
b. width 1. Bandbreite *f* *(Förderband)*; 2. Riemenbreite *f* *(Treibriemen)*

belted motor drive Antrieb *m* durch Elektromotor über Riemen
belting 1. *(Brg)* Förderband *n*, Fördergurt *m*; 2. Riemenantrieb *m*; 3. Riemenleder *n*
b. leather Treibriemenleder *n*
beltscanner Linsenkranzabtaster *m* *(mechanische Fernsehbildzerlegung)*
belvedere Aussichtsturm *m*
bench 1. Werktisch *m*, Werkbank *f*, Bank *f*; Maschinenbett *n* mit Füßen; 2. *(Bw)* Absatz *m*, Bankett *n*, Berme *f*
b. axe Handaxt *f*
b. belt conveyor *(Brg)* Strossenband *n*
b. centre *(Masch)* Spitzengerät *n* *(zur Aufnahme von Teilen zwischen festen Körnerspitzen)*
b.-cut *(Brg)* stufenförmiger Einbruch *m* *(Schachtabteufen)*
b. dog *(Holz)* Hobelbankeisen *n*
b. drill Tischbohrmaschine *f*
b. drill stand Tischbohrständer *m*
b. drilling machine Tischbohrmaschine *f*
b. filing machine Werkbankfeilmaschine *f*
b. grinder Tischschleifmaschine *f*
b. height *(Brg)* Stoßhöhe *f*
b. lathe Mechanikerdrehmaschine *f*, Tischdrehmaschine *f*
b. level Maschinenrichtwaage *f*
b. micrometer Standschraublehre *f*
b.-mounted machine Tischmaschine *f*
b. oiler Ölkanne *f*, Schmierkanne *f*
b. plate Anreißplatte *f*, Tuschierplatte *f*
b. press Tischpresse *f*
b. saw Sägemaschine *f*
b. shear Hebelblechschere *f*, Bankschere *f*, Bockschere *f*
b. stone Abziehstein *m*, Bankstein *m*, Wetzstein *m*
b. test Prüfstandversuch *m*
b. type drilling machine Tischbohrmaschine *f*
b. vice Bankschraubstock *m*; Hobelbank *f*
benched foundation *(Bw)* abgetreppte Gründung *f*, Stufenfundament *n*
benching *(Bw)* 1. Abtreppung; 2. Betonbettung *f* für eine Rohrleitung
benchwood Tischlerholz *n*
bend / to [sich] biegen, durchbiegen; verknoten
b. over umbiegen
b. up aufbiegen *(Stahlbeton, Bewehrung)*
bend 1. Abbiegung *f*, Biegung *f*, Durchbiegung *f*, Krümmung *f*, Ausbiegung *f*; 2. Biegung, Kurve *f* *(Straßenbau)*; 3. Abbiegung, Aufbiegung *f* *(Stahlbeton, Bewehrung)*; 4. *(Masch)* Bogen[rohr *n*] *m*, Knierohr *n*, Krümmer *m*, Kröpfung *f*
b. coupling *(El)* Koppelbogen *m* *(Wellenleiter)*
b. number Biegezahl *f*
b. of the fork Gabelkrümmung *f*
b.-over test Faltprobe *f*, Faltversuch *m*
b. pipe Rohrkrümmer *m*
b. point Extrempunkt *m*
b. radius Biegehalbmesser *m*
b. specimen Biegeprobe *f*
b. test Biege[falt]versuch *m*
b. test ductility Biegefähigkeit *f*
b. with ball and socket joint *(Brg)* Kugelgelenkkrümmer *m*
bendability Biegbarkeit *f*, Falzbarkeit *f*
bendable biegbar, falzbar
bender Biegeapparat *m*
bending 1. Biegen *n*, Falzen *n*; 2. Biegeumformen *n*
b. and axial load Biegung *f* mit Längskraft *(Statik)*
b. and straightening machine Biege- und Richtmaschine *f*
b. and straightening press Biegerichtpresse *f*
b. arm Biegearm *m*, Biegebalken *m*, Spannarm *m*
b. axle Biegeachse *f*
b. block *(Holz)* Biegemodell *n*
b. brittle point Kaltbiegeschlagfestigkeit *f* *(Plaste usw.)*
b. coefficient Biegegröße *f*
b. coil *(Fs)* [Elektronenstrahl-]Krümmungsspule *f*
b. couple Biegemoment *n*
b. cycles Biegewechselzahl *f*
b. device Biegeeinrichtung *f*
b. dimension Biegemaß *n* *(Beton)*
b. distance Biegungsabstand *m* *(Spitzennadel)*
b. drum *(Holz)* Biegetrommel *f*
b. elasticity Biegeelastizität *f*
b. equation Biegegleichung *f* *(Gleichung des Biegemomentes)*
b. fatigue resistance Biegeschwellfestigkeit *f*
b. fatigue strength Biegedauerfestigkeit *f*

b. fixture Biegeeinrichtung *f*
b. force Biegekraft *f*
b. furnace *(Schiff)* Spantglühofen *m*
b. height Biegungshöhe *f*
b. horse *(Holz)* Biegebock *m*
b. impact test Biegeschlagversuch *m*
b. in two planes schiefe Biegung *f* *(Statik)*
b. iron *(Holz)* Biegezange *f*
b. line Biegelinie *f*, elastische Linie *f*
b. load Biegebelastung *f*, Biegebeanspruchung *f*
b. machine Biegemaschine *f*, Biegeapparat *m*
b. moment Biegemoment *n*
b.-off press *(Met)* Abkantpresse *f*
b. plate *(Holz)* Biegeblech *n*
b. pliers Biegezange *f*
b. point for the tip of the beard *(Text)* Spitzenbiegung *f* *(Spitzennadel)*
b. press Biegepresse *f*
b. property Biegefähigkeit *f*
b. punch Biegestempel *m*
b. quality Biegefähigkeit *f*
b. radius Biegeradius *m*, Biegehalbmesser *m*
b. rigidity Biegesteifigkeit *f*
b. ring *(Holz)* Biegering *m*, Biegungsring *m*
b. roll Biegewalze *f*
b. schedule Biegeplan *m* *(Stahlbeton, Bewehrung)*
b. shackle *(Schiff)* Ankerschäkel *m*, Verbindungsschäkel *m*
b. sketch Biegeskizze *f* *(Stahlbeton)*
b. stiffness Biegesteifigkeit *f*
b. strength Biegefestigkeit *f*
b. strength under alternating loading Wechselbiegefestigkeit *f*
b. stress Biegebeanspruchung *f*, Biegespannung *f*
b.-stress durability Dauerbiegefestigkeit *f*
b. stress fatigue limit Dauerbiegefestigkeit *f*
b. test Biegeversuch *m*
b. tool Biegewerkzeug *n*
b. vibration 1. Biegeschwingung *f*; 2. Knickschwingung *f*
b. vibration test Biegeschwingungsprüfung *f*
b. wrench Biegezange *f*
bendix drive Bendixtrieb *m* *(Anlasser)*
b. starter Bendixanlasser *m*
bends 1. *(Led)* halbe Croupons *mpl*; 2. *s.* caisson disease
beneficiation Aufbereitung *f*, Anreicherung *f* *(Erze)*
Benson boiler Bensonkessel *m*
bent *(Bw)* Bock *m*, Joch *n*, Portalrahmen *m*
bent gebogen
b. bar aufgebogener (abgebogener) Bewehrungsstab *m* *(Stahlbetonbau)*
b. bolt Hakenkopfschraube *f*
b. crystal gewölbter (gebogener) Kristall *m*
b. electron gun geknicktes Strahlerzeugungssystem *n* *(Elektronenstrahloszillografie)*
b. finishing tool gebogener Schlichtmeißel *m*
b. frame eingebogener Spant *m*
b.-hole boring Herstellen *n* von Bohrungen mit gekrümmter Achse
b. in a circle kreisförmig gebogen
b.-in timbers *(Schiff)* eingebogene Spanten *npl*
b. iron *(Schiff)* Kalfatereisen *n*
b. lathe tool gebogener Drehmeißel *m*
b. lever Winkelhebel *m*, Kniehebel *m*
b. pendulum arm *(Masch)* geknickter Pendelarm *m* *(Regler)*
b. pipe Bogenrohr *n*, Krümmer *m*
b. plywood gebogenes Sperrholz *n*
b. pointer geknickter Zeiger *m*
b. rasp Bildhauerraspel *f*
b.-shank tap Gewindebohrer *m* mit gebogenem Schaft
b. stanchion Rungenbügel *m*
b.-tail[ed] carrier gebogenes Drehherz *n*
b.-up aufgebogen
b.-up Aufbiegung *f*
b. wood gebogenes Holz *n*, Biegeholz *n*
b. wood turning machine Runddrehmaschine *f* für gebogene Hölzer
bentonite *(Bw)* Bentonit[ton] *m*
benzene Benzol *n*
b. blow torch Benzollötlampe *f*
b. nucleus Benzolkern *m*, Benzolring *m*
b. recovery plant Benzolwaschanlage *f*
benzidine Benzidin *n*
benzine Benzin *n*
benzoic Benzoe-
b. acid Benzoesäure *f*
benzol Benzol *n*
b. extraction Entbenzolung *f*
b. recovery Benzolgewinnung *f*
b. recovery plant Benzolfabrik *f*
b. scrubber Benzolwäscher *m*
benzole Benzol *n*
benzosulfimide Benzoesäuresulfimid *n*, Saccharin *n*
benzoylate / to benzoylieren
benzyl cellulose Benzylzellulose *f*
Berkefeld filter Kieselgurfilter *n* nach Berkefeld
berkelium *(Ch)* Berkelium *n*
Berlin blue Berliner Blau *n*
berm *(Bw)* Absatz *m*, Bankett *n*, Berme *f*
Bermuda ring *(Schiff)* Hochtakelung *f*, Bermudatakelung *f*, Focktakelung *f*
Bernoulli distribution Bernoullische Verteilung *f*
Bernoulli's equation Bernoullische Gleichung *f*
berries Beerenobst *n*
berry transformer Drehtransformator *m*

berth *(Schiff)* 1. Koje *f*, Schlafkoje *f*; Schlafplatz *m*; Schiffskammer *f*; 2. Liegeplatz *m*, Ankerplatz *m*
b.-deck *(Schiff)* Zwischendeck *n*, Wohnraumdeck *n*
b. for fuel-oil bunkering Tankanlage *f* [für Schiffe]
berthage *(Schiff)* 1. Ankerplatz *m*; 2. Kaigebühr *f*, Dockgebühr *f*
Berthelot's calorimeter Berthelotsche Bombe *f*
berthierite Eisenantimonerz *n*, Eisenantimonglanz *m*
berthing *(Schiff)* Außenbeplankung *f* über Schergang
b. dues Kaigebühr *f*, Dockgebühr *f*
b. master Hafenmeister *m*, Hafenkapitän *m*
beryll *(Min)* Beryll *m*
beryllium Beryllium *n*
b. bronze Berylliumbronze *f*
b. glass Berylliumglas *n*
b.-moderated reactor Berylliumreaktor *m*
b. oxide Berylliumoxid *n*
beset *(Schiff)* von Treibeis eingeschlossen
Bessel functions *(Math)* Besselsche Funktionen *fpl*, Besselfunktionen *fpl*, Zylinderfunktionen *fpl*
Bessemer blow Bessemerschmelze *f*; Windfrischschmelze *f*
Bessemer converter 1. Konverter *m* für Windfrischverfahren; 2. Bessemerkonverter *m*, Bessemerbirne *f (bei saurer Zustellung)*
Bessemer foundry *(Met)* Bessemerei *f*
Bessemer iron Roheisen *n* für das Windfrischverfahren
Bessemer low carbon steel Bessemerflußstahl *m (sauer)*
Bessemer pig iron Bessemerroheisen *n (bei saurer Zustellung)*
Bessemer process 1. Windfrischverfahren *n*; 2. Bessemerverfahren *n (bei saurer Zustellung)*; 3. Thomasverfahren *n (bei basischer Zustellung)*
Bessemer steel Bessemerstahl *m*
bessemerize / to *(Met)* bessemern
best bower anchor *(Schiff)* Reservebuganker *m*
beta absorption gauge Beta-Dickenmesser *m*
b.-active beta-[radio]aktiv
b.-active sample beta-aktive Probe *f*
b.-activity Beta-Aktivität *f*
b.-backscattering *(Kern)* Beta-Rückstreuung *f*
b.-brass Beta-Messing *n*
b. counter *(Kern)* Beta-Zähler *m*
b.-counting *(Kern)* Beta-Zählung *f*
b.-decay Beta-Zerfall *m*, Beta-Umwandlung *f*
b. disintegration *s.* b.-decay
b. disintegration energy Beta-Zerfallsenergie *f*
b.-electron Beta-Elektron *n*
b.-emission Beta-Strahlung *f*
b. emitter Beta-Strahler *m*
b. gauge Beta-Dickenmesser *m*
b. particle Beta-Teilchen *n*
b.-particle beam Beta-Strahlenbündel *n*
b.-particle emission Beta-Strahlung *f*
b.-phase Beta-Phase *f*
b.-radioactive beta-radioaktive
b.-radioactive substance beta-aktive Substanz *f*, Beta-Strahler *m*
b.-radioactivity Beta-Aktivität *f*
b.-ray Beta-Teilchen *n*, Beta-Strahl *m*
b.-ray emission Beta-Strahlung *f*
b.-ray spectrometer Beta-Spektrometer *n*
b.-stable beta-stabil
b.-thickness gauge *(Kern)* Beta-Dickenmeßgerät *n*
b. transformation Beta-Umwandlung *f*
b. transition Beta-Übergang *m*
b. uranium Beta-Uran *n*
betatron Betratron *n*, Elektronenschleuder *f*
b. orbit Betatronbahn *f*
b. oscillation Betatronschwingung *f*
Bethe-Weizsäcker progressive nuclear reaction Bethe-Weizsäcker-Zyklus *m (Kernreaktionszyklus, bei dem ein Heliumkern aus vier Wasserstoffkernen erzeugt wird)*
Bethell's process Kreosot-Imprägnierverfahren *n*, Vollimprägnierverfahren *n (Holzschutz)*
better / to *(Led)* zubessern
between deck *(Schiff)* Zwischendeck *n*
b.-grind tool life Standzeit *f* des Werkzeuges
b.-the-lens shutter Zentralverschluß *m (Kamera)*
bevatron *(Kern)* Bevatron *n*
bevel / to [ab]fasen, abflächen, abkanten, abschrägen, Kanten brechen, schmiegen
bevel 1. *(Masch)* Gehrung *f*, Schräge *f*, Schiefe *f*; Schmiege *f*, Gehrungswinkel *m*; 2. Haltering *m (bei Scheinwerfergläsern)*
bevel schräg, abgefast
b. cut *(Masch)* Schrägschnitt *m*; Gehrungsschnitt *m*
b. cutter *(Holz)* Kegelfräser *m*
b. cutting Schrägschnitt *m*; Gehrungsschnitt *m*; Gehrungssägen *n*
b. differential gear Kegelraddifferential *n*, Kegelradausgleichsgetriebe *n*
b. drive gear Antriebskegelrad *n*
b. drive pinion Antriebskegelrad *n*, Kegelantriebsritzel *n*
b. driving pinion *s.* b. drive pinion
b. facing Plandrehen *n* kegeliger Stirnflächen
b. friction wheel Kegelreibrad *n*
b. gear Kegel[zahn]rad *n*, konisches Getriebe
b. gear cutter Kegelradfräser *m*
b. gear cutting machine Kegelradfräsmaschine *f*

b. gear drive Kegelradantrieb *m*
b. gear generator Kegelradwälzfräsmaschine *f*
b. gear grinding machine Kegelradschleifmaschine *f*
b. gear milling machine Kegelradfräsmaschine *f*
b. gear pinion Kegelritzel *n*
b. gear planer Kegelradhobelmaschine *f*
b. gear planing machine Kegelradhobelmaschine *f*
b. gear wheel Kegelrad *n*
b. gear with involute-tooth spiral Evolventenzahnkegelrad *n*
b. gearing Kegelrädergetriebe *n*; Kegelradverzahnung *f (als System)*
b. hub *(Kfz)* Kegelnabe *f*
b. lead kegeliger Anschnitt *m*
b. lead length Anschnittlänge *f*
b. of a beam *(Holz)* Balkenschmiege *f*
b. plug *(BMSR)* Ventilteller *m*
b. protractor Anlege[winkel]messer *m*, Winkelmaß *n*
b. rule Stellwinkel *m*; Schmiege *f*
b. sawing *(Masch)* Gehrungssägen *n*
b. steel square Winkelmaß *n*, Winkelmesser *m*
b. way Gehrmaß *n*, Schrägmaß *n*
b. wheel Kegelrad *n*
b. wheel casing Kegelradgehäuse *n*
b. wheel drive Kegelradtrieb *m*
b. wheel train Kegelradgetriebe *n*
b. wheel transmission Kegelradtrieb *m*
bevelled kegelförmig, kegelig, konisch, verjüngt; abgeschrägt, gefast, schräg, gebrochen *(Kanten)*
b. cheek *(Masch)* abgeschrägte Wange *f*
b. circular saw konische Kreissäge *f*
b. corner gefaste (gebrochene) Kante *f*
b. corner halving *(Holz)* Eckàberblattung *f* mit schrägem Schnitt
b. edge *(Masch)* abgefaste Kante *f*
b. edge square *(Masch)* Haarwinkel *m*, Präzisionswinkel *m*
b. raising cam *(Text)* innen abgeschrägter Schlauchteil *m* eines Hebers *(Schlauchklappe)*
b. roughing tooth Vorschneidezahn *m*
b. steel square schrägabgekanteter Stahlwinkel *m*
b. steel straight edge *(Masch)* Haarlineal *n*, abgefastes Stahllineal *n*
b. teeth abgeschrägte Zähne *mpl (Zahnräder)*
beveller *(Masch)* Gehrungsschere *f*
bevelling 1. Abfasen *n*, Abschrägen *n*, Brechen *n*; 2. Fase *f*
b. plane *(Masch)* Gehrungshobel *m*
b. shear Gehrungsschere *f*
beverage aerial Langdrahtantenne *f*
bewoid size *(Holz)* Bewoid-Harzleim *m*, teilverseifter Harzleim *m*
beyond cut-off ins Sperrgebiet *(Gasentladungsröhren)*
bezel genuteter Ring *m*, geneigte Schneide *f*
bi-coiling [system] *(Text)* Zweibandsystem *n*
b.-fuel system *(Flg)* Zwei-Kraftstoff-Verfahren *n (verschiedene Oktanzahlen für Start und Reise)*
b.-post cap *(El)* Zweistiftsockel *m*, Sockel *m* mit zwei Stiften
biacetic base zweisäurige Base *f*
biacetyl Diacetyl *n*
bias / to vormagnetisieren; vorspannen
bias 1. *(Math)* systematischer Fehler *m*; 2. *(El)* Gittervorspannung *f*; Vormagnetisierungsstrom *m*
b. cell *(Fmt)* Gitterbatterie *f*
b. current Vormagnetisierungsstrom *m*
b. detector Gleichrichter *m* zur Erzeugung der Regelspannung
b. in *(Masch)* Schrägzahntragen *n (nach innen)*
b. magnetization Vormagnetisierung *f*
b. out Schrägzahntragen *n (nach außen)*
b. rectifier *(Fmt)* Gittergleichrichter *m*
b. resistor Widerstand *m* zur Erzeugung der Gittervorspannung
b. tooth bearing *(Masch)* Schrägzahntragen *n*
b. voltage Vorspannung *f*, Gittervorspannung *f*
b. winding *(Elm)* Vormagnetisierungswicklung *f*
biased automatic gain-control *(Fmt, Fs)* selbsttätige verzögerte Schwundregelung *f*
b. differential protection *(El)* Längsdifferentialschutz *m*
b. rectifier *(Elm)* vorgespannter Gleichrichter *m*, Vorsatzgleichrichter *m*
b. relay *(El)* einseitig eingestelltes Relais *n*; Schutzrelais *n* mit Vormagnetisierung, Prozentrelais *n*
b. resistor Katodenwiderstand *m (Schaltelement)*
biasing grid voltage Gitterverschiebespannung *f*, Gittervorspannung *f*
b. potential *(El)* Vorspannung *f*; Sperrspannung *f*
b. transformer *(Elm)* Differentialschutzwandler *m*
b. voltage Pendelrückkopplungsspannung *f*
biatomic zweiatomig
biaxial *(Opt)* zweiachsig
b. mica weißer Glimmer *m*
b. stress ebener (zweiachsiger) Spannungszustand *m*
bib cock Wasserhahn *m*
bibasic *(Ch)* zweibasisch

bible paper *(Typ)* Bibeldruckpapier *n*, Dünndruckpapier *n*
bicable *(Bw)* Zweiseilbahn *f*, Doppelseil-Schwebebahn *f*, Zweiseil-Schwebebahn *f*
bichromate of potassium Kaliumbichromat *n*
Bickford fuse *(Brg)* Zündschnur *f*
biconcave bikonkav
b. lens bikonkave Linse *f*, Bikonkavlinse *f*
biconditional *(BMSR)* zweifach bedingt *(z. B. logische Äquivalenz)*
bicone coupling Doppelkonuskupplung *f*
biconical Doppelkonus-
b. aerial Doppelkegelantenne *f*, Doppelkonusantenne *f*
biconvex bikonvex
b. lens bikonvexe Linse *f*, Bikonvexlinse *f*
bicrohm *(El)* 10^{-9} Ohm
bicuspid zweispitzig
bicycle Fahrrad *n*
b. track Rad[fahr]weg *m*
bicyclic *(Ch)* bizyklisch,
bidirectional 1. doppelsinnig *(Peilung)*; 2. nach beiden Richtungen
b. counter *(BMSR)* Vor-Rückwärtszähler *m*
b. motor Umkehrmotor *m*
b. pulses *(Fs)* Zweirichtungs[im]pulse *mpl*
biennial zweijährig; in zweijährigem Zyklus
bifid Y-förmig
bifilar cable zweiad[e]riges Kabel *n*, Zweifachkabel *n*, Zweileiterkabel *n*, Zwillingskabel *n*
b. coil Bifilarspule *f*, Spule *f* mit doppelter Drahtwicklung
b. oscillograph Schleifenoszillograf *m*
b. suspension *(El)* Doppelfadenaufhängung *f*, Bifilaraufhängung *f*, Zweifadenaufhängung *f*
b. winding *(El)* bifilare Wicklung *f*, Bifilarwicklung *f*, Doppelfadenwicklung *f*, Zweifadenwicklung *f*, Zweidrahtwicklung *f*
bifocal lens Bifokallinse *f*, Glas *n* mit zwei Brennweiten, Zweistärkenglas *n*
bifuel heterogener Treibstoff *m*
bifurcate / to sich gabeln, sich teilen
bifurcated rivet Zweispitzniet *m*
bifurcation Gabelung *f*; Abzweigung *f*; Zweiteilung *f*
b. buoy *(Schiff)* Spaltungstonne *f*
big bearing *(Kfz)* Pleuellager *n (an der Kurbelwelle)*
b. bell *(Met)* Unterglocke *f (Gichtverschluß)*
b.-end *(Masch)* kurbelwellenseitig
b. end *(Masch)* Kurbelzapfenende *n (Schubstange)*; Pleuelfuß *m*
b. end bearing Pleuellager *n*, Pleuelkopf *m*
b. end cap Pleuellagerdeckel *m (Motor)*
b. end hole großes Pleuelstangenauge *n*
b. size *(Opt)* Großformat *n*
bight 1. *(El)* Schleife *f (Kabel oder Draht)*; Ring *m (Kabel)*; 2. *(Schiff)* Doppelpart *m (n) (beim Knoten eines Taues)*
biharmonic biharmonisch
bilateral zweiseitig; bilateral; beiderseitig
b. area track Doppelzackenschrift *f (Tonaufzeichnung)*
b. characteristic *(Fmt)* Doppelkreisrichtcharakteristik *f*
b. sense *(Fmt)* Zweiseitenempfindlichkeit *f (Peilung)*
b. tolerance Plus- und Minus-Abweichung *f*, Toleranz *f*
b. transmission beiderseitiger Achsantrieb *m*
bile acid Gallensäure *f*
bilge *(Schiff)* Bilge *f*; Bauch *m* des Schiffes; Bodenrundung *f*; Kimm *f*; Kielraum *m*, Sodraum *m*
b. board Kimmschwert *n*
b. bracket Kimmstützplatte *f (bei einfachem Schiffsboden)*
b. ceiling Kimmwegerung *f*, Bilgewegerung *f*
b. chock Schlingerkiel *m*, Kimmkiel *m*
b. cock Bilgehahn *m*, Lenzhahn *m*
b. diagonal Kimmsente *f (Spantenriß)*
b. harpin Kimmsente *f*
b. keel Kimmkiel *m*, Schlingerkiel *m*
b. keelson Kimmkielschwein *n*, Kimmstringer *m*
b. line Lenzleitung *f*
b. log *s.* b. keelson
b. main Lenzleitung *f*, Bilgeleitung *f*, Hauptlenzleitung *f*
b. pipe Lenzrohr *n*
b. pipe line Lenzleitung *f*
b. planking Kimmbeplankung *f*, Kimmwegerung *f*
b. plate Kimmplatte *f*
b. pump Lenzpumpe *f*, Bilgepumpe *f*
b. rail Kimmreling *f*
b. shore Kimmstütze *f*
b. strake Kimmgang *m*
b. stringer Kimmweger *m*, Kimmstringer *m*
b. suction Lenzleitung *f*
b. wale Kimmweger *m (Holzschiffbau)*
b. water Bilgewasser *n*
b. well Bilge[lenz]brunnen *m*
bilinear form *(Ph)* Bilinearform *f*, bilineare Form *f*
bill band Banderole *f*
b.-feed device *(Büro)* Rechnungsvorschub *m*
b. of admeasurement *(Schiff)* Meßbrief *m*
b. of lading *(Schiff)* Konnossement *n*
b. of materials Stückliste *f (mit Werkstoffangabe)*
b. of quantities *(Bw)* Massenauszug *m*, Massenermittlung *f*, Massenverzeichnis *n*
b. of tonnage Meßbrief *m (Schiff)*

billet 1. *(Met)* Barren *m*, Strang *m*, Walzblock *m*, Knüppel *m*; 2. *(Plast)* Puppe *f*, Walzpuppe *f* *(Extruder)*; 3. *(Holz)* Klotz *m*, Scheit *n*
b. chute Ausfallrutsche *f* *(Walzwerk)*
b. drilling machine Rohknüppelbohrmaschine *f*
b. driver Knüppelbohrer *m*
b. mill *(Met)* Knüppelwalzwerk *n*; Mittelstraße *f*
b. of wood Vierkantholz *n* mit einer unbearbeiteten Seite
b. roll Knüppelwalze *f*, Vorwalze *f*
b. shear Knüppelschere *f*
billiard cue (queue) Billardstock *m*, Billardqueue *n*
billiards Billard *n*
billing *(Büro)* Rechnungsschreibung *f*
b. machine *(Büro)* Fakturiermaschine *f*, Rechnungsschreibmaschine *f*
billion 1. Billion *f*; 2. *(Am)* Milliarde *f*
billy Vorspinnmaschine *f*
bimetal Bimetall *n*
b. actinometer Bimetallaktinometer *n*
b. relay Bimetallrelais *n*
b. spring Bimetallfeder *f*
bimetallic bimetallisch
b. corrosion elektrochemische Korrosion *f*
b. rectifier Trockengleichrichter *m*
b. release Bimetallauslöser *m*
b. thermostatic strip *(El)* Bimetallregler *m*, Bimetallstreifen *m*
b. trip Bimetallauslöser *m*
b. wire Bimetalldraht *m*
bimodal *(Math)* zweigipfelig
bimolecular reaction bimolekulare Reaktion *f*
bin 1. Silo *m*, Bunker *m*; 2. *(Masch)* Behälter *m*, Fülltrichter *m*, Bunker [der Zuführeinrichtung]
b. batcher Bunkerdosierapparat *m*
b. level indicator Bunkerfüllstandszeiger *m*
b. vibrator Silorüttler *m*
binary *(Am)* *(BMSR)* Binärzählstufe *f*, Zählstufe *f*
binary 1. *(Math)* binär, dual; 2. *(Met)* binär, Zweistoff-
b. adder Binäraddierer *m*, Binäraddierglied *n*, Binäraddierstufe *f*
b. alloy *(Met)* Zweistofflegierung *f*, binäre Legierung *f*
b. cell *(BMSR)* binäre Speicherzelle *f* *(Datenspeicherung)*
b. code *(BMSR)* Binärkode *m*, Dualkode *m*
b.-coded decimal code Binärkode *m* für Dezimalziffern
b. coded decimal digit *(BMSR)* binär verschlüsselte Dezimalziffer *f*
b. coded decimal notation *(BMSR)* binär verschlüsselte Dezimaldarstellung (Dezimalschreibweise) *f*
b. collision *(Ph)* Zweierstoß *m*
b. converter *(Elm)* asynchroner Einankerumformer *m*
b. counter *(BMSR)* Binärzähler *m*, Binäruntersetzer *m*, Binärzählwerk *n*
b.-decision program *(BMSR)* Binärentscheidungsprogramm *n*
b. digit 1. *(Math)* Binärziffer *f*; 2. *s.* bit 5.
b. increments *(BMSR)* binäre Inkremente *npl* *(hervorgegangen aus ternären Inkrementen durch Nullausschluß)*
b. majority-decision element *(BMSR)* binäres Majoritätsentscheidungselement *n*
b. notation *(Math, BMSR)* Binärdarstellung *f* *(von Zahlen)*, Binärschreibweise *f*, Binärsetzung *f*
b. number Binärzahl *f*, Dualzahl *f*
b. point *(Math, BMSR)* Binärkomma *n*
b. punch binäre Lochung *f*
b. scaler *(Kern)* binärer Zähler *m*, Dualzähler *m*, Zweifachuntersetzer *m*
b. scaling circuit *(El)* Elektronenschalter *m*
b. solid-state logic circuit *(BMSR)* binäre logische Festkörperschaltung *f*
b. system *(Ch)* Zweistoffsystem *n*
b.-to-decimal conversion *(BMSR)* Binär-Dezimalumsetzung *f*
b. trigger *(BMSR)* binärer Trigger *m*
b.-valued variable *(BMSR)* logische Variable *f*
b. vapour engine Dampfmaschine *f* mit kombinierten Dämpfen
b. weighing encoder *(BMSR)* Verschlüßler (Umsetzer) *m* nach dem Subtraktionsprinzip
binaural effect Raumtoneffekt *m*
binder 1. *(Plast)* Bindemittel *n*, Harzträger *m*; 2. *(Typ)* Hefter *m*; Binder *m*; Buchbinder *m*; 3. *(Bw)* Binder, Strecker *m* *(Mauerwerk)*; Bügel *m* *(Stahlbeton)*; 4. *(Lw)* Mähbinder *m*
b. cloth *(Lw)* Bindetuch *n*
b. lever *(Masch)* Klemmhebel *m*
b. of moulding sand Formsandbindemittel *n*
b. picks *(Text)* Bindeschüsse *mpl*
b. screw Klemmschraube *f*
b. twine *(Lw)* Bindegarn *n*
b. warp *(Text)* Bindekette *f*
b. yarn *(Text)* Bindegarn *n*
binder's board Buchbinderpappe *f*
bindery Buchbinderei *f*
binding 1. Binden *n*; 2. *(Kfz)* Blockieren *n* *(bei Bremsen)*; 3. *(Masch)* Festfressen *n*; Anfressen *n* der Fasen des Bohrers; Klemmen *n*; 4. *(Brg)* Anlegen *n* *(Einstellung)*
b. agent Bindemittel *n*

b. apparatus *(Lw)* Bindevorrichtung *f (Sammelpresse)*
b. attachment *(Masch)* Bindevorrichtung *f*
b. beam Zange *f (Holzbau)*
b. bolt Klemmbolzen *m*; Feststellschraube *f*
b. cloth 1. *(Lw)* Bindetuch *n*; 2. Buchbindermusselin *m*
b.edge *(Typ)* Heftkante *f*, Heftrand *m*
b. energy *(Ph, Ch)* Bindungsenergie *f*
b. force 1. Klebkraft *f*; 2. Klemmkraft *f*
b. hoop *(Holz)* Bindereif *m (Böttcherei)*
b. joint *(El)* Drahtverbindung *f*
b. margin *(Typ)* Bundsteg *m*
b. material Bindemittel *n*; Klebstoff *m*
b. muslin Buchbindermusselin *m*
b. nut *(Masch)* Gegenmutter *f*
b. pick *(Text)* Einschnittschuß *m*, Grundschuß *m*
b. point Bindepunkt *m*
b. post 1. *(Masch)* Klemmschraube *f*, Verbindungsklemme *f*; 2. *(Fmt)* Anschlußklemme *f*; Anschlußschraube *f*, Apparateklemme *f*, Lochklemme *f*
b. power Bindefähigkeit *f*; Klemmfähigkeif *f*; Klebfähigkeit *f*
b. rafter *(Holz)* Bindersparren *m (Zimmerei)*
b. ripeness *(Lw)* Bindereife *f*
b. rivet *(Masch)* Heftniet *m*
b. screw Halteschraube *f*, Klemmschraube *f*, Spannschraube *f*
b. strake Verbindungsgang *m (Holzschiff)*
b. strips Einfaßleisten *fpl*
b. tape 1. Buchbinderklebestreifen *m*; 2. Isolierband *n*
b. thread Bindegarn *n*
b. warp *s.* binder warp
b. wire Bandagendraht *m*; Bindedraht *m Stahlbeton, Bewehrungsbau)*; Packdraht *m*
b. yarn Bindegarn *n*
bineutron *(Ph)* Doppelneutron *n*
bing *(Brg)* Bergehalde *f*
binistor Binistor *m*, Halbleitertetrode *f (Halbleiterelement mit Injektor als vierte Elektrode)*
b. switching device npn-Silizium-Schaltertetrode *f (Halbleitertechnik)*
binit *(Dat)* binit *(Binärziffer)*
binnacle *(Schiff)* Kompaßhaus *n*, Nachthaus *n*
binocular binokular, beidäugig
binocular [telescope] Doppelfernrohr *n*
binode *(El)* Binode *f*; Diode-Tetrode *f*; Diode-Triode *f*
binomial binomisch, zweigliedrig
binomial Binom *n*
b. coefficient Binomialkoeffizient *m*
b. corner *(El)* Binomialwinkelsystem *n (Wellenleiter)*
b. twist *(El)* Binomialtorder *m (Hohlleiter)*
binormal *(Math)* Binormale *f*
binoxide *(Ch)* Doppeloxid *n*
bio-aeration Abwasserbelüftung *f (Gesundheitstechnik)*
b.-cell Biozelle *f*
b.-technology Biotechnologie *f (Anpassung der Maschinen an den menschlichen Organismus)*
bioamplifier Bioverstärker *m*, Verstärker *m* bioelektrischer Signale
bioassay biologische Probe *f*
biocatalyst Biokatalysator *m*
biochemical oxygen demand biochemischer Sauerstoffbedarf *m (BSB)*
biochemistry Biochemie *f*
biocides *(Chemikalien, die zur Bekämpfung schädlicher Kleinlebewesen eingesetzt werden)*
biological dose *(Kern)* biologische Dosis *f*
b. filter Tropfkörper *m (Gesundheitstechnik)*
b. half-life *(Kern)* biologische Halbwertzeit *f*
b. hazard *(Kern)* biologische Gefährdung *f*
b. hole *(Kern)* Strahlenkanal *m* für biologische Versuche
b. oxygen demand biologischer Sauerstoffbedarf *m (BSB)*
b. response *(Kern)* biologische Reaktion *f*
b. shield *(Kern)* biologischer Schirm *m*
b. waste treatment Belebtschlammverfahren *n (Abwasserreinigung)*
bionics Bionik *f*
biophysics Biophysik *f*
Biot and Savart's law *(El)* Biot-Savartsches Gesetz *n*
biotite Biotit *m*, Magnesiaeisenglimmer *m*
bipartite zweiteilig
biphase zweiphasig, Zweiphasen-
b. alternating current Zweiphasen[wechsel]-strom *m*
b. connection Zweiphasenschaltung *f*
b. current Zweiphasen[wechsel]strom *m*
b. equilibrium Zweiphasengleichgewicht *n*
b. rectifier *(El)* Doppelweggleichrichter *m*
biplane *(Flg)* Zweidecker *m*, Doppeldecker *m*
b. fuselage model Rumpf-Zweidecker *m*
biplug *(El)* Doppelstecker *m*, Zwillingsstecker *m*, zweipoliger Stecker *m*
bipod mast *(Schiff)* Zweibeinmast *m*, Bipodmast *m*
b. mounting *(Mil)* Zweibeinlafette *f*
bipolar *(El)* zweipolig, doppelpolig; Doppelpol-
b. armature *(El)* zweipoliger Anker *m*
b. electrode system bipolare Elektrodenanordnung *f*, Bipolarsystem *n*
b. motor zweipoliger (doppelpoliger) Motor *m*, Doppelpolmotor *m*
b. winding *(El)* zweipolige Wicklung *f*
biprism Biprisma *n*
bipropellant [fuel] *(Rak)* Doppeltreibstoff *m*
bipyramid Bipyramide *f*

biquadratic *(Math)* biquadratisch
biquinary code *(BMSR)* Biquinärkode *m*
b. coded decimal notation *(BMSR)* biquinär verschlüsselte Dezimaldarstellung (Dezimalschreibweise) *f*
b. notation *(BMSR)* biquinäre Darstellung (Schreibweise) *f*, Biquinärsetzung *f* *(Grundzahl abwechselnd 2 und 5)*
birch Birke[nholz *n*] *f*
bird-feed Vogelfutter *n*
b. protection Vogelschutz *m*
b. sanctuary Vogelschutzgebiet *n*
bird's eye 1. *(Brg)* gesiebter Anthrazitgrus *m*; 2. *(Holz)* Vogelauge *n*
b.-eye spot effect *(Text)* Vogelauge-Effekt *m* *(Effektbindung der Weberei)*
b.-eye view Vogelperspektive *f*, Vogelschau *f*
b.-eye weave *(Text)* Pfauenaugebindung *f*, Vogelaugebindung *f*
b. mouth / to *(Bw)* aufklauen *(Holzbau)*
b.-mouth *(Bw)* Geißfuß *m*, Klaue *f* *(Holzbau)*
birdsmouthing *(Bw)* Aufklauung *f* *(Holzbau)*
birdsnesting Nesterbildung *f* *(z. B. Feuerbuchsrohrwand)*
birefringence Doppelbrechung *f*
b. of flow Strömungsdoppelbrechung *f*
birefringent *(Opt)* doppelbrechend
Birmingham wire gauge *(Masch)* englische Drahtlehre *f*
birthplace Entstehungsort *m*
biscuit Metallschwamm *m*, Regulus *m*; Bisquit *m*
b. porcelain Biskuitporzellan *n*, unglasiertes Porzellan *n*
bisect / to zweiteilen, halbieren; sich gabeln
bisecting halbierend
b. line of an angle Winkelhalbierende *f*
bisection Halbierung *f* *(Winkel)*
bisector Halbierungslinie *f*, Mittellinie *f*; Mittelebene *f*
bisectrix *(Math)* Winkelhalbierende *f*, Bisektrix *f*, Mittellinie *f*
bisilicate Doppelsilikat *n*
Bismarck brown Bismarckbraun *n* *(Anilinbraun)*
bismuth Wismut *n*
b. fission Wismutspaltung *f*
b. litharge Wismutglätte *f*
b. metal Wismutmetall *n*
bismuthic compound Verbindung *f* des fünfwertigen Wismut
bismuthous compound Verbindung *f* des dreiwertigen Wismut
bissel axle Laufachse *f*
b. bogie Deichselgestell *n*, Laufdrehgestell *n*
b. truck Deichselgestell *n*, Laufdrehgestell *n*
b. wheel set Laufradsatz *m*
bistable multivibrator *(BMSR)* Flip-Flop-Generator *m*, bistabiler Multivibrator *m*, Eccles-Jordan-Schaltung *f*
bistoury Skalpell *n*
bistre Bister *m* *(n)* *(braune Aquarellfarbe)*
bit 1. *(Masch)* Bohrmeißel *m*; Bohrkrone *f*; Einsteckmeißel *m*; Einsatzdrehmeißel *m*; Bohreinsatz *m*, Bohrschneide *f*; 2. *(Am)* Bohrstange *f* mit Schneide, Bohrer *m*; 3. Lötkolben *m*; 4. Hobeleisen *n*; 5. *(BMSR)* *(binary digit)* Bit *n* *(Einheit des Informations- oder Nachrichteninhaltes)*
b. brace *(Masch)* Bohrwinde *f*, Windenbohrer *m*
b. for rifle barrel drilling Gewehrlaufbohrer
b. hook Meißelfanghaken *m* *(Erdöl, Erdgas)*
b. stock drill Leierspiralbohrer *m*
bitartrate *(Ch)* Hydrogentartrat *n*
bitch *(Bw)* Bauklammer *f*
bite / to 1. einhaken *(Werkzeug)*; fassen; 2. *(Led)* fassen *(Farbe)*; 3. *(Plast)* anlösen, angreifen
bite 1. *(Holz)* Sägeschnitt *m*; 2. *(Text)* Berührungsstelle *f* der Walzen
b. of rolls Walzenangriff *m*, Fassen *n* der Walzen
bitermitron *(El)* Rückwärtswellenröhre *f* *(als M-Oszillator und M-Verstärker)*
bitt *(Schiff)* Poller *m*, Beting *m* *(f)*
bitter almond oil Bittermandelöl *n*
b. earth *(Ch)* Magnesia *f*
b. principle *(Ch)* Bitterstoff *m*
b. spar *(Ch)* Bitterspat *m*, Dolomit *m*, Kalzium-Magnesium-Karbonat *n*
bitton *(Schiff)* [großer] Poller *m*
bitumen Bitumen *n*, Bergpech *n*, Erdharz *n*, Erdpech *n*
b. boiler Bitumenkocher *m*
b. cable asphaltiertes Kabel *n*
b. felt Bitumenpappe *f*, Teerpappe *f*
b. varnish Bitumenlack *m*
bituminate / to asphaltieren
bituminiferous asphalthaltig, bitumenhaltig
bituminizing Imprägnierung *f* mit Bitumen
bituminous bituminös, erdpechartig, pechhaltig
b. board *(Holz)* bituminöse Platte *f*
b. coal Backkohle *f*, Fettkohle *f*, pechhaltige Kohle *f*, Steinkohle *f*
b.-coal carbonization Steinkohlenentgasung *f*
b.-coal gas Steinkohlengas *n*
b. concrete Asphaltbeton *m*
b. felt Bitumenpappe *f*, Teerpappe *f*
b. lignite ölreiche Braunkohle *f*
b. paint Bitumenanstrich *m*
b. pavement Schwarzdecke *f* *(Straßenbau)*
b. paver finisher Schwarzdeckenfertiger *m*, Schwarzdeckeneinbaumaschine *f* *(Straßenbau)*

b. shale Brandschiefer *m*, Kohlenschiefer *m*, pechhaltiger Schiefer *m*
b. slate Porzellanschiefer *m*
biunique *(Math)* eineindeutig
bivalence *(Ch)* Zweiwertigkeit *f*
bivalent *(Ch)* zweiwertig
bivariant function generator *(BMSR)* bivarianter Funktionsgenerator *m* *(Analogrechner)*
bivariate *(Math)* mit zwei Merkmalsvariablen
bivinyl *(Ch)* Butadien *n*, Divinyl *n*
b. rubber Butadienkautschuk *m*
black / to schwärzen
b. finish brünieren
black Ruß *m*, Moorkohle *f*, Schwärze *f*
black 1. schwarz; 2. übergar *(Roheisen)*
b. after white *(Fs)* Schwarz *n* hinter Weiß, Fahnenziehen *n*
b.-and-white film Schwarzweißfilm *m*
b.-and-white transmission *(Fs)* Einfarbenübertragung *f*
b. annealing *(Met)* Schwarzglühen *n*
b. ash Rohsoda *f*, unreine Soda *f*; Schwarzschmelze *f*
b.-ash revolver *(Ch)* Drehofen *m* *(Sodaverfahren nach Leblanc)*
b. ball *(Schiff)* Ankerball *m*, Signalball *m*, Sturmball *m*
b. bat *(Geol)* Brandschiefer *m*, Kohlenschiefer *m*
b. blend Pechblende *f*, Uranerz *n*
b. body *(Ph)* schwarzer Körper *m*
b. body radiation *(Ph)* schwarze Strahlung *f*, Hohlraumstrahlung *f*
b. body temperature *(Ph)* schwarze Temperatur *f*
b. bolt industry Rohschraubenindustrie *f*, Schwarzschraubenindustrie *f*
b. book *(Schiff)* Schwarzes Buch *n* *(Lloyds Buch der Schiffsverluste)*
b. cargo *(Schiff)* Schwerölladung *f*
b. copper Rohkupfer *n*, Schwarzkupfer *n*
b. detector *(Kern)* schwarzer (total absorbierender) Detektor *m*
b. ends *(Brg)* Abbrand *m*
b. fuel peat Pechtorf *m*
b. game Schwarzwild *n*
b. gang *(sl)* Maschinenpersonal *n*
b. heart malleable cast iron Schwarz[kerntemper]guß *m*
b.-heart process amerikanisches Temperverfahren *n*
b. hematite Hartmanganerz *n*
b. japan [feiner] Schwarzlack *m*
b. japanned schwarz lackiert
b.-knotty *(Holz)* schwarzästig
b. lead Graphit *m*, Schwärze *f*
b.-lead crucible Graphittiegel *m*
b. level *(Fs)* Austastpegel *m*, Dunkelsteuerungspegel *m*; Schwarzpegel *m*
b.-level control *(Fs)* automatische Schwarzwerthaltung *f*
b.-level picture signal *(Fs)* Oberstrich *m*
b.-line paper Lichtpauspapier *n*
b. liquor *(Ch)* Ablauge *f*, Natronablauge *f*, Schwarzlauge *f*
b. locust falsche Akazie *f*
b. oil Rohschmieröl *n*
b. peak Schwarzspitze *f*
b. photo-paper Fotoeinwickelpapier *n*, schwarzes Lichtschutzpapier *n*
b. pig iron übergares Roheisen *n*
b. plate Schwarzblech *n*
b. positive paper Lichtschutzpapier *n*
b. powder *(Brg)* Schwarzpulver *n*, Sprengpulver *n*
b. powder pellets Schwarzpulvertabletten *fpl*, Sprengsalpeter *m*
b. prong *(Kern)* schwarze (schwere) Spur *f*
b. rope *(Schiff)* geteertes Tau *n*
b. screen television set Fernsehempfänger *m* mit Grauscheibe
b. sheet-iron Schwarzblech *n*
b. shortness *(Met)* Schwarzbruch *m*
b. spotter *(Fs)* Entstördiode *f*; Störschutzdiode *f*
b. squall *(Schiff)* schwarze Bö *f*
b. tin Schwarzblech *n*
b. top pavement *(Bw)* Schwarzdecke *f*
b. top paver finisher *(Bw)* Schwarzdeckenfertiger *m*, Schwarzdeckeneinbaumaschine *f*
b. wash Schwärze *f* *(Gießerei)*
blackdamp *(Brg)* matte Wetter *pl*
blacken / to schwärzen
blackening Schwärzung *f* *(Fototechnik)*
blacker-than-black Ultraschwarz-
b.-than-black pulse *(Fs)* Synchronisierimpuls *m*
b.-than-black region *(Fs)* Ultraschwarzpegel *m*
blacking 1. Schwärze *f* *(Formerei)*; 2. Schwärzen *n*
blacklistener Schwarzhörer *m* *(Radio)*
blackout *(El)* Austasten *n*
b. effect *(Fmt, Fs)* Betäubung *f*; Empfindlichkeitsverlust *m* einer Elektronenröhre nach kurzem Impuls
b. fading *(Fmt)* Tiefschwund *m*
b. headlight *(Mil)* Tarnscheinwerfer *m*
b. level *(El)* Austastpegel *m*
b. marker Dunkelpunkt *m* *(Radar)*
b. of flyback *(Fs)* Rücklaufverdunkelung *f*
b. paper Verdunkelungspapier *n*
b. pulse *(Fs)* Austastimpuls *m*, Austastsignal *n*
blacksmith Schmied *m*

b. welding Hammerschweißen *n*, Feuer[preß]schweißen *n*
blacksmiths' anvil Schmiedeamboß *m*
b. chisel Schrotmeißel *m*, Meißel *m* für Metall
b. flatter Schlichthammer *m (Schmiedewerkzeug)*
b. fuller Ballhammer *m*
b. hammer Schmiedehammer *m*
b. hardy Abschröter *m*, Abschrot *m*, Schröter *m*, Umschlageisen *n*
b. tongs Schmiedezange *f*, Stockzange *f*
b. tool Schmiedegerät *n*, Schmiedewerkzeug *n*
b. vice Schmiedeschraubstock *m*
blackstrap Melasse *f (Rohrzuckerherstellung)*
bladder Speicherblase *f*
blade 1. Blatt *n (z. B. Sägeblatt)*; 2. Schaufel *f (Turbine)*; 3. Messer *n (Fräser)*; 4. Klinge *f (Schraubenzieher)*; 5. Schneide *f*; 6. Zunge *f (Schieblehre)*; 7. Schiene *f*; 8. Band *n*; 9. Schenkel *m*; 10. Schild *m (Planierraupen usw.)*; 11. *(Typ)* Rakel *f*, Abstreichklinge *f*
b. angle Schaufelwinkel *m (Turbine)*
b. area *(Flg)* Blattfläche *f (Luftschraube)*
b. area ratio *(Schiff)* Flügelflächenverhältnis *n*
b. base Schaufelblock *m*, Schaufelfuß *m (Turbine)*
b. break down Schaufelsalat *m (Turbine)*
b. carrier Schaufelträger *m (Turbine)*
b. clearance *(Schiff)* Schraubenblattfreigang *m*
b. contact *(Opt)* Messerkontakt *m*
b. cooling Schaufelkühlung *f (Turbine)*
b. crusher *(Brg)* Flügelbrecher *m*
b. curvature Schaufelkrümmung *f (Turbine)*
b. curved throughout bis zur Austrittsstelle gekrümmte Schaufel *f (Turbine)*
b. depth *(Schiff)* Flügelbreite *f*
b. disk Laufradscheibe *f (Turbine)*
b.-edge microscope Schneidenschattenmikroskop *n*
b. efficiency *(Schiff)* Flügelwirkungsgrad *m*
b. exit angle Laufschaufelaustrittswinkel *m* β_2 *(Turbine)*
b. face 1. Schaufelfläche *f (Turbine)*; 2. Blattvorderseite *f*, Druckseite *f (Luftschraube)*
b. for crosscut saw frame Bügelsägeblatt *n*
b. grader *(Bw)* Motorstraßenhobel *m*, Straßenhobel *m*
b. grinder Messerschleifmaschine *f*, Scherenschleifmaschine *f*
b. groove Befestigungskerbe *f*, Schaufelkerbe *f (Turbine)*
b. guide Messerführung *f (Kontakte)*
b. head Schaufelkopf *m*, Schaufelspitze *f (Turbine)*
b. height Schaufellänge *f (Turbine)*
b. holder Blatthalter *m (Säge)*
b. hollow Schaufelhöhlung *f (Turbine)*
b. inlet angle Laufschaufeleintrittswinkel *m* β_1 *(Turbine)*
b. leading edge Schaufeleintrittskante *f (Turbine)*
b. life Messerschneidenstandzeit *f*
b.-like control electrode Steuersteg *m (Abstimmanzeigeröhre)*
b. locked by screws aufgeschraubtes Messer *n*
b.-locking insert Schaufelschloß *n (Turbine)*
b. loss Schaufelverlust *m (Turbine)*
b. of agitator Rührwerkflügel *m*
b. of hockey stick Schlägerschaufel *f (Eishockey)*
b. of sculls Ruderblatt *n*
b. of skate Schlittschuhkufe *f*
b. pitch Schaufelteilung *f (Turbine)*
b. pressure 1. *(Wärm)* Schaufeldruck *m (Turbine)*; 2. Sägeblattdruck *m*
b. rim Schaufelkranz *m (Turbine)*
b. ring Laufschaufelkranz *m*
b. root Schaufelfuß *m (Turbine)*
b. root fixation Schaufel[fuß]befestigung *f (Turbine)*
b. section 1. Schaufelprofil *n (Turbine)*; 2. *(Flg)* Luftschraubenprofil *n*
b. spacing Schaufelteilung *f (Turbine)*
b. speed 1. Schaufelgeschwindigkeit *f (Turbine)*; 2. *(Flg)* Luftschraubengeschwindigkeit *f*
b. stopper Endpaßstück *n*, Schlußstück *n (Turbine)*
b. tip 1. Schaufelspitze *f (Turbine)*; 2. *(Flg)* Luftschraubenblattspitze *f*
b. trailing edge Schaufelaustrittskante *f (Turbine)*
b.-type potato digger (grubber) Kartoffelrodepflug *m*
b. velocity 1. Schaufelgeschwindigkeit *f (Turbine)*; 2. *(Flg)* Luftschraubengeschwindigkeit *f*
b. wheel Laufrad *n*, Schaufelrad *n (Turbine)*
b. width 1. Schaufelbreite *f (Turbine)*; 2. Blattbreite *f (Sägeblatt)*
b. with convex tooth edge gebauchtes (gekrümmtes) Blatt *n (Säge)*
b. with prongs Schaufel *f* mit Gabelzinken *(Turbine)*
b. with round breast *(Masch)* gebauchtes (gekrümmtes) Blatt *n (Säge)*, Sägeblatt *n* mit gebauchter Zahnspitzenlinie
bladed wheel Schleuderrad *n*
blades broken off the wheel Schaufelsalat *m (Turbine)*
blading 1. Einplanieren *n*, Planieren *n (Straßenbau)*; Schaufelanordnung *f (Turbine)*
b. material Schaufelwerkstoff *m*

b. reaction Überdruckbeschaufelung *f (Turbine)*
blae bituminöser Sandstein *m (Schottland)*
blake thread Durchnähzwirn *m (Schuh)*
blank / to 1. *(Masch)* [aus]stanzen, [aus]schneiden *(Rohlinge)*; 2. *(Met)* blank machen
blank 1. *(Met)* Rohteil *n*, Rohling *m*, Rohstück *n*, rohes Werkstück *n*, Werkstückrohling *m*, Ausgangsteil *n*, rohes Arbeitsstück *n*; Formteil *n*; Platine *f*, Zuschnitt *m*; Schnitteil *n*; Blechzuschnitt *m*; 2. Rohling *(Kunststoffverarbeitung)*; 3. *(Ch)* Blindprobe *f (Analytik)*; 4. *(BMSR)* Leerstelle *f*, Leertaste *f*, Zwischenraum *m*; 5. *(Opt)* Leeraufnahme *f*
b. bar *(Fs)* Bildsynchronisierlücke *f*
b. bolt Schraubenrohling *m*
b. card *(Büro)* Leerkarte *f (z. B. Lochkarte vor der Verarbeitung)*
b. cartridge Platzpatrone *f*
b. column detector *(Büro)* Leerspaltensucher *m (z. B. bei Lochkarten)*
b. cook *(Holz)* Blindkochung *f*
b. door *(Bw)* vorgetäuschte Tür *f*
b. flange *(Wärm)* Blindflansch *m*, Deckelflansch *m*
b. hardness test Blindhärteversuch *m*
b. line *(Typ)* Blindzeile *f*
b. moulding *(Plast)* geharzte Pappe *f*
b. of a file unbehauener Feilenkörper *m*
b. offset *(Masch)* Radkörperversetzung *f*
b. out signal *(Fs)* Dunkelsteuerzeichen *n*
b. page *(Typ)* Vakat *n*, leere Seite *f*
b. sheet *(Typ)* unbedruckter (blinder) Bogen *m*, Mönchsbogen *m*
b. spot *(El, Fs)* eingebrannter Leuchtschirmfleck *m*
b. tape *s.* newtape
b. trial *(Ch)* Blindversuch *m*
b. value *(Opt)* Leerwert *m (Fotometrie)*
b. wall *(Bw)* durchgehendes Mauerwerk *n*, Mauer *f* ohne Öffnungen
b. window *(Bw)* vorgetäuschtes Fenster *n*
blanker Vorschmiedegesenk *n*
blanket 1. wollene Bettdecke *f*, Wolldecke *f*; 2. *(Bw)* Dichtungsschürze *f*; 3. *(Typ)* Drucktuch *n*, Druckfilz *m*; 4. *(Kern)* Schutzabdeckung *f*; Brutmantel *m*, Brutschild *m*, Brutzone *f*; 5. Schutzdecke *f*, Abdeckung *f (eines Schmelzbades)*
b. coating *(Am) (Led)* Gummituchstreichverfahren *n (Beschichtung)*
b. material Mantelmaterial *n*
b.-material exposure *(Kern)* Bestrahlung *f* der Brutzone
b. printing machine Deckendruckmaschine *f*
b. stitch Zierstich *m (Stickerei)*
b.-to-blanket *(Typ)* Gummi-gegen-Gummi *(gleichzeitiger Schön- und Widerdruck im Offset)*
blanketing 1. *(Fs, Fmt)* Signalüberdeckung *f*; 2. *(Schiff)* Abdecken *n*
b. clouds *(Ch)* giftige Nebelstoffe *mpl*
b. frequence Abdeckungsfrequenz *f*
b. gas *(Kern)* Schutzgas *n*
blankholder *(Masch)* Blechhalter *m*, Festhalter *m (Ziehtechnik)*
b. ring *(Masch)* Blechhaltering *m*
blankholding pressure Blechhalterdruck *m*
blanking 1. *(Fs)* Austastung *f (Strahlrücklauf in Katodenstrahlröhren)*; 2. Ausschneiden *n*, Stanzen *n*
b. die *(Masch)* Stanzmatrize *f*; Schnittwerkzeug *n*; Schnittplatte *f*
b. interval *(Fs)* Austastlücke *f*, Schwarzlücke *f*
b. level *(Fs)* Austastpegel *m*, Schwarzpegel *m*
b. out *(Fs)* Dunkelsteuerung *f*, Austastung *f*
b.-out pulse *(Fs)* Schwarzaustastimpuls *m*
b. press *(Masch)* Schneidpresse *f*, Schnittpresse *f*
b. pulse *(Fs)* Austastimpuls *m*, Austastsignal *n*; Schwarz[tast]impuls *m*; Zeilenlöschimpuls *m*
b. signal *(Fs)* Dunkelsteuerzeichen *n*, Austastsignal *n*, Schwarzsignal *n*
b. time *(Fs)* Sperrzeit *f*
b. tool Formgebewerkzeug *n*, Stanzwerkzeug *n*
b. voltage Sperrspannung *f*
blashed überröstet *(Flachs)*
blast / to *(Bw)* schießen, sprengen
blast 1. Explosionsdruck[welle *f*] *m*; Sprengung *f*; 2. Gebläse[luft *f*] *n*, Wind *m*
b. cleaning Putzstrahlen *n*
b. cupola *(Met)* Gebläsekupolofen *m*
b. furnace Blasofen *m*, Gebläse[schacht]ofen *m*, Hochofen *m*
b. furnace armour Hochofenpanzer *m*
b. furnace bear Hochofensau *f*
b. furnace bell Gichtglocke *f*
b. furnace blower Hochofengebläse *n*
b. furnace bustle pipe Hochofenwindring *m*
b. furnace casting Hochofenabstich *m*
b. furnace cement Hochofenzement *m*
b. furnace charging Hochofenbegichtung *f*, Hochofenbeschickung *f*
b. furnace charging gallery Gichtbühne *f*
b. furnace coke Hüttenkoks *m*, Hochofenkoks *m*
b. furnace dust Gichtstaub *m*
b. furnace elevator Gichtaufzug *m*, Schrägaufzug *m* am Hochofen
b. furnace foamed slag Hochofenschaumschlacke *f*
b. furnace framework Hochofengerüst *n*
b. furnace gas Gichtgas *n*, Hochofengas *n*
b. furnace gas engine Gichtgasmaschine *f*, Gichtgasmotor *m*

b. furnace gas take Gichtgasfang *m*
b. furnace gun Stichlochstopfmaschine *f* *(Hochofen)*
b. furnace hearth Hochofengestell *n*
b. furnace hopper Hochofengicht *f*
b. furnace jacket Hochofenpanzer *m*
b. furnace lid Hochofenkranz *m*
b. furnace lines Hochofenprofil *n*
b. furnace lining Hochofenfutter *n*
b. furnace process Hochofenverfahren *n*
b. furnace salamander Hochofensau *f*
b. furnace slag Hochofenschlacke *f*
b. furnace stack Hochofenschacht *m*
b. furnace stack casing Hochofenschachtpanzer *m*
b. furnace with bucket hoist Hochofen *m* mit Kübelbegichtung
b. furnace with closed front Hochofen *m* mit geschlossener Brust
b. furnace with skip hoist Hochofen *m* mit Kippgefäßbegichtung
b. furnace without appendices Hochofensumpf *m*
b. furnaceman Hochöfner *m*, Hüttenmann *m*
b. heating apparatus Winderhitzer *m*, Cowper *m*
b. hole Spreng[bohr]loch *n*
b. inlet Windform *f*
b. main Hauptwindleitung *f* *(Hochofen)*
b. pipe Windzuleitung *f*
b. pressure 1. Detonationsdruck *m*, Explosionsdruck *m*; 2. Gebläsedruck *m*, Luftdruck *m*, Winddruck *m*
b. process Windfrischen *n*
b. roasting *(Met)* Verblaseröstung *f*, Sinterröstung *f*
b.-roasting plant Sinteranlage *f*
b. scaling laws Gesetze *npl* der Expansionswelle
b. side Windseite *f* *(Hochofen)*
b. sintering process Dwight-Lloyd-Sinterverfahren *n*
b. superheater Luftüberhitzer *m*
blasted grain brandiges Korn *n*
blaster Schießhauer *m*, Schießberechtigter *m*, Sprengmeister *m*
blasting 1. Putzstrahlen *n*; 2. Sprengen *n*, Sprengung *f* *(mit Sprengstoffen)*; 3. *(Brg)* Sprengarbeit *f*, Schießarbeit *f*; 4. *(El)* Übersteuerung *f* *(Mikrofon)*
b. accessories *(Brg)* Zündmittel *npl*
b. agent Sprengmittel *n*
b. bomb Sprengbombe *f*
b. cap Sprengkapsel *f*, Zündhütchen *n*
b. cartridge Sprengpatrone *f*
b. charge Sprengladung *f*
b. fuse Zündschnur *f*
b. grit Gebläsekies *m*
b. ignition cable Sprengkabel *n*
b. machine *(Brg)* Zündmaschine *f*
b. method *(Ch)* Verblaseverfahren *n*
blaze bituminöser Sandstein *m* *(Schottland)*
bleach / to bleichen, entfärben
b. out ausbleichen
bleach Bleiche *f*, Bleichflüssigkeit *f*, Bleichmittel *n*
b. boiler *(Holz)* Kugelkocher *m*
b. manufacture Bleichlaugenbereitung *f*
b. mixing plant Bleichlaugenzubereitungsanlage *f*
bleachability Bleichbarkeit *f*
bleachable bleichbar
bleached cloth gebleichte Leinwand *f*
b. goods *(Text)* Bleichartikel *mpl*, Weißware *f*
b. kraft gebleichtes Kraftpapier *n*
b. pulp gebleichter Zellstoff (Papierfaserstoff) *m*
b. warp *(Text)* gebleichte Kette *f*
bleaching Bleiche[n *n*] *f*, Bleichung *f*, Entfärbung *f*
b. agent Bleichmittel *n*
b. assistant Bleichhilfsmittel *n*
b. bath *(Text)* Bleichflotte *f*
b. earth Bleicherde *f*
b. engine *(Holz)* Bleichholländer *m*
b. in rope form *(Text)* Strangbleicherei *f*
b. lime Bleichkalk *m*, Bleichpulver *n*
b. liquid Bleichlauge *f*, Chlorkalklösung *f*
b. of coloured goods *(Text)* Buntbleiche *f*
b. powder Bleichkalk *m*, Bleichpulver *n*; Chlorkalk *m*
b. process Ausbleichverfahren *n*
b. pulp leicht bleichbarer Zellstoff *m*
b. requirements Bleichmittelbedarf *m*
b. stabilizer Bleichstabilisator *m*
b. tower *(Holz)* Bleichturm *m*
bleed / to 1. [ab]bluten; abzapfen; aus kleinen Öffnungen abfließen; ausschwitzen; verlaufen, verwaschen; 2. *(Kfz)* entlüften *(Bremse)*
b. off 1. ablassen, ableiten, abströmen lassen, abzapfen; 2. *(Brg)* abfackeln
bleed 1. *(Ch)* Ausblühen *n*; Durchlaufen *n* der Feststoffe *(Filtration)*; 2. Bluten *n*
b. air Zusatzluft *f*, Nebenluft *f*
b. cock *(Schiff)* Leckschraube *f*
b. hole *(Flg)* kleine Luftöffnung *f*
b. plug *(Kfz)* Entlüftungsstopfen *m*
b. screw *(Kfz)* Entlüftungsschraube *f*
b. steam Anzapfdampf *m*, Entnahmedampf *m*
b.-type pneumatic relay *(BMSR)* pneumatischer Verstärker *m* mit dauerndem Luftverbrauch, abblasender pneumatischer Verstärker
bleeder 1. Ablaßventil *n* *(Hydraulik)*; 2. *(El)* Ableitwiderstand *m*; Schutzwiderstand *m*,

Entladewiderstand *m*; Vorbelastungswiderstand *m (für Gleichrichter)*
b. current Vorbelastungsstrom *m*, Anzapfstrom *m*
b. pipe *(Brg)* Fackelrohr *n*
b. resistance *(El, BMSR)* Ableitwiderstand *m*, Parallelwiderstand *m*
b. turbine Anzapfturbine *f*, Entnahmeturbine *f*
b. valve *(Masch)* Anzapfventil *n*, Ablaßventil *n*, Abblaseventil *n*
bleeding 1. *(Wärm)* Dampfentnahme *f (Anzapfung)*; 2. *(Text)* Ausbluten *n*, Bluten *n*
b. coil *(El)* Erdungsdrossel[spule] *f*, Entladespule *f*
b. of ingots Auslaufen *n* nicht erstarrter Blöcke
blemish 1. Makel *m*, Flecken *m*, Fehler *m*; 2. Speicherfehler *m*
blend / to mischen, [ver]mengen, verschmelzen; verschneiden *(Flüssigkeiten)*; *(Text)* gattieren
b. gasoline Benzin verschneiden
b. into each other *(Text)* ineinander übergehen, verlaufen, verschwimmen *(Farbschattierung)*
blend 1. Mischung *f*; Verschnitt *m*; 2. Mischgarn *n*
b. for two-strokers *(Kfz)* Zweitaktmischung *f*
b.-in button Tricktaste *f (Magnettontechnik)*
blende Zinkblende *f*
blended design *(Text)* Melangemuster *n*
b. fabric Melangegewebe *n*
b. gasoline *(Am) (Kfz)* Verschnittbenzin *n*
b. oil *(Kfz)* gemischtes (verschnittenes) Öl *n*
b. worsted Kammgarngewebe *n*
b. yarn Melangegarn *n*
blender Mischer *m*
blending agent Verschnittmittel *n*
b. chart Viskogramm *n (Erdöl)*
b. component Mischungskomponente *f*
b. feeder *(Text)* Mischwolf *m*
b. intersecting frame *(Text)* Mischstrecke *f*
b. machine *(Text)* Mischmaschine *f*
b. ratio *(Text)* Mischungsverhältnis *n*
b. room *(Text)* Mischraum *m*
b. value Mischwert *m*, Mischoktanzahl *f (Erdöl)*
b. willow *(Text)* Mischwolf *m*
blight 1. Verfall *m (von Gebäuden)*; 2. Brandpilz *m*
b. of trees Baumbrand *m*
blighted area *s.* b. district
b. district Elendsviertel *n*, sanierungsbedürftiges Gebiet *n*, Slums *pl*; überalterte Bebauung *f*
blimp *(Foto)* Kamerahülle *f*
blind / to 1. *(Brg)* verstopfen; 2. blenden
blind 1. *(Brg)* blind, nicht zutage tretend; 2. verdeckt, nicht offen, geschlossen
blind Jalousie *f*, Rollade *f*, Rollvorhang *m*, Rouleau *n*
b. alley Sackgasse *f*
b. approach beacon system Leitstrahllandeverfahren *n*
b. approach procedure *(Flg)* Blindanflugverfahren *n*
b. arch Blendbogen *m*
b. area Schattengebiet *n (Radar)*
b. beam *(Schiff)* Lukenschiebebalken *m*
b. blocking *(Typ)* Blindprägung *f*, Heißprägung *f (Buchbinderei)*
b. chintz *(Text)* Vorhangchintz *m*
b. coal Magerkohle *f*
b. drift *(Brg)* Blindstrecke *f*, Blindort *n*
b. drilled roll *(Holz)* mit Löchern versehene Blindwalze *f (Naßpressenpartie)*
b. drilling Bohren *n* ohne geologische Vorarbeiten *(Erdöl, Erdgas)*
b. end geschlossenes Ende *n (Zylinder)*, Zylinderdeckel *m* ohne Durchbruch
b.-end bore Grundbohrung *f*, Sackloch *n*
b. flange Blindflansch *m*
b. flight Blindflug *m*
b. flying Blindflug *m*
b. hatch *(Schiff)* blinde Luke *f (Zwischendeckluke ohne darüber befindliche Oberdeckluke)*
b. hole *(Masch)* Grundbohrung *f*, Sackloch *n*
b. hole thread *(Masch)* Grundlochgewinde *n*
b. landing Blindlandung *f*
b. navigation Blindflug *m*
b. nut *(Masch)* Blindmutter *f*
b. opening Blindkanal *m*, blinde Öffnung *f*
b. pass Blindstich *m (Walzwerk)*
b. plug Blindstöpsel *m*, Isolierstöpsel *m*
b. power *(Am) (El)* Blindleistung *f*
b. riser Massel *f*
b. sector 1. *(Fs)* sektorweiser Bildausfall *m*; 2. Schattensektor *m (Radar)*
b. spot *(El)* 1. tote Zone *f*, toter Punkt *m*, Wellenschatten *m (Rundfunk)*; 2. blinder Leuchtschirmfleck *m*
b. stitch Blindstich *m (Maschinennähen)*
b. tapped hole Gewindegrundbohrung *f*
b. tooling *(Typ)* Heißprägung *f (Buchbinderei)*
b. transmission Blindübertragung *f*
b. wall durchgehendes Mauerwerk *n*, Mauer *f* ohne Öffnungen
blinding 1. Blendung *f*; 2. Streuen *n* von Splitt *(Straßenbau)*
b. of the stone surface Abdecken *n* der Schleifsteinoberfläche
b. sand *(Bw)* Streusand *m*, Walzsand *m (für Schotterdecke)*

blinker *(Kfz)* Blinker *m*, Fahrtrichtungsanzeiger *m*
b. lamp Blinklampe *f*
b. signal Blinkzeichen *n*
blip Blip *m*, Echosignal *n*; Leuchtfleck *m* *(Funkmeßtechnik)*
blister 1. Blase *f*; 2. Gußblase *f*, Gasblase *f* *(Stahl)*; 3. *(Mil)* Bodenlafette *f* *(Flugzeug)*
b. copper Rohkupfer *n*, Blasenkupfer *n*
b. copper slag Schwarzkupferschlacke *f*
b. fabric *(Text)* Blasengewebe *n*
b.-forming blasenbildend; hautätzend *(Säuren usw.)*
b. of the hawse hole *(Schiff)* Wulst *m* der Ankerklüse
b. packaging Schutzhüllenverpackung *f* *(z. B. Klarsichtverpackung)*
b. steel Blasenstahl *m*, Zementstahl *m*
blistered blasig
blistering 1. Blasenbildung *f*; 2. *(Led)* Selbstspalten *n* *(Häute)*
blob Blob *n* *(Emulsion)*
b. counting *(Ph)* Clusterzählung *f*
blobby wool geschmeidige Wolle *f* mit vollem Griff
Bloch band *(Ph)* Energieband *n* *(Leitungselektronen in Festkörpern)*
Bloch interaction *(Ph)* Blochsche Wechselwirkung *f*
block / to blockieren, hemmen, sperren; arretieren, verriegeln
b. out *(Opt)* ausblenden
block 1. Flasche *f*, Kloben *m*; 2. *(Typ)* Klischee *n*, Druckstock *m*; 3. Klotz *m* *(Holz)*; Unterlegeklotz *m*; 4. Kulissenstein *m*; 5. Klemmstück *n*; 6. Meßblock *m*; 7. *(Brg)* Block *m*, Abbaublock *m*; 8. *(BMSR)* Block, Wörterblock *m*, Kästchenblock *m* *(Informationsmenge, die als Ganzes übertragen wird; oft in benachbarten Speicherzellen untergebracht)*; 9. Speicher *m*; 10. *(Eis)* Blockstrecke *f*
b. aerial Gemeinschaftsantenne *f*, Zentralantenne *f*
b. and tackle Flaschenzug *m*
b. base *(Typ)* Klischeefuß *m*
b. bender *(Typ)* Klischeebiegevorrichtung *f*
b. brake Backenbremse *f*, Klotzbremse *f*
b.-by-block fighting *(Mil)* Straßenkampf *m*
b.-caving *(Brg)* Blockbruchbau *m*
b. chain *(Masch)* Blockkette *f*, Hebezeugkette *f*
b. chain hob Blockkettenwälzfräser *m*
b.-chain sprocket Blockkettenrad *n*, Hebezeugkettenrad *n*
b. circuit *(El)* Sperrkreis *m*, Sperrschaltung *f*
b. coefficient *(Schiff)* Völligkeitsgrad *m* der Verdrängung
b. condenser *(El)* Blockkondensator *m*
b. diagram 1. *(El)* Blockschaltbild *n*; 2. *(BMSR)* Blockdiagramm *n* *(schematisch dargestellter Rechenvorgang)*; Schema *n*
b. gap *(El)* Blockabstand *m*
b. gauge Endmaß *n*, Meßblock *m*, Meßklötzchen *n*, Parallelendmaß *n*
b. gauge holder Endmaßhalter *m*
b.-in-course *(Bw)* schweres Quadermauerwerk *n*
b. letter *(Typ)* Blockschrift *f*, Druckschrift *f*
b. level Setzlibelle *f*
b.-maker *(Typ)* Klischeehersteller *m*, Chemigraf *m*
b. mica Blockglimmer *m*, Rohglimmer *m*, Naturglimmer *m*
b. of flats Wohnblock *m*, Mietshaus *n*
b. of information Wörterblock *m* *(Datenverarbeitung)*
b. pavement Pflasterdecke *f*, Steinpflaster *n* *(Straßenbau)*
b. pin *(Masch)* Blocknagel *m*
b. plan *(Bw)* Lageplan *m*
b.-polymerization Block[misch]polymerisation
b. printing *(Text)* Blockdruck *m*, Handdruck *m*, Klotzdruck *m*
b. radiator *(Kfz)* Teilblockkühler *m*
b.-room *(Bw)* Raumzelle *f*
b. schematic *(El)* Blockschaltbild *n*
b. section Blockabschnitt *m* *(Signal)*
b. set Endmaßsatz *m*, Satz *m* von Prüflehren
b. shears Stockblechschere *f*
b. signal *(Eis)* Blocksignal *n*
b. signal post *(Eis)* Blockanlage *f*; Streckenblock *m*
b. switch *(El)* Durchschalthebel *m*
b. system *(Eis)* Blockstreckensystem *n*
b. terminal *(Fmt)* Endverzweiger *m*
b. tin Blockzinn *n*, Reinzinn *n*
b.-type reamer Einbaureibahle *f*
b. tyre *(Kfz)* Blockreifen *m*
b. wood pavement Holzpflaster *n*
blockade Verstopfung *f* *(Straßenverkehr)*
blockboard *(Holz)* Verbundplatte *f*
blockbuster *(Mil)* überschwere Sprengbombe *f*
blocked impedance *(El)* Scheinwiderstand *m* bei Systemverriegelung (festgehaltenem System)
b. out ore *(Brg)* vorgerichtetes Erz *n*
blocker 1. Sperrer *m*; 2. *(Met)* Vorschmiedegesenk *n*
blockholing *(Fs)* Knäpperschießen *n*
blocking 1. Blockierung *f*, Hemmung *f*, Sperrung *f*; 2. *(Schiff)* Aufklotzung *f*, Stapelung *f*
b. action Sperrwirkung *f*
b. and embossing press *(Typ)* Vergoldepresse *f*
b. axe Bandhacke *f*, Bundaxt *f*, Zimmeraxt *f*

b. barrier Sperrschicht *f*
b. bias Sperrspannung *f*
b. brake Feststellbremse *f*
b. capacitor *(Fmt, Fs)* Blockkondensator *m*; Sperrkondensator *m*, Koppelkondensator *m*
b. circuit *(El)* Verriegelungsschaltung *f (elektronische Sperre)*
b. contact *(El)* Sperrkontakt *m*
b. current circuit Blockierungsstromkreis *m*
b. device for shuttles *(Text)* Schützenblockierung *f*
b. element *(El)* Sperrglied *n (Schutzrelais)*
b. generator *(BMSR)* Sperrschwinger *m*
b. layer *(El)* Halbleiterschicht *f*, Sperrschicht *f*
b. layer cell *(El)* Halbleiterzelle *f*
b. lever Arretierhebel *m*
b. magnet Sperrmagnet *m*
b. medium Absorptionsmittel *n*
b. of the recording *(El)* Aufnahmesperre *f*
b. oscillator *(Fs)* Kipp[spannungs]generator *m*; Sperrschwinger *m*
b. period *(El)* positive Sperrzeit *f*, Zündwinkel *m (Thyratron)*
b. signal *(Fmt)* Besetztsignal *n*, Blockierungssignal *n*
b. slide Feststellschieber *m*
b. the transmission of sound schalldämpfend
b. tube oscillator *(El)* Sperrschwinger *m*
b. valve Abschaltventil *n*, Absperrventil *n*, Haltventil *n (Hydraulik)*
b. voltage Sperrspannung *f*
b. voltage rating maximal zulässiger Spitzenwert *m* der Sperrspannung
blocklaying *(Bw)* Versetzen *n* von Quadern, Großblocksteinen u. dgl.
blockout Aussparung *f (Betonbau)*
blondin Kabelkran *m*
blood Blut *n*
b. cell Blutkörperchen *n*
b. clot Blutgerinnsel *n*
b. count 1. Anzahl *f* der Blutkörperchen; 2. Blutkörperchenzählung *f*
b. donor Blutspender *m*
b. group Blutgruppe *f*
b. picture Blutbild *n*
b. plasma Blutplasma *n*
b.-presser follower [elektronischer] Blutdruckmesser *m*
b. pressure Blutdruck *m*
b. serum Blutserum *n*
b. transfusion Blutübertragung *f*
b. vessel Ader *f*, Blutgefäß *n*
bloodstone *(Min)* Heliotrop *m*
bloom / to *(Met)* vorblocken *(Gußblöcke auf der Blockstraße)*
bloom 1. Luppe *f*, Block *m*, Rohblock *m*; Knüppel *m*; 2. *(Bw)* Ausblühung *f*, Salzausblühung *f*; 3. *(Text)* „Blume" *f (Schaum auf der Badoberfläche)*; 4. *(Fs)* Bildweichheit *f*; Überstrahlung *f*; 5. Feuchtigkeitsfilm *m*
b. reheating furnace Brammen[schweiß]-ofen *m*
b. roll *(Met)* Vorwalze *f*
b. shears Blockschere *f*
b. side *(Led)* Haarseite *f*
b. steel Luppenstahl *m*
bloomery *(Met)* Luppenfeuer *n*, Rennfeuer *n*, Rennherd *m*
b. forge Luppenherd *m*
b. hearth *(Met)* Rennfeuer *n*, Rennherd *m*
b. iron Herdfrischeisen *n*, Rennstahl *m*
b. process Rennverfahren *n*
blooming mill Blockwalzwerk *n*, Grobwalzwerk *n*; Blockwalzstraße *f*, Grobstraße *f*
b. of textured yarn „Blühen" *n*, Sich-Entfalten *n* von Strukturgarn *(nach der Verarbeitung zu textilem Flächengebilde)*
b. pass Blockkaliber *n (von Walzen)*
b. roll *(Met)* Blockwalze *f*
b. train Grobstraße *f*; Grobwalzwerk *n*
bloop / to kleben *(Film)*
blooping patch Tonkleb[e]stelle *f (Film)*
blotch printing *(Text)* Deckdruck *m*, Flächendruck *m*, großflächiges Drucken *n*
blotter test *(Kfz)* Löschblattprobe *f*, Tropfenprobe *f*
blotting paper Löschpapier *n*; Saugpapier *n*; Fließpapier *n*; Filterpapier *n*
blow / to 1. abblasen; blasen; drücken *(Förderung durch Rohrleitung)*; 2. *(El)* durchbrennen *(Sicherung)*
b. a fuse *(El)* eine Sicherung durchbrennen lassen
b. again wieder anblasen *(Hochofen)*
b. at anblasen
b. back abgasen *(Zellstoff)*
b.-by durchblasen, vorbeiblasen *(an Kolbenringen)*
b. down 1. *(Met)* ausblasen, ausgehen lassen, abblasen *(Kessel; Hochofen)*; 2. abgasen *(Zellstoff)*
b. in einblasen *(z. B. Dampf)*
b. in the furnace Hochofen anblasen
b. off ausblasen, ablassen *(Dampf)*; abblasen; zurückströmen lassen
b. out 1. *(El)* durchbrennen *(Sicherung)*; 2. ausblasen *(Dampf)*
b. over the paper Papierbahn überführen
b. through durchblasen
b. up explodieren, zerplatzen, aufplatzen aufreißen
blow 1. Schlag *m*, Stoß *m*; 2. Hub *m (Presse, Schere)*; 3. *(Met)* Frischen *n*, Frischvorgang *m*; 4. *(Met)* Schmelze *f*
b. bending test Schlagbiegeprobe *f*

b. down Abblaseverlust *m*, Abschlämmung *f* *(Dampfkessel)*
b.-down cycle zeitlicher Verlauf *m* des Abgasens *(Zellstoff)*
b.-down process Abgasprozeß *m* *(Zellstoff)*
b. forming *(Plast)* Blasen *n*, Blasverformung *f*
b. gun Spritzpistole *f*
b. heat recovery Wiedergewinnung *f* von Abgaswärme
b. hole Gaseinschluß *m*, Gußblase *f*, Gasblase *f*
b. lamp Lötlampe *f*
b. magnet *(El)* Blasmagnet *m* *(Schalter)*
b. mould Blasform *f*, Fertigform *f*
b. of the slay with open shed *(Text)* Ladenanschlag *m* bei offenem Fach
b.-off cock Ablaßhahn *m*, Luftventil *n*
b.-off pipe 1. Dampfabblaserohr *n*, Dampfabblasestutzen *m*; 2. *(Masch)* Ausblasrohr *n*, Abblasrohr *n*
b.-off valve Abblashahn *m*, Abblasventil *n*, Ablaßventil *n*, Sicherheitsventil *n*, Überdruckventil *n*
b.-off vapours Abgasbrüden *pl*
b.-out 1. Reifenpanne *f*, Platzen *n* des Reifens; 2. Ölausbruch *m*, Ausbruch *m*, Eruption *f* *(Erdöl)*; 3. Durchbrennen *n* einer Sicherung
b.-out coil *(El)* Blasspule *f*, Funkenlöschspule *f* *(Leistungsschalter)*
b.-out fuse *(El)* Schmelzsicherung *f*
b.-out valve *(Masch)* Ausblashahn *m*, Ausblasventil *n*
b. overcast Stich *m* ins Blaue
b. pipe Gebläsebrenner *m*; Schneidbrenner *m*; Schweißbrenner *m* (*s. a.* blowpipe)
b. pit Ausblasebütte *f*; Diffuseur *m*, Kochergrube *f*, Waschgrube *f*
b. room machine *(Text)* Putzereimaschine *f*
b. torch *(Am)* *s.* b. lamp

blowable verblasbar

blower 1. *(Masch)* Gebläse *n*; 2. *(Brg)* blasender Lüfter *m*
b.-assisted scavenge Gebläsespülung *f*
b. cooled engine zwangsgekühlter (durch Gebläse gekühlter) Motor *m*
b. head Gebläseförderhöhe *f*
b. kiln *(Holz)* Trockenraum *m* mit mechanischer Luftbewegung
b.-scavenged mit Gebläsespülung
b. separator Spreugebläse *n* *(Dreschmaschine)*
b. set Lüftersatz *m*, Gebläse *n*

blowing 1. Blasen *n* *(Glas)*; 2. Abblasen *n* *(Papier vom Filz mittels Luft)*
b. a steel melt Verblasen *n* einer Stahlschmelze
b. agent *(Plast)* Treibmittel *n*, Blähmittel *n*
b. current *(El)* Abschmelzstrom *m* *(Sicherung)*
b. engine Gebläsedampfmaschine *f*
b. in Anblasen *n* *(Hochofen)*
b. iron Glasmacherpfeife *f*
b. machine Glasblasmaschine *f*
b. magnet Blasmagnet *m*
b. room *(Text)* Putzerei *f*
b. through the water gauge cocks Durchblasen *n* der Wasserstandshähne
b. up the chimney Schornsteinsprengung *f*
b. ventilation *(Brg)* blasende Bewetterung *f*

blown 1. geblasen; 2. blasig; 3. durchgebrannt *(Sicherung)*
b. casting *(Met)* blasiger Guß *m*
b.-extrusion method *(Plast)* kombiniertes Strangpreß-Blasverfahren *n*
b. fuse *(El)* durchgegangene (abgeschmolzene) Sicherung *f*
b. gasket defekter Dichtungsring *m*
b. metal *(Met)* Vormetall *n*
b. oil anoxydiertes (geblasenes) Öl *n*, Blasöl *n*
b.-out hole *(Brg)* Bohrlochpfeife *f*
b.-out shot *(Brg)* Ausbläser *m*

blowpipe 1. Blasrohr *n*; Entlüftungsrohr *n*; Gebläse *n*; 2. Lötrohr *n*, Schweißbrenner *m*, Brenner *m*
b. analysis Lötrohranalyse *f*, Lötrohrprobe *f*
b. bead Schmelzperle *f*
b. brazing Flammenlötung *f*
b. flame Lötflamme *f*
b. test Lötprobe *f*, Lötversuch *m*

blue / to blau anlaufen [lassen], blauen; brünieren

blue adder *(Fs)* Blaubeimischer *m*
b. annealing Ofenglühen *n* von warmgewalzten Blechen
b. asbestos Krokydolit *m*
b. brick *(Bw)* Klinkerstein *m*, Klinkerziegel *m*
b. brittleness *(Met)* Blaubrüchigkeit *f*, Blausprödigkeit *f*
b. cotton außergewöhnlich weiße Baumwolle *f*
b. dip Blaubrenne *f* *(Galvanotechnik)*
b. filter *(Opt)* Blaufilter *n*
b.-finished blau angelassen *(Stahl)*
b. glass Kobaltglas *n*, Blauglas *n*
b. glow [blaues] Glimmlicht *n* *(in Elektronenröhren)*
b. glow oscillograph Oszillograf *m* mit Kalziumwolframatleuchtschirm
b. haze [blaues] Glimmlicht *n*
b. heat Blauhitze *f*, Blauwärme *f*
b. iron ore Vivianit *m*, Blaueisenerz *n*
b. ironstone Krokydolit *m*
b. japanned blau lackiert
b. john Fluorit *m* *(in Derbyshire)*
b. lacquer finished blau lackiert
b. metal Zinkpulver *n* *(nach einer Destillation)*
b. print Blaupause *f*

b. print paper Lichtpauspapier *n*
b.-sap fungus *(Holz)* Blau[fäule]pilz *m*
b.-short *(Met)* blaubrüchig
b. shortness Blaubrüchigkeit *f*
b. stain *(Holz)* Blaufäule *f*
b. steel plate Stahlblech *n* mit blaugrauer Walzhaut, Blaublech *n*, Stahlglanzblech *n*
b.-stone battery *(El)* Kupfer-Zink-Element *n*
b. tinged blaustichig
b. vitriol Kupfervitriol *m (n)*, Kupfersulfat *n*, blauer Vitriol *m*, blaues Vitriol *n*
blueing 1. Blauanlaufen *n*, Blauwerden *n*; Brünieren *n*; 2. *(Text)* Blaumachen *n*, Bläuen *n*
blueness on the edges *(Holz)* Kantenbläue *f*
bluette Gewebe *n* für Overalls
bluff bow 1. *(Schiff)* voller (völliger) Bug *m*; 2. vorstehende Klippe *f*
b. link Sparglied *n (Strickmaschine)*
b. sinker *(Text)* blinde Platine *f*
b. wheel *(Text)* Sparrad *n*
blunger Tonmischmaschine *f*
blunt abgestumpft, stumpf
b. end stumpfes Schiffsende *n*
b. file Stumpffeile *f*
b. start of a thread Gewindeansatz *m* ohne Abflachung
blunting Abstumpfen *n*
b. of threads Abrunden *n* von Gewinden
blur / to verwischen, unscharf machen; Fahnen ziehen *(Fernsehbild)*
blur circle *(Opt)* Zerstreuungskreis *m*, Unschärfering *m*
b. factor 1. *(El, Fmt)* Klirrfaktor *m*; 2. *(Opt)* Bewegungsunschärfe *f*
blurred image *(Opt)* unscharfes (verwaschenes) Bild *n*
b. picture *s.* b. image
b. voice *(Fmt)* verschwommene Sprache *f*
blurring *(Fs)* Fleckigkeit *f*, Undeutlichkeit *f* *(Fernsehbild)*
blush weiß *(Baumwolle)*
blushing Trübung *f*, Weißwerden *n (von Lacken)*
board / to 1. *(Bw)* mit Brettern verkleiden, verschalen; 2. besteigen *(Verkehrsmittel)*; entern *(ein Schiff)*
b. four ways *(Led)* über vier Quartiere krispeln
board 1. Brett *n*, Diele *f*, Planke *f*; 2. Pappe *f*; 3. *(El, Fmt)* Schalttafel *f*; Vermittlungsschrank *m*; 4. *(Brg)* Abbaustrecke *f*, Pfeiler *m* *(Pfeilerbau)*; 5. *(Schiff)* Schlag *m (Segelschiff)*; Bord *m (oberster Schiffsrand)*
b. coloured throughout durchgefärbte Pappe *f*
b. containing mechanical wood pulp holzhaltige Pappe *f*
b. cut Bretterschnitt *m*
b.-cuttings for composing Kartonagezuschnitte *mpl*
b. drop hammer Brettfallhammer *m*, Stangenfallhammer *m*
b. felt Naßfilz *m (Papierherstellung)*
b. foot *Volumen einer Pappe von 1 in. Dicke und 1 sq.ft. Fläche = 2,3597* dm^3
b. for pressing Ziehpappe *f*
b.-free erection *(Schiff)* bordfreier Aufbau *m*
b.-glazed Deckelsatinage *f*
b. landing light *(Flg)* Landescheinwerfer *m*
b. liner Kaschiermaschine *f*, Kartonbeklebemaschine *f*
b. machine Kartonmaschine *f*, Pappenmaschine *f*, Rundsiebmaschine *f*
b. mould Pappenform *f*, Schöpfrahmen *m* *(Papierherstellung)*
b.-mounted controller *(BMSR)* Schalttafelregler *m*
B. of Trade ohm internationales Ohm *n*
B. of Trade unit Kilowattstunde *f*
b. operator *(Flg)* Planzeichner *m*
b. plotter *s.* b. operator
b. saw Furniersäge *f*
b. sling *(Schiff)* Plattformstropp *m (zum Hieven von Säcken)*; Ladepritsche *f*
b. wall Bordwand *f (Güterwagen)*
boarded floor Bretterfußboden *m*, Dielenfußboden *m*, Holzdielenfußboden *m*
boarding 1. *(Bw)* Bretterung *f*, Brettverkleidung *f*, Dielung *f*; Verbretterung *f*; Schalung *f*, Verschalung *f*; 2. *(Text)* Trockenformen *n*; Trocknen *n* und Formen *n*; *(Strümpfe)*; 3. *(Schiff)* Kollision *f*, Schiffszusammenstoß *m*
b. bridge Auflaufbrücke *f*, Brücke *f*
b. fence Bretterzaun *m*
b. joist Dielenbalken *m*
b. ladder *(Schiff)* Fallreep *n*
boardy *(Text)* brettartig *(Griff)*
boaster Scharriereisen *n*
boasting *(Bw)* Scharrierung *f*
boat Boot *n*, Barke *f*, Kahn *m*
b. basin Bootshafen *m*
b. building yard Bootswerft *f*
b. canvas Persenning *f (Segelschiff)*
b. carrier Anhänger *m* für Bootstransport *(Landfahrzeug)*
b. cradle Bootsklampe *f*
b. deck Bootsdeck *n*
b. form Wannenform *f (Stereochemie)*
b. harbour Bootshafen *m*
b. hoist Bootsheißmaschine *f*
b. landing Bootsanlegestelle *f*, Anlandeplatz *m*
b. rail Bootsreling *f*
b. trailer Anhänger *m* für Bootstransport *(Landtransport)*

b. winch Bootswinde *f*, Bootswinsch *f*
b. yard Bootswerft *f*
boatswain Bootsmann *m*
boatwalk *(Schiff)* Bootssteg *m*
bob 1. *(Bw)* Lot *n*; 2. Rodelschlitten *m*, Bob *m*
b. run Bobbahn *f*
bobbin 1. Trommel *f*, Wickelkörper *m*; Ablaufhaspel *f*, Bandseiltrommel *f*, Flachseiltrommel *f*; 2. *(El)* Spule *f*; Spulenkörper *m*; 3. *(Text)* Klöppel *m* *(Spitze)*; Spule
b. board *(Text)* Spulenbrett *n*
b. brake *(Text)* Spulenbremse *f*
b. case *(Text)* Spulenhülse *f* *(Maschinenähen)*
b. centre rod Haltestift *m* am Spulenteller *(Strickmaschine)*
b. changer *(Text)* Spulenwechseleinrichtung *f*
b. changing *(Text)* Spulenwechsel *m*
b. core *(El)* Spulenkern *m*
b. creel *(Text)* Spulengatter *n*
b. cylinder *(Text)* Spulenwalze *f*
b. feeder *(Text)* Spulenaufstecker *m*
b. finings *(Text)* Spitzenlinderung *f*
b. frame *(Text)* Spindelbank *f*
b. frame with ten needles per inch *(Text)* Zehnpointsmaschine *f*, Zehnpunktmaschine *f*
b. holder *(Text)* Spulenträgerarm *m*
b. lace Klöppelspitze *f*
b. lace making Klöppeln *n* *(Spitze)*
b. lace work Klöppelarbeit *f*
b. lead *(Text)* Voreilen *n* der Spule
b. plate *(Text)* Trageplatte *f* für Spulenträgerarme
b. rail *(Text)* Spulenbank *f*; Spulentisch *m*
b. shaft *(Text)* Spulenträgerarm *m*
b. stand *(Text)* Spulengestell *n*
b. support shaft *(Text)* Tragewelle *f* des Spulengestells
b. support stem *(Text)* Verlängerung *f* der mittleren Tragestange des Spulengestells
b. table *(Text)* Spulengestell *n*
b. warping creel *(Text)* Schärbank *f*; Schärkanter *m*, Schärstock *m*
b. winder *(Text)* Schlußspulmaschine *f*
b. winder for cops *(Text)* Kopsspulmaschine *f*
bobbinet Tüll *m*
b. frame Bobinetmaschine *f*, Tüllmaschine *f*
b. machine Tüllmaschine *f*
bobbins for slubbing and roving *(Text)* Grobspulen *fpl*
bobby *(sl)* *s.* bobbin 2.
bobsleigh Bobsleigh *m* *(Rodelschlitten)*
bobstay *(Schiff)* Wasserstag *n*
Bode diagram *(BMSR)* Bodediagramm *n*, Frequenzgang *m* in logarithmischer Darstellung, logarithmischer Amplituden- und Phasengang *m*
bodkin Ahle *f*, Schnürnadel *f*
body [up] / to eindicken, verdicken *(Lack)*
body 1. Körper *m*, Hauptteil *n*; 2. *(Masch)* Ständer *m*, Gestell *n*; 3. Aufbau *m*, Karosserie *f*; 4. Unterteil *n* *(Schraubstock)*; 5. *(El)* Masse *f* *(meist elektrisch Null)*; 6. *(Text)* Fläche *f* zwischen den Leisten des Gewebes; Stumpen *m* *(Hut)*; Körper *(Nadelzylinder)*; Griff *m* *(des Gewebes)*; 7. Schallkörper *m*; 8. *(Mil)* Topf *m* *(Stielhandgranate)*; 9. Körper *(Farbe)*
b. apron Bleigummischürze *f* *(Strahlenschutz)*
b. capacitance *(El)* Körperkapazität *f*; Handkapazität *f*
b.-centred [cubic] raumzentriert, kubisch zentriert *(Kristallgitter)*
b.-centred [cubic] lattice kubisch raumzentriertes Gitter *n* *(Kristalle)*
b. cleaning compound *(Kfz)* Aufbaupflegemittel *n*
b. clearance Rücken *m* *(des Spiralbohrers)*
b. clearance diameter Rückendurchmesser *m* *(Spiralbohrer)*
b. contact *(El)* Körperschluß *m*
b. contact terminal Masseschlußklemme *f*
b. diameter Schaftdurchmesser *m* *(Schraube)*
b. diameter clearance Fasenhöhe *f*
b. drag Formwiderstand *m*
b. effect *(El)* Handeffekt *m*; Handkapazitätseffekt *m*
b. fabric machine Leibweitenstrickmaschine *f*
b. fittings *(Kfz)* Karosseriebeschläge *mpl*
b. of a furnace *(Met)* Ofenmassiv *n*
b. force Raumkraft *f*
b. framework *(Kfz)* Aufbaugerippe *n*, Karosseriegerippe *n*
b. length *(Met)* Ballenlänge *f* *(Walze)*
b. maker Zargenbiegemaschine *f* *(Konservendosenherstellung)*
b. of a ship Schiffskörper *m*, Schiffsrumpf *m*
b. of car Wagenkasten *m*
b. of revolution rotationssymmetrischer Körper *m*
b. of slide rule Stabkörper *m*
b. of the bearing Lagerkörper *m*
b. of the crank shaft Kurbelwellenschaft *m*
b. of the gear Zahnradkörper *m*
b. of the twist drill Schneid[en]schaft (Schneidenteil) *m* des Spiralbohrers
b. of water *(Schiff)* Wassermasse *f*
b. paper Rohpapier *n*
b. plan *(Schiff)* Spantenriß *m*, Konstruktionsspantenriß *m*
b. rib *(Masch)* Leiste *f* *(zur Spannklobenführung)*
b. section Konstruktionsspant *n*
b. sheet Tiefziehblech *n*
b. size *(Typ)* Kegel *m* *(der Schrift)*

b. size plain circular knitting machine einfonturige Leibweitenrundstrickmaschine *f*
b.-wing interference *(Flg)* gegenseitige Beeinflussung *f* von Tragflügel und Rumpf
bodying Eindickung *f (z. B. Öl)*
bodywork *(Kfz)* Aufbau *m*, Karosserie *f*
bog Wiesenmoor *n*
b. blasting Moorsprengung *f*, Sumpfsprengung *f*, Schüttsprengverfahren *n*
b. burst Moorbruch *m*
b. coal Sumpfkohle *f*
b. [iron] ore Rasen[eisen]erz *n*, Sumpfeisenstein *m*, Sumpferz *n*
boggy sumpfig
b. soil Moorboden *m*
Boghead coal Bogheadkohle *f*
bogie 1. Drehgestell *n*, Drehschemel *m*, Rollenbock *m*; Laufrad *n (bei Kettenfahrzeugen)*; 2. Fahrgestell *n*, Untergestell *n*, Unterwagen *m*; 3. *(Brg)* Wagen *m*
b. casting *(Met)* Wagenguß *m*
b. hearth furnace *(Met)* Herdwagenofen *m*
b. suspension Drehgestellaufhängung *f*
b. truck Schemelwagen *m*
Bohr atom Bohrsches Atommodell *n*
boil / to sieden; kochen, wellen, abkochen
b. blank blank kochen
b. down eindampfen, eindicken, verkochen
b. off abkochen *(Seide)*
b. over überkochen, übersieden
b. up aufsieden
boil hole *(Geol)* Frostaufbruch *m*
b.-off agent *(Ch)* Entschlichtungsmittel *n*
b.-proof bond *(Holz)* kochfeste Fuge *f*
boiled bar Puddelblock *m*
b. fur Kesselstein *m*
b.-off abgekocht; entbastet *(Seide)*
b. [linseed] oil Leinölfirnis *m*
boiler Boiler *m*, Kessel *m*, Heißwasserspeicher *m*, Dampferzeuger *m*, Wasserkocher *m*
b. acceptance Kesselabnahme *f*
b. accessories Kesselzubehör *n*
b. anti-incrustant composition Kesselsteinverhütungsmittel *n*
b.-antiscaling composition Kesselsteinverhütungsmittel *n*
b. attendance Kesselbedienung *f*
b.-attendants Kesselbedienungsmannschaft *f*, Kesselbedienungspersonal *n*
b. blow-off cock Kesselabblasehahn *m*
b. blower Kesselgebläse *n*
b. brush Kesselsteinbürste *f*
b. capacity Leistungsfähigkeit *f* des Kessels
b. casing Kesselbekleidung *f*, Kesselmantel *m*
b.-cleansing compound Kesselsteinlösemittel *n*
b.-coaling plant Kesselbekohlungsanlage *f*
b.-code Kesselbauvorschriften *fpl*
b. construction Kesselbau *m*
b. control Kesselregelung *f*
b. draft Kesselzug *m*
b. drain cock Kesselablaßhahn *m*
b. drum Kesseltrommel *f*
b. efficiency Kesselwirkungsgrad *m*
b. embrittlement Laugensprödigkeit *f*
b. end Kesselboden *m*
b. envelope Kesselmantel *m*
b. fan Kesselgebläse *n*
b.-feed water Kesselspeisewasser *n*
b.-feed-water meter Kesselspeisewassermesser *m*
b. feeder Kesselspeisepumpe *f*
b. firing Kesselfeuerung *f*
b. fitted with emulseurs Walzenkessel *m* mit Dubiauscher Rohrpumpe, Emulseurkessel *m*
b. fittings Kesselarmatur *f*, Kesselausrüstung *f*
b. flue Kesselflammrohr *n*
b. flue gas Kesselgas *n*
b. flue pass Kesselzug *m*
b. fluid Kesselsteinverhütungsmittel *n*, Mittel *n* gegen Kesselstein
b. furnace Kesselfeuerung *f*
b. grate Kesselrost *m*
b. hatch Kesselluke *f*
b.-heating surface Kesselheizfläche *f*
b. hopper Aschentrichter *m*, Aschenbunker *m*
b. horsepower Kesselleistung *f*
b. house Kesselhaus *n*
b. inspection Kesselprüfung *f*
b. inspector Kesselprüfer *m*
b. jacket Kesselverkleidung *f*
b. level Kesselsohle *f*
b. load Kesselbelastung *f*
b. maker's name plate Kesselfirmenschild *n*
b. maker's number Kesselnummer *f*, Fabriknummer *f* des Kessels
b. mountings Kesselarmaturen *fpl*
b. name plate Kesselschild *n*
b. operation Kesselbetrieb *m*
b. output Kesselleistung *f*
b. pick Kesselsteinhammer *m*, Schrupphammer *m*
b. plant Kesselanlage *f*
b. plate Kesselplatte *f*, Kesselblech *n*
b. plate eaten through by corrosion durchgerostetes Kesselblech *n*
b. pressure Kesseldruck *m*, Kesselspannung *f*
b. reamer Kesselreibahle *f*
b. rivet Kesselniet *m*
b. riveting Kesselnietung *f*
b. room *(Schiff)* Kesselraum *m*, Heizraum *m*
b. scale Kesselstein *m*, Kesselniederschlag *m*
b.-scale deposit Kesselsteinablagerung *f*
b.-scale formation Kesselsteinbildung *f*

b.-scaling hammer Kesselhammer *m*
b.-scaling tool Kesselabklopfwerkzeug *n*
b. seating Kesselfundament *n*
b. setting Kessellagerung *f*
b. shell Kesselwand *f*
b. shop Kesselbauwerkstatt *f*, Kesselbauanstalt *f*
b. shop clamp Kesselschmiedeschraubzwinge *f*
b.-supervising apparatus Kesselüberwachungsapparat *m*
b. support Kesselstütze *f*, Kesselstuhl *m*, Tragstuhl *m*
b. switchboard Kesselwarte *f*
b. tap Kesselgewindebohrer *m*
b. test pump Kesselprobierpumpe *f*
b. tube Heizrohr *n*, Kesselrohr *n*, Siederohr *n*
b. tube expander Siederohrdichtmaschine *f*
b. tube temperature Heizrohrtemperatur *f*; Siederohrtemperatur *f*
b. wall Kesselmantel *m*
b.-water feed pump Kesselspeisewasserpumpe *f*
b.-water treatment Kesselwasserreinigung *f*, Kesselwasserbehandlung *f*
b. with corrugated flue tube Wellrohrkessel *m*
b. with double steam space Doppeldampfraumkessel *m*, Kessel *m* mit zwei Dampfräumen
b. with eccentric flue Seitenrohrkessel *m*
b. with large water space Großwasserraumkessel *m*, Kessel *m* mit großem Wasserraum
b. with partly corrugated flue Halbwellrohrkessel *m*
b. with plain flue Flammrohrkessel *m* mit glattem Flammrohr
b. with ribbed flue Rippenrohrkessel *m*
b. with single steam space Einfachdampfraumkessel *m*, Kessel *m* mit einem Dampfraum
b. with small water space Kleinwasserraumkessel *m*, Kessel *m* mit kleinem Wasserraum
b. with stepped flue Stufenrohrkessel *m*
boilerhouse operation Kesselhausbetrieb *m*
boilermaker Kesselschmied *m*, Kesselhersteller *m*
boilermaker's hammer Kesselschmiedhammer *m*
b. reamer Kesselreibahle *f*, Nietlochreibahle *f*
boiling 1. Kochen *n*, Aufwallen *n*; 2. *(Text)* Abkochen *n (Seide)*
b. cooling Siedekühlung *f*
b. curve Siedekurve *f*, Siedeverlauf *m*
b. flask Kochkolben *m*, Siedekolben *m*
b. heat Siedehitze *f*
b. noise Brodeln *n*, Rauschen *n (Röhre)*
b. off Entbasten *n*
b. plate elektrische Kochplatte *f*
b. point Siedepunkt *m*, Kochpunkt *m*
b.-point curve Siedekurve *f*
b.-point elevation Siedepunkterhöhung *f*
b. process Siedebetrieb *m*
b. range Siedebereich *m*; Siedegrenze *f*
b. reactor *s.* b. water reactor
b. test Kochprobe *f*, Kochversuch *m*
b. up *(Brg)* Hochpuffen *n (Sohle)*
b. water reactor *(Kern)* Siedewasserreaktor *m*; Siedereaktor *m*; Verdampferreaktor *m*
bold *(Typ)* [halb]fett *(Schrift)*
boldface *(Typ)* Fettdruck *m*
bole 1. Stamm *m*; 2. Bolus *m (Ton)*
bollard *(Bw)* [eiserner] Poller *m*; Prellstein *m*, Radabweiser *m*; Inselpfosten *m (Straßenbau)*
b. effekt *(Text)* Pollereffekt *m*
b. pull *(Schiff)* Pfahl[trossen]zug *m*
b. with a cross-piece *(Schiff)* Kreuzpoller *m*
bolograph Bolometerdiagramm *n*
bolometer Bolometer *n*
b. bridge Bolometerbrücke *f (Meßtechnik)*
b. detection Bolometerstrahlungsmessung *f*
bolometric bolometrisch
b. correction bolometrische Korrektur *f*
bolster / to unterlegen, unterbauen
bolster 1. *(Masch)* Aufspannplatte *f*; 2. Wiege *f (Drehgestell)*; 3. *(Bw)* Sattelholz *n*; Schwarte[nbrett *n*] *f*
b. plate 1. Auflageplatte *f (Stanze)*; Aufspannplatte *f*; Grundplatte *f*; 2. *(Schiff)* Scheuerplatte *f (Decksplatte der Ankerklüse)*
b. spring Wiegenfeder *f*
bolt / to 1. verschrauben, festklemmen, festschrauben, mittels Schrauben befestigen, verriegeln, anschrauben; 2. *(Brg)* Ankerausbau einbringen, durch Ankerausbau sichern
b. down aufspannen *(Werkstück)*
b. on aufschrauben auf
bolt 1. Schraube[nbolzen *m*] *f*, Bolzen *m*, Befestigungsschraube *f*; Dorn *m*; Riegel *m*; 2. *(Brg)* Anker[stange *f*] *m*
b.-chisel Kreuzmeißel *m*, Spitzmeißel *m*
b. circle of junk ring Deckelschraubenkreis *m*
b. clipper Bolzenschneider *m*, Gewindebolzenabschneider *m*
b. cutter Bolzenschneider *m*, Bolzenschneidmaschine *f*, Schraubenschneidmaschine *f*
b. cutting machine Bolzenschneidmaschine *f*
b. die Gewindeschneidbacke *f*
b. die head Gewindeschneidkopf *m* mit Außengewinde
b. head Schraubenkopf *m*, Bolzenkopf *m*
b. heading Schraubenkopfstauchen *n*
b. hole Schraubenloch *n*, Bolzenloch *n*
b.-hole circle Schraubenlochkreis *m*
b. hook Hakenschraube *f*
b. industry Schraubenindustrie *f*
b. lever Spannhebel *m*
b. milling machine Bolzenschaftfräsmaschine

b. nut Schraubenmutter *f*
b. of a hinge Dorn *m* eines Scharniers
b. pitch circle Schraubenlochkreis *m*
b. pointer Bolzenanspitzmaschine *f*
b. stock Schraubenmaterial *n*, Schraubenstahl *m*
b. thread Bolzengewinde *n*, Schraubengewinde *n*
b. washer Unterlegscheibe *f*, Bolzenscheibe *f*
b. with head Kopfschraube *f*
bolted 1. verschraubt; 2. durchgesiebt
b. assembly Durchsteckschraubenverbindung *f*
b. connection Schraubverbindung *f*
b. flanges Flanschverschraubung *f*
b. joint Bolzenverbindung *f*, Schraubverbindung *f*, Verschraubung *f*
b. pipe joint Rohrverschraubung *f*
b. tenon vernagelter Zapfen *m*
b. to aufgeschraubt auf
b. together zusammengeschraubt
b. union Bolzenverbindung *f*
bolter Sieb *n*, Beutelsieb *n*, Mühlbeutel *m*
bolter *(Am) (Brg)* Ankerlochbohrmaschine *f*; Bohrhauer *m* für das Herstellen der Ankerbohrlöcher
bolting Verschraubung *f*; Bolzenverschraubung *f*, Verbolzung *f*
b. cloth *(Text)* Siebtuch *n*, Schablonenseide *f*, Müllergaze *f*
b. down Aufspannen *n (Werkstücke)*
b. longeron *(Rak)* Trennholm *m*
b. of veneer logs *(Holz)* Ablängen *n* der für Furniere bestimmten Stämme
b. up Verschraubung *f*
Boltzmann constant Boltzmannsche Konstante *f*, Boltzmann-Konstante *f*
Boltzmann distribution Boltzmann-Verteilung *f*
Boltzmann equation Boltzmannsche Gleichung (Stoßgleichung) *f*
bomb / to bombardieren, mit Bomben belegen
bomb Bombe *f*
b.-aiming device Bombenrichtgerät *n*
b. area Bombenabwurfgelände *n*
b.-bay Bombenschacht *m (Bombenflugzeug)*
b.-bay rack Bombenaufhängung *f*
b. calorimeter kalorimetrische Bombe *f*
b. carrier Bombenträger *m*
b. chute Bombenabwurfschacht *m*
b. crater Bombentrichter *m*
b. drop[ping] Bombenwurf *m*
b.-dropping device Bombenabwurfvorrichtung *f*
b. hit Bombeneinschlag *m*
b. panel Bombenteppich *m*
b.-proof bombensicher
b. rack Bombenaufhängevorrichtung *f*
b. release lever Bombenabwurfhebel *m*
b.-sight Bombenzielgerät *n*, Bombenvisier *n*
b.-sight equipment Bombenzielgerät *n*
b. suspension Bombenaufhängung *f*
bombard / to *(Ph)* beschießen, bestrahlen
bombarded particle *(Kern)* beschossenes (getroffenes) Teilchen *n*, Targetteilchen *n*
bombarding particle *(Kern)* einfallendes Teilchen *n*, Beschußteilchen *n*
bombardment Beschuß *m*, Beschießung *f*
b. aircraft (aviation) Bombenflugzeug *n*, Bombenfliegerkräfte *fpl*
b. by ions Ionenbeschuß *m*
b. damage *(Kern)* Schaden *m* durch Teilchenbeschuß
b. energy *(Kern)* Energie *f* der Geschoßteilchen
bomber-fighter taktisches Jagdflugzeug *n*, Jagdbomber *m*
bombing radar Bombenvisier *n*
b. raid Bombenangriff *m*
bombsight *s.* bomb-sight
bonanza *(Geol)* sehr reicher Lagerstättenteil *m*
bond / to [ver]binden, verkleben; plattieren
b. in *(Bw)* einbinden
bond 1. Band *n*; Bindemittel *n*; Lasche *f*; Verbindungsstelle *f*, Verbindungsstück *n*; Verkittung *f*, Verklebung *f*; 2. *(Ch)* Bindung *f*; 3. *(Bw)* Mauerverband *m*, Ziegelverband *m*; Verbund *m (Beton)*; 4. Schienenverbinder *m*, Brücke *f (zwischen Schienenstößen)*
b. direction Valenzrichtung *f*
b. of the wheel Bindung *f* der Schleifscheibe
b. paper Briefpapier *n*; Banknotenpapier *n*
b. strength 1. Haftfestigkeit *f*; 2. Bindungsenergie *f*, Bindungsstärke *f*
b. stress *(Bw)* Haftspannung *f*
bondability Bindungsfähigkeit *f*; Haftfähigkeit *f*
bondable bindungsfähig, haftfähig
bonded abrasive Schleifkörper *m*
b. abrasive grain gebundenes Korn *n*
b. abrasive stick Schleifleiste *f*
b. fabric *(Text)* Faservliesstoff *m*
b. post-tensioning Vorspannung *f* mit nachträglichem Verbund *(Spannbetonbau)*
b.-rubber damper Gummi[schwingungs]dämpfer *m*
b.-rubber pad *(El)* Schwingmetall *n*
b. shield picture tube Fernsehbildröhre *f* mit Sicherheitsglasscheibe
b. warehouse *(Schiff)* Freilager *n*, Tollager *n*, Tollschuppen *m*
b. yarn Klebgarn *n (mit Klebstoff statt durch Feinspinnen verfestigtes Garn)*
bonder *(Bw)* Binder *m*, Strecker *m (Mauerwerk)*
bonder *(Am) (El)* Masseverbindung *f*, Erdverbindung *f*; Erdungsschraube *f*
bonderize / to bondern *(Oberflächenvergütung)*

bonding 1. Verbinden *n*; Kleben *n*; Haften *n*; Haftung *f*; 2. *(Text)* Binden *n*, Verfestigen *n* *(zu Faservliesstoff)*
b. agent Bindemittel *n*
b. and densifying plywood Kleben *n* und Vergüten *n* von Sperrholz
b. cement Kitt *m*, Klebekitt *m*
b. clip *(El)* Abbindeklemme *f*, Beidrahtschelle *f*, Erdungsschelle *f*, Verbindungsklemme *f*
b. defect *(Opt)* Bindungsfehler *m*
b. drill Leierbohrer *m*
b. electron Valenzelektron *n*
b. material *(Ch)* Klebstoff *m*, Imprägnierungsmittel *n*
b. orbital *(Ph, Ch)* gemeinsame Elektronenbahn *f (homöopolare Bindung)*
b. ribbon *(Am)* *(El)* elektrische Bleimantelverbindung *f (zur Verhütung von Korrosion)*
b. ring *(El)* Erdungsring *m*, Erdungsschelle *f*
b. strength Bindefähigkeit *f*, Bindekraft *f*, Klebefestigkeit *f*, Leimfestigkeit *f*
b. strip *(El)* Bleimantelverbinder *m*
b. surface Verbindungsfläche *f (beim Löten)*
b. technique Bindetechnik *f (Vliesstoffherstellung)*
b. wire *(El)* Abbindedraht *m*, Bindedraht *m*
bondstone Binder *m*, Strecker *m (Mauerwerk)*
bone Knochen *m*; Gräte *f*
b. binder *(Am)* *s.* b. glue
b. black Knochenkohle *f*, Beinschwarz *n*, Knochenruß *m*, Knochenschwarz *n*
b. charcoal Knochenkohle *f*
b. coal *(Am)* Kohlenschiefer *m*, Brandschiefer *m*, bergehaltige (verwaschene) Kohle *f*
b.-dry absolut trocken, atro
b. dry pulp absolut trockener Zellstoff *m*
b. glue Knochenleim *m*
b.-grease-extracting apparatus Knochenentfettungsanlage *f*
b. lace Klöppelspitze *f*
b. oil Knochenfett *n*; Knochenöl *n*; Tieröl *n*
b. seeker *(Kern)* Knochensucher *m*
boning *(Brg)* Nivellement *n*; Visieren *n*
b. rod *(Bw)* Krücke *f*, Visiertafel *f*
Bonjean's curves *(Schiff)* Bonjeansche Kurven *fpl*, Spantinhaltskurven *fpl*
bonnet 1. Deckel *m*, Haube *f*, Kappe *f*; Schutzkorb *m*; Ventilhaube *f*; 2. Aufsatz *m (Schornstein)*; 3. *(Kfz)* Motorhaube *f*, Kühlerhaube *f*, Haube
b. clip Motorhaubenhalter *m*
b. line Motorhaubenform *f*
b. lock Motorhaubenschloß *n*
b. stay Motorhaubenstütze *f*
b. tile *(Bw)* Gratziegel *m*, Firstziegel *m*
bonney *(Geol)* Erznest *n*
b. coal schiefrige Kohle *f*
bonny *(Am)* *s.* bonney
bonus scheme Prämiensystem *n*
book / to *(Fmt)* anmelden *(Gespräch)*; vormerken [lassen]
book block Buchblock *m*
b. cloth Buchbinderleinen *n*
b. of reference Nachschlagewerk *n*
b. paper Bücherpapier *n*
b.-plate Exlibris *n*; Buchzeichen *n*
b. review Buchbesprechung *f*
b. sewing machine Fadenheftmaschine *f*
b. stack Bücherregal *n*
b. trimming machine Dreischneider *m*
b.-wrapper Buchhülle *f*
bookbinder Buchbinder *m*
bookbinder's calico *s.* buckram
bookbinding Buchbinden *n*; Buchbinderei *f*
b. fabric Einbandgewebe *n*
b. leather Buchbinderleder *n*
b. thread Heftgarn *n*, Heftfaden *m*
bookkeeping transfer *(Büro)* Umbuchung *f*
booklet Broschüre *f*
b. stitcher Broschürenheftmaschine *f*
Boolean algebra *(BMSR)* Algebra *f* der Logik, Boolesche Algebra; Schaltalgebra *f*
Boolean function *(BMSR)* logische Funktion *f*
boom / to dröhnen, schallen
b.-off *(Schiff)* mit der Stange fortbewegen; abbäumen *(Schiff seitlich abstützen)*
b. out the sails die Segel mit den Spieren aussetzen
boom 1. Ausleger *m (Baumaschinen)*; 2. Gurt *m (Fachwerkbinder)*; 3. Trägerstange *f (Turnen)*; 4. *(Schiff)* Spiere *f*; Baum *m*, Segelbaum *m*; 5. *(El, sl)* Stromabnehmerstange *f*
b. dragline Schlepplöffelbagger *m*, Schürfkübelbagger *m*, Zugschaufelbagger *m*
b. fore-sail *(Schiff)* Baumfock *f*
b. fore-sail of schooner Schonersegel *n*
b. lift *(Schiff)* Kranleine *f*; Baumtoppnant *f*, Dirk *m*
b. point *(Bw)* Auslegerspitze *f*
b. stacker *(Am)* Bandabsetzer *m (Fördertechnik)*
boomer Tieftonlautsprecher *m*
boomsman Flößer *m*
boomy speech *(El)* Sprachverzerrung *f* durch Tiefenresonanz
boon Schäbe *f*
boost / to 1. [elektrische] Spannung erhöhen; 2. aufladen; verstärken; anheben
boost 1. *(El)* Zusatzspannung *f*; 2. Verstärkung *f*, Ladedruck *m*; 3. Öldruckverstärkung *f*; Druckerhöhung *f*
b. action Druckstauwirkung *f*
b. charge *s.* quick charge
b. coil *(El)* Zündspule *f*, Verstärkerspule *f*
b. control Boostersteuerung *f*
b. curve *(Kfz)* Ladedruckkurve *f*

b. pressure *(Kfz)* Ladedruck *m*, Aufladedruck *m*
boosted feed *(Wärm)* Speisung *f* über Druckerhöhungspumpe
booster 1. *(BMSR)* Beschleuniger *m*; Verstärker *m*; Öldruckverstärker *m*; 2. *(Fmt)* Antennenverstärker *m*; 3. Hilfsmotor *m*; 4. *(El)* Zusatzdynamo *m*; 5. *(Kfz)* Aufladegebläse *n*, Servomotor *m*; 6. *(Bw)* Druckverstärker *m*, Verstärkerpumpe *f* *(Rohrleitungen)*; 7. *(Ch)* Dampfstrahlsauger *m*; 8. *(Rak)* Startbeschleuniger *m*
b. amplifier *(El)* Zusatzverstärker *m*, Booster-Verstärker *m*
b. battery *(El)* Zusatzbatterie *f*, Verstärkerbatterie *f*
b. brake Servobremse *f*
b. coil Anlaßzündspule *f*
b. diode Schalterdiode *f*
b. fan Sonderlüfter *m*
b. light Aufheller *m*
b. machine Hilfsmaschine *f*, Zusatzmaschine *f*
b. motor Hilfsmotor *m*
b. relay [pneumatischer] Leistungsverstärker *m*, Schnellverstärker *m*
b. response *s.* derivative action
b. rocket Startrakete *f*
b. station Druckverstärker *m*, Verstärkeranlage *f*, Verstärkerstation *f* *(Rohrleitungen)*
b. switch Hilfsschalter *m*, Zusatzschalter *m*
b. transformer Zusatztransformator *m* *(für Zu- und Gegenschaltung)*
boosting Verstärkung *f*, Druckerhöhung *f*
b. charge Teilladung *f*
b. voltage Zusatzspannung *f*, Verstärkungsspannung *f*
boot *(Kfz)* 1. Kofferraum *m*; 2. Staubklappe *f* *(am Bremszylinder)*
b. board Schuhpappe *f*
b. floor *(Kfz)* Kofferraumboden *m*
b. lid *(Kfz)* Kofferraumdeckel *m*, Gepäckraumklappe *f*
b.-top line *(Schiff)* Wasserpaß *m*
booth Marktverkaufsstand *m*; Jahrmarktsbude *f*,
bootlace Schnürsenkel *m*
bootstrap Bandbefehl *m* *(Rechenautomaten)*
b. amplifier Verstärker *m* mit Katodenausgang
b. circuit Schaltung *f* mit mitlaufender Ladespannung
b. integrator *(Am)* *(BMSR)* Bootstrap-Integrator *m* *(analoger Integrator)*
boracic acid *s.* boric acid
boracite *(Min)* Borazit *m*
boral *(Met)* Boral *n*
borate Borat *n*, Salz *n* der Borsäure
borated boriert, mit einer Borschicht überzogen, mit Bor bedeckt

b. graphite borierter Graphit *m*
boratofluoric acid Borfluorwasserstoffsäure *f*, Fluoborsäure *f*, Tetrafluoborsäure *f*
borax Borax *m*
b. bead Boraxperle *f*
b. glass Boraxglas *n*
bord / to paspeln
bord *(Am)* *(Brg)* Pfeiler *m* *(Pfeilerbau)*
b. and pillar method *(Brg)* Pfeilerbau[verfahren *n*] *m*
Bordeaux mixture Kupferkalkbrühe *f*, Bordeauxbrühe *f*, Bordelaiser Brühe *f*
border / to einfassen; [um]bördeln; rändeln, säumen
b. a determinant *(Math)* eine Determinante rändern
border 1. Leiste *f*; 2. *(Typ)* Zierleiste *f*, Umrandung *f*; 3. *(Text)* Salband *n*; Einfassung *f*, Randverzierung *f*; 4. *(Met)* Kalibrierringe *mpl*
b. bed Rabatte *f* *(Gartenbau usw.)*
b. scheme Kantenschema *n*
b. stone Grenzstein *m*, Begrenzungstein *m*
bordering Bördeln *n*
b. tool *(Masch)* Bördeleisen *n*
borderline Grenzlinie *f*, Grenze *f*
b. case Grenzfall *m*
b. condition Grenzbedingung *f*
b. state Grenzzustand *m*
bordroom-man *(Brg)* Streckenzimmerhauer *m* beim Pfeilerbau
bore / to [auf]bohren; lochen *(Holzbearbeitung)*
b. a well *(Brg)* eine Bohrung abteufen (herunterbringen)
b. out ausbohren
bore Bohrloch *n*; Schacht *m*; Sonde *f*; Bohrung *f*; Kaliber *n*; lichte Weite *f*
b. axis Bohrungsachse *f*
b. brush *(Mil)* Rohrreiniger *m*
b. camera Bohrlochkamera *f*
b. depth Bohrungstiefe *f*
b. diameter Bohrungsdurchmesser *m*
b. diameter limits Bohrungsdurchmessertoleranz *f*
b. hole Bohrloch *n*, Ausbohrung *f*, Bohrung *f*, Durchbohrung *f*, Loch *n*
b.-hole logging apparatus Bohrlochuntersuchungsgerät *n*
b. indicating gauge Bohrungsmeßgerät *n*, Innenmeßgerät *n*
b. measuring gauge Bohrungsmeßgerät *n*, Innenmeßgerät *n*
b. meter *s.* b. indicating gauge
b. of cylinders *(Kfz)* Zylinderbohrung *f*; Hub *m* des Zylinders
b. of pipe lichte Weite *f* eines Rohres
b. of the piston Kolbenbohrung *f*
b. pump Bohrlochpumpe *f*
b. size Bohrungsdurchmesser *m*

b. testing instrument *(Masch)* Bohrungsmesser *m*
b. tolerance Bohrungstoleranz *f*
b. tubing Bohrlochfutterrohr *n*
b. wear pattern *(Kfz)* Zylinderlaufbild *n*, Zylindertragbild *n*
b. well *(Bw)* Bohrbrunnen *m*
boreal power station Windkraftwerk *n*
bored caisson *(Bw)* Bohrpfahl *m*
b. hole Bohrung *f*, Bohrloch *n*; vorgebohrter Durchbruch *m*
b. out guide *(Wärm)* geschlossenes Führungsrohr *n*
b. pile *(Bw)* Bohrpfahl *m*
b. through hohlgebohrt
borehole *s.* bore hole
borer Bohrmaschine *f*, Bohrer *m*
boric acid Borsäure *f*, Orthoborsäure *f*
b.-acid ointment Borsalbe *f*
boring 1. Aufbohren *n*, Ausbohren *n*; Nachdrehen *n* von Bohrungen; Tiefbohrung *f*; Bohrarbeit *f*; 2. Bohrloch *n*; 3. Bohrspan *m*
b. and mortising machine for deep holes *(Holz)* Langlochbohr- und -stemmaschine *f*
b. attachment Zusatzeinrichtung *f* zum Bohren
b. bar Bohrstange *f*
b. bar cutter Bohr[stangen]messer *n*; Bohrmeißel *m*
b.-bar holder Bohrstangenhalter *m*
b. bar with micrometer adjustable cutter Feinsteinstellbohrstange *f*
b. bit Bohrmesser *n*; Bohrmeißel *m*
b. capacity Aufbohrleistung *f*, Bohrleistung *f*
b. capacity from solid größter Bohrdurchmesser *m* aus dem Vollen
b. chip Bohrspan *m*
b. collar Bohrlager *n* *(Drechslerei)*
b. cutter Bohrmesser *n*, Bohrmeißel *m*
b. depth Bohrtiefe *f*
b. fixture Bohrlehre *f*, Bohrvorrichtung *f*
b. head 1. Bohrkopf *m* *(Bohrmaschine)*; 2. Bohrsupport *m* *(Hobelmaschine)*; 3. Bohreinrichtung *f* *(Fräsmaschine)*
b. head employing two cutters Bohrkopf *m* mit zwei Bohrmeißeln
b. head for finishing cuts Fertigbohrkopf *m*
b. in lathe Ausdrehen *n*
b. jig Bohrlehre *f*, Bohrvorrichtung *f*
b. machine Aufbohrmaschine *f*, Bohrmaschine *f*, Bohrwerk *n*
b.-machine table Bohrmaschinentisch *m*, Tisch *m* einer Aufbohrmaschine
b. mill Bohrwerk *n*
b. miner *(Brg)* Bohrschrämlader *m* *(Maschine)*
b. oil Bohröl *n*
b. operation Bohrarbeit *f*, Nachdreharbeit *f*
b. position Bohrstellung *f*
b. pressure Bohrdruck *m*
b. rods Bohrgestänge *n*
b. sludge Bohrschmant *m*
b. spindle Bohrspindel *f*
b. spindle speed Bohrspindeldrehzahl *f*
b. stay Bohrsetzstock *m*
b. stay bearing Gegenlager *n*
b. tool 1. Bohrmeißel *m*; Bohrwerkzeug *n*; 2. Innendrehmeißel *m*
b. tool holder Bohrmeißelhalter *m*, Bohrkopf *m*
b. tower Bohrturm *m*
b. unit Bohreinheit *f*
borings Bohrspäne *mpl*
borizing *(Am)* Feinbohren (Genauigkeitsbohren) *n* mit Diamant- oder Hartmetallwerkzeug
Born approximation *(Ph)* Bornsche Näherung *f*
Borneo camphor *s.* borneol
borneol Borneol *n*, Borneokampfer *m*
bornite Bornit *m*, Buntkupferkies *m*
boron Bor *n*
b. carbide Borkarbid *n*
b. chamber *(Kern)* Borkammer *f*, Bortrifluorid[ionisations]kammer *f*
b. compound Borverbindung *f*
b. content Borgehalt *m*
b. counter tube *(Kern)* Borzählrohr *n* *(Zählrohr mit Borfüllung)*
b.-filled chamber *s.* b. chamber
b. hydride Borwasserstoff *m*
b. iodide Borjodid *n*
b.-lined counter *(Kern)* mit Bor ausgekleideter Zähler *m*
b.-lined pocket chamber *(Kern)* mit Bor ausgekleidete Taschenionisationskammer *f*
b. nitride Bornitrid *n*, Borstickstoff *m*
b. oxide Boroxid *n*
b. rod *(Kern)* Borkarbidstab *m*
b. steel Borstahl *m*
b. trifluoride Bortrifluorid *n*
borrow *(Bw)* Massenentnahme *f*
b. excavation Entnahmegrube *f*
b. material Füllmaterial *n* *(Straßenbau)*
b. mechanism *(Büro)* Borgmechanismus *m* *(beim Subtrahieren)*
b. pit Entnahmegrube *f*
b. source Entnahmestelle *f*, Gewinnungsstelle *f*
borrowed light Innentür *f* mit Glasfüllung
bort Bort *m*, Industriediamant *m*
Bosch injection nozzle Bosch-Einspritzdüse *f*
Bosch lubricating pump Bosch-Schmierpumpe *f*
Bosch-type jerk pump Bosch-Einspritzpumpe *f*
bosh Rast *f* *(Hochofen)*
b. angle Rastwinkel *m*
b. band Rastankerring *m*

b. casing Rastpanzer *m*
b. cooling plate Rastkühlkasten *m*
boson *(Ph)* Boson *n*
boss / to senken *(Naben)*; aussenken
boss 1. Nabe *f*, Vorsprung *m*; Nocken *m*, Nocke *f*, Auge *n*, Warze *f*; 2. *(Ch)* Muffe *f* *(Labortechnik)*; 3. *(Geol)* Stock *m*, Intrusivkörper *m*; 4. *(Schiff)* Propellernuß *f*, Wellennuß *f*
b. for foundation bolt *(Masch)* Ankernocken *m*
b. joint Nabenverbindung *f*
b. joint bolt Nabenschraube *f*
b. of guide wheel Leitradnabe *f* *(Turbine)*
b. plate *(Schiff)* Nußplatte *f*, Wellenhosenplatte *f*
bossage *(Bw)* Bossenwerk *n*
bossing 1. *(Masch)* Nabensenken *n*; 2. *(Schiff)* Wellenhose *f*
bosun *s.* boatswain
BOT-ohm *s.* Board of Trade ohm
bothway beidseitig, doppelgerichtet
b. circuit *(Fmt)* Leitung *f* für wechselseitigen Verkehr
bott Stopfen *m*
bottle / to 1. abfüllen *(in Flaschen)*; 2. sich einschnüren *(Zugversuch)*
bottle Flasche *f*
b. battery *(El)* Flaschenelement *n*
b.-blowing machine Flaschenblasmaschine *f*
b. bobbin winder Flaschenspulmaschine *f*
b. cell *s.* b. battery
b.-cleaning machine Flaschenreinigungsmaschine *f*
b.-cleaning plant Flaschenreinigungsanlage *f*
b. closure Flaschenverschluß *m*
b.-corking machine Flaschenkorkmaschine *f*
b.-filling machine Flaschen[ab]füllmaschine *f*
b. glass Flaschenglas *n*
b.-green flaschengrün
b. jack Flaschenwinde *f*
b.-moulding press Flaschenpresse *f*
b. neck 1. Flaschenhals *m*; 2. Engpaß *m* *(Produktion usw.)*
b. opener Flaschenöffner *m*
b. rinsing machine Flaschenspülmaschine *f*
b. screw Spannschraube *f*
b. shutter Flaschenverschluß *m*
b. top mould *(Met)* Flaschenhalskokille *f*
b. washer Flaschenwaschmaschine *f*
bottled gas Flaschengas *n*, Flüssiggas *n*, Speichergas *n*
bottleneck *s.* bottle neck
bottlenose oil Spermazetöl *n*, Waltran *m*
bottlery Abfüllraum *m*, Flaschenabteilung *f*
bottom / to 1. den Grund bearbeiten *(Bohrung)*; 2. mit Boden versehen; 3. grundieren
bottom 1. Boden *m*, Grund *m*; 2. Fuß *m* *(Schraubstock)*; Basis *f*; 3. *(Brg)* Liegendes *n*, Sohle *f*; 4. *(Text)* Fond *m*, Grund *m*, Untergrund *m*; 5. Schiffsboden *m*; Schiff *n*
b. and lid stamping machine *(Holz)* Boden- und Deckelstanzmaschine *f*
b. angle of floor plate *(Schiff)* Bodenspantwinkel *m*
b. anvil Bodenamboß *m*, Kasserolamboß *m*
b. anvil tool Unterteil *n*
b. bearing *(Masch)* Unterlager *n*
b. belt conveyor Untergurtförderer *m*
b. blade *(Masch)* Untermesser *n* *(Schere)*
b. blow valve Bodenventil *n*
b. board *(Text)* Platinenboden *m* *(Jacquardmaschine)*
b. boom *(Bw)* Untergurt *m* *(Fachwerkbinder)*
b. box Unterkasten *m*, unterer Formkasten *m* *(Gießerei)*
b. cage *(Text)* untere Siebtrommel *f*
b. cager *(Brg)* Anschläger *m*
b. casting *(Met)* steigender Guß *m*; Gießen *n* von unten; Gießen im Gespann
b. ceiling *(Schiff)* Bodenwegerung *f*
b. centre *s.* b. dead centre
b. charging machine Bodenbeschickungsmaschine *f*, Bodeneinsetzmaschine *f*
b. chord *(Bw)* Untergurt *m* *(Fachwerkbinder)*
b. cloth weave *(Text)* Bindung *f* der Unterware
b. cover *(Wärm)* Bodendeckel *m*, Zylinderbodendeckel *m*
b. covering Bodenbelag *m*
b. cut *(Brg)* unterer Schram *m*
b. dead centre unterer Totpunkt *m* *(Antriebsmaschinen)*
b. deposits Bodenschlamm *m* *(Wasseraufbereitung)*
b. die Untergesenk *n*, Unterstempel *m*, Unterwerkzeug *n*, Matrize *f*
b. discharge Bodenentleerung *f* *(Förderwagen)*
b. discharge skip Skip *m* mit Bodenentleerung
b. discharge waggon Bodenentleerer *m*
b. door Bodenklappe *f*
b. drafting roller untere Streckwalze *f*, Unterwalze *f*
b. drain valve Bodenentleerungsventil *n*
b.-draw cut *(Brg)* Sohllöseneinbruch *m* *(Schießarbeit)*
b. dump hatch Bodenklappe *f* eines Transportfahrzeuges
b. dump truck Fahrzeug *n* mit Bodenentleerung, Bodenentleerer *m*
b. edge Unterkante *f*
b. electrode *(El)* Bodenelektrode *f*
b. filler Balleneinlage *f* *(Schuh)*
b. fire Sohlenfeuer *n*

b. flange Untergurt *m*
b. flask Unterkasten *m (Gießerei)*
b. flue *(Wärm)* Unterzug *m*, Grundzug *m*
b. frame *(Schiff)* Bodenspant *n*
b. frequency *(Fmt, Fs)* untere Höckerfrequenz *f*
b. friction Bodenreibung *f*
b. gate 1. *(Bw)* Untertor *n (einer Schleuse)*; 2. Bodenklappe *f*
b.-half Bodenteil *n*, untere Hälfte *f*
b.-half mould untere Formhälfte *f (Gießerei)*
b.-heated bodenbeheizt
b. hide *(Led)* unterste Haut *f (Stapel)*
b. hole 1. *(Masch)* Grundbohrung *f*; 2. *(Brg)* Sohlenschußbohrloch *n*
b. in a blind hole Grund *m* einer Grundbohrung
b. land Zahnlückenbreite *f*
b. landing *(Brg)* Füllort *n*
b. leather Bodenleder *n*, Unterleder *n*
b. limit untere Grenze *f*
b. line *(BMSR)* Grundlinie *f (Zeichenerkennung)*
b. loading belt *(Brg)* Untergurtförderer *m*, Teppichförderer *m*
b. longitudinal *(Schiff)* Bodenlängsspant *n*
b. lower edge *s.* b. edge
b. mordant *(Text)* Vorbeize *f*
b. nipper plate *(Text)* Unterzangenbacke *f (Kämmaschine)*
b. of a furnace *(Met)* Ofensumpf *m*
b. of crankcase *(Kfz)* Kurbelwellenuntergehäuse *n*
b. of ditch *(Bw)* Grabensohle *f*
b. of dock *(Schiff)* Dockboden *m*
b. of drilled hole Bohrungsgrund *m*
b. of keel *(Schiff)* Unterkante *f* Kiel
b. of key Keilbauch *m*
b. of keyseat [Keil-]Nutgrund *m*
b. of shell Gehäuseboden *m*
b. of the well bore Bohrlochsohle *f*, Bohrort *n (Erdöl, Erdgas)*
b. opening Bodenöffnung *f*, Bodenklappe *f*
b. peak *(Fmt, Fs)* unterer Höcker *m*
b. planking *(Schiff)* Bodenbeplankung *f*
b. plate Grundplatte *f*, Bodenplatte *f*, Fundamentplatte *f*, Sohl[en]platte *f*
b. pouring *(Met)* steigender Guß *m*
b. pouring plate *(Met)* Gespannplatte *f*
b. pressure Bodendruck *m*, Sohldruck *m*
b. print Untergrund *m*, Vordruck *m*
b. right-hand corner untere rechte Ecke *f*
b. ring 1. hinterlegter Spannring *m*, Grundring *m*; 2. *(Brg)* Schneidschuh *m (Senkschachtverfahren)*
b. road *(Brg)* Grundstrecke *f*
b. roll Unterwalze *f*, Matrizenwalze *f*
b. runner *(Kfz)* Bodenlängsträger *m*
b. scouring machine *(Led)* Bodenglasmaschine *f*
b. season *(Led)* erste Appretur *f*, erster Glanz *m*
b. sediments Bodensatz *m*, Bodenrückstand *m*
b. shade *(Text)* Fond *m*, Untergrundfarbe *f*
b. shed *(Text)* Tieffach *n (Jaquard)*, Unterfach *n*
b. side of tooth *(Masch)* Zahnunterseite *f*
b. slide *(Masch)* Unterschieber *m*, Unterschlitten *m*
b. spacer shaft untere Verbindungsstange *f (Strickmaschine)*
b. step 1. *(Bw)* Antrittstufe *f*; 2. *(Rak)* untere (unterste) Stufe *f*
b.-stoping *(Brg)* Strossenbau *m*
b. strake *(Schiff)* Bodengang *m*
b. surface *(Flg)* Druckseite *f*; Unterseite *f*
b. swage Untergesenk *n*, Unterteil *n* eines Gesenkes
b.-tap ladle *(Met)* Stopfenpfanne *f*
b. view Unter[an]sicht *f*
b. yeast untergärige Hefe *f*
bottomer *(Brg)* Anschläger *m*
bottoming 1. *(Bw)* Packlage *f*, Unterbau *m (Straßenbau)*; 2. *(Text)* Grundierung *f*; Vorwäsche *f (Bleicherei)*; 3. *(Fmt, Fs)* Signalbegrenzung *f* durch Sättigung der Röhre; 4. Vorbeizen *n*
b. die *(Masch)* Fertigschneideisen *n*
b. reamer Grundreibahle *f*; Stirnreibahle *f*
b. tap Grund[loch]gewindebohrer *m*
bottoms *(Ch)* Bodenkörper *m*; Destillationsrückstand *m*, Rückstände *mpl*
Boucherie process *(Holz)* Boucherieverfahren *n*, Saftverdrängungsverfahren *n*
boucherizing Holzschutzverfahren *n* durch Kupfersulfatinjektion
boucherot squirrel-cage motor *s.* double squirrel-cage induction motor
bouclé *(Text)* Bouclégarn *n*; Bouclestoff *m*
b. yarn Bouclégarn *n*
boulder Felsblock *m*; gerundeter Flußstein *m*
b. clay *(Geol)* Geschiebemergel *m*
Boulton process Holzschutzverfahren *n* durch Kochen in Teeröl
Boulvin diagram *(Wärm)* Boulvinsches Diagramm *n*
bounce / to aufprallen, rückprallen; schleudern
bounce 1. Aufprall *m*; 2. *(Fs)* Zeilenzittern *n*
b. back of springs Rückfederung *f*
b. plate Prallplatte *f*
bouncing 1. Springen *n*, Abprallen *n*; 2. *(Kfz)* Anschlagen *n* der Düsennadel
b. of the trailer *(Kfz)* Schleudern *n* des Anhängers
b. pin Schlagstift *m*, Prallstift *m*

bound / to 1. begrenzen, abgrenzen; 2. rückprallen; springen; aufbäumen
bound 1. Rückprall *m*; Sprungbewegung *f*; 2. *(Math)* Schranke *f*; Grenze *f*
b.-bound transition *(Kern)* Übergang *m* zwischen gebundenen Zuständen
b. charge gebundene Ladung *f*
b. electricity Aufladung *f*; gebundene Elektrizität *f*
b.-free transition *(Ph)* Übergang *m* vom gebundenen zum freien Zustand, Übergang zum kontinuierlichen Teil des Spektrums
b. level *(Ph)* stationäres Niveau *n*
b. scattering Streuung *f* an gebundenen Atomen
b. state *(Ph)* gebundener (stationärer) Zustand *m (Zustand, in dem das Teilchen nicht frei ist)*
b. structure of atoms Atomverbindung *f*
b. vortex *(Ph)* gebundener Wirbel *m*
b. water *(Ch)* gebundenes Wasser *n*, Kristallwasser *n*
boundaries *(Schiff)* Umschottung *f*
boundary 1. Begrenzung *f*, Grenze *f*, Rand *m*; 2. *(Brg)* Markscheide *f*
b. beacon Begrenzungsbake *f*
b. beam *(Bw)* Randträger *m*
b. break-away Grenzschichtablösung *f*
b. bulkhead *(Schiff)* Begrenzungsschott *n*
b. case Grenzfall *m*
b. condition *(Math, Ph)* Grenz[flächen]bedingung *f*, Randbedingung *f*
b. density *(Ph)* Randdichte *f*
b. ditch Grenzgraben *m*
b. effect Sperrschichtfotoeffekt *m*
b. electron Randelektron *n*
b. fence Grenzzaun *m*; Einfriedung *f*
b. flow Grenzströmung *f*, Randumströmung *f*
b. friction Grenzreibung *f*
b. layer *(Ph)* Grenzschicht *f*
b.-layer cooling Grenzschichtkühlung *f*
b.-layer effect Grenzschichteffekt *m (z. B. bei Transistoren)*
b.-layer flowmeter *(BMSR)* thermischer Durchflußmesser *m*
b.-layer friction Grenzflächenreibung *f*
b.-layer separation Grenzschichtablösung *f*
b. lights Randbefeuerung *f (Flugsicherung usw.)*; Schwellenbefeuerung *f*
b. lubrication Grenz[schicht]schmierung *f*, Teilschmierung *f*
b. marker Randmarkierung *f*; Grenzfunkbake *f*
b. of shadow Schattengrenze *f*
b. pillar *(Brg)* Markscheidesicherheitspfeiler *m*, Bergfeste *f* an der Markscheide
b. plank *(Schiff)* Randplanke *f*
b. point Grenzpunkt *m*
b. post 1. Grenzpfahl *m*; 2. Zaunpfeiler *m*
b. potential *(Ph)* Grenzflächenpotential *n*
b. resistance Grenzwiderstand *m*
b. size *(Min)* Grenzkorn *n*
b. stone Grenzstein *m*
b. stress Randspannung *f (Statik)*
b. surface Grenzfläche *f*
b. turbulence Randturbulenz *f*
b. value *(Math)* Randwert *m*
b. value problem *(Math)* Randwertaufgabe *f*, Randwertproblem *n*
b. wall Einfriedungsmauer *f*
b. wavelength Grenzwellenlänge *f*
bounded *(Math)* beschränkt
b. by begrenzt durch
boundedness *(Math)* Beschränktheit *f*
boundless unbegrenzt
boundlessness Unbegrenztheit *f*
Bourdon tube Bourdonrohr *n*, Rohrfeder *f (Manometer)*
Bourdon-type pressure gauge Röhrenfedermanometer *n*
bourgeois *(Typ)* Borgis *(Schriftgröße)*
bournonite Bournonit *m*, Rädelerz *n*
bout Schar *f (Weberei)*
bovine animals Hornvieh *n*
b. race Rind *n (als Gattung)*
bow / to biegen, krümmen, wölben
b. out *(Bw)* ausbiegen
bow 1. Bogen *m*; Bügel *m*; Krümmung *f*; 2. *(Text)* verzogener Schuß *m* im Gewebe, verzogene Kette *f* im Gewebe
b. and buttock lines *(Schiff)* Schnitte *mpl (im Linienriß)*
b. chock *(Schiff)* Bugaufklotzung *f*
b.-chock mooring pipe *(Schiff)* Tauklüse *f*; runde Bugklüse *f*
b. compartment *(Schiff)* Bugraum *m*
b. compass Teilzirkel *m*, Bogenzirkel *m*
b. compasses Nullenzirkel *m*
b. grace *(Schiff)* Eisschutz *m (am Bug)*
b. gun *(Mil)* Buggeschütz *n*
b. hawse-hole *(Schiff)* Bugankerklüse *f*; Bugklüse *f*
b.-heavy *(Flg)* buglastig; kopflastig
b. instrument Nullenzirkel *m*
b. jet manoeuvring system *(Schiff)* Bugstrahl-Manövriersystem *n*
b. lines Vorschiffslinien *fpl (im Linienriß)*
b. loading *(Schiff)* Be- und Entladung *f* durch Bugöffnung
b.-net-type aerial Reusenantenne *f*
b. pen Nullenzirkel *m* mit Einsatz für Tusche
b. pencil Nullenzirkel *m* mit Einsatz für Blei
b. port Bugpforte *f*, Ladepforte *f (Ladeöffnung im Bug oder Heck eines Schiffes)*
b. propeller Bugpropeller *m*

b. spring compass Nullenzirkel *m*; Federzirkel *m*
b. string Bogensehne *f*
b. thruster drive *(Schiff)* Bugstrahlruderantrieb *m*
b. trolley *(El)* Bügelstromabnehmer *m*
b.-type collector *(El)* Bügelstromabnehmer *m*
b.-type electrode Bügelelektrode *f*
b.-type saw Bogensäge *f*
b. wave *(Schiff) (seitliche)* Bugwelle *f*
b. window rundes Erkerfenster *n*
Bowden wire Bowdenzug *m*
bower *(Schiff)* Buganker *m*
b. anchor Buganker *m*
bowk *(Brg)* Gezähekübel *m (Schachtabteufen)*
bowl 1. Schale *f*, Schüssel *f*, Napf *m*; 2. *(El)* Leuchtenschale *f*; 3. Kübel *m (Schurfkübelwagen)*
b. centrifuge *(Brg)* Vollmantelzentrifuge *f*
b. classifier *(Brg)* Schüsselklassierer *m*
b. feeder topfförmiger Vibrationsförderer *m*
b.-type resonator *(Fmt)* Hohlraumresonator *m*
bowline *(Schiff)* Bulin *f*, Buleine *f*
bowman vorderster Ruderer *m* im Boot
bowspring Bogenfeder *f*
bowsprit *(Schiff)* Bugspriet *m*
b. shroud *(Schiff)* Bugstag *n*
b.-shrouds *(Schiff)* Bugsprietwanten *pl*
bowstring arch *(Bw)* Bogen *m* mit Zugband
b. girder *(Bw)* Bogenfachwerk *n*, Fachwerkbogenträger *m*
b. truss *(Bw)* Bogenfachwerk *n*, Fachwerkbogenbinder *m*
box 1. Kasten *m*; Kiste *f (Glühen)*; Behälter *m*; Gehäuse *n*; 2. Meißelklappenträger *m*; 3. Schließfach *n (Post)*
b. all-round tipping wagon Kastenrundkipper *m*
b.-annealed *(Met)* kastengeglüht
b. annealing *(Met)* Kastenglühen *n*, Kastenglühung *f*
b. annealing furnace *(Met)* Kastenglühofen *m*, Kistenglühofen *m*
b. barrel Bohrer *m (Drechslerei)*
b. base *(Typ)* Schachtelunterteil *n*
b. beam *(Bw)* Hohlkastenträger *m*, Kastenträger *m*
b. board Kartonagepappe *f*, Schachtelkarton *m*
b. caisson *(Bw)* Senkkasten *m*
b. calf *(Led)* Boxkalf *n*
b.-car generator *(Am)* Treppenfunktionsgenerator *m*
b. casting *(Met)* Kastenguß *m (Verfahren)*
b. castings *(Met)* Kastenguß *m (Produkt)*
b. chisel Kistenbeitel *m*; Kastenmeißel *m*
b. column *(Bw)* Hohlkastenstütze *f*, Kastensäule *f*
b. column drill Kastenständerbohrmaschine *f*
b.-column model (design, type) Kastenständerausführung *f*
b. connector *(El)* Gehäuseklemme *f*
b. contact *(El)* Löschkammerkontakt *m*
b. coupling Schalenkupplung *f*
b. cover Schachteldeckel *m*; Schachteloberteil *n*
b. culvert *(Bw)* Kastendurchlaß *m*
b. dam *(Bw)* Kastenfang[e]damm *m*
b. dock U-förmiges Schwimmdock *n*, U-Dock *n*
b. drain *(Bw)* Kastendrän *m*, Kastengerinne *n*, rechteckiger Abwasserkanal *m*
b. drill jig Bohrkasten *m*
b. enamels Kartonagenglanzpapier *n*
b. end tipping wagon Kastenvorderkipper *m*
b. floating dock U-förmiges Schwimmdock *n*
b. frame assembling machine Rahmenmaschine *f*
b. fuse *(El)* Dosensicherung *f*
b. gauge Kastenkaliber *n*
b. girder *(Bw)* Hohlkastenträger *m*, Kastenträger *m*
b. gutter *(Bw)* Kastenrinne *f*, rechteckige Regenrinne *f*
b. hatchet Kistenbeil *n*
b. head Tabellenkopf *m*, Rubrik *f*
b. hole borer *(Brg)* Rollochbohrmaschine *f*
b. hook Kistenheber *m*
b. iron Bügeleisen *n*
b. jig Bohrkasten *m*
b. jointing and assembling machine Maschine *f* zum Zusammenfügen von Kisten
b. keelson *(Schiff)* Kastenkielschwein *n*
b.-kite Kastendrachen *m*
b. level *(Masch)* Dosenlibelle *f*
b.-like kastenförmig [ausgebildet]
b.-loading supply *(Text)* Schußladevorrichtung *f*, Ladevorrichtung *f*
b. loom Webstuhl *m* mit Schützenwechsel
b. making machine Kistenmaschine *f*
b. motor Kapselmotor *m*
b. nailing machine Kistennagelmaschine *f*
b. nut Überwurfmutter *f*
b. pass Flachkaliber *n (von Walzen)*
b. pasting machine Schachtelklebemaschine *f*
b. peg *(Text)* Gehäusespindel *f*
b. piston *(Wärm)* doppelwandiger (geschlossener) Kolben *m*
b. plate girder *(Bw)* Blechkastenträger *m*
b. plate tappet *(Text)* mehrteiliger zusammengesetzter Exzenter *m*
b. relay Dosenrelais *n*
b. room Kofferraum *m (einer Schiffskammer)*
b. section Kastenquerschnitt *m*, Kastenprofil *n*
b.-section bridge Vollwandbrücke *f*
b.-section frame *(Kfz)* Kastenrahmen *m*

b.-section girder kastenförmiger Hohlgußbalken *m*
b. sextant Dosensextant *m*
b. shock Kistenbretter *npl*
b. sleeper Kastenschwelle *f*
b. spanner Steckschlüssel *m*
b. stock Kistenware *f*
b. switch Dosenschalter *m*
b. table Kastentisch *m*
b. tipping wagon Kastenkippwagen *m*, Muldenkippwagen *m*
b.-tool cutter Vierkantdrehmeißel *m*
b. trawlers Fischereifahrzeuge *npl (fahren im Verband einer Flotte)*
b.-type cofferdam *(Bw)* Kastenfang[e]damm *m*
b.-type column Kastenständer *m*
b.-type delivery van *(Kfz)* Kastenwagen *m*
b.-type pedestal Kastenfuß *m*
b. veneer Schachtelspan *m*
b. veneer cutting machine Schachtelspanteilmaschine *f (Zündholz)*
b. veneer length Länge *f* des Schachtelspans *(Zündholz)*
b. wrench Ringschlüssel *m*, Hahnschlüssel *m*
boxhaul / to *(Schiff)* backhalsen, mit backen Segeln halsen
boxheart checking *(Holz)* Herzriß *m*, Kernriß *m*, Radialriß *m*
boxhole *(Brg)* Rolloch *n*, Rolle *f*
boxing *(Brg)* Kastenzimmerung *f*
boxwood Buchsbaumholz *n*
Boyle's law Boyle-Mariottesches Gesetz *n*
bra *(Math, Ph)* Diracs *(Zustandssymbol* < | *in der Quantenmechanik)*
brace / to *(Bw)* aussteifen, versteifen; unterbauen, verstreben; verschwerten
brace 1. Verbindungsklammer *f*; Aufhängeriemen *m*, Band *n*; Bohrleier *f*, Schere *f (Fräsmaschine)*; 2. Draufbohrer *m (Tischlerei)*; 3. *(Bw)* Diagonalstab *m*, Kopfband *n*, Strebe *f*; Zange *f*, Zugband *n*; Stütze *f*; 4. *(El)* Schelle *f*, Klammer *f*; Strebe *(Mast)*; 5. *(Schiff)* Brasse *f*; 6. geschweifte Klammer *f*
b. and bit *s.* b. drill
b. bit Windenbohrer *m*
b. chuck Bohrwindenbüchse *f*
b. drill Leierbohrer *m*
b. rim wrench Felgenkurbel *f*
braced verstrebt; versteift; verankert
b. arch *(Bw)* Fachwerkbogen *m*
b. beam *(Bw)* Fachwerkbinder *m*, Fachwerkträger *m*
b. girder *(Bw)* Fachwerkbinder *m*, Fachwerkträger *m*
braceless unverspannt, verspannungslos; ohne Versteifung
bracer *(Schiff)* Tragseil *n*
brachistochrone *(Math)* Brachistochrone *f*
brachyaxis *(Math)* Brachyachse *f*, Brachydiagonale *f*
brachydiagonal *(Math)* Brachyachse *f*, Brachydiagonale *f*
bracing 1. *(Bw)* Diagonalverband *m*, Dreiecksverband *m*; Aussteifung *f*, Versteifung *f*, Verschwertung *f*, Verspannung *f*, Verstrebung *f*; 2. Steife *f*, Strebe *f*; 3. Absteifring *m*
b. plate Versteifungsblech *n*
b. wire *(El)* Spanndraht *m*
Brackelsberg furnace Brackelsbergofen *m*
bracket 1. *(Math)* [eckige] Klammer *f*; 2. Befestigungsarm *m*, Halter *m*, Konsole *f*, Sockel *m*, Stütze *f*, Träger *m*, Tragkonsole *f*, Wandträger *m*, Kragarm *m*, Schelle *f*; 3. Ausleger *m (für Masten)*
b. arm *(El)* Fahrleitungsausleger *m*, Fahrdrahtausleger *m*
b. clip Steglasche *f*
b. crane Wandkran *m*, Konsolkran *m*
b. floor *(Schiff)* Stützplatten-Bodenwrange *f (beim Doppelboden)*
b. of pump Pumpenträger *m*
b. plate Befestigungsplatte *f*
b. pole Auslegermast *m (Holzmast)*
b. relation *(Ph)* Klammerrelation *f*
b. sleeper Nockenschwelle *f*
b. support *(Bw)* Tragstütze *f*
b. table Konsoltisch *m*
bracketed in [eckigen] Klammern
brackish water Brackwasser *n*
bradawl Bandbohrer *m*, Dübelbohrer *m*, Riegelbohrer *m*, Spitzbohrer *m*
Bradford breaker Bradford-Brecher *m*, Trommelsiebbrecher *m*
Bradford spinning Bradford-Spinnverfahren *n*
Bragg angle Braggscher Winkel *m*, Glanzwinkel *m*
Bragg condition Braggsche Bedingung (Formel) *f*
Bragg curve Braggsche Kurve *f*
Bragg law Braggsches Gesetz *n*
Bragg scattering Braggsche Streuung *f*
Bragg X-ray spectrometer Braggsches Röntgenspektrometer *n*
Bragstad converter *(El)* Kaskadenumformer *m*
braid / to umklöppeln; umspinnen *(Draht)*; flechten; paspeln
braid 1. Umklöppelung *f*, Umflechtung *f*; 2. Litze *f*, Leitungsschnur *f*
b. embroidery Bändchenstickerei *f*
b.-plaiting machine Bandflechtmaschine *f*
b. stitch *(Text)* Zopfstich *m*
braided umsponnen
b. asbestos cord geklöppelte Asbestschnur *f*
b. cable umflochtenes (umklöppeltes) Kabel *n*
b. conducting covering *(Fmt)* geflochtene leitende Abschirmung *f*

b. material Geflecht *n*
b. metal screen geflochtene Metallabschirmung *f*
braiding Beflechtung *f*, Bespinnung *f*, Umklöppelung *f* *(Kabel usw.)*
b. machine Flechtmaschine *f*
b. machine carrier Flechtgurtträger *m*
brail *(Schiff)* Geitau *n* *(Gaffelsegel)*
brake / to 1. [ab]bremsen; 2. brechen *(Flachs)*
brake 1. Bremse *f*; 2. Hebelstange *f*
b.-actuating rod Bremsnockenwelle *f*
b.-adjusting bearing Bremsregulierlager *n*
b.-adjusting mechanism Bremseinstellvorrichtung *f*
b. air conduit Bremsleitung *f*
b. airscrew Bremsluftschraube *f*
b. anchor pin Bremsbackenlagerbolzen *m*
b. anchor plate *(Kfz)* Bremsträger *m*
b. arm Bremshebel *m*
b. assembly Bremsanlage *f*
b. backing plate *(Kfz)* Bremstrommelrückwand *f*
b. band 1. Bremsband *n*, Bremsring *m*; 2. Spreizring *m* *(Kupplung)*
b. bar Bremszugstange *f*
b. block Bremsbock *m*, Bremsklotz *m*
b. block shoe Bremsklotzsohlen *fpl*
b. buffer Bremsanschlag *m*
b. bushing Bremsbüchse *f*
b. cable Bremsseil *n*
b. cable assembly Bremsseilzug *m*
b. cam Bremsnocken *m*
b. cam bushing Bremsnockenlager *n*
b. cam lever Bremsnockenhebel *m*
b. cam shaft Bremsnockenwelle *f*
b. chamber Bremszylinder *m*, Bremskammer *f*
b. cheek Bremsbacke *f*
b. circuit Bremskreis *m*
b. clearance Bremsspiel *n*
b. collar Bremsband *n*
b. collar bearing Halslager *n* *(Bremse)*
b.-compensating device Bremsausgleich *m*
b.-compensating lever Bremsausgleichhebel *m*
b.-compensating shaft Bremsausgleichwelle *f*
b. compensator Bremsausgleicher *m*
b. cone Bremskegel *m*
b.-contact surface Bremsfläche *f*
b. control Bremssteuerung *f*
b. counterweight lever Bremswurfhebel *m*
b. coupling Bremskupplung *f*
b. covering Bremsbelag *m*
b. cross shaft *(Kfz)* Brems[zwischen]welle *f*
b. cylinder Bremszylinder *m*
b. disk Bremsscheibe *f*
b. drag Bremsbackenandrückung *f*
b. drum Bremstrommel *f*
b. drum hub Bremstrommelnabe *f*
b. dynamometer Bremsdynamometer *n*
b. effect Bremswirkung *f*
b. factor Bremsenkennwert *m*
b. fading Bremsschwund *m* *(durch Abnutzung der Bremsbeläge)*
b. field Bremsfeld *n*, Verzögerungsfeld *n*
b. field tube (valve) Bremsfeldröhre *f*
b. fluid Bremsflüssigkeit *f*
b. fluid tank Bremsflüssigkeitsbehälter *m*
b. guard Bremsschutzkappe *f*
b. hand lever Bremshandhebel *m*
b. handle Bremshebel *m*
b. hanger Bremsgehänge *n*
b. hanger support Gehängeträger *m* *(Bremse)*
b. horsepower Bremsleistung *f*, Bremspferdestärke *f*, effektive (nutzbare) Leistung *f*, Nutzleistung *f*
b. horsepower-hour Brems-PS-Stunde *f*
b. hose Bremsschlauch *m*
b. incline *(Brg)* Bremsberg *m*
b. jaw Bremsbacke *f*
b. lever Bremshebel *m*
b.-lever guide Bremshebelführung *f*
b.-lever sector Bremssegment *n*
b. light Bremsleuchte *f*; Bremslicht *n*
b. line coupling *(Kfz)* Kupplungskopf *m*
b. liner Bremsbelag *m*, Bremsfutter *n*
b. lining Bremsbelag *m*, Bremsfutter *n*
b. linkage Bremsgestänge *n*
b. load Bremskraft *f*
b. magnet *(El)* Bremslüfter *m*; Bremsmagnet *m*
b. man Bremser *m* *(s. a.* brakeman)
b. master cylinder Hauptbremszylinder *m*
b. motor Bremslüfter[motor] *m*
b. operation Bremsbetätigung *f*
b. parachute Bremsfallschirm *m*
b. pedal Bremspedal *n*, Fußbremse *f*, Fußbremshebel *m*
b. performance Bremsleistung *f*, Bremsverhalten *n*
b. pipe Bremsleitung *f* *(bei Öldruckbremsen)*
b. piston Bremskolben *m*
b. power Bremskraft *f*, Bremsleistung *f*
b. pressure regulator Bremskraftregler *m*
b. pull rod Bremszugstange *f*
b. pulley Bremsscheibe *f*
b. quadrant Bremssegment *n*
b. ratchet Bremssperrklinke *f*
b.-rigging Bremsgestänge *n*
b. ring Bremsring *m*
b. rod Bremsspindel *f*; Bremsstange *f*
b.-rodding Bremsgestänge *n*
b. safety strap Fangbügel *m* *(Bremse)*
b. screw Bremsspindel *f*
b. screw support Bremsspindelstütze *f*
b. shoe Bremsbacke *f*, Bremsklotz *m*, Bremsschuh *m*

b.-shoe check spring Bremsbackenrückzugfeder *f*
b.-shoe holder Bremsbackenhalter *m*, Bremsklaue *f*
b.-shoe mounting bolt Bremsbackenlagerbolzen *m*
b. shoe pin bushing Bremsbackenlager *n*
b.-shoe return spring Bremsbackenrückholfeder *f*
b. solenoid Bremslüftmagnet *m*
b.-strap Bremsband *n*
b.-support bearing Bremsstützlager *n*
b. suspension Bremsgehängeträger *m*
b. test Bremsversuch *m*, Bremsprobe *f*
b. torque Bremsmoment *n*
b. valve Bremsventil *n*
b.-wheel Bremstrommel *f*
b. wire Bremsseil *n*
braked power abgebremste Leistung *f*
brakefield oscillations *(El)* Barkhausen-Schwingungen *fpl*
brakeman Bremser *m*
brakeman's cab Bremserhaus *n*
b. platform Bremserbühne *f*
b. stand Bremsstand *m*
braking 1. Bremsen *n*, Bremsung *f*; Verzögerung *f*; 2. Hemmung *f (eines Satelliten)*
b. action Bremswirkung *f*
b. area Bremsfläche *f*
b. controller Bremsschalter *m*
b. current Bremsstrom *m*
b. curve *(Rak)* Bremskurve *f*
b. distance Bremsstrecke *f*, Bremsweg *m*
b. effect Bremswirkung *f*
b. effort Bremskraft *f* am Radumfang
b. equipment Bremseinrichtung *f*
b. force Bremskraft *f*
b. heat Bremswärme *f*
b. incline *(Brg)* Bremsberg *m*
b. magnet Bremsmagnet *m*
b. period Bremsdauer *f*
b. power 1. *(Kfz)* Bremskraft *f*, Bremsvermögen *n*; 2. *(Masch)* Bremsleistung *f*, Nutzleistung *f*
b. resistance Bremswiderstand *m*
b. surface Bremsfläche *f*
b. torque Bremsmoment *n*
bran / to *(Led)* im Kleienbad beizen
bran Kleie *f (Futtermittel)*
branch / to abzweigen, verzweigen
branch 1. Zweig *m*, Abzweig[ung *f*] *m*, Verzweigung *f*; 2. Zweigstelle *f*; 3. Schenkel *m (Formstück für Rohrleitung)*, Abzweigrohr *n*
b. box *(El)* Abzweigdose *f*
b. circuit Abzweigstromkreis *m*, Anschlußkreis *m*, Nebenstromkreis *m*, Zweigstromkreis *m*
b. circuit connection Leitungsanschluß *m*
b.-circuit distribution centre Hauptverteiler[tafel *f*] *m*
b. conductor Abzweigleiter *m*, Abzweigleitung *f*
b. connection Anzapfung *f*, Anschluß *m*, Abzweigschaltung *f*, Abzweigverbindung *f*
b. exchange *(Fmt)* Nebenamt *n*, Nebenstellenanlage *f*
b. knot *(Holz)* Astknoten *m*
b. line 1. *(El)* Abzweigleitung *f*, Zweigleitung *f*; 2. Lokalbahn *f*, Nebenbahn *f*, Nebenlinie *f*, Zweigbahn *f*; 3. Verzweigungslinie *f*
b. of a curve *(Math)* Kurvenast *m*
b. of industry Industriezweig *m*
b. operation *(BMSR)* Nebenbetrieb *m (Wirkung eines Abzweigbefehls)*
b. order Abzweigbefehl *m (Programmieren von Rechenautomaten)*
b. piece *(Masch)* Abzweigstück *n*
b. pilot *(Schiff)* Revierlotse *m*
b. pipe Abzweigrohr *n*, Zweigrohr *n*, Seitenrohr *n*
b. piping Zweigrohrleitung *f*
b. point *(Math)* Verzweigungspunkt *m (Riemannsche Fläche)*
b. product Zweigprodukt *n (radioaktiver Zerfall)*
b. resistance *(El, Fmt)* Zuleitungswiderstand *m*, Zweigwiderstand *m*
b. road *(Brg)* Nebenstrecke *f*
b. switch *(El)* Abzweigschalter *m*
b. switchboard *(El)* Verteilertafel *f*
b. tee *(Masch)* T-Stück *n*, Abzweigmuffe *f*
b. terminal Abzweigklemme *f*
b. terminal line Stichbahn *f*
b. vein *(Geol)* zerschlagener Gang *m*
branched verzweigt; *(Plast)* vernetzt
branching Verzweigung *f*, Verästelung *f*
b. decay *(Kern)* Mehrfachzerfall *m*
b. dendrites *(Met)* Dendritenverästelung *f*
b. jack *(Fmt)* Mithörklinke *f*
b. network of wires Stromverzweigung *f*
b. off of current Stromverzweigung *f*
brand *(Led)* Brandzeichen *n*; Marke *f*
b. iron Brandeisen *n*
branding iron Brennstempel *m*
brank Buchweizen *m*
brasque / to *(Met)* auskleiden *(Schmelzofen)*
brass / to vermessingen
brass 1. Messing *n*; 2. *(Brg)* Eisenkies *m*, Schwefelkies *m (bergmännische Bezeichnung)*
b.-band Blasorchester *n*
b. die-casting Messingspritzguß *m*, Messingdruckguß *m*
b. ferrule *(Masch)* Messingzwinge *f*
b.-foil packing *(Wärm)* Stopfbuchse *f* aus Messingblechen

b. foundry *(Met)* Messinggießerei *f*, Gelbgießerei *f*
b. furnace Messingschmelzofen *m*
b. holder Messingfassung *f*
b. insert Messingeinsatz *m*
b. peg Messingstöpsel *m*
b. pipe Messingrohr *n*
b. plating galvanisches Vermessingen *n*
b. pressure casting Messingpreßguß *m*
b. sheet Messingblech *n*
b. slug Messingrohling *m (zum Pressen)*
b. solder Messinglot *n*
b. tube Messingrohr *n*
b. welding rod Messingschweißdraht *m*
b. wind instrument Blechblasinstrument *n*
b. wire Messingdraht *m*
brasses 1. Lagerschalenpaar *n*; 2. *s.* brass 2.
b. of the collar thrust bearing *(Masch)* Kammbuchse *f (Lager)*
b. with lubricating grooves Lagerschale *f* mit Schmiernut
brassière cloth *(Text)* Miedergewebe *n*
brassil *(Geol)* schwefelkieshaltige Kohle *f*
brassy messingartig
brassylic acid Brassylsäure *f*
brattice *(Brg)* Wetterscheider *m*
b. cloth *(Brg)* Wettertuch *n*
braunite Braunit *m*
Braun['s] oscillograph Katodenoszillograf *m*, Katodenstrahl[en]oszillograf *m*
B. tube Katoden[strahl]röhre *f*, Braunsche Röhre *f*
braze / to hartlöten
braze Hartlötung *f*
b. welding Schweißlöten *n*
brazed joint Lötverbindung *f*
b.-on tip *(Masch)* hart aufgelötetes Plättchen *n (Schneidplatte)*
b.-tip carbide tool Werkzeug *n* mit hart aufgelötetem Hartmetallplättchen
brazier 1. Klempner *m*, Messingarbeiter *m*; 2. Kokskorb *m*, Koksofen *m*; 3. Lötofen *m*
Brazil wax Carnaubawachs *n*
B. wood Brasilholz *n*
brazing Hartlötung *f*
b. alloy Hartlot *n*
b. flux Flußmittel *n* beim Löten
b. joint Lötfuge *f*
b. mixture Hartlot *n*
b. outfit Hartlötapparat *m*, Lötapparat *m*
b. solder Hartlot *n*, Messinglot *n*
b. spelter Schlaglot *n*
b. torch Hartlötkolben *m*
B.R.C. fabric *(Bw)* Bewehrungsmatte *f*, Mattenbewehrung *f (Stahlbeton; B.R.C. = British Reinforced Concrete)*
breach Bruch[stelle *f*] *m*, Einbruchstelle *f*
breadth Breite *f*, Weite *f*
b. coefficient *(El)* Sehnungsfaktor *m*, Wicklungsfaktor *m*
b. moulded *(Schiff)* Konstruktionsbreite *f*, Breite *f* auf Spanten
b. of beam Schiffsbreite *f*
b. to length *(Schiff)* Längenverhältnis *n* L:B *(Länge zu Breite)*
break / to 1. [zer]brechen; abreißen *(Faden)*; 2. *(El)* ausschalten, abschalten; unterbrechen; 3. *(Led)* strecken *(Haut)*; 4. *(Brg)* abreißen, brechen *(Hangendes)*, hereingewinnen, grob zerkleinern
b. a circuit einen Stromkreis öffnen
b. a connection eine Verbindung trennen
b. a line *(Fmt)* eine Leitung unterbrechen
b. a voyage eine Reise abbrechen
b. bulk *(Schiff)* Ladung löschen
b. down 1. *(El, Fmt)* zusammenbrechen *(Feld)*; durchschlagen *(Widerstände)*; 2. *(Brg)* abbauen, [herein]gewinnen, lösen; 3. *(Holz)* ablängen; 4. abreißen, zerreißen *(Ölfilm)*
b. free *(Schiff)* über Bord gehen, losreißen
b. ground *(Bw)* den ersten Spatenstich tun
b. loose from sich befreien; austreten aus
b. off abbrechen *(Ferngespräch usw.)*; aufhören
b. surface auftauchen *(U-Boot)*
b. through durchbrechen, durchtreten *(Bohrer)*
b. up 1. unterbrechen *(Signal)*; 2. aufteilen; unterteilen; verteilen; 3. aufspalten; 4. aufhacken, aufpicken
b. up chips Späne unterteilen
b. up into a fine mist zerstäuben; versprühen
b. up into eddies in Wirbel zerfallen
break 1. Pause *f*, Unterbrechung *f*; 2. *(El, Fmt)* Drahtbruch *m*; 3. *(Bw)* Maueraussparung *f*; Mauervorsprung *m*; 4. *(Holz)* Riß *m*, Abriß *m*; 5. *(El)* Trennstrecke *f (der Kontakte)*; Unterbrechung *f*; 6. *(Brg)* Branddamm *m*
b. arc *(El)* Abschaltlichtbogen *m*
b.-away torque *(Kfz)* Losreißmoment *n (beim Starten)*
b. bulkhead *(Schiff)* Absatzschott *n*, Frontschott *n*
b. bulkhead of raised quarter-deck Quarterdeckschott *n*, Frontschott *n* des erhöhten Quarterdecks
b. bulkhead of sunk forecastle Backschott (Endschott) *n* der versenkten Back
b. contact *(El)* Unterbrechungskontakt *m*, Ruhekontakt *m*, Öffnungskontakt *m*, Öffner *m*, Trennkontakt *m*
b. distance *(El)* Kontaktabstand *m*, Schaltstrecke *f*
b. frequency 1. *(El)* Frequenz *f* im Kennlinienknick; 2. *(BMSR)* Eckfrequenz *f*, Knickfrequenz *f*, 45°-Frequenz *f (Bodediagramm)*

b. impulse *(El)* Abschaltüberspannung *f*, Öffnungsimpuls *m*, Stoßwelle *f* beim Abschalten
b.-in operation *(Fmt)* „break-in"-Betrieb *m*
b.-in period *(Kfz)* Einlaufzeit *f*
b.-in system *(Fmt)* Sende-Empfangsumschaltung *f (Duplexverkehr)*
b. induced current Öffnungsstrom *m*, Unterbrechungsstrom *m*
b. interlock Verriegelung *f* bei Abrissen *(Papierherstellung)*
b. jack *(El, Fmt)* Abschaltklinke *f*, Trennklinke *f*, Unterbrechungsklinke *f*; Anodenschutzbuchse *f*
b. joint *(Bw)* versetzte Fugen *fpl*
b. key *(El)* Unterbrechungstaste *f*
b. lathe Plandrehmaschine *f* mit Gleitkröpfung
b. line Bruchkante *f (technische Zeichnung)*, Bruchlinie *f*
b. of vision Bildsprung *m (Augenoptik)*
b.-off phenomenon *(Rak)* Abreißvorgang *m*
b.-out *(Met)* Durchbruch *m (des Schmelzofens)*
b. point *(BMSR)* Verzweigungspunkt *m*, Zwischenstopp *m (im Laufe eines Programms)*
b.-proof bruchsicher
b. pulse *(Fs)* Öffnungs[im]puls *m*
b. sign *(Fmt)* Trennungszeichen *n*
b. spark *(El)* Abschaltfunke *m*
b.-through 1. *(Bw)* Durchbruch *m*; 2. *(Ph)* Adsorptionsschwelle *f*
b.-through of the drill Bohreraustritt *m*
b.-through of the drill point Durchtreten *n* (Durchtritt *m*) der Bohrerspitze
b.-through tank *(Mil)* Durchbruchspanzer *m*, Durchbruchskampfwagen *m*
b.-time *(El)* Ausschaltdauer *f*
breakable zerbrechlich
breakage Bruch *m*, Zerspringen *n*
breakaway *(El)* Anlaufen *n (elektrischer Zug)*
b. force *(El)* Anlaufwiderstand *m*, Losbrechwiderstand *m (elektrischer Zug)*
breakdown 1. Betriebsstörung *f*, Betriebsausfall *m*, Ausfall *m*, Störung *f*, Maschinenschaden *m*; 2. Autopanne *f*; 3. *(El)* Durchbruch *m*, Durchschlag *m (Isolatoren)*; Überschlag *m*; 4. *(Ph)* Zusammenbruch *m (elektromagnetisches Feld)*
b. bench *(El)* Durchschlagmeßtisch *m (insbesondere für Gleichrichterröhren)*
b. diode Zenerdiode *f*
b. due to overvoltage Überspannungsdurchschlag *m*
b. due to thermal instability *(Wärm)* Wärmedurchschlag *m*
b. factor *(El)* relatives Kippmoment *n*
b. field [strength] *(El)* Durchbruchsfeldstärke *f*
b. moment Abfallmoment *n*
b. of logs Einschnitt *m* der Rundhölzer
b. of service *(Fmt)* Betriebsstörung *f (z. B. Funkstörung)*
b. strength *(El)* Durchschlagfestigkeit *f*
b. test 1. *(El)* Durchschlagsprobe *f*; 2. *(Met)* Bruchprüfung *f*
b. torque Höchstmoment *n*, Kippmoment *n*
b. train *(Met)* Vorwalzstraße *f*
b. van Abschleppfahrzeug *n*, Abschleppkran *m*
b. vehicle *(Kfz)* Abschleppwagen *m*
b. voltage *(El)* Durchschlagsspannung *f*, Durchbruchspannung *f*
breaker 1. *(El)* Schalter *m*, Unterbrecher *m*; Kommutator *m (selten)*; 2. Brecher *m*, Hadernwolf *m*, Halbstoffholländer *m*, Lumpenzerreißer *m (Papierherstellung)*; 3. *(Brg)* Brecher; 4. *(Kfz)* Leinwandeinlage *f (unter der Lauffläche eines Reifens)*; Zwischenbaugewebe *n*; 5. *(Schiff)* Wasserfaß *n (Frischwasser)*; Sturzsee *f*; 6. *(Text)* Brechmaschine *f*; Appretbrecher *m*
b. arm *(El)* Unterbrecherarm *m*
b. beater Auflöseholländer *m*, Brechholländer *m*, Schlagholländer *m*
b. card *(Text)* Vorkarde *f*, Grobkrempel *f*, Vorkrempel *f*, Reißkrempel *f*
b. contact Unterbrecherkontakt *m*
b. drum Halbzeugholländerwalze *f*
b. engine Halbzeugholländer *m*
b. lever Unterbrecherhebel *m*
b. plate *(El)* 1. Lochscheibe *f*, Prallfläche *f*, Stauscheibe *f*; 2. Brecherplatte *f (Extruder)*
b. point gap *(Kfz)* Unterbrecherabstand *m*
b. press Wendepresse *f (Papiermaschine)*
b. prop *(Brg)* Reihenstempel *m*, Bruchstempel *m*
b. row *(Brg)* Bruchstempelreihe *f*
b. trap Hartkocherstofföffner *m*
b. vat Kollerbütte *f*
breakers *(Schiff)* Brecher *mpl*, Brandung *f*, Sturzsee *f*, Sturzwellen *fpl*
breaking Brechen *n*; Reißen *n*; Zerreißen *n*; Unterbrechung *f*; Trennen *n*
b. away Abreißen *n*
b. ball *(Bw)* Zertrümmerungskugel *f*
b. capacity *(El)* Ausschaltleistung *f*, Ausschaltvermögen *n*
b. contact *(El)* Unterbrecherkontakt *m*
b. current Ausschaltstrom *m*, Öffnungsstrom *m*
b.-down pass Vorstich *m (beim Walzen)*
b. edge *(Brg)* Bruchkante *f*
b. elongation *(Met)* Bruchdehnung *f*
b. ground Abbauen *n*, Hereingewinnung *f (Erzbergbau)*
b. iron *(Led)* Streckeisen *n*
b. length *(Text)* Reißlänge *f*

b. limit *(Met)* Bruchgrenze *f*, Zerreißgrenze *f*
b. load Bruchbelastung *f*, Bruchlast *f*
b. machine 1. *(Led)* Streckmaschine *f*; 2. *(Bw)* Grobzerkleinerungsmaschine *f*
b. of local calls for trunk calls Fernamtstrennung *f*
b. of the oil film Abreißen *n* des Schmierfilms
b. off *(Schiff)* Abbau *m*, Abbruch *m (Demontage)*
b. period *(El)* [gesamte] Abschaltdauer *f*
b. pin *(Masch)* Abscherstift *m*
b. point Bruchgrenze *f*, Festigkeitsgrenze *f*, Zerreißgrenze *f*
b. spark gap Abreißfunkenstrecke *f*, Unterbrecherfunkenstrecke *f*
b. speed *(Text)* Reißgeschwindigkeit *f*
b. strength Zerreißfestigkeit *f*; Reißfestigkeit *f (z. B. Baumwolle)*
b. stress Bruchbeanspruchung *f*, Bruchspannung *f*, Zerreißspannung *f*; Knickspannung *f*
b. test Zerreißversuch *m*, Bruchversuch *m*, Knickversuch *m*
b. the corners Brechen *n* der Kanten
b. the ground Aufreißen *n (Erde)*
b. up of soil Lockern *n* des Bodens
b. up of the cooling water Kühlwasserverteilung *f*
b.-up yard *(Schiff)* Abbruchwerft *f*
breakwater *(Schiff)* Wellenbrecher *m*
bream / to abbrennen *(Schiffsboden)*
breast 1. *(Bw)* Fensterbrüstung *f*; Setzstufe *f*; 2. *(Brg)* Arbeitsstoß *m*, Ortsstoß *m*; Pfeiler *m (Pfeilerbau)*
b. and bench *(Brg)* Kammerbau *m* mit strossenartigem Verhieb, stufenförmiger Festenbau *m*
b. beam *(Am)* Brustbaum *m (Weberei)*
b. board *(Brg)* Ortsbrett *n*, Zumachebrett *n (Getriebezimmerung)*
b. borer Bohrwinde *f*, Brustleier *f*
b. box Hochstau *m*; *(Text)* Stoffauflauf *m*
b. drill Brustbohrmaschine *f*
b. height Fensterbrüstung *f*
b. lining Vertäfelung *f* der Fensterbrüstung
b. machine *(Brg)* Kleinschrämmaschine *f*
b. pan *(Met)* Vorherd *m*
b. planer Blechkantenhobelmaschine *f*, Flachbesäummaschine *f*
b.-plate microphone (transmitter) Brustmikrofon *n*
b. rail *(Schiff)* Reling *f*
b. roller Brustwalze *f*
b. rope *(Schiff)* Dwarsleine *f*
b. stoping *(Brg)* Festenbau *m*
b. transmitter *(Fmt)* Brustmikrofon *n*
b. wheel mittelschlächtiges Wasserrad *n*
breasting of the beater Holländerkropf *m*
breastsummer *s.* bressummer
breather Atmungsventil *n*, Entlüfter *m*, Entlüftungsöffnung *f*, Entlüftungsstutzen *m*, Entlüftungsvorrichtung *f*
b. cap *(Kfz)* Entlüfterkappe *f*
b. line Entlüftungsleitung *f*
b. pipe Entlüfterstutzen *m*; Belüftungsrohr *n*
b. roof Atmungsdach *n (eines Tanks)*
b. vent Entlüftungsventil *n*; Schnüffelventil *n*
breathing 1. Atemtätigkeit *f*, Atmung *f*; 2. Entgasen *n*, Druckablassen *n*, Entlüften *n*, Lüften *n*
b. characteristics of an engine *(Kfz)* Gaswechseleigenschaften *fpl* eines Motors
b. vibration *(Ch)* Valenzschwingung *f* in Ringverbindungen
breccia *(Geol)* Brekzie *f*, Breccie *f*
breech Bodenstück *n (Geschütz)*
b.-block thread Sägegewinde *n*
b. cover Verschluß *m*, Verschlußkappe *f*
b. loader Hinterlader *m*
b.-loading gun Hinterlader *m*
b. mechanism Verschluß *m (Geschütz)*
breeches buoy Hosenboje *f*
b. chute *(Brg)* Hosenrutsche *f*
b. pipe Hosenrohr *n*, Rohrabzweigung *f*
b. shoot *s.* b. chute
breeder 1. Züchter *m (Tiere)*; 2. *s.* b. reactor
b. blanket *(Kern)* Brutmantel *m*
b. material *(Kern)* Brutstoff *m*, Brutmaterial *n*
b. pile *s.* b. reactor
b. reactor *(Kern)* Brutreaktor *m*, Brüter *m*
breeding *(Kern)* Brüten *n*, Brutvorgang *m*
b. animals 1. Zuchtvieh *n*; 2. Viehaufzucht *f*
b. blanket *(Kern)* Brutmantel *m*, Brutzone *f*
b. cycle *(Kern)* Brutzyklus *m*
b. gain *(Kern)* Brutgewinn *m*
b. material *(Kern)* Brutstoff *m*, Brutmaterial *n*
b. of fishes Fischzucht *f*
b. of sheep for wool Wollschafzucht *f*
b. reactor *s.* breeder reactor
b. season Brutzeit *f*; Laichzeit *f (Fisch)*
breeze 1. *(Brg)* Grus *m*, Abrieb *m*, Lösche *f*; 2. *(Bw)* Asche *f*, Koksasche *f*, Schlacke *f*; kleinstückiger Koks *m*
b. brick *(Bw)* Dübelstein *m (aus Koksasche und Zement)*
b. concrete *(Bw)* Bimsbeton *m*, Koksaschenbeton *m*, Leichtbeton *m (mit Kokszusatz)*
b. fixing brick *s.* b. brick
Bréguet spring Breguetfeder *f*
Bréguet-sprung mit Breguetspiralfeder *(versehen)*
Breit-Wigner formula *(Ph)* Breit-Wigner-Formel *f*
bremsspectrum *(Ph)* Bremsspektrum *n*

bremsstrahlung *(Ph)* Bremsstrahlung *f*
b. counter Bremsstrahlungszähler *m*
b. cross section Wirkungsquerschnitt *m* der Bremsstrahlung, Bremsstrahlungsquerschnitt *m*
b. energy Bremsstrahlungsenergie *f*
b. radiation Bremsstrahlung *f*, kontinuierliche Röntgenstrahlung *f*
bremsstrahlungsspectrum *s.* bremsspectrum
bressummer *(Bw)* Sturzbalken *m*
brevier *(Typ) (etwa)* Kolonel *f*, Mignon *f (Schriftgröße)*
brew kettle Sudkessel *m (Brauerei)*
brewer Brauer *m*
brewery Brauerei *f*
brewing barley Braugerste *f*
Brewster's angle *(Opt)* Polarisationswinkel *m*, Brewster-Winkel *m*
Brewster's bands Brewstersche Streifen *mpl*
Brewster's law Brewstersches Gesetz *n*
briar scythe Gestrüppsense *f*
b. tooth *(Holz)* Kehlenzahn *m*
brick *(Bw)* Ziegel[stein] *m*, Mauerziegel *m*, Backstein *m*
b. and clay industry grobkeramische Industrie *f*
b. arch *(Wärm)* Feuerschirm *m*
b. burning Ziegelbrennen *n*
b. carrier Ziegelkarren *m*, Ziegeltransportwagen *m*
b. chisel Steinmeißel *m*, Scharriereisen *n*
b. construction Ziegelbauweise *f*
b. cutting wire Ziegelabschneidedraht *m*
b. dust Ziegelmehl *n*
b. facing Ziegelverblendung *f*, Ziegelverkleidung *f (Mauerwerk)*
b. foundation Ziegel[werk]fundament *n*
b. joint Mauerwerkfuge *f*
b. kiln Ziegelofen *m*
b.-lined mit Ziegelausmauerung
b.-lined tunnel ziegelgefütterter Tunnel *m*
b. lining Ausmauerung *f*, Mauerausbau *m*; Ziegelausfütterung *f*, Ziegelauskleidung *f*, Ziegelausmauerung *f (Schornstein)*
b. making Ziegelherstellung *f*
b. masonry Ziegelmauerwerk *n*
b. nogging Ziegelausfachung *f* einer Trennwand
b. of clay chippings Ziegelsplittmauerstein *m*
b.-on-edge Rollziegel *m*
b.-on-edge course Rollschicht *f (Mauerwerk)*
b. paving Ziegelpflaster *n*
b. press Ziegelpresse *f*
b.-red ziegelrot
b. reservoir gemauerter Behälter *m*
b. stacker Ziegelstapler *m*
b. trowel Maurerkelle *f*, Ziegelkelle *f*
b. yard Ziegelei *f*
brickdust Ziegelmehl *n*
brickearth Ziegelton *m*
bricking Ausmauerung *f*, Mauerung *f*
b.-in Einmauerung *f*
b. with air passages Mauerung *f* mit Luftkanälen
bricklayer Maurer *m*
bricklayer's hammer Maurerhammer *m*
b. line Maurerschnur *f*
brickwork Ziegelmauerwerk *n*
b. lining Futtermauer *f*
b. of a furnace *(Met)* Zustellung *f* eines Ofens
brickworks Ziegelei *f*
bridge 1. *(El)* Meßbrücke *f*; 2. *(Bw)* Brücke *f*; 3. Wagenübergang *m*, Laufsteg *m (zwischen Wagen)*; 4. Steg *m*; 5. *(Masch)* Portal *n*
b. abutment Brückenwiderlager *n*
b. amplifier *(El)* Meßbrückenverstärker *m*
b. approach *(Bw)* Brückenauffahrt *f*, Brückenrampe *f*
b. balance *(El)* Brückenabgleich *m*
b. balance point *(El)* Brückenausgleichspunkt *m*; Nulldurchgang *m*
b.-balanced modulator Ringmodulator *m (Drahtnachrichtentechnik)*
b. bearing *(Bw)* Brückenlager *n*
b. bent Brückenjoch *n*
b. bond *(Ch)* Brückenbindung *f*
b.-carrying tank *s.* b.-laying tank
b. circuit *(El)* Brückenschaltung *f*
b.-connected *(El)* in Nebenschluß, nebengeschlossen
b.-connected rectifier *(El)* Brückengleichrichter *m*
b. connection *(El)* Brückenschaltung *f*
b.-contact *(El)* Brückenkontakt *m (Relais)*
b. control *(Schiff)* Brückenfahrstand *m*; Brückensteuerung *f (Steuerung des Schiffes von der Brücke aus)*
b. crane Brückenkran *m*, Verladebrücke *f*
b. crossing Kletterkreuzung *f*
b. deck Brückentafel *f*, Fahrbahntafel *f (Brückenbau)*
b. diplexer *(El)* Brückenweiche *f*
b. driving voltage *(El)* Brückenspeisespannung *f*
b. duplex connection *(El)* Brückenduplexschaltung *f*, Doppelbrückenschaltung *f*
b. duplex system *(Fmt)* Gegensprecheinrichtung *f* in Brückenschaltung
b. engineer Brückenbautechniker *m*
b. engineering Brückenbau *m*
b. feedback *(El)* Brückenrückkopplung *f*
b. filter Brückenfilter *n*
b. for capacity measuring *s.* capacity bridge
b. fuse *(El)* Trennsicherung *f*, Sicherung *f* in Steckdose, Lamellensicherung *f*
b.-laying tank *(Mil)* Brückenlegepanzer *m*

b. **limiter** *(BMSR)* Brückenschaltungsverstärker *m (Analogrechner)*
b. **measurement** *(El)* Brückenmessung *f*
b. **modulator** Ringmodulator *m (Nachrichtentechnik)*
b. **network** *(El)* Brückenfilter *n*; Brückenschaltung *f*; Vierpolkreuzglied *n*
b. **[of a chain]** Steg *m* [einer Kette]
b. **of the bolt cover** *(Mil)* Hülsenbrücke *f (Gewehr)*
b. **piece** 1. Brücke *f (Auflaufbrücke)*; 2. *(Schiff)* Wölbstück *n*, Bogenstück *n*, Brückenstück *n*
b. **pier** Brückenpfeiler *m*
b. **pillar** Brückenpfeiler *m*
b. **railing** Brückengeländer *n*
b. **reamer** Nietlochreibahle *f*
b. **rectifier** *(El)* Brückengleichrichter *m*, Gleichrichterbrücke *f*, Gleichrichter *m* in Graetz-Schaltung
b. **resistance** *(El)* Brückenwiderstand *m*
b. **sponson** *(Schiff)* Brückenausleger *m*, Brückennock *f*
b. **substructure** Brückenunterbau *m*
b. **superstructure** Brückenüberbau *m*
b. **supply voltage** *(El)* Brückenspeisespannung *f*
b. **transition** *(El)* Brücken[fahr]schaltung *f*
b.**-type capacitance analyzer** *(El)* Meßbrückenkapazitätsanalysator *m*
b. **voltage** *(El)* Brückenspannung *f*
b. **wall** *(Met)* Brückenkörper *m (SM-Ofen)*
b. **wing** *(Schiff)* Brückennock *f*
bridged lengthwise bearing *(Masch)* überbrücktes Längstragbild *n*
b. **profile bearing** *(Masch)* überbrücktes Profiltragbild *n*
b.**-T-network** *(BMSR)* überbrücktes T-Glied *n*
bridging 1. *(El)* Überbrückung *f*; 2. *(Bw)* Gewölbebildung *f (z. B. Schüttgüter in Silos)*; Riegel *m (Holzbau)*; 3. *(Plast)* Zusammenbacken [des Materials] *(Extruder)*
b. **amplifier** Kontrollverstärker *m*
b. **board** Schalbrett *n*
b. **capacitor** Überbrückungskondensator *m*
b. **coil** *(Fmt)* Querspule *f (Phantomschaltung)*; Parallelspule *f*
b. **jack** Parallelvielfachklinke *f*
b. **plug** Kurzschlußstecker *m*
b. **telephone** Nebenhörer *m*
bridle 1. *(El)* Spanndraht *m*; 2. Zaum *m*, Zügel *m*
b. **gear** Zaumzeug *n*
b. **slide** Zaumschnalle *f*
bridoon Trense *f*
brief dipping process Kurztauchverfahren *n (z. B. bei Holzimprägnierung)*
brigantine Schoner-Brigg *(Zweimastschiff)*
Briggsian logarithm Briggscher (dekadischer, gewöhnlicher) Logarithmus *m*
bright hell; blank
b. **adaptation** *(Opt)* Hellanpassung *f*
b. **annealed wire** blankgeglühter Draht *m*
b. **annealing** Blankglühen *n*
b.**-annealing furnace** Blankglühofen *m*
b. **bolt industry** Blankschraubenindustrie *f*
b. **coal** Glanzkohle *f*
b.**-crystalline fracture** *(Met)* grobkörnige (grobkristalline) Bruchfläche *f*
b. **dip** Glanzbrenne *f (Galvanotechnik)*
b.**-drawing** Blankziehen *n*
b.**-emitting cathode** *(El)* Hochtemperaturkatode *f*
b. **enamels** *(Foto)* Hochglanzpapier *mpl*
b.**-field condenser** *(Opt)* Hellfeldkondensor *m*
b.**-field-darkground changeover condenser** *(Opt)* Hell-Dunkelfeldkondensor *m*
b.**-field observation** *(Opt)* Beobachtung *f* im Hellfeld
b. **finished** blank
b. **intensification** *(Fs)* Hellsteuerung *f (Katodenstrahloszillografie)*
b.**-level value** *(Fs)* Weißwert *m*
b.**-line spectrum** *(Opt)* Hellinienspektrum *n*, Emissionsspektrum *n*
b.**-luster sheet** Hochglanzblech *n*
b. **orange** *(Met)* Gelbglut *f (Glühfarbe)*
b. **polished** auf Hochglanz poliert
b. **red** *(Met)* Kirschrot *n (Anlaßfarbe)*
b.**-red heat** Hellrotglühhitze *f*, Hellrotglut *f*
b. **square head set-screw** Vierkantgewindestift *m*
b. **steel** *(Met)* Blankstahl *m*
b. **stock** Brightstock *m*, hochmolekulare isoparaffinische Kohlenwasserstoffe *mpl (Mischkomponenten für Motorenöle)*
b.**-up pulse** *(Fs)* Helltastimpuls *m*
b. **work** *(Schiff)* Glanzteile *npl* an Bord
b. **zinc coating** Glanzverzinkung *f*
brighten / to *(Text)* avivieren; aufhellen
brightener Glanzzusatz *m (zu galvanischen Bädern)*
brighteners *(Ch)* [optische] Aufheller *mpl*, [optische] Bleichmittel *npl*
brightening *(Text)* Avivieren *n*; Avivage *f*
brightening agent *(Text)* Aufheller *m*
brightfield illumination Hellfeldbeleuchtung *f (Metallmikroskopie)*
brightness 1. *(Opt)* Flächenhelle *f*; Grundhelligkeit *f*, Helligkeit *f*; 2. Weißgehalt *m*; 3. *(El)* Helligkeit, Leuchtdichte *f*
b. **contrast** *(Fs)* Helligkeitskontrast *m*
b. **difference** Leuchtdichtekontrast *m*
b. **flicker** Helligkeitsflimmern *n*

b. fluctuation Helligkeitsschwankung *f*
b. knob *(Fs)* Bildhelligkeitsregler *m*
b. level Leuchtdichtewert *m*
b. of stars Helligkeit *f* der Sterne
b. of the fluorescent screen Helligkeit *f* des Bildschirms, Bildschirmhelligkeit *f*
b. of the image *(Fs)* Helligkeit *f* des Bildes
b. of the spot *(Fs)* Fleckhelligkeit *f* *(Katodenstrahlröhren)*
b. range *(Fs)* Helligkeitsbereich *m*
b. response of the eye *(Opt)* Hellempfindlichkeitsgrad *m* des Auges
b. reversion Vergilben *n* *(gebleichter Zellstoff durch Alterung)*
brilliance control Helligkeitseinregelung *f* *(Fernsehbild)*
b. modulation Helligkeitsmodulation *f*, Intensitätsmodulation *f*
brilliancy Leuchtkraft *f* *(Farbe)*
brilliant view finder *(Foto)* Brillantsucher *m*
brim Rand *m*
brimstone *(Ch)* Stangenschwefel *m*
brine / to *(Led)* [in Salzlake] einsalzen
brine Sole *f*, Lauge *f*, Salzlauge *f*, Salzlösung *f*, Salzsole *f*, Salzwasser *n*; Kühlsole *f*
b. cooler Solekühler *m*
b. leaching Salzlaugung *f*
b. pump Kühlsolepumpe *f*
b. quenching Salzhärten *n*
Brinell hardness *(Met)* Brinellhärte *f*
Brinell hardness test Brinell-Kugeldruckprüfung *f*
Brinell hardness tester Brinell-Härteprüfer *m*
Brinell number *(Met)* Brinellzahl *f*
Brinell test *(Met)* Brinellversuch *m*, Härteprüfung *f* nach Brinell
bring / to bringen, schaffen, zuführen, mitbringen, herbringen
b.-back *(Brg)* Rückbau treiben
b. clear of the work vom Werkstück abheben
b. into focus *(Opt)* fokussieren
b. into view sichtbar machen
b. off abbringen *(gestrandetes Schiff)*
b. to *(Schiff)* beidrehen
b. to rest anhalten, zum Stehen bringen, stillsetzen
b. up [to the wind] beidrehen
bringing-in Bergung *f* der Ernte
brining *(Led)* Salzen *n* in Salzlake
briquet *s.* briquette
briquettable brikettierfähig
briquette / to brikettieren, pressen
briquette Brikett *n*, Preßkohle *f*; Preßkuchen *m*, Preßling *m*, Ziegel *m*
b. crushing strength Brikettdruckfestigkeit *f*
briquetted grease *(Kfz)* Blockfett *n*
briquetting coal Brikettierkohle *f*
b. factory Brikettfabrik *f*
b. machine Brikettiermaschine *f*
b. quality Brikettierungseigenschaft *f*
brise-soleil *(Bw)* Sonnenblende *f*
bristle 1. Borste *f*; 2. *(Holz)* aufgespleißtes Ende *n* von Hackspänen
British Association Ohm Britisches Standard-Ohm *n* *(0,9886 internationale Ohm)*
British Columbian pine Douglastanne *f*
British Standard candle Britische Normalkerze *f* *(1,02 internationale Kerzen)*
British Standard Fine [screw thread] Whitworth-Feingewinde *n*
British Standard Whitworth Whitworthgewinde *n*
British Thermal Unit Britische Wärmeeinheit *f*
brittle spröde, brüchig; mürbe
b. cracking *(Met)* Sprödbruch *m*
b. fracture *(Met)* Sprödbruch *m*
b. grain *(Led)* brüchiger Narben *m*
b. lacquer *(Plast)* Reißlack *m*
b. point *(Plast)* Bruchpunkt *m*, Sprödigkeitstemperatur *f*
b. silver ore Sprödglanzerz *n*
b. while hot warmbrüchig
brittleness Brüchigkeit *f*, Sprödigkeit *f*
b. when cold Kaltbrüchigkeit *f*
b. when red hot Rotbrüchigkeit *f*
broach / to *(Masch)* räumen
broach 1. Räumwerkzeug *n*; 2. Dorn *m*; Stift *m*
b. assembly zusammengesetztes Räumwerkzeug *n*
b. carrier Räumwerkzeugträger *m*
b. handling Zubringung *f* des Räumwerkzeuges
b. holder Räumnadelhalter *m*
b. lifter cylinder Räumnadelanhebezylinder *m*
b. puller Ziehwerk *n*, Zieheinrichtung *f* *(Räumen)*
b. retriever mechanism Räumwerkzeugzubringeeinrichtung *f*
b. sharpener Räumwerkzeugschärfmaschine *f*
broached work Scharrierung *f*
broaching *(Masch)* Räumen *n*; Ausdornen *n*
b. machine Räummaschine *f*
b. press 1. Räumpresse *f*; 2. Dornpresse *f*
b. process Räumverfahren *n*
b. stroke Räumhub *m*
b. tool Räumwerkzeug *n*
broad axe Breitbeil *n*, Zimmeraxt *f*
b. cut finishing *(Masch)* Breitschlichten *n*
b. finishing tool *(Masch)* Breitschlichtmeißel
b. flange Breitflansch *m*
b. flange beam Breitflanschträger *m*
b. flange girder Breitflanschträger *m*
b. gauge Breitspur *f*
b. goods [Chemie-]Seidengewebe *n* *(von mindestens 18 [Am 12] Zoll Breite)*

b. loom carpet nahtloser gewebter Teppich *m* großer Breite
b. pendant (pennant) *(Schiff)* Toppstander *m*
b. pulse *(Fs)* Impuls *m* mit langer Laufzeit
b. stitch *(Text)* Plattstich *m*
b. strip mill Bandblechwalzwerk *n*, Streifenwalzwerk *n*, Breitbandwalzwerk *n*
b.-tailed sheep Fettschwanzschaf *n*
b. timber Breitschnittholz *n*
broadband aerial Breitbandantenne *f*
b. amplifier Breitbandverstärker *m*
b. circuit Breitbandkreis *m*
b. noise Breitbandgeräusch *n*
broadbeam measurement *(Kern)* Großfeldmessung *f*
broadcast / to 1. streuen; 2. senden *(Rundfunk)*
broadcast Rundfunkübertragung *f*; Radioübertragung *f*; Rundfunk *m*
b. band *(El)* Normalwellenbereich *m*, Rundfunkband *n*
b. frequency input Rundfunkeingang *m* *(Tonbandgerät)*
b. interference Rundfunkstörung *f*
b. listener Rundfunkhörer *m*
b. monitor Überwachungsgerät *n* (Monitor *m*) für Rundfunksendungen
b. receiver Rundfunkempfänger *m*, Radio *n*
b. reception Rundfunkempfang *m*
b. transmission Rundfunkübertragung *f*, Radioübertragung *f*
b. transmitter Rundfunksender *m*
b. writing drahtloses Fernschreiben *n*
broadcasting Senden *n*, Sendung *f*
broadcasting Radio-, Rundfunk-, Funk-
b. centre Funkhaus *n*
b. emitter Rundfunksender *m*
b. house Funkhaus *n*
b. network Rundfunknetz *n*
b. station Rundfunkstation *f*
b. studio Senderaum *m*
broadening Verbreiterung *f*
b. by damping *(Fs)* Dämpfungsverbreiterung *f (Spektrallinien)*
b. of a spectral line *(Fs)* Linienverbreiterung *f*
b. of interference lines *(Fs)* Verbreiterung *f* von Interferenzlinien
broadleaved wood Laubholz *n*
broadside Breitseite *f*
b. array *(El)* Breitstrahler *m*, Querstrahler *m*
b. format Querformat *n*
b. radiation Querstrahlung *f*
broadwall method *(Brg)* Strebbau *m* mit breitem Blick
Broca galvanometer astatisches Nadelgalvanometer *n*
brocade *(Text)* Brokat *m*
brochure Broschüre *f*
broiler *(Bw)* Backofen *m*
broiling product Rösterzeugnis *n*
broke Ausschuß *m*, Kollerstoff *m* *(Papierherstellung)*
b. beater Auflöseholländer *m*, Brechholländer *m*, Schlagholländer *m*
b. handling Ausschußaufbereitung *f*
b. pit Ausschußbütte *f*
broken gebrochen, zerbrochen; unterbrochen; [ab]gerissen
b. chip Kurzspan *m*
b. drawing-in draft *(Text)* gebrochener Einzug *m*
b. end *(Text)* Kettfadenbruch *m*
b.-end collection *(Text)* Fadenabsaugung *f*, Fadenbruchabsauganlage *f*
b. grain *(Led)* gebrochener Narben *m*
b. line 1. gestrichelte Linie *f*; 2. Knicklinie *f*
b. log *(Holz)* Bruchstück *n* *(Holzeinschlag)*
b. number gebrochene Zahl *f*
b. ore Erzklein *n*
b. pattern *(Text)* unterbrochenes Muster *n* *(Fehler)*
b. pick *(Text)* gerissener Schußfaden *m*, Schußfadenbruch *m*; Platzer *m*, Schußplatzer *m*
b. section gebrochener Schnitt *m* *(technisches Zeichnen)*
b. span saw englische Fuchsschwanzsäge *f*
b. stones Schotter *m*, Steinschlag *m*
b.-stowage *(Schiff)* Stauung *f* mit Staulücke, Staulücke *f*
b. twill *(Text)* gebrochener Köper *m*
b. warp thread *(Text)* Kettfadenbruch *m*, gebrochener Kettfaden *m*
b. wire Drahtbruch *m*
brokes *(Am)* Abfälle *mpl* bei der Verarbeitung, Ausschuß *m*
bromic acid Bromsäure *f*
b. counter *(Kern)* Bromzähler *m*
bromide paper *(Foto)* Bromsilberpapier *n*, Bromidpapier *n*
b. print Bromsilberdruck *m*
brominate / to bromieren
bromine Brom *n*
b. compound Bromverbindung *f*
b.-phenol blue Bromphenolblau *n*
b. value Bromzahl *f*
bromoacetone Bromazeton *n*
bromobenzene Brombenzol *n*
bromoil print Bromöldruck *m*
bronze / to bronzieren
bronze Bronze *f*
b.-backed bearing *(Masch)* Lager *n* mit Bronzestützschalen

b. bearing Bronzelager *n*
b.-box Bronzebezug *m*, Bronzebuchse *f*
b. foundry Bronzegießerei *f*
b. hinge Bronzegelenk *n*
b. plate Bronzeblech *n*
b. powder Bronzepulver *n*, Bronzierfarbe *f*
b.-welding Fugenlöten *n*
b. wire Bronzedraht *m*
bronziness *(Text)* Bronzeglanz *m*
bronzing Braunbeizen *n*, Brünieren *n*
b. medium *(Typ)* Bronzeunterdruckfarbe *f*
b. pickle Brünierbeize *f*
bronzite *(Min)* Bronzit *m*
brookite *(Min)* Brookit *m*
brooming Ausfasern *n*
broomy flow unruhige (divergierende) Strömung *f*
brought-forward balance Saldovortrag *m* *(Buchhaltung)*
brow 1. *(Schiff)* Laufplanke *f*, Landgang *m*, Gangway *f*; 2. *(Brg)* steil ansteigende Strecke *f*; Oberteil *m*
b. leakage *(Elm)* Stirnstreuung *f*
b. plate *(Schiff)* Lukenrampe *f*
brown coal Braunkohle *f*, Hartbraunkohle *f*
b. coal for low-temperature retort process Schwelkohle *f*
b. coat *(Bw)* mittlere Putzschicht *f* bei dreischichtigem Außenputz
b. finished *(Masch)* brüniert
b. haematite Limonit *m*, Brauneisenerz *n*
b. iron ore Braun[eisen]erz *n*, Brauneisenstein *m*, Limonit *m*
b. magnesite Magnesiumeisen *n*
b. mechanical pulp board Braunschliffpappe *f*
b. mechanical woodpulp brauner Holzschliff *m*, Braunschliff *m*
b. mixed pulp board braune Mischpappe *f*
b. ore Limonit *m*, Brauneisenerz *n*
b. paper Packpapier *n*
b. pulp ungebleichter Zellstoff *m*
b. rot *(Holz)* Braunfäule *f*
b. stock ungebleichter Holzzellstoff *m* in Breiform
Brownian movement Brownsche Bewegung *f*
Brownian particle Braunsches Teilchen *n*
browning Brünieren *n*, Braunbeizen *n*
brownstone *(Geol)* eisenhaltiger Sandstein *m*
brucine *(Ch)* Bruzin *n*
brucite *(Min)* Brucit *m*
bruising Schrotmahlen *n*, Zerschroten *n* *(Getreide)*
b. mill Schrotmühle *f*
Brunswick green Braunschweiger Grün *n*
brush / to bürsten; anstreichen, anpinseln
b. down *(Brg)* nachreißen
b. on aufstreichen, aufpinseln

brush 1. *(El)* Bürste *f*, Bürstenstromabnehmer *m*, Stromabnehmerbürste *f*; 2. *(Brg)* Versatzstrecke *f*, Strecke *f* im Alten Mann; 3. Pinsel *m*
b. adjustment *(El)* Bürstenverstellung *f*, Bürstenverschiebung *f* *(Elektromotor)*
b. atomizer *(Text)* Bürstenzerstäuber *m*
b. box *(El)* Bürstenhalter[kasten] *m*
b. carriage Bürstenträger *m* *(Lochkartenmaschinen)*
b. coater Bürstenstreichmaschine *f* *(Papierherstellung)*
b. collar Bürstenträger *m*
b. contact *(El)* Bürstenkontakt *m*; Bürstenstromabnehmer *m*
b. detector Pinseldetektor *m*
b. discharge *(Ph)* Büschelentladung *f*
b. displacement *(El)* Bürstenverschiebung *f*
b.-dyeing *(Led)* Bürstfärbung *f*
b. finger Bürstenträger *m*
b.-finished *(Text)* gerauht
b. gear *(Elm)* Bürstengestell *n*
b. heel *(Elm)* ablaufende Bürstenkante *f*
b. holder *(Elm)* Bürstenhalter *m*
b. latch opener *(Text)* Bürstenzungenöffner *m*
b. lead *(Elm)* Bürstenverschiebung *f* *(nach vorn)*, Bürstenvorschiebung *f*
b. letters *(Typ)* Pinselschrift *f*
b. lifter (lifting) *(Elm)* Bürstenabheber *m*, Bürstenabhebevorrichtung *f*
b. light *(El)* Büschellicht *n*; Glimmlicht *n*
b. method Anstrichverfahren *n* *(Papierherstellung)*
b. motor *(Elm)* Kommutatormotor *m*
b. pressure Bürstendruck *m*
b. proof *(Typ)* Bürstenabzug *m*, Fahne[nabzug *m*] *f*
b. ring Bürstenhalter *m*
b. riser Bürstenabheber *m*
b.-rocker *(Elm)* [verstellbare] Bürstenbrücke *f*
b. rod *(El)* Kontaktarmträger *m*
b. roll Bürstenwalze *f* *(Lackindustrie)*
b. scanning Pinselabtastung *f*
b. shift *(Elm)* Bürstenverschiebung *f*
b. shifting Bürstenverstellung *f*, Bürstenverschiebung *f*
b.-shifting control *(Elm)* Spannungsregelung *f* durch Bürstenverschiebung
b.-shifting motor *(Elm)* Verstellmotor *m*, Bürstenverstellmotor *m*, regelbarer Nebenschlußmotor *m*
b. spark Büschelfunken *m*, Bürstenfeuer *n*
b. spreading *(Led)* Bürstenstreichverfahren *n* *(Beschichtung)*
b. spring *(Elm)* Bürsten[druck]feder *f*; Schleiffeder *f*

b.-staining *(Text)* Trockenfärben *n*
b.-type sheet remoistener Bürstenfeuchtwerk *n*
b. wood Buschholz *n*, Reiserholz *n*; Bürstenholz *n*
b. yoke *(Elm)* [stellbare] Bürstenbrücke *f*
brushability *(Text)* Streichfähigkeit *f*
brushable streichbar, streichfähig
brushed *(Text)* gerauht
b. woollens gerauhtes Wollgewebe *n*
brushing lacquer Streichlack *m*
b. machine Tuchbürstmaschine *f*, Hutbürstapparat *m*
b. property *(Plast)* Streichfähigkeit *f*
b. roller *(Text)* Rauhwalze *f*, Bürstwalze *f*
b. wheel Rundbürste *f*; Bürstenscheibe *f*; Fiberrundbürste *f*
brushwood cooling stack Reisiggradierwerk *n*
bubble / to brodeln *(Funkempfangstechnik)*
bubble Blase *f*; Libelle *f*
b. bath Sprudeltabletten *fpl*
b. boat Schlauchboot *n*, aufblasbares Boot *n*
b. cap Bodenglocke *f*, Fraktionierbodenglocke *f*, Glocke *f*
b. cap column Glockenbodenkolonne *f*
b. cap plate Glockenboden *m*; Boden *m* einer Glockenbodenkolonne
b. cap tower Glockenbodenkolonne *f*
b. chamber *(Kern)* Blasenkammer *f*
b. control Blasenregelung *f* *(Siedewasserreaktor)*
b. flow meter Blasenströmungsmesser *m*
b. formation Bläschenbildung *f*, Blasenbildung *f*
b. glass Sprudelschauglas *n* *(Umlaufschmierung)*
b. in vial Libellenblase *f*
b. of the level Libellenblase *f*
b. plate *(Ch)* Glockenboden *m*
b. plate column Glockenbodenkolonne *f*
b. point Temperatur *f* des Siedebeginns; Blasenpunkt *m* *(Öl)*
b. point of crude oil Entgasungsbeginn *m* von Erdöl in der Lagerstätte
b. tray Glockenboden *m*, Sprudelplatte *f*
bubbler *(Ch)* Barboteur *m*; Waschflasche *f*, Gasspüler *m*
b. gauge *(BMSR)* Plubberrohr *n*, Sprudelrohr *n* *(Füllstandsmessung)*
bubbling Bläschenbildung *f*, Blasenbildung *f*
b. fluidized bed brodelnde Wirbelschicht *f*
b.-type electrode Gaselektrode *f*, Durchperlungselektrode *f*
b. washer *(Ch)* Gaswäscher *m* *(Gaswaschflasche)*
Buchholtz relay *(El)* Buchholtz-Schutz *m*, Buchholtz-Relais *n*
buck / to 1. einlaugen; brühen *(Wäsche)*; beuchen; 2. *(Brg)* brechen; 3. *(El)* *(Am)* entgegenschalten
bucket 1. Becher *m*, Bütte *f*, Kübel *m*, Ladeschaufel *f* *(Fördergeräte)*; Greiferkorb *m* *(Bagger)*; Eimer *(Bagger)*; Löffel *m*; Schöpfer *m* *(Bagger)*; 2. Laufschaufel *f*, Schaufeltasche *f* *(Turbine)*; 3. *(Masch, Wärm)* Kolben *m*; Klappenkolben *m*, Ventilkolben *m*; 4. *(El)* Auffängerelektrode *f*; 5. Adresse *f* *(Rechenautomaten)*
b. arch *(Bw)* Schleppschaufelbügel *m*
b. base Schaufelfuß *m*
b. chain Eimerkette *f*
b.-chain dredger Eimerkettenbagger *m*
b. charging *(Met)* Kübelbegichtung *f*
b. conveyor Becher[ketten]förderer *m*, Eimerkettenförderer *m*, Becherwerk *n*
b. crane Greif[er]bagger *m*
b. dredger Eimerketten[schwimm]bagger *m*
b. elevator Becheraufzug *m*, Becherelevator *m*, Becherwerk *n*, Kübelaufzug *m*
b. elevator belt Elevatorgurtband *n*
b. excavator Eimerkettenbagger *m*; Löffelbagger *m*
b. hoist Kippkübelaufzug *m*
b. inlet edge Schaufeleintrittskante *f*
b.-ladder *(Brg)* Eimerleiter *f*
b.-ladder dredger Eimerkettenschwimmbagger *m*
b.-ladder excavator Eimerkettentrockenbagger *m*
b. line Eimerkette *f*, Becherstrang *m*
b. loader Becherwerkslader *m*, Ladeschaufler *m*
b. milking aggregate Eimermelkgerät *n*
b. piston (plunger) *(El)* λ/4-Kontaktkolben *m* *(Wellenleiter)*
b. pump Baggerpumpe *f*
b. segment Laufschaufelsegment *n* *(Turbine)*
b. trap Kondensationstopf *m*
b.-type blade Becherschaufel *f*
b. wheel Schaufelrad *n* *(Bagger)*; Schöpfrad *n* *(Schaufelradbagger)*
b. wheel dredge Schaufelradbagger *m*
b. wheel excavator Schaufelradbagger *m*
b.-wheel loader *(Brg)* Schaufelradlader *m*
bucking Ablängen *n*, Holzschneiden *n*, Sägen *n*
b. bar Gegenhalter *m* *(Niet)*
b. boiler Beuchkessel *m*
b. circuit *(El)* Kompensationskreis *m*
b. coil *(El)* Kompensationsspule *f*
b. lye Beuchlauge *f*
b. saw Abkürzsäge *f*, Ablängsäge *f*, Steifsäge *f*
b. voltage *(El)* Kompensationsspannung *f*, Ausgleich[s]spannung *f*
buckle / to [aus]knicken; [aus]beulen; kippen

buckle 1. Schnalle *f*, Spange *f*; 2. Angel *f* *(Gattersäge)*
b. for eccentric adjustment Exzenterangel *f*
buckled plate Buckelblech *n*; Buckeleisen *n*, Buckelplatte *f*
buckler *(Schiff)* Klüsendeckel *m*
buckling 1. Ausbeulen *n*, Beulen *n*; 2. Ausknicken *n*, Knicken *n*; 3. *(Kern)* Flußwölbung *f*, Buckling *n*
b. coefficient *(Bw)* Knickbeiwert *m*
b. load Knickbelastung *f*, Knicklast *f*; Beullast *f*
b. strength Knickfestigkeit *f*, Ausknickungsfestigkeit *f*; Beulfestigkeit *f*
b. stress *s.* critical stress
b. test Knickversuch *m*; Beulversuch *m*
buckram *(Typ)* Buckram *m*, Steifleinen *n* *(Buchbinderleinen)*
buckshot Rehposten *m* *(besonders grobes Schrot für Jagdpatronen)*
buckskin Hirschleder *n*
b. crank loom Buckskinkurbelwebstuhl *m*
buckwheat Buchweizen *m*
b. coal Kleinkohle *f*, Nußgruskohle *f*
Bucky screen *(Ph)* Streustrahlenblende *f*
budding knife Pfropfmesser *n*
buff / to polieren, schwabbeln; abbuffen
buff Schwabbelscheibe *f*, Glanzscheibe *f*, Schwabbelschale *f*
b. bob Putzscheibe *f*, Polierscheibe *f*
buffalo 1. Büffel *m*; 2. *(Schiff)* Bugschanzkleid *n*
b. hide Büffelhaut *f*
buffer 1. Puffer *m*; Stoßfänger *m*, Stoßdämpfer *m*; Schwingungsdämpfer *m*; 2. *(BMSR)* Puffer[kreis] *m*, Pufferstufe *f*; 3. Poliermaschine *f*
b. action *(Ch)* Pufferwirkung *f*
b. amplifier Pufferverstärker *m* *(Analogrechner)*; Trennverstärker *m*
b. bar *s.* b. beam 2.
b. battery *(El)* Pufferbatterie *f*
b. beam 1. *(Brg)* Leitbaumverdickung *f*; 2. *(Eis)* Pufferbohle *f*
b. box *(Eis)* Puffergehäuse *n*
b. circuit *(El)* Pufferschaltung *f*, Pufferkreis *m*
b. condenser *(BMSR)* Löschkondensator *m* *(Stromrichter)*
b. disk *(Eis)* Pufferteller *m*
b. housing *(Eis)* Puffergehäuse *n*
b. link Schaken *fpl* zur Pufferung
b. memory *s.* b. storage
b. plunger *s.* b. rod
b. reagent *(Ch)* Puffer *m*
b. rod *(Eis)* Pufferstange *f*
b. solution *(Ch)* Pufferlösung *f*
b. spring *(Eis)* Pufferfeder *f*
b. stage *(El)* Pufferstufe *f*; Trennstufe *f*; Koppelstufe *f*, Kopplungsstufe *f*
b. stop *(Eis)* Prellbock *m*
b. storage *(BMSR)* Pufferspeicher *m*, Eingangs- oder Ausgangsspeicher *m*
b. tube *(El)* Trennröhre *f*
buffering Pufferung *f*
buffing Polierläppen *n*; Schwabbeln *n*
b. composition Poliermittel *n*
b. cone Polierdorn *m*
b. lathe Schwabbelmaschine *f*
b. machine Bimsmaschine *f*, Poliermaschine *f*, Schleifmaschine *f*, Schwabbelmaschine *f*
b. marks Verschleißspuren *fpl*
b. material Poliermittel *n*
b. slicker *(Led)* Abnarbeisen *n*
b. wheel Schwabbelscheibe *f*
bug *(Am)* *(BMSR)* Störung *f*, unsteter Fehler *m*, zeitweises Versagen *n*
buggy 1. *(Bw)* Karre[n *m*] *f*, Kippkarre[n *m*] *f*; 2. *(Brg)* *(Am)* kleiner Förderwagen *m*
b. board Spritzlederpappe *f*, Rückenlehnenkarton *m*
b. casting *(Met)* Wagenguß *m*
b. ladle *(Met)* Gießpfanne *f* mit Wagen
bugle horn Flügelhorn *n* *(Musikinstrument)*
build / to [auf]bauen, errichten, anlegen
b. in einbauen
b. up aufbauen, zusammensetzen; aufschweißen, auftragen, auftragsschweißen *(z. B. Radkränze)*
build-up 1. *(Bw)* Aufbau *m*; Stau *m*; 2. *(El)* Einschwingung *f*
b.-up factor *(Kern)* Zuwachsfaktor *m*; Aufbaufaktor *m*
b.-up heat *(Met)* Aufbauschmelze *f*
b.-up metal *(Met)* Auftragsmetall *n*
b.-up of a self-excited generator *(Elm)* Auferregung *f* eines Generators mit Selbsterregung
b.-up of current *(El)* Stromanstieg *m*
b.-up of pressure Druckaufbau *m*
b.-up time *(El)* Anklingzeit *f*, Einschwingzeit *f*, Einschwingdauer *f*; Anstiegszeit *f* *(Signal)*
builder Baufachmann *m*; Bauarbeiter *m*
builders 1. *(Ch)* Aufbaustoffe *mpl*, Gerüststoffe *mpl*, Skelettsubstanzen *fpl*; 2. Waschhilfsmittel *npl*
builder's certificate *(Schiff)* Bauschein *m*, Beilbrief *m*
b. hardware Baubeschläge *mpl*, Kleineisenzeug *n*
b. hoist Bauaufzug *m*
b. joinery Bautischlerei *f*
b. labourer Bauhilfsarbeiter *m*
b. level Baunivellier *n*
b. plant Baugeräte *npl*, Baumaschinen *fpl*
b. rotating crane Baudrehkran *m*
b. staging Baugerüst *n*
b. tape Bandmaß *n*

b. trial trip *(Schiff)* Werftprobefahrt *f*
building Gebäude *n*
b. basin *(Schiff)* Baudock *n*
b. berth Helling *f*, Helgen *m*
b. block 1. Baustein *m*; Blockstein *m*; 2. *(Masch)* Baueinheit *f*
b. block construction Konstruktion *f* nach dem Baukastensystem
b.-block machine Aufbaumaschine *f*
b. block principle Baukastenbauweise *f*, Baukastenprinzip *n*, Baukastensystem *n*
b. block system Baukastensystem *n*
b. board Leichtbauplatte *f*, Bauplatte *f*
b. brick 1. *(Bw)* Mauerziegel *m*; 2. *(Masch)* Baueinheit *f*
b. code Bauordnung *f*
b. concrete Hochbaubeton *m*
b. construction Baukonstruktion *f*; Hochbau *m*; Bauausführung *f*, Baudurchführung *f*
b. contractor Bauunternehmer *m*
b. crane Baukran *m*, Hochbaukran *m*
b. dock *(Schiff)* Baudock *n*
b. foundation Gebäudefundament *n*
b. framework *(Bw)* Skelett *n*
b. frontage Hausfront *f*
b. legislation Baugesetzgebung *f*, Baurecht *n*
b. lime Baukalk *m*
b. line Bauflucht *f*, Gebäudeflucht *f*
b. material Baustoff *m*
b. mortar Baumörtel *m*
b. operative Bauarbeiter *m*, Bauhandwerker *m*
b. paper Dachpappe *f*, Teerpappe *f*
b. pit Baugrube *f*
b. pit hoist Baugrubenaufzug *m*
b. plumber Bauklempner *m*
b. project Bauvorhaben *n*
b. regulations Baubestimmungen *fpl*, Bauvorschriften *fpl*
b. sand Bausand *m*
b. service ingenieurtechnischer Ausbau *m*; Hausinstallationen *fpl*, Gebäudeinstallationen *fpl*
b. slip Helling *f*, Helgen *m*
b. specifications Baubestimmungen *fpl*, Bauvorschriften *fpl*
b. stone Baustein *m*
b. time on the stocks *(Schiff)* Hellingbauzeit *f*
b. trade Baugewerbe *n*, Bauhandwerk *n*
b.-up process *(El)* Einschwingvorgang *m*
b.-up welding Auftragsschweißen *n*
b. vibration Gebäudeschwingung *f*
b. ways Helling *f*, Helgen *m*
b. works area Baugelände *n*; Großbaustelle *f*
built beam with keys verdübelter Balken *m*
b.-in Einbau-, eingebaut; eingespannt
b.-in aerial Einbauantenne *f*, eingebaute Antenne *f*, Gehäuseantenne *f*
b.-in check *(BMSR)* Prüfung *f* durch eingebaute Schalteinrichtungen
b.-in motor Einbaumotor *m*
b.-in radiation heating eingebaute elektrische Fußboden-, Wand- oder Deckenheizung *f*, Strahlungsheizung *f*
b.-in reactivity *(Kern)* Überschußaktivität *f*
b.-in switch *(El)* Einbauschalter *m*, eingebauter Schalter *m*
b.-in timber *(Schiff)* eingeschnittenes (gewachsenes) Spant *n*
b.-up 1. bebaut *(Gelände)*; 2. mehrteilig; zusammengesetzt; mehrlagig, mehrschichtig; 3. mit eingesetzten Zähnen
b.-up addressing *(BMSR)* zusammengesetztes Adressieren *n (Ziffernrechner)*
b.-up area *(Bw)* bebautes Gebiet *n*
b.-up beam zusammengesetzter Balken *m*
b.-up crankshaft zusammengebaute Kurbelwelle *f*
b.-up edge Aufbauschneide *f*, Schneidenaufbau *m*
b.-up frame *(Schiff)* gebautes Spant *n*
b.-up girder *(Bw)* Verbundträger *m*
b.-up mould Form *f* mit eingesetzten Matrizenteilen *(Kunststoffverarbeitung)*
b.-up roof[ing] *(Bw)* mehrlagiges Pappdach *n*
b.-up rotary switch *(El)* Paketschalter *m*
b.-up rotor *(Elm)* Scheibenrotor *m*; Scheibenläufer *m*
b.-up stack of wool *(Text)* Wollbett *n*
b.-up truss Brettbinder *m (Holzbau)*
b.-up type of face mill Messerkopf *m*
bulb 1. Ballon *m*, Birne *f*, Kolben *m*, kugelförmiges Gefäß *n*, Küvette *f*, Trog *m*; 2. Kolben *(einer Lampe)*; Glühbirne *f*, Glühlampe *f*; 3. Wurzelknolle *f*; Knolle *f*; Zwiebel *f*
b. angle Winkelwulst[stahl] *m*, Wulstwinkel[profil *n*] *m*, Winkelwulstprofil *n*
b. angle iron *s.* b. angle
b. blackening *(El)* Kolbenschwärzung *f (Elektronenröhren, Lampen)*
b. blowing machine Kolbenblasmaschine *f*
b. bow *(Schiff)* Wulstbug *m*
b. cutting machine Kolbenabschmelzmaschine *f*
b. flat *(Masch, Bw)* Flachwulst[profil *n*] *m*, Flachwulststahl *m*
b.-holder *(sl)* Fassung *f (Glühlampe)*
b.-keel yacht *(Schiff)* Wulstkieljacht *f*
b. lifter Zwiebelerntegerät *n*
b. of flame *(Ph)* Flammenkern *m*, Flammenkegel *m*
b. pipette *(Ch)* Gaspipette *f*, Kugelpipette *f*
b. plate *(Masch, Bw)* Flachwulststahl *m*
b. socket *(El)* Glühlampenfassung *f*
b. terminal *(El)* Glühlampenanschluß *m*

b. thermometer *(BMSR)* Federthermometer *n*, Tensionsthermometer *n*
bulbiform knollenartig
bulbous knollenartig
b. bow *(Schiff)* Wulstbug *m*
b. plant Zwiebelgewächs *n*, Knollenpflanze *f*
bulge / to [aus]bauchen, ausbeulen, ausbiegen, ausbuchten, stauchen; bauschen *(Faden)*
bulge Beule *f (Blech)*; Wulst *m*
bulged-in flue-tube *(Wärm)* eingedrücktes Flammrohr *n*
bulging Ausbeulen *n*, Beulen *n*; Ausbauchen *n (Umformen)*
b. die Stauchwerkzeug *n*
b.-in of the tubes Eindrücken *n* der Rohre
b. of lead blocks Bleiblockausbauchung *f (Sprengstofftest)*
b. of the boiler Verbeulung *f* des Kessels
b. test *(Met)* Aufweiteversuch *m*, Stauchversuch *m*
bulk 1. *(Schiff)* Massengut *n*, Schüttgut *n*, Schüttladung *f*; 2. Volumen *n*, Umfang *m*, Raumbedarf *m*
b. boiling Blasensiedung *f*, Blasenverdampfung *f*
b. cargo Bulkladung *f*; Schüttgut *n*; Massengut *n*
b. carrier *(Schiff)* Schüttgutfrachter *m*, Massengutfrachter *m*
b. cement loser Zement *m*
b. cement silo Zementsilo *m*, Bindemittelsilo *m*; Baustellensilo *m*
b. density 1. *(Bw)* Raumgewicht *n*; 2. *(Brg)* Schüttwichte *f*, Schüttdichte *f*
b. factor *(Ch)* Füllkonstante *f*, Verdichtungsgrad *m*
b. feed hopper Großfutterbehälter *m*, Großfuttertrichter *m*
b. lorry Tankwagen *m*
b. material Schüttgut *n*
b. meter Mengenmesser *m*
b. modulus *(Ph)* Kompressionsmodul *m*, Raummodul *m*
b. oil Ladeöl *n*, Öl *n* als Ladung *(in Tankern)*
b.-oil carrier Öltanker *m*
b.-oil carrying vessel Öltanker *m*, Öltankschiff *n*
b.-oil circuit-breaker *s.* dead-tank oil circuit-breaker
b. polymerization *(Ch)* Blockpolymerisation *f*
b. production Fertigung *f* größerer Mengen
b.-specific gravity spezifisches Schüttgewicht *n*
b. test *(Kern)* Schirmprüfung *f*, Schutzmaterialprüfung *f*
b. weight Schüttgewicht *n*
b. yarn *(Text)* Bauschgarn *n*
bulked-type yarn *(Text)* Bauschgarn *n*
bulkhead 1. *(Kfz)* Querwand *f*, Trennwand *f*; 2. *(Wärm)* Kesselboden *m*; 3. *(Schiff)* Schott *n*; 4. *(Flg)* Spant *n*; 5. Zwischenwand *f*
b. at after end of bridge Brückenendschott *n*
b. at front of bridge Brückenfrontschott *n*
b. coamsing Schottsüll *m (n)*
b. deck Schottendeck *n*
b. door Schottür *f*, Schottverschluß *m*
b. spigot Schottdurchführung *f*
b. stiffening Schottversteifung *f*
bulkheading *(Schiff)* Abschottung *f*, Umschottung *f*
bulking *(Bw)* Auflockerung *f*, Volumenvergrößerung *f (infolge Wasseraufnahme)*; Quellen *n*, Quellung *f*
b. board voluminöse Pappe *f*
b. material *(Am)* Füllstoff *m*
b. value Schüttvolumen *n*, Stampfvolumen *n*
bulky voluminös, dick, groß, massig, sperrig, unhandlich; griffig *(Gewebe)*
bull gear *(Masch)* großes Antriebszahnrad *n*, Hubscheibenrad *n*, Kulissenrad *n*
b. head 1. *(Met)* Polierwalzengerüst *n*; 2. Keil[ziegel]stein *m*
b. header *(Bw)* Binder *m* mit einer abgerundeten Vorderkante
b.-ladle Scherenpfanne *f*
b. rope *(Schiff)* Beiholer *m*
b. screen Splitterfänger *m (Papierherstellung)*
b. stretcher *(Bw)* Läufer *m* mit einer abgerundeten Kante
b. wheel 1. Bohrtrommel *f (Seilbohren, Erdöl, Erdgas)*; 2. *s.* b. gear
bulldog wrench Gasrohrschlüssel *m*, Rohr- und Schraubenschlüssel *m*
bulldoze / to *(Brg)* nachzerkleinern
bulldozer Bulldozer *m*; Fronträumer *m*, Planiergerät *n*, Planierraupe *f*
b. blade Planierschild *m*, Räumschild *m*
bulldozing Planieren *n* mit Bulldozer
bulldozing chamber Kammer *f* zum Zerkleinern des hereingeschossenen Erzes
bullet-proof kugelsicher
bullgrader Planierschild *m*
bullhead rail *(Eis)* Doppelkopfschiene *f*
bullion Barrenmetall *n*, Edelmetallbarren *m*
b. furnace Goldschmelzofen *m*, Edelmetallschmelzofen *m*
bullrope rundes Hanfantriebskabel *n*
bull's eye 1. Bullauge *n*, Rundfenster *n (kreisrundes oder elliptisches Fenster)*; kreisrunde *oder* elliptische Luke *f*; 2. Ochsenauge *n (Gaupe)*
b.-eye arch *(Bw)* Kreisbogen *m*
b.-eye window Bullauge *n*, Ochsenauge *n*, Rundfenster *n (kreisrundes oder elliptisches Fenster)*

bulwark Schanzkleid *n*, Verschanzung *f* *(im Schiffbau)*; Bohlwerk *n*, Bollwerk *n* *(im Hafenbau)*
b.-mooring pipe *(Schiff)* Verholklüse *f*
b. port *(Schiff)* Schanzkleidpforte *f*, Wasserpforte *f*
b. stair *(Schiff)* Schanzkleidreling *f*
b.-stay *(Schiff)* Relingstütze *f*, Schanzkleidstütze *f*
bumboat Proviantboot *n*, Bumboot *n*
bump *(Brg)* Gebirgsschlag *m*
b. rubber *(Kfz)* Gummipuffer *m*
b. table Rüttelherd *m*, Stoßherd *m*
b. testing *(Kfz)* Stoßprüfung *f*
bumper 1. Prellbock *m*, Puffer *m*; Stoßdämpfer *m*; 2. Rüttelformmaschine *f*; 3. *(Schiff)* Holzfender *m*
b. bar *(Kfz)* Stoßstange *f*
b. missile *(Am)* Zweistufenrakete *f*
b. spring Pufferfeder *f*
b. tooth *(Am)* Anschnittzahn *m* *(Räumwerkzeug)*
bumping Anschlagen *n*, Anstoßen *n*; Aufschlagen *n*, Aufstoßen *n*; Gegenschlagen *n*, Gegenstoßen *n*; Gegenlaufen *n* *(Drehmaschine gegen Anschlag)*
b. of the fell *(Text)* Walken *n* des Geweberandes *(im Webstuhl)*
b. post Prellbock *m*
b. table *(Brg)* Stoßherd *m*
b. tool Ausbeulwerkzeug *n*, Schlagwerkzeug *n*
bumpless transfer *(BMSR)* stoßfreie Umschaltung *f* *(Wechsel zwischen Hand- und Automatikbetrieb eines Regelkreises)*
bumpy coal *Kohle, die unter Entgasung zersplittert*
buna, Buna Buna *n (m)* *(synthetischer Kautschuk)*
bunch / to bündeln; büscheln
bunch Büschel *n*
b. discharge *(El)* Büschelentladung *f*
b. of circuits *(El)* Leitungsbündel *n*
b. of conductors *(El)* Leiterbündel *n*
b. of electrons Elektronenpaket *n*
b. of pairs *(El)* Leitungsbündel *n*
bunched gebündelt
b. circuit *(El)* Simultanleitung *f*
b. circuits *(El)* gebündelte Leitungen *fpl*
b. ions *(Ph)* Ionenpaket *n*
buncher *(El)* Bündelungsröhre *f*; Eingangsresonator *m*
b. space Einkoppelstrecke *f*; Steuerraum *m* *(Radio)*
bunching Bündelung *f*; Büschelung *f*; Zusammenballung *f*, Paket[is]ierung *f*
b. of ions *(Ph)* Ionenpaketierung *f*
b. of particles *(Ph)* Teilchenpaketierung *f*
bundle wire Ringdraht *m*, Rollendraht *m*
bundled wood Bundholz *n*
bung 1. Kohlenbunker *m* *(im Schiff)*; 2. Stopfen *m*, Spund *m*
bunghole Abstichloch *n*, Faßloch *n*, Spundloch *n*, Zapfloch *n*
b. boring machine Spundlochbohrmaschine *f*
bunker Bunker *m*; Kohlenbunker *m*
b. bulkhead *(Schiff)* Bunkerschott *n*
b. discharge wagon Bunkerentleerungswagen *m*
b. evacuating wagon Bunkerentleerungswagen *m*
b. for unwashed fines Rohfeinkohlenbunker
b. outlet Bunkerauslauf *m*
b. ring *(Schiff)* Bunkerdeckel *m*
b. train Bunkerzug *m*
bunkering boat Bunkerschiff *n*
Bunsen burner Bunsenbrenner *m*
Bunsen cell *(Ch)* Bunsensches Element *n*
Bunsen flame Bunsenflamme *f*
Bunsen gas burner *s.* Bunsen burner
Bunsen photometer Bunsen-Fotometer *n*, Fettfleckfotometer *n*
bunton *(Brg)* Einstrich *m* *(Schachtausbau)*
buoy / to *(Schiff)* austonnen *(Fahrwasser)*
buoy Boje *f*, Tonne *f* *(Seezeichen)*
b. for a shelf *(Schiff)* Untiefentonne *f*
b. layer *s.* b.-laying vessel
b.-laying vessel Seezeichenschiff *n*, Tonnenleger *m*
buoyage *(Schiff)* Betonnung *f*
buoyancy 1. *(Ph)* Auftrieb *m* *(in einer Flüssigkeit)*; 2. *(Schiff)* Schwimmfähigkeit *f*, Auftrieb[skraft *f*] *m*
b. gas balance Gasdichtemesser *m* nach dem Auftriebsprinzip
b. probe Füllstandsmesser *m* nach dem Auftriebsprinzip, Schwimmerfüllstandsmesser *m*
b. process *(Ch)* Schwimmaufbereitungsverfahren *n*
buoyant box Schwimmkasten *m*
b. force Auftrieb *m*
buoyed up mit Auftrieb, auftreibend, aufgetrieben
buoying *(Schiff)* Betonnung *f*
bur Klette *f*
b. crushing Entfernen *n* der Kletten *(aus Wolle auf mechanischem Wege)*
b. extracting *(Text)* Entkletten *n*
b. ore Klettenerz *n*
burble / to abreißen
burble point *(Ph)* Grenzschichtablösungspunkt *m*, Übergangspunkt *m*, Wirbelpunkt *m*
burden / to *(Met)* begichten, möllern
burden 1. *(El)* Bürde *f* *(des Stromwandlers)*; 2. Last *f*, Belastung *f*; Ladefähigkeit *f*; Tragkraft *f*, Tragfähigkeit *f*; 3. *(Met)* Gattierung *f*; 4. *(Geol)* Deckgebirge *n*; 5. *(Brg)* Vorgabe *f* *(Schießarbeit)*

b. balance Gattierungswaage *f* *(Hochofenbetrieb)*
b. calculation *(Met)* Möllerberechnung *f*
b. charge *(Met)* Möllergicht *f*
b. ratio *(Met)* Möllerverhältnis *n*
b. squeezer *(Met)* Luppenmühle *f*
b. yield *(Met)* Möllerausbringen *n*
burdening 1. Belastung *f*, Beschwerung *f*; 2. *(Met)* Begichtung *f*, Möllerung *f*
burdock Klette *f*
burette Bürette *f*
b. brush Bürettenbürste *f*
burgee Doppelstander *m*, Reedereiflagge *f*
burglar alarm Einbruchsalarmanlage *f*
burial ground *(Kern)* Abfallager *n*, radioaktiver Friedhof *m*, Vergrabungsstelle *f*
buried cable [verlegtes] Erdkabel *n*
b. in sand sandverweht
b. installation *(El)* Unterputzanlage *f*, UP-Anlage *f*
b. pipe erdverlegte Rohrleitung *f*
b. rock *(Geol)* von weichen Schichten überdecktes hartes Gestein *n*
b. wiring Unterputz-Leitungsverlegung *f*, Erdverlegung *f*
burin *(Typ)* Gravierstichel *m*; Grabstichel *m*
burl / to *(Text)* noppen; durchrauhen
burl 1. *(Text)* Noppe *f*, Knoten *m*; 2. Auswuchs *m*, Beule *f* *(Baum)*
burlap Sackleinen *n*
burling frame *(Text)* Noppmaschine *f*, Nopprahmen *m*
b. iron *(Text)* Noppeisen *n*, Noppzange *f*
burly *(Holz)* gemasert
burn / to 1. brennen, abfackeln; verbrennen; 2. *(El)* durchbrennen *(z. B. Lampe)*
b. in einbrennen
b. off abbrennen
b. up aufbrennen
burn Verbrennung *f*; Brand *m*; Brandstelle *f*; Brandfleck *m*
b. cut *(Brg)* Brennereinbruch *m*, Kanonenschießen *n*, Röhreneinbruch *m*
b. off Abbrand *m*
b.-off rate Abschmelzgeschwindigkeit *f* *(Schweißen)*
b.-out 1. *(Kern)* Durchbrennen *n*; 2. *(Rak)* Brennschluß *m*
b.-out altitude *(Rak)* Brennschlußhöhe *f*
b.-out lace *(Text)* Luftspitze *f*, Ätzspitze *f*
b.-out speed *(Rak)* Brennschlußgeschwindigkeit *f*
b.-out time *(Rak)* Brennschlußzeit *f*
b.-out velocity *(Rak)* Fluggeschwindigkeit *f* bei Brennschluß
b.-out weight *(Rak)* Gewicht *n* bei Brennschluß, Brennschlußmasse *f*
b.-through Durchbrennen *n* *(Schweißen)*
b.-up *(Kern)* Abbrand *m*
burned embroidery Ätzstickerei *f*
b. gases Verbrennungsgase *npl*
b. speed *(Rak)* Fluggeschwindigkeit *f* bei Brennschluß
b. spots *(Plast)* Verbrennungsmarkierungen *fpl* *(Spritzfehler)*
b.-up überbelichtet
b. velocity *(Rak)* Fluggeschwindigkeit *f* bei Brennschluß
burner Brenner *m*, Lampe *f*
b. capacity Brennerleistung *f*
b. cone *(Ch)* Brennerkegel *m*, Mischkegel *m*
b. fuel oil Heizöl *n*
b. liner Brennkammerauskleidung *f*
b. mouth Brennerdüse *f*
b. nozzle *(Rak)* Brennerdüse *f*
b. tube Brennerrohr *n* *(Lötlampe)*
burnettize / to *(Ch)* mit Zinkchlorid tränken
burning Brennen *n*, Verbrennen *n*; Oxydieren *n* durch Überhitzen; lokale Schmelzung *f*
b. brand Brennstempel *m*
b.-gas Verbrennungsgas *n*
b. glass *(Opt)* Brennglas *n*
b.-in stove Einbrennofen *m*
b.-in temperature Einbrenntemperatur *f*
b. kiln Brennofen *m*
b. lens Brennglas *n*
b. life Brenndauer *f*
b. of the grinding wheel Brennen *n* der Schleifscheibe
b. oil Brennöl *n*, Leuchtpetroleum *n*
b.-out effect Ausbrenneffekt *m*
b. point Brennpunkt *m*, Entzündungspunkt *m*, Zündpunkt *m*
b. rate Brenngeschwindigkeit *f*, Verbrennungsgeschwindigkeit *f*
b.-rate of the carbon Kohlenabbrand *m*
b. test Brennprobe *f*, Brennversuch *m*; Brennbarkeitsprobe *f* *(Fasern)*
b. time Brenndauer *f*
b. torch *(Am)* Schneidbrenner *m*
b. velocity Verbrennungsgeschwindigkeit *f*, Brenngeschwindigkeit *f*
b. voltage *(El)* Brennspannung *f*
b. zone Brennzone *f* *(eines Ofens)*
burnish / to *(Masch)* polieren, drücken, glätten, hochglanzpolieren, hochglanzschleifen, polierrollen, preßpolieren
burnisher 1. *(Masch)* Polierer[in *f*] *m*; 2. Polierstahl *m*; 3. Glättzahn *m* *(Räumwerkzeug)*
burnishing 1. *(Met)* Brünieren *n*; 2. Polierdrücken *n*, Polierrollen *n*
burnishing verdichtend
b. gear Glättrad *n*
b. lathe Prägepoliermaschine *f*
b. tooth Glättzahn *m*

burnt clay gebrannter Ton *m*
b. clay ware Grobkeramik *f*
b. cook Schwarzkochung *f (Zellstoff)*
b. deposit poröser Überzug *m (galvanisch)*
b. gases Abgas *n*, Rauchgas *n*
b. lime gebrannter Kalk *m*
b. metal überhitztes Metall *n*
b.-out *(El)* durchgebrannt *(Glühfaden einer Lampe)*
b.-out embroidery *(Text)* Ätzstickerei *f*
b. paper übertrocknetes (verbranntes) Papier *n*
b. steel verbrannter (überhitzter) Stahl *m*
b. velocity *(Rak)* Fluggeschwindigkeit *f* bei Brennschluß
burr / to *(Met, Masch)* abgraten, entgraten, putzen
burr 1. *(Met, Masch)* Grat *m*, Bart *m*; 2. *(Masch)* Grat *m*, Bohrgrat *m*, Preßgrat *m*, Walzgrat *m*; 3. Schärfrolle *f (beim Schleifstein)*
b. crusher roller Klettenbrechwalze *f*
b. extracting *s.* burring 2.
b. fork *(Holz)* Schärfrollengabel *f*
b.-free gratfrei
b. removing machine *(Met)* Abgratmaschine *f*
burred mit Grat
b. wool entklettete Wolle *f*
burring 1. *(Masch)* Entgraten *n*; Kantenbrechen *n*; 2. *(Text)* Entkletten *n*
b. chisel Abgratmeißel *m*
b. lathe Schärfrollenspindel *f*
b. machine *(Led)* Haarwaschmaschine *f*
burry *(Text)* 1. klettenhaltig; 2. flauschig
burst / to bersten, explodieren, platzen, zerreißen
burst 1. Ausbruch *m (Gas, Feuer, Wasser)*; 2. *(Kern)* Hoffmannscher Stoß *m*; Ionisationsstoß *m*; 3. *(Fs)* Schwingungszug *m (Farbfernsehtechnik)*
b. in the barrel *(Mil)* Rohrkrepierer *m*
b. pipe gebrochenes Rohr *n*, Rohrbruch *m*
b. ratio Berstindex *m*
bursting limit Berstdruck *m*, Zerreißgrenze *f*
b. pressure Berstdruck *m*
b. speed Zerreißdrehzahl *f (Festigkeit)*
b. strength Berstdruck *m*, Zerreißgrenze *f*, Berstfestigkeit *f*, Berstwiderstand *m*; Bruchfestigkeit *f*; Zerreißfestigkeit *f*
b. stress Zerreißspannung *f*, Bruchspannung *f (Festigkeit)*
bursts of errors *(BMSR)* Fehlerausbrüche *mpl*
bury / to in die Erde verlegen *(Kabel)*
bus 1. Autobus *m*, Bus *m*, Omnibus *m*; 2. *(BMSR) (Am)* Hauptweg *m*; Sammelschiene *f*
b.-bar Sammelschiene *f*
b.-bar sectionalizing switch Sammelschienenlängstrennschalter *m*
b.-bar support Sammelschienenisolator *m*
b.-bar transformer Schienen[strom]wandler *m*
b. depot Autobushof *m*, Omnibushof *m*
b. line 1. Autobuslinie *f*, Omnibuslinie *f*; 2. Fahrdraht *m*, Oberleitung *f*
b. reactor Sammelschienendrossel *f*
b. terminal 1. [Autobus-]Endstation *f*; 2. Autobushof *m*, Omnibushof *m*
bush / to 1. *(Masch)* ausbuchsen, ausbüchsen, füttern; 2. *(Bw)* scharrieren
bush 1. Buchse *f*, Büchse *f (Lager)*; Halter *m (Reibahle)*; Lagerschale *f*, Laufbuchse *f*; Muffe *f*; Rohrstutzen *m*; Zapfenlager *n*; 2. *(El)* Steckerbuchse *f*; Leitungseinführung *f*
b. chisel *(Bw)* Scharriereisen *n*
b. extractor Buchsenauszieher *m*
b. hammer *(Bw)* Scharrierhammer *m*
b.-hammered *(Bw)* scharriert
b.-harrow Buschegge *f*
b. of diaphragm Leitradbüchse *f*, Dichtungsbüchse *f (Turbine)*
b. packing *(Masch)* Ringbüchse *f*
b. planter Strauchpflanzgerät *n*
b. roller chain Buchsen[rollen]kette *f*
b. weir Buhne *f*
b.-wood Buschholz *n*
bushel[l]ing furnace Schrottofen *m (Puddelprozeß)*
bushing 1. Ausbüchsen *n*; 2. *(Masch)* Buchse *f*, Büchse *f*, Lager *n*, Laufbuchse *f*; Futterstück *n*; Muffe *f*; Durchführung *f*; Aufnahmestück *n (zum Halter für Bohrerhülsen)*; 3. *(El)* Einführung *f*, Durchführung *f*, Gerätedurchführung *f*, Geräteeinführung *f*
b. chain *(Masch)* Stiftkette *f*
b. clamp *(El)* Durchführungsklemme *f*
b. for holding work Spannbuchse *f*
b. guide *(Masch)* Führungsbuchse *f*
b. insulator Durchführungsisolator *m*, Einführungsisolator *m*, Isolierhülse *f*
b. nipple *(El)* Einführungstülle *f*, Endtülle *f*
b. on Aufbuchsen *n*
b. transformer *(El)* Durchführungswandler *m*
b.-type current transformer Stützentransformator *m*, Stützenstromwandler *m*
b. with flange Laufbuchse *f* mit Bund
b. with grease-pocket feed Laufbuchse *f* mit Fettschmierung
business machine Bürorechenmaschine *f*
buss Buise *f*, Heringsbüse *f*, Heringsbuise *f (Schiff für Heringstransport von den Fangschiffen)*
bustle-pipe *(Met)* Windringleitung *f*, Ringleitung *f (Hochofen)*
busy besetzt *(Telefon)*
b.-back jack Besetztklinke *f*, besetzte Klinke *f*

b.-back tone Besetztzeichen *n*
b.-flash signal Besetztflackerzeichen *n*
b. jack Besetztklinke *f*, besetzte Klinke *f*
b. lamp Besetztlampe *f*, Freimeldelampe *f*
butane *(Ch)* Butan *n*
b. diacid Äthandikarbonsäure *f*
butanoic acid Buttersäure *f*
butanone diacid Oxalessigsäure *f*
butt / to stumpf aneinanderfügen (gegeneinandersetzen, zusammensetzen)
butt 1. *(Brg)* normal zur Schlechtenrichtung verlaufender Stoß *m*; 2. *(Text)* Fuß *m* *(Nadel)*; 3. *(Led)* Croupon *m*, Schild *m*; 4. *(Mil)* Kolben *m (z. B. Gewehr)*
b. and butt *(Holz)* gerader (stumpfer) Stoß *m*
b. contact *(El)* Druckkontakt *m*
b. end 1. Kappenkopf *m (Schubstange)*; 2. *(Bw)* Kopfende *n*, Stirnfläche *f*
b. entry *(Brg)* 1. Grundstrecke *f*; Querstrecke *f*; 2. *(Am)* Abbaustrecke *f*
b. ingot *(Met)* kurzer Flußstahlblock *m*, Restblock *m*, Stummelblock *m*
b. joint 1. *(Masch, Holz)* Stumpfstoß *m*, Stumpfverbindung *f*; 2. *(El)* Flachkupplung *f*, Flachflanschverbindung *f (Kabel)*
b. joint graft *(Holz)* stumpfe Pfropfung *f*
b.-joint riveting *s.* b. riveting
b. mortiser Einstemmaschine *f*
b. of an ingot unteres Blockende *n*
b. plate Deckplatte *f*, Verbindungsplatte *f*
b. riveting Laschennietung *f*
b. roller *(Led)* Karrenwalze *f*
b. seal *(Opt)* Stumpfanglasung *f (Glas an Metall)*
b. seam welding Stumpfnahtschweißen *n*
b. strap Lasche *f (Nietverbindung)*, Stoßlasche *f*
b.-strap [riveted] joint Laschennietverbindung *f*
b. weld Stumpfnaht *f (Schweißen)*
b. welded stumpf geschweißt
b. welder Stumpfschweißmaschine *f*
b. welding Stumpfschweißung *f*
b. welding with pressure Stauchschweißen *n*
b. wood Stockholz *n*
butted joint *(Masch, Holz)* Stumpfstoß *m*, Stumpfverbindung *f*
butter churn Buttermaschine *f*
b.-making machine Butterfertiger *m*
b. of tin Zinnbutter *f*
butterfly Drosselklappe *f*
b. bolt Flügelschraube *f*
b. circuit *(Fs)* Schmetterlingskreis *m (cm-Wellen)*
b. cock *(Masch)* Flügelhahn *m*
b. nut Flügel[schrauben]mutter *f*
b. screw Flügelschraube *f*
b. shell *(Bw)* Schmetterlingsschale *f*, Y-Schale *f*
b. throttle spindle Drosselklappenachse *f*
b. valve Drosselklappe[nventil *n*] *f*, Drosselventil *n*, Flügelventil *n*
buttock *(Brg)* Vorgriff *m*
b. lines Schnitte *fpl (im Linienriß)*
buttocks *(Schiff)* Billen *pl*; Hinterschiffsschnitte *mpl*, Hauptschnitte *mpl (im Linienriß)*
button Knopf *m*
b.-border *(Text)* Knopfleiste *f*, Knopfblende *f*
b. breaker *(Text)* Knopfmaschine *f*
b. die Schneideisen *n* für Gewinde
b. fixing machine Knöpfesetzmaschine *f (Schuhfertigung)*
b. head Halbrundkopf *m (Niet)*
b.-head bolt Halbrundschraube *f*
b. head rivet Halbrundniet *m*
b. head rivet set Nietkopfsetzer *m* für Halbrundniete
b. head screw Halbrundkopfschraube *f*
b. of lute Lautenkamm *m*
b. sewing machine Knopfannähmaschine *f*
b. switch Druckknopfschalter *m*
b.-type blowcock Druckknopfventil *n*
buttonhole punch Knopflocheisen *n*
b. [sewing] machine Knopfloch[ausnäh]maschine *f (Schuhfertigung)*
b. spring punch *(Masch)* Knopflochzange *f*
b. stitch *(Text)* Knopflochstich *m*
buttress 1. *(Bw)* Strebepfeiler *m*; Mauerpfeiler *m*, Verstärkungspfeiler *m*; 2. *(Holz)* Wurzelanlauf *m*
b. dam *(Bw)* aufgelöste Staumauer *f*, Pfeilerstaumauer *f*
b. of battery *(Brg)* Batterieendknopf *m*
b. pack *(Brg)* Entlastungsdamm *m*
b. [screw] thread Sägengewinde *n*
b. wall Winkelstützmauer *f (Grundbau)*
buttressed dam *(Bw)* aufgelöste Staumauer *f*, Pfeilerstaumauer *f*
buttressing pier Verstärkungspfeiler *m (Mauerwerk)*
b. wall *(Bw)* aussteifende Querwand *f*
butty *(Led)* kräftig im Kern
b. boat Kohlenfördererboot *n*
b. gang *(Brg)* Kameradschaft *f*
butyl acetic acid Butylessigsäure *f*, Kapronsäure *f*
b. rubber Butylkautschuk *m*
butyric acid Buttersäure *f*
b. ester Buttersäureester *m*
buyer Einkäufer *m*; Käufer *m*, Kunde *m*
buzz-type starting ignition *(Kfz)* Summeranlaßzündung *f*
buzzer *(El)* Summer *m*, Selbstunterbrecher *m*; Zündspule *f* mit Unterbrecher

b. sound *(Fmt)* Summerton *m*
by-flow filter *(Bw)* Nebenstromfilter *n*
by frictional resistance reibschlüssig
by-gases Destillationsgase *npl (Kohle)*
by hand von Hand
by means of mittels
by-oven Nebenproduktofen *m*
by-pass 1. *(Bw)* Umgehungsstraße *f*; 2. *(Kfz)* Umleitung *f*; 3. *(BMSR)* Beipaß *m*, Umgehungsleitung *f*, Umgehungsventil *n*
by-pass capacitor *(Fmt)* Ableitkondensator *m*; Parallelkondensator *m*; Überbrückungskondensator *m*
by-pass control *(BMSR)* Beipaßregelung *f*, By-pass-Regelung *f*
by-pass jet engine Beipaß-Strahltriebwerk *n*
by-pass line 1. Umgehungsleitung *f*; 2. *(Rak)* Umwälzleitung *f*
by-pass pipe Überleitrohr *n*, Überströmrohr *n*
by-pass rectifier Nullanode *f (Stromrichter)*
by-pass road Umgehungsstraße *f*
by-pass switch *(El)* Brückenschalter *m*, Überbrückungsschalter *m*
by-pass valve Umgehungsventil *n*, Umlaufventil *n*; Steuerschieber *m (Dreifachschieber)*
by-product Nebenprodukt *n*; Abfallerzeugnis *n*, [verwertbares] Abfallprodukt *n*
by-product coke Hüttenkoks *m*, Zechenkoks *m*
by-recovery plant Nebengewinnungsanlage *f*
by the head *(Schiff)* vorlastig, kopflastig
by the stern *(Schiff)* achterlastig
by-wash Entwässerungskanal *m*, Entlastungskanal *m*
by wireless *(El)* auf drahtlosem Wege
byre Rinderstall *m*, Kuhstall *m*
byssaceous, byssoid baumwollartig; aus feinen Fäden bestehend
bytes *(BMSR)* Verarbeitungsmengen *fpl* von 1 bis 8 bit (*s. a.* bit)

C

C-battery Gitter[vorspannungs]batterie *f*
C-clamp Schraubzwinge *f*
C display C-Darstellung *f (Radar)*
C-frame mit tiefer Ausladung
C-frame press Einständerpresse *f*
C-horizon *(Geol)* C-Horizont *m*, Untergrund *m*
C-neutrons C-Neutronen *npl (Neutronen mit einer mittleren Energie zwischen 0 und 0,3 eV)*
C-power supply *(Fmt)* Gittervorspannungszusatz *m*
C process Formmaskenverfahren *n*
C-shaped traveller *(Text)* C-förmiger Läufer *m*
C-voltage Gitterspannung *f*
C-wire *(Fmt)* Meßleitung *f*; Prüfader *f*; Prüfdraht *m*
caam Blatt *n (Webstuhl)*
cab 1. Kabine *f*; 2. *(Kfz)* Fahrerhaus *n*, Führerhaus *n*; 3. *(Eis)* Führerstand *m*
c. cable *(El)* Steuerleitung *f*
c. floor *(Kfz)* Fahrerboden *m*
c. signal *(El)* Führerstandsignal *n*
c.-tyre cable *(Opt)* Gummischlauchleitung *f*
c.-tyre line *(El)* Gummischlauchleitung *f*; Gummimantelleitung *f*
c.-tyre sheathing Gummischlauchmantel *m*; Gummischlauchummantelung *f*
CAB vessel *s.* captured air bubble vessel
caber rohbehauener [Nadelholz-]Stamm *m*
cabin 1. Kabine *f (auf Fahrgastschiffen)*; Kammer *f (auf Kriegsschiffen)*; Kajüte *f*; 2. Baubüro *n (Bude)*
c. hook Türbeschlag *m*
c. liner Kajütschiff *n*, Fahrgastlinienschiff *n*
c. zone Stellbezirk *m (Signal)*
cabinet 1. Schrank *m*; 2. *(Masch)* Kastenfuß *m*; 3. *(Fmt, Fs)* [Empfänger-]Gehäuse *n*; 4. Normalformat *n* für Fotoabzüge
c. clamp Schreinerschraubzwinge *f*
c. drawing *(Math)* Kavalierprojektion *f*
c. drier *(Ch)* Trockenschrank *m*
c. file Kabinettfeile *f*, Wälzfeile *f*
c. lathe Drehmaschine *f* mit Kastenfuß
c. leg *(Masch)* Kastenfuß *m*
c. maker Schreiner *m*, Tischler *m*, Möbeltischler *m*
c. rasp *(Holz)* Tischlerraspel *f*
c. saw Tischlersteifsäge *f*
cable 1. Kabel *n*; 2. Kabeldepesche *f*, Telegramm *n*; 3. *(measure of length)* Kabellänge *f* (= *0,1 Seemeile* = *185,2 m)*; 4. Seil *n*, Drahtseil *n*, Stahlseil *n*
c. accessories Kabelgarnitur *f*
c. armouring Kabelbewehrung *f*
c. armouring machine Kabelbewehrungsmaschine *f*
c. armouring wire Kabelarmierungsdraht *m*
c. assignment record *(Am)* Belegungsplan *m (Kabel)*
c. attenuation Kabeldämpfung *f*
c. balancing Kabeladerausgleich *m*
c. bearer Kabelträger *m*, Kabelaufhänger *m*
c. binding of skis Kandaharbindung *f*
c. binding paper Kabelwickelpapier *n*
c. bond Kabelmantelverbinder *m (für durchgehende Erdung)*
c. box Kabelmuffe *f*, Kabelabzweiger *m*, Kabelkasten *m*
c. bracing Seilverspannung *f*
c. brake *(Kfz)* Seilzugbremse *f*
c. branching Kabelverzweigung *f*, Kabelabzweigung *f*
c. break Kabelbruch *m*, Kabelunterbrechung *f*
c. bridge *(Bw)* Seilbrücke *f*, Kabelbrücke *f*

c. buoy *(Schiff)* Ankerboje *f*, Vertäuboje *f*, Kabeltonne *f*
c. capacity Ladekapazität *f (Kabel)*
c. cellar Kabelkeller *m*
c. channel Kabelkanal *m*, Kabelrinne *f*
c. chock *(Schiff)* Verholklampe *f*, Leitklampe *f*
c. chute *(Bw)* Kabelhochführungsschacht *m*, Aufstiegskanal *m*
c. clamp Kabelschelle *f*, Seilklemme *f*
c. clip Kabelklemme *f*, Leitungsklemme *f*
c. code Kabelkode *m*
c. collar Kabelschelle *f*
c. compound Kabelvergußmasse *f*
c. compressor *(Schiff)* Ankerkettenkneifer *m*, Trossenkneifer *m*
c. conductor Kabelader *f*, Kabelleiter *m*
c. conduit 1. Kabelschlauch *m*, Kabelrohr *n*; 2. *(Kfz)* Seilhülle *f*
c. connecting socket Kabelkupplung *f*, Leitungskupplung *f*
c. connecting terminal Kabelanschlußklemme *f*
c. connection Kabelanschluß *m*
c. connector Leitungsverbinder *m*
c. control *(Kfz)* Seilzug[steuerung *f*] *m*
c.-controlled mit Seilsteuerung *(Schürfkübelwagen)*
c. core Kabelseele *f*; Kabelader *f*
c. coupling Kabel[steck]verbindung *f*
c. covering machine Kabelisoliermaschine *f*, Kabelumwickelmaschine *f*
c. crane Kabelkran *m*
c. current transformer Kabelstromwandler *m*
c. dismantling knife Kabelmesser *n*
c. distribution box Kabelverzweiger *m*
c. distribution head Kabelendverteiler *m*, Kabelübergangskasten *m*, Kabelendverschluß *m*, Kabelverzweiger *m*, Linienverzweiger *m*, Trennendverschluß *m*
c. draw pit Kabelschacht *m*
c. drilling bit Schlagbohrmeißel *m*
c. drum Kabeltrommel *f*, Kabelhaspel *f*
c. duct 1. *(El)* Kabelkanal *m*; 2. *(Bw)* Gleitkanal *m*, Hüllrohr *n*; Spann[glied]kanal *m* *(Spannbeton)*
c. end box Kabelendverschluß *m*
c. end piece Kabelendverschluß *m*
c. end sleeve Kabelendverschluß *m*
c. entry Kabeleinführung *f*
c. equipment Kabelzubehör *n*
c. eye Kabelschuh *m*; Kabelöse *f*
c. fan Kabelzopf *m*
c. fault Leiterbruch *m (im Kabel)*, Aderbruch *m*
c. fill Kabelauslastung *f*
c. filler Kabelmasse *f*, Kabelfüllmaterial *n*; Kabelfett *n*, Kabelvaseline *f*
c. fitting Kabelarmatur *f*; Kabelmuffe *f*; Kabelverbinder *m*
c. for communication circuit Schwachstromkabel *n*
c. gas-check point *(Fmt)* Schleuse *f (bei mit Druckgas überwachten Kabeln)*
c. gear Kabelgeschirr *n*
c. gland Kabeleinführung *f*
c. grip Kabelziehstrumpf *m*, Kabelziehkeil *m*
c. head Kabelabschluß *m*
c. housing Kabelhülle *f*, Kabelrohr *n (Spannbetonbau)*
c. interruption Kabelunterbrechung *f*
c. jack *(Schiff)* Ankerkettenhiever *m*
c. jacket Kabelmantel *m*
c.-joining box Kabelverbindungskasten *m*
c. joint Kabelmuffe *f*, Verbindungsstelle *f*; Kabellötstelle *f*
c. joint box Kabelanschlußkasten *m*, Kabelübergangskasten *m*, Kabelverbindungskasten *m*
c. jointing sleeve Abzweigmuffe *f*, Anschlußmuffe *f*, Verbindungsmuffe *f*
c. junction box Verbindungsmuffe *f*, Durchgangsmuffe *f*
c.-laid hawser *(Schiff)* Kabeltrosse *f*
c. layer Kabelleger *m*, Kabelschiff *n*; Kabelverlegemaschine *f*
c.-laying Kabelverlegung *f*; Verkabelung *f*
c.-laying detachment Kabelbautrupp *m*
c.-laying hoist Kabelverlegegewinde *n*
c.-laying ship Kabelschiff *n*, Kabelleger *m*
c. length Kabellänge *f (0,1 Seemeile = 185,2 m)*
c. lifting Kabelhochführung *f*
c. locator Kabelsuchgerät *n*
c. locker *(Schiff)* Kettenkasten *m*
c. lubricant Kabel[gleit]fett *n*
c. lug Kabelöse *f*, Kabelklemme *f*, Kabelschuh *m*
c. make-up Kabelaufbau *m*
c. manhole Kabelzweigkasten *m*
c.-operated mit Seilzugbetätigung
c. outlet Kabelausgang *m*
c. paper Kabelpapier *n*, Isolierpapier *n*
c. paraffin Kabelwachs *n*
c. pipe Kabelführung *f*
c. pit Kabelbrunnen *m*
c. pole Kabelmast *m*; Endmast *m (der Freileitung)*
c. pole line auf Masten verlegte Luftkabelleitung *f*
c. post Kabelmast *m*
c. power factor Kabelverlustwinkel *m*, Kabelleistungsfaktor *m*
c. profile Kabelführung *f*, Spanngliedführung *f (Spannbeton)*
c. protected against interference induktionsgeschütztes Kabel *n*, Kabel mit induktionsgeschützten Leitungen

c. protecting hose Kabelschutzschlauch *m*
c. protection Kabelabschirmung *f*
c. punch Kettendorn *m*
c. rack Kabelgestell *n*
c. railway Seilbahn *f*
c. reel Kabelhalter *m*; Kabelrolle *f*, Kabeltrommel *f*
c. run Kabelführung *f*, Kabelverlauf *m*, Kabelbahn *f*
c. running Kabelführung *f*, Leitungsführung *f*
c. sag Seildurchhang *m (einer Brücke)*
c. scraper *(Bw)* Seilschrapper *m*
c. screening Kabelabschirmung *f*
c. sealing compound Kabelvergußmasse *f*, Kabeldichtungsmasse *f*
c. sealing head Kabelendverschluß *m (vergossen)*
c. shaft Kabelschacht *m*
c. sheath Kabelmantel *m*
c. sheath protection factor Kabelreduktionsfaktor *m*
c. sheath splitting knife Kabelmesser *n*
c. sheathing Kabelhülle *f*, Kabelmantel *m*
c. sheave Kabelrolle *f (eines Kabellegers)*
c. shelf Kabelrost *m*
c. ship Kabelschiff *n*, Kabelleger *m*
c. sleeve Kabelmuffe *f*
c. sling Kabelaufhänger *m*
c. socket Kabelschuh *m*
c. specification Kabelpflichtenheft *n*
c. splitting pocket knife Kabelklappmesser *n*
c. stitch *(Text)* Kreuzdeckmuster *n*
c. stopper *(Schiff)* Kettenstopper *m*
c. strand Kabellitze *f*
c. stranding machine Kabelverseilmaschine *f*
c. stripper Kabelmesser *n*, Abisoliermesser *n*
c. stripping knife Kabelmesser *n*
c. subway Kabeltunnel *m (begehbar)*, Kabelstollen *m*
c. support rack Kabelendgestell *n*, Kabelrost *m*
c. suspender Kabelaufhänger *m*; Kabelhalter
c. suspension clamp Kabelhängeschelle *f*, Hängeschelle *f*
c. tank Imprägniertank *m*
c. terminal Kabelklemme *f*, Kabelanschluß *m*; Klemmkabelschuh *m*
c. terminal box Kabelendverschluß *m*; Kabelschrank *m*, Verteiler *m*
c. test Kabelprüfung *f*
c. test service Kabelmeßdienst *m*
c. testing Kabelmessung *f*
c. testing car Kabelmeßwagen *m*
c. testing van Kabelmeßwagen *m*
c. thimble Kabelschuh *m*
c. tile Kabelformstein *m*
c. tool drilling Seilbohren *n*; Seilschlagverfahren *n*
c. tool drilling equipment Seilbohranlage *f*
c. tool hole Seilbohrloch *n*
c. tool method of drilling Seilbohrverfahren *n*
c. trench Kabelgraben *m*
c. trough Kabelgraben *m*
c. troughing Kabelkanal *m*
c. tunnel Kabelkanal *m*
c. vessel Kabelschiff *n*
c. wax Kabelwachs *n*
c.-way crane Kabelkran *m*
c. winch Seilwinde *f*, Kabelwinde *f*
c. with cab-tyre sheathing *(El)* Gummischlauchleitung *f*
c. yarn Kabelgarn *n*
cabled *(Text)* mehrfach gezwirnt
c. yarn *(Text)* Mehrfachzwirn *m*
cablegram Überseetelegramm *n (durch Seekabel)*
cablet Kabeltrosse *f*
cableway Seilschwebebahn *f*
c. excavator Kabelbagger *m*
c. for handling materials Lastenseilschwebebahn *f*
c. transporting persons Personenseilschwebebahn *f*
cabling Verkabelung *f*; Verdrahtung *f*, Beschaltung *f*
c. spindle *(Text)* Kordierspindel *f*
cabochon Edelstein *m* mit mugeligem Schliff
caboose 1. *(Schiff)* Kombüse *f*, Schiffsküche *f*; 2. *(Am.Eis)* Personalwagen *m*
cabotage Küstenfahrt *f*
cabriolet Kabriolett *n*
cacao Kakaofrucht *f*, Kakaobohne *f*
c. tree Kakaobaum *m*
cacodyl reaction *(Ch)* Kakodylreaktion *f*
cacodylic acid Dimethylarsinsäure *f*, Kakodylsäure *f*
cadastral accuracy Katastergenauigkeit *f*
c. office Katasteramt *n*
c. survey Katasteraufnahme *f*
cadastre Kataster *n*
cade oil *(Ch)* Cadeöl *n*, Wacholderteeröl *n*
cadence Tonfall *m*, Kadenz *f*
c. signal *(Fmt, Fs)* Taktzeichen *n*
cadet ship Kadettenschiff *n*
cadet's training ship Kadettenschulschiff *n*
cadmia *(Met)* Ofengalmei *m*, Galmei *m*, Zinkschwamm *m*
cadmic compound Verbindung *f* des zweiwertigen Kadmiums
cadmiferous kadmiumhaltig
cadmium Kadmium *n*
c. cell *(El)* Kadmiumelement *n (Normalelement)*
c.-nickel storage battery Nickel-Kadmium-Akkumulator *m*
c.-plate / to verkadmen, kadmieren

c. plating Verkadmen *n*, Kadmieren *n*; Kadmiumüberzug *m*
c. red line rote Spektrallinie *f* des Kadmiums
c. regulator *(Kern)* Kadmiumregelstab *m*
c. rod *(Kern)* Kadmiumstab *m*
c. storage cell *(El)* Kadmiumakkumulator *m*, Kadmiumsammler *m*
c. yellow Kadmiumgelb *n*
cadmous compound Verbindung *f* des einwertigen Kadmiums
caesium Zäsium *n*
caffeine Koffein *n*, Kaffein *n*
cage 1. *(Masch)* Gehäuse *n*; Abstandshalter *m*, Käfig *m (Lager)*; 2. *(Bw, Brg)* Förderkorb *m*, Förderschale *f*; 3. *(Bw)* Bewehrungskorb *m (Stahlbeton)*; 4. Siebtrommel *f*; 5. Korb *m (in Erntemaschinen)*
c. aerial Reusenantenne *f*
c. arresting device *(Brg)* Übertreib-Schutzvorrichtung *f*
c. compartment *(Brg)* Förderturm *m*
c. effect *(Kern)* Käfigeffekt *m*
c. hoist Schachtgerüstaufzug *m*; Fahrstuhlaufzug *m*
c. hoisting *(Brg)* Gestellförderung *f*
c. of reinforcement Bewehrungskorb *m (Stahlbeton)*
c. relay Käfigrelais *n*, Scheibenrelais *n*
c. rotor *(Elm)* Kurzschlußläufer *m*, Käfigläufer *m*, Käfiganker *m*
c. seats *(Brg)* Aufsetzvorrichtung *f*
c. shoe *(Brg)* Führungsschuh *m*
c. tendent *(Brg)* Anschläger *m*
c.-type lightning conductor Käfigblitzableiter *m*
c. type of grid *(Fmt)* Käfiggitter *n*
c. winding Käfigwicklung *f*
cager *(Brg)* Anschläger *m*, Aufschieber *m*
cageway *(Brg)* Fördertrum *m*
cagework *(Text)* Durchbrucharbeit *f*
caging *(Brg)* Aufschieben *n*
cairngorm *(Min)* Rauchquarz *m*
caisson 1. *(Bw, Schiff)* Caisson *m*, Senkkasten *m*; 2. Kassette *f (Decke)*
c. disease Caissonkrankheit *f*
c. foundation Caissongründung *f*, Senkkastengründung *f*; Pfahlgründung *f*
c. pile *(Bw)* Kegelfußpfahl *m*
c. slab *(Bw)* Kassettenplatte *f*
cake / to *(Met)* [zusammen]backen, zusammenballen, sintern
c. together zusammenbacken
cake 1. Sinterkörper *m*, Preßkörper *m*, Preßling *m*; 2. *(Text)* Spinnkuchen *m*
c. copper Kupfer *n in Platten*
c.-glue *(Typ)* Tafelleim *m*
c. of clinker Schlackenkuchen *m*, Schlackenmasse *f*
c.-ore *(Met)* Kuchenerz *n*
caking Sintern *n*, Backen *n*, Festbacken *n*, Sinterung *f*
c. capacity Backfähigkeit *f*
c. coal backfähige Kohle *f*, Backkohle *f*, Fettkohle *f*
c. power Backfähigkeit *f*
c. property Backfähigkeit *f*, Backvermögen *n*, Ballungsfähigkeit *f*
calamiferous hohlstichig
calamine *(Ch, Met)* Galmei *m*, Zinkkarbonat *n*, Zinkerz *n*
calamus Federkiel *m*
c. oil Kalmusöl *n*
calc-spar Kalkspat *m*
calcar arch *(Ch)* Brennofen *m*, Frittofen *m*
calcareous kalkartig, kalkhaltig, kalkig
c. alabaster Kalkalabaster *m*
c. clay Kalkmergel *m*
c. earth Kalkerde *f*
c. gravel Kalkkies *m*
c. ore kalkiges Erz *n*
calcic kalziumhaltig, kalkartig
calcify / to verkalken
calcimine Leimfarbe *f*
calcinate / to brennen, kalzinieren
calcination *(Ch, Met)* Kalzinierung *f*, Röstung *f*, Rösten *n*, Brennen *n*
calcinator Brennofen *m*
calcine / to *(Ch, Met)* abschwelen, brennen, glühen, rösten, kalzinieren
calcined baryta Bariumoxid *n*
c. magnesia gebrannte Magnesia *f*, Magnesiumoxid *n*
c. ore Rösterz *n*
calciner Röstofen *m*
calcining plant *(Met)* Röstanlage *f*
calcite Kalkspat *m*, Kalkstein *m*
c. interferometer *(Opt)* Kalkspatinterferometer *n*
calcium Kalzium *n*
c. alginate thread Kalziumalginatfaden *m*
c. carbide Kalziumkarbid *n*
c. chloride Chlorkalzium *n*
c. cyanamide Kalkstickstoff *m*
c. fertilizer Kalkdünger *m*
c.-silicate brick Kalksandstein *m*, Kalziumsilikatstein *m*
calculability *(Math)* Berechenbarkeit *f*
calculable berechenbar, durch Rechnung lösbar
calculate from / to errechnen aus
c. the position *(Schiff)* das Besteck rechnen
calculated ceiling *(Flg)* theoretische (errechnete) Gipfelhöhe *f*
c. curve theoretische Kurve *f*
calculating *(Math)* Rechen-
c. disk Rechenscheibe *f*
c. machine Rechenmaschine *f*

c. **procedure** Rechnungsgang *m*
c. **puncher** *(Büro)* Rechenlocher *m*
c. **register** Rechenwerk *n (Buchungs- oder Fakturiermaschinen)*
calculation of floodable length *(Schiff)* Schottenrechnung *f*
c. **of length of cloth** *(Text)* Längenberechnung *f*
c. **of mixture** *(Met)* Gattierungsberechnung *f*
c. **of strength** Festigkeitsberechnung *f*
c. **of the damaged ship** *(Schiff)* Leckrechnung *f*
calculator Rechenautomat *m*, Rechenmaschine *f*; Rechner *m*
calculus *(Math)* Integralrechnung *f*; Differentialrechnung *f*, Infinitesimalrechnung *f*; höhere Analysis *f*; Kalkül *m (mathematische Logik)*
c. **of observations** Ausgleichsrechnung *f*
c. **of probability** Wahrscheinlichkeitsrechnung *f*
c. **of variations** Variationsrechnung *f*
caldron Kessel *m*, Heizkessel *m*, Kochkessel *m*
calendar-back board Kalenderrückenkarton *m*
c.**-pad bond board** Kalenderblockkarton *m*
calender / to glätten, kalandern, kalandrieren, pressen, satinieren
calender Glättmaschine *f*, Kalander *m*, Preßmaschine *f*, Satiniermaschine *f*
c.**-blackened** mit Satinierfehler
c. **box** Feuchtungswanne *f* am Kalander
c. **broke** Satinageriß *m*
c. **coating** Kalanderauftrag *m (Beschichtung)*
c.**-coloured** [am Kalander] oberflächengefärbt
c. **finish** *(Text)* Kalanderappretur *f*, Walzendekatur *f*
c.**-finished** satiniert, kalandriert
c. **for crêping** *(Text)* Kreppkalander *m*
c. **for scrooping** *(Text)* Seidenfinishkalander *m*
c. **roll[er]** Kalanderwalze *f*
c. **sheeting** *(Text)* Kalandertücher *npl*
c. **stack** Kalander *m (Papier)*
c. **with several rollers** Mehrwalzenkalander *m*
calendered kalandriert; satiniert, gestrichen
c. **sheet** *(Led)* Kalanderfolie *f*
calendering Kalandern *n*, Kalandrieren *n*, Pressen *n*, Satinieren *n*; Kalandrierung *f*
c. **roll** Kalanderwalze *f*
calf wheel *(Brg)* Fördertrommel *f (Seilbohren)*
calfeteria Kälberfütterungsanlage *f*
calfskin Kalbfell *n*
caliber *s.* calibre
calibrate / to eichen, kalibrieren, normieren; [ein]teilen, unterteilen
c. **resistances** *(El)* Widerstände abgleichen
calibrated airspeed *(Flg)* berichtigte angezeigte Eigengeschwindigkeit *f*, gemessene Luftgeschwindigkeit *f*
c. **detector** *(Kern)* geeichtes Nachweisgerät *n*
c. **focal length** Kammerkonstante *f (Fotogrammetrie)*
c. **orifice** Meßblende *f*
c. **prop** *(Brg)* Eichstempel *m*
c. **spark gap** Meßfunkenstrecke *f*
c. **vessel** Meßgefäß *n*
calibrating frequency *(El)* Eichfrequenz *f*
c. **line** *(Fmt)* Eichleitung *f (z. B. Dämpfungsbestimmung)*
c. **plot** Eichkurve *f*
c. **terminal** *(El)* Prüfklemme *f*; Prüfpunkt *m*
calibration Eichung *f*, Kalibrierung *f*, Normierung *f*; Einteilung *f*, Unterteilung *f*
c. **accuracy** Eichgenauigkeit *f*
c. **capacitor** Eichkondensator *m*
c. **card** Deviationstabelle *f*
c. **circuit** Eichkreis *m*
c. **curve** Eichkurve *f*
c. **factor** Eichfaktor *m*
c. **frequency** Eichfrequenz *f*
c. **instrument** Eichgerät *n*, Eichinstrument *n*
c. **mark** Eichstrich *m*
c. **meter** Eichzähler *m*
c. **pulse** Eichimpuls *m*
c. **table** *s.* c. card
c. **test** Eichversuch *m*
c. **voltage** Eichspannung *f*
calibre Kaliber *n*, Innendurchmesser *m*
caliche 1. weicher tonhaltiger Kalkstein *m*; 2. Salpeter *m* (*s. a.* soda nitre)
c. **wagon** Salpeterwagen *m*
calico 1. *(Typ)* Steifleinen *n*, Leinen *n*, Kaliko *m (zum Binden)*; 2. *(Text)* Druckperkal *m* minderer Qualität
c. **bellows** Kalikobalgen *m (Kamera)*
c. **weave** Leinwandbindung *f*, Kattunbindung *f*
caliduct Heißwasserheizungsrohr *n*; Dampfheizungsrohr *n*
californium *(Ch)* Kalifornium *n*
caliper / to messen *(mittels verstellbarer Lehre)*
caliper Taster[zirkel] *m*
c. **compasses** Tasterzirkel *m*
c. **for inside and outside measurements** Innen- und Außentaster *m*, Achterzirkel *m*, Doppeltaster *m*
c. **gauge** Gabellehre *f*, Rachenlehre *f*, Tast[er]lehre *f*, verstellbares Strichmaß *n*
c. **jaw** Schenkel *m (Schieblehre)*
c. **joint** Tastergelenk *n*
c. **leg** Tasterschenkel *m*
c. **of paper** Papierdicke *f*
c. **point** Tasterende *n*, Tasterspitze *f*
c. **profile recorder** Dickenmesser *m (Papier, Karton)*
c. **rule** Schieblehre *f*, Schublehre *f*, Werkstattschieblehre *f*

c. **scale** Schieblehre *f*, Schublehre *f*
c. **setting** Tastereinstellung *f*
c. **square** Schieblehre *f*, Schublehre *f*
calipering Ablehren *n*, Lehrung *f*, Messung *f* *(mittels verstellbarer Lehre)*
calipers Dickenmesser *m*, Greifzirkel *m*, Innenzirkel *m*, Krummzirkel *m*, Taster *m*, Tastzirkel *m*
calistor Kalistor *m (fotoelektronischer Vierpol)*
calk / to 1. [ab]dichten, Fugen verstopfen; 2. [durch]pausen; 3. Stollen einschrauben; (*s. a.* caulk)
calkin Stollen *m (Hufeisen)*
calking *s.* caulking
call / to *(Fmt)* anrufen
c. **at a harbour** einen Hafen anlaufen
c. **back** *(Fmt)* rückfragen
c. **up** *(Fmt)* anrufen
c. **up a subscriber** *(Fmt)* einen Teilnehmer anrufen
call Ruf *m*, Anruf *m*; Telefongespräch *n*; Signal *n*
c. **address** *(Fmt)* Rufadresse *f*
c. **announcer system** *(Fmt)* Nummernanzeigeeinrichtung *f*
c. **bell** Alarmglocke *f*; *(Fmt)* Rufwecker *m*, Anrufwecker *m*
c. **box** Fernsprechkabine *f*, Fernsprechzelle *f*, Sprechzelle *f*
c. **circuit** *(Fmt)* Dienstleitung *f*
c. **circuit selector** *(Fmt)* Dienstleitungswähler *m*
c. **control watch** *(Fmt)* Gesprächskontrolluhr *f*
c.**-count meter** Belegungszähler *m*, Gesprächszähler *m*
c. **counter** Gesprächszähler *m*
c. **detector** *(Fmt)* Anruffeststeller *m*
c. **device** *(Fmt)* Rufeinrichtung *f*
c. **distributor** *(Fmt)* Anrufverteiler *m*
c. **fee** Gesprächsgebühr *f*
c. **finder** *(Fmt)* Anrufsucher *m*
c.**-finder relay** *(Fmt)* Relaisanrufsucher *m*
c.**-in** *(Dat)* Anruf *m*, Ruf *m*
c. **indicator** *(Fmt)* Anrufklappe *f*; Fallklappentableau *n*; Rufanzeiger *m*
c. **letter** *(Fmt)* Rufzeichen *n*
c. **meter** *(Fmt)* Gesprächszähler *m*
c. **metering** *(Fmt)* Gesprächszählung *f*
c. **minute** *(Fmt)* Belegungsminute *f*
c. **number** Rufnummer *f*, Fernsprechnummer *f*; *(BMSR)* Anrufzahl *f*
c. **office** Fernsprechstelle *f*
c. **on hand** *(Fmt)* aufgespeicherter Ruf *m* *(Warteeinrichtung)*; unerledigter Anruf *m*
c. **sign** *(Fmt)* Meldeanrufzeichen *n*
c. **signal** 1. Anrufzeichen *n*; 2. Funkrufzeichen *n*
c. **signal battery** *(Fmt)* Rufbatterie *f*
c. **station** Fernsprechstelle *f*
calling *(Fmt)* Anrufen *n*, Ruf[en *f*] *m*
c. **and distress wave** *(Fmt)* Anruf- und Notwelle *f*
c. **cord** *(Fmt)* Rufschnur *f*
c. **dial** *(Fmt)* Nummernscheibe *f*, Wählscheibe *f*
c. **drop** *(Fmt)* Anrufklappe *f*
c. **frequency** *(Fmt)* Anruffrequenz *f*
c. **jack** *(Fmt)* Abfrageklinke *f*
c. **magneto** *(Fmt)* Anrufinduktor *m*
c. **plug** *(Fmt)* Rufstöpsel *m*
c. **position** *(Fmt)* Rufstellung *f*
calliper *s.* caliper
callnumber *s.* call number
calm Stille *f*, Windstille *f*, Kalme *f*
c. **belt** Kalmengürtel *m (Meteorologie)*
calm down / to nachlassen, abflauen *(Wind)*
calming agent Beruhigungsmittel *n (Stahlherstellung)*
calomel electrode Kalomelelektrode *f*
c. **half-cell** Kalomelelektrode *f*
Calor gas Flaschengas *n (Handelsname)*
caloric kalorisch; Warm-, Wärme-; thermisch
c. **heat unit** *(Ph)* Wärmeeinheit *f*
c. **power** Heizkraft *f*
c. **radiation** Wärmestrahlung *f*
c. **unit** *(Ph)* Wärmeeinheit *f*
c. **value** Heizwert *m*
calorie *(Ph)* Kalorie *f*
calorific wärmeerzeugend; Wärme-
c. **capacity** Wärmekapazität *f*
c. **effect** Heizeffekt *m*, kalorischer Effekt *m*
c. **energy** Wärmeenergie *f*, thermische Energie *f*
c. **heating value** Heizwert[zahl *f*] *m*
c. **power** Heizkraft *f*, Heizwert *m*
c. **value** Brennwert *m*, Heizwert[zahl *f*] *m*, Wärmewert *m*
calorimeter Kalorimeter *n*, Heizwertmesser *m*
c. **vessel** Kalorimetergefäß *n*
calorimetric kalorimetrisch
c. **bomb** *(Ph)* kalorimetrische Bombe *f*
c. **determination of temperature** kalorimetrische Temperaturbestimmung *f*
c. **measurement** Kalorimetermessung *f*, kalorimetrische Messung *f*, Kalorimetrierung *f*
calorized steel alitierter Stahl *m*
calorizing *(Met, Masch)* Alitieren *n*
calorstat Thermostat *m*
calory *s.* calorie
calotte Kugelkalotte *f*, Kugelschnitt *m*
calutron *(Kern)* Calutron *n* (*Cal*ifornia *U*niversity Cyclo*tron*)
calx *(Am)* Kalziumoxid *n*
calyx Sedimentrohr *n*, Schlämmrohr *n (Erdöl, Erdgas)*
c. **drill** Calyx-Bohrgerät *n (Schrotbohren)*

c.-**drilling** Schrotbohren *n*
cam 1. *(Masch)* Kurve *f*, Nocken *m*; Daumen *m*; Knagge *f*; 2. Schloßteil *n* eines Strickschlosses *(Strickmaschine)*; Platinenexzenter *m (Platine)*
c.-actuated kurvengesteuert, nockengesteuert
c. and bowl picking motion Exzenterschlag *m* durch Exzenter und Schlagwelle mit Rolle *(Webstuhl)*
c.-and-peg steering gear *(Kfz)* Schraubenganglenkung *f*, Spindellenkgetriebe *n*
c. assembly [of knitting machine] Strickschloß *n (Strickmaschine)*
c. bit Exzenterkeil *m (Wirkmaschine)*
c. bowl *(Text)* Exzenterrolle *f*
c. box 1. Nockengehäuse *n*; 2. Schloßkasten *m (Strickmaschine)*
c. brake Kniehebelbremse *f*
c. case Nockengehäuse *n*
c. contact *(El)* Nockenkontakt *m*
c. contact surface Nockenflanke *f*, Nockenbahn *f*
c. contour Nockenform *f*
c. control Nockensteuerung *f*, Kurvensteuerung *f*
c.-controlled kurvengesteuert, nockenbetätigt
c. controller *(El)* Nockenfahrschalter *m*
c. curve Nocken[erhebungs]kurve *f*
c. cutting lathe Unrunddrehmaschine *f*
c. cutting machine Nockenformfräsmaschine *f*; Kurvenfräsmaschine *f*
c. disk Nockenscheibe *f*, Daumenscheibe *f*, Kurvenscheibe *f*
c. drive Nockenantrieb *m*
c. drive gear Antriebsrad *n* für den Schloßmantel *(Strickmaschine)*
c. drum Kurventrommel *f*, Nockentrommel *f*
c. dwell Nockenerhebungszeit *f*, Nockenverweilzeit *f*
c. flank Nockenflanke *f*
c. follower Exzenterrolle *f*; Nockenstößel *m*, Kurvenstößel *m*, Ventilstößel *m*
c.-follower guide Stößelführung *f*
c. for crossing layer *(Text)* Exzenter *m* für Kreuzwindung
c. for drum racking Exzenter *m* für Schalttrommel *(Strickmaschine)*
c. for tubular knitting *(Text)* Schloßteil *n* für Schlaucharbeit
c. gear Nockensteuerung *f*, Daumenantrieb *m*, Daumensteuerung *f*
c. grinding Nockenschleifen *n*
c. groove Schloßkanal *m (Strickmaschine)*
c. lead Nockensteigung *f*
c. lever for traverse rail *(Text)* Triebhebel *m* für die Fadenführerstange
c. lever with toothed sector for cutting motion *(Text)* Schneidexzenterhebel *m* mit Zahnsektor
c. lift Nockenhub *m*
c.-lifted durch Nocken angehoben
c. lobe Nockenbuckel *m*, Nockenerhebung *f*
c.-lock nockenverriegelt
c.-lock shank nockenverriegelter Schaft *m*
c. miller Kurvenfräsmaschine *f*
c. milling attachment Kurvenfräseinrichtung *f*
c.-operated switch *(El)* Nockenschalter *m*
c. path Nockenlaufbahn *f*, Kurvenbahn *f*; Steuerkurve *f*
c. piece for open shed *(Text)* Exzenterstück *n* für Offenfach
c. planing attachment Kurvenhobeleinrichtung *f*
c. plate Schloßplatte *f (Strickmaschine)*
c. profile Nockenprofil *n*
c. raceway Schloßkanal *m (Strickmaschine)*
c.-retaining ring *(Text)* Schloßteiltragering *m*
c. ring 1. Nockenring *m*; 2. *(Text)* Schloßmantel *m*
c. ring of sinker *(Text)* Platinenexzenterring *m*
c. roller Kurvenrolle *f*
c. section *(Text)* Schloßplatte *f*
c. section ring *(Text)* aus Teilen bestehender Schloßmantel *m*
c.-share Nockenschar *n (des Pflugs)*
c. spindle Nockenwelle *f*
c. system Strickschloß *n (Strickmaschine)*
c. throw Kurvenhub *m*
c. track Nockenlaufbahn *f*, Kurvenbahn *f*; *(Text)* Schloßkanal *m*
c. travel *(Kfz)* Nockenbahn *f*
c. wheel Nockenrad *n*
camber / to wölben, biegen, krümmen
camber *(Bw)* 1. Krümmung *f*, Wölbung *f*; 2. Pfeilhöhe *f*, Stich *m*; 3. Quergefälle *n*, Querneigung *f*, Überhöhung *f (Straßenbau)*; 4. *(Schiff)* Balkenbucht *f*
c. angle *(Kfz)* Radsturzwinkel *m*
c. arch *(Bw)* scheitrechter Bogen *m*
c. of a leaf spring Sprenghöhe *f* einer Blattfeder
c. ratio *(Schiff)* Wölbungsverhältnis *n (Propeller)*
cambered face flachgewölbte Hammerbahn *f*
c. flatter Knopfamboß *m*, Pilzamboß *m*
c. girder *(Brg)* Streckenausbau *m* mit gewölbter Kappe
c. slipway *(Schiff)* Kreisbahnhelling *f*
c. standing ways *(Schiff)* Kreisstapel *m*; gekrümmte Ablaufbahn *f*, Kreisablaufbahn *f*
camel *(Schiff)* Hebeleichter *m*, Hebetank *m*
c. hair Kamelhaar *n*
camera Kamera *f*, Fotoapparat *m*, Aufnahmegerät *n*

c. **amplifier** Kameraverstärker *m*
c. **angle** Aufnahmewinkel *m*
c. **aperture** Blendenöffnung *f*
c. **attachment** Kameraansatz *m*
c. **body** Kameragehäuse *n*
c. **buckle** Filmsalat *m*
c. **crane** Kamerabühne *f*, Hebetisch *m*
c. **dolly** Kamerawagen *m*
c. **exposure** Belichtungszeit *f*
c. **extension** Kameraauszug *m*
c. **filter** Aufnahmefilter *n*
c. **frame** Kameragehäuse *n*
c. **hood** Sonnenblende *f*
c. **monitor** Kamerakontrollgerät *n*
c. **stand** Kamerastativ *n*
c. **tube** *(Fs)* Aufnahmeröhre *f*, Bildaufnahmeröhre *f*, Fernsehaufnahmeröhre *f*
camlike *(Masch)* kurvenförmig
camming *(Text)* Anordnung *f* der Strickschlösser
c. **arrangement** Steuerkurveneinrichtung *f*
camouflage Tarnung *f*
c. **material** *(Mil)* Tarnmittel *n*
campaign Reise *f (Hochofen)*
campanile Kampanile *m*, freistehender Glockenturm *m*
campeachy wood Campecheholz *n*
camphor Kampfer *m*
camphoric acid Kampfersäure *f*
c. **anhydride** Kampfersäureanhydrid *n*
camshaft Nockenwelle *f*, Steuerwelle *f*
c. **and servo-motor equipment** Nockensteuerung *f* mit Hilfsmotor
c. **bearing** Nockenwellenlager *n*
c. **casing** Nockenwellengehäuse *n*, Steuerwellengehäuse *n*
c. **chain drive** Nockenwellenkettenantrieb *m*
c. **chamber** Nockenwellengehäuse *n (Verbrennungsmotor)*
c. **controller** *(El)* Steuerschalter *m*; Nockenschalter *m*
c. **drive** Nockenwellenantrieb *m (Verbrennungsmotor)*
c. **gear** 1. Nockenwellenrad *n*; 2. *(Kfz)* Nockenwellen[schalt]werk *n*
c. **grinding machine** Nockenwellenschleifmaschine *f*
c. **roller** Nockenwellenrolle *f*
c. **sprocket** Nockenwellenkettenrad *n*
c. **switch** Nockenschalter *m*
c. **tester** Nockenwellenmeßgerät *n*
c. **testing element** Nockenwellenmeßeinrichtung *f*
c. **timing gear** *(Kfz)* Nockenwellenantriebsrad *n*
can / to 1. kapseln, umhüllen, ummanteln, einhülsen; 2. *(Am)* in Dosen einmachen *(Lebensmittel)*

can 1. Hülle *f*, Hülse *f*, Kapsel *f*; 2. Kanister *m*, Kanne *f*; 3. *(Am)* Konservendose *f*; 4. *(Fmt)* Abschirmbecher *m*; 5. *(Text)* Spinnkanne *f*; 6. Filmkassette *f*
c. **anode** Topfanode *f (Elektronenröhre)*
c. **creel** *(Text)* Kannengatter *n*
c. **drier** *(Ch)* Zylindertrockner *m*
c. **gill box** *(Text)* Nadelstabstreckwerk *n* mit Kanne; Topfstrecke *f*
c. **making machinery** *(Am)* Konservendosenherstellungsmaschinen *fpl*
c. **opener** Büchsenöffner *m*
c. **rupture** *(Kern)* Hüllenschaden *m*
c. **rupture detection** *(Kern)* Hüllenüberwachung *f*
c. **spindle** *(Text)* Topfspindel *f*
c. **spinning** *(Text)* Topfspinnen *n*
c. **spinning frame** *(Text)* Topfspinnmaschine *f*, Dosenspinnmaschine *f*
Canada balsam Kanadabalsam *m*
C. **turpentine** *s.* Canada balsam
Canadian asbestos Chrysotilasbest *m*
C. **canoe** Kanadier *m (Sportboot)*
canal 1. Kanal *m*; Rohr *n*; Schacht *m (für Abwässer)*; 2. *(künstlich angelegter)* Kanal, Binnenschiffahrtskanal *m*; Bewässerungskanal *m*
c. **construction** Kanalbau *m*
c. **cross-section** Kanalquerschnitt *m*
c. **fees** Kanalgebühren *fpl*
c. **lining** Kanalausmauerung *f*, Kanalauskleidung *f*
c. **lock** Kanalschleuse *f*
c. **lock chamber** Schleusenkammer *f*
c. **pumping station** Kanalpumpwerk *n*
c.**-ray analysis** *(Ph)* Kanalstrahlenanalyse *f*
c.**-ray discharge** *(Ph)* Kanalstrahlenentladung *f*
c.**-ray particle** *(Ph)* Kanalstrahl[teilchen *n*] *m*
c.**-ray tube** *(El)* Kanalstrahlröhre *f*
c. **rays** Kanalstrahlen *mpl*
c. **section** *(Ph)* Kanalprofil *n*
canalization Kanalisierung *f (Flußbau)*
cancel / to 1. ausstreichen, durchstreichen; löschen; 2. *(Math)* kürzen; [sich] wegheben; 3. stornieren *(Auftrag)*
c. **a number out of a fraction** *(Math)* einen Bruch durch eine Zahl kürzen
c. **out** *(Math, Ph)* [sich] aufheben, ausgleichen, [sich] wegheben
cancellated structure Netzwerkstruktur *f*
cancellation *(Math)* 1. Kürzung *f (eines Bruches)*; 2. Aufhebung *f*, Wegstreichen *n*
c. **key** *s.* cancelling key
cancelling key *(El)* Korrekturtaste *f*, Rücknahmetaste *f*, Löschtaste *f*; *(Büro)* Stornotaste *f*
c. **voltage drops** sich aufhebende Spannungsabfälle *mpl*

canch *(Geol, Brg)* Nachreißstelle *f*, Versatzstrecke *f*
candela *(Opt)* Candela *f (n)*
candelilla wax Candelillawachs *n*
candle Kerze *f*, Licht *n*
  c. coal *s.* cannel coal
  c. filter *(Ch)* Filterkerze *f*
candlewick Lampendocht *m*
cane Rohr *n*; Schilf[rohr] *n*; Zuckerrohr *n*
  c. fender *(Schiff)* Rohrgeflechtfender *m*
  c. mat Schilfmatte *f*, Rohr[stäbchen]matte *f*
  c. plaiting Rohrgeflecht *n*
  c. splitting machine Rohrspaltmaschine *f*
  c. sugar Rohrzucker *m*
canned 1. aufgenommen, aufgezeichnet; 2. *(Am)* in Blechdosen, in Dosen *(konservierte Lebensmittel)*
  c. meat Büchsenfleisch *n*
  c. milk Büchsenmilch *f*
  c. motor pump lecksichere Motorpumpe *f*
  c. music Konservenmusik *f*
  c. rotor gekapselter Rotor[läufer] *m*
  c. valve 1. *(El)* Röhre *f* mit Glaskolben und Metallumhüllung; 2. Metallröhre *f*, Röhre mit Metallkolben
cannel coal Kännelkohle *f*
cannery *(Am)* Konservenfabrik *f*
canning 1. *(Am)* Einlegen *n*; Konserven-; 2. *(sl)* Tonaufnahme *f*, Aufzeichnung *f*
  c. industry *(Am)* Konservenindustrie *f*
  c. material *(Kern)* Hülsenmaterial *n*
Cannizzaro reaction *(Ch)* Cannizzarosche Reaktion *f*
cannula Kanüle *f*
canoe stern *(Schiff)* Kanuheck *n*, Wikingerheck *n*
canon *(Typ)* 48 britische Punkte *mpl (Schriftgröße)*
canonical kanonisch
  c. distribution *(Ph)* kanonische Verteilung *f*
  c. equation *(Ph)* kanonische Gleichung *f*
  c. representation *(Ph)* kanonische Darstellung *f*
canonically conjugated quantities *(Ph)* kanonisch konjugierte Größen *fpl*
canopy *(Bw)* Kragdach *n*, Vordach *n*; Kabinendach *n*; Fallschirmkuppel *f*
  c.-switch *(El)* Führerstandsautomat *m*
cant / to kanten, kippen; abfasen
cant *(Bw)* 1. geneigte Fläche *f*, Schräge *f*, Schrägung *f*; 2. Überhöhung *f (Straßenbau)*
  c. bay *(Bw)* sechseckiger (achteckiger) Erker *m*
  c. deck *(Holz)* Kantholzboden *m*
  c. frame *(Schiff)* [stählernes] Kantspant *n*
  c. timber *(Schiff)* [hölzernes] Kantspant *n*
canted *(Bw)* abgefast, abgekantet, abgeschrägt; geneigt, schräg
  c. beam abgefaster Balken *m*, Balken mit behauenen Kanten
  c. pillar mehreckiger Pfeiler *m*
  c. wall spitz- oder stumpfwinklige Mauerecke *f*
canteen Großküche *f*, Kantine *f*, Werkküche *f*; Mensa *f*
cantilever / to *(Bw)* auskragen, vorkragen; frei vorbauen, im Freivorbau errichten *(Brückenbau)*
cantilever 1. *(Masch)* Ausleger *m*; Überhang *m*; 2. *(El)* Mastausleger *m*; 3. *(Bw)* Auskragung *f*, Kragarm *m*; Freiträger *m*, Kragträger *m*
  c. beam *(Bw)* Freiträger *m*, Kragträger *m*
  c. bridge Auslegerbrücke *f*
  c. construction Freivorbau *m (Brückenbau)*
  c. crane Auslegerkran *m*
  c. detector *(El)* Spiralfederdetektor *m*
  c. fashion *(Bw)* auskragende Bauart *f*
  c. frame *(Schiff)* Toppseitenspant *n*, Kragspant *n*
  c. load einseitig freitragende Belastung *f*
  c. method Freivorbau *m (Brückenbau)*
  c. monoplane *(Flg)* freitragender Eindecker *m*
  c. principle Freivorbau *m (Brückenbau)*
  c. retaining wall *(Bw)* Winkelstützmauer *f*
  c. roof Kragdach *n*
  c. spring *(Kfz)* Auslegerfeder *f*
  c. steps *(Bw)* eingespannte (freitragende) Stufen *fpl*
  c. undercarriage freitragendes Fahrwerk *n*
cantilevered 1. *(Bw)* ausgekragt, vorgekragt; 2. *(Brg)* vorgepfändet, freitragend vorgebaut
  c. and suspended beam *(Bw)* 1. Gelenkträger *m*, Gerberträger *m*; 2. Koppelträger *m*
  c. beam *(Bw)* Freiträger *m*, Kragträger *m*
  c. retaining wall *(Bw)* Winkelstützmauer *f*
canvas 1. Verdeckleinwand *f*; 2. Segeltuch *n*
  c. boat Faltboot *n*
  c. duct Segeltuchschlauch *m*
canyon *(Kern)* Betontunnel *m*, Lagerkammer *f*
cap / to 1. bedecken; 2. sockeln *(Elektronenröhren)*; 3. runderneuern *(Kfz-Reifen)*
cap 1. *(Masch)* Abschlußkappe *f*, Deckel *m*, Haube *f*, Kappe *f*; 2. *(Bw)* Abdeckung *f*, Aufsatz *m*, Kopfplatte *f*; 3. *(El)* Sockel *m (Lampe)*; 4. *(El, Fmt)* Anschlußkontakt *m* an der Spitze des Röhrenkolbens; 5. Kappe *(Schuh)*; 6. *(Brg)* Sprengkapsel *f (Schießarbeit)*; 7. *(Typ)* Anfangsbuchstabe *m*, Versal *m*; 8. *(Math)* Kalotte *f*; 9. Halter *m (Schneidbacke)*
  c.-and-pin [type] insulator Kappenisolator *m*, Stützisolator *m*
  c. bar *(Text)* Oberwalzenführung *f*
  c. bolt Lagerdeckelschraube *f*, Lagerbolzen *m*

c. check buckle Einschubangel *f (Gattersäge)*
c. crown Lagerdeckel *m*
c. jewel Deckstein *m (Uhr)*
c. lamp *(Brg)* Kopflampe *f*, Mützenlampe *f*, Stirnlampe *f*
c. lining Mützenfutter *n*
c. nut Kapselmutter *f*, Hutmutter *f*
c. of sinker *(Text)* oberer Abschluß *m* des Platinenringes
c.-piece *(Brg)* Kappe *f (Ausbau)*
c. rock *(Brg)* Abraum *m*
c. screw Kopfschraube *f*, Überwurfmutter *f*, Deckelschraube *f*
c. sleeve Sockelbuchse *f (Katodenstrahlröhre)*
c. spinning frame Glockenspinnmaschine *f*
c.-type insulator Kappenisolator *m*
capability Fähigkeit *f*, Leistungsfähigkeit *f*
c. of oxidation Oxydierbarkeit *f*
capable of being copied kopierfähig
c. of direct reading direkt ablesbar
c. of retaining the cutting edge schneidhaltig
capacitance *(El)* Kapazitanz *f*, Kapazität *f*, kapazitiver Blindwiderstand (Widerstand) *m*
c. altimeter kapazitiver Höhenmesser *m*
c. balancing *(Fmt)* Querausgleich *m*
c. branch kapazitiver Zweig *m*
c. bridge Kapazitätsmeßbrücke *f*
c. deviation Kapazitätsunterschied *m*
c. earth Gegengewicht *n*; kapazitive Erdung *f*
c. for shock protection Berührungsschutzkapazität *f*
c. gauge *(BMSR)* kapazitiver Geber *m*
c. meter Kapazitätsmesser *m*
c. ratio Kapazitätsverhältnis *n*
capacitive kapazitiv
c. feedback kapazitive Rückkopplung *f*
c. load kapazitive Belastung (Last) *f*
c. reactance kapazitiver Blindwiderstand *m*, Kapazitanz *f*, kapazitive Reaktanz *f*
c. sawtooth generator kapazitiver Sägezahngenerator *m*
c. shunt kapazitiver Nebenschluß *m*
c. tuning kapazitive Abstimmung *f*
c. voltage divider kapazitiver Spannungsteiler *m*
capacitively coupled kapazitiv gekoppelt
capacitor *(El)* Kondensator *m*
c. aerial Kondensatorantenne *f*
c. bank Kondensatorbatterie *f*
c. fed track circuit *(Eis)* kapazitiv gespeister Gleisstromkreis *m*
c. input filter kapazitives Eingangsfilter *n*; pi-Tiefpaß *m*; Eingangssiebschaltung *f*
c. loudspeaker Kondensatorlautsprecher *m*, elektrostatischer Lautsprecher *m*
c. motor Kondensatormotor *m*
c. of the grid circuit Gitterkreiskondensator *m*
c.-run motor Kondensatormotor *m (Kapazität wird nach Anlauf verringert)*
c. split-phase motor Kondensatormotor *m (Kondensator bleibt eingeschaltet)*
c.-start motor Motor *m* mit Kondensatoranlauf (kapazitiver Hilfsphase)
c. starting Anlassen *n* mit Hilfsphase
c. transformer kapazitiver Spannungswandler *m*
c. with indication of outer coating Kondensator *m* mit Kennzeichnung des Außenbelages
capacity 1. Kapazität *f*, Leistungsvermögen *n*, Leistungsfähigkeit *f*; Fassungsvermögen *n*; Tragfähigkeit *f*, Tragvermögen *n*; 2. *(El)* Kapazität (*s. a.* capacitance); 3. *(Masch)* Leistung *f*, Ausbringung *f*; Arbeitsbereich *m*; 4. Schluckvolumen *n (Hydraulikmotor)*; 5. Fördermenge *f (Pumpe)*; 6. *(Dat)* Speicherfähigkeit *f*, Speichervermögen *n*; maximale Stellenzahl *f*; 7. *(Schiff)* Fassungsvermögen *n*, Räumte *f*, Inhalt *m (eines Hohlraumes)*
c. balance *(El)* Kapazitätssymmetrie *f*
c. balancing *(Fmt)* Kapazitätsausgleich *m*
c. bridge Kapazitätsmeßbrücke *f*
c. cooling Verdampfungskühlung *f*
c. coupling kapazitive Kopplung *f*
c. current heating kapazitive Hochfrequenzerhitzung (Hochfrequenzerwärmung) *f*
c. drift with time *(El)* zeitliche Kapazitätsveränderung *f*
c. effect Kapazitätseffekt *m (insbesondere bei Hochantennen)*, Hochantenneneffekt *m*
c. factor *(El)* Lastfaktor *m*, Belastungsfaktor *m*
c. for chip removal *(Masch)* Schnittleistung *f*, Spanleistung *f*
c. for forming foam Schaumbildungsvermögen *n*
c. measure Hohlmaß *n*
c. meter Leistungsmeßgerät *n*, Leistungsmesser *m*
c. metering device *(El)* Kapazitätsmesser *m*, Leistungsmesser *m*
c. network *(El)* Hochpaß *m*
c. of a shop's machinery Maschinenkapazität *f* eines Betriebs
c. of a transmitter Senderleistung *f*
c. of heat *(Ph)* Wärmekapazität *f*
c. of material 1. Stangendurchlaß *m*, Werkstoffdurchlaß *m*; 2. Schnittdurchlaß *m (Holzbearbeitungsmaschine)*
c. plan *(Schiff)* Ladeplan *m*, Stauungsplan *m*
c. rating Leistungs[angabe] *f*; Leistungsberechnung *f*

c. **reactance** *(El)* kapazitiver Blindwiderstand *m*
c. **to earth (ground)** *(El)* Kapazität *f* gegen Erde, Erdkapazität *f*
c. **top** *(El)* kapazitive Kopflast *f*
cape chisel Kreuzmeißel *m*
c. **keyseating chisel** *s.* c. chisel
capillarity Kapillarität *f*; Kapillarkraft *f*; Kapillarwirkung *f*
capillary Kapillare *f*, Kapillarrohr *n*
c. **attraction** Kapillaranziehung *f*, Kapillarwirkung *f*
c. **chink** *s.* c. crack
c. **column** *(Ch)* Kapillarsäule *f*; Trennsäule *f* *(Gas-Chromatografie)*
c. **combustion tube** Verbrennungskapillare *f*
c. **constant** Kapillaritätskonstante *f*
c. **crack** Haarriß *m (Werkstoffehler)*
c. **depression** Kapillardepression *f*
c. **electrometer** Kapillarelektrometer *n*
c. **elevation** *s.* c. rise
c. **force** Kapillarkraft *f*
c. **pressure** Kapillardruck *m*
c. **rise** kapillarer Anstieg (Aufstieg) *m*, kapillare Steighöhe *f*, [kapillare] Saughöhe *f*
c. **structure** adrige Struktur *f*, Faserstruktur *f*
c. **tube** Kapillarrohr *n*, Kapillare *f*, Kapillargefäß *n*, Haarrohr *n*, Haarröhrchen *n*; Kanüle *f*
c. **tube viscosimeter** Kapillarviskosimeter *n*
c.**[-type] viscosimeter** *s.* c. tube viscosimeter
c. **water** Kapillarwasser *n*
capital *(Bw)* Kapitell *n*, Kapitäl *n*
c.**-heading** *(Typ)* Haupttitel *m*
c.**-letter** *(Typ)* Großbuchstabe *m*, Majuskel *f*
c. **ship** Großkampfschiff *n*
capitalized *(Typ)* in Versalien (Großbuchstaben) gesetzt
caplastometer Kapillar-Rheometer *n*, Viskosimeter *n*
capless lamp sockellose Glühlampe *f*
capped shell *(Mil)* Kappengeschoß *n*, Haubengeschoß *n*
capping 1. *(Bw)* Abdeckung *f*; 2. *(Am, Brg)* Deckgebirge *n*; 3. Kanisterschließen *n*
c. **beam** *(Bw)* Holm *m*
c. **mass** *(Brg)* Deckgebirge *n*
capraldehyde Kaprinaldehyd *m*
caprate Kaprat *n*, Salz *n* (Ester *m*) der Kaprinsäure
capreolate rankenförmig
capric acid Kaprinsäure *f*
caprilate Kaprylat *n*, Salz *n* (Ester *m*) der Kaprylsäure
caproaldehyde Kapronaldehyd *m*
caproate Kapronat *n*, Salz *n* (Ester *m*) der Kapronsäure
capro[n]ic acid Kapronsäure *f*
caprylaldehyde Kaprylaldehyd *m*
caprylate Kaprylat *n*, Salz *n* (Ester *m*) der Kaprylsäure
caprylic acid Kaprylsäure *f*
capsicum Kapsikum *n*, schwarzer (spanischer) Pfeffer *m*
capsize / to *(Schiff)* kentern
capstan 1. *(Masch)* Auflaufhaspel *f*, Ankerwinde *f*, Haspel *f*, Drehkreuz *n*, Spill *n*, Winde *f*; 2. *(Schiff)* Spill mit vertikaler Welle, Vertikalspill *n*; 3. Göpel *m*; 4. *(El)* Tonrolle *f (bestimmend für die Transportgeschwindigkeit; Magnettontechnik)*; 5. automatisch schaltender Revolver *m*
c. **bar** *(Schiff)* Spillspake *f*
c. **barrel** *(Schiff)* Spillkörper *m*
c. **handle** Drehkreuz *n*
c. **head** *(Schiff)* Spillkopf *m*
c. **head screw** Knebelschraube *f*, Kreuzlochschraube *f*
c. **horse** Göpelpferd *n*
c. **lathe** Revolverdrehmaschine *f* mit Revolverschlitten auf Support
c. **screw** Kreuzlochschraube *f*
c. **slide** *(Masch)* Revolverschlitten *m*
c. **spike** *(Masch)* Stellstift *m*
c. **wheel** Handkreuz *n*
c. **windlass** *(Schiff)* Ankerspill *n*
capstone Abdeckstein *m (für Kabel)*
capsular chamber Membrankapsel *f*
capsule 1. Kapsel *f*, Dose *f*, Druckdose *f*; 2. *(El)* Membrandose *f*
c. **clay** Kapselton *m (Porzellanherstellung)*
caption Bildunterschrift *f*, Legende *f*; Blickfang *m*
captive balloon Fesselballon *m*
c. **lock washer** Sicherungsscheibe *f*
capture 1. *(Kern)* Einfang *m*; 2. *(Schiff)* Aufbringung *f*, Kaptur *f*, Prise *f*
c. **area** Absorptionsfläche *f (von Antennen)*
c. **cross section** *(Kern)* Einfangquerschnitt *m*
c. **cross section for neutrons** Einfangquerschnitt *m* für Neutronen, Neutroneneinfangquerschnitt *m*
c. **efficiency** *(Kern)* Einfangausbeute *f*
c. **mean free path** *(Kern)* mittlere freie Einfangweglänge *f*
c. **probability** *(Kern)* Einfangwahrscheinlichkeit *f*
c. **radiation** *(Kern)* Einfangstrahlung *f*
c. **reaction** *(Kern)* Einfangprozeß *m*, Einfangreaktion *f*
captured air bubble vessel CAB-Schiff *n (Schiff, das sich auf den unter dem Schiffskörper austretenden Luftblasen vorwärts bewegt)*
c. **electrons** eingefangene Elektronen *npl*
car 1. Wagen *m*, Auto *n*; Fahrzeug *n*; 2. Waggon *m*; 3. *(Brg)* Förderwagen *m*, Hunt *m*; 4.

Luftschiffgondel *f*; Luftschiffführergondel *f*; 5. Förderkorb *m*, Plattform *f*; Fahrkorb *m*; Fördermittel *n*
c. aerial Wagenantenne *f*
c.-axle Wagenachse *f*
c. battery Autobatterie *f*
c. body Karosserie *f*
c.-bottom furnace *(Met)* Wagenherdofen *m*
c.-bottom hearth *(Met)* ausfahrbarer Herd *m*
c. carrier Autofrachter *m*, Autotransportschiff *n*
c. casting *(Met)* Wagenguß *m*
c. changing *(Brg)* Wagenwechsel *m*
c.-deck Wagendeck *n*, Autodeck *n* *(eines Schiffes)*
c.-door electric contact *(Kfz)* Türkontakt *m*
c. dumper Waggonkipper *m*, Wagenkipper *m*
c. ferry Wagenfähre *f*, Kraftwagenfähre *f*, Autofähre *f*, Trajektschiff *n*
c. gear Kraftwagengetriebe *n*
c.-hearth furnace *(Met)* Wagenherdofen *m*
c. heater Wagenheizer *m*
c. heating installation Fahrzeugheizungsanlage *f*
c. jack Wagenheber *m*, Wagenwinde *f*, Wagenhebebock *m*
c. launching *(Flg)* Autoschleppstart *m*
c. lift Hebebühne *f*; Wagenheber *m*
c. maintenance and overhaul shop *(Eis)* Bahnbetriebswagenwerk *n*
c. owner Kraftfahrzeugbesitzer *m*, Kraftfahrzeughalter *m*
c. park Parkplatz *m*
c. polish Lackpflegemittel *n*
c. radio Autoradio *n*, Kraftwagenempfänger *m*
c. reamer Nietlochreibahle *f*
c. shielding Autoentstörung *f*
c. spotter *(Brg)* Wagenvorziehvorrichtung *f*, Zugverholeinrichtung *f*
c. spotting *(Brg)* Zugverholung *f*, Vorziehen *n* von Wagen
c. starter *(Kfz)* Starter *m*
c. stop *(Brg)* Wagenfeststellvorrichtung *f* *(Förderkorb)*
c. switching *(Brg)* Wagenwechsel *m*
c. wired for radio mit Antenne ausgerüsteter Wagen *m*
caracole Wendeltreppe *f*
carat Karat *n*
caraway oil Kümmelöl *n*
carbamate Karbamat *n*, Salz *n* der Karbaminsäure
carbamic acid Karbaminsäure *f*
carbaminate Karbaminat *n*, Salz *n* der Karbaminsäure
carbide 1. Karbid *n*, Kalziumkarbid *n*; 2. Metallkarbid *n*; Hartmetall *n*
c. band zeilenförmige Karbideinlagerung *f*, Karbidzeile *f*
c. boring tool Hartmetallaufbohrwerkzeug *n*
c. cutter Hartmetallfräser *m*
c. cutting edge Hartmetallschneide *f*
c. cutting material Sinterkarbidschneidwerkstoff *m*; Hartmetall *n*
c. cutting tool Hartmetallwerkzeug *n*
c.-forming karbidbildend
c. grinding machine Hartmetallschleifmaschine *f*
c. insert Hartmetalleinsatz *m*; Hartmetalleinlage *f*; Hartmetalleinsatzmeißel *m*; Hartmetallzahnungseinsatz *m*
c. inserted bit Hartmetallschneide *f*, Hartmetalleinsatzmeißel *m*
c. lip Hartmetallschneide *f* *(eines Bohrers)*
c. milling Fräsen *n* mit Hartmetallwerkzeug
c. of iron Eisenkarbid *n*
c. planing tool Hartmetallhobelmeißel *m*
c. reamer Hartmetallreibahle *f*
c. soldering outfit Karbidlötapparat *m*
c. tip Hartmetallschneide *f* *(Meißel)*
c. tip bit Hartmetallschneide *f* *(Bohrkrone)*
c.-tipped hartmetallbestückt
c.-tipped gun drill Einlippenbohrer *m* [mit Hartmetallschneide]
c.-tipped tool Hartmetallwerkzeug *n*, hartmetallbestücktes Werkzeug *n*, hartmetallbestückter Meißel *m*
c. tipping Hartmetallauflage *f* *(Arbeitsvorgang)*
c. tips Hartmetallbestückung *f*
c.-to-water gas generator Azetylengasentwickler *m*, Karbideinwurfentwickler *m*
c. tool Hartmetallwerkzeug *n*; Hartmetallmeißel *m*
c. tool grinding machine Hartmetallschleifmaschine *f*
c. tool tip Hartmetallschneide *f* *(Meißel)*
c.-tooled mit Hartmetallwerkzeugen ausgerüstet
c. tooling Ausrüstung *f* mit Sinterhartmetallwerkzeugen
c. twist drill Hartmetallspiralbohrer *m*
carbithionic acid Thiokohlensäure *f*
carboatomic ring *(Ch)* Kohlenstoffring *m*
carbocoal tar Tieftemperaturteer *m*
carbocycle *(Ch)* Kohlenstoffring *m*
carbohydrate Kohlehydrat *n*
c. metabolism Kohlehydratstoffwechsel *m*
carbolic acid Karbolsäure *f*, Phenol *n*
carbolineum Karbolineum *n* *(Holzschutz)*
carbon 1. Kohlenstoff *m*; 2. Kohleelektrode *f*, Kohlestab *m*
c. arc Kohlelichtbogen *m*
c.-arc electrode Kohleelektrode *f* für elektrische Lichtbogenschweißung
c.-arc lamp Kohlebogenlampe *f*

c.-arc welding Kohlelichtbogenschweißen *n*
c. bar Kohlestab *m*
c. black Druckerschwärze *f*; Gasruß *m*, Ruß *m*
c. bleach Behandlung *f* *(Bleichen, Entfärben, Reinigen)* mit Aktivkohle
c. brick *(Bw)* Kohlenstoffstein *m*
c. brush *(Elm)* Graphitbürste *f*, Kohlebürste *f*
c. chain *(Ch)* Kohlenstoffkette *f*
c. compound *(Ch)* Kohlenstoffverbindung *f*
c.-consuming cell *(El)* Kohlenstoffelement *n*
c. contact pick-up Kohlemikrofontonabnehmer *m*
c. content [of steel] Kohlenstoffgehalt *m*; Zähgehalt *m* [von Stahl]
c. copy *(Büro)* Durchschlag *m*
c. crucible Graphittiegel *m*, Kohletiegel *m*
c. deposit Kohleablagerung *f*; Ölkohleansatz *m*
c. deposited resistor *(El)* Kohleschichtwiderstand *m*
c. deposits Ölkohle *f* *(Erdöl)*
c. dichloride Perchloräthylen *n*, Tetrachloräthylen *n*
c. dioxide Kohlendioxid *n*; Kohlensäure *f*
c. disulfide Schwefelkohlenstoff *m*
c. dust Kohlenstaub *m*
c. dust microphone Kohlenstaubmikrofon *n*
c. dust transmitter Kohlenstaubmikrofon *n*
c. electrode Kohleelektrode *f*, Kohlestab *m*
c. filament Kohlefaden *m*
c. filament lamp Kohlefadenlampe *f*
c.-film resistor Kohleschichtwiderstand *m*
c. formation Kohlebildung *f*, Ölkohlebildung *f*
c.-forming properties Verkokungsneigung *f*
c. grain Kohlegrieß *m*
c. granule microphone Kohlenkörnermikrofon *n*
c. granules Kohlekörner *npl*, Kohlegrieß *m*
c. microphone Kohlemikrofon *n*
c. mike Kohlemikrofon *n*
c. monoxide Kohlen[mon]oxid *n*
c. oxysulfide Kohlen[stoff]oxysulfid *n*
c. paper Kohlepapier *n*, Durchschlagpapier *n*
c. pickup *(Ch, Met)* Kohlenstoffaufnahme *f*
c. pile 1. *(Kern)* graphitmoderierter Reaktor *m*, Graphitreaktor *m*; 2. *(El)* Kohledrucksäule *f*, Kohlewiderstand[ssäule *f*] *m*
c. pile resistance Kohlewiderstand *m*
c.-pile voltage regulator Kohledruckspannungsregler *m*
c. regulator *(El)* Kohledruckregler *m*
c. residue Verkokungsrückstand *m*
c. residue test Conradson-Test *m*, Koksbestimmung *f* nach Conradson
c. resistor *(Am)* Massewiderstand *m* für Hochfrequenz *(Glanzkohleschicht)*
c. restoration Wiederaufkohlung *f*
c. rod Kohlestift *m*; Kohlestab *m*
c. steel Kohlenstoffstahl *m*, unlegierter Stahl *m*
c. sulfide Schwefelkohlenstoff *m*
c. switch *(El)* Vorkontaktschalter *m*
c. tetrachloride Kohlenstofftetrachlorid *n*, Tetra *n*, Tetrachlorkohlenstoff *m*, Tetrachlormethan *n*
c. tetrahalide Kohlenstofftetrahaloid *n*
c. tool steel unlegierter Werkzeugstahl *m*
c. value Maßzahl *f* für Verkokungsneigung *(z. B. Conradson-Zahl)*
carbonaceous 1. kohlenartig; kohlen[stoff]haltig, kohlig; 2. schwärzlich, verbrannt, hart
c. deposits kohleartige Rückstände *mpl*
c. ironstone *(Met)* Kohleneisenstein *m*
c. refuse *(Brg)* Vergasungsberge *pl*
c. shale *(Geol, Brg)* Brandschiefer *m*
carbonado Karbonado *m* *(Diamant)*
carbonatation *(Ch)* Karbonatbildung *f*
carbonate / to karbonisieren, mit Kohlendioxid sättigen
carbonate Karbonat *n*, Salz *n* der Kohlensäure
c. hardness Karbonathärte *f*, vorübergehende (temporäre) Härte *f*
carbonated kohlensäurehaltig
c. juice saturierter Saft *m*
carbonation juice Schlammsaft *m* *(Zuckerindustrie)*
carbonator Saturationsapparat *m* für Kohlensäure *(Zuckerindustrie)*
carbonic acid Kohlensäure *f*
c. oxide Kohlenoxid *n*
carbonization Verkokung *f*, Durchkohlung *f*; Entgasung *f*; Inkohlung *f*; Trockendestillation *f*; Verkohlung *f*
c, gas Schwelgas *n*
c. involving gas recirculation Spülgasschwelung *f*
c. of bituminous coal Steinkohlendestillation *f*
c. of lignite Braunkohlenschwelung *f*
c. of wood Holzverkohlung *f*
c. plant Kokerei[anlage] *f*, Verkokungsanlage *f*
c. zone Schwelzone *f*
carbonize / to abschwelen, ankohlen; karbonisieren, [ver]schwelen, verkohlen, verkoken
c. under vacuum abschwelen
c. under vacuum at low temperature verschwelen
carbonized bur *(Text)* verkohlte Klette *f*
c. form Durchschreibeformular *n*
c. lignite Braunkohlenkoks *m*
carbonizing Verkokung *f*, Garen *n*
c. bath *(Text)* Karbonisierflotte *f*
c. chamber 1. *(Brg)* Verkokungskammer *f*; 2. *(Text)* Karbonisierkammer *f*

c. cleavage *(Brg)* Verkokungsnaht *f*
c. drum *(Text)* Karbonisiertrommel *f*
c. flame reduzierende Flamme *f*
c. liquor *(Text)* Karbonisierlauge *f*
c. machine *(Text)* Karbonisiermaschine *f*
c. period *(Brg)* Garungszeit *f*, Verkokungszeit *f*
c. qualities Verkokungsneigung *f*
c. split *(Brg)* Verkokungsnaht *f*
c. stain *(Text)* Karbonisierfleck *m*
c. stove *(Text)* Karbonisierofen *m*
c. time *(Brg)* Garungszeit *f*, Verkokungszeit *f*
carbonyl iron Karbonyleisen *n*
c. process Karbonylverfahren *n*
c. sulfide Kohlenoxysulfid *n*
carbonylate / to *(Ch)* Karbonylgruppe einführen
carborundum Karborund[um] *n*
c. paste Einschleifpaste *f*
c. wheel Karborundschleifscheibe *f*
carbothioic acid Thiokohlensäure *f*
carboxy group Karboxylgruppe *f*
carboxylic acid Karbonsäure *f*
carboy Glasballon *m*, Ballon *m*, Korbflasche *f*, Transportflasche *f*
carburation Gemischaufbereitung *f*
carburet / to *(Ch)* karburieren, vergasen; mit Kohlenstoff verbinden
carbureter *(Am) s.* carburetter
carburetion Gemischaufbereitung *f*
carburetter Vergaser *m* *(Verbrennungsmotor)*
c. adapter Vergaserzwischenstück *n*
c.-air heater Ansaugluftvorwärmer *m*, Vergaserluftvorwärmer *m*
c.-air scoop Vergaserluftfänger *m*
c. anti-icer Vergaserfrostschutzvorrichtung *f*
c. choke Mischrohr *n*
c. control Vergasergestänge *n*
c. engine Vergasermotor *m*
c. fire Vergaserbrand *m*
c. flange Vergaseranschlußstutzen *m*
c. float Vergaserschwimmer *m*
c. float chamber Schwimmergehäuse *n* des Vergasers
c. float spindle Schwimmernadel *f*
c. hot-air door Vergaserwarmluftklappe *f*
c. induction system Vergaserlufteinlaß *m*
c. linkage Vergasergestänge *n*
c. main body Vergasergehäuse *n*
c. needle valve Brennstoffnadelventil *n* im Vergaser, Kraftstoffnadelventil *n*
c. primer Tipper *m*
c. tickler Tipper *m*
c. tuning Vergasereinstellung *f*
carburetting *(Met)* Karburierung *f* *(im SM-Ofen)*
carburettor *s.* carburetter
carburization *(Met)* Einsatzhärtung *f*; Aufkohlung *f*
c. by pig iron Aufkohlung *f* mit Roheisen
carburize / to *(Met)* aufkohlen, einsetzen *(Einsatzstahl)*, einsatzhärten; mit Kohlenstoff anreichern
carburized case Einsatzhärteschicht[dicke] *f*
carburizer Aufkohlungsmittel *n*
carburizing Aufkohlung *f*
c. agent Einsatzpulver *n*, Einsatzmedium *n*, Einsatzmittel *n*
c. box Einsatzhärtekasten *m*
c. compound Einsatzmittel *n* *(Härten)*
c. furnace Einsatzhärteofen *m*
c. operation Einsatzhärtung *f*, Einsatzhärtevorgang *m*
c. oven Einsatzhärteofen *m*
c. steel Einsatzstahl *m*
c. temperature Einsatzhärtetemperatur *f*
carcase *(Bw)* bauliche Hülle *f*, Rohbau *m*
c. work Verlegen *n* von Installationsleitungen während des Rohbaues
carcass 1. Karkasse *f*; 2. *(El) s.* frame; 3. *(Bw) s.* carcase; 4. Gestell *n*; Gerippe *n*
c. of a ship Schiffsgerippe *n*
carcassing 1. *(Bw)* bauliche Hülle *f*, Rohbau *m*; Installation *f*; 2. Fertigstellung *f* im Rohbau; 3. Installation *f* von Gasleitungen
card 1. *(Dat)* Lochkarte *f*, Karte *f*; 2. *(Text)* Karde *f*, Krempel *f*
c. blending *(Text)* Mischen *n* auf der Karde
c. bracket *(Text)* Stelleisen *n* an der Krempel
c. chain *(Text)* Krempelkette *f*
c. clothing *(Text)* Kardenbeschlag *m*, Kratzenbeschlag *m*
c. column *(Dat)* Lochkartenspalte *f*
c. counter Kartenzähler *m* *(z. B. bei Lochkartenmaschinen)*
c. cut for several weaves *(Text)* Karte *f* mit mehreren Bindungen
c. cutter *(Text)* Kartenschläger *m*, Musterschläger *m*
c. cutting *(Text)* Kartenschlagen *n*, Kartenstanzen *n*, Kartenlochen *n*
c. cutting direct from the sketch *(Text)* Kartenschlagen *n* nach der Patrone
c. cycle total transfer Summenübertragung *f* im Kartengang *(Lochkartenmaschine)*
c. cylinder *(Text)* Kartenprisma *n*
c. delivery *(Text)* Kardenbandzuführung *f*
c. doffing *(Text)* Abnehmen *n*
c. duplicator *(Dat)* Kartendoppler *m*
c. fancy Trommelputzwalze *f* *(Krempel)*
c. feed Kartenzuführung *f*, Kartenvorschub *m*
c. feeder *(Dat)* Kartenzuführer *m*
c. feeding *s.* c. feed
c. feeding apparatus *(Text)* Krempelspeiser *m*
c. file Kartei *f*

c. fitted in *(Text)* eingespannte (eingelegte) Karte *f*
c. fixer *(Text)* Kardensetzer *m*
c. flat *(Text)* Krempeldecke *f*
c. fly *(Text)* Kardenflug *m*, Krempelflug *m*
c. for interchangeable pegs *(Text)* Karte *f* für auswechselbare Stifte
c. for keeping boxes at rest *(Text)* Karte *f* für Wechselstillstand *(Kastenstillstand)*
c. for moving the box *(Text)* Karte *f* für die Schützenkastenbewegung
c. grinder *(Text)* Kardenschleifer *m*, Kardennadelrichter *m*
c. grinding *(Text)* Schleifen *n* des Krempelbeschlages
c. hole location *(Dat)* Kartenlochlage *f*
c. index Kartei *f*
c. jam Kartenstau *m (in Lochkartenmaschinen)*
c. lacer *(Text)* Kartenbinder *m*, Kartenschnürer *m*
c. lacing machine *(Text)* Kartenbindemaschine *f*
c. leather *(Text)* Kratzenleder *n*
c. mounting *(Text)* Belegen *n* der Krempel mit Kratzenbändern
c. mounting machine *(Text)* Aufziehvorrichtung *f* für Kratzenbänder
c. of warping particulars *(Text)* Schärbrief *m*, Schärzettel *m*
c. operating two picks *(Text)* Karte *f* mit zwei Bindungen
c. perforating Kartenschlagen *n*, Kartenlochen *n*, Kartenstanzen *n*
c. perforator Kartenlocher *m*, Kartenstanzer *m*
c. programmed calculator lochkartengesteuerter Rechner *m*
c. punch Kartenlocher *m*, Kartenstanzer *m*
c. punching Kartenschlagen *n*, Kartenstanzen *n*
c. punching machine Kartenschlagmaschine *f*, Kartenlochmaschine *f*
c. random access memory *(Dat)* Magnetkartenspeicher *m* mit beliebigem Zugriff
c. read punch lesender Kartenlocher *m*, Stanz-Lesegerät *n (Lochkartenmaschine)*
c. reader Kartenabfühler *m*, Kartenabtaster *m*, Kartenleser *m*, Kartenlesestation *f (Lochkartenmaschine)*
c. register of technical terms *(Büro)* Fachwortkartei *f*
c. repeating *(Text)* Kartenkopieren *n*
c. reproducing puncher *(Dat)* Kartendoppler *m*
c. reversing motion *(Text)* Kartenrückschlagvorrichtung *f*
c. roller grinding machine *(Text)* Krempelwalzenschleifmaschine *f*
c. room *(Text)* 1. Kardenraum *m*, Krempelei *f*, Karderie *f (Spinnerei)*; 2. *(Am)* Spinnereivorwerk *n*
c. saving device Kartensparеinrichtung *f*
c. saving motion *(Text)* Kartenspareinrichtung *f*
c. setter *(Text)* Kardensetzer *m*
c. setting machine *(Text)* Kratzensetzmaschine *f*
c. sheet *(Text)* Blatt *n* der Kratze
c. sliver *(Text)* Kardenband *n*
c. sliver count numbers *(Text)* Kardenbandnummer *f*
c. stacker *(Dat)* Kartenstapler *m*
c. staple *(Text)* Kratzenzahn *m*
c. stripper *(Text)* Wender *m* der Karde
c. stripping *(Text)* Krempelreinigung *f*
c. strips *(Text)* Krempelausputz *m*, Kardeneinputz *m*
c. system Karteisystem *n*
c.-to-magnetic tape converter *(Dat)* kartengesteuerter Magnetbandschreiber *m*, Lochkarte-Magnetbandumsetzer *m*
c.-to-tape converter *(Dat)* kartengesteuerter Streifenlocher *m*
c. waste *(Text)* Kardierabfall *m*
c. waste cleaning machine *(Text)* Kardenabfallreinigungsmaschine *f*
c. web *(Text)* Krempelflor *m*
c. wire *(Text)* Kratzendraht *m*
c. with automatic stripper *(Text)* Krempel *f* mit selbsttätiger Reinigung
c. with breast roller *(Text)* Krempel *f* mit Vorwalze (Vortrommel)
c. with dog's hair extractor *(Text)* Krempel *f* mit Stichelhaarausscheidevorrichtung
cardan drive Kardanantrieb *m*
c. joint Kardangelenk *n*, Kreuzgelenk *n*
c. shaft Kardanwelle *f*, Gelenkwelle *f*
Cardan's solution *(Math)* Kardanische Formel (Lösung) *f (Gleichung dritten Grades)*
cardboard Karton *m*; Kartonpapier *n*; Pappe *f*
c. binding *(Typ)* Pappeinband *m*, Ganzpappeinband *m*
c.-box factory Kartonagenfabrik *f*
c. box machinery Kartonagenmaschinen *fpl*
c. boxes Kartonagen *pl*
c. cramp plier Kartonagennietzange *f*
c. middles minderwertiger Karton *m*, Kaschierrohmaterial *n*
c. reel cutter Kartonrollenschneidmaschine *f*
c. template Lochkartenmuster *n*
c. tube winding and gumming machine Hartpapierhülsenwickel- und -klebemaschine
carded *(Text)* kardiert
c. sliver *(Text)* Kardenband *n*
c. yarn *(Text)* kardiertes Garn *n*
carder *(Text)* Krempler *m*

Cardew voltmeter *(El)* Hitzdrahtvoltmeter *n*
cardigan cam *(Text)* Fangteil *n* eines Hebers
c. lock *(Text)* Fangschloß *n*
c. rack *(Text)* Fangversatz *m*
c. stitch *(Text)* Fangmaschine *f*
cardinal number Grundzahl *f*
c. plane Grundfläche *f*, Kardinalfläche *f*
c. points *(Flg)* 1. Hauptpunkte *mpl*, Kardinalpunkte *mpl*, Grundpunkte *mpl*; 2. Himmelsrichtungen *fpl*
cardinality *(Math)* Mächtigkeit *f (Mengen)*
carding *(Text)* Krempeln *n*; Kardieren *n*
c. arm Kardierflügel *m*
c. beater Flügel *m* mit Nadelbesatz
c. comb Wollkamm *m*
c. constant Kardierkonstante *f*
c. dust Kardenstaub *m*
c. leather Kratzenleder *n*
c. machine Karde *f*; Krempel *f*
c. machine for slubbing Vorspinnkrempel *f*
c. number Kämmungszahl *f*
c. willow Krempelwolf *m*
cardioid *(Math)* Kardioide *f*, Herzkurve *f*
c. diagram herzförmiges Diagramm *n*, Herzcharakteristik *f (Antenne)*
c.-reception herzförmige Empfangscharakteristik *f (Antenne)*
Cardox blasting *(Brg)* Schießen *n* mit flüssiger Kohlensäure
careening *(Schiff)* Kielholung *f*
caret Einfügungszeichen *n*
cargo Fracht *f*, Schiffsladung *f*
c. and passenger vessel Fracht- und Fahrgastschiff *n*
c. batten *(Schiff)* Wegerungslatte *f*
c. book *(Schiff)* Ladebuch *n*
c. boom *(Am) (Schiff)* Ladebaum *m*
c. capacity Laderauminhalt *m*
c. carrying capacity Ladefähigkeit *f*
c.-carrying glider *(Mil)* Lastensegler *m*
c. chute *(Schiff)* Ladeschütte *f*
c. derrick *(Schiff)* Ladebaum *m*
c. door *(Schiff)* Ladepforte *f*
c. gear *(Schiff)* Ladegeschirr *n*
c. handling Güterumschlag *m (Laden und Löschen)*
c. handling equipment Lade- und Löschausrüstung *f*
c. handling gear Ladegeschirr *n*
c. hatch[way] Ladeluke *f*
c. hold Laderaum *m (eines Schiffes)*
c. hook Ladehaken *m* mit Abweiser
c. liner Frachtschiff *n* in der Linienfahrt, Linienfrachter *m*
c. motorship Frachtmotorschiff *n*
c. net *(Schiff)* Netzschlinge *f*, Ladungsnetz *n*
c.-passenger liner Fracht- und Fahrgastschiff *n (im Liniendienst)*
c. plan *(Schiff)* Ladeplan *m*, Stauplan *m*
c. port *(Schiff)* Ladepforte *f (im Schanzkleid)*
c. rates Frachtkosten *pl*
c. rig *(Schiff)* Ladetakelung *f*
c. runner *(Schiff)* Lastseil *n*
c. ship Frachtschiff *n*
c. space *(Schiff)* Laderäumte *f*
c. steamer Frachtdampfer *m*
c. transport airplane Transportflugzeug *n*
c. vessel Frachtschiff *n*
c. whip *(Schiff)* Ladeseil *n*
carlin *(Schiff)* Kielschwein *n*
carline Dachspriegel *m*
carmine Karmin *n*
carnallite *(Min)* Karnallit *m*
carnauba wax Karnaubawachs *n*
carnaubic acid Lignocerinsäure *f*
Carnot cycle *(Ph)* Carnotscher Kreisprozeß *m (Thermodynamik)*
carnotite *(Min)* Carnotit *m*
carotene Karotin *n*
carpenter Zimmerer *m*, Zimmermann *m*
carpenter's adze Dechsel *f* für Zimmerleute, Flachdechsel *f*
c. axe Bundaxt *f*
c. bench vice Hobelbank *f*
c. bevel Stellwinkel *m*, Schmiege *f*
c. grinder Handschleifmaschine *f*
c. hammer Latt[en]hammer *m*
c. hatchet Zimmermannsbeil *n*
c. pincers Beißzange *f*, Kantenzange *f*, Kneifzange *f*, Nagelzange *f*
c. plane Holzhobel *m*
c. roofing hammer Latt[en]hammer *m*
c. rule Zollstock *m*
c. scratch gauge Streichmaß *n*
c. shop Zimmermannswerkstatt *f*, Tischlerei *f*
c. try square of wood Anschlagholzwinkel *m*, Zimmermannsholzwinkel *m*
carpentry Zimmer[er]handwerk *n*, Zimmermannshandwerk *n*; Zimmererarbeit *f*, Zimmermannsarbeit *f*
carpet 1. *(Text)* Teppich *m*; 2. *(Bw)* Teppichbelag *m (Straßenbau)*; Straßendecke *f*
c. backing Teppich[rückseite]beschichtung *f*, Teppichleimung *f*
c. binding *(Text)* Teppichbänder *npl*
c. loom Teppich[web]stuhl *m*
c. wire pick loom Rutenteppichstuhl *m*
c. wool Teppichwolle *f*
carpeting 1. *(Text)* Bodenbelagstoffe *mpl*, Teppiche *mpl*; 2. Teppichbelag *m (Straßenbau)*
carragheen Isländisches Moos *n*
Carrara marble Carraramarmor *m*
carriage 1. Fuhrwerk *n*, Wagen *m*, Waggon *m*; 2. *(Kfz)* Fahrgestell *n*, Wagengestell *n*; 3. *(Flg)* Fahrgestell; 4. *(Büro)* Wagen;

5. *(Masch)* Bettschlitten *m*, Hauptschlitten *m*, Schlitten *m*, Support *m*, Werkzeugschlitten *m*; 6. *(Text)* fahrbarer Spindelträger *m*, Wagen; 7. *(Mil)* Lafette *f*
c. and apron unit Werkzeugschlitten *m* mit Schloßplatte
c. band arrangement *(Text)* Seilführung *f*
c. body Karosserie *f*
c. bolt Schloßschraube *f*
c. building Wagenbau *m*
c. by rail Eisenbahntransport *m*
c. cam im Schlitten montiertes Schloßteil *n* *(Strickmaschine)*
c. comb *(Text)* Führungskamm *m*, Kammschlitzbarre *f*
c. control Vorschubsteuerung *f*
c. disk *(Text)* Spulenschlittenscheibe *f*
c. drag Wagenverzug *m*
c. edger *(Masch)* Laufwagentrennsäge *f*
c. grease Wagenschmiere *f*
c. guide Schlittenführung *f*
c. return *(Büro)* Wagenrücklauf *m*
c. saddle *(Masch)* Support *m*, Hauptschlitten *m*
c. spring Wagenfeder *f*
c. travel *(Masch)* Schlittenverschiebung *f*
c. with spent bobbin *(Text)* Spulenschlitten *m* mit leerer Spule
carriageway Fahrbahn *f*, Fahrdamm *m*
c. marking Fahrbahnmarkierung *f*
c. slab Fahrbahntafel *f*
c. surfacing Fahrbahnbelag *m*
carrick bend *(Schiff)* Kreuzknoten *m*
carried around by frictional contact durch Reibschluß mitgenommen
c. by a hollow spindle mit Hohlspindellagerung
c.-over moment *(Bw)* fortgeleitetes Moment *n* *(Momentenausgleichverfahren)*
carrier 1. *(Fmt, Fs)* Frequenzträger *m*, Ladungsträger *m*, Trägerwelle *f*; 2. *(Masch)* Drehherz *n*, Mitnehmer *m*; Zwischenrad *n*; 3. Trägersubstanz *f*, Trägermaterial *n*; 4. Einspannrahmen *m*; 5. Übertrager *m*; 6. Gepäckträger *m* *(Fahrrad)*; 7. Transporter *m* *(Schiff)*
c. addition Trägerzusatz *m*
c. aeroplane Trägerflugzeug *n*
c. aircraft Trägerflugzeug *n*
c. amplitude Trägeramplitude *f*
c. analysis Trägeranalyse *f*
c. arm for pressing the bobbin on to the drum *(Text)* Gabelarm *m* zur Spulenpressung
c. art Trägerfrequenztechnik *f*
c. bar Lagerbalken *m*
c.-based fighter bordgestütztes Jagdflugzeug *n*
c. beam Trägerbündel *n*
c.-borne fighter bordgestütztes Jagdflugzeug *n*
c. channel Trägerstromkanal *m*
c. circuit Drahtfunkleitung *f*
c. colour signal *(Fs)* Farbenträgersignal *n*
c. compound Trägerverbindung *f*
c. control *(Fmt)* Trägersteuerung *f* *(Hapug-Modulation)*
c. current *(El)* Trägerstrom *m*
c. current communication Trägerfrequenznachrichtenübermittlung *f*
c. current multiple telegraphy Wechselstrommehrfachtelegrafie *f*
c.-current protection Streckenschutz *m* mit Trägerfrequenzverbindung *(Schutzrelais)*
c. current telegraphy Träger[strom]telegrafie *f*
c. current telephone communication Trägerfrequenztelefonie *f*
c. current telephony Hochfrequenztelefonie *f* auf Leitungen
c. current transmission Drahtfunk *m*
c. density Trägerdichte *f*
c. felt Tragfilz *m* *(Papier)*
c. free trägerfrei
c. free tracer trägerfreier Indikator *m*
c. frequency *(El)* Trägerfrequenz *f*
c. frequency oscillator Trägerfrequenzerzeuger *m*
c.-frequency pulse impulsmodulierte Trägerschwingung *f*
c.-frequency stability Trägerfrequenzkonstanz *f*
c.-frequency telegraphy Trägerstromtelegrafie *f*
c. frequency telephony Trägertelefonie *f*
c.-frequency voltage Trägerspannung *f*
c.-frequency wire broadcasting hochfrequenter Drahtfunk *m*, Trägerfrequenzdrahtfunk *m*
c. gas Trägergas *n*, Schleppgas *n* *(Gaschromatografie)*
c. isolating choke coil *(El)* trägersperrende Drossel *f*
c. lifetime *(Ph)* Ladungsträgerlebensdauer *f*
c. line Drahtfunk *m*
c. mobility Trägerbeweglichkeit *f*
c. noise *(Fmt)* Trägerrauschen *n*
c. of electricity Elektrizitätsträger *m*
c. operated device antinoise trägerwellengesteuerter Geräuschtöter *m*, trägergesteuerte Störsperre *f*
c. piece Zwischenbaustück *n* *(Walzwerk)*
c. plate *(Masch)* Mitnehmerscheibe *f*
c. power *(El)* Trägerleistung *f*
c. power plant Trägerstromversorgungsanlage *f*
c. rediffusion Hochfrequenzdrahtfunk *m*

c.-reference white level *(Am) (Fs)* Trägerwellenamplitude *f* für Weißmaximum; Weißpegel *m (bezogen auf den Träger)*
c. repeater Trägerstromverstärker *m*
c. rocket Trägerrakete *f*
c.-rod dog Mitnehmerfinger *m (Wirkmaschine)*
c. roller Tragrolle *f*
c. script *(Dat)* Trägerwellenschrift *f (Übersetzung einer Basisbandschrift in eine Trägerwellenschrift und umgekehrt)*
c. shift Trägerfrequenzabweichung *f*; Trägeramplitudenabweichung *f*
c. slide bar Fadenführergleitschiene *f (Strickmaschine)*
c. solution Trägerlösung *f*
c. suppression Trägerfrequenzunterdrückung *f*
c. telegraphy Hochfrequenztelegrafie *f*, Trägerfrequenztelegrafie *f*, Wechselstromtelegrafie *f*, Trägerstromtelegrafie *f*
c. telephone system Trägerstromtelefonieeinrichtung *f*
c. telephony Träger[frequenz]telefonie *f*, Trägerfrequenzsprechfunk *m*
c. telephony repeater Trägertelefonieverstärker *m*
c. terminal circuit Drahtfunkendschaltung *f*
c. transmission Übertragung *f* durch Trägerfrequenz
c. tube *(Text)* Fadenführerschlauch *m*
c. value Trägerwert *m*
c. voltage Trägerspannung *f*
c. wave *(El)* Trägerwelle *f*
c. wave oscillator Trägerwellenoszillator *m*
c. wave telephony Trägerfrequenztelefonie *f*
c. wire broadcasting hochfrequenter Drahtfunk *m*
carrion-wool Blötwolle *f*, Kadaverwolle *f*
carrot Möhre *f*, Mohrrübe *f*
carry / to 1. tragen, unterstützen, halten, aufnehmen, führen, lagern; 2. leiten *(Flüssigkeit)*; 3. führen *(Strom)*
c. current Strom führen
c. forward the balance *(Büro)* den Saldo vortragen
c. off heat Wärme ableiten
carry *(Dat)* Übertrag *m*
c. circuit *(Dat)* Übertragschaltung *f*
c. contact Zehnerübertragkontakt *m (Lochkartenmaschine)*
c. gate *(Dat)* Übertragschaltung *f*
c.-over factor *(Bw)* Fortleitungsfaktor *m (Momentenausgleichverfahren)*
c. propagation Übertragausbreitung *f*, Übertragfortpflanzung *f (Addierwerk)*
carrying agent Trägerflüssigkeit *f*
c. axle Laufachse *f*
c. bracket Stützhülse *f*
c. cable Tragseil *n*
c. capacity 1. Tragfähigkeit *f*, Tragkraft *f*, Tragvermögen *n*, Ladefähigkeit *f*; 2. Ladevermögen *n (Waggon usw.)* 3. *(El)* Strombelastbarkeit *f*, Belastungsfähigkeit *f*
c. case Transportbehälter *m*
c. clamp Kabeltragschelle *f*
c. gate *(Brg)* Hauptförderstrecke *f*
c. ion *(Ph)* Trägerion *n*
c. lee helm *(Schiff)* leegierig
c. of flags *(Schiff)* Flaggenführung *f*
c. of lights *(Schiff)* Lichterführung *f*
c. of sails *(Schiff)* Segelführung *f*
c. roller Tragrolle *f*
c. side Tragseite *f (Band)*
c. stick Tragholm *m*
c. strap Traggurt *m*
c.-trade Handelsschiffahrt *f*
c. weather helm *(Schiff)* luvgierig
cart house Remise *f*; Wagenschuppen *m*
c. load Fuder *n*, Wagenladung *f*
c. plough Karrenpflug *m*
c. pump Karrenpumpe *f*
c. sprayer Karrenspritze *f*
cartel ship Austauschschiff *n*, Parlamentärschiff *n*
Cartesian *(Math)* kartesisch *(nach Descartes)*
cartogram Diagrammkarte *f*, Kartogramm *n*
cartographer Kartograf *m*
cartography Kartografie *f*, Kartenkunde *f*
carton Karton *m*, Pappschachtel *f*
cartoon camera Trickfilmkamera *f*
c. camera table Tricktisch *m*
c. film Trickfilm *m*
cartridge 1. *(Mil)* Patrone *f*; Sprengladung *f*, Pulverladung *f*; 2. *(Kern)* Spaltstoffhülse *f*, Brennstoffhülse *f*; 3. Hüllpapier *n*
c. belt Patronengurt *m*; *(Am)* Patronentasche *f*
c. brass Messingkartusche *f*
c. case Patronenhülse *f*
c.-case deflector *(Mil)* Hülsenabweiser *m*
c.-case extractor *(Mil)* Hülsenauszieher *m*
c. fuse *(El)* Patronensicherung *f*, Sicherungspatrone *f*
c. fuse-link *(El)* Patronensicherung *f*
c. hammer *(Bw)* Bolzenschießgerät *n*
c. pouch Patronentasche *f*
c. starter Patronenstarter *m*
c.-to-cartridge film feed *(Opt)* Doppelkassettensystem *n*
c.-type brush holder *(Elm)* Bürstenhalterpatrone *f*, Patronenbürstenhalter *m*
cartwheel aerial Radantenne *f*
cartwright Stellmacher *m*, Wagner *m*
cartwright's rabbet plane iron Wangenhobeleisen *n*
carve / to schnitzen *(Holz)*
carved veneer Messerfurnier *n (Holz)*

carvel building *(Schiff)* Kraweelbau *m*
c. planking Kraweelbeplankung *f*
c. work Kraweelbau[art *f*] *m*
carver Kerbschnittwerkzeug *n*; Schnitzer *m*
carving knife Schnitzmesser[klinge *f*] *n*, Küferschnitzmesser[klinge *f*] *n*
c. machine cutter Holzfräser *m*
c. tool Kerbschnitzwerkzeug *n*, Schnitzwerkzeug *n*
cascade 1. *(El)* Kaskade *f*, Reihenschaltung *f*, Kettenschaltung *f*; 2. *(Ch, Kern)* Kaskade, Trennkaskade *f*; 3. *(Kern)* Kaskadenschauer *m*
c. amplifier Kaskadenverstärker *m*, Stufenverstärker *m*
c. arrangement Kaskadenanordnung *f*
c. battery Stufenbatterie *f*
c. carry *(Dat)* Kaskadenübertrag *m*
c. compensation *(BMSR)* Serienkompensation *f*
c. connection *(El)* Kaskadenschaltung *f*, Hintereinanderschaltung *f*, Reihenschaltung *f*
c. control *(BMSR)* Kaskadenregelung *f*, Regelung *f* durch Kaskadenschaltung *f*, Kaskadensteuerung *f*
c. converter *(Elm)* Kaskadenumformer *m*
c. culvert *(Bw)* Kaskadendurchlaß *m*, Treppendurchlaß *m*
c. cut *(Brg)* Terrassenschnitt *m*
c. decay *(Kern)* Kaskadenzerfall *m*
c. drier Kaskadentrockner *m*
c. evaporator Kaskadenverdampfer *m*
c. generator Kaskadengenerator *m*
c. jumps *(Kern)* Kaskadensprünge *mpl*
c. radiation Kaskadenstrahlung *f*
c. rectifier circuit Kaskadenschaltung *f* *(Gleichrichter)*
c. set *(Elm)* Kaskade *f*
c.-type deduster *(Brg)* Kaskadensichter *m*
c. voltage doubler Spannungsverdopplungsschaltung *f*
c. X-ray tube Kaskadenröntgenröhre *f*
cascaded carry *(Dat)* Kaskadenübertrag *m*
cascades of blades Schaufelgitter *n*
cascading Kaskadenüberschlag *m* *(bei Isolatoren)*
cascode amplifier *(El)* Kaskodenverstärker *m*
case / to 1. umhüllen, ummanteln, umkleiden; 2. *(Typ)* einhängen *(den Buchblock in die Decke)*
c. a wheel ein Rad beschlagen
c. harden *(Met)* einsatzhärten, einsetzen, oberflächenhärten, hartgießen
case 1. Gehäuse *n*, Behälter *m*, Kasten *m*; Etui *n*, Futteral *n*; 2. Einsatz[schicht *f*] *m* *(Stahl)*; 3. Fall *m*, Beispiel *n*; 4. *(Typ)* Setzkasten *m*; Buchstabe *m*; Buchdecke *f*; 5. Wachsträgerplatte *f* *(Elektrotypie)*
c. board Kistenbrett *n*
c. chisel Kistenbeitel *m*
c. depth *(Met)* Einsatztiefe *f*, Härtetiefe *f*
c.-hardenability *(Met)* Einsatzhärtbarkeit *f*
c.-hardening 1. *(Met)* Einsatzhärtung *f*, Oberflächenhärtung *f* *(im Einsatz)*; 2. *(Holz)* Verschalen *n*
c.-hardening box Einsatzkasten *m*
c.-hardening carburizer Einsatzmittel *n* *(für Einsatzhärtung)*
c.-hardening composition Einsatzmittel *n*, Härtemittel *n*
c.-hardening compound Einsatzmittel *n* *(für Einsatzhärtung)*
c.-hardening furnace Zementierofen *m*, Einsatzofen *m*, Härteofen *m* *(für Einsatzhärtung)*
c.-hardening pot Einsatztopf *m*
c.-hardening steel Einsatzstahl *m*
c. of loading Belastungsfall *m*
c. of stress Spannungszustand *m*
c. opener Kistenbrecher *m*
c. shot *(Mil)* Kartätsche *f*
c. spring Deckelfeder *f*
c. study *(Ök)* Arbeitsstudie *f*
c. temperature Gehäusetemperatur *f*
cased beam betonummantelter Träger *m*, ummantelter Stahlträger *m*
c. column betonummantelte Stütze *f*, ummantelte Stahlstütze *f*
c. motor gekapselter Motor *m*
c. pile Ortbetonpfahl *m* mit Stahlmantel
c. toe *(Led)* Steifkappe *f*
casehardening *s.* case-hardening
casein Kasein *n*
c. paint Kaseinfarbe *f*
casement 1. Flügelfenster *n*; 2. Fensterflügel *m*
c. cloth Scheibengardine *f*
c. fastener *(Bw)* Einreiber *m*; Vorreiber *m*
c. frame Fensterrahmen *m*
c. light Flügelfenster *n*
c. sash Fensterflügel *m*
c. stay Fensterfeststeller *m*, Sturmhaken *m*
c. turn *(Bw)* Einreiber *m*; Vorreiber *m*
c. type window Flügelfenster *n*
c. window Flügelfenster *n*
cash cheque Barscheck *m*
c. register Kontrollkasse *f*, Registrierkasse *f*
c. register roll Registrierkassenrolle *f*
c. register roll reader Kontrollkassenrollenleser *m*, Registrierkassenrollenleser *m*
c. tally sheet Kassenbuchungsblatt *n*
cashew nut Kaschunuß *f*
Cashmere goat Kaschmirziege *f*
C. hair Kaschmirwolle *f*
casing 1. Gehäuse *n*, Hülle *f*, Kapsel *f*, Mantel *m*; Umkleidung *f*, Ummantelung *f*; Verkleidung *f*; Hüllrohr *n*, Mantelrohr *n*; 2. *(Brg)* Verrohrung *f*; Bohrrohr *n*, Futterrohr *n*
c. bore Gehäusebohrung *f*
c. clamp Rohrschelle *f*

c. cover Gehäusedeckel *m*
c. flange Futterrohrflansch *m*
c. guide Gehäuseführung *f (Turbine)*
c. guide-bolt Gehäuseführungsbolzen *m (Turbine)*
c. machine *(Typ)* Bucheinhängemaschine *f*
c. of cock Hahngehäuse *n*
c. of fuel throwing wheel Wurfradgehäuse *n*
c. plies *(Kfz)* Gewebelagen *fpl* des Unterbaus; Mantelschichten *fpl (Reifen)*
c. rib Gehäuserippe *f (Turbine)*
c. tube Futterrohr *n*
c. wall Gehäusewand *f*
casinghead gas Bohrlochkopfgas *n (Erdgas)*
c. gasoline Rohrkopfbenzin *n*
cask schweres Faß *n*, Tonne *f*
c. body Faßrumpf *m*
c. buoy *(Schiff)* Faßboje *f*, Faßtonne *f*
c. making machine Faßbindemaschine *f*
c. manufacture 1. Faßherstellung *f*; 2. Faßfabrik *f*
c. trussing and hoop driving machine Faßzusammensetz- und Reifenaufziehmaschine *f*
c. wood Böttcherholz *n*, Faßholz *n*, Daubenholz *n*
casket Kontainer *m*, Transportbehälter *m (für radioaktive Substanzen)*
Cassegrain grating spectrograph Cassegrain-Gitterspektrograf *m*
Cassegrain reflecting telescope Cassegrain-Spiegelteleskop *n*
cassette Kassette *f*
c. method *(Bw)* Batterieverfahren *n (Fertigteilherstellung)*
cassia oil Kassiaöl *n*, Zimtöl *n*
cassiterite *(Min)* Zinnstein *m*, Kassiterit *m*
cast / to 1. *(Met)* [ver]gießen; 2. *(Bw)* betonieren; 3. lotwerfen, loten; 4. werfen
c. anchor Anker werfen
c. downhill fallend gießen
c. from the bottom steigend gießen
c. from the top fallend gießen
c. in pairs paarweise gießen
c. integral in einem Stück gießen, angießen
c. off *(Schiff)* losmachen, ablegen, loswerfen; *(Text)* abnehmen *(Maschen)*
c. on angießen
c. solid massiv gießen
c. the blades die Schaufeln eingießen *(Turbine)*
c. uphill steigend gießen
cast Guß *m*; Gußstück *n*; Wurf *m*, Werfen *n*; Gipsverband *m*
c. alloy Gußlegierung *f*
c. brass Gußmessing *n*
c. bronze Gußbronze *f*
c. centrifugally *(Bw)* geschleudert *(Schleuderbeton)*
c.-coated paper gußgestrichenes Papier *n*; Hochglanzpapier *n*, Porzellanpapier *n*
c.-coater Abgußstreichmaschine *f (Papier)*
c. coating Verbundguß[plattierung *f*] *m*
c. concrete Gußbeton *m*
c. history sheet *(Met)* Schmelzverlaufkarte *f*
c.-in eingegossen
c.-in-place am Einbauort betoniert, monolithisch hergestellt *(Ortbeton)*
c.-in-place concrete monolithischer Beton *m*, Ortbeton *m*
c.-in-place pile Ortbetonpfahl *m*
c.-in-position am Einbauort betoniert, monolithisch hergestellt
c.-in-situ am Einbauort betoniert, monolithisch hergestellt
c.-in-situ concrete monolithischer Beton *m*, Ortbeton *m*
c.-in-situ pile Ortbetonpfahl *m*
c.-in working barrel *(Masch)* eingegossener Arbeitszylinder *m*
c. insitu *s.* c.-in-situ
c. iron Gußeisen *n*
c. iron gußeisern; Gußeisen-
c.-iron assembly *(El)* Gußgruppe *f (Schalter usw.)*
c.-iron disk wheel Graugußscheibenrad *n*
c.-iron furnace front Gußeisengeschränk *n*
c. iron roll Gußeisenwalze *f*, gegossene Walze *f*
c.-iron scale Gußhaut *f*
c.-iron scrap Gußbruch *m*, Gußschrott *m*
c.-iron superheater not placed in a chamber kammerloser Gußeisenüberhitzer *m*
c. light alloy Leichtmetallguß *m*
c. magnet Gußmagnet *m*
c. metal Gußmetall *n*
c. nonferrous alloys Hartlegierungen *fpl*
c. off *(Am) (Text)* Maschenabnehmen *n*
c.-off cam *(Am) (Text)* Abschlag *m*
c.-off position *(Am)* Abschlagstellung *f (Strickmaschine)*
c.-on *(Met)* angegossen
c. resin Gießharz *n*
c. seam Gußnaht *f*, Gußgrat *m*
c. steel Gußstahl *m*, Stahl[form]guß *m*
c.-steel disk wheel Stahlgußscheibenrad *n*
c. steel jaw *(Masch)* Gußbacke *f*
c. steel wire heald *(Text)* Stahldrahtlitze *f*
c. stone Beton[werk]stein *m*, Kunststein *m*
c. structure Gußgefüge *n*
c. tooth gear Gußzahnrad *n*
c. toothed mit gegossenen Zähnen *(Verzahnung)*
c. welding *(Met)* Gießschweißen *n*
castability Gießbarkeit *f*, Vergießbarkeit *f (insbesondere Plaste)*
castable slip Gießschlicker *m*

castaway Schiffbrüchiger *m*
castellated nut Kronenmutter *f*
castellating *(Am)* Schlitzen *n* von Kronenmuttern
castellation 1. Kronierung *f*, Kronenmuttereinfräsung *f*; 2. Verzahnung *f*
caster effect Nachlauf *m*
c. knives *(Led)* Stanzmesser *npl*
casters *(Typ)* Schließrollen *fpl*
casting 1. Formgußstück *n*, Gußstück *n*; 2. Gießen *n*; 3. Hintergießen *n (Elektrotypie)*; 4. *(Bw)* Betonieren *n*
c. alloy Gußlegierung *f*
c. bay Gießhalle *f*
c. bay crane Gießereikran *m*
c. bed 1. *(Met)* Gießbett *n*; 2. *(Bw)* Matrize *f*, Betonierbett *n (Fertigteilherstellung)*
c. brush Gußputzbürste *f*
c. crack Schwindriß *m*
c. defect Gußfehler *m*
c. floor level Hüttenflur *m (Martinofenanlage)*
c. gutter Gießrinne *f*
c. ladle Gießkelle *f*, Gießlöffel *m*; Gießpfanne *f*
c. machine Gießmaschine *f*
c. model Gußmodell *n*
c. mould 1. *(Met, Plast)* Gießform *f*, Gießkasten *m*; 2. *(Bw)* Matrize *f*, Schalungsform *f (Fertigteilherstellung)*
c. out *(Math)* Restklassenprobe *f*
c. pit Gießgrube *f*, Gießstelle *f*, Gießplatz *m*
c. resin Schmelzharz *n*, Vergußharz *n*
c. roll *(Plast)* Auftragwalze *f*
c. scrap Gußschrott *m*, Gußbruch *m*
c. shoe Vorherd *m (Martinofen)*
c. skin Gußhaut *f*
c. stress Gußspannung *f*
c. wheel Gießtrommel *f*
c. yard *(Bw)* Betonierfläche *f*, Betonierplatz *m*, offenes Betonwerk *n*; Vorfertigungsplatz *m*; Vorfertigungsfläche *f*
castle nut Kronenmutter *f*
c. nut without turret Kronenmutter *f* ohne Hals
castor angle *(Kfz)* Nachlaufwinkel *m*, Vorlaufwinkel *m (der Vorderräder)*
c. oil Rizinusöl *n*
casual shirt Campinghemd *n*, Freizeithemd *n*
casualty Unfall *m*; Schiffsunfall *m*; Verlust *m*; Schiffsverlust *m*
cat / to Anker katten
cat-cracked gasoline Krackbenzin *n*
c. crusher Haferquetscher *m*
c. fall *(Schiff)* Kattläufer *m*
c. hole *(Schiff)* Heckklüse *f*, Achterklüse *f*
c. rolling machine Haferquetschmaschine *f*
c. shaft Spillwelle *f (Rotarybohren; Erdöl, Erdgas)*
c. tackle *(Schiff)* Kattakel *n*, Ankerkatt *n*
catadioptric telescope Spiegellinsenfernrohr *n*
catalan furnace *(Met)* Luppenfeuer *n*
catalogue Katalog *m*
catalysis Katalyse *f*, Kontaktwirkung *f*
catalyst Katalysator *m*, Kontaktsubstanz *f*
c. furnace Kontaktofen *m*
c. poison Katalysatorgift *n*, Kontaktgift *n*
c. support Katalysatorträger *m*, Kontaktträger *m*
catalytic katalytisch
c. action Katalyse *f*, katalytische Wirksamkeit *f*
c. combustion oxygen analyzer Wärmetönungssauerstoffanalysator *m (katalytische Verbrennung)*
c. exchange reactor *(Kern)* katalytischer Austauschreaktor *m*
catalyze / to katalysieren
catalyzed lacquers *(Plast)* säurehärtende Lacke *mpl*; säurehärtende Harze *npl*
catalyzer Katalysator *m* (*s. a.* catalyst)
catalyzing katalysierend, katalytisch wirkend
catamaran Katamaran *m*, Doppelrumpfboot *n*
catapult Katapult *m*, Schleuder *f*
c. fittings Katapultierbeschläge *mpl (Flugzeug)*
c. launcher *(Flg)* Katapultstartanlage *f*
c. launching *(Flg)* Katapultstart *m*
c. seat *(Flg)* Schleudersitz *m*
cataract lens *(Opt)* Starglas *n*
catch / to *(Masch)* auffangen, fassen, greifen, [ein]schnappen, einklinken
c. in einrasten
catch 1. Anschlag *m*; Arretierhebel *m*; Arretierung *f*; Eingriff *m*; Gesperr *n*; Klaue *f*, Mitnehmer *m*; Schließhaken *m*, Sperrhaken *m*, Sperrklinke *f*, Sperrstift *m*; Knagge *f*; 2. *(Ch)* Kühlfalle *f*, Falle *f*; 3. *(Brg)* Aufsetzvorrichtung *f*, Schachtfalle *f*
c. band *(Masch)* Fangband *n*
c. basin Auffangschale *f*, Auffangtrog *m*; Sammelbecken *n*; Auffangbecken *n*; Senkgrube *f*
c.-board Spritzbrett *n (Dreschmaschine)*
c. crop Zwischenfrucht *f*
c. drain Entwässerungsgraben *m*
c. end at the selvedge *(Text)* Fangfaden *m*
c. feeder Bewässerungsgraben *m*
c. for picking arm *(Text)* Schlägeranschlag *m*
c. hook Fanghaken *m*
c.-net Fangnetz *n*
c. of the hook *(Text)* Platinennase *f*
c. pit Senkgrube *f*; Auffangbecken *n*
c. plate *(Masch)* Mitnehmerscheibe *f*
c. prop *(Brg)* Hilfsstempel *m*

c. **ring** *(Masch)* Mitnehmerring *m*
c. **spring** Einschnappfeder *f*, Schnappfeder *f*
c. **thread** *(Text)* Fangfaden *m*, Schlingfaden *m*
c.-**water drain** Sickerkanal *m*
catcher 1. *(Schiff)* Fangboot *n*, Fänger *m*; 2. *(El)* Auskoppelungsglied *n*; Auskoppelungsfeld *n* *(Klystron)*; Auskoppelraum *m*; 3. Abnehmer *m* *(Walzwerker)*; 4. Schnäpper *m*
c. **electrode** Sammelelektrode *f*
c. **space** *(El)* Auskoppelraum *m*
catching cloth Spritzlein[e]wand *f* *(Dreschmaschine)*
catchment area Einzugsbereich *m*, Einzugsgebiet *n*, Sammelgebiet *n* *(Wasserwirtschaft)*
catchword *(Typ)* Stichwort *n*; Kustos *m* *(auf der Satzfahne)*
catechol *(Ch)* Katechin *n*, Brenzkatechin *n*
catechu Katechu *n*, Cachou *n*
category 1. Kategorie *f*; 2. *(Brg)* Fraktion *f*, Kornklasse *f*
catenary *(Math)* Kettenlinie *f*
c. **cable** Tragseil *n*
c. **trolley wire** *(El)* Kettenfahrleitung *f*, Kettenfahrdraht *m*
c. **wire** *(Fmt)* Aufhängedraht *m*, Tragdraht *m*
caterpillar 1. Raupe *f*, Raupenschlepper *m*, Gleiskettengerät *n*; 2. Raupe, Gleiskette *f*, Raupenkette *f*, Raupenfahrwerk *n*
c. **bulldozer** Bulldozer *m* mit Raupenfahrwerk
c. **chain** Raupenkette *f* *(Traktor)*
c. **crane** Raupenkran *m*
c. **drive** Gleiskettenantrieb *m*
c. **track** *(Am)* Raupenkette *f*
c. **tracks** Gleiskettenfahrwerk *n*, Raupenfahrwerk *n*
c. **tractor** Gleiskettenschlepper *m*, Raupenschlepper *m*, Raupentraktor *m*
catgut Darmsaite *f*; Katgut *n*
c. **string** Darmsaite *f*
cathead *(Schiff)* Kattdavit *m* *(n)*; Kranbalken *m*
cathedral glass Kathedralglas *n*
cathetometer *(Ph)* Kathetometer *n*
cathetron Gleichrichter *m* mit Außengittersteuerung
cathode Katode *f*
c. **anchor** Katodenhalterung *f* *(Elektronenröhre)*
c. **area** Stromeintrittszone *f*
c. **arrester** Katodenfallableiter *m*
c. **basis connection** Katodenbasisschaltung *f*
c. **bias** Katodenvorspannung *f*
c. **border** Katodensaum *m*
c. **by-pass [capacitor]** Katodenkondensator *m*
c. **cleaning** katodische Reinigung *f*
c. **compensation** Katodenkompensation *f*
c. **connecting strip** Katodenbändchen *n*
c. **contamination** Katodenverunreinigung *f*, Katodenvergiftung *f*
c. **copper** Katodenkupfer *n*, Elektrolytkupfer *n*
c.-**coupled** katodengekoppelt
c.-**coupled circuit** Anodenbasisschaltung *f*
c. **coupling** Katodenkopplung *f*
c. **dark space** Katodendunkelraum *m*, Hittorffscher (Crookesscher) Dunkelraum *m* *(Gasentladungsröhren)*
c. **deposit** Katodenniederschlag *m*
c. **disintegration** Katodenabbau *m*, Katodenzerstäubung *f*, Katodenabnutzung *f*
c. **drop** Katoden[spannungsab]fall *m*
c. **drop arrester** Katodenfallableiter *m*
c. **efficiency** katodische Stromausbeute *f*
c. **fall** *s.* c. drop
c. **fall of potential** Katoden[spannungsab]fall *m*
c. **filament** Glühkatode *f*
c. **fluorescence** Katodenfluoreszenz *f*
c. **follower** Anodenbasisstufe *f*, Katodenstufe *f*; Anodenbasisverstärker *m*, Katodenverstärker *m*; Anodenbasisschaltung *f*
c. **follower output** Katodenauskopplungsstufe *f*
c. **follower tube** Katodenverstärkerröhre *f*
c. **gate** *(Dat)* Torschaltung *f* mit Katodenausgang
c. **glow** Katoden[glimm]licht *n* *(Gasentladungsröhre)*
c.-**glow tube** Katodenlumineszenzlampe *f*
c.-**grounded circuit** *(Am)* Katodenbasisschaltung *f*
c. **heater tube** indirekt geheizte Röhre *f* *(Elektronenröhre)*
c. **heating power** Heizleistung *f* *(Elektronenröhre)*
c. **heating time** Katodenanheizzeit *f*
c. **hum** Katodenrauschen *n*
c. **keying** Katodentastung *f*
c. **layer** Katodenschicht *f*, katodische Schicht *f*
c. **leakage** Widerstand *m* zwischen Katode und Heizwendel
c. **light** *s.* c. glow
c. **liquor** Katolyt *m*, Katodenflüssigkeit *f*
c. **load circuit** Katodenbelastungskreis *m*
c. **output** Katodenausgang[sleistung *f*] *m*
c. **phase splitter** Phasenumkehrschaltung *f*
c. **pickling** katodische Beizung *f*
c. **poisoning** Katodenvergiftung *f*
c. **potential fall** Katoden[spannungsab]fall *m*
c. **pulse modulation** Katodenimpulsmodulation *f*
c. **ray** Katodenstrahl *m*, Elektronenstrahl[bündel *n*] *m*
c.-**ray accelerator** Elektronenbeschleuniger *m*

c.-ray **beam** Katodenstrahlbündel *n*
c.-ray **current** Katoden[strahl]strom *m*, Elektronen[strahl]strom *m*
c.-ray **direction finding** Peilung *f* mit oszillografischer Anzeige
c.-ray **function generator** *(Dat)* Katodenstrahlfunktionsgenerator *m*
c.-ray **furnace** Katodenstrahlofen *m*
c.-ray **indicator** magisches Auge *n*
c.-ray **memory tube** Katodenstrahlspeicherröhre *f*
c.-ray **oscillograph** Katodenstrahloszillograf *m (mit Fotografiereinrichtung)*
c.-ray **oscilloscope** Katodenstrahloszillograf *m (nur mit Betrachtungschirm)*, Elektronenstrahloszillograf *m (Oszilloskop)*
c.-ray **particle** Katodenstrahlteilchen *n*, Elektron *n*
c.-ray **pattern** Leuchtschirmbild *n*
c.-ray **picture tube** Bildwiedergaberöhre *f*
c.-ray **spot** Katodenstrahl[leucht]fleck *m*
c.-ray **television tube** Katodenstrahlfernsehröhre *f*
c.-ray **trace** Leuchtschirmbild *n (Katodenstrahlröhre)*
c.-ray **tube** Elektronenstrahlröhre *f*, Braunsche Röhre *f*, Katodenstrahlröhre *f*
c.-ray **tube display** Elektronenbild *n*, Leuchtschirmbild *n*; Anzeige *f* mit Katodenstrahlröhre
c.-ray **tube harness** Befestigungseinrichtung *f* einer Katodenstrahlröhre
c.-ray **tube screen** Bildschirm *m*, Leuchtschirm *m*
c.-ray **tube trace** Leuchtschirmbild *n*
c. **resistor** Katoden[vor]widerstand *m*
c. **return** Gitterspannungsgegenkopplung *f*
c. **saturation** Katodensättigung *f*
c. **sleeve** Emissionsschichtträger *m*; Katodenröhrchen *n*
c. **spot** Brennfleck *m*, Katodenfleck *m*
c. **sputtering** Katodenzerstäubung *f*
c. **stream** Emissionsstrom *m*, Katodenstrom *m*
c. **tail** Katodenableitung *f*

cathodic katodisch
c. **area** Stromeintrittszone *f*
c. **cleaning** katodische Reinigung *f*
c. **deposit** Katodenniederschlag *m*
c. **discharge** katodische Entladung *f*
c. **flux** Ionenstrom *m*
c. **polarization** Katodenpolarisation *f*
c. **protection** *(El)* galvanischer Korrosionsschutz *m*; Katodenschutz *m (Korrosionsschutz bei Schiffen)*; katodischer Schutz *m*; Cumberlandverfahren *n*

cathodoluminescence *(Ph)* Katodolumineszenz *f*

cathodophone *(El)* Katodofon *n*, Ionisationsmikrofon *n*

cathodophosphorescence *(Ph)* Katodophosphoreszenz *f*

catholyte *s.* catolyte

cation Kation *n*
c. **exchange** Kationenaustausch *m*
c. **exchanger** Kationenaustauscher *m*
c. **hole** Kationenloch *n*
c. **migration** Kationenwanderung *f*

cationic dye basischer Farbstoff *m*

cationics *(Ch)* kationenaktive Substanzen *fpl*

catolyte Katolyt *m*

catoptric condenser *(Opt)* Spiegelkondensor *m*
c. **objective** Spiegelobjektiv *n*

catoptrics Katoptrik *f*, Spiegeloptik *f*

cat's eye 1. Katzenauge *n*; 2. *(Min)* Augenachat *m*, Augenstein *m*
c. **whisker** *(Fmt)* Detektordrähtchen *n*; Kontaktpinsel *m*, Sucherkontakt *m*

cattle Zuchtvieh *n*; Vieh *n*
c. **barley** Futtergerste *f*
c. **breeding** Viehzucht *f*
c. **ear marker** Viehmarkierungszange *f*
c. **feeder** Futterzuführungsapparat *m*
c. **fittings** Einrichtung *f* für Viehtransportschiff
c. **foot oil** Klauenöl *n*
c. **nose plier** Bullenringlochzange *f*, Nasenlochzange *f*
c. **plague** Rinderpest *f*
c. **punch plier** Viehmarkierungszange *f*
c.**-shed** Rinderstall *m*
c. **shelter** Rinderstall *m*
c. **ship** Viehtransportschiff *n*
c. **truck** Viehwagen *m*
c. **turnip** Futterrübe *f*
c. **wagon** Viehwagen *m*
c. **weighing machine** Viehwaage *f*

catwalk 1. *(Bw)* Laufgang *m*, Laufsteg *m*; 2. *(Schiff)* Laufbrücke *f*

caul *(Holz)* Zulageblech *n*

cauldron-bottom *(Brg)* Sargdeckel *m*

caulk / to 1. *(Bw, Masch)* stemmen, dichten, einstemmen, verstemmen; 2. *(Schiff)* kalfatern, abdichten, verstemmen
c. **in the blade** die Schaufel einstemmen *(Turbine)*

caulk weld Stemmnaht *f*

caulked-in liner eingestemmte Zylinderbuchse *f*
c. **joint** Stemmfuge *f*

caulking 1. *(Bw, Masch)* Abdichten *n*, Abdichtung *f*, Verstemmen *n*; 2. *(Schiff)* Kalfatern *n*, Kalfaterung *f*
c. **chisel** Setzmeißel *m*
c. **compound** Dichtungsmasse *f*
c. **hammer** Stemmhammer *m*
c. **iron** Stemmeisen *n*
c. **joint** Stemmfuge *f*
c. **mallet** Kalfaterhammer *m*

c. strip Dichtungsstreifen *m*, Dichtungsband *n*
c. tool Stemmeißel *m*
caunch *s.* canch
cause / to verursachen, veranlassen, bewirken
c. the tool to feed Vorschub geben
cause Ursache *f*, Grund *m*
c. of error Fehlerquelle *f*
c. of trouble Fehlerquelle *f*
causeway Dammstraße *f*, Straßendamm *m*
caustic Ätzmittel *n*, Beize *f*, Beizmittel *n*
caustic ätzend, beißend, beizend, kaustisch, scharf
c. alkali Ätzalkali *n*
c. alkali in flakes Ätzalkali *n* in Schuppenform
c. curve *(Opt)* Kaustik *f*, Brennlinie *f*
c. embrittlement interkristalline Spannungsrißkorrosion *f*; Laugensprödigkeit *f*
c. fusion Ätzalkalischmelze *f*
c. lime Ätzkalk *m*, gebrannter Kalk *m*
c. line *(Opt)* Kaustik *f*, Brennlinie *f*
c. lye Ätzlauge *f*
c. pot Laugekessel *m*
c. potash Ätzkali *n*, Kaliumhydroxid *n*
c. resistant laugenbeständig
c. soda Ätznatron *n*, Natriumhydroxid *n*
c. soda cell *(El)* Alkalielement *n*
c. soda pulping liquor Natronkochlauge *f* *(Papierherstellung)*
c. stick Ätzstift *m*
c. surface *(Opt)* kaustische Fläche *f*, Brennfläche *f*
c. washing plant Laugewaschanlage *f*
causticity Ätzkraft *f*, Ätzwirkung *f*
causticizing operation Kaustifizieren *n*
c. plant Alkalisieranlage *f*
c. process Alkalisierung *f* *(z. B. Zellstoff)*
cauterization Abbeizen *n*, Beize *f*, Zerstörung *f* durch Laugeeinwirkung
cauterize / to [ab]beizen, [aus]ätzen, zerfressen
c. superficially anätzen
cautery transformer *(El)* Kaustiktransformator *m*
caution signal Langsamfahrtsignal *n*; Gefahrensignal *n*
c. zone Gefahrenzone *f*
cavalier perspective *(Math)* Kavalierperspektive *f*, Militärperspektive *f*
cave / to *(Brg)* Bruchbau betreiben, hereinwerfen, zu Bruch bauen (gehen); aushöhlen
c. in [her]einbrechen
cave 1. Höhle *f*; 2. *(Kern)* Zelle *f* *(Arbeitsraum für radioaktive Materialien)*
caved area *(Brg)* Alter Mann *m*, Bruchfeld *n*, Verbruch *m*
c. goaf *s.* caved area
c. waste *s.* caved area
cavel *(Schiff)* Belegklampe *f*, Warpklampe *f*
cavern Felshöhle *f*
cavetto *(Bw)* Hohlkehle *f*
cavil *s.* cavel
caving *(Brg)* Bruchbau *m*
c. back *(Brg)* Abbau *m* der Teilsohle über und neben dem Querbrechen *(Teilsohlenbruchbau)*
c. method *(Brg)* Bruchbau[verfahren *n*] *m*
cavitation 1. Hohlraumbildung *f*, Kavitation *f* *(Hydraulik)*; 2. Verschleiß *m* *(durch Auswaschung)*; 3. *(Schiff)* Kavitation *f*, Hohlsog *m* *(am Propeller)*
c. loss Kavitationsverlust *m* *(Hydraulik)*
c. tunnel *(Schiff)* Kavitationstunnel *m*, Kavitationstank *m*
cavity Hohlraum *m*; Gasblase *f*, Luftblase *f* *(in Gußstücken)*
c. brick Hohlziegel *m*
c. chimney doppelwandiger Schornstein *m*
c.-coupled filter hohlraumgekoppeltes Filter *n*
c. dam Hohlmauer *f* *(Talsperre)*
c. frequency meter *(Fmt)* Hohlraumfrequenzmesser *m*, Hohlraumwellenmesser *m*
c.-insulated hohlraumisoliert *(Kabel)*
c. magnetron *(El)* Radmagnetron *n*, Vielfachmagnetron *n*
c. radiation Hohlraumstrahlung *f*
c. resonator Hohlraumresonator *m*
c. wall Hohlmauer[werk *n*] *f*
c. wavemeter *s.* c. frequency meter
cawk unreiner Baryt *m*, Schwerspat *m*
cedar [wood] oil Zedernholzöl *n*
ceiling 1. *(Bw)* Decke *f*; 2. *(Schiff)* Wegerung *f*, Bodenwegerung *f*; 3. *(Flg)* Gipfelhöhe *f*
c. board *(Bw)* Dämmpappe *f*
c. countershaft *(Masch)* Deckenvorgelege[welle *f*] *n*
c. duct *(El, Fmt)* Deckendurchführung *f*
c. fan Deckenventilator *m*
c. fitting Deckenleuchte *f*
c.-hatch *(Schiff)* Wegerungsluke *f*
c. hook Deckenhaken *m*, Lampenhaken *m*
c. joist Deckenbalken *m*
c. lamp Deckenlampe *f*
c. light Deckenleuchte *f*
c. lighting Deckenbeleuchtung *f*
c. of the floor heads *(Schiff)* Kimmwegerung
c. plaster *(Bw)* Deckenputz *m*
c. plate Baldachin *m* *(Leuchten)*
c. reflecting lamp indirekte Leuchte *f*
c. rose Deckenrosette *f* *(Leuchten)*
c. rosette *s.* c. plate
c. switch Deckenschalter *m*; Zugschalter *m*
c. travelling crane Deckenlaufkran *m*
c. voltage maximale Spannung *f*, Höchstspannung *f*, Spitzenspannung *f*
c. voltage of an exciter höchste Erregerspannung *f*

ceilometer *(Flg)* fotoelektrischer Wolkenhöhenmesser *m*
celestial axis Himmelsachse *f*
c. body Himmelskörper *m*
c. globe Himmelsglobus *m*
c. guidance *(Rak)* astronomische Kurssteuerung *f*
c. mechanics Himmelsmechanik *f*
c. navigation Astronavigation *f*; astronomische Navigation *f*; astronomische Ortung *f*; astronomische Kurssteuerung *f*
celestine *(Min)* Zölestin *m*
cell 1. Behälter *m*, Dose *f*; Gaszelle *f*, Zelle *f*; Hohlraum *m*, Pore *f*; Küvette *f*; 2. *(El)* [galvanisches] Element *n*; 3. *(Dat)* Datenspeicherzelle *f*, Speicherzelle *f*
c. cavity *(Ch)* Porenraum *m*
c. compartment *(Opt, Ch)* Küvettenraum *m*
c. connector *(El)* Steg *m*, Polbrücke *f*
c. constant innerer Widerstand *m* eines Batterieelements
c. terminal *(El)* Polklemme *f*, Klemme *f* eines Elements
c. tester Zellenprüfer *m (Akkumulatoren)*
c. voltage 1. Badspannung *f (Galvanotechnik)*; 2. Zellenspannung *f (Batterie)*
cellar 1. Keller *m*; 2. *(Brg)* Vorschacht *m*
c. flap Kellerfalltür *f*
c. for shavings *(Holz)* Spänekeller *m*
celloid zellähnlich
cellophan Zellophan *n*, Zellulosefolie *f*, Glashaut *f*, Zellglas *n*
cellucotton Papierwatte *f*, Zellstoffwatte *f*
cellular 1. zellähnlich, zellenartig, zellförmig, zellig; fächerartig; 2. *(El)* gekapselt
c. board Wellpappe *f*
c. brick Lochziegel *m*
c. cofferdam *(Bw)* Zellenfang[e]damm *m*
c. concrete Porenbeton *m*, Zellenbeton *m*
c. double bottom *(Schiff)* Zellendoppelboden *m (zellartiger Doppelboden)*
c. filter *(Ch)* Zellenfilter *n*
c. flooring *(Bw)* Stahlzellendecke *f*
c. glass Schaumglas *n*
c. radiator *(Kfz)* Lamellenkühler *m*
c. structure Netz[werk]struktur *f*, zellenartige Struktur *f*, Zellenbau *m*, Zellenstruktur *f*
c. tissue Zellgewebe *n*
celluloid Zelluloid *n*, Kunsthorn *n*, Zellhorn *n*
c. templet Zelluloidschablone *f*
cellulose Zellulose *f*, Holzfaserstoff *m*, Holzzellstoff *m*, Lignose *f*, Zellstoff *m*
c. acetate Azetylzellulose *f*, Zelluloseazetat *n*
c. lacquer Zelluloselack *m*
c. nitrate Zellulosenitrat *n*, Nitrozellulose *f*
c. solution in cuprammonium hydroxide Cuoxam-Lösung *f (Chemieseideherstellung)*
cellulosics Zellulosederivate *npl*, Zelluloseplaste *mpl*
celo-navigation astronomische Navigation *f*, nautische Astronomie *f*
Celsius scale Celsiusskala *f*
cement / to 1. [ein]zementieren, [ver]kitten, [ver]kleben; 2. *(Met)* aufkohlen, einsetzen, einsatzhärten; 3. sintern
cement 1. Bindemittel *n*, Kitt *m*, Klebemittel *n*, Klebstoff *m*; 2. *(Bw)* Zement *m*; Bindemittel, Binder *m*; 3. *(Met)* Einsatz[härte]pulver *n*; 4. *(Opt)* Kitt
c.-aggregate ratio Verhältnis *n* des Zements zu den Zuschlagstoffen *(Betontechnologie)*
c.-asbestos Asbestzement *m*
c. batcher Zementdosierapparat *m*
c. batching Zementdosierung *f*
c. burning Zementbrennen *n*
c. carbon *(Met)* Zementit *m*
c. chock *(Schiff)* Zementfüllstück *n (Spantnische)*
c. clinker Zementklinker *m*
c. content Zementfaktor *m (Betontechnologie)*
c. copper Zementkupfer *n*
c. facing Zementaußenputz *m*
c. factor Zementgehalt *m*, Zementfaktor *m (Betontechnologie)*
c. fibre slab Holzzementplatte *f*, zementgebundene Faserplatte *f*
c. finish Zementputz *m (Außenputz)*; Zementestrich *m*
c. for optical elements Optikkitt *m*
c. foundation Zementfundament *n*
c. grout feiner Zementmörtel *m*; Zementschlämme *f*
c. gun Betonspritzmaschine *f*, Torkretkanone *f*
c. injection Zementeinpressung *f*, Zementinjektion *f*
c. kiln Zementofen *m*, Drehrohrofen *m*
c.-lime mortar Kalkzementmörtel *m*, verlängerter Zementmörtel *m*
c. mill Zementwerk *n*
c. mortar Zementmörtel *m*
c. pat Zementkuchen *m*
c. pipe Betonrohr *n*
c. rendering Zementaußenputz *m*
c. slurry Zementmilch *f*, Zementschlämme *f*
c. steel Zementstahl *m*
c. suspension Zementaufschlämmung *f*
c. water ratio Zement-Wasser-Verhältnis *n (Beton)*
cementation 1. Zementierung *f*; 2. *(Bw)* Bodenstabilisierung *f*, chemische Baugrundverbesserung (Bodenfestigung) *f*; 3. *(Met)* Aufkohlung *f*; Einsatzhärtung *f*, Oberflächenhärtung *f*
c. steel *(Met)* Einsatzstahl *m*
cementatory copper Zementkupfer *n*

cemented zementiert; abgebunden, verkittet; gesintert
c. carbide Sinterhartmetall *n*, Sinterkarbid *n*
c. carbide product gesintertes Hartmetall *n*
c. carbide tip HM-Plättchen *n*; Hartmetallschneidplatte *f*; Hartmetallschneide *f*
c. carbide tipped tool Werkzeug *n* mit Hartmetallschneide; hartmetallbestücktes Werkzeug
c. insert Hartmetalleinsatz[meißel] *m*
c. metal carbide Hartmetall *n*, Karbidhartmetall *n*
c. steel Zementstahl *m*
c. tungsten carbide gesintertes Wolframkarbid *n*
cementing 1. Zementierung *f*, Verkittung *f*, Verkleben *n*; Dichten *n*; 2. *(Met)* Aufkohlen *n*, Einsatzhärten *n*
c. agent 1. Dichtungsmittel *n*; 2. anorganisches Bindemittel *n*
c. in Einkitten *n (Schleifscheibe)*
c. machine Zementiermaschine *f (Schuhproduktion)*
c. metal Hilfsmetall *n (Sintern)*
c. process *(Text)* Klebebeschichtungsverfahren *n*
c. property *(Ch)* Backfähigkeit *f*
cementite Zementit *m*, Eisenkarbid *n*, Härtungskohle *f*
c.-bearing zementithaltig
cementitious verkittend
cements *(Plast)* Klebstoffkitte *mpl*
census card Zählkarte *f*, Zählblatt *n*
cent *(Kern)* Cent *m (Einheit der Reaktivität)*
center *(Am) s.* centre
centering *s.* centring
centesimal degree *(Math)* Neugrad *m*
c. graduation Neugradteilung *f*
c. minute Neuminute *f (Winkelmessung)*
c. second Neusekunde *f (Winkelmessung)*
centigrade heat unit = 453,6 cal
c. scale 100-Grad-Skale *f*, Celsius-Skala *f*
centigram Zentigramm *n*
centimetre Zentimeter *n*, cm
c.-gram-second unit CGS-Einheit *f*
c.-wave oscillator *(El)* Zentimeterwellenerzeuger *m*
c.-wave technique Zentimeterwellentechnik *f*
centimetric radar Zentimeterradar *n*
central mittig, auf Mitte; zentral; Mittel-; Mitten-
c. aerial Gemeinschaftsantenne *f*
c. aisle *(Eis)* Mittelgang *m*
c. axis Mittelachse *f*, Zentralachse *f*
c. battery Zentralbatterie *f*
c. battery signalling *(Fmt)* Rufsystem *n* mit Zentralbatterie; Zentralbatterieruf *m*
c. battery supply Zentralbatteriespeisung *f*
c. chassis lubrication *(Kfz)* Zentralschmierung *f*
c. chip disposal innere Spanabfuhr *f (Bohrer)*
c. circle Mittenkreis *m*
c. clock Zentraluhr *f*
c. collision *(Ph)* Zentralstoß *m*, zentraler Stoß *m*
c. condensing plant *(Wärm)* Zentralkondensationsanlage *f*
c. conductor *(El)* Mittelleiter *m*
c. control zentrale Regelung *f*
c. control desk Hauptsteuerpult *n*
c. control grouping zentrale Anordnung *f* der Schaltelemente
c. control room *(Fs)* Regieraum *m*
c. driven log frame *(Holz)* Hängegatter *n*
c. electric station Elektrizitätszentrale *f*
c. entrance *(Bw)* Mitteleingang *m*; *(Eis)* Mitteleinstieg *m*
c. European time mitteleuropäische Zeit *f*
c. exchange Vermittlungsamt *n*, Telefonzentrale *f*; Hauptamt *n*
c. feed Zentralvorschub *m*
c. field *(Ph)* Zentralfeld *n*
c. flux zentraler Fluß *m (der Neutronen im Reaktorzentrum)*
c. force Zentralkraft *f*
c. gate *(Met)* Eingußrohr *n (Gießtrichter)*
c. heating Zentralheizung *f*
c. heating fittings Zentralheizungsarmaturen *fpl*
c. heating plant Heizzentrale *f*; Zentralheizungsanlage *f*
c. hole Mittenbohrung *f*
c. intensity *(Opt)* Zentralintensität *f*, Intensität *f* der Linienmitte
c. lubricant distributing device (manifold) Zentralschmierapparat *m*
c. lubrication Zentralschmierung *f*
c. lubrication plunger Zentralschmierstößel *m*
c. mix[ing] plant zentrale Betonaufbereitungsanlage (Mischstation) *f*
c. monitoring position Überwachungszentrale *f*
c. motion Zentralbewegung *f*
c. office *(Fmt)* Hauptamt *n*; Zentralamt *n*
c. pile flux Neutronenfluß *m* im Reaktorzentrum, zentraler Fluß *m*
c. pintle *(Masch)* Mittelzapfen *m*
c. pivot Drehzapfen *m*
c. plane 1. *(Math)* Zentralebene *f*, Normalebene *f*; 2. Normalschnittebene *f (Schneckengetriebe usw.)*
c. position Mittelstellung *f*
c. position hook *(Text)* zur Nadelachse vorgebogener Nadelhaken *m (Zungennadel)*

c. power plant Maschinenzentrale *f*
c. producer Zentralgenerator *m*
c. reservation Mittelstreifen *m* *(Autobahn)*
c. shutter *(Foto)* Zentralverschluß *m*
c. station Kraftwerk *n*, Kraftzentrale *f*
c. tube frame *(Kfz)* Mittelrohrrahmen *m*
c. weft stop motion *(Text)* Mittelschußwächter *m*
c. wire *(El)* Kernleiter *m*
c. zero apparatus Nullinstrument *n* *(Meßgeräte)*
centralize / to zentrieren, ausmitten
centralized control Zentralsteuerung *f*
c. force-feed lubrication Zentraldruckschmierung *f*
c. grouping zentrale Anordnung *f*
c. lubricating system Zentralschmierung *f*
c. lubrication Zentralschmierung *f*
c. lubricator Zentralschmierapparat *m*; Schmierpumpe *f*
c. traffic control Fernsteuerung *f* einer Strecke *(Eisenbahnsignalwesen)*
centre / to ausmitten; zentrieren; ankörnen; zentrierbohren
c.-drill zentriersenken
c.-feed *(El)* im Mittelpunkt speisen
c. punch *(Masch)* ankörnen
centre 1. *(Math)* Mitte *f*, Mittelpunkt *m*, Zentrum *n*; 2. *(Masch)* Spitze *f*; 3. *(El)* Sternpunkt *m (Sternschaltung)*; 4. *(Bw)* Lehrgerüst *n (Bogen- und Gewölbebau)*; 5. *(BMSR)* Wirbel *m (Phasenebenendarstellung)*; 6. Zentrierbohrung *f*
c. axis *(Masch)* Bohrungsachse *f*, Mittenachse *f*
c. bar *(El)* Mittelleiter *m*
c. bed Kopfstück *n (Wirkmaschine)*
c. bit Schlangenbohrer *m*
c.-board *(Schiff)* Bootsschwert *n*, Mittelschwert *n*, Kielschwert *n*, Schwert *n*
c.-board cruiser Wanderjolle *f*
c.-board racer Binnenjolle *f*
c.-board trunk Schwertkasten *m*, Schwertbrunnen *m*
c.-board well *s.* c.-board trunk
c.-board yacht Kielschwertjacht *f*
c. bore *(Masch)* Zentrierbohrung *f*
c. buffer with top link Mittelpufferung *f* mit obenliegendem Zug
c. column rotary indexing machine Rundschaltmaschine *f* der Ringtischbauart
c. column rotary transfer machine Ringtischmaschine *f*
c. conductor *(El)* Innenleiter *m*
c. conductor rail Mittelstromschiene *f*
c. core *(El)* Kernstopfen *m*, Mittelkern *m*
c. disk Scheibe *f (Radscheibe)*
c. distance 1. Spitzenentfernung *f*, Spitzenweite *f (Drehmaschine)*; 2. Achsabstand *m (Verzahnung)*; 3. Mittenabstand *m*
c. drilling Zentriersenken *n*
c. drilling head casing Zentrierbohrkopfgehäuse *n*
c. electrode Mittelelektrode *f (Zündkerze)*
c. exchange *(Fmt)* Knotenamt *n (einer Netzgruppe)*
c. expansion Nullpunktaufweitung *f*, Mittelpunktaufweitung *f (Radartechnik)*
c. feed 1. *(El)* Mittelpunktspeisung *f*; 2. Zentralbeschickung *f*
c. feed aerial T-Antenne *f*, stromgespeister Dipol *m*
c. frequency Trägerfrequenz *f*, Bandmittenfrequenz *f*, Mittenfrequenz *f*
c. gauge Gewindelehre *f*
c. girder *(Schiff)* Mittelkielplatte *f*, Mittelträger *m*
c. grinder Spitzenschleifapparat *m*
c. grip Mitteleinspannung *f (z. B. einer Schweißelektrode)*
c. guide Innenführung *f (bei Zahnketten)*
c. guide flange Innenbord *m (Pendelrollenlager)*
c. height *(Masch)* Spitzenhöhe *f*
c. hole Zentrierbohrung *f*, Zentrierung *f*
c. hole grinding machine Zentrierbohrungsschleifmaschine *f*
c. keel *(Schiff)* Mittelkiel *m*
c. key *(Masch)* Austreiber *m* für Körnerspitze
c. knocking over surface *(Text)* mittlere Abschlagkante *f (A-Platine)*
c. lathe Spitzendrehmaschine *f*
c. lens *(Opt)* Zentrallinse *f*
c. line 1. *(Math)* Mittellinie *f*; Mittelachse *f*; Zentrale *f*; 2. Mittelachse *f*; 3. *(Masch)* Zentrale *f (Kurbelstellung)*
c. line bulkhead *(Schiff)* Mittellängsschott *n*
c. line of the connecting rod *(Masch)* Schubstangenachse *f*
c. line plane Mittschiffsebene *f*
c. loading zentrale Belastung *f*, Zentralbelastung *f (Statik)*
c. locations Lagegenauigkeit *f* der Bohrungsmittelachsen
c. mark Ankörnung *f*, Körnermarke *f*
c. of anticyclone Hochdruckzentrum *n (Meteorologie)*
c. of area Flächenmittelpunkt *m*
c. of attraction Anziehungsmittelpunkt *m*
c. of buoyancy *(Schiff)* Verdrängungsschwerpunkt *m*, Formschwerpunkt *m*, Deplacementsschwerpunkt *m*
c. of crystallization Kristallisationskeim *m*, Kristallisationskern *m*, Kristallisationszentrum *n*

c. of curvature Krümmungsmittelpunkt *m*
c. of diffraction *(Opt)* Beugungszentrum *n*
c. of distribution *(El)* Lastschwerpunkt *m*
c. of disturbance Störstelle *f (Realkristalle)*
c. of eccentric *(Masch)* Exzentermittel *n*
c. of effort *(Schiff)* Segelschwerpunkt *m*
c. of flexure Biegungszentrum *n*
c. of gravity 1. *(Ph)* Gravitationszentrum *n*; 2. Schwerpunkt *m*, Massenmittelpunkt *m*
c. of gravity of the disk Scheibenschwerpunkt *m (Turbine)*
c. of gyration 1. *(Ph)* Drehpunkt *m*, Rotationszentrum *n*; 2. *(Masch)* Schwungmassenmittelpunkt *m*
c. of inertia Massenmittelpunkt *m*, Schwerpunkt *m*
c. of inflexions *(Math)* Wendepol *m*
c. of lateral resistance *(Schiff)* Schwerpunkt *m* des Treibungswiderstands, Lateralschwerpunkt *m*
c. of machine Maschinenmitte *f*
c. of mass Massenmittelpunkt *m*, Massenzentrum *n*, Schwerpunkt *m*
c. of mass coordinate Schwerpunktskoordinate *f*
c. of mass system Schwerpunktsystem *n*
c. of movement *(Masch, Wärm)* Schubmittelpunkt *m*
c. of oscillation 1. *(Ph)* Schwingungsmittelpunkt *m*; 2. *(Masch)* Pendeldrehpunkt *m*
c. of percussion Stoßmittelpunkt *m*
c. of pressure Druck[mittel]punkt *m*, Druckzentrum *n*
c. of pressure coefficient Druckmittelpunktkoeffizient *m*
c. of rotation Rotationszentrum *n*, Drehpunkt *m*
c. of suspension Aufhängepunkt *m (Leitungen)*
c. of thrust Antriebsmittelpunkt *m*
c. of vision *(Opt)* Augenpunkt *m (Perspektive)*
c. pillar Mittelpfosten *m (Karosserie)*
c. pin Mittelstift *m (z. B. eines Röhrensokkels)*
c. pip Zentralpunkt *m (Visier)*
c. ply Mittellage *f (Sperrholz)*
c. point 1. *(Masch)* Spitze *f*; 2. *(El)* Sternpunkt *m*, Mittelpunkt *m (Sternschaltung)*
c.-point mooring *(Schiff)* Zentralverankerung *f*
c. point of bore (hole) Bohrungsmitte *f*, Lochmitte *f*
c. pole suspension *(El)* Mittelmastaufhängung *f*
c. portion Mittelteil *n*
c. position Mittelstellung *f*
c. post *(Brg)* Mittelstempel *m*
c. punch Bohrkörner *m*; Kontrollkörner *m*
c.-punch mark Körnermarke *f*; Körnung *f*
c. punching Ankörnen *n*
c. rest Lünette *f (Drehmaschine)*
c. riser Gießtrichter *m*
c. roll Mittelwalze *f*
c. section Mittelteil *n (z. B. Rakete)*
c. section strut *(Flg)* Baldachin-Strebe *f*
c. section wing *(Flg)* Flügelmittelstück *n*
c. shaft *(Text)* mittlere Tragestange *f* des Spulengestells *(Rundstuhl)*; Spindelschaft *m*
c. shaft washer *(Text)* Spindelschaftdichtungsring *m*
c. shed dobby *(Text)* Schaftmaschine *f* für Hoch- und Tieffach
c. shed Jacquard [machine] *(Text)* Hoch- und Tieffach-Jacquardmaschine *f*
c. shot *(Brg)* Einbruchschuß *m*
c. square Zentrierwinkel *m*
c. stop *(Masch)* Mittenanschlag *m*
c. support Königsstuhl *m (Drehscheibe)*
c. tap *(El)* Mittelabgriff *m*, Mittelanzapfung *f*
c. tapped *(El)* mit Mittelanzapfung, mit Mittelabgriff
c. through-plate *(Schiff)* durchlaufende Mittelkielplatte *f*
c. to centre Achsabstand *m*, Achsmaß *n*, Systemmaß *n*
c.-to-centre distance Mittenabstand *m*
c.-to-centre distance of holes Lochmittenabstand *m*
c.-to-centre line *(Bw)* Systemlinie *f*
c. track rod *(Kfz)* Mittelteil *n* einer dreiteiligen Spurstange *(Lenkung)*
c. traverse beam *(Text)* Mittelriegel *m*
c. visional ray *(Opt)* Hauptstrahl *m*
c. web Radscheibe *f*; Kern *m*, Seele *f*, Steg *m (Spiralbohrer)*
c.-well dredger Bagger *m* mit Mittelschlitz
c. wire *(El)* Seelendraht *m*, Seelenelektrode *f*
c. zero instrument Instrument *n* mit Nullpunkt in der Skalenmitte, Instrument mit symmetrischer Skala
c. zero relay *(El)* Wechselrelais *n*, polarisiertes Relais *n*

centreboarder *(Schiff)* Kielschwertjacht *f*
centreless spitzenlos
c. bar skimming machine spitzenlose Wellenschäl- und -drehmaschine *f*
c. cylindrical grinding machine spitzenlose Rundschleifmaschine *f*
c. grinding spitzenloses Schleifen *n*

centric zentrisch, mittig
c. application of force *(Mech)* zentrischer (mittiger) Kraftangriff *m*

centricleaner Zentrifugalreiniger *m (Papierherstellung)*
centrifugal zentrifugal
c. acceleration Zentrifugalbeschleunigung *f*

c. air cleaner Schleuderluftfilter *n*
c. ball *(Masch)* Schwingkugel *f*, Schwungkugel *f*
c. blast cleaning Schleuderputzstrahlen *n*
c. blower Kreiselgebläse *n*, Schleudergebläse *n*, Zentrifugalgebläse *n*
c. box *(Text)* Spinnzentrifuge *f*
c. casting 1. Schleuderguß *m*; Schleudergußstück *n*; 2. *(Bw)* Schleuderverfahren *n* *(Schleuderbeton)*
c. classifier *(Brg)* Fliehkraftklassierer *m*
c. clutch Fliehkraftkupplung *f*
c. compressor Kreiselverdichter *m*, Radialverdichter *m*
c. condenser *(Wärm)* Kreiselkondensator *m*
c. conveyor *(Bw)* Schleuderförderer *m*
c. disk atomizer Fliehkraftzerstäuber *m*
c. drier Trockenschleuder *f*, Trockenzentrifuge *f*
c. fan *(Brg)* Schleuderlüfter *m*, Zentrifugallüfter *m*
c. filter Siebschleuder *f*, Schleuderfilter *n*
c. force Fliehkraft *f*, Zentrifugalkraft *f*
c.-force governor Fliehkraftregler *m*
c.-force tachometer Fliehkrafttachometer *n*
c. governor Fliehkraftregler *m*, Zentrifugalregler *m*
c. grinder Pendelmühle *f*
c. impeller mixer Turbinenmischer *m*
c. lubrication Schleuderschmierung *f*
c. lubricator Zentrifugalschmiereinrichtung *f*
c. machine Zentrifuge *f*
c. mill Schlagmühle *f*, Schleudermühle *f*
c. moment Fliehkraftmoment *n*
c. motion Zentrifugalbewegung *f*
c. pot *(Am) (Text)* Spinnzentrifuge *f*
c. pouring method Schleudergußverfahren *n*, Schleudergießverfahren *n*
c. power Fliehkraft *f*, Schwungkraft *f*
c. pump Kreiselpumpe *f*, Schleuderpumpe *f*, Zentrifugalpumpe *f*
c.-ring-type knife-cutting machine Zentrifugal-Ring-Messerschneidemaschine *f*
c. screen Schleudersieb *n*; Zentrifugalsortierer *m*
c. scrubber Zentrifugalabscheider *m*
c. spark advance *(Kfz)* Fliehkraftzündversteller *m*
c. sprayer Zentrifugalzerstäuber *m*
c. starter *(Elm)* Fliehkraftanlasser *m*, Zentrifugalanlasser *m*
c. stower *(Brg)* Versatzschleuder *f*, Schleuderversatzmaschine *f*
c. sugar Schleuderzucker *m*
c. supercharger Schleuderlader *m*
c. switch Fliehkraftschalter *m*
c. wheel Schleuderrad *n*, Schaufelrad *n*

centrifugally cast im Schleudergußverfahren hergestellt
c. cast concrete Schleuderbeton *m*
c. controlled fliehkraftgesteuert
centrifugation Zentrifugierung *f*, Ausschleuderung *f*, Trennung *f* mit der Zentrifuge
centrifuge / to [ab]schleudern, ausschleudern, zentrifugieren
centrifuge Schleuder *f*, Zentrifuge *f*
centring 1. Einmitten *n*, Zentrieren *n*; Mitten *n*; 2. Bogenrundung *f* *(Zimmerei)*; 3. *(Bw)* Lehrgerüst *n* *(Bogen- und Gewölbebau)*
c. chuck Zentrierkopf *m* *(Bohrwinde)*
c. control *(El)* Mitteneinregelung *f*
c. device *(Masch)* 1. Zentriervorrichtung *f*; 2. Ankörnvorrichtung *f*
c. dowel Zentrierstift *m*
c. head Zentrierbohrkopf *m*
c. lathe (machine) Zentrier[dreh]maschine *f*; Anbohrmaschine *f*
c. microscope Zentriermikroskop *n*
c. operation Zentrierung *f*
c. pilot Zentrierzapfen *m*
c. pin 1. Zentrierstift *m*; 2. zentrischer Dorn *m* *(Turbine)*
c. plug Zentrierzapfen *m*
c. ring 1. Zentrierring *m*; 2. Paßring *m*
c. shoulder Zentrierleiste *f*
c. sleeve Zentrierbüchse *f*, Zentrierhülse *f*
c. taper Zentrierkegel *m*; Zentrierhülse *f*
c. tool Zentrierbohrer *m*
centripetal zentripetal
c. acceleration Zentripetalbeschleunigung *f*
c. force Zentripetalkraft *f*
c. motion Zentripetalbewegung *f*
c. turbine Radialturbine *f* mit Strömung von außen nach innen
centriscope *(Opt)* Zentriskop *n*
centriscreen Zentrifugalsortierer *m* *(Papierherstellung)*
centrode Schwerpunktskurve *f*
centroid Schwerpunkt *m*; Massenmittelpunkt *m*
centroidal distance Schwerpunktsabstand *m*, Schwerpunktsdistanz *f*
ceramic keramisch
c. bit keramischer Einsatzmeißel *m*
c. capacitor *(El)* Keramikkondensator *m*, keramischer Kondensator *m*
c. coating keramische Schutzschicht *f*
c. coil Keramikspule *f*
c. fuel [material] *(Kern)* keramischer Brennstoff *m*
c. insulator Porzellanisolator *m*
c. products keramische Erzeugnisse *npl*; Baukeramik *f*
c. tip Keramikschneidplatte *f*
c.-tipped mit Schneiden aus keramischem Werkstoff

ceramics Keramik *f*
cerargyrite Hornsilber *n*
cereals Halmfrüchte *fpl*
Cerenkov counter *(Kern)* Tscherenkow-Zähler *m*, Čerenkov-Zähler *m*
Cerenkov radiation *(Kern)* Tscherenkow-Strahlung *f*, Čerenkov-Strahlung *f*
cereous wachsartig, wächsern
ceresine wax Zeresin *n*
ceria Zer(III)-oxid *n*
ceric zer(IV)-haltig
 c. compound Verbindung *f* des vierwertigen Zers
cerite earth *(Ch)* Zeriterde *f*
cerium Zer *n*, Zerium *n*
cermet *(Am) s.* metal ceramics
cerotic acid Zerotinsäure *f*
cerotinic acid Zerotinsäure *f*
cerous zer(III)-haltig
 c. compound Verbindung *f* des dreiwertigen Zers
certainty of measurement *(Opt)* Meßsicherheit *f*
certificate of airworthiness *(Flg)* Lufttüchtigkeitsschein *m*
 c. of identity Schiffszertifikat *n*
 c. of incorporation Eintragung *f* in das Handelsregister
certified colours gesetzlich zugelassene Farbstoffe *mpl* für Nahrungsmittel
 c. copy beglaubigte Abschrift *f*
cerussite Weißbleierz *n*, Bleikarbonat *n*, Bleispat *m*, Zerussit *m*
cesium Zäsium *n*
cessation reaction *(Ch)* Kettenabbruchsreaktion *f*
cesspipe Abflußrohr *n* für Senkgrube
cesspit *s.* cesspool
cesspool Klärgrube *f*, Faulgrube *f*, Senkgrube *f*
cetane number Zetanzahl *f*
 c. value Zetanwert *m*, Zetanzahl *f*
cetene number Zetenzahl *f*
cetol Cetylalkohol *m*
cetone *(Ch)* Keton *n*
 c. acid Ketonsäure *f*
CFR-engine CFR-Motor *m* *(Kraftstoffprüfung)*
chabasite *(Min)* Chabasit *m*
chabazite *s.* chabasite
chafe / to scheuern, reiben
chafe mark *(Text)* Scheuerstelle *f*, Schleifstelle *f* *(Fehler)*
chafed abgescheuert *(Garn)*
chafer *(Kfz)* Wulstschutzstreifen *m*
chaff 1. Häcksel *m (n)*, Spreu *f*, Kaff *n*; 2. Düppel *m (Radarstörung)*
 c.-blowing pipe Spreurohr *n*
 c. bunker Spreubehälter *m*
 c. chopping knife Häckselmesser *n*
 c. cutter Häcksler *m*, Häckselbank *f*, Häckselmaschine *f*
 c.-cutting combine Mähhäcksler *m*
 c.-cutting harvester Mähhäcksler *m*
 c.-cutting knife Häckselmesser *n*
 c.-cutting machine Häcksler *m*, Häckselschneidmaschine *f*
 c. dropping *(Am)* Düppelung *f* *(Radar)*
 c. ejector Häckselauswerfer *m*, Enthäcksler *m*
 c. hose Spreuschlauch *m*
 c. pipe Spreurohr *n*
 c. sifter Spreusichter *m*
chaffer Spreusichter *m*
chafing *(Schiff)* Schamfilung *f*
 c. of the cords *(Text)* Aneinanderreihen *n* der Schnüre
 c. piece *(Schiff)* Schamfilungsleiste *f*
chain 1. Gliederkette *f*, Kette *f*; 2. *[Längenmaß = 66 feet (Gunther's chain) oder 100 feet (Ramsden engineer's chain); Längenmeßgerät]*
 c. attachment link Kettenbefestigungsglied *n*
 c. bar drawing frame *(Text)* Kettenstrecke *f*
 c. belt endlose Kette *f*
 c. bitts *(Schiff)* Kettenpoller *m*, Ankerkettenpoller *m*
 c. block Kettenflaschenzug *m*
 c. boat *(Schiff)* Ankerleichter *m*
 c. branching *(Ch)* Kettenverzweigung *f*
 c. breaking *(Ch)* Kettenabbruch *m*
 c. breast machine *(Brg)* Kettenschrämmaschine *f*
 c. bridge *(Bw)* Kettenbrücke *f*
 c. bridging *(Ch)* Brückenbildung *f* zwischen Kettenmolekülen
 c. bucket elevator Eimerkettenförderer *m*
 c. bushing Kettenhülse *f*
 c. cable *(Schiff)* Ankerkette *f*
 c. carrier Förderkette *f*; Transportkette *f*
 c. case Kettenkasten *m*
 c. coal-cutter *(Brg)* Kettenschrämmaschine *f*
 c. combination *(Ch)* Kettenverbindung *f*
 c. conformation *(Ch)* Kettenkonformation *f* *(Proteinstruktur)*
 c.-controlled carrier stop durch eine Kette gesteuerter Fadenführeraufläufer *m* *(Strickmaschine)*
 c. conveyor Kettenförderer *m*, Kettenförderband *n*
 c. coupler link Kettenverbindungsglied *n*
 c. coupling Zahnkettenkupplung *f*
 c. creeper *(Brg)* Vorziehkettenbahn *f*, Unterkettenbahn *f*
 c. cutter moulding machine Kettenfräse *f*
 c. decay (disintegration) *(Kern)* [radioaktiver] Kettenzerfall *m*, Kettenumwandlung *f*
 c.-dotted line Strichpunktlinie *f*
 c. drag Gliederschleppe *f*
 c. drill Kettenbohrapparat *m*

c. **drive** Ketten[an]trieb *m*
c.-**driven** mit Kettenantrieb
c. **dyeing** *(Text)* Kettenfärbung *f*
c. **elevator** Kettenbecherwerk *n*
c. **end link** Kettenendglied *n*
c. **filter** *(El)* Filterkette *f*
c. **fission yield** *(Kern)* Gesamtspaltproduktausbeute *f*
c. **follower** Kettenzieher *m*
c. **gearing** Kettenübersetzung *f*
c. **grate** *(Wärm)* Kettenrost *m (Wanderrost)*
c. **grate stoker** Kettenrostfeuerung *f*
c. **greasing device** Kettenschmiereinrichtung *f*
c. **grinder** *(Holz)* Stetigschleifer *m*
c. **gripper** *(Typ)* Kettengreifer *m*
c. **guard** 1. Kettenschutz *m*; 2. *(Schiff)* Geländer *n* mit Kettendurchzug
c. **guide** Kettenführung *f*
c.-**guide pulley** Kettenführungsrolle *f*, Kettenleitrolle *f*
c. **halogenation** *(Ch)* Kettenhalogenierung *f*
c. **haulage** *(Brg)* Kettenbahnförderung *f*
c. **index number** *(Math)* Kettenindexziffer *f*
c. **insulator** Kettenisolator *m*, Hängekette *f*, Isolatorkette *f*
c. **interruption** *(Ch)* Kettenabbruch *m*
c. **isomerism** *(Ch)* Kettenisomerie *f*
c. **jockey** Kettenmitnehmer *m*
c. **lease** *(Text)* kleines Kreuz *n*, Fußkreuz *n*, Gangkreuz *n*
c. **length** Kettenlänge *f*
c. **lever** Kettenhebel *m*, Kupplungshebel *m*
c. **link** Kettenglied *n*, Kettenöse *f*, Kettenschäkel *m*
c. **locker** 1. *(Masch)* Kettenschloß *n*; 2. *(Schiff)* Kettenkasten *m*, Kettenkammer *f*, Decksklüse *f*
c. **loom** Schaft[web]stuhl *m*
c. **method** *(Math)* Kettenrechnung *f*
c. **molecule** Kettenmolekül *n*
c. **nose cutting plier** Kettenzange *f*
c. **of cards** *(Text)* Kartenkette *f*
c. **of lags** *(Text)* gelenkig verbundene Holzkarten *fpl*
c. **of traverse rail** *(Text)* Fadenführerkette *f*
c. **pin** Kettenbolzen *m*
c. **pipe tong** Kettenrohrzange *f*
c. **pipe vice** Kettenrohrspanner *m*
c. **pipe wrench** Kettenrohrzange *f*
c. **pitch** Kettenteilung *f*
c. **propagator** *(Ch)* Kettenträger *m (Reaktionskinetik)*
c. **pulley** Kettenrolle *f*
c.-**reacting amount** *(Kern)* kritische Masse *f*
c.-**reacting pile** kritischer Reaktor *m*
c. **reaction** Kettenreaktion *f*
c.-**reaction pile** *s.* nuclear reactor
c. **relative** *(Math)* Kettenzahl *f*
c. **rim** Kettenkranz *m*
c. **roller** Ketten[lauf]rolle *f*
c. **roller bearing** Kettenrollenlager *n*
c. **rule** *(Math)* Kettenregel *f*
c. **saw** Kettensäge *f*
c. **screen grubber** Kettensiebroder *m*
c. **screw brake** Kettenspindelbremse *f*
c. **sequence** *(Ch)* Kettensequenz *f (Proteinstruktur)*
c. **sling** Kettenschlinge *f*
c. **snip** Kettensäge *f*
c.-**spaced rows of rivets** Kettennietung *f*
c.-**stitch technique** *(Text)* Nähwirktechnologie *f (Malimo-Verfahren)*
c. **stopper** *(Ch)* Kettenabbruch bewirkendes Agens *n*
c. **stopping** *(Ch)* Kettenabbruch *m*
c. **stud** Kettenbolzen *m*
c. **tape** Meßkette *f*
c. **tensioner** Kettenspanner *m*
c. **termination** *(Ch)* Kettenabbruch *m*
c. **terminator** *(Ch)* Kettenabbruch bewirkendes Agens *n*
c. **tightener** Kettenspanner *m*
c. **tube cleaner** *(Wärm)* Röhrenreinigungskette *f*, Kettenröhrenreiniger *m*
c. **twisting** *(Text)* Andrehen *n* der Kette
c.-**type float gauge** Pegelmesser *m*
c.-**type manure distributor (spreader)** Kettendüngerstreuer *m*
c.-**type table planing machine** *(Holz)* Kettentischhobelmaschine *f*
c. **velocity** Kettengeschwindigkeit *f*
c. **wheel** Kettenrad *n*, Kettenrolle *f*
c. **width** Kettenbreite *f*
c. **winding** *(El)* Doppelschichtkorbwicklung *f*
chainette *(Math)* Kettenlinie *f*
chaining *(Text)* Aufziehen *n* der Kette
c. **with rectangular offsets** Orthogonalaufnahme *f (Vermessung)*
chainman *(Brg)* Markscheidergehilfe *m*
chair back bending machine Stuhllehnenbiegemaschine *f*
c. **form** *(Ch)* Sesselform *f (Stereochemie)*
c.-**making industry** Stuhlindustrie *f*
c. **rail** *(Bw)* Schutzleiste *f (an Wänden angebracht zum Schutz gegen Stuhllehnen)*
chalcanthite *(Min)* Chalkanthit *m*; Kupfervitriol *m*
chalcopyrite Eisenkupferkies *m*, Gelbkupfererz *n*, Kupferkies *m*, Kupferpyrit *m*
chalcostibite Kupferantimonglanz *m*
chalk / to mit Kreide schreiben (zeichnen)
c. **up** *(Typ)* [durch]pausen *(manuell)*
chalk Kreide *f*
c. **cliff** Kreidefelsen *m*
c. **fog** Kalkschleier *m*
c. **line** 1. Kreidestrich *m*; 2. Kreideschnur *f*

c. marl Kreidemergel *m*
c. test for cracks Kalkmilchprobe *f*
c. transfer-paper *(Typ)* Umdruckpapier *n*
chalking *(Plast)* Ausschwitzen *n*, Beschlag[en *n*] *m (eines Preßlings)*
chalky kreideartig, kreidehaltig, kreidig
c. clay Mergelton *m*
challenger Kennungsabfragegerät *n (Radar)*
chalybeate eisenähnlich, eisenartig, eisenhaltig
chalybite Spateisenstein *m*
chamber 1. Raum *m*, Kammer *f*; 2. *(Brg)* Abbaukammer *f*, Abbaustrecke *f*; 3. Kammer *(z. B. Sprengkammer)*
c. acid Kammer[schwefel]säure *f*
c. angle Kammerwinkel *m (Ophthalmologie)*
c. blasting *(Brg)* Kammersprengung *f (Schießarbeit)*
c. crystals Bleikammerkristalle *mpl*
c.-depth indicator Kammertiefenmesser *m (Ophthalmologie)*
c. envelope Kesselmantel *m*
c. filter-press Kammerfilterpresse *f*
c. furnace Kammerofen *m*
c. of a pump *(Wärm)* Pumpenstiefel *m*
c. of beam *(Schiff)* Balkenbucht *f*
c. oven Kammerofen *m*
c. pressure *(Ch)* Ofendruck *m*
c. process Kammerverfahren *n (Schwefelsäure)*
c. superheater *(Wärm)* Kammerüberhitzer *m*
c. wall Kesselmantel *m*
c. working Kammerbau[verfahren *n*] *m (insbesondere Erz- und Salzbergbau)*; Pfeilerbau[verfahren *n*] *m (insbesondere Kohlenbergbau)*
Chamber of Commerce Handelskammer *f*
Chamber of Industry and Commerce Industrie- und Handelskammer *f*
chambered hole *(Brg)* gekesseltes Bohrloch *n*
chambering *(Brg)* Auskesseln *n* eines Bohrlochs
chamfer / to 1. *(Masch)* fasen, abkanten, abschrägen, anfasen, kegelig abkanten, Kanten brechen; 2. *(Holz)* abfasen, auskehlen
c. edges *(Masch)* Kanten brechen
chamfer 1. *(Masch)* Abfasung, Abschrägung *f*, Fase *f*, Schrägkante *f*; Anschnitt *m (Gewindebohrer)*; 2. *(Holz)* Fase, Kehle *f*, Schrägkante *f*; 3. *(Bw)* Fase *(meistens 45°)*; Abschrägung, Schräge *f*; 4. *(Opt)* Facette *f*
c. angle Abschrägungswinkel *m*
c. at the corner of a cutter tooth Stumpfung *f*
c. cutting edge *(Am)* Anschnittschneide *f*
c. length *(Am)* Anschnittlänge *f*
c. strip Dreikantleiste *f*
chamfered abgefast, abgekantet, abgesetzt, abgeschrägt; verjüngt; gebrochen
c. core konischer Kern *m*
c. edge 1. Anschnitt *m (Gewindeschneidbacke)*; 2. *(Holz)* Fase *f*
c. end Kegelansatz *m*
c. joint *(Masch)* abgefaste Fuge *f*
c. point Kegelansatz *m (Schraube)*
c. teeth abgeschrägte Zähne *mpl*
chamfering 1. Abfasen *n*, Anfasen *n*, Fasen *n*; 2. Abschrägung *f*; Abkantung *f*
c. attachment Abkantvorrichtung *f*
c. off *(Holz)* Abfasen *n*
chammy Sämischleder *n*
chamois dressing *(Led)* Sämischgerbung *f*
c. [leather] Gemsleder *n*, Sämischleder *n*, Chamoisleder *n*
champion lode *(Geol)* Hauptgang *m*
chance coincidence *(Math, Ph)* zufällige Koinzidenz *f*
c. of error Fehlermöglichkeit *f*, Fehlerwahrscheinlichkeit *f*
c. process *(Brg)* Sandschwimmverfahren *n*
c. rate *(Kern)* Zählrate *f* der Zufallsimpulse
change / to *(Masch)* auswechseln *(Werkzeuge)*; umschalten *(Vorschübe, Drehzahlen usw.)*; schalten *(Ventile)*
c. back to sich rückverwandeln in
c. colour sich verfärben
c. matter to energy Materie in Energie umwandeln
c. rotation of cutter Drehrichtung des Fräskopfes ändern
c. the tool set-up *(Masch)* umrüsten
change Wechsel *m*, Veränderung *f*, Abänderung *f*, Verwandlung *f*
c.-box slay *(Text)* Wechsellade *f*
c. collar Wechselmuffe *f*
c. collet Wechselhülse *f*
c. gear Umsteckrad *n*, Wechselrad *n*
c. gear calculation Wechselräderberechnung *f*
c. gear quadrant Wechselräderschere *f*
c. gear train Wechselrädergetriebe *n*
c. gear wheel Wechselrad *n*
c. gears Wechselräder *npl*; Wechselrädergetriebe *n*
c. in lignin content Änderung *f* der Aufschlußzahl *(Zelluloseherstellung)*
c. lever Stellhebel *m*; Umschalthebel *m*
c. mixer *(Ch)* Mischer *m* mit auswechselbarem Behälter
c. of colour Farbumschlag *m*, Farbänderung *f*, Farbwechsel *m*
c. of direction Richtungswechsel *m*
c. of length Längenänderung *f*
c. of pitch Luftschraubenverstellung *f*
c. of polarity Polaritätswechsel *m*; Polwechsel *m*
c. of shift Schichtwechsel *m (Arbeitszeit)*
c. of slags Schlackenwechsel *m (Roheisenfrischverfahren)*

c. of solubility Löslichkeitsänderung *f*
c. of state *(Ph)* Zustandsänderung *f*
c. of style Stilwandel *m (z. B. Mode)*
c. of trim *(Schiff)* Trimmänderung *f*, Tauchungsänderung *f*
c. of unused velocity into pressure Rückverwandlung *f* der nicht aufgebrauchten Geschwindigkeit in Druck *(Turbine)*
c. of voltage 1. Spannungswechsel *m*, Spannungsveränderung *f*; 2. Umspannung *f*
c. of weave *(Text)* Bindungswechsel *m*
c.-over 1. Umrüsten *n*; 2. Übergang *m*; 3. Überblendung *f*
c.-over contact [unit] *(El)* Wechselkontakt *m*
c.-over contactor *(El)* Umschaltschütz *n*
c.-over jet *(Kfz)* Übergangsdüse *f*
c.-over slide switch for tape speeds Bandgeschwindigkeitenschiebeumschalter *m*
c.-over switch *(El)* Wechselschalter *m*, Umschalter *m*
c.-over time *(Masch)* Umrüstzeit *f*
c.-over to Übergang *m* auf
c.-over tube *(Opt)* Wechseltubus *m*
c.-over valve *(Masch)* Umschaltventil *n*
c. point *(Met)* Haltepunkt *m*, Umwandlungspunkt *m*
c.-pole induction motor polumschaltbarer Asynchronmotor *m*
c.-pole motor polumschaltbarer Motor *m*
c.-speed drive Stufengetriebe *n*
c.-speed gear Schaltgetriebe *n*
c.-speed geared drive Stufenrädergetriebe *n*
c.-speed lever Schalthebel *m*, Getriebeschalthebel *m*
c.-speed lever knuckle Schalthebelfinger *m (Schaltgetriebe)*
c.-speed motor Regelmotor *m*, Stellmotor *m*, Motor *m* mit regelbarer Drehzahl
c. tune switch *(El)* Wellenbereichsschalter *m*, Wellenumschalter *m*
c.-valve box *(Masch)* Wechselventilkasten *m*
c. wheel Wechselrad *n*

changeover *s.* change-over

changes in the parameters of the orbit *(Astr)* Veränderungen *fpl* der Parameter der Flugbahn, Veränderungen der Flugbahnwerte

changing-over *(El)* Umschaltung *f*
c. rolls Walzenwechsel *m*
c. speed lever *(Kfz)* Schalthebel *m*

channel / to [aus]kehlen, aussparen, riffeln *(Holzbearbeitung)*

channel 1. Kanal *m*; Rohr *n*; Schacht *m*; Auskehlung *f*, Riefe *f*, Rille *f*, Rinne *f*; 2. *(künstlicher)* Kanal, Fahrrinne *f*; Hafeneinfahrt *f*; 3. Kanal *(Strömungslehre)*; 4. Fernsehkanal *m*; 5. Flugkanal *m*; 6. Nachrichtenkanal *m*; Übertragungskanal *m*; 7. *(Dat)* Speicher[kanal] *m*; 8. U-Profil *n*, U-Stahl *m*
c. allocation 1. *(Fs)* Kanaleinteilung *f*; 2. *(Fmt)* Wellenverteilung *f*
c. amplifier *(Fs)* Kanalverstärker *m*, KV
c. buoy *(Schiff)* Fahrwassertonne *f*
c. capacity *(Dat)* Kanalkapazität *f*
c. filter Kanalblende *f*; Kanalfilter *n*, Trennfilter *n*
c. identification letter *(Flg)* Kanalkennbuchstabe *m*
c. iron U-Eisen *n*, U-Stahl *m*
c. sample *(Brg)* Schlitzprobe *f*
c. section Doppel-T-Profil *n*, U-Profil *n*, U-Querschnitt *m*
c.-section cross member Querträger *m* mit U-Profil
c.-section rail U-Schiene *f*
c. selector *(Fmt, Fs)* Kanalwähler *m*
c. separation *(El)* Kanaltrennung *f*; Trennschärfe *f*
c. spacing *(Fmt, Fs)* Kanalabstand *m*
c. spin *(Ph)* Kanalspin *m*
c. steel *s.* c. iron
c. width *(Fs)* Kanal[band]breite *f*
c. working *(Fmt)* Mehrfachausnutzung *f*

channeling *s.* channelling

channelled kanneliert
c. plate Riffelblech *n*
c. roller Riffelwalze *f*

channelling 1. Auskehlen *n*, Auskehlung *f*, nutenförmige Aussparung *f*, Riefenbildung *f*; Riffeln *n*; 2. *(Kern)* Kanaleffekt *m*, Kanalwirkung *f*
c. effect factor *(Kern)* Kanalfaktor *m*
c. fluidized bed durchbrochene Wirbelschicht

channellized transmitter *(El)* Sender *m* mit Simultankanälen

chaplet *(Met)* Kernbock *m*, Kernnagel *m*, Kernstütze *f*

char / to 1. [an]kohlen, brennen, schmoren, verkohlen, verkoken; 2. *(Bw)* scharrieren

char Pflanzenkohle *f*, Tierkohle *f*
c.-dust bin Kohlenstaubbunker *m*

character *(BMSR, Dat)* Schriftzeichen *n (einschließlich Zahlzeichen oder deren in einer Maschinensprache verschlüsselte Darstellung)*
c. counter Zeichenzähler *m*, Zeichenwählvorrichtung *f*
c. light Kennfeuer *n (Navigation)*
c. of gas Gasbeschaffenheit *f*
c. of surface Oberflächenart *f*, Oberflächenzustand *m*, Oberflächenbeschaffenheit *f*
c. reader *(Dat)* automatischer Schriftzeichenleser *m (mit optischen, elektrischen oder magnetischen Vorrichtungen)*
c. reading *(Dat)* Zeichenerkennung *f*, Zeichenlesen *n*
c. recognition *(Dat)* selbsttätige Zeichenerkennung (Schriftzeichenerkennung) *f*

c. translator *(Dat)* Zeichenzuordner *m*

characteristic 1. *(El)* Kennlinie *f (insbesondere Elektronenröhren)*; 2. *(Masch)* Kenngröße *f*; 3. *(Math)* Kennziffer *f (eines Logarithmus)*

c. admittance *(El)* reziproker Wert *m* des Wellenwiderstandes einer Leitung

c. angular difference of phases *(El)* Phasenfolgewinkel *m*

c. curve Kennlinie *f*, Charakteristik *f*

c. curve of a vacuum tube *(El)* Röhrenkennlinie *f*

c. curve tracer *(BMSR)* Kennlinienschreiber *m*

c. data Kenndaten *pl*

c. energy Eigenenergie *f*

c. factor Kennzahl *f*

c. film curve Schwärzungskurve *f (Fotografie)*

c. frequency *(Ph)* Eigenfrequenz *f*

c. function *(Ph)* Eigenfunktion *f*

c. impedance *(El)* Wellenwiderstand *m (insbesondere Vierpole und Hohlleiter)*

c. number Kennzahl *f*, Kenngröße *f*, Eigenwert *m*

c. parameters *(Ph)* Zustandsparameter *mpl*

c. radiation *(Ph)* Eigenstrahlung *f*, charakteristische Strahlung *f*

c. rotation Eigendrehbewegung *f*

c. surface *(El)* Kennliniengebirge *n (insbesondere Elektronenröhren)*

c. temperature Eigentemperatur *f*

c. time Eigenschwingungsdauer *f (Schwingungsglied)*

c. time delay *(Kern)* charakteristische Verzögerung *f*

c. value Kennwert *m*

c. vector Eigenvektor *m*

c. vibration Eigenschwingung *f*

c. wave impedance Feldwellenwiderstand *m (Hohlleiter)*

c. X-radiation charakteristische Röntgenstrahlung *f*

characteristics 1. Charakteristiken *fpl (Differentialgleichungen)*; 2. *(Ph, Ch)* Kenndaten *pl*, Kennziffern *fpl*

c. curves *(El)* Kennlinien *fpl*

characterization factor UOP-Faktor *m (Charakterisierung von Kohlenwasserstoffgemischen)*

charactron Charactron *(Firmenbezeichnung einer Elektronenstrahlröhre zur Bildzeichendarstellung)*

charcoal Holzkohle *f*, Aktivkohle *f*

c. bar Schweißeisenluppe *f*

c. blast furnace mit Holzkohle betriebener Hochofen *m*

c. bottom Löschboden *m*

c. hearth *(Met)* Rennfeuer *n*

c.-hearth iron Frischfeuereisen *n*, Herdfrischstahl *m*

c. hearth process *(Met)* Holzkohlenfrischverfahren *n*, Herdfrischverfahren *n*

c. hearth steel Herdfrischstahl *m*

c. kiln Holzkohlenmeiler *m*, Kohlenmeiler *m*, Meiler[ofen] *m*

c. pig iron Holzkohlenroheisen *n*

c. wood Meilerholz *n*

charge / to 1. [be]laden; belasten, beschweren; 2. füllen, speisen; 3. *(Met)* begichten *(Hochofen)*; einsetzen, beschicken, chargieren *(SM-Ofen)*; 4. *(El)* [auf]laden *(z. B. eine Batterie)*; zuschalten; 5. *(Ök)* anrechnen *(Kosten)*, in Rechnung stellen; 6. *(Text)* beschicken *(Bad)*

c. a battery eine Batterie laden

c. the furnace *(Met)* [be]gichten

c. up *(El)* [sich] aufladen

c. with belasten mit

charge 1. Ladung *f*; Last *f*; 2. Füllung *f*; 3. *(Met)* Charge *f*, Beschickungsmaterial *n*; 4. *(El)* Ladung *f*; 5. *(Ök)* angerechnete (anzurechnende, berechnete, zu berechnende) Kosten (Unkosten) *pl*, Gebühren *fpl*

c. carrier *(Ph)* Ladungsträger *m*

c. carrier diffusion Ladungsträgerdiffusion *f*

c. carrier injection Ladungsträgerinjektion *f*

c. conservation Erhaltung *f* der Ladung

c. control lamp Ladestromkontrollampe *f*

c. density Ladungsdichte *f*

c. dissipation Kriechentladung *f*

c. distribution Ladungsverteilung *f*

c. end Beschickungsseite *f*

c. exchange Ladungsaustausch *m*; Umladung *f*

c.-exchange scattering *(Ph)* Austauschstreuung *f*, Streuung *f* mit Umladung

c. hoist Ladewinde *f*, Ladeaufzug *m*

c. independence Ladungsunabhängigkeit *f*

c. indicator 1. Spannungsprüfer *m*; 2. Ladungsanzeige *f (an Blitzgeräten)*

c. invariance Ladungsunveränderlichkeit *f*

c. level indicator *(Met)* Gichtsonde *f*, Möllersonde *f*, Teufenanzeiger *m*

c.-mass ratio *(Ph)* spezifische Ladung *f*

c. material *(Kern)* Ausgangsmaterial *n*, Spaltstoff *m*

c. multiplication Ladungsvervielfachung *f*

c. number Ladungszahl *f*

c. of a capacitor Belegung (Ladung) *f* eines Kondensators

c. of a nucleus Kernladung *f*

c. of the electron Elektronenladung *f*, elektrische Elementarladung *f*, Ladung *f* des Elektrons

c.-over *(El)* Wechselschalter *m*

c. production *(Ph)* Erzeugung *f* von Ladungsträgern

c. pump *(Kfz)* Ladepumpe *f*

c.-rate Ladestrom *m*

c. spin *(Ph)* Ladungsspin *m*
c.-storage tube Ladungsspeicherröhre *f* *(Informationsspeicherung)*
c. table Beschickungstisch *m*
c. tester Batterieprüfer *m*, Ladeprüfer *m*
c.-to-mass ratio *(Ph)* Ladung-Masse-Verhältnis *n*
c. transfer Ladungsübertragung *f*
chargeable gebührenpflichtig
c. indicator *(Fmt)* Gesprächsuhr *f*
c. minutes Gebührenminuten *fpl*, Gesprächsminuten *fpl*
charged [auf]geladen
c. conductor stromführender Leiter *m*
c. particle geladenes Teilchen *n*
charger Ladungseinrichtung *f*, Ladegerät *n* *(für eine Batterie)*
c. of threshing machine Einleger *m* in Dreschmaschine
c.-reader Lade- und Meßgerät *(für Taschenionisationskammern)*
charging 1. *(El)* Ladung *f*, Aufladung *f*; 2. Füllung *f*; Beschickung *f*, Begichten *n*; 3. *(Brg)* Laden *n* *(Bohrloch)*, Einbringen *n* der Ladung
c. area Beschickungsfläche *f*, Ladefläche *f*
c. barrow Chargierwagen *m*, Begichtungswagen *m*
c. basket Gichtkübel *m*
c. box Einsetzmulde *f*, Beschickungsmulde *f*
c. bridge *(Schiff)* Verladebrücke *f*
c. bucket Gichtkübel *m*
c. by batches satzweise Beschickung *f*
c. capacitor Ladekondensator *m*
c. car *(Brg)* Füllwagen *m*
c. chute *(Wärm)* Schüttkanal *m*
c. circuit Ladestromkreis *m*
c. crane Chargierkran *m*, Begichtungskran *m*, Beschickungskran *m*
c. current Ladestrom *m*
c. device Begichtungseinrichtung *f*, Begichtungsvorrichtung *f*, Einsetzvorrichtung *f*
c.-discharging unit *(Kern)* Be- und Entladungsvorrichtung *f*
c. door 1. Chargieröffnung *f*, Beschickungsöffnung *f*; 2. Füllklappe *f* *(Rostfeuerung)*
c. face Beschickungsseite *f*, Aufgabeseite *f*
c. floor Begichtungsbühne *f*, Chargierbühne *f*, Einsetzbühne *f*, Gichtbühne *f*, Kocherboden *m*, Setzboden *m*
c. generator Lademaschine *f*
c. hole *(Brg)* Fülloch *n*, Füllschacht *m*
c. hopper Fülltrichter *m*, Füllvorrichtung *f*
c. impulse Ladestromstoß *m*
c. level Setzboden *m* *(Kupolofen)*
c. machine Chargiermaschine *f*; Beschikkungsmaschine *f*
c. of a surface Oberflächenbedeckung *f*, Oberflächenbeladung *f*
c. of the grate Rostbeschickung *f* *(Kessel)*
c. operation Begichtung *f*
c. peel Chargierlöffel *m*, Einsetzlöffel *m*
c. period *(Masch)* Ansaughub *m*, Einlaßhub *m*
c. plant Beschickungsanlage *f*
c. platform Gichtbühne *f*
c. rate Stromaufnahme[wert *m*] *f*, Ladewert *m*
c. rectifier Ladegleichrichter *m*
c. resistance Aufladewiderstand *m*, Ladewiderstand *m*
c. scales Gattierungswaage *f*
c. sequence Gichtenfolge *f* *(im Hochofen)*
c. set 1. Ladeaggregat *n*, Ladegerät *n*; 2. Gasgenerator *m* *(Kompressor und Brennkammer einer Gasturbine)*
c. side 1. Einsetzseite *f* *(des Ofens)*; 2. Beschickungsseite *f*
c. side of harvester Einlaufseite *f* der Erntemaschine
c. spoon Chargierlöffel *m*, Einsetzlöffel *m*
c. switchboard Ladeschalttafel *f*
c. time constant Aufladezeitkonstante *f*, Ladezeitkonstante *f*
c. tube *(Kern)* Laderohr *n*, Ladekanal *m*
c. turbine Laderturbine *f*
c. voltage Ladespannung *f*
Charpy impact testing machine Charpyscher Pendelschlaghammer *m* *(Werkstoffprüfung)*
Charpy test Kerbschlagversuch *m* nach Charpy *(Werkstoffprüfung)*
charred leather Lederkohle *f*
c. slop Schlempekohle *f*
c. sponge Schwammkohle *f*
charring Ankohlen *n*, Brennen *n*, Verkohlen *n*, Verkohlung *f*
c. plant Verkohlungsanlage *f*
chart / to aufzeichnen
chart 1. Karte *f*, Tabelle *f*, Plan *m*, Diagramm *n*; 2. *(Schiff)* Seekarte *f*
c. datum *(Schiff)* Kartennull *f*
c. drive *(BMSR)* Diagrammtransportwerk *n*; Uhrwerk *n*
c. matrix *(Math)* Darstellungsmatrix *f*
c.-recording instrument Registrierinstrument *n*
c. speed Geschwindigkeit *f* des Registrierstreifens
charter / to *(Schiff)* chartern, heuern, befrachten
charter broker *(Schiff)* Befrachtungsmakler *m*
c. party Frachtvertrag *m*, Charterpartie *f*
charthouse *(Schiff)* Kartenhaus *n*
charwal bar Schweißeisenluppe *f*
chase / to *(Masch)* [gewinde]strehlen
chase 1. *(Typ)* Schließrahmen *m*, Formrahmen *m*; 2. *(Bw)* Mauerschlitz *m*; eingemeißelte

Rinne *f* zur Rohrverlegung, Graben *m* für Rohrverlegung
chased work getriebene Arbeit *f*; Punzenarbeit *f*; Ziselierung *f*
chaser *(Masch)* Gewindestrehler *m*, Strehler *m*, Gewindestahl *m*; Schneidbacke *f (Schneidkopf für Innengewinde)*
c. die Strehlerbacke *f*
c. die stock Strehlerkluppe *f*
c. tooth Strehlerzahn *m*
chasing Strehlen *n*
c. jig Leitapparat *m* zum Gewindestrehlen
c. tool Gewindestrehler *m*, Gewindestahl *m*
chassis Chassis *n*, Fahrgestell *n*
c. frame Fahrgestellrahmen *m*
c. grease Abschmierfett *n*
chassisless rahmenlos, chassislos
c. construction rahmenlose Bauart *f*, selbsttragende Karosserie *f*
chatee Jutegewebe *n (indisch)*
chatter / to rattern
chatter Rattern *n*
c.-free ratterfrei
c. mark *(Masch)* Rattermarke *f*
chattering Rattern *n*
chatterless ratterfrei
chatty ore fein eingesprengtes Erz *n*
chavage *(Kern, Ch)* Abbau *m* von Molekülketten
check / to 1. kontrollieren; nachrechnen; [nach]-prüfen, überprüfen; überwachen; 2. nachmessen; reißen *(Gußstücke, Holz)*; 3. arretieren, hemmen
c. a tube die Emission einer Röhre messen, eine Röhre prüfen
c. for eccentricity auf Schlag prüfen
c. ultrasonically mit Ultraschallgerät prüfen
check 1. Kontrolle *f*, Nachrechnung *f*, Prüfung *f*; Nachmessung *f*, Überprüfung *f*; 2. *(Met)* Haarriß *m*; 3. Arretierfeder *f*; 4. *(Text)* kariertes Muster *n*
c. analysis Kontrollanalyse *f*
c. band *(Text)* Gegenseil *n (Selfaktor)*
c. clock Stechuhr *f*
c. digit *(Dat)* Kontrollziffer *f*, Prüfbit *n*, Prüfziffer *f (zusätzliche mitgeführte Ziffer zum Prüfen)*
c. flange Anlaufflansch *m*
c. flight Probeflug *m*
c. gauge Prüflehre *f*, Prüfmaß *n*
c. key Kontrolltaste *f*
c. list Prüftabelle *f*
c. loom *(Text)* Wechselstuhl *m*
c. nut Gegenmutter *f*, Kontermutter *f*
c. of balances carried forward Vortragskontrolle *f (Buchungsmaschine)*
c.-out point Kontrollstelle *f* der Bezahlung *(z. B. in einem Selbstbedienungsladen)*
c. plate *(Masch)* Anlaufscheibe *f*; Sperrplatte *f*
c. plug Paßdorn *m*, Prüflehrdorn *m*
c. problem Prüfaufgabe *f*
c. rail Zwangsschiene *f (Radlenker)*
c. ring Ansatzring *m*, Anschlagring *m*
c. room Prüfraum *m*, Prüfstand *m*
c. screw Begrenzungsschraube *f*, Stellschraube *f*, Halteschraube *f*
c. solution *(Math, BMSR)* [unabhängige] Kontrollösung *f*
c. valve 1. Absperrventil *n*; Regulierventil *n*; Rückschlagklappe *f*, Rückschlag[klappen]-ventil *n*, Rückströmventil *n*, Steuerventil *n*; 2. *(El)* Sperrkondensator *m*
c.-valve suction pump Membranpumpe *f*
checker Gitter *n*, Gitterraum *m (SM-Ofen)*
checkerwork 1. Gittermauerwerk *n*; 2. *(Text)* kariertes Muster *n*
checking 1. *(Bw)* Rißbildung *f*, Rissebildung *f (Anstrichtechnik)*; 2. Überwachung *f*, Prüfen *n*, Nachmessung *f*
c. bollard *(Schiff)* Verholpoller *m*, Landpoller *m*
c. equipment Überwachungseinrichtung *f*
checkout *(BMSR, Dat)* Durchprüfung *f*, Kontrolle *f*, Überprüfung *f*
cheek 1. Wange *f (Kurbelwelle)*; 2. Bettwange *f (Drehmaschine)*; 3. *(Text)* Seitenwand *f* des Schlitzes zur Lagerung der Zunge *(Zungennadel)*; 4. *(Led)* Kopf *m (im zugerichteten Leder)*
c. board Seitenschalung *f (Betonbau)*
c. brake Backenbremse *f*
c. of bellows Balgenboden *m*
c. ring of bridle Trensenring *m*
cheese 1. Zylinderreflektor *m (Antenne)*; 2. *(Text)* Kreuzspule *f*
c. cone *s.* cheese 2.
c. head screw Zylinder[holz]schraube *f*
chelate compound Chelatverbindung *f*, Scherenverbindung *f (Komplexchemie)*
chelating agent Chelatbildung bewirkendes Agens *n*
chelation Chelatbildung *f*
chemi-groundwood process chemisches Holzschliffverfahren *n*
chemical Chemikalie *f*
chemical chemisch
c. agent Reagens *n*
c. atomic mass unit chemische Atommasseneinheit *f*
c. attack chemischer Angriff *m*
c. balance Analysenwaage *f*, chemische Waage *f*
c. binding effect Einfluß *m* der chemischen Bindung
c. board holzfreier Karton *m*
c. composition measurements Analysenmeßtechnik *f*

c. **compound** chemische Verbindung *f*
c. **cotton** 1. reine Zellulose *f*; 2. *(Am)* Baumwollinters *pl (gereinigt, gebleicht, in Ballen von 115 lb)*
c. **debarking** chemische Holzentrindung *f*
c. **engineer** Chemieingenieur *m*; Verfahrenstechniker *m*
c. **engineering** chemische Technik (Technologie, Verfahrenstechnik) *f*
c. **finishing** *(Text)* Ausrüstung durch chemische Verfahren
c. **impurity** Fremdatom *n*
c. **metallurgy** Eisenhüttenchemie *f*
c. **milling** Formätzen *n*
c. **precipitation** chemische Fällung *f*
c. **processing plant (equipment)** Chemieanlagen *fpl*
c. **protection [of radiation]** *(Kern)* chemischer Strahlenschutz *m*
c. **pulp** Zellulose *f*; Zellstoff *m*
c. **recovery cycle** chemisches Wiedergewinnungsverfahren *n*; *(Holz)* Wiedergewinnungsverfahren von Aufschlußmitteln, Dampf und Beiprodukten aus Schwarzlauge *(Kraftprozeß)*
c. **rectifier** Elektrolytgleichrichter *m*, elektrolytischer Gleichrichter *m*
c. **refrigeration** Kühlung *f* mit Kältemitteln
c. **reprocessing** *(Kern)* chemische Aufarbeitung *f*
c. **resistant** chemikalienfest
c. **rocket** chemische Rakete *f*
c. **seasoning** chemische Trocknung *f*
c. **tracer** *(Kern)* chemischer Indikator *m*
c. **warfare** chemische Kriegführung *f*, Krieg *m* mit chemischen Kampfstoffen
chemically combined water chemisch gebundenes Wasser *n*
c. **equivalent quantities** chemisch äquivalente Mengen *fpl*
c. **pure** chemisch rein
chemichromatography Fällungschromatografie *f*
chemiluminescence Chemilumineszenz *f*
chemist Chemiker *m*; Apotheker *m*, Drogist *m*
chemistry Chemie *f*
c. **of colloids** Kolloidchemie *f*
chemolysis chemische Zersetzung *f*
chenille [yarn] Chenille[garn *n*] *f*
cheque counterfoil *(Büro)* Scheckkontrollabschnitt *m*
chequer / to kar[r]ieren
chequer-board effect schachbrettartiges Aussehen *n*
c.-board pattern Schachbrettmuster *n*
c.-board weave *(Text)* Bindung *f* für Würfelmuster
chequerdot handles *(Masch)* karierte Schenkel *mpl (Zange)*
chequered jaw Backe *f* mit gerauhten Greifflächen *(Zange)*
c. **plate** Riffelblech *n*
c. **sheet** Riffelblech *n*
chequerwork *(Bw)* Schachbrettmuster *n*
cherry / to ausfräsen *(Gesenk)*
cherry Gesenkfräser *m*, Kugelfräser *m*
c. **coal** Sinterkohle *f*
c.-stoner Kirschenentkerner *m*
cherrying Konvexfräsen *n* in Gesenken; Konvex- und Konkavfräsen *n*; Kugelfräsen *n*
c. **operation** Ausfräsvorgang *m* auf einer Gesenkfräsmaschine
chert *(Geol)* Hornstein *m*
chessylite *(Min)* Azurit *m*, Chessylith *m*, Kupferlasur *f*
chest set Brustmikrofon *n*
chestnut Kastanie[nholz *n*] *f*
c. **coal** Nußkohle *f*
chevilled *(Text)* chevilliert
chevilling Chevillieren *n (Seide)*
chevron packing *(Masch)* „V“-Dichtung *f*
c.-type sealing ring Nutringmanschette *f*
chicken-roost Hühnerstall *m*
c.-wire glass *(Plast)* Zelldrahtglas *n*
chief designer Chefkonstrukteur *m*, Konstruktionsleiter *m*
c. **engineer** technischer Leiter (Direktor) *m*; Chefingenieur *m*
c. **flying instructor** erster Fluglehrer *m*
c. **frame** *(Schiff)* Richtspant *n*
c. **mate** *(Schiff)* Obermaat *m*; erster Offizier *m*; erster Steuermann *m*
c. **of locomotive operations** Lokdienstleiter *m*
c. **officer** *(Schiff)* erster Offizier *m*
c. **surveyor** *(Brg)* Hauptmarkscheider *m*
c. **valence** *(Ch)* Hauptvalenz *f*
chill / to 1. [ab]schrecken; 2. in Kokillen gießen
chill 1. Abschreckung *f*; Kälte *f*; 2. Kokille *f*, Schalengußform *f*
c. **casting** Schalenhartguß *m*, Hartguß *m*, Kokillenguß *m*
c. **casting machine** Kokillengußmaschine *f*
c. **mould** Schlacken[guß]form *f*
c. **point** Erstarrungspunkt *m*, Kristallisationsbeginn *m*
c. **roll** Kühlwalze *f (Folienherstellung)*
c. **testing** Härtetiefenuntersuchung *f (Werkstoffprüfung)*
chilled cast iron Schalenhartguß *m*; Schalenhartgußstück *n*
c. **goods** Kühlgut *n*
c. **layer** *(Met)* Schreckschicht *f*
c. **roll** Hartgußwalze *f*
chilling Abschrecken *n*, Abschreckung *f*
c. **device** Kühlvorrichtung *f*
c. **layer** *(Met)* Schreckschicht *f*

c. roller *(Typ)* Kühlzylinder *m*
chillroom Kältelagerraum *m*
chimes Glockenspiel[werk] *n*, Spielwerk *n*
chiming clock Spieluhr *f*
c. gear Spielwerk *n*
chimney Schornstein *m*, Kamin *m*, Esse *f*, Schlot *m*
c. base Schornsteinsockel *m*
c. brickwork Schornsteinmauerwerk *n*
c. cap Schornsteinkopf *m*
c. capital Schornsteinkapitäl *n*
c. casing Schornsteinmantel *m*
c. cooler Kaminkühler *m*
c. cooler with natural draught selbstventilierender Kaminkühler *m*
c. cowl Schornsteinaufsatz *m*, Schornsteinhaube *f*
c. damper Essenschieber *m*
c. draught Schornsteinzug *m*, Kaminzug *m*
c. duty Schornsteinleistung *f*
c. fire Schornsteinbrand *m*
c. flue Esse *f*, Schornstein *m*; Rauchrohr *n*, Schornsteinzug *m*; Schornsteinfuchs *m*
c. foot Schornsteinfuß *m*
c. head Schornsteinkopf *m*
c. lining Schornsteinfutter *n*
c. pot Schornsteinaufsatz *m*, Schornsteinkappe *f*, Schornsteinhaube *f*, Schornsteinkrone *f*
c. shaft Schornsteinschaft *m*, Schornsteinsäule *f*, Kaminsäule *f*
c. stack Esse *f*, Kamin *m*, Schlot *m*, Schornstein *m*; Schornsteingruppe *f*
c. step Trommelabsatz *m*, Schornsteinabsatz *m*
c. superheater Schornsteinüberhitzer *m*
c.-sweep Schornsteinfeger *m*
c. with fire clay lining Schornstein *m* mit Schamottefutter
c. with stepped lining Schornstein *m* mit Etagenfutter
china Porzellan *n*
c. clay Kaolin *n*, Porzellanerde *f*
c.-grass Ramie *f*, Chinagras *n*
c. wood oil chinesisches Holzöl *n*, Tungöl *n*
chinaware Porzellan *n*
chinchona *s.* cinchona
chine boat *(Schiff)* Knickspantboot *n*
c. construction *(Schiff)* Knickspantbauweise *f*
c. piece *(Schiff)* Kimmweger *m*
c.-type frame Knickspant *n*
Chinese oil Zimtöl *n*, Kassiaöl *n*
C. red Chromrot *n*
chinic acid Chinasäure *f*
chink / to [auf]reißen, [sich] spalten
chink Bruch *m*; Sprung *m*, Riß *m*
chintz paper Kattundruckpapier *n*
chip / to 1. meißeln, schroten; 2. abblättern; 3. ausbrechen

chip 1. Span *m (Holz, Metall)*; 2. *(Kern)* Kernbruchstück *n*, Spaltprodukt *n*
c. axe *(Holz)* Dünnbeil *n*
c. bin Hackspänebehälter *m*
c. board *s.* chipboard
c. breaker Spanbrecher[nut *f*] *m*
c.-breaker flute (groove) Spanbrechernut *f*
c.-breaking nick Spanbrechernut *f*
c.-breaking shoulder *(Masch)* Spanformstufe *f*
c. cavity Spannut *f*
c. charger Hackbecken *n*, Spänesilo *m*
c. chatter *(Masch)* Spanrattern *n*, Spanschwingung *f*
c. chopper Spanbrecher *m*
c. chute 1. *(Holz)* Schnitzelrutschrinne *f*; 2. *(Masch)* Spänerutsche *f*, Spanleitblech *n*
c. clearance 1. Spanabfuhr *f*; 2. Spanraum *m*
c. clogging Spänestauung *f*
c. congestion Spänestauung *f*
c. conveyor Spänetransporteinrichtung *f*, Späneförderer *m*
c. cooking Hackschnitzelkochung *f (Zellulose-herstellung)*
c. cross section Spanquerschnitt *m*
c. curl Spanlocke *f*
c. curling Aufrollen *n* des Spans
c. deflector Spanführung *f*
c. deformation Spanstauchung *f*
c. disposal Spanabfuhr *f*
c. disposal space Spänedurchlaß *m*
c. distribution Spanaufschüttung *f*
c. flow Spanfluß *m*
c. flow type lathe bed spangerechtes Drehmaschinenbett *n*
c. formation Spanbildung *f*
c.-forming spanend
c. groove Spannut *f*
c. guard Spanschirm *m*
c. jamming Spanstauung *f*
c. loft Hackschnitzelsilo *m*
c. packer Füllapparat *m* für Hackschitzel
c. packing *(Holz)* Schnitzelfülldichte *f*
c. pan Spänefangschale *f*
c. process Schnitzelverfahren *n (Chemie-faserherstellung)*
c.-producing spanend
c. relief *(Masch)* Ausspänen *n*
c. removal *(Masch)* 1. Spanabführung *f*; 2. Zerspanung *f*, Spanabnahme *f*; 3. Spanleistung *f*; abgenommenes Volumen *n*
c. screen Sortiertrommel *f* für Hackspäne
c. screening Sortieren *n* von Hackspänen
c. space Spanraum *m*, Spankammer *f*
c. tray Spänefang *m*, Spänekasten *m*, Späneschale *f*

c. upset Spanstauchung *f*
chipboard 1. Span[holz]platte *f*; 2. Graupappe *f*
chipless spanlos
c. deformation (forming) spanlose Formgebung *f*; spanlose Formung *f*
chipper 1. Hackmaschine *f*, Holzhackmaschine *f*; 2. *(Masch)* Putzbeitel *m*
chipping 1. Zerkleinern *n*; 2. *(Masch)* Meißelarbeit *f*
c. chisel Flachmeißel *m*
c. hammer Meißelhammer *m*
c. knife Kabelmesser *n*
c. of the cutting edge Ausbrechen *n* (Ausbruch *m*) der Schneide
chippings *(Bw)* Splitt *m*, Steinsplitt *m (insbesondere Feinsplitt)*
chips 1. Schnitzel *npl*, Abfall *m*, Späne *mpl*, Flocken *fpl*, Granulat *n*; 2. *(Bw)* Splitt *m*, Steinsplitt *m*
chirp Zirpen *n (Funkempfang)*
chisel / to meißeln; stemmen
c. off abmeißeln; abstemmen, behauen
c. out the foundation bolt hole das Ankerloch ausstemmen
chisel Meißel *m*, Beitel *m*, Schneide *f*, Stemmeisen *n*
c. angle Querschneidenwinkel *m (Spiralbohrer)*
c. cut Meißelhieb *m*
c. edge Querschneide *f (Spiralbohrer)*
c. handle Beitelheft *n*
c. point Querschneide *f (Spiralbohrer)*
c. share Meißelschar *n*
chiselled work Punzenarbeit *f*, getriebene Arbeit *f*; Getriebenes *n*
chive turning machine *(Holz)* Spunddrehbank *f*
chloanthite *(Min)* Weißnickelkies *m*, Chloanthit *m*
chloracetic acid Chloressigsäure *f*
chlorargyrite Hornsilber *n*
chlorate / to chlorieren
chlorate Chlorat *n*, Salz *n* der Chlorsäure
c. liquor Bleichlauge *f*
chlorauric acid Gold(III)-chlorwasserstoffsäure *f*
chlorazotic acid Königswasser *n*
chlorhydric acid Chlorwasserstoffsäure *f*, Salzsäure *f*
chloric acid Chlorsäure *f*
chloride Chlorid *n*, Salz *n* der Chlorwasserstoffsäure
c. of lime Chlorkalk *m*
chloridize / to chlorieren
chlorinate / to chlorieren; chloren
chlorinated chloriert; gechlort
c. lime Kalziumhypochlorit *n*, Chlorkalk *m*
c. rubber Chlorkautschuk *m*
c. wool gechlorte Wolle *f*
chlorination plant Chlorieranlage *f (z. B. Leichtmetallveredlung)*
c. stage Chlorierung *f*
chlorinator Chlorierungsgefäß *n*
chlorine Chlor *n*
c. fastness *(Text)* Chlorechtheit *f*
c. in chain *(Ch)* Kettenchlor *n*
c. logging Chlorlog *n (Erdöl, Erdgas)*
c.-peroxide bleach Chlorperoxidbleiche *f*
c. retention Chlorrückhaltevermögen *n*
chlorinity Chlorgehalt *m*
chlorinolysis Chlorierung *f* unter Aufspaltung der Kohlenwasserstoffkette
chlorite Chlorit *n*, Salz *n* der chlorigen Säure
c. schist Chlorschiefer *m*
chlorizate Chlorierungsprodukt *n*
chlorobenzene Chlorbenzol *n*
chlorohydrocarbons Chlorkohlenwasserstoffe *mpl*
chloromethylation Chlormethylierung *f*, Einführung *f* der Chlormethylgruppe
chloronitric acid Königswasser *n*; Nitrosylchlorid *n*
chloroplatinic acid Hexachloroplatin(IV)-säure *f*
chlorosulphonation Chlorsulfonierung *f*, Umsetzung mit Sulfonylchlorid
chlorosulphonic acid Chlorsulfonsäure *f*
chlorous chlorhaltig, chlorig
c. acid chlorige Säure *f*
chock / to verkeilen; auf Holzkeile aufsetzen
c. up verkeilen, mit Holzkeilen befestigen
chock 1. Füllstoff *m*; 2. *(Brg)* Bruchkantenstempel *m*, Kasten *m*; 3. *(Schiff)* Klotz *m*, Aufklotzung *f*
c. damp *(Brg)* Nachschwaden *m*
c. release *(Brg)* Auslösebalken *m (Ausbau)*
chokage Drosselung *f*; Verstopfung *f*
choke / to 1. [sich] festklemmen, klemmen; [sich] verstopfen, stauen; 2. drosseln; sperren
c. out drosseln, unterdrücken
choke 1. *(El)* Drossel[spule] *f*; 2. Luftklappe *f*, Vergaserlufttrichter *m*; 3. Drosselklappe *f*
c. capacitance coupling *(Fmt)* L.C.-Kopplung *f*
c. coil *(El)* Drosselspule *f*
c. connector *(El)* Drosselflansch *m*
c. control Drosselklappe *f*
c.-control knob Luftklappenknopf *m*
c. coupled amplifier *(Fmt)* Drosselverstärker *m*
c. coupling *(El)* Drosselkupplung *f (insbesondere Wellenleiter)*
c. filter *(El)* Drosselkette *f*
c.-input filter Filter *n* mit induktivem Eingang
c. joint *(Fmt)* Drosselverbindung *f*

c. modulation *(El)* Anodenspannungsmodulation *f*, Heising-Modulation *f*, Parallelröhrenmodulation *f*
c. piston *(El)* kapazitiver Kurzschlußkolben *m*, Drosselkolben *m (insbesondere Wellenleiter)*
c. plate *(Wärm)* Entgasungsplatte *f*
c. plunger *s.* c. piston
c. throttle lever *(Kfz)* Handgashebel *m*
c. tube *(Kfz)* Drosselleitung *f*, Drosselkanal *m*; Lufttrichter *m*, Mischrohr *n (Vergaser)*
c. valve *(Kfz)* Drosselventil *n*
chokedamp Nachschwaden *m*
choker valve *(Masch)* Drosselventil *n*
choking 1. *(Plast)* Spritzfehler *m*, Einschnürung *f*, Verstopfung *f*; 2. *(El)* Verdrosselung *f*; 3. *(Kfz)* Drosselung *f*
c. coil *(El)* Drosselspule *f*; Drossel *f*
c. effect *(Fmt)* Drosselwirkung *f*
c. of the chips Spanstauung *f*
c. resistance *(El)* Drosselwiderstand *m*
c. with swarf Verstopfen *n* mit Spänen
choleic acid Choleinsäure *f*
cholesterol Cholesterin *n*
cholic acid Cholsäure *f*
choline Cholin *n*
chop / to [zer]hacken; abschroten; zerschnitzeln
c.-off abkappen, abköpfen
chop Holzschraubstockbacke *f*; Zange *f*
c.-filled plastic *(Plast)* Schnitzelmasse *f*
chopped *(El)* zerhackt; *(Holz)* gespalten
c. direct current zerhackter Gleichstrom *m*
c.-filled plastic *(Plast)* Schnitzelmasse *f*
c. mat *(Ch, Plast)* Faservlies *n*, Glasseidenmatte *f*
c. strands *(Plast)* Stapelglasseide *f*
c. straw Krummstroh *n*
c. straw press Krummstrohpresse *f*
c. up *(Fmt)* zerhackt *(z. B. Sprache bei Fernsprechübertragungen)*
chopper 1. *(Masch)* Hackmesser *n*, Hackbeil *n*; 2. Futterklinge *f*, Häckselmesser *n*; Schneidemaschine *f*; 3. *(Kern)* Zerhacker *m*, Neutronenmonochromator *m*; 4. *(BMSR, Dat)* elektromechanischer Modulator *m*, Zerhacker *m*
c. amplifier *(El)* Zerhackerverstärker *m*
c. bar controller *(Am) (BMSR)* Tastbügelregler *m*, Fallbügelregler *m*
c. disk 1. Zerhackerscheibe *f (z. B. Lochscheibe)*; 2. *(Fmt)* Scheibenunterbrecher *m*
c. knife Häckselklinge *f*
chopping *(Fmt)* Amplitudenbegrenzung *f*
c. bench Schneidlade *f*
c. board Futterbank *f*; Schneidlade *f*
c. knife Futterklinge *f*, Häckselmesser *n*
c. machine Fleischwolf *m*
choppy sea *(Schiff)* kabbelige See *f*
chord 1. *(Math)* Sehne *f*; 2. Saite *f (Musikinstrumente)*; 3. *(Bw)* Gurt *m (Fachwerkbinder, Bögen usw.)*; 4. Tönung *f (Farbstoffe)*
c. length *(Math)* Sehnenlänge *f*
c. winding *(Elm)* Sehnenwicklung *f*
c. wire Spanndraht *m*
chordal addendum *(Masch)* Zahnkopfhöhe *f (an der Sehne gemessen)*
c. dimension Sehnenmaß *n*
c. dimension over two *(or more)* **teeth** Zahnweitenmaß *n*
c. pitch *(Masch)* Rollkreisteilung *f*
c. thickness *(Masch)* Zahndicke *f (an der Sehne gemessen)*
c. thickness of the pitch circle Zahndickensehnenmaß *n*
christening of a ship *(Schiff)* Schiffstaufe *f*
chroma control Farbtonregler *m (Farbfernsehen)*
chromalizing Inchromieren *n*, Chromieren *n*
chromate Chromat *n*, Salz *n* der Chromsäure
c. discharge *(Text)* Chromatätze *f*
chromatic aberration *(Opt)* chromatische Aberration *f*, Farbabweichung *f*, Farb[en]fehler *m*, chromatischer Fehler *m*
c. aberration of position Farb[en]ortsfehler *m*
c. bleeding Farbsaum *m*
c. coma *(Opt)* Farbquerfehler *m*
c. difference of magnification *(Opt)* Farbenvergrößerungsfehler *m*, chromatische Vergrößerungsdifferenz *f*, Farbenunterschied *m* der Vergrößerung
c. purity Farbenreinheit *f*
c. selection Farbauszug *m*
c. sensitivity Farbenempfindlichkeit *f*
c. spectrum *(Opt)* Farbenspektrum *n*
c. tone scale chromatische Tonskala *f (Musik)*
c. variation of spherical aberration *(Opt)* chromatische Abweichung *f* der sphärischen Aberration
chromaticity *(Fs)* Farbton *m*
c. coordinates Farbwertanteile *mpl*
c. diagram Farbtondiagramm *n*
chromatics *(Ph)* Farbenlehre *f*
chromatism *(Opt)* chromatischer Fehler *m*, Farbzerstreuung *f*, Farbfehler *m*
chromatogram Chromatogramm *n*
chromatographic chromatografisch
c. analysis chromatografische [Adsorptions-] Analyse *f*
chromatography Chromatografie *f*
chromatology Farbenlehre *f*
chromatometry *(Ph)* Farbmessung *f*
chromatosulfuric acid Chromschwefelsäure *f*
chrome 1. *(Ch)* Chrom *n*; 2. *(Led)* Chromleder *n*
c. alum Chromalaun *m*

c. **azo dye** Chromazofarbstoff *m*
c. **brick** Chromitstein *m*
c. **complex dye** Chromkomplexfarbstoff *m*
c. **dye** Chromfarbstoff *m*, Chromier[ungs]-farbstoff *m*, Beizenfarbstoff *m*
c.-**dye printing paste** *(Text)* Chromfarbenteig *m*
c. **green** Chromgrün *n*
c. **iron ore** Chromeisenstein *m*; Chromit *m*
c. **leather** Chromleder *n*
c. **magnesite brick** Chrommagnesitstein *m*
c. **mordant dye** *(Text)* Chrombeizfarbstoff *m*
c.-**plated** *s.* chromium-plated
c.-**plating** *s.* chromium-plating
c. **red** Chromrot *n*
c. **side leather** chromgares Rindleder *n*
c. **side upper leather** chromgares Rindoberleder *n*
c. **sole leather** Chromsohlleder *n*
c. **sole leather splits** Chromsohlspaltleder *n*
c. **tannage** *(Led)* Chromgerbung *f*; Chromgerberei *f*
c. **tanned** *(Led)* chromgar
c.-**tanned leather** Chromleder *n*
c. **tanner** Chromgerber *m*
c. **tanning** Chromgerbung *f*
c. **tanning extract** Chromgerbeextrakt *m*
c. **top dye** *(Text)* Nachchromierungsfarbstoff *m*
c.-**topped** *(Text)* nachchromiert
c.-**tungsten-steel** Chrom-Wolframstahl *m*
c. **yellow paste** *(Text)* Chromgelb *n* im Teig
chromic chromhaltig
c. **acid** Chromsäure *f*
c. **anhydride** Chromsäureanhydrid *n*
c. **compound** Verbindung *f* des sechswertigen Chroms
c. **oxychloride** Chromylchlorid *n*
chromiferous chromhaltig
chrominance *(Fs)* Chrominanz *f*, Farbigkeit *f*
c. **band** Farbfrequenzband *n*
c. **channel** Farbträgerkanal *m*
c. **component** Chrominanzkomponente *f*
c. **information** Chrominanzinformation *f*
c. **modulator** Farbmodulator *m*
c. **subcarrier** Farbhilfsträger *m*
chromite Chromit *m*
c. **brick** *s.* chrome brick
chromium Chrom *n*
c. **compound** Verbindung *f* des dreiwertigen Chroms
c.-**layer filter** *(Opt)* Chromfilter *n*
c.-**plated** galvanisch verchromt
c. **plating** galvanische Verchromung *f*
c. **steel** Chromstahl *m*
chromize / to verchromen; [in]chromieren
chromizing Verchromen *n*; Inchromieren *n*, Chromieren *n*
chromo typography Farbenbuchdruck *m*
chromoboard Chromokarton *m*
chromodizing Chromatieren *n*
chromolithography Chromolithografie *f*, Farbensteindruck *m*
chromometer Kolorimeter *n*
chromometry Kolorimetrie *f*
chromophore Farbträger *m*; chromophore Gruppe *f*; Chromophor *m*
chromoscope Farbfernsehbildröhre *f*
chromosome deficiency *(Kern)* Chromosomenstückausfall *m (Strahlenschaden)*
chromosphere *(Astr)* Chromosphäre *f*
chromous compound Verbindung *f* des zweiwertigen Chroms
chronic exposure *(Kern)* chronische Strahlenbelastung *f*; Dauerbelastung *f*
chronograph Zeitschreiber *m*, Chronograf *m*
chronometer Chronometer *n*, Marinechronometer *n*
c. **escapement** Chronometerhemmung *f*
chronometric tachometer *(BMSR)* Stich-Drehzahlmesser *m*, Umdrehungszähler *m*
chronometry Chronometrie *f*, Zeitmessung *f*
chronoscope *(BMSR)* Kurzzeitmesser *m*
chrysoberyl *(Min)* Chrysoberyll *m*
chrysocolla *(Min)* Chrysokoll *m*; Kieselkupfer *n*
chrysotile *(Min)* Chrysotyl *m (Asbest)*
chuck / to einspannen, [fest]spannen; aufspannen *(Werkzeuge)*
chuck Spannfutter *n*, Spannvorrichtung *f*, Futter *n*, Klemmfutter *n*, Spannwerkzeug *n (Werkzeugmaschine)*; Bohrwindenbüchse *f*
c. **capacity** Spannfähigkeit *f*, Spannweite *f*
c. **closing device** Spanneinrichtung *f* des Futters
c. **collet** Einsatzhülse *f*, Reduziereinsatz *m*
c. **groove** Spann-Nute *f*
c. **jaw** Spannbacke *f* des Futters
c. **jig** Aufspannvorrichtung *f*
c. **lathe** Planscheibendrehmaschine *f*; Drehmaschine *f* mit Futteraufspannung
c. **lever** Spannhebel *m*
chuckable im Futter spannbar
chucker *(Am) (Masch)* Futterautomat *m*
chucking *(Masch)* Einspannen *n*, Festspannen *n*, Spannen *n* im Futter
c. **appliance** Spannkopf *m*, Spannvorrichtung *f (Werkzeugmaschine)*
c. **automatic** *(Masch)* Futterautomat *m*
c. **capacity** Spannbereich *m*, Spannweite *f*
c. **device** Einspannvorrichtung *f*, Spannvorrichtung *f*
c. **effect** Spannkraft *f*
c. **machine** 1. Maschine *f* mit Spannfutter; 2. Futterautomat *m*
c. **or holding equipment** Spannmittel *npl*
c. **power** Spannkraft *f*

c. reamer Maschinenreibahle *f*
c. tool Spann[werk]zeug *n*
c. wedge Spannkeil *m*
c. work *(Masch)* Futterarbeit *f*
chunk *(Brg)* nachzuzerkleinernder Mineralbrocken *m*
churn / to Latex herstellen; Schaum schlagen; buttern; schütteln, umrühren
churn 1. Erdbohrer *m*; Kneter *m*, Rührwerk *n*; 2. Butterfaß *n*
c. dasher Butterstößel *m*
c. drill *(Brg)* Stoßbohrmaschine *f*
c. drilling *(Brg)* Stoßbohren *n*, stoßendes Bohren *n*
c. staff Butterstößel *m*
churner Stoßbohrer *m*
churning 1. Schäumen *n*; Buttern *n*; Rühren *n*; 2. Waten *n* [im Öl] *(Zahnrad)*
c. barrel Butterfaß *n*
c. machine Kirnvorrichtung *f*
c. vat Butterfaß *n*
chute / to rutschen, schurren *(über eine Rutsche fördern)*
chute 1. [abschüssige] Rinne *f*, Rutsche *f*, Schurre *f*, Schütte *f*, Schüttrinne *f*; 2. Schüttgosse *f* *(Rostfeuerung)*; 3. *(Brg)* Rolloch *n*, Rolle *f*; 4. Sammelbehälter *m*; 5. *(Masch)* Zuführrutsche *f*; Leitblech *n*
c. borer *(Brg)* Rollochbohrmaschine *f*
c. box Rutschenkasten *m*
c. compartment *(Brg)* Bergetrum *m* *(Aufhauen)*
c. feeder Fülltrichter *m*
c. feeding *(Masch)* Leitschienenzuführung *f*; Rutschzuführung *f*
c.-gate *(Brg)* Rollochverschluß *m*
c. grate *(Wärm)* Schüttrost *m*
c. hopper Rutschentrichter *m*
c. raise *(Brg)* Förderüberhauen *n*, Förderaufhauen *n*, Förderberg *m*
chuter Gleitbahn *f*, Gleitrinne *f*, Ablaufrinne *f*
chutes and ducts Rinnen *fpl* und Leitungen *fpl*
cicatrization swelling *(Holz)* Vernarbungsrindenwuchs *m*
cicatrize / to narben
cigar lighter Zigarrenanzünder *m*
cigarette lighter Zigarettenanzünder *m*
cimented *s.* cemented
cinchona Chinarinde *f*
c. bark *s.* cinchona
cinder Schlacke *f*; Zunder *m*
c. concrete block Schlackenstein *m*
c. dump Schlackenhalde *f*
c. heat Entzunderungswärme *f*
c. notch Schlackenstich *m* *(am Ofen)*
c. paste Schwärze *f*, Schlichte *f*
c. path Aschenbahn *f* *(Laufbahn)*
c.-pit Aschekasten *m*, Ascheraum *m*, Ascheloch *n*
c. pot Schlackenpfanne *f*, Schlackenbehälter *m*, Krätzebehälter *m*
c. spout Schlackenrinne *f*
c. yard Schlackenhalde *f*
ciné camera Amateurfilmkamera *f*, Schmalfilmkamera *f*
c. film Schmalfilm *m*
c. projection Kinoprojektion *f*
c. projection lens Kino-Projektionsobjektiv *n*
c. projector Kinoprojektor *m*
c. recording lens Kino-Aufnahmeobjektiv *n*
cinema Kino *n*, Lichtspieltheater *n*
c. organ Kinoorgel *f*
cinematograph film Film *m* *(Kino)*
cineol Zineol *n*
cineradiography Kino-Röntgenografie *f*
cinnabar Zinnober *m*
cinnamaldehyde Zimtaldehyd *m*
cinnamamide Zimtsäureamid *n*
cinnamic acid Zimtsäure *f*
cinnamo nitrile Zimtsäurenitril *n*
cinnamon bark Zimtrinde *f*, Zimtrohr *n*
c. flower Zimtblüte *f*
cipher / to 1. [aus]rechnen; 2. verschlüsseln, chiffrieren
cipher Zahl *f*, Ziffer *f*; Chiffre *f*
c. key Geheimwort *n*
circle / to kreisen
c. around kreisen um, umkreisen
c. in an elliptical orbit *(Astr)* eine Ellipse beschreiben
c. in its orbit *(Astr)* in seiner Bahn kreisen *(Satellit)*
c. the earth die Erde umkreisen
c. the moon den Mond umkreisen
circle Kreis *m*
c. c coefficient *(Schiff)* C-im-Kreis-Koeffizient *m* *(Vergleichskoeffizient für Schiffsgeschwindigkeiten)*
c. coefficient *(Fmt)* Streufaktor *m*; *(Elm)* Schlüpfungskoeffizient *m*
c. cutter *(Masch)* Kreisbohrer *m*
c. cutting Rundschneiden *n*
c. cutting attachment Kreistrennvorrichtung *f*
c. minute Kreisminute *f*
c. of confusion *(Ph)* Unschärfenkreis *m*, Zerstreuungskreis *m*
c. of contact Wälzkreis *m* *(Verzahnung)*
c. of convergence *(Math)* Konvergenzkreis *m*
c. of curvature *(Math)* Krümmungskreis *m*
c. of diffusion *(Ph)* Streukreis *m*, Zerstreuungskreis *m*
c. of holes *(Masch)* Lochkreis *m*
c. polarimeter *(Opt)* Kreispolarimeter *n*
c. test *(Schiff)* Drehkreisversuch *m*

circling *(Met)* Rondenschneiden *n*
circlip *(Masch)* Sprengring *m*, Ringfeder *f*, Seegersicherung *f*
circuit 1. *(El)* Stromkreis *m*; Schaltung *f*; 2. *(Flg)* Platzrunde *f*; 3. *(Math)* geschlossener Linienzug *m*; 4. Kreislauf *m (Hydraulik)*
c. **analyzer** *(Dat)* Netzwerkanalysator *m*, Schaltungsanalysator *m*; Netzmodell *n*
c. **angle** Phasenwinkel *m (Stromrichterschaltung)*
c. **arrangement** *(El)* Schaltung[sanordnung] *f*
c. **blue-print** *(Fmt, Fs)* Schaltpause *f*
c. **break** *(El)* Leitungsbruch *m*, Unterbrechung *f (der Leitung)*
c. **breaker** Ausschalter *m*, Trennschalter *m*; Leistungsschalter *m*
c.**-breaker oil** Schalteröl *n*
c. **breaking** *(El)* Abschaltung *f*, Ausschaltung *f*
c.**-breaking capacity** Abschaltleistung *f*, Abschaltvermögen *n*, Ausschaltvermögen *n*
c.**-breaking transient** Ausgleichsvorgang *m*; Einschwingen *n (beim Ausschalten)*
c. **breaking transient current** Ausschalt[strom]stoß *m*
c. **closer** Stromschließer *m*; Einschalter *m*
c. **closing** Einschaltung *f*, Einschaltvorgang *m*, Stromschluß *m*
c.**-closing connection** Arbeitsstromschaltung *f (Relais)*
c. **controller** *(El)* Schaltwalze *f*, Walzenschalter *m*
c. **damping** Kreisdämpfung *f*
c. **diagram** Stromkreisbild *n*, Stromkreisdiagramm *n*; Schaltbild *n*, Schaltplan *m*, Schaltschema *n*; Verdrahtungsschema *n*
c. **element** Schalt[ungs]element *n*
c. **indicator** Strom[an]zeiger *m*, Leitungsprüfer *m*
c. **logic** *(BMSR)* Schaltungslogik *f*
c. **noise** Kreisrauschen *n*, thermisches Rauschen *n* im elektrischen Kreis
c. **noise meter** Geräuschspannungsmesser *m*, Geräuschspannungszeiger *m*
c. **of holding coil** Haltestromkreis *m*
c. **opening** Öffnen *n* des Stromkreises; Abschaltung (Ausschaltung) *f* eines Kreises
c.**-opening connection** Ruhestromschaltung *f (Relais)*
c. **schematic** *(El)* Schaltungsschema *n*
c. **stability** Frequenzkonstanz *f* [der Schaltung]
c. **survey** *(Kern)* Kreislaufüberwachung *f*
c. **tester** *(El)* Leitungsprüfer *m*
c. **voltage** *(El)* Leiterspannung *f (zwischen den Phasen)*
c. **wiring** *(El)* Schaltung *f*; Verdrahtung *f*
c. **wiring capacity** Schaltungskapazität *f*
circuital vector field *(El)* Wirbelfeld *n*
circuitry Schaltungstechnik *f*; Schaltung *f*, Verdrahtung *f*
circular kreisförmig; rund, kreisrund; Kreis-, Rund-; zirkulierend, umlaufend, wiederkehrend
c. **acceleration** Drehbeschleunigung *f*
c. **aerial** Rundrahmenantenne *f*
c. **air grid** *(Wärm)* Rosette *f*
c. **air grid wheel** *(Wärm)* Rosettenrädchen *n*
c. **aperture** Kreisöffnung *f*
c. **arc** *(Math)* Kreisbogen *m*
c. **arc method** *(Bw)* Geländebruchuntersuchung *f (Grundbau)*
c. **band** *(Wärm)* Ringleiste *f*
c. **bending apparatus** Rundbiegevorrichtung *f*
c. **blade** rundgebogene Klinge *f (Blechschere)*; Rundmesser *n*
c. **blank** *(Met)* Ronde *f*
c. **box lathe** *(Text)* Revolverlade *f*
c. **box loom** Webstuhl *m* mit Revolverwechsel, Revolverwebstuhl *m*
c. **box motion** *(Text)* Drehkastenwechsel *m*; Revolverwechsel *m*
c. **box underpick loom** Unterschlagrevolverwebstuhl *m*
c. **breaking machine** Rundbrechmaschine *f (Flachs)*
c. **broach** Formwälzwerkzeug *n*
c. **brush** Bürstenwalze *f*; Rundbürste *f*; Fiberrundbürste *f*
c. **cam** Schlauchteil *m* eines Hebers *(Strickmaschine)*
c.**-chart recorder** *(BMSR)* Kreisblattschreiber *m*
c. **chuck** Rundfutter *n*
c. **column** Rundsäule *f*
c. **comber** *(Text)* Rundkämmaschine *f*
c. **combing machine** *(Text)* Rundkämmmaschine *f*
c. **cone** Kreiskegel *m*
c. **course** Kreisbewegung *f*, Kreislauf *m*
c. **cross section** Kreisquerschnitt *m*
c. **cut** Rundschnitt *m*
c. **cut-off saw** Trennkreissäge *f*
c. **cut-off saw blade** Trennkreissägeblatt *n*
c. **cylinder** Kreiszylinder *m*
c. **cylindrical valve** Drehkolbenschieber *m*
c.**-deck table** *(Brg)* Rundherd *m*
c. **dividing machine** Kreisteilmaschine *f*
c. **dividing table** Rundtisch *m* mit Teileinrichtung
c. **division** Kreisteilung *f*, Winkelteilung *f*
c. **dovetailing saw** Hobelnutsäge *f*
c. **electric wave** zylindersymmetrische elektromagnetische Welle *f*
c. **file** *(Masch)* Feilscheibe *f*
c. **flanging press** *(Met)* Kümpelpresse *f*

c. fore sight *(Mil)* Kreiskorn *n*
c. forming tool Formscheibenstahl *m*
c. frequency Kreisfrequenz *f*
c. functions Kreisfunktionen *fpl*
c. furnace Ringofen *m*
c. gear shaving cutter Schabrad *n* für Zahnradfertigung
c. glass cutter Rundschneider *m*, Glasschneider *m* für Rundschnitt
c. graduation Kreis[ein]teilung *f*
c. grate *(Wärm)* Kreisrost *m*
c. grinding Rundschleifen *n*; Rundschliff *m*
c. grinding machine Rundschleifmaschine *f*
c. groove Ringnut *f*
c. indexing table Rundschalttisch *m*
c. knit Rundstrickware *f*
c. knit hosiery Rundstrickstrumpfware *f*
c.-knitted rundgestrickt
c. knitted fabric Rundstrickware *f*
c. knitting machine Rundstrickmaschine *f*, Rundstuhl *m*
c. knitting machine with one set of needles einfonturige Rundstrickmaschine *f*
c. laminations *(Elm)* Rundbleche *npl (des Ankers)*
c. land Führungsfase *f*
c. latch needle machine Rundstrickmaschine *f*
c. linking machine *(Text)* Rundkettelmaschine *f*
c. links-and-links knitting machine Links-Links-Rundstrickmaschine *f*
c. loom Rundwebstuhl *m*
c. loom with gripper Greiferrundwebmaschine *f*
c. magnetic chuck Rundfutter *n*
c. measure Bogenmaß *n*
c. metal cutting saw Metallkreissäge *f*
c. milling attachment Rundfräseinrichtung *f*
c. motion Kreisbewegung *f*, Drehbewegung *f*
c. movement Kreisbewegung *f*, Drehbewegung *f*
c. nozzle runde Düse *f*
c. oil groove umlaufende Schmiernut *f*
c. orbit Kreisbahn *f*, kreisförmige Umlaufbahn *f*
c. orbit speed Kreisbahngeschwindigkeit *f*
c. path Kreisbahn *f*, kreisförmige Bahn *f*
c. pattern *(Text)* Kreismuster[ung *f*] *n*
c. pitch Wälzkreisteilung *f (Zahnrad)*
c. planing attachment Rundhobeleinrichtung *f*
c. polarization *(Fs)* Zirkularpolarisation *f*
c. profiling *(Masch)* Rundkopieren *n*
c. punching comb *(Text)* kreisförmiger Vorstechkamm *m*
c. radio-beacon *(Schiff)* Kreisfunkbake *f*, Kreisfunkfeuer *n*
c. rib knitting machine Rundstrickrippmaschine *f*
c. ring Kreisring *m*
c. ring sector Kreisringsektor *m*
c. rip saw Spaltkreissäge *f*
c. saw Kreissäge *f*
c. saw bench Tischkreissäge *f*
c. saw blade Kreissägeblatt *n*
c. saw for cutting wood Holzkreissäge *f*
c. saw for edging boards Besäumsäge *f*
c. saw for parallel sawing Parallelsaumkreissäge *f*
c. saw for squaring logs Kantholzkreissäge *f*
c. sawing machine Kreissäge[maschine] *f*
c. scale Kreisskale *f*
c. scanning Kreisabtastung *f*
c. screen *(Typ)* Drehraster *m*, Rundraster *m*
c.-section mit kreisförmigem Querschnitt
c. section *s.* c. cross section
c. sector *(Math)* Kreisausschnitt *m*, Sektor *m*
c. segment *(Math)* Kreisabschnitt *m*, Segment
c. shaft *(Brg)* Rundschacht *m*, Schacht *m* mit runder Schachtscheibe
c. shape Kreisform *f*; Ronde *f*
c. shaped blade *(Masch)* rundgebogene Klinge *f*
c. shear[s *pl*] *(Masch)* Kreisschere *f*
c. slide valve Schieber *m* mit kreisförmiger Schieberplatte
c. slotted nut Nutmutter *f*
c. string border machine *(Text)* Rundleistenmaschine *f*
c. syphon *(Wärm)* Trompetenrohr *n*
c. table Rundtisch *m (Werkzeugmaschine)*
c. table with indexing attachment Rundtisch *m* mit Teileinrichtung
c. thickness *(Masch)* Zahndicke *f (Länge der Zahnkreisteilbogen)*
c. trajectory Kreisbahn *f*
c. vault *(Bw)* Tonnengewölbe *n*
c. velocity Kreisbahngeschwindigkeit *f*
c. vibratory screen Kreisschwingsieb *n*
c. vortex Kreiswirbel *m*
c. warp knitting machine Rundkettenwirkmaschine *f*
c. warp loom *(Text)* Rundkettenstuhl *m*
c. washer *(Text)* Rundwaschmaschine *f*
c. wave Kreiswelle *f*
c. weft knitting machine Rundkulierwirkmaschine *f*
c. weld Rund[schweiß]naht *f*
c. woven fabric Rundgewebe *n*
circularly polarized *(Opt)* zirkular polarisiert
circulate / to zirkulieren, kreisen, umlaufen; umwälzen, im Umlauf sein
circulating current 1. *(Ph)* Kreisstrom *m*; 2. *(Fmt)* Schwingkreisstrom *m (z. B. Sendekreis)*; 3. *s.* eddy current

c. currents *(El)* vagabundierende Ströme *mpl*
c. digester Umwälzkocher *m*
c. fan Umwälzlüfter *m*
c. fuel reactor *(Kern)* Kreislaufreaktor *m* *(Reaktor mit umlaufendem Spaltstoff)*
c. lubrication Umlaufschmierung *f*
c. lubrication [system] using force feed Umlaufdruckschmierung *f*
c. memory *(BMSR)* Umlaufspeicher *m (in einer Laufzeitvorrichtung umlaufende Information)*
c. motion Zirkulationsbewegung *f*
c.-oil system Ölumlaufsystem *n*
c. pump Umlaufpumpe *f*; Umwälzpumpe *f*
c. reactor *s.* c. fuel reactor
c. signal umlaufendes Signal *n*
c. system of lubrication Umlaufschmierung *f*
c. water Umlaufwasser *n*
c. water mains Kühlrohrleitung *f*
c. water pump Kühlwasserpumpe *f*
circulation 1. Zirkulation *f*, Kreisbewegung *f*, Kreislauf *m*; Umlauf *m (z. B. eines Schmiermittels)*; 2. *(Math)* Rundintegral *n*
c. area *(Bw)* Verkehrsfläche *f (Fläche der Eingangshallen, Treppenhäuser, Flure usw.)*
c. boiler *(Wärm)* Umlaufkessel *m*, Zirkulationskessel *m*
c. channel *(Bw)* Umströmkanal *m*
c. cook[ing] Umwälzkochung *f*
c. feed water heater *(Wärm)* Zirkulationsvorwärmer *m*
c. flow Zirkulationsströmung *f*
c. gas Spülgas *n*
c. loop Umlaufleitung *f*
c. of current Stromlauf *m*
c. of the lubricant Umlauf *m* des Schmiermittels
c. oiling Umlaufschmierung *f*
c. pump Umlaufpumpe *f*
c. tar *(Brg)* Spülteer *m*
c. theorem *(Math)* Zirkulationssatz *m*
circumcentre *(Math)* Umkreismittelpunkt *m*
circumcircle *(Math)* Umkreis *m*
circumcone *(Math)* umbeschriebener Kegel *m*
circumference Kreisumfang *m*, Umfang *m*
c. of reel *(Text)* Haspelumfang *m*
c. pitch *(Masch)* Umfangsteilung *f*
circumferential peripheral, Umfangs-
c. force Umfangskraft *f*
c. groove Ringnut *f*
c. length of clearance space Spaltlänge *f*, Umfanglänge *f* des Spaltes *(Turbine)*
c. pressure Umfangdruck *m*, Umfangkraft *f* *(Turbine)*
c. prestressing Ringvorspannung *f (Spannbetonbau)*
c. seam welding machine Rundnahtschweißmaschine *f*
c. speed Umfangsgeschwindigkeit *f*, Periphergeschwindigkeit *f*
c. stress *(Mech)* Umfangsspannung *f*
c. surface Mantelfläche *f*
c. velocity Umfangsgeschwindigkeit *f*
circumfluence Umströmung *f*
circumfluent umflutend, fließend um, umströmend, umgebend
circumscribe / to *(einen Kreis)* umschreiben
circumscribed umschrieben, abgegrenzt
c. circle *(Math)* Umkreis *m*, umbeschriebener Kreis *m*
circumsphere *(Math)* umbeschriebene Kugel *f*
cire perdue process *(Met)* Wachsausschmelzverfahren *n*
cis-addition *(Ch)* Anlagerungsreaktion *f*, als deren Ergebnis eine cis-Verbindung entsteht
c.-compound cis-Verbindung *f*
cissing Blasenbildung *f (Ablösen n des Farbanstriches oder Lackes)*
cissoid *(Math)* Zissoide *f*, Efeulinie *f*
cistern 1. *(Bw)* Flüssigkeitsbehälter *m*, Tank *m*, Wasserspeicher *m*, Zisterne *f*; Spülkasten *m*; 2. Hafen *m*, Wanne *f (Glasherstellung)*
c. barometer Gefäßbarometer *n*
c. float Spülkastenschwimmer *m*
citizen radio ziviler UKW-Verkehr *m*
citizens radio service Jedermann-Funksprechverkehr *m*
citrate Zitrat *n*, Salz *n* der Zitronensäure
citric acid Zitronensäure *f*
c. soluble zitronensäurelöslich
citrico dehydrogenase Zitronensäuredehydrase *f*
citridic acid Akonitsäure *f*
citrine zitronenfarben
citron yellow Zinkgelb *n*
citrus fruit Zitrusfrucht *f*
city water Leitungswasser *n*
civet *(Ch)* Zibet *m*
civetone *(Ch)* Zibeton *n*
civic centre kommunales Verwaltungszentrum *n*
c. planning Städtebau *m*, Stadtplanung *f*
civil engineer Bauingenieur *m*
c. engineering Bauingenieurwesen *n*, Ingenieurbau[wesen *n*] *m*; Bautechnik *f*
c. engineering structures Ingenieurbauten *pl*
Civil Court Zivilgericht *n*
civilian power type nuclear reactor *(Kern)* Kernreaktor *m* für Energieerzeugung (Energieversorgung)
clack valve Klappenventil *n*, Scharnierventil *n*
clad / to plattieren; umhüllen, überziehen, verkleiden
clad 1. plattiert; umhüllt, überzogen, verkleidet; 2. *(Kern)* abgeschirmt; 3. *s.* clothed
c. material Verbundwerkstoff *m*, plattierter Werkstoff *m*

cladding 1. *(Bw)* Verkleidung *f*, Verschalung *f*; 2. *(Met)* Plattierung *f*; Plattierschicht *f*; 3. *(Kern)* Umhüllung *f* *(Brennstoffelement)*
c. by rolling Walzplattieren *n*
c. material 1. Plattierungsmaterial *n*, Auflagewerkstoff *m*; 2. *(Kern)* Hüllenmaterial *n* *(für Spaltstoffelemente)*
c. process *(Met, Masch)* Plattierverfahren *n*
clagging Anhaften *n*
claggy *(Brg)* Nachfallpacken *n*
claim 1. *(Brg)* Grubenfeld *n*; 2. Patentanspruch *m*
claimant Patentanmelder *m*
Claisen flask Claisen-Kolben *m*
clam nut Schloßmutter *f*
c. shell markings Muschelbruch *m*
clamp / to befestigen, einspannen, festklemmen, festspannen; [auf]spannen *(Werkstücke)*; verriegeln
c. in place festklemmen
c. in position einspannen; festspannen; feststellen
c. rigidly fest einspannen
clamp 1. Klammer *f*, Klampe *f*, Krampe *f*, Schelle *f*; Schraubzwinge *f*, Zwinge *f*; Bügel *m*; Greifbacke *f*, Greifhaken *m*, Haken *m*; 2. Klammer *(chemische Labortechnik)*; 3. Klemmstück *n* *(Schraubstockbefestigung)*; 4. Rohrschelle *f* *(Rohrverlegung)*; 5. Spanneisen *n*
c. attachment Klemmschelle *f*
c. bar 1. Klemmstange *f*; 2. Preßbalken *m* *(Papierschneidemaschine)*
c. bed Rundführung *f* *(Fräsmaschine)*
c. bolt Spannschraube *f*
c. circuit Blockierschaltung *f*
c. collar Klemmring *m*, Verschlußring *m*; Spannhülse *f*, Spannring *m*
c. coupling Klemmkupplung *f*, Schalenkupplung *f*
c. dog Spannkloben *m*
c. handle Klemmhebel *m*
c.-hub pulley Riemenscheibe *f* mit geteilter Nabe für Klemmbefestigung
c. iron *(Bw)* Bauklammer *f*, Klammer *f*
c. kiln Feld[brenn]ofen *m* *(Ziegelherstellung)*
c. knob Klemmknopf *m*
c. lever Klemmhebel *m*
c. nut Spannmutter *f*
c.-on skates Klammerschlittschuhe *mpl*, Klemmschlittschuhe *mpl*
c. plate Spannplatte *f*, Spannscheibe *f*
c. ring Klemmring *m*
c.-type tool holder Klemmhalter *m*
clamper *(Fs)* Schaltung *f* zur Zurückführung des Schwarzpegels auf den Sollwert
clamping 1. Aufspannen *n*; Befestigung *f*, Einspannen *n*; Einspannung *f*, Festklemmen *n*, Festspannung *f*; Klemmung *f*; 2. *(Fs)* Schwarzwertrückgewinnung *f*; Signalpegelblockierung *f*; Verriegelung *f*
c. accessories Spannzeuge *npl*; Aufspannwerkzeuge *npl*
c. angle *(Masch)* Aufspannwinkel *m*
c. area *(Masch)* Aufspannfläche *f*
c. arrangement *(Masch)* Spannvorrichtung *f*
c. bar *s.* clamp bar
c. bar head Klemmknebel *m*
c. block *(Masch)* Spanneisen *n* *(Vorrichtung)*
c. bolt Befestigungsbolzen *m*, Befestigungsschraube *f*, Druckbolzen *m*, Halteschraube *f*, Klemmbolzen *m*
c. bush[ing] *(Masch)* Klemmbüchse *f*, Spannbüchse *f*
c. chuck Einspannklaue *f*, Einspannkopf *m*, Klemmfutter *n*
c. circuit 1. *(Dat)* Klemmschaltung *f*, Blokkierschaltung *f*; 2. *(Fs)* Null-Linienfesthalteschaltung *f*; Schwarzsteuerungskreis *m*, Schwarzwertrückgewinnungsschaltung *f*
c. collar *(Masch)* Schelle *f*
c. cone Spannkegel *m*; Klemmkegel *m*, Reibkegel *m*
c. device 1. *(Masch)* Aufspannvorrichtung *f*, Festspannvorrichtung *f*; 2. *(Flg)* Feststellvorrichtung *f*; 3. *(Fs)* [elektronische] Verriegelung *f*
c. diode *(Fs)* Klemmdiode *f*, Schwarzsteuerdiode *f*
c. face *(Masch)* Aufspannfläche *f*
c. head Klemmknopf *m* *(insbesondere optischer Geräte)*
c. in place Festklemmen *n*
c. jaw Klemmbacke *f*, Spannbacke *f*
c. lever 1. Spannhebel *m*; 2. Klemmhebel *m* *(insbesondere optischer Geräte)*
c. mechanism Klemmung *f*
c. member *(Masch)* Spannteil *n*
c. nut Haltemutter *f*, Klemmutter *f*, Überwurfmutter *f*
c. pad Druckstück *n*
c. plate Spannplatte *f*
c. range Spannbereich *m* *(Schraubstock)*
c. ring Klemmring *m*, Schelle *f*, Spannring *m*; Gegenring *m* *(Turbine)*
c. screw Klemmschraube *f*
c. sleeve Klemmuffe *f*, Klemmbüchse *f*, Spannbüchse *f*
c. slot Spannschlitz *m* *(Tisch)*
c. spring Federklammer *f*
c. stud Spannbolzen *m*
c. surface Aufspannfläche *f*, Spannfläche *f*
c. table Spanntisch *m*
c. tool Spann[werk]zeug *n*
clamshell bucket *(Bw)* Greifer *m*, Greifkorb *m*
c. grab Zweischalengreifbagger *m*

Clapeyron's theorem Clapeyronsche Dreimomentengleichung *f*
clapper 1. Klöppel *m (Glocke)*; 2. *(Masch)* Meißelklappe *f*
c. block *(Masch)* Meißelklappe *f*
c. box *(Masch)* Klappenträger *m*, Meißelklappenträger *m*
clarain *(Min)* Clarit *m*, Mattkohle *f*
clarification Reinigung *f*; Raffination *f*; Läuterung *f*; Klärung *f*
c. of sewage Abwässerklärung *f*
c. plant Kläranlage *f (Gesundheitstechnik)*
c. tank Klärbecken *n*, Klärbehälter *m*
clarified juice Dünnsaft *m (Zuckerindustrie)*
clarifier Klärer *m*; Reiniger *m*; Klärapparat *m*; Klärmittel *n*; Absetzgefäß *n*
clarify / to [ab]klären; läutern; reinigen
c. the dirty waste water schmutziges Abwasser klären
clarifying basin Absetzbecken *n*; Klärbecken *n*
clarinet Klarinette *f*
clarionet Klarinette *f*
clashing of gears Klirren *n* [von Zahnrädern], Geklirr *n* [von Zahnrädern]
clasp / to zuhaken, einhaken
clasp nut Schloßmutter *f*
class Klasse *f*
c.-A amplifier *(Fmt)* A-Verstärker *m*
c.-A connection *(Fmt)* A-Schaltung *f*
c.-A operation *(Fmt)* A-Betrieb *m*
c.-AB amplifier *(Fmt)* AB-Verstärker *m*
c.-B amplifier *(Fmt)* B-Verstärker *m*
c.-B modulation *(Fmt)* B-Modulation *f*
c.-C amplifier *(Fmt)* C-Verstärker *m*
c. frequency *(Math)* Besetzungszahl *f*; Klassenhäufigkeit *f*
c. interval *(Math)* Klassenbreite *f*, Klassengröße *f*
c. of accuracy Genauigkeitsklasse *f*
c. of coal Kohlensorte *f*, Kohlenart *f*
c. of fit Passung[sgüte] *f*
c.-room film Lehrfilm *m*
c. sum *(Math)* Vereinigungsklasse *f*, Klassensumme *f*
classes of star magnitudes Größenklassen *fpl* der Sterne
classical system *s.* non-quantized system
c. theory of fields *(Ph)* klassische Feldtheorie
classification Einordnung *f*; Einteilung *f*; Sortierung *f*; Klassierung *f*
c. keyboard Fachwählertastatur *f (Lochkartenmaschinen)*
c. of fits *(Masch)* Passungssystem *n*
c. of ores Klassierung *f* der Erze
classifier Klassierer *m (insbesondere Erze)*
classify / to klassifizieren, [ein]gliedern, [ein]ordnen, einteilen, einstufen; klassieren *(sortieren durch Sieben)*, sichten, sortieren
classifying cylinder Sortierzylinder *m*
c. screen 1. Klassiersieb *n (in der Chemie und für technische Produkte)*; 2. Sortiersieb *n (insbesondere für landwirtschaftliche Produkte)*
clastic *(Geol)* brüchig, klastisch
Clausius-Clapeyron equation Clausius-Clapeyronsche Gleichung *f*
Clausius' cycle Clausiusscher Kreisprozeß *m (Thermodynamik)*
claw 1. Haken *m*, Klaue *f (Zange)*, Kralle *f*, Nagelheber *m*, Pratze *f*, Schere *f*, Tatze *f*, Greifdorneisen *n*; 2. Greifer *m*
c. bar Nagelklaue *f*
c. belt fastener Riemenkralle *f*
c. clutch Klauenkupplung *f*
c. coupling Klauenkupplung *f*
c. hammer Klauenhammer *m*
c. hatchet Klauenbeil *n*
c. of the hammer Klaue *f* eines Hammers
c. trussing machine Reifenaufziehmaschine *f*
c.-type fastener Krallenverbinder *m (Riemen)*
c.-type nail hammer Zimmererhammer *m*
clay / to *(Brg)* verletten
clay Ton *m*; Bleicherde *f*; Lehm *m*; Mergel *m*
c. bank *(Geol)* Tonschicht *f*
c. body Tonscherbe *f*; Tonsubstanz *f*
c.-bond fire-clay Tonerdeschamotte *f*
c. brick Tonziegel[stein] *m*
c. contacting Kontaktreinigung *f* mit Hilfe von Ton *(Erdgas, Erdöl)*
c. course *(Geol)* Lettenbesteg *m*, Salband *n*
c. filler Kaolinfüllstoff *m*
c. floor Lehmestrich *m*, Lehmfußboden *m*
c. gouge *(Geol)* Lettenbesteg *m*, Salband *n*
c. gun Stichlochstopfmaschine *f (z. B. SM-Ofen)*
c. marl Tonmergel *m*
c. mill Tonknetmaschine *f*
c. pigeon Wurftaube *f (für Jagdzwecke)*
c. pipe Tonrohr *n*, Steinzeugrohr *n*
c. pit Tongrube *f*
c. pot Tontiegel *m*
c. preparation Tonaufbereitung *f*
c. products Keramikerzeugnisse *npl*, Tonwaren *fpl*
c. slate Tonschiefer *m*
c. slurry Tonschlempe *f*
c. stemming *(Geol)* Lettenbesatz *m*
c. tile Tonziegel *m*; Tonplatte *f*; Dachziegel *m (aus gebranntem Ton)*
c. tile filler Deckenfüllkörper *m (aus gebranntem Ton)*
c. treatment Bleicherdebehandlung *f*
c. ware Keramikerzeugnisse *npl*, Tonwaren *fpl*
c. wash Tonwasser *n*, Tonschlämme *f (Formerei)*
c. water *s.* c. wash
c. working Tonabbau *m*, Tongewinnung *f*

clayey 1. lehmhaltig, tonartig, tonhaltig, tonig; 2. lettenhaltig, lettig
c. matter tonige (tonerdehaltige) Bestandteile *mpl*
c. shale Tonschiefer *m*
clean / to 1. reinigen, säubern, putzen; 2. *(Lw)* ausmisten *(Stall)*; 3. *(Brg)* sortieren
c. castings Gußteile putzen
c. off the cask das Faß abdrehen (glätten) *(Küfnerei)*
c. ship *(Schiff)* Reinschiff machen
clean rein, sauber; gereinigt, gesäubert; klar, scharf, deutlich
c. bomb *(Kern)* „saubere" Bombe *f (H-Bombe mit geringem radioaktivem Fallout)*
c. break *(El)* funkenfreie Unterbrechung *f*
c.-cut separation *(Ch)* destillative Trennung *f* in enge Fraktionen
c.-cut thread sauberes Gewinde *n*
c.-grained *(Led)* narbenrein
c.-lined ship *(Schiff)* Schiff *n* mit glatten, gut strakenden Linien
c.-out hole Reinigungsöffnung *f*
c. reactor *(Kern)* jungfräulicher Reaktor *m*
c. split *(Holz)* Spiegelkluft *f*
c. tape *(Typ)* korrigiertes Lochband *n*
c. timber astreines (fehlerfreies) Bauholz *n*
c. wood astreines Holz *n*
cleaned coal gewaschene Kohle *f*, Reinkohle *f*
c. fines *(Brg)* angereicherte Feinstkohle *f*
c. from rust entrostet
c. wire dekapierter Draht *m*
cleaner 1. [chemische] Reinigung[sanstalt] *f*; 2. Filter *n*; Reiniger *m*, Reinigungsmittel *n*; 3. *(Lw)* Abstreifer *m (an Erntemaschinen)*; 4. *(Brg)* Nachschäumzelle *f*
cleaner's naphtha *(Ch)* Ligroin *n (Siedegrenze 138 bis 205 °C; zur chemischen Reinigung)*
cleaning 1. Putzen *n*, Reinigen *n*, Säubern *n*; 2. Reinigen *n*, Entfetten *n (Galvanotechnik)*; 3. Ausmisten *n (des Stalles)*; 4. *(Brg)* Sortierung *f*; Abräumen *n*
c. agent Reinigungsmittel *n*
c. and dressing Putzen *n (Gießerei)*
c. basin Reinigungsbecken *n*, Reinigungsbassin *n*
c. brush Reinigungspinsel *m*
c. cloth Putztuch *n*
c. cotton waste Putzwolle *f*
c. doctor *(Text)* Walzenreinigungsrakel *f*
c. door Reinigungstür *f (z. B. einer Feuerung)*
c. efficiency *(Text)* Reinigungsfähigkeit *f*, Reinigungswirksamkeit *f*
c. grid Putzrost *m*
c. hole Putzöffnung *f*, Putzloch *n*
c. leather Putzleder *n*
c. machine Putzmaschine *f*
c. magnet Auslöschmagnet *m*
c. material Reinigungsmittel *n*, Putzmittel *n*
c. needle *(Masch)* Düsennadel *f*
c. of coal Kohlenaufbereitung *f*
c. of gas Gasreinigung *f*, Gaswäsche *f*
c. of the stable Ausmisten *n* des Stalles
c. plant *(Brg)* Aufbereitungsanlage *f*, Wäsche *f*
c. pole piece Auslöschmagnet *m*
c. rag Putzlappen *m*
c. room Putzerei *f*
c. solution Waschlösung *f*
c. spirit Fleckenentfernungsmittel *n*
c. table 1. *(Brg)* Herd *m*; 2. *(Met)* Putztisch *m (Gießerei)*
cleanliness Reinheitsgrad *m*; Reinlichkeit *f*
c. of the cotton fibre Reinheitsgrad *m* der Baumwollfaser
cleanse / to reinigen, säubern; von Schmutz oder Fett befreien; spülen
c. by fumigation *(Holz)* ausbrennen
cleanser Putzer *m*; Putzmittel *n*
cleansing agent Reinigungsmittel *n*
c. blower *(Wärm)* Durchblasreiniger *m*
c. cream Reinigungscreme *f*, Abschminkcreme *f (Kosmetik)*
c. solvent Ligroin *n* zur chemischen Reinigung
c. tissue Moptextilien *pl*, textiles Mopmaterial *n (zum Reinigen)*; Gesichtswatte *f (Kosmetik)*
cleanup work *(Brg)* Reinigungsarbeiten *fpl (auf dem freigelegten Flöz)*; Putzen *n (des Kohlenhangenden)*
clear / to 1. hinterdrehen, hinterschleifen, räumen; freiarbeiten; abheben *(Werkzeug vom Werkstück)*; 2. *(El)* entstören; trennen; 3. *(Dat)* freigeben; löschen; räumen *(Datenspeicher usw.)*; 4. *(Bw)* [be]räumen, wegräumen; 5. *(Brg)* wegfüllen; räumen
c. a port *(Schiff)* aus einem Hafen auslaufen
c. a ship *(Schiff)* Ladung löschen
c. away abbacken *(Tische in der Schiffsmesse)*
c. by grinding hinterschleifen
c. goods abfertigen *(Güter zollamtlich)*
c. the chips ausspänen
c. the ground *(Lw)* Land urbarmachen
c. the hole of chips die Späne aus der Bohrung entfernen (fördern)
c. the loop *(Text)* einschließen *(Strickerei)*
c. the messtable abbacken *(Tische in der Schiffsmesse)*
c. the shade *(Text)* Ton aufhellen, avivieren
clear 1. hell, klar, licht; durchsichtig; 2. *(Ök)* unkostenfrei; 3. berührungsfrei
c. ahead *(Schiff)* klar voraus
c. air gust *(Flg)* unbehinderte Luftbö *f*
c. astern *(Schiff)* klar achteraus
c. channel Sendekanal *m* für 10 bis 50 kW *(nach FCC)*
c. distance lichte Weite *f*, lichter Abstand *m*

c. **glass** Klarglas *n*
c. **height** lichte Höhe *f*
c. **mesh** [lichte] Maschenweite *f (eines Siebes)*
c. **of** ohne Berührung mit; mit Abstand von, getrennt von
c. **shed** *(Text)* schräges (reines, offenes) Fach *n*
c. **sheet** *(Kfz)* Klarscheibe *f*; Klarglasscheibe *f*
c. **signal** Freisignal *n*
c. **space** freier Raum *m*
c. **span** lichte Weite *f*
c. **span building** stützenfreies Gebäude *n*
c. **stuff** fehlerfreies Bauholz *n*
c. **switch** Freigabeschalter *m*
c.**-text sorter reader** Klarschriftsortierleser *m*, Klartextsortierleser *m*
c. **width** lichte Breite (Weite) *f*
c. **wood** astreines Holz *n*

clearance 1. Zwischenraum *m*; Spielraum *m*; Abstand *m*; Spiel *n*; Luft *f*; 2. Freiwinkel *m (Werkzeugschneide)*; 3. Kopfspiel *n (Zahnspiel)*; 4. schädlicher Raum *m (Dampfzylinder)*; 5. Spalt *m (Turbine)*; 6. lichter Raum; lichte Höhe *f*, Durchgangshöhe *f*, Durchfahrtshöhe *f*; Begrenzung *f (Durchfahrt)*; 7. Abbruch *m*, Beseitigung *f*; Aufräumung *f*, Wegräumung *f*; Sanierung *f (Städtebau)*; 8. Lichtung *f (im Wald)*
c. **angle** *(Masch)* Freiwinkel *m*
c. **certificate** *(Schiff)* Klarierungspapier *n*
c. **diameter** Spaltdurchmesser *m (Turbine)*
c. **face** *(Masch)* Freifläche *f*
c. **fit** *(Masch)* Spielsitz *m*
c. **for approach** *(Flg)* Freigabe *f* für den Anflug
c. **gauge for pantographs** *(El)* Lichtraumprofil *n* für Stromabnehmer
c. **height** Durchgangshöhe *f*, Durchfahrtshöhe *f*; Kopffreiheit *f*, Kopfhöhe *f*
c. **leakage** Spaltundichtigkeit *f (Turbine)*
c. **limit** Durchfahrtsbegrenzung *f*, Durchfahrtsprofil *n*
c. **loss** Spaltverlust *m (Turbine)*
c. **of bridge** Durchfahrthöhe *f* [der Brücke]
c. **of freight** Güterabfertigung *f (Zoll usw.)*
c. **opening** lichter Querschnitt *m*; Durchflußquerschnitt *m*
c. **papers** Zollpapiere *pl*
c. **pressure** Spaltdruck *m*, Zwischendruck *m (Turbine)*
c. **space** *(Masch)* Spielraum *m*, Hohlraum *m*; Zwischenraum *m*; Toleranzfeld *n*
c. **volume** Kompressionsvolumen *n*, Kompressionsraum *m*, Verdichtungsraum *m (Verbrennungsmotor)*

clearer 1. *(Brg)* Hauer *m*; 2. Räumer *m*; 3. *(Text)* Wendewalze *f*, Wender *m*
c. **board** *(Text)* Filzwalzenbelag *m*
c. **cloth** Putztuch *n (zum Putzen von Spinnmaschinen)*
c. **roller** *(Text)* Plüschwalze *f (Kämmaschine)*

clearing 1. *(Masch)* Hinterschliff *m*; Abheben *n*; 2. *(Fmt)* Trennung *f*, Unterbrechung *f*; Fehlerbeseitigung *f*; Reinigung *f (der Röhre von Gasresten)*; 3. *(BMSR)* Freigabe *f*; Löschung *f*; 4. Räumung *f (von Baugelände)*; 5. *(Brg)* Räumungsarbeiten *fpl*; Beräumung *f*; 6. Aushau *m*, Roden *n (Wald)*
c. **agreement** Verrechnungsabkommen *n (Außenhandel)*
c. **and baling press** Räum- und Sammelpresse *f*
c. **basin** Klärbecken *n*, Klärteich *m*
c. **bath** *(Ch)* Klärbad *n*
c. **cam** Heber *m (Strickmaschine)*; Austriebsexzenter *m*
c. **circuit** *(El)* Blockstromkreis *m*
c. **current** *(Fmt)* Schlußzeichenstrom *m*
c. **device** *(Dat)* automatische Null[ein]-stellung *f*
c. **element** *(Text)* Austriebsteil *m*
c. **hoe** Rodehacke *f*
c. **implement** Räumer *m*
c. **key** Schlußtaste *f (Telefonie)*
c. **lamp** *(Fmt)* Schlußlampe *f*
c. **lever** *(Dat)* Löschhebel *m*
c. **machine** Räumer *m*
c. **needle** Räumspieß *m*
c. **of a call** *(Fmt)* Aufheben *n* einer Verbindung
c. **of a fault** *(Fmt)* Störungsbeseitigung *f*
c. **of fields** Räumen *n* der Felder
c. **of forests** Ausholzen *n*, Roden *n (des Waldes)*
c. **of the yarn** *(Text)* Entfernen *n* von Garnverunreinigungen
c. **plow** Hobelpflug *m*
c. **poker** Räumspieß *m*
c. **port** *(Schiff)* Wasserpforte *f*, Schanzkleidpforte *f*
c. **position** *(Text)* Einschlußstellung *f (Zungennadel)*; Austriebsstellung *f*
c. **press** Aufsammelpresse *f*, Räumpresse *f*
c. **pulse** *(Fmt)* Schlußzeichen *n*; Überwachungszeichen *n*
c. **relay** *(Fmt)* Schluß[zeichen]relais *n*; Überwachungsrelais *n*
c. **signal** 1. *(Fmt)* Schlußzeichen *n*; 2. Räumungssignal *n*
c. **stoker** Räumspieß *m*
c. **stripper** Ausdünnstriegel *m*, Striegel *m*
c. **time** *(El)* [Gesamt-]Abschaltdauer *f*, Ausschaltdauer *f*

clearness 1. *(Opt)* Schärfe *f*; 2. *(Ak)* Deutlichkeit *f*, Klarheit *f*; 3. *(Fmt)* Verständlichkeit *f*
c. **of sound** Wiedergabebrillanz *f*

c. of tuning *(Fmt)* Abstimmschärfe *f*
cleat 1. Bügel *m*; 2. *(Bw)* Knagge *f (Holzbau)*; 3. *(Geol)* Schlechte *f*, Ablösungsfläche *f*, Schichtfläche *f*; 4. *(Schiff)* Klampe *f (zum Belegen)*
c. insulator *(El)* Klemmisolator *m*
cleating ring Verkleidungsring *m*
cleavability Spaltbarkeit *f (Kristalle, Mineralien)*
cleavable spaltbar
cleavage 1. Spalt *m*, Riß *m*, Sprung *m*; 2. Spaltung *f*, Abspaltung *f*, Aufspaltung *f*; [ebene] Spaltbarkeit *f (Kristalle, Mineralien usw.)*; 3. *(Brg)* Schlechte *f*
c. brittleness interkristalline Brüchigkeit (Sprödigkeit) *f*, Spaltbrüchigkeit *f*
c. face Spaltebene *f*, Spaltfläche *f (Kristalle)*
c. of crystals Spaltbarkeit *f* von Kristallen
c. plane Spaltebene *f*, Spaltfläche *f*; Bruchfläche *f (Kristalle)*
c. rhombohedron Spaltrhomboeder *n*
c. strength Spaltfestigkeit *f*
c. surface Spaltfläche *f*
cleave / to spalten; [auf]schlitzen
cleave saw *(Holz)* Bohlensäge *f*
cleaved *(Geol)* zerklüftet
cleaver 1. Spaltmesser *n*; 2. Baumaxt *f*, Baumhacke *f*
cleaving test Spaltversuch *m*
cleft Sprung *m*, Riß *m*; Kluft *f*, Spalte *f*
c. billet *(Holz)* Kloben *m*
c. timber Spaltholz *n*
c. weld Kehlschweißung *f*
clench / to 1. festklammern, zusammenpressen; 2. kalt nieten; 3. abdichten
clerk of works Baukontrolleur *m*
clevis 1. Gabelkopf *m (am Bremsseil)*; 2. *(Bw)* Schäkel *m*
c. type connector Einhängelasche *f*
click / to 1. knacken, klicken; 2. einrasten
click 1. Sperrhaken *m*, Sperrkegel *m*, Sperrklinke *f*, Sperrstift *m*; 2. *(Fmt)* Knacken *n*, Knackton *m*
c. filter *(Fmt)* Filter *n* zur Unterdrückung der Knackgeräusche, Rastfilter *n*
c.-in spring Einrastfeder *f*
c. spring Einrastfeder *f*, Sperrfeder *f*
c. stop Rastblende *f*
c.-stop device Einrastvorrichtung *f*, Rastvorrichtung *f*
clicking board Zuschneidebrett *n (Schuhfertigung)*
c. department (room) Zuschneiderei *f (Schuhfertigung)*
clickless *(Fmt)* knackgeräuschfrei
clicks *(Fmt)* Knackgeräusche *npl*, Knackstörung *f*
clickwork *(Masch)* Gesperr *n*
climate investigation Klimaprüfung *f*
climatic chamber Klimakammer *f*
c. conditions Klima *n*, klimatische Verhältnisse *npl*
c. strain klimatische Beanspruchung *f*
climatology Klimatologie *f*, Klimakunde *f*
Climax boiler *(Wärm)* Climaxkessel *m*, Leiterkessel *m*
climb / to klettern; ranken *(Pflanzen)*; steigen; Höhe gewinnen *(Flugzeug)*
c. up the chimney den Schornstein befahren
climb *(Flg)* Steigflug *m*
c. and dive competition *(Flg)* Steig- und Sturzflug-Wettbewerb *m*
c.-cut milling Gleichlauffräsen *n*
c. hobbing Gleichlaufwälzfräsen *n*
c. milling Gleichlauffräsen *n*
climbers Steigeisen *npl (für Telegrafenmaste usw.)*
climbfeed method *(Masch)* Gleichlauffräsverfahren *n*
climbing ability *(Kfz)* Steigvermögen *n*
c. boots Bergkletterschuhe *mpl*
c. capacity Kletterfähigkeit *f*
c. crane Kletterkran *m*
c. crossing Kletterkreuzung *f*
c. curve Auflaufkurve *f*
c. force Steigkraft *f*
c. form[work] Kletterschalung *f (Betonbau)*
c. gradient befahrbare Steigung *f*
c. path Aufstiegsbahn *f (einer Rakete)*
c. pole Kletterstange *f*
c. power 1. Steigflugleistung *f*; 2. Kletterfähigkeit *f*, Klettervermögen *n*
c. program *(Rak)* Umlenkung *f*, Umlenkprogramm *n*
c. shuttering Kletterschalung *f (Betonbau)*
c. speed *(Flg, Rak)* Steiggeschwindigkeit *f*, Aufstiegsgeschwindigkeit *f*
c. switch Kletterweiche *f*
c. turn gezogene Kurve *f*, Kurve im Steigflug
c. up the chimney Befahren *n* des Schornsteins
clinch / to 1. festklammern; einschnappen; 2. kalt nieten
c. a rivet head einen Nietkopf stauchen
clinch *(Kfz)* Felgenhorn *n*
c.-bolt Nietbolzen *m*
c. nut Nietmutter *f*
clincher band *(Kfz)* Wulstband *n*
c. rim Wulstfelge *f*
c. tyre Wulstreifen *m*
cling to / to haften an
clinic boat Klinikboot *n*, Lazarettschiff *n*
clinker / to *(Bw)* 1. klinkern; 2. sintern *(Zement)*
clinker *(Bw)* 1. Klinker[stein] *m*, Klinkerziegel *m*; 2. Sinterschlacke *f*

c. **brick** Klinker[stein] *m*, Klinkerziegel *m*
c. **built** Klinker[bau]-, in Klinkerbauweise; geklinkert
c. **built boat** Klinkerboot *n*, geklinkertes Boot *n*
c. **grate** *(Wärm)* Schlackenrost *m*
clinkering temperature Sintertemperatur *f* *(Zementherstellung)*
c. **zone** Sinterzone *f* *(Zementherstellung)*
clinoclasite Strahlerz *n*, Strahlenkupfer *n*
clinometer Inklinometer *n*, Klinometer *n*, Neigungsmesser *m*
clip / to *(El)* begrenzen, beschneiden, verkürzen *(Amplitude)*
c. **the fleece close to the skin** Vlies an der Haut glatt abscheren
clip 1. Büroklammer *f*, Klammer *f*; Klemme *f*, Krampe *f*, Lasche *f*, Schelle *f*; Spannbügel *m*; 2. *(El)* Kabelschuh *m (insbesondere bei Kraftfahrzeugen)*
c. **bolt** Hakenschraube *f*
c.-**ring** Klemm[en]ring *m*
c.-**spot loom** *s.* swivel loom
c. **stenter** *(Text)* Spannrahmen-Trockenmaschine *f* mit Kluppen
clipped connection Klemmverbindung *f*
clipper 1. *(Fmt, Fs)* Abkapper *m*, Clipper *m*, Amplitudenbegrenzer *m*, Begrenzer *m*; 2. *(Schiff)* Klipper *m*, Schnellsegler *m*; 3. Beißzange *f*, Schneidzange *f*
c. **circuit** Amplitudenbegrenzungskreis *m*, Abschneideschaltung *f*
c. **stage** *(El)* Impulsabschneidstufe *f*
c. **tube** *(Fs)* 1. Trennröhre *f* für Bild und Impuls; 2. Abschneideröhre *f*, Schwellenröhre *f*
clipping Abkappen *n*, Abschneiden *n*, Begrenzung *f (Funktechnik)*
c. **machine** Beschneidemaschine *f*
c. **of words** Wortverstümmelung *f* *(Telegrafie)*
cloak-room Garderobe *f*, Kleiderablage *f*
cloanthite *s.* chloanthite
clock / to *(Zeit)* stoppen, Zeit nehmen
clock 1. Uhr *f*; 2. Taktgeber *m*, Zeitgeber *m* *(in Gestalt von Impulsen)*
c. **case** Uhrgehäuse *n*
c.-**controlled** uhrgesteuert
c. **drive** Uhrwerk *n*
c. **for sidereal time** Sternzeituhr *f*
c. **frequency** *(BMSR)* Taktgeberfrequenz *f*, Zeitgeberfrequenz *f*
c. **hour figure of the vector group** *(El)* Schaltgruppen-Numerierung *f* nach dem Zifferblatt
c. **impulse** *(Dat)* Taktimpuls *m*
c. **installation** Uhrenanlage *f*
c. **movement** Uhrwerk *n*
c.-**pulse gate** *(Dat)* Impuls-Synchronisierschaltung *f*
c.-**pulse generator** *(Dat)* Taktgeber *m*
c. **rate** Taktgebermaß *n*, Zeitgebermaß *n*
c. **relay** *s.* time-switch
c. **spring** Uhrfeder *f*
c.-**spring motor** Federmotor *m*; Uhrwerk *n*
c. **switch** *(El)* Schaltuhr *f*
c. **thermostat** *(BMSR)* Zeitplan-Temperaturregler *m*
c. **track** *(BMSR)* Taktgeberspur *f*, Zeitgeberspur *f*
clockwise rechtsdrehend, im Uhrzeigersinn
c. **rotation** Rechtslauf *m*; Rotation *f* im Uhrzeigersinn
c. **running** Rechtslauf *m*
clockwork fuse *(Mil)* Uhrwerkzünder *m*
clod 1. Ackerscholle *f*, Erdscholle *f*; 2. *(Brg)* Nachfallpacken *m*
c. **breaker** Schollenbrecher *m*
c. **crusher** Schollenbrecher *m*
clog / to 1. sich festhaken *(Kreissäge)*; sich festsetzen *(Werkzeug)*; [sich] verstopfen, sich zustopfen *(Schleifscheibe)*; sich zusammenballen, stauen *(Späne)*; 2. verkleben, verschmutzen *(Verbrennungsmotor, Gasturbine)*
c. **up** sich stauen, verstopfen
clog pack *(Brg)* Holzkasten *m*, Holzschrank *m*
clogging of piston rings *(Kfz)* Verkleben (Verschmutzen) *n* der Kolbenringe
c. **of the chips** Spanstauung *f*
close / to schließen, zumachen; beenden
c. **a circuit** einen Stromkreis schließen, einschalten
c. **down** 1. schließen, stillegen, stillsetzen *(Betriebe)*; 2. beenden *(eine Sendung)*
c.-**in** sich *(einem Ziel)* [an]nähern; losjagen, lossteuern *(auf ein Ziel)*
c. **off** abketteln *(Strickerei)*
close eingeschlossener Raum *m*
close 1. nahe, benachbart; vertraut; 2. dicht, undurchdringlich *(Nebel)*; 3. bedrückend, stickig *(Luft)*; 4. eng; eng anliegend *(Hüllen, Kleider)*
c. **annealing** *(Met)* Kastenglühung *f*, Kistenglühung *f*
c.-**boarded fence** Bretterzaun *m*
c. **by the wind** *(Schiff)* hart am Wind
c. **ceiling** *(Schiff)* feste Wegerung *f*
c. **centre holes** Bohrungen *fpl* mit geringem Mittenabstand
c. **collision** *(Ph)* enger Zusammenstoß *m*
c. **contact glue** Kontaktkleber *m*
c. **coupling** *(El)* feste Kopplung *f*
c. **cribbing** *(Brg)* Vollschrotausbau *m*, Vollschrotzimmerung *f*
c.-**cut separation** *(Ch)* destillative Trennung *f* in enge Fraktionen

c. down 1. Stillegung *f (Betriebe)*; 2. Sendeschluß *m (Radio, Fernsehen)*
c. driving fit [Johansson system] *(Masch)* enger Treibsitz *m* [System Johansson]
c. grain *(Led)* dichter (geschlossener) Narben *m*
c.-grained *(Met)* feinkörnig, dicht *(Guß)*; engrining *(Holz)*
c.-grained cast iron Feinkorngußeisen *n*
c.-grained pig iron feinkörniges Roheisen *n*
c. inshore *(Schiff)* dicht unter Land
c. joint Schweißverbindung *f* mit gegenseitiger Berührung der zu verbindenden Teile
c. limits enge Maßgrenzen *fpl*
c. meshed engmaschig
c.-packed code *(Dat)* dicht gepackter Kode *m*
c.-packed lattice dichtgepacktes Gitter *n* *(Kristallgitter)*
c.-packed structure Dichtpackung *f*, [dichteste] Kugelpackung *f (Kristallgitter)*
c. quarter drill [for use in corners] Eckenbohrmaschine *f*
c. quarter situation *(Am) s.* nearest approach
c. quote *(Typ)* Abführungszeichen *n*
c.-range disturbing effect *(Fmt)* Nahstörung *f*
c.-reconnaissance aircraft Nahaufklärungsflugzeug *n*, Nahaufklärer *m*
c. riveting Dichtnietung *f*
c. running fit *(Masch)* enger Laufsitz *m*
c. scanning *(Fmt, Fs)* Feinabtastung *f*
c.-shed *(Text)* Geschlossenfach- *(Jacquard)*
c. spacing kleinporiges (dichtes) Gefüge *n*
c. stitch *(Text)* Knopflochstich *m*
c. stowing anchor *(Schiff)* Patentanker *m*
c.-support tank Nahkampfpanzer *m*, Panzerkampfwagen *m* für Nahkampfunterstützung
c.-talking sensitivity *(El)* Empfindlichkeit *f* [bei naher Besprechung], Übertragungsfaktor *m (Mikrofon)*
c.-tolerance engtoleriert
c. tolerance bolt Paßschraube *f*
c.-up Nahaufnahme *f*, Großaufnahme *f (Fotografie, Film, Fernsehen)*
c.-up camera Kamera *f* für Nahaufnahmen (Großaufnahmen)
c.-up [sectional] view vergrößerter Bildausschnitt *m*
c.-up shot *(Opt)* Nahaufnahme *f*, Großaufnahme *f*
c. weld Schweißverbindung *f* mit gegenseitiger Berührung der zu verbindenden Teile
closed 1. [ab]geschlossen; 2. angebogen *(Feder)*
c. aerial geschlossene Antenne *f*, Rahmenantenne *f*
c. ashpit furnace *(Wärm)* Unterwindfeuerung *f*
c. assembly *(Plast)* geschlossene Verleimung *f*
c. barge *(Schiff)* Kastenschute *f*
c. block *(El)* geblockte (blockierte) Blockstrecke *f*
c.-centre valve *(BMSR)* Steuerventil *n* mit Überlappung in der Ruhelage
c. chain with constrained movement *(Wärm)* zwangsläufig geschlossene Kette *f*
c. circuit *(El)* geschlossener Stromkreis *m*; Ruhestromkreis *m*
c.-circuit alarm device Ruhestromalarmgerät *n*
c.-circuit arrangement Ringschaltung *f*
c.-circuit battery Ruhestrombatterie *f*
c.-circuit condition Einschaltzustand *m*
c.-circuit connection Ruhestromschaltung *f*
c.-circuit contact Ruhekontakt *m*, Öffner *m*
c.-circuit current Ruhestrom *m*
c.-circuit current principle Ruhestromprinzip *n*
c.-circuit system Ruhestromanlage *f*
c.-circuit television Fernsehen *n* im Kurzschlußverfahren
c.-circuit ventilated machine Maschine *f* mit Umlaufkühlung
c.-circuit voltage Arbeitsspannung *f*
c.-circuit working Ruhestrombetrieb *m*
c. coil winding [in sich] geschlossene Wicklung *f*
c. core transformer Transformator *m* mit geschlossenem Kern
c.-cup flash point Flammpunkt *m* im geschlossenen Tiegel *(Erdöl)*
c.-cup tester *(Ch)* Flammpunktprüfer *m* mit geschlossenem Tiegel
c. cycle 1. vollständiger Arbeitszyklus *m*; 2. Kreisprozeß *m (Thermodynamik)*
c. dock *(Schiff)* Flutbecken *n*, Schleusendock *n*
c.-end impedance *(El)* Kurzschlußimpedanz *f*, Kurzschlußwiderstand *m*
c.-end line *(El)* [am Ende] kurzgeschlossene Leitung *f*
c. end of sleeper Kappe *f (Schwelle)*
c. eye bolt Augenschraube *f*
c. forced-draught stokehold mit Raumdruck gefahrener Kesselraum *m*
c. form geschlossene Form *f (Lösung von mathematischen Problemen)*
c. frame geschlossener Rahmen (Rechteckrahmen) *m*
c. groove geschlossenes Kaliber *n (an Walzen)*
c. hearth Schmelzherd *m* mit Verschlußdeckel
c. housing geschlossener Ständer *m (Walzwerk)*
c. interval *(Math)* abgeschlossenes Intervall *n*, abgeschlossene Intervallmenge *f*

c. **joint** Schweißverbindung *f* mit Berührung der zu verbindenden Teile
c. **kinematic chain** *(Masch)* geschlossene kinematische Kette *f*
c.**-loop control** *(BMSR)* Regelung *f* [mittels Regelkreises]
c.**-loop control system** Regelkreis *m*, Regelungssystem *n* [mit geschlossenem Kreis]
c.**-loop frequency response** *(BMSR)* Frequenzgang *m* des geschlossenen Regelkreises
c.**-loop radiator** *(Fs)* Rahmenstrahler *m*
c.**-loop series street lighting system** Reihenlampenbeleuchtung *f*, Straßenbeleuchtung *f* mit Reihenlampen
c.**-loop system** *s.* c.-loop control system
c. **pair** formschlüssige Paarung *f*
c. **pass** geschlossenes Walzenkaliber *n*
c. **race-track** *(Kern)* geschlossener Racetrack *m (Synchrotron)*
c. **railing** *(Schiff)* Schanzkleid *n (geschlossene Reling)*
c. **resonator** *(El)* Hohlraumschwingkreis *m*, Topfkreis *m*
c. **shell** *(Ph)* abgeschlossene (vollbesetzte) Schale *f (Atomelektronen)*
c.**-shell atom** Atom *n* mit abgeschlossener Schale, Atom mit Edelgaskonfiguration
c. **shop** *(Dat)* geschlossener Betrieb *m (nicht im eigenen Betrieb erfolgte Programmierung)*
c. **stub** *(Fmt)* Kurzschlußstichleitung *f*
c. **subroutine** *(Dat)* geschlossenes Teilprogramm (Unterprogramm) *n*
c. **surface** *(Math)* geschlossene Fläche *f*
c. **television** *(Fs)* Betriebsfernsehen *n*
c. **washery** *(Brg)* Wäsche *f* mit geschlossenem Wasserkreislauf
c. **weld** Schweißverbindung *f* mit Berührung der zu verbindenden Teile
closely coiled mit eng aneinanderliegenden Windungen
c. **measured** genau bemessen
c.**-spaced** mit geringen Mittenabständen (Abständen), dicht aneinander
closer 1. *(Masch)* Spannkopf *m*, Reduzierhülse *f*; 2. Riemchen *n (Ziegelmauerwerk)*
closet 1. Kleiderkammer *f*; Kleiderschrank *m*; Schrank *m*; 2. Abstellkammer *f*, Nebengelaß *n*; 3. *s.* water closet
closing coil *(El)* Einschaltspule *f*
c. **down** 1. *(Ök)* Betriebsstillegung *f*; 2. *(Fmt)* Beendigung *f* einer Sendung
c. **error** *(Math)* Abschlußfehler *m*, Endfehler *m*
c. **handle** Schließgriff *m*
c. **machine** *(Text)* Verseilmaschine *f*
c. **of the chain** *(Wärm)* Kettenschluß *m*
c. **of the port** *(Wärm)* Kanalschluß *m*
c. **plate** *(Wärm)* Abschlußplatte *f*, Verschlußplatte *f*
c. **rate** 1. *(Ök)* Betriebsstillegungsrate *f*; 2. Endgeschwindigkeit *f*; 3. Annäherungsgeschwindigkeit *f (an das Ziel)*
c. **room** Stepperei *f (Schuh)*
c. **rope** Schließseil *n (Greifer)*
c. **speed** 1. Schlußgeschwindigkeit *f*, Abschlußgeschwindigkeit *f*, Endgeschwindigkeit *f*; 2. Annäherungsgeschwindigkeit *f (an das Ziel)*
c. **switch** *(El)* Einschalter *m*
c. **the shed** *(Text)* Fachschluß *m*
c.**-up** 1. Annäherung *f (an das Ziel)*; 2. Formen *n* des Schließkopfes *(Nieten)*
c. **velocity** Annäherungsgeschwindigkeit *f (an das Ziel)*
c. **wire lead** *(Text)* Schließdrahtblei *n*
closure 1. *(Math)* abgeschlossene Hülle *f*; 2. *(Bw)* Verschluß *m*; 3. *(Brg)* Querschnittsverminderung *f (Strecke)*
clot / to *(Ch)* sich zusammenballen, gerinnen
clot Klumpen *m*; Gerinnsel *n*; Pfropfen *m*
cloth 1. *(Text)* Tuch *n*; [schwerer] Stoff *m*, Zeug *n*; 2. *(Typ)* Einbandgewebe *n*; Leinen *n*
c. **abrasive disk** Schleifleinenscheibe *f*
c. **beam** *(Text)* Warenbaum *m*
c.**-centred board** *(Holz)* Triplexkarton *m* mit Stoffmittellage
c. **cleaning machine** *(Text)* Gewebeputzmaschine *f*
c. **covering** Tuchbespannung *f*
c. **face** Gewebeoberseite *f*
c.**-faced paper** *(Typ)* Papyrolin *n*
c. **fall-out** *(Text)* Absprengen *n* des Stoffes *(Strickerei)*
c. **fall-out detector** *(Text)* Absprengabsteller *m*
c. **filter** Tuchfilter *n*
c. **manufacture** Tuchindustrie *f*
c. **printer** Tuchdrucker *m*, Textildrucker *m*
c. **roll-up** *(Text)* Stoffaufrollvorrichtung *f*
c. **roller** *(Text)* Warenzylinder *m*; Warenbaum *m (Webstuhl)*
c. **roller apron** *(Text)* Zuführtuch *n* des Warenzylinders (Warenbaums)
c. **shearing machine** Tuchschermaschine *f*
c. **shrinker** Tuchdekateur *m*
c. **spreader arm** Stoffbreithalterarm *m*
c. **tester** Gewebeprüfapparat *m*
c. **weaving loom** Tuchwebstuhl *m*
clothing Beschlag *m (Trommel)*
c. **industry** Bekleidungsindustrie *f*
c. **leather** Bekleidungsleder *n*
clothoid *(Math)* Klothoide *f*
cloud 1. Wolke *f*; 2. *(Fs)* Schatteneffekt *m*
c. **bank** Wolkenwand *f*, Wolkenbank *f*
c. **base** untere Wolkengrenze *f*
c.**-burst** Wolkenbruch *m*, Regenschauer
c. **ceiling** Wolkenhöhe *f*

c. ceilometer Wolkenhöhenmesser *m*
c. chamber *(Ph)* Nebelkammer *f*, Wilsonkammer *f*
c.-chamber method Nebelkammermethode *f*
c.-chamber tracks Nebelkammerspuren *fpl*
c. clutter Wolkenechos *npl*
c. discharge Wolkenentladung *f*
c. of electrons Elektronenwolke *f*, Elektronenschwarm *m*
c. point Trübungspunkt *m*, Temperatur *f* der beginnenden Paraffinausscheidung *(Erdöl)*
c. pulse *(El)* Raumladungsimpuls *m*
c. soaring *(Flg)* Wolkensegeln *n*
c.-track *(Ph)* Nebelspur *f (Wilsonkammer)*
c. trails Nebelspuren *fpl (Wilsonkammer)*
cloudiness Trübung *f*
cloudy fabric *(Text)* moirierter Stoff *m*
clout / to schienen; beschlagen; flicken
cloven aufgespalten
c. billet *(Holz)* Kloben *m*
c. timber Nutzspaltholz *n*, Spaltware *f*
clover *(Lw)* Klee *m*
c. leaf 1. Kleeblatt *n*; 2. Kleeblattkreuzung *f (Autobahn)*
c. leaf aerial Kleeblattantenne *f*
c. rasper Kleereiber *m*
c. rasping machine Kleereiber *m*
c. seed broadcaster Klee-Breitsämaschine *f*
c. seeder Kleesägerät *n*
c. thresher Kleedrescher *m*
c. threshing machine Kleedrescher *m*
cloves oil *(Ch)* Nelkenöl *n*
club / to zusammenschließen; verkeilen
c.-haul *(Schiff)* wenden mit Hilfe des Leeankers
club-foot *(Schiff)* Wulstbug *m*
cluster 1. Büschel *n*, Schwarm *m*; Anhäufung *f*; Häufung *f*; 2. Sternhaufen *m*
c. count *(Ph)* Clusterzählung *f*
c. gears *(Masch)* Schieberäderblock *m*
c. mill *(Met)* Vielwalzengerüst *n*
c. of atoms Atomhaufen *m*
c. of flames *(Wärm)* Flammenbüschel *n*
c. of incendiary bombs *(Mil)* Brandbombenbündel *n*
c. plate Stirnplatte *f*; Spindellagerplatte *f (Bohrmaschine)*
c. rocket launcher *(Mil)* Raketengeschütz *n* mit mehreren Läufen
c. structure *(Ph)* Büschelstruktur *f*
clustered column *(Bw)* Bündelpfeiler *m*
c. errors *(Dat)* Fehlerhäufungen *fpl*
clutch/to *(Masch)* einrücken, schalten *(Kupplung)*
clutch *(Masch)* [aus- und einrückbare] Kupplung *f*, Schaltkupplung *f*
c.-actuating fork *(Masch)* Gabelstück *n*
c.-and-brake steering Kupplungssteuerung *f (Raupenfahrzeuge)*
c. bellhousing *(Kfz)* Kupplungsglocke *f*, Kupplungsgehäuse *n*
c. bolt Kupplungsbolzen *m*
c. brake Kupplungsbremse *f*
c. cam Kupplungsdaumen *m*
c. case Kupplungsgehäuse *n*
c. collar Kupplungsring *m*
c. cone Kupplungskegel *m*
c. control lever cable Kupplungskabel *n*
c. cover Kupplungsdeckel *m*
c. disk Kupplungsscheibe *f*
c. dog Kupplungsklaue *f*
c. drive plate Kupplungstreibscheibe *f*
c. facing Kupplungsbelag *m*
c. fading Abnutzung *f* des Kupplungsbelages, Kupplungsschwund *m*
c. fork Kupplungs[ausrück]gabel *f*
c. guide bearing Kupplungsführungslager *n*
c. head Schraubenkopf *m* mit Innenführung für einen Schlüssel in Sonderform
c. housing Kupplungsgehäuse *n*
c. keyway Mitnehmernut *f (Frässpindel)*
c. knob *(Masch)* Griff *m* für Längs- und Planzug
c. lining Kupplungsbelag *m*
c. linkage Kupplungsgelenk *n*, Kupplungsgestänge *n*
c. member Kupplungshälfte *f*
c. operating device Kupplungsgestänge *n*
c. operating shaft Kupplungsausrückwelle *f*
c. pedal Kupplungspedal *n*
c. pedal clearance Kupplungsspiel *n*
c. plate Kupplungsscheibe *f*
c. pressure plate Kupplungsdruckscheibe *f*, Druckplatte *f*
c. pressure spring Kupplungsfeder *f*
c. release bearing Ausrücklager *n (Kupplung)*
c. release [fork] lever Ausrückhebel *m (Kupplung)*
c. release plate Ausrückplatte *f (Kupplung)*
c. release shaft Ausrückwelle *f (Kupplung)*
c. release sleeve Kupplungsausrückmuffe *f*
c. release yoke Ausrückgabel *f (Kupplung)*
c. segment Kupplungslamelle *f*
c. shaft Kupplungswelle *f*
c. sleeve Kupplungsmuffe *f*
c.-sleeve guide Kupplungsmuffenführung *f*
c. slip Kupplungsschlupf *m*
c. spindle *(El)* Wickelteller *m (Bandgerät)*
c. spring Kupplungsfeder *f*
c.-throwout collar Kupplungsdrucklager *n*
c.-throwout fork Kupplungsausrückgabel *f*
c. thrust bearing Kupplungsdrucklager *n*
c. thrust plate Kupplungsdruckscheibe *f*
c. thrust spring Kupplungsdruckfeder *f*
c. withdrawal shaft Kupplungsauslösewelle *f*, Kupplungsausrückwelle *f*
clutching Schaltung *f (Kupplung)*

clutter 1. Störungszeichen *n (Radar)*; 2. Trübung *f (Radar)*
c. signal *s.* clutter 1.
clystron *(Fs)* Klystron[röhre *f*] *n*
co-applicant Mitanmelder *m (Patentwesen)*
co-axial *s.* coaxial
co-channel interference *(Fs)* Störung *f* aus demselben Kanal
co-channel station *(Fs)* Sendestelle *f* im gleichen Kanal
co-channel transmitter Gleichkanalsender *m*
co-condensation *(Ch)* Mischkondensation *f*
co-driver Beifahrer *m*
co-ordinated insulation *(El)* abgestufte Isolation *f*
co-ordination of insulation *(El)* Stufung *f* der Isolierstoffe, Abstufung *f* der Isolierung, Festlegung *f* des Isolationspegels
co-phase component *(El)* gleichphasige Komponente *f*, Wirkkomponente *f*
co-reactant *(Ch)* Reaktionspartner *m*
coach Eisenbahnwagen *m*
c. leather Wagenbezugleder *n*
c. screw Vierkantkopfschraube *f (für Holz)*
coachwork *(Kfz)* Karosserie *f*, Wagenaufbau *m*
coagulable erstarrungsfähig, gerinnbar
coagulant Gerinnungsmittel *n*, Koagulierungsmittel *n*, Verdickungsmittel *n*
coagulate / to koagulieren, gerinnen, verdicken
coagulate Koagulat *n*, Koagulierungsprodukt *n*
coagulating bath Fällbad *n (Chemiefaserherstellung)*
coagulation Koagulation *f*, Ausflocken *n*, Flockung *f*
c. spinning *(Text)* Naßspinnen *n*, Lösungsspinnen *n*
coagulum *(Ch)* Koagulat *n*, Koagel *n*, Gerinnsel *n*
coal / to *(Schiff)* Kohlen einnehmen, bunkern
coal Kohle *f*; Steinkohle *f*
c. apples Kugelkohle *f*
c.-bearing formation Kohlengebirge *n*
c.-bearing rock Kohlengebirge *n*
c. briquette Kohlenbrikett *n*
c. bunker *(Schiff)* Kohlenbunker *m*
c. burnt per unit grate area *(Wärm)* Rostbelastung *f*, Rostflächenbeanspruchung *f*
c. burst *(Brg)* Kohlenausbruch *m*
c. carbonization Kohlenentgasung *f*; Steinkohlenschwelung *f*
c. carbonization coking Steinkohlenverkokung *f*
c.-carbonizing plant Kokerei *f*
c. carrier Kohlenschiff *n*, Kohlentransporter *m*
c. consumption Kohlenverbrauch *m*
c. consumption trial trip *(Schiff)* Kohlenmeßfahrt *f*
c. crusher Kohlezerkleinerer *m*
c. crushing Kohlezerkleinerung *f*
c.-cutter *(Brg)* Schrämmaschine *f*
c.-cutter cable Schrämleitung *f (Kabel)*
c.-cutter chain Schrämkette *f*
c. drill Kohlenbohrer *m*, Kohlenbohrmaschine
c.-dust engine Kohlenstaubmotor *m*
c.-dust furnace Kohlenstaubfeuerung *f*
c.-fuelled kohlengeheizt
c. gas Kokereigas *n*, Stadtgas *n*, Steinkohlengas *n*
c. hoist *(Schiff)* Kohlenwinde *f*
c. hopper *(Wärm)* Fülltrichter *m*, Schütttrichter *m*, Kohletrichter *m*, Schüttkasten *m*
c. jetty *(Schiff)* Kohlenpier *m*
c. lighter *(Schiff)* Kohlenprahm *m*, Kohlenleichter *m*
c. liquefaction Kohleverflüssigung *f*
c. measures *(Geol)* Kohlenlager *n*, Kohlenschichten *fpl*
c. mining explosive *(Brg)* Wettersprengstoff *m*, Sicherheitssprengstoff *m*
c. outburst *(Brg)* plötzlicher Kohlenausbruch *m*
c. planer *(Brg)* Kohlenhobel *m*
c. planing schälende Kohlengewinnung *f*
c. screening and grading plant Kohlenaufbereitungsanlage *f*
c. shaft *(Brg)* Förderschacht *m*
c. shift *(Brg)* Gewinnungsschicht *f*
c. slack Kohlengrus *m*
c. tar Steinkohlenteer *m*
c.-tar asphaltum Teerpech *n*
c.-tar dye[stuff] Teerfarbstoff *m*; synthetischer Farbstoff *m*
c.-tar naphtha Solventnaphtha *n*
c.-tar oil Steinkohlenteeröl *n*, Teeröl *n*
c.-tar pitch Steinkohlenteerpech *n*
c. transformation Kohleveredlung *f*
c. types Kohlenarten *fpl*
coalescence 1. Schmelzfluß *m*; 2. Sinterung *f*
coalification Inkohlung *f*
coaling *(Schiff)* Kohlenübernahme *f*; *(Eis)* Bekohlung *f*
c. facilities Bekohlungsanlage *f*
c. plant *(Ch)* Bekohlungsanlage *f*
c. ship Bekohlungsschiff *n*
coalite tar Tieftemperaturteer *m*
coaming *(Schiff)* Lukensüll *m (n)*, Süll *m (n)*
c. level window Süllfenster *n*
c. stay Lukensüllstütze *f*
c. stiffener Süllversteifungsprofil *n*
coarse grob[stückig], großstückig *(Kohle usw.)*; roh *(Schrauben usw.)*
c. adjustment Grobeinstellung *f*, Grobverstellung *f*
c. aggregate Grobzuschlag[stoffe *mpl*] *m (Beton)*

c. coal Stückkohle *f*
c.-coal jig *(Brg)* Grobkornsetzmaschine *f*
c. control rod *(Kern)* Grobregelstab *m*
c. crushing Grobzerkleinerung *f*; Vorbrechen *n*
c. cut Grobhieb *m*, grober Hieb *m* *(Feile)*
c.-cut file Grobfeile *f*, Schruppfeile *f*, Vorfeile *f*
c.-feed series *(Masch)* Grobvorschubreihe *f*
c. feel Rauhgriffigkeit *f* *(Stoffe, Leder usw.)*
c. file Vorfeile *f*, Schruppfeile *f*
c.-fine action *(BMSR)* Grob-Fein-Regelung *f*
c. fodder *(Lw)* Rauhfutter *n*
c. forage *(Lw)* Rauhfutter *n*
c. fraction *(Brg)* Grobgut *n*
c. grain annealing *(Met)* Grobkornglühen *n*
c.-grained grobkörnig; grobkristallin; grobfaserig
c.-grained structure Grobkorngefüge *n*, Grobkornstruktur *f*
c. gravel *(Bw)* Grobkies *m*
c.-mesh weitmaschig
c. motion Grobtrieb *m* *(optische und feinmechanische Geräte)*
c. of screen Siebüberlauf *m*, Siebgrobes *n*, Siebrückstand *m*
c. ore furnace *(Met)* Groberzofen *m*
c. pitch steilgängig *(Gewinde)*
c.-pitch blade Sägeblatt *n* mit grober Zahnteilung
c. pitch of needles *(Text)* grobe Teilung *f*, grober Nadelstich *m*
c.-pitch thread grobgängiges (steilgängiges) Gewinde *n*, Grobgewinde *n*
c. positioning Grobpositionierung *f*; Groblageeinstellung *f*
c. roving *(Text)* grobes Vorgarn *n*
c. salt for cattle Viehsalz *n*
c. sand Grobsand *m*; Grieß *m*
c. scanning *(Fs)* Grobabtastung *f*
c. screen 1. Gräpelsieb *n*, Gröpelsieb *n*; 2. *(Typ)* Grobraster *m*
c. screw Steilgewindespindel *f*
c. screw thread Grobgewinde *n*
c. screw thread series Grobgewindereihe *f*
c.-seed wheel Grobsaatrad *n* *(einer Sämaschine)*
c. setting Grobeinstellung *f*
c. stuff *(Bw)* Unterputz *m*
c. thread Grobgewinde *n*, Steilgewinde *n* *(Gewinde mit geringer Gangzahl je Zoll)*
c. threading attachment Steilgewindeschneideinrichtung *f*
c.-tooth mit grober Zahnteilung
c. tuning *(El)* Grobabstimmung *f*
c.-weft fabric grobschüssiges Gewebe *n*
coarsely crystalline grobkristallin
coarseness of file teeth Feilenhiebteilung *f*
coarsening Vergröberung *f*

coast / to 1. *(Masch)* auslaufen, ausrollen; im Leerlauf fahren; weiterlaufen; 2. *(BMSR)* frei auslaufen *(Folgeregler mit Dreipunktglied)*
coast Küste *f*, Ufer *n*
c. guard *(Schiff)* Küstenwache *f*; *(Am)* Wasserschutzpolizei *f*
c. guard station *(Am)* Rettungsbootstation *f* an Land
c. light Küstenleuchtfeuer *n*
c. lighting Küstenbefeuerung *f*
c. steamer Küstendampfer *m*
c. works Küstenschutzbauten *pl*, Seebauten *pl*
coastal barge Küstenlastkahn *m*
c. cargo-liner Küstenfrachter *m* *(im Liniendienst)*
c. defence craft Vorpostenboot *n*, Streifenboot *n*
c. deviation Küsteneffekt *m* *(Radar)*
c. direction finding station Küstenpeilstelle *f*
c. engineering Küstenschutz[bau] *m*, Seebau *m*
c. fishing Küstenfischerei *f*
c. fortification *(Mil)* Küstenbefestigung *f*
c. motor ship Küstenmotorschiff *n*
c. navigation Küstenschiffahrt *f*, terrestrische Navigation *f*
c. pack Küstenpackeis *n*
c. refraction Küstenbrechung *f* *(Radar)*
c. vessel Küstenschiff *n*
c. works Küstenschutzbauten *pl*, Seebauten *pl*
coaster Küstenschiff *n*
coasting 1. *(El)* Auslauf *m*; Fahren *n* mit Freilauf; 2. *(Schiff)* Küstenfahrt *f*
c. barge *(Schiff)* Seeleichter *m*
c. vessel Küstenschiff *n*
coastwise entlang der Küste, längs der Küste
c. fortification *(Mil)* Küstenbefestigung *f*
c. steamer Küstendampfer *m*
c. trade Küstenfahrt *f*
c. trade master Kapitän *m* auf kleiner Fahrt
coat / to [an]streichen, bestreichen; lackieren; überziehen; bekleiden, belegen, beschichten; beplanken *(Baustoffe)*; kaschieren *(Papier)*
coat 1. *(Masch)* Anstrich *m*, Auftrag *m*, Belag *m*, Schutzschicht *f*, Schicht *f*, Überzug *m*; 2. *(Text)* Mantel *m*; 3. Überform *f* *(Gießerei)*; 4. Straßendecke *f*
c. of paint Farbanstrich *m*, Farbüberzug *m*
coated abrasive Schleifgewebe *n*, Schmirgelleinwand *f*
c. capillary überzogene Kapillare *f* *(imprägnierte Kapillarsäule; Gas-Chromatografie)*
c. cathode pastierte Katode *f*
c. chippings Mischsplitt *m* *(Straßenbau)*
c. electrode Mantelelektrode *f*, umhüllte Elektrode *f*

c. fabric beschichteter Stoff *m*, beschichtetes Gewebe *n*; ein- oder beidseitig mit Harz überzogenes Gewebe, Lackgewebe *n*
c. film *(Plast)* gegossener Film *m*
c. paper *(Typ)* gestrichenes Papier *n*; Kunstdruckpapier *n*
c. side *(Typ)* Schichtseite *f*, beschichtete Seite *f (des Papiers)*
c. tape Schichtband *n*, Schichtstreifen *m (z. B. Magnetband)*
c. web beschichtete Bahn *f (Textilien)*
c. welding rod Tauchelektrode *f (Schweißen)*
coater Streichmaschine *f*; Beschichtungsanlage *f*; Auftragswerk *n*
coating 1. Schutzschicht *f*; Belag *m*; Umhüllung *f*; Umkleidung *f*; Überzug *m*; 2. Beschichtung *f*, Auskleidung *f*, Oberflächenimprägnierung *f*; 3. Streichmasse *f (Papier)*; 4. *(Schiff)* Anstreichen *n*; Anstrich *m*; Schicht *f*, Überzug *m*
c. conductivity *(El)* Leitfähigkeit *f* des Überzugs, Überzugsleitfähigkeit *f*
c. machine Maschine *f* für Oberflächenbehandlung, Streichmaschine *f*; Beschichtungsanlage *f*; Auftragswerk *n (Papierherstellung)*
c. material Deckmaterial *n*; Überzugsmaterial *n*
c. of a condenser *(El)* Kondensatorbelag *m*, Kondensatorbelegung *f*
c. of fabrics with rubber *(Text)* Gummieren *n* von Geweben
c. pan *(Am)* Tränkwanne *f (Plastherstellung)*
c. press Streichpresse *f*, Überziehpresse *f (Papierherstellung)*
c. process Oberflächenbehandlung[sverfahren *n*] *f*
c. resin Lackharz *n*
c. varnish Imprägnierlack *m*, Überzugslack *m*
coax *s.* coaxial cable
coaxial gleichachsig, koaxial
c. aerial Koaxialantenne *f*
c. cable *(El)* konzentrisches (koaxiales) Kabel *n*, Koax[ial]kabel *n*
c. circuit *(El)* konzentrische Leitung *f*
c. connector *(El)* konzentrischer Stecker *m*, Koaxialanschluß *m*
c. diode Koaxialdiode *f*
c. electrode system konzentrische Elektrodenanordnung *f*
c. line *(El)* Koaxialleitung *f*, konzentrische (koaxiale) Leitung *f*
c. loudspeaker Koaxiallautsprecher *m*
c. pair *(El)* konzentrische Doppelleitung *f*, Koaxialleitung *f*
c. solenoid *(El)* Zylinderspule *f*
c. transistor *(El)* Koaxialtransistor *m*
c. transmission line *(El)* konzentrische Speiseleitung (Übertragungsleitung) *f*
cob / to *(Brg)* klauben, scheiden *(von Hand)*
cob 1. ungebrannter Ziegel *m*; 2. Gemenge *n* von Lehm, Kies und Stroh
c. wall Lehmwand *f*
cobalt Kobalt *n*
c. bloom Kobaltblüte *f* (*s. a.* erythrite)
c. blue *(Ch)* Kobaltblau *n*, Thenards Blau *n*
c. bomb Kobaltbombe *f*, H-Bombe *f* mit Kobaltmantel
c. glance Kobaltglanz *m*
c. green *(Ch)* Kobaltzinkat *n*, Rinmanns Grün *n*
c. high speed steel kobalthaltiger Schnellarbeitsstahl *m*
c. pyrites Kobaltkies *m*
c. steel Kobaltstahl *m*
cobaltic compound Verbindung *f* des dreiwertigen Kobalts
cobaltiferous kobalthaltig
cobaltite Kobaltglanz *m*
cobaltous compound Verbindung *f* des zweiwertigen Kobalts
cobble Feldstein *m*, Kopfstein *m (Straßenpflasterung)*
cobbler's suture *(Text)* Naht *f* mit zwei Fäden
cobbles *(Brg)* Knabbelkohle *f*
cobs *(Fs)* glockenähnliche Bildverzerrung *f*
cocaine Kokain *n*
cochineal Koschenille *f*, Cochenille *f (roter Farbstoff)*
cochleoid *(Math)* Kochleoide *f*, Schraubenkurve *f*
Cochran patent vertical multitubular boiler *(Wärm)* stehender Vielröhrenkessel *m (Patent Cochran)*
cock 1. *(Masch)* Hahn[ventil *n*] *m*, Absperrglied *n*, Absperrorgan *n*; 2. kleiner Schober *m*
c. loft *(Bw)* Spitzboden *m*
c. nail *(Masch)* Hahnschraube *f*
c. plug Hahnkegel *m*, Hahnkücken *n*
c. stopper Stöpselhahn *m*
c. wrench Hahnschlüssel *m*
cockboat Beiboot *n*
Cockcroft generator *(Ph)* Kaskadengenerator *m (zur Beschleunigung von Teilchen)*
cockering *(Brg)* Polygonausbau *m*, Vieleckausbau *m (Holz)*
cockermeg *(Brg)* Strebe *f (Ausbau)*
cocking Aufstellen *n* der Heuhaufen
c. lever Spannhebel *m*
c. piece Schlößchen *n (Gewehr)*
c. spring Spannfeder *f*
cockpit 1. *(Flg)* Führerraum *m*, Kabine *f*, Kanzel *f*, Kockpit *f*; Führersitz *m*, Pilotensitz *m*; 2. *(Schiff)* Plicht *f*
c. enclosure *(Flg)* Führersitzverkleidung *f*
coco-nut fibre Kokosfaser *f*
cocoon 1. Kokon *m*; 2. *(Kern)* Abziehfarbe *f*
c. spinning *(Text)* Spinnen *n* des Kokons

cod end *(Schiff)* Stert *m*, Netzstert *m (eines Scherbrettnetzes)*
c. oil Tran *m*
CODAN *s.* carrier operated device antinoise
code / to kodieren, verschlüsseln
code Kode *m*, Schlüssel *m*, Verschlüsselung *f*
c. beacon *(Flg)* Kodebake *f*, verschlüsselte Bake *f*
c. book *(Schiff)* Signalbuch *n*
c. conversion Kodeumsetzung *f*; Umkodierung *f (Telegrafie)*
c. converter Kodeumsetzer *m*, Kodeumwandler *m*
c.-disk Kodierschablone *f*, Kodierscheibe *f (Analog-digital-Umsetzer)*
c. language Geheimsprache *f*
c. letter Kodezeichen *n*
c. letters *(Fmt)* Rufzeichen *npl*
c. light Kennfeuer *n (Navigation)*
c. of practice *(Bw)* Ausführungsbestimmung *f*, Richtlinie *f*
c. pennant *(Schiff)* Antwortflagge *f*
c. punching *(Dat)* Schlüssellochung *f*
c. reception Telegrafieempfang *m*
c. selector *(Fmt)* Amts[gruppen]wähler *m*
c. sender *(Fmt)* Zeichengeber *m*
c. signal Morsezeichen *n*; Telegrafiezeichen *n*
c. system Verschlüsselungssystem *n*
c. tape recorder Morsestreifenlocher *m*
c. translation Dekodierung *f*, Entschlüsselung *f*
c. translator Kodewandler *m*
c. void *(Büro)* „ohne Kode" *(Tastenbezeichnung)*
Code of Civil Procedure Zivilprozeßordnung *f*
coded character kodiertes (verschlüsseltes) Schriftzeichen *n*
c. decimal digit verschlüsselte Dezimalziffer *f*
c. decimal notation verschlüsselte Dezimaldarstellung (Dezimalschreibweise) *f*
c. program kodiertes (verschlüsseltes) Programm *n*
coder 1. *(Dat)* Kodierer *m*, Kodiervorrichtung *f*, Verschlüßler *m*, Verschlüsselungsvorrichtung *f*; 2. *(Fmt, Fs)* Übersetzer *m*, Umrechner *m*, Umsetzer *m*
coding *(Dat)* Kodieren *n*, Verschlüsseln *n*; [kodierte] Befehlsfolge *f*
c. line Befehlszeile *f*
c. method Kennzeichnungsverfahren *n*
coefficient Koeffizient *m*; Beiwert *m*; Zahl *f*
c. of absorption of radiant heat 1. *(Ph)* Wärmeabsorptionskoeffizient *m*; 2. *(Wärm)* Einstrahlungszahl *f*, Einstrahlungskoeffizient *m*
c. of adhesion Adhäsionskoeffizient *m*; Haftreibzahl *f*
c. of amplification *(El)* Verstärkungsfaktor *m*
c. of apparent expansion *(Wärm)* relativer Ausdehnungskoeffizient *m*
c. of charge *(El)* Ladekoeffizient *m*; Ladungskoeffizient *m*
c. of compressibility Steifezahl *f*, Steifeziffer *f (Bodenmechanik)*
c. of conductivity *(Ph)* Leitfähigkeitsfaktor *m*; Leitungskoeffizient *m (Elektrizität, Wärme)*
c. of coupling *(El)* Kopplungsgrad *m*, Kopplungsfaktor *m*, Kopplungskoeffizient *m*
c. of cubical expansion kubischer (räumlicher) Ausdehnungskoeffizient *m (Wärmeausdehnung)*
c. of cyclic variation *(Masch, Wärm)* Ungleichförmigkeitsgrad *m*, Ungleichförmigkeitsverhältnis *n*, Gleichförmigkeitsgrad *m*
c. of diffusion *(Ph)* Diffusionskoeffizient *m*
c. of dispersion *(El)* Zerstreuungskoeffizient *m*
c. of dissociation Dissoziationsgrad *m*
c. of drag Widerstandsbeiwert *m*
c. of earthing *(El)* Erdungsziffer *f*
c. of elasticity E-Modul *m*, Elastizitätsmodul *m*
c. of electrolyte dissociation Dissoziationsgrad *m*
c. of elongation Dehnungskoeffizient *m*
c. of expansion 1. *(Math)* Entwicklungskoeffizient *m (bei Reihenentwicklung)*; 2. Ausdehnungskoeffizient *m*, Ausdehnungszahl *f*
c. of extinction *(Opt)* Extinktionskoeffizient *m*
c. of fineness *(Schiff)* Völligkeitsgrad *m*
c. of fineness of midship section Völligkeitsgrad *m* des Hauptspantes
c. of friction Reibungskoeffizient *m*, Reibungsbeiwert *m*, Reibungszahl *f*
c. of harmonic distortion *(Fmt)* Klirrfaktor *m*
c. of heat transmission (transition, transfer) Wärmedurchgangszahl *f*, Wärmeübergangszahl *f*, Wärmeübertragungszahl *f*
c. of impact Stoßzahl *f*
c. of leakage *(Wärm)* Undichtigkeitszahl *f*, Undichtigkeitskoeffizient *m*
c. of linear expansion linearer Ausdehnungskoeffizient *m (Wärmeausdehnung)*
c. of losses *(Masch, Wärm)* Verlustzahl *f*, Verlustkoeffizient *m*
c. of magnification Vergrößerungszahl *f*
c. of mutual induction *(El)* Gegeninduktion[skoeffizient *m*] *f*
c. of negative feedback Gegenkopplungsfaktor *m*
c. of non-linear distortion Klirrfaktor *m (Verstärkerröhre)*
c. of performance Nutzeffekt *m*
c. of permeability Durchlässigkeitskoeffizient *m*, Durchlässigkeitszahl *f (Bodenmechanik)*

c. of reaction *(El)* Rückkopplungsfaktor *m*; Rückwirkungsfaktor *m*
c. of refraction Brechungskoeffizient *m*, Brechungsexponent *m*, Refraktionskoeffizient *m*
c. of rigidity Schubelastizitätsmaß *n*; Steifezahl *f*, Steifheitszahl *f*
c. of self-induction *(El)* Selbstinduktion[skoeffizient *m*] *f*
c. of shearing *(Ph, Masch)* Schubzahl *f*
c. of sliding friction Gleitreibungszahl *f*, Reibungszahl *f* für Gleitreibung
c. of solubility Lösungskoeffizient *m*
c. of sound damping Schalldämmzahl *f*
c. of surface expansion Oberflächenausdehnungskoeffizient *m*
c. of swelling Schwellzahl *f*, Schwellziffer *f* *(Bodenmechanik)*
c. of thermal conduction Wärmeleitfähigkeit[szahl] *f*, Wärmeleitungskoeffizient *m*
c. of thermal expansion Wärmeausdehnungskoeffizient *m*, Wärmeausdehnungszahl *f*
c. of transmission *(Fmt, El)* Durchlaßfaktor *m*
c. of variation Variationskoeffizient *m*
c. of viscosity Viskositätskoeffizient *m*, Viskositätskonstante *f*
c. potentiometer Koeffizientenpotentiometer *n (Analogrechner)*
coelostat Zölostat *m*
coercive force Koerzitivkraft *f (Magnetismus)*
coffee mill squeezer *(Met)* Luppenmühle *f (Puddelbetrieb)*
coffer 1. Kasten *m*, Kiste *f*; 2. Schleusenkammer *f (Kanal)*
cofferdam Caisson *m*; Wand *f* zum Abhalten von Wasser
c. bulkhead *(Schiff)* Kofferdammschott *n*
coffered ceiling *(Bw)* Kassettendecke *f*
coffering *(Brg)* wasserdichter Ausbau *m*
coffin *(Kern)* 1. Kontainer *m*, Transportbehälter *m (für radioaktive Stoffe)*; 2. Sarg *m*
c. finish Sarglack *m*
cog / to *(Elm)* rucken *(ungleichmäßige Drehung eines Motors oder Generators infolge der Ankernutung)*
c. down *(Met)* herunterwalzen *(Blöcke)*; vorblocken; vorschmieden
cog 1. *(Masch)* [eingesetzter] Zahn *m (eines Zahnrads)*; Knagge *f*, Nase *f*; 2. *(Brg)* Holzkasten *m*, Holzschrank *m*; Pfeiler *m*, Versatz *m*
c. tooth Radzahn *m*, [eingesetzter] Zahn *m* eines Zahnrads
c.-wheel 1. *(Masch)* Zahnrad *n (Zähne aus Holz)*, Kammrad *n*, Kron[en]rad *n*; 2. *(Elm)* Polrad *n*
cogged *(Masch)* gezahnt, verzahnt
c. ingot *(Met)* Vorblock *m*, vorgewalzter Block *m*
cogger *(Brg)* Versatzarbeiter *m*
cogging 1. *(Bw)* Aufkämmung *f*, Verkämmung *f (Holzbau)*; 2. *(Met)* Vorblocken *n*
c.-down roll *(Met)* Streckwalze *f*, Vorwalze *f*
c.-down stand *(Met)* Vorwalzgerüst *n*
c. mill Blockwalzwerk *n*, Vorwalzwerk *n*
c. pass *(Met)* Blockkaliber *n*, Vorstich *m*, Vorstreckkaliber *n*
c. roll *(Met)* Blockwalze *f*, Streckwalze *f*
c. stand *(Met)* Vorgerüst *n*
c. train *(Met)* Blockstraße *f*, Vor[walz]straße *f*
cogrediency *(Math)* Kogredienz *f*
cogredient *(Math)* kogredient
cohere / to 1. haften; zusammenhalten; 2. *(El)* fritten
c. to each other aneinanderhaften
coherence 1. Kohärenz *f*, Zusammenhang *m*, Zusammenhalt *m*; 2. *(El)* Frittung *f*
coherent detection phasenabhängige Gleichrichtung *f*
c. oscillator kohärenter (durchstimmbarer) Oszillator *m*, Oszillator mit fester Phasenbeziehung
c. scattering kohärente Streuung *f*
coherer *(Fmt)* Fritter *m*, Kohärer *m*
c. action Fritterwirkung *f*
c. current Fritterstrom *m*
c. resistance Fritterwiderstand *m*
cohesion Kohäsion *f*
c. force Kohäsionskraft *f*
c. pressure Binnendruck *m*, Kohäsionsdruck *m*
cohesionless soil *(Bw)* nichtbindiger (rolliger) Boden *m (Grundbau)*
cohesive resistance *(Met)* Trennwiderstand *m*
c. soil *(Bw)* bindiger Boden *m (Grundbau)*
coil / to 1. sich rollen, sich werfen; 2. *(Feder)* wickeln
c. load pupinisieren, bespulen
c. up abspulen
coil 1. Rohrschlange *f*, Rohrspirale *f*, Schlangenrohr *n*, Wendel *f*; 2. *(El)* Spule *f*, Wendel; 3. Windung *f (Feder)*
c. aerial Rahmenantenne *f*; Spulenantenne *f*
c. amplification factor Gütefaktor *m* einer Spule
c. armature Ankerspule *f*
c. assembly Spulensatz *m*
c. base Spulenkörper *m*
c. capacitance Spulenkapazität *f*
c. chain Rundgliederkette *f*
c. constant Spulengüte *f*
c.-driven loudspeaker Tauchspulenlautsprecher *m*
c. end Spulenkopf *m*
c. former Spulenkörper *m*, Spulenträger *m*
c. ignition Spulenzündung *f*, Abreißzündung *f (Verbrennungsmotor)*

c. **impedance** Spulenimpedanz *f*, Spulenwiderstand *m*
c. **jacket** Umkapselung *f* der Spulen
c. **leads** Spulenanschlußdrähte *mpl*
c.-**loaded** spulenbelastet, pupinisiert *(Kabel)*
c.-**loaded cable** Pupinkabel *n*
c.-**loaded circuit** Pupinleitung *f*
c. **loading** Spulenbelastung *f*, Pupinisierung *f*, induktive Belastung *f (von Kabeln)*
c. **of spring** Federwindung *f*
c. **of wire** Drahtspule *f*
c. **of yarn** *(Text)* Fadenwindung *f*, Garnwindung *f*
c. **pitch** Nutenschritt *m*, Spulenweite *f* in Nutteilungen
c. **Q** Gütefaktor *m* (Güte *f*) der Spule, Spulengüte *f*
c. **quality** *s.* c. Q
c. **reactance** Spulenblindwiderstand *m*
c. **resistance** Spulenwiderstand *m*
c. **spring** Schneckenfeder *f*, Schraubenfeder *f*, Spiralfeder *f*
c. **switch** Spulenschalter *m*, Wellenumschalter *m*
c. **tap** Spulenabgriff *m*, Spulenanzapfung *f*
c.-**type milk cooler** Kühlschlange *f* für Milch
c. **valve** Schlangenventil *n (Luftreifen)*
c. **varnish** Spulenlack *m*, Isolierlack *m*
c. **weight** Bundgewicht *n (Walz- oder Ziehgut)*
c. **winder** Spulenwickelmaschine *f*
c. **winding head** Spulenwickelkopf *m*
c. **winding machine** Spulenwickelmaschine *f*
c. **winding set** Spulenwickelgerät *n*
c. **wire** Wicklungsdraht *m*, Spulendraht *m*
coiled coil Doppelwendel *f*
c. **coil filament** Doppelwendel[leucht]draht *m*
c. **coil lamp** Doppelwendellampe *f*
c. **expansion pipe** *(Wärm)* Ausgleichschlangenrohr *n*
c. **filament** *(El)* gewendelter Leuchtdraht (Glühfaden) *m*
c. **filament lamp** Wendeldrahtlampe *f*
c. **loops** *(Text)* Riegelmaschen *fpl*
c. **piston ring** *(Wärm)* schraubenförmiger Liderring *m*, Schraubenringspanner *m*
c. **torsion spring** gewundene Drehfeder *f*
c. **wire[s** *pl***]** Rollendraht *m*
coiler *(Text)* Drehkanne *f*; Drehtopf-Vorrichtung *f*
c. **calender roll** Glättwalze *f* der Drehkanne
coin / to prägen; münzen
coin Geldstück *n*, Münze *f*
c.-**automat telescope** Automatenfernrohr *n*, Münzfernrohr *n*
c. **collector** Münzfernsprecher *m*
c.-**freed television** Fernsehsystem *n* mit Gebührenautomat
c.-**operated machine** Münzautomat *m*
coinage alloy Münzlegierung *f*
coincide / to koinzidieren, übereinstimmen, zusammenfallen, zur Deckung kommen
coincidence 1. Deckung *f*, Koinzidenz *f*, Übereinstimmung *f*; Zusammenfallen *n*; 2. Zufall *m*, zufälliges Zusammenfallen
c. **chamber** *(Ph)* Koinzidenzkammer *f*
c. **circuit** *(Ph)* Koinzidenzschaltung *f*
c. **counter** *(Ph)* Koinzidenzzähler *m*
c. **gate** *(BMSR)* Koinzidenztor *n*
c. **level** *(Opt)* Koinzidenzlibelle *f*
c. **of phases** Phasengleichheit *f*, Phasensymmetrie *f*
c. **scintillation spectrometer** Koinzidenz-Szintillationsspektrometer *n*
c. **setting** *(Opt)* Koinzidenzeinstellung *f*
coincident current core storage Koinzidenzstrom-Kernspeicherung *f (Datenspeicherung in Magnetkernen)*
c. **current read-out** Koinzidenzstromauslesen *n (z. B. aus einem Magnetkernspeicher)*
c. **drive selection** koinzidente Steuerungswahl *f (Magnetkernspeicher)*
coining machine Prägepresse *f*
c. **press** Kalibrierpresse *f*; Prägepresse *f*; Prägestanze *f*
coir [fibre] Kokosfaser *f*
c. **mat** Kokosmatte *f*
coke / to verkoken, verkohlen
coke Koks *m*
c. **bench** Koksplatz *m*, Koksrampe *f*
c. **bench side** Koks[lösch]seite *f*
c. **breeze** Koksgrus *m*, Koksstaub *m*
c. **chamber** Kokskammer *f*
c. **charge** Koksbrand *m*
c. **cleavage** Koksnaht *f*
c. **cutter** Koksbrecher *m*
c. **extinguishing car** Kokslöschwagen *m*
c. **fired furnace** Koksofen *m*
c. **guide** Koksführungsschild *m*
c. **guide car** Koks[kuchen]führungswagen *m*
c. **man** Kokereiarbeiter *m*
c. **nuts** Brechkoks *m*
c. **oven charging car** Koksofenfüllwagen *m*
c.-**oven coke** Hüttenkoks *m*
c.-**oven gas** Kokereigas *n*, Destillationsgas *n*
c. **oven plant** Kokerei[anlage] *f*
c. **parting** *(Brg)* Koksnaht *f*
c. **per charge** Satzkoks *m*, Kokssatz *m* *(Kupolofenbetrieb)*
c. **pusher** Koksausdrückmaschine *f*
c. **pushing machine** Koksausdrückmaschine *f*
c. **quenching car** Kokslöschwagen *m*
c. **quenching tower** Kokslöschturm *m*
c. **ram** Koksausdrückstange *f*
c. **wharf** Koksrampe *f*, Abwurframpe *f*
cokery Kokerei *f*
coking Verkokung *f*

c. capacity Backfähigkeit *f*
c. chamber Kokskammer *f*; Vorraum *m (bei Halbgasfeuerungen)*
c. coal Kokskohle *f*
c. coal bunker Kokskohlenturm *m*
c. period Garungszeit *f*
c. plant Kokerei *f*
c. plate *(Wärm)* Schürplatte *f*, Vorstellplatte *f (Rostfeuerung)*
c. power Kokungsvermögen *n*
c. smalls Koksfeinkohle *f*
c. stoker with coal feed *(Wärm)* Verkokungsfeuerung *f* mit Kohlenvorschub
c. time Kokungszeit *f*, Verkokungszeit *f*, Garungszeit *f*
coky centre *(Met)* Fadenlunker *m*
Colcrete *(Bw)* Prepaktbeton *m*
cold blast / to kaltblasen
c. cast kalt (matt) vergießen
c. draw *(Met)* kaltziehen
c. drive kaltschlagen *(Niete)*
c.-forge kaltschmieden
c.-form kaltverformen
c. press kaltpressen
c. roll kaltwalzen
c. saw kaltsägen
c. straighten kaltrichten
c. strain kaltstrecken
c. vulcanize kaltvulkanisieren
c.-work harden kalthärten, kaltverfestigen
cold 1. kalt; 2. *(Kern)* aktivitätsfrei, nicht [radio]aktiv; unterthermisch
c. aging *(Plast)* Kalthärtung *f*
c. area *(Kern)* nicht radioaktiver Bereich *m*, aktivitätsfreier Raum *m*
c.-bath method *(Text)* Kaltfärbeverfahren *n*
c. bend[ing] test Kaltbiegeprobe *f*, Kaltbiegeversuch *m*, Kaltfaltversuch *m*
c. blast furnace *(Met)* Kaltwindofen *m*
c.-blast pig iron kaltgeblasenes Roheisen *n*
c. blocking *(Typ)* Kaltprägung *f*
c. blowing *(Met)* Kaltblasen *n*
c. bonderizing *(Met)* Kaltbonderverfahren *n*
c. brittleness Kaltsprödigkeit *f*
c. casting Kaltguß *m*
c. cathode kalte Katode *f*, Kaltkatode *f*
c. chisel Kaltmeißel *m*, Hartmeißel *m*
c. circular saw Kaltkreissäge *f*
c.-coiled kaltgewickelt *(Feder)*
c. composition *(Typ)* Nichtmetall-Satz *m*, Kaltsatz *m*
c. crack *(Met)* Spannungsriß *m (im abkühlenden Guß)*, Kaltriß *m*
c. cupping Kaltziehen *n*; Kalttiefziehen *n*
c. curing Kaltvulkanisation *f*
c.-cut varnish Kaltansatzlack *m*
c. cutter Kaltschrotmeißel *m*
c. deformation Kaltverformung *f*, Kaltumformung *f*
c. die [block] *(Met)* Kaltmatrize *f*
c. drawing 1. Kaltziehen *n (Metalle)*; 2. Kaltverstrecken *n (Chemiefaserherstellung)*
c. drawing bench *(Met)* Kaltziehbank *f*
c.-drawing tool *(Met)* Kaltziehmatrize *f*
c.-drawn kaltgezogen
c.-drawn oil kaltgepreßtes (kaltgeschlagenes) Öl *n*
c.-drawn steel Blankstahl *m*, kaltgezogener Stahl *m*
c.-dyeing colour *(Text)* Kaltfärber *m*
c. electron langsames Elektron *n*
c. electronic discharge *(El)* Dunkelentladung *f*
c. emission *(El)* Feldemission *f (von Elektronen)*
c. extrusion Kaltfließpressen *n*
c. forging Kaltschmieden *n*; Kaltstauchen *n*, Kaltpressen *n*
c. forging machine Kaltstauchmaschine *f*
c. forming Kaltumformen *n*; Kaltformgebung *f*; Kaltverformung *f*
c. galvanizing galvanische Verzinkung *f*
c. glue Kaltleim *m*
c.-headed screw Schraube *f* mit kaltangestauchtem Kopf
c. header Kaltstauchautomat *m*
c. heading Kaltstauchen *n (Schraubenköpfe)*
c. heading machine Kaltstauchmaschine *f*, Kaltstauchautomat *m*
c. immersion method *(Holz)* Kaltbadtauchverfahren *n*
c. impact forging Kaltschlagschmieden *n*
c. junction Vergleichsstelle *f*, kalte Lötstelle *f (Thermoelement)*
c.-junction compensation Kompensation *f* der Vergleichsstellentemperatur *(Thermoelement)*
c. laboratory *(Kern)* aktivitätsfreies Laboratorium *n*
c. lap 1. Walzgrat *m*; 2. Kaltschweißstelle *f*
c. neutron *(Ph)* langsames (kaltes, unterthermisches) Neutron *n*
c. pilger rolling *(Met)* Kaltpilgern *n*
c.-plasticity Kaltformbarkeit *f*, Kaltformvermögen *n*
c. plug *(Kfz)* Zündkerze *f* mit hohem Wärmewert
c. pressboard Kaltpreßpappe *f*
c.-pressed forging Preßteil *n*
c. pressing 1. Kaltdrücken *n*, Kaltpressen *n*; 2. *(Holz)* Kaltverleimung *f*
c. protective *(Masch)* Kälteschutzmittel *n*
c. reactivity *(Kern)* Kaltreaktivität *f*
c. reactor kalter (jungfräulicher) Reaktor *m*
c. reduction mill Kaltwalzwerk *n*

c. refining Kaltveredlung *f*
c.-resisting kältebeständig, kältefest
c. riveting Kaltnietung *f*
c. roll Kaltwalze *f*, Dressierwalze *f*
c.-rolled kaltgewalzt
c.-rolled section *(Met)* Kaltprofil *n*
c.-rolled strip Kaltband *n*, kaltgewalztes Band *n*
c. rolling mill Kaltwalzwerk *n*
c. run Blindversuch *m (Reaktor)*
c. sate Kaltschrotmeißel *m*
c. saw Kaltsäge *f*
c.-sawing machine Kaltsäge[maschine] *f*
c.-set ink *(Typ)* Kalttrockenfarbe *f*
c. sett Kaltschrotmeißel *m*
c.-setting kaltabbindend, kalthärtend
c.-short kaltbrüchig
c. shortness Kaltbrüchigkeit *f*
c. shut Kaltschweißstelle *f*
c.-start lamp Kaltstartlampe *f*
c.-starting *(Kfz)* Kaltstart *m*
c.-starting aid Kaltstarthilfe *f*
c. state Kaltgang *m (Hochofen)*
c. steam cylinder *(Wärm)* Kaltdampfzylinder *m*
c. storage 1. Kühlhaus *n*, Kühlraum *m*; 2. Kühlhausaufbewahrung *f*, Kühlraumaufbewahrung *f*
c. storage hold *(Schiff)* Kühlraum *m*
c. store Kühlhalle *f*, Kühlhaus *n*, Kühlraum *m*
c. straining Kaltbeanspruchung *f*, Kaltverformung *f*, Kaltstrecken *n*
c. strengthening Kaltverfestigung *f*
c. strip Kaltband *n*
c. swaging Kaltgesenkpressen *n*, Kaltgesenkschlagen *n*
c. swaging machine Kaltgesenkdrückmaschine *f*
c.-sweat *(Led)* Kaltwasserschwitze *f*
c. test Kälteprüfung *f*, Kälteprobe *f*; Kühlversuch *m*
c.-test thermometer Kältethermometer *n*
c. top email Kaltemail *n*
c. treatment Kältebehandlung *f*
c. trim Kaltabgraten *n*
c. twisting Kaltverwinden *n*, Kaltverdrehen *n*
c. upsetting *(Met)* Kaltstauchen *n*
c. upsetting die Kaltstauchmatrize *f*
c. waste inaktiver Abfall *m (Reaktorbetrieb)*
c. water jacket Kühlwassermantel *m*
c. water milling *(Text)* Kaltwasserwalke *f*
c. water retting *(Text)* Kaltwasserröste *f*
c. water spinning frame *(Text)* Halbnaßspinnmaschine *f*
c. wave 1. Kältewelle *f (Wetter)*; 2. Kaltwelle *f (Kosmetik)*
c.-waxed board [mit Eiswasser] abgeschreckter Wachskarton *m*
c. welding Kaltpreßschweißen *n*
c. wetting agent *(Text)* Kaltnetzer *m*, Kaltnetzmittel *n*
c.-work hardening Kalthärtung *f*, Kaltverfestigung *f (Kaltverformung)*
c.-worked kaltgereckt, kaltverformt
c. working *(Met)* 1. Rohgang *m (Hochofen)*; 2. Kaltverformung *f*; Kaltbearbeitung *f*
c.-wound kaltgewickelt
cole-seed oil Rüböl *n*
colemanite *(Min)* Colemanit *m*
Colgrout *(Bw)* kolloidaler Zementmörtel *m (Prepaktbeton)*
collapse / to 1. *(Bw)* einstürzen, zusammenbrechen; 2. *(Brg)* zubruchgehen, hereinbrechen; 3. zusammensacken; flach werden *(Luftreifen usw.)*
collapse Einsturz *m*; Zusammenbruch *m*; Zusammensacken *n*
c. load Traglast *f*; Bruchlast *f*
collapsibility Zerdrückbarkeit *f*; Zerfallseigenschaften *fpl*
collapsible aerial ausziehbare (zusammenlegbare) Antenne *f*
c. boat Faltboot *n*, Klappboot *n*
c. container *(Schiff)* zusammenlegbarer Behälter *m*
c. dinghy Faltboot *n*
c. mast *(Schiff)* Klappmast *m*, klappbarer Mast *m*
c. reel *(Text)* Fallhaspel *m*
c. rubber boat Faltboot *n*
c. top *(Kfz)* Faltdach *n*, Klappverdeck *n*
c. tripod zusammenschiebbares *(dreibeiniges)* Stativ *n*
c. tube Tube *f*
collapsing stress Knickspannung *f*, Knickbelastung *f*
c. test Knickversuch *m*, Ausknickungsversuch *m*
collar 1. Bund *m*, Hals *m*, Manschette *f*, Ring *m*, Stellring *m*, Wulst *m*, Hülse *f*, Lagerschale *f*, Zapfenlager *n*; Abstandsring *m*; Skalenring *m*; 2. *(Text)* Kragen *m (Hemd usw.)*; 3. Kehlbalken *m (Holzbau)*; 4. *(Brg)* Schachtmündung *f*
c. beam Kehlbalken *m (Holzbau)*
c.-beam roof Kehlbalkendach *n (Holzbau)*
c. bearing *(Masch)* Halslager *n*, Kammlager *n*
c. burner *(Ch)* Ringbrenner *m*
c. coupling *(Masch)* Muffenschloß *n*
c. end bearing *(Masch)* Bundlager *n*, Lager *n* mit Bundringen
c. head screw Bundschraube *f*
c. leather Kummetleder *n*
c. nut *(Masch)* Achsmutter *f*, Bundmutter *f*

c. **of axle journal** *(Eis)* Bund *m (Achsschenkel)*
c. **of piston rod** Kolbenstangenbund *m*
c. **roof** Kehlbalkendach *n (Holzbau)*
c. **screw** Bundschraube *f*, Vierkantschraube *f* mit Bund
c.**-shaped** *(Masch)* bundartig, kragenartig, ringartig, wulstartig
c. **shim** Zwischenring *m*
c. **socket** Muffenhülse *f*
c. **step bearing** Ringspurlager *n*
c. **stud** Bolzen *m* mit Bund
c. **thrust bearing** Kamm[druck]lager *n*
collate / to 1. *(BMSR, Büro)* ordnen, vergleichen, zuordnen *(höhere Stufe des Sortierens)*; mischen; 2. *(Typ)* kollationieren
collateral chain *(Kern)* zugehörige Zerfallskette *f*
c. **family** *(Kern)* zugehörige Zerfallsfamilie *f*
collation *(BMSR)* Konjunktion *f (logischer Operator)*
collator Zuordner *m*; Mischer *m (z. B. Lochkartenmischer)*
c. **counting device** *(Büro)* Zählvorrichtung *f* am Lochkartenmischer
collect / to auflesen, sammeln; auffangen, einfangen *(z. B. Elektronen)*; abnehmen, entnehmen *(elektrischen Strom)*
collecting anode Auffanganode *f*
c. **bar** *(El)* Sammelschiene *f*, Hauptsammelschiene *f*
c. **cable** Sammelkabel *n*
c. **conveyor** Entnahmeförderband *n*
c. **electrode** Auffangelektrode *f*, Sammelelektrode *f*
c. **flue** *(Wärm)* Sammelfuchs *m*
c. **grid** *(El, Fmt)* Fanggitter *n*
c. **lens** *(Opt)* Sammellinse *f*; Kollektor *m*
c. **main** 1. *(El)* Sammelleitung *f*; 2. *(Text)* Sammelkanal *m*
c. **pipe** *(Text)* Sammelrohr *n*
c. **pond** *(Bw)* Sammelbecken *n*
c. **ring** *(Elm)* Schleifring *m*
c. **roller table** Sammelrollgang *m*
c. **screen** Auffangschirm *m*
c. **tank** Sammelbecken *n*
c. **zone** *(Fmt)* Fangbereich *m*, Einfangbereich *m*; Mitnahmebereich *m*
collective freight train Sammelgüterzug *m*
c. **lens** *(Opt)* Sammellinse *f*
c. **number** *(Fmt)* Sammelnummer *f*
c. **shipment** *(Schiff)* Sammelladung *f*
collector 1. *(Elm)* Schleifring *m*, Kollektor *m*, Kommutator *m*; Stromabnehmer *m (elektrische Bahn)*; 2. *(Bw)* Kollektor, Sammelkanal *m*, Sammler *m*
c. **bar** 1. Kollektorlamelle *f*, Kollektorsegment *n*, Kommutatorlamelle *f*, Kommutatorsegment *n*; 2. Stromabnehmerstange *f*
c. **bow** Stromabnehmerbügel *m*
c. **breakdown** Kollektordurchbruch *m*
c. **brush** Kollektorbürste *f*, Kommutatorbürste *f*
c. **cutoff current** Kollektorreststrom *m (Halbleitertechnik)*
c. **electrode** Kollektorelektrode *f*
c. **junction** Kollektorrandschicht *f (Halbleitertechnik)*
c. **motor** Bürstenmotor *m*, Kommutatormotor *m*, Kollektormotor *m*
c. **peak voltage** Kollektorspitzenspannung *f (Halbleitertechnik)*
c. **plate** *(Ph)* Auffangelektrode *f*; Auffanganode *f (Elektronen, negative Ionen)*
c. **plates** *(El)* Kontaktstäbe *mpl*, Kontaktplatten *fpl*
c. **ring** Kommutatorring *m*, Kollektor[ring] *m*, Schleifring *m*
c. **road** *(Bw)* Zubringestraße *f*
c. **saturation voltage** Kniespannung *f*; Sättigungsspannung *f*
c. **segment** Kollektorlamelle *f*, Kollektorsegment *n*, Kollektorstreifen *m*, Kommutatorlamelle *f*, Kommutatorsegment *n*
c.**-shoe gear** Stromabnehmer *m* für [oberirdische] Stromleitung
c. **terminal** Kollektoranschluß *m*
c.**-type motor** Bürstenmotor *m*, Kommutatormotor *m*, Kollektormotor *m*
c. **voltage** Kollektorspannung *f*
collet 1. Klemmhülse *f*, Klemmring *m*, Schraubhülse *f*, Schneideisenkapsel *f*; 2. Spannzange *f*; 3. Patrone *f*, Spannpatrone *f*; 4. Reduziereinsatz *m*
c. **attachment** Spannzangeneinrichtung *f*
c. **bar chuck** Keilspannfutter *n*
c. **capacity** Werkstoffdurchlaß *m* durch die Zange
c. **chuck** Zangenspannfutter *n*
c. **chucking** Zangenspannung *f*
c. **chucking attachment** Spannpatroneneinrichtung *f*
c. **for die** Schneideisenkapsel *f*
c. **gripping** Patronenspannung *f*, Zangenspannung *f*
c. **jaw** Spannbacke *f*
c. **pad** Patroneneinsatz *m (Spannpatrone)*
c.**-type sleeve** Kegelklemmhülse *f*
collide / to zusammenstoßen
collier 1. *(Brg)* Bergmann *m*, Bergarbeiter *m*, Grubenarbeiter *m*; 2. *(Schiff)* Kohlendampfer *m*, Kohlenschiff *n*, Kohlenfahrer *m*
colliery *(Brg)* Grube *f*, Kohlengrube *f*, Zeche *f (nur Steinkohlenbergbau)*
c. **air conditioner** Grubenwetterkühler *m*
collimated beam *(Ph)* ausgeblendetes Strahlenbündel *n*

collimating slit *(Opt)* Spaltblende *f*
collimation *(Opt)* Kollimation *f*
collimator *(Opt)* Kollimator *m*
collinear *(Math)* kollinear
  c. array *(El)* Dipolreihe *f*, Dipolzeilenantenne *f*, Dipolzeile *f*
collineation *s.* collimation
collineatory *(Math)* kollinear
colliquation Schmelzen *n*, Schmelzung *f*
collision Zusammenstoß *m*, Stoß *m*, Kollision *f*, Anprall *m*
  c. bulkhead *(Schiff)* Kollisionsschott *n*
  c. cross section *(Ph)* Stoßquerschnitt *m*
  c. damping *(Ph)* Stoßdämpfung *f*, Dämpfung *f* durch Stöße *(Spektrallinienbreite)*
  c. drill *(Schiff)* Schottenmanöver *n*
  c. excitation *(Ph)* Stoßanregung *f (von Gasmolekülen)*
  c. frequency *(Ph)* Stoßzahl *f* [pro Zeiteinheit]
  c. integral *(Ph)* Stoßintegral *n*
  c. ionization *(Ph)* Stoßionisation *f*
  c. mat *(Schiff)* Leckmatte *f*, Stopfmatte *f*, Kollisionsmatte *f*
  c. mean free path *(Ph)* mittlere freie Stoßweglänge *f*
  c. number *(Ph)* Stoßzahl *f*
  c. of the first kind *(Ph)* Stoß *m* erster Art
  c. of the second kind *(Ph)* Stoß *m* zweiter Art
  c. probability *(Ph)* Zusammenstoßwahrscheinlichkeit *f*
collisional damping *(Ph)* Stoßdämpfung *f*, Dämpfung *f* durch Stöße
  c. excitation *(Ph)* Stoßanregung *f*
  c. multiplication *(Ph)* Stoßvervielfachung *f*
collodion Kollodium[wolle *f*] *n*
  c. cotton Kollodiumwolle *f*, Schießbaumwolle *f*
colloid Kolloid *n*
  c. mill Kolloidmühle *f*
  c. rectifier Kolloidgleichrichter *m*
colloidal *(Ch)* kolloidal
  c. chemistry Kolloidchemie *f*
  c. electrolyte kolloidaler Elektrolyt *m*
  c. filament *(El)* gespritzter Glühfaden (Leuchtfaden) *m*
  c. graphite Kolloidgraphit *m*
  c. solution kolloidale Lösung *f*
  c. state *(Ch, Ph)* kolloid[al]er Zustand *m*
colonnade *(Bw)* Säulengang *m*
colophony Kolophonium *n*, Geigenharz *n*
color *(Am) s.* colour
colorant *(Am) s.* colouring agent
coloration *(Typ)* Kolorierung *f*
Colorcrete *(Bw)* farbiger, schnell abbindender Zement *m (Handelsname)*
colorimeter Kolorimeter *n*, Farb[en]messer *m*
colorimetric kolorimetrisch
  c. pyrometer *(Ph, Met)* Farbpyrometer *n*
colorimetry Kolorimetrie *f*, Farb[en]messung *f*
colour / to färben
  c. again auffärben
  c. in the drum *(Led)* im Walkfaß färben
  c. in the tray *(Led)* in der Mulde färben
colour Farbe *f*, Farbton *m*
  c. agglutinant Farbbindemittel *n*
  c. analysis Farbenanalyse *f*, Farbenzerlegung *f*
  c. and weave effect *(Text)* Farb[en]- und Bindungseffekt *m*
  c. answer print Farbmusterkopie *f*
  c. atlas Farbatlas *m*
  c. balance Farbgleichgewicht *n*
  c. bar generator Farbbalkengenerator *m (Farbfernsehtechnik)*
  c. bar pattern Farbbalkenmuster *n (Farbfernsehtechnik)*
  c.-blind farbenblind
  c.-blindness Farbenblindheit *f*
  c. break-up Farbenzerlegung *f*
  c. burst Farbsynchronsignal *n*
  c. burst pedestal *(Fs)* Farbsynchronisationsaustastung *f*
  c. carrier reference *(Fs)* Farbenträgerbezugswert *m*
  c. centre Farbzentrum *n*
  c. change *(Ch)* Farbumschlag *m*
  c. chart *(Opt)* Farbatlas *m*
  c. check print Farbmusterkopie *f*
  c. coder Farbkoder *m*
  c. collotype *(Typ)* Farb[en]lichtdruck *m*
  c. compensation Farbausgleich *m*
  c. component Farbkomponente *f*; Farbauszug *m (Farbfilm)*
  c. contamination Farbwertverschiebung *f (Farbfernsehtechnik)*
  c. decoder Farbdekoder *m*
  c. difference signal *(Fs)* Farbdifferenzsignal *n*
  c. dilution Farbsättigungsabnahme *f*
  c. discharge *(Text)* Buntsätze *f*
  c. doctor *(Text)* Farbabstreichmesser *n*, Farbrakel *f*
  c. emissivity Farbaussendefähigkeit *f*
  c. equation *(Ph)* Farbengleichung *f*
  c. fast *(Text)* farbbeständig, lichtecht
  c. fastness *(Text)* Farbbeständigkeit *f*, Farbechtheit *f*
  c. fidelity Farbtreue *f (von Farbfilmen)*
  c. field Farbteilbild *n*
  c. film Farbfilm *m*
  c. filter Farbfilter *n*
  c. flicker Farbenflimmern *n*
  c. frame Farbenraster *m*
  c. free *(Opt)* achromatisch

c. fringe *(Opt)* Farbsaum *m*
c. fringing *(Fs)* Farbenfransen *n*
c. furnisher *(Text)* Farb[auftrag]walze *f*, Speisewalze *f* für Farbe
c. grinding mill Farbmühle *f*, Farbreib[e]-mühle *f*
c. indicator Farbindikator *m*
c. killer Farbensperre *f*, Farbträgerunterdrücker *m (Farbfernsehtechnik)*
c. kinescope Farb[fernseh]bildröhre *f*
c. match[ing] Farbangleichung *f*, Farbabstimmung *f*
c. mixture Farbenmischung *f*
c. perception Farbempfindung *f*, Farbwahrnehmung *f*
c. phase alternation *(Fs)* Farbenphasenumtastung *f*, Wechsel *m* der Farbenphase
c. photography Farbenfotografie *f*
c. picture screen Farb[fernseh]bildschirm *m*
c. picture tube Farb[fernseh]bildröhre *f*
c. pits *(Led)* Farbengang *m*
c. print *(Typ)* Farbdruck *m*
c. print film Positivfarbfilm *m*
c. printing *(Typ)* Farbendruck[verfahren *n*] *m*
c. range *(Text)* Farbskala *f*, Farbreihe *f*
c. record Farbauszug *m (Farbfilm)*
c. registration Farbenüberdeckung *f (Farbfernsehtechnik)*
c. rendering Farbwiedergabe *f*
c. rendition Farbwiedergabe *f*
c. retention [property] Farbtonechtheit *f*
c. sampling frequency (rate) *(Fs)* Farbschaltfrequenz *f*
c. saturation Vollfarbigkeit *f*
c. scanner *(BMSR)* Farbenabtaster *m (ermöglicht nach Farben getrennte Negative)*
c. scintillation *(Ph)* Farbflimmern *n*
c. separation Farbtrennung *f*
c. shade Farbschattierung *f*, Farbton *m*
c. stability Lichtechtheit *f*, Farbbeständigkeit *f*
c. stimulus Farbreiz *m*
c. stimulus specification Farbvalenz *f (Beleuchtung)*
c. subcarrier Farbenzwischenträger *m (Farbfernsehtechnik)*
c. superimposition Farbenüberdeckung *f (Farbfernsehtechnik)*
c. suspender *(Led)* Farbengang *m*
c. synchronization signal *(Fs)* Farbsynchronisationssignal *n*
c. synchronizer Farbsynchronisationsstufe *f (Farbfernsehtechnik)*
c. television Farbfernsehen *n*
c. temperature *(Ph)* Farbtemperatur *f*
c. tinge Farbstich *m*
c. trough Chassis *n (Zeugdruck)*
c. tuning Farbabstimmung *f*
c. value Farbwert *m*
c. value rendition Tonwertwiedergabe *f* von Farben
c. yield Farbausbeute *f*
coloured gefärbt
c.-all through massegefärbt
c. bleach goods *(Text)* Buntbleichartikel *mpl*
c. discharge print *(Text)* Buntätzdruck *m*
c. overprint resist *(Text)* Überdruckbuntreserve *f*
c. plate *(Typ)* Farbkunstdruck *m*
c. pre-print resist *(Text)* Vordruckbuntreserve *f*
c. resist under vat printing *(Text)* Buntreserve *f* mit Küpenfarbenüberdruck
c. tannin discharge printing *(Text)* Tanninbuntätzdruck *m*
c. thread for identification *(El)* [farbiger] Kennfaden *m (bei Kabeln)*
c. tracer thread *(El)* [farbiger] Kennfaden *m (Kabel)*
c. wax pencil Fettstift *m*, Signierstift *m*
colourfast farbecht
colouring Farbgebung *f*, Kolorierung *f*
c. agent Färbemittel *n*
c. machine Buntpapiermaschine *f*
c. matter aufgetragene Farbe *f*
c. pit *(Led)* Angerbefarbe *f*
c. strength *(Text)* Anfärbevermögen *n*, Farbkraft *f*
colourless farblos
colours *(Schiff)* Flagge *f*
Colpitts circuit *(El)* kapazitive Dreipunktschaltung *f*
Colpitts oscillator circuit *(El)* Dreipunktoszillator *m*
colposcope *(Opt)* Kolposkop *n*
colposcopy *(Opt)* Kolposkopie *f*
colstaff *(Masch)* Wuchtbaum *m*, Hebebaum *m*
coltainer *s.* collapsible container
colter [of plough] Pflugeisen *n*, Pflugmesser *n*
columbate Niobat *n*
columbic compound Verbindung *f* des fünfwertigen Niob[ium]s
columbite Kolumbit *n*, Niobit *n*
columbium Niob *n*, Niobium *n*
columbous compound Verbindung *f* des dreiwertigen Niob[ium]s
column 1. *(Bw)* Pfeiler *m*, Säule *f*; Stiel *m*, Stütze *f*; 2. *(Masch)* Ständer *m*; 3. *(Ch, Ph)* Kolonne *f*, Säule *f*, Turm *m*; 4. *(Math)* Spalte *f*, Kolonne *f (Matrix, Lochkarte)*; 5. *(Typ)* Spalte *f*, Setzspalte *f*, Kolumne *f*, Kolonne *f*; 6. Knickstab *m*; 7. Vertikalstab *m*, Füllstab *m*
c. analogy Säulenanalogie *f (Statik)*
c.-and-knee type milling machine Konsolfräsmaschine *f*

c. base Maschinenfuß *m*
c. cap *(Ch)* Kolonnenkopf *m*, Säulenkappe *f*, Säulenkopf *m*
c. clamp Säulenklemmung *f*
c. drilling machine Säulenbohrmaschine *f*
c. galley *(Typ)* Spaltenschiff *n*
c. guide *(Masch)* Säulenführung *f (Pressen, Schnitt- und Stanzwerkzeuge)*
c. indicating device *(Büro)* Spaltenanzeiger *m (Lochkarte)*
c. matter *(Typ)* gespaltener Satz *m*
c.-mounted drill Säulenbohrmaschine *f*
c. movement *(Schiff)* Fahrt *f* in Kiellinie
c. of adsorbent Adsorptionssäule *f*
c. of box section Kastenständer *m*
c. of liquid Flüssigkeitssäule *f*
c. of mercury *(Ph)* Quecksilbersäule *f*
c. packing *(Ch)* Säulenfüllmaterial *n*, Säulenfüllung *f*
c. scrubber *(Ch)* Turmwäscher *m*
c. splits *(Büro)* Absplitthebel *mpl (Druckwerksplittung z. B. bei Tabelliermaschinen)*
c. still *(Ch)* Destilliersäule *f*
c. strength *(Met)* Knickfestigkeit *f*
c. stress *(Met)* Knickspannung *f (Festigkeit)*
c. transformer Säulentransformator *m*
c. tray *(Ch)* Kolonnenboden *m*
c.-type transformer Säulentransformator *m*
c. washer *(Ch)* Skrubber *m*, Sprühwäscher *m*
c. ways *(Masch)* Führungsbahnen *fpl* am Ständer
columnar structure *(Met)* Stengelgefüge *n*
colure Kolur *m (Großkreis)*
colza oil Rapsöl *n*, Rüböl *n*
coma *(Opt)* Koma *n*, Asymmetriefehler *m*
comb / to kämmen
comb *(Text)* Kamm *m*, Riet *n*, Abschlagzahn *m*
c. cylinder *(Text)* Kämmtrommel *f*
c. filter *(Fmt)* Filter *n* mit kammförmigem Frequenzspektrum
c. plate *(Text)* Fräsblech *n (Raschelmaschine)*
c. segment *(Text)* Kreiskamm *m*, Kammsegment *n (Kämmaschine)*
c.-type gauge Kammlehre *f*, Grenzgewinderachenlehre *f*
combed sliver *(Text)* gekämmtes Band *n*; Kammzug[band *n*] *m*
c. stucco *(Bw)* Kammputz *m*
comber *(Text)* Kämmaschine *f*
c. board Harnischbrett *n*
c. board hole Schnürbrettloch *n*
c. fly Kammflug *m*
c. lap Kämmwickel *m*, Bandwickel *m*
c. lap preparation Bandwickelvorbereitung *f*
c. sliver Kammzug *m*
c. strippings Kämmereiausputz *m*
c. waste Kämmling *m*
c. waste sliver Kämmlingsband *n*

combination Verbindung *f*, Vereinigung *f*, Zusammensetzung *f*, Kombination *f*
c. band Kombinationsbande *f (Spektroskopie)*
c. bar *(Fmt)* Wählerschiene *f*
c. bass control and switch for incorporated speaker *(El)* Tieftonregler *m* mit Schalter für eingebauten Lautsprecher
c. bevel Doppelschmiege *f*
c. board mehrlagig gegautschter Karton *m*
c. boiler Verbundkessel *m*, zusammengesetzter Kessel *m*; kombinierter Kessel (Dampfkessel) *m*
c. centre drill Zentrierbohrer *m* mit beiderseitig abgesetztem Schneidenteil
c. centre drill and countersink Zentrierbohrer *m* mit Schutzsenkung
c. chuck kombiniertes *(zentrisch und exzentrisch spannendes)* Drehmaschinenspannfutter *n*
c. column *(Bw)* Verbundsäule *f (Formstahl und Beton)*
c. cutter Messerkopf *m*
c. distributing frame *(Fmt)* kombinierter Haupt- und Zwischenverteiler *m*
c. framing *(Schiff)* kombiniertes Spantensystem *n*
c. frequency 1. Kombinationsfrequenz *f (insbesondere Atom- und Molekülspektren)*; 2. Schwebungsfrequenz *f (Überlagerung von Schwingungen oder Wellen)*
c. gauge block *(Masch)* Parallelendmaß *n*
c. heat *(Ch)* Bildungswärme *f*
c. light *(Schiff)* Leuchtfeuer *n* mit wechselnden Farben, Wechselfeuer *n*
c. oscillator detector *(El)* Mischstufe *f*
c. pipe and nut (screw) wrench [kombinierter] Rohr- und Schraubenschlüssel *m*, Gasrohrschraubenschlüssel *m*
c. plier Kombi[nations]zange *f*
c. protractor Kombinationswinkelmesser *m*
c. reaction *(Ch)* Anlagerungsreaktion *f*
c. rig kombiniertes Seil- und Drehbohrgerät *n (Erdöl, Erdgas)*
c. saw Kreissäge *f* für Längs-, Quer- und Gehrungsschnitt
c. square Universalwinkel *m*, Kombinationswinkel *m*
c. switch Kombinationsschalter *m*
c. system *(Schiff)* kombiniertes Spantensystem *n (Längs- und Querspantensystem)*
c. tannage *(Led)* gemischte Gerbung *f*, Mischgerbung *f*
c. tone Kombinationston *m*; Schwebungston *m*
c. tool Programmwerkzeug *n*
c. tool block Mehrfachmeißelhalter *m*
c. tool plate holder Vielmeißelaufspannplatte *f*

c. **weaves** *(Text)* zusammengesetzte Bindungen *fpl*
c. **wrench** Universalschlüssel *m*; kombinierter Rohr- und Schraubenschlüssel *m*, Gasrohrschraubenschlüssel *m*
combinational switching circuits *(BMSR)* kombinatorische Schaltkreise *mpl*, Verknüpfungsschaltungen *fpl*
combinatorial *(Math)* kombinatorisch, Kombinations-
c. **analysis (theory)** *(Math)* Kombinatorik *f*
combinatory *s.* combinatorial
combine / to kombinieren; [sich] miteinander verbinden
combine Mähdrescher *m*
c. **drill** Kombinationsdrillmaschine *f*, Kombinationsreihensämaschine *f*
c. **harvester** [kombinierter] Mähdrescher *m*
combined verbunden; gebunden
c. **aerial** Gemeinschaftsantenne *f*
c. **bending and axial load** Biegung *f* mit Längskraft *(Statik)*
c. **board** kaschierter Karton *m*
c. **cable** gemischtpaariges Kabel *n*
c. **carbon** *(Ch)* gebundener Kohlenstoff *m*
c. **check and stop valve** *(Wärm)* Speisekopf *m*
c. **conductivity** *(El)* Kombinationsleitwert *m*, Mittelwert *m* des spezifischen Leitwertes
c. **counterbore and reamer** Senkreibahle *f*
c. **coupling** *(Fmt)* gemischte Kopplung *f*
c. **curve** *(Wärm)* Kurvenzug *m*
c. **direct and bending stresses** Spannungen *fpl* aus Längskraft und Biegung *(Statik)*
c. **drill and countersink** Zentriersenker *m*
c. **field weakening** *(Elm)* gemischte Feldschwächung *f*
c. **flasher and tail lamp** *(Kfz)* Blink-Schlußleuchte *f*
c. **flow turbine** vereinigte Axial- und Radialturbine *f*
c. **footing** *(Bw)* gemeinsames Fundament *n* mehrerer Stützen
c. **fuse and cut-out** *(El)* Trennsicherung *f*
c. **gas and coal furnace** *(Wärm)* gemischte (kombinierte) Feuerung *f*
c. **grain-fertilizer drill** Korn- und Düngerdrillmaschine *f*
c. **greasing agent** *(Text)* Verbundschmälze *f*
c. **horizontal and vertical steam engine** kombinierte liegende und stehende Dampfmaschine *f*
c. **impulse and reaction turbine** kombinierte Aktions- und Reaktionsturbine *f*
c. **load** zusammengesetzte Last *f (Statik)*
c. **method** *(Brg)* kombiniertes Abbauverfahren *n*
c. **milling and drilling machine** Fräs- und Bohrmaschine *f*
c. **milling and slotting machine** Fräs- und Stoßmaschine *f*
c. **paver** Mischerfertiger *m (Straßenbau)*
c. **plastic** Verbund[preß]stoff *m*
c. **sawing and filing machine** kombinierte Säge- und Feilmaschine *f*
c. **scutching and breaking machine** kombinierte Schwing- und Brechmaschine *f (Flachs)*
c. **sewer** Mischwasserkanal *m*, Mischwassersammler *m (kommunale Entwässerung)*
c. **steam pipe isolating and stop valve** *(Fmt)* vereinigtes Rohrbruch- und Absperrventil *n*
c. **stop and tail lamp** *(Kfz)* Brems-Schlußleuchte *f*
c. **stop-tail-number plate lamp** *(Kfz)* Brems-Schluß-Kennzeichenleuchte *f*
c. **stresses** zusammengesetzte Spannungen *fpl (Statik)*
c. **stroke** Gesamthub *m (bei Gegenkolben)*
c. **television-telephone service** Fernsehgegensprechen *n*; Fernsehsprechdienst *m*
c. **truck-conveyor system** *(Brg)* kombinierter Zug-Bandbetrieb *m*
c. **twill** *(Text)* Mehrgratköper *m*
c. **water** *(Ch)* chemisch gebundenes Wasser *n*, Konstitutionswasser *n*
c. **water tube and fire tube boiler** *(Wärm)* Wasserrohr-Heizrohr-Kessel *m*
c. **welding machine** Universalschweißmaschine *f*
combing *(Text)* Kämmen *n*
c. **machine** Kämmaschine *f*, Kammstuhl *m*
c. **wool** Kammwolle *f*
combining equivalent *(Ch)* Äquivalentgewicht *n*, Verbindungsgewicht *n*
c. **nozzle** Mischdüse *f*
c. **weight** *(Ch)* Äquivalentgewicht *n*, Verbindungsgewicht *n*
combustibility Verbrennbarkeit *f*, Brennbarkeit *f*
combustible Brennstoff *m*
combustible brennbar
c. **gas** Heizgas *n*, Brenngas *n*
c. **matter** brennbare Bestandteile *mpl*
combustibles Brennstoffe *mpl*
combustion Verbrennung *f*
c. **adjustment** Brennereinstellung *f*
c. **air** Verbrennungsluft *f (Verbrennungsmotor)*
c. **analysis** *(Ch)* organische Elementaranalyse *f*, Verbrennungsanalyse *f*
c. **boat** *(Ch)* Glühschiffchen *n*, Verbrennungsschiffchen *n*
c. **bomb** *(Ph, Ch)* Explosionskalorimeter *n*
c. **chamber** 1. *(Kfz)* Verbrennungsraum *m*; 2. *(Rak, Wärm)* Brennkammer *f*; 3. *(Holz)* Rauchkammer *f*

c. **chamber liner** *(Rak)* Brennkammereinsatz *m*
c. **controls** Flammenüberwachung *f*, Brennerüberwachung *f*
c. **cowl** *(Rak)* Brennrohr *n*
c. **cutoff** *(Rak)* Brennschluß *m*
c. **cutoff frequency** Brennschlußfrequenz *f*
c. **cutoff method** Abschaltverfahren *n*
c. **cutoff signal** Brennschlußkommando *n*, Brennschlußsignal *n*
c. **cutoff system** Brennschlußanlage *f*
c. **efficiency** Verbrennungswirkungsgrad *m*
c. **engine** Verbrennungskraftmaschine *f*
c. **gases** Rauchgase *npl*, Verbrennungsgase *npl*, Abgase *npl*
c. **heat** Verbrennungswärme *f*
c. **indicator** Rauchgasmesser *m*
c. **pressure** Verbrennungsdruck *m*
c. **product** Verbrennungsprodukt *n*
c. **space** Verbrennungsraum *m*, Brennraum *m*
c. **temperature** Verbrennungstemperatur *f*
c. **termination** *(Rak)* Brennschluß *m*
c. **test stand** *(Rak)* Brennstand *m*
c. **time** *(Rak)* Brenndauer *f*
c. **tube furnace** *(Ch)* Verbrennungsofen *m* *(Elementaranalyse)*
c.**-type oxygen analyzer** Wärmetönungs-Sauerstoffanalysator *m* *(Verbrennung an hocherhitztem Draht)*
c. **unit cooling line** *(Rak)* Brennkammerkühlrohr *n*
c. **velocity** Brenngeschwindigkeit *f*
c. **zone** Verbrennungszone *f*, Brennzone *f*
combustor Brennkammer *f*
come / to kommen; sich entwickeln
c. **down** *(Flg)* heruntergehen
c. **down from aloft** *(Schiff)* niederentern
c. **down in a spin** *(Flg)* abtrudeln
c. **down through the overcast** *(Flg)* [Wolkendecke] durchstoßen
c. **in** einlaufen, einkommen *(Schiff in den Hafen)*
c. **into collision with** stoßen auf, aufeinandertreffen
c. **into effect** wirksam werden
c. **into land** *(Flg)* zur Landung ansetzen
c. **loose** sich lockern
c. **off** abarbeiten, abbringen, flottmachen *(Schiff)*
c. **out to the day** *(Geol)* ausbeißen, ausstreichen
c. **to shore** anlegen *(Schiff)*
c. **up** *(Brg)* ausfahren
Comecon Rat *m* für gegenseitige Wirtschaftshilfe
comet *(Astr)* Komet *m*, Schweifstern *m*
c. **finder** *(Opt)* Kometensucher *m*
c. **trace** *(Astr)* Kometenschweif *m*
cometary orbit *(Astr)* Kometenbahn *f*
coming into force Inkrafttreten *n* *(eines Gesetzes)*
c. **into step** *(El)* Intrittfallen *n* *(einer Synchronmaschine)*
comma-shape kommaförmig
command 1. *(BMSR)* Befehl *m* *(in Rechenautomaten)*; 2. Aufgabewert *m*, Sollwert *m*
c. **guidance** *(Rak)* Kommandolenkung *f*, Fernlenkung *f*
commencement Beginn *m*; Einleitung *f* *(z. B. eines elektrischen Vorganges)*
c. **chip thickness** *(Masch)* Spandicke *f* am Anfang des Schnittes
commencing course *(Text)* Anschlagreihe *f*
commensurable 1. *(Math)* kommensurabel, einen gemeinsamen Teiler besitzend; 2. *(Ph)* kommensurabel, im gleichen Maß meßbar, dimensionsgleich
commercial handelsüblich; geschäftlich; wirtschaftlich; großtechnisch
c. **airplane** Verkehrsflugzeug *n*
c. **art** Gebrauchsgrafik *f*
c. **broadcast** Werbefunk *m*
c. **buildings** Büro- und Geschäftsbauten *pl*
c. **colours** *(Text)* nicht wasch- oder lichtechte Farben *fpl*
c. **company** Handelsgesellschaft *f*
c. **copper** Handelskupfer *n*
c. **data processing** kaufmännische (kommerzielle) Datenverarbeitung *f*
c. **efficiency** Ausnutzungsfaktor *m*, Gesamtwirkungsgrad *m*, wirtschaftlicher Nutzen *m*
c. **enterprise** Erwerbsgesellschaft *f*
c. **forest** Wirtschaftswald *m*
c. **grade** Handelsqualität *f*, handelsübliche Qualität *f*
c. **grades of coal** handelsübliche Kohlensorten *fpl*
c. **intercourse** Handelsverkehr *m*
c. **jig** *(Brg)* Betriebssetzmaschine *f*
c. **length** Handelslänge *f*
c. **material** handelsüblicher (technischer) Werkstoff *m*
c. **moisture regain** handelsüblich festgelegter Feuchtigkeitsgehalt *m*
c. **navigation** Handelsschiffahrt *f*
c. **navy** Handelsflotte *f*, Handelsmarine *f*
c. **port** Handelshafen *m*
c. **production** *(Ch)* [groß]technische Herstellung (Produktion) *f* *(Gegensatz: labormäßige Herstellung)*
c. **radiotelegraphy** kommerzielle Telegrafie *f*
c. **register** Handelsregister *n*
c.**-scale plant** Betriebsanlage *f*
c. **sewage** Gewerbeabwässer *npl*, gewerbliche Abwässer *npl*
c. **speed** *(Schiff)* Dienstgeschwindigkeit *f*

c. standard 1. handelsüblicher Standard *m*, handelsübliche Norm *f*; 2. technische Liefer- und Gütebestimmung *f*
c. tank Produktionsbad *n* *(Galvanotechnik)*
c. television Werbefernsehen *n*
c. transmitting station Werbefunksender *m*
c. vehicle *(Kfz)* Nutzfahrzeug *n*
c. voltage Netzspannung *f*, Spannung *f* im öffentlichen Netz
Commercial Code Handelsgesetzbuch *n*
commercially available im Handel erhältlich
c. pure technisch rein *(Chemikalien)*
commercials Fernseh-Werbefilme *mpl*
comminute / to zerkleinern, zermahlen, zerstükkeln, pulverisieren
comminution [mechanische] Zerkleinerung *f*, Vermahlung *f*; Pulverisierung *f*; Brechen *n*
commission / to ernennen, beauftragen, bevollmächtigen
commission Ernennungsurkunde *f*; Bevollmächtigungsurkunde *f*
Committee for Mutual Economic Co-operation Rat *m* für gegenseitige Wirtschaftshilfe, RGW
common gemein[sam]
c. aerial Gemeinschaftsantenne *f*
c. base current gain cutoff frequency Grenzfrequenz *f* der Stromverstärkung in Basisschaltung *(Transistoren)*
c. battery *(El)* gemeinsame Batterie *f*; Zentralbatterie *f*
c. battery telephone set Fernsprechapparat *m* mit Zentralbatterie
c. battery working Zentralbatteriebetrieb *m*
c. board *(Typ)* Schrenz[pappe *f*] *m*
c. brick Hintermauerungsstein *m*, Mauerziegel *m*
c. caliper square Werkstattschieblehre *f*
c. channel interference *(Fmt)* Gleichkanalstörung *f*
c. chord *(Ak)* Dreiklang *m*
c. conductor *(Am)* *(El)* Null-Leiter *m*
c. cost *(Ök)* Gemeinkosten *pl*
c. denominator *(Math)* gemeinsamer Nenner
c. drive gemeinsamer Antrieb *m*
c. emitter current gain cutoff frequency Grenzfrequenz *f* der Stromverstärkung in Emitterschaltung *(Transistoren)*
c. emitter frequency for unity current gain Frequenz *f* der Einheitsstromverstärkung in Emitterschaltung *(Transistoren)*
c. fraction *(Math)* gemeiner Bruch *m*
c.-frequency broadcasting Gleichwellenrundfunk *m*
c.-frequency working *(Fmt)* Gleichwellenbetrieb *m*, Gleichwellenfunk *m*
c. hook *(Text)* rundgebogener Haken *m* *(Zungennadel)*
c. iron pyrites Schwefelkies *m*
c. language *(Dat)* Maschinensprache *f*
c. logarithm gewöhnlicher (gemeiner) Logarithmus *m*, Briggscher Logarithmus, dekadischer Logarithmus
c.-mode rejection *(El, BMSR)* Unterdrükkung *f* gleichphasiger Signale, Gleichtaktunterdrückung *f* *(Differentialverstärker)*
c.-mode rejection quotient *(El)* Diskriminationsfaktor *m* *(Differentialverstärker)*
c.-mode signal *(El)* gleichphasiges Signal *n*, Gleichtaktsignal *n*
c. osier Korbweide *f*
c. position of hook *(Text)* Nadelhaken *m* in der Linie der Auflagefläche der Nadel angeordnet *(Zungennadel)*
c. practice übliches Verfahren *n*
c. reactance *(El)* gegenseitige Induktion *f*
c. reactive element *(El)* gemeinsamer Blindwiderstand *m*; Kopplungswiderstand *m*
c. salt Kochsalz *n*
c. timbers Handelshölzer *pl*
c. tooth gerader Zahn *m* *(Säge)*, Dreieckszahn *m*
c. trunc gemeinsames Leitungsbündel *n*
c. workman Lohnarbeiter *m*
communal gemeinschaftlich, Gemeinschafts-
c. aerial Gemeinschaftsantenne *f*
c. reception Blockempfang *m* *(Radio)*
communicate / to 1. *(Fmt)* in Verkehr stehen, verkehren; mitteilen; 2. in Verbindung stehen, kommunizieren
communicating tubes *(Ph)* kommunizierende Röhren *fpl*
communication 1. Nachricht *f*; 2. Nachrichtenwesen *n*; Sprechverkehr *m*; Schreibverkehr *m*; 3. Verbindung *f*, Kommunikation *f*
c. art Nachrichtentechnik *f*
c. band *(El)* Übertragungsband *n*
c. by train radio Funkverständigung *f* im Zug
c. channel *(Fmt)* Übertragungskanal *m*, Kanal *m*
c. code Verkehrskode *m*
c. counter Gesprächszähler *m* *(Telefonie)*
c. engineer Nachrichteningenieur *m*, Fernmeldeingenieur *m*
c. engineering Fernmeldetechnik *f*, Nachrichtentechnik *f*
c. network Nachrichtennetz *n*
c. receiver Nachrichtenempfänger *m*; Betriebsempfänger *m*
c. satellite Nachrichtensatellit *m*
c. system Nachrichtennetz *n*, Nachrichtenverbindungssystem *n*
communications Fernmeldewesen *n*; Nachrichtenwesen *n*
community aerial Gemeinschaftsantenne *f*

c. automatic exchange automatisches Ortsfernsprechamt *n*
c. centre *(Bw)* Gemeinschaftszentrum *n*; Wohngebietszentrum *n*
c. planning *(Bw)* Siedlungsplanung *f*
c. television Gemeinschaftsfernsehempfang *m*
c. unit Wohnkomplex *m*
commutate / to kommutieren, vertauschen, umpolen
c. the current den Strom wenden (umpolen)
commutating angle *(Elm)* Kommutierungswinkel *m*
c. condenser Löschkondensator *m* *(Stromrichter)*
c. field *(Elm)* Wendefeld *n*, Kommutierungsfeld *n*
c. pole *(Elm)* Kommutierungspol *m*, Wendepol *m*, Kompensationspol *m*, Ausgleich[s]pol *m*, Hilfspol *m*
c. variable transformer *(Am)* *s.* variable ratio transformer
c. winding *(Elm)* Wendefeldwicklung *f*
commutation *(Elm)* Kommutierung *f*, Kommutation *f*, Vertauschung *f*, Stromwendung *f*
c. factor Kommutationsfaktor *m*
c. failure Kippen *n* *(der Kommutierung)*
c. field Kommutierungsfeld *n*, Wendefeld *n*
c. machine Kommutatormaschine *f*, Kollektormaschine *f*
c. margin Löschwinkel *m*, Kommutierungswinkel *m*
c. pole Kommutierungspol *m*, Wendepol *m*, Kompensationspol *m*, Ausgleich[s]pol *m*, Hilfspol *m*
commutative law *(Math)* Kommutativgesetz *n*
commutator *(El)* Stromwender *m*, Kommutator *m*, Kollektor *m*
c. armature Kommutatoranker *m*
c. bar Kommutatorlamelle *f*, Kommutatorsegment *n*, Kollektorlamelle *f*, Kollektorsegment *n*, Kollektorstab *m*, Kollektorstreifen *m*
c. brush Kollektorbürste *f*, Kommutatorbürste *f*
c. bush Kommutatorbuchse *f*, Kommutatorbüchse *f*, Kollektorbuchse *f*, Kollektorbüchse *f*, Stromsammlerbuchse *f*, Stromsammlerbüchse *f*
c. design Kommutatorkonstruktion *f*, Kollektorkonstruktion *f*, Stromwenderkonstruktion *f*
c. end Kommutatorseite *f*
c. end ring Kommutatorendring *m*, Kollektorendring *m*, Stromsammlerendring *m*, Stromsammlerschlußring *m*
c. grinder Kollektorschleifmaschine *f*
c. hub *s.* c. bush
c. induction motor Kommutatorinduktionsmotor *m*
c. lug Stromwenderfahne *f*, Kommutatorfahne *f*, Kollektorfahne *f*
c. machine Kommutatormaschine *f*, Kollektormaschine *f*
c. mica Kollektorglimmer *m*
c. motor Kommutatormotor *m*, Stromwender *m*, Kollektormotor *m*
c. motor meter Kommutatormotorzähler *m*
c. phase advancer Kommutatorphasenschieber *m*
c. pitch Kommutatorschritt *m*, Kollektorschritt *m*, Stromwenderschritt *m*
c. polishing paper Kommutatorschleifpapier *n*, Kollektorschleifpapier *n*
c. rectifier mechanischer Gleichrichter *m*, Kontaktgleichrichter *m*, Kommutatorgleichrichter *m*
c. ripple frequency Kommutierungsfrequenz *f*
c. riser *s.* c. lug
c. section *s.* c. segment
c. segment Kommutatorlamelle *f*, Kommutatorsegment *n*, Kollektorlamelle *f*, Kollektorsegment *n*, Kollektorstab *m*, Kollektorstreifen m
c. shell *s.* c. bush
c. shell insulation Kollektorisolierhülse *f*, Kommutatorisolierhülse *f*
c. sleeve *s.* c. bush
c. tag *s.* c. lug
c. voltage Kommutatorspannung *f*, Kollektorspannung *f*
commuter 1. Einpendler *m* *(Straßenverkehr)*; 2. Verkehrsboot *n*, Tourenboot *n*
compact / to verdichten, preßformen
compact Preßteil *n* aus Pulver; Preßling *m*; Formteil *n*; Halbzeug *n*
compact 1. *(Math)* kompakt, dicht *(Mengen)*; 2. *(Met)* dicht *(Guß)*; 3. gedrungen, lederhaft, voll *(Fell, Haut)*
c. bars *(Holz)* Derbstangen *fpl*
c. car *(Kfz)* Wagen *m* *jeden Karosserietyps, konstruiert für leichtes Parken, leichte Bedienung, niedrige Unterhaltungskosten und geringen Brennstoffverbrauch*
c. design gedrungene Bauform *f*
c. material *(Ch)* Verbundstoff *m*
c. powder Kompaktpuder *m* *(Kosmetik)*
c. wood Derbholz *n*
compacted earth fill foundation *(Bw)* Sandpolstergründung *f*
compactibility Verdichtbarkeit *f*, Verdichtungsfähigkeit *f*
compacting Verdichtung *f*
c. beam *(Bw)* Stampfbohle *f*
c. of soil Verfestigung *f* des Bodens
compaction Verdichtung *f*, [mechanische] Bodenverdichtung *f*

c. roller *(Bw)* Verdichtungswalze *f*
compactor *(Bw)* Verdichtungsgerät *n*
compactron *(El)* Kompaktron *n (12-Stift-Vielfachsystem-Verbundröhre in den USA)*
COMPANDER *s.* compressor-expander
companding *(Fs)* Pressung-Dehnung *f* [der Dynamik], Dynamikregelung *f*
companion hatch *(Schiff)* Kajütenluke *f*, Niedergangsluke *f*
c. ladder *(Schiff)* Kajütstreppe *f*, Niedergangstreppe *f*
c. matrix *(Math)* Begleitmatrix *f*
c. part *(Masch)* Gegenstück *n*
c.-way *(Schiff)* Kajütsniedergang *m*, Niedergang *m*
comparative measurement Unterschiedmessen *n*; Vergleichsmessung *f*; mittelbares Messen *n*
c. measuring instrument Vergleichsmeßgerät *n*
comparator 1. *(Dat)* Komparator *m*, Vergleicher *m*, Vergleichseinrichtung *f*; Vergleichsvorrichtung *f (insbesondere in Analogrechnern zum Ausführen der Amplitudenwahl)*; 2. *(Masch)* Komparator, Vergleichsmesser *m*; 3. *(El)* Differenzverstärker *m*
c. gauge Mikrotastgerät *n*
c. principle *(Opt)* Komparatorprinzip *n*
c. resistance *(El)* Vergleichswiderstand *m*
comparing dimensions Unterschiedmessen *n*
comparison electrode Bezugselektrode *f*, Vergleichselektrode *f*
c. element *(BMSR)* Vergleichsglied *n*
c. eyepiece *(Opt)* Vergleichsokular *n*
c. lamp Vergleichslampe *f*
c. microscope Vergleichsmikroskop *n*
c. spectrum Vergleichsspektrum *n*
c. standard Normal *n*, Vergleichsnormale *f*
c. test Vergleichsprüfung *f*
c. theorem *(Math)* Vergleichssatz *m*
comparometer Vergleichsmesser *m*
comparoscope Oberflächenprüfgerät *n*
compartment 1. Abteil *n*; 2. Abteilung *f (im Schiff)*; 3. *(Bw)* Tasche *f (Bunker)*; Zelle *f (Silo)*; 4. *(Brg)* Trum *m*; 5. *(Brg)* Bett *n (Setzmaschine)*
c. furnace *(Ch)* Kammerofen *m*
c. kiln *(Holz)* Trockenkammer *f* mit abschnittweiser Beladung
c. mill *(Ch)* Mehrkammermühle *f*
compass 1. Kompaß *m*; 2. *(Am) (Bw)* Zirkel *m*; 3. *(Opt)* Bussole *f*
c. bearing *(Schiff)* Kompaßpeilung *f*
c. brick *(Bw)* Keilziegel *m*
c. bridge *(Schiff)* Peildeck *n*
c. card Kompaßrose *f*, Windrose *f*
c. declination Kompaßabweichung *f*
c. flat *(Schiff)* Peildeck *n*
c. heading Kompaßkurs *m*
c. plane *(Holz)* Bogenhobel *m*, Rundhobel *m*; Schiffshobel *m*
c. platform *(Schiff)* Peildeck *n*
c. pole *(Schiff)* Kompaßsäule *f*
c. rafter *(Bw)* gekrümmter Sparren *m*
c. roof *(Bw)* halbkreisförmiges Dach *n*, Runddach *n*, Tonnendach *n*
c. run *(Schiff)* Kompensierungsfahrt *f*
c. saw Stichsäge *f*, Durchbruchsäge *f*, Lochsäge *f*
c. timber Krummholz *n*
c. traverse *(Opt)* Bussolenzug *m*
c. variation *(Flg)* Kompaßabweichung *f*
c. window rundes Erkerfenster *n*
c. with pen[cil] point *s.* compasses with pen[cil] point
compasses Zirkel *m*
c. bow instrument Zirkel *m*; Nullenzirkel *m*
c. with pen[cil] point Zirkel *m* mit Einsatz
compatibility Kompatibilität *f (Farb- und Schwarzweiß-Fernsehempfang)*
c. condition *(Math, Ph)* Verträglichkeitsbedingung *f*
c. of the skin Hautverträglichkeit *f (Chemikalien)*
compatible colour television austauschbares Farbfernsehverfahren *n*
c. dyes mischbare Farbstoffe *mpl*
compensate / to 1. ausgleichen, kompensieren, neutralisieren; entzerren *(Bild- und Tonübertragung)*; 2. abfinden, entschädigen *(Patentwesen)*
c. for damages einen Schaden abfinden *(Patentwesen)*
compensated amplifier Kompensationsverstärker *m*
c. circuit Ausgleichskreis *m*, Ausgleichsleitung *f*
c. control *(Am) (BMSR)* Nachführregelung *f*
c. ionization chamber *(Ph)* kompensierte Ionisationskammer *f*, Kompensationskammer *f*
c. motor Motor *m* mit Kompensationswicklung
c. pendulum Kompensationspendel *n*
c. repulsion motor kompensierter Repulsionsmotor *m*
c. scan *(Fmt)* kompensierte Abtastung *f*
c. semiconductor Kompensationshalbleiter *m*
c. squelch kompensierte Geräuschsperre *f*
c. volume control *(El)* kompensierte (physiologische) Lautstärkeregelung *f*
c. winding *(El)* Kompensationswicklung *f*, Ausgleich[s]wicklung *f*
compensating charge Pufferladung *f (Batterie)*
c. circuit Kompensationsschaltung *f*, Ausgleichsschaltung *f*
c. coil Kompensationsspule *f*, Ausgleichsspule *f*

c. device 1. *(Wärm)* Ausgleichvorrichtung *f*; 2. *(Masch)* Korrektureinrichtung *f* *(Lehrenbohren)*
c. eyepiece *(Opt)* Kompensationsokular *n*
c. feedwheel *(Text)* ausgleichendes konisches Fournisseurrad *n*
c. fore-and-aft magnet *(Schiff)* Ausgleichslängsmagnet *m*
c. gear *(Masch)* Ausgleichsrad *n*
c. jet Ausgleichdüse *f* *(Verbrennungsmotor)*
c. loop with swivel joint *(Wärm)* Ausgleichschleife *f* mit drehbarem Flansch
c. magnet Ausgleich[s]magnet *m*, Kompensationsmagnet *m*
c. magnetometer Kompensationsmagnetometer *n*
c. microphone Kompensationsmikrofon *n*
c. network *(BMSR, Dat)* Kompensationsschaltung *f*, Korrekturschaltung *f*; Korrekturglied *n*
c. pole *(Elm)* Ausgleich[s]pol *m*, Hilfspol *m*, Wendepol *m*
c. reservoir *(Bw)* Ausgleichbecken *n*; Ausgleichbehälter *m*
c. resistance *(El)* 1. Kompensationswiderstand *m*, Ausgleich[s]widerstand *m*; 2. Vorschaltwiderstand *m* *(Meßtechnik)*
c. resistor *(El)* 1. Kompensationswiderstand *m*, Ausgleich[s]widerstand *m*; 2. Vorschaltwiderstand *m*
c. surface *(Flg)* Ausgleichflosse *f*
c. tank Ausgleichbehälter *m*
c. triode *(El)* Kompensationstriode *f*
c. voltage *(El)* Kompensationsspannung *f*, Ausgleichsspannung *f*
c. weight Ausgleichsgewicht *n*
c. winding *(El)* Kompensationswicklung *f*, Ausgleich[s]wicklung *f*
compensation 1. Ausgleich *m*, Kompensation *f*; Entzerrung *f* *(Bild- und Tonübertragung)*; 2. Abfindung *f*; Entschädigung *f*
c. circuit *(Fs, Fmt)* Ausgleichsstromkreis *m*
c. developer Ausgleichsentwickler *m* *(Filme)*
c. device Ausgleichseinrichtung *f* *(Lehrenbohren)*
c. for play *(Masch)* Spielausgleich *m*
c. for wear Ausgleich *m* der Abnutzung
c. gear Differentialgetriebe *n*, Ausgleichsgetriebe *n*
c. magnet Kompensationsmagnet *m*
c. method Ausgleichsverfahren *n*
c. of errors Fehlerausgleich *m*
compensator *(Masch)* 1. Kompensator *m* *(Turbine)*; 2. Ausgleichseinrichtung *f* *(Lehrenbohren)*
c. jack Ausgleichsstrebe *f*
c. reservoir Ausgleichbehälter *m*
c. starter Anlaßtransformator *m*, Transformatoranlasser *m*
c. winding Kompensationswicklung *f*, Ausgleich[s]wicklung *f*
compensatory leads Ausgleichsleitungen *fpl*, Kompensationsleitungen *fpl*
compensed instrument transformer *(Am)* *(Elm)* kompensierter Meßwandler *m*
compete / to konkurrieren, im Wettbewerb stehen
competitive konkurrenzfähig; wettbewerbsfähig
competitor Konkurrent *m*
compilation 1. Anhäufung *f*; 2. *(Büro)* Kompilation *f* *(Programmierung)*
compiler *(Dat)* Kompiler *m*, Selbstprogrammierer *m*
compiling routine *(Dat)* Selbstprogrammierung *f*
complaint section Beschwerdestelle *f*
complement / to ergänzen, komplementieren; vervollständigen
complement 1. Ergänzung *f*, Komplement *n*, Vervollständigung *f*; Bestückung *f* *(insbes. mit Röhren bei elektr. Geräten)*; 2. *(Schiff)* vollständige Schiffsmannschaft *f*, komplette Besatzung *f*
c. on b *(BMSR)* b-Komplement *n* *(b bedeutet Grundzahl zu einer Potenz)*
c. on (b-1) *(BMSR)* (b-1)-Komplement *n* *([b-1] bedeutet die um 1 verminderte Grundzahl einer Potenz)*
complementary komplementär, Komplementär-; ergänzend, Ergänzungs-
c. angle *(Math)* Ergänzungswinkel *m* *(zu 90°)*, Komplementwinkel *m*
c. angle of the bevel lead angle *(Masch)* Anschnittwinkel *m*
c. angle of the chisel edge angle *(Masch)* Querschneidenwinkel ψ *m*
c. colour Komplementärfarbe *f*
c. functions *(BMSR)* komplementäre Funktionen *fpl*
c. rectifier Hilfsgleichrichter *m*
c. transistor Komplementärtransistor *m*
complete / to vervollständigen; beenden, fertigmachen, fertigstellen
c. a call *(Fmt)* ein Gespräch zustande bringen, eine Verbindung herstellen
c. a circuit einen Stromkreis schließen
c. a revolution around the earth *(Astr)* einmal um die Erde fliegen
c. a turn *(Flg)* Kurve beenden
c. the apprenticeship auslernen
complete vollständig; beendet, fertiggestellt
c. carry vollständiger Übertrag *m*
c. circuit geschlossener Stromkreis *m*
c. circuit diagram *(El)* Gesamtschaltbild *n*
c. cycle 1. [vollständiger] Arbeitszyklus *m* *(insbesondere Taktstraße)*; 2. [geschlossener] Kreisprozeß *m* *(Thermodynamik)*

c. **diagram** *(Wärm)* Gesamtdiagramm *n*
c. **miscibility** vollständige Mischbarkeit *f*
c. **modulation** *(Fmt)* vollständige Modulation (Aussteuerung) *f*
c. **operation** *(Dat)* Volloperation *f*
c. **overhaul** Generalüberholung *f*
c. **rotation** volle Umdrehung *f*
c. **subsidence** *(Brg)* Vollsenkung *f*
c. **superstructure** *(Schiff)* durchlaufende Aufbauten *pl*
c. **superstructure vessel** durchlaufendes Aufbauschiff *n*
completed draft 1. fertiger Entwurf *m*; 2. *(Text)* vollständige Patrone *f*
c. **shell** *(Ph)* abgeschlossene (vollbesetzte, aufgefüllte) Schale *f (Elektronen)*
completely ionized gas vollionisiertes Gas *n*, vollständig ionisiertes Gas
c. **wet surface** vollkommen benetzte Fläche *f*
completement 1. *s.* completion; 2. Ergänzungsstück *n*; Gegenstück *n*; Gegenlehre *f*
completing circuit Schließstromkreis *m*
completion 1. *(Bw)* Ausbau *m*; 2. Fertigstellung *f*, Komplettierung *f*; 3. fertiggestellter Neubau *m (Schiff)*
c. **date** Fertigstellungstermin *m*
complex 1. *(Math)* komplex; 2. komplex, zusammengesetzt; vielseitig, vielschichtig, kompliziert
c. **admittance** *(El)* komplexer Scheinleitwert *m*
c. **alloy** Vielstofflegierung *f*
c. **display** zusammengesetztes Bild *n (Radar)*
c. **ether** gemischter Äther *m*
c. **fraction** *(Math)* Doppelbruch *m*
c. **impedance** *(El)* komplexer Scheinwiderstand *m*
c. **ions** *(Ph)* Komplexionen *npl*
c. **number** *(Math)* komplexe Zahl *f*
c. **of figures** *(Math)* Zahlenkomplex *m*
c. **oscillation** *(Ph)* zusammengesetzte Schwingung *f*
c. **plane** *(Math)* komplexe Ebene *f*, komplexe Zahlenebene *f*, Ebene der komplexen Zahlen
c. **plane analyzer** *(Dat)* Gerät *n* zur Vektorzerlegung
c. **radiation** *(Ph)* zusammengesetzte Strahlung *f*, Mischstrahlung *f*
c. **sine wave** *(Ph)* zusammengesetzte Sinuswelle *f*
c. **sound** *(Ak)* Tongemisch *n*, zusammengesetzter Klang *m*
c. **spectrum** linienreiches Spektrum *n*
c. **waves** 1. zusammengesetzte Wellen *fpl (insbesondere Elektrizität)*; 2. *(Math, Ph)* komplexe Wellen[felder *npl*] *fpl*
complexing *(Ch)* Komplexbildung *f*
c. **agent** *(Ch)* Komplexbildner *m*
complimentary copy *(Typ)* Freiexemplar *n*
comply with the requirements of law / to die gesetzlichen Anforderungen erfüllen
compo 1. *s.* composition; 2. Zementmörtel *m*
component 1. *(Math)* Komponente *f (eines Tensors, insbesondere eines Vektors)*; 2. *(Ph)* Komponente *(Tensorgrößen; zusammengesetzte Prozesse)*; 3. *(Masch)* Maschinenteil *n*, Teil *n*, Baugruppe *f*; Werkstück *n*; 4. Komponente; Teilkraft *f*
c. **at right angles to plane of blade motion** Seitenkomponente *f* des Bahndruckes *(Turbine)*
c. **clamping** Werkstückspannung *f*
c. **current** *(El)* Teilstrom *m*
c. **dimension** Werkstückabmessung *f*
c. **drawing** Werkstückzeichnung *f*, Stückzeichnung *f*
c. **efficiency** Teilwirkungsgrad *m*
c. **equalizer network** Entzerrer-Baustein *m (Tonbandgerät)*
c. **force** Einzelkraft *f*; Kraftkomponente *f*; Teilkraft *f (Statik)*
c. **frequency** Frequenzkomponente *f*
c. **layout** Bauplan *m (insbesondere elektrischer Geräte)*
c. **motion** Bewegungskomponente *f*
c. **of a force** Kraftkomponente *f*
c. **of compressive force** Druckkomponente *f*
c. **of the cutting force** Schnittkraftkomponente *f*
c. **part** Bestandteil *m*; Einzelteil *n*, Maschinenteil *n*, Stückteil *n*; Ersatzteil *n*
c. **picture** Teilbild *n (Farbbild; stereoskopisches Bild)*
c. **voltage** Spannungskomponente *f*
components of strain *(Ph)* Komponenten *fpl* des Deformationstensors, Formänderungskomponenten *fpl*
c. **of stress** *(Ph)* Komponenten *fpl* des Spannungstensors, Spannungskomponenten *fpl*
compose / to 1. zusammensetzen; 2. *(Typ)* setzen; 3. komponieren *(Musik)*
composing frame *(Typ)* Setzpult *n*
c. **machine** Setzmaschine *f*
c. **room** Setzerei *f*
c. **stick** Winkelhaken *m (des Setzers)*
composite *(Plast)* Verbundstoff *m*
composite zusammengesetzt; Verbund-
c. **action** *(Bw)* Verbundwirkung *f*
c. **beam** *(Bw)* Verbundträger *m*
c. **bearing** Verbundlager *n*
c. **board** Verbundplatte *f*
c. **board including veneer-faced particle** furnierbeplankte Spanplatte *f*
c. **boiler** Abgaskessel *m* mit Zusatzfeuerung
c. **building construction** Verbundbau[weise *f*] *m*
c. **cable** gemischtpaariges Kabel *n*

c. cast roll Verbundgußwalze *f*
c. casting Verbundguß *m*
c. check of errors Sammelfehlerprüfung *f* *(Zahnrad)*
c. circuit *(Fmt)* Simultanschaltung *f*
c. colour signal zusammengesetztes Farbsignal *n*
c. column Verbundsäule *f* *(Formstahl und Beton)*
c. conductor *(El)* Zweimetalleiter *m*
c. construction 1. Verbundbau[weise *f*] *m*; 2. *(Schiff)* Kompositbauweise *f*, Gemischtbauweise *f* *(Holz und Stahl)*
c. dike *(Geol)* zusammengesetzter Gesteinsgang *m*
c. error Sammelfehler *m* *(Zahnrad)*
c. excitation *(Elm)* Doppelschlußerregung *f*, Verbunderregung *f*, Kompounderregung *f*, gemischte Erregung *f*
c. function *(Math)* Funktion *f* einer Funktion; Funktional *n*
c. girder Verbundträger *m*
c. loading *(Fmt)* Viererpupinisierung *f*
c. metal Verbundmetall *n*; Zwiemetall *n*
c. number *(Math)* in Faktoren zerlegbare Zahl *f*
c. panel Verbundplatte *f*
c. plate Mehrschichten-Überzug *m*, Verbundüberzug *m* *(Galvanotechnik)*
c. rocket Mehrstufenrakete *f*, mehrstufige Rakete *f*
c. roof truss Fachwerkbinder *m* aus mehreren Baustoffen
c. shot Kombinationsaufnahme *f* *(Film)*
c. signal *(Fs)* Signalgemisch *n* *(Bild- und Synchronsignal)*
c. steel Verbundstahl *m*
c. stress zusammengesetzte Spannung *f*
c. structure Verbundkonstruktion *f*
c. synchronization signal zusammengesetztes Synchronisiersignal *n*
c. test Sammelfehlerprüfung *f* *(Zahnrad)*
c. truss Fachwerkbinder *m* aus mehreren Baustoffen
c. vessel Kompositschiff *n*
composited circuit *(Fmt)* mehrfach ausgenützte Leitung *f*; Simultanschaltung *f*
composition 1. *(Ph)* Zusammensetzung *f* *(Kräfte usw.)*; 2. *(Ch)* Aufbau *m*, Zusammensetzung *f* *(Materialien)*; 3. *(Typ)* Satz *m*
c. flooring fugenloser Fußbodenbelag *m*, Steinholzestrich *m*
c. of forces *(Mech)* Zusammensetzung *f* von Kräften
c. resistor *(Fmt)* Massewiderstand *m*
c. roller *(Typ)* Massewalze *f*
compositor *(Typ)* Setzer *m*
compost Kompost *m*
compound / to mischen; kompoundieren
compound 1. [chemische] Verbindung *f*; 2. Masse *f*, Verbindung *f*, Vergußmasse *f*, Kabelmasse *f*
compound 1. zusammengesetzt; 2. *(El)* Doppelschluß-; Verbund-
c. alternator kompoundierter Wechselstromgenerator *m*
c. brush *(Elm)* Metallkohlebürste *f*
c. casting Verbundguß *m*
c. catenary suspension line Verbundkettenfahrleitung *f*
c. cloth mehrfaches Gewebe *n*
c. coil Doppelspule *f*, Verbundspule *f*
c. coke oven Verbundkoksofen *m*
c. connection *(Elm)* Verbundschaltung *f*
c. crystal Zwillingskristall *m*
c. cupola Kupolofen *m* mit erweitertem Herd
c. ether gemischter Äther *m*
c. excitation Doppelschlußerregung *f*, Verbunderregung *f*, Kompounderregung *f*, gemischte Erregung *f*
c. fraction *(Math)* Doppelbruch *m*
c. frame *(Schiff)* gebautes Spant *n*
c. generator Doppelschlußgenerator *m*, Verbundgenerator *m*
c. gland cock Hahn *m* mit Verbundstopfbuchse
c. glass Schichtglas *n*, Sicherheitsglas *n*, Verbundglas *n*
c. glass tube Verbundglasröhre *f*
c. indexing *(Masch)* Verbundteilen *n*
c. interest *(Ök)* Zinseszins *m*
c. load zusammengesetzte Last *f*
c. locomotive Verbundlokomotive *f*
c. machinery Aggregat *n* *(insbesondere von Raketen)*
c. magnet *(El)* Lamellenmagnet *m*, Blättermagnet *m*, Magnetbündel *n*, kombinierter (zusammengesetzter) Magnet *m*
c. mechanism zusammengesetztes Getriebe *n*
c. microscope zusammengesetztes Mikroskop *n*
c. mine *(Brg)* Verbundzeche *f*
c. motor *s.* c.-wound motor
c. navigation Koppelnavigation *f*
c. nucleus *(Kern)* Zwischenkern *m*, Kompoundkern *m*
c. oscillation zusammengesetzte Schwingung *f*
c. pendulum zusammengesetztes Pendel *n*, Mehrfachpendel *n*
c. portable steam engine Verbundlokomobile *f*
c. rest Kreuzschieber *m*; Kreuzsupport *m*
c. rest slide Kreuzschlitten *m*, Oberschlitten *m* *(Drehmaschine)*

c. **shaft** *(Brg)* Seigerschacht *m*, dessen unteres Ende tonnlägig fortgesetzt ist *(mit einer gemeinsamen Fördereinrichtung)*
c. **signal** *(Fmt) (aus mehreren Frequenzen)* zusammengesetztes Zeichen *n*
c. **slide** Kreuzschlitten *m*, Oberschlitten *m (Drehmaschine)*
c. **spring** geschichtete Blattfeder *f*
c. **steam engine** Verbunddampfmaschine *f*, Zweifachexpansionsdampfmaschine *f*
c. **steel** Duplexstahl *m*, Verbundstahl *m*
c. **stress** zusammengesetzte Spannung (Beanspruchung) *f*
c. **strip** Bimetallstreifen *m*
c. **table** Kreuztisch *m (Werkzeugmaschine)*
c. **tie-up** *(Text)* mehrfache Schnürung *f*
c. **tool** Verbundwerkzeug *n*
c. **train** 1. doppelte Wechselräderübersetzung *f*, 2. Gliederzug *m*
c. **transformer** Kompoundtransformator *m*, Verbundtransformator *m*
c. **turbine** Verbundturbine *f*
c.-**type of workholding table** Kreuzaufspanntisch *m*
c. **winding** *(El)* Kompoundwicklung *f*, Doppel[be]wicklung *f*, Doppelschlußwicklung *f*, gemischte Wicklung *f*, Verbundwicklung *f*
c.-**wound current transformer** Stromwandler *m* mit Zusatzmagnetisierung
c.-**wound motor** Doppelschlußmotor *m*, Kompoundmotor *m*, Verbundmotor *m*
compounding Kompoundierung *f*; Versatz *m*; Versetzung *f*
c. **coil** *(Elm)* Kompoundwicklung *f*
compreg *s.* compregnated wood
compregnated wood kunstharzimprägniertes und gepreßtes Holz *n*, Kunstharzpreßholz *n*, Kunstholz *n*, Lignofol *n*, Preßholz *n*
comprehensive insurance *(Kfz)* Kaskoversicherung *f*
compress / to komprimieren, verdichten, zusammendrücken
c.-**form** druckumformen
compressed komprimiert, verdichtet, zusammengedrückt
c. **air** Druckluft *f*, Preßluft *f*
c.-**air blast cleaning** Druckluftputzstrahlen *n*
c.-**air bottle** Druckluftflasche *f*, Preßluftflasche *f*
c.-**air brake** *(Kfz)* Druckluftbremse *f*
c.-**air brake cylinder** Bremsluftbehälter *m*
c.-**air caisson** Luftsenkkasten *m*
c.-**air cleaner** Druckluftfilter *n*
c.-**air cleaning** Druckwindreinigung *f (Getreide)*
c.-**air distributor** Druckluftverteiler *m*
c.-**air engine** Preßluftmotor *m*
c.-**air equilibrator** Luftausgleicher *m*
c.-**air hammer** Preßlufthammer *m*
c.-**air hose** Druckluftschlauch *m*
c.-**air lamp** *(Brg)* Preßluftleuchte *f*
c.-**air line** Druckluftleitung *f*
c.-**air locomotive** *(Brg)* Preßluftlokomotive *f*
c.-**air motor** Preßluftantrieb *m*, Preßluftmotor *m*
c.-**air motor hand drill** Drucklufthandbohrmaschine *f*
c.-**air operated** druckluftbetätigt
c.-**air pipe fitting** Druckluftarmatur *f*
c.-**air plant** Druckluftanlage *f*, Preßluftanlage *f*
c.-**air reservoir** Druckluftbehälter *m*
c.-**air shift cylinder** *(Kfz)* Druckluftschaltzylinder *m*
c.-**air tank** Druckluftbehälter *m*
c.-**air tool** Preßluftwerkzeug *n*
c.-**air tunnel** *(Flg)* Überdruckwindkanal *m*
c. **asphalt** Stampfasphalt *m*
c. **coal** Stampfkohle *f*
c. **coal cake** *(Brg)* Stampfkuchen *m*
c. **fibreboard** vorgepreßte Faserplatte *f*
c. **gas** *(Ch)* Flaschengas *n*; komprimiertes Gas *n*
c.-**gas accumulator** hydropneumatischer Speicher *m*
c.-**gas circuit breaker** *(El)* Druckgasschalter *m*
c. **spring** Feder *f* mit Vorspannung
c. **wood laminate flooring surface** Fußbodenoberfläche *f* aus verpreßtem Holz
compressibility Kompressibilität *f (physikalische Größe)*; Zusammendrückbarkeit *f*, Komprimierbarkeit *f*
c. **burble** *(Flg)* Wirbelbildung *f* bei Überschallgeschwindigkeit infolge Verdichtungsstoßes
compressible kompressibel, zusammendrückbar; komprimierbar
compression 1. Kompression *f*, Verdichtung *f*; Zusammendrücken *n*; 2. Druck *m*; 3. *(Fs)* Dynamikkompression *f*; Pressung *f*
c. **and eddy losses** Umlagerungsverlust[e *mpl*] *m (Turbine)*
c. **area** Druckquerschnitt *m*, Druckfläche *f*
c. **bar** Druckstab *m (Stahlbeton)*
c. **bars** Druckbewehrung *f*, Druckstäbe *mpl (Stahlbeton)*
c. **boom** Druckgurt *m (Fachwerk)*
c. **cable** Druckkabel *n*
c. **chamber** *(Kfz)* Kompressionsraum *m*
c. **chord** Druckgurt *m (Fachwerk)*
c. **coil spring** Schraubendruckfeder *f*
c. **column** *(Bw)* Druckstütze *f*
c. **condenser** Quetschkondensator *m*
c. **coupling** *(Masch)* 1. Klemmkupplung *f*; Schalenkupplung *f*; 2. Klemmverbindung *f* für Wellen

c. curve Verdichtungskurve *f*
c. cycle Verdichtungstakt *m*, Verdichtungshub *m (Verbrennungsmotor)*
c. fibre Druckfaser *f (eines Trägers)*
c. gauge Kompressionsdruckprüfer *m*
c. heat sealing *(Plast)* Heißsiegeln *n*
c. ignition Selbstzündung *f* durch Verdichtung *(Diesel)*
c.-ignition engine Dieselmotor *m*, Motor *m* mit Selbstzündung (Verdichtungszündung)
c. line *s.* c. curve
c. liquefying Kompressionsverflüssigung *f*, Verflüssigung *f* durch Komprimieren
c. load Kompressionsdruck *m*, Verdichtungsdruck *m*
c. loss Kompressionsverlust *m (Turbine)*
c. member *(Bw)* Druckglied *n*, Druckstab *m (Fachwerk)*
c. mould Preßform *f*
c.-moulded wood-base item druckverformter Gegenstand *m* auf Holzbasis *(Preßteil)*
c. moulding Formpressen *n*, Pressen *n*
c. of speech *(Dat)* Fortlassen *n* (Unterdrükkung *f*) von unwesentlichen Wörtern eines gesprochenen Textes
c. packing ineinandergreifende Dichtungsringe *mpl*
c. pressure Kompressionsdruck *m*, Verdichtungsdruck *m*
c. ratio Verdichtungsverhältnis *n*, Kompression *f (Verbrennungsmotor)*
c. reducer Verdichtungsverminderer *m*
c. refrigeration Kompressionskühlung *f*, Kühlung *f* durch Kompression
c. reinforcement Druckbewehrung *f (Stahlbeton)*
c. ring Kolbenring *m*, Kompressionsring *m*, Verdichtungsring *m (Verbrennungsmotor)*
c. riveter Preßnietmaschine *f*
c. shock *(Flg)* Verdichtungsstoß *m (Überschreiten der Schallgeschwindigkeit)*
c. spring Druckfeder *f*
c. stage Verdichtungsstufe *f*
c. strapping Bandeisenverpackung *f (Papierbogen gleichen Formats)*
c. strength 1. Druckfestigkeit *f*; Druckspannung *f*; 2. Stauchfestigkeit *f (Papier)*
c. stroke Kompressionstakt *m*; Kompressionshub *m*; Verdichtungshub *m*
c.-swirl Druckwirbelung *f*
c. test Druckversuch *m*
c. test specimen Druckprobe *f*
c.-type refrigerator Kompressorkühlschrank *m*, Kompressionskühlschrank *m*, Kühlschrank *m* mit Kompressor
c. valve jet Düse *f (beim Stoßdämpfer)*
c. volume *(Kfz)* Verdichtungsraum *m*
c. welding Preßschweißung *f*
c. wood Druckholz *n*
c. zone Druckzone *f*
compressive amplitude Druckamplitude *f*
c. clamp Federspanneisen *n*
c. force Druckkraft *f*
c. load Drucklast *f*, Druckbelastung *f*, Druckbeanspruchung *f*
c. oil cup Preßöler *m*
c. reinforcement Druckbewehrung *f (Stahlbeton)*
c. strength Druckfestigkeit *f*
c. strength of a cube Würfeldruckfestigkeit *f*
c. stress Druckbeanspruchung *f*, Druckspannung *f*
c. yield point Quetschgrenze *f*
c. zone Druckzone *f*
compressor 1. Verdichter *m*; Kompressor *m*; 2. Lader *m*; 3. *(Schiff)* Kettenkneifer *m*, Kneifstopper *m*, Zwischendeckstopper *m*
c. blade Kompressorenschaufel *f*, Verdichterschaufel *f*
c.-expander Dynamik-Kompressions- und -Expansionsgerät *n*
c. inlet Gebläseeinlauf *m*
c. pulsation limit *(Wärm)* Pumpgrenze *f*
c. shaft Kompressorwelle *f*
c. turbine Verdichterturbine *f*
c. valve Verdichterventil *n*
comprise / to einschließen, enthalten, umfassen
Compton absorption Compton-Absorption *f*
Compton collision Compton-Stoß *m*
Compton recoil Compton-Rückstoß *m*
Compton scattering Compton-Streuung *f*
Compton shift Compton-Verschiebung *f*
compulsory checking zwangsläufig wirkende Überwachung *f (Eisenbahnsignalwesen)*
c. landing *(Flg)* Pflichtlandung *f*
c. pilotage *(Schiff)* Lotsenzwang *m*
c. standard verbindliche Norm *f*, verbindlicher Standard *m*
c. working of a patent (of patents) Erzwingung *f* der Patentausübung, obligatorische Patentausübung *f*
computability *(Math)* Berechenbarkeit *f*
computable berechenbar
computation Ausrechnen *n*; Berechnung *f*; [maschinelle] Rechentechnik *f*
c. centre Rechenzentrum *n*
c. table Rechentafel *f*
computational rechnerisch, Rechen-
compute / to [be]rechnen, errechnen, ausrechnen
computer Rechenanlage *f*, Rechenmaschine *f*, Rechenautomat *m (meist im Sinne von programmgesteuert)*
c. code Rechnerkode *m*
c.-controlled durch Rechengerät gesteuert
c. for guidance into line of sight *(Rak)* Einlenkrechner *m*

c. operation Rechenautomatenbetrieb *m*, [programmierter] Rechnerbetrieb *m*
c. process control Steuerung *f* [von Produktionsanlagen bzw. des Produktionsablaufes] durch Prozeßrechner
c. variable Rechnervariable *f* *(Analogrechner: abhängige, dargestellte Variable)*
computing centre Rechenzentrum *n*, Rechenstelle *f*
c. circuit *(BMSR)* Rechenschaltung *f*
c. machine Rechenmaschine *f*
c. machinery Rechenanlage *f*
c. network *(BMSR, Dat)* Rechenschaltung *f*, Rechennetzwerk *n*
con rod *(Kfz)* Pleuelstange *f*
concatenated connection *(El)* Kaskadenschaltung *f*
c. motor Kaskadenmotor *m*
c. system *s.* concatenated connection
concatenation Hintereinanderschaltung *f*, Kaskadenschaltung *f*
concave konkav
c.-back mit hohlem Rücken
c. cathode Hohlkatode *f*
c. cutter Halbkreisformfräser *m*
c. cutter for milling half circles konkaver Halbkreisformfräser *m*
c. fillet weld Hohlkehlnaht *f*
c. grinding Hohlschleifen *n*, Hohlschliff *m*
c. ground hohlgeschliffen, mit Hohlschliff
c. lens Konkavlinse *f*, Zerstreuungslinse *f*
c. mirror Hohlspiegel *m*, Konkavspiegel *m*
c. mirror cathode Hohlspiegelkatode *f*
c. mould[ing] Hohlkehle *f*
c. reflector Hohlspiegel *m*
c. spade Hohlspaten *m*, Hohlspatel *m*
c. surface Hohlfläche *f*
c. tapered hohlkegelig
c. tile Nonne *f* *(Dachziegel)*
c. weld Hohl[schweiß]naht *f*
concavity Austiefung *f*, Konkavität *f*, Rundhöhlung *f*
concavo-convex *(Opt)* konkavkonvex
c.-convex lens Konkavkonvexlinse *f*
concealed verdeckt; verborgen
c. lighting indirekte Beleuchtung *f*
c. motion *(Ph)* verborgene Bewegung *f* *(kinetische Theorie der Wärme)*
c. wiring *(El)* Unterputzleitung *f*
concentrate / to anreichern, konzentrieren
concentrate *(Ch)* Konzentrat *n*
concentrated capacitance punktförmig verteilte Kapazität *f*
c. load Einzellast *f*, konzentrierte Last *f*, Punktlast *f* *(Statik)*
c. nuclear fuel *(Kern)* konzentrierter (angereicherter) Spaltstoff *m*
c. winding *(El)* konzentrierte Wicklung *f*, Spulenwicklung *f*
concentrating coil *(El)* Bündelungsspule *f*, Fokussier[ungs]spule *f*
c. reflector Punktstrahler *m*, Richtstrahler *m*; Tiefstrahler *m*
concentration 1. Bündelung *f*; Fokussierung *f*; 2. Konzentration *f*, Anreicherung *f*; Eindicken *n*; 3. Sättigungsgrad *m*
c. cell *(El)* Konzentrationselement *n*
c. coil Sammelspule *f*
c. key *(Fmt)* Platzschalter *m*
c. of ions Ionenkonzentration *f*
c. of the electron beam Elektronenstrahlbündelung *f*
c. of the feeding liquor *(Text)* Nachsatzkonzentration *f*
c. of workings *(Brg)* Abbaukonzentration *f*, Betriebskonzentration *f*
c. overvoltage konzentrationsbedingte Überspannung *f*, Konzentrationspolarisation *f*
c. plant Eindickanlage *f*
c. position *(Fmt)* Sammelplatz *m*
c. switch Sammelschalter *m*
concentrator 1. *(El)* Konzentrator *m*; 2. *(Fmt)* Zentralumschalter *m*; 3. *(Brg)* Aufbereitungsanlage *f*
concentric konzentrisch, zentriert, zentrisch; gleichachsig, mittig
c. braiding *(Fmt)* Beklöppelung *f*, Umflechtung *f*
c. bushes *(Wärm)* konzentrische Lagerschalen *fpl*
c. cable konzentrisches (koaxiales) Kabel *n*, Koaxialkabel *n*
c. chuck *(Masch)* Universalfutter *n*
c. electrode system konzentrische Elektrodenanordnung *f*
c. load Axiallast *f*, mittige Last *f* *(Statik)*
c. pipe transmission line konzentrische Speiseleitung *f*
c. plug and socket *(El)* konzentrische Steckverbindung *f*
c. running *(Masch)* [genauer] Rundlauf *m*
c. winding[s] *(El)* konzentrische Wicklung (Röhrenwicklung, Spulenwicklung, Zylinderwicklung) *f*
concentricity 1. Gleichachsigkeit *f*, Mittigkeit *f*, Zentrizität *f*; 2. Rundlauf *m*
c. tester Rundlaufmeßgerät *n*; Schlagfehlerprüfgerät *n*
concertina Akkordeon *n* (in kleiner Ausführung), Handharmonika *f*, Harmonika *f*, Ziehharmonika *f*
c. cover ziehharmonikaähnlicher Schutz *m*
c. folding *(Typ)* Zickzackfalzung *f*
c.-type covers of the ways *(Masch)* harmonikaähnliche Gleitbahnabdeckung *f*

c.-type ducting *(Brg)* Faltlutte *f*, zusammenschiebbare Lutte *f*
c. walls *pl* Faltenbalg *m*
conchoid *(Math)* Konchoide *f*, Muschellinie *f*
conchoidal muschelig, schneckenlinienförmig
c. motion *(Wärm)* Konchoidenlenker *m*
conclude a contract / to einen Vertrag [ab]-schließen
conclusion Schluß[folgerung *f*] *m*; Folgerung *f*
concrete / to Beton einbringen, betonieren
c. in einbetonieren
concrete Beton *m*
c. additive Betonzusatz *m*
c. admixture Betonzusatz *m*
c. aggregate Betonzuschläge *mpl*, Zuschlagstoffe *mpl*
c. block Betonblock *m*, Betonstein *m*
c.-block walling Betonformsteinausbau *m*
c. breaker Betonbrecher *m*; Betonaufbruchhammer *m*
c. breaking machine *s.* c. breaker
c. casing Betonummantelung *f*
c. cast in position (situ) Ortbeton *m*
c. channel Betonrinne *f*; Betonrinnstein *m*
c. chimney cooler Betonkaminkühler *m*
c. compaction Betonverdichtung *f*
c. compactor Betonverdichter *m*
c. construction Betonbau *m*, Stahlbetonbau *m*
c. cover Betondeckung *f (Stahlbeton)*
c. cutter *s.* c. saw
c. drill Steinbohrer *m* für Beton
c.-encased betonummantelt
c. engineering Betonbau *m*, Stahlbetonbau *m*
c. envelope Betonmantel *m (Reaktor)*
c. extruding machine Betongleitfertiger *m*, Gleitfertiger *m*
c. factory Betonwerk *n*
c. finisher Betonfertiger *m*
c. foundation Betonfundament *n*
c. grinder Betonschleifmaschine *f*
c. gun Betonspritzmaschine *f*, Torkretkanone *f*
c. hardener Betonhärtungsmittel *n*
c. hull Beton-Schiffskörper *m (Schiffskörper aus Beton)*
c. lining Betonausbau *m*
c. masonry Betonmauerwerk *n*
c. mix Betonmischung *f*, Frischbeton *m*
c. mixer Betonmischer *m*, Betonmischmaschine *f*
c. pancake column *(Brg)* künstlicher Pfeiler *m* aus Betonscheiben
c. paver Betonstraßenfertiger *m*
c. pile Betonpfahl *m (Grundbau)*
c. pipe Betonrohr *n*
c. placement Betoneinbringung *f*, Betonieren *n*
c. placement funnel Betontrichter *m*
c. placing Betoneinbringung *f*, Betonieren *n*
c. plank Betonbohle *f*
c. plant *(Am)* Betonwerk *n*
c. product Betonware *f*, Fertigteil *n*
c. proportion Betonmischungsverhältnis *n*
c. protection Betondeckung *f (Stahlbeton)*
c. rammer Betonstampfer *m*
c. reinforcement Bewehrung *f (Stahlbeton)*
c. road Betonstraße *f*
c. road finisher Betonstraßenfertiger *m*
c.-rolling technique Betonwalzverfahren *n*
c. roofing tile Betondachstein *m*
c. sand Betonsand *m*
c. saw Betonsäge *f*; Fugensäge *f (Straßenbau)*
c. ship Betonschiff *n (Schiff aus Beton)*
c. sleeper Beton[eisenbahn]schwelle *f*
c. spreader Betonverteiler *m*
c. strength Betonfestigkeit *f*
c. stress Betonspannung *f*
c. tamper Betonstampfer *m*, Erdstampfer *m*
c. texture Betongefüge *n*
c. topping Aufbeton *m*, Betondeckschicht *f*
c. vibrator Betonrüttler *m*
c. waterproofing compound Dichtungsmittel *n*
c. work Betonarbeiten *fpl*
concreting Betoneinbringung *f*, Betonieren *n*
concretor Beton[fach]arbeiter *m*, Betonierer *m*
concurrency logic *(BMSR)* Gleichzeitigkeitslogik *f*
concurrent 1. gleichlaufend, parallel, gleichzeitig; 2. *(Math)* durch ein und denselben Punkt gehend
c. boiler *(Wärm)* Gleichstromkessel *m*
c. centrifuge *(Kern)* Durchflußzentrifuge *f*, Durchlaufzentrifuge *f (Gaszentrifuge)*
c. forces zusammenwirkende Kräfte *fpl*
c. operation *(BMSR)* simultaner Operationsablauf *m*
c. shaft sinking and equipping *(Brg)* gleichzeitiges Abteufen *n* und Einrichten *n*
concussion Erschütterung *f*
c.-free erschütterungsfrei
condemn a vessel / to ein Schiff einziehen
condensability Verdichtbarkeit *f*, Kondensierbarkeit *f*
condensable verdichtbar, kondensierbar
condensate Kondensat *n*, Niederschlag *m*, Kondensationsprodukt *n*
c. collector Kondensatsammelbehälter *m*
c. water Niederschlagswasser *n*
condensating plant Kondensationsanlage *f*
condensation Kondensierung *f*, Verdichtung *f*; Zusammenziehung *f*; Niederschlag *m*
c. by injection Einspritzkondensation *f*
c. by mixing Mischkondensation *f*
c. level (point) Kondensationspunkt *m*
c. polymer *(Plast)* Polykondensat *n*
c. resin Kondensationsharz *n*

c. **runoff** Kondensatablauf *m*
c. **space** Kondensationsraum *m*
c. **tank** Kondensatbehälter *m*
c. **trail** *(Flg)* Kondensstreifen *m*
c. **trap** Kondensatsammelgefäß *n*
c. **water** Kondenswasser *n*
condense / to [sich] verdichten, kondensieren, [sich] niederschlagen
condensed steam kondensierter Dampf *m*
c.**-steam pump** Kondensatpumpe *f*
c.**-water discharge** Kondenswasseraustritt *m*
c. **yarn** *(Text)* Zweizylindergarn *n*
condenser 1. Kühler *m (Destillation)*; 2. *(El)* Kondensator *m* (*s. a.* capacitor); 3. Kondensator *m*, Verflüssiger *m*; 4. *(Opt)* Kondensor *m*
c. **aerial** Kondensatorantenne *f*; Netzantenne *f*
c. **armature** *(El)* Kondensatorbelag *m*, Kondensatorbelegung *f*
c. **bank** *(El)* Kondensatorenbank *f*, Kondensatorengruppe *f*; Kondensator[en]batterie *f*
c. **battery** *(El)* Kondensator[en]batterie *f*
c. **bobbin** *(Text)* Vorgarnspule *f*
c. **bushing** *(El)* Kondensatordurchführung *f*
c. **card with tape divider** *(Text)* Vorspinnkrempel *f*, Kontinuekrempel *f* mit Riemchenflorteiler
c. **chamber** *(Wärm)* Kondensatorkammer *f*
c. **circuit** *(El)* Kondensatorkreis *m*
c. **coating** *(Fmt)* Kondensatorbelag *m*, Kondensatorbelegung *f*
c. **coil** *(Wärm)* Kühlschlange *f*
c. **combined with the turbine** in die Turbine eingebauter Kondensator *m*
c. **constant** *(El)* Kondensatorgüte *f*
c. **cotton yarn** Baumwollstreichgarn *n*
c. **coupler** *(El)* kapazitiver Spannungsteiler *m*
c. **dielectric** Kondensatordielektrikum *n*
c. **discharge** *(El)* Kondensatorentladung *f*
c. **disk** *(Am) s.* c. plate
c. **electrolyte** Kondensatorelektrolyt *m*
c. **electroscope** Kondensatorelektroskop *n*
c. **extraction pump** Kondensatpumpe *f*
c. **jacket** *(Wärm)* Kühlermantel *m*
c. **lens** Kondensorlinse *f*
c. **load** *(El)* kapazitive Belastung (Last) *f*
c. **loudspeaker** Kondensatorlautsprecher *m*
c. **microphone** elektrostatisches Mikrofon *n*, Kondensatormikrofon *n*
c. **pick-up** kapazitiver Tonabnehmer *m*
c. **piping** *(Wärm)* Kondensatorrohrleitung *f*
c. **plate** *(El)* Kondensatorplatte *f*, Kondensatorscheibe *f*
c. **pressure** *(Wärm)* Kondensatordruck *m*
c. **receiver** *(Wärm)* Kondensationsgefäß *n*, Kondensatorkasten *m*
c. **reel** *(Fmt)* Kondensatorwickel *m*
c. **spindle** *(El)* Kondensatorachse *f*
c. **spinning** *(Text)* Zweizylinderspinnerei *f*
c. **system** *(Text)* Streichgarnverfahren *n*
c. **transmitter** Kondensatormikrofon *n*
c. **tube** *(Wärm)* Kondensatorrohr *n*
c. **tube plate** *(Wärm)* Kondensatorrohrboden
c. **tubing** *(Wärm)* Kondensatorberohrung *f*
c. **type terminal** *(El)* Kondensatordurchführung *f*; Nagelsche Klemme *f*
c. **voltage** Kondensatorspannung *f*
c. **water** Kondensationswasser *n*
c. **with spiral condensing tube** Kühler *m* mit Schlangenrohr
c. **with spiral tube sealed to body** Kühler *m* mit eingeschmolzenem Schlangenrohr
condensing coil Kühlschlange *f*
c. **engine** Kondensationsmaschine *f*
c. **lens** *(Opt)* Sammellinse *f*
c. **pump** Kondensationspumpe *f*
c. **steam engine** Kondensationsdampfmaschine *f*
c. **steam turbine** Kondensationsturbine *f*
c. **vessel** *(Wärm)* Niederschlaggefäß *n*
condensive *s.* capacitive
condensor *s.* condenser
condition / to klimatisieren; konditionieren; härten; anpassen; abbinden
condition 1. Bedingung *f*; 2. Zustand *m*; 3. Erhaltungsgrad *m (Münzen)*
c. **of acceptance** Übernahmebedingung *f*
c. **of equilibrium** *(Ph)* 1. Gleichgewichtsbedingung *f (im Zustandsdiagramm)*; 2. Gleichgewichtszustand *m*
c. **of gas** Gasbeschaffenheit *f*
c. **of insulation** Isolationszustand *m*
c. **of sale** *(Am)* Liefer[ungs]bedingung *f*
c. **of surface** Oberflächenzustand *m*, Oberflächenbeschaffenheit *f*
c. **of the cylinder liner surface** *(Kfz)* Zylindertragbild *n*
c. **signal light** *(Rak)* Zustandslampe *f*
c. **under which flow takes place** Strömungsvorgang *m (Turbine)*
conditional *(BMSR)* bedingt
c. **breakpoint instruction** *(BMSR)* bedingter Haltbefehl (Stoppbefehl) *m*
c. **jump** *(BMSR)* bedingter Sprung *m (Programmierung)*
c. **transfer of control** *(BMSR)* bedingter Sprung *m (Programmierung)*
conditioning 1. Klimatisieren *n*; 2. Abbinden *n*; Härten *n*, Verputzen *n*, Zubereiten *n*; 3. *(Text)* Konditionieren *n*
c. **cabinet** Klimaschrank *m*
c. **loss** *(Text)* Konditionierverlust *m*
c. **network** Anpassungsnetzwerk *n*
c. **period** *(Plast)* Nachhärtungsfrist *f*
c. **stimulus** elektrischer Vorreiz *m*
conduct / to 1. leiten *(Strom)*; führen; 2. abhalten *(Lehrgang)*

conductance *(El)* Steilheit *f (Röhre)*; Leitwert *m*, Konduktanz *f*, Wirkleitwert *m*; Realteil *m* des Scheinleitwertes
c.-bridge hydrometer Leitfähigkeitsdichtemesser *m*
c. characteristic Leitwertcharakteristik *f*
c. relay Konduktanzrelais *n*
conductibility *s.* conductivity
conductible leitfähig
conducting leitend *(z. B. elektrischen Strom)*, leitfähig
c. capacity Leitfähigkeit *f*
c. direction Durchlaßrichtung *f (Stromrichter)*
c. electrons Leitungselektronen *npl*
c. layer leitende Schicht *f*
c. path Strombahn *f*
c. period *(El)* Brennzeit *f*; Durchlaßdauer *f*; Stromflußwinkel *m*
c. power Leitvermögen *n*, Leitungskraft *f*
c. process *(El)* Stromleitung *f*
c. salts Leitsalze *npl*, Elektrolyte *mpl*
c. tissue Leitgewebe *n (Holz)*
c. wire *(El)* Leitungsdraht *m*
conduction *(Ph)* Leitung *f*; Leitfähigkeit *f*
c. band *(Ph, El)* Leitungsband *n*
c. current Leitungsstrom *m*
c. direction *(El)* Durchlaßrichtung *f*
c. electron Leitungselektron *n*; Valenzelektron *n*
c. heating surface Leitungsheizfläche *f*
c. of heat Wärmeleitung *f*
c. time *(El)* Brenndauer *f (Stromrichter)*
conductive leitend, leitfähig
c. coupling *(El)* galvanische Kopplung *f*
c. material elektrischer Leiter *m*, leitendes Material *n*
c. pencil *(Büro)* elektrisch leitender Stift *m*
c. resistance *(El)* Leitwiderstand *m*
conductivity Leitfähigkeit *f*, Leitvermögen *n*
c. bridge *(El)* Kohlrausch-Brücke *f*
c. cell *(Ch, El)* Leitfähigkeitsmeßzelle *f*
c. electron *s.* conduction electron
c. modulation transistor Transistor *m* mit veränderlicher Leitfähigkeit *f*
c. phenomena *(Ph)* Leitungserscheinungen *fpl*
c. test *(El)* Leitfähigkeitsprüfung *f*; Fehlerbestimmung *f* aus dem Spannungsabfall
c.-type moisture meter Leitfähigkeits-Feuchtemesser *m*
c. water *(Ch)* Leitfähigkeitswasser *n*
conductometer Leitwertmesser *m*
conductor 1. *(elektrischer)* Leiter *m*, Stromleiter *m*; 2. Ableiter *m*; Abführrohr *n*; 3. Schaffner *m*; Dirigent *m*
c. break cut-out *(El)* Leiterbruchausschalter *m*, Drahtbruchausschalter *m*, Leiterbruchwächter *m*, Drahtbruchwächter *m*
c. constant *(El)* Leitungskonstante *f*
c. element *(El)* Teilleiter *m*
c. joint *(El)* Leitungsverbinder *m*; Leiterverbinder *m*
c. lead Stromzuführung *f*, Zuleitung *f*
c. of a cable Kabelader *f*
c. of heat Wärmeleiter *m*
c. pipe Leitrohrtour *f*, Standrohr *n (Erdöl)*
c. rail Stromschiene *f*
c. rail anchor Stromschienenverankerung *f*
c. rail system Stromschienensystem *n*
c. size *(El)* Leiterdicke *f*
conductor's ticket punch Lochzange *f (Schaffner)*
conduit 1. Abzug[kanal] *m*; Leitung[srohr *n*] *f*, Röhre *f*, Rohr[leitung *f*] *n*; 2. *(El)* Kabelkanal *m*, Kabelrohr *n*; Isolierrohr *n*, I-Rohr *n*, Schutzrohr *n*
c. bend Normalbogen *m (I-Rohr)*, Rohrbogen *m*, Bogenstück *n*, Krümmer *m*
c. bender Rohrbiegemaschine *f*
c. bending pliers I-Rohrbiegezange *f*, Biegezange *f*
c. bilge tank *(Schiff)* begehbarer Bilgetank *m*
c. box *(El)* Klemmkasten *m*, Verteilerdose *f*
c. clip *(El)* I-Rohrschelle *f*
c. conductor rail system *(El)* Schlitzkanalstromschiene *f*, Unterleitung *f*
c. coupling *(El)* Muffe *f (von Schutzrohren)*
c. elbow Knierohr *n*, Kniestück *n*; Bogen-Winkelstück *n (von Schutzrohren)*
c. fittings Rohrleitungszubehör *n*
c. rack *(El)* Registerschelle *f*, Sammelschelle *f*
c. supporting clip *(El)* Hängeschelle *f (für I-Rohr)*
c. tee T-Stück *n (von Schutzrohren)*
c. wire *(El)* Rohrdraht *m*, Einziehdraht *m*
c. wiring *(El)* Verlegung *f* in Isolierrohr
cone 1. *(Math, Ph)* Kegel *m*, Konus *m*; Lichtkonus *m*; 2. Zapfen *m (Nadelbäume)*; 3. *(Text)* Spule *f*
c. aerial Kegelantenne *f*, Konusantenne *f*, Schirmantenne *f*
c. anchor Keilanker *m*, Konusanker *m*
c. apex to back Abstand *m* des Kegelscheitels bis zur Rückenfläche
c. apex to crown Abstand *m* des Kegelscheitels bis zur äußeren Kante des Kopfkegels
c.-baffle classifier *(Ch)* Kegelfilter *n*, Umlenksichter *m*
c. bearing *(Masch)* Kegellager *n*
c. belt Keilriemen *m*
c. block Ausrichtkeil *m*
c. brake Kegelbremse *f*
c. breaker Einweichtrommel *f*, Knetmühle *f*, Papierzerfaserer *m*
c. centre *(Masch)* Kegelscheitel *m*, Kegelspitze *f*
c. chucking ring *(Holz)* Einsatzring *m (Faß)*

c. **clutch** Kegel[reibungs]kupplung *f*
c. **crusher** Kegelbrecher *m*, Kreiselbrecher *m*
c. **diaphragm** Konusmembran[e] *f*
c. **distance** äußere Teilkegellänge *f (Zahnrad)*; Spitzenentfernung *f*
c. **drive** Stufenscheibenantrieb *m*
c. **friction clutch** Kegelreibungskupplung *f*
c. **gear** Globoidzahnrad *n*
c.**-hardness test** Kegelhärteprüfung *f (Werkstoffprüfung)*
c. **head lathe** Stufenscheibendrehmaschine *f*
c. **head rivet** Kegelniet *m*
c. **impeller** Kegelkreiselmischer *m*, Konuskreiselmischer *m*
c. **indentation test** Kegeldruckprobe *f*
c. **mill** Trichtermühle *f*
c. **nut** Konusmutter *f*
c. **of fire** *(Mil)* Geschoßgarbe *f*
c. **of friction** Reibungskegel *m*
c. **of light** Lichtkegel *m*
c. **of rays** Strahlenkegel *m*, Strahlenbündel *n*
c. **penetration test** Kegeleindringversuch *m (Werkstoffprüfung)*
c. **pin** Haltestift *m* am Spulenteller *(Strickmaschine)*
c. **plate** 1. *(Text)* Spulenteller *m*; 2. Brille *f (Drechslerei)*
c. **plate bar** *(Text)* Tragestange *f* des Spulentellers
c. **point** Kegelansatz *m*; Ansatzspitze *f*; Kegelzapfen *m*
c. **point screw** Schraube *f* mit Ansatzspitze
c.**-point setscrew** Druckschraube *f* mit Spitze
c. **pulley** kegeliges Antriebsrad *n*, Kegel[riemen]scheibe *f*, Stufenrad *n*, Stufenscheibe *f*
c. **pulley drive** Stufenscheibenantrieb *m*
c. **pulley drive with back gear** Stufenscheibenantrieb *m* mit Zahnrädervorgelege
c.**-pulley transmission** Riemengetriebe *n*
c. **shaped** kegelig, kegelförmig, konisch
c.**-shaped needle** *(Text)* kegelförmig zulaufende Nadel *f*
c. **shell** Kegelschale *f*
c.**-thrust test** Kegeldruckprobe *f (Härteprüfung)*
c.**-type crusher** *s.* c. crusher
c.**-type face milling cutter** Messerkopf *m* mit schräg eingesetzten Messern
c. **valve** Kegelventil *n*, konisches Ventil *n*
c. **warping machine** *(Text)* Konusschärmaschine *f*
c. **winder** *(Text)* Kreuzspulmaschine *f*
c. **worm** Globoidschnecke *f (Verzahnung)*
cones Zäpfchen *npl (Retina)*
confectioning Konfektionierung *f*
confer / to konferieren; übertragen
c. **a right on somebody** jemandem ein Recht verleihen
c. **the authority on somebody** jemanden bevollmächtigen
conference call *(Fmt)* Sammelgespräch *n*, Konferenzgespräch *n*
c. **calling equipment** *(Fmt)* Rundgesprächseinrichtung *f*, Sammelgesprächseinrichtung *f*
c. **circuit** *(Fmt)* Sammelverbindung *f*
c. **connection** *(Fmt)* Konferenzschaltung *f*, Rundgesprächsanlage *f*, Sammelschaltung *f*
confidence Sicherheit *f (Statistik)*
c. **interval** Sicherheitsbereich *m*, Sicherheitsintervall *n*; Vertrauensbereich *m (Statistik)*
c. **limit** Vertrauensgrenze *f*, Sicherheitsschwelle *f (Statistik)*
configurated glass Kathedralglas *n*
configuration *(Ph, Ch)* Bildung *f*, Konfiguration *f*, Strukturschema *n*, Zusammensetzung *f*
c. **control** Konfigurationsregelung *f*, Regelung *f* durch Änderung der Brennstoffelementanordnung *(Reaktor)*
c. **interaction** Konfigurationswechselwirkung *f (Quantentheorie der chemischen Bindung)*
confined begrenzt, beschränkt, behindert; geschlossen *(Flüssigkeit)*
c. **compression test** *(Bw)* Druckversuch *m* mit behinderter Querdehnung *(Bodenmechanik)*
c. **compressive strength** *(Bw)* Druckfestigkeit *f* bei behinderter Querdehnung *(Bodenmechanik)*
c. **fluid** eingeschlossenes Medium *n*
c. **pitting** zusammenfließende grübchenförmige Abzehrungen *fpl (Kesselkorrosion)*
confocal *(Math, Ph)* konfokal, mit gemeinsamem Brennpunkt
conform to standards / to den Standards (Normen) entsprechen
conformal mapping *(Math)* konforme Abbildung *f*
confused sea wilde (durcheinanderlaufende) See *f*
confusion *(Opt)* Unschärfe *f*
c. **reflector** Störreflektor *m*
c. **region** Verwaschungsgebiet *n (Radar)*
congeal / to *(Ch)* erstarren, gerinnen, koagulieren
congealing of a melt Erstarrung *f* einer Schmelze, Einfrieren *n*
c. **point** *(Ch)* Erstarrungspunkt *m*, Koagulierungspunkt *m*
congealment Gerinnen *n*
congelation Erstarrung *f*; Gerinnen *n*
congeries Komplex *m*; Anhäufung *f*
congestion meter *(Fmt)* Überlastungszähler *m*
conglomerate *(Geol)* Konglomerat *n*, Trümmergestein *n*
congruence Kongruenz *f*
congruent deckungsgleich, kongruent
conic section Kegelschnitt *m*

conical kegelig, kegelförmig, konisch
c. aerial Kegelantenne *f*, Konusantenne *f*
c. bearing *(Masch)* Spitzenlager *n*
c. buoy *(Schiff)* Spitzboje *f*, Kegeltonne *f*
c. classifier *(Brg)* Klassierspitze *f*
c. clutch Kegelkupplung *f*, Reibungskegelkupplung *f*
c. cock Kegelhahn *m*
c. coil spring Kegelfeder *f*
c. dense-medium bath *(Brg)* Konussinkscheider *m (Aufbereitung)*
c. diaphragm Konusmembran[e] *f*
c. drum kegelförmige Trommel *f (Turbine)*
c.-drum hoist *(Brg)* Fördermaschine *f* mit konischer Trommel
c. ended screw Gewindestift *m* mit Kegelansatz
c. feedwheel *(Text)* konisches ausgleichendes Fournisseurrad *n*
c. frustum Kegelstumpf *m*
c. hopper *(Brg)* Spitztrichter *m*
c. indentation hardness Kegeldruckhärte *f*
c. mixer *(Ch)* Doppelkonus-Trommelmischer *m*
c. pivot Spitzzapfen *m*
c. refiner *(Holz)* Kegelrefiner *m*, Kegelmühle *f*
c.-roller bearing Kegelrollenlager *n*
c. seat[ing] konischer Sitz *m*
c. settling tank *(Brg)* Klärspitze *f*
c. [spiral] spring Kegelfeder *f*
c. valve Kegelventil *n*
c. wheel Kegelrad *n*
c. widening kegelförmige (konische) Erweiterung *f (z. B. Düse)*
conicity Kegeligkeit *f*; Verjüngung *f*
conicoid *(Math)* Fläche *f* zweiter Ordnung
coniferous wood Nadelholz *n*, Weichholz *n*
conjugate / to *(Masch)* paaren
conjugate 1. formschlüssig; 2. *(Math)* konjugiert
c. axis Nebenachse *f (einer Hyperbel)*
c. complex konjugiert-komplex
c. complex number konjugiert-komplexe Zahl *f*
c. impedances konjugiert-komplexe Impedanzen *fpl* (Scheinwiderstände *mpl*)
c. transfer coefficient (constant) konjugiert-komplexes Übertragungsmaß *n*
conjugated 1. zugeordnet; 2. konjugiert
c. double bonds *(Ch)* konjugierte Doppelbindung *f*
conjunction Konjunktion *f (logischer Operator)*
conjunctive normal form *(BMSR)* konjunktive Normalform *f (Aussagenlogik)*
connate water *(Geol)* fossiles Wasser *n*
connect / to *(El)* [ein]schalten; koppeln; verbinden
c. ahead vorschalten
c. directly galvanisch verbinden
c. in parallel nebeneinanderschalten, parallelschalten, nebeneinander (parallel) verbinden
c. in series hintereinanderschalten, in Reihe (Serie) schalten, hintereinander (in Reihe, in Serie) verbinden
c. in series parallel gemischt schalten (verbinden)
c. through durchschalten
c. to earth erden, an Erde legen, mit Erde verbinden
c. to frame mit dem Gehäuse verbinden, an Masse legen
c. up as in diagram nach Schaltbild schalten
connected load *(El)* Anschlußwert *m*
c. wheel turbine Einwellenturbine *f*
connecting block *(El)* Anschlußleiste *f*, Klemmleiste *f*, Schaltleiste *f*
c. bolt Verbindungsschraube *f*
c. box *(El)* Verbindungsmuffe *f*, Anschlußdose *f*
c. branch *(Wärm)* Anschlußstutzen *m*
c. cable *(El)* Anschlußkabel *n*, Verbindungskabel *n*
c. circuit Verbindungsstromkreis *m*
c. clamp *(Kfz)* Anschlußstück *n (z. B. am Tachometer)*
c. contact *(El)* Anschlußkontakt *m*
c. cord *(El)* Verbindungsschnur *f*
c. coupling Anschlußkupplung *f*
c. curve Übergangsbogen *m*; Übergangskurve *f*
c. eye Angriffsauge *n (leichte Weiche)*
c. flex *(El)* Verbindungsschnur *f*
c. jack *(Fmt)* Verbindungsklinke *f*
c. lead *(El)* Anschlußleitung *f*
c. line *s.* c. lead
c. link Bindeglied *n*, Verbindungsglied *n*, Verbindungsgelenk *n*
c. link plate Seitenlasche *f* des Verbindungsgliedes *(Rollenkette)*
c. longeron *(Rak)* Trennholm *m*
c. lug *(Masch)* Verbindungsstift *m*
c. passage 1. Verbindungsgang *m*; Verbindungsweg *m*; 2. *(Brg)* Verbindungsstrecke *f*
c. piece Mittelstück *n*, Verbindungsstück *n*
c. pin *(Masch)* Haltebolzen *m*
c. pipe Anschlußrohr *n*, Verbindungsleitung *f*, Verbindungsrohr *n*
c. plate Verbindungsblech *n*
c. plug Verbindungsstöpsel *m*
c. rack *(Fmt)* Verbindungsgestell *n*
c. relay Durchschaltrelais *n*
c. rivet Anschlußniet *m*
c. rod Gestänge *n*, Pleuel[stange *f*] *m*; Schubstange *f*; Verbindungsstange *f*; Druckstelze *f (Presse, Schere)*; Koppel *f*
c.-rod angle Pleuelstangenausschlag *m*

c.-rod assembly Kurbelantrieb *m*
c.-rod bearing Pleuellager *n*
c.-rod blade Pleuelschaft *m*
c.-rod bolt Pleuelschraube *f*
c.-rod head Schubstangenkopf *m*, Pleuel[stangen]kopf *m*
c. rod of brake-lever *(Kfz)* Bremszugstange *f*
c. rod of four-wheel steering *(Kfz)* Koppelstange *f*
c. rod of steering assembly Lenkverbindungsstange *f*
c.-rod pressure Schub[stangen]kraft *f*
c.-rod shank Pleuelschaft *m*
c.-rod strap Pleuelstangenbügel *m*
c. set *(El)* Anschlußgerät *n*
c. shaft *(Masch)* Verbindungswelle *f*
c. socket *(El)* Anschlußbuchse *f*
c. strip *(Fmt)* Anschlußstreifen *m*; Klemm[en]leiste *f*
c. terminal *(El)* Anschlußklemme *f*
c. through *(Fmt)* Durchschaltung *f*
c.-up *(Fmt)* Beschaltung *f*; Verdrahtung *f*
c. wire *(El)* Anschlußdraht *m*, Verbindungsleitung *f*, Anschlußleitung *f*, Rangierdraht *m*, Schaltdraht *m*
c. zone *(Rak)* Trennstelle *f*

connection *(El)* Verbindung *f*; Schaltung *f*; Anschluß *m*; Motorklemme *f*
c. angle *(Masch)* Befestigungswinkel *m*
c. block *(El)* Anschlußleiste *f*, Klemm[en]leiste *f*
c. box *(El)* Anschlußdose *f*, Schaltdose *f*, Abzweigdose *f*
c. contact *(El)* Anschlußkontakt *m*
c. diagram *(El)* Anschlußplan *m*, Schaltplan *m*
c. in opposition *(El)* Gegenschaltung *f*
c. nut Überwurfmutter *f*
c. of a recorder Schreiberanschluß *m*
c. of rim segments Kranzverbindung *f (des Schwungrades)*
c. plate *(El)* Anschlußplatte *f*, Klemm[en]platte *f*
c. sleeve Verbindungsmuffe *f*
c. zone Trennstelle *f*

connective *(BMSR)* Verknüpfung[szeichen *n*] *f (logisch)*
c. operation *(BMSR)* Verknüpfungsoperation *f (logisch)*

connectivity Zusammenhang *m*

connector *(El)* Anschlußklemme *f*, Klemme *f*, Verbindungsklemme *f*; Stecker *m*, Gerätestecker *m*; Steckverbindung *f*; Kopplungsstück *n (Wellenleiter)*; *(Bw)* Dübel *m*, Industriedübel *m (Holzbau)*
c. bar Verbindungsschiene *f (Batterie)*
c. socket Steckdose *f*

connexion *s.* connection

conning Schiffsführung *f (durch Kommandos an den Rudergänger)*
c. tower hatch Turmluk *n (U-Boot)*

conode Glockenkurve *f*, Konode *f (insbesondere Darstellung von Phasengleichgewichten)*

conoidal shell Konoidschale *f*

conrod *(Masch)* Pleuel[stange *f*] *m*

consecutive aufeinanderfolgend

consequent pole *(Elm)* Folgepol *m*
c. reaction *(Ch)* Folgereaktion *f*

conservations laws *(Ph)* Erhaltungssätze *mpl*
c. of angular momentum Erhaltung *f* des Drehimpulses, Drehimpulserhaltung *f*
c. of charge Erhaltung *f* der Ladung, Ladungserhaltung *f*
c. of energy Erhaltung *f* der Energie, Energieerhaltung *f*
c. of momentum Erhaltung *f* des Impulses, Impulserhaltung *f*
c. of parity Erhaltung *f* der Parität, Paritätserhaltung *f*

conservative flux *(Ph)* quellenfreier Fluß *m*
c. system ungedämpftes System *n*

conservatory 1. Konservatorium *n*; 2. Gewächshaus *n*

consist of / to bestehen aus

consistence Konsistenz *f (Beton usw.)*
c. control Stoffdichteüberwachung *f*
c. regulator Stoffdichteregler *m*

consistency Konsistenz *f (Beton usw.)*; Verträglichkeit *f*
c. index Konsistenzzahl *f (Beton; Bodenmechanik)*
c. of equations *(Math)* Verträglichkeit (Widerspruchsfreiheit) *f* von Gleichungen
c. of reading *(BMSR)* Reproduzierbarkeit *f* der Anzeige

console *(BMSR)* Bedienungspult *n*
c. receiver Radiotruhe *f*; Fernsehtruhe *f*

consolidation Festigung *f*, Verdichtung *f*, Verfestigung *f*; Bodenfestigung *f*, Bodenverdichtung *f*, Bodenverfestigung *f*

consolute temperature *(Ch)* kritische Lösungstemperatur *f*

consonance *(Ak)* Gleichklang *m*

constancy Konstanz *f*
c. of pitch *(El)* Tonkonstanz *f*

constant Konstante *f*

constant konstant, unveränderlich
c. amplitude konstante Amplitude *f*
c.-capacity motor Hydraulikmotor *m* mit unveränderlichem Schluckvolumen
c. current *(El)* konstanter Strom *m*, Dauerstrom *m*
c.-current charging Ladung *f (der Batterie)* bei konstantem Strom
c.-current dynamo *(Kfz)* stromregelnde Lichtmaschine *f*

c.-current equivalent circuit Kurzschluß-ersatzschaltbild *n*
c.-current modulation *(El)* Anodenspannungsmodulation *f*, Heising-Modulation *f*
c.-current transformer Konstantstrom-Transformator *m*, Transformator *m* für gleichbleibenden Strom
c.-delivery pump nichtverstellbare Pumpe *f*, Konstantförderpumpe *f*
c.-displacement pump Konstantförderpumpe *f*
c. duplicate production *(Masch)* Massenerzeugung *f*, Massenfertigung *f*
c. error systematischer Fehler *m*
c. excitation Konstanterregung *f*
c. expansion Konstantdehnung *f*
c. field Gleichfeld *n*
c. for the material Werkstoffkonstante *f*
c. force konstante Kraft *f*
c. frequency drive *(BMSR)* Frequenzsteuerung *f (Uhren usw.)*
c. level carburettor *(Kfz)* Schwimmervergaser *m*
c. load Dauerlast *f*, Dauerbelastung *f*
c.-mesh gear Getriebe *n* mit ständigem Eingriff
c. multiplier Konstantenmultiplikator *m*, Konstantenmultiplizierer *m (Analogrechner)*
c. of a meter *(El)* Zählerkonstante *f*
c. of armature *(Elm)* Ankerfestwert *m*, Ankerkonstante *f*, Rotorkonstante *f*
c. of gravitation *(Mech)* Gravitationskonstante *f*
c. of measuring instrument Instrumentkonstante *f*
c. of radio-active transformation *(Kern)* Zerfallskonstante *f*
c. potential generator *(Elm)* Konstantspannungsgenerator *m*
c. potential transformer Transformator *m* mit konstanter Primärspannung
c. power *(Kfz)* Dauerleistung *f (Motor)*
c. pressure combustion Gleichdruckverbrennung *f*
c. pressure gas turbine Gleichdruckgasturbine *f*
c. pressure line Isobare *f*, Linie *f* gleichen Druckes
c. screen-grid voltage feste Schirmgitterspannung *f*
c. speed 1. konstante Drehzahl *f (z. B. eines Motors)*; 2. gleichbleibende Geschwindigkeit *f*
c.-speed control Drehzahlstabilisierung *f*
c.-speed drive *(Masch)* Antrieb *m* mit gleichbleibender Geschwindigkeit (Drehzahl)
c.-speed floating control *(BMSR)* Zweilaufregelung *f*
c.-speed governor Isodromregler *m*
c.-speed motor Motor *m* mit konstanter Drehzahl
c.-temperature oven *(Am)* Thermostat *m*
c.-tension dynamo Gleichspannungsdynamo *m*
c. time lag stromunabhängige Verzögerung *f*
c.-torque mit gleichbleibendem Drehmoment
c.-velocity joint *(Kfz)* Doppelgelenk *n*
c. voltage Dauerspannung *f*; konstante (gleichbleibende) Spannung *f*
c.-voltage cell konstantes Element *n (Batterie)*
c.-voltage charging Ladung *f* bei konstanter Spannung
c.-voltage control dynamo *(Kfz)* spannungsregelnde Lichtmaschine *f*
c.-voltage equivalent circuit Leerlaufersatzschaltbild *n*
c.-voltage modulation Vorröhrenmodulation *f*
c.-voltage regulator Stabilisator *m*, Spannungsregler *m*, Spannungskonstanthalter *m*
c. volume combustion *(Rak)* Gleichraumverbrennung *f*
c. volume gas turbine Explosionsturbine *f*
constellation figures projector *(Opt)* Sternbildprojektor *m*
constituent Gefügebestandteil *m*; Komponente *f*
constitution diagram Zustandsschaubild *n*
c. of surface agents *(Text)* Tensidstruktur *f*
constitutional change Zustandsänderung *f*, Gefügeänderung *f*, Gefügeumwandlung *f*
c. diagram Zustandsdiagramm *n*, Zustandsschaubild *n*
c. formula *(Ch)* Strukturformel *f*
constrained behindert, gehemmt, eingeschränkt; erzwungen, Zwangs-; eingespannt *(Balken)*
c. current operation *(Elm)* Betrieb *m* mit erzwungener Stromform; Betrieb mit erzwungener Erregung
c. movement 1. *(Ph)* Bewegung *f* mit Nebenbedingungen, [an Nebenbedingungen] gebundene Bewegung, nichtfreie Bewegung; 2. *(Masch)* Bewegungszwang *m*, Zwanglauf *m*
c. oscillation erzwungene Schwingung *f*, Zwangsschwingung *f*, fremderregte (aufgedrückte) Schwingung
c. system *(Mech)* [an Nebenbedingungen] gebundenes mechanisches System *n*
c. vibration *s.* c. oscillation
constraint 1. Beschränkung *f*, Behinderung *f*; 2. Einspannung *f (Balken)*; 3. Beschränkung *(lineares Programmieren)*
constriction Einschnürung *f*, Verengung *f*, Verjüngung *f*; Zusammenschnürung *f*
c. ratio *(Rak)* Klemmung *f*
construct / to bauen; ausführen; konstruieren
c. a triangle ein Dreieck konstruieren
constructible konstruierbar

construction Konstruktion *f*, Bau *m*, Aufbau *m*, Bauart *f*, Bauweise *f*, Bauausführung *f*, Gestaltung *f*; Bildaufbau *m*
c. car Werkstattwagen *m*
c. drawing Konstruktionszeichnung *f*
c. engineer Bauingenieur *m*
c. engineering Bauwesen *n*; Bautechnik *f*
c. equipment Baugeräte *npl*, Baumaschinen *fpl*
c. glass Bauglas *n*
c. joint Arbeitsfuge *f (Betonbau)*
c. lines of brickwork Steinzeichnung *f*, Profillinie *f (Hochofen usw.)*
c. machinery Baumaschinen *fpl*
c. material Baustoff *m*
c. mixer Baumischer *m*
c. of weave *(Text)* Bindeart *f*
c. plan Eisenplan *m*, Eisenlängsschnitt *m (im Schiffbau)*
c. plant *s.* c. machinery
c. project Bauvorhaben *n*
c. schedule Bauablaufplan *m*
c. site Bauplatz *m*, Baustelle *f*
c.-site crane Baustellenkran *m*
c. timber Bauholz *n*
c. way Bauzufahrt *f*
constructional characteristic Konstruktionsmerkmal *n*
c. feature Bauart *f*, Konstruktionsmerkmal *n*
c. gluing Montageleimung *f*, Verkleben *n* von Konstruktionsteilen
c. length Baulänge *f*
c. section Bauspant *n*
c. sketch Konstruktionsentwurf *m*, Konstruktionsskizze *f*
c. steel Baustahl *m*
c. steelwork Stahlbau *m*
constructor's number Baunummer *f*, Werknummer *f*
consumable electrode abschmelzbare Elektrode *f*
consume / to verbrauchen
consumer goods Verbrauchsgüter *npl*, Konsumgüter *npl*
consumer's cable Hausanschlußkabel *n*
c. terminals Verbraucher[anschluß]klemmen *fpl*; Ausgangsklemmen *fpl* des Zählers
consumption 1. Verbrauch *m (Kraftstoff)*; 2. Bedarf *m (Leistung)*
c. figures Verbrauchsziffern *fpl*
contact / to 1. berühren; andrücken, anlegen, drücken; tragen *(Gewindeflanken)*; 2. *(Flg)* aufsetzen *(bei der Landung)*
contact *(El)* Kontakt *m*, Berührung *f*, leitende Verbindung *f*; Kontakt[stück *n*], Schaltstück *n*, Pol *m*
c. acid Kontakt[schwefel]säure *f*
c. adhesive Haftkleber *m*
c. adjustment Kontakteinstellung *f*
c. angle 1. *(El)* Kontaktwinkel *m*; 2. Umschlingungswinkel *m (Riemenantrieb)*
c. area 1. Berührungsfläche *f*, Kontaktfläche *f*; 2. Traganteil *m (Oberflächengüte)*
c. arm 1. Meßbügel *m*; 2. *(Kfz)* Unterbrecherhebel *m*
c. assemblies *(El)* Kontaktfedersätze *mpl*
c. bank Kontaktbank *f*, Kontaktreihe *f*
c. bead *(El)* Kontaktperle *f*
c. bed Tropfkörper *m (Gesundheitstechnik)*
c. between two lines *(El)* Leitungsberührung *f*
c. blade Kontaktmesser *n (eines Schalters)*
c. box Kontaktdose *f*
c. breaker Unterbrecher *m*, Ausschalter *m*
c. breaker cam Unterbrechernocken *m*, Kontaktnocken *m*
c.-breaker contact Unterbrecherkontakt *m*
c.-breaker gap Unterbrecherkontaktabstand *m*, Öffnungsspalt *m (Zündverteiler)*
c.-breaker point Unterbrecherkontakt *m*
c. breaking device Abschaltvorrichtung *f*
c. bridge Kontaktbrücke *f*
c. brush *(El)* Kontaktbürste *f*
c. bush *(Elm)* Kontakthülse *f*
c. carbon *(El)* Kontaktkohle *f*
c. carriage *(El)* Kontaktschlitten *m*
c. centre *(Kfz)* Auflagepunkt *m (Reifen)*
c. clearance *(El)* Kontaktabstand *m*, Kontaktweite *f*
c. collecting bar *(El)* Kontakt[sammel]schiene *f*; Sammelschiene *f* mit Abgriffen; *(selten)* Fahrstromschiene *f*
c. conductor *s.* c. electrode
c. current breaking capacity *(El)* Ausschaltvermögen *n* des einzelnen Kontaktes
c. deposit *(Geol)* Kontaktlagerstätte *f*, Kontaktablagerung *f*
c. detector *(El)* Kontaktdetektor *m*
c. distance *(El)* Kontaktabstand *m*; Schaltstrecke *f (zwischen den Kontakten)*
c. distillation *(Ch)* Destillation *f* über Adsorptionsmittel, Kontaktdestillation *f*
c. drop *(El)* Kontakt-Spannungsabfall *m*
c. electricity Berührungselektrizität *f*, Kontaktelektrizität *f*
c. electrode Kontaktelektrode *f*; Eisenpulverelektrode *f*, Hochleistungselektrode *f*
c. electromotive force Kontakt-EMK *f*
c. emf *s.* c. electromotive force
c. face Berührungsfläche *f*; Meßfläche *f*; Ansatz *m*, Anschlag *m*, Tastfläche *f*
c. fault *(El)* Kontaktfehler *m*
c. file Kontaktfeile *f*
c. filtration *(Ch)* Filtration *f* durch eine Adsorptionsschicht, Kontaktfiltration *f*
c. finger *(El)* Kontaktfinger *m*

c. fluid *(El)* Kontaktflüssigkeit *f*
c. force *(Met)* Anpreßkraft *f*
c. friction Kontaktreibung *f*
c. gap *(El)* Kontaktabstand *m*; Kontaktöffnung *f*
c. head *(BMSR)* Meßkopf *m*
c. heating surface *(Wärm)* Berührungsheizfläche *f*, Kontaktheizfläche *f*
c. image Direktaufnahme *f (Röntgenfotografie)*
c. insecticide Kontaktinsektizid *n*; Kontaktgift *n*, Berührungsgift *n*
c. jaw Einspannbacke *f (Schweißen)*
c. laminating *(Plast)* Schichtstoffherstellung *f* unter niedrigem Druck
c. lens *(Opt)* Haftschale *f*, Haftglas *n*
c. lever Kontakthebel *m*
c. lights *(Flg)* Start- und Landebahnfeuer *n*, Start- und Landebahnbefeuerung *f*
c. line 1. Berührungslinie *f*; 2. Fahrdraht *m*, Fahrleitung *f*
c. loss *(El, Fmt)* Übergangsverlust *m*
c. maker *(El)* Schaltstück *n*, Kontaktstück *n*
c. making *(El)* Kontaktgabe *f*
c. making clock Schaltuhr *f*
c. making voltmeter Kontaktvoltmeter *n*
c. material Kontaktwerkstoff *m*
c. mechanism Schaltglied *n*
c. member *(El)* Kontaktstück *n (Relais)*; Schaltstück *n (Schalter)*
c. metal Kontaktmetall *n*; metallischer Kontaktwerkstoff *m*
c. microphone Kontaktmikrofon *n*
c. moulding *(Plast)* Niederdruckpreßverfahren *n*
c. noise *(El, Fmt)* Kontaktrauschen *n*; Kratzgeräusch *n*
c. of belt and pulley Umschlingung *f*
c. on line *(El)* Drahtberührung *f*
c. path *(El)* Kontaktbahn *f*, Kontaktstrecke *f*
c. pin *(El)* Sockelstift *m*, Kontaktstift *m*, Steckerstift *m*
c. piston *(El)* Kontaktkolben *m*
c. plane Berührungsebene *f*
c. plate *(El)* Bodenkontakt *m*, Kontaktplättchen *n (Lampe)*
c. plating *(El)* Kontaktplattierung *f*
c. plug *(El)* Kontaktstöpsel *m*
c. plunger *s.* c. piston
c. point 1. Berührungspunkt *m*; 2. Prüfstelle *f*; Meßspitze *f*; 3. Unterbrecherkontakt *m*
c. poison Berührungsgift *n*, Kontaktgift *n*
c. potential *(El)* Berührungspotential *n*, Kontaktpotential *n*
c. potential barrier Kontaktpotentialwall *m (Halbleitertechnik)*
c. potential difference *(El)* Kontaktspannung *f*, Kontaktpotentialdifferenz *f*
c. pressure *(Masch)* Anpreßdruck *m*; Meßdruck *m*; Anpressung *f*
c. pressure moulding *(Plast)* Niederdruckpreßverfahren *n*
c. print Kontaktabzug *m*, Kontaktkopie *f (Fotografie)*
c. radiation therapy Nahbestrahlungstherapie *f*; Kontakttherapie *f*
c. rail Stromschiene *f*
c. rail gap Stromschienentrennung *f*, Stromschienenabstand *m*
c. rail gauge Stromschienenlichtraumprofil *n*
c. ratio *(Masch)* Überdeckungsgrad *m*
c. rectifier Sperrschichtgleichrichter *m*
c. resin *(Plast)* Harz *n* für das Niederdruckpreßverfahren
c. resistance *(El)* Übergangswiderstand *m*, Kontaktwiderstand *m*
c. ring *(El)* Kontaktring *m*
c. rock *(Geol)* Kontaktgestein *n*
c. roller 1. *(El)* Kontaktrolle *f*; 2. *(Masch)* Führungsrolle *f*
c. rolling test Wälzprüfung *f (von Zahnrädern)*
c. rule Anlagelineal *n*
c. screen *(Typ)* Kontaktraster *m*
c. screw *(El)* Kontaktschraube *f*
c. series *(El)* Spannungsreihe *f*
c. set *(El)* Kontaktsatz *m*
c. shoe *s.* c. slipper
c. slipper Stromabnehmerlöffel *m*
c. spraying *(Plast)* Niederdruckspritzverfahren *n*
c. spring Kontaktfeder *f*
c. strips Schleifleisten *fpl (Stromabnehmer)*
c. stud *(El, Fmt)* Kontaktknopf *m*, Kontaktklotz *m*, Schaltstück *n*
c. surface Kontaktfläche *f*, Berührungsfläche *f*
c. tension *s.* c. potential
c. thermography Kontaktthermografie *f (Lumineszenzverfahren zur Temperaturmessung)*
c. tip *(Opt)* Meßhütchen *n*
c. to earth Erdschluß *m*
c. touch *(El)* Kontaktberührung *f*
c. travel Kontaktweg *m*, Schaltweg *m*, Schaltstrecke *f*
c. tube Schaltröhre *f*, Umschaltröhre *f*
c. voltage Berührungsspannung *f*, Kontaktspannung *f*
c. voltmeter Kontaktvoltmeter *n*
c. wear Kontaktabnutzung *f*
c. wheel Andrückscheibe *f*
c. winding *(Elm)* Schablonenwicklung *f*
c. wire *(El)* Fahrdraht *m*
c. with a liquid *(El)* flüssiger Kontakt *m*
contacting Kontaktbehandlung *f (Erdöl, Erdgas)*
c. plate *(Ch)* Kontaktboden *m*
contactless berührungsfrei

contactor 1. Relais *n*, Tastrelais *n*; [elektromechanischer] Kontaktgeber *m*; Schaltschütz *n*; Schützenschalter *m*; 2. *(Ch)* Destillationsturm *m*, Extraktionsturm *m*
c. control *(Am) (BMSR)* 1. Mehrpunktregelung *f*; 2. Schützensteuerung *f*
c. control circuit Steuerstromkreis *m*
c. controller *(El)* Schützensteuerwalze *f*
c. equipment *(El)* Schützensteuerung *f*
c. starter *(Elm)* Anlaßschütz *n*
contain / to enthalten
contained plastic deformation eingeschränkte plastische Verformung *f*
container Behälter *m*, Transportgefäß *n*
c. board Schachtelpappe *f*, Behälterpappe *f*
c. ship Behälterschiff *n*
c. vessel Behälterschiff *n*
containerized cargo *(Schiff)* Fracht (Ladung) *f* in Behältern
containing cell Akkumulatorenglas *n*, Sammlergefäß *n*
c. one core einadrig
c. two cores zweiadrig
c. uranium uranhaltig
c. vessel Behälterschiff *n*
containment *(Masch)* Kapselung *f*
c. vessel *(Kern)* Sicherheitsbehälter *m*
contaminant Verseuchungsstoff *m*, Kontaminationsmittel *n*
contaminate / to verschmutzen, verunreinigen; vergiften; [radioaktiv] verseuchen
contaminated iron *(Met)* verunreinigtes Eisen *n*
contamination *(Ch)* Kontamination *f*, Vergiftung *f*; Verschmutzung *f*; Verseuchung *f*; Verunreinigung *f*
c. hazard *(Kern)* [radioaktive] Verseuchungsgefahr *f*
content Gehalt *m*, Inhalt *m*
contents Inhalt *m*; Speicherinhalt *m (Elektronenrechner)*
c. gauge Ölstandanzeiger *m*
context *(BMSR)* Zusammenhang *m (logisch)*
contiguity Aneinanderstoßen *n*, Aneinanderliegen *n*, Aneinandergrenzen *n*
contiguous benachbart, aneinanderliegend, angrenzend, berührend
continental horsepower Pferdestärke *f*; PS
continued fortlaufend; Ketten-
c. fraction *(Math)* Kettenbruch *m*
continuity *(Math)* Stetigkeit *f*; Kontinuität *f*
c. control *(Fs)* Einsatzkontrolle *f*
c. equation *(Ph)* Kontinuitätsgleichung *f*
c. limit *(Ph)* Stetigkeitsgrenze *f*
c. preserving contact unterbrechungsloser Kontakt *m*, Schleifkontakt *m*
c. reinforcement Bewehrungsstahl *m* im Bereich negativer Momente zur Herstellung der Durchlaufwirkung über Stützen usw. *(Stahlbetonbau)*
c. rod Zulageeisen *n*, Zusatzstab *m*; Bewehrungsstab *m* im Bereich negativer Momente zur Herstellung der Durchlaufwirkung über Stützen usw. *(Stahlbetonbau)*
c. steel Bewehrungsstahl *m* im Bereich negativer Momente zur Herstellung der Durchlaufwirkung über Stützen usw. *(Stahlbetonbau)*
c. test Prüfung *f* auf Durchgang *(Kabel, Spulen usw.)*
continuous 1. fortlaufend, ununterbrochen, pausenlos, kontinuierlich; 2. *(Math)* stetig; 3. *(Text)* endlos *(Faden)*
c. absorption *(Ph)* kontinuierliche Absorption *f*, Absorption im Kontinuum
c. absorption coefficient *(Ph)* kontinuierlicher Absorptionskoeffizient *m*
c. action stetige Regelung *f*
c. annealing furnace *(Met)* Durchziehglühofen *m*, Durchlaufglühofen *m*
c. articulated beam Gelenkträger *m*, Gerberträger *m*
c. beam Durchlaufträger *m*, Mehrfeldträger *m*
c. beater Stetigholländer *m (Papierherstellung)*
c. bleaching Kontinue-Bleichen *n (Textilveredlung)*
c. bulk centrifuge *(Ch)* Schälzentrifuge *f*
c. casting Stranggießen *n*
c. casting method *(Met)* Stranggußverfahren *n*
c. charge furnace Schüttrostfeuerung *f*, Füllschachtfeuerung *f*
c. charging grate Schüttrost *m (Feuerung)*
c. chip Fließspan *m*
c. chip material langspanender Werkstoff *m*
c. chip with built-up edge Scherspan *m*
c. concrete mixer kontinuierlich arbeitender Betonmischer *m*
c. control stetige Regelung *f*
c. controller Stetigregler *m*
c. conveyor Stetigförderer *m*
c. core drilling rig fortlaufend kernende Bohranlage *f (Tiefbohrtechnik)*
c. cross-feed veneer splicer *(Holz)* kontinuierlich arbeitende Furnierfugen-Verleimmaschine *f* mit Quervorschub
c. curly chip *(Masch)* Lockenspan *m*, Wendelspan *m*
c. current *(El)* Gleichstrom *m*; stetiger Strom *m*; Dauerstrom *m*
c.-current armature Gleichstromanker *m*
c.-current circuit Gleichstromkreis *m*
c.-current component Gleichstromkomponente *f*
c.-current converter Gleichstromumformer *m*
c.-current excitation Gleichstromerregung *f*

c.-current interrupter Gleichstromunterbrecher *m*
c.-current line Gleichstromleitung *f*
c.-current meter Gleichstromzähler *m*
c.-current motor Gleichstrommotor *m*
c.-current plant Gleichstromanlage *f*
c.-current system Gleichstromsystem *n*
c.-current traction motor Gleichstrom-Bahnmotor *m*
c.-current transformer Gleichstromtransformator *m*
c.-current voltage Gleich[strom]spannung *f*
c.-current voltmeter Gleichspannungsmesser *m*
c. dash signal *(Flg)* Dauerstrich *m*
c. design *(Text)* fortlaufendes Muster *n*, Kontinuemuster *n*
c. diagram fortlaufendes Diagramm *n*
c. discharge *(El)* Dauerentladung *f*
c. discharge heating furnace *(Met)* Durchstoßofen *m*
c. double helical teeth echte Pfeilverzahnung *f*
c. duty Dauerbetrieb *m (mit gleichbleibender Belastung)*
c. dyeing machine *(Text)* Kontinue-Färbeapparat *m*
c. earth Dauererdschluß *m*
c. electrode Dauerelektrode *f*
c. filament yarn Chemieseide *f*
c. fire *(Mil)* Dauerfeuer *n*
c. firing *(Wärm)* ununterbrochene Beschikkung *f*
c. floor *(Schiff)* durchlaufende Bodenwrange *f*
c. fluidized bed *(Ch)* stetig arbeitende Wirbelschicht *f*
c. footing Streifenfundament *n*
c. form guide *(Büro)* Endlosformularführung *f*, Leporellopapierführung *f*
c. frame *(Bw)* Durchlaufrahmen *m*, Mehrfeldrahmen *m*
c. furnace *(Met)* Durchlaufofen *m*, Fließofen *m*
c. girder Durchlaufträger *m*, Mehrfeldträger *m*, durchlaufender Unterzug *m*
c. grinder Stetigschleifer *m*
c. grinding Durchlaufschleifen *n*
c. ground Dauererdschluß *m*
c. high-power caterpillar grinder stetiger Großkraftschleifer *m*, Großstetigschleifer *m*
c. indicator Indikator *m* für fortlaufende Diagramme
c. irradiation kontinuierliche Einstrahlung (Bestrahlung) *f*
c. line Vollinie *f*
c. load 1. Dauerlast *f*, Dauerbelastung *f*; 2. gleichmäßig verteilte Last *f*
c. loading gleichmäßige Bespulung (Pupinisierung) *f (von Kabeln)*, Krarupisierung *f (Kabel)*
c. low-power caterpillar grinder Kleinstetigschleifer *m*
c. mixer kontinuierlich arbeitender Betonmischer *m*
c. normalizing furnace *(Met)* Durchlaufofen *m* zum Normalglühen
c. open-width washer *(Text)* Kontinue-Breitwaschmaschine *f*
c. operation 1. Dauerbetrieb *m*; Fließarbeit *f*, Fließbetrieb *m*; 2. Normallauf *m*, ununterbrochener Lauf *m*
c. oscillation ungedämpfte (kontinuierliche) Schwingung *f*
c. output Dauerleistung *f*
c. pattern *(Text)* durchlaufendes Muster *n*
c. plywood scarf jointing kontinuierliche Sperrholzschäftung *f*
c. power consumption Dauerleistungsaufnahme *f*; kontinuierliche (andauernde) Leistungsaufnahme *f*
c. power output Dauerleistung[sabgabe] *f*; kontinuierliche (andauernde) Leistungsabgabe *f*
c. pusher-type furnace *(Met)* Durchstoßofen *m*, Tunnelofen *m*
c. rating[s] *(Elm, Masch)* [Nenn-]Dauerleistung *f*, [Nenn-]Dauerbelastung *f*
c. recorder *(BMSR)* Linienschreiber *m*
c. repetition work Dauerbetrieb *m*
c. rheostat stetig regelbarer Widerstand *m*
c.-roll reader Leser *m* für durchlaufende Rollen *(z. B. von Kontrollkassen)*
c. roller dryer kontinuierlicher Walzentrockner *m*
c. roller dryer with multi-deck arrangement kontinuierlicher Walzentrockner *m* mit Mehretagenanordnung
c. rolling method Betonwalzverfahren *n*
c. rolling mill kontinuierliche Walzstraße *f*
c. rope drive *(Wärm)* Kreisseiltrieb *m*
c. service Dauerbetrieb *m*
c. shaft durchlaufende Welle *f*
c. shock test *(El)* Durchschlagprüfung *f*
c. slab *(Bw)* Durchlaufplatte *f*, Mehrfeldplatte *f*
c. sound Dauerton *m*
c. spectrum kontinuierliches Spektrum *n*
c. speed Motordrehzahl *f* bei Dauerleistung
c. spinning *(Text)* Drosselspinnerei *f*
c. spinning system *(Text)* kontinuierliches Spinnverfahren *n*, Kontinue-Spinnverfahren *n*
c. stress Dauerbeanspruchung *f*
c. supply ununterbrochene Versorgung (Zufuhr) *f*
c. test Prüfung *f* bei Dauerentladung *(Batterie)*

c. tone Dauerton *m*
c. tractive effort *s.* tractive effort at continuous rating
c.-type furnace *(Met)* Stoßofen *m*, Tunnelofen *m*
c.-type mixer *(Bw)* kontinuierlich arbeitender Mischer *m*
c. use Dauerbetrieb *m*
c. vat dyeing *(Text)* Färben *n* aus der Kontinuekufe
c. voltage Gleich[strom]spannung *f*
c. wave *(Ph)* ungedämpfte Welle *f*
c. wave signal *(Flg)* Dauerstrich *m*
c. wing *(Flg)* durchlaufender Flügel *m*
c. work Dauerbetrieb *m*
c. X-ray radiation Bremsstrahlung *f*, kontinuierliche Röntgenstrahlung *f*
continuously adjustable inductor stetig regelbare Induktivität *f*, Variometer *n*
c. adjustable resistor stetig regelbarer Widerstand *m*
c. loaded mit gleichmäßig verteilter Belastung
c. loaded cable Krarupkabel *n*
c.-running duty *(Elm)* Dauerbetrieb *m* *(mit veränderlicher Belastung)*
c.-running duty with intermittent loading *(Elm)* durchlaufender Betrieb *m* *(mit aussetzender Belastung)*
c.-sensitive [cloud] chamber *(Kern)* kontinuierliche Nebelkammer *f*
c. variable stetig (stufenlos) einstellbar, kontinuierlich regelbar
continuum *(Ph, Math)* Kontinuum *n*
c. of numbers Zahlenkontinuum *n*
contort / to verdrehen
contour / to im Umriß fräsen, profilieren, zweidimensional nachformen
contour Form *f*, Profil *n*, Umriß *m*, Außenlinie *f*, Umrißlinie *f*, Umgrenzungslinie *f*
c. boring Umrißbohren *n*
c. controls Umrißnachformsteuerung *f*
c. facing Plannachformdrehen *n*
c. forming Formdrehen *n*; Nachformen *n*
c. grinding Formschleifen *n*, Unrundschleifen *n*
c. lighting *(El)* Konturbeleuchtung *f*
c. line Höhenlinie *f*
c. map Höhenlinienkarte *f*
c. milling Formfräsen *n*
c. of file Feilenform *f*
c. of the keel *(Schiff)* Kielform *f*
c. of tooth Zahnform *f*
c. projector *(Opt)* Profilprojektor *m*
c. saw Kopiersäge *f*
c. sawing attachment Umrißschnitteinrichtung *f*
c. turning Nachformdrehen *n*, Kopierdrehen *n*
contouring 1. Nachformen *n*; 2. Bahnsteuerung *f* *(Werkzeugmaschinensteuerung)*
c. lathe Nachformdrehmaschine *f*
c. machine Kopier[werkzeug]maschine *f*
c. operation Nachformdreharbeit *f*
c. tool Formmeißel *m*
contra coater Gegenlaufstreichmaschine *f* *(Papierherstellung)*
c.-flow Gegenstrom *m*
c.-guide rudder *(Schiff)* Kontraruder *n*, Leitflächenruder *n*
c.-propeller *(Schiff)* Gegenschraube *f*, Kontrapropeller *m*
c.-rotating propellers *(Flg)* gegenläufige Luftschrauben *fpl*
contrabass Kontrabaß *m*, Baßgeige *f*
contract / to sich zusammenziehen, [zusammen]-schrumpfen, schwinden, sich verengen
contract Vertrag *m*, vertragliche Abmachung *f*
c. of employment Arbeitsvertrag *m*
c. of sale Kaufvertrag *m*
c. work Akkordarbeit *f*
contracted passage between blades verengter Schaufelkanal *m* *(Turbine)*
contractible zusammenziehbar
contractile zusammenziehbar
contracting-band cutch Klemmbandkupplung *f*
c. company Baubetrieb *m*, Bauunternehmen *n*
c. firm Baubetrieb *m*, Baufirma *f*
c. jet kontrahierter Strahl *m*
c. nozzle sich verengende Düse *f*, verengte Düse
c. segment *(Masch)* zusammenziehendes Segment *n*
contraction 1. Kontraktion *f*, Schrumpfung *f*, Zusammenziehung *f*; 2. Einschnürung *f* *(Zugversuch)*; 3. Verengung *f* *(Rohr usw.)*
c. allowance Schrumpfzugabe *f*
c. cavity *(Met)* Lunker *m*
c. crack *(Met)* Schrumpfriß *m*
c. joint *(Bw)* Dehn[ungs]fuge *f*
c. of a mould Saugen (Nachsaugen, Fallen) *n* einer Form
c. of jet Strahlkontraktion *f* *(Turbine)*
c. of the gas current *(Kält)* Einziehung *f* des Gasstromes
c. rule Schwindmaßstab *m*
c. strain *(Ph, Met)* 1. Schrumpfdeformation *f*; 2. Schrumpfspannung *f*, Schrumpfbeanspruchung *f* *(Betriebsausdruck)*
c. stress *(Ph, Met)* Schrumpfspannung *f*
contractor Hauptauftragnehmer *m*, Auftragnehmer *m*, Bauunternehmer *m*; Baubetrieb *m*, Bauunternehmen *n*
contractor's plant Baumaschinen *fpl*, Baugeräte *npl*
c. yard Bauhof *m*
contragredience *(Math)* Kontragredienz *f*

contragredient *(Math)* kontragredient
contrast / to Kontrast bilden (machen), kontrastieren
contrast Kontrast *m*; Dynamik *f* *(Rundfunk)*
  c. amplifier *(Fs)* Kontrastverstärker *m*
  c. etching Kornflächenätzung *f* *(Metallografie)*
  c. grade Härtegrad *m* *(Fotopapier)*
  c. of brightness Helligkeitskontrast *m*, Leuchtdichtekontrast *m*
  c. range 1. *(Fs)* Kontrastumfang *m*, Helligkeitsbereich *m*; 2. *(Foto)* Schwärzungsbereich *m*
  c. reduction Kontrastverminderung *f*; Kontrastherabsetzung *f*
  c. screen Kontrastfilter *n*
  c. stop *(Opt)* Kontrastblende *f*
  c. threshold Kontrastschwelle *f*
contrasting print Kontrastdruck *m*
  c. shade *(Text)* abstechender Farbton *m*
contravariance *(Math)* Kontravarianz *f*
contravariant *(Math)* kontravariant
contrivance Instrument *n*, Apparat *m*, Anlage *f*, Einrichtung *f*
control / to 1. kontrollieren, überwachen; 2. regeln, regulieren; schalten; steuern; 3. bekämpfen; verhüten *(Seuchen usw.)*
control 1. Kontrolle *f*, Überwachung *f*; 2. Regulierung *f*, Regelung *f*, Steuerung *f*; 3. Regler *m*; 4. Bekämpfung *f*, Verhütung *f* *(von Seuchen usw.)*
  c. accuracy Regelgenauigkeit *f*
  c. action Steuerwirkung *f*, Steuerung *f*
  c. algorithm *(BMSR)* Steuer[ungs]algorithmus *m*
  c. area *(BMSR)* Regelfläche *f* *(Zeitintegral der Regelabweichung)*
  c. assembly Regeleinrichtung *f*, Regelanlage *f*
  c. band *s.* proportional band
  c. bars Steuerbrücke *f* *(Buchungsautomaten)*
  c. board 1. Schalttafel *f*; Überwachungstafel *f*; 2. *(Eis)* Führerpult *n*
  c. box Schaltkasten *m*
  c. break Gruppenunterbrechung *f* *(Lochkartenmaschinen)*
  c. brushes obere Bürsten *fpl* *(Lochkartenmaschinen)*
  c. cabinet Schaltschrank *m*
  c. cable Steuerkabel *n*
  c. car Führergondel *f* *(Luftschiff)*
  c. centre 1. Kontrollraum *m*, Kontrollzentrum *n*; Meßwarte *f*; 2. *(Mil)* Gefechtsstand *m*, Leitstelle *f*
  c. characteristic 1. Regelkennlinie *f*, Steuerkennlinie *f*; 2. Gitter-Anodenspannungskennlinie *f*; 3. Zündkennlinie *f*
  c. chart Kontrollkarte *f*
  c. circuit 1. Steuerkreis *m*; 2. Stellstromkreis *m* *(Signalwesen)*
  c.-circuit transformer Steuerkreisumspanner *m*
  c. circuits *(Dat)* Steuerwerk *n*
  c. coil Steuerwicklung *f*, Richtspule *f*
  c. column Steuersäule *f*, Schaltsäule *f*
  c. console 1. Bedienungspult *n*, Mischpult *n*, Schaltpult *n*; 2. *(Schiff)* Steuerstand *m*
  c. criterion *(BMSR)* Gütekriterium *n*, Kriterium *n* der Regelgüte
  c. crystal Steuerquarz *m*
  c. current Regelstrom *m*, Steuerstrom *m*
  c. cut-out switch Steuerstromunterbrecher *m*
  c. desk Steuerpult *n*, Schaltpult *n*, Bedienungspult *n*; Regiepult *n* *(Radio, Fernsehen)*
  c. differential transmitter *(BMSR)* Drehmelderdifferentialgeber *m*
  c. effect Steuerwirkung *f*
  c. electrode Lichtsteuerelektrode *f*; Steuer[ungs]elektrode *f*; Wehnelt-Zylinder *m*
  c. element Bedienteil *n*, Bedienelement *n*; Verstellelement *n*
  c. engineering Regelungstechnik *f*
  c. error *(Flg)* Leitfehler *m*
  c. fin *(Rak)* Leitwert *m*
  c. gear Steuergerät *n*, Steuerung[sgerät *n*] *f*; Schaltwerk *n*; Schaltgeräte *npl*
  c. grid *(El)* Steuergitter *n*
  c. grid of a variable mu-tube *(El)* Regelgitter *n*, Steuergitter *n* einer Regelröhre
  c.-grid plate transconductance Steilheit *f* *(Regelröhren)*
  c. hysteresis *(BMSR, El)* Steuerhysteresis *f*
  c. index setting *s.* desired value
  c. input *(BMSR)* Aufgabewert *m*, Sollwert *m*
  c. instrument *s.* automatic controller
  c. instruments Kontrollinstrumente *npl*
  c. knob Bedienungsknopf *m*; Betätigungsknopf *m*
  c. lamp 1. Überwachungslampe *f* *(Signal)*; 2. *(Kfz)* Kontrolleuchte *f*
  c. lever 1. *(Masch)* Schalthebel *m*, Steuerhebel *m*; Bedienungshebel *m*; 2. Stellhebel *m*, Stelltaste *f*; Stellknopf *m*
  c. limit Kontrollgrenze *f*
  c. linkage *(Kfz)* Steuergestänge *n*
  c. loop Regelkreis *m*, Steuerstromkreis *m*
  c. loudspeaker Kontrollautsprecher *m*, Abhörlautsprecher *m*
  c. magnet Richtmagnet *m*
  c. map *(Flg)* Leitkarte *f*

c. medium *(BMSR)* Stellgröße *f*
c. member *(Masch)* Bedien[ungs]organ *n*, Schaltorgan *n*
c. mode Typ *m* der Regelung, Regelungsgesetz *n*
c. motor Stellmotor *m*, Regelmotor *m*
c. movement Schaltbewegung *f*
c. of solar activity Sonnenüberwachung *f*
c. panel 1. Schalttafel *f*, Steuerschalttafel *f*, Steuertafel *f*, Warte *f*; 2. *(Büro)* Schalttafel *f*, Steckertafel *f*; 3. *(Eis)* Führerpult *n*
c. piston solenoid *(Rak)* Steuerkolbentauchspule *f*
c. plunger Steuerkolben *m*
c. point Meßstelle *f*
c. point adjustment Sollwerteinstellung *f*; Sollwerteinsteller *m*, Sollwertgeber *m*
c.-point setting Sollwerteinstellung *f*
c. points of known position and elevation Festpunkte *mpl* bei Triangulation
c. post Führungsstelle *f*, Leitstelle *f*
c. power Steuerleistung *f*
c. precision *(BMSR)* Reproduzierbarkeit *f* der Regelgröße
c. pressure Steuerdruck *m*
c. quantity Regelgröße *f*
c. rack *(Kfz)* Regelstange *f*
c. range Regelbereich *m*; Steuerhub *m*
c. ratio 1. Führungsübertragungsfunktion *f* des geschlossenen Regelkreises; 2. Steuerfaktor *m (Gasentladungsröhren)*
c. register *(Dat)* Steuerbefehlsspeicher *m*, Steuerbefehlsregister *n*
c. relay 1. Steuerrelais *n*, Steuerschütz *n*; 2. Kontrollrelais *n*
c. response Schaltreaktion *f*
c. rod Regelstab *m (Kernreaktor)*
c. rod calibration Eichung *f* des Regelstabes
c. rod efficiency function Steuer[stab]kennlinie *f*
c.-rod opening Kanal *m* (Öffnung *f*) für den Regelstab
c. room *(BMSR)* Meßwarte *f*
c. section *(BMSR)* Leitwerk *n*
c. sequence *(Dat)* Befehlsfolge *f*
c. setting *s.* desired value
c. slip Kontrollstreifen *m*
c. spindle *(Kfz)* Reglerspindel *f*
c. stage Steuerstufe *f*
c. station Meßwarte *f*; Leitstation *f*, Leitstelle *f*
c. stick *(Flg)* Steuerknüppel *m*
c. surface Ruder *n*, Steuerfläche *f (Flugkörper, Schiff)*
c. switch Steuerschalter *m*; Fahrschalter *m*
c. system Regeleinrichtung *f*, Regelanlage *f*, Reglersystem *n*; Regelkreis *m*; Steuerkette *f*; Steuerungssystem *n (z. B. Raketen)*
c. take-up *(BMSR)* Ausgleichsschlaufe *f (Förderband)*
c. technology Regelungstechnik *f*
c. timer *(BMSR)* Zeitplangeber *m*, Programmgeber *m*
c. transformer Meßwandler *m*, Steuertransformator *m*
c. transmitter 1. Steuersender *m*; 2. *(BMSR)* Drehmeldergeber *m*, Geber *m*
c. unit *s.* controlling unit
c. valve 1. *(Masch)* Regelschieber *m*, Regelventil *n*, Steuerschieber *m*, Steuerventil *n*; 2. *(Fmt, Fs)* Regelröhre *f*, Steuerröhre *f*
c. vane *(Rak)* Druckstück *n*
c. voltage Betätigungsspannung *f*, Steuerspannung *f*
c. winding *(Elm)* Steuerwicklung *f*
c. wire *(Kfz)* Betätigungsseil *n*
c. wiring Steuerleitungen *fpl*
c. word *(BMSR)* Leitzelle *f (besondere Informationsspeicherzelle)*
c. worm Schaltschnecke *f*
c. zone Regelbereich *m*
controllability 1. Regelbarkeit *f*; 2. Steuerbarkeit *f (Schiffe, Flugzeuge, Raketen)*
controllable-pitch propeller *(Schiff)* Verstellpropeller *m*
c. vane Verstelleitschaufel *f (Turbine)*
controlled geregelt, gesteuert
c. anti-aircraft missile gelenkte Fla-Rakete *f*
c.-atmosphere furnace brazing Schutzgaslötung *f*
c. avalanche effect gesteuerter Lawineneffekt *m (in Halbleitervorrichtungen)*
c. by a single lever mit Einhebelsteuerung *f*
c. carriage *(Text)* zwangsläufig geführter Wagen *m*
c. carrier *(El)* [modulationsabhängig] gesteuerter Träger *m*
c. centring *(Opt)* Zwangszentrierung *f*
c. chain reaction *(Kern)* gesteuerte Kettenreaktion *f*
c. condition *(BMSR)* Regelgröße *f*
c. drilling gerichtetes Bohren *n*, Richtbohren *n*, Schrägbohren *n (Erdöl, Erdgas)*
c. gravity stowing *(Brg)* Fließversatz *m*
c. medium *s.* c. condition
c. member *(BMSR)* Regelstrecke *f*
c. mosaic *(Opt)* Bildplan *m (Fotogrammetrie)*
c. operation [elektrisch] gesteuerter Betrieb *m*
c. projectile gelenkte (gesteuerte) Rakete *f*
c. rocket gelenkte (gesteuerte) Rakete *f*
c. spark gesteuerter Funke *m*
c. steam supply kontrollierte (regelmäßig eingestellte) Dampfzufuhr *f*

c. switch Steuerschalter *m*, steuerbare Siliziumzelle *f*, steuerbarer Gleichrichter *m*, Vierschichtentriode *f*
c. system Regelstrecke *f*
c. transmitter Sender *m* mit Steuerstufe
c. variable *s.* c. condition
c.-volume pump Regelpumpe *f*
controller 1. *(Fmt, Fs)* Verstärkungsregler *m*, Regler *m*, Regelgerät *n*; 2. Fahrschalter *m* *(Fahrzeuge, Krane)*; 3. *(Masch)* Steuerwalze *f*, Walzenanlasser *m*; Regelorgan *n*; 4. *(Schiff)* Kettenstopper *m*, Ankerkettenstopper *m* (*s. a.* automatic controller)
c. barrel Schaltwalze *f*
c. cylinder Schaltwalze *f*
c. drum Schaltwalze *f*
c. finger Kontaktfinger *m*
c. lag *(BMSR)* Eigenverzögerung *f* des Reglers
controlling Regelung *f*; Kontrolle *f*
c. by wireless drahtlose Steuerung *f*
c. cam Steuerkurve *f*
c. contact Überwachungskontakt *m (Signal)*
c. current Regelstrom *m*, Steuerstrom *m*
c. magnet Richtmagnet *m*, Stellmagnet *m*
c. means *(BMSR)* Regler *mpl*
c. member *s.* c. unit
c. torque Einstellmoment *n (eines Instrumentes)*
c. unit Regler *m*, Regelgerät *n*
c. valve Dampfeinlaßventil *n*
controls *(BMSR)* Einrichtungen *fpl* zur Regelung und Steuerung; Steuerungen *fpl*; Bedien[ungs]elemente *npl*; Rechenknöpfe *mpl*, Rechnungsknöpfe *mpl*; Bedienungsorgane *npl*; Steuergeräte *npl*, Schaltgeräte *npl*
conurbation Ballungsgebiet *n (Siedlungsplanung)*
convect/to übertragen, überführen *(z. B. Wärme)*
convection Konvektion *f*, Übertragung *f*
c. current *(Ph)* Konvektionsstrom *m*
c. heater *(Bw)* Konvektionsheizkörper *m*
c. of heat Wärmekonvektion *f*, Wärmeübertragung *f*
c. potential *(Ph)* Konvektionspotential *n*
convective discharge *(Elm)* elektrostatischer Wind *m*, Konvektionsentladung *f*
convector Konvektor *m*, Wärmeleitplatte *f* *(Heizung)*
convenience outlet Haushaltgerätesteckdose *f*
c. receptacle Gerätesteckdose *f*, Steckdose *f*
conventional cut *(Masch)* Gegenlauffrässchnitt *m*
c. fuel konventioneller Brennstoff *m (Gegensatz: Kernbrennstoff)*
c. herringbone teeth unechte Pfeilverzahnung *f*
c. milling Gegenlauffräsen *n*
c. teeth *s.* c. herringbone teeth
converge/to konvergieren, zusammenlaufen; sich verengen
convergence 1. *(Math)* Konvergenz *f*; 2. Annäherung *f* von Hangendem und Liegendem *(Gebirgsmechanik)*
c. circuit *(Math)* Konvergenzkreis *m*
c. theorem *(Math)* Konvergenzsatz *m*
convergent 1. *(Math, Ph)* konvergent, konvergierend; 2. sich verengend, verjüngend
c. beam *(Ph)* konvergentes Strahlenbündel *n*
c. beam radiation *(Ph)* Konvergenzbestrahlung *f*
c. sequence *(Ph)* konvergente Folge *f*
converging 1. *(Math, Ph)* konvergent, konvergierend; 2. sich verengend, verjüngend
c.-diverging doppeltrichterförmig
c. lens Sammellinse *f*
conversation *(Fmt)* Gespräch *n*
c. counter Gesprächszähler *m*
c. meter Gesprächszähler *m*
c. time Gesprächszeit *f*
converse piezo-electric effect reziproker piezoelektrischer Effekt *m*, Elektrostriktion *f*
conversion 1. Umwandlung *f*, Verwandlung *f*, Überführung *f*; 2. Umformung *f*; 3. Umrechnung *f (in ein anderes Maßsystem)*; 4. Umbau *m (eines Schiffes für andere Zwecke)*; 5. Umwandlung *f*, Umwicklung *f (elektrischer Geräte)*; Mischung *f (Elektronenröhren)*; 6. *(BMSR)* Konvertierung *f*, Umsetzung *f*; Umkodierung *f*
c. angle Umsetzungswinkel *m*
c. coating Beschichten *n* außerhalb der Maschine
c. coefficient Umrechnungskoeffizient *m*, Umrechnungsfaktor *m*
c. conductance *(El)* Konversionssteilheit *f*, Mischsteilheit *f*, Überlagerungssteilheit *f*
c. constant *(Ph)* mechanisches Wärmeäquivalent *n*
c. detector *(Fmt, Fs)* Mischstufe *f*
c. factor *(Math, Ph)* Umrechnungsfaktor *m*
c. gain *(Fmt, Fs)* Mischverstärkung *f*
c. loss *(Fmt, Fs)* Mischdämpfung *f*
c. mutual conductance *(Fmt, Fs)* Mischsteilheit *f*
c. of energy *(Mech)* Energieumwandlung *f*
c. of potential energy into kinetic energy Umwandlung *f* potentieller Energie in kinetische Energie; Umwandlung der Spannungsenergie in Bewegungsenergie *(Turbine)*
c. ratio Konversionsverhältnis *n*, Umrechnungsverhältnis *n*; Umwandlungsverhältnis *n*
c. table Umrechnungstabelle *f*
c. transconductance *(El)* Konversionssteilheit *f*, Mischsteilheit *f*, Überlagerungssteilheit *f*

convert / to 1. umwandeln, verändern; umbauen; 2. umrechnen *(in ein anderes Maßsystem)*; 3. *(Elm)* umformen, umspannen
c. back zurückverwandeln
c. into power in Energie umsetzen
c. the steam pressure into velocity Dampfdruck in Geschwindigkeit umsetzen *(Turbine)*
c. to umrechnen in
convert Deckel *m*, Verschlußklappe *f*
converted deviation umgeformte Regelabweichung *f*
c. timber Schnittholz *n*
converter 1. *(Met)* Konverter *m*, Bessemerbirne *f*, Frischbirne *f*, Birne *f*; Windfrischapparat *m*; 2. *(Plast)* Verarbeiter *m*; 3. *(El)* Umformer *m*, Wandler *m*; selbstschwingende Mischröhre *f* *(Elektronenröhre)*; Frequenzbandumsetzer *m*; Mischstufe *f*; 4. *(Dat)* Konverter, Wandler, Umsetzer *m*, Umwandler *m*; 5. *(Kern)* Konverter, Konversionsreaktor *m*
c. belly *(Met)* Konverterbauch *m*
c. blast box *(Met)* Konverterwindkasten *m*
c. charge *(Met)* Konvertereinsatz *m*
c. exciter *(El)* Erregerumformer *m*
c. iron *(Met)* Flußstahl *m*
c. lining *(Met)* Konverterfutter *n*
c. matte Feinstein *m* *(Nickelgewinnung)*
c. tube *s.* c. valve
c. valve *(Fmt, Fs)* Mischröhre *f*; Umkehrröhre *f*
c. with differential winding *(Elm)* Sparumformer *m*
convertible Wagen *m* mit zurückklappbarem Verdeck
converting 1. *(Ph)* Konvertieren *n*, Konversion *f*; 2. *(Met)* Windfrischen *n*, Verblasen *n* in flüssigem Zustand
c. furnace Zementierofen *m*
c. station *(Elm)* Umspannwerk *n*, Umformerwerk *n*
c. vessel *(Met)* Frischbirne *f*
convertor *s.* converter
convex 1. *(Math, Opt)* konvex; 2. gebaucht, erhaben
c.-concave lens Konvexkonkavlinse *f*
c. fillet weld Wölbkehlnaht *f* *(Schweißtechnik)*
c. grinding Balligschliff *m*
c. mirror Konvexspiegel *m*
c. tile Mönch *m* *(Dachziegel)*
c. turning attachment Balligdreheinrichtung *f*
c. weld Wölbnaht *f* *(Schweißtechnik)*
convexity Balligkeit *f*
convexo-concave *(Opt)* konvexkonkav
convey / to [be]fördern *(nicht Personen)*
c. current *(El)* Strom zuführen (abführen)
conveyance 1. Förderung *f*, Transport *m*; 2. *(Brg)* Fördergefäß *n*, Fördermittel *n*
c. of coal Anfuhr *f* der Kohlen
c. of steam *(Wärm)* Dampfüberleitung *f*, Dampfüberführung *f*
conveyer *(Am)* *s.* conveyor
conveying belt Förderband *n*, Transportband *n*
c. chute Förderrinne *f*, Förderrutsche *f*
c. dryer Bandtrockner *m*
c. equipment Förderanlage *f*
c. machinery Fördermaschinen *fpl*
c. plant Förderanlage *f*, Fördereinrichtung *f*
conveyor Förderer *m*, Stetigförderer *m*, Zubringer *m*
c. advancing Rücken (Vorrücken) *n* des Förderers
c. belt Förderband *n*, Fördergurt *m*
c.-belt drying machine Bandtrockner *m*
c. belting Förderband *n*, Fördergurt *m*
c. bridge *(Bw)* Förderbrücke *f*; *(Brg)* Bandbrücke *f*
c. chain Becherkette *f*
c. drying machine Bandtrockner *m*
c. line Fließband *n*, Montageband *n*
c. line assembly Fließbandmontage *f*
c. roller *(Typ)* Bogenführungstrommel *f*
c. screw Förderschnecke *f*, Transportschnecke *f*
c. structure *(Brg)* Förderbandgerüst *n*
c. worm Förderschnecke *f*
convolution 1. *(Math)* Faltung *f* *(bei Integralgleichungen)*; 2. Einfaltung *f*; Einrollung *f*; 3. Gang *m*
c. [of cotton fibre] korkenzieherartige Windung *f* [der Baumwollfaser]
c. theorem *(Math)* Faltungssatz *m*
convoy guard *(Mil)* Geleitschutz *m*
cooked überentwickelt *(Film)*
cooking acid Kochsäure *f*
c. chart Kochschema *n*
c. cycle Kreislauf *m* der Kochung, Kochperiode *f*
c. liquor Kochlauge *f*
c. without liquor circulation Kochung *f* ohne Umwälzsystem
cool / to [ab]kühlen
c. back rückkühlen
c. down abkühlen, erkalten; kaltschüren, abtempern
coolant Kühlflüssigkeit *f*, Kühlmittel *n*, Kühlstoff *m*; Wärmeträger *m*
c. circulation Kühlmittelumlauf *m*
c. container Kühlmittelbehälter *m*
c. exit temperature Kühlstoffaustrittstemperatur *f*
c. feed Kühlmittelzuführung *f*
c. filter Kühlmittelfilter *n*
c. flow Kühlmittelfluß *m*
c. for grinding Schleifflüssigkeit *f*
c. gutter Kühlmittelrinne *f*

c. **make-up** *(Kern)* Kühlmittelaufbereitung *f*
c. **mist** Kühlnebel *m*
c. **nozzle** Kühlmitteldüse *f*
c. **pump** Kühlmittelpumpe *f*
c. **reservoir** Kühlmittel[flüssigkeits]behälter *m*
c. **supply** Kühlmittelversorgung *f*; Schmiermittelzufuhr *f*
c. **system** Kühlsystem *n*
c. **tank** Kühlmittelbehälter *m*
c. **temperature gauge** Kühlstofftemperaturmesser *m*
c. **tray** Fangschale *f* für Tropföl
cooler 1. Kühler *m*, Kühlapparat *m*; 2. *s.* cooling floor
c. **crystallizer** Kühlmaische *f*
Coolidge tube Coolidgeröhre *f (älterer Typ einer Röntgenröhre)*
cooling Abkühlung *f*, Kühlung *f*, Erkaltung *f*
c. **action** Kühlwirkung *f*
c. **agent** Kühlmittel *n*, Abkühlmittel *n*, Kältemittel *n*, Kälteträger *m*
c. **air** Kühlluft *f*
c. **attachment** Kühlmitteleinrichtung *f*
c. **baffle** Kühlband *n*, Kühlluftleitblech *n*, Luftleitblech *n*
c. **belt** Kühlband *n (z. B. für Glas)*
c. **block** Kühlerblock *m*
c. **blower** Kühlluftgebläse *n*
c. **by air** Luftkühlung *f*
c. **by evaporation** Verdampfungskühlung *f*
c. **by injection** Einspritzkühlung *f*
c. **by means of a cooling tower** *(Wärm)* Kaminkühlung *f*
c. **by means of circulating water** Wasserumlaufkühlung *f*
c. **by outspread surface** Flächenkühlung *f*
c. **cap** Kühlhaube *f*
c. **chamber** Kühlkammer *f*; *(Text)* Kühlmansarde *f*
c. **coil** Kühlschlange *f*; Kühlwindung *f*
c. **conduit** Kühlgraben *m*
c. **conveyor** Kühlband *n*
c. **crack** *(Met)* Schrumpfriß *m*
c. **crystallizer with stirring device** Kühlrührmaische *f*
c. **curve** Abkühlungskurve *f*, Abkühlungsverlauf *m*
c. **disk** Kühlplatte *f*
c. **down** Abkühlen *n*
c. **drag** der zur Kühlung dienende aerodynamische Widerstandsanteil *m*, Kühl[ungs]-widerstand *m*
c. **ducts** *(Elm)* Kühlschlitze *mpl*
c. **due to expansion** Entspannungsabkühlung *f*
c. **effect** Kühlwirkung *f*
c. **fan** Kühlventilator *m*, Lüfter *m*
c.**-fan increasing gear** Lüfterübersetzungsgetriebe *n*
c. **fin** Kühllamelle *f*, Kühlrippe *f*
c. **flange** Kühlrippe *f*
c. **floor** Kühlstock[werk *n*] *m*
c. **guide vane** Kühlluftleitblech *n*
c. **jacket** Kühlmantel *m*
c. **liquid** Kühlflüssigkeit *f*
c. **of electrodes** Elektrodenkühlung *f*
c. **of piston** Kolbenkühlung *f*
c. **of the commutator** Kommutatorkühlung *f*, Kollektorkühlung *f*
c. **oil** Kühlöl *n*
c. **period** Abkühl[ungs]zeit *f*
c. **pipe** Kühlkanal *m*, Kühlleitung *f*
c. **plant** Kühlanlage *f*, Kühlwerk *n*
c. **plant for low temperature** Tiefkühlanlage *f*
c. **plate box** Kühlnische *f*
c. **pond** Kühlteich *m*, Kühlbassin *n*, Kühlbecken *n*
c.**-power loss** für Kühlzwecke angewendeter Leistungsanteil *m*, Kühlverlustleistung *f*
c. **rate** Abkühl[ungs]geschwindigkeit *f*
c. **rib** Kühllamelle *f*, Kühlrippe *f*
c. **stack** Berieselungsanlage *f (Kühlturm)*; Gradierwerk *n*
c. **stack with fan** Ventilatorgradierwerk *n*
c. **surface** Abkühlungsfläche *f*, Kühlfläche *f*
c. **surface of the cylinder cover** *(Wärm)* abkühlende Oberfläche *f* eines Zylinderdeckels
c. **tank** Kühlgefäß *n*
c. **tower** Gradierwerk *n*; Kühlturm *m*; Kühlwerk *n*
c. **trap** Kühlfalle *f (Vakuumtechnik)*
c. **wall** Kühlwandung *f*
c. **water** Kühlwasser *n*
c.**-water circulating pump** Kühlwasserpumpe *f*
c.**-water circulation** Kühlwasserumlauf *m*
c. **water discharge** Kühlwasserablauf *m*
c.**-water discharge pipe** Kühlwasserableitung *f*
c.**-water failure** Kühlwasserausfall *m*
c.**-water outlet** Kühlwasseraustritt *m*
c. **water pipe** Kühlwasserleitung *f*
c.**-water tank** Kühlwasserbehälter *m*
cooltainer *(Schiff)* Kühlbehälter *m (Transportbehälter)*
coop *s.* Cooper-Hewitt lamp
cooper Böttcher *m*, Küfer *m*, Faßbinder *m*
Cooper-Hewitt lamp Quecksilberdampflampe *f*
Cooper-Hewitt rectifier Quecksilberdampfgleichrichter *m*
cooperage Böttcherei *f*
cooperate / to zusammenarbeiten; zusammenwirken
cooperation Zusammenarbeit *f*

Cooperative Fuel Research Comittee-engine CFR-Motor *m (Kraftstoffprüfung)*
cooper's adze Küferdechsel *f*
coordinate Koordinate *f*
c. bond *(Ch)* Koordinationsbindung *f*
c. boring machine Koordinatenbohrmaschine *f*
c. drilling Koordinatenbohren *n*
c. drilling table Koordinatenbohrtisch *m*
c. lattice Koordinatenraster *m*
c. measuring instrument Koordinatenmeßgerät *n*
c. measuring table Koordinatenmeßtisch *m*
c. paper Netzpapier *n (Millimeterpapier, Logarithmenpapier usw.)*
c. stage Koordinatenmeßtisch *m*
c. table Koordinatenmeßtisch *m*
c. trihedral *(Math)* Koordinatendreibein *n*
coordinately linked *(Ch)* durch Koordinationsbindung gebunden
coordinates of the state system Landeskoordinaten *fpl*
coordination number *(Ch)* Koordinationszahl *f*
c. of insulation *(El)* Koordination *f* der Isolation, Isolationskoordination *f (Festlegung des Isolationspegels)*
cop *(Text)* Kops *m*, Kötzer *m*; Spule *f*
c. dyed *(Text)* im Kops gefärbt
copaiba balsam Kopaivabalsam *m*
copal[s] Kopal[e *pl*] *m*
cope *(Met)* Oberform *f*, obere Formhälfte *f (Formerei)*
cophasal *(El)* gleichphasig
coping Mauerabdeckung *f*, Mauerabschluß *m*, Mauerkrone *f*
c. saw Bogensäge *f*, Laubsäge *f*
c. saw blade Laubsägeblatt *n*
c. saw frame Laubsägebogen *m*
coplanar *(Math)* komplanar, in derselben Ebene, koplanar
copolymer *(Plast)* Mischpolymerisat *n*
copolymerization *(Plast)* Mischpolymerisation *f*
copper / to [ver]kupfern
c. plate verkupfern
copper Kupfer *n*
c. aftertreatment *(Text)* Nachkupfern *n*
c. alloy Kupferlegierung *f*
c. anode Kupferanode *f*
c. ashes Kupferhammerschlag *m*
c. bit Lötkolben[spitze *f*] *m*
c. block *(Typ)* Kupferklischee *n*
c. brazing mixture Schlaglot *n*
c. brush *(Elm)* Kupferbürste *f*
c. cable Kupferkabel *n*
c. cake Kupfer *n* in Platten (Tafeln)
c.-chrome aftertreatment *(Text)* Kupfer-Chrom-Nachbehandlung *f*
c.-clad steel conductor Stahlkupferleiter *m*
c.-clad wire Kupfermanteldraht *m*
c. damper *(Elm)* Wirbelstromdämpfer *m*
c. deposit Kupferlagerstätte *f*
c. engraver *(Typ)* Kupferstecher *m*
c. factor *(Fmt)* Füllfaktor *m*
c. finish *(Text)* Kalandern *n* in Kupferwalzen
c. foil Kupferfolie *f*; Blattkupfer *n*
c. gasket Kupferdichtung *f*
c. glance *(Min)* Kupferglanz *m*
c. loading *(El)* Kupferbeanspruchung *f*, Kupferbelastung *f*
c. loss with direct current Gleichstromverluste *mpl*
c. losses *(El)* Kupferverluste *mpl*
c. nickel *(Min)* Rotnickelkies *m*
c. number Kupferzahl *f (Abbaubestimmung für Zellulose)*
c. oxide Kupferoxid *n*
c. oxide cell *(El)* Kupferoxydulzelle *f*, Kupronelement *n*
c. oxide rectifier *(Elm)* Kupferoxydulgleichrichter *m*
c. packing *(Wärm)* Kupfereinlage *f*
c.-plated carbon *(Elm)* Kupferkohle *f*, verkupferte Kohle *f*
c. plated wire verkupferter (kupferplatierter) Draht *m*
c. plating Verkupfern *n*
c. pyrites *(Min)* Kupferkies *m*
c. rain Streukupfer *n*, Sprühkupfer *n*
c. rayon Kupferseide *f (Zelluloseregeneratseide)*
c. roller Musterzylinder *m* aus Kupfer
c. sheet Kupferblech *n*; Blattkupfer *n*
c. sponge Kupferschwamm *m*
c. strand Kupferlitze *f*
c. strip earth conductor *(El)* Banderder *m*
c. sulphate Kupfersulfat *n*, Kupfervitriol *n*
c. voltameter Kupfervoltameter *n*
c. works Kupferhütte *f*
copperas Eisenvitriol *n*, Eisen(II)-sulfat *n*
coppered block *(Text)* kupferverstärkter Model *m*
c. carbon *s.* copper-plated carbon
coppering *(Met)* Verkupfern *n*
copping Aufwinden *n*
coprecipitation *(Ch)* Mitfällung *f*
c. method Mitfällungsmethode *f*
coproduct *(Ch)* Nebenprodukt *n*, Beiprodukt *n*
coprosterol *(Ch)* Koprosterin *n*
copulate / to kopulieren
copunctal kopunktal, durch denselben Punkt gehend
copy / to nachformen, kopieren; pausen
c.-plane nachformhobeln
c.-turn kopierdrehen
copy Kopie *f*; Exemplar *n*

c.-holder *(Typ)* Tenakel *n*, Vorlagenhalter *m*, Manuskripthalter *m*
c. milling machine Kopierfräsmaschine *f*, Schablonenfräsmaschine *f*
c. paper Durchschlagpapier *n*, Durchschreibpapier *n*
c.-planing attachment Einrichtung *f* zum Nachformhobeln
c.-proof *(Typ)* Korrekturabzug *m*, Bürstenabzug *m*
c. saddle Nachformschlitten *m*
c. tape Kopieband *n*, Kopiestreifen *m (z. B. Lochstreifen, Magnetstreifen)*
c. turning Kopierdrehen *n*
c.-turning attachment Kopierdreheinrichtung *f*
c. turning lathe Kopierdrehmaschine *f*, Nachformdrehmaschine *f*
copying Nachformen *n*, Kopieren *n*, Pausen *n*
c. apparatus *(Opt)* Lichtpauseinrichtung *f*, Kopiereinrichtung *f*
c. attachment Kopiereinrichtung *f*, Nachformeinrichtung *f*
c. cycle Nachformzyklus *m*
c. device Kopiereinrichtung *f*
c. lathe Fassonniermaschine *f*, Kopier[dreh]-maschine *f*, Schablonendrehmaschine *f*, Nachformdrehmaschine *f*
c. master Kopierbezugsstück *n*
c. motion Nachformbewegung *f*
c. saddle Kopierschlitten *m*
c. slide *s.* c. saddle
copyright in effects Musterschutz *m*
coral *(Geol)* Koralle *f*
c. agate *(Min)* Korallenachat *m*
c. rag Korallenkalk *m*
c. reef Korallenriff *n*
corbel / to *(Bw)* 1. auskragen, vorkragen; 2. einen Schornstein ziehen *(Schornsteinmauerwerk)*
corbel *(Bw)* Auskragen *n*, Auskragung *f*, Konsole *f*, Kragstein *m*
c. beam Freiträger *m*, Kragträger *m*
corbelling *(Bw)* Auskragung *f*, Auskragen *n*, Vorkragen *n*
cord / to verseilen
cord 1. Schnur *f*, Bindfaden *m*, Kordel *f*; Strick *m*, *dünnes* Tau *n*; 2. *(Text)* Mehrfachzwirn *m*, Kord *m*; 3. *(El)* Verbindungsschnur *f*, Leitungsschnur *f*, Anschlußschnur *f*; 4. *(Fmt)* Vermittlungsschnur *f*; 5. *(Typ)* Bund *m*; 6. *(amerikanische Volumeneinheit für Holz: 1 cd = 128 cu.ft. = 3,625 m³)*
c. and plug Steckerschnur *f*
c. circuit 1. Schnurschaltung *f*, Schrankstromkreis *m*; 2. Steckdosenstromkreis *m (Haushalt)*
c. circuit repeater *(Fmt)* Schnurverstärker *m*
c. covering *(Wärm)* Isolierschnur *f*, Wärmeschutzschnur *f*
c. drive Seiltrieb *m (zur Betätigung von Skalenzeigern)*
c. eye *(El)* Schnuröse *f*
c. fastener Zugentlastung *f (für Geräteschnur)*
c. guide *(Wärm)* Schnurführung *f (Isolierung)*
c. maker Seiler *m*
c. packing *(Wärm)* Schnurpackung *f*
c. pair *(El)* Schnurpaar *n*
c. ply *(Kfz)* Kordgewebelage *f (Reifen)*
c. protecting means Schnurschutz *m (z. B. Spirale)*
c. pulley Schnurrolle *f*
c. pulp Kordzellstoff *m*
c. repeater Schnurverstärker *m*
c. rigger *(Fmt)* Schnurrolle *f*
c. rim *(Wärm)* Schnurkranz *m*
c. roller *(Fmt)* Schnurrolle *f*
c. switch *(El)* Schnurschalter *m (Schalter in Verbindungsschnur)*
c. terminal *(El)* Schnuröse *f*
c. test *(El)* Schnurprüfung *f*
c. tightener *(Fmt)* Schnurspanner *m*
c. tyre *(Kfz)* Kordreifen *m*
c. tyre fabric Reifenkord *m*
cordage Seilerware *f*, Seilwerk *n*, Seilereierzeugnisse *npl*
c. fibre Faser *f* zur Herstellung von Seilerwaren
corded timber Schichtnutzholz *n*
c. weave *(Text)* gerippte Bindung *f*
cordieriolite *s.* cordierite
cordierite Cordierit *m*
cordless board *(Fmt)* schnurloser Schrank *m*
c.-radio schnurloser Rundfunkempfänger *m*
corduroy Kordsamt *m*
cordwood Brennholz *n*
core / to mit einem Kern versehen *(Guß)*
core 1. Kern *m (Gewinde)*; Seele *f (Seil)*; 2. Gußkern *m*; Kern; 3. *(El)* Ader *f*, Leiter *m (Kabel)*; Kern *(Spule)*; 4. *(Kern)* aktive Zone *f*, Spaltzone *f*; 5. *(Holz)* Mittellage *f*; 6. Obstgehäuse *n*; 7. Bohrkern *m*; 8. Düsendorn *m*
c. baking *(Met)* Kerntrocknung *f*
c. bar *(Masch)* Kernspindel *f*, Kernstange *f*
c. barrel Kernrohr *n (Erdöl, Erdgas)*
c. binder *(Met)* Kernbinder *m*
c. blower *(Met)* Kernblasmaschine *f*
c. board Kernbrett *n*, Kernschablone *f*
c. boring Kerndrehbohren *n*
c. box Kernkasten *m*
c. casting Hohlguß *m*, Gießen *n* mit Kern
c. catcher Kernfänger *m*
c. conductor *(El)* Innenleiter *m*, Leiterseele *f*
c. diameter Kerndurchmesser *m (Gewinde, Spule usw.)*

c. diameter snap gauge Kernrachenlehre *f*
c. drill Aufbohrer *m* für vorgegossene Löcher
c. drill bit Bohrkrone *f*
c. eccentricity *(Holz)* exzentrisches Herz *n*
c. ejecting hammer *(Met)* Kernausstoßhammer *m*
c. flow *(Ph)* Kernströmung *f*
c.-fuel exposure Bestrahlung *f* des Kernbrennstoffs in der aktiven Zone
c. hole pilot Kernlochzapfen *m*
c. hole plug Verschlußpfropfen *m*, Verschlußputzen *m*
c. knockout *(Met)* Kernausdrücker *m* *(Gießerei)*
c. lamination *(El)* Kernblech *n*
c. lifting Kernziehen *n* *(Erdöl, Erdgas)*
c. losses *(El)* Eisenverluste *mpl*
c.-making machine Kernformmaschine *f* *(Gießerei)*
c. mark Kernmarke *f*
c. memory *(BMSR)* Kernspeicher *m*, Magnetkernspeicher *m*
c. mixture *(Met)* Kernmasse *f*
c. mould *(Met)* Kernform *f*
c. moulder Kernmacher *m*
c. moulding Kernmacherei *f*, Kernformerei *f*
c. nail Kernnagel *m*
c. of a coil *(El)* Spulenkern *m*
c. of a shaft Wellenseele *f*
c. of projectile *(Mil)* Geschoßkern *m*
c. of the thread *(Text)* Innere *n* des Fadens, innerer Fadenanteil *m*, Fadenseele *f*
c. of the vortex *(Ph)* Wirbelkern *m*
c. oil Kernöl *n*
c. paper Kernpapier *n*
c. plate *(El)* Kernblech *n*
c. print Kernmarke *f*
c. receiving barrel inneres Kernrohr *n* *(Erdöl, Erdgas)*
c. recovery Kerngewinn *m* *(Erdöl, Erdgas)*
c. refining *(Met)* Kernrückfeinung *f*
c. sample Kernprobe *f* *(Erdöl, Erdgas)*
c. sand Kernsand *m*
c. setting gauge Kerneinstellehre *f*
c. shooter Kerngießmaschine *f*
c. slab *(Bw)* Langlochplatte *f*, Lochplatte *f*; Hohlplatte *f*
c. snap gauge Kernrachenlehre *f*
c. store *s.* c. memory
c. tank *(Kern)* Reaktorgefäß *n*
c. thread *(Text)* Kernfaden *m*, Seelenfaden *m*
c. transformer *(El)* Kerntransformator *m*
c. type induction furnace Induktionsofen *m* mit Eisenkern
c.-type transformer *(Elm)* Kerntransformator *m*
c. yarn *(Text)* Garn *n* mit Kernfaden (Seelenfaden); Kerngarn *n*

cored carbon *(Elm)* Dochtkohle *f*
c. electrode Seelenelektrode *f*
c. hole vorgegossene Bohrung *f*
c. panel *(Bw)* Hohlplatte *f*, Langlochplatte *f*, Lochplatte *f*; Hohltafel *f*
c. part Kerngußstück *n*
coreless induction furnace Induktionstiegelofen *m*, kernloser Induktionsofen *m*, Hochfrequenzofen *m*
c. wattmeter [eisenloser] Drehspulleistungsmesser *m*
coremaker Kernmacher *m*
coring 1. Kernen *n* *(Erdöl, Erdgas)*; 2. Kristallseigerung *f*; 3. Kernmachen *n* *(Gießerei)*
cork / to verkorken, zukorken, zupfropfen, zustöpseln
cork Korkrinde *f*; Kork[en] *m*, Stöpsel *m*
c. board Korkplatte *f*; Pantoffelholz *n*
c. dust Korkschrot *m*
c. insulation Korkisolation *f*
c. oak Korkeiche *f*
c. tile Korkformstück *n*
c. tree Korkeiche *f*
Cork lace *(Text)* irische Spitze *f*
corkscrew Kork[en]zieher *m*
c. aerial Wendelantenne *f*
c. divider Drehbahnteiler *m*, rotierender Halmteiler *m*
c. drawing-in draft *(Text)* Mehrfacheinzug *m*
c. motion *(Schiff)* Schlingerbewegung *f*
c. rule *(Ph)* Korkzieherregel *f*, Dreifinger-Regel *f* *(elektromagnetische Induktion)*
c. stair[case] Wendeltreppe *f*
corkscrewing Gierschwingungen *fpl* *(eines Schiffes)*, Schlingerbewegungen *fpl*
Corliss engine Corliss-Maschine *f*, Corliss-Dampfmaschine *f*
Corliss valve Corliss-Ventil *n*
corn Korn *n*, Getreide *n*; *(Am auch)* Mais *m*
c.-bin Futterkasten *m*
c.-chest Futterkasten *m*
c.-cob *(Am)* Maiskolben *m*
c.-for cattle Futterkorn *n*
c.-house Scheune *f*, Scheuer *f*
c.-husk *(Am)* Maishülse *f*
c.-loft Kornboden *m* *(Scheune)*
c.-mill Getreidemühle *f*
c.-stalk Getreidehalm *m*, Getreidestengel *m*; *(Am auch)* Maisstengel *m*
corneal lens *(Opt)* Cornealschale *f*, Hornhautlinse *f*
c. microscope Hornhautmikroskop *n*
corned leather genarbtes Leder *n*
corner Ecke *f*; Kante *f*
c. boat Knickspantboot *n*
c. bracing *(Bw)* Eckaussteifung *f*
c. buffers *(Eis)* Doppelpuffer *m*
c. bumper *(Kfz)* Eckenstoßfänger *m*

c. clamp Kantenzwinge *f (Holzbearbeitung)*
c. cutting machine *(Led)* Eckenschneidemaschine *f (Schuh)*
c. drill Eckbohrmaschine *f*, Winkelbohrmaschine *f*
c. drilling machine Eckbohrmaschine *f*, Winkelbohrmaschine *f*
c. frequency *(BMSR)* Eckfrequenz *f*, Knickfrequenz *f*, 45°-Frequenz *f (Bodediagramm)*
c. joint 1. Eckverbindung *f (Holz)*; 2. Eckstoß *m (Schweißstoß)*
c. locking *(Holz)* gerade Zinke *f*
c. locking machine Zinkenfräsmaschine *f*
c. mark Eckwasserzeichen *n*
c. point Knick *m (Kurve)*
c. post *(Brg)* Bolzen *m (Bolzenschrotausbau)*
c. pulley Umlenkrolle *f*
c. reflector *(Fmt)* Winkelreflektor *m*
c. riveting hammer Eckniethammer *m*
c.-rounding cutter Abrundfräser *m*; Radiusfräser *m*
c. tenon jointing *(Holz)* Eckverzapfung *f*
c. tool gebogener Schlichtmeißel *m*
c.-up mit umgebogener Ecke
c. valve Eckventil *n*
c. vane Umlenkschaufel *f*
c.-working machine *(Typ)* Eckenbearbeitungsmaschine *f*
cornered kantig, winklig
cornering characteristics *(Kfz)* Kurvenlage *f*
c. stability *(Kfz)* Kurvenfestigkeit *f*, Neigungsstabilität *f*
cornice *(Bw)* Gesims *n*, Gurtgesims *n*, Kranzgesims *n*, Sims *m (n)*
c. level Gesimshöhe *f*
Cornish boiler Cornwallkessel *m*, Einflammrohrkessel *m*
Cornish method of underhand stoping *(Brg)* gelöster Strossenbau *m*
Cornish multitubular boiler Flammrohrkessel *m* [mit vorgehenden Heizröhren], Heizröhrenkessel *m* mit zylindrischer Feuerbüchse und vorgehenden Heizröhren, Feuerbüchsenkessel *m*
Cornu prism Cornusches Prisma *n*
corollary Folgerung[ssatz *m*] *f*
corona 1. *(Astr)* Korona *f*; 2. *(El)* Korona[entladung] *f*
c. brushing Sprühentladung *f*
c. discharge Koronaentladung *f*, Sprühentladung *f*
c. effect Koronaeffekt *m*, Koronaentladung *f*
c. loss Koronaverluste *mpl*, Glimmverluste *mpl*
c. power loss *s.* c. loss
c. tube Glimmentladungsröhre *f*, Koronaentladungsröhre *f*
coronagraph *(Opt)* Koronograt *m*
coronal ray Koronastrahl *m*
c. streamer Koronastrahl *m*
corporate body juristische Person *f*
corpuscle *(Ph)* Korpuskel *f*, Materieteilchen *n*, Teilchen *n*
corpuscular anode rays *(Ph)* Anodenstrahlen *mpl*
c. emission *(Ph)* Teilchenemission *f*, Korpuskularemission *f*
c. radiation *(Ph)* Teilchenstrahlung *f*, Korpuskularstrahlung *f*
c. rays Korpuskularstrahlen *mpl*
c. theory of light Korpuskulartheorie *f* des Lichtes
correct / to berichtigen, korrigieren, verbessern
c. deviation *(Rak)* auslenken
c. distortion *(Fmt)* entzerren
c. previous positional inaccuracy in a hole ein versetztes Loch ausbohren (auf Mitte bringen)
correct korrekt, richtig
c. application of cutter *(Holz)* rechter Anschliff *m*
c. dimensions / of maßgerecht
c. scale maßstäblich, maßstabgerecht
c. size / of maßgerecht
correcting *(El)* Entzerrung *f*
c. apparatus Entzerrer *m*
c. circuit Entzerrschaltung *f*
c. coil *(Fs)* Korrektionsspule *f*
c. current Gleichlaufimpuls *m*, Gleichlaufzeichen *n*, Regelstrom *m*
c. impulse *(Fmt)* Gleichlaufimpuls *m*
c. member *s.* regulating unit
c. network Ausgleichsschaltung *f*; Entzerrerschaltung *f*
c. repeater entzerrender Verstärker *m*
c. signal Berichtigungszeichen *n*
correction Korrektion *f*, Verbesserung *f*, Berichtigung *f*
c. angle Vorhaltewinkel *m (Navigation)*
c. bar Ausgleichslineal *n*
c. by addition *(Rak)* Zusatzberichtigung *f*, Zusatzkorrektur *f*
c. factor Korrekturfaktor *m*, Korrektionsfaktor *m*
c. for covered specimens *(Opt)* Deckglaskorrektion *f (Mikroskopie)*
c. network Ausgleichsschaltung *f*, Entzerrerschaltung *f*
c. of amplitudes Amplitudenentzerrung *f*
c. of distortion *(El)* Entzerrung *f*
c. of errors *(BMSR)* Fehlerberichtigung *f*
c. of phase Phasenentzerrung *f*
c. of the drilling Nacharbeiten *n* des Bohranschnitts
c. relay Gleichlaufrelais *n*
c. signal *(Rak)* Korrekturkommando *n*

c. strip Ausgleichlineal *n*
c. term *(Ph)* Korrektionsglied *n*
corrective action *s.* control action
c. factor *(Opt)* Korrekturfaktor *m*
c. filter *(Opt)* Korrektionsfilter *n*
c. network Ausgleichsschaltung *f*; Entzerrer[schaltung *f*] *m*
corrector *(BMSR)* Entzerrer *m*
c. circuit Korrektionskreis *m*; Entzerrer *m*
c. relay Gleichlaufrelais *n*
correlate / to korrelieren, in Wechselbeziehung setzen, zuordnen
correlated control system Mehrfachregelung *f*, vermaschte Regelung *f*
correlation Korrelation *f*; Wechselbeziehung *f*; gegenseitige Abhängigkeit *f*, kausaler Zusammenhang *m*; Verkettung *f*
c. computer Korrelationsrechenautomat *m*, Korrelationsrechner *m*
correspond / to entsprechen; korrespondieren, zugehörig sein
correspondence envelope Briefhülle *f*, Briefumschlag *m*
c. system Wechselsprechanlage *f*
corresponding angles *(Math)* Gegenwinkel *mpl*
c. dummy piston korrespondierender Ausgleichskolben *m (Turbine)*
c. wires korrespondierende Drähte *mpl*, Korrespondierende *pl (bei Wechselschaltungen)*
corridor *(Flg)* Flugschneise *f*
c. carriage *(Eis)* Durchgangswagen *m*
corrode / to ätzen, korrodieren, rosten; anätzen, anrosten; verrosten, zerfressen, verschleißen
corrodent aggressives Mittel *n*, Korrosionsmittel *n*
corrodibility Korrodierbarkeit *f*; Korrosionsneigung *f*, Rostneigung *f*
corrodible korrodierbar
corroding korrodierbar, angreifbar
c. agent aggressives Mittel *n*, Korrosionsmittel *n*
corrosion Korrosion *f*; Rostbildung *f*
c. behaviour Korrosionsverhalten *n*
c. cell Korrosionselement *n*
c. fatigue Korrosionsermüdung *f*, Ermüdungskorrosion *f*
c. pit Rostnarbe *f*
c. preventive korrosionshindernd
c. proof korrosionsbeständig, korrosionsfest
c. protection Korrosionsschutz *m*
c. resistance Korrosionsfestigkeit *f*
c.-resistant korrosionsbeständig, korrosionsfest
c. resisting quality Korrosionsbeständigkeit *f*
corrosive Korrosionsmittel *n*, Abbeizmittel *n*, Beize *f*
corrosive aggressiv, korrodierend; korrodierbar
c. action Korrosionsvorgang *m*
c. attack Korrosionswirkung *f*
corrugated gewellt, geriffelt
c. board Wellpappe *f*
c. cable sheathing Balgenmantel *m (Kabel)*
c. cardboard Wellpappe *f*
c. expansion pipe *(Wärm)* Ausgleichswellrohr *n*
c. fibre board Well[faser]pappe *f*
c. furnace Wellflammrohr *n*
c. gasket Welldichtung *f*
c. insulator *(El)* Rillenisolator *m*
c. iron [sheet] Wellblech *n*
c. metal pipe Wellblechrohr *n*
c. mould *(Met)* gewellte Kokille *f*
c. packing ring gewellter Dichtungsring *m*
c. pipe Wellrohr *n*
c. plate Wellblech *n*
c. plate capacitor *(Fmt)* wellengerader Kondensator *m*
c. profile Sägezahnprofil *n*
c. seal *(Masch)* Welldichtung *f*
c. sheet [metal] Wellblech *n*
c. shell plating *(Schiff)* gewellte Außenhaut *f*
c. steel bar *(Brg)* Wellstahlkappe *f*
c. steel bulkhead *(Schiff)* Wellschott *n*
c. vessel Schiff *n* mit gewellter Außenhaut
corrugating board Wellpappe *f*
c. medium Wellpappenmittellage *f*
corrugation Riffelung *f*; Wellung *f*
corset cloth Miedergewebe *n*
cortex Außenschale *f*; Rinde *f*
corundum Aluminiumoxid *n*, Korund *m*
c. brick *(Bw)* Korundstein *m*
cosecant, cosec *(Math)* Kosekans *m*, cosec, $\cos^{-1}$
coseparation *(Kern)* Mitfällung *f*, Trennung *f* mit Trägersubstanz
coset *(Math)* Nebengruppe *f*, Nebenklasse *f*; Restklasse *f*
cosettes Zuckerrübenschnitzel *npl*
cosine curve kosinusförmige Kurve *f*, Kosinuskurve *f*
c. governor *(Wärm)* Kosinusregler *m*
c. integral Integralkosinus *m*
c. law [Lambertsches] Kosinusgesetz *n*
cosinusoid *(Math)* Kosinuslinie *f*
coslettize / to *(Met)* phosphatieren
cosmetics Kosmetika *npl*, Schönheitspflegemittel *npl*
cosmic abundance kosmische Häufigkeit *f (von chemischen Elementen)*
c. dust kosmischer Staub *m*
c. expansion Expansion *f* des Weltalls
c. law *(Rak)* Raumrecht *n*
c. radiation Höhenstrahlung *f*, kosmische Strahlung *f*
c. ray background kosmischer Strahlungshintergrund (Strahlungsuntergrund) *m*, Höhenstrahlungsuntergrund *m*

c. ray burst Ionisationsstoß *m*, Hoffmannscher Stoß *m*
c. ray decay electron Zerfallselektron *n* der kosmischen Strahlung
c. ray intensity Höhenstrahlungsintensität *f*, Intensität *f* der kosmischen Strahlung
c. ray meson Höhenstrahlmeson *n*
c. ray shower Schauer *m* der kosmischen Strahlung, Höhenstrahlungsschauer *m*
c. ray star Höhenstrahlungsstern *m*, durch kosmische Strahlung erzeugter Zertrümmerungsstern *m*
c. rays Höhenstrahlung *f*, kosmische Strahlung *f*
c. space Weltraum *m*
c. system Weltsystem *n*
cosmogony Kosmogonie *f*
cosmography Kosmografie *f*
cosmology Kosmologie *f*
cosmonautics Kosmonautik *f*
cosmotron Cosmotron *n (Protonen-Synchrotron)*
cost Kosten *pl*, Selbstkosten *pl*; Preis *m*
c. accountant Kalkulator *m*, Kostenrechner *m*
c. accounting Kostenrechnung *f*, Kalkulation *f*, betriebliches Rechnungswesen *n*, Selbstkostenrechnung *f*, Selbstkostenkalkulation *f*
c. accounting department Selbstkostenberechnungsabteilung *f*
c. and method analysis technisch-wirtschaftliche Analyse *f*
c. centre Kostenstelle *f*
c. comparison Kostenvergleich *m*
c. cutting programm Kostensenkungsprogramm *n*
c. department Kalkulationsabteilung *f*
c. estimation Kostenanschlag *m*
c. of attendance Wartungskosten *pl*, Bedienungskosten *pl*
c. of construction Baukosten *pl*, Herstellungskosten *pl*, Konstruktionskosten *pl*
c. of erection Montagekosten *pl*
c. of installation Installationskosten *pl*, Anlagekosten *pl*
c. of labour Lohnkosten *pl*
c. of maintenance Wartungskosten *pl*; Unterhaltungskosten *pl*
c. of management Verwaltungskosten *pl*
c. of marketing Vertriebskosten *pl*
c. of production Herstellungskosten *pl*
c. of steam production *(Wärm)* Dampfherstellungskosten *pl*
c. of upkeep Unterhaltungskosten *pl*
c. records Kostenbeleg *m*
c. reduction Kostensenkung *f*, Einsparung *f*, Reduktion *f* der Selbstkosten
c.-saving kostensparend
costean / to *(Brg)* schürfen
costeaning ditch *(Geol)* Schürfgraben *m*, Schurfgraben *m*
cot Schwingbett *n (Schiffsbett)*
cotangent Kotangens *m*
Cotes's spiral (Math) Ährenkurve *f (Cotesche Spiralen)*
cottage piano Pianino *n*
c. steamer *(Text)* Dämpfkasten *m*, Kastendämpfer *m*, Runddämpfer *m*
cotter / to *(Masch)* versplinten
cotter *(Masch)* Querkeil *m*, Nutenkeil *m*, Keil *m*, Splint *m*, Dorn *m*, Vorsteckkeil *m*
c. and keyseat (keyway) milling machine Keilloch- und -nutenfräsmaschine *f*
c. bolt Bolzen *m* mit Splint, Vorsteckbolzen *m*
c. driver Keiltreiber *m*
c. file dickflache Feile *f*
c. joint Splintverbindung *f*
c. key *(Masch)* Querkeil *m*
c. lock *(Masch)* Splintsicherung *f*
c. mill Langlochfräser *m*
c. pin Splint *m*, Quersplint *m*
c. pin extractor Splintenzieher *m*
c. pin hole Splintloch *n*
c. slot Keilloch *n*
c. slot milling machine Keillochfräsmaschine *f*
c. way Keilloch *n* für Querkeil
cottered joint Querkeilverbindung *f*
c. pin versplinteter Bolzen *m*
cottering Keilverbindung *f*, Verkeilung *f*
cotton Baumwolle *f*
c. and wool mixture Halbwolle *f*
c. bale Baumwollballen *m*
c. baling plant Baumwoll[ballen]preßanlage *f*
c. binder *(El)* [farbiger] Kennfaden *m (Kabel)*
c. blending Mischen *n* der Baumwolle *(vor dem Be- und Verarbeiten)*
c. braid hose baumwollarmierter Schlauch *m*
c. cake Baumwollsaat[öl]kuchen *m*
c.-carding process Baumwollkarden *n*
c. combed yarn gekämmtes Baumwollgarn *n*
c. combing machine Baumwollkämmaschine *f*
c. cord Baumwollschnur *f*
c. count Feinheit *f* des Baumwollgarns
c.-covered wire baumwollisolierter (baumwollumsponnener) Draht *m*
c. fabric Baumwollgewebe *n*
c. feeder Baumwollspeiser *m*
c. fibre Baumwollfaden *m*, Baumwollfaser *f*
c. gin Baumwollentkörnungsmaschine *f*, Baumwollkratze *f*
c. growing Baumwollanbau *m*
c. growing area Baumwollanpflanzfläche *f*
c. insulation Baumwollisolation *f*, Baumwollisolierung *f*
c. linters Baumwollinters *pl*, Baumwollabfälle *mpl*

c. linters pulp Baumwollzellstoff *m*
c. mill Baumwollspinnerei *f*
c. opener Baumwollraspel *f*
c. packing Baumwollpackung *f*
c. picker Baumwollpflückmaschine *f*, Baumwollpflücker *m*
c. pile fabric Florgewebe *n* aus Baumwolle
c. plant Baumwollstrauch *m*
c. printing *(Text)* Kattundruck *m*, Zeugdruck *m*
c.-roving frame tenter Baumwollgarnspinner *m*
c. shag Baumwollplüsch *m*
c. shredder Baumwollraspel *f*
c. speeder tenter Baumwollfeinstrecker *m*
c. spinner Baumwollspinner[in *f*] *m*
c. spinning Baumwollspinnerei *f*
c. stalk Baumwollstengel *m*
c. thread *s.* c. twine
c. threader Baumwollüberspinner *m*
c. twine Baumwollgarn *n*, Baumwollzwirn *m*
c. wadding Watte *f*
c. waste 1. Baumwollabfall *m*, Baumwollabgang *m*; 2. Putzwolle *f*
c. waste spinning Baumwollabfallspinnerei *f*
c. weaving 1. Baumwollweben *n*; 2. Baumwollweberei *f*
c. winding Baumwollhaspelei *f*
c. wool Rohbaumwolle *f*
cottonized bast fibre Flockenbast *m*
c. flax Flockenflachs *m*
cottonseed hulls Baumwollsamenkapselschalen *fpl*
c. oil Baumwollsamenöl *n*, Cottonöl *n*
couch / to gautschen, ausdrücken
couch Gautsche *f*; Auftrag *m*, Auflage *f*, Schicht *f*
c. box Saugkasten *m* der Gautsche
c. pit Gautschbruchbütte *f*
c. pit recovery Gautschbruchrückgewinnung *f*
c. press Gautschpresse *f*
c. stool Gautschstuhl *m*
coucher Gautscher *m*, Papierstreicher *m*
couching Ausbreiten *n* der Gerste *(Mälzerei)*
Coudé Grating Spectrograph Coudé-Gitterspektrograf *m*
coulomb Coulomb *n*, Amperesekunde *f*
Coulomb energy Coulomb-Energie *f*
Coulomb friction trockene (Coulombsche) Reibung *f*
Coulomb interaction Coulombsche Wechselwirkung *f*
Coulomb scattering Coulomb-Streuung *f*
Coulomb's law Coulombsches Gesetz *n*
coulometer Coulometer *n*, Voltameter *n*
coulometric titration *(Ch)* coulometrische Titration *f*
coulometry Coulometrie *f* *(elektrochemisches Analysenverfahren)*
coulter Schar *n*; Pflugmesser *n*, Pflugsech *n*
c. beam *(Lw)* Pflugsechholz *n*
c. point *(Lw)* Pflugsechklinge *f*
coumalic acid Kumalinsäure *f*
coumaric acid Kumarsäure *f*
coumarin Kumarin *n*, Lakton *n* der Kumarinsäure
coumarinic acid Kumarinsäure *f*
coumarone *(Ch)* Kumaron *n*
Council for Mutual Economic Aid Rat *m* für gegenseitige Wirtschaftshilfe
count / to zählen
count 1. Zählung *f*, Registrierung *f* *(z. B. von Impulsen)*; 2. *(Text)* Feinheit[sgrad *m*] *f* *(Garn)*
c.-by-n-circuit *(Dat)* Impulsuntersetzer *m* 1:n
c.-controlled operation *(Kern)* Betrieb *m* mit Impulsvorwahl *(Zählgerät)*
c. of cloth *(Text)* Feinheit *f* des Gewebes
c. of delivered sliver *(Text)* Bandendnummer *f*, Bandfeinteil *m* *(Strecken)*
c. of feed *(Text)* Eingangsnummer *f* *(Strecken)*
c. of yarn *(Text)* Garnnummer *f*
c. range Nummernbereich *m* *(Garn)*
c.-up count-down chain *(BMSR, Büro)* Aufwärts-Abwärts-Zählkette *f*
countability *(Math)* Abzählbarkeit *f* *(Menge)*
countable abzählbar *(Menge)*
counter 1. *(Kern)* Zähler *m*, Zählrohr *n*, Zählgerät *n*, Zählapparat *m*; 2. Zählwerk *n* *(z. B. einer Rechenmaschine)*; 3. *(Wärm)* Hubzähler *m*; 4. Ladentisch *m*; Anrichte *f*, Arbeitsfläche *f* *(Küche)*; 5. Kappe *f*, Hinterkappe *f* *(Schuh)*; 6. *(Text)* Steigungszahl *f* *(Atlasbindung)*
counter entgegengesetzt, gegen; Gegen-
c. accelerator *(Büro)* Zählerbeschleuniger *m* *(für elektromechanische Zähler)*
c. arrangement *(Kern)* Zähleranordnung *f*, Zählrohranordnung *f*
c.-blow hammer Gegenschlaghammer *m*
c. buffing machine *(Led)* Kappenschleifmaschine *f*
c. capacity *(BMSR, Büro)* Zählerkapazität *f*
c. circuit Zählrohrschaltung *f*
c. control *(BMSR, Büro)* Zählerkontrolle *f*
c.-currently nach dem Gegenstromprinzip *(Absorption)*
c. dead time *(Kern)* Zählrohrtotzeit *f*
c. die Oberstempel *m*
c. discharge *(Kern)* Zählrohrentladung *f*
c.-emf Gegen-EMK *f*
c. force Gegenkraft *f*
c. gas *(Kern)* Zählrohr[füll]gas *n*
c. light Gegenlicht *n*
c. list exit *(Büro)* Zählerausgang *m* für Posten
c. position *(BMSR, Büro)* Zählerstelle *f*
c.-pressure Gegen[halte]druck *m*
c.-propeller *(Schiff)* Gegenpropeller *m*

c. quench circuit *(Kern)* Zählrohrlöschkreis
c. reel Abreißrolle *f*
c. resolving time *(BMSR)* Zählerauflösungszeit *f*
c. stage *(Fs)* Addierstufe *f*
c. time-lag *(BMSR)* Zählzeitverzögerung *f* *(statistisch)*
c. tube *(Kern)* Zählrohr *n*
c.-wheel printer *(Dat)* Ziffernraddrucker *m*
counteract / to gegenwirken
counteracting entgegenwirkend; gegeneinanderarbeitend
c. force Gegenkraft *f*
c. winding *(El)* Differentialwicklung *f*
counterbalance / to *(Masch, El)* ausbalancieren, auswuchten, durch Gegengewicht ausgleichen
counterbalance *(Masch, El)* Gegengewicht *n*
counterbalanced mit Gegengewicht
counterbattery fire Artilleriebekämpfung *f*
counterbinding *(Text)* Gegenbindung *f*
counterbore / to [zylindrisch] aufsenken, zylindrische Bohrungen aussenken
counterbore 1. Senker *m* [zum Aussenken zylindrischer Vertiefungen]; 2. Zylindersenkung *f*, zylindrische Senkbohrung *f*
c. with interchangeable blades (cutters) and guides Zapfensenker *m* mit auswechselbaren Messern und auswechselbaren Führungszapfen
c. with removable cutters (blades) Messerstangensenker *m*
c. with spiral flutes Spiralsenker *m*
counterboring [zylindrisches] Aussenken *n*
c. tool Senkwerkzeug *n*
c. tool with pilot Zapfensenker *m*
counterceiling *(Bw)* untergehängte Decke *f*
countercell *(El)* gegengeschaltete Sammlerzelle *f*, Gegenzelle *f*
counterclaim Gegenforderung *f*
counterclockwise entgegengesetzt dem Uhrzeigersinn
countercurrent Gegenstrom *m*
c. boiler Gegenstromkessel *m*
c. braking *(Elm, El)* Gegenstrombremsung *f*
c. classifier *(Brg)* Gegenstromklassierer *m*
c. condenser *(Wärm)* Gegenstromkondensator *m*
c. flow drying *(Text)* Gegenstromtrocknung *f*
c. heater *(Wärm)* Gegenstromvorwärmer *m*
c. method Gegenstrommethode *f*
c. pipe cooler Gegenstromröhrenkühler *m*
c. principle Gegenstromverfahren *n*
c. surface condenser *(Wärm)* Oberflächengegenstromkondensator *m*
countereffect Gegenwirkung *f*
counterelectrode *(Fs)* Gegenelektrode *f*
counterelectromotive force Gegen-EMK *f*, gegenelektromotorische Kraft *f*

c. force cell Gegenzelle *f*
counterfaller *(Text)* Gegenschläger *m*
counterflow *(Ph)* Gegenstrom *m*
c. condensation Gegenstromkondensation *f*
c. cooling tower Gegenstromkühlturm *m*
c. heat exchanger Gegenstromwärme[aus]tauscher *m*
counterflush drilling rig Gegenstrombohranlage *f (Erdöl, Erdgas)*
counterfoil *(Büro)* Kontrollabschnitt *m*, Kontrollschein *m*, Quittung *f*
counterforce Gegenkraft *f*
counterfort *(Bw)* Strebepfeiler *m*; Mauerpfeiler *m*, Verstärkungspfeiler *m*
c. wall Winkelstützmauer *f*
counterguide link *(Wärm)* Gegenlenker *m*
counterhatching Kreuzschraffur *f*
counterlathing *(Bw)* Putzleisten *fpl*
counterlight Gegenlicht *n*
countermark / to gegenstempeln *(Münze)*
countermark Gegenstempel *m (Münze)*
countermeasure Gegenmaßnahme *f*
counterpart Gegenstück *n*
counterpoise 1. *(Masch)* Gegengewicht *n*; 2. *(El)* künstliche Erde *f*, Gegengewicht
counterpressure roller *(Text)* Gegendruckwalze *f*
counterproof *(Typ)* Gegenabdruck *m*
counterpunch Gegenstanze *f*, Gegenstempel *m*
counterreaction *(Ch)* Rückreaktion *f*, Umkehrreaktion *f*
counterregulation Gegenregelung *f (insbesondere Turbine)*
counterrevolving air screw gegenläufige Luftschraube *f*
counterrotation Gegenlauf *m*, Gegendrehung *f*
countershaft *(Masch)* Vorgelege[welle *f*] *n*
c. drive gear Vorgelegetriebrad *n*
c. pulley Vorgelegescheibe *f*
countersink / to *(Masch)* [aus]senken, [kegelig] aufsenken, versenken
countersink 1. Senker *m* zum Aussenken kegeliger Vertiefungen; Spitzsenker *m*; 2. kegelige Senkbohrung *f*
c. bit Holzversenkbohrer *m*
countersinking Senkung *f*, [kegelige] Aussenkung *f*
c. cutter Spitzsenker *m*
counterslope *(Bw)* Gegenböschung *f*
countersteam brake Gegendampfbremse *f*
counterstern *(Schiff)* elliptisches Heck *n*, Gillungsheck *n*
countersunk bolt versenkter Bolzen *m*
c. head Senkkopf *m (Niet)*
c. head rivet Senkniet *m*
c. oval head Linsensenkkopf *m*
c. recess Schutzsenkung *f*
c. rivet Senkniet *m*

c. rivet head Nietsenkkopf *m*
c. screw Senkschraube *f*
counterturning screws *(Schiff)* gegenläufige Schrauben *fpl*
countervoltage Gegenspannung *f*
countervortex Gegenwirbel *m*
counterweight Gegengewicht *n*, Ausgleichgewicht *n*
c. hoisting Förderung *f* mit Gegengewicht
c. pendant Zugpendellampe *f*
counterweighted crankshaft Kurbelwelle *f* mit Gegengewicht
counting Zählen *n*, Zählung *f*
c. circuit Impulszählschaltung *f*; Zählkreis *m*, Zählschaltung *f*
c. device Zählwerk *n*, Zähluhr *f*
c. glass *(Text)* Fadenzähler *m*
c. mechanism Zählvorrichtung *f*
c. mechanism of a meter *s.* register of a meter
c. rate Zählgeschwindigkeit *f*
c.-rate meter 1. *(BMSR)* Zählgeschwindigkeitsmesser *m*; 2. *(Kern)* Ratemeter *n*, Mittelwertmesser *m*, Impulsdichtemesser *m*, Integrator *m*
c. rate voltage characteristic Impuls-Spannungs-Charakteristik *f*
c. relay *(Fmt)* Zählerrelais *n*
country beam Fernlicht *n*
c. fair Landwirtschaftsausstellung *f*
c. rock Nebengestein *n*, Begleitgestein *n*
c. satellite exchange *(Fmt)* ländliches Unteramt *n*
county road Landstraße *f*, Straße *f* 2. Ordnung
c. surveyor Bezirksstraßenmeister *m*
couple / to [an]koppeln; kuppeln
c. back rückkoppeln
couple 1. Paar *n*; 2. Kräftepaar *n*; 3. Plattenpaar *n (der Batterie)*; 4. *(sl) s.* thermocouple
c. of forces Kräftepaar *n*
c. of instantaneous rotations momentanes Rotationspaar *n*
c. of rotations Rotationspaar *n*
c. roof *(Bw)* Sparrendach *n*
coupled circuit *(El)* angekoppelter Kreis *m*
c. equations gekoppelte Gleichungen *fpl*
c. oscillation gekoppelte Schwingung *f*
c. rigidly *(Fmt)* phasenstarr verkoppelt
c. running *(El)* Vielfachsteuerung *f*
c. together zusammengekuppelt
c. wheel pair on axle *(Eis)* Kuppelradsatz *m*
coupler 1. Gerätesteckvorrichtung *f*, Steckkupplung *f (zwischen Verlängerungsschnur und Geräteschnur)*; 2. Verbindungsmuffe *f (Stahlrohrgerüst)*; 3. *(Am) s.* sleeve
c. capacitor Kopplungskondensator *m*; Koppelkondensator *m*
c.-on *(Brg)* Wagenkuppler *m*
c. loop *(El)* Koppelschleife *f*, Kopplungsschleife *f*
c. network *(El)* Anpassungsvierpol *m*
c. nut Überwurfmutter *f*; Spannschloß *n*
c. pin Verbindungsstift *m*, Verschlußbolzen *m*
c. point Kuppelstelle *f*, Einkuppelstelle *f*
c. pressure Anpressungsdruck *m*
c. reaction Kupplungsreaktion *f (Farbenchemie)*
c. resistance Kopplungswiderstand *m*
c. ring Spannring *m (an Zangen)*
c. rod Kupplungsstange *f*, Kuppelstange *f*
c. screw Kupplungsschraube *f*, Verbindungsschraube *f*
c. sleeve Kupplungsmuffe *f*, Verbindungsmuffe *f*
c. tappet Auslöseknagge *f*
c. transformer *(El)* Anpassungstransformator *m*, Kopplungstransformator *m*; Übertrager *m*
c. wave Kopplungswelle *f*
c. with internal valve *(Kfz)* Kupplungskopf *m* mit Ventil
c. with pin *(Kfz)* Kupplungskopf *m* mit Stift
c. plug Gerätestecker *m*, Anschlußstecker *m*, Kupplungsstecker *m*
c. socket Gerätesteckdose *f*, Geräteanschluß *m*, Kupplungs[steck]dose *f*
c. terminal strip Lüsterklemme *f*; Registerklemme *f*, Reihenklemme *f*, Klemm[en]streifen *m*
coupling 1. *(Masch)* Kopplung *f*, Kupplung *f*, Verbindung *f*; Kupplungsstück *n*, Verbindungsstück *n*; Muffe *f*; Schlauchverbindung *f*; 2. *(El)* Ankopplung *f*, Kopplung *f*, Verbindung *f*
c. attenuation Kopplungsdämpfung *f*, Koppeldämpfung *f*
c. bolt Kuppelbolzen *m*, Kupplungsbolzen *m*
c. box Kupplungsmuffe *f*, Durchgangsmuffe *f*
c. capacitor *(El)* Kopplungskondensator *m*
c. capacity *(El)* Kopplungskapazität *f*
c. chain Kuppelkette *f*, Zugkette *f*
c. circuit Anpassungskreis *m*; Antennenanpassungskreis *m*
c. coefficient *(El)* Kopplungsfaktor *m*, Koppelfaktor *m*, Kopplungsbeiwert *m*
c. condenser Kopplungskondensator *m*, Koppelkondensator *m*
c. disk Kupplungsscheibe *f*
c. draw hook Zughaken *m*
c. element Kopplungselement *n*
c. end Kupplungsseite *f*
c. energy Kopplungsenergie *f*
c. factor *s.* c. coefficient
c. flange Kupplungsflansch *m*
c. hook Kuppelhaken *m*
c. impedance *(BMSR)* Kopplungswiderstand *m (eines Verzögerungsgliedes)*

course 1. Bahn *f*, Lauf *m*, Weg *m*; Kurs *m* *(Flugzeuge, Schiffe)*; 2. *(Bw)* Lage *f*, Reihe *f*, Schicht *f*; 3. Lehrgang *m*; 4. *(Geol)* Erzgang *m*; 5. *(Text)* Maschenreihe *f*
c. and distance calculator Kurs- und Abstandsrechner *m*
c.-line computer *(Flg)* Flugwegrechner *m*
c. of ore *(Geol)* Gangstreichen *n*
c. of the thread *(Text)* Fadenlauf *m*
c. participant Lehrgangsteilnehmer *m*
c. recorder *(Schiff)* Kursschreiber *m*
c. setting device *(Flg)* Kursgeber *m*
c. signal *(Fmt)* Leitton *m*; Kurszeichen *n*
Court of Admiralty Seegericht *n*, Seeamt *n*
Court of Arbitration Schiedsgericht *n*
Court of Inquiry 1. Seeamt *n*; 2. Seeamtsverhandlung *f*
covalence *s.* covalency
covalency *(Ch)* Kovalenz *f*
covalent bond *(Ch)* Kovalenz[bindung] *f*, Atombindung *f*, Elektronen[paar]bindung *f*, kovalente (homöopolare, unpolare) Bindung *f*
c. compound *(Ch)* kovalente (homöopolare, unpolare) Verbindung *f*
c. link[age] *s.* c. bond
covariance *(Math)* Kovarianz *f*
covariant *(Math)* kovariant
cove *(Bw)* ausgerundetes Ecksel *n*, Hohlkehle *f* zwischen Decke (Fußboden) und Wand
covellite Covellin *m*
cover / to verdecken, bedecken, zudecken; verkleiden; überziehen *(mit Farbe)*; beschlagen; auskleiden; abdecken *(Boot)*
c. a range *(Fmt)* einen Bereich umfassen (versorgen); einen Bereich überstreichen
c. the top of the boiler *(Wärm)* den Kessel isolieren
c. with non-conducting material isolieren, mit Wärmeschutzmasse verkleiden
cover 1. Kappe *f*, Schutzhaube *f*; Kapsel *f*, Hülle *f*, Deckel *m*, Verschlußdeckel *m*; 2. *(Bw)* Abdeckplatte *f*, Abdeckung *f*, Verkleidung *f*; 3. *(Typ)* Buchdeckel *m*, Einband *m*; Buchdecke *f*; Umschlag *m*; 4. Betondeckung *f* *(Stahlbeton)*; 5. *(Brg)* Abraum[schichten *fpl*] *m*, Deckgebirge *n*
c. board Verzugsbrett *n* *(Holz)*; Verzugsplatte *f* *(Beton)*
c. brick Abdeckstein *m* *(Kabel)*
c. caving *(Brg)* Zubruchgehen *n*, Zubruchbauen *n* des Hangenden
c. cloth Überzugsgewebe *n*
c. coat Deckmasse *f*; Überzug *m*
c. disk Deckscheibe *f*
c. dressing Deckelverkleidung *f*
c. factor *(Text)* Deckfaktor *m*, Dichtefaktor *m* *(Gewebe)*
c. fillet *(Bw)* Abdeckleiste *f*, Fugenleiste *f*
c. flange Deckelflansch *m*
c. glass Deckglas *n*
c. lid Verschlußdeckel *m*
c. nut Hutmutter *f*
c. of casing Gehäusedeckel *m* *(Turbine usw.)*
c. of the layers *(Led)* Lohhut *m*
c. paper Deckpapier *n*, Umschlagpapier *n*
c. plate Deckel *m*; Deckplatte *f*; Lasche *f*
c. plate lock Deckelsperre *f*
c.-printing Überdruck *m*; Überdrucken *n*
c. screwing Deckelverschraubung *f*
c. strap *(Bw)* Lasche *f*
c. strip *(Bw)* Deckleiste *f*
c. valve *(Masch)* Deckelschieber *m*
coverage 1. Reichweite *f* *(Radio)*; abgetastetes Gebiet *n*, Auffaßbereich *m* *(Radar)*; 2. Betondeckung *f* *(Stahlbeton)*; Deckung *f* *(Dach)*
c. of a receiver Reichweite *f* eines Empfängers
covered abgedeckt *(Bereich)*; ummantelt, umhüllt *(Kabel)*; umspannt, umklöppelt *(Draht usw.)*; isoliert
c. machine for rubber threads Gummiumspinnmaschine *f*
c. material Bespannstoff *m*, Bezugmaterial *n* *(Innendekoration)*; Deckwerkstoff *m*
c. metal Deckmetall *n*
c. moving in the opposite direction of the wire points *(Text)* sich entgegengesetzt zur Zahnrichtung bewegender Beschlag *m*
c. needle *(Text)* Decknadel *f*
c. of non-conducting material *(Wärm)* Wärmeschutzhülle *f*, Isolierhülle *f*
c. of seat Außenbezug *m* des Sitzes
c. paper *(Typ)* Bezugspapier *n*, Deckpapier *n*
c. plate 1. Abdeckplatte *f*; 2. Lasche *f*
c. power Deckkraft *f* *(Farbe)*
c. signal Deckungssignal *n*
c. strip Lasche *f*
c. the guide-blade disk Abdecken *n* des Leitapparates *(Turbine)*
c. area überdeckte Fläche *f* *(Messe)*
c. cable umhülltes Kabel *n*
c. electrode Mantelelektrode *f*, ummantelte (umwickelte) Elektrode *f*
c. port opening *(Wärm)* überdeckte Kanalmündung *f*
c. way abgedeckte (verdecktliegende) Führungsbahn *f*
c. wire bespannter (isolierter, ummantelter) Draht *m*, Manteldraht *m*
covering 1. Ummantelung *f*, Umhüllung *f*, Außenmantel *m* *(Kabel)*; 2. Belag *m*, Deckschicht *f*, Überzug *m*; Dach[ein]deckung *f*, Dachhaut *f*
c. board *(Schiff)* Schandeck *n*, Schandeckel *m*
c. colour Deckfarbe *f*
c. knife Abdruckblech *n* *(Strickmaschine)*

c. thread *(Text)* Deckfaden *m*
c. tilt Abdeckplane *f*
covibrate / to mitschwingen
covolume Kovolumen *n*
cow beet Futterrübe *f*
c. catcher Bahnräumer *m*
cowl 1. Schornsteinaufsatz *m*, Schornsteinhaube *f*; Rohraufsatz *m*; Rohrhaube *f (Entlüftungsrohre)*; 2. *(Schiff)* Lüfterkopf *m*; 3. *(Kfz)* Stirnwand *f*; 4. Motorhaube *f*
c. flap *(Masch)* Kühlluftregelklappe *f*
c. support *(Kfz)* Stirnwandstütze *f*
c. ventilator Haubenventilator *m*
cowling Abdeckung *f* des Flugzeugmotors
Cowper [hot-blast] stove Cowperscher Winderhitzer *m*, Cowper *m*
cowshed Kuhstall *m*, Rinderstall *m*
c. pipeline milkers Melkrohrsystem *n*
coxswain Steuermann *m*, Bootführer *m*
coxswain's seat Steuersitz *m*
coyote-hole *(Brg)* Sprengkammer *f*
c.-hole blasting *(Brg)* Kammersprengung *f*
coyoting *(Am) (Brg)* Weitungsbau *m*, unsystematischer Abbau *m*
crab Laufkatze *f*
c. angle *(Flg)* Luvwinkel *m*, Vorhaltewinkel *m*
c. bolt durchgehender Ankerbolzen *m*
c. connection Katzfahrschaltung *f (Laufkatze)*
c. winch Handkabelwinde *f*, Vorgelegehaspel *f*
crabbing *(Text)* Einbrennen *n*, Krabben *n*
c. jack *(Text)* Brennbock *m*
c. liquor *(Text)* Brennflotte *f*
c. machine *(Text)* Krabbmaschine *f*, Einbrennmaschine *f*
crack / to 1. [auf]brechen, spalten; 2. reißen, Risse bilden; [zer]springen; 3. kracken *(Erdöl)*; 4. *(Brg)* knallen *(Stempel)*
crack Bruch *m*, Riß *m*, Spalt *m*, Sprung *m*
c. arrestor Rißstopper *m*, Rissefänger *m*
c. detection Rißprüfung *f*
c. detector Rißprüfer *m*
c. in the cooling jacket Kühlmantelriß *m*
c. liner Mammutschiff *n*; hervorragendes Schiff *n*
c. wave Sprungwelle *f*
cracked 1. gebrochen, gerissen, mit Rißbildung; gespalten; 2. *(Led)* brüchig *(Narben)*
c. concrete gerissener Beton *m*, Beton mit gerissener Zugzone *(Stahlbeton)*
c. gasoline Krackbenzin *n*, gekracktes Benzin *n*, Spaltbenzin *n*
c. state Zustand *m* II *(Stahlbeton mit gerissener Zugzone)*
cracker Brechwalze *f*
crackiness *(Led)* Narbenbrüchigkeit *f*
cracking 1. Reißen, Rißbildung *f*, Rissebildung *f*; 2. Knacken *n*, Zerknacken *n*; Zertrümmern *n*, Zerbrechen *n*; Zerspringen *n*; 3. Aufspaltung *f*, Krackung *f*, thermische Spaltung *f (Erdöl)*
c. distillation Krackdestillation *f*, Spaltungsdestillation *f (Erdöl)*
c. plant *(Ch)* Spaltanlage *f*
crackle Rißmuster *n* in Glasur
crackled lacquer Reißlack *m*
crackling *(El)* Kratzgeräusche *npl*; Nebengeräusche *npl*
c. sound of tin Zinngeschrei *n*, Kreischen *n* des Zinns
cracky rissig
cradle 1. Wiege *f (Wälzfräsen)*; Wälztrommel *f*; 2. Wiege *(Drehgestell)*; 3. *(Bw)* Hängebühne *f*, Hängekorb *m*; 4. Gabel *f (Fernsprechapparat)*; 5. *(Schiff)* Stapelschlitten *m*; Schleppwagen *m (zum Aufholen eines Schiffes)*; 6. *(Text)* Laufbügel *m (Jaquardstuhl)*
c. angle Wiegenwinkel *m (Verzahnung)*
c. car Kippwagen *m*
c. member Führungssegment *n*
c. mounting 1. *(Kfz)* Spannbandbefestigung *f*, Sattelbefestigung *f*; 2. *(Mil)* Wiegenlafette *f*
c. of engine *(Kfz)* Motorblock *m*
c. roll change gear Wechselrad *n* für Wiegenrollung
c. side tipping wagon Wiegenkipper *m*
c. switch *(El)* Gabel[um]schalter *m*
c. test roll Wiegenprüfrollung *f*
c. tip wagon Wiegenkipper *m*
craft 1. berufliches (handwerkliches) Können *n*; 2. Beruf *m*, Handwerk[sberuf *m*] *n*; 3. [kleines] Schiff *n*
c. training Berufsausbildung *f*, Facharbeiterausbildung *f*
crafting method handwerkliche Methode *f*
craftsman Facharbeiter *m*, Handwerker *m*
cramp [iron] Bauklammer *f*, Klammer *f*, Krampe *f*; Klemme *f*; Bankhaken *m*
crampons of climbing boots Bergschuhsteigeisen *npl*
cranage *(Schiff)* Krangeld *n*, Krangebühr *f*
crane Kran *m*
c. beam Kranbahnträger *m*
c. boom *(Am)* Kranausleger *m*
c. cab Krankabine *f*, Kranführerkorb *m*
c. excavator Baggerkran *m*
c. girder Kran[bahn]träger *m*
c. hook Kranlasthaken *m*
c. jib Kranausleger *m*
c. ladle Kranpfanne *f*
c. magnet Hubmagnet *m*
c. navvy Hochlöffelbagger *m*, Löffelhochbagger *m*
c. pillar Kransäule *f*, Kranständer *m*
c. rail Kran[bahn]schiene *f*
c. skip Kranförderkübel *m*

c. **trolley** Kranlaufkatze *f*, Laufkatze *f*
c. **way** Kran[lauf]bahn *f*
craneage Krananlage *f*
crank / to 1. *(Masch)* [ver]kröpfen, beugen, biegen; 2. kurbeln; 3. anwerfen, andrehen, ankurbeln *(Motor)*
crank 1. Kurbel *f*; Kulisse *f*; 2. Andrehkurbel *f*, Anwerfkurbel *f* *(Motor)*; 3. Kurbel, Handgriff *m*
c. **and flywheel pump** Kurbelgetriebeschwungradpumpe *f*
c. **and rocker** Kurbelschwinge *f*
c. **angle** Kurbeldrehwinkel *m*
c. **arm** 1. *(Masch)* Kurbelarm *m*; 2. *(Text)* Ladenschere *f*, Ladenzugstange *f*
c. **assembly** Kurbelgestänge *n*, Kurbeltrieb *m*
c. **axle** *s.* crankshaft
c. **bearing** Kurbel[wellen]lager *n*, Hauptlager *n*
c. **bearing pedestal** Kurbellagerbock *m*
c. **bolt** Kurbelbolzen *m*
c. **boss** Kurbelnabe *f*
c. **brace** Brustleier *f*, Brustbohrer *m*
c. **cam** Kurbelscheibe *f*
c. **capstan** *(Schiff)* Gangspill *n* mit Handkurbelantrieb
c. **chamber** Kurbelgehäuse *n*, Kurbelkammer *f*, Kurbelkasten *m*
c. **cheek** Kurbelschenkel *m*, Kurbel[kröpfungs]-wange *f*
c. **circle** Bahnkreis *m*, Kurbelkreis *m*
c.**-connecting link** Kurbelkupplung *f*
c. **cylinder** Kurbelzylinder *m*
c. **disk** Kurbelblatt *n*, Kurbelscheibe *f*
c. **displacement** Kurbelversetzung *f*
c. **drawing press** Kurbelziehpresse *f*
c. **drive** Kurbel[an]trieb *m*
c. **drive assembly** Kurbeltriebwerk *n*
c. **driving mechanism** Kurbelgetriebe *n*; Kurbelschleifentrieb *m*
c. **effort diagram** Drehkraftdiagrammlinie *f*, Tangentialdruckdiagramm *n*
c. **end** Kurbelkopf *m*
c. **eye** Kurbelwellenauge *n*
c. **gear** *(Masch)* Kulissenrad *n*; Hubscheibe *f*
c. **guard** Kurbelschutzhaube *f*
c. **guide** Kurbelschleife *f*
c. **handle** Andrehkurbel *f*; Handkurbel *f*, Kurbel *f*; Kurbelgriff *m*; Kurbelschlüssel *m*
c. **inductor** *(El, Elm)* Kurbelinduktor *m*
c. **latch** Kurbelriegel *m*
c. **lever** Doppelhebel *m*
c. **loom** *(Text)* Kurbelstuhl *m*
c. **mechanism** Kurbeltrieb *m*, Kurbelgetriebe *n*; Kurbelschleifenantrieb *m*
c. **mechanism extrusion press** *(Brg)* Schubkurbelpresse *f*
c.**-operated [side] window** *(Kfz)* Kurbelfenster *n*
c. **path** Kurbelweg *m*
c. **pin** *s.* crankpin
c. **pit** Kurbelwanne *f*
c. **planer** *(Holz)* Hobelmaschine *f* mit Kurbelantrieb
c. **press** Kurbelpresse *f*
c. **quadrilateral with converging links** konisches Kurbelviereck *n*, viergliedrige konische Kurbelkette *f*
c. **quadrilateral with isosceles members** gleichschenkliges Kurbelgetriebe *n*, gleichschenkliger Kurbeltrieb *m*
c. **race** Kurbelbilge *f*; Kurbeltrog *m*
c. **radius** Kurbelradius *m*, Hebelarm *m* der Kurbel
c.**-rotation angle** Kurbeldrehwinkel *m*
c. **shaft** *s.* crankshaft
c. **shaking screen** Kurbelschwingsieb *n*
c. **start motor** Anwurfmotor *m*
c. **starter** Kurbelwellenanlasser *m*
c. **stock** *(Led)* Kurbelwalke *f*
c. **stroke** Kurbelhub *m*
c. **throw** Kurbel[wellen]kröpfung *f*; Kurbelhub *m*
c.**-type folding press** Kurbelabkantpresse *f*
c. **web** Kurbelblatt *n*, Kurbelscheibe *f*, Kurbelwange *f*, Hubwange *f*
c. **windlass** Kurbelhaspel *f* *(m)*
c. **with drag link** Gegenkurbel *f*
crankcase Kurbelgehäuse *n*, Kurbelkasten *m*, Kurbelwanne *f*
c. **bleeder** Kurbelgehäuseentlüfter *m*
c. **bottom** Kurbelwanne *f*, Ölwanne *f*
c. **bottom half** Kurbelgehäuseunterteil *n*
c. **breather** Kurbelgehäuseentlüfter *m*, Kurbelkastenentlüfter *m*
c. **breather pipe** Kurbelgehäuseentlüftung[s-rohr *n*] *f*
c. **compression** Kurbelgehäuseverdichtung *f*, Vorverdichtung *f* *(Zweitakter)*
c. **dilution** Kurbelgehäuse-Ölverdünnung *f*
c. **door** Kurbelgehäusefenster *n*
c. **drain recommendation** *(Kfz)* Ölwechselempfehlung *f*
c. **drainage periods** *(Kfz)* Ölwechselzeiten *fpl*
c. **scavenging** Kurbelgehäusespülung *f*, Kurbelkastenspülung *f*
c. **sump** Ölwanne *f*, Kurbelwanne *f*, Motorwanne *f*
c. **supercharger** Kurbelkastenvorverdichter *m*
c. **top half** Kurbelgehäuseoberteil *n*
c. **ventilation system** Kurbelgehäuseentlüftung *f*
cranked gebogen; gekröpft

c. lever Gelenkhebel *m*
c. portion of shaft Kurbelkröpfung *f*
c. shaft gekröpfte Kurbelwelle *f*
c. wing *(Flg)* Knickflügel *m*
cranking handle Handandrehkurbel *f*
c. jaw Andrehklaue *f*
c. motor Anlasser *m*, Anlaßmotor *m*, Starter *m*
c. speed Anwerfdrehzahl *f*
c. torque Anwerfdrehmoment *n*
crankless engine kurbelwellenloser Motor *m*
crankness Rankheit *f (eines Schiffes)*
crankpin Kurbelzapfen *m*; Hubzapfen *m (Kurbelwelle)*
c. bearing Kurbelzapfenlager *n*
c. clamping screw Kurbelwellenklemmbolzen *m*
c. end Kurbelzapfenende *n*
c. grinder Kurbelzapfenschleifmaschine *f*
c. level Kurbelzapfenrichtwaage *f*
cranks moving in opposite directions *(Masch)* gegenläufige Kurbeln *fpl*, Gegendrehungskurbeln *fpl*
c. rotating in the same direction *(Masch)* gleichläufige Kurbeln *fpl*
crankshaft Kurbelwelle *f*
c. bearing Kurbel[wellen]lager *n*, Hauptlager *n*
c. bearing-end of frame Rahmenkopf *m*, Lagerbalkenkopf *m*
c. counterweight Kurbelwellengegengewicht *n*
c. deflection Kurbelwellendurchbiegung *f*
c. degree Kurbelwellengrade *mpl (Ventilsteuerung)*
c. drill Tieflochspiralbohrer *m*
c. drill having heavy web Tieflochspiralbohrer *m* mit verstärktem Kern
c. drive Kurbeltrieb *m*
c. eccentricity Schlag *m* der Kurbelwelle
c. extension Kurbelwellenverlängerung *f*
c. gear wheel Kurbelwellenrad *n*
c. grinding machine Kurbelwellenschleifmaschine *f*
c. journal Kurbelwellenende *n*, Kurbelwellenschaft *m*
c. lathe Kurbelwellendrehmaschine *f*
c. milling machine Kurbelwellenfräsmaschine *f*
c. pin Kurbelzapfen *m*
c. position Kurbelstellung *f*
c. shoulder Kurbelwellenbund *m*
c. throw Kurbelwellenkröpfung *f*
c. timing gear Kurbelwellenrad *n*
c. torsional vibration Kurbelwellendrehschwingung *f*
c. vibrations Kurbelwellenschwingungen *fpl*
crash 1. Krach *m*; Zusammenstoß *m*; Bruch *m*; 2. Grobleinen *n*
c. accelerometer Beschleunigungsmesser *m* für Luftfahrtzwecke
c. barrier Leitschiene *f*, Sicherheitsgeländer *n (Straßenbau)*
c. boat Fliegerrettungsboot *n*
c. helmet Sturzhelm *m*
c. landing Bruchlandung *f*
crashes lang andauernde starke atmosphärische Störungen *fpl (Rundfunk)*
crater / to auskolken
crater 1. *(Geol)* Krater *m*; 2. Auskolkung *f*, Kolkung *f*; 3. Mulde *f (Schweißtechnik)*
c. mountain *(Astr)* Ringgebirge *n (des Mondes)*
c.-resistant kolkfest
cratering Auskolken *n*, Auskolkung *f*, Kolkbildung *f*
crawler 1. Raupenschlepper *m*, Gleiskettenzugmaschine *f*, Kettenschlepper *m*; 2. Gleiskette *f*, Raupenband *n*, Raupenkette *f*
c. chain Raupenkette *f*
c. crane Raupenkran *m*, Gleiskettenkran *m*
c. drive Raupen[ketten]antrieb *m*, Gleiskettenantrieb *m*
c. link Raupenkettenglied *n*
c. mounted Raupen-, auf Raupenfahrwerk, auf Raupenfahrgestell
c.-mounted loader *(Brg)* Laderaupe *f*
c. track Kettenfahrwerk *n*, Raupe[nkette] *f*
c. tractor Kettenschlepper *m*, Raupenschlepper *m*
c. tractor crane Raupenschlepperkran *m*, Gleiskettenschlepperkran *m*
c. tractor truck wheel Raupenschlepprad *n*
c.-type crane Raupen[ketten]kran *m*
c.-type vehicle Kettenfahrzeug *n*
crawling *(Fmt)* Intrittkommen *n (mit einer Unterfrequenz)*
c. speed Kriechgang *m (insbesondere des Traktors)*
craze mill *(Met)* Zinnpochwerk *n*
crazing 1. Brandriß *m (in Kokillen)*; 2. Haarriß[bildung *f*] *m*
creaking of tin Zinnschrei *m*
cream / to entrahmen *(Milch)*
cream heater Rahmerhitzer *m (Butterei)*
c. of milk Rahm *m*, Sahne *f*
c. of tartar *(Ch)* Kaliumbitartrat *n*, Weinstein *m*
c. separator Entrahmer *m*, Milchzentrifuge *f*, Separator *m*
creamery Molkerei[geschäft *n*] *f*
crease 1. *(Text)* Bügelfalte *f*, Falte *f*, Kniff *m*; Knitter[falte *f*] *m*; 2. *(Masch)* Sicke *f*
c. angle *(Text)* Knitterwinkel *m*
c. formed in the washer *(Text)* Waschfalte *f*
c. mark *(Text)* Faltspur *f*, Faltstelle *f*
c.-proof knitterfest

c. recovery *(Text)* Knittererholung *f*
c. recovery angle *(Text)* Knittererholungswinkel *m*
c. resistance *(Text)* Knitterwiderstandsvermögen *n*; Knitterfestigkeit *f*
c.-resistant finish *(Text)* Knitterarmausrüstung *f*
c.-retention *(Text)* Faltenbeständigkeit *f*
creaser *(Typ)* Nutmaschine *f*; Rillmaschine *f*
creasing die Sickeneisen *n*, Sickenform *f*
c. property *(Text)* Knittereigenschaft *f*
c. resistance *(Text)* Knitterfestigkeit *f*
c. test Knitterversuch *m*, Knitterprobe *f (Papier)*
c. tool Sickenzug *m*
create / to [er]schaffen; hervorbringen, produzieren
c. a magnetic field ein Magnetfeld aufbauen
c. patterns by the warp *(Text)* aus der Kette mustern
creatine *(Ch)* Kreatin *n*, Methylguanidinessigsäure *f*
creatinine *(Ch)* Kreatinin *n*, Anhydrid *n* der Methylguanidinessigsäure
creative work schöpferische Arbeit *f*
crèche Kinderkrippe *f*
credit symbol *(Büro)* Habenzeichen *n*
Creed printing telegraph system Creed-Fernschreiber *m*
creel *(Text)* Aufsteckgatter *n*, Spulenständer *m*; Spulengestell *n*; Gatter *n*
c. board *(Text)* Aufsteckbrett *n*
c. bobbin *(Text)* Ablaufspule *f*
c. frame *(Text)* Aufsteckrahmen *m*
c. in three heights for double roving dreistöckiges Gatter *n* für doppeltes Vorgespinst
c. peg Spulenständerstift *m*
c. table Spulentisch *m*
creel the bobbins / to *(Text)* Spulen aufstecken
creep / to 1. *(Met, Plast)* fließen, kriechen *(Festigkeit)*; 2. *(El)* kriechen *(Strom)*; 3. *(Brg)* quellen, heben *(Sohle)*; 4. wandern *(Schiene)*; 5. ranken
c. out ausschlüpfen *(Tiere aus Eiern)*
creep 1. *(Met, Plast)* Fließen *n*, Kriechen *n (Festigkeit)*; 2. *(El)* Kriechen *(Strom)*
c. action *(Masch)* Dehnungsschlupf *m (Riementrieb)*
c. behaviour Dauerstandverhalten *n*, Kriechverhalten *n*
c. feed Kriechgang *m*; Langsamvorschub *m*
c. leakage Kriechentladung *f*, Kriechstromverlust *m*
c. limit Dauerdehngrenze *f*, Kriechgrenze *f*; Dauer[stand]festigkeit *f*, Kriechfestigkeit *f*
c. motion Kriechgang *m*
c. rate Kriechgeschwindigkeit *f*
c. rate stress curve Zeit-Dehnungskurve *f*
c. resistance Kriechfestigkeit *f*, Dauerstandfestigkeit *f*
c. resistance limit *s.* c. limit
c. speed 1. Kriechgeschwindigkeit *f*; 2. *(Masch)* Schleichganggeschwindigkeit *f*
c. strain Kriechdehnung *f*
c. strain-time curve Zeit-Dehnungskurve *f*
c. strength Kriechfestigkeit *f*, Dauer[stand]festigkeit *f*
c. strength depending on time Zeitstandfestigkeit *f*
c. test Dauerstandversuch *m*; *(El auch:)* Dauerprüfung *f* auf Kriechstrom
c. testing machine Zeitstandprüfmaschine *f*
creepage *(El)* 1. Kriechen *n*; 2. Kriechstrom *m*
c. distance Kriechstrecke *f*
c. path Kriechweg *m*
creeper 1. *(Masch)* endlose Kette *f*; endloses Beförderungstuch *n (Textilmaschinen)*; 2. Kletterpflanze *f*, Rankengewächs *n*
c. tractor Raupenschlepper *m*
creeping Kriechen *n*, langsame Vertikalbewegung *f*
c. current Kriechstrom *m*
c. distance *(El)* Kriechstrecke *f*, Kriechweg *m*
c. movement Kriechen *n*, Kriechbewegung *f*
c. speed Geschwindigkeit *f* des Kriechganges; Kriechgang[geschwindigkeit *f*] *m*
c. stimulus *s.* accumulating stimulus
creosol Kreosol *n*
creosote / to mit Kreosot (Teeröl) tränken
creosote Kreosot *n*
c. oil Kreosotöl *n*
creosoting Imprägnieren *n* mit Kreosot
crepe *s.* crêpe
crêpe *(Text)* Flor *m*, Krepp *m*
c. paper Kreppapier *n*
c. rubber Kreppgummi *m*
c. tissue paper Seidenkreppapier *n*
c. yarn Kreppgarn *n*
crescent 1. *(Math)* Kreiszweieck *n*; 2. *(Masch)* sichelförmiges Widerlager *n*
c. type spanner verstellbarer Schraubenschlüssel *m*
cresol dye Kresolfarbe *f*
c. resin Kresolharz *n*
crest 1. *(Math, Ph)* Scheitelwert *m*, Schwingungsbauch *m*, Wellenberg *m*; 2. *(Bw)* Krone *f*, Kuppe *f*, Scheitel *m*; 3. *(Masch)* Gewindespitze *f*, Zahnkopf *m*; 4. *(Geol)* Scheitel *(Falte)*
c. clearance *(Masch)* Kopfspiel *n (Verzahnung)*; Spitzenspiel *n (Gewinde)*
c. factor *(El)* Scheitelfaktor *m*; Spitzenwert *m*
c. line *(El)* Scheitellinie *f*
c. of thread Spitze *f* eines Gewindes; Gewindespitze *f*
c. of tooth Zahnspitze *f*; Zahnoberteil *n*

c. template Spitzenlehre *f* *(Gewinde)*
c. truncation *(Masch)* Gewindespitzenabrundung *f*
c. value *(Ph, El)* Scheitelwert *m*, Spitzenwert *m*
c. voltage Scheitelspannung *f*, Scheitelwert *m* der Spannung
c. voltmeter Scheitel[wert]spannungsmesser *m*, Spitzenspannungsmesser *m*, Spitzenspannungsvoltmeter *n*, Voltmeter *n* mit Ablesung des Scheitelwertes
cresting *(Bw)* Aufstand *m* *(z. B. an Firstziegeln)*
cresylate *(Ch)* Kresylat *n*
cresylic acid Kresolsäure *f*, technisches Kresol *n*
c. resin Kresolharz *n*
cretacious Kreide-; kreidehaltig; aus Kreide
cretonne *(Text)* bedruckter Möbelkattun *m*, Kretonne *m* *(f)*
crevice Spalt *m*, Riß *m*
crew 1. Brigade *f*, Trupp *m*; Besatzung *f*; 2. Mannschaft *f*, Schiffsmannschaft *f*; 3. Flugzeugbesatzung *f*, Flugpersonal *n*; 4. Zugbegleitmannschaft *f*, Zugpersonal *n*
c. compartment Abteil *n* für Zugpersonal
crew's bunks *(Schiff)* Mannschaftsschlafraum *m*
c. mess-room *(Schiff)* Mannschaftsmesse *f*
c. quarter *(Schiff)* Mannschaftsraum *m*, Logisraum *m*
c. space *(Schiff)* Mannschaftsraum *m*, Logisraum *m*
crib / to *(Brg)* ausbauen, verzimmern
crib Krippe *f*; Futterkrippe *f*; Holzunterbau *m*; Holzfundament *n*
c. chute *(Brg)* in Holz ausgebautes Rolloch *n*
c.-ring *(Brg)* Schachtgeviert *n*, Geviert *n*
cribbing 1. *(Brg)* Bolzenschrotausbau *m*, Bolzenschrotzimmerung *f*; 2. *(Schiff)* Stapelung *f* *(auf Helling)*, Kielstapelung *f*
cribwork *(Brg)* Schachtgeviert *n*, Geviert *n*
cricket bat Kricketschläger *m*
criminal case Strafprozeß *m*
c. procedure Strafverfahren *n*
Criminal Law Strafgesetz *n*
crimp / to 1. *(Text)* falten, kräuseln; 2. sicken
crimp 1. *(Text)* Kräuselung *f*; Falte *f*; 2. *(Schiff)* Heuerbaas *m*
c. of the warp Kettkräuselung *f*, Kräuseln *n* der Kettfäden *(Fehler)*
c. retentivity *(Text)* Fähigkeit *f*, die Kräuselung aufrechtzuerhalten
c. warp Faltenkette *f*
crimped yarn Kräuselgarn *n*
crimper *(Brg)* Würgezange *f* *(Schießarbeit)*
crimping 1. *(Text)* Kräuseln *n*, Kräuselung *f*; 2. Sicken *n*,
c. arc of the fibre Kräuselungsbogen *m* der Faser
c. machine 1. *(Text)* Kräuselmaschine *f*; 2. Sickenmaschine *f*
crimpy wool fibre gekräuselte Wollfaser *f*
crinkle crêpe Baumwollkräuseldruckgewebe *n*
c. varnish Kräusellack *m*
crippled leap-frog test *(Am)* *(Dat)* Teildurchprüfung *f*
crippling Ausknicken *n*, Knicken *n*; Ausbeulen *n*, Beulen *n*
c. load Knicklast *f*
c. test Knickversuch *m*
crispening Umrißversteilerung *f* *(Fernsehbild)*
criss-cross in Längs- und Querrichtung
criss-crossing kreuzweises Aufeinanderlegen (Aufbringen) *n*
crit *(Kern)* kritische Masse *f* [ohne Reflektor]
criterion Kriterium *n*, Gradmesser *m*
c. of convergence *(Math)* Konvergenzkriterium *n*
critical kritisch
c. absorption *(Ph)* kritische Absorption *f*
c. altitude Gleichdruckhöhe *f*
c. amount *(Kern)* kritische Masse (Menge) *f*
c. angle *(Opt)* Grenzwinkel *m* *(Brechung)*
c. anode spacing kritischer Anodenabstand *m*
c. area of extraction *(Brg)* Vollfläche *f*
c. behaviour *(Kern)* kritisches Verhalten *n*
c. build-up resistance kritischer Widerstand *m* für die Selbsterregung (Auferregung)
c. build-up speed kritische Drehzahl *f* für die Selbsterregung (Auferregung) *f*
c. case *(Ph)* Grenzfall *m*
c. condition *(Ph)* Grenzbedingung *f*; kritische Bedingung *f*
c. controlling current of a cryotron *(BMSR)* kritischer Steuerstrom *m* eines Cryotrons
c. cooling rate kritische Abkühlungsgeschwindigkeit *f*
c. coupling *(El)* kritische Kopplung *f*
c. current *(El)* Grenzstromstärke *f*
c. docking draft *(Schiff)* kritischer Tiefgang *m* beim Docken *(Stabilität gleich Null)*
c. focusing *(Opt)* Scharfeinstellung *f*
c. frequency Grenzfrequenz *f* *(z. B. Filter)*; kritische Frequenz *f*
c. grid current Gitterzündstrom *m*, Gitterstrom *m* im Augenblick der Zündung
c. grid voltage Gitterzündspannung *f* *(Gasentladungsröhren)*
c. head kritisches Gefälle *n* *(Talsperre)*
c. interval *(Met)* kritisches Intervall *n*, Haltepunktsdauer *f*, Haltezeit *f*
c. load kritische Belastung (Last) *f*, Knicklast *f*, Beullast *f*
c. mass *(Kern)* kritische Masse *f* *(Kettenreaktion)*

c. path method *(Bw)* CPM-Verfahren *n*, Methode *f* des „kritischen Wegs" *(Netzwerkplanung)*
c. point 1. *(Ph)* kritischer Punkt *m (Zustandsdiagramm)*; 2. *(Met)* Umwandlungspunkt *m*, Haltepunkt *m (Phasendiagramm)*
c. pressure at the outlet opening kritischer Mündungsdruck *m (Turbine)*
c. range *(Met)* kritisches Intervall *n*, Umwandlungsbereich *m*; Haltezeit *f*, Haltepunktsdauer *f*
c. rate of cooling kritische Abkühlungsgeschwindigkeit *f*
c. reactor *(Kern)* kritischer Reaktor *m*
c. resistance *(El)* Grenzwiderstand *m*; kritischer Widerstand *m*
c. self-excitation kritische (volle) Selbsterregung *f*
c. speed of rotation *(Büro)* kritische Drehzahl *f*
c. state *(Kern)* kritischer Zustand *m (eines Reaktors)*
c. stress kritische Spannung *f*, Knickspannung *f*
c. temperature kritische Temperatur *f*, Umwandlungstemperatur *f*
c. temperature range *(Met)* kritisches Intervall *n*; Umwandlungsbereich *m*; Haltezeit *f*, Haltepunktsdauer *f*
c. value kritischer Wert *m*
c. velocity *(Ph)* Grenzgeschwindigkeit *f*
c. voltage *(El)* Grenzspannung *f*
c.-voltage parabola Magnetfeldröhrenkennlinie *f*
c. wavelength *(El)* Grenzwellenlänge *f*
crizzling Haarrißbildung *f* auf Glas
crochet Häkelarbeit *f*
c. file Gabelfeile *f*
c. hook Häkelnadel *f*
c. yarn Häkelgarn *n*
crock-resistant *(Text)* reibecht
crockery Topfwaren *fpl*; Haushaltkeramik *f*
crocking *(Text)* Abreiben *n*
crockmeter *(Text)* Gerät *n* zur Messung der Reibechtheit
crocodile clip *(El)* Krokodilklemme *f*
c. shear *(Masch)* Hebelschere *f*
crocoisite *s.* crocoite
crocoite Krokoit *m*, Rotbleierz *n*
Crompton's mule frame *(Led)* Schemelschaftmaschine *f*
crook Knieholz *n*, Krummholz *n*
crooked 1. *(Holz)* gekrümmt, nichtschnürig; verzogen; krumm; 2. *(Math)* ungerade
Crookes' dark space *(Ph)* Crookesscher Dunkelraum *m*
crooking of the tubes *(Wärm)* Krummziehen *n* der Rohre
crop / to 1. abschneiden, kurzschneiden; scheren; kupieren; 2. einsammeln, ernten; 3. besäen, bepflanzen
c. out *(Geol)* ausbeißen, ausstreichen
c. the nap of a fabric *(Text)* Gewebe kahlscheren
crop 1. Getreideernte *f*; Ernteertrag *m*; 2. Tierhaut *f* für Sohlenleder; 3. Abschnitt *m*; 4. *(Brg)* Ausbiß *m*, Ausgehendes *n*
c. failure Mißernte *f*
cropped piece Spaltstück *n*
cropping *(Brg)* Ausbiß *m*, Ausgehendes *n*
c. and shearing machine *(Text)* Putz- und Schermaschine *f*
c. machine *(Text)* Tuchschermaschine *f*
c. punch Schneidstempel *m*; Beschneidestempel *m*
Crosby clip Seilklemme *f (gerillte Lasche und U-förmige Schraube mit zwei Muttern)*
cross / to 1. kreuzen; überqueren *(Straße)*; 2. durchkreuzen, durchqueren; *(sich)* schneiden; 3. verschränken *(Riemen)*
c.-multiply *(Math)* über Kreuz multiplizieren
cross 1. Kreuzrohrstück *n*; 2. *(El)* Berührung *f*, Leitungsberührung *f*; Querschluß *m*; 3. *(Typ)* Kreuzmaß *n*
c.-adit *(Brg)* Querschlag *m*
c. adjustment *(Masch)* Querverstellung *f*
c. ampere-turns Querfeldamperewindungen *fpl*
c. arm Ausleger *m*; Querbalken *m* der Einständerhobelmaschine; Querarm *m*, Traverse *f*; Querträger *m*
c.-band *s.* crossband
c. bar 1. *(Bw, Masch)* Ausleger *m*; Querriegel *m*, Querstab *m*, Querstange *f*, Querstück *n*, Querträger *m*, Traverse *f*; 2. *(Masch, Wärm)* Spanneisen *n*; 3. *(Typ)* Mittelsteg *m (des Schließrahmens)*; 4. Sprunglatte *f*; Torlatte *f*
c.-bar selector *(El)* Crossbarwähler *m*, Koordinatenschalter *m*
c.-bar switch Kreuzschienenschalter *m (Matrixschalter)*
c.-bar transformer *(Elm)* Jochwandler *m*
c.-beam *(Bw, Masch)* Ausleger *m*; Holm *m*, Querbalken *m*, Querträger *m*, Traverse *f*
c. bearing Kreuzpeilung *f*, Doppelpeilung *f (Radar)*
c.-beater mill *(Ch)* Schlagkreuzmühle *f*
c.-bedding Kreuzschichtung *f*
c. bit Kreuzbohrer *m*
c.-bitt *(Schiff)* Kreuzpoller *m*
c. bombardments *(Kern)* gekreuzte Kernreaktionen *fpl*, Kreuzbeschuß *m*
c. brace Querstrebe *f*
c. bracing 1. *(Bw)* Queraussteifung *f*, Querverband *m*, Querverstrebung *f*; 2. *(Flg)* Diagonalverspannung *f*, Kreuzverspannung *f*

c. bracket Querriegel *m*
c.-breaking strength Durchbiegungsfestigkeit *f*, Querbiegefestigkeit *f*
c.-brushing machine *(Text)* Querbürstmaschine *f*
c. bulkhead *(Schiff)* Querschott *n*
c. bunker *(Schiff)* Querbunker *m*
c. call *(Fmt)* Rückruf *m*
c. channel interference *(El)* Kreuzmodulationsstörung *f*
c.-channel steamer Kanaldampfer *m*
c.-channel vessel Kanalschiff *n*
c.-checking system *(Büro)* Lochstreifenprüfungssystem *n* durch querverlaufendes Prüfen, Querprüfungssystem *n*
c. coil Kreuzspule *f*
c.-coil aerial Kreuzantenne *f*
c. conduction *(El)* Kreuzleitung *f*
c.-connecting board *(Fmt)* Verteilertafel *f* *(für Stöpselbetrieb)*; Kreuzschienenverteiler *m*
c.-connecting terminal *(Am)* *(Fmt)* Überführungskasten *m*
c.-connection *(El)* Querschaltung *f*; Querverbindung *f*
c. conveyor Querförderer *m* *(Sammelpresse)*
c.-country capacity *(Kfz)* Geländegängigkeit *f*
c.-country car *(Kfz)* Geländewagen *m*
c.-country mill Zickzacktrio *n* *(Walzstraße)*
c.-country pipeline Überlandrohrleitung *f*
c.-country properties *(Kfz)* Geländegängigkeit *f*
c.-country racing ski Langlaufski *m*
c.-country tyre *(Kfz)* Geländereifen *m*
c.-country vehicle *(Mil)* Geländefahrzeug *n*
c. coupling *(Fmt)* Übersprechkopplung *f*
c.-curve diagram *(Schiff)* Schottquerkurven *fpl*
c.-cut 1. *(Holz)* Hirnschnitt *m*, Querschnitt *m* *(Sägen)*; 2. *(Brg)* Querschlag *m*, Querstrecke *f*
c.-cut chisel Kreuzmeißel *m*
c.-cut frame saw Quergatter *n* *(Sägewerk)*
c.-cut hammer Schrämhammer *m*
c.-cut method *(Brg)* Querbau *m*
c.-cut method in ascending order aufwärtsgeführter Querbau *m*
c.-cut method in descending order abwärtsgeführter Querbau *m*
c.-cut saw Quersäge *f*, Säge *f* für Querschnitt
c.-cut saw set Schränkzange *f* für Quersägen
c. cutter 1. *(Typ)* Querschneider *m*; 2. *(Brg)* Schrämmaschine *f*
c.-cutting *(Holz)* Abkürzen *n*, Quersägen *n*, Querschneiden *n*
c.-cutting chisel Kreuzmeißel *m*
c.-cutting machine *(Text)* Querschermaschine *f*
c. direction Querrichtung *f*
c.-drain Querrinne *f*
c.-draught carburettor *(Kfz)* Flachstromvergaser *m*
c.-dyeing *(Text)* Überfärben *n*, Überfärbung *f*
c.-dyeing in acid liquor saures Überfärben *n*
c.-face *(Brg)* Querort *n*
c.-fall *(Bw)* Quergefälle *n*, Querneigung *f* *(Straßenbau)*
c. feed Quervorschub *m*, Planvorschub *m*; Planzug *m*
c.-feed ignition Querstromzündung *f*
c.-feed mechanism Plantrieb *m* *(Drehmaschine)*
c.-feed screw Plan[vorschub]spindel *f*
c.-feed stop Anschlag *m* für Planvorschub
c. field *(Elm)* Querfeld *n*
c. fire *(El)* Induktionsstörung *f*
c.-flow heat exchanger Querstromwärme[aus]tauscher *m*
c. fold *(Typ)* Querfalz *m*
c. frame Querrahmen *m*
c. girder *(Bw)* Querträger *m*
c. girth *(Masch)* Querhaupt *n*, Traverse *f*
c.-grain quer zur Faserrichtung
c. grain Querfaser *f*; Hirnseite *f*
c.-grain leaf Hirnholzplatte *f*
c.-grain veneer Querfurnier *n*
c.-grained 1. *(Holz)* Hirn gegen Hirn geschnitten; 2. quer zur Faser
c.-grained stuff Hirnholz *n*
c.-hair[s] *(Opt)* Fadenkreuz *n*, Strichkreuz *n*
c.-handle *(Masch)* Handkreuz *n*
c.-hatch kreuzweise (schräg) schraffieren
c. hatching Kreuzschraffierung *f*, Kreuzschraffur *f*
c. head 1. *(Typ)* Zwischentitel *m*, Zwischentext *m*; 2. Querkopfziehform *f* *(Strangpresse)*
c. heading *(Brg)* Querort *n*, Querstrecke *f*
c. hole Querbohrung *f*
c.-hole nut Kreuzlochmutter *f*
c. induction *(Fmt)* Nebensprechen *n*, Übersprechen *n*
c. jaw Kreuzschnabel *m* *(Schiebelehre)*
c.-jet circuit breaker Querstromschalter *m* [ölarmer] Leistungsschalter *m* mit Querbesprühung
c. joint *(El)* Kreuzkupplung *f*, Kreuzverbindung *f*; Kreuzstutzen *m*; Kreuzverzapfung *f*
c. jointer *(El)* Kreuzverbinder *m*; Querverbinder *m*
c.-keyed coupling Kreuzscheibenkupplung *f*
c.-knurl kreuzrändeln
c.-knurled head Kreuzrändelkopf *m* *(Schraube)*
c.-knurling Kreuzrändeln *n*, Kreuzrändelung *f*
c.-laid rope Kreuzschlagseil *n*
c. lathe *(Text)* Quertritt *m*
c.-lay rope Kreuzschlagseil *n*

c. level *(Opt)* Kreuzlibelle *f*
c. link *(Ch)* Querverbindung *f*; Vernetzung *f*
c.-linkage *(Plast)* Vernetzung *f*
c. member *(Bw, Masch)* Querträger *m (Rahmen)*
c. modulation *(Fmt)* Kreuzmodulation *f*, Quermodulation *f*
c. movement Querbewegung *f*; Planbewegung *f (Drehen)*
c. neutralization *(El)* Kreuzneutralisation *f*, Gegentaktneutralisation *f*
c.-over *(El)* [erster] Bündelknoten *m (Elektronenoptik)*; Überkreuzung *f (von Leitungen)*
c.-over bend Übergangsbogen *m (Rohrleitung)*
c.-over clamp *(El)* Kreuzungsklemme *f (Freileitung)*
c.-over design Kreuzplan *m*
c.-over frequency 1. *(Ak)* Übergangsfrequenz *f*; 2. *(El)* Überschneidungsfrequenz *f*
c.-over valve *(Masch)* Wechselventil *n*, Umschaltventil *n*
c. pane *(Masch)* Kreuzschlaghammer *m*; Vorschlagpinne *f*
c.-peen sledge Kreuzschlaghammer *m*
c. piece 1. Kreuzstück *n*; 2. *(Schiff)* Querholz *n*
c. pin Kreuzzapfen *m*
c. pointer Kreuzzeiger *m (am Meßinstrument)*
c.-pointer instrument Kreuzzeigerinstrument *n*
c. polarized kreuzpolarisiert
c. position Kreuzstellung *f*
c. product *(Math)* Vektorprodukt *n*, vektorielles Produkt *n*
c. profiling *(Masch)* Plankopieren *n*
c. rail Querbalken *m (Werkzeugmaschine)*
c. raising *(Text)* Querrauhen *n*
c. ratio *(Math)* Doppelverhältnis *n*
c. reel *(Text)* Kreuzweife *f*
c. reeling *(Text)* Kreuzhaspelung *f*
c. reference Kreuzverweisung *f (in Büchern)*
c. rib Querrippe *f*
c.-ribbed querverrippt; mit Querrippen
c.-ribbed rubbing leather *(Text)* kreuzgerippte Nitschelhose *f*
c.-ribbon feed *(Text)* Bandübertragung *f* mit Kreuzung
c. ring Kreuzring *m*
c.-road Kreuzweg *m*, Querstraße *f*
c. rolling Schrägwalzwerk *n*
c. scale *(Masch)* Strichplatte *f*
c. seam Kreuznaht *f*
c. section 1. [geometrischer] Querschnitt *m*; 2. Wirkungsquerschnitt *m*
c. section of fracture Bruchquerschnitt *m*
c. section of port *(Wärm)* Kanalquerschnitt *m*
c. section under tension *(Mech)* Zugquerschnitt *m*
c.-sectional area Querschnittsfläche *f*
c.-sectional view *(Masch)* Querriß *m*, Schnittzeichnung *f*
c. shaft *(Masch)* Querwelle *f*
c.-shaped hole Kreuzloch *n*
c.-shaped link *(Masch)* Kreuzlenker *m*
c. shed *(Text)* Kreuzfach *n*
c.-sill *(Eis)* Schwelle *f*
c. sleeper Querverbindung *f*
c. sleeve Kreuzmuffe *f*
c. slide 1. Querschlitten *m*; Planschlitten *m*; Quersupport *m*; Unterschieber *m*; 2. Querbalken *m*
c. slide rest Plansupport *m*, Quersupport *m*
c.-slide screw Planspindel *f*
c. slot Kreuznut *f*
c.-span Querdraht *m (Fahrleitung)*
c. spider-line Fadenkreuz *n*
c.-stick *(Typ)* Kreuzsteg *m*
c. stiffening Querversteifung *f*
c.-stitch *(Text)* Kreuzstich *m*
c. stop Anschlag *m* für Planvorschub, Queranschlag *m*, Plananschlag *m*
c. strip Querleiste *f*
c. stud Kreuzstrebe *f*
c. switch *(Eis)* Kreuzweiche *f*
c. talk *(Fmt)* Übersprechen *n*
c.-talk attenuation Nebensprechdämpfung *f*
c.-talk coupling Nebensprechkopplung *f*
c.-talk meter Nebensprechdämpfungsmesser *m*
c.-talk proof nebensprechfrei
c.-talk trouble Übersprechstörung *f*, Kreuzmodulation *f*
c. thread Kreuzgewinde *n*
c.-tie 1. Querhalter *m*; Querstrebe *f (am Mast)*; 2. Schwelle *f (Gleisbau)*; 3. *(Text)* Kreuzschnürung *f (Jacquard)*
c. tieing *(Masch)* Fachwerkverband *m*, Dreieckverband *m*
c. timber Achsriegel *m*
c. track on turn-table Fahrrille *f (Drehscheibe)*
c. tube boiler Quersieder[rohrkessel] *m*
c. twill *(Text)* Kreuzköper *m*
c.-type folded dipole *(Fmt)* Kreuzfaltdipol *m*
c. vault *(Bw)* Kreuzgewebe *n*
c. walk Fußgängerüberweg *m*
c.-wall Querwand *f*
c. weaving Gazeweberei *f*
c. wheel *(Masch)* Malteserkreuz *n*
c. wind *(Flg)* Seitenwind *m*
c.-wind landing Seitenwindlandung *f*, Landung *f* mit Seitenwind
c.-wind take-off Seitenwindstart *m*, Start *m* mit Seitenwind
c.-winged guided missile *(Rak)* Kreuzflügler *m*

c.-wire micrometer Fadenkreuzmikrometer *n*
c.-wire sight *(Mil)* Fadenkreuzvisier *n*
c.-wound bobbin *(Text)* Kreuzspule *f*
Cross method *(Bw)* Cross-Verfahren *n*, Momentenausgleichsverfahren *n*, Methode *f* wiederholter Momentenverteilung *(Statik)*
crossband *(Holz)* Absperrlage *f*, Absperrfurnier *n*
c.-twist thread linksgedrehtes Garn *n*
crossbanded veneered panels abgesperrte und furnierte Paneele *npl*; abgesperrte Holzplatte *f*
crossbanding *(Holz)* kreuzweise Absperrung *f*
crosscut Kreuzhieb *m* (*s. a.* cross-cut)
crossed gekreuzt
c.-axes angle Achskreuzwinkel *m (Zahnradschaben)*
c.-axes point Achskreuzpunkt *m (Zahnradschaben)*
c.-axes shaving Schaben *n* mit gekreuzten Achsen
c.-belt drive gekreuzter Riementrieb *m*, Kreuzriementrieb *m*
c.-coil aerial Kreuzrahmenantenne *f*
c.-coil device *(El)* Kreuzspulgerät *n*
c.-coil instrument *(El)* Kreuzspulgerät *n*, Kreuzspulinstrument *n*
c.-coil measuring instrument *(El)* Kreuzspulmeßgerät *n*
c. coils *(Elm)* Kreuzspulen *fpl*, übereinandergreifende Spulen *fpl*
c. crank mechanism *(Masch)* durchschlagendes Kurbelgetriebe *n*; überschlagendes (gekreuztes) Gelenkviereck *n*
c.-field machine *s.* amplidyne
c.-fields [electron-beam] multiplier *(Dat)* Kreuzfeldmultiplikator *m*
c. harness verkreuzter Harnisch *m*
c. helical gears Schraubenräder *npl*; Schraubtrieb *m*
c.-loop aerial Kreuzrahmenantenne *f*
c. nicols *(Opt)* gekreuzte Nicolsche Prismen *npl*
c. threads Fadenkreuz *n*
crossfire 1. *(Mil)* Kreuzfeuer *n*; 2. *(El)* Induktion *f* durch Telegrafie
crossfooting *(BMSR, Büro)* Querrechnen *n*
crosshead 1. Kreuzkopf *m*; Querhaupt *n*, Traverse *f*; 2. Traverse, Querträger *m*; Spanntraverse *f (Spannbeton)*; 3. Führungsschlitten *m (Abteufkübel)*
c. beam Kolbenstangentraverse *f*; Querhaupt *n*
c. bearing 1. Kreuzkopf[zapfen]lager *n*; 2. Kreuzkopfgleitplatte *f*
c. cap Kreuzkopf[lager]deckel *m*
c. end Kreuzkopfende *n (der Schubstange)*
c. end of the piston rod Kreuzkopfende *n* der Kolbenstange
c. engine Kreuzkopfmaschine *f*
c. for four slide bars viergleisiger Kreuzkopf *m*
c. guide Kreuzkopfführung *f*, Geradführung *f*
c. pin Kreuzkopfzapfen *m*, Kreuzkopfbolzen *m*, Gabelzapfen *m*
c.-pin bearing Kreuzkopfzapfenlager *n*
c. piston Kolben *m* mit Kreuzkopfführung, Kreuzkopfkolben *m*
c. shoe Gleitschuh *m*, Gleitbacke *f*
c.-type engine Kreuzkopfmaschine *f*
crossing 1. *(Bw)* Kreuzung *f*, Straßenkreuzung *f*; 2. *(El)* Linienkreuzung *f*, Überführung *f*, Vertauschung *f (von Leitungen)*; 3. Herzstück *n (Weiche)*; 4. *(Text)* Fadenkreuz *n*
c. file Vogelzungenfeile *f*
c.-over clamp Kreuzklemme *f*
c. products of strains *(Holz)* Provenienzkreuzungen *fpl*
c. rod *(Text)* Kreuzschiene *f*
c. track *(Eis)* Kreuzgleis *n*
c. valve Kreuzventil *n*
c. warp *(Text)* Dreherkette *f*
crosslinking *(Ch)* Querverzweigung *f*, Vernetzung *f*
crossover *s.* cross-over
crosstie *s.* cross-tie
crosswall *s.* cross-wall
crosswinder dyeing tube Kreuzspulfärbehülse *f (Textilveredlung)*
crosswise in Querrichtung
c. movement *(Masch)* Querbewegung *f*, Querverschiebung *f*, Planzug *m*
crotch Gabelung *f*, Rohrverzweigung *f*
crotonaldehyde Krotonaldehyd *m*
crotonic acid Krotonsäure *f*
crotonyl alcohol Krotylalkohol *m*
crow-fly distance Luftlinienentfernung *f*
crowbar Brechstange *f*
crowd / to stauen, füllen, vollstopfen; abdrängen
crowd brake Vorschubbremse *f (Bagger)*
c. noise Saalgeräusch *n (Rundfunkübertragung)*
c. shovel Hochlöffel[bagger] *m*
crowding 1. Füllen *n*, Füllvorgang *m (Hochlöffelbagger)*; 2. Grabvorgang *m*, Vorschub *m (Löffelbagger)*
c. of the chips Spanstauung *f*
crown 1. *(Bw)* Krone *f*, Scheitel *m*, Spitze *f*, Schlußstein *m*; 2. *(Schiff)* Ankerkreuz *n*; 3. *s.* c. glass
c. block Turmrolle *f (Seilbohren)*
c. dimension Balligkeit[smaße *npl*] *f (einer Walze)*
c. face pulley ballige Riemenscheibe *f*
c. gear Kron[en]rad *n*, [erzeugendes] Planrad *n*, Tellerrad *n*
c. glass *(Opt)* Kronglas *n*
c. hinge Scheitelgelenk *n*

c. lapping Balligläppen *n*
c. nut Kronenmutter *f*
c. of a press Pressenoberteil *n*
c. of furnace Feuerungsdecke *f*, Feuerraumdecke *f*
c. of tank *(Schiff)* Wasserballasttankdecke *f*, Doppelbodendecke *f*
c.-piece *(Brg)* Kappe *f (Ausbau)*
c. plate Deckplatte *f*
c. ring *(Kfz)* Starterzahnkranz *m*
c. runner *(Brg)* Firstenläufer *m (Ausbau)*
c. section *(Brg)* Firstensegment *n (Ausbau)*
c. wheel 1. Kronenrad *n*, Planrad *n*; 2. *(Kfz)* großes Differentialantriebskegelrad *n*, Tellerrad *n*
crowned ballig
c.-face pulley ballig gedrehte Riemenscheibe *f*
c. gear balligtragendes Zahnrad *n*
c. gears Balligverzahnung *f*
c. grinding Balligschliff *m*
crowning 1. *(Masch)* Balligkeit *f*; 2. *(Brg)* Heben *n*, Quellen *n (Streckensohle)*; 3. Herstellen *n* einer balligtragenden Fläche *(Zahnrad)*
c. attachment *(Masch)* Vorrichtung *f* zur Herstellung einer balligtragenden Fläche *(z. B. Balligdrehvorrichtung)*
c. of rolls Walzenbombierung *f*
c. pulley ballige Riemenscheibe *f*
crow's nest *(Schiff)* Ausguck *m*, Krähennest *n*, Mastkorb *m*
croze *(Holz)* Kröse *f*, Kimme *f*
crozzle verbrannter Ziegel *m (verformt und blau)*
crucible Schmelztiegel *m*, Tiegel *m*
c. charge Tiegeleinsatz *m*
c. furnace Tiegelofen *m*
c.-lifting tongs Tiegelziehzange *f*
c. of an induction furnace Schmelzrinne *f* eines Induktionsofens
c. steel Tiegelstahl *m*
c. tongs Tiegelzange *f*
cruciform kreuzförmig
c. bollard *(Schiff)* Kreuzpoller *m*
c. edges gekreuzte Schneiden *fpl (Kreuzmeißel)*
c. twins kreuzartig verwachsene Kristalle *mpl*, Durchkreuzungszwillinge *mpl*
crude roh
c. ammonia liquor Rohgaswasser *n (Kohleveredlung)*
c. antimony Rohantimon *n*
c. board *(Holz)* Rohpappe *f*
c. bottom Bodensatz *m* nach der Destillation von Rohöl *(Erdöl)*
c. flash tower Erdöldestillationskolonne *f*
c. gas Rohgas *n*, ungereinigtes Gas *n*
c. iron Roheisen *n*
c. metal Rohmetall *n*
c. oil Erdöl *n*, Rohöl *n*
c. ore Fördererz *n*, Roherz *n*
c. paraffin Rohparaffin *n*
c. product Vorprodukt *n (Kohleveredlung)*
c. rubber Rohgummi *m*
c. salt Rohsalz *n*
c. sewage Rohabwasser *n*, unbehandeltes Abwasser *n*
c. steel Rohstahl *m*
c. tar Dickteer *m*, Rohteer *m*
cruiser *(Schiff)* Kreuzer *m (gedecktes Fahrzeug mit Wohneinrichtung)*
c. stern *(Schiff)* Kreuzerheck *n*
cruising altitude *(Flg)* Reiseflughöhe *f* über NN
c. range *(Flg)* Aktionsradius *m*
c. speed *(Flg, Schiff)* Reisegeschwindigkeit *f*, Dauergeschwindigkeit *f*
crumble / to [zer]krümeln; zerbröckeln *n*; abbröckeln *(Schneide)*
crumbling harrow Krümelegge *f*
c. roller Krümelwalze *f*
crumbly bröckelnd, bröcklig
crunode *(Math)* Doppelpunkt *m* [einer Kurve] *(Kreuzungspunkt)*
crupper leather Schwanzriemenleder *n*
c. turnback Rückenriemen *m* des Pferdes
crush / to *(Ch, Masch)* 1. [grob]mahlen; brechen, [grob]zerkleinern, kollern, zerdrücken, zerquetschen; 2. zerknittern, zerknüllen *(Papier usw.)*; stauchen *(Festigkeitsprüfung für Papier)*
c.-form abrichten mit Profilrolle *(Schleifscheibe)*
crush dressing *s.* c.-form dressing
c.-form dressing Abrichten *n* mit Profilrolle *(Schleifscheibe)*
c.-formed mit Profilrolle abgerichtet *(Schleifscheibe)*
c. forming Abrichten *n* mit Profilrolle *(Schleifscheibe)*
c. resistance 1. Beständigkeit *f* gegen Zerdrücken, Druckfestigkeit *f*; 2. Knitterfestigkeit *f*
c. resistant knitterbeständig; druckfest
crushed brick Ziegelsplitt *m*
c. brick aggregate concrete Ziegelsplittbeton *m*
c. coarse aggregate zerkleinerter Grobzuschlagstoff *m*
c. coke Brechkoks *m*
c. electrodes Elektrodenmehl *n*
c. gravel Betonsplitt *m*; Brechkies *m*
c. rock Schotter *m*
c. stone Grobsplitt *m*; Schotter *m*
c. sugar cane Bagasse *f (Zuckerindustrie)*
crusher Brecher *m*, Brechwerk *n*, Grobzerkleinerungsmaschine *f*; Quetsche *f*, Schlagmühle *f*, Vorbrecher *m*

c. block *(Brg)* Quetschholz *n*, Kopfholz *n*
c.-drier *(Brg)* Mahltrockner *m*
c. roll Profilrolle *f (Schleifscheibe)*
c. stower *(Brg)* Blasversatzmaschine *f* mit angebautem Bergebrecher
crushing 1. Brechen *n*, Quetschen *n*, Kollern *n*, Zerdrücken *n*, Stauchen *n*, Zerknittern *n*, Zerknüllen *n* (*s. a.* to crush); 2. Bruch *m (durch Stauchen)*
c. and screening plant Brech- und Siebanlage *f*
c. cone Brechkegel *m*
c. load Bruchlast *f (Statik)*
c. machine Reibwolf *m*
c. mill Brechwalzwerk *n*, Grobzerkleinerungsmühle *f*; Quetschwalzwerk *n*
c. plant Brechanlage *f*
c. plate Quetschplatte *f*
c. ring Brechring *m*
c. roll Brechwalze *f*
c. roller Quetschwalze *f (Walzstraße)*
c. segment Quetschsegment *n*
c. strength Bruchfestigkeit *f*, Druckfestigkeit *f (Statik)*
c. stress Bruchspannung *f (Statik)*
c. test Druckversuch *m (Statik)*
c. test specimen Druckprobe[körper *m*] *f*
c. yield point *(Met)* Quetschgrenze *f*
crust / to verkrusten
crust Straßendecke *f*
c. leather Leder *n* in der Kruste
c. of scale *(Met)* Gußhaut *f*
crutch stake *(Led)* Stollkrücke *f*
crutcher Seifenmischer *m*; Mischer *m*
cry of tin Zinnschrei *m*
cryochemistry Tieftemperaturchemie *f*, Kältechemie *f*
cryogen Kältemischung *f*, Kühlmischung *f*
cryogenic kälteerzeugend
cryolite Kryolith *m*, Natriumaluminiumhexafluorid *n*
cryosar Kryosar *m (Zweielektrodenbauelement aus Germanium in flüssigem Helium: Steuervorrichtung mit negativem Widerstand)*
cryoscopic kryoskopisch
cryosistor Kryosistor *m (dreipoliges Halbleiterelement; wirkt wie ein Thyratron)*
cryostat 1. *(Masch)* Kälteregler *m*; 2. *(BMSR)* Kryostat *m (dient zum Messen der Halbleitereigenschaften bei Tiefsttemperaturen)*
cryotron *(BMSR)* Kryotron *n (supraleitende Schaltvorrichtung mit elektromagnetischer Steuerspule)*
cryptocrystalline kryptokristallin
crystal Kristall *m*
c. activity *(Ph)* Kristallaktivität *f*
c. analysis Kristallstrukturuntersuchung *f*
c. angle Kristallwinkel *m*
c. anomaly Kristallanomalie *f*, Anomalie *f* der Kristalle
c. axis Kristallachse *f*
c. blank Rohkristall *m*
c. boundary Korngrenze *f (Metall)*
c. capacity *(El)* Parallelkapazität *f* des Quarzes
c. case Quarzgehäuse *n*
c. chemistry Kristallchemie *f*
c. circuit Schwingquarzschaltung *f*
c. clock Quarzuhr *f*
c. control *(Fmt)* Quarzsteuerung *f*
c. controlled *(Fmt)* quarzgesteuert
c.-controlled clock Quarzuhr *f*
c.-controlled transmitter *(El)* kristallgesteuerter (quarzgesteuerter) Sender *m*
c. counter *(Kern)* Kristallzähler *m*
c. defect Kristallbaufehler *m*
c. detector Kristalldetektor *m*
c. diffraction *(Ph)* Beugung *f* an Kristallen, Kristallbeugung *f*
c. drive *(Fmt)* Quarzsteuerung *f*
c. edge Kristallkante *f*
c. effects *(Ph)* Einfluß *m* des Kristallgitters
c. element Quarzelement *n*
c. face Kristallebene *f*, Kristallfläche *f*
c. filter *(El)* Kristallfilter *n*, Quarzfilter *n*
c. formation Kristallbildung *f*
c. glass Kristallglas *n*
c. grain Kristallkorn *n*, Kristallit *m*
c. grating *(Ph)* Kristallgitter *n (Beugung)*
c. growing Kristallzüchtung *f*
c. growth Kristallwachstum *n*; Kristallzüchtung *f*
c. holder *(El)* Quarzfassung *f*, Kristallhalter[ung *f*] *m*
c. junction line Kristallgrenzlinie *f*
c. lattice Kristallgitter *n*, Raumgitter *n*
c. lattice imperfection Kristallbaufehler *m*
c. lattice spacing Kristallgitterabstand *m*
c.-lattice vibration Kristallgitterschwingung *f*
c. loudspeaker piezoelektrischer Lautsprecher
c. microphone Kristallmikrofon *n*, piezoelektrisches Mikrofon *n*, Quarzmikrofon *n*
c. nucleus Kristallkeim *m*
c. orientation Kristallrichtung *f*
c.-oriented kristallorientiert
c. oscillator Quarzoszillator *m*
c. phasing Kristallfilterneutralisation *f*
c. physics Kristallphysik *f*
c. pick-up piezoelektrischer Tonabnehmer *m*
c. probe-head Schallkopf *m*, Ultraschallkopf *m (Werkstoffprüfung)*
c. pulling Kristallziehverfahren *n*
c. quartz window Kristallquarzfenster *n*, Quarzfenster *n*
c. receiver Detektorempfänger *m*; Detektorglied *n*, Kristalldetektor *m*

c. resonator *(Fmt)* Schwingquarz *m*, Quarzschwinger *m*
c. skeleton Kristallskelett *n*
c.-stabilized quarzstabilisiert
c. structure Kristallstruktur *f*, Kristall[auf]bau *m*
c. texture Kristallanordnung *f*, Textur *f*
c.-tuned quarzgesteuert
c. video rectifier *(Fs)* Kristallbildgleichrichter *m*
crystalliform kristallförmig, kristallin
crystalline kristallin, kristallinisch
c. acid Eisessig *m*
c. aggregation Kristallaggregat *n*
c. clinker kristallinische Schlacke *f*
c. deposit kristallinischer Niederschlag *m*
c. force Kristallisationsvermögen *n*
c. form Kristallform *f*
c. growth Kristallwachstum *n*
c. lens Kristallinse *f*
c. solid solution Mischkristall *m*
c. state kristallinischer Zustand *m*
crystallite Dendrit *m*, Kristallgerippe *n*, Kristallit *m*
crystallizable kristallisierbar, kristallisationsfähig
crystallization Kristallisation *f*, Kristallbildung *f*
crystallize / to kristallisieren
crystallizing dish Kristallisierschale *f*
c. pan *(Ch)* Kristallisiergefäß *n*
crystallographic kristallografisch
c. growth [reguläres] Kristallwachstum *n*
c. plane Kristallebene *f*
crystallography Kristallografie *f*
crystalloid kristallähnlich, kristalloid
cubage Berechnung *f* des Inhalts; Inhalt *m*; Volumen *n*
cubature *(Math)* Kubatur *f*, Volumberechnung *f*
cube / to in die dritte Potenz erheben
cube *(Math)* 1. Kubus *m*, Würfel *m*; 2. dritte Potenz *f*
c. root *(Math)* Kubikwurzel *f*
c.-shaped würfelförmig
c. spar Anhydrit *m*, Karstenit *m*, wasserfreies Kalziumsulfat *n*
c. strength Würfelfestigkeit *f* *(Betonprüfung)*
cubic *(Math)* Gleichung (Kurve) *f* dritten Grades
cubic kubisch; Raum-, Würfel-
c. body centred kubisch raumzentriert *(Kristallgitter)*
c. capacity *(Kfz)* Hubraum *m*, Hubvolumen *n*, Kubikinhalt *m*
c. centimetre Kubikzentimeter *n*
c. dilatation Volumendilatation *f*
c. equation Gleichung *f* dritten Grades
c. face centred kubisch flächenzentriert *(Kristallgitter)*
c. lattice kubisches Gitter *n* *(Raumgitter)*
c. metre Kubikmeter *n*, Raummeter *n*
c. space lattice kubisches Gitter *n* *(Raumgitter)*
c. test block Versuchskörper *m* in Form eines Würfels
c. yardage *(Bw)* umbauter Raum *m*; *(Brg)* Ausbruchsleistung *f* *(in Kubikyards)*
cubical kubisch, würfelförmig, würf[e]lig; dritten Grades
c. contents Kubikinhalt *m*, Rauminhalt *m*; umbauter Raum *m*
cubically centred kubisch zentriert *(Kristallgitter)*
cubicle 1. *(Bw)* Kabine *f*, Zelle *f*; 2. *(El)* Schaltschrank *m*, Schaltzelle *f*
cubit Elle *f* *(altes Längenmaß)*
cuboid *(Math)* Quader *m*
cuddy *(Schiff)* Plicht *f*
cue 1. *(Fs)* Regiesignal *n*, Signal *n* *(für Studio usw.)*; Stichwort *n*; 2. Queue *n* *(Billardstock)*
cul-de-sac Sackgasse *f*
cull *(Am)* *(Ch)* Abfall[stutzen] *m*
c. timber Ausschußware *f* *(Schnittholz)*
cullet Glasbruch *m*, Glasscherben *fpl*
culm Schmiedegrus *m*, Steinkohlengrus *m*
c. coal Anthrazit-Rohfeinkohle *f*
culmiferous plants Halmpflanzen *fpl*
culmination *(Astr)* Kulmination *f*
c. of a star Durchgang *m* eines Fixsterns durch den Meridian, Kulmination *f* eines Fixsterns
c. point Kulminationspunkt *m*, Scheitelpunkt *m*, höchster Stand *m*
cultivate / to urbarmachen *(Land)*
cultivated plants Kulturpflanzen *fpl*
c. soil Ackerboden *m*
cultivating ridger Häufelpflug *m*
cultivation Anbau *m*; Pflege *f*
c. of soil Bodenbearbeitung *f*, Landbestellung *f*
cultivator 1. Züchter *m* *(Pflanzen)*; 2. Hackpflug *m*, Grubber *m*, Lockerpflug *m*
culture broth Kulturlösung *f*, Nährlösung *f* *(Biochemie)*
c. medium Kulturmittel *n*, Nährboden *m* *(Biochemie)*
culvert *(Bw)* Durchlaß[kanal] *m*, Wasserdurchlaß *m*
cumarone resin Kumaronharz *n*
Cumberland protection Cumberlandverfahren *n*, katodischer Schutz *m* *(gegen Korrosion)*
cumulative kumulativ, aufgehäuft; Summen-
c. compound excitation *(Elm)* Mitverbunderregung *f*
c. compound motor *(Elm)* Schlupfmotor *m*
c. compound winding *(Elm)* Doppelschlußwicklung *f*, Kompoundwicklung *f*, mitkompoundierende Wicklung *f*
c. compounded *(Elm)* in Serie kompoundiert

c. detection Gitter[strom]gleichrichtung *f*, Audiongleichrichtung *f*
c. dose *(Kern)* zeitliche Gesamtdosis *f*
c. double bonds *(Ch)* kumulierte Doppelbindungen *fpl*
c. effect Gesamtwirkung *f*
c. error Summenfehler *m*
c. frequency Summenhäufigkeit *f*
c. grid detection Gitter[strom]gleichrichtung *f*, Audiongleichrichtung *f*
c. grid rectifier gittergesteuerter Gleichrichter *m*
c. line Summenlinie *f*
c. rectification Audiongleichrichtung *f*, Gitter[strom]gleichrichtung *f*
cumylic acid Isopropylbenzoesäure *f*
cuneate *s.* cun[e]iform
cun[e]iform keilförmig
cunit Holzmaß *n (100 Festkubikfuß ungeschälten Holzes = 2,83 fm)*
cup / to 1. auskolken; 2. tiefziehen
cup 1. *(Masch)* Becher *m*, Kalotte *f*, Schale *f*; Außenring *m (Kegelrollenlager)*; Hohlform *f*, Hohlgefäß *n*; 2. *(Holz)* deformierte Krümmung *f* quer zur Maserung
c.-and-cone bearing Kegelkugellager *n*
c. anemometer Schalenkreuzanemometer *n*
c. bolt *(Masch)* Pfannenbolzen *m*
c. flow figure *(Plast)* Prüfbecherfließzahl *f*
c. grease Staufferfett *n*
c. grinding wheel Topf[schleif]scheibe *f*
c. head Halbrundkopf *m*
c. insulator *(El)* Glockenisolator *m*
c. packing *(Masch)* Topfmanschette *f*
c. point screw Schraube *f* mit Ringschneide
c. seal *(Masch)* Kolbenmanschette *f*, Napfmanschette *f*
c. shake *(Holz)* Kernschäle *f*, Ringschäle *f*, Ringriß *m*, Schälriß *m*
c.-shaken *(Holz)* ringschälig
c.-shaky *(Holz)* ringschälig
c.-shaped topfförmig
c.-shaped capacitor Hütchenkondensator *m*
c.-shaped cutter *(Masch)* Topffräser *m*
c. spring Tellerfeder *f*
c. test 1. *(Met)* Löffelprobe *f*, Schöpfprobe *f*; 2. Näpfchenziehversuch *m*
c. valve Glockenventil *n*, halbkugelförmiges Ventil *n*, Kronenventil *n*
c. wheel Topfschleifscheibe *f*
c. winding machine *(Text)* Trichterspulmaschine *f*
cupel / to [ab]treiben *(Silbergewinnung)*
cupel 1. *(Ch)* Glühschale *f*, Probetiegel *m*, Scheidekapelle *f*; 2. Treibofen *m (Silbergewinnung)*
c. mould *(Met)* Kapell[en]form *f*
cupellation *(Ch, Met)* Kapellenprobe *f*; Kupellation *f*; Treibarbeit *f*
c. furnace *(Met)* Treibofen *m*
cupelling *(Met)* Treiben *n*, Kupellieren *n*
cupola 1. Kuppel *f (verglast)*; 2. Kupolofen *m*
c. charging Kupolofenbegichtung *f*
c. furnace Kupolofen *m*
c. lining Kupolofenausfütterung *f*
c. mixture Kupolofengattierung *f*
c. receiver Kupolofenvorherd *m*
c. shaft Kupolofenschacht *m*
c.-type gunmount *(Mil)* Kuppellafette *f*
cupped mit Lunkern, mit inneren Hohlräumen
cupping *(Met)* 1. Kümpeln *n*; 2. Tiefziehen *n*
c. ductility value *(Met)* Tiefungswert *m*, Ziehwert *m*
c. process Tiefziehverfahren *n*
c. test Tiefungsprüfung *f*, Näpfchenziehversuch *m (Werkstoffprüfung)*
cupra silk Kupferseide *f*
cuprammonium rayon *(Text)* Kupferseide *f*
cuprate Kuprat *n*, Tetrahydroxykuprat(III) *n*
c. silk Kupferseide *f*
cupreous kupferhaltig, kupferartig
cupric chloride Kupfer(II)-chlorid *n*
c. compound Verbindung *f* des zweiwertigen Kupfers
c. subacetate basisches Kupferazetat *n*
c. subcarbonate basisches Kupferkarbonat *n*
cupriferous kupferhaltig
c. pyrite Kupferkies *m*
cuprite Rotkupfererz *n*
cupro-nickel alloy Kupfer-Nickel-Legierung *f*
c.-uranite Torbernit *m*
cuprocompound Verbindung *f* des einwertigen Kupfers
cuprous acetylide Kupferazetylid *n*
c. compound Verbindung *f* des einwertigen Kupfers
c. oxide Kupfer(I)-oxid *n*
curative agent *(Led)* Konservierungsmittel *n*
curb *(Bw)* Bordeinfassung *f*, Bordkante *f*, Bordschwelle *f*, Bordstein *m*, Rinnstein *m*, Schrammbord *m*, Schrammschwelle *f*
c. fender Bordschwelle *f*, Schutzschwelle *f*
c. ring Keilkranz *m (Schachtausbau)*
c. roof gebrochenes Dach *n*, Mansard[en]-dach *n*
curbing Schachtausbau *m* mit Schrotzimmerung
curbstone Bordstein *m*, Randstein *m*, Rinnstein *m*
curcuma paper Kurkumapapier *n*
curd soap Kernseife *f*
curdled milk Dickmilch *f*
curdling Eindickung *f*; Festwerden *n*
curds *(Ch)* Gerinnsel *n*; käsiger Niederschlag *m*
curdy precipitation *(Ch)* käsiger Niederschlag *m*

cure / to 1. *(Plast)* [aus]härten; 2. *(Text)* entwässern, trocknen *(z. B. mit Kunstharz ausgerüstete Gewebe)*; 3. *(Led)* konservieren; 4. nachbehandeln *(Beton)*
cure 1. *(Plast)* Aushärten *n*, Aushärtung *f*; 2. *(Text)* Entwässerung *f*, Trocknung *f*; 3. *(Led)* Konservierung *f*; 4. Nachbehandlung *f* *(Beton)*; 5. Fermentieren *n*
cured weight *(Led)* Salzgewicht *n*, salzfrei verwogen
curie *(Ph)* Curie *n* *(Maßeinheit)*
Curie constant Curie-Konstante *f*
Curie point Curie-Punkt *m*
Curie temperature Curie-Temperatur *f*
curing *s.* cure
c. chamber *(Text)* Trockenanlage *f* hoher Wärmegrade
c. machine *(Text)* Kondensationsmaschine *f*
c. mat Abdeckmatte *f* *(Betonnachbehandlung)*
curium Curium *n* *(chemisches Element)*
curl / to kräuseln; sich kräuseln (rollen, werfen); aufrollen, locken *(Span)*; bördeln
c. up sich rollen (werfen, abheben)
curl 1. Schlinge *f* *(Draht, Faden usw.)*; 2. Rotation *f* *(Strömung)*
c. field Wirbelfeld *n*
c. of a vector field *(Math)* Rotor *m* eines Vektorfeldes
c. yarn *(Text)* Bouclégarn *n*
curlator Wirbelsichter *m*
curled chip *(Masch)* aufgerollter Span *m*, Lockenspan *m*
c. spot *(Holz)* Flader *f*, Maser *f*
curling 1. Aufrollen *n*; Kräuselung *f*; 2. *(Text)* Verdrehungsneigung *f* *(von Schmaltextilien)*; 3. Einrollen *n* *(Blechkanten)*; Anrollen *n*
c. cut 1. Schälanschnitt *m* *(Bohrer)*; 2. Schälschnitt *m* *(Holzspan)*
c. tool Rollwerkzeug *n*
curly bracket geschweifte Klammer *f*
c. wool fibre gekräuselte Wollfaser *f*
currector *(Am)* *(El)* Stromstabilisierdiode *f*
current laufend, gültig, gegenwärtig
current 1. Fluß *m*, Strom *m* *(Strömung)*; 2. [elektrischer] Strom; Stromstärke *f*; 3. *(Bw)* Neigung *f*
c. amplification Stromverstärkung *f*
c. amplifier Stromverstärker *m*
c. antinode Strombauch *m*
c. balance Stromwaage *f*, Amperewaage *f*; Stromabgleich *m*
c. beat Stromschwebung *f*
c.-bedding *(Geol)* Kreuzschichtung *f*, Diagonalschichtung *f*
c. branch Stromzweig *m*
c. branching Stromverzweigung *f*
c. capacity Strombelastbarkeit *f*
c.-carrying stromführend, unter Strom stehend
c.-carrying capacity Strombelastbarkeit *f*; Kontaktdauerstrom *m* *(z. B. Relais)*
c. changer [ruhender] Umformer *m*
c. circuit Strompfad *m*, Stromkreis *m*
c. coil Stromspule *f*
c. collecting bus-bar Stromsammelschiene *f*
c. collector Stromabnehmer *m*
c. combination *(Fmt)* Stromschrittfolge *f*
c. conduction Stromleitung *f*
c. consumer Stromabnehmer *m*, Stromverbraucher *m*
c. consumption Stromaufnahme *f*; Stromverbrauch *m*, Stromentnahme *f*
c. control Stromregelung *f*
c. control dynamo (generator) *(Kfz)* stromregelnde Lichtmaschine *f*
c. controlled stromgesteuert, stromabhängig
c. delivery Stromabgabe *f*
c. demand Strombedarf *m*
c. density Stromdichte *f*; *(Kern)* Neutronenstromdichte *f*
c. density per solid angle Stromdichte *f* je Raumwinkeleinheit; Richtstrahlwert *m*
c.-dependent stromabhängig
c. detector Leitungsprüfer *m*
c. direction indicator Polaritätsanzeiger *m*
c. directional relay Stromrichtungsrelais *n*, Richtungsrelais *n*
c. displacement Stromverdrängung *f*
c. distribution Stromverteilung *f*
c. divider Stromteiler *m*
c. division Stromverteilung *f*
c. drain Stromentnahme *f*
c. efficiency Stromausbeute *f*; Wirkungsgrad *m* des Stromes
c. failure Stromausfall *m*
c.-fed stromgespeist
c. feed Stromspeisung *f*, Stromzuführung *f*
c. feedback Stromgegenkopplung *f*; Stromrückkopplung *f*
c. flight plan gültiger Flugplan *m*
c. flow Stromfluß *m*, Stromdurchgang *m*
c. fluctuation Stromschwankung *f*
c. for lighting purposes Lichtstrom *m*
c. for power Kraftstrom *m*
c. force *s.* c. forcing
c. forcing Felderhöhung *f*
c. generation Stromerzeugung *f*
c. illumination characteristic Licht-Fotostromkennlinie *f*
c. impulse Stromimpuls *m*, Stromstoß *m*
c. indicator Am[pere]meter *n*, Amperemesser *m*, Strom[an]zeiger *m*, Strom[stärke]messer *m*
c. input Stromaufnahme *f*
c. intensity Stromstärke *f*
c. lead Stromzuführung *f*

c. **leakage** Kriechstrom *m*, Streustrom *m*
c. **leakage path** Kriechweg *m*, Kriechstrecke *f*
c. **limitation** Strombegrenzung *f*
c. **limiter** Strombegrenzer *m*, Strombegrenzungsschalter *m*, Überstromwächter *m*; Strombegrenzungsrelais *n*; *(selten)* Sicherung *f*
c.**-limiting coil** Strombegrenzungsdrossel *f*
c.**-limiting device** Strombegrenzer *m*
c.**-limiting fuse** Strombegrenzungssicherung *f*, strombegrenzende Sicherung *f*
c.**-limiting impedance** Begrenzungswiderstand *m*
c.**-limiting inductor** Strombegrenzungsdrossel *f*, Kurzschlußdrosselspule *f*
c.**-limiting reactor** Kurzschlußdrossel *f*, Strombegrenzungsdrossel *f*
c.**-limiting resistor** Strombegrenzungswiderstand *m*
c. **load[ing]** Strombelastung *f*
c. **loop** Strombauch *m*
c. **loss** Stromverlust *m*
c. **meter** Stromzähler *m*, Zähler *m*
c. **node** Stromknoten *m*
c. **noise** Stromrauschen *n*
c. **of air** Luftstrom *m*, Luftströmung *f*
c. **of high amperage** Starkstrom *m*
c.**-operated** strombetätigt, stromgesteuert
c. **output** Stromabgabe *f*
c. **passage** Stromdurchgang *m*, Stromdurchfluß *m*
c. **path** Stromverlauf *m*, Stromweg *m*, Strompfad *m*
c. **per unit cross-section** Stromdichte *f*
c. **phase** Stromphase *f*
c. **phase-balance protection** Stromdifferentialschutz *m*, Differentialphasenschutz *m*
c. **protection** Stromschutz *m*
c. **pulse** Stromstoß *m*
c. **range** Strommeßbereich *m*
c. **rating** Nennwertbemessung *f* der Stromstärke; Nennstromstärke *f*
c. **ratio** Stromverstärkungsfaktor *m*; Stromübersetzung *f*, Übersetzung[sfaktor *m*] *f* *(Stromwandler)*
c. **recorder** Stromschreiber *m*
c. **rectifier** Stromgleichrichter *m*
c. **regulator** Stromregler *m*
c. **regulator tube** Stromregelröhre *f*
c. **relay** Stromrelais *n*
c. **requirement** Strombedarf *m*
c. **resonance** Stromresonanz *f*
c.**-responsive element** Ansprechglied *n*; Schmelzleiter *m* *(Sicherung)*
c. **reversal** Stromumkehr *f*, Stromwendung *f*
c. **reverser** Polwechselschalter *m*, Wendeschalter *m*
c. **rush** Stromstoß *m* *(Schaltvorgänge)*
c. **sensitiveness (sensitivity)** Stromempfindlichkeit *f*
c. **source** Stromquelle *f*, Elektrizitätsquelle *f*, Spannungsquelle *f*
c. **square meter** Stromquadratmeßgerät *n*, Effektivstrommesser *m*
c. **strength** Stromstärke *f*
c. **strength in amperes** Stromstärke *f* in Ampere
c. **supply** Stromversorgung *f*, Stromzuführung *f*
c. **surge** Stromstoß *m*
c. **tap** Mehrfachstecker *m*
c. **telemeter** strombetätigtes Fernmeßgerät *n*; Fernamperemeter *n*
c. **transformer** Stromwandler *m*, Stromtransformator *m*, Stromumformer *m*
c. **undulation** periodische Stromschwankung *f*
c. **unit** Stromeinheit *f*
c. **variation** Stromschwankung *f*; Stromänderung *f*
c. **velocity** Stromgeschwindigkeit *f*
c.**-voltage characteristic** Entladungscharakteristik *f*, Strom-Spannungskennlinie *f*
c.**-voltage control** *(Kfz)* Strom-Spannungsregelung *f* *(Lichtmaschine)*
c. **volume** *s.* c. strength
c. **wave** Stromwelle *f*
c. **wave form** Stromwellenform *f*
c.**-wavelength characteristic** Charakteristik (Kurve) *f* der spektralen Empfindlichkeit *(Fotozelle)*
c. **zero** Stromnulldurchgang *m*
currentless conductor stromloser Leiter *m*
currier *(Led)* Lederzurichter *m*
currier's laying steel *(Led)* Legstahl *m*
currying Zurichten *n* *(Leder)*
cursor 1. Läufer *m* *(Rechenschieber)*; Reiter *m*; Schieber *m*; 2. *(El)* Gleitkontakt *m*, Schiebekontakt *m*, Schleifkontakt *m*
curtain 1. Vorhang *m*, Gardine *f*; 2. *(Brg)* Wettertuch *n*; 3. *(Kern)* Abschirmfolie *f*, Neutronenfänger *m*
c. **aerial** Fächerantenne *f*
c. **array [of aerials]** Antennenwand *f*
c. **board** Gardinenleiste *f*
c. **wall** vorgehängte Wand *f*, Vorhangwand *f*
curtate verkürzt, reduziert
curtilage Hausgrundstück *n*
curvature 1. *(Math)* Krümmung *f*; 2. *(Masch, Opt)* Wölbung *f*
c. **of field** *(Opt)* Bildfeldwölbung *f*
c. **of orbit** Bahnkrümmung *f*
c. **of swan neck** *(Text)* Kulissenkrümmung *f*
c. **of the stitching thread** *(Text)* Biegung *f* des anbindenden Fadens
curve 1. *(Math)* Kurve *f*; 2. Biegung *f*, Kurve *(Straßenbau)*

c. follower *(Dat)* Kurvenabtaster *m*
c.-going *(Brg)* kurvengängig *(Fördermittel)*
c. in space Raumkurve *f*
c. of admissible lengths *(Schiff)* Schottkurve *f*
c. of areas of cross-sections *(Schiff)* Spantflächenkurve *f*, Verdrängungskurve *f*
c. of bulkhead spacing *(Schiff)* Schottkurve *f* *(Schottabstände)*
c. of buoyancy *(Schiff)* Auftriebskurve *f*, Formschwerpunktkurve *f*
c. of deflection Biegungslinie *f*
c. of equal velocity *(Ph)* Isotache *f*
c. of flight path *(Flg)* Flugbahnlinie *f*
c. of flotations *(Schiff)* Wasserlinienflächenkurve *f*
c. of horizontal distribution *(El)* horizontales Polardiagramm *n*, Kennlinie *f* der horizontalen Verteilung
c. of intensity distribution *(El)* Lichtstärkeverteilungskurve *f* *(einer Lampe oder Leuchte)*
c. of magnetization *(Elm)* Magnetisierungskurve *f*
c. of normal magnetization *(El)* Kommutierungskurve *f*
c. of pressure distribution Druckverteilungskurve *f*
c. of pursuit *(Math)* Hundekurve *f*, Verfolgungskurve *f*
c. of spectral distribution [spektrale] Energieverteilungskurve *f*
c. of statical stability *(Schiff)* Hebelarmkurve *f*
c. of track Bahnkurve *f*
c. of vertical distribution *(Fmt)* vertikales Polardiagramm *n*
c. plotter Kurvenschreiber *m* *(Analogrechner)*
c. roller Kurvenrolle *f*
c. shear *(Masch)* Kurvenschere *f*
c. showing the time as a function of the course Zeit-Weg-Kurve *f*
c. work Schweifarbeit *f* *(Holzbearbeitung)*
curved gekrümmt; gebaucht, geschweift; hohlflächig
c. back gebauchter Rücken *m*
c. bearing scraper Löffelschaber *m*
c. blade 1. gebauchtes Sägeblatt *n*; 2. gekrümmte Schaufel *f* *(Turbine)*
c. breast gekrümmte Zahnspitzenlinie *f*
c. cam surface effecting circular motion *(Wärm)* kreisförmiger Schub *m*
c. guides Bogenführung *f*, Kurvenführung *f*
c. guiding mechanism *(Text)* Kurvenführung *f*
c. hole *(Brg)* gekrümmtes Bohrloch *n*
c. jib *(Brg)* Profilschrämarm *m*, Formschrämarm *m*, gebogener Schrämarm *m*
c. needle nose plier gebogene Justierzange *f*
c. piston-path diagram *(Wärm)* bogenförmiges Kolben-Weg-Diagramm *n*
c. stereo *(Typ)* Rundstereo *n*
c. timber Krummholz *n*
c.-tooth kurvenverzahnt
c. tooth file Feile *f* mit Bogenhieb
c.-tooth type bevel gear Kegelrad *n* mit Bogenverzahnung
c. twill *(Text)* krummer Köper *m*
c.-wedge mechanism Geradschubkurvengetriebe *n*
curvilinear motion (movement) krummlinige Bewegung *f*
c. slotting machine Senkrechtstoßmaschine *f* zum Bearbeiten von Eisenbahnrädern
curving Schweifsägen *n* *(Holzbearbeitung)*
curvometer Kurvenmesser *m*
cushion / to [ab]federn, dämpfen; auspolstern
cushion 1. Auflage *f*, Kissen *n*, Polster *n*, Puffer *m*; 2. Gummipolster *n* des Zwischenbaus *(Kfz-Reifen)*; 3. Ziehkissen *n* *(Presse)*; 4. *s.* c. block
c. block *(Bw)* Rammhaube *f* *(Pfahlgründung)*
c. cam Führungsteil *n* unterhalb der Senkerspitze *(Strickmaschine)*
c. craft Luftkissenfahrzeug *n*
c. face Aufschlagfläche *f*, Stoßfläche *f*
c.-geared drive gefederter Zahnradantrieb *m*
c. idler *(Brg)* gefederte Rolle *f*
c. lace Klöppelspitze *f*
c. spring Dämpfungsfeder *f*
c. tyre *(Kfz)* hochelastischer Reifen *m*
cushioncraft *s.* cushion craft
cushioned gefedert, weich, stoßfrei; mit Pufferung
c. blasting *(Brg)* Hohlraumschießen *n*
c. socket *(Fmt)* federnde Röhrenfassung *f*
cushioning Dämpfung *f*, Federung *f*
c. capacity *(Kfz)* Federungs- und Dämpfungsfähigkeit *f* *(Reifen)*
c. effect Federungsvermögen *n*
cusp Spitzpunkt *m*
cuspidal point Rückkehrpunkt *m*
custodian Treuhänder *m*
custom boat *(Schiff)* in Einzelkonstruktion gebautes Boot *n*
customary in the trade handelsüblich
customer 1. Kunde *m*, Käufer *m*; 2. Abnehmer *m*, Verbraucher *m* *(z. B. von elektrischer Energie)*; 3. Teilnehmer *m* *(Fernsprechverkehr)*
customs cruiser *(Schiff)* Zollkreuzer *m*
c. cutter Zollkutter *m*, Zollschiff *n*
cut / to 1. *(Masch)* [zer]schneiden *(Draht usw.)*; beschneiden *(Blech)*; spanend bearbeiten; fräsen; scheren; [auf]hauen *(Feilen)*; ziehen *(Nuten)*; schneiden *(Zahnrad)*; verzahnen *(Räder)*; 2. *(Holz)* ablängen, kürzen, schneiden; 3. *(Text)* zuschneiden; 4. *(Brg)* schrämen, Einbruch herstellen, schlitzen, schneiden

c. a connection *(Fmt)* eine Verbindung trennen
c. a thread ein Gewinde schneiden
c. away 1. abschneiden, wegschneiden; 2. *(Schiff)* kappen; 3. *(Brg)* wegspitzen
c. back verschneiden, versetzen *(Erdöl)*
c. down 1. zerschneiden, zerkleinern; 2. fehlerhaftes Papier auf kleineres Format schneiden
c. down the voltage die Spannung reduzieren
c. free *(Masch)* freischneiden, hinterdrehen
c. in *(Masch)* einschneiden, einstechen
c. into chips zerspanen
c. off 1. abstechen, trennen; abheben *(Späne)*; kürzen *(Holzbearbeitung)*; 2. abschalten, ausschalten, sperren, trennen, unterbrechen; beschneiden *(Frequenz)*; 3. abkappen, abköpfen; 4. abschneiden
c.-off to length ablängen
c. out 1. *(El)* ausschalten; 2. *(Masch)* ausrücken
c. production cost die Fertigungskosten (Erzeugungskosten) senken
c. the circuit den Strom unterbrechen
c. timber [ab]holzen *(Wald)*
c. up [ab]sägen

cut 1. Schnitt *m*; Span *m*; Spantiefe *f*; 2. Hieb *m*; 3. *(Brg)* Einbruch *m (Schießarbeit)*; Schram *m*; 4. *(Bw)* Abtrag *m*; Durchstich *m*; Einschnitt *m (Erdbau)*; 5. Schnitt *(Film)*; 6. *(Fs)* scharfe Überblendung *f*
c.-and-fill stoping *(Brg)* Firstenbau *m* mit Versatz
c.-away centre abgeflächte Spitze *f*
c.-away forefoot *(Schiff)* runder Stevenlauf *m*, stark auflaufender Steven *m*
c.-away model Modell *n* im Schnitt, Schnittmodell *n*
c.-away stern *(Schiff)* Schlepperheck *n*
c.-back *(Ch)* Verschnitt-
c.-back bitumen Verschnittbitumen *n*
c.-back product mit Destillaten verschnittene Destillationsrückstände *mpl*
c. coke Brechkoks *m*
c.-down milling Gleichlauffräsen *n*
c. edge of cloth ausgerissener Geweberand *m*
c. film Planfilm *m*
c. goods *(Text)* Meterware *f*
c. height *(Brg)* Schrämdicke *f*
c.-in current rush Einschaltstromstoß *m*
c.-in delay Einschaltverzögerung *f*
c.-in portion Aussparung *f*
c.-in speed Einschaltgeschwindigkeit *f*
c. into chips zerspant
c. loop *(Text)* aufgeschnittene Flornoppe *f*
c. mark *(Text)* Längenzeichen *n*
c. micro-section *(Met)* Schliffbild *n*
c.-nippers *(Masch)* Vorschneider *m*
c.-off *(Fmt, Fs)* gesperrt
c.-off 1. *(Ph)* Abschneidung *f (Teil eines Spektrums)*; 2. Abkneifer *m*; 3. *(Fmt, Fs)* Unterbrechung *f*; 4. *(Wärm)* Füllung[sgrad *m*] *f*; 5. *(Rak)* Brennschluß *m*
c.-off angle 1. abgestumpfter Winkel *m*; 2. *(Rak)* Brennschlußwinkel *m*
c.-off attenuator Hohlrohrabschwächer *m (Wellenleiter)*
c.-off battery Abschaltbatterie *f*
c.-off bias Gittersperrspannung *f*
c.-off burr Abstechgrat *m*
c.-off cock Absperrhahn *m*
c.-off coupling *(Masch)* Schaltkupplung *f*
c.-off current Ausschaltspitzenstrom *m*
c.-off end Abstichseite *f*
c.-off evaluation receiver *(Rak)* Brennschluß[signal]empfänger *m*
c.-off evaluation truck *(Rak)* Brennschlußempfängerwagen *m*
c.-off frequency *(Ph, El)* Abschneidefrequenz *f*; Grenzfrequenz *f*
c.-off grid potential Gittersperrspannung *f*
c.-off ground installation *(Rak)* Brennschlußsignalanlage *f*
c.-off jack *(Am)* *s.* disconnecting jack
c.-off length Abschnittlänge *f*; Trennlänge *f*
c.-off limiting Begrenzung *f* durch Gitterstromeinsatz
c.-off of a transistor Sättigung *f* eines Transistors *(wichtig bei Verwendung des Transistors zum Schalten)*
c.-off relay Trennrelais *n*, Abschaltrelais *n*; Löschrelais *n*
c.-off rest *(Masch)* Abstechsupport *m*
c.-off saw Trennsäge *f*
c.-off signal modulator *(Rak)* Brennschlußsignalgeber *m*, Kommandogeber *m*
c.-off signal receiver *(Rak)* Brennschluß[signal]empfänger *m*, Kommandoempfänger *m*
c.-off signal test *(Rak)* Brennschluß[signal]rückmeldung *f*, Kommandorückmeldung *f*
c.-off signal transmission *(Rak)* Brennschluß[signal]gabe *f*
c.-off signal transmitter *(Rak)* Brennschluß[signal]sender *m*, Kommandosender *m*
c.-off signal transmitter output stage *(Rak)* Kommandosenderendstufe *f*
c.-off tool Stechmeißel *m*
c.-off tube Sperröhre *f*, Trennröhre *f*
c.-off value *(Rak)* Brennschlußgröße *f*
c.-off valve 1. *(Masch)* Abschlußventil *n*; 2. Sperröhre *f*, Trennröhre *f*
c.-off velocity *(Rak)* Brennschlußgeschwindigkeit *f*
c.-off voltage Einsatzspannung *f*, Entladegrenzspannung *f*; *(Am)* Endspannung *f*

c.-off waveguide Wellenleiter *m* unterhalb der Grenzfrequenz
c.-off wavelength Grenzwellenlänge *f*
c.-out Auslöser *m*, Ausschalter *m*, Trenner *m*, Trennschalter *m*; Ausschaltung *f*; Sicherung *f*
c.-out box Sicherungskasten *m*
c.-out portion Auskehlung *f*
c.-out relay Reglerschalter *m*
c. per tooth *(Masch)* Schnittiefe *f* je Zahn
c. pile carpet Tournaivelourteppich *m*
c. plane iron Lochhobeleisen *n*
c. plush geschnittener Plüsch *m*
c. point *(Brg)* Trennschnitt *m*
c. selvedge eingerissene Gewebekante *f*
c. shot *(Brg)* Einbruchschuß *m*
c. skin *(Led)* verschnittenes Fell *n*
c. staple geschnittener Stapel *m*
c. surface Schnittfläche *f (technische Zeichnung)*
c. temple roller *(Text)* Walzenbreithalter *m* ohne Nadelspitzen
c.-thread tap Gewindebohrer *m* mit geschnittenem Gewinde
c.-through *(Brg)* Durchhieb *m*
c.-through relay Durchschalterelais *n*, Fortschalterelais *n*
c. work *(Text)* Doppeldurchbrucharbeit *f*
cutability Schneidfähigkeit *f*; Spanbarkeit *f*
cuticle remover Nagelhautentferner *m*
cutlery Stahlwaren *fpl*
c. steel Messerstahl *m*
cutoff *s.* cut-off
cutoffs *(Holz)* Kürzungen *fpl*
cutter 1. *(Masch)* Fräser *m*; Schneidwerkzeug *n*; Einsatzdrehmeißel *m*; Messer *n*; Zange *f*; Hobeleisen *n*; 2. *(Typ)* Schneidemaschine *f*; Beschneider *m*; 3. Schneidemaschine, Schnittmaschine *f*; 4. *(Holz)* Querschneider *m*; 5. *(Brg)* Hauer *m*; 6. spaltbarer Ziegelstein *m*; 7. Kutter *m (Fischereifahrzeug)*; 8. *(Text)* Zuschneider[in *f*] *m*
c. arbor Fräs[er]dorn *m*
c. arbor collar Fräs[er]dornbund *m*
c. bar 1. Messerbalken *m*; 2. *(Masch)* Messer *n*
c. beam Messerbalken *m*
c. bit Drehzahn *m*
c. block spindle Messerspindel *f*, Messerwelle *f (Hobel)*
c. block with curved knives Messerkopf *m* mit gekrümmten Messern *(Hobelmaschine)*
c. block with hollow faces Messerkopf *m* mit aufliegenden Messern *(Hobelmaschine)*
c. body Fräserkörper *m*
c. clearance gauge Fräserprüflehre *f (Hinterschliff)*
c. cradle Fräserwiege *f*
c. disk Messerscheibe *f*
c. drive Fräsermitnahme *f*
c. for cornering Radiusfräser *m* für Ecken
c. for cutting sprockets Zahnformfräser *m* für Zahnkettenräder
c. for fluting countersinks Senkernutenfräser *m*
c. for fluting reamers Reibahlennutenfräser *m*
c. for fluting twist drills Spiralbohrernutenfräser *m*
c. for gear shaping Schneidrad *n (Wälzstoßen)*
c. for grooving taps Gewindebohrernutenfräser *m*
c. for milling half circles Halbkreisformfräser *m*, Nutenfräser *m*
c. for milling wrenches Schlüsselfräser *m* für Schraubenschlüssel
c. for stub-arbor mounting Fräser *m* zum Aufstecken auf fliegend angeordnetem Dorn
c. gang Fräsersatz *m*, Satzfräser *m*
c. grinder Fräserschleifmaschine *f*
c. head Messerkopf *m*; Fräskopf *m*; Frässpindelstock *m*; Hobelmesser *n*
c. head grinder Messerschleifapparat *m*
c. hole Fräserbohrung *f*
c. lift Fräserabhebung *f*
c. loader Schrämlader *m*, Schrämlademaschine *f*, Kohlenkombine *f*
c. pick Schrämmeißel *m*
c. plough Schrämhobel *m*
c. relief Fräserabhebung *f*
c. relieving attachment Fräser-Hinterdrehapparat *m*
c. retraction Fräserabhebung *f*
c. rig *(Schiff)* Kuttertakelung *f*
c. saddle Frässchlitten *m*
c.-scraper Schrämschrapper *m*
c. shank Messerkopfzapfen *m (Hobelmaschine)*
c. sharpening machine Fräserschleifmaschine *f*
c. spindle Fräs[er]welle *f*
c.-suction dredger Saug[e]bagger *m* mit Schneidkopf
c. sweep Nutenauslauf *m*
c. testing fixture Fräser[schleif]lehre *f*
c. tooth contour tester Fräser[schleif]lehre *f*
c. track *(Brg)* Schrämgasse *f*
c. wheel Schneidrädchen *n*
cutting 1. Schneidarbeit *f*, Schneiden *n*, spanende Bearbeitung *f*, Spanen *n*, Zerspanung *f*, Schnitt *m (als Arbeitsvorgang)*; 2. *(Bw)* Abtrag *m*, Abtragen *n*, Einschnitt *m (Erdbau)*; Stemmen *n (Holzbearbeitung)*
c. ability Schneidfähigkeit *f*, Schnittleistung *f*, Zerspanungsleistung *f*
c. ability of the grinding wheel Griffigkeit *f* der Schleifscheibe

c. across the grain Querschnitt *m*
c. action Schneidwirkung *f*
c. angle Schnittwinkel *m*
c. appliance Schneidvorrichtung *f*
c. area 1. *(Masch)* Schneidzone *f*; 2. *(Holz)* Hiebfläche *f*
c. attachment Schneidvorrichtung *f*
c. bench Filmschneidetisch *m*
c. blade Messer *n (Fräser)*; Bohrmesser *n*
c. blocks *(Led)* Stanzklötze *mpl*
c. blowpipe Schneidbrenner *m*
c. burner Schneidbrenner *m*
c. capacity Schneidfähigkeit *f*; Schnittleistung *f*, Spanleistung *f*
c. chain gear *(Brg)* Schrämkopf *m*, Schrämkettenantrieb *m*
c. compasses Schneidzirkel *m*
c. components Komponenten *fpl* der Schnittkräfte, Schnittkraftkomponenten *fpl*
c. compound Kühlmittel *n*; Kühlöl *n*
c. coolant Kühlmittel *n*
c. depth Spantiefe *f*; Schnittiefe *f*
c. die Schnittmatrize *f*; Schnittstempel *m*, Stanzmesser *n*, Stanzwerkzeug *n*; Schneidbacke *f (Gewindeschneidkopf für Außengewinde)*
c. disk *(Brg)* Schrämscheibe *f*
c. down Gleichlauffräsen *n*
c. down wood Abholzen *n (des Waldes)*
c. drum *(Brg)* Schrämwalze *f*
c. duty vorgeschriebene Schnittleistung *f*
c. edge 1. Schneide *f*, Schneidkante *f*; 2. Senkschuh *m*, Schneidschuh *m (Senkschachtverfahren)*
c. edge at the periphery of the [twist] drill Fasenschneide *f*
c. edge cutting across the grain *(Holz)* quer zur Faser liegende Schneidkante *f*
c. edge cutting obliquely to the grain *(Holz)* schräg gegen die Faser liegende Schneidkante *f*
c.-edge geometry Geometrie *f* der Schneide
c. edge inclined in a forward direction *(Holz)* nach vorn überhängende Schneidkante *f*
c. edge on the feed side Hauptschneide *f*
c. edge parallel to the grain *(Holz)* gleichlaufend mit der Faser liegende Schneidkante *f*
c.-edge wear Schneidenverschleiß *m*; Schneidenabnutzung *f*
c.-edge wear land Verschleißfase *f* an der Schneide
c. edges at the end Anschnittschneiden *fpl (Bohrer)*
c. efficiency Zerspanungsleistung *f*
c. face Spanfläche *f*
c. facet Schneidfacette *f*
c. fluid Schneidflüssigkeit *f*
c. fly tool *(Masch)* Schlag[form]messer *n*
c. force Schnittkraft *f*
c. gear Schneidwerk *n*
c. head 1. Schneidenkopf *m*; Bohrkopf *m*; 2. Schneidbrennerkopf *m*
c. head for boring long bores Tieflochbohrkopf *m*
c.-in Einstechen *n*, Einstich *m*
c.-in profile *(Holz)* Profilschnitt *m*
c.-in speed Einschaltdrehzahl *f*
c. internal threads Innengewindeschneiden *n*
c. iron 1. Pflugsech *n*; 2. Hobeleisen *n*
c. jack *(Text)* Schneideplatine *f*
c. jib *(Brg)* Schrämarm *m*, Schrämausleger *m*
c. knife Schnitzmesser *n*; Trennmesser *n*
c. life Standzeit *f (Zerspanungswerkzeuge)*
c. lip Lippe *f*, Schneide *f (Bohrer)*
c. load Schnittkraft *f*, Schnittbelastung *f*
c. lubricant Schneidöl *n*, Kühlmittel *n*, Kühlöl *n*
c. machine 1. Abschneidemaschine *f*, Schneidmaschine *f*; 2. Beschneidmaschine *f (Papier)*, Papierschneider *m*; 3. *(Text)* Zuschneidemaschine *f*
c. material Schneidstoff *m*
c. metal Schneidmetall *n*
c. nippers Zange *f*, Vorschneider *m*
c. of asparagus Spargelstechen *n*
c. of rectangular spoke tenons *(Holz)* Fräsen *n* der flachen Speichenzapfen
c. of teeth Verzahnung *f*, Schneiden *n* von Zahnrädern
c. of the pile tubes *(Text)* Aufschneiden *n* der Florschläuche
c.-off Abstechen *n*
c.-off a rope Seiltrennen *n*
c.-off lathe Abstechdrehmaschine *f*
c.-off machine Abstechmaschine *f*
c.-off operation Trennarbeit *f*
c.-off process Trennverfahren *n*
c.-off steam Dampfabsperrung *f*
c.-off the steam admission Abschneiden *n* der Dampfeinströmung
c.-off tool Stechmeißel *m*; Abstechwerkzeug *n*
c.-off wheel Trennschleifscheibe *f*
c. oil Kühlöl *n*, Kühlmittel *n*, Schneidöl *n*
c. operation spanende Bearbeitung *f*
c.-out machine *(Text)* Zuschneidemaschine *f*
c.-out saw Trennsäge *f*
c.-out stope *(Brg)* unterster Stoß *m (Firstenstoßbau)*
c. parallel to the grain *(Holz)* Längsschnitt *m*
c. path Schnittweg *m*
c. periphery of the grinding wheel Schleifscheibenumfläche *f*
c. perpendicular to the grain *(Holz)* Querschnitt *m*
c.-plane line Linie *f* zur Angabe der Schnittebene *(Zeichnung)*

c. **pliers** Drahtzange *f*, Schneidzange *f*
c. **power** Schneidfähigkeit *f*; Schnittleistung *f*, Spanleistung *f*
c. **press** Schnittpresse *f*
c. **pressure** Arbeitsdruck *m*, Schnittdruck *m*, Schnittkraft *f*
c. **property** Schneidfähigkeit *f*, Zerspanbarkeit *f*
c. **punch** Schnittstempel *m*
c. **quality** 1. Schneidfähigkeit *f*; 2. Zerspanungseigenschaft *f*
c. **rake** Spanwinkel γ *m (Spiralbohrer)*
c. **shaping** spanende Formung *f*
c. **site** *(Masch)* Schnittstelle *f*
c. **solution** *(Masch)* Kühlmittel *n*
c. **speed** *(Masch)* Schnittgeschwindigkeit *f*
c. **speed chart** Schnittgeschwindigkeitstabelle *f*
c. **stroke** Arbeitshub *m*
c. **the thread** Gewindeschneiden *n*
c. **tip** *(Masch)* Schneide *f*; Schneidplatte *f*
c. **tool** Schneid[werk]zeug *n*; Meißel *m*; Schnittwerkzeug *n*
c.**-tool edge** Werkzeugschneide *f*
c.**-tool life** *(Masch)* Standzeit *f*
c.**-tool tip** *(Masch)* Plättchen *n (Schneidwerkzeuge)*
c. **tooth** Schneidzahn *m*, Sägezahn *m*
c. **torch** Schneidbrenner *m*
c. **up** *(Masch)* Gegenlauffräsen *n*
c. **with the grain** *(Holz)* Längsschnitt *m (Arbeitsvorgang)*
cuttings 1. *(Masch)* Schneidspäne *mpl*, Späne *mpl*; 2. *(Holz)* Beschneideabfälle *mpl*; 3. *(Brg)* Schrämklein *n*; 4. *(Text)* Abfall *m (Schneiderei)*
c. **from a bore** Bohrklein *n*, Bohrgut *n*, Bohrschmant *m (Erdöl, Erdgas)*
cuttler *(Text)* Täfler *m*
cuttling frame *(Text)* Ablegevorrichtung *f*
CX-lamp UV-Lampe *f*
cyanate / to zyanieren, die CN-Gruppe einführen
cyanate zyansauer
cyanate Zyanat *n*, Ester *m* (Salz *n*) der Zyansäure
cyanic acid Zyansäure *f*
cyanidation Zyan[id]laugung *f*, Zyanidprozeß *m*, Zyanidverfahren *n*
cyanide / to einsatzhärten, [im Zyanbad] härten
c. **harden** *s.* to cyanide
cyanide Zyanid *n*, Salz *n* der Zyanwasserstoffsäure
cyaniding Zyaneinsatzhärtung *f*, Zyanbadhärtung *f*
cyanoacetic acid Zyanessigsäure *f*
cyanogen Zyan *n*, Dizyan *n*, Oxalsäuredinitril *n*
c. **halide** Zyanhalogenid *n*, Halogenzyan *n*
cyanotrichite Kupfersamterz *n*
cyanotype Zyanotypie *f (Blaupause)*
cyanuric acid Zyanursäure *f*
cybernation Steuerungstechnik *f*
cybernetic kybernetisch
cybernetics *(Dat)* Kybernetik *f*
cyclanes Zykloparaffine *npl*, Naphthene *npl*
cycle / to 1. radfahren; 2. *(BMSR)* umlaufen; periodisch wiederholen
cycle 1. Kreisprozeß *m (Thermodynamik)*; 2. *(Masch, BMSR)* Arbeitsablauf *m*, Arbeitstakt *m*, Arbeitszyklus *m*, Doppelhub *m*, Gang *m*, Umlauf *m*; Spiel *n*, Zyklus *m*; 3. *(El)* Periode *f*, einzelne Schwingung *f*; 4. Fahrrad *n*
c. **car** *(Kfz)* Kleinstwagen *m*; Dreiradwagen *m*
c. **compound** *(Ch)* zyklische Verbindung *f*, Ringverbindung *f*
c. **criterion** *(Dat)* Umlaufzahl *f*, Spielzahl *f*
c. **delay** *(Büro)* Gangverzögerung *f*
c. **index** *(BMSR)* Spielzahlanzeiger *m*, Zyklusindex *m*
c. **of charges** Gichtenfolge *f (im Hochofen)*
c. **of stress** Lastspiel *n*
c. **path** Rad[fahr]weg *m*
c. **reset** *(Dat)* Rückstellung *f (des Spielzeigers)*
c. **selector** Frequenzwähler *m*
c. **setting** Einstellung (Voreinstellung) *f* des Arbeitsablaufes
c. **spanner** Fahrradschlüssel *m*
c. **time** 1. *(Masch)* Gesamtstückzeit *f*; Taktzeit *f*; 2. *(Brg)* Fahrspieldauer *f*, Förderspieldauer *f*
c. **track** Rad[fahr]weg *m*
cycles per second Schwingungen *fpl* je Sekunde; Hertz *n*, Hz
cyclic 1. zyklisiert, ringförmig; 2. periodisch; schwingend; wechselnd
c. **accelerator** *(Kern)* Zirkularbeschleuniger *m*, Ringbeschleuniger *m*, Kreisbeschleuniger *m*
c. **admittance** *(El)* Drehfeldadmittanz *f*
c. **compound** *(Ch)* zyklische Verbindung *f*, Ringverbindung *f*
c. **efficiency** Wirkungsgrad *m* des Kreisprozesses *(Wärmekraftmaschine)*
c. **impedance** *(El)* Drehfeldimpedanz *f*
c. **irrotational motion** *(Ph)* zirkulatorische Potentialströmung *f*
c. **loading** Dauerschwingbeanspruchung *f*
c. **magnetization** 1. *(Ph)* zyklische Magnetisierung *f*; 2. *(El)* Ummagnetisierung *f*
c. **memory** *(BMSR)* periodischer Speicher *m (z. B. Magnettrommel als periodischer Speicher)*
c. **reactance** *(El)* Drehfeldreaktanz *f*
c. **shift** zyklische Verschiebung *f*
c. **stress** Schwingspannung *f*
c. **work** *(Bw)* Taktfertigung *f*, Taktverfahren *n*; zyklische Fertigung *f*

cycling 1. *(Masch)* Taktablauf *m*; 2. Schwingung *f*; Dauerschwingbeanspruchung *f*; Wechselbeanspruchung *f*; Schwellbeanspruchung *f*; 3. Radfahren *n*
cyclization *(Ch)* Zyklisierung *f*, Ringbildung *f*
cyclo-polymerization zyklisierende Polymerisation *f*, Polymerisation *f* unter Ringbildung
cyclograph *(El)* Polar-Oszillograf *m*
cycloid *(Math)* Zykloide *f*, Radkurve *f*, Radlinie *f*
cycloidal kreisläufig
c. curve zyklische Rollkurve (Kurve) *f*; Zykloide *f*
c. orbit Zykloidenbahn *f*
c. path *s.* c. orbit
c. pendulum *(Ph)* Zykloidenpendel *n*
c. teeth Zykloidenverzahnung *f*
c. tooth profile Zykloidenprofil *n*
cyclometer *(BMSR)* Umdrehungszählwerk *n*, Zyklometer *n*
cyclometric *(Math)* Kreismessungs-; zyklometrisch; trigonometrisch
c. function zyklometrische (trigonometrische) Funktion *f*
cyclometry *(Math)* Kreismessung *f*, Zyklometrie *f*
cyclone 1. Zyklon *m (Meteorologie)*; 2. *(Ch)* Staubabscheider *m*, Zyklon *m*
c. clarifier Klärzyklon *m (Aufbereitung)*
c. classifier Klassierzyklon *m (Aufbereitung)*
c. firing *(Met)* Zyklonfeuerung *f*
c. re-washing plant Zyklonnachwäsche *f (Aufbereitung)*
c. separator Waschzyklon *m*, Sortierzyklon *m*; Zyklonabscheider *m (Aufbereitung)*
c. thickener Eindickzyklon *m (Aufbereitung)*
c.-type evaporator *(Holz)* Zyklonverdampfer *m*
c. warning Wirbelsturmwarnung *f*
c. washer Waschzyklon *m*, Sortierzyklon *m (Aufbereitung)*
cyclopean masonry *(Bw)* Zyklopenmauerwerk *n*
cyclotomic *(Math)* Kreisteilungs-
cyclotomy *(Math)* Kreisteilung *f*
cyclotron *(Kern)* Zyklotron *n*
cylinder 1. Zylinder *m*, Walze *f*, Rolle *f*, Trommel *f*; 2. Laufbuchse *f*; 3. Stahlflasche *f (für Gase)*; 4. Kolbenbohrung *f*
c. barrel Zylinderlaufbuchse *f*, Zylindermantel *m*
c. base Zylinderflansch *m*
c. bit Genauigkeitsspiralbohrer *m*
c. blanket *(Typ)* Drucktuch *n*
c. block *(Kfz)* Zylinderblock *m*
c. block boring machine Gehäuseausbohrmaschine *f*
c. block production line Zylinderblockfertigungsstraße *f*
c. board Rundsiebmaschinenpappe *f*
c. bolt Deckelschraube *f*
c. bore Zylinderbohrung *f*, innerer Zylinderdurchmesser *m*, lichte Weite *f* des Zylinders
c. boring machine Zylinderbohrmaschine *f*
c. boring mill Zylinderbohrwerk *n*
c. cam Mantelkurve *f*
c. cam section Zylinderschloßsystem *n*
c. capacity *(Kfz)* Zylinderinhalt *m*, Zylindervolumen *n*
c. clearance schädlicher Raum *m (einer Kolbenmaschine)*
c. cooling fin (rib) Zylinderkühlrippe *f*
c. cover Zylinderdeckel *m*, abnehmbarer Zylinderboden *m*, Gehäusedeckel *m (Turbine)*
c. covering *(Text)* Walzenbeschlag *m*
c. cutter *(Holz)* runde Messerwelle *f (Hobelmaschine)*
c. diameter Zylinderdurchmesser *m*, Zylinderweite *f*
c. displacement *(Kfz)* Zylinderinhalt *m*
c. dressing *(Typ)* Zylinderaufzug *m*
c.-dried paper maschinengetrocknetes Papier *n*
c. fittings Zylinderausrüstung *f*, Zylinderarmatur *f*, Zylindergarnitur *f*
c. foot Zylinderstütze *f*, Zylinder[trag]fuß *m*, Zylinderfußgestell *n*
c. for drop valves Ventilzylinder *m*
c. gas Flaschengas *n*
c. head 1. Zylinderkopf *m*, Zylinderdeckel *m*; 2. *(Wärm)* Kesselboden *m*
c.-head bolt Zylinderkopfschraube *f*
c.-head casing Zylinderkopfgehäuse *n*
c.-head cover Zylinderkopfdeckel *m*
c.-head gasket Zylinderkopfdichtung *f*, Zylinderdichtungsring *m*
c.-head ring Zylinderkopfring *m*
c.-head stud Zylinderkopfschraube *f*
c. holding down bolt Zylinderbefestigungsschraube *f*
c. in swing *(Text)* Prismahängelade *f*
c. jacket Zylindermantel *m*, Kühlwassermantel *m*
c. lagging Zylinderverschalung *f*, Zylinderverkleidung *f*, Zylinderschutzmantel *m*
c. liner Arbeitszylinder *m*, Zylinderlaufbuchse *f*, Einsatzzylinder *m*, Zylindereinsatz *m*
c. machine *s.* c. press
c. mixer Trommelmischer *m*
c. mould Rundsiebzylinder *m*
c. needle loop expanded gespreizte Zylindermaschine *f*
c. needle revolving fabric machine einfonturige Rundstrickmaschine *f* mit sich drehendem Zylinder
c. needle wale Zylindernadelstäbchen *n*
c. of portable engine Lokomobilzylinder *m*
c. oil Zylinder[schmier]öl *n*

c. opener combined with beater and lap machine *(Text)* Trommelöffner *m* mit Schläger und Wickelvorrichtung
c. output *(Kfz)* Zylinderleistung *f*
c. port Zylinderschlitz *m (Zweitaktmotor)*
c. press Schnellpresse *f*
c. printing Walzendruck *m*
c. printing machine *(Text)* Rouleaudruckmaschine *f*
c. roller *(Masch)* Zylinderwalze *f*
c. sanding machine Zylinderschleifmaschine *f (für Holzbearbeitung)*
c. shape Zylinderform *f*
c. singeing machine *(Text)* Zylindersenge *f*
c. sizing machine *(Text)* Trommelschlichtmaschine *f*
c. support *(Wärm)* Zylinderstützung *f*, Zylinderauflagerung *f*
c. surface Ballenoberfläche *f (einer Walze)*
c. tooth *(Text)* Trommelnase *f (Trommelöffner)*
c. type chopper *(Lw)* Trommelhäcksler *m*
c. vat Rundsieb *n (Kartonmaschine)*
c. vat machine Rundsiebmaschine *f (Kartonherstellung)*
c. verge Zylinderoberkante *f*
c. wall Zylinderwand[ung] *f*
c. wheel Schleifring *m*
cylinders arranged in a vee *(Kfz)* Zylinderanordnung *f* in V-Form
c. in line *(Kfz)* Zylinder *mpl* in Reihe
c. in radial arrangement *(Kfz)* Zylinder *mpl* in Sternform
c. in V-arrangement *(Kfz)* Zylinder *mpl* in V-Form
cylindrical zylindrisch, walzenförmig
c. battery Stabbatterie *f*
c. bearing Zylinderrollenlager *n*
c. boiler *(Wärm)* zylindrischer Kessel *m*, Zylinderkessel *m*
c. boiler with feed water heater Walzenkessel *m* mit Vorwärmer
c. buoy *(Schiff)* stumpfe Tonne *f*
c. cam milling attachment Mantelkurven-Fräsvorrichtung *f*
c. cathode Rundkatode *f (Elektronenröhren)*
c. coil *(El)* Zylinderspule *f*
c. controller *(El)* Walzenfahrschalter *m*
c. cutter Walzenfräser *m*
c. die thread rolling Gewinderollen *n*
c.-drum hoist *(Brg)* Fördermaschine *f* mit zylindrischer Trommel
c. fit Rundpassung *f*
c. function *(Math)* Besselfunktion *f*, Zylinderfunktion *f*
c. gear differential Stirnraddifferential *n*
c. grate *(Wärm)* Walzenrost *m*
c. grinder Rundschleifmaschine *f*
c. grinding Rundschleifen *n*
c. grinding attachment Rundschleifeinrichtung *f*
c. grinding machine Außenrundschleifmaschine *f*
c. guide *(Wärm)* Rundführung *f*
c. land Rundschliffase *f*
c. lap Läppdorn *m*
c. lapping Rundläppen *n*
c. lens Zylinderlinse *f*
c. magnet Topfmagnet *m*
c. margin Rundschliffase *f*
c. milling cutter Walzenfräser *m*
c. part [of the reamer] Zylinderteil *n*
c. pin Zylinderstift *m*
c. portion [of the reamer] Zylinderteil *n*
c. roller bearing Zylinderrollenlager *n*
c. roller thrust bearing Axial-Zylinderrollenlager *n*
c. rotary kiln Drehrohrofen *m*
c. rotary valve Drehkolbenventil *n*
c. rotor *(Elm)* Trommelläufer *m*, Trommelanker *m*
c. rough grinding Rundschruppschleifen *n*
c. screen Rundsieb *n*
c. screw gear zylindrisches Schraubrad *n*
c. shell Zylinderschale *f*, Tonnenschale *f*
c. slab mill Walzenfräser *m*
c. slide valve Rohrschieber *m*; Kolbenschieber *m*
c. spectacle lens astigmatisches Brillenglas *n*
c. spiral spring zylindrische Schraubenfeder *f*
c. thread zylindrisches Gewinde *n*
c. turning Langdrehen *n*
c. winding *(El)* Zylinderwicklung *f*
c. worm Zylinderschnecke *f*, zylindrische Schnecke *f*
cylindrically ground rundgeschliffen, mit Rundschliff
cylindro-conical drum hoist *(Brg)* Fördermaschine *f* mit zylindro-konischer Trommel
cymbal Becken *n*, Schallbecken *n (Musikinstrument)*
cymene *(Ch)* Zymol *n*
cymophane Cymophan *m*, Chrysoberyll *m*
cynometer *(El)* Wellenmesser *m*
cynoscope *(El)* Wellenanzeiger *m*
cypress Zypresse[nholz *n*] *f*
cyprian vitriol Kupfersulfat-Zinksulfat-Doppelsalz *n*
cysteic acid Zysteinsäure *f*
cysteine Zystein *n*
cystine Zystin *n*
cystoscope Zystoskop *n*

D

D-bit Langlochbohrer *m*, Tieflochbohrer *m*
dabber *(Typ)* Filzwalze *f*
dabbing brush Einschlagbürste *f*
dabrey Sammelschale *f* für Balata (Kautschuk)
dag Graphitstaub *m (Schmiermittelkomponente)*
dagger *(Schiff)* Stopper *m (beim Stapellauf)*
d. board *(Schiff)* kurzes Schwert *n*
d. function *s.* Peirce function
d. mast *(Schiff)* Steckmast *m*
d. shore *(Schiff)* Schlagbettungsstrebe *f*; Stopper *m (beim Stapellauf)*
dairy Molkerei *f*; Milchwirtschaft *f*
d. barn Milchviehstall *m*
d. cattle Milchkühe *fpl*
d.-farming Milchwirtschaft *f*
d.-house Milchkammer *f*
d. machine Molkereimaschine *f*
dais *(Bw)* Podium *n*; Laufsteg *m*
Dall tube *(BMSR)* Dallrohr *n (Durchflußmesser)*
dam / to stauen; eindämmen
d. up rückstauen
dam 1. *(Bw)* Staudamm *m*; Staumauer *f*; Talsperre *f*; 2. Wehr *n (am Holzschleifer)*
d. lake Talsperrensee *m*
d. lip Wehrüberlauf *m (am Holzschleifer)*
d. plate Schlackenblech *n*
damage Beschädigung *f*, Schaden *m*; Schädigung *f (Werkstoff)*; *(Schiff)* Havarie *f*
d.-control splinter and stuffing box *(Schiff)* Leckstopfbeutel *m*
d. curve Schadenslinie *f*
d. due to mining Bergschaden *m*
d. due to singeing *(Text)* Sengschaden *m*
d. survey Havarieuntersuchung *f*
damaged beschädigt; *(Schiff)* havariert, leck
d. condition schadhafter Zustand *m*; *(Schiff)* lecker Zustand
damaging of the fibre Faserangriff *m*, Faserschädigung *f*
damascene / to damaszieren
damask weave *(Text)* Damastbindung *f*
damaskeen / to *s.* damascene / to
damaskeen Damaszierung *f*
dammar Dammar[harz] *n (natürliches Weichharz)*
damming Absperrung *f*
d.-up Rückstau *m*
damp / to 1. *(El) (zeitlich)* dämpfen; 2. anfeuchten, befeuchten
d. out unterdrücken *(Schwingungen)*; abdämpfen
damp 1. Grubengas *n (Steinkohlenbergbau)*; 2. Feuchtigkeit *f*
d.-proof feuchtigkeitsbeständig; feuchtigkeitsgeschützt; feuchtigkeitsisolierend
d.-proof course *(Bw)* Sperrschicht *f*
d.-proof membrane *(Bw)* Dichtungshaut *f*
d.-proofing *(Bw)* Feuchtigkeitsschutz *m*; Abdichtung *f*, Sperrung *f*
damped-down blast furnace gedämpfter Hochofen *m*
d. harmonic system gedämpfter harmonischer Schwinger *m*
d. moulding sand feuchter Formsand *m*
d. oscillation gedämpfte Schwingung *f*
d. periodic instrument gedämpft schwingendes Instrument *n*
d. pointer action Zeigerdämpfung *f (Meßinstrumente)*
d. sinusoidal quantity gedämpfte Sinusgröße *f*
d. vibration gedämpfte Schwingung *f*
d. wave gedämpfte Welle *f*
dampen / to befeuchten, anfeuchten, benetzen; dämpfen
damper 1. Dämpfer *m*, Schalldämpfer *m*, Tondämpfer *m*; 2. *(Masch)* Drossel[klappe] *f*, Zugklappe *f*; Schieber *m*, Rauchgasschieber *m*; 3. *(Brg)* Ablenkseil *n (Schachtförderung mit Führungsseilen)*
d. chain *(Wärm)* Schieberkette *f*
d. diode Dämpfungsdiode *f*
d. guide Schieberführung *f*
d. pit Schiebergrube *f*
d. valve Dämpfungsventil *n*
d. winding *(El)* Dämpferwicklung *f*, Dämpfungswicklung *f*
damping 1. Dämpfung *f*; 2. Anfeuchtung
d. coefficient Abklingkonstante *f*; Dämpfungsfaktor *m*
d. constant Dämpfungskonstante *f*
d. constant of a cable Verlustkonstante *f* eines Kabels
d. current Dämpfungsstrom *m*
d. diode Dämpfungsdiode *f*
d. down of the coke Kokslöschen *n*
d. factor Dämpfungsfaktor *m*
d. fin Dämpfungsflosse *f (Schwingungsdämpfer eines Schiffes)*
d. machine *(Text)* Anfeuchtemaschine *f*, Netzapparat *m*
d. machine with water spray *(Text)* Anfeuchtemaschine *f* mit Wasserzerstäuber
d. magnet Dämpfungsmagnet *m*, Bremsmagnet *m*
d. mica Dämpfungsglimmer *m (Elektronenröhrenherstellung)*
d. of vibration Schwingungsdämpfung *f*
d. period Beruhigungszeit *f (Meßinstrumente)*
d. ratio *(BMSR)* Dämpfungsfaktor *m*, relative Dämpfung *f*
d. reduction Dämpfungsverminderung *f*
d. resistor Dämpfungswiderstand *m*
d. roller *(Typ)* Anfeuchtwalze *f*, Wischwalze *f*

d. rope *(Brg)* Ablenkseil *n (Schachtförderung mit Führungsseilen)*
d. system Feuchtwerk *n (Drucken)*
d. time Einstellzeit *f (Meßinstrumente)*
d. torque Dämpfungsmoment *n*
d. tube Dämpfungszylinder *m*
d. water *(Typ)* Wischwasser *n*, Feuchtwasser *n*
d. winding Dämpferwicklung *f*, Dämpfungswicklung *f*
damstone *(Met)* Wallstein *m (Hochofen)*
dan buoy *(Schiff)* Markierungsboje *f*, Schwimmboje *f*
dandy loom halbmechanischer Webstuhl *m*
d. roll[er] *(Typ)* Vordruckwalze *f*; Wasserzeichenwalze *f*, Egoutteur *m*
danger Gefahr *f*
d. arrow Hochspannungswarnpfeil *m*
d. buoy *(Schiff)* Warnungsboje *f*
d. dose *(Kern)* Gefährdungsdosis *f*
d. light optisches Gefahrensignal *n*
d. lock Haltsperre *f (Signal)*
d. money *(Ök)* Gefahrenzulage *f*
d. of icing Vereisungsgefahr *f*
d. of nosing-over Überschlaggefahr *f (bei der Landung)*
d. of skidding Rutschgefahr *f (bei der Landung)*
d. position Haltlage *f (Signal)*
d. report Gefahrenmeldung *f*
d. signal Alarmsignal *n*, Gefahrensignal *n*
d. zone 1. bestrichener Raum *m (Schießen)*; 2. Gefahrenzone *f*
dangerous cross section gefährdeter Querschnitt *m (Festigkeit)*
d. materials class Gefahrenklasse *f (Chemikalien)*
Daniell cell Daniell-Element *n*
danlayer *(Schiff)* Bojenboot *n*
Darcy's law Darcysches Gesetz *n (Bodenmechanik)*
dark dunkel, finster
d. adaptation *(Opt)* Dunkeladaptation *f (Auge)*
d. and light spots Schatteneffekt *m (Ikonoskop)*
d. band Kompensationsstreifen *m (Polarisationsmikroskopie)*
d. burn Ermüdung *f* des Leuchtstoffes
d. current Dunkelstrom *m (lichtelektrische Zelle)*
d. discharge *(El)* Dunkelentladung *f*, stille Entladung *f*
d. field *(Opt)* Dunkelfeld *n*
d.-field condenser Dunkelfeldkondensor *m*
d.-field observation Dunkelfeldbeobachtung *f*
d. ground *(Opt)* Dunkelfeld *n*
d. intensification Dunkelsteuerung *f (Katodenstrahloszillografie)*
d.-line spectrum Absorptionsspektrum *n*
d. nebula Dunkelnebel *m*
d. oils Dunkelöle *npl*, nichtraffinierte Schmieröle *npl*
d. pulse Dunkel[strom]impuls *m*
d. radiation *(Ph)* Dunkelstrahlung *f*
d. red silver ore dunkles Rotgiltigerz *n*
d. resistance Dunkelwiderstand *m (lichtelektrische Zelle)*
d. slide Filmkassette *f*, Kassette *f*
d. space Dunkelraum *m*
d.-straw *(Met)* dunkelgelb *(Anlaßfarbe)*
d. substitute schwarzer Faktis *m (Lackindustrie)*
d. trace screen Leuchtschirm *m* einer Farbschriftröhre (Dunkelschriftröhre)
d. trace tube Dunkelschriftröhre *f*, Farbschriftröhre *f*, Skiatron *n*
darken / to abdunkeln; schwärzen
darkening agent *(Text)* Abdunkelungsmittel *n*
darkroom Dunkelkammer *f*
d. camera *(Typ)* Dunkelraumkamera *f*
d. illumination Dunkelkammerbeleuchtung *f*
darn / to *(Text)* ausnähen; mit einem stopfartigen Stich nähen *(Stickerei)*; stopfen
darner *(Text)* Stopfmeister *m*
darning yarn Stopfgarn *n*
D'Arsonval galvanometer Drehspulinstrument *n*, Drehspulgalvanometer *n*
darting flame Stichflamme *f*
dash / to *(Math)* stricheln; mit einem oberen Strich (′) versehen
dash 1. *(Math)* oberer [senkrechter] Strich *m* (′); 2. [waagerechter] Strich *m (Morsealphabet)*
d. and dot line Strichpunktlinie *f*, strichpunktierte Linie *f*
d.-dot mode *(BMSR)* Strichpunktverfahren *n*
d.-dotted line Strichpunktlinie *f*, strichpunktierte Linie *f*
d. line gestrichelte Linie *f*, Strichlinie *f*
dashboard Armaturenbrett *n*; Instrumentenbrett *n*; Spritzbrett *n (Traktor)*
d. gear change *(Kfz)* vorgesetzte Schaltung *f*
dashed line *s.* dash line
dashers of churner Flossen *fpl* der Buttermaschine
dashpot Bremszylinder *m*, Dämpfungszylinder *m*, Puffer *m*, *(hydraulischer)* Stoßdämpfer *m*
d. action *(Masch)* dämpfende Wirkung *f*
d. carburettor *(Kfz)* Vergaser *m* mit Beschleunigerpumpe *f*
d. cylinder Dämpfungszylinder *m*
d. pump *(Kfz)* Beschleunigerpumpe *f (Vergaser)*
d. pump carburettor *(Kfz)* Vergaser *m* mit Beschleunigerpumpe
d.-type shock absorber hydraulischer Stoßdämpfer *m*
dasymeter Dasymeter *n*, Gasdichtemesser *m*

data *(BMSR, Büro)* Daten *pl*, Angaben *fpl*; Kenngrößen *fpl*
d. acquisition Datenerfassung *f*, Datengewinnung *f*
d. acquisition system Datenerfassungssystem *n*
d.-biased read-out device Auslesevorrichtung *f* mit Datenvorgabe
d. bit Datenbit *n (im binären System Null oder Eins)*
d. block Datenblock *m (die Zeichen sind in Blocks gruppiert)*
d. dump Informationsverlust *m*
d. gathering Datenerfassung *f*, Datengewinnung *f*
d. handling system Datenverarbeitungssystem *n (zum Koordinieren, Berechnen, Tabellieren)*
d. logger Datenerfassungs- und -aufzeichnungsanlage *f*, Daten[mit]schreiber *m*; Datenspeicher *m*
d. logging Registrierung (Speicherung) *f* von Daten
d. logging equipment Meßwertschreibeinrichtung *f (auf Schiffen bei Meßfahrten)*
d.-logging system Betriebskontrollanlage *f*, BKA
d. processing Datenverarbeitung *f*
d. processor Datenverarbeiter *m*, Datenverarbeitungsgerät *n*
d. recorder Meßwertschreiber *m (auf Schiffen bei Meßfahrten)*
d. reduction Datenreduktion *f*
d. sampling system Datenabtastsystem *n*
d. transaction posting unit Datenbuchungsmaschine *f* für Umsätze *(Materiallager)*
d. transmission system Meßwertübertragungssystem *n*, Datenübertragungssystem *n*
d. transmitter Meßwertsender *m*
d. transmitter-receiver Datensende- und -empfangsgerät *n*
d. word format Datenwortformat *n (Ziffernrechner)*
date line Datumsgrenze *f*, Frist *f*; Fristablauf *m*
d.-time group *(Schiff)* Kennummer *f* eines Signals
datum gegebene Größe *f*; Bezugsgröße *f*
d. face *(Masch)* Bezugsfläche *f*
d. horizon *(Geol)* Leithorizont *m*
d. level Grundebene *f*, Ausgangsebene *f*; Bezugshöhe *f*, Höhenmarke *f*
d. line *(BMSR)* Bezugslinie *f*
d. plane 1. *(Bw)* Bezugsebene *f*, Bezugsfläche *f*; 2. *(Masch)* Teilebene *f* einer Zahnstange
d. point Bezugspunkt *m*
d. surface Bezugsfläche *f (z. B. Oberflächenrauhigkeit)*
daub / to [grob] auftragen, bestreichen, beschmieren, zuschmieren
d. with ashes and lime *(Led)* anschwöden
daub Auftrag *m*, Bewurf *m*, Schicht *f*; *(Led)* Grund *m*
daughter *(Kern)* Tochtersubstanz *f*, Folgeprodukt *n*, Zerfallsprodukt *n*
d. activity Tochteraktivität *f*
d. atom Folgeatom *n*, Tochteratom *n*
d. decay Tochterzerfall *m*
d. element Folgeelement *n*, Tochterelement *n*; Zerfallsprodukt *n*
d. fraction Tochterfraktion *f*, Anteil *m* der Tochtersubstanz
d. missile zweite Raketenstufe *f (einer Zweistufenrakete)*
d. nucleus Folgekern *m*, Tochterkern *m*
davit *(Schiff)* Davit *m (n)*, Schiffsgalgen *m*; Ankerdavit *m*
d. craft Davitboot *n*, Seitenboot *n*
davy Sicherheitslampe *f*, Wetterlampe *f*
day *(Brg)* Tagesoberfläche *f*
d.-and-night transfer key *(Fmt)* Nacht-[alarm]schalter *m*
d.-drift *(Brg)* Stollen *m*
d. frequency Tagesfrequenz *f (Rundfunk)*
d. load *(El)* Tagesbelastung *f*, Belastung *f* am Tage
d. range Tagesreichweite *f*
d. shift *(Brg)* Mittagsschicht *f*
d. turn *(Brg)* Mittagsschicht *f*
d. wave Tageswelle *f (Rundfunk)*
daylight developing box Tageslichtentwicklungsdose *f (Fotografie)*
d. development Tageslichtentwicklung *f*
d. loading spool Tageslichtspule *f (Fotografie)*
d. mining *(Brg)* Tagebau *m*
d. reconnaissance *(Mil)* Tagaufklärung *f*
daylighting natürliche Beleuchtung *f*, Tagesbeleuchtung *f*
days under fire *(Brg)* Feuertage *mpl*
day's work *(Schiff)* Etmal *n (Navigationsarbeit über 24 Stunden)*
daywork grinding Schleifen *n* im Zeitlohn
d. joint Arbeitsfuge *f (Betonbau)*
dazzle / to blenden
dazzle-free blendungsfrei *(Beleuchtung)*
d. painting *(Schiff)* Tarnanstrich *m*, Täuschungsanstrich *m*
D.C., d.c. *s.* direct current
de-acceleration Verzögerung *f*
de-acidification Entsäuerung *f*
de-aerating feed-water heater Entgasungsspeisewasservorwärmer *m*
de-aeration Entlüftung *f*; Entgasung *f*
de-aerator Entlüftungseinrichtung *f*, Entgaser *m*
de-aired water entlüftetes Wasser *n*
de-awning machine Entgranner *m (Dreschmaschinen)*

de Broglie wave *(Ph)* Materiewelle *f*, de-Broglie-Welle *f*
de-dust / to entstauben
de-emphasis *(Fmt, Fs)* Dämpfung *f* der hohen Frequenzen *(empfängerseitig)*; Deemphasis *f*; Vorentzerrung *f*
de-emulsification Demulgierung *f*
de-energize / to 1. *(BMSR, Büro)* ab[er]regen; 2. *(El)* auslösen, zum Abfall bringen *(Relais)*; ausschalten, abschalten
de-excitation *(Ph)* Abregung *f*, Übergang *m* in den Grundzustand
de-excitation photon *(Kern)* Kern-Gamma-Quant *n*
de-icer Enteisungsanlage *f*
de-icing Enteisen *n (z. B. Windschutzscheibe)*, Enteisung *f*
de-icing device Enteisungsanlage *f*
de-inking Entfärben *n (Aufbereitung von Altpapier)*
de-inking solution Entfärbelösung *f*, Entfärber *m*
de-ionizing grid *(El)* Deionisierungsgitter *n*
de-ionizing potential Löschspannung *f (Gasentladung)*
de-ionizing time Löschzeit *f (Gasentladung)*
de-laminating *(Holz)* Aufspalten *n*
de-railer Entgleisungsschuh *m*
de-railing block Entgleisungsschuh *m*
de-rust / to entrosten
de-rusting Entrosten *n*
de-scale / to entzundern
de-scaling *(Met)* Entzundern *n*
de-scrambler Entzerrer *m (z. B. beim Münzfernsehen)*
de-screen / to *(Opt)* entrastern
de-screening *(Opt)* Entrasterung *f*
de-shaler *(Brg)* Bergeaustrag *m*
deaccentuator *(Fs)* Vorentzerrungs-Schaltung *f (FM-Technik)*
deacetylation Abspaltung *f* der Azetylgruppe
deacidify / to *(Ch)* entsäuern
deacidized tar entsäuerter Raffinationsabfall *m (Erdöl)*
deactivation *(Ch)* Desaktivierung *f*, Entaktivierung *f*
dead 1. spannungslos, stromlos; 2. absolut, konstant; 3. unbeweglich, nicht drehend; 4. Eigen-
d. air *(Brg)* matte Wetter *pl*
d. annealing *(Met)* Weichglühen *n*
d. band *s.* d. zone
d.-beat aperiodisch, überschwingfrei
d.-beat ammeter aperiodisches Amperemeter *n*
d. burn / to *(Met)* totbrennen
d. central genau mittig, genau auf Mitte
d. centre 1. *(Masch)* feste Körnerspitze (Spitze) *f*; 2. *(Wärm)* Totpunkt *m*, Umkehrpunkt *m*
d.-centre buffer Halbrundpuffer *m*
d.-centre ignition *(Kfz)* Totpunktzündung *f*
d.-centre position *(Kfz)* Totpunktlage *f*, Totlage[nstellung] *f*
d. conductor *(El)* stromloser Leiter *m*
d. contact *(El)* Leerkontakt *m*
d. cotton unreife (nicht voll entwickelte) Baumwolle *f*
d. dip / to mattbrennen
d.-drawn wire biegungsfreier gezogener Draht *m*
d. earth *(Am) (El)* Erdschluß *m*
d. end *(Brg)* Blindort *n*
d.-end effect *(Fmt)* Nebenresonanz *f*
d.-end pressure regulator *(Am) (BMSR)* Abströmregler *m*
d.-end street Sackgasse *f*
d.-end switch Schalter *m* mit Kurzschlußkontakten
d.-end tower *(Elm)* Abspannmast *m*, Endmast *m (Stahlmast)*
d.-eye *(Schiff)* Kausche *f*
d. flat *(Schiff)* Hauptspant *n*
d. freight Ballast *m*
d. ground 1. *(Geol)* taubes Gestein *n*; 2. *(Am) (El)* vollkommener Erdschluß *m*; 3. gedeckter Raum *m (bei Beschuß)*
d.-harrowing Toteggen *n* des Ackers
d. head verlorener Kopf *m (eines Gußstückes)*
d. hole *(Brg)* Pfeife *f (Schießarbeit)*
d. knot Aststummel *m*
d. length accuracy maßgenaue Länge *f*
d. level *(Schiff)* Nullpegel *m*
d. line *(Fmt)* tote Leitung *f*
d. load Eigengewicht *n*, Eigenlast *f*, tote Last *f*; ständige Last; Dauerbelastung *f*, konstante Belastung *f*
d. loss *(Ök)* Totalverlust *m*
d. man 1. *(Bw)* Ankerblock *m*; rückwärtige Verankerung *f (Grundbau)*; 2. *(Am) (El)* Abspannpfahl *m*, Abspannmast *m (am Haus)*
d.-man brake *(Eis)* Totmann-Bremse *f*
d.-man's-handle Sicherheitseinrichtung *f*
d. match *(sl) (Text)* genaue Übereinstimmung *f*
d. matter *(Typ)* Ablegesatz *m*
d. pass Blindkaliber *n (Walzwerk)*
d. plate Entgasungsplatte *f*, Schürplatte *f*, Vorstellplatte *f (Rostfeuerung)*
d. plate for inclined grate Schrägrostvorstellplatte *f*
d. point *(Wärm)* Totpunkt *m*; Hubwechselpunkt *m*
d. point of eccentric *(Wärm)* Exzentertotlage *f*

d. position Totpunkt *m*, Endlage *f (Kolben)*
d. reckoning *(Schiff)* gegißtes Besteck *n*, Besteckrechnung *f*, Gissung *f*
d.-reckoning navigation *(Schiff)* Koppelnavigation *f*
d. reckoning position *(Schiff)* gegißtes Besteck *n*, Koppelort *m*
d. rise *(Schiff)* Aufkimmung *f*
d. roasting Totrösten *n*
d. roll Blindwalze *f (Walzwerk)*
d. round genau rund
d. shore *(Bw)* Stempel *m (Unterfangung)*
d. short-circuit direkter (vollständiger) Kurzschluß *m*
d.-smooth sea spiegelglatte See *f*
d.-soft *(Met)* totweich
d.-soft annealing Weichglühen *n*, Totweichglühen *n*
d. steel *(Met)* ausgegarter (beruhigter) Stahl *m*
d. stick *(Flg)* stehende Luftschraube *f*
d. stock *(Ök)* unverkäufliche Ware *f*; totes Inventar *n*
d. studio *(Fmt, Fs)* schalltotes Studio *n*
d. tailstock centre Reitnagel *m*
d.-tank oil circuit-breaker Ölkesselschalter *m*, Kesselölschalter *m*, isolierter Ölschalter *m*
d.-tie *(Am) (El)* Abspannbund *m (Freileitung)*
d. time *(BMSR)* Totzeit *f*, Laufzeit *f*
d. travel *(Kfz)* Leerweg *m*, toter Gang *m*
d. wall *(Bw)* blinde Mauer *f*
d. water 1. stehendes Wasser *n*; 2. Sog *m*, Totwasser *n (hinter einem Schiff)*, Kielwasser *n*
d. weight 1. *(Bw)* Eigenmasse *f*, Eigenlast *f*; 2. *(Kfz)* Eigenmasse, Leergewicht *n*; 3. Schwergut *n (in der Schiffahrt)*
d. weight all told Tragfähigkeit *f (eines Schiffes: Ladung, Personen, Kohle, Öl, Wasser, Proviant usw.)*
d. weight capacity *(Schiff)* Tragfähigkeit *f*, Gesamtzuladung *f*, Ladevermögen *n*
d. weight cargo Schwergut *n*, Gewichtsfracht *f (Ladung, nach Gewicht berechnet)*
d. weight cargo factor Tragfähigkeitsfaktor *m (Faktor × Registertonnen)*
d. weight carrying capacity *(Schiff)* Tragfähigkeit *f*, Gesamtzuladung *f*, Ladevermögen *n (effektive Ladung einschließlich Bunkervorräte, Ballast und Proviant)*
d. weight / displacement coefficient *(Schiff)* Tragfähigkeitskoeffizient *m (Tragfähigkeit/Verdrängung)*
d. weight loading capacity *(Schiff)* Tragfähigkeit *f*, Ladefähigkeit *f (effektive Nutzladung ausschließlich Bunker, Ballast und Proviant)*
d.-weight piston gauge *(BMSR)* gewichtsbelastetes Kolbenmanometer *n*
d. weight tester *s.* d.-weight piston gauge
d. weight tonnage *(Schiff)* Frachtraum *m*, Frachttonnage *f (in t)*; Ladetonnage *f*
d. window *(Bw)* Scheinfenster *n*, vorgetäuschtes Fenster *n*
d. wood *(Lw)* Abholz *n*, Reiserholz *n*; *(Schiff)* Totholz *n*
d. wool *(Text)* Sterblingswolle *f*
d. work 1. *(Brg)* Gesteinsarbeit *f*; unproduktive Arbeit *f*; 2. totes Werk *n*, Oberschiff *n*, Oberwerk *n (Überwasserteil eines Schiffes)*
deadbeat aperiodisch, schwingungsfrei *(Meßinstrumente)*; gedämpft
deaden / to 1. dämpfen, abfangen, aufnehmen *(Kraft)*; 2. *(Text)* abmatten
deadener 1. Schalldämpfer *m*; 2. Schwingungsdämpfer *m (bei Leitungen)*
deadening *(Bw)* Schalldämmung *f*, Schalldämpfung *f*, Schallschluckung *f*; Aufnahme *f*
d. of noises Dämpfung *f* von Geräuschen
d. shock 1. Stoßdämpfen *n*; 2. stoßdämpfend
deadfreight *(Schiff)* Fehlfracht *f (vereinbarte, vom Befrachter nicht angelieferte Fracht)*
deadhead *(Am)* blinder Passagier *m*
deadline Endtermin *m*
deadlock [vollständige] Stockung *f*
deadly nightshade root Belladonnawurzel *f*
deadrise boat Knickspantboot *n*
deads *(Geol)* taubes Gestein *n*
deadstock *s.* dead stock
deadweight *s.* dead weight
deaf aid Hörapparat *m*
deal 1. *(Ök)* Geschäft[sabschluß *m*] *n*, Abkommen *n*, Vertrag *m*; 2. *(Bw)* Nadelholz *n*; Bohle *f*, Brett *n*, Diele *f (aus Nadelholz)*
dealer Maschinenhändler *m*, Händler *m*
dealkylation Dealkylierung *f*, Entalkylierung *f*
deamination Desamidierung *f*, Desaminierung *f*
dean 1. Dekan *m*; 2. *(Brg)* Blindort *n*
deasphaltizing Entasphaltieren *n*
debacle 1. Eisaufbruch *m (in Flüssen)*; 2. Hochwasserflut *f* mit Gesteinsmassen
debark / to *s.* disembark / to
debasement Qualitätsminderung *f*
debenzolize / to entbenzolen
debenzolizing Entbenzolung *f*
debit note Lastschriftzettel *m*
deblooming Entscheinen *n (Farbstoffbehandlung von Ölen zur Entfernung des blauen Scheins)*
debris 1. Trümmer *pl*; 2. *(Geol)* Geröll *n*, Geschiebe *n*; 3. *(Brg)* Bergeklein *n*, Bohrklein *n*, Haufwerk *n*, hereingeschossenes Gestein *n*, Schrämklein *n*
debromination Bromabspaltung *f*, Entbromierung *f*
debugging *(Am) (BMSR)* Störbeseitigung *f*, Fehlerbeseitigung *f*
deburr / to *(Met)* entgraten
deburrer Abgratmaschine *f*, Entgratmaschine *f*

deburring 1. Entgraten *n*; 2. Anfasen *n*
d. machine Abgratmaschine *f*, Entgratmaschine *f*
debus / to absteigen, aussteigen *(Motorfahrzeug)*; abladen
debye Debye *n*
Debye-Scherrer method Debye-Scherrer-Verfahren *n (Röntgenfeinstrukturverfahren)*
decade Dekade *f*
d. connection *(BMSR, El)* Dekadenschaltung *f*
d. counter tube *(BMSR, El)* Dekadenzählröhre *f*, dekadische Zählröhre *f*, Dezimalzählröhre *f*
d. resistance Dekadenwiderstand *m*
d. scaler *(BMSR)* dekadischer Untersetzer *m*; Dekadenzähler *m*
decadic dekadisch; Dezimal-
decagon *(Math)* Zehneck *n*
decahedron Dekaeder *n*
decahydronaphthalene Dekalin *n*, Dekahydronaphthalin *n*
decalcification Entkalkung *f*
decalcify / to entkalken
decalescence Wärmeaufnahme *f*; Abschrecken *n*, Ablöschen *n*
d. point Haltepunkt *m* bei Erwärmung
decan / to *(Kern)* entkapseln
decant / to *(Ch)* abgießen, abfließen, dekantieren; abfüllen, umfüllen
decantation under pressure *(Brg)* Druckscheidung *f*
decanter *(Ch)* Abklärgefäß *n*, Dekantiergefäß *n*
decarbonization Entkohlung *f*, Kohlenstoffentziehung *f*
decarbonize / to entkohlen; Ölkohle entfernen
decarboxylation Dekarboxylierung *f*
decarburate / to *s.* decarburize / to
decarburization Entkohlung *f*, Kohlenstoffentziehung *f*
decarburize / to entkohlen, frischen
decarburizing agent Entkohlungsmittel *n*
d. plant Entkohlungsanlage *f*
d. process *(Met)* Frischverfahren *n*, Frischprozeß *m (Glühfrischen)*
decascaler *(BMSR)* Dekadenzähler *m*, Dekadenuntersetzer *m*
decate / to *(Text)* dekatieren
decating blanket *(Text)* Dekatiertuch *n*
d. machine *(Text)* Dekatiermaschine *f*
decationizing *(Ch)* Kationenabspaltung *f*
decatize / to *(Text)* dekatieren
d. with dry steam trockendekatieren
decatizing *(Text)* Dekatur *f*
d. machine *(Wärm)* Kesseldekatiervorrichtung *f*
decay / to 1. abnehmen, abklingen, abfallen, geringer werden; 2. [ver]faulen, [ver]modern *(Holz, Pflanzen)*; 3. verderben *(Lebensmittel)*; 4. zerfallen *(z. B. Atomkerne)*
decay 1. Abnahme *f*, Abklingen *n*, Abfall[en *n*] *m*; 2. Verfaulung *f*; 3. Verderb *m*; 4. Zerfall *m*
d. coefficient 1. *(Ph, El)* Dämpfungsfaktor *m*, Abkling[ungs]konstante *f*; 2. *(Kern)* Zerfallskonstante *f*
d. constant 1. *(Ph, El)* Abkling[ungs]konstante *f*; 2. *(Kern)* Zerfallskonstante *f*
d. electron *(Kern)* Zerfallselektron *n*
d. energy *(Kern)* Zerfallsenergie *f*
d. family *(Kern)* radioaktive Zerfallsreihe *f*
d. luminescence *(Ph)* Zerfallsleuchten *n*
d. of luminescence *(Ph)* Abklingen *n* der Lumineszenz
d. of vorticity *(Ph)* Wirbeldiffusion *f*
d. probability *(Kern)* Zerfallswahrscheinlichkeit *f*
d. product *(Kern)* Zerfallsprodukt *n*
d. series *(Kern)* Zerfallsreihe *f*
d. time 1. *(Fmt, Fs)* Abfallzeit *f (Impulstechnik)*; Abklingzeit *f (Schwingungen usw.)*; 2. *(Kern)* Zerfallszeit *f*
decaying Verfaulen *n*, Verderben *n*
decaying *(Kern)* instabil; zerfallend
d. current abklingender Strom *m*; Ausschwingstrom *m*
d. current impulses abklingende Stromimpulse *mpl*
decelerate / to 1. *(Ph)* verlangsamen, verzögern; 2. *(Kfz)* Gas wegnehmen, Geschwindigkeit herabsetzen, bremsen
decelerating electrode *(Ph)* Bremselektrode *f*
deceleration Verlangsamung *f*, Verzögerung *f*; Verminderung (Herabsetzung) *f* der Geschwindigkeit, Bremsung *f*
d. parachute Bremsfallschirm *m*
decendioic acid Oktylendikarbonsäure *f*
decenoic acid Decylensäure *f*
decentralization of cores *(El)* Aderverlagerung *f*
dechlorination Chlorabspaltung *f*, Entchlorierung *f*
dechuck / to *(Masch)* entspannen
decibel Dezibel *n (akustische Maßeinheit)*
d.-meter Dämpfungsmesser *m*, Pegelmesser *m*
decimal dezimal
decimal Dezimalbruch *m*
d. balance Dezimalwaage *f*
d. candle *(Opt)* Bougie-Kerze *f (Einheit)*
d. digit Dezimalziffer *f*
d. fraction Dezimalbruch *m*
d. notation Dezimaldarstellung *f*, Dezimalschreibweise *f*
d. place Dezimalstelle *f*
d. point *(Math)* Dezimalpunkt *m (anstelle des Kommas im Deutschen)*

d.-to-binary conversion *(BMSR)* Dezimal-Binärumsetzung *f*, Dezimal-Binärkonvertierung *f*, Dezimal-Binärumwandlung *f*
decimetric wave Dezimeterwelle *f*
decinormal *(Ch)* zehntelnormal
decipher / to dekodieren, entschlüsseln, dechiffrieren
decision element *(Dat)* Entscheidungselement *n*, Entscheidungsgerät *n*
d. making *(BMSR, Dat)* Entscheidungtreffen *n*
d. procedure *(BMSR, Dat)* Entscheidungsverfahren *n*
deck *(Schiff)* Deck *n*
d. area Decksfläche *f (auf dem Schiff)*
d. auxiliaries Deckshilfsmaschinen *fpl*
d. beam Decksbalken *m*
d.-beam clamp Decksbalkenweger *m*
d. bridge *(Bw)* Brücke *f* mit obenliegender Fahrbahn
d. cabin Kajütsaufbau *m*
d. cargo Decksladung *f*, Deckslast *f*
d. composition Decksbelag *m*
d. control Decksfahrstand *m (elektrische Steuerung von Deck aus)*
d. crane Bordkran *m*
d. drain Speigatt *n*
d. erection Decksaufbau *m*
d. fighter bordgestütztes Jagdflugzeug *n*
d. flat *(Schiff)* Decksbelag *m*
d. gang Schauerleute *pl (die an Deck arbeiten)*
d. hands Deckspersonal *n*
d. hook Deckswrange *f*
d. interceptor bordgestütztes Abfangjagdflugzeug *n*
d. load Deckslast *f*, Decksladung *f*
d. log Deckslog[buch] *n*
d. longitudinal Deckslängsbalken *m*
d. machinery Decks[hilfs]maschinen *fpl*
d. of cards *(Am)* Stoß *m* Karten
d. pillar Decksstütze *f*
d. planking Decksbeplankung *f (Holz)*
d. planks Decksplanken *fpl*
d. plating Decksbeplattung *f (Stahl)*
d. ponds Fischsortierkästen *mpl*, Fischsortierabteile *npl (an Deck eines Fischdampfers)*
d. seam Decksnaht *f*
d. slab *(Bw)* Fahrbahnplatte *f (Brücke)*
d. stops Decksruderstopper *m (Ruderausschlag)*
d. stringer Decks[balken]stringer *m*
d. stringer plate Decks[balken]stringerplatte *f*
d. strongback Deckslängsträger *m*
d. transom Heckbalken *m*, Deckswrange *f*
decker Entwässerungsmaschine *f*, Eindickzylinder *m*, Eindickmaschine *f* für Schliff und Zellstoff *(Papierherstellung)*
deckered eingedickt *(Schliff)*, entwässert *(Zellstoff)*
decking 1. *(Schiff)* Decksbelag *m*; 2. *(Bw)* Abdeckung *f*; Deckenschalung *f*; Balkenlage *f*; Brückentafel *f*, Fahrbahntafel *f (Brückenbau)*; Flachdach *n*; 3. *(Brg)* Aufschieben *n*
deckle Schöpfrahmen *m (Papierherstellung)*
d. box formation Gießverfahren *n (Papierherstellung)*
d. box with travelling screen Gießkasten *m* mit beweglichem Sieb *(Papierherstellung)*
d. doctor Abstreifer *m*, Abstreifschaber *m (Papierherstellung)*
d. edge Büttenrand *m (Papier)*
d.-edged board Büttenkarton *m (Papier)*
d.-edged paper Büttenrandpapier *n*
d. frame Gießkastenrahmen *m (Papierherstellung)*
d. strap Deckelriemen *m (Papierherstellung)*
declaration of validity Gültigkeitserklärung *f*
declare / to erklären, verkünden; aussagen, deklarieren
d. a patent invalid ein Patent für nichtig erklären
d. a protective right extinct ein Schutzrecht für erloschen erklären
declination Deklination *f*; Neigung *f*; Abweichung *f*
d. of a compass Mißweisung *f*
declined geneigt, abfallend, schräg
declinometer Deklinationsnadel *f*
declivitous geneigt, abschüssig
declivity geneigte Stellung *f*
d. of standing ways *(Schiff)* Hellingbahnneigung *f*, Stapellaufbahnneigung *f*
declutch / to auskuppeln, ausrücken, entkuppeln
declutching Lösen *n* der Kupplung; Auskuppeln *n*
decoction *(Ch)* Dekokt *n*, Sud *m*, Abkochung *f*
d. of lustre *(Led)* Glanzabkochung *f*
decode / to *(Dat)* entschlüsseln, dekodieren, dechiffrieren
decoder 1. Entschlüßler *m*, Dekodierer *m*; 2. *s.* digital-to-analog converter
decoding Entschlüsselung *f*, Impulsentschlüsselung *f*
decohere / to entfritten
decoic acid Kaprinsäure *f*
decoiler Abwickeleinrichtung *f*
decoiling cradle Abwickeleinrichtung *f*
decolorize / to entfärben *(Glas)*
decolorizer Entfärbungsmittel *n*
decolour / to entfärben; bleichen
decolourant Bleichmittel *n*, Entfärbungsmittel *n*
decolourate / to bleichen, entfärben
decolourizing charcoal Entfärbungskohle *f*
decomposability 1. *(Math, Ph)* Zerlegbarkeit *f*; 2. *(Ch)* Zersetzbarkeit *f*
decompose / to 1. *(Math, Ph)* zerlegen, zergliedern; zerfallen; 2. *(Ch)* zersetzen, abbauen; auflösen; zerfallen

decomposition 1. *(Math, Ph)* Zerlegung *f*; Zerfall *m*; 2. *(Ch)* Zersetzung *f*, Dissoziation *f*; Auflösung *f*; Zerfall *m*; Umsetzung *f (Sprengstoff)*
d. into partial fractions *(Math)* Partialbruchzerlegung *f*
d. potential *(El)* Zersetzungsspannung *f*
d. voltage *(El)* Zersetzungsspannung *f*
decompounded motor *(Elm)* Gegenverbundmotor *m*, Gegenkompoundmotor *m*
decompression Kompressionsverminderung *f*
decontaminate / to entgiften; entseuchen
decontaminating agent Entseuchungsmittel *n*; Entgiftungsmittel *n*
decontamination *(Kern)* Entseuchung *f*, Dekontamination *f*
d. factor Entseuchungsgrad *m*, Entseuchungsfaktor *m*
d. index [logarithmischer] Entseuchungsfaktor *m*, Entseuchungsindex *m*
d. vessel Abklingbehälter *m*, Abkühlbecken *n*
decopperate / to entkupfern
decorative bond Zierverband *m (Mauerwerk)*
d. concrete *(Bw)* Sichtbeton *m*
decorator Dekorateur *m*; Maler *m*
decorticated entrindet, geschält *(Holz)*
decortication Entholzung *f (Faser)*
decouple / to *(Mech)* auskuppeln, ausrücken, entkuppeln; *(El)* entkoppeln
decoupling Entkopplung *f*
d. capacitor *(El)* Entkoppelungskondensator *m*
d. circuit *(El)* Entkoppelkreis *m*, Entkopplungskreis *m*, Trennkreis *m*
d. resistance *(El)* Entkopplungswiderstand *m (physikalische Größe)*
d. resistor *(El)* Entkopplungswiderstand *m (Schaltelement)*
decoy vessel *(Mil)* U-Boot-Falle *f*
decrease / to abnehmen, absinken, sich vermindern
d. to zero *(Ph)* auf Null abfallen (absinken)
decrease Abnahme *f*; Absinken *n*; Verminderung *f*
d. of current Stromabnahme *f*, Stromrückgang *m*, Stromabfall *m*
d. of gain *(Fmt, Fs)* Verstärkungsabfall *m*
d. of intensity *(Ph)* Intensitätsabfall *m*, Absinken *n* der Intensität
d. of output 1. Leistungsrückgang *m (Maschinen)*; 2. Produktionsrückgang *m*
d. of pressure Druckabfall *m*
d. of the cross-sectional area Verjüngung *f*, Querschnittsverjüngung *f*
d. of weight Gewichtsabnahme *f*, Gewichtsverminderung *f*
decrement Abnahme *f*; Stufe *f (Verminderung)*
decrements of variables *(BMSR)* Dekremente *npl* der Variablen
decremeter *(BMSR)* Dämpfungsmesser *m*
decrepitate / to *(Ch)* dekrepitieren, bersten, prasseln
deculator Entlüfter *m (zwischen Sortierer und Einlaufkasten der Papiermaschine)*
decuple / to verzehnfachen
decuple zehnfach
dedendum *(Masch)* Zahnfußhöhe *f*
d. angle *(Masch)* Zahnfußwinkel *m*, Fußwinkel *m (Verzahnung)*
d. circle *(Masch)* Fußkreis *m (Zahnrad)*
deduce / to *(Math)* ableiten
deducibility *(Math)* Ableitbarkeit *f*, Deduzierbarkeit *f*
deducible ableitbar
deduction Ableitung *f*
deductive durch Ableitung; deduktiv
dedust / to entstauben
dee *(Kern)* Dee *n*, Duant *m (Zyklotron)*
dee gap Duantenabstand *m (Zyklotron)*
deep beam *(Bw)* wandartiger Träger *m*
d. boring Tiefbohren *n*
d. bracket floor *(Schiff)* Stützplattenbodenwrange *f (im Doppelboden)*
d. cultivator *(Lw)* Tieflockerer *m*
d.-cut / to *(Holz)* sägen, zerlegen
d. dimension picture *(Fs)* durchgezeichnetes Bild *n*
d. diving submarine Tiefseetauchboot *n*
d. draw / to tiefziehen
d.-drawing *(Masch)* Tiefziehen *n*
d.-drawing press Tiefziehpresse *f*
d.-drawing [quality] sheet[s] Tiefziehblech *n*
d.-drawing strip steel Tiefziehbandstahl *m*
d.-drawing test Tiefziehversuch *m*, Tiefungsversuch *m*
d.-drawing testing device Tiefungsprüfvorrichtung *f*, Tiefungsprüfgerät *n*
d.-drawing vessel tiefgehendes Schiff *n*
d.-drill / to tiefbohren
d.-etch / to tiefätzen
d. floor hohe Bodenwrange *(an den Schiffsenden)*
d. foundation *(Bw)* Tief[en]gründung *f*
d. frame *(Schiff)* Hochspant *n*
d.-freeze trawler Tiefkühltrawler *m*
d.-freezing plant Tiefkühlanlage *f*
d. girder *(Bw)* wandartiger Träger *m*
d. groove ball bearing *(Masch)* Schulterkugellager *n*
d.-hole boring Tieflochaufbohren *n*
d.-hole boring machine Tieflochbohrmaschine *f*
d.-hole drill 1. Tieflochbohrer *m*, Langlochbohrer *m*; 2. Tiefbohrer *m (Erdöl usw.)*; 3. *(Brg)* Langlochbohrhammer *m*
d.-hole drilling 1. Tieflochbohren *n*; 2. Tiefbohren *n*; 3. *(Brg)* Langlochbohren *n*

d.-hole drilling machine Tieflochbohrmaschine *f*
d.-hole job Tieflochbohrarbeit *f*
d.-hole tooling Werkzeuge *npl* zum Tieflochbohren
d. knurled head screw hohe Rändelschraube *f*
d.-level garage Tiefgarage *f*
d.-level mine *(Brg)* tiefe Grube *f*
d.-mined coal im Tiefbau gewonnene Kohle *f*
d. penetration electrode Tiefeinbrandelektrode *f*
d.-pile carpet hochfloriger Teppich *m*
d. pool *(Schiff)* Kolk *m (ausgewaschene Vertiefung)*
d. reach blacksmith's vice Tiefspannschmiedeschraubstock *m*
d. reach clamp *(Masch)* Tief[en]spanner *m*
d.-sea fishery Hochseefischerei *f*
d.-sea navigation Hochseeschiffahrt *f*
d.-sea port Hochseehafen *m*
d.-sea trawler Hochseefischdampfer *m*, Hochseetrawler *m*
d.-sea wave Tiefseewelle *f*
d. tank *(Schiff)* Hochtank *m*, Tieftank *m*
d. test Suchbohrung *f (nach tieferen Horizonten in einem Ölfeld)*
d. throat[ed] press *(Masch)* Einständerpresse *f* mit tiefer Ausladung
d. waterline *(Schiff)* Tiefladelinie *f*
d. well *(Bw)* Tiefbrunnen *m*
d.-well pump Tiefbrunnenpumpe *f*
deepen / to 1. vertiefen, tiefer machen; 2. *(Brg)* weiterteufen
deepening 1. Vertiefung *f*; 2. *(Brg)* Weiterteufen *n*
deeply contrasting kontrastreich; mit starken Kontrasten
deerskin Wildfell *n*
d. leather Wildleder *n*
defecant Saftreinigungsmittel *n (Zuckerindustrie)*
defecate / to läutern, reinigen, klären, scheiden
defecator *(Ch)* Scheidepfanne *f*
defect 1. Fehler *m*, Mangel *m*, Materialschaden *m*; 2. Maschinenschaden *m*
d. detection Fehlersuche *f*
d. electron Defektelektron *n*; Elektronenfehlstelle *f (Halbleiter)*
d.-free störungsfrei, ohne Defekt, ohne Maschinenschaden, ohne Materialschaden
d. in casting Gußfehler *m*
d. in insulation *(El)* Isolierfehler *m*, Isolationsfehler *m*, Isolationsstörung *f*
d. in material Werkstoffehler *m*
d. in weaving Webfehler *m*
defective fehlerhaft, mangelhaft, schadhaft
d. angles *(Holz)* Wahnkante *f*, Waldkante *f*
d. contact Wackelkontakt *m*
d. paper Ausschußpapier *n*
d. parts fehlerhafte Teile *npl*, Ausschuß *m*
d. ringing *(Fmt)* mangelhafter Ruf *m*, Rufstörung *f*
defectively sighted *(Opt)* fehlsichtig
defence boom *(Schiff)* Schutzspiere *f*, Abweiser *m*
deferring relay *(El)* Zeitrelais *n*
defibratability Aufschlagbarkeit *f (Überführen trockenen Papierzellstoffs in wäßrige Fasersuspension)*
defibrator Aufschlaggerät *n*, Defibrator *m*, Zerfaserer *m*
defibre / to zerfasern *(Zellstoff)*
defibrinate / to defibrinieren, entfasern
defibrinated von Fasern befreit, entfasert
defibring Zerfaserung *f*
deficiency Fehler *m*, Mangel *m*, Manko *n*, Unzulänglichkeit *f*
d. of sight *(Opt)* Sehfehler *m*
definite *(Math)* bestimmt; definit
d. direction of lay *(Masch)* geordnet-rilliger Oberflächencharakter *m*
d. form definite Form *f (Algebra)*
d. integral bestimmtes Integral *n*
d. metric definite Metrik *f (Geometrie)*
d. time-lag circuit breaking *(El)* begrenzt verzögerte Auslösung *f*
definition 1. Definition *f*, Begriffsbestimmung *f*; 2. *(Opt)* Definition *f*, Schärfe *f (Bild)*
d. chart *(Fs)* Probebild *n*, Testbild *n*
deflagrate / to deflagrieren, explosionsartig verbrennen, verpuffen; abbrennen; aufflammen, entflammen; mit heller Flamme verbrennen
deflagration Deflagration *f*, Verpuffung *f*
deflake / to entstippen *(Holz)*
deflaker *(Holz)* Entstipper *m*
deflate / to Luft ablassen *(Reifen)*
deflect / to ablenken; abbiegen; durchfedern; abweichen; ausschlagen
deflecting blade Ablenkschaufel *f*, Umlenkschaufel *f (Turbine)*
d. coil *(El)* Ablenkspule *f*
d. electrode Ablenkelektrode *f*
d. prism *(Opt)* Ablenkprisma *n*
d. sheave Ablenkscheibe *f*
d. system Ablenksystem *n*
d. torque *(El)* Meßmoment *n (eines Instrumentes)*
d. vane anemometer Klappen-Strömungsmesser *m*
d. voltage Ablenkspannung *f*
d. wall Prallfläche *f*
d. yoke Ablenkspule *f*
deflection Ablenkung *f*; Abbiegung *f*; Durchbiegung *f*; Aufbäumung *f*; Abweichung *f*; Ausschlag *m (Zeiger)*; Auslenkung *f*

d. coefficient *(Fs)* Ablenkfaktor *m (reziproke Ablenkempfindlichkeit)*
d. coil Ablenkspule *f*
d. current Ablenkstrom *m*
d. defocusing Defokussierung (Fleckunschärfe) *f* bei Ablenkung *(Elektronenoptik)*
d. electrode Ablenkelektrode *f*
d. factor *(Fs)* Ablenkfaktor *m (reziproke Ablenkempfindlichkeit)*
d. field Ablenkfeld *n*
d. multiplier *s.* crossed-fields multiplier
d. of spar *(Flg)* Holmdurchbiegung *f*
d. of spring Durchfederung *f*, Federweg *m*
d. of the needle Nadelausschlag *m (z. B. eines Meßinstrumentes)*
d. of the plumb line Lotabweichung *f*
d. plate Ablenkplatte *f*
d. potentiometer *(El, BMSR)* Ausschlagkompensator *m*; Lindeck-Rothe-Kompensator *m*
d. pressure Ablenkungsdruck *m (Turbine)*
d. pulley Umlenkrolle *f*, Umlenkscheibe *f*
d. sensitivity Ablenkempfindlichkeit *f (Meßinstrumente)*
d. setting *(Flg)* Klappenstellung *f (Landeklappen)*
d. shoulder *(Am) (Masch)* Spanleitstufe *f*
d. surface *(Math, Ph)* Biegungsfläche *f*
d. test Biegeversuch *m*
d. theory *(Bw)* Theorie *f* 2. Ordnung *(Statik)*
deflectometer Durchbiegungsmesser *m*
deflector 1. *(Ph)* Ablenkvorrichtung *f*; Auslenkvorrichtung *f (Teilchenbeschleuniger)*; 2. *(Masch)* Ablenker *m*, Ablenkblech *n*, Ablenkfläche *f*; Ablenkplatte *f*; Leitblech *n*, Leitfläche *f*; Nase *f*; Prallschirm *m*; 3. *(Wärm)* Prallklappe *f*; 4. Streuscheibe *f (Lampe)*
d. guide Leitblech *n*
d. plate Spritzblech *n*
d. vane Ablenkblech *n*, Leitblech *n*
deflexion *s.* deflection
deflocculated entflockt; pulverisiert, zerstäubt
deflocculation Ausflockung *f*, Entflockung *f*; Zerstäubung *f*
defluorinate / to entfluorieren, Fluor abspalten
defoamer *s.* defoaming agent
defoaming agent Demulgator *m*, Schaumbrecher *m*; Antischaummittel *n*
defocus / to defokussieren
defocusing Defokussierung *f*, Entbündeln *n*, Entbündelung *f*, Streuung *f*
d. lens Zerstreuungslinse *f*
defocussing *s.* defocusing
defoliation Entlaubung *f*
deforest / to abholzen
deform / to deformieren, verspannen, werfen; verformen
deformability Verformbarkeit *f*, Formänderungsfähigkeit *f*
deformable verformbar
deformation Umformung *f*, Verformung *f*, Deformation *f*, Formung *f*, Formgebung *f*
d. curve Deformationskurve *f*
d. energy Deformationsarbeit *f*, Deformationsenergie *f*
d. resistance Formänderungswiderstand *m*
d. vibrations Deformationsschwingungen *fpl*, Verformungsschwingungen *fpl*
d. work Formänderungsarbeit *f*, Umformarbeit *f*
deformed bar *(Bw)* Betonformstahl *m*, Rippenstahl *m*
defroster *(Kfz)* Enteiser *m*, Entfroster *m*
defrosting *(Kfz)* Abtauen *n*, Entfrosten *n*, Enteisen *n (Windschutzscheibe)*
degas / to entgasen
degasification Entgasung *f*
degasify / to entgasen
degasifying Entgasung *f*
degating Enttrichtern *n*
degauss / to entmagnetisieren
degaussing Entmagnetisierung *f*
d. coils on board *(Schiff)* magnetische Eigensicherung *f*
degeneracy Ausartung *f*; Entartung *f*
degenerate / to ausarten; entarten
degenerate ausgeartet; entartet
d. conic section entarteter Kegelschnitt *m*
d. eigenvalue *(Math, Ph)* entarteter (mehrzähliger) Eigenwert *m*
d. gas entartetes Gas *n (Quanteneffekt)*
d. level *(Ph)* entartetes (mehrzähliges) Niveau *n*, entartetes (mehrzähliges) Energieniveau *n*
d. state entarteter Zustand *m (Quantentheorie)*
degenerated nucleus *(Kern)* entarteter Kern *m*
degenerative *(El)* gegengekoppelt
d. amplifier gegengekoppelter Verstärker *m*
d. feedback *(BMSR)* Gegenkopplung *f*
d. voltage Gegenkoppelspannung *f*
degradation 1. *(Ch)* Abbau[reaktion *f*] *m*, Degradation *f*; 2. *(Brg)* Abrieb[bildung *f*] *m*; 3. stufenförmige Verwitterung; 4. *(Opt)* Abschattierung *f (Spektrum)*; 5. *(Kern)* Energieabnahme *f*, Energieverlust *m (durch Stöße)*; 6. Qualitätsverlust *m*
d. of energy *(Ph)* Abwertung *f* der Energie *(2. Hauptsatz der Thermodynamik)*
d. product Abrieb *m (Kohle)*
degrain / to *(Led)* Narben abstoßen
degras Degras *m (n)*, Lederfett *n*, Lederschmiere *f*
d. acid Wollfettsäure *f*
degreasant Entfettungsmittel *n*
degrease / to entfetten, entölen
degreased wool entschweißte Wolle *f*
degreasing agent Entfettungsmittel *n*, Entschweißungsmittel *n*, schmutzlösendes Mittel *n*

degree *(Ph, Math)* Grad *m*
d. Baumé *(Ch)* Baumégrad *m*
d. centigrade Grad *m* Celsius, °C
d. Kelvin Kelvingrad *m*, °K
d. of accuracy Genauigkeitsgrad *m*
d. of achromatic correction *(Opt)* Achromasiekorrektionsgrad *m*
d. of admission Füllungsgrad *m (Verbrennungsmotor)*
d. of advance *(Schiff)* Fortschrittsgrad *m (beim Propeller)*
d. of angle *(Math)* Winkelgrad *m*
d. of arc Bogengrad *m*
d. of attenuation *(Ph)* Schwächungsgrad *m*
d. of blackness Schwärzegrad *m*; Schwärzungsgrad *m*
d. of brightness Helligkeitsstufe *f*
d. of coarseness Gröbe *f*, Feinheit *f*
d. of condensation *(Ch)* Verdichtungsgrad *m*
d. of current rectification *(El)* Gleichrichtungsfaktor *m*
d. of deformation Verformungsgrad *m*
d. of dissociation Dissoziationsgrad *m*
d. of disturbance *(El)* Störungsgrad *m*
d. of draft Verzugshöhe *f*, Verzugsgröße *f*
d. of expansion Expansionsgrad *m*
d. of feeding *(Wärm)* Füllungsgrad *m*
d. of fineness Feinheitsgrad *m*
d. of flatness Ebenheitsgrad *m*
d. of freedom Freiheitsgrad *m (Mechanik)*
d. of hardness Härtegrad *m*, Grad *m* der Härte
d. of hardness of the water Härtegrad *m* des Wassers
d. of incline Überhöhung *f*
d. of indeterminacy *(Bw)* Grad *m* der statischen Unbestimmtheit
d. of insensitiveness Unempfindlichkeitsgrad *m*
d. of interference *(El)* Störungsgrad *m*
d. of ionization Ionisationsgrad *m*, Ionisierungsgrad *m*
d. of irregularity of the governor *(Wärm)* Ungleichförmigkeitsgrad (Stabilitätsgrad) *m* des Reglers
d. of modulation Modulationsgrad *m*, Aussteuerungsgrad *m*
d. of openness *(Text)* Öffnungsgrad *m (von Fasern)*
d. of oxidation Oxydationsgrad *m*, Oxydationsstufe *f*
d. of polarization *(Ph)* Polarisationsgrad *m (transversale Wellen)*
d. of probability Wahrscheinlichkeitsgrad *m*
d. of purity Reinheitsgrad *m*
d. of reaction Reaktionsgrad *m*, Überdruckgrad *m (Turbine)*
d. of reset Rückstellgrad *m*
d. of saturation Sättigungsgrad *m*
d. of screening *(El)* Abschirmfaktor *m*, Schirmfaktor *m*
d. of shrinkage Schwindmaß *n*; Schrumpfmaß *n*
d. of sizing Leimungsgrad *m (des Holzes)*
d. of slope Überhöhung *f*
d. of smearing *(Plast)* Weichheitsgrad *m*
d. of statical indeterminacy *(Bw)* Grad *m* der statischen Unbestimmtheit
d. of tanning *(Led)* Durchgerbezahl *f*
d. of taper *(Masch)* Kegelneigung *f*; Verjüngung *f*
d. of terrain sloping Neigungswinkel *m* des Geländes
d. of uniformity Gleichförmigkeitsgrad *m*, Ungleichförmigkeitsgrad *m*
d. of viscosity Viskositätsgrad *m*
degrees below zero Kältegrade *mpl*
d. centigrade Grad Celsius, °C
degum / to 1. *(Text)* abziehen, degummieren, souplieren, entschälen, entbasten *(Seide)*; 2. *(Ch)* raffinieren *(Öl)*
degumming of oil Ölhydrierung *f*, Ölraffination *f*
dehumidification Entfeuchtung *f*
dehumidifier Trockenmittel *n*; Lufttrockner *m (Klimatisierung)*
dehumidify / to Feuchtigkeit entziehen, [aus]-trocknen
dehydrate / to *(Ch)* dehydratisieren, Wasser entziehen, entwässern, kristallwasserfrei machen
dehydrated vegetables Trockengemüse *n*
dehydration Dehydratisieren *n*
dehydrator *(Lw)* Trocknungsanlage *f*
dehydrogenate / to dehydrieren, Wasserstoff entziehen
dehydrogenize / to *s.* dehydrogenate / to
dehydromucic acid Dehydroschleimsäure *f*
deion circuit-breaker *(El)* Deion-Schalter *m*, Deionisationsschalter *m*
deionization *(Ph)* Entionisierung *f*, Deionisierung *f*
d. rate Geschwindigkeit *f* der Entionisierung *(Gasentladungsröhren)*
d. time 1. Entionisierungszeit *f*; 2. Löschzeit *f (Gasentladung)*
deionize / to entionisieren
dekatron Dekatron *n (Elektronenröhre)*
del *(Math)* Nabla *n*, Nabla-Operator *m*
delaminate / to ablösen, spalten
delamination *(Plast)* Schichtentrennung *f*, Aufspaltung *f*, Schichtspaltung *f*
delay *(El)* Verzögerung *f*; Gruppenlaufzeit *f*, Laufzeit *f*; *(Fmt)* Wartezeit *f*
d.-action verzögertes Ansprechen *n*
d.-action exploder *(Brg)* Zeitzünder *m*
d.-action fuse 1. *(El)* träge Sicherung *f*; 2. *(Mil)* Verzögerungszünder *m*, Zeitzünder *m*

d. angle Zündverzögerungswinkel *m (eines Stromrichters)*
d. bias verzögerte Regelspannung (Vorspannung) *f*
d. cable *s.* d. line
d. coincidence circuit *(BMSR)* Verzögerungskoinzidenzschaltung *f*, Laufzeitkoinzidenzschaltung *f*
d. distortion in data systems *(BMSR)* Laufzeitverzerrung (Verzögerungsverzerrung) *f* in Datensystemen
d. electric detonator Zeitzünder *m (für Sprengungen)*
d. equalizer *(El)* Laufzeitentzerrer *m*
d. fuse 1. *(El)* träge Sicherung; 2. *(Mil)* Verzögerungszünder *m*, Zeitzünder *m*
d. line *(El)* Verzögerungsleitung *f*, Verzögerungskabel *n*, Laufzeitkette *f*
d.-line member *(BMSR)* Laufzeitglied *n*, Verzögerungsglied *n*
d. network Verzögerungsglied *n*, Verzögerungskette *f*
d. of ignition Zündverzug *m (Verbrennungsmotor)*
d. pattern *(Brg)* Zündfolge *f*
d. period Zündverzug *m (Verbrennungsmotor)*
d. time *(BMSR)* Verzugszeit *f*
d. time characteristic *(El)* Laufzeitcharakteristik *f*
d. working *(Fmt)* Rückrufbetrieb *m*; Wartezeitbetrieb *m*

delayed automatic volume control verzögerte automatische Lautstärkeregelung *f*, verzögerte Schwundregelung *f*
d. coincidence *(Kern)* verzögerte Koinzidenz *f*
d. filling *(Brg)* nach dem Abbau eingebrachter Versatz *m*
d. neutrons verzögerte Neutronen *npl (Kernspaltung)*
d. ringing *(Fmt)* verzögerter Ruf *m*
d. test *(El)* Kapazitätsprüfung *f* nach Lagerung

delayer *(Rak)* Verzögerungsmittel *n*

deletion Streichung *f*, Tilgung *f*
d. mark *(Typ)* Streichungszeichen *n*, Deleatur *n*

delicate empfindlich *(Instrumente)*
d. relay empfindliches Relais *n*

delignification Delignifizierung *f*, Entfernung *f* von Lignin, Entholzung *f*

delime / to entkalken

delinquent accounts *(BMSR)* offene Posten *mpl*

deliver / to 1. liefern; 2. fördern *(Pumpe)*

delivery 1. Lieferung *f*; 2. Fördermenge *f*, Förderung *f (Pumpe)*
d. box *(Text)* Ablieferungskasten *m*
d. cart Fuhrmannswagen *m*
d. chamber *(Wärm)* Druckkammer *f*
d. date Liefertermin *m*
d. gate *(Typ)* Auslegerrechen *m*, Rechenableger *m*, Rechenausleger *m*
d. head Förderhöhe *f (Pumpe)*
d. lattice *(Text)* Abführlattentuch *n*
d. line Druckleitung *f*
d. loss Verlust *m* an Fördermenge *(Pumpe)*
d. mechanism *(Text)* Ablegevorrichtung *f*
d. of energy Energieabgabe *f*, Energielieferung *f*
d. of heat Wärmeabgabe *f*
d. plunger *(Kfz)* Förderkolben *m*
d. port Auslaßschlitz *m*
d. pressure Förderdruck *m (Flüssigkeit)*
d. quantity Fördermenge *f (Pumpe)*
d. rate 1. Fördergeschwindigkeit *f (Pumpe)*; 2. Abliefergeschwindigkeit *f (Textilmaschinen)*
d. reversal Umkehr *f* der Förderrichtung *(Pumpe)*
d. rolls *(Text)* Ablieferungswalzen *fpl*
d. side 1. Auslaufseite *f (Walzwerk)*; 2. Austrittseite *f*, Druckseite *f (Pumpe)*
d. speed Walzaustrittsgeschwindigkeit *f*
d. stroke Auslaßhub *m*
d. system Auslegersystem *n (Drucken)*
d. table *(Text)* Abführtisch *m*
d. tap Ausströmungsöffnung *f (Pumpe)*
d. time Lieferzeit *f*
d. valve Auslaßventil *n (Pumpe)*
d. van [geschlossener] Lieferwagen *m*, Kastenwagen *m*

Delon rectifier Delon-Gleichrichter *m*, mechanischer Gleichrichter *m*

delta Delta *n*; Deltasymbol *n*
d.-connected *(El)* im Dreieck geschaltet, dreieckgeschaltet
d. connection *(El)* Dreieckschaltung *f*
d. function *(Math)* [Diracsche] Deltafunktion *f*
d. metal Deltametall *n*
d. modulation Deltamodulation *f*
d. ray *(Kern)* Deltastrahl *m*, Deltateilchen *n*, Deltaelektron *n*
d.-star connection *(El)* Dreiecksternschaltung *f*
d. voltage verkettete Spannung *f*, Dreieckspannung *f*
d. winding *(El)* Dreieckswicklung *f*
d. wing fighter Deltajäger *m*, Jagdflugzeug *n* mit dreieckigen Tragflächen
d.-Y connection *s.* d.-star connection
d.-zigzag connection *(El)* Dreieckzickzackschaltung *f*

deltoid deltaförmig, dreieckig

deltoid Deltoid *n (Geometrie)*
d. dodecahedron Deltoiddodekaeder *n (Kristallografie)*

d. icositetrahedron Deltoidikositetraeder *n* *(Kristallografie)*
delusterant Mattierungsmittel *n*
delustre / to *(Text)* mattieren; Glanz entfernen
delustring *(Text)* Mattieren *n*
d. in the piece *(Text)* Stückmattierung *f*
demagnetization Entmagnetisierung *f*
d. by continued reversals Entmagnetisierung *f* durch [wiederholte] Ummagnetisierung
d. curve Entmagnetisierungskurve *f*
d. factor Entmagnetisierungsfaktor *m*
demagnetize / to entmagnetisieren
demagnetizer Entmagnetisierer *m*
demagnetizing coil Entmagnetisierungsspule *f*
d. field entmagnetisierendes Feld *n*
d. reactance Gegenreaktanz *f*
d. turns gegenmagnetisierende Windungen *fpl*
demand 1. Bedarf *m*; 2. Belastung *f* *(elektrischer Maschinen)*
d. attachment Maximumanzeiger *m* *(eines Zählers)*
d. indicator Rufanzeiger *m*
d. limiter Strombegrenzer *m*
d. meter Spitzenzähler *m*, Maximumzähler *m*
d. service *(Fmt)* wartezeitloser Ferndienst *m* *(mit Vorwärtsaufbau)*
d. setter Leistungsvorwählschalter *m*
d. working exchange *(Fmt)* Schnell[verkehrs]amt *n*
demarcation potential *s.* injury potential
demesh / to *(Masch)* ausrücken, außer Eingriff bringen *(Getriebe)*
demethylate / to entmethylieren
demi-sec spinning frame Halbnaßspinnmaschine *f*
demijohn Korbflasche *f*, Ballon *m*, Glasballon *m*
demirror / to *(Opt)* entspiegeln
demirroring *(Opt)* Entspiegelung *f*
demisting *(Kfz)* Verhindern *n* des Beschlagens *(Windschutzscheibe)*
demodulate / to entmodulieren, demodulieren, entmodeln
demodulation Entmodelung *f*, Demodulation *f*
demodulator *(El, BMSR)* Demodulator *m*, Hochfrequenzgleichrichter *m*, Detektor *m*
d. circuit Demodulatorschaltung *f*
demography Bevölkerungsstatistik *f*
demolition *(Bw)* Abbruch *m*, Abriß *m*
d. bomb Sprengbombe *f*
d. chamber Sprengkammer *f* *(Brücke)*
d. squad *(Mil)* Sprengtrupp *m*
d. works *(Schiff)* Abwrackbetrieb *m*
demonstrability *(Math)* Beweisbarkeit *f*
demonstration Vorführung *f* *(Messe)*
d. eye *(Opt)* Demonstrationsauge *n*
d. eyepiece *(Opt)* Demonstrationsokular *n*
d. vessel Vorführungsschiff *n*
demould / to entformen, entschalen *(Betonbau)*
demoulding Entformung *f*, Entschalung *f* *(Betonbau)*
demount / to demontieren, abbauen, abmontieren; abbrechen, ausbauen; abnehmen
demountable demontierbar, zerlegbar, abnehmbar, abmontierbar
d. tube (valve) zerlegbare Senderöhre *f*
demounting Demontage *f*, Abbau *m*
demulsify / to *(Ch)* demulgieren, brechen *(Emulsion)*
demurrage *(Schiff)* Liegegebühr *f*; Überliegetage *mpl*
denaturant *(Ch)* Denaturierungsmittel *n*; Vergällungsmittel *n*
denaturated nuclear fuel denaturierter Kernbrennstoff *m*
denaturation Denaturierung *f*; Vergällung *f*
denature / to denaturieren; vergällen; ungenießbar machen
dendritic dendritisch
denial for non-payment *(Fmt)* Sperre *f* eines Anschlusses
denier Denier *n*, den *(Garnnummer)*
denitration Denitrierung *f*, Entstickung *f*, Entfernen *n* des Stickstoffs
denitrifying bacteria Denitrifikationsbakterien *fpl*
denominate benannt
d. number *(Math)* benannte Zahl *f*
denominator Nenner *m*
dense dicht
d. concrete Schwerbeton *m*
d. medium *(Brg)* Schwertrübe *f*
d.-medium bath Kastensinkscheider *m*
d.-medium cyclone Schwerflüssigkeitszyklon
d.-medium jig Schwerflüssigkeitssetzmaschine *f*
d.-medium process Schwerflüssigkeitsverfahren *n*
d.-medium washer Sinkscheider *m*
d.-medium washing Schwerflüssigkeitssortierung *f*
d.-medium washing drum Trommelsinkscheider *m*
d. structure dichtes Gefüge *n*
densener *(Met)* Verdichtungsanlage *f*
denseness Dichtheit *f*
densimeter Densimeter *n*, Dichtemesser *m*, Dichtigkeitsmesser *m*; Aräometer *n*
densimetry Densimetrie *f*, Dichtebestimmung *f*
densitometer Densitometer *n*, Schwärzungsmesser *m*
densitometry *(Opt, Kern)* Densitometrie *f*, [fotografische] Dichtemessung *f*, Schwärzungsmessung *f*
density 1. *(Ph)* Dichte *f*; 2. *(Opt)* Schwärzung[sdichte] *f*; 3. Bebauungsdichte *f*, Wohndichte *f* *(Siedlungsplanung)*

d. curve *(Opt, Kern)* Schwärzungskurve *f*
d. distribution *(Ph)* Dichteverteilung *f*; *(Opt)* Schwärzungsverteilung *f*, Schwärzungsverlauf *m*
d. matrix [Diracsche] Dichtematrix *f*
d. modulation Dichtemodulation *f* *(Elektronenstrahl)*
d. of charge *(Ph)* Ladungsdichte *f*
d. of cloth *(Text)* Gewebedichte *f*
d. of electron current Elektronenstromdichte *f*
d. of electrons [räumliche] Elektronendichte *f*
d. of energy Energiedichte *f*
d. of growth *(Lw)* Bestandsdichte *f*
d. of probability Wahrscheinlichkeitsdichte *f*
d. of source distribution *(Ph)* Quelldichte *f*
d. of volume charge *(El)* Raumladungsdichte *f*, Raumdichte *f* der elektrischen Ladung
d. range *(Opt)* Schwärzungsbereich *m*
d. value *(Opt)* Helligkeitswert *m*
dent 1. *(Masch)* Beule *f*, Einschnitt *m*, Kerbe *f*; 2. *(Text)* Rohröffnung *f*
d. of the reed *(Text)* Webblattzahn *m*, Rietzahn *m*, Rohr *n*
dental gold Zahngold *n*
denticulation *(Bw)* Verzahnung *f*
denting Verbeulung *f*
d. machine *(Text)* Blattstechmaschine *f*; Blatteinziehmaschine *f* *(Webereihilfsmaschine)*
dentifrice Zahnpflegemittel *n*
denuder *(El)* Zersetzungszelle *f*, Amalgamzersetzer *m*
denumerability *(Math)* Abzählbarkeit *f* *(einer Menge)*
denumerable *(Math)* abzählbar
denumerate / to *(Math)* abzählen
denumeration *(Math)* Abzählung *f*
deny patentability / to die Patentfähigkeit aberkennen
deodorant Desodorierungsmittel *n*, Geruchsbeseitiger *m*
deodorizing Desodorierung *f*, Geruchsbeseitigung *f*
deoxidation Desoxydation *f*, Sauerstoffentfernung *f*
d. alloy Desoxydationslegierung *f*
deoxidize / to *(Met)* desoxydieren, beruhigen, feinen
depart from the port / to aus dem Hafen auslaufen
departure 1. Abfahrt *f*, Auslaufen *n*; 2. *s.* deviation
dependability Sicherheit *f*, Zuverlässigkeit *f*
dependable in service betriebssicher
dependence Abhängigkeit *f*
d. of frequency Frequenzabhängigkeit *f*
dependency area Abhängigkeitsgebiet *n*
dependent time-lag relay abhängiges Zeitrelais *n*
d. variable *(Math)* abhängige Veränderliche (Variable) *f*
dephased *(El)* außer Phase, phasenverschoben
dephenolizer *(Ch)* Phenolabscheider *m*
dephlegmator Dephlegmator *m*, Rücklaufkondensator *m*; Fraktionierkolonne *f*
dephosphorization *(Met)* Entphosphorung *f*
d. of pig iron Roheisenentphosphorung *f*
depickle / to *(Led)* entpickeln
depict / to bildlich darstellen; abbilden *(optisch)*
depilatory Depiliermittel *n*, Haarentfernungsmittel *n*
depitching Entpechen *n*
deplete / to 1. *(Led)* entquellen; 2. erschöpfen, ausbeuten
depleted *(Geol)* erschöpft
d. fuel *(Kern)* abgereicherter (verarmter) Spaltstoff *m*; ausgebrannter (erschöpfter) Spaltstoff
depletion Abreicherung *f*, Verarmung *f*; Ausbrennen *n*, Erschöpfung *f* *(Kernbrennstoff)*
d. layer Verarmungsrandschicht *f*; Sperrschicht *f* *(Halbleiter)*
deplistor *(BMSR)* Deplistor *m* *(Halbleiterschaltvorrichtung: Nutzbarmachung der Sperrschicht)*
deployment base Ausweichflugplatz *m*
depolished glass Mattglas *n* *(Leuchten)*
deposit / to 1. *(Ch)* ausscheiden, [sich] niederschlagen, [sich] absetzen; 2. *(Geol)* ablagern; 3. aufschweißen, auftragen
d. hardchromium hartverchromen
deposit 1. *(Ch)* Niederschlag *m*, Absatz *m*, Schicht *f*, *(galvanische)* Abscheidung *f*; 2. *(Geol)* Ablagerung *f*; Lagerstätte *f*; 3. Auftrag *m* *(Schweißen)*
d. metal Auftragsmetall *n* *(Schweißen)*
d. of moisture Schweißwasserbildung *f*, Schwitzwasserbildung *f*
d. of primary magmatic origin *(Geol)* magmatogene Lagerstätte *f*, Lagerstätte der magmatischen Folge
d. of soot Rußabscheidung *f*
d. of the cooling surface *(Wärm)* Belegung *f* der Kühlfläche
d.-welding Auftragschweißung *f*
deposited carbon resistor *(Fmt, Fs)* Kohleschichtwiderstand *m*
depositing dock *(Schiff)* Absetzdock *n*, Fingerdock *n*
d.-out tank Entmetallisierungsbad *n*
deposition 1. *(Ch)* Abscheidung *f*, Absatz *m*, Niederschlag *m*; 2. *(Geol)* Ablagerung *f*
d. of copper galvanische Verkupferung *f*
d. welding Auftragschweißung *f*
depot ship Mutterschiff *n*, Vorratsschiff *n*
d. tug kleiner Hafenschlepper *m*

depreciate / to abschreiben *(Wert)*
depreciation Abschreibung *f*; Wertverminderung *f*
d. cost Abnutzungsaufwand *m*
depress / to niederdrücken, senken, erniedrigen
depressant Drücker *m (Flotation)*
depressed strand gekrümmtes Spannglied *n (Spannbeton)*
depressing agent *(Brg)* Drücker *m (Flotation)*
depression Absenkung *f*; Vertiefung *f*; Druckverminderung *f*; Tief *n*
d. curve *(Ph)* Absenkungskurve *f*
d. of freezing point *(Ph)* Gefrierpunktserniedrigung *f*
depth 1. Tiefe *f*; 2. *(Brg)* Teufe *f*, Tiefe
d. accuracy of a bore Bohrtiefengenauigkeit *f*
d. adjustment Tiefeneinstellung *f*
d. bomb Wasserbombe *f*
d. charge Wasserbombe *f*
d. control *(Lw)* Tiefeneinstellung *f (Pflug usw.)*; Tiefenbegrenzer *m (Traktor)*; *(Masch)* Bohrtiefenkontrolle *f*
d. control stop *(Masch)* Tiefenanschlag *m*
d. cultivator *(Lw)* Tieflockerer *m*, Tiefgrubber *m*
d. dial Bohrtiefenrundskala *f*
d. drilling operation Tieflochbohrarbeit *f*
d. feed *(Masch)* Senkrechtvorschub *m*, Tiefenvorschub *m*
d. gauge 1. *(Masch)* Tiefen[schieb]lehre *f*, Tiefenmaß *n*, Tiefenmesser *m*, Tiefentaster *m*; 2. *(Bw)* Pegel *m*
d. gauge micrometer Tiefenmaß *n*
d. gauge with dial indicator Meßuhrtiefenlehre *f*
d. indicator Pegel *m*; Teufenanzeiger *m*
d.-measuring appliance Tiefenmeßgerät *n*
d.-measuring device *(Masch)* Bohrtiefenmeßeinrichtung *f*; Bohrtiefenbegrenzung *f*
d. micrometer Tiefenmaß *n*; Meßschraube *f* für Tiefenmessungen
d. moulded *(Schiff)* Seitenhöhe *f*
d. of aileron *(Flg)* Querrudertiefe *f*
d. of beam *(Bw)* Trägerhöhe *f*
d. of body clearance *(Masch)* Fasenhöhe *f*
d. of bore Bohrtiefe *f*
d. of building Gebäudebreite *f*
d. of burn Einbrandtiefe *f*
d. of case *(Met)* Einsatztiefe *f*, Härtetiefe *f*
d. of chill *(Met)* Abschrecktiefe *f*
d. of column *(Typ)* Satzspiegelhöhe *f*, Spaltenhöhe *f*, Kolumnenhöhe *f*
d. of cut 1. Frästiefe *f*, Schnittiefe *f*, Spantiefe *f*; 2. Zahnhöhe *f (Fräser)*; 3. Schleiftiefe *f*
d. of definition *(Opt)* Schärfentiefe *f*
d. of drilling Bohrtiefe *f*
d. of engagement Eingriffstiefe *f (Zahnräder)*; Tragtiefe *f*, Überdeckung *f (Gewinde)*
d. of fairway Fahrwassertiefe *f*
d. of focus *(Opt)* Schärfentiefe *f*
d. of fuel layer *(Wärm)* Schichthöhe *f*, Schütthöhe *f*, Höhe *f* der Schüttung
d. of gap *(Masch)* Ausladung *f*
d. of girder *(Bw)* Trägerhöhe *f*
d. of hardening Härtetiefe *f*
d. of hardness *(Met)* Härtetiefe *f*
d. of hole Hülsenlänge *f (Kegellehre)*
d. of hole drilled (bored) Bohrungstiefe *f*
d. of immersion Eintauchtiefe *f*
d. of modulation *(Fmt)* Aussteuerungsgrad *m*, Modulationsgrad *m*
d. of packing *(Masch)* Dichtungstiefe *f*
d. of passage Kanaltiefe *f (Turbine)*
d. of penetration 1. Eindringtiefe *f*; 2. *(Met)* Einbrandtiefe *f*
d. of pin Zinkenlänge *f*
d. of reed *(Text)* Blatthöhe *f*
d. of round *(Brg)* Abschlaglänge *f*
d. of teeth Zahntiefe *f (Sägeblatt)*
d. of the drilling step Bohrstufentiefe *f*
d. of thread Gewindetiefe *f*
d. of thread engagement Gewindeeingriffstiefe *f*
d. of throat Ausladung *f*
d. of tooth Zahnhöhe *f*
d. plough *(Lw)* Tiefpflug *m*
d. recorder *(Schiff)* Echograf *m*
d. regulation *(Lw)* Tiefgangregulierung *f*
d. resolution Tiefenauflösung *f (Fotografie)*
d. slide gauge Tiefenmaß *n*
d. sounder Echolot *n*
d. stop Tiefenanschlag *m*
d. tormentor *(Lw)* Tieflockerer *m*
d. trip Tiefenauslösung *f*
deputy *(Brg)* Steiger *m*
deputy's district *(Brg)* Steigerabteilung *f*, Steigerrevier *n*
derail / to entgleisen
derailment Entgleisung *f*
d. alarm Entgleisungsmeldeanlage *f*
derby *(Met)* Regulus *m*, König *m*
d. doubler *(Text)* Bandwickelmaschine *f*
d. red Chromrot *n*
derelict treibendes Wrack *n*, verlassenes Schiff *n*
deresination *(Ch)* Entharzung *f*
derivation 1. *(Math)* Ableitung *f*, Herleitung *f*, Beweis *m*; Differenzierung *f*; 2. Abweichung *f*
derivative *(Math)* abgeleitet
derivative 1. *(Math)* abgeleitete Funktion *f*, Ableitung *f (einer Funktion)*, Differentialquotient *m*; Abgeleitete *f*; 2. *(Ch)* Derivat *n*
d. action *(BMSR)* Differentialquotienteneinfluß *m*

d. action time *(BMSR)* Vorhaltezeit *f*, Beschleunigungszeit *f*
d. control differenzierend wirkende Regelung *f*, D-Regelung *f*
d. control factor *(BMSR)* Differentialzeit *f*
d. equalization *(BMSR)* Kompensation *f* durch Vorhalt, Vorhaltstabilisierung *f*
d. following the fluid *(Ph)* substantielle Ableitung *f*
d. restriction D-Drossel *f*, Vorhaltdrossel *f* *(pneumatische Regler)*
d. time constant *s.* d. action time
derive / to *(Math)* ableiten, herleiten, beweisen
derived circuit *(El, Fmt)* Ableitkreis *m*, Abzweigungskreis *m*
d. unit *(Ph)* abgeleitete Einheit *f*
d. weave *(Text)* abgeleitete Bindung *f*
derrick 1. *(Bw)* Derrick[kran] *m*; Mastenkran *m*; 2. *(Schiff)* Ladebaum *m*; Auslegerwippkran *m*; 3. Bohrturm *m*
d. brace Bohrturmstrebe *f*
d. cornice Bohrturmkranz *m*
d. crane Derrick[kran] *m*; Bohrturmkran *m*
d. floor Bohrturmflur *m*
d. girders Bindebalken *mpl* des Bohrturmes, Bohrturmquerträger *mpl*
d. grillage Bohrturmrost *m*
d. mast *(Schiff)* Lademast *m*
d. post *(Schiff)* Ladepfosten *m*
d. span *(Schiff)* Ladebaumhanger *m*
d. wagon *(Eis)* Kranwagen *m*
desalination *(Ch)* Entsalzung *f*
desamidation *(Ch)* Desaminierung *f*
descale / to entzundern
descend / to sich abwärts bewegen, niedergehen; abspringen *(mit Fallschirm)*
descendant *(Kern)* Folgeprodukt *n*; Tochterelement *n*
descending piping fallende Rohrleitung *f*
descensional ventilation *(Brg)* abfallende Wetterführung *f*, fallende Bewetterung *f*
descent 1. *(Flg)* Fallschirmabsprung *m*; Höhenaufgabe *f* *(beim Flug)*; 2. *(Brg)* Anfahrt *f*, Einfahrt *f*
d. of piston Kolbenniedergang *m*
d. velocity Sinkgeschwindigkeit *f*
describing function *(BMSR)* Beschreibungsfunktion *f*, harmonische Balance (Linearisierung) *f*
descriptive geometry darstellende Geometrie *f*
descriptor *(Dat)* Deskriptor *m* *(Dokumentation)*
deseam / to *(Met)* [brenn]putzen *(Blöcke)*
desensitization Desensibilisierung *f*
desensitizer Desensibilisator *m*
desert [rocket] proving establishment Raketenversuchsgelände *n* in der Wüste
desiccant Trockenmittel *n*, wasserentziehendes Mittel *n*
desiccate / to trocknen, Wasser entziehen
desiccation Austrocknung *f*; Heißlufttrocknung *f* *(Holz)*
desiccator *(Ch)* Exsikkator *m*
d. cartridge *(Opt)* Trockenpatrone *f*
d. plate *(Ch)* Exsikkatoreinsatz *m*
design / to konstruieren; entwerfen; auslegen *(Maschinen)*, bemessen, berechnen *(Bauteile)*
d. rolls Walzen kalibrieren
design 1. Bauart *f*, Bauweise *f*, Konstruktion *f*, Ausführung *f*, Aufbau *m*, Gestaltung *f*, Form *f*; 2. Entwurf *m*; Bemessung *f*, Berechnung *f* *(Bauteile)*; 3. *(Text)* Muster *n* *(auf Stoffen)*; Entwurf *(Modegestaltung)*
d. application [in] Konstruktionsanwendung *f* [bei]
d. assumptions *(Bw)* Berechnungsannahmen *fpl*
d. bureau Konstruktionsbüro *n*
d. calculation 1. *(Bw)* statische Berechnung *f*; 2. *(Masch)* Festigkeitsberechnung *f*
d. department Konstruktionsabteilung *f*
d. engineer Konstrukteur *m*
d. feature Konstruktionsmerkmal *n*
d. frequency *(El)* Nennfrequenz *f*
d. load *(Bw)* Lastannahme *f*; zulässige Last *f*
d. method Bemessungsverfahren *n*, Berechnungsverfahren *n* *(Bauteile)*
d. of experiments Planen *n* von Versuchen
d. of gauges Lehrenbau *m*
d. of jigs and fixtures Vorrichtungsbau *m*
d. of jigs, fixtures, and gauges Vorrichtungs- und Lehrenbau *m*
d. office Konstruktionsbüro *n*
d. on the building-block principle Aufbau *m* nach Baukastensystem
d. paper 1. Zeichenpapier *n*; 2. *(Text)* Patronenpapier *n*
d. pressure Betriebsdruck *m*
d. sheet *(Masch)* Konstruktionsentwurf *m*
d. sinker top knitting machine Musterrundstrickmaschine *f* mit Einschließplatinen
d. speed Ausbaugeschwindigkeit *f* *(Straßenbau)*
d. strength Gestaltsfestigkeit *f*
d. stress *(Bw)* zulässige Spannung *f*
d. table Bemessungstafel *f* *(z. B. Stahlbetonbau)*
designating symbol Kurzzeichen *n*
designation of radio-electric waves Bezeichnung *f* der Funkwellen
d. strip Bezeichnungsstreifen *m* *(Kabel)*
designed displacement *(Schiff)* Konstruktionswasserverdrängung *f*
d. draft *(Am)* *s.* d. draught
d. draught *(Schiff)* Konstruktionstiefgang *m*, Berechnungstiefgang *m*, Konstruktionstiefe *f*, Berechnungstiefe *f*, Tauchtiefe *f* *(eines Überwasserschiffes)*

d. section *(Schiff)* Konstruktionsspant *n*
d. waterline *(Schiff)* Konstruktionswasserlinie *f*, KWL., CWL.
designer Zeichner *m*; Entwurfsbearbeiter *m*; Konstrukteur *m*; Modegestalter *m*; Dessinateur *m*
designer's needle *(Text)* Ausnehmenadel *f*
designing Konstruieren *n*, Konstruktion *f*; Entwerfen *n*, Entwurf *m*
d. on the building-block principle Aufbau *m* nach Baukastensystem
desilverization Entsilberung *f*
desired bearing under full load *(Masch)* erwünschtes Tragbild *n* unter voller Belastung
d. dimension Sollmaß *n*
d. position Sollstellung *f*
d. value Sollwert *m*, Aufgabewert *m*
desize / to *(Text)* entschlichten
desizing *(Text)* Entschlichten *n*, Entschlichtung *f*
desk Pult *n*, Tisch *m*; Schaltpult *n*
d. calculator Bürorechenmaschine *f*, Tischrechenmaschine *f*
d. control Pultsteuerung *f*
d. instrument Tischgerät *n*; Schaltpult[einbau]gerät *n*
d. stand set *(Am)* Fernsprechtischapparat *m* mit festem Mikrofon
deslag / to entschlacken
desliming *(Brg)* Entschlämmung *f*
desmut / to entschmutzen, reinigen
desorb / to desorbieren
desorption Desorption *f*
destabilization Entstabilisierung *f*, Verschlechterung *f* der Stabilität
destaticizer Mittel *n* zur Beseitigung oder Verminderung elektrostatischer Aufladung
destrengthening annealing Spannungsfreiglühen *n*
destructibility Zerstörbarkeit *f*
destructible zerstörbar
destructibleness Zerstörbarkeit *f*
destructive zerstörend
d. distillation Entgasung *f*, trockene Destillation *f*
d. fire Schadenfeuer *n*
d. reading destruktives (zerstörendes) Lesen *n* *(Löschen des gespeicherten Dateninhaltes beim Lesen von Magnetkernspeichern)*
d. test nicht zerstörungsfreie Prüfung *f*; Prüfung unter Zerstörung des Prüfkörpers
destructor Müllverbrennungsofen *m*
desugarization Entzuckerung *f*
desulf . . . *(Am)* *s.* desulph . . .
desulphurate / to entschwefeln
desulphur[iz]ation Entschwefelung *f*
desulphurize / to entschwefeln
detach / to abnehmen, ausbauen, entfernen, loslösen
detachable abnehmbar
d. bit abnehmbare (aufsteckbare) Krone *f* *(eines Bohrers)*
d.-blade propeller Propeller *m* mit abnehmbaren Flügeln
d. rim abnehmbare Felge *f*
d. wheel abnehmbares Rad *n*
detached house alleinstehendes Haus *n*, Einzelhaus *n*
d. keelson *(Schiff)* Trägerkielschwein *n* *(auf die Bodenwrangen aufgesetztes Kielschwein)*
d. superstructure *(Schiff)* freistehender Aufbau (Decksaufbau) *m*, alleinstehender Aufbau (Decksaufbau)
detaching device for taper joint drill bits Bohrkronenabschlagvorrichtung *f*
d. mill Lösewalzwerk *n*
d. roller *(Text)* Abreißwalze *f* *(Kämmaschine)*
detachment of electrons Elektronenablösung *f*, Freimachung *f* von Elektronen
detail Detail *n*, Einzelheit *f*
d. contrast *(Fs)* Detailkontrast *m*, Bildeinzelheitenkontrast *m*
d. contrast ratio *(Fs)* Kontrastverhältnis *n* der Bildeinzelheiten, Einzelheitenkontrastverhältnis *n*
d. drawing Detailzeichnung *f*, Teilzeichnung *f*
d. of machine Maschinenelement *n*, Maschinenbauteil *n*
d. part Einzelteil *n*
detailer Detailzeichner *m*; Teilkonstrukteur *m*
detailing Detailbestimmung *f*; Detailauswertung *f*
d. by polar coordinates polare Detailvermessung *f*
d. in lines linienweise Detailauswertung *f*
detar / to entteeren
detarring Entteerung *f*
detect / to entdecken, auffinden; demodulieren; nachweisen
detecting chamber *(Kern)* Nachweiskammer *f*
d. element 1. Meßfühler *m*, Meßwertgeber *m*; 2. *(Kern)* Nachweisglied *n*
d. instrument Indikator *m*
detection 1. Nachweis *m*, Nachweisen *n*, Beobachten *n*, Feststellung *f*, Aufspüren *n*; 2. Aufnahme *f*, Registrierung *f*, Aufzeichnung *f*; 3. HF-Gleichrichtung *f*; Demodulation *f*; 4. Entwicklung *f* *(Chromatografie)*
d. instrument *s.* detector
d. limit *(Ch)* Empfindlichkeitsgrenze *f*; Nachweisgrenze *f* *(Analytik)*
d. of radiation Strahlennachweis *m*
d. of trace elements Spurenanalyse *f*, Spurennachweis *m*
d. sensitivity Nachweisempfindlichkeit *f*
d. time Nachweiszeit *f*

detector 1. Meßzelle *f*, Meßfühler *m*; 2. Suchgerät *n*, Spürgerät *n*; 3. *(El)* Detektor *m*, Empfangsgleichrichter *m*, Hochfrequenzdemodulator *m*
d. balanced bias Entstörschaltung *f* gegen Radarstörungen
d. coefficient Gleichrichterwirkungsgrad *m*
d. contact Überwachungskontakt *m*
d. of inactive accounts *(Büro)* Sucher *m* für unbewegte Konten
d. reaction coupling Audionrückkopplung *f*
d. receiver Detektorempfänger *m*
d. receiving set Detektorempfänger *m*
d. valve Detektorröhre *f*
d. wire *(Text)* Abstellnadel *f*
detension / to *(Ph)* entspannen
detent 1. Raste *f*, Auslöser *m*, Klinke *f*; 2. *(Kfz)* Synchron-Körper *m*; 3. Arretierung *f*
d. lever Auslösehebel *m*
d. pin Auslösestift *m*, Anschlagstift *m*
d. spring Arretierungsfeder *f*
detention reservoir Rückhaltebecken *n*
d. time Verweilzeit *f*
detergency Waschwirkung *f*; Reinigungsvermögen *n*
detergent reinigend, ablösend
detergent [oberflächenaktives] Reinigungsmittel (Lösungsmittel) *n*; Waschmittel *n*
d. characteristics Wascheigenschaften *fpl*
d. for the coloured wash Buntwaschmittel *n*
determinable bestimmbar, feststellbar, meßbar
determinant *(Math)* Determinante *f*
d. theorem Determinantensatz *m*
determinate [statisch] bestimmt
determination Bestimmung *f*, Ermittlung *f*; Messung *f*
d. of elongation by extensometer Feindehnungsmessung *f*
d. of errors Fehlerbestimmung *f*
d. of heating value Heizwertbestimmung *f*
d. of suitable tolerances Tolerieren *n*
d. of tanning matter *(Led)* Gerbstoffbestimmung *f*
d. of the swelling value *(Led)* Quellwertbestimmung *f*
detin / to entzinnen
detonating characteristics Klopfeigenschaft[en *pl*] *f* *(Kraftstoffe)*
d. charge Zündladung *f*
d. fuse detonierende Zündschnur *f*
detonator Sprengkapsel *f*, Zündkapsel *f*, Zünder *m*, Initialzünder *m*
detour Umleitung *f*, Umweg *m*
detoxicate / to entgiften
detoxi[fi]cation Entgiftung *f*
d. of gas Gasentgiftung *f*
detrital sediments Geröllsedimente *npl*
detritus 1. Geröll *n*, Trümmermasse *f*; 2. Sinkstoffe *mpl*
d. chamber Kläranlage *f*, Absetzbecken *n*
d. pit Kläranlage *f*
d. tank Absetzbecken *n*
detuning Verstimmen *n*, Verstimmung *f*
d. measurement Verstimmungsmessung *f*
deuterium Deuterium *n*
d.-impregnated emulsion *(Kern)* mit Deuterium getränkte Emulsion *f*
d.-moderated pile deuteriummoderierter Reaktor *m*, Schwerwasserreaktor *m*
deuteron Deut[er]on *n*
develop / to 1. *(Math)* abwickeln *(Oberfläche eines Körpers in die Ebene)*; 2. entwickeln; 3. *(Brg)* ausrichten, vorrichten
d. by steam *(Text)* durch Dämpfen fixieren
developability *(Math)* Abwickelbarkeit *f*
developable *(Math)* abwickelbar
d. hull form abwickelbare Schiffskörperform *f*
d. surface abwickelbare Fläche *f*
developed dimension *(Masch)* Fertigmaß *n*
d. dyeing *(Text)* Entwicklungsfärbung *f*
d. dye[stuff] Entwicklungsfarbstoff *m*
d. shell plating *(Schiff)* Außenhautabwicklung *f*; abgewickelte Außenhaut *f*
d. turbulence vollständig entwickelte Turbulenz *f*
d. V-port plug Laternenventilkegel *m*
d. view Abwicklung *f* *(geometrische Darstellung)*
developer 1. Entwickler[substanz *f*] *m* *(Fotografie)*; 2. *(Schiff)* Abwickler *m* *(Schnürbodenabwickler)*
developing agent Entwicklersubstanz *f* *(Fotografie)*
d. bath Entwicklerbad *n*, Entwicklungsbad *n*
d. by air-passage *(Text)* Luftgangentwicklung *f*
d. corrosion korrosionsfördernd
d. dye Entwicklungsfarbstoff *m*
d. liquor *(Text)* Kupplungsflotte *f*
d. paper Kunstlichtpapier *n* *(Fotografie)*
d. tank Entwicklertank *m* *(Fotografie)*
development 1. *(Math)* Abwicklung *f* *(Oberfläche eines Körpers in die Ebene)*; 2. Entwicklung *f*; 3. *(Bw)* Bebauung *f*, Erschließung *f*; 4. *(Brg)* Aufschluß *m*, Ausrichtung *f*
d. by oxidation oxydative Entwicklung *f*
d. department Entwicklungsabteilung *f*
d. of heat Wärmeentwicklung *f*
d. opening *(Brg)* Ausrichtungsbau *m*, Ausrichtungsstrecke *f*
d. raise *(Brg)* Überhauen *n* zur Vorrichtung einer neuen Kammer
d. road *(Brg)* Vorrichtungsstrecke *f*
d. rock *(Brg)* Ausrichtungsberge *pl*

d. **stoping** *(Brg)* Förderung *f* aus der Vorrichtung
d. **work** *(Brg)* Aus- und Vorrichtungsarbeiten *fpl*
developmental work Entwicklungsarbeit *f*
deviate / to abweichen
d. **from the true course** verlaufen *(Bohrer)*
deviation 1. Ablenkung *f*, Abweichung *f* [vom Sollwert]; Ausschlag *m*; 2. *(Masch)* Abmaß *n*; 3. *(BMSR)* [momentane] Regelabweichung *f*, [momentane] Abweichung *f* vom Sollwert; 4. Abweichung *f* von der Lotrechten *(Tiefbohren; Erdöl, Erdgas)*; 5. *(Opt, Fs)* Auswanderung *f*, Bildauswanderung *f*
d. **angle** Ablenkungswinkel *m*
d. **buoy** *(Schiff)* Kompensierungsboje *f*
d. **distortion** Verzerrung *f* durch Frequenzabweichung *(FM-Empfänger)*
d. **indicator** Abweichungsanzeiger *m*
d. **of the bubble** Libellenausschlag *m*
d. **ratio** 1. Abweichungsverhältnis *n*, dynamischer Regelfaktor *m*; 2. Frequenzhub *m*
d. **recorder** Abweichungsanzeiger *m*
device Gerät *n*, Apparat *m*, Vorrichtung *f*
d. **for guidance to the line of sight** *(Rak)* Einlenkgerät *n*
devices for holding work Spannzeuge *npl*
devil tragbarer Lötofen *m (Klempner)*
d. **gas line** *(Brg)* Abgasleitung *f*
devilling Aufkratzen *n* des Putzes
dew point Taupunkt *m (Luftfeuchtigkeit)*
d.**-point indicator** Taupunktmesser *m*, Taupunktanzeiger *m*
d.**-point meter** Taupunktmesser *m*, Taupunktmeßgerät *n*
d. **retting** Tauröste *f*, Landröste *f (Flachs)*
dewater / to entwässern, trockenlegen
dewatering Entwässerung *f*, Trockenlegung *f*; Wasserhaltung *f*; Sümpfung *f*
d. **screening trough** Entwässerungssiebrinne *f*
d. **trough** Entwässerungsrinne *f*
dewax / to entwachsen
dewaxed wachsfrei; entwachst
dewaxing Entparaffinieren *n*
dewing *s.* damping
dewooled sheepskin entwolltes Schaffell *n*
dextrin Dextrin *n*
dextrinate / to in Dextrin umwandeln
dextro-compound [optisch] rechtsdrehende Verbindung *f*
dextrogyrate rechtsdrehend *(optische Aktivität)*
dextrogyric rechtsdrehend *(optische Aktivität)*
dextrorotatory rechtsdrehend *(optische Aktivität)*
dextrorse *(Math)* rechtsgewunden, rechtsgängig, rechtswendig
dextrorsum *(Math)* rechtsgewundene Kurve *f*
dextrose Dextrose *f*
dezinc / to entzinken
dezincification Entzinkung *f*
D. F. ... *s.* direction finding ...
diacetic acid Azetessigsäure *f*
d. **ester** Azetessigester *m*
diaceto-succinic ester Diazetylbernsteinsäureester *m*
diachylon [plaster] Bleipflaster *n (Verbandsmaterial)*
diacid zweisäurig
diacoustics Schallbrechungslehre *f*
diacritical diakritisch
diagnostic routine *(BMSR)* Diagnoseprogramm *n*, Fehlersuchprogramm *n*
diagonal Diagonale *f*
diagonal diagonal
d. **astigmatism** *(Opt)* Diagonalastigmatismus *m*
d. **bond** Schränkverband *m*, Stromverband *m (Mauerwerk)*
d. **brace** *(Bw)* Diagonalstab *m*, Diagonalstrebe *f*, Strebe *f*
d. **bracing** *(Bw)* Dreiecksverband *m*, Kreuzverstrebung *f*, Verschwertung *f*
d. **bulkhead** *(Schiff)* diagonalgebautes Holzschott *n*
d. **corner locking** schräge Zinke *f*
d. **course** Schränkschicht *f*, Stromschicht *f (Mauerwerk)*
d. **cross brace** *(Bw)* Diagonalstab *m*, Kreuzstrebe *f*, Strebe *f (einer Verschwertung)*
d. **cross bracing** Kreuzverstrebung *f*, Verschwertung *f*
d. **cross ribbing over full length of bed** *(Masch)* Diagonalverrippung *f* über volle Bettlänge
d. **cut[ting]** Schrägschnitt *m*; Gehrungsschnitt *m*
d. **eyepiece** Zenitokular *n*, gebrochenes Okular *n*
d. **grain** Diagonalschnitt *m*
d. **joint** *(Holz)* Stoß *m* auf Gehrung
d. **member** *(Bw)* Diagonalstab *m*, Diagonalstrebe *f*, Strebe *f*
d. **rib with irregular floats** *(Text)* Adriabindung *f* mit ungleichen Flottungen
d. **shaving** Diagonalschaben *n*
d. **strut** *(Bw)* Diagonalstab *m (Druck)*; Druckstrebe *f*
d. **system** *(Brg)* Dreiecksanordnung *f (Ausbau)*
d. **tension** schiefe Hauptzugspannung *f (Statik)*
d. **tie** *(Bw)* Diagonalstab *m (Zug)*; Zugstrebe *f*
d. **tieing** Fachwerkverband *m*, Dreieckverband *m*
diagram 1. Bild *n*; Diagramm *n*; Schaubild *n*, Schema *n*; Plan *m*; 2. *(Dat)* Rechenschema *n*
d. **board** *(BMSR)* Blindschaltbild *n*

d. curve Diagrammlinie *f*, Diagrammkurve *f*
d. loop Diagrammschleife *f*
d. of connections Schaltbild *n*
d. of equilibrium Zustandsschaubild *n*, Zustandsdiagramm *n*
d. of stages Stufenplan *m* *(Turbine)*
d. of the blade wheel Abwicklung *f* des Schaufelrades *(Turbine)*
d. of work Arbeitsdiagramm *n*
d. panel optische Meldetafel *f*
diagrammatic plan of pipes *(Schiff)* Rohrleitungsplan *m*
d. symbol *(El)* Schaltzeichen *n*
diagrid Gitterrost *m*
dial / to 1. *(Fmt)* [an]wählen; Nummernscheibe drehen; anrufen; 2. *(Masch)* wählen; Skalenscheibe drehen
d. through *(Fmt)* durchwählen
dial 1. Einstellscheibe *f*, Drehscheibe *f*; Rundskala *f*, Skalenscheibe *f*, Scheibe *f*, Zifferblatt *n*; 2. *(Fmt)* Wählscheibe *f*, Nummernscheibe *f*; 3. *(Lw)* Abteilerspitze *f*; 4. *(Text)* Rippscheibe *f*
d. barometer Zeigerbarometer *n*
d. bench gauge Meßtisch *m* mit Meßuhr, Feinmeßvorrichtung *f*
d. cam section *(Text)* Rippschloß-System *n*
d. central office *(Fmt)* Wählamt *n*
d. depth gauge Meßuhrtiefenmaß *n*
d. drive Skalenantrieb *m*
d. exchange *(Fmt)* Wählvermittlung *f*
d. feed Drehtellerzuführung *f*, Revolvertellerzuführung *f*
d. feed press Presse *f* mit Drehtellerzuführung, Revolverpresse *f*
d. gauge Meßuhr *f*
d. gauge caliper Tastermeßuhr *f*
d. gauge instrument Meßuhrgerät *n*
d. hum *(Fmt)* Amtszeichen *n*
d. illumination Skalenbeleuchtung *f*
d. indicator 1. Gewindeuhr *f*, Meßuhr *f*; Ableseskala *f*; 2. Ablesevergrößerungsgerät *n*
d. indicator micrometer Mikrometermeßuhr *f*
d. intercommunicating system *(Am)* *(Fmt)* Wählsystem *n*
d. key *(Fmt)* Wählerscheibenkontakt *m*, Nummernscheibenkontakt *m*
d. lamp Skalenlampe *f*
d. needle wale *(Text)* Rippnadelstäbchen *n*
d.-operated mit Wahlscheibeneinstellung; von Skalenscheibe aus eingestellt
d. pointer Skalenzeiger *m*
d. pulse Wählimpuls *m*
d. reading Skalenablesung *f*, Skalenanzeige *f*
d. scale Ringskala *f*, Kreisskala *f*
d. service *(Fmt)* Selbstwählbetrieb *m*
d.-set feeding *(Masch)* Wahlscheibeneinstellung *f*
d. system *(Fmt)* Wählsystem *n*
d. telegraph Zeigertelegraf *m*
d. test Skalenprüfung *f*
d. thickness gauge Meßuhrdickenmesser *m*
d. thickness indicator Meßuhrdickenmesser *m*
d. toll circuit *(Am)* Fernwahlleitung *f*
d. tone *(Fmt)* Amtszeichen *n*
d.-type machine 1. Maschine *f* mit Wahlscheibeneinstellung; 2. *(Am)* Rundschalttischmaschine *f*
dialing *s.* dialling
dialkylate *(Ch)* dialkylierte Verbindung *f*
dialling *(Fmt)* Wählen *n*
d. battery Wählbatterie *f*
d. effect Wählgeräusch *n*
d. impulse Wählimpuls *m*
d. tone Amtsfreizeichen *n*
dialyze / to dialysieren, durch Dialyse trennen
dialyzer Dialysator *m*, Dialisierzelle *f*
diamagnetic diamagnetisch
d. substances diamagnetische Stoffe *mpl*
diamagnetism Diamagnetismus *m*
diameter Durchmesser *m*
d. class *(Holz)* Dickeklasse *f*
d. load coefficient *(Schiff)* Durchmesserbelastungsgrad *m* *(Propeller)*
d. of a molecule Moleküldurchmesser *m*
d. of armature *(Elm)* Ankerdurchmesser *m*, Rotordurchmesser *m*
d. of ball impression Kalottendurchmesser *m*
d. of bore Bohrungsdurchmesser *m*
d. of grinding wheel Schleifscheibendurchmesser *m*
d. of head Bahngröße *f* *(Hammer)*
d. of pilot head Führungszapfendurchmesser *m*
d. of recess reduzierter Schaftdurchmesser *m* *(Werkzeug)*
d. of steam jet Strahldicke *f*, Strahlstärke *f* *(Dampfturbine)*
d. of the pitch circle Lochkreisdurchmesser *m*
d. of the root circle Fußkreisdurchmesser *m*
d. of the turning circle Drehkreisdurchmesser *m*
d. of tooth Zahndurchmesser *m*
d. of wheel Schleifscheibendurchmesser *m*
d. of wire Drahtdicke *f*
d. tolerance Durchmessertoleranz *f*
diametral Durchmesser-
d. clearance 1. Lagerspiel *n*; 2. Freifläche *f* am Umfang *(Spiralbohrer)*
d. pitch Modulkehrwert *m* *(Zähnezahl je Zoll Teilkreisdurchmesser)*
d. plane *(Masch)* Stirnebene *f*

d. voltage Durchmesserspannung *f*
d. voltage of a polyphase system größte verkettete Spannung *f*
d. winding Durchmesserwicklung *f*
diamide *(Ch)* Diamid *n*
diamine dyes Diaminfarbstoffe *mpl*
diammonium phosphate sekundäres Ammoniumphosphat *n*
diamond 1. Diamant *m*; 2. *(Typ) (etwa)* Diamant *f (4½ britische Punkte; Schriftgröße)*
d. aerial Rautenantenne *f*, Rhombusantenne *f*
d. bit Diamantbohrkrone *f*
d. boring Genauigkeitsbohren *n* mit Diamant- oder Hartmetallaufbohrwerkzeug
d. coil Ankerformspule *f*
d. core drilling Diamantkernbohren *n (Erdöl, Erdgas)*
d. crossing *(Eis)* spitzwinklige Gleiskreuzung *f*
d. cut *(Brg)* Pyramideneinbruch *m*
d. dresser Abrichtdiamant *m*, Abdrehdiamant *m*
d. dressing device Diamantabzieheinrichtung *f*
d. drill Diamantenbohrmaschine *f*, Diamantbohrer *m*
d. drill bit Diamantkrone *f*, Diamantbohrer *m*
d. drilling Diamantbohren *n*, Bohren *n* mit Diamantkronen
d. dust Diamantstaub *m*
d. file Rautenfeile *f*
d. grinding wheel *s.* d. wheel
d. impregnated mit Diamantstaub durchsetzt
d. knurling Kordeln *n*
d. nib Diamantpatrone *f*
d. particle Diamantsplitter *m*
d. pattern Kreuzmuster *n*, Kordel *f*
d. penetrator hardness *(Met)* Vickershärte *f*
d. point Diamantspitze *f*
d. point drill for concrete, brick, and stone Steinmeißelbohrer *m*
d. powder Diamantstaub *m*
d. pyramid hardness Diamantpyramidhärte *f*, Vickershärte *f*
d.-shaped rautenförmig; kordiert
d.-shaped aerial Rautenantenne *f*, Rhombusantenne *f*
d.-shaped knurling Kordeln *n*
d. structure Diamantstruktur *f*
d. tie up *(Text)* Spitzschnürung *f*
d. tipped diamantbestückt
d. tool Diamantwerkzeug *n*
d. wheel Diamantschleifscheibe *f*
d. wheel dresser Abrichtdiamant *m*, Formdiamant *m*; Diamantabrichter *m*
d. winding Gleichspulenwicklung *f*
dianodal *(Math)* durch einen Knoten gehend *(Kurve)*
diaphanceity Transparenz *f*, Durchsichtigkeit *f*
diaphanometer Diaphanometer *n*, Lichtdurchlässigkeitsprüfer *m*
diaphanous transparent, durchscheinend
diaphragm 1. Membran[e] *f*, Diaphragma *n*, federnde Platte *f*, Schwingwand *f*, Zwischenwand *f*, Zwischenboden *m*; 2. Lichtblende *f*, Lochblende *f*, Blende *f*; 3. *(Bw)* Aussteifungsträger *m*, Querbalken *m*, Querträger *m*, Querriegel *m*; Querwand *f*; Scheibe *f*, wandartiger Träger *m*; 4. Leit[rad]scheibe *f*, Druckstufenscheidewand *f (Turbine)*
d. accumulator Membranakkumulator *m*
d. action *(Bw)* Scheibenwirkung *f*
d. beam *(Bw)* Aussteifungsträger *m*, Querriegel *m*; Querbalken *m*, Querträger *m*, wandartiger Träger *m*
d.-box level detector Membranschalter *m*
d. displacement Membranauslenkung *f*
d. expansion joint Membranausgleichvorrichtung *f*
d. loudspeaker Membranlautsprecher *m*
d. motor Membranantrieb *m*, Membran[stell]motor *m*
d. oscillation Membranschwingung *f*
d. plane Blendenebene *f (Foto)*
d. pressure gauge Membranmanometer *n*
d. pulsator Membranpulsator *m (Melkmaschine)*
d. pump Membranpumpe *f*
d. return spring Membranrückholfeder *f*
d.-ring [mode] filter Membranringmodusfilter *n*, Membranringwellentypfilter *n (Wellenleiter)*
d. stress Membranspannung *f*
d.-type pump Membranpumpe *f*
diapositive Diapositiv *n*
diathermal wärmedurchlässig
diathermancy Wärmedurchlässigkeit *f*
diathermanous diatherman, wärmedurchlässig
diathermic wärmedurchlässig
diatomaceous earth Diatomeenerde *f*, Kieselgur *f*
diatomic zweiatomig
d. molecule zweiatomiges Molekül *n*
diatomite Kieselgur *f*
diatoms Diatomeen *fpl*, Kieselalgen *fpl*
diatonic tone scale diatonische Tonskala *f*
diazo compound Diazokörper *m*, Diazoverbindung *f*
d. fast dye Diazoechtfarbstoff *m*
diazotize / to diazotieren
diazotized dyestuff Diazofarbstoff *m*
diazotizing Diazotieren *n*, Diazotierung *f*
dibasic *(Ch)* zweibasisch
dibble Pflanzstock *m*, Pflanzholz *n*, Setzholz *n*, Steckholz *n*
dibbling Dibbeln *n*, Pflanzlochen *n*; Häufchensaat *f*

d. machine Dibbelmaschine *f*
d. tool Pflanzlochwerkzeug *n*, Pflanzstock *m*
dice / to in Würfel schneiden
dice Würfel *m*
dichroism *(Opt)* Dichroismus *m*, Zweifarbigkeit *f*
dichromate Dichromat *n*
dicing machine Plättchenschneidemaschine *f* *(Halbleiterfertigung)*
dickite *(Min)* Dickit *m*
dictaphone Diktiermaschine *f*
d.-type system Diktieranlage *f*
dictating machine Diktiergerät *n*, Diktiermaschine *f*
dictation machine Diktiergerät *n*, Diktiermaschine *f*
die / to 1. erlöschen; 2. aussetzen *(Verbrennungsmotor)*
d. away nachlassen, abflauen *(Wind)*; abklingen *(Schwingung)*
d.-cast spritzgießen, durch Spritzgießverfahren herstellen
d. down abklingen, ausschwingen
d. out abklingen, ausklingen
d. sink ausfräsen, nachformfräsen
die 1. Schnittwerkzeug *n*; Stanzwerkzeug *n*; Unterwerkzeug *n*; Schnittplatte *f*; 2. Gesenk *n*; Form *f*; Hohlform *f*; Matrize *f*; 3. Schneideisen *n*; 4. Lochstempel *m*, Stanzstempel *m* *(Lochkarten)*; 5. Ziehdüse *f*, Zieheisen *n*, Ziehtrichter *m* *(Draht)*; 6. Ziehring *m* *(Tiefziehen)*
d. and mould copy milling machine Gesenk- und Formennachformfräser *m*
d. and tap holder Schneideisen- und Gewindebohrerhalter *m*
d. away time Abklingzeit *f*
d. block 1. Preßform *f*; 2. Gesenk *n*
d. body Schneidkopfgehäuse *n*
d. burnishing Polierdrücken *n*, Preßpolieren *n*
d. cast Spritzgußstück *n*
d.-cast metal Spritzguß *m*
d. casting 1. Spritzguß *m*, Dauerformguß *m*, Druckguß *m*, Preßguß *m*, Kokillenguß *m*; 2. Dauerformgußstück *n*
d.-casting alloy Spritzgußlegierung *f*
d.-casting die Spritz[guß]form *f*
d.-casting machine Spritzgußmaschine *f*
d.-casting practice Spritzgußtechnik *f*
d.-casting zinc alloy Zinkdruckgußlegierung *f*
d. cavity Gesenkgravur *f*
d. channel Düsenkanal *m* *(Extruder)*
d. chaser Schneideisengewindestrehler *m*
d. clearance Stempelspiel *n* *(Schnittwerkzeug)*
d. cushion Ziehkissen *n* *(Presse)*
d. feeding Schneidkopfvorschub *m*
d. filing machine Matrizenfeilmaschine *f*
d. forging Gesenkschmieden *n*
d. forging die Gesenk *n*
d. forging hammer Gesenkschmiedehammer *m*
d.-formed im Gesenk geformt
d.-formed part 1. Preßteil *n*; 2. Gesenkschmiedestück *n*
d. grinding fixture Backenschleifvorrichtung *f*
d. head Gewindeschneidkopf *m*, Schneideisenkopf *m*; Strehlerbackenkopf *m*
d. holder Matrizenhalter *m*, Schneideisenhalter *m*
d. hole Ziehdüse *f*
d.-life Standzeit *f*
d. load meter Zugkraftmesser *m*
d. making Schnittbau *m*
d. mill Gesenkfräser *m*
d. milling cutter Gesenkfräser *m*
d. plate Zieheisen *n*, Ziehdüse *f* *(Draht)*
d. press Gesenkpresse *f*
d.-pressed part Gesenkpreßteil *n*
d. ring Ziehring *m* *(Presse)*
d. set Schnitt *m*
d. shop Gesenkmacherei *f*
d. sinker 1. Gesenkfräsmaschine *f*; 2. *(Typ)* Stempelschneider *m*
d. sinker's file Werkzeugfeile *f*, Werkstattfeile *f*, Gesenkmacherfeile *f*
d. sinking Ausfräsung *f*, Gesenkfräsen *n*, Raumnachformfräsen *n*
d. sinking machine Nachformfräsmaschine *f*, Gesenkkopiermaschine *f*, Matrizenfräsmaschine *f*
d. stamp *(Typ)* Trockenstempel *m*, Trockensiegel *n*
d. stamping Preßteil *n*
d. stamping press Gesenkpresse *f*
d. steel Matrizenstahl *m*
d. stock Gewindeschneidkluppe *f*, Schneideisenhalter *m*
d.-stock chaser Schneideisenstrehler *m*
d.-stock die Schneideisen *n* *(für Kluppe)*
d. tap Gewindebohrer *m*, Schneideisengewindebohrer *m*
d. typing press Kalteinsenkpresse *f*
d. wear Gesenkverschleiß *m*
d. work Gesenkschmieden *n*
dieing press Umformautomat *m* mit Unterantrieb
dielectric Dielektrikum *n*, Nichtleiter *m*
dielectric dielektrisch
d. absorption dielektrische Absorption *f*
d. breakdown dielektrischer Durchschlag *m*, Spannungsdurchschlag *m*
d. constant Dielektrizitätskonstante *f*
d. current dielektrischer Strom *m*, Polarisationsstrom *m*
d. displacement dielektrische Verschiebung *f* *(Maxwell)*; elektrische Erregung *f* im Medium
d. drier Hochfrequenztrockner *m*

d. **fatigue** dielektrische Nachwirkung *f*
d. **fluid** Dielektrikum *n*
d. **guide** dielektrischer Leiter *m*
d. **heating** dielektrische Erwärmung *f*, Hochfrequenzheizung *f*, HF-Heizung *f*
d. **lens** elektrostatische (dielektrische) Linse *f*
d. **loss** dielektrischer Verlust *m*
d. **loss angle** Verlustwinkel *m*
d. **polarization** dielektrische Polarisation *f*
d. **resistance** dielektrischer Widerstand *m*, Isolationswiderstand *m*
d. **rigidity** Spannungsfestigkeit *f*, dielektrische Festigkeit *f*
d. **strength** Durchschlagsfestigkeit *f*
d. **stress** elektrische Beanspruchung *f (eines Isolators)*
d. **test** Isolationsprüfung *f*; Prüfung *f* der Durchschlagsfestigkeit; *(El)* Spannungsprüfung *f*
d. **test voltage** Prüfspannung *f*
d. **viscosity** dielektrische Nachwirkung *f (Hysteresis)*
dielectrical dielektrisch
dielectronic recombination Zweielektronenrekombination *f*
diemaker's reamer Schälreibahle *f*
diene *(Ch)* Dien *n*
dienes with independent double bonds Diene *npl* mit isolierten Doppelbindungen
Diesel compressor Dieselkompressoranlage *f*
Diesel-driven dieselgetrieben, mit Dieselantrieb
Diesel-driven generating set Dieselaggregat *n*
Diesel-electric dieselelektrisch
Diesel-electric drive dieselelektrischer Antrieb *m*
Diesel-electric generating set Dieselaggregat *n*
Diesel-electric ship Schiff *n* mit dieselelektrischem Antrieb, dieselelektrisches Schiff *n*
Diesel engine Dieselmotor *m*
Diesel engine oil Dieselschmieröl *n*
Diesel engine vehicle Dieselfahrzeug *n*
Diesel engine with pre-combustion chamber Vorkammermotor *m*
Diesel engine with solid injection kompressorloser Dieselmotor *m*
Diesel-engined road vehicle Dieselstraßenfahrzeug *n*
Diesel fuel Dieselkraftstoff *m*
Diesel hammer Dieselpfahlramme *f*
Diesel-hydraulic dieselhydraulisch
Diesel locomotion Dieselzugförderung *f*
Diesel-powered *s.* Diesel-driven
Diesel railcar Dieseltriebwagen *m*
Diesel roller Dieselstraßenwalze *f*
diethylene glycol Diäthylenglykol *n*
difference Differenz *f*, Unterschied *m*, Abweichung *f*
d. **amplifier** Differenzverstärker *m*, Differentialverstärker *m*
d. **depth of modulation** Modulationstiefenabstand *m*
d. **equation** Differenzengleichung *f*
d. **frequency** Differenzfrequenz *f*
d. **in draught** Zugdifferenz *f*, Zugunterschied *m (z. B. bei Öfen)*
d. **in light intensity** Lichtunterschied *m*
d. **in the moisture contents** *(Brg)* Wassergehaltsspanne *f*
d. **number** *s.* neutron excess
d. **of level** Niveauunterschied *m*
d. **of path** Gangunterschied *m (Strahlen)*
d. **of phase** Phasenunterschied *m*, Phasenverschiebung *f*
d. **of potential** Spannungsunterschied *m*, Potentialdifferenz *f*
d. **of pressure** Druckdifferenz *f*, Druckunterschied *m*, Druckstufe *f*
d. **tone** Differenzton *m*, Schwebungston *m*
different-sized verschieden groß, von verschiedener Größe
differentiability *(Math)* Differenzierbarkeit *f*, Ableitbarkeit *f*
differentiable differenzierbar, ableitbar
differential 1. *(Math)* Differential *n*; 2. *(Kfz)* Ausgleichsgetriebe *n*, Differential; 3. *(Am) (Dat)* Vergleichsglied *n*, Vergleichsstelle *f*; 4. *(Am) (BMSR)* Unempfindlichkeitsbereich *m*, tote Zone *f*
differential differential, Differential-
d. **accumulator** Druckluftdifferentialkolbenakkumulator *m*
d. **aileron** *(Flg)* Differentialquerruder *n*
d. **aileron control** *(Flg)* Querrudersteuerung *f*
d. **amplifier** Differentialverstärker *m*, Differenzverstärker *m (Analogrechner)*
d. **analyzer** Differentialanalysator *m*, Differentialgleichungslöser *m (Analogrechner)*
d. **assembly** *(Kfz)* Ausgleichsgetriebe *n*
d. **axle drive bevel gear** *(Kfz)* Differentialantriebskegelrad *n*
d. **band brake** *(Masch)* Differentialbandbremse *f*
d. **bevel pinion** *(Kfz)* Ausgleichskegelrad *n*, kleines Differentialantriebskegelrad *n*
d. **booster** *(Elm)* Zusatzmaschine *f* mit Differentialerregung
d. **brake** *(Kfz)* Differentialbremse *f*
d. **bridge** Differential[meß]brücke *f*
d. **calculus** Differentialrechnung *f*
d. **calorimeter** Differentialkalorimeter *n*
d. **capacitor** Differentialkondensator *m*
d. **carrier** *(Kfz)* Hinterachsbrücke *f*, Hinterachsträger *m*; Differentiallagerkasten *m*; Differentialgehäusevorderteil *n*, Ritzelträger *m* am Differential
d. **case (casing)** *(Kfz)* Ausgleichgehäuse *n*, Differentialgehäuse *n*

d. chain block Differentialflaschenzug *m*
d. coefficient Differentialquotient *m*, abgeleitete Funktion *f*, Ableitung *f*
d. compound winding Gegenkompoundwicklung *f*, Kompoundwicklung *f* mit entgegengesetzter Stromrichtung
d. compound-wound motor Gegenverbundmotor *m*
d. compounded gegenkompoundiert
d. compounding Gegenkompoundierung *f*
d. condenser Differentialkondensator *m*
d. connection Vergleichsschaltung *f*, Differenzschaltung *f*
d. current Differenzstrom *m*
d. cylinder Differentialzylinder *m*
d. decay method Isotopenbestimmung *f* nach unterschiedlichen Halbwertzeiten
d. diagram *(Wärm)* Differenzdiagramm *n*
d. draught gauge *(Wärm)* Differenzzugmesser *m*
d. drive *(Kfz)* Differentialantrieb *m*
d. duplex system *(Fmt)* Differentialgegensprechsystem *n*
d. equation Differentialgleichung *f*
d. excitation Gegenverbunderregung *f*, Differentialerregung *f*; Eigenerregung *f*
d. fading selektiver Schwund (Wellenschwund)
d. flowmeter *(Am)* Differenzdruckdurchflußmesser *m*
d. gap *(BMSR)* 1. tote Zone *f*, Unempfindlichkeitsbereich *m*; 2. *(Am)* Hysterese *f (Zweipunktregelung)*
d. gear Ausgleichgetriebe *n*, Differential[getriebe] *n*; Kompensator *m (Turbine)*
d. gear with bevel wheels Kegelraddifferential *n*, Kegelradausgleichgetriebe *n*
d. gearbox Ausgleichgetriebe[kasten *m*] *n*
d. indexing Differentialteilen *n*
d. indexing head Differentialteilgerät *n*, Teilkopf *m* für Differentialteilen
d. indicator Differenzindikator *m*
d. lock *(Kfz)* Ausgleichsperre *f*, Differentialsperre *f*
d. manometer Differentialmanometer *n*
d. master gear *(Kfz)* [großes] Differentialantriebskegelrad *n*
d. measurement Differenzmessung *f*, Unterschiedsmessung *f*
d. measuring instrument Differenzmesser *m*
d. mechanism Differentialgetriebe *n*
d. modulation meter Brückenmodulationsmesser *m*
d. pawl Differentialsperre *f*
d. pinion Ausgleichkegelrad *n*, Ausgleichritzel *n*, Planetenrad *n (Ausgleichgetriebe)*; kleines Differentialantriebskegelrad *n*
d. pinion shaft *(Kfz)* Ausgleichradachse *f*, Differentialachse *f*

d. piston gauge Differentialkolbenmanometer *n*
d. pressure Wirkdruck *m*, Differenzdruck *m*
d. pressure gauge Differentialmanometer *n*
d. pulley Differentialkettenzug *m*, Differentialflaschenzug *m*
d. quantity 1. Differenz[en]größe *f*, Größendifferenz *f*; 2. Infinitesimalgröße *f*; unendlich kleine Größe *f*
d. relay Differentialrelais *n*; Fehlerrelais *n*
d. repeating coil *(Fmt)* Differentialübertrager *m*
d. resistance innerer Röhrenwiderstand *m*
d. satellite pinion Trabantenrad *n (Planetengetriebe)*
d. settlement ungleichmäßiges Setzen *n*, ungleichmäßige Setzung *f (erhärtende Baustoffe)*
d. shunt motor Reihenschlußmotor *m* mit Gegennebenschlußerregung
d. side gear Hinterachswellenrad *n*
d. slide wire Schleifdraht *m (Meßbrücke)*
d. spider *(Kfz)* Ausgleichstern *m*, Differentialstern *m*
d. spider pinion *(Kfz)* Ausgleichkegelrad *n*, Planetenrad *n*; Zwischenrad *n*
d. spur gear *(Kfz)* Ausgleichstirnrad *n*, Differentialstirnrad *n*
d. star piece *(Kfz)* Ausgleichstern *m*, Differentialstern *m*
d. tackle Differentialflaschenzug *m*
d. thermal analysis Differentialthermoanalyse *f*
d. thermal pyrosynthesis *(Min)* Differentialthermopyrosynthese *f*
d. thermostat Differenztemperaturregler *m*
d. tone Differenzton *m*, Schwingungston *m*
d. tooth spacing ungleiche Zahnteilung *f (Räumwerkzeug)*
d. transformer Differentialtransformator *m*, Differentialübertrager *m*
d. winding Differentialwicklung *f*
d. winding motion Differentiallaufwindebewegung *f (Spinnmaschine)*
d. wound motor Kompoundmotor *m*
d. Zener resistance Durchbruchswiderstand *m (Halbleitertechnik)*
differentiate / to 1. *(Math)* differenzieren, ableiten; 2. unterscheiden
differentiating circuit Differenzierglied *n (Differentiation von Impulsen, insbesondere Umwandeln eines Rechteckimpulses in einen Nadelimpuls)*
differentiation *(Math)* Differentiation *f*, Ableitung *f*
differentiator *(Dat)* Differenziergerät *n*, Differenziereinrichtung *f*
difficult intersections *(Text)* schwer ausspringende Bindungen *fpl*

d. to get at schwer zugänglich
d. to split *(Holz)* schwer spaltbar
diffract / to beugen
diffraction *(Ph)* Beugung *f (Wellen)*
d. angle Beugungswinkel *m*
d. curve Beugungskurve *f*
d. fringes Beugungsstreifen *mpl*
d. grating Beugungsgitter *n*, Diffraktionsgitter *n*
d. image Beugungsbild *n*
d. microscopy Diffraktionsmikroskopie *f*
d. of electrons Elektronenbeugung *f*
d. of light Beugung *f* des Lichtes
d. pattern Beugungsbild *n*, Beugungsfigur *f*
d. peak Beugungsmaximum *n*
d. ring Beugungsring *m*
d. scattering Diffraktionsstreuung *f*, Beugungsstreuung *f*
d. spectrograph Beugungsspektrograf *m*
d. spectrum Beugungsspektrum *n*, Gitterspektrum *n*
diffuse / to *(Ph)* [ein]diffundieren, einwandern
diffuse *(Ph)* unscharf, diffus; zerstreut, gestreut
d.-background radiation [diffuse] Hintergrundradiostrahlung *f (Radioteleskopie)*
d. reflection diffuse Reflexion *f*
d. reflection photometer Diffusionsfotometer *n*
d. series diffuse Serie *f*, erste Nebenserie *f (Spektrum)*
d. sound diffuser Schall *m*
d. transmission gestreute (diffuse) Durchlassung *f (Licht)*
diffused-alloy-power transistor diffusionslegierter Leistungstransistor *m*
d. junction Diffusionsschicht *f (Halbleiter)*
d. light Streulicht *n*, diffuses Licht *n*
d. lighting gestreute (diffuse) Beleuchtung *f*
d. transistor Diffusionstransistor *m*
diffuser 1. *(Ph)* [licht]streuender Körper *m*, Streukörper *m (insbesondere Lichttechnik)*; 2. *(Masch)* Diffuser *m (umgekehrte Düse)*; 3. Luftverteiler *m (Lüftungstechnik)*; 4. *(Kfz)* Ladeleitrad *n (Lader)*; 5. *(Flg)* Landeleitdraht *m*
d. angle Erweiterungswinkel *m*
d. jet Zerstäuberdüse *f*
d. vane Diffusorschaufel *f*
diffusing glass Streuglas *n*, Mattglas *n*; Kathedralglas *n*
d. screen Mattscheibe *f*; Streuschirm *m (Leuchten)*
diffusiometer Diffusionsmesser *m*
diffusion Diffusion *f*; Zerstreuung *f*
d. approximation Diffusionsnäherung *f*
d. barrier *(Kern)* poröse Wand *f*
d. by wire *(Fmt)* Drahtrundspruch *m*
d. capacitance Diffusionskapazität *f (Halbleiter)*
d. chamber *(Ph)* Diffusionskammer *f*
d. cloud chamber *(Kern)* Diffusionsnebelkammer *f*
d. constant Diffusionskonstante *f*
d. cooling Diffusionskühlung *f*
d. current density Diffusionsstromdichte *f*
d. equation Diffusionsgleichung *f*
d. extrapolation distance *(Kern)* extrapolierte Diffusionslänge *f*
d. factor Streuvermögen *n (Lichttechnik)*
d. flow Diffusionsströmung *f*, Diffusionsstrom *m*
d. kernel Diffusionskern *m*
d. mean free path mittlere [freie] Diffusionsweglänge *f (Neutronen)*
d. moulding Niederdruckpreßverfahren *n* mit unvorbereitetem Füllstoff
d. plant Gasdiffusionsanlage *f*, Diffusionsanlage *f (Isotopentrennung)*
d. process Diffusionsverfahren *n*, Osmoseverfahren *n*; Pastenverfahren *n*
d. pump Diffusionspumpe *f*
d. rate Diffusionsgeschwindigkeit *f*
d. separator (separating column) Diffusionstrennkolonne *f*, Trennrohr *n (Isotopentrennung)*
d.-type cloud chamber *(Kern)* Diffusionsnebelkammer *f*
diffusivity Diffusionsvermögen *n*; Diffusität *f*
diffusor *s.* diffuser
dig / to graben; mit Spaten ausschachten; roden
d. in einhaken *(Meißel)*
digest / to 1. *(Ch)* digerieren, einweichen, plastifizieren, versetzen; aufschließen *(z. B. Holz)*; 2. faulen, sich zersetzen *(Gesundheitstechnik)*
digester Autoklav *m*, Digestor *m*, Dämpfer *m*; Dampfkochtopf *m*; Zellstoffkocher *m*
d. house Kocherei *f*, Kocherhaus *n*
d. lining Kocherauskleidung *f*
d. test Kochlaugenprüfung *f*
digestion 1. *(Ch)* Digerieren *n*; Aufschluß *m*, Aufschließen *n (z. B. Holz)*; 2. Faulung *f*, Schlammfaulung *f (Gesundheitstechnik)*
d. bottle *(Ch)* Digerierflasche *f*
d. chamber Schlammfaulraum *m (Gesundheitstechnik)*
d. of emulsion *(Ch)* Reifung *f* der Emulsion
d. of sewage sludge Faulen *n* von Abwasserschlamm
d. tank Faulraum *m (Emscherbrunnen)*
digestive poison Fraßgift *n (Schädlingsbekämpfung)*
digestor *s.* digester
digger 1. Roder *m*; Kartoffelroder *m*; 2. Bagger *m*, Löffelbagger *m*

digging 1. Schürfen *n*, Ausgraben *n*, Graben *n*, Roden *n*; 2. Ausbaggern *n*; Grabvorgang *m* beim Baggern; Ausschachten *n* mit Spaten
d. and riddling machine Vorratsroder *m*
d. cycle Arbeitsspiel *n (Löffelbagger)*
d. fork Grabgabel *f*
d.-in [of the drill] Einhaken *n* des Bohrers
d. machine Grabemaschine *f*
d. motion Grabbewegung *f*
d. plough Grabepflug *m*
d. point Baggerstelle *f*
d. shovel Schaufel *f*, Schippe *f*
digit 1. *(Math)* einstellige Zahl *f*, Ziffer *f*; 2. *(Dat)* Stelle *f*; 3. *(Fmt, Fs)* Stufe *f*; Wahlstufe *f*; Ziffer *f*
d. absorber *(Fmt)* Impulsvernichter *m*, Stromstoßunterdrücker *m*
d. absorbing selector *(Fmt)* Nummernschlukker *m*
d. place (position) *(Dat)* Ziffernstelle *f*
d. punching *(Dat)* Ziffernlochung *f*
d. selector *(Fmt)* Direktorwähler *m*
digital *(Math, Dat)* digital, ziffernmäßig; Digital-, Ziffern-
d. automation Automatisierung *f* mit Hilfe digitaler Rechengeräte, digitale Automatisierung
d. computer Digitalrechner *m*, Ziffernrechenautomat *m*, Ziffernrechenmaschine *f (programmgesteuert)*
d. differential analyzer digitaler Differentialgleichungslöser *m*, Digital-Differentialanalysator *m*, DDA
d. display Zahlenwertanzeige *f*
d. sampled-data system digitales Stichprobendatensystem *n*
d. speech communication system *(Dat)* digitales Sprachverständigungssystem *n*
d.-to-analog converter *(Dat)* Digital-Analog-Umsetzer *m*, DA-Umsetzer *m*
digitize / to *(Dat)* digital (ziffernmäßig) darstellen, digitieren
digitizer *(Dat)* Quantisierer *m*
diglycerol Diglyzerin *n*
dihalogenated zweifach halogeniert (halogensubstituiert)
dihedral 1. *s.* dihedron; 2. *(Flg)* V-Form *f*, V-Stellung *f (Tragflächen)*
dihedral *(Math)* zweiflächig, von zwei Flächen gebildet
d. angle Flächenwinkel *m*
dihedron *(Math)* Zweiflach *n*, Zweiflächner *m*
diheptal base 14poliger Stiftsockel *m (für Katodenstrahlröhren)*
dihexahedral dihexaedrisch
dihydric *(Ch)* zweibasisch, zweiwertig
dike Damm *m*, Deich *m*; Trockenmauer *f*
d. burst Deichbruch *m*

dilapidated baufällig, verfallen
dilapidation Verfall *m (von Gebäuden usw.)*
dilatability Ausdehnungsvermögen *n*, Dehnungsvermögen *n*, Ausdehnarbeit *f*, Dehnbarkeit *f*
dilatation Ausdehnung *f*, Längung *f*
d. interval *(Bw)* Dehn[ungs]fuge *f*
dilational waves *(Ak)* Dehn[ungs]wellen *fpl*
dilatometer Dehnungsmesser *m*, Dilatometer *n*
diluent *(Ch)* Verdünnungsmittel *n*, Streckungsmittel *n*, Verschnittmittel *n* für Lösungsmittel
dilute medium *(Brg)* Dünntrübe *f*
diluted phase fluid bed Flugstaubbett *n*
diluting ratio *(Ch)* Verdünnungsfaktor *m*
d. roller Erweichwalze *f (Papierherstellung)*
dim / to 1. verdunkeln; trüben; 2. *(Text)* abmatten; 3. abblenden *(Kino)*
d. lights abblenden *(Scheinwerfer)*
dim dunkel; trübe
d.-out cap Abblendkappe *f (Scheinwerfer)*
dimension / to bemaßen
dimension 1. Dimension *f*; 2. Abmessung *f*, Maß *n*, Größe *f*, Ausmaß *n*
d. figure Maßzahl *f (technisches Zeichnen)*
d. letter Maßbuchstabe *m (technisches Zeichnen)*
d. line Maßlinie *f (technisches Zeichnen)*
d. specification Maßzahl *f (technisches Zeichnen)*
d. stock Dimensionsstockholz *n*
d. timber Dimensionsholz *n*
d. with tolerance toleriertes Maß *n*
d. without tolerance nichttoleriertes Maß *n*
dimensional accuracy Maßgenauigkeit *f*
d. analysis *(Ph)* Dimensionsanalyse *f*
d. co-ordination *(Bw)* Maßordnung *f*
d. control *(Opt)* Meßsteuerung *f*
d. error Maßfehler *m*
d. limit Grenzmaß *n*
d. restorability *(Text) Fähigkeit eines Gewebes, nach Schrumpfung oder Strecken in die Ausgangsmaße zurückzukehren*
d. stability Maßbeständigkeit *f*, Maßhaltigkeit *f*; Formbeständigkeit *f (von Geweben während des Tragens)*
d. stability under heat Wärmefestigkeit *f*, Warmfestigkeit *f*
d. tolerance Toleranz *f*, Maßtoleranz *f*
d. variation Abmaß *n*; Maßabweichung *f*
dimensionally stable maßhaltig; formbeständig
dimensioned bemaßt
d. drawing Maßzeichnung *f*
d. sketch Maßskizze *f*
dimensioning Bemaßung *f*, Dimensionierung *f*
dimensionless dimensionslos
dimensions of the bore Bohrungsabmessungen *fpl*
dimethyl phthalate Dimethylphthalat *n*, Phthalsäuredimethylester *m*

dimethylaniline Dimethylanilin *n*
dimethylglyoxime Dimethylglyoxim *n*
dimetric dimetrisch
diminishing pipe Reduzierstück *n (Rohrleitungsbau)*
diminution of section Querschnittsverminderung *f (beim Zerreißversuch)*
dimmed light Abblendlicht *n*, abgeblendetes Licht *n*
dimmer Verdunklungswiderstand *m*, Abblendwiderstand *m*
d. switch 1. Verdunklungsschalter *m*; 2. *(Kfz)* Abblendschalter *m*
d. transformer Verdunklungstransformator *m*
dimming device Verdunklungsschiebewiderstand *m*
d. filament *(Kfz)* Abblendfaden *m*
d. resistor *(El)* Verdunklungswiderstand *m*
d. switch 1. Verdunklungsschalter *m*; 2. *(Kfz)* Abblendschalter *m*
dimple Sicke *f*, Vertiefung *f*; Anbohrung *f*
dinas rock brick Dinasstein *m*
dineric interface Grenzfläche *f* flüssig/flüssig
dineutron Doppelneutron *n (kurzzeitige Kombination zweier Neutronen)*
dinghy Ding[h]i *n*
dinging *(Bw)* Bürstenputz *m*, Pinselputz *m*
d. hammer Ausbeulhammer *m*, Blechrichthammer *m*
dingot *(Met)* Regulus *m*
dining car *(Eis)* Speisewagen *m*
d. compartment Speiseabteil *n (im Zug)*
dink Ding[h]i *n*
dint / to *(Brg)* nachnehmen *(Sohle)*
diode Diode *f*, Zweipolröhre *f*
d.-capacitor store Diodenkondensatorspeicher *m (Datenspeicherung)*
d. characteristic Diodencharakteristik *f*
d. circuit Diodenschaltung *f*
d. detection Diodengleichrichtung *f*
d. detector Diodengleichrichter *m*
d. gate *(Dat)* Dioden-Torschaltung *f*
d. limiter Diodenbegrenzer *m*
d. load resistance Diodenaußenwiderstand *m*
d. rectification Diodengleichrichtung *f*
d. rectifier Diodengleichrichter *m*
d.-resistor-logic *(Dat)* logische Schaltung *f* mit Dioden und Widerständen
d.-transistor-logic Dioden-Transistor-Logik *f*, Dioden-Transistoren-Schaltkreistechnik *f*, DT-Schaltkreistechnik *f*
d. voltage Diodenspannung *f*
dionic water tester Leitfähigkeitsmesser *m (für Wasser)*
diophantine *(Math)* diophantisch
dioptase *(Min)* Dioptas *m*
dioptre *(Opt)* Diopter *m*
d. scale Dioptrienteilung *f*
dioptric *(Opt)* Dioptrie *f*
dioptrics Dioptrik *f*, Theorie *f* der Lichtbrechung
dioptry Dioptrie *f*
dip / to eintauchen; untertauchen
d. the light [Licht] abblenden
dip 1. *(Math)* Einschnitt *m*, Senkung *f (in einer Kurve)*; Einsattelung *f (in einer Fläche)*; 2. *(Ph)* Inklination *f (Erdmagnetismus)*; 3. *(Schiff)* Kimmtiefe *f*; 4. *(Geol)* Einfallen *n*, Fallen *n*, Einsenkung *f*; Neigung *f*; Einfallende *f*, einfallende Strecke *f*; 5. Eintauchflüssigkeit *f*; 6. Durchhang *m (Leitung)*
d. brazing Tauchlöten *n*
d.-coated electrode getauchte Elektrode *f*
d. coater Tauchstreichmaschine *f (Papierherstellung)*
d.-coating Beschichten *n* durch Tauchen, Heißtauchen *n* von Überzügen
d.-dyeing Tauchfärben *n (Tauchverfahren beim Färben von Wirk- und Strickwaren)*
d.-filament *(Kfz)* Abblendfaden *m*
d.-galvanizing Feuerverzinken *n*
d.-moulding *(Plast)* Heißtauchen *n*; Herstellung *f* von Tauchartikeln
d. needle Inklinationsnadel *f (Erdmagnetismus)*
d. of the horizon *(Schiff)* Kimmtiefe *f*
d. plating Eintauchplattierung *f*
d. soldering Tauchlötung *f*
d. stick Peilstab *m*; Öl[stand]meßstab *m*
d.-stick level indicator Ölstand[s]kontrolle *f*
d.-stick reading Peilstabablesung *f*; Ölstandablesung *f*
d. stock *s.* d. stick
d. switch *(Kfz)* Abblendschalter *m*
d. tape method Tauchmeßverfahren *n*
d. working *(Brg)* Unterwerksbau *m*
diphase zweiphasig
diphenylamine Diphenylamin *n*
diplex Diplex *n*
d. modulation Doppelmodulation *f*
d. operation *(Fmt)* Doppelsprechbetrieb *m*, Gegensprechbetrieb *m*
diplexer Frequenzweiche *f*, Diplexer *m*
dipole Dipol *m*
d. aerial Dipol[antenne *f*] *m*
d. field Dipolfeld *n*
d. gamma-ray transition *(Kern)* Dipol-Gamma-Übergang *m*
d. layer Dipolschicht *f*
d. radiation Dipolstrahlung *f*
d. transition Dipolübergang *m*
dipped electrode Tauchelektrode *f*
Dippel's oil Tieröl *n*, Dippelöl *n*
dipper *(Kfz)* 1. Ölschöpfer *m*; 2. Abblendschalter *m*; 3. Nase *f* des Pleuelkopfes; 4. Baggerlöffel *m*

d. boom Löffelstiel *m (Bagger)*
d. bucket Löffel *m (Bagger)*
d. dredger Schöpfbagger *m*, Löffelbagger *m*, Schaufelbagger *m*
d. handle Löffelstiel *m (Bagger)*
d. ladle Löffel *m* des Löffelbaggers
d. relay Tauchkontaktrelais *n*
dipping 1. Eintauchen *n*; 2. Tauchbad *n*
d. drum Eintauchtrommel *f*, Tauchtrommel *f*
d. method Tauchverfahren *n*
d. needle *s.* dip needle
d. process Tauchprozeß *m*, Tauchverfahren *n*
d. refractometer Eintauchrefraktometer *n*
d. roller *(Text)* Tauchwalze *f*
d. thermometer Tauchthermometer *n*
d. varnish Tauchlack *m*
d. vat Schöpfbütte *f*, Tauchbütte *f (Papierherstellung)*
diproton *(Ph)* Doppelproton *n*
Dirac interaction *(Ph)* Diracsche Wechselwirkung *f*
Dirac waves Dirac-Wellen *fpl*
dirch *(El)* Wellenfalle *f (Wellenleiter)*
direct / to 1. *(Ök)* leiten, verwalten; 2. *(Ph)* richten; lenken
d. a beam einen Strahl richten
d. teem *(Met)* fallend gießen
direct direkt, unmittelbar, unvermittelt; gerade
d. acting 1. *(BMSR)* direkt *(Pneumatik; steigender Eingangsgröße entspricht steigende Ausgangsgröße)*; 2. direkt betätigt
d.-acting recording instrument *(BMSR)* direkt betätigter Schreiber *m*; direkt betätigtes Registrierinstrument *n*
d.-acting steam engine direkt wirkende Dampfmaschine *f*
d.-acting steam pump direkt wirkende Dampfpumpe *f*, unmittelbar angetriebene Dampfpumpe
d.-acting valve *(BMSR)* mit steigendem Steuerdruck schließendes Ventil *n*
d. arc furnace Héroult-Ofen *m*
d. axis *(Elm)* Längsfeldachse *f*, Polachse *f*, magnetische Längsachse *f*
d.-axis component of a magnetomotive force Längsdurchflutung *f* einer Synchronmaschine
d.-axis component of armature current Längsfeldkomponente *f* des Ankerstromes
d.-axis component of the electromotive force Längs-EMK *f*
d.-axis component of the voltage Längsspannung *f*
d.-axis subtransient electromotive force subtransitorische Längs-EMK *f*
d.-axis subtransient impedance subtransitorische Längsimpedanz *f*
d.-axis subtransient reactance Subtransientreaktanz *f* des Längsfeldes
d.-axis synchronous impedance synchrone Längsimpedanz *f*
d.-axis synchronous reactance synchrone Längsfeldreaktanz *f*, Synchronreaktanz *f*
d.-axis transient impedance transitorische Längsimpedanz *f*
d.-axis transient reactance Transientreaktanz *f* des Längsfeldes
d.-axis voltage Längsfeldspannung *f*, Längsfeld-EMK *f*, Haupt-EMK *f*
d. bilge pump *(Schiff)* Bilgelenzpumpe *f*
d. call *(Fmt)* direkte (unmittelbare) Verbindung *f*
d. chill casting *(Met)* Stranggießen *n*
d.-connected motor-driven pump Pumpenaggregat *n*
d.-connected steering gear *(Schiff)* Rudermaschine *f* mit direktem Antrieb
d. connection 1. *(Fmt)* direkte (unmittelbare) Verbindung *f*; 2. galvanische Verbindung
d. control Regelung *f* ohne Hilfsenergie, direkte Regelung
d.-coupled direkt (galvanisch) gekoppelt
d.-coupled amplifier Gleichstromverstärker *m*
d.-coupled transistor logic circuitry *(Dat)* direkt gekoppelte transistorisierte logische Schaltung *f*
d. coupling direkte (galvanische) Kopplung *f*
d. current, D.C., d.c. Gleichstrom *m*
d.c.-alternating current converter (motor generator) Gleichstrom-Wechselstrom-Umformer *m*, Wechselrichter *m*
d.c. ammeter Gleichstrommesser *m*, Amperemeter *n*
d.c. amplifier Gleichstromverstärker *m*
d.c. analogue computer Gleichspannungsanalogrechner *m*
d.c. armature Gleichstromanker *m*
d.c. bell Gleichstromwecker *m*, Wecker *m* mit Selbstunterbrechung
d.c. brake lifting magnet Gleichstrommagnetbremslüfter *m*
d.c. component Gleichstromkomponente *f*, Gleichwert *m*
d.c. conductance Gleichstromleitvermögen *n*, Gleichstromleitwert *m*
d.c. control grid current Steuergittergleichstrom *m*
d.c. distribution system Gleichstromnetz *n*
d.c. drive Gleichstromantrieb *m*
d.c. excitation Gleichstromerregung *f*
d.c. kilowatt-hour meter Gleichstromzähler *m*
d.c. level Gleichstrompegel *m*
d.c. mains Gleichstromnetz *n*

d.c. measurement Gleichstrommessung *f*
d.c. measuring transductor Gleichstrommeßtransduktor *m*, Gleichstromwandler *m*
d.c. meter Gleichstromzähler *m*
d.c. motor Gleichstrom[elektro]motor *m*
d.c. motor-generator Gleichstromumformer *m*
d.c. network Gleichstromnetz *n*
d.c. operated receiver Gleichstromnetzempfänger *m*
d.c. plant Gleichstromanlage *f*
d.c. plate current Anodengleichstrom *m*
d.c. potential Gleichspannungspotential *n*
d.c. power Gleichstromleistung *f*
d.c. resistance Gleichstromwiderstand *m*, Ohmscher Widerstand *m*
d.c. restorer Nullpunktfesthalteschaltung *f*
d.c. saturation Gleichstromvormagnetisierung *f*
d.c. series wound motor Gleichstromreihenschlußmotor *m*, Hauptstrommotor *m*
d.c. source Gleichstromquelle *f*
d.c. supply Gleichstromspeisung *f*
d.c. transformer Gleichstrom[meß]wandler *m*
d.c. voltage Gleichspannung *f*
d.c. working voltage Gleichstrombetriebsspannung *f*
direct dialling *(Fmt)* direkte Wahl *f*
d. drive unmittelbarer Antrieb *m*
d.-drive engine Achsmotor *m*
d.-drive motor Achsmotor *m*
d. dye[stuff] direktziehender Farbstoff *m*, Direktfarbstoff *m*
d. e.m.f. Gleichstrom-EMK *f*
d. exchange line *(Fmt)* Amtsleitung *f*, Haupt[anschluß]leitung *f*
d. gas primary cooling direkte Gasvorkühlung *f (Kohleveredlung)*
d. grid current Gittergleichstrom *m*
d.-heated cathode direkt geheizte Katode *f*
d. heating Heizung *f* mittels Heizkörper *(im Gegensatz zur Warmluftheizung)*
d. hit *(Mil)* Volltreffer *m*
d. impact gerader Stoß *m*
d. indicator reading direkte Ablesung *f* vom Meßgerät
d. injection *(Kfz)* Zylindereinspritzung *f*
d. interelectrode capacitance Teilkapazität *f* zwischen Elektroden
d. iron process *(Met)* Rennverfahren *n*
d. lighting direkte Beleuchtung *f*
d. measurement unmittelbares Messen *n*
d. motor drive Einzelantrieb *m*
d. noise amplifier geräuschmodulierter Sender *m* für Radarstörungen
d. on-line starter *s.* d. switching starter
d. pick-up Originalaufnahme *f*, Originalabtastung *f*
d. play-back unmittelbares Abspielen *n* nach der Aufnahme *(Tonband usw.)*
d. potential Gleichspannung *f*
d. potential difference Gleichstrompotentialdifferenz *f*, Gleichspannung *f*
d. printing *(Text)* Direktdruck *m*
d. ray *(Fmt, Fs)* Bodenstrahl *m*, Bodenwelle *f*
d. reading direkt anzeigend, mit direkter (unmittelbarer) Ablesung *(Meßgeräte)*
d.-reading frequency meter Frequenzmesser *m* mit direkter Ablesung, Zeigerfrequenzmesser *m*
d.-reading instrument direkt anzeigendes Instrument *n*, geeichtes Skalengerät *n*
d.-reading transmission level indicator Pegel[an]zeiger *m*
d. recording Direktschreibung *f (Bildfunk)*
d. reflection gerichtete Reflexion (Spiegelung) *f*
d.-relation telemeter *(BMSR)* Proportionalfernmeßgerät *n*
d. reversing engine direkt umsteuerbarer Motor *m*
d. rope haulage *(Brg)* Einseilförderung *f*
d. selection direkte Wahl *f*
d. shear test Druckversuch *m* mit behinderter Seitenausdehnung *(Bodenmechanik)*
d. stress and bending Längskraft *f* mit Biegung *(Statik)*
d. switching starter Anlaßschalter *m*
d. transmission 1. direkte Übertragung *f*; 2. gerichtete Durchlassung *f (Licht)*
d. trip[ping] direkte Auslösung *f (Impulstechnik)*
d. trunk connection line *(El)* Fernanschlußleitung *f*
d. vacuum tube current Vakuumröhrengleichstrom *m*
d. view tube *(Fs)* Direktsichtbildröhre *f*
d.-vision prism geradsichtiges Prisma *n*, Geradsichtprisma *n*, Amici-Prisma *n*
d.-vision spectroscope geradsichtiges Spektroskop *n*
d.-vision tube Bildröhre *f* für Direktbetrachtung
d. voltage Gleichspannung *f*
d. wave direkter Strahl *m*, direkte Welle *f (Antenne)*
d. wire direkte Leitung *f*
directed gerichtet, orientiert
d. anti-aircraft missile (rocket) gelenkte Fla-Rakete *f*
d. beam gerichteter Strahl *m*
directedness Gerichtetheit *f*
directing effect of a group *(Ch)* dirigierende Wirkung *f* eines Substituenten
d. force Richtkraft *f*, Direktionskraft *f*, Rückstellkraft *f*

d. magnet Richtmagnet *m*
d. pulse Steuerstromstoß *m*
d. station 1. *(Fmt)* Leitfunkstelle *f*; 2. *(Flg)* Leitstation *f*
direction Richtung *f*; Richtungssinn *m*, Orientierung *f*; Führung *f*, Steuerung *f*
d.-bearing signal Peilsignal *n*
d. commutator Richtungsschalter *m*
d. cosine Richtungskosinus *m*
d. distortion Richtungsverzerrung *f*
d. finder Funkortungsgerät *n*, Funkpeiler *m*, Peilgerät *n*
d.-finder error Funkfehlweisung *f*
d.-finder evaluating station Peilauswertungsstelle *f*
d.-finder signal Peilsignal *n*
d.-finder system Peilanlage *f*
d. finding Funkortung *f*, Funkpeilung *f*
d.-finding bearing Funkpeilung *f*
d.-finding loop Peilrahmen *m*
d.-finding map Peilkarte *f*
d.-finding on board Eigenpeilung *f*
d.-finding receiver Funkpeilempfänger *m*
d.-finding transmitter Peilsender *m*
d.-finding wireless Peilfunk *m*
d.-indicating switch *(Kfz)* Winkerschalter *m*
d. indicator *(Kfz)* Fahrtrichtungsanzeiger *m* *(Winker oder Blinker)*
d.-indicator arm *(Kfz)* Winker *m*
d.-indicator control lamp *(Kfz)* Blinkerkontrolleuchte *f*, Winkerkontrolleuchte *f*
d.-indicator lamp *(Kfz)* Blinkleuchte *f*
d. of beam incidence Strahleinfallsrichtung *f*
d. of current Stromrichtung *f*
d. of cutting Schnittrichtung *f*
d. of dip *(Geol)* Fallrichtung *f*
d. of draw *(Masch)* Zugrichtung *f*
d. of flow Strömungsrichtung *f*
d. of force Kraftrichtung *f*
d. of growth Wuchsrichtung *f* *(Holz)*
d. of hand Gangrichtung *f* *(Fräser)*; Drehrichtung *f*
d. of helix Steigungssinn *m* *(Zahnrad)*
d. of lines of force Kraftlinienrichtung *f*
d. of mining *(Brg)* Abbaurichtung *f*
d. of motion (movement) Bewegungsrichtung *f*; Fahrtrichtung *f*
d. of outflow Ausflußrichtung *f*
d. of predominant surface pattern Richtungsverlauf *m* der Rauheit *(Oberflächenbeschaffenheit)*
d. of propagation *(Ph)* Fortpflanzungsrichtung *f*
d. of reading off *(Text)* Ableserichtung *f*
d. of rotation Drehsinn *m*, Umlaufrichtung *f*
d. of running Laufrichtung *f*
d. of scanning Abtastrichtung *f*
d. of spindle rotation Spindeldrehrichtung *f*
d. of strata *(Geol)* Schichtstreichen *n*
d. of the travel of the slide valve *(Masch)* Schieberbewegungsrichtung *f*, Schieberschubrichtung *f*
d. of the velocity component Richtungssinn *m* der Geschwindigkeitskomponente
d. of transverse power feed *(Masch)* Plandrehrichtung *f*
d. of travel *(Kfz)* Fahrtrichtung *f*, Bewegungsrichtung *f*
d. of traverse Bewegungsrichtung *f*
d. of twist *(Text)* Drehrichtung *f*, Richtung *f* der Zwirndrehung
d. of vibration Schwingungsrichtung *f*
d. switch Richtungsschalter *m*
directional 1. in Bewegungsrichtung; gerichtet; Richt-; 2. *(Masch)* sinnfällig
d. accuracy Richtungsgenauigkeit *f*
d. aerial Richtantenne *f*
d. beam Richtstrahl *m*
d. beam transmitter Richt[strahl]sender *m*, gerichteter Sender *m*
d. change Richtungsänderung *f*
d. characteristic Richtcharakteristik *f*, Richtkennfläche *f* *(Antenne, Mikrofon, Zählrohr usw.)*
d. circuit-breaker Rückstrom[aus]schalter *m*
d. control 1. *(Masch)* sinnfälliges Schalten *n*; 2. Richtungskontrolle *f*
d.-control four-way valve Umsteuervierwegeschieber *m*
d. control lever sinnfällig angeordneter Schalthebel *m*
d. control valve Umsteuerschieber *m*
d. controls sinnfällige Schaltelemente *npl*
d. counter *(Ph)* richtungsabhängiger Zähler *m*, richtungsabhängiges Zählrohr *n*
d. coupler Richt[ungs]koppler *m* *(Wellenleiter)*
d. current protection gerichteter Stromschutz *m*
d. diagram Richtcharakteristik *f* *(Antenne)*
d. distortion Richtungsfehler *m*
d. distribution Richtungsverteilung *f*
d. drilling gerichtetes Bohren *n*, Richtbohren *n*; Schrägbohren *n* *(Erdöl, Erdgas)*
d. element Richtungsglied *n* *(Relais)*
d. error Richtungsfehler *m*
d. field *(Math)* Richtungsfeld *n*
d. gyro Kreiselkompaß *m*
d. gyroscope Richtkreisel *m*
d. homing gerichtete Eigenpeilung *f* *(Radar)*
d. interlocking *(Eis)* Richtungsblockierung *f*, Ausschluß *m* von Fahrten entgegengesetzter Richtung
d. light *(Schiff, Flg)* Richtfeuer *n*
d. lighting gerichtetes Licht *n* *(Beleuchtungstechnik)*
d. loop Peilrahmen *m*

d. measurement Richtungsprüfung *f*
d. operation richtungsabhängiges Arbeiten *n* *(Relais)*
d. phase changer *s.* d. phase shifter
d. phase shifter Richtphasenschieber *m*
d. power protection Leistungsrichtungsschutz *m*
d. quantization *(Ph)* Richtungsquantelung *f*
d. radio Peilfunk *m*
d. radio beacon Richtfunkfeuer *n*
d. radio communication Richtfunkverkehr *m*
d. radio link Richtfunkstrecke *f*
d. receiver Richtempfänger *m*, Richtmikrofon *n*
d. relay Richtungsrelais *n*, Rückstromrelais *n*
d. resistance Richtwiderstand *m*
d. response pattern Richtcharakteristik *f*, Richtkennfläche *f*; Richtungsempfindlichkeit[sfläche] *f*
d. transmission Richtfunk *m*
d. transmitter Richtfunksender *m*, Richt[strahl]sender *m*, gerichteter Sender *m*
d. tripping magnet Rückstromauslöser *m*
d. valve *(Masch)* Umsteuerschieber *m*
d. vane Leitschaufel *f (Turbine)*
directionality Anisotropie *f*, Richtungsabhängigkeit *f*; Richtwirkung *f*
directive gerichtet
d. aerial Richtantenne *f*, Richtstrahler *m*
d. aerial system Richtantennenanlage *f*, Richtstrahleranlage *f*
d. efficiency Richtwirkung *f* *(einer Antenne)*
d. feed *s.* directional coupler
d. force Richtkraft *f*, Richtvermögen *n*
d. gain Antennen[richtungs]gewinn *m*, Richtungsgewinn *m* einer Antenne
d. reception Richtempfang *m*
d. wave Richtwelle *f*
directivity Richtvermögen *n*, Richtwirkung *f* *(einer Antenne)*
directly controlled equipment Fahrschaltersteuerung *f*
d. indicating amplifier Anzeigeverstärker *m*
d. reversible Diesel engine direkt umsteuerbarer Dieselmotor *m*
director 1. Direktor *m*, Produktionsleiter *m*; Filmregisseur *m*; 2. Interpolator *m (Werkzeugmaschinensteuerung)*; Nummernspeicher *m*, Register *n*, Speicher *m*; Übersetzer *m*, Umrechner *m*; 3. Direktor, Führungsantenne *f*, Leitantenne *f*
d. general Generaldirektor *m*
directory Telefonbuch *n*
directrix *(Math)* Leitkurve *f*; Leitlinie *f*, Direktrix *f*

dirigible lenkbar
dirt 1. Dreck *m*, Schmutz *m*; Erde *f*; 2. *(Am)* Land *n*, Erdboden *m*; 3. *(Brg)* Berge *pl*, Abgänge *mpl*
d. band Bergemittel *n*, Gesteinsmittel *n*
d. bed Bergeschicht *f*, Bergemittel *n*
d. count Schmutzpunktzahl *f (Papierherstellung)*
d. descending pipe Bergefalleitung *f*
d. heap Bergehalde *f*
d. moving Erdbewegung *f*, Bodenförderung *f*
d. separator Schmutzabscheider *m (Papierherstellung)*
dirtec Stoffreiniger *m (Papierherstellung)*
dirty schmutzig, verunreinigt; erdig; fehlerhaft *(Druck)*
d. bomb *(Kern)* „schmutzige“ Bombe *f*
d.-clothes side of pitbath *(Brg)* Schwarzseite *f (Kaue)*
d. coal ballastreiche Kohle *f*
d. oil tank Schmutzöltank *m*
d. ship schmutziges Schiff *n (Tanker für schmutzige Öle, wie Rohöl, Kesselöl usw.)*
d. water Abwasser *n*
disaccharose *(Ch)* Biose *f*, Disaccharose *f*
disaggregation Zerlegung *f* in Bestandteile
disappearing fiddles *(Schiff)* herunterklappbare Schlingerleisten *fpl (Eßtisch)*
d. filament pyrometer [optisches] Glühfadenpyrometer *n*, [optisches] Heizfadenpyrometer *n*
disassemble / to zerlegen, abbauen, demontieren, abmontieren, auseinandernehmen
disassembly Zerlegung *f*, Demontage *f*, Ausbau *m*, Abbau *m*
disazo dye Disazofarbstoff *m*
disbursement accounts Hafenkosten *pl (für ein Schiff)*
disbursements Schiffsunkosten *pl*
disc *s.* disk
discard / to wegwerfen, ablegen
discard *(Brg)* Berge *pl*, Abgänge *mpl*
discernibility Sichtbarkeit *f*, Erkennbarkeit *f*, Wahrnehmbarkeit *f*; Unterscheidbarkeit *f*
discernible sichtbar, erkennbar, wahrnehmbar; unterscheidbar
discharge / to 1. *(El)* entladen; 2. *(Schiff)* [Ladung] löschen; 3. *(Eis)* entladen *(Fracht)*; entleeren *(Waggons)*; 4. ablassen, auslassen; ausströmen, ausfließen *(Gase, Flüssigkeiten)*; 5. *(Ök)* entlassen *(Personal)*; abmustern *(Schiffsmannschaft)*; 6. *(Text)* ätzen
d. steam Dampf auslassen
discharge 1. [elektrische] Entladung *f*; 2. Löschen *n (Schiffsfracht)*; 3. *(Eis)* Entladen *n (Fracht)*; Entleeren *n (Waggons)*; 4. Ablassen *n*; Ausströmen *n*, Ausfließen *n (Gase, Flüssigkeiten)*; 5. *(Ök)* Entlassung *f*; Abmusterung *f*

(Schiffsbesatzung); 6. Beaufschlagung *f*; 7. Verdrängung *f*, Fördermenge *f*
d. afterglow Entladungsnachglimmen *n*
d. capacity 1. *(El)* Entladefähigkeit *f*, Entladekapazität *f*, Entladeleistung *f*, Entladungskapazität *f*; 2. Ableitvermögen *n* *(Schaltgeräte)*; 3. Förderleistung *f* *(Pumpe)*
d. channel *(El)* Entladungskanal *m*
d. circuit Entlade[strom]kreis *m*, Entladungs[strom]kreis *m*
d. coefficient Ausflußzahl *f*, Durchflußzahl *f* *(Durchfluß von Gasen und Flüssigkeiten)*
d. coil Entladungsdrossel *f*; Erdungsdrossel *f*
d. current Ableitstrom *m*, Entladungsstrom *m*
d. delay Entladeverzug *m*
d. end of the furnace Ziehende *n* eines Ofens, Ziehherd *m* *(Walzwerk)*
d. frequency Entladungsfrequenz *f*
d. gap Entladungsstrecke *f*, Funkenstrecke *f*
d. gate *(Bw)* Entleerungsöffnung *f*
d. header Luftauslaßrohr *n*; Löschverteilerstutzen *m* *(an Deck von Tankern)*
d. hopper Schütttrichter *m*
d. jet *(Kfz)* Einspritzdüse *f*
d. lag Entladeverzug *m*
d. lamp Entladungslampe *f* *(Sammelbegriff)*; Glimmlampe *f*; Mischlichtlampe *f*
d.-line pressure Förderdruck *m* *(Pumpe)*
d. nozzle 1. *(Kfz)* Zapfhahn *m*; 2. *(Flg)* Schubdüse *f* *(Turbo-Antrieb)*
d. of clear water Reinwasserabfluß *m*
d. of cooling water Kühlwasseraustritt *m*
d. of the fuel *(Kern)* Ausbringen *n* des Brennstoffs, Ausfahren *n* der Brennstoffelemente
d. path Entladungsstrecke *f*, Entladungsweg *m*
d. pattern *(Text)* Ätzmuster *n*
d. pipe friction Druckrohrreibung *f*
d. pipeline Druckleitung *f*
d. piping Abflußleitung *f*
d. plasma Gasentladungsplasma *n*
d. port Austrittskammer *f*
d. pressure Förderdruck *m* *(Pumpe)*
d. printing *(Text)* Ätzdruck *m*
d. rate 1. Fördermenge *f* *(Pumpe usw.)*; 2. Entladedauer *f* *(Batterie)*
d. register Jalousieauslaß *m* *(Heizungs- und Lüftungstechnik)*
d. resistor Entladewiderstand *m*
d. space Entladungsraum *m*
d. spark Entladungsfunke *m*, Abreißfunke *m*
d. switch Batterieschalter *m*, Zellenschalter *m*
d. tube Entladungsröhre *f*, Gasentladungsröhre *f*; Nullode *f*
d. valve Druckventil *n*
d. voltage Entladespannung *f*
discharged water *(Wärm)* Ablaufwasser *n*
discharger Widerstandszünder *m*; Parallelfunkenstrecke *f*
discharging arch Entlastungsbogen *m* *(Mauerwerk)*
d. berth Löschplatz *m* *(eines Schiffes)*
d. chute Entladungsrutsche *f*
d. current Entladestrom *m*
d. funnel Ablauftrichter *m*
d. gear *(Schiff)* Löschgeschirr *n*, Ladegeschirr *n*
d. overside *(Schiff)* Außenbordlöschung *f*
d. point Entladestelle *f*
d. resistor Entladewiderstand *m*; Erdungswiderstand *m*
disciplinary fine Ordnungsstrafe *f*
discolour / to verfärben
discomposition Atomumlagerung *f* [durch Kernstoß] *(im Gitterverband)*
disconnect / to 1. *(El)* trennen; ausschalten, abschalten, unterbrechen; 2. *(Masch)* auslösen *(Kupplung)*; ausrücken
d. a call *(Fmt)* eine Verbindung trennen (aufheben)
disconnect relay Trennrelais *n*
d. signal *(Fmt)* Schlußzeichen *n*
disconnectable *(El)* trennbar, abschaltbar
d. coupling lösbare Verbindung *f* *(Rohr)*
disconnected grid *(Fs)* freies (offenes) Gitter *n*
disconnecting blade [bewegbares] Schaltstück *n*, Trennmesser *n* *(Schalter)*
d. contact [Trenn-]Schaltstück *n*, Abreißkontakt *m*
d. fuse *s.* fuse disconnecting switch
d. jack Trennklinke *f*, Unterbrechungsklinke *f*
d. lever Ausrücker *m*, Ausrückhebel *m*
d. link Trennlasche *f*; Trennsicherung *f*; Schaltstück *n*
d. switch Ausschalter *m*; Trennschalter *m*
d. switch reverser Fahrtwende- und Motortrennschalter *m*
disconnection 1. Abschaltung *f*, Ausschaltung *f*, Trennung *f*, Unterbrechung *f* *(einer Leitung)*; 2. *(Masch)* Auslösung *f*, Ausrückung *f* *(Kupplung)*
d. fault Störung *f* durch Drahtbruch
d. noise Unterbrechungsgeräusch *n*
discontinuity Unstetigkeit *f*; Unstetigkeitsstelle *f*; Sprung *m*; Unregelmäßigkeit *f*
d. wave *(Ph)* Stoßwelle *f*, Unstetigkeitswelle *f*
discontinuous unterbrochen, nicht fortlaufend, nicht kontinuierlich; stoßartig; unstetig
d. chip Bruchspan *m*; Reißspan *m*
d. control unstetige Regelung *f*, Abtastregelung *f*
d. current lückender Strom *m* *(Stromrichter)*
d. function unstetige Funktion *f*
d. wave Wellenpaket *n*, Wellenzug *m*
discontinuum Diskontinuum *n*

discord *(Ak)* Dissonanz *f*, Mißklang *m*
discordance *(Geol)* Schichtungsdiskordanz *f*
discount *(Ök)* Rabatt *m*, Vergütung *f*
d. meter Vergütungszähler *m*, Rückvergütungszähler *m*
discover / to 1. entdecken; 2. *(Brg)* erschürfen
discovery Entdeckung *f*
d. of petroleum Fündigkeit *f* auf Erdöl, Entdeckung *f* von Erdöl
d. well Entdeckungsbohrung *f*, Erschließungssonde *f*, erste fündige Bohrung *f*, Fundbohrung *f*
discrete *(Ph)* diskret *(Teilchen)*
d. alpha-rays diskrete Alpha-Strahlung *f*
d. particles diskrete Teilchen *npl*
d. spectrum diskretes Spektrum *n*; Linienspektrum *n*
discriminant *(Math)* Diskriminante *f*
d. curve Diskriminantenkurve *f*
discriminating circuit-breaker *(El)* Rückstrom[aus]schalter *m*, Differentialausschalter *m*
d. cut-out Rückstromautomat *m*
d. element Meßglied *n*
d. protective system Selektivschutzsystem *n*
d. relay Wählerrelais *n*
d. selector Dienstwähler *m*, Hilfswähler *m*, Mitlaufwerk *n*, Mitlaufwähler *m*
discrimination bedingter Sprung *m (elektronische Rechner)*
discriminator *(El)* Diskriminator *m*, Frequenzdemodulator *m*
d. stage Frequenzmodulatorstufe *f*
discriminatory logic *(BMSR, Dat)* unterscheidende Logik *f*
discus Diskus *m*
disembark / to landen; ausschiffen; aussteigen
disengage / to 1. ausschalten, stillsetzen; auslösen, entkuppeln; ausrücken; ausklinken; 2. *(El)* trennen *(Kontakte usw.)*
d. the clutch *(Kfz)* auskuppeln
disengaged line *(Fmt)* freie Leitung *f*
disengagement 1. Ausschaltung *f*, Stillsetzung *f*; Auslösung *f*, Lösen *n*; Entkuppeln *n*; Ausrükken *n*; Ausklinken *n*; 2. Trennen *n (Kontakte)*
d. by hand Handauslösung *f*
disengaging arbor Ausrückwelle *f*
d. coupling Schaltkupplung *f*
d. sleeve *(Kfz)* Ausrückmuffe *f*
disentangle / to entwirren; *(Schiff)* klarieren *(eine Leine)*
dish / to *(Met)* 1. kümpeln; 2. tiefziehen
dish drier Tellertrockner *m*
d. [grinding] wheel Tellerschleifscheibe *f*
dished 1. hohlflächig, hohlgeschliffen, nach innen gewölbt; 2. erhaben; nach außen gewölbt; 3. mit abgerundeter Ecke
d. boiler end gewölbter Kesselboden *m*
d. cover nach außen gewölbter Deckel *m*
d. electrode Schalenelektrode *f*
d. metal slitting saw seitlich hohlgeschliffene Metallkreissäge *f*
d. plate Buckelblech *n*, Buckeleisen *n*, Buckelplatte *f*
d. sleeper *(Eis)* Fischbauchschwelle *f*
dishing 1. Hohlschleifen *n*; 2. *(Met)* Kümpeln *n*; 3. *(Met)* Tiefziehen *n*
d. press Kümpelpresse *f*
dishwasher Geschirrspülmaschine *f*
dishwashing agent Geschirrspülmittel *n*
d. machine Geschirrspülmaschine *f*
disinfectant Desinfektionsmittel *n*
disinflate / to Luft entweichen lassen
disintegrate / to 1. abbauen, desintegrieren; 2. *(Kern)* zerfallen
disintegrating cut Trennschnitt *m*
d. mill Hartzerkleinerungsmaschine *f*
disintegration 1. *(Kern)* Zerfall *m*, Umwandlung *f*; 2. *(Ch)* Aufschluß *m*; 3. Zerstäubung *f (der Katode)*; 4. *(Holz)* Aufschlagen *n*, Zerfasern *n*; Mahlung *f*
d. chain *(Kern)* Zerfallsreihe *f*, Zerfallskette *f*
d. electron Zerfallselektron *n*
d. of atom Atomzerfall *m*, Kernzerfall *m*
d. of filament *(El)* Fadenzerstäubung *f*
d. of the nucleus Kernzerfall *m*
d. probability *(Kern)* Zerfallswahrscheinlichkeit *f*
d. rate *(Kern)* Zerfallsgeschwindigkeit *f*
d. series *(Kern)* Zerfallsreihe *f*
disintegrator Schlag[korb]mühle *f*, Schleudermühle *f*, Zerkleinerer *m*
disjunction Disjunktion *f*, einschließendes ODER *n (logischer Operator)*
disjunctive normal form *(Dat)* disjunktive Normalform *f (Aussagenlogik)*
disk 1. Scheibe *f*, Platte *f*; Stammblatt *n*; 2. Schallplatte *f*; 3. Scheibenrad *n (Turbine)*; 4. Diskus *m*
d. anode Telleranode *f*, Lochscheibenanode *f (Elektronenröhren)*
d. area ratio *(Schiff)* Flächenverhältnis *n (abgewickelte Flügelfläche zur vollen Schraubenkreisfläche)*
d. armature *(Elm)* Scheibenanker *m*
d. attenuator Scheibenabschwächer *m (Hohlleiter)*
d. attrition mill Scheibenmühle *f*
d. auger Tellerbohrer *m (zur Bodenuntersuchung)*
d. bar for dumb-bells Scheibenstange *f*
d. bar for weight lifting Scheibenstange *f*
d. bearing Scheibenlager *n*
d. bit Scheibenmeißel *m*
d. boss Scheibennabe *f*
d. brake 1. Scheibenbremse *f*; 2. *(Elm, El)* Wirbelstrombremse *f*

d. cam Scheibenkurve *f*
d. chipper *(Holz)* Hackrotor *m*
d. chuck Scheibenfutter *n*
d. clutch Scheibenkupplung *f*
d. coil Flachspule *f*; Scheibenwicklung *f*
d. commutator Scheibenkollektor *m*, Flachkommutator *m*
d. coulter *(Lw)* Scheibendrillschar *n*; Rundsech *n*, Scheibensech *n*
d. crank Kurbelscheibe *f*
d. crusher Scheibenmühle *f*, Scheibenbrecher *m*
d. discharger rotierende Funkenstrecke *f*, Scheibenfunkenstrecke *f*
d. drill *(Am)* *(Lw)* Scheibenschardrillmaschine *f*
d. drop *(Fmt)* Fallklappe *f*
d. drying *(Holz)* Scheibentrocknung *f* *(Späne)*
d. electrometer Scheibenelektrometer *n*
d. feeder Tellerspeiser *m*, Tellerbeschicker *m*, Tellerzuteiler *m*, Telleraufgeber *m*
d. filter *(Kfz)* Spaltfilter *n*
d. friction gear Diskusgetriebe *n*
d. grinder Abrichtschleifmaschine *f*
d. harrow *(Lw)* Scheibenegge *f*
d. impeller Scheibenkreiselmischer *m*
d. indicator Scheibenzähler *m*
d. insulator Scheibenisolator *m*
d. joint Scheibengelenk *n*
d.-loading bar-bell Hantel *f*, Scheibenhantel *f*
d. mill Stiftmühle *f*; Refiner *m* *(Papierherstellung)*
d. of a clutch *(Kfz)* Kupplungslamelle *f*
d. of wire screen *(Opt)* Netzobjektträger *m*
d. pig Scheibeneisen *n*
d. piston Scheibenkolben *m*
d. piston blower Kreiskolbengebläse *n*
d. plate drawing frame *(Text)* Scheibenwalzstrecke *f*
d. plough Scheibenpflug *m*
d. plough scuffling Schälpflügen *n*
d. record Schallplatte *f*
d. recording Schallplattenaufnahme *f*
d. refiner Scheibenrefiner *m* *(Papierherstellung)*
d. rotary valve Kreisschieberventil *n*
d. sander Scheibenschleifmaschine *f* *(Papierherstellung)*
d. scanner *(Fs)* Bildabtastscheibe *f*
d. screen Scheibenrost *m*, Grobrost *m*
d. seal diode Scheibendiode *f*
d. seal valve Scheibenröhre *f*
d.-sealed triode Scheibentriode *f*
d.-shaped anode Scheibenanode *f*
d. shutter *(Opt)* Scheibenblende *f*, Scheibenverschluß *m*
d. spring Tellerfeder *f*
d. trimmer Scheibentrimmer *m*
d. turbine Scheibenturbine *f*
d.-type condenser Scheibenkondensator *m*
d.-type dumb-bell Scheibenhantel *f*
d.-type electrometer Scheibenelektrometer *n*
d.-type fertilizer sprayer Tellerdüngerstreuer *m*
d.-type knife cutting machine [Scheibentyp-]Messerschneidemaschine *f* *(Holzzerkleinerung)*
d.-type rotary selector Kreisschieberumsteuerventil *n*
d. valve Tellerventil *n*, Plattenventil *n*, Scheibenventil *n*
d. wheel Scheibenrad *n*
d. winding Scheibenwicklung *f*
d. with balancing holes Scheibe *f* mit Ausgleichlöchern *(Turbine)*
disks running in opposite directions gegenläufige (entgegengesetzt laufende) Räder *npl* *(Turbine)*
dislocation Versetzung *f* *(Kristall)*
d. line Versetzungsgrenze *f* *(Realkristalle)*
dislodge electrons / to Elektronen befreien
dismantle / to demontieren, abmontieren, abbauen, auseinandernehmen, abbrechen, ausbauen
dismantled ship abgetakeltes Schiff *n*, Blockschiff *n*
dismantling Demontage *f*, Abbau *m*, Ausbau *m*, Auseinandernehmen *n*
dismast / to *(Schiff)* entmasten
dismember / to auseinandernehmen
dismount / to abnehmen; absteigen
dismounting Abbau *m* *(Demontage)*
disorder Fehlordnung *f* *(Kristalle)*
d. energy Fehlordnungsenergie *f*
d. scattering *(Ph)* diffuse (ungeordnete) Streuung *f*, Fehlordnungsstreuung *f*
disordered fehlgeordnet *(Kristalle)*
d. crystal gestörter Kristall *m*
d. motion *(Ph)* ungeordnete (regellose) Bewegung *f*
disparity 1. *(Ph)* Disparität *f*; 2. *(BMSR)* Disparität, Ungleichheit *f*, Unpaarigkeit *f*
dispatching import goods *(Schiff)* Versendung *f* von Importgütern *(nach dem Inland)*
dispense / to *(Kfz)* abgeben, auslitern
dispenser cathode Vorratskatode *f*
dispensing vehicle Tankwagen *m*
disperge / to dispergieren
dispersal dispers, feinverteilt
d. colour Dispersionsfarbstoff *m*
d. landing field *(Flg)* Ausweichlandeplatz *m*
d. plant Zweigbetrieb *m*
dispersant Dispersionsmittel *n*
disperse / to dispergieren, zerstreuen, zerteilen, verteilen
disperse dispers
d. dye Dispersionsfarbstoff *m*
dispersed state Verteilungszustand *m*

dispersing agent Dispergiermittel *n*
d. cone *(Ph)* Streuungskegel *m*
d. lens Streuungslinse *f*
dispersion 1. *(Math)* Dispersion *f*, Streuung *f* *(Statistik)*; 2. *(Ph)* Dispersion *(Wellenausbreitung in Medien)*; Zerstreuung *f*; 3. *(Ch)* Dispersion, Verteilung *f*
d. coefficient 1. *(Ph)* Dispersionskoeffizient *m*; 2. *(El)* Streukoeffizient *m*
d. current Streustrom *m*
d. forces Dispersionskräfte *fpl*
d. formula *(Opt)* Dispersionsformel *f*
d. hardening Ausscheidungshärtung *f*, Dispersionshärtung *f*, Aushärtung *f*
d. lens Zerstreuungslinse *f*
d. line Dispersionslinie *f*
d. loss Streuungsverlust *m* *(Turbine)*
d. of heat Wärmestreuung *f*
d. of light Dispersion *f* des Lichtes, optische Dispersion
d. relation Dispersionsrelation *f*, Dispersionsgesetz *n*
d. theory Dispersionstheorie *f*
dispersive medium dispergierendes Medium *n*
d. power Dispersionsvermögen *n*, Streu[ungs]vermögen *n*
displace / to 1. verschieben; 2. verdrängen *(Flüssigkeit)*; 3. fördern *(Pumpe)*; 4. auslenken
displaceable lock *(Eis)* auffahrbare Verriegelung
displaced fluid verdrängte Flüssigkeit *f*
d. vector verschobener Vektor *m*
displacement 1. *(Math)* Verschiebung *f*, Translation *f*; 2. *(Ph)* Umlagerung *f*, Verschiebung *f* *(Kristallgitterstörung)*; 3. *(Schiff)* Wasserverdrängung *f*; Werkstoffverdrängung *f*; 4. Förderung *f* *(Pumpe)*; Fördermenge *f*; Fördergut *n*; 5. *(Fs)* Strahlverschiebung *f*, Unschärfe *f*
d. collision *(Ph)* Umlagerungsstoß *m*
d. current Verschiebungsstrom *m*
d. curve *(Schiff)* Verdrängungskurve *f*
d. factor Verschiebungsfaktor *m*
d. gyro *(BMSR)* Lagekreisel *m*
d. law Verschiebungsgesetz *n*
d. of current Stromverdrängung *f*
d. of cylinder *(Kfz)* Zylinderinhalt *m*
d. of equilibrium Gleichgewichtsverschiebung *f*
d. of phase Phasenverschiebung *f*
d. of piston *(Kfz)* Hubraum *m*, Hubvolumen *n*
d. of spectral lines Verschiebung *f* der Spektrallinien, Spektrallinienverschiebung *f*
d. of support *(Bw)* Auflagerverschiebung *f*; Stützensenkung *f*
d. of the needle board *(Text)* Nadelbrettverschiebung *f*
d. series elektrochemische Spannungsreihe *f*
d. ship Verdrängungsschiff *n*
d. supercharger *(Kfz)* Drehkolbenlader *m*, Kolbenlader *m*
d. ton *(Schiff)* Verdrängungstonne *f*
d.-type controller *(BMSR)* Wegvergleichsregler *m*
d. vector *(Math)* Verschiebungsvektor *m*
displacer Kapselgebläse *n*, Verdrängergebläse *n* *(Verbrennungsmotor)*
d. piston Verdrängerkolben *m* *(Verbrennungsmotor)*
displacing 1. Verdrängen *n*; 2. Versetzen *n*, Verschieben *n* *(in Kristallen)*
d. limit Schubgrenze *f*
display / to 1. [optisch] anzeigen, optisch ausgeben; 2. *(El, Fs)* darstellen, wiedergeben; 3. *(Typ)* auszeichnen *(Manuskript)*
display 1. [optische] Anzeige *f*; 2. *(El, Fs)* Bildaufzeichnung *f*, Darstellung *f*, Wiedergabe *f*, Oszillogramm *n*; 3. *(Typ)* Auszeichnung *f* *(Manuskript)*
d. board *(Flg)* Luftlagekarte *f*
d. box Vitrine *f*, Schaukasten *m*
d. chart *(Flg)* Luftlagekarte *f*
d. lighting Schaufensterbeleuchtung *f*
d. of solutions Darstellung *f* der Aufgabenlösungen *(Analogrechner)*
d. panel *(Fmt)* Lampenfeld *n* für Rufanzeige
d. types *(Typ)* Auszeichnungsschrift *f*
d. unit 1. [optische] Anzeigeeinrichtung *f*, Sichtgerät *n*; 2. Sichtgerät *(Radar)*
d. window Schaufenster *n*
displayable darstellbar, ausstellbar
disposable container Wegwerfbehälter *m*
d. output *(Brg)* verwertbare Förderung *f*
disposal plant *(Kern)* Abfallbeseitigungsanlage *f*
dispose / to anordnen
d. of beseitigen
disposition of the tolerance zone Lage *f* des Toleranzfeldes *(zur Nullinie)*
disregard / to vernachlässigen, außer acht lassen
disruption Zerreißen *n*
disruptive breakdown *(El)* Durchschlag[en *n*] *m*
d. discharge *(El)* Durchschlag *m*, Durchbruch *m*
d. force Zerreißkraft *f*
d. power Durchschlagskraft *f*
d. strength Durchschlagfestigkeit *f* *(dielektrische)*
d. voltage Durchschlagspannung *f*
dissecting change-over condenser *(Opt)* Präparierwechselkondensor *m*
d. magnifier Präparierlupe *f*
d. microscope Präpariermikroskop *n*
d. utensils Präparierbesteck *n*
dissector *(Fs)* Zerleger *m*, Bildzerleger *m*
d. tube Aufnahmeröhre *f* [nach Farnsworth]
disseminated *(Geol)* eingesprengt

dissertation Dissertation *f*
dissimilar unähnlich; ungleichnamig
d. fractions *(Math)* ungleichnamige Brüche *mpl*
dissipation 1. *(Ph)* Dissipation *f*, Zerstreuung *f*; 2. *(El)* Verlustleistung *f*
d. limit Anodenverlustleistungsgrenze *f*; Belastungsgrenze *f*
d. of energy Energiedissipation *f*; Energieverlust *m*
d. of heat Wärmeableitung *f*, Wärmeabstrahlung *f*, Wärmedissipation *f*
dissipationless line *(El)* verlustfreie Leitung *f*
dissipative force *(Ph)* dissipative Kraft *f*
d. system *(Fmt, Fs)* Abstrahlvorrichtung *f*
dissociation Dissoziation *f*, Aufspaltung *f*, Zerlegung *f*
d. energy Dissoziationsenergie *f*
d. equilibrium Dissoziationsgleichgewicht *n*
d. potential Dissoziationspotential *n*
dissociative capture dissoziative Anlagerung *f*, dissoziativer Einfang *m (Spezialfall der Ionen-Rekombination)*
d. recombination dissoziative Rekombination (Wiedervereinigung) *f*
dissolution *(Ch)* Auflösung *f*, Auflösen *n*
dissolvant *(Ch)* Lösungsmittel *n*
dissolve out / to herauslösen
dissolver Stofflöser *m (Zellstoffherstellung)*
dissolving chest Auflösebehälter *m (Zellstoffherstellung)*
d. pulp Viskosezellstoff *m*
d. salt Solutionssalz *n*
dissonance Dissonanz *f*
dissymmetrical asymmetrisch
d. transducer asymmetrischer Wandler *m*
dissymmetry Asymmetrie *f*
distance 1. Weg *m*; 2. Abstand *m*, Entfernung *f*; 3. *(Math)* Strecke *f*
d. between centres *(Masch)* größte Drehlänge *f*; Spitzenabstand *m*
d. between end rows of bolt-holes Seitenlänge *f* von Schraubenmitte bis Schraubenmitte
d. between rolls Walzenabstand *m*
d. between side rows of bolt holes *(Masch)* Lochreihenabstand *m*
d. between the crank and its dead centre *(Masch)* Abstand *m* der Kurbel von ihrer Totlage
d. control Fernsteuerung *f*, Fernregelung *f*; Fernlenkung *f*
d.-controlled ship Fernlenkschiff *n*, ferngelenktes Schiff *n*
d. covered *(Ph)* zurückgelegte Strecke *f*; Weg *m*
d. gauge Entfernungsmesser *m*, Wegzähler *m*
d. judgement Entfernungsschätzung *f*
d. measurement Abstandsmessung *f*, Streckenmessung *f*
d. measuring Streckenmessung *f*
d.-measuring equipment Entfernungsmeßeinrichtung *f*
d.-measuring interrogator *(Flg)* DME-Abfragesender *m*
d.-measuring responder *(Flg)* Antwortbake *f* beim Entfernungsmeßverfahren DME
d.-measuring transponder *(Flg)* Wiederholerbake *f* beim Entfernungsmeßverfahren DME
d. of centre of gravity from mouth of venturi *(Rak)* Schwerpunktabstand *m*
d. of the valve from the centre position *(Wärm)* Schieberausschlag *m*, Schieberauslenkung *f*, Ausschlag *m* des Schiebers aus der Mittellage
d. of transmission *(Fmt, Fs)* Reichweite *f*
d. off-course Flugwegabweichung *f*
d. piece 1. *(Masch)* Distanzstück *n*, Zwischenlage *f*; Zwischenscheibe *f*; 2. *(Bw)* Abstandshalter *m*; 3. *(Wärm)* Trennungszunge *f*; 4. Klebstift *m (Relais)*
d. plate Distanzplatte *f*, Distanzscheibe *f*
d. protection Distanzschutz *m*
d. relay Distanzrelais *n*
d. ring *(Masch)* Distanzring *m*, Zwischenring *m*, Abstandsring *m*
d. signal *(Eis)* Vorsignal *n*
d. to go verbleibende Flugstrecke *f*
d. to travel verbleibende Flugstrecke *f*
d. tube *(Masch)* Abstandshülse *f*
d. velocity lag *(BMSR)* Laufzeit *f*, reine Totzeit *f*
distant control *s.* remote control
d. control gear Fernsteuerungsanlage *f*
d. ignition Fernzündung *f (Sprengungen)*
d. office *(Am) (Fmt)* Gegenamt *n*
d.-reading Fernmessungs-
d.-reading instrument Instrument *n* mit Fernanzeige
d.-reading tachometer Ferndrehzahlmesser *m*
d.-reading thermometer Fernthermometer *n*
d. reception *(Fmt, Fs)* Fernempfang *m*
d. regulation Fernregelung *f*
d. signal *(Eis)* Vorsignal *n*
d. station reception *(Fmt, Fs)* Fernempfang *m*
d. terminal *(Fmt)* Gegenamt *n*
d. trade *(Schiff)* große (lange) Fahrt *f*
d. zone Fernzone *f*
distemper / to mit Leimfarbe streichen
distemper Leimfarbe *f*
distempering Leimfarbenanstrich *m*
distension Ausdehnung *f*, Dehnung *f*, Streckung *f*
disthene *(Min)* Disthen *m*, Kyanit *m*

distil / to destillieren
distillate Destillat *n*
distillation Destillation *f*
d. chamber Vorraum *m (bei Halbgasfeuerungen)*
d. characteristics Destillationsverhalten *n*, Siedeverhalten *n*, Siedeverlauf *m*
d. column Destillationskolonne *f*
d. flask Destillierkolben *m*, Siedekolben *m*
d. range Siedebereich *m*
d. tail Destillationsnachlauf *m*
d. under vacuum Verschwelung *f*, Vakuumdestillation *f*
distilled heat Abwärme *f*
d. oil abgetriebenes Öl *n*
d. water destilliertes Wasser *n*
distiller Destillierapparat *m*; Brüdenkondensator *m*
distiller's grains Treber *pl (Bierbrauen)*
d. yeast Weinhefe *f*
distinctive note Klangcharakter *m (Musikinstrumente; Elektroakustik)*
distinguishing letter Kennbuchstabe *m*
d. signal Erkennungssignal *n*
distort / to 1. *(El)* verzerren, stören; 2. *(Opt)* verzeichnen, verzerren *(Abbildung)*; 3. *(Masch)* verbiegen; verziehen; verdrehen; verwerfen
distorted waves Wellen *fpl* verzerrter Kurvenform; nichtsinusförmige Wellen
distorting of the grate Werfen *n* des Rostes *(Feuerungstechnik)*
distortion 1. *(El)* Verzerrung *f (Übertragung, Wiedergabe)*; 2. *(Opt)* Verzeichnung *f*, Verzerrung *f (Abbildung)*; 3. *(Masch)* Verbiegung *f*; Verziehung *f*, Verzug *m*; Verdrehung *f*; Verwerfen *n*
d. analyzer Klirranalysator *m*
d. bridge Klirrfaktormeßbrücke *f*, Oberwellenmeßgerät *n*
d. by welding Schweißverzug *m*
d. circuit Verzerrer *m*; Verzerrerschaltung *f*
d. constant *(El)* Klirrfaktor *m*
d.-correcting device Ausgleichsvorrichtung *f*, Ausgleichsbestandteile *mpl (z. B. Fernsprechverstärker)*
d. due to feedback *(Fmt)* Rückkopplungsverzerrung *f*
d. effect Verzerrungseffekt *m*
d. factor *(El)* Oberwellengehalt *m*, Klirrfaktor *m*
d. factor meter Klirrfaktormeßgerät *n*
d. factor of a static converter Verzerrungsfaktor *m* eines Stromrichters
d.-free verzerrungsfrei
d. of image Bildverzerrung *f*
d. of the field *(Ph)* Feldverzerrung *f*
d. on hardening Härteverzug *m*, Verwerfen *n* durch Härten
d. value *(Opt)* Verzeichnungswert *m*
distortionless *(Opt)* verzeichnungsfrei
distress *(Schiff)* Notlage *f*
d. at sea Seenot *f*
d. call Notruf *m*, Notzeichen *n*, Seenotruf *m*
d. cargo Eilzustellungsladung *f*
d. flag Notflagge *f*
d. frequency Seenotfrequenz *f*
d. light Notsignalfeuer *n*
d. wavelength Seenotwellenlänge *f*
distribute / to 1. verteilen; 2. *(Typ)* ablegen *(Satz)*; 3. verreiben *(Farbe)*
distributed verteilt
d. amplifier *(Fs)* Kettenverstärker *m*
d. capacitance *(El)* verteilte Kapazität *f*
d. capacitance of a coil Spulenkapazität *f*
d. constant verteilte elektrische Größe *f (Wellenleiter)*
d. load Streckenlast *f*, verteilte Last *f*, Flächenlast *f (Statik)*
distributing bar *(Typ)* Ablegestange *f (Setzmaschine)*
d. box *(El)* Verteilerkasten *m*, Verteilungsdose *f*, Verteilungskasten *m*
d. bus-bar Verteiler[sammel]schiene *f*
d. cable Verteilungskabel *n*, Zweigkabel *n*
d. edge of the slide valve *(Wärm)* steuernde Schieberkante *f*
d. edge of the steam port *(Fmt)* steuernde Kanalkante *f*
d. factor *s.* distribution factor
d. frame *(El)* Verteilergestell *n*; Verteiler *m*
d. main[s] Verteil[ungs]leitung *f (Elektrizität, Wasser)*
d. network Verteil[ungs]netz *n*
d. pillar *(El)* Verteilersäule *f*
d. rollers Farbreibewalzen *fpl*
d. valve chest *(Wärm)* Steuerungsventilkasten *m*
d. valve motion Steuerbewegung *f (Dampf)*
d. wire Schaltdraht *m*
distribution Verteilung *f*
d. board Verteiler[schalt]tafel *f*, Stromverteilungstafel *f*
d. box Abzweigdose *f*; Verteilerkasten *m*, Verteilungsdose *f*
d. bucket Gichtkübel *m (Hochofen)*
d. cable Verteilungskabel *n*, Zweigkabel *n*
d. case *(Typ)* Ablegekasten *m (für Satz)*
d. centre *(El)* Hauptverteilung *f*
d. curve *(Math, Ph)* Verteilungskurve *f*
d. cutout *(El)* Verteilungsausschalter *m*
d. diagram 1. *(Math)* Verteilungsdiagramm *n (Statistik)*; 2. *(Wärm)* Steuerungsdiagramm *n*
d. factor Verteilungszahl *f (Statik, Momentenausgleichsverfahren)*
d. feeder *(El)* Verteilungsleitung *f*

d. fluctuations 1. *(Math, Ph)* Verteilungsschwankungen *fpl*; 2. *(El)* Verteilungsrauschen *n*
d. frame *(El)* Verteiler[gestell *n*] *m*, Leitungsverteiler *m*
d. fuse board Verteilersicherungstafel *f*
d. network 1. *(El)* Leitungsnetz *n*, Verteil[ungs]netz *n*; 2. Drahtfunknetz *n*
d. of atmospheric pressure Luftdruckverteilung *f*
d. of charge *(Ph, El)* Ladungsverteilung *f*
d. of current Stromverteilung *f*
d. of electricity Elektrizitätsversorgung *f*, Stromversorgung *f*
d. of electrons Elektronenverteilung *f*
d. of energy Energieverteilung *f*
d. of load Lastverteilung *f*; Gewichtsverteilung *f*
d. of strain Spannungsverteilung *f* *(Turbine; Betriebsausdruck)*
d. of stress[es] *(Ph)* Spannungsverteilung *f*
d. of velocity Geschwindigkeitsverteilung *f*
d. panel Schalttafel *f*
d. steel Querbewehrung *f*, Verteilerstahl *m* *(Stahlbeton)*
d. switchboard Verteilungsschalttafel *f*; Umschalteinrichtung *f*
d. system *(Bw)* Verteilernetz *n*, Verteilungsnetz *n* *(Versorgungsanlagen)*
d. transformer Netzumspanner *m*, Netztransformator *m*
d. tube Saatleitungsrohr *n* *(Drillmaschine)*
distributive law [of algebra] Distributivgesetz *n*
distributor 1. *(Kfz)* Verteiler *m*, Zündverteiler *m*; 2. Verteiler *m* *(Straßenbaumaschine)*; 3. Leitapparat *m* *(Turbine)*
d. arm *(Kfz)* Verteilerfinger *m*
d. bar *(Typ)* Ablegestange *f* *(Setzmaschine)*
d. box *s.* distribution box
d. cam *(Kfz)* Verteilernocken *m*, Zündverteilernocken *m*
d. cap *(Kfz)* Verteilerkappe *f* *(Zündverteiler)*
d. carriage *(Typ)* Ablegeschlitten *m* *(Setzmaschine)*
d. coil *(Kfz)* Zündspule *f*
d. contact *(Kfz)* Unterbrecherkontakt *m*
d. head *(Kfz)* Verteilerkopf *m*
d. pole *(El)* Verteilungsmast *m*
d. roller *(Typ)* Farbverreibewalze *f*, Reibwalze *f*
d. rotor arm *(Kfz)* Verteilerfinger *m*
d. switch Verteilerschalter *m*, Mehrwegeschalter *m*
d. valve Ausgleichsventil *n*, Verteilerventil *n*
district 1. Abteilung *f*; Gebiet *n*; Bezirk *m*; 2. *(Brg)* Revier *n*
d. boss *(Am) (Brg)* Abteilungssteiger *m*
d. central heating Fernheizung *f*
d. exchange *(Fmt)* Endamt *n*
d. heating Fernheizung *f*
d. surveyor *(Brg)* Reviermarkscheider *m*
d. transmitter Bezirkssender *m*
disturb / to stören; erregen
disturb pulse Störimpuls *m*
disturbance 1. *(BMSR)* Störgröße *f*; 2. [atmosphärische] Störung *f*; Funkstörung *f*; Betriebsstörung *f*; 3. *(Geol)* Störung *f*, Dislokation *f*, Gebirgsstörung *f*, Schichtenstörung *f*; 4. Erregung *f*
d. function *(Math)* Störfunktion *f*
disturbation theory *s.* perturbation theory
disturbed 1. gestört *(z. B. Rundfunkempfang)*; 2. *(Geol)* gestört, verworfen
d. orbit *(Ph)* gestörte Bahn *f*
disturbing current Störstrom *m*
d. noises Störgeräusche *npl*
d. voltage Störspannung *f*
disulphuric acid Dischwefelsäure *f*, Pyroschwefelsäure *f*
disused working *(Brg)* abgeworfener Grubenbau *m*
ditch / to 1. Graben ziehen; 2. *(Flg)* notwassern
ditch Graben *m*
d. cleaner Grabenräumer *m*, Grabenreinigungsgerät *n*
d. digger Grabenbagger *m*; Tieflöffelbagger *m*
d. digging Grabenaushub *m*
ditcher Grabenziehmaschine *f*, Grabenbagger *m*; Tieflöffelbagger *m*
ditching 1. Grabenaushub *m*; 2. *(Flg)* Notwasserung *f*
dithiocarbonic acid Dithiokohlensäure *f*
dithionic acid Dithionsäure *f*
diurnal range *(Schiff)* täglicher Tidenhub *m*
d. variation tägliche Schwankung *f*; tägliche Variation *f*
div div *(Vektoranalysis)*
divalent zweiwertig
dive / to [ein]tauchen; untertauchen; stürzen auf
dive bomber Sturzbomber *m*
d. brake Sturz[flug]bremse *f*
d. flap Sturzflugbremse *f*
d. logger Taucherboot *n*
diver Taucher *m*
diverge / to auseinanderlaufen, divergieren, sich erweitern, zerstreuen
divergence 1. Divergenz *f* *(Reihe; Vektorfunktion)*; 2. Ausschlag *m* *(Zeiger)*
divergent divergent, divergierend *(Reihe; Integral)*
d. jet guiding divergente Strahlführung *f* *(Turbine)*
diverging lens Zerstreuungslinse *f*, negative Linse *f*

diversion 1. Ablenkung *f*; 2. Umleitung *f* *(Straßenverkehr, Flüsse usw.)*
diversity receiver Diversity-Empfänger *m*, Mehrfachempfänger *m* zum Fadingausgleich
d. receiving system *(El)* Mehrfachempfang *m*
d. reception *(El)* Mehrfachempfang *m*, Diversity-Empfang *m*
divert / to 1. ablenken; 2. umleiten *(Straßenverkehr)*
diverted time *(Ök)* bezahlte Ausfallzeit *f*
diverter Ableiter *m*, Überspannungsableiter *m*
divide / to 1. dividieren, teilen; 2. *(Masch)* teilen
divided by [geteilt] durch
d. circular groove unterteilte Ringnut *f*
d.-conductor protection *(El)* Spaltleiterschutz *m*
d. deflection of the steam jet doppelte (zweiseitige) Strahlablenkung *f (Turbine)*
d. iron-core *(El)* unterteilter Eisenkern *m*
d. sliver delivery *(Text)* Teilbandstrecke *f*
divider 1. Verteiler *m*; 2. Dividierer *m*, Teiler *m (Analogrechner)*; 3. *(Brg)* Einstrich *m (Schachtausbau)*; 4. *s.* dividers
d. apparatus Teilkopf *m*
d. leg Zirkelbein *n*, Zirkelschenkel *m*
d. point Zirkelspitze *f*
d. with quadrant Bogenspitzzirkel *m*
dividers Spitz[en]zirkel *m*, Stechzirkel *m*, Teilzirkel *m*; Kreisreißer *m*
dividing 1. *(Math)* Teilen *n*, Teilung *f*, Dividieren *n*; 2. *(Masch)* Teilen, Teilung
d. apparatus (attachment) Teilapparat *m*, Teileinrichtung *f*
d. box *(Fmt)* Abzweigkasten *m*; Verzweigungsmuffe *f*, Abzweigmuffe *f*
d. cam Brille *f (Wirkmaschine)*; Verteilschloßteil *n (Strickmaschine)*
d. comb *(Text)* Riedkamm *m*; Ried *n*
d. device for calculating punch Divisionsvorrichtung *f* für Rechenlocher
d. edge Mittelgrat *m (Turbine)*
d. error Teilfehler *m (Teilkopf)*
d. filter Trennfilter *n*
d. head *(Masch, Opt)* Teilkopf *m*
d.-head plate Teilscheibe *f*
d.-head sector Teilkopfzeigerpaar *n*
d. line Trennlinie *f*; Teillinie *f*
d. network *(El)* Frequenzweiche *f*
d. plate Teilscheibe *f*
d. point Teilnadel *f*, Teilspitze *f*
d. roller *(Text)* Teilwalze *f*
d. ruler Teillineal *n*
d. strip Trennstreifen *m (Straßenbau)*
d. thread *(Text)* Fitzfaden *m*
d. wall Trennwand *f*
d. worm Teilschnecke *f*
diving bell Taucherglocke *f*
d.-bell barge Taucherglockenprahm *m*
d. board Sprungbrett *n (Schwimmsport)*
d. boat Taucherboot *n*
d. tank Tauchtank *m (U-Boot)*
d. tower Sprungturm *m (Schwimmsport)*
divining rod Wünschelrute *f*
divinyl Butadien *n*, Divinyl *n*
divisibility Teilbarkeit *f*
divisible [unter]teilbar
division 1. Aufteilung *f*, Verteilung *f*; 2. *(Math)* Division *f*; 3. Skalenintervall *n*; Skalenstrich *m*, Teilstrich *m (Meßinstrumente)*
d. line Skalenstrich *m*, Teilstrich *m (Meßinstrumente)*
d. of labour Arbeitsteilung *f (zwischen Ländern)*
d. of the card web *(Text)* Krempelflorteilung *f*
d. of velocity Geschwindigkeitsverteilung *f (Turbine)*
d. plate *(Masch)* Teilscheibe *f*
d. sign Divisionszeichen *n*
d. testing device *(Masch, Opt)* Teilkopfprüfgerät *n*
d. wall *(Bw)* Trennwand *f*
divisor *(Math)* Teiler *m*
d. chain Teilerkette *f*
dobby *(Text)* Schaftmaschine *f*
d. design Muster *n* für die Schaftmaschine
d. loom Schaft[web]stuhl *m*
d. weaving Weben *n* auf der Schaftmaschine
docimasy Probierkunde *f*
dock / to *(Schiff)* docken
d. out ausdocken
dock Dock *n*, Schiffsdock *n*, Schiffsliegeplatz *m (zwischen zwei Pieren)*
d. caisson Docktor *n*, Dockcaisson *m*
d. fee Dockgebühr *f*
d. floor Docksohle *f*
d. gate Docktor *n*, Schleusentor *n*, Stemmtor *n*
d. hand Hafenarbeiter *m*, Dockarbeiter *m (ungelernt)*
d. pontoon Dockponton *m*, Dockverschlußponton *m*
d. section Docksektion *f*
d. shore Dockstütze *f*
d. sill Dockhaupt *n*, Dockdrempel *m*, Docksüll *m (n)*
d. trial Standprobe *f*, Stehprobe *f*
d. tug Dockschlepper *m*
d. wing-structure Dockseitenkasten *m*
dockage *(Schiff)* Docken *n*; Dockgebühr *f*, Dockgeld *n*
docker Hafenarbeiter *m*, Dockarbeiter *m*, Schauermann *m*
docking *(Schiff)* Eindockung *f*, Einlaufen *n* ins Dock

d. keel Dockkiel *m*
d. television Fernsehen *n* für die Dockfahrt
docks Hafenanlage *f*
dockside travelling crane *(Schiff)* Kai-Laufkran *m*
dockyard Schiffswerft *f*, Werft *f*; Marinewerft *f*
doctor blade 1. *(Typ)* Rakel *f*, Farbabstreicher *m*; 2. *(Text)* Abstreifmesser *n*, Rakel[messer *n*] *f*, Schabermesser *n*
d. blade inking device *(Typ)* Rakelfarbwerk *n*
d. finishing machine *(Text)* Rakelappreturapparat *m*
d. kiss coating *(Led)* Rakelverfahren *n* mit Walzenauftrag *(Beschichtung)*
d. knife Rakel *f*
d. roll Abstreifwalze *f*, Walzenauftragmaschine *f*
d. roller *(Typ)* Rakelwalze *f*
d. ruler Rakelführer *m*, Rakellineal *n*
document box *(Eis)* Begleitkartenkasten *m*
d. film *(Opt)* Dokumentenfilm *m*
documentary film Dokumentarfilm *m*
dodecahedral zwölfflächig
dodecahedron Dodekaeder *n*
dodger *(Schiff)* Brückenkleid *n*, Schutzkleid *n*
doff / to *(Text)* abziehen *(Spinnen)*; ablegen, abnehmen
d. the web *(Text)* Flor abkämmen
doffer *(Text)* Abzieher *m*; Abnehmer *m*; Peigneur *m*; Abnehmerkratze *f*
d. clearer roller Abnehmerputzwalze *f*
d. comb Hacker *m*
d. covering Abnehmerbeschlag *m (Karde)*
d. filleting Kratzenbeschlag *m* der Abnehmerwalze
d. roller Abnehmerwalze *f*
d. system Abnehmersystem *n*
doffing Abzug *m (Spinnerei)*
dog 1. Aufspannfrosch *m*, Drehherz *n*, Mitnehmer *m*; Kamm *m*; Klaue *f*, Klinke *f*, Sperrklinke *f*; Kloben *m*, Knagge *f*; Kropfeisen *n*; Anschlag *m*; Bauklammer *f*; 2. *(Brg)* Hunt *m*, Förderwagen *m*
d.-actuated anschlagbetätigt, nockenbetätigt
d.-and-chain *(Brg)* Raubvorrichtung *f*, Silvester *m*
d. clutch Klauenkupplung *f*
d.-controlled anschlagbetätigt, nockenbetätigt
d. dung *(Led)* Hundekotbeize *f*
d. hole *(Brg)* Fahrweg *m*, Durchhieb *m*
d.-house 1. *(Met)* Hundehütte *f*; 2. *(Schiff)* Hundehaus *n (kleiner Aufbau über der Niedergangsöffnung)*
d. plate Mitnehmerscheibe *f*
d. shore *(Schiff)* Schlagbettungsstrebe *f*, Schlittenständer *m*
d. spike Schienennagel *m (Gleisbau)*
d. tooth Mitnehmerzahn *m*
d. vice Spannkluppe *f (Holzbearbeitung)*
dogs Mitnehmer *m*, Ziehzange *f*, Ziehwagen *m*
doll buggy *(Fs)* Kamerawagen *m*
dollar *(Kern)* Dollar *m (Reaktivitätseinheit)*
dollie *(Schiff)* [kleiner] Poller *m (am Kai)*
dollop auf Deck überkommende See *f*
dolly / to gegenhalten, vorhalten *(Nieten)*
dolly Gegenhalter *m (Nieten)*
d. tub Rührbottich *m*, Rührfaß *n*
dolomite *(Min)* Dolomit *m*
d. brick Dolomitziegel *m*
d. lining *(Met)* Dolomitzustellung *f*
dolphin 1. *(Bw)* Dalbe *f*; 2. *(Schiff)* Dückdalben *m*, Ankerboje *f*
d. pile *(Bw)* Dalbenpfahl *m*
domain 1. Bereich *m*, Bezirk *m*; 2. *(Ph)* Elementarbereich *m*; ferromagnetischer Bereich *m*; Domäne *f*
d. of definition Definitionsbereich *m*
dome *(Bw)* Kuppel *f*
d. lamp Deckenlampe *f*
d.-shaped insulator Glockenisolator *m*
d. steam pipe Dampfsammelrohr *n*
domed gewölbt
d. nut *(Masch)* Hutmutter *f*
d. roof Kuppeldach *n*
domestic appliances Haushaltgeräte *npl*
d. coal Hausbrandkohle *f*
d. engineering Installationstechnik *f*
d. hard soap Haushaltseife *f*
d. heating system Heizungsanlage *f*
d. hide *(Led)* Zahmhaut *f*
d. installation Haus[halts]installation *f*, Wohnungsinstallation *f*
d. sewage Haushaltsabwasser *n*
d. waste Müll *m*, Haushaltsabfälle *mpl*
domical roof Kuppeldach *n*
dominant fault line *(Geol)* vorherrschende Verwerfungslinie *f*
d. mode Grundtyp *m*, Grundmodus *m*, Wellenform *f* mit der untersten Grenzfrequenz *(Wellenleiter)*
d. wave *(El)* Grundwelle *f*
d. wavelength dominierende (effektive) Wellenlänge *f*
doming Wölbung *f*; Einbeulung *f*
domino board biegbare Pappe *f*
donator *s.* donor
donkey boiler Hilfskessel *m*
d. engine *(Schiff)* Hilfsmaschine *f*
d. pump Dampf[speise]pumpe *f*
donkeyman *(Schiff)* Donkeymann *m*, Hilfskesselwärter *m*
Donnan potential Donnan-Potential *n (Elektrobiologie)*
donor Donator *m (Störstelle im Festkörper mit Abgabemöglichkeit für ein Elektron)*

d. impurity Donatorverunreinigung *f*
d. level Donatorniveau *n*
d. migration Donatorenwanderung *f*
don't answer *(Am) (Fmt)* Nichtbeantwortung *f (des Anrufes)*; keine Antwort *f (des angerufenen Teilnehmers)*
don't care terms *(BMSR, Dat)* überflüssige Glieder *npl (in Gleichungen)*
donut *(Ph)* 1. Ringkammer *f*, ringförmige Vakuumkammer *f*; 2. Flußkonverter *m (Neutronenfluß)*
dook 1. Dübel *m*; 2. *(Brg)* Bremsberg *m*
door Tür *f*
d. bell Hausglocke *f*
d. case *s.* d. frame
d. catch *(Kfz)* Türanschlag *m*, Türnase *f*
d. check selbsttätiger Türschließer *m*
d. closing device Türschließanlage *f*
d. contact Türkontakt *m*
d. contact interrupter Türkontaktschalter *m*
d. extracting device Türabhebevorrichtung *f*
d. extractor Türabheber *m*
d. frame Türrahmen *m*, Türzarge *f*
d. framework Türrahmen *m*, Türzarge *f*
d. furniture Türbeschläge *mpl*
d. guide Türführung *f*
d. handle Türdrücker *m*
d. hinge Türscharnier *n*
d. hinge pillar Türscharniersäule *f*
d. jamb Türpfosten *m*
d. knob transformer *(El)* Keulenwandler *m*
d. knob tube *(Fmt)* Knopfröhre *f*, Tellerröhre *f*
d. latch Türverriegelung *f*
d. lifter Türabheber *m*
d. lifting machine Türabhebemaschine *f*
d. lock cylinder Türschließzylinder *m*
d. lock pillar Türschloßsäule *f*
d. operating cylinder Türbetätigungszylinder
d. panel Türverkleidung *f*
d. post Türsäule *f*
d. pull handle Türziehgriff *m*
d. sill Türschwelle *f*
d. skin Türschale *f*
d. strip *s.* d. weather strip
d. swing Türaufschlag *m*
d. switch Kulissenschalter *m*, Tür[kontakt]-schalter *m*
d. weather strip Türdichtungsstreifen *m*
d. winch Türwinde *f*
doormat Fußabtreter *m*
doorway Tordurchfahrt *f*, Torweg *m*
dope / to 1. *(Ph)* dotieren *(d. h. Fremdatome in einen Festkörper einbauen, z. B. zur Veränderung der elektrischen Eigenschaften eines Halbleiters oder zur Verbesserung katalytischer Eigenschaften)*; 2. Additive beigeben *(Öl)*; 3. mit Zusätzen versehen
dope 1. Wirkstoff *m*, Zusatzmittel *n*; Firnis *m*; Flugzeuglack *m*; Klebelack *m*; Filmlösung *f*; 2. *(Kfz)* Kraftstoffzusatz *m*; 3. Additiv *n (Öl)*; 4. Abbindeverzögerer *m (Zement)*
d. dyeing *(Am)* Spinnfärbung *f*
doped gasoline mit Zusätzen versehenes Benzin *n*, verbleites Benzin
d. junction dotierte (gedopte) Schicht *f (Halbleiter)*
doping *(Ch, Ph)* Dopen *n (Dotieren)*; Dotierung *f*, Einbau *m* von Fremdatomen; Beifügung *f* von Zusätzen
Doppler broadening Doppler-Verbreiterung *f (Spektrallinien)*
Doppler shift *(Ph)* Doppler-Verschiebung *f*
Doppler width Doppler-Breite *f (Spektrallinien)*
dormer [window] *(Bw)* Dachgaupe *f*
dormitory town Schlafstadt *f (Siedlungsplanung)*
dorsal fin *(Flg)* Seitenflosse *f*
Dortmund tank Dortmunder Brunnen *m*
dosage Dosis *f*; Dosierung *f*; Dosismessung *f*
d. measurement Dosismessung *f*
d. meter Dosimeter *n*
d. rate Dosisleistung *f*
dose / to dosieren
dose Dosis *f*
d. fractionation Dosisfraktionierung *f*
d.-meter *s.* dosimeter
d. of ore Erzgicht *f*, Erzsatz *m*
d. of radiation Strahlungsdosis *f*
d. reduction factor Dosisreduzierungsfaktor *m*
dosimeter Dosimeter *n*
dosimetry Dosimetrie *f*; Strahlen[schutz]messung *f*
d. probe Strahlenmeßkopf *m*, Strahlenmeßsonde *f*
dosing and mixing system Meß- und Mischsystem *n*
d. tank Dosierbehälter *m*
dot 1. *(Fmt)* Punkt *m*, Morsepunkt *m*; 2. *(Fs)* Punkt *m*, Fleck *m*
d. frequency 1. Telegrafierfrequenz *f*; 2. *(Fs)* Punktfrequenz *f*
d. interlacing *(Fs)* Punktsprungverfahren *n*; Zwischenpunktabtastung *f*
d. product of two vectors Skalarprodukt (skalares Produkt) *n* zweier Vektoren
d. screen *(Typ)* Punktraster *m*
d. sequential system Punktfolgeverfahren *n*, Farbwechsel[verfahren *n*] *m* in Punktfolge *(Farbfernsehen)*
dotted line punktierte Linie *f*
dotting mark *(Masch)* Körnerpunkt *m*, Körnermarke *f*
double / to 1. verdoppeln; 2. *(Met)* doppeln, falten; 3. *(Text)* doppeln, dublieren, fachen *(Spinnerei)*

d. over [doppelte] Falze umlegen
d. up *(Brg)* verstärken *(Ausbau)*
double 1. Dublett[e *f*] *n*; 2. Double *n* *(Film)*; 3. *(Typ)* Hochzeit *f*
double doppelt; Doppel-
d.-acting doppeltwirkend
d.-acting barring gear doppeltwirkendes Schaltwerk *n* *(Schwungrad)*
d.-acting brake doppeltwirkende Bremse *f*
d.-acting cylinder 1. doppeltwirkender Zylinder *m*; 2. wechselseitig beaufschlagter Arbeitszylinder *m*
d.-acting hammer Oberdruckhammer *m*
d.-acting shock absorber doppeltwirkender Stoßdämpfer *m*
d.-acting steam engine doppeltwirkende Dampfmaschine *f*
d. action Doppelwirkung *f*
d.-action drawing press doppeltwirkende Tiefziehpresse *f*
d.-action fuse *(Mil)* Doppelzünder *m*
d.-action harp Pedalharfe *f*
d. admission piston valve Kolbenschieber *m* mit doppelter Einströmung *(Dampfmaschine)*
d. aerial Doppelantenne *f*, Zwillingsantenne *f*, Antennenpaar *n*
d.-amplitude-peak Spitze-zu-Spitze-Amplitude *f*, ss-Amplitude *f*
d. anastigmat *(Opt)* Doppelanastigmat *m*
d. apron Doppelriemchen *n* *(Spinnerei)*
d. arm semaphore (signal) *(Eis)* doppelflügeliges (zweiflügeliges) Signal *n*
d. armature *(Elm)* Doppelanker *m*
d.-armed zweiarmig *(Hebel)*
d.-armed traveller *(Text)* Doppelgabelläufer *m*
d.-aspect signal *(Eis)* Zweibegriffssignal *n*
d. back-gear Zweifachvorgelege *n*
d. back-geared head Spindelkasten *m* mit zweifachem Rädervorgelege
d.-bass große Baßgeige *f*, Kontrabaß *m*
d. bassoon Kontrafagott *n*
d. bath Doppelbad *n*
d.-bath dyeing *(Text)* Zweibadfärbung *f*
d.-bath process *(Led)* Zweibadgerbung *f*
d. battery switch *(El)* Doppelzellenschalter *m*
d.-beam frame *(Wärm)* Doppelbalkenbett *n*
d.-beam radar system Radar *n* mit zwei Richtstrahlen
d. bearing Doppellager *n*
d. bell insulator Doppelglocke[nisolator *m*] *f*
d. bell-shaped insulator Doppelglocke[nisolator *m*] *f*
d. bevel *(Masch)* doppelter Anschnitt *m*
d.-bevel butt weld K-Naht *f* *(Schweißtechnik)*
d.-biased doppelt vorgespannt
d.-bladed mit zwei Meißeln (Messern)
d.-block brake Zweibackenbremse *f*
d.-blow cold header Zweistufenstauchautomat *m*
d. boiler Wasserbad *n*
d. bond *(Ch)* Doppelbindung *f*
d. bottom Doppelboden *m*
d.-bottom tank *(Schiff)* Doppelbodentank *m*
d.-bowed boat Doppelender *m* *(Boot)*
d.-break circuit-breaker Schalter *m* mit doppelter Trennstrecke
d.-break jack *(Fmt)* Doppelunterbrechungsklinke *f*
d.-breasted overcoat zweireihiger Mantel *m*
d. bridge Doppel[meß]brücke *f*, Thomsonbrücke *f*
d. bucket zweiseitige Schaufel *f*, Doppelschaufel *f* *(Turbine)*
d.-bottom cells *(Schiff)* Doppelbodenzellen *fpl*
d. cage rotor *(Elm)* Doppelstabläufer *m*, Doppelkäfigläufer *m*
d. caliper Doppeltaster *m*
d. cap insulator Doppelkappenisolator *m*
d. carbon-arc lamp Doppel[kohlen]bogenlampe *f*, Zwillingsbogenlampe *f*
d. carbon-arc welding Zerener-Verfahren *n*
d. card *(Text)* Doppeltrommelkrempel *f*, Doppelkarde *f*
d.-casing machine *(Elm)* Maschine *f* mit Mantelkühlung
d. catenary construction *s.* d. catenary suspension line
d. catenary suspension *(Fmt)* Doppeltragdrahtaufhängung *f*
d. catenary suspension line *(El)* Dreieck-Kettenfahrleitung *f*, Kettenfahrleitung *f* mit doppeltem Tragseil
d. chain scraper conveyor Doppelkettenkratzförderer *m*
d.-chamfered nut Mutter *f* mit Fase
d.-charged *(Kern)* doppeltgeladen
d. chopper Doppelbeil *n*, Doppelhacke *f*
d. circuit receiver Zweikreisempfänger *m*
d. cleat Doppelklemme *f*
d. clip Doppellasche *f*
d. clip wool Zweischurwolle *f*
d. cloth Doppelgewebe *n*
d.-cloth weave Doppelgewebebindung *f*
d.-cloth weaving loom Doppelwebstuhl *m*
d. cloth with stuffing threads Doppelgewebe *n* mit Ausfüllungsfäden
d.-coated film Zweischicht[en]film *m*
d. cock with half-couplings Doppelhahn *m* mit Kupplung
d. coil relay zweispuliges Relais *n*
d. coil spring doppeltgewundene Indikatorfeder *f*
d. column *(Bw)* Doppelstütze *f*, Doppelsäule *f* *(z. B. an Dehnungsfugen)*

d.-column construction Doppelständerbauart *f*
d.-column jig borer Lehrenbohrwerk *n* in Zweiständerausführung
d.-column mill Zweiständerdrehwerk *n*
d.-column planer Doppelständerhobelmaschine *f*
d.-column plano milling machine Zweiständerbettfräsmaschine *f*, Portalfräsmaschine *f*
d.-column vertical turning and boring machine Zweiständerkarusselldrehmaschine *f*
d. commutator motor Doppelkommutatormotor *m*, Doppelkollektormotor *m*
d.-compounded engine Doppelverbundmaschine *f*
d. concave lens bikonkave Linse *f*, Bikonkavlinse *f*
d.-conductor cable zweiadriges Kabel *n*
d.-conductor cord Zweileiterschnur *f*, Doppelschnur *f*, zweiadrige Schnur *f*
d.-cone aerial Doppelkonusantenne *f*
d.-cone blender Doppelkonusmischer *m*, Trommelmischer *m*
d.-cone clamping coupling Doppelkonusklemmkupplung *f*, Sellerskupplung *f*, Hülsenkupplung *f*
d.-cone impeller Doppelkonuskreiselmischer *m*
d.-cone pulley zweistufige Riemenscheibe *f*
d.-cone spring Doppelkegelfeder *f*
d. connector Doppelklemme *f*
d. contact Doppelkontakt *m*
d.-contact wire system doppelter Fahrdraht *m*, Zwillingsfahrleitung *f*
d. control Doppelsteuerung *f*
d.-convex lens bikonvexe Linse *f*, Bikonvexlinse *f*
d.-cotton covered *(El)* mit doppelter Baumwollumspinnung, doppelt baumwollumsponnen *(Draht)*
d. cover *(Wärm)* doppelwandiger Boden *m*
d. coverage Doppeldeckung *f (Dach)*
d.-covered butt joint *(Bw)* zweischnittiger Laschenstoß *m*
d.-crank press Doppelkurbelpresse *f*; Kurbelpresse *f* mit zweifach gekröpfter Welle
d.-crank toggle drawing press Kniehebelbreitziehpresse *f*
d.-cross track *(Eis)* Kreuzgleis *n (Drehscheibe)*
d. crossed-loop aerial Doppelkreuzrahmenantenne *f*
d.-cup insulator Doppelglocke[nisolator *m*] *f*
d. current *(Fmt)* Doppelstrom *m*
d. current cable code Doppelstromtelegrafie *f*
d.-current signalling Doppelstromsystem *n (Telegrafie)*
d.-current working Doppelstrombetrieb *m*
d. cusp *(Math)* Selbstberührungspunkt *m*
d.-cut *(Masch)* zweiseitig schneidend
d. cut Doppelhieb *m*, Kreuzhieb *m (Feilen)*
d.-cut file Doppelhiebfeile *f*, Kreuzhiebfeile *f*
d.-cutting *(Masch)* für Vor- und Rückwärtsschnitt
d.-cylinder dobby loom *(Text)* Zweiprismenschaftmaschine *f*
d.-cylinder jacquard Zweiprismenjacquardmaschine *f*
d.-cylinder knitting machine Doppelzylinderstrickmaschine *f*
d. cylindrical boiler Doppelwalzenkessel *m*, Doppelzylinderkessel *m*, Kessel *m* mit Unterkessel
d. dash *(Fmt)* Doppelstrich *m (Trennungszeichen)*
d.-deck bridge *(Bw)* Doppelstockbrücke *f*, Etagenbrücke *f*
d.-deck bunk *(Schiff)* Doppelkoje *f (zwei Kojen übereinander)*
d.-deck car *(Eis)* Doppelstockwagen *m*, Zweistockwagen *m*
d.-deck motor Asynchronmotor *m* mit Anlaß- und Betriebswicklung
d.-deck screen *(Brg)* Doppeldeckersieb *n*
d.-deck tool holder Aufbauwerkzeughalter *m*
d. decker *(Eis)* Doppelstockwagen *m*, Zweistockwagen *m*
d. deflection Doppelausschlag *m*
d. deflection of the steam jet doppelte (zweiseitige) Strahlablenkung *f (Turbine)*
d.-delta connection *(Elm)* Doppeldreieckschaltung *f*
d. detection receiver Superhetempfänger *m*
d. detector reception *(Fmt)* Überlagerungsempfang *m*
d. diaphragm draught gauge *(Wärm)* Kapselfederzugmesser *m*
d. diode Doppeldiode *f*, Duodiode *f*
d.-diode output pentode Doppeldiode-Endpentode *f*
d.-dipping *(Kfz)* Abblenden *n* beider Scheinwerfer
d.-dipping headlamp Abblendlampe *f*, Biluxlampe *f*
d. disk attrition mill Doppelscheibenschleifmühle *f (Holzzerkleinerung)*
d. distributor *(Typ)* Doppelableger *m (Setzmaschine)*
d. doffer card *(Text)* Doppelabnehmerkrempel *f*
d. dog *(Masch)* Doppelsperrklinke *f*
d. drop frame *(Kfz)* doppelt gekröpfter Rahmen *m*
d.-drum cutter loader *(Brg)* Doppelwalzenschrämlader *m (für Abbau und Laden bei Berg- und Talfahrt)*
d.-edged zweischneidig

d.-end clamp Aufspannklaue *f*
d.-end control *(El)* doppelseitiger Führerstand *m*
d.-end drill Bohrer *m* mit beidseitig abgesetztem Schneidenteil
d.-end end mill doppelseitiger Schaftfräser *m*
d.-ended doppelseitig; doppelmäulig; zweischneidig
d.-ended boat Spitzgattboot *n*
d.-ended boiler Doppelenderkessel *m*
d.-ended boiler with common combustion chamber Doppelenderkessel *m* mit gemeinsamer Flammkammer
d.-ended boiler with separate combustion chambers Doppelenderkessel *m* mit getrennten Flammkammern
d.-ended boring-bar cutter doppelseitig arbeitendes Bohrmesser *n*
d.-ended cord Schnur *f* (Verbindungskabel *n*) mit zwei Steckern, Schnur (Verbindungskabel) mit zwei Stöpseln
d.-ended cutter *(Masch)* zweischneidiges Bohrstangenmesser *n*, beiderseitig schneidendes Bohrmesser *n*; zweiseitiger Fräser *m*
d.-ended firm-joint outside caliper *(Masch)* Doppeltaster *m*
d.-ended plate doppelt herausgeführte Anodenleitung *f*
d.-ended saw file Doppelsägenfeile *f*
d.-ended Scotch marine boiler Doppelenderschiffskessel *m*
d.-ended seaming machine Boden- und Dekkelauffalzmaschine *f*
d.-ended spanner Doppel[maul]schraubenschlüssel *m*
d.-ended tube lathe doppelte Rohrdrehmaschine *f*
d.-ended vessel Schiff *n* mit Propellern vorn und hinten *(z. B. Fähre)*
d. ender 1. *(Schiff)* Spitzgattboot *n*, Doppelender *m*; 2. Doppelender *(Kessel)*
d. english *(Typ)* 28 britische Punkte *mpl* *(Schriftgröße)*
d. entry *(Brg)* Strecke *f* mit Begleitort
d. expansion compound steam engine Verbunddampfmaschine *f* mit doppelter Expansion, Zweifachexpansionsdampfmaschine *f*
d. exposure Doppelbelichtung *f*
d.-exposure prevention Doppelbelichtungssperre *f*
d. face hammer doppelbahniger Hammer *m*
d.-faced doppelseitig, zweiseitig
d.-faced corrugated board doppelgedeckte Wellpappe *f*
d. feather-edged board an den Seiten dünn auslaufende Pappe *f*
d.-filament bulb Doppelfadenlampe *f*, Zweifadenlampe *f*
d. fillet weld Doppelkehlnaht *f*
d. fired furnace Zweizonenofen *m*
d.-flange hub *(Kfz)* Zweiflanschnabe *f*
d.-flanged doppelflanschig
d.-flanged bobbin *(Text)* Scheibenhülse *f*; Flanschenhülse *f*
d.-flanged butt joint Bördelstoß *m*
d. flexible Doppelschnur *f*, Zweileiterschnur *f*, zweiad[e]rige Schnur *f*
d. flow turbine Hoch- und Niederdruckturbine *f* mit entgegengerichteter Dampfströmung
d. folder *(Typ)* Zweibruchfalzapparat *m*, Kreuzbruchfalzer *m*
d. frame *(Schiff)* Doppelspant *n*
d. frequency changing doppelte Frequenzüberlagerung *f*
d. full-fillet lap weld zweiseitiger Überlappstoß *m* mit Flachkehlnaht *(Schweißen)*
d. furnace *(Met)* Doppelpuddelofen *m*
d.-furrow plough Zweischarpflug *m*
d. gang tuning condenser Zweifachabstimmkondensator *m*
d.-geared *(Masch)* mit Vorgelege
d. girder Zwillingsträger *m*
d.-glazed casement *(Bw)* Verbundfenster *n*
d. glazing doppelte Verglasung *f*
d. guide *(Masch)* doppelseitige Führung *f*
d.-head flat border machine zweiköpfige Leistenflachstrickmaschine *f*
d.-head needle *(Text)* Links-Links-Nadel *f*
d.-head porcupine drawing *(Text)* Doppelnadelwalzenstrecke *f*
d.-head rail Doppelkopfschiene *f*
d.-head socket wrench doppelseitiger Steckschlüssel *m*
d.-head wrench Doppelschraubenschlüssel *m*
d.-headed rail Doppelkopfschiene *f*
d.-header boiler with circulating device Zweikammerkessel *m* mit Umlaufvorrichtung
d.-header boiler with cylindrical headers Zweikammerkessel *m* mit zylindrischen Wasserkammern
d.-helical gear Stirnrad *n* mit Pfeilverzahnung
d.-helical gearing Pfeilverzahnung *f*, Winkelverzahnung *f*
d.-helical teeth Doppelschrägzähne *mpl*; Pfeilverzahnung *f*
d.-helical wheel Pfeilrad *n*
d. heterodyne receiver Doppelüberlagerungsempfänger *m*
d.-hinged arm *(Kern)* Doppelgelenkarm *m* *(Manipulator)*
d.-hook hoist block Zwillingsflaschenzug *m*
d.-hook spanner Doppelhakenschlüssel *m*
d. horizontal bars Doppelspannreck *n*
d. horizontal grate *(Wärm)* Doppelplanrost *m*

d.-housing *(Masch)* Doppelständer-, Zweiständer-
d.-hulled boat Zweihüllenboot *n*
d. humped doppelhöckerig, zweihöckerig *(z. B. Resonanzkurven)*
d. ignition *(Kfz)* Doppelzündung *f (Magnet und Batterie)*
d. image *(Fs)* Echo[bild] *n*, Geist[erbild *n*] *m*
d.-image ocular Doppelbildokular *n*
d. in-and-out stitch *(Text)* Doppelkettenstich *m*
d. insulator spindle Isolatordoppelstütze *f*
d. integral Doppelintegral *n*
d.-interruption jack *(Fmt)* Doppelunterbrechungsklinke *f*
d.-J butt weld Doppel-J-Naht *f (Schweißtechnik)*
d.-jack flat knitting machine Flachstrickmaschine *f* mit zwei Platinenreihen und einem Nadelbett
d.-jack frame Extradoppelfeinbank *f*, Extrafeinflyer *m*
d.-jack links-links flat knitting machine Links-Links-Flachstrickmaschine *f* mit zwei Platinenreihen im hinteren Nadelbett
d.-jack plane Doppelschrupphobel *m*
d.-jaw chuck Zweibackenfutter *n*
d.-jet carburetter Doppeldüsenvergaser *m*
d.-lap machine Zweischeibenläppmaschine *f*
d. layer *(Ph, El)* Doppelschicht *f*
d. layer winding *(Elm)* zweilagige Wicklung *f*, Zweischichtwicklung *f*
d. lean-to roof Schmetterlingsdach *n*, V-förmig nach innen geneigtes Dach *n*
d. letter Ligatur *f*
d.-lever contact breaker zweipoliger Ausschalter *m*
d.-lift dobby *(Text)* Schaftmaschine *f* für Hoch- und Tieffach
d.-lift jacquard [machine] Doppelhubjacquardmaschine *f*
d. limit *(Math)* Doppellimes *m*
d. limiter *(Fs)* Auftastimpulskreis *m*, Ausblendstufe *f*
d. linkage *s.* d. bond
d. linkage leakage *(El)* doppelt verkettete Streuung *f*
d. liquor circulation *(Text)* zweiseitige Flottenzirkulation *f*
d. liquor flow *(Text)* doppelter Flottenweg *m*
d.-lock carriage zweisystemiger Schlitten *m (Flachstrickmaschine)*
d.-lock flat knitting machine zweisystemige Flachstrickmaschine *f*
d. log frame *(Holz)* Doppelbundgatter *n*
d.-make contact Doppel[arbeits]kontakt *m*
d.-make relay Doppelkontaktrelais *n*, Relais *n* mit zwei Schließkontakten
d. mandrel method [of alignment testing] Zweidornverfahren *n* [der Fluchtungsprüfung]
d. modulation Doppelmodulation *f*
d. modulator Modulator *m* mit Trägerunterdrückung
d. motion mixer gegenläufiges Rührwerk *n*
d. needle machine Doppelnadelmaschine *f (zur Herstellung von Nadelfilzen)*
d. note amplifier zweistufiger Tonfrequenzverstärker *m*
d. open-end wrench Doppelschraubenschlüssel *m*
d. output pentode Doppelpentoden-Endröhre *f*
d. packing *(Brg)* Mitnahme *f* von Doppeldämmen auf beiden Streckenseiten
d. paper Duplopapier *n*
d. parallel crank Doppelparallelkurbel *f*
d. parallel crank mechanism Doppelparallelkurbelmechanismus *m*
d. partition Hohlmauertrennwand *f*, zweischalige Trennwand *f*
d. passage of light doppelter Strahlendurchgang *m*
d. pattern attachment *(Text)* Doppelmustereinrichtung *f*
d.-peak resonance curve zweispitzige Resonanzkurve *f*, Resonanzkurve mit zwei Maxima (Spitzen)
d. pendulum Doppelpendel *n*
d. petticoat insulator Doppelglockenisolator *m*, Zweimantelisolator *m*
d. phantom circuit Achterschaltung *f*, Superphantomschaltung *f*
d. picker Zunahmefinger *m (Rundstrickmaschine)*
d. picks *(Text)* Doppelschuß *m (Fehler)*
d. pile *(Text)* Doppelflor *m*
d.-pin *(Masch)* Doppelzapfen *m*
d.-pipe stock Duplex-Schneidkluppe *f*
d.-piston engine Doppelkolbenmotor *m*
d.-pitch system Verbundteilungssystem *n (Verzahnung)*
d.-plate clutch Zweischeibenkupplung *f*
d. ploughing *(Lw)* Zwiebrache *f*
d. plug Doppelstecker *m*
d. plug and jack Zweifachstecker *m*
d.-plugged cord Schnur *f* (Verbindungskabel *n*) mit zwei Steckern
d.-ply belt Doppelriemen *m*
d. point *(Math)* Doppelpunkt *m (Kurve)*
d. pole *(El)* zweipolig, doppelpolig
d. pole Doppelmast *m (für elektrische Leitungen)*
d. pole change-over switch zweipoliger Umschalter *m*
d. pole double-throw switch zweipoliger Umschalter *m*

d. pole keying Doppeltastung *f*
d.-ported control valve Doppelsitzregelventil *n*
d. prismatic square Doppelwinkelprisma *n*
d. punch and blank column detection device *(BMSR)* Doppel- und Fehllochaussuchvorrichtung *f*
d. ram Doppelstößel *m*
d.-range instrument Meßgerät *n* für zwei Meßbereiche
d.-rate meter Doppeltarifzähler *m*, Zweitarifzähler *m*
d.-reduction gear Doppeluntersetzungsgetriebe *n*
d. reed suspension doppelte Blattfederaufhängung *f*
d.-refracting *(Opt)* doppelbrechend
d. refraction *(Opt)* Doppelbrechung *f*
d. reinforced *s.* doubly reinforced
d. resistance box *(El)* gleichläufiger Doppelkurbelwiderstand *m*
d. reverberatory furnace Doppelflammofen *m*
d. rib warp goods *(Text)* Fangkettenware *f*
d.-ribbed inner ring roller bearing innenbordgeführtes Rollenlager *n*
d.-riveted doppelreihig genietet
d.-riveted joint zweireihige Nietverbindung *f*
d. riveting Zweireihennietung *f*, Doppelreihennietung *f*
d.-roll cam Doppelwälzkurve *f*
d.-roll crusher Zweiwalzenbrecher *m*
d.-roll cutting Verzahnen *n* mit doppelter Wälzung
d.-roller chain Zweifachrollenkette *f*
d. root *(Math)* Doppelwurzel *f*
d.-row doppelreihig, zweireihig *(Lager)*
d.-row ball bearing zweireihiges Kugellager *n*
d.-row lighting *(Flg)* Doppelreihenbefeuerung *f*
d. safety-locking filling doppelt gesicherte Bemesserung *f (Holländer)*
d. safety valve Doppelsicherheitsventil *n*, doppeltes Sicherheitsventil *n*
d. safety valve with alarm Doppelsicherheitsventil *n* mit Signalvorrichtung
d. safety valve with high lift Hochhubdoppelsicherheitsventil *n*
d. salt Doppelsalz *n*
d. sampling doppelte Stichprobenprüfung *f (Qualitätskontrolle, Statistik)*
d. saw-tooth system *(Brg)* Doppelsägeblattanordnung *f (Ausbau)*
d. scale Doppelskala *f (Meßinstrumente)*
d. scanning Doppelabtastung *f (Film)*
d. scull Doppelzweier *m (Sportboot)*
d. scutcher *(Text)* doppelte Schlagmaschine *f*
d.-seated control valve Doppelsitzregelventil *n*
d. series expansion *(Math)* Doppelreihenentwicklung *f*, Entwicklung *f* in eine Doppelreihe
d. set of arms Doppelarmkreuz *n (Schwungrad)*
d. shear *(Bw)* zweischnittig
d. shed insulator Doppelglockenisolator *m*
d. shell insulator Zweimantelisolator *m*
d.-shielded antifriction bearing Wälzlager *n* mit zwei Deckscheiben
d. shoe brake Zweibackenbremse *f*, Doppelbackenbremse *f*
d. sideband Zweiseitenband *n (Rundfunk)*
d. sideband transmission Zweiseitenbandübertragung *f*
d.-sided eccentric press Zweiständerexzenterpresse *f*
d.-sided four-crank press Vierpunkt-Doppelständerkurbelpresse *f*
d.-sided plush doppelfloriger Plüsch *m*
d.-silk-covered wire doppeltseidenisolierter Draht *m*
d.-sleeve valve engine Doppelschiebermotor *m*
d. slide broaching machine Doppelstößelräummaschine *f*, Zwillingsräummaschine *f*
d. slit Doppelspalt *m*
d. sliver *(Text)* Doppelband *n*
d. spindle shaper Doppelfräsmaschine *f (Holzbearbeitung)*
d. spinning wheel Doppelspinnrad *n*, zweispuliges Spinnrad *n*
d. spiral coil Spule *f* mit bifilarer Wicklung; bifilare Wicklung *f*
d. spiral gear Pfeilrad *n*
d.-spot tuning Zweipunktabstimmung *f (Funkempfangstechnik)*
d. spreading *(Plast)* beiderseitiges Auftragen *n*
d. squirrel cage Doppelkäfig[läufer] *m*
d. squirrel-cage induction motor Doppelkäfiginduktionsmotor *m*
d. squirrel-cage rotor Doppelkäfiganker *m*, Doppelkäfigläufer *m*, Doppelnutläufer *m*
d. squirrel-cage winding Doppelkäfigwicklung *f*
d.-stage air-pump zweistufige Luftpumpe *f*
d. standard *(Masch)* Doppelständer *m*
d. star connection *(El)* Doppelsternschaltung *f*
d.-start thread doppelgängiges Gewinde *n*
d. stave cross cut saw *(Holz)* Daubenkürzsäge *f*
d.-stitched satin *(Text)* doppelbindiger Atlas *m*
d. straining *(El)* Doppelabspannung *f*

d.-strand braid *(Text)* zweiflechtige Tresse *f*
d.-strand chain Doppelkette *f*, Zweifachkette *f*
d.-strap [riveted] butt joint doppelseitige Laschennietung *f*
d.-stream amplifier Doppelstrahlröhre *f* *(Wanderwellenverstärker)*
d.-strength paste *(Text)* Doppelteig *m*
d.-stretch articles *(Text)* Zweizugware *f*
d. stroke Doppelhub *m*
d. suction pump Pumpe *f* mit doppelseitigem Einlauf
d. superheterodyne reception Doppelüberlagerungsempfang *m*
d.-swing mechanism Doppelschwinggetriebe *n*, Doppelschwinge *f*
d.-system flat knitting machine zweisystemige Flachstrickmaschine *f*
d.-system power-driven flat jacquard knitting machine zweisystemige Motorjacquardflachstrickmaschine *f*
d.-system V-type knitting machine zweisystemige Rechts-Rechts-Flachstrickmaschine *f*
d.-T-beam Doppel-T-Träger *m*, I-Träger *m*
d. tabbing of mill saw Doppelleiste *f* für Gattersäge
d. tackle block Flaschenzug *m* mit zwei Rollen
d.-tandem engine Doppelreihenmotor *m*
d.-tariff meter Doppeltarifzähler *m*
d. tee joint Kreuzstoß *m* *(Schweißen)*
d. tenon head Doppelzapfenschneidkopf *m*
d. terminal Doppelklemme *f*
d. thread zweigängiges (doppelgängiges) Gewinde *n*
d.-threaded doppelgängig
d.-threaded screw Doppelgewindeschraube *f*
d.-throat Venturi tube *(Kfz)* Doppelsaugrohr
d.-throw contact *s.* two-way contact
d.-throw disconnecting switch *(Fmt)* Trennumschalter *m*
d.-throw pump Zweikurbelpumpe *f*
d.-throw switch Umschalter *m*, Wechselschalter *m*, Zweiweg[um]schalter *m*
d. thrust ball bearing zweiseitig wirkendes Axialrillenkugellager *n*
d.-thrust bearing Wechsellager *n*
d. tie *(Text)* doppelte Bindung *f*
d. tonguing and grooving machine *(Holz)* doppelte Spundmaschine *f*
d. track bed *(Masch)* Zweibahnenbett *n*
d.-tracked zweigleisig
d. triode Doppeltriode *f*, Duotriode *f*
d. truck wagon *(Eis)* Doppeltruckwagen *m*
d.-tube core-barrel Doppelkernrohr *n* *(Erdöl, Erdgas)*
d. tube injector Doppelinjektor *m*, Universalinjektor *m* mit zwei Düsensätzen
d.-tuned *(Fmt, Fs)* mit zwei Einzelkreisen abgestimmt, zweifach abgestimmt
d.-tuned amplifier Zweikreisverstärker *m*
d.-tuned circuit type discriminator Frequenzdemodulator *m* mit abgestimmten Einzelkreisen
d. turbine with the blade wheels running in opposite directions gegenläufige Doppelturbine *f*
d. turning zweiseitige Bogenführung *f* *(einzelnes Umblättern bei der Papiersortierung)*
d. twister Ringzwirnmaschine *f* für Seiden- oder Rayongarne
d. two-housing mill Doppelduowalzwerk *n*
d. type Ligatur *f*
d. U butt weld Doppel-U-Naht *f* *(Schweißen)*
d.-unit panel *(Brg)* zweiflügeliger Streb *m*
d. V butt weld X-Naht *f* *(Schweißen)*
d.-V engine *(Kfz)* Fächermotor *m*
d.-V joint X-Stoß *m*, Doppel-V-Stoß *m* *(Schweißen)*
d. velvet Doppelsamt *m*
d. velvet loom Doppelsamtwebstuhl *m*
d. vibration volle Periode *f*, Vollwelle *f*, zwei Halbwellen *fpl*
d.-voltage generator Generator *m* für zwei Spannungen
d.-walled doppelwandig, hohlwandig
d.-warp linings Futterstoffe *mpl* aus Zweifachkettgarn
d. warp loom Doppelketten[web]stuhl *m*
d.-way connection Zweiwegschaltung *f*
d. way launching Stapellauf *m* auf zwei Schlitten, Zweibahnablauf *m*
d. weaves stitched to form the patterns musterartig verbundenes oder durchstepptes Doppelgewebe *n*
d.-weft jacquard velvet zweispuliger Jacquardsamt *m*
d.-welded butt joint Stumpfnaht *f* mit Gegenschweißung
d.-wheel knurl holder Kordierhalter *m*
d.-wheel knurling Kordeln *n*
d.-wheel knurling tool Kordelwerkzeug *n*, Kordierwerkzeug *n*
d. white patent-coated board gegautschter Triplexkarton *m*
d. width chain Zweifachkette *f*, Zwillingskette *f*
d.-winding armature *(Elm)* Stufenanker *m*
d.-wing door zweiflügelige Tür *f*
d.-wing plough Gruppenpflug *m* *(Grabenräumung)*
d.-wire system Zweileitersystem *n*, Zweidrahtsystem *n*
d. worsted card *(Text)* Kammgarndoppelkrempel *f*
d.-wound *(El)* bifilar gewickelt

d.-wound coil Bifilarspule *f*; Spule *f* mit doppelter Drahtwicklung

d.-wound relay Relais *n* mit zwei Wicklungen

d.-woven pile fabric Doppelsamt *m*

doubled on one another *(Text)* umeinandergezwirnt

d. warp *(Text)* Zwirnkette *f*

doubler 1. *(El)* Frequenzverdoppler *m*; 2. *(Text)* Dubliermaschine *f*

doublet 1. *(Opt, Kern)* Dublett *n*, Doppellinie *f (Spektrum)*; 2. *(El)* [Elementar-]Dipol *m*; 3. *(Opt)* zweiteilige Linse *f*; zweilinsiges Objektiv *n*

d. aerial Dipolantenne *f*

d. objective zweilinsiges Objektiv *n*

d. separation *(Opt, Kern)* Dublettaufspaltung *f*

doubling 1. Verdopplung *f*; 2. *(Met)* Doppelung *f*; 3. *(Text)* Dublierung *f*

d. machine *(Text)* Fachmaschine *f*

d. test *(Met)* Faltversuch *m*

d. time *(Kern)* Verdopplungszeit *f*

d. winder *(Text)* Dubliermaschine *f*

d. winding frame *(Text)* Dubliermaschine *f*

doubly bonded *(Ch)* durch Doppelbindung gebunden

d. fed motor Repulsionsmotor *m*

d. fed polyphase shunt commutator motor ständergespeister (statorgespeister) Mehrphasen-Nebenschlußmotor *m*

d. reinforced doppelt bewehrt *(Stahlbeton)*

doubtful operation zweifelhaftes Arbeiten *n (z. B. Relais)*

dough Teig *m*; teigartige Masse *f*

d. moulding Teigpreßverfahren *n*

doughnut *(Kern)* Flußumwandler *m*, Flußkonverter *m*, Ringkammer *f*, Ringröhre *f*

Douglas fir Douglastanne *f*

doup weave *s.* doupe weave

doupe weave *(Text)* Dreherbindung *f*

douplex brake Duplexbremse *f*

d. wheel cylinder Duplexradzylinder *m*

douse / to 1. in Flüssigkeit werfen, abschrecken; 2. einholen; herablassen; schließen; abschalten

dove cot Taubenhaus *n*, Taubenschlag *m*

dovetail / to 1. Schwalbenschwanznuten einarbeiten *(Metall)*; 2. *(Holz)* [ein]schwalben, [ver]zinken

dovetail 1. Zinkung *f*; Schwalbenschwanzzinkung *f (Holzverbindung)*; 2. Schwalbenschwanz[führung *f*] *m (Metall)*

d. cutter Schwalbenschwanznutenfräser *m*; Zinkenfräser *m*

d. groove Schwalbenschwanznut *f*

d. guide Schwalbenschwanzführung *f*

d. halving Schwalbenschwanzblatt *n*

d. hole Schwalbenschwanzeinschnitt *m*

d. joint Schwalbenschwanzverbindung *f*, Schwalbenschwanzzinkung *f*

d. key Schwalbenschwanzkeil *m*

d. keyway schwalbenschwanzförmig ausgebildete Keilnut *f*

d. plane Federhobel *m*; Grathobel *m*

d. plate *(Schiff)* Stevenbeschlag *m*, Stevenschwalbe *f*

d. saw Zinkensäge *f*

d. slideway Schwalbenschwanzführung *f*

d. tenon Kammzapfen *m*

d.-way schwalbenschwanzförmige Keilnut *f*; Schwalbenschwanzführung *f*

dovetailed schwalbenschwanzförmig; gezinkt, mit Schwalbenschwanzverzapfung

d. beam verdübelter Balken *m* mit Schwalbenschwanz

d. board gezinktes Brett *n*

d. disk rim hinterdrehter Scheibenrand *m (Turbine)*

d. overlapping blade schwalbenschwanzförmig übergreifende Schaufel *f (Turbine)*

d. root of blade schwalbenschwanzförmiger Schaufelfuß *m (Turbine)*

d. to *(Masch)* in Schwalbenschwanzführung an

dovetailing 1. *(Holz)* Verzinkung *f*, Zinkung *f*, Zinkenverbindung *f*; 2. Schwalbenschwanzverbindung *f*

d. cutter Winkelstirnfräser *m*; Schwalbenschwanznutenfräser *m*, Zinkenfräser *m (Holzbearbeitung)*

d. machine Zinkenschneidmaschine *f*, Zinkenfräse *f*

d. machine with circular saw Zinkensäge *f*

dowel / to 1. mit Zapfen versehen; 2. mit Paßstift verbinden, verstiften; 3. dübeln

dowel 1. Dübel *m*; 2. Paßstift *m*; 3. Führungsschraube *f*, Pressonstift *m*

d. bit Dübelbohrer *m*

d. iron Dübeleisen *n*

d. pin Dübel *m*, Federbolzen *m*, Herzbolzen *m*; Paßstift *m*, Spannstift *m*, Stehbolzen *m*, Zapfen *m*, Zylinderzapfen *m*

dowelled gedübelt, verstiftet

d. beam verdübelter Balken *m*

dowelling Verdübelung *f*, Verstiftung *f*

dowk unreiner Schwerspat *m*

dowlas *(Text)* Lederleinen *n*

down Daune *f*

down by the head *(Schiff)* kopflastig, vorlastig

d. by the stern *(Schiff)* achterlastig, steuerlastig

d.-cut milling Gleichlauffräsen *n*

d.-draft sintering Saugzugsinterung *f*

d.-draught carburetter *(Kfz)* Fallstromvergaser *m*

d.-draught furnace Feuerung *f* mit umgekehrter Flamme
d.-dredge Tiefschnittbagger *m*
d.-hand welding Abwärtsschweißen *n*
d.-hill brake Talfahrtbremse *f*
d. hole abwärts gerichtetes Bohrloch *n*
d. lead Antennenableitung *f*
d.-lead insulator [Antennen-]Durchführungsisolator *m*, Einführungsisolator *m*
d. milling Gleichlauffräsen *n*
d. pipe Fallrohr *n*
d.-spout Fallrohr *n*
d. storage Untergrundspeicherung *f*
d.-stream talwärts, stromabwärts *(Binnenschifferei)*
d.-stroke Abwärtshub *m*
d.-the-hole drill *(Brg)* Tauchbohrhammer *m*
d. timber Windfallholz *n*
d. time Nebenzeit *f*; Maschinenausfallzeit *f*, Stillstandzeit *f*, Wartezeit *f*; Abschaltzeit *f*; *(Dat)* Ausfallzeit *f*, Totzeit *f*
d. time required to change tools, d. time required for tool change für Werkzeugwechsel erforderliche Zeit *f*
downcast shaft *(Brg)* Einziehschacht *m*, einziehender Wetterschacht *m*
downdraft carburetter *(Kfz)* Fallstromvergaser *m*
downfeed Tiefenzustellung *f*
d. screw *(Masch)* Tiefenzustellspindel *f*
d. slide Meißelschlitten *m*, Senkrechtvorschubschlitten *m*, Vertikalschieber *m*, Werkzeughalter *m (Waagerechtstoßmaschine)*
downgate Trichter *m (Gußstück)*
downpipe *s.* down pipe
downproof finish *(Text)* daunendichte Appretur *f*
downstream flußabwärts, stromab, talwärts; in Richtung der Strömung
d. pressure *(BMSR)* Druck *m* hinter dem Ventil
downtake [type of] superheater hinter dem Kessel eingebauter Überhitzer *m*
downthrown fault *(Geol)* Sprung *m* ins Liegende
downward modulation *(Fs)* Negativmodulation *f*, subtraktive Modulation *f*
d. sloping abfallend
d. stroke Niedergang *m*
d. welding Schweißung *f* mit Abwärtsführung
downwind landing *(Flg)* Rückenwindlandung *f*
dowser 1. Wünschelrutengänger *m*; 2. Feuerschutzschirm *m (Kino)*
dowsing Wünschelrutengängerei *f*
d. rod Wünschelrute *f*
doze Fäulnis *f (Holz)*
dozer Planierraupe *f*, Planierschild *m*, Planiergerät *n*, Frontraümer *m*, Räumschild *m*
draft / to 1. zeichnen, entwerfen; 2. *(Text)* patronieren
draft 1. Entwurfszeichnung *f*, Riß *m*, Skizze *f*; 2. Zug *m*, Luftzug *m*; 3. Zug *(z. B. Drahtziehmaschine)*; 4. Konizität *f*, Verjüngung *f* *(Gußmodell)*; 5. *(Text)* Patrone *f*, Schnürmuster *n*, Verzug; Strecke *f*; 6. *(Am) s.* draught
d. agreement Vertragsentwurf *m*
d. angle Aushebewinkel *m (Gußteil)*
d. gear Zugvorrichtung *f*
d. roll Zugwalze *f*
d. specification Vorstandard *m*, Normblattentwurf *m*, Vornorm *f*
d. standard Vornorm *f*, Vorstandard *m*
d. standard specification Normenentwurf *m*, Vorstandard *m*
drafting Strecken *n*, Verziehen *n (Spinnerei)*
d. area Verzugszone *f*, Verzugsbereich *m* *(Spinnerei)*
d. board Reißbrett *n*, Zeichenbrett *n*
d. instruments Reißzeug *n*
d. machine Zeichenmaschine *f*
d. office Zeichenbüro *n*
d. paper Patronenpapier *n*
d. room Zeichensaal *m*
d. symbol for surface roughness Oberflächenzeichen *n* [zur Angabe der geforderten Oberflächengüte]
d. table Zeichentisch *m*
draftsman [technischer] Zeichner *m*
drag 1. *(Flg, Rak)* aerodynamischer Widerstand *m*; Luftwiderstand *m*; 2. *(Schiff)* konstruktive Steuerlastigkeit *f*; Kielneigung *f* *(bei nicht horizontalem Kiel)*; 3. *(Met)* untere Formhälfte *f* *(Formerei)*; 4. *(Lw)* [schwere] Egge *f*; Schleppe *f*; 5. *(Bw)* Planierschleppe *f*; 6. Schürfkübel *m*; 7. *(Masch)* Zwischenschneide *f*
d. acceleration Widerstandsbeschleunigung *f*
d. aerial Schleppantenne *f*
d. anchor *(Schiff)* Dragganker *m*, Dregganker *m*, Schleppanker *m*; Suchanker *m*; Bremsanker *m (Stapellauf)*
d. bar Kupplungsstange *f*; Zugstange *f*
d. beam *(Lw)* Anhängebalken *m*; Anhängeschiene *f*; Schleppbalken *m*, Eggenbalken *m*
d. bit Blattmeißel *m*
d. bolt Kuppelbolzen *m*
d. bucket Schöpfbecher *m*
d. chain Hemmkette *f*, Schleppkette *f*, Schurfkette *f*, Zugkette *f*
d. cup *(BMSR)* Glocke *f (Ferrarismaschinenläufer)*
d.-cup motor Ferrarismotor *m*
d.-cup rotor Glockenläufer *m (Ferrarismaschine)*
d.-cup tachometer Ferraristachodynamo *m*, Ferraristachometermaschine *f*

d. cut *(Brg)* Löseneinbruch *m*
d. disk flowmeter *(BMSR)* Klappendurchflußmesser *m*, Strömungswächter *m*
d. gyro Schleppkreisel *m*
d. hook Kuppelhaken *m*, Ziehhaken *m*
d.-in Eintrag *m*, eingeschleppte Lösung *f* *(Galvanotechnik)*
d. link 1. *(Masch)* Schleppglied *n*; 2. *(Kfz)* Lenkspurhebel *m*, Lenkstange *f*
d.-link conveyor Trogkettenförderer *m*
d.-link mechanism Doppelkurbelmechanismus *m*
d.-net *(Schiff)* Schleppnetz *n*
d. of the engine Reibungswiderstand *m* des Motors
d.-out Austrag *m*, herausgeschleppte Lösung *f* *(Galvanotechnik)*
d.-over mill Umführwalzgerüst *n*
d. plough Schlepppflug *m*
d. rail Anhängeschiene *f*
d.-ring Bördelring *m*
d. rod *(Kfz)* Lenkschubstange *f*
d. scraper and loader Schrapplader *m*
d. shovel Tieflöffel[bagger] *m*
d. stop motion *(Text)* Zupfabsteller *m*
d. suction dredger *(Schiff)* Schleppsaugkopfbagger *m*
d. surface *(Schiff)* Vorderfläche *f (z. B. eines Propellers)*
d.-torque tachometer Wirbelstromtachometer *n*
d.-torque viscosimeter Rotationsviskosimeter *n*
d. triangle *(Lw)* Schleppdreieck *n*
d. type classifier *(Am) s.* scraper-type classifier
dragger Schleppnetzfischerboot *n (mit Scherbrettern)*
dragging Ziehen *n*, Schleppen *n*; Abschleppen *n*
d. back of the cutting edge *(Masch)* drückender Schneidenrücken *m*
dragline Schürfkübelbagger *m*, Schlepplöffelbagger *m*, Zugschaufelbagger *m*
d. bucket Schürfkübel *m*, Schlepplöffel *m*, Zugschaufel *f*, Schleppschaufel *f*
d. excavator Schürfkübelbagger *m*, Schlepplöffelbagger *m*, Zugschaufelbagger *m*, Schleppschaufelbagger *m*
d. scraper *s.* dragline
dragon Drachenboot *n*
d. beam *(Bw)* Diagonalstichbalken *m*, Stichbalken *m* unter einem Gratsparren *(Holzbau)*
drain / to 1. dränen, dränieren, entwässern; 2. ablassen, zurückfließen lassen; entleeren; 3. abtropfen lassen
d. away the charge *(El)* Ladung abführen
d. gas from the reservoir Gas aus der Lagerstätte abzapfen
d. off ablassen; abstechen
drain 1. Drän[rohr *n*] *m*; Abflußgraben *m*, Entwässerungsgraben *m*; 2. Abfluß *m*, Ablaß *m*, Rückfluß *m*
d.-and-spray screen *(Brg)* Enttrübungs- und Abbrausesieb *n*
d. board Abtropfbrett *n (Küche)*
d. box Ausgußkasten *m*
d. cock 1. Abflußhahn *m*, Ablaßhahn *m*, Auslaufhahn *m*, Entleerungshahn *m*; 2. Entwässerungshahn *m*, Wasserabflußhahn *m*, Kondenshahn *m*; 3. Abschlämmhahn *m (am Kfz-Kühler)*
d. coil Erdungsdrossel[spule] *f*, Entladespule *f*
d. hole *(Schiff)* Bodenventil *n*, Wasserablaßloch *n*
d. pipe 1. Abziehrohr *n*, Dränrohr *n*, Dränleitung *f*, Entwässerungsleitung *f*, Entwässerungsrohr *n*; 2. Abflußrohr *n*, Abwasserrohr *n*; Fallrohr *n*; 3. Ablaufleitung *f*, Abführleitung *f*; Sickerrohr *n*; 4. *(Kfz)* Leckölleitung *f*
d. plug Ablaßpfropfen *m*, Ablaßschraube *f*, Ablaßstopfen *m*, Entleerungsschraube *f*
d. tank Kondensatsammelbehälter *m*
d. tap Ablaßhahn *m*, Abflußhahn *m*
d. tile Dränrohr *n*, Sickerrohr *n*
d.-trap Geruchverschluß *m*
d. valve Abflußventil *n*, Ablaßhahn *m*, Abzapfhahn *m*, Entleerungsventil *n*; Lenzschieber *m*
drainage 1. Dränage *f*, Drän[ier]ung *f*, Entwässerung *f*, Kanalisation *f*; Bodenentwässerung *f*, Trockenlegung *f*; 2. *(Brg)* Entwässerung *f*, Wasserhaltung *f*, Wasserhebung *f*, Wasserlosung *f*, Wasserlösung *f*; 3. Ablaß *m*
d. adit *(Brg)* Wasser[haltungs]stollen *m*
d. area 1. Einzugsgebiet *n*, Entwässerungsgebiet *n*, Niederschlagsgebiet *n*, Sammelgebiet *n*; 2. Siebnummer *f (Papierherstellung)*
d. coil Erdungsdrossel[spule] *f*, Entladespule *f*; Saugdrossel *f*
d. district Einzugsgebiet *n*, Entwässerungsgebiet *n*, Niederschlagsgebiet *n*, Sammelgebiet *n*
d. ditch 1. Dränggraben *m*, Entwässerungsgraben *m*; Vorfluter *m*; 2. *(Brg)* Wasserrösche *f*
d. of the superheater Überhitzerentwässerung *f*, Entwässerung *f* des Überhitzers
d. pit Abflußgrube *f*; Entwässerungsschacht *m*
d. screen *(Brg)* Entwässerungssieb *n*
d. station Pumpstation *f*
d. trench Drängraben *m*, Entwässerungsgraben *m*

d. tunnel Wasserstollen *m*
drainer Abtropfpfanne *f*, Stoffkasten *m*; Absetzbehälter *m*, Abtropfbehälter *m*; Absetzbütte *f*, Eindickbütte *f (Papierherstellung)*
d. conveyor Entwässerungsförderband *n*
draining Ablassen *n*; Entwässerung *f*, Trockenlegung *f*; Wasserhaltung *f*
d. board Abtropfbrett *n (Spülbecken)*
d. channel Abflußgraben *m*, Drängraben *m*; Abzugsgraben *m*, Dränstrang *m*
d. device Entwässerungsvorrichtung *f*
d. ditch Abflußgraben *m*, Abzugsgraben *m*, Drängraben *m*
d.-off voltage Absaugspannung *f*
d. plant Entwässerungsanlage *f*; Vorflutanlage *f*
d. plough Abziehpflug *m*, Dränpflug *m*
d. rack Ablaufgestell *n*, Trockengestell *n (Labortechnik)*
d. rate Entwässerungsgeschwindigkeit *f*
d. tank Absetzbehälter *m*, Stoffkasten *m (Papierherstellung)*
d. the cylinder Entwässerung *f* des Arbeitszylinders
d. transformer Saugtransformator *m*
d. trench Abflußgraben *m*, Abzugsgraben *m*, Drängraben *m*
drainway *(Brg)* Sumpfstrecke *f*
dram system Garnnumerierung *f (Seide)*
drape forming *(Plast)* Streckformverfahren *n*
d. point Umlenkstelle *f (Spannbetonbau)*
d. retention *(Text)* Formbeständigkeit *f* der Drapierung
drapery Tuchmacherei *f*
d. fabric Deko[rations]stoff *m*
draught / to zeichnen, entwerfen
draught 1. Luftzug *m*, Zugluft *f*; Schornsteinzug *m*; 2. *(Schiff)* Tiefgang *m*, Tauchung *f*; 3. *(Masch)* Gesenkschräge *f*
d. aft Tiefgang *m* achtern
d. beer Faßbier *n*
d. by forging Querschnittsverminderung *f* durch Schmieden
d. by rolling Querschnittsverminderung *f* durch Walzen
d. check plate *(Wärm)* Zugabsperrklappe *f* für Flammrohre
d. designed *(Schiff)* Konstruktionstiefgang *m*
d. distributing box *(Wärm)* Windverteilungskasten *m*
d. effect of the cooling tower *(Wärm)* Zugwirkung *f* des Kamins
d. gauge Zugmesser *m*
d. horse Zugpferd *n*
d. in chimneys Schornsteinzug *m*
d. lobby Windfang *m*
d.-marks *(Schiff)* Ahmings *fpl*, Tiefgangsmarken *fpl*
d. of pattern Anzug *m*, Verjüngung *f (eines Modells)*
d. over keel blocks Wassertiefe *f* über Pallen
d.-proof zugfrei; zugsicher
d. regulating register Rosette *f*
d. regulating wheel Rosettenrädchen *n*
d. regulation Regelung *f* des Zuges
d. regulator Zugregler *m*
d. retarder Zugverzögerer *m*
d. scale *(Schiff)* Tiefgangsskala *f*
draughting board Reißbrett *n*, Zeichenbrett *n*
d. table Zeichentisch *m*
draughtsman [technischer] Zeichner *m*
dravite Dravit *m (Turmalin)*
draw / to 1. zeichnen, eine Zeichnung anfertigen (herstellen); 2. ziehen; abziehen; verziehen; 3. *(Met)* anlassen *(Stahl)*; 4. *(El)* entnehmen, abnehmen, aufnehmen *(Strom)*
d. a circle einen Kreis beschreiben
d. a line eine [gerade] Linie ziehen
d. and double the card slivers before combing Krempelbänder vor dem Kämmen strecken und doppeln
d. current from Strom entnehmen aus, Strom aufnehmen von; Strom entziehen aus (von)
d. deep tiefziehen
d.-file mit der Feile abziehen
d. home anziehen, hineinziehen
d. in 1. einzeichnen; 2. einziehen; 3. ansaugen *(Pumpe)*
d. in a cable Kabel einziehen
d. off 1. abfüllen, abziehen *(Flüssigkeiten)*; 2. *(Brg)* rauben *(Ausbau)*; 3. abstechen *(Hochofen)*
d. off liquid Flüssigkeit ablassen
d. out 1. ausziehen *(Zeichnung)*; 2. herausziehen; 3. ausziehen, ausdehnen; strecken; recken, ausschmieden
d. out threads Fäden ziehen
d. the fires Feuer herausreißen *(Kessel)*
d. the sliver *(Text)* das Band verziehen
d. the temper anlassen *(Stahl)*
d. the temper when grinding beim Schleifen ausglühen, die Härte verlieren
d. up *(Brg)* fördern *(Schacht)*
d. wire Draht ziehen
draw 1. Zug *m*; 2. *(Plast)* Füllraumkonizität *f*
d. beam *(Led)* Stollbock *m*, Stollpfahl *m*
d. bench Ziehbank *f*, Ziehmaschine *f*
d. bolt *(Masch)* Spannbolzen *m*
d. box *(Text)* Lieferwalze *f*, Abzugswalze *f*; Lieferzylinder *m*, Abzugszylinder *m*
d.-box underclearer Streckwerksunterwalzenreiniger *m*, untere Putzwalzen *fpl* im Streckwerk
d. bridge *s.* drawbridge

d. button Zugknopf *m*
d. cam *(Text)* Kulierexzenter *m*
d. crack *(Met)* Spannungsriß *m*
d.-cut *(Brg)* Dreieckseinbruch *m* *(Schießarbeit)*
d.-cut type keyseater Keilnutenziehmaschine *f*
d.-die *(Met)* Ziehmatrize *f*
d. frame *(Text)* Strecke *f*, Streckwerk *n*; Streckmaschine *f*
d. frame gearing Streckentriebwerk *n*
d. head Ziehkopf *m* *(Räummaschine)*
d.-head support Schlitten *m* *(Räummaschine)*
d. hook Zughaken *m*
d.-in arbor Spanndorn *m*
d.-in attachment Spannzange *f*
d.-in bolt 1. Anzugschraube *f*; 2. Spannschraube *f*
d.-in collet auf Zug arbeitende Spannpatrone *f*
d.-in roller Einzugswalze *f*
d.-in screw thread Anzugsgewinde *n*
d.-in spindle Festspannschraube *f*
d.-in tube Spannrohr *n*
d.-in type arbor Anzugdorn *m*
d. knife Zugmesser *n*
d. lever Zughebel *m*
d. loom *(Text)* Kegelstuhl *m*
d.-off cock Ablaßhahn *m*, Abzapfhahn *m*
d.-off roller *(Text)* Gehäuse *n* der Abzugswalze
d.-out radiator ausziehbarer Kühler *m*
d.-out track *(Eis)* Stichgleis *n*
d. plate *(Met)* Zieheisen *n*, Ziehdüse *f* *(Draht)*; Ziehlochplatte *f*
d. punch Ziehstempel *m*
d. ring Ziehring *m*
d. rod Zugstange *f*
d. rod tension Spann[stangen]zug *m*
d. rope *(Schiff)* Codleine *f*
d. slate *(Brg)* Nachfallpacken *m*, Nachfalldach *n*
d. sliver *s.* drawn sliver
d. spindle Spannschraube *f*
d. spring Zugfeder *f*
d. tempering Anlassen *n* *(Stahl)*
d. thread Trennfaden *m* *(Strickerei)*
d. tongs *(El)* Seilspannklemme *f*, Froschklemme *f*, Spannfrosch *m*
d. tube *(Opt)* Ausziehtubus *m*
d.-twister *(Text)* Streckzwirnmaschine *f*
d. vice *(El)* Seilspannklemme *f*, Spannfrosch *m*
d. well Ziehbrunnen *m*
d. winch Zugwinde *f*
d. wire *(Fmt)* Einziehdraht *m*
d. work Hebewerk *n* *(Rotarybohren; Erdöl, Erdgas)*
drawability *(Met)* Ziehfähigkeit *f*
drawback bolt Anzugschraube *f* *(Frässpindel)*
d. collet auf Zug wirkende Spannpatrone *f*
d. spring Arretierfeder *f*
drawbar Anzugstange *f*, Zugeisen *n*, Zugstange *f*
d. pull *(Am)* Zugkraft *f*
drawbolt *s.* drawback bolt
drawbridge Zugbrücke *f*
d. signal Zugbrückensignal *n*
drawer 1. Schubkasten *m*, Schublade *f*; 2. *(Text)* Streckwerk *n*; 3. *(Brg)* Schlepper *m*
d. box Schiebeschachtel *f*
d. for dressing knives *(Holz)* Abziehmesserkasten *m*
d.-type panel Geräteeinschub *m*
drawing 1. Riß *m*, Zeichnung *f*; 2. *(Met)* Ziehen *n*; 3. Anlassen *n* *(Stahl)*; 4. Strecken *n* *(Spinnerei)*
d. and stretching press Streckziehpresse *f*
d. apart of the warp *(Text)* Auseinanderziehen *n* der Kette
d. awl Riemenahle *f*
d. bead Ziehwulst *m*
d. board Reißbrett *n*, Zeichenbrett *n*
d. cage Förderkorb *m*, Fördergestell *n*, Förderschale *f*
d. cardboard Zeichenkarton *m*
d. cloth Pausleinen *n*
d. curve Kurvenlineal *n*
d. cushion *(Met)* Ziehkissen *n*
d. cut ziehender Schnitt *m*
d. depth *(Met)* Ziehtiefe *f*
d. device Ausziehvorrichtung *f*, Ziehvorrichtung *f*
d. die Ziehmatrize *f*, Ziehstein *m*, Ziehwerkzeug *n*
d. down Ausziehen *n*; Ausschmieden *n*
d. frame 1. *(Text)* Strecke *f*, Streckwerk *n*; 2. Förderkorb *m*, Fördergestell *n*, Förderschale *f*
d. frame for maximum draft Höchstverzugsstreckwerk *n*
d. frame with divided sliver delivery Teilbandstrecke *f*
d. frame with x lines of rollers x-Walzenstrecke *f*
d. furnace *(Met)* Anlaßofen *m*
d. grade *(Met)* Tiefziehgüte *f*
d. grease Ziehfett *n*
d. hoop *(El)* Ziehband *n*, Fädelband *n*
d.-in 1. Einziehen *n*, Einzug *m* *(Weberei)*; 2. Hineinziehen *n*
d.-in band Einzugsseil *n* *(Selfaktor)*
d.-in frame Ketteneinziehstuhl *m*
d.-in hook Einziehhäkchen *n*, Einziehnadel *f*
d.-in machine [Webketten-]Einziehmaschine *f* *(Weberei)*
d.-in of the reamer into the hole Hereinziehen *n* [der Reibahle in Bohrung]

d.-in the reed *(Text)* Blattstechen *n*
d.-in the warp *(Text)* Einziehen *n* der Kette
d.-in the yarn on the warp weaving machine *(Text)* Einziehen *n* der Kette auf die Kettenwirkmaschine
d.-in width *(Text)* Einzugbreite *f*
d. ink Ausziehtusche *f*; Zeichentusche *f*
d. instruments Reißzeug *n*
d. knife Schälmesser *n (Furnier)*
d. limit Ziehgrenze *f*
d. loft Schnürboden *m (einer Werft)*
d. machine Zeichenmaschine *f*
d. mirror *(Opt)* Zeichenspiegel *m*
d. needle *(Text)* Ziehnadel *f*
d. of needles *(Text)* Nadelzug *m*
d.-off band *(Text)* Abzugsband *n*
d. off the flue gases Absaugen *n* der Rauchgase
d. off the water Wasserableitung *f*
d. office Konstruktionsbüro *n*; Zeichenbüro *n*
d. out Recken *n*, Strecken *n (Schmiedetechnik)*
d.-out phenomenon *(El)* Zieherscheinung *f*
d. paper Zeichenpapier *n*
d. pen Reißfeder *f*, Ziehfeder *f*
d. pencil Zeichenstift *m*
d. pin Reißnagel *m*, Reißzwecke *f*, Reißbrettstift *m*
d. plant Zieherei *f*
d. plate Zieheisen *n (Drahtherstellung)*
d. press Ziehpresse *f*
d. pump ziehende Pumpe *f*
d. roller *(Text)* Streckwalze *f*, Streckzylinder *m*, Ausziehwalze *f*, Verzugswalze *f*
d. room Zieherei *f*
d. set Reißzeug *n*
d. shaft Förderschacht *m*
d. shown partly in section Teilschnitt *m*
d. sliver *(Text)* Streck[en]band *n*
d. table Zeichentisch *m*
d. the staple *(Text)* Stapelziehen *n*
d. the warp yarn into the harness *(Text)* Einzug *m* des Kettfadens in den Schaft
d. thin wire Feindrahtzug *m*
d. tool *(Masch)* Ziehring *m*, Ziehwerkzeug *n*
drawn-in eingezogen; eingefahren
d. plough Anhängepflug *m*
d. sliver *(Text)* Streckenband *n*
d. steel bar gezogener Stabstahl *m*
d. [true] to scale maßstäblich gezeichnet
drawrock *(Am) (Brg)* zum Nachfall neigendes Gestein *n (unmittelbares Hangendes)*
drawtwister *(Text)* Streckzwirnmaschine *f*
dredge / to [aus]baggern
dredge 1. *(Am)* Naßbagger *m*, Schwimmbagger *m*; 2. *(Schiff)* Schleppnetz *n*; 3. *(Kern)* Aufschlämmung *f*, Schlamm *m*, Trübe *f*
d. conveyor Tauchband *n*
dredger Naßbagger *m*, Schwimmbagger *m*
d. excavator Trockenbagger *m*
d. track level Baggerplanum *n*
d. trailing cable Baggerschleppkabel *n*
dredging Naßbaggern *n*, Naßbaggerung *f*, Unterwasserbaggern *n*
d. boat *(Schiff)* Baggerprahm *m*
d. craft Naßbagger *m*, Bagger *m*
d. cutter Baggerschneidkopf *m*
d. pontoon Baggerponton *m*
d. pump Baggerpumpe *f*
d. scoop Baggereimer *m*; Baggerlöffel *m*; Schürfkübel *m*
d. spoil Baggergut *n*
dredgings Baggergut *n*
dregs Bodensatz *m*, Trester *pl*
dress / to 1. nach[be]arbeiten; [aus]richten, schlichten, zurichten; abrichten, abziehen *(Schleifkörper)*; 2. besäumen, zurichten *(Holz)*; behauen *(Stein)*; 3. appretieren, schlichten; 4. *(Typ)* fertigmachen *(Satz)*; 5. verbinden *(Wände)*; 6. einölen *(Leder)*; 7. dekorieren *(Schaufenster)*; 8. bekleiden, ankleiden, anziehen; 9. zubereiten
d. with formed crusher roll mit Profilrolle abrichten
dress-fabric circular knitting machine Rundstrickmaschine *f* für Oberbekleidungsstoffe
d. hook Hakenschraube *f*, Halbösenschraube *f*
d. preserver (shield) *(Text)* Schweißblatt *n*
dressed lumber Kantholz *n*
d. ore angereichertes (aufbereitetes) Erz *n*
d. sawn wood ausgeformtes (geputztes) Schnittholz *n*
d. skin Fell *n*
d. stone behauener Bruchstein *m*
d. timber Kantholz *n*
dresser 1. Anrichte *f*, Büfett *n (Küche)*; 2. *(Typ)* Fertigmacher *m (Setzerei)*; 3. Beizer *m (Saatgutbereitung)*; 4. Abrichter *m*; Abrichteinrichtung *f (Schleifscheibe)*
dressing 1. Nachbearbeitung *f*, Schlichten *n*; Abrichten *n (Schleifscheibe)*; 2. *(Met)* Formschlichte *f*; Kokillenausbruch *m*; 3. Aufbereitung *f (Erze)*; 4. *(Led)* Zurichten *n*; Appretur *f*
d. diamond Abrichtdiamant *m (Schleifscheibe)*
d. fixture Abrichteinrichtung *f (Schleifscheibe)*
d. frame *(Text)* Schlichtmaschine *f*, Appretiermaschine *f*
d. hammer 1. *(Masch)* Pritschhammer *m*; 2. *(Brg)* Scheidebeil *n*
d. of cable Vorbereitung *f* des Kabelendes
d. of grain Beizen *n* des Getreides
d. of seeding grain Saatgutaufbereitung *f*
d. off Gußputzen *n*
d. room Ankleidezimmer *n*, Umkleideraum *m*
d. shop *(Met)* Putzerei *f*, Gußputzerei *f*

d. the grinding wheel Abziehen *n* der Schleifscheibe
d. the grinding wheel at an angle Winkelabrichten *n* der Schleifscheibe
d. tool Abrichtwerkzeug *n*
dribbling Tropfen *n*, Nachtropfen *n*
dried fruit Dörrobst *n*, Backobst *n*
drier 1. Trockenapparat *m*, Trockner *m*; Trokkenofen *m*; 2. Sikkativ *n*, Trockenmittel *n*, Trockenstoff *m*, Trockner *m*
d. furnace Trockenofen *m*
d. journal *(Brg)* Trocknerzapfen *m*
drift / to 1. schwimmen; treiben, abgetrieben werden; triften; 2. *(El)* weglaufen *(Frequenz)*
d. out austreiben
drift 1. *(Rak)* Abdrängung *f*, Abtrift *f*; 2. *(Flg)* Drift[komponente] *f*, Horizontalkomponente *f* des aerodynamischen Druckes; 3. *(Dat)* Drift, Nullpunktwanderung *f*; Driftfehler *m*; 4. *(El)* Abweichung *f*, Drift, Weglaufen *n (Frequenzen)*; 5. Schlupf *m (Akustik)*; 6. Abweichung *f* von der Lotrechten *(Tiefbohren)*; 7. *(Bw)* Stollen *m*; 8. *(Brg)* Gesteinsstrecke *f (allgemein)*, Richtstrecke *f (insbesondere)*, Strecke *f (allgemein)*; 9. Austreiber *m*; Ausweitdorn *m*; 10. *s. a.* offset
d. advance *(Brg)* Vortriebsgeschwindigkeit *f*
d. along the lattice Elektronenwanderung *f* durch den Kristallgitterraum
d. anchor Driftanker *m*
d. angle *(Rak)* Abdrängungswinkel *m*, Abtriftwinkel *m*, Seitentriftwinkel *m*
d.-band *(Geol)* dünnes Bergemittel *n*
d. boat Logger *m*
d. bottle Flaschenpost *f*, Treibflasche *f*
d. correction 1. *(Ph)* Driftkorrektion *f*; 2. *(Schiff)* Abdriftkorrektur *f*, Abtriftberichtigung *f*
d. error *(BMSR)* Driftfehler *m (Analogrechner)*
d. hammer Treibhammer *m*
d. indicator *(Schiff)* Abtriftanzeiger *m*, Abtriftmesser *m*
d. key *(Masch)* Austreiber *m*
d. landing-gear *(Flg)* Seitenwindfahrwerk *n*
d. mining Stollenbergbau *m*
d. net Treibnetz *n (Fischerei)*
d. percentage Driftgrad *m*, Grad *m* der Abweichung
d. pin Keiltreiber *m*, Splintentreiber *m*
d. punch Durchschläger *m*, Durchschlaghammer *m*, Durchtreiber *m*
d. rail *(Schiff)* zerbrochene Gangreling *f*
d. sail Treibsegel *n*
d. set *(Brg)* Türstock *m*
d.-slicing *(Brg)* Querbruchbau *m* mit Türstockausbau in den Querbrechen
d. slot *(Masch)* Schlitz *m* zum Einschieben des Austreibers
d. space Laufraum *m (Laufzeitröhren, Hohlraumresonatoren)*
d. speed *(Ph, Kern)* Driftgeschwindigkeit *f*, [mittlere] Wander[ungs]geschwindigkeit *f*; mittlere Durchdringungsgeschwindigkeit *f*
d. test Aufweitversuch *m*, Aufdornversuch *m*
d. tube Elektronenlaufzeitröhre *f*, Laufzeitröhre *f*
d. tunnel Laufraumelektrode *f*
d. velocity Abtriftgeschwindigkeit *f*
drifter 1. *(Schiff)* Logger *m*, Treibnetzfischereifahrzeug *n*; 2. *(Brg)* Bohrhammer *m (für Vortrieb)*
drifting *(Brg)* Streckenvortrieb *m (in Gestein)*, Auffahrungsarbeiten *fpl*, Streckenauffahren *n*
d. out Austreiben *n (Keil)*
drill / to 1. *(Masch, Brg)* bohren *(aus dem vollen Werkstoff)*; 2. *(Lw)* drillen, säen
d. a well eine Bohrung abteufen (herunterbringen) *(Erdöl, Erdgas)*
d. holes at an angle schräg bohren
d. off centre exzentrisch bohren
d. round konzentrisch bohren
d. true lehrenhaltig bohren
drill 1. Bohrer *m*, Bohrmaschine *f*; 2. *(Lw)* Sämaschine *f*
d. axis Bohrerachse *f*
d. barge Bohrinsel *f*
d. bit 1. Langlochbohrer *m*, Tieflochbohrer *m*; 2. Bohrmeißel *m*; Schneidenkopf *m*; 3. Kernbohrkrone *f (Tiefbohren)*; 4. Bohrstück *n*
d. blank Bohrerrohling *m*
d. breakage Bohrerbruch *m*
d. bush[ing] Bohrbuchse *f*
d. capacity from solid größter Bohrdurchmesser *m* aus dem Vollen
d. carriage Bohr[unter]wagen *m*
d. casing Futterrohre *npl*, die als Bohrgestänge verwendet werden und als Verrohrung im Bohrloch verbleiben
d.-chamfer-spot face [subland] tool Programmwerkzeug *n* zum Bohren, Brechen der Kante und Plansenken
d. changing Bohrerwechsel *m*
d. chuck 1. *(Masch)* Bohrfutter *n (Spiralbohrer)*; 2. *(Holz)* Drilldocke *f*
d. chuck key *(Masch)* Futterschlüssel *m*
d. chuck with three jaws Dreibackenbohrfutter *n*
d. collar Schwerstange *f*, Bohrstange *f (Tiefbohren)*
d. column Bohrsäule *f (Tiefbohren)*
d. coulter Drillschar *n*
d. cuttings Bohrklein *n (Tiefbohren)*
d. depth indicator *(Masch)* Bohrtiefenanzeiger *m*
d. diameter Bohrerdurchmesser *m*

d. drift Bohreraustreiber *m*
d. driver Kegelklemmhülse *f*
d. extension Bohrerverlängerung *f*
d. failure Bohrerbruch *m*
d. feed rate Bohrvorschub[größe *f*] *m*
d. flute Bohrernut *f*
d. follows a straight line / the der Bohrer *m* verläuft nicht
d. footage Bohrleistung *f* gemessen in ft. *(Tiefbohren)*
d. gauge Bohrerlehre *f*
d. grinding gauge Spiralbohrerschleiflehre *f*
d. grinding machine Bohrerschleifmaschine *f*
d. guide bush Bohr[erführungs]buchse *f*
d. hammer Bohrhammer *m*
d. harrow Furchenegge *f*, Saategge *f*
d. head Bohrkopf *m*; Spindelstock *m (Bohrmaschine)*
d.-hole Bohrloch *n*, Bohrung *f*; Aufbohrung *f*; Ausbohrung *f*
d. jig Bohrvorrichtung *f*
d.-jig bush[ing] Bohrbüchse *f*
d. life 1. Bohrerstandzeit *f (minutes between regrinds)*; 2. Standmenge *f* des Bohrers *(holes between regrindings; yards of hole)*
d. man *(Brg)* Bohrhauer *m*
d. mounting Bohrgerüst *n*, Bohrgestell *n (Tiefbohren)*
d. pipe Bohrgestänge *n*, Gestängerohr *n*, Hohlgestänge *n (Tiefbohren)*
d. pipe slips Bohrrohrfangkeile *mpl (Tiefbohren)*
d. pipe string Bohrgestängezug *m (Tiefbohren)*
d. plough *(Lw)* Säpflug *m*; Drillmaschine *f*, Reihensämaschine *f*
d. point Bohrerspitze *f*
d. point angle Bohrerspitzenwinkel *m*
d. point gauge Spiralbohrerschleiflehre *f*
d. point grinding Spiralbohrernachschliff *m*
d.-point leading edges Hauptschneiden *fpl* an der Bohrerspitze
d. point sharpening machine Spiralbohrerspitzenschleifmaschine *f*
d. point thinner Spiralbohrerausspitzmaschine *f*
d. pointing *(Am)* Ausspitzen *n* des Bohrers
d. press *(Am)* Säulenbohrmaschine *f*
d. return Bohrerrücklauf *m*
d. revolution Bohrerumdrehung *f*
d. rigidity Bohrersteifigkeit *f*, Bohrerstarrheit *f*
d. rod Bohrgestänge *n*, Bohrstange *f (Tiefbohren)*
d. runner Bohrarbeiter *m*, Bohrhauer *m*
d.-share *(Lw)* Drillschar *n*
d. sharpener Bohrerschärfmaschine *f*
d. ship *(Schiff)* Schulschiff *n*
d. sleeve Bohrerhülse *f*, Kegelhülse *f*, Reduzierhülse *f*
d. socket Bohrfutterkegel *m*, Einsatzhülse *f*, Kegelhülse *f*, Reduziereinsatz *m*
d. spade *(Lw)* Lochspaten *m*
d. spindle Bohrspindel *f*
d. spindle sleeve Bohrspindelhülse *f*
d. stand Bohrständer *m (Handbohrmaschine)*
d. steel Bohrstahl *m*, Bohrer *m*
d. stem Schwerstange *f*, Bohrstange *f (Tiefbohren)*
d. stem with bit Bohrzeug *n*, Bohrgarnitur *f (Tiefbohren)*
d. strength Bohrerfestigkeit *f*
d. table Bohrtisch *m*
d. tang Bohrerzapfen *m*, Austreiberlappen *m* am Bohrer
d. thinning Bohrerausspitzen *n*
d. tip Schneidenkopf *m* am Bohrer
d. tower Bohrturm *m*
d. tube Bohrrohr *n*
d. unit *(Masch)* Bohreinheit *f*
d. upsetting machine Bohrerstauchmaschine *f*
d. with increase twist Spiralbohrer *m* mit veränderlichem Drall; Bohrer *m* mit verschiedener Steigung der Nuten, Bohrer mit verschiedenen Spanwinkeln
d. with interior oil channel Spiralbohrer *m* mit innerer Ölzuführung
d. with split point grind Bohrer *m* mit geschlitzter Querschneide
drillability Bohrbarkeit *f*
drillable bohrbar
drilled hole 1. Bohrloch *n*, Bohrung *f*; 2. vorgebohrtes Loch *n*
d.-in caisson *(Bw)* Bohrpfahl *m*
d. jig Bohrlehre *f*
d. rows *(Lw)* Drillreihen *fpl*
d. seed Drillsaat *f*, Reihensaat *f*
driller 1. Bohrhauer *m (Tiefbohren)*; 2. *(Am)* Bohrmaschine *f*
drilling 1. Bohren *n*; Bohrtechnik *f*; 2. Bohrspan *m*
d. apparatus Bohreinrichtung *f*
d. appliance Bohrgerät *n*
d. attachment Bohreinrichtung *f*
d. barge *(Schiff)* Bohrschute *f*
d., boring, milling, and tapping machine Bohr-, Fräs-, Gewindeschneidmaschine *f*
d. box *(Lw)* Drillkasten *m*, Säkasten *m*
d. cable Bohrseil *n (Tiefbohren)*
d. campaign Bohrprogramm *n (Tiefbohren)*
d. capacity Bohrleistung *f*; maximaler Bohrdurchmesser *m*
d. crew Bohrmannschaft *f (Tiefbohren)*
d. engineering Bohrtechnik *f*
d. feed Bohrvorschub *m*
d. fixture Spiralbohrerspanneinrichtung *f*

d. fluid Bohrtrübe *f (Tiefbohren)*; Bohröl *n*
d. from the solid Bohren *n* aus dem vollen
d. hammer Minenhammer *m*, Schlegel *m*
d. head Bohr[spindel]schlitten *m*; Bohrkopf *m*
d. head clamping lever Hebel *m* zum Klemmen des Bohrschlittens
d. jar Rutschschere *f (Tiefbohren)*
d. jig Bohrvorrichtung *f*
d. job Bohrarbeit *f*
d. line Bohrseil *n (Tiefbohren)*
d. machine *(Lw)* Drillmaschine *f*, Reihensämaschine *f*
d.-machine spindle Bohrspindel *f*
d. mud Bohrschlamm *m*, Bohrtrübe *f (Tiefbohren)*
d. mud additives Bohrspülungszusatzmittel *npl*
d. of microscopic holes Mikrobohren *n*
d. oil Bohröl *n*
d. operation Bohrarbeit *f*
d. pattern Bohrschema *n*
d. position Bohrstellung *f*
d. pressure Bohrdruck *m*
d. rate Bohrfortschritt *m*, Bohrgeschwindigkeit *f*
d. rig Bohranlage *f*, Bohrgestell *n*, Bohrkran *m (Tiefbohren)*
d. rod *s.* drill rod
d. saddle Bohrschlitten *m (Radialbohrmaschine)*
d. seed box Drillsaatkasten *m*
d. speed Bohrfortschritt *m*, Bohrgeschwindigkeit *f*
d. string Bohrzeug *n*, Bohrgarnitur *f (Tiefbohren)*
d. swivel Spülkopf *m (Tiefbohren)*
d. template Bohrschablone *f*
d. tool Bohrwerkzeug *n*
d. torque Drehmoment *n* beim Bohren ins volle
d. tower Bohrturm *m*
d. water Spülwasser *n (Tiefbohren)*
d. well abteufende (im Abteufen befindliche) Bohrung *f (Tiefbohren)*
drillings Bohrspäne *mpl*
drip / to [herab]tropfen, tröpfeln, träufeln, triefen
drip *(Bw)* Wasserablaufrinne *f*; Wasserschenkel *m (Fenster)*
d. cap Tropfnase *f*, Wasserablauf *m*
d.-dry tropftrocken
d.-dry finish Tropf-Trocken-Ausrüstung *f*, Tropfnaß-Trocken-Ausrüstung *f (Textilveredlung)*
d.-feed lubrication Tropf[öl]schmierung *f*
d.-feed lubricator Tropföler *m*, Öltropfapparat *m*
d. furnace Tropffeuerung *f*
d. moulding *(Kfz)* Regenrinne *f*
d. nozzle Tropfdüse *f*
d. oiler Tropföler *m*
d. pan Abtropfblech *n*
d. plug Ablaufstopfen *m*
d.-proof tropfwasserdicht, tropfwassersicher; tropfwassergeschützt
d.-proof protected tropfwassergeschützt
d.-proof screen-protected *(Elm)* gegen Tropfwasser und Berührung geschützt
d.-tight tropfwasserdicht
dripping cooling plant Rieselkühler *m*
d. oil Tropföl *n*
d. water Tropfwasser *n*
dripstone *(Bw)* Wasserschenkel *m*
drive / to 1. [an]treiben *(Maschine)*; mitnehmen *(Werkzeug)*; schlagen *(Nieten)*; einschlagen *(Nägel)*; [ein]rammen *(Pfähle)*; 2. *(Kfz)* fahren, führen, lenken *(Wagen)*; 3. *(El)* [aus]steuern *(Röhre, Verstärker)*; 4. *(BMSR)* erregen, steuern, treiben; 5. *(Brg)* auffahren, treiben *(Strecke)*
d. hard weit aussteuern *(Elektronenröhren)*
d. home eintreiben *(Keil)*
d. in [hin]eintreiben; einpressen *(Buchsen)*
d. in a screw einschrauben
d. on cold kalt aufpressen
d. out heraustreiben; [her]auspressen *(Buchsen)*
d. the work mitnehmen *(z. B. Drehmaschine)*
drive 1. Antrieb *m*, Trieb *m*; 2. *(Kfz)* Fahrt *f*
d. assembly Antrieb *m*
d. centre *(Masch)* Mitnehmerspitze *f*
d. control Antriebsregelung *f*
d. drum Antriebstrommel *f*
d. element Antriebselement *n*, Antriebsorgan *n*
d.-fitted key Treibkeil *m*
d.-fitted to eingetrieben in
d. flange Antriebsflansch *m*
d. magnet Antriebsmagnet *m (z. B. automatisches Telefon)*
d. of selector Wählerantrieb *m (Telefon)*
d.-on-drive-off carrier *(Schiff)* „Drive-on-drive-off"-Frachter *m (Frachter mit Auffahrtsrampe)*
d.-on-drive-off layout Auf- und Abfahrtsvorrichtung *f (auf einem Schiff)*
d.-on system Auffahrtsystem *n (Frachter)*
d. oscillation *(El)* Steuerschwingung *f*
d. pin Mitnehmerstift *m*
d. pin punch zylindrischer Durchtreiber *m*
d. pinion of the differential *(Kfz)* Ausgleichkegelrad *n*, Planetenrad *n*
d. plate Mitnehmerscheibe *f*
d. pulley Antriebsrolle *f*, Antriebsscheibe *f*, Treibrolle *f*
d. shaft Antriebswelle *f*
d.-shaft spigot *(Kfz)* Antriebswellenstumpf *m*

d.-shaft stub *(Kfz)* Antriebswellenstumpf *m*
d.-shaft tube *(Kfz)* Gelenkwellenrohr *n*
d. sheave Treibscheibe *f*
d. side *(Masch)* Treibflanke *f*, Zugflanke *f* *(eines Zahnes)*
d. slot Mitnehmernut *f*
d. winding *(El)* Steuerwicklung *f* *(Magnetkernwicklung)*
drivehead *(Brg)* Antriebsstation *f*, Antriebsseite *f*
driven getrieben; mitgenommen
d. aerial [direkt]gespeiste Antenne *f*
d. gear getriebenes Rad *n*; Getriebenes *n*; Abtriebsrad *n*
d. head Schließkopf *m* *(Niet)*
d. link Abtriebsglied *n* *(Getriebelehre)*
d. member Abtriebsglied *n*
d. pile *(Bw)* Rammpfahl *m*
d. plate Kupplungslamelle *f* *(Traktor)*
d. shaft getriebene Welle *f*
d. side Abtrieb *m*
d. sweep gesteuerter Kippvorgang *m* *(z. B. einer Katodenstrahlröhre)*
d. time base gesteuerte Zeitbasis *f* *(z. B. einer Katodenstrahlröhre)*
d. transmitter fremdgesteuerter (extern gesteuerter) Sender *m*
driver 1. Mitnehmer *m* *(Werkzeug)*; Drehherz *n*; Mitnehmerbolzen *m*; 2. treibendes Rad *n*, Antriebsrad *n*; 3. Antriebsglied *n* *(Getriebelehre)*; 4. Antriebswelle *f*, treibende Welle *f*; 5. *(Kfz)* Wagenführer *m*, Fahrer *m*, Chauffeur *m*; 6. *(Fmt, Fs)* Steuersender *m*
d. motor Antriebsmotor *m*
d. pin Mitnehmerstift *m*
d. plate Mitnehmerscheibe *f*
d. stage *(El)* Treiber[stufe *f*] *m*
d. transformer Treibertransformator *m*
d. tube Treiberröhre *f*
d. valve Treiberröhre *f*
driver's cabin Führerstand *m*
d. compartment Fahrerraum *m*
d. control valve Führerbremsventil *n*
d. lamp Mannschaftslampe *f*, Grubenlampe *f*
d. seat Fahrersitz *m*
driveway Auffahrt *f*, Zufahrt *f*; Garagenauffahrt *f*
driving 1. Fahren *n*; 2. *(El)* Steuerung *f*, Ansteuerung *f* *(Senderstufen)*; Aussteuern *n* *(Elektronenröhren)*; 3. Einrammen *n* *(von Pfählen)*; 4. *(Brg)* Auffahren *n* von Flözstrecken *(in der Lagerstätte)*; 5. *(Masch)* Antrieb *m*
d. agent Antriebsorgan *n* *(z. B. Schalter)*
d. axle Treibachse *f*, Triebachse *f*, Antriebswelle *f*
d. band Führungsring *m* *(Granate)*
d. belt Treibriemen *m*, Transmissionsriemen *m*
d. by steam Dampfantrieb *m*; Dampfbetrieb *m*
d. cam Antriebsnocken *m*
d. cap *(Bw)* Rammhaube *f*
d. chain Antriebskette *f*, Treibkette *f*
d. chain wheel Antriebskettenrad *n*
d. circuit Antriebsstromkreis *m*
d. clutch Antriebskupplung *f*
d. coil Ansprechspule *f*, Erregerspule *f*
d. collar Mitnehmerring *m*
d.-cone pulley Antriebsstufenscheibe *f*
d. disk Antriebsscheibe *f*, Mitnehmerscheibe *f*
d. dog Anlasserklaue *f*, Antriebsklaue *f*, Mitnehmer *m*, Drehherz *n*
d. effect of inertia *(Masch)* treibende Wirkung *f* der Massenkräfte (Trägheitskräfte)
d. element Antriebselement *n*, Antriebsorgan *n*
d. end *(Masch)* Antriebsseite *f*
d. face 1. *(Flg)* Druckseite *f* *(Luftschraube)*; 2. *(Schiff)* Druckfläche *f* *(Propeller)*
d. fit *(Masch)* Übergangspassung *f*
d. force Antriebskraft *f*
d. formula Rammformel *f* *(Pfahlgründungen)*
d. frequency *(Fmt, Fs)* Steuerfrequenz *f*
d. gear Antriebsrad *n*, Triebrad *n*; Treiber *m*; Laufwerk *n*
d.-in the working barrel *(Wärm)* Einpressen (Eintreiben) *n* des Arbeitszylinders
d. instructor Fahrlehrer *m*
d.-key transmission Ziehkeilgetriebe *n*
d. lamp *(Kfz)* Scheinwerfer *m*
d. licence Fahrerlaubnis *f*
d. light *(Kfz)* Fernlicht *n*
d. lug Mitnehmerzapfen *m*, Mitnehmernase *f*
d. magnet Antriebsmagnet *m*, Schaltmagnet *m*
d. mechanism *(Masch)* Triebwerk *n*
d. medium Zugkraftglied *n*, Zugkraftorgan *n*
d. member Antriebsteil *n*, Trieborgan *n*; Antriebsglied *n*
d. mirror *(Kfz)* Rück[blick]spiegel *m*
d. motor Antriebsmotor *m*
d. motor [of a drilling machine] Bohrmotor *m*
d. part Antriebsteil *n*, Triebwerksteil *n*
d. pawl Stoßklinke *f* *(z. B. Telefonie)*
d. pin Antriebsbolzen *m*, Mitnehmerbolzen *m*, Mitnehmerstift *m*, Treibzapfen *m*
d. pinion Antriebsritzel *n*, antreibendes Zahnrad *n*, Ritzel *n*
d. piston Treibkolben *m*
d. plate Mitnehmerscheibe *f*
d.-point admittance Eingangsscheinleitwert *m*; Zweipoladmittanz *f*, Zweipolscheinleitwert *m*

d.-point function *(BMSR, Dat)* Zweipolfunktion *f*
d.-point impedance *(BMSR, Dat)* Zweipolimpedanz *f*, Zweipolscheinwiderstand *m*; Eingangsimpedanz *f*
d. power *(Fmt)* Antriebsleistung *f*, Steuerleistung *f*, Treiberstufenleistung *f*
d. pulley Antriebs[riemen]scheibe *f*, Scheibenschwungrad *n*, Treibrolle *f*, Treibscheibe *f*
d. pulse *(BMSR)* Steuerimpuls *m*, Treiberimpuls *m*
d. rain *(Bw)* Schlagregen *m*
d. rate 1. Durchsatzgeschwindigkeit *f (Hochofen)*; 2. *(Brg)* Vortriebsgeschwindigkeit *f*, Leistung *f* beim Streckenauffahren
d. rod Antriebsstange *f*, Pleuelstange *f*, Schubstange *f*
d. roller Transportrolle *f*, Treibrolle *f*
d. shaft Antriebswelle *f*, antreibende Welle *f*, Transmissionswelle *f*; Hauptwelle *f*
d.-shaft bushing Antriebswellenbüchse *f*
d.-shaft coupling Antriebswellenkupplung *f*
d. side *(Masch)* Zugtrum *m (Riemen)*
d. signal *(BMSR)* Steuersignal *n*, Treibersignal *n*
d. slot Mitnehmernut *f*; Quernut *f* zur Mitnahme
d. sprocket Antriebskettenrad *n*, Kettentriebsrad *n*
d. steam engine Betriebsdampfmaschine *f*, Antriebsdampfmaschine *f*
d. string Antriebseil *n*; Skalenseil *n (elektrische Geräte)*
d. tang *s.* tang
d. the work Mitnahme *f* [des Werkstücks]
d. voltage Steuerspannung *f*
d. wheel Triebrad *n*
drizzle Nieseln *n*, Nieselregen *m*, Sprühregen *m*
drone 1. Dröhnen *n*; 2. funkgesteuertes Wasserfahrzeug *n*; 3. Fernlenkflugzeug *n*
droop 1. *(Fs)* Abfall *m* [des Impulsdaches]; 2. *s.* offset
d. control *s.* proportional control
d. governor *s.* proportional acting controller
drop / to 1. fallen [lassen]; abwerfen; sinken; abnehmen; 2. tropfen, tröpfeln; sickern; lecken; 3. abfallen, sinken *(z. B. Spannung, Geschwindigkeit)*
d. a boat astern ein Boot achtern aussetzen *(zu Wasser bringen)*
d. a perpendicular to a line *(Math)* auf eine Gerade das Lot fällen
d. forge [im Gesenk] schmieden
d.-in einrasten
d. off the nose-cone *(Rak)* den Schutzkegel abwerfen
d. out abfallen *(Relais)*

d. out of engagement (gear, mesh) lösen, außer Eingriff bringen
d. the anchor den Anker fallen lassen
d. to zero auf Null abfallen
drop 1. Fall *m*; 2. Tropfen *m*; 3. Abfall *m (z. B. Spannung, Geschwindigkeit)*; 4. Klappe *f*, Fallklappe *f (Telefonie)*; 5. Fallhammer *m*, Riemenfallhammer *m*; 6. *(Bw)* Pilzkopf *m*, Säulenkopf *m*
d. annunciator *(El)* Fallklappenanlage *f*
d. arm *(Kfz)* Lenkstockhebel *m*, Lenkspurhebel *m*
d. ball Fallhammer *m*, Fallbär *m*
d.-ball-viscosimeter Kugelfallviskosimeter *n*
d.-base rim *(Kfz)* Tiefbettfelge *f*
d.-bottom bucket *(Met)* Setzkübel *m*
d.-bottom door *(Eis)* Bodenklappe *f*
d.-bottom skip Skip *m* mit Bodenentleerung
d. box *(Text)* Steigkasten *m*, Fallkasten *m*, Hubkasten *m*
d.-box loom Steigkastenwechselwebstuhl *m*
d.-box motion Steigkastenwechsel *m*
d. calling *(Fmt)* Klappenruf *m*
d.-centre rim *(Kfz)* Tiefbettfelge *f*
d. deck *(Schiff)* tiefgesetztes Deck *n*
d.-feed lubrication (oiling) Tropfschmierung *f*
d. feed oiler Tropföler *m*
d.-finger auxiliary stop motion *(Text)* Hilfsabstellvorrichtung *f* mit Fühlerfinger
d. forge Gesenkschmiede *f*
d.-forged gesenkgeschmiedet, im Gesenk geschmiedet
d.-forged wrench Gesenkschraubenschlüssel *m*
d. forging 1. Gesenkschmieden *n*; 2. Gesenkschmiedestück *n*
d. forging component Gesenkschmiedestück *n*
d. forging die Gesenk *n*, Schmiedegesenk *n*
d. forging press Gesenkschmiedepresse *f*
d. forging shop Gesenkschmiede *f*
d. frame *(Kfz)* Niederrahmen *m*
d. hammer Fallhammer *m*, Rammbär *m*, Freifallbär *m*
d. hammer die Fallhammergesenk *n*
d. hanger *(Masch)* Hängelager *n*
d. hardness test Fallhärteprüfung *f*
d.-in pin Einrastbolzen *m*, Rastzapfen *m*
d.-in winding *(Elm)* Träufelwicklung *f*
d. indicator *(Fmt)* Fallklappe *f*, Fallscheibe *f*
d.-indicator panel *(Fmt)* Klappenschrank *m*
d.-indicator shutter *(Fmt)* Anrufklappe *f*
d. keel *(Schiff)* Hängekiel *m*, Steckschwert *n*
d. lid Klappdeckel *m*
d. of potential 1. *(Ph)* Potentialabfall *m*; 2. *(El)* Spannungsabfall *m*, Spannungsabnahme *f*

d. of pressure Druckabnahme *f*; Druckgefälle *n (Turbine)*
d.-out current Abfallstrom *m (Relais)*
d.-out fuse Sicherung *f* mit Ausschaltung, abfallende Sicherung
d.-out time Abfallzeit *f (Relais)*
d.-out value Abfallwert *m (Relais)*
d. panel Pilzkopf *m*, Säulenkopf *m (einer Pilzdecke; Stahlbetonbau)*
d. pin *(Text)* Abstellnadel *f*
d. point Tropfpunkt *m*
d. shaft *(Brg)* Senkschacht *m*
d. sheet *(Brg)* Wettertuch *n*
d. shunt of a track circuit für den Abfall des Gleisrelais zulässiger Achswiderstand *m*
d. shutter *(Fmt)* Anrufklappe *f*
d. stamp Fallhammer *m*
d. stamping Preßling *m (s. a.* d. forging)
d. stamping shop Gesenkschmiede *f*
d. stanchion *(Eis)* umklappbare Runge *f*
d. stowing pipeline *(Brg)* Bergefalleitung *f*
d. switchboard *(Fmt)* Klappenschrank *m*
d. tank *(Rak)* Abwurfbehälter *m*, Zusatzbehälter *m*
d. test *(Schiff)* Fallprobe *f (Anker)*
d. type switchboard *(Fmt)* Klappenschrank *m*
d. valve *(Kfz)* hängendes Ventil *n*
d. window Fallfenster *n*
d.-worm [mechanism] Fallschnecke *f*
drope-in tape loading slot Bandeinlegeschlitz *m*
droppable fuel tank abwerfbarer Brennstoffbehälter *m*
d. tank *(Rak, Flg)* Zusatzbehälter *m*, Abwurfbehälter *m*
dropped axle gekröpfte Achse *f*
d. ceiling *(Bw)* untergehängte Decke *f*
d. frame *(Kfz)* gekröpfter Rahmen *m*
d.-out knot herausgefallener Ast *m*
d. panel Pilzkopf *m*, Säulenkopf *m (einer Pilzdecke; Stahlbetonbau)*
d. stitch *(Text)* Fallmasche *f*
dropper 1. *(Text)* Fadenreiter *m*; 2. *(Geol)* Nebengang *m*, Seitentrum *m*; 3. Tropfröhrchen *n*
dropping Fallen *n*, Abfallen *n*, Sinken *n*, Abnehmen *n*
d. angle Vorhaltewinkel *m (Bombenabwurf)*
d. ball test Kugelfallprüfung *f*
d. electrode Tropfelektrode *f (Polarografie)*
d. glass *(Ch)* Bürette *f*
d. point Tropfpunkt *m*, Schmelzpunkt *m (Schmierfett)*
d. resistor Reduzierwiderstand *m*; Vorwiderstand *m*
d. step *(Brg)* Falltreppe *f*
dropwise tropfenweise
dross Gekrätz *n*, Krätze *f*, Schlacke *f*, Garschaum *m*
d. lost Krätzeverlust *m*, Schlackeverlust *m*
drossing oven Abstrichofen *m*
drossy schlackenhaltig, schlackig
d. coal schwefelkieshaltige Kohle *f*
drought Dürre *f*, Regenmangel *m*
drove 1. Bewässerungskanal *m*; 2. Meißel *m*; 3. Herde *f*
d. cattle Herdenvieh *n*
drown / to 1. übertönen, überdecken *(Elektroakustik)*; 2. *(Brg)* ersaufen
drum 1. Trommel *f*; Zylinder *m*; Walze *f*; Kanister *m*; 2. Kabeltrommel *f*; 3. *(Am)* Schaltwalze *f*, Steuerwalze *f*; 4. Trommel, Radtrommel *f (Turbine)*
d. armature *(Elm)* Doppel-T-Anker *m*, Trommelanker *m*
d. barker *(Holz)* Entrindungstrommel *f*
d. basket *(Lw)* Trommelkorb *m (z. B. einer Sortiermaschine)*
d. bearing Trommellager *n*
d. brake Trommelbremse *f*
d. cam Mantelkurve *f*; Kurventrommel *f*
d.-chart recorder Trommelschreiber *m*
d. controller Walzenfahrschalter *m*
d.-curb *(Brg)* Schachtring *m*
d. dryer Walzentrockner *m*
d. filter Trommelfilter *n*
d. governor Flachregler *m*
d. head Trommelfell *n*
d. information assembler and dispatcher *(BMSR)* Rufnummernspeicher *m* und -geber *m*
d. memory Trommelspeicher *m*, Magnettrommelspeicher *m*
d. mixer Freifallmischer *m*, Trommelmischer *m*
d.-net-type aerial Reusenantenne *f*
d. rotor *(Elm)* Trommelläufer *m*
d. scanner *(Fs)* Trommelabtaster *m*
d. screen Trommelsieb *n*
d. shaft Trommelwelle *f*; Hebewerkstrommelwelle *f (Rotarybohren)*
d.-shaped recorder Trommelschreiber *m*
d.-shaped scale Trommelskala *f*
d. shutter *(Opt)* Trommelblende *f*, Zylinderverschluß *m*
d. starter Anlaßwalze *f*; Schaltwalzenanlasser *m*, Walzenanlasser *m*
d. stick Trommelstock *m*, Trommelschlägel *m*, Schlägel *m*
d. store *(BMSR)* Magnettrommelspeicher *m*
d.-stuff / to *(Text)* im Faß schmieren
d. surface Trommelfläche *f*, trommelförmige Oberfläche *f*
d. switch Walzen[fahr]schalter *m*, Rastenschalter *m*

d. tanning Faßgerbung *f*
d. test *(Brg)* Trommelversuch *m (auf Abrieb)*
d. tumbler Walkfaß *n*
d. turbine Trommelturbine *f*
d.-type chronograph Walzenchronograf *m*
d.-type controller Steuerwalze *f*
d.-type hoist Trommelfördermaschine *f*
d.-type magnetic separator Trommelmagnetscheider *m*
d.-type milling machine Trommelfräsmaschine *f*
d.-type retort furnace *(Met)* Tonnenretortenofen *m*
d.-type transfer machine Trommelmaschine *f*
d. water-meter Trommelwassermesser *m*
d. weir Trommelwehr *n*
d. winding Trommelwicklung *f*
d.-wound armature *(Elm)* Trommelanker *m*
drumming *(Brg)* Abklopfen *n* der Firste
Drummond light Drummondsches Kalklicht *n*
drunkenness Taumelfehler *m (Gewindegänge)*
dry / to trocknen
d.-dock [ein]docken
d. in eintrocknen
d. off *(Holz)* austrocknen
d. out austrocknen, vertrocknen
d. up eintrocknen
dry 1. trocken; 2. übergar *(Metall)*; 3. nicht aktiv
d. ash-free wasser- und aschefrei *(z. B. Kohle)*
d. back boiler Flammrohrkessel *m* mit an der Rückseite nicht wasserberührter Wendekammer
d. barometer Trockenbarometer *n*
d. batch Trockenmischung *f (Betontechnologie)*
d. battery Trockenbatterie *f*
d. bone ore Smithsonit *m*
d. box *(Kern)* Arbeitskammer *f*, Schutzkammer *f* mit eingebauten Handschuhen, Handschuhbox *f*
d. brined *(Led)* in Salzlake gelaugt und getrocknet
d. broke *(Holz)* Trockenausschuß *m*
d.-bulb temperature Trockentemperatur *f*
d.-bulb thermometer Trockenthermometer *n*
d. cargo vessel Trockenfrachter *m*
d. cell Trockenbatterie *f*, Trockenelement *n*
d.-cell battery Trockenbatterie *f*
d.-chemical fire extinguisher Trocken[feuer]löscher *m*
d. chips Trockenschnitzel *npl (Zuckerrüben usw.)*
d.-cleanable *(Text)* trocken reinigungsfähig, chemisch reinigungsfähig
d. cleaning 1. *(Text)* chemische Reinigung *f*; 2. *(Brg)* Trockensortierung *f*
d. cleaning solvent chemisches Reinigungsmittel *n*, Trockenreinigungsmittel *n*
d.-cleaning table *(Brg)* Luftherd *m*
d. condenser *(Wärm)* trockener Kondensator *m*, Fallrohrkondensator *m*, Gefällekondensator *m*
d. construction Montagebauweise *f*, Trockenbauweise *f*
d. core cable Kabel *n* mit Luftraumisolierung, Luftraumkabel *n*, Papierhohlkernkabel *n*
d. decating *(Text)* Trockendekatur *f*
d. deposition nichtaktiver Niederschlag *m*
d.-development Trockenentwicklung *f*
d. diode *(Am)* Halbleiterdiode *f*, Trockengleichrichter *m*
d.-disk clutch Trockenkupplung *f*
d.-disk joint Trockengelenk *n*
d.-disk rectifier Plattengleichrichter *m*
d. distillation Trockendestillation *f*
d. dock Trockendock *n*
d. dressing *(Lw)* Trockenbeizen *n*
d. drilling Trockenbohren *n*
d. drum Dampfsammler *m (Etagenkessel)*
d. end Trockenpartie *f (Papiermaschine)*
d. extrusion *(Plast)* Schlauchspritzen *n* ohne Anwendung von Lösungsmitteln, Strangpressen *n*
d. feeding Trockenfütterung *f*
d. finishing *(Text)* Trockenappretur *f*
d. friction Trockenreibung *f*
d. goods *(Am)* Textilien *pl*
d. grinding Trockenschleifen *n*, Trockenschliff *m*; Trockenmahlung *f*
d. harbour *(Schiff)* Strandhafen *m*
d. hole söhliges (aufwärts gerichtetes) Bohrloch *n*
d. ice Trockeneis *n*
d. iron siliziumarmes Roheisen *n*
d. joint kalte (schlechte) Lötstelle *f*
d. kiln Trockenkammer *f*
d. liner *(Kfz)* trockene Laufbuchse *f*
d. masonry Trockenmauerwerk *n*
d. matter Trockensubstanz *f*
d. milk Trockenmilch *f*
d. mineral-matter-free wasser- und mineralstofffrei *(Erdöl)*
d. offset *(Typ)* indirekter Hochdruck *m*, Trockenoffset *m*
d.-packing wire netting *(Brg)* Versatzmatte *f*, Versatzgeflecht *n*
d. parting Trockenscheidung *f*
d. pile Stabbatterie *f*
d. pipe Dampfsammelrohr *n*
d. pitch Hartpech *n*
d.-plate clutch *(Kfz)* Trocken[scheiben]kupplung *f*

d.-plate rectifier Trockengleichrichter *m*
d.-point engraving *(Typ)* Kaltnadelradierung *f*
d. process of ore dressing Erzaufbereitung *f* auf trockenem Wege
d. quencher Trockenlöscher *m*
d. quenching Trockenlöschung *f*
d. raising *(Text)* Trockenrauhen *n*
d. rectifier Trockengleichrichter *m*
d. rot Hausschwamm *m*, Holzschwamm *m*; Trockenfäule *f*
d. rot fungus *(Bw)* echter Hausschwamm *m*, Tränenschwamm *m*
d. rubble wall Trockenmauer[werk *n*] *f*
d. running Trockenlauf *m (Lager)*
d.-sand mould Trockensandform *f*
d. sand moulding Trockensandformen *n*
d.-saturated steam trocken gesättigter Dampf *m*, Sattdampf *m*
d. season hide Haut *f* der trockenen Jahreszeit *(Leder)*
d. slag nichtschmelzende Schlacke *f*
d. sludge Trockenschlamm *m (Gesundheitstechnik)*
d. spinning Trockenspinnen *n*
d. spinning frame Trockenspinnmaschine *f*
d. spinning method Trockenspinnverfahren *n*
d. spot *(Plast)* Fehlstelle *f*, Luftstelle *f*
d. steam trocken gesättigter Dampf *m*
d.-stores carrier Trockenfrachter *m (im Gegensatz zu Tanker)*
d. strength *(Text)* Trockenfestigkeit *f*
d.-sump lubrication *(Kfz)* Trockensumpfschmierung *f*
d. tack *(Plast)* Trockenklebrigkeit *f*
d. treatment Trockenaufbereitung *f*
d. twisting *(Text)* Trockenzwirnen *n*
d. wall[ing] Trockenmauer[werk *n*] *f*
d. weight Leergewicht *n*, Trockengewicht *n*
d. well Fehlbohrung *f*, trockene Sonde *f (Erdöl, Erdgas)*
dryer Trockenapparat *m*, Trockner *m*
d. canvas Baumwolltrockenfilz *m*
d. felt tightener Trockenfilzspanner *m*
drying agent Trockenmittel *n*
d. air Trockenluft *f*
d. barn Trockenscheune *f*, Trocknungsscheune *f*
d. bin Trockenbunker *m*
d. by circulating air Umlauftrocknung *f*
d. cabinet Trockenschrank *m*
d. chamber 1. Trockenkammer *f*, Trockenschrank *m*; 2. *(Text)* Trockenhänge *f*; 3. *(Lw)* Darre *f*
d. chamber with steam condensing apparatus Trockenraum *m* mit Dampfniederschlagvorrichtung
d. drum Trommeltrockner *m*
d. end *(Typ)* Trockenpartie *f*
d. from the surface *(Holz)* Trocknung *f* von außen nach innen
d. in the log Rundholztrocknung *f*
d. kiln Trockenkammer *f (z. B. für Holz)*
d. loft Trockenboden *m*
d. on frames *(Led)* Trocknung *f* im Rahmen
d. press 1. Trockenpresse *f*; 2. *(Led)* Abwelkpresse *f*
d. room 1. Trockenraum *m*, Trockenkammer *f*; 2. *(Lw)* Darre *f*
d. section 1. *(Text)* Trockenteil *n*; 2. *(Holz)* Trockenpartie *f*; 3. *(Brg)* Trockendienst *m*
d. shrinkage Trocknungsschrumpfen *n (Beton)*
d. stove Trockenofen *m*
d. tunnel Trockendarre *f*, Trockenkanal *m (z. B. für Holz)*
dryness Trockenheit *f*, Dürre *f*
d. of steam Dampftrockenheit *f*
dual dual, doppelt; Doppel-, Zwei-
d.-action telescopic shock absorber *(Kfz)* doppeltwirkender Teleskopstoßdämpfer *m*
d. amplification Reflexverstärkung *f (Funktechnik)*
d. belt sander *(Holz)* Doppelbandschleifmaschine *f*
d. card *(Büro)* Verbundkarte *f (Lochkartenmaschinen)*
d. carriageway road Autobahn *f*, Fernverkehrsstraße *f* mit zwei getrennten Fahrbahnen
d. control *(Rak)* Doppelsteuerung *f*
d. cycle reactor *(Kern)* Zweidruckreaktor *m*
d. decay *(Kern)* dualer Zerfall *m*
d. diversity Zweifach-Diversity *f (Funktechnik)*
d.-drive mit doppeltem Spindelantrieb (Antrieb)
d. feed tape carriage *(Büro)* doppelter Lochstreifenvorschub *m*
d.-fuel engine Zweistoffmotor *m*
d.-fuel operation Zweistoffbetrieb *m*, Betrieb *m* mit zwei Brennstoffsorten
d. grip chuck *(Masch)* Doppelspannfutter *n*
d. ignition Doppelzündung *f*
d.-lane zweispurig *(Straßenbau)*
d.-lane road zweispurige Fahrbahn *f*
d.-lateral coil *(El)* Wabenspule *f*, Honigwabenspule *f*
d. modulation Doppelmodulation *f*
d. pressure hydraulic system hydraulisches Doppeldrucksystem *n*
d. printing *(BMSR, Büro)* Zweifachschreibung *f*
d. producer Zwillingsfördersonde *f (Erdöl, Erdgas)*
d.-purpose machine Doppelzweckmaschine *f*

d. ram broaching machine Doppelstößelräummaschine *f*, Zwillingsräummaschine *f*
d. range variable capacitor (condenser) Doppeldrehkondensator *m*
d. receiver Reflexempfänger *m (Funktechnik)*
d.-trace oscillograph Zweistrahloszillograf *m*
d. tyre *(Kfz)* Doppelreifen *m*, Zwillingsreifen *m*
d. vertical control surface *(Flg)* doppeltes Seitenleitwerk *n*
duality *(Ph)* Dualität *f*
d. principle Dualitätsprinzip *n*
Duane and Hunt's law Duane-Huntsches Gesetz *n*
duant *(Kern)* Duante *f*
dub / to 1. *(Holz)* dechseln; 2. *(Led)* einfetten; 3. synchronisieren *(Film)*
dub 1. *(El)* Tonbandkopie *f*, Kopie *f*; 2. Pfuscher *m (Druckerei)*
dubbing 1. Lederfett *n*; 2. Schnitt *m*, Tonmischung *f (Tonband)*; 3. Synchronisierung *f (Film)*; 4. Ausfüllen *n*
duck 1. *(Lw)* Ente *f*; 2. *(Text)* Segeltuch *n*
duckbill Stoßschaufellader *m*, Entenschnabellader *m*
d. loader *s.* duckbill
d.-nose pliers Entenschnabelzange *f*
d. pliers Weberzange *f*
d. wiring plier Entenschnabelzange *f*
duckboard Laufbohle *f*, Laufsteg *m (z. B. für Dachdecker)*
duckfoot cultivator *(Lw)* Grubber *m* mit Gänsefußscharen
duct 1. Kanal *m*, Lüftungskanal *m*, Schacht *m (Lüftung, Heizung)*; Leitungskanal *m (Elektroinstallation)*; Spann[glied]kanal *m (Spannbeton)*; 2. *(El)* Kabelrohr *n*, Leitungsrohr *n*; 3. *(Masch)* Kühlluftführung *f*; 4. *(Typ)* Farbkasten *m (Druckmaschine)*
d. capacitor Durchführungskondensator *m*
d. cleaner Rohrbürste *f*
d. keel *(Schiff)* Tunnelkiel *m*, Kastenkiel *m*
d. route Mehrlochkabelkanal *m*
d. width *(Fmt)* Kanalbreite *f*
ducter Widerstandsmeßgerät (Ohmmeter) *n* für kleine Widerstände
ductile dehnbar, duktil, streckbar; plastisch, [kalt] verformbar
d. cast iron Gußeisen *n* mit Kugelgraphit
d. fracture Scherbruch *m*
ductility Dehnvermögen *n*, Dehnbarkeit *f*, Duktilität *f*, Ziehvermögen *n*, Streckbarkeit *f*; Plastizität *f*, Verformbarkeit *f*
d. at elevated temperatures Warmbildsamkeit *f*, Warmverformbarkeit *f*
d. limit Dehn[ungs]grenze *f*
ducting *(Bw)* Kanalführung *f*, System *n* der Leitungskanäle
d. of the air Kühlluftführung *f* in Kanälen
ductor *(Typ)* Rakel *f*, Abstreichmesser *n*
d. key Farbklotz *m*
d. roller Farbwalze *f*, Duktor *m*
dud Blindgänger *m*
Duddel oscillograph Drehspuloszillograf *m*, Schleifenoszillograf *m*
due date *(Am)* Fälligkeitstag *m*, Fälligkeitstermin *m*
duff / to fälschen
duff Feinkohle *f*
dug well Handbrunnen *m*
dulcitol *(Ch)* Dulcit *n*
dull / to 1. abstumpfen; stumpf werden; 2. mattieren; matt werden
dull 1. glanzlos, matt *(Oberfläche)*; matt, trübe *(Farbe, Licht)*; 2. dumpf *(Töne)*; 3. abgestumpft, stumpf *(Werkzeuge)*
d. black Mattschwarz *n*
d. clear varnish Hartmattlack *m*
d. coal Mattkohle *f*
d. decating *(Text)* Mattdekatur *f*
d. edge Baumkante *f (am Brett)*
d.-edged beam baumkantiger (waldkantiger) Balken *m*
d. emitter Kaltkatode *f*, Sparkatode *f*
d. emitter valve Röhre *f* mit Sparkatode
d. finish Mattanstrich *m*
d.-finish lacquer Schleiflack *m*
d.-finish sheet Mattblech *n*
d. gloss Mattglanz *m*
d. iron mattes Eisen *n*
d. polish Mattierung *f*
d. rayon mattierte Viskoseseide *f*
d. red heat Dunkelrotglut *f*, Dunkelrothitze *f*
dulling 1. Stumpfwerden *n*, Abstumpfen *n (Werkzeuge)*; 2. Mattieren *n*
d. agent Mattierungsmittel *n*
dullness 1. Glanzlosigkeit *f*, Trübe *f (Oberfläche; Farbe, Licht)*; 2. Dumpfheit *f (Töne)*; 3. Stumpfheit *f (Werkzeuge)*
dumb aerial verstimmte Antenne *f*
d. barge Schute *f* ohne Eigenantrieb, Schleppkahn *m*, Treckschute *f*
d.-bell Hantel *f (Sportgerät)*
d.-bell disk Hantelscheibe *f*
d.-bell model *(Kern)* Hantelmodell *n*
d. chalder Ruderträger *m (beim Holzschiff)*
d. vessel *(Schiff)* Fahrzeug *n* ohne Eigenantrieb
dummy 1. *(Typ)* Attrappe *f*; Blindband *m*; 2. Gesenkschmiederohling *m*; 3. Ponton *m*
d. aerial Ersatzantenne *f*, künstliche Antenne *f*
d. aerial resistor Antennenersatzwiderstand *m*
d. bar Anfahrstück *n (Stranggießen)*
d. barge Anlegerponton *m*, Anlegerfloß *n*

d. coil Hilfsspule *f*
d. coupling *(Eis)* Leerkupplung *f*
d. funnel blinder Schornstein *m*
d. fuse *(El)* Blindsicherung *f*
d. order *(Dat)* fiktiver Befehl *m*, Scheinbefehl *m*
d. piston Ausgleichkolben *m*, Druckausgleichkolben *m*, Entlastungskolben *m*, Gegendruckkolben *m*
d. plug 1. *(El)* Blindstecker *m*; 2. *(Masch)* Blindstopfen *m*, Blindstöpsel *m*
d. resistor Lastausgleichswiderstand *m*
d. rivet Heftniet *m*, Füllniet *m*
d. road *(Brg)* Blindort *n*
d.-road packing Blindortversatz *m*
d. roll Blindwalze *f*, Schleppwalze *f*
d. turn *s.* idle turn
d. valve *(El)* Blindröhre *f*, nichtausgenützte Röhre *f*
dump / to abkippen, entladen, kippen; 2. *(BMSR)* abwerfen *(Datenspeicherinhalt)*
dump 1. *(Geol)* Bergehalde *f*; 2. *(Fmt)* Schnellablaß *m*
d. barge Klappschute *f*, Klappenprahm *m*
d. body *(Kfz)* Kippkastenkarosserie *f*
d. car Kippwagen *m*
d. check *(Dat)* Abwurfkontrolle *f*
d. grate Kipprost *m (Kessel)*
d. pit Stoffgrube *f (Papierherstellung)*
d. scow *(Schiff)* Abfallschute *f*, Abfallprahm *m*
d. site Schuttabladeplatz *m*
d. truck Kipper *m*, Kippfahrzeug *n*; Kippwagen *m*
d. valve Ablaßventil *n*, Sicherheitsventil *n*
dumped zusammengepreßt *(Ballen)*
d. materials *(Eis)* Schüttgut *n*
dumper Vorderkipper *m*, Dumper *m*; Muldenkipper *m*, Autoschütter *m*
dumping boards *(Schiff)* Bodenwegerung *f (nur unter Luken)*
d. car Kipplore *f*, Muldenkippwagen *m*
d. cart Schüttkarren *m*
d. chute Entladungsrutsche *f*
d. grate Kipprost *m*, Klapprost *m*
d. ground 1. Lagerplatz *m*; Abladeplatz *m*
d. height Schütthöhe *f*
d. mechanism Kippvorrichtung *f (am Kfz)*
d. place Lagerplatz *m*, Holzlager *n*
d. reach *(Brg)* Schüttweite *f*
dung Stalldung *m*, Mist *m*
d. fork Mistgabel *f*
d. removal aggregate Entmistungsgerät *n*
d. spreader Miststreuer *m*, Stalldungstreuer *m*
d.-water Jauche *f*
d.-water cistern Jauchegrube *f*
dunking Eintauchen *n*
dunnage *(Schiff)* Stauhölzer *npl*, Garnier *n*
d. wood Garnierholz *n*, Stauholz *n*
duo-clad metal Verbundguß *m*
d.-core loudspeaker Doppelkonuslautsprecher *m*
d.-lateral coil Wabenspule *f*
d. mill Duowalzwerk *n*, Zwillingswalzwerk *n*
d. rolls Duowalzen *fpl*, Zwillingswalzen *fpl*
duodecagon Zwölfeck *n*
duodecimal system Zwölfersystem *n*, Duodezimalsystem *n*
duodiode Doppeldiode *f*, Duodiode *f*
d. rectifier Zweifachdiodengleichrichter *m*
duodynatron *(El)* Doppelfelddynatronschaltung *f*, Duodynatron *n*
dupe Kopie *f*, Duplikat *n (Film, Tonband)*
d. negative Duplikatnegativ *n (Film)*
d. positive Duplikatpositiv *n (Film)*
duplet *(Ph)* Dublett *n*
duplex *(Fmt)* Gegensprechverkehr *m*, Duplexverkehr *m*
d. artificial circuit *(Fmt)* Duplexnachbildung *f*
d. bearing Doppellager *n*, Lagerpaar *n*
d. beating engine Doppelholländer *m (Papierherstellung)*
d. cable Zweileiterkabel *n*
d. carburetter Doppelvergaser *m*
d. channel Duplexkanal *m*; Gegensprechkanal *m*
d. circuit Duplexleitung *f*; Gegensprechleitung *f*; Duplexschaltung *f*; Gegensprechschaltung *f*
d. communication Duplexverbindung *f*; Gegensprechverbindung *f*
d. cord Doppelschnur *f*, Zweileiterschnur *f*, zweiad[e]rige Schnur *f*
d. cutter Längs- und Querschneidemaschine *f*
d. diode Duodiode *f* mit Doppelkatode (zwei getrennten Katoden)
d. grinder Zwillingsschleifmaschine *f*
d. halftone *(Typ)* Duplexautotypie *f*
d. ignition *(Kfz)* Doppelzündung *f*
d. lathe Doppeldrehmaschine *f*
d. microscope Doppelmikroskop *n*
d. oil trap Doppelölabscheider *m*
d. operation *(Fmt)* Duplexbetrieb *m*
d. phone connection Doppelsprechschaltung *f*, Gegensprechschaltung *f*
d. planing machine Doppelhobelmaschine *f*
d. print[ing] *(Text)* beiderseitiger (zweiseitiger) Druck *m*, beiderseitiger Farbauftrag *m*, Drucken *n* auf beiden Seiten
d. printing machine *(Text)* Duplexdruckapparat *m*, Maschine *f* für zweiseitigen Druck
d. process Duplexverfahren *n (Bessemer- und SM-Verfahren)*
d. pruning saw Astsäge *f*
d. pump Duplexpumpe *f*

d. screen printer *(Text)* Duplexfilmdruckmaschine *f*
d. set *(Fmt)* Gegensprechsatz *m*
d. spiral-type reamer kreuzverzahnte Reibahle *f*
d. star connection *(El)* Doppelsternschaltung *f*
d. steam engine Zwillingsdampfmaschine *f*
d. steel Duplexstahl *m*
d. tapping machine zweispindlige Gewindeschneidmaschine *f*
d. telephony Gegensprechverkehr *m*, Duplexsprechverkehr *m*
d.-type filter Doppelfilter *n*
d. variable-area track Doppelzackenschrift *f* *(Tonfilm)*
d. winding *(El)* Doppelwicklung *f*, zweifach geschlossene Wicklung *f* *(Anker)*
duplexer *(Fmt)* Simultanantenne *f* *(Senden/Empfang)*; Sende-Empfangs-Weiche *f* *(Radar)*
duplexing 1. *(Fmt)* Gegensprechanordnung *f*, Viereranordnung *f*; 2. *(Met)* Duplexschmelzverfahren *n*
duplicate / to 1. kopieren, nachformen; reproduzieren; in Serie fertigen; nachbauen *(Geräte, Maschinen)*; 2. *(Typ)* vervielfältigen, abziehen; 3. kopieren *(Film)*
d. in three dimensions *(Masch)* dreidimensional nachformen, raumnachformen
duplicate Kopie *f*; Duplikat *n*; Abzug *m*; Reproduktion *f*
d. boring Serienbohren *n*
d. part Massenteil *n*
d. plate *(Typ)* Duplikatplatte *f*
d. production Großserienfertigung *f*, Reihenfertigung *f*, Serienherstellung *f*
d. turning Nachformdrehen *n*
d. work holding fixture Doppelspannvorrichtung *f*
duplicating Nachformen *n*; Reproduzieren *n*
d. attachment Nachformeinrichtung *f*
d. lathe Nachformdrehmaschine *f*
d. paper Saugpost *f*, Abzugpapier *n*, Vervielfältigungspapier *n*
d. rack *(Büro)* Lochkartenbett *n* *(Hauptkarten)*
d. stencil Wachsmatrize *f*
duplication 1. Kopieren *n*; Nachformung *f*; Reproduktion *f*; 2. *(Typ)* Vervielfältigung *f*
d. check *(Büro)* Duplikatvergleich *m*, Zwillingsprüfung *f*
duplicator 1. *(Typ)* Abziehapparat *m*, Vervielfältigungsapparat *m*; 2. *(Dat)* Doppler *m* *(Lochkarten)*; 3. Nachformeinrichtung *f* *(Metallbearbeitung)*
duplicity *(Math)* Zweiwertigkeit *f*
duprene Chloroprenkautschuk *m*, Dupren *n*
Dupuis boiler Dupuiskessel *m*, stehender Heizröhrenkessel *m* mit vorderem (liegendem) Walzenkessel
durability Haltbarkeit *f*, Dauerfestigkeit *f*, Lebensdauer *f*, Standfestigkeit *f*; Alterungsbeständigkeit *f*, Beständigkeit *f*, Dauerhaftigkeit *f*
d. of the tool Standzeit (Lebensdauer) *f* des Werkzeugs
durable finish *(Text)* Dauerausrüstung *f*, Permanentappretur *f*
d. pleat *(Text)* Dauerfalte *f*
durain Durit *m*, Mattkohle *f*
duraluminium Duraluminium *n*, Dur[al] *n*
duration Dauer *f*, Zeitdauer *f*
d. of a call in minutes *(Fmt)* Gesprächsminuten *fpl*
d. of a pulse *(Fs)* Impulsdauer *f*
d. of ascent *(Rak)* Steigzeit *f*
d. of charge Ladezeit *f*, Ladedauer *f* *(Sammler)*
d. of combustion *(Wärm)* Brenndauer *f*, Brennzeit *f*
d. of contact Berührungszeit *f*, Kontaktdauer *f*
d. of powered flight *(Rak)* Dauer *f* des Antriebsfluges
d. of short-circuit Kurzschlußdauer *f*
d. of the free vibrations *(El)* Ausschwingzeit *f*
d. of thrust *(Rak)* Antriebsdauer *f*
d. time modulation Impulslängenmodulation *f*
durobronze Hartbronze *f*
dust / to abstauben, entstauben, abwischen, ausstäuben
d. with bark mit Lohe bestreuen *(Leder)*
dust 1. Staub *m*; 2. Haushaltabfälle *mpl*; Müll *m*; 3. Blütenstaub *m*
d. aspiration Staubabsaugung *f*
d. bars *(Text)* Abfallrost *m*
d. bin Mülleimer *m*, Müllkasten *m*, Mülltonne *f*
d. binding oil Reißöl *n*, Staubbindeöl *n*
d. brush Staubpinsel *m*
d. cage filter Siebtrommelfilter *n*
d. cap Staubkappe *f*
d. car Müllabfuhrwagen *m*
d. catcher Staubabscheider *m*
d. coal Kohlestaub *m*
d. collecting equipment Staubabsaugeeinrichtung *f*
d. collector Staubsichter *m*, Staubabscheider *m*
d. core *(El)* Eisenpulverkern *m*, Massekern *m*
d. cover *(Typ)* Schutzumschlag *m*
d. exclusion lip *(Masch)* Staubschutzlippe *f*
d. exhaust Staubabsaugung *f*

d. exhaust system Staubabsaugeeinrichtung *f*
d. exhauster Staubabsaugeeinrichtung *f*
d. exposure *(Brg)* Staubbelastung *f*
d. extraction equipment Staubabsaugeeinrichtung *f*
d. guard Staubschutz *m*
d. iron core *(El)* Hochfrequenzeisenkern *m*, Pulverkern *m*
d.-laden air staubhaltige Luft *f*
d. man Müllabfuhrarbeiter *m*
d. ore mulmiges Erz *n*
d. precipitation Staubniederschlag[ung *f*] *m*
d.-proof staubdicht, staubgeschützt, staubsicher
d.-protective cover Staubschutzhülle *f*
d. removal Entstauben *n*, Entstaubung *f*
d. remover Staubentferner *m*, Staubsammler *m*, Zyklon *m*
d.-resistant staubbeständig
d. respirator Staubmaske *f*
d. separator Staubabscheider *m*
d. shield Staubschutz *m*
d. suppression Staubbekämpfung *f*
d.-tight staubdicht, staubsicher
d. washer Staubschutzring *m*

duster 1. Stäubegerät *n*, Zerstäuber *m (Pflanzenschutz)*; 2. Abfallabscheider *m*, Staubsammler *m*; 3. Staublappen *m*
dustiness *(Brg)* Staubdichte *f*
dusting Einstauben *n*; Abstauben *n*
d. roller *(Typ)* Abstaubwalze *f*
dusty *(Brg)* staubhaltig, staubbeladen
Dutch bond Kreuzverband *m (Mauerwerk)*
Dutch timber holländische Balken *mpl*
Dutch tongs Froschklemme *f*
duty 1. Beanspruchung *f (Maschine, Werkzeug)*; Leistung *f*, Auslastung *f (Maschine)*; 2. *(El, Fmt)* Betriebsart *f*; 3. Zoll *m*
d. cycle *(BMSR)* Arbeitszyklus *m*, Arbeitsspiel *n*
d. factor 1. Arbeitsphase *f*, Belastungszeitverhältnis *n (z. B. bei Elektronenröhren)*; 2. Tastverhältnis *n (Radar)*
d.-free gebührenfrei
d. ratio Tastverhältnis *n (Radar)*
d. signal boat Kurierboot *n*
duvetyn[e] *(Text)* Duvetine *m*
dwarf signal *(Eis)* Zwergsignal *n*
d. wood krummfaseriges Holz *n*
dwell *(Masch)* Bewegungspause *f*, Stillstand[sdauer *f*] *m*, Verzögerung *f*
d. cam Verzögerungsnocke *f*
d. for table index Stillstandsdauer *f* für Tischschaltung
d. of cam Nockenerhebungszeit *f*, Verweilzeit *f*
d. period Haltedauer *f*; Stillstandsdauer *f*

dwelling Bewegungspause *f*, Stillstand *m*, Verzögerung *f*
d. house Wohngebäude *n*
d. unit Wohn[ungs]einheit *f*
dyad *(Math)* Dyade *f*
dyadic system dyadisches System *n*, Zweiersystem *n*
dye / to *(Text)* färben
d. again nachfärben
d. at the boil kochend färben
d. completely ausfärben
d. during milling in der Walke färben
d. fast [shades] echt färben
d. in a neutral bath neutral färben
d. in a single bath einbadig färben
d. in the dye back auf der Kufe färben
d. in the hose im Schlauch färben
d. in the loose wool in der Wolle färben
d. in the package spulenfärben, im Garnkörper färben
d. in the piece im Stück färben
d. in the size in der Schlichte färben
d. in the yarn im Garn färben
d. to pattern nach Muster färben
d. under the surface unter der Flotte färben

dye Farbe *f*; Farbstoff *m*, Färbemittel *n*
d. absorption Farbaufnahme *f*
d. absorptive farbaufnahmefähig
d. affinity Anfärbbarkeit *f*, Farbaufnahmefähigkeit *f*, Farbstoffaffinität *f*
d. base Farbbase *f*
d. bath Farbflotte *f*, Färbebad *n*
d. dispersing agent Farbstoffdispergiermittel *n*, Farbverteilungsmittel *n*
d.-forming farbbildend
d. in powder form pulverförmiger Farbstoff *m*
d. liquor Färbeflotte *f*, Färbeflüssigkeit *f*
d. paste Farbpaste *f*, Farbteig *m*
d. remover Farbentferner *m*
d. solution Flotte *f*
d. solvent Farbstofflösemittel *n*, Farbstofflöser *m*
d. vessel with time and temperature control Kufe *f* mit Färbedauer- und Temperaturmeßvorrichtung
d. warp beam Färbe[kett]baum *m*

dyed in the yarn stranggefärbt
dyeing apparatus Färbeapparat *m (bewegte Farbflotte)*
d. in equal depth gleichmäßiges Durchfärben *n*
d. in the hank Strangfärben *n*
d. in the loose state Färben *n* in der Flocke
d. machine Färbemaschine *f (bewegtes Färbegut)*
d. of the stock Anfärben *n* des Papierstoffes

d. on a mordant Färben *n* auf Vorbeize
d. on the padder Foulardfärberei *f*
d. pad Färbefoulard *m*
d. tube perforierter Garnträger *m*, Färbehülse *f*
d. under pressure Hochdruckfärberei *f*
d. with sulphur dyes Schwefelfärbung *f*
dyeline paper *(Typ)* Lichtpauspapier *n*
dyer Färber *m*
dyer's rocket Färberwau *m*
dyestuff Farbstoff *m*
dyewood Farbholz *n*
dying away *(Ak)* Ausschwingen *n*
d. down time Ausschwingzeit *f*
d.-out Ausschwingen *n*
d.-out constant Ausschwingkonstante *f*
d.-out process Ausschwingvorgang *m*
d.-out time Ausschwingdauer *f*, Abklingzeit *f*
dyke 1. Damm *m*, Deich *m*; Trockenmauer *f*; 2. *(Geol)* Intrusivgang *m*, Gesteinsgang *m*
dynaforming Explosionsumformen *n* mit Wasser als Druckübertragungsmittel
dynamic dynamisch
d. air-speed indicator Staudruckmesser *m*
d. balance *(Mech, Flg)* dynamisches Gleichgewicht *n*, Massenausgleich *m*; Schleuderfreiheit *f*
d. balance test Schleuderprüfung *f*
d. balancer Ausgleich[s]masse *f*, Ausgleich[s]gewicht *n*
d. balancing dynamische Ausgleichung *f*; dynamische Auswuchtung *f*
d. balancing rig Auswuchtmaschine *f*
d. ball-impact test Kugelschlagprüfung *f* *(Materialprüfung)*
d. ball-impact testing apparatus Kugelschlaggerät *n* *(Materialprüfung)*
d. behaviour dynamisches Verhalten *n*
d. braking *(Elm)* Gegenstrombremsung *f*, Kurzschlußbremsung *f*, Widerstandsbremsung *f*
d. characteristic *(El)* Arbeitskennlinie *f*, dynamische Charakteristik (Kennlinie) *f*
d. compaction *(Bw)* Rüttelverdichtung *f*, Vibrationsverdichtung *f*
d. condenser *(Elm)* Synchronphasenschieber *m*
d. condenser electrometer Meßverstärker *m* mit Schwingkondensator
d. convergence *(Ph)* dynamische Bündelung *f*
d. electricity dynamische (strömende) Elektrizität *f*
d. equilibrium dynamisches Gleichgewicht *n* *(z. B. zwischen Emission und Absorption)*
d. equilibrium ratio dynamisches Gleichgewichtsverhältnis *n*
d. expander Dynamikentzerrer *m*
d. impedance Impedanz *f* bei Parallelresonanz
d. lateral pressure dynamischer Seitendruck *m*
d. load dynamischer Lastwert *m*
d. luminous sensitivity Wechsellichtempfindlichkeit *f* *(Fotozellen usw.)*
d. memory *(Dat)* dynamischer Speicher *m*
d. multiplier *(Ph)* Pendelvervielfacher *m*, dynamischer Vervielfacher *m*
d. plate resistance innerer Widerstand *m* *(der Elektronenröhre)*, Röhrenarbeitswiderstand *m*
d. pressure Staudruck *m*
d. range Lautstärkeumfang *m* *(Elektroakustik)*
d. resistance *(El, Fmt)* Arbeitswiderstand *m*, dynamischer Widerstand *m*, Resonanzwiderstand *m*
d. storage *(Dat)* dynamische Speicherung *f*
d. strength Schwingungsfestigkeit *f*, dynamische Festigkeit *f*
d. stress schwingende (dynamische) Belastung *f*
d. subroutine *(Dat)* dynamisches Teilprogramm (Unterprogramm) *n*
d. viscosity *(Ph)* dynamische Zähigkeit *f*
dynamical similitude *(Ph)* dynamische Ähnlichkeit *f*
dynamicizer *(Dat)* Parallelserienwandler *m*
dynamics Dynamik *f*
dynamite Dynamit *n*, Sprengstoff *m*
dynamited [über]beschwert *(Seide)*
dynamo Dynamo *m*; Gleichstromgenerator *m*, Gleichstrommaschine *f*; *(Kfz)* Lichtmaschine *f*
d. armature Dynamoanker *m*
d. battery ignition unit *(Kfz)* Licht-Batterie-Zünder *m*
d.-electric machine Maschine *f* mit Elektromagnet
d. machine Dynamomaschine *f*
d. magneto *(Kfz)* Licht-Magnet-Zünder *m*
d. pressure *(Kfz)* Maschinenspannung *f* *(Lichtmaschine)*
d. sheet Dynamoblech *n*
d. with ground return *(Kfz)* Licht *n* mit Masserückleitung
d. with insulated return *(Kfz)* Licht *n* mit isolierter Rückleitung
dynamometer *(Ph)* Kraftmesser *m*, Dynamometer *n*, Federzugwaage *f*; *(El)* Kraftmesser; Leistungsmesser *m*; Drehmomentmesser *m*
d.-type meter movement *(El)* Repulsionsmeßwerk *n* *(Abart des Kreuzspuleninstrumentes für Wechselgrößen)*
d.-type wattmeter Drehspulleistungsmesser *m*, Drehspulwattmeter *n*
dynamotor Drehumformer *m* *(Motor und Generator)*

Dynapak high energy rate forming Umformen *n* mit Dynapak-Schnellschlagpresse
Dynapak machine Dynapak-Schnellschlagpresse *f*
Dynapak process Dynapak-Verfahren *n* *(Umformtechnik)*
dynatron Dynatron *n*
d. characteristic *s.* d. effect
d. effect Dynatronwirkung *f*
d. oscillator Oszillator *m* mit negativem Widerstand
dyne Dyn *n* *(Krafteinheit)*
dyotron Dyotron *n* *(cm-Wellen-Oszillatorröhre)*
dysprosia Dysprosiumoxid *n*
dysprosium Dysprosium *n*

E

E-layer E-Schicht *f* *(Ionosphäre)*
eagre Flutwelle *f* in Untiefen
ear 1. Lasche *f*, Öse *f*; Auge *n* *(Säge)*; 2. *(Lw)* Ähre *f*; 3. Zipfel *m*
e. cap *(El)* Hör[er]muschel *f*
e. lifting device *(Lw)* Ährenheber *m*
earing *(Met)* Zipfelbildung *f*
earlier stage *(Rak)* untere Stufe *f*
early-bearing prop *(Brg)* frühtragender Stempel *m*
e. ignition Frühzündung *f*, Vorzündung *f*
e. shift Frühschicht *f*
e. spark Frühzündung *f*; Vorzündung *f*
e. warning Frühwarnung *f*, Fernwarnung *f* *(Funkmeßtechnik)*
e. warning system Vorwarnsystem *n* *(Radar)*
e. wood Frühholz *n*
earmark / to bezeichnen, kennzeichnen, markieren; vormerken
earnings Arbeitslohn *m*, Einkommen *n*, Verdienst *m*
e. from exports Exporteinnahmen *fpl*
earphone Kopfhörer *m*
earth / to erden, an Erde legen
earth 1. Erde *f*, Erdboden *m*; 2. *(El)* Erdschluß *m*
e. arrester Erdleitungskurzschließer *m*
e. auger Erdbohrer *m* *(z. B. für Bodenuntersuchungen)*
e. bar Erdungsschiene *f*
e. boring bit Erdbohrer *m*
e. boring machine Erdbohrer *m*
e. capacitance Erdkapazität *f*, Kapazität *f* gegen Erde
e. circuit Erdstromkreis *m*, Erdschleife *f*
e. coil Erdungsdrossel[spule] *f*, Erdungsspule *f*, Entladespule *f*
e. conduction telegraphy Erdtelegrafie *f*
e. conductivity [spezifische] Leitfähigkeit *f* des Erdbodens
e. connection Erdleitung *f*, Erdverbindung *f*, Erdung[sleitung] *f*
e. connection point Erdungspunkt *m*, Erdungsanschluß[stelle *f*] *m*
e. current 1. [natürlicher] Erdstrom *m*; 2. Erd[schluß]strom *m*
e.-current equalizer Erd[schluß]stromausgleicher *m*
e.-current prospecting Erdstromschürfung *f*
e. detector Erdschlußanzeiger *m*, Erdschlußprüfer *m*
e. displacement Bodenbewegung *f*, Erdbewegung *f*; Erdbewegungsarbeiten *fpl*
e. electrode Masseelektrode *f* *(Zündkerze)*
e. equalizer conductor Antennenabgleichelement *n*, Abgleichelement *n*
e. fault Erdschluß *m*
e.-fault protection Erdschlußschutz *m*
e. fill Erdaufschüttung *f*
e. fill dam Erd[schütt]damm *m*
e. free *(Am)* erdschlußfrei; erdschlußsicher
e. induction compass *(Flg)* Induktionskompaß *m*
e. inductor Erdinduktor *m*
e. lead Erddraht *m*; Erd[ungs]leitung *f*, Erd[ungs]leiter *m*
e. leakage [schleichender] Erdschluß *m*
e. leakage current Erdschlußstrom *m*, Kriechstrom *m* gegen Erde
e. leakage indicator Erdschlußanzeiger *m*, Erdschlußprüfer *m*
e. leakage meter Erdschlußmesser *m*
e. leakage relay Erdschlußrelais *n*
e.-leakage trip Erdschlußautomat *m*, Heinisch-Riedl-Schutzschalter *m*
e. location Erdschlußbestimmung *f*
e. magnetism Erdmagnetismus *m*
e. metal Erdmetall *n*
e. movement Bodenbewegung *f*
e.-phantom circuit Viererkreis *m* mit Erdrückführung *(Fernsprechtechnik)*
e. plate Erdungsplatte *f*
e. potential Erdpotential *n*, Nullpotential *n*
e. potential of radio-frequency hochfrequenzmäßiges Erdpotential *n*
e. pressure Bodendruck *m*, Erddruck *m*
e. protective sheath geerdete Abschirmung *f*, geerdeter Schirm *m*
e. resistance Erdausbreitungswiderstand *m*
e. resistance meter Erdungsmesser *m*
e. return Erdrückleitung *f*
e. return brush Erdungsbürste *f*
e. return circuit eindrähtige Leitung *f*, Einfachleitung *f*, Einzelleitung *f*, Stromkreis *m* mit Erdrückleitung

e. screen geerdete Schirmwicklung *f*, Gegengewicht *n*
e. sheath geerdeter Mantel (Schirm) *m*
e. silo Kartoffelmiete *f*
e.-space service Erde-Weltraum-Funkdienst *m*
e. station *(Flg, Astr)* Erdfunkstelle *f*
e. switch Erd[ungs]schalter *m*
e. system of an aerial Antennenerdleitung *f*
e. terminal Erdanschlußklemme *f*, Erd[ungs]-klemme *f*, geerdete Klemme *f*
e. wax Erdwachs *n*, Bergwachs *n*, Ozokerit *m*
e. wire Erd[ungs]draht *m*, Erdleiter *m*, Erdleitung *f*
earthed geerdet
e. neutral geerdeter Sternpunkt *m*, Nullpunkt *m*
e. neutral conductor geerdeter Mittelleiter *m*, [geerdeter] Nulleiter *m*, Nulleiter mit Stützerder
e. neutral system System *n* mit geerdetem Mittelpunkt
e. system geerdetes Netz (System) *n*
e. wire geerdeter Draht *m*, Erdseil *n*
earthenware Steingut *n*
earthing Erdung *f*
e. auto-transformer Erdungsdrossel *f (Bauchtransformator)*
e. bush Erd[ungs]buchse *f*
e. choke coil Erdungsdrossel[spule] *f*, Entladespule *f*
e. coil Erdungsdrossel[spule] *f*, Entladespule *f*
e. collar Erdungsring *m*
e. contact Erdungskontakt *m*
e. contact-type plug Schukostecker *m*
e. core Erdungsader *f*
e. device Erdungsanlage *f*
e. inductor Erdungsdrossel *f*
e. isolator Erdungstrenn[schalt]er *m*, Erdungsschalter *m*
e. lead *(Kfz)* Massekabel *n*
e. reactor Erdungsdrossel *f*
e. resistance Erdungswiderstand *m*
e. strip Banderder *m*
e. switch Erdungsschalter *m*
e. terminal Erdanschlußklemme *f*, Erd[ungs]-klemme *f*, geerdete Klemme *f*
earthmoving Erdbewegung[sarbeiten *fpl*] *f*
earthquake Erdbeben *n*
earth's attraction Erdanziehung *f*
e. crust Erdrinde *f*
e. magnetic field erdmagnetisches Feld *n*, Erdmagnetfeld *n*
e. motion Erdbewegung *f*
e. oblateness Erdabplattung *f*
e. orbit Erdbahn *f*
e. shadow Erdschatten *m*
e. surface Erdoberfläche *f*
earthworks Erdarbeiten *fpl*, Erdbau[arbeiten *fpl*] *m*
earthy brown coal Weichbraunkohle *f*, Lignit *m*
e. dilution of ore Gangart *f*, Ganggestein *n*
ease / to 1. entlasten, erleichtern; lockern; 2. lösen *(Klemme usw.)*; 3. abrunden, ausrunden
e. a vessel *(Schiff)* aufkommen *(mit dem Ruder nachgeben)*
ease of care Leichtigkeit (Bequemlichkeit) *f* der Pflege; leichte (bequeme) Pflege *f*
e. of handling Leichtigkeit (Bequemlichkeit) *f* der Handhabung; leichte (bequeme) Handhabung *f*
e. of maintenance Leichtigkeit (Bequemlichkeit) *f* der Wartung; leichte (bequeme) Wartung *f*
e. of operation Leichtigkeit (Bequemlichkeit) *f* der Bedienung; leichte (bequeme) Bedienbarkeit *f*
e. of steering *(Kfz)* Leichtgängigkeit *f* der Steuerung, leichtgängige Steuerung *f*
easel *(Typ)* Staffelei *f*
easement curve Übergangsbogen *m*, Übergangskurve *f (Straßenbau)*
easer *(Brg)* Kranzschuß *m*
easily shifted gear *(Kfz)* leicht schaltbarer Gang *m*
e. split *(Holz)* leicht spaltend
easing 1. Entlastung *f*; Lockerung *f*; 2. Lösen *n (Klemme usw.)*; 3. Abrunden *n*, Ausrunden *n*
e. at the point Kopfkürzung *f (Verzahnung)*
e. of work Arbeitserleichterung *f*
East-Indian geranium oil Palmarosaöl *n*
easy bilge *(Schiff)* runde Kimm *f*
e.-care *(Text)* pflegeleicht, leicht zu pflegen
e. draught *(Schiff)* geringer Tiefgang *m*
e. machining leicht zerspanbar
e. roll *(Schiff)* leichtes Schlingern *n*
e. speed mäßige Geschwindigkeit *f*
e. start jet Startdüse *f*
e. starting arbor *(Masch)* Aufsteckhalter *m* mit Abdrückmutter
e. steerability *(Schiff)* Wendigkeit *f*
e. steerer wendiges Schiff *n*
e. to shape bildsam, leicht verformbar
e. to use handlich
e. to view übersichtlich
e.-trimmer Leichttrimmer *m (Kohlendampfer)*
e.-view übersichtlich
e.-wear *(Text)* trageleicht
e. working vessel im Seegang ruhig liegendes Schiff *n*
eat / to fressen *(Säure an Metall)*
e. out herausfressen
eave[s] Dachrinne *f*, Dachtraufe *f*, Traufe *f*
eaves *(Schiff)* Wetterdach *n*

e. **board** *(Bw)* Traufbohle *f*
e. **gutter** Dachrinne *f*
ebb [tide] Ebbe *f*
ebb and flood / to ab- und zuströmen
e. **and flood** Ebbe *f* und Flut *f*
ebonite Ebonit *n*, Hartgummi *m*
e. **insulator** Ebonitisolator *m*, Hartgummiisolator *m*
e. **plug** Hartgummistopfen *m*, Hartgummiverschluß *m*
ebony Ebenholz *n*
ebullition Aufwallung *f*
eccentric Exzenter *m*, Scheibenkurbel *f*
eccentric ausmittig, außermittig, exzentrisch
e. **action of force** außermittige Krafteinwirkung *f*
e. **bearing** Exzenterlager *n*
e. **bore** Außermittebohrung *f*
e. **buckle** Exzenterangel *f (Gattersäge)*
e. **cam** Exzenter *m*
e. **chuck** Außermittespannfutter *n*
e. **chucking** Außermittespannen *n*
e. **circle** 1. *(Math)* Scheitelkreis *m*; 2. *(Masch)* Exzenterkreis *m*; 3. Exzenterrille *f*
e. **collar strap** Exzenterbügel *m*
e. **crank mechanism** exzentrisches (geschränktes) Schubkurbelgetriebe *n*
e. **disk** Exzenterscheibe *f*, Kröpfscheibe *f*
e. **drive** Exzenterantrieb *m*
e. **driven feed pump** Exzenterspeisepumpe *f*, Speisepumpe *f* mit Exzenterantrieb
e. **hoop** Exzenterring *m*
e. **housing** Exzenterhülse *f*
e. **lathe** Unrunddrehmaschine *f*
e. **lever** Exzenterhebel *m*
e. **motion** exzentrische Bewegung *f*; Exzenterbewegung *f*
e. **pin** Exzenterzapfen *m*
e. **press** Exzenterpresse *f*
e. **rod** Schwingenstange *f*
e. **shaft** Exzenterwelle *f*
e. **shaft press** Exzenterpresse *f*
e. **sheave** Exzenterscheibe *f*
e. **sleeve** Exzenterhülse *f*
e. **tang** Exzenterangel *f*
e. **turning** Außermittedrehen *n*
e. **turning lathe** *(Holz)* Passigdrehmaschine *f*
e. **vibrating screen** Exzenterschwingsieb *n*
e. **wheel** Exzenterrad *n*
eccentrically außermittig, außerachsig, exzentrisch
eccentricity 1. Außermittigkeit *f*; Exzentrizität *f*; exzentrische Lage *f*; Außermittemaß *n*; 2. Schlag *m (Rad, Welle)*
e. **of crankshaft** Schlag *m* der Kurbelwelle
e. **of rest** Exzentrizität *f* der Ruhe
e. **of rotation** Exzentrizität *f* der Drehung
e. **of the disk** Scheibenexzentrizität *f*
e. **signals** *(Fmt)* Mehrfachzeichen *npl*, Doppelzeichen *npl*
Eccles-Jordan-circuit *(Fs)* Eccles-Jordan-Schaltung *f*; Flip-Flop-[Kipp-]Schaltung *f*
Eccles-Jordan trigger Flip-Flop-Auslöser *m*, Eccles-Jordan-Schalter *m*
echelon Staffelung *f*
e. **aerial** Antenne *f* mit gestaffelten Strahlern
e. **grating** [Michelsonsches] Stufengitter *n*; Echelon *m*
e. **lens** Stufenlinse *f*
e. **working** *(Fmt)* Staffelbetrieb *m*
echo / to nachhallen, widerhallen
echo 1. Nachhall *m*, Widerhall *m*, Echo *n*; 2. *(Fs)* Geist[erbild *n*] *m*
e. **attenuation measuring set** Echomesser *m*
e. **box** Echobox *f*, Echohohlraumresonator *m* *(Mikrowellen)*; Impulstastgerät *n*, Prüfechogerät *n (Radar)*
e. **cancellation** *(Fs)* Echounterdrückung *f*
e. **checking** *(Dat)* Schleifenprobe *f*
e. **current attenuation** Echodämpfung *f*
e. **depth sounding** Echolotung *f (Tiefenbestimmung)*
e. **image** *(Fs)* Doppelbild *n*, Geist[erbild *n*] *m*
e. **interval** *(Fmt, Fs)* Echolaufzeit *f*
e. **killer** *(Fs)* Echosperre *f*
e. **matching** Echoabgleichung *f*
e. **ranging** Entfernungsmessung *f (Radar)*
e. **sounder** [akustisches] Echolot *n*
e. **sounding** [akustische] Echolotung *f*
e. **suppressor** *(Fmt)* Echosperre *f*
e. **transient time** *(Fmt, Fs)* Echolaufzeit *f*
e. **trouble** Echostörung *f*
e. **wave** Echowelle *f*
economic planning Wirtschaftsplanung *f*
e. **unit** wirtschaftliche Einheit *f*
economical grate *(Wärm)* Sparrost *m*
e. **mixture** *(Flg)* Sparfluggemisch *n*
e. **transformer** Spartransformator *m*, Sparwandler *m*, Autotransformator *m*, Einspulentransformator *m*
economics 1. Wirtschaftslehre *f*, Volkswirtschaftslehre *f*, politische Ökonomie; 2. Einsparungen *fpl*
economizer *(Wärm)* Abgasvorwärmer *m*, Rauchgasvorwärmer *m*, Röhrenvorwärmer *m* mit Heizung durch Abgase, Ekonomiser *m*, Speisewasservorwärmer *m*, Luftvorwärmer *m*
e. **jet** Spardüse *f*
economizing circuit *(El)* Sparschaltung *f*
economy Wirtschaft[ssystem *n*] *f*; Wirtschaftlichkeit *f*, Sparsamkeit *f*
e. **coil** Spartransformator *m*
e. **filament** Sparkatode *f*
e. **of operation** Wirtschaftlichkeit *f* im Betrieb
e. **of the working of the boiler** Wirtschaftlichkeit *f* des Kesselbetriebes

e. resistance Sparwiderstand *m*; Haltewiderstand *m*, Vorwiderstand *m* der Haltespule *(Relais)*
e. transformer Spartransformator *m*, Sparwandler *m*, Autotransformator *m*, Einspulentransformator *m*
eddy / to wirbeln
eddy Wirbel *m*, Strudel *m*
e. current 1. *(El)* Wirbelstrom *m*, Foucault-Strom *m*; 2. turbulente Strömung *f (Hydraulik)*
e.-current brake Wirbelstrombremse *f*
e.-current braking Wirbelstrombremsung *f*
e.-current heating method Wirbelstromglühverfahren *n*
e.-current loss Wirbelstromverlust[e *pl*] *m*
e. diffusion *(Ph, Kern)* turbulenzüberlagerte Diffusion *f*
e. flow Wirbelströmung *f (Hydrodynamik)*
e. formation Wirbelbildung *f*
e. loss Wirbelverlust *m (Turbine)*
e. space Wirbelraum *m (Turbine)*
eddying Wirbelung *f*, Wirbelbewegung *f*
eddying wirbelnd
Edeleanu process Edeleanu-Verfahren *n*
edge / to 1. *(Masch)* schärfen; 2. *(Bw)* abkanten, eine Kante abschrägen
e. away *(Geol)* auskeilen
edge 1. *(Masch)* Kante *f*, Rand *m*, Schneide *f*, Schneidkante *f*; 2. *(Typ)* Rand, Kante; Schnitt *m*; 3. Körperkante *f (Zeichnen)*
e. beam *(Bw)* Randbalken *m*, Randträger *m*
e. bottom *(Schiff)* Doppelboden *m* mit Bilge, abgeknickter Boden *m*
e. break *(Met)* Kantenriß *m*
e. chipping *(Masch)* Ausbrechen *n* der Schneide
e. crack Querriß *m*, Kantenriß *m*
e. damping *(Ak)* Randdämpfung *f*
e. definition *(Fs)* Bildschärfe *f*
e. dislocation Stufenversetzung *f (Kristallphysik)*
e. disturbance Randstörung *f (Statik)*
e. effect Kanteneffekt *m*
e. fillet weld Ecknaht *f (Schweißen)*
e. filtration Saugfiltration *f*
e. flare Randaufbruch *m (Fernsehbild)*
e. fog Randschleier *m (Film)*
e. fracture Kantenriß *m*
e. frequency *(BMSR)* Grenzfrequenz *f*, Eckfrequenz *f*
e. gilding *(Typ)* Schnittvergoldung *f*, Goldschnittmachen *n*
e. grain *(Holz)* Spiegelhochschnitt *m*
e. hide *(Led)* Eckhaut *f (Stapel)*
e. interference Kanteneingriff *m (Verzahnung)*
e. joint Parallelstoß *m*
e. jointing veneer Furnierfügen *n*
e. lace braid *(Text)* Spitzenband *n*
e. life Standzeit *f (Werkzeug)*
e. loading Kantenpressung *f (Verzahnung)*
e. mill Kollergang *m*, Kollermühle *f*, Läufermühle *f*, Erzmühle *f*
e. moulder Kantenkehlmaschine *f (Holzbearbeitung)*
e. of a crystal Kristallkante *f*
e. of rise *(Wärm)* Staukante *f (Kondensator)*
e. of space Schwelle *f* des kosmischen Raumes
e. pattern *(Text)* Bordürenmuster *n*
e. perturbation Randstörung *f (Statik)*
e. planer Blechkantenhobelmaschine *f*
e. preparation Fugenvorbereitung *f*
e. pressure *(Bw)* Kantenpressung *f*
e.-punched card Randlochkarte *f*
e.-punched card reader Randlochkartenleser *m*
e. radius Radius *m* der Kantenabrundung
e. raising *(Masch)* Bördeln *n*
e. roller mill *(Text)* Hanfreibe *f*
e. runner Kollergang *m*
e.-runner mill Kollermühle *f*, Läufermühle *f*, Pfannenmühle *f*
e. steepness Flankensteilheit *f (z. B. eines elektrischen Impulses)*
e. strength Schneidhaltigkeit *f*, Schnitthaltigkeit *f*
e. stress Randspannung *f (Statik)*
e. strip 1. Randstreifen *m (Straßenbau)*; 2. *(Holz)* Umleimer *m*
e.-type filter *(Kfz)* Kantenfilter *n*, Spaltfilter *n*
e. watermark *(Holz)* Eckwasserzeichen *n*
e. wear Verschleiß *m* an der Werkzeugkante
e. weld Stirnnaht *f*, Stirnstoß *m (Schweißen)*
e. winding *(Elm)* Hochkantwicklung *f*
edged pincers Kantenzange *f*
edger 1. Besäumkreissäge *f*; 2. Kantenhobelmaschine *f*
edgeways *s.* edgewise
edgewise hochkant[ig]
edging 1. *(Bw)* Kantenschutzleiste *f*; Randeinfassung *f*; 2. Besäumen *n (Holzbearbeitung)*; 3. *(Holz)* Säumling *m*
e. mill stand Stauchgerüst *n*
e. of bloom Kanten *n* eines Blockes
e. pass Stauchstich *m*, Stauchkaliber *n*
e. plane Bestoßhobel *m (Holzbearbeitung)*
e. roll Stauchwalze *f*
Edison accumulator Edison-Akkumulator *m*
Edison cell Edison-Akkumulator *m*
Edison lamp-holder Edison-Fassung *f*, E-Fassung *f*
Edison screw cap Edison-Sockel *m*, Schraubsockel *m*

Edison screw holder Edison-Fassung *f*
Edison thread Edison-Gewinde *n*
edit / to 1. *(BMSR)* redigieren *(z. B. Neuanordnung von Information)*; 2. schneiden *(Film)*
editing Montage *f (z. B. von Tonbändern)*
e. of the tape Bandschnitt *m*, Schnitt *m (Tonbandaufnahme)*
e. recorder Tonmontagegerät *n*
e. table Schneidetisch *m (Film)*
e. tape recorder Tonmontagegerät *n*
e. of the tape Bandschnitt *m*, Schnitt *m (Tonbandaufnahme)*
edition *(Typ)* Auflage *f (von Büchern)*
e. printing Auflagendruck *m*
editorial Leitartikel *m*
e.d.p. system *s.* electronic data-processing system
education Ausbildung *f*, Schulung *f*
educational film Lehrfilm *m*
e. television Schulfernsehen *n*
eduction Abzug *m (Dampf, Gase usw.)*
e. valve Dampfabgangsventil *n*
eel basket Aalkorb *m*
effect / to bewirken, hervorrufen, leisten
effect Wirkung *f*, Resultat *n*; Effekt *m*
e. of attraction Anziehungswirkung *f*
e. of contrast Kontrastwirkung *f*
e. of gravity Schwerewirkung *f*, Schwerkraftwirkung *f*
e. of light Lichteinwirkung *f*
e. of scattering Streuwirkung *f*; Streueffekt *m*
e. of weight Gewichtswirkung *f*
e. twist Effektzwirn *m*, Zierzwirn *m*
e. yarn Effektgarn *n*
effective effektiv, tatsächlich, wirksam
e. aerodynamic atmosphere ausnutzbarer Bereich *m* aerodynamischen Auftriebs
e. alternating current resistance Echtwiderstand *m*, Wechselstromwirkwiderstand *m*
e. area 1. effektive Antennenfläche *f*; 2. Arbeitsfläche *f (Dampfdiagramm)*; 3. *s.* e. surface
e. atomic number effektive Kernladungszahl *f*
e. bunching angle *(El)* effektiver Laufwinkel *m*
e. call *(Fmt)* ausgeführte (zustande gekommene) Verbindung *f*
e. capacity 1. Nutzinhalt *m*; 2. Nutzleistung *f (Maschinen)*; 3. *(El)* wirksame Kapazität *f*
e. centre Wirkungszentrum *n*
e. coil federnde Windung *f*
e. collision cross-section effektiver Stoßquerschnitt *m*, [effektiver] Wirkungsquerschnitt *m*, [effektiver] Wirkungsquerschnitt für Stoß *(insbesondere Gasentladungsröhren)*
e. conductance Wirkleitwert *m (Wechselstrom)*
e. conductivity effektive Leitfähigkeit *f*, Wechselstromleitfähigkeit *f*
e. cross section effektiver Wirkungsquerschnitt *m*
e. current Effektivstrom *m*, Effektivwert *m* des Stromes
e. cut-off frequency effektive Grenzfrequenz *f*
e. depth Nutzhöhe *f (Stahlbeton)*
e. depth of atmosphere for aerodynamic lift ausnutzbarer Bereich *m* aerodynamischen Auftriebs
e. diameter 1. Innendurchmesser *m*, lichte Weite *f*, Nennweite *f*; 2. Flankendurchmesser *m (Gewinde)*
e. dose *(Kern)* Wirkungsdosis *f*
e. electron mass effektive Elektronenmasse *f*
e. face width aktive Zahnbreite *f*
e. field ratio *(El)* Erregergrad *m*
e. flange width mitwirkende Plattenbreite *f (Stahlbetonbau)*
e. gap width effektive Spaltbreite *f (Magnetkopf)*
e. half-life effektive Halbwertzeit *f*
e. height wirksame Höhe *f*, Effektivhöhe *f (z. B. von Antennen)*
e. horse-power Nutzpferdestärke *f*, Nutz-PS *f*, effektive Pferdestärke *f*, PS_e; Nutzleistung *f*
e. input capacitance dynamische Gitterkapazität *f (Elektronenröhre)*
e. length Knicklänge *f (Statik)*
e. length of spinning bath Spinnbadstrecke *f*
e. mass effektive (wirksame) Masse *f*
e. multiplication factor *(Kern)* effektiver Vermehrungsfaktor *m*
e. output effektive (abgegebene) Leistung *f*
e. part of scale [effektiver] Meßbereich *m*
e. percentage modulation effektiver Modulationsgrad *m*
e. power Nutzleistung *f*, effektive Leistung *f*; Wirkleistung *f*
e. power output effektive (abgegebene) Leistung *f*; *(El)* effektive Ausgangsleistung *f*
e. pressure *(Kfz)* Arbeitsdruck *m*
e. principle of pesticides Wirkstoff *m* der Schädlingsbekämpfungsmittel
e. range 1. effektive Reichweite *f*; 2. *(BMSR)* Nutzmeßbereich *m*
e. reactance Blindwiderstand *m*, Reaktanz *f*
e. resistance Wirkwiderstand *m*
e. resistivity spezifischer Wechselstromwiderstand *m*
e. screen cut-point *(Brg)* Kornscheide *f*, Trennkorngröße *f*
e. slab width mitwirkende Plattenbreite *f (Stahlbetonbau)*
e. surface wirksame Oberfläche *f (Adsorption, Katalyse usw.)*

e. transmission equivalent *(Fmt)* Bezugsdämpfung *f* eines Übertragungssystems, Nutzdämpfung *f*
e. tyre roll radius wirksamer Reifenrollradius *m*
e. value Effektivwert *m*, quadratischer Mittelwert *m*
e. voltage Effektivspannung *f*
e. width mitwirkende Plattenbreite *f* *(Stahlbetonbau)*
e. work *(Ph)* Nutzarbeit *f*
effectiveness Leistungsfähigkeit *f*, Wirksamkeit *f*, Wirkungsgrad *m*
effervesce / to schäumen, aufwallen, aufbrausen
effervescent steel unberuhigter Stahl *m*
efficiency 1. Nutzeffekt *m*, Nutzleistung *f*, Wirkungsgrad *m*; Leistungsfähigkeit *f*; 2. Ausbeute *f* *(z. B. an Neutronen)*
e. circuit *(Fs)* Sparschaltung *f*
e. department Leistungskontrollabteilung *f*
e. diode Spardiode *f*
e. factor 1. Gütegrad *m*; Wirkungsgrad *m*; 2. Mikrofonübertragungsfaktor *m*
e. in open water *(Schiff)* Wirkungsgrad *m* des freifahrenden Propellers
e. of a luminous source Lichtausbeute *f*
e. of counting *(Kern)* Zählausbeute *f*, Ansprechwahrscheinlichkeit *f* *(eines Zählrohrs)*
e. of metal removal Spanungsleistung *f*
e. of rectification Gleichrichterwirkungsgrad *m*
e. of the combustion Verbrennungswirkungsgrad *m*
e. of the non-conducting covering [thermischer] Isolierungswirkungsgrad *m*
e. of the stoking Feuerungswirkungsgrad *m*
e. test Leistungsprüfung *f*; Wirkungsgradbestimmung *f*
efficient leistungsfähig, wirksam, wirkungsvoll; mit hohem Wirkungsgrad
effloresce / to ausblühen *(Mauerwerk, Beton usw.)*
efflorescence Ausblühung *f*, Auswitterung *f* *(z. B. Beton)*
effluence Ausfließen *n*, Ausströmen *n*
effluent Ausströmung *f*, Ausfluß *m*, abgehender Strom *m*; Abwasser *n*
e. tester Abwasserprüfgerät *n*
e. water Abwasser *n*
efflux Ausfluß *m*, Ausströmung *f*; Ausflußmenge *f*; Abflußmenge *f*
e. angle Austrittwinkel *m*
e. velocity Austrittsgeschwindigkeit *f*, Ausströmgeschwindigkeit *f*
e. viscosimeter Ausflußviskosimeter *n*
effusiometer Ausströmungsmesser *m*, Effusiometer *n*
effusion Ausströmung *f*, Effusion *f*
e. chamber *(Kern)* Effusionskammer *f*
e. cooling Effusionskühlung *f*
egg *(Ch)* Birne *f*, Blase *f*
e. calipers Ellipsenzirkel *m*
e. coal Eierbrikett *n*
e. hand grenade Eierhandgranate *f*
e. insulator Nußisolator *m*, Eiisolator *m*, Isolierei *n*
e.-shaped eiförmig, oval
e.-shaped hand grenade Eierhandgranate *f*
e.-shell finish halbglänzend
eggette [eiförmiges] Brikett *n*, Eierbrikett *n*
eggs Rogen *m* *(Fisch)*
e.h.v. . . . *s.* extra-high voltage . . .
Ehringhaus composite plate *(Opt)* Kompensationsplatte *f* nach Ehringhaus
eiconagon Zwanzigeck *n*
eigenfunction *(Math)* Eigenfunktion *f*
eigenoscillation *(Mech)* Eigenschwingung *f*
eigenvalue *(Math)* Eigenwert *m*
e. spectrum Eigenwertspektrum *n*
eight-angled achteckig
e.-cam machine *(Text)* Achtschloßmaschine *f*
e.-digit word generator *(BMSR)* achtstelliger Wortgenerator *m*
e.-electron shell *(Kern)* Achterschale *f*, L-Schale *f*
e.-film reel Achterspule *f* *(früher von Kurzwellenamateuren verwendet)*
e.-fold achtzählig
e.-harness *(Text)* achtbindig
e.-hr. working day Achtstundentag *m*
e. line pica *(Typ)* 96 britische Punkte *mpl* *(Schriftgröße)*
e.-lock flat knitting machine Achtschloßflachstrickmaschine *f*
e.-lock machine *(Text)* Achtschloßmaschine *f*
e.-oared shell Achter *m* *(Sportrennboot)*
e.-sided achtseitig
e.-spindle automatic Achtspindelautomat *m*
e. to pica *(Typ)* *(etwa)* Achtelcicero *f*, $1\frac{1}{2}$ britische Punkte *mpl* *(Schriftgröße)*
e.-valued achtwertig
eiking *(Schiff)* Bugfüllstück *n*
eikonometer *(Opt)* Ikonometer *n*
Einstein-de Haas effect Einstein-de-Haas-Effekt *m*, gyromagnetischer Effekt *m*
Einstein equation *(Ph)* Einsteinsche Gleichung *f*
Einstein relation Einsteinsche Beziehung *f* *(Fotoeffekt)*
Einstein shift *(Ph)* Einsteinsche Spektralverschiebung *f*
einsteinium *(Ch)* Einsteinium *n*
Einstein's equivalence principle *(Ph)* 1. Einsteinsches Äquivalenzprinzip *n* *(Äquivalenz von Beschleunigung und Gravitation)*; 2. *(inkorrekte Bezeichnung für)* [Einsteinsche] Energie-Masse-Beziehung *f*

Einstein's law of photochemical equivalence Einsteinsches Äquivalenzgesetz *n*
eject / to 1. emittieren, ausstoßen *(Teilchen)*; 2. auslösen, herausschießen, herausschlagen *(Teilchen)*; 3. *(Masch)* ausstoßen, auswerfen
ejection 1. Ausstoß *m*, Auswurf *m*; 2. *(Plast)* Entformung *f*, Herausnehmen *n* aus der Preßform
e. curve of chaff-cutter Auswurfbogen *m* der Häckselmaschine
e. fixture *(Masch)* Auswerfeinrichtung *f*
e. of chaff Auswerfen *n* des Häcksels
e. of particles Emission *f* von Teilchen; Partikelausbruch *m (Astrophysik)*
e. pipe of chaff-cutter Auswurfrohr *n* der Häckselmaschine
ejector 1. Auswerfer *m*, Ejektor *m*; 2. Ausstoßhebel *m (Setzmaschine)*; 3. Strahlsaugapparat *m (Turbine)*
e. bar Auswerferschiene *f*
e. condenser *(Wärm)* Strahlkondensator *m*
e. mechanism Auswerfeinrichtung *f*
e. pin Ausrückstift *m*
eking *(Schiff)* Bugfüllstück *n*
elaeostearic acid Eläostearinsäure *f*
elaidic acid Elaidinsäure *f*
elastic Gummiband *n*
elastic elastisch, federnd
e. aftereffect elastische Nachwirkung *f*
e. axis Elastizitätsachse *f (Kristalle, anisotrope Festkörper)*
e. bending elastische Biegung *f*, Durchbiegung *f* im elastischen Bereich
e. bond elastische Bindung *f*
e. boots Zugstiefel *mpl*
e. collision elastischer Stoß *m (Zusammenstoß von zwei Teilchen)*
e. constant Elastizitätskonstante *f*
e. curve Biegelinie *f*, elastische Linie *f*
e. deformation elastische Verformung (Deformation) *f*
e. design Berechnung *f* nach der Elastizitätstheorie, n-Verfahren *n (Stahlbetontheorie)*
e. diaphragm *(Rak)* Federungsblech *n*
e. force 1. *(Ph)* elastische (Hookesche) Kraft *f*; 2. *(Masch)* Federkraft *f*, Spannkraft *f*
e. foundation elastische Bettung (Unterlage) *f*
e. halfspace elastischer Halbraum *m*
e. hysteresis Elastizitätshysterese *f*
e. impact elastischer Stoß *m*
e. lag *(Ph)* elastisches Nachkriechen *n*, elastische Nachwirkung *f*
e. limit Elastizitätsgrenze *f*
e. line *s.* e. curve
e. modulus Elastizitätsmodul *m*
e. nozzle flap federnde Düsenzunge *f (Turbine)*
e. recoil *(Ph)* elastischer Rückstoß *m*
e. recoverable stretch *(Text)* elastische Dehnung *f*
e. ring elastischer Kolbenring *m*
e. scattering *(Ph)* elastische Streuung *f*
e. spindle *(Text)* Hängespindel *f*
e. spring Sprungfeder *f*
e. strain elastische Dehnung *f*
e. support *(Bw)* elastische Bettung (Lagerung) *f*
e. supports elastische Stützung *f (Statik)*
e. theory Elastizitätstheorie *f*
e. yarn Elastikgarn *n*
elasticity Elastizität *f*; Federkraft *f*, Spannkraft *f*
e. of bending Biegeelastizität *f*
e. of compression Druckelastizität *f*
e. of elongation Zugelastizität *f*
e. of flexure Biegeelastizität *f*
e. of shearing Schubelastizität *f*
e. of torsion Drehungselastizität *f*
elasto-plastic elastoplastisch
elastomer Elastomer *n*
elastomeric plastics elastomere (gummiartige) Kunststoffe *mpl*, Elaste *mpl*
elastometer Elastizitätsmesser *m*
elastostatics Elastostatik *f*, [statische] Elastizitätstheorie *f*
elat[e]rometer Dampfmesser *m*
elbow Knie[stück] *n*, Krümmer *m*, Winkelstück *n*
e. bed plate Grundwerk *n* mit V-förmig gewinkelten Messern *(Holzbearbeitung)*
e. flowmeter Krümmerdurchflußmesser *m*
e. forming machine Knierohrbiegemaschine *f*
e. joint Kniegelenk *n*, Knieverbindung *f*, Krümmerverbindung *f*
e. lever Winkelhebel *m*, Kniehebel *m*
e. piece Eckstück *n*, Kniestück *n*
e. pipe Kniestück *n*, Bogenrohr *n*, Schenkelrohr *n*
e. pipe with foot Ellbogenrohr *n* mit Fuß
e. telescope Winkelfernrohr *n*
e. tube plier Eckrohrzange *f*
elbowed abgewinkelt
electret Elektret *n*
electric elektrisch (*s. a.* electrical)
e. accounting machine Tabelliermaschine *f*
e.-acoustic index elektro-akustisches Verhältnis *n* eines Empfangssystems
e. alarm elektrische Alarmanlage *f*, elektrisches Läutewerk *n*
e. alarm pressure gauge *(Wärm)* Signalmanometer *n*
e. arc Lichtbogen *m*; elektrische Bogenentladung *f*
e. arc drilling Lichtbogenbohren *n*
e. arc furnace Lichtbogenofen *m (für Elektrostahlerzeugung)*

e. arc heating Lichtbogenheizung *f*
e. arc lamp Bogenlampe *f*
e. arc welding Lichtbogenschweißung *f*
e. attraction elektrostatische Anziehung *f*
e. axis elektrische Achse *f*, X-Achse *f (eines Kristalls)*
e. baggage and freight truck Elektrokarren *m*
e. baking oven elektrischer Backofen *m*, Elektrobackofen *m*
e. battery truck Elektrokarren *m*
e. battery vehicle Batteriefahrzeug *n*
e. bell elektrische Klingel *f*
e. bench drill Elektrotischbohrmaschine *f*
e. blanket elektrische Heizdecke *f*
e. blast furnace Elektrohochofen *m*, elektrischer Hochofen *m*
e. blasting cap elektrischer Zünder *m*
e. boiler elektrischer Kocher *m*, Elektrokocher *m*, Elektrodampfkessel *m*
e. boiling pan elektrischer Kochkessel *m*
e. boiling plate elektrische Kochplatte *f*, Elektrokochplatte *f*
e. breeze elektrischer Wind *m*, Ionenwind *m*
e. bulb Glühlampe *f*
e. calamine Hemimorphit *m*
e. capacity altimeter kapazitiver Höhenmesser *m*
e. car Elektromobil *n*
e. carburizing furnace elektrischer Einsatzofen *m*
e. carpet Heizteppich *m*
e. cash-register elektrische Registrierkasse *f*
e. central station Kraftzentrale *f*
e. chain block Elektrokettenzug *m*
e. charge elektrische Ladung *f*
e. circuit Stromkreis *m*
e. clock with reserve power elektrische Uhr *f* mit Gangreserve
e. clutch *(Kfz)* Elektrokupplung *f*
e. conductance Leitwert *m*, elektrische Leitfähigkeit *f*
e. conductivity [elektrische] Leitfähigkeit *f*, [elektrisches] Leitvermögen *n*
e. control elektrische Regelung *f*
e. convector elektrischer Konvektionsofen *m*
e. cooker elektrischer Haushaltsherd *m*, Elektrohaushaltsherd *m*
e. cooking oven elektrischer Brat- und Backofen *m*, Elektrobrat- und -backofen *m*
e. counting mechanism [elektrisch betriebenes] Zählwerk *n*
e. crane Elektrokran *m*
e. cushion elektrisches Heizkissen *n*
e. deep frier elektrische Fritürepfanne *f*
e. deflection elektrische Ablenkung *f*
e. dehydrator elektrischer Dörrschrank *m*, Elektrodörrschrank *m*
e. delay line elektrische Verzögerungsleitung *f*, Laufzeitglied *n*
e. demand Energiebedarf *m*
e. depth finder elektrisches Echolot *n*, elektrischer Höhenmesser *m*
e. dipol elektrischer Dipol *m*
e. dipol moment elektrisches Dipolmoment *n*
e. dipol radiation elektrische Dipolstrahlung *f*
e. discharge elektrische Entladung *f*
e. discharge forming elektro-hydraulische Umformung *f*
e. discharge lamp Gasentladungslampe *f*
e. discharge machine Elektroerosionsmaschine *f*
e. discharge machining elektroerosive Bearbeitung *f*
e. displacement elektrische Verschiebung *f*
e. distance thermometer elektrisches Fernthermometer *n*
e. door operator elektrischer Türöffner *m*
e. double layer Dipolschicht *f*
e. driller Elektrohandbohrmaschine *f*
e. drive Elektroantrieb *m*
e. drying machine elektrischer Wringer *m*, Zentrifuge *f*; Trockenschleuder *f*
e. dumbwaiter *(Am)* elektrischer Gepäckaufzug (Küchenaufzug, Speiseaufzug) *m*
e. dynamometer Pendelmaschine *f*
e. energy Elektroenergie *f*, elektrische Energie *f*
e. excitation elektrische Anregung (Erregung) *f*
e. fence elektrischer Zaun (Weidezaun) *m*
e. fence system elektrische Zaunanlagen *fpl*
e. field elektrisches Feld *n*
e. field intensity *s.* e. field strength
e. field strength elektrische Feldstärke *f*
e. filter Elektrofilter *n*, elektrisches Filter *n*
e. fire elektrischer Strahlofen *m*, Heizsonne *f*
e. fire alarm Feuermelder *m*
e. fire bowl Heizsonnenreflektor *m*
e. floor polisher elektrische Bohnermaschine *f*
e. flux Verschiebungsfluß *m*, elektrischer Fluß *m (Flächenintegral der dielektrischen Verschiebung ϑ)*
e. flux density elektrische Flußdichte (Kraftlinienzahl) *f*
e. freight truck Elektrokarren *m*
e. frier elektrische Bratpfanne *f*
e. furnace Elektroofen *m*
e. furnace melting Elektrostahlschmelzen *n*
e. furnace steel Elektrostahl *m*
e. generating station Elektrizitätswerk *n*, Kleinkraftwerk *n*
e. generator Generator *m*, Stromerzeuger *m*, Dynamo[maschine *f*] *m*
e. grill Elektrogrill *m*
e. geyser Durchlauferhitzer *m*

e. hand drill of the breast type Elektro-Handbohrmaschine *f* mit Brustschild
e. hand saw Elektrohandsäge *f*
e. hearth furnace Elektroherdofen *m*
e. heater Elektroofen *m*, elektrischer Heizkörper *m*
e. hoist Elektro[seil]zug *m*
e. hooter *(Kfz)* elektrische Hupe *f*
e. horn *(Kfz)* Boschhorn *n*
e. hot table elektrischer Wärmetisch *m*
e. hotplate elektrische Kochplatte *f*
e. ignition elektrische Zündung *f*
e. incubator elektrische Brutmaschine *f*
e. instrument elektrisches Meßinstrument *n*
e. iron elektrisches Bügeleisen *n*
e. ironing machine elektrische Bügelmaschine *f*
e. ironing press elektrische Kleiderpresse *f*
e. jar Leydener Flasche *f*
e. kettle Elektrokocher *m*, elektrischer Kocher *m*
e. lamp bulb Glühlampe *f*
e. leakage tester Erdschlußprüfer *m*, Leckstrommesser *m*
e. lens spectrometer Massenspektrometer *n* mit elektrostatischen Linsen
e. lifting magnet Elektrohebemagnet *m*
e. light elektrisches Licht *n*
e. light appliance elektrisches Beleuchtungsgerät *n*
e. lighting mains Lichtleitung *f*
e. line of force elektrische Kraftlinie *f*
e. lock Sperrmagnet *m*
e. locomotive elektrische Lokomotive *f*
e. machine Elektrisiermaschine *f*
e. mains Netz *n*, Stromnetz *n*
e. melting furnace Elektroschmelzofen *m*
e. metal hochwertiges HF-Eisen *n*
e. metal snips Elektrohandblechschere *f*
e. meter Zähler *m*, Stromzähler *m*
e. milker elektrische Melkmaschine *f*
e. motor Elektromotor *m*
e. motor car Elektromobil *n*
e. multipole radiation elektrische Multipolstrahlung *f*
e. network reciprocity theorem Reziprozitätssatz *m* für elektrische Netze
e. output elektrische Leistung *f*
e. oven elektrischer Ofen *m*, Elektroofen *m*
e. panel heater elektrischer Flächenheizkörper *m*
e. parts list Schaltteilliste *f*
e. phonograph recorder elektrischer Plattenschneider *m*
e. pig Elektroroheisen *n*
e. plant elektrische Anlage *f*, Elektroanlage *f*
e. platform truck Elektrokarren *m*
e. polarity elektrische Polarität *f*
e. potential elektrisches Potential *n*
e. potential difference elektrische Potentialdifferenz *f*
e. power elektrische Leistung *f*
e. power cable Energiekabel *n*, Kraft[strom]kabel *n*
e. power pylon Überlandleitungsmast *m*
e. power station Kraftwerk *n*
e. precipitation *(Ch)* Fällung *f* im elektrischen Feld
e. preselector gear elektrischer Gangvorwähler *m*
e. pressure elektrische Spannung *f*
e. printing punch *(Büro)* Magnetschreiblocher *m*
e. punch *(Büro)* Magnetlocher *m*
e. quadrupole radiation elektrische Quadrupolstrahlung *f*
e. quantity Elektrizitätsmenge *f*
e. radiator elektrischer Strahlungsofen *m*
e. rail clamp elektrische Querbalkenklemmung *f (Hobelmaschine, Bettfräsmaschine)*
e. rammer Elektrostampfer *m*
e. range *(Am) s.* e. cooker
e. razor Trockenrasierapparat *m*, Trockenrasierer *m*
e. recorder of a round elektrische Wächterkontrollanlage *f*
e. residuum elektrischer Rückstand *m*, elektrisches Residuum *n*
e. rocking-type furnace direkter Lichtbogenofen *m*
e. safety lamp elektrische Grubenlampe *f*
e. screening elektrische Abschirmung *f*
e. separator 1. *(Ch)* Elektrofilter *n*; 2. *(Lw)* elektrische Zentrifuge *f*
e. shaft furnace Elektrohochofen *m*, Elektroschachtofen *m*
e. shock elektrischer Schlag *m*
e. shocker Elektrisierapparat *m*
e. sleep Elektronarkose *f*
e. smelting furnace Elektroeinschmelzofen *m*
e. soldering iron Elektrolötkolben *m*
e. steam raiser Elektrodampfkessel *m*
e. steel Elektrostahl *m*
e. steel furnace Elektrostahlofen *m*
e. steel process Elektrostahlverfahren *n*
e. steel works Elektrostahlwerk *n*
e. strength Durchschlagsfestigkeit *f*, Spannungsfestigkeit *f*
e. stress dielektrische Beanspruchung *f*
e. supply Stromversorgung *f*
e. supply failure Stromausfall *m*
e. supply line Stromnetz *n*, Netz *n*; Zuleitung *f*
e. supply meter Strom[verbrauchs]zähler *m*
e. surface resistance Oberflächenwiderstand *m*

e. susceptibility elektrische Suszeptibilität *f*
e. table cooker elektrische Tischkochplatte *f*
e. tamper Elektrostampfer *m*
e. tape Isolierband *n*
e. thermometer elektrisches Temperaturmeßgerät *n*
e. torch [elektrische] Taschenlampe *f*
e. traction elektrische Zugförderung *f*
e. tractor Elektrozugmaschine *f*, Elektroschlepper *m*
e. tramway elektrische Straßenbahn *f*
e. transducer elektrische Übertragungsschaltung *f*; Vierpol *m*
e. trolley Elektrolaufkatze *f*
e. truck Elektrokarren *m*
e. typewriter elektrische Schreibmaschine *f*
e. water heater Warmwasserbereiter *m*
e. welding Elektroschweißung *f*
e. wind elektrischer Wind *m*, Ionenwind *m*, konvektive Entladung *f*
e. windscreen wiper *(Kfz)* elektrischer Scheibenwischer *m*
e. wire rope block Elektroseilzug *m*
electrical elektrisch (*s. a.* electric)
e. accessories elektrisches Zubehör *n*
e. accommodation elektrische Akkommodation *f*
e. anesthesia Elektroanästhesie *f*, elektrische Narkose *f*
e. angle Phasenwinkel *m*
e. appliances Elektrogeräte *npl*
e. bore-hole prospecting Erdwiderstandsbohrschürfung *f*
e. communication art [elektrische] Nachrichtentechnik *f*; Fernmeldetechnik *f*
e. control board Geräteplatte *f*
e. controls elektrische Schaltgeräte *npl*
e. coupler elektrische Kupplung *f*
e. depression bar Druckschienenkontakt *m*, Schienenkontakt *m*, Hebelkontakt *m*
e. double layer Dipolschicht *f*
e. efficiency elektrischer Wirkungsgrad *m* *(einer Anlage)*; Stromausbeute *f*
e. elementary quantum *s.* elementary charge
e. engineer Elektrotechniker *m*; Elektroingenieur *m*
e. engineering Elektrotechnik *f*
e. engineering industry Elektromaschinenbau *m*
e. equipment elektrische Einrichtung *f*
e. equivalent of heat elektrisches Wärmeäquivalent *n*
e. fence *s.* electric fence
e. fitter Elektroinstallateur *m*, Elektriker *m*
e. flux *s.* electric flux
e. forming elektrisches Formieren *n*
e. fusion welding process elektrisches Widerstandsschmelzschweißen *n*
e. image elektrisches Ladungsbild *n*; Spiegelbild *n* der Ladung
e. insulating paper Isolierpapier *n* für elektrotechnische Zwecke, Kabelpapier *n*
e. insulation value Isolationswiderstand *m*
e. interlock post *(Eis)* elektrische Stellwerkanlage *f*
e. interlocking elektrische Sperre (Verriegelung) *f*
e. leakage Erdschluß *m*
e. load Strombelastung *f*
e. load capacity Strombelastbarkeit *f*
e. mains plug-in unit Netzeinschub *m*
e. measuring instrument elektrisches Meßinstrument *n (für elektrische Größen)*
e. moisture meter elektrischer Wassergehaltmesser *m*
e. power elektrische Leistung *f*
e. power engineering Starkstromtechnik *f*
e. precipitation elektrische Staubabscheidung *f*
e. precipitator EGR-Anlage *f*
e. quantity elektrische Größe *f*
e. resistance elektrischer Widerstand *m*
e. resistance fusion welding process elektrisches Widerstandsschmelzschweißen *n*
e. salinometer elektrischer Salzgehaltmesser *m*
e. sensing *(BMSR)* elektrische Abfühlung *f* *(Lochkarten)*
e. signal box elektrisches Stellwerk *n*
e. supply Stromzuführung *f*; Stromversorgung *f*
e. surface prospecting elektrische Oberflächenschürfung *f*
e. technology Elektrotechnik *f*
e. thermometer elektrisches Temperaturmeßgerät *n*; Kontaktthermometer *n*
e. time-distribution system Zentraluhrenanlage *f*
e. transmission elektrische Kraftübertragung *f*
e. welding Elektroschweißen *n*
electrically controlled clock elektrisch geregelte Uhr *f*
e. driven elektrisch betrieben, mit elektrischem Antrieb
e. heated pad elektrisches Heizkissen *n*
e. heated thermocouple *s.* thermal converter
e. measuring instrument elektrisches Meßgerät *n (für nichtelektrische Größen)*
e. operated chuck Elektrospannfutter *n*
e. wound clock Uhr *f* mit elektrischem Aufzug
electrician Elektriker *m*, Elektroinstallateur *m*
electricity Elektrizität *f*
e. by friction Reibungselektrizität *f*
e. meter Elektrizitätszähler *m*, Stromverbrauchszähler *m*

e. of fibres Faserelektrizität *f*
e. of opposite kind ungleichnamige Elektrizität *f*
e. of opposite polarity ungleichnamige Elektrizität *f*
e. of opposite sign ungleichnamige Elektrizität *f*
e. of precipitations Niederschlagselektrizität *f*
e. of same kind gleichnamige Elektrizität *f*
e. of same sign gleichnamige Elektrizität *f*
e. supply Stromversorgung *f*, Elektrizitätsversorgung *f*
electrifiable elektrisierbar
electrification Elektrifizierung *f*; Aufladung *f*
electrify / to aufladen; elektrifizieren
electrize / to elektrisieren
electro brighten / to elektrolytisch glänzen (beizen)
e. extraction elektrische Extraktion (Gewinnung) *f*
e. gild / to galvanisch vergolden
electroacoustic elektroakustisch
e. force factor elektroakustischer Übertragungsfaktor *m*
e. four-pole network elektroakustischer Vierpol *m*
e. transducer [elektroakustischer] Wandler *m*
electroacoustics Elektroakustik *f*
electroaffinity *(Ch)* Elektroaffinität *f*
electroanalysis *(Ch)* Elektroanalyse *f*, elektrolytische Analyse *f*
electroballistics Elektroballistik *f*
electrobiology Elektrobiologie *f*
electrobioscopy Elektrobioskopie *f*
electroblock *(Typ)* Galvano *n*, Prägegalvano *n*
electrocapillarity Elektrokapillarität *f*
electrocapillary current Kapillarstrom *m*
e. curve Elektrokapillarkurve *f*
electrocardiogram Elektrokardiogramm *n*, Ekg *n*, EKG *n*
electrocardiography Elektrokardiografie *f*
electrocautery Elektrokauter *m*, Glühkauter *m*
electrochemical elektrochemisch; elektrolytisch
e. condenser Elektrolytkondensator *m*
e. dissociation elektrolytische Dissoziation *f*
e. migration [elektrochemische] Ionenwanderung *f*
e. oxidation elektrochemische Oxydation *f*
e. reduction elektrochemische Reduktion *f*
e. series galvanische (elektrochemische) Spannungsreihe *f*
e. valve elektrochemisches Ventil *n*, Elektrolytgleichrichter *m*
electrochemistry Elektrochemie *f*
electrocoagulation *(Ch, Ph)* Elektrokoagulation *f*
electrocoating *s.* electroplating
electrocorrosion Korrosion *f* durch Lokalströme
electrocratic elektrokratisch
electrode Elektrode *f*
e. adjusting Elektrodenjustierung *f*
e. adjusting gear Elektrodennachsteller *m*, Elektrodenverstellwerk *n*
e. adjustment Elektrodenjustierung *f*
e. admittance Elektrodenadmittanz *f*, Elektrodenscheinleitwert *m*
e. arm Elektrodenausleger *m (E-Ofen)*
e. bias Elektrodenvorspannung *f*
e. boiler Elektrodampfkessel *m*
e. cable detector Kabelsuchgerät *n*, Kabelordnungsgerät *n*
e. clamp Elektrodenfassung *f*, Elektrodenhalter *m*
e. compound Elektrodenkitt *m*
e. conductance Gittersteilheit *f*
e. control Elektrodensteuerung *f*
e.-current averaging time Stromflußwinkel *m*
e. differential resistance Widerstandskomponente *f* der Elektrodenimpedanz
e. dissipation Elektrodenverlust[leistung *f*] *m*; Gitterverlust[leistung *f*] *m*
e. drop Spannungsabfall *m* an der Elektrode
e.-hearth furnace *(direkter)* Lichtbogenofen *m*
e. holder Elektrodenhalter *m (z. B. am Schmelzofen)*
e. impedance Elektrodenimpedanz *f*, Elektrodenscheinwiderstand *m (Elektronenröhren)*; innerer Widerstand *m (Entladungsröhren)*
e. jib Elektrodenträger *m*
e. mast Elektrodenfahrsäule *f* *(im E-Ofen)*
e. point Elektrodenspitze *f*
e. potential Elektrodenpotential *n*
e. prong Elektrodenspanner *m*
e. radiator Kühlflügel *m*, Elektrodenkühllamelle *f*
e. reactance Elektrodenblindwiderstand *m*, Elektrodenreaktanz *f*
e. resistance Elektrodenresistenz *f*, Elektrodenwirkwiderstand *m*
e. salt bath Elektrodensalzbad *n*
e. separation Elektrodenabstand *m*
e. shaper Elektrodenspitzmaschine *f*
e. shell Elektrodenmantel *m*
e. short-circuit Elektrodenschluß *m*
e. spacing Elektrodenabstand *m*
e. spatter Spritzen *n* einer Elektrode
e. sticking Kleben *n* der Elektrode *(Schweißen)*
e. support Elektrodenhalter *m*, Elektrodenträger *m*
e. surface Elektrodenfläche *f*
e. susceptance Elektrodenblindleitwert *m*

e. terminal Elektrodenklemme *f*; Polklemme *f* *(Batterie)*
electrodeless discharge elektrodenlose Entladung *f*
electrodeposit / to galvanisieren
electrodeposit galvanischer Überzug *m*, galvanisch (elektrolytisch) aufgebrachte Schicht *f*
electrodeposited galvanisch gefällt
e. metallic paper packing galvanische Metallpapierpackung *f*
electrodeposition elektrolytische Abscheidung *f*, galvanischer Niederschlag *m*; galvanische Metallisierung *f*, Galvanisierung *f*; Galvanotechnik *f*
electrodesiccation *(Ch)* elektrische Trocknung *f*
electrodetarrer Elektroteerscheider *m*
electrodiagnosis Elektrodiagnose *f*
electrodiagnostics Elektrodiagnostik *f*
electrodialysis *(Ch)* Elektrodialyse *f*
electrodissolution anodische Korrosion *f*; elektrolytische Auflösung *f*; galvanische Zersetzung *f*
electrodynamic elektrodynamisch
e. balance Stromwaage *f*
e. loudspeaker elektrodynamischer Lautsprecher *m*
e. loudspeaker clutch *(BMSR)* Tauchspulkupplung *f*
e. pick-up elektrodynamischer Tonabnehmer *m*
e. vibration pick-up elektrodynamischer Schwingungsgeber (Schwingungsmesser) *m* *(Messung mechanischer Schwingungen)*
electrodynamics Elektrodynamik *f*
electrodynamometer Elektrodynamometer *n*
electroencephalograph Elektroenzephalograf *m*
electroendoscopy Elektroendoskopie *f*
electroengraving elektrolytische Ätzung *f*
electroerosion Elektroerosion *f*
e. metal working process Bearbeitung *f* durch Elektroerosion
electroetch / to galvanisch ätzen
electroextraction elektrolytische Gewinnung *f* *(von Metall)*
electrofacing Elektroplattierung *f*
electroforming 1. Verformung *f* durch Funkenentladung; 2. Galvanoplastik *f*
electrogalvanize / to galvanisch verzinken
electrogalvanizing galvanische Verzinkung *f*
electrogilding galvanische Vergoldung *f*
electrograph Elektrograf *m*
electrographic ink leitfähige Tinte *f*
e. pen Graphitstift *m*
electrography Elektrografie *f*
electrohydraulic elektrohydraulisch, hydroelektrisch
e. forming hydroelektrisches Umformen *n*
e. method of high-velocity forging hydroelektrisches Verfahren *n* der Hochleistungsumformung
electrokinetic elektrokinetisch
electrokinetics Elektrokinetik *f*
electrolier Kronleuchter *m*, Lüster *m*
electrolimit gauge elektrisches Toleranzmeßgerät *n*
electrolysis Elektrolyse *f*
e. in the dry way Schmelzflußelektrolyse *f*
electrolyte Elektrolyt *m*, Lauge *f*
e. dissociation constant *(Ch)* Dissoziationskonstante *f*
electrolytic elektrolytisch
e. analysis elektrolytische Analyse *f*, Elektroanalyse *f*
e. bath Elektrolysebad *n*
e. capacitor Elektrolytkondensator *m*
e. cleaning elektrolytische Reinigung *f*
e. copper elektrolytisches Kupfer *n*, Elektrolytkupfer *n*
e. deposition elektrolytische Abscheidung *f*
e. detector Elektrolytdetektor *m*, Schlömilch-Zelle *f*
e. dissociation elektrolytische Dissoziation *f*
e. interrupter Elektrolytunterbrecher *m*, elektrolytischer Unterbrecher *m*
e. iron Elektrolyteisen *n*
e. lead Elektrolytblei *n*
e. lightning arrester Elektrolytableiter *m*, elektrolytischer Überspannungsableiter *m*
e. meter Elektrolytzähler *m*
e. oxidation process Aloxidverfahren *n*, Eloxalverfahren *n*
e. parting elektrolytische Metallscheidung *f*
e. polarization elektrolytische Polarisation *f*
e. polisher elektrolytisches Poliergerät (Blankätzgerät) *n*
e. polishing elektrolytisches Polieren *n*
e. rectifier Elektrolytgleichrichter *m*
e. refining elektrolytische Raffination *f*
e. separation of isotopes elektrolytische Isotopentrennung *f*
e. slime Elektrolyseschlamm *m*
e. solution voltage Lösungsspannung *f*
e. surge arrester Elektrolytableiter *m* *(für Überspannungen)*
e. tank elektrolytischer Trog *m*
e. valve Elektrolytgleichrichter *m*, Elektrolytventil *n*
e. water-tester [elektrolytischer] Speisewasserprüfer *m*
e. zinc Elektrolytzink *n*
electrolytical pickling bath elektrolytisches Beizbad *n*
electrolytically oxidized eloxiert
electrolyze / to elektrolysieren, durch Elektrolyse zersetzen

electrolyzer Elektrolyseur *m*
electromagnet Elektromagnet *m*
electromagnetic elektromagnetisch
e. chuck elektromagnetische Spannplatte *f*
e. clutch elektromagnetische Kupplung *f*
e. constant elektromagnetische Konstante *f*; Lichtgeschwindigkeit *f*
e. damping Wirbelstromdämpfung *f*
e. field elektromagnetisches Feld *n*
e. flowmeter induktiver Durchflußmesser *m*, Durchflußmesser nach dem Induktionsprinzip
e. holding device elektromagnetisches Spannfutter *n*
e. hum pickup elektromagnetische Brummschleife *f (z. B. Radio)*
e. induction elektromagnetische Induktion *f*
e. instrument Dreheiseninstrument *n*
e. lens magnetische Linse *f (Elektronenoptik)*
e. linkage elektromagnetische Verkettung *f*
e. mass elektromagnetische Masse *f*
e. mass separator elektromagnetischer Massentrenner *m*
e. method elektromagnetisches Verfahren *n*
e. radiation elektromagnetische Strahlung *f*
e. receiver elektromagnetischer Fernhörer *m (mit Eisenanker)*
e. retroaction induktive Rückkopplung *f*
e. separation elektromagnetische Trennung *f*
e. separation of isotopes elektromagnetische Isotopentrennung *f*
e. separation plant elektromagnetische Isotopentrennanlage *f*
e. separator elektromagnetischer Isotopentrenner *m*; elektromagnetischer Abscheider *m (Aufbereitung)*
e. shoe brake elektromagnetische Schienenbremse *f*
e. theory of light elektromagnetische Lichttheorie *f*
e. unit elektromagnetische Einheit *f*
e. wave elektromagnetische Welle *f*
electromagnetics *s.* electromagnetism
electromagnetism Elektromagnetismus *m*
electromatic dropper stop motion elektrische Abstellvorrichtung *f* mit Fühlerfinger
e. needle detector stop motion *(Text)* elektrische Abstellvorrichtung *f* gegen Nadelbruch und Aufhocken
e. yarn stop motion *(Text)* [auf den Faden wirkende] elektrische Abstellvorrichtung *f*
electromechanical elektromechanisch
e. coupling impedance *s.* e. force factor
e. force factor elektromechanischer Umwandlungsfaktor *m*
e. repeater *(BMSR)* elektromechanischer Melder *m*
e. signal box elektromechanisches Stellwerk *n*
e. transducer elektromechanischer Wandler *m*
electromechanics Elektromechanik *f*
electromedical elektromedizinisch
electrometallization Elektrometallisierung *f*, Herstellung *f* metallischer Überzüge auf nichtleitendem Material
electrometallurgy Elektrometallurgie *f*
electrometer Elektrometer *n*
e. valve Elektrometerröhre *f*
electrometric titration elektrometrische Maßanalyse *f*, Elektrometrie *f*
electromobile Elektromobil *n*; Elektrokarren *m*
electromotive elektromotorisch
e. force elektromotorische Kraft *f*, EMK *f*
e. series elektrochemische (galvanische) Spannungsreihe *f*
electromotor Elektromotor *m*, elektrischer Motor *m*
electromyogram Elektromyogramm *n*
electron Elektron *n*
e. accelerator Elektronenbeschleuniger *m*
e. accumulation Elektronenstauung *f*
e. affinity Elektronenaffinität *f*; Austrittsarbeit *f*; Elektronenaustrittsarbeit *f*
e. attachment Elektronenanlagerung *f*, Anlagerung *f* von Elektronen
e. avalanche Elektronenlawine *f*
e. background Elektronenuntergrund *m*, Elektronenrauschen *n*
e. beam Elektronenstrahl *m*, Katodenstrahl *m*; Katodenstrahlbündel *n*
e.-beam scanner Elektronenstrahlabtaster *m*
e.-beam tube (valve) Braunsche Röhre *f*; Elektronenstrahlröhre *f*, Katodenstrahlröhre *f*
e. bombardement Elektronenbeschuß *m*
e. brain Elektronengehirn *n*
e. camera elektronischer Bildfänger *m*, Elektronenkamera *f*
e. capture Elektroneneinfang *m*
e. catch-up region Elektronenüberholungsgebiet *n*
e. charge Elektronenladung *f*
e. charge-mass ratio spezifische Ladung *f* des Elektrons
e. cloud Elektronenwolke *f*; Elektronenhülle *f*
e. collection Absaugen (Auffangen) *n* von Elektronen
e. conduction Elektronenleitung *f*
e. conductor Elektronenleiter *m*
e. counter Elektronenzähler *m*
e.-coupled elektronengekoppelt
e. coupling 1. Elektronenkopplung *f*; 2. Atombindung *f*, Elektronenbindung *f*, homöopolare (kovalente) Bindung *f*, Kovalenzbindung *f*
e. current Elektronenstrom *m*
e. density [räumliche] Elektronendichte *f*, Elektronenbelegung *f*

e. density modulation Modulation *f* der Elektronendichte
e. detachment Elektronenablösung *f*, Ablösung (Abtrennung) *f* von Elektronen
e. diffraction Elektronenbeugung *f*
e. diffraction pattern Elektronenbeugungsbild *n*, Elektronenbeugungsaufnahme *f*
e. discharge Elektronenentladung *f*
e. discharge tube 1. Elektronenentladungsrohr *n*; 2. Röhre *f*, Elektronenröhre *f*
e. donor Elektronendonator *m*
e. drift Elektronenwanderung *f*
e. emission Elektronenemission *f*, Elektronenaussendung *f*, Elektronenabgabe *f*
e.-emission area Elektronenabgabefläche *f*
e.-emissive elektronenemittierend
e. emitter Elektronenquelle *f*; Elektronenstrahler *m*
e.-emitting area Elektronenabgabefläche *f*
e. evaporation Elektronenverdampfung *f*, glühelektrische Elektronenemission *f*
e. focusing Elektronenfokussierung *f*; Elektronenbündelung *f*
e. gas Elektronengas *n*
e. gas degeneracy Elektronengasentartung *f*, Gasentartung *f*
e. gun Elektronenkanone *f*, Elektronenspritze *f*, Elektronenstrahlerzeuger *m*
e. image elektronenoptisches Abbild *n*, Elektronenbild *n*
e. image dissector Elektronenbildzerleger *m*
e. impact Elektronenstoß *m*
e. impact ion source Elektronenstoßionenquelle *f*
e. injector Elektroneneinspritzer *m*, Elektroneninjektor *m*
e. jet instrument Elektronenstrahlgerät *n*
e. jump Elektronensprung *m*, Elektronenübergang *m*
e. lens Elektronenlinse *f*
e. level Elektronenniveau *n*
e. liberation Elektronenablösung *f*
e. linear accelerator Linearbeschleuniger *m* für Elektronen, linearer Elektronenbeschleuniger *m*
e. metal Elektron *n*, Elektronmetall *n* *(Magnesiumlegierung)*
e. microscope Elektronenmikroskop *n*
e. microscopy Elektronenmikroskopie *f*
e. migration Elektronenwanderung *f*
e. mirror Elektronenreflektor *m*
e. multiplier Elektronenvervielfacher *m*, Sekundärelektronenvervielfacher *m*
e. multiplier phototube Fotozellenverstärker *m*
e. multiplier tube Elektronenvervielfacher *m*, Sekundärelektronenvervielfacher *m*
e.-optical elektronenoptisch
e. optics Elektronenoptik *f*
e. orbit Elektronenbahn *f* *(Bohrsches Atommodell)*
e. oscillations Laufzeitschwingungen *fpl* *(z. B. Magnetron)*
e. oscillator Laufzeitgenerator *m*
e. over-take region Elektronenüberholungsgebiet *n*
e. pair Elektronenpaar *n*
e.-pair bond homöopolare (kovalente) Bindung *f*
e. path Elektronenbahn *f*
e.-permeable elektronendurchlässig
e.-photon component Elektronen-Photonen-Komponente *f* *(kosmische Strahlung)*
e. physics Elektronenphysik *f*
e. plasma Elektronenplasma *n*
e. position Elektronenort *m*
e.-ray indicator tube Abstimmanzeigeröhre *f*, magisches Auge *n*, magischer Strich *m*
e. ray tube Elektronenstrahlröhre *f*
e. rest mass Ruhemasse *f* des Elektrons
e.-scan microscope Elektronenrastermikroskop *n*
e.-scanning beam Elektronenabtaststrahl *m*
e.-scanning pencil Elektronenabtaststrahl *m*
e. scattering Elektronenstreuung *f*
e. screening Kernabschirmung *f*, Abschirmung *f* des Atomkerns durch Elektronenhülle
e. sheath Elektronenschicht *f*
e. shell Elektronenhülle *f*, Elektronenschale *f*
e.-shell structure Aufbau *m* der Elektronenhülle
e. site Elektronenplatz *m*
e. source Elektronenquelle *f*
e. spectrograph Elektronenspektrograf *m*
e. spectroscopy Elektronenspektroskopie *f*
e. spin Elektronenspin *m*
e. spot Elektronenfleck *m* *(z. B. Leuchtschirm)*
e. state Elektronenzustand *m* *(Atome, Moleküle)*
e. stream Elektronenströmung *f*
e.-stream potential Bündel[ungs]potential *n*
e. structure of atoms Aufbau *m* der Atomhülle
e. trajectory Elektronenbahn *f*
e. transfer Elektronenübergang *m* *(Festkörper)*
e. transit time Elektronenlaufzeit *f*
e. transition Elektronenübergang *m*, Elektronensprung *m*
e. transition probability Elektronenübergangswahrscheinlichkeit *f*
e.-transmitting elektronendurchlässig
e. trap Elektronenfänger *m*
e. trapping Elektroneneinfang *m*
e. tube Elektronenröhre *f*

e. tube rack Röhrengestell *n*
e. tube tester Röhrenprüfgerät *n*
e. valve Elektronenröhre *f*
e. velocity Elektronengeschwindigkeit *f*
e.-volt Elektronenvolt *n*, eV
electronegative elektrisch negativ, elektronegativ
electronic elektronisch, Elektronen-
e. autopilot *(Flg, Rak)* elektronisch gesteuerter Kursgeber *m*
e. band spectrum Elektronenbandspektrum *n*
e. beam Elektronenstrahl *m*
e. brain Elektronengehirn *n*, elektronisches Rechengerät *n*
e. calculating punch elektronischer Rechenlocher *m*
e. calculator elektronischer Kleinrechner *m*, elektronisches Rechengerät *n*
e. charge Elementarladung *f*
e. chopper lichtelektrischer Zerhacker *m*
e. collision Elektronenstoß *m*
e. colour correction *(Typ)* elektronische Farbkorrektur *f*
e. computer elektronische Rechenanlage *f*, elektronische Rechenmaschine *f* *(programmgesteuert)*
e. conductivity Elektronenleitfähigkeit *f*
e. conductivity control elektronische Regelung *f* der Leitfähigkeit *(Lösungen)*
e. configuration Elektronenkonfiguration *f* *(Atome, Moleküle)*
e. contact rectifier Sperrschichtgleichrichter *m*
e. control elektronische Regelung (Steuerung) *f*, Regelung (Steuerung) mit Elektronenröhren
e.-controlled elektronisch gesteuert
e. core Elektronenrumpf *m (Atome)*
e. current Elektronenstrom *m*
e. data-processing machine elektronische Datenverarbeitungsmaschine *f*
e. data-processing system elektronisches Datenverarbeitungssystem *n*, elektronische Datenverarbeitungsanlage *f*
e. delay storage automatic calculator elektronisches Rechengerät *n* auf Speicherprinzip
e. diffraction grating Elektronenbeugungsgitter *n*
e. editor elektronische Dokumentationsmaschine *f*
e. engineering Elektronik *f*
e. engraving *(Typ)* elektronische Gravur *f*
e. engraving machine *(Typ)* elektronische Graviermaschine *f*
e. equilibrium Elektronengleichgewicht *n*
e. exposure control *(Typ)* Belichtungsautomat *m*
e. flash Elektronenblitz *m*
e. flashlamp *(Opt)* Blitzröhre *f*
e. flight simulator elektronischer Flugsimulator *m (Radar)*
e. formula Elektronenformel *f (Atome, Moleküle)*
e. frequency meter Röhrenfrequenzmesser *m*
e. heating induktive Hochfrequenzerhitzung *f*, Hochfrequenzheizung *f*
e. helmsman [elektronischer] Ruderautomat *m*
e. impact Elektronenstoß *m*
e. instrument elektronisches Meßinstrument *n*; elektronisches Gerät *n*
e. interpreter Übersetzungsmaschine *f*
e. key *(Fmt)* Schnelltaste *f*; vollautomatische Taste *f*
e. keying elektronisches (vollautomatisches) Tasten *n*
e. lens Elektronenlinse *f*
e. level *(Ph)* Elektronenterm *m*, Elektronenniveau *n*
e. marker generator elektronischer Markengeber *m*
e. micrometer Röhrenmikrometer *n*
e. mine detector elektronisches Minensuchgerät *n*
e. mobility Elektronenbeweglichkeit *f*
e. multiplier elektronischer Multiplikator *m* *(Rechenmaschine)*
e. office *(sl)* Datenverarbeitungsanlage *f*, kommerzieller Rechner *m*
e. peak-reading voltmeter elektronisches Voltmeter *n* mit Ablesung der Spitzenspannung
e. position indicator Bildschirm *m (Radar)*
e. print reader elektronischer Druckschriftleser *m*
e. programme control elektronische Programmsteuerung *f*
e. projection apparatus elektronenoptisches Abbildungsgerät *n*
e. punch elektronisch lochendes Rechengerät *n*
e. punch card machine elektronischer Locher *m*
e. quenching circuit elektronischer Löschkreis *m*, elektronische Löschschaltung *f*
e. random number generator elektronisches Gerät *n* zur Erzeugung von Zufallszahlen
e. reader elektronisches Lesegerät *n*; elektronische Briefsortiermaschine *f*
e. rectifier Röhrengleichrichter *m*
e. roughness tester elektronischer Rauhigkeitsmesser *m*
e. semiconductor Elektronenhalbleiter *m*
e. sewing HF-Schweißung *f (von Kunststoffen)*
e. shell Elektronenschale *f*

e. speed control elektronische Drehzahlregelung *f*, Drehzahlregelung mit Elektronenröhren
e. sweep circuit Kippschaltung *f*, Ablenkschaltung *f*
e. switch Schaltröhre *f*, Elektronenschalter *m*
e. tachometer elektronischer Drehzahlmesser *m*, elektronisches Tachometer *n*
e. television Katodenstrahlfernsehen *n*
e. temperature recorder elektronischer Temperaturschreiber *m*
e. thickness tester elektronisches Schichtdickenmeßgerät *n*
e. timer elektronischer Zeitschalter *m*
e. torch Plasmabrenner *m*
e. tuning elektronische Abstimmung *f*
e. tuning figure elektronisches Verstimmungsmaß *n (Oszillatoren)*
e. weighing equipment elektronische Wiegeanlage *f*
e. welding control elektronische Schweißzeitbegrenzung *f*
electronical elektronisch
electronically elektronisch
e. regulated elektronisch geregelt, röhrengeregelt
e. tuned elektronisch abgestimmt
electronicam Kombination *f* aus Fernseh- und Filmkamera
electronics Elektronik *f*
electronogenic elektronenerzeugend
electrooptical shutter elektrooptischer Verschluß *m*
electrooptics Elektrooptik *f*
electroosmosis Elektroosmose *f*
electroparting elektrolytische Scheidung (Trennung) *f*
electropercussive welding Widerstandsstoßschweißung *f (Nichteisenmetalle)*
electrophoresis Elektrophorese *f*
electrophotographic printing *(Typ)* elektrofotografische Wiedergabe *f*
electrophotography *(Typ)* Elektrofotografie *f*
electroplate / to elektroplattieren, galvanisieren, elektrolytisch metallisieren
electroplate galvanischer Überzug *m*, Deckmetall *n*
electroplating Elektroplattieren *n*, Elektroplattierung *f*, Galvanisieren *n*, Galvanisierung *f*; Galvanotechnik *f*, Galvanostegie *f*
e. shop Galvanisierwerkstatt *f*
electropneumatic elektropneumatisch
e. contactor Druckluftschütz *n* mit Magnetsteuerung
e. controller Druckluftsteuerschalter *m*; elektropneumatischer Programmschalter *m*
e. regulator Pumpenselbstschalter *m*
e. signal box *(Eis)* Druckluftstellwerk *n*
e. valve elektrisch betätigtes Druckluftventil *n*, elektropneumatisches Ventil *n*
electropolishing elektrolytisches Polieren *n*, Elektropolieren *n*
electropositive elektropositiv, elektrisch positiv
electrorefining elektrolytische Raffination *f*
electroscope Elektroskop *n*
electrosherardizing Elektrosherardisierung *f (Oberflächenvergütung)*
electrostatic elektrostatisch
e. actuator Hilfselektrode *f (Akustik)*
e. charge of textile fibres elektrostatisches Aufladen *n* von Textilfasern
e. coupling elektrostatische (kapazitive) Kopplung *f*
e. detarrer elektrischer Teerabscheider *m*
e. displacement elektrostatische Verschiebung *f*
e. energy-storage spot welder Speicherschweißgerät *n*
e. feedback kapazitive Rückkopplung *f*
e. field elektrostatisches Feld *n*, Influenzfeld *n*
e. filter Elektrofilter *n*
e. flocking elektrostatisches Beflocken *n*
e. focusing elektrostatische Fokussierung (Strahlenbündelung) *f (Elektronenoptik)*
e. force elektrostatische Kraft *f*
e. generator elektrostatischer Generator *m*
e. hum pickup elektrostatische Brummschleife *f*, Antibrummkondensator *m (Radio)*
e. induction [elektrische] Influenz *f*
e. instrument elektrostatisches Meßinstrument *n*
e. ion microscope Feldionenmikroskop *n*
e. lens elektrostatische Linse *f*
e. loudspeaker Kondensatorlautsprecher *m*
e. microphone Kondensatormikrofon *n*
e. microscope Elektronenmikroskop *n* mit elektrostatischen Linsen
e. precipitation elektrostatische Abscheidung (Fällung) *f*
e. precipitator Elektrofilter *n (Gasreinigung)*
e. printing *(Typ)* elektrostatischer Druck *m*
e. repulsion *(Ph)* elektrostatische Abstoßung *f*
e. retentive force elektrostatische Klebkraft *f*
e. screen elektrostatischer Schirm *m*, Faraday-Käfig *m*
e. screening elektrostatische Abschirmung *f*
e. shielding elektrostatische Abschirmung *f*
e. spraying elektrostatisches Spritzen *n*
e. storage elektrostatische Speicherung *f (Datenspeicherung)*
e. storage tube Speicherröhre *f (Datenspeicherung)*
e. time base generator Ablenkgenerator *m* für elektrostatische Ablenkung
e. transmitter Kondensatormikrofon *n*
e. unit elektrostatische Einheit *f*

electrostatical *s.* electrostatic
electrostatics Elektrostatik *f*
electrostenolysis Elektrostenolyse *f*
electrostriction Elektrostriktion *f*
electrosyntonic switch durch Tonfrequenz ferngesteuerter Schalter *m*
electrotechnical elektrotechnisch
electrotechnician Elektrotechniker *m*
electrotechnics Elektrotechnik *f*
electrotellurograph Erdstrommeßgerät *n*
electrotherapeutics Elektrotherapeutik *f*
electrothermal elektrothermisch
e. instrument Hitzdrahtinstrument *n*
e. relay elektrothermisches Relais *n*
electrothermic elektrothermisch
e. release elektrothermischer Auslöser *m*
e. tensioning elektrothermisches Vorspannen *n (Spannbetonbau)*
electrothermics Elektrothermie *f*, Elektrowärmelehre *f*
electrotinning galvanische Verzinnung *f*
electrotype / to galvanografisch abbilden
electrotype Galvano[plastik *f*] *n*; galvanografische Abbildung *f*
electrotyper Galvanoplastiker *m*
electrotyping Elektrotypie *f*
electrotypograph elektrische Setzmaschine *f*
electrovalency *(Ch)* Elektrovalenz *f*, heteropolare Valenz *f*
e. bond *(Ch)* Elektrovalenzbindung *f*, heteropolare Bindung *f*
electrowinning elektrolytische Metallgewinnung *f*
element Element *n*, Grundstoff *m*; *(Masch)* Baugruppe *f*, Bauteil *n*; Bestandteil *m*, Organ *n*; *(El)* Bauelement *n*; Schaltelement *n*; Plattenpaket *n (Batterie)*; *(Flg)* Flugzeugpaar *n*
e. former Heizleiterträger *m*, Widerstandsträger *m (Heizkörper)*
e. jacket *(Kern)* Brennstoffelement[en]hülse *f*
e. of a cone *(Math)* Erzeugende (Mantellinie) *f* eines Kegels
e. spacing Strahlerabstand *m (Antennentechnik)*
e. transformation Elementumwandlung *f*
elemental elementar, primär, einfach
e. area *(Fs)* Bildpunkt *m*
elementary elementar
e. aerial Antennenelement *n*, Strahlerelement *n*
e. analysis Elementaranalyse *f*
e. area Flächenelement *n*
e. cell Elementarzelle *f (Kristalle)*
e. charge Elementarladung *f*,
e. circuit *(El)* Grundschaltung *f*
e. circuit diagram Prinzipschaltbild *n*
e. interval Schritt *m (z. B. Telegrafie)*
e. particle Elementarteilchen *n*
e. process Elementarprozeß *m*
e. quantum Elementarquantum *n*
e. sone einfacher Klang (Ton) *m*
e. unit 1. Grundeinheit *f (Maßsystem)*; 2. *(Ak)* Elementarintervall *n*
e. wave *(Opt)* Elementarwelle *f (Huygenssches Prinzip)*
e. weave *(Text)* Grundbindung *f*
elements of the orbit Bahnelemente *npl*
e. of the pair Getriebeelemente *npl (Getriebelehre)*
elemi gum Elemiharz *n*
elephant boiler Bouilleurkessel *m*, Walzenkessel *m* mit Sieder, Siederkessel *m*
e. machine tool *(Am)* Schwerstwerkzeugmaschine *f*
elevate / to [an]heben, erhöhen
elevated aerial Hochantenne *f*
e. frame Hochrahmen *m*
e. highway Hochstraße *f*
e. projection kotierte Projektion *f (technisches Zeichnen)*
e. railway Hochbahn *f*; S-Bahn *f*, Stadtbahn *f*
e. road[way] Hochstraße *f*
e. tank Hochbehälter *m*, Hochreservoir *n*
e. temperature hardness Warmhärte *f*, Härte *f* bei erhöhten Temperaturen
e. temperature test Warmversuch *m*
elevating 1. Heben *n*; Erhöhen *n*; 2. Hubbewegung *f (Bagger)*
e. arm radial drilling machine Radialbohrmaschine *f* mit Höhenverstellung
e. control Höhenverstellung *f*
e. gear Hochstellvorrichtung *f*
e. mechanism Hubwerk *n*; Höhenrichtmaschine *f*
e. motion Hubbewegung *f*; Senkrechtverstellbewegung *f*
e. motor Hubmotor *m (Radialbohrer)*
e. nut Hubspindelmutter *f*
e. platform Hubplattform *f*
e. screw Hubspindel *f*; Gewindespindel *f* für Vertikalverstellung
elevation 1. Höhe *f* über Normalnull; Höhe über Meeresspiegel; 2. Steighöhe *f (Kapillaren)*; 3. Anhebung *f*, Hochhub *m*; 4. Aufriß *m*, Riß *m*, Ansicht *f (technisches Zeichnen)*; 5. Fassade *f*
e. angle Neigungswinkel *m*; Erhebungswinkel *m (Radio)*; Erhöhungswinkel *m (Ballistik)*
e. difference *(Flg)* Höhenstufung *f*, Tiefenstufung *f*
e. head Lagehöhe *f*, Ortshöhe *f*
e. of boiling point Siedepunktserhöhung *f*
e. setter Höhenrichtkanonier *m*
e.-tracking telescope Höhenrichtfernrohr *n*
elevator 1. *(Masch)* Hebewerk *n*, Aufzug *m*; 2. *(Flg)* Höhenruder *n*, Höhensteuer *n*; 3. *(Bw)* Aufzug, Fahrstuhl *m*, Lift *m*;

4. *(Masch)* Anhebeeinrichtung *f*; 5. Elevator *m*, Höhenförderer *m*, Becherwerk *n*
e. **assembly** *(Flg)* Höhenleitwerk *n*
e. **control** *(Flg)* Höhensteuerung *f*
e. **dredge[r]** Eimerkettenbagger *m*
e. **[for elevating work]** *(Masch)* Teileheber *m*
e. **rope** Aufzugseil *n*
e. **servo motor** *(Flg)* Höhenrudermaschine *f*
e. **shaft** Aufzugschacht *m*
e. **suction pipe** Saugrohr *n* zum Wasserheber
e. **unit** *(Flg)* Höhenleitwerk *n*
e. **well** Aufzugschacht *m*, Fahrstuhlschacht *m*
eleven-unit code Elfschrittalphabet *n (Telegrafie)*
e.**-year [sunspot] cycle** elfjährige Sonnenfleckenperiode *f*
eleventh step contact *(Fmt)* Durchdrehkontakt *m*, W11-Kontakt *m*
eliminant *(Math)* Resultierende *f*, Resultante *f (eines Gleichungssystems)*
eliminate / to 1. *(Math)* eliminieren; 2. ausscheiden, wegnehmen, überflüssig machen
e. **a failure** einen Defekt (Schaden) beheben, eine Störung beseitigen
e. **carbon** entkohlen
e. **defective components** schadhafte Teile aussondern
e. **heat** Wärme abführen
elimination 1. *(Math)* Elimination *f*; 2. *(Ch)* Abführung *f*, Absonderung *f*, Abspaltung *f*, Ausscheidung *f*
e. **of air from the steam pipes** Entlüftung *f* der Dampfrohrleitung
e. **of defects** Fehlerbeseitigung *f*; Schadenbehebung *f*
e. **of radio interference** Funkentstörung *f*
elinvar Elinvar *n (thermoelastisch stabile Legierung aus Fe, Ni, Cr, W, Mn)*
eliquation *(Met)* Herausschmelzen *n*, Seigerung *f*
elk hide Elchhaut *f*, Elchleder *n*
ell Seitenflügel *m* eines Gebäudes *(im rechten Winkel)*
ell L-förmig
e.**-beam** *(Bw)* einseitiger Plattenbalken *m*
ellipse Ellipse *f*
ellipsograph Ellipsograf *m*, Ellipsenzirkel *m*, Ellipsenzeichner *m*
ellipsoid Ellipsoid *n*
e. **of revolution** Rotationsellipsoid *n*
ellipsoidal ähnlich einem Ellipsoid; elliptisch *(nur in Verbindung mit Koordinaten)*
elliptic elliptisch
e. **trammel** Ellipsenzeichner *m*, Ellipsenzirkel *m*
elliptical elliptisch
e. **arch** Korbbogen *m (Mauerwerk)*
e. **field** elliptisches Drehfeld *n*
e. **polarization** elliptische Polarisation *f*
e. **rotating field** elliptisches Drehfeld *n*
e. **stern** *(Schiff)* elliptisches Heck *n*, Rundheck *n*, rundes Heck
elliptically polarized elliptisch polarisiert
ellipticity Abplattung *f (eines Rotationsellipsoides)*
e. **coefficient (ratio)** *(Math)* Halbachsenverhältnis *n (Ellipse)*
elm Bergrüster *f*, Feldrüster *f*, Flatterrüster *f*
elongate / to 1. verlängern, dehnen; 2. *(Met)* strecken, längen, recken
e. **excessively** überdehnen
elongated hole Langloch *n*
elongating slots Langlochfräsen *n*
elongation 1. Längenzunahme *f*, Längung *f*, Verlängerung *f*; 2. Bruchdehnung *f*
e. **at break** Bruchdehnung *f*
e. **limit** Dehnungsgrenze *f*
e. **line** Dehnungslinie *f*
e. **rupture** Bruchdehnung *f*
elpidite *(Min)* Elpidit *m*
eluate / to *(Ch)* eluieren, herauslösen
eluate *(Ch)* Eluat *n*
eluent *(Ch)* Elutionsmittel *n*
elusive transients *(BMSR)* flüchtige (nicht zu erfassende) Augenblickswerte *mpl*
elution *(Ch)* Eluierung *f*, Elution *f*, Auslaugung *f*, Auswaschung *f*
e. **analysis** Elutionsmethode *f*
elutriate / to ausschlämmen, [ab]schlämmen, abschwemmen
elutriating funnel Schlämmtrichter *m*
elutriation Abschlämmung *f*, Ausspülung *f*, Auswaschung *f*, Schlämmung *f*; Schlammspülverfahren *n*
e. **analysis** Schlämmanalyse *f (im aufsteigenden Wasserstrom)*
elutriator Schlämmgerät *n*, Schlämmapparat *m*
Elverson oscilloscope Lichtblitzstroboskop *n*
eman Eman *n (Einheit der Konzentration der Radioaktivität)*
emanating power *(Ph)* Emaniervermögen *n*
emanation 1. Ausstrahlung *f*; Ausfluß *m*; 2. *(Kern)* Emanation *f*, Ausströmung *f (radioaktiver Gase)*; 3. *(Ch)* Emanation, Radon *n (chemisches Element)*
embankment Böschung *f*; Damm *m*, Erddamm *m*; Eisenbahndamm *m*; Uferanlage *f*, Uferbefestigung *f*, Kai *m*; Aufschüttung *f*
embargo Ausfuhrverbot *n*; Hafensperre *f (über ein Schiff)*
embark / to sich einschiffen, an Bord gehen
embattlemented *(Bw)* mit Zinnen versehen
embayment *(Schiff)* Bucht *f*, Golf *m*
embed / to 1. *(Masch)* einlagern, einschließen; umhüllen; einlassen; einbauen; 2. *(Math)* einbetten *(Geometrie)*
e. **in concrete** einbetonieren

embedded in eingebettet in, eingelagert in, eingelassen in, gekapselt in
e. rectifier gekapselter Gleichrichter *m*
e. temperature detector eingebautes Thermoelement *n*; eingebaute Widerstandsspule *f*
embedding material Einbettungsmasse *f*
embellish / to schmücken
ember glimmende Asche (Kohle) *f*
emboss / to bossieren, [hohl]prägen, pressen; narbenpressen *(Leder)*
embossed erhaben, geprägt
e. crêpe *(Text)* geprägter Krepp *m*
e. hardboard geprägte Hartplatte *f*
e. leather [narben]gepreßtes Leder *n*, Preßleder *n*
e. map Reliefkarte *f (Landkarte)*
e. plate Buckelblech *n*, Buckelplatte *f*
e. print Prägedruck *m*
e. sheet Prägefolie *f*; *(Led)* genarbte Folie *f*
e. wiring [aufgeprägte] Verdrahtung *f*; Prägung *f (gedruckte Schaltung)*
embosser *(Fmt)* Reliefschreiber *m*
embossing 1. Hohlprägung *f*; 2. *(Text)* Prägung *f*, Relief *n*, erhabene Arbeit *f*; 3. Netzraster *m*, Raster *m (Farbfilm)*; 4. Gaufrage *f*
e. calender 1. *(Text)* Prägekalander *m*, Profilkalander *m*; 2. *(Typ)* Prägekalander, Gaufrierkalander *m*
e. die Prägewerkzeug *n*
e. hammer Treibhammer *m*; Bossierhammer *m*
e. iron *s.* e. hammer
e. machine 1. Prägepresse *f*; 2. Gaufriermaschine *f (Papierherstellung)*
e. plate *(Led)* Narbenplatte *f*
e. press Hohlprägepresse *f*, Prägepresse *f*
e. press for hot process Heißprägepresse *f*
e. roller *(Led)* Chagrinierrolle *f*, Narbenpreßrolle *f*
embossment Warze *f (Buckelschweißung)*
embranchment Abzweigung *f*, Zweig *m*; Gabelung *f*
e. frame *(El)* Leitungsverteiler *m*
embrasure *(Bw)* abgeschrägte Tür- oder Fensterleibung *f*; Schießscharte *f*
embrittle / to spröde werden, verspröden
embrittlement Versprödung *f*, Brüchigwerden *n*
embrittling influence versprödender Einfluß *m*
embroider / to sticken
embroidered goods *(Text)* lanziertes Gewebe *n*
embroidery lathe *(Text)* Broschierlade *f*
e. loom Broschierwebstuhl *m*
emerald 1. *(Min)* Smaragd *m*, grasgrüner Beryll *m*; 2. *(Typ) (etwa)* Nonpareille *f (Schriftgröße)*
e. copper Dioptas *m*
e. green Schweinfurter Grün *n*, gemischtes Kupferarsenitazetat *n*
e. nickel Nickelsmaragd *m*
emerge / to austreten *(z. B. Licht)*
e. from the metal aus dem Metall austreten *(Elektronen)*
e. into eintreten in
emergency Notfall *m*
e. aerial Behelfsantenne *f*; Ersatzantenne *f*; Hilfsantenne *f*
e. air valve *(Eis)* Notbremse *f*
e. airport Ausweichflughafen *m*
e. battery Ersatzbatterie *f*; Notbatterie *f*
e. boat Unfallboot *n (an Bord stets ausgeschwungen gefahren)*
e. brake Notbremse *f*; Sicherheitsbremse *f*
e. cable 1. Hilfskabel *n*; Baukabel *n*; 2. Hilfstragseil *n*
e. call Unfallanruf *m*, Unfallmeldung *f*; Überfallanruf *m*, Überfallmeldung *f (Fernsprechwesen)*
e. connection *(El)* Notschaltung *f*
e. cutoff *(Rak)* Notbrennschluß *m*
e. cutout Notausschalter *m*, Notausschaltung *f*
e. danger lamp *(Eis)* Nothaltlampe *f (Signal)*
e. disconnection *(Eis)* Notauflösung *f (Stellwerk)*
e. door 1. Notausgang *m (Gebäude)*; Notausstieg *m (Verkehrsmittel)*; 2. Brandtür *f*
e. exchange Ausweichvermittlung *f (Fernsprechwesen)*
e. exit Notausgang *m (Gebäude)*; Notausstieg *m (Verkehrsmittel)*
e. fire pump Notfeuerlöschpumpe *f*, Reservefeuerlöschpumpe *f*
e. generator Notstromgenerator *m*
e. governor Sicherheitsregler *m*
e. lighting Notbeleuchtung *f*, Hilfsbeleuchtung *f*; *(Flg)* Notbefeuerung *f*
e. lighting system Notbeleuchtungsanlage *f*
e. outfit Notausrüstung *f*
e. power plant Notstromanlage *f*
e. ratings für Notfälle festgesetzte Leistung *f*
e. rod *(Kern)* Abschaltstab *m*, Sicherheitsstab *m*, Sicherungsstab *m*
e. route Hilfsweg *m (Fernsprechtechnik)*
e. shut-down Notabschaltung *f*, Schnellabschaltung *f*, Sicherheitsabschaltung *f (Reaktor)*
e. shut-off rod Abschaltstab *m*, Sicherheitsstab *m (Reaktor)*
e. slaughter *(Lw)* Notschlachtung *f*
e. speed *(Schiff)* Höchstgeschwindigkeit *f (im Notfall)*
e. stop Notausschalter *m (Fahrstuhl)*
e. stop valve Schnellschlußventil *n*
e. switch Not[brems]schalter *m*
e. tiller *(Schiff)* Reservepinne *f*
e. train stop Notbremse *f*
e. transmitter Notsender *m*

e. tripping Notausschaltung *f*
e. turn *(Schiff)* Ausweichmanöver *n*
e. valve Notabsperrventil *n*, Sicherheitsventil *n*
emergent austretend; Austritts-
e. direction Austrittsrichtung *f*
e. stem correction Korrektur *f* für den aus der Flüssigkeit herausragenden Teil des Quecksilberfadens *(Thermometrie)*
emerizing *(Text)* Velourieren *n*
emery Schmirgel *m*
e. base-paper Schmirgelrohpapier *n*
e. cake Schmirgel-Öl-Gemisch *n*
e. cloth Schmirgelleinen *n*
e. dust Schmirgelpulver *n*
e. grinding machine Schmirgelschleifmaschine *f*
e. paper Schmirgelpapier *n*
e. paste Schmirgel-Öl-Gemisch *n*
e. wheel Schmirgelscheibe *f*
e. wheel dresser Abdrehvorrichtung *f* *(für Schleifscheiben)*
emigrant ship Auswanderungsschiff *n*, Auswandererschiff *n*
emission Emission *f*, Ausstrahlung *f*, Strahlenaussendung *f*, Strahlung *f*
e. analysis Emissionsanalyse *f*
e. band *(Ph)* Emissionsbande *f*
e. characteristic *(El)* Emissionskennlinie *f*
e. current Emissionsstrom *m*
e. efficiency Elektronenausbeute *f*; Heizmaß *n (Röhren mit Glühkatode)*
e. limitation Emissionsbegrenzung *f*, Sättigung *f (Elektronenröhren)*
e. line *(Opt)* Emissionslinie *f*
e. of electrons Elektronenemission *f*
e. of heat Wärmeausstrahlung *f*, Wärmeabstrahlung *f*
e. of light Ausstrahlung *f* [von Licht]; Lichtaussendung *f*, Leuchten *n*
e. of particles Teilchenemission *f*
e. of quanta Quantenemission *f*
e. of sound Schallabgabe *f*, Schallaussendung *f*
e. of steam Dampfausströmung *f*
e. of waves Wellenaussendung *f*
e. probability *(Ph)* Emissionswahrscheinlichkeit *f*
e. reflector Reflektorstrahler *m (Antennentechnik)*
e. saturation echte Sättigung *f*, Katodensättigung *f*
e. spectrum Emissionsspektrum *n*
e. work Austrittsarbeit *f* *(Fotoelektronen)*
emissive ausstrahlend
e. power 1. Emissionsvermögen *n*; 2. Strahlungsleistung *f*
emissivity Ausstrahl[ungs]vermögen *n*, Emissionsvermögen *n*, Emissionsfähigkeit *f*; Elektronenemission *f*, Elektronenabgabe *f*
e. of a thermal radiator Emissionsgrad *m* (Emissionsvermögen *n*) eines Temperaturstrahlers
emit / to aussenden, ausstrahlen, emittieren
e. heat Wärme ausstrahlen (abstrahlen)
e. light Licht aussenden, leuchten
e. particles Teilchen emittieren
e. radiation ausstrahlen, Strahlung emittieren
e. sound Schall aussenden; tönen; schallen
e. steam Dampf ausströmen [lassen]
e. waves Wellen aussenden
emitron Emitron *n (englischer Spezialtyp des Ikonoskops)*
e. camera Ikonoskop *n*
emittance Ausstrahlungsvermögen *n*; spezifische Ausstrahlung *f*
emitted emittiert
e. particle *(Ph)* emittiertes Teilchen *n*
emitter 1. *(Ph)* Strahler *m*, Strahlenquelle *f*; 2. *(Kern)* radioaktiver Strahler; 3. *(El)* Emissionselektrode *f*, Emitter *m (Transistoren)*
e. barrier Emittersperre *f*
e. bypass capacitor Emitterkondensator *m*
e. contact Emitteranschluß *m*
e. cutoff current Emitterreststrom *m*
e. follower Emitterfolger *m*, Emitterverstärker *m*
e. junction Emitterschicht *f*, Emitterübergang *m*
e. region Emitterzone *f*
e. resistance [innerer] Emitterwiderstand *m*
e. resistor äußerer Emitterwiderstand *m*
emitting aerial Sendeantenne *f*
e. area Strahlungsfläche *f*, Emissionsfläche *f*
e. electron Leuchtelektron *n*
e. layer Emissionsschicht *f*
e. nucleon Leuchtnukleon *n*
e. power Emissionsvermögen *n*
e. surface *s.* e. area
emmetropia Normalsichtigkeit *f*
emmetropic normalsichtig
emmonite *(Min)* Emmonit *m*
emmonsite *(Min)* Emmonsit *m*
emollient *(Ch)* Erweichungsmittel *n*, Weichmacher *m*
e. cream Hautnährcreme *f*, Wirkcreme *f*
empennage *(Flg)* Leitwerk *n*
emphasing *s.* emphasizing
emphasis Anhebung (Betonung, Bevorzugung, Hervorhebung) *f* bestimmter Tonfrequenzen *(Elektroakustik)*
e. on basstones Tiefenhervorhebung *f* *(Radio)*
emphasizer Frequenzanheber *m (Radio)*

emphasizing Anheben (Betonen, Bevorzugen, Hervorheben) *n* bestimmter Tonfrequenzen *(Elektroakustik)*
empire cloth *(Plast)* Ölstoff *m*
Empire *(Typ)* Empireformat *n* *(6½ × 4½ in.)*
empirical empirisch
e. equation empirische Gleichung *f*
e. fact Erfahrungstatsache *f*
e. formula empirische Formel *f*
e. value Erfahrungswert *m*
emplecite *(Min)* Emplezit *m*, Emplektit *m*
employ / to 1. beschäftigen; 2. verwenden, anwenden
e. work rotation *(Masch)* das Werkstück die Umlaufbewegung ausführen lassen
empties *(Brg)* Leerwagen *mpl*
empty / to [ent]leeren
e. a boiler einen Kessel ablassen
empty leer
e. band unbesetztes Energieband *n* *(Festkörper, Halbleiter)*
e.-cell process *(Holz)* Spartränkverfahren *n*
e.-frame prevention Leerbelichtungssperre *f* *(Fotografie)*
e. journey *(Brg)* Leer[wagen]zug *m*
e. lap *(Text)* leerer Wickel *m*
e. run *(Eis)* Leerfahrt *f*
e. running weight *(Eis)* Leergewicht *n*
e. slot *(Elm)* Blindnut *f*, unbesetzte (unbewickelte) Nut *f*
e. stroke *(Büro)* Leerzug *m*
e. weight Leermasse *f*
emptying Entleeren *n*, Entladung *f* *(Wagen)*
e. poker *(Lw)* Räumspieß *m*
Emscher tank *(Bw)* Emscherbrunnen *m*
emulsibility Emulgierbarkeit *f*, Emulgierfähigkeit *f*
emulsifiability Emulgierbarkeit *f*, Emulgierfähigkeit *f*
emulsifiable emulgierfähig, emulsionsfähig
emulsification Emulgierung *f*, Emulsionsbildung *f*
emulsifier Emulgator *m*, Emulsionsbildner *m*, Emulgierzusatz *m*; Emulgiermaschine *f*
emulsify / to emulgieren
emulsifying Emulgierung *f*
e. agent Emulgator *m*, Emulgiermittel *n*
emulsion 1. Emulsion *f*; 2. *(Led)* Nahrung *f* *(Weißgerbung)*
e. base Schichtträger *m* *(Film)*
e. carrier Schichtträger *m* *(Film)*
e. face of film Schichtseite *f* des Films
e. paint Emulsionsfarbe *f*
e. polymerization Emulsionspolymerisation *f*
e. side of film Schichtseite *f* des Films
e. track Schichtspur *f*, Teilchenspur *f* in der Fotoschichtspur
emulsionize / to emulgieren
en bloc casting Blockgußstück *n*
en cabochon mit mugeligem Schliff
en route marker beacon Flugstreckenmarkierungsfunkfeuer *n*
enamel / to emaillieren, glasieren
enamel Email[lack *m*] *n*, Glasur *f*; Lackfarbe *f*; Kunstharzlack *m*, Schmelz *m*
e. bead Boraxperle *f*, Phosphorsalzperle *f* *(Analytik)*
e. bonded cotton mit Lack getränkte Baumwolle *f* *(für Kabelisolierung)*
e. colour Schmelzfarbe *f*, Emailfarbe *f*
e. foil Emailbelag *m*
e.-insulated wire Lackdraht *m*, emaillierter Draht *m*
e. paint Isolierlack *m*; Lackfarbe *f*
e. paper Glanzpapier *n*
e. pearl Schmelzperle *f*
e. remover Nagellackentferner *m*
e. stripping Entemaillierung *f*
e. varnish Emaillack *m*
e.-wire Lackdraht *m*, emaillierter Draht *m*
enamelled emailliert, glasiert
e. board Hochglanzstreichkarton *m*
e. brick Glasurstein *m*
e. cloth Lederimitation *f*; Wachstuch *n*
e. copper [wire] Kupferlackdraht *m*
e. hides *(Led)* Lackvachetten *fpl*
e. paper Kunstdruckpapier *n*; Hochglanzpapier *n*
e. wire Lackdraht *m*, emaillierter Draht *m*
enamelling Emaillieren *n*, Glasieren *n*
e. furnace Emaillierofen *m*
enanthaldehyde Önanthaldehyd *m*, Önanthal *n*, Heptanal *n*
enanthic acid Önanthsäure *f*
enantiomorphous enantiomorph
enantiotropic enantiotrop
enargite Enargit *m* *(Kupfererz)*
encapsulate / to [ein]kapseln, einhülsen, einschließen
encapsulated fuel element *(Kern)* eingehülstes (eingeschlossenes) Brennstoffelement *n*
e. source eingeschlossenes radioaktives Präparat *n*; eingeschlossene Strahlenquelle *f*
encase / to einkapseln, einschließen, mit Gehäuse versehen
encased eingekapselt, ummantelt, umhüllt
e. capacitor Becherkondensator *m*
e. knot eingewachsener Ast *m*
e. magnet Mantelmagnet *m*
e. steelwork betonummantelte Stahlbauteile *npl*
encasing Ummantelung *f*
e. tube Hüllrohr *n*
encircle / to umschreiben; einkreisen
enclose / to [ein]kapseln; einschließen, umgrenzen

e. in brackets einklammern
enclosed verdeckt, geschlossen
e. arc lamp Dauerbrandbogenlampe *f*
e. blower Kapselgebläse *n*
e. discharge safety valve Sicherheitsventil *n* mit Dampfabblasestutzen, geschlossenes Sicherheitsventil
e. fuse Sicherungspatrone *f*, Patronensicherung *f*
e. gear gekapseltes Getriebe *n*
e. in an oil-tight case in öldichter Verkleidung
enclosing rock Nebengestein *n*
enclosure 1. *(Masch)* Gehäuse *n*, Hülle *f*, Kapsel *f*, Umhüllung *f*; 2. *(Kern)* abgeschlossener Raum *m*; 3. Einfriedung *f*, Umfriedung *f*, Umzäunung *f*, Kabine *f*, Gehege *n*; 4. *(Math)* abgeschlossene Hülle; 5. *(Met)* Einschluß *m*
encode / to kodieren, verschlüsseln, chiffrieren
encoder 1. Verschlüßler *m*, Verschlüsselungsgerät *n*; 2. *s.* analog-to-digital converter
e. matrix Verschlüsselungsmatrix *f*
encoding Verschlüsseln *n*
encounter Zusammenstoß *m*
encrustation Krustenbildung *f*
end 1. Beendigung *f*, Ende *n*, Schluß *m*; 2. *(Masch)* Stirnfläche *f*, Stirnseite *f*; Ansatz *m* *(Stiftschraube)*; Nase *f* *(Spindel)*; 3. *(Text)* einzelner Kettfaden *m*; 4. *(Brg)* Ort *n*, Ortsstoß *m*
e. anchorage Endverankerung *f* *(Spannbeton)*
e.-anchored reinforcement endverankerte Spannbewehrung *f* *(Spannbeton)*
e.-and-end lease *(Text)* Kopfkreuz *n*
e.-around carry *(BMSR)* Endübertrag *m*, Rück[lauf]übertrag *m*, Umlaufübertrag *m* *(bei komplementierenden Addierschaltungen)*
e. bander Rollenpackmaschine *f* *(Papier)*
e. bearing Auflager *n*
e. bearing capacity Spitzentragfähigkeit *f* *(Pfahlgründungen)*
e.-bearing pile Aufstandspfahl *m*, stehender Pfahl *m* *(Grundbau)*
e. bell *(Elm)* Lagerschild *m*, Motorschild *m*
e. block Endmaß *n*, Parallelendmaß *n*
e. block accuracy Endmaßgenauigkeit *f*
e. block set Endmaßsatz *m*
e. breakage *(Text)* Fadenbruch *m*
e. breakage rate *(Text)* Fadenbruchzahl *f*
e. bulkhead *(Schiff)* Endschott *n*, Abschlußschott *n*; Frontschott *n*, Absatzschott *n*
e.-centred basisflächenzentriert *(Kristalle)*
e. clearance Axialspiel *n*, Längsspiel *n*
e. coaming *(Schiff)* Quersüll *m* *(n)*
e. cogging End[ver]kämmung *f* *(Zimmerei)*
e. collar Anschlagring *m*, Bundring *m*
e. column Gegenständer *m*
e. condition Randbedingung *f*
e. connection *(Elm)* Stirnverbindung *f* *(Anker)*
e. contact Endkontakt *m*
e.-cut turning tool Stechdrehmeißel *m*
e.-cutting in axialer Richtung schneidend, mit der Stirnschneide schneidend
e.-cutting edge Stirnschneide *f*; Stirnzahn *m*
e.-cutting nipper Vorschneider *m*
e.-cutting plier Vorschneider *m*, Vorschneidezange *f*
e.-cutting reamer Stirnreibahle *f*
e. door Stirnklappe *f*
e. double bonds *(Ch)* endständige Doppelbindungen *fpl*
e.-dump truck Hinterkipper *m*; Rückwärtskipper *m*
e. entrance *(Eis)* Endeinstieg *m* *(Personenwagen)*
e. face Planfläche *f*, Stirnseite *f*
e. face mill Walzenstirnfräser *m*
e. facing Abflächen *n* *(von Stirnseiten)*
e. fe[e]d aerial L-Antenne *f*, spannungsgespeister Lambda-2-Dipol *m*
e. filling piece Endpaßstück *n*, Schlußstück *n*
e. fixing Endeinspannung *f* *(Balken)*; Fußeinspannung *f* *(Säulen)*
e.-frame *(Elm)* Lagerschild *m*
e. grip Endeinspannung *f* *(Schweißelektrode)*
e. impedance Abschlußwiderstand *m*, Endimpedanz *f*
e. journal Stirnzapfen *m*
e. launching *(Schiff)* Längsablauf *m*, Längsstapellauf *m*
e. leakage *(Elm)* Stirnstreuung *f*
e. load Axialbelastung *f*
e. measure Endmaß *n*, Parallelendmaß *n*
e.-measure jig borer Endmaßlehrenbohrmaschine *f*
e. measuring block Endmaß *n*, Parallelendmaß *n*
e. measuring disk rundes Prüfmaß *n* *(Schraublehre)*
e. measuring rod stabförmiges Prüfmaß *n* *(Schraublehre)*
e. measuring rod with spherical ends Kugelendmaß *n*
e. mill Schaftfräser *m*; Fingerfräser *m*
e. milling Stirnflächenfräsen *n*; Fräsen *n* mit Fingerfräser
e. milling cutter Schaftfräser *m*
e. moment *(Bw)* Einspannmoment *n*
e. movement Axialbewegung *f*
e. of a gallery *(Brg)* Ort *n*, Ortsstoß *m*
e. of pole piece Polende *n*
e. of pulsing signal Ende *n* des Zifferübertragungszeichens *(Trägerfrequenztechnik)*
e. of stroke Hubende *n*

e. of transmission Sendeschluß *m (Radio, Fernsehen)*
e.-of-word character Wortendezeichen *n*
e.-on 1. *(Schiff)* mit dem Bug voraus; 2. *(Am, Brg)* Abbau *m* mit senkrecht zu den Schlechten gestellter Abbaufront
e.-on directional aerial (array) Längs[richt]strahler *m*
e.-paper *(Typ)* Vorsatz[papier] *n*
e.-papering machine Vor[satz]klebemaschine *f*
e. pier Endpfeiler *m*
e. plate 1. *(Masch)* Abschlußdeckel *m*, Abschlußwand *f*; Seiten[begrenzungs]platte *f*; 2. Lagerschild *m*; 3. *(Brg)* kurzes Joch *n (Bolzenschrotausbau)*; 4. *(Schiff)* Endschott *n (eines Doppelbodens)*; *(Schiff)* Kesselboden *m*
e. play Axialspiel *n*, Längsspiel *n*, toter Gang *m*, Leerlauf *m*
e. point *(Math)* Randpunkt *m*
e.-point analyzer *(Am)* Analysenmeßgerät *n*
e.-point control *(BMSR)* Prozeßregelung *f* auf Grund von Analysen der Ausgangsdaten
e.-point energy *(Kern)* Grenzenergie *f*, Endenergie *f*
e.-point gasoline Benzin *n* mit definierten Siedegrenzen
e.-point of reaction Endpunkt *m* der Reaktion
e. pole *(El)* Abspannmast *m*; Endgestänge *n*, Endmast *m*
e. pressure Axialdruck *m*
e. product 1. Fertigerzeugnis *n*; 2. *(Rak)* Endprodukt *n*, Endkern *m (Zerfall)*; 3. Verbrennungsprodukt *n (Kernreaktor)*
e. quench test *(Met)* Stirnabschreckversuch *m*
e. relieving *(Masch)* axiales Hinterdrehen *n*
e. restraint 1. Eindeinspannung *f (Balken)*; Fußeinspannung *f (Säulen)*; 2. Einspannung *f* der Kabelenden
e. ring 1. Abschlußring *m*, Bundring *m*; Deckring *m (Dichtung)*; 2. *(Elm)* Kommutatorring *m*
e. scale value maximaler Skalenwert *m*
e. seal Endverschluß *m*
e. shield Lagerschild *m*, Seitenplatte *f*
e. sizing installation Kalibervorrichtung *f (Walzwerk)*
e. standard Endmaß *n*, Parallelendmaß *n*
e. stone Deckstein *m (Uhr)*
e. stop Endanschlag *m*, Längsanschlag *m*
e. stopping of chain *(Ch)* Kettenabbruch *m*
e. support Gegenlagerständer *m*; Kopfquerträger *m*; Lünettenständer *m*
e. surface *(Masch)* Stirnfläche *f*
e. tank *(Schiff)* Endluftkasten *m (Auftriebstanks an den beiden Enden von Rettungsbooten)*
e. thread Gewindeauslauf *m*
e. thrust Axialschub *m*
e.-thrust bearing Axiallager *n*
e. tipping box car Muldenvorderkipper *m*
e. tipping lorry Hinterkipper[-LKW] *m*, Muldenhinterkipper *m*
e. tipping wagon Vorderkipper *m*
e. tippler *(Brg)* Kopfwipper *m*
e.-to-end von einem Ende zum anderen
e.-to-end-grain joints *(Holz)* Hirnholzverbindung *f*
e.-to-end measurement Streckenmessung *f*
e.-to-end scavenging *(Kfz)* Gleichstromspülung *f*
e.-to-end signalling *(Fmt)* Signalisierung *f* ohne Wiederholung der Signale
e. tooth Stirnzahn *m (Schaftfräser)*
e. tower *s.* straining tower
e. trip Endauslösung *f*
e.-use performance Gebrauchsverhalten *n*
e. view Seitenansicht *f*
e. wall Giebelwand *f*; Kopfwand *f*, Stirnwand *f*
e.-weld / to stirnseitig verschweißen
e. welding Stumpfschweißung *f*
e. winding *(Elm)* Spulenkopf *m*, Stirnverbindung *f*
e.-window counter *(Kern)* Endfensterzählrohr *n*, Fensterzählrohr *n*
endbell Lagerschild *m*
endfire aerial Endfire-Antenne *f (Spezialtyp einer Yagiantenne)*
ending fitting *(El)* Abspannklemme *f*
endless endlos
e. belt endloses Band *n*, endloser Riemen *m*, Transmissionsband *n*, Transmissionsriemen *m*
e.-belt flexible coupling elastische Kupplung *f* mit endlosem Lederriemen
e.-belt screen Bandsieb *n*, Siebband *n*
e. chain haulage Kettenförderbahn *f*
e. conveyor endloses Förderband *n*, endloser Förderer *m*
e. grate Bandrost *m*
e.-rope haulage *(Brg)* endlose Seilförderung *f*, Förderung *f* mit geschlossenem Seil
e. saw Bandsäge *f*
e. saw-blade Bandsägeblatt *n*
e. winding geschlossene Wicklung *f*
endo-methylene group Methylenbrücke *f* innerhalb von Ringverbindungen
endocarp innere Fruchthülle *f*
endocyclic bond *(Ch)* intraannulare Bindung *f*
endodyne *(El)* Selbstüberlagerer *m*; Überlagerungsschwebungsempfänger *m*
endoscope Endoskop *n*
endosmose *(Ch)* Endosmose *f*
endothermal *s.* endothermic
endothermic endotherm, wärmeaufnehmend, wärmebindend

e. **reaction** endotherme Reaktion *f*
ends down beim Weben gerissene Kettfäden *mpl*
endurance 1. Lastspielzahl *f*; 2. Dauer[schwing]-festigkeit *f*; 3. Standzeit *f (Werkzeug)*
e. **bending strength** Biegewechselfestigkeit *f*
e. **crack** Ermüdungsriß *m*, Dauerriß *m*
e. **failure** Dauerbruch *m*
e. **fracture** Dauerbruch *m*
e. **limit** Dauer[schwing]festigkeit *f*
e. **ratio** Ermüdungsverhältnis *n*
e. **run** Dauerlauf *m (Maschinentest)*
e. **strength** Dauerfestigkeit *f*
e. **tensile strength** Dauerzugfestigkeit *f*
e. **test** Dauerprüfung *f*, Dauerversuch *m*
endwise axial, in Achsrichtung, in Richtung der Stirnseite
energetic energiereich, hochenergetisch
e. **event** energiereicher Stoß *m*
e. **particle** energiereiches Teilchen *n*, Teilchen hoher Energie
e. **recoil nucleus** *(Kern)* energiereicher Rückstoßkern *m*
energetics Energiewirtschaft *f*
energize / to verstärken; Energie zuführen; erregen *(Relais)*
energized 1. erregt; 2. energiereich
e. **speaker** elektrodynamischer Lautsprecher *m*
energizer Erreger *m*
energizing circuit Erreger[strom]kreis *m*, Erregungs[strom]kreis *m*
energy Energie *f*; Arbeitsvermögen *n*; Arbeit *f*
e. **absorbed** verbrauchte Energie *f*, aufgenommene Leistung *f*
e. **absorption** Energieaufnahme *f*, Energieabsorption *f*
e. **accumulator** Energiespeicher *m*
e. **amplification** Leistungsverstärkung *f*
e. **analyzer** Energieanalysator *m*
e. **balance** Energiebilanz *f*, Energiegleichgewicht *n*; Energiesatz *m*
e. **band** *(Ph)* Energieband *n*
e. **build-up factor** *(Kern)* Energiezuwachsfaktor *m*
e. **carrier** Energieträger *m*
e. **component** *(El)* Wirkkomponente *f*
e. **conservation** Energieerhaltung *f*, Erhaltung *f* der Energie
e. **consumed** aufgebrauchte Energie *f*
e. **consumption** Energieverbrauch *m*
e. **content** Energieinhalt *m*
e. **decrement** Energieverlust *m*
e. **demand** Energiebedarf *m*
e. **density** Energiedichte *f*
e. **density of radiation** räumliche Strahlungs[energie]dichte *f*, Strahlungs[energie]-dichte
e. **density of sound** Schalldichte *f*
e. **dependence** Energieabhängigkeit *f*
e. **dissipation** Energiezerstreuung *f*, Dissipation (Zerstreuung) *f* der Energie
e. **dissipator** Energievernichter *m*
e. **distribution** Energieverteilung *f*
e. **efficiency** energetischer Wirkungsgrad *m*
e. **equivalent** Energieäquivalent *n*
e. **exchange** Energieaustausch *m*
e.**-exchange time** Zeitdauer *f* für den Energieaustausch, Austauschzeit *f* der Energie
e. **expended** Arbeitsaufwand *m*
e. **flow** Energiefluß *m*, Energiestromdichte *f*
e. **flow per unit area** Energiestromdichte *f*
e. **flux** Energiestromdichte *f*
e. **gap** Energielücke *f*; Bandabstand *m*, verbotenes Band *n (Halbleiter)*
e. **generation** Energieerzeugung *f*
e. **input** zugeführte Energie *f*
e. **jump** Energiesprung *m*
e. **level** Energieniveau *n*, Term *m*
e.**-level diagram** Termschema *n*, Niveauschema *n (Atome, Kerne usw.)*
e.**-level spacing** Energieniveauabstand *m*, Termabstand *m*
e. **loss** Energieverlust *m*
e. **meter** Strom[verbrauchs]zähler *m*, Wattstundenzähler *m*
e. **migration** Energiewanderung *f*
e. **needed** Kraftbedarf *m*
e. **of absolute zero** innere Energie *f* am absoluten Nullpunkt
e. **of activation** Aktivierungsenergie *f*
e. **of deformation** Formänderungsenergie *f*
e. **of dislocation** Versetzungsenergie *f*
e. **of dissociation** Dissoziationsenergie *f*
e. **of fission** *(Kern)* Spaltungsenergie *f*
e. **of flow** Strömungsenergie *f*, Ausströmungsenergie *f (Turbine)*
e. **of impact** *(Mil)* Auftreffwucht *f*
e. **of light** Lichtenergie *f*
e. **of motion** Bewegungsenergie *f*, kinetische Energie *f*; Wucht *f (im Zusammenhang mit Stoß)*
e. **of rotation** Rotationsenergie *f*
e. **of volume** Volumenenergie *f*
e. **operator** Energieoperator *m*
e. **output** Energieabgabe *f*, Energieleistung *f*
e. **put in** zugeführte Leistung *f*
e. **quantization** Energiequantelung *f*
e. **quantum** Energiequant *n*
e. **requirement** Energiebedarf *m*
e. **sensitiveness** Energieempfindlichkeit *f*
e. **shell** Energieschale *f*
e. **source** Energiequelle *f*
e. **storage** Energiespeicherung *f*
e. **storage device** Energiespeicher *m*
e. **storage spot welder** Impulsschweißgerät *n*

e. **stored up in flywheel** Schwungenergie *f*
e. **supply** Energiezufuhr *f*, Energieversorgung *f*
e. **surface** Energiefläche *f*
e. **tensor** Energie-[Impuls-]Tensor *m*
e. **theorem** Energiesatz *m*
e. **thickness** Energiedichte *f*
e. **transfer** Energieübertragung *f*
e. **transformation** Energieumwandlung *f*
e. **transport** Energietransport *m*
e. **unit** Energieeinheit *f*
e. **value** Energiewert *m*
e. **width** Energiebreite *f*
e. **yield** Energieausbeute *f*

engage / to eingreifen, in Eingriff kommen (bringen); einrasten, einrücken; zuschalten, einschalten; kuppeln
e. **needles** *(Text)* Nadeln betätigen
e. **the clutch** [ein]kuppeln, [ein]schalten
e. **with** *(Masch)* im Eingriff sein mit, kämmen mit
e. **with a contact** *(El)* einen Kontakt berühren

engaged 1. *(Masch)* im Eingriff [befindlich]; 2. besetzt, belegt *(Telefonleitung)*; 3. eingerastet; eingeschaltet
e. **condition** *(Fmt)* Besetztfall *m*
e. **in local call** *(Fmt)* ortsbesetzt
e. **jack** *(Fmt)* Besetztklinke *f*, besetzte Klinke *f*
e. **lamp** *(Fmt)* Besetztlampe *f*
e. **on trunk call** *(Fmt)* fernbesetzt
e. **signal** *(Fmt)* Besetztflackerzeichen *n*

engagement 1. Einrücken *n*, Einstellen *n*; 2. Eingriff *m (Zahnrad)*; 3. *(Fmt)* Einklinken *n*
e. **factor** Überdeckungsgrad *m (Verzahnung)*

engaging 1. eingreifend, im Eingriff befindlich, kämmend; 2. Einschalten *n*
e. **dog** Mitnehmerklaue *f*
e. **gear** Einrückeinrichtung *f*
e. **gears** miteinander im Eingriff stehendes Räderpaar *n*, gepaarte Räder *npl*
e. **lever** Einrückhebel *m*, Rasthebel *m*
e. **of stop lock** *(Eis)* Einfallen *n* der Haltsperre
e. **piece** Mitnehmer *m*
e. **plane** Einrückfläche *f (Klauen- bzw. Zahnkupplung)*; Reibfläche *f (Reibungskupplung)*
e. **surface** 1. Überdeckung *f (Verzahnung)*; 2. *s.* e. plane

engine Wärmekraftmaschine *f*, Kraftmaschine *f*; Maschine *f*; Motor *m*; Triebwerk *n*; Lokomotive *f*; Dampfmaschine *f*
e. **attendant** Maschinenwärter *m*
e. **base** Maschinensockel *m*; Motorsockel *m*
e. **bearer** Motorbock *m*, Motorträger *m*
e. **bed** Maschinenfundament *n*; Motorbock *m*, Motorfundament *n*
e. **bedplate** Maschinenrahmen *m*; Motortragplatte *f*, Motorgrundplatte *f*
e. **bonnet** Motorhaube *f*
e. **bracket** Motorkonsole *f*
e. **breakdown** Motorpanne *f*
e. **capacity** Hubraum *m*
e. **clutch** Motorkupplung *f*
e. **compartment** Maschinenkammer *f*, Maschinenraum *m*; Motorraum *m*
e. **control platform** *(Schiff)* Maschinenleitstand *m*, Maschinenfahrstand *m*
e. **controls** *(Schiff)* Maschinentelegraf *m*, Maschinenleiteinrichtung *f*, Maschinensteuerwerk *n*
e. **cover** Motorhaube *f*
e. **cowling** *(Am)* Motorhaube *f*, Motorverkleidung *f*, Motorverschalung *f*
e. **currently in production** Serienmotor *m*
e. **data** Maschinenkenndaten *pl*; Motorkenndaten *pl*
e. **details** Elemente *npl* der Kraftmaschinen
e. **detonation limit** Klopfgrenze *f (Kraftstoff)*
e.**-driven** motorbetrieben, Motor-
e.**-driven pump** Motorpumpe *f*
e.**-driven welding set** Motor-Schweißaggregat *n*
e. **driver** 1. Maschinist *m (Dampfmaschine)*; 2. Lokomotivführer *m*
e. **fitter** Maschinenschlosser *m*
e. **for superheated steam** Heißdampfmaschine *f*
e. **foundation** Maschinenfundament *n*, Maschinenunterbau *m*
e. **gauge** *(BMSR)* Indikator *m*
e. **governor** Drehzahlregler *m*, Kraftmaschinenregler *m*
e. **grease** Maschinenfett *n*
e. **guard plate** Motorschutzblech *n*
e. **gun mounting** *(Mil)* Motorlafette *f*
e. **hatch** *(Schiff)* Maschinenluke *f*
e. **heater** Motorvorwärmer *m*
e. **hood** Motorhaube *f*
e.**-hour indicator** Betriebsstundenzähler *m*
e. **hours** Motorbetriebsstunden *fpl*
e. **humming** *(Kfz)* Motorbrummen *n*
e. **lathe [having a lead screw]** Zug- und Leitspindeldrehmaschine *f*
e. **line** Fertigungslinie *f* für Motoren
e. **logbook** *(Flg)* Motorflugbuch *n*
e. **mechanic** Autoschlosser *m*; Motorenwart *m*
e. **mount** Motorblock *m*, Motorlager *n*
e. **mounting** Motoraufhängung *f*, Motoreinbau *m*
e. **oil** Motorenöl *n*
e. **operator** Maschinist *m*
e. **output per 1000 cc. cylinder capacity** *(Kfz)* Literleistung *f* des Motors

e. overhaul Motorüberholung *f*
e. performance Betriebsverhalten *n* des Motors; Motorleistung *f*
e. plane *(Brg)* Haspelberg *m*
e. plate Motorschild *n*
e. priming Einspritzen *n* von Anlaßkraftstoff in den Motor
e. revolutions Motordrehzahl *f*
e. room *(Schiff)* Maschinenraum *m*; Dieselraum *m* *(U-Boot)*; Maschinenkammer *f* *(Torpedo)*
e.-room auxiliaries Maschinenraumhilfsmaschinen *fpl*
e. room complement Maschinenpersonal *n*
e. room crane Maschinenraumkran *m (Laufkatze)*
e. room layout Maschinenraumanordnung *f*
e. room staff Maschinenraumpersonal *n*
e.-room telegraph Maschinentelegraf *m*
e. seating Maschinenfundament *n*, Maschinenlagerung *f*
e. shaft Maschinenwelle *f*
e. shed Lokomotivschuppen *m*, Lok-Schuppen *m*
e. sizing Stoffleimung *f*
e. skylight *(Schiff)* Maschinenoberlicht *n*
e. speed Motordrehzahl *f*
e.-speed indicator Drehzahlmesser *m*, Tachometer *n*
e. starter Startmotor *m*, Anlaßmotor *m*
e.-test certificate Motorprüfschein *m*
e.-test log Motorabnahmeprotokoll *n*
e. timing *(Kfz)* Motorsteuerung *f*
e. transmission Getriebe *n*
e. trouble Motorstörung *f*, Motorschaden *m*
e. with side-by-side valve arrangement *(Kfz)* untengesteuerter Motor *m*
engineer / to gestalten, technisch ausführen
engineer 1. Metallarbeiter *m*; Maschinenbauer *m*; 2. Techniker *m*; Ingenieur *m*; 3. Pionier *m*; 4. *(Am)* Lokomotivführer *m*
e. tank Pionierpanzerkampfwagen *m*
e. troops Pioniertruppen *fpl*
engineered fibre *s.* man-made fibre
engineering Technik *f*
engineering technisch
e. alloys Konstruktionslegierungen *fpl*
e. and design department Konstruktionsbüro *n*
c. aspects Konstruktionsrichtlinien *fpl*
e. brick Hartbrandstein *m*, Klinkerziegel *m*
e. clinker Hartbrandstein *m*, Klinkerziegel *m*
e. construction Maschinenbau *m*
e. cybernetics technische Kybernetik *f*
e. department Konstruktionsbüro *n*
e. design technische Zeichnung *f*
e. drawing technisches Zeichnen *n*
e. education 1. technisches Bildungswesen *n*; 2. technisches Schulwesen *n*
e. fair technische Messe *f*
e. industry Maschinenbau[industrie *f*] *m*
e. material technischer Werkstoff *m*
e. measurement technische Messung *f*
e. process technisches Verfahren *n*
e. properties technische Eigenschaften *fpl*
e. service technischer Dienst *m*
e. structure Ingenieurbauwerk *n*
e. structures Ingenieurbauten *pl*
e. subject technisches Fach *n*
e. test reactor technischer Versuchsreaktor *m*
e. thermodynamics technische Thermodynamik *f*
engineer's field book *(Opt)* Feldbuch *n*
e. hammer Schlosserhammer *m*, Bankhammer *m*
e. square Anschlagwinkel *m*
e. wrench Gabel[schrauben]schlüssel *m*, Maulschlüssel *m*
engineman Maschinist *m*; Lokführer *m*
Engler degree Grad *m* Engler, Viskositätsangabe *f* nach Engler
Engler distilling flask Englerkolben *m*
Engler visco[si]meter Englerscher Zähigkeitsmesser *m*, Englersches Viskosimeter *n*
English *(Typ)* 14 britische Punkte *mpl (Schriftgröße)*
English bond Blockverband *m (Mauerwerk)*
E. cross bond Kreuzverband *m (Mauerwerk)*
E. finish halbglänzend *(Papier)*
E. flute Blockflöte *f*
E. graduation Zollteilung *f*
E. mount *(Opt)* englische Montierung *f*
E. pitch Zollteilung *f (Verzahnung)*
E. thread Zollgewinde *n*
engobe Engobe *f*, Begußmasse *f*, Angußmasse *f* *(für keramische Erzeugnisse)*
engorgement of a blast furnace Hängen *n* eines Hochofens
engrave / to stechen, [ein]gravieren *(in Metall)*; schneiden *(in Holz)*
e. on copper Kupfer stechen
engraved screen *(Typ)* Gravurraster *m*
engraver Graveur *m*, Stecher *m*
engraving Gravierung *f*, Stich *m*; Gravur *f*
e. machine Graviermaschine *f*
e. needle Gravierstichel *m*
enhance / to vergrößern, steigern *(Genauigkeit)*
enhanced lines hervortretende Linien *fpl* *(Spektroskopie)*
enhancement *(Kern)* Anreicherung *f*, Verstärkung *f*
enimic form *(Ch)* Enimin-Form *f*
enimization *(Ch)* tautomeres Gleichgewicht *n* der Enimin- und Enamin-Form

enlarge / to erweitern, vergrößern; weiten; aufweiten; ausbohren
enlargement Vergrößerung *f*; Erweiterung *f*; Aufweitung *f*
enlarger *(Opt)* Vergrößerungsapparat *m*
enlarging Erweitern *n*; Aufweiten *n*; Ausbohren *n*; Vergrößern *n*
e. apparatus *(Opt)* Vergrößerungsapparat *m*
e. focusing magnifier *(Opt)* Projektionsbildlupe *f*
e. of sound base by stereo-box Basisverbreiterung *f* durch Raumklangbox
enolization *(Ch)* Enolisierung *f*
enquiry *(Fmt)* Anfrage *f*, Rückfrage *f*
enrich / to anreichern; konzentrieren
enriched angereichert; konzentriert
e. fuel angereicherter Kernbrennstoff (Spaltstoff) *m*
e. fuel reactor angereicherter Reaktor *m*
e. output Ausbeute *f* an angereichertem Material
e. reactor angereicherter Reaktor *m*
e. slag angereicherte Schlacke *f*
enrichment Anreicherung *f*, Konzentration *f*
e. jet Kompensatordüse *f*, Ausgleichdüse *f*
e. section Abtriebsäule *f (Destillation)*
enrol / to immatrikulieren
ensign staff *(Schiff)* Flaggenstock *m (Heck)*
ensilage / to silieren
ensilage Silieren *n*, Silolagerung *f*; Silage *f*
enstatite *(Min)* Enstatit *m*
entangled straw *(Lw)* Wirrstroh *n (Bündel)*
e. wires verwickelte (verwirrte) Drähte *mpl*
enter / to 1. *(Masch)* einsetzen; einführen; 2. *(Dat)* eingeben, einschreiben *(Informationen)*; 3. anlaufen, einlaufen *(Hafen)*; 4. einströmen
e. a circuit *(Fmt)* in eine Verbindung eintreten; mithören
e. a curve *(Kfz)* in die Kurve gehen
e. a hole in ein Loch einrasten
e. a port in einen Hafen einlaufen
e. appeal Beschwerde führen
e. into the result in das Resultat eingehen
e. the cut *(Masch)* anschneiden, Spanabnahme beginnen
e. the line *(Fmt)* in die Leitung kommen
e. the work off centre verlaufen *(Bohrer)*
entering Einführen *n*, Einführung *f*; Eintritt *m*
e. angle Eintrittswinkel *m*, Einzugswinkel *m*
e. edge *(Elm)* auflaufende Bürstenkante *f*, Vorderkante *f* der Bürste
e. end Anschnitt *m*, Anschnittseite *f*
e. ladder *(Schiff)* Fallreepstreppe *f*
e. plan *(Text)* Schema *n* für das Einziehen der Kettfäden
e. port *(Schiff)* Fallreepspforte *f*
e. rope *(Schiff)* Fallreepstau *n*
e. tap Vorschneider *m (Gewindebohrer)*
e. tooth Eingriffszahn *m*
entertainment Instandhaltung *f*
enthalpy Enthalpie *f*, Wärmeinhalt *m*
entrain / to mitreißen, mitführen
entrained air 1. Lufteinschluß *m (z. B. Beton)*; 2. *(Met)* Falschluft *f*
e. oil mitgerissenes Öl *n*
entraining station Verladebahnhof *m*
entrainment Mitführung *f*, Mitschleppen *n*, Mitreißen *n*
e. separator *(Ch)* Absetzgefäß *n*, Falle *f*, Kühlfalle *f*
entrance 1. *(Bw)* Eingang *m*, Zugang *m*; 2. *(Ph)* Eintritt *m*; 3. Vorschiffsschärfe *f*, Scharf *n* des Vorschiffes *(in der Wasserlinie)*
e. hatch Einstiegluke *f*
e. into the nozzle Eintritt *m* in die Düse *(Turbine)*
e. lock *(Schiff)* Schleusenkammer *f*, Kammerschleuse *f*
e. of cooling water Kühlwassereintritt *m*
e. port *(Opt)* Eintrittsluke *f*
e. pupil *(Opt)* Eintrittspupille *f*
e. slit *(Opt)* Eintrittsspalt *m*
e. window *(Opt)* Eintrittsluke *f*
entrap / to [ein]fangen; einschließen
entrapped mitgerissen
e. air [bubble] Lufteinschluß *m*; Luftblase *f*
entrapping Einschluß *m*
entrepôt port Bestimmungshafen *m*
entresol Halbgeschoß *n*, Mezzanin *n*, Zwischengeschoß *n*
entropy Entropie *f*
e. diagram *(Wärm)* Entropiediagramm *n*, T-S-Diagramm *n*
e. of activation *(Ch)* Aktivierungsentropie *f*
e. principle Entropiesatz *m*, 2. Hauptsatz *m* der Thermodynamik
entry 1. *(Dat)* Eingabe *f*, Einschreibung *f*; 2. *(Schiff)* Klarierung *f*; 3. Abbaustrecke *f*, Fahrstrecke *f*; Förderstrecke *f*; Strecke *f*; Pfeilerstrecke *f*; 4. Eintritt *m*; Einströmen *n*
e. hub Eingangsbuchse *f*
e. pressure Eintrittsdruck *m*
e. stump *(Brg)* Streckenpfeiler *m*
enumerability Abzählbarkeit *f (einer Menge)*
enumerable abzählbar
enumerated einzeln genannt (aufgezählt, aufgeführt)
envelop / to einhüllen, umhüllen; ummanteln; umfassen; einschließen
envelope 1. *(Math)* Einhüllende *f*, Hüllkurve *f*, Umhüllende *f*; 2. Umhüllung *f*, Ummantelung *f*, Umkleidung *f*, Verkleidung *f*; 3. Röhrenkolben *m*, Kolben *m*; 4. Ballonhülle *f*; 5. Briefumschlag *m*
e. delay [time] Gruppenlaufzeit *f*

e. distortion Modulationsverzerrung *f*, Modulationsklirrfaktor *m*
e. velocity Gruppengeschwindigkeit *f*
enveloping surface *(Math)* Hüllfläche *f*
envenomation Korrosion *f* durch Oberflächenberührung; Schädigung *f* einer Kunststoffoberfläche
environment Umwelt *f*
environmental activity *(Kern)* Aktivität *f* der Umgebung
e. factor Einfluß *m* der Umgebung, Umwelteinfluß *m*
e. influence Umwelteinfluß *m*
e. temperature Umgebungstemperatur *f*
enzymatic enzymatisch, fermentativ
enzyme Ferment *n*, Enzym *n*
enzymic enzymatisch, enzymisch
enzymosis Enzymwirkung *f*
eocene *(Geol)* Eozän *n* *(Periode)*
eosin Eosin *n*
E.P. lubricant *(Kfz)* Hochdruckschmiermittel *n*, Höchstdruckschmiermittel *n*
epicadmium neutrons Epikadmiumneutronen *npl*
epicarp *(Lw)* äußere Fruchthülle *f*
epicenter *(Am)* *s.* epicentre
epicentre Epizentrum *n*, Oberflächenmittelpunkt *m*
epicentrum *s.* epicentre
epicycle Epizykel *m*, Epikreis *m*
epicyclic band brake Schlingbandbremse *f*
e. gear Planetengetriebe *n*, Umlaufgetriebe *n*
epicycloid *(Math)* Epizykloide *f*
epicycloidal motion epizyklische Bewegung *f*
epidiascope Epidiaskop *n*
epididymite *(Min)* Epididymit *m*
epidote *(Min)* Epidot *m*
epigenite *(Min)* Epigenit *m*
epilatory Epiliermittel *n*, Haarentfernungsmittel *n*
epinephrine Adrenalin *n*
episcope Episkop *n*
episcopic illumination Auflichtbeleuchtung *f*
e. projection Epiprojektion *f*
epistemological erkenntnistheoretisch
epistemology Erkenntnistheorie *f*
epitaxy *(Krist)* Epitaxie *f*
epithermal *(Kern)* epithermisch
e. energy epithermische Energie *f*
e. neutron epithermisches Neutron *n*
e. neutron yield Ausbeute *f* an epithermischen Neutronen
e. reactor epithermischer Reaktor *m*
epitrochoid Epitrochoide *f* *(Geometrie)*
epoxide Epoxyd *n*
e. resin Epoxy[d]harz *n*, Epoxydkunstharz *n*
epoxy casting resin Epoxydgußharz *n*
e. resin Äthoxylinharz *n*, Epoxy[d]harz *n*
Epsom salts Bittersalz *n*, Magnesiumsulfat *n*
Epstein hysteresis tester Epstein-Gerät *n*, Epstein-Rahmen *m*
equal / to gleichen, gleich [groß] sein
e. zero gleich Null sein
equal gleich
e. angle gleichschenkliges Winkelprofil *n*
e. area projection flächentreue Abbildung *f*
e.-armed balance Balkenwaage *f*
e.-energy particles energiegleiche Teilchen *npl*
e.-energy white *(Fs)* Idealweiß *n*
e.-laid rope *(Brg)* Gleichschlagseil *n*
e. leg angle gleichschenkeliger Winkelstahl *m*, gleichschenkeliges L-Profil *n*
e.-length multi-unit code Einheitenkode *m*
e. percentage control valve gleichprozentiges Regelventil *n*, Regelventil mit logarithmischer Kennlinie
e.-phase *(El)* gleichphasig, konphas
e. settlement *(Bw)* gleichmäßige Setzung *f*
e.-sided angle steel gleichschenkeliger Winkelstahl *m*, gleichschenkeliges L-Profil *n*
e. sign *(Math)* Gleichheitszeichen *n*
e.-signal white *(Fs)* Mittelweiß *n*
e. triangles *(Math)* flächengleiche Dreiecke *npl*
equality Gleichheit *f*
e. of brightness (contrast) photometer Kontrastfotometer *n*
e. sign *(Math)* Gleichheitszeichen *n*
equalization 1. *(Masch)* Abgleichung *f*, Ausgleichung *f*; 2. *(El)* Entzerrung *f*; 3. *(BMSR)* Stabilisierung *f*
e. of levels *(Fmt)* Pegelausgleich *m*
e. of temperature Temperaturausgleich *m*
e. passage *(Wärm)* Ausgleichkanal *m*
equalize / to 1. *(Masch)* abgleichen, ausgleichen; gleichmachen, gleichstellen; 2. *(El)* entzerren
equalized circuit *(El)* entzerrte Leitung *f*
e. delay line *(Dat)* Verzögerungsleitung *f* mit Phasenkorrektur
equalizer 1. *(El, Fmt, Fs)* Entzerrer *m*, Frequenzkorrekturfilter *n*; 2. *(BMSR)* Kompensationsglied *n*, Stabilisierungsglied *n*
e. network *(El)* Entzerrerschaltung *f*
e. pulse Ausgleichsimpuls *m* *(Fernsehnorm)*
e. switch Ausgleichschalter *m*
equalizing *(Masch)* Ausgleichszustellung *f*
e. buffer *(Eis)* Balancierpuffer *m*
e. charge Ausgleichladung *f* *(Batterie)*
e. connection *(El)* Ausgleichschaltung *f*
e. current Ausgleich[s]strom *m*; Entladungsstrom *m* *(Kondensatoren)*
e. gear Ausgleichgetriebe *n*
e. network 1. *(El)* Entzerrerschaltung *f*; Entzerrerkette *f*; 2. *(BMSR)* Kompensationsschaltung *f*, Korrekturschaltung *f*

e. **pipe** Ausgleichrohr *n*
e. **pulse** Ausgleichimpuls *m (Fernsehnorm)*
e. **repeater** *(Fmt)* entzerrender Verstärker *m*
e. **tank** Ausgleichbehälter *m*
e. **valve** Ausgleichventil *n*
e. **voltage** Kompensationsspannung *f*, Ausgleich[s]spannung *f*
equalling needle file Flachstumpffeile *f*
equally spaced in gleichem Abstand [befindlich], abstandsgleich
equate / to *(Math)* gleichsetzen
equation *(Math)* Gleichung *f*
e. **of continuity** Kontinuitätsgleichung *f*, differentieller Erhaltungssatz *m*
e. **of first degree** Gleichung *f* ersten Grades
e. **of moment** Momentengleichung *f*
e. **of motion** Bewegungsgleichung *f*
e. **of oscillation** Schwingungsgleichung *f*
e. **of state** Zustandsgleichung *f*
e. **of time** *(Ph)* Zeitgleichung *f (Beobachtung)*
e. **solver** *(Dat)* Rechner *m* zur Auflösung linearer Gleichungssysteme, Lineargleichungslöser *m*, Polynomrechner *m*
equations of equilibrium Gleichgewichtsbedingungen *fpl (Statik)*
equator Äquator *m*
equatorial äquatorial
e. **air** äquatoriale Luftmassen *fpl*
e. **mount** *(Opt)* parallaktische Montierung *f*
equi-angled gleichwinklig
e.**-energy spectrum** energiegleiches Spektrum *n*
e.**-radial aerial** Rundstrahler *m*; ungerichtete Antenne *f*
equiangular gleichwinklig; winkeltreu
e. **spiral** logarithmische Spirale *f*
equiaxed gleichachsig
equicontinuous gleichmäßig stetig *(Funktion)*
equidirectional gleichgerichtet, gleichsinnig
equidistant abstandsgleich, gleichweit entfernt, äquidistant
equidistant Äquidistante *f*
equifrequent conductor *(El)* mitschwingender Leiter *m*
equigranular gleichkörnig
equilateral gleichseitig
e. **hyperbola** gleichseitige Hyperbel *f*
e. **triangle** gleichseitiges Dreieck *n*
equilaterality Gleichseitigkeit *f*
equilibrate / to 1. ausbalancieren, ins Gleichgewicht bringen; auswuchten; 2. abgleichen *(z. B. Dunkelstrom einer Fotozelle)*
equilibration Abgleichung *f*; Ausbalancieren *n*; Auswiegen *n*; Auswuchten *n*
equilibrium Gleichgewicht *n*; Ausgleich *m*; Ruhelage *f*
e. **concentration** Gleichgewichtskonzentration *f*
e. **conditions** Gleichgewichtsbedingungen *fpl*
e. **constant** Gleichgewichtskonstante *f*, Massenwirkungskonstante *f*
e. **curve** Gleichgewichtsdruckkurve *f*
e. **depletion** *(Kern)* Gleichgewichtserschöpfung *f*
e. **diagram** Gleichgewichtsdiagramm *n*, Gleichgewichtsschaubild *n*
e. **distribution** Gleichgewichtsverteilung *f*
e. **electrode potential** Gleichgewichtspotential *n* einer Elektrode
e. **of forces** Kräftegleichgewicht *n*
e. **of growth and decay** radioaktives Gleichgewicht *n*
e. **reaction potential** statisches Gleichgewichtspotential *n* der Reaktion
e. **state** Gleichgewichtszustand *m*
e. **temperature** 1. Beharrungstemperatur *f*; 2. *(Met)* Umwandlungstemperatur *f (Härtetechnik)*
e. **value** Gleichgewichtswert *m*
e. **vapour pressure** Dampfdruck *m* im Gleichgewichtszustand
equip / to ausrüsten, ausstatten, einrichten, versehen [mit]
equipartition Gleichverteilung *f*
e. **theorem** Gleichverteilungssatz *m (statistische Mechanik)*
equiphase surface Fläche *f* gleicher Phase
e. **zone** Überschneidungsgebiet *n*, Zone *f* gleicher Phase *(Funksignale)*
equipment Ausrüstung *f*; Zubehör *n*, Einrichtung *f*, Ausstattung *f*; Gerät *n*, Apparatur *f*; Gerätesatz *m*
e. **box** *(El)* Gerätekasten *m*
e. **casting** Präzisionsguß *m*
e. **fleet** *(Bw)* Gerätepark *m*, Maschinenpark *m*
e. **for tropical climates** Tropenausrüstung *f*
e. **maker** Gerätebauer *m*
e. **tonnage** *(Schiff)* Tonnengehalt *m*
equipoise Gleichgewicht *n*
equipollent *(Math)* gleichmächtig; gleichwertig
equipotent gleichmächtig
equipotential äquipotentiell; Äquipotential-
e. **cathode** Äquipotentialkatode *f*
e. **connection** Ausgleichsleiter *m*, Ausgleichsverbindung *f*; Ausgleichsschaltung *f*, Spannungsausgleichschaltung *f*
e. **curve** *(Geol)* Äquipotentiallinie *f*
e. **line** Äquipotentiallinie *f*
e. **surface** Äquipotentialfläche *f*, Niveaufläche *f*
equisetic acid Akonitsäure *f*
equisignal line Leitstrahllinie *f (Navigation)*
e. **zone** Leitstrahlsektor *m (Navigation)*
equivalence Äquivalenz *f*, Gleichwertigkeit *f*
e. **law** Äquivalenzgesetz *n*
e. **theorem** *(Ph)* Äquivalenztheorem *n*

equivalent Äquivalent *n*
equivalent äquivalent, gleichwertig
e. air speed *(Flg)* äquivalente Eigengeschwindigkeit *f*
e. binary digit *(Dat)* äquivalente Binärstellenzahl *f*
e. circuit Ersatzschaltung *f*, Ersatz[strom]-kreis *m*
e. concentration *(Ch)* Äquivalenzkonzentration *f*
e. conductance Äquivalentleitwert *m*
e. conductivity Äquivalentleitfähigkeit *f*
e. control circuit resistance Äquivalentwiderstand *m* des Steuerkreises
e. current Äquivalentstrom *m*
e. dilution *(Ch)* Äquivalentverdünnung *f*
e. disturbing current äquivalenter Störstrom *m*
e. disturbing voltage äquivalente Störspannung *f*
e. electron Ersatzelektron *n*
e. focal length *(Opt)* Äquivalentbrennweite *f*
e. grid-leak resistance äquivalenter Gitter-[ableit]widerstand *m (Rauschspannung einer Röhre)*
e. grid-noise resistance äquivalenter Gitterrauschwiderstand *m*
e. logarithmic [damping] decrement scheinbares logarithmisches Dämpfungsdekrement *n*
e. loudness subjektive Lautstärke *f (Elektroakustik)*
e. network *(El)* Ersatzschaltung *f*
e. noise input *(El)* äquivalentes Eingangsrauschen *n*
e. noise resistance *(El)* Rausch[ersatz]widerstand *m*
e. noise side-band input äquivalente Seitenbandrauschleistung *f*
e. number of teeth *(Masch)* Ersatzzähnezahl *f*
e. reactance *(El)* wirksamer Blindwiderstand *m*
e. resistance Ersatzwiderstand *m*
e. shaft-horsepower Wellenvergleichsleistung *f*
e. sideband input äquivalente Seitenbandrauschleistung *f*
e. source circuit Rauschquellenersatzschaltbild *n*
e. sparking distance äquivalente Funkenschlagweite *f*
e. stopping power *(Kern)* Bremsäquivalent *n*
e. thickness *(Kern)* äquivalente Schichtdicke *f*
e. weight *(Ch)* Äquivalentgewicht *n*
E.R. crane *s.* engine room crane
eradiation Strahlung *f*, Ausstrahlung *f*
erasable memory löschbarer Speicher *m (Datenspeicherung)*
e. storage löschbare Speicherung *f*
e. store löschbarer Speicher *m (Datenspeicherung)*
erase / to 1. löschen *(gespeicherte Information, Tonband)*; 2. radieren
erase amplifier *(Dat)* Löschverstärker *m*
e. cut-out key Löschsperre *f*
e. head Löschkopf *m*
eraser gate *(Dat)* Löschtorschaltung *f*
erasing Löschung *f*
e. head Löschkopf *m*
e. shield Radierschablone *f*
e. speed Löschgeschwindigkeit *f*
erasure *(Fmt)* Irrung *f*
e. signal Irrungszeichen *n*
erbia Erbiumoxid *n*
erbium Erbium *n*
e. oxide Erbinerde *f*, Erbiumoxid *n*
erect / to aufbauen, aufstellen, aufrichten, errichten, montieren, setzen
e. a perpendicular *(Math)* eine Senkrechte errichten
erect image aufrechtstehendes Bild *n*
erecting crane Montagekran *m*
e. derrick Montagederrick *m*
e. frame *(Rak)* Kipprahmen *m*
e. prism Umkehrprisma *n*
e. shop Montierwerkstatt *f*; Montagehalle *f*
e. tool Montagewerkzeug *n*
erection 1. Aufstellung *f*; Errichtung *f*; Montage *f*; Stahlbau *m*; Zusammenbau *m*; Aufstellung *f*; 2. *(Schiff)* Aufbau *m*, Decksaufbau
e. building Montagehalle *f*
e. crew Montagebrigade *f*, Montagekolonne *f*
e. drawing Montagezeichnung *f*
e. of overhead lines *(El)* Anlage *f* von Freileitungen, Freileitungsmontage *f*
e. pit Montageplatz *m (Bagger)*
e. sag Montagedurchhang *m (Luftkabel)*
e. schedule Montageplan *m*
e. shop Montagewerkstätte *f*
e. site Baustelle *f (Stahlbau)*
erector Monteur *m*; Stahlbaumonteur *m*
erg Erg *n (Einheit für Arbeit, Energie und Wärmemenge: 1 erg =* 10^{-7} *J)*
ergodic ergodisch
e. process ergodischer Prozeß *m*
e. theory Ergodentheorie *f*
ergon *(Ch)* Ergon *n*, Zellwirkstoff *m*
ergosterol Ergosterin *n*
ergot Mutterkorn *n (Getreide)*
Erichsen cupping test Tiefungsversuch (Einbeulversuch) *m* nach Erichsen, Erichsen-Tiefung *f*
Erichsen test Erichsen-Versuch *m*
Ericsson [line] selector *(Fmt)* Ericsson-Kulissenwähler *m*, Ericsson-Wähler *m*; Kugelwähler *m* [nach Ericsson]

Ericsson separating filter *(Fmt)* Ericsson-Weiche *f*
eriometer *(Text)* Wolldickenmesser *m*
eriskop Eriskop *n (Handelsname; Spezialtyp einer französischen Bildaufnahmeröhre)*
Erlagol process *(El)* Erlagol-Verfahren *n (britisches Silberpreßverfahren für gedruckte Schaltungen)*
Erlenmeyer flask Erlenmeyerkolben *m*
erode / to erodieren, anfressen, zerfressen; auskolken
erosion Erosion *f*; Kolkbildung *f*
e. by slags Schlackenangriff *m*
e. control Erosionsschutz *m*
erosive erodierend
errata Druckfehler *mpl*; Druckfehlerberichtigung *f*
erratic regellos, sprunghaft, ungleichmäßig
e. block *(Geol)* Findling *m*
e. firing unregelmäßiges Zünden *n (Verbrennungsmotor)*
e. value Streuwert *m*
erratum Druckfehler *m*; Schreibfehler *m*
erroneous fehlerhaft
error 1. *(Math)* Fehler *m*, Abweichung *f*; 2. *(Am) (BMSR)* Regelabweichung *f*; Fehler *(Folgeregler)*
e. correcting code *(Dat)* fehlerkorrigierender Kode *m*
e. corrector *(Am) s.* controller
e. curve Fehlerkurve *f*
e.-detecting code *(Dat)* fehlersuchender (selbstprüfender) Kode *m*
e. detection Fehlersuche *f*
e. distribution Fehlerverteilung *f*
e. distribution curve Fehlerkurve *f*
e. due to friction Reibungsfehler *m*
e. estimation Fehlerabschätzung *f*
e.-free fehlerfrei *(Messung)*
e. in bearing Peilfehler *m*, Fehlweisung *f*
e. in card punching Falschlochung *f*
e. in fix *(Schiff)* Besteckfehler *m*
e. in grinding Schleiffehler *m*
e. in indication Anzeigefehler *m (Meßinstrument)*
e. in latitude *(Schiff)* Breitenfehler *m*
e. in measurement Meßfehler *m*
e. in observation Beobachtungsfehler *m*
e. in operation Bedienungsfehler *m*
e. in reading Ablesefehler *m (Meßinstrumente)*
e. in rounding off *(Math)* Abrundungsfehler *m*
e. in tooth spacing *(Masch)* Teilungsfehler *m*
e. latitude Fehler[band]breite *f*
e. measuring device *(BMSR)* Vergleichsglied *n*, Vergleicher *m*
e. of measurement Meßfehler *m*
e. of observation Beobachtungsfehler *m*
e. of pitch *(Masch)* Steigungsfehler *m*
e. of reading Ablesefehler *m*
e. propagation *(Math)* Fehlerfortpflanzung *f*
e. range *(BMSR)* Störbereich *m*
e.-ratio *(BMSR)* Regelfaktor *m*
e.-sensing *s.* e. detection
e. signal Fehlersignal *n*; *(Fmt)* Korrektionssignal *n*
e. source Fehlerquelle *f*
erubescite 1. Buntkupferkies *m*; 2. *s.* bornite
erucic acid Erukasäure *f*
eruginous copper mit Grünspan beschlagenes Kupfer *n*
eruption Eruption *f*, Ausbruch *m (eines Vulkans)*
eruptive [rock] Eruptivgestein *n*
erythrite *(Min)* Erythrin *m*
erythritol *(Ch)* Erythrit *n*
Esaki diode Tunneldiode *f*
escalator Rolltreppe *f*
escape / to ausströmen, austreten, entweichen, abströmen
e. terrestrial gravity dem Gravitationsfeld der Erde entkommen
escape 1. Ausströmen *n (Gas)*; Entweichen *n (Gas, Teilchen)*; 2. *(Masch)* Ablauf *m (Späne)*
e. hatch Notausstieg *m*; *(Schiff)* Notluke *f*, Sicherheitsluke *f*
e. lung *(Am) (Brg)* Atemgerät *n*
e. of chips Späneabfluß *m*, Späneabfuhr *f*
e. of neutrons Neutronenverlust *m*
e. of steam Dampfentweichung *f*, Dampfaustritt *m*
e. probability Entkommwahrscheinlichkeit *f (Neutronen)*
e. shaft *(Brg)* Fluchtschacht *m*, Notschacht *m*, Rettungsschacht *m*
e. velocity 1. Ausströmungsgeschwindigkeit *f (Gas)*; 2. zweite kosmische Geschwindigkeit *f*
escapement 1. Hemmung *f (Uhrwerk)*; 2. Ankerhub *m (Relais)*; 3. Ausströmung *f*
e. file Präzisionsnadelfeile *f*, Nadelfeile *f*
e. mechanism Rücklaufhemmung *f (z. B. Elektrizitätszähler)*
e. wheel Gangrad *n (Uhrwerk)*
escort fighter Begleitjagdflugzeug *n*
e. sloop *(Schiff)* Flottenbegleiter *m*
escort[ing] vessel Begleitschiff *n*
escribe / to *(Math)* anschreiben *(einen Kreis dem Dreieck)*
escribed circle *(Math)* Ankreis *m*
escutcheon Schlüsselschild *n*; Namenschild *n (am Schiff)*
espagnolette bolt Espagnolettenverschluß *m*, Drehstangenverschluß *m*
espalier Spalier *n (Lattengerüst)*
esparto grass Espartogras *n*
e. paper Alfapapier *n*

e. pulp Espartozellstoff *m*
ess pipe S-Rohr *n*, S-Stück *n*
essaying vessel *(Ch)* Probiergeschirr *n*
essence Essenz *f*
e. of mirbane Mirbanöl *n*, Nitrobenzol *n*
essential wichtig; Haupt-
e. equation Hauptgleichung *f*
e. oil ätherisches Öl *n*
e. reaction Grundreaktion *f*
establish / to herstellen, bewerkstelligen; einrichten; bestimmen, feststellen; bilden
e. a call (connection) *(Fmt)* eine Verbindung herstellen
ester Ester *m*
e. gum an der Luft härtender Harzester *m*, Estergummi *m*, Harzester *m*
e. interchange Umesterung *f*
e. number Esterzahl *f*
e. resin *s.* e. gum
esterifiable veresterbar
esterification Esterbildung *f*, Veresterung *f*
esterify / to verestern
estimable [ab]schätzbar, bestimmbar
estimate / to [ab]schätzen; veranschlagen, überschlagen
estimate Schätzung *f*, Überschlag *m*, Voranschlag *m*; Kostenschätzung *f*, Kostenüberschlag *m*, Kosten[vor]anschlag *m*
e. calculation Überschlagsrechnung *f*
estimated time of arrival voraussichtliche Ankunftszeit *f*
estimation 1. *(Math)* Abschätzung *f*; 2. *(Ch)* Bestimmung *f*, Beurteilung *f*; 3. Schätzen *n*; Kalkulation *f*, Kostenanschlag *m*
estimator Kalkulator *m*, Kostenplaner *m*
estron Azetatzellulosefaser *f*
estuary Ästuarium *n*, Mündungsgebiet *n* eines Flusses, Flußmündung *f*, Gezeitenflußmündung *f*
e. cable Flußkabel *n*
e. harbour Stromhafen *m*, Mündungshafen *m* *(den Gezeiten unterliegend)*
e. trading *(Schiff)* Revierfahrt *f*
e. tug *(Schiff)* Revierschlepper *m*
etalon Etalon *m*
etch / to [ab]ätzen, anätzen, beizen; radieren
e.-polish ätzpolieren
etch figure Ätzfigur *f* *(Kristall)*
e. test Ätzverfahren *n*, metallografische Untersuchung *f*
etchant Ätzmittel *n*, Beize *f*
etcher *(Typ)* Ätzmaschine *f*
etching 1. Ätzen *n*, Beizen *n*; 2. Radierung *f* *(Kunst)*
e. agent Ätzmittel *n*
e. bands Ätzstreifung *f*
e. bath Ätzbad *n*
e. figure Ätzfigur *f*
e. fluid Ätzflüssigkeit *f*
e. ground Ätzgrund *m*
e. machine *(Typ)* Ätzmaschine *f*
e. pit Ätzvertiefung *f*, Ätzgrübchen *n*, Ätzgrube *f*
e. solution Ätzlösung *f*, Ätzwasser *n*
ethane Äthan *n*
e. diacid Oxalsäure *f*
e. dicarboxylic acid Äthandikarbonsäure *f*, Bernsteinsäure *f*
ethanedioic acid Oxalsäure *f*
ethanol Äthylalkohol *m*
ether Äther *m*
e. theory of light *(Ph)* Äthertheorie *f* des Lichtes *(historisch)*
ethereal ätherisch
ethereous ätherisch
etherify / to Äther bilden, veräthern
etherizing Ausäthern *n*, Ausschütteln *n* mit Äther
ethide Äthyl-Verbindung *f* eines Metalles
ethionic acid Äthansulfonsäure
ethoxide Äthylat *n*
ethoxylene resins Äthoxylinharze *npl*, Epoxydharze *npl*
ethyl acetate Äthylazetat *n*, Essig[säureäthyl]ester *m*
e. alcohol Äthanol *n*, Äthylalkohol *m*
e. cellulose Äthylzellulose *f*
e. ether [einfacher] Äther *m*, Diäthyläther *m*
e. hydrosulphide Äthylmerkaptan *n*
e. mustard oil Äthylsenföl *n*, Äthylester *m* der Isothiozyansäure, Äthylisothiozyanat *n*
e. oxide Diäthyläther *m*
e. tin Zinntetraäthyl *n*
ethylate / to äthylieren
ethylate Äthylat *n*, Äthylierungsprodukt *n*
ethylene Äthylen *n*
e. benzene Äthylenbenzol *n*, Styrol *n*
e. epoxide Äthylenoxid *n*
ethylenic bond Äthylenbindung *f*, Doppelbindung *f*
ethylized fuel Treibstoff *m* mit Bleitetraäthylzusatz
ethynylation Äthinylierung *f*, Einführung *f* der Kohlenstoffdreifachbindung
eucalyptole Eukalyptol *n*
eucalyptus Eukalyptusbaum *m*
e. oil Eukalyptusöl *n*
euchlorine Euchlorin *n*
euchroite *(Min)* Euchroit *m*
euclase *(Min)* Euklas *m*
Euclidian euklidisch
eucolloid Eukolloid *n*
eudialyte *(Min)* Eudialyt *m*
eudiometer Eudiometer *n* *(Gase)*
eugenol Eugenol *n*
Euler's constant Eulersche Konstante *f*

Euler's crippling load Eulersche Knicklast *f*
Euler's formula Eulersche Formel *f*
Euler's formulae for columns Eulersche Knickformeln *fpl*
euphonium Baßtuba *f*
Eureka Eureka-Legierung *f (Widerstandsdrähte)*
European fox Rotfuchs *m*
E. outside-contact base Europasockel *m*, Topfsockel *m (Elektronenröhren)*
europia Europiumoxid *n*
europium Europium *n*
eustatic eustatisch
eutectic Eutektikum *n*
eutectic eutektisch
e. alloy eutektische Verbindung *f*
e. cast iron eutektisches Gußeisen *n*
e. system Eutektikum *n*
eutectoid Eutektoid *n*
evacuate / to auspumpen, evakuieren, luftleer machen (saugen)
evacuating machine Evakuiervorrichtung *f*, Vakuumpumpanlage *f*, Pumpstand *m*
e. plant *s.* e. machine
evacuation Evakuieren *n*, Evakuierung *f*
evaluate / to auswerten
evaluation Auswertung *f*
e. receiver Vermessungsstelle *f*
evanescent mode räumlich gedämpfter Schwingungstyp *m*, unterkritischer Modus *m (Wellenleiter)*
e. waveguide Wellenleiter *m* mit unterkritischer Betriebsfrequenz; (*s. a.* cut-off waveguide)
evaporate / to verdampfen, eindampfen; abdampfen; verdunsten; abrauchen; entweichen
e. on aufdampfen auf
e. to dryness zur Trockne eindampfen
evaporated milk 1. Milchpulver *n*, Trockenmilch *f*; 2. Büchsenmilch *f*
evaporating capacity Verdampfungsfähigkeit *f*, Verdampfungsvermögen *n*
e. dish Abdampfschale *f*
e. point Verdampfungspunkt *m*
e. temperature Verdampfungstemperatur *f*
e. unit Verdampferanlage *f*, Verdampfungsanlage *f*
evaporation Verdampfung *f*; Verdunstung *f*; Eindampfung *f*
e. cathode Aufdampfkatode *f*
e. coating Aufdampfen *n (Überzüge)*
e. metre Verdunstungsmesser *m*
e. of electrons Elektronenverdampfung *f (thermische Emission)*
e. of liquids Flüssigkeitsverdampfung *f*
e. of the nucleus Kernverdampfung *f*
e. per sq.m. of heating surface *(Wärm)* spezifische Dampfleistung *f*, Verdampfung *f* je m^2 Heizfläche
e. section Verdampferteil *m*
e. surface Verdampfungsheizfläche *f*
e. test Verdampfungsversuch *m*
e. theory Kernverdampfungstheorie *f*
evaporative Verdunstungs-; Verdampfungs-
e. capacity Verdampfungsfähigkeit *f*; Dampfleistung *f*
e. cooling Verdampfungskühlung *f*; Verdunstungskühlung *f*
e. power Verdampfungsleistung *f*
e. surface condenser Oberflächenrieselkondensator *m*
evaporator Verdampfer *m*, Verdampfungsgefäß *n*, Verdampfungsapparat *m*, Abdampfvorrichtung *f*, Evaporator *m*, Eindampfer *m*
e. coil Verdampferschlange *f*
e. feed pump Verdampferspeisepumpe *f*
e. heating coil Verdampferschlange *f*
e. section Verdampferteil *m*
e. tube Siederohr *n*
evaporimeter Verdunstungsmesser *m*
evaporization Abdampfung *f*, Abdampfen *n*, Verdampfen *n*
evaporize / to abdampfen, eindampfen, verdampfen
evaporizer Verdampfer *m*
even / to abgleichen, glätten
e. out ausgleichen
even 1. gleichmäßig, einheitlich; 2. eben *(Fläche)*; 3. *(Math)* gerade, geradzahlig; ohne Rest *(teilbar)*
e.-even gerade-gerade, gg *(z. B. Kern)*
e.-even nucleus Gerade-gerade-Kern *m*, gg-Kern *m*
e. fracture *(Met)* feinkörnige Bruchfläche *f*
e. function gerade (symmetrische) Funktion *f*
e.-grained gleichmäßig gekörnt
e. harmonic geradzahlige Oberschwingung (Oberwelle) *f*
e. keel *(Schiff)* gleichlastig
e.-line interlace *(Fs)* geradzahliger Zeilensprung *m*
e. mass gerade Massenzahl *f*
e. number gerade Zahl *f*
e. number of flutes gerade Nutenzahl *f*
e.-numbered geradzahlig
e.-odd gerade-ungerade, gu *(z. B. Kern)*
e.-odd nucleus Gerade-ungerade-Kern *m*, gu-Kern *m*
e.-order harmonics geradzahlige Oberschwingungen (Oberwellen) *fpl*
e. parity *(Ph)* gerade (positive) Parität *f*
e. rotational quantum number gerade Rotationsquantenzahl *f*
e.-running crimp *(Text)* regelmäßige Kräuselung *f*
e.-sided twill *(Text)* gleichseitiger Köper *m*
e. state *(Ph)* gerader Zustand *m*

e. stitches *(Text)* Maschen *fpl* gleicher Länge
evener cylinder *(Text)* Abstreichwalze *f*
e. roller with receding spikes *(Text)* Abstreifregler *m* mit zurücktretenden Stiften
evenly gleichmäßig
e.-napped surface *(Text)* Oberfläche *f* mit gleichmäßigem Gewebeflor
e.-shared load gleichmäßig verteilte Last *f*
e.-spaced abstandsgleich
e.-waved wool fibre wellentreue Wollfaser *f*
evenness Ebenheit *f*, Glätte *f*, Glattheit *f*
e. of the wool fibre Treue *f* der Wollfaser
everdur Everdur *n*, Everdurmetall *n (Cu–Si–Mn-Legierung)*
everready case *(Opt)* Bereitschaftsbehälter *m*
evidence Beweismaterial *n*, Sachverhalt *m*, Tatbestand *m*
e. of damage Tatbestandsaufnahme *f (bezüglich eines Schadens)*
evoke a volt drop / to einen Spannungsabfall hervorrufen
evolute Evolute *f*, Krümmungsmittelpunktskurve *f*
e. connection *(Elm)* Evolventenstirnverbindung *f*
e. winding *(Elm)* Evolventenwicklung *f*
evolution 1. Entwicklung *f*, Evolution *f*; 2. *(Math)* Wurzelziehen *n*, Radizieren *n*; Evolution *f*
e. of gases Gasentwicklung *f*
e. of heat 1. Wärmetönung *f*; 2. Wärmeentwicklung *f*
evolve / to entwickeln
e. gas Gas entwickeln, gasen
evolvent Abwicklungskurve *f*, Evolvente *f*
ewe Zuchtschaf *n*
ewe's wool Mutterwolle *f*
ex stock vom Lager
ex work ab Werk
exact exakt, genau, scharf
exacting demand hoher Anspruch *m (z. B. an Maßhaltigkeit)*
exactly drawn genau gezogen *(Draht)*
e. elliptical motion *(Masch)* genauer Ellipsenlenker *m*
exactness Genauigkeit *f*, Schärfe *f*
exalted carrier angehobener Träger *m (Funktechnik)*
examination Untersuchung *f*, Prüfung *f*, Überprüfung *f*, Beobachtung *f*; Durchsicht *f*
e. in dissolved state Prüfung *f* auf nassem Wege
e. in dry state Prüfung *f* auf trockenem Wege
e. of line *(El)* Leitungsüberprüfung *f*, Linienrevision *f*
e. of materials Werkstoffprüfung *f*
e. of structure Gefügeuntersuchung *f*, Strukturuntersuchung *f*
e. requirements Prüfungsanforderungen *fpl*
examine / to untersuchen, [über]prüfen
e. a line *(El)* eine Leitung prüfen
e. by photometry fotometrieren, fotometrisch untersuchen
examiner 1. Prüfer *m (Patent)*; Warenprüfer *m*; 2. *(Text)* Stoffbeschauer *m*; 3. *(Brg)* Sicherheitssteiger *m*, Wettersteiger *m*
excavate / to 1. ausschachten, ausbaggern, Boden ausheben; 2. *(Brg)* abbauen, ausfördern
excavated cross-section *(Brg)* Ausbruchquerschnitt *m*
e. diameter *(Brg)* Ausbruchdurchmesser *m*
e. material Baggergut *n*
excavation 1. *(Bw)* Ausschachtung *f*, Bodenaushub *m*, Erdaushub *m*; Baggerung *f*; 2. Baugrube *f*; 3. *(Brg)* Grubenraum *m*
excavator 1. Bagger *m*, Trockenbagger *m*, Erdbagger *m*; 2. *(Lw)* Grabmaschine *f*
e. crane Baggerkran *m*, Kranbagger *m*
e. hoist Baggerwinde *f*
e. on caterpillars Raupenbagger *m*
e. track Baggergleis *n*
e. track level Baggerplanum *n*
exceed / to überschreiten, übersteigen, übertreffen
excelsior *(Am)* Holzwolle *f*
e. hook *(Text)* eckig gebogener Nadelhaken *m (Zungennadel)*
excenter *(Math)* Ankreismittelpunkt *m (des Dreiecks)*
excentric Exzenter *m*
excentric exzentrisch
e. clamp *(El)* Froschklemme *f*
e. disk Exzenterscheibe *f*
e. motion Exzenterbewegung *f*
except *(Dat)* JEDOCH-NICHT, UND-NICHT *(Schaltungen)*
e. gate JEDOCH-NICHT-Schaltung *f* (Tor *n*), UND-NICHT-Schaltung (Tor)
excess Überschuß *m*
e. absorption Überschußabsorption *f*
e. air Luftüberschuß *m (z. B. Verbrennungsmotor)*
e.-air factor Luftüberschußzahl *f*
e.-air meter Luftüberschußmesser *m*
e. and total meter *(El)* Gesamt- und Spitzenzähler *m*
e. base Basenüberschuß *m*
e. blast-furnace gas burner *(Met)* Gichtgasfackel *f*
e. conductor Überschußhalbleiter *m (n-Typ)*
e. consumption Mehrverbrauch *m*, Überverbrauch *m*
e. consumption meter Überverbrauchszähler *m*
e. current Überstrom *m*
e. current release Überstromauslösung *f*

e. electron überschüssiges Elektron *n*, Überschußelektron *n*
e. filling *(Brg)* Fremdbergeversatz *m*
e. fuel Kraftstoffüberschuß *m*
e.-gas burning Abfackeln *n*, Abbrennen *n* der Hochofengase
e. length Überlänge *f*, Überstand *m*; Übermaß *n (Langholz)*
e. load Überlast *f*
e. meter Überverbrauchszähler *m*, Spitzenzähler *m*
e. multiplication factor *(Kern)* Überschuß[vermehrungs]faktor *m (Vermehrungsfaktor minus 1)*
e. of air Luftüberschuß *m*
e. of hatchway *(Schiff)* Lukenübermaß *n*, Lukenüberschuß *m*, Übermaß *n* an Luken
e. of heat Wärmeüberschuß *m*
e. of work *(Wärm)* Arbeitsüberschuß *m*, Überschuß *m* an Arbeit *(Dampfdiagramm)*
e. potential *(El)* Überpotential *n*
e. pressure Überdruck *m*
e.-pressure regulator Überdruckregler *m*
e.-pressure valve Überdruckventil *n*
e. reactivity *(Kern)* Überschußreaktivität *f*
e. reverse current Überstrom *m* in Sperrichtung
e. speed *(Elm)* Schleuderdrehzahl *f*
e.-three-code *(Dat)* Drei[er]-Exzeß-Kode *m*, Drei-Überschuß-Kode *m*, Plus-Drei-Kode *m*, Stibitz-Kode *m*
e. voltage Überspannung *f*
e.-voltage preventer Überspannungsschutz *m*, Überspannungswächter *m*
e.-voltage wave Überspannungswelle *f*
excessive übermäßig; überschüssig
e. compensation Überkompensierung *f*; Überregulierung *f*
e. pressure Überdruck *m*
e. regulation Überregulierung *f*
e. stress Überbeanspruchung *f (Festigkeit)*
exchange / to 1. *(Ph)* austauschen; 2. *(Masch)* auswechseln; 3. *(Dat)* umspeichern
exchange 1. *(Ph)* Austausch *m*; 2. *(Masch)* Auswechslung *f*; 3. *(Fmt)* Amt *n*, Vermittlungsamt *n*; 4. *(Dat)* Umspeicherung *f*
e. area *(Fmt)* Anschlußbereich *m*, Gebührenzone *f*
e. battery *(Fmt)* Amtsbatterie *f*
e. cable Ortskabel *n*
e. call Amtsanruf *m*
e. call key Amtsanrufschlüssel *m*
e. dial private branch *(Am)* Privatnebenstellenanlage *f*
e. energy Austauschenergie *f*
e. force Austauschkraft *f*
e. interaction Austauschwechselwirkung *f*
e. jack Amtsklinke *f*
e. line *(Fmt)* Amtsleitung *f*
e. number Amtsnummer *f*
e. of electrons Elektronenaustausch *m*
e. of experience Erfahrungsaustausch *m*
e. of goods Warenaustausch *m*
e. of heat Wärmeaustausch *m*
e. of ions Ionenaustausch *m*
e. of oil Ölwechsel *m*
e. operator Austauschoperator *m*
e. potential Austauschpotential *n*
e. private branch *(Am)* Privatnebenstellenanlage *f*
e. reaction *(Ch)* Austauschreaktion *f*
e. reactor Austauschreaktor *m*
e. tone *(Fmt)* Amtszeichen *n*
exchangeability Auswechselbarkeit *f*, Austauschbarkeit *f*
exchangeable auswechselbar, austauschbar, ersetzbar
exchanger Austauscher *m*, Wärme[aus]tauscher *m*
excircle *(Math)* Ankreis *m*, angeschriebener Kreis *m*
excitability Anregungsvermögen *n*, Anregungsfähigkeit *f*; Reizbarkeit *f*; Erregbarkeit *f*
excitation 1. Anregung *f (Energieniveau)*; 2. *(Elm)* Erregung *f*
e. anode Erregeranode *f*, Halteanode *f*; Zündanode *f*; Hilfsanode *f*
e. band *(Ph, El)* Anregungsband *n*
e. by collision *(Ph)* Stoßanregung *f*
e. circuit *(El)* Erregerkreis *m*
e. coil Erregerspule *f*
e. collision Anregungsstoß *m*
e. conditions *(Ph)* Anregungsbedingungen *fpl*
e. cross section *(Ph)* Anregungsquerschnitt *m*, Wirkungsquerschnitt *m* für Anregung
e. current Erregerstrom *m*
e. drive *(El)* Gitteranregung *f*, Gittersteuerung *f*
e. electrode Zündelektrode *f*
e. energy Anregungsenergie *f*
e. losses *(Elm)* Erregerverluste *mpl*
e. luminescence Anregungsleuchten *n*
e. of a gas Anregung *f* eines Gases
e. of an aerial Antennenerregung *f*
e. of oscillations Schwingungsanfachung *f*, Schwingungserregung *f*
e. output *(El)* Erregerleistung *f*
e. potential Anregungspotential *n*, Anregungsspannung *f*
e. set *(Elm)* Erreger[maschinen]satz *m*, Erregergruppe *f*
e. time *(Ph)* Anregungszeit *f*
e. voltage Anregungsspannung *f*
e. winding *(Elm)* Erregerwicklung *f*, Feldwicklung *f*
excite / to 1. *(Ph)* anregen; 2. *(El)* erregen

e. **oscillations** Schwingungen erregen (erzeugen)
excited 1. *(Ph)* angeregt; 2. *(El)* erregt
e. **atom** angeregtes Atom *n*
e.-**atom density** Dichte *f* der angeregten Atome *(eines Gases)*
e.-**field loudspeaker** Lautsprecher *m* mit Fremderregung
e. **level** *(Ph)* angeregtes Niveau *n*
e. **nuclear state** angeregter Kernzustand *m*
e. **nucleon** Nukleon *n* im angeregten Zustand
e. **nucleus** angeregter Kern *m*
e. **state** angeregter Zustand *m*, Anregungszustand *m*
exciter 1. *(Elm)* Erreger[maschine *f*] *m*; 2. *(Fmt)* Steuersender *m*; 3. Strahler *m (Antenne)*
e. **build-up time constant** *(Elm)* Zeitkonstante *f* der Auferregung
e. **capacity** Erregerleistung *f*
e. **circuit** *(Elm)* Erregerkreis *m*
e. **coil** Erregerspule *f*, Erregerwicklung *f*
e. **field rheostat** *(Elm)* Erregerfeldwiderstand *m*, Feldregler *m*
e. **lamp** Spaltbeleuchtungslampe *f*, Tonlampe *f (Tonfilm)*
e. **light-source** Erregerlichtquelle *f*
e. **rectifier** Erregergleichrichter *m*
e. **resistance** Erregerwiderstand *m*, Feldregler *m*
e. **response** *(Elm)* Änderungsverhältnis *n* der Erregerspannung
e. **set** *(Elm)* Erreger[maschinen]satz *m*, Erregergruppe *f*
e. **stage** *(El)* Steuerstufe *f*
e. **turbine** Erregerturbine *f*
exciting anode Erregeranode *f*, Zündanode *f*
e. **circuit** Erregerkreis *m*
e. **coil** Erregerspule *f*, Erregerwicklung *f*
e. **current** Erregerstrom *m*
e. **dynamo** Erregermaschine *f*
e. **field** Erregerfeld *n*
e. **generator** Erregermaschine *f*
e. **spark gap** Erregerfunkenstrecke *f*
e. **voltage** Erregerspannung *f*
exciton *(Ph)* Exciton *n*
excitor *s.* exciter
excitron *(El)* Excitron *n*
e. **valve** Excitronröhre *f*
exclude / to abschließen *(Luft)*
exclusion Abschluß *m*, Ausschluß *m*; Entfernung *f*, Ausschaltung *f*
e. **of air** Luftabschluß *m*
e. **principle** *(Ph)* Ausschließungsprinzip *n*, Pauli-Prinzip *n*
exclusive OR *(Dat)* ausschließendes ODER *n (logischer Operator)*; Antivalenz *f*; ENTWEDER-ODER *n*
e. **right** Alleinrecht *n*
excursion *(Masch)* Auslenkung *f*, Auswanderung *f*
e. **of the stylus** Auslenkamplitude *f* der Nadel *(Tonabnehmer)*
e. **vessel** Vergnügungsschiff *n*, Ausflugsschiff *n*, Bäderschiff *n*
executive decision Unternehmungsentscheidung *f*
e. **power** Staatsgewalt *f*
e. **routine** *(BMSR)* Leitprogramm *n*, Steuerprogramm *n*
e. **work** Verwaltungsarbeit *f*
exempt from fees gebührenfrei
exemption Ausnahme[genehmigung] *f*
e. **from charges (fees)** Gebührenbefreiung *f*
exercise yard *(Lw)* Auslauf *m (Stallbauten)*
exert pressure / to Druck ausüben
exfoliate / to abblättern, aufblättern
exfoliation 1. Abblätterung *f (einer Oberfläche)*; 2. *(Text)* Auflösen *n* eines Fadens in seine Bestandteile
exhalation Ausdünstung *f*, Dunst *m*, Dampf *m*, Brodem *m*
exhaust / to 1. ablassen; ausblasen; auspuffen; ausschieben; 2. luftleer machen, auspumpen, evakuieren; 3. erschöpfen *(Vorräte)*; 4. abströmen lassen; absaugen
exhaust Ausschieben *n (Verbrennungsmotor)*; Auspuff *m*
e. **air** Abgas *n*, Abluft *f*
e. **back pressure** Auspuffgegendruck *m*
e. **chamber** Austrittskammer *f (Turbine)*
e. **coefficient** Ausflußkoeffizient *m (Turbine)*
e. **connecting branch** Austrittstutzen *m*, Auslaßstutzen *m*, Abdampfstutzen *m*
e. **connection** *s.* e. connecting branch
e. **curve** Ausströmungslinie *f*
e. **depression** *(Schiff)* Heckmulde *f*
e.-**driven supercharger** Abgasturbolader *m*
e. **duct** Abgasleitung *f*
e. **edge** Auslaßkante *f*
e. **elbow** Auspuffkrümmer *m*
e. **fan** Entlüfter *m*, Exhaustor *m*, Sauggebläse *n*, Sauglüfter *m*
e. **feed heater** mit Abdampf geheizter Vorwärmer *m*
e.-**fired calorifier** Abgaskessel *m*, Abhitzekessel *m*
e.-**flame damper** *(Kfz)* Auspuffflammendämpfer *m*
e. **fumes** Auspuffgase *npl*, Qualm *m*
e. **gas** Abgas *n*; Gichtgas *n*
e.-**gas analyzer** Abgasprüfgerät *n*
e.-**gas-fired boiler** Abgaskessel *m*
e.-**gas flow** Abgasstrom *m*
e.-**gas turbine** Abgasturbine *f*
e. **gas turbine-driven supercharger** Abgasturbolader *m*

e.-gas turbo-charger Abgasturbolader *m*
e. gases Verbrennungsgase *npl*
e.-gear mechanism Auslaßsteuerung *f*
e. head Auspufftopf *m*, Auspuffstutzen *m*
e. heat Abwärme *f*
e. heat boiler Abhitzekessel *m*, Abgaskessel *m*
e.-heat de-icer *(Flg)* Abgasenteisung[sanlage] *f*
e. hood Abzugshaube *f*
e. hopper bale opener *(Text)* Kastenballenöffner *m* mit Staubabzug
e.-jacketed carburetter Vergaser *m* mit Auspuffheizmantel
e. jet Abgasdüse *f*
e. lap *(Wärm)* Auslaßüberdeckung *f*, innere Überdeckung *f*
e. lead 1. *(Wärm)* Vorausströmung *f*, Voraustritt *m*; 2. *(Schiff)* Abgasleitung *f*
e. manifold Abgassammelleitung *f*; Auspuffsammelrohr *n*, Auspuffkrümmer *m*
e. manifold gasket Auspuffdichtung *f*
e. muffler Auspuffschalldämpfer *m*, Auspufftopf *m*
e.-operated air heating *(Kfz)* Abgasheizung *f*
e. passage Auspuffkanal *m*, Auslaßkanal *m*, Abgaskanal *m*; Dampfausströmkanal *m*; Ausströmhals *m (Turbine)*
e. period Dampfausströmperiode *f*; Ausflußzeit *f (Turbine)*
e. pipe Abzugsrohr *n*, Entlüftungsrohr *n (Lüftungstechnik)*; Auspuffrohr *n*
e. piping Auspuffleitung *f*
e. port 1. Auspufföffnung *f*; 2. Dampfaustrittsöffnung *f*, Dampfausströmungsöffnung *f*; 3. Pumploch *n (Vakuumröhren)*; 4. Austrittsschlitz *m*
e.-port piston Kolben *m* eines 2-Takt-Gegenkolbenmotors, der die Auspuffschlitze steuert
e.-port side *(Wärm)* Dampfauslaßseite *f*
e.-pulse supercharging *(Kfz)* Abgasimpulsaufladung *f*
e. reheat *(Flg)* Nachverbrennung *f (Turboantrieb)*
e. resistance Ausflußwiderstand *m (Turbine)*
e. silencer Auspuffschalldämpfer *m*, Auspufftopf *m*
e. stack Auspuffkrümmer *m*, Auspuffrohr *n*, Auspuffstutzen *m*
e. steam Abdampf *m*
e.-steam passage Dampfaustrittskanal *m*
e.-steam pipe Dampfaustrittsrohr *n*, Dampfabblasrohr *n*, Dampfauslaßrohr *n*
e.-steam utilizing plant Abdampfverwertungsanlage *f*
e.-steam preheater Abdampfvorwärmer *m*
e.-steam turbine Abdampfturbine *f*
e. stroke 1. Ausschubtakt *m*; 2. Auspufftakt *m*, Auspuffhub *m (Verbrennungsmotor)*
e. temperature Abgastemperatur *f*, Auspufftemperatur *f*
e. tube Pumprohr *n*, Pumpstengel *m*, Pumpstutzen *m (Vakuumröhren)*
e. turbine Abdampfturbine *f*, Abgasturbine *f*
e.-turbo charger Abgasturbolader *m*
e.-turbo-driven blower Abgasturbogebläse *n*
e.-turbo supercharger Abgasturbolader *m*
e.-turbo supercharging Abgasturboaufladung *f*
e.-type turbo blower Abgasturbogebläse *n*
e. valve Auspuffventil *n*; Dampfauslaßventil *n*; Druckventil *n*
e.-valve chest Auslaßventilgehäuse *n*
e.-valve head Auslaßventilteller *m*
e.-valve lid Auslaßventilabschlußdeckel *m (Turbine)*
e. vane *(Rak)* Strahlruder *n*
e. velocity *(Rak)* Austrittsgeschwindigkeit *f*, Ausströmgeschwindigkeit *f*
e. ventilation 1. Zwangsentlüftung *f*; 2. *(Brg)* saugende Bewetterung *f*
e. ventilator Sauglüfter *m*, Ablüftejektor *m*, Ablüfterkopf *m*
e. whistle *(Eis)* Auspuffpfeife *f*
exhausted 1. erschöpft, verbraucht *(Batterie)*; 2. *(Brg)* völlig abgebaut
e. coal-pit abgebaute Kohlenzeche *f*
e. of air luftleer gepumpt
exhauster Entlüfter *m*, Exhaustor *m*; Luftsauger *m*, Sauggebläse *n*, Saugventilator *m*
exhaustion Evakuierung *f*; Absaugen *n*; Herausziehen *n*
e. machine *(Am)* s. evacuating machine
e. of battery Erschöpfung *f* der Batterie
exhibit / to ausstellen
exhibit 1. Ausstellungsstück *n*; 2. Beweisstück *n*
exhibition Ausstellung *f*
e. area Ausstellungsgelände *n*
e. building Ausstellungsgebäude *n*
e. ground Ausstellungsgelände *n*
e. hall Ausstellungshalle *f*; Messehalle *f*
e. pavilion Ausstellungshalle *f*
exhibitor Aussteller *m*
exit 1. Ausgang *m*; 2. *(Masch)* Auslaß[öffnung *f*] *m*
e. angle Austrittswinkel *m (z. B. Turbinenschaufel)*
e. coefficient Entweichungskoeffizient *m (z. B. für Gase)*
e. diameter Austrittsdurchmesser *m (z. B. Strahltriebwerk)*
e. dose *(Kern)* Austrittsdosis *f*
e. gas Abgas *n*; Gichtgas *n*
e. hub *(Büro)* Ausgangsbuchse *f*
e. losses Austrittsverluste *mpl*

e. port *(Opt)* Austrittsluke *f*
e. portal *(Kern)* Austrittsfeld *n*
e. potential Austrittspotential *n (Elektronenröhren)*
e. pupil *(Opt)* Austrittspupille *f*
e. road Ausfallstraße *f*
e. side 1. Ausgangsseite *f*; 2. Austrittsseite *f (Hochofen)*
e. side of the machine Maschinenausgangsseite *f*
e. slit *(Opt, Kern)* Austrittsspalt *m*
e. velocity Austrittsgeschwindigkeit *f*
e. window *(Opt)* Austrittsluke *f*
exitelite Antimonblüte *f*, Weißspießglanzerz *n*
exo-form Exokonfiguration *f*
exoelectron emission Exoelektronenemission *f*
exoelectrons Exoelektronen *npl*
exoenergetic *s.* exothermic
exoenergic *s.* exothermic
exoergic *s.* exothermic
exograph Röntgenaufnahme *f*
exophthalmometer *(Opt)* Exophthalmometer *n*
exothermal *s.* exothermic
exothermic exotherm, energieabgebend, wärme[ab]gebend, wärmeentwickelnd
e. reaction exotherme Reaktion *f*
exotic hide Wildhaut *f*
expand / to 1. *(Math)* entwickeln *(in Reihe)*; 2. *(Ph)* expandieren, sich ausdehnen; sich ausbreiten *(Wellen)*; 3. *(Masch)* [aus]dehnen; erweitern, ausweiten *(Loch)*; auftreiben, aufweiten *(Rohre)*; spreizen; 4. *(Plast)* [auf]schäumen
e. in[to] a series *(Math)* in eine Reihe entwickeln
e. in ascending powers *(Math)* nach steigenden Potenzen entwickeln
e. the nozzle section die Düsen aufweiten *(Turbine)*
expanded blast-furnace slag Hüttenbims *m (Baustoff)*
e. centre plan display Mittelpunktaufweitung *f (Radar)*
e. clay Blähton *m (Baustoff)*
e. contrast *(Fs)* Kontrastvergrößerung *f*, Kontrastverbesserung *f*
e. metal Streckmetall *n*
e. metal forming press Streckmetallpresse *f*
e. plastic Schaumstoff *m*
e. plastic film Schaumstoffschicht *f*
e. position indicator Radarschirm *m* mit Mittelpunktaufweitung
e. rubber Schaumgummi *m*
e. shale Bläh[schiefer]ton *m*, aufgeblähter Tonschiefer *m (Baustoff)*
e. shell plating *(Schiff)* abgewickelte Außenhaut *f*
e. slag Hüttenbims *m (Baustoff)*
e. slate Bläh[schiefer]ton *m*, aufgeblähter Tonschiefer *m (Baustoff)*
e. steel Streckstahl *m*
e. sweep *(Am) (El)* kompensierte Abtastung *f*
expander 1. *(Fmt)* Dynamikdehner *m*; 2. *(Text)* Ausbreiter *m*, Ausbreitemaschine *f*; 3. Spreizmechanismus *m (Werkzeug)*
expandibility *(Plast)* Schaumfähigkeit *f*
expanding Ausbauchen *n (Umformen)*
e. agent *(Plast)* Treibmittel *n*, Porenbildner *m*
e. band clutch (coupling) Spreizringkupplung *f*
e. block type cutter mounting Spreizfassung *f* [in Bohrstange]
e. cement Quellzement *m*
e. cone Spreizkonus *m*
e. mandrel Spreizdorn *m*
e. mill Aufweitewalzwerk *n (Rohre)*
e. nozzle Expansionsdüse *f (Turbine)*
e. pipe Ausgleichsrohr *n*, Expansionsrohr *n*
e. reamer Reibahle *f* mit Spreizkörpereinstellung; Spreizreibahle *f*; Dehnreibahle *f*; nachstellbare geschlitzte Reibahle
e. washer Abdichtungsring *m*
e. wave *(Ph)* sich ausbreitende Welle *f*
e. work arbor Spreizdorn *m*
expansion 1. *(Math)* Reihenentwicklung *f*; 2. *(Ph)* Expansion *f*; Ausdehnung *f*; 3. Auseinanderspreizen *n (Werkzeug)*; 4. *(El)* Zeitbasisdehnung *f*, Dynamikdehnung *f*
e. and condensing steam engine Dampfmaschine *f* mit Expansion und Kondensation
e. arbor *(Masch)* Spreizdorn *m*
e. bearing *(Bw)* bewegliches Auflager *n*
e. bellow Wellrohrausdehnungsstück *n*
e. bend Ausgleichbogen *m*, Dehnungsschlaufe *f (Rohrleitungen)*
e. by heat Wärme[aus]dehnung *f*
e. chamber 1. *(Kfz)* Ausdehnungsraum *m*, Entspannungsraum *m (Schalldämpfer)*; 2. *(Ph)* Nebelkammer *f*, Wilsonkammer *f*
e. circuit-breaker Expansionsschalter *m*
e. coupling Ausdehnungskupplung *f*, längsbewegliche Kupplung *f*
e. crack Dehnungsriß *m*, Ausdehnungsriß *m*, Wärmedehnungsriß *m*
e. curve *(Wärm)* Expansionslinie *f*
e. due to heat Wärme[aus]dehnung *f*
e. fit 1. Einschrumpfverbindung *f*; 2. Dehnpreßpassung *f*
e.-fit into / to einpressen in *(bei Dehnpreßpassung)*
e. gear Expansionsapparat *m (Regler)*
e. hand reamer verstellbare Handreibahle *f*
e. instrument Hitzdrahtinstrument *n*
e. joint *(Bw)* Dehn[ungs]fuge *f*; *(Eis)* Stoßfuge *f*

e. **liquefying** Expansionsverflüssigung *f*
e. **loop with rotary flange** *(Wärm)* Ausgleichschleife *f* mit drehbarem Flansch
e. **loop with spherical joint** *(Wärm)* Ausgleichschleife *f* mit Kugelgelenk
e. **nozzle** Expansionsdüse *f*
e. **of the grate bars** *(Wärm)* Rostausdehnung *f*
e. **phase** Expansionsphase *f*
e. **pipe** Ausgleichsrohr *n*, Expansionsrohr *n*
e. **pipe joint** Ausdehnungsrohrverbindung *f*
e. **ratio** Expansionsverhältnis *n*, Druckverhältnis *n*
e. **reamer** Reibahle *f* mit Spreizkörpereinstellung; Spreizreibahle *f*; Dehnreibahle *f*; nachstellbare geschlitzte Reibahle
e. **ring** Ausdehnungsring *m*, Spannring *m*
e.-sleeve [roof-]bolt *(Brg)* Spreizhülsenanker *m*
e. **slide valve** Expansionsschieber *m*
e. **sliding block** Expansionsgleitbacke *f*
e. **stage** Expansionsstufe *f*
e. **steam engine** Dampfmaschine *f* mit Expansion, Expansionsdampfmaschine *f*
e. **stress** Ausdehnungsspannung *f*
e. **stroke** *(Kfz)* Ausdehnungstakt *m*; Arbeitstakt *m*
e. **stuffing box** *(Wärm)* Ausgleichstopfbüchse *f*
e. **tank** Ausdehnungsgefäß *n* *(Heizungstechnik)*
e. **test** *(Met)* Aufweiteversuch *m*
e. **theorem** Entwicklungssatz *m*
e. **tong** Fernbedienungszange *f*, Zangenmanipulator *m*
e. **valve** Entspannungsventil *n*, Expansionsventil *n*
e. **vessel** Ausdehnungsgefäß *n*
expansive cement Quellzement *m*
e. **force** Ausdehnungskraft *f*
expansivity Ausdehnungsvermögen *n*
expectation value *(Math, Ph)* Erwartungswert *m*
expedance *(El)* negative Impedanz *f*
expedor phase transformer Phasenschieber *m* *(Induktionsmotor)*
expelled particle *(Ph)* emittiertes Teilchen *n*
expenditure Aufwand *m*; Kostenaufwand *m*
experience Erfahrung *f*; Sachkenntnis *f*
experiment Experiment *n*, Versuch *m*; Prüfung *f*
experimental experimentell, Versuchs-
e. **alloy** Versuchslegierung *f*
e. **animal** Versuchstier *n*
e. **approach** experimentelle Annäherung *f*
e. **area** Versuchsgelände *n*
e. **arrangement** Versuchsanordnung *f*; Meßanordnung *f*
e. **boiling water reactor** Versuchssiedewasserreaktor *m*
e. **breeder reactor** Versuchsbrutreaktor *m*
e. **building** Versuchsbau *m*, Experimentalbau *m*
e. **check** experimentelle Nachuntersuchung *f*
e. **chemistry** Experimentalchemie *f*
e. **data** Versuchsangaben *fpl*
e. **direction** Versuchsvorschrift *f*
e. **drilling** Versuchsbohrung *f* *(z. B. Erdöl)*
e. **engine** Versuchsmaschine *f*; Versuchsmotor *m*
e. **error** Versuchsfehler *m*
e. **facilities** Versuchseinrichtungen *fpl*
e. **gallery** *(Brg)* Versuchsstrecke *f*
e. **hole** *(Kern)* Experimentierkanal *m*, Versuchskanal *m*
e. **house** Experimentalhaus *n*, Versuchshaus *n*
e. **lathe** Versuchsdrehmaschine *f*
e. **material** Untersuchungsmaterial *n*
e. **physics** Experimentalphysik *f*
e. **plant** Versuchsanlage *f*
e. **scale** Versuchsmaßstab *m*
e. **set-up** Versuchsanordnung *f*; Meßanordnung *f*
e. **state** Versuchsstadium *n*
e. **station** Versuchsfunkstelle *f*
e. **vessel** *(Schiff)* Versuchsfahrzeug *n*
experimenter Experimentator *m*; Experimentalphysiker *m*
expert erfahren, fachkundig, sachkundig; sachgemäß
expert Sachverständiger *m*, Experte *m*, Fachmann *m*, Spezialist *m*
e. **knowledge** Fachkenntnis *f*, Sachkenntnis *f*
e. **on** Fachmann *m* für
e. **opinion** Meinung *f* der Fachleute; [fachmännisches] Gutachten *n*
expert's report Sachverständigengutachten *n*
expiration of patent Patentablauf *m*, Erlöschen *n* des Patent[e]s
expired cost abgeschriebener Kostenanteil *m*
e. **time** abgelaufene (verstrichene) Zeit *f*
explement Ergänzungswinkel *m* *(zu 360°)*
explicit explizit *(Funktion)*
e. **solution** *(Math)* explizite Lösung *f*
explode / to explodieren, bersten, zerknallen, platzen, zerspringen, verpuffen, detonieren
exploded view Darstellung *f* in auseinandergezogener Anordnung, auseinandergezogene Darstellung
exploder *(Brg)* Zündmaschine *f*, elektrische Zündvorrichtung *f*
exploit / to *(Brg)* ausbeuten, gewinnen, abbauen
exploitation 1. *(Brg)* Abbau *m*, Ausbeutung *f*, Gewinnung *f*; 2. Nutzung *f*
e. **drilling** Gewinnungsbohrung *f*
e. **factor** Ausnutzungsfaktor *m*

e. of a duplexed line *(Fmt)* Gegensprechbetrieb *m*
e. of soil Nutzung *f* des Bodens
exploration Untersuchung *f*; *(Bw)* Bodenuntersuchung *f*; *(Brg)* Schürfung *f*
e. of space Weltraumforschung *f*
e. of the ionosphere Ionosphärenforschung *f*
exploratory boring Suchbohrung *f*, Untersuchungsbohrung *f (Erdöl, Erdgas)*
e. drift *(Brg)* Erkundungsstrecke *f*
e. drilling Aufschlußbohren *n*, Explorationsbohren *n*, Suchbohren *n*, Untersuchungsbohren *n*
e. rocket Rakete *f* zur Erforschung der oberen Schichten der Atmosphäre; meteorologische Rakete
exploring coil *(El)* Prüfspule *f*, Suchspule *f*
e. spot *(Fs)* Abtastlichtfleck *m*
explosibility Explosionsfähigkeit *f*, Explodierbarkeit *f*; Explosionsneigung *f*
explosion Explosion *f*, Zerknall *m*, Knall *m*
e. centre Detonationszentrum *n*
e. chamber *(El)* Löschkammer *f (Unterbrecher)*
e. crater Explosionskrater *m*, Sprengkrater *m*, Explosionstrichter *m*
e. hazard Explosionsgefahr *f*
e. of fire damp *(Brg)* Schlagwetterexplosion *f*
e. pressure Explosionsdruck *m*
e.-proof explosionssicher, explosionsgeschützt; schlagwettergeschützt
e.-proof luminaire *(Am) s.* flameproof lighting fitting
e.-proof transformer Transformator *m* mit Explosionsschutz
e. protection Explosionsschutz *m*
e. rivet Sprengniet *m*
e. stroke *(Kfz)* Arbeitshub *m*, Arbeitstakt *m*
e. trial Versuchsexplosion *f*
explosive explosiv
explosive Sprengstoff *m*, Sprengmittel *n*
e. action Sprengwirkung *f*
e.-cartridge fastening tool Bolzenschießgerät *n*
e. charge Sprengladung *f*
e. combustion engine Explosionsmotor *m*, Verbrennungsmotor *m*
e. cotton Schießbaumwolle *f*
e. effect Explosionswirkung *f*
e. forming Explosionsumformung *f*, Explosivumformen *n*, Hochenergieumformen *n*
e. limits Explosionsgrenzen *fpl*
e. rivet Sprengniet *m*
e. store Sprengstofflager *n*
explosively formed durch Explosivverfahren umgeformt
explosiveness Explodierbarkeit *f*
explosives Explosivstoffe *mpl*

e. canister Sprengstoffbehälter *m*
e. drilling Sprengbohren *n (Erdöl, Erdgas)*
exponent Exponent *m*
e. of expansion Expansionsexponent *m*
exponential exponentiell, Exponential-
e. absorption exponentielle Absorption *f*
e. approximation *(Math)* exponentielle Näherung *f*
e. characteristic Exponentialkennlinie *f*
e. curve Exponentialkurve *f*
e. decay exponentieller Abfall *m*
e.-horn loudspeaker Pilzlautsprecher *m*
e. law Exponentialgesetz *n*
e. line Exponentialkurve *f*
e. pentode Bremsgitterregelröhre *f*
e. pile *s.* e. reactor
e. reactor Exponentialreaktor *m*
e. series *(Math)* Potenzreihe *f*
e. well *(Ph)* Exponentialtopf *m*
export Export *m*, Ausfuhr *f*
e. delivery time Exportlieferzeit *f*
e. order Exportauftrag *m*
e. version Exportausführung *f*
exported article Exporterzeugnis *n*
exporter Exporteur *m*
exports Außenhandel *m*
expose / to 1. aussetzen *(Witterung, Strahlung)*; 2. belichten *(Film)*
e. on the surface auf Putz verlegen *(Leitung)*
e. to radiation exponieren; bestrahlen; belichten
e. to ultrasonic waves [mit Ultraschallwellen] beschallen
e. to weather der Witterung aussetzen
exposed 1. exponiert; freigelegt, offen, ungeschützt; 2. strahlenexponiert, bestrahlt
e. aggregate concrete Sichtbeton *m* mit freigelegten Zuschlagstoffen
e. concrete Sichtbeton *m*
e. deck *(Schiff)* freiliegendes Deck *n*
e. installation *(El)* gefährdete (offene) Anlage *f*
e. weather side *(Flg)* Luvseite *f*
e. wiring *(El)* Verlegung *f* auf Putz, A.P.-Installation *f*
exposing 1. Freilegen *n (Kohlenflöz)*; 2. Belichten *n (Film)*
exposure 1. Belichtung *f (Film)*; 2. Strahlenexponierung *f*, Exponierung *f*
e. cavity *(Kern)* Bestrahlungszelle *f*
e. counter *(Opt)* Aufnahmezähler *m (Kamera)*
e. field Bestrahlungsfeld *n*
e. hole *(Kern)* Bestrahlungskanal *m*
e. latitude Belichtungsspielraum *m*
e. meter Belichtungsmesser *m*
e. rate Bestrahlungsstärke *f*
e. time Belichtungszeit *f*
e. time table Belichtungstabelle *f*

e. timer Belichtungsuhr *f*; Belichtungsmesser *m*
e. to light Belichtung *f*
e. to rays Bestrahlung *f (Elektromedizin)*
e. to ultrasonic waves Beschallung *f*, Ultraschallbestrahlung *f*
express call *(Fmt)* dringendes Gespräch *n*
e. highway Autobahn *f*
e. liner Schnelldampfer *m*, Schnellmotorschiff *n*
e. telephone service *(Fmt)* Schnellverkehr *m*
e. traffic Schnellverkehr *m (Straßenverkehr)*
e. train Schnellzug *m*
e. way Autobahn *f*, Schnellverkehrsstraße *f*
expressible ausdrückbar
expressing tolerances Tolerieren *n*
expression *(Math)* Ausdruck *m*
e. in brackets Klammerausdruck *m*
expulsion-fuse *(El)* Löschrohrsicherung *f*, Ausblassicherung *f*
e.-type arrester *(El)* Löschrohrableiter *m*
exradius *(Math)* Ankreisradius *m*
exsiccate / to [aus]trocknen
exsiccation Trocknung *f*, Austrocknung *f*
exsiccator Exsikkator *m*
extend / to [aus]dehnen
e. a call *(Fmt)* ein Gespräch verlängern
e. a line eine Leitung (Linie) verlängern, eine Leitung (Linie) weiterführen
e. the landing gear *(Flg)* das Fahrwerk ausfahren
e. the range [of measurement] den Meßbereich erweitern
extended langgestreckt
e. centre line *(Flg)* Anfluggrundlinie *f*
e. electron beam breites Elektronenbündel *n*
e. furnace with horizontal grate Planrostvorfeuerung *f*
e. mean value theorem *(Math)* erweiterter Mittelwertsatz *m*
e. point transformation *(Math)* erweiterte Punkttransformation *f*
e. time test Dauerversuch *m*, Langzeitversuch *m*
extender Füllmittel *n*, Streckmittel *n (Papier)*; Beschwerungsmittel *n*, Verschnittmittel *n (Kunststoffe)*
extensibility Dehnbarkeit *f*, Streckbarkeit *f*
extension 1. Dehnung *f*; Verlängerung *f*, Ausdehnung *f (in Länge oder Zeit)*; 2. *(Fmt)* Nebenstelle *f*; 3. Verlängerung *f (eines Maschinenteils)*; 4. *(Met)* verlorener Kopf *m (beim Blockgießen)*
e. bar Verlängerungsschiene *f*
e. circuit Verlängerungsleitung *f*; Nebenstellenleitung *f*
e. cord Verlängerungsschnur *f*
e. crack Dehnungsriß *m*
e. drill socket Kegelhülse *f* in verlängerter Form
e. for dividers Zirkelverlängerung *f*
e. funnel Verlängerungstrichter *m*
e. furnace Vorfeuerung *f*, Außenfeuerung *f*
e. line 1. Nebenschlußleitung *f*; 2. Maßhilfslinie *f (Zeichnung)*
e. loudspeaker Nebenlautsprecher *m*, zweiter Lautsprecher *m*
e. rod Verlängerungsstange *f*; Verlängerung *f*; Einsatz *m*
e. set *(Fmt)* Nebenstellenapparat *m*
e. shaft Verlängerungswelle *f*
e. spring Zugfeder *f*
e. spring eye Aufhängeöse *f* der Zugfeder
e. tong Fernbedienungszange *f*
e. wire Ausgleichsleitung *f (Thermoelement)*
extensions of the wing *(Flg)* Außenflügel *m*
extensity Extensität *f*
extensive umfassend, ausgedehnt
e. air shower ausgedehnter Luftschauer *m*, Auger-Schauer *m (kosmische Strahlung)*
e. mechanization weitgehende Mechanisierung *f*
extensometer Dehnungsmesser *m (Werkstoffprüfung)*
extent Ausdehnung *f*, Ausmaß *n*, Umfang *m*; Grad *m*, Größe *f*; Maß *n*
e. of delignification Aufschlußgrad *m (Zellstoff)*
e. of shrinkage Schwindmaß *n*
e. of tolerance Toleranzfeld *n*
exterior Außenansicht *f*; Außenseite *f*
e. aerial Außenantenne *f*
e. angle Außenwinkel *m*
e. cutter Außenschere *f (Webstuhl)*
e. face *(Holz)* Bundseite *f*, Seitenfläche *f*
e. lighting Außenbeleuchtung *f*, Beleuchtung *f* im Freien
e. lining Außenverkleidung *f*
e. paint Außenanstrich *m*; Außenanstrichfarbe *f*
e. sheathing board Platte *f* für Außenzwecke
e. work *(Bw)* Außenarbeiten *fpl*, Rohbauarbeiten *fpl*
exterminants Schädlingsgifte *npl*
extermination of pests Schädlingsbekämpfung *f*
external Außen-; außenseitig; äußerlich, oberflächlich; Fremd-; fremd
e. aileron *(Flg)* angesetztes Querruder *n*
e. amplification äußere Verstärkung *f*
e. broaching machine Außenräummaschine *f*
e. caliper gauge Außentaster *m*, Wellenlehre *f*
e. circuit äußerer Stromkreis *m*
e. clutch *(Eis)* Fahrkupplung *f*

e. conductor äußerer Leiter *m*, leitender Mantel *m*, Mantelleiter *m*
e. connection Außenanschluß *m (z. B. Fernsehbildröhre)*
e. contact base Außenkontaktsockel *m*
e.-contracting [shoe] brake Außenbackenbremse *f*
e. control Fernsteuerung *f*
e. corrosion Außenkorrosion *f*
e. cylindrical grinding machine Außenrundschleifmaschine *f*
e. cylindrical lapping Außenrundläppen *n*
e. diameter Außendurchmesser *m*
e. dimension Außenmaß *n*
e. electrode Außenelektrode *f*
e. excitation Fremderregung *f*
e. exhaust collector ring *(Rak)* Außensammler *m*
e. exposure Fremdbestrahlung *f*, äußere Bestrahlung *f*
e. fan kiln Trockenkammer *f* mit Außenventilator
e. flue *(Wärm)* Außenzug *m*
e. force äußere Kraft *f*
e. furnace with step grate Treppenrostvorfeuerung *f*
e. gear Außenrad *n*
e. grinding Außenschleifen *n*
e. gripping Außenspannung *f (Spannvorrichtung)*
e. guidance *(Rak)* Fernsteuerung *f*
e. holding Außenspannung *f (Spannvorrichtung)*
e. induction Fremdinduktion *f*
e.-internal teeth lock washer federnde Zahnscheibe *f* mit Innen- und Außenzähnen
e. irradiation Fremdbestrahlung *f*, äußere Bestrahlung *f*
e. jaw Außenbacke *f*
e. limit gauge Grenzrachenlehre *f*
e. measurement Außenmessung *f*
e. member Außen[paß]teil *n*
e. memory Außenspeicher *m*, Fremdspeicher *m*; Ausgabespeicher *m*
e. micrometer Bügelmeßschraube *f*; Meßschraube *f* für Außenmessungen
e. modulation Fremdmodulation *f*
e. photoelectric effect äußerer fotoelektrischer Effekt *m*, lichtelektrische Elektronenemission *f*
e. pole armature *(Elm)* Außenpolanker *m*
e. pole frame *(Elm)* Außenpolgestell *n*
e. Q Lastgüte *f*, Güte *f* bei Belastung
e. radius Abrundung *f*
e. regulating unit *s.* regulating unit
e. scrap Fremdschrott *m*
e. screw cutting tool Gewindedrehmeißel *m (Außengewinde)*
e. screw thread Außengewinde *n*, Bolzengewinde *n*
e. series gap [äußere] Vorfunkenstrecke *f (des Ableiters)*
e. serration Kerbzahnwellenprofil *n*
e. shoe brake Außenbackenbremse *f*
e. spline[s] Keilwellenprofil *n*
e. spur gear außenverzahntes Stirnrad *n*
e. surface Außenfläche *f*
e. taper Vollkegel *m*
e. taper cylindrical gauge Rachenlehre *f*, Kettenlehre *f*
e. teeth lock washer federnde Zahnscheibe *f* mit Außenzähnen
e. thermometer Außenthermometer *n*
e. thread Außengewinde *n*, Bolzengewinde *n*
e. thread chaser Außengewindestrehler *m*
e. thread grinding Außengewindeschleifen *n*
e. threading Außengewindeschneiden *n*
e. threading operation Außengewindeschneidarbeit *f*
e. [threading] tool Außen[gewinde]drehmeißel *m*
e.-toothed mit Außenverzahnung
e. toothing Außenverzahnung *f*
e. vibrator *(Bw)* Außenrüttler *m*, Schalungsrüttler *m*
e. voltage äußere Spannung *f*; Fremdspannung *f*
e. wall Außenmauer *f*, Außenwand *f*
externally acting brake Außenbremse *f*
e. finned pipe Rippenrohr *n*
e. fired gas furnace Gasunterfeuerung *f*
e. grinding cylindrical parts Außenrundschleifen *n*
e. heated arc Lichtbogen *m* mit fremdbeheizter Katode
e. mounted radiator freifahrender Kühler *m*
extinction *(Opt)* Auslöschung *f*, Löschung *f*; Extinktion *f*
e. coefficient Extinktionskoeffizient *m*
e. potential Löschspannung *f*
e. recorder *(Opt)* Extinktionsregistriergerät *n*
e. voltage Löschspannung *f*
extinguish / to löschen
extinguisher Löscher *m*
extinguishing medium Löschmittel *n*
e. voltage Löschspannung *f*
extra extra, zusätzlich
e. anode Zusatzanode *f*
e.-atmospheric außeratmosphärisch
e.-axial rays außeraxiale Strahlen *mpl*
e. bold *(Typ)* fett *(Schrift)*
e. bright hochglanzpoliert
e. charge Preisaufschlag *m*, Mehrpreis *m*
e. cost Mehrkosten *pl*
e. dense [barium] crown glass Schwerstkronglas *n*

e. dense flint glass Schwerflintglas *n*
e.-excitation Übererregung *f*
e. expenditure Mehraufwand *m*
e. fine thread Feingewinde *n*
e. heavy crown glass Ultrakronglas *n*
e.-high frequency Höchstfrequenz *f*
e.-high pressure discharge Höchstdruckentladung *f*
e.-high tension Höchstspannung *f*
e.-high vacuum Höchstvakuum *n*
e.-high voltage Höchstspannung *f*
e.-high voltage insulation Höchstspannungsisolation *f*
e.-high voltage line Höchstspannungsleitung *f*
e.-light loading besonders leichte Bespulung (Pupinisierung) *f*
e. lightly loaded sehr leicht bespult
e.-long length Sonderlänge *f (Bohrer)*
e.-low voltage lighting Kleinspannungsbeleuchtung *f*
e. otter board Höhenscherbrett *n (Fischerei)*
e. pay Lohnzuschuß *m*, Gehaltszuschuß *m*, Zuschuß *m*
e. power Zusatzleistung *f*
e.-pure substance Reinststoff *m*
e. time allowance *(Ök)* Zeitzugabe *f (für Erschwernisse usw.)*
e. work Mehrarbeit *f*
extract / to 1. ausziehen, entziehen, extrahieren; 2. *(Ch)* auslaugen, entziehen, extrahieren; 3. *(Brg)* abbauen, gewinnen
e. a root *(Math)* eine Wurzel ziehen
e. a voltage *(El)* eine Spannung abnehmen
e. gas entgasen
extract Extrakt *m*, Auszug *m*
e. printing *(Text)* Ätzdruck *m*
e. tanning *(Led)* Extraktgerbung *f*
extractable ausziehbar
extractant Extraktionsmittel *n*
extracting 1. *(Ch)* Extrahieren *n*; 2. *(Math)* Radizieren *n*; 3. *(Met)* Treiben *n (Erze)*
e. electrode Absaugelektrode *f*
e. roots *(Math)* Wurzelziehen *n*
extraction 1. *(Ch)* Extrahierung *f*, Extraktion *f*, Auslaugung *f*; 2. *(Wärm)* Dampfentnahme *f*; 3. *(Brg)* Verhieb *m*, Abbau *m*, Gewinnung *f*
e. bleeder valve Entnahmeventil *n*
e. by fusion *(Met)* Röstreduktionsarbeit *f*
e. flask *(Ch)* Extraktionskolben *m*
e. of copper Kupfergewinnung *f*
e. of dust Entstaubung *f*
e. of roots *(Math)* Radizieren *n*, Wurzelziehen *n*
e. of the flax fibre Gewinnung *f* der Flachsfaser
e. potential Absaugspannung *f*
e. process Gewinnungsverfahren *n*
e. valve Entnahmeventil *n (Dampf)*
extractive Extraktionsmittel *n*
extractor Auszieher *m*, Ausziehvorrichtung *f*; Abziehvorrichtung *f (z. B. für Schleifscheibe)*; Auswerfer *m*
e.-type centrifuge Siebschleuder *f*
extrados *(Bw)* Bogenrücken *m*
extradosal burst *(Brg)* Gebirgsschlag *m* aus dem Festen
extraneous field *(Ph)* äußeres Feld *n*; *(El)* Fremdfeld *n*; Streufeld *n*
extranuclear außerhalb des Kerns
extraordinary ray *(Opt)* außerordentlicher Strahl *m (Doppelbrechung)*
extrapolate / to *(Math)* extrapolieren
extrapolation *(Math)* Extrapolation *f*
extras 1. Extrazubehör *n*, Sonderzubehör *n*; 2. Zusatzleistungen *fpl*
extraterrestrial außerterrestrisch, extraterrestrisch, außerhalb der Erdatmosphäre
e. disturbances außerirdische Störungen *fpl (Radio usw.)*
extreme backward position hintere Endlage *f*
e. breadth [over all] *(Schiff)* größte Breite *f*, Breite über alles
e. fibre Randfaser *f (eines Festkörpers)*
e. fibre stress Randspannung *f*
e. forward position vordere Endlage *f*
e. length [over all] *(Schiff)* größte Länge *f*, Länge über alles
e. pressure Hochdruck *m*
e. pressure dopes Hochdruckzusätze *mpl*
e. pressure lubricant Hochdruckschmieröl *n*
e. ultraviolet äußerstes (extremes) Ultraviolett *n*
e. value Extremwert *m*, Spitzenwert *m*
extremely fine vorzüglich *(Erhaltungsgrad einer Münze)*
extremes Extremwerte *mpl*
extrinsic conduction Störstellenleitung *f (Halbleiter)*
e. properties durch Störstellenleitung bedingte Eigenschaften *fpl (Halbleiter)*
e. semiconductor Störstellenhalbleiter *m*, Fehlstellenhalbleiter *m*
extrudability Verarbeitbarkeit *f* auf der Strangpresse
extrude / to fließpressen; strangpressen; extrudieren
extruded stranggepreßt
e. coating electrode Preßmantelelektrode *f*
e. sections gespritzte (stranggepreßte) Profile *npl*
e. shape Strangpreßprofil *n*
e. strand gezogener Faden (Strang) *m*, stranggepreßter Faden
extruder Spritzmaschine *f*, Strangpresse *f*
e. particle board stranggepreßte Spanplatte *f*

e. for metal Metallabscheider *m*
e. head Spritzkopf *m*
extruding Strangpressen *n*; Fließpressen *n*; Extrudieren *n*
e. die Spritzstempel *m*, Preßstempel *m*, Preßform *f*
e. press Strangpresse *f*
extrusion 1. Strangpressen *n*, Spritzen *n*; 2. gespritztes Profil *n*; Strangpreßling *m*; Strangpreßprofil *n*; Faden *m*
e. coating Extruderbefilmen *n*; Kaschieren *n* mit Schneckenfütterung *(Folienherstellung)*; Spritzbefilmen *n*
e. compound Strangpreßmischung *f*
e. die Preßform *f*; Preßstempel *m*, Spritzstempel *m*
e. laminating Aufgießen *n (Beschichtung)*
e. moulding Strangpressen *n*; Strangpreßverfahren *n*
e. of metals Strangpressen *n* von Metallen; Fließpressen *n* von Metallen
e. press Strangpresse *f*
e. process Fließpreßverfahren *n*, Strangpreßverfahren *n*
extrusive rock Ergußgestein *n*
exudate / to ausscheiden, ausschwitzen; abscheiden, ausblühen
exudation Ausscheidung *f*, Ausschwitzung *f*, Exsudat *n*
exude / to *s.* exudate / to
eye 1. *(Masch)* Auge *n*; Öhr *n*, Öse *f*; 2. *(Text)* Auge *(Helfe)*; Rille *f (Hakennadel)*; Zasche *f (Spitzennadel)*; 3. Schauloch *n*
e. ball Augapfel *m*
e. bolt *(Masch)* Augenschraube *f*; Ringbolzen *m*, Ösenbolzen *m*
e.-brow *(Schiff)* Regen[schutz]leiste *f (über einem Bullauge)*
e.-catcher Blickfang *m*
e.-cup Okularmuschel *f*
e. hook Ösenhaken *m*
e. level Augenhöhe *f*
e. nut Ringmutter *f*
e. of fork Gabelöse *f*, Gabelauge *n*
e.-sight Schauloch *n*
eyebrow pencil Augenbrauenstift *m*
eyehole Schauloch *n*
eyelet 1. Durchziehloch *n*, Öse *f*, Lötöse *f*; 2. Gatt[chen] *n (Schnürloch im Segel)*
e. plier *s.* e. set
e. set Ösenzange *f*
e.-stitch pattern Lochmaschenmuster *n (Strickerei)*
eyeletting machine Ösenmaschine *f (Schuhproduktion)*
eyepiece Okular *n*
e. cup Okularmuschel *f*
e. diaphragm Okularblende *f*
e. micrometer Okularmikrometer *n*
e. revolver Okularrevolver *m (Mikroskop)*
e. screw micrometer Okularschraubenmikrometer *n*
eyes of a ship Vorschiff *n*, Schiffsaugen *npl (der vorderste Teil des Wetterdecks am Schiffsbug)*

F

F-layer F-Schicht *f (Ionosphäre)*
f-number *(Opt)* Blendenzahl *f*
fabric 1. *(Text)* Stoff *m*, Gewebe *n*; 2. *(Bw)* Konstruktion *f*, Skelett *n (Gebäude)*; Rohbau *m*; 3. Aufbau *m*, Bau *m*, Zusammensetzung *f*; 4. Erzeugnis *n*, Fabrikat *n*
f. at full width *(Text)* Gewebe *n* in Stückbreite
f.-backed *(Led)* mit Gewebe kaschiert
f. base plastic Gewebekunststoff *m*
f. bearing *(Masch)* Preßstofflager *n*
f. belt Faserstoffriemen *m*, Textilriemen *m*
f. bonding with foam Schaumstofflaminieren *n*
f. clippings Gewebeschnitzel *npl*
f. coating Kunstharzbeschichtung *f*, Kunstharzüberzug *m*
f. collector Staubfilter *n*, Tuchfilter *n*
f. covering Stoffbespannung *f*
f.-filled mit Gewebeeinlage
f. filler Gewebefüllstoff *m*
f.-foam laminates Schaumstofflaminate *npl*
f. insulation Gewebeisolierung *f*, Baumwollisolierung *f*
f. ply Gewebeeinlage *f*
f. printing Stoffdruck *m*, Textildruck *m*, Zeugdruck *m*
f. reinforced hose Schlauch *m* mit Gewebeeinlage
f. reinforced laminate Hartgewebe *n*
f. reinforcement Bewehrungsmatte *f*, Mattenbewehrung *f*; Streckstahlbewehrung *f (Stahl-, beton)*
fabricability Verarbeitungsfähigkeit *f*
fabricate / to 1. fertigen, formen; fügen; 2. vorfertigen *(Stahlbau)*; 3. *(Masch)* bearbeiten, [weiter]verarbeiten
fabricated cutter Messerkopf *m*
f. ship fabrikmäßig (in Sektionen) gebautes Schiff *n*
f. steel plate design geschweißte Stahlblechkonstruktion *f*
f. stern frame *(Schiff)* Walzblechhintersteven *m*
f. structural parts Hochbauelemente *npl*
fabricating method Fertigungsart *f*, Fabrikationsmethode *f*

fabrication 1. Fertigung *f*; Formung *f*; Fügung *f*; 2. Vorfertigung *f (Stahlbau)*, Werkstattvorfertigung *f*; 3. Bearbeitung *f*, Verarbeitung *f*; 4. Formkörper *m*; Erzeugnis *n*
fabricator *(Bw)* Stahlbaufirma *f*; Stahlbauer *m*
f. in plastics Kunststoffverarbeiter *m*
fabrics printing Stoffdruck *m*
fabroid gear Hartgewebezahnrad *n*
Fabry and Pérot interferometer Interferometer *n* nach Fabry-Pérot
façade Fassade *f*, Vorderansicht *f*, Vorderfront *f*
f. decoration Fassadenschmuck *m*
face / to 1. plandrehen, planfräsen, stirnen; 2. *(Bw)* verblenden, verkleiden, verputzen, verschalen; belegen; 3. schlichten; einpudern *(Gießerei)*
f. a mould *(Met)* eine Gießform mit Kohle bestäuben
f. down plandrehen
f.-grind planschleifen
f. to length ablängen
face 1. *(Math)* Fläche *f*, Oberfläche *f*; 2. Bahn *f (Hammer)*; 3. Stirnfläche *f*, Vorderfläche *f*; 4. Spanfläche *f (Werkzeug)*; 5. *(Bw)* Außenseite *f*, Sichtfläche *f*, Vorderseite *f*; 6. Oberseite *f*, rechte Seite *f*, Vorderseite *f (Gewebe)*; 7. *(Holz)* Decklage *f*; 8. *(Brg)* Abbaustoß *m*, Gewinnungsstoß *m*, Ortsbrust *f*, Ortsstoß *m*, Stoß *m (Tiefbau)*; Strosse *f (Tagebau)*; 9. Kopfflanke *f (Zahnrad)*; 10. Zifferblatt *n*; 11. *(Typ)* Schnitt *m (des Buchstaben)*, Schriftbild *n*
f. advance *(Brg)* Abbaufortschritt *m*, Verhiebsfortschritt *m*
f. angle 1. *(Math)* Winkel *m* an der Spitze; 2. Stirnwinkel *m (Verzahnung)*; Kegelwinkel *m*; 3. Spanwinkel *m (Räumwerkzeug)*
f. angle of blank Radkörperstirnwinkel *m*
f. angle radius Zahngrundrundung *f* zur Spanfläche *(Räumwerkzeug)*
f. apex Stirnkegelscheitel *m (Verzahnung)*
f. boards *(Brg)* Ortsbretter *npl*
f. boss *(Am) (Brg)* Brigadier *m*; Ortsältester *m*
f. cam *(Masch)* Plankurve *f*
f.-centered flächenzentriert *(Kristallgitter)*
f.-centered cubic kubisch flächenzentriert *(Kristalle)*
f.-centered cubic lattice kubisch flächenzentriertes Gitter *n*
f.-centered lattice flächenzentriertes Gitter *n*
f. collapse *(Brg)* Strebbruch *m*
f. concrete Sichtbeton *m*, Vorsatzbeton *m*
f. cone Kopfkegel *m (Zahnrad)*
f. contact ratio Überdeckung[sgrad *m*] *f*, Überlappung[sgrad *m*] *f (Verzahnung)*
f. conveying *(Brg)* Abbauförderung *f*
f. conveyor *(Brg)* Abbaufördermittel *n*
f. crew *(Brg)* Ortsbelegschaft *f*
f. cutting edges Stirnschneiden *fpl (Stirnfräser)*
f. diameter Schirmdurchmesser *m (Katodenstrahlröhre)*
f. driver Stirnmitnehmer *m (Stirnfräser)*
f.-entry *(Brg)* Abbaustrecke *f* senkrecht zu den Schlechten
f. finish *(Text)* Oberflächenappretur *f*
f. gear Kron[en]rad *n*, Planrad *n*, Zahnscheibe *f*
f.-gear guard Stirnradschutzkappe *f*
f. grinder Planschleifmaschine *f*
f. grinding Stirnschleifen *n*; Planschleifen *n*; Flächenschleifen *n*
f. grinding attachment Planschleifeinrichtung *f*
f. hardening Oberflächenhärtung *f*
f. joint *(Bw)* sichtbare Fuge *f*
f. lathe Plandrehmaschine *f*
f. length Zahnbreite *f*; Ballenlänge *f (Walze)*
f.-lift Modernisierung *f*, Renovierung *f*, Verjüngungskur *f (Fassadenerneuerung)*
f. man *(Brg)* Hauer *m*
f. mill Stirnfräser *m*
f. milling Flächenfräsen *n*, Stirnfräsen *n*, Stirnen *n*
f. milling and centring machine Abflächund Zentriermaschine *f*
f. milling cutter Stirnfräser *m*, Messerkopf *m*
f. milling cutter grinder Messerkopfschleifmaschine *f*
f. mix Sichtbeton *m*, Vorsatzbeton *m*
f. mould Schablone *f*
f.-mounting motor Flanschmotor *m (stirnseitige Befestigung)*
f. of a rim Kranzfläche *f (Rad)*
f. of drum Trommelstirnfläche *f (Turbine)*
f. of gateway *(Brg)* Ortsstoß *m*
f. of hammer Hammerbahn *f*
f. of joint Dichtungsleiste *f*
f. of packing rings Anlagefläche *f* der Dichtungsringe
f. of propeller blade *(Schiff)* Druckfläche *f* eines Propellerflügels
f. of the plane Hobelsohle *f*
f. of the tooth Spanfläche *f (Fräserzahn)*
f. of the weld Oberfläche *f* der Schweißnaht
f. of thread Gewindeflanke *f*
f. of tooth Stirnfläche *f*, Kopfseite *f*, Kopfflanke *f (Zahnrad)*
f. of work *(Brg)* Ort *n*
f.-on *(Am) (Brg)* Abbau *m* mit parallel zu den Schlechten gestellter Abbaufront
f. page *(Typ)* Vorderseite *f*
f.-parallel cut Orientierung II *f (Kristalle)*
f.-perpendicular cut Orientierung I *f (Kristalle)*

f. pick *(Text)* Oberschuß *m*
f. piece Endspreize *f*, Stirnspreize *f (Grabenverbau)*
f. pillar *(Brg)* Ortspfeiler *m*
f. plate 1. Frontplatte *f*; Stirnblech *n*; 2. Planscheibe *f (Drehmaschine)*; 3. Schirmträger *m (einer Katodenstrahlröhre)*
f. plate clamping slot Planscheibenspannschlitz *m*
f. plate controller Flachbahnsteuerschalter *m*
f. plate dog Planscheibenbacke *f*
f. plate jaw Planscheibenklaue *f*
f. plate of chuck Aufspannplatte *f*
f. plate starter *(Elm)* Flachbahnanlasser *m*
f. positioning *(Brg)* Strebstellung *f*
f.-ring connection *(Elm)* Stirnverbindung *f (Anker)*
f.-ring stray loss *(Elm)* Stirnstreuverlust *m (Anker)*
f. sharpening Brustschleifen *n*
f. shield Schutzmaske *f*, Gesichtsmaske *f*; Schweißerschutzschild *m*
f. shovel Hochlöffel *m (Bagger)*
f. side Vorderseite *f*
f. slips *(Brg)* gegen den Stoß einfallende Schichtflächen *fpl*
f. support *(Brg)* Strebausbau *m*, Ausbau *m* im Abbau
f.-type motor Flanschmotor *m (stirnseitige Befestigung)*
f.-type tailstock Reitstockplanscheibe *f*
f. view Vorderansicht *f*
f. waling Endspreize *f*, Stirnspreize *f (Grabenverbau)*
f. wall Frontwand *f*, Verblendmauer *f*
f. warp *(Text)* Oberkette *f*
f. weft *(Text)* Oberschuß *m*
f. weft thread Oberschußfaden *m*
f. wheel Kron[en]rad *n*, Tellerrad *n*
f. width 1. Zahnbreite *f*; 2. Kranzbreite *f*; 3. Breite *f* der oberen Auflagefläche *(Hobel)*
f. wire Obersieb *n (Papierherstellung)*
facebrick Vormauerziegel *m*, Verblendstein *m*
faced brickwork Verblendmauerwerk *n*
f. mould geglättete Form *f (Gießerei)*
f. with clinkers mit Klinker ausgelegt, mit Klinkerverkleidung
f. with lead mit Blei ausgeschlagen *(Säureschutz)*
f. with leather lederbezogen
faceplate Stirnblech *n*; Planscheibe *f*
facet Facette *f*; Begrenzungsfläche *f*, Seitenfläche *f*; Kristallfläche *f*
facework bündige Mauerseite *f*, Verblendmauerwerk *n*
facia Schriftfeld *n* über einem Schaufenster
f. board Sturzbrett *n* über einem Schaufenster; Firmenschild *n*
f. panel Armaturenbrett *n*
facial mask Gesichtsmaske *f*, *(kosmetische)* Maske *f*
facilitate / to erleichtern
facilities Einrichtungen *fpl*
facility Leichtigkeit *f*
facing 1. Planen *n*; Plandrehen *n*, Flächenfräsen *n*; 2. *(Bw)* Putzschicht *f*; Verkleidung *f*, Schutzschicht *f*, Sichtfläche *f*; Verblendung *f*; 3. Auflage *f*; Plattierschicht *f*, Deckschicht *f*; 4. Belag *m*
f. a mould Einpudern (Einstäuben) *n* einer Form
f. and boring head Plan- und Ausdrehkopf *m*
f. attachment Plandreheinrichtung *f*
f. board Blendholz *n*, Bekleidungsbrett *n*
f. bond Läuferverband *m* *(Mauerwerk)*
f. brick Verblender[ziegel] *m*, Vormauerziegel *m*
f. concrete Sichtbeton *m*, Vorsatzbeton *m*
f. cut Planschnitt *m*
f. cutter Stirnfräser *m*
f. head Planwerkzeugträger *m*; Plandrehsupport *m*
f. lathe Plandrehmaschine *f*, Kopfdrehmaschine *f*
f. material 1. Deckmaterial *n*; Belagmaterial *n*; 2. *(Met)* Schlichte *f (Glätten einer Gießform)*
f. of leather Lederbelag *m*; Belederung *f*
f. operation *(Masch)* Planarbeit *f*
f. page *(Typ)* Vorderseite *f*
f. plate Verkleidungsblech *n*
f. rail *(Bw)* Anlaufschiene *f*
f. sand feiner Formsand *m*, Modellsand *m (Gießerei)*
f. slab *(Bw)* Verkleidungsplatte *f*
f. slide Plandrehsupport *m*
f. stone Fassadenstein *m*; Naturstein *m* für Wandverkleidung; Verkleidungsplatte *f*
f. stop *(Masch)* Plananschlag *m*
f. tool Plandrehmeißel *m*; Plansenker *m*; Stirnsenker *m*
f. with leather Belederung *f*
facsimile Bildübertragung *f*, Bildtelegrafie *f*
f. broadcast station Bildsender *m*
f. posting machine *(Büro)* Postenumdrucker *m*
f. printing Faksimiledruck *m*
f. radio Bildfunk *m*
f. receiver Bildempfänger *m*
f. telegram Bildtelegramm *n*
f. telegraph Bildtelegraf *m*
f. telegraphy Bildtelegrafie *f*
f. transient distortion Ausschwingungsverzerrung *f (Bildübertragung)*

f. **transmission** Bildübertragung *f*; Faksimile-Übertragung *f* *(Übertragung von Zeitungsseiten mit UKW oder KW)*
factice Faktis *m*, geschwefeltes Öl *n*, Ölkautschuk *m*
factor / to *(Math)* in Faktoren zerlegen
f. **out** ausklammern
factor Faktor *m*, Koeffizient *m*, Einflußgröße *f*, Zahl *f*
f. **of assurance** Prüfspannungsverhältnis *n* *(Meßgeräte)*
f. **of correction** Korrekturfaktor *m*
f. **of ignorance** Unsicherheitsfaktor *m*
f. **of induction** Induktionskoeffizient *m*
f. **of merit** Gütefaktor *m* *(Stromkonstante eines Galvanometers bei 1 Ohm Widerstand und 10 s Schwingungsdauer)*
f. **of quality** *(El)* Kreisgüte *f* *(Parallelschwingkreis oder Spule)*
f. **of safety** Sicherheitsfaktor *m*, Sicherheitskoeffizient *m*
f. **of uncertainty** Unsicherheitsfaktor *m*
f. **of utilization** *(Bw)* Ausnutzungsgrad *m* *(umgekehrter Sicherheitsfaktor, d. h. Verhältnis der zulässigen Spannung zur Bruchfestigkeit)*
factorable *(Math)* in Faktoren zerlegbar
factorial *(Math)* Fakultät *f* *(Zeichen: !)* *(als nachgestellter Operator)*
factorial *(Math)* Faktoren-
f. **experiment** Faktorenversuch *m*
f. **magnification** Eigenvergrößerung *f*
factoring *s.* factorization
factorization *(Math)* Faktorenzerlegung *f*, Zerlegung *f* in Faktoren
factorize / to *(Math)* in Faktoren zerlegen
factory Betrieb *m*; Fabrik *f*; Werk *n*
f. **accident** Betriebsunfall *m*
f. **building** Fabrikgebäude *n*, Industriebau *m*
f. **canteen** Werkskantine *f*
f. **chimney** Fabrikschornstein *m*, Industrieschornstein *m*
f. **cost** Gestehungskosten *pl*, Herstellungskosten *pl*
f. **equipment** Betriebsausrüstung *f*
f.**-finished** fabrikfertig, vorgefertigt
f. **hall** Fabrikhalle *f*, Werkhalle *f*
f. **hand** Fabrikarbeiter *m*
f. **length** Herstellungslänge *f* *(z. B. Kabel)*
f.**-made** fabrikmäßig [hergestellt]
f. **medical centre** Betriebsambulanz *f*, Betriebspoliklinik *f*
f.**-new** fabrikneu
f. **overhead cost** allgemeine Betriebsunkosten *pl*
f. **plant** Fabrikanlage *f*
f. **precasting** Vorfertigung *f* im Betonwerk
f. **regulation** Arbeitsordnung *f* *(im Betrieb)*
f. **regulations** Arbeitsschutzbestimmungen *fpl*
f. **serial number** Werknummer *f*
f.**-set at** vom Werk eingestellt auf *(z. B. Ventil auf bestimmten Druck)*
f. **ship** Fabrikschiff *n* *(Fischerei)*
f. **standard** Werksstandard *m*, Werksnorm *f*
f.**-tested** im Werk geprüft
f. **trawler** Fabriktrawler *m*, Fischereifang- und Verarbeitungsschiff *n*
f. **worker** Fabrikarbeiter *m*
fade / to 1. verschießen, verblassen *(Farbe)*; 2. schwinden, Schwund haben *(Funkwellen)*
f. **away** abklingen *(Schall)*
f. **down** abblenden, ausblenden *(Fernsehen, Film, Ton)*
f. **in** 1. aufblenden *(Fernsehen, Film, Ton)*; 2. [nach Schwund] wiederkommen *(Funktechnik)*
f. **out** 1. abblenden, ausblenden *(Fernsehen, Film, Ton)*; 2. schwinden, Schwund haben *(Funkwellen)*
f. **out and in** überblenden *(Fernsehen, Film, Ton)*
f. **over** überblenden *(Fernsehen, Film, Ton)*
f. **up** *(Schwarz- bzw. Weißwert-)*Pegel anheben und halten *(Fernsehtechnik)*
f. **up and down** *(Fs)* überblenden
fade Schwund *m*, Fading *n* *(Funkwellen)* *(s. a.* fading*)*
f.**-in** 1. Aufblenden *n*, Aufblendung *f* *(Fernsehen, Film, Ton)*; 2. Wiederkommen *n* [nach Schwund] *(Funktechnik)*
f.**-out** 1. Ausblenden *n*, Ausblendung *f* *(Fernsehen, Film, Ton)*; 2. Totalschwund *m* *(Funkwellen)*
f.**-over** [weiche] Überblendung *f* *(Fernsehen, Film, Ton)*
f.**-reducing aerial** schwundmindernde Antenne *f*
fadeometer *(Text)* Echtheitsmesser *m*, Fadeometer *n*
fader Bildüberblender *m*, Tonüberblender *m*, Überblender *m*, Überblendungseinrichtung *f*
fading 1. *(El)* Fading *n*, Funkschwund *m*, Intensitätsabfall *m*, Intensitätsausfall *m*; 2. Überblendung *f* *(Fernsehen, Film, Ton)*; 3. Abnutzung *f*, Schwund *m* *(Kupplung, Bremsen)*
f. **area** Nahschwundzone *f*
f. **by absorption** Absorptionsschwund *m* *(Funktechnik)*
f. **by interference** Interferenzschwund *m* *(Funktechnik)*
f. **by polarization** Polarisationsschwund *m* *(Funktechnik)*
f. **control** 1. automatische Schwundregelung *f*; 2. Überblendungseinrichtung *f* *(Fernsehen, Film, Ton)*
f. **effect** Fadingeffekt *m*, Schwunderscheinung *f*

f. hexode Fadinghexode *f*, Viergitterregelröhre *f*
f. hollow Wellenschwundloch *n* *(Wellenausbreitung)*
f. outage Schwundausfall *m*, Ausfall *m* durch Fading
f. rate Schwundhäufigkeit *f*
f. reduction aerial schwundmindernde Antenne *f*
f. trouble Schwundstörung *f*
faecal . . . *s.* fecal . . .
faeces *s.* fecal matter
fag end aufgedrehtes Seilende *n*
fag[g]ot 1. *(Bw)* Faschine *f*; 2. *(Met)* paketierter Schrott *m*
f. steel Bundstahl *m*
f. wood Bundholz *n*, Brennreisig *n*
fag[g]ot[t]ing 1. *(Bw)* Faschinenbau *m*; Faschinenpackwerk *n*; 2. *(Met)* Paketieren *n* *(Schweißstahlerzeugung)*; 3. Paketierung *f*; Faschinenverbauung *f*
f. wire Faschinendraht *m*
fahlerz *(Min)* Fahlerz *n*
Fahrenheit scale Fahrenheit-Skala *f*
Fahrenheit thermometer Fahrenheit-Thermometer *n*
faience Fayence *f*; Steingut *n*
fail / to versagen; mißlingen; mißraten; fehlerhaft sein; abbrechen; sich unzulässig verformen; geschädigt werden
f. to pass nicht bestehen *(Abnahmeprüfung, Probe, Test)*, den Anforderungen nicht genügen
fail-safe schädigungssicher; bruchsicher
failing strain Bruchdehnung *f*
f. stress Bruchspannung *f*
failure 1. Schädigung *f*; Bruch *m*; plastische Verformung *f* *(Werkstoff)*; 2. Versager *m*; ungeeigneter Werkstoff *m*; 3. Versagen *n*; Störung *f*; Aussetzen *n*; Panne *f*
f. indication Störungsanzeige *f*
f. load Bruchlast *f*
f. of oscillation[s] Aussetzen *n* der Schwingungen
f. rate Ausfallrate *f*
f. recovery Fehlerbeseitigung *f*, Störungsbeseitigung *f*
f. strain Bruchdehnung *f*
f. stress Bruchspannung *f*
faint crimp *(Text)* schlichte Kräuselung *f*
f. headed *(Schiff)* leegierig
faintly acid schwach sauer
fair / to verkleiden *(z. B. stromlinienförmig)*
fair *(Bw)* bündig, glatt
fair Messe *f*
f. catalogue Messekatalog *m*
f. copy Reinschrift *f*
f. face bündige Mauerseite *f*, bündiges Mauerwerk *n*
f.-faced *(Bw)* bündig, glatt
f.-faced plaster Glattputz *m*
f.-lead *(Schiff)* Verholklampe *f*
f. ship Ausstellungsschiff *n*
Fairbairn boiler Fairbairn-Kessel *m*, Flammrohrheizröhrenkessel *m* mit mittlerer Flammkammer
fairing 1. *(Flg, Schiff)* Verkleidung *f*; Verkleiden *n*; Ausrunden *n*; 2. *(Schiff)* Ausstraken *n*, Ausrichten *n* *(von Bauteilen auf der Helling)*
fairline Straklinie *f* *(des Schiffskörpers)*
fairness *(Schiff)* Straken *n*, Geradheit *f*, Ausgerichtetsein *n*
fairway Fahrrinne *f*, Fahrwasser *n*
f. span Durchfahrtsöffnung *f*, Schiffsdurchfahrt *f* *(Brücke)*
fake Windung *f* eines aufgerollten Taus; Tauwindung *f*
falciform sichelförmig
fall / to 1. fallen; 2. [ab]fallen, sinken *(Spannung)*; 3. *(Text)* verfallen *(in der Beize)*; 4. absprengen; abdrücken
f. apart zerrieseln *(z. B. Schlacke)*
f. astern *(Schiff)* nach achtern absacken, abfallen
f. behind zurückbleiben
f. below a minimum of weniger betragen als; unterschreiten
f. down einstürzen, zusammenbrechen, zusammenfallen
f. in einfallen, einstürzen
f. off *(Flg)* abkippen, abrutschen
f. off to abfallen auf
f. out of synchronism asynchron werden, aus dem Synchronismus fallen
f. to pieces in Stücke zerfallen
f. to ruin verfallen *(Gebäude)*
fall 1. Sturz *m*, Fall *m*; 2. Abfall *m*, Sinken *n*, Absinken *n* *(z. B. der Spannung)*; 3. *(Brg)* Bruch *m* *(Streb)*; 4. *(Geol)* Fallen *n*, Einfallen *n*; 5. *(Lw)* Niederschlag[smenge *f*] *m*; 6. *(Bw)* Gefälle *n*, Neigung *f*
f.-away Trennung *f* *(Raketenstufen)*
f.-back circuit *(Am)* *(El)* Hilfskreis *m*; Reservekreis *m*
f. block Unterflasche *f* *(Flaschenzug)*
f. boom *s.* hatch boom
f. hammer Gewichtsbär *m*, Freifallramme *f*
f. in tension *(El)* Abnahme *f* der Spannung, Spannungsabfall *m*
f. of gain *(El)* Verstärkungsabfall *m*
f. of ground *(Brg)* Steinfall *m*
f. of potential Potentialabfall *m*, Spannungsabfall *m*
f. of pressure Druckabfall *m*
f. of roof *(Brg)* Steinfall *m*
f. of scaffold Stürzen (Fallen) *n* der Gicht *(Hochofen)*

f. **of temperature** Temperaturabfall *m*
f.-off Abfall *m*, Absinken *n (z. B. der Spannung)*
f.-out Fallout *m*, radioaktiver Niederschlag *m*
f.-out detector 1. *(Kern)* Falloutdetektor *m*; 2. *(Text)* Lochabsteller *m*
f.-out detector spring Kontaktfeder *f* des Lochabstellers
f.-out shelter Schutzraum *m* gegen radioaktiven Niederschlag
f. pendant Zugpendel *n (Leuchte)*
f. pipe Fallrohr *n*
f. plate *(Text)* Fallblech *n (Raschelmaschine)*
f. rime Abfalldauer *f*, Ansprechdauer *f (Relais)*
f. rope Hubseil *n*, Lastseil *n*
f. time 1. *(Ph)* Fallzeit *f*; Abfall[s]zeit *f*; 2. *(El)* Abstieg[s]zeit *f (Impulstechnik)*
f. wool *(Am)* im Herbst geschorene Wolle *f*
fallen hide *(Led)* Sterbling *m*
faller *(Text)* Gillstab *m*, Nadelstab *m*
f. box *(Led)* Broschierwürfel *m*, Walkwürfel *m*
f. stock *(Led)* Kurbelwalke *f*
f. wire *(Text)* Garnführer *m*
falling Fallen *n*, Stürzen *n*
f. ball test Kugelfallprobe *f*
f. ball viscosimeter Kugelfallviskosimeter *n*
f. characteristic [ab]fallende Kennlinie *f*
f. gradient Gefälle *n*
f.-off in power Leistungsabfall *m*
f. out of phase (step, synchronism) *(Elm)* Außertrittfallen *n*
f. solution *(Led)* Pickelbad *n*
f. speed 1. Fallgeschwindigkeit *f*; Abnahmegeschwindigkeit *f*; 2. abnehmende Geschwindigkeit *f*
f. track Ablaufgleis *n (Eisenbahnbau)*
f. weight Hammerbär *m*
f. weight test *(Met)* Fallprobe *f*, Schlagversuch *m*
fallow Brachland *n*, Brache *f*
f. crop Hackfrucht *f*
f. ground Brachland *n*, Brache *f*
f. land Brachland *n*, Brache *f*
f.-plough Brachpflug *m*
f. season Brachzeit *f*
f. soil Brachland *n*, Brache *f*
false falsch, nachgemacht
f. bearing *(Bw)* nicht satt aufliegend
f. bearing falsche Peilung *f*
f. bottom 1. Doppelboden *m*, Zwischenboden *m*; 2. *(Text)* Siebboden *m*
f. call *(Fmt)* Fehlanruf *m*
f. ceiling *(Bw)* Einschubdecke *f*; untergehängte Decke *f*
f. core Keilstück *n*, Kernstück *n (Formerei)*
f. cover *(Wärm)* Blenddeckel *m*, Blinddeckel *m*
f. floor *(Bw)* Blindboden *m*, Zwischendecke *f*
f. header halber Ziegel *m*
f. hip-roof Krüppelwalmdach *n*
f. jaw Zangeneinsatz *m*, Einsatzbacke *f*; Aufsatzspannbacke *f (für Drehfutter)*
f. keel loser Kiel *m*, Loskiel *m*, achteres Ballastkielholz *n (Jacht)*
f. lining Zwischenschacht *m (Hochofenfutter)*
f. pass blindes Kaliber *n (Walzwerk)*
f. post *(Schiff)* Außenachtersteven *m*, loser Hintersteven *m*
f. rib *(Flg)* Nebenrippe *f*
f. set 1. vorzeitiges Abbinden *n (Zement)*; 2. *(Brg)* Hilfstürstock *m*
f. spar *(Flg)* Zwischenholm *m*
f. start call *(Fmt)* Fehlanruf *m*
f. stem *(Schiff)* Außensteven *m*, Scheg[g] *n*
f. twill *(Text)* Scheinköper *m*
f. twist *(Text)* falscher (vorübergehender) Draht *m*
f.-twist process *(Text)* Falschdrahtverfahren *n*
f.-twist technique *(Text)* Falschdrahtverfahren *n*
f. twisting frame Falschzwirnmaschine *f*
f. twisting spindle Falschdrahtspindel *f*
falsework Schalung *f*, Schal[ungs]gerüst *n (Betonbau)*
f. structure Lehrgerüst *n*
falt boat Faltboot *n*, Klappboot *n*
famatinite *(Min)* Famatinit *m*
family *(Math)* Schar *f*; Büschel *n*
f. of characteristics Kennlinienfeld *n*, Kennlinienschar *f*
f. of curves Kurvenschar *f*
f. of plane surfaces Ebenenschar *f*
f. of surfaces Flächenschar *f*
fan / to [an]fächeln; anfachen, anblasen *(Feuer)*
f. out 1. sich fächerförmig ausbreiten *(z. B. Radiowellen)*; 2. *(El)* [aus]spreizen, wachsen *(Akkumulatorenplatten)*; 3. aufspleißen *(Kabel)*
fan 1. Gebläse *n*, Lüfter *m*, Ventilator *m*; 2. Amateur *m*; Bastler *m*; Liebhaber *m*
f. aerial Fächerantenne *f*
f. anchorage Besenverankerung *f*, Fächerverankerung *f (Spannbetonbau)*
f. beacon *(Schiff)* Fächerfunkfeuer *n*, Fächermarkierungsbake *f*
f. belt Lüfterkeilriemen *m*, Ventilatorriemen *m*
f.-belt idler (tensioner) Ventilatorriemenspannrolle *f*
f. blade Lüfterflügel *m*, Ventilatorflügel *m*, Ventilatorschaufel *f*
f. blast Ventilatorwind *m*
f. blower Ventilatorgebläse *n*
f. bracket Lüfterbock *m*
f. brake *(Flg)* Bremsluftschraube *f*

f. **casing** Ventilatorgehäuse *n*
f.-cooled luftgekühlt, gebläsegekühlt
f. cowl *(Kfz)* Lüfterhaube *f*
f. cut *(Brg)* Fächereinbruch *m*
f. damper Zugklappe *f (Ventilator)*
f. drift Lüfterkanal *m*, Saugkanal *m*, Abzugskanal *m*
f. drive shaft Lüfterwelle *f*
f. fixed shaft Lüfterachse *f*
f.-forced heater Heizlüfter *m*; Heißluftdusche *f*
f. guard *(Bw)* Auslegergerüst *n*, Auslegerrüstung *f*
f. hub Lüfternabe *f*
f. light Türoberlicht *n*; Fächerfenster *n*; Fensterrose *f*
f. marker *s.* 1. f. beacon; 2. f. marker aerial
f. marker aerial Fächerfunkfeuerantenne *f*
f. motor Lüftermotor *m*, Ventilatormotor *m*
f. oven Heißlufttrockenschrank *m*
f. propeller Ventilatorrad *n*
f.-shaped fächerförmig
f.-shaped aerial Fächerantenne *f*
f. shroud *(Kfz)* Lüfterhaube *f*
f.-tailed burner Rundstrahlbrenner *m*
f.-type air cooling Gebläseluftkühlung *f*
f. wheel anemograph Windradanemograf *m*
fancy cloth gemustertes Gewebe *n*, Bildgewebe *n*
f. cutting machine *(Text)* Effektschermaschine *f*
f. doubling frame Effektzwirnmaschine *f*
f. letters Zierschrift *f*, Kunstschrift *f*
f. lighting Effektbeleuchtung *f*
f. paper Ausstattungspapier *n*, Buntpapier *n*, Dekorationspapier *n*, Luxuspapier *n*, Maserpapier *n*, Phantasiepapier *n*
f. soap Toilettenseife *f*
f. thread Zierfaden *m*
f. twist Effektzwirn *m*, Zierzwirn *m*
f. twisting Effektzwirnen *n*
f. weaves *(Text)* Phantasiebindungen *fpl*
f. weaving Bildweberei *f*
f. yarn Effektgarn *n*, Ziergarn *n*
fanned-beam aerial Biberschwanzantenne *f*
fanning corn Schwingen *n (Getreide)*
f. of the steam jet Streuung (Zersplitterung) *f* des Dampfstrahls *(Turbine)*
f. strip Kamm *m (am Lötösenstreifen)*
fantail 1. Verbindungszug *m (Siemens-Martin-Ofen)*; 2. *(Schiff)* Hecküberhang *m*
far-end cross-talk *(Fmt)* Gegennebensprechen *n*
f. infrared fernes Infrarot *n*
f. infrared radiation ferne Infrarotstrahlung *f*
f.-off selectivity Weitabtrennschärfe *f (Radio)*
f. ultraviolet fernes Ultraviolett *n*
f. ultraviolet radiation ferne Ultraviolettstrahlung *f*
farad *(El)* Farad *n (Einheit der Kapazität)*
faraday Faraday-Konstante *f*, Faradaysche Zahl (Ladung, Äquivalentladung) *f*
Faraday cage Faradayscher Käfig *m*, Faraday-Käfig *m*
Faraday constant *s.* faraday
Faraday cylinder Faradayscher Becher *m*
Faraday dark space Faradayscher Dunkelraum *m (Gasentladung)*
Faraday ice pail Faradayscher Becher *m*
Faraday's law Induktionsgesetz *n*
Faraday's laws of electrolysis Faradaysche Gesetze *npl*
Faraday's space *s.* Faraday dark space
faradism Faradisation *f (Elektromedizin)*
faradization Faradisation *f (Elektromedizin)*
faradize / to faradisieren, elektrisieren *(Elektromedizin)*
faradmeter Kapazitätsmesser *m*
farm Bauernhof *m*; Gehöft *n*, Hof *m*
f.-building construction landwirtschaftliches Bauen *n*
f. buildings landwirtschaftliche Gebäude *npl*
f. labourer Feldarbeiter *m*, Landarbeiter *m*
f. land Acker *m*
f. machinery Landmaschinen *fpl*
f. mechanization Mechanisierung *f (eines landwirtschaftlichen Betriebes)*
f. tractor Ackerschlepper *m*
farmhand Feldarbeiter *m*, Landarbeiter *m*
farmhouse Bauernhaus *n*
farming Ackerbau *m*, Feldbau *m*
f. attachment Zusatzgerät *n*, Anhängegerät *n*
f. implement Anhängevorrichtung *f*
f. implements Ackergeräte *npl*, landwirtschaftliche Geräte *npl*
f. of sheep for wool Wollschafzucht *f*
f. roller Ackerwalze *f*, Feldwalze *f*
f. tool Werkzeug *n* für Bodenbearbeitung
f. tractor Traktor *m* für die Landwirtschaft, Ackerschlepper *m*
f. trailer Anhänger *m*
farmyard Bauernhof *m*, Wirtschaftshof *m*
f. work Innenwirtschaft *f*
Farnsworth image dissector *s.* Farnsworth tube
Farnsworth tube *(Fs)* Dissektorröhre *f* nach Farnsworth, Farnsworthscher Bildzerleger *m*, Farnsworth-Bildröhre *f*
farrier Hufschmied *m*
farrier's pincers Huf[beschlag]zange *f*
f. shoeing hammer Hufhammer *m*
fascia 1. Seitenansicht *f* eines Trägers oder Balkens; 2. Schriftfeld *n* über einem Schaufenster; 3. Instrumentenbrett *n*
f. board Gesimsbrett *n*, Simsbrett *n*, Traufbohle *f*, Traufbrett *n*
fascine Faschine *f*

f. barrier wall Faschinendamm *m*
f. work Faschinenbau *m*; Faschinenpackwerk *n*
fashion design *(Text)* Modegestaltung *f*; modisches Dessin *n*
f. piece *(Schiff)* Randsomholz *n*
f. plate stem *(Schiff)* Platten[vor]steven *m*
fashionable make-up modischer Zuschnitt *m*
fashioning Formgebung *f*
fast 1. schnell *(Verkehrsmittel)*; 2. fest[sitzend]; dauerhaft; 3. echt *(Farbe)*
f.-access storage *(Dat)* Schnellspeicher *m*
f.-acting schnellansprechend *(z. B. Relais)*
f.-action circuit breaker Schnellschalter *m*
f.-action switch Schnellschalter *m*
f. activation cross section Aktivierungsquerschnitt *m* für schnelle Neutronen
f.-and-loose pulleys Fest- und Losscheiben *fpl*, Last- und Leerscheiben *fpl*
f. boat Schnellboot *n*
f. breeder schneller Brutreaktor *m*
f.-charging current Schnelladestrom *m*
f. colour *(Text)* licht- und waschechte Farbe *f*
f. cosmic ray neutron schnelles Neutron *n* der kosmischen Strahlung
f.-curing moulding compound *(Plast)* Schnellpreßmasse *f*
f. detector *(Ph, Kern)* Detektor *m* mit hohem [zeitlichem] Auflösungsvermögen
f. driving 1. schnelles Fahren *n*; 2. beschleunigter Betrieb *m (z. B. eines Hochofens)*
f. dye Echtfarbstoff *m*
f. effect *s.* f. fission effect
f. fission *(Kern)* schnelle Spaltung *f*, Schnellspaltung *f*
f. fission effect Einfluß *m* der schnellen Neutronen; Schnellspaltungseffekt *m (Neutronenvervielfachung)*
f. forward Schnellvorlauf *m (Tonband)*
f. going clock (watch) vorgehende Uhr *f*
f. green Benzylgrün *n*, Neptungrün *n*
f. lens lichtstarkes Objektiv *n*
f. milling red Echtbordeaux *n*, Tuchrot *n*
f. mordant dye Echtbeizenfarbstoff *m*
f. motion 1. *(Masch)* Schnellauf *m*, Schnellgang *m*; 2. Zeitraffung *f (Film)*
f. multiplication factor Vermehrungsfaktor *m* für schnelle Neutronen
f. neutron fission Spaltung *f* durch schnelle Neutronen
f. neutron fission cross section Spalt[ungs]querschnitt *m* für schnelle Neutronen
f. neutron selector Geschwindigkeitsselektor (Monochromator) *m* für schnelle Neutronen
f. neutron shield Abschirmung *f* gegen schnelle Neutronen
f.-operating relay Schnellschaltrelais *n*
f. photolens lichtstarkes Fotoobjektiv *n*
f. plough *(Brg)* Schnellhobel *m*, Anbauhobel *m*, Löbbehobel *m*
f. pulley Festscheibe *f*, Lastscheibe *f*, feste Rolle *f*
f.-pulse amplifier Breitbandimpulsverstärker
f. reed *(Text)* fester Kamm *m*
f. reed loom Webstuhl *m* mit feststehendem Blatt
f.-release relay Schnelltrennrelais *n*
f.-response schnellansprechend *(z. B. Relais)*
f.-return / to im Eilgang zurücklaufen
f. return *(Masch)* Eilrücklauf *m*
f. rewind Schnellrücklauf *m (Tonband, Film)*
f. rod insertion *(Kern)* Schnelleinfahren *n* von Sicherheitsstäben
f. runner Schnelläufer *m*
f.-running schnellaufend
f. sailer Schnellsegler *m*
f. setting *(Plast)* schnellhärtend
f. sheet feststehender Fensterrahmen *m*
f. speed machine Schnelläufer *m*, schnellaufende Maschine *f*
f.-spiral engspiralig
f. steamer Schnelldampfer *m*, Eilschiff *n*
f. stuff *(Am)* röscher Stoff *m*
f. time gain control Verstärkungsregelung *f* mit geringer Zeitverzögerung *(besonders Automatik)*
f. to acid milling *(Text)* säurewalkecht
f. to bleaching bleichecht
f. to boiling soapsuds seifenkochecht
f. to brightening *(Text)* avivierecht
f. to kier-boiling *(Text)* beuchecht
f. to light lichtecht
f. to lye laugenecht
f. to perspiration schweißecht
f. to rubbing *(Text)* reibecht
f. to sea water meerwasserecht
f. to sizing *(Text)* schlichtecht
f. to solvents lösungsmittelbeständig
f. to stoving *(Text)* schwefelecht
f. to washing waschecht
f. to weather witterungsecht
f. to wet treatment *(Text)* naßecht
f. traverse *(Masch)* Eilgang *m*
f. yellow Echtgelb *n*
fasten / to anbringen, anheften, anstecken; befestigen, festbinden, festmachen; verspannen
f. with bolts verbolzen; verschrauben
f. with screws festschrauben
fastener 1. Befestigungsmittel *n*; 2. Verbindungsklemme *f*; Halter *m*; 3. Verschlußeinrichtung *f*; Riemenverbinder *m*
fastening 1. Befestigungsmittel *n*; 2. Befestigung *f*; Verspannung *f*; 3. Verblockung *f*; Verriegelung *f*; 4. Verbindungselement *n*; lösbare Verbindung *f*

f. device 1. Befestigungsvorrichtung *f*; 2. Verbindungselement *n*; lösbare Verbindung *f*
f. lug Befestigungslappen *m (Turbine)*
f. of nozzles Düsenbefestigung *f (Turbine)*
f. plate *(Masch)* Druckplatte *f*
f. ring Haltering *m*
f. screw Befestigungsschraube *f*
f. screw thread Befestigungsgewinde *n*
f. the blades by riveting rings Schaufelbefestigung *f* durch Nietung *(Turbine)*
fastness 1. Schnelligkeit *f*, Schnelle *f*; 2. Echtheit *f (Farben)*
f. demanded Echtheitsanforderung *f*
f. rating Echtheitsgrad *m*, Echtheitswert *m*
f. test Echtheitsprüfung *f*
f. to bleaching Bleichechtheit *f*
f. to chlorinated water Echtheit *f* gegen Schwimmbadwasser (gechlortes Wasser)
f. to cold washing Kaltwaschechtheit *f*
f. to cross-dyeing Überfärbeechtheit *f*
f. to dry cleaning Echtheit *f* gegen chemisches Reinigen
f. to light Lichtechtheit *f*
f. to rubbing Reibechtheit *f*
f. to swimming pool water Echtheit *f* gegen Schwimmbadwasser (gechlortes Wasser)
f. to washing Waschechtheit *f*
f. to wet pressing Naßbügelechtheit *f*
fat 1. fett[ig], ölhaltig, fetthaltig; 2. ergiebig, reich[haltig]
fat 1. Fett *n*, Schmierfett *n*; 2. klebriger Ton *m*
f. coal Fettkohle *f*, bituminöse Kohle *f*
f. dyes fettlösliche Farbstoffe *mpl*, Fettfarben *fpl*
f. extraction apparatus Entfettungsvorrichtung *f (Wolle)*
f.-faced fett *(Schrift)*
f. ignition spark satter Zündfunke *m*
f.-like fettartig
f. lime Fettkalk *m*, Weißkalk *m*
f. liquor / to fettlickern, degrassieren
f. liquor *(Led)* Fettlicker *m*
f. lute Kitt *m* auf Ölbasis
f. sand fetter Sand *m*
f. spark satter Zündfunke *m*
f. splitting *(Ch)* fettspaltend
f.-tailed sheep Fettschwanzschaf *n*
f. wrinkle *(Led)* Mastfalte *f*
fatal accident tödlicher Unfall *m*
fathom / to *(Schiff)* abfaden, nach Faden loten, ausloten
fathom *(Schiff)* Faden *m (Tiefenmaß: = 6 ft ~ 1,80 m)*
f. line Lotungslinie *f*, Tiefenlinie *f*
fathometer Tiefenmesser *m*, Echolot *n*
fatigue / to dynamisch beanspruchen; ermüden
fatigue 1. Ermüdung[serscheinung] *f*; 2. Dauerhaltbarkeit *f*
f. bend[ing] test Dauerbiegeversuch *m (Werkstoffprüfung)*
f. bend[ing] testing machine Dauerbiegemaschine *f*
f. crack Daueranriß *m*; Dauerbruch *m*
f. failure Ermüdungsbruch *m*, Dauerbruch *m*
f. fracture *s.* f. failure
f. impact test Dauerschlagprüfung *f*
f. life Lebensdauer *f*
f. limit Dauer[schwing]festigkeit *f*
f. limit test Dauerfestigkeitsversuch *m*
f. limit under reversed tension-compression stresses Zug-Druck-Wechselfestigkeit *f*
f.-loaded dauerschwingungsbeansprucht
f. loading Dauerschwingungsbeanspruchung *f*
f. notch factor Kerbwirkungszahl *f*
f. of materials Werkstoffermüdung *f*
f. point Dauerschwingfestigkeit *f*; Ermüdungsgrenze *f*
f. range Ermüdungsgrenze *f*
f. resisting ermüdungsbeständig
f. strength 1. *(indefinite number of cycles)* Dauerschwingfestigkeit; 2. *(stated number of cycles)* Zeitschwingfestigkeit *f*
f. strength for completely reversed stress Wechselfestigkeit *f*
f. strength reduction factor Kerbwirkungszahl *f*
f. strength under pulsating stresses Schwellfestigkeit *f*
f. strength under pulsating tensile stresses Zugschwellfestigkeit *f*
f. test Dauer[schwing]versuch *m*, Dauerprüfung *f*
f. test diagram Dauerfestigkeitsschaubild *n*, Grenzspannungsdiagramm *n*
f. testing Dauerschwingversuch *m*
f. testing machine Schwingprüfmaschine *f*
f. yield limit Dauerdehngrenze *f*
fatten / to mästen; ansetzen; fett werden
fattened cattle Mastvieh *n*
fattening 1. Mästung *f (Vieh)*; 2. *(Met)* Aufbauchung *f*
fatty fettartig, fetthaltig, fettig
f. acid Fettsäure *f*
f. alcohol Alkohol *m* der Fettreihe, Fettalkohol *m*
f. degeneration Verfettung *f (Vieh)*
f. oil Fettöl *n*, fettes Öl *n*
f. series *(Ch)* Fettreihe *f*
f. spew *(Led)* Fettausschlag *m*
faucet Hahn *m*, Absperrglied *n*, Absperrorgan *n*; Zapfen *m*; Spund *m*, Absperrhahn *m*; Wasserhahn *m*
f. plug Ansteckspund *m*
f. screw Zapfenschraube *f*

fault 1. Mangel *m*; Fehler *m*; Defekt *m*; 2. *(El)* Störung *f*; Störstelle *f*; Isolationsfehler *m*; 3. *(Geol)* Verwerfung *f*, Verwurf *m*, Sprung *m*, Wechsel *m*
f. clearance Störungsbeseitigung *f*
f. clearing service Störungsdienst *m*
f. current Erdschlußstrom *m*; Fehlerstrom *m*
f. finding Fehlersuche *f*
f. heave *(Geol)* Sprungweite *f*
f. in material Werkstoffehler *m*
f. in weaving Webfehler *m*
f. indication Fehleranzeige *f (mit Alarmeinrichtung)*
f. line 1. *(Geol)* Verwerfungslinie *f*; 2. *(El)* gestörte Leitung *f*
f. localization Fehlereingrenzung *f*, Störungseingrenzung *f (Kabel, Leitung)*
f. location Fehlereingrenzung *f*, Störungseingrenzung *f (Kabel, Leitung)*
f. on a line Leitungsfehler *m*, Störung *f* auf einer Leitung
f. plane *(Geol)* Verwerfungsebene *f*
f. report Störungsmeldung *f*, Störungsbericht *m*
f. resistance Fehlerwiderstand *m*
f. section Fehlerstrecke *f (Kabel)*; Störungsstelle *f (Fernsprechtechnik)*
f. selection *(Fmt)* verzerrte (fehlerhafte) Nummernwahl *f*
f. space *(Geol)* Verwurfzone *f*
f. throw *(Geol)* Sprunghöhe *f*
f. throwing Auslösung *f* durch künstlichen Fehler *(Relais)*
f.-trap Ölfalle *f* an einer Verwerfung *(Erdöl)*
faultage *(Geol)* Sprungbildung *f*
faulted *(Geol)* gestört, verworfen
f. overfold *(Geol)* Bruchfalte *f*
faulting *(Geol)* Sprungbildung *f*
faultless fehlerfrei
faultsman *(Fmt)* Störungssucher *m*
faulty 1. fehlerhaft, schadhaft; 2. gestört *(elektrische Leitung)*; 3. falsch
f. annealing Verglühung *f (falsches Glühen von Metall)*
f. branch *(Fmt)* gestörte (fehlerhafte) Abzweigung *f*, gestörter (fehlerhafter) Zweig *m*
f. card *(Büro)* fehlerhafte Karte *f*
f. chucking Fehlspannung *f*, Verspannung *f (von Werkstücken)*
f. circuit fehlerhafter (gestörter) Stromkreis *m*
f. control *(Masch)* Fehlschaltung *f*
f. design Fehlkonstruktion *f*; Konstruktionsfehler *m*
f. dimension *(Masch)* Maßfehler *m*
f. formation of the fibre *(Text)* mangelhafte Fadenbildung *f*
f. gripping Fehlspannung *f*, Verspannung *f (von Werkstücken)*
f. insulation schadhafte Isolation *f*
f. machining Bearbeitungsfehler *m*
f. measurement Fehlmessung *f*
f. pitch *(Masch, Fs)* Teilungsfehler *m*
f. section Fehlerstrecke *f (Fernsprechleitung)*
f. selection *(Fmt)* Fehlwahl *f*, Falschwahl *f*
f. soldered joint fehlerhafte (kalte) Lötstelle *f*
f. spot control Einstellfehler *m (des Strahlpunkts) (Katodenstrahlröhren, Elektronenmikroskopie)*
f. switching 1. Fehlschaltung *f*; 2. Fehlerbetrieb *m (des Schalters)*
f. wiring Verdrahtungsfehler *m*
Faure plate *s.* pasted plate
fayalite *(Min)* Fayalit *m*
faying surface Paßfläche *f*
feasibility [technische] Durchführbarkeit *f*
feasible ausführbar, [technisch] durchführbar
feather / to federn; sich federartig bewegen; mit Federn versehen
f. edge / to zuschärfen
feather 1. Feder[keil *m*] *f*; Paßfeder *f*; Gleitfeder *f*; Nase *f*; 2. künstliches Echo *n (Radar)*
f. edge scharfe (zugeschärfte) Kante *f*
f.-edge brick Keilziegel *m*
f.-edge file Schwertfeile *f*
f.-edged keilartig, scharfkantig
f.-edged board Büttenkarton *m*
f.-edged brick Keilziegel *m*
f.-edged coping schräge Mauerabdeckung *f*
f. head Tellerkopf *m (Schraube)*
f. joint tongue Federkeil *m (Tischlerei)*
f. key Gleitfeder *f*; Paßfeder *f*
f. plucking machines Federrupfmaschinen *fpl*
featherbed idler gefederte Rolle *f*
feathered aufgekeilt
f. pitch *(Flg)* Segelstellung *f*
f. shot copper feingranuliertes Kupfer *n*
feathering ausgelaufener Tintenstrich *m (auf ungeleimtem Papier)*
f. position *(Am)* Leerstellung *f (bei Schalterkontakten)*
f. propeller Verstellpropeller *m*
f. screw propeller *(Schiff)* Verstellpropeller *m*, Umkehrpropeller *m*
feathers *(Met)* Gußnaht *f*
featherstitch / to im Zickzack nähen, zickzack nähen
featherstitch *(Text)* Hexenstich *m*
featherweight paper Federleichtpapier *n*
feature / to darstellen, charakterisieren
feature Merkmal *n*, Hauptzug *m*, Charakteristik *f*
f. film Spielfilm *m*
f. of construction Konstruktionsmerkmal *n*
fecal matter Fäkalien *pl*, Kotstoffe *mpl*
feces *s.* fecal matter
fee Gebühr *f*

f.-television Fernsehsystem *n* mit Gebührenautomat
feeble schwach *(z. B. ankommende Signale)*
f. current cable Schwachstromkabel *n*
feed / to 1. zustellen; vorschieben; verschieben; Vorschub geben *(Werkzeug)*; 2. zuleiten; einfüllen; zuführen *(Flüssigkeit)*; 3. beschicken; 4. speisen *(Dampf)*; 5. *(Typ)* anlegen *(Papier)*; 6. *(El)* anlegen *(Spannung)*; 7. füttern; sich ernähren; grasen
f. a boiler einen Kessel speisen
f. a furnace einen Ofen beschicken
f. a voltage eine Spannung anlegen
f. by hand von Hand beistellen (vorschieben)
f. current Strom einspeisen
f. current to mit Strom speisen
f. in 1. einspeisen, zuführen; 2. einlegen *(Tonband)*
feed 1. Beschickung *f (Ofen)*; Speisung *f (Kessel)*; 2. Beschickungsgut *n*, Aufgabegut *n*; Materialzuführung *f*; 3. *(Masch)* Vorschub *m*; Vortrieb *m*; Zuführer *m*, Zuführeinrichtung *f*
f. adjustment Vorschubregelung *f*
f.-back *s.* feedback
f. bin Schüttrumpf *m*
f. block Speisestutzen *m (am Dampfkessel)*
f. board *(Typ)* Anlegetisch *m (Papier)*
f. box 1. *(Met)* Eingußkasten *m*, Aufgabekasten *m*; 2. *(Masch)* Vorschubgetriebekasten *m*
f. box of Norton type Norton-Vorschubgetriebe *n*
f. cable Speisekabel *n*
f. cam Vorschubnocken *m*
f. canal Zuleitungskanal *m*; Speisekanal *m*
f. change Vorschubwechsel *m*
f. change box Vorschubwechselräderkasten *m*
f. change gear Vorschubwechselrad *n*
f. change lever Vorschubwechselhebel *m*
f. check Speisewasserabsperrung *f (Dampfkessel)*
f. check valve Speiseventil *n (Dampfkessel)*
f. chute 1. *(Met)* Fülltrichter *m*; 2. Schüttrinne *f*, Schüttrutsche *f*; Förderrutsche *f*; Förderrinne *f*; 3. Leitschienenzuführung *f*
f. circuit Speise[strom]kreis *m*; Erregerkreis *m (Relais)*
f. clutch Vorschubkupplung *f*
f. cock Speisehahn *m*; Füllhahn *m (Dampfkessel)*
f. coil Speisedrossel *f*; Speisespule *f*
f. control Vorschubeinstellung *f*
f. control lever Vorschubhebel *m*
f. control unit *(Büro)* Auswahleinheit *f (Kartenmischer)*
f. controller Speisewasserbetätigungsvorrichtung *f (Dampfkessel)*
f. conveyor 1. Zubringer *m*, Beschicker *m*; 2. *(Lw)* Futterzuführungsapparat *m*
f. current Speisestrom *m*
f.-cut-off device Vorschubabschalteinrichtung *f*
f. de-aerator Speisewasserentlüfter *m*; Entgaser *m*, Entlüfter *m*
f. delivery pipe Speisedruckrohr *n*, Druckrohr *n* der Speisepumpe
f. depth gauge Bohrtiefenbegrenzung *f*
f. dial Vorschubskala *f*; Vorschubwähler *m*; Vorschubeinstellscheibe *f*
f. drive Vorschubantrieb *m*; Nebenantrieb *m*
f. end Beschickungsseite *f*, Einfahrseite *f (Ofen)*
f. engaging lever Vorschubeinrückhebel *m*
f. filter Speisewasserfilter *n (Dampfkessel)*
f. force Vorschubkraft *f*
f. gear Vorschubrad *n (Fräsmaschine)*
f. gear [box] Vorschubgetriebe *n*
f. gearbox Vorschubgetriebekasten *m*
f. handle Vorschubhandhebel *m*; Gefühlshebel *m*
f. heater Speisewasservorwärmer *m (Dampfkessel)*
f. heater with double tubes Speisewasservorwärmer *m* mit Doppelröhren
f. heater with exhaust in contact Mischungsvorwärmer *m*
f. heater with ribbed tubes Rippenröhrenvorwärmer *m*
f. heater with U-tubes Vorwärmer *m* mit gebogenen Röhren
f. heating plant Vorwärmeranlage *f*
f. hole Führungsloch *n*, Transportloch *n*, Vorschubloch *n*
f. hopper Aufgabetrichter *m*, Fülltrichter *m*, Beschickungstrichter *m*, Einfülltrichter *m*, Schütttrichter *m*
f. impedance *(El)* Begrenzungswiderstand *m*
f. injector Injektor *m*, Kesselspeisepumpe *f*
f. leg *(Brg)* Bohrhammerstütze *f*, Bohrknecht *m*
f. lever Vorschubhebel *m*
f. line 1. Zuleitung[srohr *n*] *f*; Speiseleitung *f*; Zulaufleitung *f*; 2. *(Rak)* Langrohrleitung *f*
f. magazine Zuführungsmagazin *n*
f. main Hauptleitung *f*
f. make-up *(Wärm)* Speisewassererzeuger *m*
f. marks Vorschubriefen *fpl*
f. material Ausgangsmaterial *n*; Beschickungsmaterial *n*
f. mechanism Vorschubgetriebe *n*
f. motion Vorschubbewegung *f*
f. motor Vorschubgetriebemotor *m*
f. movement Vorschubbewegung *f*
f. of frame Bildvorschub *m*, Bildschaltung *f (Film)*

f.-out of the honing tool Zustellung *f* der Honsteine
f. pawl Vorschubklinke *f*, Vorschubstift *m*
f. pin Vorschubstift *m*
f. pipe Speiserohr *n*, Zuleitung[srohr *n*] *f*
f. plate Aufgabeteller *m*, Streuteller *m*
f. point *(El)* Speisepunkt *m*
f. preheater Speisewasservorwärmer *m*
f. press Stanzautomat *m*
f. pressure Vorschubdruck *m*
f. pressure in sawing Arbeitsdruck *m* beim Sägen
f. pulper *(Lw)* Futtermuser *m*
f. pump 1. *(Wärm)* Speise[wasser]pumpe *f*; 2. *(Kfz)* Förderpumpe *f*, Kraftstoffpumpe *f*
f.-pump with flywheel *(Wärm)* Schwungradspeisepumpe *f*
f. rack Vorschubzahnstange *f*
f. ratchet wheel Vorschubklinkenrad *n*; Schaltrad *n*
f. reel 1. Abwickelspule *f* *(Film, Tonband)*; 2. *(BMSR, Büro)* Zuführungsrolle *f*, Zuführungsspule *f*
f.-regulating conveyor Dosierband *n*
f.-regulating valve *(Wärm)* Speiseregulierventil *n*
f. regulation 1. *(Masch)* Vorschubregulierung *f*; 2. Speisewasserregulierung *f* *(Dampfkessel)*; 3. Dosierung *f* des Aufgabegutes
f. regulator 1. *(Masch)* Vorschubregler *m*; 2. Speiseregler *m*, Verstellschieber *m* *(Dampfkessel)*
f. release Vorschubausschaltung *f*
f.-relief valve *(Wärm)* Sicherheitsventil *n* der Speisepumpe
f. reverse Vorschubumsteuerung *f*
f. reverse lever Vorschubumsteuerungshebel *m*
f. ribbon *(Text)* Zuführungsband *n*
f. rod Vorschubwelle *f*; Zugspindel *f* *(Drehmaschine)*
f. roll Zuführwalze *f*, Vorschubrolle *f*; Einzugsrolle *f* *(Draht)*
f. roller 1. Treibrolle *f*, Ständerrolle *f*; 2. *(Wärm)* Speisewalze *f*; 3. *(Text)* Einzugszylinder *m*; 4. Zuführrolle *f*, Transportrolle *f*
f. roller table Zufuhrrollgang *m* *(Walzwerk)*
f. screw 1. Förderschnecke *f*; 2. Zustellspindel *f*; Vorschub[gewinde]spindel *f*; 3. Zugspindel *f*
f. section *(Ch, Met)* Beschickungszone *f*
f. shredder *(Lw)* Futterreißer *m*
f. slide Vorschubschlitten *m*
f. spool Abwickelspule *f* *(Film, Tonband)*
f. sprocket Vorwickelrolle *f*, Vorwickeltrommel *f* *(Film)*
f. stock *(Ch, Met)* Ausgangsmaterial *n*, Vormaterial *n* *(für metallurgische Bearbeitung)*
f. suppression Vorschubunterbrechung *f*
f. tank 1. Speisewassertank *m* *(Dampfkessel)*; 2. Beschickungsbunker *m*
f.-through capacitor Durchführungskondensator *m*
f.-through ceramicon keramischer Durchführungskondensator *m*
f. train Vorschubgetriebe *n*
f. trip Vorschubauslösung *f*; Tiefenauslösung *f* *(Bohrer)*
f. trip dial Skala *f* für Einstellung der Tiefenauslösung
f. valve Speiseventil *n*; Vorschubventil *n*
f. water Speisewasser *n* *(Dampfkessel)*
f.-water analysis Speisewasseruntersuchung *f*
f.-water conditioning Speisewasseraufbereitung *f*
f.-water container Speisewasserbehälter *m*
f.-water control Speisewasserregelung *f*
f.-water de-aerating Speisewasserentlüftung *f*
f.-water generator Speisewassererzeuger *m*
f.-water heater Speisewasservorwärmer *m*, Anwärmer *m*
f.-water heating by steam bled at six points 6stufige Speisewasservorwärmung *f* durch Anzapfdampf
f.-water preparation Speisewasseraufbereitung *f*
f.-water preparation plant Speisewasseraufbereitungsanlage *f*
f.-water purification Speisewasserreinigung *f*
f.-water purifier Speisewasserreiniger *m*
f.-water purifying plant Speisewasseraufbereitungsanlage *f*
f.-water surface heater Speisewasseroberflächenvorwärmer *m*
f.-water tank Speisewassertank *m*, Speisewasserbehälter *m*
f. weighing Massedosierung *f*, Gewichtsdosierung *f*
f. wheel 1. *(Text)* Fournisseurrad *n*; 2. *(Masch)* Zustellhandrad *n*
f. winch Zubringerhaspel *f*
f. wire Zuleitungsdraht *m*, Zuleitungskabel *n*
f. worm shaft Vorschubschneckenwelle *f* *(Drehmaschine)*
f. zone *(Ch, Met)* Beschickungszone *f*
feedback / to rückkoppeln
feedback *(El)* Rückkopplung *f*, Rückwirkung *f*; *(BMSR)* Rückführung *f*, Rückkopplung *f*, Rückmeldung *f*
f. admittance Rückwirkungsleitwert *m*, Rückwärtsscheinleitwert *m* *(Anode-Gitter)*
f. amplifier rückgekoppelter Verstärker *m*
f. circuit Rückkopplungskreis *m*; Rückkopplungsschaltung *f*

f. coil Rückkopplungsspule *f*
f. control Rückkopplungsregelung *f*, selbsttätige Regelung *f*
f. control system *(BMSR)* Regelkreis *m*
f. element *(BMSR)* Rückführglied *n*
f. limiter Rückkopplungsbegrenzer *m*
f. loop Rückkopplungsschleife *f*, Rückkopplungsweg *m*
f. magnetron oscillator Rückkopplungsmagnetron *n*
f. oscillator Rückkopplungsgenerator *m*
f. path Rückkopplungsweg *m*
f. size control *(BMSR)* Meßsteuerung *f*
f.-stabilized amplifier rückgekoppelter Verstärker *m*
f. suppressor Rückkopplungssperre *f*
f. system *s.* f. control system
f. transfer resistance Koppelwiderstand *m*, Rückkopplungswiderstand *m*
f. transmitter rückgekoppelter Sender *m*
f. variable *(BMSR)* Rückführgröße *f*
feeder 1. *(Bw)* Aufgabevorrichtung *f*, Aufgeber *m*, Beschickungsvorrichtung *f*; 2. *(Masch)* Vorschubeinrichtung *f*, Werkstückzuführeinrichtung *f*; 3. *(BMSR)* Zubringer *m*; 4. *(El)* Speiseleitung *f*, Speisekabel *n*; Zuleitung *f*; 5. *(Met)* Einguß *m*, Gießaufsatz *m*; 6. Wassergraben *m*; Vorfluter *m*; 7. *(Geol)* Nebengang *m*; 8. *(Typ)* Bogenanleger *m*, Bogeneinleger *m*, Anleger *m*
f. air service *(Flg)* Zubringerdienst *m*, Zubringerluftverkehr *m*
f. bar Sammelspeiseschiene *f*
f. bus *s.* f. bar
f. cable Speisekabel *n*, Zuführungskabel *n*
f. car *(Brg)* Aufgabewagen *m*
f. casting Spritzguß *m*
f. circuit Speisestromkreis *m*
f. clamp Stromklemme *f*
f.-conveyor-elevator Stetigförderer *m*
f. die Werkzeug *n* für stufenweises Formen eines Werkstückes
f. distribution centre Speiseleitungsverteiler *m*
f. head Gießkopf *m*, verlorener Kopf *m (Gießerei)*
f. hopper *(Text)* Kastenspeiser *m*
f. junction point *(El)* Speisepunkt *m*
f. line Speiseleitung *f*, Zuführungsleitung *f*
f. protection Speiseleitungsschutz *m*
f. reactor *(El)* Speisedrossel *f*
f. road Zubringerstraße *f*
f. scale *(Ch, Met)* Beschickungswaage *f*
f. skip hoist Beschickungsaufzug *m*
f. switch Speise[leitungs]schalter *m*
f. transformer Speisetransformator *m*
f. valve Speiseventil *n*
f. voltmeter Leitungsvoltmeter *n*

feedforward *(BMSR)* Aufschaltung *f*, positive Rückkopplung *f*
f. control Störgrößenaufschaltung *f*; Steuerung *f*
f. transfer function Übertragungsfunktion *f* des Vorwärtszweiges
feeding 1. Speisung *f*, Zuführung *f*, Zuleitung *f*; 2. Werkstückzuführung *f*; Vorschub *m*; 3. Nachspeisung *f (Gießen)*
f. apron *(Text)* Einführtuch *n*, Zuführtuch *n*
f. attachment *(BMSR)* Zuführungseinrichtung *f*
f. barley Futtergerste *f*
f. beet Futterrübe *f*
f. by bridges *(El)* Brückenspeisung *f*
f. cable Speisekabel *n*
f. chute Zuführschrägrinne *f*
f. circuit *(El)* Speisebrücke *f*
f. collet Vorschubpatrone *f*, Vorschubzange *f*
f. device 1. Vorschubeinrichtung *f*, Vorschubmechanismus *m*; 2. Eingebeeinrichtung *f (Werkstücke)*; 3. Füllvorrichtung *f*, Speisevorrichtung *f*, Zuführungsvorrichtung *f*
f. elevator Schrägförderer *m*, Aufgabebecherwerk *n*; Aufgabeelevator *m*
f. finger Vorschubzange *f*
f. gate *(Met)* Gußtrichter *m*
f. gear Nachschubvorrichtung *f (z. B. Bogenlampe)*
f. head *(Met)* verlorener Kopf *m (am Gußstück)*, Aufsatzkopf *m*
f. herbs Futterkräuter *npl*
f. hopper 1. *(Text)* Speisekasten *m*; 2. *(Bw)* Füllrumpf *m*, Fülltrichter *m*
f. line Speiseleitung *f*, Speiserohr *n*
f. mark *(Typ)* Anlegemarke *f*, Paßmarke *f*
f. mechanism Vorschubgetriebe *n*
f. of cut wood by rollers Walzenzuführung *f* geschnittener Hölzer
f. plant Futterpflanze *f*
f. point *(El)* Speisepunkt *m*
f. potatoes Futterkartoffeln *fpl*
f. rate Zustellbetrag *m*
f. roller 1. *(Met)* Speisewalze *f*, Einziehwalze *f*, Zuführwalze *f*; 2. *(Text)* Lieferzylinder *m*, Lieferwalze *f*
f. roller table Zuführrollgang *m*
f. rye Futterroggen *m*
f. side of harvester Einlaufseite *f* der Erntemaschine
f. trough Beschickungsmulde *f*, Schüttgosse *f (Rostfeuerung)*
f. wheel to work Schleifscheibenzustellung *f*
feedwater . . . *s.* feed-water . . .
feel / to 1. fühlen; tasten; 2. *(Brg)* abklopfen, untersuchen
feel *(Text)* Griff *m*, Griffigkeit *f*
feeler Taster *m*, Taststift *m*, Meßdorn *m*

f. **gauge** Blechlehre *f*; Dicktenlehre *f*, Dicktenschablone *f*; Fühllehre *f*, Spion *m*
f. **pin** Taster *m*, Taststift *m*
f. **pin of the dial gauge** Meßstift *m*; Taststift *m* der Meßuhr
f. **switch** Prüfschalter *m*
f. **wire** Tasterdraht *m*
feet head of liquid Höhe *f* der Flüssigkeitssäule in Fuß
Fehling's solution *(Ch)* Fehlingsche Lösung *f*
feint attack *(Mil)* Scheinangriff *m*
f. **plug** *(El)* Leerstecker *m*
fel-bovinum *(Typ)* Ochsengalle *f*
feldspar Feldspat *m*
feldspathic feldspatartig, feldspathaltig
f. **stoneware** Halbporzellan *n*
feldspathoid *(Min)* Feldspatvertreter *m*
fell / to fällen *(Holz)*
fell 1. Fell *n*; Balg *m*; 2. gefällte Holzmenge *f*
f. **of the cloth** Warenrand *m (im Webstuhl)*
felling *(Holz)* Fällen *n*, Einschlag *m*
f. **axe** Baumaxt *f*, Baumhacke *f*
f. **rotation** Hiebfolge *f*, Hiebumlauf *m*
fellings of wood Holzeinschlag *m*
fellmonger Fellzubereiter *m*
felloe Radfelge *f*
f. **bending machine** Radfelgenbiegemaschine *f*
f. **rounder** Felgenabrunder *m*
f. **wood** Felgenholz *n*
fellow Facharbeiter *m*, Geselle *m*
f. **inventor** Miterfinder *m*
f. **station** *(El)* Gegenstation *f*; Gegenstelle *f*
f. **worker** Arbeitskollege *m*; Mitarbeiter *m*
felly *s.* felloe
felsitic *(Geol)* felsitisch
felspar *s.* feldspar
felspathic *s.* feldspathic
felt / to verfilzen
felt 1. Filz *m*; 2. Baupappe *f*
f. **board** Filzpappe *f*
f. **calender** Filzkalander *m*
f. **conditioner** Filzinstandhalter *m*, Filzreiniger *m*
f. **disk** Filzscheibe *f*
f. **filter** Filzfilter *n*
f. **finish** Filzprägung *f*
f. **guide** Filzleitwalze *f*
f. **hob** Filzpolierscheibe *f*
f. **layer** Filz[unter]lage *f*
f. **napping machine** Filzrauhmaschine *f*
f. **packing ring** Filzdichtungsring *m*
f. **pad** 1. Filzunterlage *f*; 2. Filzröllchen *n* *(Tonbandgerät)*
f. **pad lubricator** Filzöler *m*
f. **polishing wheel** Polierfilzscheibe *f*
f. **roofing** Pappdach *n*
f. **rubbing pad** Filzscheibe *f* zum Glätten
f. **seal** Filzdichtung *f*
f. **seam mark** Führungsstreifen *m*, Laufstreifen *m*, Richtstreifen *m*
f. **side** Filzseite *f (Papier)*
f. **strip** Filzstreifen *m*
f. **suction roll** Filzsauger *m*, Filzsaugwalze *f* *(Papierherstellung)*
f. **washer** Filz[unterleg]scheibe *f*
f.**-wheel polishing** Filzen *n*
felter 1. Filzer *m*, Filzmaschine *f*, Walker *m* *(Hutherstellung)*; 2. *(Holz)* Verfilzer *m*, Verfilzungsmaschine *f*
felting Filzbildung *f*, Verfilzen *n*
f. **machine** Walkmaschine *f (Hutherstellung)*
f. **roller** Walkholz *n (Hutherstellung)*
female centre Hohlkörnerspitze *f*
f. **cone** Innenkonus *m*
f. **coupling** Innengewindekupplung *f*
f. **die** Matrize *f*, Gesenk *n*
f. **die part** Matrize *f*, Unterwerkzeug *n*, Schnittplatte *f*
f. **form** Hohlform *f*
f. **form block** Hohlgesenk *n*, Matrize *f*
f. **gauge** Lehrring *m*, Kaliberring *m*
f. **half** Gegenstück *n (hohl)*
f. **part** Außenteil *n*; aufnehmendes Teil *n*
f. **piece** aufnehmendes Teil *n*, Gegenstück *n*; Matrize *f*
f. **plug** Mutterstecker *m*
f. **screw** Hohlschraube *f*; Schraubenmutter *f*
f. **screw thread** Innengewinde *n*, Muttergewinde *n*
f. **spline** Keilnut *f*
f. **spline gauge** Keilbahnlehrring *m*
f. **taper** Innenkegel *m*
f. **taper gauge** Lehrhülse *f*
f. **thread** Innengewinde *n*, Muttergewinde *n*
f. **type of piston rod** hohle Kolbenstange *f*
femic *(Min)* femisch
femic femisches Mineral *n*; Magnesium-Eisen-Silikat *n*
femmer schlank, dünn, schwach
fence 1. Zaun *m*, Gitter *n*, Schutzgitter *n*; 2. *(Masch)* Anschlagbacke *f*, Führungsbacke *f*, Führungslineal *n*
f. **wire** Maschendraht *m*, Gitterdraht *m*
f. **wire stretcher** Zaundrahtspanner *m*
fencing Einzäunung *f*, Umfriedung *f*
fend off / to *(Schiff)* abfendern
fender 1. Fender *m*, Prellpfahl *m*, Schutzschwelle *f*, Schwelle *f*; 2. *(Kfz)* Kotblech *n*, Schutzblech *n*, Stoßfänger *m*; 3. *(Fl)* Schutzkappe *f*
f. **beam** Prellbalken *m*, Prellbock *m*
f. **guard** *(Schiff)* Scheuerleiste *f*
f. **pile (pole)** Prellpfahl *m*, Reibepfahl *m*, Streichpfahl *m*
f. **post** *(Bw)* Radabweiser *m*
f. **stock** Stoßstange *f*

fenestra *(Bw)* Öffnung *f* in der Außenwand
fenestration 1. Fensteranordnung *f*, Fensteraufteilung *f*; 2. Einrichtung *f* für [regelbare] Zuführung von Tageslicht
fennel oil Fenchelöl *n*
fent Stoffreststück *n*
ferberite *(Min)* Ferberit *m*
fergusonite *(Min)* Fergusonit *m*
Fermat's last theorem *(Math)* großer Fermatscher Satz *m*
Fermat's theorem *(Math)* kleiner Fermatscher Satz *m*
ferment / to fermentieren; gären
ferment Ferment *n*; Gärstoff *m*, Gärungserreger *m*
fermentability Fermentationsfähigkeit *f*; Gär[ungs]fähigkeit *f*, Vergärbarkeit *f*
fermentable fermentierbar; gärbar, gärfähig
fermentation Fermentation *f*; Gärung *f*, Gären *n*, Gärverfahren *n*
f. flask Gärkolben *m*
f. in open tubs offene Bottichgärung *f*
f. intermediate Gärungszwischenprodukt *n*
f. method Faulverfahren *n*
f. putrefaction Gärfaulverfahren *n*
f. room Gärkeller *m*, Gärraum *m*
fermentative fermentativ; gärungserregend
f. test Gärkraftbestimmung *f*
fermented fermentiert
fermenter Gärbottich *m*, Gärkessel *m*
fermenting in the pile Fermentieren *n* im Stapel
f. room *s.* fermentation room
f. trough Faulbütte *f (Papierherstellung)*
fermentive fermentativ; gärungserzeugend
Fermi age *(Kern)* Fermi-Alter *n*
Fermi analysis *(Kern)* Fermi-Analyse *f*
Fermi characteristic-energy level Fermi-Kante *f (Energiespektrum)*
Fermi constant *(Kern)* Fermi-Konstante *f*
Fermi-Dirac-Sommerfeld distribution law Fermi-Dirac-Sommerfeld-Geschwindigkeitsverteilungsgesetz *n*, FDS-Gesetz *n*
Fermi potential *(Kern)* Fermi-Potential *n*
Fermi spectrum *(Kern)* Fermi-Spektrum *n*
Fermi statistics *(Kern)* Fermi-Statistik *f*
Fermi temperature *(Kern)* Fermi-Temperatur *f*
Fermi theory *(Kern)* Fermische Theorie *f* des Beta-Zerfalls
fermion *(Ph)* Fermion *n*, Fermi-Teilchen *n*, Teilchen *n* mit ungeradzahligem Spin
fermium Fermium *n*
fern Farn[kraut *n*] *m*
Ferraris motor Ferraris-Motor *m (Zweiphaseninduktionsmotor)*
Ferraris wattmeter Drehfeldleistungsmesser *m*, Ferraris-Instrument *n*
ferrate Ferrat *n*
ferrated tartar Eisenkaliumtartrat *n*
ferreous *s.* ferrous
ferri-electric ferrielektrisch
ferric Eisen(III)-, Ferri-
f. chloride Eisen(III)-chlorid *n*
f. compound Verbindung *f* des dreiwertigen Eisens
f. ferrocyanide Ferriferrozyanid *n*, Berliner Blau *n*
f. induction Magnetisierung *f*, magnetische Induktion *f*
f. oxide Eisen(III)-oxid *n*
ferrichloric acid Ferrichlorwasserstoff *m*
ferricyanic compound Ferrizyanverbindung *f*
ferricyanide Ferrizyanid *n*
f. of iron Eisen(II)-ferrizyanid *n*; Eisen(III)-ferrizyanid *n*
f. of potassium Kaliumferrizyanid *n*, rotes Blutlaugensalz *n*
ferriferous eisenhaltig; eisenschüssig
f. manganese ore Eisenmanganerz *n*
ferrite Ferrit *m (Alphaeisen in fester Lösung)*
f. aerial Ferritantenne *f*
f. brittleness Blausprödigkeit *f (Stahl)*
f.-carbide aggregate Ferrit-Zementit-Gemenge *n*
f.-carbide aggregate structure Ferrit-Zementit-Mischgefüge *n*
f. core Ferritkern *m (Datenspeicherung, Schaltelement)*
f. duplexer Ferritduplexer *m*, Ferritumschalter *m (z. B. Radar)*
f. envelope *(Met)* Ferrithof *m*
f. grain *(Met)* Eisenkorn *n*
f. isolator Ferritrichtungsisolator *m*, Ferritrichtungsleitung *f (Mikrowellen)*
f. lattice *s.* f. network
f. memory Ferrit[kern]speicher *m*
f. network Ferritnetzwerk *n (Metallografie)*
f. rotator Ferritdreher *m (Mikrowellen)*
f. steel ferritischer Stahl *m*
ferritic ferritisch
f. ductile iron ferritisches Kugelgraphitgußeisen *n*
ferritize *(Met)* Glühen *n (bis zur Bildung von ferritischem Gefüge)*
ferro alloy Ferrolegierung *f*
f.-aluminium Ferroaluminium *n*, Aluminiumeisen *n*
f.-cement Armozement *m*, Ferrozement *m*
f.-chromium Ferrochrom *n*
f.-coke Ferrokoks *m*
f.-compound Verbindung *f* des zweiwertigen Eisens
f.-concrete *s.* reinforced concrete
f.-manganese Ferromangan *n*, Manganeisen *n*
f.-molybdenum Ferromolybdän *n*

f.-prussiate paper Blaupauspapier *n*
f.-silicon Ferrosilizium *n*
f.-titanium Ferrotitan *n*
f.-tungsten Ferrowolfram *n*
f.-uranium Ferrouran *n*
f.-vanadium Ferrovanadin *n*
ferrocarbide brick Ferrokarbidstein *m*
ferroclad brick blechummantelter Stein *m*
ferrod radiator system Ferritstabstrahlungssystem *n*
ferrodynamic instrument ferrodynamisches Meßinstrument *n*
ferroelectric ferroelektrisch
ferroelectric ferroelektrische Substanz *f*, Ferroelektrikum *n*
ferroginous eisenhaltig
ferromagnetic ferromagnetische Substanz *f*, Ferromagnetikum *n*
ferromagnetic ferromagnetisch
f. circuit Eisenkreis *m*, ferromagnetischer Kreis *m*
f. powder Magnetpulver *n*
ferromagnetics Ferromagnetik *f*
ferromagnetism Ferromagnetismus *m*
ferropentacarbonyl Eisenpentakarbonyl *n*
ferroresonant switching circuits ferroresonante Schaltkreise *mpl*
ferroso-ferric oxide Eisen(II, III)-oxid *n*
ferrostatical pressure ferrostatischer Druck *m*
ferrous eisenhaltig; Eisen(II)-, Ferro-
f. compound Verbindung *f* des zweiwertigen Eisens
f. metallurgy Eisenhüttenkunde *f*
f. oxide Ferrooxid *n*, Eisen(II)-oxid *n*
f. scrap Eisenschrott *m*
ferrugineous *s.* ferruginous
ferruginous 1. rostfarben; 2. eisenhaltig, mit Eisenzusätzen; eisenschüssig
f. calamine Eisenzinkspat *m*
f. earth Eisenerde *f*
ferrule / to mit Zwinge versehen
ferrule Eisenband *n*, Zwinge *f*; Endring *m*, Ring *m*
f. terminal *(El)* Anschlußhülse *f*
f. with bolster Wulstzwinge *f*
ferruled mit Zwinge *(versehen)*
ferry / to übersetzen; überqueren *(Strom)*; überführen
ferry Fähre *f*, Fährschiff *n*, Fährboot *n*
f. boat *s.* ferry
f. bridge 1. *(Schiff)* Fähranleger *m*, Anlegeponton *m*; 2. Trajekt *m*, fliegende Brücke *f*
f. punt Fährprahm *m*
f. rack Fährbett *n*
f. rope Fährseil *n*
f. service Fährdienst *m*
f. terminal Fährhafen *m*
ferrying flight *(Flg)* Überführungsflug *m*
fertile 1. *(Kern)* brutfähig, in Spaltstoff umwandelbar; 2. fruchtbar; fett *(Boden)*
f. material Brutstoff *m (umwandlungsfähiges Kernmaterial)*
fertility *(Kern)* Brütbarkeit *f*, Brutfähigkeit *f*
fertilization *(Lw)* Fruchtbarmachung *f*
fertilizer Düngemittel *n*, Dünger *m*, Mineraldünger *m*
f. broadcaster Düngerstreuer *m*
f. drill-lever Düngerdrillhebel *m*
f. dumping cart Düngerkippwagen *m*
f. grade ammonium nitrate Sprengstoff *m (selbsthergestellt aus Ammonsalpeter in Düngergüte)*
f. requirements Düngerbedarf *m*
f. spreader Düngerstreumaschine *f*
f. value Dungwert *m (von Abfällen)*
fertilizing Düngen *n*
f. capacity Düngevermögen *n*
f. drill Düngelanze *f*
f. lime Düngekalk *m*
ferul aldehyde Koniferylaldehyd *m*
festoon *(Bw)* Girlande *f*
f. bulb Soffittenlampe *f*
f. cap *(El)* Hängesockel *m*
f. drier *(Ch)* Girlandentrockner *m*
f. lamp Soffittenlampe *f*, Röhrenlampe *f*, Linienlampe *f*
f. lighting Soffittenbeleuchtung *f*
fettle / to *(Met)* ausbessern *(Ofenausmauerung)*; putzen *(Guß)*; säubern
fettling 1. Gußputzen *n*; 2. Ausbessern *n*, Nacharbeiten *n*
f. cost Ausbesserungskosten *pl (Ofen)*
f. material Flickmasse *f (Ofenausbesserung)*
f. room Putzerei *f (für Gußstücke)*
fiber *(Am) s.* fibre
fiberize / to *(Am)* zerfasern *(Holz)*
fibre Faser *f*, Fiber *f*
f. [building] board Faserplatte *f*, Leichtbauplatte *f*
f. core Hanfseele *f (eines Drahtseils)*
f. flash-drying of pulp Blitztrocknung *f* von Zellstoff
f. flax Faserlein *m*
f. fleece chain-stitch machine Faservliesnähwirkmaschine *f*
f. flow Faserverlauf *m (Metallografie)*
f. gasket Fiberdichtung *f*
f. glass Fiberglas *n*; Glas[faser]gespinst *n*
f. glass reinforced laminates Glasfaserschichtstoffe *mpl*
f. glass reinforced plastics Glasfaserkunststoffe *mpl*
f. glass yarn Garn *n* aus Glasfaser
f. grease fadenziehendes Fett *n*
f. matrix Fasergefüge *n (Holz)*
f. metallurgie Fasermetallurgie *f*

f. of flax Flachs *m*; Flachsfaser *f*
f. optics Faseroptik *f*
f. paper Vulkanfiber *f*
f. pattern Fasermuster *n*, Faserstruktur *f*
f. pulley Faserstoffriemenscheibe *f*, Preßholzriemenscheibe *f*
f. recovery Faserrückgewinnung *f*
f. saturation point Fasersättigungspunkt *m*
f. sheet Vulkanfiberplatte *f*
f. slab Faserplatte *f*, Leichtbauplatte *f*
f. stress Faserspannung *f*, Drehspannung *f*
f. structure Faserstruktur *f*; Walzstruktur *f*
f. suspension Fadenaufhängung *f* *(physikalischer Geräte)*
fibriform faserförmig, fas[e]rig
fibrillation *(Opt)* Faserung *f*, Faserbildung *f*
fibring Faserung *f*
fibrinous faserstoffartig; fibrinhaltig
fibroin Fibroin *n*
fibrolite *(Min)* Fibrolith *m*, Sillimanit *m*
fibrous fas[e]rig
f. and metallic packing *(Masch)* armierte Weichpackung *f*
f. artificial leather Faserkunstleder *n*
f. blende *(Min)* Strahlenblende *f*
f. cassiterite Holzzinnerz *n*
f. coal Faserkohle *f*
f. crystal *(Min)* Trichit *m*
f.-felted board faserverfilzte Platte *f*
f. fracture fas[e]riger Bruch *m*, fas[e]rige Bruchfläche *f*
f. glass Glasfasern *fpl*; Fiberglas *n*
f. insulation Faser[stoff]isolierung *f*
f. limestone Faserkalk *m*
f. structure fas[e]rige Struktur *f*, Faserstruktur *f*
f. suspension Fadenaufhängung *f* *(physikalischer Geräte)*
fibrousness fas[e]rige Beschaffenheit *f* *(eines Stoffes)*
fictile tönern, irden
f. art Kunsttöpferei *f*
fictitious fiktiv, Schein-
f. bar Ersatzstab *m* *(Statik)*
f. boundary scheinbare Grenze *f*, fiktive Begrenzung *f*
f. force Scheinkraft *f*
fiddle *(Schiff)* Schlingerleiste *f* *(z. B. am Tisch)*
f. bow *(Schiff)* Klipperbug *m*
f. bow drill Fiedelbohrer *m*
fiddley *(Schiff)* Schornsteinumbau *m*; Kesselschacht *m*
f. hatch *(Schiff)* Schornsteinluke *f*
fidelity Wiedergabegüte *f*, Wiedergabetreue *f* *(Elektroakustik)*
f. curve Wiedergabekurve *f*
fiducial Bezugs-; Vergleichs-
f. limit Grenzwahrscheinlichkeit *f*
f. mark Bildrahmenmarke *f* *(Fotogrammetrie)*
field 1. *(Lw)* Feld *n*; Acker *m*; 2. *(Ph, El)* Feld; Prüffeld *n*; 3. Feld; Körper *m*; Bereich *m*, Gebiet *n*; 4. *(Bw)* Baustelle *f*; Gelände *n*; Baustrecke *f*, Strecke *f* *(Leitungsbau)*; 5. *(Dat)* Gruppe *f* *(Information)*; Lochkartenfeld *n*; 6. *(Fs)* Raster[bild *n*] *m*, Teilbild *n* *(Zeilensprungverfahren)*; 7. *(Brg)* Revier *n*; 8. Fachgebiet *n*
f. ammeter Feldamperemeter *n*, Streckenamperemeter *n*
f.-astronomical observation astronomische Anschlußmessung *f*
f. bend *(Fs)* Rasterkorrektur *f*, Teilbildkorrektion *f*, Trapezentzerrung *f*
f. blackout (blanking) period *(Fs)* Teilbildaustastperiode *f*
f. book Feldbuch *n* *(Vermessung)*
f. breaking switch Magnet[feld]ausschalter *m*, Feldtrennschalter *m*, Feldtrenner *m*
f. brush Ackerbürste *f*
f. coil Feldwicklung *f*, Feldspule *f*; Polwicklung *f*
f. concrete Baustellenbeton *m*
f. connection Baustellenverbindung *f* *(Montagebau)*
f. control *(BMSR)* Feldsteuerung *f*
f. controlled motor feldgesteuerter Motor *m*, Motor mit Feldsteuerung
f. controller *(BMSR)* Feldregler *m*, in der Anlage montierter Regler *m*
f. current Erregerstrom *m*, Feldspulenstrom *m*
f. diaphragm Sehfeldblende *f*
f. distortion 1. *(Ph, El)* Feldverzerrung *f*; 2. *(Fs)* Bildverzerrung *f*
f. distribution *(Ph, El)* Feldverteilung *f*
f. distribution of the aperture Feldverteilung *f* der wirksamen Öffnung (Strahlungsfläche) *(cm-Wellen-Antennen)*
f. diverter rheostat *(Elm)* Feldregulierparallelwiderstand *m*
f. divider *(Fs)* Rasterfrequenzteiler *m*, Teilbildfrequenzteiler *m*
f. drag Ackerschleppe *f*, Ackerschlichte *f*
f. drain Dräne *f*, Dränrohr *n*
f. drilling Bohren *n* innerhalb eines Ölfeldes
f. duration *(Fs)* Teilbilddauer *f*
f. effect tetrode Feldeffekttetrode *f* *(vierpoliges Halbleiterelement)*
f. effect transistor Feld[effekt]transistor *m*
f. electron discharge Feldelektronenentladung *f*
f. emission Feld[elektronen]emission *f*, kalte Emission *f* *(Elektronenröhren)*
f. emission microscope Feldelektronenmikroskop *n*

f. energy Feldenergie *f*
f. erection Aufbau *m* am Aufstellungsort
f. excitation Felderregung *f*
f.-extension well Anschlußbohrung *f*, Bohrung *f* zur Untersuchung der Ausdehnung eines Ölfeldes, Erweiterungsbohrung *f*
f. forge Feldschmiede *f*
f.-free emission current feldfreier Emissionsstrom *m (z. B. Langmuir-Katode)*
f. frequency *(Fs)* Bildwechselfrequenz *f* *(Teilbild)*, Rasterfrequenz *f*
f. frequency control *(Fs)* Bildwechselfrequenzregelung *f*, Rasterfrequenzregelung *f*
f. glass[es] *(Opt)* Feldstecher *m*
f. grid space Gitterraum *m (Laufzeitröhre)*
f. installation *(Rak)* Einsatzanlage *f*
f. intensity Feldstärke *f*
f. iron loss *(Elm)* Ständereisenverlust *m*
f. laboratory Baustellenlabor *n*
f. lens *(Opt)* Feldlinse *f*
f. line construction *(Fmt)* Feldleitungsbau *m*
f. linearity control *(Fs)* Linearitätsregelung *f*, Regelung *f* der Rasterlinearität
f. magnet Feldmagnet *m*, Poljoch *n*
f. moisture equivalent natürlicher Sättigungsgrad *m (Bodenmechanik)*
f. monitoring tube *(Fs)* Rasterkontrollröhre *f*, Teilbildkontrollröhre *f*
f. monitoring unit 1. *(El)* Betriebsprüfgerät *n*; 2. *(Fs)* Rasterkontrollgerät *n*
f. neutralizing coil Abschirmspule *f (z. B. Fernsehen)*
f. newly cultivated *(Lw)* Reutland *n*
f. of a particle *(Ph)* Partikelfeld *n*
f. of activity Arbeitsgebiet *n*
f. of application Anwendungsgebiet *n*, Verwendungsbereich *m*
f. of forces Kraft[linien]feld *n*, Kräftefeld *n*
f. of gravity Schwerefeld *n*
f. of observation Gewichtsfeld *n*
f. of pressure gradient Druckgradientenfeld *n*
f. of research Forschungsgebiet *n*
f. of uses Anwendungsgebiet *n*
f. of view Blickfeld *n*, Gesichtsfeld *n*, Sehfeld *n*
f.-of-view number *(Opt)* Feldzahl *f*
f. of vision Blickfeld *n*, Gesichtsfeld *n*
f. operator Feldoperator *m*
f. particle Feldteilchen *n*
f. pattern Feldstärke[n]diagramm *n*, Strahlungsdiagramm *n (Antennenausstrahlung)*
f. pick-up *(Fs)* Außenaufnahme *f*
f.-pole *(Elm)* Feldpol *m*, Spulenkasten *m*
f. quantities *(Ph)* Feldgrößen *fpl*
f. quantization Feldquantisierung *f*, Feldquantelung *f*
f. quantum Feldquant *n*
f. radio station Kleinfunkstelle *f*
f. rail Feldbahnschiene *f*
f. ratio *(Elm)* Erregungsgrad *m*
f. recording Außenaufnahme *f (Tonband)*
f. rectifier Feldgleichrichter *m*
f. regulator [selbsttätiger] Feldregler *m*, [selbsttätiger] Magnetfeldregler *m*
f. rheostat Feldregler *m*, Magnetfeldregler *m*
f. rivet Baustellenniet *m*, Montageniet *m*
f. rod *(Opt)* Fluchtstab *m*
f. roller *(Lw)* Feldwalze *f*
f. scan *(Fs)* Teilbildabtastung *f*
f. selection Lochfeldsteuerung *f*
f. sequential system Rasterfolgeverfahren *n*, Rasterwechselverfahren *n*, Zeitfolgeverfahren *n* *(Farbfernsehen)*
f. shoes Rennschuhe *mpl*, Laufschuhe *mpl*
f. shunting Feldschwächung *f* durch Nebenschluß
f. simultaneous system Simultanverfahren *n* *(Farbfernsehen)*
f. spider *(Elm)* Polrad *n*, Polstern *m*
f. spool Feldspule *f*; Feldspulenkasten *m*, Spulenkasten *m*
f. stone Feldstein *m*
f. stop Sehfeldblende *f*
f. strength Feldstärke *f*, Feldintensität *f*
f. strength distribution Feldstärkeverlauf *m*, Feldstärkeverteilung *f*
f. superintendent Bauleiter *m*; Oberbauleiter *m*
f. supervision Baustellenüberwachung *f*
f. suppressor *(Elm)* Feldschwächungseinrichtung *f*; Feldschwächungsschalter *m*
f. sweep *(Fs)* Teilbildabtastung *f*
f. switch Feldschalter *m*
f. switchboard *(Fmt)* Feldklappenschrank *m*
f. synchronizing [im]pulse (signal) Bildsynchronisierungsimpuls *m*, V-Impuls *m*
f. tapping *s.* f. weakening by tapping
f. telephone Feldtelefon *n*
f. television camera Fernsehkamera *f* für Außenaufnahmen, Freilichtfernsehkamera *f*
f. test Baustellenprüfung *f*
f. theory *(Ph)* Feldtheorie *f*
f. tile *(Am)* *s.* f. drain
f. tilt *(Fs)* Rasterentzerrung *f*, Trapezentzerrung *f*
f.-to-noise ratio Verhältnis *n* von Nutzfeldstärke (Signalstärke) zur Störfeldstärke, Störfeldabstand *m*
f. tripod Feldstativ *n (Vermessung)*
f. use Prüfzweck *m*
f. weakening Feldschwächung *f*
f.-weakening apparatus Feldschwächungsschalter *m*
f. weakening by tapping Feldschwächung *f* durch Anzapfung

f. weakening ratio Feldschwächungsverhältnis *n*, relative Feldschwächung *f*
f. weld Baustellenschweißung *f*
f. winding Erregerwicklung *f*, Feldwicklung *f*; Feldspule *f*
f. work Feldarbeit *f (Vermessung)*
f. worker Feldarbeiter *m (Vermessung)*
Field boiler *(Wärm)* Field-Kessel *m*
fieldistor Fieldistor *m*, Feld[effekt]transistor *m*
fiery 1. feurig; explosiv; 2. *(Brg)* schlagwetterführend
f. fracture *(Met)* feinkörniger Bruch *m*
f. mine *(Brg)* Schlagwettergrube *f*
fifth-wheel load *(Kfz)* Aufsattellast *f*, Aufsatteldruck *m*
f.-wheel steering *(Kfz)* Drehschemellenkung *f*
fig tree Feigenbaum *m*
fighter air division Jagdfliegerdivision *f*
f.-bomber Jagdbombenflugzeug *n*
f. control Führung *f*, Leitung *f (Jagdflugzeuge)*
f. control room Jägerleitstelle *f*
f. direction Jägerleitung *f (Jagdflugzeuge)*
f. direction post Jägerleitstelle *f*
f. plane Jagdflugzeug *n*
figure / to [zeichnerisch] darstellen; [be]rechnen
f. out ausrechnen; Resultat zahlenmäßig angeben
figure 1. Ziffer *f*, Zahl *f*; Figur *f*; 2. *(Typ)* Abbildung *f*; 3. *(Holz)* Maserung *f*, Zeichnung *f*
f.-eight pattern *(El)* Achterfigur *f*, Achterdiagramm *n*, Doppel-Kosinus-Charakteristik *f*, Doppelkreischarakteristik *f*
f.-eight winding *(El)* Achterwicklung *f*
f.-head *(Schiff)* Galionsfigur *f*, Bugfigur *f*
f. of eight Dipolcharakteristik *f (Antennenemission)*
f.-of-eight characteristic Lemniskatenkennlinie *f*
f.-of-eight suture *(Text)* Achternaht *f*
f. of merit 1. *(BMSR)* Leistungszahl *f*, Gütezahl *f*; 2. *(El)* S/C-Verhältnis *n (Verhältnis der Steilheit zu Eingangs- und Ausgangskapazität; Elektronenröhre)*; 3. Parallelgüte *f (Kristalle)*
f. pointing drum *(Fmt)* Zahlentrommel *f*
f. shift[ing] Zahlenwechsel *m (Fernsprechtechnik)*; Zahlenumschaltung *f (Telegrafie)*
f. skating Eiskunstlauf *m*
f. space *(Text)* Figurstelle *f*
figured fabric Bildgewebe *n*
f. glass Ornamentglas *n*, Profilglas *n*
f. iron Profileisen *n*
f. ribs figurierter Rips *m*
f. veneer gemasertes Furnier *n*
f. weft backed fabric *(Text)* gemusterter Schußdouble *m*
f. woven fabric gemustertes Gewebe *n*
figuring end *(Text)* Musterfaden *m*, musterbildender Faden *m*
f. harness *(Text)* Musterharnisch *m*, Figurenharnisch *m*
f. shafts *(Text)* Mustergeschirr *n*
f. weft *(Text)* Figurschuß *m*
filament 1. Glühfaden *m*, Leuchtfaden *m (Glühlampe)*; Heizfaden *m (Elektronenröhre)*; 2. *(Text)* Einzelfaden *m* [unbegrenzter Länge], Endlosfaden *m*; 3. Faser *f (Balken)*
f. accumulator Heizakkumulator *m*
f. activity test Röhrenprüfung *f*
f. battery Heizbatterie *f*
f. break Glühfadenbruch *m*; Heizfadenbruch *m*
f. cathode direkt beheizte Katode *f*
f. choke Heizdrossel *f*
f. circuit Heizstromkreis *m (Elektronenröhre)*
f. count of yarn Nummer *f* des Endlosfadens
f. current Heizstrom *m (Elektronenröhre)*
f.-current consumption Heizstromverbrauch *m*
f. efficiency Glühfadenwirkungsgrad *m (Glühlampe)*; Heizfadenwirkungsgrad *m (Elektronenröhre)*
f. electrometer Fadenelektrometer *n*
f. emission Glühkatodenemission *f*, Heizfadenemission *f*
f.-forming substance fadenbildende Substanz *f*
f. generator Heizgenerator *m (für Elektronenröhre)*
f. heating circuit Heizstromkreis *m*
f. heating transformer Heiztransformator *m (für Elektronenröhre)*
f. holder Glühdrahthalter *m*, Glühfadenträger *m (Glühlampe)*; Heizfadenträger *m (Elektronenröhre)*
f. input Heizleistung *f (Elektronenröhre)*
f. instrument Faden[meß]instrument *n*
f. lamp Glühlampe *f*
f. power Heizleistung *f (Elektronenröhre)*
f. rheostat Heizwiderstand *m*
f. supply Heizfadenspeisung *f*
f. supply battery Heizbatterie *f*
f. supply transformer Heiztransformator *m*
f. support Heizfadenträger *m (Elektronenröhre)*; Glasgestell *n*, Glühdrahtträger *m (Glühlampe)*
f. terminal Heizfadenanschluß *m*
f. thermometer Fadenthermometer *n*
f.-type cathode direkt geheizte Katode *f*
f. voltage Heizspannung *f (Elektronenröhre)*
f. wattage Heizleistung *f* in W
f. winding Heizwicklung *f (Transformator)*
f. yarn aus Einzelfäden gedrehtes Garn *n*
filamentary fadenförmig
f. cathode Fadenkatode *f*

f. molecule Fadenmolekül *n*
f. rectifying tube direkt geheizte Gleichrichterröhre *f*
f. transistor Fadentransistor *m*
filar micrometer Fadenmikrometer *n*
file / to 1. [aus]feilen; 2. einordnen; ablegen, abheften *(z. B. Akten)*
f. away abfeilen
file 1. Feile *f*; 2. Aktendeckel *m*, Heftmappe *f*, Pultordner *m*, Schnellhefter *m*; Kartei *f*; Archiv *n*
f. blade Feilenblatt *n*
f. blank Feilenrohling *m*
f. card 1. Archivkarte *f*, Nachweiskarte *f*; Karte *f*, Zettel *m*; 2. Drahtbürste *f* aus Kratzenstoff, Feilenbürste *f*
f. cleaner Feilenbürste *f*
f. copy Belegexemplar *n*
f. cut Feilenhieb *m*
f. cutting machine Feilenhaumaschine *f*
f. cutting tool Feilenhaue *f*
f. dust Feilstaub *m*
f. handle Feilenheft *n*
f. hard feilenhart
f. hardness testing Anfeilhärteprüfung *f*
f.-punch Locher *m*
f. reference Aktenzeichen *n*
f. scratch test Anfeilhärteprüfung *f*
f.-searching machine Aktensuchmaschine *f*, Karteisuchmaschine *f*
f. tang Feilenangel *f*
f. with oblique edges abgeschrägte Feile *f*
f. with tang Angelfeile *f*
filed patent angemeldetes Patent *n*
filiform fadenförmig, in Drahtform
filigree Filigran *n*
filing 1. Feilen *n*, Feilarbeit *f*; Feilspan *m*; 2. Einordnen *n*; Ablegen *n*; Abheften *n (z. B. Akten)*
f. and sawing machine Feil- und Sägemaschine *f*
f. cabinet Karteikasten *m*; Aktenschrank *m*
f. card Karteikarte *f*
f. dust Feilstaub *m*
f. machine Feilmaschine *f*
f. method Registrierverfahren *n (Bürotechnik)*
f. the tooth gullet Ausschleifen *n* des Zahngrundes *(Sägeblatt)*
f. time *(Fmt)* Anmeldezeit *f* eines Gespräches
filings Feilstaub *m*, Feilspäne *f*
fill / to 1. füllen; 2. *(Met)* eingießen *(z. B. Roheisen)*; 3. hinterfüllen, verfüllen *(Erdbau)*; 4. *(Brg)* versetzen
f. in hinterfüllen, verfüllen, zuschütten *(Erdbau)*
f. the tan pit with liquors *(Led)* Grube abtränken
f. up auffüllen, nachfüllen *(z. B. Batterie)*
fill *(Bw)* Anschüttung *f*, Aufschüttung *f*, Auftrag *m*, Schüttung *f*; Auftragsmaterial *n*, Füllboden *m*, Schüttmaterial *n*
f. checking Füllstandsmessung *f*
f. chute *(Brg)* Versatzrolloch *n*, Versatzrolle *f*
f. raise *(Brg)* Versatzrolloch *n*, Versatzrolle *f*
filled [ab]gefüllt
f. band vollbesetztes Energieband *n (Halbleiter)*
f. board Mehrlagenpappe *f*
f. chip-board kaschierte Schrenzpappe *f*
f. electronic shell abgeschlossene (vollbesetzte) Elektronenschale *f*
f. flat-back stope *(Brg)* söhliger Firstenstoß *m (Firstenstoßbau mit Versatz)*
f. ground aufgeschütteter Boden *m*, Aufschüttung *f*, Auftrag *m*, Bodenauftrag *m (Erdbau)*
f. joint ausgegossene Kabelmuffe *f*
f. pulp-board minderwertiger Dreilagenkarton *m*
f. rill stope *(Brg)* Firstenstoß *m* mit schräggestellter Firste *(Firstenstoßbau mit Versatz)*
f. section rail Vollschiene *f*
f. stope versetzter Abbau *m (Erzbergbau)*
f. stoping Abbauverfahren *n* mit Versatz *(Erzbergbau)*
f.-system thermometer Kapillarthermometer *n*, Ausdehnungsthermometer *n*
f. up versetzt *(Filz)*
filler 1. Füllmasse *f*, Füllmaterial *n*, Füllmittel *n*, Füllstoff *m*; Abfüllmaschine *f*; 2. *(Plast)* Harzträger *m*, Porenfüller *m*, Streckmittel *n*; 3. *(Brg)* Füllkörper *m*; 4. *(Masch)* Einsatz *m*; 5. Füllkanne *f*; 6. *(Bw)* Füllholz *n*, Futterstück *n*; 7. *(Typ)* Füller *m (z. B. einer Zeitung)*; 8. Zusatzwerkstoff *m (Schweißen)*; Lot *n (Löten)*
f. block *(Bw)* Füllkörper *m*, Hohlkörper *m*
f. board Kaschierrohpappe *f*
f. brick Besatzstein *m (Siemens-Martin-Ofen)*
f. cap Einfüllkappe *f*, Einfüllverschluß *m*; Füllstutzen *m*
f. gas Füllgas *n*
f. joist *(Bw)* Stichbalken *m*, Deckenträger *m*
f.-joist floor Stahlträgerdecke *f* mit Beton- oder Ziegelausfachung
f. material 1. Füllmaterial *n*, Zusatzstoff *m*; 2. *s.* filler 8.
f. metal Aufschweißlegierung *f*; Zusatzmetall *n (Schweißen)*
f. neck Einfüllstutzen *m*, Füllstutzen *m*; Betankungsstutzen *m*
f.-neck coupling Betankungskupplung *f*
f. plate Füllplatte *f*
f. rod Zusatzdraht *m (Schweißen)*, Schweißdraht *m*, Schweißstab *m*; Füllungsstab *m*
f.-speck *(Plast)* Materialfehler *m*, vom Füllstoff herrührende Fehlstelle *f*

f. tube Einfüllstutzen *m* [für Kraftstoff]
f. valve Betankungsventil *n*
f. wire Zusatzdraht *m*
fillet / to auskehlen
fillet *(Masch)* 1. Hohlkehle *f*, Kehle *f*; Kehlnaht *f (Schweißung)*; Ausrundung *f*; Schaftausrundung *f (Niet)*; 2. *(Bw)* Deckleiste *f*, Leiste *f*, Saumleiste *f*; Kehlleiste *f*; 3. *(Flg)* Verkleidungsübergang *m (Rumpf zu Tragfläche)*
f. curve *(Masch)* Ausrundungskurve *f*
f. gauge Radiuslehre *f*; Radienschablone *f*; Konkav- und Konvexlehre *f*
f. moulding machine Leistenhobelmaschine *f*
f. of the gullet Zahngrundrundung *f*
f. plane Filethobel *m (Tischlerei)*
f. radius Rundungshalbmesser *m*
f. strip *(Bw)* Deckleiste *f*, Kehlleiste *f*
f. weld Kehlnaht *f*; Kehlnahtschweißung *f*
filleted ausgerundet
filleting Ausrunden *n*; Ausrundung *f*; Verkleidung *f*
filletster plane *s.* fillister plane
filling 1. Aufschüttung *f*, Bodenauftrag *m*, Füllung *f*, Schüttung *f*; 2. *(Met)* Vollgießen *n (Kokille)*; Begichtung *f (Hochofen)*; 3. Rieselanlage *f (Kühlturm)*; 4. *(Text)* Schuß *m*; Füllmittel *n*; 5. Bemesserung *f* am Holländer; 6. *(Brg)* Wegfüllen *n*; Bergeversatz *m*, Versatz *m*
f.-backed fabric *(Text)* Schußdouble *m*
f. bogie Füllwagen *m (Kokerei)*
f. brick Füllstein *m*
f. compound Füllmasse *f*; Ausgußmasse *f*
f. float *(Text)* Schußflottung *f*
f. fork *(Text)* Schußgabel *f*
f. frame 1. *(Met)* Aufsatzrahmen *m (Formerei)*; 2. *(Schiff)* Füllspant *n*
f. funnel Einfülltrichter *m*, Fülltrichter *m*
f.-futtock *(Schiff)* Hanger *m (für Püttings)*
f. hole Fülloch *n*, Einfüllöffnung *f*
f. hose Zapfschlauch *m*
f.-in cam Führungsschloßteil *n* [, das den Nadeln keine Bewegung erteilt] *(Strickmaschine)*
f.-in material Zusatzwerkstoff *m*
f.-in of the meshes *(Text)* Zellenfüllung *f (geknüpfter Netzstoff)*
f.-in work Ergänzungsarbeit *f (an Landkarten)*
f. knife Spachtel *f*
f. knit fabric *(Text)* Kulierware *f*
f. mark Füllstrich *m (Gefäße)*
f. notch Füllnut *f*
f. of tap-hole Schließen *n* der Abstichöffnung, Schließen des Stichlochs
f. pattern *(Text)* Schußfolgemuster *n*
f. pick *(Text)* Schuß *m*
f. pick fabric *(Text)* Schußsamt *m*
f. plate *(Bw)* Futterblech *n*
f. 5-ply *(Text)* Schuß *m* fünffach
f. point *(Schiff)* Tankhof *m*
f. pressure Fülldruck *m*
f. puttey Spachtelkitt *m*
f. raise *(Brg)* Versatzüberhauen *n*
f. rod *(Bw)* Futterleiste *f*
f. slot bearing Lager *n* mit Füllnut
f. spout Füllstutzen *m*
f. station 1. Tankstelle *f*, Zapfstelle *f*; 2. *(Brg)* Füllort *n*
f.-station equipment Tankstellenausrüstung *f*
f. stop motion *(Text)* Schußwächtergabel *f*
f.-timber *(Schiff)* Füllspant *n*
f. trowel *(Bw)* Fugenkelle *f*
f. with mud Verschlammen *n (des Bodens)*
f. with sand Versandung *f (des Bodens)*
fillings 1. *(Ch)* Füllkörper *mpl*; 2. *(Schiff)* Füllhölzer *npl*
fillister 1. Kittfalz *m (Fenster)*; 2. Kittfalzhobel *m*, Grathobel *m*
f.-head screw Zylinderschraube *f*
f. plane Grathobel *m*
f. plane iron Grathobeleisen *n*
film Film *m*; Feinfolie *f*; dünner Überzug (Belag) *m*, dünne Schutzschicht *f*; Häutchen *n*
f. badge *(Kern)* Filmplakette *f (Strahlenschutz)*, Filmdosimeter *n*
f. base Schichtträger *m*, Blankfilm *m*
f. camera Filmkamera *f*, Kamera *f*
f. car Aufnahmewagen *m*
f. cartridge Filmspule *f*, Filmpatrone *f*
f. cassette Filmkassette *f*
f. cement Filmkitt *m*
f. channel Filmbahn *f*, Führungsbahn *f*
f. characteristic *(Opt)* Schwärzungskurve *f*
f. composer *(Typ)* Lichtsetzmaschine *f*
f. composing *(Typ)* Lichtsatz *m*
f. copying machine Filmkopiermaschine *f*
f. course *(Am)* Lehrfilm *m*
f. cut-out Durchschlagssicherung *f*, flinke Feinsicherung *f*
f. demonstration Filmvorführung *f*
f. developing device Filmentwicklungsgerät *n*
f. distribution schleierförmige Verteilung *f (Kondensator)*
f. dosimeter *(Kern)* Filmdosimeter *n (Strahlenschutz)*
f.-drive mechanism Filmlaufwerk *n*
f. drying drum Filmtrockentrommel *f*
f. dublicating Filmdoppeln *n*
f. editing Filmschneiden *n*
f. emulsion Filmemulsion *f*, Filmschicht *f*
f. feed Filmtransport *m*, Filmvorschub *m*
f. gate Filmfenster *n*, Filmrahmen *m*; Bildfenster *n*
f. grader Kopiermeister *m*
f. grain Filmkorn *n*

f. **grain noise** Kornrauschen *n (Film)*
f. **holder** Filmrahmen *m*
f. **lamination** *(Typ)* Glanzfolienkaschierung *f*
f. **library** Filmarchiv *n*
f. **magazine** Filmkassette *f*
f. **magazine roll** Filmvorratsspule *f*
f. **memory** Filmspeicher *m (Datenspeicherung)*
f. **of lubricant** *(Masch)* Schmierfilm *m*
f. **of oxide** Oxidhäutchen *n*
f. **pick-up** Filmabtaster *m*, Filmgeber *m (Fernsehwiedergabe von Filmaufnahmen)*
f. **processing** Filmbearbeitung *f*
f. **processing tank** Entwicklerdose *f*
f. **projection** Filmvorführung *f*
f. **projector** Filmprojektor *m*
f. **pulling mechanism** Durchzugslaufwerk *n (Filmprojektor)*
f. **rack** Filmrahmen *m*
f. **recorder** [Tonfilm-]Schallaufzeichner *m*
f. **reel** Filmrolle *f*
f. **resistor** Schichtwiderstand *m*
f.**-rewinding outfit** Umspuleinrichtung *f (Filmprojektion)*
f. **ring** *(Kern)* Fingerring *m* mit Filmeinlage *(Strahlenschutz)*
f. **scan disk** Ringlochscheibe *f*
f. **scanner** 1. Filmgeber *m*; 2. *(Fs)* Filmabtaster *m*
f. **scanning** Filmabtastung *f*, Filmübertragung *f*
f. **size** Filmformat *n*
f.**-speed indicator** Filmtachometer *n*
f. **splicer** Klebeapparat *m*, Filmkleber *m*
f. **spool** Filmspule *f*, Filmkassette *f*
f. **stock** 1. Rohfilm *m*; 2. Aufnahmematerial *n*
f. **strip transport** Bildbandführung *f*
f.**-take-up reel (spool)** Filmaufwickelspule *f*, Filmrolle *f*
f. **television** Fernsehkino *n*, Fernsehen *n* mit Zwischenfilm
f. **track** Filmbahn *f*, Führungskanal *m (Projektor)*
f. **transmitter** Fernkinosender *m*
f. **type carbon resistor** Kohleschichtwiderstand *m*
f. **viewer** Durchleuchtungskasten *m (für Filme)*
f.**-winding mechanism** Filmaufwickelvorrichtung *f*
filmsetter *(Typ)* Lichtsetzmaschine *f*
filmsetting *(Typ)* Lichtsatz *m*
f. **machine** *(Typ)* Lichtsetzmaschine *f*
filter / to 1. *(Ch)* filtern, filtrieren, [durch]seihen; 2. *(El)* sieben; 3. *(Masch)* durchschlagen
f. **by suction** *(Ch)* abnutschen
f. **in** *(Brg)* einsickern *(Wasser)*
f. **off** *(Ch)* abfiltrieren
filter 1. *(Ch, Opt)* Filter *n*; 2. *(El)* Filter, Siebglied *n*, Siebkette *f*, elektrische Weiche *f*; 3. *(Masch)* Durchlaß *m*, Durchschlag *m*
f. **aid** *(Ch)* Filtrationshilfsmittel *n*
f. **attenuation band** *(El)* Filtersperrbereich *m*, Sperrbereich *m*
f. **bag** Filtertasche *f*
f. **bed** Filterbecken *n*, Filterbett *n*, Filterschicht *f*, Tropfkörper *m (Gesundheitstechnik)*
f. **block** Tropfkörperbrocken *m (Gesundheitstechnik)*
f. **board** dickes Filterpapier *n*
f. **bowl** *(Kfz)* Filtertopf *m*
f. **cake** Filterkuchen *m*, Preßkuchen *m*
f. **capacitor** Siebkondensator *m*, Glättungskondensator *m*
f. **capacity** Siebkapazität *f*
f. **cartridge** *(Kfz)* Filterpatrone *f*
f. **case** Filtergehäuse *n*
f. **cell** *(Ch)* Filtertasche *f*
f. **cell unit** Zellenfiltereinsatz *m*
f. **chain** *(El)* Siebkette *f*
f. **choke** *(El)* Siebdrossel *f*, Glättungsdrossel[spule] *f*
f. **circuit** Siebkreis *m*, Siebschaltung *f*, Glättungsschaltung *f*
f. **cloth** Filtertuch *n*, Preßtuch *n*, Tuchfilter *n*
f. **coil** Siebspule *f*
f. **condenser** *s.* f. capacitor
f. **cone** Filterkegel *m*, Filtereinlage *f*
f. **core** *(Ch)* Siebkern *m*
f.**-coupled** *(El)* bandfiltergekoppelt
f. **discrimination** *(El)* Trennschärfe *f* eines Filters
f. **drain** Filterrohr *n*, Sickerrohr *n*
f. **element** *(Kfz)* Filtereinsatz *m*, Filterelement *n*, Filterpaket *n*
f.**-element container** Filtereinsatztopf *m*
f. **feed heater** *(Wärm)* Filtervorwärmer *m*
f. **filling** Filtereinsatz *m*
f. **flask** Filtrierflasche *f*, Filtrierkolben *m*, Saugkolben *m*
f. **for band elimination** Bandsperrfilter *n*
f. **gauze** Filtergaze *f*
f. **gravel** Filterkies *m*
f. **insert** Filtereinsatz *m (Gasmaske)*
f. **lens** Filterlinse *f*
f. **liquor** Filtrat *n*
f. **medium** Filtermittel *n*, Filterstoff *m*
f. **network** *(El)* Siebschaltung *f*
f. **pad** *(Kfz)* Filterkörper *m*
f. **paper** Filterpapier *n*, Filtrierpapier *n*
f. **paper analysis** Tüpfelanalyse *f* auf Filterpapier
f. **pass** *(El)* Lochpaß *m*
f. **pass band** Durchlaßbandbreite *f*
f. **polarizer** *(Opt)* Filterpolarisator *m*

f. press Filterpresse *f*
f. pulp *(Ch)* Filterbrei *m*
f. pump *(Ch)* Wasserstrahlpumpe *f*
f. range Durchlässigkeitsbereich *m* eines Filters
f. reactor *(El)* Siebdrossel *f*, Siebspule *f*
f. screen Filtertuch *n*, Filtergewebe *n*, Filterstoff *m*
f. section *(El)* Siebglied *n*
f. self rescuer *(Brg)* Filterselbstretter *m*
f. slot Sickerschlitz *m*
f. spout Filterstutzen *m*
f. stand Filterstativ *n*, Filtriergestell *n*, Filtrierstativ *n*
f. stop band *(El)* Bandsperre *f*, Filtersperrbereich *m*
f. transmission band *(El)* Durchlaßfrequenzbereich *m*
f. transmittance *(Opt)* Filterdurchlässigkeit *f*
f. unit Filteranlage *f*, Filtereinsatz *m*
f. well Filterbrunnen *m*, Großfilterbrunnen *m*, Schachtbrunnen *m*
filterability Filtrierbarkeit *f*, Fließfähigkeit *f*
f. at low temperatures Kältefließfähigkeit *f* *(z. B. von Dieselkraftstoff)*
filterable filtrierbar
filtering 1. *(Ch)* Filtrierung *f*, Filtration *f*; 2. *(El)* Siebung *f*, Filterung *f*
f. action *(Ch)* Filterwirkung *f*, Ausscheidungseffekt *m*
f. basin Filterbecken *n*, Filtertank *m*, Klärbecken *n (Gesundheitstechnik)*
f. circuit *(El)* Filterkreis *m*
f. medium Filterstoff *m*, Filtermittel *n*
f. of power leads Siebung *f* der Netzspannung
f. of used oil Regenerierung *f* von Altöl
f. property Filtrierbarkeit *f*
filtrate / to filtrieren
filtrate Filtrat *n*
filtration Filterung *f*, Filtrierung *f*
f. chamber *(Wärm)* Filterraum *m*
filty *s.* fire damp
fin / to rippen, mit Rippen versehen
fin 1. Grat *m*; Gußnaht *f*; Preßnaht *f*; Nase *f*; 2. Kühlrippe *f*, Rippe *f*; 3. *(Flg, Rak)* Flosse *f*, Stabilisierungsfläche *f*
f.-adjusting instrument Flossenverstellgerät *n*
f. adjustment Flossenverstellung *f*
f. aerial Flossenantenne *f (an Flugkörpern)*
f. keel *(Schiff)* Flossenkiel *m*
f. motor Rudermaschine *f*
f. plate *(Schiff)* Flossenplatte *f*, Stabilisierungsplatte *f*
f.-stabilized missile Flossengeschoß *n*
f.-stabilized rocket flügelstabilisierte Rakete *f*
f. trimming Flossenverstellung *f*
f. tube Rippenrohr *n*
final End-, Fertig-; endgültig
f. acceptance Schlußabnahme *f*
f. addition *(Met)* Fertigzuschlag *m*
f. altitude *(Flg)* Mindestsicherheitshöhe *f*
f. amplification Endverstärkung *f*
f. amplifier Endverstärker *m*; Fernleitungsendverstärker *m*
f. amplifier stage Endverstärkerstufe *f*
f. annealing Fertigglühung *f*
f. anode Endanode *f*
f. approach *(Flg)* Endanflug *m*
f. assembling Endmontage *f*
f. blowing *(Met)* Fertigblasen *n (Windfrischen)*
f. boiling point Siedeendpunkt *m*, oberer Siedepunkt *m*
f. coat of paint Endanstrich *m*
f. condition Endzustand *m*
f. connector *(Fmt)* Leitungswähler *m*
f. control element Stellglied *n*, Stellmotor *m* (*s. a.* regulating unit)
f. control unit *s.* regulating unit
f. copy Reinschrift *f*
f. current Endstrom *m*, Strom *m* im [stationären] Endzustand
f. diameter Enddurchmesser *m*; Drehkreisdurchmesser *m*
f. disposal *(Kern)* endgültige Beseitigung *f*
f. drawn sliver *(Text)* Endstreckenband *n*
f. drive Abtrieb *m*, Achsantrieb *m*, Radantrieb *m*, Hinterradantrieb *m*, Hinterachsantrieb *m*
f.-drive pinion Endantriebsritzel *n*
f.-drive unit Hinterachsantrieb *m*
f. erection Endmontage *f*
f. etching Scharfätzung *f*, Tiefätzung *f*
f. examination Endkontrolle *f*, Schlußkontrolle *f*; Abnahmekontrolle *f*; Endabnahme *f*
f. examination of apprentices Lehrabschlußprüfung *f*
f. gear reduction in the rear axle *(Kfz)* Hinterachsübersetzung *f*
f. grade Fundamentsohle *f*
f. grading Feinplanierung *f*
f. hardening *(Met)* Schlußhärten *n*
f. heating wall *(Brg)* Endheizwand *f*
f. inspection Abnahmekontrolle *f*; Endkontrolle *f*, Schlußkontrolle *f*; Bauabnahme *f*
f. limit switch Endausschalter *m*
f. liquor Endlauge *f*
f. magnification Endvergrößerung *f*
f. mass *(Rak)* Endmasse *f*, Masse *f* am Ende des aktiven Flugabschnitts; Masse der letzten Raketenstufe
f. measurement Abnahmemessung *f*
f. moisture Endfeuchtigkeit *f*
f. pass Schlichtstich *m*, Polierstich *m (einer Walzstraße)*

f. **payload** Nutzlast *f* der letzten Raketenstufe
f. **position** Endlage *f*, Endstellung *f*
f. **pressure** Enddruck *m*
f. **pressure of expansion** *(Wärm)* Kompressionsendspannung *f*
f. **prestress** Endvorspannkraft *f*, Endvorspannung *f (Spannbetonbau)*
f. **prestressing force** Endvorspannkraft *f*, Endvorspannung *f (Spannbetonbau)*
f. **print** Theaterkopie *f (Film)*
f. **product** Enderzeugnis *n*, Fertigerzeugnis *n*
f. **proof** *(Typ)* Aushängebogen *m*, Revisionsbogen *m*
f. **rejection** endgültige Ablehnung *f*, Verwerfung *f*
f. **resistance** Endwiderstand *m*, Abschlußwiderstand *m*
f. **result** Endresultat *n*
f. **saturation** 1. Endsättigung *f*; 2. *(Ch)* Nachsaturation *f*
f. **selector** *(Fmt)* Leitungswähler *m*
f. **separation** Endsortierung *f*, Nachsortierung *f*
f. **set** Erstarrungsende *n (Beton)*
f. **setting time** Zeitspanne *f* bis zum Erstarrungsende *(Beton)*
f. **settling basin (tank)** Nachklärbecken *n (Gesundheitstechnik)*
f. **sintering** Fertigsintern *n (Metallpulver)*
f. **size** Fertigmaß *n*
f. **slag** Reduktionsschlacke *f (Elektrometallurgie)*
f. **speed** Endgeschwindigkeit *f*
f. **spinning** Feinspinnen *n*, Ausspinnen *n*
f. **stage** 1. *(El)* Endstufe *f*; 2. *(Rak)* obere (letzte) Stufe *f*
f. **stage valve** Endröhre *f*
f. **state** *(Ph)* Endzustand *m*
f. **step** *(Rak)* obere (letzte) Stufe *f*
f. **strength** Endfestigkeit *f*
f. **structure** *(Met)* Endgefüge *n (nach abgeschlossener Wärmebehandlung)*
f. **terminal-stopping device** *(El)* Endausschalter *m (Automatik)*
f. **test** Abschlußprüfung *f*; Endabnahme *f*, Endprüfung *f*
f. **thermal state** thermischer Endzustand *m*
f. **tightening** Nachziehen *n (Schrauben usw.)*
f. **vacuum** Endvakuum *n*
f. **value** Endwert *m*
f. **velocity** Endgeschwindigkeit *f*
f. **voltage** [stationäre] Endspannung *f*
f. **well log** endgültiges Bohrprofil *n (Erdöl, Erdgas)*
financial difficulties Zahlungsschwierigkeiten *fpl*
f. **year** Rechnungsjahr *n*
finback whale oil Finnfischtran *m*
find / to finden; suchen; ermitteln, aufsuchen
f. **out** ermitteln
f. **out a trouble** *(El)* eine Störung suchen
f. **the square root** *(Math)* die Quadratwurzel suchen
f. **the target** *(Mil)* einschießen
finder 1. *(Opt)* Sucher *m*; 2. *(Fmt)* Sucher, Suchschalter *m*
f. **circle** Finderteilkreis *m (Spektroskopie)*
f. **lens** Sucherobjektiv *n*
f. **selector** *(Fmt)* Anrufschalter *m*
f. **switch** *(Fmt)* Anrufsucher *m*, Suchwähler *m*
finding Finden *n*; Suchen *n*; Ermitteln *n*, Aufsuchen *n*; Befund *m*
f. **action** *(Fmt)* Freisuchen *n*, Freiwählen *n*
f. **place** Fundort *m*
f. **telescope** Sucherfernrohr *n*
findings Untersuchungsergebnisse *npl*
f. **of the court** Gerichtsspruch *m*
fine / to 1. klären, schönen *(Bier, Wein)*; 2. *(Led)* gerben; 3. *(Met)* frischen, feinen, raffinieren; 4. mit Geldstrafe belegen
fine fein, rein, verfeinert; dünn
f. **adjusting screw** Feineinstellschraube *f*
f. **adjustment** Feineinstellung *f*, Feinabgleich *m*
f. **aggregate** Feinzuschlag *m (Beton: < 3/16"; Schwarzdecken: < 1/8")*; Feinsand *m*, Feinsplitt *m*
f. **board** Feinpappe *f*
f. **borer** Feinbohrwerk *n*
f. **boring** Fein[st]bohren *n*
f.**-boring machine** Feinbohrmaschine *f*
f.**-boring tool** Feinbohrmesser *n*
f. **chippings** *(Bw)* Feinsplitt *m*
f. **coal** Feinkohle *f*, Staubkohle *f*
f. **coke** Koksgrus *m*
f. **control** Feineinstellung *f*, Feinregelung *f*
f. **control rod** *(Kern)* Feinregelstab *m*
f. **crack** Haarriß *m*
f. **crusher** Feinbrecher *m*, Nachbrecher *m*
f. **crushing** Feinbrechen *n*, Nachbrechen *n*
f.**-detail resolution** *(Opt)* Feinauflösung *f*; Durchzeichnung *f*
f. **dimension planing** Abrichthobeln *n*
f. **dispersion** Vernebelung *f*
f. **drive** *(Opt)* Feinstelltrieb *m*
f. **dust** Flugstaub *m*
f.**-edged** feinschneidig *(Werkzeuge)*
f. **feed** Feinvorschub *m*
f.**-fibrous** feinfaserig
f. **file** Schlichtfeile *f*
f. **finishing** Feinbearbeitung *f*
f. **fraction** *(Brg)* Feingut *n*
f. **glazing** Hochglanzpolieren *n*
f. **gold** Feingold *n*
f.**-grading** *(Bw)* Feinplanierung *f*
f.**-grain** 1. feinkörnig; Feinkorn-; 2. zart *(Narben im Leder)*

f.-grain *(Met)* Feinkorn *n (Gefüge)*
f. grain emulsion Feinkornemulsion *f*
f.-grained feinkörnig
f.-grained fracture feinkörniger Bruch *m*, feinkörnige Bruchfläche *f*
f.-grained iron Feinkorneisen *n*
f. gravel Feinkies *m*; Grus *m*; Splitt *m*
f. grinding 1. Feinstschleifen *n*, Polierschleifen *n*, Schleifpolieren *n*; 2. Feinmahlen *n*, Feinmahlung *f*, Feinzerkleinerung *f*, Nachmahlen *n*
f. ground feingemahlen
f. hand feed Handfeinvorschub *m*
f. hole drilling Feinbohren *n*
f. lace re-embroidered with metallic thread *(Text)* Filigranarbeit *f* mit Metallfäden
f. machining Feinschlichtbearbeitung *f*
f. mechanics Feinmechanik *f*
f. mesh[ed] engmaschig, feinmaschig
f. metallics *(Met)* Metallpartikel *fpl*
f. micrometer adjustment Feineinstellung *f*
f. ore Feinerz *n*
f. pearlite *(Met)* feinstreifiger Perlit *m*
f.-pitch mit enger Zahnteilung
f.-pitch system amerikanisches Zahnteilungssystem *n*
f.-pitch thread Feingewinde *n*
f. plane *(Bw)* Putzhobel *m*
f.-pointed flame Stichflamme *f*
f. polishing Vorfeinpolieren *n*
f. pored feinporig, mikroporig
f. raggings *(Met)* Erzpartikel *fpl (grobe Sandkorngröße)*
f. reduction crusher Granulator *m*, Feinbrecher *m (Zerkleinerung auf Korngrößen unter 30 mm)*
f. regulating knob Feinstellknopf *m*
f. regulation Feinregelung *f*
f. rib circular knitting machine Feinrippenrundstrickmaschine *f*
f. roving *(Text)* feines Vorgarn *n*
f. roving frame Doppelfeinbank *f*, Doppelfeinflyer *m*, Extrafeinflyer *m*
f. rubble Feinschotter *m*
f. sand Feinsand *m*
f. saw Feinsäge *f*
f. scanning Feinabtastung *f*
f. scrap leichter Schrott *m*
f. screen Feinrost *m*, Feinsieb *n (Aufbereitung)*
f. screw thread Feingewinde *n*
f. screw thread series Feingewindereihe *f*
f. seed wheel Feinsaatrad *n*
f. seeds Sämereien *pl*
f. segmental chip Feinspan *m*
f. setting Feineinstellung *f*
f. sight *(Mil)* Feinkorn *n*
f. silver Feinsilber *n*, Kapellensilber *n*
f. small coal Feinförderkohle *f*
f. small duff gewaschene Feinkohle *f*
f. smudge Feinkohle *f*
f. spinning Feinspinnen *n*, Ausspinnen *n*
f. spinning frame Feinspinnmaschine *f*, Fertigspinnmaschine *f*
f. steel Edelstahl *m*
f. steel wire drawing Stahlfeinzug *m*
f.-step feinstufig
f. structure Feinstruktur *f*, Feingefüge *n*
f. structure anomaly Feinstrukturanomalie *f*
f. structure constant Feinstrukturkonstante *f*
f. structure term Feinstrukturterm *m (Spektroskopie)*
f. thread Feingewinde *n*
f. threaded mit Feingewinde [versehen]
f. tooth blade for web saws Absatzsägeblatt *n*
f. tooth milling cutter Walzenfräser *m* mit kleiner Zahnteilung
f. tooth web saw Absatzsäge *f*
f. tuning *(El)* Feinabstimmung *f*
f. weave mesh fein eingestelltes Gewebe *n*
f.-wire fuse Fein[draht]sicherung *f*
finely broken fein zerrieben; gut unterteilt
f. crystalline feinkristallin
f. deposited feinverteilt *(Niederschlag)*
f. divided feinverteilt, feinzerteilt; staubförmig
f. ground feingemahlen
f. ground grain *(Brg)* Kunstfeinkorn *n*
fineness 1. *(Text)* Feinheit *f (Fasern allgemein)*; Denier *n (Seide)*; 2. Titer *m*; 3. *(Met)* Feingehalt *m*; 4. Völligkeit *f (eines Schiffes)*
f. modulus Feinheitszahl *f (Beton)*
f. of an alloy Feingehalt *m* einer Legierung
f. of gold standard Feingehalt *m* des Goldes
f. of grain Kornfeinheit *f*
f. of scanning *(Fs)* Rasterfeinheit *f*
f. of the cotton fibre Feinheit *f* der Baumwollfaser
f. ratio *(Flg)* Schlankheitsverhältnis *n*
finery *(Met)* Frischhütte *f*
fines 1. Feinerz *n*, pulverförmiges Erz *n*; 2. *(Bw)* Feinstoffe *mpl*, Feinteile *npl*; 3. Feinkohle *f*; 4. Feinstoff *m*, Nullfaser *f*; Schlammanteil *m (Papierherstellung)*; 5. Späne *mpl*
finest grain Feinstkorn *n*
f. grain briquette Feinstkornbrikett *n*, F4-Brikett *n*
f. wire Feinstdraht *m*, Haardraht *m*
finger 1. Finger *m*; Daumen *m*; Greifer *m*; 2. Flügel *m (Flügelpumpe)*
f.-action tool *(Kern)* Fingergreifer *m (Manipulator)*
f. bars on the grinder seitliche Abdichtungsplatten *fpl* am Schleifer *(Holzschliff)*
f. contact *(El)* Fingerkontakt *m*

f. guard Fingerschutzeinrichtung *f*
f. nut Flügelmutter *f*
f. ring Haltering *m (Schraublehre)*
f. stop *(Fmt)* Fingeranschlag *m*
f.-tip control Tippschaltung *f*
f.-tip switch Tippschalter *m*
fingering wool feste Handstrickwolle *f*
fingernail test Fingernagelprobe *f*
fining Raffination *f*; Reinigung *f*; Frischen *n*; Feinen *n*
f. agent Klärmittel *n*, Aufheller *m*, Schönungsmittel *n*
f. forge *(Met)* Frischfeuer *n*, Luppenfeuer *n*
f. hearth *(Met)* Frischherd *m*
f. in hearth Herdfrischen *n*
f. period *(Met)* Oxydationsperiode *f*
f. process *(Met)* Frischverfahren *n*, Frischprozeß *m*, Frischarbeit *f*
finish / to 1. fertigstellen, beenden; 2. *(Masch)* fertigbearbeiten; schlichten; aufarbeiten; 3. *(Text)* appretieren, ausrüsten, veredeln; 4. ausputzen *(Schuhe)*; 5. *(Bw)* [ver]putzen; 6. feinmahlen *(Papier)*; 7. richten *(Elektrotypie)*
f. bright hochglanzpolieren
f. fair *(Bw)* glätten
f. grind fertig schleifen; nachschleifen
f.-machine schlichten; fertig bearbeiten
f. ream fertig reiben
f.-size *(Masch)* kalibrieren; auf Endmaß bringen
f. to gauge auf genaues Maß bringen (fertigen); kalibrieren
f. to size auf Fertigmaß (Endmaß) bearbeiten; fertig schleifen; fertig schlichten; fertig polieren
f.-turn schlichten; fertig drehen
finish 1. Oberflächengüte *f*, Oberflächenbeschaffenheit *f*; Ausführung *f*, Ausrüstung *f*; Aussehen *n*; 2. Oberflächenschutz *m*; Deckschicht *f*; 3. *(Text)* Appretur *f*, Ausrüstung *f*
f. accuracy Bearbeitungsgenauigkeit *f*; Schleifgenauigkeit *f (optische Systeme, Kristalle)*
f. allowance *(Masch)* Schlichtzugabe *f*
f. anneal[ing] Schlußglühung *f*
f. boring Fertigbohren *n*; Nachbohren *n*
f.-boring operation Fertigbohrarbeit *f*
f. breaking machine *(Text)* Appreturbrechmaschine *f*
f. coat *(Bw)* Oberputz *m*; Deckanstrich *m*
f. cut Fertigschliff *m*, Feinschliff *m (optisches Glas)*
f. draft *(Text)* Endverzug *m*
f. draw punch Fertigzugstempel *m*
f. for increasing the body *(Text)* Erschwerungsappretur *f*
f. forged genaugeschmiedet
f. forging Fertigschmieden *n*
f. grinding Fertigschliff *m*
f. machining Schlichten *n*; Fertigbearbeitung *f*
f. mark Schlichtzeichen *n*; Oberflächen[bearbeitungs]zeichen *n*
f. of setting Abbindeende *n*, Ende *n* des Abbindevorgangs *(Beton)*
f. of the cut *(Masch)* Ausschneiden *n*
f. planing *(Masch)* Schlichthobeln *n*
f. polishing Fertigfeinpolieren *n*
f. quality Oberflächengüte *f*
f. reaming with flute reamer Fertigreiben *n* mit Maschinenreibahle
f. refining *(Met)* Fertigfrischen *n*
f. rolling Fertigwalzen *n*
f. smoothing *(Bw)* Glätten *n*
f. stamping Fertigschmieden *n*
f. turning Fertigdrehen *n*
f.-turning tool Schlichtdrehmeißel *m*
finished article Enderzeugnis *n*, Endprodukt *n*, Fertigerzeugnis *n*, Fertigprodukt *n*, Fertigware *f*
f. black plate Schwarzblech *n*
f. forging Endform *f (Schmiedestück)*
f. paper ausgerüstetes Papier *n*
f. part Fertigteil *n*
f. parts store Fertigteillager *n*
f. product Enderzeugnis *n*, Endprodukt *n*, Fertigerzeugnis *n*, Fertigprodukt *n*, Fertigware *f*
f. rolling mill products fertige Walzerzeugnisse *npl*
f. section Fertigprofil *n*
f. size Fertigmaß *n*
f. width *(Text)* Fertigbreite *f*
finisher 1. Fertigmacher *m*, Ausarbeiter *m*; 2. *(Bw)* Deckenfertiger *m*, Straßenfertiger *m*; 3. *(Masch)* Fertiggesenk *n*; 4. *(Met)* Polierwalze *f (Walzwerk)*; 5. Feinzeugholländer *m*
f. card *(Text)* Feinkrempel *f*, Feinkarde *f*
finishing 1. Fertigstellung *f*; 2. Fertigbearbeitung *f*; Feinbearbeiten *n*, Schlichten *n*; 3. *(Text)* Appretur *f*, Ausrüstung *f*, Veredlung *f*; 4. *(Met)* Oberflächenbehandlung *f*; Hochglanzpolieren *n*
f. allowance Schlichtzugabe *f*
f. and decatizing machine *(Text)* Finishdekatiermaschine *f*
f. beam *(Bw)* Glättbohle *f*
f. bit Fertigbohrmeißel *m*, Endbohrmeißel *m*
f. block Feinzug *m (Drahtzieherei)*
f. brush *(Led)* Ausreibebürste *f*
f. burn Garbrand *m (keramische Industrie)*
f. card *(Text)* Feinkarde *f*
f. chip *(Holz)* Putzspan *m*
f. coat *(Bw)* Deckanstrich *m*; Oberputz *m*
f. coat of varnish *(Holz)* Deckpolitur *f*
f. composition Hochglanzpoliermittel *n*
f. compound 1. Appreturmasse *f*; 2. Schleifpaste *f*

f. cut Fertigschnitt *m*, Schlichtschnitt *m*; Feinhieb *m (Feile)*
f. cutter Fertigfräser *m*, Nachfräser *m*, Schlichtfräser *m*
f. die 1. Fertiggesenk *n (Gesenkschmieden)*; 2. Endstein *m (Ziehen)*
f. draft *(Text)* Endverzug *m*
f. edger Fertigstaucher *m (Walzwerk)*
f. end Kalibrierteil *n (Räumnadel)*
f. file Schlichtfeile *f*
f. furnace *(Met)* Fertigfrischofen *m*, Fertigschmelzofen *m*, Feinungsofen *m*
f. groove Polierkaliber *n*, Fertigkaliber *n (Walzwerk)*
f. hand taper reamer Handkegelreibahle *f* zum Fertigreiben
f. head Schlichtkopf *m*
f. hob Fertigwälzfräser *m*
f. holes Nacharbeiten *n* von Bohrungen
f. kiln Glattofen *m (keramische Industrie)*
f. knife *(Holz)* Abziehmesser *n*, Putzmesser *n*
f. lathe Feinbearbeitungsdrehmaschine *f*
f. lathe tool Schlichtdrehmeißel *m*
f. layer *(Bw)* Verschleißschicht *f*, Deckschicht *f*, oberste Schicht *f*
f. line *(Bw)* Fertigungsbahn *f*
f. machine 1. *(Bw)* Fertiger *m*; Deckenfertiger *m*, Straßenfertiger *m*; 2. Feinmühle *f (Papier)*
f. mill Fertigwalzwerk *n*; Fertig[walz]straße *f*; Feinwalzstraße *f*
f. mill train Fertigwalzstraße *f*
f. mould Fertigform *f (z. B. Flaschenherstellung)*
f. operation Feinbearbeitungsverfahren *n*
f. paint Deckanstrich *m*; Deckfarbe *f*
f. pass 1. Fertigzug *m (Ziehen)*; 2. Fertigstich *m*, Polierstich *m*, Schlichtstich *m*; Schlußkaliber *n (Walzwerk)*
f. process *(Masch)* Schlichtverfahren *n*
f. reamer Fertigreibahle *f*
f. roasting *(Met)* Garrösten *n*
f. roll Endwalze *f*, Fertigwalze *f*; Polierwalze *f*
f. roll train Fertigwalzstraße *f*
f. rolls Fertig[walz]gerüst *n*
f. room Ausrüstungssaal *m*
f. screed *(Bw)* Glättbohle *f*
f. shop *(Met)* Zurichterei *f*
f. stand Polierständer *m (Walzwerk)*
f. stand of rolls Fertig[walz]gerüst *n*
f. tap Fertigschneider *m (Gewinde)*
f. taper pin reamer Fertigkegelreibahle *f*
f. taper reamer Fertigreibahle *f* für kegelige Bohrungen
f. temperature Abbiegetemperatur *f*
f. tool Schlichtmeißel *m*; Schlichtwerkzeug *n*
f. tooth Schlichtzahn *m*; Kalibrierzahn *m (Räumnadel)*
f. trades *(Bw)* Ausbaugewerke *npl*
f. train Fertigstraße *f*, Fertigstrecke *f (Walzwerk)*
f. trowel Glätter *m*, Glättkelle *f*
f. varnish Lacküberzug *m*, Überzugslack *m*
f. work Schlichtarbeit *f (Spanen)*
finite 1. *(Math)* endlich; 2. begrenzt, [von] endlicher Abmessung (Länge)
f. automaton *(BMSR)* Automat *m* mit Endzuständen
f. game *(Dat)* endliches Spiel *n*
f. length endliche Länge *f*
f. life-time Neutronenlebensdauer *f* im endlich ausgedehnten Medium
f. number *(Math)* endliche Zahl *f*
f. stretch endliche Dehnung *f*
finiteness Endlichkeit *f*
finity Endlichkeit *f*
Fink truss *(Bw)* Dreieckfachwerkbinder *m*, Dreieckbinder *m* System Fink, Fink-Binder *m*
finless gratlos *(Walzwerkerzeugnisse)*; nahtlos *(Gußstücke)*
f. rocket Rakete *f* ohne Stabilisierungsflossen
f. three-stage rocket Dreistufenrakete *f* ohne Stabilisierungsflossen
finned *(Masch)* gerippt
f. pipe Rippenrohr *n*
f. projectile *s.* f. rocket
f. rocket flügelstabilisierte Rakete *f*
finning 1. Gratbildung *f (Walzerzeugnisse)*; Nahtbildung *f (Gußstücke)*; 2. *(Flg)* Verrippung *f*
Finsen lamp Finsen-Lampe *f*
fintube Rippenrohr *n*
fiordvessel Fördeschiff *n*, Fjordschiff *n*
fioryte *(Min)* Geyserit *m*, Kieselsinter *m*
fir Tanne *f*; Tannenholz *n*
f. apple Tannenzapfen *m*
f. ball *s.* f. apple
f. cone *s.* f. apple
f. needle Tannennadel *f*
f. oil Tannennadelöl *n*
f. resin Tannenharz *n*
f.-tree crystallite *(Met)* Dendritenkristall *m*, Tannenbaumkristall *m*
f.-tree form *(Masch)* Tannenbaumprofil *n*
f.-tree like *(Met)* dendritisch, tannenbaumartig *(Gefüge)*
f.-tree root Tannenbaumfuß *m (Turbinenschaufel)*
f.-tree root fixing Tannenbaumfußbefestigung *f*
f.-tree structure *(Met)* Dendritenstruktur *f*, Tannenbaumstruktur *f*
f. wood Tannenholz *n*
fire / to 1. heizen *(Dampfkessel)*; 2. zünden; 3. brennen *(Keramik)*; 4. *(Am)* auslösen; ausklinken; fristlos entlassen *(Betriebsausdruck)*; 5. abfeuern; abschießen

f.-gild [feuer]vergolden
f. in lane *(Mil)* sturmreif schießen
f.-tin [feuer]verzinnen
fire Feuer *n*
f.-alarm Feuermelder *m*
f.-arms Handfeuerwaffen *fpl*
f. assay Brandprobe *f*
f. bar 1. *(Wärm)* Roststab *m*; 2. *(El)* Keramikstab *m* für Heizstrahler
f. bar of the trough grate Muldenroststab *m*
f. bar spacing Rostspaltweite *f*
f. boat Feuerlöschboot *n*, Spritzenboot *n*
f. bomb Brandbombe *f*
f. boss *(Am)* *(Brg)* Fahrhauer *m*, Fahrsteiger *m*
f. box Feuerungsraum *m*, Feuerkammer *f*, Flammkammer *f*, Feuerbuchse *f*, Heizkammer *f*
f.-box bottom flange Feuerbuchsbodenring *m*
f.-box shell Feuerbuchsmantel *m*
f. brick Schamottestein *m*, Schamotteziegel *m*
f. bridge *(Wärm)* Feuerbrücke *f*
f. brigade Feuerwehr *f*
f. bronze finish[ed] goldbronziert
f. bulkhead *(Schiff)* Feuerschott *n*, Rauchschott *n*
f. chamber Verbrennungsraum *m*
f. clay feuerfester Ton *m*, Schamotte[erde] *f*
f.-clay brick Schamotteziegel *m*
f.-clay crucible Schamottetiegel *m*, Tontiegel *m*
f.-clay lining Schamotteauskleidung *f*, Schamotteausfütterung *f*
f.-clay mortar Schamottemörtel *m*
f. combatting Brandbekämpfung *f*
f. control *(Mil)* Feuerleitung *f*
f.-control device Feuerleitgerät *n*
f.-control director Kommandogerät *n* *(Flak)*
f.-control equipment Feuerleitgerät *n*
f.-control instrument *s.* f.-control equipment
f.-control post Feuerleitzentrale *f*, Feuerleitstelle *f*
f.-control radar Feuerleitradar *n*, Radar *n* zur Feuerleitung
f.-control set Feuerleitgerät *n*, Geschützrichtgerät *n*
f.-control station Feuerleitstation *f*, Geschützrichtstation *f*; Feuerleitzentrale *f*
f. crack Haarriß *m* in Glasuren
f. crack of roll Warmriß *m* einer Walze
f. cracking Brandrissigkeit *f*
f. dam Branddamm *m*
f. damage Brandschaden *m*
f. damp Grubengas *n*, schlagende Wetter *pl*
f. damp drainage Methanabsaugung *f*
f. damp indicator Schlagwetteranzeiger *m*
f. damp proof schlagwettergeschützt
f. detecting arrangement Feuermeldeanlage *f*
f. detection device [elektronisches] Feuermeldegerät *n*
f. director *(Mil)* Feuerleitgerät *n*, Kommandogerät *n*
f.-division wall Brandmauer *f*
f. door Feuertür *f*, Heiztür *f*, Ofentür *f*; Brandtür *f*
f. door frame Heiztürrahmen *m*
f. door handle Heiztürgriff *m*
f. door latch Heiztürverschluß *m*
f. drencher Feuerlöschgerät *n*, Handfeuerlöscher *m*
f. engine Brandspritze *f*, Feuerspritze *f*
f. escape Feuerleiter *f*, Nottreppe *f*
f. extinguisher Handfeuerlöscher *m*, Feuerlöscher *m*, Feuerlöschgerät *n*
f.-extinguishing equipment Feuerlöschanlage *f*
f. fighter Feuerwehrmann *m*
f. fighting Feuerbekämpfung *f*, Brandbekämpfung *f*
f. fighting by foam systems Schaumlöschverfahren *n*
f. fighting pump Feuer[lösch]pumpe *f*, Feuer[lösch]spritze *f*
f. float Feuerlöschboot *n*
f. for effect *(Mil)* Wirkungsschießen *n*, Wirkungsfeuer *n*
f.-gilded feuervergoldet
f. gilding Feuervergoldung *f*
f. grate Feuerrost *m*
f. hazard Brandgefahr *f*
f. hole Schürloch *n*
f. hose Feuerwehrschlauch *m*, Feuerspritzenschlauch *m*
f. hunter Brandwache *f*
f. insurance Feuerversicherung *f*
f.-lighter Feueranzünder *m*
f. load Brandlast *f*
f. main Feuerlöschhauptleitung *f*
f.-monitor Feuerlöschboot *n*
f. net Sprungnetz *n*
f.-place 1. Kamin *m*, Herd *m*; 2. Heizraum *m*
f. plug Feuerhahn *m*; Hydrant *m*
f. point Entflamm[ungs]punkt *m*; Brennpunkt *m* *(Öl)*
f. policy Feuerversicherungspolice *f*
f. pond Feuerlöschteich *m*
f. prevention Feuerverhütung *f*
f.-proof brandfest, feuerbeständig, feuerfest, feuersicher
f.-proof brick feuerfester Stein (Ziegel) *m*
f.-proof clothing Feuerschutzanzug *m*; Feuerschutzanzüge *mpl*
f.-proof lining feuerfeste Auskleidung *f*
f.-proof mortar Feuerkitt *m*
f.-proof wall Brandmauer *f*
f. proofing Feuerschutz *m*, Brandschutz *m*
f. protection Brandschutz *m*

f. refining *(Met)* Raffinieren *n* auf trockenem Wege
f. resistance Feuerbeständigkeit *f*, Feuerwiderstandsfähigkeit *f*
f.-resistant feuerbeständig, flammwidrig
f.-resistant finish *(Text)* Flammschutzausrüstung *f*, Flammschutzimprägnierung *f*
f.-resistant test *(Text)* Feuerbeständigkeitsprüfung *f*
f.-resisting feuerbeständig, flammwidrig; feuerhemmend; schwer entflammbar *(Öl)*
f.-resisting bulkhead Feuerschott *n* *(feuersicheres Schott aus Metall)*
f.-resisting wall Brandmauer *f*
f.-resistive feuerbeständig; feuerhemmend; schwer entflammbar
f.-retardant feuerdämmend, feuerhemmend
f.-retardant agent Brennbarkeitsverzögerer *m*
f.-retarding bulkhead Feuer[hemm]schott *n*
f.-retarding paint Feuerschutzanstrich *m*
f. risk Brandgefahr *f*
f. runner Brandwache *f*
f. screen bulkhead Feuerschirmschott *n*
f. shrinkage *(Met)* Feuerschwindung *f*
f. sprinkler *(Bw)* Sprinkleranlage *f*
f.-stopping feuerdämmend, feuerhemmend
f. stroke Arbeitshub *m*, Explosionshub *m* *(Verbrennungsmotor)*
f. test Brandversuch *m*
f.-tinned feuerverzinnt
f.-tinning Feuerverzinnen *n*
f. tongs Tiegelzange *f*
f. top Gichtfeuer *n* *(Hochofen)*
f. tube Heizrohr *n*, Rauchrohr *n*
f.-tube boiler Flammrohrkessel *m*, Heizrohrkessel *m*, Rauchrohrkessel *m*; Feuerrohrdampfkessel *m*
f.-tube heating surface Röhrenheizfläche *f*
f. wall Brandmauer *f*; Brandschott *n*
f. welding Feuerschweißen *n*
firebox *s.* fire box
fireclay *s.* fire clay
firecracker welding *(Hafergut method)* Elin-Hafergut-Schweißverfahren *n*
firecrackers Schwärmer *mpl* *(Feuerwerk)*
fired lime Branntkalk *m*, gebrannter Kalk *m*
f. with beheizt mit
firedamp *s.* fire damp
fireman 1. Heizer *m*; 2. Sprengmeister *m*; 3. Sicherheitsinspektor *m*; 4. Feuerwehrmann *m*
fireman's axe Feuerwehraxt *f*, Feuerwehrbeil *n*
firemen Löschmannschaft *f*
firemen's quarters *(Schiff)* Heizerlogis *n*
fireplace *s.* fire place
fireproof *s.* fire proof
firer *(Brg)* Schießmeister *m*
firestone Feuerstein *m*, Flint *m*
firewall *s.* fire wall
firewood Brennholz *n*
f. factory Brennholzspalterei *f*
f. truck Scheitholzwagen *m*
firing 1. Zündung *f*, Zünden *n*; Anzünden *n*; 2. Abfeuern *n*; Abschießen *n*; 3. Brennen *n* *(Keramik)*; 4. Heizen *n*, Beheizen *n*, Feuern *n*, Feuerung *f*; Heizung *f*; 5. Überhitzen *n* *(eines Lagers)*; 6. *(Brg)* Schießen *n*, Abtun *n* *(von Schüssen)*
f. apparatus Zündmaschine *f* *(für Sprengungen)*
f. area Schießstand *m*
f. base *(Rak)* Startplatz *m*
f. bricks Ziegelbrennen *n*
f. cone Hitzdaube *f*
f. days *(Brg)* Feuertage *mpl*
f. duration *(Rak)* Brennzeit *f*
f. kiln Brennofen *m*
f. lever *(Mil)* Abzugshebel *m*
f. order Zündfolge *f* *(Verbrennungsmotor)*
f. pad *(Rak)* Starttisch *m*
f. period *(Wärm)* Beschickungsperiode *f*
f. pin *(Mil)* Schlagbolzen *m*
f. platform *(Rak)* Startplattform *f*
f. point Zündzeitpunkt *m*
f. position Anschlag *m* *(Schießen)*
f. potential Zündspannung *f*
f. range Schußentfernung *f*
f. site *(Rak)* Startplatz *m*
f. stroke Arbeitshub *m*, Explosionshub *m* *(Verbrennungsmotor)*
f. technique Feuerführung *f*; Feuerbedienung *f*
f. time 1. Zündzeit *f* *(z. B. Gasentladungsröhre)*; 2. Brennzeit *f* *(z. B. Rakete)*
f. tools Heizgeräte *npl*
f. top-centre oberer Totpunkt *m* *(vor Arbeitshub; Verbrennungsmotor)*
f. torque Zünddrehmoment *n* *(Wärmekraftmaschine)*
f.-up Anfeuern (Anheizen) *n* des Kessels
f. voltage Zündspannung *f*
firm Firma *f*
firm fest, dauerhaft, solid
f. grained *(Led)* festnarbig
f. ground fester Boden *m* *(Erdboden)*
f. jaw fester Schenkel *m*
f. joint dividers Spitzzirkel *m* mit Genaugelenk
f. joint inside caliper Innentaster *m* mit Schraubenscharnier
f. joint outside caliper Außentaster *m* mit Schraubenscharnier
f. knot *(Holz)* eingewachsener Ast *m*
f. line *(Bw)* Vollinie *f*
f. offer festes Angebot *n*
firmer chisel Stechbeitel *m*

firmly secured eingespannt
firmness Festigkeit *f*
firm's symbol Firmenzeichen *n*
firring *(Bw)* Aufschiebling *m*, Futterholz *n*, Dachverschalung *f*, Unterfutterung *f*
first accelerator Voranode *f*, Sauganode *f*
f. aid erste Hilfe *f*
f. aid box Rettungskasten *m*, Verbandskasten *m*
f. aid dressing Verbandspäckchen *n*
f. anode erste Anode *f*, Sauganode *f* *(Katodenstrahlröhre)*
f. bit Anfangsbohrer *m*, Vorbohrer *m*
f. breaker *(Text)* Vorkrempel *f*, Reißkrempel *f*
f. cabinet *(Rak)* Vorgestell *n*
f. circuit noise Rauschen *n* des ersten Kreises (Schwingkreises)
f.-class erstklassig, erster Klasse
f. class *(Eis)* erste Klasse *f*
f.-class conductor *(El)* Leiter *m* erster Klasse
f.-class ore verhüttungsfähiges Erz *n*
f.-class radio telegraph licence Funker-Zeugnis *n* 1. Klasse
f.-class workmanship Wertarbeit *f*
f. coat Unterputz *m*
f. coat of paint Grundanstrich *m*, Grundierung *f*
f. collision *(Ph)* Erststoß *m*, erster Stoß *m*
f. collision correction Erststoßkorrektion *f*
f. cost Gestehungskosten *pl*; Anschaffungskosten *pl*; Anlagekosten *pl*; Herstellungskosten *pl*
f. curvature of a space curve erste Krümmung *f* einer Raumkurve
f. cut Anschnitt *m (Gewindeschneiden)*
f. degree [skin] burn Verbrennung *f* ersten Grades
f. detector HF-Gleichrichter *m*
f. draw Vorzug *m (Tiefziehen)*
f. drawing frame *(Text)* Grobstrecke *f*
f. etching Vorätzung *f*
f. finished section Vorschlichtprofil *n (Walzen)*
f. finishing pass Vorpolierstich *m (Walzen)*
f. finishing roll Vorschlichtwalze *f (Walzwerk)*
f. focusing lens Vorsammellinse *f*
f. form print *(Typ)* Schöndruck *m*
f. gear *(Kfz)* erster Gang *m*
f.-grade hochwertig, höchstwertig
f. growth Urwaldbestand *m*
f. harmonic Grundschwingung *f*; Grundfrequenz *f*
f. image *(Opt)* [erstes] Zwischenbild *n*
f. interlace *(Fs)* erster Zeilensprung *m*
f. law of thermodynamics erster Hauptsatz *m* der Thermodynamik
f. loading coil section *(El)* Anlauflänge *f (Länge bis zur ersten Spule)*
f. marking signal Vorsignal *n (Impulstechnik)*
f. moment [of area] statisches Moment *n*
f.-order system *(BMSR)* System (Verzögerungsglied) *n* 1. Ordnung
f.-order transition Phasenumwandlung *f* erster Ordnung
f. paint Grundfarbe *f (Anstrich)*
f. pass Anstich *m (Walzwerk)*
f. pickling Vorbeizen *n*
f. printing *(Typ)* Schöndruck *m*
f. proof Vorauskorrektur *f*, Hauskorrektur *f*
f. propellant valve *(Rak)* Brennstoffvorventil *n*
f. quadrant angle Winkel *m* im ersten Quadranten
f. quality pulp Gutstoff *m (Papierherstellung)*
f. quantum number Hauptquantenzahl *f*
f.-rate erstklassig, hochfein
f. refined copper rohgares Kupfer *n*
f. resonating frequency Resonanzgrundfrequenz *f*
f. run slag Reduktionsschlacke *f (Zinngewinnung)*
f. runnings Destillationsvorlauf *m*, Vorlauf *m*
f. slag Frischschlacke *f*
f.-smelting pig iron Vorschmelzeisen *n*
f. stage erste Raketenstufe *f*
f. stage engine Triebwerk *n* der ersten Raketenstufe
f. tap Vorschneider *m (Gewinde)*
f.-time operation erstmalige Inbetriebsetzung *f*
f. Townsend discharge primäre Townsend-Entladung *f*
f. upsetter *(Masch)* Vorstauchstempel *m*
f. voltage range Spannungsbereich *m* unter 30 V, unterster Spannungsbereich
f. weight *(Brg)* erster Setzdruck *m (Ausbau)*
Fischer-Tropsch process *(Ch)* Fischer-Tropsch-Verfahren *n*
fish / to 1. fischen; Fischfang treiben; 2. *(Bw, Masch)* verlaschen; verstärken; 3. *(El)* durchfädeln; durchziehen *(Drähte)*
fish 1. Fremdkörper *m (im Gußstück)*; 2. Antennenendgewicht *n (Schleppantenne)*
f.-bar Lasche *f*
f.-beam *(Bw)* Fischbauchträger *m*
f.-bellied fischbauchig, wellig, gewellt
f.-bellied girder *(Bw)* Fischbauchträger *m*
f.-bellied grate bar *(Wärm)* fischbauchförmiger Roststab *m*
f.-bladder Fischblase *f*; Schwimmblase *f*
f.-bolt Laschenschraube *f*, Laschenbolzen *m*
f.-bone *(Text)* Fischgratmuster *n*
f.-bone aerial Tannenbaumantenne *f*
f. boom Anker[aufwinde]baum *m*
f.-design *s.* f.-bone

f. eye Materialfehler *m*; Flockenriß *m*; Knötchen *n*; ringförmiger Schweißfehler *m*; Kalklunker *m (Putz)*
f.-factory ship Fischverarbeitungsschiff *n*, Fischfabrikschiff *n*
f. glue Fischleim *m*
f. grid *(Bw)* Fischrechen *m*
f. hold Fisch[lade]raum *m*
f. hook Angelhaken *m*
f. joint Laschenstoß *m*, Verlaschung *f*
f. ladder Fischtreppe *f*, Fischleiter *f*, Fischschleuse *f*, Fischpaß *m*
f. locating equipment Fischortungsgerät *n*, Fischortungsausrüstung *f*
f. oil Tran *m*, Fischöl *n*
f. paper dünne Vulkanfiber *f*; Isolierpapier *n*
f. pass *s.* f. ladder
f. plate *(Bw)* Lasche *f*, Stoßlasche *f*, Zuglasche *f*; *(Eis)* Schienenlasche *f*; *(Brg)* Klammerlasche *f (Ausbau)*
f.-pole [probe] Angelsonde *f (amerikanisches Strahlungsmeßgerät)*
f. pond Fischteich *m*
f. screen *(Bw)* Fischrechen *m*
f.-stocking pond Setzteich *m (für Fische)*
f.-tail *(Bw)* auseinandergebogene Enden *npl*; Anker *m*
f.-tail bit Fischschwanzmeißel *m*
f.-tail bolt *(Bw)* Ankerschraube *f*
f.-tail kneader Fischschwanzkneter *m*
f.-tail twin Schwalbenschwanzzwilling *m (Kristallografie)*
f. tape *(El)* Fädelband *n (zum Einziehen von Drähten)*
f.-track *(Kern)* kurze Nebelspur *f*
f.-wire *(El)*
Einziehdraht *m*, Durchziehdraht *m*
fishbelly girder *(Bw)* Fischbauchträger *m*
fished beam Balken *m* mit Seitenverstärkungen
f. joint Laschenstoß *m*, Laschenverbindung *f*
fisher Fischer *m*
fisherman 1. Fischer *m*; 2. Fischereifahrzeug *n*
fishery Fischerei *f*
f. patrol boat Fischereischutzboot *n*
f.-protecting vessel Fischereischutzboot *n*
fishing 1. Fischerei *f*; Fischen *n*; Angeln *n*; 2. Laschenstoß *m*, Laschenverbindung *f*, Verlaschung *f*
f. beam verlaschter Balken *m*
f. boat Fischerboot *n*
f. box *(El)* Abzweigdose *f*
f. cutter Fischkutter *m*
f. fleet Fischereiflotte *f*
f. gear Fischfanggerät *n*
f. grab Bohrfänger *m (Tiefbohrtechnik)*
f. grounds Fischgründe *mpl*
f. line Angelleine *f*
f. rod Angelrute *f*
f. season Fangzeit *f (Fischerei)*
f. smack Fischkutter *m* mit Bünn
f. tools Fangwerkzeug *n*, Fanggerät *n (Tiefbohrtechnik)*
f. vessel Fischereifahrzeug *n*
fishtail *s.* fish-tail
fishtailing *(Flg)* Abbremsen *n*
fishway *s.* fish ladder
fissel / to *(El)* leise zischen
fissile *(Kern)* spaltbar; spaltfähig
f. fuel Kernbrennstoff *m*
f. material Spaltmaterial *n*, spaltbares Material *n*
f. nucleus spaltbarer Kern *m*
fissility *(Kern)* Spaltbarkeit *f*
fission Kernspaltung *f*
f. bomb Spaltbombe *f*
f. capture Spalteinfang *m*
f. chain Spaltproduktreihe *f*
f. chain reaction Kettenspaltung *f*, Spaltungskettenreaktion *f*
f. chamber Spaltungskammer *f*
f. counter Spaltungszähler *m*
f. cross section Spaltquerschnitt *m*, Wirkungsquerschnitt *m* für Kernspaltung
f. energy Spalt[ungs]energie *f*
f. fragments Kernbruchstücke *npl*; Kerntrümmer *pl*, Spaltprodukte *npl*
f. gas Spaltgas *n*, gasförmiges Spaltprodukt *n*
f. heat Spaltungswärme *f*
f. hole Spaltkanal *m*, Reaktorkanal *m* für Bestrahlungsversuche mit Spaltneutronen
f. neutron Spaltneutron *n*
f. neutron energy Spaltneutronenenergie *f*
f. poison Reaktorgift *n*, Spaltgift *n*
f. process Spaltprozeß *m*, Spaltungsreaktion *f*
f. product Spaltprodukt *n*
f. product activity Spaltproduktaktivität *f*
f. product of uranium Uranspaltprodukt *n*
f. product poisoning Spaltproduktvergiftung *f*
f. product refining plant Aufarbeitungsanlage *f* für Spaltprodukte
f. rate Spaltungshäufigkeit *f*
f. reaction Spaltungsreaktion *f*, Spaltprozeß *m*
f. reactor Spaltungsreaktor *m*
f. spectrum Energiespektrum *n* der Spaltneutronen, Spalt[neutronen]spektrum *n*
f. threshold Spaltschwelle *f*, Energieschwelle *f* für Kernspaltung
f. yield Spalt[produkt]ausbeute *f*
fissionability *(Kern)* Spaltbarkeit *f*
fissionable *(Am) s.* fissile
fissuration Rißbildung *f*
fissure 1. Riß *m*, Spalt *m*, Spalte *f*, Sprung *m*; 2. *(Geol)* Kluft *f*, Bruchspalte *f*
f. occupation *(Geol)* Spaltausfüllung *f*
f. on hardening Härteriß *m*

f. vein *(Geol)* Lagergang *m*
fissured zerklüftet
fissuring Rißbildung *f*; Spaltbildung *f*
fist Hinweis[zeichen *n*] *m*, Verweisung *f*
f. ball Faustball *m*
fistuca Pfahlrammer *m*
fit / to [ein]passen, zusammenpassen; passen; anbringen, montieren
f. by scraping durch Schaben aufpassen, einschaben
f. in einpassen
f. on aufpassen
f. one another zueinander passen
f. out ausrüsten, ausstatten
f. snugly *(Masch)* fest sitzen
f. to anschließen an; anbauen an
f. together zusammenpassen; montieren
f. up zurichten, einrichten
f. with a shunt *(El)* shunten, [mit einem Widerstand] überbrücken
fit Passung *f*; Sitz *m*; Gütegrad *m (Passung)*
fit geeignet, fertig, passend
f. for flying duties flugtauglich
f. for service afloat *(Schiff)* borddienstfähig
f. of thread Gewindepassung *f*
f. tolerance Paßtoleranz *f*
f.-up *(Bw)* Fertigungsschalung *f*, wiederverwendungsfähige Schalung *f*; Schal[ungs]tafel *f (Betonbau)*
fitch 1. zubereitetes Iltisfell *n*; 2. kleiner Malerpinsel *m (zum Kantenziehen)*
fitchering 1. Feststecken *n (Bohrarbeit)*; 2. Kantenziehen *n*
fitments Ausstattungsgegenstände *mpl*; Einbaumöbel *npl*
fitted bolt Paßbolzen *m*, Paßschraube *f*, eingepaßte Schraube *f*
f. cupboard Einbauschrank *m*, eingebauter Schrank *m*
f. in eingepaßt
f. kitchen Einbauküche *f*
f. surface Paßfläche *f*
f. with versehen mit, ausgerüstet mit
fitter Monteur *m*; Motorenschlosser *m*; Maschinenschlosser *m*; Triebwerkschlosser *m*; Spengler *m*; Installateur *m*
fitter's shop Schlosserei *f*; Montagehalle *f*; Montagewerkstatt *f*
fitting 1. Paßarbeit *f*, Nacharbeit *f*; Anbringung *f*; Aufbau *m*, Montage[arbeit] *f*; Ausrüstung *f*; 2. Armatur *f*, Fassung *f*, Formstück *n*, Rohrverbindungsstück *n*, Verbindungsstück *n*; 3. Beleuchtungskörper *m*, Leuchte *f*
f. allowance *(Masch)* Einpaßzugabe *f*
f. assembly Einpaßarbeit *f*
f. berth Ausrüstungsdock *n*, Ausrüstungsbecken *n (Werft)*
f. block Paßstein *m*
f. for lubrication Schmiernippel *m*
f. in Paßarbeit *f*
f. instruction Montageanweisung *f*
f. key Paßfeder *f*
f. member Paßteil *n*
f. of spectacles Brillenanpassung *f*
f.-out basin Ausrüstungsbassin *n*, Ausrüstungsbecken *n (Werft)*
f.-out berth Ausrüstungsdock *n*, Ausrüstungsbecken *n (Werft)*
f.-out shore Ausrüstungskai *m (Werft)*
f. part Paßteil *n*
f. piece Paßstück *n*
f. pin Paßstift *m*
f. shop Schlosserei *f*; Montagewerkstatt *f*, Montagehalle *f*
f. sleeve Montagehülse *f (Dichtungen)*
f. surface Paßfläche *f*
f. wrench Paßschlüssel *m*
fittings Armaturen *fpl*, Beschläge *mpl*, Zubehörteile *npl*
f. and fixtures *s.* fixtures and fittings
five-beam cathode ray tube Fünfstrahlröhre *f*
f.-bearing crankshaft fünfmal gelagerte Kurbelwelle *f*
f.-centred arch *(Bw)* Korbbogen *m* aus fünf Leierpunkten (Mittelpunkten)
f.-cornered pointed hand reamer fünfkantiger Aufreiber *m*
f.-digit fünfstellig
f.-electrode tube (valve) Fünfpolröhre *f*
f.-figure fünfstellig
f.-hole punching die Fünflochmatrize *f (Telegrafie)*
f.-key transmitter Fünftastengeber *m (Telegrafie)*
f.-level punched tape fünfteiliger Lochstreifen *m*
f. line pica *(Typ) (etwa)* fünf Cicero *(60 britische Punkte; Schriftgröße)*
f.-masted bark Fünfmastbark *f (Fock-, Groß-, Mittel-, Achter- und Besanmast)*
f.-masted fore-and-aft schooner Fünfmastgaffelschoner *m*
f.-masted full-rigged vessel Fünfmastvollschiff *n*
f.-membered fünfgliedrig, fünfteilig
f.-membered ring *(Ch)* Fünfring *m*
f.-minute vernier 5-Minuten-Nonius *m*
f.-nines *(Met)* Fünfneuner- *(Metall mit einer Reinheit von 99,999 Prozent)*
f.-pin base fünfpoliger Sockel *m*, Fünfstift-[stecker]sockel *m (z. B. Elektronenröhren)*
f.-point jack *(Fmt)* fünfpolige Klinke *f*
f.-point switch fünfpoliger Schalter *m*, Schalter *m* mit fünf Ausgängen
f.-prong base *s.* f.-pin base
f.-sided turret Fünfkantrevolverkopf *m*

f.-spindle automatic Fünfspindelautomat *m*
f.-unit alphabet (code) Fünferalphabet *n* *(Telegrafie)*
f.-valve superhet Fünfröhrensuper *m*
f.-wire system *(El)* Fünfleitersystem *n*
fix / to 1. *(Masch)* anbringen, befestigen, festmachen, festspannen; aufspannen *(Werkstücke)*; einspannen *(Werkzeuge)*; 2. *(Met)* einsetzen *(Kerne in die Gießform)*; 3. *(Ch)* binden; 4. fixieren *(Film)*
f. a link ein Glied gestellfest machen
f. in einbinden *(Mauerwerk)*
f. together zusammenfügen, zusammensetzen
f. up instandsetzen, reparieren
fixation 1. Fixierung *f*; Befestigung *f*; Einspannung *f*; Endeinspannung *f*; *(Bw)* Einspannungsgrad *m*; 2. Fixierung *f (Film)*
fixative Fixativ *n*; Fixiermittel *n (Farbe, Kreide)*
fixature Fixativ *n (Haar)*
fixed 1. ortsfest, stationär; fest, unbeweglich, unverschiebbar; 2. fixiert *(Film)*; 3. gestellfest *(Getriebelehre)*
f. aerial Festantenne *f*
f. and flashing light *(Schiff)* Mischfeuer *n*, Festfeuer *n* mit Blitzen
f. and loose pulley feste und lose Riemenscheibe *f*
f. armature *(Elm)* Festanker *m*; Primäranker *m*
f. ash *(Brg)* gebundene Asche *f*
f. back-iron feststehendes Putzmesser *n*
f. bar Gestellglied *n (Getriebelehre)*
f. battery ortsfeste (stationäre, fixe) Batterie *f*
f. beam eingespannter Balken (Träger) *m*
f. bearing festes Auflager *n*
f. bed 1. Festbett *n*, stationäre Schicht *f*; 2. ruhendes Festbett, Ruheschüttung *f (Wirbelschichtverfahren)*
f. bed process Prozeß *m* mit fest angeordnetem Katalysator
f. bed type miller Planfräsmaschine *f*
f. bias feste Gittervorspannung *f*
f. blade Leitschaufel *f (Turbine)*
f.-blade propeller *(Flg)* Starrluftschraube *f*
f. bobbin *(Text)* Schleifspule *f*
f. caliper gauge Loch- und Tasterlehre *f*
f. caliper jaw Meßschenkel *m*, Meßschnabel *m*, fester Schenkel *m (Schiebelehre)*
f. capacitor Festkondensator *m*, fester (unveränderlicher) Kondensator *m*
f. carbon gebundener Kohlenstoff *m*
f. cathode spot fixierter Katodenfleck *m*
f. charge Grundgebühr *f*, Festbetrag *m*
f. column eingespannte Säule (Stütze) *f*; eingespannter Stiel *m (Rahmen)*
f. condenser *s.* f. capacitor
f. contact *(El)* festes Schaltstück *n*
f. coupling starre Kupplung *f*
f.-cycle operation *(Dat)* Taktgeberbetrieb *m*, Zeitgeberbetrieb *m*
f. datum *(Bw)* Festpunkt *m*, Richtpunkt *m*
f. discharger feste (unveränderliche) Funkenstrecke *f (Lichtbogenlampe)*
f.-drum-type rotary drier Rotationstrommeltrockner *m* mit stationärem Gehäuse
f.-echo elimination Festzeichenbeseitigung *f (Radar)*
f. emitter fester Sender *m*
f. end 1. Gegenzapfen *m*, Fixpunkt *m*; 2. eingespanntes Trägerende *n*; 3. Amboß *m (fester Meßanschlag einer Schraublehre)*
f.-end arch eingespannter Bogen *m*
f.-end girder eingespannter Träger *m*
f.-end moment Einspann[ungs]moment *n (Statik)*
f. flange fester Flansch *m*
f.-focus mit konstanter Brennweite
f. frame eingespannter Rahmen *m*
f. frequency feste (konstante) Frequenz *f*
f. frequency transmitter Sender *m* mit fester (konstanter) Frequenz
f. fulcrum fester Drehpunkt *m*
f. gauge feste Lehre *f*, Grenzlehre *f*
f. grate *(Wärm)* fester Rost *m*
f. grid bias feste Gittervorspannung *f*
f. grizzly *(Brg)* Steilsieb *n*
f. gun starre Kanone *f*; starres Maschinengewehr *n*
f.-hammer mill *(Brg)* Schlagkreuzmühle *f*
f.-head blowpipe Schweißbrenner *m* ohne Regelung
f.-head T-square Reißschiene *f* mit festem Kopf
f. hub feste Nabe *f*
f. jaw fester Schenkel *m*, Meßschnabel *m* *(Schieblehre)*; feststehende Backe *f (Schraubstock)*
f. junction of thermocouple kalte Lötstelle *f* eines Thermoelementes
f. light *(Schiff)* Festfeuer *n*, festes Feuer *n*
f.-line posting *(Büro)* feste Zeileneinstellung *f* *(z. B. beim Buchen)*
f. link 1. Gestell *n (Getriebelehre)*; 2. *s.* f. bar
f. load Eigenmasse *f*, Eigenlast *f*
f. net *(Schiff)* stehendes Netz *n*, Stellnetz *n*
f. nozzle Festdrossel *f*, Vordrossel *f (pneumatischer Regler)*
f. oil nichtflüchtiges Öl *n*; Fettöl *n*
f. open-hearth furnace feststehender Siemens-Martin-Ofen *m*
f. packing festliegende Dichtung *f*
f. path of instantaneous centre feste (ruhende) Polbahn *f*
f.-pitch airscrew (propeller) *(Flg)* Starrluftschraube *f*
f. plane *(Flg)* Flosse *f*

f. plane iron Furniermesser *n*, Stahllamelle *f* *(Holzbearbeitung)*
f. plates Statorplatten *fpl* *(Drehkondensator)*
f. point 1. *(Math, Opt)* Festpunkt *m* *(Abbildung)*; 2. *(Ph)* Fixpunkt *m*, Fundamentalpunkt *m* *(z. B. Temperaturskala)*; 3. *(Dat)* festes Komma *n*
f.-point calculation *(Dat)* Festkommarechnung *f*, Festkommabetrieb *m*
f.-point levelling *(Opt)* Festpunktnivellement *n*
f.-point net *(Math, Opt)* Festpunktnetz *n*
f.-point representation *(Dat)* Festkommadarstellung *f*
f. pole *(Bw)* Standmast *m*
f. price Festpreis *m*
f. pulley Festrolle *f*, feste Rolle *f*
f. range mark generator Eichgenerator *m*, Markierungsgenerator *m*
f. rear square sight *(Mil)* Balkenvisier *n*
f. resistance Festwiderstand *m*, unveränderlicher Widerstand *m*
f. resistor Festwiderstand *m*
f. rigid to fest an
f. rim feste Felge *f*
f.-size gauge Maßlehre *f*
f. stay (steady) feststehender Setzstock *m*
f. stop Festanschlag *m*
f. tail surface *(Flg)* Leitwerkflosse *f*
f. tailplane *(Flg)* Höhenflosse *f*
f. target rejection filter Selektivfilter *n* zur Festzielunterdrückung *(Radar)*
f. timber *(Schiff)* gewachsenes (festes, eingeschnittenes) Spant *n*
f.-time call Ferngespräch *n* zu bestimmter Zeit, Festzeitgespräch *n*
f.-time lag unabhängige (konstante) Zeitverzögerung *f*
f. vessel Standgefäß *n*
f. voltage Festspannung *f*
f. ways Ablaufbahn *f* *(Helling)*; Stapellaufbahn *f*
f. weight festgesetztes Gewicht *n*
f. wheel festes Rad *n*
f.-wire resistance Drahtfestwiderstand *m*
fixedness Ortsunbeweglichkeit *f*
fixer Fixierbad *n*; Fixiermittel *n*
fixing 1. Aufspannen *n*; Aufspannung *f*; Einspannung *f*; Befestigung *f*; 2. Einsetzen *n* *(Kern)*; 3. Binden *n*; 4. Fixieren *n* *(Film)*; 5. *s.* position finding
f. accessories Befestigungsmittel *npl*
f. agent Fixiermittel *n*
f. and toning bath Tonfixierbad *n*
f. bath Fixierbad *n*
f. block *(Bw)* Dübelstein *m*
f. bolt Befestigungsschraube *f*, Befestigungsriegel *m*
f. brick Dübelstein *m*
f. by directional [wireless] telegraphy Standortbestimmung *f* mittels Richtungstelegrafie
f. catch Befestigungsriegel *m*
f. clip Befestigungsschelle *f*; Halteklammer *f*
f. flange Aufspannflansch *m*
f. key Verbindungskeil *m*
f. lug *(Opt)* Spannauge *n*
f. moment Einspannmoment *n* *(Statik)*
f. of insulator Isolatorbefestigung *f*
f. of pipes Rohrbefestigung *f*
f. plate Aufspannplatte *f*
f. position *(Schiff)* Anbringen *n* des Bestecks *(in der Karte)*
f. ring Stellring *m*
f. salt Fixiersalz *n*
f. screw Befestigungsschraube *f*
f. tape Befestigungsband *n*
fixity 1. Einspannung *f* *(Statik)*; 2. Beständigkeit *f*; Verflüchtigungsbeständigkeit *f*
fixture / to ausrüsten *(mit Vorrichtungen)*
fixture 1. Vorrichtung *f*; Spannvorrichtung *f*; 2. Installationsobjekt *n*
fixtures and fittings nichtbewegliche (festgemachte) Teile *npl* der Wohnungseinrichtung *(Decken- und Wandleuchten, Gardinen usw.)*
fizelyite *(Min)* Fizelyit *m*
fizz / to *s.* fizzle / to
fizzing Zischen *n*
fizzle / to [schwach] zischen, summen, gluckern, sprudeln
f. out langsam ausgehen, [er]sterben; versagen *(Feuer, Licht, Spannung usw.)*
flabelliform barley Fächergerste *f*
flag / to 1. Flaggensignale geben; mit Flaggen signalisieren; 2. mit Platten belegen *(Gehwege)*; mit Fliesen belegen, fliesen *(Fußböden, Wände)*
flag 1. Flagge *f*; Signalflagge *f*; 2. *(Bw)* Fliese *f*, Steinplatte *f*; 3. Gegenlichtblende *f* *(Fernsehkamera)*
f. discrimination *(Schiff)* Flaggenunterscheidung *f*, Flaggendiskriminierung *f*
f. halliard *(Schiff)* Flaggleine *f*
f. officer Flaggoffizier *m*
f. ore plattiges Erz *n*
f. pole *s.* flagpole
f. ship Flaggschiff *n*
f. station *(Eis)* Bedarfshaltestelle *f*
flagged mit Platten belegt *(Gehweg)*
flagging Plattenbelag *m* *(Gehweg)*
flagpole 1. Fahnenstange *f*; 2. *(Fs)* Balken *m* *(im Testbild)*
flags of convenience *(Schiff)* bequeme (billige) Flaggen *fpl* *(Flaggen von Ländern, die für den ausländischen Schiffsreeder günstige Steuerbedingungen bieten)*

flagstaff Fahnenmast *m*
flagstone Natursteinplatte *f*; Gehwegplatte *f*
f. of the furnace Ofenbodenstein *m*; Bodenstein *m*
flake / to abblättern, abplatzen, abschuppen
flake 1. Flocke *f*, Schuppe *f*; 2. Gerüst *n*; 3. Flachspan *m*
f. board Flachspanplatte *f*
f. caustic Ätzkali *n* in Schuppenform
f. formation Flockenbildung *f*
f. graphite Flockengraphit *m*, Schuppengraphit *m*
f.-graphite iron Gußeisen *n* mit in Streifen abgeschiedenem Graphit
f. mica *(Min)* Schuppenglimmer *m*
f. soap Seifenflocken *fpl*
f. white Bleiweiß *n*
flakes Stippen *fpl*
flakiness Flockigkeit *f*
flaking Abblättern *n*, Abblätterung *f*, Schälen *n*; Faltenbildung *f (Tiefziehen)*
flaky graphite Flockengraphit *m*, flockiger Graphit *m*
flamard flammgehärtet
flambé mit schillernder Glasur *(Keramik)*
flame / to 1. flammen, lodern *(Feuer)*; 2. flammen, sengen, schönen *(Farbleder)*
f. out ausflammen
f. up aufflammen, auflodern
flame Flamme *f*
f. adjustment Flammeneinstellung *f*
f. arc *(El)* Flammbogen *m*
f. arc lamp Effektbogenlampe *f*
f. baffling Flammenführung *f (z. B. im Brenner)*
f. bonded foamback flammgebundenes Laminat *n*
f. bonder Flammenbindevorrichtung *f (zum Laminieren)*
f. chipping Brennputzen *n*
f. cleaning Flammstrahlen *n*
f. cone Flammenkegel *m*
f. control Flammenführung *f (Lötlampe)*
f. cored carbon lamp Flammenbogenlampe *f*
f. cutter Schneidbrenner *m*
f. cutting Brennschneiden *n*
f. damper *(Rak)* Muffel *f*
f. descaling Brennputzen *n*
f. failure alarm *(BMSR)* Flammenwächter *m*
f. flue Flammenzug *m*
f. formation Flammenbildung *f*
f. furnace Flammofen *m*
f. gun *s.* f.-spray gun
f.-hardened flammengehärtet
f. hardening *(Met)* Brennhärten *n*, Flammenhärten *n*, Oberflächenhärtung *f*
f.-hardening blowpipe Härtebrenner *m*
f.-hardening machine Brennhärtemaschine *f*, Flammenhärtemaschine *f*
f.-hardening torch *(Am)* *s.* f.-hardening blowpipe
f. jet Flammenstrahl *m*
f. photometer Flammenfotometer *n*
f. photometry Flammenfotometrie *f*
f. plating *(Met)* Flammplattieren *n*
f.-proof *s.* flameproof
f. propagation Flammenfortpflanzung *f*
f. protection *(Brg)* Schlagwetterschutz *m*
f. reaction *(Ch)* Flammenfärbungsreaktion *f*
f. resistance Flammbeständigkeit *f*; Schwerbrennbarkeit *f*
f.-resistant nicht entflammbar, flamm[en]-beständig; flamm[en]sicher
f. retardant *(Bw)* feuerdämmend, feuerhemmend
f. safety lamp *(Brg)* Wetterlampe *f*
f. scarfing Brennputzen *n*, Flämmen *n*
f. spectrography Flammenspektrografie *f*
f. spectrum Flammenspektrum *n*
f.-spray gun Flammspritzpistole *f*
f. spraying *(Plast)* Flammspritzen *n*, Wärmespritzen *n*; *(Led)* Aufdüsen *n*, Aufspritzen *n*
f. thrower *(Mil)* Flammenwerfer *m*
f. throwing tank Flammenwerferpanzer *m*
f. trap *(Kfz)* Flammenrückschlagsicherung *f*, Flammenschutz *m*, Rückschlagtopf *m*
f. travel Flammenausbreitung *f*, Flammenweg *m*
f. tube Flammrohr *n*
flamed burls *(Text)* geflammte Noppen *fpl*
flameholder Flammenhalter *m*
flameproof *(Brg)* schlagwettergeschützt, schlagwettersicher
f. detonator schlagwettersichere Kapsel *f*
f. enclosure Schlagwetterschutzkapselung *f*
f. lighting fitting schlagwettergeschützte Leuchte *f*
f. switch schlagwettergeschützter Schalter *m*
flameproofing agent Flammenschutzmittel *n*
flamestat *(BMSR)* Flammenwächter *m*
flaming arc Flammbogen *m*, Effekt[licht]bogen *m*
f. arc lamp Effektbogenlampe *f*
f. furnace Flammofen *m*
flammability Brennbarkeit *f*, Entflammbarkeit *f*
f. test *(Text)* Flammfestigkeitsprüfung *f*
flammable [ent]flammbar
flange / to flanschen, bördeln
flange 1. Flansch *m*; Muffenstück *n*; Kranzstück *n*; Bund *m*; vorstehender Rand *m*; Steg *m*; Spurkranz *m*; 2. Steinscheibe *f (Holzbearbeitung)*
f. bolt Flanschbolzen *m*, Flanschenschraube *f*, Bundbolzen *m*
f. bushing Bundbuchse *f*
f.-connected angeflanscht

f. coupling Flanschkupplung *f*, Flanschverbindung *f*
f. edge Bördelrand *m*
f. edge joint Stirnstoß *m (Schweißtechnik)*
f.-face coupling Scheibenkupplung *f*
f.-guided rollers bordgeführte Wälzkörper *mpl*
f. hub Flanschnabe *f*
f. joint Flanschverbindung *f*; Flanschenverschraubung *f*; Bördelnaht *f*; Bördelstoß *m*
f. leather *(Masch)* Hutmanschette *f*
f.-mounted angeflanscht
f.-mounted motor Flanschmotor
f. mounting Flanschbefestigung *f*
f.-mounting motor Flanschmotor *m*, Anbaumotor *m*
f. nut Bundmutter *f*
f. of a beam Trägerflansch *m*
f. of casing Gehäuseflansch *m (z. B. Turbine)*
f. of coupling Wellenkupplungsflansch *m*, Kupplungsscheibe *f*
f. of distance piece Laternenflansch *m*
f. of gland Brillenfutter *n*
f. of piston rod Kolbenstangenflansch *m*
f. of the circular guide Rundführungsflansch *m*
f. rail Breitflanschschiene *f*, Breitfußschiene *f*
f. tool Flanschwerkzeug *n*
f.-type motor Flanschmotor *m*
f.-welding seam Bördelschweißnaht *f*
f. with feet Fußflansch *m*
f. with packing grooves Flansch *m* mit Dichtungsrillen
f. with plain face glatter Rohrflansch *m*
f. with spherical seat Gelenkflansch *m*
flanged angeflanscht; aufgebördelt
f. bearing Flanschlager *n*
f. branch Gegenflansch *m*
f. bush Flanschbuchse *f*
f. coupling Flanschverbindung *f*
f. coupling joint Flanschverbindung *f*
f.-edge type of weld Bördelnaht *f (Schweißtechnik)*
f. fittings Flanschfittings *npl*
f. girder I-Träger *m*, Doppel-T-Träger *m*
f. idler roller Umlenkrolle *f*
f. joint Flanschverbindung *f (Rohrleitungsbau)*
f. joint with conical ring Flanschdichtung *f* durch eingelegten konischen Ring
f. joint with tongue and groove Flanschdichtung *f* durch Feder und Nut
f. motor Flanschmotor *m*, Anbaumotor *m*
f. pipe Flanschrohr *n*
f. rail Breitflanschschiene *f*, Breitfußschiene *f*
f. seam Flanschnaht *f*, Bördelnaht *f*
f. seam riveting Bördelnietung *f*
f. sheet Bördelblech *n*; gebördeltes Blech *n*
f. socket Rohransatz *m*, Rohrstutzen *m*
f. spindle nose Spindelkopf *m* mit Spindelflansch
flanger Bördelvorrichtung *f*
flanging Umbördeln *n*, Flanschen *n*
f. machine Bördelmaschine *f*
f. test Bördelprobe *f*
flank 1. Fußflanke *f (Zahnrad)*; 2. Freifläche *f (Schneide)*; 3. Seitenfläche *f*; Giebelseite *f*; 4. Flanke *f (Gewinde)*
f. abrasion Freiflächenverschleiß *m*
f. clearance Flankenspiel *n*
f. diameter Flankendurchmesser *m (Gewinde)*
f. dispersion *(Elm)* Streufluß *m*
f. of a thread Flanke *f* eines Gewindes
f. of the drill point Hinterschliffläche *f*
f. of tooth Fußflanke *f*
f. profile Flankenform *f (Gewinde)*
f. wall Giebelmauer *f*, Giebelwand *f*
f. wear land *(Masch)* Freiflächenverschleißfase *f*
flanked by flankiert von
flanking hole *(Brg)* schräg in den Stoß gebohrtes waagerechtes Loch *n*
flanning Anschlag *m*, Leibung *f (Mauerwerk)*
flap / to flattern; schlagen; klappen
flap 1. Meißelklappe *f (Hobelmaschine)*; 2. Klappe *f*
f. actuation *(Flg)* Klappenbetätigung *f*
f. attenuator Streifenabschwächer *m (Wellenleiter)*
f. guide Klappenführung *f*
f. hinge Klappenscharnier *n*
f. shutter *(Opt)* Klappenverschluß *m*
f. valve Klapp[en]ventil *n*
f.-valve pump Klappen[luft]pumpe *f*
flapper Prallplatte *f (pneumatischer Verstärker)*
f. box *(Masch)* Klappenträger *m*, Meißelklappenträger *m*
flapping 1. Schlagen *n (z. B. des Förderseils oder eines Riemens)*; 2. Verblasen *n (Kupferraffination)*
flare / to 1. aufflammen; aufleuchten *(Licht)*; abfackeln; 2. aufweiten *(Rohre)*
f. gas Gas abfackeln
f. outward *(Masch)* nach außen ausweichen
f. up aufflackern
flare 1. *(Mil)* Leuchtsignal *n*; 2. Eruption *f (Sonne)*; 3. *(El)* Fußrohr *n*, Tellerrohr *n (Glühlampe)*; 4. *(Schiff)* Ausfall *m* [der Schiffsseiten am Bug], Seitenausfall *m* der Bordwand, Ausbauchung *f*
f. angle *(El)* Öffnungswinkel *m (Antenne)*
f. out bow *(Schiff)* ausfallender Steven *m*, ausfallendes Vorschiff *n*, springender Bug *m*
f.-path *(Flg)* befeuerte Landebahn *f*
f. point *(Mil)* Leuchtspitze *f (Torpedo)*

f. spot *(Fs)* Linsenlichtfleck *m*

flared radiating guide *(El)* Hohlleiter *m* mit Leitblechen

f. tube of rear axle *(Kfz)* Hinterachstrichter *m*

flaring cup wheel kegelige Topfschleifscheibe *f*

f. test *(Am)* Aufweitprobe *f (Rohre)*

flash / to 1. aufleuchten *(Licht)*; blinken *(Signale)*; feuern *(elektrische Maschinen)*; zucken *(Blitz)*; 2. entgraten, abgraten

f. a warning *(mittels Lampe)* Warnsignal geben

f. back zurückschlagen *(Flamme)*

f. off abschmelzen

f. over überschlagen *(Spannung)*; überspringen *(Funken)*

f. trim *(Met)* entgraten

f. up aufflammen

flash 1. Aufblitzen *n (Licht)*; 2. Szintillationsblitz *m (Auftreffen von Teilchen auf Schirm)*; 3. *(Masch)* Grat *m*, Walzgrat *m*; Preßgrat *m*; überfließender Stoff *m*; 4. erhärteter Leimwulst *m*, der aus den Leimfugen gequollen ist

f. annealing *(Met)* Stoßglühen *n*

f.-back Flammenrückschlag *m (Brenner)*

f.-back arrestor Flammensperre *f*

f.-barrier *(El)* Schutzwand *f*, Rundfeuerschutz *m*

f. boiler Kessel *m* für Augenblicksverdampfung

f.-butt weld Abbrennstumpfschweißnaht *f*

f.-butt welding Abbrennstumpfschweißung *f*

f. chamber Raum *m* zur Aufnahme überfließenden Materials *(Plastverarbeitung)*

f. column Verdampfungskolonne *f (Turmdestillation)*

f. current Kurzschlußstrom *m (Batterie)*

f. distillation Turmdestillation *f (Erdöl; Destillation, bei der das Flüssigkeitsgemisch über Siedepunkt unter Druck erhitzt wird und durch Verminderung des Drucks verdampft)*

f. drier 1. Schnelltrockner *m*; 2. *(Brg)* Schwebetrockner *m*

f. figures *(Met)* Glühspuk *m*

f. gun Blitzleuchte *f (Fotografie)*

f. gutter Gratmulde *f*

f. hider 1. *(Wärm)* Feuerdämpfer *m*; 2. *(Mil)* Mündungsfeuerdämpfer *m*

f. lamp Blitzlichtlampe *f*; Stroboskoplampe *f*

f.-light 1. Blitzlicht *n (Fotografie)*; 2. Taschenlampe *f*; 3. Blinkfeuer *n*; Blinkgerät *n*

f.-light battery Taschenlampenbatterie *f*

f.-light photo[graph] Blitzlichtaufnahme *f*

f. light unit Blitzgerät *n (Fotografie)*

f. line Gratkante *f*, Gratlinie *f*

f. magnetization Stoßmagnetisierung *f (Werkzeugprüfung)*

f. of light Lichtblitz *m*

f.-over *(El)* Funkenüberschlag *m*, Überschlag *m*

f.-over test Überschlagsprüfung *f*

f.-over voltage Überschlagsspannung *f*

f. overflow Herausfließen *n* der überschüssigen Preßmasse *(Plastverarbeitung)*

f. period Blinkfolge *f (Blinkfeuer)*

f. plate Schnellüberzug *m (Galvanotechnik)*

f. point Flammpunkt *m*, Entflammungspunkt *m*

f. point closed cup Flammpunkt *m* im geschlossenen Tiegel

f. point open cup Flammpunkt *m* im offenen Tiegel

f. point tester Flammpunktprüfer *m*

f. set Schnellabbinden *n (Zement)*

f.-steam cycle Kreisprozeß *m*, bei dem siedendes Wasser außerhalb der Heizfläche nach Drosselung plötzlich verdampft

f. suppressor *(Elm)* Überschlaglöscher *m*

f. test 1. Entflammungsprobe *f*; 2. *(Masch)* Stoßprüfung *f*

f. tower Verdampfungskolonne *f (Erdöl; Turmdestillation)*

f. trimmer Abgratmaschine *f*

f. trimming Abgraten *n*

f.-trimming press Abgratpresse *f*

f. tube Blitzröhre *f (z. B. Fotografie)*

f.-up *(Kern)* Aufblitzen *n*, Hervorschießen *n*

f. welding Abbrennschweißung *f*

flasher 1. automatische Schaltvorrichtung *f* für Leuchtreklame; 2. [automatischer] Blinkzeichengeber *m (Navigation)*; 3. Düppel *m (Radarstörung)*

f. motor *(Kfz)* Blinkmotor *m*

f. unit *(Kfz)* Blinkgeber *m*

flashing 1. Aufblitzen *n*, Aufleuchten *n*, Feuern *n*, Zucken *n*; 2. Entgraten *n*; Abfall *m (beim Entgraten)*; 3. Abdeckblech *n*, Kehlblech *n*, Schürze *f*; 4. Zunderflecken *m (auf Feinblech)*; 5. *(Text)* Schußstreifen *m (Gewebefehler)*; 6. Durchschleusen *n (Schiff)*; 7. *s.* f. test

f. beacon *(Schiff)* Blinkfeuer *n*

f. direction indicator lamp *(Kfz)* Blinker *m*

f. light Blinkfeuer *n*

f.-light direction indicator *(Kfz)* Blinker *m*

f.-off 1. Abtreiben (Verdampfen) *n* leichter Fraktionen *(Erdöl)*; 2. *(Met)* Abschmelzen *n*

f.-over 1. Funkenüberschlag *m*; 2. *(Elm)* Rundfeuer *n*

f. rate Lichtblitzfolge *f*

f. relay Flackerrelais *n*

f. signal 1. Blinksignal *n*; 2. *(Fmt)* Flackerzeichen *n*

f. test 1. Hochfrequenzvakuumprüfung *f* von Elektronenröhren; 2. *(Fmt)* Besetztprüfung *f* mit Flackerzeichen

f. trafficator *(Kfz)* Blinker *m*

flashlight *s.* flash-light
flashograph Amplitudenanzeiger *m*
flashometer Lichtblitzanalysator *m*
flashover *s.* flash-over
flask 1. *(Ch)* Flasche *f*, Kolben *m*, Flakon *m*; 2. *(Met)* Formkasten *m*
f. annealing *(Met)* Kastenglühen *n*
f. bar *(Met)* Formkastenschore *f*, Formkastentraverse *f*
f. board Formkastenbrett *n*
f. moulding Kastenformerei *f*
f. pin Formkastenführungsstift *m*
f. with flange Formkasten *m* mit Sandrippe
flat / to glätten
f. the grain *(Led)* ausrecken, aussetzen, ausstoßen
flat 1. [angeschliffene] Fläche *f*; Abflachung *f* *(Gewinde)*; Mitnahmefläche *f*; Zwischenschneide *f*; 2. Etagenwohnung *f*, Wohnung *f*, Hausetage *f*; 3. *(Text)* Kardendeckel *m*; 4. Niederung *f*; 5. Untiefe *f*; 6. *(Schiff)* Plattformdeck *n*
flat 1. plan; flach; eben, plano; 2. ungefalzt *(Druckbogen)*; 3. kontrastarm; 4. matt, stumpf *(Farbe)*; 5. platt *(Reifen)*; 6. erschöpft *(Batterie)*
f. angle gestreckter Winkel *m*
f. antitank mine Tellermine *f*
f. anvil Flachamboß *m*
f. arch scheitrechter Bogen *m* *(Mauerwerk)*
f. back needle Nadel *f*, deren Nadelhaken in Linie der Auflagefläche angeordnet ist *(Strickerei)*
f.-back stope *(Brg)* Firstenstoß *m* parallel zur Sohlenstrecke
f.-bar Flacheisen *n*, Flachstahl *m*
f.-bar copper Flachkupfer *n*
f.-bar iron *s.* f.-bar
f.-bar knitting machine Flachstrickmaschine *f*
f.-bar steel *s.* f.-bar
f.-base rim *(Kfz)* Flachbettfelge *f*
f. bastard rasp flachspitze Bastardraspel *f*
f. bearing *(Masch)* Gleitbahn *f*, Führungsbahn *f*; Gleitlager *n* mit ebenen Flächen; Lagerung *f* für hin- und hergehende Bewegung
f. bearing cap *(Masch)* Lagerdeckelplatte *f*, Deckelplatte *f*
f.-bed knitting machine Flachstrickmaschine *f*
f.-bed letterpress *(Typ)* Flachform[en]buchdruck *m*
f.-bed machine *(Typ)* Flachformmaschine *f*
f.-bed press Flachbügelpresse *f*
f.-bed printing *(Typ)* Flachdruck *m*
f.-bed rotary press Flachdruckrotationsmaschine *f*
f. belt Flachriemen *m*
f. belt drive Flachriementrieb *m*
f. bench rule *(Masch)* Bankmaßstab *m*
f. billet grobes Flacheisen *n* *(Walzwerk)*
f. bloom *(Am)* *(Met)* Rohbramme *f*
f. boat *(Schiff)* Zille *f*
f.-bottom rail Breitflanschschiene *f*, Breitfußschiene *f* *(Eisenbahnbau)*
f.-bottom self-discharging wagon Flachbodenselbstentlader *m*
f. bottomed flask *(Ch)* Standkolben *m*
f. bottomed vessel Plattbodenschiff *n*, flachbodiges Schiff *n*
f. box Flachsauger *m*
f. bulb iron Flachwulststahl *m*
f. cable 1. Bandkabel *n*; Flachkabel *n*; 2. *(Brg)* Flachseil *n*, Bandseil *n*
f. car Plattformwagen *m*, Rungenwagen *m*
f. card *(Text)* Deckelkarde *f*
f. chisel Flachmeißel *m*
f. coat Grundanstrich *m*
f. coil Flachspule *f*, Scheibenspule *f*
f. coil instrument Flachspul[meß]instrument *n*
f. cold chisel Kaltmeißel *m*
f. colour Grundfarbe *f*
f. conical cheese *(Text)* Sonnenspule *f*
f. construction Wohnungsbau *m*, Massenwohnungsbau *m*
f. contact Flachkontakt *m*
f. copper [bar] Flachkupfer *n*
f. counter *(Schiff)* Spiegelheck *n*
f. covering *(Text)* Deckelbeschlag *m* *(Karde)*
f. crush Streifenstauchprüfung *f* *(Papier)*
f.-cut *(Holz)* Fladerschnitt *m*, Flachschnitt *m*
f. design paper *(Text)* Schalpatrone *f*
f. die Flachbacke *f* *(Gewindewalzen)*
f.-die forging Freiformschmieden *n*
f.-die thread rolling Gewindewalzen *n*
f.-die thread rolling machine Gewindewalzmaschine *f*
f. double head border machine zweiköpfige Leistenflachstrickmaschine *f*
f. drill Spitzbohrer *m*
f. drill for microscopic holes Kleinst[spitz]-bohrer *m*
f. drill with inserted cutting edges Messerspitzbohrer *m*
f. driving tang Austreiberlappen *m* *(Werkzeug)*
f. drying and sizing machine *(Text)* Plantrockenschlichtmaschine *f*
f.-ended tube Katodenstrahlröhre *f* mit ebener Schirmfläche
f. engine Motor *m* mit liegendem Zylinder
f. fabric Flachstrickware *f*
f.-face pulley Flachscheibe *f*
f.-faced flachbahnig *(Werkzeug)*
f. faced die Gleitbacke *f*, Rollbacke *f* *(Gewindewalzen)*

f. faced thread rolling die Gewindewalzstahlbacke *f*
f. file Ansatzfeile *f*, Flach[spitz]feile *f*
f. film flache Folie *f*
f. finish Mattlack *m*
f. finishing planing tool Breitschlichthobelmeißel *m*
f. frame knitting Flachstrickerei *f*
f. furrow freshly turned Schälfurche *f*
f. gain *(Fmt)* geradlinige Verstärkung *f*
f. gauge Flachlehre *f*; Schablone *f*
f. glass Flachglas *n*, Fensterglas *n*
f. grain *(Holz)* Flachschnitt *m*
f. grate Planrost *m*
f. grinding Flachschleifen *n*
f. guide *(Wärm)* Ebenführung *f*, ebene Schlittenführung *f*; Flachführung *f*
f. guideway Flachbahn *f*
f. handle flaches Heft *n (Säge)*
f. harvester binder Flachmähbinder *m*
f. head rivet Flachkopfniet *m*
f.-head screw Senkschraube *f*
f.-headed bolt Flachkopfbolzen *m*, Zylinderschraube *f*
f.-headed counter sunk bolt Senkschraube *f*
f. hearth type mixer *(Met)* Flachherdmischer
f. internal limit gauge Grenzflachlehre *f*
f. iron Flacheisen *n*
f. jaw Planschnabel *m*
f. joint bündige Mauerwerksfuge *f*, Schnittfuge *f*
f. key Flachkeil *m*
f. knitting frame with loop forming sinkers *(Text)* Flachkulierstuhl *m*
f. knitting machine Flachstrickmaschine *f*
f. lacquer Mattlack *m*
f. lap *(ebene)* Läppscheibe *f*
f. lapping machine Flachschläppmaschine *f*
f. lapping [operation] Flachläppen *n*
f. limit plug gauge Grenzflachlehre *f*
f. low-lying coast Flachküste *f*
f. material Flachmaterial *n*
f. milled tooth file Flachstumpffräserfeile *f*
f. mirror Planspiegel *m*
f.-needle bed Nadelbett *n* einer Flachstrickmaschine
f.-needle file Flachspitznadelfeile *f*
f.-nose bullet *(Mil)* Flachkopfgeschoß *n*
f. nose plier Flachzange *f (zum Drahtbiegen)*
f. packing Flachpackung *f (Abdichtung)*
f. paint coat Grundanstrich *m*
f. paper Planopapier *n*
f. path *(Rak)* gestreckte Flugbahn *f*
f. picture *(Fs)* kontrastarmes Bild *n*
f. plain taper key Flachkeil *m*
f. plate control valve Flachschiebersteuerventil *n*
f. plate keel *(Schiff)* Flachkiel *m*
f.-platen press Flachplattenpresse *f*
f.-platen pressed board flachgepreßte Platte *f*
f. pliers Flachzange *f*
f. point screw Schraube *f* mit Kegelansatz, Schraube ohne Kuppe
f.-pointed bullet *(Mil)* Flachkopfgeschoß *n*
f. purl knitting machine Links-Links-Flachstrickmaschine *f*
f. rasp Flachspitzraspel *f*
f. rate Pauschalgebühr *f (Wasser, Telefon usw.)*
f. response [curve] *(El)* abgeflachte Durchlaßkurve (Übertragungskurve) *f*, flache Einhöckerkurve *f*
f. ring armature *(Elm)* Flachringanker *m*
f. roll glatte Rolle *f (Walzwerk)*
f. roof Flachdach *n*
f. rope Flachseil *n*, Bandseil *n*
f. saddle key Flachkeil *m*
f. sawn *(Holz)* im Tangentialschnitt
f. scraper Flachschaber *m*
f. screen Langsieb *n*; Plansieb *n*; Plansortierer *m*
f.-section jet burner Flachstrahlbrenner *m*
f. setting *(Text)* Einstellung *f* des Kardendeckels
f.-sheet deliverer *(Typ)* Planoausleger *m*, Planoableger *m*
f. slab Pilzdecke *f (Stahlbetonbau)*
f. slide valve Flachschieber *m*
f. slideway Flachführung[sbahn] *f*
f. spin[ning] *(Flg)* Flachtrudeln *n*
f. spiral superheating tube Überhitzerspiralrohr *n*
f. spring Blattfeder *f*, Flachfeder *f*
f. steel Flachstahl *m (Breite 12 bis 150 mm, Dicke ≧ 5 mm)*; Breitflachstahl *m (Breite > 150 mm, Dicke ≧ 3 mm)*
f. steel sleeper *(Eis)* Flacheisenschwelle *f*
f. stern *(Schiff)* plattes Heck *n*
f. stripping apparatus *(Text)* Deckelputzapparat *m*
f. strips *(Text)* Deckelausputz *m*
f. superheater coil Überhitzerflachrohr *n*
f. surface plane Fläche *f*
f. tail Biberschwanz *m (Dachziegel)*
f. tang Lappen *m*, Mitnahmefläche *f*
f. taper gauge Kegelflachlehre *f*
f. thread Flachgewinde *n*
f. tongs Flachzange *f*
f.-top guide Flachbahn *f*
f.-top resonance peak abgeflachte Resonanzspitze *f*
f.-top response [curve] *(El)* flacher Kurvenverlauf *m (insbesondere Bandfilter)*
f.-topped combustion chamber *(Kfz)* Verbrennungsraum *m* mit flachem Boden

f.-topped piston Flachkolben *m*, Kolben *m* mit flachem Boden
f. track Flachführung *f*, Flachbahn *f*
f.-trajectory fire *(Mil)* Flachfeuer *n*
f.-trajectory gun Flachbahngeschütz *n*
f. trowel *(Bw)* Spachtel *m*
f. truss Parallelfachwerk *n*, parallelgurtiger Fachwerkbinder *m*
f. tube *(Fs)* Flachröhre *f*
f. tuned *(Fmt)* unscharf abgestimmt
f. tuning *(El)* Grobabstimmung *f*, unscharfe (unselektive) Abstimmung *f*; flache (weiche) Abstimmung
f. turret Flachtischrevolver *m*
f. turret lathe Drehmaschine *f* mit Flachtischrevolverkopf
f.-type relay Flachrelais *n*
f.-type thimble ionization chamber *(Kern)* flache Fingerhut-Ionisationskammer *f*
f. varnish Mattlack *m*
f. warp knitting machine Flachkettenwirkmaschine *f*
f. washer Beilagsscheibe *f*
f. waste *(Text)* Deckelausputz *m*
f. weft knitting machine Flachkulierwirkmaschine *f*
f. weld Flachnaht *f (Schweißen)*
f. wire Flachdraht *m*
f. wood bench rule Bankmaßstab *m*
flatbed . . . *s.* flat-bed . . .
flatlet Kleinstwohnung *f*
flatness 1. Ebenheit *f*, Flachheit *f*; 2. *(Flg)* Rasanz *f (Flugbahn)*; 3. Mattheit *f*, Stumpfheit *f (Farbe)*; 4. *(Fs)* Kontrastlosigkeit *f*
flats Flachstahl *m*
flatstitch embroidery Plattstichstickerei *f*
flatted abgeflacht, abgeplattet, geebnet, egalisiert
f. plank vollkantiges Brett *n*
flatten / to abflachen, abplatten, ebnen, egalisieren, flach machen; plattdrücken *(Presse)*; breiten *(Walzwerk)*
f. down flach machen
f. out plattdrücken; ausbeulen; ausschweben *(Flugzeug)*
flattened abgeflacht, abgeplattet, geebnet
f. field wölbungsfreies Feld *n*
f. spectrum geebnetes Spektrum *n*
f. strand rope Flachlitzenseil *n*
f. strand winding rope Flachlitzenförderseil *n*
f. thread abgeflachtes Gewinde *n*
f. zone *(Kern)* Bereich *m* ausgeglichener Flußverteilung
flattening Abflachung *f*; Verflachung *f*; Abplattung *f*
f. out *(Flg)* Abfangen *n*
flatter / to flattern *(Ventile)*
flatter *(Mech)* Planierhammer *m*; Spannhammer *m*
flatting agent *(Am)* *(Plast)* Mattierungsmittel *n*
f. coat Wachsfarbenanstrich *m*
f. varnish Spachtellack *m*
flattop . . . *s.* flat-top . . .
flatwise flach[liegend]
f. wound flach gewickelt
flatwork ironer Bügelmaschine *f* für Haushaltleinen
flaunching Betonabdeckung *f (z. B. am Schornstein)*
flavone Flavon *n*
flavour Aroma *n*, Duftstoff *m*; Würze *f*
flavouring agent Aromastoff *m*; Geschmacksstoff *m*
flaw Fehler *m*; Materialfehler *m*; Defekt *m*; Gußblase *f*; Riß *m*; Sprung *m*
f. detection Fehlersuche *f*
f. detection sensitivity Fehlernachweisempfindlichkeit *f*
f. detector Defektoskop *n*
f. echo Fehlerecho *n (Materialprüfung)*
flawed geplatzt, rissig, gerissen, gespalten; blasig *(Glas)*; aschfleckig *(Porzellan)*
flawless fehlerfrei, fehlerlos
flawy *s.* flawed
flax 1. Lein *m (Pflanze)*; 2. *(Am)* Flachs *m (Faser)*
f. breaking and dressing machine Flachsbrech- und -reinigungsmaschine *f*
f. breaking machine Flachsbrechmaschine *f*
f. comber Flachshechelgerät *n*
f. cutting machine Flachsschneidemaschine *f*, Flachsreißmaschine *f*
f. dressing aggregate Flachsreinigungsgerät *n*
f. drying bunker Flachsdörre *f*, Flachsdarre *f*
f. drying kiln Flachsdarrofen *m*
f.-field Leinacker *m*
f.-hackler *s.* f.-heckler
f. hackling machine Flachshechelmaschine *f*
f. heckler Flachshechler *m*
f. puller Flachsraufmaschine *f*
f. pulling Leinraufen *n*, Flachsraufen *n*
f. pulling machine Leinraufmaschine *f*, Flachsraufmaschine *f*
f. rettery Flachsröste *f*
f. retting Flachsrösten *n*
f. ribbon Flachsfaserband *n*
f. rippling comb Flachsriffelkamm *m*
f. seed Leinsamen *m*
f. seed oil Leinöl *n*
f. shives Flachsschäben *fpl*
f. sliver Flachsfaserband *n*
f. stack Flachshocke *f*
f. tow Flachswerg *n*
f. tow finisher card with drawing head Feinkrempel *f* für Flachswerg mit Streckkopf

f. tow preparing Flachswergaufbereitung *f*
flay / to *(Led)* abhäuten
flaying Abzug *m (Schlachtung)*
fleece 1. Vlies *n*, Schaffell *n* [mit Wolle]; 2. Vliesstoff *m*
f. coming from the card Kardenvlies *n*
f. technique Vliestechnik *f*
f. wool Vlieswolle *f*
fleeced flaumig, flauschig *(Gewebeoberseite)*
fleet flach; nicht tief
f. angle *(Brg)* Seilablenkungswinkel *m*
f. supply ship Troßschiff *n*
flench / to zerlegen *(Wal)*; abhäuten
flensing deck Flensdeck *n*, Schlachtdeck *n (auf einem Walfangmutterschiff)*
flesh outside *(Led)* Fleischseite *f* außen *(im Bündel)*
f. side Aasseite *f*, Fleischseite *f (Riemen)*
f. split Fleischspalt *m*
fleshing *(Led)* Entfleischung *f*
f. beam Schabebaum *m*
f. knife Schabeisen *n*, Gerbermesser *n*
f. machine Entfleischmaschine *f*
fleshings Aas *n*; Scherfleisch *n*
Flettner balance *(Flg)* Flettner-Hilfsruder *n*
fletton [brick] in Südengland hergestellter, hauptsächlich als Hintermauerungsstein verwendeter rosafarbener Tonziegel *m*
Flewelling circuit *(El)* Pendelrückkopplungsschaltung *f*, Flewelling-Schaltung *f*
flex / to [sich] biegen; durchbiegen; beugen
flex *(sl) s.* flexible cord
flexibility 1. Biegsamkeit *f*, Federungsvermögen *n*, Flexibilität *f*; 2. vielseitige Anpassungsfähigkeit *f*
flexible 1. biegsam, elastisch, federnd, flexibel; 2. anpassungsfähig, umstellbar *(z. B. Maschinen auf veränderte Produktion)*
f. armoured cable bewehrtes biegsames Kabel *n*
f. axle Schwingachse *f*
f. belt sander Bandschleifmaschine *f*
f. binding *(Typ)* Klebebindung *f*
f. boring machine Gelenkbohrmaschine *f (Holzbearbeitung)*
f. cable biegsames (flexibles) Kabel *n*, biegsame (flexible) Leitung *f*; Geräte[anschluß]schnur *f*; Lampenschnur *f*
f. cargo container flexibler Ladungsbehälter *m*
f. chrome splits Chromsohlspaltleder *n*
f. conductor *(El)* biegsamer Leiter *m*
f. connection *(Kfz)* Verbindungsschlauch *m*
f. contact federnder Kontakt *m*
f. container *(Schiff)* flexibler Behälter *m (schwimmender langgestreckter Ölbehälter aus Kunststoff; Ölwurst)*
f. cord Anschlußschnur *f*, flexible Leitung *f*, Litze *f*, Litzenleitung *f*, Schnurleitung *f*, Verbindungsleitung *f*; Telefonschnur *f*
f. coupling Ausgleichskupplung *f*; Kreuzgelenkkupplung *f*
f. die einstellbares (offenes) Schneideisen *n*
f. disk *(Kfz)* Hardyscheibe *f*
f. door Falttür *f*, Harmonikatür *f*
f.-drive screen *(Brg)* Freischwingsieb *n (Aufbereitung)*
f. drive shaft *(Kfz)* biegsame Welle *f (Tachometer)*
f. fastening elastische Befestigung *f*; elastische Aufhängung *f (z. B. Fahrdraht)*
f. joint Kardangelenk *n*, elastisches Gelenk *n*
f. lead *s.* f. cord
f. lead connector Kupplung *f*, Schnurverbinder *m*
f. membrane clutch elastische Kupplung *f*
f. metal hose Metallschlauch *m*
f. metal tubing biegsame Rohrleitung *f*
f. oil barge *(Schiff)* flexibler Ölleichter *m (Ölwurst)*
f. pavement elastische Decke *f (Straßenbau)*
f. pigtail *(El)* Anschlußlitze *f*, Schaltlitze *f*, Rangierlitze *f*
f. pipe Metallschlauch *m*, biegsame Rohrleitung *f*
f. plastic gummiartiger (weicher) Kunststoff *m*
f. relief plate *(Typ)* flexible Buchdruckplatte *f*
f. roller bearing Federrollenlager *n*
f. rule Bandmaß *n*
f. shaft Biegewelle *f*, biegsame Welle *f*
f. shaft grinder Elektroschleifmaschine *f* mit Biegewelle
f. sheet weiche Folie *f*
f. sleeve federnde Buchse *f*
f. steel rule Stahlbandmaß *n*
f. varnished tubing *(El)* Lackschlauch *m*, Isolierschlauch *m*
f. waveguide biegsamer Wellenleiter *m*
f. wire *(El)* Litze *f*, Litzendraht *m*
fleximer flooring Spachtelboden *m*
flexing *(Geol)* Faltung *f*
f. life Lebensdauer *f* bei Dauerbiegebeanspruchung, Dauerbiegefestigkeit *f (z. B. Plaste)*
f. machine Knittermaschine *f*
flexographic [printing] press Anilindruckmaschine *f*
flexural cracking Biegerißbildung *f*
f. load Biegelast *f*; Biegebeanspruchung *f*
f. member Biegeglied *n*, Biegestab *m*
f. mode of vibration Biegeschwingung *f*, Biegungsschwingung *f (z. B. Schwingquarz)*
f. moment Biegemoment *n*
f. rigidity *s.* f. stiffness
f. stiffness Biegesteifigkeit *f*
f. strength Biegefestigkeit *f*
f. stress Biegespannung *f*

f. tensile stress Biegezugspannung *f*
f. test Biegeprüfung *f*, Biegeversuch *m*
f. wave Biegungswelle *f*
flexure Durchbiegung *f*; Biegung *f*, Biegbarkeit *f*
f. modulus Biegemodul *m*, Biegefestigkeitszahl *f*
flicker / to flackern; funkeln, flimmern
f. away verflackern
f. up aufflackern
flicker effect Funkeleffekt *m*, Flackereffekt *m* *(Elektronik)*; *(Opt)* Flimmereffekt *m*
f. fading Flackerschwund *m*
f.-free flimmerfrei
f.-free reproduction flimmerfreie Wiedergabe *f*
f. frequency Flackerfrequenz *f*, Flimmerfrequenz *f*
f. method *(Opt)* Flimmerverfahren *n*
f. noise *(Fmt)* Flackerrauschen *n*
f. photometer Flimmerfotometer *n*
flickering Flackern *n*, Geflacker *n*; Flimmern *n*
f. lamp Flackerlampe *f*
f. signal Flackersignal *n*, Flackerzeichen *n* *(z. B. Fernsprechtechnik)*
flier Stufe *f (Treppe)*
fliers Außentreppe *f*, Freitreppe *f*; Treppenlauf
flight 1. Flug *m*; Flugbereich *m*; 2. Treppenlauf *m*, Freitreppe *f*; 3. Reichweite *f*; 4. Gang *m (einer Schraubenlinie)*; 5. Flügel *m (einer Mischerschnecke)*
f. altitude Flughöhe *f*
f. axis Flugbahnachse *f*
f. computation Flugberechnung *f*
f. control instruments Flugüberwachungsgeräte *npl*
f. conveyor Kratzkettenförderer *m*, Mitnehmerförderer *m*
f. deck *(Schiff)* Flugdeck *n*, Abflugdeck *n*
f.-dynamic pressure Flugstaudruck *m*
f. engineer Bordmechaniker *m*
f. evaluation Flugauswertung *f*
f. information Flugdaten *pl*
f. information centre Flugsicherungs[informations]zentrale *f*, FS-Informationszentrale *f*
f. information region Flugsicherungsinformationsgebiet *n*
f. information service Flugsicherungs[informations]dienst *m*, FS-Informationsdienst *m*
f. instrument Flugüberwachungsgerät *n*
f. lane Flugschneise *f*
f. loader *(Brg)* Schrämmaschine *f*, die bei der Talfahrt als Ladegerät arbeitet
f. log *(Flg)* 1. Flugbuch *n*; 2. Flugwegschreiber *m*
f. of stairs Treppe[nlauf *m*] *f*
f. path *(Flg)* Flugweg *m*, Flugroute *f*; *(Rak)* Flugbahn *f*
f. path angle Gleitwinkel *m*, Steigwinkel *m*; Flugbahn[neigungs]winkel *m*
f.-path computer Flugwegrechner *m (Flugzeuge)*; Flugbahnrechner *m (Raketen)*
f.-path plotter 1. *(Flg)* Flugschreiber *m*; 2. *(Rak)* Flugbahnzeichner *m*
f.-path recorder *s.* f.-path plotter
f.-path speed 1. *(Rak)* Bahngeschwindigkeit *f*; 2. *(Flg)* Flugbahngeschwindigkeit *f*
f. plotter *s.* f.-path plotter
f. recorder *s.* f.-path plotter
f. simulator Flugnachahm[ungs]gerät *n*
f. time Flugzeit *f*, Flugdauer *f*
f. track tatsächlicher (aufgezeichneter) Flugweg *m*
f. tracking *(Rak)* Flugbahnverfolgung *f*
flights on the mixer tube Gänge *mpl* des Mischrohres
flimsy *(sl) (Typ)* Durchschlagpapier *n*
fling / to schleudern, schnellen
flinger Schleuder *f*
flint 1. Feuerstein *m*, Flint[stein] *m*; 2. *(Opt)* *s.* f. glass
f.-dry *(Led)* volltrocken
f. glass *(Opt)* Flintglas *n*, Kristallglas *n*; Quarzglas *n*
f. glass prism Flintglasprisma *n*
f.-like steinhart
f. ware *s.* stone ware
flinty kieselhaltig
f. earth Kieselerde *f*
f. ground kiesiger Boden *m*, Kiesgrund *m*
flip / to [um]kippen *(Multivibrator)*
flip-flap telegraph Klipp-Klapp-Telegraf *m*
f. flop Kippschaltung *f*, Flip-Flop-Schaltung *f*
f.-flop circuit Flip-Flop-Schaltung *f*, [bistabile] Kippschaltung *f*, Flip-Flop-Generator[kreis] *m*, bistabiler Multivibrator[kreis] *m*
f.-flop oscillator Flip-Flop-Generator *m*; bistabiler Multivibrator *m*
flipper *(Flg)* Tiefenruder *n*, Höhenruder *n*
flitch cut *(Holz)* Herzschnitt *m*, Gatterschnitt *m*
f. plate *(Bw)* Lasche *f*
f. veneer Flitschfurnier *n*
flitched beam verdübelter Balken *m (Holzbau)*
flitches Balken *mpl* ohne Herz, in der Mitte aufgetrennte Balken, Zwalpen *pl*
f. in the veneer Flitschholz *n*
flitting Talfahrt *f*, Umsetzen *n (Schrämmaschine)*
float / to 1. schwimmen; treiben *(auf dem Wasser)*; 2. pendeln; beweglich sein; 3. *(Brg, Min)* flotieren, aufschwimmen *(Aufbereitung)*
f. down a river abflößen
float 1. Schwimmer *m*, Schwimmkörper *m*; Schwebekörper *m (Rotameter)*; 2. einhiebige Feile *f*; 3. Aufziehbrett *n*; Reibebrett *n*; 4. *(Text)* Flott[ier]ung *f*; 5. Milchkarren *m*

f.-and-sink analysis *(Brg)* Wichteanalyse *f*
f.-and-sink apparatus *(Brg)* Sinkscheider *m*
f.-and-sink test *(Brg)* Schwimm- und Sinkversuch *m*
f. chamber *(Kfz)* Schwimmergehäuse *n*
f. charger Ladegerät *n* für Pufferbetrieb, Pufferlader *m* für Batterie
f.-control Schwimmerregelung *f*
f.-controlled durch Schwimmer geregelt (betätigt); schwimmergesteuert
f.-controlled valve Schwimmerventil *n*
f.-cut file einhiebige Feile *f*
f. gauge Wasserstandanzeiger *m* mit Schwimmer; Schwimmermesser *m*
f. holder *(Masch)* Pendeleinsatz[hülse *f*] *m*
f. needle *(Kfz)* Schwimmernadel *f*
f.-needle valve *(Kfz)* Schwimmernadelventil *n*
f. position Nichtstrickstellung *f* *(Strickmaschine)*
f. switch *(El)* Schwimmerschalter *m*
f. test Schwimmprobe *f* *(Viskositätsbestimmung)*
f.-type densitometer Schwimmerdichtemesser *m*, Auftriebsdichtemesser *m*
f.-type manometer Schwimmermanometer *n*
f. valve Schwimmerventil *n*
floatability Flotierbarkeit *f*; Schwimmfähigkeit *f*
floatage 1. schwimmende (treibende) Masse *f*; 2. Auftrieb *m*; 3. *s.* floating
floated battery *(El)* Pufferbatterie *f*
f. concrete Gußbeton *m*
f. gyro Schwimmerkreisel *m*
f. timber Floßholz *n*
f.-type gyroscope Schwimmerkreisel *m*
f. wheel Wasserschaufel *f*
floating 1. Schwimmen *n*; Auftrieb *m*; 2. *(Holz)* Flößen *n*, Triften *n*; 3. Glätten *n* *(Beton)*; 4. *(El)* Pufferbetrieb *m*
floating 1. pendelnd, frei beweglich; schwimmend [gelagert]; Pendel-; 2. *(El)* ungeerdet
f. action *(BMSR)* integrierende Regelung *f*
f. adapter Pendelhülse *f*
f. address *(Dat)* symbolische Adresse *f*, Pseudo-Adresse *f*
f. axle *(Kfz)* Schwingachse *f*, fliegende Achse *f*
f. battery Notstrombatterie *f*
f.-battery operation Notstrombatteriebetrieb *m*
f. beacon *(Schiff)* Treibbake *f*
f. blubber factory Walfangmutterschiff *n*
f. body Schwimmkörper *m*
f. box *(Bw)* Schwimmkasten *m*
f. brick Schwimmstein *m*
f. bridge Pontonbrücke *f*, Schwimmbrücke *f*
f. bulb *(Wärm)* Tauchglocke *f*
f. bush *(Masch)* Pendelhalter *m*, pendelnder Halter *m*
f. channel Triftkanal *m*
f. collet Pendeleinsatzhülse *f*
f. control *s.* integral control
f. controller Integralregler *m*, I-Regler *m*
f. controller action I-Regelung *f*
f. crane Pontonkran *m*, Schwimmkran *m*
f. dock Schwimmdock *n*
f. dredger Schwimmbagger *m*, Naßbagger *m*
f. drilling platform schwimmende Bohrstation *f* *(Ölbohrinsel)*
f. dry-dock Schwimmdock *n*
f. fish factory Fischereifabrikschiff *n*
f. floor schwimmender Estrich (Fußbodenbelag) *m*
f. grid *(El)* offenes Gitter *n*
f. gudgeon pin *(Kfz)* schwimmender Kolbenbolzen *m*
f. holder Pendeleinsatz[hülse *f*] *m*; Pendelfutter *n*; Pendelmeißelhalter *m*
f. ice Treibeis *n*
f. knife coating *(Led)* Luftrakelstreichverfahren *n* *(Beschichtung)*
f. knife roll coating *(Led)* Luftrakelverfahren *n* mit Walzenauftragung *(Beschichtung)*
f. light Leuchtboje *f*; Leuchtschiff *n*, Feuerschiff *n*
f.-line *(Schiff)* Wasserlinie *f*
f. magnetic head *(BMSR)* schwebender Magnetkopf *m* *(auf Luftpolster)*
f. matter Schwimmstoff *m*
f. meadow Wasserwiese *f*
f. mine *(Mil)* Treibmine *f*
f. mount *(Masch)* pendelnde Lagerung *f*
f. movement *(Masch)* Pendelbewegung *f*
f. neutral nichtgeerdeter Nulleiter (Mittelleiter) *m*
f. packing *(Wärm)* bewegliche Dichtung *f*
f. paper ungeleimtes Papier *n*, Löschpapier *n*
f. particles *(Ph, Ch)* Schwebeteilchen *npl*
f. pile schwebender Pfahl *m* *(Grundbau)*
f. pile-driver Schwimmramme *f*
f. point *(BMSR)* gleitendes Komma *n*
f.-point calculation Gleitkommarechnung *f* *(Elektronenrechner)*
f.-point representation halblogarithmische Zahlenschreibung (Zahldarstellung) *f*
f.-point routine Gleitkommaprogramm *n* *(Elektronenrechner)*
f. position *(Masch)* Schwimmlage *f*
f. rate Laufgeschwindigkeit *f* *(I-Regler)*
f. reamer Pendelreibahle *f*
f. reamer holder Reibahlenpendelhülse *f*
f. reinforced concrete box Stahlbetonschwimmkasten *m*
f. roof Schwimmdach *n* *(Öltank)*
f. shaft Schwingachse *f*
f. soap Schwimmseife *f*
f. speed *(BMSR)* Laufgeschwindigkeit *f* *(I-Regler)*

f. **stage** 1. Gleittisch *m (ohne mechanische Führungselemente)*; 2. *(Schiff)* Kalfatfloß *n*
f. threads *(Text)* flottliegende Fäden *mpl*
f. timber Floßholz *n*
f. tool holder pendelnder Werkzeughalter *m*, Pendelhalter *m*, Pendelfutter *n*
f. trickle [battery] Stromausgleichbatterie *f*
f.-type chuck Pendeleinsatz[hülse *f*] *m*
f.-type holder Pendelhalter *m (Reibahle)*
f. whale-oil factory Walfangmutterschiff *n*
floats *(Text)* flottierende Fäden *mpl*
flocculant flockig, flockenartig
flocculate / to *(Ch)* [aus]flocken, koagulieren
flocculating agent Flockungsmittel *n*
flocculation *(Ch)* Flockenbildung *f*; Ausflockung *f*, Flockung *f*
f. test Flocktest *m*
flocculator Flockenbildner *m*, Flockungsmittel *n*
floccule Flocke *f*
flocculent flockenartig, flockig
flocculent *s.* flocculator
flock Flocke *f*; Wollflocke *f*; Flockenbündel *n*
f. carpet Flockteppich *m*
f. dot fabric Flockpunktgewebe *n*
f. fibre Flockfaser *f (zum elektrostatischen Beflocken)*
f. of sheep Schafherde *f*
f. pile Flockflor *m (auf Flockteppich)*
f. print Flockdruck *m*
f.-printing Flockdrucken *n*
flocked carpet Flockteppich *m*
f. yarn Flockfaden *m*
flocking fabric by electrostatic method Gewebebeflockung *f* durch elektrostatisches Verfahren
f. fabric by spray method Gewebebeflockung *f* durch Sprühverfahren
floe Treibeis *n (Stück)*
f. berg schwimmender Eisberg *m*
flog / to peitschen *(Pferd)*; auswerfen *(Angelleine)*; grob zuhauen *(Holz)*
flong *(Typ)* Mater *f*, Matrize *f*; Matrizenpappe *f*
flood / to 1. überfluten, überströmen; berieseln; 2. *(Brg)* ersaufen *(Grube)*
f. with oil mit Öl berieseln
flood Hochwasser *n*
f.-anchor *(Schiff)* Flutanker *m*
f. bridge Flutbrücke *f*
f. car Kokslöschwagen *m*
f. control Hochwasserschutz *m*
f.-control levee Hochwasserschutzdeich *m*
f.-control storage Hochwasserspeicherung *f*
f. dam Hochwasserschutztalsperre *f*
f. discharge Flutwelle *f*, Hochwasserwelle *f*; Hochwasserentlastung *f*
f. district Überschwemmungsgebiet *n*
f. gate Flutschleuse *f*, Fluttor *n*, Schleusentor *n*
f. levee Hochwasserschutzdeich *m*
f. light Flutlicht *n*; Flutlichtscheinwerfer *m*, Scheinwerferlicht *n*
f.-light plant Flutlichtanlage *f*
f.-light scanning *(Fs)* Flutlichtabtastung *f*
f.-lighting Anstrahlung *f*, Flutlichtbeleuchtung *f*
f.-lit angestrahlt, durch Flutlicht erhellt, im Scheinwerferlicht
f. lubrication Eintauchschmierung *f*
f. measuring post *(Bw)* Pegelmeßstelle *f*
f.-oiling Berieselung *f* mit Öl, Ölberieselung *f*
f. projection *(Fs)* Abtasten *n* bei voller Beleuchtung des Objekts
f. protection Hochwasserschutz *m*
f. protection works Hochwasserschutzbauten *pl*
f. pump *(Schiff)* Leckpumpe *f*
f. storage basin (reservoir) Hochwasserspeicher *m*
f. stream Flut[strömung] *f*
f.-tide Flutzeit *f*
f. tide gate Fluttor *n*
f. warning Hochwasseralarm *m*
f. water Hochwasser *n*
floodable lengths curve *(Schiff)* Schottkurve *f*, Kurve *f* der flutbaren Längen
flooding 1. Überflutung *f*; 2. *(Brg)* Ersaufen *n (Grube)*
f. calculation *(Schiff)* Schottrechnung *f*, Leckrechnung *f*
f. curve *(Schiff)* Schottkurve *f*
f. of the carburettor *(Kfz)* Fluten (Überlaufen) *n* des Vergasers
f. point *(Ch)* Überflutungspunkt *m (Füllkörperkolonne)*, Verflüssigungspunkt *m*
f. water Flutwasser *n*; Leckwasser *n*
floor / to 1. den Fußboden (Boden, Estrich) herstellen; pflastern; 2. ausbreiten [auf Tenne] *(Malz)*
floor 1. Etage *f*, Geschoß *n*, Stock[werk *n*] *m*; 2. Fußboden *m*; Boden *m (Meer)*; Sohle *f (z. B. Tunnel)*; Tenne *f (Scheune)*; 3. *(Schiff) (stählerne)* Bodenwrange *f*, Bodenstück *n*; 4. *(Brg)* Liegendes *n*, Sohle *f*
f. arch *(Brg)* Sohlengewölbe *n*
f. area *(Bw)* Grundfläche *f*, Wohnfläche *f*
f. bar *(Brg)* Schwelle *f*, Sohlenholz *n*
f. beam Hauptbalken *m (Balkendecke)*, Deckenbalken *m*, Deckenträger *m*
f. board Diele *f*, Fußbodenbrett *n*
f.-bolting *(Brg)* Verankerung *f* der Sohle, Verankerung des Liegenden
f. contact *(El)* Fuß[boden]schalter *m*, Bodenkontakt *m*
f. covering Fußbodenbelag *m*
f.-frame *(Kfz)* Bodenrahmen *m*
f. guide Führung *f* für Schiebetür

f. **gunmount** Bodenlafette *f (Flugzeug)*
f. **height** *(Kfz)* Pritschenhöhe *f*
f. **jack** Heber *m*, Rangierheber *m*
f. **joist** Deckenbalken *m*, Deckenträger *m*
f. **lamp** Stehlampe *f*
f. **level** *(Kfz)* Bodenhöhe *f*
f. **lift** *(Brg)* Quellen *n* der Sohle, Sohlenhebung *f*
f. **mat** *(Kfz)* Bodenmatte *f*
f. **moulding** *(Met)* Herdformerei *f*
f. **mounting** Aufstellen *n* auf dem Fußboden
f. **oil** Fußbodenöl *n*
f. **panel** Deckenplatte *f*
f. **paving** Fußbodenbelag *m* [aus Platten]
f. **plan** Etagengrundriß *m*
f. **plaster** Estrichgips *m*
f. **plate** 1. *(Masch)* fest fundamentierte Spannplatte *f*; Grundplatte *f*; 2. *(Schiff)* Flurplatte *f*, *(stählerne)* Bodenwrange *f*
f.-plate of magazine Kastenboden *m (Gewehr)*
f.-plate-type horizontal boring mill Plattenbohrwerk *n*
f. **polish** Bohnerwachs *n*; Fußbodenpflegemittel *n*
f. **polisher** [elektrische] Bohnermaschine *f*
f. **rest** Hilfsständer *m (Einständerhobelmaschine)*
f. **rider** *(Schiff)* Innenbodenwrange *f*
f. **sand** *(Met)* Altsand *m (Formerei)*
f. **sanding machine** Fußbodenschleifmaschine *f*
f. **slab** Deckenplatte *f*
f. **space** Grundfläche *f*, Nutzfläche *f*; Bodenfläche *f (Büroräume, Fabrikhallen, Ausstellungshallen usw.)*
f. **space required** *(Masch)* Platzbedarf *m (Bodenfläche)*
f. **standard** Stehlampe *f*
f. **stop** Türschwelle *f*
f. **switch** Fußschalter *m*
f. **timber** *(Schiff)* [hölzerne] Bodenwrange *f*, Bauchstück *n*
f.-to-floor height Geschoßhöhe *f*
f.-to-floor time *(Masch)* Boden-zu-Boden-Zeit *f*
f. **topping** Fußbodenbelag *m*
f.-type horizontal boring and facing machine Plattenbohrwerk *n*
f.-type horizontal boring and milling machine with travelling column Waagerecht-bohr- und Fräswerk *n* mit fester Aufspannplatte und waagerecht verschiebbarem Ständer
f.-type [horizontal boring] machine with cross-sliding column Grundplattenmaschine *f* mit kreuzbeweglichem Ständer, Grundplattenmaschine mit Kreuzbett
f.-type turning machine Großdrehmaschine *f* ohne Bett
f. **wax** Bohnerwachs *n*
flooring Fußboden *m*; Dielung *f*; Fußbodenbelag *m*
f. **light** [begehbares] Oberlicht *n*
f. **off** *(Schiff)* Stauen *n* der untersten Ladung
f. **tile** Fußbodenfliese *f*, Fußbodenplatte *f*
floors *(Brg, Geol)* Stockwerk *n*
flop / to plumpsen, hin und her schlagen (schwingen), auf und nieder schwingen (schlagen)
f. **back** *(El)* zurückkippen *(Multivibrator)*
flop gate Umstellklappe *f (Bunker)*
flopper Falte *f (auf Blechen)*
floreated *(Bw)* blumenverziert, mit Blumenmustern verziert
floret *(Typ)* Vignette *f*
floriculture Blumenzucht *f*
florist Blumenzüchter[in *f*] *m*; Blumenverkäufer[in *f*] *m*
floss Rohseidenfaden *m*
flotation 1. *(Ch, Brg)* Flotation *f*, Naßaufbereitung *f*, Schwemmaufbereitung *f*, Schwemmung *f*, Schwemmverfahren *n*, Schwimmaufbereitung *f*; 2. Ausflockung *f*, Flockung *f (Gesundheitstechnik)*; 3. Schwimmen *n*
f. **agent** Flotationsmittel *n*
f. **cell** Flotationszelle *f*
f. **concentration** Schwimmaufbereitung *f*
f. **liquid** Flotationstrübe *f*
f. **plant** Schwimmaufbereitungsanlage *f*
f. **process** Flotation *f*, Schwimmaufbereitung
flour / to zerstäuben, mahlen *(zu Mehl)*
flour Mehl *n*; Staub *m*
f. **abrasive** gepulvertes Schleifmittel *n*
f. **binder** mehliger Binder *m (Gießen)*
f. **core** Mehlkern *m (Gießen)*
f. **emery** feines Schmirgelpulver *n*
f. **gold** Flittergold *n*, Staubgold *n*
f. **mill** Getreidemühle *f*
f. **of emery** Schmirgelstaub *m*
f. **paste** Mehlkleister *m*
f. **spar** *s.* fluorite
flow / to rinnen *(Flüssigkeiten)*; fließen *(Flüssigkeiten, Werkstoffe)*; strömen *(Flüssigkeiten, Gase)*; kriechen *(Werkstoffe)*
f. **against** anströmen
f. **around** umspülen; umströmen
f. **away** verlaufen *(Wasser)*
f. **back** zurückfluten
f. **down** herablaufen *(Flüssigkeiten)*
f. **from** entströmen
f. **in** einlaufen, einfließen, zufließen *(Flüssigkeiten)*; einströmen *(Gase, Flüssigkeiten)*
f. **off** abfließen, ausfließen, ausströmen
f. **out** ausfließen, ausströmen
f. **over** überfließen *(Gefäß)*; überströmen *(Turbine)*

f. through durchfließen
f. together zusammenfließen
flow Fließen *n*, Fluß *m*; Strömung *f*, Strom *m*; Durchfluß *m*, Durchsatz *m*; Förderstrom *m*
f. area Fließquerschnitt *m*
f. around fuselage *(Flg)* Rumpfumströmung *f*
f. assembly Durchflußgefäß *n* *(kontinuierliche pH-Messung)*
f. at subsonic velocity Unterschallströmung *f*
f. at supersonic velocity Überschallströmung *f*
f. baffle Strömungsleitblech *n*
f. bean Eruptionsdüse *f (Tiefbohrung)*
f. behaviour Fließverhalten *n*
f. birefringence Strömungsdoppelbrechung *f*
f. capacity Durchflußkapazität *f* *(Rohrleitungen)*
f. characteristic *(BMSR)* Flußkennlinie *f*; Durchflußkennlinie *f (Ventil)*
f. chart 1. Strömungsdiagramm *n*; 2. *(BMSR)* Ablaufschema *n*, Flußdiagramm *n*, Flußschema *n (Programmablauf)*; Befehlsschema *n* *(elektronische Rechner)*
f. cleavage *(Geol)* Druckschieferung *f*
f. coefficient Durchflußzahl *f*; Durchflußkoeffizient *m*
f. conditions Strömungsverhältnisse *npl*
f. control *(BMSR)* kontinuierliche Mengenregelung *f*
f. control valve *s.* f. controller
f. controller *(BMSR)* Mengenregelungsventil
f. diagram 1. Strömungsdiagramm *n*; 2. *(BMSR)* Ablaufschema *n*, Flußdiagramm *n*; Flußschema *n*
f. drag Strömungswiderstand *m*
f. energy Strömungsenergie *f*
f. field *(Ph)* Strömungsfeld *n*
f. free from turbulence (vortices) wirbelfreie Strömung *f*
f. gate *(Met)* Steiger *m (Gießerei)*
f. gauge *s.* flow meter
f. governor Mengenregler *m*
f. hardening *(Met)* Kaltverfestigung *f*
f. heater Durchflußerhitzer *m*
f.-in pressure Einströmdruck *m*
f. indicator Durchflußanzeiger *m*, Strömungswächter *m*
f. limit Fließgrenze *f (Metall, Keramik)*
f. line 1. Fließband *n (Montage)*; Fließreihe *f (Fertigung)*; Fertigungslinie *f*; 2. Vorlaufleitung *f*; 3. *(Ph)* Strömungslinie *f*, Stromlinie *f*; 4. Gleitlinie *f (Struktur)*
f.-line assembly Fließbandmontage *f*
f.-line principle Fließfertigung *f*
f.-line production Fließ[band]fertigung *f*, Fließbandproduktion *f*
f. lines *(Met)* [Lüderssche] Fließfiguren *fpl*
f. machine Strömungsmaschine *f*
f. medium Flußmittel *n*
f. meter Durchfluß[mengen]messer *m*, Strömungsmesser *m*
f.-meter manometer Differenzdruckmanometer *n*
f. mixer Durchflußmischer *m*, Fließmischer *m*
f.-moulding *(Plast)* Preßspritzen *n*
f. of chips Spanfluß *m*
f. of coolant Kühlmittelzufluß *m*
f. of diffusing molecules Diffusionsstrom *m*
f. of distribution *(Holz)* Stoffverteilung *f*
f. of energy Energiestrom *m*
f. of heat Wärmestrom *m*
f. of mass *(Ph)* Massestrom *m*, Massetransport *m*, Diffusion *f*
f. of materials Werkstofffluß *m*
f. of metal waste Spanfluß *m*
f. of momentum *(Ph)* Impulsstrom *m*
f. of oil Öldurchfluß *m*; Ölzufluß *m*
f. of the pulp *(Holz)* Stofffluß *m*
f. pattern 1. *(Ph)* Strömungsbild *n*; 2. *(BMSR)* Ablaufschema *n*, Fließdiagramm *n*
f.-point *(Met)* Fließpunkt *m*, Fließgrenze *f*, Fließtemperatur *f*
f. pressing Fließpressen *n*
f. pressure Fließdruck *m*
f. process Fließverfahren *n*
f. property *(Met)* Fließfähigkeit *f*; Fließverhalten *n*
f. rate Strömungsmenge *f*; Strömungsgeschwindigkeit *f*
f. recorder Mengenschreiber *m*
f. regulator *(BMSR)* Durchflußregler *m*, Mengenregler *m*
f. resistance 1. Strömungswiderstand *m*; 2. Fließfestigkeit *f*
f. separation Abreißen (Sichablösen) *n* der Strömung *(Stromlinien)*
f. sheet 1. *(Ch)* Fließbild *n*, Fließschema *n*; 2. *(BMSR)* Arbeitsablaufkarte *f*; 3. *(Geol)* Gleitschicht *f*
f. split Strömungsteilung *f*
f. stowing *(Brg)* Fließversatz *m*
f. test Fließprobe *f*; Ausbreitversuch *m (Betonprüfung)*
f.-through Durchfluß *m*
f.-through treatment *(Met)* Durchflußbehandlung *f*
f.-through type cell Durchflußküvette *f*
f.-type pH electrode pH-Durchflußmeßelektrode *f*
f.-type pressure regulator kontinuierlich arbeitender Druckregler *m*
f. velocity Strömungsgeschwindigkeit *f*
f. volume Durchflußmenge *f*
f. washer Durchlaufwaschmaschine *f*

f. washing machine *s.* f. washer
f. welding Gießschweißen *n*
f. when an acceleration potential exists Strömung *f* im Feld eines Beschleunigungspotentials
f. without loss verlustlose Strömung *f*
flowability *(Plast)* Fließfähigkeit *f*; Viskositätsgrad *m*
flowable fließfähig; strömfähig
flowage *(Geol)* Fließbewegung *f*
flower bed Blumenbeet *n*
f. box Blumenkasten *m*
f.-lifting tool Ausheber *m (Gärtnerei)*
f. of cotton plant Baumwollblüte *f*
f. oil ätherisches Öl *n*
f. plot Blumenbeet *n*, Rabatte *f*
flowering season Blütezeit *f*
flowers of sulphur Schwefelblüte *f*
f. of zinc *(Met)* Zinkblumen *fpl*
flowing Fließen *n*; Strömen *n*; Kriechen *n*; Strömung *f*; Strömungsmenge *f*; Flut *f*
f. life Dauer *f* des freien Ausflusses *(Erdöl; Bohrung)*
f. well Eruptionssonde *f*, Eruptivsonde *f*; frei ausfließende Bohrung *f*, selbstausfließende Bohrung *f (Tiefbohrung)*
flowsolder method (processing) Fließlötverfahren *n*
fluate *(Ch)* Fluat *n*
flucan *(Geol)* Salband *n*, Lettenbesteg *m*
fluctuate / to fluktuieren, [zeitlich] schwanken, schwingen
fluctuating current *(El)* schwankender Strom *m*
f. load Dauerschwingbeanspruchung *f*
f. power *(El)* Schwingleistung *f (Mehrphasensystem)*
f. voltage *(El)* schwankende Spannung *f*
fluctuation [zeitliche oder statistische] Schwankung *f*; Fluktuation *f*
f. in mains voltage Netzspannungsschwankung *f*
f. in temperature Temperaturschwankung *f*
f. in the parameters of the orbit Schwankung *f* der Flugbahnparameter (Flugbahnwerte)
f. noise *(El)* statistisches Rauschen *n*
f. of pressure Druckschwankung *f*
f. of sensitivity Empfindlichkeitsschwankung *f*
flue 1. Abzugskanal *m*, Abzugsrohr *n*, Zug *m*; Rauchkanal *m*, Rauchrohr *n*, Fuchs *m*; 2. Flammrohr *n (Dampferzeuger)*
f. ash Flugasche *f*
f. boiler Flammrohrkessel *m*
f. bridge Fuchsbrücke *f*
f. cleaner Kesselrohrreiniger *m*
f. damper *(Wärm)* Zugabsperrklappe *f* für Flammrohre
f. dust Flugstaub *m*; Flugasche *f*; *(Met)* Gichtstaub *m*
f. dust catcher Flugaschenfang *m*
f. dust collector *s.* f. dust catcher
f. end Fuchsende *n (des Ofens)*
f. furnace Flammrohrfeuerung *f*
f. gas Verbrennungsgas *n*, Abgas *n*, Rauchgas *n*, Feuergas *n*
f. gas analyzer Rauchgasanalysator *m*
f. gas collector Rauchgasreiniger *m*
f. gas explosion Rauchgasexplosion *f*, Gasexplosion *f* im Feuerzug
f. gas pump Rauchgaspumpe *f*
f. gas tester Rauchgasprüfer *m*
f. gas utilization Abhitzeverwertung *f*
f. heating surface Flammrohrheizfläche *f*
f. hole Schürloch *n*
f. loss Fuchsverlust *m*
f. opening Fuchsöffnung *f (des Ofens)*
f. pipe 1. Abzugskanal *m*, Abzugsrohr *n*, Wrasenrohr *n*; 2. Lippenpfeife *f*, Labialpfeife *f*
f. with downward draught absteigender Zug *m*
f. with upward draught aufsteigender Zug *m*
fluellite *(Min)* Fluellit *m*
fluff / to 1. *(Led)* bimsen, dolieren, schleifen; 2. *(Met)* auflockern *(Formsand)*
fluffing machine Bimsmaschine *f*, Doliermaschine *f*, Schleifmaschine *f*
f. wheel Bimswalze *f*, Dolierrad *n*, Schleifzylinder *m*
fluid Flüssigkeit *f*, Fließendes *n*, fließendes Medium *n (Flüssigkeit oder Gas)*
fluid [dünn]flüssig; geschmolzen; fließend; pneumatisch
f. agitation *(Ch)* Rühren *n* fließender Medien
f. amplifier *(Am) (BMSR) Oberbegriff für hydraulische und pneumatische Verstärker*
f. bed Fließbett *n*; Wirbelschicht *f*
f.-bed process Wirbelschichtverfahren *n*
f. brake Wasserwirbelbremse *f (Motorenprüfstand)*
f.-carrying duct Durchflußrohr *n*
f. catalyst process Prozeß *m* mit bewegtem Katalysator, Wirbelschichtverfahren *n*
f. circuit Flüssigkeitskreislauf *m*
f. clutch Flüssigkeitskupplung *f*
f. container Flüssigkeitsbehälter *m*
f. coupling Flüssigkeitskupplung *f*; hydraulische Kupplung *f*
f. de-icing Flüssigkeitsenteisung *f*
f. displacement Flüssigkeitsverdrängung *f*
f. distribution Strömungsverteilung *f*
f. distributor *(Kfz)* Flüssigkeitsverteiler *m*
f. drive 1. hydraulischer Antrieb *m*; 2. *s.* f. gear
f. driven mit hydraulischem Antrieb

f. dynamics Strömungslehre *f*; Flüssigkeitsdynamik *f*
f.-feed milling machine Fräsmaschine *f* mit hydraulischem Vorschub
f. film zusammenhängende Flüssigkeitsschicht *f*
f. film lubrication Fließschmierung *f*; Filmschmierung *f* *(Gleitlager)*
f. flow 1. Strömungslehre *f*; 2. Strömung *f*
f.-flow engine Strömungsmaschine *f*
f.-flow measurement Durchflußmessung *f*
f.-flow velocity Strömungsgeschwindigkeit *f* *(Flüssigkeiten)*
f. flywheel dynamisches Flüssigkeitsgetriebe *n*, Drehmomentwandler *m*, hydrodynamisches Getriebe *n*, hydrodynamische Kupplung *f*, Strömungsgetriebe *n*, Turbogetriebe *n*
f. friction Flüssigkeitsreibung *f*
f. friction damping Flüssigkeitsdämpfung *f*, hydraulische Dämpfung *f*
f. gear Flüssigkeitsgetriebe *n*, hydraulisches Getriebe *n*, hydraulischer Transformator (Wandler) *m*
f. kinetic machine Strömungsmaschine *f*
f. level Flüssigkeitsspiegel *m*
f. level gauge Flüssigkeits[stand]anzeiger *m*
f. manometer Flüssigkeitsmanometer *n*, U-Rohr-Manometer *n*
f. mechanics Hydromechanik *f*
f. medium Hydraulikmedium *n*; Druckmittel *n*; Fördermedium *n*
f. motor drilling Bohren *n* mittels durch die Spülung angetriebener Motoren *(Erdöl, Erdgas)*
f. mud Spültrübe *f* *(für Spülbohrungen)*
f.-power line Hydraulikleitung *f*
f.-power motor Hydraulikmotor *m*
f.-powered mit hydraulischer Hilfsenergie
f. pressure hydrostatischer Druck *m*
f. pressure[-type] gear hydrostatisches Getriebe *n*, hydrostatischer Transformator (Wandler) *m*
f. pump Flüssigkeitspumpe *f*
f. seal Flüssigkeitsabdichtung *f*; hydraulischer Verschluß *m*
f. starter *(Elm)* hydraulischer Anlasser *m*
f. stream Strömung *f*; Flüssigkeitsstrom *m*
f.-tight flüssigkeitsdicht
f.-to-fluid cold extrusion Kaltfließpressen *n* von durch hohen Druck duktil gemachten Werkstoffen
f. torque converter hydraulischer Drehmomentwandler *m*
f. transmission hydraulische Kraftübertragung *f*; Flüssigkeitskupplung *f*, Strömungsgetriebe *n*
fluidity Fluidität *f*, Fließfähigkeit *f*; Viskosität *f*; Formfüllungsvermögen *n*, Gießbarkeit *f*
fluidization Wirbelschichttechnik *f*
fluidized bed *(Ch, Kern)* Fließbett *n*, Wanderschicht *f*, Wirbelbett *n*, Wirbelschicht *f*
f. bed reactor Wirbelschichtreaktor *m*
f.-fuel reactor Reaktor *m* mit quasiflüssigem (suspendiertem) Brennstoff
f. operation Wirbelschichtverfahren *n*
f. pressure Druck *m* eines fließenden Mediums
f.-solid reactor Fließbettreaktor *m*
fluidizing point Wirbelpunkt *m* *(Wirbelschichtverfahren)*
f. process Wirbelschichtverfahren *n*, Fließverfahren *n*
fluke *(Brg)* Schlammlöffel *m*
f. of the anchor *(Schiff)* Ankerpflug *m*, Ankerflunke *f*, Ankerflügel *m*
flume 1. Venturikanal *m*, Meßkanal *m* *(Durchflußmessung)*; 2. *(Brg)* Wasserrösche *f*, Gefluder *n*, Gefluter *n*; 3. Betongleitblech *n*
f. end trogförmiges Schiffsende *n* *(Bug oder Heck)*
fluoboric acid Borfluorwasserstoffsäure *f*
fluorate / to *(Ch)* fluorieren
fluorcarbon Fluorkohlenwasserstoff *m*, fluorsubstituierter Kohlenwasserstoff *m*
fluorene *(Ch)* Fluoren *n*
fluoresce / to fluoreszieren
fluorescence Fluoreszenz *f*
fluorescent fluoreszierend, fluoreszent
f. bulb Leuchtstoffröhre *f*
f. dye Fluoreszenzfarbstoff *m*
f. dye staining Fluorochromierung *f*
f. ink *(Typ)* Fluoreszenzdruckfarbe *f*, Leuchtdruckfarbe *f*
f. lamp Leuchtstofflampe *f*, Leuchtstoffröhre *f*
f. material Leuchtmasse *f*
f. paint Leuchtfarbe *f*
f. penetrant Fluoreszenzlösung *f* *(zerstörungsfreie Werkstoffprüfung)*
f. radiation Fluoreszenzstrahlung *f*
f. screen Leuchtschirm *m*
f.-screen scanning Leuchtschirmabtastung *f*
f. substance Fluoreszenzstoff *m*, Leuchtstoff *m*
f. tube Leucht[stoff]röhre *f*
fluorhydric acid Fluorwasserstoffsäure *f*, Flußsäure *f*
fluoric acid Flußsäure *f*
fluoridate / to fluorieren
fluoride Fluorsalz *n*, Fluorid *n*
fluorimeter *(Ch)* Fluormesser *m*
fluorinate / to fluorieren
fluorine Fluor *n*
fluorite Flußspat *m*
fluoro-aliphatic compound fluorsubstituierte aliphatische Verbindung *f*
fluorogen fluoreszenzerregender Stoff *m*
fluorogenic fluoreszenzerregend

fluorography *(Am)* Schirmbildaufnahme *f (Medizintechnik)*
fluorohydrocarbon Fluorkohlenwasserstoff *m*, fluorsubstituierter Kohlenwasserstoff *m*
fluorolubricant Schmiermittel *n* auf Basis fluorierter Kohlenwasserstoffe
fluorometer *(Ph)* Fluorometer *n*, Fluoreszenzmesser *m*
fluoroscope Röntgenbildschirm *m*
fluoroscopy Röntgendurchleuchtung *f*
fluorothene Fluoräthylen, Tetrafluoräthylen *n*
fluorspar Flußspat *m*
fluosilicate Fluorsilikat *n*
fluosilicic acid Fluorkieselsäure *f*, Kieselfluorwasserstoffsäure *f*
flurry leichte Brise *f*, Windstoß *m*, Bö *f*
flush / to sprengen, spülen; einschlämmen, einspülen; einebnen; einfluchten
f. away wegspülen
f. out ausschwemmen, ausspülen, auswässern
flush 1. Wasserstrahl *m*; 2. Spüler *m (Abort)*; 3. Flottierung *f*
flush 1. *(Masch, Bw)* abgeglichen, bündig, fluchteben, fluchtgerecht, glatt; eingelassen; 2. *(Brg)* satt anliegend *(Kappe)*
f. bolt *(Bw)* Riegel *m*
f. boring Spülbohren *n (Tiefbohrungen)*
f. box Spülkasten *m*
f. coating Fließbeschichtung *f*
f. deck vessel Glattdeckschiff *n*
f.-decked mit Glattdeck
f. diaphragm *(BMSR)* Flachmembrane *f*
f. door Tür *f* mit glattem Blatt
f.-encased lock eingelassenes Schloß *n*
f.-fitting hatch cover *(Schiff)* glatt mit dem Deck abschließender Lukendeckel *m*
f. gas *(Brg)* Spülgas *n*
f. head 1. Senkkopf *m (Niet)*; 2. Spülkopf *m (bei Bohrungen)*
f. headlight *(Kfz)* eingelassener Scheinwerfer *m*
f. joint *(Holz)* bündige Fuge *f*
f.-mounted eingebaut, eingelassen, versenkt
f.-mounted switch Frontplattenschalter *m*
f.-mounting instrument Einbauinstrument *n*
f.-mounting meter *(El)* Einbauinstrument *n*, versenkter Zähler *m*
f. pipe Spülrohr *n*
f. rivet Senkniet *m*
f. socket Unterputzsteckdose *f*
f. surface glatte Oberfläche *f*
f. switch Unterputzschalter *m*, Einbauschalter *m*
f. toilet Spülklosett *n*
f.-type instrument Einbau[meß]instrument *n*, versenkbar einzubauendes Meßinstrument *n*
f.-type switch Unterputzschalter *m*, Einbauschalter *m*
f. water *(Brg)* Transportwasser *n*
f. weld *(eingeebnete)* Flachnaht *f*
f. with bündig mit
f. wood screw Senkholzschraube *f*
flushed colours Direktfarben *fpl*
flusher Straßensprengwagen *m*
flushing 1. Spülung *f*; Wasserspülung *f*; Bespülung *f*; Einschlämmen *n (z. B. einen Pfahl)*; 2. *(Brg)* Spülung *f*, Spülversatz *m (Bohrloch)*
f. dredger *(Brg)* Schwemmbagger *m*, Spülbagger *m*
f. dump Spülkippe *f*
f. hole *(Met)* Schlacken[stich]loch *n*, Schlackenstich *m*
f. method *(Plast)* Verdrängen *n* des Wassers in Farbpasten durch Bindemittel
f. oil Spülöl *n*
f. tank Spülkasten *m*
f. water Nachspülwasser *n*
flute / to 1. kannelieren, auskehlen; 2. nuten; riefen, riffeln
flute 1. Riefe *f*, Rille *f*, Hohlkehle *f*; 2. Spannut *f (Schneidwerkzeug)*; Stollen *m (Gewindebohrer)*; Spankammer *f (Fräser)*; 3. Welle *f (Wellpappe)*; 4. Flöte *f (Musikinstrument)*
f. edge *(Masch)* Umfangszahn *m*
f. helix angle Drallsteigungswinkel *m*
f. length Spirallänge *f (Spiralbohrer)*; Nutenlänge *f*; Schneidenlänge *f (Bohrer)*
f. miller Nutenfräsmaschine *f*
f. milling Nutenfräsen *n*
f. pipe Lippenpfeife *f*
f. reaming Fertigreiben *n* mit Maschinenreibahle
f. recorder Flötenbaß *m (Orgel)*
f. stop Flötenregister *n*, Flötenwerk *n*, Flötenzug *m (Orgel)*
fluted 1. gerieft, geriffelt, gerillt, kanneliert; 2. *(Holz)* spannrückig
f. bar iron Hohlkanteisen *n*
f. cutter Fräser *m* mit gefrästen Zähnen
f. feed roller geriffelte Speisewalze *f (z. B. Textilmaschinen)*
f. handle gerillter Griff *m*
f. roller *(Lw)* Riffelwalze *f*
f. segment Riffelsegment *n (z. B. Kämmmaschine)*
f. socket head cap-screw Schraube *f* mit Innenriffelung
f. tooth gefräster Zahn *m*
f. type machine bit Maschinenholzspiralbohrer *m*
fluting 1. Nuten *n*; Nutenfräsen *n*; 2. Auskehlung *f*, Kannelierung *f*; Riefelung *f*; Riffelung *f*; 3. *(Text)* Kräuseln *n* und Rüschen *n (Besatz)*
f. attachment Kannelierapparat *m (Holzbearbeitung)*

f. cutter Spannutfräser *m*
f. machine *(Lw)* Riffelmaschine *f*
f. plane Hohlkehlhobel *m*, Kehlhobel *m*, Nuthobel *m*, Rinnenhobel *m*
f. tool Riffelstahl *m*
flutter / to flattern, zittern; schwanken *(Zeiger an Meßinstrumenten)*
flutter *(El)* [schnelle] Tonhöhenschwankungen *fpl*; Flattereffekt *m*
f. echo Flatterecho *n*
f. effect *(Fs)* Flattereffekt *m*
f. of piston rings Flattern *n* der Kolbenringe
f. phenomenon Flattererscheinung *f*
flux / to verschlacken
flux 1. Fluß *m (Kraftlinien)*; 2. *(Ch, Met)* Flußmittel *n*; Schmelzmittel *n*; Schmelzfluß *m*; Zuschlag *m*; 3. Löt[fluß]mittel *n*
f. and reflux Ebbe *f* und Flut *f*
f.-cored soldering wire Lötdraht *m* mit Flußmittelkern, Kombinationslötdraht *m*
f.-cored wire solder Lötdraht *m* mit Flußmittelkern, Kombinationslötdraht *m*
f. density [magnetische] Kraftliniendichte *f*
f. distribution [magnetische] Kraftlinienverteilung *f*; Flußverteilung *f (Neutronen)*
f.-gate magnetometer Luftspaltmagnetometer *n*
f. interlinking [magnetische] Flußverkettung *f*
f. leakage [magnetische] Kraftlinienstreuung *f*
f. leakage field [magnetisches] Streufeld *n*
f. linkage [magnetische] Flußverkettung *f*; induktive Kopplung *f*
f. linking a coil Spulenfluß *m*
f. linking a turn Windungsfluß *m*
f. meter Fluxmeter *n*, Flußmesser *m*, Kriechgalvanometer *n*
f. of heat Wärmefluß *m*
f. of lines of force Kraftlinienfluß *m*
f. of luminous energy Lichtstrom *m*
f. of radiation Strahlungsfluß *m*
f. of vorticity Wirbelfluß *m*
f. oil Flußmittel *n*, Stellöl *n*
f. residue Flußmittelrest *m*; Lötmittelrest *m*
f. reversal Flußumkehr *f*
f. shunting *(Elm)* Feldschwächung *f*
f. time integraler Fluß *m*, Zeitintegral *n* des Neutronenflusses
fluxed electrode Tauchelektrode *f* [mit Flußmittel]
fluxing 1. *(Plast)* Erweichen *n*, Herabsetzen *n* des Schmelzpunktes, Plastifizieren *n*, Weichmachen *n*; 2. *(Met)* Zusetzen *n* von Flußmitteln
f. ore Zuschlagerz *n*
f. power Verschlackungsfähigkeit *f*
f. temperature Schmelztemperatur *f*
fluxion *(Math)* Differential *n*
fluxless soldering Löten *n* ohne Flußmittel
fluxmeter *s.* flux meter
fluxograph Flußdichteschreiber *m*, Fluxograf *m*
fly / to fliegen; auflassen
f. the flag *(Schiff)* Flagge führen
f. with tail wind *(Flg)* mit dem Wind fliegen
fly 1. *(Text)* Flug *m (Abfall)*; 2. Flugstrecke *f*; 3. *(Typ)* Ausleger *m*
f. ash Flugasche *f*
f.-away oil Motorrostschutzmittel *n*
f.-back *s.* flyback
f.-ball *s.* flyball
f. cutter Schlagmesser *n*
f. cutter for holes in thin metal Blechausstechbohrer *m*
f. cutter in stub bar fliegendes Ausdrehwerkzeug *n*
f. cutting Schlagzahnfräsen *n*
f. doubler Flügelzwirnmaschine *f*
f. frame drawing rollers *(Text)* Streckwerk *n* der Spulenbank
f. leaf *(Typ)* Vorsatz[blatt] *n*, Vorsatzpapier *n*
f. machine Flügelspinnmaschine *f*
f. net Moskitonetz *n*
f. nut Flügelmutter *f*
f.-over Überführung *f (insbesondere bei Straßenkreuzungen und über Autobahn)*
f. press Spindelpresse *f*
f. rock *(Brg)* weggeschleudertes Haufwerk *n (Schießarbeit)*
f. shedding *(Text)* Faserflugbildung *f*, Flugbildung *f*
f. shuttle *s.* flying shuttle
f. shuttle lathe *(Text)* Schnellade *f*
f.-spring *(El)* Relaisfeder *f* [für erste Kontaktgabe]
f. twister Flügelzwirnmaschine *f*
f.-under Unterführung *f*
f. weight Fliehgewicht *n*
flyback *(Fs)* Rücklauf *m*, Zeilenrücklauf *m*
f. blanking Rücklaufaustastung *f*
f. extra high tension generator Rücklaufhochspannungsgenerator *m*
f. speed Rücklaufgeschwindigkeit *f*, Rückkippgeschwindigkeit *f*
f. supression Rücklaufverdunkelung *f*
flyball Schwungkörper *m*
f. governor Fliehkraftregler *m*
f. integrator Kugelintegrator *m*
flyer Flügelspinnmaschine *f*; Fleier *m*, Flyer *m*, Flügel *m*
f. frame *s.* flyer
f. frame with differential motion Differentialflyer *m*
f. head Flügelkopf *m*
f. lathe Federlade *f*
f. spindle Flügelspindel *f*
f. spinning frame machine *s.* flyer

f. twisting frame Flügelzwirnmaschine *f*
f. with presser arm Flügel *m* mit Preßfinger
f. yarn Flügelgarn *n*
flying Fliegen *n*; Flugwesen *n*, Fliegerei *f*
f. altitude Flughöhe *f*
f. apart of the grinding wheel Bersten *n* der Schleifscheibe
f. boat Flugboot *n*
f. boat base Wasserflughafen *m*
f. bond *s.* monk bond
f. bridge *(Schiff)* Laufbrücke *f*, fliegende Brücke *f*, Verbindungsbrücke *f*
f. buttress *(Bw)* Strebepfeiler *m*
f. control equipment Flugüberwachungsgerät *n*
f. missile Flügelrakete *f*
f.-off deck Flugdeck *n*, Abflugdeck *n* *(Flugzeugträger)*
f. saucer fliegende Untertasse *f*
f. scaffold hängendes Gerüst *n*
f. shears fliegende Schere *f*
f. shuttle *(Text)* Schnellschützen *m*
f. sounder *(Schiff)* Tieflotungsapparat *m*
f. speed Fluggeschwindigkeit *f*
f. spot wandernder Lichtpunkt *m*
f. spot method of scanning Lichtpunktabtastung *f*
f. spot scanner Lichtpunktabtaster *m*
f. spot store *(BMSR)* halbpermanenter Lichtpunktspeicher *m*
f. suit Fliegerschutzanzug *m*
f. time Flugzeit *f*
f. tuition Flugunterricht *m*
f. weather Flugwetter *n*
f.-wing design *(Flg)* Nurflügelbauart *f*
flyweight governor Fliehkraftregler *m*, Zentrifugal[kraft]regler *m*
flywheel Schwungrad *n*; Schwungscheibe *f* *(Kupplung)*
f. action Schwungwirkung *f*
f. arranged between two bearings beiderseitig gestütztes Schwungrad *n*
f. blower Schwungradlüfter *m*
f. brake Schwungradbremse *f*
f. calculation Schwungradberechnung *f*
f. casing Schwungradgehäuse *n*
f. converter *(Elm)* Schwungradumformer *m*
f. dowelling Schwungradverklammerung *f*
f. drive Schwungradantrieb *m*
f. effect *(Fmt)* Schwungradeffekt *m* *(Resonanzschwingkreis)*
f. explosion Schwungradexplosion *f*
f. force Fliehkraft *f*
f. generator *(Elm)* Schwungradgenerator *m*
f. governor Schwungradregler *m*
f. in quarters vierteiliges Schwungrad *n*
f. key Schwungradkeil *m*
f. magneto *(Kfz)* Schwungradmagnetzünder *m*
f. pit Schwungradgrube *f*
f. pulley Schwungradriemenscheibe *f*, als Schwungrad dienende Riemenscheibe *f*, als Riemenscheibe ausgeführtes Schwungrad *n*
f. starter-generator ignition unit *(Kfz)* Schwungradlichtanlaßzünder *m*
f. synchronization *(Fs)* Schwungradsynchronisation *f*
f. timing mark *(Kfz)* Schwungradmarke *f*
f. toothing Schwungradverzahnung *f*
f. with starting gear *(Kfz)* Schwungrad *n* *n* mit Anlasserzahnkranz
f. with starting ring gear *(Kfz)* Schwungrad *n mit Anlasserzahnkranz*
foam / to schäumen
f. onto *(Text)* aufschäumen
f. over überschäumen
foam Schaum *m*
f.-back carpet Teppich *m* mit Schaumstoffrückseite
f.-backed knitted wear laminierte Wirkbekleidung *f*
f.-backed textiles mit Schaumstoff laminierte Textilien *pl*
f. backing Laminieren *n* mit Schaumstoff
f. bubble Schaumblase *f*
f. collapse *(Plast)* ungenügendes Schäumen *n*
f. concrete Schaumbeton *m*
f. control agent Schaumregulierungsmittel *n*
f. drainage Schaumentwässerung *f*
f. extinguisher Schaum[feuer]löscher *m*
f. film *(Text)* Schaumlamelle *f*
f. fire-engine Schaumlöschwagen *m*
f. generator Schaumerzeuger *m*
f. glue Schaumkleber *m*
f.-in-place material *(Plast)* im Fertigteil geschäumtes Material *n*
f.-in-place moulding *(Plast)* Schäumen *n* im Fertigteil
f. inhibitor Schaumdämpfungsmittel *n*
f. laminate Schaumstoff *m*
f. lamination Schaumstofflaminieren *n*
f.-like backing *(Text)* Kaschieren *n* mit Schaumstoff (schaumstoffähnlichem Material)
f.-mat drying Schaummattentrocknung *f* *(Verfahren zur Herstellung von Instantpulvern)*
f. mist for vermin destruction Schaumnebel *m* zur Ungezieferbekämpfung
f. overblow *(Plast)* übermäßiges Schäumen *n*
f. rubber Schaumgummi *m*
f. skimmer Schaumabstreifer *m*
f. trap Schaumfang *m*
f. type [fire] extinguisher Schaum[feuer]löscher *m*
f. yarn Schaumfaden *m*, Faden *m* aus Schaumstoff

foamback Schaumstoffrückseite *f* von Textilien
foambacks mit Schaumstoff kaschierte Textilien *fpl*
foamclay Schaumton *m*
foamed geschäumt, Schaum-
f. blast-furnace slag Hüttenbims *m*, Schaumschlacke *f*
f. concrete Schaumbeton *m*
f. plastic Schaumstoff *m*
f. rubber Schaumgummi *m*
f. slag Hüttenbims *m*, Schaumschlacke *f*
foamer Schaumbildner *m*
foaming Schäumen *n*, Aufschäumen *n*, Schaumbildung *f*
f. agent Schaumbildner *m*, Schaummittel *n*, Treibmittel *n*
f. in place *(Plast)* Herstellung *f* von Schaumstoff in der Form
f. tendency Neigung *f* zur Schaumbildung, Schaumneigung *f*
foamy schaumig
fo'c'sle, foc's'le *s.* forecastle
focal fokal; Fokal-, Brenn[punkt]-
f. aperture Blendenöffnung *f*
f. axis Brennachse *f*
f. chord Parameter *m (eines Kegelschnittes)*
f. curve Brennlinie *f*
f. distance Brennpunktabstand *m*
f. length Brennweite *f*, Fokallänge *f*; Brennpunktabstand *m*
f. line Brennlinie *f*
f. plane Brenn[punkt]ebene *f*, Fokalebene *f*
f.-plane frame Anlegerahmen *m (Fotogrammetrie)*
f.-plane shutter Schlitzverschluß *m*
f. point Brennpunkt *m*, Fokus *m*
f. ray Brennstrahl *m*
f. shutter Schlitzverschluß *m (Kamera)*
f. spot Brennfleck *m*, Fokus *m*
f.-spot characteristic Brennfleckcharakteristik *f*
foci of the ellipse Brennpunkte *mpl* der Ellipse
focimeter *s.* focometer
focometer Brennweitenmesser *m*, Fokometer *n*
focus / to fokussieren, bündeln
f. the headlamps *(Kfz)* die Scheinwerfer einstellen
focus 1. Brennpunkt *m*; Fokus *m*; 2. *(Fs)* Scharfeinstellung *f*; 3. *(BMSR)* Strudel *m (Phasenebenendarstellung)*; 4. Schmelzzone *f (Hochofen)*
f. anode *(Fs)* Fokussieranode *f*
f. control Scharfeinstellung *f*; Schärferegler *m*
f. displacement Fokusverlagerung *f*
f.-to-film distance Fokus-Filmabstand *m*
focusable fokussierbar
f. eyepiece stellbares Okular *n*
focused for infinity auf unendlich eingestellt
f. spot Brennfleck *m*
focusing Fokussierung *f*; Bündelung *f*
f. coil Fokussier[ungs]spule *f*; elektromagnetische Linse *f*
f. electrode Bündelungselektrode *f*, elektrostatische Linse *f*
f. field Fokussierungsfeld *n*; Fokuseinstellungsfeld *n*
f. lens Fokussierlinse *f*, Sammellinse *f*; Einstellupe *f*
f. magnet Fokussier[ungs]magnet *m*
f. microscope Einstellmikroskop *n*
f. screen Einstellschirm *m*; Mattscheibe *f (Kamera)*
f. shot Fokussieraufnahme *f*
f. voltage Fokussierungsspannung *f*
focussing *s.* focusing
fodder / to füttern *(Vieh)*
fodder Viehfutter *n*
f. barley Futtergerste *f*
f. chopping bench Futterschneidebank *f*
f. plant Futterpflanze *f*
f. processing Futterbehandlung *f*
fog 1. Nebel *m*; 2. *(Opt, Fs)* Schleier *m*
f. band Nebelschwaden *m*
f. cooling *(Kern)* Direktkühlung *f (mit einem Hochdruckdampf-Wassertröpfchen-Gemisch; Leistungsreaktoren)*
f. density *(Opt)* Schleierdichte *f*, Schleierschwärzung *f*
f. formation Nebelbildung *f*
f. horn Nebelhorn *n*
f. lamp Nebellampe *f*; *(Kfz)* Nebelscheinwerfer *m*
f. level *(Opt)* Schleier[dichte *f*] *m*
f. light *(Kfz)* Nebelscheinwerfer *m*
f. prevention *(Opt)* Schleierschutz *m*
f. return Nebelrückstrahlung *f (Radar)*
f. tracks Nebelspuren *fpl*
f. value Schleierwert *m*
fogged condition Verschleierung *f (fotografischer Film)*
foil Folie *f*, Blatt *n*, Blättchen *n*
f. detector *(Kern)* Aktivierungsfolie *f*
f. electrometer Blattelektrometer *n*
f. gauge *s.* f. strain gauge
f.-jamming Verdüppelung *f (Radar)*
f. metal Metallfolie *f*
f. mill Folienwalzwerk *n*
f. of metal Metallfolie *f*
f. strain gauge Dehn[ungs]meßstreifen *m*
f. tin Zinnfolie *f*, Stanniol *n*
fold / to falten; abkanten *(Blech)*; falzen
f. down herunterklappen, herabklappen
f. in *(Typ)* einfalzen
f. up *(Flg)* einziehen *(Fahrwerk)*; zusammenfalten *(Tragfläche)*
f. upwards hochklappen, aufklappen

fold 1. Falte *f*; 2. Falz *m (Blech, Papier, Holz)*; 3. Glied *n (Gliedermaßstab)*
f.-back *(Plast)* Raupenbildung *f* des Kunststoffs in der Spritzform
f.-fault *(Geol)* Bruchfalte *f*
f.-over *(Am) (Fs)* Geisterbild *n*
foldability Falzfähigkeit *f*, Faltbarkeit *f*
foldable 1. zusammenklappbar; 2. falzbar, falzfähig
foldableness *s.* foldability
folded [zusammen]gefaltet; *(Met)* abgekantet *(Blech)*
f. belt *(Brg)* Faltenband *n*
f. dipole antenna Faltdipolantenne *f*
f.-horn loudspeaker Faltenlautsprecher *m*
f. plate *(Bw)* Faltwerk *n*
f.-plate roof Faltwerkdach *n*
f. yarn Zwirn *m*
folder 1. Aktendeckel *m*, Ordner *m*, Schnellhefter *m*; 2. Faltprospekt *m*, Broschüre *f*; 3. Falzbein *n (Buchbinderei)*; 4. *(Text)* Täfler *m*
folding 1. *(Math, Geol)* Faltung *f*; 2. Falten *n (Papier)*; Falzen *n (Blech, Papier)*; Knicken *n (Leder)*; 3. Abkanten *n (Blech)*; 4. Doppeln *n (Bleche)*
folding zusammenklappbar, zusammenlegbar
f. and glueing machine Falz- und Leimmaschine *f (Buchbinderei)*
f. blade (Falzschwert *n (Buchbinderei)*
f. boat Faltboot *n*
f.-bone *s.* f. stick
f. bow *(Kfz)* Faltspriegel *m*
f. box Faltschachtel *f*
f. boxboard Faltschachtelkarton *m*
f. camera Klappkamera *f*
f. chair Klappstuhl *m*
f. climber Klappsteigeisen *n*
f. crank umlegbare Kurbel *f*
f. door Flügeltür *f*; Harmonikatür *f*
f. endurance Faltbeständigkeit *f (Papier)*; Knickfestigkeit *f (Leder)*
f. handle Einschlagheft *n*; umlegbare (versenkbare) Kurbel *f (z. B. Bandmaß)*
f. hood *(Kfz)* Faltdach *n*, Klappverdeck *n*
f. machine 1. *(Masch)* Abkantmaschine *f*; 2. *(Text)* Dubliermaschine *f*, Gewebefaltmaschine *f*, Legemaschine *f*; 3. *(Led)* Buggmaschine *f*, Sturzmaschine *f*
f. magnifier *(Opt)* Einschlaglupe *f*
f. magnifying glass Einschlaglupe *f*
f. map zusammenlegbare Karte *f (Landkarte, Stadtplan usw.)*
f. mount Einschlagfassung *f*
f. plier Deckzange *f*
f. press Abkantmaschine *f*, Abkantpresse *f*
f. resistance Falzwiderstand *m (Blech)*; Knickwiderstand *m (Leder)*
f. rule Gliedermaßstab *m*, Zollstock *m*
f. seat Klappsitz *m*, Klappstuhl *m*
f. sight Klappvisier *n (Gewehr)*
f. steel rule Gliedermaßstab *m* aus Stahl
f. step Klappfußtritt *m*
f. stick *(Typ)* Falzbein *n*
f. strength Faltfestigkeit *f (Papier)*; Falzfestigkeit *f (Blech, Papier)*; Knickfestigkeit *f (Leder)*
f. table Klapptisch *m*
f. test *(Met)* Faltversuch *m*, Falzversuch *m*
f. top *(Kfz)* Verdeck *n*, Klappverdeck *n*
f.-top base Verdecklager *n*
f.-top bow Verdeckspriegel *m*
f.-top clamp Verdeckverschluß *m*
f.-top cover Verdeckhülle *f*
f.-top frame Verdeckrahmen *m*
f.-top structure Verdeckgestell *n*
f. trail-spade Klappsporn *m (Geschütz)*
f. undercarriage *(Flg)* Einziehfahrwerk *n*
f. wedges Doppelkeil *m*
foliaceous blätt[e]rig; Blatt-, Blätter-
f. spar Blätterspat *m*
f. tree Laubbaum *m*
foliage Laub[werk] *n*
f. plant Blattpflanze *f*
foliate / to mit Folie belegen
foliated blätt[e]rig
f. gypsum Schiefergips *m*
f. texture blättrige Struktur *f (Festkörper)*
foliation *(Geol)* Schichtung *f*, schichtenförmige Lagerung *f*
f. cleavage *(Geol)* Schieferung *f*
folio *(Typ)* Folio[blatt] *n*; Bogen *m*
f.-signature Bogensignatur *f*
f. volume Foliant *m*
folium 1. Blatt *n*; Lamelle *f*; 2. *(Math)* Blattkurve *f*
follow / to folgen, mitgehen; abtasten
f. directions for use Gebrauchsanweisung befolgen
f. up nachlaufen; verfolgen *(Lauf)*
f. working instructions Bedienungsvorschriften befolgen
follow Folge-, Mitlauf-, Nachlauf-; Folgen *n*
f. current Folgestrom *m*
f.-focus device mitlaufender Sucher *m (Filmkamera)*
f. motor Nachlaufmotor *m*
f.-on current nachfolgender Strom *m*, Folgestrom *m*
f.-on tool Folgewerkzeug *n*
f. rest *(Masch)* mitgehende Lünette *f*
f.-up control Nachlaufregelung *f*
f.-up device Nachführung[seinrichtung] *f (z. B. astronomisches Fernrohr)*
f.-up gear *(Schiff)* Fernsteuerungsanlage *f*
f.-up invoice Zweitschriftrechnung *f*

f.-up mechanism Folgeregler *m*, Nachlaufregler *m*
f.-up motor Nachlaufmotor *m*, Folgemotor *m*
f.-up potentiometer Nachlaufpotentiometer *n* *(Analogrechner)*
f.-up pressure gleichbleibender (gleichgehaltener) Druck *m*
follower 1. Folgeregler *m*, Nachlaufregler *m*; 2. *(El)* Folgestufe *f*, Verstärker *m*; 3. Leitbacke *f* *(Gewindeleitapparat)*; 4. *(Kfz)* Stößel *m*; 5. *(Am)* *(Mil)* Zubringer *m* *(z. B. Gewehr)*; 6. Tastnase *f*; Tastrolle *f* *(Nachformen)*; 7. Abtriebsrad *n*; getriebenes Rad *n*; 8. *(Wärm)* Kolbendeckel *m*, Deckelring *m*, Kolbendecke *f*, Winkelring *m*
f. control Folgeregelung *f*
f. rest mitlaufender Setzstock *m*
f. roll Tastrolle *f*
f. roller Nachformwalze *f*
f. tooth Nachschneidezahn *m* *(Säge)*
following rest mitgehender Setzstock *m*
font *(Typ)* Schriftkegel *m*
food Essen *n*; Nahrungsmittel *npl*; Proviant *m*
f. container Essenbehälter *m*
f. industry Nahrungsmittelindustrie *f*
f. investigation Lebensmitteluntersuchung *f*
f. irradiation reactor Reaktor *m* für Nahrungsmittelbestrahlung
f. poison Fraßgift *n* *(Schädlingsbekämpfung)*
f. preservatives Konservierungsmittel *npl*
f. ship Proviantschiff *n*
foodstuff Lebensmittel *n[pl]*, Nahrungsmittel *n[pl]*, Proviant *m*
f. chemist Nahrungsmittelchemiker *m*
f. processing Lebensmittelverarbeitung *f*
f. refractometer Lebensmittelrefraktometer *n*
fool-proof betriebssicher, mißgriffsicher; narrensicher
fool's gold *s.* iron pyrites
foolscap paper Kanzleipapier *n*
foot 1. Fuß *m* *(Maßeinheit)* *(1 ft. = 12 in = 30,48 cm)*; 2. *(Fmt)* Röhrenfuß *m*, Sockel *m*; 3. *(Brg)* Liegendes *n*, Sohle *f*
f.-and-mouth-disease *(Lw)* Maul- und Klauenseuche *f*
f. bearing *(Masch)* Stützlager *n*, Spurlager *n*
f. block *(Brg)* Grundschwelle *f*, Sohlenholz *n* *(Ausbau)*
f. brake Fußbremse *f*
f.-bridge Fußgängerbrücke *f*
f.-candle Footcandle *(englische Einheit der Beleuchtungsstärke; 1 fc. = 1 lm/sq.ft. = 10,76 lx)*
f.-candle meter Beleuchtungsstärkemesser *m*
f. control Fußbedienung *f*, Fußschaltung *f*
f. dip switch *(Kfz)* Fußabblendschalter *m*
f.-gear Fußbekleidung *f*; *(meist:)* Schuhwerk *n*
f. lambert Footlambert *(englische Einheit der Leuchtdichte)*
f. lever Fußhebel *m*, Tritthebel *m*
f.-mounted motor Fußmotor *m*
f.-note Fußnote *f*
f. of bearing Lagerfuß *m*
f. of frame Rahmenfuß *m*
f. of the spoke Blattzapfen *m* *(Stellmacherei)*
f.-operated switch Fußschalter *m*, Tretschalter *m*
f. operation Fußantrieb *m*, Fußbetrieb *m*
f.-path Fußweg *m*, Gehweg *m*; Feldweg *m*
f. pedal Fußhebel *m*
f.-plate *(Eis)* Führerstand *m*
f.-plate of ski Schiplatte *f*, Fußplatte *f*
f.-poundal *englische Einheit der Arbeit; 1 ft.pdl. = 0,042 Nm = 4,297 mp*
f. per minute *englische Einheit der Geschwindigkeit; 1 ft./min = 0,508 cm/s*
f.-pound Footpound *(englische Einheit der mechanischen Energie; 1 ft.lb. = 1,356 J = 0,138 kpm)*
f. per second squared *englische Einheit der Beschleunigung; 1 ft./s² = 30,48 cm/s²*
f.-pump Fußpumpe *f*
f. rest Fußstütze *f*
f.-road Bürgersteig *m*, Fuß[gänger]weg *m*
f.-rule Zollstab *m* *(ein Fuß lang)*
f. stall *(Bw)* Postament *n*, Sockel *m*
f. starter switch *(Kfz)* Fußanlaßschalter *m*
f. switch Fußschalter *m*, Pedalschalter *m*, Tretschalter *m*
f.-ton *englische Einheit der Arbeit; 1 ft.ton = 0,3097 Mpm*
f. treadle Fußtritt *m*, Tretschemel *m*, Tritt *m*
f. valve Bodenventil *n*
f.-way Bürgersteig *m*, Fußweg *m*, Gehweg *m*
f.-wear Fußbekleidung *f*; *(meist)* Schuhwerk *n*
footage 1. Filmlänge *f* *(Kinofilm)*; 2. Länge *f* *(in Fuß)*
f. capacity Filmfassungsvermögen *n*
f. counter Filmlängenmesser *m*, Filmuhr *f*
football bladder Fußballblase *f*
f. boots Fußballschuhe *mpl*
footboard *(Eis)* Trittbrett *n* *(zum Ein- und Aussteigen)*
footbrake *s.* foot brake
footfall Fußtritte *mpl* *(Geräusch)*
f. sound attenuation Schalldämpfung *f* der Fußtritte, Trittschalldämpfung *f*
footing 1. Flachgründung *f*, Fundament *n*; Sockel *m*; 2. Untergrund *m*; Auflage *f*; Stützung *f*
f. machine *(Text)* Fußmaschine *f*
footless lamp sockellose Glühlampe *f*
f. valve *(Fmt, Fs)* sockellose Röhre *f*
footstep bearing Fußlager *n*, Spurlager *n*
f. pillow Spurplattenträger *m* *(Turbine)*
footstock Reitstock *m*
f. centre Reitstockspitze *f*

footwall *(Brg)* Liegendes *n*, Sohle *f*
f. drift Sohlenstrecke *f*
footwarmer Fußwärmer *m*
forage Viehfutter *n*
f. harvester Futter[mittel]erntemaschine *f*
f. loft Futterkammer *f*, Futterboden *m*
f. plant Futterpflanze *f*
f. shed Futterkammer *f*
f. shredder Futterreißer *m*
forbidden band verbotenes Energieband *n* *(Halbleiter)*
f. region verbotener Energiebereich *m* *(Halbleiter)*
f. transition verbotener Übergang *m* *(Atome, Kerne)*
force / to 1. *(Masch)* drücken, pressen, treiben, zwängen, zwingen; 2. *(Ph)* erzwingen; 3. *(BMSR)* eingreifen *(in den Programmablauf)*
f. against anpressen, andrücken, gegendrücken
f. down niederdrücken, herunterdrücken, herabdrücken
f. in einpressen
f. into hineintreiben
f.-land *(Flg)* notlanden
f. on aufpressen; aufziehen *(Kupplung)*
f. open aufbrechen
f. out [her]auspressen
f. through durchzwängen [durch]; durchdrücken [durch]
force 1. Kraft *f*; 2. Wucht *f*, Gewalt *f*; 3. Stärke *f* *(Magnetismus)*; 4. *(Bw)* Schnittkraft *f* *(Statik)*
f. along the path Bahnkraft *f*, Tangentialkraft *f*
f. arm Kraftarm *m*
f. arrangement Kräfteanordnung *f*
f. balance Kräfteausgleich *m*; Kraftvergleich *m*
f.-balance controller *(BMSR)* Kraftvergleichsregler *m*
f.-balance flowmeter Kraftvergleichsdurchflußmesser *m*
f. component Kraftkomponente *f*
f. couple Kräftepaar *n*
f. diagram Kräftediagramm *n*, Kräfteplan *m*
f. equation Kraftgleichung *f*
f. equilibrium Kräftegleichgewicht *n*
f.-feed circulating lubrication Druckumlaufschmierung *f*
f.-feed lubrication Druckschmierung *f*
f.-feed oiler Drucköler *m*, Ölspritzkanne *f*
f. field Kräftefeld *n*
f. fit Preßpassung *f*
f. function Kräftefunktion *f*
f. lubricated druckgeschmiert
f. moment Drehmoment *n*
f. of attraction Anziehungskraft *f*
f. of friction Reibungskraft *f*
f. of gravity Schwerkraft *f*
f. of inertia Trägheitskraft *f*
f. of percussion Stoßkraft *f*
f. of repulsion Abstoßungskraft *f*, abstoßende Kraft *f*
f. of sea Seegangstärke *f*
f. parallel to the path Bahnkraft *f*, Tangentialkraft *f*
f. parallelogram Kräfteparallelogramm *n*
f. polygon Krafteck *n*, Kraftpolygon *n*
f. pump Druckpumpe *f*
f.-stacked diaphragm controller *(BMSR)* Stapelregler *m* *(pneumatischer Reglertyp)*
f. triangle Kräftedreieck *n*
f. vector Kraftvektor *m*
forced erzwungen, Zwang[s]-, Druck-
f. absorption erzwungene Absorption *f*
f.-air cooled druckluftgekühlt
f.-air-cooled oil-immersed transformer Öltransformator *m* mit Anblasung; Öltransformator mit Lüfterkühlung
f.-air cooling Druckluftkühlung *f*, Umlaufluftkühlung *f*
f.-air oven Trockenvorrichtung *f* mit Luftumwälzung
f.-air system Umwälzverfahren *n*; Zwangslüftung *f*
f.-air ventilation Druckluftkühlung *f* *(Maschinen)*
f. circulation Zwangsumlauf *m*
f.-circulation blower Umwälzgebläse *n*
f.-circulation boiler Zwangsumlaufkessel *m*
f.-circulation cooling *(Kfz)* Umlaufkühlung *f*, Wasserpumpenkühlung *f*
f. coding *(Dat)* Minimalsuchzeitprogrammierung *f*, Schnellprogrammierung *f*
f. commutation *(Elm)* Zwangskommutierung *f*, erzwungene Kommutierung *f*
f. crystallization erzwungene Kristallisation *f*
f. double refraction *(Opt)* erzwungene Doppelbrechung *f*
f. draft *s.* f. draught
f. draught künstlicher Zug *m*; Druckzug *m*; Zwangslüftung *f*
f.-draught blower Druckzugventilator *m*, Ventilator *m* für Unterwindfeuerung
f.-draught cooling *(Kfz)* Gebläsekühlung *f*
f.-draught fan Druckzuggebläse *n*
f.-draught furnace Unterwindfeuerung *f*; unterwindgefeuerter Ofen *m*
f.-draught furnace with steam jet blower underneath the grate Dampfstrahlunterwindfeuerung *f*
f.-draught kiln *(Holz)* Trockenkammer *f* mit künstlichem Zug
f. drop lock Sperrmagnet *m* mit mechanischer Abstützung *(Relais)*
f. emission *(Ph)* erzwungene Emission *f*

f.-feed lubrication Druckölung *f*, Druck[umlauf]schmierung *f*, Zwangsschmierung *f*
f. fit Preßpassung *f*
f.-induction engine Auflademotor *m*, Gebläsemotor *m*, Kompressormotor *m*
f. landing *(Flg)* Notlandung *f*
f. lubrication Druckschmierung *f*, Preßölschmierung *f*
f. motion erzwungene Bewegung *f*
f. oscillation erzwungene (fremderregte, aufgedrückte) Schwingung *f*, Zwangsschwingung *f*
f.-ventilated motor *(Elm)* fremdbelüfteter Motor *m*, Motor mit Fremdbelüftung
f. ventilation 1. Fremdbelüftung *f*, Druckluftkühlung *f* *(elektrische Maschinen usw.)*; 2. *(Brg)* künstliche Bewetterung *f*
f. vibration erzwungene (fremderregte, aufgedrückte) Schwingung *f*, Zwangsschwingung *f*
forceps Zange *f*, Probierzange *f*, Pinzette *f*
forces 1. *s.* force; 2. *(Am)* Streitkräfte *fpl*
f. of molecular attraction [inter]molekulare Anziehungskräfte *fpl*
f. of molecular interaction intermolekulare Kräfte *fpl*
forcible kräftig; zwang[s]läufig
forcibly actuated zwang[s]läufig betätigt
forcing hose Druckschlauch *m*
f. on Aufpressen *n* auf
f. out Auspressen *n* [aus]
f. screw Druckschraube *f*; Abdrückschraube *f*
f. ventilation *(Brg)* blasende Bewetterung *f*
ford Furt *f*
fording capacity *(Kfz)* Watfähigkeit *f*
fore and aft längsschiffs
f. and aft bridge *(Schiff)* Verbindungsbrücke *f*, fliegende Brücke *f*, Laufbrücke *f*
f.-and-aft gangway Laufbrücke *f* *(Tanker)*
f.-and-aft line Mittschiffslinie *f*
f.-and-aft moorings Bug- und Heckvertäuung *f*
f. and aft schooner Gaffelschoner *m* *(Segelschiff mit zwei Masten)*
f.-and-afters *(Schiff)* längslaufende Bauteile *npl* *(z. B. Längsspanten)*
f.-blow *(Met)* Vorblasen *n*
f.-blow period Vorblaseperiode *f*
f.-cabin Vorschiffkabine *f*
f.-masher *(Ch)* Maischbottich *m*, Maischkessel *m*
f. part Vorderteil *n*
f. pole *(Brg)* Getriebepfahl *m*
f. poppet *(Schiff)* vorderes Schlittendrehlager *n*, vordere Schlittenstütze *f* *(eines Ablaufschlittens beim Stapellauf)*
f. shift Frühschicht *f*
f.-tank *(Schiff)* Wasserballast *m* vorn
f.-woman Vorarbeiterin *f*
forebody Vorschiff *n*
foreboom of a schooner Schonersegelbaum *m*
forecarriage of plow Pflugkarren *m*
forecast Vorhersage *f* *(z. B. Wetter)*
forecasting Vorausberechnung *f*
forecastle *(Schiff)* Vorderdeck *n*; Back *f* *(Decksaufbau)*
f. bulkhead Backschott *n*
f. deck Backdeck *n*
forecourt *(Bw)* Vorhof *m*
foreedge *(Typ)* Buchschnitt *m*; Beschnittkante *f*, Beschnittrand *m*
f. painting Verzieren *n* des Buchschnitts mit Farbe
foreend of running ways *(Schiff)* Läufervorkante *f*, Schlittenvorkante *f* *(Stapellauf)*
foreframe *(Schiff)* Vorderspant *n*
foreground Vordergrund *m*
forehand welding *(Am)* Nachlinks-Schweißen *n*
forehead *(Brg)* Streckenort *n*, Feldort *n*
forehearth *(Met)* Vorherd *m*
forehold *(Schiff)* vorderer Laderaum *m*
foreign atom Fremdatom *n*
f. body Fremdkörper *m*
f. ion Fremdion *n*
f. matter Fremdstoffe *mpl*
f.-metal impurities Begleitmetalle *npl*
f. order Auslandsauftrag *m*
f. particle Fremdkörper *m*
f. to the lattice gitterfremd *(Kristalle)*
f. trade Außenhandel *m*
forelift *(Schiff)* Focktoppnant *f*
forelock Splint *m*, Vorsteckstift *m*
foreman Brigadier *m*, Vorarbeiter *m*; *(Brg)* Steiger *m*; *(Typ)* Faktor *m*
f. driller Bohrmeister *m*
foreman's office Betriebsbüro *n*
foremarker *(Flg)* Vor-Einflugzeichen *n*
foremast *(Schiff)* Fockmast *m*, Vortopp *m*
f.-sail Focksegel *n*
forensic chemistry forensische (gerichtliche) Chemie *f*
forepoling *(Brg)* Getriebszimmerung *f*, Abtreibezimmerung *f*
forepoppet *(Schiff)* vordere Schlittenstütze *f* *(Stapellauf)*
forepump Vorvakuumpumpe *f*
foresail Focksegel *n*
forescattering *s.* forward scattering
foreshore Vorland *n*
foreshorten / to verkürzen *(in Perspektive)*
foresight *(Mil)* Korn *n* *(Visier)*
forest Hochwald *m*, Wald *m*, Forst *m*
f. clearing saw Durchforstungssäge *f*
f. destroyers Forstschädlinge *mpl*
f. economy Forstwirtschaft *f*
f. fire Waldbrand *m*
f. lane Schneise *f*
f. pests Forstschädlinge *mpl*

f. pruning shears Durchforstungsschere *f*
f. service Forstverwaltung *f*
f. year Forstwirtschaftsjahr *n*
forestalling switch Quittungsschalter *m*
forester 1. Waldarbeiter *m*; Förster *m*; 2. Baumschere *f*, Durchforstungsschere *f*, Heckenschere *f*; 3. Waldvogel *m*; Waldtier *n*
forestry Forstwesen *n*; Forstkunde *f*; Forstkultur *f*, Forstwirtschaft *f*; Waldgelände *n*
forewarn / to vorwarnen
forfeiture of a right Verlust *m* eines Rechtes
forge / to 1. [aus]schmieden; 2. fälschen, falschmünzen
f. from a bar von der Stange schmieden
f. from a billet vom Stück schmieden
f.-on anschmieden
forge Schmiede[werkstatt] *f*; Schmiedefeuer *n*
f. alloy Schmiedelegierung *f*
f. cinder Hammerschlag *m*, Schmiedezunder *m*
f. coal Schmiedekohle *f*
f. hammer Schmiedehammer *m*
f. hearth Luppenfeuer *n*, Frischfeuer *n*
f. iron schmiedbares Eisen *n*, Schmiedeeisen *n*
f. pig iron Puddelroheisen *n*
f. scale Hammerschlag *m*, Zunder *m*
f. welding Feuerschweißen *n*
forgeability Schmiedbarkeit *f*
forgeable schmiedbar
f. alloy Knetlegierung *f*
forged geschmiedet
f. piece Schmiedestück *n*
f. solid with aufgeschmiedet auf, angeschmiedet an
f. steel roll geschmiedete Walze *f*
forger Fälscher *m*
forging Hämmern *n*, Schmieden *n*; Schmiedestück *n*
f. brass Schmiedemessing *n*
f. coal Schmiedekohle *f*
f. crack Schmiederiß *m*
f. die Schmiedegesenk *n*
f. die cavity Gesenkhohlraum *m*
f. furnace Schmiedeofen *m*
f.-grade steel nicht härtbarer Stahl *m*
f. hammer Schmiedehammer *m*
f. ingot Schmiedeblock *m*
f. machine Schmiedemaschine *f*
f. practice Schmiedetechnik *f*
f. press Schmiedepresse *f*
f. roll Schmiedewalze *f*; Reckwalze *f*
f. scale Schmiedezunder *m*
f. strain Schmiedespannung *f*
f. temperature Schmiedetemperatur *f*, Schmiedehitze *f*
f. test Schmiedeprobe *f*
fork / to sich gabeln, sich verzweigen
fork Gabel *f*; Bügel *m (z. B. Lastrolle)*
f. axle Gabelachse *f*
f. beam Gabelholm *m*
f. coil Gabelspule *f*
f. connection Sechsphasengabelschaltung *f*
f.-lift truck Gabelstapler *m (Hubstapler)*
f. mount Gabelmontierung *f*
f. of the connecting rod Schubstangengabelung *f*
f. opening lichte Gabelweite *f*
f. stacking truck Gabelstapler *m*
f. tedder Gabelheuwender *m*
f. truck Gabelstapler *m*
f. with bearings for crosshead pin Gabel *f* mit Stirnzapfenlagern
f. with crosshead pin Gabel *f* mit Gabelzapfen
forked gabelförmig, gegabelt; verzweigt
f. chain *(Ch)* verzweigte Kette *f*
f. clamp Gabelklammer *f*
f. coil *(BMSR)* Gabelspule *f*
f.-coil eddy current test Induktionsprüfverfahren *n* mit Gabelspule
f. connecting-rod Gabelpleuel *m*
f. crosshead gabelförmiger (gegabelter) Kreuzkopf *m*
f. hoe *(Lw)* zweizinkige Hacke *f*
f. lever Gabelhebel *m*, Klauenkurbel *f*, Zangenhebel *m*
f. lightning Linienblitz *m*
f. shaft *(Lw)* Gabeldeichsel *f*
f. sleeve Schiebestück *n (Getriebe)*
f. steering arm *(Kfz)* Gabelhebel *m*
f. strut Gabelstiel *m*; Y-Strebe *f*
f. terminal (tongue) *(El)* offener Kabelschuh *m*
forks *(Brg)* Stempel *m*
form / to 1. bilden; profilieren; fassonieren, [um]formen; 2. einschalen *(Beton)*
f. a compound eine Verbindung eingehen
f. an image *(Opt)* abbilden
f. crystals Kristalle bilden
f. the croze *(Holz)* Kimmung einschneiden
f. to umarbeiten auf
form 1. Vordruck *m*, Formular *n*; 2. Form *f*, Schalung[sform] *f (Beton)*; 3. Modell *n (Elektrotypie)*; 4. Bank *f (ohne Lehne)*; 5. Schulklasse *f*
f. and punch shaping machine Form- und Stempelhobelmaschine *f*
f. copying planer Kopierhobelmaschine *f*
f. cutter Formfräser *m*, Profilfräser *m*
f. cutter for gear teeth Zahnformfräser *m*
f. drag Formwiderstand *m (Strömungslehre)*
f. factor *(Ph, El)* Formfaktor *m*
f. factor indicator Formfaktormeßbrücke *f*
f. feeding device Formularzuführungsvorrichtung *f*, Formularzuführungseinrichtung *f*
f. gauge Formlehre *f*

f. grinder Profilschleifscheibe *f*
f. grinding Formschleifen *n*
f.-ground profilgeschliffen
f. liner Schalungseinlage *f (Sichtbetonherstellung)*
f. lining Schalungsauskleidung *f (Sichtbetonherstellung)*
f. milling method Formfräsverfahren *n*
f. of energy Energieform *f*
f. of oscillation Schwingungsform *f (Schallerzeuger)*
f. of trajectory Bahnverlauf *m*
f. oil Schalungsöl *n (Betonbau)*
f. panel Schal[ungs]tafel *f (Betonbau)*
f. plate Leitlineal *n*, Schablone *f*; Fassonschiene *f*
f. reamer Profilreibahle *f*
f. relieved cutter Fräser *m* mit hinterdrehten Zähnen, Formfräser *m*
f. removal Ausschalung *f*
f. setting Formlegen *n (Brikettpresse)*
f. strength Gestaltfestigkeit *f*
f. stripping Ausschalen *n*, Entschalung *f (Betonbau)*
f. tool Formmeißel *m*
f. tool gauge Formlehre *f*
f. tracer Nachformfühler *m*, Taster *m*
f. trueing device Formabrichteinrichtung *f*
f. turning Formdrehen *n*
f. turning lathe Formdrehmaschine *f*
f. turning tool Formdrehmeißel *m*
f.-vibrated concrete Rüttelbeton *m*
f. vibrator Schalungsrüttler *m (Betonbau)*
f. work 1. *(Masch)* Fassonarbeit *f*; 2. Form *f*, Schalung[sform] *f (Betonbau)*
f. work lining Schalungsauskleidung *f (Sichtbetonherstellung)*
formability Formbarkeit *f*, Formungsfähigkeit *f*; Nachformbarkeit *f*
formable [ver]formbar; umformbar
formaldehyde Formaldehyd *m*
formalin Formalin *n*
formate 1. Format *n (Buch)*; 2. *(Ch)* Formiat *n*, Ester *m* (Salz *n*) der Ameisensäure
f. cutting process Formwälzmethode *f (Zahnrad)*
formation 1. Bildung *f*; 2. *(Geol)* Formation *f*, Schichtengruppe *f*; 3. *(El)* Formieren *n*, Formierung *f (Sammlerplatten)*; 4. *(Bw)* Feinplanum *n*, Planum *n*
f. heat *(Ch)* Bildungswärme *f*
f. level *(Bw)* Feinplanumshöhe *f*; Fundamentsohle *f*, Gründungssohle *f*
f. of a blow pipe flame Stichflammenbildung *f*
f. of a neck Einschnürung *f*
f. of a wedge-shaped oil film Schmierkeilbildung *f*
f. of bubbles Blasenbildung *f*; Gasen *n*
f. of cavities Hohlraumbildung *f*
f. of cracks Rißbildung *f*
f. of craters Kraterbildung *f*
f. of crystal layers Schichtkristallbildung *f*
f. of crystalline grain Kristallkornbildung *f*
f. of distortion wedges Zipfelbildung *f (an Blechen)*
f. of eddies Wirbelbildung *f*
f. of fog Nebelbildung *f*
f. of hydrogen Wasserstoffentwicklung *f (Elektrolyse)*
f. of nuclei Keimbildung *f (Kristallisation)*
f. of oxygen Sauerstoffentwicklung *f (Elektrolyse)*
f. of pairs *(Kern)* Paarbildung *f*; *(Fs)* Paarigkeit *f*
f. of scale *(Met)* Zunderbildung *f*
f. of scum Schaumbildung *f*
f. of sheet *(Holz)* Blattbildung *f*
f. of stitch *(Text)* Maschenbildung *f*
f. of the structure *(Met)* Gefüge[aus]bildung *f*
f. of wrinkles Faltenbildung *f*
f. take-off *(Flg)* Verbandsstart *m*
f. time Aufbauzeit *f (Funkenentladung)*
f. voltage Formierungsspannung *f (Sammler)*
forme *(Typ)* Druckform *f*
f. reglet *(Typ)* Formatreglette *f*
f. roller *(Typ)* Auftragwalze *f*
formed coil Ankerformspule *f*, Formspule *f*, Schablonenspule *f*
f. crusher roll Profilrolle *f* mit eingearbeitetem Profil *(Schleifen)*
f. part Formteil *n*
f. plate *(El)* formierte Platte *f (Sammler)*
f. tooth cutter Fräser *m* mit hinterdrehten Zähnen
f. turning tool Formdrehmeißel *m*
f. wheel Profil[schleif]scheibe *f*
former 1. *(Masch)* Kopierlineal *n*, Kopierschablone *f*; 2. *(Elm)* Wickelschablone *f*; 3. Dorn *m (Biegeprüfung für Stahl)*
f. plate 1. Formlineal *n (Schleifmaschine)*; 2. Leitlineal *n (Drehmaschine)*
f. strip *(Holz)* Formatleiste *f*
f. template *s.* former
f.-wound coil *(Elm)* Ankerformspule *f*, Formspule *f*, Schablonenspule *f*
formiate Formiat *n*, Ester *m* (Salz *n*) der Ameisensäure
formic acid Ameisensäure *f*
f. aldehyde Formaldehyd *m*
Formica „Formica"-Kunststoffplatten *fpl (Handelsname)*
forming 1. Formgebung *f*, Umformung *f*; 2. Einschalung *f*, Schalungsbau *m (Betonbau)*
f. a wedge-shaped oil film Schmierkeilbildung *f*

f. attachment Fassondreheinrichtung *f*, Kopier[fräs]vorrichtung *f*
f. bell *(Met)* Ziehtrichter *m*
f. board 1. Kabelformbrett *n*; 2. Siebtisch *m*
f. die Formwerkzeug *n*; Ziehgesenk *n*; Stanzwerkzeug *n*
f. edge *(Met)* Biegekante *f*
f. force Umformkraft *f*; Formänderungskraft *f*
f. pass *(Met)* Profilstich *m*
f. process Formverfahren *n (Zahnrad)*
f. resistance Formänderungswiderstand *m*
f. technique Umformverfahren *n*
f. turret lathe Formrevolverdrehmaschine *f*
formula *(Math)* Formel *f*, Gleichung *f*; *(Ph)* Formel, Gesetz *n*; *(Ch)* Formel *(Moleküle)*
f. for altitude Höhenformel *f*
f. for guidance to the line of sight *(Rak)* Einlenkgesetz *n*
f. sign Formelzeichen *n*
f. translator Formelübersetzer *m (Programmierung)*
formwork Schalgerüst *n*; Schalung *f*
formylate / to *(Ch)* Formylgruppe einführen, formylieren
forsterite *(Min)* Forsterit *m*
forte stop Forteregister *n*, Fortezug *m (Orgel)*
forth vorwärts
fortuitous zufällig; unregelmäßig
f. distortion *(Fmt)* unregelmäßige Verzerrung *f*
f. finding Zufallsergebnis *n*
f. result *s.* f. finding
forward / to weiterschicken; aufgeben, befördern
forward acting regulation *(BMSR)* Vorwärtsregelung *f*
f.-backward counter *(Büro)* Vorwärts-Rückwärts-Zähler *m*
f. balance Saldovortrag *m*
f. body plan Vorschiffsspantenriß *m*
f. clutch Vorwärtskupplung *f*
f. cockpit *(Schiff)* Vorderplicht *f*
f. conductance Durchlaßleitwert *m*, Flußleitwert *m (Halbleiter)*
f. control chassis *(Kfz)* Frontlenker *m*
f. current Durchlaßstrom *m*, Strom *m* in Durchlaßrichtung
f. direction Durchlaßrichtung *f (Gleichrichter, Transistoren)*
f.-driving pulley Vorlaufscheibe *f*
f.-facing rocket Bremsrakete *f*
f. flow angle Stromflußwinkel *m*
f. gear *(Kfz)* Vorwärtsgang *m*
f. impedance Vorwärtsimpedanz *f*
f. lead of the brushes *(Elm)* Bürstenvorverstellung *f*, Bürstenvorschub *m*
f. motion *(Masch)* Vorlauf *m*, Vorwärtsgang *m*; Vorwärtsbewegung *f*
f. of the beam *(Schiff)* vorlicher als dwars
f. pass Hingang *m (Walzgut)*
f. piston stroke Kolbenhingang *m*
f. quarter Vorschiff *n*
f. resistance Durchlaßwiderstand *m*
f. rib Vorschiffsspant *n (Boot)*
f. running *(Masch)* Vorlauf *m*; Vorwärtsgang *m*
f. scattering *(Ph)* Vorwärtsstreuung *f*
f. scattering collision Stoß *m* mit kleinem Streuwinkel
f. shovel Hochlöffelbagger *m*
f. slip[ping] Voreilen *n (beim Walzprozeß)*
f. speed 1. *(Kfz)* Vorwärtsgang *m*; 2. *(Flg)* Geradeausgeschwindigkeit *f*
f. square Vorplatz *m (auf Fahrgastschiffen)*
f. stroke 1. Kolbenhingang *m*, Kolbenvorlauf *m*; 2. *(Fs)* Hinlaufperiode *f*, Vorlaufperiode *f*
f. stroke guide surface Vorwärtsgleitbahn *f*, Vorwärtsgleitfläche *f (Dampfzylinder)*
f. stroke standard Vorwärtsständer *m (Dampfmaschine)*
f. surge current Stromstoß *m (z. B. Halbleiterdiode)*
f. tail group *(Flg)* Kopfleitwerk *n*
f. thrust *(Flg)* Vortrieb *m*
f. transadmittance Übertragungsleitwert *m (Elektronik)*
f. travel *(Masch)* Vorwärtsgang *m*
f. voltage Durchlaßspannung *f*, Spannung *f* in Flußrichtung
forwarder [of goods] Spediteur *m*
forwarding agency Spedition *f*
f. agent Spediteur *m*
f. import goods *(Schiff)* Versenden *n* von Importgütern *(nach dem Inland)*
fossil botany Paläobotanik *f*
f. coal Steinkohle *f*
f. meal *(Min)* Infusorienerde *f*, Kieselgur *f*
f. wax *(Min)* Ozokerit *m*, Erdwachs *n*
fossilism Paläontologie *f*
Foster-Seely discriminator Diskriminator *m* nach Foster und Seely, Riegger-Kreis *m (Demodulator für frequenzmodulierte Spannungen)*
fothering means *(Schiff)* Leckstopfmittel *npl*
Föttinger coupling Flüssigkeitskupplung *f*; Föttinger-Wandler *m*
Föttinger transmission *s.* Föttinger coupling
Foucault current *s.* eddy current
Foucault gyroscope Foucault-Kreisel *m*
Foucalt's pendulum experiment Foucaultscher Pendelversuch *m*
foul / to 1. beschmutzen, verschmutzen; 2. verrammeln, verstopfen, blockieren
foul unsauber, schmutzig; bewachsen; schlecht
f. air *(Brg)* schlechte Wetter *pl*
f. anchor *(Schiff)* unklarer Anker *m*

f. bottom 1. schlechter Ankergrund *m*; 2. bewachsener Schiffsboden *m*
f. electrolyte verbrauchter Elektrolyt *m*
f. gas 1. Erdgas *n* mit Schwefelwasserstoffgehalt; 2. Abgas *n*
f. ground *(Schiff)* schlechter Ankergrund *m*
f. pit Schlagwettergrube *f*
f. tide Gegenströmung *f*, gegenlaufender Gezeitenstrom *m*
f. water Abwasser *n*, Schmutzwasser *n*
f. water line Abwasserleitung *f*
f. weather schlechtes Wetter *n*

fouling 1. Verschmutzung *f*; 2. Bewuchs *m (am Schiff)*; 3. Blockierung *f*; Verstopfung *f*

found / to 1. [be]gründen; 2. *(Bw)* gründen, fundieren; unterbauen; 3. *(Masch)* fundieren; 4. *(Met)* gießen; 5. blankschmelzen, reinschmelzen *(Glas)*

foundation 1. Gründung *f*; Stiftung *f*; 2. *(Bw)* Fundament *n*, Gründung *(Bauwerke)*; Unterbau *m (Straßenbau)*; 3. *(Masch)* Fundament, Unterbau; Bettung *f*
f. base Fundamentsockel *m*, Fundamentsohle *f*
f. block *(Masch)* Fundamentblock *m*, Fundamentklotz *m*; *(Bw)* Einzelfundament *n*, Säulenfundament *n*
f. bolt Ankerbolzen *m*, Ankerschraube *f*; Fundamentschraube *f*, Steinschraube *f*
f. bolt hole Ankerloch *n*
f. bolt secured with a cotter Splintanker *m*, Anker *m* mit Keilverschluß
f. bolt with hammerhead Anker *m* mit Hammerkopf
f. cylinder *(Bw)* Zylinderpfahl *m*
f. engineering Grundbau *m*
f. for bottom of the main flue *(Wärm)* Fuchssohle *f*
f. girder Fundamentträger *m*
f. level *(Bw)* Fundamentsohle *f*, Gründungssohle *f*
f. make-up Fondcreme *f (Kosmetik)*
f. pit Baugrube *f*
f. plan Fundamentzeichnung *f*
f. plate 1. *(Bw)* Fundamentplatte *f*, Grundplatte *f*, Gründungsplatte *f*; 2. *(Masch)* Ankerplatte *f*; 3. *(Schiff)* Fundamentplatte, Kielschweinplatte *f*
f. pressure *(Bw)* Bodendruck *m*, Bodenpressung *f (Baugrundmechanik)*
f. raft *(Bw)* Plattenfundament *n*, Plattengründung *f*; Fundamentplatte *f*, Grundplatte *f*, Gründungsplatte *f*
f. ring *(Eis)* Bodenring *m (Feuerbuchse)*
f. slab *(Bw)* Fundamentplatte *f*, Grundplatte *f*, Gründungsplatte *f*
f. slide rail *(Masch)* Fundamentschiene *f*
f. stone *(Bw)* Grundstein *m*
f. wall Grundmauer *f*

founder / to zusammenbrechen; einfallen; untergehen

founder 1. Gründer *m*, Begründer *m*; Stifter *m*; 2. *(Met)* Gießer *m*; *(Typ)* Schriftgießer *m*; Schmelzer *m (Glas)*

foundering *(Schiff)* Untergehen *n*, Sinken *n*

founding *(Met)* Gießen *n*, Guß *m*; Gießereiwesen *n*; Reinschmelzen *n (Glas)*

foundry 1. *(Met)* Gießerei *f*; 2. *(Typ)* Stereotypie *f*, Schriftgießerei *f*; 3. Schmelzerei *f (Glas)*
f. air-furnace Gießereiflammofen *m*
f. cleaning room Gußputzerei *f*
f. coke Gießereikoks *m*
f. cupola Gießereikupolofen *m*
f. facing graphite Gießereigraphit *m*, Formgraphit *m*
f. flask Formkasten *m*
f. furnace Gießofen *m*
f. ladle Gießpfanne *f*
f.-man 1. Gießer *m*; 2. Schmelzer *m (Glas)*; 3. *(Typ)* Stereotypeur *m*
f. moulding machine Gießereiformmaschine *f*
f. pattern Gußmodell *n*, Gießmodell *n*
f. pig iron Gießereiroheisen *n*
f. pit Gieß[erei]grube *f*
f. properties Gießeigenschaften *fpl*
f. resin Harz *n* für Gießereikerne
f. sand Formsand *m*
f. stove Trockenofen *m*

foundryman *s.* foundry-man

fount *(Typ)* Schriftablage *f*

fountain pen Füllfederhalter *m*
f. solution (water) *(Typ)* Wischwasser[lösung *f*] *n*

four Vierer *m*, Vier *f*
f.-address operation *(BMSR)* Vier-Adressenbetrieb *m*, Vier-Adreßbetrieb *m*
f.-arm network *(El)* Vierpol *m*
f.-axial vierachsig
f.-bar Viergelenk-
f.-bar linkage Gelenkviereck *n*, Viergelenkgetriebe *n*
f.-bar motion with axes of rotation parallel, linked quadrilateral zylindrisches Kurbelviereck (Gelenkviereck) *n*, zylindrische Kurbelkette *f*
f.-bar motion with axes of rotation tending to a point konisches Kurbelviereck *n*, viergliedrige konische Kurbelkette *f*
f.-bar type of crosshead viergleisiger Kreuzkopf *m*
f.-barreled anti-aircraft gun Vierlingsflak *f*
f.-bladed vierflügelig *(z. B. Propellerschraube)*; mit vier Messern *(Messerkopf)*
f.-block brake Vierklotzbremse *f*
f.-channel Vierkanal-

f.-channel stereophonic film Vierkanalstereofoniefilm *m*
f.-channel stereophonic sound Vierkanalton *m*
f.-colour automatic loom *(Text)* Vierfarbenspulenwechselautomat *m*
f.-component 1. vierkomponentig *(z. B. Viererverktor)*; 2. *(Masch)* vierteilig
f.-component alloy quaternäre Legierung *f*
f.-core cable vieradriges Kabel *n*, Vierleiterkabel *n*
f.-cornered viereckig
f.-course radio-range Vierkursleitstrahlbake *f*, Bake *f* für A-N-System
f. cutter planing and moulding machine vierseitige Hobel- und Kehlmaschine *f (Holzbearbeitung)*
f.-cycle engine Viertaktmotor *m*
f.-cylinder engine Vierzylindermotor *m*
f.-cylinder triple expansion steam engine Dreifachexpansionsdampfmaschine *f* mit geteiltem Niederdruck
f.-cylinder turbine viergehäusige (vierzylindrische) Turbine *f*
f.-decked vessel Vierdeckschiff *n*
f.-edged vierkantig
f.-electrode valve Tetrode *f*, Vierpolröhre *f*
f.-end twill *(Text)* vierbindiger Köper *m*
f.-faced turret Vierfachrevolverkopf *m*
f.-figure vierstellig
f.-figure number vierstellige Zahl *f*
f.-flute drill Vierschneidenbohrer *m*
f.-fluted viernutig; mit vier Spannuten
f.-fold vierfach; vierzählig
f.-frequency diplex system Vierfrequenzdiplexsystem *n*
f.-handle die stock vierarmige Gewindeschneidkluppe *f*
f. high rolling mill Doppelduostraße *f*
f.-high rolling stand Vierwalzengerüst *n*
f.-jaw chuck Vierbackenfutter *n*
f.-jaw independent chuck exzentrisch spannendes Vierbackenfutter *n*
f.-leaf transposed double faced twill *(Text)* vierbindiger versetzter zweiseitiger Köper *m*
f.-leaf twill *(Text)* vierbindiger Köper *m*
f.-leaved vierflügelig *(Fenster)*
f. line brevier *(Typ) (etwa)* Doppelmittel *n (32 britische Punkte; Schriftgröße)*
f.-lipped end mill Vierschneider *m (Langlochfräser)*
f.-mast-barque *(Schiff)* Viermast-Bark *f (Fock-, Groß-, Achter- und Besanmast)*
f.-mast ship Viermast-Schiff *n (Fock-, Groß-, Achter- und Kreuzmast)*
f.-necked flask *(Ch)* Vierhalskolben *m*
f. nines zinc *(Met)* Vierneunerzink *n*
f.-part alloy quaternäre Legierung *f*
f.-part bearing vierteiliges Lager *n*
f.-party line *(Fmt)* Gemeinschaftsleitung *f* für vier Anschlüsse
f. phase vierphasig
f.-piece timber set *(Brg)* Geviertzimmerung *f*, Viergespann *n*
f.-point contact bearing Vierpunktlager *n*
f.-point mounting Vierpunktaufhängung *f*, Vierpunktbefestigung *f*
f.-point suspension Vierpunktsystem *n*
f.-pointed vierschneidig
f.-polar vierpolig
f.-pole Vierpol *m*
f.-pole vierpolig
f. pole [circuit] network Vierpol[schaltung *f*] *m*
f.-position mit vier Arbeitsstellen (Stellungen)
f.-post drawing press Säulenziehpresse *f*
f.-quadrant multiplier Vierquadrantmultiplizierer *m (Analogrechner)*
f.-roller mill Vierwalzenwerk *n*
f.-rules calculator Vierartenrechner *m*, Vierspeziesrechner *m*
f.-sided vierseitig
f.-speed drive Vierganggetriebe *n*
f.-speed gearbox Vierganggetriebe *n*
f.-spindle automatic Vierspindelautomat *m*
f.-spindle automatic chucking machine Vierspindelfutterautomat *m*
f.-spindle automatic lathe Vierspindelautomat *m*
f.-spindle bar auto Vierspindelstangenautomat *m*
f.-spindle drilling machine Vierspindelbohrmaschine *f*
f.-spindle semi-automatic [lathe] Vierspindeldrehhalbautomat *m*
f. spring peg with shoulder *(Text)* Vierfederspindel *f* mit Ansatz
f.-stacker Vierschornsteinschiff *n*
f. stage rocket vierstufige Rakete *f*
f.-step vierstufig
f. step rocket vierstufige Rakete *f*
f.-stranded rope vierschäftiges Seil *n*
f.-stroke cycle Viertaktverfahren *n*, Viertaktlauf *m*
f.-stroke engine Viertaktmotor *m*, Viertakter *m*
f.-tensor *(Math)* Vierertensor *m*
f.-terminal network Vierpol[schaltung *f*] *m*
f.-terminal transistor vierteiliger Transistor *m*, Vierelementtransistor *m*
f.-throw vierhübig *(Kurbelwelle)*
f. to pica *(Typ) (etwa)* Brillant *f (3 britische Punkte; Schriftgröße)*
f. track Viertelspur *f (Tonband)*
f.-vector *(Math)* Vierervektor *m*

f.-way cock Vierwegehahn *m*
f.-way drilling machine Vierwegebohrmaschine *f*
f.-way jack *(Fmt)* vierteilige Klinke *f*
f.-way pipe Vierwegestück *n*
f.-way rim wrench Kreuzschlüssel *m*
f.-way tap Vierwegehahn *m*
f.-way tool block (post) Vierfachmeißelhalter *m*
f. wheel brake Vierradbremse *f*
f.-wheel drive Vierradantrieb *m*, Geländeantrieb *m*
f.-wheel steer arrangement *(Kfz)* Vierradsteuerung *f*
f.-wire balanced line *(El)* symmetrische Vierdrahtleitung *f*
f.-wire repeater *(Fmt)* Vierdrahtverstärker *m*
f.-wire terminating set *(Fmt)* Gabelschaltung *f*; Vierdrahtgabel *f*
f.-zone heat control Heizzylinder *m* mit vier Temperaturzonen
Fourdrinier Langsiebpapiermaschine *f*
Fourdrinier section Langsiebpartie *f*
Fourdrinier sheet auf dem Langsieb gebildetes Blatt *n*
Fourdrinier type washer Langsiebwäscher *m*
Fourdrinier wire Langsieb *n*, Maschinensieb *n*
Fourdrinier wire weaving loom Drahtwebmaschine *f* für Feingewebe
Fourier analysis *(Math)* Fourier-Analyse *f*, Fouriersche Zerlegung *f*
Fourier decomposition Fouriersche Zerlegung *f*
Fourier expansion Fouriersche Reihenentwicklung *f*, Fourier-Entwicklung *f*
Fourier integral Fourier-Integral *n*, Fouriersches Integral *n*
Fourier series Fourier-Reihe *f*
Fourier synthesizer Gerät *n* zur Fourier-Synthese
Fourier transform Fourier-Transformierte *f*
fourth power equation Gleichung *f* vierten Grades
f. stage vierte Raketenstufe (Stufe) *f*
fovea Netzhaut *f (Auge)*
f. centralis Netzhautgrube *f*
fowl Geflügel *n*
f.-house Hühnerstall *m*
f.-piece Vogelflinte *f*
fowlerite *(Min)* Fowlerit *m*
fox Fuchs *m*; Fuchsfell *n*
f.-tail wedging Zapfenverkeilung *f*
foxed *s.* foxy
foxhole *(Mil)* Schützenloch *n*
foxy moderfleckig, stockfleckig, stockig
fraction 1. *(Math)* Bruch *m*; 2. Korngröße *f*, Körnung *f*
f. stroke Bruchstrich *m*
fractional abundance relative (prozentuale) Häufigkeit *f*
f. analysis *(Ch)* Siedeanalyse *f*
f. cooling-power loss Anteil *m* des Leistungsaufwandes für die Kühlung
f. crystallization fraktionierte Kristallisation *f*
f. detuning Feinabstimmung *f (Radio)*
f. distillation fraktionierte Destillation *f*, Fraktionierung *f*
f. distillation flask Destillierkolben *m*
f. horsepower mit weniger als 1 PS
f. horsepower motor Klein[st]motor *m*
f. isotopic abundance relative (prozentuale) Isotopenhäufigkeit *f*
f. line *(Math)* Bruchstrich *m*
f. load *(Elm)* Teilbelastung *f*, Unterbelastung *f*
f. loss 1. relativer (prozentualer) Verlust *m*; 2. Verlust je Stufe, Verlustanteil *m*
f. mistuning Feinverstimmung *f (Radio)*
f. number *(Math)* Bruch[zahl *f*] *m*
f. operation[s] Bruchrechnung *f*
f. package aufteilbare Verpackung *f*
f. part *(Math)* Bruchteil *m*
f. pitch Teilschritt *m*
f.-pitch winding Sehnenwicklung *f*, Wicklung *f* mit verkürztem Wicklungsschritt
f.-slot winding Bruchlochwicklung *f*
f. washing Stufenwäsche *f (Papierindustrie)*
f. yield *(Ch, Kern)* Ausbeute *f* je Stufe
fractionate / to fraktionieren, fraktioniert destillieren
fractionating column Dephlegmator *m*, Destillieraufsatz *m*, Fraktionieraufsatz *m*, Fraktionierkolonne *f*
f. tower *s.* f. column
fractionation Fraktionierung *f*
fractionator *(Holz)* Rollsichter *m*
fracture / to brechen
fracture Bruch *m (Werkstoff; Knochen)*; Bruchfläche *f*
f. across the grains intrakristalliner Bruch *m*
f. area Bruchfläche *f*
f. at the flange due to bad flanging Bördelriß *m* durch schlechtes Umbördeln
f. limit Bruchgrenze *f*, Zerreißgrenze *f*
f. load Bruchlast *f*; Bruchbelastung *f*
f. of a stand pipe Stutzenbruch *m*
f. of rolls Walzenbruch *m*
f. of the rim Kranzbruch *m (Schwungrad)*
f.-proof bruchfest
f. strain Bruchdehnung *f*
f. stress Bruchspannung *f*
fractured ground *(Brg)* zerrüttetes Gebirge *n*
f. zone *(Geol)* Ruschelzone *f*
fracturing *(Geol)* Bruchbildung *f*, Zerklüftung *f*
fragile brüchig *(Metalle usw.)*; zerbrechlich *(Glas)*

fragility Brüchigkeit *f*, Sprödigkeit *f (Metalle)*; Zerbrechlichkeit *f (Glas)*
fragment 1. Bruchstück *n*, Spaltstück *n*; 2. *(Kern)* Kernbruchstück *n*, Spaltprodukt *n*
fragmentary trümmerartig
f. rocks *(Geol)* Trümmergesteine *npl*
fragmentate / to zertrümmern
fragmentation 1. *(Brg)* Zerkleinerung *f (Schießarbeit)*; 2. *(Kern)* Kernexplosion *f*, Kernzertrümmerung *f*; 3. Brucherscheinung *f (in Kristallen)*
f. bomb *(Mil)* Splitterbombe *f*
f. ring Splitterring *m (Handgranate)*
fragments *(Geol)* Trümmer *pl*
fragrance Duft[stoff] *m*
Frahm frequency meter *(El)* Frahmscher Zungenfrequenzmesser *m*
frail zerbrechlich
frame / to 1. *(Bw)* Bauteile zu einem Tragwerk zusammenfügen; 2. *(Led)* speilen; 3. einrahmen; 4. verzimmern
f. the patent claims die Patentansprüche abfassen
frame 1. *(Masch)* Gestell *n*, Rahmen *m*; Bügel *m (z. B. Säge)*; Gehäuse *n (Maschinen)*; Trog *m (Schleifstein)*; 2. *(Bw)* Fachwerk *n*; Skelett *n*; Stabwerk *n*; Tragwerk *n*; Zarge *f*; 3. Scherenständer *m*, Pressenständer *m*; Pressenkörper *m*; 4. *(Opt)* einzelnes Bild *n* eines Films; 5. *(Fs)* Bild *n*; Teilbild *n*, Raster *m*; *(Am)* Vollbild *n*; 6. *(Brg)* Ausbaugeviert *n (Schachtausbau)*; 7. *(Schiff)* Spant *n (stählern)*; Bauspant *n*
f. aerial Rahmenantenne *f*
f. amplitude control *(Fs)* Bildamplitudenregler *m*, Rasterregler *m*; Bildamplitudenregelung *f*, Rasterregelung *f*
f. and panel construction Fertigteilbauweise *f* mit Skelett und Ausfachungsplatten
f. angle *(Schiff)* Spantwinkel *m*
f. bend teilbildfrequente Störkompensation *f*, Rasterkorrektur *f (Fernsehkamera)*
f. bending machine Siebränderbiegemaschine *f*
f.-bending slabs *(Schiff)* Spantenbiegeplan *m*, Biegeplan *m*
f. blanking *(Fs)* Teilbildaustastung *f*, Rasteraustastung *f*
f. blanking period *(Fs)* Rasteraustastungsperiode *f*
f. centre rest *(Kfz)* Rahmenbrille *f*
f. coil 1. *(El)* Scheibenspule *f*; 2. *(Fs)* Bildeinstellspule *f*
f. coil direction finder Rahmenpeiler *m*
f. construction Skelettbau[weise *f*] *m*
f. counter *(Opt)* Bildzählwerk *n*
f. crane Bockkran *m*, Torkran *m*, Portalkran *m*
f. cross member Rahmenquerträger *m*
f. deflection *(Fs)* Rasterablenkung *f*
f. direction finder Drehrahmenpeiler *m*
f. distortion *(Fs)* Rasterverzerrung *f*
f. divider *(Fs)* Frequenzteiler *m (Rasterfrequenz)*, Rasterfrequenzteiler *m*
f. drop *(Kfz)* Rahmenkröpfung *f*
f. duration *(Fs)* Teilbilddauer *f*, Rasterdauer *f*; *(meist:)* Halbbilddauer *f*
f. extension *(Kfz)* Rahmenverlängerung *f*
f. flange Rahmenflansch *m*
f. floor *(Schiff)* Bodenwrange *f*
f. flyback *(Fs)* Bildrücklauf *m*
f. fork *(Kfz)* Rahmengabel *f*
f. frequency *(Fs)* 1. Bild[wechsel]frequenz *f*; 2. Rasterfrequenz *f*
f. girder Rahmenbalken *m*; Fachwerkträger *m*
f. grid Spanngitter *n (Elektronenröhren)*
f. height *(Fs)* Bildhöhe *f*
f.-jib *(Brg)* Rahmenschrämarm *m*
f.-leakage protection Gestellschlußschutz *m (Relais)*; *(Elm)* Gehäuseschlußschutz *m*
f. level *(Opt)* Rahmenlibelle *f*
f. line noise Bildstrichgeräusch *n (Film)*
f. linearity control *(Fs)* Regelung *f* der Rasterlinearität
f. liner *(Schiff)* Spantfüllstreifen *m*
f. lines *(Schiff)* Spantenriß *m*
f.-lock grid structure Spanngitter *n (Elektronenröhren)*
f. loom Webstuhl *m*
f. member Fachwerkstab *m*
f. monitoring tube *(Fs)* Rasterkontrollröhre *f*
f. of plate *(El)* Plattenrahmen *m*, Rahmenplatte *f (Sammler)*
f. of spectacles Brillenfassung *f*
f. pulse *(Fs)* Bildwechselimpuls *m*; Rasterimpuls *m*
f. repetition rate *(Fs)* 1. Rasterwechselfrequenz *f*; 2. Bildwechselfrequenz *f*
f. ring *(Schiff)* Spantrahmen *m*
f. saw 1. Bundgatter *n*; Gattersäge *f*; 2. Spannsäge *f*, Stellsäge *f*, Stoßsäge *f*
f. scan *(Fs)* Bildabtastung *f*, Rasterabtastung *f*, Teilbildabtastung *f*; *(meist:)* Halbbildabtastung *f*
f. scan circuit *(Fs)* Bildkippschaltung *f*
f. sequential system *(Fs)* Rasterfolgeverfahren *n*, Rasterwechselverfahren *n*, Zeitfolgeverfahren *n*
f. side bar Längsträger *m*, Rahmenholm *m*
f. simultaneous system *(Fs)* Rastersimultanverfahren *n*
f. size *(Fs)* Bildgröße *f*, Bildformat *n*
f. spacing *(Schiff)* Spantabstand *m*
f. stations *(Schiff)* Spantenriß *m*
f. suppression *(Fs)* Bildaustastung *f (am Ende eines Rasters)*; Rasteraustastung *f*

f. suppression period *(Fs)* Rasteraustastungsperiode *f*
f. suspended motor Gestellmotor *m*, Elektromotor *m* mit Rahmenaufhängung
f. sweep unit *(Fs)* Bildkippgerät *n*
f. synchronizing impulse *(Fs)* Bildgleichlaufimpuls *m*, Rastersynchronisier[ungs]impuls *m*
f. tilt *(Fs)* Rasterentzerrung *f*, Trapezentzerrung *f*
f. timber *(Schiff)* Spantholz *n*
f. timbering *(Brg)* Türstockausbau *m*
f. time base *(Fs)* Bildablenkgerät *n*, Rasterzeitbasis *f*
f. time-base oscillation *(Fs)* Bildkippschwingung *f*
f. trussing *(Kfz)* Rahmenunterzug *m*
f. view-finder *(Opt)* Rahmensucher *m*
f. wall Fachwerkwand *f*
f. with circular end bettförmiger Rahmen *m*
f. with circular guide Rundführungsrahmen *m*
framed construction *(Bw)* Skelettbau *m*
f. cross-cut saw Bügelsäge *f*, Strecksäge *f*
f. girder Fachwerkbinder *m*, Fachwerkträger *m*
f. hand saw Strecksäge *f*
f. hand saw with stretcher gefaßte Säge *f*
f. pruning saw Baumsäge *f*
f. roof abgebundenes Dach *n*, Zimmermannsdach *n*; Fachwerkdach *n*
f. ship in Spanten stehendes Schiff *n*
f. winding *(Brg)* Gestellförderung *f*
frameless construction *(Kfz)* selbsttragende Konstruktion *f (Karosserie)*, rahmenlose Bauart *f*
framework *(Bw)* Fachwerk *n*; Skelett *n*; Tragwerk *n*; Stabwerk *n*; *(Flg)* Gerippe *n*; *(Schiff)* Spanten *npl*
f. house Fachwerkhaus *n*
framing 1. Zusammenbau *m*; Bauen *n*; 2. Einrahmen *n (Dias)*; 3. *(Opt)* Bildstricheinstellung *f (Kinofilm)*; 4. *(Fs)* Bildausschnitt *m*; Bildeinstellung *f*, Höhe-Breite-Einstellung *f (Bild)*; 5. *(Schiff)* Spantwerk *n*, Bespantung *f*
f. body plan *(Schiff)* Bauspantenriß *m*
f. condition *(Opt)* Bildstand *m*
f. control *(Fs)* Bild[einstell]regler *m*, Bildfangregler *m*; Bild[einstell]regelung *f*
f. depth *(Schiff)* Seitenhöhe *f* für Spanten
f. knob *(Fs)* Bildeinstellknopf *m*
f. mask Bildmaske *f*
f. oscillator *(Fs)* Rasterkippgenerator *m*
f. timber Verbandholz *n*
Francis turbine Francis-Turbine *f*
Franki pile *(Bw)* Franki-Pfahl *m (Grundbau)*
Franklin-type aspherical cataract bifocal lens asphärisches Zweistärken-Starglas *n*
franklinite *(Min)* Franklinit *m*
fraught animal Zugtier *n*, Zugvieh *n*
Fraunhofer diffraction Fraunhofersche Beugung *f*
Fraunhofer lines Fraunhofersche Absorptionslinien *fpl*, Fraunhofer-Linien *fpl (Sonnenspektrum)*
Fraunhofer region Fraunhofersches Gebiet *n (Beugung)*
freak value Zufallswert *m (Messung)*
free / to 1. befreien, entlasten, frei machen; 2. *(Schiff)* lenzen, lenzpumpen
f. from branches *(Holz)* entasten
free 1. frei *(Atome, Elektronen, Ionen usw.)*; 2. unbesetzt *(Energieniveaus; Fernsprechleitung)*; 3. *(Ch)* ungebunden
f.-air ionization chamber offene Luftionisationskammer *f*
f.-air [over]pressure Überdruck *m* in freier Atmosphäre
f.-air thermometer Außenthermometer *n*
f.-air transformer Außentransformator *m*
f. balloon 1. voller Fadenballon *m (Spinnerei)*; 2. Freiballon *m*
f. bend test Freibiegeversuch *m (Schweißnaht)*
f.-blown glass freihandgeblasenes Glas *n*
f.-burning coal Flammkohle *f*
f. call *(Fmt)* gebührenfreier Anruf *m*
f. cantilevered construction Freivorbau *m (Brückenbau)*
f. carbon *(Ch)* freier Kohlenstoff *m*
f. copy Freiexemplar *n*
f. current operation Betrieb *m* mit natürlicher Erregung *(Transduktor)*
f. cutting *(Met)* [gut] zerspanbar
f. cutting alloy Automatenlegierung *f*
f. cutting brass Automatenmessing *n*
f. cutting property Zerspanbarkeit *f*
f. cutting steel Automatenstahl *m*
f.-electron model Modell *n* des freien Elektrons
f. electron theory Sommerfeldsche Elektronentheorie *f* der Metalle
f. energy freie Energie *f (Thermodynamik)*
f. face *(Brg)* freie Fläche *f*, Freifläche *f (Schießarbeit)*
f. fall freier Fall *m*
f. falling cutter Freifallbohrer *m (Tiefbohrungen)*
f. falling lathe *(Text)* freifallende Lade *f*
f. fatty matter *(Ch)* Unverseifbares *n*
f. fit Gleitsitz *m*; Feinpassung *f*, Schlichtpassung *f (z. B. Verschraubung)*
f.-flight rocket ungesteuerte (ungelenkte) Rakete *f*
f.-flight trajectory *(Rak)* Freiflugbahn *f*
f. floating collet Pendeleinsatz[hülse *f*] *m*
f.-floating plunger freifliegender Kolben *m*

f. flowing property gute Fließfähigkeit *f*
f.-fly rocket ungesteuerte (ungelenkte) Rakete *f*
f.-free transition *(Kern)* Übergang *m* zwischen freien Zuständen, Frei-frei-Übergang *m*
f. from burrs gratfrei
f. from close range fading *(Fmt)* nahschwundfrei
f. from contradiction widerspruchsfrei
f. from defects fehlerfrei
f. from distortion verzerrungsfrei *(Bild, Tonwiedergabe)*; verzeichnungsfrei *(Bild)*; klirrfrei *(Tonwiedergabe)*
f. from disturbance stör[ungs]frei
f. from eddies wirbelfrei
f. from flicker flimmerfrei
f. from impurities ohne Beimischungen, beimischungsfrei
f. from interference *(El)* stör[ungs]frei
f. from knots *(Holz)* astfrei, astrein
f. from noise *(El)* rauschfrei
f. from phase shift *(El)* phasenrein
f. from play spielfrei
f. from scale zunderfrei
f. from slag 1. schlackenfrei; 2. zunderfrei
f. from stress *(Mech)* spannungslos
f. from tension *(Mech)* spannungslos
f. from undercut unterschnittfrei
f. from vibration schwingungsfrei
f. from vortices wirbelfrei
f. from wear verschleißfrei
f. grid offenes Gitter *n (Elektronenröhren)*
f. hearth method *(Met)* offenes Herdverfahren *n*
f. heat *(Ch)* freie Wärme *f*
f. height of blade freitragende Schaufelhöhe *f (Turbine)*
f. impedance Eingangsimpedanz *f* bei unbelastetem Ausgang
f. ionization chamber *(Kern)* offene Ionisationskammer *f*
f. jet turbine Freistrahlturbine *f*
f. line *(Fmt)* unbesetzte Leitung *f*
f.-line signal Freizeichen *n*
f.-machining property Zerspanbarkeit *f*, Spanbarkeit *f*
f. moisture *(Geol)* Oberflächenwasser *n*
f. molecule diffusion Knudsen-Strömung *f*
f. motion *(Masch)* Leerlauf *m*
f. neutron freies Neutron *n*
f. of charge 1. gebührenfrei, unentgeltlich; 2. *(El)* ladungsfrei
f. of duty steuerfrei *(Waren)*
f. of losses verlustfrei, verlustlos
f. of ripples *(El)* welligkeitsfrei
f. of singing *(Fmt)* pfeiffrei
f. oscillation *(El)* freie (selbststeuernde) Schwingung *f*
f. path *(Kern)* freie Weglänge *f*
f. path length *(Kern)* freie Weglänge *f*
f.-piston compressor Freikolbenverdichter *m*
f.-piston engine Freikolbenmaschine *f*, Flugkolbenmaschine *f*
f.-piston engine compressor Freikolbenverdichter *m*
f.-piston gas generator turbine Freikolbenvergaserturbinenanlage *f*
f.-piston gas turbine plant Freikolbengasturbinenanlage *f*
f.-piston gasifier Freikolbentreibgaserzeuger *m*
f.-piston gasifier turbine unit Turbinenanlage *f* mit Freikolbengaserzeuger
f. piston gauge Freikolbenmanometer *n*
f.-piston-type engine Freikolbenmotor *m*, Flugkolbenmotor *m*
f.-piston-type gas turbine Freikolbengasturbine *f*
f. port Freihafen *m*
f. pulp *(Holz)* röscher Schliff *m*
f. radical *(Ch)* freies Radikal *n*
f. rocket ungesteuerte (ungelenkte) Rakete *f*
f.-running *(Fs)* Freilaufsystem *n*
f.-running freilaufend *(Maschinen, Motore)*; freifließend *(Sand, Dünger, Mehl usw.)*
f.-running frequency *(sl) (Fmt, Fs)* Eigenfrequenz *f*
f.-running multivibrator *(Dat)* frei (selbständig) kippender Multivibrator *m*, frei (selbständig) laufender Multivibrator
f. section of the steam port *(Wärm)* freier Kanalquerschnitt *m*
f. settling classifier Freifallklassierer *m*
f. source *(Kern)* nichtabgeschirmte (offene) Quelle *f*
f. space 1. leerer (freier) Raum *m*; Vakuum *n*; 2. *(Masch)* Spiel *n*
f. space charge freie Raumladung *f*
f. space charge wave freie Raumladungswelle *f*
f. space diagram Freiraumausbreitungsdiagramm *n (Antennenstrahlung)*
f. space propagation Ausbreitung *f* im leeren Raum
f. space velocity Vakuumgeschwindigkeit *f*
f. space wave Raumwelle *f*
f.-standing freistehend
f. stuff rösch gemahlener Zellstoff *m*
f. surface freie Oberfläche *f*
f. swinging loudspeaker Freischwingerlautsprecher *m*
f. to align and centre itself pendelnd angeordnet *(Läppspindel)*
f. to float pendelnd angeordnet
f. to move frei beweglich
f. to slide axially in Achsrichtung verschiebbar

f. **trajectory** *(Rak)* Freiflugbahn *f*
f. **travel** Spiel *n (Kupplung)*
f. **valence** *(Ch)* freie Valenz *f*
f. **vibration** freie (selbststeuernde) Schwingung *f*
f. **water** *(Ch)* nicht gebundenes Wasser *n*
f. **wave** *(Fmt)* freie (ungehindert fortschreitende) Welle *f*
f.**-wheel** *s.* freewheel
f.**-wheeling** freilaufend
f.**-wheeling clutch** Freilaufkupplung *f*
f.**-wheeling rectifier** Nullanode *f (Stromrichter)*
freeboard *(Schiff)* Freibord *m*
f. **depth** Freibordhöhe *f (gemessen von Oberkante Kiel bis Decksstrich)*
f. **draught** Freibordtiefgang *m*
freedom Freiheit *f*
f. **from brittle fracture** *(Met)* Trennbruchsicherheit *f*
f. **from distortion** Verzerrungsfreiheit *f (Bild, Tonwiedergabe)*; Verzeichnungsfreiheit *f (Bild)*
f. **from disturbance[s]** Stör[ungs]freiheit *f*
f. **from phase shift** *(El)* Phasenreinheit *f*
f. **from secondary lobes** *(El)* Nebenzipfelfreiheit *f*
f. **of motion** Bewegungsfreiheit *f*
freehand drawing 1. Freihandzeichnen *n*; 2. *s.* f. sketching
f. **grinding** Handschleifarbeit *f*
f. **line** Freihandlinie *f*
f. **representation** *s.* f. sketching
f. **sketching** freihändige Darstellung *f*
freeing-port *(Schiff)* Wasserpforte *f*, Schanzkleidpforte *f*
freely supported beweglich gelagert, frei aufliegend
f. **suspended** frei aufgehängt
freeness Mahl[ungs]grad *m (Papier und Zellstoff)*
f. **degree of grinding** Schmierigkeitsgrad *m*
f. **index** Mahlgrad *m*
f. **tester** Mahlgradprüfer *m*
freeway kreuzungsfreie Schnellverkehrsstraße *f*
freewheel 1. Freilauf *m*; 2. Freilaufkupplung *f*
f. **brake roller** Freilaufklemmrolle *f*
f. **brake roller cage** Freilaufklemmrollenkäfig *m*
f. **change** Freilaufschaltung *f*
f. **housing** Freilaufgehäuse *n*
f. **inner ring** Freilaufinnenring *m*
f. **lock** Freilaufsperre *f*
f. **outer ring** Freilaufaußenring *m*
f. **overrunning clutch** Freilaufüberholkupplung *f*
f. **turbine** *(Am)* Zweiwellenturbine *f*
freeze / to 1. gefrieren, [ein]frieren; erstarren; tiefkühlen; frosten; 2. *(Masch)* sich festklemmen; 3. kleben [bleiben], haften [bleiben] *(z. B. Relais)*
f. **in** einfrieren
f. **out** ausfrieren
f. **up** zufrieren
freeze-drying Gefriertrocknung *f*
f. **out** *(Fmt)* Anschlußsperrung *f*
f.**-proof** frostbeständig
f.**-thaw-stable** beständig gegen Frost und Tauwasser *(z. B. Lack)*
freezer Eisgenerator *m*, Eiserzeuger *m*
f. **stern trawler** Tiefkühlhecktrawler *m*
freezing 1. Gefrieren *n*; Erstarren *n*; 2. *(Masch)* Sichfestklemmen *n*; 3. Klebenbleiben *n (z. B. Relais)*; 4. Frost *m*
f. **curve** Erstarrungsdiagramm *n*
f. **hole** Gefrierloch *n*
f. **method** Gefrierverfahren *n*
f. **mixture** Kältemischung *f*, Kältegemisch *n*
f. **of a melt** Erstarren (Einfrieren) *n* einer Schmelze
f. **plant** Gefrieranlage *f*
f. **point** 1. Nullpunkt *m (Thermometer)*; 2. Gefrierpunkt *m*; Erstarrungspunkt *m (Glas, Metall, Plaste)*
f. **point depression** Gefrierpunktserniedrigung *f*
f. **preventive** Frostschutzmittel *n*
f. **range** Erstarrungsbereich *m*
f. **shaft** *(Brg)* Gefrierschacht *m*
f. **shaft sinking** Gefrierverfahren *n (Schachtabteufen)*
f. **shrinkage** Erstarrungsschrumpfung *f*
f. **temperature** Erstarrungspunkt *m*
freibergite *(Min)* Freibergit *m*
freight / to beladen; verfrachten
freight Fracht *f*; Last *f*; Güter *npl*
f. **airplane** Güterflugzeug *n*
f. **bill** Frachtbrief *m*
f. **boat** Lastkahn *m*
f. **car** *(Am) (Eis)* Güterwagen *m*
f. **clearance** Güterabfertigung *f*
f. **declaration** Frachtbrief *m*
f. **elevator** Lastenaufzug *m*
f. **handling** *(Schiff)* Güterumschlag *m (Laden und Löschen)*
f. **locomotive** Güterzuglokomotive *f*
f. **office** Güterabfertigungsstelle *f*
f. **platform** Güterverladeplatz *m*
f. **rate** Frachtsatz *m*
f. **ship** Frachtschiff *n*
f. **space** Frachtraum *m*
f. **station** Güterbahnhof *m*
f. **steamer** Frachtdampfer *m*
f. **ton** *(Schiff)* Maßtonne *f (40 ft³ oder 2240 lbs)*
f. **traffic** Güterverkehr *m*
f. **yard** Güterbahnhof *m*

freighter 1. Frachtschiff *n*, Frachter *m*; 2. Spediteur *m*
French boiler Bouilleurkessel *m*, Walzenkessel *m* mit Sieder, Siederkessel *m*
French calorie *s.* kilocalorie
French chalk *(Am)* Talkum *n*
French comb *(Text)* Heilmannsche Kämmmaschine *f*
French curve Kurvenlineal *n*
French door Fenstertür *f*, französisches Fenster *n*
French grey grauscheckiges Papier *n*
French horn Horn *n*, Waldhorn *n*
French polish Möbelpolitur *f*
French sewn *(Typ)* mit Heftmaschine geholländert, mit französischer Broschur
French spun nach dem Kontinentalsystem versponnenes Kammgarn *n*
French wheat Hirse *f*
French white Bleiweiß *n*, Silberweiß *n*
French window Fenstertür *f*, französisches Fenster *n*
Freon Freon *n*, Frigen *n (Difluordichlormethan; Kältemittel)*
frequency 1. Häufigkeit *f (Statistik)*; 2. Frequenz *f*, Schwingungszahl *f*
f. accuracy Frequenzgenauigkeit *f (für Messungen)*
f. allocation (assignment) Frequenzzuteilung *f*, Wellenzuteilung *f*
f. band Frequenzband *n*
f. band of emission Emissionsfrequenzband *n*
f. band width Frequenzbandbreite *f*
f. bridge *(El)* Frequenzmeßbrücke *f*
f. change Frequenzwechsel *m (Radio)*; Frequenzgleiten *n (z. B. Magnetron)*
f.-change oscillator Frequenzwandler *m*, Hilfsoszillator *m*
f. changer Frequenzumsetzer *m*, Frequenzwandler *m*, Wechselumrichter *m (elektrische Maschinen)*; Mischstufe *f (Elektronik)*
f.-changer crystal Kristallmodulator *m*
f.-changer valve Mischröhre *f*
f. changing receiver Transponierungsempfänger *m*
f. changing stage Mischstufe *f*
f. characteristic Frequenzgang *m*
f. check Frequenzkontrolle *f*
f. composition Frequenzgemisch *n*; Frequenzspektrum *n*
f. constancy Frequenzkonstanz *f*
f. control 1. Frequenzstabilisierung *f*; 2. *(El)* Frequenzregelung *f*, Regelung *f* durch Änderung der Frequenz
f. control limits Nachstimmbereich *m*, Stabilisierungsgrenzen *fpl*
f. converter 1. *(Elm)* Frequenzwandler *m*, Frequenztransformator *m*, Periodenumformer *m*; 2. Mischröhre *f*; Mischstufe *f*
f. converter grid Mischgitter *n*, Signalgitter *n* am Mischrohr
f. counter 1. Frequenzmesser *m* [nach Impulszählverfahren]; 2. Häufigkeitszähler *m (statistische Gütekontrolle)*
f. curve 1. Häufigkeitskurve *f (Statistik)*; 2. Frequenzgang *m*, Frequenzverlauf *m*
f. demultiplication Frequenzerniedrigung *f*, Frequenzteilung *f*
f. departure *s.* f. deviation
f.-dependent frequenzabhängig
f. deviation Frequenzabweichung *f*; [statischer] Frequenzhub *m (Frequenzmodulation)*
f. diagram Häufigkeitsdiagramm *n (Statistik)*
f. difference detector Diskriminator *m*, Frequenzdifferenzdetektor *m*
f. discrimination Frequenzauflösung *f*, Frequenzunterscheidung *f*
f. discriminator Diskriminator *m*; Demodulator *m* für Frequenzmodulation
f. distortion [Amplituden-]Frequenzverzerrung *f*, Dämpfungsverzerrung *f*, Verzerrung *f* 2. Art
f. distribution 1. Häufigkeitsverteilung *f (Statistik)*; 2. Frequenzverteilung *f*
f. divider Frequenzteiler *m*
f. division Frequenzteilung *f*
f.-division multiplexing *(BMSR)* Frequenzschachtelungsverfahren *n*
f. doubler Frequenz[ver]doppler *m*
f. doubling Frequenzverdopp[e]lung *f*
f. drift Frequenzwanderung *f*
f. error 1. Häufigkeitsfehler *m*; 2. Frequenzabweichung *f*
f. factor Häufigkeitsfaktor *m (Statistik)*
f. filter Frequenzfilter *n*
f. fluctuation 1. Häufigkeitsschwankung *f*; 2. Frequenzschwankung *f*
f. for comparison Vergleichsfrequenz *f*
f.-gain Frequenzentdämpfung *f*
f. generating unit Frequenzsatz *m (Telegrafie)*
f. generator Frequenzgenerator *m*
f. holding range *(Fmt)* Haltebereich *m (automatische Frequenzstabilisierung)*
f. indicator Frequenzanzeiger *m*, Frequenzmesser *m*
f. interlacing Frequenzschachtelung *f*; Frequenzverschränkung *f*
f. inversion 1. Frequenzumkehrung *f*; 2. Frequenzumsetzung *f (Sprachverschlüsselung)*
f. limit of equalization *(Fmt)* Entzerrungsgrenze *f*
f. maintenance Frequenzkonstanthaltung *f*
f. match Frequenzanpassung *f*
f. measuring bridge Frequenzmeßbrücke *f*
f. memory Frequenzspeicher *m*
f. meter Frequenzmesser *m*, Wellenmesser *m*

f.-modulated frequenzmoduliert
f. modulation Frequenzmodulation *f*, FM
f. modulator Frequenzmodulator *m*
f. monitor Frequenzkontrollgerät *n*
f. multiplexing Frequenzschachtelung *f*
f. multiplication Frequenzvervielfachung *f*
f. multiplier Frequenzvervielfacher *m*
f. multiplier clystron Frequenzvervielfachungsklystron *n*
f. of calls Gesprächsfrequenz *f*
f. of errors Fehlerhäufigkeit *f*
f. of excitation Erregerfrequenz *f*
f. of load cycles Lastspielfrequenz *f*
f. of oscillations Schwingungsfrequenz *f*
f. of recurrence Wiederholungsfrequenz *f*
f. of resonance Resonanzfrequenz *f*
f. of vibration Schwingungsfrequenz *f*; Schüttelfrequenz *f*
f. overlap Frequenzüberlappung *f*; Frequenzschachtelung *f*
f. pulling *(Fmt)* Frequenzziehen *n*, Mitziehen *n* der Frequenz, Mitzieheffekt *m*
f. range Frequenzbereich *m*, Frequenzband *n*
f.-range of equalization Entzerrungsbereich *m*
f.-range of radio interference Störbreite *f* *(Frequenzbandbreite einer Störung)*
f.-range switch Frequenzbereichumschalter *m*
f. regulator Frequenzregler *m*
f. response 1. *(BMSR)* Frequenzgang *m*; 2. Übertragungsfrequenzgang *m*, Frequenzwiedergabe *f* *(Elektroakustik)*
f. response of attenuation Frequenzgang *m* der Dämpfung, Dämpfungsgang *m*
f. selectivity Frequenztrennschärfe *f*, Trennschärfe *f*
f. sensitive relay Frequenzrelais *n*
f. separation Frequenzabstand *m*
f. shift 1. [ungewollte] Frequenzänderung *f*; 2. Tasthub *m* *(Telegrafie)*
f. shift keying Frequenzumtastung *f*
f. shift signalling Frequenzmodulation *f* *(Telegrafie)*
f. sifter Frequenzsieb *n*, Wellenschlucker *m*
f. sliding Frequenzgleiten *n*, Inkonstanz *f* der Frequenz
f. spacing Frequenzabstand *m*
f. spectrum Frequenzspektrum *n*; Frequenzgemisch *n*
f. splitting *(Fmt)* Entstehung *f* von Parasitärfrequenzen
f. stability Frequenzkonstanz *f*, Frequenzstabilität *f*
f. standard Frequenznormal *n*
f. sub-standard Frequenzhilfsnormal *n*
f. submultiplication Frequenzteilung *f*
f. sweep Frequenzhub *m* *(Wobbelgenerator)*
f. sweeping Wobbeln *n*
f. swing Frequenzhub *m* *(bei Frequenzmodulation)*
f. synthesizer Frequenzgenerator *m* mit Frequenzsynthese
f. table Häufigkeitstabelle *f*
f. telemetering Frequenzfernmeßverfahren *n*
f. tolerance zulässige Frequenzabweichung *f*
f. transfer unit Frequenzmeßwandler *m*
f. transformer Frequenztransformator *m*, Frequenzumformer *m*, Frequenzwandler *m*, Periodenumformer *m*
f. translation Frequenztransponierung *f*
f. tripler Frequenzverdreifacher *m*
f. used in radiolocation Ortungsfrequenz *f*
f. variation Frequenzänderung *f*, Frequenzschwankung *f*
f. variation method Bandbreiteverfahren *n* *(Messungen an Oszillatorkreisen)*
f. wobbler Frequenzwobbler *m*
fresh frisch; grün *(Formsand)*
f. air *(Brg)* Frischwetter *pl*, frische Wetter *pl*
f. air heater Frischluftheizanlage *f*, Frischluftheizung *f*
f. air heating Frischluftheizung *f*
f. air inlet Frischluftzufuhr *f*
f. air respirator Frischluftgerät *n*
f. air supply Frischluftzufuhr *f*
f. batch *s.* f. concrete
f. cleavage frische Bruchfläche *f*
f. concrete Frischbeton *m*
f. from the forest waldfrisch
f.-keeping wrapping Frischhaltepackung *f*
f. lining Neuauskleidung *f*, Neuzustellung *f* *(Schmelzofen)*
f. liquor Frischlauge *f*
f. milk Frischmilch *f*
f. sand Neusand *m*, Frischsand *m* *(Formerei)*
f. water Süßwasser *n*, Flußwasser *n*, Frischwasser *n*
f.-water cooling Süßwasserkühlung *f*
f.-water [storage] tank Süßwasserbehälter *m*, Frischwassertank *m*
freshly repeated frisch hergestellt
f. turned flat furrow Schälfurche *f*
Fresnel diffraction *(Opt)* Fresnelsche Beugung *f*
Fresnel diffraction fringe Fresnelscher Beugungssaum *m*
Fresnel pattern Fresnelsches Beugungsbild *n*
Fresnel prism *(Opt)* Fresnelsches Prisma (Doppelprisma, Zwillingsprisma) *n*
Fresnel region Fresnelsche Beugungszone *f*
fret / to fressen *(Korrosion)*
fret saw Dekoupiersäge *f*; Schweifsäge *f*; Laubsäge *f*
f.-saw machine Schweifsägemaschine *f*
f.-saw timber Laubsägeholz *n*
f.-saw work Laubsägearbeit *f*

fretting Fressen *n (Korrosion)*
f. corrosion Reibkorrosion *f*
f. damage Freßschaden *m*
friability Brüchigkeit *f*, Mürbheit *f*
friable bröck[e]lig, mürbe; gebräch *(Kohle, Gestein usw.)*
f. brown-coal mulmige Braunkohle *f*
f. iron ore mulmiges Eisenerz *n*
friction Reibung *f*, Friktion *f*
f. action Friktionswirkung *f*; Bremswirkung *f*
f. and rack locomotive Reibungs- und Zahnradlokomotive *f*
f. band Friktionsband *n*; Spreizband *n (Kupplung)*
f. between dry surfaces trockene Reibung *f*
f. between lubricated surfaces Schmiermittelreibung *f*
f. board gewalzte Graupappe *f*, Stanzpappe *f*
f. brake Friktionsbremse *f*, Reibungsbremse *f*
f. circle *(Mech)* Reibungskreis *m*
f. clutch Reibungskupplung *f*; im Betrieb schaltbare Kupplung *f*
f. clutch sleeve Reibungskupplungsmuffe *f*
f. coefficient Reibungskoeffizient *m*, Reibungszahl *f*
f. cone Bremskegel *m*
f. contact Reibungskontakt *m*, Wälzkontakt *m*
f. countershaft Reibvorgelege *n*
f. disk Friktionsscheibe *f*
f. disk clutch *(im Betrieb schaltbare)* Scheibenkupplung *f*
f. drive Friktionsantrieb *m*, Reib[scheiben]-antrieb *m*
f. driving roll Schleppwalze *f*
f. drum Reibtrommel *f*
f. facing Friktionsbelag *m*, Reibbelag *m*
f. factor Widerstandsziffer *f*; Reibungsfaktor *m*
f. forging press Schmiedepresse *f* mit Friktionsantrieb
f. formula *(Schiff)* Mitstromformel *f*
f.-gear drive Reibradtrieb *m*
f. gearing Reibgetriebe *n*
f. glazing Friktionsglättung *f*, Satinieren *n* mit Friktion
f. grip Reibschluß *m*
f. grip bolt *(Bw)* hochfest vorgespannte Schraube *f*, HV-Schraube *f*
f. grip bolting gleitfeste Schraubenverbindung *f*, HV-Schraubenverbindung *f*
f. head Verlusthöhe *f*, Druckverlust *m* in der Leitung durch Reibung
f. headstock Spindelkasten *m* mit Friktionskupplung
f. heating Reibungserwärmung *f*
f. horsepower *s.* frictional horsepower
f. law Reibungsgesetz *n*
f. loss Reibungsverlust *m*, Energieverlust *m* durch Reibung
f. machine Reibungselektrisiermaschine *f*
f. material Friktionsmaterial *n*, Reibungsmaterial *n (Bremsbelag)*
f. of motion Bewegungsreibung *f*
f. of repose (rest) Haftreibung *f*
f. pad Bremskissen *n (Scheibenbremse)*
f. pawl Klemmkegel *m*, Reibkegel *m*
f. pile Reibungspfahl *m (Grundbau)*
f. resistance Reibungswiderstand *m*, Friktionswiderstand *m*
f. saw Reibsäge *f*
f. sawing Reibsägen *n*
f. screw press Reibradspindelpresse *f*, Friktionsspindelpresse *f*, Reibtriebpresse *f*
f. shock absorber Reibungsstoßdämpfer *m*
f.-socket Fanghülse *f (Erdöl, Erdgas)*
f. stress Reibspannung *f*
f. surface Reibungsfläche *f*
f. tape *(Am)* Isolierband *n*
f. tapping device Reibungsgewindeschneideinrichtung *f*
f. torque *s.* frictional torque
f. turbine Friktionsturbine *f*, Reibungsturbine *f*
f.-type hoist *(Brg)* Treibscheibenfördermaschine *f*
f.-type viscosimeter Kapillarviskosimeter *n*
f. welding Reibungsschweißen *n*
f. wheel Friktionsrad *n*, Friktionsrolle *f*, Reibrad *n*, Treibscheibe *f*
frictional Reib[ungs]-
f. contact method *(Masch)* Anreibeverfahren *n*
f. countershaft Reibungsvorgelege *n*
f. electric machine Elektrisiermaschine *f*
f. electricity Reibungselektrizität *f*, Triboelektrizität *f*
f. force Reibungskraft *f*
f. gearing Reibungsgetriebe *n*, Friktionsgetriebe *n*
f. heat Reibungswärme *f*
f. horsepower *(Kfz)* Reibungsleistung *f*, Reibungs-PS
f. losses Reibungsverluste *mpl*
f. oxidation Reiboxydation *f*
f. properties Gleiteigenschaften *fpl (z. B. von Gleitlagern)*
f. resistance Reibungswiderstand *m*; Reibhemmung *f*
f. resistance transmission Reibschlußverbindung *f*
f. surface Reibungsoberfläche *f*
f. torque Friktions[dreh]moment *n*, Reibungs[dreh]moment *n*, Drehmoment *n* der Reibungskräfte
f. work Reibungsarbeit *f*
frictionless reibungslos, reibungsfrei
frieze *(Bw, Text)* Fries *m*
frigate Fregatte *f*

frige *(sl)* Kühlschrank *m*
frigid *(sehr, eisig oder künstlich)* kalt
  f. zone kalte Zone *f*
frigorific kälteerzeugend; Kälte-
  f. mixture Kältemischung *f*
  f. unit Kälteeinheit *f*
frigorimeter Kältemesser *m*
frill / to sich kräuseln *(Gelatineschicht auf Fotoplatte)*
frilling Kräuseln *n* der Schicht *(auf Fotoplatte)*
frills Kräuselung *f*
fringe Saum *m*, Rand *m*; Streifen *m*
  f. area 1. Randgebiet *n*, Randzone *f (z. B. Städteplanung)*; 2. *(Opt, El)* Interferenzgebiet *n*; 3. *(Fs)* Grenzzone *f*, Grenzreichweite *f*
  f. crystals Stengelkristalle *mpl*
  f. effect 1. *(El)* Randeffekt *m (Streufeld z. B. an Kondensator- oder Ablenkplatten)*; 2. *(Fs)* Fransenbildung *f*; Fransenwirkung *f*
  f. howl *(El)* Grenzrauschen *n (Radio)*
  f. of superposition *(Met)* Überlagerungsstreifen *m*
  f. of the spectrum Randstreifen *m* des Spektrums
  f. pattern Streifenbild *n*; Beugungsbild *n*; Interferenzbild *n*
  f. stitch *(Text)* Fransenlegung *f*, Fransenbindung *f*
fringes Streifen *mpl (Beugung, Interferenz)*; Fransen *fpl (Fernsehbild)*
fringing field *(El)* Randfeld *n*, Streufeld *n*
frit / to fritten, schmelzen, sintern, zusammenbacken
frit *(Ch)* Fritte *f*
  f. furnace Fritteofen *m (Glasindustrie)*
fritted glass filter Fritte *f*, Glasfilterfritte *f*
fritted glazes gefrittete Glasuren *fpl*
  f. hearth bottom *(Met)* Sinterherd *m (SM-Ofen)*
fritting Zusammenbacken *n*, Zusammenfritten *n*, Zusammensintern *n*
friz[z] / to kräuseln; frisieren; aufkratzen
  f. off the grain *(Led)* Narben spalten *(Sämischgerberei)*
frog 1. *(Eis)* Herzstück *n (Weiche)*; 2. *(Masch)* Drehmaschinenherz *n*
  f. and switch planer Weichenhobelmaschine *f*
  f.-clip Froschklemme *f*
  f.-leg winding *(El)* Froschbeinwicklung *f*
  f. rammer *(Bw)* Explosionsstampfer *m*
frogman's flipper Schwimmflosse *f*
frogsuit *(Kern)* Hautanzug *m*, Schutzanzug *m (Strahlenschutz)*
front 1. Vorderseite *f*; Stirnseite *f (Maschinen)*; 2. Front *f (Wellenausbreitung)*; 3. Brust *f (Hochofen)*
  f. abutment pressure *(Brg)* voreilender Kämpferdruck *m*
  f. axle *(Kfz)* Vorderachse *f*
  f. axle beam Vorderachskörper *m*
  f. axle bracing rod Vorderachsstrebe *f*
  f. axle drive Vorderradantrieb *m*
  f. axle fork Vorderachsgabel *f*
  f. axle pivot pin Achsbolzen *m*
  f. axle shaft Vorderachswelle *f*
  f. axle suspension Vorderachsaufhängung *f*
  f. brace Gegenhalterschere *f*
  f. brake *(Kfz)* Vorderradbremse *f*
  f. bumper *(Kfz)* Frontstoßstange *f*, vordere Stoßstange *f*
  f. clearance *(Masch)* [vorderer] Freiwinkel *m (Schneidwerkzeug)*
  f. compartment *(Rak)* Gerätespitze *f*
  f. connection *(El)* vorderseitiger Anschluß *m*
  f. crossmember of frame *(Rak)* vorderer Kopfträger *m*
  f. cutting edge Stirnschneide *f*; Nebenschneide *f*; Stirnzahn *m*
  f. cylinder cover Stirnwand *f* des Zylinders, Zylinderstirnwand *f*
  f. drive *(Kfz)* Vorderradantrieb *m*, Frontantrieb *m*
  f. edge Vorderkante *f*
  f. elevation Vorderansicht *f*; Stirnansicht *f*
  f. end 1. *(Masch)* Stirnseite *f*; 2. Verarbeitungsstelle *f (Glas)*
  f.-end equipment Baggergeräte *npl (auswechselbare Geräte für Universalbagger)*
  f.-end loader Front[schaufel]lader *m*, Frontschaufler *m*, Ladeschaufel *f*, Schaufellader *m*
  f.-end of the frame *(Wärm)* Rahmenkopf *m*, Lagerbalkenkopf *m*
  f.-end plate of boiler Kesselstirnwand *f*
  f. face Vorderseite *f*, Stirnseite *f*
  f. flap *(Typ)* vordere Klappe *f (eines Schutzumschlages)*
  f. focal plane *(Opt)* vordere Brennebene *f*
  f. gap Nutzspalt *m (eines Magnetkopfwandlers)*
  f. guiding *(Brg)* Kopfführung *f*, Stirnführung *f*
  f. hanger lens *(Opt)* Nahvorhänger *m*
  f. hub Vorderradnabe *f*
  f. joint Stirnstoß *m (Schweißen)*
  f. ledge Scheuerleiste *f*
  f. lens Vorsatzlinse *f*; Frontlinse *f*
  f. lighting Fassadenbeleuchtung *f*
  f.-loading truck Frontlader *m*
  f. lock *(Text)* vorderes Strickschloß *n*
  f. mill table Rollengang *m (vor einer Walze)*
  f. milling tool Stirnfräser *m*
  f. needle bar *(Text)* vordere Nadelbarre *f*
  f.-operated frontbedient; mit Frontantrieb
  f.-operated lathe Frontdrehmaschine *f*
  f. panel 1. *(El)* Frontplatte *f (elektrische Geräte)*; 2. *(Kfz)* Vorderwand *f*

f.-panel control *(El)* Einstellung (Regelung) *f* an der Frontplatte, Einstellung (Regelung) an der Vorderseite
f.-panel frame *(Kfz)* Vorderwandrahmen *m*
f. pitch kollektorseitiger Wicklungsschritt *m*
f. porch *(Fs)* Vorderschwarzschalter *m*, vorderer Schwarzschalter *m*, vordere Schwarztreppe *f*
f. power take-off Rumpfzapfwelle *f*
f. projection Aufprojektion *f (darstellende Geometrie)*
f. rest *(Masch)* Vordersupport *m*
f. roller bearing *(Text)* Vorderzylinderlager *n*
f. roller nip *(Text)* Vorderzylinderklemmpunkt *m (Streckwerk)*
f. scanning Vorwärtsabtastung *f*
f. seat Vordersitz *m*
f. section *(Rak)* Behälterspitze *f*
f. share Vorschar *n* des Pfluges
f. side Stirnseite *f*, Vorderseite *f*
f.-side projection Aufprojektion *f (darstellende Geometrie)*
f. sight Korn *n (Visier; Flinte, Gewehr)*
f. spar *(Flg)* Vorderholm *m*
f. spring *(Kfz)* Vorderfeder *f*
f. spring hanger Vorderfederbock *m*
f. spring support Vorderfederstütze *f*
f. stairs Freitreppe *f*
f. steering *(Kfz)* Vorderradlenkung *f*
f. steering device Vordersteuer *n*
f. stub-shaft Rumpfzapfwelle *f*
f. surface Stirnfläche *f*
f. suspension *(Kfz)* Vorderradaufhängung *f*
f. teeth *(Masch)* Stirnzähne *mpl*
f. tipper *(Kfz)* Vorderkipper *m*, Frontkipper *m*
f.-to-back drawing-in *(Text)* Einzug *m* von vorn nach hinten
f.-to-back (rear) ratio Vor-Rück-Verhältnis *n (Antennentechnik)*; Ausblendungsverhältnis *n (bei Richtstrahlern)*
f. track Spurweite *f* vorn
f.-tyre Vorderradreifen *m*
f. upright *(Masch)* Vorderständer *m*
f. view Aufriß *m (technisches Zeichnen)*; Vorderansicht *f*
f. wall of firebox *(Wärm)* Feuerbuchsrohrwand *f*
f. wave *(Ph)* Kopfwelle *f*
f. wheel Vorderrad *n*
f. wheel brake *(Kfz)* Vorderradbremse *f*
f. wheel drive *(Kfz)* Frontantrieb *m*, Vorderradantrieb *m*
f. wheel set *(Eis)* Vorderradsatz *m*
f. wing *(Kfz)* Vorderkotflügel *m*, vorderer Kotflügel *m*
frontage line Gebäudeflucht *f*, Fluchtlinie *f*
frontal frontal; Vorder-, Stirn-
f. analysis Frontanalyse *f (Chromatografie)*
f. area Stirnfläche *f*, Vorderfläche *f*
f. edge Vorderkante *f*
f. lathe Frontdrehmaschine *f*
f. plane Frontalebene *f*
f. screen Vordersieb *n*
f. section Frontalschnitt *m*
frontispiece 1. *(Typ)* Titelseite *f*, erste Seite *f*; Titelbild *n*; 2. Giebelseite *f*, Front[seite] *f*
frost / to matt ätzen, mattieren *(Glas)*; auftrocknen unter Bildung von Kristall- und Eisblumenmustern *(Lackindustrie)*
frost 1. Frost *m*; 2. Flitter *m (Glasindustrie)*
f. action Frosteinwirkung *f*
f. blanket Frostschutzschicht *f (Straßenbau)*
f.-cleft eisklüftig *(Holz)*
f. crack Frostriß *m*
f. heave *(Bw)* Frosthebung *f*
f. layer Frostschutzschicht *f (Straßenbau)*
f. line Frostgrenze *f (Grundbau)*
f. penetration depth Frosttiefe *f (Grundbau)*
f.-proof frostbeständig, frostsicher
f. protection Frostschutz *m*
f. resistance Frostbeständigkeit *f*
f.-resistant frostbeständig
f. rib *(Holz)* Frostleiste *f (Fehler)*
f. scaling Frostabblätterung *f*
frosted finish Eisblumenlack *m*
f. glass Mattglas *n*
f. lamp Mattglaslampe *f*, mattierte Lampe *f*
f. silver Mattsilber *n*
f. varnish Eisblumenlack *m*
frostiness Mattigkeit *f*, Glanzlosigkeit *f (Metalloberfläche)*
frosting Mattätzen *n*, Mattierung *f*; Glasur *f*
f. agent Mattsalz *n (Glasindustrie)*
froth / to [auf]schäumen
f. out herausschäumen
froth Schaum *m*, Gischt *m*
f. breaker Schaumbrecher *m (Flotation)*
f. cell *(Brg, Min)* Flotationszelle *f*
f.-contact heat exchanger *(Ch)* Schaumkontaktwärme[aus]tauscher *m*
f.-destroying nozzle Schaumzerstörungsdüse *f*
f. fines *(Brg, Min)* Flotationskonzentrat *n*
f. flotation Flotation *f*, Schaumschwimmverfahren *n*, Schaum[schwimm]aufbereitung *f*
f. killer Schaumbrecher *m (Flotation)*
f. rubber Schaumgummi *m*, Zellgummi *m*
f. skimmer Schaumabheber *m*, Schaumabstreifer *m*
f. stabilizer Schaumstabilisierer *m (Flotation)*
frother Schäumer *m*, Schaumbildner *m (Flotation)*
f. cell Flotationszelle *f*
frothing Schäumen *n*; Schaumbildung *f*
f. agent Schäumer *m*, Schaumbildner *m*

f. technique (*Text*) Aufschäumtechnik *f*
frozen eingefroren, erstarrt, fest
f. casing festgewordene Verrohrung *f* (*Erdöl, Erdgas*)
f.-in degrees of freedom (*Ph*) eingefrorene Freiheitsgrade *mpl*
f.-in field lines eingefrorene Feldlinien *fpl* (*Supraleitung*)
f. lacquer Eisblumenlack *m*
f. slag erstarrte Schlacke *f*
fructification (*Lw*) Fruchtbarmachung *f*
fructose Fruchtzucker *m*, Fruktose *f*
fruit Obst *n*; Südfrucht *f*; Südfrüchte *fpl*; Frucht *f*
f. brandy Obstbranntwein *m*
f.-carrier Obsttransportschiff *n*, Fruchttransportschiff *n*
f.-crop Obsternte *f*
f. culture Obstbau *m*
f. drier Dörrapparat *m* (*für Obst*)
f. grove Obstplantage *f*
f. growing Obstbau *m*
f. juice Fruchtsaft *m*
f. peel Obstschale *f*
f. plantation Obstpflanzung *f*, Obstplantage *f*
f. press Fruchtpresse *f*, Kelter *f*
f.-pressing machine Obstkeltermaschine *f*
f. ship Obsttransportschiff *n*, Fruchttransportschiff *n*
f. sugar Fruchtzucker *m*, Fruktose *f*
f. to be peeled Schälobst *n*
f. tree Obstbaum *m*
fruiter Obstzüchter *m*
frustum (*Math*) Stumpf *m*
f. of a cone Kegelstumpf *m*
f. of a pyramid Pyramidenstumpf *m*, abgestumpfte Pyramide *f*
frying (*El*) Knistern *n*, Kratzgeräusche *npl*, Nebengeräusche *npl*
fuchsine (*Ch*) Fuchsin *n*, Rosanilin *n*
f. acid Fuchsinsäure *f*
fuchsite (*Min*) Fuchsit *m*
fuel / to tanken, Brennstoff einnehmen (aufnehmen)
fuel 1. Brennstoff *m* (*Oberbegriff*); 2. Kraftstoff *m*, Treibstoff *m* (*Vergaser, Diesel, Turbine*); 3. (*Rak*) Treibstoff; 4. Heizstoff *m* (*Heizgas, Kohle, Heizöl usw.*); 5. Spaltstoff *m* (*Kernreaktor*)
f.-air mixture Kraftstoff-Luft-Gemisch *n*
f.-air ratio Kraftstoff-Luft-Verhältnis *n*
f. alcohol Brennspiritus *m*
f. assembly (*Kern*) Brennstoffanordnung *f* (*im Reaktor*); Brennstoffkassette *f*
f.-atomizer Kraftstoffzerstäuber *m*
f. bed Brennstoffschicht *f*
f. boost pump Kraftstoffüberladungspumpe *f* (*Strahlantrieb*)
f.-burning engine Brennkraftmaschine *f*, Verbrennungskraftmaschine *f*
f. by-pass regulator Brennstoffnebenwegregler *m*
f. capacity Fassungsvermögen *n* des Brennstoffbehälters (Kraftstoffbehälters)
f. charge (*Kern*) Reaktorbeschickung *f*
f. consumption Brennstoffverbrauch *m*; Kraftstoffverbrauch *m*; Treibstoffverbrauch *m*
f.-consumption gauge (indicator) Brennstoffverbrauchsmesser *m*; Kraftstoffverbrauchsmesser *m*
f. cutoff valve Brennstoffabsperrschieber *m*, Kraftstoffabsperrschieber *m*
f. cycle (*Kern*) Brennstoffzyklus *m*, Brennstoffkreislauf *m* (*Reaktor*)
f. deaerator Kraftstoffentlüfter *m*
f. delivery Brennstoffzufuhr *f*; Kraftstoffzufuhr *f*
f. depletion (*Kern*) Spaltstofferschöpfung *f*
f. diffusion pump Treibmittelpumpe *f* (*Vakuumtechnik*)
f. dope Antiklopfmittel *n*
f. doubling time (*Kern*) Spaltstoffverdopplungszeit *f* (*Brutreaktor*)
f. dump Brennstofflager *n*; Kraftstofflager *n*, Treibstofflager *n*
f. dust Brennstaub *m*
f. economizer Brennstoffsparvorrichtung *f*
f. economy 1. Brennstoffwirtschaft *f*; 2. Brennstoffersparnis *f*; Kraftstoffersparnis *f*
f. element (*Kern*) Brenn[stoff]element *n*
f. element can (*Kern*) Brennstoffelementhülse *f*
f. engineering Brennstofftechnik *f*; Feuerungskunde *f*
f. enrichment (*Kern*) Spaltstoffanreicherung *f* (*Brutreaktor*)
f. feed Brennstoffzuführung *f*
f. feed pump Brennstofförderpumpe *f*
f. feeder Brennstoffspeisevorrichtung *f*
f. filler cap Kraftstoffüllstutzen *m*
f. filter Kraftstoffilter *n*
f.-fired mit Brennstoff geheizt
f.-flow gauge Brennstoffverbrauchsmesser *m*
f. gas Brenngas *n*; Treibgas *n*; Heizgas *n*
f. gauge Kraftstoffstandanzeiger *m*; Kraftstoffuhr *f*; Brennstoffmesser *m*; Kraftstoffvorratsanzeiger *m*
f. heating 1. Kraftstoffheizung *f*; 2. Ölheizung *f*
f. inhibitor Antiklopfmittel *n*, Oktanzahlverbesserer *m*
f. injection Kraftstoffeinspritzung *f*
f.-injection engine Einspritzmotor *m*
f. injection nozzle Brennstoffdüse *f*; Einspritzdüse *f*
f. injection pump Kraftstoffeinspritzpumpe *f*

f. injection system Kraftstoffeinspritzanlage *f*
f. injection valve Kraftstoffventil *n*, Einspritzventil *n*
f. injector Kraftstoffeinspritzer *m*, Einspritzdüse *f*
f. injector pump Kraftstoffeinspritzpumpe *f*
f. inventory *(Kern)* Brennstoffbestand *m* *(Reaktorbetrieb)*
f. jet Kraftstoffstrahl *m*; Kraftstoffdüse *f*
f. jettisoning *(Flg)* Treibstoffschnellentleerung *f*
f. jettisoning gear Treibstoffschnellablaß *m*
f. level Brennstoffspiegel *m*, Brennstoffvorrat *m*; Kraftstoffspiegel *m*, Kraftstoffvorrat *m*
f.-level gauge Brennstoffvorratsmesser *m*, Brennstoffuhr *f*; Kraftstoffvorratsmesser *m*
f. lifetime *(Kern)* Brennstofflebensdauer *f*
f. lift pump Kraftstoffsaugpumpe *f*
f. line Brennstoffleitung *f*; Kraftstoffleitung *f*
f. loading *(Schiff)* Ölübernahme *f*, Treibstoffübernahme *f*
f. make-up *(Kern)* Brennstoffnachfüllung *f*; Brennstoffauffrischung *f*
f. manifold Kraftstoffleitung *f*, Brennstoffleitung *f*
f. material *(Kern)* Brennstoff *m*, Spaltstoff *m*
f. meter Brennstoff[meß]uhr *f*; Kraftstoff[meß]uhr *f*
f.-metering nozzle Kraftstoffmeßdüse *f*
f. mileage *(Kfz)* Kraftstoffergiebigkeit *f (auf Meilen bezogen)*
f. mist *(Kfz)* Kraftstoffnebel *m*
f. mixed with air Brennstoff-Luft-Gemisch *n*
f. mixture *(Kern)* Brennstoffgemisch *n*, Spaltstoffgemisch *n*
f. oil 1. Heizöl *n*; 2. Treiböl *n*, Dieselöl *n*
f. oil additive Treibölzusatz[mittel *n*] *m*
f.-oil supply tank Brennstoffvorratsbehälter *m*
f.-oil valve Brennstoffventil *n*
f. pipe Brennstoffleitung *f*; Kraftstoffleitung *f*
f. piping *s.* f. pipe
f. poisoning *(Kern)* Brennstoffvergiftung *f*
f. pump Kraftstoff[förder]pumpe *f*; Brennstoff[förder]pumpe *f*
f. pump diaphragm Förderpumpenmembran *f*, Kraftstoffpumpenmembran *f*
f. pump timing Förderpumpeneinstellung *f*, Kraftstoffpumpeneinstellung *f*
f. rate Brennstoffverbrauch *m*; Kraftstoffverbrauch *m*; Treibstoffverbrauch *m*
f. ratio Heizwertzahl *f*
f.-receiving chamber *(Wärm)* Füllrumpf *m*
f.-reclaiming tank Kraftstoffrücklaufbehälter
f. reconditioning *(Kern)* Brennstoffaufarbeitung *f*
f. regeneration *s.* f. reconditioning
f. replenishment *(Kern)* Brennstoffnachfüllung *f*
f. reprocessing *s.* f. reconditioning
f. reprocessing plant *(Kern)* Brennstoffaufarbeitungsanlage *f*, Spaltstoffaufbereitungsanlage *f*
f. research Kraftstofforschung *f*; Kraftstoffuntersuchung *f*
f. research engine Kraftstoffuntersuchungsmotor *m*
f. reservoir Brennstoffbehälter *m*; Kraftstoffbehälter *m*; Treibstoffbehälter *m*
f.-return line *(Rak)* Umwälzleitung *f*
f. rod *(Kern)* Brennstoffstab *m*
f. sensitivity of an engine Kraftstoffempfindlichkeit *f* eines Motors
f. service pump Brennstofftagespumpe *f*, Brennstoffförderpumpe *f*
f. sheathing *(Kern)* Brennstoffumhüllung *f*
f. shortage Brennstoffknappheit *f*
f. slug *(Kern)* Brennstoffblock *m*
f. solution *(Kern)* Spaltstofflösung *f*, Brennstofflösung *f*
f. sprayer Brennstoffzerstäuber *m*
f. strainer Brennstoffilter *n*; Kraftstoffreiniger *m*
f.-structure ratio Kraftstoffgewichtsverhältnis *n*
f. supply Brennstoffzufuhr *f*; Kraftstoffzufuhr *f*
f. supply at full load *(Kfz)* Vollastfördermenge *f*
f. supply gauge (indicator) Kraftstoffvorratsanzeiger *m*; Kraftstoffvorratsmesser *m*
f. supply line (pipe) Brennstoff[zuführungs]leitung *f*
f. supply pump *(Kfz)* Förderpumpe *f*
f. system Kraftstoffanlage *f*
f. tank Brennstoffbehälter *m*; Benzintank *m*, Kraftstoffbehälter *m*, Treibstoffbehälter *m*
f. testing engine Kraftstoffuntersuchungsmotor *m*
f. throughput *(Kern)* Brennstoffdurchsatz *m*
f.-throwing wheel *(Wärm)* Wurfrad *n*, umlaufende Wurfschaufel *f*
f. transfer pump Brennstofförderpumpe *f*, Brennstoffumwälzpumpe *f*
f. trap Kraftstoffabschneider *m*
f. utilization Brennstoffausnutzung *f*
f. value Heizwert *m*
f. valve Brennstoffventil *n*; Kraftstoffhahn *m*
f. vaporizer Kraftstoffverdampfer *m*
f. vapour lock Dampfblasenbildung *f* in der Kraftstoffleitung
f. vapour reflux Rückdiffusion *f (Vakuumtechnik)*
f.-weight ratio Kraftstoffgewichtsverhältnis *n*
f. wharf Ölhafen *m*
f. wood Brennholz *n*
fueling *s.* fuelling

fuelling Auftankung *f*, Betankung *f*; Brennstoffübernahme *f*, Treibstoffübernahme *f*
f. attendant Tankwart *m*
f. facilities Betankungseinrichtungen *fpl*
f. pump *(Schiff)* Übernahmepumpe *f*
fugacity Fugazität *f*, Aktivität *f* in der Gasphase, Flüchtigkeit *f (Thermodynamik)*
fugitive dye *(Text)* unechter Farbstoff *m*
fulcrum Drehpunkt *m*, Gelenkpunkt *m*, Stützpunkt *m*
f. pin Drehbolzen *m*, Drehzapfen *m*
f. stud Gelenkzapfen *m*
fulcrumed lever zweiarmiger (drehbar gelagerter) Hebel *m*
fulfil / to genügen *(Bedingung)*; erfüllen *(Plan)*
f. the requirements den Anforderungen (Forderungen) genügen
fulguration 1. *(Met)* Blick[en *n*] *m (Treiben des Silbers)*; 2. Elektrosikkation *f (Elektromedizin)*
full / to *(Am) (Text)* walken
full 1. voll; 2. *(Masch)* mit 1/64 Zoll Übermaß
f. adder *(Dat)* Volladdierer *m*, Volladdierwerk *n (binär)*
f. admission *(Wärm)* volle Beaufschlagung *f (Turbine)*; Vorfüllung *f*
f. admission turbine Vollturbine *f*
f. advance *(Kfz)* Frühzündung *f*, Vorzündung *f*
f. and down Höchstbefrachtung *f (ein Schiff so beladen, daß die Höchstmenge an Frachttonnen erzielt wird)*
f. annealing Ausglühen *n*, Hochglühen *n*
f. automatic Vollautomat *m*
f. automatic lathe Drehautomat *m*
f.-automatic processing vollautomatische Verarbeitung *f*
f.-automatic working *(Fmt)* Selbstwählbetrieb *m*
f. band [voll]besetztes Energieband *n (Halbleiter)*
f.-bodied ship völliges Schiff *n*
f. bulb Vollwulststahl *m*
f. can *(Text)* gefüllte (volle) Spinnkanne *f*
f. cantilever voll freitragend
f. caving *(Brg)* Totalbruchbau *m*
f.-cell process *(Holz)* Volltränkverfahren *n*
f. circle Vollkreis *m*, Ganzkreis *m*
f.-circle rotation Drehung *f* um 360°
f. comprehensive insurance Vollkaskoversicherung *f*
f. countersunk head Senkkopf *m*
f. curve ausgezogene Kurve *f*
f. deflection Endausschlag *m (Meßinstrument)*
f. depth Zahnhöhe *f*
f.-depth of jaws Spannbackenhöhe *f*
f.-depth system *(Masch)* Normalverzahnungssystem *n*
f.-depth teeth Hochverzahnung *f*
f. dial service *(Am) (Fmt)* Vollwählbetrieb *m*
f. diameter Außendurchmesser *m (z. B. Innengewinde)*
f. dog screw Schraube *f* mit Zapfenende
f.-duplex circuit *(Fmt)* Gegenschreibschaltung *f (Telegrafie)*
f.-duplex operation *(Fmt)* 1. Gegensprechen *n (Telefonie)*; 2. Gegenschreiben *n (Telegrafie)*
f.-edged vollkantig
f. employment Vollbeschäftigung *f*
f. ended ship Schiff *n* mit völligen Enden
f. excursion *s.* f. scale travel
f. face *(Typ)* vollfett
f.-fashioned *(Text)* 1. formgerecht *(Stricken)*; 2. Cotton-
f.-fashioned hosiery machine *(Text)* Cottonmaschine *f*
f. fillet Flachkehlnaht *f (Schweißen)*
f. fillet weld Flachkehlnaht *f (Schweißen)*
f.-flow filter *(Kfz)* Hauptstromfilter *n*
f. gantry crane Volltorkran *m*, Vollportalkran *m*
f. hardening Durchhärten *n*, Aushärten *n*; Durchhärtung *f*, Aushärtung *f*
f. headlight beam *(Kfz)* Fernlicht *n*
f. journey *(Brg)* Voll[wagen]zug *m*
f.-length durchlaufend
f.-length feature film abendfüllender Spielfilm *m*
f. lift Vollhub *m*
f. line Vollinie *f*
f. load Vollast *f*; Vollbelastung *f*; Volleistung *f*
f.-load current Vollaststrom *m*
f.-load displacement *(Schiff)* Ladeverdrängung *f*
f.-load efficiency Vollastwirkungsgrad *m*
f.-load excitation *(Elm)* Vollasterregung *f*
f.-load field *(Elm)* Vollastfeld *n*
f.-load loss *(Elm)* Vollastverlust *m*
f.-load meter adjustment *(El)* Einstellvorrichtung *f* für den Nennlastabgleich *(eines Zählers)*
f.-load needle *(Kfz)* Vollastnadel *f*
f.-load torque Vollastdrehmoment *n*
f.-load voltage Vollastspannung *f*
f.-logarithmic doppeltlogarithmisch *(Zeichenpapier)*
f.-magnetic controller *(BMSR)* vollmagnetischer Fahrschalter (Programmschalter) *m*
f. measure *(Holz)* vollmaßig nach dem Trocknen
f. modulation *(El)* Vollaussteuerung *f*
f.-pitch winding *(Elm)* Durchmesserwicklung *f*
f. power volle Leistung *f*
f.-pressure Volldruck-
f.-pressure altitude Volldruckhöhe *f*
f.-pressure line Volldrucklinie *f*

f. professor ordentlicher Professor *m*
f. profile Vollumriß *m*
f.-profile tooth hob Grundprofilzahnfräser *m*
f. radiator *(Ph)* schwarzer Strahler *m*
f. range deflection Endausschlag *m*, voller Zeigerausschlag *m (Meßinstrument)*
f.-red gold hochrotes Gold *n*
f.-revolving excavator (shovel) Schwenkbagger *m*
f.-rigged ship Vollschiff *n (Dreimastschiff; Fock-, Groß- und Kreuzmast)*
f. rudder *(Schiff)* hart Ruder
f. satellite exchange *(Fmt)* Teilamt *n* ohne Überbrückungsverkehr
f. scale 1. natürliche Größe *f*; 2. *(BMSR)* Aussteuerungsbereich *m (Analogrechner)*; nomineller maximaler Wert *m*; Vollaussteuerung *f*
f. scale maßstäblich; mit Maßstab 1:1
f.-scale deflection Endausschlag *m*, Vollausschlag *m*, voller Zeigerausschlag *m (Meßinstrument)*
f.-scale reading Meßbereichendwert *m (auf der Skala)*
f.-scale test Großversuch *m (im Gegensatz zum Modellversuch)*
f. scale travel Endausschlag *m (Meßinstrument)*
f. scantling vessel Volldeckschiff *n*, Volldecker *m*
f. section Vollschnitt *m (technisches Zeichnen)*
f. section of the steam port *(Wärm)* voller Kanalquerschnitt *m*
f. set *(Brg)* Geviertzimmerung *f*
f. shadow Vollschatten *m*
f. sight *(Mil)* Vollkorn *n*
f. size 1. Vollmaß *n*, Vollformat *n*; 2. *(Typ)* Planoformat *n*
f.-size drawing Zeichnung *f* in natürlicher Größe, Zeichnung im Maßstab 1:1
f.-size test zerstörungsfreie Werkstoffprüfung *f*
f.-size trimmed *(Typ)* auf Format beschnitten *(Papier usw.)*
f. speed ahead *(Schiff)* volle Fahrt voraus
f. sphere Vollkugel *f*
f. strength *(Mil)* Etatstärke *f*
f. supply turbine Vollturbine *f*
f. throttle *(Kfz)* Vollgas *n*
f. throttle altitude *(Flg)* Volleistungshöhe *f*
f.-time apprentice instructor hauptberuflicher Lehrausbilder *m*
f. tub *(Brg)* Vollwagen *m*
f. universal radial drilling machine Universalauslegerbohrmaschine *f*
f.-view illustration Gesamtansicht *f*
f.-view window Fenster *n* mit unbeschränkter Sicht, Vollsichtfenster *n*
f.-wave circuit Zweiwegschaltung *f*
f.-wave dipole Ganzwellendipol *m*
f.-wave rectification Zweiweggleichrichtung *f*
f.-wave rectifier Zweiweggleichrichter *m*, Doppelweggleichrichter *m*, Vollweggleichrichter *m*
f. width rinsing *(Text)* Breitspülen *n*
f. width temple *(Text)* Vollbreithalter *m*, Ganzbreitenbreithalter *m*, über die ganze Gewebebreite wirkender Breithalter *m*
fuller 1. Ballhammer *m*, Rundbahnhammer *m*; Abschreckwerkzeug *n*; Rollgesenk *n*; 2. *(Text)* Tuchwalker *m*
fuller's earth Bleicherde *f*, Fullererde *f*, Walkerde *f*
fulling *(Text)* Walken *n*, Tuchwalke[n *n*] *f*
f. board Walkbrett *n*
f. cube Broschierwürfel *m*, Walkwürfel *m*
f. mill Walkmühle *f*
fullness 1. Kernigkeit *f (Leder)*; 2. Völligkeit *f (Schiff)*
f. of tone Klangfülle *f*
fulls *(Brg)* Vollwagen *mpl*
fully voll, ganz
f. aged *(Met)* vollausgehärtet
f. automatic vollselbsttätig; vollautomatisch *(automatisch)*
f.-automatic machine Vollautomat *m*
f. carbonized völlig gar *(Koks)*
f. energized erregt für das Ansprechen *(Relais)*
f.-floating axle *(Kfz)* vollfliegende Achse *f*, Steckachse *f*
f.-floating suspension *(Kfz)* Vollschwingaufhängung *f*
f. heat-treated vergütet
f. intermeshed network *(El)* Maschennetz *n*
f. ionized gas vollionisiertes (vollständig ionisiertes) Gas *n*
f. killed steel beruhigter Stahl *m*
f. mechanized *(Mil)* vollmotorisiert
f. modulated *(Fmt)* durchmoduliert
f. restrained *(Bw)* voll eingespannt; beidseitig eingespannt *(Träger)*
f. seated *(Masch)* voll aufliegend
f.-trained voll ausgebildet, mit abgeschlossener Ausbildung
f. versatile excavator Universalbagger *m*
fulminate / to zur Explosion bringen
fulminate Fulminat *n*, Salz *n* der Knallsäure
fulminating cotton Schießbaumwolle *f*
f. damps *(Brg)* schlagende Wetter *pl*
f. gold goldsaures Ammonium *n*, Knallgold *n*
f. mercury Knallquecksilber *n*, Quecksilberfulminat *n*
f. silver Knallsilber *n*, Silberfulminat *n*
fulminic acid Knallsäure *f*
fulness *s.* fullness
fumaric acid Fumarsäure *f*

fume / to gasen, rauchen; dampfen; qualmen; räuchern
  f. away verdunsten
  f.-off verpuffen
fume Abgas *n*, Rauch *m*; Dampf *m*; Dunst *m*
  f. constituents Rauchgasbestandteile *mpl*
  f. cupboard Abzug[sschrank] *m (Labor)*
  f. extraction equipment Gasabsaugungsanlage *f*
  f. fading *(Text)* Entfärben (Verschießen) *n* in Abgasatmosphäre
  f.-proof rauchgasfest; gasdicht, rauchdicht
  f. scrubber *(Ch)* Nebelwäscher *m*
fumes *(Brg)* Schießschwaden *mpl*
fumigate / to ausräuchern; desinfizieren
  f. with sulphur ausschwefeln *(Stall usw.)*
fumigating candle Räucherkerze *f*
fumigation Ausgasung *f*, räuchernde Desinfektion *f*
fuming rauchend
  f. nitric acid rauchende Salpetersäure *f*
  f. off Verpuffung *f*
fumivorous rauchverzehrend
function / to arbeiten; funktionieren
function Arbeitsweise *f*, Tätigkeit *f*, Zweck *m*, Aufgabe *f*; *(Math)* Funktion *f*
  f. generating potentiometer nichtlinearer Spannungsteiler *m (Analogrechner)*
  f. generator Funktionsgeber *m*, Funktionsgenerator *m (Analogrechner)*
  f. multiplier Funktionsmultiplizierer *m (Analogrechner)*
  f. plotter Kurvenschreiber *m*
  f. potentiometer Funktionspotentiometer *n (Analogrechner)*
  f. switch Funktionsschalter *m (Analogrechner)*
  f. table *(Math)* Funktionstabelle *f*, Funktionstafel *f*
  f. translator *(BMSR)* Funktionszuordner *m*
functional *(Math)* Funktional *n*, Funktion *f* einer Funktion, zusammengesetzte Funktion
functional Funktions-; funktionell
  f. analysis Funktionalanalysis *f*
  f. circuit logische Schaltung *f*; Dekodierschaltung *f*
  f. diagram logisches Diagramm *n (Elektronenrechner)*
  f. equation Funktionalgleichung *f*
  f. gauge Paarungslehre *f*
  f. gauging 1. Paarungslehren *n*; 2. Funktionsprüfung *f*
  f. matrix Funktionalmatrix *f*
  f. symbol *(Math)* Funktionssymbol *n*; logisches Symbol *n (Elektronenrechner)*
functionality Funktionstüchtigkeit *f*
functions of angles Winkelfunktionen *fpl*
functor *(Math)* Funktor *m*
fundamental grundlegend, grundsätzlich, ursprünglich
  f. arithmetic Grundrechnungsart *f*
  f. bridge equation *(El)* allgemeine Gleichung *f* der Meßbrücke
  f. circle Grundkreis *m (Zahnrad)*
  f. colours Grundfarben *fpl*
  f. component distortion Amplitudenverzerrung *f*
  f. constant Grundkonstante *f*
  f. crystal Grundschwingungsquarz *m*
  f. equation of reaction propulsion Raketengrundgleichung *f*
  f. equation of rocket motion Raketengrundgleichung *f*
  f. frequency Grundfrequenz *f*
  f.-frequency attenuation Grundwellendämpfung *f*
  f. harmonic oscillation Grundschwingung *f*
  f. knowledge Grundwissen *n*
  f. lattice Grundgitter *n (Kristalle)*
  f. law Grundgesetz *n*
  f. mode Grundtyp *m (Welle)*
  f. mode of vibration Grundschwingung *f (Schallsender)*
  f. operation Grundrechnungsart *f*
  f. oscillation Grundschwingung *f*, Fundamentalschwingung *f*
  f. particles Elementarteilchen *npl*
  f. principle Grundprinzip *n*, Axiom *n*
  f. research Grundlagenforschung *f*, Erkundungsforschung *f*
  f. rule Grundregel *f*
  f. science Grundwissenschaft *f*
  f. series Grundreihe *f*
  f. solution *(Math)* Fundamentallösung *f*, Grundlösung *f*
  f. sound (tone) Grundton *m*
  f. subject Grundlagenfach *n*
  f. theorem Fundamentalsatz *m*
  f. unit Grundeinheit *f*
  f. vibration Grundschwingung *f*, Fundamentalschwingung *f*
  f. wave Bodenwelle *f*, Grundwelle *f*
  f. wave frequency Grundwellenfrequenz *f*
  f. wave of voltage Grundschwingung *f (einer Wechselspannung)*
  f. zero-sequence Grundwellenblindleistung *f* der Nullkomponente
fundamentals Grundlagen *fpl (einer Wissenschaft)*
fundus Fundus *m*
  f. oculi Augenhintergrund *m*
  f. of the eye Augenhintergrund *m*
fungi-proof beständig gegen Pilzbildung
fungicide chemisches Mittel *n* gegen Schimmelbildung und Bakterienbildung, Fungizid *n*, Konservierungsmittel *n*, Pilzvertilgungsmittel *n*

funginertness Unempfänglichkeit *f* für Schimmelbildung
fungus Pilz *m*, Schimmelpilz *m*, Schwamm *m*; *(Bw)* Hausschwamm *m*
f. attack *(Holz)* Pilzbefall *m*
funicular 1. Seil-, Ketten-, Band-; 2. faserig
f. polygon Seileck *n*, Seilpolygon *n*, Seilzug *m*
f. railway Seilbahn *f*
f. traction Seilförderung *f*
funnel 1. Trichter *m*, Fülltrichter *m*; 2. Schornstein *m*, Kamin *m*, Schlot *m*; 3. Lichtschacht *m*; Lüftungsschacht *m*
f. aerial Trichterantenne *f*, trichterförmige Antenne *f*
f.-base temperature Temperatur *f* am Schornsteinfuchs
f. for auxiliary boiler Hilfskesselschornstein *m*
f. mark *(Schiff)* Schornsteinabzeichen *n*
f. opening Schornsteinluke *f*
f. shaft Schornsteinschacht *m*
f.-shaped trichterförmig
f.-shaped aerial Trichterantenne *f*, trichterförmige Antenne *f*
f.-shaped crater *(Brg)* Wurftrichter *m (Schießarbeit)*
f. stand Filtriergestell *n*
f. stay *(Schiff)* Schornsteinstag *n*
f. type straight oil cup Trichteröler *m*
f. with spark arrester Schornstein *m* mit Funkenfängersieb
fur / to sich mit Kesselstein bedecken
f. the boiler den Kessel ausklopfen
fur 1. Pelzwerk *n*, Rauchware *f*, Rauchwerk *n*; 2. Kesselstein *m*
f. fabric pelzartiges Gewebe *n*, Pelzstoff *m*
f.-farm Pelzfarm *f*
f. imitation Pelzimitation *f*, Kunstpelz *m*
f. skin Pelzfell *n*
furaldehyde Furfurol *n*
furan resin Furanharz *n*
furcation Gabelung *f*
furfur Kopfschuppen *fpl*
furfural Furfural *n*, Furfurol *n*
furfurane Furan *n*
furl the sails / to Segel bergen
furlong englische Achtelmeile *f*
furnace 1. Ofen *m*, Industrieofen *m*; 2. Feuerraum *m*, Feuerung *f*
f. addition *(Met)* Ofenzuschlag *m*; Schmelzzuschlag *m*
f. arch Feuer[ungs]gewölbe *n*
f. atmosphere Ofenatmosphäre *f*, Schutzatmosphäre *f*
f. attendant Ofenwärter *m*
f. bear Ofenbär *m*, Ofensau *f*, Bodensau *f*
f. blast Hochofenwind *m*
f. bleeder Gichtgasfackel *f*
f. body Ofengestell *n*
f. bottom Ofensohle *f*
f. brazing Ofenlöten *n*
f. brick lining Ofenausmauerung *f*
f. bridge Feuerungswehr *n*
f. burdening Möllerung *f (Hochofen)*
f. campaign *(Met)* Ofenreise *f*, Ofengang *m*
f. capacity Ofenleistung[svermögen *n*] *f*
f. casting Herdguß *m*
f. charge Ofenbeschickung *f*, Charge *f*
f. charging machine Chargierkran *m*, Chargiermaschine *f*
f. cinder Ofenschlacke *f*
f. coke Hochofenkoks *m*
f. construction Ofenbau *m*
f. control Ofenregelung *f*, Ofenüberwachung *f*
f. control station Ofenwarte *f*
f. cooling Abkühlung *f* im Ofen
f. crown Ofengewölbe *n*
f. designer Ofenbauer *m*
f. flue Ofen[abzugs]kanal *m*
f.-flue superheater Flammrohrüberhitzer *m*
f. for bright tempering Blankglühofen *m*
f. for coal Kohlefeuerung *f*
f. for tan bark and saw dust Lohe- und Sägespänefeuerung *f*
f. forge Schmiedeofen *m*
f. front Ofengeschränk *n*
f. front with double door Doppelgeschränk *n*
f. front with single door einfaches Geschränk *n*
f. gas Rauchgas *n*
f. gas circulation drier Umlauftrockner *m*
f. grate Ofenrost *m*
f. in front of the boiler Vorfeuerung *f*
f. lid Ofendeckel *m*
f. life Standzeit *f* eines Ofens
f. lines Ofenprofil *n*
f. lining 1. *(Met)* Ofenfutter *n*; 2. *(Wärm)* Feuerungsauskleidung *f*
f. operation Ofenbetrieb *m*, Ofenführung *f*
f. operator Ofenwärter *m*
f. pig Ofenbär *m*, Ofensau *f*, Bodensau *f*
f. pusher *(Met)* Blockdrücker *m*
f. refractories Material[ien *pl*] *n* für Ofenausmauerung
f. returns Ofenrücklauf *m*
f. roof 1. *(Met)* Ofengewölbe *n*; 2. *(Wärm)* Feuerungsdecke *f*, Feuerraumdecke *f*
f. run Ofenreise *f*
f. scale Ofenzunder *m*
f. seat Ofensohle *f*, Ofenboden *m*
f. shaft Ofenschacht *m*
f. soldering Ofenlöten *n*
f. sow Ofensau *f*, Bodensau *f*, Ofenbär *m*
f. temperature Ofentemperatur *f*
f. top Gicht *f (Hochofen)*
f.-top bell Gichtglocke *f (Hochofen)*

f. transformer Ofentransformator *m*
f. trouble Ofenstörung *f*
f. under the boiler Unterfeuerung *f*
f. with adjustable feeding device Regulierschüttfeuerung *f*
f. with auxiliary fire bridge Wehrfeuerung *f*
f. with basket grate Korbrostfeuerung *f*
f. with combustion chamber Kammerfeuerung *f*
f. with cylinder grate Walzenrostfeuerung *f*
f. with direct heating surface Feuerung *f* mit unmittelbarer Heizfläche
f. with external inclined grate Schrägrostvorfeuerung *f*
f. with forced draught Preßluftfeuerung *f*; Druckluftfeuerung *f*
f. with grate-bars cooled by spraying water on them Rieselrostfeuerung *f*
f. with horizontal grate Planrostfeuerung *f*
f. with horizontal grate underneath the boiler Planrostunterfeuerung *f*
f. with inclined grate Schrägrostfeuerung *f*
f. with inclined grate underneath the boiler Schrägrostunterfeuerung *f*
f. with inclined grate with back-draught arranged in the interior of the boiler Schrägrostinnenfeuerung *f* mit Rückbrennung (Gegenfeuer), Tenbrink-Feuerung *f*
f. with internal horizontal grate Planrostinnenfeuerung *f*
f. with internal inclined grate Schrägrostinnenfeuerung *f*
f. with knee-shaped grate bars Knierostfeuerung *f*
f. with movable grate-bars Schüttelrostfeuerung *f*
f. with multi-stage grate Staffelrostfeuerung *f*, Etagenrostfeuerung *f*
f. with over-grate blower Feuerung *f* mit Oberluftzuführung
f. with ridge grate Muldenfeuerung *f*; Dachrostfeuerung *f*
f. with steam jets above grate Dampfschleierfeuerung *f*
f. with stepped grate Treppenrostfeuerung *f*
f. with stepped grate arranged underneath the boiler Treppenrostunterfeuerung *f*
f. with travelling grate Wander[rost]feuerung *f*
f. with trough grate Muldenrostfeuerung *f*
f. zone Ofenzone *f*
furnish / to ausrüsten, ausstatten; möblieren; liefern
furnish Eintrag *m*, Holländerfüllung *f*, Mahlgut *n (Papierherstellung)*
furnishings Ausstattungsgegenstände *mpl*, Einrichtungsgegenstände *mpl*
furniture 1. Möbel *n[pl]*; 2. Beschläge *mpl*; Zubehör *n*; 3. *(Typ)* Blindmaterial *n*, Stege *mpl*; 4. *(Schiff)* Takelung *f* und Ausrüstung *f*
f. factory Möbeltischlerei *f*
f. grubs Holzwürmer *mpl*
f. hides Möbelvachetten *pl*
furoate Ester *m* (Salz *n*) der Furankarbonsäure
furoic acid Brenzschleimsäure *f*, Furankarbonsäure *f*
furol Furfural *n*, Furfurol *n*
furonic acid Furylessigsäure *f*
furred 1. verbrettert; 2. mit Kalkansatz
furrier Kürschner *m*
furring Verbretterung *f*
furrow 1. Ackerfurche *f*, Saatfurche *f*; 2. *(Bw)* Hohlkehle *f*; 3. *(Brg)* Rinne *f*
f.-width Furchenbreite *f*
furrowing *(Lw)* Furchenziehen *n*
further development Weiterentwicklung *f*
f. processing Weiterverarbeitung *f*
f. sinking *(Brg)* Weiterteufen *n*
f. treatment Weiterverarbeitung *f*
furtherance Erschwerniszulage *f*
fury-fighter Jagdflugzeug *n*
fusain Faserkohle *f*, fasrige Steinkohle *f*, Fusit *m*
fusarc electrode Netzmantelelektrode *f*
f. process *s.* f. welding process
f. welding process Netzmantelelektrodenschweißen *n*
fuse / to 1. durchbrennen, schmelzen *(Sicherung)*; 2. schmelzschweißen; zusammenschweißen
fuse 1. Schmelzeinsatz *m*, Schmelzsicherung *f*, Sicherung *f*, Stromsicherung *f*; 2. Zündschnur *f*; Zünder *m*
f. alarm device Sicherungsüberwachungseinrichtung *f*
f. alloy niedrigschmelzende Legierung *f*
f.-base Fassung *f* (Sockel *m*, Unterteil *n*) der Sicherung
f. board Sicherungstafel *f*
f. box Sicherungskasten *m*
f. cabinet *s.* f. box
f. carrier Sicherungsgriff *m (Rohrpatronensicherung)*; Schraubkopf *m (Stöpselsicherung)*
f. cord Zündschnur *f*
f. cut-out Ausschalter *m* mit Sicherung; Schmelzsicherung *f*
f. disconnecting switch Trennsicherung *f*, Sicherungstrenner *m*, Trenner *m* mit Sicherung
f. disk Trennsäge *f*; zahnloses Kreissägeblatt *n*
f. element Schmelzleiter *m*
f. filler Löschmittel *n (Sicherung)*
f. holder Sicherungshalter *m*; Schraubkappe *f*; Stöpselkopf *m*
f. lighter *(Brg)* Zündfackel *f*, Zündstäbchen *n*, Zündstift *m*
f. link Schmelzeinsatz *m (Sicherung)*; Schmelzstreifen *m*

f. muffler *(El)* Sicherungsauspuff *m*, Sicherungsblasrohr *n*
f. plug Sicherungsstöpsel *m*, Stöpselkopf *m*
f. protection *(El)* Absicherung *f*
f. screw plug Schraubeinsatzsicherungspatrone *f*
f. setter *(Mil)* Zünderstellmaschine *f*
f.-setting *(Mil)* Zünderstellung *f*
f.-setting machine Zünderstellmaschine *f*
f. strip Sicherungsstreifen *m*, Schmelzstreifen
f. switch Sicherungsschalter *m*
f. tongs Sicherungszange *f*
f. tube Löschrohr *n* [der Sicherung]
f. with enclosed fuse element geschlossene Sicherung *f*
fused 1. [ab]gesichert; 2. schmelzflüssig *(Metall)*
f. corundum Elektrokorund *m*
f. electrolyte Schmelze *f*
f. electrolyte cell Hochtemperaturelement *n*
f. hearth bottom Sinterherd *m (Martinofen)*
f. junction Rekristallisationsschicht *f (Halbleiter)*
f. nucleus *(Ch)* kondensiertes Ringsystem *n*
f. quartz geschmolzener Quarz *m*, Quarzglas *n*
f. silica Quarzgut *n*
f. switch [Last-]Schalter *m* mit Sicherung kombiniert
fusehead Zündsatz *m*
fusel oil Fuselöl *n*
fuselage *(Flg)* Rumpf *m*
f. drag Rumpfwiderstand *m*
f. of steel-tubing framework Stahlrohrrumpf *m*
f. strut Rumpfstrebe *f*
fusibility Schmelzbarkeit *f*
fusible schmelzbar, schmelzend
f. alloy leicht schmelzbare Legierung *f*, Schmelzlegierung *f*
f. arc-welding electrode Abschmelzelektrode
f. cone *(Wärm)* Brennkegel *m*, Schmelzkegel *m*, Segerkegel *m*
f. cut-out Schmelzsicherung *f*
f. element Schmelzleiter *m*
f. glass Emailglas *n*
f. metal Schmelzmetall *n*, Schnellot *n*
f. plug Schmelzstift *m*, Schmelzstöpsel *m (Temperaturbestimmung)*
fusing 1. Schmelzen *n*; Durchbrennen *n*; 2. Schmelzschweißen *n*
f. additions Schmelzzusätze *mpl*, Schmelzzuschläge *mpl*
f. burner Schneidbrenner *m*
f. conductor *(El)* Schmelzleiter *m*
f. current Abschmelzstrom *m*, Ansprechstrom *m (Sicherung)*
f. disk Reibsäge *f*
f. of contacts Festbrennen *n* von Kontakten, Kontaktverschweißen *n*
f. point Schmelzpunkt *m*
f. temperature Schmelztemperatur *f*, Erweichungstemperatur *f*
f. time Abschmelzdauer *f*, Schmelzdauer *f*; Schmelzzeit *f*
fusion 1. Schmelzung *f*; Verschmelzung *f*; Schmelzvorgang *m*; 2. *(Kern)* Fusion *f*, Kernverschmelzung *f*; 3. homogene Vermischung *f*, Homogenisierung *f (Plastverarbeitung)*
f. bomb Thermonuklearbombe *f*, Kernfusionsbombe *f*
f.-cast blocks aus der Schmelze gegossene Steine *mpl*
f. cone Segerkegel *m*
f. curve Schmelzkurve *f*
f. frequency *(Opt)* Verschmelzungsfrequenz *f*
f. heat Schmelzwärme *f*
f. laminating method Schweißlaminierverfahren *n (Plastverarbeitung)*
f. mixture Aufschlußmischung *f*
f. point Schmelzpunkt *m*; Erweichungspunkt *m*
f. process 1. Aufschlußverfahren *n*, Schmelzverfahren *n*; 2. Kernverschmelzung *f*
f. thermit process Schmelzschweißverfahren *n*
f.-type weapon thermonukleare Waffe *f*
f. weld 1. Schmelzschweißung *f*; 2. Schmelzschweißnaht *f*
f. weldable schmelzschweißbar
f. welding Schmelzschweißen *n*
f. with alkali Alkaliaufschluß *m*, Alkalischmelze *f*
fustic Fustikholz *n*, Fisetholz *n*, Visetholz *n*
fusty schimmelig
futtock *(Schiff)* Auflanger *m (zweites Spant neben dem eigentlichen Spant)*
f. frame *(Schiff)* Holzspant *n*
fuze *s.* fuse
fuzziness Unschärfe *f*, Verschwommenheit *f*
fuzzy 1. verschwommen; verwaschen, unscharf *(Bild)*; 2. *(Text)* unerwünscht haarig *(Gewebeoberseite)*

G

g-coefficient Spannungskoeffizient *m (Piezoelektrizität)*
g-factor *(Ph)* Landéscher Aufspaltungsfaktor *m (Atomspektren)*
gab Haken *m*, Gabel *f*
gabbard scaffold Baugerüst *n (aus Kanthölzern)*
gabbro *(Geol)* Gabbro *m (grobkörniges Tiefengestein)*
gabers scaffold *s.* gabbard scaffold
gable Giebel *m*

g. **bottom** *(Masch)* gewölbter Zahngrund *m*
g. **end** Giebelwand *f*, Brandgiebel *m*
g.**-ended roof** Satteldach *n*, Giebeldach *n*
g. **roof** Giebeldach *n*, Satteldach *n*
g.**-type roof** *s.* g. roof
gabled dormer window Dacherker *m*
g. **roof** *s.* gable roof
gablet *(Bw)* Zwerggiebel *m*
gad *(Brg)* Fimmel *m*, Bergeisen *n*
gadget Vorrichtung *f*, Gerät *n*
gadolinia Gadoliniumoxid *n*
gadolinite Gadolinerde *f*, Ytterbit *m*
gadolinium Gadolinium *n*
g. **oxide** Gadolinerde *f*, Gadoliniumoxid *n*
gaff *(Schiff)* Gaffel *f*, Fischhaken *m*
g. **sail** Gaffelsegel *n*
gaffer Chefelektriker *m*, Beleuchter *m*
gag Knebel *m*
g. **press** Richtpresse *f*, Stempelrichtpresse *f*
gage *(Am) s.* gauge
gagger Sandhaken *m (Formerei)*
gahnite *(Min)* Gahnit *m*
gain / to 1. gewinnen, aufnehmen; 2. Nuten schneiden *(Holz)*
g. **energy** Energie aufnehmen
g. **in time** vorgehen *(Uhr)*
gain *(El, Fmt, Fs)* Entdämpfung *f*; Gewinn *m* *(Antenne)*; Leistungsgewinn *m*; Verstärkung[sfaktor *m*] *f*, Verstärkungsgrad *m*, Verstärkungsmaß *n (Verstärker)*
g. **adjusting amplifier** Verstärker *m* mit automatischer Lautstärkebegrenzung
g.**-band merit** Bandbreite *f* bei Verstärkung
g. **booster** Übertragersatz *m*; Zündübertrager *m*
g. **constant** Verstärkungsfaktor *m*, Amplitudenfaktor *m*
g. **control** Verstärkungsregelung *f*
g. **controller** Verstärkungsregler *m*
g. **cross-over frequency** *(Fmt)* Amplitudenschnittfrequenz *f*; Frequenz *f*, bei der die Verstärkung den Wert 1 einnimmt
g. **in sensitivity** Empfindlichkeitsgewinn *m*
g. **in time** Vorlauf *m (Uhr)*
g. **margin** Verstärkungsrand *m*, Verstärkungsvorrat *m*; Amplitudenrand *m*, Amplitudenvorrat *m*; Verstärkungssicherheitsabstand *m* gegen Selbsterregung
g. **of amplification** Verstärkungsgrad *m*
g. **of antenna** Antennengewinn *m*
g. **of heat** Wärmeaufnahme *f*
g. **of the stage** Stufenverstärkung *f*
g. **of work** Arbeitsgewinn *m (Dampfdiagramm)*
g. **per stage** Stufengewinn *m*; Stufenverstärkung *f*
g.**-phase plot** Amplituden-Phasen-Diagramm *n*
g. **sensitivity control** selektive Dämpfung *f*
gainer Nutmaschine *f (Holzbearbeitung)*
gaining head Nutenschneidkopf *m (Holzbearbeitung)*
gal Gal *n (1 cm/s^2, Einheit der Beschleunigung, bes. in der Geophysik)*
galactic *(Astr)* galaktisch
g. **nebula** *(Astr)* Milchstraßennebel *m*
g. **noise** *(Fmt)* galaktisches Rauschen *n* *(extraterrestrische Strahlung)*
g. **system** *(Astr)* Milchstraßensystem *n*
galactometer Galaktometer *n*, Milchwaage *f*
galactonic acid Galaktonsäure *f*
galacturonic acid Galakturonsäure *f*
galaxy Milchstraße[nsystem *n*] *f*
gale Sturm *m (von Windstärke 8)*
g. **warning service** *(Schiff)* Sturmwarnungsdienst *m*
galena Bleiglanz *m*, Galenit *m*, Grauerz *n*
g. **detector** *(El)* Bleiglanzdetektor *m*
galenical preparation galenisches Präparat *n*
galenoid bleiglanzähnlich
Galilean principle of relativity Galileisches Relativitätsprinzip *n*
Galilean telescope Galileisches Fernrohr *n*
Galilean transformation Galilei-Transformation *f*
gall / to 1. scheuern; reiben; wundreiben *(Pferd)*; 2. sich festklemmen; sich festfressen
gall-fly *(Lw)* Gallmücke *f*, Gallwespe *f*
g. **mite** Gallmilbe *f*
g. **wasp** Gallwespe *f*
gallant rail *(Schiff)* Obergeländer *n (auf dem Schanzkleid)*
gallate Gallat *n*, Gallussäureester *m*
gallery 1. Empore *f*, Galerie *f*, Rang *m*; 2. *(Brg)* Stollen *m*, Tunnel *m*; 3. *(Brg)* Strecke *f*; 4. *(El)* [begehbarer] Kabeltunnel *m*
g. **block** Außenganghaus *n*, Laubenganghaus *n*
g. **burner** Galeriebrenner *m (Glastechnik)*
g. **driving** *(Bw)* Stollenbau *m*
g. **end** *(Brg)* Ortsstoß *m*, Streckenstoß *m*
g. **face** *(Brg)* Streckenstoß *m*
g. **roof** *(Brg)* Streckenfirste *f*
gallet Steinsplitter *m*
galley 1. *(Schiff)* Küche *f*, Kombüse *f*; Galeere *f*; 2. *(Flg)* Bordküche *f*; 3. *(Typ)* Setzschiff *n*
g. **furnace** *(Ch)* Galeerenofen *m*
g. **proof** *(Typ)* Fahne[nabzug *m*] *f*
g. **range** *(Schiff)* Kombüsenherd *m*
g. **stroke** Galeerenschlag *m (ein Ruderstil)*
gallic 1. galliumhaltig; Gallium(III)-; 2. Gallus-
g. **acid** Gallussäure *f*
g. **compound** Verbindung *f* des dreiwertigen Galliums
gallinaceous birds Hühnervögel *mpl*

galling 1. Scheuern *n*; Reiben *n*; 2. Festfressen *n*; Verklemmen *n*; 3. Abnutzung *f* durch Reiben
g. marks Reibspuren *fpl (Beschädigung von Metalloberflächen)*
gallium Gallium *n*
g. arsenide tunnel diode Galliumarsenidtunneldiode *f (über 10000 MHz)*
gallon Gallone *f (1. englische Volumeneinheit für Flüssigkeiten und trockene Substanzen: 1 gal. = 4,546 dm^3; 2. amerikanische Volumeneinheit für Flüssigkeiten: 1 gal. = 3,785 dm^3)*
gallonic acid Galloninsäure *f*
galloping *(Am)* Stottern *n (eines Motors infolge zu fetten Gemisches)*
gallous compound Verbindung *f* des zweiwertigen Galliums
gallows Galgen *m (Baumgalgen)*
Galtonian board Galtonsches Brett *n*, Zufallsbrett *n*
Galtonian curve Galtonsche Kurve *f (Statistik)*
galvanic galvanisch
g. battery galvanische Batterie *f*, Primärbatterie *f*
g. cell galvanisches Element *n*, galvanische Kette *f*
g. corrosion elektrochemische Korrosion *f*
g. current galvanischer Strom *m*
g. electricity galvanische Elektrizität *f*
g. metal colouring galvanische Metallanfärbung *f*
g. pile Voltasche Säule *f*
galvanism Galvanismus *m (Elektromedizin)*
galvanize / to verzinken; feuerverzinken *(Eisen, Kupfer)*
galvanized by spraying [method] spritzverzinkt
g. sheet-iron verzinktes Eisenblech *n*
g. wire verzinkter Draht *m*
galvanizing Verzinkung *f*; Feuerverzinkung *f*
g. bath Verzinkungsbad *n*
g. by dipping Verzinken *n* im Tauchverfahren
g. plant Verzinkerei *f*
galvanocautery *(El)* Galvanokaustik *f*
galvanometer Galvanometer *n*, Stromanzeiger *m*; Strommesser *m*
g. arm Meßdiagonale *f (Meßbrücke)*
g. mirror Galvanometerspiegel *m*
g. oscillograph Schleifenoszillograf *m*
g. spot Lichtanzeige *f* eines Spiegelgalvanometers, Stellung *f* des Lichtzeigers auf der Skala
g. with moving magnet Nadelgalvanometer *n*
galvanoplastics Galvanostegie *f*, Galvanoplastik *f*
galvanoscope Galvanoskop *n*
gambrel Aufhängevorrichtung *f*
game 1. Wild *n*; 2. *(Dat)* Spiel *n (mit vorgeschriebenen Folgezuständen)*
g. bird Jagdvogel *m*
g. keeper Wildhüter *m*, Heger *m*
g. law Jagdgesetz *n*
g.-playing machine Maschine *f*, die als Spielpartner automatisch eine Partie *(z. B. Schach, Dame)* gegen einen anderen Partner spielt
g. preserve Wildbahn *f*
gamma *(Fs)* Gamma *n*
g. acid γ-Säure *f*, 2-Amino-8-naphthol-6-sulfonsäure *f (Farbenchemie)*
g.-active gammaaktiv
g. activity Gamma-Aktivität *f*
g. cross section Wirkungsquerschnitt *m* für Gamma-Strahlen
g.-decay Gamma-Zerfall *m*
g.-discriminating counter gammaempfindlicher Zähler *m*
g. flux Gamma-Fluß *m*
g. iron Gamma-Eisen *n*, γ-Eisen *n*
g. lifetime Gamma-Lebensdauer *f*
g. meter Gamma-Strahlungsmesser *m*
g. phase Gamma-Phase *f*
g. quantum Gamma-Quant *n*
g. radiation Gamma-Strahlung *f*, Gamma-Emission *f*
g.-radiation meter Gamma-Strahlungsmesser *m*, Gammameter *n*
g. radiator Gamma-Strahler *m*
g.-radioactive gammaaktiv
g. radioactivity Gamma-Aktivität *f*
g.-radiography Gamma-Radiografie *f*, Gammagrafie *f*
g. ray Gamma-Strahl *m*
g.-ray cascade Gamma-Kaskade *f*
g.-ray counter Gamma-Zähler *m*
g.-ray emission Gamma-Strahlung *f*, Gamma-Emission *f*
g.-ray material testing Gamma-Defektoskopie *f*, [zerstörungsfreie] Werkstoffprüfung *f* mittels Gamma-Strahlen
g.-ray materiology *s.* g.-ray material testing
g.-ray photon *s.* g.-ray quantum
g.-ray quantum Gamma-Quant *n*
g.-ray shield Gamma-Strahlenschutz *m*
g.-ray source Gamma-Strahlenquelle *f*, Gamma-Quelle *f*, Gamma-Präparat *n*
g.-ray thickness gauging Dickenmessung *f* mit Gamma-Strahlen
g.-ray transition Gamma-Übergang *m*
g.-ray yield Gamma-Strahlenausbeute *f*
g. sensitivity Gamma-Empfindlichkeit *f*
g. solid solution Gamma-Mischkristall *m*
g. space Gibbsscher Phasenraum *m*
g. spectroscopy Gamma-Spektroskopie *f*
g. unit *(Fs)* Gamma-Korrekturgerät *n*

gammagraphy *s.* gamma-radiography
gamut *(El)* Stimmumfang *m*; Tonbereich *m*
gang / to zu einem Satz zusammenstellen, zusammensetzen *(Fräser)*
g. up in einem Satz aufspannen
gang 1. Arbeitsgruppe *f*, Brigade *f*, Bautrupp *m*, Kolonne *f*, Mannschaft *f*; Rotte *f*; 2. Gruppe *f*, Satz *m*, Serie *f (Werkzeuge)*
g.-board *(Schiff)* Laufplanke *f*, Laufbrett *n*, Steg *m*
g. boss *(Am) s.* g. foreman
g. capacitor Mehrfach[dreh]kondensator *m*, Mehrgangkondensator *m*
g. car *(Eis)* Rottenwagen *m*
g. control *(Fmt)* Gleichlaufsteuerung *f*
g. cutters Gruppenfräser *mpl*
g. drill Reihenbohrmaschine *f*
g. drilling Reihenbohren *n*
g. foreman Bautruppführer *m*; Brigadier *m*; Kolonnenführer *m*; Rottenführer *m*
g. mandrel Drehspanndorn *m*
g. milling cutter *s.* g. cutters
g. punch *(Dat)* Blocklocher *m*, Schnellocher *m*, Schnellstanzer *m (Lochkartenmaschinen)*
g. puncher *(Dat)* Kartendoppler *m*
g. saw Bundgatter *n*; Vollsäge *f*; Gattersäge *f*; Spaltgatter *n*
g.-saw blade Vollgattersägeblatt *n*
g.-spindle drilling machine Reihenbohrmaschine *f*
g. stacker Gruppeneinschubvorrichtung *f (Kühlofen für Glas)*
g. summary punch *(Dat)* Summenlocher *m*, Summenstanzer *m (Lochkartenmaschinen)*
g. switch *(El)* Gruppenschalter *m*, Paketschalter *m*
g.-tool Meißelsatz *m*; Werkzeugsatz *m*
g.-tool operation Mehrmeißelarbeit *f*
g.-tool[ing] setup Mehrmeißelaufspannung *f*
g.-type multiple-spindle drilling machine Mehrspindelbohrmaschine *f* mit Anordnung der Spindeln in Reihe
g.-type tool holder Mehrmeißelhalter *m*
ganged capacitor Mehrfach[dreh]kondensator *m*
g. circuits gleichlaufende Mehrfachabstimmkreise *mpl*
g. tuning Einknopfabstimmung *f*
ganging 1. *(Masch)* Satzaufspannung *f*; 2. *(El, Fmt)* Gleichlauf *m*, Abgleich *m*, Superabgleich *m*
gangue Gangart *f*, Ganggestein *n (Erzbergbau)*
g. material Gangmasse *f*
g. rock Ganggestein *n*
gangway 1. Gang *m*, Laufgang *m*; Laufbrücke *f*, Laufsteg *m*; 2. *(Schiff)* Laufplanken *fpl*, Laufbrücke *f*; Auffahrtsklappe *f*; Fallreep *n*; 3. *(Brg)* Pfeilerstrecke *f (Pfeilerbau)*; Feldstrecke *f*, Fahrstrecke *f*; 5. Wagenübergang *m (in D-Zügen)*
g. bellows Faltenbalg *m (Wagenübergang)*
g. port *(Schiff)* Fallreepspforte *f*
ganister *(Geol)* Ganister *m (englischer Quarzit)*
g. brick *(Bw)* Dinasstein *m*
ganja Indischer Hanf *m (Haschisch, Marihuana)*
ganjah *s.* ganja
gannister *s.* ganister
ganomalite *(Min)* Ganomalith *m*
ganophyllite *(Min)* Ganophillit *m*
gantry Gerüst *n*; Krangerüst *n*
g. lathe Prismendrehmaschine *f*
g. post Pendelstütze *f*, Pendelsäule *f*, Pendelpfeiler *m*
g. rail Kranbahnschiene *f*
gap 1. Öffnung *f*, Spalt *m*, Lücke *f*, Zwischenraum *m*, Spiel[raum *m*] *n*, Fuge *f*; 2. *(Geol)* Kluft *f*, Spalte *f*; 3. *(Masch)* Kröpfung *f*; Ausladung *f*; Rachen *m (Lehre)*; Maulhöhe *f (Nietmaschine)*; 4. *(El)* Lücke *f*, Luftspalt *m (z. B. bei Magnetköpfen)*; Loch *n*, Leerstelle *f (Halbleiter)*; Minimum *n* im Polardiagramm *(Radar)*; 5. *(Ch)* Siedelücke *f*
g. admittance Spaltleitwert *m (einer Entladungsstrecke)*
g. arrester Überspannungsableiter *m*
g. bed gekröpftes Bett *n*
g. clearance Luftspaltbreite *f*, Spaltbreite *f*
g. depth Spalttiefe *f (Magnetkopf)*
g. distance of electrodes Elektrodenabstand *m*
g. field Spaltfeld *n (Magnetkopf)*
g. filler Hilfsantenne *f (Radar)*
g.-filling adhesive Kitt *m* zum Ausfüllen größerer Klebefugen, Klebekitt *m*
g. frame eccentric press Einständerexzenterpresse *f*
g. gauge Rachenlehre *f*
g. lathe Drehmaschine *f* mit herausnehmbarer Brücke
g. limit gauge Passungslehre *f*, Grenzrachenlehre *f*
g. of the joint *(Holz)* Zwischenraum *m (Stoßfuge)*
g. reluctance [magnetischer] Luftspaltwiderstand *m*
g. section Schutzstrecke *f (Fahrdraht)*
g. testing *(El)* Überschlagsprobe *f*
g.-type filter Spaltfilter *n*
g. width Spaltbreite *f (Magnetkopf)*
garage / to in die Garage stellen
garage Garage *f*
garbage Küchenabfälle *mpl*, Müll *m*, Kehricht *m*
g. can Abfallbehälter *m*
g. car Müll[abfuhr]wagen *m*
g. chute Müllschlucker *m*

g. disintegrator Müllzerkleinerer *m*
g. disposal Müllbeseitigung *f*
g. dressing Müllverwertung *f*
g. incineration Müllverbrennung *f*
g. lighter *(Schiff)* Abfallprahm *m*
g. pit Müllabladeplatz *m*
g. scow *(Schiff)* Abfallschute *f*, Abfallprahm *m*
garble / to verstümmeln *(Telegramm)*
garbled message verstümmelte Funkmeldung *f*
g. telephony Geheimtelefonie *f*
garboard plank *(Schiff)* Kielplanke *f*
g. plate Kielplatte *f*
g. strake Kielgang *m*
garden fence Gartenzaun *m*
g. furniture Gartenmöbel *npl*
g. hose Gartenschlauch *m*, Wasserschlauch *m*
g. house Wohnlaube *f*
g. mould Gartenerde *f*
g. pot Gießkanne *f*
g. roller Gartenwalze *f*
g. spray Gartenspritze *f*
g. syringe Gartenspritze *f*
gardener Gärtner *m*
gardener's shears Gartenschere *f*
gardening Gartenarbeit *f*; Gartenbau *m*
g. knife Gartenmesser *n*
gardenroller Gartenwalze *f*
gargoyle *(Bw)* Wasserspeier *m*
garland 1. *(Schiff)* Taukragen *m*; 2. *(Brg)* Wasserring *m*, Wasserrinne *f (Schachtausbau)*
garment industry Bekleidungsindustrie *f*; Konfektion *f*
garmenture Bekleidung *f (allgemeine Bezeichnung für alle Kleidung)*
garnet *(Min)* Granat *m*
g. rock Granatfels *m*
garnett clothing card *(Text)* Garnettkrempel *f*, Krempel *f* mit Sägezahndrahtbeschlag
g. machine *(Text)* Garnette *f*
garnetting *(Text)* Krempeln *n*
garnierite *(Min)* Garnierit *m*; Nickelsilikat *n*
garret / to verblenden; Steine in Fugen einsetzen
garret 1. Boden *m*, Dachboden *m*; Dachkammer *f*, Dachstube *f*; Speicher *m*
g. room Bodenkammer *f*
g. story Dachgeschoß *n*
g. window Bodenfenster *n*, Dachluke *f*
garter spring endlose Schraubenfeder *f (Radialdruck)*
garth begrünter Innenhof *m*
Garton lightning arrester Blitzableiter *m* nach Garton, Blitzableiter mit Spule
gas / to 1. gasen, Gas abgeben (entwickeln) *(z. B. Sammler)*; 2. *(Text)* gasieren, abflammen, sengen
g.-weld autogenschweißen
gas 1. Gas *n*; Grubengas *n*, Methan *n*; 2. *(Am)* s. petrol
g.-absorbing agent Gasabsorptionsmittel *n*
g. absorption Gasabsorption *f*
g. accumulation Gasansammlung *f*
g. amplification Gasverstärkung *f*
g. analysis Gasanalyse *f*
g. analyzer Gasprüfer *m*
g. and water supply Gas- und Wasserversorgung *f*
g. angle Strahlwinkel *m (Gasturbine)*
g. attack *(Mil)* Gasangriff *m*
g. bag Gassack *m*
g. balance Gaswaage *f*
g. bath *(Ch)* pneumatische Wanne *f*
g. black Gasruß *m*
g. blanketing *(Ch)* Abdeckung *f* mit einer Gasschicht, Überschichtung *f* mit Gas
g.-blast circuit-breaker Druckgasschalter *m*, Gasstromschalter *m*
g. bleb Gasblase *f*
g. bleeder *(Brg)* Fackelrohr *n*
g. blow torch Gasgebläselampe *f*, Lötbrenner *m*, Lötrohr *n*
g. blower Gasgebläse *n*
g. blowpipe Lötbrenner *m*
g. bottle Gasflasche *f*
g. bubble Gasbläschen *n*, Gaseinschluß *m (z. B. im Glas)*
g. buoy Leuchttonne *f*, Leuchtboje *f*, Gastonne *f*
g. burner Gasbrenner *m*
g. cap *(Brg)* Flammenaureole *f (Wetterlampe)*
g. carbon Retortenkohle *f*
g. carburizing *(Met)* Gasaufkohlung *f*; Gaseinsatzhärtung *f*, Gaszementieren *n*
g. case hardening Gaseinsatzhärtung *f*
g. cavity Gasblase *f*, Gaseinschluß *m (im Guß)*
g. cell 1. Gasküvette *f*; 2. [elektrisches] Gaselement *n*, [elektrische] Gaszelle *f*
g. chamber kiln Gaskammerofen *m*
g. chromatography Gaschromatografie *f*
g. circulator gasbeheizter Warmwasserspender *m*
g. coal Gaskohle *f (26 bis 32% flüchtige Bestandteile)*
g. cock Gashahn *m*
g. coke Gaskoks *m*
g. collector Gassammler *m*
g. concrete Gasbeton *m*
g. conditioning Gasreinigung *f*
g. conduit Gasleitung *f*
g. constant *(Ph)* [allgemeine, universelle] Gaskonstante *f (ideale Gase)*
g. containing gasführend *(Rohrleitungen)*
g.-contaminated surface gasbehaftete (mit Gas verunreinigte) Oberfläche *f*

g. converting Gaswandlung *f*, Gasumformung *f*
g. cooker Gaskocher *m*
g. cooking stove Gaskochherd *m*
g.-cooled gasgekühlt
g. cooling Gaskühlung *f*
g. counter *(Kern)* Gaszählrohr *n*, Gaszähler *m*
g.-cut mud mit Erdgas durchsetzter Bohrschlamm *m*
g. cutting Brennschneiden *n*; autogenes Trennen *n* (*s. a.* flame cutting)
g. cylinder Druckgasflasche *f*, Gasbombe *f*
g. density Gasdichte *f*
g. detector Gasspürgerät *n*
g. developing agent Treibmittel *n* für Schaumstoffe
g. discharge Gasentladung *f*
g. discharge chamber Gasentladungskammer *f*
g. discharge gap Gasentladungsstrecke *f*
g. discharge gauge Glimmentladungsmanometer *n*
g. discharge lamp Gasentladungslampe *f*, Leuchtstofflampe *f*
g. discharge protector Gasentladungsableiter *m*
g. discharge relay Gas[entladungs]relais *n*
g. discharge tube Glimmlampe *f*, Glimmröhre *f*
g. dispersoid Aerosol *n*
g.-driven engine blower Gasgebläse *n*
g. ducting Gasleitung *f*
g.-electric benzinelektrisch
g. emission *(Brg)* Ausgasung *f*
g. engine Gasmotor *m*
g. engineer Gastechniker *m*
g. envelope Gashülle *f*
g. equation Gasgleichung *f*
g. equilibrium Gasgleichgewicht *n*
g. eruption Gasausbruch *m* *(Erdgas)*
g. escape Gasausbruch *m*, Gasentweichung *f*
g. evolution Ausgasung *f*; *(Ch)* Gasentwicklung *f*
g.-exhaust pipe Abgasleitung *f*
g. exhausting during charging *(Brg)* Füllgasabsaugung *f*
g. exit angle Strahlaustrittswinkel *m (Gasturbine)*
g. expansion thermometer Gas[ausdehnungs]thermometer *n*
g. expansion turbine Gasspannungsturbine *f*
g. fading *(Text)* Entfärben (Verschießen) *n* in Abgasatmosphäre
g. field Naturgasfeld *n*, Gasfeld *n*
g.-filled gasgefüllt, mit Gasfüllung
g.-filled rectifier Gasentladungsventil *n*, Ionenventil *n*
g.-filled thermometer Gas[ausdehnungs]thermometer *n*
g.-filled triode gasgefüllte Triode *f*, Gas[entladungs]relais *n*, Gastriode *f*
g.-filled tungsten lamp gasgefüllte Wolframlampe *f*
g.-fired gasbeheizt, mit Gasheizung
g.-fired air furnace Gasflammofen *m*
g.-fired calcining kiln Gasröstofen *m*
g.-fired furnace Gasfeuerung *f*
g.-fired muffle furnace gasbeheizter Muffelofen *m*
g. firing Gasfeuerung *f*
g. fitter Rohrleger *m*, Klempner *m*
g. flow Gasstrom *m*, Gasströmung *f*
g.-flow counter Gasdurchflußzähler *m*
g. flue Gasabzug *m*, Abgaskanal *m*
g. focusing Ionenstrahlfokussierung *f* durch Gasfüllung
g. formation Gasbildung *f*, Gasentwicklung *f*
g.-freeing equipment *(Schiff)* Entgasungsanlage *f (Tanker)*
g. fuel Heizgas *n*
g.-fuelled gasbeheizt
g. furnace Gasofen *m*
g. fusion welding Gasschmelzschweißen *n*
g. gallery *(Brg)* Meistergang *m*, Begehkanal *m*
g. gap *s.* g. discharge gap
g. gauge Manometer *n*
g. generating bottle Gasentwickler *m*, Gasentwicklungsflasche *f*
g. generator Gaserzeuger *m*; Gasentwickler *m*
g. geyser Gasbadeofen *m*
g. grid Gasnetz *n*
g. gun oven *(Brg)* kopfbeheizter Ofen *m*
g. heating system Gasheizung *f*
g. helmet Gasmaske *f* mit Respirator
g. holder Gasbehälter *m*, Gassammler *m*
g. hole *(Met)* Gaseinschluß *m*, Gasblase *f*
g. illumination Gasbeleuchtung *f*
g. indicator Gasdruckmesser *m*
g. inflation Gasfüllung *f (Ballons usw.)*
g. inlet Gasanschluß *m*, Gaseinlaßleitung *f*
g. inlet angle Strahleintrittswinkel *m (Gasturbine)*
g. input well Gaseinpreßbohrung *f*, Gaseinpreßsonde *f*
g. ionization Gasionisation *f*
g. law *(Ph)* Gasgesetz *n*
g.-leak indicator Gasanzeiger *m*
g. leakage Gasaustritt *m* an undichter Stelle
g. lever *(Kfz)* Gashebel *m*
g. lift Druckgasförderverfahren *n (Erdgas)*
g.-light paper Chlorbromsilber-Fotopapier *n (Gaslichtpapier)*
g. lighter Gasanzünder *m*
g. lighting Gasbeleuchtung *f*
g. liquor Ammoniakwasser *n*

g.-loaded druckgasbelastet
g. main Gasleitung *f*, Vorlage *f*
g. make Gasausbeute *f*
g. making Gaserzeugung *f*
g. man *(Brg)* Wettersteiger *m*
g. mantle Glühkörper *m*
g. mask Gasmaske *f*
g. meter Gasmesser *m*, Gaszähler *m*, Gasuhr *f*
g. mixing chamber *(Wärm)* Mischraum *m*, Gasmischungskammer *f*
g. molecule Gasmolekül *n*
g. motor *s.* g. engine
g. occlusion Gaseinschluß *m (in Metall)*
g. oil Gasöl *n*, Dieselöl *n*, Treiböl *n*
g. operation Gasbetrieb *m*
g. outburst *(Brg)* [plötzlicher] Gasausbruch *m*
g. oven Gasofen *m*
g. passage Gasdurchgang *m*; Durchgasung *f* *(des Ofens)*
g. phototube gasgefüllte Fotozelle *f*
g. pilot burner Sparbrenner *m*, Sparflamme *f*
g. pipe Gasrohr *n*
g. pipe stock Gaskluppe *f*
g. pipe tap Gasgewindebohrer *m*
g. pipe thread Rohrgewinde *n*
g. pipe vice clamp Gasrohrklemmfutter *n*
g. piping Gasleitung *f*
g. plant for long-distance supply Gasfernversorgungsanlage *f*
g. pliers Gasrohrzange *f*
g. pocket *(Met)* Gaspore *f*, Gaslunker *m*
g. poisoning Gasvergiftung *f*
g.-powered gasbetrieben, gasangetrieben
g. pressure Gasdruck *m*
g.-pressure cable Druckgaskabel *n*
g.-pressure relay Buchholz-Relais *n*
g. processing plant Gasaufbereitungsanlage *f*; Gaserzeugungsanlage *f*, Gaswerk *n*
g. producer Gasgenerator *m*, Gaserzeuger *m*
g. producing well Gasfördersonde *f*
g.-proof gasdicht
g. protective apparatus *(Brg)* Gasschutzgerät *n*
g. puddling *(Met)* Gas[flammofen]frischen *n*
g. purification Gasreinigung *f*, Gaswaschen *n*; Gasaufbereitung *f*
g. purifier Gasreiniger *m*
g. purifying medium Gasreinigungsmasse *f*
g. quenching *(Met)* Abblasen *n* mit kalten Gasen
g. ratio Vakuumfaktor *m (Elektronenröhren)*
g. reforming plant Gasspaltanlage *f*
g. reheating Rauchgaszwischenüberhitzung *f*
g. relay Ionenrelais *n*, Gastriode *f*
g.-reverberatory furnace Gasflammofen *m*
g. ring 1. ringförmiger Gaskocher *m*; 2. Kolbenring *m*; Dichtungskragen *m*
g.-scintillation counter *(Kern)* Gasszintillationszähler *m*
g. scrubber Gaswäscher *m*, Skrubber *m*
g. scrubbing Gasberieselung *f*, Gaswaschung *f*, Gaswäsche *f*
g. seal Gasverschluß *m*
g. seepage Erdgasausbiß *m*
g. separator Gasabscheider *m*
g.-shielded welding Schutzgasschweißung *f*
g. singeing *(Text)* Gassengen *n*
g. singeing machine Gassenge *f*, Gassengmaschine *f*
g. slack Gasgruskohle *f*
g. sluice valve Gasschieber *m*
g. soldering iron Gaslötkolben *m*
g. stepping tube gasgefülltes Zählrohr *n* *(Elektronenstrom wird durch Raumladewirkung von Elektrode zu Elektrode weitergeschaltet)*
g. stove Gasofen *m (für Heizung)*
g. tank Gasbehälter *m*, Gasometer *m*
g. tension Gasspannung *f*
g. testing *(Brg)* Ableuchten *n*
g. thermometer Gas[ausdehnungs]thermometer *n*
g. thread Rohrgewinde *n*
g.-tight gasdicht
g. tongs Gasrohrzange *f*
g. torch Gasschweißbrenner *m*
g. torch welding Autogenschweißen *n*; Gasschmelzschweißen *n*
g. transmission Gas[weiter]leitung *f*; Gasförderung *f*
g. trap Gasschleuse *f*
g. tube gasgefüllte Röhre *f*
g. tube switch Thyratron *n*; Stromtor *n*
g. turbine Gasturbine *f*
g. turbine combustor Gasturbinenbrennkammer *f*
g.-turbo-alternator Gasturbinen[wechselstrom]generator *m*
g. vane *(Rak)* Strahlruder *n*
g.-vapour generator Gasdampferzeuger *m*
g. voltameter Knallgasvoltameter *n*, Gasvoltameter *n*
g. warfare chemischer Krieg *m*
g. washer Skrubber *m*
g. washing bottle *(Ch)* Gaswaschflasche *f*
g. welding Gasschweißen *n*; Gasschweißung *f*; Autogenschweißung *f*
g. welding apparatus Autogen[schweiß]gerät *n*
g. well Gasbohrung *f*, Gasquelle *f*
g. works Gaswerk *n*, Gasanstalt *f*
g. X-ray tube gasgefüllte (weiche) Röntgenröhre *f*
g. yield Gasausbeute *f*, Gasertrag *m*
gaseous gasartig; gasförmig
g. atmosphere Gasatmosphäre *f*

g. compound gasförmige Verbindung *f*
g. conduction lamp Glimmlampe *f*
g. diffusion Gasdiffusion *f*
g. discharge Gasentladung *f*
g. discharge counter tube Gasentladungszählrohr *n*
g. discharge lamp Gasentladungslampe *f*, Leuchtröhre *f*
g. fuel gasförmiger Brennstoff *m*
g. medium Gasatmosphäre *f*
g. mine Schlagwettergrube *f*
g. mixture Gasgemisch *n*
g. products of combustion Abgas *n*, Rauchgas *n*
g. products of incomplete combustion Schwelgase *npl*
g. state Gaszustand *m*, gasförmiger Zustand *m*
gash / to *(Masch)* Zahnlücken vorfräsen
gash 1. Lücke *f*; Nut *m*; Spalt *m*; Schlitz *m*; 2. *(Masch)* Zahnlücke *f*; Spanlücke *f*, Schnittnut *m*
gasholder Gasbehälter *m*
gashouse *(Am)* Gasanstalt *f*, Gaswerk *n*
gasifiable vergasbar
gasification Gaserzeugung *f*; Vergasung *f*
g. in place *(Brg)* Untertagevergasung *f*, UTV *f*
g. of coal Kohlevergasung *f*
g. speed Vergasungsgeschwindigkeit *f*
g. under pressure Druckvergasung *f*
gasifier *(Ch)* Vergaser *m*
gasiform gasförmig
gasify / to vergasen
gasket Dichtung *f*; Dichtungsring *m*; Manschette *f*; Packung *f*; Unterlagscheibe *f*
g. board Dichtungspappe *f*
g. coating Düsenlackierverfahren *n*
g. ring Dicht[ungs]ring *m*
g.-sealing washer Dichtung *f*
gasketed abgedichtet
gaskin Dichtungsstrick *m*, Hanfseil *n*
gasolene *(Am)* *s.* gasoline
gasoline *(Am)* Benzin *n*
gasometer Gasbehälter *m*, Gasometer *m*
gasser Gasfördersonde *f*
gassing 1. Gasen *n*, Gasentwicklung *f*, Blasenbildung *f*; 2. *(Text)* Absengen *n*, Sengen *n*
g. accumulator *(El)* kochender Akkumulator *m*
gassy gashaltig, vergast
g. mine schlagwettergefährdete Grube *f*, Schlagwettergrube *f*
g. tube *(El)* schlecht evakuierte Röhre *f*
gastaldite *(Min)* Gastaldit *m*
gate / to *(Fs)* austasten, Strahl sperren; einblenden
gate 1. *(Bw)* Tor *n*; Eingangstür *f*; 2. Gatter *n* *(Säge)*; 3. *(Plast)* Angußsteg *m*; Anschnitt *m*; 4. *(Met)* Eingießöffnung *f*; Gießtrichter *m*; Zulauf *m*; 5. Absperrventil *n*; Schieber *m* *(Hydraulik)*; 6. *(Dat)* Tor[schaltung *f*] *n*, Ventilschaltung *f*; 7. *(Fs)* Auftastimpulskreis *m*; Ausblendstufe *f*; 8. Projektionsfenster *n* *(Film)*; 9. Schleuse[ntor *n*] *f*; 10. *(Brg)* Strecke *f*; 11. *(Holz)* Schaumlatte *f*, Spalt *m* des Stoffauflaufes
g. amplifier *(Fs)* Auslöseimpulsverstärker *m*
g. angle *(El)* Sättigungswinkel *m*
g. chamber *(Bw)* Verschlußkammer *f*
g. change *(Kfz)* Kulissenschaltung *f*
g. circuit 1. *(Dat)* Torschaltung *f (UND-Tor, ODER-Tor)*, Durchlaßschaltung *f*; 2. *(Fs)* Auftastschaltung *f*
g. dock Hafendock *n*, Hafenschleuse *f*
g. end *(Brg)* Streckenende *n*, Ortsstoß *m*
g. end box Streckenschalter *m*
g. end road Flözstrecke *f*
g. filter *(El)* Sperrfilter *n*
g. generator *(Fs)* Auftastgenerator *m*
g.-house Pförtnerhaus *n*
g. logic *(Dat)* Torschaltungslogik *f*, Durchlaßlogik *f*, Gitterschaltungslogik *f*
g. of mould *(Met)* Formenanschnitt *m*, Eingußtrichter *m*
g. opening *(Met)* Trichterkopf *m*, Trichtermündung *f*
g. paddle mixer *(Ch)* Gitterrührer *m*
g. pulse *(Fs)* Auftastimpuls *m*
g. road *(Brg)* Abbaustrecke *f*
g. roll coater Eingußwalzenstreichmaschine *f* *(Papierherstellung)*
g. shears geschlossene Blechschere *f*
g. structure Verschlußbauwerk *n*
g. time *(Dat)* Öffnungszeit *f (Torschaltung)*
g. valve Absperrschieber *m*, Absperrventil *n*, Durchgangsschieber *m*, Durchlaßventil *n*, Schieber *m*; Stromtorröhre *f*
g. vessel Torschiff *n*, Schwimmtor *n (für Schleusen, Trockendocks usw.)*
g. voltage 1. Torsteuerspannung *f*, Steuerspannung *f* eines elektronischen Tores, Gattersteuerspannung *f*; 2. *(Fs)* Sperrspannung *f*, Verriegelungsspannung *f*
g.-width *(Fs)* Öffnungszeit *f*
gated beam tube Doppelstromtor-Strahlsteuerungsröhre *f*, Strahlsteuerungsröhre *f* mit Doppelstromtor
g. decade counter *(Dat)* Stromtordekade *f*
g. multivibrator *(El)* Univibrator *m*
gatekeeper Schrankenwärter *m*
gateroad *(Brg)* Abbaustrecke *f*
gateside pack *(Brg)* Streckendamm *m*, Bergemauer *f*
gateway 1. Tor[durchfahrt *f*] *n*, Torweg *m*; 2. *(Brg)* Abbaustrecke *f*

gather / to [auf]sammeln; sich ansammeln; sich zusammenballen *(Wolken)*; lesen, pflücken, ernten
gathering Rollen *n (Drehen einer Stange zwischen Gesenken)*
g. anode Auffanganode *f*, Fanganode *f*, Sammelanode *f*
g. electrode Auffangelektrode *f*, Fangelektrode *f*, Sammelelektrode *f*
g. ground Einzugsgebiet *n*
g. machine *(Typ)* Zusammentragmaschine *f*
g. of water Wasserfassung *f*
g. plough *(Brg)* Räumpflug *m*
gating 1. *(BMSR, El)* Ausblenden *n*, Ausblendung *f*, Austasten *n*, Austastung *f (Zeitsignale)*; Impulssperrung *f*, Strahlsperrung *f*; 2. *(Met)* Eingießöffnung *f*, Gießrinne *f*, Gießtrichter *m*; 3. Ausrichten *n* von Webstuhlteilen
g. circuit *s.* gate circuit
g. impulse *s.* gating pulse
g. multivibrator monostabiler Multivibrator *m*
g. pulse 1. Torimpuls *m (Torschaltung)*; 2. Auftastimpuls *m (Auftastimpulskreis)*
gauffer calender *(Text)* Gaufrierkalander *m*, Prägekalander *m*
gauge / to ablehren, abmessen, ausmessen; eichen, kalibrieren *(Instrumente, Skalen)*; beurteilen *(Güte)*; 2. dosieren, zumessen *(z. B. Betontechnologie)*
gauge 1. Lehre *f*, Lehrwerkzeug *n*; Normalmaß *n*; Eichmaß *n*; Bezugsmaß *n*; Meßelement *n*, Meßuhr *f*; Prüfkopf *m*; Schablone *f*; Meßstreifen *m*; 2. Drahtdurchmesser *m*, Drahtdicke *f*; Drahtlehre *f*; Blechlehre *f*; 3. Spurweite *f (Gleis)*; 4. *(Text)* Gauge *(Wirkwaren)*; Spindelteilung *f (Spinnmaschine)*; Warendichte *f (Gewebe)*; Feinheit *f* von Nadeln und Platinen *(Stricknadeln)*; Nadelanzahl *f (auf einer Länge von 1½ in. = 38,1 mm)*; 5. *(Ph)* Eichung *f*; 6. *(Schiff)* Tiefgang *m*, Pegel *m*
g. blade Meßlineal *n*
g. block 1. *(BMSR)* Endmaß *n*, Parallelendmaß *n*; 2. *(Text)* Vorrichtblock *m*
g.-block interferometer *(Opt)* Interferenzkomparator *m*
g. box Abmeßkasten *m (Betonherstellung)*
g. changing displaceable wheels *(Eis)* Lösräder *npl (Spurwechsel)*
g.-cock Wasserstandhahn *m*, Hahn *m* am Wasserstandanzeiger
g. cylinder Prüfzylinder *m*
g. designer (engineer) *(Masch)* Lehrenbauer *m*
g. factor Empfindlichkeitsfaktor *m*, *K*-Faktor *m (Dehnungsmeßstreifen)*
g. for metallic coatings Metallschichtdikkenmesser *m*
g. glass Wasserstandsanzeiger *m*; Standglas *n*
g. glass protector and reflector Reflexionswasserstandglas *n*
g. glass protector frame *(Wärm)* Wasserstandschutzrahmen *m*
g. in millimeter Millimeterlehre *f*
g. invariance Eichinvarianz *f*
g. length Meßlänge *f*, Meßstrecke *f*
g. maker Lehrenbauer *m*
g. mark Eichstrich *m*; Füllstrich *m*; Meßmarke *f*
g. of sheet Blechdicke *f*
g. of welding wire Schweißdrahtdicke *f*
g. of wire Drahtdicke *f*; Drahtlehre *f*
g.-plate Zieheisen *n*; Drahtzieheisen *n*
g. pressure Überdruck *m*, auf Atmosphäre bezogener Druck *m*
g. ring Paßring *m*
g.-rod *(Schiff)* Peilstock *m*
g.-rod of a pump Pumpenpeilstock *m*
g. transformation Eichtransformation *f*; Umeichung *f*
g. with two steps Grenzlehre *f*; Austauschlehre *f*
gauging Lehren *n*, Prüfen *n* der Einhaltung gegebener Maße; maßliches Prüfen *n*
g. circuit Meßschaltung *f*
g. cylinder Meßzylinder *m*
g. equipment Meßeinrichtung *f*
g. instrument Lehr[en]werkzeug *n*, Lehrgerät *n*, Meßgerät *n*
g. length Meßlänge *f*
g. method Lehrmeßverfahren *n*
g. of ships Schiffseichung *f*, Schiffsausmessung *f (meistens bei Binnenschiffen)*
g. pin Meßbolzen *m*, Meßzapfen *m*, Backe *f*, Tastbolzen *m*
g. position Meßstellung *f*
g. practice Meßtechnik *f*
g. radius Prüfradius *m*
g. roller Meßrolle *f*
g. section *s.* g. station
g. station Meßstelle *f*, Meßstation *f*
g. surface Meßfläche *f*
g. the cutter block Ablehren *n* des Messerkopfes
gauntlet *(Kern)* [arm]langer Gummihandschuh *m*, Stulpenhandschuh *m* aus Gummi
gauntree *s.* gantry
gauntry *s.* gantry
gauss Gauß *n (veraltete Maßeinheit der magnetischen Flußdichte oder der magnetischen Induktion)*
Gauss error distribution curve Gaußsche Fehlerverteilungskurve *f*
Gauss error integral Gaußsches Fehlerintegral *n*
Gaussian curve Gaußsche Fehlerkurve *f*

Gaussian eyepiece Gaußsches Okular *n*
Gaussian kernel Gaußscher Integralkern *m*
Gaussian well Gaußscher Potentialtopf *m*
gauze Gaze *f*
g. filter Siebfilter *n*; Gazefilter *n*
g. weave *(Text)* Dreherbindung *f*
g. winding loom Gazewebstuhl *m*
Gay-Lussac tower Gay-Lussac-Turm *m*
Gay-Lussac's law Gay-Lussacsches Gesetz *n*
gaylussite *(Min)* Gaylussit *m*
gazebo Aussichtspavillon *m*
gear / to einrücken; zum Eingriff bringen; mit Zahnrädern verbinden; eingreifen; ineinander greifen
g. into eingreifen *(Zahnräder usw.)*
g. out ausrücken *(Zahnräder)*
g. together ineinander eingreifen, in Eingriff sein, miteinander kämmen
gear 1. *(Masch)* Zahnrad *n*; 2. Gerät *n*; Ausrüstung *f*; 3. *(Kfz)* Getriebe *n*; Gang *m*; 3. *(Schiff)* Geschirr *n (Ladegeschirr)*
g. being machined Werkrad *n*
g. blank Zahnradrohling *m*; unbearbeiteter Radkörper *m*; Werkrad *n*
g. body Zahnradkörper *m*
g. box Getriebe[gehäuse] *n*; Getriebekasten *m*; Räderkasten *m (Drehmaschine)*; Zahnradkasten *m*; Schaltgetriebe *n*
g. box case cap Getriebegehäusedeckel *m*
g. box casing Getriebegehäuse *n*
g. box flange Getriebeflansch *m*
g. box shaft Getriebewelle *f*
g. brake Getriebebremse *f*
g. broaching machine Zahnradräummaschine *f*
g. case Getriebegehäuse *n*, Getriebekasten *m*
g. chain Zahnkette *f*
g. chamfering cutter Zahnkantenabrundfräser *m*
g. change *(Kfz)* Gangschaltung *f*
g.-change box Wechselgetriebe *n*
g.-change diagram Gangschaltschema *n*, Ganganordnungsschaubild *n*
g. changing Gangwechsel *m*
g. cutter Zahnradfräser *m*
g. cutting Zahnradschneiden *n*; Verzahnen *n*
g. cutting machine Verzahnungsmaschine *f*, Zahnradfräsmaschine *f*
g. cutting tool Verzahnwerkzeug *n*
g. differential Differentialgetriebe *n*
g. drive Zahnradgetriebe *n*, Zahnradantrieb *m*; Rädertrieb *m*
g.-driven face plate Zahnkranzplanscheibe *f*
g.-driven fan Ventilator *m* mit Zahnradantrieb *m*
g.-driven supercharger *(Kfz)* Getriebelader *m*
g. error Verzahnungsfehler *m*
g. fatigue-testing machine Zahnraddauerprüfmaschine *f*
g. finishing Feinbearbeitung *f* von Zahnflanken; Zahnradfeinbearbeitung *f*
g.-flushing unit *(Kfz)* Getriebespüler *m*, Getriebespülapparat *m*
g. forming tool *s.* g. cutting tool
g. gauge Zahnmeßschraublehre *f*
g. generating machine [Zahnrad-]Wälzfräsmaschine *f*
g. generating shaper cutter [Zahnrad-] Schneidrad *n*
g. generating tool *s.* g. cutting tool
g. generator Wälzfräsmaschine *f*, Wälzstoßmaschine *f (Verzahnung)*
g. grinding machine Zahnflankenschleifmaschine *f*, Zahnradschleifmaschine *f*
g. head *(Brg)* Schrämkopf *m*
g. hob Zahnradwälzfräser *m*
g. hobber Zahnradwälzfräsmaschine *f*
g. hobbing Zahnradfräsen *n* im Wälzverfahren; Zahnradwälzfräsen *n*
g. hobbing machine Zahnradwälzfräsmaschine *f*
g. honing Zahnradhonen *n*
g. honing machine Zahnradhonmaschine *f*
g. hub Zahnradnabe *f*
g.-lapping Zahnradläppen *n*
g. lapping machine Zahnradläppmaschine *f*
g. lever Schalthebel *m*
g. mechanism Getriebe *n*, Zahnradgetriebe *n*
g. milling Zahnradfräsen *n*
g. milling cutter Zahnrad[form]fräser *m*
g. milling machine Zahnradfräsmaschine *f*
g. noise characteristics Geräuschverhalten *n* der Zahnräder
g. planer Wälzhobelmaschine *f* für Zahnräder; Zahnradwälzhobelmaschine *f*
g. planing Zahnradwälzhobeln *n*
g. pump Zahnradpumpe *f*
g. quadrant Räderschere *f*
g. rack Zahnstange *f*
g. ratio 1. Zähnezahlverhältnis *n*; 2. Getriebesteigung *f*; Übersetzungsverhältnis *n*
g. reduction Räderübersetzung *f* [ins Langsame]
g. reduction set Zahnradgetriebe *n* [für Übersetzung ins Langsame]
g. rim Zahnradkranz *m*
g. ring *(Kfz)* Starterzahnkranz *m*
g. roughing cutter Zahnradvorfräser *m*
g. run testing machine Zahnradlaufprüfmaschine *f*
g. scroll chuck Universalfutter *n*
g. selector mechanism *(Kfz)* Gangwähler *m*
g. shaft Getriebewelle *f*, Radwelle *f*, Transmissionswelle *f*
g. shaper Wälzstoßmaschine *f*

g.-shaper cutter Schneidrad *n (Wälzstoßen)*
g. shaping Wälzstoßen *n*
g.-shaping cutter Zahnradschneidrad *n*
g.-shaping machine Wälzstoßmaschine *f (für Zahnräder)*
g.-shaping process Wälzstoßverfahren *n*
g. shaving Zahnradschaben *n*
g.-shaving machine Zahnradschabmaschine *f*
g.-shift bar Schaltstange *f*
g.-shift control mechanism Gangschaltung *f*
g.-shift cover Schaltdeckel *m*
g.-shift diagram Gangschaltbild *n*
g.-shift dome Schaltturm *m*
g.-shift lever Radverstellhebel *m (Räderkasten)*; Schalthebel *m*, Handschalthebel *m*
g.-shift-lever shaft Schaltwelle *f*
g.-shift lug Schaltbock *m*
g.-shift mechanism Gangschaltung *f*
g.-shift pattern Ganganordnung *f*
g.-shift rail Schaltschiene *f*
g.-shift sleeve Schaltmuffe *f*
g.-shifter fork Schaltgabel *f*, Schaltklaue *f (Getriebe)*
g. size Räderabmessungen *fpl*
g. sleeve Zahnradbüchse *f*
g. stocking cutter Zahnformvorfräser *m*
g.-testing device Zahnradprüfgerät *n*
g.-testing machine Zahnradprüfmaschine *f*
g. tooth Zahnradzahn *m*
g. tooth caliper Zahnmeßlehre *f*
g. tooth cutter Verzahnungsfräser *m*
g. tooth flank Zahnflanke *f*
g. tooth gauge Zahnmeßlehre *f*
g. tooth milling machine Zahnfräsmaschine *f*
g. tooth number Zahnradzähnezahl *f*
g. tooth outline Zahnprofil *n*
g. tooth pressure Zahndruck *m*
g. tooth profile Zahnprofil *n*
g. tooth rounding and deburring machine Zahnkantenabrund- und -entgratmaschine *f*
g. tooth type clutch Zahnradkupplung *f*
g. tooth vernier [caliper] Zahnmeßschieblehre *f*
g. train 1. Getriebezug *m*, Rädergetriebe *n*; 2. *(Kfz)* Kraftübertragungsweg *m*
g. transmission Räderübersetzung *f*, Zahnradübersetzung *f*
g. transmission ratio [Zahnrad-]Übersetzungsverhältnis *n*
g.-type coupling Zahnkupplung *f*
g.-type pump Zahnradpumpe *f*
g. unit Getriebeaggregat *n*, Getriebebaugruppe *f*, Getriebeeinheit *f*
g. wheel Zahnrad *n*
g. wheel engagement Radeingriff *m*
g. wheel transmission Rädertrieb *m*
g. with dog clutch Klauenrad *n*
g. with localized tooth bearing balligtragendes Zahnrad *n*
g. with minimum number of teeth *(to avoid undercutting)* Grenzrad *n (Zahnrad mit Grenzzähnezahl)*

geared 1. in Eingriff; 2. mit Rädergetriebe; zwangsläufig; mit Zahnrädern; mit Übersetzung durch Zahnräder
g. coupling Zahnkupplung *f*
g. down übersetzt ins Langsame
g. drilling machine Bohrmaschine *f* mit Antriebskopf mit Rädergetriebe
g. drive zwangsläufiger Antrieb *m*, Zahnradantrieb *m*
g. engine *(Flg)* Getriebemotor *m*
g. feed Vorschub *m* durch Zahnräder
g. feed pump Speisepumpe *f* mit Zahnradantrieb
g. head Räderspindelkasten *m*; Antriebskopf *m* mit Rädergetriebe
g. headstock Räderspindelstock *m*
g. oil engine Dieselmotor *m* mit Übersetzungsgetriebe [ins Langsame]
g. pump Zahnradpumpe *f*
g. reducing press Räderziehpresse *f*
g. rim Zahnkranz *m (Spannfutter)*
g. screw chuck Kranzspannfutter *n*
g. speed reducer Reduziergetriebe *n*
g. steam turbine Getriebedampfturbine *f*
g. to durch (über) Zahnräder verbunden mit
g. turbine Getriebeturbine *f*

gearing Getriebe *n*; Zahnradantrieb *m*; Zahnradgetriebe *n*; Zahnradübersetzung *f*; Eingreifen *n*, Eingriff *m*, Kämmen *n (Zahnräder)*; Verzahnung *f*
g. down Übersetzen *n* ins Langsame
g. of the fly frame *(Text)* Spulenbankgetriebe *n*
g. unit Getriebeaggregat *n*, Getriebeeinheit *f*
g. up Übersetzen *n* [ins Schnelle]

gearless räderfrei, räderlos

gears Räderwerk *n*; Getrieberäder *npl*, Zahnräder *npl*
g. on nonparallel and intersecting axes Schraubräder *npl*
g. on parallel and intersecting axes Wälzgetriebe *n*
g. on parallel or intersecting axes Wälzräder *npl*
g. operating on crossed axes Schraubgetriebe *n*

geat *s.* gate

Gee-system Gee-Verfahren *n (Funknavigation)*

gehlenite *(Min)* Gehlenit *m*

Geiger counter *(Ph)* Geiger-Zähler *m*, Geiger-Müller-Zählrohr *n*

Geiger-Muller counter tube Geiger-Müller-Zählrohr *n*, GM-Zählrohr *n*

Geiger-Muller tube *s.* Geiger-Muller counter tube
Geiger region Auslösebereich *m (Geiger-Zähler)*
Geiger threshold Auslösespannung *f*, Einsatzspannung *f (Geiger-Zähler)*
geikielite *(Min)* Geikielit *m*
Geissler discharge *(El)* Geißlersche Entladung *f*
Geissler tube Geißler-Röhre *f*
gel / to erstarren, gelieren
gel *(Ch)* Gel *n*
  g.-coat finish gelartiger Überzug *m*
  g. lacquer Tauchlösung *f* für die Herstellung bleibender Überzüge
gelatin *(Am) s.* gelatine
gelatinate / to gelatinieren, gelieren
gelatine Gelatine *f*
  g. dynamite Sprenggelatine *f*, Gelatinedynamit *n*, Sprenggummi *m*
  g. glue Gelatineleim *m*
  g. roller *(Typ)* Gelatinewalze *f*
gelatinform state Gelzustand *m*, gallertartiger Zustand *m*
gelatinize / to gelatinieren, Gel bilden, gelieren
gelatinizing agent Gelatinierungsmittel *n*
gelatinous gallertartig, gelartig, gelatinös
gelation Gelatinierung *f*, Gelbildung *f*
gelignite *s.* gelatine dynamite
gelling 1. Gelierung *f (z. B. Öl)*; 2. Quellung *f*
  g. agent gelbildendes Mittel *n*
gem [echter] Edelstein *m*
  g.-stone [echter] Edelstein *m*
gemming machine *(Led)* Gemmaschine *f*
gemmology Edelsteinkunde *f*
gemmy fibre Jute[hanf *m*] *f*
genapping *(Text)* Absengen *n*, Sengen *n*
genemotor *(Am)* Einankermotorgenerator *m*, Einankerumformer *m*
general allgemein, generell
  g. account Sachkonto *n*
  g. arrangement plan Generalplan *m*, Übersichtsplan *m (eines Schiffes)*
  g. assembly drawing Zusammenstellungszeichnung *f*
  g. average große Havarie *f*
  g. call *(Fmt)* Anruf *m* an alle
  g. calling frequency allgemeine Ruffrequenz *f*
  g. cargo *(Schiff)* Stückgut[ladung *f*] *n*, gemischte Ladung *f*
  g. case Allgemeinfall *m*
  g. case of strain räumlicher (dreiachsiger) Formänderungszustand *m*
  g. case of stresses räumlicher (dreiachsiger) Spannungszustand *m*
  g. coverage *(Fmt)* Allwellenempfang *m*
  g. diffused lighting gleichförmige Beleuchtung *f*
  g. drawing Übersichtszeichnung *f*
  g. ellipsoid dreiachsiges Ellipsoid *n*
  g. foreman Meister *m*, Polier *m*; Oberpolier *m*; Bauführer *m*
  g. integral allgemeines Integral *n (Differentialgleichungen)*
  g. layout Gesamtanordnung *f*
  g. maintenance work Instandhaltungsarbeiten *fpl*
  g. operating costs Betriebsgemeinkosten *pl*
  g. overhaul Generalüberholung *f*
  g. plan Gesamtübersicht *f*, Situationsplan *m*; (*s. a.* g. arrangement plan)
  g. post office Hauptpostamt *n*
  g. pumping arrangement *(Schiff)* Rohrleitungsplan *m*
  g. purpose aircraft Mehrzweckflugzeug *n*
  g. purpose beater Ganzzeugholländer *m (Papierherstellung)*
  g.-purpose cargo vessel Mehrzweckfrachtschiff *n*, Mehrzweckfrachter *m*
  g.-purpose computer Allzweckrechner *m*, Universalrechner *m*
  g.-purpose diode Allzweckdiode *f*, Universaldiode *f*
  g.-purpose machine Vielzweckmaschine *f*
  g.-purpose manipulator *(Kern)* Universalmanipulator *m*
  g.-purpose plough Allzweckpflug *m*
  g.-purpose radar Allzweckradar *n*, Universalradar *n*
  g.-purpose vehicle Mehrzweckfahrzeug *n*
  g. reactor equation allgemeine Reaktorgleichung *f*
  g. routine *(Dat)* allgemeines Programm *n*
  g. sketch Übersichtsskizze *f*
  g. theory of relativity allgemeine Relativitätstheorie *f*
  g. trader Mehrzweckfrachter *m*; Stückgutfrachter *m*
  g. trend *(Geol)* Hauptstreichen *n*, allgemeines Streichen *n*
  g.-use switch *(El)* Universalschalter *m*, normaler Schalter *m*
  g. view Gesamtansicht *f*, Übersichtsbild *n*
generalized *(Ph)* generalisiert *(Lagrangesche Form der Mechanik)*
generate / to 1. erzeugen *(z. B. Strom)*; 2. wälzfräsen; verzahnen nach Abwälzverfahren
generated cut Hüllschnitt *m (Verzahnung)*
  g. heat Bildungswärme *f*
  g. surface *(Math)* Mantelfläche *f*
generating erzeugend
generating Wälzfräsen *n*; Wälzstoßen *n*
  g. bevel gear planer (planing machine) Kegelradwälzstoßmaschine *f*
  g. cam Wälzkurve *f*; Abwälzkurve *f (Verzahnung)*

g. circle Rollkreis *m (Zahnrad)*
g. cutter Schneidrad *n*, Stoßrad *n*
g. electrostatic voltmeter Feldmühle *f*; rotierendes Voltmeter *n*
g. function *(Math)* erzeugende Funktion *f*
g. gear planing machine Wälzhobelmaschine *f*
g. gear shaper Wälzstoßmaschine *f* [für Zahnräder]
g. line *(Math)* Erzeugende *f*, Mantellinie *f*
g. machine Wälzfräsmaschine *f*
g. method Wälzverfahren *n*
g. motion Wälzbewegung *f*
g. pitch circle Erzeugungswälzkreis *m (Zahnrad)*
g. pitch diameter Wälzkreisdurchmesser *m (Erzeugungswälzkreis)*
g. plant *(El)* Stromversorgungsanlage *f*
g. process Wälzverfahren *n (Zahnrad)*
g. set Generatorsatz *m*, Generatorgruppe *f*
g. shaping Wälzstoßen *n*
g. station *(El)* Elektrizitätswerk *n*, Kraftwerk *n*
g. station auxiliary power Eigenbedarfsleistung *f (des Kraftwerkes)*
g. station capacity installierte Kraftwerksleistung *f*
g. station efficiency Kraftwerkswirkungsgrad *m*
g. unit *(El)* Generator[satz] *m*
g. voltmeter Schwingungsvoltmeter *n*
generation 1. Erzeugung *f*; 2. Verzahnung *f* durch Wälzverfahren
g. grinding Abwälzschleifen *n*
g. of gears by rack Zahnradwälzbohren *n*
g. of heat Wärmeentwicklung *f*, Wärmeerzeugung *f*
g. of teeth Verzahnung *f* nach dem Wälzverfahren
generator 1. Generator *m*, Stromerzeuger *m*; 2. Wälzfräsmaschine *f*, Wälzstoßmaschine *f*; 3. *(Dat)* Kodeerzeuger *m*; 4. *(Fmt)* Induktor *m*
g. call *(Fmt)* Induktoranruf *m*
g. for battery charging Ladegenerator *m*, Ladedynamo *m*, Lademaschine *f*, Dynamo *m* für Akkumulatorentladung
g. gas Generatorgas *n*
g. hum *(Elm)* Maschinengeräusch *n*
g. machine Kesseldampfmaschine *f*
g. of sound Schallerzeuger *m*
g. room Kraftzentrale *f*
g. signalling *(Fmt)* Ruf *m* mit Handinduktor
g. surface *(Met)* Heizfläche *f*, beheizte Fläche *f*
generatrix *(Math)* Erzeugende *f*
Geneva drive Malteserkreuztrieb *m*
Geneva motion Malteserkreuzbewegung *f*
Geneva movement *s.* Geneva motion
Geneva stop Malteserkreuz *n*
Geneva wheel Malteserkreuzscheibe *f*
gentianic acid Gentiansäure *f*
gentle schonend, mild; leicht, mäßig
g. bank *(Flg)* leichte Querlage *f*
g. breeze leichte Brise *f*
g. flow leichte Strömung *f*
genuine echt, rein, unverfälscht; wirklich
g. part Original[ersatz]teil *n*
g. signal echtes Signal *n*
g. silk echte Seide *f*
g. soap ungefüllte Seife *f*
geocentric geozentrisch
geochemical geochemisch
geochemistry Geochemie *f*
geocronite *(Min)* Geochronit *m*
geode *(Min)* Druse *f*, Erzdruse *f*
geodesic geodätisch
geodesic *(Math)* Geodätische *f*
geodesy [höhere] Geodäsie *f*, Erdvermessung *f*
geodetic geodätisch; Vermessungs-
g. head *(Geol)* geodätisches Gefälle *n*
g. station Vermessungsamt *n*
g. surveying höhere Geodäsie *f*
geodetical *s.* geodetic
geodynamics Geodynamik *f*
geoelectric geoelektrisch, erdelektrisch
geoelectricity Erdelektrizität *f*
geognosy Geognosie *f*
geographer Geograf *m*
geographic geografisch
g. range Tragweite *f (eines Leuchtfeuers)*
geographical geografisch
g. latitude geografische Breite *f*
g. longitude geografische Länge *f*
g. panel *(El)* Gleisbildstellwerk *n*
geography Geografie *f*
geohydrology Hydrogeologie *f*
geoid Geoid *n*
geologic *s.* geological
geological geologisch
g. age determination geologische Altersbestimmung *f*
g. clock 1. [geologische] Zeittafel *f*; 2. geologische Uhr *f (radioaktive Mineralien)*
g. epochs geologische Perioden *fpl*
g. history Erdgeschichte *f*
g. map geologische Karte *f (Landkarte)*
geologist Geologe *m*
geologist's hammer Geologenhammer *m*
geology Geologie *f*
geomagnetic geomagnetisch
g. anomaly erdmagnetische (geomagnetische) Anomalie *f*
g. field Erdmagnetfeld *n*, erdmagnetisches (geomagnetisches) Feld *n*
g. field intensity erdmagnetische Feldstärke *f*
g. storm geomagnetischer Sturm *m*

geomagnetism Erdmagnetismus *m*, Geomagnetismus *m*
geomechanics Geomechanik *f*
geometer Geometer *m*
geometric *s.* geometrical
geometrical geometrisch
  g. accuracy Formgenauigkeit *f*
  g. chord Flügelsehne *f*
  g. inaccuracy Formungenauigkeit *f*
  g. locus *(Math)* geometrischer Ort *m*
  g. luminosity geometrische Lichtstärke *f*
  g. mean *(Math)* geometrisches Mittel *n*
  g. optics geometrische Optik *f*, Strahlenoptik *f*
  g. progression *(Math)* geometrische Folge (Progression) *f*
  g. proposition geometrischer Satz *m*
  g. series geometrische Reihe *f*
geometrically accurate formgenau
  g. similar geometrisch ähnlich
  g. true formgenau
geometrician Geometer *m*
geometry 1. Geometrie *f*; 2. Formtoleranz *f*
  g. of kinematics Bewegungsgeometrie *f*, geometrische Bewegungslehre *f*
  g. of solids Stereometrie *f*
  g. of the cutting edge Schneidengeometrie *f*
geophysical geophysikalisch
  g. prospecting geophysikalische Schürfung *f*
geophysics Geophysik *f*
geosyncline Geosynklinale *f*
geothermal degree of depth geothermische Tiefenstufe *f*
germ Keim[ling] *m*
  g. carrier Bazillenträger *m*
  g. disease Keimlingskrankheit *f*
  g. warfare bakteriologischer Krieg *m*, bakteriologische Kriegführung *f*
German flute Querflöte *f*, große Flöte *f*
German jack plane with double iron *(Holz)* Doppelhobel *m*
German representation deutsche Darstellungsweise *f (technisches Zeichnen)*
German silver Argentan *n*, Neusilber *n*, Alpaka *n*
German spring flesher *(Led)* Scherdegen *m*
German Standards Committee Deutscher Normenausschuß *m*
German stitch *(Text)* Kreuzstich *m*
German system of spinning *(Text)* deutsches Spinnverfahren *n*
germanate Germanat *n*, Salz *n* der Germaniumsäure
germanic compound Verbindung *f* des vierwertigen Germanium
germanite *(Min)* Germanit *m*
germanium Germanium *n*
  g. diode Germaniumdiode *f*
  g. dioxide Germaniumdioxid *n*
  g. junction rectifier Germaniumflächengleichrichter *m*
  g. p-n diode Ge-pn-Diode *f*
germanous compound Verbindung *f* des zweiwertigen Germanium
germantown grobes 4faches Strickgarn *n*
germicidal antiseptisch, keimtötend
germicide Keimvernichtungsmittel *n*, bakterizides (keimtötendes) Mittel *n*
germinating apparatus *(Lw)* Auskeimungsapparatur *f*
  g. capacity *s.* g. power
  g. power *(Lw)* Keimfähigkeit *f*
germination 1. *(Lw)* Keimung *f*; 2. *(Ch)* Keimbildung *f (Kristalle)*
  g. test Keimprobe *f*
germproof keimfrei
gersdorffite Arsennickelglanz *m*, Nickelarsenkies *m*
get / to gewinnen, abbauen, hereingewinnen, verhauen *(Kohle)*
  g. a ship aback Schiff zum Achterauslaufen bekommen
  g. afloat abbringen *(ein gestrandetes Schiff)*
  g. away *s.* g. afloat
  g. barren *(Geol)* vertauben
  g. free frei bekommen, losbekommen *(Schiff)*
  g. going in Gang bringen (setzen)
  g. jammed sich festklemmen, klemmen
  g. off *(Schiff)* losbekommen *(festsitzendes Fahrzeug)*, abbringen *(ein gestrandetes Schiff)*
  g. off the rails entgleisen
  g. out of commission ausfallen *(Flugzeugtriebwerk)*
  g. rusty anrosten; verrosten
  g. stuck stecken bleiben *(Fahrzeuge)*; sich festklemmen
  g. up steam Dampf aufmachen
get-away power *(Kfz)* Anzugsvermögen *n*
  g.-away speed *(Flg)* Abhebegeschwindigkeit *f*
getter Getter *m*, Fangstoff *m (Elektronenröhren)*
  g. fastener Getterhalter *m*; Getterpfanne *f*; Gettertasche *f (Elektronenröhren)*
gettering Getterverdampfung *f*, Getterung *f*
  g. agent Fangstoff *m*, Getter *m*
geyser Warmwasserdurchlauferhitzer *m*
geyserite *(Min)* Geyserit *m*
ghee *(Ch) (indische)* Schmelzbutter *f*
ghost Schattenstreifen *m*
  g. image *(Fs)* Geisterbild *n*
  g. images *(Opt)* Geister *mpl (insbesondere im Spektrum)*
  g. line *(Met)* Längsstreifen *m*, Seigerungsstreifen *m*
  g. working *(Fmt)* Phantombetrieb *m*
ghosts *(Fs)* Geisterbilder *npl*
giant aerial Großantenne *f*

g. jazz drum Riesenfelltrommel *f*
g. molecule Riesenmolekül *n*
g. tyre Überballonreifen *m*; LKW-Reifen *m*
gib Führungsleiste *f*, Nachstelleiste *f*; Hakenkeil *m*, Nasenkeil *m*, Keilbeilage *f*; Ausleger *m* *(Kran)*
g. and cotter stellbare Keilverbindung *f*
g. and cotter end Kappenkopf *m* *(Schubstange)*
g. head Keilnase *f*
g. head flat saddle key Nasenflachkeil *m*
g. head hollow saddle key Nasenhochkeil *m*
g.-head key Nasenkeil *m*
g.-head taper key angeschrägter Nasenkeil *m*
g. strip Keilleiste *f*; Nachstelleiste *f*
gibbed mit Leiste
gibbet Kranbalken *m*
Gibbs' function Enthalpie *f*
Gibbs-Helmholtz equation Gibbs-Helmholtzsche Gleichung *f*
Gibbs' phase rule Gibbssches Phasengesetz *n*, Gibbssche Phasenregel *f*
gibbsite *(Min)* Gibbsit *m*; Hydrargillit *m*
gig Gigboot *n*
g. end bearing *(Eis)* hinteres Lager *n* *(Kuppel- oder Treibstange)*
gigging *(Text)* Rauhen *n*
gild / to vergolden
g. with gold leaf blattvergolden
gilder's aqua fortis *(Ch)* Goldscheidewasser *n*
gilding Vergoldung *f*, Vergolden *n*
g. metal unechtes Blattgold *n*; Goldtombak *m*
g. press *(Typ)* Vergolderpresse *f*
g. with gold leaf Blattvergoldung *f*
gill / to *(Text)* hecheln
gill 1. *(Masch)* Luftregelklappe *f*; 2. *(Text)* Hechelapparat *m*, Igel *m*; Gillspinnmaschine *f*; 3. Rippe *f* *(eines Heizkörpers)*
g. bar *(Text)* Gillstab *m*, Kammstab *m*, Nadelstab *m*
g. box *(Text)* Nadelstabstrecke *f*; Einfachhechelfeldstrecke *f*
g. box minder Rohstrecker *m* *(Spinnerei)*
g. box with double set of fallers Doppelnadler *m*, Doppelhechelfeldstrecke *f*
g.-netter Logger *m*, Treibnetzfischereischiff *n*
gilled pipe Lamellenrohr *n*, Rippenrohr *n* *(z. B. Heizung)*
g. radiator Lamellenkühler *m*
g. tube Lamellenrohr *n*, Rippenrohr *n* *(z. B. Heizung)*
gilling *(Text)* Kardieren *n*
gilsonite *(Min)* Gilsonit *m*
gilt vergoldet
g. copper vergoldetes Kupfer *n*
g. edge *(Typ)* Goldschnitt *m*
gimbal Kardanring *m*
g. mounting *s.* g. suspension
g. suspension kardanische Aufhängung *f*
g.-type mounting *s.* g. suspension
gimlet Holzbohrer *m*, Nagelbohrer *m*
g. point Bohrerspitze *f*, konischer Schaft *m*; Schraubenende *n* mit Schneidspitze
gin / to entkörnen, egrenieren *(Baumwolle)*
gin 1. Aufzug *m*; Hebegerät *n*; 2. *(Text)* Egreniermaschine *f*; 3. Trommel *f*; Scheibe *f*
g. block Baurolle *f*
g. pole 1. Rüstbaum *m*; Standmast *m* *(Derrickkran)*; 2. Hilfsmast *m* *(Antennenanlage)*; 3. *(Lw)* Göpelbaum *m*
g. tackle *(Schiff)* Hebezeug *n*
ginger oil Ingweröl *n*
ginging Ausmauerung *f*
ginning Entkörnen *n*, Egrenieren *n* *(Baumwolle)*
Giorgi units Giorgisches Maßsystem *n*
gipsy capstan *(Schiff)* Verholspill *n*
girder Balken *m*, Träger *m*, Tragbalken *m*; Unterzug *m*
g. and section mill Formeisenwalzwerk *n*
g. beam Trägerbalken *m*
g. bridge Balkenbrücke *f*
g. grillage *(Bw)* Trägerrost *m* *(Gründungen)*
g. rolling mill Trägerwalzwerk *n*
g. section Trägerprofil *n*
girdle / to 1. [um]gürten; 2. *(Holz)* ein ringförmiges Stück Rinde ausschneiden, damit der Baum abstirbt
Girod furnace *(Met)* Girod-Ofen *m*, Gleichstromlichtbogenofen *m*
girt *s.* girder
girth Gurt *m*, Gürtel *m*; Tragriemen *m*
g. of frame *(Schiff)* Spantumfang *m* *(Bauspant)*
g. of section *(Schiff)* Spantumfang *m* *(Konstruktionsspant)*
g. seam Rundnaht *f* *(Nieten)*
git 1. *(Met)* Gießtrichter *m*; 2. *s.* gate
g. cutter Eingußabschneider *m*
g. mould Einlaufkokille *f*
give / to [an]geben; [an]zeigen; ergeben; erteilen, leisten, verabreichen; übermitteln; sich werfen *(Holz)*
g. a cut with the saw einsägen
g. a slope abschüssig machen
g. back energy Energie zurückliefern, rückspeisen
g. expert opinion on something ein Gutachten über etwas abgeben, etwas begutachten
g. off abgeben; locker werden, sich lockern *(Schraube)*
g. off heat Wärme abgeben
g. off light Licht ausstrahlen
g. out power Leistung abgeben
g. rise to bewirken, veranlassen
g. rotary motion umlaufen lassen

g. the blades an angle by grinding den Schaufelwinkel anschleifen *(Turbine)*
g. up *(Brg)* auflassen, aufgeben
given clearance *(Masch)* hinterschliffen
g.-off heat abgegebene Wärme *f*
glacé kid Glacéleder *n*
glacial acetic acid Eisessig *m*
g. dye Eisfarbe *f*, Eisfarbstoff *m*
g. polish Hochglanzpolitur *f*, Spiegelpolitur *f*
glaciate / to vereisen, vergletschern
glaciation Vereisung *f*, Vergletscherung *f*
glacier Gletscher *m*
g. berg Gletscherberg *m (Eisberg)*
glaciology Gletscherlehre *f*
glade Aushau *m (Wald)*; Lichtung *f*
glance Glanz *m*
g. coal Glanzkohle *f*
g. cobalt *(Min)* Glanzkobalt *m*
glancing angle Glanzwinkel *m (Kristallanalyse)*
g. incidence streifender Einfall *m (Strahlen)*
gland Stopfbuchsenbrille *f*, Stopfbuchsendeckel *m (Dampfmaschine)*
g. cock Hahn *m* mit Stopfbuchse
g. condenser Sperrdampfkondensator *m*
g. leak-off Wrasenleitung *f (Stopfbuchsensperrdampf)*
g. lining Brillenfutter *n (Stopfbuchsenbrille)*
g. nut Dichtungsmuffe *f*
g.-ring Flansch[spann]ring *m*
g.-sealing service air Stopfbuchsensperrluft *f*
g.-sealing service steam Stopfbuchsensperrdampf *m*
g.-shaped ring stopfbuchsenartig eingreifender Ring *m*
g. stud Stopfbuchsenschraube *f*
glandless ohne Stopfbuchse, stopfbuchslos
glare / to hell scheinen, grell leuchten, blenden
g. up aufleuchten
glare 1. blendendes Licht *n*; Blendlicht *n*; Blendwirkung *f*; 2. *(Kfz)* Reflexion *f* von nasser Straße
g. effect Blendwirkung *f*
g.-reducing glass Blendschutzglas *n*
glareless blendungsfrei
g. light blendungsfreies Licht *n*
glarimeter Glanzmesser *m*
glaring colour grelle (schreiende) Farbe *f*
glascrete Glas[stahl]beton *m*
glass / to emaillieren
glass Glas *n*
g. batch Glasgemenge *n (für Glasherstellung)*
g. bath Glasbad *n*
g. bead Glasperle *f*
g. blank Glaspreßling *m*, Preßling *m*
g. block Glasbaustein *m*, Glasziegel *m*
g. blower Glasbläser *m*
g. blowing Glasblasen *n*
g. brick Glasbaustein *m*, Glasziegel *m*
g. bubbler Waschflasche *f*
g. bulb Glaskolben *m (Glühlampen)*
g.-bulb forming machine Kolbenblasmaschine *f*
g.-bulb rectifier *(El)* Glaskolbenventil *n*, Glasgleichrichter *m*
g. ceramic keramisches Glas *n*
g.-clear glasklar
g. cock Glashahn *m*
g. concrete Glas[stahl]beton *m*
g. constituent Glasbestandteil *m*
g. cover Deckglas *n (Mikroskop)*
g. crack Glasfehler *m*, Sprung *m*
g. cutter 1. Glasschneider *m (Werkzeug)*, Glaserdiamant *m*; 2. Glasschneider *m (Beruf)*, Zuschneider *m (Flachglas)*
g.-cutting shop Glasschneiderei *f*
g.-cutting tool Glasschneidwerkzeug *n*, Glaserdiamant *m*
g.-cutting wheel Glasschleifrad *n*
g.-cutting [work]shop Glasschneiderei *f*
g. defect Glasfehler *m*
g. depth Glasstand *m (im Ofen)*
g. disk memory Glasplattenspeicher *m (permanenter Datenspeicher)*
g. drill Glasbohrer *m*
g. electrode Glashalbzelle *f*
g.-enclosed verglast
g.-enclosed fuse *(El)* Röhrchensicherung *f*, Feinsicherung *f*
g. etching Glasätzung *f*
g. fabric Glasgewebe *n*
g. factory Glashütte *f*
g. fibre Glasfaser *f*
g.-fibre cloth Glasfasergewebe *n*
g.-fibre reinforced glasfaserverstärkt; glasfaserbewehrt
g.-fibre reinforced plastics glasfaserverstärkter Kunststoff (Plast) *m*
g.-fibre starved areas (regions) glas[faser]-arme Stellen *fpl*
g. fibres textiler Glasfaserstoff *m*
g. filaments *s.* g. fibres
g. film Glasfolie *f*
g. fleece Glas[faser]vlies *n*
g. former Glasbildner *m*, Glasformer *m*
g. forming Glasformung *f*; Glasverarbeitung *f*
g. fragment Glassplitter *m*
g. frit Glasfritte *f*
g. funnel Glastrichter *m*
g. gilding Glasvergoldung *f*
g.-grinding machine Glasschleifmaschine *f*
g.-grinding [work]shop Glasschleiferei *f*
g. guard Schutzscheibe *f*
g.-hard glashart
g.-house 1. Glashütte *f*; 2. Gewächshaus *n*, Treibhaus *n*
g. jar Glasbehälter *m*, Glasgefäß *n*

g. level Glasstand *m (im Ofen)*
g.-lined steel glasierter Stahl *m*
g. machinery Glasbearbeitungsmaschinen *fpl*, Glasverarbeitungsmaschinen *fpl*
g. maker Glasmacher *m*
g.-maker's bench Glasmacherbank *f*
g. making Glasfertigung *f*, Glasherstellung *f*
g. melting furnace Glasschmelzofen *m*
g. melting pot Glashafen *m*
g. mould Glasform *f*
g. muddiness Glastrübung *f*
g. painter Glasmaler *m*
g. painting Glasmalerei *f*
g. pane Glasscheibe *f*
g. panel Glasplatte *f*
g. paper Glaspapier *n*, Schmirgelpapier *n*
g. pinch Quetschfuß *m (Elektronenröhre)*
g.-plate capacitor Glasplattenhochspannungskondensator *m*
g. pliers Glasbrecherzange *f*
g. roof Glasdach *n*
g.-roofed mit Glasdach
g. scale Glasmaßstab *m*
g. screen 1. *(Bw)* Glastrennwand *f*; 2. *(Typ)* Glasraster *m*
g. silvering Glasversilberung *f*
g. slicker *(Led)* Glasierglas *n*
g. stain Farbbeize *f* für Glas
g. staining Glasmalerei *f (z. B. von Kirchenfenstern)*
g. stopper Glasstopfen *m*
g. tank Glasofen *m*
g. tank crown Glas[schmelz]wanne *f*
g. tank furnace Glas[schmelz]wanne *f*
g. thread Glasfaden *m*
g. tile Glasziegel *m*, Fliese *f*
g.-to-metal seal Glas-Metall-Verschmelzung *f*
g. tube Glasrohr *n*, Glasröhre *f*; Röhrenlibelle *f*
g. tube fuse Glasröhrchensicherung *f*, Feinsicherung *f*
g. tube protector *(Wärm)* Schutzglas *n*
g. ware Glaswaren *fpl*
g. wool Glaswolle *f*
g. worker Glashüttenarbeiter *m*
g. working Glasverarbeitung *f*
g. works Glasfabrik *f*, Glashütte *f*
glassine Glassin *n*, Pergamin *n (Papier)*
glassing jack *s.* g. machine
g. machine *(Led)* Glasiermaschine *f*
glassy glasartig, glasig
g. lustre *(Min)* Glasglanz *m*
g. silver ore Schlackenerz *n*
Glauber's salt Glaubersalz *n*
glaze / to 1. verglasen; 2. glasieren, lasieren; überglasen, engobieren; 3. satinieren; kalandrieren *(Papier)*; 4. lüstrieren, mit Glanz pressen *(Stoffe)*; satinieren *(Leder)*
g. over überglasen
glaze glasige Oberfläche *f*; Überglasung *f*; Glasur *f (z. B. Keramik)*; Lasur *f (z. B. Ölgemälde)*
g. baking Glasurbrand *m*
glazed brick Glasurstein *m*, Glasurziegel *m*
g. door 1. verglaste Tür *f*; 2. Fenstertür *f*, französisches Fenster *n*
g. horse *(Led)* Roßchevreaux *n*
g. paper Hochglanzpapier *n*, Glacépapier *n*, Glanzpapier *n*, satiniertes Papier *n*
g. tracing paper Transparentpapier *n*
g. wire Emaildraht *m*; Glanzdraht *m*
glazer Satinierer *m (Papierherstellung)*
glazier Glaser *m*
glazier's diamond Glaserdiamant *m*, Glasschneider *m*
g. hammer Glaserhammer *m*
g. putty Glaserkitt *m*
g. shop Glaserei *f*
glazing 1. Glasieren *n*; Verglasen *n*; 2. Feuerpolieren *n (von geschliffenem Glas)*; 3. Glaserarbeiten *fpl*; Verglasung *f*; 4. Glänzigwerden *n* des Schleifmittelbelags
g. calender Glättkalander *m*, Glanzkalander *m*; Satinierkalander *m*
g. compound Fensterkitt *m*, Glaserkitt *m*
g. glass *(Led)* Glas *n* zum Glanzstoßen
g. machine Glanzmaschine *f*, Satiniermaschine *f*
g. material glasartige Kunststoffe *mpl*
g. mill Rundschmelzmaschine *f (Elektronenröhrenherstellung)*
g. of the grinding wheel Verglasen *n* der Schleifscheibe
g. roller *(Led)* Glanzstoßzylinder *m*
gleam / to glänzen, blinken, schimmern
gleam Glanz *m*, Schein *m*, Schimmer *m*
g. of light Lichtschimmer *m*
g. of silver Silberblick *m (Bleientsilberung)*
gleaning *(Lw)* Ährenlesen *n*, Nachlese *f*
glebe Erzstufe *f*
glide / to 1. *(Met)* gleiten; 2. *(Flg)* fliegen ohne Motor
g. down herabgleiten
glide 1. Gleiten *n*; 2. *(Flg)* Gleitflug *m*
g. mechanism 1. Gleitvorrichtung *f (z. B. Plattenwechsler)*; 2. *(Met)* Gleitvorgang *m*
g.-over door *(Bw)* Kipptor *n*, Kipptür *f*
g. path *(Flg)* Gleitweg *m*; *(Rak)* Gleitbahn *f*
g. path angle Gleitwinkel *m*
g. path beacon Gleitstrahlbake *f*
g. path beam Gleitwegleitstrahl *m*
g. path transmitter Gleitwegsender *m*
g. plane Gleitebene *f*
g.-reflection plane Gleitspiegelebene *f*
g. trajectory *(Rak)* Gleitbahn *f*

glider 1. Segelflugzeug *n*; 2. Gleitboot *n*; 3. Gleitvorrichtung *f* *(z. B. für Plattenwechsler)*
g.-towing Segelflugzeugschlepp *m*
gliding 1. Gleitung *f* *(z. B. Kristalle)*; 2. *(Flg)* Segelfliegen *n*
g. boat Gleitboot *n*
g. movement Gleitbewegung *f*
g. plane Gleitebene *f* *(Kristall)*
g. reflection Gleitspiegelung *f*
g. scale bewegliche Skala *f*
glim lamp Glimmlampe *f*
glimmer / to glimmen; schimmern; flimmern
glimmer Schimmer *m*; Flimmern *n*
glisette *(Math)* Gleitkurve *f*
glitter / to glitzern
global missile Globalrakete *f*
g. radiation Globalstrahlung *f*
g. rocket Globalrakete *f*
globate kugelförmig
globe 1. Globus *m*; Kugel *f*; kugelförmiger Körper *m*; 2. *(El)* Leuchtenglocke *f*; [große] Glühlampe *f*
g. body control valve *(BMSR)* Durchgangsregelventil *n*
g. buoy *(Schiff)* Kugelbakentonne *f*
g. caliper Kugeltaster *m*
g. joint Kugelgelenk *n*
g. lightning Kugelblitz *m*
g. photometer Kugelfotometer *n*
g.-shaped kugelförmig
g. strain insulator Ei[er]isolator *m*
g. valve Durchgangsventil *n*, Kugelventil *n*, Kugelschieber *m*, Tellerventil *n*
globoid gear Globoidgetriebe *n*
g. worm Globoidschnecke *f*
globoidal worm gear Globoidschneckentrieb *m*
globosity Kugelgestalt *f*
globular kugelig, kugelförmig; Kugel-
g. discharge *(El)* geschichtete Entladung *f*
g. fender *(Schiff)* Kugelfender *m*
g. lightning Kugelblitz *m*
g. perlite körniger (kugelförmiger) Perlit *m*
g. structure *(Met)* globoidales Gefüge *n*
globularize / to *(Met)* Kugelgraphit bilden [lassen]
glonoin *(Am)* Nitroglyzerin *n*
gloria hole *s.* beam hole
gloriole Aureole *f*
glory hole 1. Anwärmeofen *m*, Auftriebofen *m*, Einbrennofen *m* *(z. B. für Glas)*; 2. *(Brg)* Trichter *m* *(Trichterbau)*
g.-hole method untertägiger Trichterbau *m*
g.-hole work Trichterbau *m* *(Tagebau)*
gloss Glanz *m*, Glasur *f*, Glätte *f*, Politur *f*
g. finishing paint Glanzdeckanstrichfarbe *f*
g. ink Glanzdruckfarbe *f*
g. oil *(Am)* Harttrockenglanzöl *n* *(Lösung von Kolophonium in Benzin oder Terpentinöl)*
g. paint Glanzlack *m*
g. retention Glanzbeständigkeit *f*, Dauerhaftigkeit *f* des Glanzes
glossimeter *s.* glossmeter
glossmeter Glanzmesser *m*
glossy glänzend, blank
g.-black bituminous coal Glanzkohle *f*
g. ink Glanztinte *f*
g. paper Glanzpapier *n*
glost fire Glasurbrand *m*
glove box 1. Handschuhkasten *m*; 2. *(Kern)* Schutzkasten *m* mit eingebauten Handschuhen
g. compartment Handschuhfach *n*
g. leather Handschuhleder *n*
Glover tower Gloverturm *m* *(Bleikammerverfahren)*
glover's wool Gerberwolle *f*
glow / to glühen, glimmen
g. away verglimmen
glow Glut *f*; Glühen *n*, Glimmen *n*
g. counting tube *(BMSR)* Glimmzählröhre *f*
g. discharge Glimmentladung *f*
g.-discharge cathode Glimmkatode *f*
g.-discharge lamp Glimmlampe *f*
g.-discharge rectifier Glimmentladungsventil *n*
g.-discharge tube Glimmlampe *f*
g.-discharge tube audio generator Glimmlampensummer *m*
g.-discharge tube voltage divider Glimmspannungsteiler *m*
g. gap divider Glimmspannungsteiler *m*
g. lamp Glimmlampe *f*
g. lamp for switch Schalterglimmlampe *f*
g. lamp rectifier Glimmlichtgleichrichter *m*
g. plug *(Kfz)* Glühkerze *f*
g. plug filament Glühspirale *f*
g. plug switch Glühanlaßschalter *m*
g. potential Glimmpotential *n*, Zündspannung *f*
g. rectifier Glimmlichtgleichrichter *m*
g. regulator Glimmröhrenstabilisator *m* *(Spannung)*
g. temperature Glühtemperatur *f*
g. tube Glimm[entladungs]röhre *f*
g. tube fuse Feinsicherung *f*
g. wire Glühdraht *m*
glower Glühkörper *m*
glowing glühend
glowing Glühen *n*
g. arc Glimmbogen *m*
g. filament Glühfaden *m*
g. heat Glühtemperatur *f*, Glühhitze *f*
g. lamp Glühlampe *f*
g. red rotglühend
g. white weißglühend

glucina Berylliumoxid *n*, Beryllerde *f*
glucinum *s.* beryllium
gluconate Ester *m* (Salz *n*) der Glukonsäure, Glukonat *n*
gluconic acid Glukonsäure *f*
glucose Glukose *f*, Glykose *f*, Traubenzucker *m*
glucoside Glykosid *n*
glucuronic acid Glukuronsäure *f*
glue / to [ver]leimen, [ver]kleben
  g. on aufkleben
  g. together zusammenleimen
glue Leim *m*
  g. applicator *(Typ)* Anleimmaschine *f*, Leimauftragmaschine *f*
  g. boiler Leimsieder *m*
  g. clamp Leimzwinge *f*
  g. heater Leimkocher *m*
  g. heating apparatus Leimkocher *m*
  g. in pearls Perlleim *m*
  g. joint Leimfuge *f*
  g.-line shear test Leimfugenscherfestigkeitsprüfung *f*
  g. pot Leimtopf *m*
  g. press Leimknecht *m*, Leimzwinge *f*
  g. priming Leimgrundierung *f*
  g. putty Leimkitt *m*
  g. smearing machine *s.* g. spreading machine
  g. spread aufgetragene Leimmenge *f*
  g. spreading machine Leimauftragmaschine *f*
  g. stock Leimleder *n*
glued joint 1. Leimfuge *f*; 2. Klebeverbindung *f*
  g. laminated construction verleimte Schichtkonstruktion *f*
  g. laminated wood verleimtes Schichtholz *n*
glueing *s.* gluing
gluing Verleimung *f*; Kleben *n*
  g. clamp spindle *(Holz)* Furnierbockspindel *f*
  g. cramp Leimknecht *m*, Leimzwinge *f*
  g. cylinder Leimauftragwalze *f*
  g. edge veneer joint Furnierfugenleimmaschine *f*
  g. machine *(Typ)* Anleimmaschine *f*
  g. of metals Metallklebetechnik *f*
  g. practice Verleimungstechnik *f*; Klebetechnik *f*
  g. roller Leimauftragwalze *f*
glulam lamellenverleimt *(Holzbau)*
gluside Benzoesäuresulfimid *n*, Saccharin *n*
glutamic acid Glutaminsäure *f*
glutamine Glutamin *n*, Amid *n* der α-Aminoglutarsäure
glutarate Ester *m* (Salz *n*) der Glutarsäure
glutaric acid Glutarsäure *f*
glutinate / to verkleben, verkleistern
glutinous klebrig
glycerin[e] Glyzerin *n*
glycerol Glyzerin *n*
  g. trinitrate Glyzerintrinitrat *n*, Trinitroglyzerin *n*
glycine Glyzin *n*, Glykokoll *n*, Aminoessigsäure *f*
glycocholic acid Glykocholsäure *f*
glycolate / to zu Glykol umsetzen
glycollic acid Glykolsäure *f*
glyoxal[ic] acid *s.* glyoxylic acid
glyoxylic acid Glyoxylsäure *f*
glyptal resin Glyptalharz *n*
gneiss Gneis *m*
go / to gehen, fahren, sich fortbewegen; sich erstrecken; funktionieren
  g. about *(Schiff)* wenden, über Stag gehen
  g. aloft *(Schiff)* aufentern, in die Takelage klettern
  g. alongside *(Schiff)* anlegen
  g. astern *(Schiff)* rückwärts fahren, über Steuer gehen, achterauslaufen
  g. down 1. *(Brg)* einfahren; 2. sich vermindern; 3. versinken, untergehen *(in Wasser)*; 4. niedergehen *(z. B. Flugzeug)*
  g. in a boat Kahn fahren
  g. into a spin *(Flg)* abtrudeln
  g. into conduction beginnen Strom durchzulassen, leitend werden, Strom durchlassen
  g. into low gear *(Kfz)* den ersten Gang einschalten
  g. off tune verstimmt werden, sich verstimmen
  g. on angehen, aufleuchten
  g. out ausgehen *(elektrisches Licht)*
  g. out of the line *(Fmt)* aus der Leitung gehen
  g. through a hysteresis loop eine Hystereseschleife durchlaufen
  g. through a training eine Ausbildung absolvieren
  g. to pieces zu Bruch gehen
  g. to press in Druck gehen
go-and-return line *(El)* Hin- und Rückleitung *f*
  g.-around *(Flg)* Durchstarten *n*
  g.-devil Reinigungsbürste *f*
  g.-devil plane *(Brg)* Bremsberg *m*
  g.-end Gutseite *f (Lehre)*
  g.-gauge Gutlehre *f*
  g.-getter *(Rak)* Steuerung *f*, Leitsystem *n*, Lenksystem *n*
  g.-limit Gutgrenze *f (Lehre)*
  g.-no-go radiation detector Strahlungsüberwachungsgerät *n (das bei bestimmter Dosisleistung anspricht)*
  g.-not-go testing Gut-Schlecht-Prüfung *f*
  g.-plug screw gauge Gutgewindelehrdorn *m*
  g.-ring [gauge] Gutgewindering *m*
  g.-side *s.* g.-end

g.-thread roller Gutrolle *f (Gewinderachenlehre)*
goaf *(Brg)* alter Mann *m*
g. side *(Brg)* Versatzseite *f*
g.-side versatzseitig
g.-side road Abbaustrecke *f* mit einseitig anstehender Kohle
g. stower *(Brg)* Versatzmaschine *f*
goal Fußballtor *n*, Tor *n*
g. post Torpfosten *m*
g.-post system *(Brg)* Rahmenausbau *m*
g.-seeking system *(Dat)* zielstrebiges System *n*
goat skin coat Ziegenledermantel *m*
gob *(Brg)* alter Mann *m*
g. fire Grubenbrand *m*
g.-road Strecke *f* im alten Mann
g. stuff *(Brg)* Versatzberge *pl*
gobbed-up blast furnace verstopfter Hochofen *m*
gobbing *(Brg)* Versetzen *n*, Versatzarbeit *f*; Bergeversatz *m*
gobbler Truthahn *m*
goblet Pokal *m*, Kelch *m*, Kelchglas *n*
gobo *(sl)* 1. Gegenlichtblende *f (Kamera)*; 2. seitliche Schallabschirmung *f (Mikrofon)*
godown *(Am, sl)* Kaischuppen *m (in asiatischen Häfen)*
goethite *(Min)* Goethit *m*, Nadeleisenerz *n*
goffer / to kräuseln, fälteln, plissieren
goffering calender Gaufrierkalander *m (Papierherstellung)*
goggles Schutzbrille *f*
going Stufenbreite *f (Treppe)*
g. about Wenden *n (z. B. eines Schiffes)*
gold-coat / to vergolden; feuervergolden
g.-plate *(galvanisch)* vergolden
gold Gold *n*
g. amalgam Goldamalgam *n*
g. assay Goldprobe *f*
g.-bearing goldführend *(Fluß)*; goldhaltig *(Erz)*
g.-beater Goldschläger *m*
g. beating Ausschlagen *n* von Gold [zu Blattgold]
g. blocking *(Typ)* Preßvergoldung *f*, Prägevergoldung *f*
g.-bonded diode Golddrahtdiode *f*
g. bronze Goldbronze *f*
g. bullion 1. Goldbarren *m*; 2. goldhaltige Legierung *f*; Rohgold *n*
g.-coated mit Goldüberzug; vergoldet
g. coating 1. Vergolden *n*; Feuervergolden *n*; 2. [Steuergitter-]Goldschicht *f (Elektronenröhren)*
g. coin Goldmünze *f*
g. content Goldgehalt *m*
g. deposit *(Min)* Goldseife *f*
g. dredging Goldbaggern *n*
g. dust Goldstaub *m*
g. filigree threads Goldgespinst *n*
g. finder Goldsucher *m*
g. foil Blattgold *n*; Goldfolie *f*
g. fulminate Knallgold *n*
g. hydrochloric acid Tetrachlorogoldsäure *f*
g. in bars ungemünztes Gold *n*
g. ingot Goldbarren *m*, Barrengold *n*
g.-lace maker *(Text)* Goldbortenwirker *m*
g. leaf Blattgold *n*
g.-leaf electroscope Blattelektroskop *n (mit Goldfolie)*
g. litharge *(Ch)* Goldglätte *f*
g. mine Goldgrube *f*
g. nugget Goldklumpen *m*
g. number Goldzahl *f (Kolloidchemie)*
g. parting Goldscheidung *f*
g.-placer *(Geol)* Goldseife *f*
g.-plated contacts vergoldete Kontakte *mpl*
g.-plated metal Golddoublé *n*, Doublé *n*
g.-plating *(galvanisches)* Vergolden *n*
g. quartz Goldkies *m*
g. refinery Goldscheideanstalt *f*
g. ruby glass Goldrubinglas *n*
g. size Goldleim *m*
g.-stamping works Goldprägeanstalt *f*, Goldprägerei *f*
g. standard Goldwährung *f*
g. stripping Entgoldung *f*
g. vein Goldader *f*
g. washer Goldwäscher *m*, Goldsucher *m*
Goldberg wedge *(Opt)* Goldbergkeil *m*, Graukeil *m*
golden goldartig, golden
g. lustre Goldglanz *m*
g. section *(Math)* goldener Schnitt *m*
goldsmith Goldschmied *m*
Goliath cap Goliath-Sockel *m (Lampe)*
Goliath Edison screw cap Goliath-Sockel *m*
gon Neugrad *m*, Gon *n*
gondola car *(Am) (Eis)* Niederbordwagen *m*
g. wall *(Am)* Bordwand *f (Güterwagen)*
gong buoy *(Schiff)* Gongtonne *f*, Gongboje *f*
goniometer Goniometer *n*, Winkelmesser *m*
g. eyepiece Winkelmeßokular *n*
g. ocular *s.* g. eyepiece
goniometric goniometrisch, winkelmessend
g. instrument Winkelmeßinstrument *n*
goniometry Goniometrie *f*, Winkelmessung *f*
goniophotometer Winkelfotometer *n (Fotometer zur Messung des Reflexionsverhaltens)*
gonioscope Gonioskop *n*
Gooch crucible Goochtiegel *m*
Gooch filter *s.* Gooch crucible
good 1. gut; echt; haltbar; tauglich; 2. *(Met)* gar *(Schmelze)*
g. anchorage ankerfester Grund *m*

g. board *(Schiff)* Hulebug *m*, Holebug *m*, Streckbug *m*; Schlagbug *m*
g. conductor *(El)* guter Leiter *m*
g. earth gute Erdung (Erdverbindung) *f*
g. ignition quality Zündwilligkeit *f*
g. road *(Schiff)* geschlossene Reede *f*
g. service area *(Fmt, Fs)* Bereich *m* guten Empfangs
goods Gut *n*, Güter *npl*; Waren *fpl*
g. department *(Eis)* Güterabfertigung *f*
g. lift Lastenaufzug *m*
g. locomotive Güterzuglokomotive *f*
g. office *(Eis)* Güterabfertigung *f*
g. station Güterbahnhof *m*
g. traffic Güterverkehr *m*
g. train Güterzug *m*
g. wagon Güterwagen *m*
goose-neck S-Bogen *m*, Schwanenhals *m (Rohr)*
g. neck claw bar Nagel[zieh]eisen *n*, Nagelzieher *m*
g. neck finishing tool *(Am)* gekröpfter Schlichtmeißel *m*
g.-necked S-förmig; gekröpft
gooseneck *s.* goose-neck
gopher hole *(Brg)* kleine Minenkammer *f*
g.-hole blasting Sprengen *n* mit einem kleinen ausgekesselten Bohrloch
gophering 1. Sprengen *n* mit einem kleinen ausgekesselten Bohrloch; 2. *(Am)* Weitungsbau *m*, unsystematischer Abbau *m*
gore *(Text)* Zwickel *m*
gorge *(Geol)* Schlucht *f*, Engpaß *m*
goslarite *(Min)* Goslarit *m*
gossan *(Geol)* eiserner Hut *m*
Gott's method kapazitives Kabelfehlerortungsverfahren *n (Meßwesen)*
goudron Asphaltteer *m*, Goudron *m (n)*
gouge Hohlbeitel *m*, Kehlbeitel *m*; Bohrlöffel *m*; Schaumlöffel *m*; 2. *(Holz)* Drechslerröhre *f*; 3. *(Geol)* Salband *n*, Lettenbesteg *m*
g. bit Erdbohrer *m*
govern / to regeln, regulieren, steuern
governer *s.* governor
governing Regelung *f*, Regulierung *f*
g. by cutting out individual nozzles Einzelregelung *f (Turbine)*
g. by cutting out nozzles Düsenregelung *f (Turbine)*
g. by throttling the steam supply Drosselregelung *f*, Regelung *f* durch Dampfdrosselung *(Turbine)*
g. impulse *(Fs)* Gleichlaufimpuls *m*
government call *(Fmt)* Staatsgespräch *n*
g. priority call dringendes Staatsgespräch *n*
g. shipyard Staatswerft *f*
g. vessel Behördenschiff *n*, Regierungsfahrzeug *n*
governor Regler *m*; *(speziell:)* Drehzahlregler *m*
g. balance weight Reglergewicht *n*
g. ball Reglerkugel *f*, Schwungkugel *f*
g. bush Reglerbüchse *f (Turbine)*
g. casing Reglergehäuse *n*
g. collar Reglermuffe *f*
g. cone Reglerkegel *m*
g. cover 1. *(Kfz)* Reglerhaube *f*; 2. Reglerdeckel *m (Turbine)*
g. dancing Reglertanzen *n*
g. deflection Reglerausschlag *m*
g. frame Reglerkorb *m (Turbine)*
g. friction Eigenreibung *f* des Reglers
g. gear[ing] Reglergetriebe *n*
g. hood Regleraufsatz *m (Turbine)*
g. housing Reglergehäuse *n*
g. influence Reglereinwirkung *f*
g. piston valve Reglerschieber *m (Turbine)*
g. position Reglerstellung *f*
g. relief valve Überdruckventil *n*
g. rodding Reguliergestänge *n (Regler)*
g. sensitivity Empfindlichkeitsgrad *m* eines Reglers
g. setting Regler[ein]stellung *f*
g. shaft Reglerwelle *f*
g. sleeve Reglermuffe *f*, Reglerhülse *f*
g. slide valve Reglerschieber *m*
g. spindle Regulatorachse *f*, Reglerspindel *f*
g. tap pen Regulierstößel *m*
g. valve Reglerventil *n*, Regelventil *n*, Düsenventil *n (Turbine)*
g. with crossed arms Regler *m* mit gekreuzten Stangen
g. with fixed centre of oscillation Regler *m* mit Kreuzschleife
g. with inverted suspension Regler *m* mit umgekehrter Pendelaufhängung
g. with open links Regler *m* mit Schubkurbelgetriebe
g. with rhombic suspension Regler *m* mit rhombischer Aufhängung
g. with sleeve Muffenregler *m*, Kegelpendelregler *m*
GP-computer *s.* general-purpose computer
grab Baggerkorb *m*, Greifer *m*, Greifkorb *m*, Greifkübel *m*
g. bucket *s.* grab
g. crane Greiferkran *m*
g. dredge Greifbagger *m (z. B. für Molenbau)*
g. magnet Greifmagnet *m*
grabbing Anfressen *n* der Fasen des Bohrers
g. crane Greifbagger *m*
gradation 1. Abstufung *f*, Gradation *f*; Stufenfolge *f*; 2. *(Opt)* Helligkeitsabstufung *f*; Schwärzungsabstufung *f*; 3. Kornabstufung *f (Beton)*
g. of tones *(Opt)* Abstufung *f* der Helligkeitsunterschiede

g. photometer Stufenfotometer *n* *(Spektrografie)*

grade / to 1. abstufen, staffeln; einstufen, einteilen, klassieren; 2. *(Ch, Bw)* sortieren, trennen; 3. kreuzen *(Zucht)*

grade 1. Güte[grad *m*] *f*; Qualität *f*; 2. Härtestufe *f*; Güteklasse *f*; 3. Gefälle *n*, Neigung *f*; 4. Kreuzung *f (Zucht)*; 5. Neugrad *m*; 6. Flur *f*, Terrain *n*

g.-bottomed launch Gleitboot *n* mit gestaffeltem Boden

g.-builder Planierraupe *f*; Trassenschäler *m*

g. crossing [ebener] Bahnübergang *m*, Schienenübergang *m*

g. level Geländehöhe *f*, OK-Terrain *n*, Terrainhöhe *f*

g. of accuracy Qualität *f*, Toleranzstufe *f*

g. of accuracy of the machine tool Herstellungsgenauigkeit *f* der Werkzeugmaschine

g. of concrete Betongüte *f*

g. of cut Hiebnummer *f (Feile)*

g. of indeterminacy Grad *m* der statischen Unbestimmtheit

g. of ore Metallgehalt *m (Erz)*

g. of pig iron Roheisensorte *f*

g. of steel Stahlgüte *f*

g. of the grinding wheel Schleifscheibenhärte *f*

g. of tolerance Toleranzstufe *f*, Qualität *f*

g. separation Straßenkreuzung *f* (Bahnübergang *m*) in getrennten Ebenen

graded 1. abgestuft; 2. sortiert

g. coal Nußkohle *f*

g. filter 1. *(El)* abgestuftes Filter *n (Radio)*; 2. Kiessandfilter *n* mit abgestuften Korngrößen

g. insulation abgestufte Isolation *f*

g. junction abgestufter (allmählicher) Übergang *m (Transistoren)*

g. seal Verschmelzstück *n (Glas-Metall)*, Zwischenstück *n* mit abgestuftem Ausdehnungskoeffizienten

grader 1. Sortiermaschine *f*; 2. Planiergerät *n*; Straßenhobel *m*; Bodenhobel *m*, Erdhobel *m*

grades of fits Arten *fpl* der Passungen

gradient 1. *(Math)* Gradient *m*; 2. Steigung *f*; Gefälle *n*; Neigung *f (Böschung)*

g. angle Steigungswinkel *m (einer Straße)*

g. brake *(Eis)* Gefällebremse *f*

g. density filter *(Opt)* Verlauffilter *n*

g. meter Steigungsmesser *m*

g. of track *(Eis)* Gleissteigung *f*

grading 1. Abstufung *f*, Staffelung *f*; Einstufung *f*, Einteilung *f*, Klassierung *f*; Korngrößentrennung *f*; 2. *(Ch, Bw)* Sortierung *f*, Trennung *f*; 3. Kornabstufung *f (Beton)*; 4. Klassierung *(Wolle)*

g. analysis Siebanalyse *f*

g. apparatus Sortierapparat *m*

g. coefficient Körnungsziffer *f*

g. curve Sieblinie *f*, Siebkurve *f*

g. cylinder Sortierzylinder *m*

g. diagram *(Fmt)* Stafflungsplan *m*

g. drum Sortiertrommel *f*

g. efficiency Siebtrennungsgrad *m*

g. jigging screen Klassierrüttelsieb *n*

g. of groups *(Fmt)* Gruppenstaffelung *f*, Staffelung *f* von Gruppen

g. of ores Klassieren *n* der Erze

g. screen Klassiersieb *n*

g. test Siebprobe *f*

gradual allmählich; stufenweise, schrittweise

g. tempering of steel gebrochenes (stufenweises) Stahlhärten *n*

g. transition stufenweiser Übergang *m*

g. variation schrittweise Änderung *f*

graduate / to 1. mit Skalenteilung versehen; eichen, kalibrieren; 2. den ersten akademischen Grad erwerben, Diplom erwerben

graduate Graduierter *m (Inhaber eines akademischen Grades)*

g.-acting control *(Am) s.* proportional control

g. apprentice Praktikant *m (von Hochschulen)*

graduated mit Gradeinteilung

g. arc *(Math)* Gradbogen *m*, Limbus *m*

g. burette Meßbürette *f*

g. circle Teilkreis *m*

g. collar Buchse *f* mit Maßeinteilung (Gradeinteilung), Skalenring *m*

g. cylinder Meßzylinder *m*

g. depth gauge Bohrtiefenskala *f*

g. dial *s.* g. disk

g. disk Skalenscheibe *f*, Teilscheibe *f*

g. drum Meßtrommel *f* mit Meßstrichen

g. drum with vernier Meßtrommel *f* mit Nonius

g. flask Meßkolben *m*

g. glass circle Glasteilkreis *m*

g. in degrees mit Gradeinteilung

g. jug Meßbecher *m*

g. measuring tube graduiertes Meßrohr *n*

g. rate gestaffelte Gebühr *f*, Staffelgebühr *f*

g. ring Skalenring *m*

g. scale [geeichte] Meßskala *f*

g. slide-wire kalibrierter Schleifdraht *m (Potentiometer)*

g. steel straight edge Strichmaßstab *m*

g. straight edge Meßschiene *f*

g. swivel base Drehscheibe *f* mit Gradeinteilung

g. vessel Meßgefäß *n*

graduating Teilen *n (in Grade)*

graduation 1. Abstufung *f*, Graduierung *f*; Gradeinteilung *f*; 2. *(Ch)* Gradierung *f*, Abdampfung *f*; 3. Erwerbung *f* des ersten akademischen Grades *(Erwerbung eines Diploms)*;

4. Skalenteilung *f*; Teilstrich *m*; Teilungsabstand *m*
g. in degrees Grad[ein]teilung *f*
g. in inches Zollteilung *f*
g. in millimeters Millimeterteilung *f*
g. mark Skalenstrich *m*, Teilstrich *m*
g. of resistance *(El)* Widerstandsabstufung *f*
g. of the brine Gradierung *f* der Salzsole
g. of tonal intensities *(Ak)* Tonabstufung *f*
graduator 1. *(El)* Verflacherschaltung *f*; 2. Gradierwerk *n*
Graetz rectifier Gleichrichter *m* in Graetz-Schaltung
graft / to pfropfen, veredeln *(Bäume)*
graft-hybrid Pfropfbastard *m*
g. polymerization *(Ch)* Pfropfpolymerisation
grafted tree Pfröpfling *m*
grafting knife Okuliermesser *n*, Pfropfmesser *n*
g. scion Pfropfreis *n*
g. slit Pfropfschnitt *m*, Pfropfspalt *m*
g. twig Pfropfreis *n*
graftling Pfröpfling *m*
grain / to adern, masern; körnen, granieren
g. up *(Led)* aufkrausen
grain 1. *(Met)* Korn *n*; Kristall *m*; 2. *(Opt)* Korn *(fotografischer Film)*; 3. Getreide *n*, Korn, Samenkorn *n*; 4. *(Holz)* Faserstruktur *f*, Maserung *f*; 5. *(Led)* Narbe *f*; 6. *(Text)* parallel zur Leiste laufende Fäden *mpl* im Material; 7. Körnung *f (Schleifkörper)*; 8. Grain *m (Massemaß für Edelmetalle und Medikamente)*
g. auger Kornförderschnecke *f*, Schneckenkornförderer *m*
g. awn Getreidegranne *f*
g. boundary Korngrenze *f (Kristalle)*
g. boundary attack Korngrenzenangriff *m*
g. boundary barriers Korngrenzensperrschichten *fpl (Halbleiter)*
g. boundary breakdown Korngrenzenzerfall *m*
g. boundary crack Korngrenzenriß *m*, intrakristalliner Riß *m*
g. boundary transistor Korngrenzentransistor *m*
g. bulkhead *(Schiff)* Getreideschott *n*, Kornschott *n*
g.-catching cloth cover Fangtuch *n (Dreschmaschine)*
g. classification Korngrößeneinteilung *f*
g. coarsening *(Met)* Kornvergröberung *f*
g. colony Kristallgebilde *n*
g. count Korngrößenbestimmung *f (durch Zählung)*
g. defect Narbenfehler *m (Leder)*
g. deformation *(Met)* Kornverzerrung *f*
g. direction *(Holz)* Faserrichtung *f*, Maserrichtung *f*; *(Typ)* Laufrichtung *f (des Papiers)*
g. drier Getreidetrockner *m*, Körnertrockner
g. drill Drillmaschine *f*
g. ear Kornähre *f*
g. feeder *(Schiff)* Getreidefeeder *m*, Getreide-[nachfüll]schacht *m*
g. fittings *(Schiff)* Getreide[lade]einrichtungen *fpl*
g. growth Kristallwachstum *n*; Kornwachstum *n*
g. harvester Mähbinder *m*
g. hygroscope Getreidefeuchtigkeitsmesser *m*
g. loft Kornboden *m (Scheune)*
g. mixture concentrates *(Lw)* Kraftfutter *n*
g. refinement *(Met)* Kornverfeinerung *f*
g. roll *(Met)* ungehärtete Walze *f*
g. roller *(Led)* Narbenrolle *f*
g. screen Getreidesieb *n*
g. silo Kornspeicher *m*
g. size Korngröße *f*, Körnung *f*
g. size distribution Korngrößenverteilung *f (z. B. Metalle)*
g. size number Körnungsnummer *f*
g. sorting machine Trieur *m (Getreidesortierung)*
g. spacing Gefüge *n* der Schleifscheibe
g. spike Kornähre *f*
g. split *(Led)* Narbenspalt *m*
grained 1. körnig *(Film)*; 2. genarbt *(Leder)*; 3. gemasert *(Holz)*
graining 1. Körnung *f (Film)*; 2. *(Holz)* Maserung *f*; 3. Granulierung *f*
g. board *(Led)* Krispelholz *n*
g. calender Granierkalander *m*
g. machine *(Led)* Krispelmaschine *f*
gram Gramm *n*
g. atom Grammatom *n*
g. atomic weight *s.* g. atom
g. calorie Grammkalorie *f*, kleine Kalorie *f*
g. equivalent Grammäquivalent *n*
g. force *(Am)* Pond *n*
g. molecule Mol *n*, Grammolekül *n*
grammatite *(Min)* Grammatit *m*
gramme *s.* gram
g. ring armature *(Elm)* Flachringanker *m*
g. weight Pond *n*
gramophone Grammofon *n*, Plattenspieler *m*
g. amplifier Schallplattenverstärker *m*
g. attachment Plattenspielerzusatz *m*, Tonabnehmerzusatz *m*
g. disk Schallplatte *f*
g. pick-up Tonabnehmer *m*
g. record Schallplatte *f*
g.-record manufacture Schallplattenherstellung *f*
granary Getreidespeicher *m*; Kornkammer *f*
g. weevil Kornkäfer *m*
granatohedron Rhombendodekaeder *n (Kristallografie)*

grand average Mittel *n* von Mittelwerten
g. jazz drum Riesenfelltrommel *f*
g. piano Konzertflügel *m*; Flügel *m*
g. total *(Math)* Endsumme *f*, Gesamtsumme *f*
grandstand *(Bw)* Tribüne[nbau *m*] *f*
grange Scheune *f*, Scheuer *f*
graniform kornähnlich
granite Granit *m*
g.-block paving Granitplattenpflaster *n*
g. drill Hartgesteinsbohrer *m*
g. insulator Granitisolator *m*
granitic granitisch
g. insulator Granitisolator *m*
granitoid granitähnlich
granodize / to Stahloberflächen phosphatieren *(nach dem Granodine-Verfahren)*
grant / to bewilligen, gewähren, erteilen, zugestehen
g. a licence under a patent eine Lizenz auf ein Patent erteilen
g. a patent claim einen Patentanspruch gewähren
grant 1. Stipendium *n*; 2. Verleihung *f*, Konzession *f*
g. of licence Lizenzerteilung *f*
g. of patent claim Gewährung *f* eines Patentanspruchs
grantee of a licence Lizenznehmer *m*
granular körnig
g.-crystalline grobkristallin
g. flux Schweißpulver *n*
g. fracture *(Met)* körniger (kristallinischer) Bruch *m*
g. fusing flux Schweißpulver *n*
g. microphone Körnermikrofon *n*
g. noise Eigenrauschen *n (von Röhren oder Widerständen)*
g. size Korngröße *f*
g. soil *(Bw)* nichtbindiger (rolliger) Boden *m*
g. surface structure körnige Oberflächenstruktur *f*
granularity Körnigkeit *f*; *(BMSR)* Stufigkeit *f*, begrenztes Auflösungsvermögen *n*, Unstetigkeit *f*
granulate / to *(Met)* granulieren, zerkleinern
granulate depth fertilization Granulattiefdüngung *f*
granulated carbon Kohlegrieß *m*, Kohlekörner *npl*, granulierte Kohle *f*
g. metal Granalien *pl*
g. screen *(Typ)* Kornraster *m*
g. welding composition Schlackenpulver *n (UP-Schweißung)*
granulating crusher Granuliermühle *f*, Kornmühle *f*
g. machine Pulverkörnmaschine *f*
g. point Kornbildungspunkt *m*
g. spout Granulierrinne *f (Hochofen)*
granulation Granulation *f*, Körnung *f*, Kornbildung *f*
g. of pitch Körnungspunkt *m*, Kristallbildungspunkt *m*
g. of slags *(Met)* Schlackengranulierung *f*
g. size Körnung *f*
granulator 1. Feinbrecher *m*, Granulator *m*; 2. Zuckertrockner *m*
granule Körnchen *n*; Kornspan *m*
granules 1. Körnchen *npl*; 2. Granulat *n*; 3. Kohlekörner *npl (Mikrofon)*
granulitic körnig
grape Weinbeere *f*
g. juice Traubensaft *m*
g. sugar Traubenzucker *m*, Dextrose *f*
graph / to grafisch darstellen
graph Diagramm *n*, Schaubild *n*, grafische Darstellung *f*; Bild *n*
g. follower *(Dat)* Kurvenabtaster *m*
g. of a function Funktionsbild *n*, Kurve *f*
g. paper Koordinatenpapier *n (gewöhnlich mit Zollteilung)*
g. paper with millimeter squares Millimeterpapier *n*
g. plotter Kurvenschreiber *m*
g. reader Diagrammleser *m*, Kurvenleser *m (die Koordinaten werden an einen Rechenautomaten gegeben)*
grapher Schreiber *m*, Schreibvorrichtung *f*
graphic grafisch
g. formula *(Ch)* Formelbild *n*
g. instrument Registriermeßgerät *n*
g. method of determination grafisches Verfahren *n* zur Bestimmung, grafische Bestimmung *f*
g. panel *(BMSR)* Betriebsschaltbild *n*, Blindschaltbild *n*
g. recorder Schreiber *m*, Schreibvorrichtung *f*
g. representation grafische Darstellung *f*
graphical *s.* graphic
graphite / to graphitieren
graphite Graphit *m*
g.-blacking *(Met)* Formenschwärzung *f* mit Graphit
g. column *(Kern)* thermische Säule *f*
g. low-energy experimental pile Graphitversuchsreaktor *m* niedriger Energie
g. lubrication Graphitschmierung *f*
g. lubricator Graphitschmierapparat *m*
g. melting pot Graphitschmelztiegel *m*
g.-moderated *(Kern)* graphitmoderiert, graphitgebremst
g. pyrometer Graphitpyrometer *n*, Graphitthermometer *n*
g. resistor Graphitwiderstand *m*
g.-treated graphitiert
graphited grease graphitiertes Fett *n*

graphitic carbon Kohlenstoff *m* in Form von Graphit, graphitischer Kohlenstoff
g. cast iron graphitreiches (kohlenstoffreiches) Gußeisen *n*
g. pig iron graues Roheisen *n*
graphitiferous graphithaltig
graphitization Graphit[aus]bildung *f*, Graphitierung *f*, Temperkohleabscheidung *f*
grapple / to *(Schiff)* entern; verankern
grapple 1. Mehrschalengreifer *m* *(Bagger)*; 2. *(Schiff)* Enterhaken *m*
g. dredger Greifbagger *m*
g. hook Greiferhaken *m*
grappler Steigeisen *n*, Klettereisen *n*
grappling Verankerung *f*
g. anchor *(Schiff)* Draggen *m*, Dragganker *m*
g. iron Enterhaken *m*
grasper Greifer *m*
grass bleaching Rasenbleiche *f*
g.-covered vergrast
g. mowing machine Grasmäher *m*
g. plot Rasenplatz *m*
g. pulp Papierfaserstoff *m* aus einkeimblättrigen Pflanzen
grate / to vergittern
grate Gitter *n*, Rost *m*; Siebrost *m*
g. area Rost[ober]fläche *f*
g. bar Roststab *m*
g. bar support Roststabträger *m*, Querträger *m*
g.-chain Rostkette *f*
g.-chain carriage Kettenrostwagen *m*
g.-chain drum Rostkettentrommel *f*
g. consisting of thin bars Blattrost *m*
g. cooling Rostkühlung *f*
g. firing Rostfeuerung *f*
g. in one section Ganzrost *m*
g. opening Rostfuge *f*, Rostspalt *m*
g. plate Rostplatte *f*
g. residues Feuerungsrückstände *mpl*
g. section Rostfeld *n*
g. sluice Rostschieber *m*
g. step Rosttreppe *f*
g. stoker Rostbeschickungsanlage *f*
g. surface Rostfläche *f*
g. with sprinkling of the fire bars Rieselrost *m*, Rost *m* mit Wasserberieselung der Roststäbe
grated hatch *(Schiff)* Lukengräting *f*
grater Raspel *f*
graticule Meßgitter *n*, Raster *m*, Strichgitter *n*, Strichplatte *f*
grating 1. *(Bw)* Gitter[werk] *n*; Vergitterung *f*; 2. *(Opt)* Beugungsgitter *n*, Gitter; 3. *(Fs)* Bildraster *m*; 4. *(Schiff)* Gräting *f*
g. constant *(Fs)* Gitterkonstante *f*
g. converter *(El)* Gitterumformer *m*, Gitterwandler *m*
g. line *(Opt)* Gitterstrich *m*
g. over stokehold *(Schiff)* Kesselraumgräting *f*
g. reflector Rasterreflektor *m*
g. replica *(Opt)* Gitterkopie *f*
g. space Gitterkonstante *f*
g. spectrograph Gitterspektrograf *m*
g. spectroscope Gitterspektroskop *n*
g. spectrum Beugungsspektrum *n*, Gitterspektrum *n*
grave-yard *(Kern)* Abfallager *n*; radioaktiver Friedhof *m*; Vergrabungsstelle *f*
gravel / to beschottern *(Straße)*; bekiesen *(Dach)*
gravel Geröll *n*, Kies *m*, Schotter *m*
g. aggregate Kieszuschlag[stoff] *m*
g. asphalt Kiesasphalt *m*
g. deposite Kiesvorkommen *n*
g. dredger Kiesschwimmbagger *m*, Kiesnaßbagger *m*
g. excavator Kiestrockenbagger *m*
g. fraction Kieskörnung *f*
g. pit Kiesgrube *f*
g. plant Kiesaufbereitungsanlage *f*
g. road Kiesstraße *f*
gravelling *(Bw)* Beschotterung *f*
gravelly kiesartig, kieselig, kieshaltig
g. soil Kiesboden *m*
graver Stichel *m*
graveyard *s.* grave-yard
gravimeter *(Geol)* Gravimeter *n*, [statischer] Schweremesser *m*
gravimetric gewichtsanalytisch, gravimetrisch
g. analysis Gewichtsanalyse *f*, gravimetrische Analyse *f*
g. batching Massedosierung *f*
g. factor Umrechnungsfaktor *m* in der Gravimetrie
g. flotation Schwerflüssigkeitssinkscheidung *f*
g. measuring *(Geol)* Gravimetermessung *f*
g. method Masseverfahren *n*
gravimetrical *s.* gravimetric
gravimetry gravimetrische Bestimmung *f*, Gravimetrie *f*, Gewichtsanalyse *f*
graving dock *(Schiff)* Trockendock *n*
gravitate / to gravitieren, im Schwerefeld kreisen; durch Schwerkraft fortbewegen, durch eigene Schwere fließen
gravitation Gravitation *f*, Massenanziehung *f*
g. force Massenanziehungskraft *f*, Schwerkraft *f*
gravitational schwer; Schwere-, Schwerkrafts-, Gravitations-
g. acceleration Gravitationsbeschleunigung *f*
g. constant 1. Gravitationskonstante *f*, Schwerebeschleunigung *f*; 2. Fallbeschleunigung *f*, Erdbeschleunigung *f*
g. field Gravitationsfeld *n*; Schwerefeld *n*
g. force Gravitationskraft *f*, Schwerkraft *f*
g. mass Gravitationsmasse *f*, schwere Masse *f*

g. method of prospecting *(Brg)* gravimetrische Lagerstättenforschung *f*
g. potential Gravitationspotential *n*
g. segregation *(Met) (gewöhnliche)* Blockseigerung *f*
g. shift of spectral lines Gravitationsverschiebung *f* der Spektrallinien
gravity Schwere *f*; Schwerkraft *f*, Erdanziehung *f*
g. acceleration Schwerebeschleunigung *f*, Schwerkraftbeschleunigung *f*, Fallbeschleunigung *f*, Erdbeschleunigung *f*
g. anomaly Schwereanomalie *f*
g. axis Schwerpunktsachse *f*
g. balance Gravitationsdrehwaage *f*
g. bucket conveyor Schwerkraftbecherwerk *n*
g. casting *(Met)* Standguß *m*
g. change Schwereschwankung *f*
g. chute Schrägrutsche *f*
g. chute feed press Presse *f* mit Zuführung durch Schrägrutsche
g. classifier Schwerkraftklassierer *m*
g. concentration Dichteanreicherung *f*; Schwerkraftsaufbereitung *f*; naßmechanische Aufbereitung *f*
g.-controlled instrument Meßinstrument *n* mit rein gravitativer Rückstellkraft *f*
g. conveyor Wuchtförderer *m*
g. dam *(Bw)* Gewichtsmauer *f*, Schwergewichtsmauer *f*
g. davit *(Schiff)* Schwerkraftdavit *m (n)*
g. die casting Schwerkraftkokillenguß *m*
g. discharge furnace *(Met)* Durchstoßofen *m*
g. dressing *(Brg)* Schwerkraftaufbereitung *f*
g. drive Schwerkraftantrieb *m*
g. feed 1. *(Bw, Ch)* Freifallbeschickung *f*, Materialzuführung *f* auf Grund der Schwerkraft, Materialzuführung durch Gefälle; 2. Schnittdruck *m* durch Gewicht des Sägebügels
g.-feed casting Schwerkraftguß *m*
g.-feed lubrication *(Kfz)* Trockensumpfschmierung *f*
g.-feed tank 1. *(Kfz)* Fallbehälter *m*, Falltank *m*; 2. *(Wärm)* Speisegefäß *n* zum Speisen ohne Pumpe, Speiseflasche *f*
g. field Schwerefeld *n*
g.-free schwerefrei, schwerelos
g.-free space schwerefreier Raum *m*
g. haulage *(Brg)* Gefälleförderung *f*
g. mixer Freifallmischer *m*
g. observation Schwerebeobachtung *f*
g.-operated durch Schwerkraft betrieben
g. plane *(Brg)* Bremsberg *m*
g. prospecting *(Geol)* Schweremessung *f*
g. retaining wall *(Bw)* Schwergewichtsmauer *f*, Stützmauer *f*
g. road *(Brg)* Bremsbergstrecke *f*
g. sand casting Schwerkraftsandguß *m*
g. screen *(Brg)* Schrägsieb *n*, Siebfege *f*
g. segregation *(Met)* Schwerkraftseigerung *f*, Gravitationsseigerung *f*
g. shunting incline Ablaufberg *m (Rangierbetrieb)*
g. spindle *(Text)* Hängespindel *f*
g. switch Hakenschalter *m*
g. tank Hochtank *m*, Falltank *m*
g. tram *(Brg)* Bremsberg *m*
g. voltameter Gewichtsvoltameter *n*, Massenvoltameter *n*
g. wall *(Bw)* Gewichtsmauer *f*, Schwergewichtsmauer *f*
gravure *(Typ)* Tiefdruck *m*
g. cells Tiefdrucknäpfchen *npl*
g. paper Tiefdruckpapier *n*
g. press *s.* g. printing press
g. printer Kupferdrucker *m*; Tiefdrucker *m*
g. printing press Tiefdruckmaschine *f*, Kupfertiefdruckpresse *f*
g. rotary printing Rotationstiefdruck *m*
g. screen Tiefdruckraster *m*
gravy boat *(Schiff)* Soßenschüssel *f*, Sauciere *f*, Tunkenschüssel *f (Küchenschüssel)*
gray *(Am) s.* grey
Gray code *(Dat)* Gray-Kode *m*, reflektierter Binärkode *m*
graze / to 1. leicht berühren, streifen; 2. grasen, weiden
grazing animals Weidevieh *n*
g. incidence *(Opt)* streifender Lichteinfall *m*
grease / to schmieren; einfetten
grease Fett *n*; Schmierfett *n*; Staufferfett *n*
g. bearing Schmierlager *n*
g. box Schmierbüchse *f*
g. cup Staufferbüchse *f*
g.-drawn wire blankgezogener Draht *m*
g. extracting apparatus *(Text)* Entschweißvorrichtung *f*
g. gun Fettspritze *f*, Fettpistole *f*; Schmierpresse *f*, Fettpresse *f*
g.-lubricated mit Fettschmierung
g. lubricating point Fettschmierstelle *f*
g. lubrication Fettschmierung *f*
g. nipple Schmierkopf *m* für Fettschmierung, Schmiernippel *m*
g. packed and sealed mit versiegelter Dauerschmierpackung
g.-proof fettdicht
g.-proof paper Pergamentersatzpapier *n*, Pergamin *n*; Butterbrotpapier *n*
g. renewal period Fettwechselperiode *f*
g. resistant paper fettdichtes Papier *n*
g. scum *(Text)* Wollfettschlamm *m*
g. separator 1. Fettabscheider *m*; 2. *(Wärm)* Dampfentöler *m*

g. sleeve wheel set *(Eis)* Fetthülsenradsatz *m*
g. solvent Fettlösemittel *n*
g.-spot photometer Fettfleckfotometer *n*, Bunsenfotometer *n*
g. trap Fettfang *m*
greaser Fettbüchse *f*
greasy fettig, fettfleckig
g. quartz milchiger Quarz *m*
g. wool Schmutzwolle *f*, [rohe] Schweißwolle *f*
Great Lakes Seaway *(Schiff)* Große-Seen-Fahrt *f*
Great Lakes Service *(Schiff)* Große-Seen-Fahrt *f*
Great Lakes Vessel Große-Seen-Schiff *n (ein Schiff, das von Europa nach den Großen Seen und auf diesen selbst fährt)*
great primer *(Typ) (etwa)* Tertia *f*, 18 britische Punkte *mpl (Schriftgröße)*
greatest common divisor (measure) *(Math)* größter gemeinsamer Teiler *m*, Hauptnenner *m*
greave / to abbrennen *(Schiffsboden)*
Grecian aerial Bruce-Antenne *f*
G. saw Rebensäge *f*
G.-type aerial Bruce-Antenne *f*
green 1. grün; 2. frisch, neu; 3. ungebrannt; ungesintert; 4. naß
g. adder *(Fs)* Grünbeimischer *m*
g. belt Grüngürtel *m (Stadtplanung)*
g. ceramics ungebrannte Töpferware *f*
g. coal Rohkohle *f*, frische Kohle *f*
g. concrete Frischbeton *m*
g. copper ore *(Min)* Malachit *m*
g. core process *(Met)* Grünkernverfahren *n*
g.-flesh / to *(Led)* entfleischen
g. fodder Grünfutter *n*, Futtergras *n*
g.-fodder cutter Grünfutterschneider *m*
g. goose Junggans *f*
g. hide *(Led)* ungesalzene grüne Haut *f*
g. highs *(Fs)* Grünhöhe *f*
g. liquor *(Ch)* 1. Abfallauge *f* beim Natronzellstoffprozeß; 2. Frischlauge *f*
g. lows *(Fs)* Grüntiefe *f*
g. manuring Gründüngung *f*
g. ore Roherz *n*
g. salted glue stock gesalzenes Leimleder *n*
g. sand grüner (nasser) Sand *m*, Grüngußsand *m*, Neusand *m (Formerei)*
g. sand mould Grünsandform *f*, Naßgußform *f*
g. sand moulding Naßsandformen *n*
g. sea *(Schiff)* Sturzsee *f*
g. strength Grünstandfestigkeit *f (Form)*
g. vitriol Eisenvitriol *n*, Ferrosulfat *n*
g. weight Grüngewicht *n (Haut)*
g. wood frisches (grünes) Holz *n*, Frischholz *n*
Green function *(Math)* Greensche Funktion *f*
greenery *(Bw)* Grünanlagen *fpl*
greenhouse Gewächshaus *n*, Treibhaus *n*
greening lacquers Patinierfarben *fpl*
greenockide Kadmiumblende *f*
greens Rasenplatz *m*
Green's economizer *(Wärm)* Greenscher Abgasvorwärmer (Ekonomiser) *m*
Greenwich mean time mittlere Greenwichzeit *f*
greige cloth *(Text)* Rohgewebe *n*
g. goods Rohware *f*
grenade Handgranate *f*; Gewehrgranate *f*
Grenet cell *(El)* Bichromatelement *n (nach Grenet)*
grey grau
g. antimony *(Min)* Antimonglanz *m*, Antimonit *m*
g. antimony ore Grauspießglanzerz *n*
g. board Graupappe *f*, Hadernpappe *f*
g. body *(Ph)* grauer (nicht selektiver) Strahler *m*
g. body radiation *(Ph)* graue Strahlung *f*
g. cast iron Grauguß *m*
g. cast iron casting Graugußstück *n*
g. cylinder felt grauer Rundsiebfilz *m*
g. diffuser *(Opt)* neutral (grau) streuender Körper *m*
g. filter Graufilter *n*
g. fog Grauschleier *m (Film)*
g. iron scrap Gußbruch *m*, Gußschrott *m*
g. pig iron graues Roheisen *n*
g. radiation *(Ph)* graue Strahlung *f*
g. scale recording *(BMSR)* Graustufungsaufzeichnung *f*; Halbtonaufzeichnung *f*
g.-shade Grauton *m*
g. warp *(Text)* Rohkette *f*
g. wedge Graukeil *m*
g. width *(Text)* Rohwarenbreite *f*
greyhound Greyhound *m (schnelles Segelschiff)*
grid 1. Gitter *n (Elektronenröhre)*; 2. Verbundnetz *n (Kraftstrom)*; 3. *(Math)* Koordinatennetz *n*; 4. Raster *m*, Gitternetz *n*; 5. *(Bw)* Systemliniengitter *n*, Systemliniennetz *n*
g. action Gittersteuerung *f*
g. alternating current voltage Gitterwechselspannung *f*
g. base Gittersperrspannung *f*; Steuergittereinsatzspannung *f*; Aussteuerungsbereich *m*
g. bias Gittervorspannung *f*
g. bias battery Gitterbatterie *f*
g. bias control Gittervorspannungsregelung *f*
g. bias generator Gittervorspannungserzeuger *m*
g. bias modulation Gittervorspannungsmodulation *f*
g. bias resistance Katodenwiderstand *m (zur Erzeugung der Gittervorspannung)*

g. bias voltage Gittervorspannung *f*
g. blocking Gitterblockierung *f* durch statische Aufladung, Gitterverriegelung *f*
g. blocking capacitor Gitterblockkondensator *m*
g. branch Gitterkreis *m*, Gitterzweig *m*
g. cap Gitteranschluß[kappe *f*] *m*, Gitterkappe *f*
g. capacitor Gitterkondensator *m*
g.-cathode capacitance Gitterkatodenkapazität *f*
g. characteristic Gitterstrom-Gitterspannungs-Kennlinie *f*
g. circuit Gitterkreis *m*
g.-circuit rectification Gitter[strom]gleichrichtung *f*, Audiongleichrichtung *f*
g. clip Gitteranschluß *m*, Gitterkappe *f*, Gitterclip *m*
g. control Gittersteuerung *f*
g. controlled rectifier gittergesteuerter Gleichrichter *m*
g. cooler *(Wärm)* Rostkühler *m*
g. coupling Gitterkreiskopplung *f*
g. current Gitterstrom *m*
g. current characteristic Gitterstromkennlinie *f*
g. current detection Audiongleichrichtung *f*, Gittergleichrichtung *f*
g. current modulation Gitterstrommodulation *f*
g. current point Gitterstromeinsatzpunkt *m*
g. current swing Gitterstromaussteuerung *f*
g. cut-off Gitterblockierung *f*
g. cut-off voltage Steuergittereinsatzspannung *f*
g. detection Gitter[strom]gleichrichtung *f*, Audiongleichrichtung *f*
g. detector Audion *n*, Gittergleichrichter *m*
g.-dip meter Absorptionswellenmesser *m*
g.-dip oscillator Grid-dip-Oszillator *m*
g. door Auswurfklappe *f (Baumwollöffner)*
g.-driving power Gittersteuerleistung *f*, Steuerleistung *f*
g. electrode Gitterelektrode *f (Akku)*
g. emission Gitteremission *f*
g. extinguishing Gittersperrung *f*, Gitterabschaltung *f (Stromrichter)*
g. faults Gitterstörungen *fpl*
g.-filament circuit Gittereingangskreis *m*
g. focusing Gitterfokussierung *f*
g. gas Ferngas *n*
g. gas line Ferngasleitung *f*
g. glow tube gittergesteuerte Glimmröhre *f*, Gasentladungsrelais *n*, Glimmrelais *n*
g. holes Rasterlöcher *npl*
g. hum Gitterbrummen *n*
g. inductance coil Gitterdrossel *f*
g. keying Gittertastung *f*
g. leak Gitterableitung *f*, Gitterableitwiderstand *m*
g. leak bias Gitterableitvorspannung *f*, [durch Abfall am Gitterableitwiderstand bewirkte] Gittervorspannung *f*
g. leak condenser Gitter[ableit]kondensator *m*
g. leak detection Audiongleichrichtung *f*, Gitter[strom]gleichrichtung *f*
g. leak detector Audion[gleichrichter *m*] *n*, Gitteraudion *n*
g. leak rectification Gitter[strom]gleichrichtung *f*, Audiongleichrichtung *f*
g. leak resistance Gitterableitwiderstand *m*
g. leak voltmeter Röhrenvoltmeter *n* mit Gittergleichrichtung
g. limiting Gitterstrombegrenzung *f*
g. line Rasterlinie *f*; Systemlinie *f*
g. mesh Gittermasche *f*
g. micrometer Netzmikrometer *n*
g. modulation Gittermodulation *f*
g.-noise resistance Gitterrauschwiderstand *m*
g. pin Gitterstift *m*
g. pitch Ganghöhe (Steigung) *f* des Gitters; Ganghöhe der Gitterdrähte
g. plate Gitterplatte *f (Sammler)*
g.-plate capacitance Gitteranodenkapazität *f*
g.-plate [trans]conductance Steilheit *f (Elektronenröhren)*
g. point Gitterpunkt *m*
g. polarization voltage Gittervorspannung *f*
g. potential Gitterpotential *n*
g. potential anode-current characteristic Anodenstrom-Gitterspannungs-Kennlinie *f*
g. priming voltage Gittervorspannung *f*
g. ratio Rasterverhältnis *n*
g. rectification Audiongleichrichtung *f*, Gittergleichrichtung *f*
g. referencing *(Flg)* Gradnetzmeldeverfahren *n*
g. resistor Gitterdämpfungswiderstand *m*, Gittervorwiderstand *m*
g. return Gitterrückleitung *f*
g. shielding can Gitterabschirmhaube *f*
g. space Gitterraum *m*
g. spacing Rasterteilung *f*
g. square Planquadrat *n (Navigation)*
g. stopper Gitterschutzwiderstand *m*; Gittersperrkreis *m (Radio)*
g. sweep doppelter Gitteraussteuerungsbereich *m*
g. system Verbundnetz *n*
g. tank Gitterschwingkreis *m*
g. therapy Rastertherapie *f*
g.-to-plate capacitance Anodengitterkapazität *f*, Durchgriffskapazität *f*
g. transparency *(El)* Durchgriff *m*
g. tray column *(Ch)* Siebbodenkolonne *f*
g. valve Gitterschieber *m*

g. voltage swing Gitteramplitude *f*, Gitterspannungsschwankung *f*
g. winding Gitterspule *f*
g. wire Richtdraht *m (Flugkompaß)*
gridded chamber Gitterionisationskammer *f*
griddle Rost *m*, Rätter *m*
gridiron *(Schiff)* Balkenrost *m*, Kielbank *f*, Trockengräting *f*, Rostdock *n*
grief stem Mitnehmerstange *f (Rotarybohren)*
grillage *(Bw)* Fundamentrost *m*; Gründungsrost *m*
g. foundation *s.* grillage
grille *(Bw)* Gitter *n*, Rost *m*; Jalousie *f*
g.-type microphone Gittermikrofon *n*
grind / to 1. [zer]mahlen, zerkleinern; 2. schleifen; 3. abreiben *(Farben)*; 4. radieren *(Reifen)*; 5. knirschen *(Getriebe)*; 6. schärfen, wetzen *(Sense, Sichel)*
g. across the grain *(Holz)* querschleifen
g. and mix kollern *(Mischkollergang)*
g. bright blankschleifen
g. concentric rundschleifen
g. cylindrical rundschleifen
g. down abschleifen
g. glass mattschleifen
g. in einschleifen
g. into a powder *s.* g. to powder
g. off abschleifen
g. rough vorschleifen
g. splines Keilflanken schleifen
g. to powder zu Pulver mahlen
g. to size auf genaues Maß schleifen
g. two parallel surfaces simultaneously planparallelschleifen
g. valves Ventile einschleifen
g. with emery [ab]schmirgeln
g. wood across the grain *(Holz)* querschleifen
grind Schliff *m*; Nachschliff *m*
g. fineness of paint Vermahlungsgrad *m* eines Anstrichstoffes
grindability 1. Schleiffähigkeit *f*; 2. Mahlbarkeit *f*, Zermahlbarkeit *f*
grindable 1. schleiffähig; 2. zermahlbar
grinder 1. Schleifmaschine *f*, Schleifer *m*; 2. atmosphärische Störung *f*; 3. Mahlstein *m*
g. case Schleifergehäuse *n*
grinders Brodelstörung *f (Funkempfang)*
grindery plant Schleiferei *f*
grinding 1. Schleifarbeit *f*, Schleifen *n*, Schliff *m*; Schärfen *n*; 2. Zermahlen *n*, Zerkleinern *n*
g. accuracy to size maßgenaues Schleifen *n*
g. allowance Schleifaufmaß *n*; Schleifzugabe *f*
g. angular work Schrägschleifen *n*
g. area Schleifzone *f*
g. attachment Schleifeinrichtung *f*, Schleifgerät *n*
g. ball Mahlkugel *f*
g. bodies Mahlkörper *mpl*
g. brush *(Text)* Verstreichbürste *f*
g. check Schleifriß *m*
g. clearance Schleifwinkel *m*
g. compound Einschleifpaste *f*, Schleifpaste *f*
g. concave surfaces Konkavschleifen *n*
g. convex surfaces Konvexschleifen *n*
g. crack Schleifriß *m*
g. cylindrical work Rundschleifen *n*
g. disk *s.* g. wheel
g. drum Mahltrommel *f*
g. error Schleiffehler *m*
g. fineness Mahlfeinheit *f*
g. fixture Schleifeinrichtung *f*, Schleifgerät *n*, Schleifvorrichtung *f*
g. gauge Schleiflehre *f*
g. head Schleifkopf *m*
g. heat Schleifwärme *f*
g.-in Einschleifen *n*
g.-in paste Einschleifpaste *f*
g. length Schleiflänge *f*
g. life *(Am)* Standzeit *f*
g. machine Schleifmaschine *f*
g. machine for plane irons Hobelmesserschleifmaschine *f*
g. machine with swivelling wheel head Schleifmaschine *f* mit schwenkbarem Schleifkopf
g. material 1. Schleifmittel *n*; 2. Schleifrohstoff *m*
g. oil Schleiföl *n*
g. operation 1. Schleifvorgang *m*, Schleifarbeit *f*; 2. Mahlgang *m*, Mahlarbeit *f*
g. outfit Schleifausrüstung *f*
g. path Mahlgang *m*
g. pedestal Schleifbock *m*
g. pencil Schleifer *m (Werkzeug)*
g. plant Mahlanlage *f*
g. practice Mahltechnik *f*
g. property Vermahlungsfähigkeit *f*
g. quality Schliffgüte *f*
g. raw material Schleifrohstoff *m*
g. resin dispergiertes Harz *n*, Organosol *n*
g. sand Schleifsand *m*
g. slideways Führungsbahnschleifen *n*
g. sludge Schleifschlamm *m*
g. stock Mahlgut *n*
g. stone Schleifstein *m*, Wetzstein *m*
g. swarf Schleifschmant *m*
g. table Schleiftisch *m*
g. test Reibprobe *f*; Schleifprobe *f*
g. throat Schleifspalt *m*
g. time Schleifzeit *f*
g. wheel Schleifscheibe *f*, Schleifkörper *m*
g. wheel arbor Aufspanndorn *m* für Schleifkörper *m*
g. wheel balance Wuchtzustand *m* der Schleifscheibe

g. **wheel bond** Schleifscheibenbindung *f*
g. **wheel chuck** Schleifscheibenfutter *n*
g. **wheel crushing** Einrollverfahren *n*
g. **wheel dresser** Schleifscheibenabrichter *m*, Abrichteinrichtung *f*; Abrichtwerkzeug *n*
g. **wheel dresser capable of being used for shaping angles** Winkelabrichteinrichtung *f*
g. **wheel dresser for shaping radii** Radienabrichteinrichtung *f*
g. **wheel flange** Schleifscheibenflansch *m*
g. **wheel guard** Schleifscheibenschutzhaube *f*
g. **wheel head** Schleifkopf *m*
g. **wheel in vitrified bond** Schleifkörper *m* in keramischer Bindung
g. **wheel mount** Schleifscheibenaufnahme *f*, Schleifscheibenbefestigung *f*
g. **wheel slide** Schleifscheibensupport *m*
g. **wheel speed** Scheibendrehzahl *f (Schleifen)*
g. **wheel spindle** Schleifspindel *f*
g. **wheel spindle axis** Schleifspindelachse *f*
g. **wheel spindle bearing** Schleifspindellagerung *f*
g. **wheel spindle head** Schleifspindelstock *m*
g. **wheel truing attachment** Schleifscheibenabrichteinrichtung *f*
g. **work parallel** Planparallelschleifen *n*
grindings Schleifspäne *mpl*
grindstone Naturschleifstein *m*, Schleifstein *m*
grinning Rißbildung *f (im Putz)*
grip / to fassen, greifen, klemmen; einspannen, verspannen *(Spannbacken)*
g. **the selvedges** *(Text)* Leisten einklemmen
grip 1. Griff *m*; Heft *n*; Schaft *m*; Schenkel *m*; 2. Klemmlänge *f (Niet)*; 3. Reibschluß *m*; 4. Kabelziehstrumpf *m*, Kabelziehschlauch *m*; 5. *(Text)* Griffigkeit *f*
g. **bar** Betonformstahl *m*, Rippenstahl *m*
g. **car** *(Brg)* Förderwagen *m* mit Mitnehmer
g. **end** Griffstück *n (Manipulator)*
g. **feed** Greiferzuführung *f*, Zangenzuführung *f*
g. **holder** Spannkopf *m*
g. **pattern** Griffmuster *n*
g. **screw** Faßschraube *f*
g. **wrench** Blitzrohrzange *f*
gripe lashing *(Schiff)* Bootszurring *n (am Davit)*
g. **of the stem** *(Schiff)* Vorstevenunterteil *n (Übergang vom Kiel zum Vorsteven)*
griping *(Schiff)* luvgierig
gripper carriage Einspannwagen *m (Metallbearbeitung)*
g. **disk** Greiferscheibe *f*
g. **feed** Zangenvorschub *m (Presse)*
g. **fork** Mitnehmergabel *f*, Seilgabel *f (Seilbahn)*
g. **loom** Greiferwebstuhl *m*
g. **rail** Greiferschienen *fpl (Vorschubeinrichtung)*
gripping Einspannung *f*, Festspannung *f*; Greifen *n*; Fassen *n*
g. **action** Spannwirkung *f*
g. **appliance** Spannvorrichtung *f*, Spannwerkzeug *n*
g. **cam** Spannkurve *f*
g. **capacity** Spannkraft *f (einer Spannvorrichtung)*
g. **collet** Spannpatrone *f*
g. **cone** Spannkegel *m*
g. **device** Einspannvorrichtung *f (Zerreißprüfung)*; Spannkopf *m (Werkzeugmaschinen)*
g. **effect** Spannwirkung *f*, Spannkraft *f*; *(Text)* Klemmwirkung *f (Streckwalzen)*
g. **jaw** Spannbacke *f (Werkstoffprüfung)*
g. **lever** Spannhebel *m*
g. **mechanism** Spannvorrichtung *f*
g. **member** Spannteil *n*
g. **of rolling piece** *(Met)* Angreifen (Fassen) *n* der Walzen *(Walzgut)*
g. **of the fibres between the rollers** *(Text)* Klemmung *f* der Fasern zwischen den Walzen *(Strecken)*
g. **plate** Spannplatte *f*
g. **pliers** Haltezange *f*
g. **position** Spannlage *f*, Spannstelle *f*
g. **power** Spannkraft *f*
g. **surface** Grifffläche *f*
g. **yoke** Spannbrücke *f*
grist Mahlgut *n*
grit / to bekiesen; Sand (Splitt) streuen
grit 1. *(Bw)* Grieß *m*, Grobsand *m*, Kies *m*, Splitt *m*; 2. *(Masch)* Abrieb *m*, Schleifstaub *m*
g. **blasting** Sanden *n*; Putzstrahlen *n* mit Strahlkies
g. **chamber** Sandfang *m*
g. **size** Körnung *f*
grits Schleifstaub *m*
gritter Sandstreugerät *n*
grittiness Sandhaltigkeit *f*
gritty 1. kiesartig, kiesig, körnig, sandartig, sandig; 2. körnig, spröde *(Metall)*
grizzley *(Am)* *s.* grizzly
grizzly *(Brg)* Roststab *m*, Sieb[rost]stab *m*
g. **bar** *s.* grizzly
g. **chamber** Rostkammer *f*
g. **feeder** Aufgabestangenrost *m*
g. **level** Rostsohle *f*
g. **rolls** Walzenrost *m*
grog gebrannter Ton *m*, Schamotte *f*
g. **mortar** Schamottemörtel *m*
groin 1. *(Bw)* Grat *m*, Rippe *f (Gewölbe)*; 2. *(Schiff; Am)* Buhne *f*
groined vault Kreuzgewölbe *n*
g. **vaulting** Kreuzgewölbe *n*
grommet 1. Gummiring[dichtung *f*] *m*, Gummi-

dichtungsring *m*; 2. *(El)* isolierte Durchführung *f*, Durchführungshülse *f*, Durchführungstülle *f*; 3. Unterlegscheibe *f*; 4. Öse *f*

groove / to nuten; kalibrieren; auskehlen; kerben; riefen; riffeln; kehlen

g. and tongue *(Holz)* spunden

groove 1. Nut *m*, Nute *f*; Kerbe *f*; Kehle *f*; Furche *f*; Lücke *f*; Raste *f*; Riefe *f*; 2. Fuge *f* *(Schweißen)*; 3. Rille *f* *(Schallplatte)*; 4. Bodenrille *f*

g. and tongue 1. *(Bw)* Nut *f* und Feder *f*; 2. *(Holz)* Spund *m*

g. angle 1. Öffnungswinkel *m* der Schweißfuge; 2. Rillenwinkel *m* *(Schallplatte)*

g. face Rillenoberfläche *f* *(Schallplatte)*

g. joint Stumpfnaht *f*

g. jumping Überspringen *n* [von Rillen] *(Schallplatten)*

g. milling machine *(Holz)* Langlochfräsmaschine *f*, Nutenfräsmaschine *f*

g. of insulator Isolatorenhals *m*

g. of lay for the reed *(Holz)* Blattfuge *f*

g. of ski Führungsrille *f*

g. opening *(Met)* Kaliberöffnung *f*

g. parallelism with side Axialschlag *m* *(Wälzlager)*

g. pulley [for V-belts] Keilriemenscheibe *f*

g. roll *(Met)* Kaliberwalze *f*

g. shape Rillenquerschnitt *m* *(Schallplatte)*

g. spacing Rillenabstand *m*, Füllgrad *m* *(Schallplatte)*

g. under the rim of the disk Eindrehung *f* des Scheibenrandes *(Turbine)*

g. wear *(Met)* Kaliberverschleiß *m*

grooved [aus]gekehlt, gefurcht, gekerbt, genutet, gerieft, geriffelt, gerillt

g. and tongued gespundet

g. ball bearing Rillenkugellager *n*

g. cylinder *(Met)* Kaliberwalze *f*

g. pin Kerbstift *m*

g. piston ring *(Kfz)* Schlitzring *m*

g. pulley Keilriemenscheibe *f*

g. roll profilierte Walze *f*

g.-shank drill Bohrer *m* mit genutetem Schaft

g. trolley wire Profilfahrdraht *m*, Rillenfahrdraht *m*

grooves for anchoring white metal to liner bush Verklammerungen *fpl* für Lagerausgüsse

grooving 1. *(Masch)* Auskehlung *f*, Riefenbildung *f*, Riefung *f*; Nutung *f*; 2. rillenförmige Abzehrungen *fpl* *(Kesselkorrosion)*

g. and tonguing machine Spundmaschine *f*

g. and tonguing of boards Spunden *n* von Brettern

g. cutter Nut[en]fräser *m*

g. cutter block *(Holz)* Schlitzscheibe *f*

g. head Nutenschneidkopf *m*

g. plane Falzhobel *m*, Nuthobel *m*, Spundhobel *m*

g. saw Nutkreissäge *f*

gross 1. brutto; 2. grob, roh

g. amount Bruttobetrag *m*

g. calorific value Verbrennungswärme *f*, oberer Heizwert *m*

g. efficiency Gesamtwirkungsgrad *m*

g. floor area Gesamtfläche *f*

g. heat rate Bruttowärmeleistung *f*

g. heat value oberer Heizwert *m*

g. income Bruttoeinkommen *n*

g. knot grober Ast *m*

g. output Bruttoleistung *f*; Rohförderung *f*

g. profit Bruttogewinn *m*

g. reaction *(Ch)* Bruttoreaktion *f*

g. receipts Bruttoeinnahmen *fpl*, Roheinnahmen *fpl*

g. register ton *(Schiff)* Bruttoregistertonne *f*

g. structure analysis Grobstrukturanalyse *f*

g. tonnage Bruttoraumgehalt *m*, Bruttotonnengehalt *m*

g. wage Bruttolohn *m*

g. weight 1. Bruttomasse *f*; 2. *(Kfz, Flg)* Gesamtmasse *f*; 3. *(Text)* Rohmasse *f*

grossular *(Min)* Grossular *m*, grüner Granat *m*

ground / to 1. *(Schiff)* auflaufen, auf Grund geraten; 2. *(Flg)* am Aufstieg hindern; 3. *(Am)* *(El)* erden; 4. grundieren

g. a line eine Leitung erden

ground geschliffen; geschärft; gemahlen

ground 1. Grund *m*, Boden *m*; 2. Grund *(logisch)*; 3. *(Am)* *(El)* Erde *f*; 4. *(Brg)* Gebirge *n*, Gestein *n*

g. absorption *(El)* Dämpfung *f* der Bodenwelle, Bodenabsorption *f*

g. aerial Erdantenne *f*

g.-air communication *(Flg)* Boden-Bord-Verkehr *m*

g.-air radiotelephony *(Flg)* Boden-Bord-Funksprechverkehr *m*

g. angle shot Aufnahme *f* aus der Froschperspektive *(Fotografie, Film)*

g. anti-aircraft control unit Flakzielgerät *n* *(Radar)*

g.-attack aircraft Schlachtflugzeug *n*

g. auger Erdbohrer *m*

g. back-scattering Bodenrückstreuung *f* *(Radiowellen)*

g. balancing *(El)* Erdausgleich *m*

g. basic slag Thomas[schlacken]mehl *n*

g. boosting Bodenaufladung *f*

g.-breaking erster Spatenstich *m* *(symbolischer Beginn von Bauarbeiten)*

g. bus wire Erdsammelleitung *f*

g. cable Erdkabel *n*, Untergrundkabel *n*, unterirdisches (unterirdisch verlegtes) Kabel *n*

g. chain *(Schiff)* Mooringkette *f*

g. chisel Bodenmeißel *m*
g. circular rundgeschliffen
g. clamp Erdanschlußklemme *f*, Erd[ungs]-klemme *f*, geerdete Klemme *f*
g. clearance *(Kfz)* Bodenfreiheit *f*, Bauchfreiheit *f*
g.-clearance indicator Funkecholot *n*, Radarhöhenmesser *m*
g. clearance of the frame *(Kfz)* Rahmenhöhe *f*
g. clutter Bodenecho *n*
g. coat[ing] Grundanstrich *m*, Grundierung *f*
g. consolidation Baugrundverdichtung *f*, Bodenverfestigung *f*
g. control box *(Rak)* Ventilkasten *m*, Steuerkasten *m*
g. control landing *(Flg)* [vom Boden geleitete] Blindlandung *f*
g.-controlled approach *(Flg)* 1. Bodenpeilanlage *f*; 2. bodenseitig geleiteter Anflug *m*, Funkmeßlandung *f*, Radarlandeverfahren *n*; bodengesteuerte Radaranlage *f* für Blindlandungen; Blindlandesystem *n* mit Boden-Bord-Sprechfunk
g.-controlled interception *(Mil)* Flugabwehr *f* durch vom Boden geleitete Jäger; bodengesteuertes Radarleitgerät *n (Leitverfahren)*
g. controlled landing *(Flg)* vom Boden überwachte Landung *f*; Blindlandesystem *n* mit Boden-Bord-Sprechfunk
g. controller *(Flg)* Bodenfunkstelle *f*
g. crew *(Flg)* Bodenpersonal *n*
g. damp Bodenfeuchtigkeit *f*
g. detector Erdschlußanzeiger *m*
g. direction finder Bodenpeiler *m*
g. direction finding [einfache] Fremdpeilung *f*
g. distance Horizontalabstand *m*
g. electrode Masseelektrode *f*
g. equalizer conductor Antennenabgleichelement *n*
g. equalizer inductor Abgleichspule *f*, Ausgleichspule *f (Antennen)*
g. equipment *(Rak)* Bodenanlage *f*
g. fault Störung *f* durch Erdschluß
g.-fault neutralizer Erdschlußlöschvorrichtung *f*
g. floor Erdgeschoß *n*
g. fog Bodennebel *m*
g. friction Bodenreibung *f*
g. frost Bodenfrost *m*
g. glass mattiertes Glas *n*; Mattscheibe *f*
g.-glass stop cock Schliffhahn *m (Bürette)*
g.-glass stopper eingeschliffener Glasstopfen *m*, Schliffstopfen *m*
g.-glass tap Schliffhahn *m (Bürette)*
g.-grip tyre Geländereifen *m*
g.-hog *(Brg)* Gegengewicht *n* im Bremsberg
g. humidity Bodenfeuchtigkeit *f*
g. ice Grundeis *n*
g. idling condition *(Flg)* Leerlaufbetriebszustand *m*
g. imperfections schlechte Erdungsverhältnisse *npl*
g.-in eingeschliffen
g. indication Erdschlußanzeige *f*
g. indicator Erdschlußanzeiger *m*
g. installation 1. *(Rak)* Bodenanlage *f*; 2. Erdungsanlage *f*
g. joint eingeschliffene Verbindung *f*
g. joint female part Schliffhülse *f*
g. joint male part Schliffkern *m*
g.-joist Schwellenholz *n*
g. keel *(Schiff)* Scheuerkiel *m*
g. lead Erdungsleitung *f*
g. leak Verlust *m* durch Erdschluß, Erdschluß *m*
g. leak[age] on two phases Doppelerdschluß *m*
g. level 1. *(Bw)* Geländehöhe *f*, OK-Terrain *n*, Terrainhöhe *f*; 2. *(Ph)* Grundterm *m*, Energie *f* des Grundzustandes
g. line Grundlinie *f*
g. location Erdschlußbestimmung *f*
g. loop Erdrückleitungsweg *m*, Erdungsschleife *f (bei Analogrechnern Vereinigung der Rückleitungen an einer Stelle)*
g.-mass *(Geol)* Matrix *f*, Grundmasse *f*
g. mat 1. Turnmatte *f*; 2. Erd[ungs]-drahtnetz *n (Antenne)*
g. mine *(Mil)* Grundmine *f*
g. moraine *(Geol)* Grundmoräne *f*
g.-mounted winder Flurfördermaschine *f*
g. movement Bodenbewegung *f*
g. noise *(El)* Eigenrauschen *n*, Grundgeräusch *n*
g. of the channel *(Schiff)* Kanalsohle *f*
g. plan Grundriß *m*
g. plane 1. Horizontalebene *f*; 2. Grundhobel *m*
g. plane iron Grundhobeleisen *n*
g. plate 1. *(El)* Erdungsplatte *f*; 2. *(Bw)* Bodenschwelle *f*, Schwelle *f*
g. position finding Fremdortung *f*
g. position indicator Ortungsgerät *n (Radar)*; Standortanzeiger *m* ohne Höhenkorrektur
g. pressure Bodendruck *m*, Bodenpressung *f*; Flächenpressung *f*
g. projection plane Grundrißebene *f*
g. radar Bodenradar *n*, Bodenfunkmeßgerät *n*
g. radio beacon Bodenfunkfeuer *n*
g. radio station Bodenfunkstelle *f*
g. range *(Fmt)* horizontale Reichweite *f*, Reichweite *f* am Erdboden
g. ray *(Rak)* Bodenstrahl *m*
g. reflection Bodenreflexion *f*
g. relay Erdschlußrelais *n*

g. resistance Erdungswiderstand *m*
g. return 1. *(El)* Erdrückleitung *f*; 2. Bodenecho *n (Radar)*
g. return current Erdrückstrom *m*
g. sample Metallschliff *m*
g. slag Schlackenmehl *n*
g. sluicing Schwemmarbeit *f* mittels Kanälen *(Seifenbergbau)*
g. state Grundzustand *m (Atom, Molekül)*
g. state disintegration [radioaktiver] Zerfall *m* in den Grundzustand
g. station Bodenstation *f*
g. station direction finder Bodenpeiler *m*
g. strap *(El)* Erdungslasche *f*
g. structure beneath the foundation (building) Baugrund *m*
g. surveillance radar Rundsuchradar[anlage *f*] *n*
g. survey Geländeaufnahme *f*
g. swell *(Schiff)* Grunddünung *f*
g. switch Erd[ungs]schalter *m*
g. system Erdungsanlage *f*
g. tackle *(Schiff)* Ankergeschirr *n*, Grundgeschirr *n*
g. tank Tiefbehälter *m*
g. terminal Erdanschlußklemme *f*, Erd[ungs]-klemme *f*, geerdete Klemme *f*
g. texture Grundgewebe *n*
g.-thread tap Gewindebohrer *m* mit geschliffenem Gewinde
g.-to-airplane communication Funkverkehr *m* vom Boden zum Flugzeug
g.-to-fit auf Passung geschliffen
g. transformer Erdungstransformator *m*, Erdungswandler *m*, Erdungsdrossel *f*
g. warp *(Text)* Einschnittkette *f*, Grundkette *f*
g. water Grundwasser *n*
g.-water bottom Grundwassersohle *f*
g.-water level Grundwasserspiegel *m*
g.-water lowering Grundwasserabsenkung *f*
g.-water pressure Grundwasserdruck *m*
g.-water table Grundwasserspiegel *m*
g. wave *(Fmt)* Bodenwelle *f*
g. ways *(Schiff)* Ablaufbahn[gründung] *f (Helling)*
g. wire *(El)* Erdleitung *f*
g. wood Holzschliff *m*
g.-wood pulp Holzstoff *m*
g. zero Hypozentrum *n (Erdbeben)*
groundage Hafenliegegebühr *f*, Ankergebühr *f*
grounded 1. geerdet; 2. *(Flg)* am Aufstieg gehindert; 3. aufgelaufen; 4. grundiert
g.-anode circuit Anodenbasisschaltung *f*
g.-base circuit Basisschaltung *f (Transistor)*
g.-cathode amplifier Katodenbasisverstärker *m*
g.-cathode circuit Katodenbasisschaltung *f*
g.-collector circuit Kollektorbasisschaltung *f*
g.-emitter circuit Emitterschaltung *f (Transistor)*
g. end geerdete Seite *f*
g. grid *(El)* Gitterbasis *f*
g.-grid amplifier Gitterbasisschaltung *f*, Gitterbasisverstärker *m*
g.-grid circuit Gitterbasisschaltung *f*
g.-grid triode Gitterbasistriode *f*
g. pole geerdeter Pol *m*
grounding 1. *(El)* Erden *n*; 2. *(Flg)* Festhalten *n* am Boden; 3. *(Schiff)* Auflaufen *n*; 4. Grundierung *f*
g. checkup *(Flg)* Masseschlußprüfung *f*
g.-choke coil Erdungsdrossel[spule] *f*, Entladespule *f*
g. coil *s.* g.-choke coil
g. collar Erdungsring *m*
g. contact Erdungskontakt *m*
g. device Erdungsanlage *f*
g. keel *(Schiff)* Scheuerkiel *m*
g. plane iron Falzhobeleisen *n*
g. ring Erdungsring *m*
g. switch Erd[ungs]schalter *m*
g. terminal Erdanschlußklemme *f*, Erd[ungs]-klemme *f*, geerdete Klemme *f*
groundmass Grundmasse *f*
grounds Gelände *n*
groundwork Erdarbeit *f*; Fundament *n*, Gründung *f*
group / to gruppieren, [in Gruppen] anordnen, zusammenstellen
group 1. *(Math)* Gruppe *f*; 2. *(Geol)* Formation *f*
g. amplifier Gruppenverstärker *m*
g. casting Gespannguß *m*
g. centre *(Fmt)* Verteilerfernamt *n*
g. codes for prescribed error patterns *(BMSR)* Gruppenkodes *mpl* für vorgeschriebene Fehlermuster
g. control change Gruppenwechsel *m (Lochkartenmaschinen)*
g. delay *(Ph)* Gruppenverzögerung *f*; Laufzeitverlängerung *f*
g. delay characteristic Laufzeitentzerrung *f*
g. delay-frequency distortion Laufzeitverzerrung *f*
g. delay meter Gruppenlaufzeitmesser *m*
g. delay time Gruppenlaufzeit *f*
g. distribution frame Gruppenverteiler *m (Trägertelefonie)*
g. distributor *(Fmt)* Gruppenverteiler *m*
g. drive *(Masch)* Gruppenantrieb *m*
g. flashing light *(Flg)* Blinkkennfeuer *n*
g. frequency *(Ph)* Gruppenfrequenz *f*
g. fusing *(El)* Gruppenabsicherung *f*
g. indication Gruppenanzeige *f (Lochkartenmaschinen)*

g. indication cycle Gruppenanzeigegang *m (Lochkartenmaschinen)*
g. indication elimination Gruppenanzeigeelimination *f (Lochkartenmaschinen)*
g. interchangeability Gruppenaustauschbarkeit *f*
g. modulator *(El)* Gruppenumsetzer *m*
g. of beddings *(Geol)* Schichtgruppe *f*
g. of conductors Adernbündel *n (Kabel)*
g. of contacts Kontaktkranz *m (Fernsprechwähler)*
g. of curves *(Math)* Kurvenschar *f*
g. of incoming lines (trunks) *(Fmt)* ankommendes Leitungsbündel *n*, Zubringerbündel *n*
g. of lines Leitungsbündel *n*
g. of nozzles Düsengruppe *f (Turbine)*
g. of outgoing lines (trunks) *(Fmt)* abgehendes Leitungsbündel *n*
g. of parts Baugruppe *f*
g. of plates Plattensatz *m*
g. of speeds Drehzahlreihe *f*
g. of stages Stufengruppe *f (Turbine)*
g. of strata *(Geol)* Schichtkomplex *m*
g. of threads *(Text)* Schärgang *m*
g. precipitant *(Ch)* Gruppenfällungsreagens *n*
g. retardation Gruppenlaufzeit *f (Radiowellen)*
g. selector *(Fmt)* Gruppenwähler *m*
g. selector for no-delay service Gruppenwähler *m* für Schnellverkehr
g. selector for trunk traffic Gruppenwähler *m* für Fernverkehr
g. sorting device Leitkartensortiervorrichtung *f*
g. switch Gruppenschalter *m*
g. technology Gruppenbearbeitung *f (Mitrofanowmethode)*
g. teeming stool *(Met)* Gespannplatte *f*
g. theory *(Math)* Gruppentheorie *f*
g. translator *(El)* Gruppenumsetzer *m*
g. velocity Gruppengeschwindigkeit *f*
g. wave delay frequency distortion Laufzeitverzerrung *f (Radio)*
grout / to vergießen, untergießen *(mit Zement)*; [ver]gipsen
g. in einpressen *(Mörtel)*
g. the steam engine die Dampfmaschine untergießen
g. up auspressen *(Spannbeton)*
g. with cement mit Zement ausgießen
grout Auspreßmörtel *m*, Einpreßmörtel *m*; feiner (dünner) Mörtel *m*; Vergußmaterial *n*
grouting Einpressen *n* von Zementmörtel; Vermörteln *n*, Vermörtelung *f*; Umgießen *n (Maschine mit Zement)*; *(Brg)* Verfestigen *n*, Versteinen *n*, Zementieren *n*
g. compound Vergießmasse *f*
g. hole Eingußloch *n (Turbine)*
g. of the ground Zementieren *n*, Versteinen *n (des Untergrundes durch Einpressen von Zementmilch)*
g. with liquid cement Zementeinpressung *f*
grow / to [an]wachsen, zunehmen; *(Lw)* züchten; keimen
g. cold erkalten, kalt werden
g. dark dunkel werden, dunkeln
g. hot heiß werden
g. mouldy schimmlig werden, Schimmel ansetzen
g. rampant wuchern *(Pflanzen)*
g. rich sich anreichern
g. rusty einrosten
g. warm warm werden, sich erwärmen
growable kultivierbar *(Nutzpflanzen)*
grower Pflanzenzüchter *m*
growing 1. *(Lw)* Wachsen *n*; 2. *(Ph)* Züchten *n (Kristalle)*
g. crop [sprießende] Saat *f*
g. of crystals Kristallzüchtung *f*
growler *(El)* Ankerprüfgerät *n*
grown timber *(Schiff)* gewachsenes Spant *n*
growth 1. Zunahme *f*, Anwachsen *n*; 2. *(Ph)* Wachstum *n (Kristalle)*; 3. *(Lw)* Anbau *m*
g. condition Wachstumsbedingung *f*
g. curve 1. Wachstumskurve *f*, Zunahmekurve *f (Statistik)*; 2. Nachbildungskurve *f (Kernreaktor)*
g. factor Wachstumsanreger *m*, Wachstumsfaktor *m (Kristalle)*
g. layer *(Holz)* Wachstumsschicht *f*
g. of cast iron Wachsen *n* von Gußeisen
g. of trees Baumbestand *m*
g. per cent Zuwachsprozent *n*
g. ring Wachstumsring *m*; Jahresring *m (Baum)*
g. substance Wuchsstoff *m*
groyne / to Buhnen bauen
groyne Buhne *f*
grub / to 1. ausjäten, ausreuten, ausroden; 2. auflockern, graben *(Erde)*
grub-axe Rodehaue *f*
g. screw Madenschraube *f*
grubber Ausroder *m*, Grubber *m*, Grupper *m*, Reuter *m*, Roder *m*, Vorratsroder *m*; Kultivator *m*
grubbing Ausjäten *n*, Ausroden *n*; Kultivieren *n*
g. hoe Hacke *f*, Rodehaue *f*
g. machine Vorratsroder *m*
g. mattock Reuthaue *f*
g. screen Abreutersieb *n*
grubstone mortar Grundmörtel *m*
gruenerite *(Min)* Grünerit *m*
guaiac resin Guajakharz *n*
guanajuatite *(Min)* Selenwismutglanz *m*
guanidine Guanidin *n*
g. acetic acid Guanidylessigsäure *f*

guano Guano *m*, Vogeldünger *m*
g. drill Guanostreuer *m*
guanosine Guanosin *n*, Guaninribosid *n*
guanylic acid Guanidin-riboso-phosphorsäure *f*, Guanylsäure *f*
guarantee / to Garantie geben, garantieren, gewährleisten
guaranties of performance Leistungsgarantien *fpl*
guaranty Garantie *f*, Gewähr[leistung] *f*
g. card Garantieschein *m*
g. terms Garantiebedingungen *fpl*
g. test Garantieversuch *m*
guard 1. Schutzeinrichtung *f*; Schutzblech *n*; Schutzkorb *m*; 2. Wächter *m*; Schaffner *m*
g. band *(Fmt, Fs)* Schutzfrequenzband *n*, Sicherheitsabstand *m (zwischen zwei Kanälen)*
g. bar Schutzstange *f*
g. cam Sicherungsschloßteil *n (Strickmaschine)*
g. circuit Schutzschaltung *f*; Überwachungskreis *m (Empfangstechnik)*
g. cradle Schutznetz *n (unter Hochspannungsleitung)*
g. grille Schutzgitter *n*
g. plate Schutzblech *n*; Anlaufscheibe *f*, Gegenplatte *f*
g. position Sparstelle *f (Elektronenrechner)*
g. rail 1. Schutzgeländer *n*, Geländer *n*, Schutzschiene *f*; 2. *(Eis)* Zwangsschiene *f (Radlenker)*; Balkenschiene *f*, Anschlagschiene *f*
g. relay Halterelais *n*, Schutzrelais *n*, Sperrrelais *n*
g. ring 1. Schutzring *m*; Haltering *m*; 2. *(El)* ringförmige Hilfselektrode *f*, [elektrostatischer] Schutzring *(Meßkondensator)*
g. ring condenser Schutzringkondensator *m*
g. socket Schutztülle *f*
g. wire Fangdraht *m (Starkstromleitung)*; Schutzdraht *m (Meßgerät)*
guarded field ionization chamber Faßkammer *f*
guardrail *s.* guard rail
guardstone Eckstein *m (bei Toreinfahrten)*
gudgeon Metallbolzen *m*; Stift *m*; Drehzapfen *m*; Stirnzapfen *m*; Kreuzkopfzapfen *m*
g. pin *(Kfz)* Kolbenbolzen *m*, Kolbenstift *m*
g.-pin bearing Kolbenbolzenlager *n*
g.-pin bushing Kolbenbolzenbüchse *f*
g.-pin extractor Kolbenbolzenauszieher *m*
g.-pin hole Kolbenbolzenauge *n*, Bolzenauge *n*
g.-pin puller Kolbenbolzenauszieher *m*
gudmundite *(Min)* Gudmundit *m*
guejarite *(Min)* Guejarit *m*, Wolfsbergit *m (Kupferantimonglanz)*
guess warp *(Schiff)* Warpleine *f*, Verholleine *f*, Vertäuleine *f*
guhr Infusorienerde *f*, Kieselgur *f*

guidance 1. Lenkung *f*, Führung *f*; 2. Fernlenkung *f*, Fernsteuerung *f (Flugkörper)*
g. by radio Funkführung *f*, Funklenkung *f*, Funksteuerung *f*
g. counsellor *(Am)* Berufsberater *m*
g. of the saw blade Führung *f* des Sägeblattes
g. to the line of sight *(Rak)* Einlenkung *f*, Einlenkvorgang *m*
guide / to führen, leiten
guide 1. Führung *f*; 2. Führungsschaft *m*; Führungszapfen *m*; Lineal *n* zur Werkstückführung; 3. Kenngröße *f*; Richtwert *m*; Maß *n*; Ausgangsgröße *f*; 4. *(El)* Hohlleiter *m*; Wellenleiter *m*; 5. Spurlatte *f*, Leitbaum *m (Schachtförderung)*
g. baffle Leitblech *n (Destillationstechnik)*
g. bar Führungsschiene *f*, Gleitschiene *f*
g. beam Leitstrahl *m*
g.-beam aerial Leitstrahlantenne *f*
g.-beam receiver Leitstrahlbordgerät *n*
g.-beam superposition *(Rak)* Leitstrahlaufschaltung *f*
g.-beam unit Leitstrahlgerät *n*
g. bearing Führungslager *n*
g. blade Leitschaufel *f (Turbine)*
g. blade disk Leit[rad]scheibe *f*, Druckstufenscheidewand *f (Turbine)*
g. blade segment Leitschaufelsegment *n*, Leitschaufelsektor *m (Turbine)*
g. block Kreuzkopfschleifer *m*, Gleitklotz *m*, Schlitten *m*, Führungsschlitten *m*
g. bolt Führungsbolzen *m*; Abhebestift *m*
g. box Führungsbuchse *f*
g. bush[ing] Führungsbuchse *f*; Bohrbuchse *f*
g. cable Leitkabel *n*, Lotsenkabel *n*, Führungskabel *n*
g. cam Führungsnocken *m*
g. cap *(Wärm)* Lenkungskappe *f*
g. casing *(Rak)* Führungsbock *m*
g. channel 1. Kantrinne *f (Walzwerk)*; 2. Ringkanal *m*, Umführungskanal *m (Turbine)*
g. claw Führungsklaue *f*
g. comb *(Fmt)* Führungskamm *m*
g. cylinder Führungszylinder *m*
g. disk Leitrad *n (Turbine)*
g. drum Führungstrommel *f (Dreschmaschine)*
g. fillet Führungsleiste *f*
g. finger Führungsfinger *m*; Kopierfinger *m*; Zunge *f*
g. flange Bord *m (Rollenlager)*
g. fork *(El)* Führungsbügel *m*
g. fossil *(Geol)* Leitfossil *n*
g. groove Führungsrille *f*; Leitnut *f*
g. jaw Führungsbacke *f*
g. link Führungskettenglied *n (Zahnkette)*
g. lug Führungswarze *f*
g. mast *(Bw)* Führungsmast *m*

g.-number Leitzahl *f*
g. of the tail piston eingleisige Kolbenstangenführung *f*
g. opening Rahmenfenster *n*
g. passage Leitkanal *m*
g. path Führung[srille] *f*
g. piece Führungskörper *m*; Führungsstück *n*
g. pile Richtpfahl *m (Grundbau)*
g. pin *(Masch)* Führungsstift *m*, Führungszapfen *m*; Kopierstift *m*; *(Met)* Abhebestift *m (Formkasten)*
g. plate *(Masch)* Führungsleiste *f*, Führungsschiene *f*; Kopierschiene *f*; *(Kfz)* Gleitplatte *f*
g. pulley *(Masch)* Führungsrolle *f*, Leitrolle *f*, Umlenkrolle *f*
g. pulley of the band saw Bandführungsrolle *f*
g. rail Führungsschiene *f*, Leitschiene *f*, Laufschiene *f*
g. ring Führungsring *m*, Stellring *m*; Klauenring *m*
g. rod Führungsstange *f*, Leitstange *f*
g. roller Führungsrolle *f*, Führungswalze *f*, Leitrolle *f*; Umlenkrolle *f*
g. shoe Führungsschuh *m (Förderkorb)*
g. sleeve Führungsbuchse *f*, Führungsmuffe *f*
g. surface Führungsfläche *f*, Führungsbahn *f*
g. vane Laufschaufel *f*, Leitschaufel *f*, Umlenkschaufel *f*
g.-vane ring Leitkranz *m*
g. wall *(Bw)* Leitmauer *f*, Leitwand *f*
g. wave impedance Feld[wellen]widerstand *m (Hohlleiter)*
g. wavelength Leiterwellenlänge *f*
g. way cover Führungsbahnenschutz *m*
g. wheel Leitrad *n*, Leitrolle *f*
g. wire *s.* single-wire transmission line
guided aircraft missile ferngelenktes Flugzeuggeschoß *n*
g. missile ferngelenkter Flugkörper *m*
g. missile angle Aggregatwinkel *m*
g. rocket gesteuerte (gelenkte) Rakete *f*
g. wave leitungsgebundene Welle *f*
g.-wave propagation leitergebundene Wellenausbreitung *f*
g. weapon Fernlenkwaffe *f*
g. weapons establishment Forschungsinstitut *n* für ferngelenkte Geschosse
guider *(Opt, Ph)* Leitrohr *n*
guideway Führung[sbahn] *f*, Führungsfläche *f*
guiding bar Leitlineal *n*
g. cable Leitkabel *n*, Lotsenkabel *n*, Führungskabel *n*
g. centre *(Ph)* Gyrationszentrum *n*
g. disk Führungsscheibchen *n*
g. face of blade führende Schaufelfläche *f (Turbine)*
g. in paths Führung *f* in Bahnen
g. ocular head *(Opt)* Pointierungsokularkopf *m*
g. of the hot gases Heizgasführung *f*
g. pulley *s.* guide pulley
g. stud *(Rak)* Führungsbeschlag *m*
g. surface for sliding members Führungsfläche *f* für gleitende Teile
g. telescope Leitfernrohr *n*
Guillery impact testing machine Guillery-Hammer *m*
guillotine *(Typ, Opt)* Schneidemaschine *f*
g. attenuator Fähnchenabschwächer *m*, Tauchteiler *m (Hohlleiterwellen)*
g. door *(Brg)* Falltür *f (Skip)*
g. plate shear *s.* g. shear
g. shear Blechtafelschere *f*, Tafelblechschere *f*
g. shearing machine Blechtafelschere *f*, Tafelschere *f*
g. veneer cutting machine schwere Furnierfügemaschine *f (Holzbearbeitung)*
guipure lace *(Text)* Luftspitze *f*
guitar Gitarre *f*
gulch gold Seifengold *n*
Guldin's rule Guldinsche Regel *f*
gull wing *(Flg)* Knickflügel *m*
gullet Zahngrund *m*; Zahnlücke *f*; Rundung *f* am Zahnfuß
g. tooth Kransägezahn *m (Sägeblatt)*
g. tooth system Wolfszahnung *f*
gully Rinne *f*, Wasserablauf *m*; Einlaufschacht *m*, Gully *m*, Straßeneinlauf *m*
gulonic acid Gulonsäure *f*
gum / to 1. gummieren, mit Klebstoff (Gummilösung) bestreichen; 2. gummiartig werden, klebrig werden; verharzen, verpichen
g.-up abreiben, abschleifen; verharzen
gum 1. Klebstoff *m*, Gummilösung *f*; 2. Gummiharz *n*, Pflanzengummi *m*; 3. Abdampfrückstand *m*; Harzrückstand *m (Öl)*; 4. Kohlenklein *n*
g. arabic Gummiarabikum *n*, Akaziengummi *m*
g. elastic Gummielastikum *n*
g. formation Verharzung *f*
g. inhibitor *(Ch)* Stabilisator *m*
g. mastic Mastix *m*
g. pocket *(Holz)* Harzgalle *f*
g. printing *(Typ)* Gummidruck *m*
g. printing varnish Gummidrucklack *m*
g. resin *(Holz)* Schleimharz *n*
g. rosin Zapfkolophonium *n*
g. spirit of turpentine Terpentinöl *n*
g. sugar Arabinose *f*
g. test Abdampfprobe *f (Erdöl)*
g. tragacanth Tragantgummi *m*
g. waste *(Text)* unentbasteter Seidenabfall *m*
gumbo *(Am) (Bw)* hochplastischer zäher Ton *m*

gumenite *(Min)* Uranogummit *m*, Gummit *m*, Gummierz *n*
gummed paper gummiertes Papier *n*
g. tape Klebstreifen *m*
g. with old oil mit Ölrückständen behaftet
gummer 1. Schärfmaschine *f (Holzbearbeitung)*; 2. *(Brg)* Schrämkleinräumer *m*
gumming[s] *(Brg)* Schrämklein *n*
gummy harzartig; klebrig
gumthrower *(Brg)* Schrämkleinräumer *m*
gun 1. Pistole *f*, Presse *f*, Spritze *f*; 2. *(Mil)* Gewehr *n*; Kanone *f*; *(Am)* Revolver *m*; 3. Jagdflinte *f*; 4. *(Ch)* Druckgefäß *n*, Druckrohr *n*; 5. *s.* electron gun
g. base Geschützbettung *f*
g. bore brush *(Mil)* Rohrreiniger *m*
g.-boring cutter head Tieflochbohrkopf *m*
g.-camera Foto-MG *n*, Schußkamera *f*
g.-carriage Lafette *f*
g. cotton Schieß[baum]wolle *f*
g. drill Langlochbohrer *m*; Tieflochbohrer *m*; Spindelbohrer *m*
g.-drill work Tieflochbohren *n (mit Kanonen- oder Spindelbohrer)*
g.-fire control radar (system) Feuerleitradar *n*
g. harpoon Harpune *f*
g. metal Geschützbronze *f*; Lagerbronze *f*
g. mount[ing] Lafette *f*
g. pointer Richtschütze *m*
g. reamer Tieflochreibahle *f*
g. ring Maschinengewehrdrehkranz *m*, MG-Drehkranz *m*
g. tank Geschützpanzerkampfwagen *m*
gunboat 1. *(Mil)* Kanonenboot *n*; 2. *(Brg)* Kippkübel *m*, Skip *m (für tonnlägige Schächte)*
gundalow *(Schiff)* flachgehender Binnenleichter *m (mit stumpfem Bug)*
gunhouse Geschützturm *m (Kriegsschiff)*
gunite / to torkretieren
gunite Spritzbeton *m*, Torkretbeton *m*, Torkretputz *m*
guniting Torkretieren *n*, Torkretierung *f*
gunja *s.* ganja
gunjah *s.* ganja
gunk *(Am)* Polyesterglasfasermasse *f*
g.-moulding *(Am)* Teigpreßverfahren *n*
gunned concrete Spritzbeton *m*, Torkretbeton *m*
gunner *(Am)* Richtschütze *m*
gunner's cockpit Maschinengewehrkanzel *f*, MG-Kanzel *f*
gunnery fleet spotter Schiffsartilleriebeobachtungsflugzeug *n*
g. manual Schießvorschrift *f*
gunny fibre Jute[hanf *m*] *f*
gunpowder Schwarzpulver *n*, Schießpulver *n*
gunwale Schandeck[el *m*] *n*, Kante *f* zwischen Schergang und Deckstringer *(Stahlschiff)*

g. tank *(Schiff)* Toppseitentank *m*
gurgoyle *(Bw)* Wasserspeier *m*
gush / to ausbrechen; hervorbrechen; quellen; sprudeln *(Wasser)*
g. over überwallen
gusher Geysir *m*, Geiser *m*
g. of natural gas Erdgasquelle *f*
gusset 1. Knotenblech *n*; Stützblech *n*; 2. *(Brg)* Stempel *m*
g. plate *s.* gusset 1.
gust Bö *f*, Stoßwind *m*
g. meter Böenmesser *m*
gusty böig
g. wind böiger Wind *m*
gut Fahrrinne *f*, Gatt *n*, Seegatt *n (enger Kanal)*
g.-string Darmsaite *f*
guts for sausages Wurstdarm *m*, Wurstdärme *mpl*
guttapercha Guttapercha *f (n)*
g. insulation Guttaperchaisolation *f*
g. mastic Guttaperchakitt *m*
gutter 1. Dachrinne *f*, Ablaufrinne *f*; Gosse *f*, Rinnstein *m*, Straßenrinne *f*; 2. *(Brg)* Wasserrösche *f*; 3. *(Met)* Gießrinne *f*
g. board Traufbohle *f*, Traufbrett *n*
g. channel Gosse *f*, Rinnstein *m*, Straßenrinne *f*
g. pipe Fallrohr *n*
g.-stick *(Typ)* Bundsteg *m*; Kreuzsteg *m*
guttering Regenrinne *f*; Dachrinne *f*, Abflußrinne *f*
guy / to abspannen, verankern, verspannen *(Maste)*
guy Abspannung *f*; Abspannseil *n*, Ankerseil *n*, Ankerdraht *m*; Mastanker *m*; (*s. a.* g. rope)
g. cable Abspannseil *n*, Ankerseil *n*
g. derrick abgespannter Derrickkran *m*
g. rope Abspannseil *n*, Ankerseil *n*, Halteseil *n*, Pardune *f*, Drahtseil *n*; Spannseil *n*
g. wire Abspanndraht *m*; Spanndraht *m*
guyed abgespannt, seilverankert *(Mast)*
guying Verankern *n* mit Drahtseilen
gym outfit Turnkleidung *f*
g. shoes Turnschuhe *mpl*
g. shorts Turnhose *f*
g. suit Turnanzug *m*
gymnasium Turnhalle *f*, Sporthalle *f*; Sportschule *f*
gymnastic bank Turnbank *f*
g. hand apparatus Handgeräte *npl (Turnen)*
g. mat Matte *f*, Turnmatte *f*
gypseous gipsartig
gypsiferous gipshaltig
gypsum Kalziumsulfat *n*, Gips *m*
g. board Gipsplatte *f*
g. calcining Gipsbrennen *n*
g. cast Gipsabdruck *m*
g. furnace Gipsbrennofen *m*

g. **industry** Gipsindustrie *f*
g. **mould** Gipsform *f*
g. **rock** Rohgips *m*
gyrate / to taumeln; kreiseln
gyrating mass Schwungmasse *f*
gyration Kreisbewegung *f*, kreisförmige Taumelbewegung *f*
g. **compass** Kreiselkompaß *m*
gyratory Kreiselbrecher *m*
g. **breaker** *s.* g. crusher
g. **crusher** Kegelbrecher *m*; *(Met)* Walzenbrecher *m*
g. **screen** Planrätter *m*, Plansichter *m*
gyro-axis Kreiselachse *f*
g. **bearing** korrigierte Peilung *f*
g. **blower** Kreiselradgebläse *n*
g.-**compass** Kreiselkompaß *m*
g. **flowmeter** gyrometrischer Durchflußmesser *m*
g.-**frequency** Gyrationsfrequenz *f*
g. **integrator** Integrierkreisel *m*
g. **mixer** Kreiselmischer *m*
g. **stabilizer** Schiffskreisel *m*
g. **vertical** Kreisellot *n*, Kreiselvertikale *f*
gyrobus *(El)* Gyrobus *m*
gyromagnetic gyromagnetisch
g. **compass** Kreiselmagnetkompaß *m*
g. **effect** *(Ph)* gyromagnetischer Effekt *m*, Einstein-de-Haas-Effekt *m*
g. **frequency** gyromagnetische Frequenz *f*
g. **ratio** gyromagnetisches Verhältnis *n*
gyrometer Gyrometer *n*, Flüssigkeitstachometer *n*
gyrometric direction finder Kompaßpeiler *m*
gyropilot *(Flg)* Selbststeuergerät *n*
g. **control** Selbststeuerung *f*
gyroscope Kreisel[gerät *n*] *m*, Gyroskop *n*
g. **pendulum** Kreiselpendel *n*
gyroscopic action Kreiselwirkung *f*
g. **model** *(Ph)* Kreiselmodell *n*
g. **moment** Kreiselmoment *n*
gyrostatic compass Kreiselkompaß *m*
g. **moment** *(Ph)* Kreiselmoment *n*
gyrostatics Gyrostatik *f*, Kreisellehre *f*
gyrosyne compass Erdinduktionskompaß *m*

H

H-amplifier Horizontalverstärker *m* *(Katodenstrahlröhre)*
H and D curve *(Opt)* Gradationskurve *f*, Schwärzungskurve *f*
H-armature *(Elm)* Doppel-T-Anker *m*, H-Anker *m*
H-bar control *(Fs)* Horizontal[balken]regler *m*
H-beam *(Bw)* Doppel-T-Träger *m*, H-Träger *m*; Breitflanschträger *m*
H-bend H-Bogen *m*, H-Krümmer *m* *(Wellenleiter)*
H-bomb Wasserstoffbombe *f*
H-circuit H-Leitung *f*
H-display H-Darstellung *f* *(Radar)*
H-fixture *(El)* Mastverbindung *f* in H-Form
H-girder *(Bw)* Doppel-T-Träger *m*, H-Träger *m*; Breitflanschträger *m*
H-mode H-Typ *m* *(Wellenleiter)*
H-navigation *(Flg)* Hypermeßverfahren *n*
H-network *(El)* symmetrisches T-Glied *n*, Vierpol *m* in H-Schaltung
H-paper H-Papier *n*, Höchstädter Papier *n* *(Kabelisolierung)*
H-parameter equivalent circuit *(El)* Ersatzschaltung *f* mit H-Parametern (Hybridparametern)
H-particle *(Ph)* H-Teilchen *n*, Wasserstoffteilchen *n*
H-plane H-Ebene *f* *(Wellenleiter)*
H-plane half-power width Halbwertsbreite *f* der H-Ebene
H-plane T[ee] junction H-Verzweiger *m*
H-pole *(El)* Doppelmast *m*, H-Gestänge *n*, H-Mast *m*
H-pole fixture Mastverbinder *m* in H-Form, Verbinder *m* für Doppelmast
H-scanner Abtaster *m* vom H-Typ *(Radar)*
H-scope *s.* H-type oscilloscope
H-section 1. *(Bw)* H-Querschnitt *m*; Breitflanschprofil *n*; 2. *(El)* H-Glied *n*, Doppel-T-Glied *n* *(Schaltung)*
H-section steel Doppel-T-Stahl *m*
H-steel Doppel-T-Stahl *m*
H-type cable H-Kabel *n*, Höchstädter-Kabel *n*
H-type core *(El)* H-Kern *m*
H-type engine H-Motor *m*
H-type oscilloscope Anzeigegerät *n* [vom] Typ H *(Radar)*
haar feuchter Nebel *m* *(englische Ostküste im Sommer)*
Haber-Bosch process Haber-Bosch-Verfahren *n* *(Salpetersäureherstellung)*
Haber process *s.* Haber-Bosch process
habit Habitus *m* *(Kristall)*
h. **cloth** englische Woll- und Kammgarnstoffe *mpl* feiner Qualität
h. **effect** Habituswirkung *f*
habitat Habitat *n* *(Tiere, Pflanzen)*
hachure / to schraffieren, schattieren
hachure Schraffierung *f*, Schattierung *f* *(technisches Zeichnen)*
h. **line** Schraffierungslinie *f*
hack / to 1. aufrauhen; zerhacken; kappen; 2. Ziegel aufschichten zum Trocknen
hack 1. Haue *f*; 2. Ziegeltrockengerüst *n*
h. **knife** Kabelmesser *n*
h. **saw** Bügelsäge *f*

h.-saw blade Bügelsägeblatt *n*
h.-saw frame Metallsägebügel *m*
h.-saw machine maschinelle Bügelsäge *f*
hacking Bruchsteinmauer *f*, unregelmäßiges Schichtenmauerwerk *n*
hackle / to *(Text)* hecheln, kämmen
hackle chain Hechelkette *f*
hackled flax gekämmter Flachs *m*, Hechelflachs *m*
hackling bench Hechelstuhl *m (Flachs)*
h. block Hechelbrett *n*
h. machine Hechelmaschine *f*
h.-threshing machine *(Am)* Stiftendreschmaschine *f*
h. tow Hechelwerg *n*
hacksaw *s.* hack saw
haddocker Schiff *n* für Schellfischfang
hade / to *(Brg)* einfallen
hade *(Brg)* Neigung *f (Störungsfläche)*
hading against the dip *(Geol)* widersinnig fallend
h. shaft *(Brg)* tonnlägiger Schacht *m*
h. with the dip *(Geol)* rechtsinnig fallend
haematite *(Min)* Hämatit *m*
h. iron Hämatiteisen *n*
h. pig iron Hämatit *n*
hafnium *(Ch)* Hafnium *n*
haft Heft *n*, Griff *m*
Haidinger sheaf [of interference fringes] *(Opt)* Haidingersches Büschel *n*
hailsquall Hagelschauer *m*
hailstorm Hagelschauer *m*
hair Haar *n*
h. bleaches Haarbleichmittel *npl*
h. cloth Haartuch *n*
h. colourings Haarfärbemittel *npl*
h. conditioner Frisierhilfsmittel *n*
h. crack Haarriß *m (Glas, Metall)*
h. drier Heißluftdusche *f*
h.-grooming preparations Frisierhilfsmittel *npl*
h. hygrometer Haarhygrometer *n*
h-line 1. aus Haar[en] gemachtes Seil *n*; 2. *s.* hairline
h. mortar *(Bw)* Haarkalkmörtel *m*
h. outside *(Led)* Haarseite *f* außen *(im Bündel)*
h.-pin *s.* hairpin
h.-slippiness *(Led)* Haarlässigkeit *f*
h.-space *(Typ)* Haarspatium *n*
h. spring *s.* hairspring
h. straighteners Haarentkräuselungsmittel *npl*
h. tonic Haartonikum *n*, Haarwasser *n*
haircloth Haar[einlage]stoff *m*
hairline feiner Strich *m*, Skalenstrich *m*, Strichmarke *f*
h. crack *s.* hair crack
h. gauge block Strichendmaß *n*
h. stripe *(Text)* Nadelstreifen *m*
hairpin bend Haarnadelkurve *f (Straße)*
h. cathode Haarnadelkatode *f*
h. shaped haarnadelförmig
hairspring Spiralfeder *f (Meßinstrument, Uhr)*
halation Halobildung *f*, Lichthofbildung *f (Filmaufnahme; Katodenstrahlröhre)*
half halb
h. adder Halbaddierer *m*
h. and half *(Am) (Brg)* Abbau *m* mit 45° zu den Schlechten gestellter Abbaufront
h. angle Halbwinkel *m*
h. angle of thread halber Teilflankenwinkel *m*
h.-axle Halbachse *f*
h.-back lighting Seitenbeleuchtung *f*
h.-balk Halbholz *n*
h. bat halber Stein *m*, Kopf *m (Ziegel)*
h. beam *(Schiff)* Kurzbalken *m (neben der Luke)*
h.-breadth Halbwertsbreite *f (Spektrallinien)*
h.-breadth plan *(Schiff)* Wasserlinienriß *m*, Wasserpaßriß *m*, Grundriß *m*
h.-brick wall *(Bw)* Halbsteinwand *f*, halbsteinige (halbsteindicke) Wand *f*
h.-calf *(Typ)* Halbleder *n*
h.-cell *(El)* Halbzelle *f (Sammler)*
h. circle Halbkreis *m*
h.-cloth *(Typ)* Halbgewebe *n*
h.-countersunk head Linsensenkkopf *m*
h.-coupling Kupplungshälfte *f*
h.-course *(Brg)* Abbau *m* mit 45° zum Streichen gestelltem Stoß
h.-crossed belt drive Halbkreuzriementrieb *m*, halb geschränkter Riementrieb *m*
h.-cycle Halbperiode *f*
h. davit *(Schiff)* Einzeldavit *m (n)*
h.-digit delay *(Dat)* Halbzifferverzögerung *f*
h.-divided oven *(Brg)* halbgeteilter Ofen *m*
h.-dog screw Schraube *f* mit Ansatzkuppe
h.-dry spinning frame Halbnaßspinnmaschine *f*
h.-duplex operation (system) Halbduplexbetrieb *m (Oberbegriff)*; bedingtes Gegensprechen *n (Radiotelefonie)*; Wechselschreiben *n (Fernschreibtechnik)*
h.-duplex teletype Wechselschreiber *m*
h.-element *(El)* Halbzelle *f (Sammler)*
h.-elliptic spring Halbelliptikfeder *f*
h.-enclosed halbgeschlossen, halboffen
h. fine mittelfein, halbfein
h. finished product Halbzeug *n*, Halbfabrikat *n*
h. gap halbe Spaltbreite *f*
h.-gas fired furnace Halbgasfeuerung *f*
h.-hard halbhart, mittelhart
h.-hose Kniestrumpf *m*

h.-integer *(Math)* halbganze Zahl *f*
h.-integer total angular momentum *(Ph)* halbzahliger Gesamtdrehimpuls *m*
h.-integral spin *(Ph)* halbzahliger Spin *m*
h. intensity width *(Ph)* Halbwertsbreite *f*
h. joint gerades Blatt *n (Holzbau)*
h.-landing Zwischenpodest *n (m)*
h. life Halbwertzeit *f*
h.-life period *s.* h. life
h.-line *(Math)* Strahl *m*
h.-linen *(Typ)* Halbleinen *n*
h.-log Halbholz *n*
h. mast Fallmast *m*
h. measure *(Typ)* gespaltener Satz *m*, Schmalsatz *m*
h. nut Schloßmutter *f*
h. nut lever Mutterschloßhebel *m (Leitspindel)*
h.-on *(Am) (Brg)* Abbau *m* mit 45° zu den Schlechten gestellter Abbaufront
h. port *(Schiff)* halbe Stückpfortenklappe *f*
h.-power point Halbwertspunkt *m*, Bandgrenze *f (Funktechnik, Impulstechnik)*; *(Am)* Halbwertbreite *f (Antennenausstrahlung)*
h.-power width [of a radiation lobe] *(Am)* Halbwertbreite *f (Antennenausstrahlung)*
h. quartered auf dem Halbspiegel
h. rip saw englische Fuchsschwanzsäge *f*
h. roll *(Met)* halbkalibrierte Walze *f*
h.-rotary hand pump *(Schiff)* Hand[flügel]-pumpe *f*
h. round *s.* h. round bar
h. round bar *(Bw)* Halbrundstahl *m*, Segmentstahl *m*; Segmenteisen *n*
h.-round chisel Nutenmeißel *m*
h.-round file Halbrundfeile *f*
h.-round scraper Löffelschaber *m*
h.-round wood Breitholz *n*, Halbholz *n*, hochkantiges (halbrundes) Holz *n*
h.-scoop Greiferschale *f*
h. section 1. Halbschnitt *m (technisches Zeichnen)*; 2. Halbholz *n*
h. select current *(BMSR)* Halbwählstrom *m (eine Hälfte der Koinzidenz)*
h.-select pulse *(BMSR)* Halbwählimpuls *m*
h. set *(Brg)* halber Türstock *m*, Handweiser *m*
h.-shade *(Ph)* Halbschatten *m*, Penumbra *f*
h.-shadow *s.* h.-shade
h. shaft *(Kfz)* Hinterachswelle *f*, Hinterradwelle *f*, Seitenwelle *f*
h. shift register *(BMSR)* Halbschieberegister *n*
h.-side milling cutter Scheibenfräser *m* mit einseitigen Umfangszähnen
h.-sinusoid halbe Sinuswelle *f*
h.-sized halbgeleimt *(Papier)*
h.-space *(Math)* Halbraum *m*
h.-spherical halbrund
h. stock Halbzeug *n (Papierherstellung)*
h.-stock Hollander beater Halbzeugholländer *m (Papierherstellung)*
h.-stuff Papierfaserstoff *m (Halbstoff)*
h. sweep *(Lw)* einseitiges Gänsefußschar *n*
h.-thickness *s.* h.-value thickness
h.-thread angle *(Masch)* halber Flankenwinkel *m*
h. tide *(Schiff)* Mitte *f* der Gezeiten
h.-timbered building Fachwerkhaus *n*
h.-title *(Typ)* Schmutztitel *m*
h.-tone *(Ak)* Halbton *m*
h.-tone block *(Typ)* Rasterätzung *f*, Rasterklischee *n*, Autotypie *f*
h.-tone characteristic Halbtoncharakteristik *f*
h.-tone dots *(Typ)* Rasterpunkte *mpl*
h.-tone picture Halbtonbild *n*
h.-tone process image *(Fs)* gerastertes Bild *n*
h.-tone screen *(Typ)* Bildraster *m*, Autotypieraster *m*
h.-tone storage *(BMSR)* Halbtonspeicherung *f*
h.-track 1. Halbspur *f (Magnetband)*; 2. Teilspur *f*, Teilaufzeichnung *f (Film)*; 3. Anbauraupe *f*, Ansteckraupe *f (Traktor)*
h.-track motorcycle *(Mil)* Kettenkrad *n*
h.-track recorder Doppelspur[tonband]-gerät *n*
h.-value Halbwert *m*
h.-value layer Halbwertschicht *f (Absorption, radioaktiver Zerfall usw.)*
h.-value thickness Halbwertdicke *f*
h.-wave Halbwelle *f*
h.-wave aerial Halbwellenantenne *f*
h.-wave dipole Halbwellendipol *m*; Hertzscher Dipol *m*
h.-wave[length] plate *(Opt)* Halbwellenlängenplatte *f*, $\frac{\lambda}{2}$-Plättchen *n*
h.-wave rectification Einweggleichrichtung *f*, Halbwellengleichrichtung *f*
h.-wave rectifier Einweggleichrichter *m*, Halbwellengleichrichter *m*
h.-wave rectifier circuit Einweggleichrichterschaltung *f*
h.-width Halbwert[s]breite *f (Spektrallinien)*
h.-write pulse Halbschreibimpuls *m*, Rückschreibimpuls *m (Magnetkernspeicher)*
h.-wrought material *(Met)* Halbzeug *n*
halibuter Schiff *n* für Heilbuttfang
halide Halogenid *n*, Salz *n* einer Halogenwasserstoffsäure
halite Steinsalz *n*
hall Diele *f*, Flur *m*, Vorhalle *f*; Halle *f (Messe)*
h. exhibition space überdeckte Ausstellungsfläche *f*
h. noise Saalgeräusch *n*
h. set Flurgarderobe *f*

h. space überdeckte Fläche *f (Messe)*
Hall coefficient (constant) Hall-Koeffizient *m*, Hall-Konstante *f (Hall-Effekt)*
Hall effect *(Ph, El)* Hall-Effekt *m*
Hall rule Hallsche Regel *f (Bewegungsregel) (Hall-Effekt)*
Hall signal Hall-Spannung *f (Hall-Effekt)*
halloysite *(Min)* Halloysit *m*
halo Halo *m*, Lichthof *m*; *(Opt)* Hof *m*, Strahlenring *m*; Schwärzungsring *m (fotografisches Negativ)*
h. disturbance lichthofartige Störung *f (Fotografie)*
halocarbon Halogenkohlenwasserstoff *m*, halogensubstituierter Kohlenwasserstoff *m*
h. plastics aus perhalogenierten Verbindungen hergestellte Kunststoffe *mpl*
halochromism Halochromie *f*
halogen carrier Halogenübertrager *m*
h. counter *(Ph, Kern)* Halogenzählrohr *n*
h.-quenched counter *(Ph, Kern)* Halogenzähler *m*, Halogenzählrohr *n*
h. quenching addition Halogenlöschzusatz *m*
halogenate / to halogenieren
halogenize / to halogenieren
halogens Halogene *npl*
haloid Halogenid *n*, Haloid *n*
halter Halfter *f*
h.-rope Halfterriemen *m*
halting of the steam engine Hinken *n* der Dampfmaschine
halve / to 1. *(Math)* halbieren; 2. anblatten, überblatten, verblatten *(Holzbau)*
halved joint gerades Blatt *n (Holzbau)*
halyard Aufzugsseil *n*
ham Funkamateur *m*
h.-receiver Funkamateurempfänger *m*
hambergite *(Min)* Hambergit *m*
Hamilton function *(Mech)* Hamilton-Funktion *f*
Hamilton operator Hamilton-Operator *m (Quantenmechanik)*
Hamiltonian *s.* Hamilton operator
hammer / to *(Met)* hämmern, [aus]schmieden
h.-dress aushämmern
h. even aushämmern
h.-forge freiformschmieden
h. into hineinschlagen in
h. out glatthämmern
hammer Hammer *m*; Ramme *f*; Pfahlramme *f*, Rammbär *m*, Rammhammer *m*; Fäustel *m*
h. and chisel (wedge) *(Brg)* Schlägel *m* und Eisen *n*
h. blow Hammerschlag *m*
h. break *(El)* Hammerunterbrecher *m*, Wagnerscher Hammer *m*
h.-break spark coil Funkeninduktor *m* mit Hammerunterbrecher
h.-crusher Hammerbrecher *m*; Hammermühle *f*
h. drill Bohrhammer *m*
h. enamel Hammerschlaglack *m*
h. eye Hammerauge *n*, Hammerloch *n*
h. face Hammerbahn *f*, Hammerfläche *f*
h. forging Freiformschmieden *n*
h. gear Schlagwerk *n*
h.-hardening Kalthämmern *n*
h. mill 1. *(Met)* Hammerwerk *n*; 2. Hammermühle *f*
h. of bell Glockenklöppel *m*
h. pipe Schmiedekreuz *n*
h. ram Schlagkolben *m (Drucklufthammer)*
h. scale Hammerschlag *m*
h.-scales varnish Hammerschlaglack *m*
h. spark coil Hammerinduktor *m*, Funkeninduktor *m* mit Hammerunterbrecher
h. split *(Büro)* Hammersplitthebel *m*
h. strap Hammerfeder *f*
h. throwing cage Hammerwurfgitter *n*
h. welding Hammerschweißen *n*
hammering Hämmern *n*; Schmieden *n*, Ausschmieden *n*, Reckschmieden *n*
hammerlock *(Büro)* Hammersperre *f*
Hamming-code *(Dat)* Hamming-Kode *m (fehlererkennende Kodeform)*
hammock Hängematte *f*
Hammond organ Hammond-Orgel *f*
hand / to übergeben, [über]reichen; einhändigen; ergreifen
h.-clean *(Brg)* klauben, lesen
h.-pick *(Brg)* aushalten, klauben, lesen
h.-stuff mit der Hand schmieren
hand 1. Zeiger *m (Uhr)*; 2. Richtung *f*; Drehrichtung *f*; Sinn *m*; Schnittrichtung *f*; 3. *(Text)* Griff *m*
h. actuation Handbetätigung *f*
h. adjustment Handverstellung *f*
h.-and-foot counter *(Kern)* Zählrohr *n* zur Prüfung von Händen und Füßen
h. anvil Bankamboß *m*
h. bar *(Bw)* Handlauf *m*
h. barrow Handkarre *f*, Schiebkarren *m*, Schubkarre *f*, Stoßkarren *m*; Tragbahre *f*
h.-blocked *(Text)* handgedruckt
h.-blocking *(Typ)* Hand[blind]prägung *f*
h. boring Freihandbohren *n*
h. brake Handbremse *f*
h. brake handle Handbremszuggriff *m*
h. brake lever Handbremshebel *m*
h. brush Handfeger *m*
h. calculator Handrechenmaschine *f*
h. capacity effect *(Fmt)* Wirkung *f* der Handkapazität
h. capstan Handkreuz *n*
h. car *(Eis)* Draisine *f*
h. card *(Text)* Handkratze *f*, Handkardätsche *f*

h. cart *(Lw)* Handkarre *f*
h. chain tooth saw Kettenzugsäge *f*
h. charging Handbeschickung *f*
h. chaser Hand[gewinde]strehler *m*
h. composition *(Typ)* Handsatz *m*
h. control Handsteuerung *f*; Schaltung *f* von Hand
h.-controlled handgesteuert
h. crank Andrehkurbel *f*; Handkurbel *f*
h. die hob Handbackengewindebohrer *m*
h. drill [Kurbel-]Handbohrmaschine *f*
h. drill point Bohrer[einsatz] *m* für Handdrillbohrer
h. drilling Handbohren *n*
h.-drilling appliance Handbohrgerät *n*
h. drilling unit Handbohrmaschine *f*
h.-dry handtrocken
h.-embroidered handgestickt *(z. B. Plauener Nähspitze)*
h. fed handbeschickt *(z. B. SM-Ofen)*
h.-fed push-bench Tischsäge *f* mit Handvorschub *(Holzbearbeitung)*
h. feed Handvorschub *m*
h. feed punch Handlocher *m*
h. file Flachstumpffeile *f*, Handfeile *f*
h. filling Handabfüllung *f*
h. fine adjustment for feeding wheel to work Handfeinzustellung *f* der Schleifscheibe
h. fine feed Handfeinzustellung *f* der Schleifscheibe
h. firing Handfeuerung *f*, Feuerung *f* mit Handbeschickung
h. flag Winkflagge *f*
h. flat frame Handflachstrickmaschine *f*
h. flat knitting machine Handflachstrickmaschine *f*
h. forged handgeschmiedet, freiformgeschmiedet
h. forging Handschmieden *n*; Freiformschmieden *n*
h. gag Handknebel *m*
h. gear Handsteuerung *f*
h. generator Handdynamo *m*, Kurbelinduktor *m*
h. grinder Handschleifmaschine *f*
h.-grip Handgriff *m*
h. groover Falzmeißel *m (Holzbearbeitung)*
h. grooving and tonguing plane Nut- und Federhobel *m (Holzbearbeitung)*
h. hack saw Handbügelsäge *f*
h.-held drill Handbohrmaschine *f*
h. hydrometer Spindelaräometer *n*
h. jack *(Kfz)* Handwinde *f*, Wagenheber *m*
h. knit handgestrickt
h.-knitting yarn Handstrickgarn *n*
h. knurl holder *(Masch)* Rändelhalter *m*
h. ladle Gießlöffel *m*, Handgießpfanne *f*
h. ladling Gießen *n* mit Gießlöffel
h. lamp Handleuchte *f*; Handlampe *f (mit Verlängerungsschnur)*
h. lapping Handläppen *n*
h. lead line *(Schiff)* Handlotleine *f*
h. lever Bedienungshebel *m*, Griffhebel *m*, Handhebel *m*
h.-lever bearing Lagerbock *m*
h.-lever control Handhebelsteuerung *f*
h.-lever cross shaft *(Kfz)* Handhebelwelle *f*
h.-lever operated handhebelbetätigt
h.-lever rod bender Betoneisenbieger *m* mit Handbetätigung
h.-lever shears Hebelblechschere *f*
h. line *(Schiff)* Rettungsleine *f*
h. log *(Schiff)* Handlog *n*
h. loom Handwebstuhl *m*
h. loom weaver Handweber *m*
h. loom weaving Handweberei *f*
h. loomed *s.* h. woven
h.-made brick Handstrichziegel *m*
h.-made glass mundgeblasenes Glas *n*
h.-made lace *(Text)* Handspitze *f*
h.-made paper Handbütten[papier] *n*, echtes Büttenpapier *n*, handgeschöpftes Papier *n*
h. magnifier Handlupe *f*
h. metal punch Plombierzange *f*
h. miller Handfräsmaschine *f*
h. mortising machine Handstechmaschine *f* *(Holzbearbeitung)*
h. moulding Handformen *n*, Handformerei *f*
h. mucking *(Brg)* Schaufelarbeit *f*, Wegfüllen *n* von Hand
h. nut Rändelmutter *f*
h. nut tap Handmutternbohrer *m*
h. of a reamer cut Schnittrichtung *f* einer Reibahle
h. of clock Uhrzeiger *m*
h. of coils Windungsrichtung *f (Feder)*
h. of cut Schnittrichtung *f*
h. of helix Drallrichtung *f*
h. of rotation Drehrichtung *f*; Drehsinn *m*
h. of [the] spiral Drallrichtung *f*
h. of welding Schweißrichtung *f*
h. oiler Ölkanne *f*
h.-operated handbedient, handbetätigt, handbetrieben; Hand-
h.-operated capstan Handspill *n*
h.-operated chuck Handspannfutter *n*
h.-operated exchange *(Fmt)* Handamt *n*, Handvermittlung *f*
h.-operated fire extinguisher Handfeuerlöscher *m*
h.-operated pump Pumpe *f* für (mit) Handbetrieb
h. operation Handbetätigung *f*; Betätigung *f* von Hand
h. ophthalmoscope Handaugenspiegel *m*

h. pick *(Geol)* leichte Hacke *f* für Bemusterungszwecke
h.-picked dirt *(Brg)* Klaubeberge *pl*, Leseberge *pl*
h. plane Handhobel *m*
h. planer Abrichtmaschine *f*
h. planer and jointer Handabrichthobel- und Fügemaschine *f*
h. power winch Bauwinde *f*, Zahnstangenwinde *f*
h.-powered handbetrieben
h. primer *(Kfz)* Handpumpenvorrichtung *f*
h. puddling *(Met)* Handpuddeln *n*
h. pull *(Typ)* Handabzug *m*
h. radar tragbares Radargerät *n*
h. rail Handlauf *m*; Haltegriff *m*; Haltestange *f*
h.-rail of stairs (steps) Treppengeländer *n*
h. rails Maschinengeländer *n*, Schutzgeländer *n*
h. ramming machine Handstampfmaschine *f* *(Formerei)*
h. reamer Handreibahle *f*
h. reamer for parallel holes Zylinderhandreibahle *f*
h. reamer with straight squared shank Handreibahle *f* mit zylindrischem Schaft und Vierkant
h. receiver Hörer *m (Fernsprechapparat)*
h.-regulation Handregelung *f*
h. rejector Handabweiser *m (Schutzeinrichtung)*
h. rest Handauflage *f*
h. riveting Handhammernieten *n*
h.-rubbed von Hand verrieben
h. saw Handsäge *f*
h. saw file Säge[n]feile *f*, Schärffeile *f*
h. scraper Handschrapper *m*
h. screw press Handspindelpresse *f*
h. seed drill Handsämaschine *f*
h.-set handgesetzt
h. set *(Fmt, El)* Handapparat *m*; Mikrotelefon *n*
h.-set type-matter *(Typ)* Handsatz *m*
h. setting 1. Handeinstellung *f*; Einstellung *f* von Hand; 2. *(Typ)* Handsatz *m*
h. shield Schweißerschutzschild *m*
h. sketch Skizze *f*, Freihandzeichnung *f*
h. slit lamp Handspaltleuchte *f*
h. sowing machine Handsämaschine *f*
h.-specimen *(Min)* Handstück *n*
h. sprayer Handspritze *f (Schädlingsbekämpfung)*
h. stake *(Led)* Handstollen *m*
h. swinging Anreißen *n*, Anwerfen *n (Motor)*
h. switch Handschalter *m*, handbetätigter Schalter *m*
h. tally [register] Handstückzähler *m*
h. tap Handgewindebohrer *m*
h. taper reamer Handkegelreibahle *f*
h. tool Handwerkzeug *n*
h. tramming *(Brg)* Schlepperförderung *f*
h. vice Handfeilkloben *m*; Handschraubstock *m*
h. welding Handschweißen *n*
h. wheel Handrad *n*, Kurbelrad *n*
h. wheel for wheel feed Beistellhandrad *n* [der Schleifscheibe]
h. winch Handwinde *f*
h. winding Handwicklung *f*
h. woven handgewebt
handball Handballspiel *n*
handbook Handbuch *n*, Nachschlagewerk *n*
handiness *(Schiff)* Manövrierfähigkeit *f*, Wendigkeit *f*
handle / to 1. handhaben, behandeln; bedienen; 2. fordern; transportieren; zubringen; 3. bearbeiten
h. a frequency range einen Frequenzbereich verarbeiten
h. current Strom führen *(Kabel)*
handle 1. Heft *n*, Handgriff *m*, Griff *m*, Schaft *m*, Stiel *m*; Schenkel *m*; 2. *(Text)* Griff *m*
h. bar for rim wrench Drehstange *f* für Radmutterschlüssel
h. end Griffangel *f*
h. of the plane Hobelnase *f*
h. type fuse *(El)* Griffsicherung *f*
handled mit Stiel, mit Griff
handler Gerbgrube *f*
handling 1. Handhabung *f*; Behandlung *f*; Bedienung *f*; 2. Förderung *f*, Transport *m*; Zubringung *f*; 3. Bearbeitung *f*
h. appliances Umschlagseinrichtungen *fpl (eines Hafens)*
h. device Fördereinrichtung *f*
h. engineering Fördertechnik *f*
h. equipment 1. Hebe- und Fördergeräte *npl*, Hebe- und Fördermittel *npl*; 2. Werkstückzuführeinrichtung *f*
h. gear Transportgeräte *npl*; Fördermittel *npl*
h. hook Transportbügel *m*, Transportschlaufe *f (Betonfertigteile)*
h. period *(Masch)* Griffzeit *f*
h. plant Förderanlage *f*; Transportanlage *f*
h. time *s.* h. period
handmould Schöpfrahmen *m (Herstellung von Büttenpapier)*
hands of clock Uhrzeiger *mpl*
handspike *(El, Fs)* Tastspitze *f*
handwraps Bandagen *fpl (Boxsport)*
handwriting reader *(Dat)* [automatischer] Handschriftleser *m*
handy boat wendiges Boot *n*
h.-talkie tragbares Sende-Empfangsgerät *n*

hang / to [auf]hängen; hängen bleiben *(in der Schwebe)*
h. from gimbals kardanisch [auf]hängen
h. on *(Fmt)* am Apparat bleiben
h. up 1. *(Fmt)* [Hörer] auflegen, [Hörer] anhängen, abhängen; 2. *(Brg)* hängenbleiben *(Erz oder Berge in Rollen)*
h. wallpaper tapezieren
hang-down Hängelager *n*
hangar Flugzeughalle *f*, Hangar *m*
h. deck *(Schiff)* Hangardeck *n*, Hallendeck *n*
hanger 1. Hängestab *m*; 2. *(Bw)* Hängelager *n*, Lagerträger *m*; 3. Aufhänger *m (Luftkabel)*
h. bar Montageeisen *n (Stahlbeton)*
h.-on *(Brg)* Anschläger *m*
hanging 1. Aufhängung *f*; 2. *(Brg)* Hangendes *n*, Dach *n*, Firste *f*
h. compass *(Schiff)* Kajütkompaß *m*, Hängekompaß *m*
h. mat *(Schiff)* Schamfielungsmatte *f*
h. steps *(Bw)* eingespannte Stufen *fpl*, Kragtreppe *f*
h. truss *(Bw)* Hängewerk *n*
h. wall *(Brg)* Hangendes *n*, Dach *n*, Firste *f*
hangings *pl* Tapeten *fpl*; Tapetenrohpapier *n*
hangover *(Fs)* Auftreten *n* von Fahnen (Streifen) an Stellen starker Helligkeitsunterschiede
h. time *(El)* Nachwirkzeit *f (Schaltelemente mit Trägheit, z. B. sprachbetätigte Schalter)*; Teilsperrzeit *f (Echounterdrückung)*
hank *(Text)* Strähn *m*, Strähne *f*
h. sizing machine Strangschlichtmaschine *f*
h. winding Strangspulen *n*
h. yarn Stranggarn *n*
haphazard arrangement *(Met)* regellose Anordnung *f (Kristallgitter)*
haplite *s.* aplite
harbor *(Am) s.* harbour
harbour Hafen *m (nur Hafenbereich, der aus den Schiffsliegeplätzen, Molen, Piers usw. besteht)*
h. authority Hafenbehörde *f*
h. bar Hafendamm *m*
h. basin Hafenbecken *n*
h. dues Hafengebühr *f*
h. engineering Hafenbau *m*
h. establishment Hafenzeit *f*
h. facilities Hafeneinrichtung *f*
h. icebreaker Hafeneisbrecher *m*
h. launch Hafenbarkasse *f*
h. light Hafenfeuer *n*, Molenfeuer *n*
h. lighter Hafenleichter *m*
h. master Hafenmeister *m*, Hafeninspektor *m*
h. of refuge Nothafen *m*
h. pilot Hafenlotse *m*
h. railway Hafenbahn *f*
h. routine *(normaler)* Hafenbetrieb *m*
h. service Hafenbetrieb *m*, Hafendienst *m*
h. tender Hafenbarkasse *f*, Hafentender *m*
h. tug Hafenschlepper *m*
hard 1. hart; 2. hart, durchdringend, energiereich *(Strahlen)*; 3. schwierig, schwer
h.-a-port hart Backbord
h.-a-starboard hart Steuerbord
h. all through durchgehärtet
h. alloy Hartmetall *n*
h.-and-fast rule Faustregel *f*
h. anodic coating Harteloxierung *f*
h. bilge scharfe Kimm *f (bei flachbodigen Schiffen)*
h. board *s.* hardboard
h. brass Hartmessing *n*
h. brick Hartbrandstein *m*
h. brown coal Hartbraunkohle *f*, Braunkohle *f*
h. carbide Sintermetall *n*
h.-centred steel Stahl *m* mit hartem Kern
h. chrome plating Hartverchromung *f*
h. coal Steinkohle *f*; *(Am)* Anthrazit *m*, Magerkohle *f*
h.-coal briquette Steinkohlenbrikett *n*
h. coating Harteloxierung *f*
h. component harte (durchdringende) Komponente *f (kosmische Strahlung)*
h. copper Hartkupfer *n*, hartgezogenes Kupfer *n*
h. copy schwer lesbare Kopie *f (z. B. der Lochstreifen eines Schreiblochers im Gegensatz zur Klartextkopie)*
h. core 1. „hard core" *(Struktur der Elementarteilchen)*; 2. *(Bw)* Packlage *f*
h. cured *(Led)* salzhart *(Haut)*
h. drawn *(Met)* hartgezogen
h.-drawn brass Hartmessing *n*
h.-drawn copper Hartkupfer *n*, hartgezogenes Kupfer *n*
h. face / to 1. auftragschweißen; 2. bestükken, Stahlplättchen auflöten; 3. panzern
h.-faced valve *(Kfz)* oberflächenvergütetes Ventil *n*
h.-facing Auftragschweißen *n (einer harten Oberflächenschicht)*
h.-facing alloy Aufschweißlegierung *f*
h. fibre Hartfaser *f*
h. flow geringe Fließfähigkeit *f*, schlechter Fluß *m*
h.-galvanized hartverzinkt
h. glass Hartglas *n*
h. head Härtling *m (Legierung)*
h. heading *(Brg)* Gesteinsstrecke *f*
h. image *(Fs)* hartes (übersteuertes) Bild *n*
h. lead Hartblei *n*
h.-luck pilot *(sl) (Flg)* Bruchpilot *m*
h. metal Hartmetall *n*, Sinterhartmetall *n*
h.-metal cutting bit Hartmetallschneide *f*
h.-nickel plating Hartvernickelung *f*

h. nut to crack schwieriges (schwer lösbares) Problem *n*
h. paper Hartpapier *n*
h. pitch Hartpech *n*
h. plating Hartverchromen *n*
h. porcelain Hartporzellan *n*
h. pulp schwer bleichbarer Stoff *m*
h. radiation harte (durchdringende, energiereiche) Strahlung *f*
h. rays Hartstrahlen *mpl (kurzwellige Röntgenstrahlen)*
h. rolled hart (kalt) gewalzt
h. rubber Hartgummi *m*
h. soil schwerer Boden *m*
h. solder Hartlot *n*, Schlaglot *n*, Messinglot *n*
h. soldering Hartlöten *n*
h. spot *(Met)* harte Stelle *f*, harter Einschluß *m*
h. surfacing 1. Auftragschweißen *n*; 2. Oberflächenhärten *n*
h. tap *(Met)* eingefrorenes Abstichloch *n*
h. temper *(Met)* Vollhärte *f*
h. to get at schwer zugänglich
h. to hold schwer zu spannen, schwer spannbar
h. to machine schwer spanbar
h. to melt schwer schmelzbar
h. to reach schwer erreichbar
h. to read schwer lesbar
h. tube rectifier Hochvakuumgleichrichter *m*
h.-twist[ed] stark gezwirnt, hartgezwirnt, stark überdreht *(Garn)*
h. usage rauhe Behandlung *f*
h.-vacuum lamp Hochvakuumlampe *f*
h. valve *(El)* harte Röhre *f*, Hochvakuumröhre *f*
h.-valve pulser Hochvakuumpulsmodulator *m (Radar)*
h. water hartes (kalkhaltiges) Wasser *n*
h. wax Hartwachs *n*
h. wear strapazierfähig *(z. B. textile Stoffe)*
h. wearing verschleißbeständig, verschleißfest
h. wheat Hartweizen *m*
h. wiring *(El)* festverlegte Installation *f*
h. wood fibre slab Hartholzfaserplatte *f*
hardboard 1. Hartpappe *f*; 2. Hart[faser]-platte *f*
h.-faced plywood mit harter Platte beplanktes Sperrholz *n*
h.-type particle board hartplattenartige Spanplatte *f*
harden / to 1. härten *(Stahl usw.)*; 2. erhärten, hart werden *(Beton usw.)*; abbinden
h. by dispersion (precipitation) *(Met)* aushärten, dispersionshärten
h. thoroughly (throughout) durchhärten
hardenability 1. Härtbarkeit *f*; Härteannahme *f*
h. by aging Aushärtbarkeit *f*
hardenable härtbar
hardened gehärtet
hardener Härtemittel *n*, Härter *m*
hardening 1. Härten *n*, Härtung *f*; 2. Erhärten *n*, Hartwerden *n*, Verhärtung *f*; Abbindung *f*
h. acceleration Härtebeschleunigung *f*
h. and heat-treating plant Härte- und Vergüteanlage *f*
h. and tempering Vergütung *f (Stahl)*
h. bath Härtebad *n*
h. by dispersion Aushärtung *f*, Dispersionshärtung *f*
h. by high-frequency current Hochfrequenzhärtung *f*
h. by immersion Tauchhärten *n*
h. by precipitation *s.* h. by dispersion
h. compound Härtepulver *n*
h. constituent Härtebildner *m*
h. crack Härteriß *m*
h. crucible Härtetiegel *m*
h. distortion Härteverzug *m (Deformation)*
h. fixing bath Härtefixierbad *n (Film)*
h. flaw Härteriß *m*
h. furnace Härteofen *m*
h. of X-ray tube Hartwerden *n* der Röntgenröhre
h. plant Härteanlage *f*; Härterei *f*
h. salt *(Met)* Härtesalz *n*
h. shop Härterei *f*, Vergüterei *f*
h. shrinkage Härteschrumpfung *f*
h. strain Härtespannung *f*
h. under cold work Kaltverfestigung *f*
hardness 1. Härte *f*; 2. Evakuierungsgrad *m*
h. drop tester Fallhärteprüfer *m*
h. gauge Härtemesser *m*
h. grading 1. Härteabstufung *f*; 2. Härtebestimmung *f*
h. increase Verfestigung *f*, Verhärtung *f*
h. indicator Härtemesser *m*
h. of radiation Härte (Kurzwelligkeit) *f* der Strahlung, Strahlenhärte *f*
h. of valves *(El)* Güte *f* des Röhrenvakuums
h. penetration *(Met)* Härtetiefe *f*
h. penetration by quenching Abschreckhärtetiefe *f*
h. scale [Mohssche] Härteskala *f*
h. test Härteprüfung *f*
h. tester Härteprüfgerät *n*, Härteprüfmaschine *f*
h. testing machine *s.* h. tester
hardpan tragfähige Schicht *f*, tragfähiger Untergrund *m (Grundbau)*
hardpressed fibreboard hartverpreßte Faserplatte *f*
hards *(Text)* Schäbe *f*

hardware 1. Eisenwaren *fpl*, Stahlwaren *fpl*; 2. Beschläge *mpl*; Kleineisenzeug *n*; 3. *(Am)* Bauelemente *npl*, Bausteine *mpl (der Elektrotechnik)*
h. representation *(Büro)* maschineneigene Darstellung (Sprache) *f*
hardwearing verschleißfest
hardwood Hartholz *n*, Laubholz *n*
h. pulp Papierfaserstoff *m* aus Laubholz
hardy Amboßschröter *m*
Hardy Cross method Cross-Verfahren *n*, Momentenausgleichverfahren *n (Statik)*
harmful ingredients schädliche Beimischungen *fpl*
harmless ungefährlich, nicht schädlich
harmonic 1. *(Math, Ph)* harmonisch; 2. *(El)* Oberwellen-
harmonic [harmonische] Oberschwingung *f*; Oberwelle *f*
h. aerial auf Oberwellen erregte Antenne *f*
h. analysis 1. *(Math, Ph)* Fourier-Analyse *f*, harmonische Analyse *f*; 2. *(El)* Oberwellenanalyse *f*
h. analyzer 1. *(Math, Ph)* Analysator *m*, Fourier-Analysator *m*; 2. *(El)* Oberwellenanalysator *m*
h. cam *(Kfz)* harmonischer Nocken *m*
h. components harmonische Teilschwingungen *fpl*; Oberwellen *fpl*
h. content Anteil *m* an Oberschwingungen; Oberwellenanteil *m*
h. current Sinusstrom *m*, sinusförmiger Strom *m*
h. detector Klirrfaktormeßbrücke *f*
h. distortion characteristic Charakteristik *f* der nichtlinearen Verzerrung, Klirrfaktorcharakteristik *f*
h. distortion factor Klirrfaktor *m*, Klirrkoeffizient *m*
h. emission *(El)* Oberwellenemission *f*; Oberwellenleistung *f*
h. filter *(El)* Oberwellenfilter *n*
h. function *(Math)* harmonische Funktion *f*
h. generator Frequenzvervielfacher *m*
h. interference Oberwellenstörung *f (Radio)*
h. mean *(Math)* harmonisches Mittel *n*, harmonischer Mittelwert *m*
h. mode crystal Oberschwingungsquarz *m*
h. mode resonator auf Oberwelle abgestimmter Hohlraumresonator *m*
h. motion harmonische Bewegung *f*, Sinusschwingung *f*
h. multiple telegraph Mehrfachtelegraf *m* mit abgestimmten Wechselströmen
h. number Ordnungszahl *f* der Oberschwingung (Oberwelle)
h. of fundamental sound harmonische Oberschwingung *f*
h. oscillation 1. *(Ph, El)* harmonische (sinusförmige) Schwingung; 2. *(El)* Oberschwingung *f*
h. oscillator 1. *(Ph)* harmonischer Oszillator *m*; 2. *(El)* Oberwellenoszillator *m*
h. power Oberwellenleistung *f*, Verzerrungsleistung *f*
h. response *(BMSR)* Frequenzverhalten *n*, Frequenzgang *m*
h. response diagram Ortskurve *f* des Frequenzganges
h. ringing *s.* h. selective ringing
h. ripple Oberwellengehalt *m*
h. selective ringing *(Fmt)* Frequenz[an]ruf *m*, Anruf *m* mit zugeteilter Frequenz
h. series *(Math)* harmonische Reihe *f*
h. suppression circuit Sendersiebkreis *m*
h. suppressor Oberwellensperrfilter *n*, Oberwellensperrkreis *m*
h. system of working *(Brg)* harmonischer Abbau *m*
h. telegraphy *(Fmt)* Tonfrequenztelegrafie *f*, Wechselstromtelegrafie *f*
h. trap Oberwellenfilter *n*
h. wave harmonische (sinusförmige) Welle *f*
harmonics harmonische Oberschwingung *f*; Oberwellen *fpl*
harness 1. *(Text)* Webgeschirr *n*, Harnisch *m*; 2. Pferdegeschirr *n*
h. bar Harnischstab *m*
h. chain Geschirrkette *f*
h. cords Harnischschnüre *fpl*
h. eye Litzenauge *n*
h. frame Geschirrhalter *m*
h. leather Blankleder *n*, Geschirrleder *n*
h. loom Schaft[web]stuhl *m*
h. mounting Schnürung *f*
h. reed Chorbrett *n*, Harnischbrett *n*
h. room Schirrkammer *f*
h. tie Gallierung *f*
h. wire Geschirrdraht *m*, Schaftaufhänger *m*
harnessing Anschirren *n (Pferde)*
harp-shape shackle *(Schiff)* Buchtschäkel *m*
harpoon Harpune *f*
h. gun Harpunenkanone *f*, Harpunengeschütz *n*
h. line Harpunenschnur *f*
harpsichord Cembalo *n*
harrow / to eggen
harrow Egge *f*
h.-bar Eggenbalken *m*
h. disk Eggenscheibe *f*
h. for animal traction Gespannegge *f*
h. sledge Eggenschlitten *m*
h. spike Eggenzinken *m*
harrowing the ground *(Lw)* Aufreißen *n* der Erde
harsh hart, rauh, scharf *(Ton)*; hart *(Bild)*

h. concrete erdfeuchter (steifer) Beton *m*
h. flame harte Flamme *f*
h.-grained *(Led)* rauhnarbig
Hartinger coincidence refractometer *(Opt)* Koinzidenzrefraktometer *n* nach Hartinger
Hartley circuit *(Fmt)* Dreipunktschwingkreis *m* *(mit Spulenanzapfung)*, Hartley-Oszillator *m*
Hartmann oscillator Hartmann-Generator *m* *(Ultraschall)*
harvest / to ernten
harvest Ernte *f*
h. wagon Erntewagen *m*
harvested cropland *(Am)* Erntefläche *f*
harvester Erntemaschine *f*
h. thresher Mähdrescher *m*
h. threshing box Dreschkasten *m* *(Mähdrescher)*
h. threshing machine Mähdrescher *m*
harvesting machine Erntemaschine *f*
h. rake Ernterechen *m*
h. van (wagon) Erntewagen *m*
h. yields Ernteertrag *m*
hash *(Am)* *(Fmt)* Rauschstörung *f* *(Zerhackerbetrieb)*
hasp 1. Haspe *f*, Haspen *m*; 2. *(Text)* 3600 Yards *(Leinen- oder Jutegarn)*
hat *(Masch)* Hut *m*, Kappe *f*
h. leather [packing] Hutmanschette *f* *(Dichtung)*
h. orifice *(Am)* *(BMSR)* zylindrische Blende *f*
h. press Schichtstoffverformungspresse *f*
h. press mould Form *f* mit Matrize, Patrize und Gummibeutel *(Plastverarbeitung)*
h. shaped ferrule *(Masch)* Hutzwinge *f*
h.-type condenser *(El)* Hütchenkondensator *m*
hatch / to 1. schraffieren; 2. *(Lw)* brüten
hatch Klappe *f*, Luke *f*; Durchreiche *f*, Durchreichfenster *n*
h. beam *(Schiff)* Lukenbalken *m*, Scherstock *m*
h. boat Lukenboot *n*
h. boom *(Schiff)* Lukenbaum *m*
h. coaming *(Schiff)* Lukensüll *n*
h. cover Lukendeckel *m*
h. covering Lukenbezug *m*
h. grating Lukengräting *f*
h. rest Lukendeckelauflage *f*
h. securing bar Lukensicherungsriegel *m*
h.-side girder Lukenlängsträger *m*
h. stanchion Lukengeländerstütze *f*
h. wedge Lukenschalkkeil *m*
hatched 1. schraffiert; 2. *(Lw)* ausgebrütet
hatchel Hechel *f* *(Flachsaufbereitung)*
hatchery Brutanstalt *f*
hatchet Beil *n*, Handbeil *n*
h. soldering copper Hammerlötkolben *m*
h. with hammer end and claw Klauenbeil *n*
hatching 1. Schraffierung *f*, Schraffur *f*; 2. *(Lw)* Brüten *n*; 3. *(Brg)* selbsttätige Bremsvorrichtung *f*
hatchway *(Schiff)* Luke *f*
h. beam Lukenschiebebalken *m*, Schiebebalken *m*
h. tie plate Lukenstringer *m*
haul / to 1. befördern, transportieren; 2. *(Brg)* fördern
haulage 1. Gütertransport *m*, Transport *m*; 2. *(Brg)* Streckenförderung *f*, Förderung *f*
h. capacity Förderleistung *f* *(von Förderern)*
h. clip Schlepphaken *m*
h. drift Förderstrecke *f*
h. gear Windwerk *n* *(Schrämmaschine)*
h. level Fördersohle *f*
h. man Schlepper *m*
h. plane Bremsberg *m*
h. road Förderstrecke *f*
h. rope Förderseil *n*
h. vehicle Transportwagen *m*
h. way Förderstrecke *f*
hauler *(Brg)* Schlepper *m*, Fördermann *m*
haulier *s.* hauler
hauling *(Brg)* Fördern *n*, Schleppen *n*
h. bridge Förderbrücke *f*
h. cable Förderseil *n*
h. engine Fördermaschine *f*
h. load *(Eis)* Anhängelast *f*
h. of wood Holzabfuhr *f*
h. off buoy *(Schiff)* Verholboje *f*
h. power Zugleistung *f* *(Traktor usw.)*
h. power at drawhook *(Eis)* Hakenzugkraft *f*
h. slip *(Schiff)* Aufschlepphelling *f*, Slip *m*, Aufschleppe *f*
h. system *(Schiff)* Netzeinholsystem *n*
h.-up slip *(Schiff)* Aufschlepphelling *f*, Aufschleppe *f*
haulway *(Brg)* Förderstrecke *f*
haunch 1. Bogenschenkel *m*; Gewölbeschenkel *m*; 2. Auflagerschräge *f*, Schräge *f*, Voute *f* *(Stahlbeton)*
haunched beam Träger *m* mit Auflagerschrägen, Voutenbalken *m* *(Stahlbeton)*
hausmannite *(Min)* Schwarzmanganerz *n*, Hausmannit *m*
haven Hafen *m* *(für kleine Fahrzeuge)*
having a taper on both sides beiderseits kegelig
h. close spacing kleinporig
h. front teeth stirnverzahnt
h. good ignition quality zündwillig
h. poor ignition quality zündträg
h. wide spacing großporig
Hawkins cell *(El)* Zinkeisenelement *n*, Hawkins-Element *n*
hawse *(Schiff)* Klüse *f*

h.-bag Klüssack *m*, Stopfbeutel *m (für Ankerklüsen)*
h. bolster Ankerklüsenwulst *m (f)*
h.-buckler Klüsendeckel *m*
h. hole Ankerklüse *f*, Klüse *f*, Klüsengatt *n*
h. hook Klüsenband *n*, Ankerklüsenband *n*
h.-piece Klüsenholz *n*
h. pipe Ankerklüse *f*, Klüse *f*, Klüsengatt *n*
h.-timber Klüsholz *n*
hawser *(Schiff)* Trosse *f*
h. port Verholklüse *f*, Vertäuklüse *f*
h. reel Trossenrolle *f*, Trossenwinde *f*
h. rudder Trossenruder *n (Notruder)*
hay-baler *(Lw)* Heuballenpresse *f*
h.-baling press Heuballenpresse *f*
h.-band Heuseil *n*
h.-band machine Heudrehmaschine *f*
h.-band twister Heu[seil]drehmaschine *f*
h.-barn Heuschuppen *m*
h.-bay Heubanse *f*
h.-cock kleiner Heuschober *m*, Heuhaufen *m*
h.-crop geerntetes Heu *n*; Heuernte *f*
h.-fork Heugabel *f*
h.-knife Heumesser *n*
h.-loader Heulader *m*
h.-loft Heuboden *m*
h.-maker Heuwendemaschine *f*
h.-making Heuernte *f*
h.-moving Heuernte *f*
h.-rack Heuraufe *f (Stall)*
h. rick *s.* h. stack
h. stack Heuhaufen *m*
h.-stacker Schieberechen *m*, Schobersetzvorrichtung *f*
h.-tedder Heuwender *m*
h.-trailer Heuwagen *m*, Futterwagen *m*
h.-turning machine Heuwendemaschine *f*
h.-wire wild verlegt *(z. B. Kabel)*
haylage *(Lw)* Futterportion *f*
hazard Gefahr *f*, Gefährdung *f*; Strahlengefahr *f*
h. to life Lebensgefahr *f*
haze [atmosphärischer] Dunst *m*, Schleier *m*
haziness *(Fs)* Unschärfe *f*
hazle *(Geol)* fester schiefriger Sandstein *m*
hazy dunstig, trübe
h. picture *(Am) (Fs)* unscharfes Bild *n*
head / to *(Masch)* anstauchen *(Köpfe)*
h. a cask ausböden, [ver]bodmen *(Küfnerei)*
h. for *(Schiff)* ansteuern
h. out auffahren, vortreiben, vorhauen
head 1. Spindelstock *m*; Stößelkopf *m (Waagerechtstoßmaschine)*; Support *m (Hobelmaschine)*; 2. Blatt *n*; Kopf *m*; Messerkopf *m*; Vorschneider *m (Werkzeug)*; 3. Schieber *m*; Steg *m*; 4. Druckhöhe *f*; Fallhöhe *f*, Gefälle *n*; 5. Vorschiff *n*, Vorderteil *n* eines Schiffes; Schiffsabort *m*; 6. *(Brg)* Fördergerüst *n*; 7. Sturz *m (Mauerwerksöffnung)*; Rähm *m (Holzbau)*; 8. Kesselboden *m*; 9. *(Typ)* Titel *m*, Überschrift *f*
h. amplifier Bildverstärker *m*, Videovorverstärker *m (Fernsehkamera)*; Vorverstärker *m* im Kondensatormikrofon
h.-and-bench-tunnelling *(Brg)* Streckenauffahren *n* mit strossenartiger Arbeitsweise
h. and tunnel construction Stollen- und Tunnelbau *m*
h. band 1. Kopf[hörer]bügel *m*; 2. *(Typ)* Kapitalband *n*; Kapitälchen *n*
h.-band lamp Stirnleuchte *f*
h. board *(Brg)* Anpfahl *m*
h. bolster *(Eis)* Kopfpolster *n (Sitze)*
h. box Papiermaschinenbütte *f*
h. centre Spindelkopfspitze *f*
h. clearance lichte Höhe *f*
h.-coil-type fuse *(Fmt)* rücklötbare Sicherung *f*
h. compressor Bodenpresse *f*; Bodendübelpresse *f (Küfnerei)*
h. control Steigungsführung *f (Gewindeschneidbacke)*
h. cord Kopfhörerschnur *f*
h. dressing machine Bodenabrichtmaschine *f (Küfnerei)*
h. efficiency Kopfwirkungsgrad *m (Magnettontechnik)*
h. end 1. Kopfende *n (1. Anfang eines neuen Gewebestückes auf dem Webstuhl; 2. Muster einer größeren Partie)*; 2. *(Brg)* Ort *n*, Abbaustoß *m*
h. end of piston stangenfreie Kolbenseite *f*
h. fast *(Schiff)* Bugfestmacher *m*
h. feed thresher *(Am)* Stiftendreschmaschine *f*
h. for side extrusion *(Plast)* Querspritzkopf *m*
h. foreman Werkleiter *m*
h. frame *(Brg)* Förderturm *m*, Fördergerüst *n*
h. gap Magnetkopfspalt *m*
h. gate *(Brg)* Kopfstrecke *f*
h. gear 1. Kopfhörer *m*, Kopftelefon *n*; 2. *(Lw)* Zaumzeug *n*; 3. *(Schiff)* Bugankergeschirr *n*; 4. *(Brg)* Fördergerüst *n*, Förderturm *m*; 5. Kopfbekleidung *f*
h. gear pulley *(Brg)* Seilscheibe *f*
h. house Schachtgebäude *n*
h. jointing machine Bodenfügemaschine *f (Küfnerei)*
h. lamp *s.* headlamp
h. lease *(Text)* Kopfkreuz *n*
h. light *s.* headlight
h. loss [hydraulischer] Druckabfall *m*, Verlusthöhe *f*
h. meter Druckhöhenmesser *m*, Zuflußmesser *m*
h. mirror *(Büro)* Kopfspiegel *m (Magnetbandtechnik)*

h. of a gallery *(Brg)* Ort[sstoß *m*] *n*
h. of cask Faßboden *m*
h. of delivery Förderhöhe *f*
h. of department Abteilungsleiter *m*
h. of grate bar Roststabkopf *m (Feuerung)*
h. of institute Institutsdirektor *m*
h. of press *(Met)* Preßkopf *m*
h. of screw Schraubenkopf *m*
h. of suction Saughöhe *f*
h. of water Wasserhöhe *f*, Wassersäule *f (Druckhöhe)*
h.-office [building] Hauptverwaltungsgebäude *n*
h.-on collision *(Ph)* zentraler Stoß *m*; Zentralstoß *m*
h.-on crash *(frontaler)* Zusammenstoß *m (von zwei Fahrzeugen)*
h.-on view Seitenansicht *f*; Kopfansicht *f*, Stirnansicht *f*
h. piece 1. *(Brg)* Kappe *f*; 2. *(Typ)* Kopfleiste *f*; 3. *(Bw)* Rähm *m (Holzbau)*; 4. Bodenbrett *n (Küfnerei)*
h. planing and jointing machine Hobel- und Fügemaschine *f*
h. planing machine Bodenhobelmaschine *f*
h. plate *(Bw)* Rähm *m (Holzbau)*
h. protection Kopfschutz *m*
h. receiver Kopfhörer *m*
h. rest *(Kfz)* Kopfstütze *f*
h. rope *(Brg)* Oberseil *n*, Förderseil *n*
h. runner *(Bw)* Rähm *m (Holzbau)*
h. sail Vorsegel *n*, Fock *f*
h. sea Bugsee *f*, Gegensee *f*
h. set Kopfhörer *m (Radio usw.)*
h. sheets *(Schiff)* Vorschote *fpl*
h. shield Kopfschutz *m*
h. stick *(Typ)* Kopfsteg *m*
h. stone Eckstein *m (Fundament)*
h. straightening machine Abrichthobelmaschine *f (Küfnerei)*
h.-surface configuration *(Büro)* Aufzeichnungsstrukturen *fpl (Magnetbandtechnik)*
h. tank Hochbehälter *m*
h. tin Feinzinn *n*
h.-to-coating separation *(Büro)* Magnetkopf-Belag-Abstand *m*
h. turf Kopfrasen *m*
h. wall Flügelmauer *f (Brücken)*; Stirnmauer *f (Durchlässe)*
h. wave *(Ph)* Kopfwelle *f*
h. wheel *(Brg)* Seilscheibe *f*
h. wind Gegenwind *m*
headed screw Kopfschraube *f*
header 1. *(Masch)* Stauchstempel *m*; Nietkopfsetzer *m*; 2. *(Met)* verlorener Kopf *m (beim Gießen)*; 3. *(Bw)* Sturz *m (Mauerwerksöffnung)*; Binder[stein] *m*, Kopfstein *m*, Strecker *m (Ziegelmauerwerk)*; 4. *(Brg)* Kerb- und Schrämmaschine *f*, Vortriebsmaschine *f*; Kopfschußbohrloch *n*; 5. Falltank *m*; 6. Ährenmäher *m*
h. auger Schneckenförderer *m*
h. bond Streckerverband *m (Ziegelmauerwerk)*
h. die Gesenk *n* zum Schraubenkopfstauchen
h. plate Rohrboden *m*, Rohrwand *f*
headgear *s.* head gear
heading 1. *(Typ)* Titel *m*, Überschrift *f*; 2. Stauchen *n (Schraubenköpfe)*; 3. Kurs *m*, Richtung *f*; 4. Anfang *m* und Ende *n* eines Gewebestückes; 5. *(Bw)* Stollen *m*; 6. *(Brg)* Abbaustrecke *f*, Strecke *f*, Vorrichtungsstrecke *f*; 7. *(Masch)* Anstauchen *n*
h. correction Kursverbesserung *f*
h. course Binderschicht *f*, Streckerschicht *f (Ziegelmauerwerk)*
h. die *s.* header die
h. displacement Kursabweichung *f*, Kursversetzung *f*
h. indication Vorausanzeige *f (Radar)*
h. machine *(Brg)* Streckenauffahrmaschine *f*, Streckenvortriebsmaschine *f (nur in Kohle)*
h. marker Bordpeiler *m*
h. of paper rolls Verkleben *n* des Kopfes von Papierrollen
h. set *(Am)* Nietkopfsetzer *m*
h. slide Stauchschlitten *m*
h. through *(Brg)* Durchörterung *f*
h. tool Preßstempel *m*
h. upwards *(Brg)* schwebende Strecke *f*
headlamp *(Kfz)* Scheinwerfer *m*
h. bulb Fernlichtlampe *f*, Scheinwerferlampe *f*
h. socket Scheinwerferstütze *f*
h. support tie rod Scheinwerferverbindungsstange *f*
headlands ungepflügter Ackerrain *m*, Vorgewende *n (Feldbestellung)*
headledge Lukenquersüll *n (Holzschiff)*
headless rivet kopfloser Niet *m*
h. screw Madenschraube *f*
headlight 1. *(Kfz)* Fernlicht *n*, Scheinwerfer[licht *n*] *m*; 2. *(Flg)* in die Flügel eingebaute Radarantenne *f*
h. dimmer *(Kfz)* Scheinwerferabblender *m*
h. main beam indicator *(Kfz)* Fernlichtkontrollampe *f*
h. shield *(Kfz)* Abblendkappe *f*
headline *(Typ)* Schlagzeile *f*
headphone Kopfhörer *m*
h. bow Kopfhörerbügel *m*
h. connection Kopfhöreranschluß *m*
h. level Kopfhörerlautstärke *f*
h. reception Kopfhörerempfang *m*
headphones Kopfhörer *mpl*
headrail *(Schiff)* Galionsreling *f*

headroom Bauhöhe *f* *(Betriebsanlage)*; Kopfhöhe *f*, lichte Höhe *f*
heads tank Vorlaufbehälter *m* *(Destillation)*
headset kopfbefestigter Fernsprechapparat *m*; Kopffernhörer *m*
headsheave *(Brg)* Seilscheibe *f*
headspace Freiraum *m*, Leerraum *m* *(in einer Flasche)*
headstock 1. Spindelstock *m* *(Drehmaschine)*; 2. *(Text)* Triebstock *m*, Triebwerk *n*
h. gearing Hauptgetriebe *n*, Spindelstockgetriebe *n*
h. spindle Drehspindel *f* *(Drehmaschine)*
headway *(Schiff)* Fahrt[geschwindigkeit] *f*
heald Helfe *f* *(Weblitze)*
h. braiding machine Litzenflechtmaschine *f*
h. calculation Helfenberechnung *f*, Litzenberechnung *f*
h. eye Litzenauge *n*
h.-mail Litzenauge *n*
h. shaft Geschirr *n* *(Schaft am Webstuhl)*
healing Dachdecken *n*
health hazard Gesundheitsgefährdung *f*
h. measurement Strahlenschutzmessung *f*
h. physicist Strahlenschutzphysiker *m*, Strahlenschutzbeauftragter *m*
h. physics Strahlenschutzphysik *f*
h. precaution Strahlenschutz *m*
heap / to [an]häufen; aufhäufen
heap 1. *(Lw)* Haufen *m*, Hocke *f*, Stapel *m*; 2. *(Brg)* Halde *f*
h. keeper Haldenarbeiter *m*
h. of logs Schichtholz *n*
h. threshing Hockendrusch *m*
heapable goods Schüttgut *n*
heaped concrete Schüttbeton *m*
heaping Aufschüttung *f*, Schüttung *f*
heapstead 1. Zwischenbühne *f*, Hängebank *f* *(Schacht)*; 2. Schachtgebäude *n*
hearing 1. Gehör *n*; Hören *n*; 2. Gerichtsverhandlung *f*
h. aid Hörhilfe *f*; Hörgerät *n* für Schwerhörige
h. of witnesses Beweisaufnahme *f*
heart cam Herznocken *m*, herzförmiger Nocken *m*
h. check Kernriß *m* *(Holz)*
h. cut Mittelfraktion *f* *(Destillation)*
h.-rotten *(Holz)* kernfaul
h. shake Herzriß *m*, Kernriß *m*
h. shaky kernrissig
h.-shaped diagram Kardioidenkennlinie *f* *(Antennentechnik)*
hearth Herd[fläche *f*] *m*; Ofenraum *m*; Nietwärmofen *m*; Gestell *n* *(Hochofen)*
h. accretion Herdansatz *m*
h. area Herdfläche *f*
h. area output Herdflächenleistung *f*
h. bottom Bodenstein *m* *(Hochofen)*
h. casing Gestellpanzer *m* *(Hochofen)*
h. electrode *(Met)* Bodenelektrode *f* *(E-Ofen)*
h.-fettling Herdfutter *n*, Herdfütterung *f* *(Puddelofen)*
h. furnace *(Wärm)* Herdfeuerung *f*
h. jacket Gestellmantel *m* *(Hochofen)*
h. level Herdsohle *f*
h. load Gestellbelastung *f* *(Hochofen)*
h. moulding Herdformerei *f*
h. refining *(Met)* Herdfrischen *n*
h.-type furnace Herdofen *m*
heartshake *s.* heart shake
heartwood Kernholz *n*
heary-media jig *(Am)* *(Brg)* Schwerflüssigkeitssetzmaschine *f*
heat / to 1. erwärmen, erhitzen *(einen Körper)*; 2. heizen *(einen Raum)*
h. evenly gleichmäßig erhitzen
h. glue Leim kochen
h.-press *(Met)* warmpressen, heißpressen
h. to redness rotglühen
h.-treat härten, vergüten *(Stahl)*
h. up anwärmen
heat 1. Wärme *f*; Hitze *f*; 2. *(Met)* Schmelze *f*; Charge *f*; Schmelzung *f*
h.-absorbing wärmeaufnehmend
h.-absorbing filter Wärmeschutzfilter *n*
h.-absorbing power Wärmeschluckvermögen *n*
h.-absorbing surface Wärmeaufnahmefläche *f*
h. absorption Wärmeabsorption *f*; Wärmeaufnahme *f*, Wärmebindung *f*; Wärmeentzug *m*
h.-absorption capacity Wärmeaufnahmefähigkeit *f*
h. absorptivity Wärmeaufnahmevermögen *n*
h.-abstracting characteristic Wärmeableitfähigkeit *f*
h. abstraction Wärmeentzug *m*; Wärmeableitung *f*
h.-accumulator Wärmeakkumulator *m*
h.-affected wärmebeeinflußt
h.-ag[e]ing Wärmealterung *f*
h. balance Wärmebilanz *f*, Wärmehaushalt *m*
h. barrier *(Rak)* Hitzegrenze *f*
h. bonding Warmbindung *f*
h. capacity Wärmeinhalt *m*
h. change Wärmetönung *f* *(chemische Reaktion)*
h. check Wärmeriß *m*
h. circulation Wärmeumlauf *m*
h.-cleaning Flammensäuberung *f*, Heißreinigung *f* *(Glas)*
h. coil Hitzdrahtspule *f*; Feinsicherungselement *n*, Feinsicherungspatrone *f*
h. coil fuse Feinsicherung *f*
h. colour Glühfarbe *f*; Anlaßfarbe *f*

h. compensation Wärmeausgleich *m*
h. compensator *s.* h. exchanger
h. conduction Wärmeleitung *f*
h. conduction error Wärmeableitungsfehler *m*
h. conductivity Wärmeleitfähigkeit *f*, Wärmeleitvermögen *n*
h.-conductivity pressure gauge Wärmeleitungsmanometer *n*, Wärmeleitungsvakuummeter *n*
h. conductor Wärmeleiter *m*
h. consuming wärmeverbrauchend, wärmeverzehrend; endotherm
h. consumption Wärmeverbrauch *m*
h. content Wärmeinhalt *m*
h. control Temperaturregelung *f*
h. convection Wärmeumwälzung *f*, Wärmekonvektion *f*
h. crack *(Met)* Härtungsriß *m*, Wärmeriß *m*
h. curing Warmbehandlung *f (Beton)*
h. damper Hitzedämpfer *m*
h.-deformable thermoplastisch
h. dilatation Wärmedehnung *f*
h. dilatometer Wärmedehnungsmesser *m*
h.-dissipating wärmeabgebend, wärmeableitend
h. dissipating element wärmeableitendes Element *n (z. B. Transistorkühlung)*
h. dissipation Abführung *f* der Wärme; Wärmeableitung *f*
h. dissipator Wärmeableiter *m*
h. distortion point *(Plast)* Wärmefestigkeitsgrenze *f*
h. drop Wärmegefälle *n*, Wärmeabfall *m*
h. economy Wärmewirtschaft *f*
h. effect Wärmetönung *f (chemische Reaktion)*
h. element *(Rak)* Ofentopf *m*
h. emission 1. *(Ph)* Emission *f* von Wärmestrahlung; 2. Wärmeabgabe *f* [durch Strahlung]
h.-emitting surface 1. *(Ph)* Abstrahlfläche *f*; 2. *(Wärm)* Wärmeabgabefläche *f*
h. energy Wärmeenergie *f*
h. engine Wärmekraftmaschine *f*
h. engineering Wärmetechnik *f*
h. equalization *(Wärm)* Wärmeausgleich *m*
h. equivalent *(Ph)* Wärmeäquivalent *n*
h. evolution Wärmeentwicklung *f*
h. exchanger Wärme[aus]tauscher *m*
h. expansion Wärmeausdehnung *f*
h. flash Wärmeblitz *m*
h. flow Wärmefluß *m*, Wärmestrom *m*
h. flush *(Kern)* Wärmestoß *m*
h. fusion process *(Text)* Wärmeschmelzverfahren *n*
h. generated in grinding Schleifwärme *f*
h. generation Wärmeerzeugung *f*
h. governor Thermostat *m*
h. input Wärmezufuhr *f*
h.-insulated gegen Wärmeabgabe (Wärmezufuhr) geschützt
h.-insulating wärmedämmend
h.-insulating board Wärmedämmplatte *f*
h.-insulating casing Wärmeschutzpanzer *m*
h.-insulating material Wärmeschutzstoff *m*, Wärmeisoliermittel *n*
h. insulation Wärme[schutz]isolierung *f*
h. insulator Wärmeisolator *m*, Wärmeschutz[masse *f*] *m*
h. jacket Heizmantel *m*
h.-laminated *(Text)* durch Wärmeschmelzverfahren laminiert
h. laminating machine Heißlaminiermaschine *f*
h. load Wärmebelastung *f*
h. loss Wärmeverlust *m*
h. losses due to excess of air Wärmeverlust *m* durch Luftüberschuß
h. losses due to formation of soot Wärmeverlust *m* durch Rußbildung
h. losses due to incomplete combustion Wärmeverlust *m* durch unvollkommene Verbrennung
h. losses due to unburnt gases Wärmeverlust *m* durch unverbrannte Gase
h. meter Wärmemengenmesser *m*
h. motion Wärmebewegung *f*
h. motor *s.* h. engine
h. of absorption Absorptionswärme *f*
h. of activation Aktivierungswärme *f*
h. of coagulation *(Ch)* Koagulationswärme *f*
h. of combustion Verbrennungswärme *f*
h. of compression Verdichtungswärme *f*, Kompressionswärme *f*
h. of condensation Kondensationswärme *f*
h. of dissociation Dissoziationswärme *f*, Zerfallswärme *f*
h. of evaporation Verdampfungswärme *f*
h. of formation *(Ph, Ch)* Bildungswärme *f*
h. of fusion Schmelzwärme *f*
h. of fusion per atomic weight *(Ch)* Atomschmelzwärme *f*
h. of hydration Hydrationswärme *f*; Abbindewärme *f*
h. of radiation Strahlungswärme *f*
h. of radioactivity radioaktive Zerfallswärme *f*
h. of reaction *(Ch)* Reaktionswärme *f*, Wärmetönung *f*
h. of recalescence *(Ch)* Umwandlungswärme *f*
h. of solidification Kristallisationswärme *f*
h. of solution *(Ch)* Lösungswärme *f*
h. of steam Dampfwärme *f*
h. of transformation *(Ph, Ch)* Umwandlungswärme *f*
h. of transition Übergangswärme *f*

h. of wetting Benetzungswärme *f*
h.-on ratio *(El)* Schaltverhältnis *n (Thermostat)*
h. output Wärmeabgabe *f*, Wärmeleistung *f*
h. peak *(Kern)* örtliche Wärmequellenspitze *f*
h. producer *(Wärm)* Wärmebildner *m*
h. production Wärmeerzeugung *f*
h.-proof hitzebeständig, wärmebeständig
h. protection Wärmeschutz[maßnahmen *fpl*] *m*
h. pump Wärmepumpe *f*
h. pyrometer Strahlungspyrometer *n*
h. radiation Wärme[aus]strahlung *f*, Temperaturstrahlung *f*
h. radiator Wärmestrahler *m*; Kühlschelle *f*; Wärmeableiter *m (Leistungstransistoren)*
h. received zugeführte Wärme *f*
h. recovery Wärmerückgewinnung *f*
h. refining *(Met)* Vergüten *n*
h. reflection Wärmerückstrahlung *f*
h. release Wärmefreisetzung *f*
h. requirement Wärmebedarf *m*
h. resistence Hitzebeständigkeit *f*; Wärmefestigkeit *f*
h.-resisting hitzebeständig; wärmefest
h.-resisting glass feuerfestes Glas *n*
h. retention Wärmehaltung *f*
h. run *(Elm)* Heißlaufprobe *f*
h. screen Wärmeschutz[schirm] *m*
h. sealable heißsiegelfähig
h.-sealing 1. *(Ch)* Heißsiegeln *n*, Heißverkleben *n*, Heißverschweißen *n*; 2. *(Typ)* Anschmelzen *n*, Schweißen *n (des Buchblocks)*
h. sensitive wärmeempfindlich
h.-set [finish] Fixieren *n*
h.-set ink *(Typ)* Heißtrockenfarbe *f*
h. setting *(Plast)* hitzehärtbar
h. setting *(Wärm)* Heißformen *n*; Hitzefixierung *f*
h. shield Wärmeschutz[schirm] *m*; Strahlschutz *m (einer Katode)*
h. shock Temperatursturz *m*
h. shrinking Aufschrumpfen *n*
h. sink *(Ph)* Wärmesenke *f*; Wärmeableitung *f (Halbleitertechnik)*
h.-solvent tape sealing *(Plast)* Heißsiegeln *n* mit Lösemitteln
h. source Wärmequelle *f*
h. stability Wärmefestigkeit *f*
h. stabilizer Wärmestabilisierungsmittel *n*
h. station Fernheizwerk *n*
h. storage Wärmespeicherung *f*
h.-strengthened glass gehärtetes Glas *n*
h. stress Wärmespannung *f*
h. supply Wärmezufuhr *f*
h.-tensioning thermisches Vorspannen *n (Spannbeton)*
h. throughput Wärmedurchsatz *m*
h. tint Anlaßfarbe *f*, Anlauffarbe *f (Stahl)*
h. tinting Anlassen *n (Stahl)*
h. tone Wärmetönung *f (chemische Reaktion)*
h. transfer Wärmeübertragung *f*, Wärmedurchgang *m*, Wärmeübergang *m*
h.-transfer agent Wärmeträger *m*, Wärmeübertragungsmittel *n*
h.-transfer coefficient Wärmeübergangszahl *f*, Wärmedurchgangszahl *f*
h.-transfer cycle Kühlkreislauf *m*
h.-transfer factor Wärmeübergangszahl *f*, Wärmedurchgangszahl *f*
h.-transfer medium Wärmeübertrager *m*; Kühlmittel *n*
h.-transfer rate Wärmeübergangsgeschwindigkeit *f*
h. transformation Wärmeumsatz *m*
h. transmission *s.* h. transfer
h. treatability Wärmevergütbarkeit *f (Stahl)*
h.-treatable härtbar, vergütbar *(Stahl)*
h.-treatable steel Vergütungsstahl *m*
h.-treated steel vergüteter Stahl *m*; wärmebehandelter Stahl
h. treating Wärmebehandlung *f*, Vergütung *f (Stahl)*
h.-treating department Härterei *f*, Vergüterei *f (Stahl)*
h.-treating engineer Härteingenieur *m*
h.-treating furnace Härteofen *m (Stahl)*
h.-treating plant Vergütungsanlage *f (Stahl)*
h.-treating property Vergütbarkeit *f (Stahl)*
h. treatment Wärmebehandlung *f*; Warmbehandlung *f*
h.-treatment bath Wärmebehandlungsbad *n*
h.-treatment crack *(Met)* Härteriß *m*
h.-treatment steel Vergütungsstahl *m*
h. unit Wärmeeinheit *f*
h. utilization Wärmeausnutzung *f*
h. value 1. Heizwert *m*; 2. Wärmewert *m (Zündkerze)*
h. vibration Wärmeschwingung *f*
h. wave Hitzewelle *f*
heatable stage heizbarer Tisch *m (Mikroskop)*
heated air *(Met, Ch)* Heißwind *m*, erhitzter Gebläsewind *m*
h. press roll heizbare Preßwalze *f*
h. steam überhitzter Dampf *m*
heater 1. Heizkörper *m*, Heizapparat *m*; Erhitzer *m*; 2. *(El)* Heizfaden *m (indirekt geheizte Katode)*
h. by-pass Heizfadenüberbrückungskondensator *m*
h.-cathode leakage Stromübergang *m* zwischen Heizfaden und Katode
h.-cathode leakage current Heizfaden-Katoden-Verluststrom *m*
h.-cathode-type valve indirekt geheizte Röhre *f*

h. circuit Heizstromkreis *m (indirekt geheizte Katode)*
h. coil Heizspule *f*; Heizwendel *f (Elektronenröhren)*
h. element 1. *(El)* Heizelement *n*; 2. Heizkörper *m*
h. fan Heizgebläse *n*
h. filament Heizfaden *m*
h. filament current Heizfadenstrom *m*
h. of carburetter air *(Kfz)* Ansaugluftvorwärmer *m*, Vergaserluftvorwärmer *m*
h. plug *(Kfz)* Glühkerze *f*
h.-plug control Glühüberwacher *m*
h.-plug resistor Glühkerzenwiderstand *m*
h.-plug starting switch Glühanlaßschalter *m*
h. spiral *s.* h. coil
h. starter switch *(Kfz)* Glühanlaßschalter *m*
h. supply Heizstromquelle *f (indirekt geheizte Katode)*
h. switch *(Kfz)* Glühschalter *m*
h. transformer Heiztransformator *m*
h.-type cathode indirekt geheizte Katode *f*
h.-type tube Röhre *f* mit indirekt geheizter Katode, indirekt geheizte Röhre
h. voltage Heizfadenspannung *f*
h. winding *(Elm)* Heizwicklung *f*
heather mix *(Text)* Mischung *f* vielfarbiger Fasern zur Herstellung eines gemischten oder gesprenkelten Garnes
heating 1. Erhitzen *n*, Erwärmen *n*; 2. Beheizung *f*, Heizung *f*
h. and ventilating system Heizungs- und Lüftungsanlage *f*
h. area Heizfläche *f*
h. by circulating air Umluftheizung *f*
h. by gas Gasheizung *f*
h. by infrared-radiation Infrarotstrahlungsheizung *f*
h. by waste gases Vorwärmung *f* durch Abgase
h. capacity Heizfähigkeit *f*
h. chamber *(Ch)* Heizkammer *f*, Heizschrank *m*
h. circuit Heizstromkreis *m*
h. coil 1. *(El)* Heizspirale *f*, Heizspule *f*; 2. Heizschlange *f (Heizungstechnik)*
h. conductor Heizleiter *m*
h. crack *(Met)* Warmriß *m*; Härteriß *m*
h. current Heizstrom *m*
h. duct Heizkanal *m*
h. effect Heizeffekt *m*, kalorischer Effekt *m*
h. effect indicator Heizeffektmesser *m*
h. element *(El)* Heizelement *n*
h. equipment Heizanlage *f*
h. flue 1. Feuerungszug *m*; 2. *(Brg)* Heizzug *m*, Heizkanal *m*
h. flue casting *(Brg)* Heizzugverschluß *m*
h. furnace *(Met)* Glühofen *m*; Warmhalteofen
h. gas Heizgas *n*, Brenngas *n*, technisches Gas *n*
h. hood Heizhaube *f*
h. of bearing Warmlaufen (Heißlaufen) *n* eines Lagers
h. of receiver Aufnehmerheizung *f*
h. of superheater Heizung *f* der Dampfüberhitzer; Überhitzerheizung *f*
h. of the cylinder cover Deckelheizung *f*
h. of the cylinder walls Mantelheizung *f*, Heizung *f* des Zylindermantels
h. pad Heizkissen *n*
h. panel Heizplatte *f*, Heiztafel *f (Heizungstechnik)*
h. pit *(Met)* Tiefofen *m*; Wärmegrube *f*
h. plug *(Kfz)* Glühkerze *f*
h. power Heizkraft *f*
h.-power machine Abwärmekraftmaschine *f*
h. resistor *(El)* Heizleiter *m*, Heizwiderstand *m*
h. sample *(Ch)* Glühprobe *f*
h. shock Wärmestoß *m*
h. surface Heizfläche *f*
h. surface of the feed heater Vorwärmerheizfläche *f*
h. test *(Met)* Glühprobe *f*
h. time *(Met)* Wärmzeit *f*
h. transformer Heiztransformator *m (Radio)*
h. tube Kesselrohr *n*
h.-up period 1. Aufheizperiode *f*, Aufheizzeit *f (Elektronenröhren)*; 2. *(Wärm)* Anwärmperiode *f*
h.-up schedule Temperplan *m (Metall, Glas)*
h.-up time Aufheizzeit *f*
h. value Heizwert *m*
h. value per unit weight Heizwert *m* pro Masseeinheit, spezifischer Heizwert, Heizwertzahl *f*
h. wire *(El)* Heizdraht *m*
h. with exhaust in contact *(Wärm)* Mischungsvorwärmung *f*
heatronic moulding Verfahren *n* der Hochfrequenzvorheizung von Kunststoffen
heave / to 1. *(Schiff)* hieven, hochwinden; 2. *(Brg)* quellen, heben *(Sohle)*
h. and set *(Schiff)* stampfen
h. to *(Schiff)* beidrehen
heave 1. Auf- und Abbewegung *f (eines Schiffes)*; 2. Auftrieb *m*, Hebung *f (Grundbau)*; 3. *(Geol)* Sprungweite *f (einer Verwerfung)*
h. of the sea Seegang *m*
h. the lead *(Schiff)* Lotwerfen *n*
heaver *(Lw)* Hebebaum *m*
heavily laden ship schwerbeladenes Schiff *n*
h. loaded stark belastet
h. plated stark verchromt
h. ribbed stark verrippt

h. silver-plated stark versilbert
heaving 1. Hieven *n*; Hochwinden *n*; 2. *(Schiff)* Tauchschwingungen *fpl*; 3. *(Brg)* Hebung *f*; Quellung *f (Sohle)*
h. line *(Schiff)* Wurfleine *f*, Hievleine *f*
Heaviside layer Heaviside-Schicht *f*
Heaviside unit-function [Heavisidesche] Sprungfunktion *f*
Heaviside's operational calculus Heavisidesche Operatorenrechnung *f*
heavy schwer; massiv; solid; schwerflüssig *(Öl)*
h.-aggregate concrete Schwer[st]beton *m*
h.-aggregate shield Schwerbetonabschirmung *f (Reaktorbau)*
h. anode massive Anode *f*, Vollanode *f*
h.-armature relay Relais *n* mit schwerem Anker
h. bedded *(Geol)* dickbankig
h.-bodied zähflüssig
h. boom *(Schiff)* Schwergutbaum *m*, Schwerlastbaum *m*
h. broken ground *(Brg)* druckhaftes Gebirge *n*
h. cargo schwere Ladung *f*, Schwergut *n (Güter, die das Schiff bis zur Lademarke abladen, es jedoch raummäßig nicht ausfüllen)*
h. cargo-handling gear *(Schiff)* Schwergutgeschirr *n*
h. cargo ship Schwergutfrachter *m*
h. ceramics Grobkeramik *f*
h. clay industry grobkeramische Industrie *f*
h. concrete Schwer[st]beton *m*
h. crown glass Schwerkronglas *n*
h. current Starkstrom *m*
h. current engineering Starkstromtechnik *f*
h. cut *(Masch)* Kraftschnitt *m*; kräftiger Schnitt *m*
h. derrick *(Schiff)* Schwergutbaum *m*, Schwerlastbaum *m*
h. drive fit *(Masch)* enger Treibsitz *m*
h. duck loom Schwergewebe[web]stuhl *m*
h. duty Hochleistung *f*
h. duty 1. Hochleistungs-; 2. hoch belastbar
h.-duty armoured conveyor schwerer gepanzerter Stegkettenförderer *m*
h. duty detergent Grobwaschmittel *n*
h. duty drilling and boring Hochleistungsbohren *n*
h.-duty drilling machine Hammerbohrmaschine *f*
h.-duty lathe Großdrehmaschine *f*; Hochleistungsdrehmaschine *f*; Schwerdrehmaschine *f*
h.-duty lorry Schwerlastwagen *m*
h. duty machine Hochleistungsmaschine *f*
h. duty oils HD-Öle *npl (Öle, die durch Zusätze speziell den Anforderungen in Dieselmotoren angepaßt sind)*
h.-duty power unit Hochleistungstriebwerk *n*
h.-duty roll turning lathe Hochleistungswellendrehmaschine *f*
h. duty screwdriver Kraftschraubenzieher *m*
h.-duty switch Leistungsschalter *m*
h.-duty trailer Schwerlastanhänger *m*
h. engineering industry Schwermaschinenbau *m*
h. fabric Schwergewebe *n*
h. flint glass Schwerflintglas *n*
h. flywheel Massenschwungrad *n*
h. force and shrink fit Längsquerpreßpassung *f*
h. fuel Masut *n*, zähflüssiges Heizöl *n*
h. gauge *(Met)* schweres Walzkaliber *n*
h.-gauge line *(El)* dickdrähtige Leitung *f*
h.-gauge wire dicker Draht *m*
h. gymnastics Geräteturnen *n*
h. hours verkehrsreiche Zeit *f*
h. hydrocarbons schwere Kohlenwasserstoffe *mpl*
h. hydrogen schwerer Wasserstoff *m*, Deuterium *n*
h. industry Schwerindustrie *f*
h. keying fit *(Masch)* enger Keilsitz *m*
h. lift ship Schwergutfrachter *m*
h. line fette (dick gezeichnete) Linie *f*
h. liquid Schwerflüssigkeit *f*
h.-liquid separation Schwerflüssigkeitsaufbereitung *f*
h.-load capacity Hochbelastbarkeit *f*
h.-load wire-wound resistor hochbelastbarer Drahtwiderstand *m*
h. loaded cable schwer bespultes (pupinisiertes) Kabel *n*
h. machinery Schwermaschinen *fpl*
h. mass schwere Masse *f*
h. material *(Brg)* Beschwerungsstoff *m*
h.-media separation *(Am)* Schwerflüssigkeitsverfahren *n (Aufbereitung)*
h.-media separator *(Am)* Sinkscheider *m (Aufbereitung)*
h. medium Schwertrübe *f (Aufbereitung)*
h. meson *(Ph)* schweres Meson *n*
h. metal Schwermetall *n*
h.-metal casting Schwermetallguß *m*
h. naphtha Schwerbenzin *n*
h. nucleus *(Ph)* schwerer Kern *m*
h. oil Schweröl *n*
h.-oil carburetter Schwerölvergaser *m*
h.-oil engine Schwerölmotor *m*
h. particle *(Ph)* schweres Teilchen *n*
h. petrol Schwerbenzin *n*
h. plate Grobblech *n*
h. punching bag Sandsack *m (Boxen)*
h. rocker plough Rodepflug *m (Forst)*
h. sailer *(Schiff)* schlechter Segler *m*
h. scrap Blockschrott *m*, schwerer Schrott *m*
h. sections schwerer Formstahl *m*

h. service C clamp Schraubzwinge *f* für schwere Arbeiten
h. solid line dicke Vollinie *f (Zeichnung, grafische Darstellung)*
h. spar Schwerspat *m*, Baryt *m*
h. starting duty Schweranlauf *m (Motor)*
h. timber Starkholz *n*
h. traffic starker Verkehr *m*
h. type *(Typ)* fette Schrift *f*
h. water *(Ph, Ch)* schweres Wasser *n*, Schwerwasser *n*
h.-water moderated reactor Schwerwasserreaktor *m*
h.-water plant Anlage *f* zur Erzeugung schweren Wassers, Schwerwassererzeugungsanlage *f*
h.-weight concrete Schwer[st]beton *m*
h. wire dicker Draht *m*
h.-working ship schwer bedienbares Schiff *n*
heckle Hechel *f*
hectograph / to hektografieren, mit Hektograf vervielfältigen
hectograph Hektograf *m*
h. pad Hektografenplatte *f*
hectolitre Hektoliter *n*
hectowatt Hektowatt *n*
h. hour Hektowattstunde *f*
heddle *(Am)* Litze *f (Weben)*
h. eye *(Text)* Mittelauge *n (Litze)*; *(Am)* Litzenauge *n*
h. hook Einziehnadel *f*, Einziehhaken *m*
hedenbergite *(Min)* Hedenbergit *m*
hedge-bill Gartenmesser *n*
h. clipping shears Heckenschere *f*
h. cutter Heckenschneider *m*
h.-hop / to *(Flg)* tieffliegen
h. hopper *(Flg)* Tiefflieger *m*
h. trimmer Heckenbeschneidemaschine *f*
hedgehog drawing *(Text)* Nadelwalzenstrecke *f*, Igelstrecke *f*
h. transformer *(El)* Igeltransformator *m*
heel / to krängen
h. over *(Schiff)* überholen, krängen, sich überlegen
heel 1. *(Text)* Ferse *f (Strümpfe)*; 2. *(Led)* Schuhabsatz *m*; 3. Schleifsohle *f (Pflug)*; 4. Hacke *f (Ruder)*; Krängung *f (eines Schiffes)*; 5. *(Masch)* Auflage *f*; Schneidenrücken *m*; Rückenkante *f (Spiralbohrer)*; Fase *f*; Rückenfläche *f*; 6. Absatz *m (Stützmauer)*
h.-board *(Kfz)* Auflagebrücke *f* für Rücksitze
h. building machine *(Led)* Absatzvorbaumaschine *f*
h. leather Absatzleder *n (Schuhe)*
h. of bead *(Kfz)* Wulstferse *f*
h. of coal Kohlenbein *n*
h. of stern frame *(Schiff)* Hinterstevenhacke *f*
h. of stope *(Brg)* unterer Stoß *m (Strossenbau)*; oberer Stoß *(Firstenstoßbau)*
h. of tooth Schneidenrücken *m (Fräser)*
h. tool Drehhaken *m*, Falzhaken *m*, Schrothaken *m (Drechslerei)*
heeling Krängung *f (eines Schiffes)*
h. angle Krängungswinkel *m*
h. error *(Schiff)* Krängungsfehler *m (des Kompasses)*
h. machine *(Text)* Fersenmaschine *f*
h. tank *(Schiff)* Krängungstank *m*
heer *(Text)* 600 yards *(Jute)*
Hefner candle Hefnerkerze *f*
Hefner lamp Hefnerlampe *f*
heifer Jungkuh *f*, Kalbe *f*
height Höhe *f*
h. above sea level Höhe *f* über Meeresspiegel
h. adjustment Höhenverstellung *f*; Senkrechtverstellung *f*
h. change-over switch Höhenumschalter *m (Radar)*
h. control *(Fs)* Höheneinstellung *f*, Einstellung *f* der Bildhöhe
h. determination Höhenbestimmung *f*
h. equivalent to a theoretical plate (stage) *(Kern)* Trennstufenhöhe *f*, Bodenwert *m*
h. finder Höhenmesser *m*
h.-finding instrument Höhenmesser *m*
h. gain *(El)* Höhenzunahme *f* [der Feldstärke]
h. gauge 1. Höhenmaß[stab *m*] *n*; 2. Höhenanreißlehre *f*, Höhenreißer *m*, Reißmaß *n*
h. indicator Höhenmesser *m*, Altimeter *n*
h. marker Höhenzielmarke *f (Vermessung)*
h. measuring Höhenmessung *f*
h. of ascent Steighöhe *f (in Kapillaren)*
h. of bosh Rasthöhe *f (Hochofen)*
h. of centres *(Masch)* Spitzenhöhe *f*
h. of coal line Füllhöhe *f* der Kohle *(Koksofen)*
h. of damming Stauhöhe *f*
h. of growth Wuchshöhe *f*
h. of high water Hochwasserstand *m*
h. of low water Niedrigwasserstand *m*
h. of rise 1. Steighöhe *f (in Kapillaren)*; 2. *(Bw)* Stauhöhe *f*
h. of the heald *(Text)* Sprung *m* der Helfe
h. of the lathe centres Drehmaschinenspitzenhöhe *f*
h. of water Druckhöhe *f* des Wassers
h. of water column Wasserbarometerhöhe *f*, Druck *m* in cm Wassersäule
h. of waves *s.* wave amplitude
Heisenberg force Heisenberg-Kraft *f*
Heisenberg picture Heisenberg-Bild *n*
Heisenberg relations [Heisenbergsche] Unbestimmtheitsrelationen *fpl*

Heisenberg representation Heisenberg-Darstellung *f*
Heisenberg theory [Heisenbergsche] nichtlineare Spinortheorie *f (Elementarteilchen)*
Heisenberg's approximation Heisenbergsche Näherung *f*
Heisenberg's principle of indeterminacy [Heisenbergsche] Unbestimmtheitsrelation *f*
Heisenberg's uncertainty principle [Heisenbergsche] Unbestimmtheitsrelation *f*
Heising modulation Absorptionsmodulation *f*, Heising-Modulation *f*
Heizmann's plate superheater Plattenüberhitzer *m*, plattenförmiger Rauchrohrüberhitzer *m*, Heizmannscher Überhitzer *m*
heliarc welding process Heliarcschweißverfahren *n*
helical Schrägzahnrad *n*
helical schraubenförmig; wendelförmig; gedrallt; schraubig; schneckenförmig
  h. aerial Spiralantenne *f*, Wendelantenne *f*
  h. blower Schrauben[rad]gebläse *n*, Schneckengebläse *n*
  h. coil compression spring Schraubendruckfeder *f*
  h. core of the nozzle sprayer Heizspirale *f* des Düsenzerstäubers
  h. curve Schraubenlinie *f*
  h. element *(El)* Spiralheizelement *n*
  h. extension spring Zugfeder *f*
  h. fan *s.* h. blower
  h. flute Drallnut *f*; Wendelnut *f*
  h. gear Schrägzahnstirnrad *n*
  h. gear train Schrägzahnstirnradgetriebe *n*
  h. gear with helical splines Schrägrad *n* mit Schraubennuten
  h. geared schrägverzahnt, mit Schrägverzahnung
  h. gearing Schrägverzahnung *f*
  h. groove Drallnut *f*; Wendelnut *f*
  h. lens mount Schneckengangfassung *f*
  h. line Schraubengewindelinie *f*
  h. milling cutter Fräser *m* mit spiralig gebogenen Schneidkanten
  h. motion Schraubenbewegung *f*; Schraubvorschub *m*
  h. packing ring schraubenförmiger Dichtungsring *m*
  h. potentiometer mehrgängiges Potentiometer *n*, Spiralpotentiometer *n*
  h. rack Zahnstange *f* mit Schrägverzahnung
  h. reinforcement Spiralbewehrung *f (Stahlbeton)*
  h. reversing bucket schraubenförmige Umkehrschaufel *f (Turbine)*
  h. scanning Schraubenlinienabtastung *f*
  h. spring [zylindrische] Schraubenfeder *f*, Spiralfeder *f*
  h. spur gear Schrägzahnstirnrad *n*
  h. stair[case] Wendeltreppe *f*
  h. superheating tube *(Wärm)* Überhitzerschraubenrohr *n*
  h. surface *(Math)* Schraubenfläche *f*
  h. tape Spiralband *n*
  h. taper pin reamer Kegelreibahle *f* mit Spiralzähnen für Stiftlöcher, Kegelschälreibahle *f*
  h. teeth Schrägzähne *mpl*; Schrägverzahnung *f*
  h. tooth *(Masch)* Schraubzahn *m*, Schrägzahn *m*
  h.-toothed spur gear Schrägzahnstirnrad *n*
  h. tube spiralförmiges Rohr *n*, Wendelrohr *n*
  h. twist spring schraubenförmige Drehfeder *f*
helically toothed wheel schrägverzahntes Rad *n*
helicoid Schraubenfläche *f*
helicoid schraubenförmig; schneckenförmig; spiralförmig
  h. parabola parabolische Spirale *f*
helicoidal *s.* helicoid
  h. motion Schraubenbewegung *f*
helicometer Helikometer *n (Spiralmessung)*
helicopter Hubschrauber *m*
heliocentric heliozentrisch
heliodor *(Min)* Heliodor *m*
heliography Lichtdruck *m*; Halbtonätzung *f*
heliogravure Heliogravüre *f*; Fotogravüre *f*
heliothermometer Heliothermometer *n (Astrophysik)*
heliotrope 1. *(Min)* Heliotrop *m*; 2. *(Astr)* Sonnenspiegel *m*, Heliotrop *n*
helipot Helipot *(Firmenbezeichnung für mehrgängiges Potentiometer)*
helium Helium *n*
  h. arc welding Heliarcschweißen *n*
  h. nucleus Heliumkern *m*
helix 1. *(Math)* Schraubenlinie *f*; 2. *(Masch)* Wendel *f*; Drall *m*; 3. Förderschraube *f (Lochbohrer für Holzbearbeitung)*; 4. *(Fs)* Verzögerungsleitung *f*
  h. angle Spiralwinkel *m*; Drallsteigungswinkel *m*, Steigungswinkel *m*; Spanwinkel *m*; Schrägungswinkel *m (Zahnrad)*
  h. angle tester Schrägungswinkelmeßgerät *n*
  h. line Schraubenlinie *f*
  h.-tooth cutter Fräser *m* mit Spiralverzahnung, spiralverzahnter Fräser
  h.-type plunger Schrägkantenkolben *m*
  h.-type variable capacitor Wendelkondensator *m (Spezialtyp eines abstimmbaren Kondensators)*
  h. waveguide Wendelhohlleiter *m*, wendelförmiger Wellenleiter *m*
helm *(Schiff)* Ruder *n (Steuer)*
  h. angle indicator Ruderzeigeranlage *f*
  h. indicator Ruderanzeiger *m*

helmet 1. Rammhaube *f*, Schlaghaube *f (Rammpfahl)*; 2. *(El)* Funkerhaube *f*
h. crew *(Brg)* Rettungsmannschaft *f*
Helmholtz galvanometer Tangentenbussole *f* [nach Helmholtz]
helmsman Steuermann *m*
help material Hilfsmaterial *n*
helve Stiel *m (Beil oder Axt)*
hem *(Text)* Saum *m*
hematin Hämatin *n*
hematite *(Am)* Eisenglanz *m*, Hämatit *m*, Roteisenstein *m*
h. ore Hämatiterz *n*
h. pig iron Hämatitroheisen *n*
hemeralopia Tagblindheit *f*
hemeralopic tagblind
hemianopsia Halbseitenblindheit *f*, Hemianopsie *f*
hemianopsy spectacles Hemianopikerbrille *f*
hemicolloid Hemikolloid *n*
hemihedral halbflächig *(Kristall)*
h. crystal Halbflächner *m (Kristall)*
hemihedrism Hemiedrie *f (Kristalle)*
hemihedron Halbflächner *m (Kristall)*
hemihedry Hemiedrie *f (Kristalle)*
hemimorphism Hemimorphie *f (Kristalle)*
hemimorphite Hemimorphit *m*; Kieselzinkerz *n*
hemin *(Ch)* Hämin *n*
hemisphere Halbkugel *f*, Hemisphäre *f*
h. combustion chamber Kugelbrennraum *m*
hemispherical halbkugelförmig, halbkugelig
h. bearing Kugelpfannenlager *n*
hemitrope Zwillingskristall *m*
hemitropic hemitrop *(Kristall)*
hemlock fir Schierlingstanne *f*
hemoglobin Hämoglobin *n*
hemp Hanf *m*
h. centre Hanfseele *f (Seil)*
h. cord Hanfschnur *f*
h. core Hanfseele *f (Seil)*
h. drawing frame *(Text)* Hanfstrecke *f*
h. farming Hanfanbau *m*
h. hackling machine with automatic screwers Hanfhechelmaschine *f* mit selbsttätiger Umspannung
h. knifing machine Hanfschneidemaschine *f*, Hanfzerreißmaschine *f*
h. packing *(Masch)* Hanfdichtung *f*, Hanfpackung *f*; Hanfumwicklung *f*
h. paper Hanfpapier *n*
h. rope Hanfseil *n*, Hanfstrick *m*
h. softening mill *(Text)* Hanfreibe *f*
h. strick *(Text)* Hanfriste *f*
h. tow Hanfwerg *n*
h. tow preparing Hanfwergaufbereitung *f*
h. twist Hanfzopf *m*
Hempel pipette Hempel-Bürette *f*
hen or dove dung bate Beize *f* aus Hühner- oder Taubenmist *(Ledergerbung)*
hendecane *(Ch)* Undekan *n*
hendecanoic acid Undekansäure *f*
hendecene *(Ch)* Undezen *n*
henequen Henequen *m (als Pflanze)*; Henequen *n (als Faser)*, Sisalhanf *m*
heniquen *s.* henequen
henry *(El)* Henry *n (Einheit der Induktivität)*
heptacyclic compound *(Ch)* Siebenringverbindung *f*
heptagon Siebeneck *n*
heptagonal siebeneckig
heptahedral siebenflächig
heptaldehyde Heptaldehyd *m*, Önanthaldehyd *m*
heptane Heptan *n*
h. diacid Pentandikarbonsäure *f*
heptavalent siebenwertig
heptene Hepten *n*
heptode *(El)* Heptode *f*, Siebenpolröhre *f*
herbage Kräuter *npl*, Gräser *npl*
herbicide Herbizid *n*, Unkrautvertilgungsmittel *n*
herd Viehherde *f*
h. of cattle Rinderherde *f*
herf technique Hochenergieumformverfahren *n*, Hochleistungsumformverfahren *n*
Herman „Herman" *(Bezeichnung einer Lernmaschine)*
hermaphrodite caliper Tastzirkel *m*
h. tank *(Mil)* Räderraupenkampfwagen *m*, Räderraupenpanzer *m*
hermetic hermetisch, luftdicht
h. seal luftdichter Abschluß (Verschluß) *m*
hermetically-closed *s.* h.-sealed
h.-sealed luftdicht abgeschlossen
hermitian *(Math)* hermitisch
herpolhode cone Rastpolkegel *m*
herring drifter Heringslogger *m*, Treibnetzfischer *m*
h. smack Heringskutter *m*
herringbone fischgrätenartig
herringbone *(Text)* Fischgrätenmuster *n*, Fischgrätenmusterung *f*
h. bond *(Bw)* Schränkschicht[verband *m*] *f*
h. gear Pfeil[stirn]rad *n*, Stirnrad *n* mit Pfeilverzahnung
h. gear cutter Pfeilzahnradfräser *m*
h. gear pump Zahnradpumpe *f* mit Pfeilverzahnung
h. gear reducer Pfeilraduntersetzungsgetriebe *n*
h. gearing Pfeilverzahnung *f*; Getriebe *n* mit Pfeilverzahnung (Winkelverzahnung)
h. [milking] parlour Fischgrätenmelkstand *m*
h. profile Pfeilprofil *n (Reifen)*

h. rack Zahnstange *f* mit Pfeilverzahnung
h. stitch *(Text)* Hexenstich *m*
h. tooth Pfeilzahn *m*, Winkelzahn *m*
h. wheel Pfeilrad *n*
hertz Hertz *n (Einheit der Frequenz; 1 Hz = 60 U/min)*
Hertzian oscillator Hertzscher Oszillator (Dipol) *m*
hesitation roll *(Flg)* gesteuerte langsame Rolle *f*
Hess hammer lamp *(Opt)* Hammerlampe *f* nach Heß
Hessian Hessian *n*; Sackleinen *n*
H. cloth *s.* hessian
hessite *(Min)* Hessit *m*
Hess's law Heßscher Satz *m*
hetero-polymerization Mischpolimerisation *f*
heterochromatic verschiedenfarbig, heterochrom
heterocyclic heterozyklisch *(chemische Verbindung)*
heterodyne / to *(El)* überlagern
heterodyne amplifier Schwebungsverstärker *m*
h. bridge Schwebungsmeßbrücke *f*
h. code reception Überlagerungstelegrafieempfang *m*
h. detector Mischstufe *f*
h. filter 9-kHz-Sperre *f*
h. frequency Überlagerungsfrequenz *f*
h. frequency meter Frequenzmesser *m* nach dem Überlagerungsprinzip, Überlagerungsfrequenzmesser *m*
h. interference Überlagerungspfeifen *n*
h. note Schwebungston *m*, Überlagerungston *m*
h. oscillator erster Überlagerer *m* im Superhet; Schwebungssummer *m*; Überlagerungssender *m*
h. receiver Überlagerungsempfänger *m (Radio)*
h. reception Überlagerungsempfang *m (Radio)*
h. tone Schwebungston *m*, Überlagerungston *m*
h. wave Überlagerungswelle *f*
h. wave analyzer Überlagerungsanalysator *m*
h. wavemeter Interferenzwellenmesser *m*; Überlagerungswellenmesser *m*; Wellen[längen]messer *m* nach dem Überlagerungsprinzip
heterogeneity Heterogeneität *f*, Ungleichartigkeit *f*
heterogeneous heterogen, ungleichartig
heteromorphic heteromorph
heteropolar 1. *(Ch)* heteropolar; 2. *(El)* wechselpolig
h. bond *(Ch)* heteropolare Bindung *f*, Ionenbindung *f*
h. field magnet Wechselpolfeldmagnet *m*
h. generator Wechselpolgenerator *m*
h. molecule Ionenmolekül *n*
heterostatic circuit heterostatische Schaltung *f*, Quadrantenschaltung *f*
h. electrometer heterostatisches Elektrometer *n*
heterotopic *(Kern)* heterotop, nichtisotopisch
heterotropic *s.* anisotropic
heulandite *(Min)* Heulandit *m*
Heusinger valve gear *(Eis)* Heusinger-Steuerung *f*
Heusler alloy Heuslersche Legierung *f*
hew / to behauen *(Steine, Holz)*; fällen *(Bäume)*
hewer Hauer *m*, Bergmann *m*
hewing 1. *(Brg)* Unterschrämen *n*; 2. Behauen *n (Steine)*; 3. Fällen *n (Bäume)*
hewn-squared timber auf Vierkant behauenes Holz *n*
h. timber behauenes Holz *n*
hex / to ein Sechskant fräsen
hex Sechskant *n*
h. head screw Sechskantschraube *f*
hexacid *(Ch)* sechssäurig
hexacyclic compound *(Ch)* Sechsringverbindung *f*
hexad sechswertiges Element *n*
hexaedron Hexaeder *n*, Würfel *m (Kristalle)*
hexagon sechseckig
hexagon 1. Sechseck *n*; 2. Sechskantprofil *n*, Sechskantstahl *m*
h. bar steel Sechskantprofil *n*, Sechskantstahl *m*
h. box wrench Sechskantsteckschlüssel *m*
h. castle nut Kronenmutter *f*
h. connection *(El)* Sechseckschaltung *f*
h. die Sechskantschneidmutter *f*
h. head Sechskantkopf *m (einer Schraube)*
h.-head bolt Sechskantschraube *f*
h.-head coach screw Sechskantholzschraube *f*
h.-head deck screw Sechskantholzschraube *f*
h.-head screw Sechskantschraube *f*
h.-head wood screw Sechskantholzschraube *f*
h. lock nut Sechskantgegenmutter *f*
h. nut Sechskantmutter *f*
h. ring spanner Sechskantringschlüssel *m*
h. slotted nut Sechskantkronenmutter *f*
h. socket Sechskanteinsatz *m (Steckschlüssel)*
h.-socket wrench Sechskantsteckschlüssel *m*
h. stock Sechskantmaterial *n*
h. turret Sechskantrevolverkopf *m*, Sechskantsternrevolverkopf *m*
h. voltage verkettete Sechsphasenspannung *f*
hexagonal sechseckig; sechsseitig
h. bolt Sechskantschraube *f*
h. head Sechskantkopf *m (einer Schraube)*
h. nut Sechskantmutter *f*
h. screw Sechskantschraube *f*
h. socket Sechskanteinsatz *m*
h. socket-head bolt Zylinderschraube *f* mit Innensechskant, Innensechskantschraube *f*

h. socket-head cap-screw Innensechskantschraube *f*
h. turret Sechskantrevolverkopf *m*
hexagonally centred hexagonal zentriert *(Kristallgitter)*
hexagrid valve Oktode *f*, Sechsgitterröhre *f*, Achtpolröhre *f*
hexahedral hexaedrisch, sechsflächig
hexahedron Hexaeder *n*, Sechsflach *n*
hexane *(Ch)* Hexan *n*
hexaphase sechsphasig
hexavalent sechswertig
hexitol *(Ch)* Hexit *n*
hexode Hexode *f*, Sechspolröhre *f*
HF-, H.F.-, h-f- *s.* high-frequency-
Hi-boy *(Am)* Warmluftheizung *f* mit unterem Luftaustritt
hiatus Wärmeausdehnungsfuge *f* *(Ofenbau)*
hibernate / to überwintern
hibernating animals Winterschläfer *mpl (Tiere)*
hibschite *(Min)* Hibschit *m*
hicap Kondensator *m* mit hoher Dielektrizitätskonstante
hidden verdeckt; versteckt; unsichtbar; latent
h. conductor unter Putz verlegter Leiter *m*
h. edge unsichtbare Körperkante *f*
h. energy latente Energie *f*
h. line Linie *f* für unsichtbare Kanten und Umrisse, unsichtbare Körperkante *f (technische Zeichnung)*
hide / to verdecken; verstecken
hide Haut *f*, Balg *m*
h. and skin of pachydermata Dickhäuterfell *n*
h. glue Hautleim *m*
hiding power Deckfähigkeit *f*, Deckkraft *f*, Deckvermögen *n (Anstrichfarbe)*
high 1. hoch, groß; stark; 2. hoch, hell *(Stimme)*; 3. mächtig *(z. B. Flöz)*; 4. reich *(z. B. an Beimischungen)*
h.-accuracy instrument Präzisions[meß]-instrument *n*
h.-accuracy wave meter *(El)* Präzisionswellenmesser *m*
h. activity *(Kern)* hohe („heiße“) Energie *f*
h. aerial Hochantenne *f*
h.-alloy material hochlegierter Werkstoff *m*
h.-alloy steel hochlegierter Stahl *m*
h.-altitude fighter Höhenjagdflugzeug *n*
h.-altitude passage *(Astr)* hoher Durchgang *m*
h.-altitude research rocket Rakete *f* zur Erforschung der oberen Schichten der Atmosphäre
h.-altitude rocket Höhenrakete *f*
h.-alumina cement Tonerdezement *m*
h.-aluminous tonerdereich
h.-amplitude detector Kraftdetektor *m*, Leistungsdetektor *m*
h.-angle belt conveyor *(Brg)* Steilband *n*
h.-back-resistance diode hochsperrende Diode *f*
h.-beam indicator lamp *(Kfz)* Fernlichtkontrolleuchte *f*
h.-bituminous lignite Schwelkohle *f*
h. boiled stark kochend *(Stahlbad)*
h.-boiled pulp weichgekochter Zellstoff *m*
h.-boiling hochsiedend
h. boost *(Am) (Fs)* Hochfrequenzausgleich *m*
h. brass Gelbguß *m*
h.-bulk yarn Hochbauschgarn *n*
h. butt needle *(Text)* Hochfußnadel *f*
h. calcium lime fetter Kalk *m*, Weißkalk *m*
h. capacity mine car *(Brg)* Großraumwagen *m*
h.-carbon hochgekohlt; zähreich
h. class fit Edelpassung *f*
h.-compression engine hochkomprimierter (hochverdichteter) Motor *m*, Motor mit hoher Kompression
h. conductivity diode gut leitende Diode *f*
h.-contrast kontrastreich
h. contrast image (picture) kontrastreiches Bild *n*
h.-current discharge Starkstromentladung *f*
h. curved crimp hochbogige Kräuselung *f*
h.-cut Hochschnitt *m (Getreide)*
h. definition television hochzeiliges Fernsehverfahren *n*
h.-density bleaching Dickstoffbleiche *f*
h.-density concrete Schwerbeton *m*
h.-density recording Registrierung *f* mit hoher Informationsdichte
h.-division index plate Teilscheibe *f* für weiten Teilbereich
h.-draft mechanism *(Text)* Hochverzugsstreckwerk *n*
h. duty Hochleistung *f*
h.-duty grey iron Sondergrauguß *m*
h.-early strength cement frühhochfester Zement *m*
h.-efficiency Hochleistungs-
h.-energy hochenergetisch, energiereich; hart, durchdringend *(z. B. Röntgenstrahlen)*
h.-energy collision energiereicher (hochenergetischer) Stoß *m*
h.-energy component *(Ph)* durchdringende Komponente *f (z. B. Höhenstrahlung)*
h.-energy level *(Ph)* hohes Energieniveau *n*
h.-energy neutron energiereiches Neutron *n*
h.-energy physics Physik *f* hoher Energien
h.-energy radiation energiereiche (harte) Strahlung *f*
h. energy rate forming Hochenergieumformen *n*, Hochleistungsumformen *n*
h.-energy state *(Ph)* hoher Energiezustand *m*
h.-expansion cement Quellzement *m*

h. explosive brisanter Sprengstoff *m*
h. explosive shell Brisanzgranate *f*, Sprenggranate *f*
h. face Hochschnitt *m (Tagebau)*
h.-feed *(Masch)* mit großem Vorschub
h. fidelity hohe Wiedergabegüte *f (Elektroakustik)*
h.-field-emission arc *(El)* Feld[emissions]-bogen *m*
h.-fin hochgerippt
h. finish besonders glatte Oberfläche *f*, Hochglanz *m*
h. flash point drier Trockner *m* mit hoher Temperatur
h.-flash reactor Reaktor *m* mit hohem Neutronenfluß
h.-flux core [aktive] Zone *f* mit hohem Neutronenfluß
h.-freeboarded *(Schiff)* mit großem Freibord, hochbordig
h. frequency Hochfrequenz *f*
h.-frequency 1. hochfrequent; 2. Hochfrequenz-, Kurzwellen-
h.-frequency absorber Schallschlucker *m* für hohe Frequenzen
h.-frequency aerial Kurzwellenantenne *f*
h.-frequency alternator Hochfrequenzgenerator *m*
h.-frequency amplification Hochfrequenzverstärkung *f*
h.-frequency amplifier Hochfrequenzverstärker *m*
h.-frequency biasing Hochfrequenzvormagnetisierung *f*
h.-frequency bridge Hochfrequenz[meß]-brücke *f*
h.-frequency carrier current telephone channel Hochfrequenzsprechkanal *m*, Hochfrequenzsprechweg *m*
h.-frequency choke Hochfrequenzdrossel *f*, Hochfrequenzsperre *f*
h.-frequency circuit Hochfrequenz[strom]-kreis *m*
h.-frequency compensation Höhenanhebung *f*
h.-frequency cutoff Hochfrequenzabschneidung *f*; obere Grenzfrequenz *f*, Maximalfrequenz *f*
h.-frequency direction finding Hochfrequenzpeilverfahren *n*
h.-frequency drying Hochfrequenztrocknung *f*
h.-frequency [electrical] engineering Hochfrequenztechnik *f*
h.-frequency heating Hochfrequenztrocknung *f*
h.-frequency induction furnace kernloser Induktionsofen *m*, Hochfrequenzofen *m*
h.-frequency induction hardening Hochfrequenzinduktionshärten *n*
h.-frequency induction heat brazing Hochfrequenzlöten *n*
h.-frequency mass spectrometer Hochfrequenzmassenspektrometer *n*
h.-frequency power Hochfrequenzleistung *f*
h.-frequency range Kurzwellenbereich *m (Radio)*
h.-frequency resistance Hochfrequenzwiderstand *m (physikalische Größe)*
h.-frequency resistor Hochfrequenzwiderstand *m (Schaltelement)*
h.-frequency response Hochfrequenzverhalten *n*, Verhalten *n* gegen Hochfrequenz
h.-frequency screen *(Brg)* Vibratorsiebmaschine *f*
h.-frequency sound Ultraschall *m*
h.-frequency sounder Hochfrequenzlot *n*
h.-frequency speaker Hochtonlautsprecher *m*
h.-frequency stage Hochfrequenzstufe *f*
h.-frequency vacuum gauge Hochfrequenzvakuumprüfer *m*
h.-frequency wave Kurzwelle *f*
h.-furnace cement Hochofenzement *m*, Hüttenzement *m*
h.-gain *(El)* mit hoher Verstärkung; mit hohem Wirkungsgrad
h.-gamma tube Fernsehröhre *f* mit besonders guter Gradation
h.-glazed coloured art-paper farbiges Glacépapier *n*
h. gloss lamination *(Typ)* Hochglanzbeschichtung *f*; Hochglanzkaschierung *f*
h.-grade hochwertig; reichhaltig; hochangereichert
h.-grade casting Qualitätsguß *m*, hochwertiger Guß *m*
h.-grade fuel *(Kern)* hochangereicherter Brennstoff *m*
h.-grade iron Qualitätseisen *n*
h.-grade ore reichhaltiges Erz *n*
h.-grade paper Papier *n* bester Güteklasse
h.-grade steel Edelstahl *m*, Qualitätsstahl *m*
h.-grade zinc Feinzink *n*
h.-helix drill engspiraliger Bohrer *m*, Bohrer mit enger Steigung
h.-impedance mit hoher Impedanz; hochohmig
h.-impedance relay Relais *n* mit hoher Impedanz, hochinduktives Relais, Drosselrelais *n*
h.-impedance winding hochohmige Wicklung *f*
h.-insulating fibre hochgradig isolierende Faser *f*
h.-intensity carbon arc Hochstromkohlebogen *m*, Flammen[kohle]bogen *m*

h.-internal resistance valve hochohmige Röhre *f*
h. inverse voltage rectifier Gleichrichter *m* mit hoher Sperrspannung
h. level *(Ph)* hohes Energieniveau *n*
h.-level camshaft *(Kfz)* obenliegende Nockenwelle *f*
h.-level modulation *(El)* Endstufenmodulation *f*
h.-lift slabbing mill Hochhubbrammenwalzwerk *n*
h.-lift truck Hubstapler *m*
h.-light brightness Spitzenhelligkeit *f*
h. lights hellste Stellen *fpl (Fernsehbild, fotografisches Bild)*; Glanzlichter *npl*, Glanzpunkte *mpl*
h.-load capacity *(Elm)* Hochbelastbarkeit *f*
h.-low level control *(BMSR)* Zweipunktniveauregelung *f*
h. lustre Hochglanz *m*
h. manganese steel Manganhartstahl *m*
h.-melting schwerschmelzbar, hochschmelzend
h. melting-point alloy hochschmelzende Legierung *f*
h.-melting waxes Hartparaffine *npl*
h. mirror-finished sheet Hochglanzblech *n*
h.-mu valve steile Röhre *f*, Röhre mit hohem (großem) Verstärkungsfaktor
h.-note response *(El)* Hochtonwiedergabe *f*
h.-octane gasoline hochoktaniges (klopffestes) Benzin *n*
h. ohmic hochohmig
h.-pass filter *(El)* Hochpaßfilter *n*
h.-pass selective circuit Niederfrequenzsperrkreis *m*
h.-percentage hochprozentig
h.-performance leistungsfähig
h.-pitched note hoher Ton *m*
h. plate-resistance tube Röhre *f* mit hohem Innenwiderstand
h. polish Hochglanz *m*
h.-potential capacitor Hochspannungskondensator *m*
h. potential line Hochspannungsleitung *f (über 750 V)*
h.-power broadcasting station Großrundfunksender *m*
h.-power choke modulator Antennenmodulationsdrossel *f*
h.-power loudspeaker Hochleistungslautsprecher *m*, Großlautsprecher *m*
h.-power modulation *(El)* Modulation *f* mit hohem Pegel, Modulation in der Endstufe
h.-power motor Hochleistungsmotor *m*
h.-power objective *(Opt)* starkes Objektiv *n*
h.-power plant Großkraftwerk *n*
h.-power reactor Hochleistungsreaktor *m*
h.-power signal generator Leistungsmeßsender *m*
h.-power test generator Leistungsmeßsender *m*
h.-power three-pocket grinder Dreipressengroßkraftschleifer *m*
h.-power transformer Hochleistungstransformator *m*
h.-power transmission line Starkstromleitung *f*
h.-power transmitter Großsender *m*
h.-power water boiler Hochleistungskochwasserreaktor *m*
h.-powered engine Hochleistungsmotor *m*
h.-powered motor starker Motor *m*, Motor großer Leistung[saufnahme]
h.-precision levelling Nivellement *n* hoher Präzision
h.-pressure Hochdruck *m*
h.-pressure 1. Hochdruck-; 2. *(El, sl)* Hochspannung[s]-
h.-pressure arc discharge Hochdruckbogenentladung *f*
h.-pressure atomizer Hochdruckzerstäuber *m*
h.-pressure baler *(Lw)* Hochdruckpresse *f*, Ballenpresse *f*
h. pressure blower Hochdruckgebläse *n*
h.-pressure boiler Hochdruckkessel *m*
h.-pressure chamber Hochdruckkammer *f*
h.-pressure cloud chamber *(Ph)* Hochdrucknebelkammer *f*
h.-pressure counter *(Kern)* Hochdruckzählrohr *n*
h.-pressure cover *(Wärm)* Hochdruckdeckel *m*
h.-pressure crank Hochdruckkurbel *f (Dampfmaschine)*
h.-pressure cruising turbine Hochdruckmarschturbine *f*
h.-pressure cylinder Hochdruckzylinder *m (Dampfmaschine)*
h.-pressure die casting Druckguß *m*, Hochdruckgußverfahren *n*
h.-pressure disk Hochdruckrad *n (Turbine)*
h.-pressure flow pressing Hochdruckfließpressen *n*
h.-pressure forming Hochdruckverformung *f*
h.-pressure gasification Druckvergasung *f*
h.-pressure grease gun Druckschmierapparat *m*
h.-pressure lamp Hochdrucklampe *f*
h.-pressure lubrication Hochdruckschmierung *f*
h.-pressure mercury vapour lamp Quecksilberhochdrucklampe *f*
h. pressure pipe line Hochdruckleitung *f*
h.-pressure pump Hochdruckpumpe *f*

h.-pressure ram filters Hochdruckfilterpresse *f*
h.-pressure reactor Hochdruckreaktor *m*
h.-pressure stage Hochdruckstufe *f (Turbine)*
h.-pressure steam curing Autoklavbehandlung *f*, Hochdruckdampferhärtung *f (Betonbau)*
h.-pressure steam engine Hochdruckdampfmaschine *f*
h.-pressure stream Hochdruckstrahl *m*
h.-pressure suit *(Flg)* Druckanzug *m*
h.-pressure tube of sprayer Hochstrahlrohr *n* des Spritzapparates
h.-pressure turbine Hochdruckturbine *f*
h.-pressure tyre *(Kfz)* Hochdruckreifen *m*
h.-pressure vessel Drucktank *m*, Hochdruckbehälter *m*
h.-quality sheet Qualitätsblech *n*
h.-quality steel Edelstahl *m*
h.-radiation field starkes Strahlungsfeld *n*
h. range *(Kfz)* schnelle (hohe) Übersetzung *f*, Straßengang *m*
h. rank hoch inkohlt, hochwertig *(Kohle)*
h.-ratio transformer Transformator (Übertrager) *m* mit hohem Übersetzungsverhältnis
h.-resistance hochohmig
h.-resistance winding hochohmige Wicklung *f*
h. resistive hochohmig
h. resolution hohe Auflösung *f*
h.-resolution hochauflösend, mit großem Auflösungsvermögen
h.-resolution detector Nachweisgerät *n* mit großem Auflösungsvermögen
h.-road Chaussee *f*, Fernverkehrsstraße *f*, Landstraße *f*
h. sea schwere See *f*, Wellengang *m* bestimmter Höhe
h. seas schiffbares Gewässer *n*, offenes Meer *n (außerhalb der Dreimeilenzone oder anderer Territorialgewässer)*; hohe (offene) See *f*, weites Meer, Hochsee *f*
h.-sensitive gas analyzer Gasspurenanalysator *m*
h.-sided goods wagon *(Eis)* Hochbordwagen *m*
h.-silicon pig iron hochsiliziertes Roheisen *n*
h. sinker *(Text)* Hochfuß-Links-Links-Platine
h.-slope valve Röhre *f* großer Steilheit, [sehr] steile Röhre
h.-slump concrete [hoch]plastischer Beton *m*
h.-speed schnell[um]laufend, mit hoher Drehzahl; mit hoher Geschwindigkeit
h.-speed arc suppression relay Schnellabschaltrelais *n*
h.-speed attachment Schnellaufspindel *f*
h.-speed automatic transmitter Schnellmorsesender *m*
h.-speed auxiliary spindle Schnellaufspindel *f*
h.-speed boat Schnellboot *n*
h.-speed boring head Schnellbohrgerät *n*
h.-speed camera Zeitlupenkamera *f*
h.-speed carbide tool machining Schnellbearbeitung *f* mit Hartmetallwerkzeugen
h.-speed centrifuge Ultrazentrifuge *f*
h.-speed circuit breaker Schnellschalter *m*
h.-speed circuit protection Schnellabschaltung *f*
h.-speed commercial airplane Schnellverkehrsflugzeug *n*
h.-speed computer Schnellrechner *m*
h.-speed computing machine *s.* h.-speed computer
h.-speed cutter 1. Schnellfräser *m*; Schnellstahlmeißel *m*; 2. Schnellschneidemaschine *f*
h.-speed cutting tool Schnellstahlmeißel *m*; Schnellfräser *m*
h.-speed draw frame Hochleistungsstrecke *f (Spinnerei)*
h.-speed drill Hochleistungsbohrer *m*
h.-speed drilling attachment Schnellbohreinrichtung *f*
h.-speed drilling machine Schnellaufbohrmaschine *f*
h.-speed drilling method (operation) Schnellbohrverfahren *n*
h.-speed electron schnelles Elektron *n*
h.-speed engine hochtouriger Motor *m*; Motor mit hoher Drehzahl
h.-speed excitation system *(El)* Schnellerregung *f*, Stoßerregung *f*
h.-speed feed beschleunigte Zuführung *f*, Schnellzuführung *f*
h.-speed flash tube Lichtblitzröhre *f*
h.-speed grinding Schnellschleifen *n*
h.-speed guillotine *(Holz)* Schnellschneidemaschine *f*, Schnellquerschneider *m*
h.-speed jigging screen *(Brg)* Schnellschwingsieb *n*
h.-speed lathe Schnelldrehmaschine *f*
h.-speed loom Schnelläuferwebstuhl *m*
h.-speed machining Schnellbearbeitung *f*; Schnellzerspanung *f*
h.-speed machining technique Schnellzerspanungsverfahren *n*
h.-speed magnetic amplifier schnellansprechender magnetischer Verstärker *m*
h.-speed memory Schnellspeicher *m*
h.-speed milling attachment Schnellfräseinrichtung *f*, Schnellfräskopf *m*
h.-speed motor-driven drilling machine Schnellaufbohrmaschine *f*
h.-speed moving coil recorder Drehspulenschnellschreiber *m*
h.-speed nozzle große Düse *f*, Hauptdüse *f*

h.-speed planer Schnellhobelmaschine *f*
h.-speed press Schnellpresse *f*
h.-speed printer *(BMSR)* Schnelldrucker *m* *(Ausgabevorrichtung)*
h.-speed production plain circular knitting machine einfonturige Hochleistungsrundstrickmaschine *f*
h.-speed readout Schnelleseeinrichtung *f*
h.-speed recorder Schnellschreiber *m*
h.-speed regulator Schnellregler *m*
h.-speed relay schnellansprechendes Relais *n*, Schnell[schalt]relais *n*
h.-speed roughing wheel Schnellschrupp-[schleif]scheibe *f*
h.-speed runabout schnelles Verkehrsboot *n*
h.-speed scanning aerial schnellrotierende Radarantenne *f*
h.-speed screw slotter Schnellschraubenschlitzmaschine *f*
h.-speed selector *(Fmt)* schnellaufender Wähler *m*
h.-speed snagging wheel Schnellschrupp-[schleif]scheibe *f*
h.-speed spark camera Funkzeitlupenkamera *f*
h.-speed spark photograph Funkzeitlupenaufnahme *f*
h.-speed spindle Schnellaufspindel *f*
h.-speed steam engine schnellaufende Dampfmaschine *f*
h.-speed steel Schnell[arbeits]stahl *m*
h.-speed steel reamer Schnellstahlreibahle *f*
h.-speed switching *(BMSR)* Hochleistungsschalten *n* *(meist elektronisch)*
h.-speed tape punch *(Büro)* Hochleistungsstreifenlocher *m*
h.-speed tape transmitter *(Fmt)* Lochstreifenschnellsender *m*
h.-speed telecommunication engineering Schnellfernmeldetechnik *f*, Schnellnachrichtentechnik *f*
h.-speed telegraph Schnelltelegraf *m*
h.-speed teleprinter Hochleistungsfernschreiber *m*, Fernschreiber *m* hoher Geschwindigkeit
h.-speed tool Schnell[dreh]meißel *m*
h.-speed tricot machine *(Text)* Schnelläuferkettenstuhl *m*
h.-speed turning Schnelldrehen *n*
h.-speed two-colour press *(Typ)* Zweifarbenschnellpresse *f*
h.-speed xerographic printer *(Büro)* xerografischer Schnelldrucker *m*
h.-spliced heel Hochferse *f* *(Strumpf)*
h.-strength alloy hoch[zug]feste Legierung *f*
h.-strength bolt hochfest vorgespannte Schraube *f*, HV-Schraube *f*
h.-strength cast iron Edelguß *m*
h.-strength friction grip bolt hochfest vorgespannte Schraube *f*, HV-Schraube *f*
h.-strength steel hochfester Stahl *m*
h.-stretch rubber core Gummikern *m* *(für Elastikgarn)*
h.-temperature warmfest; hochtemperaturbeständig
h.-temperature carbonization *(Brg)* Entgasung *f*, Verkokung *f*
h.-temperature characteristics Heißverhalten *n*, Warmverhalten *n*; Warmfestigkeit
h.-temperature coal tar Steinkohlenhochtemperaturteer *m*
h.-temperature coking Hochtemperaturverkokung *f*
h.-temperature grinding Warmschleifen *n*
h.-temperature quenching *(Met)* Warmbadhärtung *f*
h.-temperature resistance 1. Warmfestigkeit *f*; 2. *(El)* Widerstand *m* bei hoher Temperatur; Arbeitswiderstand *m* *(z. B. Glühfaden)*
h.-temperature resistant warmfest; hochtemperaturbeständig
h.-temperature resistant alloy warmfeste Legierung *f*
h.-temperature stability (strength) Warmfestigkeit *f*
h.-temperature veneer drying Furnierhochtemperaturtrocknung *f*
h.-tenacity mit hoher Reißfestigkeit
h.-tenacity viscose filament hochfeste Viskoseseide *f*
h.-tenacity yarn hochfestes Garn *n*
h.-tensile hoch[zug]fest
h.-tensile bolt hochfest vorgespannte Schraube *f*, HV-Schraube *f*
h.-tensile bolting gleitfeste Schraubenverbindung *f*, HV-Schraubenverbindung *f*
h.-tensile grip bolt hochfest vorgespannte Schraube *f*, HV-Schraube *f*
h.-tensile prestressed bolt hochfest vorgespannte Schraube *f*, HV-Schraube *f*
h.-tensile steel hochzugfester Stahl *m*; hochwertiger Baustahl *m*
h. tension *(El)* Hochspannung *f*
h.-tension battery Hochspannungsbatterie *f*; Anodenbatterie *f* *(Radio)*
h.-tension coupling circuit *(Fmt)* Hochspannungsübertrager *m*
h.-tension direct current hochgespannter Gleichstrom *m*
h.-tension dynamo Hochspannungserzeuger *m*, Hochspannungsgenerator *m*
h.-tension electric detonator *(Brg)* Spaltzünder *m*
h.-tension end Hochspannungsseite *f*
h.-tension keying Anodenstromtastung *f*
h.-tension line Hochspannungsleitung *f*
h.-tension magneto [kleine] Hochspannungsmaschine *f*

h.-tension mains Hochspannungsnetz *n*
h.-tension power unit Hochspannungsnetzgerät *n*
h.-tension probe *(BMSR)* Hochspannungsmeßkopf *m*
h.-tension protected hochspannungssicher
h.-tension supply [unit] Anodenspannungsnetzteil *n (Radio)*
h. tension tester Hochspannungsprüfgerät *n*
h.-tension transmission line Hochspannungsleitung *f*
h.-tension voltage *(El)* Hochspannung *f*
h.-test cast iron Hochleistungsgußeisen *n*
h. tide Hochwasser *n*
h. trajectory fire *(Mil)* Steilfeuer *n*
h.-transmission *(Opt)* besonders stark lichtdurchlässig
h. vacuum Hochvakuum *n*
h.-vacuum coating plant Hochvakuumbedampfungsapparatur *f*
h.-vacuum gauge Druckmesser *m* für hohes Vakuum
h.-vacuum measuring instrument Hochvakuummeßgerät *n*
h.-vacuum melting Hochvakuumschmelzen *n*
h.-vacuum melting furnace Hochvakuumofen *m*
h.-vacuum oil diffusion pump Hochvakuumöldiffusionspumpe *f*
h.-vacuum pumping system Hochvakuumerzeuger *m*
h.-vacuum technique Hochvakuumtechnik *f*
h.-vacuum thermionic rectifier Hochvakuumgleichrichter *m*
h.-vacuum valve 1. Hochvakuumventil *n*; 2. *(El)* Hochvakuumröhre *f*, harte Röhre *f*
h.-valued resistor hochohmiger Widerstand *m*
h.-velocity forging *(Am)* Hochleistungsumformen *n*
h.-velocity turning Schnelldrehen *n*
h.-volatile hochflüchtig *(z. B. Kohle)*
h. voltage Hochspannung *f* (*s. a.* h. tension)
h. voltage Hochspannungs-
h.-voltage accelerator Hochspannungsbeschleuniger *m*
h.-voltage condenser Hochspannungskondensator *m*
h.-voltage d.c. transmission Hochspannungsgleichstromübertragung *f*
h.-voltage electrode Hochspannungselektrode *f*
h.-voltage generator Hochspannungsgenerator *m*, Hochvoltgenerator *m*
h.-voltage-heater tube Röhre *f* mit Hochspannungsheizfaden
h.-voltage supply Hochspannungsversorgung *f*
h.-voltage switchgear Hochspannungsschaltgerät *n*; Hochspannungsschaltanlage *f*
h. warp loom *(Text)* Hautelisse[web]stuhl *m*
h. water Hochwasser *n*
h.-water level Hochwasserstand *m*
h.-water mark Hochwassermarke *f*, Flutmarke *f*, Flutlinie *f*
h.-water neap Nipphochwasser *n*
h.-water stand Hochwasserstillstand *m*
h.-wing monoplane *(Flg)* Schulterdecker *m*
higher chain products künstliche Transurane *npl (Reaktorprodukte)*
h. harmonics [harmonische] Oberschwingungen *fpl*; Oberwellen *fpl*
h. limiting filter *(Fmt)* Sperrkette *f*, Tiefpaß
h. pair höheres Elementenpaar *n (Getriebelehre)*
highlight *(Typ)* Spitzlicht *n*, Hochlicht *n*
highly alloyed hochlegiert
h. collimated beam *(Ph)* stark ausgeblendeter Strahl *m*
h. concentrated hochkonzentriert
h. damped instrument aperiodisches Meßinstrument *n*
h. directional aerial Antenne *f* mit starker Richtwirkung
h.-disperse feindispers
h.-dispersive prism *(Opt)* Prisma *n* mit großer Dispersion, dispersionsstarkes Prisma
h. evacuated hochevakuiert
h. excited *(Ph)* hochangeregt
h. excited level hochangeregter Zustand *m*
h. fluid dünnflüssig
h. ionized hochionisiert
h. magnifying *(Opt)* stark vergrößernd
h. polished auf Hochglanz poliert
h. radioactive stark radioaktiv, „heiß", hochaktiv
h. saturated *(Ch, Ph)* hochgesättigt
h. sensitive hochempfindlich
h. stressed hochbeansprucht
h. tensioned straff gespannt *(z. B. Treibriemen)*
h. viscous stark zähflüssig, hochviskos
highs *(Am) (El)* hohe Frequenzen *fpl*
highwall Abbaustoß *m (Tagebau)*
highway 1. Chaussee *f*, Fernverkehrsstraße *f*, Landstraße *f*, 1. Ordnung; 2. Hauptweg *m (elektronische Rechner)*
h. bridge Straßenbrücke *f*
h. code Straßenverkehrsordnung *f*
h. construction Straßenbau *m*
h. engineer Straßenbauingenieur *m*
h. engineering Straßenbau *m*
h. radiotelephony Landstraßenfunk[telefonie *f*] *m*
h. safety Straßenverkehrssicherheit *f*
Hilbert space *(Math)* Hilbert-Raum *m*
hill climbing *(Dat)* Feststellen *n*, ob man (z. B. beim Schachspiel) dem gesteckten Ziel näher ist

h.-climbing ability *(Kfz)* Steigfähigkeit *f*, Steigvermögen *n*
h. gear *(Kfz)* Berganggetriebe *n*
h.-side up-current Hangaufwind *m* *(Segelfliegerei)*
hilling *(Lw)* Häufeln *n*
h. share Häufelschar *n*
h.-up Aufhäufeln *n* [der Kartoffeln]
Hindley worm Globoidschnecke *f*
hinge / to mit Scharnier versehen; gelenkig befestigen; schwenken [um]; anlenken; um einen Bolzen drehen
hinge Drehachse *f*, Gelenk *n*; Scharnier *n*; Angel *f*, Türangel *f*
h. band Scharnierband *n*
h. bolt Scharnierriegel *m*, Scharnierstift *m*, Überwurfmutter *f*; Gelenkbolzen *m*
h. coupling Gelenkkupplung *f*
h. fitting Gelenkanschluß *m*; Drehgelenk *n*; Drehpunkt *m*
h. hasp Scharnierriegel *m*
h. lid Klappdeckel *m*
h. pin Federbolzen *m*, Kippachse *f*, Lagerstift *m*, Scharnierstift *m*, Drehstift *m*
h. pivot Türangel *f*
h. plate Schloßplatte *f*
h. rod Gelenkstange *f*
h. teeth scharnierartige Zähne *mpl*
hinged gelenkig [angeschlossen], gelenkig gelagert (befestigt); schwenkbar; in einem Gelenk drehbar; schrägstellbar; kippbar; aufkippbar; aufklappbar
h. bar *(Brg)* Gelenkkappe *f*
h. bearing Gelenk *n*, Kipplager *n*
h.-body car Kippwagen *m*
h. bolt Gelenkschraube *f*, Klappschraube *f*
h. bucket Gelenkbecher *m*, Klappkübel *m*
h. bulkhead door Hängetür *f*, Klapptür *f*
h. cantilever *(El)* Drehausleger *m* *(an Masten)*
h. cover Klappdeckel *m*, Verschlußklappe *f*; Scharnierdeckel *m*
h. davits *(Schiff)* Barkunen *pl*, Klappdavits *mpl (npl)*
h. frame Klapprahmen *m*
h. funnel Klappschornstein *m*, umlegbarer Schornstein *m*
h. joint Gelenk *n*
h. loading ramp Schwingbühne *f*
h. mast Klappmast *m*
h. pier Pendelpfeiler *m*, Pendelstütze *f* *(Brückenbau)*
h. side-swinging door seitlich ausschwingende Klappe *f*
h. steel bar Gelenkstahlkappe *f*
h. tie bar Gelenkstange *f*, Zugstrebe *f*
h. valve Drehklappe *f*
h. vice Gelenkschraubstock *m*
h. window Klappfenster *n*
hinny Maulesel *m*
hip *(Bw)* Walm *m*
h. gable Walm *m*
h. jack rafter Walmschifter *m*
h. rafter Gratsparren *m*
h. roof Walmdach *n*
h. tile Gratziegel *m*
h. truss Trapez[fachwerk]binder *m*
hipped plate Faltwerk *n* *(Stahlbeton)*
h. roof Walmdach *n*
hippuric acid Hippursäure *f*, Benzoylglyzin *n*, Benzoylglykokoll *n*
hire / to 1. mieten, pachten; 2. leihen *(Geld)*; 3. *(Schiff)* anheuern
h. a vessel *(Schiff)* befrachten, in Fracht nehmen
hire 1. Miete *f*; 2. Arbeitslohn *m*; 3. *(Schiff)* Heuer *f*
hiss / to zischen
hiss *(Fmt)* Zischen *n*
h. effect Zischeffekt *m*
h. silencer Rauschdämpfer *m*, Zischtondämpfer *m*
hissing *(Fmt)* Zischen *n*
h. of the valve *(Wärm)* Pfeifen *n* des Schiebers
histamine *(Ch)* Histamin *n*
histidine *(Ch)* Histidin *n*
histogram grafisch-tabellarische Darstellung *f*; Säulenschaubild *n* *(Statik)*
hit / to [auf]prallen; [zusammen]stoßen; treffen
hit 1. *(Mil)* Treffer *m*; 2. Stoß *m*; Hieb *m*
hitch / to einhaken, ankuppeln, festmachen
hitch 1. Haken *m*; Schnellkupplung *f*; Knoten *m*; 2. *(Brg)* Bühnloch *n*; 3. *(Geol)* kleine (geringfügige) Verwerfung *f*
hitchback Spannstelle *f*
hitcher *(Brg)* Anschläger *m*
hitching Anhängevorrichtung *f*, Kupplung *f*
h. device Schnellkupplung *f*
Hittorf number *(Ch)* Hittorfsche Überführungszahl *f*
hive Bienenstock *m*
hoar frost Reif *m*, Rauhreif *m*
h. frost line Sublimationskurve *f*
hoarding Bauzaun *m*, Bretterzaun *m*
hob / to 1. wälzfräsen; verzahnen nach dem Abwälzverfahren; 2. kalt einsenken; prägen
hob 1. Wälzfräser *m*; 2. Prägestempel *m*
h. arbor Fräsdorn *m*
h. cutter Abwälzfräser *m*
h.-cutter testing machine Wälzfräserprüfgerät *n*
h. grinding (sharpening) machine Wälzfräserscharfschleifmaschine *f*
h. speed Wälzfräserdrehzahl *f*
h. spindle Wälzfräserspindel *f*

h. top *(Met)* Blockaufsatz *m*, Gießaufsatz *m*, verlorener Kopf *m*
hobbed gear wälzgefrästes Rad *n*
h. part Wälzfrästeil *n*
hobber Wälzfräsmaschine *f*
h. barge with bottom doors *(Schiff)* Klappschute *f*
h. barge with bottom doors and for reclamation purposes *(Schiff)* Klapp- und Spülprahm *m*
hobbing Wälzfräsen *n*, Fräsen *n* im Wälzverfahren
h. cutter Wälzfräser *m*
h. faults Wälzfehler *mpl*
h. head Fräskopf *m*
h. machine Wälzfräsmaschine *f*
h. method Wälzverfahren *n*
h. press Kalteinsenkpresse *f*
h. process Wälzverfahren *n*
hobble skirt *(Text)* Humpelrock *m*
hobbyist Bastler *m*
hobnail Hufnagel *m*
hockey stick Hockeyschläger *m*
hod Mörteltragekasten *m*; Ziegeltragekasten *m*, Tragmulde *f*
hodograph Hodograf *m*
hoe / to [auf]hacken
hoe 1. Hacke *f*, Haue *f*; 2. Grabenbagger *m*, Tieflöffelbagger *m*
h. dipper *s.* backhoe dipper
Hoechstaedter cable Höchstädter Kabel *n*, H-Kabel *n* *(Hochspannungskabel)*
hoeing *(Lw)* Hacken *n*
h. blade Hackmesser *n* *(Hackmaschine)*
h. up Umhacken *n*
hog / to nach oben krümmen
h. back *(Brg)* einfallen
hog-backed girder Halbparabelträger *m*
h. frame *(Schiff)* Seitenstringer *m*
h. house *(Am)* Schweinestall *m*
h. pea Futtererbse *f*
h. piece *(Schiff)* Innenkiel *m*
h. trough Schweinetrog *m*, Futterkrippe *f*
hogged fuel zerkleinertes Brennholz *n*
hogger-pump *(Brg)* Schachtpumpe *f*
hoggin tonhaltiger Kiessand *m*
hogging *(Schiff)* Kielbruch *m*, Kielaufbuchtung *f*
h. and sagging Kielaufbuchtung *f* und -durchsenkung *f* *(im Seegang)*
h. cut *(sl)* Schnitt *m* mit großer Spantiefe
h. of the flues *(Wärm)* Aufbäumen *n* der Flammrohre durch ungleichmäßige Ausdehnung
hog's hair brush Schweineborstenpinsel *m*
hogskin Schweinsleder *n*
hoick / to hochreißen
h. off vom Boden wegnehmen *(Flugzeugstart)*
hoist / to 1. [an]heben; [hoch]winden; 2. *(Brg)* fördern, treiben, zutage heben; 3. *(Schiff)* heißen, hissen
h. out a boat ein Boot aus dem Wasser heben
hoist Aufzug *m*, Fahrstuhl *m*; Flaschenzug *m*; Haspelwinde *f*, Hebewerk *n*, Hebezeug *n*; Kranlaufwinde *f*, Winde *f*; Haspel *f*
h. brake Hubbremse *f*
h. chain Hubkette *f*
h. frame Fördergerüst *n*
h. man Fördermaschinist *m*
h. motor Fördermaschinenmotor *m*
h. operator's control Fördermaschinenbedienungspult *n*
h. room Fördermaschinengebäude *n*
h. rope Windwerksseil *n* *(Bagger)*
h. trolley Laufkatze *f* für Hebezeuge
hoisting *(Brg)* Heben *n*, Fördern *n*, Treiben *n*; Schachtfördern *n*
h. and conveying Fördertechnik *f*
h. apparatus Hebezeug *n*
h. barrel Seiltrommel *f*
h. block Flaschenzugsblock *m*
h. cable Hubseil *n*, Lastseil *n*; Förderseil *n*
h. cage *(Brg)* Förderkorb *m*
h. capacity Schachtförderleistung *f*, Schachtkapazität *f*
h. compartment Fördertrum *n*
h. crab Bockwinde *f*
h. device Hebevorrichtung *f*, Hubvorrichtung *f*
h. drum Seiltrommel *f*
h. engine Fördermaschine *f*; Hubmotor *m*
h. equipment Hebezeug *n*
h. gear *(Schiff)* Anschlagmittel *n* *(Zusatzgerät im Kranumschlag)*; Hebezeug *n*; Hubwerk *n*
h. line Hubseil *n*, Lastseil *n*; Förderseil *n*
h. machine Hebezeug *n*
h. machinery Hebezeuge *npl*
h. machinery and conveyors Fördermaschinen *fpl*
h. machinery construction Hebezeugbau *m*
h. power Hubkraft *f*
h. rope Hubseil *n*, Lastseil *n*, Förderseil *n*
h. sheave Seilscheibe *f*
h. signal Schachtsignal *n*
h. sling *(Schiff)* Heißschlinge *f*
h. speed Hubgeschwindigkeit *f*, Fördergeschwindigkeit *f*; Seilgeschwindigkeit *f*
h. steam engine Lasthebedampfmaschine *f*
h. tackle Flaschenzug *m*
h. winch Heißwinde *f*, Heißwinsch *f*, Aufzugswinde *f*
hold / to 1. [fest]halten; aufnehmen; [auf]spannen; 2. gelten, gültig sein; 3. *(Büro)* [besetzt] halten

h. a call *(Fmt)* eine Anmeldung zurückstellen
h. a circuit *(Fmt)* eine Leitung besetzt halten
h. a keen cutting edge schneidhaltig sein
h. a tolerance eine Toleranz einhalten
h. against gegenhalten
h. drills Bohrer spannen
h. frequency *(El)* Frequenz fahren (halten)
h. in place festhalten
h. limits to Toleranzen einhalten bis zu
h. on vorhalten *(Nietung)*
h. size maßhaltig bleiben
h. the line *(Fmt)* in der Leitung bleiben
h. together zusammenhalten

hold *s.* cargo hold
h.-back Zurückhalten *n*, Aufhalten *n*
h.-back agent Rückhalteträger *m*
h.-back carrier Rückhalteträger *m*
h. bulkhead *(Schiff)* Laderaumschott *n*
h. ceiling *(Schiff)* Laderaumwegerung *f*, Raumwegerung *f*
h. circuit Festhalteschaltung *f* *(Analogierechner)*
h. condition Haltezustand *m* *(Analogierechner)*
h. control Stabilitätsregler *m* *(Regler zur Einstellung der Ablenkeigenfrequenz)*; Kippfrequenzregelung *f*
h. current Haltestrom *m*
h.-down Niederhalter *m*; Blechhalter *m*
h. frame *(Schiff)* Raumspant *n*
h.-off Abfangen *n* *(Flugzeug)*
h. on Gegenhalter *m*, Setzkopfeisen *n*
h. stanchion *(Schiff)* Raumstütze *f*
h. time *(Dat)* Speicherzeit *f*
h. up 1. Stauung *f* des Materials; 2. Materialeinsatz *m* *(Isotopentrennung)*
h.-up of the wastes *(Kern)* Lagerung *f* der Abfälle

holder Fassung *f*, Halter[ung *f*] *m*; Klemme *f*; Sockel *m*; Träger *m*
h. for shell reamers Aufsteckhalter *m* für Aufsteckreibahlen
h. for taper shank tools Einsatzhülse *f* *(für Werkzeuge mit Kegelschaft)*
h.-on Gegenhalter *m*, Vorhalter *m* *(Nietung)*
h. sinker *(Text)* Gegenhalterstift *m*

holderbat Schelle *f* *(für Abflußrohr)*

holdfast Zwinge *f*, Klammer *f*

holding Halten *n*; Aufnahme *f*; Spannen *n* *(Werkstück, Werkzeug)*
h. beacon *(Flg)* Wartefunkfeuer *n*
h. beam *(Dat)* Regenerierstrahl *m*, Haltestrahl *m* *(Speicherröhre)*
h. brake Gefällebremsung *f* *(elektrischer Zug)*
h. braking effort Bremskraft *f* bei Gefällebremsung
h. capacity *(Masch)* Spannbereich *m*
h. circuit Haltestromkreis *m* *(Relais)*
h. coil Haltespule *f* *(Relais)*
h. crucible Warmhaltetiegel *m*
h. device Spannvorrichtung *f*, Haltevorrichtung *f*
h. devices Spannzeuge *npl*
h.-down Niederhalter *m* *(Blechschere)*
h.-down bolt Befestigungsschraube *f*; Ankerbolzen *m*; Ankerschraube *f*; Fundamentbolzen *m*
h.-down stud Befestigungsbolzen *m*, Befestigungsstiftschraube *f*
h. effect *(Fmt)* Mitnahme *f* *(Schwingkreis)*
h. fixture Werkstückspannvorrichtung *f*, Spannzeug *n*
h. from the inside Innenspannen *n*
h. furnace Warmhalteofen *m*
h. ground Ankergrund *m*
h. key Sperrschalter *m*, Halteschalter *m*
h. neck Nase *f*
h.-on magnet Haltemagnet *m*; Magnetanker *m*
h. pad Nase *f*
h. pattern *(Flg)* Warteschleife *f*
h. pawl Sperrklinke *f*
h. relay Halterelais *n*
h. screw Halteschraube *f*
h. strap Spanneisen *n*
h. time 1. *(Ch)* Verweilzeit *f*; 2. *(Fmt)* Belegungsdauer *f*; 3. Haltezeit *f*
h. to close limits toleranzhaltig
h. to close limits Einhalten *n* enger Toleranzen
h. tool Spannzeug *n*; Aufspannwerkzeug *n*
h. trigger Stopper *m* für Stapellaufschlitten
h.-up Gegenhalten *n* *(Nieten)*
h.-up hammer Gegenhalter *m* *(Nieten)*
h. voltage *(El)* Haltespannung *f*
h. winding Haltewicklung *f* *(Relais)*
h. wire *(Fmt)* Prüfdraht *m*
h. wire current Meßstrom *m*

holdsmen *(Schiff)* Stauereiarbeiter *mpl*, Schauerleute *pl* *(im Laderaum)*

hole / to 1. durchbrechen, vorbohren; bohren *(Dachschiefer)*; 2. *(Brg)* Durchhieb herstellen, durchbauen, durchfahren, durchörtern

hole 1. Loch *n*; Durchbruch *m*; 2. *(Masch)* Bohrung *f*; 3. Bohrloch *n*; Sprengloch *n*; 4. Defektelektron *n*, Loch *(Halbleitertechnik)*; 5. tote Zone *f* *(Radar)*
h. absorption Absorption *f* durch Defektelektronen
h. bottom Bohrlochtiefstes *n*, Bohrlochsohle *f* *(Erdöl, Erdgas)*
h. centre distance *(Masch)* Bohrungsmittenabstand *m*
h. circle *(Masch)* Lochkreis *m*

h. collar Bohrlochmund *m (Erdöl, Erdgas)*
h. conduction Löcherleitung *f*, Defekt[elektronen]leitung *f*
h. conductivity Defektleitung *f*
h. cutter Ausschneidbohrer *m*; Blechbohrer *m*
h. cutting shear Lochschere *f*
h. density Löcherdichte *f (Halbleitertechnik)*
h. depth *(Masch)* Bohrungstiefe *f*; Lochtiefe *f*
h. depth control Bohrtiefenkontrolle *f*
h. dia[meter] Bohrungsdurchmesser *m*
h. digger Pflanzlocher *m*, Erdbohrer *m*
h. disk Lochscheibe *f*
h. drum Lochtrommel *f*
h. electron Mangelelektron *n (Halbleitertechnik)*
h. exit Bohrungsdurchbruch *m*
h. for pin Stiftloch *n*
h. gauge Innenmeßgerät *n (Bohrung)*
h. injection Lochwanderung *f (Halbleitertechnik)*
h. lapping Innenrundläppen *n*
h. layout *(Masch)* Bohrbild *n*
h.-location positioning accuracy Lagegenauigkeit *f* der Bohrung
h. measuring gauge Innenmeßgerät *n (Bohrung)*
h. pattern *(Masch)* Bohrbild *n*
h. punching Lochen *n*
h. punching machine for patterns Modell-, Loch- und Setzmaschine *f (Schuhfertigung)*
h. size 1. Bohrungsdurchmesser *m*; 2. Bohrungsabmessungen *fpl*
h. space Bohrlochabstand *m*
h. spacing *(Masch)* Festlegung *f* der Mittenabstände
h. storage effect Diffusionskapazität *f (Halbleitertechnik)*
h. theory *(Ph)* Löchertheorie *f*
h.-type conductivity Fehlstellenleitung *f (Halbleitertechnik)*
h. wall Bohrungswand[ung] *f*
h. with shoulder Ansatzbohrung *f*
holes in the trap board *(Text)* Hebebrettbohrungen *fpl*
h., plain and tapped Bohrungen *fpl* ohne und mit Gewinde
holing 1. *(Brg)* Durchschlag *m*, Durchbauen *n*, Durchfahren *n*, Durchhieb *m*, Durchörtern *n*; Einbruch *m (in der Kohle)*; 2. Durchbrechen *n*
h. prop *(Brg)* Spreizstempel *m*
h. through *s.* holing
holland finish *(Text)* Halbbleiche *f (Baumwolle, Leinen)*
hollander Holländer *m*
hollow / to auskehlen *(Holzbearbeitung)*
h. drill *(Masch)* hohlbohren
h. out aushöhlen, eine Vertiefung ausarbeiten
hollow 1. Höhlung *f*, Mulde *f*, Rinne *f*, Vertiefung *f*; 2. Einsattelung *f (z. B. Bandfilter)*
hollow hohl
h. 'air cooled grate bar *(Wärm)* Hohlroststab *m* mit Luftkühlung
h. anode tube Hohlanodenröhre *f*
h. ball Hohlkugel *f*
h. bit Hohlmeißel *m*
h.-bit tongs *(Masch)* Hohlmaulzange *f*
h. blade Hohlschaufel *f*
h. block *(Bw)* Füllkörper *m*, Hohlkörper *m*, Hohlstein *m*
h.-block floor Stahlbetonrippendecke *f*
h. bolt Hohlschraube *f*
h. bored hohlgebohrt
h. boring bit *(Holz)* Hohlbohrer *m*
h. brick *(Bw)* Füllkörper *m*, Hohlkörper *m*, Hohlstein *m*
h.-cast cover doppelwandiger Boden *m*, Hohlgußdeckel *m*
h.-cast piston hohlgegossener (hohler) Kolben *m*
h. casting Hohlguß *m*
h. cathode Hohlkatode *f*
h. centre *(Masch)* Hohlkörnerspitze *f*
h.-charge ammunition Hohlladungsmunition *f*
h.-charge projectile Hohlkopfgeschoß *n*, Hohlladungsgeschoß *n*, Hohlraumgeschoß *n*
h. concrete Einkornbeton *m*
h. conductor *(El)* Hohlrohr *n*, Hohlleitung *f*, Hohlseil *n*
h. cylinder Hohlzylinder *m*, Rohr *n (Stereometrie)*
h. drill *(Masch)* Hohlbohrer *m*
h. drill steel Hohlbohrgestänge *n (Erdöl, Erdgas)*
h. drill stem Hohlbohrgestänge *n (Erdöl, Erdgas)*
h. ended vessel Schiff *n* mit hohlen Wasserlinien
h. filament Hohlseide *f*
h. forging Warmlochen *n*
h. glass Hohlglas *n*
h. glass prism Hohlprisma *n*
h. grain *(Led)* losnarbig
h. grate bar *(Wärm)* Hohlroststab *m*
h. grinding Hohlschliff *m*
h. grinding attachment Hohlschleifeinrichtung *f*
h.-ground mit Hohlschliff
h. ground slotting saw seitlich hohlgeschliffener Schlitzfräser *m*
h. head screw Hohlkopfschraube *f*
h. key Hohlkeil *m*; Rundschlüssel *m*
h. knot hohler Ast *m*
h. man-made fibres Spinnmasse *f* mit Lufteinschlüssen

h. mandrel Hohlmandrill *m*, Hohldorn *m*
h.-mandrel lathe Hohlspindeldrehmaschine *f*
h. mill Fräsring *m*; Hohlfräser *m*; Bohrring *m*
h. milling tool Hohlfräser *m*
h. mortising chisel Hohlstichel *m*
h. pin Hohlbolzen *m*
h. prism Hohlprisma *n*
h. punch Locheisen *n*, Lochstanze *f*
h. resonator *(El)* Hohlraumresonator *m*
h. rivet Hohlniet *m*
h.-rod clamp Gestängerohrklammer *f*
h. roll Hohlwalze *f*
h. saddle key Hohlkeil *m*
h. set-screw Innenvierkantschraube *f*; Innensechskantschraube *f*
h. shaft Hohlwelle *f*
h. shaft motor drive Ankerhohlwellenantrieb *m*
h. sphere Hohlkugel *f*
h. spindle Hohlspindel *f*
h. steel shaft Stahlhohlwelle *f*
h. stick *(Typ)* Hohlsteg *m*
h. stranded conductor *(El)* Hohlseil *n*
h. tile *(Bw)* Füllkörper *m*, Hohlkörper *m*, Hohlstein *m*
h.-tile floor *(Bw)* Füllkörperdecke *f*, Hohlkörperdecke *f*
h. trunnion Windzapfen *m (Bessemerbirne)*
h. vortex *(Ph)* Hohlwirbel *m*
h.-walled hohlwandig
h. water cooled grate bar *(Wärm)* Hohlroststab *m* mit Wasserkühlung
hollowed iron Hohleisen *n*
hollowing Auskehlung *f*
holmium *(Ch)* Holmium *n*
holocrystalline vollkristallin
holohedral vollflächig, holoedrisch
holohedrism Holoedrie *f*, Vollflächigkeit *f*
holohedron Holoeder *m*, Vollflächner *m*
holomorphic holomorph
holonomic holonom
holster Walzenständer *m*
homaloid ebener Raum *m*
home 1. Heim[stätte *f*] *n*; 2. *(Brg)* Ausfahrt *f*
h. amateur Radioamateur *m*
h. building Wohnungsbau *m*
h. cinema Heimkino *n*
h. constructor Funkbastler *m*, Radioamateur *m*; Bastler *m*
h. country Inland *n*
h. freight *(Schiff)* Rückfracht *f*
h.-grown wood inländisches Holz *n*
h. interview method Haus-zu-Haus-Befragung *f*, Haus-zu-Haus-Umfrage *f*
h. lot *(sl)* Filmstudio *n*
h.-made transmitter Amateursender *m*, selbstgebauter Sender *m*
h. market Inlandsmarkt *m*
h. organ Harmonium *n*
h. port Heimathafen *m*
h. position Nullstellung *f (z. B. Schalter)*; Ausgangsstellung *f (z. B. Wähler)*
h.-produced product Inlandserzeugnis *n*
h. record Mitlesestreifen *m (Fernschreiber)*
h. scrap Umlaufschrott *m*
h. sewage treatment plant Grundstückskläranlage *f*
h. signal *(Eis)* Einfahrtssignal *n*
h. study course Fernstudium *n*
h. trade 1. *(Schiff)* kleine Fahrt *f*, Inlandsfahrt *f*; 2. Binnenhandel *m*
h. trade ship Binnenschiff *n*, Inlandsschiff *n*, Schiff *n* in der Inlandsfahrt
h. weaving Hausweberei *f*, Heimweberei *f*
homeomorphic homöomorph
homeopolar homöopolar
homeward bound auf der Heimreise (Rückreise) *(befindliches Schiff)*
h. trip Heimreise *f*
homilite *(Min)* Homilit *m*
homing 1. *(Wärm)* Zielsuchen *n (Flugkörper)*; 2. Eigenpeilung *f*, Zielauffindung *f (Radar)*
h. action *(Fmt)* Rücklauf *m*
h. apparatus *(Flg)* Zielfluggerät *n*
h. beacon *(Flg)* Zielflugbake *f*
h. beacon light Ansteuerungsfeuer *n*
h. device *(Flg)* Ziel[an]fluggerät *n*
h. guidance Zielflugverfahren *n*, Zielsuchlenkung *f*
homocentric homozentrisch
homochromatic gleichfarbig, isochrom
homochrome *s.* homochromatic
homodyne *(El)* Trägerfrequenzüberlagerer *m*
h. reception Homodynempfang *m*, Synchronempfang *m*
homogeneity Homogenität *f*
homogeneous homogen, gleichartig
h. field homogenes (gleichförmiges) Feld *n*
h. line *(El)* homogene Leitung *f*
h. oil immersion homogene Ölimmersion *f*
h. reactor experiment homogener Versuchsreaktor *m*
h. thorium reactor homogener Thoriumreaktor *m*
homogeneousness Gleichartigkeit *f*, Homogenität *f*
homogenize / to 1. homogenisieren, gleichförmig machen; 2. diffusionsglühen
homogenizer *(Text)* Mischapparat *m*
homogenizing Homogenisierungsglühen *n*; Diffusionsglühen *n*
homohedral homoedrisch *(Kristalle)*
homologous homolog, übereinstimmend
h. series homologe Reihe *f*
homologue Homolog *n*

homologues of benzene Schwerbenzol *n* *(Homologe des Benzols, die als industrielle Lösungsmittel verwendet werden)*
homopolar homöopolar
h. bond homöopolare Bindung *f*, Atombindung *f*, Elektronen[austausch]bindung *f*, Kovalenzbindung *f*
h. component *s.* zero-sequence component
h. field magnet Gleichpolfeldmagnet *m*
h. generator Schenkelpolgenerator *m*, Gleichpolgenerator *m*
h. machine *(El)* Unipolarmaschine *f*
homothetic *(Math)* ähnlich liegend; ähnlich und ähnlich liegend; perspektivisch
hone / to 1. wetzen, abziehen *(Messer)*; 2. ziehschleifen, honen
hone 1. Schleifstein *m*, Wetzstein *m*; 2. Ziehschleifwerkzeug *n*, Honahle *f*
h.-abrading Ziehschleifen *n*, Honen *n*
honey Honig *m*
h. centrifuge Honig[waben]schleuder *f*
h. combing zahllose grübchenförmige Abzehrungen *fpl (Kesselkorrosion)*
h. separator Honig[waben]schleuder *f*
h. stone *(Min)* Honigstein *m*, Mellit *m*
honeycomb 1. Wabe *f*; Wabenstruktur *f*; 2. Honigwabenbaustoff *m*
honeycomb wabenförmig
h. coil *(El)* Honig[waben]spule *f*
h. core *(Bw)* Wabenkern *m*
h. laminate *(Plast)* Schichtstoff *m* mit wabenförmigem Kern
h. pattern *(Text)* Bienenwabenmuster *n*
h. radiator Wabenkühler *m*, Zellenkühler *m*
h. structure 1. *(Ch)* Wabenstruktur *f*; 2. *(Masch)* Wabenbauelement *n*; Wabenbaukonstruktion *f*
h. weave *(Text)* Waffelbindung *f*
honeycombing *(Bw)* wabenförmige Rissebildung
Honigmann method *(Brg)* Honigmann-Schachtbohrverfahren *n*
honing 1. Wetzen *n*, Abziehen *n (Messer)*; 2. Honen *n*; Ziehschleifen *n*; Langhubhonen *n*
h. fixture Honvorrichtung *f*
h. machine Honmaschine *f*, Ziehschleifmaschine *f*
h. stick Schleifleiste *f*; Ziehschleifstein *m*
h. stone Honstein *m*; Ziehschleifstein *m*
h. stone holder Honsteinhalter *m*
h. tool Honahle *f*, Ziehschleifahle *f*
h. tool with six sticks sechssteinige Honahle *f*
hood 1. *(Ch)* Abzug[sschrank] *m*; 2. Abzugshaube *f*, Decke *f*, Haube *f*, Kappe *f*, Verdeck *n*; 3. Anodenschutzkopf *m*; 4. *(Kfz) s.* bonnet
h. catch Haubenschloß *n*
h. covering *(Kfz)* Verdeckbezug *m*
h. end *(Schiff)* Plankenende *n* *(Vor- oder Hintersteven)*
hooded headlight *(Mil)* Tarnscheinwerfer *m*
h. target mit Elektronenschutzkopf
hoof cutting pincers Hufschneidezange *f*
hoofed animals Hufgänger *mpl*, Hufftüßler *mpl*, Huftiere *npl*
hook / to 1. angeln; 2. abbiegen, krümmen; 3. haken *(Bohrer)*
h. up anspannen
hook 1. Haken *m*, Lasthaken *m (eines Krans)*; 2. *(Text)* Haken, Platine *f (Jacquardstuhl)*
h. and eye coupling Seilschloß *n*
h. bolt Hakenbolzen *m*
h. end Hakenende *n (eines Kettengliedes)*
h.-link chain Hakenkette *f*
h. nail Hakennagel *m*
h. pliers Hakenzange *f*
h. screw Schraubhaken *m*
h. setting machine Agraffenmaschine *f* *(Schuhfertigung)*
h. shank taper tap Gewindefertigschneider *m* für Automaten *(mit langem gebogenem Schaft)*
h. spanner Hakenschlüssel *m*
h. spring ring Hakensprengring *m*
h. stick *(El)* Schaltstange *f*
h. switch Haken[um]schalter *m*
h. tool Hakenmeißel *m*
h. tooth Hakenzahn *m*
h.-type bottom block Hakenflasche *f* *(Flaschenzug)*
h.-up wire *(El)* Schaltdraht *m*, Rangierdraht *m*
Hookean solid ideal-elastischer Körper *m*
hooked hakenförmig
h. bar Bewehrungsstab *m* mit Endhaken *(Stahlbeton)*
h. key for brasses Lagerschalenschlüssel *m*
h. peg *(Text)* Einfederspindel *f* mit Nase, Hakenspindel *f*, Nasenspindel *f*
h. rug Knüpfteppich *m*
h. spanner *s.* h. wrench
h. tongs Schmiedezange *f*
h. wrench Hakenschlüssel *m*
hooker *(Schiff)* Hukboot *n*, Angelleinenfahrzeug *n*
Hooke's coupling Kardangelenk *n*
Hooke's law Hookesches Gesetz *n*
hooking *(Masch)* Schlagen *n (des Fräsers)*
h. into the work Einhaken *n* des Bohrers; Anfressen *n* der Fasen des Bohrers
hoop / to 1. abbinden, bereifen, binden *(Küfnerei)*; 2. beringen
h. a cask ein Faß binden (aufsetzen, aufschlagen)
hoop Bügel *m*; Daube *f*; Reif[en] *m*; Faßband *n*, Faßreifen *m*
h. driving machine Reifenauftreibmaschine *f*, Reifenaufdrückmaschine *f*
h. drop relay Fallbügelrelais *n*

h. button *(Kfz)* Horndruckknopf *m*
h. comb Hornkamm *m*
h. gap *(El)* Hörnerfunkenstrecke *f*
h.-gap arrester Hörnerableiter *m*
h.-gap switch Hörnerschalter *m*
h.-like hornähnlich
h. loudspeaker Trichterlautsprecher *m*
h. radiator *(El)* Hornstrahler *m*
h.-rimmed spectacles Hornbrille *f*
h. ring *(Kfz)* Horndruckring *m*
h. socket Hornbüchse *f (Erdöl, Erdgas; Fangwerkzeug)*
h. spoon Hornlöffel *m*
h. stake Sperrhaken *m* für Klempner
h. switch *(El)* Hörnerschalter *m*
h. timber *(Schiff)* Mittelheckspant *n*, Hinterstevenmittelteil *n*
h.-type loudspeaker Trichterlautsprecher *m*
h.-type radiator *(El)* Hornstrahler *m*
hornbeam Weißbuche *f*
hornblende Hornblende *f*
h. slate *(Min)* Hornblendeschiefer *m*
hornblock *(Eis)* Achslagerführung *f*, Führungsgabel *f*
horned cattle Rindvieh *n*
Horner's method *(Math)* Hornersches Schema *n*
hornfels *(Geol)* Hornfels *m*
hornplate *(Eis)* Achshalter *m*, Führungsgabel *f*
h. bogie *(Eis)* Achshalterdrehgestell *n*
hornsilver Hornsilber *n*
hornslate *(Min)* Hornblendeschiefer *m*
horse 1. Pferd *n*; 2. Bock *m*, Gerüstbock *m*
h. bean Futterbohne *f*, Saubohne *f*
h. box car Stallwagen *m*
h.-breeder Pferdezüchter *m*
h.-drawn cultivator Gespanngrubber *m*, Pferdehacke *f*
h.-drawn duster Gespannstäuber *m (Pflanzenschutz)*
h.-flesh ore Buntkupferkies *m*
h.-heading *(Brg)* Vorpfänden *n*
h. hide leather Roßleder *n*
h. hoe Pferdehacke *f*
h. mower Grasmähmaschine *f*
h. path Treidelpfad *m*
h. power *s.* horsepower
h.-radish Meerrettich *m*
h. rake Pferderechen *m*
h. rasp Hufraspe *f*
h. shoe *s.* horseshoe
h. team Pferdegespann *n*
h. traction Pferdezug *m*
horsehair Roßhaar *n*
h. stuff Roßhaargewebe *n*
horsepower Pferdestärke *f*, PS *(englische Leistungseinheit; 1 h.p. = 1,014 PS)*
h. delivered by the pump Pumpenleistung *f*
h.-hour Pferdestärkenstunde *f*, PSh *(englische Einheit für mechanische Arbeit; 1 h.p.hr. = 1,014 PSh)*
h. input Leistungsaufnahme *f*
h. loss Leistungsverlust *m*
h. output Leistung *f* in PS, PS-Leistung *f*
h. rating Leistung *f* in PS
horseshoe 1. Hufeisen *n*; 2. *(Brg)* Stahlausbau *m* in Hufeisenform
h. filament Haarnadelglühfaden *m*
h. magnet Hufeisenmagnet *m*, hufeisenförmiger Magnet *m*
h. mixer *(Ch)* Ankerrührer *m*
horticultural tractor Gartentraktor *m*, Einachstraktor *m*
horticulture Gartenbau *m*
hose / to mit Schlauch abspritzen
hose 1. Schlauch *m*; 2. Muffe *f*; 3. Strümpfe *mpl*
h. clamp Schlauchklemme *f*
h. clip Schlauchklemme *f*
h. connection Schlauchverbindung *f*, Schlauchanschluß *m*
h. coupling Schlauchkupplung *f*
h.-coupling thread Schlauchkupplungsgewinde *n*
h. for compressed air Preßluftschlauch *m*
h. joint dip Schlauchschelle *f*
h. line Schlauchleitung *f*
h. nipple Schlauchanschlußstück *n*, Schlauchverschraubung *f*
h. nozzle Strahlrohr *n (Feuerwehr)*
h.-proof strahlwassergeschützt *(Maschine)*
h. reel Schlauchwinde *f*, Schlauchhaspel *f*
h. strap Schlauchriemen *m*
h. woven round with wire Panzerschlauch *m*
hosiery Strumpf- und Trikotwaren *fpl*
h. anklet top Strickrand *m* an Knöchelsocken
h. knitting Strumpfwirkerei *f*
hospital clothing Krankenhauskleidung *f*
h. ship Lazarettschiff *n*
h. switch Notstromschalter *m*, Notbeleuchtungsschalter *m*
hot 1. warm; heiß; glühend; 2. hochangeregt; 3. hochaktiv, stark radioaktiv
h.-acting solvent method heißes Lösemittelverfahren *n (Herstellung von Zellulosederivatkunststoffen)*
h. activity *(Kern)* hohe Aktivität *f*
h. air Heißluft *f*
h.-air apparatus Heißluftdusche *f*
h.-air blast 1. *(Rak)* Düsenstrahl *m*; 2. Heißwind *m*
h.-air bridge Heißluftfeuerbrücke *f*
h.-air burner Warmluftbrenner *m*
h.-air cabinet *(Ch)* Trockenschrank *m*

h.-air cupola furnace *(Met)* Heißwindkupolofen *m*
h.-air de-icer Heißluftenteiser *m*
h.-air device Heißlufteinrichtung *f*, Warmlufteinrichtung *f*; Heißluftdusche *f*
h.-air drawing furnace Heißluftanlaßofen *m*
h.-air drying Warmlufttrocknung *f*
h.-air engine Heißluftmotor *m*
h.-air heating Warmluftheizung *f*; Luftheizung *f*
h.-air main Heißwindleitung *f (Hochofen)*
h.-air seasoning künstliche Trocknung *f*
h.-air setting stenter *(Text)* Heißluftfixierrahmen *m*
h.-air sizing machine *(Text)* Schlichtmaschine *f* mit Heißlufttrocknung
h.-air supplier Heißluftdusche *f*
h.-air turbine Heißluftturbine *f*
h.-air valve Heißluftschieber *m*
h. and cold bath process *(Holz)* Heiß-Kalt-Verfahren *n*
h. area heiße Zone *f (z. B. im Reaktor)*
h. atom *(Kern)* hoch angeregtes Atom *n*
h. band grease *(Masch)* Heißlagerfett *n*
h. band instrument Hitzbandinstrument *n*
h. bearing *(Masch)* heißgelaufenes Lager *n*
h. bed 1. *(Lw)* Treibbeet *n*, Frühbeet *n*; 2. *(Masch)* Warmbett *n (Lager)*
h. bend test Warmbiegeversuch *m (Festigkeit)*
h. bending Warmbiegen *n*
h. blast Heißwind *m*; vorgewärmter Gebläsewind *m*
h.-blast apparatus Winderhitzer *m*, Cowper *m (Hochofen)*
h.-blast furnace *(Met)* Heißwindofen *m*
h.-blast pig iron heißerblasenes Roheisen *n*
h. blast stove Winderhitzer *m*, Cowper *m*
h. blow / to *(Met)* warmblasen
h. body glühender Körper *m*
h. brittle warmbrüchig, rotbrüchig
h. brittleness Warmsprödigkeit *f*
h.-bulb Glühkopf *m*
h.-bulb engine Glühkopfmotor *m*
h.-bulb ignition *(Kfz)* Glühkopfzündung *f*
h. cabinet Trockenschrank *m*
h. cathode Glühkatode *f*, Heizkatode *f*
h.-cathode discharge Glühkatodenentladung *f*
h.-cathode gas-filled valve Stromtor *n*, Thyratron *n*
h.-cathode lamp Glühkatodenlampe *f*
h.-cathode mercury-vapour rectifier Glühkatodenquecksilberdampfgleichrichter *m*
h.-cathode rectifier Glühkatodengleichrichter *m*
h.-cathode valve Glühkatodenröhre *f*, Glühkatodenventil *n*
h. cave *(Kern)* heiße Zelle *f*
h. channel effect *(Kern)* Kanalüberhitzung *f*
h. chisel Warmschrotmeißel *m*
h. circular saw Warmkreissäge *f*
h.-coiled warmgewickelt *(Feder)*
h. coiling Warmwickeln *n*
h. coke glühender Koks *m*
h. cooling *(Met)* Heißkühlung *f*
h. crack Warmriß *m*
h. critical *(Kern)* heißkritisch
h. die Warmgesenk *n*
h.-dip coating Heißtauchen *n*, Überziehen *n* durch Heißtauchen
h.-dip galvanize / to feuerverzinken, tauchverzinken
h.-dip galvanizing bath Zinkbadofen *m*
h.-dip silver-plating Feuerversilberung *f*
h.-dip solution Tauchlösung *f* für die Herstellung bleibender Überzüge
h. dipped tin plate feuerverzinntes Blech *n*, Weißblech *n*
h. draw / to warmziehen
h. drawing 1. Heißverstrecken *n (Chemiefaserherstellung)*; 2. *(Met)* Warmziehen *n*
h. drawing bench Warmziehbank *f*
h.-drawn warmgezogen; heißverstreckt
h.-ductility Warmbildsamkeit *f*
h. embrittlement Warmversprödung *f*
h. flue *(Text)* Heißluftmansarde *f*, Heißlufttrockenmaschine *f*, Hotflue
h. forging Warmschmieden *n*
h. forging die Warmarbeitsgesenk *n*
h. forging press Schmiedepresse *f*
h. forming Warmumformung *f*
h.-galvanization Feuerverzinkung *f*
h.-galvanize / to feuerverzinken
h. gases *(Rak)* Verbrennungsgase *npl*
h. gilding Feuervergoldung *f*
h. groove Warmkaliber *n*
h.-ground pulp Warmschliff *m (Papierherstellung)*
h. hardness Warmhärte *f*
h.-house Gewächshaus *n*, Treibhaus *n*
h. junction Schweißperle *f*
h. ladle crane Gießkran *m*
h. material *(Kern)* hochradioaktives Material *n*, stark radioaktive Substanz *f*
h.-melt coating *(Holz)* Schmelzstreichüberzugverfahren *n*
h. metal ladle Roheisenpfanne *f*
h. metal mixer Roheisenmischer *m*
h. mould centrifugal casting Kokillenschleuderguß *m*
h. neck grease *(Masch)* Heißwalzenfett *n*
h. pad Heizkissen *n*
h. penetration test *(Plast)* Wärmedruckprüfung *f*
h. plate Wärmeplatte *f*; Heizplatte *f*, Kochplatte *f*

h. plate mill Warmblechstraße *f*
h. platen press Heißplattenpresse *f*
h. plug Zündkerze *f* mit niedrigem Wärmewert
h. press / to 1. *(Met)* warmpressen; 2. *(Text)* dekatieren, Preßglanz entfernen
h. press Leimpresse *f*
h. press plate Heißplattenpresse *f*
h. pressing *(Met)* Warmpressen *n*
h. pressure welding Feuerpreßschweißen *n*
h. quenching Warmbadhärtung *f*; Stufenhärtung *f*
h.-reactor loading Reaktorbeschickung *f* ohne Betriebsunterbrechung
h. refining Heißveredlung *f*
h. resistance Warmwiderstand *m*
h.-rivet / to warmnieten
h. riveting Warmnieten *n*
h.-rolled warmgewalzt
h. rolling *(Met)* Warmwalzen *n*
h. rolling mill Warmwalzwerk *n*
h. rolling train Warmstraße *f*
h. runner moulding *(Plast)* Spritzgußverfahren *n* mit geheizten Stegen
h. saw Heiß[eisen]säge *f*, Warmsäge *f*
h. sawing Warmsägen *n*
h. sawing machine Warmsäge *f*
h. sealing Heißsiegelverfahren *n*, Heißversiegeln *n*
h.-sealing heißsiegelfähig
h. setting adhesive warmabbindender Klebstoff *m*
h. short warmbrüchig
h.-short crack Warmriß *m*
h. shortness Warmbrüchigkeit *f*
h.-solder / to heißlöten
h.-speed landing *(Flg)* Landung *f* mit großer Geschwindigkeit
h.-spelter galvanizing Feuerverzinken *n*
h. spot 1. *(Kern)* örtliche Überhitzung *f* (Ansammlung *f* von Radioisotopen); 2. *(Met)* Aufblasestelle *f*
h. stamping press Warmgesenkpresse *f*
h.-start lamp Glühstartlampe *f*, Warmstartlampe *f*
h. strain / to *(Met)* warmrecken
h. strength Warmfestigkeit *f*
h.-strip instrument Hitzbandinstrument *n*
h. strip mill Warmbandstraße *f*
h.-stuff / to *(Led)* einbrennen *(zurichten)*
h. swaging machine Warmgesenkdrückmaschine *f*
h. tar Heißteer *m*
h. tear *(Met)* Wärmeriß *m*
h.-tensile test Warmzerreißversuch *m*
h. test Warmversuch *m*
h. tin / to feuerverzinnen
h. tinning Feuerverzinnung *f*
h. transfer foil *(Typ)* Aufbügelfolie *f*
h. trim / to warmentgraten
h. trimming die Warmabgratwerkzeug *n*
h. upset forging Warmstauchen *n*
h. vulcanization Heißvulkanisation *f*
h. waste hochaktiver Abfall *m*
h.-water heating system Warmwasserheizungsanlage *f*
h. water injector Warmwasserinjektor *m*
h.-water supply Warmwasserversorgung *f*
h. well *(Wärm)* Heißwasserbehälter *m*
h. wire Hitzdraht *m*
h. wire altitude meter Hitzdrahthöhenmesser *m*
h. wire ammeter Hitzdrahtamperemeter *n*
h. wire anemometer Hitzdrahtanemometer *n*
h. wire cautery Hitzdrahtausbrennung *f*, Glühkaustik *f*
h. wire expansion instrument Hitzdrahtinstrument *n*
h. wire gas analyzer Wärmeleitfähigkeitsgasanalysator *m*
h. wire instrument Hitzdrahtinstrument *n*
h. wire microphone *s.* thermal microphone
h. wire pressure gauge Hitzdrahtmanometer *n*, Pirani-Manometer *n*
h. wire relay Thermorelais *n*
h. wire voltmeter Hitzdrahtvoltmeter *n*
h. work / to warmverformen
h. workability Warmumformbarkeit *f*
h. workable warm bearbeitbar (umformbar)
h. working Warmumformung *f*
h.-working brass Schmiedemessing *n*
h.-wound warmgewickelt *(z. B. Federn)*
hotbox heißgelaufenes Lager *n*
hotplate *s.* hot plate
hour Stunde *f*, h
h. axis *(Opt)* Stundenachse *f*
h.-axis drive Stundenantrieb *m*
h. circle *(Opt)* Stundenkreis *m*
h.-counter Stundenzähler *m*
h.-glass worm Globoidschnecke *f*
h.-glass worm-and-sector steering gear Globoidschneckenlenkung *f*
h. meter Zeitzähler *m*
hourly earning Stundenlohn *m*
h. output Stundenleistung *f*
h. plate Zifferblatt *n*
house / to einbauen; unterbringen; aufnehmen
house Haus *n*
h. cable *(Fmt)* Innenkabel *n*
h. coal Hausbrandkohle *f*
h. coaming *(Schiff)* Deckshaussüll *m (n)*
h. connection line *(El)* Hausanschluß[leitung *f*] *m*
h. flag Reedereiflagge *f*, Hausflagge *f*
h. lead-in *(El)* Hauseinführung *f*

h. pole *(El)* Dachabspanngestänge *n*, Dachabspannstange *f*
h. service *(El)* Hausanschluß[leitung *f*] *m*
h. service meter *(El)* Haushaltszähler *m*
h. sewage Abwässer *npl*
h. telephone Hausfernsprecher *m*
houseboat Deckhausboot *n*, Wohnschiff *n*, Hausboot *n*
housebreaker Abbrucharbeiter *m*
housed in a common body in einem Gehäuse untergebracht
household article Haushaltsgegenstand *m*
h. coal Hausbrandkohle *f*
h. detergent Haushaltswaschmittel *n*
housing 1. Wohnungsbau *m*; Wohnungswesen *n*; 2. Aufnahme *f*; 3. *(Masch)* Gehäuse *n*; Ständer *m*; Stütze *f*
h. construction Wohnungsbau *m*
h. density Wohndichte *f*
h. estate Wohnkomplex *m*; Wohnsiedlung *f*
h. project Wohnungsbauvorhaben *n*
h. scheme Wohnungsbauvorhaben *n*
h. stopper *(Schiff)* Kettenstopper *m*, Kettenklaue *f*
hovel 1. Schuppen *m*; 2. *(Brg)* Kaue *f*
hover Schirmglucke *f (Geflügelzucht)*
hovercraft *(Schiff)* Luftkissenfahrzeug *n*
hovering flight Schwebeflug *m (Hubschrauber)*
h. vessel *(Am)* Schmuggelschiff *n*
hovership *s.* hovercraft
howitzer *(Mil)* Haubitze *f*
howl Heulton *m (durch akustische oder elektrische Rückkopplung)*
howler *(Fmt, El)* Heuler *m*, Summer *m*
howling buoy *(Schiff)* Heultonne *f*, Heulboje *f*
hoy Lastboot *n*
HSS reamer Schnellstahlreibahle *f*
hub 1. Nabe *f*; Radnabe *f*; 2. *(Brg)* Nivellierlatte *f*; 3. Spule *f (für Magnetbänder)*; 4. *(Büro)* Buchse *f*
h. bolt Nabenschraube *f*
h. bushing Radnabenbuchse *f*
h. cap Radnabendeckel *m*, Radkappe *f*, Nabenkappe *f*
h. cover Nabendeckel *m*
h. fairing Nabenverkleidung *f*
h.-hole Nabenbohrung *f*
h. keyway Naben[keil]nut *f*
h. nut Nabenmutter *f*
h. spider Nabenstern *m*
h. with clutch cam Nabe *f* mit Kupplungsnocken
hubnerite *(Min)* Hübnerit *m*
huckaback weave *(Text)* Gerstenkornbindung *f*
hudge *(Brg)* Förderkübel *m*
hue 1. Farbton *m (des Lichtes der Lampe)*; 2. *(Fs)* Farbwert *m*; 3. *(Text)* Farbschattierung *f*, Nuance *f*, Schattierung *f*, Stich *m*
huffler Flußlotse *m*
hug the land / to sich dicht unter der Küste halten
Hughes bit Dreirollenmeißel *m (Tiefbohrungen)*
Hughes printing telegraph *(Fmt)* Hughes-Typendrucker *m*
hulk Hulk *m (Speicherschiff)*
hull 1. Hülse *f*, Samenhülse *f*; 2. Schiffsrumpf *m*, Schiffskörper *m*; 3. Luftschiffgerippe *n*
h. gunner's periscope Kugelzielfernrohr *n (Panzer)*
h. of centre section *(Rak)* Mittelteilzelle *f*
h. performance Leistungswerte *mpl* des Schiffes *(ohne Maschine)*
h. resistance *(Schiff)* Formwiderstand *m*
hulling apparatus *(Lw)* Enthülser *m*
hum / to brummen; summen *(Funktechnik)*
hum Summen *n*, Brummen *n*; Brummton *m*; Brummspannung *f*
h.-bucker Entbrummspule *f*
h.-bucking coil Spule *f* zur Brummunterdrückung *(elektrodynamischer Lautsprecher)*
h.-dinger *(Fmt)* Entbrummer *m*
h. eliminator Brummfilter *n*, Netzfilter *n*, Störungsfilter *n*, Entbrummer *m*
h. modulation Brummodulation *f*
h. pick-up Brummeinstreuung *f*
h. potentiometer *(Fmt)* Entbrummer *m*
h. reduction factor Siebfaktor *m*
h. ripple *(Fmt)* Brummwelligkeit *f*
h. voltage Brummspannung *f*
human attendance Wartung *f*
h. torpedo *(Mil)* Einmanntorpedo *m*
humic acid Huminsäure *f*, Humussäure *f*
h. coal Humuskohle *f*
humid feucht
h. analysis *(Ch)* Naßanalyse *f*, Analyse *f* auf nassem Wege
humidification Befeuchtung *f*
humidifier Befeuchter *m*, Anfeuchter *m*
humidity Feuchtigkeit *f*
h. control Luftbefeuchtung *f*, Regelung *f* der Luftfeuchtigkeit *(Klimatechnik)*
h. meter Feuchtigkeitsmesser *m*
h. of [the] air Luftfeuchtigkeit *f*
h. protection Feuchtigkeitsschutz *m*
hummer *(Fmt)* Schnarre *f*, Summer *m*
humming *(Fmt)* Summen *n*
h. sound Summerzeichen *n*, Summerton *m*
hump 1. Buckel *m*, Maximum *n*; 2. Ablaufberg *m*; 3. Buckel *(in der Straße)*
h. in brightness *(Fs)* Helligkeitssprung *m*
h. locomotive Ablaufberglokomotive *f*
humpy höckerig
humus Humuserde *f*
hunch pit Schlackengrube *f*
hung ceiling *(Bw)* untergehängte Decke *f*

hunt / to 1. jagen; 2. pendeln, schwingen *(instabiler Regelkreis)*; 3. gieren; 4. *(Fmt)* suchen
h. over a bank *(Fmt)* freiwählen
h. over a complete level *(Fmt)* durchdrehen
hunter *(Fmt)* Wähler *m* mit freier Wahl
hunting 1. Jagen *n*, Jagd *f*; 2. Pendeln *n*, Pendelschwingung *f* *(Regler)*; 3. Gieren *n*; 4. *(Fmt)* Anrufsuchen *n*; 5. Nachlauf *m*, Selbstregelung *f* *(Elektronenrechner)*
h. action *(Fmt)* Freiwahl *f*, Freisuchen *n*
h. device *(Fmt)* Abtastvorrichtung *f*
h. link Ausgleichglied *n* der Getriebekette
h. operation *(Fmt)* Freiwahl *f*
h. selector *(Fmt)* Freiwähler *m*
h. spoon rolling mill pendelndes Löffelwalzwerk *n*, Löffelwalzmaschine *f*
h. switch *(Fmt)* Abgreifer *m*
h. tooth überholender Zahn *m*
hurdle 1. *(Brg)* Gitter *n*; 2. Hürde *f*, Horde *f*; 3. Hürde *(Sport)*; 4. Faschine *f*, Reisigbündel *n*
h. race Hürdenlauf *m* *(Sport)*
h.-type scrubber *(Brg)* Hordenwascher *m*
h. washer *(Brg)* Hordenwascher *m*
h. work Flechtwerk *n*, Geflecht *n*
hurdling Hürdenlauf *m* *(Sport)*
hurricane-deck ship Sturmdeckschiff *n*
h. lamp Sturmlaterne *f*
h. warning Wirbelsturmwarnung *f*
hurrier *(Brg)* Schlepper *m*, Fördermann *m*
husk / to enthülsen, schälen
husk Hülse *f*, Schale *f*, Schote *f*
husker Maisentliescher *m*, Schäler *m*
h. shredder Bröckler *m*, Futterreißer *m*
husking feeder Schäleinleger *m*
h. of legumes Schälen *n* der Hülsenfrüchte
husky boat starkgebautes Boot *n*
hustle / to beschleunigen, treiben, drängen
hut Kate *f*; Hütte *f*; Baracke *f*
h. camp Barackenlager *n*
hutch *(Brg)* Förderwagen *m*, Grubenwagen *m*, Hunt *m*
hutching *(Brg)* Setzwäsche *f* *(Aufbereitung)*
Huygenian eyepiece *(Opt)* Huygens-Okular *n*
Huygen's eyepiece *(Opt)* Huygens-Okular *n*
h.v. *s.* high voltage
hyacinth *(Min)* Hyazinth *m*
hyaline glasig, glasartig
hyalite *(Min)* Hyalit *m*, Glasopal *m*
hybrid amplifier *(Fmt)* Vierdrahtverstärker *m*
h. base petroleum gemischtes Rohöl *n*, Hybridrohöl *n*
h. coil Ausgleichübertrager *m*, Brückenübertrager *m*, Differentialtransformator *m*, Differentialübertrager *m*, Sprechspule *f*
h. coil termination circuit *(Fmt)* Gabelschaltung *f*
h. computing elements Hybridrechenelemente *npl* *(Kombination von digitalen und analogen Rechenelementen)*
h. coupler Hybridrichtungskoppler *m* *(Wellenleiter)*
h. electromagnetic wave gemischte elektromagnetische Welle *f* *(Wellenleiter)*
h. ion Zwitterion *n*
h. junction Hybride *f* *(Wellenleiter)*
h. orbitals *(Ch)* Bastardbahnen *fpl*
h.-pi equivalent circuit Ersatzschaltung *f* nach Giacoletto *(Halbleitertechnik)*
h. ring Ringhybride *f* *(Wellenleiter)*
h. rocks *(Geol)* Mischgesteine *npl*
h. set *(Fmt)* Gabel[schaltung] *f*
h. transformer Ausgleichübertrager *m*; dreispuliger Differentialübertrager *m*; Sprechspule *f*
hybridization Kreuzen *n* *(v. Pflanzen od. Tieren)*
hydantoic acid Hydantoinsäure *f*
hydracid Wasserstoffsäure *f*
hydrant Hydrant *m*
hydrapulper *(Holz)* Hydrastoffauflöser *m*
hydrargyrum Quecksilber *n*
hydrate / to 1. *(Ch)* hydratisieren *(Wasser anlagern)*; 2. abbinden *(Zement)*
hydrate Hydrat *n*
hydrated 1. *(Ch)* hydratisiert; wasserhaltig; 2. gelöscht *(Kalk)*
h. cement Zementstein *m*
h. lime gelöschter Kalk *m*, Kalkhydrat *n*, Löschkalk *m*
hydration Hydratisierung *f*, Hydratation *f*
h. of ions Ionenhydratation *f*
h. water Hydratwasser *n*
hydraulic hydraulisch; Flüssigkeits-
h. accumulator Druckwasserspeicher *m*, hydraulischer Akkumulator *m*
h. automobile jack hydraulischer Wagenheber *m*
h. barking *(Holz)* hydraulische Naßentrindung *f*
h. binder *(Bw)* hydraulischer Binder *m*, hydraulisches Bindemittel *n*
h. boiler test Abpressen *n* des Kessels
h. brake hydraulische Bremse *f*
h. buffer Flüssigkeitsrücklaufbremse *f*, hydraulische Rücklaufbremse *f*
h. capsule Flüssigkeitsdruckmeßdose *f*
h. circuit Flüssigkeitskreislauf *m*; Treibmittelkreis *m*
h. classification *(Brg)* Naßklassierung *f*
h. compressor Wasserdruckkompressor *m*
h. computer hydraulischer Rechner *m*
h. control valve hydraulisch betätigtes Ventil *n*
h. copying device hydraulische Nachformeinrichtung *f*

h. coupling hydraulische Kupplung *f*, Flüssigkeitskupplung *f*; Strömungskupplung *f*
h. cylinder hydraulischer Arbeitszylinder *m*
h. damper Flüssigkeitsdämpfer *m*
h. depth filter *(BMSR)* hydraulisches Kartuschenfilter *n*
h. detent hydraulische Vorsteuerung *f*
h. dip *(Brg)* Hakentasse *f*
h. dredge[r] Saugbagger *m*, Pumpenbagger *m*
h. drive Flüssigkeitsgetriebe *n*; hydraulischer Antrieb *m*
h. duplicator hydraulische Nachformeinrichtung *f*
h. efficiency hydraulischer Wirkungsgrad *m*
h. engineer 1. Wasserbauingenieur *m*; 2. Hydraulikingenieur *m*
h. engineering Wasserbau *m*; Fluß- und Seebau *m*
h. fill eingeschlämmte (eingespülte) Erdmassen *fpl*
h. filling 1. *(Bw)* Sandeinspülung *f*; 2. *(Brg)* Spülversatz *m*
h. fluid Treibmittel *n*; Druckflüssigkeit *f*
h. fluid stream Treibstrom *m*
h. force pump Kolbenpumpe *f*
h. four-way valve hydraulischer Steuerzylinder *m* mit Vierkantensteuerung
h. gear change Druckölschaltung *f*
h. governor hydraulischer Regler *m*; Vorsteuerpumpe *f*
h. gun Wasserkanone *f*
h. jack Öldruckpresse *f*; hydraulischer Wagenheber *m*; Druckhebewinde *f*; Hydraulikkolben *m*
h. leather Dichtungsleder *n*; Lederdichtung *f*
h. lime hydraulischer Kalk *m*, Wasserkalk *m*
h. line Hydraulikleitung *f*
h. load-cell Flüssigkeitsdruckmeßdose *f*
h. loading mechanism hydraulische Werkstückzuführung *f*
h. logic hydraulische Logik *f*
h. loss hydraulischer Verlust *m*
h. main Hydraulikleitung *f*
h. mining *(Brg)* hydraulische Gewinnung *f*
h. motor 1. Wasserkraftmaschine *f*; 2. *(Masch)* Hydraulikmotor *m*, Flüssigkeitsmotor *m*
h. oil Treiböl *n*; Hydrauliköl *n*
h. organ Wasserorgel *f (Musikinstrument)*
h. package[-type unit] hydraulisches Aggregat *n*
h. peeler log debarker *(Holz)* hydraulischer Entrinder *m*
h. planer Hobelmaschine *f* mit Hydraulikantrieb
h. power pack Hydraulikaggregat *n*
h. power station Wasserkraftwerk *n*
h. power system hydraulisches Getriebe *n*, Hydraulik *f*
h. press hydraulische Presse *f*; Plungerpresse *f*
h. pressure Wasserdruck *m*
h. propeller *(Schiff)* Reaktionspropeller *m*
h. pump 1. Flüssigkeitspumpe *f*; 2. *(Masch)* Druckölpumpe *f*
h. radius hydraulischer Radius *m*, Profilradius *m*
h. recoil brake Flüssigkeitsrücklaufbremse *f*, hydraulische Rücklaufbremse *f*
h. shock absorber hydraulischer Stoßdämpfer *m*
h. space filter *(BMSR)* hydraulisches Spaltfilter *n*
h. stabilizer hydraulischer Stabilisator *m*
h. stop-light switch *(Kfz)* Bremslichtöldruckschalter *m*
h. stowage *(Brg)* Spülversatz *m*
h. stripping *(Brg)* hydromechanisierte Beseitigung *f* des Abraums
h. structure Wasserbauwerk *n*
h. structures Wasserbauten *pl*
h. system hydraulisches System *n*, Hydraulik *f*
h. tank Hydraulikölbehälter *m*
h. torque converter hydraulischer Drehmomentwandler *m*
h. transmission hydraulisches Getriebe *n*
h. turbine Wasserturbine *f*
h. veneering press with steam plates hydraulische Dampfplattenfurnierpresse *f*
hydraulically actuated hydraulisch betätigt
h.-assisted steering *(Kfz)* Lenkung *f* mit hydraulischer Lenkhilfe, hydraulische Servolenkung *f*
h. coupled hydraulisch gekuppelt
h. driven mit Flüssigkeitsantrieb, hydraulisch angetrieben
h. operated hydraulisch betätigt
h.-operated support leg *(Brg)* hydraulische Bohrstütze *f*
hydraulicking *(Brg)* hydraulischer Abbau *m (nur Tagebau)*
hydraulics Hydraulik *f*
hydrazide Hydrazid *n*
hydrazine Hydrazin *n*
hydrazoate Azid *n*, Salz *n* der Stickstoffwasserstoffsäure
hydrazobenzene Hydrazobenzol *n*
hydrazoic acid Stickstoffwasserstoffsäure *f*
hydride Hydrid *n*
hydro-extractor Trockenschleuder *f*, Schleuder *f*, Zentrifuge *f*
h.-leg *(Brg)* hydraulische Bohrstütze *f*
h.-mechanic brake system *(Kfz)* hydromechanische Bremsanlage *f*

h.-pneumatic accumulator pneumohydraulischer Flüssigkeitsspeicher *m*, Druckluftakkumulator *m*
hydroaromatic hydroaromatisch
hydrobromic acid Bromwasserstoffsäure *f*
hydrobromination Anlagerung *f* von Bromwasserstoff
hydrocarbon Kohlenwasserstoff *m*
h. black Ruß *m*
h. group analysis Ringanalyse *f*
h. pavement Schwarzdecke *f* *(Straßenbau)*
hydrocarbonaceous kohlenwasserstoffhaltig
hydrochloric salzsauer
h. acid Chlorwasserstoffsäure *f*, Salzsäure *f*
hydrochlorination Anlagerung *f* von Chlorwasserstoff
hydrocinnamic acid Hydrozimtsäure *f*
hydroclassifying hydraulisches Klassieren *n*, Sichten *n* mit Hilfe von Wasser
hydroconic hull form hydrokonische Schiffskörperform *f*
hydrocyanic acid Blausäure *f*, Zyanwasserstoffsäure *f*
hydrocycle tank Wasserumlaufbütte *f*
hydrodynamic hydrodynamisch
h. lubrication *(Kfz)* Vollschmierung *f*, vollflüssige Schmierung *f*
h. paradox *(Ph)* hydrodynamisches Paradoxon *n*
hydrodynamical log *(Schiff)* Staudrucklog *n*, Stevenlog *n*
h. speedometer *(Schiff)* Staudruckfahrtmesser *m*
hydrodynamics Hydrodynamik *f*
hydroelectric cell *(El)* Naßelement *n*
h. generating set *(Elm)* Wasserkraft[maschinen]satz *m*, hydraulische Generatorgruppe *f*
h. generating station Wasserkraftwerk *n*
h. power station Wasserkraftwerk *n*
hydroextract / to [aus]schleudern, zentrifugieren
hydroextracting cage Schleuderkorb *m*
hydroferricyanic acid Hexacyanoeisen(III)-säure *f*
hydroferrocyanic acid Hexacyanoeisen(II)-säure *f*
hydrofining Aufarbeitung *f* durch Hydrierung, Hydrofining *n*
hydrofluoric acid Fluorwasserstoffsäure *f*, Flußsäure *f*
hydrofluosilicic acid Flukieselsäure *f*, Kieselfluorwasserstoffsäure *f*
hydrofoil [boat] Tragflächenboot *n*, Tragflügelboot *n*
h. rudder *(Schiff)* Stromlinienruder *n*
h. seacraft *(Am)* Tragflächenboot *n*
h. section *(Schiff)* Tragflügelprofil *n*
hydroforming 1. Hydroforming *n*, katalytisches Verfahren *n* zur Veredlung von Benzinen; 2. hydraulisches Tiefziehen *n* ohne Gleitmittel
hydrogen Wasserstoff *m*
h. atom Wasserstoffatom *n*
h. bomb Wasserstoffbombe *f*
h. bond Wasserstoffbindung *f*
h. bridge *(Ch)* Wasserstoffbrücke *f*
h. brittleness Beizsprödigkeit *f*, Wasserstoffsprödigkeit *f*
h. bromide Bromwasserstoff *m*
h. chloride Chlorwasserstoff *m*
h. compound Wasserstoffverbindung *f*
h. cyanide Zyanwasserstoff *m*
h. cylinder Wasserstoffflasche *f*
h. discharge lamp Wasserstoffentladungslampe *f*
h. electrode Wasserstoffelektrode *f*
h. fluoride Fluorwasserstoff *m*
h. gas electrode Wasserstoffelektrode *f*
h. generator Wasserstofferzeuger *m*
h. iodide Jodwasserstoff *m*
h. ion Wasserstoffion *n*
h. ion concentration Wasserstoffionenkonzentration *f*
h. iron resistance *(El)* Eisenwasserstoffwiderstand *m*
h. line Wasserstofflinie *f* *(Spektrum)*
h. meter Wasserstoffzähler *m*
h. nucleus Wasserstoffkern *m*
h. peroxide Wasserstoff[su]peroxid *n*
h.-resisting wasserstoffbeständig *(Stahl)*
h. unsoundness *(Met)* Wasserstoffkrankheit *f* *(Kupfer)*
hydrogenate / to hydrieren *(Wasserstoff anlagern)*
hydrogenated fat gehärtetes Fett *n*
hydrogenation Hydrierung *f*
hydrogenic ion wasserstoffähnliches Ion *n*
hydrogenize / to hydrieren *(Wasserstoff anlagern)*
hydrographic map Seekarte *f*
h. survey ship Vermessungsschiff *n* für hydrografische Messungen
hydroiodic acid Jodwasserstoffsäure *f*
hydrology Hydrologie *f*, Wasserkunde *f*, Gewässerkunde *f*
hydrolysis Hydrolyse *f*
h. of wood Holzverzuckerung *f*
hydrolyze / to hydrolysieren
hydromagnetic compression wave hydromagnetische Kompressionswelle *f*
h. shock wave hydromagnetische Stoßwelle *f*
hydromagnetics Hydromagnetik *f*
hydromechanical hydromechanisch
hydromechanics Hydromechanik *f*
hydrometallurgy Naßmetallurgie *f*
hydrometer Aräometer *n*, Tauchwaage *f*, Hydrometer *n*, Säuremesser *m*

hydrometry Wassermessung *f*
hydromotor Wasserkraftmotor *m*
hydronics *(Am)* Warmwasserheizung *f*
hydrophilic hydrophil, wasserbindend, wasseranziehend
hydrophobic hydrophob, wasserabstoßend
hydrophone Unterwasserhorchgerät *n*, Hydrofon *n*
hydroplane Gleitboot *n*, Stufenboot *n*; Wasserflugzeug *n*
hydroquinone Hydrochinon *n*
hydroskimmer Hydroskimmer *m*, Luftkissenfahrzeug *n* *(nicht voll flugfähiges Schwebefahrzeug)*
hydrosol Hydrosol *n*, wäßrige kolloide Lösung *f*
hydrospark forming hydroelektrisches Umformverfahren *n* durch Funkenbombe
hydrostatic hydrostatisch
  h. curves Formkurven *fpl (eines Schiffes)*
  h. head hydrostatische Förderhöhe *f*
  h. paradox hydrostatisches Paradoxon *n*
  h. test Prüfversuch *m* zur Beurteilung der Beständigkeit eines Gewebes gegen Wasserdurchdringung
  h. winch hydraulische Winde *f*
hydrostatical stability *(Schiff)* Formstabilität *f*
hydrostatics Hydrostatik *f*
hydrosulphate Hydrogensulfat *n*, saures Sulfat *n*
hydrosulphite Hydrogensulfit *n*, Bisulfit *n*
hydrosulphurous acid Sulfoxylsäure *f*
hydrous wasserhaltig, wäßrig
  h. salt kristallwasserhaltiges Salz *n*
hydrovalve Hydraulikventil *n*
hydroxamic acid Hydroxaminsäure *f*
hydroxide Hydroxid *n*
hydroxy acid Oxysäure *f*
hydroxyacetic acid Glykolsäure *f*, Oxyessigsäure *f*
hydroxydibasic acid zweibasische Oxysäure *f*
hydroxypropionic acid Oxypropionsäure *f*, Milchsäure *f*
hydroxyquinone Oxychinon *n*
hydroxysuccinic acid Oxybernsteinsäure *f*, Äpfelsäure *f*
hydroxytribasic acid dreibasische Oxysäure *f*
hydrozincite *(Min)* Hydrozinkit *m*, Zinkblüte *f*, Zinkspat *m*
hyetograph Regenschreiber *m*
hyetometer Regenmesser *m*
hygrometer Hygrometer *n*, Feuchtigkeitsmesser *m*
hygrometric conditions *(Holz)* Feuchtigkeitsbedingungen *fpl*
  h. grade Feuchtigkeitsgrad *m*
hygronom Dampffeuchtemesser *m*
hygroscope Hygroskop *n*, Feuchtigkeitsanzeiger *m*, Feuchtigkeitsmesser *m*
hygroscopic hygroskopisch, wasseraufnehmend, wasseraufsaugend, feuchtigkeitsempfindlich
hygroscopicity Hygroskopizität *f*, Wasseraufnahmevermögen *n*
hypar *(Am)* *s.* hyperbolic paraboloid
hyperbola Hyperbel *f*
hyperbolic hyperbolisch; Hyperbel-
  h. angle hyperbolischer Winkel *m*
  h. cosecant hyperbolischer Kosekans *m*
  h. cosine hyperbolischer Kosinus *m*
  h. cotangent hyperbolischer Kotangens *m*
  h. curvature hyperbolische Krümmung *f*
  h. disk profile hyperboloidisches Scheibenprofil *n* *(Turbine)*
  h. function Hyperbelfunktion *f*
  h. geometry hyperbolische Geometrie *f*
  h. inverter Hyperbelinversor *m*
  h. logarithm natürlicher (hyperbolischer, Neperscher) Logarithmus *m*
  h. motion Hyperbelbewegung *f*
  h. orbit Hyperbelbahn *f*
  h. paraboloid *(Bw)* Hyperbelparaboloid *n*; hyperbolische Paraboloidschale *f*, hyperbolisches Paraboloiddach *n*
  h. path Hyperbelbahn *f*
  h. plane hyperbolische Ebene *f*
  h. position finding Hyperbelortung *f* *(Radar)*
  h. radio navigation Hyperbelnavigation *f*
  h. secans hyperbolischer Sekans *m*
  h. sine Hyperbelsinus *m*, hyperbolischer Sinus *m*
  h. tangent hyperbolischer Tangens *m*
hyperbolical wheel Hyperbolidenrad *n*, Hyperbelrad *n*, hyperbolisches Schraubenrad *(Verzahnung)*
hyperboloid Hyperboloid *n*
  h. gear Hyperboloidrad *n*
  h. of revolution Rotationshyperboloid *n*
hypercomplex hyperkomplex
hyperconductivity Supraleitung *f*, Überkonduktivität *f*
hypercritical überkritisch
hypereutectic *(Met)* hypereutektisch, übereutektisch
hypereutectoid *(Met)* Übereutektoid *n* *(Stahl)*
hyperfine hyperfein, Hyperfein-
  h. structure Hyperfeinstruktur *f*
hyperfrequency *(El)* Höchstfrequenz *f*
hyperon Hyperon *n*
hypersensitive überempfindlich
hypersonic speed Überschallgeschwindigkeit *f*
hypersynchronous übersynchron
hypnotic Schlafmittel *n*, Hypnotikum *n*
hypo *s.* sodium thiosulphate
hypobromite Hypobromit *n*
hypobromous acid hypobromige (unterbromige) Säure *f*

hypochlorite Hypochlorit *n*
h. stage oxydierende Hypochloritbehandlung *f* *(Zellstoffbleiche)*
h. sweetening Hypochloritraffination *f*
hypochlorous acid hypochlorige (unterchlorige) Säure *f*
hypocycle drive Umlaufrädergetriebe *n*
hypocycloid Hypozykloide *f*
hypocycloidal motion hypozyklische Bewegung *f*
hypodermic product Subkutanpräparat *n*
h. syringe Injektionsspritze *f*
hypoeutectic untereutektisch
hypoid gear Hypoidgetriebe *n*, achsversetztes Getriebe *n*; Hypoidkegelrad *n*
h. gearing Hypoidgetriebe *n*
h. oil Hypoidöl *n* *(Hochdruckschmiermittel)*
hypoiodous acid hypojodige (unterjodige) Säure *f*
hyponitrite Hyponitrit *n*
hyponitrous acid hyposalpetrige (untersalpetrige) Säure *f*
hypophosphate Hypophosphat *n*, Diphosphat *n*
hypophosphite Hypophosphit *n*
hypophosphoric acid Hypophosphorsäure *f*, Unterphosphorsäure *f*
hypophosphorous acid hypophosphorige (unterphosphorige) Säure *f*
hyposulphate Dithionat *n*
hyposulphite Dithionit *n*
hyposulphuric acid Dithionsäure *f*
hyposulphurous acid dithionige Säure *f*
hyposynchronous untersynchron
hypotenuse Hypotenuse *f*
hypothesis Hypothese *f*; Annahme *f*; Satz *m*
hypothetical reference circuit fiktiver Bezugskreis *m* *(Sendetechnik)*
hypsograph Pegelschreiber *m*
hypsometer Pegelmesser *m*, Pegelzeiger *m*
hysteresis Hysterese *f*, Hysteresis *f*; Satz *m*
h. coefficient Hysteresekoeffizient *m*, Hysteresefaktor *m*, Hysterese[verlust]zahl *f*
h. curve Hysteresekurve *f*, Hystereseschleife *f*, Hysteresiskurve *f*, Hysteresisschleife *f*
h. cycle *s.* h. curve
h. factor Hysteresebeiwert *m*, Hysteresefaktor *m*
h. loop[ing] *s.* h. curve
h. loss Hystereseverlust *m*
h. loss coefficient Hysteresekoeffizient *m*, Hysteresefaktor *m*, Hysterese[verlust]zahl *f*
h. meter Hysteresemesser *m*
hysteretic constant Hysteresiskonstante *f*, Steinmetzkonstante *f*
h. drag Hysteresismoment *n*
h. loss Hysteresisverlust *m*, Ummagnetisierungsverlust *m*

I

I-aerial I-Antenne *f*
I-beam Doppel-T-Träger *m*, I-Träger *m*
I-bracket I-Stütze *f*
I-display I-Darstellung *f* *(Radar)*
I-girder Doppel-T-Träger *m*, I-Träger *m*
I-groove weld I-Naht *f*
I-head engine *(Kfz)* Motor *m* mit hängenden Ventilen
I-pin I-Stütze *f*
I-section Doppel-T-Profil *n*, Doppel-T-Querschnitt *m*, I-Profil *n*, I-Querschnitt *m*
i-type semiconductor Eigenhalbleiter *m*, i-Halbleiter *m*
iatrochemistry Iatrochemie *f* *(historisch)*
iatron storage tube Iatronspeicherröhre *f*
IC engine *s.* internal combustion engine
ice Eis *n*
i. accretion Eisansatz *m*
i. adze Eispickel *m*
i. age Eiszeit *f*
i. anchor Eisanker *m*
i.-apron Eisbrecher *m* *(an einer Brücke)*
i. axe Eispickel *m*
i. boat Eisboot *n*; Segelschlitten *m*
i.-breaker Eisbrecher *m* *(Schiff und an einer Brücke)*
i.-breaker stem Eisbrechersteven *m*
i.-breaking power Eisbrecherleistung *f*
i. calorimeter Eiskalorimeter *n*
i. can Eiszelle *f*
i. cave Eistor *n*
i. coating Eisbelag *m*, Eisüberzug *m*
i. colour Eisfarbe *f*
i. concrete Eisbeton *m*
i.-crushing machine Eiszerkleinerungsmaschine *f*
i. crystal Eiskristall *m*
i.-cutter *(Schiff)* Eissporn *m*, Eisschneider *m* *(Schutz des Ruders bei Rückwärtsfahrt)*
i.-doubling *(Schiff)* Eishaut *f*, Eisdopplung *f*, Eisbug *m*, Bugverstärkung *f* *(gegen Eispressung)*
i. drag *(Schiff)* Eisdraggen *m* *(Eisanker)*
i. drill Eistrift *f*
i. dye *(Ch)* Eisfarbe *f*, Eisfarbstoff *m*
i.-eliminating system Enteisungsanlage *f*
i. field Eisfeld *n*
i. fin *(Schiff)* Eisabweiser *m* *(Propellerschutz)*
i. floe Eisscholle *f*
i. formation Eisbildung *f*
i.-freeing device Enteisungsanlage *f*
i. guard Enteiser *m*; Eisbrecher *m*
i. hockey stick Eishockeyschläger *m*
i. indicator Vereisungsanzeiger *m*
i.-lining *(Schiff)* Eishaut *f*, Eisdopplung *f*, Eisbug *m*, Bugverstärkung *f* *(gegen Eispressung)*

i. **machine** Eiserzeugungsmaschine *f*
i. **making** Kunsteisherstellung *f*
i. **pack** Eisfeld *n*
i. **patron vessel** Eiswachschiff *n*
i. **pick** Eispickel *m*
i. **piton** Bergsteigereishaken *m*
i. **point** Eispunkt *m*
i. **pressure** Eisdruck *m*
i. **spade** Eissporn *m (am Geschütz)*
i. **spar** *(Min)* Eisspat *m*
i. **strengthening** Eisverstärkung *f*
i. **wall** *(Brg)* Frostkörper *m*, Frostwand *f (Gefrierverfahren)*
i. **warning unit** Vereisungsanzeiger *m*
i.**-yachting** Eissegeln *n*
iceberg Eisberg *m*
icebound vessel eingefrorenes (eingeeistes) Schiff *n*, Schiff in Eisnot
icebreaker *(Schiff)* Eisbrecher *m*
Iceland moss Isländisches Moos *n*
Iceland spar isländischer Doppelspat *m*
icer Fruchteiserzeuger *m*
icewater Eiswasser *n*
ichnograph Grundriß *m*
ichthyocol Fischleim *m*
icicle shaped eiszapfenförmig
icing Vereisung *f*
i. **sugar** Puderzucker *m*, Staubzucker *m*
iconoscope *(Fs)* Ikonoskop *n*, Speicherröhre *f*
icosahedral zwanzigflächig
icosahedron Ikosaeder *n*, Zwanzigflächner *m*
icositetrahedron *(Math)* Ikositetraeder *n*, Vierundzwanzigflächner *m*
ideal ideal, vollkommen; theoretisch
i. **condition** Idealbedingung *f*
i. **crystal** Idealkristall *m*
i. **cycle** idealer Kreisprozeß *m*
i. **dielectric** ideales Dielektrikum *n*, absoluter (vollständiger) Nichtleiter *m*
i. **efficiency** idealer Wirkungsgrad *m*
i. **gas** ideales Gas *n*
i. **modulation** ideale Modulation *f*, IM
i. **no-load d.c. voltage** ideelle Gleichspannung *f (Stromrichter)*
i. **separation factor** *(Kern)* theoretischer Trennfaktor *m*
i. **simple process factor** *(Kern)* elementarer Trennfaktor *m* einer Stufe
i. **solution** *(Ph)* ideale Lösung *f*
i. **value** *s.* desired value
idealized characteristic idealisierte Kennlinie *f*
ideally imperfect ideal fehlgeordnet *(z. B. Kristalle)*
idempotent idempotent *(Algorithmus)*
identical *(Math)* identisch
identification 1. Identifizierung *f*; 2. *(Math)* Gleichsetzung *f*; 3. Kennung *f (Signal)*
i. **beacon** 1. Kennfeuer *n (Luftfahrt)*; 2. Kennbake *f (Radar)*
i. **friend or foe** Freund-Feind-Kennung *f*, Kennungsgerät *n* Freund oder Feind
i. **light** Kennfeuer *n (Luftfahrt)*
i. **marking** Kennung *f (Radartechnik)*
i. **plate** *(Kfz)* Typenschild *n*
identify / to identifizieren; bestimmen; nachweisen
identifying signal Kennungssignal *n*
identity *(Math)* Identität *f*; Einheitselement *n*; Einheitsoperator *m*
i. **matrix** Einheitsmatrix *f*
i. **measures** *(Schiff)* Vermessungsmaße *npl*
idioblastic *(Geol)* idioblastisch
idiochromatic idiochromatisch
idioelectric idioelektrisch
idiomorphic idiomorph
idiophanous idiophan *(Kristalle)*
idiostatic idiostatisch
i. **electrometer** idiostatisches Elektrometer *n*
i. **method** idiostatische Schaltung *f*
iditol *(Ch)* Iditol *n*
idle / to leerlaufen, mitlaufen
idle frei; untätig, in Ruhe befindlich, stillstehend
i. **adjusting screw** Leerlaufbegrenzungsschraube *f*
i. **air adjusting screw** Leerlaufluftschraube *f*
i. **capacity cost** Kostenaufwand *m* für ungenützte Kapazität
i. **component** *(El)* Blindkomponente *f*
i. **current** Blindstrom *m*, Leerlaufstrom *m*, wattloser Strom *m*
i.**-current connection** Ruhestromschaltung *f (Relais)*
i.**-current meter** Blindstrommesser *m*, Blindstromamperemeter *n*
i.**-current wattmeter** Blindleistungsmesser *m*
i. **gear** Leergang *m*, Leerlauf *m*
i. **indicating signal** *(Fmt)* Freilampe *f*
i. **jack** *(Fmt)* freie Klinke *f*
i. **jet** Leerlaufdüse *f*
i. **line** *(Fmt)* freie Leitung *f*
i. **machine time** Nebenzeit *f*; Verlustzeit *f*; Stillstandszeit *f*
i. **motion** Leerlauf *m*
i. **movement** toter Weg *m*
i. **period** 1. Griffzeit *f*, Nebenzeit *f*; Sperrzeit *f (Elektronik)*
i. **position** Ruhestellung *f (z. B. einer Maschine)*
i. **power** *(El)* Blindleistung *f*
i. **pulley** Leitrolle *f*
i. **return stroke** *(Masch)* Leerrücklauf *m*
i. **roll** Blindwalze *f*, Schleppwalze *f*
i.**-run of locomotive** Leerfahrt *f* der Lokomotive
i. **running** Leerlauf *m*

i. **running speed** Leerlaufdrehzahl *f*
i. **segment** Blindsegment *n (Dreschtrommel)*
i. **ship** Schiff *n* ohne Ladung (Auftrag)
i. **spindle** leerlaufende Spindel *f*
i. **station** *(Masch)* Leerstation *f*
i. **stroke** Leertakt *m (Verbrennungsmotor)*
i. **time** Nebenzeit *f*; Verlustzeit *f*; Stillstandszeit *f*
i. **travel** Überlauf *m* des Hobelmeißels *(Holzbearbeitung)*
i. **turn** tote (unbelegte) Windung *f (Wicklung)*
i. **watt-amperes** *(El)* Blindstrom *m*, Blindleistung *f*
i. **wheel** Zwischenrad *n*
idler Riemenspannrolle *f*, Spannrolle *f*; Laufrolle *f*; Leitrolle *f*; Umlenkrolle *f*; Zwischenrad *n*; angetriebenes Teil *n*
i. **arm** *(Kfz)* Lenkzwischenhebel *m*
i. **drum** Leerlauftrommel *f*
i. **gear** Zwischen[zahn]rad *n*
i. **impeller** angetriebene Spindel *f (Schraubenpumpe)*
i. **pulley** Spannrolle *f*; Umlenkrolle *f*; Stützrolle *f*
i. **shaft** *(Kfz)* Vorgelegeachse *f*
i. **sprocket** Spannkettenrad *n*, leerlaufendes Kettenrad *n*, Zwischenkettenrad *n*, Ausgleichszahnrad *n*
i. **wheel** Zwischenrad *n*
idling Leerlauf *m*
i. **flow** Leerlaufdurchsatz *m*
i. **jet** Leerlaufdüse *f*
i. **of a selector** *(Fmt)* Leerlauf (Nachlauf) *m* eines Wählers
i. **of locomotive** Leerfahrt *f* der Lokomotive
i. **speed** Leerlaufdrehzahl *f*
i. **time** Stillstandszeit *f*
idocrase *(Min)* Idokras *m*
idrialite *(Min)* Quecksilberbranderz *n*
IF-AND-ONLY-IF *(Dat)* WENN-UND-NUR-WENN, GENAU-DANN-WENN *(logischer Operator)*
IF-THEN *(Dat)* WENN-SO *(logischer Operator)*
igneous feurig; glühend; vulkanisch, eruptiv
i. **electrolysis** Schmelzflußelektrolyse *f*
i. **gneiss** *(Geol)* Orthogneis *m*
i. **rock** Ergußgestein *n*, Erstarrungsgestein *n*, Eruptivgestein *n*
i. **veins** *(Geol)* Eruptivgänge *mpl*
ignescent stone *(Min)* Feuerstein *m*
ignitability Entzündbarkeit *f*
ignitable zündfähig
ignite / to [ent]zünden, entflammen
igniter 1. Zünder *m*; Zündapparat *m*; Zündelektrode *f*; Zündmittel *n*; Schlagpatrone *f*; 2. Anlasser *m*
i. **chamber** *(Rak)* Vorbrennkammer *f*
i. **plug** Anlaßzündkerze *f (Turboantrieb)*
ignitibility Zündempfindlichkeit *f*
igniting Entzündung *f*; Zünd-
i. **anode** Erregeranode *f*; Zündanode *f*, Hilfsanode *f*
i. **apparatus** Zündapparat *m*
i. **current** Zündstrom *m*
i. **device** Zündeinrichtung *f*
i. **flame** Zündflamme *f*
i. **mixture** Zündmittel *n*, Zündgemisch *n*
i. **point** Zündpunkt *m*
i. **spark** Zündfunke *m*
ignition Initiierung *f*, Entzündung *f*, Zündung *f*; Verbrennung *f*
i. **accelerator** *(Kfz)* Zündbeschleuniger *m*
i. **advance** Vorzündung *f (Motor)*
i. **aid** *(Kfz)* Starthilfe *f*, Zündhilfsmittel *n*
i. **anode** Zündanode *f*, Erregeranode *f*, Hilfsanode *f*
i. **by incandescence** Glühzündung *f*
i. **by spark** Funkenzündung *f*
i. **cable** Zündkabel *n*
i. **cam** Unterbrechernocken *m*
i. **cartridge** Zündpatrone *f*
i. **characteristic** Zündkennlinie *f*
i. **circuit** Zündstromkreis *m*
i. **coil** Zündspule *f*
i. **control** Zündzeitverstellung *f*
i. **control additive** Kraftstoffzusatz *m* zur Verhinderung von Bleiablagerungen
i. **current** Zündstrom *m*
i. **cycle** Verbrennungshub *m*
i. **delay** Zündverzug *m*
i. **device** Zündvorrichtung *f*
i. **distributing wire** Zündkabel *n*
i. **distributor** Zündverteiler *m*, Verteiler *m*
i.-distributor shaft Zündverteilerwelle *f*
i. **electrode** Zündelektrode *f*
i. **equality** Zündgleichmäßigkeit *f*
i. **flame** Zündflamme *f*
i. **impuls** Zündimpuls *m*
i. **interference** Zündstörung *f*
i. **key** Zündschlüssel *m*
i. **lag** Zündverzug *m*
i. **lead** Zündleitung *f*
i. **lock** Zündschloß *n*
i. **magneto** Zündmagnet *m*
i. **noise** *(El)* Zündstörungen *fpl (Empfang)*
i. **order** Zündfolge *f*
i. **peak** Zündspitze *f*
i. **pellet** Zündkirsche *f*
i. **performance** Zündwilligkeit *f*
i. **period** Zünddauer *f*
i. **pin** Zündstift *m*
i. **plug** Zündkerze *f*
i. **point** Brennpunkt *m*, Flammpunkt *m*, Zünd[zeit]punkt *m*
i. **power** Zündfähigkeit *f*

i. properties Zündwilligkeit *f*
i. quality Zündeigenschaft *f*
i. range Zündbereich *m*, Zündgrenzen *fpl*
i. screening Funkabschirmung *f* an der Zündeinrichtung
i. shield Entstörkappe *f*
i. spark Zündfunke *m*
i. spring Zündfeder *f*
i. stroke Verbrennungshub *m*
i. switch Zünd[anlaß]schalter *m*
i. temperature Entzündungstemperatur *f*, Zündtemperatur *f*
i. test Glühprobe *f*, Zündprobe *f*
i. timing Zünd[punkt]verstellung *f*
i.-timing adjustment Zündzeitpunktverstellung *f*
i.-timing range Zündverstellbereich *m*
i. transformer Zündtransformator *m*
i. tube test *(Ch)* Glührohrprobe *f*
i. velocity Zündgeschwindigkeit *f*
i. voltage Zündspannung *f*
i. wire Zündkabel *n*
ignitor Zündstift *m*, Ignitor *m (Stromrichter)*
i. control Zündstiftsteuerung *f (Stromrichter)*
i. current Zündelektrodenstrom *m*
ignitron Ignitron *n*, Gleichrichter *m* mit Zündstift, Zündstiftröhre *f*
ignoble metal unedles Metall *n*
ignore *(Fs)* Leerzeichen *n*
ignored conductor freier Leiter *m*; Leiter mit schwebendem Potential
ikonophone Fernsprechanlage *f* mit Gegenfernsehen
Ilgner system *(El, Masch)* Ilgner-Antrieb *m*, Ilgner-Umformer *m*
ill-designed unzweckmäßig konstruiert
i.-fitting schlecht passend
i.-smelling übelriechend
illegal measure gesetzwidrige Maßnahme *f (Patentwesen)*
illium Nickellegierung *f*
illuminance Beleuchtungsstärke *f*
illuminant Beleuchtungsmittel *n*; Leuchtkörper *m*; *(Fs)* Weißpunkt *m*, Illuminant *m*
illuminate / to beleuchten
illuminated dial *(El)* Leuchtskala *f*
i. magnifier Leuchtlupe *f*
i. push button Leuchtdrucktaste *f*
illuminating Beleuchtung *f*
i. apparatus Beleuchtungsgerät *n*
i. cone Beleuchtungskegel *m*
i. gas Stadtgas *n*
i. path of rays Beleuchtungsstrahlengang *m*
i. pencil Lichtkegel *m*, Verfolgungsscheinwerfer *m*
i. power Lichtstärke *f*; Leuchtkraft *f*
i. rocket Leuchtrakete *f*
illumination 1. Ausleuchtung *f*; Beleuchtung *f*; Belichtung *f*; 2. Beleuchtungsstärke *f (in einem Punkte einer Fläche)*
i. darkening Saalverdunkelung *f*
i. engineer Beleuchtungsingenieur *m*; Lichttechniker *m*
i. engineering Beleuchtungstechnik *f*, Lichttechnik *f*
i. meter Beleuchtungsmesser *m*
i. of deck *(Schiff)* Decksbeleuchtung *f*
i. of the image *(Fs)* Bildhelligkeit *f*
i. photometer Beleuchtungsmesser *m*
illuminometer Beleuchtungsmesser *m*
illustration Abbildung *f*; Ansicht *f*
image / to abbilden
image *(optisch entworfenes)* Bild *n*, Abbild[ung *f*] *n*
i. aerial Scheinantenne *f*, imaginäre Antenne *f*
i. amplifier Bildverstärker[einrichtung *f*] *m*
i. area Bildausschnitt *m*, Bildfläche *f*
i. attenuation Dämpfungsmaß *n*
i. attenuation factor *(Am)* Vierpoldämpfungsfaktor *m*
i. brightness Bildhelligkeit *f*
i. carrier *(Fs)* Bildträger *m*
i. clearness *(Fs)* Bildschärfe *f*
i. contraction *(Fs)* Bildschrumpfung *f*
i. control Bildkontrolle *f*
i. converter Bildwandler *m*
i. converter tube Bildwandlerröhre *f*
i. defect Abbildungsfehler *m*
i. definition Bildschärfe *f*
i. dissector *(Fs)* Bildzerleger *m*, Dissektorröhre *f*, Bildsondenröhre *f*
i. dissector tube Bildzerlegerröhre *f*, Bildwandlerröhre *f*
i. distance Bildweite *f*
i. distortion Bildverzerrung *f*
i. drift *(Fs)* Bildschaukeln *n*
i.-erecting prism Umkehrprisma *n*
i.-erecting set Umkehrsatz *m*
i.-erecting tube bildaufrichtender Tubus *m*
i. fault Abbildungsfehler *m*
i.-flattening device Bildebnungseinrichtung *f*
i. flyback *(Fs)* Bildrücklauf *m*
i. formation Bildentstehung *f*
i.-forming system abbildendes System *n*
i. frequency Bild[wechsel]frequenz *f*, Bildzahl *f* je Sekunde; Spiegelfrequenz *f (Superhet)*
i.-frequency rejection ratio Spiegelfrequenzverwerfungsgrad *m*
i. height *(Fs)* Bildhöhe *f*
i. iconoscope Image-Ikonoskop *n*, Superikonoskop *n*, Zwischenbildikonoskop *n*
i. impedance stoßfreier Abschlußwiderstand *m*, Kennwiderstand *m (Vierpoltheorie)*
i. intensification Bildverstärkung *f*
i. intensifier Bildverstärker *m*
i. intercept Bildweite *f*
i. interference Videofrequenzstörung *f*

i. intermediate frequency Bildzwischenfrequenz *f*
i. inversion Bildumkehr *f*
i. line *(Fs)* Bildzeile *f*
i. output *(Fs)* Bildzeitbasis *f*
i. pattern Ladungsbild *n (elektrische Abbildung einer Helligkeitsverteilung mittels Elementarkondensatoren)*
i. phase-change coefficient Vierpolwinkelmaß *n*
i. phase factor Vierpolphasenfaktor *m*
i. plane Bildebene *f*
i. plate Bildelektrode *f*
i. point Bildpunkt *m*
i. power amplifier *(Fs)* Bildendverstärker *m*, Leistungsverstärker *m*
i. ratio Spiegelfrequenzsicherheit *f* der Fernsehbildfrequenz, Spiegelverhältnis *n*, Verhältnis *n* von Spiegelfrequenz zu Nutzfrequenz
i. reception Bildempfang *m*
i. rejection *(Fs)* Spiegelfrequenzunterdrückung *f*, Unterdrückung *f* der Bildträgerfrequenz
i. reproduction Bildwiedergabe *f*
i. response *(Fs)* Spiegelfrequenzempfang *m*, Spiegelfrequenzempfindlichkeit *f*; Bildwiedergabe *f*
i. retention Bildkonservierung *f*
i. reversing prism Umkehrprisma *n*
i. reversion Bildumkehrung *f*
i. sector Bildausschnitt *m*
i. sharpness Bildschärfe *f*
i. shift Bildverschiebung *f*
i. signal frequency Bildfrequenz *f*
i. space Bildraum *m*
i. storing tube Bildspeicherröhre *f*
i. suppression *(Fs)* Spiegelfrequenzunterdrückung *f*
i. sweep frequency Bildkippfrequenz *f*
i. transfer coefficient Vierpolübertragungsmaß *n*, Wellenübertragungsmaß *n*
i. transfer constant Vierpolübertragungsmaß *n*, Wellenübertragungsmaß *n*
i. transmission Bildübertragung *f*
i. tube Bild[wiedergabe]röhre *f*
i. tumbling *(Opt)* Bildsturz *m*
i.-viewing tube Bildwandlerröhre *f*
i. voltage Bildspannung *f*
imaginary imaginär, gedacht, ideell
i. axis Nebenachse *f (Parabel)*
i. field gedachtes (fiktives) Feld *n*
i. number imaginäre Zahl *f*
i. part Imaginärteil *m*
i. unit imaginäre Einheit *f*
imagination Abbildung *f*
imaging defect *(Opt)* Abbildungsfehler *m*
i. optics Abbildungsoptik *f*
i. path of rays Abbildungsstrahlengang *m*
imbalance Unwucht *f*
imbibe / to quellen
imbibition Quellung *f*; Imprägnierung *f*
imbricated *(Bw)* schuppenartig, übereinandergreifend, überlappend
i. winding *(Elm)* Faßwicklung *f*, Korbwicklung *f*
Imhoff tank Imhoff-Brunnen *m*
imide *(Ch)* Imid *n*
imitate / to nachahmen, nachbilden, imitieren
i. the grain of wood Holzmaserung nachahmen
imitated line *(El)* Kabelnachbildung *f*, Leitungsnachbildung *f*
imitation Imitation *f*, Nachahmung *f*
i. fur *(Text)* künstlicher Pelz *m*
i. leather Kunstleder *n*, Lederimitation *f*
i. material insulated wire kunststoffisolierter Draht *m*
i. nap yarn falsches Noppengarn *n*
i. yarn Imitatgarn *n*
immaterial stofffrei
immeasurable unmeßbar; unermeßlich
immediate access store Speicher *m* mit unmittelbarem Zugriff *(Datenspeicherung)*
i. appreciation percentage *(Fmt)* momentaner Verständlichkeitsgrad *m*
i.-bearing prop *(Brg)* soforttragender Stempel *m*
i. ringing *(Fmt)* direkter Ruf *m*
i. roof *(Brg)* unmittelbares Hangendes *n*, Dachschichten *fpl*
immerse / to [ein]tauchen, tränken
immersed area of midship section *(Schiff)* eingetauchte Hauptspantfläche *f*
i. rheostat Widerstand *m* im Wasserbad, Tauchwiderstand *m*, Wasserwiderstand *m*
i. surface *(Schiff)* benetzte Oberfläche *f*
immersion Eintauchen *n*, Untertauchen *n*, Versenken *n*
i. cap Eintauchkappe *f*
i. heater [elektrischer] Tauchsieder *m*
i. milk cooler Milchtauchkühler *m*
i. objective Immersionsobjektiv *n*
i. oil Immersionsöl *n*
i. plating Eintauchplattierung *f*, Ansieden *n (Galvanotechnik)*
i. pump Tauchpumpe *f*
i. roller *(Text)* Tauchwalze *f*
i. scale *(Am) (Schiff)* Tiefgangsskala *f*
i. soldering bath Tauchlötbad *n*
i. vibrator *(Bw)* Flaschenrüttler *m*, Innenrüttler *m*, Tauchrüttler *m*
immiscibility Nichtmischbarkeit *f*
immiscible nicht mischbar
immovable unbewegbar
immune to unempfindlich gegen
i. to shock stoßunempfindlich
i. to vibration erschütterungsunempfindlich

immunity Immunität *f*, Widerstandsfähigkeit *f*
i. from disturbance Störfestigkeit *f*
i. to corrosion Korrosionsbeständigkeit *f*
immunization Passivierung *f*
immure / to einmauern
impact / to aufschlagen, aufprallen, auftreffen
i. extrude spritzen *(Tiefziehverfahren)*
i. mould schlagpressen
impact Stoß *m*; Anstoß *m*; Schlag *m*; Zusammenstoß *m*, Berührung *f*
i. acceleration process Anstoßprozeß *m*
i. ball hardness Fallhärte *f* *(Kugelfallversuch)*
i. bending strength Schlagbiegefestigkeit *f*, Schlagbiegezähigkeit *f*
i. bending test Schlagbiegeversuch *m*
i. broadening Stoßverbreiterung *f*
i. buckling test Schlagknickversuch *m*
i. coefficient Stoßzahl *f*
i. crusher Schlagbrecher *m*
i. crushing Stoßzerkleinerung *f*
i.-die forging Gesenkschmieden *n*
i. effect Stoßwirkung *f*, Schlagwirkung *f*
i. electron Stoßelektron *n*
i. emission *(El)* Sekundäremission *f*
i. excitation Stoßerregung *f*
i. extrusion Schlagfließpressen *n*
i. extrusion press Fließdruckpresse *f*
i. fatigue limit Dauerschlagfestigkeit *f*
i. fatigue test Dauerschlagversuch *m*
i. fatigue testing machine Dauerschlagprüfmaschine *f*; Dauerprüfschlagwerk *n*
i. force Stoßkraft *f* *(Festigkeit)*
i. fuse Aufschlagzünder *m*, Kontaktzünder *m*
i. hardness Schlaghärte *f* *(Festigkeit)*
i. hardness test Schlaghärteprüfung *f*
i. hardness tester Schlaghärteprüfer *m*
i. ionization Stoßionisation *f*
i. load Stoßbelastung *f*, Stoßbeanspruchung *f*
i. machine Pendelschlagwerk *n*, Schlagwerk *n*
i. mill Prallmühle *f*
i. mixer Turbomischer *m*
i. moulding Schlagpressen *n*
i. neutron Anstoßneutron *n*
i. noise Trittschall *m*; Körperschall *m*
i. of radiation Einfall *m* von Strahlung
i. oscillator Unterwasserschallsender *m*
i. parameter Stoßparameter *m*
i.-proof schlagfest
i. properties Zähigkeitseigenschaften *fpl*
i. pulling test Schlagzerreißversuch *m*
i. pulverizer Stoßmühle *f*, Schlagmühle *f*
i. radiation Stoßstrahlung *f*
i. resistance Schlagfestigkeit *f*, Stoßfestigkeit *f*
i. riveting Hammernieten *n*
i. screen Vibrationssieb *n*; Prallsieb *n*
i. sound Trittschall *m*; Körperschall *m*
i. spring Schlagfeder *f*
i. strength Kerb[schlag]festigkeit *f*, Schlagfestigkeit *f*; Stoßfestigkeit *f*
i. stress Schlagbeanspruchung *f*, Stoßbeanspruchung *f*
i. styrene materials stoßfeste Styrol-Butadien-Mischungen *fpl*
i. tearing test Schlagzerreißversuch *m*
i. tensile stress Schlag-Zug-Beanspruchung *f*
i. test Kerbschlagversuch *m*, Schlagversuch *m*; Stoßversuch *m*
i. testing machine Pendelschlagwerk *n*; Pendelschlagmaschine *f*
i.-torsion test Schlagdrehversuch *m*
i. tube *(Ph)* Staurohr *n*, Pitotrohr *n*
i. value Kerb[schlag]zähigkeit *f*
i. velocity Auftreffgeschwindigkeit *f*; Stoßgeschwindigkeit *f*
i. wave Stoßwelle *f*, Kopfwelle *f*
i. wheel mixer *(Ch)* Turbomischer *m*
impacting Beaufschlagung *f* *(Elektronen)*
impactor crusher *(Brg)* Schlagbrecher *m*
impacts of micrometeorites Mikrometeoritenstöße *mpl*
impair / to beeinträchtigen, beschädigen
impart / to erteilen, ausüben
i. pressure to unter Druck setzen, Druck ausüben auf
impeach / to bemängeln, beschuldigen, tadeln
i. the trust-worthiness of a witness die Glaubwürdigkeit eines Zeugen anfechten
impedance *(El)* Scheinwiderstand *m*, Impedanz *f*
i. angle Phasenwinkel *m*
i. bond 1. Drosselverbinder *m*; verdrosselter Anschluß *m*; 2. Drosselstoß *m* *(bei Isolierschienen)*
i. bridge Impedanzmeßbrücke *f*, Scheinwiderstands[meß]brücke *f*
i. characteristic Impedanzkurve *f*
i. coil Drossel[spule] *f*
i. compensator Entzerrer *m*, Impedanzkorrektor *m*, Scheinwiderstandsangleicher *m*
i. converter Impedanzwandler *m*
i. corrector Entzerrer *m*, Impedanzkorrektor *m*
i. coupled drosselgekoppelt
i. coupled amplifier Drosselverstärker *m*
i. factor Verhältnis *n* des Scheinwiderstandes zum Wirkwiderstand einer Schaltung
i. losses stromabhängige Verluste *mpl*
i. match Anpassungswiderstand *m*
i. matching Impedanzanpassung *f*, Scheinwiderstandsanpassung *f*
i. matching condition Anpassungsverhältnis *n*
i. measuring unit Scheinwiderstandsmesser *m*
i. network Scheinwiderstandskettenleiter *m*
i. protective system Distanzschutz *m*, Impedanzschutz *m*

i. relay Distanzrelais *n*
i. transformation Widerstandstransformation *f*
i. voltage Gesamtspannungsverluste *mpl*, Gesamtspannungsabfall *m*
impede / to arretieren, bremsen, festhalten
i. the taking of bearings das Peilen verhindern
impeder *(El)* Begrenzungsdrossel *f*, Drossel *f*
impeller 1. Flügelrad *n*, Gebläserad *n*, Kreiselrad *n*; Leitrad *n*; Schaufelrad *n*; Windrad *n*; Impeller *m*; Meßkreisel *m*; Lüfterrad *n*; Schleuderrad *n*; 2. Kreiselmischer *m*, Wirbelmischer *m*, Schnellrührer *m*; 3. Antriebsrad *n*; Förderkörper *m*; Verdränger *m (Pumpe)*
i. blade Laufradflügel *m*, Windflügel *m*
i. breaker Schneidmühle *f*
i. channel Laufschaufelkanal *m*
i. gear treibendes Rad *n*
i. of pump Pumpenflügelrad *n*
i. shaft Laufradwelle *f*
i.-type log *(Schiff)* Propellerlog *n*
i. wheel grubber *(Lw)* Schleuderradroder *m*
impelling force Antriebskraft *f*
impending plastic flow unmittelbar bevorstehendes plastisches Fließen *n*
impenetrable paint Isolierfarbe *f*
imperfect unvollkommen, unzulänglich, minderwertig, unzureichend, unvollständig
i. combustion unvollkommene Verbrennung *f*
i. contact schlechter Kontakt *m*
i. contact joint Wackelkontakt *m*
i. crystal realer Kristall *m*, Realkristall *m*
i. dielectric nichtverlustfreies (unvollkommenes) Dielektrikum *n*, fehlerhafter (unvollkommener) Isolierstoff *m*
i. gas reales Gas *n*
i. magnetic circuit nichtgeschlossener magnetischer Kreis *m*
i. resonance unvollkommene Resonanz *f*
i. sheet *(Typ)* Fehlbogen *m*
i. tuning *(Fmt)* unscharfe Abstimmung *f*
imperfection Fehler *m*, Mangel *m*; Fehlstelle *f*, Fehlordnung *f (im Kristallgitter)*, Störstelle *f*
i. of the scale Maßstabfehler *m*
i. of workmanship Herstellungsungenauigkeiten *fpl*
imperforate nichtperforiert
impermeability Undurchlässigkeit *f*
impermeable undurchdringlich, undurchlässig
impervious undurchdringlich; dicht *(gegen Gas oder Wasser)*
i. to heat wärmeundurchlässig
i. to X-rays röntgenstrahlenundurchlässig
imperviousness Undurchlässigkeit *f*
impetus Impuls *m*; Anstoß *m*
impinge / to [auf]prallen, aufschlagen, [an]stoßen, zusammenstoßen, treffen
impingement attack Tropfenschlagkavitation *f*
implacement Eindringung *f*
implant *(Kern)* Einlage *f*, Implantation *f*
implement 1. Möbelstück *n*; Kleidungsstück *n*; 2. Instrument *n*, Werkzeug *n*, Gerät *n*
i. beam *(Lw)* Anhängebalken *m*
i. carrier *(Lw)* Geräteträger *m*
i. carrying tractor *(Lw)* Geräteträger *m*
implements for tillage and seedbed preparation *(Lw)* Bodenbearbeitungsgeräte *npl*
implication Implikation *f (logischer Operator)*
implicit *(Math)* implizit, nicht explizit
i. function implizite Funktion *f*
implode / to implodieren *(Bildröhre)*
implosion Einsturz *m*, Implosion *f*
i. safety factor *(Fs)* Implosionsfestigkeit *f*
imponderable fluid schwerelose Flüssigkeit *f (Äthertheorie)*
import / to importieren, einführen
import Import *m*, Einfuhr *f*
i. duty Einfuhrzoll *m*
i. steamer *(Schiff)* Eingangsdampfer *m*
importance function *(Math)* Einflußfunktion *f*
imported dirt *(Brg)* Fremdberge *pl*
i. stowing *(Brg)* Fremdversatz *m*
i. wood ausländisches Holz *n*, Importholz *n*
impose / to 1. auferlegen, besteuern; beeindrucken; 2. *(Typ)* ausschießen
i. on aufbringen auf
i. stresses (a load) on beanspruchen
imposed regulation *(BMSR)* übergeordnete Regelung *f*
i. voltage *(El)* eingeprägte Spannung *f*
imposing surface *(Typ)* Metteurtisch *m*; Schließplatte *f*
impost Kämpfer[stein] *m (Bogen, Gewölbe)*
impounding Stau *m (Wasser)*
impreg *(Am)* mit Phenolharz imprägniertes Preßholz *n*
impregnate / to imprägnieren, sättigen, tränken
i. with tallow Talg einlassen
impregnated fabric getränktes Gewebe *n*; Lackgewebebahn *f*
i. insulating material imprägniertes Isoliermaterial *n*
i. paper getränktes Papier *n*; Lackpapier *n*
i. sheet reinforcement for laminates Lackgewebebahnen *fpl*, Lackpapierbahnen *fpl*
i. sinter-alloy Sintertränklegierung *f*, getränkter Sinterwerkstoff *m*
impregnating Holzimprägnierung *f*
i. boiler Imprägnierkessel *m*, Tränkkessel *m*
i. matter Imprägniermasse *f*, Tränkmasse *f*
i. solution Imprägnierlösung *f*, Tränklösung *f*
i. varnish Imprägnierlack *m*, Tränklack *m*
i. with resin Verharzung *f*
impregnation Imprägnierung *f*, Tränkung *f*
i. compound Imprägnierungsmasse *f*, Tränkmasse *f*

i. **liquid** Tränkflüssigkeit *f*
i. **time** Tränkungsdauer *f*
impress / to 1. einprägen; eindrücken; 2. *(El)* anlegen *(Spannung)*
impressed electromotive force *(El)* eingeprägte elektromotorische Kraft *f*
i. **voltage** eingeprägte (angelegte) Spannung *f*
impression Eindruck *m*; Kalotte[nfläche] *f*; Vertiefung *f*
i. **die** Preßgesenk *n*
i. **moulding** *(Plast)* Pressen *n* mit niedrigem Druck, Pressen ohne Druck
impressional force *(Typ)* Druckkraft *f*
impressor Eindringkörper *m (Härteprüfung)*
imprint 1. *(Typ)* Impressum *n*, Druckvermerk *m*; Aufdruck *m*; Eindruck *m*; 2. *(Met)* Eindruck *(Materialprüfung)*
improper fraction *(Math)* unechter Bruch *m*
i. **integral** uneigentliches Integral *n*
i. **navigation** fehlerhafte Navigation *f*
improve / to verbessern
improvement Verbesserung *f*
i. **cutting** Durchhieb *m*, Durchlichtung *f* *(Forst)*
i. **in quality** Qualitätsverbesserung *f*
i. **patent** Verbesserungspatent *n*
i. **suggestion** Verbesserungsvorschlag *m*
i. **threshold** *(El)* Gewinn *m* an Rauschabstand
improving Vorraffinieren *n (Blei)*
i. **furnace** Vorraffinierofen *m*
impulse Stoß *m*, Impuls *m*; Antrieb *m*; Stromstoß *m*; Stoßkraft *f*
i. **action** *(Fmt)* Nummernwahl *f*
i. **amplification** Impulsverstärkung *f*
i. **amplitude** Impulshöhe *f*
i. **breakdown** Stoßdurchschlag *m*
i. **circuit** Stoß[strom]kreis *m*, Impuls[strom]kreis *m*
i. **clock** stromstoßgeschaltete Uhr *f*
i. **contact** Wischkontakt *m*, Impulskontakt *m* *(Relais)*
i. **control** Impulssteuerung *f*
i. **counter** Impulszähler *m*
i. **coupling** Schnapper *m*
i. **detector** Impulsdemodulator *m*
i. **distortion** *(El)* Impulsverformung *f*
i.**-driven clock** impulsgesteuerte Uhr *f*, Uhr mit Stromstoßfortsteuerung
i. **duration** Impulsdauer *f*, Impulslänge *f*
i. **excitation** Stoßerregung *f*
i. **flashover** Stoßüberschlag *m*
i. **flashover volt-time characteristic** Stoßspannungscharakteristik *f*
i. **function** Sprungfunktion *f*, Stoßfunktion *f*
i. **generator** Stoßgenerator *m*, Impulsgenerator *m*
i. **governed oscillator** impulssynchronisierter Oszillator *m*
i. **inertia** Überschlagverzögerung *f*, Durchschlagverzögerung *f*, Ansprechverzögerung *f* *(einer Funkenstrecke)*
i. **level** Stoßspannungsfestigkeit *f*; Stoß[spannungs]pegel *m*
i. **loading** dynamische Beanspruchung (Belastung) *f*
i. **machine** Impulsgeber *m*; Stromstoßgeber *m*
i. **magnetization** Impulsmagnetisierung *f*
i. **meter** Impulszähler *m*
i. **of recoil** Rückstoßimpuls *m*
i. **of the force** Impuls *m* der Kraft
i. **period** Impulsabstand *m*
i. **piston** Auslöskolben *m*
i. **reaction turbine** Aktions-Reaktionsturbine *f*
i. **regenerator** Impulsentzerrer *m*
i. **relay** Stromstoßrelais *n*, Impulsrelais *n*
i. **repeater** Impulsübertrager *m*
i. **response** 1. Stoßverhalten *n*, Verhalten *n* gegen Stoßlast (Stoßspannung); 2. Impulsantwort *f*, Impulsübergangsfunktion *f*
i. **sender** Impulsgeber *m*
i. **signal** Impulssignal *n*
i. **spark-over test** Stoßüberschlagsprüfung *f*
i. **start** Startstrom[stoß] *m*
i. **starter** *(Flg)* Abschnappkupplung *f*
i. **stepping** *(Fmt)* Nummernwahl *f*
i. **summator** Impulszähler *m*
i. **technique** Impulstechnik *f*
i. **telegraphy** Impulstelegrafie *f*
i. **train** Impulsfolge *f*
i. **transformer** Impulstransformator *m*, Impulsübertrager *m*
i. **turbine** Druckturbine *f*, Gleichdruckturbine *f*, Aktionsturbine *f*
i. **turbine with pressure stages** mehrstufige Druckturbine *f*, Druckturbine mit Druckabstufung
i. **turbine with velocity stages** Druckturbine *f* mit Geschwindigkeitsstufen
i.**-type blading** Gleichdruckbeschaufelung *f* *(Turbine)*
i.**-type telemeter** Impulsfernmeßgerät *n*
i. **voltage** Stoßspannung *f*
i. **wheel** Aktionsrad *n (Turbine)*
i. **width** Impulslänge *f*, Impulsdauer *f*
i. **withstanding voltage** Stehstoßspannung *f*
impulser Impulsgenerator *m*
impulsing Impulsgabe *f*
i. **circuit** Stoßkreis *m*, Impulsstromkreis *m*
impulsion *(Schiff)* Antrieb *m (Impuls)*
impulsive loading dynamische Beanspruchung (Belastung) *f*
i. **solar noise** Kurz[zeit]störung *f*; Mögel-Dellinger-Effekt *m*
impurification Verunreinigung *f*

impurity Fremdkörper *m*, Verunreinigung *f*; Begleitelement *n*
i. activation energy Aktivierungsenergie *f* infolge Verunreinigung *(Halbleitertechnik)*
i. concentration Fremdstoffkonzentration *f*
i. ion Fremdion *n*
i. photoconductivity Störstellenfotoleitfähigkeit *f*
i. semiconductor Störstellenhalbleiter *m*
i. spot Störstelle *f (Halbleitertechnik)*
in-and-out travel *(Masch)* Ein- und Ausfahren *n*
in bulk lose, als Schüttgut
in contact im Eingriff befindlich, kämmend; mit Berührung
in-cut method Verfahren *n* des Gleichlauffräsens
in dead centre genau mittig
in echelon *(Bw)* gestaffelt
in-feed *(Masch)* Zustellung *f*; Einstechbewegung *f*
in-feed grinding Einstechschleifen *n*
in-feed lever Zustellhebel *m*
in-feed movement Zustellbewegung *f*
in fine steps feinstufig
in gear in Eingriff, eingerückt *(Verzahnung)*
in geometrical progression geometrisch gestuft
in good balance unwuchtfrei
in layers geschichtet
in line fluchtend
in-line automatic transfer machine Taktstraße *f* als Geradförderer
in-line data processing schritthaltende Datenverarbeitung *f*
in-line drilling Reihenbohren *n*
in-line engine Reihenmotor *m*
in-line machine Geradförderer *m (Taktstraße)*
in-line multi Reihenbohrmaschine *f*
in-line multibank engine Reihenmotor *m* mit mehreren Zylinderreihen
in-line multiple spindle drilling machine Reihenbohrmaschine *f*
in-line system *(Brg)* geradlinige Anordnung *f (Ausbau)*
in-line transfer line for drilling, reaming, tapping Bohrstraße *f*
in-line transfer machine Taktstraße *f (Geradeausförderer)*
in longitudinal direction in Längsrichtung
in one operation in einem Arbeitsgang
in one setting of the work in einer Werkstückaufspannung
in opposite directions gegenläufig
in opposition *(El)* in Gegenphase
in-out connection *(Büro)* Steckverbindung *f*
in pairs paarweise
in-phase *(El)* gleichphasig (*s. a.* inphase)
in-pile innerhalb (im Innern) des Reaktors
in-pile test Bestrahlungsversuch *m* innerhalb des Reaktors
in place in situ
in-plant *(Am)* innerbetrieblich
in-plant handling innerbetrieblicher Transport *m*
in-process material Zwischenprodukt *n*
in-process work storage *(Masch)* Vorratsbildung *f* im Zwischenlager verketteter Maschinen
in quadrature um 90° phasenverschoben
in-reactor *s.* in-pile
in series *(El)* hintereinander, in Reihe, in Serie
in service in Betrieb
in-situ am Ort, an Ort und Stelle
in-situ concrete monolithischer Beton *m*, Ortbeton *m*
in stock auf Vorrat, vorrätig
in tabular form in Tabellenform
in tandem hintereinander
in the absence of forces kräftefrei, in kräftefreiem Zustand
in-the-mill drying *(Brg)* Mahltrocknung *f*
in the same direction gleichsinnig
in vacuo *(Opt)* im Vakuum
inaccessible unzugänglich
inaccuracy Ungenauigkeit *f*
i. of lead Steigungsfehler *m (Leitspindel)*
inaccurate ungenau
i. tuning unsaubere Abstimmung *f (z. B. Radio)*
inactivate / to inaktivieren, unwirksam machen
inactivated inaktiviert
inactive träge, inaktiv, wirkungslos
i. neutral *(BMSR)* Unempfindlichkeitsbereich *m*, tote Zone *f*
i. vessel Schiff *n* außer Betrieb
inactivity Passivität *f*, Trägheit *f*
inadequate unzureichend, unzulänglich
inarticulate *(Fmt)* undeutlich *(Sprache)*
inaudibility Unhörbarkeit *f*
inaudible unhörbar
inboard *(Schiff)* binnenbords, innenbords
i. cargo Raumladung *f (im Gegensatz zu Decksladung)*
i. elevation Längsschnitt *m*
i. motor Innenbordmotor *m*
i. profile Längsschnitt *m*, Längsansicht *f*
i. profile plan Längsschnitt *m*, Längsansicht *f*
inbound bearing Anflugpeilung *f*, Zielpeilung *f*
i. freight *(Schiff)* Rückfracht *f*
inbreeding Inzucht *f*
inbuilt eingebaut
inbye in Richtung der Grubenbaue, feldwärts
incandesce / to weißglühend machen; weißglühend werden
incandescence Weißglut *f*; Weißglühen *n*
incandescent [weiß]glühend; auf Weißglut erhitzt
i. bulb Glühlampe *f*

i. cathode Glühkatode *f*
i. electric lamp Glühlampe *f*
i. gas mantle Glühkörper *m*
i. heat Weißglühhitze *f*
i. lamp Glühlampe *f*
i. lighting Glühlampenbeleuchtung *f*
i. mantle Glühkörper *m*
incapsulate / to einkapseln
incasement Einschluß *m*
incendiary bomb Brandbombe *f*
i. shell Brandgranate *f*
incentre Inkreismittelpunkt *m (beim Dreieck)*
incentive bonus Leistungszulage *f*
inch Zoll *m (25,4 mm)*
i. mast *(Schiff)* Mastholz *n*
i.-measure screw thread Zollgewinde *n*
i. of water *(englische Druckeinheit; 1 in. water = 25,4 mm WS)*
i.-ounce *(englische Einheit der mechanischen Energie; 1 in.oz. = 7,201 pcm)*
i.-pound *(englische Einheit der mechanischen Energie; 1 in.lb. = 1,15 · 10^{-2} kpm)*
i. rule Zollmaß *n*, Zollstock *m*
inching 1. *(El)* Tastbetrieb *m*, Tippbetrieb *m*; 2. *(Masch)* Feinverschiebung *f*; Langsambewegung *f*; Zentimetereinstellung *f*
i. button Knopf *m* für Feinverstellung
i. control Langsamsteuerung *f*; Taststeuerung *f*, Schrittsteuerung *f*
i. starter Kriechanlasser *m*, Anlasser *m* mit Tippbetrieb
i. travel Langsamgang *m*
incidence 1. *(Rak)* Anstellwinkel *m*; 2. Einfallen *n (eines Strahles)*
i. angle *(Rak)* Anstellwinkel *m*
i. of faults *(Büro)* Störanfälligkeit *f*, Auftreten *n* von Störungen
i. point Auftreffpunkt *m*
i. wire Diagonaldraht *m*, Verspannung *f*
incident einfallend, auftreffend
i. direction Einfallsrichtung *f*
i. light einfallendes (auftreffendes) Licht *n*; Auflicht *n*
i.-light condenser Auflichtkondensor *m*
i.-light microscope Auflichtmikroskop *n*
i. primary electron auftreffendes Primärelektron *n*
i. ray einfallender Strahl *m*
i. wave Einfallwelle *f*, einfallende Welle *f*; einlaufende (ankommende) Welle
incidentals Kleinmaterial *n*, Zubehör *n*
incinerate / to veraschen; kalzinieren
incineration Veraschung *f*; Müllverbrennung *f* *(Gesundheitstechnik)*
i. dish *(Ch)* Veraschungsschale *f*
incinerator Müllbrenner *m*, Müllverbrennungsanlage *f (Gesundheitstechnik)*
incipient combustion träge Verbrennung *f*
i. crack Anriß *m*
i. decay *(Holz)* Anfangsstadium *n* der Fäulnis
i. knocking Klopfgrenze *f (Erdöl)*
i. plastic flow beginnendes plastisches Fließen *n*
incircle *(Math)* Inkreis *m*, eingeschriebener Kreis *m*
incise / to schnitzen; gravieren
inclinable neigbar, schrägstellbar
i. press neigbare Presse *f*
inclination Gradiente *f*, Neigung *f*, Steigung *f*; Hanglage *f*; Tonnlage *f*; Inklination *f (Magnetnadel)*
i. balance Neigungswaage *f*
i. of grate Rostneigung *f*
i. of orbit Neigung *f* der Bahnebene *(gegen die Ekliptik)*
i. of standing ways *(Schiff)* Ablaufbahnneigung *f*, Stapellaufbahnneigung *f*
i. of the cutting edge *(Masch)* Neigungswinkel *m*
i. of the disk Schrägstellung *f* der Scheibe *(Turbine)*
i. to crack Rißneigung *f*, Rißanfälligkeit *f*
inclinator *(Kfz)* Rückenlehnenverstellvorrichtung *f (Fahrersitz)*
inclinatorium Inklinationsnadel *f*; Inklinatorium *n*
inclinatory needle Inklinationsnadel *f*
incline / to [sich] neigen; einfallen *(Strecke)*
incline 1. *(Bw)* Gradiente *f*, Neigung *f*, Steigung *f*; schiefe Ebene *f*, Schrägebene *f*; 2. *(Brg)* flache (einfallende) Strecke *f*, Berg *m*, Flaches *n*, tonnlägiger Schacht *m*
i. attendant *(Brg)* Bremsbergführer *m*, Haspelzieher *m*
i. shaft *(Brg)* tonnlägiger Schacht *m*, Schrägschacht *m*
inclined 1. einfallend, geneigt, schief; 2. *(Brg)* flach, geneigt, schräg, tonnlägig
i. ascent *(Rak)* Schrägschuß *m*
i. ball mill Schrägkugelmühle *f*
i. bar aufgebogener Stab *m (Stahlbeton)*; Schrägstab *m*
i. bed *(Brg)* einfallendes Flöz *n*
i. blade position schiefe Schaufelstellung *f* *(Turbine)*
i. bucket elevator Schrägbecherwerk *n*
i. carbon arc Lampe *f* mit schrägstehenden Kohlen
i. chamber oven Schrägkammerofen *m*
i. conveyor Schrägbecherwerk *n*, Schrägförderer *m*
i. cut-and-fill stope *(Brg)* Firstenstoß *m* mit schräggestellter Firste *(Firstenstoßbau mit Versatz)*
i. drawing rollers *(Text)* geneigtes Streckwerk *n*

i. furnace hoist Schrägaufzug *m*, schräger Gichtaufzug *m (Hochofen)*
i. grate Schrägrost *m*
i. grate bar Schrägroststab *m*
i. hoist Schrägaufzug *m*
i. joint schräge Teilfuge *f*, schräger Schlitz *m*
i. lift Schrägaufzug *m*, schräger Gichtaufzug *m (Hochofen)*
i. overhead contact line windschiefe Kettenfahrleitung *f*
i. piling *(Brg)* schräges Anstecken *n*
i. plane 1. schiefe (geneigte) Ebene, Schrägebene *f*; 2. *(Brg)* Bremsberg *m*
i. plane-[gravity]shunting *(Eis)* Bremsbergrangieren *n*
i. polygonal grate bar Schrägpolygonroststab *m*
i. raise *(Brg)* Schrägaufhauen *n*, Berg *m*
i. roller bearing Kegelrollenlager *n*, Schräglager *n*
i. screen Schrägsieb *n*
i. shaft *(Brg)* tonnlägiger Schacht *m*, Schrägschacht *m*
i. steam engine schrägliegende Dampfmaschine *f*
i. throw schiefer Wurf *m*
i. top-slicing *(Brg)* Querbruchbau *m* mit schräggestelltem Abbaustoß
i. track Schrägbahn *f (Hochenergiephysik)*
i. trough Zuführschrägrinne *f*
i. tube *(Opt)* Schrägtubus *m*
i. tube evaporator Schrägrohrverdampfer *m*
i. tube manometer Schrägrohrmanometer *n*
i. zonal bunker Zonenschrägbunker *m*
inclining experiment *(Schiff)* Krängungsversuch *m*
inclinometer 1. *(Schiff)* Krängungsmesser *m*, Neigungsmesser *m*; Inklinationsmesser *m*; 2. Inklinationsnadel *f*
inclose / to *(Fmt, Fs)* abschirmen, abschließen
include / to einschließen, enthalten
included angle eingeschlossener Winkel *m*; Kegelwinkel *m*; Flankenwinkel *m*; Spitzenwinkel *m*
i. angle of the drill tip Bohrerspitzenwinkel *m*
i. angle of the [mill] tooth Fräserwinkel *m*
i. cutting angle Bohrerspitzenwinkel *m*
i. draught fan Saugzuggebläse *n*
i. plan angle Spitzenwinkel *m*
i. spaces *(Schiff)* vermessene Räume *mpl (in die Bruttovermessung einbezogene Räume)*
i. tooth angle Keilwinkel *m*
inclusion Einschluß *m*, Fremdkörper *m (im Grundmetall)*
i. theorem *(Math)* Einschließungssatz *m*
inclusive OR einschließendes ODER *n (logischer Operator)*
incoherence *(Math)* Inkohärenz *f*
incoherent knot Durchfallast *m*, herausfallender Ast *m*
incombustibility Unverbrennbarkeit *f*
incombustible unverbrennbar
incombustibles nicht brennbare Ladung *f (Güter npl)*
incoming *(Fmt)* ankommend
i. call ankommendes Gespräch *n*
i. circuit ankommende Leitung (Verbindung) *f*, Eingangsleitung *f*
i. frequency Eingangsfrequenz *f*
i. junction line ankommende Amtsverbindung *f*
i. position Eingangsplatz *m*, Verbindungsplatz *m*
i. power Empfangsleistung *f*
i. selector Eingangswähler *m*
i. signal Empfangssignal *n*
i. trunk call ankommendes Ferngespräch *n*
incommensurable unmeßbar, unvergleichbar
i. number Primzahl *f*
incompatibility *(Ph)* Unverträglichkeit *f*; Unvereinbarkeit *f*
incomplete unvollkommen, unvollständig
i. circuit offener Stromkreis *m*
i. combustion unvollkommene Verbrennung *f*
i. convergence Teilkonvergenz *f*
i. fusion Schweißrissigkeit *f*
i. lubrication Teilschmierung *f*
i. reaction unvollständig verlaufende Reaktion *f*
incompressibility Unzusammendrückbarkeit *f*
incompressible inkompressibel, unzusammendrückbar
i. fluid inkompressible Flüssigkeit *f*
Inconel Inconel *(Handelsname; chromreiche Stahllegierung)*
inconsistent unbeständig, unstet
inconstancy Inkonstanz *f*
incorporate / to einbauen, unterbringen; inkorporieren
incorporation Einbau *m*; Inkorporierung *f*
incorrect measurement Fehlmessung *f*
incorrodible nichtrostend, nicht korrodierbar
increase / to vergrößern; verstärken; wachsen
i. by electro-deposition elektrolytisch (galvanisch) verstärken
i. the gain *(El)* die Verstärkung erhöhen
increase Vergrößerung *f*; Zunahme *f*; Zuwachs *m*
i. in grain size Kornvergrößerung *f*
i. in hardness Verfestigung *f*
i. in pay Lohnerhöhung *f*
i. in temperature *s.* i. of temperature
i. in velocity Geschwindigkeitszunahme *f*
i. of a vibration *(Flg)* Aufschaukeln *n*

i. of admission *(Wärm)* Füllungsvergrößerung *f*
i. of grain size Kornwachstum *n*
i. of output Leistungssteigerung *f*
i. of resistance Widerstandszunahme *f*
i. of slope Versteilerung *f (Vergrößern der Flankensteilheit von Signalen)*
i. of temperature Temperaturanstieg *m*, Temperaturerhöhung *f*, Temperaturzunahme *f*
i. of the cross-sectional area *(Wärm)* Verdickung *f (des Schaftes)*
i. of voltage *(El)* Spannungsanstieg *m*
i. of volume Vergrößerung *f* des Volumens, Volumenzunahme *f*
increased efficiency Mehrleistung *f*
increaser Erweiterungsstück *n*, Reduzierstück *n* *(Fittings)*
increasing Vergrößern *n*, Vermehren *n*
i. balloon length *(Text)* wachsende Länge *f* des Fadenballons
i. pitch *(Schiff)* zunehmende Steigung *f (beim Propeller)*
i. ratio Übersetzungsverhältnis *n*
increment Zunahme *f*, Zuwachs *m*; Sprung *m*, Maßsprung *m*, Stufensprung *m*; Laststufe *f*; Inkrement *n*
i. cut reamer Reibahle *f* mit ungleicher Zahnteilung
i. load Lastzunahme *f*
i. of face advance *(Brg)* Verhiebsbreite *f*
i. of roll *(Schiff)* Inkrement *n* der Rollschwingung
i. of speed Geschwindigkeitszunahme *f*
incremental digital computer digitaler Differentialgleichungslöser *m*
i. digital transducer *(Dat)* Zuwachsmeßumsetzer *m*
i. duplex *(Fmt)* Additionsduplex *n*
i. hysteresis loss differentieller Hysteresisverlust *m*
i. inductance Induktivität *f* für überlagerten [kleinen] Wechselstrom, Wechselstrominduktivität *f*; zusätzliche Induktion *f*
i. magnetizing force differentielle magnetische Feldstärke *f (Scheitelwert)*
i. permeability differentielle Permeabilität *f*
incrementer Telegrafenrelais *n*
increments Stufung *f*
incrush of water *(Brg)* Wassereinbruch *m*
incrust / to verkrusten, inkrustieren
incrustation 1. Inkrustierung *f*; Verkrustung *f*; Kruste *f*, Belag *m*; 2. Kesselstein *m*
i. of boiler fur Kesselsteinansatz *m*
i. preventings Kesselsteinlösung *f*
incubate / to [be]brüten, entwickeln
incubation period Brutzeit *f*, Entwicklungsperiode *f*, Inkubationszeit *f*
incubator *(Lw)* Brutmaschine *f*, Inkubator *m*
incunabula *(Typ)* Frühdruck *m*, Wiegendruck *m*, Inkunabel *f*
indanthrene Indanthrenfarbstoff *m*
indecomposable unzerlegbar
indefinable undefinierbar
indefinite unbegrenzt
i. integral unbestimmtes Integral *n*
indefiniteness Unbestimmtheit *f*
indeflagrable unverbrennbar
indelible dokumentenecht *(Tinte)*
i. pencil Tintenstift *m*
indent / to [ein]kerben; auszacken
indent Kerbe *f*, Einkerbung *f*, Einschnitt *m*; Zacke *f*
indentation 1. Zahnung *f*; zahnungsartige Rauhigkeit *f (Oberflächengüte)*; Eindruck *m*, Einprägung *f*; Rille *f*, Vertiefung *f*; 2. *(Typ)* Einzug *m*
i. cup *(Met)* Eindruckkalotte *f (Kugelhärteprüfung)*
i. hardness Eindruckhärte *f*, Kugeldruckhärte *f*, Eindringhärte *f*
i. hardness test with ball indentor Kugeldruckhärteprüfung *f*
i. of the ball Kugeleindruck *m*, Kalotte *f* *(Härteprüfung)*
i. test *(Met)* 1. Tiefungsprobe *f*; 2. Kugeldruckversuch *m*
indented gedimpelt, gekerbt; verzahnt
i. chisel Zahneisen *n*
i. joint Verzahnung *f (Holzbau)*
indenter *s.* indentor
indentor *(Met)* Eindringkörper *m (Härteprüfung)*
indenumerable *(Math)* nicht abzählbar, unabzählbar
independence Unabhängigkeit *f*
i. of frequency *(El, Fmt, Fs)* Frequenzunabhängigkeit *f*
i. of temperature Temperaturunabhängigkeit *f*
independent unabhängig
i. adjustment Einzelverstellung *f*
i. boiler feeds *(Wärm)* voneinander unabhängige Speisevorrichtungen *fpl*
i. chimney freistehender Schornstein *m*
i. condenser *(von der Turbine)* getrennter Kondensator *m*
i. feeder *(El)* direkte Speiseleitung *f*
i. fission yield *(Kern)* primäre Spaltausbeute *f*
i. four-jaw chuck Planscheibe *f*
i. front suspension *(Kfz)* Einzelradaufhängung *f*
i. of frequency frequenzunabhängig
i. of time zeitunabhängig
i. of voltage fluctuation spannungsunabhängig

i. piece *(Schiff)* Vorstevenrückenstück *n*
i. sideband generation Erzeugung *f* unabhängiger Seitenbänder
i. sideband transmitter Sender *m* mit unabhängiger Seitenbandübertragung
i. spring Schwingachse *f*
i. springing *(Kfz)* Einzelradaufhängung *f*
i. steam engine freistehende Dampfmaschine *f*
i. suspension Einzel[rad]aufhängung *f*; Einzelabfederung *f*
i. time control örtliche Synchronisierung *f* *(Meßtechnik)*
i. time-lag relay unabhängiges Zeitrelais *n*
i. tower freistehender Turm *m*
i. tracks *(El)* unabhängig gespeiste Gleise *npl*
i. trip Hilfsstromauslösung *f*
i. variable unabhängige Variable (Veränderliche) *f*
indestructibility Unzerstörbarkeit *f* *(Materie)*
indestructible unverschleißbar, unzerstörbar, unzerbrechlich
indeterminacy Unbestimmtheit *f*
i. principle Unbestimmtheitsrelation *f*
indeterminate unbestimmt
index / to 1. *(Masch)* teilen; schalten *(Revolverkopf)*; 2. mit Index versehen
i. around rundschalten
i. to the next position *(Masch)* weiterschalten *(Revolver)*
index 1. Anzeiger *m*; 2. alphabetisches Verzeichnis *n*; 3. Beiwert *m*; Index *m*; Kennziffer *f*; Faktor *m*; 4. Exponent *m*
i. cam *(Masch)* Teilungskurve *f*
i. card Karteikarte *f*
i. card punch Karteikerbzange *f*
i. change gear Teilwechselrad *n*
i. circle Lochkreis *m* der Teileinrichtung
i. correction Korrektur *f* des systematischen Fehlers, Korrektur des Gerätefehlers
i. crank Teilkurbel *f* *(Werkzeugmaschine)*
i. cutting machine *(Typ)* Registerschneidemaschine *f*
i. disk Teilscheibe *f*
i. dot Einstellmarke *f*
i. error Indexfehler *m*, Instrumentenfehler *m*; Teilfehler *m*
i. fossil Leitfossil *n*
i. gear Teilzahnrad *n*
i. gear hob Schaltradwälzfräser *m*
i. graduation Teilstriche *mpl*
i. head *(Masch)* Teilkopf *m*
i. lever Schalthebel *m*
i. mineral Leitmineral *n*
i. of a root *(Math)* Wurzelexponent *m*
i. of modulation *(El)* Modulationsfaktor *m*
i. of polytropic compression Polytropenexponent *m* der Verdichtung
i. of refraction Brechungsindex *m*, Brech[ungs]zahl *f*
i. plate Rastenscheibe *f*, Teilscheibe *f*
i. register 1. Indexregister *n* *(Ziffernrechner)*; 2. Adressenregister *n*
i. ring Teilring *m*
i. setting *s.* desired value
i. spindle Teilspindel *f* *(Teilkopf)*
i. strip Registrierstreifen *m*
i. worm *(Masch)* Teilschnecke *f*
i. worm shaft Teilschneckenwelle *f*
indexed roller *(Wärm)* Ableserolle *f*
indexing *(Masch)* 1. Teilen *n*, Teilung *f*; 2. Schalten *n*, Schaltung *f* *(Revolver)*
indexing *(Masch)* schaltbar *(Revolver)*; mit Teilbewegung
i. accuracy Teilgenauigkeit *f*; Schaltgenauigkeit *f*
i. attachment Teileinrichtung *f*, Teilgerät *n*, Teilkopf *m*
i. bolt Sperrbolzen *m*
i. disk Ablesescheibe *f*; Teilscheibe *f*
i. error Teilfehler *m*
i. fixture Teileinrichtung *f*
i. handle Teilkurbelgriff *m*
i. hatch Indexklinke *f*; Indexraststift *m*
i. head Teilkopf *m*
i. method Teilverfahren *n*
i. movement Teilbewegung *f* *(Teilen)*; Schaltvorgang *m* *(Automat)*
i. notch Teilscheibenraste *f*
i. position Schaltstellung *f* *(Revolverkopf)*
i. register 1. Indexregister *n* *(Ziffernrechner)*; 2. Adressenregister *n*
i. rotary table Rundschalttisch *m*, Rundtisch *m* mit Teileinrichtung
i. spindle Teilkopfspindel *f*
i. the turret Schalten *n* des Revolverkopfes
i. time Schaltzeit *f*
i. worm Schaltschnecke *f*
India ink *s.* Indian ink
India paper Dünndruckpapier *n*
India rubber Kautschuk *m*; Radiergummi *m*
India-rubber cable Gummikabel *n*
India-rubber wire Gummidraht *m*; Gummikabel *n*
Indian clubs Turnkeulen *fpl* *(Gymnastikgeräte)*
Indian corn Mais *m*
Indian ink Zeichentusche *f*, Ausziehtusche *f*
Indian moth Mugaspinner *m*
Indian red *(Min)* Eisenglanz *m*, roter Hämatit *m*
Indian type loom indischer Teppichwebstuhl *m*
indicate / to anzeigen; indizieren; anzeichnen
indicated efficiency indizierter Wirkungsgrad *m*
i. horsepower angegebene (indizierte) Pferdestärke *f*
i. output indizierte Leistung *f*
i. work indizierte Arbeit *f* *(Dampfdiagramm)*

i. work done by the steam indizierte Dampfarbeit *f*
indicating apparatus Indikator *m*, Nachweisgerät *n*
i. contactor Meßfühlhebel *m*
i. controller anzeigender Regler *m*, Regler mit Meßinstrument
i. device Anzeigevorrichtung *f*; Abstimmanzeiger *m*
i. dial Skalascheibe *f*, Skalenscheibe *f*
i. dimensions Bemaßung *f*
i. direction finder Sichtpeiler *m*
i. fuse Sicherung *f* mit Unterbrechungsmelder
i. gauge Lehre *f* mit anzeigenden Prüfmitteln
i. instrument Anzeigegerät *n*; Zeigerinstrument *n*; Ablesegerät *n*; anzeigendes Meßgerät *n*
i. light Anzeigelampe *f*, Kontrollampe *f*
i. methane detector *(Brg)* Schlagwetteranzeiger *m*
i. micrometer Feinzeigermeßschraube *f*
i. pen Tintenschreiber *m*
i. plug gauge Passimeter *n*
i. pressure gauge Zeigermanometer *n*
i. range Anzeigebereich *m*
i. screw-thread calipers Fühlhebelgewindeschraublehre *f*
i. snap gauge Passimeter *n*
i. specific material[s] werkstoffkennzeichnend
i. specific shape[s] formkennzeichnend
i. switch Meldeschalter *m*, Quittungsschalter *m*
indication Anzeichen *n*, Zeichen *n*; Anzeige *f* *(Meßgerät)*
i. error Fehlanzeige *f*, Mißweisung *f* *(Radar)*
i. of direction Richtungsanzeige *f*
i. of tuning Abstimmanzeige *f*
i. of vacuum Vakuummeteranzeige *f*
indicator 1. Anzeigeinstrument *n*, Anzeiger *m*, Ablesegerät *n*; Sichtgerät *n*; Fühlhebel *m*; Messer *m*, Meßgerät *n*; Sekundärgerät *n*; 2. Indikator *m (Kraftmaschine)*
i. balance Zeigerwaage *f*
i. bell *(Fmt)* Fallscheibenwecker *m*
i. board *(Fmt)* Klappenschrank *m*
i. boss *(Wärm)* Indikatornocken *m*
i. card Indikatorpapier *n*, Diagrammpapier *n*
i. chart Indikatordiagramm *n*
i. clip Skalenreiter *m*
i. cock *(Wärm)* Indikatorhahn *m*
i. connection Indikatoranschluß *m*, Indikatorstutzen *m*
i. cord *(Wärm)* Indikatorschnur *f*
i. diagram Indikatordiagramm *n*, Indikatorschaubild *n*, Leistungsdiagramm *n*
i. drive Indikatorantrieb *m*
i. for continuous diagrams *(Wärm)* Indikator *m* für fortlaufende Diagramme
i. gauge Anzeige[meß]gerät *n*; Anzeigelehre *f*
i. grid *(El)* Anzeigegitter *n*
i. lag *(Am)* Anzeigeträgheit *f*
i. lamp Signallicht *n*, Signallampe *f*
i. lead Signalleitung *f*
i. light Kontrollampe *f*
i. of inclination Neigungsanzeiger *m*
i. panel Anzeigetafel *f*, Meldetafel *f*
i. paper Indikatorpapier *n*
i. pencil Indikatorschreibstift *m*
i. piston Indikatorkolben *m*
i. plug Indikatorschraube *f*, Verschlußschraube *f*, Verschlußstück *n*
i. range Indikatorbereich *m*
i. reading Zeigerablesung *f*
i. sheet Indikatorpapier *n*
i. system Meldeanlage *f*
i. test *(Wärm)* Indikatorversuch *m*
i. with external spring Kaltfederindikator *m*, Indikator *m* mit außenliegender Feder
i. with internal feeler Innenmeßfühlhebel *m*
i. with internal spring Warmfederindikator *m*, Indikator *m* mit innenliegender Feder
indicatrix *(Math)* Indikatrix *f*
i. of diffusion *(El)* Streuindikatrix *f*
indices Indizes *mpl*
indicial admittance *(El)* Kennleitwert *m*
i. equation *(Math)* determinierende Gleichung *f*
indicolite *(Min)* Indigolith *m*
indifference Indifferenz *f*
indifferent indifferent; reaktionslos, neutral; unempfindlich
i. electrode neutrale (indifferente) Elektrode *f*
i. equilibrium indifferentes Gleichgewicht *n*
indigenous timber einheimisches Holz *n*
indigo Indigo *m (n)*
i. blue Indigoblau *n*
i. carmine Indigotin *n*, Indigokarmin *n*
i. composition Indigotinktur *f*
i. copper *(Min)* Kovellin *m*, Covellin *m* *(Kupferindig[o])*
i. print Indigodruck *m*
i. red Indigorot *n*
i. white Indig[o]weiß *n*
indigoid indigoider Farbstoff *m*
indigolite *s.* indicolite
indirect mittelbar; indirekt
i. address indirekte Adresse *f (Ziffernrechner)*, Adresse von der eigentlichen Adresse
i. arc furnace Strahlungsofen *m*, indirekter Lichtbogenofen *m*
i. arc heating indirekte Lichtbogenheizung *f*
i. control Regelung *f* ohne Hilfsenergie; indirekte Regelung; Regelung mit Ersatzregelgröße

i. costs Gemeinkosten *pl*
i. coupling indirekte Kopplung *f*
i. fittings indirekte Leuchten (Lampen) *fpl*
i. heating indirekte Heizung *f (Elektronenröhren)*
i. illumination indirekte (mittelbare) Beleuchtung *f*
i.-injection indirekte Einspritzung *f*, Einspritzung in Ansaugleitung
i. interlocking mittelbare Sperre (Verriegelung) *f*
i. letterpress *(Typ)* indirekter Hochdruck *m*
i. lighting indirekte Beleuchtung *f*
i. scanning *(Fs)* indirekte Abtastung *f*
i. transmission *(Fs)* indirekte Bildübertragung *f*; Zwischenfilmübertragung *f*
i. tripping *(El)* mittelbare Auslösung *f*, Sekundärauslösung *f*
i. tripping circuit control indirekte (mittelbare) Auslösung *f*
i. wave *(El)* Raumwelle *f*, reflektierte Welle *f*
indirectly controlled variable indirekt geregelte Größe *f (Größe, die von der Regelgröße abhängt, aber nicht direkt gemessen wird)*
i. fed aerial mittelbar gespeiste Antenne *f*, strahlungsgekoppelte Antenne
i. heated cathode indirekt geheizte Katode *f*
indirubin *s.* indigo red
indistinct undeutlich, verschwommen, unklar *(z. B. Stimme, Farbe)*
i. selectivity unscharfe Trennung *f (Radio)*
indistorted unverzerrt
indium Indium *n*
individual einzeln; Einzel-
i. call *(Fmt)* Einzelanruf *m*
i. case Einzelfall *m*
i. contactor equipment *(Elm, El)* Einzelschützensteuerung *f*
i. control Einzelsteuerung *f*
i. decision making under uncertainty *(Dat)* individuelles Treffen *n* von Entscheidungen bei Ungewißheit
i. drive Einzelantrieb *m*
i. efficiency Einzelwirkungsgrad *m*
i. error Einzelfehler *m (Zahnrad)*
i. footing Einzelfundament *n (für Säulen)*
i. front-wheel suspension Einzelradaufhängung *f*
i. line *(Am) (Fmt)* Einzelanschluß *m*; Teilnehmer[anschluß]leitung *f*
i. line subscriber *(Am) (Fmt)* Hauptanschlußteilnehmer *m*
i. load Einzellast *f*
i. measurement Einzelmessung *f*
i. motor drive Einzelantrieb *m*
i. part Einzelteil *n*
i. part production time Stückzeit *f*
i. pitch Einzelteilung *f (Stirnrad)*
i. production Einzelfertigung *f*
i. production time Stückzeit *f*
i. setting Einzeleinstellung *f*
i. spot welding Einzelpunktschweißen *n*
i. steering control Einzelsteuerung *f*
i. tooth Einzelzahn *m (Zahnrad)*
i. velocity Einzelgeschwindigkeit *f*
individually powered mit Einzelantrieb
indivisibility Unteilbarkeit *f*
indivisible unteilbar
i. load unteilbare Last *f*
indoor aerial Innen[raum]antenne *f*, Zimmerantenne *f*
i. furniture Inneneinrichtung *f*
i. illumination Innenbeleuchtung *f*
i. lighting Innenbeleuchtung *f*
i. mounting Innenaufstellung *f*
i. shot Innenaufnahme *f (F*[illegible]*)*
i.[-type] transformer Haustransformator *m*
i. wiring *(El)* Inneninstallation *f*, Hausinstallation *f*
indophenol Indophenol *n*
indraft *s.* indraught
indraught *(Flg)* Saugluft *f*
induce / to induzieren
induced induziert; künstlich
i. activity *s.* i. radioactivity
i. charge induzierte Ladung *f*
i. cleavage *(Brg)* Vorzerklüftung *f*
i. conductivity induzierte Leitfähigkeit *f*
i. current induzierter Strom *m*, Induktionsstrom *m*
i. draught Saugzug *m*
i. draught fan Saugzugventilator *m*
i. electricity induzierte elektrische Ladung *f*; Induktionselektrizität *f*, induzierte Elektrizität *f*
i. electromotive force induzierte elektromotorische Kraft *f*
i. emission induzierte (erzwungene) Emission *f*
i. field *(El)* erregtes Feld *n*
i. fission *(Kern)* künstliche Spaltung *f*
i. needle *(El)* Weicheisennadel *f*
i. noise *(Am)* Starkstromgeräusch *n*; Störung *f* durch Starkstrom
i. nuclear reaction künstliche Kernreaktion *f*
i. radioactivity künstliche Radioaktivität *f*; induzierte Aktivität *f*
belüftung *f*, Selbstkühlung *f*
i. voltage induzierte Spannung *f*, Sekundärspannung *f*
inducement Reizmittel *n*
inducer Einlaufteil *n (Strömungsmaschine)*
inductance Induktivität *f*
i. balance Induktivitätssymmetrie *f*
i. bridge Induktivitätsmeßbrücke *f*
i. coil Drossel[spule] *f*; Induktivität *f*
i. ventilation künstliche Belüftung *f*, Eigen-

i. coupling induktive Kopplung *f*
i. load *(El)* induktive Belastung (Last) *f*
i. machine Influenzgenerator *m*
i. meter Induktivitätsmesser *m*
i. unbalance Induktivitätsunsymmetrie *f*
i. voltage drop induktiver Spannungsabfall *m*
induction Induktion *f*
i. accelerator 1. *(El)* Induktionsbeschleuniger *m*; 2. *(Kern)* Elektronenschleuder *f*
i. air filter *(Kfz)* Ansaugluftfilter *n*
i. air-swirl *(Kfz)* Wirbelung *f* durch Ansaugluft
i. brazing Induktions[hart]löten *n*
i. coil Funkeninduktor *m*, Induktionsspule *f*
i. compass Erdinduktorkompaß *m*, Induktionskompaß *m*, magnetelektrischer Kompaß m
i. current Induktionsstrom *m*
i. drag Kraftwirkung *f* durch induzierte Wirbelströme
i. field Induktionsfeld *n*
i. flux Induktionsfluß *m*
i. furnace *(Met)* Induktionsofen *m*
i. generator Asynchrongenerator *m*
i.-hardened induktionsgehärtet
i. hardening Induktionshärtung *f*
i. hardening machine Induktionshärtemaschine *f*
i. heating 1. Induktionserwärmung *f*, Induktionsheizung *f*, induktive Heizung *f*; 2. *(Met)* induktive Wärmebehandlung *f*
i. in air [gap] Luftinduktion *f*
i. in iron Eiseninduktion *f*, Induktion *f* in Eisen
i. instrument Drehfeldinstrument *n*
i. lamp elektrodenlose Glimmlampe *f*, Hochfrequenzglimmlampe *f*
i. loudspeaker Induktionslautsprecher *m*
i. machine Influenzmaschine *f*; Asynchronmaschine *f*
i. manifold Ansaug[sammel]leitung *f*, Einlaßkrümmer *m*
i. meter Ferrariszähler *m*, Induktionszähler *m*
i. motor Induktionsmotor *m*, Asynchronmotor *m*
i. motor meter Induktionszähler *m*
i. phase shifter Drehphasenschieber *m*, Drehtransformator *m*
i. pipe Ansaugrohr *n*, Einlaßrohr *n*
i. port Ansaugkanal *m*, Einlaßkanal *m*, Saugkanal *m*
i. pressure Ansaugdruck *m*
i. regulator Phasenschieber *m*, Phasentransformator *m*
i. relay Ferrarisrelais *n*, Induktionsrelais *n*
i. soldering Induktionslöten *n*
i. stroke Ansaughub *m*, Einlaßhub *m*
i. surface hardening Oberflächenhärtung *f* mit induktiver Erwärmung, Induktionsoberflächenhärten *n*
i. system Ansaugleitung *f*, Einlaßleitung *f*; Ansaugsystem *n*, Ansaugeinrichtung *f*
i. tachometer Ferraristachodynamo *m*, Ferraristachometermaschine *f*; Wechselstromtachodynamo *m*, Wechselstromtachometermaschine *f*
i. temperature Einlaßtemperatur *f*, Ansaugtemperatur *f*
i. top Induktionskreisel *m*
i.-type instrument Induktionsinstrument *n*
i. voltage regulator Induktions[spannungs]-regler *m*
i. wattmeter Drehfeldwattmeter *n*, Drehfeldleistungsmesser *m*
i. welding Induktionsschweißung *f*
i. winding Erregerwicklung *f*, Feldwicklung *f*, Magnetwicklung *f*
inductionfree induktionsfrei, induktionslos
inductionless *s.* inductionfree
inductive induktiv; Induktions-
i. action Induktionswirkung *f*
i. and capacitive reactance Blindwiderstand *m*
i. capacity Dielektrizitätskonstante *f*, D.K.
i. coupling induktive Kopplung *f*
i. disturbance induktive Störung *f*
i. drop induktiver Spannungsabfall *m*
i. element for solid-state circuits Festkörperinduktivität *f*
i. feedback induktive Rückkopplung *f*
i. heating Induktionsbeheizung *f*
i. impedance induktiver Scheinwiderstand *m*, Scheinwiderstand mit induktiver Komponente
i. interference Induktionsstörung *f*
i. leak *(El)* Drosselerdung *f*
i. load induktive Belastung *f*
i. potentiometer induktives Potentiometer *n*, Spulenpotentiometer *n*
i. reactance induktiver Blindwiderstand *m*, Induktanz *f*, induktive Reaktanz *f*
i. resistance induktiver Widerstand *m*
i. rise induktiver Spannungsanstieg *m (bei kapazitiver Last)*
i. shunt induktiver Nebenschluß (Shunt) *m*
i. transmitter induktiv gekoppelter Sender *m*
i. voltage drop induktiver Spannungsabfall *m*
inductively coupled induktiv gekoppelt
inductivity Induktivität *f*, Induktionsvermögen *n*
inductometer *(Fmt)* Variometer *n*
inductor 1. Drossel[spule] *f*; 2. *(Am) (El)* Induktivität *f (Schaltelement)*; Induktor *m*
i. alternator Gleichpolmaschine *f*, Induktormaschine *f*
i. circuit Induktorstromkreis *m*, Läuferstromkreis *m*

i. **current** Erregerstrom *m*, Feldstrom *m*, Induktorstrom *m*
i. **generator** Induktormaschine *f*
i. **loudspeaker** magnetischer Lautsprecher *m*, Freischwinger *m*
i. **voltage** Induktorspannung *f*, Läuferspannung *f*, Feldspannung *f*, Erregerspannung *f*
i. **winding** Läuferwicklung *f*
inductosyn Inductosyn *m* *(Firmenbezeichnung einer induktiven Längenmeßeinrichtung)*
induline Indulin *n*
indurate / to verhärten
indurated verhärtet
industrial industriell; großtechnisch
i. **art** Kunstgewerbe *n*
i. **building** Industriegebäude *n*
i. **disease** Berufskrankheit *f*
i. **design** Industrieformgebung *f*
i. **diamond** Industriediamant *m*
i. **electric tractor** Elektrozugmaschine *f*, Elektroschlepper *m*
i. **exhibition** Industrieausstellung *f*
i. **fluoroscopy** Materialdurchleuchtung *f*
i. **frequency** Netzfrequenz *f*; Normalfrequenz *f*, technische Frequenz *f*
i. **furnace** Industrieofen *m*
i. **grade benzene** Lösungsbenzol *n*
i. **haulage** Werktransport *m*
i. **heating engineer** Wärmetechniker *m*
i. **loader** Fabriklader *m*, Industrielader *m*
i. **locomotive** Industrielokomotive *f*, Werkslokomotive *f*
i. **measurement** technische Messung *f*
i. **methylated spirit** Brennspiritus *m*
i. **noise** Industrielärm *m*
i. **nuclear power** Kernenergie *f* für industrielle Zwecke
i. **plant** Industrieanlage *f*
i. **pollution** industrielle Verunreinigung *f*
i. **processing** *(industrielles)* Großverfahren *n*
i. **product** Industrieerzeugnis *n*
i. **resin** technisches Harz *n*
i. **scale manufacture** Massenherstellung *f*; großtechnische Herstellung *f*
i. **sewage** Industrieabwässer *npl*
i. **spectroscope** Industriespektroskop *n*
i. **split wood** Nutzscheitholz *n*
i. **telescope** technisches Fernrohr *n*
i. **television** industrielles Fernsehen *n*; Betriebsfernsehen *n*
i. **tractor** Zugmaschine *f*
i.**-type nuclear reactor** Kernreaktor *m* für industrielle Zwecke
i. **waste** Industrieabfälle *mpl*
i. **waste water** Industrieabwässer *npl*
i. **water** Betriebswasser *n*
i. **wood** Industrieholz *n*, Werkholz *n*
industrialization Industrialisierung *f*
industrialized building method industrielle Bauweise *f*
i. **construction method** industrielle Bauweise *f*
i. **housing** industrieller Wohnungsbau *m*
industries fair Industriemesse *f*
industry Industrie *f*; Gewerbe *n*
i. **furnace** Industrieofen *m*
i. **sponsored research** Vertragsforschung *f*
inedible nicht eßbar, nicht genießbar, ungenießbar
ineffective wirkungslos
i. **call** *(Fmt)* nicht zustande gekommene Verbindung *f*, nicht ausgeführte Verbindung
inefficiency Unwirtschaftlichkeit *f*
inefficient in operation leistungsunfähig
inelastic unelastisch
i. **collision** unelastischer Stoß *m*
i. **impact** unelastischer Stoß *m*
inelasticity Unelastizität *f*
ineligible dienstuntauglich
inequal ungleich
inequality Ungleichheit *f*; Abweichung *f*
inequation Ungleichung *f*
inequilateral ungleichseitig
inequivalent ungleichwertig
inert inert; träge, passiv; indifferent
i.**-air transformator** Schutzgastransformator *m*
i. **electrode** indifferente (neutrale) Elektrode *f*
i. **gas** 1. Schutzgas *n*, inertes Gas *n* *(Schweißen)*; 2. Edelgas *n*
i.**-gas arc welding** Schutzgaslichtbogenschweißen *n*
i.**-gas shielded** mit Schutzgas
i.**-gas-shielded consumable-electrode welding** Schutzgasschweißung *f* mit Edelgasen und abschmelzenden Elektroden
i.**-gas-shielded non-consumable-electrode welding** Schutzgasschweißung *f* mit Edelgasen und nichtabschmelzenden Elektroden
i.**-gas-shielded welding** Schutzgaslichtbogenschweißung *f*
i. **mass** träge Masse *f*
i. **soil** *(Lw)* träger Boden *m*
i. **speed** *(Eis)* Beharrungsgeschwindigkeit *f*
i. **zone** *s.* dead zone
inertia Trägheit[skraft] *f*, Beharrungskraft *f*, Beharrungsvermögen *n*
i. **effect** Beharrungswirkung *f*; Beharrungsmoment *n*, Trägheitsmoment *n*
i. **ellipsoid** Trägheitsellipsoid *n*
i. **force** Trägheitskraft *f*, Massenkraft *f*
i. **governor** Beharrungsregler *m*
i. **mass** Beharrungsmasse *f* *(Regler)*, Schwungmasse *f*
i. **moment** Trägheitsmoment *n*
i. **of the outflowing steam** Trägheitswiderstand *m* des ausströmenden Dampfes *(Turbine)*

i. **of the steam** Massenwiderstand *m* des Dampfes *(Turbine)*
i. **starter** Schwungkraftanlasser *m*
inertial flight *(Rak)* Trägheitsflug *m*
i. **force** Trägheitskraft *f*
i. **guidance** *(Rak)* Trägheitslenkung *f* (*s. a.* i. navigation)
i. **motion** *(Mech)* Trägheitsbewegung *f*
i. **navigation** Trägheitsnavigation *f*, Trägheitsortung *f*
inertness Massenwiderstand *m*; Trägheit *f*
inerts content Ballastgehalt *m*
inexact ungenau
inexpensive billig, nicht kostspielig, nicht aufwendig
inexpert unfachgemäß
inexploitable *(Geol)* abbauunwürdig, unbauwürdig
inexplosive nicht explodierend
inextensible undehnbar; dehnungslos
infant food Kindernährmittel *n*, Kindernahrung[smittel *n*] *f*
infantry *(Mil)* Infanterie *f*
i. **tank** Infanteriebegleitpanzer *m*
i. **weapon** Infanteriewaffe *f*
infeed Zuführung *f*; Beschickungsseite *f*
i. **gear cutting machine** Radeinstechmaschine *f*
i. **grinding** spitzenloses Einstechschleifen *n*
i. **method of centreless grinding** Einstechverfahren *n* beim spitzenlosen Schleifen, spitzenloses Einstechschleifen *n*
i. **process** Einstechverfahren *n*
infeeding *(Masch)* Zustellung *f*
i. **the grinding wheel** Schleifscheibenzustellung *f*
infer / to ableiten, folgern, schließen
inference Folgerung *f*, Schluß *m (logisch)*
inferential analysis schlußfolgernde Analyse *f*
inferior minderwertig; tiefer, niedriger; tief gesetzt; innerer *(Planet)*
i. **concrete** Unterbeton *m*
i. **figure** tiefstehende Zahl *f (in einer Formel)*
i. **goods** Ausschußware *f*
i. **purlin** *(Bw)* Fußpfette *f*
i. **quality** minderwertige Qualität *f*
i. **wood** Blindholz *n*, Holz *n* geringer Güte
inferred zero *(El)* unterdrückter Nullpunkt *m*
infertile unfruchtbar
infertility Unfruchtbarkeit *f*
infestation Befall *m*, Verseuchung *f*, Zerstörung *f* durch Schimmelbildung
infill *(Bw)* Hinterfüllung *f (Erdbau)*
infiller block *(Bw)* Futterholz *n*
infilling 1. *(Bw)* Ausfachung *f*; 2. *(Geol)* Gang[aus]füllung *f*
infiltration Infiltration *f*, Eindringen *n*, Einsickern *n*, Versickern *n*
i. **gallery** *(Brg)* Sickerstollen *m*
i. **water** Sickerwasser *n*
infinite *(Math)* unendlich [groß]; unbegrenzt
i. **adjacent channel rejection** *(Fmt)* völlige Auslöschung *f* benachbarter Störer *(bei Empfängern)*
i. **conductivity** unendlich große Leitfähigkeit *f*
i. **game** *(Dat)* unendliches Spiel *n*
i. **lattice** *(Ph)* unendlich ausgedehntes Gitter *n*, unbegrenztes Gitter
i. **length of connecting-rod** unendliche Schubstangenlänge *f*
i. **life-time** Neutronenlebenszeit *f* in unendlich ausgedehnten Medien
i. **multiplication constant** *(Kern)* Vermehrungsfaktor *m* der unendlich ausgedehnten Anordnung, Vermehrungsfaktor für unendlich ausgedehntes Medium
i. **series** unendliche Reihe *f*
i.**-turn potentiometer** Umlaufpotentiometer *n*
infinitely distant unendlich fern
i. **variable** stufenlos veränderlich
i. **variable speed control** stufenlose Drehzahlregelung *f*
i. **variable speed motor** feinregelbarer (stufenlos regelbarer) Motor *m*, Motor mit stufenloser Regelung, Motor mit vollem Drehzahlregelungsbereich
i. **variable speed transmission** stufenloses Getriebe *n*
infinitesimal infinitesimal, unendlich [klein]
infinitesimal Infinitesimalgröße *f*, unendlich kleine Größe *f*
i. **calculus** Infinitesimalrechnung *f*
infinitesimally small unendlich klein
infinitude Unendlichkeit *f*
infinity 1. Unendlichkeit *f*; 2. unendliche Zahl *f*
i. **adjustment** *(Opt)* Unendlicheinstellung *f*
i. **plug** *(Fmt)* Unterbrechungsstöpsel *m*, Stöpselunterbrecher *m*
inflame / to entflammen, entzünden
inflammability Entflammbarkeit *f*, Entzündlichkeit *f*; Zündwilligkeit *f*
inflammable entzündbar; feuergefährlich; zündwillig
i. **cargo** feuergefährliche Ladung *f*
i. **coal** Flammkohle *f*
i. **matter** Zündstoff *m*
inflammation Entzündung *f*; Entflammung *f*
i. **temperature** Entzündungstemperatur *f*; Zündtemperatur *f*
inflatable aufblasbar, aufpumpbar
i. **frame work** *(Text)* Stützschlauchstruktur *f*
i. **liferaft** aufblasbares Rettungsfloß *n*
i. **rubber dinghy** aufblasbares Schlauchboot *n*
inflate / to aufblasen, aufweiten; aufblähen; *(Kfz)* aufpumpen *(Reifen)*

inflation Aufpumpen *n*; Aufblasen *n*; Füllung *f* mit Gas
i. pressure *(Kfz)* Reifen[füll]druck *m*
i. valve *(Kfz)* Luftschlauchventil *n*
inflator Luftpumpe *f (für Kfz-Reifen)*
inflect / to biegen
inflection Krümmung *f (einer Kurve)*
i. point Wendepunkt *m*, Knickpunkt *m (einer Kurve)*
i.-point emission current Sättigungsstrom *m*, Grenzstrom *m*
inflectional *(Math)* Wende- *(z. B. Tangente)*
inflexible unbiegbar, steif
inflow Einfluß *m*, Einströmen *n*; Zufluß *m*
i. of the tide *(Schiff)* Auflaufen *n* der Tide
influence Einfluß *m*, Beeinflussung *f*; elektrische Influenz *f*
i. line Einflußlinie *f (Statik)*
i. machine Influenzmaschine *f*
i. surface Einflußfläche *f (Statik)*
influent einfließend, einströmend
influent Nebenfluß *m*; Zustrom *m*, Zufluß *m*
influx Einfluß *m*, Einströmen *n*, Zufluß *m*, Zustrom *m*
information Information *f*, Nachricht *f*; Nachrichtenmenge *f*; zugeleitete Angaben *fpl*; Daten *pl*
i. bit *(Dat)* Informationsbit *n*
i. board *(Fmt)* Auskunftsplatz *m*
i. booth Informationsstand *m (Messe)*
i. content Informationsgehalt *m*
i. density Informationsdichte *f*
i. flow Informationsfluß *m*
i. gate *(Dat)* Informationstor *n*
i. line *(Fmt)* Bescheidleitung *f*
i. processing cell Informationsverarbeitungszelle *f*, künstliches Neuron *n*
i. processing language Information verarbeitende Sprache *f*
i. rate Informationsmenge *f* je Zeiteinheit
i.-read-wire Informationslesedraht *m*
i. retrieval Erschließung *f* technischer und wissenschaftlicher Information, Informationsbeschaffung *f*, Informationserschließung *f*
i. searching Informationsrecherche *f*
i. source Informationsquelle *f*, Nachrichtenquelle *f*
i. storage density Informationsspeicherungsdichte *f*
i. technique Informationstechnik *f*
i. theory Informationstheorie *f*
i. word Informationswort *n*
infra acoustic infraakustisch
i.-acoustic frequency Infraschallfrequenz *f*, unhörbar tiefe Frequenz *f*
i.-acoustic telegraphy *(Fmt)* Unterlagerungstelegrafie *f*, UT
i. red *s.* infrared
infrablack ultraschwarz
i. condition Ultraschwärzung *f*
Infraphone *(Am)* Infrarottelefon *n*
infrared infrarot, ultrarot
infrared Infrarot *n*
i. detector Infrarotdetektor *m*
i. drier Infrarottrockner *m*
i. drying Infrarottrocknung *f*
i. drying duct *(Text)* Infrarottrockenstrecke *f*
i. emission *(Ph)* Infrarotemission *f*
i. filter *(Opt)* Infrarotfilter *n*
i. heating Infrarotheizung *f*
i. homing Infrarotzielsuchverfahren *n (Flugkörper)*
i. homing system Infrarotzielsuchgerät *n (Flugkörper)*
i. lamp Infrarotlampe *f*, Infrarotstrahler *m*
i. radiation Infrarotstrahlung *f*
i. radiator Infrarotstrahler *m*
i. range and direction detection Infrarotortung *f*
i. rays Infrarotstrahlen *mpl*
i. receiver Infrarotempfänger *m*
i. reconnaissance *(Mil)* Infrarotaufklärung *f*
i. response Infrarotempfindlichkeit *f*
i. spectrophotometer Infrarotspektralfotometer *n*
i. spectroscopy Infrarotspektroskopie *f*
i. transmittancy Infrarotdurchlässigkeit *f*
infrasonic untertonfrequent
i.-frequency range Infraschallbereich *m*
i. vibrational waves Infraschall *m*
i. wave Unterschallwelle *f*
infrasound Infraschall *m*
infringement *(Fmt, Fs)* Seitenbandstörung *f*
i. of copyright Urheberrechtsverletzung *f*
i. of patent Patentverletzung *f*
infusible unschmelzbar, unauflösbar; schwer schmelzbar
infusion tube *(Brg)* Tränkrohr *n*
infusorial earth Diatomeenerde *f*, Infusorienerde *f*, Kieselgur *f*
ingate Anschnitt *m*; Gießtrichter *m (Formen)*
i.-plot *(Brg)* Füllort *n*
i. with side channels angeschnittener Einguß *m*
ingenious sinnreich; talentvoll; praktisch
ingoing air Ansaugluft *f (Verbrennungsmotor)*
ingot *(Met)* Block *m*, Gußblock *m*, Rohblock *m*, Barren *m*; Massel *f*
i. and billet shear Platinen- und Knüppelschere *f*
i. buggy Blockwagen *m*
i. butt Blockrest *m*, Stummel *m*
i. cast steel Flußstahl *m*
i. charging bogie Blockeinsetzwagen *m*
i. charging carriage Blockeinsetzwagen *m*
i. charging crane Blockeinsetzkran *m*

i. **charging truck** Blockwagen *m*
i. **chariot** Blockwagen *m*
i. **chill** Blockform *f*
i. **conveying roll** Blocktransportvorrichtung *f*, Vorrolle *f*
i. **dog** Blockzange *f*
i. **dressing shop** Blockzurichterei *f*
i. **gripper** Blockgreifer *m*
i. **heating furnace** Blockwärmofen *m*
i. **iron** Flußstahl *m*
i. **metal** Blockmetall *n*
i. **mould** Blockform *f*, Blockkokille *f*, Kokille *f*; Barrenform *f*
i. **mould blackening** Kokillenschlichte *f*, Kokillenlack *m*
i. **mould casting** Kokillenguß *m*
i. **mould scrap** Kokillenbruch *m*
i. **mould stool** Kokillenuntersatz *m*
i. **mould varnish** Kokillenschlichte *f*, Kokillenlack *m*
i. **parting machine** Blockabstechmaschine *f*
i. **planer** Blockhobelmaschine *f*
i. **pusher** Blockausdrücker *m*
i. **reheating furnace** Blockwärmofen *m*
i. **scum** Blockschaum *m*
i. **shears** Blockschere *f*
i. **silver** Barrensilber *n*
i. **slab** Rohbramme *f*
i. **slab mould** Brammenform *f*
i. **slicing lathe** Gußblockabstechdrehmaschine *f*
i. **steel** Flußstahl *m*
i. **stirrup** Blockzange *f*
i. **stripper** Blockabstreifer *m*
i. **stripping crane** Stripperkran *m*
i. **tilter** Blockkipper *m*
i. **tipper** Blockkipper *m*
i. **tipping device** Blockkipper *m*
i. **tongs** Blockzange *f*
i. **withdrawing device** Blockauszieher *m*
ingotism *(Met)* Blockseigerung *f*
ingrafting Aufpfropfen *n*, Einpfropfen *n (Gartenbau)*
ingrain faser- oder garngefärbt, vor dem Stricken gefärbt *(Garn)*
i. **colour** Azofarbe *f*
i. **wool** vor dem Spinnen gefärbte Wolle *f*
ingredient Bestandteil *m*, Ingrediens *n*
ingress Eindringen *n*
inhabitable bewohnbar
inherent Eigen-; zugehörig, innewohnend
i. **ash** gebundene Asche *f*
i. **balanced** *(Fmt)* in sich abgeglichen, in sich symmetrisch
i. **colour** Eigenfarbe *f*
i. **density** *(Ph)* Eigendichte *f*
i. **distortion** 1. Eigenklirrfaktor *m (Meßinstrument)*; 2. *(Fmt)* Apparateverzerrung *f*
i. **filtration** *(Kern)* Eigenfilterung *f*
i. **losses** Eigenverluste *mpl*
i. **noise** *(Fmt)* Eigenrauschen *n*
i. **regulation** *(El)* Selbstregelung *f (bei Lastabwurf)*; Ausgleich *m*, Selbstausgleich *m (Regelstrecke)*; Spannungsanstieg *m* bei Abwurf der Nennlast
i. **resistance** innerer Widerstand *m*, Eigenwiderstand *m*
i. **safe** eigensicher, in sich sicher *(Reaktor)*
i. **stability** Eigenfestigkeit *f*, Eigenstabilität *f*
i. **tension** Eigenspannung *f*
i. **value** Eigenwert *m*
inherited error *(Dat)* mitgeschleppter Fehler *m*
inhibit / to entgegenwirken, hemmen, hindern, inhibieren, unterbinden, verhindern, verzögern
inhibit winding *(BMSR)* Verhinderungswicklung *f*
i. **wire** *(BMSR)* Blockierdraht *m*
inhibition circuit Inhibitschaltung *f*, Verhinderungsschaltung *f (logisch)*
i. **of convection** Unterdrückung *f* der Konvektion
inhibitor 1. *(Ch)* Hemmstoff *m*, Inhibitor *m*; Stabilisator *m*; 2. Sparbeize *f*, Beizzusatz *m*
inhibitory-gate JEDOCH-NICHT-Schaltung *f*, JEDOCH-NICHT-Tor *n*, UND-NICHT-Schaltung *f*, UND-NICHT-Tor *n*
inhomogeneity 1. *(Ch)* Inhomogenität *f*, [stoffliche] Ungleichheit *f*, Ungleichmäßigkeit *f*; 2. *(Math)* Inhomogenität, Heterogenität *f*, Ungleichartigkeit *f*, Ungleichförmigkeit *f*
inhomogeneous inhomogen, heterogen, ungleichartig, ungleichförmig
i. **magnetic field** inhomogenes Magnetfeld *n*
inhour *(Kern)* Inhour *f*, reziproke Stunde *f*
initial *(Typ)* großer Anfangsbuchstabe *m*, Initial *n*, Majuskel *f*; Monogramm *n*
initial Anfangs-; anfänglich; Ausgangs-; ursprünglich
i. **abscisse** Anfangsabszisse *f*
i. **activity** *(Ph)* Initialaktivität *f*, Ausgangsaktivität *f*
i. **amplitude** Anfangsamplitude *f*
i. **approach** *(Flg)* Anflugbeginn *m*
i. **approach altitude** Platzanflughöhe *f*
i. **argument** Anfangswert *m*, Anfangsgröße *f (Vektor)*
i. **boiling point** Anfangssiedepunkt *m*, Siedebeginn *m*
i. **brace** *(Math)* *(geschweifte)* Klammer *f* auf
i. **capacitance** Anfangskapazität *f*
i. **climb** *(Flg)* Anfangssteigung *f*
i. **concentration** Anfangskonzentration *f*, Ausgangskonzentration *f*
i. **condition** *(BMSR)* Anfangsbedingung *f*
i. **convergence** *(Brg)* Vorabsenkung *f*

i. cost Anlagekosten *pl*, Gestehungskosten *pl*, Anschaffungskosten *pl*
i. cracking Anriß *m*
i. creep primäres Kriechen *n*, Anfangskriechen *n*
i. cross section Anfangsquerschnitt *m*
i. current Anfangsstrom *m*, Anlaufstrom *m*; Ruhestrom *m*
i. current intensity Einschaltstromstärke *f*
i. curvature ursprüngliche Krümmung *f (Statik)*
i. detonating agent Initialsprengstoff *m*
i. direction *(Opt)* Anfangsrichtung *f*
i. equation *(Math)* Ausgangsgleichung *f*
i. grid voltage Gittervorspannung *f*
i. hardening *(Bw)* Anfangserhärtung *f*; Erstarrungsbeginn *m*
i. harvest Holzeinschlag *m*
i. high velocity scan *(Fs)* Bildabtastung *f* mit hoher Anfangszerlegungsgeschwindigkeit
i. inverse voltage Sprungspannung *f (Elektronik)*
i. ionization *(Ph)* Primärionisation *f*
i. load *(Met)* Vorlast *f (Rockwell)*
i. loading Anfangsbelastung *f*
i. magnetization *(Ph)* Erstmagnetisierung *f*
i. magnetization curve jungfräuliche Magnetisierungskurve *f*
i. mass *(Rak)* Startmasse *f*
i. metacentre *(Schiff)* Anfangsmetazentrum *n*
i. nucleus Kristallisationskeim *m*, Kristallisationskern *m*
i. operation of locomotive Einfahren *n* der Lokomotive
i. ordinate *(Math)* Anfangsordinate *f*
i. output test Frischprüfung *f (Batterie)*
i. path Anfangsbahn *f*
i. period *(Am) (Fmt)* Gesprächseinheit *f*
i. permeability Anfangspermeabilität *f*
i. point Anfangspunkt *m*; Ausgangspunkt *m*
i. point of expansion Expansionsbeginn *m*
i. position Ausgangsstellung *f*, Nulleinstellung *f*; *(Eis)* Grundstellung *f (Signale)*
i. potential *(El)* Vorspannung *f*
i. pressure 1. *(Kfz)* Anfangsdruck *m (Lader)*; 2. *(Wärm)* Eintrittdampfspannung *f*
i. prestress[ing] Anfangsvorspannung *f (Spannbeton)*
i. product Ausgangserzeugnis *n*, Vorprodukt *n*
i. rate of climb *(Flg)* Steiggeschwindigkeit *f* in Bodennähe
i. rating Anfangsleistung *f*
i. rating of well Anfangsförderung *f (Erdöl, Erdgas)*
i. reactant *(Ch)* Ausgangsstoff *m*
i. salary Anfangsgehalt *n*
i. set *(Bw)* Abbindebeginn *m*, Erstarrungsbeginn *m*
i. setting time Zeitspanne *f* bis zum Erstarrungsbeginn *(Abbindeversuch, Beton)*
i. slackness Anfangsspiel *n (Lager)*
i. speed Ausgangsdrehzahl *f*; Anfangsgeschwindigkeit *f*
i. stability *(Schiff)* Anfangsstabilität *f*
i. state Anfangszustand *m*
i. strain *(sl) s.* i. tension
i. stresses Eigenspannungen *fpl*
i. tank pressure *(Rak)* Tankvordruck *m*
i. temperature Anfangstemperatur *f*, Initialtemperatur *f*
i. tension Vorspannung *f*
i. thrust *(Rak)* Startschub *m*
i. time constant Anfangszeitkonstante *f*
i. trajectory *(Rak)* Anfangsbahn *f*
i. transient *(El)* Einschwingvorgang *m*
i. twist Anfangsdrall *m*
i. value Anfangswert *m*
i. velocity Anfangsgeschwindigkeit *f*
i. velocity current Anlaufstrom *m*
i. voltage *(El)* Anfangsspannung *f*, Anlaufspannung *f*
i. voltage response of an exciter Anfangserregungsgeschwindigkeit *f*
i. vortex *(Rak)* Anfahrwirbel *m*, Anfangswirbel *m*
i. weight *(Rak)* Startmasse *f*
i. work *(Brg)* Aufschlußarbeiten *fpl (Tagebau)*
initiate / to initiieren, [ent]zünden; in Gang setzen; *(einen Vorgang)* einleiten
initiating electron Primärelektron *n*, auslösendes Elektron *n*
i. particle *(Kern)* Primärteilchen *n*, Primäres *n*; erzeugendes (auslösendes) Teilchen *n*
i. pulse Auslöseimpuls *m*, Startimpuls *m*
i. signal Auslösesignal *n*, Auslösezeichen *n*
initiation Inbetriebnahme *f*
i. of current *(El)* Beginn *m* des Stromflusses
i. time Anfahrzeit *f*
initiator Initialzünder *m*, Initiator *m* einer Kettenreaktion
inject / to 1. einspritzen, Injektion vornehmen, injizieren, impfen; zuführen; 2. *(Bw)* verpressen, einpressen *(Zementsuspension)*; 3. *(El)* anlegen *(Spannung)*; 4. *(Kern)* einschießen, einschleusen *(Teilchenbeschleuniger)*
i. a voltage eine Spannung anlegen (einführen, zuführen)
i. on the grid *(El)* dem Gitter zuführen
inject concrete Einpreßbeton *m*
injected voltage *(El)* zugeführte Spannung *f*
injecting device *(Bw)* Injiziergerät *n*
injection 1. Einspritzung *f*, Einspritzen *n (Verbrennungsmotor)*; 2. Spritzen *n*; Spritzvorgang

m; 3. Injektion *f (Beton)*; 4. *(Kern)* Einschießen *n*, Einschleusen *n*, Injektion *f*; 5. Holztränkung *f*
i. advance *(Kfz)* Einspritzvoreilung *f*, Voreinspritzung *f*
i. advance device *(Kfz)* Einspritzversteller *m*
i.-advance mechanism *(Kfz)* Zündzeitpunktverstelleinrichtung *f*
i. condenser Einspritzkondensator *m*
i. control hub *(Kfz)* Spritzverstellernabe *f*
i. cylinder *(Plast)* Massezylinder *m*
i. engine Einspritzmotor *m*
i. equipment *(Kfz)* Einspritzanlage *f*
i. grid Mischgitter *n (Elektronenröhren)*
i. lag *(Kfz)* Einspritzverzögerung *f*
i. moment *(Kern)* Einschußmoment *m*
i. mould *(Plast)* Spritzgußform *f*, Spritzgußwerkzeug *n*
i. moulded part *(Plast)* Spritzteil *n*
i. moulding *(Plast)* Spritzgußverfahren *n*, Spritzen *n*
i. moulding compound Spritzgußmasse *f*
i. moulding defect Spritzfehler *m*
i. moulding machine Spritzgußmaschine *f*
i. moulding nozzle Spritzdüse *f*
i. moulding of zinc Zinkspritzguß *m*, Zinkdruckguß *m*
i. moulding practice Spritzgußtechnik *f*
i. nozzle Einspritzdüse *f (Diesel)*
i. optics *(Ph)* Einschußoptik *f*
i. period Einspritzdauer *f*, Öffnungsdauer *f*, Ventilerhebungszeit *f*
i. pipe Einspritzrohr *n*
i. point *(Kern)* Einschußstelle *f*
i. port *(Kfz)* Einspritzöffnung *f*
i. press Einspritzpresse *f*
i. pressure *(Kfz)* Einspritzdruck *m*, Öffnungsdruck *m*
i. pulse *(Ph)* Einschußimpuls *m*
i. pump Einspritzpumpe *f (Diesel)*
i. pump upper housing Einspritzpumpenoberteil *n*
i. spring Stößelfeder *f (Düsenhalter)*
i. timing Einspritzverstellung *f (Diesel)*
i. timing collar Spritzverstellmuffe *f*
i. timing device Einspritzverstelleinrichtung *f*, Spritzversteller *m*
i. timing gear Einspritzverstelleinrichtung *f*
i. timing mechanism Spritzversteller *m*
i. valve Einspritzventil *n*; Druckventil *n (Einspritzpumpe)*
i. water Einspritzwasser *n*, Kondenswasser *n*
i. well Einpreßsonde *f (Erdöl, Erdgas)*
injector Injektor *m*, Dampfstrahlpumpe *f*; Spritze *f*; Einspritzdüse *f*
i. check valve *(Eis)* Injektorrückschlagventil *n*
i. feed pipe *(Eis)* Injektordruckrohr *n*
i. filter Einspritzventilfilter *n*
i. grid Mischgitter *n (bei Röhren)*
i. needle Düsennadel *f (Einspritzdüse)*
i. overflow valve *(Eis)* Injektorschlabberventil *n*
i. valve Einspritzventil *n*, Brennstoffventil *n*
injure / to beeinträchtigen, beschädigen
injurious forest insects Forstschädlinge *mpl*
injury due to frost Frostschaden *m*
i. potential *(El)* Verletzungspotential *n*, Demarkationspotential *n*
ink / to einfärben, Farbe auftragen; mit Tusche ausziehen *(technische Zeichnung)*
i. the form schwärzen, Schwärze auftragen
ink 1. Tinte *f*; 2. Tusche *f*; 3. *(Plast)* Farbe *f*; *(Typ)* Druckfarbe *f*
i. acceptance *(Typ)* Farbannahme *f*
i. agitator *(Typ)* Farbrührwerk *n*
i. ball Farbballen *m*
i. block Farbläufer *m*
i. disk Farbrädchen *n*
i. drawing Tuschezeichnung *f*
i. duct *(Typ)* Farbkasten *m*
i. feathering Auslaufen *n* der Tinte *(auf Papier)*
i. grinding machine Farbenreibmaschine *f*
i. misting *(Typ)* Stäuben *n* der Druckfarbe
i. pad Stempelkissen *n*
i. powder Tintenpulver *n*
i. recorder *(Fmt)* Tintenschreiber *m*
i. reducer *(Typ)* Verdünnungsmittel *n (für Druckfarben)*
i. release *(Typ)* Farbabgabe *f*
i. ribbon Farbband *n*
i. ribbon feed Farbbandvorschub *m*
i. ribbon reversal Farbbandwechsel *m*
i. roller Farbwalze *f*
i. transfer *(Typ)* Farbübertragung *f*
i. writer Farbschreiber *m*
inker 1. Farbrolle *f*, Farbwalze *f*, Auftragwalze *f*; 2. *(Fmt)* Tintenschreiber *m*, Farbschreiber *m*
inking Einfärben *n*, Farbauftragung *f*
i. apparatus *(Typ)* Farbwerk *n*
i. compass Zirkel *m* mit Reißfeder
i. cylinder Farbauftragwalze *f*
i. device Farbwerk *n*
i. disk Farbrädchen *n*
i. noise *(Dat)* Farbauftragrauschen *n (Zeichenerkennung)*
i. roller Farbrädchen *n*; Farb[auftrag]walze *f*
i. system Tintenschreibeinrichtung *f*, Farbwerk *n (an Registriergeräten)*
i. wheel Farbrädchen *n*
inkle loom *s.* Indian type loom
inlaid border eingelegte Kante *f*
i. floor Parkett[fußboden *m*] *n*
i. work eingelegte Arbeit *f*, Einlegearbeit *f*
inland landeinwärts

i. **harbour** Binnenhafen *m*
i. **lake scow** Binnenseekahn *m*
i. **navigation** Binnenschiffahrt *f*
i. **sea** Binnenmeer *n*
i. **trade** 1. Binnenhandel *m*; 2. Binnenschiffahrt *f*
i. **traffic** Inlandsverkehr *m*
i. **vessel** Binnenschiff *n*, Binnenfahrzeug *n*
i. **waters** Binnengewässer *npl*
i. **waterway** Binnenschiffahrtsweg *m*
i. **waterway barge** Flußkahn *m*
Inland Rules Binnenwasserstraßenordnung *f*
inlay / to einlegen, furnieren *(Holz)*
inlay Holz *n* zum Einlegen, Intarsie *f*
inlaying Einlegearbeit *f*, furnierte Arbeit *f*, Intarsienarbeit *f*
inleakage Einsaugen *n*; Einströmen *n*, Eintritt *m*
inlet 1. Einlaß *m*, Einlauf *m*; Einlaufseite *f*; 2. Einführung[söffnung] *f*, Durchlaß *m*; 3. *(El)* Zuführung *f*; 4. Beaufschlagung *f (Turbine)*; 5. [kleine] Bucht *f*
i. **air** Ansaugluft *f*; angesaugte Luft *f (Verbrennungsmotor)*
i. **angle** Eintrittswinkel *m (für Turbinenschaufel)*
i. **cam** Einlaßnocken *m*
i. **casing** Eintrittsgehäuse *n*
i. **chamber** Saugkammer *f (Pumpe)*
i. **channel** Einlaufrinne *f*, Zulaufrinne *f*
i. **connecting branch** Eintrittsstutzen *m*, Einlaßstutzen *m*, Frischdampfstutzen *m*
i. **connection** Einlaufstutzen *m*
i. **cross section** Eintrittsquerschnitt *m*
i. **diagram** Eintrittdreieck *n (Turbine)*
i. **edge** Eintrittkante *f (Turbine)*
i. **elbow** Ansaugkrümmer *m*
i. **fitting** Einfüllansatz *m*
i. **flange** Einlaßöffnungsflansch *m*
i. **funnel** Einlauftrichter *m*, Einführungspfeife *f*
i. **governor** Einlaßsteuerung *f*
i. **header** Eintrittsvorlage *f*
i. **housing** Leitschaufelträger *m*
i. **insulator** Einführungsisolator *m*
i. **manifold** Ansaugleitung *f*, Einlaßkrümmer *m*; Einlaßleitung *f*
i. **manifold gasket** Ansaugkrümmerdichtung *f*
i. **nipple** Einführungsstutzen *m*
i. **pipe** Einlaufrohr *n*, Einlaßrohr *n*, Zuleitungsrohr *n*
i.-pipe connection Eintrittsstutzen *m*
i. **pipe for circulating water** Kühlwasserzufuhrrohr *n*
i. **piping** Einlaßleitung *f*
i. **plug** Zündkerze *f* an Einlaßseite
i. **port** Einlaßkanal *m*, Einlaßöffnung *f*, Einlaßschlitz *m*; Einspritzöffnung *f*
i. **pressure** Ansaugdruck *m*
i. **reservoir** Einlaufbecken *n (Wasserbau)*
i. **shaft** Einlaufschacht *m*
i. **side** Saugseite *f (Pumpe)*
i. **strainer** Eingußsieb *n*
i. **stroke** Saughub *m*
i. **structure** Einlaufbauwerk *n (Wasserbau)*
i. **temperature** Einlaßtemperatur *f*, Eintrittstemperatur *f*
i. **valve** Ansaugventil *n*, Einlaßventil *n*, Einströmventil *n*, Eintrittsventil *n*
i. **valve chest** Einlaßventilgehäuse *n*
i. **velocity** Einlaßgeschwindigkeit *f (Turbine)*
innavigability Unschiffbarkeit *f*
innavigable nicht schiffbar, nicht mit Schiffen befahrbar
inner inner; Innen-
i. **bark** Bast *m*
i. **bearing** *(Masch)* Innenlager *n*
i. **bremsstrahlung** *(Kern)* innere Bremsstrahlung *f*
i. **column** Innenrohr *n*, Innensäule *f (Radialbohrer)*
i. **conductor** Innenleiter *m (Drehstrom)*
i. **cone** Innenkonus *m*
i. **cone of a flame** Flammenkern *m*
i. **core** Innenkern *m*
i. **core of a shaft** Wellenseele *f*
i. **cowling** Innenhaube *f (Sternflugmotor)*
i. **dead centre** oberer Totpunkt *m*
i. **diameter** Innendurchmesser *m*
i. **edge** Innenkante *f*
i. **electron** inneres Elektron *n*, Elektron einer inneren Schale
i. **end-paper** *(Typ)* fliegendes Vorsatz *n*
i. **harbour** Binnenhafen *m*
i. **jib** *(Schiff)* Innenklüver *m*, Binnenklüver *m*
i. **layer** *(El)* Innenlage *f (einer Spule)*
i. **marker** *(Flg)* Platzeinflugzeichen *n (ILS-Verfahren)*
i. **marker signal** Haupteinflugzeichen *n*
i. **member** 1. Innenteil *n (Passung)*; 2. Innenring *m (Freilauf)*
i. **orbit** *(Kern)* Innenbahn *f*, innere Bahn (Kreisbahn) *f*
i. **pole alternator** Innenpolgenerator *m*
i. **pole armature** *(Elm)* Innenpolanker *m*
i. **port** Binnenhafen *m*
i. **post** *(Schiff)* Binnenachtersteven *m*, Innensteven *m*
i. **quantum number** innere Quantenzahl *f*
i. **race** Innenlaufbahn *f*, Innenring *m (Lager)*
i. **radiation-zone** innerer Strahlungsgürtel *m (der Erde)*
i. **ring** 1. Innenring *m (Lager)*; 2. versenkt liegender Gleitring *m*
i. **self-inductance** *(El)* innere Induktivität *f*, Eigeninduktivität *f*

i. shell Innenschale *f*, innere Schale *f*; innere Elektronenschale *f*
i. shell electron *s.* i. electron
i. span Innenweite *f*
i. spring ring *(Masch)* hinterlegter Spannring *m*, Grundring *m*
i. stern-post *(Schiff)* Binnenachtersteven *m*
i. strake *(Schiff)* anliegender Gang *m*
i. surface Innenfläche *f*
i. surface of a cylinder Bohrungsmantel *m*
i. total reflection *(Opt)* innere Totalreflexion *f*
i. tube Luftschlauch *m (Bereifung)*
i.-tube valve fitting Schlauchventilbrücke *f*
i.-tube valve insert Schlauchventileinsatz *m*
i.-tube valve with dust cap Schlauchventil *n* mit Staubklappe
i. waterway *(Schiff)* inneres Leibholz *n*, innerer Wassergang *m*
i. winding *(El)* innere Wicklung *f (einer Spule)*
i. work function innere Austrittsarbeit *f*
innovation Neuerung *f*
innovator Neuerer *m*
innumerable *(Math)* unzählbar, unzählig
inoculant Graugußzusatz *m*, Impfungszusatz *m*
inoculate / to 1. *(Met)* impfen; 2. aufpfropfen, einpfropfen, okulieren *(Gartenbau)*
inoculation 1. Impfung *f (Kristalle)*; 2. Aufpfropfen *n*, Einpfropfen *n*, Okulierung *f (Gartenbau)*
i. of plants Veredlung *f* der Pflanzen
inodorous geruchfrei
inoperative ausgeschaltet, abgeschaltet, außer Betrieb, unwirksam
i. pass *(Met)* blindes (totes) Kaliber *n*
inordinate regellos
inorganic anorganisch
i. acid anorganische Säure *f*
i. bond anorganische Bindung *f*
i. chemistry anorganische Chemie *f*
i. matter anorganische Bestandteile *mpl*
inosine acid Inosinsäure *f*
inositol *(Ch)* Inosit *n*
inoxidable nicht oxydierbar, unoxydierbar, oxydationsfest
inoxidizable *s.* inoxidable
inphase *(El)* gleichphasig, phasengleich
i. amplifier Gleichtaktverstärker *m*
i. arrangement Gleichtaktschaltung *f*
i. component *(Fs)* gleichphasige Komponente *f*, Inphase-Komponente *f*; Wirkkomponente *f*
i. current gleichphasiger Strom *m*, Wirkstrom *m*
i. opposition gegenphasig
i. quadrature *(El)* [um 90 Grad] phasenverschoben
i. recording Gleichtaktaufnahme *f*
i. rejection *s.* common-mode rejection
i. signal *s.* common-mode signal
i. voltage gleichphasige Spannung *f*, Wirkspannung *f*
input 1. Eingangsleistung *f*, Leistungsaufnahme *f*; zugeführte Energie *f*; Eingangswert *m*; Aufwand *m*; 2. *(BMSR, Dat)* Eingangssignal *n*, Eingangsgröße *f*, Eingangsinformation *f*; 3. *(BMSR)* Eingabe *f*; Eingabevorrichtung *f*
i. admittance Eingangsleitwert *m*, Eingangsadmittanz *f*, Eingangs[schein]leitwert *m*, Gitterableitung *f (Kehrwert des Gittereingangswiderstandes)*
i. amplifier Eingangsverstärker *m*, Vorverstärker *m*
i. attenuator Eingangsabschwächer *m*
i. block *(Dat)* Eingangsspeicher *m*, Eingabespeicher *m*
i. capacitance Eingangskapazität *f (Elektronenröhre)*; Belastungskapazität *f (Voltmeter)*
i. capacitor Eingangskondensator *m*; Ladekondensator *m*
i. circuit Eingangs[strom]kreis *m*, Eingangsschaltung *f*
i. conductance Eingangs[wirk]leitwert *m*
i. connection Eingangsschaltung *f*
i. control Vorregelung *f*
i. current Eingangsstrom *m*
i. damping Antennendämpfung *f*, Eingangsdämpfung *f*
i. end *(El)* Eingang *m*; Eingangsklemmen *fpl*; Eingangsseite *f*, Sendeseite *f*
i. equipment *(Fs)* Eingabegerät *n*, Eingabemaschine *f (vom Rechner getrennt)*, Eingabeausrüstung *f*
i. filter Eingangsfilter *n*
i. function *(El)* Steuerverlauf *m*
i. gap Eingangsspalt *m*
i. impedance Eingangs[schein]widerstand *m*, Eingangsimpedanz *f*
i. letter *(Dat)* Eingangsbuchstabe *m*
i. level Eingangspegel *m*
i. means Sollwertumformer *m*, Sollwertgeber *m (Regler)*
i. member *(Masch)* Antriebsglied *n*
i. noise *(El)* Eingangsrauschen *n*
i. of dry wood Holztrockengewicht *n*
i.-output analysis *(Dat)* Einsatz-Ausstoßanalyse *f*
i. power *(El)* Eingangsleistung *f*
i. preamplifier Eingangsvorverstärker *m*
i. reactor *(Fmt)* Eingangsdrossel *f*
i. repeating coil *(El)* Vorübertrager *m*
i. resistance Eingangs[wirk]widerstand *m*
i. resonator Eingangsresonator *m*, Steuerraum *m*
i. routine Eingabeprogramm *n (sogenanntes Leseprogramm)*
i. sensitivity Eingangsempfindlichkeit *f*

i. shaft Antriebswelle *f*; treibende Welle *f (Getriebe)*
i. shaft extension Antriebswellenstumpf *m*
i. signal Stellgröße *f*, Eingangssignal *n*, Eingangsspannung *f*
i. slit Eintrittsspalt *m*
i. speed Eingangsdrehzahl *f*; antriebsseitige Drehzahl *f*; Motordrehzahl *f (Getriebe)*
i. stage *(El)* erste Stufe *f*, Eingangsstufe *f*
i. storage Eingabespeicher *m*
i. terminal Eingangsklemme *f*, Stromklemme *f*
i. time constant *(El)* Eingangszeitkonstante *f*
i. torque Eingangsmoment *n*
i. transformer Eingangstransformator *m*, Eingangsübertrager *m*, Vorübertrager *m*
i. tube *(El)* Eingangsröhre *f*, Vorröhre *f*
i. unit Aufnahmegerät *n*, Aufnahmeteil *n*, Eingabe *f (im Rechner eingebaut)*
i. variable Eingangsgröße *f (beim Regler)*
i. voltage Eingangsspannung *f*
inquiry call *(Fmt)* Auskunftsgespräch *n*
i. signal *(Fmt)* Suchzeichen *n*
inradius *(Math)* Inkreisradius *m*
inrigged racing boat Dollenrennboot *n*
inrush *(El) (plötzliches)* Einströmen *n*, Lasteinbruch *m*
i. load Überlastung *f* durch Einschaltspitze *(Glühlampen)*
inscribe / to einschreiben, markieren, einhauen; beschriften
inscribed angle *(Math)* Peripheriewinkel *m*
i. circle Inkreis *m*, eingeschriebener Kreis *m*
insecticidal insektenvernichtend
insecticide Insektizid *n*, Schädlingsbekämpfungsmittel *n*, Insektenvertilgungsmittel *n*
insensitive unempfindlich
i. interval *(Fmt, Fs)* Totzeit *f*, Sperrzeit *f*
i. time *s.* dead time
inseparable unscheidbar; untrennbar
insert / to einrasten; einsetzen, einlegen; aufstecken *(Wechselräder)*; zwischenschalten
i. anodes Anoden einhängen
i. resistance Widerstand einschalten
i. the tape *(El)* Band einlegen
insert 1. Einlage *f*; Einsatz[stück *n*] *m*; 2. Innenteil *n (Passung)*; 3. Einsteckmeißel *m*; Schneideneinsatz *m (Werkzeuge)*
i. bit Einsatzbohrer *m*, Bohrer *m* mit Einsätzen
i. earphone *(El)* Einsteckhörer *m*
i. socket *(El)* Einbaufassung *f*
i. stake Einsteckamboß *m*
i. tube 1. Einschubrohr *n*; 2. Einsatzröhre *f*
inserted eingesetzt, eingelegt, aufgesteckt; dazwischengeschaltet
i. blade Messer *n (z. B. Fräser)*
i. blade cutter Messerkopf *m*
i. blade reamer Reibahle *f* mit eingesetzten Messern
i. bridge piece eingesetzter Steg *m (Turbine)*
i. ceiling *(Bw)* untergehängte Decke *f*, Zwischendecke *f*
i. cover *(Masch)* Einsatzdeckel *m*
i. crankpin eingesetzter Kurbelzapfen *m*
i. joint eingelassene Verbindung *f*
i. journal eingesetzter Zapfen *m*
i. peg with spring-catch *(Text)* Steckspindel *f* mit Federhaltung
i. stanchion *(Eis)* Einsteckrunge *f*
i. stuffing box *(Masch)* Stopfbüchseneinsatz *m*, angeschraubter Stopfbüchsenbehälter *m*, Einsatzstopfbüchse *f*
i. teeth eingesetzte Einzelzähne *mpl*
i. tool Einsteckmeißel *m*
i. tooth 1. *(Masch)* eingesetzter Zahn *m*; 2. *(Text)* eingesetzter Abschlagzahn *m (Nadelbett)*
i. tooth blade Sägeblatt *n* mit eingesetzten Zähnen
i. tooth face milling cutter Messerkopf *m*
i.-tooth saw blade Sägeblatt *n* mit eingesetzten Zähnen
i. tooth section Schneideneinsatz *m (Räumwerkzeug)*
insertion Aufnahme *f*; Einsatz *m*, Einsetzen *n*
i. attenuation *(El)* Einfügungsdämpfung *f*
i. device *(BMSR, Büro)* Einzugsvorrichtung *f*
i. gain *(El)* Einfügungsgewinn *m*
i. humidity controller Kanalfeuchteregler *m*
i. loss *(Fmt)* Einfügungsverlust *m*; zusätzliche Dämpfung *f*
i. of electrons *(Kern)* Elektroneneinbau *m*
i. of guide disks Zwischenschaltung *f* von Leiträdern *(Turbine)*
i. phase change *(Fmt)* Phasenänderung *f* durch Einfügung eines Gliedes
i. thermostat Kanaltemperaturregler *m*
inset / to beilegen
inset 1. *(Brg)* Füllort *n*; 2. *(Typ)* Beilage *f*, Einlage *f*, Einschaltblatt *n*
i. frame *(Brg)* Schachtstuhl *m*
i. stake Einsteckamboß *m*
inshore am Ufer, dicht am Land, nach dem Land zu, küsteneinwärts
i. fisheries Küstenfischerei *f*
i. navigation Binnenschiffahrt *f*, Nahfahrt *f*
inside in, innen, innerhalb; Innen-
inside Innenseite *f* eines Körpers
i. aerial Innenantenne *f*, Zimmerantenne *f*
i. cable Innenkabel *n*
i. caliper Innentaster *m*, Lochtaster *m*; Lochzirkel *m*
i. chaser Innengewindestrehler *m*
i. cheek brake *(Eis)* Innenbackenbremse *f*
i. conductor Innenleiter *m (Drehstrom)*

i. **diameter** Innendurchmesser *m*, lichte Weite *f*; Gewindekerndurchmesser *m*
i. **diameter of spring** Federinnendurchmesser *m*
i. **dimension** lichter Durchmesser *m*; Innendurchmesser *m*
i. **door panel** Türinnenblech *n*
i. **edge** Innenkante *f*
i. **firm-joint caliper** Innentaster *m*
i.**-flange tubbing ring** *(Brg)* deutscher Tübbing *m*
i. **focusing device** *(Opt)* Innenfokussiereinrichtung *f*
i. **foreman** Obersteiger *m*
i. **frosting** *(Opt)* Innenmattieren *n*
i. **hand brake** *(Kfz)* Innenhandbremse *f*
i. **height** lichte Höhe *f*
i. **labour** *(sl)* Untertagearbeit *f*
i. **lap** *(Wärm)* Auslaßüberdeckung *f*, innere Überdeckung *f*
i. **link** *(Masch)* Innenglied *n*
i. **measure** Innenmaß *n*
i. **micrometer** Stichmaß *n*
i. **painting** Innenanstrich *m*
i. **panel** Innenblech *n*
i. **plant** *(Am)* Fernsprechinneneinrichtung *f*
i. **roof lining** *(Kfz)* Himmel *m*
i. **shoe brake** *(Kfz)* Innenbackenbremse *f*
i. **slope** *(Brg)* Innenböschung *f*
i. **spring-joint caliper** Innenfedertaster *m*
i. **strake** *(Schiff)* anliegender Gang *m*, Kraweelgang *m*, Binnengang *m*
i. **surface** Innenfläche *f*
i. **taper** Innenkonus *m*
i. **tube scraper** Bleirohraufreiber *m*
i. **wall** Innenwand *f*
i. **width** Innenmaß *n*, lichte Weite *f*, Innendurchmesser *m*
i. **width of a frame** *(Holz)* lichte Rahmenweite *f*, Stammdurchgang *m (Gatter)*
i. **work** Arbeitsverrichtungen *fpl* während der Maschinenlaufzeit
insignificant *(Math)* ohne Zahlenwert
insinkability Unsinkbarkeit *f*
insitu *s.* in-situ
insolate / to besonnen, der Sonnenstrahlung aussetzen
insolation Besonnung *f*, Sonnenbestrahlung *f*
insole Brandsohle *f (Schuh)*
i. **chanelling machine** Brandsohlenrißmaschine *f*
i. **leather** Brandsohlleder *n*
i. **tacking machine** Brandsohlenheftmaschine *f*
insolubility Unlöslichkeit *f*
insoluble *(Ch)* Unlösliches *n*
insoluble unlöslich; unlösbar
i. **in water** wasserunlöslich
i. **matter** Unlösliches *n*
insolvent zahlungsunfähig
inspect / to kontrollieren *(Gütekontrolle)*; [nach]-prüfen, überprüfen, überwachen
inspection Gütekontrolle *f*; Überprüfung *f*, Überwachung *f*; Revision *f*
i. **and quality control** Gütekontrolle *f*
i. **bore** Kontrollbohrung *f*
i. **car** *(Eis)* Werkstattwagen *m*
i. **clap** Schauklappe *f*
i. **department** Gütekontrolle *f*
i. **door** Reinigungsöffnung *f*, Kontrollöffnung *f*
i. **equipment** Prüfeinrichtung *f*
i. **eye** Kontrollöffnung *f*
i. **fixture** Prüfeinrichtung *f*
i. **gallery** Beobachtungsstollen *m*
i. **gauge** Abnahmelehre *f*
i. **glass** Schauglas *n*
i. **hole** Schauloch *n*
i. **lamp** Handleuchte *f*, Ableuchtlampe *f*
i. **launch** *(Schiff)* Bereisungsboot *n*
i. **of damage** Havarieuntersuchung *f*
i. **of individual [gear] errors** Einzelfehlerprüfung *f (Zahnrad)*
i. **of materials** Werkstoffprüfung *f*
i. **of the boiler** *(Wärm)* Befahren *n* des Kessels
i. **panel** Schauglas *n*, Sichtglas *n*
i. **pit** *(Kfz)* Reparaturgrube *f*
i. **procedure** Prüfverfahren *n*
i. **sheet** Prüf[ungs]protokoll *n*
i. **station** *(Masch)* Prüfstation *f (Taktstraße)*
i. **test** Abnahmeprüfung *f*, Werkstattprüfung *f*
i. **trolley** *(Eis)* Bahnmeisterwagen *m*, Draisine *f*
inspector Gütekontrolleur *m*
inspector's launch *(Schiff)* Bereisungsboot *n*
insphere *(Math)* eingeschriebene Kugel *f*
inspissate / to einkochen; eindicken; verdicken *(Flüssigkeit)*
inspissated eingedickt, dickflüssig
inspissation Verdickung *f (Flüssigkeit durch Verdampfen)*; Verdickungsmittel *n*
i. **of oil** Ölverdichtung *f*
instability Instabilität *f*; Inkonstanz *f*; Labilität *f*
instable labil, unstabil, instabil
i. **equilibrium** labiles (instabiles) Gleichgewicht *n*
install / to einbauen, installieren, montieren, verlegen, errichten, aufstellen, anbringen
installation Einbau *m*, Verlegung *f*, Anlage *f*, Aufstellung *f*, Einrichtung *f*, Montage *f*; Betriebsanlage *f*; Installationselemente *npl*
i. **drawing** Einbauplan *m*
i. **engineer** Montageingenieur *m*
i. **into automated production line** Einbau *m* in Taktstraße

i. **material** Installationsmaterial *n*
i. **microscope** Einbaumikroskop *n*
i. **of a subscriber's telephone** *(Fmt)* Einrichtung *f* (Anlegen *n*) eines Teilnehmeranschlusses
i. **switch** Lichtschalter *m*; Kleinschalter *m*, Normalschalter *m*
i. **work** Installierung *f*
installment Teillieferung *f*; Teilzahlung *f*
i. **payment** Abzahlung *f*, Rate *f*, Teilzahlung *f*
instant Zeitpunkt *m*, Moment *m*, Augenblick *m*
i.**-heating valve** Röhre *f* mit Schnellheizkatode
instantaneous augenblicklich; Augenblicks-; momentan; Momentan-; verzögerungsfrei, unverzögert
i. **acceleration** Momentanbeschleunigung *f*
i. **acoustic intensity per unit area** momentane Schallintensität *f*
i. **automatic gain control** selbsttätiger Pegelregler *m (Radar)*
i. **axis** Momentanachse *f*
i. **axis of rotation** Momentanachse *f* [der Bewegung]
i. **centre** Momentanzentrum *n*, Pol *m*, Doppelpunkt *m*, Rollpunkt *m (Kinematik)*
i. **centre of rotation** Momentanzentrum *n*
i. **connection** *(El)* Momentschaltung *f*
i. **current** Augenblickswert (Momentwert) *m* des Stromes; Augenblicksstrom *m*
i. **detonator** Momentzünder *m*
i. **deviation control** *(El)* unverzögerte Phasenhubregelung *f*
i. **electric detonator** Momentzünder *m*
i. **exposure** Momentaufnahme *f*
i. **force** Momentankraft *f*
i. **frequency** Momentanfrequenz *f*, momentane Frequenz *f*
i. **fuse** flinke (unverzögerte) Sicherung *f*; Momentzünder *m*
i. **locking** Momentfeststellung *f (Schiebelehre)*
i. **motion** Momentanbewegung *f*
i. **photograph** Momentaufnahme *f*
i. **power** Augenblicksleistung *f*, Momentanleistung *f*
i. **power output** momentane Ausgangsleistung *f*
i. **relay** Momentrelais *n*, Relais *n* mit Schnellauslösung
i. **release** Momentauslöser *m*
i. **rotation** Momentandrehung *f*
i. **sampling** momentane Analyse *f*
i. **short-circuit current of a transformer** dynamischer Grenzstrom *m* eines Stromwandlers
i. **shutter** Momentverschluß *m*
i. **sound energy density** momentane Schallenergiedichte *f*
i. **sound pressure** momentaner Schalldruck *m*
i. **value** Augenblickswert *m*, Momentanwert *m*
i. **velocity** augenblickliche Geschwindigkeit *f*, Momentangeschwindigkeit *f*
i. **voltage** Augenblicksspannung *f*, Momentanspannung *f*
i. **water-heater** Warmwasserspender *m*, Durchlauferhitzer *m*
instantaneously operating unverzögert
instanter winch *(Schiff)* Hangerwinde *f*
institute Institut *n*, Anstalt *f*
i. **of technology** *(Am)* Technische Hochschule *f*
instruct / to ausbilden, anlernen; unterrichten, belehren
instruction 1. Anweisung *f*, Vorschrift *f*; 2. *(Dat)* Befehl *m*, Instruktion *f*
i. **book** Bedienungsanweisung *f*, Gebrauchsanweisung *f*, Betriebsanweisung *f*
i. **code** Befehlskode *m*
i. **counter** Befehlszähler *m*
i. **for assembly** Montageanweisung *f*
i. **for maintenance** Wartungsvorschrift *f*
i. **for the [inspection and] testing** Prüfanweisungen *fpl*
i. **leaflet** Lehrbrief *m*
i. **model** Anschauungsmodell *n*, Lehrmodell *n*
i. **sheet** Arbeitsplan *m*
i. **to operator** *(Masch)* Bedienungsanweisung *f*
instructional course Einführungslehrgang *m*
instructions for use Gebrauchsanweisung *f*
instrument Instrument *n*, Gerät *n*, Apparat *m*, Einrichtung *f*
i. **approach procedure** *(Flg)* Blindanflugverfahren *n*
i. **assembly** Armaturenbrett *n*
i. **autotransformer** Meßwandler *m* in Sparschaltung
i. **bay** Geräteraum *m*
i. **board** Armaturenbrett *n*
i.**-carrying rocket** Instrumente tragende Rakete *f*
i. **characteristic** Gerätekennwert *m*
i. **compartment** Geräteraum *m*
i. **cord** *(El)* Anschlußschnur *f*
i. **department** Betriebskontrollabteilung *f*
i. **engineer** Gerätebauer *m*
i. **equipment** Instrumentenausstattung *f*
i. **error** Instrumentenfehler *m*, Gerätefehler *m*
i. **flight** Blindflug *m*, Schlechtwetterflug *m*
i. **flight training** Blindflugschulung *f*
i. **flying** Blindflug *m*, Schlechtwetterflug *m*
i. **for measuring** Meßgerät *n*
i. **hole** Meßkanal *m*
i. **industries** Gerätebauindustrie *f*
i. **lamp** Skalen[beleuchtungs]lampe *f*
i. **landing** Blindlandung *f*

i. landing system Instrumentenlandesystem *n*, Blindlandesystem *n* mit Eigenpeilung
i. lay-out Instrumentenanordnung *f*
i. lead[s] [Apparat-]Zuleitung *f*; Meßschnur *f*; Meßleitung *f*
i. lighting Instrumentenbeleuchtung *f*
i. multiplier *(El)* äußerer Vorwiderstand *m*
i. panel Schalttafel *f*, Instrumententafel *f*; Armaturenbrett *n*
i. panel equipment Instrumententafelausrüstung *f*
i. panel lamp Instrumentenleuchte *f*
i. panel support Instrumententafelstütze *f*
i. parameter Gerätekennwert *m*
i. pointer Instrumentenzeiger *m*
i. range Periodenmeßbereich *m*
i. runway Blindfluglandebahn *f*, Instrumentenlandebahn *f*
i. shunt *(El)* Nebenwiderstand *m*
i. switch Instrumentenschalter *m*, Meßbereichumschalter *m*
i. table Instrumententisch *m*
i. transformer Meßwandler *m*, Meßtransformator *m*
i. with locking device Fallbügelinstrument *n*
i. with optical pointer Lichtmarkeninstrument *n*
instrumental error Instrumentenfehler *m*
i. flight time Instrumentenflugzeit *f*
i. straggling instrumentell bedingte Streuung *f*
instrumentation Instrumentierung *f*, Ausstattung *f* mit Betriebskontrollgeräten
i. section Geräteraum *m*
instrumented rocket Instrumente tragende Rakete *f*
insubmersible unsinkbar *(Schiff)*
insulance Isolationswiderstand *m*
insulant 1. Dämmstoff *m (Schall, Wärme)*; Sperrstoff *m (Feuchtigkeit)*; 2. *(El)* Isolierstoff *m*
insular steamer Inseldampfer *m*
insulate / to 1. dämmen *(Schall, Wärme)*; sperren *(Feuchtigkeit)*; 2. *(El)* isolieren
insulated bolt *(Fs)* Isolierklemme *f*
i. brush isolierte Bürste *f*
i. bushing Durchführungsisolator *m*
i. cable isoliertes Kabel *n*
i. circuit ungeerdeter Stromkreis *m*
i. conductor isolierter Leiter *m*
i. contact isolierter Kontakt *m*
i. hold *(Schiff)* Kühlladeraum *m*, isolierter Laderaum *m*
i. instrument transformer Isolierwandler *m*
i. middle wire isolierter (nicht geerdeter) Mittelleiter *m*
i. neutral isolierter Mittelleiter (Nulleiter, Sternpunkt) *m*
i. return system isolierte Rückleitung *f*; doppelpolige Speisung *f*
i. sleeving Isolierschlauch *m*
i. support Stützisolator *m*, Stütze[r *m*] *f*
i. system isoliertes (ungeerdetes) System *n*
i. tongs Isolierzange *f*
insulating Isolier-
i. asphalt felting Asphaltisolierpappe *f*
i. base Isoliersockel *m*, Isolierfuß *m*
i. bead Isolierperle *f*
i. bed Isolierschicht *f*
i. board Isolierpappe *f*; Dämmplatte *f*
i. bottle Thermosflasche *f*
i. bush Isolierbuchse *f*, Isolierbüchse *f*, Isolatormuffe *f*
i. capacity Isolierfähigkeit *f*, Isoliervermögen *n*, Isolationsvermögen *n*; dielektrische Festigkeit *f*
i. cardboard Isolierpappe *f*, Isolierkarton *m*
i. case Isoliergehäuse *n*
i. chair Isolierschemel *m*
i. clamp Isolierklemme *f*, Isolationsklemme *f*
i. cloth Isoliertuch *n*
i. composition *s.* i. compound
i. compound Isoliermittel *n*, Isoliermasse *f*, Isolationsmasse *f*; Vergußmasse *f*, Ausgußmasse *f*; Kabelmasse *f*
i. conduit Isolationskanal *m*
i. course Dämmschicht *f (Schall, Wärme)*
i. disk Isolierscheibe *f*
i. felt Isolierfilz *m*
i. ferrule *(El)* Isolierring *m*
i. fishplate Isolierlasche *f*
i. glove Gummihandschuh *m*
i. intermediate layer isolierende Zwischenlage *f*
i. joint *(El)* Isoliermuffe *f*, Isolierstoß *m*
i. key Isolierkeil *m*
i. lacquer Isolierlack *m*, Isolierfirnis *m*
i. layer 1. Isolierschicht *f*, Isolationsschicht *f*; 2. Dämmschicht *f (Schall, Wärme)*
i. liquid Isolierflüssigkeit *f*
i. mat Isolierteppich *m*, Isoliermatte *f*
i. material elektrischer Isolierstoff *m*; Dämmstoff *m*
i. medium Isoliermittel *n*, Isolierstoff *m*
i. oil Isolieröl *n*, Transformatorenöl *n*
i. paint Isolierfarbe *f*
i. paper Isolationspapier *n*, Isolierpapier *n*
i. part Isolierteil *n*
i. paste Isolationsmasse *f*, Isoliermasse *f*
i. plate Isolierplatte *f*
i. plug Isolierstöpsel *m*, Blindstöpsel *m*
i. ply Isolierschicht *f*
i. power Isoliervermögen *n*, Isolationsvermögen *n*
i. property 1. elektrisches Isoliervermögen *n*, isolierende Eigenschaft *f*; 2. Dämmeigenschaft *f*, Dämmfähigkeit *f (Schall, Wärme)*

i. resistance Isolationswiderstand *m*
i. ring Isolierring *m*
i. sheet Isolierfolie *f*
i. silk Isolierseide *f*
i. slab Isolierplatte *f*
i. sleeve Isolierrohr *n*
i. stand Isolierstand *m*
i. stool Isolierschemel *m*
i. strength Isolierfestigkeit *f*
i. substance Isolationsmaterial *n*, Isoliermaterial *n*, Isolationsmittel *n*, Isolierstoff *m*; Dämmstoff *m (Schall, Wärme)*
i. support Isolierfuß *m*
i. tape Isolierband *n*
i. testing apparatus Isolationsprüfapparat *m*, Isolationsprüfer *m*, Isolationsmeßapparat *m*, Isolationsmesser *m*; Isolationsmeßgerät *n*, Megger *m*, Megohmmesser *m*, Megohmmeter *n*
i. tube Isolierrohr *n*, I-Rohr *n*
i. varnish Isolierlack *m*, Isolierfirnis *m*
insulation 1. Dämmung *f (Schall, Wärme)*; Sperrung *f (Feuchtigkeit)*; 2. *(El)* Isolierung *f*
i. against ground Isolation *f* gegen Erde
i. compound Isolationsmasse *f*, Isoliermasse *f*
i. current Isolationsstrom *m*
i. defect Isolierfehler *m*, Isolationsfehler *m*, Isolationsstörung *f*
i. detector Isolationsprüfapparat *m*, Isolationsprüfer *m*, Isolationsmeßapparat *m*, Isolationsmesser *m*, Isolationsmeßgerät *n*, Megger *m*, Megohmmesser *m*, Megohmmeter *n*
i. failure *s.* i. fault
i. fault Isolationsfehler *m*, Isolierfehler *m*, Isolationsstörung *f*
i. for cold Kälteschutzisolierung *f*
i. hum Isolationsbrumm *m*
i. indicator Isolationsprüfer *m*
i. loss Isolationsverlust *m*
i. material Isoliermaterial *n*; Dämmaterial *n*
i. measurement Isolationsprüfung *f*, Isolationsmessung *f*
i. measuring device Isolationsmeßeinrichtung *f*
i. resistance Isolationswiderstand *m*
i. sheath Isolierüberzug *m*, Kabelschlauch *m*
i. strength Isolationsfestigkeit *f*
i. tape Isolierband *n*
i. test Isolationsprüfung *f*, Isolationsmessung *f*
i. tester Isolationsprüfer *m* (*s. a.* i. detector)
i. testing apparatus *s.* i. tester
i. transformer Isolationstransformator *m*
insulativity Isolierfähigkeit *f*; spezifischer Widerstand *m*; Würfelwiderstand *m*
insulator 1. Isolator *m*; Isolierstoff *m*; Isolierkörper *m*; 2. Dämmstoff *m*
i. bracket Isolatorstütze *f*
i. cap Isolatordach *n*, Isolatorkappe *f*
i. chain Isolator[en]kette *f*
i. groove Isolatorenhals *m*
i. pin Isolatorenklöppel *m*; Isolator[en]stütze *f*
i. string Isolatorenkette *f*
i. support Isolator[en]stütze *f*
i.-type transformer Topfwandler *m*
insurance of a ship Kaskoversicherung *f (eines Schiffes)*
i. on hull and equipment Kaskoversicherung *f (eines Schiffes)*
insusceptible [of, to] unempfänglich [für], unempfindlich [für]
i. to aging *(Met)* alterungsunempfindlich
intaglio / to *(Typ)* tiefätzen
intaglio printing *(Typ)* Tiefdruck *m*
i. printing press Tiefdruckpresse *f*
intake Einlaß *m*, Eintritt *m*, Einlaufseite *f*, Aufgabeseite *f*; Einlauf *m*; Ansaugen *n*; Zulauf *m*; Zuführungsrohr *n*; Einströmungsöffnung *f*
i. air 1. *(Kfz)* Ansaugluft *f*; 2. *(Brg)* einziehende Wetter *pl*
i. construction Einlaufbauwerk *n (Wasserbau)*
i. elbow Ansaugkrümmer *m*
i. manifold Ansaugkrümmer *m*
i. pipe Ansaugrohr *n*
i. pressure Zulaufdruck *m (Pumpe)*
i. shaft *(Brg)* einziehender Schacht *m*
i. side of a pump Saugseite *f* einer Pumpe
i. silencer Ansauggeräuschdämpfer *m*
i. stack Ansaugstutzen *m*
i. stroke *(Kfz)* Ansaughub *m*, Saughub *m*, Ladehub *m*
i. structure Einlaufbauwerk *n (Wasserbau)*
i. tube Einlaßkanal *m*; Ansaugstutzen *m*
i. valve Einlaßventil *n*
intarsia Einlegearbeit *f*, Intarsia *f*, Intarsie *f*
i. plain knitting machine einbettige Intarsiaflachstrickmaschine *f*
integer ganze Zahl *f*
integer ganzzahlig
i.-slot winding *(Elm)* Ganzlochwicklung *f*
i. solution *(Math)* ganzzahlige Lösung *f*
integrable integrabel, integrierbar
integral Integral *n*
i. action Integralwirkung *f (Regler)*
i. action controller Integralregler *m*, I-Regler *m*
i. action time *(BMSR)* Nachstellzeit *f*
i. action zone *s.* control zone
i. bellow Integralbalg *m*, I-Wellrohr *n (pneumatischer Regler)*
i. body and chassis unit *(Kfz)* selbsttragende Karosserie *f*
i. calculus Integralrechnung *f*
i. construction selbsttragende Konstruktion *f (Karosserie)*

i. control *(BMSR)* Integralregelung *f*, I-Regelung *f*
i. controller *(BMSR)* Integralregler *m*, I-Regler *m*
i. equalization *(BMSR)* Kompensation *f* durch Integration
i. equalizer *(BMSR)* Kompensationsglied *n* mit I-Wirkung, integrierendes Kompensationsglied
i. equation Integralgleichung *f*
i. multiple ganzzahliges Vielfaches *n*
i. representation *(Math)* Integraldarstellung *f*
i. restriction *(BMSR)* Integraldrossel *f*, I-Drossel *f (pneumatischer Regler)*
i. sign Integral[zeichen] *n*
i. time constant *s.* i. action time
i. transform *(Math)* Integraltransformierte *f*
i. with aus einem Stück mit, zusammenhängend mit, fest an
integrally cast in einem Stück gegossen
integralness Ganzzahligkeit *f*
integrate / to integrieren
integrated circuit zu einer Baugruppe zusammengefaßte Schaltelemente *npl*
i. data processing ganzheitliche (integrierte) Datenverarbeitung *f*
i.-demand meter Spitzenzähler *m (elektrische Leistung)*
i. marine steam power plant kompakte Dampfschiffsmaschinenanlage *f*
i. tank *(Schiff)* Festtank *m*
integrating amplifier Integrationsverstärker *m*, Summierverstärker *m*
i. circuit *(Fs)* Integrationskreis *m*
i. device Zählwerk *n*
i. divider *(Fs)* integrierender Frequenzteiler *m*
i. factor integrierender Faktor *m*
i. gyroscope *(Dat)* Integrierkreisel *m*
i. indicator *(Wärm)* planimetrierender (totalisierender) Indikator *m*
i. instrument *(El)* [zeitlich] integrierendes Meßinstrument *n*, [zeitlich] integrierendes Instrument *n*, Zähler *m*
i. meter Zähler *m*
i. motor *(Dat)* Integrationsmotor *m*
i. photometer *(El)* Lichtstrommesser *m*, integrierendes Fotometer *n*
i. wattmeter *(El)* zählender Leistungsmesser *m*
integration Integration *f*, Zusammenfassung *f*
i. by parts partielle Integration *f*
integrator Integrator *m*, Integrierglied *n*; Integrationsverstärker *m*
integrity Ganzheit *f*
integro-differential equation Integrodifferentialgleichung *f*
intelligence Nachricht *f*; Nachrichtengehalt *m*
i. aircraft Aufklärungsflugzeug *n*
i. bandwidth *(Fmt)* Nachrichtenbandbreite *f*
i. of the sea marks *(Schiff)* Seekennung *f*
i. signal Nutzsignal *n*
intelligibility Deutlichkeit *f*, Verständlichkeit *f (Sprache)*
i. of phrases *(Fmt)* Satzverständlichkeit *f*
i. of words *(Fmt)* Wortverständlichkeit *f*
intended trajectory *(Rak)* vorausberechnete (gewünschte) Flugbahn *f*
intensification Verstärkung *f*, Intensivierung *f*; *(Fs)* Helltastung *f*
i. modulation Helligkeitsmodulation *f*
intensifier Verstärker *m*
i. electrode Nachbeschleunigungselektrode *f (Katodenstrahlröhren)*
intensify / to verstärken, intensivieren
intensifying bath Verstärkungsbad *n*
i. factor Verstärkungsfaktor *m*
i. ring Nachbeschleunigungselektrode *f*
i. screen *(Kern)* Verstärkerfolie *f*
intensimeter Intensimeter *n*
intensitivity to microphonics *(El)* Klingfestigkeit *f*
intensity Stärkegrad *m*, Intensität *f*; spezifische (auf Fläche bezogene) Kraft *f*
i. anomaly *(Ph)* Intensitätsanomalie *f*
i. control Intensitätsregelung *f*
i. decrease Intensitätsschwächung *f*
i. distribution Intensitätsverteilung *f*
i. level Schall[druck]pegel *m*
i. measurement Intensitätsmessung *f*
i. meter Intensitätsmesser *m*
i. modulation Helligkeitsmodulation *f*, Intensitätsmodulation *f (Katodenstrahloszillografie)*
i. of activation Aktivierungsenergie *f*
i. of colour Farbtiefe *f*, Farbintensität *f*
i. of draught Zugstärke *f (z. B. Ventilator)*
i. of echo Echointensität *f*
i. of exposure Belichtungsstärke *f*
i. of field Feldstärke *f*
i. of illumination Beleuchtungsstärke *f*
i. of light Helligkeit *f*
i. of noise Geräuschstärke *f*
i. of radiation Strahlungsintensität *f*, Strahl[ungs]stärke *f*
i. of sound Schallintensität *f*, Schallstärke *f*
i. of traffic Verkehrsdichte *f*
i. of X-rays Röntgenstrahlenintensität *f*
i. time variations zeitliche Schwankungen *fpl* der kosmischen Strahlung
intensive cooling Tiefkühlung *f*
inter-diffusion gegenseitige Diffusion *f*
i.-electrode capacitance (capacity) *(Ph)* Elektrodenkapazität *f*
i.-engaging ineinandergreifend
i.-operational zwischen den Arbeitsgängen
i.-row crop Zwischenreihenfrucht *f*

i.-row cultivation Zwischenreihenkultivierung *f*
i.-turbine duct Verbindungsleitung (Überströmleitung) *f* zwischen zwei Turbinen
interaccelerator Zwischenbeschleuniger *m*
interact / to in Wechselwirkung stehen; einwirken
interacting particles *(Kern)* in Wechselwirkung stehende Teilchen *npl*, Wechselwirkungsteilchen *npl*
interaction Wechselwirkung *f*, Austauschvorgang *m*, gegenseitige Beeinflussung *f* [der Reglereinstellungen], Rückwirkung *f*, Überlagerung *f*, Zusammenwirken *n*
i. cross section Wechselwirkungsquerschnitt *m*, Reaktionsquerschnitt *m*
i. energy Wechselwirkungsenergie *f*
i. factor *(Fmt)* Wechselwirkungsfaktor *m* *(Übertrager)*
i. force Wechselwirkungskraft *f*
i. impedance *(BMSR)* Kopplungswiderstand *m (eines Verzögerungsgliedes)*
i. law Wechselwirkungsgesetz *n*
i. loss *(Fmt)* Wechselwirkungsverlust *m* *(Übertrager)*
i. range Wechselwirkungsbereich *m*
i. space Koppelraum *m*, Wechselwirkungsraum *m*
i. time Wechselwirkungszeit *f*
interatomic interatomar, zwischenatomar
i. distance Atomabstand *m*
i. forces interatomare Kräfte *fpl*
interbedded eingebettet, zwischengelagert
interbedding *(Geol)* Zwischenlagerung *f*
interblend / to vermischen
intercalation Einschaltung *f*
intercarrier demodulator *(Fs)* Intercarrier-Demodulator *m*, Differenzträgerdemodulator *m*
i. interference ratio *(Fs)* Intercarrier-Störabstand *m*
i. sound system Differenzträgerverfahren *n*, Intercarrier-Verfahren *n*, Zwischenträgerverfahren *n*
intercell connector Zellenbrücke *f (Batterie)*
intercept / to abfangen, auffangen; begrenzen; abhören, abschneiden
intercept *(Math)* Abschnitt *m*
i. form Achsenabschnittsform *f*, Achsenschnittform *f (Geradengleichung)*
i. post *(Fmt)* Abhörstelle *f*
i. rocket Fliegerabwehrrakete *f*, Fla-Rakete *f*
i. screen Abfangschirm *m*
i. station *(Fmt)* Abhörstelle *f*
interception activity *(Fmt)* Abhörtätigkeit *f*
i. circuit *(Fmt)* Fangeinrichtung *f*, Halteeinrichtung *f*
i. service Abhördienst *m*, Horchfunk *m* *(Radar)*
interceptor 1. Sammler *m*; Abschneider *m*; Auffänger *m*; 2. Geruchverschluß *m (Gesundheitstechnik)*; 3. Abfangjagdflugzeug *n*
i. grid Fanggitter *n*
interchange / to auswechseln, austauschen, vertauschen
interchange 1. Vertauschung *f*; Austausch *m*, Auswechslung *f*; 2. Abfahrt *f*, Auffahrt *f*, Ausfahrt *f*, Einfahrt *f (Autobahn)*
i. effect *(Kern)* Austauscheffekt *m*
i. of heat Wärmeaustausch *m*
i. of work Arbeitsaustausch *m*
interchangeability Austauschbarkeit *f*, Auswechselbarkeit *f*, Auswechslung *f*
i. of constructional elements Austauschbau *m*
interchangeable austauschbar; auswechselbar; ersetzbar
i. gear Wechselrad *n*
i. lenses auswechselbare Optik *f*, Austauschobjektiv *n*
i. manufacture *(Masch)* Austauschbau *m*
interchanger Austauscher *m*
interchannel modulation *(Fs)* Kreuzmodulation *f*
i. spacing *(Fs)* Kanalabstand *m*
intercoastal navigation Küstenfahrt *f*, mittlere Fahrt *f*
i. service Küstendienst *m*
i. trade Zwischenküstenfahrt *f*
intercom system Gegen[laut]sprechanlage *f*
intercombination Interkombination *f (Röntgenspektrum)*
i. line Interkombinationslinie *f*
intercommunication *(Fmt)* Wechselverkehr *m*, Zwischenverkehr *m*, Wechselsprechanlage *f*
i. channel Funkquerverbindung *f*
i. plant *(Fmt)* Linienwähleranlage *f*
i. speaker *(Fmt)* Sprechstelle *f*
i. system Wechselverkehrsanlage *f*, Wechselsprechanlage *f*; Gegensprechanlage *f*; Linienwähleranlage *f*
interconnect / to zusammenschalten, zwischenschalten; verbinden; miteinander verbinden
interconnected verkettet, vermascht *(Netz)*; eingeschaltet, zwischengeschaltet
i. controls *(BMSR)* vermaschte Regelungen *fpl* (Regelkreise *mpl*)
i. star connection Zickzackschaltung *f*
i. two-phase system *(El)* verkettetes Zweiphasensystem *n*, Zweiphasen-Dreileitersystem *n*
i. voltage *(El)* verkettete Spannung *f*
i. with in Abhängigkeitsschaltung, verblockt mit
i. Y-connection *(El)* Zickzackschaltung *f*
interconnecting bar Verbindungsstab *m*
i. duct Verbindungsleitung *f*
i. mains *(El)* vermaschtes Netz *n*

interconnection Zwischenverbindung *f*; Zwischenglied *n*; Zwischenschaltung *f*; Verkettung *f*
i. of phases Phasenverkettung *f*
i. system Verbundnetz *n*
interconnector *(El)* Ausgleichsleitung *f*
intercontinental ballistic missile interkontinentale ballistische Rakete *f*, interkontinentales Fernlenkgeschoß *n*
interconversion *(Dat)* Umkodierung *f* *(Umsetzung aus einem Kodesystem in ein anderes)*; Umwandlung *f*
interconvertible umsetzbar, umwandelbar
intercooler Zwischenkühler *m*
intercooling Zwischenkühlung *f*
intercostal floor *(Schiff)* durchschnittene Bodenwrange *f*
i. girder *(Schiff)* Zwischenplattenunterzug *m*, Interkostalträger *m*
i. keelson-plate *(Schiff)* interkostale Mittelplatte *f*
intercropping Zwischenfruchtbau *m*
intercrystalline interkristallin, zwischenkristallin
i. corrosion interkristalline Korrosion *f*
i. failure interkristalliner Bruch *m*, Korngrenzenbruch *m*, Bruch entlang der Korngrenze
interdependence *(Math)* gegenseitige Abhängigkeit *f*
i. of signals *(Eis)* Signalabhängigkeit *f*
interdependent *(Math)* gegenseitig abhängig
interdialling *(Fmt)* Zwischenwahl *f*
interdot-flicker Zwischenpunktflimmern *n* *(Farbfernsehen)*
interelectrode capacitance Zwischenelektrodenkapazität *f*
i. capacitancy Zwischenelektrodenkapazität *f*
i. distance Elektrodenabstand *m*
i. gap Elektrodenabstand *m*
interest calculation Zinsenberechnung *f*
interesterification *(Ch)* Umesterung *f*
interface Grenzfläche *f*, Berührungsfläche *f*, Grenzschicht *f*, Übergangsfläche *f*
i. level Trennschicht *f*, Niveau *n* der spezifisch schwereren zweier nicht mischbarer Flüssigkeiten
i. potential Grenzflächenpotential *n*
i. resistance Zwischenschichtwiderstand *m*
interfacial Grenzflächen-
i. angle Grenzflächenwinkel *m*
i. diffusion Grenzflächendiffusion *f*
i. energy Grenzflächenenergie *f*
i. surface energy Grenzflächenenergie *f*
i. tensiometer Grenzschichtspannungsmesser *m*
i. tension *(Ch)* Grenzflächenspannung *f*
interfere / to *(Ph)* interferieren
i. with klemmen, stören, beeinträchtigen
interference 1. *(Ph)* Interferenz *f*; 2. *(El)* Interferenz, Überlagerung *f* *(einer Störung)*; 3. Abmaß *n*, Übermaß *n*; Störung *f*; Behinderung *f*; Beeinflussung *f*
i. apparatus Interferenzgerät *n*
i. area Stör[ungs]gebiet *n* *(Funktechnik)*; Verwirrungsgebiet *n* *(bei Gleichwellenbetrieb)*
i. band Interferenzstreifen *m*
i. blanker Stör[ungs]sperre *f*
i. by other stations Störung *f* durch andere Sender
i. causing frequency Störfrequenz *f*
i. colours Interferenzfarben *fpl*
i. comparator *(Opt)* Interferenzkomparator *m*
i. condenser Störschutzkondensator *m*
i. current Störstrom *m*
i. drag Interferenzwiderstand *m*
i.-eliminating condenser Störschutzkondensator *m*
i. elimination *(El)* Entstörung *f*
i. eliminator Störschutzanlage *f*, Interferenzfilter *n*
i. factor Störfaktor *m*
i. field Störfeld *n*
i. field strength Störfeldstärke *f*
i. figure *(Opt)* Interferenzbild *n*
i. filter Interferenzfilter *n*
i. fit Preßpassung *f*
i.-free störungsfrei
i. fringes Interferenzstreifen *mpl*
i. fringes of equal inclination Interferenzstreifen *mpl* gleicher Neigung
i. guard band *(El)* Störschutzband *n*
i. inverter *(Fs)* Entstördiode *f*
i. level Störniveau *n*, Störpegel *m*, Störspiegel *m*, Störhöhe *f*, Geräuschniveau *n*, Geräuschpegel *m*, Geräuschspiegel *m*
i. line *(Ph)* Interferenzstreifen *m*; Interferenzlinie *f*, Störungslinie *f*
i. location Störungssuche *f*
i. locator Störungssuchgerät *n*
i. microscope Interferenzmikroskop *n*
i. microscopy Interferenzmikroskopie *f*
i. pattern Moiréstörung *f* *(Fernsehempfangstechnik)*; Interferenzbild *n*
i. protection *(El)* Störschutz *m*
i. spectrometer Interferenzspektrometer *n*
i. spot Interferenzfleck *m*
i. suppression *(El)* Entstörung *f*
i. suppression capacitor Entstörungskondensator *m*
i. suppressor Entstörer *m*
i.-type fading *(Fmt)* Nahschwund *m*
interferential action Interferenzwirkung *f*
interfering störend; Stör-
i. effect Störwirkung *f*
i. elements *(Ch)* störende Elemente *npl* *(Analyse)*

i. **frequency** Störfrequenz *f*
i. **station** Störsender *m*
i. **transmitter** Störsender *m*
interferometer *(Opt)* Interferometer *n*, Interferenzmesser *m*
interferometry Interferenzmeßverfahren *n*
interferric space *(El)* Luftspalt *m*
interfolder *(Typ)* Leporellofalzmaschine *f*
interfusion Verschmelzung *f*
intergalactic travel Flug *m* im intergalaktischen Raum
interglacial period Interglazialzeit *f*, Zwischeneiszeit *f*
intergranular intergranular, interkristallin
i. **corrosion** *(Met)* interkristalline Korrosion *f*, Korngrenzenkorrosion *f*
intergrind with / to vermahlen mit
intergrowth Verwachsung *f (von Erzen, Kristallen)*
interior innere[r]; inwendig; verdeckt; Innen-
i. **angle** *(Math)* Innenwinkel *m*
i. **decoration** Innengestaltung *f*, Raumgestaltung *f*
i. **design** Innenarchitektur *f*
i. **diameter** Innendurchmesser *m*
i. **finish board** Platten *fpl* zur Innenausstattung
i. **lighting** Innenbeleuchtung *f*
i. **navigation** Binnenschiffahrt *f*
i. **of a rocket** Raketeninneres *n*
i. **paint** Innenanstrichfarbe *f*
i. **plastering** Innenputz *m*
i. **space** Innenraum *m*
i. **view** Innenansicht *f*, Innenperspektive *f*
i. **work** Ausbau *m*, Innenausbau *m*
interlace / to 1. verflechten, verschlingen; vernetzen, ineinanderschachteln; 2. verflechten, verschränken *(z. B. verschränkte Speicherzellennumerierung)*; 3. *(Text)* abbinden, verweben
i. **tightly** *(Text)* eng abbinden
interlaced roots Wurzelwerk *n (eines Baumes)*
i. **scanning** *(Fs)* Zeilensprungverfahren *n*
interlacing *(Ch)* Vernetzung *f*
i. **agent** *(Ch)* Netzmittel *n*
i. **of warp and weft threads** *(Text)* Verweben *n* von Kett- und Schußfäden
interlaminar bonding *(Plast)* Schichtverband *m*
i. **strength** *(Plast)* Spaltfestigkeit *f*
interlayer Zwischenschicht *f*; Zwischenlage *f*; Einlageschicht *f*
interlayered zwischengelagert; eingelagert
interleave / to *(Typ)* durchschießen *(z. B. mit leeren Blättern)*
interleave paper *s.* interleaving paper
interleaving paper *(Typ)* Durchschußpapier *n*
interline / to 1. *(Typ)* durchschießen; 2. *(Text)* mit Zwischenfutter versehen
interline flicker *(Fs)* Zeilenflimmern *n*
interlinear space *(Typ)* Durchschuß *m*
interlingual *(Dat)* zwischensprachlich
interlining *(Text)* 1. Zwischenfutter *n*; 2. Einlagestoff *m*
interlink / to verketten
interlinked verkettet
i. **current** verketteter Strom *m*
i. **electromagnetic circuit** elektromagnetische Verkettung *f*
interlock / to blockieren, verriegeln, verblocken; ineinandergreifen; kuppeln
interlock Verblockung *f*, Blockierung *f*, Sperrung *f*, Verriegelung *f*; Abhängigkeitsschaltung *f*
i. **ball** *(Kfz)* Riegelkugel *f*
i. **circuit** Verriegelungsstromkreis *m*, Verriegelungsschaltung *f*
i. **device** Riegel *m*
i. **plug** Riegelstopfen *m*
i. **spring** Riegelfeder *f*
interlocked verblockt, verriegelt; gekuppelt; zusammengeschlossen; mit Abhängigkeitsschaltung
i. **multiplets** *(Opt, Kern)* verkettete Multipletts *npl*
i. **type of flexible metal hose** gefalzter, biegsamer Metallschlauch *m*
interlocking Sperre *f*, Verriegelung *f*; Ineinandergreifen *n*
interlocking ineinandergreifend, verriegelnd
i. **circuit** *(El)* Verriegelung *f (der Schaltung)*
i. **control** *(Masch)* Abhängigkeitsschaltung *f*
i. **cutter** gekuppelter Fräser *m*
i. **frame** *(Eis)* Signal- und Weichenstellwerk *n*
i. **grate** Verschlußgitter *n*
i. **plant** *(Eis)* Stellwerksanlage *f (Signal)*
i. **sinker cams** *(Text)* ineinandergreifende Platinenexzenter *mpl*
i. **switch** *(El)* Verriegelungsschalter *m*
i. **switchgroup** *(El)* Verriegelungsschalter *m*
i. **tile** *(Bw)* Falzziegel *m*
i. **tower** *(Eis)* Stellwerksturm *m*
intermediary intermediär
i. **product** Halbfertigprodukt *n*, Zwischenprodukt *n*
i. **section** Zwischenausschnitt *m*
intermediate Zwischenkern *m*, Zwischenglied *n*
intermediate intermediär [gebildet], zwischenliegend, Zwischen-, Mittel-; mittelschnell *(Neutronen)*
i. **alloy** Zwischenlegierung *f*, Vorlegierung *f*
i. **amplifier** Zwischenverstärker *m*
i. **anchoring** Zwischenverankerung *f*
i. **annealing** *(Met)* Zwischenglühen *n*
i. **approach** *(Flg)* Zwischenanflug *m*
i. **audio frequency** Tonzwischenfrequenz *f*
i. **axis** Zwischenachse *f*
i. **bearing** Zwischenlager *n*
i. **bearing support** Zwischentraglager *n*

i. bottom Zwischenboden *m*
i. cable Verbindungskabel *n*, Zwischenkabel *n*; Kabelzwischenstück *n*
i. calender Feuchtglätter *m (zwischen zwei Trockenpartien)*
i. ceiling Zwischendecke *f*
i. circuit *(El)* Zwischenkreis *m*
i. circuit reception Zwischenkreisempfang *m*
i. connecting shaft Zwischenwelle *f (Kreuzgelenk)*
i. cooler Zwischenkühler *m*
i. cooling agent Zwischenkühlmittel *n*
i. coupling *(Kern)* intermediäre Kopplung *f*
i. crop Zwischenfrucht *f*, Zwischenernte *f*
i. crushing Mittelzerkleinerung *f*, Zwischenzerkleinerung *f*
i. density fibreboard *(Holz)* mitteldichte (halbharte) Faserplatte *f*
i. depiction *(Opt)* Zwischenabbildung *f*
i. distribution frame *(Fmt)* Zwischenverteilergestell *n*
i. draft *(Text)* Zwischenverzug *m*
i. drawing frame *(Text)* Mittelstrecke *f*
i. drive Zwischengetriebe *n*
i. energy mittlere Energie *f*
i. exchange *(Fmt)* Zwischenamt *n*
i. fibreboard *(Holz)* mitteldichte (halbharte) Faserplatte *f*
i.-film method *(Fs)* Zwischenfilmverfahren *n*
i.-film transmitter *(Fs)* Zwischenfilmsender *m*
i. flange Zwischenflansch *m*
i. frame 1. *(Schiff)* Zwischenspant *n*; 2. *(Text)* Mittelbank *f*, Mittelflyer *m*
i. frequency Zwischenfrequenz *f*
i. frequency amplifier Zwischenfrequenzverstärker *m*
i. frequency bandpass filter Zwischenfrequenzbildwandler *m*
i. frequency circuit Zwischenfrequenzkreis *m*
i. frequency rejection Zwischenfrequenzdurchschlagssicherheit *f*
i. frequency rejector Zwischenfrequenzsperre *f*
i. frequency section Zwischenfrequenzteil *m*
i. frequency stage Zwischenfrequenzstufe *f*
i. frequency strip Zwischenfrequenzteil *n*
i. frequency transformation Zwischenfrequenztransformation *f*, ZF-Transformation *f*
i. frequency transformer Zwischenfrequenztransformator *m*, Zwischenfrequenzübertrager *m*
i. frequency tube Zwischenröhre *f*
i. fuse *(El)* Zwischensicherung *f*
i. gate Zwischentor *n*
i. gear Vorgelege *n*, Zwischenrad *n*
i. gear wheel Vorgelege *n*, Zwischenrad *n*
i. gearing Zahnradvorgelege *n*, Zwischengetriebe *n*
i. girder *(Bw)* Zwischenträger *m*
i. grade Zwischenstadium *n*
i. guide blade Zwischenleitschaufel *f (Turbine)*
i. heat exchanger Zwischenwärme[aus]tauscher *m*
i. image *(Opt)* Zwischenbild *n*
i. image plane Zwischenbildebene *f*
i. layer Zwischenschicht *f*
i. layer of paper Papierzwischenlage *f*
i. level *(Brg)* Zwischensohle *f*, Teilsohle *f*
i. member Zwischenträger *m*, Zwischenglied *n*, Zwischenorgan *n*; Zwischenstück *n*
i. neutron mittelschnelles (intermediäres) Neutron *n*
i. office *(Fmt)* Zwischenamt *n*
i. optic Zwischenoptik *f*
i. orbit *(Rak)* Zwischenbahn *f*
i. pack *(Brg)* Mitteldamm *m (Versatz)*
i. phase *(Ph)* intermediäre Phase *f*
i. pier *(Bw)* Zwischenpfeiler *m*
i. pillar *(Bw)* Zwischenpfeiler *m*
i. port Zwischenhafen *m*, Durchgangshafen *m*
i. position Mittelstellung *f*
i. power amplifier Verstärker *m* mit mittlerer Leistung
i. preparation *(Ch)* Zwischenstufe *f*; Herstellung *f* von Zwischenprodukten, Zwischenstufe *f* in der Herstellung
i. pressure Mitteldruck *m*
i.-pressure cruising turbine Mitteldruckmarschturbine *f*
i. pressure cylinder Mitteldruckzylinder *m (Dampfmaschine)*
i. pressure diagram Mitteldruckdiagramm *n*
i. product Zwischenprodukt *n*
i. quality Zwischensorte *f*
i.-range rocket Mittelstreckenrakete *f*, Rakete *f* mit mittlerer Reichweite
i. reactor intermediärer (mittelschneller) Reaktor *m*
i. region *(Kern)* mittlerer Bereich *m*
i. repeater *(El)* Zwischenverstärker *m*
i. roll stand *(Met)* Mittelgerüst *n*
i. rolling mill *(Met)* Zwischenwalzwerk *n*
i. saddle Zwischensupport *m (Revolverdrehmaschine)*
i. selector *(Fmt)* Gruppenwähler *m*, GW
i. shaft Laufwelle *f*, Zwischenwelle *f*; Vorgelegewelle *f*
i. site Zwischengitterlage *f*
i. size Zwischenmaß *n*, Zwischengröße *f*
i. sleeper *(Eis)* Mittelschwelle *f*
i. slide valve Mitteldruckschieber *m*
i. softening *(Met)* Zwischenglühung *f*
i. stage Zwischenstadium *n*
i. state Zwischenzustand *m*
i. storage Zwischenspeicherung *f (von Daten)*

i. story Zwischengeschoß *n*
i. structure Zwischenstufengefüge *n*, Bainit *m*
i. subcarrier Zwischenhilfsträger *m (z. B. Farbfernsehtechnik)*
i. superheater Zwischenüberhitzer *m*
i. superheating Zwischenüberhitzung *f*
i. switch Zwischenschalter *m*; Kreuzschalter *m*
i. tapping *(Wärm)* Zwischendampfentnahme *f*
i.-temperature setting adhesive bei mittlerer Temperatur abbindender Klebstoff *m*
i. timing gear Zwischenrad *n*
i. track Zwischengleis *n*; Mittelfahrbahn *f*
i. transverse frame Zwischenring *m*
i. value Zwischenwert *m*
i. vessel Fracht- und Fahrgastschiff *n*
i. wall Zwischenwand *f*
i. waves *(Fmt)* Grenzwellen *fpl*
i. wheel Zwischenrad *n*
intermesh / to kämmen *(Zahnräder)*
intermeshing im Eingriff [befindlich] *(Zahnräder)*
i. rotors *(Flg)* ineinanderkämmende Rotoren *mpl (beim Hubschrauber)*
intermetallic intermetallisch
i. compound intermetallische (halbleitende) Verbindung *f*
i. semi-conductor intermetallischer Halbleiter *m*
intermingled *(Min)* durchwachsen
intermittent diskontinuierlich; absatzweise, stoßweise; ruckweise; periodisch, zyklisch; unterbrochen; intermittierend; aussetzend; zeitweise
i. arc *(Opt)* Abreißbogen *m*
i. arc producer Abreißbogenerzeuger *m*
i. board machine Handpappenmaschine *f*
i. contact Wackelkontakt *m*; zeitweise Leitungsberührung *f*
i. control *s.* discontinuous control
i. current intermittierender Strom *m*
i. cutting unterbrochene Spanung *f*
i. defect zeitweiser Defekt (Ausfall) *m*
i. direct current intermittierender Gleichstrom *m*, zerhackter pulsierender Gleichstrom *m*
i. discharge aussetzende Entladung *f*
i. disconnection zeitweise Unterbrechung *f (der Leitung)*
i. drive schrittweiser (ruckweiser) Transport *m*
i. duty *(Elm)* aussetzender Betrieb *m (mit während eines Spiels gleichbleibender Belastung)*
i. earth *(El)* aussetzender Erdschluß *m*
i. fault *s.* i. earth
i. feed *(Masch)* Sprungvorschub *m*
i. feed milling Sprungtischfräsen *n*
i. film feed Schrittschaltung *f* des Films
i. film motion Schrittschaltung *f* des Films
i. firing unterbrochene Beschickung *f (des Ofens)*
i. fringe stitch *(Text)* Wechselfranse *f*
i. interference *(El)* zeitweise Störung *f*
i. light Gleichtaktfeuer *n*, unterbrochenes Leuchtfeuer *n*, Blinklicht *n*
i. load aussetzende (pulsierende) Belastung *f*, Aussetzlast *f*
i. movement schrittweise Bewegung *f*
i. noise unterbrochenes Rauschen *n*
i. operation *(Elm)* Aussetzbetrieb *m*, aussetzender Betrieb *m*
i. periodic duty *(Elm)* periodisch aussetzender Betrieb *m (mit gleichbleibender Belastung)*
i. process absatzweiser (diskontinuierlicher) Betrieb *m*
i. production Aussetzbetrieb *m*; diskontinuierliche Fertigung *f*
i. service *(Ök)* aussetzender Betrieb *m* mit gleichbleibender Belastung
i. shock load stoßweise Belastung *f*
i. table feed Sprungtischschaltung *f (Frästisch)*
i. test Prüfung *f* bei aussetzender Entladung *(Batterie)*
i. weld unterbrochene Schweißnaht *f*
i. welding Schrittschweißung *f*
intermitter Pumpe *f* mit unterbrochener Leistung
intermix / to durchmischen, vermischen
intermixing Vermischung *f*
intermodulation *(El)* Zwischenmodulation *f*
i. distortion Verzerrung *f* durch Zwischenmodulation
i. frequency Differenzton *m*
intermolecular intermolekular, zwischenmolekular
i. cross linking *(Ch)* intermolekulare Vernetzung *f*
i. forces intermolekulare Kräfte *fpl*
i. rearrangement *(Ch)* intermolekulare Umlagerung *f*
internal 1. innerlich; Innen-; 2. inländisch
i. absorption coefficient *(Kern)* Selbstabsorptionskoeffizient *m*
i. adsorption *(Ph)* innere Adsorption *f*
i. aerial Innenantenne *f*, Zimmerantenne *f*
i. air lines Inlandsflugstrecken *fpl*
i. annual shake *(Holz)* Ringschäle *f*
i. ballast *(Schiff)* Innenballast *m*
i. base resistance Basisinnenwiderstand *m (Halbleitertechnik)*
i. bevel gear Innenkegelrad *n*
i. broach Innenräumwerkzeug *n*, Räumnadel *f*
i. broaching Innenräumen *n*
i. broaching machine Innenräummaschine *f*
i. cable Innenkabel *n*

i. caliper gauge Bohrungslehre *f*, Innentaster *m*
i. capacitance Wicklungskapazität *f (Spulen)*; Elektrodenkapazität *f (Röhren)*
i. capacities of tubes Röhrenkapazitäten *fpl*
i. characteristic innere Kennlinie *f*
i. charge *(Fmt)* Inlandsgebühren *fpl*
i. circuit innerer Stromkreis *m*
i. combustion engine Verbrennungsmotor *m*
i. combustion rammer *(Bw)* Explosionsramme *f*
i. combustion turbine Verbrennungsturbine *f*, Gasturbine *f (mit innerer Verbrennung und offenem Kreislauf)*
i. conductor *(El)* Innenleiter *m*
i. consumption Eigenverbrauch *m*
i. conversion *(Kern)* innere Konversion (Umwandlung) *f*
i. conversion coefficient Konversionskoeffizient *m*, Koeffizient *m* der inneren Umwandlung
i. conversion electron Konversionselektron *n*
i. conversion probability Konversionswahrscheinlichkeit *f*
i. coolant Kühlmittel *n* für direkte Kühlung
i. corrosion *(Wärm)* innere Abzehrung *f*, Innenkorrosion *f*
i. crack *(Met)* Innenriß *m*
i. cylindrical gauge Bohrungslehre *f*, Lochlehre *f*
i. cylindrical grinding machine Innenrundschleifmaschine *f*
i. cylindrical lapping Innenrundläppen *n*
i. damping Eigendämpfung *f*, innere Dämpfung *f*
i. diameter Innendurchmesser *m*, lichte Weite *f*
i. dimension Innenmaß *n*
i. discharge innere Beaufschlagung *f (Pumpe)*
i. dose *(Kern)* Dosis *f* für innere Bestrahlung
i. electrode Innenelektrode *f*
i. energy Energiegehalt *m*
i.-expanding clutch Spreizbandkupplung *f*, Innenbackenkupplung *f*
i.-expanding shoe brake Innenbackenbremse *f*
i. exposure *(Kern)* innere Bestrahlung *f*
i. fan kiln Trockenkammer *f* mit Innenventilator
i. faultsman *(Fmt)* Störungssucher *m* im Innendienst
i. field alternator Innenpolwechselstromgenerator *m*
i. finish Ausbau *m*, Innenausbau *m*
i. firebox Flammrohrkessel *m*
i. firebox boiler Kessel *m* mit Innenfeuerung
i. firing Innenfeuerung *f*
i. fissure innerer Riß *m*
i. flange eingezogener Flansch *m*
i. flue *(Wärm)* Innenzug *m*
i.-flue boiler Feuerbuchskessel *m*
i. force Schnittkraft *f (Statik)*; innere Kraft *f*
i. friction Eigenreibung *f*; innere Reibung *f*, Viskosität *f*
i. friction factor Koeffizient *m* der inneren Reibung, dynamische Zähigkeit *f*, Viskosität *f*
i. furnace Innenfeuerung *f*
i. furnace boiler Flammrohrkessel *m*
i. gear Hohlrad *n*, Innenrad *n*, innenverzahntes Rad *n*; Zahnkranz *m*
i. gear shaving machine Innenverzahnungsschabemaschine *f*
i. gearing Innenverzahnung *f*
i. grinder Innenschleifmaschine *f*
i. grinding Innenschleifen *n*, Innenschliff *m*
i. grinding attachment Innenschleifeinrichtung *f*
i. grinding machine Innenschleifmaschine *f*
i. grinding of cylindrical holes Ausschleifen *n* zylindrischer Bohrungen
i. gripping *(Masch)* Innenspannung *f*
i. groove Auskehlung *f*
i. hardening stress *(Met)* Härtungsspannung *f*, Härtespannung *f*
i. heating Selbst[auf]heizung *f*; Innenfeuerung *f*
i. holding *(Masch)* Innenspannung *f*
i. inductance Eigeninduktivität *f*
i. input resistance *(El)* Gitterarbeitswiderstand *m*
i. installation Inneneinrichtung *f*
i. irradiation *(Kern)* innere Bestrahlung *f*
i. jaw *(Masch)* Innenbacke *f*
i. kinetic energy innere kinetische Energie *f*, innere Bewegungsenergie *f*
i. lapping machine Innenläppmaschine *f*
i. limit gauge Grenzlehrdorn *m*, Toleranzlochlehre *f*
i. losses *(El)* Eigenverluste *mpl*
i. measurement Innenmessung *f*
i. member Innen[paß]teil *n*
i. memory innerer (eingebauter) Speicher *m (für Daten)*
i. micrometer [Feinmeß-]Schraublehre *f* für Innenmessungen
i. milling attachment Innenfräsapparat *m*
i. mixer *(Ch)* Banburry-Mischer *m*, Innenmischer *m*
i. navigation Binnenschiffahrt *f*
i. noise Eigenrauschen *n*
i. pair Konversionselektronenpaar *n*
i. photoeffect innerer Fotoeffekt *m*, Halbleiterfotoeffekt *m*
i. photoelectric effect *s.* i. photoeffect
i. poison against vermin Fraßgift *n* gegen Schädlinge

i. **pole** Innenpol *m*
i. **pressure** Innendruck *m*; innerer Druck *m*
i. **profile** Innenprofil *n*
i. **radius** Ausrundung *f*
i. **resistance** *(El)* Innenwiderstand *m*, innerer Widerstand *m*
i. **rotation** Eigenrotation *f*; innere Rotation *f*
i. **screw-cutting tool** Innengewindemeißel *m*
i. **screw gauge** Gewindelehrdorn *m*
i. **screw thread** Innengewinde *n*, Muttergewinde *n*
i. **self-inductance** *(El)* innere Selbstinduktion *f*
i. **serration** Kerbzahnnabenprofil *n*
i. **shoe** Innenbacke *f*, Innenbremsbacke *f*
i. **shoe brake** Innenbackenbremse *f*
i. **slackness** Luft *f*, Spiel *n (Lager)*
i. **spline[s]** Keilnabenprofil *n*
i. **spring ring** hinterlegter Spannring *m*, Grundring *m*
i. **springing** Innenfederung *f*
i. **spud vibrator** *(Bw)* Flaschenrüttler *m*, Innenrüttler *m*, Tauchrüttler *m*
i. **spur gear** Innenzahnkranz *m*
i. **store** *s.* i. memory
i. **stress** Eigenspannung *f*; Restspannung *f*; innere Spannung *f*
i. **supercharger** *(Masch)* Innenvorverdichter *m*
i. **surface** Innenfläche *f*
i. **taper** Hohlkegel *m*
i. **taper cylindrical gauge** Kegellehrdorn *m*
i. **teeth lock washer** federnde Zahnscheibe *f* mit Innenzähnen
i. **tension** Eigenspannung *f*
i. **thermometer** Innenthermometer *n*
i. **thread** Innengewinde *n*, Muttergewinde *n*
i. **thread chaser** Innengewindestrehler *m*
i. **thread gauge** Gewindelehrdorn *m* [für Innengewinde]
i. **thread grinding** Innengewindeschleifen *n*
i. **thread milling attachment** Innengewindefräsapparat *m*
i. **threading** Innengewindeschneiden *n*
i. **threading operation** Innengewindeschneidarbeit *f*
i. **toothed** mit Innenverzahnung
i. **toothing** Innenverzahnung *f*
i. **traffic** innerbetrieblicher Verkehr *m*
i. **trouble** Eigenstörung *f*
i. **turning tool** Innendrehmeißel *m*
i. **upset drill pipe** innenverstärktes Bohrgestänge *n*
i. **vibration** Innenrüttlung *f (Betontechnologie)*
i. **vibrator** *(Bw)* Flaschenrüttler *m*, Innenrüttler *m*, Tauchrüttler *m*
i. **viscosity** *(Ch)* Eigenviskosität *f*
i. **voltage** *(El)* innere Spannung *f*
i. **volume** Eigenvolumen *n*
i. **volume drop** innerer Spannungsabfall *m*
i. **wiring** *(El)* Innenverdrahtung *f*
i. **wrenching bolt** Innensechskantschraube *f*
internally fired furnace Innenfeuerung *f*
i. **frosted** innenmattiert *(Beleuchtungskörper)*
i. **geared wheel** Zahnrad *n* mit Innenverzahnung
i. **ribbed** *(Masch)* mit Innenverrippung
i. **sprung** innen gefedert
i. **threaded** mit Innengewinde
international international; *(Fmt)* zwischenstaatlich
i. **air law** internationales Luftrecht *n*
i. **ampere** internationales Ampere *n (= 0,99985 absolute Ampere)*
i. **call** *(Fmt)* Auslandsgespräch *n*
i. **calling frequency** *(Fmt)* internationale Ruffrequenz *f*
i. **candle** internationale Kerze *f (= 1,019 Candela)*
i. **date line** Datumsgrenze *f*
i. **distress frequency** internationale Seenotfrequenz *f*
i. **exhibition** Weltausstellung *f*
i. **line** *(Fmt)* Auslandsleitung *f*
i. **metric measures** metrisches Maßsystem *n*
i. **monitoring of radio** internationale Funküberwachung *f*
i. **monitoring system** internationales Überwachungssystem *n (Funk)*
i. **ohm** internationales Ohm *n (= 1,0049 absolute Ohm)*
i. **prototype meter** Urmeter *n*
i. **radio silence** internationale Funkstille *f*
i. **radium standard** internationaler Radiumstandard *m*
i. **sight-testing chart** *(Opt)* internationale Sehprobentafel *f*
i. **spelling table** internationale Buchstabiertafel *f*
i. **standard metric thread** metrisches Gewinde *n*
i. **telegraphic convention** Welttelegrafenvertrag *m*
i. **telegraphic union** Welttelegrafenverein *m*
i. **telephone circuit** zwischenstaatliche Fernsprechleitung *f*
i. **terminal office** zwischenstaatliches Fernamt *n*
i. **titre** *(Ch)* Dezimaltiter *m*
i. **trade** Welthandel *m*; Weltmarkt *m*
i. **traffic** zwischenstaatlicher Verkehr *m*
i. **transit exchange** *(Fmt)* zwischenstaatliches Durchgangsamt *n*
i. **unit** internationale Einheit *f*

i. unit for measurement internationale Maßeinheit *f*
i. volt internationales Volt *n* *(= 1,00034 absolute Volt)*
internet traffic *(Fmt)* Querverkehr *m*
internodal distance Knotenabstand *m*
internode of oscillations Schwingungsbauch *m*
internuclear internuklear, zwischen den Kernen
i. distance Kernabstand *m*
interocular distance *(Opt)* Augenabstand *m*
interoffice trunk *(Fmt)* Amtsverbindungsleitung *f*
interpenetrate / to sich gegenseitig durchdringen
interpenetration gegenseitige Durchdringung *f*
i. twin Durchdringungszwilling *m (Kristall)*
interphase Zwischenphase *f*; Phasengrenze *f*, Phasengrenzfläche *f*
i. reactor *(El)* Ausgleichsdrossel *f*, Saugdrossel *f*
i. transformer Phasendrehtransformator *m*; Phasenschieber *m*; Saugdrossel *f (Stromrichter)*
interphone Haustelefon *n*, Haussprechanlage *f*
interplanar crystal spacing Netzebenenabstand *m*, Abstand *m* der Kristallebenen
i. spacing Ebenenabstand *m*, Abstand *m* zwischen zwei Ebenen
interplanetary interplanetar
i. aviation [interplanetare] Raumfahrt *f*, Kosmonautik *f*
i. communications interplanetarische Verbindungen *fpl*
i. flight interplanetarer Flug *m*
i. navigation interplanetare Raumnavigation *f*
i. radio contact interplanetare Funkverbindung *f*
i. space interplanetarer Raum *m*
i. trip interplanetarer Flug *m*
interpolate / to *(Math)* interpolieren
interpolation Interpolation *f*; Einbeziehen *n*
i. resistance Interpolationswiderstand *m*
interpolator *(Dat)* Interpoliereinrichtung *f*
interpole *(El)* Zwischenpol *m*, Wendepol *m*
interpolymer *(Plast)* Mischpolymerisat *n*
interposition *(El)* Zwischenschaltung *f*
interpret / to deuten; auswerten *(Statistik)*; lesen *(Zeichen, Zeichnung)*
i. drawings Zeichnungen lesen
interpretative routine *(Dat)* interpretierendes (interpretatives) Programm *n*, Zuordnerprogramm *n*, Zuordnungsprogramm *n*
interpreter 1. Dolmetscher *m*; 2. *(Dat)* Zuordner *m*, interpretierende Einrichtung *f*; Zuordnungsprogramm *n*; Zuordnerkode *m*
interpreting routine *s.* interpretative routine
interpupillary distance *(Opt)* Augenabstand *m*
i. distance gauge Augenabstandsmesser *m*
interrogate / to abfragen *(z. B. Magnetkernspeicher)*
interrogate winding *(BMSR)* Abfragewicklung *f*
interrogating pulse Abfrageimpuls *m (Magnetkernspeicher)*
interrogation Abfragung *f (Radar)*
i. frequency Abfragefrequenz *f*
i. range Abfrageweite *f*
interrogator *(Fs)* Abfragesender *m*
i.-responder Frage-Antwort-Radargerät *n*, Abfragesender *m* mit Empfänger
i.-responder transponder unit kombiniertes Frage-Antwort-Relaisgerät *n (Radar)*
interrupt / to 1. *(El)* unterbrechen; trennen; 2. abreißen *(Lichtbogen)*; 3. *(Geol)* sich verschieben *(Schichten)*
i. a connection *(Fmt)* eine Verbindung trennen
interrupted unterbrochen
i. arc Abreiß[licht]bogen *m*
i. continuous wave telegrafenmodulierte ungedämpfte Welle *f*; Dauerstrich *m* mit Anodenwechselspannung, getastete Trägerwelle *f*
i. direct current zerhackter Gleichstrom *m*
i. quench[ing] Stufenhärten *n*, gestuftes Härten *n*
i. waves *(El)* periodisch zerhackte Wellen *fpl*
interrupter Ausschalter *m*, Unterbrecher *m*
i. cam *(Fmt)* Unterbrechernocke *f*
i. shaft *(El)* Schaltwelle *f*
i. spark Unterbrecherfunke *m*
i. spring Unterbrecherfeder *f*
interrupting Unterbrechen *n*
i. capacity *(El)* Ausschaltleistung *f*
i. time *(El)* Abschaltzeit *f*
interruption Unterbrechung *f*
i. of current Stromabschaltung *f*
i. of traffic Verkehrsunterbrechung *f*
interruptive capacity *(El)* Abschaltleistung *f*
interruptor Unterbrecher *m*
intersect / to *(Math)* durchdringen; sich schneiden; überschneiden
intersecting angle Schnittwinkel *m*
i. frame *(Text)* Doppelnadelstabstrecke *f*
i. gill box *(Text)* Doppelnadler *m*, Doppelhechelfeldstrecke *f*
i. gill drawing frame *(Text)* Doppelnadelstabstrecke *f*, Doppelhechelfeldstrecke *f*
i. plane Schnittebene *f*
i. point Schnittpunkt *m*
intersection 1. Durchdringung *f (technisches Zeichnen)*; Durchdringungslinie *f*; Schnittpunkt *m*; Schnittlinie *f*; Stoßstelle *f*; Durchschnitt *m*; 2. Straßenknoten *m*, Straßenkreuzung *f*, Kreuzung *f*; 3. *(Brg)* Durchörterung *f*
i. angle Schnittwinkel *m*

i. **line** Schnittlinie *f*
i. **point** Schnittpunkt *m*
intersetting packing *(Masch)* ineinandergreifende Packungsringe *mpl*
interspace Zwischenraum *m*
intersperse / to durchsetzen [mit]
interstage *(El)* Zwischenstufe *f*
i. **network** *(BMSR, Dat)* Koppelnetzwerk *n*, Kopplungsschaltung *f*
i. **pickling** *(Met)* Zwischenbeizen *n*
i. **reservoir** *(El)* Vorvakuumbehälter *m*
i. **transformer** *(Elm)* Zwischentransformator *m*
interstellar interstellar, kosmisch
i. **aviation** interstellare Raumfahrt *f*, Weltraumschiffahrt *f*, Kosmonautik *f*
i. **craft** Weltraumschiff *n*
i. **dust** kosmischer Staub *m*
i. **flight** interstellarer Flug *m*, [interstellare] Raumfahrt *f*, Weltraumfahrt *f*
i. **matter** interstellare Materie *f*
i. **navigation** [interstellare] Raumnavigation *f*
interstice Zwischenraum *m*, Spalt *m*, Fuge *f*; Riß *m*; Pore *f*
interstitial Zwischengitteratom *n*
interstitial interstitiell, in (mit) Zwischenräumen; Zwischen[raum]-
i. **atom** Zwischengitteratom *n*
i. **compound** *(Ch)* nichtstöchiometrische Verbindung *f*
i. **measurement** *(Opt)* Lückenmessung *f*
i. **solid solution** Einlagerungsmischkristall *m*
i. **water** Porenwasser *n*
interstratification *(Geol)* Zwischenlagerung *f*
interstratified *(Geol)* streifig
intertrack pulse time displacement *(BMSR)* Zwischenspurimpulszeitverschiebung *f*
intertripping *(El)* Mitnahmeschaltung *f (Relais)*
interval Intervall *n*; Zeitspanne *f*, Zwischenzeit *f*; Strecke *f*; Abstand *m*; Zwischenraum *m*
i. **error** Pausenfehler *m*
i. **in transmission** Sendepause *f*
i. **of graduation** *(Masch)* Teilungsintervall *n*
i. **of integration** Integrationsbereich *m*
i.**-selector circuit** *(Fs)* Auftastkreis *m*; Koinzidenzkreis *m*, Zeitdiskriminatorkreis *m*
i. **signal** Pausenzeichen *n*
intervalve coupling *(El)* Röhrenkopplung *f*, Stufenkopplung *f*
i. **space** Saugkammer *f*
i. **timer** Schaltuhr *f*
intervening bulkhead *(Schiff)* Trennschott *n*
i. **size** *(Masch)* Zwischenmaß *n*
interweave / to verflechten, verweben
interwinding capacitance *(El)* Wicklungskapazität *f (Spule)*
i. **floating take-up** *(Text)* schwingender Warenabzug *m*
intimate innig; eng
i. **blending** *(Text)* gutes Mischen *n (von Fasern)*
i. **contact** inniger Kontakt *m*
i. **mixture** inniges Gemisch *n*
intolerant tree Lichtholzart *f*
intraatomic inneratomar, intraatomar
intracavitary irradiation intrakavitäre Bestrahlung *f*, Innenraumbestrahlung *f*
intracrystalline intrakristallin, durch Einzelkristalle
intrados Bogenleibung *f*, Gewölbeleibung *f*; Unterseite *f*, Untersicht *f (Bogen, Gewölbe usw.)*
i. **surface** Leibungsfläche *f*
intradosal burst *(Brg)* Gebirgsschlag *m* aus der Zerrüttungszone
intramolecular intramolekular, innermolekular
i. **forces** innermolekulare Kräfte *fpl*
i. **motions** intramolekulare Bewegungen *fpl*
i. **rearrangement** innermolekulare Umlagerung (Umordnung) *f*
intranuclear intranuklear, innernuklear
intratomic *s.* intraatomic
intravitam fluorescent dye staining *(Opt)* Vitalfluorochromierung *f*
intricacy Kompliziertheit *f*, Schwierigkeit *f*, Verwickeltheit *f*
intricate kompliziert, verwickelt, schwierig
i. **moulded part** Preßteil *n* von komplizierter Formgebung, schwieriges Preßteil
intrinsic innerlich; eigentlich; Eigen-; wesentlich; zugehörig, innewohnend
i. **angular momentum** Eigendrehmoment *n*, Spin *m*
i. **brilliance (brilliancy, brightness)** Flächenhelligkeit *f*, Flächenhelle *f*
i. **conduction** Eigenleitung *f (Halbleiter)*
i. **conductivity** Eigenleitung *f (Halbleiter)*
i. **coordinates** natürliche Koordinaten *fpl*
i. **curve** Mohrsche Umhüllungskurve *f (Gebirgsmechanik)*
i. **density** *(Ph)* Intrinsic-Dichte *f*
i. **energy** innere Energie *f*; Eigenenergie *f*
i. **error** Grundfehler *m* eines Meßinstrumentes
i. **layer** eigenleitende Sperrschicht *f (des Transistors)*
i. **magnetic moment** magnetisches Eigenmoment *n*
i. **permeability** *(El)* Suszeptibilität *f*
i. **pressure** Binnendruck *m*, Kohäsionsdruck *m*
i. **radiance** Strahlungsvermögen *n*
i. **safety** Eigensicherheit *f*
i. **semiconductor** Halbleiter *m* mit Eigenleitung, Eigenhalbleiter *m*
i. **value** Eigenwert *m*

i. viscosity Strukturviskosität *f*, grundmolare Viskositätszahl *f*, Grundviskosität *f*, Eigenviskosität *f*
intrinsically safe eigensicher
introduce / to einführen; einziehen, einlegen *(Papier)*; einsetzen, hineinführen
introduction Einführung *f*, Einleitung *f*
i. of lye Laugenzuführung *f*
introductory lecture Einführungsvorlesung *f*
intrude / to *(Geol)* intrudieren, eindringen
intrusion 1. *(Geol)* Intrusion *f*; 2. *(Met)* Durchfressen *n (Ofenauskleidung)*
i. grout Einpreßmörtel *m*
i. mortar Einpreßmörtel *m*
intrusive rock *(Geol)* Intrusivgestein *n*
i. sheet *(Geol)* Intrusivlager *n*, Lagergang *m*, Sill *m*
i. vein *(Geol)* Eruptivgang *m*
inunction Einölung *f*
inundate / to überfluten, überschwemmen
inundation Überflutung *f*, Überschwemmung *f*, Hochwasser *n*
invalid ungültig *(z. B. Lehrsatz)*
invalidity Ungültigkeit *f*
invar *(Met)* Invar *n*
invariable unveränderlich, gleichbleibend, invariabel
i. plane unveränderliche Ebene *f*
i. voltage *(El)* gleichbleibende Spannung *f*
invariance *(Math)* Invarianz *f*
invariant *(Math)* Invariante *f*
invariant *(Math)* invariant
i. theory Invariantentheorie *f*
invariants of strain Invarianten *fpl* des Deformationszustandes
i. of stress Invarianten *fpl* des Spannungszustandes
invent / to erfinden
invention Erfindung *f*
inventor Erfinder *m*
inventor's right Erfinderrecht *n*
inventory Stückliste *f*, Verzeichnis *n*; Inventar *n*; Inventur *f*
i. and material accounting *(Büro)* Inventur- und Materialabrechnung *f*
i. control *(Dat)* Vorratslenkung *f*, Regelung *f* von Speichervorräten
i. machine Inventurmaschine *f*
i. verification Inventurprüfung *f*
inverse Kehrwert *m*; Umgekehrte *f*
inverse entgegengesetzt, entgegengerichtet, umgekehrt, verkehrt; reziprok
i. amplification factor Durchgriff *m*, Anodendurchgriff *m (Elektronenröhre)*
i. anode voltage Sperrspannung *f (Anode)*
i. beta transformation *(Ph)* inverse Beta-Umwandlung *f*
i. circular zyklometrisch
i. cosecant *(Math)* Arkuskosekans *m*
i. cosine *(Math)* Arkuskosinus *m*
i. cotangent *(Math)* Arkuskotangens *m*
i. current Sperrstrom *m (Elektronenröhren, Halbleiter)*
i. current method Gegenstrommethode *f*
i. current ratio Rückwärtsübersetzung *f (Stromübersetzung; Vierpole)*
i. derivative controller Nachhalter *m (pneumatischer Reglerbaustein)*
i. direction Sperrichtung *f (Elektronenröhren, Halbleiter)*
i. electrode current Katodenstrom *m*
i. feedback Gegenkopplung *f*, negative Rückkopplung *f*
i. feedback circuit Gegenkopplungsschaltung *f*
i. function Umkehrfunktion *f*, reziproke Funktion *f*
i. grid current positiver Gitterstrom *m*
i. leakage current *(El)* Sperrstrom *m*
i. logarithm Numerus *m*
i. matrix Kehrmatrix *f*
i. modulation Gegenmodulation *f*
i. multiplet *(Ph)* verkehrtes Multiplett *n*
i. of a point [with respect to a circle] inverser Punkt *m* [bezogen auf einen Kreis]
i.-parallel connection Antiparallelschaltung *f*
i. peak voltage Spitzensperrspannung *f*
i. period Sperrzeit *f (bei negativer Anodenspannung; Stromrichter)*
i. process Umkehrprozeß *m*
i. proportional umgekehrt proportional
i. ratio Kehrwert *m*
i. rectifier Wechselrichter *m*
i. resonance Stromresonanz *f*, Parallelresonanz *f*, Antiresonanz *f*
i. secant *(Math)* Arkussekans *m*
i. segregation *(Met)* umgekehrte Blockseigerung *f*
i. sine *(Math)* Arkussinus *m*
i. slope Schaltkonstante *f (Ferrite)*
i. speed motor Motor *m* mit Hauptstromverhalten
i. square law quadratisches Abstandsgesetz *n*
i. tangens *(Math)* Arkustangens *m*
i. time gegenläufiges Zeitintervall *n*
i. time-lag abhängig verzögert *(z. B. Schalter)*
i. time-lag relay reziprok abhängiges Zeitrelais *n*
i. time-lag relay with definite minimum begrenzt abhängiges Zeitrelais *n*
i. transformation *(Ph)* umgekehrte Änderung *f*
i. trigonometric function *(Math)* Arkusfunktion *f*
i. voltage *(El)* Sperrspannung *f*, Rückspannung *f*

i. voltage ratio Rückwärtsübersetzung *f (Vierpol)*
inversely proportional umgekehrt proportional
inversion Umkehrung *f*; Inversion *f*
i. circuit *(Dat)* Inversionsschaltung *f*, NICHT-UND-Schaltung *f (logisch)*
i. doubling Inversionsverdopplung *f*
i. formula Umkehrformel *f*
i. integral Umkehrintegral *n*
i. layer Inversionsschicht *f*
i. of spectral lines *(Opt)* Selbstumkehr *f* von Spektrallinien
i. of the cycle Umkehrung *f* des Kreisprozesses
i. spectrum Umkehrspektrum *n*, Inversionsspektrum *n*
i. temperature Inversionstemperatur *f*, thermischer Umkehrpunkt *m*
i. theorem Umkehrsatz *m*
i. velocity Inversionsgeschwindigkeit *f*
invert / to umwenden; umkehren
invert 1. *(Bw)* Kanalsohle *f*; Grundplatte *f*, Sohlplatte *f*; 2. *(Brg)* Sohlenbogen *m (Ausbau)*
i. sugar Invertzucker *m*
invertase *(Ch)* Invertase *f*
inverted umgekehrt, verkehrt; kopfstehend
i. arch foundation verkehrtes Gewölbe *n (Grundbau)*
i. converter Wechselrichter *m*
i. elliptical motion umgekehrter Ellipsenlenker *m*
i. engine *(Kfz)* Motor *m* mit hängenden Ventilen, hängender Motor
i. exclusive OR umgekehrtes ausschließendes ODER *n (logisch)*
i. flat slab foundation *(Bw)* Pilzgrundplatte *f*
i. function Umkehrfunktion *f*
i. image umgekehrtes Bild *n*
i. in-line engine Reihenmotor *m* mit hängenden Zylindern
i. machine rotorgespeiste (läufergespeiste) Maschine *f*
i. microscope umgekehrtes Mikroskop *n*
i. rectifier Wechselrichter *m*
i. repulsion motor Repulsionsmotor *m* mit kurzgeschlossenem Feld
i. siphon *(Bw)* Düker *m*
i. speech *(Fmt)* verschlüsselte Sprache *f*
i. suspension umgekehrte Aufhängung *f*
i. tooth chain Zahnkette *f*
i. V-bottom boat Seeschlitten *m (amerikanischer Gleitboottyp)*
i. V-engine V-Motor *m* mit hängenden Zylindern
i. vee-engine *s.* i. V-engine
i. vee-guide *(Masch)* Prismenführung *f*
inverter 1. *(Elm)* Wechselrichter[gruppe *f*] *m*; 2. Inverter *m*, Vorzeichenumkehrverstärker *m*, Vorzeicheninverter *m*, Vorzeichenumkehrer *m (Analogrechner)*
invertible *(Math)* umkehrbar
inverting *(Math)* Umkehrung *f*
i. amplifier Vorzeichenumkehrverstärker *m (Analogrechner)*
i. circuit *s.* inverter 2.
i. mirror Umkehrspiegel *m*
i. prism Umkehrprisma *n*
invertor *(El)* Wechselrichter *m*; *(Fmt)* Inverter *m*
investigate / to erforschen, untersuchen, ermitteln
investigation Erforschung *f*, Untersuchung *f*, Ermittlung *f*
investment casting *(Met)* 1. Gießen *n* mit verlorener Gießform, Modellausschmelzgießverfahren *n*; 2. Genauigkeitsgußstück *n*
i. pattern Ausschmelzmodell *n*
inviscid dünnflüssig
invisibility Unsichtbarkeit *f*
invisible unsichtbar
invoice Rechnung *f*; Lieferschein *m*
i. form Rechnungsvordruck *m*, Abrechnungsvordruck *m*
i. writing Rechnungsausstellung *f*, Ausstellung *f* einer Faktura
invoicing machine *(Büro)* Fakturiermaschine *f*
involutary *(Math)* Faltungs- *(z. B. Integral)*
involute evolventenförmig gekrümmt, evolventisch
involute Evolvente *f (Zahnrad)*
i. connection *(Elm)* Evolventenstirnverbindung *f*
i. curve Evolvente *f*, Fadenlinie *f*
i. function Evolventenfunktion *f*
i. gear Evolventenzahnrad *n*, Zahnrad *n* mit Evolventenverzahnung
i. gear cutter Zahnformfräser *m* für Evolventenverzahnung
i. gear stocking cutter Zahnformvorfräser *m* für Evolventenzahnräder
i. gear teeth Evolventenverzahnung *f*
i. gearing Evolventenverzahnung *f*
i. helical gear Evolventenschrägzahnrad *n*
i. pinion *(Am)* Evolventenritzel *n*, Ritzel *n* mit Evolventenverzahnung
i. serration Evolventenkerbverzahnung *f*
i. splines Evolventenkeilverzahnung *f*, Zahnnabenprofil (Zahnwellenprofil) *n* mit Evolventenflanken
i. teeth Evolventenverzahnung *f*
i. tester Evolventenprüfgerät *n*
i. tooth *(Masch)* Evolventenzahn *m*
i. tooth hob Wälzfräser *m* für Evolventenprofile
i. tooth profile Evolventenprofil *n*
i. tooth profile worm hob Wälzfräser *m* für Schneckenräder mit Evolventenverzahnung

i. toothing Evolventenverzahnung *f*
involution *(Math)* Potenzierung *f*
involutometry Evolventengeometrie *f*
involve / to *(Math)* potenzieren
inwale *(Schiff)* Dollbord *m*, Wegerung *f*
inward batter Verjüngung *f*
i. bulging Einbeulung *f*, Einbauchung *f*
inwardly inclined bearing *(Masch)* nach innen geneigtes Lager *n*
inyoite *(Min)* Inyoit *m*
iodate / to jodieren
iodate Jodat *n*, Salz *n* der Jodsäure
iodic acid Jodsäure *f*
iodide Jodid *n*
i. process Van-Arkel-Prozeß *m*
iodiferous jodhaltig
iodinate / to jodieren
iodine Jod *n*
i. chloride Jodchlorid *n*, Jodtrichlorid *n*
i. flask Jodierungskolben *m*, Jodzahlkolben *m*
i. number Jodzahl *f*
iodizate / to jodieren
iodized starch Jodstärke *f*
iodoacetic acid Jodessigsäure *f*
iodoform Jodoform *n*
iodometry Iodometrie *f*
iodous acid hypojodige (unterjodige) Säure *f*
iolite *(Min)* Kordierit *m*
ion Ion *n* (*Zusammensetzungen s. a. unter* ionic)
i. acceleration Ionenbeschleunigung *f*
i. accelerator Ionenbeschleuniger *m*
i. acceptor Ionenakzeptor *m*
i. activity Ionenaktivität *f*
i. avalanche Ionenlawine *f*; Trägerlawine *f*
i. beam Ionenstrahl *m*, Ionenbündel *n*
i. beam scanning [massenspektrometrische] Ionenstrahlanalyse *f*
i. burn[ing] *(Fs)* Ionen[brenn]fleck *m*
i. cloud Ionenwolke *f*
i. cluster Ionenschwarm *m*, Ionencluster *m*
i. concentration Ionenkonzentration *f*
i. conduction Ionenleitung *f*
i. core Ionenrumpf *m*
i. counter Ionenzähler *m*, Ionenzählrohr *n*
i. current Ionengitterstrom *m*
i. deflection Ablenkung *f* des Ionenstrahlbündels
i. density Ionendichte *f*, Ionenkonzentration *f*
i. detector Ionensonde *f*
i.-dipole bond Ionenbindung *f*, heteropolare (ionogene) Bindung *f*
i. dose Ionendosis *f*, Ladungsdosis *f*
i. emission Ionenemission *f*
i. exchange Ionenaustausch *m*
i.-exchange column Ionenaustauschersäule *f*
i.-exchange resin Ionenaustausch[er]harz *n*
i.-exchange separation Trennung *f* durch Ionenaustausch
i.-exchange unit Ionenaustauscher *m*
i. exchanger Ionenaustauscher *m*
i.-free ionenfrei *(z. B. Dampf)*
i. gauge Ionenmanometer *n*, Ionisationsmanometer *n*
i. gun Ionenkanone *f*, Ionenquelle *f*
i. hose Ionenschlauch *m*
i. impact Ionenstoß *m*
i. impact ionization Ionenstoßionisation *f*
i. lattice Ionengitter *n*
i. loss Ionenverlust *m*
i. meter Ionisationsdosimeter *n*
i. microscope Ionenmikroskop *n*
i. migration Ionenwanderung *f*
i. mobility Ionenbeweglichkeit *f*
i. motor Ionenmotor *m*, Ionenantrieb *m*
i. movement Ionenbewegung *f*
i. optics Ionenoptik *f*
i. orbit Ionenbahn *f*
i. pair Ionenpaar *n*
i.-pair yield Ionenpaarausbeute *f* *(Ionisierung)*
i. physics Ionenphysik *f*
i. production Ionenerzeugung *f*
i. retarding field Ionenbremsfeld *n*
i. rocket Ionenrakete *f*
i. source Ionenquelle *f*
i. spectrum Ionenspektrum *n*
i. spot *(Fs)* Ionen[brenn]fleck *m*
i. strength Ionenstärke *f*
i. transfer *(El)* Iontophorese *f*, Ionophorese *f*
i. trap Ionenfalle *f*
i. trap magnet Magnet *m* für Ionenfalle
i. tube Ionenröhre *f*; Ionenrohr *n*
i. yield Ionenausbeute *f*
Ionian, Ionic *(Bw)* ionisch
Ionian order ionische Säulenanordnung *f*
ionic Ionen- (*s. a. unter* ion)
i. atmosphere *s.* ion cloud
i. bond Ionenbindung *f*, heteropolare (ionogene) Bindung *f*
i. centrifuge Ionenzentrifuge *f*
i. charge Ionenladung *f*
i. cleavage Ionenspaltung *f*
i. collision Ionenstoß *m*
i. compound *s.* i. bond
i. conduction Ionenleitung *f*
i. conductivity Ionenleitfähigkeit *f*
i. conductor Ionenleiter *m*
i. core Ionenrumpf *m*
i. crystal Ionenkristall *m*
i. current *s.* ion current
i. discharge device *(El)* Ionenableiter *m*
i. equilibrium Ionengleichgewicht *n*
i.-heated cathode *(El)* Selbstaufheizkatode *f*
i. lattice Ionengitter *n* *(Kristalle)*
i. link (linkage) *s.* i. bond
i. medication *s.* ion transfer

i. **mobility** Ionenbeweglichkeit *f*
i. **optics** Ionenoptik *f*
i. **product** Ionenprodukt *n*
i. **propulsion** Ionenantrieb *m (für Raumfahrzeuge)*
i. **pump** Ionenpumpe *f*
i. **quantimeter** Ionendosismesser *m*
i. **rays** Ionenstrahlen *mpl*
i. **reaction** Ionenreaktion *f*
i. **replacement** *s.* ion exchange
i. **semiconductor** Ionenhalbleiter *m*
i. **separator** Ionenzentrifuge *f*
i. **strength** Ionenstärke *f*
i. **valency** Ionenvalenz *f*, Elektrovalenz *f*, heteropolare Valenz *f*
i. **valve** *(El)* Gasentladungsröhre *f*, gasgefüllte Röhre *f*, Ionenrohr *n*; Ionenventil *n*
i. **velocity** Ionengeschwindigkeit *f*
ionium *(Ch)* Ionium *n*
ionizability Ionisierbarkeit *f*
ionizable ionisierbar
ionization Ionisation *f*, Ionisierung *f*
i. **chamber** Ionisationskammer *f*
i. **chamber region** Ionisationskammerbereich *m*
i. **coefficient** *s.* i. constant
i. **constant** Ionisierungskonstante *f*
i. **core** Ionisationskern *m*
i. **counter** Ionisationszähler *m*
i. **cross section** Ionisierungsquerschnitt *m*
i. **current** Ionisationsstrom *m*, Ionisierungsstrom *m*
i. **density** Ionisationsdichte *f*
i. **dose** Ionisationsdosis *f*
i. **dose meter** Ionisationsdosimeter *n*
i. **function** Ionisierungsfunktion *f*
i. **gap** Ionisationsspalt *m*, Ionisationsöffnung *f*
i. **gauge** Ionisationsmanometer *n*
i. **glow** Elektronenstoßleuchten *n*
i. **impact** Ionisationsstoß *m*
i. **kernel** *s.* i. core
i. **loss** Ionisierungsverlust *m*, Ionisationsverlust *m*
i. **number** Ionisierungszahl *f*
i. **potential** Ionisierungsspannung *f*, Ionisationsspannung *f*, Ionisierungsarbeit *f*, Ionisationsenergie *f*, Ionisationspotential *n*
i. **power** Ionisierungsvermögen *n*
i. **pressure gauge** Ionisationsmanometer *n*
i. **probability** Ionisierungswahrscheinlichkeit *f*
i. **pulse** Ionisationsstoß *m*, Ionisationsimpuls *m*
i. **pump** *(Ph)* Ionisationspumpe *f*
i. **rate** Ionisationsgeschwindigkeit *f*
i. **state** Ionisationszustand *m*
i. **stopping** Ionisationsbremsung *f*
i. **tension** Ionisationsspannung *f*
i. **time** Ionisationszeit *f*, Ionisierungszeit *f*
i. **track** Ionisierungsspur *f*
i. **vacuum gauge** Ionisationsvakuummeter *n*
i. **yield** Ionisierungsausbeute *f*
ionize / to ionisieren
ionized atom Atomion *n*, ionisiertes Atom *n*
i. **layer** ionisierte Schicht *f*, Ionisationsschicht *f*
ionizing *s.* ionization
i. **capacity** *s.* i. power
i. **collision** Ionisierungsstoß *m*
i. **electrode** Ionisierungselektrode *f*
i. **energy** Ionisierungsenergie *f*, Ionisierungsarbeit *f*
i. **particle** ionisierendes Teilchen *n*
i. **power** Ionisierungsvermögen *n*, Ionisationsvermögen *n*, Ionisierfähigkeit *f*
i. **radiation** ionisierende Strahlung *f*
i. **rays** Ionisierungsstrahlen *mpl*
i. **time** Ionisationszeit *f*
ionogenic ionogen
ionoluminescence Ionolumineszenz *f*
ionometer Ionometer *n (Röntgenstrahlenintensitätsmesser)*
ionometric ionometrisch
i. **measurement method** ionometrisches Meßverfahren *n*
ionone *(Ch)* Jonon *n*, Ionon *n*
ionophoresis *(Ph)* Ionophorese *f*
ionosphere Ionosphäre *f*
i. **disturbance** ionosphärische Störung *f*
i. **echo** ionosphärisches Echo *n*
ionospheric path *(Fmt)* Ausbreitungsweg *m* in der Ionosphäre
i. **wave** Raumwelle *f (Funk)*
ionotherapie *s.* ion transfer
iontophoresis *s.* ion transfer
ipot *s.* inductive potentiometer
IR drop *(El)* ohmscher Spannungsabfall *m*
iridesce / to irisieren, schillern
iridescence Irisieren *n*, Schillern *n*
iridescent irisierend, schillernd; regenbogenfarbig
i. **colour** Schillerfarbe *f*, schillernde Farbe *f*
i. **fabric** *(Text)* irisierendes Gewebe *n*
i. **printing** *(Typ)* Irisdruck *m*
iridic compound Verbindung *f* des vierwertigen Iridium
iridium Iridium *n*
irido-compound Verbindung *f* des zweiwertigen Iridium
iridous compound Verbindung *f* des dreiwertigen Iridium
irignite *(Min)* Irignit *m*
irinite *(Min)* Irinit *m*
iris diaphragm *(Opt)* Irisblende *f*
irisation Irisieren *n*

iron / to 1. mit Eisen beschlagen; 2. bügeln
i. out plätten *(Furnier)*
iron 1. Eisen *n*; 2. Messer *n*; Kolben *m (Löten)*; 3. Bügel *m*; 4. Bügeleisen *n*
i. acetate Eisenazetat *n*
i. acetate liquor Eisenazetatlösung *f (Farbstoff)*
i. alloy Eisenlegierung *f*
i. anode Eisenanode *f*
i. arsenite Eisenarsenit *n*
i. bark *(Schiff)* Eisenholz *n*
i. bars Stabeisen *n*
i. black Eisenlack *m*
i. blade of plough *(Lw)* Pflugzunge *f*
i. blast furnace Eisenhochofen *m*
i. bromide Eisenbromit *n*, Bromeisen *n*
i. bound coast felsige Küste *f*
i. carbide Eisenkarbid *n*
i.-carbon alloy Eisen-Kohlenstoff-Legierung *f*
i.-carbon diagram Eisen-Kohlenstoff-Diagramm *n*
i. carbonyl Motyl *n*, Eisenpentakarbonyl *n*
i. casting 1. Eisenguß *m*; 2. Eisenguß[stück *n*] *m*
i. cell *(El)* Eisenelement *n*
i. cement Eisenkitt *m*
i.-cementite diagram Eisen-Eisenkarbid-Diagramm *n*
i. charge Eisengicht *f (Hochofen)*
i. chloride Eisenchlorid *n*
i. cinder Hochofenschlacke *f*
i.-clad gußgekapselt, eisenbewehrt, eisengepanzert
i.-clad switchgear eisengekapseltes (gußgekapseltes) Schaltgerät *n*
i.-clad transformer Manteltransformator *m*, Panzertransformator *m*, gepanzerter Transformator *m*
i.-constantan couple Eisen-Konstantan-Thermoelement *n*
i. core *(El)* Eisenkern *m*
i.-core armature Weicheisenanker *m*
i.-core coil Eisen[kern]spule *f*
i.-core inductance Eisenkerninduktivität *f*
i.-core transformer Eisen[kern]transformator *m*
i.-cored *(El)* mit Eisenkern versehen *(z. B. Spule)*
i.-cored choke coil Eisendrossel *f*
i.-cored coil Eisenkernspule *f*
i. cutters Blechschere *f*; Eisenschere *f*
i. cutting saw Eisensäge *f*
i. deposition Eisenniederschlag *m*
i. disk mill *(Ch)* Stiftmühle *f*
i. dross Gußschlacke *f*; Hochofenschlacke *f*
i.-dust core *(Elm)* Eisenpulverkern *m*, Eisenstaubkern *m*
i. dust slug *s.* i.-dust core
i. electrode Eisenelektrode *f*
i. extraction Enteisenung *f (Wasseraufbereitung)*
i. filings Eisenfeilspäne *mpl*
i. fluoride Eisenfluorid *n*
i. founder Eisengießer *m*
i. foundry Eisengießerei *f*
i.-free eisenfrei
i. furniture *(Typ)* Eisensteg *m*
i. glance *(Min)* Eisenglanz *m (Hämatit)*
i.-graphite diagram Eisen-Graphit-Diagramm *n*
i.-iron carbide diagram Eisen-Eisenkarbid-Diagramm *n*
i. jacket Eisenmantel *m*, Eisenumhüllung *f*
i. keel *(Schiff)* Eisenkiel *m*
i. kentledge *(Schiff)* Ballasteisen *n*
i.-lead matte *(Met)* Bleieisenstein *m*
i. lode *(Geol)* Eisenader *f*, Eisengang *m*
i. losses *(El)* Eisenverluste *mpl*
i.-lung eiserne Lunge *f (Medizintechnik)*
i. matte *(Min)* Eisenstein *m*
i.-matted runway *(Flg)* Stahlplattenstartbahn *f*
i. melt Eisenschmelze *f*
i. metallurgist Eisenhüttenmann *m*
i. meteorite Eisenmeteorit *m*
i. mica *(Min)* Eisenglimmer *m*
i. mill Eisenhütte *f*
i. mine Eisenerzbergwerk *n*
i. mixer Roheisenmischer *m*
i. modulator Eisenmodulator *m*
i.-mongery *s.* ironmongery
i. mould 1. *(Met)* Kokille *f*; 2. Rostfleck *m*
i. moving receiver *(Fmt)* elektromagnetischer Fernhörer *m*
i.-nickel accumulator Edisonakkumulator *m*
i.-nickel alloy Eisennickellegierung *f*
i. nitrate Eisennitrat *n*
i. notch Stich *m*, Abstichloch *n*, Eisenabstich *m (Hochofen)*
i. ochre Eisenmennige *n*
i. ore Eisenerz *n*
i. ore briquetting Eisenerzbrikettierung *f*
i. ore cement Eisenerzzement *m*
i. ore deposit Eisenerzlager *n*
i. ore dressing Eisenerzaufbereitung *f*
i. ore mining Eisenerzbergbau *m*
i. ore smelting Eisenerzverhüttung *f*
i. ore wash Eisenerzwäsche *f*
i. oxalate Eisenoxalat *n*
i. oxide Eisenoxid *n*
i.-oxide paint Eisen[oxid]farbe *f*
i. phosphide Phosphoreisen *n*, Eisenphosphid *n*
i. pig Eisenmassel *f*
i. plane Eisenhobel *m*
i. plate Stahlblech *n*

i. plate chimney Blechschornstein *m*
i. Portland cement Eisenportlandzement *m*
i. powder Eisenpulver *n*
i.-powder core *(Ph)* Massenkern *m*
i. protosulphate Eisen(II)-sulfat *n*, Ferrosulfat *n*
i. pyrites Eisenkies *m*, Schwefelkies *m*, Pyrit *m*
i. removal Enteisenung *f* *(Wasseraufbereitung)*
i. ribbon Stahlband *n*
i. runner *(Met)* Roheisenabstichrinne *f*
i.-saturated eisengesättigt
i. scale Eisenzunder *m*, Eisensinter *m*, Eisenhammerschlag *m*
i. scrap Eisenschrott *m*
i. smelting Eisenverhüttung *f*
i. spoke shave eiserner Schabhobel *m (mit Klinge)*
i.-stone *(Min)* Eisenerz *n*, Eisenstein *m*
i. sulphate Eisensulfat *n*, Eisenvitriol *n*
i. sulphide Eisensulfid *n*, Schwefeleisen *n*
i. tap Eisenabstich *m (Hochofen)*
i. tungstate Eisenwolframat *n*
i.-vane instrument Dreheiseninstrument *n*
i. wire card clothing *(Text)* Eisendrahtkratzenbeschlag *m*
i. wire-core coil Drahtkernspule *f*
i. wood Eisenholz *n*
i. work Schmiedearbeit *f*
i. work black Eisenlack *m*
i. works Eisenwerk *n*
i. yoke Eisenkern *m*
ironclad *s.* iron-clad
irone *(Ch)* Iron *n*
ironing Bügeln *n*
i. board Plättbrett *n*
i. press Bügelpresse *f*
ironmongery Beschläge *mpl*, Eisenwaren *fpl*, Kleineisenzeug *n*
ironwork *s.* ironmongery
irradiance Bestrahlungsstärke *f* *(W/cm²)*
irradiancy *s.* irradiance
irradiate / to bestrahlen, exponieren, beschießen *(mit Strahlen)*
irradiated bestrahlt, exponiert
i. fuel bestrahlter Brennstoff (Spaltstoff) *m*
i. fuel element bestrahltes Brennstoffelement *n*
irradiation *(Ph)* Exponierung *f*, Bestrahlung *f*; Bestrahlungsstärke *f*; Belichtung *f*; Einstrahlung *f*
i. arrangement Bestrahlungsanordnung *f*
i. damage Bestrahlungsschaden *m*; Strahlungsschaden *m*
i. death Strahlungstod *m*
i. hole Bestrahlungskanal *m*
i. microscope Durchstrahlungsmikroskop *n*
i. object Bestrahlungsobjekt *n*
i. source Strahlenquelle *f*, Strahler *m*
i. time Exponierungszeit *f*
i. unit Bestrahlungseinheit *f*
irradiator Strahler *m*, Strahlenquelle *f*
irrathene *(Plast)* bestrahltes Polyäthylen *n*
irrational *(Math)* irrational
i. number irrationale Zahl *f*
i. quantity irrationale Größe *f*
irredundant nicht redundant
irrefrangible nicht brechbar
irregular unregelmäßig; anomal; ungleich; ungleichseitig
i. curve Kurvenlineal *n*
i. distortion *(Fmt)* unregelmäßige Verzerrung *f*
i. fracture ungleichmäßiges Bruchgefüge *n*
i. line *(El)* inhomogene Leitung *f*
i. motion ungeordnete (regellose) Bewegung *f*
i. nebula *(Ph)* diffuser Nebel *m*
i. roof *(Brg)* gebräches Hangendes *n*
i. selection *(Fmt)* Fehlwahl *f*
i. settlement ungleichmäßige Setzung *f*
i. spacing ungleiche Teilung *f (Reibahle)*
i. spacing of teeth ungleichmäßige Zahnteilung *f*
irregularity 1. Unregelmäßigkeit *f*, Ungleichheit *f*; 2. Fehler *m (Material)*; 3. Sprungstelle *f*, Störstelle *f (Hohlleiter)*
irregularly spaced mit nicht gleich großen Abständen, mit ungleicher Teilung
i. waved wool fibre *(Text)* wellenuntreues Haar *n*
irreparable nicht reparierbar
irreplaceable unersetzbar, nicht auswechselbar
irresolvable nicht zerlegbar; nicht [auf]lösbar
irrespirable air *(Brg)* nicht atembare Wetter *pl*
irreversibility Irreversibilität *f*; Unumkehrbarkeit *f*
irreversible irreversibel; nicht umkehrbar; selbstsperrend
i. deformation bleibende Formänderung *f*
i. element *(BMSR)* rückwirkungsfreies Glied *n*
i. process irreversibler Prozeß (Vorgang) *m*
irrigate / to bewässern, berieseln
irrigated field Rieselfeld *n*
irrigation Berieselung *f*; Bewässerung *f*, Bodenbewässerung *f*
i. canal Bewässerungskanal *m*
i. dam Bewässerungstalsperre *f*
i. ditch Bewässerungsgraben *m*
i. pipe system Bewässerungsrohrsystem *n*
i. plant Bewässerungsanlage *f*
i. sluice Bewässerungsschleuse *f*
i. storage Bewässerungsspeicherung *f*
i. works Bewässerungsanlagen *fpl*
irrigational field Rieselfeld *n*
irritant Reizmittel *n*

i. gas ätzender Kampfstoff *m*, Reizgas *n*
i. poison Reizgift *n*
irrotational drehungsfrei, drallfrei; wirbelfrei
i. flow drallfreie (wirbelfreie) Strömung *f*
i. magnetic field wirbelfreies Magnetfeld *n*
i. motion wirbelfreie Strömung *f*, Potentialströmung *f*
isallobar Isallobare *f (Meteorologie)*
isallotherm Isallotherme *f (Meteorologie)*
isatin *(Ch)* Isatin *n*
isenthalpic process *(Ph)* isenthalpische Zustandsänderung *f*
isentropic isentropisch, von gleicher Entropie
isentropic Isentrope *f*
isherwood framing *(Schiff)* Längsspantenbauart *f*, Isherwoodbauart *f*
island effect *(El)* Inselbildung *f*
iso-octane *(Ch)* Isooktan *n*
iso-pentane *(Ch)* Isopentan *n*
isobare 1. *(Kern)* Isobar *n*; 2. Isobare *f (Meteorologie)*
isobaric isobar, bei konstantem Druck
i. atom kinds isobare Atomarten *fpl*
i. process isobarer Vorgang *m*
i. surface isobare Fläche *f*
isobase *(Geol)* Isobase *f*
isobront Isobronte *f (Meteorologie)*
isobutane *(Ch)* Isobutan *n*, Methylpropan *n*
isobutyl alcohol *(Ch)* Isobutylalkohol *m*
isobutylene *(Ch)* Isobutylen *n*
isocandela curve *(El)* Kurve *f* gleicher Lichtstärke
isochoric *(Ph)* isochor *(bei konstantem Volumen)*
i. process *(Ph)* isochorer Vorgang *m*
isochromatic farbgleich, isochromatisch
isochrone Isochrone *f*, Gleichzeitenkurve *f*
isochronous isochron, gleichzeitig
i. control Isochronregelung *f*
i. governor *(BMSR)* PI-Drehzahlregler, PI-Regler *m*, Proportional-Integral-Drehzahlregler *m*, Proportional-Integral-Regler *m*
i. scanning *(Fs)* isochrone Abtastung *f*
isoclinal line *(Ph)* Isokline *f*
isocount *(Ph)* Isoimpulskurve *f (Linie gleicher Zählrate)*
isocyanate plastic Isozyanatplast *m*
isodisperse isodispers
isodose *(Ph)* Isodosis *f*
i. curve Isodosenkurve *f*
i. recorder Isodosenschreiber *m*
i. surface Isodosenfläche *f*
isodynamic curve Isodyname *f*, isodynamische Kurve *f*
i. line *s.* i. curve
i. process isodynamische Zustandsänderung *f*
isoelectric point isoelektrischer Punkt *m*
isoelectronic ions isoelektronische Ionen *npl*
i. sequence isoelektronische Atome *npl*
isoenergetic process isoenergetische Zustandsänderung *f*
isogeothermal surface *(Ph)* Isogeotherme *f*
isogonal gleichwinklig; konform, winkeltreu *(Abbildung)*
isogonality Gleichwinkligkeit *f*; Winkeltreue *f (einer Abbildung)*
isogonic *s.* isogonal
isogonous *s.* isogonal
isogony Isogonie *f (Kristalle)*
isohaline *(Ch)* Isohaline *f*
isokurtic *(Math)* normal gekrümmt *(Häufigkeitskurven)*
isokurtosis *(Math)* normale Krümmung *f (Häufigkeitskurve)*
isolate / to 1. einen Stromkreis unterbrechen, trennen, abschalten; 2. rein darstellen *(Stoff)*, absondern; 3. kapseln *(Lärm)*
isolated vereinzelt, einzeln dastehend
i. chimney freistehender Schornstein *m*
i. pitting vereinzelte grübchenförmige Abzehrungen *fpl (Kesselkorrosion)*
isolating capacitor Sperrkondensator *m*, Trennkondensator *m*
i. stage Trennstufe *f*
i. switch Trenn[schutz]schalter *m*
i. transformer Trenntransformator *m*, Schutztransformator *m*; Isolierwandler *m*
i. valve 1. *(Masch)* Absperrventil *n*, Rohrbruchventil *n*; 2. *(El)* Trennröhre *f*
isolation 1. *(El)* Trennung *f*, Abschaltung *f*; 2. *(Ch)* Reindarstellung *f*; Absonderung *f*; 3. Kapselung *f (Lärmerzeuger)*
i. felt Isolierpappe *f*
i. of beets Vereinzeln (Verziehen) *n* der Rüben
isolator Trennschalter *m*, Trenner *m*; Richtleiter *m*, Richtungsisolator *m*
i. stage Trennstufe *f (Radio)*
isolux line *(Am) s.* isophot curve
isomagnetic isomagnetisch
isomer *(Ch, Ph)* Isomer *n*
isomeric *(Ch, Ph)* isomer
i. branching isomere Verzweigung *f*
i. heat Isomerisationswärme *f*
i. level isomeres Niveau *n*
i. nuclei isomere Kerne *mpl*
i. state isomerer Zustand *m*
i. transition isomerer Übergang *m*
isomeride *(Ch)* Isomeride *f*
isomerism *(Ph)* Isomerie *f*
isomerizate / to *(Ch)* isomerisieren
isometric *(Math)* isometrisch, maßgleich
i. drawing isometrische Darstellung *f*
i. projection *(Bw)* Isometrie *f*
i. system reguläres (kubisches) System *n (Kristalle)*

isometry *(Math)* Isometrie *f*, Maßgleichheit *f*
isomorphic *(Math)* isomorph, strukturgleich
isomorphous isomorph *(Kristalle)*
i. replacement isomorphe Substitution *f*
isopentanizing Verschnitt *m* mit Isopentan, Zusatz *m* von Isopentan
isophot *(Ph)* Isophote *f*
i. curve *(El)* Kurve *f* gleicher Beleuchtungsstärke
isopolymorphic isopolymorph *(Kristalle)*
isopotential *äquipotential, isopotential*
i. process *(Ph)* isopotentielle Zustandsänderung *f*
isoprene *(Ch)* Isopren *n*, Methylbutadien *n*
i. rubber Isoprenkautschuk *m*
isopulse *s.* i. curve
i. curve Isoimpulskurve *f*
isopycnic isopyknisch
i. curve Isochore *f*
i. surface *(Ph)* isopyknische Fläche *f*
isorad *(Ph)* Isoradlinie *f*
isosceles *(Math)* gleichschenklig
i. crank mechanism gleichschenkliges Kurbelgetriebe *n*, gleichschenkliger Kurbeltrieb *m*
i. drag link gleichschenklige Drehkurbel *f*
isostasy *(Ph)* Isostasie *f*
isostatic *(Ph)* isostatisch
isosteric *(Ph, Ch)* isoster
i. molecule isosteres Molekül *n*
i. surface isostere Fläche *f*
isotherm Isotherme *f*, Temperaturgleiche *f*, Wärmegleiche *f*
isothermal isotherm
i. curve *s.* isotherm
i. expansion isotherme Expansion *f*
i. line Isotherme *f*
i. surface isotherme Fläche *f*
isotonic solution *(Ch)* isosmotische (isotonische) Lösung *f*
isotope Isotop *n*
isotope isotop[isch]; Isotopie-, Isotopen- (*s. a. unter* isotopic)
i. analysis Isotopenanalyse *f*
i. analyzer Isotopenanalysator *m*
i. application Isotopenanwendung *f*
i. chemistry Isotopenchemie *f*
i. composition Isotopenzusammensetzung *f*
i. concentration Isotopenanreicherung *f*
i. container Isotopenbehälter *m*
i. dilution Isotopenverdünnung *f*
i. dilution method Isotopenverdünnungsmethode *f*
i. dosimetry Isotopendosimetrie *f*
i. enriched isotopenangereichert
i. exchange Isotopenaustausch *m*
i. fractionation Isotopentrennung *f*
i. number Isotopennummer *f*
i. rule Isotopenregel *f*
i. sluice Isotopenschleuse *f (Reaktortechnik)*
i. structure Isotopiestruktur *f (Hyperfeinstruktur für Elemente)*
isotopic isotop[isch]; Isotopie-, Isotopen- (*s. a. unter* isotope)
i. abundance Isotopenhäufigkeit *f*
i. composition Isotopenzusammensetzung *f*, isotopische Zusammensetzung *f*
i. dating Altersbestimmung *f* mittels Isotopen
i. enrichment Isotopenanreicherung *f*
i. equilibrium Isotopengleichgewicht *n*
i. exchange Isotopenaustausch *m*
i. indicator Indikatorisotop *n*, Leitisotop *n*
i. mixture Isotopengemisch *n*
i. ratio Isotopenverhältnis *n*
i. spin Isospin *m*, Isotopenspin *m*, isotoper Spin *m*
i. spin selection Isospinauswahl *f*
i. target Isotopenauffänger *m*
i. thickness gauge Flächengewichtsmeßgerät *n*
i. weight Isotopenmasse *f*
isotopy shift *(Opt)* Isotopieverschiebung *f*
isotropic isotrop[isch]
i. distortion *(Opt)* isotrope Verzeichnung *f*
i. distribution *(Opt)* isotrope Verteilung *f*
i. radiator *(Fmt, El)* isotroper Strahler *m*; Kugelstrahler *m*
isotropy *(Opt)* Isotropie *f*
issue / to in Umlauf bringen, in Kurs setzen *(Münzen)*
i. from ausströmen aus
issue Ausfluß *m*, Austritt *m*; Ausgang *m*; Mündung *f*
itabirite *(Min)* Eisenglimmerschiefer *m*, Itabirit *m*
itaconic acid Itakonsäure *f*
italics Schrägschrift *f*, Kursivschrift *f*
item 1. *(Dat)* Gegenstand *m*; 2. Einzelheit *f*, Punkt *m*; 3. Position *f*; Posten *m*
itemize / to spezifizieren
itemized costs Einzelkosten *pl*
iterate / to *(Math)* iterieren
iterate *(Math)* Iterierte *f*
iterated fission expectation (probability) *(Kern)* iterierte (asymptotische) Spalterwartung *f*
iteration *(Math)* Iteration *f*; Iterieren *n*; Wiederholung *f*
i. method Iterationsverfahren *n*
iterative attenuation [coefficient] *(El)* Kettendämpfung *f*
i. attenuation factor *(El)* Kettendämpfungsfaktor *m*
i. impedance Kettenwiderstand *m*, Kettenimpedanz *f*, Wellenwiderstand *m*
i. matrix Kettenmatrix *f*
i. network Kettenleiter *m (Vierpole)*

i. program loops *(BMSR)* iterative Programmschleifen *fpl*
i. propagation constant Kettenübertragungsmaß *n (Vierpole)*
ivory Elfenbein *n*
i. black Elfenbeinschwarz *n*
i. paper Bristolpapier *n*, Elfenbeinpapier *n*
Izod figure Schlagzahl *f (Izod-Prüfverfahren)*
Izod test Schlagversuch *m (nach Izod)*

J

jaboridine *(Ch)* Jaboridin *n*
jaborine *(Ch)* Jaborin *n*
jacaranda wood Palisanderholz *n*
jacinth *s.* hyacinth
jack / to [an]heben *(mechanisch)*; [hoch]winden; [ver]rücken *(mechanisch)*; unterbauen
j.-knife *(Brg)* zusammenbrechen, zerknicken *(Holzausbauten)*
j. off abwinden; abwickeln
j. up 1. hochwinden; aufbocken; 2. abhobeln, abstoßen, abschruppen *(Holz)*
jack 1. Hebebock *m*, Stützbock *m*, Heber *m*, Hochstellbock *m*; Schraubenwinde *f*, Winde *f*; Wagenheber *m*; Presse *f*; 2. *(Fmt)* Klinke *f*, Schaltklinke *f*; 3. Spannpresse *f (Spannbeton)*; 4. *(Schiff)* Gösch *f*, Bugflagge *f*; 5. Kolben *m*
j. arch scheitrechter Bogen *m (Mauerwerk)*
j. barrel *(Fmt)* Klinkenhülse *f*
j. board *(Fmt)* Klinkenbrett *n*; Stöpselbrett *n*
j. body *(Fmt)* Klinkenkörper *m*
j. bolt Abdrückschraube *f*
j. box *(Fmt)* Klinkengehäuse *n*
j. cam adjusting screw *(Text)* Einstellschraube *f* des Platinenexzenters
j. field *(Fmt)* Klinkenfeld *n*
j. for busy tone *(Fmt)* Besetztklinke *f*, besetzte Klinke *f*
j. frame *(Text)* Doppelfeinbank *f*, Doppelfeinflyer *m*
j. hammer *(Brg)* leichter Bohrhammer *m*
j. ladder *(Schiff)* Lotsenleiter *f*
j. lamp Stecklampe *f*
j. panel *(Fmt)* Klinkenbrett *n*, Klinkenfeld *n*
j. pin *(Schiff)* Belegnagel *m*
j. plane Langhobel *m*, Rauhhobel *m*, Schrupphobel *m*
j. plane knife Schrupphobeleisen *n*
j. plug *(Fmt)* Klinkenstöpsel *m*
j. rafter *(Bw)* Schifter *m*; Gratschifter *m*, Walmschifter *m*
j. roll Förderhaspel *f*
j. screw Hebespindel *f*, Hubspindel *f*; Schraubenwinde *f*
j. shaft 1. *(Met)* Zwischenspindel *f (eines Walzwerkes)*; 2. *(Masch)* Vorgelegewelle *f*
j. socket Hebestutzen *m*
j. stay *(Flg)* Längsstrebe *f*
j. switch *(Fmt)* Knebelschalter *m*
j.-tensioned mittels Spannpresse vorgespannt *(Spannbeton)*
j. tensioning Vorspannen *n* mittels Spannpresse *(Spannbeton)*
j. vault Kappengewölbe *n (Mauerwerk)*
jackass *(Schiff)* Leckstopfbeutel *m*, Klüsenpfropfen *m*
jacket / to verkleiden, umhüllen, ummanteln
jacket 1. Mantel *m*; Umhüllung *f*, Hülle *f*, Umkleidung *f*, Verkleidung *f*, Ummantelung *f*, Gehäuse *n*; 2. *(Typ)* Schutzumschlag *m*; 3. *(Am) (El)* Gehäuse *n*, Papphülse *f*, Batteriekarton *m*
j. cooling Mantelkühlung *f*
j. furnace Mantelofen *m*
j. heating Mantelheizung *f*
j. of projectile Geschoßmantel *m*
j. steam Heizdampf *m*
j. tube Mantelrohr *n*
j. water Kühlwasser *n*; Mantelkondensat *n (Dampfmaschine)*
jacketed ummantelt
j. carburetter Heizmantelvergaser *m*
j. pan Doppelkessel *m*
j. pipe Heizmantelrohr *n*
j. vessel ummanteltes Gefäß *n*
j. wall Doppelwandung *f*
jacketing Verkleidung *f*, Umhüllung *f*, Ummantelung *f*
jackhammer *(Brg)* leichter Bohrhammer *m*
jacking *(Bw)* 1. Anheben *n*, Heben *n*; Verrücken *n*, Verrückung *f*; 2. Spannen *n (Spannbetonbau)*
j. block Spannblock *m (Spannbeton)*
jackknife Seemannsmesser *n*, Taschenmesser *n*
j.-mounted telescope Scherenfernrohr *n*
Jacobian *(Math)* Funktionaldeterminante *f*
Jacobian function Jakobische Funktion *f*
Jacob's ladder Jakobsleiter *f*, Seefallreep *n*
Jacquard attachment *(Text)* Jacquardeinrichtung *f*
Jacquard bottom board Jacquardplatinenboden *m*
Jacquard card lacer Jacquardkartenbindemaschine *f*
Jacquard card puncher Jacquardkartenschläger *m*
Jacquard lifting knives Jacquardhebemesser *npl*
Jacquard circular knitting machine Jacquardrundstrickmaschine *f*
Jacquard comberboard Jacquardschnurbrett *n*
Jacquard cylinder Kartenprisma *n*
Jacquard chain loom Jacquardkettenstuhl *m*
Jacquard loom Jacquardmaschine *f*; Jacquardstuhl *m*

Jacquard machine Jacquardmaschine *f*, Jacquardstuhl *m*
Jacquard machine with extra fine pitch Jacquardmaschine *f* mit feinfeiner Teilung
Jacquard needle-board Jacquardnadelbrett *n*
Jacquard needle-box Jacquardnadelgehäuse *n*
Jacquard pattern chain Jacquardkette *f* mit Mustergliedern
Jacquard repeating machine Jacquardkopiermaschine *f*
Jacquard tie Jacquardschnürung *f*
jade *(Min)* Jade[it] *m*, Nephrit *m*
jadeite *s.* jade
jag / to 1. einkerben; 2. *(Holz)* einzapfen; 3. *(Brg)* ausblatten, verstufen
jag Zacke[n *m*] *f*; Felszacken *m*; Einschnitt *m*, Kerbschnitt *m*, Kerbe *f*
jagged ausgebrochen, gezackt, gekerbt
j. bit Zahnbohrer *m*
j. casing gezahntes Rohr *n*
j. fracture zackiger Bruch *m*
jaggedness Auszackung *f*
jaggy *s.* jagged
jakobsite *(Min)* Jakobsit *m*
jalousy Jalousie *f*
jam / to 1. blockieren, [fest]fressen, klemmen, steckenbleiben, stocken; 2. sich verklemmen *(Bohrer, Werkzeuge)*; 3. sich stauen *(Späne)*; 4. *(Am)* *(Fmt, Fs)* durch Störsignale funktionsunfähig machen, absichtlich stören
j. a receiver *(Fmt)* einen Empfänger zustopfen
j. in einzwängen; hineinklemmen
jam 1. Klemmung *f*, Verstopfung *f*; Störung *f*, Stockung *f*; 2. Marmelade *f*, Konfitüre *f*; 3. *(Mil)* Ladehemmung *f*
j. nut Gegenmutter *f*, Kontermutter *f*
jamb *(Bw)* Gewände *n*, Leibung *f*, Pfosten *m*, Türpfosten *m*, Einfassung *f* *(Tür, Fenster)*
j. liner *s.* j. lining
j. lining Türfutter *n*
jambstone *(Bw)* Eckpfeiler *m*; Gewände *n*, Gewändstein *m*
Jamin refractor Jaminscher Interferenzrefraktor *m*
jammed controls *(Flg)* blockierte Ruder *npl*
jammer Störsender *m*, Störfunkstelle *f* *(für absichtliche Störungen)*
jamming 1. Klemmen *n*, Festklemmen *n*, Blockieren *n*, Festfressen *n*, Stocken *n*; Stauung *f*; 2. Störung *f* *(des Funkverkehrs durch andere Sender)*
j. frequency Störfrequenz *f*
j. of the piping due to expansion *(Wärm)* Zwängung *f* der Dampfrohrleitung durch Ausdehnung
j. of the piston Klemmen (Ecken) *n* des Kolbens
j. station Störsender *m*
j. transmitter Störsender *m*
Janus green Diazingrün *n*
Janus switch *(Fmt)* Janusschalter *m*
Jap silk Japanseide *f*, Japon *m*
japan / to lackieren; japanisieren
japan Japanlack *m*; Lederlack *m*
j. camphor Kampfer *m*
j. earth Katechin *n*, Katechu *n*, Kachugummi *m* *(Farbstoff)*
j. lacquer Einbrennlack *m*, Japanlack *m*, Zaponlack *m*
j. silk Japanseide *f*, Japon *m*
japanned leather Lackleder *n*, Glanzleder *n*
japanner Lackierer *m*
japanning 1. Emaillelackierung *f*, Lackierung *f*; 2. Lacklederherstellung *f*
j. works Lackierfabrik *f*
japonic acid Japonsäure *f*
jar / to rattern, vibrieren; rütteln; verdichten *(Formsand durch Rütteln)*
jar 1. Standgefäß *n*, Flasche *f*, Gefäß *n*; Becherglas *n*, Glaskolben *m*, Glaszylinder *m*; Krug *m*; Tiegel *m*; 2. *(El)* Zellengefäß *n*, Zellenkasten *m* *(Batterie)*; 3. Rutschschere *f*, Bohrschere *f* *(Seilbohren)*
j. mill Kugelmühle *f*
j. moulding machine Rüttelformmaschine *f*
j. ramming method *(Met)* Rüttelverfahren *n* *(Form)*
j. ramming rollover moulding machine *(Met)* Umrollrüttelformmaschine *f*
j. rubber Gummiring *m* *(für Einweckgläser)*
jargon *(Min)* Zirkon *m*
jarosite *(Min)* Jarosit *m*
jarring Rütteln *n*; Erschütterung *f*
j. machine *(Met)* Rüttelformmaschine *f*
j. moulding machine *(Met)* Rüttelformmaschine *f*
j. plate Rütteltisch *m*
j. pressure moulding machine *(Met)* Rüttelpreßformmaschine *f*
j. pressure turn-over moulding machine *(Met)* Rüttelpreßwendeplattenformmaschine *f*
j. screen Rüttelsieb *n*
j. table Rütteltisch *m*
j. valve Rüttelventil *n*
jars *s.* jar 3.
jasper 1. *(Min)* Jaspis *m*; 2. *s.* jasperated china
j.-opal *(Min)* Jaspisopal *m*
jasperated china Jaspisporzellan *n*
jaune brillant *(Ch)* Kadmiumgelb *n*
Javel water *(Ch)* Eau de Javelle *n* *(f)*, Bleichwasser *n*, Javellesche Lauge *f*
javelin Speer *m* *(Wurfspeer)*
j.-shaped fuel rod *(Kern)* speerförmiger Brennstoffstab *m*

jaw Backe *f*; Rachen *m*; Maul *n*; Klaue *f*; Kloben *m*; Schnabel *m*
j. bolt Backenschraubenbolzen *m (Parallelschraubstock)*
j. breaker Backenbrecher *m*
j. chuck Backenfutter *n*
j. clutch Backenkupplung *f*
j. coupling Klauenkupplung *f*
j. crusher Backenbrecher *m*
j. guide Backenführung *f*
j. of a gaff *(Schiff)* Gaffelklaue *f*, Gaffelschuh *m*
j. spanner Maulschlüssel *m*
j. stone-crusher Backenbrecher *m*
j. vice Schraubstock *m*
jaws opening Maulweite *f*, Backenweite *f*, Backenöffnung *f*
jazz percussion instruments Jazzschlagzeug *n*
jeep 1. Jeep *m (geländegängiger Allzweckwagen)*; 2. *(Fs)* Drahtfernsehsystem *n*
jeer *(Schiff)* Rahtakel *n*; Schwertakel *n (einer Unterrah)*; Fall *n (einer Unterrah)*
jell / to eindicken, gelieren
jelly Gallerte *f*, Gel[ee] *n*; Sülze *f*
j.-like gallertartig
jemmy [kleine] Brechstange *f*
Jena glass Jenaer Glas *n*
jenny 1. Laufkatze *f*; 2. Generator *m*; 3. Zirkel *m* mit gebogenen Schenkeln
jerk / to rütteln, stoßen, rucken
jerk Ruck *m*, ruckartige Bewegung *f*
j.-free ruckfrei
j. movement ruckweise Bewegung *f*
j. pump Einspritzpumpe *f (Dieselmotor)*
jerking of the steam engine Tanzen (Zucken) *n* der Dampfmaschine
jerry can Benzinkanister *m*
j. man *(Brg)* Zimmerhauer *m*
jersey circular knitting machine einfonturige Trikotrundstrickmaschine *f*
jet / to herausspritzen; hervorsprudeln; mit Druckwasser spülen; einspülen; spülbohren
jet 1. Strahl *m*; Düse *f*, Brennstoffdüse *f*; Brenner *m*; Sattstrahl *m*; Spinndüse *f*; *(Rak)* Düsenstrahl *m*; Strahltriebwerk *n*; 2. *(Min)* Jet *m*, Gagat *m*, dichte Braunkohle *f*; 3. Jet[t] *m (n) (zu Schmuck verarbeiteter Gagat)*
j. action Strahlwirkung *f (Turbine)*
j. agitator Strahlmischer *m*
j.-assisted take-off *(Flg)* Start *m* mit Düsenstarthilfe
j.-black tiefschwarz, pechschwarz
j. burner Strahlbrenner *m*
j. carburetter Düsenvergaser *m*, Zerstäubungsvergaser *m*
j. coal Pechkohle *f*, Gagatkohle *f*
j. condensation Strahlkondensation *f*
j. condenser Einspritzkondensator *m*
j. control Strahlsteuerung *f*
j. deflection propeller Strahlablenkungspropeller *m*
j. deflector Strahlablenker *m*; Schurre *f*
j. engine Strahltriebwerk *n*
j. engine fitter Triebwerkmechaniker *m*
j. flame Stichflamme *f*
j. foam monitor Schaumstrahlmonitor *m (Feuerlöscher)*
j. fuel Düsenkraftstoff *m*
j. gases ausströmende Gase *npl (Raketenantrieb)*
j. guiding Strahlführung *f (Turbine)*
j.-like distribution *(Wärm)* strahlenförmige Verteilung *(Kondensator)*
j. liner Verkehrsflugzeug *n* mit Strahltriebwerk
j. lubrication Spritzschmierung *f*
j. mill Strahlmühle *f*
j. moulding Spritzverfahren *n* für hitzehärtbare Massen
j. needle Düsennadel *f*
j. nozzle Strahldüse *f*
j. of a liquid Flüssigkeitsstrahl *m*
j. of flame Flammenstrahl *m*
j. of fuel Kraftstoffstrahl *m*
j. of oxygen Sauerstoffstrahl *m*
j. oil-foam separator Schmierstoffentschäumer *m*
j. oscillograph *(El)* Strahloszillograf *m*
j. piercing Flammenstrahlbohren *n*, Thermobohren *n (Erdöl, Erdgas)*
j. piercing lance Thermobohrer *m*
j.-pipe regulator Strahlrohrregler *m*
j.-pipe valve Strahlrohrverstärker *m*
j. plane Düsenflugzeug *n*, Strahltriebflugzeug *n*
j. power plant *(Flg)* Strahltriebwerk *n*
j. propelled mit Strahlantrieb
j.-propelled airplane Düsenflugzeug *n*, Strahltriebflugzeug *n*
j.-propelled vessel Düsenschiff *n*, Schiff *n* mit Düsenantrieb
j. propulsion Strahl[an]trieb *m*; Hydromotorantrieb *m*
j.-propulsion airplane Düsenflugzeug *n*, Strahltriebflugzeug *n*
j.-propulsion engine Heißluftstrahltriebwerk *n*
j.-propulsion method Heißluftstrahlverfahren *n*
j. pump Strahlpumpe *f*
j. relay Strahlrohrverstärker *m*
j. setting Düseneinstellung *f*
j. size Düsenkaliber *n*
j. spraying Strahlzerstäubung *f*
j. vane *(Rak)* Strahlruder *n*
j. velocity Austrittsgeschwindigkeit *f*, Ausströmgeschwindigkeit *f*

j. **wave rectifier** Quecksilberstrahlgleichrichter *m*
jetcrete Spritzbeton *m*, Torkretbeton *m*, Torkretputz *m*
jetsam angeschwemmtes Strandgut *n*
jetting *(Bw)* Einspülen *n*, Spülen *n*; Spülbohren *n*
jettison / to über Bord werfen
jettison Überbordwerfen *n (zur Erleichterung eines Schiffes)*, Ballastwurf *m*
j. **tank** Zusatzbehälter *m*, Abwurfbehälter *m*
j. **valve** Schnellablaßventil *n*
jettisonable tank *(Rak)* Abwurfbehälter *m*, Zusatzbehälter *m (Kraftstoff)*
jettisoning Abwurf *m (z. B. der Flugzeugkanzel)*
jetty Hafendamm *m*, Mole *f*, Buhne *f*, Wellenbrecher *m (Seebau)*
jewel / to schmücken; Uhrsteine einsetzen
jewel 1. Juwel *n*; 2. Uhrstein *m*
j. **bearing** 1. Edelsteinlager *n*; 2. Uhrsteinlager *n*
j. **hole** Lochstein *m (Uhr)*
jewelled mit Uhrsteinen [versehen]
jew's harp *(Schiff)* Ankerschäkel *m*
jib / to scheuen; ausweichen; ablehnen
j. **in** *(Brg)* anschneiden *(Flöz)*
jib 1. Ausleger *m*; Auslegerarm *m*, Kragarm *m*, Kranbalken *m*; 2. *(Schiff)* Klüver *m*, Fock *f*; 3. *(Brg)* Schrämarm *m*, Schrämausleger *m*
j. **barrow** Schubkarre *f (ohne Seitenwände)*
j. **boom** *(Schiff)* Klüverbaum *m*, Fockbaum *m*
j. **crane** Auslegerkran *m*, Drehkran *m*, Schwenkkran *m*
j. **door** Geheimtür *f*, Tapetentür *f*
j. **head** 1. *(Brg)* Schrämkopf *m*; 2. *(Schiff)* Klüverkopf *m*; 3. Auslegerspitze *(Kran)*
j. **turret** *(Brg)* Schrämkopf *m*
jibbing-in *(Brg)* Unterschrämen *n*
jig / to *(Brg)* setzen, waschen, durchsetzen
j. **bore** Lehren bohren
jig 1. *(Masch) (werkzeugsteuernde)* Vorrichtung *f*; Spannvorrichtung *f*; Bohrschablone *f*; 2. *(Text)* Jigger *m*; 3. *(Brg)* Setzmaschine *f*, Setzkasten *m*, Vorklassiersieb *n (Aufbereitung)*
j. **and fixture design** Vorrichtungskonstruktion *f*
j.**-back ropeway** Pendelseilbahn *f*
j. **bed** Setzbett *n*
j. **borer** Lehrenbohrwerk *n*
j. **borer operator** Lehrenbohrer *m*
j. **borer with uprights and cross rail** Portallehrenbohrwerk *n*
j. **boring** Lehrenbohren *n*
j. **boring machine** Lehrenbohrmaschine *f*, Lehrenbohrwerk *n*
j. **boring operation** Lehrenbohrarbeit *f*
j. **box** *(Brg)* Setzkasten *m (Aufbereitung)*
j. **builder** Vorrichtungsbauer *m*
j. **bush** *s.* j. bushing
j. **bushing** Bohrbuchse *f*
j. **department** Vorrichtungsbau *m (Abteilung)*
j. **mill** Lehrenfräsmaschine *f*
j. **plate** Bohrplatte *f (Vorrichtung)*
j. **reamer** Reibahle *f* mit Führungsschaft
j. **saw** Laubsäge *f*, Rahmenspaltsäge *f*
j. **saw machine** Schweifsägemaschine *f*
j. **screen** Setzsieb *n*
j. **shaken tray** Schüttelrutsche *f*
j. **stenter** *(Text)* Changierspannrahmentrockenmaschine *f*
jigger 1. Töpferscheibe *f*; 2. Antennentransformator *m*; Kopplungstransformator *m*; Selbstinduktionsspule *f (des geschlossenen Schwingungskreises)*
j. **conveyor** Schüttelrutsche *f*
j. **mast** *(Schiff)* Besanmast *m*
j. **sail** *(Schiff)* Treiber[segel *n*] *m*, Besansegel *n*, Begiensegel *n*, Kreuzsegel *n*
jigging 1. Vorrichtungsbau *m*; 2. *(Brg)* Setzarbeit *f (Aufbereitung)*
j. **chute** Schüttelrutsche *f*
j. **conveyor** Schüttelrutsche *f*
j. **screen** Schüttelsieb *n*, Schwingsieb *n (Aufbereitung)*
jigs and fixtures *(Masch)* Vorrichtungen *fpl*
jim-crow Schienenbiegeeinrichtung *f*; Brechstange *f*
jimmy kurze Brechstange *f*
jingle Maschinentelegrafenglocke *f*
jink Kupplung *f*
jinny *(Brg)* 1. schwebende Strecke *f*; 2. Kohlenwippe *f*
jitter *(El)* Jitter *m*, Signalunbeständigkeit *f*; Synchronisationsfehler *m*; Zeitmodulation *f*; Zittern *n*
job Anstellung *f*; Arbeit *f*; Arbeitsstück *n*; Serie *f*
j. **analysis** Arbeitsstudie *f*
j. **captain** Brigadier *m*, Vorarbeiter *m*
j. **composition** *(Am)* Akzidenzsatz *m*
j. **diary** Bautagebuch *n*
j.**-mix[ed] concrete** auf der Baustelle gemischter Beton *m*, Baustellenbeton *m*
j. **number** Auftragsnummer *f*
j. **order costing** Kostenrechnungssystem *n* für auftragsweise Fertigung
j. **record** Bautagebuch *n*
j. **shop** Betrieb *m* mit *(auftragsweiser)* Einzelteilfertigung
j. **site** *(Am)* Baustelle *f*
j. **specification** Berufsbild *n*
jobbers' length drill Spiralbohrer *m* mit Zylinderschaft
jobbing *(auftragsweise)* Einzelteilfertigung *f*
j. **[printing] press** Akzidenzdruckmaschine *f*, Werkdruckmaschine *f*

j. sheet *(Met)* Mittelblech *n*
jobsite *(Am)* Baustelle *f*
jockey 1. *(Fmt)* Reiter *m*; 2. selbsttätige Auslösung *f* für Förderkübel
j. pulley Leitrolle *f*, Spannrolle *f*
j. sprocket Spannkettenrad *n*
jog / to stoßen; langsam bewegen; rütteln
jog 1. Absatz *m*, Sprung *m*; 2. Rütteln *n*; 3. Haltepunkt *m (einer Kurve)*
jogger *(Typ)* Bogenschüttelmaschine *f*, Bogenaufstoßmaschine *f*, Bogenglattstoßmaschine *f*
jogging Rucken *n*; Feineinstellung *f*
joggle / to *(Masch)* stauchverschränken; verzahnen
joggle Nut *f* und Feder *f*; Verschränkung *f*; Dübel *m*
j. beam verzahnter Balken *m*
j. shackle *(Schiff)* Vertäuschäkel *m*
joggled butt joint Stoßfuge *f*
Johnson noise *(El, Fmt)* Röhrenrauschen *n*, Schroteffekt *m*, Grieß *m*
join / to 1. fügen, verbinden, zusammenfügen, stoßen; 2. *(El)* verbinden, anschließen, zusammenschalten
j. by casting anschmelzen, angießen, verbinden durch Gießen
j. by cogging verkämmen *(Zimmerei)*
j. by grooves spunden *(Zimmerei)*
j. by mortising fest einfügen, zusammenfügen
j. capacitively kapazitiv zusammenschalten
j. in series *(El)* in Reihe schalten
j. together aneinanderfügen
j. up by twisting *(Text)* andrehen
j. with adhesives kleben
joined lath *(Holz)* Bindelatte *f*
joiner Tischler *m*, Schreiner *m*; Bautischler *m*, Bauschreiner *m*
j. bulkhead *(Schiff)* Kammerschott *n*, leichtes Schott *n*
joiner's bench Hobelbank *f*
j. glue Tischlerleim *m*
j. putty Holzkitt *m*
joinery 1. Tischlerei *f*, Schreinerei *f*; Bautischlerei *f*, Bauschreinerei *f*, Werktischlerei *f*; 2. Tischlerarbeit *f*; Tischlerhandwerk *n*; 3. Holzbauteile *npl*
joining 1. Fügen *n*; Binden *n*, Abbinden *n*; 2. *(Holz)* Einfalzen *n*, Einfügen *n*, Einblatten *n*
j. beam *(Holz)* Bundbalken *m*
j. by rabbets Einblattung *f*, Einfalzung *f*, Einfügung *f*
j. by scarfs Einblattung *f*, Einfalzung *f*, Einfügung *f*
j. metal Zusatzwerkstoff *m (Schweißen)*
j. of timbers Holzverbindung *f*
j. shackle *(Schiff)* Verbindungsschäkel *m*, Ankerkettenschäkel *m*
j. the warp threads *(Text)* Anlängen *n* der Kette
joint / to [aus]fugen *(Mauer)*; vermuffen; verlaschen; verbinden; anlenken, durch Gelenk verbinden
joint Fuge *f*; Stoß *m*; Klebefuge *f*, Leimfuge *f*; Verbindungselement *n (unlösbares)*; Verbindung *f*; Verbindungsstelle *f*; Knotenpunkt *m*, Gelenk *n*; Teilfuge *f*; Lötstelle *f*; Gelenkstück *n*; Glied *n*; Naht *f*; Abdichtung *f*; Schweißstoß *m*; *(Typ) (geklebter, verstärkter)* Falz *m*; *(Geol)* Absonderungsfläche *f*
j. aging time *(Plast)* Nachhärtungszeit *f* zur Erreichung der endgültigen Festigkeit
j. assembly Kreuzgelenk *n*, Kardangelenk *n*
j. binding Drahtverbindung *f*
j. bolt Gelenkbolzen *m*
j. box *(El)* Abzweigdose *f*, Verbindungsdose *f*
j. cast iron pipe Muffengußrohr *n*
j. cement Fugenkitt *m*, Fugenmörtel *m*
j. cleaner Fugenreiniger *m*
j. concrete Fugenbeton *m*
j. conditioning time *(Plast)* Nachhärtungsfrist *f* zur Erreichung der endgültigen Festigkeit
j. coupling Gelenkkupplung *f*
j. cutter Fugensäge *f (Straßenbau)*
j. end Gelenkkopf *m*
j. face Fugenwandung *f*
j. filling compound Fugenvergußmasse *f*, Fugenkitt *m*
j. grease Dichtungsfett *n*
j. guide Gelenkführung *f*
j. hook *(Typ)* Winkelhaken *m*
j. in reinforcement *(Bw)* Bewehrungsstoß *m*
j. inventor Miterfinder *m*
j. leakage Fugenundichtigkeit *f (z. B. Dampfkessel)*
j. lever Kniehebel *m*
j. lining Fugeneinlage *f*
j. mortar Fugenmörtel *m*
j. of rails Schienenstoß *m*
j. of rope Seilschloß *n*
j. of rupture Bruchfuge *f*
j. owner *(Schiff)* Mitreeder *m*
j. packing Flanschendichtung *f*, Flanschenpackung *f*, Flanschenliderung *f*; Fugendichtung *f*
j. piece Verbindungsplatte *f*; Lasche *f*
j. pin Drehachse *f*, Gelenkbolzen *m*
j. plane *(Geol)* Kluftfläche *f*
j. plate Anschlußblech *n*, Verbindungsplatte *f*
j.-pliers Scharnierzange *f*
j. property Gütergemeinschaft *f*
j. raker *(Bw)* Fugenreißer *m*
j. resistance *(El)* Muffenwiderstand *m*; kombinierter Widerstand *m*
j. ring Dichtungsring *m*, Dichtungsscheibe *f*

j. rubber Dichtungsgummi *m*
j. sealing compound Fugenvergußmasse *f*
j. sleeve Verbindungsmuffe *f*, Durchgangsmuffe *f*
j. strut Gelenkstrebe *f*
j. work Gemeinschaftsarbeit *f*
jointed 1. verbunden; 2. gegliedert; gelenkig; 3. *(Geol)* zerklüftet
j. coupling Gelenkkupplung *f*
j. cross-shaft Pendelachse *f*, Schwingachse *f*
j. cross-shaft axle Pendelachse *f*
j. fork spanner Gelenkstiftschlüssel *m*
jointer 1. *(El)* Kabellöter *m*; 2. *(Holz)* Glatthobel *m*, Schlichthobel *m*; Stoßbank *f*; Fügebank *f (Küfnerei)*
jointer's tent Lötzelt *n*
jointing 1. Verbinden *n*; Vermuffen *n*; Ausfügen *n*; 2. Dichtung[smaterial *n*] *f*
j. compound Dicht[ungs]masse *f*
j. cutter Fügefräser *m (Holzbearbeitung)*
j. edge Kleberand *m*; Fugenfläche *f (Holz)*
j. machine Fügemaschine *f (Holzbearbeitung)*
j. mark Bundzeichen *n (Zimmerei)*
j. material Dichtungsmaterial *n*
j. of pipes Rohrverbindung *f*
j. plane Fügebank *f*, Stoßbank *f (Holzbearbeitung)*
j. ring Dichtungsband *n*
j. sleeve Hülsenverbinder *m*, Verbindungshülse *f*
j. solution Dichtmittel *n*
j. surface Fugenfläche *f*
j. the boards Fügen *n* der Bretter
j. tube Überwurfmuffe *f*
jointless fugenlos
joist Balken *m*, Träger *m*; Deckenbalken *m*, Deckenträger *m*; Walzträger *m*
j. bearing Balkenauflage *f*
j. end Balkenkopf *m*
j. floor Stahlsteindecke *f*
j. rolling mill Trägerwalzwerk *n*
jolly boat Jolle *f*; Ding[h]i *n*
j. boat sailing Jollensegeln *n*
j. cruiser Jollenkreuzer *m*
jolt / to stauchen; rütteln; schleudern *(Bohrer)*
j. by means of soldering anlöten
jolt Stoß *m*, Rütteln *n*
j. crete *s.* jolted concrete
j. moulding machine Rüttelformmaschine *f*
j. piston *(Masch)* Rüttelkolben *m*
j. ramming Rütteln *n*, Verdichten *n (Formsand)*
j. squeeze machine Rüttelpreßformmaschine *f*
j. squeezer Rüttelverdichter *m (Formsand)*
j. table Rütteltisch *m (Formerei)*
j. valve *(Masch)* Rüttelventil *n*
jolted concrete Rüttelbeton *m*
Jordan mill *(Am)* Kegelstoffmühle *f*
Jordan plug Jordankegelbemesserung *f*
Jordan refiner Jordanmühle *f*
joule Joule *n*, Wattsekunde *f (Einheit für Arbeit, Energie und Wärmemenge)*
Joule-Kelvin effect *s.* Joule-Thomson effect
Joule-Thomson effect Joule-Thomson-Effekt *m*
Joule's heat Joulesche Wärme *f*
Joule's heat loss Joulescher Wärmeverlust *m*
Joule's law Joulesches Gesetz *n*
journal / to *(Masch)* lagern
journal 1. Zapfen *m*; Wellenzapfen *m*; Achszapfen *m*; 2. Fachzeitschrift *f*
j. bearing Zapfenlager *n*, Wellenzapfenlager *n*, Gleitlager *n*
j.-bearing bushing Gleitlagerbüchse *f*
j. box Lagerbüchse *f*
j. friction Zapfenreibung *f*
j. load *s.* j. loading
j. loading Radialbelastung *f*
j. of the crank shaft Lagerhals *m* der Kurbelwelle
j. pin Lagerzapfen *m*
j. pressure Zapfendruck *m*
j. voucher Buchungsbeleg *m*, Journalbeleg *m*
j. with collars Halszapfen *m*
journey 1. *(Brg)* Seilfahrt *f*; Wagenzug *m*, Zugspiel *n*; 2. Münzgewicht *n*
journeyman Facharbeiter *m*; Geselle *m*
j. patent Facharbeiterbrief *m*, Gesellenbrief *m*
jow Hanfseele *f (eines Seiles)*
jowling *(Brg)* Abklopfen *n* der Firste
joystick *(sl)* Steuersäule *f*, Steuerknüppel *m*
jucten Juchtenleder *n*
judas Guckloch *n*, Spion *m*
judicial decree Gerichtsbeschluß *m*
juice 1. Saft *m*; 2. *(sl)* Strom *m*, Elektrizität *f*, „Saft" *m (Strom)*
j. boiler Kochapparat *m*
j. catcher Saftfänger *m*
j. decantation Saftabsetzung *f*
j.-gang *(sl)* Elektrikerbrigade *f (Filmstudio)*
j. pump Saftheber *m*
juicer *(sl)* Elektriker *m (Filmstudio)*
jumbo *(Bw)* fahrbares Tunnelgerüst *n*; Bohrwagen *m*; fahrbares Gerüst *n*
j. boom *(Schiff)* Stagfockbaum *m*, Schwergutladebaum *m*
jump / to [über]springen, überschlagen; [an]stauchen *(Metall)*
j. back zurückschnellen
j. back to zurückfallen auf *(z. B. Spannung)*
j. into the groove einrasten
j. weld stumpfschweißen
jump Sprung *m (Elektronen)*; Überschlag *m (Funke)*; Sprungbefehl *m (bei elektronischen Rechnern)*; *(Math)* Unstetigkeit *f*

j. feed *(Masch)* Sprungschaltung *f*, Sprung[tisch]vorschub *m*
j. in brightness *(Fs)* Helligkeitssprung *m*
j. in potential Potentialsprung *m*
j. in temperature Temperatursprung *m*
j. instruction *(Dat)* Sprungbefehl *m*
j. order *(Dat)* Sprungbefehl *m*
j. roughing mill Springduowalzwerk *n*
j. spark Überspringfunke *m*
j.-spark coil Zündspule *f*, Induktionsspule *f*
j.-spark ignition Hochspannungszündung *f*
j. weld T-Stoß *m*; Stumpfschweißung *f*
jumped sheet buckliges (ungleich dickes) Blech *n*
jumper / to *(sl)* verbinden *(durch Schaltdrähte)*
jumper 1. *(Brg)* Stoßbohrer *m*; 2. *(El)* Kurzschlußbrücke *f*; Schaltdraht *m*, Überbrückungskabel *n*
j. cable Überbrückungskabel *n*, Schienenverbindungskabel *n*
j. connection Schaltdrahtverbindung *f*
j. field Schaltfeld *n*
j. receptacle Kupplungsdose *f*
j. stay compass *(Schiff)* Springstagkompaß *m* *(über Deck)*
j. wire Schaltdraht *m*; Verteilerschaltdraht *m*
jumpering *(El)* Verbindung *f* *(durch Schaltdraht)*
jumping 1. Stauchen *n*; 2. *(Fs)* Tanzeffekt *m*
j. drilling *(Brg)* Stauchbohren *n*
j. ski Sprungschi *m*
j. spark Überschlagsfunke *m*
j. stand Hochsprunggerät *n*
j. up Anstauchen *n*
junction 1. Abzweigung *f*, Verzweigung *f*, Einmündung *f*; Knoten[punkt] *m*; 2. Straßenkreuzung *f*, Streckenabzweigung *f*; 3. *(Eis)* Knotenpunkt; 4. Stoß[stelle *f*] *m* *(eines Kessels)*; 5. Lötstelle *f*, Lötverbindung *f*; 6. *(El)* Anschluß *m* *(Leitungen)*; Flächenkontakt *m*, Übergangszone *f*, Halbleiterübergang *m*
j. box Anschlußdose *f*, Abzweigdose *f*, Verteilerdose *f*, Verbindungsdose *f*, Kabelkasten *m*
j. buoy *(Schiff)* Kreuztonne *f*
j. cable flexibles Verbindungskabel *n*
j. centre *(Fmt)* Knotenamt *n*
j. circuit *(Fmt)* Amtsverbindung *f*
j. cord Verbindungsschnur *f*
j. dialling *(Fmt)* Fernwahl *f*
j. diode Flächendiode *f*
j. in a pipeline Rohrverzweigung *f*
j. jack *(Fmt)* Verbindungsklinke *f*, Multipelklinke *f*, Vielfachklinke *f*
j. key *(Fmt)* Nachbaramtstaste *f*
j. line *(El)* Querverbindungsleitung *f*; Abzweigleitung *f*; Amtsverbindung *f*
j. loss *(Fmt)* Verbindungsverlust *m*
j. of four plates Vierplattenstoß *m* *(Kessel)*
j. of three plates Dreiplattenstoß *m* *(Kessel)*
j. of veins *(Brg)* Gangkreuz *n*
j. plate Knotenblech *n*
j. point Knotenpunkt *m*; Verbindungspunkt *m*
j. rail Anschlußschiene *f* *(Drehscheibe)*
j. selector *(Fmt)* Amtswähler *m*
j. service *(Fmt)* Nahverkehr *m*
j. station Abzweigbahnhof *m*
j. temperature Sperrschichttemperatur *f*
j. terminal Anschlußklemme *f*
j. tetrode Flächentetrode *f*, Transistorflächentetrode *f*
j. transistor Flächentransistor *m*
juncture Vereinigungspunkt *m*; Zusammentreffen *n*; Verbindungspunkt *m*
j. point Knotenpunkt *m*; Verbindungspunkt *m*
juniper gum Wacholderharz *n*
junk Schrott *m*, Altmaterial *n*
j. ring *(Wärm)* Kolbendeckel *m*, Deckelring *m*, Kolbendecke *f*, Winkelring *m*
j. ring pin Kolbendeckelschraube *f*
jura *(Geol)* Jura *m*
j. formation Juraformation *f*
j. limestone Jurakalk *m*
jury anchor *(Schiff)* Behelfsanker *m*, Notanker *m*
j. mast Notmast *m*, Behelfsmast *m*
j. rig Nottakelage *f*
j. rudder Notruder *n*
justification *(Typ)* Ausschließen *n*, Ausschluß *m*
justified tape *(Typ)* ausgeschlossener Lochstreifen *m*
justify / to *(Typ)* ausschließen *(den Satz)*; justieren *(die Mater)*
jut Auskragung *f*; hervorstehender Punkt *m*
jute Jute *f*, Jutehanf *m*, ostindischer Hanf *m*
j. fibre Jutefaser *f*, Jute *f*
j. insulated cable Kabel *n* mit Juteisolierung
j. loom Jutewebstuhl *m*
j.-root hackling machine *(Text)* Schnippmaschine *f*
j. softener Jutequetschmaschine *f*
j. softener roller Jutequetschwalze *f*
j. thread Jutefaden *m*
j. tow spinning Jutewerggarnspinnerei *f*, Jutehedegarnspinnerei *f*
j. waste picker Juteabfallklauber *m*
j. weaving Juteweberei *f*
j. yarn Jutegarn *n*
jutty Erker *m*; Anbau *m*
juxtaposition *(Min)* Aneinanderlagerung *f*; Juxtaposition *f*
j. twins *(Min)* Berührungszwillinge *mpl*; Juxtapositionszwillinge *mpl*

K

K-absorption limit *(Kern)* K-Absorptionskante *f*
K-acid *(Ch)* K-Säure *f*
K-binding energy Bindungsenergie *f* eines K-Elektrons
K-capture *(Kern)* K-Einfang *m*
K-conversion *(Kern)* K-Umwandlung *f*
K degrees Kelvingrade *mpl*
K display K-Darstellung *f (Radar)*
K-electron K-Elektron *n*, Elektron *n* der K-Schale
K-electron capture *s.* K-capture
K-emission *(Kern)* K-Strahlung *f*
K-shell *(Kern)* K-Schale *f*
K-truss *(Bw)* K-Fachwerk *n*
K X-ray radiation K-Röntgenstrahlung *f*
kaemmererite *(Min)* Kämmererit *m*
kainite *(Min)* Kainit *m*
kalinite *(Min)* Kalinit *m*
kaolin Kaolin *n*
kaolinite *(Min)* Kaolinit *m*
Kaplan water turbine Kaplan-Turbine *f (Wasserpropellerturbine)*
kapok Kapok *m*
k. gin *s.* k. ginning machine
k. ginning machine Kapokentkörnungsmaschine *f*
k. tree Baumwollbaum *m*
k. twirling machine Kapokentkörnungsmaschine *f*
Kapp coefficient *(El)* Kapp-Faktor *m*
Kapp-type phase advancer *(El)* Kapp-Phasenschieber *m*
karat *s.* carat
Karlsbad twins Karlsbader Zwillinge *mpl (Kristalle)*
Karolus cell Karolus-Zelle *f*, fotoelektrische Zelle *f*
Karolus system Karolus-Fernsehverfahren *n*
Kassel yellow *(Ch)* Kasseler Gelb *n*
kata-thermometer Katathermometer *n*
katharometer Wärmeleitfähigkeitsmesser *m* für Gase
kathetron Gasentladungsröhre *f* mit Außengitter
kauri *s.* k. pine
k. gum Kauriharz *n*
k. pine Kaurifichte *f*, Kopalfichte *f*
kayak Kajak *m (Boot)*
kedge *s.* k. anchor
k. anchor *(Schiff)* Warpanker *m*, Verholanker *m*, Schleppanker *m*
k. buoy Verholboje *f*, Verholtonne *f*
kedger *(Schiff)* Schleppanker *m*
keel *(Schiff)* Kiel *m*
k. block Kielpalle *f*, Kielstapel *m*, Stapelklotz *m*, Kielklotz *m*
k. boat Kielboot *n*
k.-laying Kiellegung *f*
k. line Kiellinie *f*, Schiffsmittenlinie *f*, Mittschiffslinie *f*
k. shoe Loskiel *m*, Kielschuh *m*
k. shore Kielstütze *f*
k. siding Kielhöhe *f*
k.-track ram Kielblockramme *f*, Kielstapelramme *f*
keelage Kielgeld *n*, Hafengebühren *fpl*
keeler Kieljacht *f*
keelover / to kentern
keelover kieloben
keelson *(Schiff)* Kielschwein *n*, Binnenkiel *m*
k. timber Kielschwein *n*
keen scharf
k.-edged scharfkantig; mit scharfer Schneide
keenness Schärfe *f*
keep / to [er]halten, festhalten, behalten; lagern, aufbewahren, konservieren; verhindern, verhüten
k. a relay ein Relais unter Spannung halten
k. aligned führen
k. dry trocken aufbewahren; vor Nässe schützen
k. in fork unter Kontrolle halten
k. in order instandhalten
k. in shape instandhalten
k. in step *(Elm)* synchron bleiben
k. off [from the shore] *(Schiff)* von Land abhalten
k. station *(Schiff)* Position halten
k. steam *(Wärm)* Dampf halten
k. the course Kurs halten
k. the dead-reckoning *(Schiff)* koppeln
k. the land aboard *(Schiff)* sich unter Land halten, in Landsicht bleiben
k. the log Logbuch (Tagebuch) führen
k. the offing *(Schiff)* frei von Land halten, außerhalb der Dreimeilenzone bleiben
k. the reckoning *(Schiff)* mitkoppeln
k. the sea die See halten *(außerhalb der Dreimeilenzone bleiben)*
k. to the shore *(Schiff)* unter Land bleiben
k. track spuren
k. under water unter Wasser lagern
k. upright nicht stürzen
k. watch *(Fmt)* besetzt halten, hörbereit sein *(Empfänger)*
k. within hail *(Schiff)* sich in Hörweite halten
keep-alive electrode Erregeranode *f*, Halteanode *f*
k. plate *(Schiff)* Befestigungsplatte *f*
keeper 1. Halteeinrichtung *f*, Halter *m*; 2. Anker *m (Magnet)*; 3. Getter *m (Elektronenröhren)*; 4. Wächter *m*, Wärter *m*; 5. erster Mann *m (am Hochofenabstich)*
keeping Verwahrung *f*
k. constant Konstanthaltung *f*

k. of cattle Viehhaltung *f*
k. qualities Haltbarkeit *f* von Waren
keeps Aufsetzvorrichtung *f (Förderkorb)*
keg *(Schiff)* Bootsfaß *n (unter 10 Gallonen)*
k. buoy *(Schiff)* Faßtonne *f*
Kell factor *(Fs)* Kell-Faktor *m*
Kellog switch *(El)* Kellogschalter *m*
kelly Mitnehmerstange *f (Rotarybohren)*
Kelvin balance *(El)* Stromwaage *f* nach Kelvin
Kelvin bridge *s.* Kelvin double bridge
Kelvin double bridge *(El)* Doppelbrücke *f*
Kelvin effect Kelvin-Effekt *m*, Thomson-Effekt *m*
Kelvin electrometer Quadrantenelektrometer *n* nach Kelvin
Kelvin's absolute scale of temperature absolute thermodynamische Temperaturskale *f*
kenaf Kenafhanf *m*
Kennedy key Tangentkeil *m*
Kennelly-Heaviside layer Kennelly-Heaviside-Schicht *f*
kenotron Hochvakuumgleichrichterröhre *f*, Glühkatodengleichrichterröhre *f*, Kenotron *n*
kentledge *(Schiff)* Ballasteisen *n*
k. goods *(Schiff)* Ballastgüter *npl (schwere, tief unten gestaute Güter)*
kep *(Brg)* Aufsetzvorrichtung *f*, Schachtfalle *f*
Keplerian law Keplersches Gesetz *n*
Kepler's laws Keplersche Gesetze *npl*
keramic *s.* ceramic
keramics *s.* ceramics
kerb Bordeinfassung *f*, Bordkante *f*, Bordschwelle *f*, Schrammschwelle *f*
k. feeler *(Kfz)* Bordsteinfühler *m*, Reifenwächter *m*
k.-stone Bordstein *m*, Randstein *m*, Rinnstein *m*
kerf Sägeschnitt *m*, Sägenute *f*, Schnittfuge *f*; Kerb *m*; Abschnittseite *f*; Einschnitt *m*; *(Brg)* Schlitz *m (senkrechter Schram)*
kerfing *(Brg)* Schlitzen *n*
kermesite *(Min)* Kermesit *m*, Antimonblende *f*, Rotspießglanz *m*, Spießglanzblende *f*
kerned letter *(Typ)* unterschnittener Buchstabe *m*
kernel 1. Kern *m*, Zellkern *m*; 2. Kern *(einer Integralgleichung)*
k. oil Kernöl *n*
k. roasting Kernrösten *n (Erze)*
kernite *(Min)* Kernit *m*
kerosene Kerosin *n*, Leuchtöl *n*, Leuchtpetroleum *n*, Paraffinöl *n*
k. blow torch Petroleumlötlampe *f*
k. engine Motor *m* für Traktorenkraftstoff, Petroleummotor *m*
k. oil Petroleum *n*
k. shale *(Geol)* Ölschiefer *m*
k. tank Petroleumtank *m*
Kerr effect *(Opt)* Kerr-Effekt *m*
Kerr electro-optic shutter *(Opt)* elektrooptischer Verschluß *m* nach Kerr
ketch Ketsch *f (Segelschiff)*
k.-rigged barge Besanewer *m*
ketene *(Ch)* Keten *n*
keto-enolic tautomerism *(Ch)* Keto-Enol-Tautomerie *f*
ketogenesis *(Ch)* Ketogenese *f*, Bildung *f* von Azetonkörpern
ketol Ketoalkohol *m*, Ketol *n*
ketone *(Ch)* Keton *n*
k. musk *(Ch)* Ketonmoschus *m*
ketonic acid Ketosäure *f*
k. oxygen Karbonylsauerstoff *m*
kettle / to verkochen *(Lackindustrie)*
kettle 1. Kessel *m*, Siedekessel *m*, Kochkessel *m*, Wasserkessel *m*; 2. *(Text)* Kesselfärbemaschine *f*
k. bottom Flachbodenboot *n*, Plattbodenboot *n*, Plätte *f*, Kahn *m*
k.-drum Kesselpauke *f*
k.-drum stick Paukenschlegel *m*
k. furnace Schalenfeuerung *f*
kevel *(Schiff)* Kreuzklampe *f*
k. head *(Schiff)* Kreuzpoller *m (aus Holz)*
key / to 1. verdübeln; verkeilen; aufkeilen; verzahnen; 2. *(Fmt)* tasten *(schreiben)*; 3. stimmen *(Instrument)*
key 1. *(Masch)* Längskeil *m*, Feder *f*; Schlüssel *m (Futter)*; 2. *(Bw)* Dübel *m*, Verdübelung *f*; Keil *m*; Taster *m (Telegraf)*; Taste *f (Schreibmaschine)*; Zeichenerklärung *f*; 3. *(Fmt)* Chiffrierschlüssel *m*; 4. *(Fs)* Gradation *f*; 5. Verschlußklappe *f*, Klappe *f (Musikinstrument)*
k. and slot *(Masch)* Feder *f* und Nut *f*
k. bed *(Geol)* Leitschicht *f*
k. bit Schlüsselbart *m*
k. blank Rohschlüssel *m*
k. bolt Keilbolzen *m*
k. brick Schlußstein *m (Mauerwerksbogen)*
k. buckle Keilangel *f (Gattersäge)*
k.-buoy Anfangsboje *f*, Anfangstonne *f (für ein Fischnetz)*
k. button Druckknopf *m*, Taste *f*
k. clicks *(Fmt)* Tastgeräusche *npl*
k. controlled continuous wave getastete ungedämpfte Welle *f*
k.-day Stichtag *m*
k.-desk Klaviaturschrank *m (Orgel)*
k. drive Mitnahme *f* durch Keil
k. file Schlüsselfeile *f*
k. for accelerated forward speed Schnellvorlauftaste *f*
k. for printing totals *(Büro)* Summendrucktaste *f*
k. fossile *(Geol)* Leitfossil *n*

k. groove Keilnut *f*
k.-grooving machine Keilnutmaschine *f*
k. head Keilnase *f*
k. holder *(El)* Schalterfassung *f*, Hahnfassung *f*
k. industry Grundindustrie *f*, Schlüsselindustrie *f*
k. joint Keilverbindung *f*
k. letter Kodeschlüssel *m*
k. lever Tastenhebel *m*, Tastatur *f*
k. light Hauptlicht *n (Studiolampe)*
k. metering Tastenzählung *f*
k. note *(Ak)* Grundton *m*
k. of cock Hahnschlüssel *m*
k. of flute Klappe *f* der Flöte
k. pipe Schlüsselrohr *n*
k. plan Übersichtsplan *m*
k. plate *(Masch)* Keilplatte *f*
k. punch Tastenlocher *m*, Handlocher *m (für Lochkarten)*
k. punch machine Locher *m*
k. ring Schlüsselbund *n*
k. rod *(Typ)* Auslösestab *m (Setzmaschine)*
k.-root Keil *m (Keilwelle)*
k. selector switch Tastenwahlschalter *m*
k. set Tastensatz *m*
k.-slot Wellennut *f*, Keilnut *f*, Welle *f*
k.-slot gauge Keilnutlehre *f*
k. speed *(Fmt)* Handtempo *n*
k. stem *(Büro)* Tastenstange *f*, Tastenstengel *m*
k.-stone mill *(Led)* Glockenmühle *f*
k. substance *(Am)* Ausgangsstoff *m*
k. thump *(Fmt)* Tastgeräusch *n*
k. to signs Zeichenerklärung *f*, Zeichenerläuterung *f*
k. well Einblasesonde *f*, Einpreßbohrung *f*
k. word Stichwort *n*; Kennwort *n*; Schlagwort *n (unter dem Dokumente zu finden sind)*
keyboard Tastatur *f*, Tastenfeld *n*; Klaviatur *f*; Einstellwerk *n*
k. contacts Tastenschalterkontakte *mpl*
k. lock *(Fmt)* Tastensperre *f*
k. perforator Tastenlocher *m*; Morsetastenlocher *m*
k. selection *(Fmt)* Wahl *f* mittels Tastatur
k. transmitter *(Fmt)* Tastengeber *m*
keyed verkeilt; getastet, durch Tastung; verdübelt
k. automatic gain control unverzögerte Schwundregelung *f*
k. beam verdübelter Balken *m (Holzbau)*
k. bugle Klapphorn *n*
k. continuous waves getastete ungedämpfte Wellen *fpl*
k. joint Keilverbindung *f*, Hohlfuge *f*
k. modulated waves getastete tonüberlagerte ungedämpfte Wellen *fpl*
k. to aufgekeilt auf
k. waves getastete Wellen *fpl*
keyer Tastgerät *n (zur Erzeugung des Tastimpulses; Radar)*; Abtrennstufe *f*; Impulszentrale *f*
keygroove / to Keilnuten ziehen
keyhole Schlüsselloch *n*
k. saw Lochsäge *f*, Stichsäge *f*
keying 1. *(Masch)* Verkeilen *n*, Verkeilung *f*, Keilbefestigung *f*; 2. *(Bw)* Verzahnung *f*; 3. *(Fmt)* Tastung *f*
k. and reading *(Fmt)* Eintasten *n* und Abtasten *n*
k. circuit Tastkreis *m*
k. cycle Tastzeit *f (Radar)*
k. filter Tastfilter *n*; Vorfilter *n*
k. period Tastzeit *f (Radar)*
k. relay Tastrelais *n*
k. space Tastpause *f*, Tastzwischenraum *m*
k. speed Tastgeschwindigkeit *f*, Eintastgeschwindigkeit *f*
k. system Tastanordnung *f*
k. wave *(Fmt)* Zeichenwelle *f*
keyless chuck schlüsselloses Futter *n*
k. ringing *(Fmt)* selbsttätiger Ruf *m*
keys Tastatur *f*, Klaviatur *f*
keyseat / to Keilnuten einarbeiten
keyseat Keilnut *f*
k. cutter Schlitzfräser *m (für Nuten)*
k. in hub Nabennut *f*
k. in shaft Wellennut *f*
keyseated mit Längskeilverbindung
keyseater Keilnutenziehmaschine *f*
keyseating chisel Nutenmeißel *m*
k. machine working on the draw-cut principle Keilnutenziehmaschine *f*
keystone Schlußstein *m (Bogen, Gewölbe)*
k. correction *(Fs)* Trapezentzerrung *f*
k. distortion *(Fs)* Trapezfehler *m*
k.-shaped trapezförmig
keyway / to nuten; Nuten ziehen
keyway Nabennut *f*, Keilnut *f* in Nabe; Keillängsnut *f*
k. broaching Keilnutenräumen *n*
k. cutting *s.* k. milling
k. file Räumfeile *f*
k. gauge Keillehre *f*
k. milling Keilnutenfräsen *n*
k. milling attachment Keilnutenfräsapparat *m*
k. milling cutter Keilnutenfräser *m*
k. milling machine Keilnutenfräsmaschine *f*
k. root Nutgrund *m*
k. seating machine Keilnutenmaschine *f*
k. shaper Nuthobelmaschine *f*
k. tool Keilnutenstoßmeißel *m*
keywaying Keilnutenfräsen *n*, Nutenziehen *n*, Wellennuten *n*
kibble / to schroten

kibble *(Brg)* Abteufkübel *m*, Förderkübel *m*
kibbling mill Schrotmühle *f*
kick / to stoßen *(mit Fuß)*; antreten; kickstarten
kick 1. Stoß *m*, Schlag *m*; Schlagen *n*; 2. Schwingung *f (Bohrseil)*
k.-back 1. *(Kfz)* Zurückdrehen *n (des Lenkrades)*; 2. *(El)* Rücklauf *m (des Abtaststrahls)*
k.-back dump *(Brg)* Kopfwipper *m*
k.-back power supply *(Fs)* Hochspannungsgewinn *m* aus dem Zeilenrücklauf
k.-back pulse *(Fs)* Rückschlagimpuls *m*
k. sorter Impulshöhenanalysator *m*
k. stamp Schmiedehammer *m*
k.-starter Kickstarter *m*
kicking coil *(El)* Drosselspule *f*
kickling Umwicklung *f* von Kabeln mit alten Seilen *(Abscheuerungsschutz)*
kickout *(Masch)* Auslösung *f*, Auslöseeinrichtung *f*
kicksorting technique Impulshöhenanalyse *f*
kid Jungziege *f*
k. crosses in Kreuzform zusammengenähte Zickelfelle *npl*
k. leather works Kidgerberei *f*
Kidderminster Kidderminsterteppich *m (dreiteiliger Teppich ohne Flor)*
kidney ore Erzniere *f*, Nierenerz *n*
k.-shaped nierenförmig
kier *(Text)* Beuchfaß *n*, Bleichkessel *m*
k.-boiling *(Text)* Beuchen *n*
kieselgu[h]r Kieselgur *f*
kieserite *(Min)* Kieserit *m*
kieve *(Brg)* Schüttelkasten *m (Aufbereitung)*
kill / to 1. *(Met)* abstehen lassen, beruhigen; 2. *(Brg)* niederschlagen *(Flotationsschaum bei Aufbereitung)*; unwirksam machen
k. lime Kalk totbrennen
killas Tonschiefer *m*
killed acid Lötwasser *n*
k. spirits of salts Lötwasser *n*
k. steel beruhigt vergossener Stahl *m*
killer *(Ch)* Lumineszenzgift *n*; neutralisierendes Mittel *n*
killick kleiner Bootsanker *m*, Muringstein *m*
killing agent *(Met)* Beruhigungsmittel *n*
k. period Ausgarzeit *f (Stahlschmelze)*
kiln / to [aus]darren, trocknen
kiln Darrofen *m*, Röstofen *m*; Schachtofen *m*; Brennofen *m*; Trockenraum *m*; Kalkofen *m*; Meiler *m*
k. apparatus Trockenkammergerät *n (für Holz)*
k.-brick feuerfester Stein *m*, Ofenbaustein *m*
k.-burning Meilerverkohlung *f (Holz)*
k. control Steuerung *f* der Trockenkammer
k.-dried künstlich getrocknet, ofengetrocknet *(z. B. Holz)*
k.-dry weight Darrgewicht *n*
k. drying of lumber *(Holz)* Trockenraumtrocknung *f*, technische Trocknung *f*, Kammertrocknung *f*
k.-drying schedule Trockenkammerarbeitsrichtlinie *f*
k.-hole Ofenloch *n*
k. operation Trockenkammerbetrieb *m*, Arbeit[sweise] *f* der Trockenkammer
k. scum Ofenschaum *m*
k.-seasoning Kammertrocknung *f*
k. sterilization Sterilisation *f* in der Trockenkammer
k. with overlying beds Etagenofen *m*
kilning künstliche Holztrocknung *f*
k. drum Darrtrommel *f*
kilo-electron-volt Kiloelektronenvolt *n* *(1 keV = 1,602 · 10^{-16} J)*
kiloampere Kiloampere *n*
kilocalorie Kilokalorie *f*
kilocurie Kilocurie *n*
kilocycle Kilohertz *n (Einheit der Frequenz; 1 kHz = 10^3 Hz)*
kilodyne Kilodyn *n (1 kdyn = 10^{-2} N)*
kilogram *(Am)* s. kilogramme
kilogramme Kilogramm *n (Einheit der Masse)*
k. atom Kilogrammatom *n*
k. calorie s. kilocalorie
k. force Kilopond *n (1 kgf = 1 kp = 9,80665 N)*
k. mass Kilogramm *n*
k. metre Meterkilopond *n*, Kilopondmeter *n*
k. weight Kilopond *n*
kilohertz s. kilocycle
kilojoule Kilojoule *n (1 kJ = 1 kWs)*
kiloliter *(Am)* s. kilolitre
kilolitre Kiloliter *n*
kilometer *(Am)* s. kilometre
kilometre Kilometer *n*
kilometric fault *(El)* Abstandskurzschluß *m*
k. waves *(El)* Langwellen *fpl (1000 bis 10000 m)*
kiloohm Kiloohm *n*
kilopond Kilopond *n (1 kp = 9,80665 N)*
kilovar Kilovar *n (1 kvar = 10^3 W)*
kilovolt Kilovolt *n*
kilovoltampere Kilovoltampere *n (Einheit der Leistung nur zur Angabe von elektrischen Scheinleistungen)*
k.-hour meter kVAh-Zähler *m*, Scheinverbrauchszähler *m*
k. reactive Blindkilovoltampere *n*
kilowatt Kilowatt *n*
k.-hour Kilowattstunde *f*
k.-hour meter Kilowattstundenzähler *m*
kind Art *f*, Sorte *f*; Güte *f*; Gattung *f*
k. of current Stromart *f*
k. of drive Antriebsart *f*
k. of surface Oberflächenart *f*, Oberflächenzustand *m*, Oberflächenbeschaffenheit *f*

kindle / to anzünden, entzünden
kindle curve *(Schiff)* gut strakende Kurve *f*
k. movements of a vessel sanfte Schiffsbewegungen *fpl*
kindling temperature Entzündungstemperatur *f*
kinematic kinematisch
k. chain kinematische Kette *f*
k. viscosity kinematische Viskosität (Zähigkeit) *f*
kinematics Kinematik *f*; Getriebelehre *f*
kinemograph *(Masch)* Kinemograf *m*
kinemometer Kinomometer *n*
kineophone Kineofon *n*
kineoptoscope Kineoptoskop *n*
kinescope *(Fs)* Bild[wiedergabe]röhre *f*
k. recording Filmaufnahme *f* eines Fernsehbildes, gefilmtes Fernsehprogramm *n*
kinetic kinetisch
k. control system *(BMSR)* Regelung *f* von Bewegungsgrößen
k. energy kinetische Energie *f*, Bewegungsenergie *f*, Wucht *f*
k. engineer Kinetiker *m*
k. friction Bewegungsreibung *f*
k. quantity Bewegungsgröße *f*
k. theory of fluids kinetische Theorie *f* der Flüssigkeiten
k. theory of gases kinetische Gastheorie *f*
k. theory of heat kinetische Theorie *f* der Wärme
kinetics Kinetik *f*
k. of development Entwicklungskinetik *f*
king beam *(Schiff)* Lukenscherstock *m*
k. bolt 1. Mittelzapfen *m (Drehscheibe)*; 2. *(Am, Kfz)* Achsschenkelbolzen *m*
k. bridge *(Schiff)* Portalmastenbrücke *f*, Ladepfostenverbindungsbrücke *f*
k. closer Dreiviertelziegel *m*
k. journal Drehzapfen *m*
k. pin 1. Mittelzapfen *m (Drehscheibe)*; 2. *(Kfz)* Achsschenkelbolzen *m*, Drehzapfen *m*, Vorderachszapfen *m*
k.-pin angle *(Kfz)* Radspreizungswinkel *m*
k.-pin bearing Achsschenkelbolzenlager *n*
k.-pin bush Achsbolzenbuchse *f*
k.-pin inclination Achszapfensturz *m*, Vorderradspreizung *f*
k.-pin steering assembly Achsschenkellenkung *f*
k. plank *(Schiff)* Mittschiffsdecksplanke *f*; Fischplanke *f*
k. post 1. Hängesäule *f (Holzbau)*; 2. *(Schiff)* Ladepfosten *m*; Schwergutmast *m*
k.-post truss[ed beam] einfaches Hängewerk *n (Holzbau)*
k.-size cartridge *(Brg)* lange Patrone *f*
k. spoke *(Schiff)* Königsspeiche *f (beim Steuerrad des Mittschiffsruders)*
king's hook *(Brg)* Königsstange *f*
kink / to 1. sich verknoten; 2. *(El)* in Schleifen legen *(Leitungen)*
kink Kink *f*, Schleife *f*
k. instability *(Mech, Ph)* Kinkinstabilität *f*, Knickinstabilität *f*
kinking 1. Knickung *f*; Schleifenbildung *f*; 2. *(Met)* Faltenbildung (Buckelbildung) *f* im Werkstoffgefüge
kinks Werkstattkniffe *mpl*
kiosk booth öffentliche Fernsprechstelle *f*
Kipp's apparatus *(Ch)* Kippscher Apparat *m*
Kirchhoff adder *(Dat)* analoge Addierschaltung *f*
Kirchhoff approximation Kirchhoffsche Näherung *f*
Kirchhoff's law of networks Kirchhoffsches Verzweigungsgesetz *n*, Kirchhoffsche Regeln *fpl* für Stromverzweigung
Kirchhoff's radiation law Kirchhoffsches Strahlungsgesetz *n*
kirner Handbohrer *m*
Kirschner beater Kirschner-Schläger *m*, Kirschner-Flügel *m (Spinnerei)*
kirver *(Brg)* Schrämhauer *m*
kirving *(Brg)* Schrämen *n*; Kerbe *f*, Schlitz *m*
kish Garschaum *m*, Eisenschaum *m (Graphit)*
k. graphite Garschaumgraphit *m*
kit Werkzeugkasten *m*; Ausrüstung *f*
kitchenette Kochnische *f*; Teeküche *f*
kite Drachen *m*
k. aerial Drachenantenne *f*
k. balloon sloop *(Schiff)* Drachenballonslup *f (Fahrzeug zum Schleppen eines Drachenballons)*
k. drag *(Schiff)* Dragganker *m*, Drachenseeanker *m*
k. spar *(Flg)* Drachenholm *m*
klaxon *(Kfz)* elektrische Hupe *f*
Kleine floor *(Bw)* Kleinesche Decke *f*, Stahlsteindecke *f*
klinker *s.* clinker
klinometre Klinometer *n*, Neigungsmesser *m*
klippe Klippe *f*
klockmannite *(Min)* Klockmannit *m*
klop *(Abkürzung für)* Chlorpikrin *n*, Trichlornitromethan *n (chemischer Kampfstoff)*
klydonograph *(El)* Klydonograf *m*
klystron Klystron *n*
k. amplifier Klystronverstärker *m*
k. control grid Klystronsteuergitter *n*
k. generator Klystrongenerator *m*
k. oscillator Klystronoszillator *m*
knag Astknorren *m*, Knoten *m*
knaggy knorrig, knotig *(Ast)*
knap / to spalten *(Stein)*
knapped gespalten *(Stein)*
knapsack Rucksack *m*, Tornister *m*
k. boiler *(Wärm)* Tornisterkessel *m*

k. mistblower tragbares Nebelgerät *n*, Schulternebelgerät *n (Sprühgerät zur Schädlingsbekämpfung)*
k. transmitter Tornisterfunkgerät *n*
knar Knorren *m*, Knoten *m (Ast)*
knead / to kneten
kneadable knetbar, plastisch
kneader Kneter *m*, Knetmaschine *f*
kneading machine Knetwerk *n*, Knetmaschine *f*
k.-trough Backmulde *f*
knee Knie[stück] *n (Rohr)*; Konsole *f*, Winkeltisch *m*
k. and column-type milling machine Konsolfräsmaschine *f*
k. brace Kopfband *n (Holzbau)*
k. breeches Kniehose *f*
k. clamping Konsolklemmung *f*
k. elevating screw Konsolhubspindel *f*
k. joint Kniekupplung *f*; durchschlagendes Doppelkurbelgetriebe *n*, umlaufende (rotierende) Doppelkurbel *f*
k. of the curve Kurvenkrümmung *f (Röhrenkennlinie)*; Sättigungsknick *m (Elektronenröhren)*
k.-saddle way Konsolführungsbahn *f*
k. table Konsoltisch *m*
k.-timber Knieholz *n*
k.-to-column way Ständerführungsbahn *f* der Konsolfräsmaschine
k.-type automatic automatische Konsolfräsmaschine *f*
knib Knoten *m (Seide)*
knife Klinge *f*, Messer *n*; Schneide *f*
k. barking machine Messerschälmaschine *f (Holzentrindung)*
k. beam Messerbalken *m (einer Mähmaschine)*
k. blade Messerklinge *f*; Klinge *f*
k. blade switch Hebelschalter *m*, Messerschalter *m*
k. coating Beschichten *n* mit Messer (Rakel), Rakelstreichverfahren *n*
k.-cut gemessert *(Furnier)*
k.-cutting machine Messerschneidemaschine *f*
k. edge Haarmeßkante *f*; Meßschneide *f*, Schneide *f*
k.-edge bearing *(Am)* Schneidenlager *n*, Zentrierschneide *f*
k.-edge contact tip Meßhütchen *n* mit Schneide
k.-edge file Messerfeile *f*
k.-edge relay Schneidenankerrelais *n*
k.-edge suspension Lagerung *f* auf Schneiden
k.-folder *(Typ)* Schwertfalzmaschine *f*
k. grinder Messerschleifmaschine *f (Holzbearbeitung)*
k. holder Messerträger *m*
k.-lifting distance Mähmesserhub *m*
k. plug Messerstecker *m*
k. pruning saw Messersäge *f (Gartenbau)*
k. rest *(Mil)* spanischer Reiter *m*
k. setting attachment Messereinstellapparat *m (Holzbearbeitung)*
k. supports Messerstütze *f*
k. switch Messerschalter *m*, Schalter *m* mit Messerkontakten; Hebelschalter *m*
k. veneer Messerfurnier *n*
k.-veneer cutting machine Messermaschine *f (Furnier)*
knifer coater Rakelstreichmaschine *f*
knifing *(Met)* Walzen *n* mit geteiltem Kaliber
k. glaze Spachtelmasse *f*
knit / to stricken, wirken
knit goods Strickware *f*
knitted cloth Wirkstoff *m*, gewirkter Stoff *m*
k. material gewirkte Ware *f*, Wirkware *f*
k. suits gewirkte (gestrickte) Kleider *npl*
knitting butt Strickfuß *m*
k. cam Strickheber *m*
k. frame Wirkstuhl *m*
k. level Strickhöhe *f*
k. machine Strickmaschine *f (mit Zungennadeln)*; Wirkmaschine *f (mit Hakennadeln)*
k. needle Stricknadel *f*
k. wool Strickwolle *f*, Strick[woll]garn *n*
knitwear Strickbekleidung *f*
knob Knopf *m*, Kugelknopf *m*, Drehknopf *m*
knock / to aufschlagen, prallen, stoßen, treffen, zusammenstoßen; klopfen *(Motor)*; flattern *(Ventil)*
k. off 1. abhauen, abschlagen; 2. auslösen *(Sekundärelektronen)*
k. out ausstoßen, herausschießen *(Teilchen)*
knock characteristics Klopfeigenschaften *fpl (Kraftstoff)*
k. inhibitor Klopfbremse *f*, Antiklopfmittel *n (Kraftstoff)*
k. intensity Klopfstärke *f (Kraftstoff)*
k.-off Auslösung *f*; Auswerfer *m*
k.-off sinker *(Text)* Abschlagplatine *f*
k. out Auswerfer *m*
k.-out electrons ausgelöste Elektronen *npl*
k.-out key Austreiber *m*
k.-out lever Ausrückhebel *m*
k. promoter Klopfförderer *m (Kraftstoff)*
k. property Klopfeigenschaft *f (Kraftstoff)*
k. rating Klopffestigkeit *f*, Oktanzahl[bestimmung] *f (Kraftstoff)*
k.-reference fuel Vergleichskraftstoff *m* zur Bestimmung der Klopffestigkeit
k.-resistance Klopffestigkeit *f (Kraftstoff)*
k. suppressor Antiklopfmittel *n*; Klopfbremse *f (Kraftstoff)*
knockabout slupgetakeltes, halbgedecktes Kielboot *n* für offene Gewässer

knocked-on atom Anstoßatom *n*, angestoßenes Atom *n*, Rückstoßatom *n*
k.-on electron Anstoßelektron *n*, Delta-Elektron *n*, Delta-Strahl *m*
knocker *(Brg)* Signalhammer *m*
k. line *(Brg)* Signaldraht *m*
knocking 1. Klopfen *n (Verbrennungsmotor)*; 2. *(Brg)* Stückerz *n*
k. characteristic *(Kfz)* Klopfeigenschaft *f*
k.-out Herausstoßen *n*; Atomumlagerung *f (im Gitterverband)*
k. pump schlagende Pumpe *f*
knockover bit *(Text)* Abschlagplatine *f*
knockproof klopffest *(Kraftstoff)*
Knoop hardness test Härteprüfung *f* nach Knoop
knop Faserknötchen *n*
k. yarn Noppengarn *n*
knopped twist Knotenzwirn *m*
knorl Astknorren *m*
knot / to 1. verknoten; 2. Äste flicken
knot 1. Ast *m*, Knorren *m*; 2. Knoten *m (Verknüpfung)*; 3. Knoten *(sm/h)*
k. borer Astbohrer *m*
k. chipper Ästehackmaschine *f*
k. cord *(Text)* Schnurplatine *f*
k. defect durch einen Ast verursachter Fehler *m*, Astfehler *m*
k. hole Astloch *n*
k. tenacity Knotenfestigkeit *f*
k. within a meteor stream Meteorschwarm *m*
knotless astlos, astfrei
knotted knorrig
k. yarn *s.* knop yarn
knotter 1. *(Lw)* Knüpfapparat *m (Sammelpresse)*; 2. Astfänger *m*, Knotenfänger *m*, Splitterfang *m (Holzschliffherstellung)*
k. beak *(Lw)* Knüpferschnabel *m (Sammelpresse)*
knottiness Astigkeit *f*
knotting 1. *(Am)* Spachtelmasse *f*; 2. Grobstoffsortierung *f*, Herauslesen *n* von Ästen *(Holzschliffherstellung)*
k. apparatus *(Lw)* Knüpfapparat *m (Sammelpresse)*
knotty firewood Knorrholz *n*
k. wood Knorrholz *n*
know-how Betriebserfahrung *f*, Fachkenntnisse *fpl*, technisches Wissen *n*
knub Astknorren *m*
knuckle 1. Gelenk *n*, Kardandrehzapfen *m*; 2. Knick *m*, Heckknick *m (bei Rundgattschiffen)*
k. joint Gabelgelenk *n*; Kniegelenk *n*; Kniehebel *m*
k. joint embossing press Kniehebelmatrizenpresse *f*
k. line Knicklinie *f (beim Schiffsheck)*
k. mast *(Schiff)* Klappmast *m*
k. pin Anlenkbolzen *m*, Kuppelbolzen *m*
k. post *(Brg)* Haspelpfosten *m*
k. strake *(Schiff)* Knickgang *m*, Plattengang *m*
k. thread Rundgewinde *n*
knurl / to 1. kordieren, kordeln; 2. rändeln
k. straight-lined patterns rändeln
k. with double-wheel knurling tool kordieren, kordeln
knurl 1. Kordelrädchen *n*; 2. Rändelrädchen *n*
knurled handle Kordelgriff *m*; Rändelgriff *m*
k. head Rändelkopf *m*
k.-head screw Rändelschraube *f*
k. knob Kordelgriff *m*; Rändelknopf *m*
k. nut Kordelmutter *f*; Rändelmutter *f*
k. screw Kordelschraube *f*; Rändelschraube *f*
knurling Rändel *n*
k. having the ridges at right angles Kreuzrändel *n*
k. roller Rändelrädchen *n*
k. straight-lined patterns Rändeln *n*
k. tool Kordelwerkzeug *n*; Rändelwerkzeug *n*
k. tool with two wheels Kordelwerkzeug *n*
k. wheel Kordelrädchen *n*
Koepe wheel Treibscheibe *f*, Koepescheibe *f (Schachtförderung)*
Kort nozzle *(Schiff)* Kortdüse *f*
kraft bleaching Kraftzellstoffbleiche *f*
k. board *s.* k. paper
k. cord *(Text)* Hartkord *m*
k. paper Kraft[pack]papier *n*; Hartpapier *n*
k. pulp Kraftzellstoff *m*
kraftliner Wellpappendeckenlage *f*
Kramer control *(Elm)* Drehzahlregelung *f* mit Hintermotor
Krarup cable *(Fmt)* Krarup-Kabel *n*
Krarup winding *(Fmt)* Krarup-Wicklung *f*
krarupize / to *(Fmt)* krarupisieren
krypton Krypton *n*
Kundt's tube Kundtsches Rohr *n*
kurtosis *(Math)* Exzeß *m (Häufigkeitskurve)*, Steilheit *f*, Krümmung *f*, Wölbung *f*
kyanize / to kyanisieren *(Holz)*
kymograph Kymograf *m*
kynurenic acid Kynurensäure *f*

L

L-binding energy Bindungsenergie *f* eines L-Elektrons
L-capture *(Kern)* L-Einfang *m*
L cathode L-Katode *f*
L-conversion *(Kern)* L-Umwandlung *f*
L display L-Darstellung *f*
L-electron L-Elektron *n*
L-electron capture *s.* L-capture

L-handled socket wrench Rohrsteckschlüssel *m*
L-head engine Motor *m* mit stehenden Ventilen, seitengesteuerter Motor
L-level *(Kern)* L-Niveau *n*
L-network Vierpol *m* in L-Schaltung, L-Schaltung *f*
L-shaped aerial L-Antenne *f*
L-shell Achterschale *f*, L-Schale *f*
L X-rays L-Röntgenstrahlung *f*
La Cour's converter Kaskadenumformer *m*
lab *(Am)* *s.* laboratory
label / to etikettieren, beschriften; bezeichnen, markieren
label 1. Etikett *n*; 2. *(Bw)* Wasserschenkel *m*; 3. *(Dat)* Kennimpuls *m*, Kennzeichen *n*; Markierung *f*
l. cloth stark geschlichtetes Baumwollgewebe *n* für Etiketten
l. puncher Etikettenausstanzmaschine *f*
labelled *(radioaktiv oder stabilisotop)* markiert, mit einem Leitisotop markiert
l. atom markiertes Atom *n*, Leitisotop *n*
l. compound *(radioaktiv)* markierte Verbindung *f*
labelling *(Kern)* Markierung *f*
labile labil
l. equilibrium labiles Gleichgewicht *n*
labor *(Am)* *s.* labour
laboratory 1. Labor[atorium] *n*; 2. Ofenherdraum *m*, Erhitzungskammer *f*, Brennraum *m*
l. assistant Laborant *m*
l. bench Laboratoriumstisch *m*
l. clothing Laborkleidung *f*
l. course Praktikum *n*
l. exercise[s] Laborübungen *fpl*
l. interferometer Laborinterferometer *n*
l. scale Versuchsmaßstab *m*
l. size Laboratoriumsmaßstab *m*
l. technician Labor[atoriums]techniker *m*
l. test Labor[atoriums]versuch *m*, Kleinversuch *m*, Prüfstandversuch *m*
l. testing Labor[atoriums]prüfung *f*
l. trial *s.* l. test
labour / to 1. bearbeiten, bebauen; 2. *(Schiff)* arbeiten, schlingern, stampfen
l. the engine *(sl)* den Motor schinden
labour Arbeit *f*
l.-aiding equipment Hilfseinrichtung *f*
l. camp Arbeiterwohnlager *n*
l. conditions Arbeitsbedingungen *fpl*
l. consumption *(Brg)* Arbeitsaufwand *m*, Schichtenaufwand *m*
l. efficiency Arbeitsleistung *f*
l. force Arbeitskräfte[anzahl *f*] *fpl*, Belegschaft *f*
l. piracy Abwerbung *f* von Facharbeitern
l. power Arbeitskräfte[anzahl *f*] *fpl*, Belegschaft *f*
l. productivity Arbeitsproduktivität *f*
l. safety regulations Arbeitsschutzanordnungen *fpl*
l.-saving arbeitsparend
l. shortage Arbeitskräftemangel *m*
labourer [ungelernter] Arbeiter *m*
labouring of the engine *(sl)* Schinden (schweres Laufen) *n* des Motors
labrador *(Min)* Labrador *m*
labradorite *(Min)* Labradorit *m*
labyrinth box Labyrinthbüchse *f*
l. collector Labyrinthabscheider *m*
l. gland Labyrinthdichtung *f*
l. packing Labyrinthdichtung *f*
l. piston Labyrinthkolben *m* *(Dichtung)*
l. protection Labyrinthdichtung *f*
l. seal *s.* l. sealing
l. sealing Labyrinthdichtung *f*
l. washer Labyrinthring *m*
lac Stocklack *m*, Imprägnierharz *n*, Lack *m*
l.-lac Lackfirnislack *m*
l. smearing machine Lackiermaschine *f* für Schichtstoffe
lacca *s.* shellac
lace / to 1. [ver]binden; schnüren; 2. mit Spitzen besetzen; 3. versetzen *(mit Alkohol)*
l. up *(Brg)* verschalen
lace Litze *f*, Schnur *f*; Schnürsenkel *m*; Spitze *f*
l. boots Schnürstiefel *mpl*
l. maker Klöppler[in *f*] *m*
l. pillow Klöppelkissen *n*
l. weaving machine Spitzenwebmaschine *f*
l.-worked button Posamentenknopf *m*
lacing 1. Binden *n*, Verbinden *n*; Verbinder *m*; 2. Vergitterung *f*, Diagonalen *fpl* *(Stahlbau)*; 3. *(Brg)* Verschalung *f*, Verzug *m*
l. board Kabelformbrett *n*
l. loom Posamentenwebstuhl *m*
l. of cables Abbinden *n* von Kabelbäumen
l. pattern *(Brg)* Anordnung *f* der Schrämmeißel in der Kette
l. thread Trennfaden *m*
l. twine Bindegarn *n* *(für Kabel)*
l. wire Versteifungsdraht *m*
lack / to ermangeln, nicht haben, nicht besitzen
lack Mangel *m*, Fehlen *n*, Nichtvorhandensein *n*
l. in water Wassermangel *m*
l. of balance Unwucht *f*
l. of definition Unschärfe *f*
l. of oxygen Sauerstoffmangel *m*
l. of roundness Unrundheit *f*
l. of space Platzmangel *m*, Raummangel *m*
l. of uniformity Ungleichmäßigkeit *f*
lacklustre glanzlos, matt
lacquer / to lackieren, lacken, firnissen
lacquer Lack *m*, Lackfarbe *f*, Anstrichstoff *m*, Firnis *m*
l. dope Zaponlack *m*

l. enamel Farblack *m*, pigmentierter Lack *m*
l. finish Farblacküberzug *m*, Lackierung *f*
l. master Lackschallplatte *f*
l. print Lackdruck *m*
l. sealer schnelltrocknender Isolierlack *m*
l.-solvent Lacklösungsmittel *n*
lacquered lackiert
l. leather Lackleder *n*
l. wire Lackdraht *m*
lacquerer Lackierer *m*
lacquering furnace Lackierofen *m*
lactate Laktat *n*, Salz *n* der Milchsäure
lactic acid Milchsäure *f*
l. bacteria Milchsäurebakterien *fpl*
l. fermentation Milchsäuregärung *f*
lactide *(Ch)* Laktid *n*
lactometer *s.* galactometer
lactone *(Ch)* Lakton *n*
lactose Laktose *f*, Milchzucker *m*
lactylic anhydride Milchsäureanhydrid *n*
ladder 1. Leiter *f*; Treppe *f (im Schiff)*; 2. Laufmasche *f*; 3. *(Brg)* Fahrt *f*; 4. Eimerleiter *f (Bagger)*
l. attenuator *(El)* Ketten[spannungs]teiler *m*
l. beam Leiterbaum *m*, Leiterholm *m*
l. dredge Eimerketten[naß]bagger *m*
l. dredger *s.* l. dredge
l. hoist gantry Leiterbock *m (beim Bagger)*
l. logic *(Dat)* Leiterlogik *f*
l. network *(El)* Kettenleiter *m*, symmetrisches Netzwerk *n*
l. pole Leiterstange *f*
l.-proof *(Text)* maschenfest
l. road *(Brg)* Fahrtrum *n*
l. rope *(Schiff)* Fallreep *n*
l. round Leitersprosse *f*
l. rung Leitersprosse *f*
l. scaffold Leitergerüst *n*
l. scaffolding *s.* l. scaffold
l. sollar *(Brg)* Ruhebühne *f*
l. step Leitersprosse *f*, Treppenstufe *f*
l.-type filter *(El)* Abzweigfilter *n*
l.-type trenching machine Eimerkettengrabenbagger *m*
l. vein *(Geol)* Leitergang *m*
ladderway 1. *(Schiff)* Niedergangsluke *f*, Treppenöffnung *f* im Deck; 2. *(Brg)* Fahr[ten]trum *n*
laddic *s.* ladder logic
lade / to [be]laden, belasten
laden beladen, belastet
lading Ladung *f*, Beladung *f*, Beladen *n*; Schiffsladung *f*
l. port Ladungshafen *m*, Verladehafen *m*
ladle / to [aus]schöpfen
ladle *(Met)* Pfanne *f*, Gießpfanne *f*; Kelle *f*, Schöpflöffel *m*, Gießlöffel *m*
l. additions Pfannenzusätze *mpl*
l. analysis Pfannenprobe *f*, Schöpfprobe *f*
l. bail Gießpfannengehänge *n*
l. barrow Gießpfannenwagen *m*
l. bowl Pfannentiegel *m*
l. capacity Pfanneninhalt *m*
l. car Pfannenwagen *m*
l. crane Gießpfannenkran *m*, Gußpfannenkran *m*, Gießereikran *m*
l. guide Pfannenführung *f*
l. lining Gießpfannenauskleidung *f*, Pfannenauskleidung *f*
l. lip Pfannenschnauze *f*
l. pit Schlackengrube *f*
l. plug Pfannenpfropfen *m*
l. pouring time Pfannengießzeit *f*
l. sample Schöpfprobe *f*
l. scull *s.* l. skull
l. shank Pfannengabel *f*
l. skull Pfannenbär *m*, Pfannenkruste *f*
l. test Pfannenprobe *f*
l. truck 1. Gießpfannenwagen *m*; 2. *(Eis)* Transportkübelwagen *m*
ladleman *(Met)* Gießer *m*; Pfannenführer *m*
lady's bicycle Damenfahrrad *n*
laevogyrate linksdrehend
l. quartz Linksquarz *m*
laevogyration Linksdrehung *f*
laevogyrous *s.* laevogyrate
laevoisomer linksdrehendes Isomer(es) *n*
laevopolarization Linkspolarisation *f*
laevorotation Linksdrehung *f*
laevorotatory linksdrehend
lag / to 1. *(El)* zurückbleiben, nacheilen, sich verzögern, nachhinken; 2. isolieren *(mit Dämmstoff)*; verschalen; verkleiden; ummanteln
l. behind nacheilen
lag 1. *(El)* Nacheilung *f*, Nacheilen *n (des Phasenwinkels)*; 2. Verzögerung *f (Relais)*; Zurückbleiben *n*; Verzug *m*; Nachhinken *n*; 3. Dämmstoff *m*
l. circuit Verzögerungsschaltung *f*
l. coefficient *s.* time constant
l. network Verzögerungsschaltung *f*, Schaltung *f* mit Phasenrückdrehung
l. of a valve Ventilverzögerung *f*
l. of control *(Flg)* Nachbleiben *n* der Steuerung
l. of ignition *(Kfz)* Zündverzug *m*
l. screw Ankerbolzen *m*, Ankerschraube *f*, Fundamentschraube *f*, Holzschraube *f* mit Rundkopf
l. screw gimlet pointer Holzschraubenanspitzmaschine *f*
l. screw wrench Schraubenschlüssel *m* für Holzschrauben
l. winding *(Fmt)* Schleifenwicklung *f*
lagan *(Schiff)* Wrackgut *n*, versenktes Gut *n*

lagged-demand meter *(El)* Spitzenzähler *m* mit Zeitverzögerung
l. hose verkleideter Schlauch *m*
l. pipe wärmeisoliertes Rohr *n*
lagging 1. Hinken *n*, Nacheilung *f*, Verzögerung *f*; 2. Isolierung *f*, Rohrisolierung *f*; Wärmeschutzverkleidung *f*, Verkleidung *f* *(Kessel)*; 3. *(Brg)* Verzug *m*, Verschalung *f*
l. behind of the steam engine Hinken *n* der Dampfmaschine
l. crankshaft *(Kfz)* Schleppkurbel *f*
l. current *(El)* nacheilender Strom *m*
l. load induktive Belastung (Last) *f*
l. of phase Phasennacheilung *f*
l. of the tide Verzögerung *f* der Gezeiten
l. power-factor nacheilender (induktiver) Leistungsfaktor *m*
l. ring Verkleidungsring *m*
Lagrange function Lagrange-Funktion *f*
Lagrange's polynomial Lagrangesches Polynom *n*
Lagrangian [function] *s.* Lagrange function
lagre Lager *n (Glas)*
laid deck *(Schiff)* Holzdeck *n (mit schmalen Planken)*
l. length Einbaulänge *f*
l. lines Wasserlinien *fpl (Egoutteurrippung)*
l. paper geripptes Papier *n*
l.-up ship außer Dienst gestelltes Schiff *n*, aufgelegtes Schiff
l.-up vehicle stillgelegtes Fahrzeug *n*
laitance Zementmilch *f*, Zementschlämme *f*, Zementschlempe *f*
lake 1. See *m*; 2. gedeckter Farbstoff *m*, Farblack *m*, Lackfarbe *f*; organisches Pigment *n*
l. iron ore Raseneisenerz *n*, See-Erz *n*
l. steamer Seendampfer *m*, Dampfer *m* vom Typ der Große-Seen-Schiffe *(Maschine achtern)*
Lalande cell Lalande-Element *n (Zink-Eisen-Primärelement)*
lamb Jungschaf *n*, Lamm *n*
Lamb shift Bethe-Effekt *m*, Lambshift *m (H-Spektrum)*
lambert Lambert *n (Einheit der Leuchtdichte; 1 la = 0,318 cd/cm²)*
Lambert's law Lambertsches Gesetz *n*
lambswool Lammwolle *f*
lamella Lamelle *f*, Schicht *f*; Gleitlamelle *f*; Plättchen *n*, Streifen *m (Gefüge)*
lamellar lamellar, geschichtet, blättchenförmig, blättrig, streifenförmig
l. magnet Lamellenmagnet *m*, Blättermagnet *m*
l. perlite lamellarer Perlit *m*
l. prop *(Brg)* Lamellenstempel *m*
l. structure lamellare Struktur *f*
lamina Lamelle *f*; Schicht *f*; Folie *f*; Blättchen *n*; Plättchen *n*; Furnierblatt *n*, dünne Holzlage *f*
l. of wood Dickte *f*, dünne Holzlage *f*
laminable streckbar, walzbar
laminar boundary layer *(Strm)* laminare Grenzschicht *f*
l. flow laminare Strömung *f*, Schichtenströmung *f*, Laminarströmung *f*
l. friction laminare Reibung *f*
l. motion laminare Bewegung *f*
l. path of current *(El)* flächenförmige Strombahn *f*
laminate / to schichten, Schichtstoff herstellen, beschichten, kaschieren, laminieren, in Blätter teilen, lamellieren, unterteilen; Folien auswalzen (schlagen)
l. to laminieren mit
l. with foam back mit Schaumstoff laminieren
laminate *(Am)* Schicht[preß]stoff *m*; Schichtplatte *f*
l. moulding Herstellung *f* geschichteter Formteile
laminated geschichtet, geblättert, laminiert, lamelliert, mehrlagig, mehrschichtig, aus Schichtstoff; Mehrschichten-
l. aerial geschichtete Antenne *f (UKW-Antenne)*
l. board Schichtstoffplattenmaterial *n*
l. brush switch *(El)* Bürstenschalter *m*
l. contact lamellierter Kontakt *m*
l. core 1. *(El)* Blechkern *m*; 2. *(Holz)* Stäbchenmittellage *f*
l. core rotor Läufer (Rotor) *m* mit Blechkern
l. fabric Schichtstoff *m*; Hartgewebe *n*
l. flow Schichtenströmung *f*
l. fracture lamellenartiger (lagenartiger) Bruch *m*
l. fuse Lamellensicherung *f*, Streifensicherung *f*
l. glass Verbundglas *n*, Schichtglas *n*, Sicherheitsglas *n*
l.-iron core Eisenblätterkern *m*, Eisenblechkern *m (Zündspule)*
l. lumber Schichtholz *n*
l. magnet Lamellenmagnet *m*, Blättermagnet *m*
l. metal geschichtetes (mehrschichtiges) Metall *n*
l. mica Spaltglimmer *m*
l. moulding Formpreßling *m*, Formstück *n* aus Schichtstoff
l. paper kaschiertes Papier *n*, Schichtenpapier *n*, Hartpapier *n*
l. paper board Platte *f* aus Papierschichten
l. plastics Schicht[preß]stoff *m*
l. sheet Schichtstoffbahn *f*, Schichtstoffplattenmaterial *n*
l. sheet[-iron] plate Blechpaket *n*
l. ski Schichtenschi *m*
l. spring geschichtete Blattfeder *f*

l. **structure** Schichtgefüge *n*
l. **tin** Walzzinn *n*
l. **wood** Schichtholz *n*, verleimte Bretter *npl*
laminating *(Led)* Aufkaschieren *n* einer fertigen Schicht
l. **glue** Sperrholzleim *m*
l. **machine** Kaschiermaschine *f*
l. **mill** Platinenwalzwerk *n*
l. **sheet** Schichtfolie *f*
l. **wax** Kaschierwachs *n*
lamination 1. Dopplung *f*, Schichtung *f (von Blechen beim Walzen)*, Walzdopplung *f*; Walzfaser *f*; 2. Blechschnitt *m*, Dynamoblech *n*, Lamelle *f*, lamelliertes Blech *n*, Transformatorenblech *n*; 3. *(Geol)* Schichtung *f*
l. **by quilting** *(Text)* Laminieren *n* durch Aufnähen
l. **coating** *(Led)* Kaschieren *n* mit Film bei Austritt aus der Breitschlitzdüse
lamp Lampe *f*; *(Brg)* Geleucht *n*; Leuchte *f*
l. **bank** Lampenbrett *n*
l. **base** Leuchtenfuß *m*, Lampensockel *m*
l. **basin** Leuchtenschale *f*
l. **black** Lampenruß *m*
l. **bracket** Lampenhalter *m*, Lampenträger *m*
l. **bulb** Lampenkolben *m*, Birne *f*, Glühbirne *f*
l. **cabin** *(Brg)* Lampenstube *f*, Lampenkammer *f*
l. **cap** Lampensockel *m*, Sockel *m*
l. **capping cement** Sockelkleber *m*
l. **circuit** Lampenstromkreis *m*, Lichtstromkreis *m*
l. **guard** Lampenschutzkorb *m*
l. **holder** Lampenfassung *f*
l.-holder plug Fassungsstecker *m*
l.-holder switch Schalter *m* an der Lampenfassung
l. **paneel** *(Fmt)* Lampenfeld *n*
l. **post** Laternenmast *m*
l. **receptacle** Lichtsteckdose *f*
l. **repeater** Meldelämpchen *n (Eisenbahnsignalwesen)*
l. **resistance** Lampenwiderstand *m*
l. **room** *(Brg)* Lampenstube *f*, Lampenkammer *f*
l. **shade** Lampenschirm *m*
l. **socket** Lampenfassung *f*
l. **with firedamp detector** *(Brg)* Verbundleuchte *f*
lanarkite *(Min)* Lanarkit *m (Bleisulfat)*
lance *(Text)* lanziertes Gewebe *n*
lancet Lanzette *f*
l. **arch** *(Bw)* überhöhter Spitzbogen *m*
land / to 1. anlanden, anlegen *(mit dem Schiff)*; ausschiffen; 2. *(Flg)* landen
l. **on water** wassern *(Seeflugzeug)*
land 1. Land *n*; Feld *n*; 2. Führungsfase *f*, Umfläche *f (Spiralbohrer)*; 3. hervorstehende Fläche *f (Nuten)*; tragende Fläche; 4. Feld *(im gezogenen Lauf einer Schußwaffe)*; 5. *(Plast)* Formtrennfuge *f*
l. **chain** Meßkette *f*
l. **clearance** Freifräsung *f* am Umfang *(Bohrer)*
l. **direction finding station** Landpeilstation *f (Radar)*
l. **evaluation** Bodennutzung *f*
l. **ground circular** Rundschliffase *f*
l. **levelling** Landplanierung *f*
l. **line** Überlandleitung *f*
l.-mark Grenzstein *m*
l. **packer** Ackerwalze *f*
l. **reclamation** Neulandgewinnung *f*, Landgewinnung *f*
l. **register** Kataster *n*
l. **relief** Freiwinkel *m*, Hinterschliff *m* der Fase
l. **roller** Ackerwalze *f*
l. **station** Landfunkstelle *f*
l. **surveying** Feldmessung *f*, Landvermessung *f*
l. **surveyor** Feldmesser *m*, Landvermesser *m*, Vermessungsingenieur *m*, Geodät *m*
l. **turbine** Landturbine *f*; stationäre Turbine *f*
l. **wear** Verschleißmarkenbreite *f (Werkzeug)*
l. **width** Fasenbreite *f*
l. **work** Landbetrieb *m*, stationärer Betrieb *m*
landed casing [im Bohrloch] abgesetzte Verrohrung *f*
lander *(Brg)* Anschläger *m*
landing 1. Landung *f*; 2. Podest *n*, Treppenabsatz *m*, Treppenpodest *n*; 3. *(Brg)* Hängebank *f*, Ruhebühne *f*; 4. *(Fs)* Ungleichmäßigkeit *f* der Signalamplitude
l. **angle** *(Flg)* Landewinkel *m*
l. **attitude** Landeflugzustand *m*
l. **beam beacon** Einflugzeichensender *m*, Landungsbake *f (Luftnavigation)*
l. **bridge** *(Schiff)* Anlandebrücke *f*
l. **cable** Haltetrosse *f*
l. **chair** Aufsetzvorrichtung *f (Förderkorb)*
l. **craft** Landungsboot *n*
l. **deck** Landedeck *n*, Flugdeck *n (Flugzeugträger)*
l. **dogs** *(Brg)* Aufsetzvorrichtung *f*
l. **dues** Landungsgebühren *fpl*
l. **field** Rollbahn *f*
l. **gear** *(Flg)* Fahrgestell *n*, Fahrwerk *n*
l. **gear cushioning** Fahrwerkabfederung *f*
l. **light** Landefeuer *n*, Landelicht *n*
l. **pier** Anlegebrücke *f*, Landungsbrücke *f*
l. **place** Anlandeplatz *m*, Anlegeplatz *m*, Anlegestelle *f*
l. **push-button** Fahrstuhlstockwerksdruckknopf *m*

l. radio beacon Landebake *f (Flugnavigation)*
l. rocket Landungsrakete *f*
l. rope Haltetrosse *f*
l. runway Landebahn *f*
l. speed Landegeschwindigkeit *f*
l. stage 1. Landungsbrücke *f*, Bootssteg *m*; 2. *(Brg)* Zwischenbühne *f*, Sturzbühne *f*
l. strip Landebahn *f*
l. switch Treppenschalter *m (automatische Treppenlichtanlage)*
landline teletype Fernschreiben *n* über Draht
landmark Ortungspunkt *m (Radar)*
landsale Landverkauf *m*
landscape *(Typ)* Querformat *n*
l. architect Landschaftsarchitekt *m*, Landschaftsgestalter *m*, Landschaftsplaner *m*
l. architecture Freiflächenplanung *f*, Landschaftsarchitektur *f*, Landschaftsgestaltung *f*
l. gardening Landschaftsgartenbau *m*, Freiflächengartenbau *m*
l. planning Landschaftsarchitektur *f*, Landschaftsgestaltung *f*, Landschaftsplanung *f*
landscaping Freiflächenplanung *f*, Landschaftsgestaltung *f*
landslide Erdrutsch *m*
landslip Erdrutsch *m*
lane 1. Gasse *f*; 2. *(Schiff)* Fahrroute *f*, Dampferweg *m*; 3. Fahrbahn *f*, Fahrspur *f (Straße)*; Fertigungsbahn *f (Betonwerk)*
l. marker Fahrspurmarkierung *f*, Straßenmarkierung *f*
l. marking *s.* l. marker
l. of approach *(Flg)* Landeschneise *f*
langley Flugzeuglandungsinsel *f*
Langmuir effect Langmuir-Effekt *m*
Langmuir probe Langmuir-Sonde *f*
Lang's lay rope Gleichschlagseil *n*, Längsschlagseil *n*
language Informationsdarstellung *f (für Datenverarbeitungsmaschinen)*
lanolin Lanolin *n*
lantern Laterne *f*
l. light Dachlaterne *f*
l. picture Diapositivbild *n*
l. pinion Triebstockkranz *m*
l. slide Diapositiv *n*
l. slide lecture Lichtbildervortrag *m*
l. slide scanning Diapositivabtastung *f*
l. wheel Triebstockzahnrad *n*
lanthana Lanthanoxid *n*
lanthanides Lanthaniden *npl*
lanthanite *(Min)* Lanthanit *m*
lanthanum Lanthan *n*
l. oxide Lanthanoxid *n*
lanyard stopper *(Schiff)* Taustopper *m*
lap / to 1. *(Masch)* läppen; 2. überlappen [lassen]; bewickeln, umwickeln; 3. *(Holz)* überblatten
l. weld überlappt schweißen
lap 1. Überlappung *f*, Überdeckung *f*; Falte *f (Oberflächenfehler)*; 2. Läppwerkzeug *n*; 3. *(Text)* Wickel *m*, Legung *f*, Schlagmaschinenwickel *m*
l. brazing Überlapplötung *f*
l. dissolve *(Fs)* weiche Überblendung *f (zweier Filmszenen)*
l. dovetail verdeckter Schwalbenschwanz *m*
l. drawing frame *(Text)* Wickelstrecke *f*, Kehrstrecke *f*
l. drum *(Text)* Pelztrommel *f*, Vliestrommel *f*
l.-edge *(Masch)* Lappenkante *f*
l.-edge fracture *(Wärm)* Kantenriß *m*
l.-ended piston ring Kolbenring *m* mit überlapptem Stoß
l. feeding *(Text)* Speisen *n* der Wickel
l. for holes *(zylindrische)* Lapphülse *f*
l. joint Überdeckungsstoß *m*, Überlappstoß *m*; Überlapp[t]klebung *f*
l.-joint riveting Überlappungsnietung *f*
l. material Läppmittel *n*
l. of fibre *(Text)* Faserwickel *m*
l. plate *(ebene)* Läppscheibe *f*
l. polishing Wischpolieren *n*
l. roller *(Text)* Abwickelwalze *f*, Wickelwalze *f*
l.-solvent sealing Überlapptsiegeln *n* mit Lösemitteln
l.-time Anzugszeit *f*, Trocknungsbeginn *m (Lackindustrie)*
l. weld Überlapp[schweiß]naht *f*
l. welding Überlappungsschweißung *f*, überlappendes Schweißen *n*
l. winding *(El)* Schleifenwicklung *f*, Scheitelwicklung *f*, überlappte Wicklung *f*
lapel Revers *n*
l. microphone Knopflochmikrofon *n*
lapicide Steinmetz *m*
lapidary Edelsteinschleifer *m*
l. trade Schmucksteinindustrie *f*
lapis *(Min)* Lapis *m*
l. lazuli Lapislazuli *m*, Lazurstein *m*, Lasurstein *m*
Laplace transformation Laplace-Transformation *f*
Laplacian Laplace-Operator *m*
Laplacian equation Laplacesche Gleichung *f*
Laporte parity rule *(Kern)* Laportesche Regel *f*
lappage Überlappung *f*
lapped geläppt; überlappt
l. connection Stoßüberdeckung *f*
l. splice Überdeckungsstoß *m*
l. winding *s.* lap winding
lappet *(Text)* Fadenführerklappe *f*
l. eye Fadenführeröse *f*
l. loom Lappetstuhl *m*, Stickereiwebstuhl *m*
l. weaving Lappetstickerei *f*
lapping 1. Läppen *n*; 2. Überdeckung *f*, Überlappung *f*

latex Latex *m*
l. coat Latexanstrich *m*
l. foam rubber Schaumgummi *m*
lath Latte *f*, Leiste *f*; *(Brg)* Getriebepfahl *m*
l.-fence Lattenzaun *m*
l. grating Lattenrost *m*
l. nail Lattenstift *m*
lathe Drehmaschine *f*
l. apron Schloßplatte *f*
l. arbor Drehdorn *m*
l. bed Drehmaschinenbett *n*
l. carriage Drehmaschinensupport *m*
l. carrier Drehherz *n*
l. centre Drehmaschinenspitze *f*
l. chuck Drehmaschinenfutter *n*
l. chuck flange Futterflansch *m*
l. dog Drehherz *n*
l. driving Drehmaschinenantrieb *m*; *(Text)* Ladenantrieb *m*
l. driving carrier Drehherz *n*
l. for bearing races Wälzlagerringdrehmaschine *f*
l. for metal spinning Drückbank *f*
l. lead screw Drehmaschinenleitspindel *f*
l. mandrel Dreh[maschinenspann]dorn *m*
l. operator Dreher *m*
l. spindle Drehspindel *f*
l. sword *(Text)* Ladenfuß *m*, Ladenarm *m*
l. tool Drehmeißel *m*
l. work Dreharbeit *f*; Drehteil *n (Werkstück)*
lather / to schäumen
lather Seifenschaum *m*
l. booster Schaumverbesserer *m*
l. value Schaumwert *m*
lathering Schäumen *n (Seifenwasser)*
lathing Putzträger *m*; Latten *fpl*; Annageln *n* von Latten
lathwork Putzträger *m*
latices *s.* latex
latitude *(geografische)* Breite *f*; Breitengrad *m*
l. effect Breiteneffekt *m*
l. of density Schwärzungsumfang *m*
latten brass Messingblech *n*
lattice / to vergittern
lattice 1. Kristallgitter *n*, Gitter *n*; 2. *(Kern)* Brennstoffgitter *n*; 3. *(Bw)* Fachwerk *n*, Flechtwerk *n*, Gitterwerk *n*
l. arrangement Gitteranordnung *f*
l. array Gitteranordnung *f*
l. atom Gitteratom *n*
l. binding energy Bindungsenergie *f (im Kristallgitter)*
l. boom *(Schiff)* Gitterladebaum *m (für Schwergut)*
l. bulkhead *(Schiff)* Lattenschott *n*, Gitterschott *n*
l. circuit *(El)* Kettenleiter *m*
l. coil *(El)* Gitterspule *f*
l. conductance Gitterleitfähigkeit *f (Festkörper)*
l. constant Gitterkonstante *f*
l. defect Gitterfehlstelle *f*, Gitterfehler *m*, Störstelle *f*
l. dimensions Gitterabmessungen *fpl*
l. dislocation Gitterstörung *f*, Störung *f* des Kristallgitterbaus; Versetzung *f*
l. distortion *s.* l. defect
l. drum *(Text)* Lattentrommel *f*
l. element Gitterbaustein *m*
l. energy *s.* l. binding energy
l. expansion Gitterdehnung *f*
l. fence Lattenzaun *m*
l. filter *(El)* Brückenfilter *n*
l. fracture gitterartiger Bruch *m*
l. girder Fachwerkbinder *m*, Gitterfachwerk *n*, Gitterträger *m*
l. imperfection *s.* l. defect
l. impurity Gitterfehlstelle *f*
l. mast Gittermast *m (Stahlbau)*
l. matrix Gittermatrix *f*
l. network Kettenleiter *m*, Kreuzglied *n*
l. parameter Gitterkonstante *f*
l. plane Gitterebene *f*, Netzebene *f*
l. point Gitterpunkt *m*; Gitterplatz *m*
l. spacing Gitterkonstante *f*; Gitterabstand *m*
l. structure Gitteraufbau *m*, Gitterstruktur *f*
l. tower Gittermast *m (Stahlmast)*
l. transition Gitterübergang *m (Halbleiter)*
l.-type filter *s.* l. filter
l.-type girder Fachwerkbinder *m*, Gitterfachwerk *n*, Gitterträger *m*
l. vacancy Gitterfehlstelle *f*
l. vibration Gitterschwingung *f*
l. void Gitterloch *n*
l. winding *(El)* Faßwicklung *f*, Korbwicklung *f*
l. window Gitterfenster *n*
l. work Gitterwerk *n*; Fachwerk *n*; Lattengerüst *n*
latticed gitterartig
l. column Gitterstütze *f*
l. girder Gitterwerksträger *m*
l. girder construction Gitterwerk *n*
latus rectum Parameter *m (von Kegelschnitten)*
laubanite *(Min)* Laubanit *m*
Laue diffraction pattern Laue-Beugungsdiagramm *n*, Laue-Diagramm *n*
Laue interference Laue-Interferenz *f*
Laue method Laue-Verfahren *n*
Laue pattern *s.* Laue diffraction pattern
Laue's diagram *s.* Laue diffraction pattern
laughing gas Lachgas *n*
laumontite *(Min)* Laumontit *m*
launch / to starten *(z. B. Rakete)*; vom Stapel lassen *(ein Schiff)*

launch 1. Stapellauf *m*; 2. *(Rak)* Start *m*; 3. Beiboot *n*, Pinasse *f*, Gleitboot *n*, Barkasse *f*, Großboot *n*
l. control car *(Rak)* Feuerleitwagen *m*
l. of the slipway *(Schiff)* Vorhelling *f*
l. slip *(Schiff)* Helling *f*, Ablaufbahn *f*
launcher *(Rak)* Abschußrampe *f*, Startrampe *f*
l. loader *(Rak)* Transportladefahrzeug *n*
launching 1. *(Rak)* Start[en *n*] *m*; 2. *(Schiff)* Stapellauf *m*; 3. Hohlleiterspeisung *f* *(mit Koaxialkabel)*
l. alongside the equator Raketenstart *m* längs des Äquators
l. angle *(Rak)* Startwinkel *m*
l. arrangement Aussetzvorrichtung *f* *(für Rettungsboote)*
l. base *(Rak)* Startplatz *m*
l. carriage *(Rak)* Startwagen *m*
l. cradle 1. *(Rak)* Starttisch *m*; 2. *(Schiff)* Ablaufschlitten *m (Stapellauf)*
l. difficulty *(Rak)* Anfahrschwierigkeit *f*, Startschwierigkeit *f*
l. drags Stapellaufschleppketten *fpl*, Bremsketten *fpl*
l. emplacement Startaufbau *m*, Startplattform *f*
l. friction Reibung *f* beim Stapellauf
l. grease Stapellaufschmiere *f*
l. installation Raketenstarteinrichtung *f*, Startanlage *f*, Startvorrichtung *f*
l. of a satellite Satellitenstart *m*, Start *m* eines Satelliten
l. pad *(Rak)* Starttisch *m*
l. platform *(Rak)* Startplattform *f*, Startaufbau *m*
l. poppet Schlittenstütze *f (Stapellauf)*
l. pressure Flächendruck (Schlittendruck) *m* beim Stapellauf
l. rail *(Rak)* Startschiene *f*, Startgestell *n*
l. rocket Startrakete *f*
l. site *(Rak)* Startplatz *m*
l. speed Ablaufgeschwindigkeit *f (Stapellauf)*
l. table *(Rak)* Starttisch *m*
l. track *(Rak)* Startgestell *n*, Startschiene *f*
l. trigger Stopphebel *m (Stapellauf)*
l. vehicle Raketenträger *m*, fahrbare Raketenstartanlage *f*
l. ways *(Schiff)* Ablaufbahn *f (Helling)*
l. wedge Stapellaufkeil *m*, Schlittenkeil *m*
l. weight 1. *(Rak)* Startmasse *f*; 2. *(Schiff)* Ablaufmasse *f*, Stapellaufmasse *f*
launder / to waschen
launder Rinne *f*; Gerinne *n*; Gefluter *n*, Gefluder *n (Aufbereitung)*
launderproof *(Text)* beständig gegen normales Waschen
laundry Wäscherei *f*, Waschhaus *n*; Wäsche *f*
l. equipment Wäschereiausrüstung *f*
laundryman Wäscher *m*, Wäschereiarbeiter *m*
lauraldehyde Laurinaldehyd *m*
laurate Laurat *n*, Laurinsäureester *m*
lauric acid Laurinsäure *f*
lava Lava *f*
l. flow Eruptivgestein *n*, vulkanisches Gestein *n*, Vulkanit *m*
l. pumice Bimssteinlava *f*
law Gesetz *n*; Naturgesetz *n*; Satz *m*, Hauptsatz *m*, Lehrsatz *m*; Prinzip *n*, Regel[n *fpl*] *f*
l. of absorption Absorptionsgesetz *n*
l. of action and reaction Prinzip *n* von Wirkung und Gegenwirkung
l. of attraction Anziehungsgesetz *n*
l. of causality Kausalgesetz *n*
l. of conservation Erhaltungssatz *m*
l. of conservation of angular momentum Gesetz *n* von der Erhaltung des Drehimpulses
l. of conservation of matter Satz *m* von der Erhaltung der Materie
l. of conservation of momentum Prinzip *n* von der Erhaltung der Bewegungsgröße, Impulserhaltungssatz *m*
l. of constant angles Gesetz *n* der Winkelkonstanz
l. of constant proportions Gesetz *n* der konstanten Proportionen (Gewichtsverhältnisse)
l. of cooling Abkühlungsgesetz *n (Newton)*
l. of corresponding states Gesetz *n* der übereinstimmenden Zustände
l. of cosines 1. *(Math)* Kosinussatz *m*; 2. *(Kern)* Kosinusgesetz *n (Strahlung)*
l. of definite proportions *s.* l. of constant proportions
l. of dilution *(Ch)* Verdünnungsgesetz *n*
l. of distribution Verteilungsgesetz *n*
l. of distribution of velocities Geschwindigkeitsverteilungsgesetz *n*
l. of Dulong and Petit Dulong-Petitsche Regel *f*
l. of elasticity Elastizitätsgesetz *n*
l. of electric networks Kirchhoffscher Verzweigungssatz *m*
l. of electromagnetic induction Faradaysches Gesetz *n*
l. of electrostatic attraction Coulombsches Gesetz *n*, Gesetz der elektrostatischen Anziehung
l. of elongation Dehnungsgesetz *n*
l. of equipartition of energy Energiegleichverteilungssatz *m*
l. of errors Fehlergesetz *n*
l. of falling bodies Fallgesetz *n*
l. of gearing Verzahnungsgesetz *n*
l. of gravitation Gravitationsgesetz *n*, Fallgesetz *n*
l. of Guldberg and Waage *s.* l. of mass action

l. **of induction** Induktionsgesetz *n*
l. **of inertia** Trägheitsgesetz *n*
l. **of isothermal radiation** Temperaturstrahlungsgesetz *n*
l. **of lever** Hebelgesetz *n*
l. **of mass action** Massenwirkungsgesetz *n*
l. **of mines** Berggesetz *n*, Bergrecht *n*
l. **of mixtures** Mischungsgesetz *n*
l. **of moment of momentum** Drehimpuls[erhaltungs]satz *m*
l. **of motion** Bewegungsgesetz *n*
l. **of multiple proportions** Gesetz *n* der multiplen Proportionen
l. **of nature** Naturgesetz *n*
l. **of networks** Verzweigungsgesetz *n*
l. **of optical absorption** optisches Absorptionsgesetz *n*
l. **of partial pressures** Gesetz *n* der Partialdrücke *(Dalton)*
l. **of pendulum motion** Pendelgesetz *n*
l. **of photochemical equivalence** Einsteinsches Äquivalenzgesetz *n*
l. **of probability** Wahrscheinlichkeitsgesetz *n*
l. **of radioactive decay** *s.* l. of radioactive disintegration
l. **of radioactive disintegration** Gesetz *n* des radioaktiven Zerfalls, [radioaktives] Zerfallsgesetz *n*
l. **of rationality** Rationalitätsgesetz *n*
l. **of reflection** Reflexionsgesetz *n*
l. **of refraction** Brechungsgesetz *n*
l. **of relativity** Relativitätsgesetz *n*
l. **of sequence** Gesetz *n* der Serie
l. **of signs** Vorzeichenregel *f*
l. **of sine** Sinussatz *m*
l. **of the conservation of mass** Erhaltungssatz *m* der Masse
l. **of the mean** Mittelwertsatz *m*
l. **of the propagation of errors** Fehlerfortpflanzungsgesetz *n*
l. **of thermodynamic similitude** Gesetz *n* der thermodynamischen Ähnlichkeit
l. **of thermodynamics** Hauptsatz *m* der Thermodynamik
l. **of volumes** Gay-Lussacsches Gesetz *n*

lawn 1. [gepflegter] Rasen *m*, Rasenplatz *m*; 2. Sieb *n (mit Seideeinsatz)*
l. **mower** Rasenmäher *m*
l. **sprinkler** Rasensprenger *m*

lawsonite *(Min)* Lawsonit *m*, farbloser Lievrit *m*

lay / to 1. legen, setzen, stellen; 2. [ver]legen *(Kabel)*
l. **away** *(Led)* versenken, versetzen
l. **away in grease** *(Led)* geschmiertes Leder ablagern lassen
l. **bar** freilegen
l. **claim to something** einen Anspruch auf etwas erheben
l. **down** auf Stapel legen
l. **in sawdust** in feuchte Sägespäne legen *(Chromleder)*
l. **off a bearing on the chart** *(Schiff)* Peilung in die Karte eintragen
l. **off a course in the chart** *(Schiff)* Kurs auf der Karte absetzen
l. **on** *(Typ)* anlegen *(Bogen)*
l. **on the stocks** auf Stapel legen
l. **out** anordnen; *(Am)* anreißen, aufreißen; trassieren *(Straßenbau)*; auslegen, verlegen *(Kabel)*
l. **out cables** Kabel legen
l. **the course** *(Schiff)* Kurs absetzen
l. **the grain** *(Led)* satinieren
l. **the keel** *(Schiff)* Kiel legen
l. **to** *(Schiff)* beidrehen
l. **up** auflegen, abrüsten *(ein Schiff)*

lay 1. Oberflächenzeichnung *f*; Richtungsverlauf *m* der Rauheit; Strich *m*; 2. Schlag[länge *f*] *m (Drahtseil)*; 3. *(Typ)* Anlage *f*; 4. *(Text)* Ladenschere *f*, Ladenzugstange *f*
l. **away pit** *(Led)* Versenkgrube *f*
l.**-by** 1. Liegeplatz *m (am Kanal für Schiffe)*; 2. Parkplatz *m*, Rastplatz *m (Autobahn)*
l. **days** Liegezeit *f (von Schiffen)*, Hafenliegezeit *f*
l. **edge** *(Typ)* Anlagekante *f*, Anlegekante *f*
l. **lathe** Lade *f*, Weblade *f*
l. **mark** *(Typ)* Anlegemarke *f*, Fröschchen *n*
l. **system** *(Am) (Brg)* Magazinbau *m*, Speicherbau *m*
l.**-up time** Auflegezeit *f*, Aufliegezeit *f (eines Schiffes)*

layer Schicht *f*, Lage *f*; *(Led)* Versatz *m*
l. **board** Trennpappe *f*
l. **lattice** Schichtengitter *n*
l.**-like** schichtweise, schichtähnlich
l. **of air** Luftmantel *m*, Luftschicht *f*
l. **of fibres** Faserschicht *f*
l. **of insulation** Isolationsschicht *f*, Isolierschicht *f*
l. **of non-conducting material** Isolierschicht *f*
l. **of tin foil** Stanniolbelag *m*
l. **rectifier** Sperrschichtgleichrichter *m*
l. **thickness** Schichtdicke *f*
l. **traversed by stream of water** wasserdurchrieselte Schicht *f*
l. **winding** Lagenwicklung *f*, Schichtenwicklung *f*

layering mechanism for the weft threads Schußfadenlegeeinrichtung *f (Vliesstoffe)*

laying Verlegung *f (von Rohren)*
l.**-in of cores** Einlegen *n* der Kerne *(Formerei)*
l. **off** *(Schiff)* Schnürbodenarbeit *f*, Abschnüren *n*
l. **on the course** *(Schiff)* Aufkurslegen *n*
l. **out** *(Am)* Anreißen *n*

l. out cables Kabellegen *n*
l.-out plate Anreißplatte *f*
l. the keel *(Schiff)* Kiellegen *n*
layout 1. Lageplan *m*, Anordnung *f*; Gruppierung *f*; Grundriß *m*; 2. Satzspiegel *m*; 3. Schaltungsanordnung *f*
l. drawing Grundrißzeichnung *f*
l. fluid *(Am)* Anreißflüssigkeit *f*
l. of pile *(Holz)* Stapelanlage *f*
l. of spare parts Ersatzteilplan *m*
l. plan Lageplan *m*
layshaft Vorgelegewelle *f*; Zwischenwelle *f*
l. gear Vorgelegerad *n*
l. gear cluster Vorgelegezahnradblock *m*
lazulite *(Min)* Lazulith *m*, Blauspat *m*
lazy jack *(Schiff)* Fangleine *f*
LD process *(Met)* LD-Verfahren *n*, Linz-Donawitz-Verfahren *n*, Sauerstoffaufblasverfahren *n*
Le-Chatellier-Braun principle Prinzip *n* von Braun und Le Chatellier
lea *(Text)* 1. Gebinde *n*; 2. Längeneinheit *f* für Garne *(Baumwolle: 109,7 m; Flachs: 274,2 m)*
l. strength Gebindefestigkeit *f*
leach / to [aus]laugen; herauslösen
leach pit *(Led)* Auslaugbottich *m*
leached zone *(Geol)* Auslaugungszone *f*
leaching agent Laugungsmittel *n*, Laugemittel *n*
lead / to *(Aussprache: li:d)* leiten, führen; voreilen
l. back zurückleiten, zurückführen
l. in einführen *(eine Leitung)*
l. off abführen
lead 1. Zuleitung *f*; Leitung[skabel *n*] *f*, Stromleitung *f*, Stromleiter *m*; 2. Voreilen *n*, Voreilung *f (Phase)*; 3. Steigung *f*, Ganghöhe *f (Gewinde)*; 4. *(Kfz)* Vorlauf *m (Vorderräder)*; 5. *(Text)* Nadelhalterung *f*; 6. *(Brg)* Zündkabel *n*
l. angle 1. *(90° – helix angle)* Steigungswinkel *β m*; 2. Voreilwinkel *m*; 3. Anschnittwinkel *m (Gewinde)*; 4. Schrägungswinkel *m (Zahnrad)*
l. bar *(Masch)* Leitlineal *n*
l. chart *(Mil)* Planfeuertafel *f (Flak)*
l. control Steigungsführung *f*
l. controller D-Regler *m*
l. error Steigungsfehler *m*
l. hole *(Am)* Vorbohrung *f*
l.-in Einführung *f*; Zuleitung *f*; Antenneneinführung *f*
l.-in cable Einführungskabel *n*, Zuführungskabel *n*
l.-in insulator Einführungsisolator *m*, Eingangsisolator *m*
l.-in spiral Einlaufrille *f (Schallplatte)*
l.-in wire Stromzuführungsdraht *m*; Durchführung *f*; Antennenzuleitung *f*
l. inductance Induktivität *f* der Zuführungen, Zuleitungsinduktivität *f*
l.-lag network *(BMSR)* PID-Schaltung *f*, Schaltung *f* mit Phasenvordrehung und Phasenrückdrehung
l. network Vorhaltschaltung *f*, Schaltung *f* mit Phasenvordrehung
l. of a thread Gewindeganghöhe *f*
l. of brushes Bürstenverschiebung *f*
l. of helix Nutensteigung *f*, Spiralsteigung *f*
l. of thread Gewindesteigung *f*
l. resistance *(El)* Widerstand *m* der Zuleitungen, Zuleitungswiderstand *m*
l. screw Leitspindel *f*
l. screw nut Leitspindelmutter *f*
l. screw tester Leitspindelsteigungsprüfer *m*
l. tester Steigungsprüfer *m*
l.-through bushing *(Fmt)* Durchführung *f*
l.-through capacitor Durchführungskondensator *m*
l. wire 1. Meßdraht *m (Zahnradmessung)*; 2. *(El)* Anschlußleitung *f*
lead / to *(Aussprache: led)* verbleien, mit Blei auskleiden (füttern)
l.-coat verbleien
l. fuel Kraftstoff verbleien
l. harden im Bleibad härten
lead 1. Blei *n*; 2. Lot *n*, Senkblei *n*; 3. *(Typ)* Durchschuß *m*
l. accumulator Bleiakkumulator *m*, Bleisammler *m*
l. acetate Bleiazetat *n*, Bleizucker *m*
l. anode Bleianode *f*
l. antimon[i]ate Bleiantimonat *n*, Neapelgelb *n*
l. arsenate Arsenblei *n*, Bleiarsenat *n*
l. ashes Bleischaum *m*, Bleiasche *f*
l. azide Bleiazid *n*
l.-base bearing metal Bleilagermetall *n*
l. bath Bleibad *n (zum Verbleien)*
l.-bath furnace Bleibadofen *m*
l.-bath patenting Bleibadpatentierung *f*
l. blast-furnace Bleischachtofen *m*
l. bronze Bleibronze *f*
l. bullion Rohblei *n*
l. burner Bleilöter *m*
l. cable Blei[mantel]kabel *n*, Kabel *n* mit Bleimantel
l. carbonate Bleikarbonat *n*; Bleispat *m*
l. case *(Typ)* Durchschußkasten *m*, Zeugkasten *m*
l. caster Bleigießer *m*
l. caulking chisel Bleistemmer *m*
l. cell Bleiakkumulator *m*
l. chamber process Bleikammerverfahren *n*
l. chloride Bleichlorid *n*
l. chromate Bleichromat *n*, chromsaures Blei *n*, Chromgelb *n*

l.-coated feuerverbleit
l. coating 1. Bleimantel *m*; 2. Feuerverbleien *n*
l.-covered cable Blei[mantel]kabel *n*, Kabel *n* mit Bleimantel
l. covered rubber cable Gummibleikabel *n*
l. covering Bleimantel *m*
l. crystal Bleikristall *n*, Bleiglas *n*, Kristallglas *n*
l. cutter Bleischneider *m*
l. deposit Bleischlamm *m (im Akku)*
l. dross Bleikrätze *f*, Bleischaum *m*, Bleiasche *f*
l. foil Bleifolie *f*
l. furnace Bleiofen *m*
l. furniture *(Typ)* Bleisteg *m*
l. glance *(Min)* Bleiglanz *m*
l. glass *s.* l. crystal
l. grip Bleiklemme *f (für Kabelbefestigungen)*
l. hammer Bleihammer *m*
l. horn *(Mil)* Bleikappe *f (Mine)*
l. jaw Bleibacke *f*
l. ladle Bleilöffel *m*
l.-like bleiartig
l.-lined wooden vat *(Text)* mit Blei ausgeschlagener Holzbottich *m*
l. lining Bleiauskleidung *f*
l. matte *(Met)* Bleistein *m*
l. ore Bleierz *n*
l. oxide Bleioxid *n*, Bleiglätte *f*
l. oxide red Mennige *f*
l. packing Bleidichtung *f*
l. pig Bleiblock *m*, Bleibarren *m*, Bleimassel *f*
l. pipe Blei[leitungs]rohr *n*
l. plating galvanisches Verbleien *n*
l. poisoning Bleivergiftung *f*
l. quenching Abschrecken *n* in Blei
l. refinery Bleiraffinerie *f*
l. response of a fuel Bleiempfindlichkeit *f* eines Kraftstoffes
l.-rubber apron *(Kern)* Bleischürze *f*
l. seal Bleiplombe *f*, Plombe *f*
l. sealing plier Plombenzange *f*, Plombierzange *f*
l. seat Bleidichtung *f*
l. sheathed cable Blei[mantel]kabel *n*, Kabel *n* mit Bleimantel
l. sheathing Bleimantel *m*
l.-shielded mit Blei abgeschirmt (gepanzert)
l. shielding Bleiabschirmung *f*
l. silicate Bleisilikat *n*; Bleiglas *n*
l. skim *(Met)* Abstrichblei *n*
l. slag Bleischlacke *f*
l. sludge Bleischlamm *m (im Akku)*
l. smelting Bleiverhüttung *f*
l. smelting plant Bleihütte *f*
l. storage battery Bleiakkumulator *m*
l. subacetate Bleiessig *m*
l. suboxide Bleiasche *f*
l. sugar Bleizucker *m*, essigsaures Blei *n*
l. sulphate Bleisulfat *n*
l. sulphide Bleisulfid *n*, Schwefelblei *n*
l. susceptibility Bleiempfindlichkeit *f*
l.-tin solder Lötzinn *n*
l. tube cutter Rohrschneider *m* für Bleirohre
l. tungstate Bleiwolframat *n*
l. vein *(Brg)* Bleiader *f*, Bleigang *m*
l. works Bleihütte *f*
leaded fuel verbleiter Kraftstoff *m*
l. gasoline *(Am)* verbleites Benzin *n*, Bleibenzin *n*
l. matter *(Typ)* durchschossener Satz *m*
leaden bleiern
leader Gewinde[leit]patrone *f*
l. cable Leitkabel *n (Funknavigation)*
l. tape Vorspannband *n (zum Magnetband)*
l. tooth Vorschneidezahn *m (Sägeblatt)*
leading 1. Voreil-; Vorder-; Leit-; Führungs-; 2. voreilend; leitend
l. axle Vorderachse *f*, Leitachse *f*
l. cable Leitkabel *n*, Lotsenkabel *n*, Führungskabel *n*
l. current voreilender Strom *m*
l. cutting edge *(Am) (Masch)* Hauptschneide *f*
l. down Herabführung *f (Antenne)*
l. eccentric der Kurbel vorauseilender Exzenter *m*
l. edge 1. Vorderkante *f*, Eintrittskante *f*; 2. Anschnittschneide *f*; 3. Vorderflanke *f (des Impulses)*
l. edge [of the land] Fasenschneide *f (Spiralbohrer)*
l. edge pulse time Anstiegszeit *f* der Vorderflanke eines Impulses
l. end Vorderende *n*, Vorderseite *f*
l. face Druckfläche *f (eines Propellerflügels)*
l.-in cable Einführungskabel *n*, Zuführungskabel *n*
l.-in wire Stromzuführung *f*
l. light Leitfeuer *n*, Richtfeuer *n*, Kursfeuer *n*
l. load kapazitive Belastung *f*
l. of phase Phasenvoreilung *f*
l. particulars Hauptdaten *pl*
l. pass section Anstichquerschnitt *m*
l. phase voreilende Phase *f*
l. pole vordere Polkante *f*; voreilender Pol *m*
l. pole horn auflaufende Polkante *f*
l. power factor kapazitiver Leistungsfaktor *m*
l. shoe *(Kfz)* Auflaufbacke *f (Bremse)*
l. wire *(Brg)* Zünderdraht *m*, Schießleitung *f*
l. zeros *(Büro)* führende Nullen *fpl (Kommafestlegung)*
leadsman *(Schiff)* Lotgast *m*

leady matte *(Met)* Bleistein *m*
l. spelter *(Met)* bleiisches Bodenzink *n*
leaf 1. Blatt *n (Feder; Baum)*; 2. Folie *f*; 3. Schale *f (Hohlmauerwerk)*; 4. Triebstock *m*; 5. Brückenflügel *m*, Torflügel *m*
l. aluminium Blattaluminium *n*, Aluminiumfolie *f*
l. bug Blattwanze *f*
l. electroscope Blattelektroskop *n*
l. fibre Blattfaser *f*
l. gilding Vergoldung *f* mit Blattgold
l. gold Blattgold *n*
l. metal Metallfolie *f*
l. powder Blättchenpulver *n*
l. spring Blattfeder *f*
l. valve Klappenventil *n*
leafy wood Laubholz *n*; Laubwald *m*
leak / to 1. ableiten *(Strom)*; kriechen *(Strom)*; durchschlagen; streuen; 2. lecken, undicht sein, durchsickern
l. back zurücksickern
l. off *(El)* abfließen
leak 1. *(El)* Ableitung *f*, Streuung *f*; 2. Undichtigkeit *f*, Leck[stelle *f*] *n*, undichte Stelle *f*
l. circuit Erdungskreis *m*
l. coil *(El)* Überbrückungsspule *f*
l. detection Lecksuche *f*, Suche *f* nach undichten Stellen
l. detector Lecksucher *m*, Leckfinder *m*
l. impedance *(El)* Parallelimpedanz *f*, Querimpedanz *f*
l. of air Falschluft *f*, Luftaustritt *m*
l.-proof dicht, abgedichtet, nicht leckend, lecksicher, durchschlagsicher
l. rate Leck[öl]menge *f*
l. resistance *(El)* Ableitungswiderstand *m*, Nebenschlußwiderstand *m*, Querwiderstand *m*
leakage 1. Undichtigkeit *f*, Leck[age *f*] *n*, Leckstelle *f*, Austritt *m*; 2. Abfließen *n*, Ausfließen *n*, Entweichen *n*, Ausströmen *n*, Lecken *n*; 3. Leckverlust *m*; Leckmenge *f*; 4. *(Kern)* Neutronenverlust *m*, Neutronenausfluß *m*; 5. *(Masch)* Ölaustritt *m*; 6. *(El)* Ableitung *f*, Isolationsfehler *m*, Isolationsverlust *m*, Ableitverlust *m*, Kriechverlust *m*; Streuung *f (Magnet)*; 7. *s.* l. radiation
l. airgap Streuspalt *m (Magnet)*
l. coefficient Streufaktor *m*
l. coupling Streukopplung *f*
l. current Kriechstrom *m*, Ableitstrom *m*, Fehlerstrom *m*, Streustrom *m*
l. currents vagabundierende Ströme *mpl*
l. detector Isolationsmesser *m*
l. distance Kriechweg *m*, Kriechstrecke *f*
l. field Streufeld *n*
l. flux Streufluß *m*
l. inductance Streuinduktivität *f*
l. induction Streuinduktion *f*
l. line *(El)* Streulinie *f*
l. loss 1. *(El)* Ableitungsverlust *m*; 2. Leckverlust *m (z. B. Öl)*
l. path 1. *(El)* Kriechstrecke *f*; 2. *(Kern)* Sickerstrahlenweg *m*, Leckweg *m*
l. radiation 1. Streustrahlung *f (Sendetechnik)*; 2. *(Kern)* Sickerstrahlung *f*, Leckstrahlung *f*
l. rate Leckverlust *m*; Abflußgeschwindigkeit *f*
l. reactance Streureaktanz *f*
l. relay Erdschlußrelais *n*
l. resistance Isolationsoberflächenwiderstand *m*, Ableitungswiderstand *m*
l. resistor *(El)* Ableitwiderstand *m*
l. return line Rückleitung *f* für die Leckflüssigkeit
l. voltage Streuspannung *f*
leakiness *(Masch)* Undichtigkeit *f*, Undichtheit *f*
leaking of the flange joint *(Wärm)* Tropfen *n* der Flanschverbindung
l. path Kriechweg *m*, Kriechstrecke *f*
leakless lecksicher
leakproof lecksicher, dicht
leaky undicht
l. area Leckstelle *f*
l. grid detection Gitter[strom]gleichrichtung *f*, Audiongleichrichtung *f*
l. grid rectification *s.* l. grid detection
lean 1. arm *(Erz)*; 2. mager *(Kohle)*; 3. scharf *(Schiffsform)*
l. bow scharfer Bug *m*
l. clay Magerungsmittel *n*, Magerungsmaterial *n*
l. coal Magerkohle *f*, Mergelkohle *f*
l. coke Magerkoks *m*
l. concrete Magerbeton *m*
l. fuel-air mixture mageres Kraftstoff-Luft-Gemisch *n*
l. gas Armgas *n*, geringwertiges Gas *n*, Schwachgas *n*
l.-gas firing Schwachgasheizung *f*
l. mix Magerbeton *m*
l. mixture Spargemisch *n (Treibstoffgemisch)*
l. ore armes Erz *n*, Magererz *n*
l.-to roof Pultdach *n*
leaning matter Magerungsmittel *n*
leap-frog test *(Am)* Durchprüfung *f (Rechner)*
lear *s.* lehr
l. board *s.* layer board
learner-driver Fahrschüler *m*
learning automaton lernender Automat *m*, Lernmaschine *f*
learsyn *(BMSR)* Learsyn *(Firmenbezeichnung einer Drehmeldertype)*
lease *(Text)* Fadenkreuz *n*
l. pegs Kreuznägel *mpl*
l. pin Kreuznagel *m*, Schranknagel *m*

l. **rod** Kreuzrute *f*, Kreuzschiene *f*; Schärholz *n*
l. **sticks** Kreuzstäbe *mpl*
leased circuit *(Fmt)* Mietleitung *f*
leasing system *(Text)* Fadenkreuzvorrichtung *f*
least action principle Maupertuissches Prinzip *n*, Prinzip der kleinsten Wirkung
l. **common multiple** kleinstes gemeinsames Vielfaches *n*
l. **count** kleinste ablesbare Teilung *f*
l. **significant digit** *(Dat)* gültige (Zahlenwert besitzende) Ziffer *f* der niedrigsten Ordnung, am wenigsten wichtige Angabe *f*
leat Mühlgraben *m*; Wasserzuführung *f*
leather Leder *n*
l. **belt** Lederriemen *m (Antrieb)*; Ledergurt *m (Sicherung)*; Ledergürtel *m*
l. **board** Lederpappe *f*
l. **bonding adhesive** Lederkleber *m*
l. **case** Lederetui *n*
l. **cement** Lederkitt *m*
l. **cloth** Kunstleder *n*
l. **coupling** Lederkupplung *f*
l.**-covered** lederbezogen
l.**-covered wood polishing wheel** Pließtscheibe *f*
l. **cup** Ledermanschettendichtung *f*
l. **detaching roller** belederte Abreißwalze *f (Kämmaschine)*
l. **dressing** Zurichten *n* des Leders
l.**-faced** mit Lederbelag
l. **gasket** Dichtungsscheibe *f (aus Leder)*
l. **gauge** Vorrichtung *f* zum Messen des Leders
l. **glove** Lederhandschuh *m*
l. **glue** Lederleim *m*, Büffelleim *m*
l. **grease** Lederfett *n*
l. **hammer** Klopfhammer *m (Schuhproduktion)*
l. **hose** *(Text)* Nitschelhose *f*
l.**-link coupling** elastische Kupplung *f* mit Lederelementen
l. **packing** Lederdichtung *f*, Ledermanschette *f*
l. **substitute** Lederersatz *m*
l. **varnish** Lederlack *m*
l. **washer** lederne Dichtungsscheibe *f*
l. **wax** Lederwichse *f*
leatherette Kunstleder *n*, Lederimitation *f*
leathering Lederdichtung *f*, Ledermanschette *f*
leatheroid Leatheroid *n*; Vulkanfiber *f*
leatherpulp board Lederfaserpappe *f*
leave / to verlassen, zurücklassen; aufhören mit, unterlassen; fortgehen, abreisen, abfahren
l. **a cut** auslaufen *(Fräser)*
l. **a port** *(Schiff)* aus einem Hafen auslaufen
l. **the pass** *(Met)* austreten *(ein Knüppel aus den Walzen)*
Leavers lace machine *(Text)* Spitzenmaschine *f*
leaves for thickness gauge Fühlerlehrplättchen *npl*
leaving energy Austrittsenergie *f*
l. **loss** Austrittsverlust *m*
l. **up** *(Brg)* Anbauen *n (z. B. Kohlenbank)*
l. **velocity** Austrittsgeschwindigkeit *f*
Leblanc connection *(El)* Leblancsche Schaltung *f*
Leblanc phase advancer *(El, Masch)* Leblancscher Phasenschieber *m*
Leblanc process *(Ch)* Leblanc-Verfahren *n*
Lecher wave Lecher-Welle *f*
Lecher wires Lecher-Leitung *f*
Leclanché battery *s.* Leclanché cell
Leclanché cell Leclanché-Element *n*, Trockenelement *n*
lecture / to vortragen; Vorträge (Vorlesungen) halten
l. **on** lesen über *(Hochschule)*
lecture Vorlesung *f*
l. **theatre** Hörsaal *m*
ledeburite *(Met)* Ledeburit *m*
ledge Absatz *m*; Gesims *n*, Sims *m*; Berme *f (Erdbau)*; Leiste *f*
l. **elevator** *(Lw)* Leistenelevator *m*
l. **matter** *(Geol)* Gangfüllung *f*
ledger Riegel *m*
l. **beam** *s.* ledger
l. **card** Kontokarte *f*
l. **paper** Bücherpapier *n*, Registerpapier *n*
l. **total** Abschlußbetrag *m*
l. **wall** *(Brg)* Liegendes *n*, Sohle *f*
lee-board *(Schiff)* Seitenschwert *n (bei kleinen Fahrzeugen)*
l.**-boarder** Seitenschwertboot *n*
leer *s.* lehr
leeway *(Schiff)* Abtrift *f*, Leeweg *m*
left and right reversed *(Opt)* seitenverkehrt
l.**-bent tool** gebogener linker Meißel *m*
l.**-hand** linksläufig; Links-
l.**-hand cut[ting]** linksschneidend
l. **hand cylinder bed sewing machine** Linksarmmaschine *f (Schuhproduktion)*
l.**-hand helical** mit Linksdrall
l.**-hand milling cutter** linksschneidender Fräser *m*, Walzenfräser *m* mit Rechtsspirale
l.**-hand nut** Mutter *f* mit Linksgewinde
l.**-hand rule** Linke-Hand-Regel *f*
l.**-hand spindle** *(El)* linker Wickelteller *m*
l.**-hand steering** *(Kfz)* Linkslenkung *f*
l.**-hand switch** *(Eis)* Linksweiche *f*
l.**-hand thread** Linksgewinde *n*
l.**-hand traffic** Linksverkehr *m*
l.**-hand twill** *(Text)* S-Grat-Köper *m*
l.**-hand twist** Linksdrall *m*
l.**-handed** 1. *(Masch)* linksgängig; 2. *(Opt)* linksdrehend
l.**-handed coordinate system** *(Math)* linkshändiges Koordinatensystem *n*, Linkssystem *n*
l.**-handed propeller** linksgängige Schiffsschraube *f*

l.-lay rope linksgängiges Seil *n*
l. side view linke Seitenansicht *f*
leftward welding Nachlinks-Schweißen *n*
leg 1. Schenkel *m (Winkel)*; Fuß *m (Instrument)*; Teilstrecke *f*; 2. *(Math)* Kathete *f*; 3. *(Geol)* Flügel *m (Falte)*; 4. *(Brg)* Stempel *m*, Bein *n*; 5. *(Bw)* Stiel *m (Rahmen)*
l. length Strumpflänge *f*
l. of a bridge *(El)* Brückenzweig *m*
l. of a circuit Außenleiter *m (Drehstromkreis)*
l. of angle iron Schenkel *m* eines Winkeleisens
l. of standard *(Wärm)* Ständerbein *n*
legal address Wohnsitz *m*
l. claim Rechtsanspruch *m*
l. entity juristische Person *f*
l. procedure Rechtsweg *m*
l. provisions relating to labour safety and accident prevention gesetzliche Bestimmungen *fpl* über Arbeits- und Unfallschutz
l. redress Rechtsmittel *n*
l. resolution Gerichtsbeschluß *m*
l. title Rechtsanspruch *m*
legalized copy beglaubigte (rechtsgültige) Abschrift *f*
legally binding rechtskräftig
l. inoperative rechtsunwirksam
l. operative rechtswirksam
legend Legende *f*, Zeichenerklärung *f*, Bildtext *m*, Beschriftung *f (Zeichnung)*
leggings Gamaschen *fpl*
legible lesbar
legumes Hülsenfrüchte *fpl*
leguminous fruits Hülsenfrüchte *fpl*
lehr Kühlofen *m (zum Entspannungsglühen von Glas)*
leimtype Leimdruck *m*
lemma *(Math)* Hilfssatz *m*, Lemma *n*
lemniscate *(Math)* Lemniskate *f*, Schlingenkurve *f*
lend / to [ver]leihen; gewähren
l. itself to a definition gut definierbar sein
lending library Leihbibliothek *f*
length Länge *f*; Strecke *f*; Dauer *f*
l. adjustment Längeneinstellung *f*
l. between centres Spitzenweite *f*, größte Drehlänge *f*
l. between perpendiculars *(Schiff)* Konstruktionslänge *f*, Länge *f* zwischen den Loten
l. compensation Längenausgleich *m*
l. constant *(Schiff)* Längenkonstante *f*, Schlankheitsgrad *m*
l. feed stop *(Masch)* Längsanschlag *m*
l. gauge Strichlehre *f*
l. graduation Längeneinteilung *f*
l. measuring comparator Längenmeßkomparator *m*
l. measuring instrument Längenmeßgerät *n*
l. of a brake path *(Kfz)* Bremsweg *m*
l. of action Eingriffstrecke *f (Zahnrad)*
l. of beaming *(Text)* Schärlänge *f*
l. of break *(El)* Gesamtschaltstrecke *f*
l. of broaching cut Räumlänge *f*
l. of chamfer *(Masch)* Anschnittlänge *f*
l. of commutating pole *(El)* Wendepollänge *f*
l. of conversation *(Fmt)* Gesprächsdauer *f*
l. of cut Schnittlänge *f*
l. of engagement Einschraublänge *f*, Eingriffslänge *f*
l. of entrance *(Schiff)* Einlauflänge *f*, Zuschärfungslänge *f*
l. of exposure Belichtungsdauer *f*, Expositionsdauer *f*
l. of grip Klemmlänge *f*
l. of jib Ausladung *f (eines Kranes)*
l. of lay Schlaglänge *f (Seil)*
l. of optical path optische Weglänge *f*
l. of packing Dichtungslänge *f*
l. of production run Größe *f* der Serie
l. of shafting Wellenstrang *m*
l. of shoulder Schaftlänge *f*
l. of span Stützweite *f*
l. of square Vierkantlänge *f*
l. of straight-line motion Geradführungslänge *f (Getriebelehre)*
l. of stroke Hublänge *f (eines Kolbens)*
l. of taper Kegellänge *f*
l. of the approach cut in milling Anschnittweg *m* des Fräsers
l. of the piece *(Text)* Warenlänge *f*
l. of the piston *(Masch)* Kolbenlänge *f*, Kolbenhöhe *f*
l. of thread contact Einschraubtiefe *f*
l. of time of table dwell Haltezeit *f* der Tischumkehrung
l. of toothed portion verzahnte Länge *f (Sägeblatt)*
l. on designed waterline *(Schiff)* Länge *f* der Konstruktionswasserlinie
l. overall Länge *f* über alles, größte Länge
l.-preserving längentreu *(Abbildung)*
l. shrinkage Längenschrumpfung *f*
l. standard Längennormal *n*, Längennormale *f*
l. stop *(Masch)* Längsanschlag *m*
l. subjected to bending Biegelänge *f*
l. unit Längeneinheit *f*
lengthen / to 1. verlängern, ausdehnen; 2. *(Met)* längen, strecken, recken; 3. anstücken
lengthening 1. Verlängerung *f*, Längenzuwachs *m*; 2. *(Met)* Strecken *n*, Streckung *f*
lengthsman Straßenwärter *m*
lengthwise in Längsrichtung
l. bulkhead *(Schiff)* Längsschott *n*
l. of the grain in Faserrichtung

leno Dreherbindung *f (Weberei)*
l. loom Webstuhl *m* für Drehergewebe
l. weave Dreherbindung *f*
lens Linse *f*; Objektiv *n*; Lupe *f*
l. aerial Linsenantenne *f*
l. aperture Objektivöffnung *f*
l. barrel Objektivfassung *f*
l. chain Linsenkette *f*
l. combination Linsensystem *n*
l. component Linsenglied *n*
l. condenser Linsenkondensor *m*
l. coverage Gesichtsfeld *n*.
l. distortion Linsenverzeichnung *f*
l. equation Linsengleichung *f*
l. error Linsenfehler *m*
l. field Linsenfeld *n*
l.-glass Linsenglas *n (Scheinwerfer)*
l. grinding Linsenschleifen *n*
l. hood Kamerakappe *f*, Sonnenblende *f*
l. mask *(Fs)* Schattenmaske *f*
l. mounting Linsenfassung *f*
l. screen Linsenschirm *m (Film)*
l. shield Sonnenblende *f*
l. spectrograph Linsenspektrograf *m*
l. spectrometer Linsenspektrometer *n*
l. stop *(Opt)* Blende *f*
l. system Optik *f*, Linsensystem *n*
l. voltage *(El)* Linsenspannung *f*
lensatic sight *(Opt)* Richtglas *n*
lenticle Linsenscheibe *f*, Linse *f*
lenticular lentikulär, linsenartig, linsenförmig
l. intercalation *(Geol)* linsenförmige Einlagerung *f*
l. profile *(Opt)* Linsenprofil *n*
l. vein *(Geol)* Linsengang *m*
lenticulated film Farbrasterfilm *m*, Rasterfilm *m*
lenticulation Linsen[farb]raster *m*
lentil Linse *f (Frucht)*
Lentz valve gear Lentz-Steuerung *f*
Lenz skiascopy disk *(Opt)* Skiaskopierscheibe *f* nach Lenz
Lenz's law Lenzsche Regel *f*
Leonard control Leonard-Schaltung *f*
leonite *(Min)* Leonit *m*
lepidolite *(Min)* Lepidolith *m*
lepidomelane *(Min)* Lepidomelan *m*
leptokurtic überhöht *(Häufigkeitskurve)*
leptokurtosis Überhöhung *f (Häufigkeitskurve)*
lepton *(Kern)* Lepton *n*
lerbachite *(Min)* Lerbachit *m*
less significant digit *(Dat)* gültige (Zahlenwert besitzende) Ziffer *f* der niedrigeren Ordnung
lessen / to vermindern; verringern
let / to lassen
l. down 1. hinablassen; 2. anlassen *(Stahl)*
l. drop abtropfen lassen
l. escape entweichen lassen
l. in einlassen, einbetten
l. into concrete einbetonieren
l.-off 1. ablassen, abzapfen, ableiten; 2. *(Text)* abbäumen
l. to freight verfrachten *(ein Schiff vermieten)*
let-go nachträglich eintretende Fehlstelle *f* in Schichtstoff
l.-off motion *(Text)* Abbäumen *n*
lethal tödlich, letal
l. amount *s.* l. dose
l. dose Letaldosis *f*, tödliche Dosis *f*
letter / to mit Buchstaben bezeichnen; beschriften; stempeln
l. a drawing eine Zeichnung beschriften
letter Zeichen *n*, Buchstabe *m*; Brief *m*
l. balance Briefwaage *f*
l. box Briefkasten *m*
l. case *(Typ)* Schriftkasten *m*, Setzkasten *m*
l. file Briefordner *m*, Schnellhefter *m*, Heftmappe *f*
l. foundry Schriftgießerei *f*
l. metal Schriftmetall *n*, Schriftblei *n*
l.-reading machine *(Dat)* Buchstabenlesemaschine *f*
l. series buchstabengekennzeichnete Reihe *f*
l. spacing *(Typ)* Sperren *n*, Spationieren *n*
l. transmission *(Fmt)* Zeichenübertragung *f*
lettered buchstabengekennzeichnet, beschriftet
lettergram Brieftelegramm *n*
lettering Beschriftung *f*; Buchstabenkennzeichnung *f*
l. guide Schriftschablone *f*
l. of drawings Zeichnungsbeschriftung *f*
l. panel Beschriftungsplättchen *n*
letterpress *(Typ)* 1. Kopiermaschine *f*, Kopierpresse *f*; 2. Abzug *m*
l. colour printing Farbenbuchdruck *m*
l. imitating gravure Buchdruck *m* mit Tiefdruckwirkung
l. paper Buchdruckpapier *n*
l. printing Hochdruck *m*, Buchdruck *m*
l. simulating gravure *s.* l. imitating gravure
letters case *(Fmt)* Buchstabenreihe *f*
l. patent Patentbrief *m*, Patenturkunde *f*
letterset *(Typ)* indirekter Hochdruck *m*
letting down Anlassen *n (Stahl)*
leucine *(Ch)* Leuzin *n*
leuco Leukoverbindung *f*, Leukobase *f*
l. dye Leukofarbstoff *m*
levee Damm *m*, Uferdamm *m*, Deich *m*, Stromdamm *m*
level / to 1. abziehen, ausgleichen; glätten; ebnen, planieren; 2. nivellieren, auswiegen, in Waage bringen, abgleichen; [aus]richten; 3. richten *(Bleche)*; 4. senkrecht stellen
l. off verflachen; flach werden; abflachen
l. out 1. ausspachteln; 2. einebnen
level eben, gleichmäßig, in Waage, horizontal; *(Brg)* söhlig

level 1. Pegel *m*, Niveau *n*, Geländehöhe *f*, Höhe *f*; Höhenlage *f*; 2. *(Bw)* Richtwaage *f*; Richtscheit *n*, Richtlatte *f*; Nivelliergerät *n*, Nivellierinstrument *n*; Spiegel *m (Flüssigkeit)*; 3. *(Brg)* Sohlstrecke *f*, Wasserstrecke *f*; 4. Horizont *m*, Horizontalebene *f*
l.-boss *(Brg)* Steiger *m*
l. characteristic Pegelkennlinie *f*; flache Kennlinie *f*
l. compensation Pegelausgleich *m*
l. compound excitation Verbunderregung *f* für gleichbleibende Spannung, Flachverbunderregung *f (Generator)*
l. control Pegelkontrolle *f*; Standregelung *f*
l. controller Pegelregler *m*
l. course streichender Grubenbau *m*
l. crossing Gleiskreuzung *f*, Bahnübergang *m*, Planübergang *m*
l. crossing signal Wegübergangssignal *n*
l. crossing without gates unbeschrankter Bahnübergang *m*
l. density Niveaudichte *f*, Dichte *f* des Energieniveaus
l. detector Füllmengenanzeiger *m*
l. distribution *(Kern)* Energieniveauverteilung *f*
l. dyeing *(Text)* Egalfärbung *f*, Gleichmäßigfärben *n*, Egalisieren *n*
l. equalization Pegelausgleich *m*
l. flesh glatte Fleischseite *f (Leder)*
l. flight Horizontalflug *m*
l. generator Pegelsender *m*, Pegelgeber *m*
l. glass Libelle *f (Wasserwaage)*
l. groove weld HV-Naht *f*, Halb-V-Naht *f*
l. indicator 1. Füllstandsmesser *m*, Füllstandszeiger *m*, Niveaustandszeiger *m*; 2. Aussteuerungsmesser *m (Radar)*
l. interval *(Brg)* Sohlenabstand *m*
l. line Höhenlinie *f*
l.-luffing crane Wippkran *m*
l.-luffing deck crane *(Schiff)* Einziehdeckskran *m (mit waagerechtem Lastweg)*
l. measurement Pegelmessung *f*
l. measuring set Pegelmesser *m*
l. meter Pegelmesser *m*
l. of confidence statistische Sicherheit *f*
l. of filling Füllstand *m*
l. of foundation *(Bw)* Gründungshöhe *f*
l. of radioactivity Gehalt *m* an Radioaktivität
l. of reference *(Ph)* Bezugsniveau *n*
l. of the water-mark post Pegelstand *m*
l. of underground water Grundwasserspiegel *m*
l. pillar Schwebe *f (zwischen Abbauraum und Streckensohle)*
l. recorder Pegelschreiber *m*
l. road fuel consumption Kraftstoffnormverbrauch *m*
l. rod Nivellierlatte *f*
l. spacing *(Kern)* Niveauabstand *m*, Niveaudistanz *f*
l. square mit Setzwaage vereinigtes Winkelmaß *n*
l. switch Füllstandsschalter *m*
l. transportation Waagerechtförderung *f*
l. with auf gleicher Höhe wie
leveling *s.* levelling
leveller Planierstange *f*; Nivellierinstrument *n*; Richtmaschine *f*
l. door Planiertür *f*, Planierloch *n*
levelling 1. Höhenaufnahme *f*, Höhenbestimmung *f*, Höhenmessung *f*; Nivellement *n*, Nivellieren *n*, Nivellierung *f*, Horizontierung *f*; 2. Abgleichen *n*, Abziehen *n*, Ausgleichen *n*, Ebnen *n*, Einebnen *n*, Glätten *n*, Planieren *n*, Planierung *f*
l. agent *(Plast)* Verlaufmittel *n*
l. base Dreifuß *m (am Vermessungsgerät)*
l. board Abstreichlineal *n (Formerei)*
l. bottle *(Masch)* Niveauflasche *f*
l. course *(Bw)* Ausgleichsschicht *f*
l. door *(Brg)* Planiertür *f*
l. indicator Dosenlibelle *f*
l. instrument Nivellierinstrument *n*
l. layer *(Bw)* Ausgleichsschicht *f*
l. midway between two points Nivellieren *n* aus der Mitte
l. nut Justiermutter *f*
l. of sheets Richten *n* von Blechen
l. pole Nivellierlatte *f*
l. rod Nivellierlatte *f*
l. screw Nivellierschraube *f*
l. staff Nivellierlatte *f*
l. straight edge Tuschierlineal *n*
l. surface Richtfläche *f*
l. survey Nivellement *n*
l. underground markscheiderisches Nivellieren *n*
l. up Ausrichtung *f*
lever Hebel *m*, Knebel *m*, Arm *m*, Kurbel *f*, Schwengel *m*; Hebestange *f*; Schwinge *f (Getriebelehre)*; Dachwippe *f (Walzwerk)*
l. arm 1. Hebelarm *m*; 2. Hebelstange *f*
l. barring gear Schaltwerk *n* mit Hebel
l. bearing Kulissenlagerung *f*
l. car Hebeldraisine *f*
l. change Stellhebel *m*
l. collar Gleitring *m (Regler)*
l. control Hebelschaltung *f*
l. feed Hebelvorschub *m*
l. fulcrum Hebeldrehpunkt *m*
l. inspection trolley Hebeldraisine *f*
l. jack Lastwinde *f*, Hebelwinde *f*
l. key *(Fmt)* Kippschalter *m*
l.-operated hebelbetätigt
l.-operated chuck Hebelspannfutter *n*

l. principle Hebelgesetz *n*
l. punch Hebellochstanze *f*
l. ratchet Hebelknarre *f*
l. rod Hebelstange *f*
l. safety valve Sicherheitsventil *n* mit Hebelbelastung, Hebelsicherheitsventil *n*
l. shear Hebelschere *f*
l. stand *(Eis)* Stellbock *m (Signal)*
l. switch Hebelschalter *m*
l. switch box *(Eis)* Stellbock *m (Signal)*
l. transmission Hebelübertragung *f*
l.-type indicator mechanischer Fernanzeiger *m*
l. with equal arms gleicharmiger Hebel *m*
l. with pawl Klinkhebel *m*
leverage 1. Gestänge *n*; 2. Hebelwirkung *f*; Hebelübersetzung *f*, Hebelübertragung *f*; Hebelwerk *n (Walzstraße)*
l. ratio Hebelverhältnis *n*
levigate / to schlämmen; zerreiben; glätten
levigation Ausschlämmung *f*; Pulverisierung *f*
levo . . . *(Am) s.* laevo . . .
levulinic acid Lävulinsäure *f*
lewis Steinwolf *m*
l. bolt Ankerschraube *f*, Steinschraube *f*
Lewis equation Lewis-Gleichung *f (Verzahnung)*
lewisite 1. *(Min)* Lewisit *m*; 2. *(Ch)* Lewisit *n*, Chlorvinyldichlorarsin *n (chemischer Kampfstoff)*
Leyden jar Leydener Flasche *f*
liability claim Haftpflichtanspruch *m*
l. to interferences Störanfälligkeit *f*
l. to recourse Regreßpflicht *f*
liable for damages schadenersatzpflichtig
lias *(Geol)* Lias *m (f)*, Liasformation *f*
l. stone Liaskalk *m*, Liasstein *m*
liberate / to 1. loslösen, abscheiden; entwickeln; befreien *(z. B. Gas aus einer Verbindung)*; 2. freisetzen, entbinden *(z. B. Wärme)*
liberation of electrons Elektronenauslösung *f*
l. of heat Freiwerden *n* von Wärme
liberator Bleibad *n*
l. cell Bleibad *n*
l. tank Entmetallisierungsbad *n*, Elektrolytreinigungsbad *n (z. B. Entkupferungsbad)*
libethenite *(Min)* Libethenit *m*
librarian Bibliothekar[in *f*] *m*
library 1. Bibliothek *f*, Bücherei *f*; 2. *(Dat)* Programmbibliothek *f*, Befehlsbibliothek *f*, Programmothek *f*
l. routine *s.* l. subroutine
l. subroutine *(Dat)* Bibliotheksprogramm *n*
licence Lizenz *f*, Zulassung *f*; Bewilligung *f*, Genehmigung *f*, Konzession *f*; Zulassungsschein *m*
l. agreement Lizenzvereinbarung *f*
l. fee Lizenzgebühr *f*
l. number *(Am) s.* l. plate
l. plate *(Kfz)* Nummernschild *n*, polizeiliches Kennzeichen *n*
l. plate light *(Kfz)* Kennzeichenbeleuchtung *f*
license / to Lizenz erteilen; zulassen
licensed under patents Lizenzbau *m* auf Grund von Patentrechten
licensee Konzessionsinhaber *m*, Lizenzinhaber *m*, Lizenznehmer *m*, Lizenzträger *m*
Lichtenberg figures *(El)* Lichtenbergsche Figuren *fpl*
licorice Lakritze *f*, Süßholzsirup *m*
lid 1. Deckel *m*; Verschlußdeckel *m*, Klappe *f*; 2. Ofenschließstein *m*, Ofenschließer *m*; 3. *(Brg)* Quetschholz *n*, Kopfholz *n*
l. arrester Deckelraste *f (Tonbandgerät)*
l. catch Deckelverschluß *m*
l. of blast furnace Hochofenkranz *m*
l. of stack Schornsteinhaube *f*
lie / to 1. liegen, sich legen; 2. *(Schiff)* beidrehen, vor Anker liegen
l. about beigedreht liegen *(Schiff)*
lie detector Lügendetektor *m*
Lieben-Reiss relay *(El)* Lieben-Röhre *f*
life Lebensdauer *f (Geräte, Werkzeuge)*; Standzeit *f (Werkzeuge; Ventile)*; Brenndauer *f (Elektronenröhren; Glühbirnen)*
l. belt Rettungsgürtel *m*, Schwimmweste *f*; Rettungsring *m*
l. between regrinds Standzeit *f* [zwischen zwei Nachschliffen]
l.-boat Rettungsboot *n*
l.-boat service Rettungsdienst *m*; Küstenrettungsdienst *m*
l. broadcast Originalsendung *f*
l. buoy Rettungsboje *f*; Rettungsring *m*
l. car tankähnliches Rettungsschiff *n*
l. expectancy *(voraussichtliche)* Lebensdauer *f*
l. float Rettungsfloß *n*
l. jacket Rettungsgürtel *m*, Schwimmweste *f*
l. of a cutting tool (edge) between regrinds Standzeit *f*
l. preserver Schwimmweste *f*; Rettungsgerät *n*; Rettungsring *m*
l. raft Rettungsfloß *n*
l. restoring apparatus Wiederbelebungsapparat *m*, Wiederbelebungsgerät *n*
l. ring Rettungsring *m*
l. rocket Rettungsrakete *f*
l.-saving equipment Rettungsausrüstung *f*
l.-stock wagon Viehwagen *m*
l.-time *s.* lifetime
l. zero *(BMSR)* lebendiger Nullpunkt *m (Signalübertragung)*
lifeguard service Rettungsdienst *m*; Küstenrettungsdienst *m*
lifetime Lebensdauer *f (z. B. radioaktiver Isotope)*; Betriebsdauer *f*; Lastspielzahl *f*

l. of an artificial satellite Lebensdauer *f* eines [künstlichen] Satelliten
lift / to 1. abheben, anheben, [hoch]heben; abnehmen; aufbäumen; 2. *(Brg)* quellen, heben
l. a chip einen Span abnehmen
l. clear of abheben von
l. from abheben von; abfedern von
l. off 1. [her]ausheben *(Modelle)*; 2. *(Fmt)* abheben *(Hörer)*
l. out herausnehmen
l. the pattern Modell herausheben *(Formguß)*
l. up aufrichten, hochrichten; hochheben
lift 1. Aufbäumen *n*; 2. Aufzug *m*, Fahrstuhl *m*, Lift *m*; 3. *(Brg)* Strosse *f (Tagebau)*; Sohle *f*; 4. *(Masch)* Hub[höhe *f*] *m*; Abhebung *f*, Lüftung *f (Meißel)*; 5. Auftrieb[skraft *f*] *m*; Sogkraft *f*; 6. Einbauschicht *f*, Lage *f (Betonbau)*; 7. *(Fs)* Schwarzabhebung *f*; 8. *(Brg)* Grundstrecke *f*, Sohlstrecke *f*
l. adjusting device Hubverstellung *f*
l. barrow Hubkarren *m*
l. bridge Hebebrücke *f*, Hubbrücke *f*
l. cam Anhebekurve *f*
l. car annunciator Stockwerksanzeiger *m (Fahrstuhl)*
l. coefficient Auftriebsbeiwert *m*
l. component Auftriebskraft *f*
l. controller Fahrstuhlsteuerschalter *m*
l. decrement Auftriebsverlust *m*
l. distribution Auftriebsverteilung *f*
l.-drag ratio *(Flg)* Gleitverhältnis *n*
l. force Auftriebskraft *f*
l. landing signal lamp Stockwerksanzeiger *m (Fahrstuhl)*
l. lock Schleuse *f*; Schiffshebewerk *n*
l. of mowing knives Mähmesserhub *m*
l. of sleeve *(Masch)* Muffenhub *m*
l. of the ring rail Ringbankhub[höhe *f*] *m*
l. shaft Fahrstuhlschacht *m*
l. slab *(Bw)* Hubplatte *f*
l.-slab method *(Bw)* Geschoßhebeverfahren *n*, Hubplattenverfahren *n*, Lift-slab-Verfahren *n*
l. spring Hubfeder *f*
l. truck Hubkarren *m*
lifter 1. Heber *m*; Wagenheber *m*; 2. Sandhaken *m (Formerei)*; 3. *(Text)* Ausheber *m*; 4. *(Brg)* Heberschuß *m (Schießarbeit)*
l. end *(Text)* Platinennase *f*
l.-hole *(Brg)* Heberloch *n (Schießarbeit)*
l. nose *(Text)* Platinennase *f*
lifting 1. Anheben *n*; Befördern *n* in senkrechter (schräger) Richtung; Lastenheben *n*; Hubbewegung *f*; Abhebung *f (Meißel)*; 2. *(Lw)* Roden *n*
l. appliance Hebezeug *n*
l. at the bow Austauchen *n* des Vorschiffs
l. barrier Schlagbaumschranke *f*
l. blocks Überhebevorrichtung *f (Walzwerk)*
l. cog Nocken *m*
l. crane Hebekran *m*
l. device Abhebevorrichtung *f*
l. dock Hebedock *n*
l. dogs Patentgreifer *m*
l. electromagnet Elektrohebemagnet *m*, *(elektrischer)* Lasthebemagnet *m*, Hubmagnet *m*
l. equipment Hebezeug *n*
l. eye Hebeöse *f*
l. flank Auflaufkurve *f (Nocken)*
l. gate Schlagbaumschranke *f*
l. gear Hebezeug *n*, Hubwerk *n*
l. guard *(Brg)* Schachtgatter *n*
l. heald *(Text)* Hebelitze *f*
l. hook Lasthaken *m*; Transporthaken *m*, Transportschlaufe *f*
l. jack Hebebock *m*, Wagenheber *m*
l. loop Transportschlaufe *f*
l. magnet *s.* l. electromagnet
l. of beets *(Lw)* Roden *n* der Rüben
l. pawl Hebeklinke *f*
l. pin Abhebestift *m*
l. platform Hebebühne *f*, Hubplattform *f*
l. pontoon *(Schiff)* Hebeponton *m*, Kamel *n*
l. power Hubkraft *f*; Tragkraft *f*
l. rod *(Brg)* Königsstange *f*
l. roller Hebewalze *f (am Walzgerüst)*
l. screw Abdrückschraube *f*, Hebeschraube *f*, Hubschnecke *f*
l. service Förderbetrieb *m*
l. tackle Hebezug *m*; Flaschenzug *m*; Aufzug *m*; Anschlagmittel *n*, Gehänge *n*
l. vat Aufzugskübel *m*
l. vessel Hebeschiff *n*, Dockschiff *n (zum Heben gesunkener Schiffe)*
ligature *(Typ)* Ligatur *f*, Doppelbuchstabe *m*
light / to 1. [be]leuchten; 2. entzünden; anzünden, anbrennen; 3. *(Schiff, Flg)* befeuern
l. a tube *(El)* eine Röhre heizen
light 1. hell; 2. leicht; 3. *(Typ)* mager *(Schrift)*
light 1. Licht *n*; Lichtschein *m*; Strahler *m*; 2. *(Brg)* Geleucht *n*; 3. *(Schiff, Flg)* Feuer *n*, Leuchtfeuer *n*; 4. Fenster *n*
l. aberration Lichtaberration *f*
l.-absorbing lichtabsorbierend, lichtschlukkend
l. absorption Lichtabsorption *f*, Lichtaufsaugung *f*
l. action Lichtwirkung *f*
l. adaptation *(Opt)* Helladaptation *f*
l. alloy Leichtmetall *n*, Leichtmetallegierung *f*
l.-alloy housing Leichtmetallgehäuse *n*
l.-alloy piston Leichtmetallkolben *m*
l.-alloy superstructure Leichtmetallaufbau *m*
l. barrier Lichtschranke *f*
l. beacon Leuchtbake *f*
l. beam Lichtstrahl *m*

l.-beam galvanometer Spiegelgalvanometer *n*, Lichtzeigergalvanometer *n*
l. boat unbemanntes Feuerschiff *n*
l.-bodied dünnflüssig
l. building board Leichtbauplatte *f*
l. bulb Glühlampe *f*, Lampe *f*
l. buoy Leuchtboje *f*, Leuchttonne *f*
l. cable Lichtleitung *f*
l. cargo leichte Ladung *f*, Leichtgut *n (Güter, die die Laderäume eines Schiffes räumlich ausfüllen, es jedoch nicht auf Lademarke bringen)*
l. centre length Lichtpunktabstand *m*
l. characteristic optische Kennlinie *f*
l. coal Gaskohle *f*
l. cone Lichtkegel *m*
l. connection Lampenschaltung *f*
l. current 1. Schwachstrom *m*; 2. lichtelektrischer Strom *m*, Fotostrom *m*
l. current engineering Schwachstromtechnik *f*
l. diffusing screen Weichzeichner *m (Film)*
l. diffusion Lichtstreuung *f*
l. draught *(Schiff)* Leertiefgang *m*
l. drive fit *(Masch)* leichter Treibsitz *m*
l. duty detergent Feinwaschmittel *n*
l. duty model leichte Ausführung *f*
l. duty part gering beanspruchter Teil *m*
l. duty scraper conveyor leichter Stegkettenförderer *m*
l. dynamo Lichtmaschine *f*
l. efficiency Lichtausbeute *f*
l. emissivity spezifische Lichtausstrahlung *f*
l. energy Energie *f* des Lichtes
l. fast lichtecht
l. fitting Beleuchtungskörper *m*, Leuchte *f*
l. float Leuchtfloß *n*
l. fluctuation Lichtschwankung *f*
l. flux Lichtstrom *m*
l. frequency Frequenz *f* des Lichtes, Lichtfrequenz *f*
l. gap Lichtspalt *m*
l. ground illumination Hellfeldbeleuchtung *f (Mikroskopie)*
l.-handed *(Schiff)* leicht bemannt
l. hydrogen leichter Wasserstoff *m*, Protium *n*
l. integrating timer Belichtungsautomat *m*
l. intensity Lichtstärke *f*
l. line dünne (feine) Linie *f*
l. magnet Lichtmagnet *m*
l. meson *(Kern)* leichtes Meson *n*
l. metal Leichtmetall *n*
l. metal hammer Ausbeulhammer *m*, Leichtmetallhammer *m*
l. meter Belichtungsmesser *m*
l. mixer *(Opt)* Strahlenvereiniger *m*
l. modulation Lichtmodulation *f*, Lichtsteuerung *f*
l.-negative lichtnegativ
l.-oil Leichtöl *n*
l.-optical microscope Lichtmikroskop *n*
l.-optical microscopy Lichtmikroskopie *f*
l. output Lichtausbeute *f*
l. output ratio of a fitting Leuchtenwirkungsgrad *m*
l. particle leichtes Elementarteilchen *n*, Lepton *n*
l. path Strahlengang *m*
l. pencil Licht[strahlen]bündel *n*
l. petroleum Petroläther *m*
l. plate Mittelblech *n*
l.-positive lichtpositiv
l. pressure Lichtdruck *m*
l. push switch Lichtschubschalter *m*
l. quantum Lichtquant *n*, Photon *n*
l. radiation Lichtstrahlung *f*
l. rail Feldbahnschiene *f*
l. rail motor tractor Kleinlokomotive *f (elektrischer Zug)*
l. railway Schmalspurbahn *f*, Feldbahn *f*
l. railway engine Feldbahnlokomotive *f*
l. railway track Feldbahngleis *n*
l. ray Lichtstrahl *m*
l. refractive lichtbrechend
l. relay fotoelektrisches Relais *n*
l. resistance Lichtwiderstand *m*
l. resonator Leuchtresonator *m*
l. running fit *(Masch)* leichter Laufsitz *m (System Johansson)*
l. scattering Lichtstreuung *f*, Streuung *f* des Lichtes
l.-section rolling mill Stabstahlwalzwerk *n*
l.-sectioning tester Lichtschnittmeßgerät *n*
l. sector Leuchtfeuersektor *m*, Leuchtwinkel *m*
l.-sensitive lichtempfindlich
l.-shaft Lichtschacht *m*
l. ship unbeladenes Schiff *n*
l. signal optisches Zeichen *n*, Lichtzeichen *n*, Lichtsignal *n*
l. slit Lichtspalt *m*
l.-slit method Lichtschnittverfahren *n*
l. socket Lichtsteckdose *f*
l. solid line dünne Vollinie *f*
l. source Lichtquelle *f*
l. spindle switch Lichtdrehschalter *m*
l. spiral of teeth schwache Spiralverzahnung *f*
l. spot Lichtfleck *m*, Leuchtfleck *m*
l.-spot scanner Leuchtfleckabtaster *m*, Lichtpunktabtaster *m*
l. stimulus Lichtreiz *m*
l. switch Lichtschalter *m*
l.-tight lichtundurchlässig
l. track Lichtspur *f*
l. transformer Lichttransformator *m*, Beleuchtungstransformator *m*, Transformator *m* für Beleuchtungszwecke

l. transmission Lichtdurchlässigkeit *f*
l. trap Lichtschleuse *f*
l. velocity measurement Lichtgeschwindigkeitsmessung *f*
l.-vessel Feuerschiff *n*
l. vibration Lichtschwingung *f*
l.-voltaic cell elektrolytisches Fotoelement *n*
l. water leichtes (gewöhnliches) Wasser *n*, Normalwasser *n*, Wasser, das kein Deuterium und Tritium enthält
l. wave Lichtwelle *f*
l.-weight leicht; in Leichtbauweise
l.-weight aggregate Leichtzuschlag[stoff] *m (Beton)*
l.-weight aggregate concrete Leichtzuschlagbeton *m*
l.-weight concrete Leichtbeton *m*
l.-weight concrete aggregate Leichtzuschlag[stoff] *m (Beton)*
l. yield Lichtausbeute *f*
lightboat Feuerschiff *n (unbemannt)*
lighted buoy befeuerte Tonne *f*, Leuchttonne *f*, Leuchtboje *f*
lighten / to *(Schiff)* leichtern, löschen
lightened silver Blicksilber *n*
lightening 1. Aufhellen *n (Anstrichtechnik)*; 2. *(Schiff)* Leichtern *n*, Ableichterung *f*
lighter 1. Anzünder *m*, Feuerzeug *n*; 2. Leichter *m*, Lastschiff *n*, Schleppschiff *n*, Schute *f*
l. flint Feuerstein *m*
l.-than-air craft Luftschiff *n (leichter als Luft)*
lighterman Leichterführer *m*; Kahnschiffer *m*
Lightgow tunnel *(Schiff)* Kavitationstank *m (in einer Schleppversuchsanstalt)*
lighthouse Leuchtturm *m*, Feuerturm *m*
l. tender Leuchtturmtender *m*, Tonnenleger *m*
lighthouses *(Schiff)* Befeuerung *f*
lighting 1. Beleuchtung *f*; Lichttechnik *f*; 2. *(Schiff, Flg)* Befeuerung *f*; 3. Zünden *n*
l. battery Lichtbatterie *f*
l. cable Lichtkabel *n*, Lampenkabel *n*
l. circuit Lichtstromkreis *m*, Lampenstromkreis *m*
l. density Leuchtdichte *f*
l. engineering Beleuchtungstechnik *f*
l. equipment Beleuchtungseinrichtung *f*, Beleuchtungsanlage *f*, Lichtanlage *f*
l. feeder Lichtleitung *f*
l. fitting Beleuchtungskörper *m*, Leuchte *f*
l. fixture Beleuchtungskörper *m*, Leuchte *f*
l. flex Lichtanschlußschnur *f*
l. gas Leuchtgas *n*, Stadtgas *n*
l. installation Lichtanlage *f*, Beleuchtungsanlage *f*
l. load Lampenlast *f*
l. of mines Grubenbeleuchtung *f*
l. of streets Straßenbeleuchtung *f*, öffentliche Beleuchtung *f*
l. outlet *(El)* Brennstelle *f*
l. power Leuchtkraft *f*
l. regulations Beleuchtungsvorschriften *fpl*
l. set Lichtanlage *f*
l. switch Lichtschalter *m*
l. system Beleuchtungsanlage *f*, Beleuchtungssystem *n*
l. transformer Lichttransformator *m*, Beleuchtungstransformator *m*, Transformator *m* für Beleuchtungszwecke
lightness 1. Leichtheit *f*, geringe Masse; 2. Helligkeit *f*
lightning Blitz *m*
l. arrester Überspannungsableiter *m*
l. call *(Fmt)* Blitzgespräch *n*
l. conductor Blitzableiter *m*
l. generator Stoßgenerator *m*
l. proofing Blitzschutz *m*
l. protection Blitzschutz *m*
l. protective system Blitzschutz *m*
l. rod Blitzableiter *m*
l. surge Gewitterüberspannung *f*
l. switch Antennenschalter *m*, Blitzschutzschalter *m*
lightship Feuerschiff *n*, Leuchtschiff *n*
lignification Verholzung *f*
lignin Lignin *n*, Holzstoff *m*
lignite Weichbraunkohle *f*, Braunkohle *f*, Lignit *m*
l. beds Braunkohlenvorkommen *npl*
l. briquette Braunkohlenbrikett *n*
l. carbonization plant Braunkohlenschwelerei *f*
l. coal mine Braunkohlengrube *f*
l. coal tar Braunkohlenteer *m*
l. coke Braunkohlenkoks *m*
l. coke plant Braunkohlenkokerei *f*
l. dressing plant Braunkohlenaufbereitungsanlage *f*
l.-fired furnace Braunkohlenfeuerung *f*
l. low temperature coke Braunkohlenschwelkoks *m*
l. mining Braunkohlenbergbau *m*
l. open cut Braunkohlentagebau *m*
l. wax Montanwachs *n*
lignitic braunkohlenhaltig, lignitisch
lignitiferous braunkohlenhaltig
ligno cellulose Holzzellstoff *m*, Holzfaserstoff *m*
lignoceric acid Lignocerinsäure *f*
lignum vitae Guajakholz *n*
ligroin *(Ch)* Ligroin *n*
like currents gleichgerichtete Ströme *mpl*
l. poles gleichnamige Pole *mpl*
l. signs gleiche Vorzeichen *npl*
limb 1. Schenkel *m*, Gradbogen *m (Meßinstrument)*; 2. *(Geol)* Flügel *m (Falte)*

l. darkening Randverdunkelung *f*
l. of a magnet Magnetkern *m*
limber Lafette *f*
l. boards *(Schiff)* Bilgeabdeckung *f*, Bilgewegerung *f*
l. strake *(Schiff)* Kielwegerungsplanke *f*
limbing Entästen *n (Holz)*
limbwood Äste *mpl*
lime / to 1. kalken, mit Kalk behandeln; mit Kalk düngen; 2. *(Led)* [ein]schwöden, kälken, äschern
l.-coat kalken
l.-wash tünchen, weißen *(mit Kalk)*
lime 1. [gebrannter] Kalk *m*; Kalziumoxid *n*; 2. *(Led)* Äscher *m*; 3. Linde[nholz *n*] *f*
l. accumulation Kalkanreicherung *f (des Bodens)*
l. agitator Kalkrührwerk *n*
l. and hair mortar Haarkalkmörtel *m*
l. blast *(Led)* Kalkflecken *m*
l. bloom Ausblühung *f*, Kalkausblühung *f (Betonfehler)*
l. boil *(Met)* Schlackenfrischreaktion *f (Martin-Verfahren)*
l. burning Kalkbrennen *n*
l. cement Zementkalk *m*, hydraulischer Kalk *m*
l. concrete Kalkbeton *m*
l. cream *(Led)* Kalkbrei *m*, Schwödebrei *m*
l. crusher Kalkmühle *f*
l. crust Kalkansatz *m*
l. dressing Kalkdüngung *f*
l. dust Kalkstaub *m*
l. fertilization Kalkdüngung *f*
l. floor Estrich *m*
l. kiln Kalkofen *m*
l. light *s.* limelight
l. lye Kalklauge *f*
l. marl Kalkmergel *m*
l. milk Kalkmilch *f*
l. mortar Kalkmörtel *m*
l. paint Tünche *f*
l. paste Kalkbrei *m*
l. pit *(Led)* Äscher *m*, Äschergrube *f*, Kalkäscher *m*
l. plant Kalkwerk *n*
l. plaster *(Bw)* Kalkputz *m*
l. reel *(Led)* Aufschlaghaspel *f* für den Äscher
l. sand Kalksand *m*
l.-sand brick Kalksandstein *m*, Kalksandziegel *m (Mauerstein)*
l. sandstone Kalksandstein *m*
l. set Kalkelend *n (Hochofen)*
l. slag *(Met)* Kalkschlacke *f*
l. slaking drum Kalklöschtrommel *f*
l. slaking plant Kalklöschanlage *f*
l. soap Kalkseife *f*
l. softening process Kalkverfahren *n (Wasseraufbereitung)*
l. soil *(Lw)* Kalkboden *m*
l. spar Kalkspat *m*
l. spreading attachment *(Lw)* Kalkstreugerät *n*
l. stone *s.* limestone
l. suspension Kalksuspension *f*
l.-tree Lindenbaum *m*
l.-wash Kalktünche *f*
l. water Kalkwasser *n*
l. whiting Kalktünche *f*
l. works Kalkwerk *n*
l. yard *(Led)* Äscherwerkstatt *f*
limelight Kalklicht *n*, Drummondsches Licht *n*
limen elektrischer Schwellenwert *m*
limer Kalktüncher *m*
limestone Kalkstein *m*
l. addition Kalksteinzuschlag *m*
l. gravel Kalksteinkies *m*
l. masonry Kalksteinmauerwerk *n*
l. pit Kalksteingrube *f*
l. quarry Kalksteinbruch *m*
l. shale Kalkschiefer *m*
liming *(Met)* Kälken *n (Draht)*; *(Led)* Äscher *m*
l. in warm liquor Sohlederäscherung *f* in warmem Wasser
l. tub *(Led)* Äscherfaß *n*
limit / to begrenzen, beschränken
l. to beschränken auf
limit 1. Grenze *f*; Grenzmaß *n*; 2. *(Brg)* Markscheide *f*; 3. Toleranz *f*, Abmaß *n (oberes und unteres)*
l. angle Grenzwinkel *m*
l. blade Grenzschaufel *f (Turbine)*
l. bridge Toleranzmeßbrücke *f*
l. case Grenzfall *m*
l. control *(BMSR)* Grenzwertüberwachung *f*
l. curve of critical state *(Wärm)* Grenzkurve *f*
l. cycle Grenzzyklus *m*, Grenzkurve *f (Phasenebenendarstellung)*
l. design Traglastverfahren *n (Stahlbetonberechnung)*
l. finger Toleranzanzeiger *m*
l. frequency Grenzfrequenz *f*
l. gap gauge Grenzrachenlehre *f*
l. gauge Grenzlehre *f*, Maßlehre *f*
l. gauging Maßlehren *n*
l. indicator Toleranzanzeiger *m*; Grenzzeiger *m*
l. load Grenzbeanspruchung *f*
l. mark Grenzmarkierung *f*
l. of a sequence *(Math)* Grenzwert *m* einer Folge
l. of accuracy Genauigkeitsgrenze *f*; Toleranzgenauigkeit *f*
l. of audibility Hörgrenze *f*, Grenze *f* des Hörbereiches

l. of detection Nachweisgrenze *f*
l. of dissociation Dissoziationsgrenze *f*
l. of elasticity Elastizitätsgrenze *f*
l. of error Fehlergrenze *f*
l. of ignitability Zündgrenze *f*
l. of perceptibility Wahrnehmbarkeitsgrenze *f*
l. of power Leistungsgrenze *f*
l. of proportionality Proportionalitätsgrenze *f*
l. of resistance Festigkeitsgrenze *f*
l. of scattering Streu[ungs]grenze *f*
l. of sensibility Empfindlichkeitsgrenze *f*
l. of size Grenzmaß *n*
l. of stability Stabilitätsgrenze *f*
l. plug gauge Grenzlehrdorn *m*
l. pointer Toleranzanzeiger *m*
l. screw plug gauge Grenzgewindelehrdorn *m*
l. speed Grenzdrehzahl *f*
l. stop Endanschlag *m*
l. switch End[lagen]schalter *m*, Grenzschalter *m*, Begrenzungsschalter *m*
l. thread gauge Grenzgewindelehre *f*
l. value Grenzwert *m*
limitation Begrenzung *f*, Einschränkung *f*
l. of amplitude Amplitudenbegrenzung *f*
l. of frequency band Frequenz[band]begrenzung *f*
l. of interference Störbegrenzung *f*
limited begrenzt
l. liability company Gesellschaft *f* mit beschränkter Haftung
l. partnership Kommanditgesellschaft *f*
limiter Begrenzer *m*
l. diode Begrenzerdiode *f*
l. stage Begrenzerstufe *f*
limiting *(Math)* Grenz-, Häufungs-
l. amplifier Begrenzungsverstärker *m*
l. angle Grenzwinkel *m*
l. case Grenzfall *m*
l. characteristic Begrenzungskennlinie *f*
l. charge Grenzladung *f*
l. circuit Begrenzerschaltung *f*
l. condition Grenzbedingung *f*
l. creep stress Kriechgrenze *f*, Dauerstandfestigkeit *f*
l. current Grenzstrom *m*
l. dimension Grenzmaß *n*
l. frequency Grenzfrequenz *f*, Abschneidefrequenz *f*
l. load Grenzbelastung *f*
l. no-damage current größter Haltestrom *m* *(einer Sicherung)*
l. output Grenzleistung *f*
l. ratings Grenzdaten *pl (Elektronenröhren)*
l. sensitivity Grenzempfindlichkeit *f*
l. size Grenzmaß *n*
l. speed Grenzgeschwindigkeit *f*
l. stage *(BMSR)* Begrenzerschaltung *f*, Begrenzerstufe *f*
l. stress *(Mech)* Grenzspannung *f*, höchstzulässige Spannung *f*
l. temperature Grenztemperatur *f*
l. velocity Grenzgeschwindigkeit *f*
l. wavelength Grenzwellenlänge *f*
limitless unbegrenzt
limits and fits Toleranzen *fpl* und Passungen *fpl*
l. of accuracy required erforderliche Arbeitsgüte *f*
limmers Gabeldeichsel *f*
limonite Brauneisenerz *n*, Raseneisenerz *n*, Limonit *m*
limp schlaff *(Seil)*
l. cover *(Typ)* flexibler Einband *m*
limy *(Ch)* kalkartig, kalkig
l. ore kalkiges Erz *n*
linage 1. Zeilenzahl *f*; 2. Zeilenhonorar *n*
linarite *(Min)* Linarit *m*, Bleilasur *f*
Linde process Linde-Verfahren *n (Kältetechnik)*
line / to 1. [ab]füttern, ausfüttern, ausgießen, auskleiden, ausschlagen, überziehen, umkleiden, verkleiden; 2. ausmauern *(Hochofen)*; 3. ausbauen *(Bergbau)*; 4. zustellen, auslegen, kaschieren; 5. einstellen, abfluchten *(Maschinen)*
l. a bearing *(Masch)* ein Lager ausfüttern
l. ahead *(Schiff)* in Kiellinie fahren
l. out abschnüren *(Schiffslinien)*; abfluchten; anreißen
l. the rolls Walzen einstellen (zustellen)
l. up ausrichten *(eine Linie)*, abfluchten
line 1. Linie *f*, Gerade *f*; Zeile *f*; Reihe *f*; Strich *m*; 2. Schnur *f*; Mauerschnur *f*; 3. Leitung *f*; Rohrleitung *f*; Strang *m*; 4. Bahnlinie *f*, Bahnstrecke *f*
l.-a-time printer *(Dat)* Zeilendrucker *m*
l.-a-time printing *(Dat)* Paralleladrucken *n*
l. absorption coefficient *(Opt)* Linienabsorptionskoeffizient *m*
l. advance *(Fs)* Zeilenabstand *m*
l. amplifier Kabelendverstärker *m (Rundfunksender)*, Leitungsverstärker *m*
l. amplitude *(Fs)* Zeilenamplitude *f*
l. amplitude control Bildbreitenregelung *f*; Zeilenbreitenregler *m*
l. assembly work Fließmontage *f*, Bandmontage *f*
l.-at-a-time printer *(Dat)* Zeilendrucker *m*
l. attenuation Leitungsdämpfung *f*; Leitungsverlust *m*
l. balance Leitungssymmetrie *f*, Leitungsentzerrer *m*
l. balance converter Symmetrierglied *n*, Symmetriertopf *m*
l. bank *(Fmt)* Leitungskontaktbank *f*
l. bar Leitungsschiene *f*
l. blanking *(Fs)* Zeilenaustastung *f*
l. blanking level Zeilenaustastpegel *m*

l. block *(Typ)* Strichätzung *f*, Strichklischee *n*
l.-break relay Drahtbruchrelais *n*
l. breaker Netzschalter *m*; Fahrleitungsschalter *m*
l.-by-line zeilenweise
l.-by-line analysis *(Fs)* zeilenweise Zerlegung *f*
l.-by-line milling Zeilenfräsen *n*
l.-by-line scanning *(Masch)* Zeilenabtasten *n*
l. capacitance Leitungskapazität *f*
l. centre *(Opt)* Linienmitte *f*
l. circuit Freileitungskreis *m*
l. circuit-breaker Überstromschalter *m*
l.-clear lock *(Eis)* Fahrsperre *f*
l.-clear position *(Eis)* Fahrstellung *f*
l. constant Leitungskonstante *f*
l. construction Telegrafenbau *m*
l. contact Linienberührung *f (Zahnräder)*
l. contactor Überstromschalter *m*
l. contradictory to Stokes' rule *(Opt)* antistokessche Linie *f*
l.-controlled leitungsgesteuert
l. cord Netzanschlußschnur *f*
l. crawl Farbenzeilenflimmern *n (Farbfernsehtechnik)*
l. current Netzstrom *m*, Leitungsstrom *m*
l. divider *(Fs)* Zeilenfrequenzteiler *m*
l. drawing Strichzeichnung *f*
l. drop Spannungs[ab]fall *m* auf der Leitung
l. duration *(Fs)* Zeilendauer *f*
l. equalizer *(Fmt)* Leitungsentzerrer *m*
l. etching Strichätzung *f*
l. failure Netzausfall *m*; Leitungsstörung *f*
l. feed *(Fs)* Zeilenvorschub *m*
l. finder *(Fmt)* Vorwähler *m*, Anrufsucher *m*
l.-finder relay *(Fmt)* Relaisanrufsucher *m*
l. flicker Farbenzeilenflimmern *n (Farbfernsehtechnik)*
l. flickering *s.* l. flicker
l. flow production Fließbandfertigung *f*
l. flyback pulse *(Fs)* Zeilenrücklaufimpuls *m*
l. focus Strichfokus *m*
l.-free television zeilenfreies Fernsehen *n*
l. frequency 1. Netzfrequenz *f*; 2. *(Fs)* Zeilenfrequenz *f*
l. gauge *(Typ)* Typometer *n*
l. gauge block Strichendmaß *n*
l. graduation Strichteilung *f*
l.-image converter Zeilenbildwandler *m*
l. in trouble gestörte Leitung *f*
l. indicating course of the piston *(Masch)* Kolbenweglinie *f*
l. inspection trolley *(Eis)* Draisine *f*
l. insulator Freileitungsisolator *m*
l. integral Kurvenintegral *n*, Linienintegral *n*
l. interlacing *(Fs)* Zeilensprung *m*
l. irregularity Leitungsinhomogenität *f*
l. jump scanning *(Fs)* Zeilensprungzerlegung *f*, Zeilensprungabtastung *f*
l. leather Zügelleder *n*
l.-lengthener Wellenstrecker *m (Wellenleitertechnik)*
l. linking element Verkettungseinrichtung *f*
l. load Linienlast *f*, Streckenlast *f (Statik)*
l. loss Leitungsdämpfung *f*; Leitungsverlust *m*
l. marker Straßenmarkierungsgerät *n*, Straßenmarkierungsmaschine *f*
l. monitor *(Fmt)* Leitungsüberwachungsgerät *n*
l. monitoring tube *(Fs)* Zeilenkontrollröhre *f*
l. noise *(El)* Leitungsrauschen *n*
l. of a whirl *(Strm)* Wirbellinie *f*
l. of action Eingriffslinie *f (Zahnrad)*; Wirkungslinie *f (Kraft)*; Angriffslinie *f*
l. of apsides Apsidenlinie *f*
l. of centres Mittellinie *f*
l. of centres of the force system Zentrallinie *f* des Kräftesystems
l. of contact Berührungslinie *f*
l. of demarcation Grenzlinie *f*, Trennungslinie *f*, Demarkationslinie *f*
l. of diagram Diagrammlinie *f*, Diagrammkurve *f*
l. of dip *(Geol)* Fallinie *f*
l. of direction Richtungslinie *f*
l. of electric field strength Linie *f* der elektrischen Feldstärke
l. of electric force elektrische Kraftlinie *f*
l. of equal steam quantity *(Wärm)* Linie *f* gleicher Dampfmenge, Linie gleichen Mischungsverhältnisses
l. of face *(Brg)* streichende Baulänge *f*
l. of faulting *(Geol)* Bruchlinie *f*
l. of flow *(Strm)* Stromlinie *f*, Strömungslinie *f*
l. of force Kraftlinie *f*, Feldlinie *f*
l. of graduation Skalenstrich *m*, Teilstrich *m*
l. of intensity of magnetic field Linie *f* der magnetischen Feldstärke
l. of least resistance *(Brg)* Linie *f* des geringsten Widerstandes *(Schießarbeit)*
l. of lode *(Geol)* Gangstreichen *n*
l. of magnetic force magnetische Kraftlinie *f*
l. of pipes Rohrstrang *m*
l. of resistance Widerstandslinie *f*
l. of segregation *(Met)* Seigerungszone *f*
l. of shafting Wellenstrang *m (Riementrieb)*
l. of sight *(Mil)* Visierlinie *f*, Ziellinie *f*
l. of solidification *(Met)* Erstarrungslinie *f*
l. of strike *(Geol)* Streichlinie *f*
l. of tubing Rohrleitung *f*
l. on the grating Gitterstrich *m (Kristalle)*
l. operated mit Netzanschluß versehen
l. output *(Fs)* Zeilenzeitbasis *f*

l. output pentode *(Fs)* Horizontal[ablenkung]-endpentode *f*
l. pattern *(Dat)* Schriftprobe *f (Zeichenerkennung)*
l. printer *(Dat)* Zeilendrucker *m*
l. radio Drahtfunk *m*
l. radio amplifier Drahtfunkverstärker *m*
l. radio station Drahtfunksender *m*
l. regulator Netzspannungsregler *m*
l. relay Linienrelais *n*
l. scan circuit *(Fs)* Zeilenkippschaltung *f*
l. scanning *(Fs)* Zeilenabtastung *f*
l. scanning frequency Zeilenfrequenz *f*
l. scratches *(Fmt)* Kratzgeräusche *npl (in Leitungen)*
l. screen *(Typ)* Strichraster *m*, Linienraster *m*
l. segment *(Math)* Strecke *f*
l. selector 1. *(Fmt)* Leitungssucher *m*, Leitungswähler *m*; 2. *(Fs)* Zeilenwähler *m*
l. sequential system Zeilenfolgesystem *n (Farbfernsehtechnik)*
l. setup *(Masch)* Aufspannung *f* in Reihe, Reihenaufspannung *f*
l. shaft Transmission[swelle] *f*; *(Wärm)* Laufwelle *f*, Zwischenwelle *f*; *(Schiff)* Hauptwelle *f*
l. shaft bearing *(Masch)* Zwischenlager *n*
l. shafting Transmission *f*
l. signal *(Eis)* Liniensignal *n*; *(Fmt)* Rufzeichen *n*
l. simulator *(El)* künstliche Leitung *f*
l. source Linienquelle *f*
l. spacing *(Fs)* Zeilenabstand *m*
l. spectrum Linienspektrum *n*
l. starting unmittelbares Einschalten *n (Kurzschlußläufermotor)*
l. synchronization *(Fs)* Zeilensynchronisation *f*
l. synchronizing pulse *(Fs)* Zeilensynchronisationsimpuls *m*
l. tangent to Tangente *f* an
l. telegraphy Drahttelegrafie *f*
l. terminals Phasenklemmen *fpl*, Strangklemmen *fpl (einer E-Maschine)*
l. termination Ende *n* der Leitung, Leitungsabschluß *m*
l. tester Fadenzähler *m*
l.-throwing rocket *(Schiff)* Rettungsrakete *f*, Leinenwurfrakete *f*
l. time base *(Fs)* Zeilenkippgerät *n*
l.-to-earth fault Erdschluß *m*
l.-to-line fault Leitungsberührung *f*, Schluß *m* zwischen Phasen
l.-to-line voltage Phasenspannung *f*
l. transformer 1. *(Fs)* Zeilentransformator *m*, Zeilenübertrager *m*; 2. *(Fmt)* Leitungsübertrager *m*
l. trap netzseitige HF-Sperre *f*; Wellenschlukker *m*, Wellenfalle *f*; Trägerfrequenzsperre *f (Kopplungskondensator)*
l. variation Netzspannungsschwankung *f*
l. voltage Netzspannung *f*; Leiterspannung *f*, Spannung *f* auf der Leitung
l. voltage of a polyphase system verkettete Spannung *f*, Leiterspannung *f*
l. voltage regulator Netz[spannungs]regler *m*
lineal foot Fuß *m (Längenmaß)*
linear linear; eindimensional; geradlinig; mit geradliniger Bewegung
l. absorption coefficient linearer Absorptionskoeffizient *m*
l. acceleration Linearbeschleunigung *f*
l. accelerator Linearbeschleuniger *m*
l. accuracy Axialität *f*
l. actuator Schubantrieb *m*, Stellantrieb *m* mit Linearbewegung
l. admission lead *(Wärm)* Voreilen *n* für den Eintritt, äußeres lineares Voreilen
l. amplification lineare Verstärkung *f*
l. amplifier linearer Verstärker *m*
l. array *(Am) (Fs)* Dipolreihe *f*, Dipolzeile *f*
l. bolometer Linearbolometer *n*
l. decrement lineare Amplitudenabnahme *f*
l. detection lineare Gleichrichtung *f*
l. detector linearer Detektor *m*
l. dilatation einachsige (lineare) Dehnung *f*
l. dimension Längenmaß *n*
l. dispersion Lineardispersion *f*
l. distortion lineare Verzerrung *f*
l. dividing machine Längenteilmaschine *f*
l. electric constant Leitungsbelag *m*
l. electron accelerator Linearelektronenbeschleuniger *m*
l. equation lineare Gleichung *f (Gleichung 1. Grades)*
l. exhaust lead *(Wärm)* Voreilen *n* für den Austritt, inneres lineares Voreilen
l. expansion Längsdehnung *f*, lineare Ausdehnung *f*
l. expansion coefficient linearer Ausdehnungskoeffizient *m*
l. extension Längenausdehnung *f*, lineare Ausdehnung *f*
l. hysteresis Ummagnetisierung *f*, Wechselmagnetisierung *f*
l. lead *(Wärm)* lineares Voreilen *n*; Voreröffnen *n*; Voröffnungsstrecke *f*
l. loading lineare Belastung *f*
l. machining *(Masch)* Spanabnahme *f* mit geradliniger Bewegung
l. magnification lineare Vergrößerung *f*
l. measurement Längenmessung *f*
l. measuring instrument Längenmeßgerät *n*
l. modulation verzerrungsfreie (lineare, g[e]radlinige) Modulation *f*
l. molecule Fadenmolekül *n*
l. moment of inertia [lineares] Trägheitsmoment *n*

l. programming lineares Programmieren *n*; Linearprogrammierung *f*; lineare Optimierung *f*
l. pulse amplifier linearer Impulsverstärker *m*
l. relationship lineare Beziehung *f*
l. response detector *(Kern)* Detektor *m* mit linearer Charakteristik
l. scanning *(Fs)* zeilenweises Abtasten *n*
l. speed Geschwindigkeit *f*
l. stopping power lineares Bremsvermögen *n*
l. sweep *(Fs)* lineare Zeitablenkung *f*
l. velocity lineare Geschwindigkeit *f*
linearity G[e]radlinigkeit *f*, Linearität *f*
l. control *(Fs)* Linearitätsregelung *f* *(Steuerung)*
linearly polarized linear polarisiert
lineation Linierung *f*
lined fabric kaschiertes Gewebe *n*
l. with refractory material feuerfest ausgemauert
lineman *(Am)* Freileitungsmonteur *m*
linen Leinen *n*; Wäsche *f*
l. cloth Leinwand *f*
l.-disinfection plant Wäschedesinfektionsanlage *f*
l. goods Weißwaren *fpl*
l.-grained paper leinengeprägtes Papier *n*
l. room Wäschezimmer *n*
liner 1. Unterlegscheibe *f*; Zylinderlaufbuchse *f*; Unterlegstreifen *m*; Buchse *f*; Deckmantel *n*; Ausguß *m*; Auskleidung *f*; 2. Kaschierpapier *n*, Deckpapier *n*; 3. Liner *m*, Filterrohr *n*; 4. *(Rak)* Feuerhaut *f*; 5. Tourendampfer *m*, Linienfahrer *m*, Linienschiff *n*
l. bushing Buchse *f* zur Aufnahme von Einsteckbohrbuchsen
l. company Linienreederei *f*
l. condition *(Kfz)* Zylinderlaufbild *n*
l. of bearing Lagerausguß *m*
l. of diaphragm Leitradbüchse *f*, Dichtungsbuchse *f* *(Turbine)*
l. out Anreißer *m*
l. trade Linienschiffahrt *f*, Seeschiffahrt *f*, Handelsschiffahrt *f*
lines *(Schiff)* Linienriß *m*, Konstruktionsriß *m*
l. drawing *(Schiff)* Linienriß *m*
l. of blast furnace Hochofenprofil *n*
l. of magnetic flux magnetische Flußlinien *fpl*
l. of magnetic force magnetische Kraftlinien *fpl*
l. of maximal shearing stresses Schubspannungslinien *fpl*
l. plan *(Schiff)* Linienriß *m*
linesetter *(Typ)* Zeilensetzmaschine *f*, Linotype *f*
lineshaft drive Transmissionsantrieb *m*
lingerie Leinenwaren *fpl*; Damenunterwäsche *f*
lingo *(Text)* Anhangeisen *n*
linguistic analysis linguistische Analyse *f*, Sprachanalyse *f* *(automatisches Übersetzen)*
linhay vorn offener Schuppen *m*
lining 1. Belag *m*, Bremsbelag *m*; 2. Auskleidung *f*; Ausmauerung *f*; Ausfütterung *f*, Futter *n*; Futterstoff *m*; 3. Kaschierung *f*; 4. Verbau *m* *(Grabenbau)*; 5. *(Brg)* Ausbau *m*; 6. *(Met)* Zustellung *f*; 7. *(Brg)* markscheiderische Vermessung *f*; 8. Einstellung *f* *(Walzen)*
l. brick Futterstein *m*, Futterziegel *m*
l. calender *(Typ)* Kaschierkalander *m*, Beschichtungskalander *m*
l. fabric Futterstoff *m*
l. felt Isolierfilz *m* *(Rohre)*
l. leather Futterleder *n*
l. mass Futtermasse *f*, Zustellungsmasse *f* *(Ofen)*
l. metal Lagermetall *n*
l. of a bearing Lager[schalen]ausguß *m*
l. of a furnace Zustellung *f* eines Ofens
l. of rolls Walzeneinstellung *f*, Walzenanstellung *f*
l. of the bearing Lager[schalen]ausguß *m*
l. out Anreißen *n*
l. paper Kaschierpapier *n*
l. pipes Futterrohre *npl*
l. up Ausrichten *n*
link / to verketten
link 1. Glied *n* *(Getriebelehre)*; 2. Gelenk *n*; Verbindungsglied *n*, Verbindungsstange *f*; Kurbelschwinge *f*, Krampe *f*; 3. Bindung *f*; 4. Bügel *m* *(Stahlbetonbau)*; 5. *(Fmt)* Kanal *m*; Verbindung *f*; 6. *amerikanische Längeneinheit: 1 li = 20,117 cm*
l. bar Gelenkkappe *f*
l. belt Gliederriemen *m*, Gliederband *n*
l.-belt chain Gliederkette *f*
l.-belt chain drive Zahnkettentrieb *m*
l. box Gelenkmuffe *f*
l. bracket Kulissenlager *n*, Schwingenlager *n*
l. chain Gelenkkette *f*, Gliederkette *f*
l. coil *(Fmt)* Zwischenkreisspule *f*
l. drive Kulissenantrieb *m*
l. fuse *(Am)* Lamellensicherung *f*, Streifensicherung *f*
l. heald *(Text)* Schlinghelfe *f*
l. line Fertigungskette *f*
l. motion Kulissensteuerung *f*
l. pin Anlenkbolzen *m*, Nebenpleuelzapfen *m*, Kettenbolzen *m*
l. polygon Seilpolygon *n*, Seilzug *m*
l. rod Gelenkstange *f*; Nebenpleuel *m*, Kulissenstange *f*
l. tooth saw Kettensäge *f*
l. transmitter Sender *m* einer Relaisstrecke
l.-up Verkettung *f*
Link trainer Link-Trainer *m*, Flugtrainer *m*
linkability *(Masch)* Verkettbarkeit *f*

linkable *(Masch)* verkettungsfähig, verkettbar
linkage 1. Gelenkgetriebe *n*; 2. [chemische] Bindung *f*; 3. *(Kern)* Kette *f*; Verbindung *f*; 4. Aufhängung *f (Ackergeräte an den Traktor)*
l. brake *(Kfz)* Gestängebremse *f*
l. of farming tools Aufhängung *f* der Ackergeräte
linked barges Schleppzug *m (von Schleppkähnen)*
l. reaction *(Ch)* gekoppelte Reaktion *f*
l. switch *(El)* Kupplungsschalter *m*
linking Verkettung *f*
l. block *(Bw)* Verbindungsbau *m*, Verbindungstrakt *m*
l. machine *(Text)* Kettelmaschine *f*
l. seam Kettelnaht *f*
l. together Verkettung *f*
links-and-links stitch *(Text)* Perlstich *m*
lino *s.* linoleum
l. cut *(Typ)* Linolschnitt *m*
linoleic acid Linolsäure *f*
linolenic acid Linolensäure *f*
linoleum Linoleum *n*
linotype *(Typ)* Zeilensetzmaschine *f*, Linotypesetzmaschine *f*
l. slug *(Typ)* Linotypezeile *f*
linseed Leinsamen *m*
l. flax Saatlein *m*
l. oil Lein[samen]öl *n*
l. plant Lein *m*
lint-free fusselfrei
lintel Sturz *m*; Fenstersturz *m*, Türsturz *m*; Sturzriegel *m*, Sturzträger *m*
l. girder Tragkranz *m*, Tragring *m (des Hochofens)*
linters *(Text)* Linters *pl*
lintol *s.* lintel
lip 1. Rand *m*, Kante *f*; 2. Hauptschneide *f (am Spiralbohrer)*, Schneidlippe *f*; 3. Gießschnauze *f*
l. angle Keilwinkel β *m (am Spiralbohrer)*
l. clearance Hinterschliff *m*, Freiwinkel *m (Spiralbohrer)*
l. of the twist drill Schneidkante *f* am Spiralbohrer
l. of tyre Ansatz *m* am Radreifen
l.-poured ladle *(Met)* Kipppfanne *f*
l. relief angle Freiwinkel α *m (am Spiralbohrer)*
l.-type oil seal Lippendichtung *f*, Wellendichtung *f*, Simmering *m*
lipid[e] Fettkörper *m*, Lipoid *n*
lipolysis Fettspaltung *f*, Lipolyse *f*
liquate / to *(Met)* [aus]seigern; darren *(Kupfer)*
liquate *(Met)* Seigerrückstand *m*
liquated copper Darrkupfer *n*
liquating furnace *(Met)* Seigerofen *m*
liquation *(Met)* Seigerung *f*, Ausseigerung *f*; Entmischen *n*
l. furnace Seigerofen *m*
l. hearth Seigerherd *m*
l. pan Seigerpfanne *f*
l. process Seigerung *f*, Seigervorgang *m*, Seigerarbeit *f*
l. residue Seigerrückstand *m*
l. slag Seigerschlacke *f*
liquefacient Verflüssigungsmittel *n*
liquefaction Verflüssigung *f*
l. of gas Gasverflüssigung *f*
liquefied gas verflüssigtes Gas *n*
l. natural gas tanker Flüssiggastanker *m*
l. petroleum gas Flüssiggas *n*
l. petroleum gas tanker Flüssiggastanker *m*
liquefy / to verflüssigen, flüssig machen, schmelzen, zerlassen
liquefying cleansing cream Reinigungsemulsion *f*
liquid Flüssigkeit *f*
liquid flüssig, Flüssigkeits-
l. absorption Flüssigkeitsaufnahme *f*
l. air 1. flüssige Luft *f*; 2. *(Brg)* Sprengluft *f*
l. alloy geschmolzene Legierung *f*, Schmelze *f*
l. ammonia Salmiakgeist *m*, flüssiges Ammoniak *n*
l. carrier Trägerflüssigkeit *f*
l. catalyst flüssiger Katalysator *m*
l. cell Flüssigkeitsküvette *f*
l. column Flüssigkeitssäule *f*
l.-column gauge Flüssigkeitsmanometer *n*, U-Rohr-Manometer *n*
l. condition of aggregation flüssiger Aggregatzustand *m*
l. condition of matter *s.* l. condition of aggregation
l. container Flüssigkeitsbehälter *m*
l.-containing structure Flüssigkeitsbehälter *m*
l. controller Flüssigkeitsregler *m*
l. cooling Flüssigkeitskühlung *f*
l. counter *(Kern)* Flüssigkeitszähler *m*
l. coupling Flüssigkeitskupplung *f*
l. damping Flüssigkeitsdämpfung *f*
l. dentrifice Mundwasser *n*
l. drop Flüssigkeitstropfen *m*, Tropfen *m*
l.-drop behaviour Tröpfchenverhalten *n*
l.-drop model Tröpfchenmodell *n*
l.-drop model of the nucleus Tröpfchenmodell *n* des Atomkerns
l. dung Jauche *f*
l. engine Flüssigkeitsraketentriebwerk *n*
l. filament *(Strm)* Stromfaden *m*
l. firing Verfeuerung *f* flüssiger Brennstoffe
l. friction Flüssigkeitsreibung *f*
l. fuel flüssiger Brennstoff *m*; Spaltstoffflüssigkeit *f*
l. fuel and oxidant motor Flüssigkeitsraketentriebwerk *n*

l.-fuel furnace with nozzle spraying Flüssigkeitsfeuerung *f* mit Rohrzerstäubung
l.-fuel reactor Reaktor *m* mit flüssigem Brennstoff, Flüssigkeitsreaktor *m*
l.-fuel rocket engine Flüssigkeitsraketentriebwerk *n*
l. incendiary bomb Flüssigkeitsbrandbombe *f*
l. insulation *(Am) s.* insulating liquid
l. jet Flüssigkeitsstrahl *m*
l.-junction potential Potentialsprung *m* zwischen zwei Elektrolyten, Kontaktpotential *n* zwischen Flüssigkeiten
l. level Flüssigkeitsspiegel *m*
l. level gauge Füllstandmeßgerät *n*, Flüssigkeitsstandmesser *m*
l. level switch *s.* float switch
l.-limit curve *(Wärm)* Flüssigkeitsgrenzkurve *f*
l. manure Jauche *f*
l. manure cistern Jauchegrube *f*
l. manure pump system Jauchenpumpsystem *n*
l. measure Hohlmaß *n*; Volumenmaß *n* für Flüssigkeit
l. metal Flüssigmetall *n*, geschmolzenes Metall *n*
l.-metal circulation Flüssigmetallumlauf *m* *(Reaktorkühlung)*
l.-metal fuel *(Kern)* Flüssigmetallbrennstoff *m*
l.-metal fuel[led] reactor Flüssigmetallreaktor *m*, Reaktor *m* mit Flüssigmetallbrennstoff
l.-moderated reactor Flüssigkeitsreaktor *m*, Reaktor *m* mit flüssigem Moderator
l. motor Flüssigkeitsraketentriebwerk *n*
l. operated hydraulisch betrieben (gesteuert), mit hydraulischer Hilfsenergie
l. packing *(Masch)* Flüssigkeitsabschluß *m*
l. paraffin Paraffinöl *n*, flüssiges Paraffin *n*
l. phase hydrogenation Sumpfphasenhydrierung *f*
l. polishing Hydropolieren *n*
l. powered *s.* l. operated
l. propellant flüssiger Treibstoff *m*
l.-propellant motor Flüssigkeitsraketentriebwerk *n*
l.-propellant rocket Flüssigkeitsrakete *f*
l.-propellant rocket engine Flüssigkeitsraketentriebwerk *n*
l.-propellant unit Flüssigkeitsraketentriebwerk *n*
l. purification *(Brg)* Naßreinigung *f*
l.-quenched cut-out *(El)* Ölsicherung *f*
l. resistance Flüssigkeitswiderstand *m*
l. rocket Flüssigkeitsrakete *f*; Flüssigkeitsraketentriebwerk *n*
l.-rocket engine Flüssigkeitsraketentriebwerk *n*
l.-rocket unit Flüssigkeitsraketentriebwerk *n*
l.-sample counter *(Kern)* Flüssigkeitszähler *m*
l. seal Trennflüssigkeit *f*, Sperrflüssigkeit *f* *(Schutz gegen aggressive Medien)*
l. starter Flüssigkeitsanlasser *m*
l. state flüssiger Zustand *m*
l. state of aggregation flüssiger Aggregatzustand *m*
l. state of matter *s.* l. state of aggregation
l.-tight flüssigkeitsdicht
l.-tight joint flüssigkeitsdichte Verbindung *f*
liquids carrier Schiff *n* für flüssige Ladung
liquidus *(Met)* Liquiduslinie *f*
liquor Flüssigkeit *f*; Lauge[nflüssigkeit] *f*; Laugenbad *n*
l. tanning Gerben *n* in Lohbrühe
l. well *(Brg)* Gaswassergrube *f*
liroconite *(Min)* Linsenerz *n*, Liroconit *m*
Lissajous' figures *(Ph)* Lissajoussche Figuren *fpl*
lisseuse *(Text)* Lisseuse *f*, Kammzugplättmaschine *f*
list / to 1. eintragen *(in eine Liste)*; 2. *(Schiff)* Schlagseite haben, überliegen, krängen
list 1. Liste *f*, Verzeichnis *n*; 2. *(Schiff)* Schlagseite *f*, Überliegen *n*, Krängung *f*; 3. *(Text)* Webkante *f*, Salleiste *f*
l. of circuit elements Schaltteilliste *f*
l. of lights Leuchtfeuerverzeichnis *n*
l. of parts Stückliste *f*
l. of scantlings *(Schiff)* Baubesteck *n*
listen / to [an]hören, zuhören, horchen
l. in on a circuit *(Fmt)* sich in eine Leitung einschalten
listener Rundfunkhörer *m*
listening button *(Fmt)* Mithörtaste *f*
l. frequency *(Fmt)* Hörfrequenz *f*, Wachfrequenz *f*
l. jack *(Fmt)* Mithörklinke *f*
l. loudspeaker Kontrollautsprecher *m*, Abhörlautsprecher *m*
l. set *(Mil)* Horchgerät *n*
l. test *(Fmt)* Verständlichkeitsprüfung *f*
listing *(Text)* Webkante *f*, Salleiste *f*
l. calculator druckende Rechenmaschine *f*
l. heavily *(Schiff)* mit schwerer Schlagseite
liter *(Am) s.* litre
litharge Bleiglätte *f*, Blei[mon]oxid *n*
lithia Lithiumoxid *n*
lithium Lithium *n*
l. base grease *(Kfz)* Lithiumfett *n*
litho screen *(Typ)* Offsetraster *m*
l. stone *(Typ)* Druckstein *m*, Lithografiestein *m*
lithochrome *s.* l. printing process
l. printing process *(Typ)* Öldruckverfahren *n*
lithographic lithografisch
l. colour Plakatfarbe *f*
l. paper Lithografenpapier *n*

l. stone Lithografenkalk *m*, Lithografenschiefer *m*
lithography Lithografie *f*
litmus Lackmus *n*
l.-paper Lackmuspapier *n*
litre Liter *n*
litter feste Abfallstoffe *mpl*, Kehricht *m*; Schüttstroh *n*
little tapering vollholzig
litzendraht wire HF-Litze *f*, Litzendraht *m*
livable bewohnbar *(Haus)*; ertragbar *(Klima)*
live 1. *(El)* stromführend, spannungsführend; 2. *(Masch)* mitlaufend, drehend
l. axle Antriebsachse *f*
l. cartridge scharfe Patrone *f*
l. centre *(Masch)* mitlaufende Körnerspitze *f*
l. conductor spannungsführender (stromführender) Leiter *m*
l. fence Hecke *f*
l.-front switchboard Schalttafel *f* mit spannungsführenden Teilen auf der Vorderseite
l. knot gesunder (eingewachsener) Ast *m*
l. load Auflast *f*, Nutzlast *f*, Verkehrslast *f*
l. matter *(Am) (Typ)* Stehsatz *m*
l. ore Schütterz *n*
l. part spannungsführender Teil *m*
l. pass Arbeitsstich *m*, Arbeitskaliber *n*
l. projectile scharfes Geschoß *n*
l.-roller table Abfuhrrollgang *m*
l. soaking pit beheizter Tiefofen *m*
l. steam Kesseldampf *m*, Frischdampf *m*
l. steam evaporator frischdampfbeheizter Verdampfer *m*
l. stock Viehbestand *m*
l. wire unter Spannung stehende Leitung *f*
l. work lebendes Werk *n (Schiff unter Wasserlinie)*
liveliness Heftigkeit *f (eines Reaktionsablaufes)*
lively boat starkes (kräftiges) Boot *n*
liver-boiling plant *s.* l. plant
l. oil Lebertran *m*
l. plant Trankochanlage *f*, Lebertrankochanlage *f (auf einem Fischereifahrzeug)*
l. pyrites *(Min)* Leberkies *m*
livering Gelieren *n*, Verdicken *n*
livery Amtstracht *f*, Livree *f*
livestock farming Viehhaltung *f*
living caravan Wohnwagen *m*
l. density Wohndichte *f*
l. hut Wohnbaracke *f*
l. knot eingewachsener Ast *m*
l. quarters Wohnunterkunft *f*
l. room Wohnzimmer *n*
l. space Wohnfläche *f*
l. unit Wohnungseinheit *f*
lixiviate / to [aus]laugen, extrahieren
lixiviation Auslaugung *f*
lixivium Absud *m*, Auszug *m*, Elixier *n*
Lloyd's Register of Shipping Lloyds Schiffsregister *n*
Lloyd's Rules and Regulations *(Schiff)* Lloyds Bauvorschrift *f*
L.N.G. tanker *s.* liquefied natural gas tanker
load / to 1. belasten, beanspruchen; aufbringen; einführen; zuführen; beschicken; auflegen; laden; 2. *(Brg)* [weg]laden, [weg]füllen
l. a cable *(Fmt)* ein Kabel pupinisieren
load Beanspruchung *f*, Belastung *f*, Last *f*; Beladung *f*; Ladung *f*; Beschickung *f*, Charge *f*, Füllmaterial *n*, Füllung *f*
l.-and-fire mechanism federgespannter Auslösemechanismus *m*
l. and unload mechanism Zu- und Abführeinrichtung *f (Werkzeugmaschine)*
l.-bearing tragend
l.-bearing capacity Tragfähigkeit *f*, Tragkraft *f (Ausbau)*
l.-bearing capacity of soils Tragfähigkeit *f* des Baugrundes
l. brake *(Kfz)* Bremsbelastung *f*
l.-break current rating *(El)* Nennlastausschaltstrom *m*
l.-break cutout *(El)* Lastausschalter *m*
l. capacity 1. *(Bw)* aufnehmbare Last *f*, Tragfähigkeit *f*, Tragkraft *f*, Tragvermögen *n*, Belastbarkeit *f*; 2. *(Masch)* Leistungsaufnahme *f*
l. capacity of the tyre cover *(Kfz)* Tragfähigkeit *f* der Luftreifendecke
l. carrying tragend
l.-carrying ability Tragfähigkeit *f*
l.-carrying capacity 1. aufnehmbare Last *f*, Tragfähigkeit *f*, Tragkraft *f*, Tragvermögen *n*; 2. *(El)* Belastbarkeit *f*, Strombelastung *f*
l.-carrying capacity of a bearing *(Masch)* Belastbarkeit *f* eines Lagers
l.-carrying properties Tragfähigkeit *f*, Druckaufnahmefähigkeit *f (eines Ölfilms)*
l. cell Druckmeßdose *f*
l. characteristic *(El)* Belastungskennlinie *f*, Lastkennlinie *f*
l. circuit Belastungs[strom]kreis *m*, Entnahmekreis *m*, Lastkreis *m*, Verbraucherkreis *m*
l.-circuit power input Eingangsleistung *f* des Belastungskreises
l. current Belastungsstrom *m*
l. cycle Lastspiel *n*
l. deflection curve for spring Federkennlinie *f*, Federdiagramm *n*
l.-deflection diagram Durchbiegungsdiagramm *n*
l. diagram Belastungsdiagramm *n*, Belastungsschaubild *n*
l. distribution *(El)* Lastverteilung *f*
l. draught *(Schiff)* Ladetiefgang *m*
l. error Belastungsfehler *m (Analogrechner)*

l.-extension diagram *(Met)* Zerreißdiagramm *n*
l. factor Lastfaktor *m*; Sicherheitsbeiwert *m* *(Traglastverfahren)*
l.-factor method n-freies Bemessungsverfahren *n*, Traglastverfahren *n* *(Stahlbetonberechnung)*
l. fluctuation Lastschwankung *f*, Belastungsänderung *f*
l. governing characteristic *(BMSR)* Netzkennlinie *f* *(Netzregelung)*
l. governor Leistungsregler *m*
l. hook Lasthaken *m*
l. impedance *(El)* Belastungsimpedanz *f*
l. increment Lastzunahme *f*; Laststufe *f*
l. level Beanspruchungshöhe *f*
l. levelling relay Lastbegrenzungsrelais *n*
l. limit Grenzbelastung *f*
l.-limiting resistor Strombegrenzungswiderstand *m*
l. line 1. *(Schiff)* Lade[wasser]linie *f*, Tiefladelinie *f*; 2. *(El)* Widerstandsgerade *f*, Belastungskennlinie *f*
l.-line mark *(Schiff)* Freibordmarke *f*, Lademarke *f*, Tieflademarke *f*
l. losses stromabhängige Verluste *mpl*
l. moment Lastmoment *n*
l. on an axle Achsdruck *m*, Achslast *f*
l. on the valve Ventilbelastung *f*
l. peak *(El)* Belastungsspitze *f*
l. position Laststellung *f*
l. rating Lastwert *m*
l. ratio control *(El)* Lastregelung *f*
l. resistance Belastungswiderstand *m*
l. reversal *(Eis)* Lastwechsel *m* *(Bremse)*
l. rheostat *(El)* Belastungswiderstand *m*
l. scheme Belastungsfall *m*, Lastfall *m*; Belastungsbild *n*, Lastbild *n*
l. sensitivity Lastabhängigkeit *f*, Lastempfindlichkeit *f*, Störverhalten *n* bei Laststörungen
l.-sharing concept Lastverteilungsprinzip *n*
l.-shedding *(Brg)* ruckweises Nachgeben *n* *(Ausbau)*
l. spitter *(Brg)* Zündrakete *f*
l. static error P-Abweichung *f*, Proportionalabweichung *f*, Regelabweichung *f* unter Störgrößeneinwirkung
l.-tensioning Spannen *n* mittels Gewichten *(Spannbeton)*
l. test Belastungsversuch *m*; Lastprobe *f*, Probebelastung *f*
l. test machine Belastungsprüfmaschine *f*
l. variable *(BMSR)* Störgröße *f*, Laststörung *f*
l. voltage *(El)* Ausgangsspannung *f*, Spannung *f* unter Last, Arbeitsspannung *f*
l. water-line *(Schiff)* Ladewasserlinie *f*
l.-water-line coefficient *(Schiff)* Wasserlinienvölligkeitsgrad *m*

loadability Belastbarkeit *f*
loaded aerial belastete Antenne *f*
l. cable bespultes Kabel *n*
l. circuit *(El)* pupinisierter (bespulter) Kreis *m*
l. impedance *(El)* Eingangsimpedanz *f* bei Sollabschluß
l. mat harzbesprühte Matte *f*
l. Q *(Fmt)* Belastungsgüte *f*
l. roving harzbesprühte Matte *f*
l. tub *(Brg)* Vollwagen *m*
loader 1. Ladegerät *n*, Lademaschine *f*, Lader *m*, Verladegerät *n*; 2. *(Mil)* Ladeschütze *m*; 3. *(Masch)* Ladeeinrichtung *f*, Magazin *n*
l. combine Mähhlader *m*
l. gate *(Brg)* Ladestrecke *f*
loading 1. Laden *n*, Aufladen *n*, Beladen *n*; 2. Werkstückzuführung *f*; 3. Verschmieren *n* der Schleifscheibe; 4. Beanspruchung *f*, Belastung *f*; 5. Einlegen *n* *(Film)*; 6. *(Text)* Beschweren *n*; Erschweren *n*, Erschwerung *f*; 7. Füllmaterial *n*, Füllstoff *m*, Beschwerungsmaterial *n*
l. agent Beschwerungsmittel *n*, Zusatz *m*
l. agents on fillers Appreturkörper *mpl* *(Papier)*
l. and unloading device Lade- und Entladeeinrichtung *f*
l. and unloading station *(Masch)* Spann- und Entspannstation *f*
l. and unloading time Ein- und Ausspannzeit *f* *(Werkzeugmaschine)*
l. berth Ladeplatz *m* *(für Schiffe)*
l. brake *(Kfz)* Auflaufbremse *f*
l. bridge Verladebrücke *f*
l. capacity Tragfähigkeit *f*, Tragkraft *f*; Ladefähigkeit *f*
l. case Lastfall *m*, Belastungsfall *m*
l. cavity Füllraum *m* *(Kunststoffverarbeitung)*
l. chamber Füllraum *m* *(Kunststoffverarbeitung)*
l. chute Verladeschurre *f*
l. coil *(El)* Pupinspule *f*
l. components *(Masch)* Aufsetzen (Einfüllen, Zuführen) *n* von Werkstücken
l. cycle Lastspiel *n*, Belastungsverlauf *m*
l. device *(Masch)* Ladeeinrichtung *f*, Zuführungseinrichtung *f*, Einlegeeinrichtung *f*
l. diagram Belastungsbild *n*, Lastbild *n*
l. equipment Ladeeinrichtung *f*, Beschickungseinrichtung *f*, Werkstückzubringungseinrichtung *f*
l. gauge Lademaß *n*
l. gear Ladegeschirr *n*
l. grab Verladegreifer *m*
l. harvester Mählader *m*
l. hatch Dachluke *f*, Ladeluke *f*
l. hopper Fülltrichter *m*, Ladetrichter *m*, Schüttrumpf *m*; Verladetrichter *m*

l. **implement** Verladegerät *n*
l. **in bulk** *(Schiff)* Verladung *f* als Massengut
l. **pigment** Pigmentfüllstoff *m*
l. **platform** Laderampe *f*, Rampe *f*, Verladebrücke *f*
l. **pocket** Fülltasche *f (Skipförderung)*
l. **point** 1. *(Ch)* Staupunkt *m (Füllkörperkolonne)*; 2. *(Am)* Ladeplatz *m*, Ladestelle *f*
l. **position** *(Masch)* Ladestellung *f*, Aufspannstellung *f*
l. **ramp** Verladebrücke *f*
l. **resistance** *(El)* Ballast[widerstand] *m*
l. **shovel** Frontlader *m*; Ladeschaufel *f*; Schaufellader *m*; Schaufler *m*
l. **slot** Füllnut *f*
l. **station** 1. Ladepunkt *m*, Beladestelle *f*, Ladestelle *f*; 2. *(Masch)* Ladestation *f*, Spannstation *f (Taktstraße)*
l. **the machine** Beschickung *f* der Maschine
l. **the work on** Auflegen *n* des Werkstücks auf
l. **time** Umspannzeit *f (Werkstück, Werkzeug)*; Ladedauer *f*, Ladezeit *f*; Belastungszeit *f*
l. **velocity** *(Ch)* Gasdurchsatz *m (Absorption)*

loadstone *s.* lodestone
loaf / to *(Masch)* leerlaufen
loam Lehm *m*, Letten *m*, Ziegelton *m*, Tonmergel *m*
l. **board** Drehbrett *n*
l. **casting** *(Met)* Lehmguß *m*, Gießen *n* in Lehmform
l. **cutter** Lehmschneider *m*
l. **luting** Lehmdichtung *f*
l. **mortar** Lehmmörtel *m*
l. **mould** Lehmform *f*
l. **moulding** Lehmformerei *f*
l. **pit** Lehmgrube *f*
l. **pugger** Lehmkneter *m*

loaming *(Geol)* Sichern *n*
loamy lehmig; lehmhaltig
l. **marl** Lehmmergel *m*
l. **sand** *(Met)* Klebsand *m*, Formsand *m*
l. **soil** lehmiger Boden *m*

loanable ausleihbar
lobby Diele *f*, Vestibül *n*, Vorhalle *f*
lobe 1. Flügel *m*, Lappen *m*; 2. Strahlungskeule *f*, Keule *f*; Zipfel *m (Antennentechnik)*
l. **of cam** Nase *f* des Nockens, Nocken *m*
l.**-type compressor** *(Kfz)* Kapselgebläse *n*, Rollkolbenkompressor *m*
l. **width** Strahlungslappenbreite *f (Antennentechnik)*

local lokal, örtlich
l. **action** örtliche (lokale) Korrosion *f*
l. **area** *(Fmt)* Ortsverkehrsbereich *m*
l. **battery** Ortsbatterie *f*, Lokalbatterie *f*
l. **cable** Ortskabel *n*
l. **call** Ortsgespräch *n*
l. **cell** Lokalelement *n*
l. **channel** *(Fmt)* Sendekanal *m* bis 250 W *(nach FCC)*
l. **echo attenuation** Nahechodämpfung *f (Radar)*
l. **exchange network** *(Fmt)* Ortsnetz *n*
l. **forestry administration** Forstamt *n*
l. **hardening** Teilhärtung *f*
l. **irradiation** *(Kern)* Lokalbestrahlung *f*
l. **lighting** Arbeitsplatzbeleuchtung *f*
l. **line** Ortsleitung *f*
l. **line selector** Ortsleitungswähler *m*
l. **message** *(Am)* Ortsgespräch *n*
l. **oscillator** Empfangsoszillator *m*, Überlagerungsoszillator *m*, Empfängeroszillator *m*; *(El)* Hilfsoszillator *m*
l. **position** *(Fmt)* Ortsplatz *m*
l. **power** lokale (örtliche) Leistung *f*
l. **railway** Lokalbahn *f*
l. **range** *(El)* Nahbereich *m*
l. **receiving** Ortsempfang *m*
l. **reception** Ortsempfang *m*
l. **sale** *(Brg)* Landabsatz *m*, örtlicher Absatz *m*
l. **sender** Ortssender *m*; Bezirkssender *m*
l. **sidereal time** örtliche Sternenzeit *f*
l. **singing** *(Fmt)* Pfeifen *n (Selbsterregung eines Verstärkers)*
l. **station** Ortssender *m*
l. **telephone network** Ortsleitungsnetz *n*
l. **time** Ortszeit *f*
l. **transit** Nahverkehr *m*, Ortsverkehr *m*
l. **velocity of current** *(Schiff)* Anströmgeschwindigkeit *f*

locality 1. Ort *m*, Örtlichkeit *f*; 2. *(Geol)* Lagerstätte *f*
localization Lokalisieren *n*. Lokalisation *f*, Eingrenzung *f* von Fehlerquellen
l. **of defects** Fehlerbestimmung *f*
l. **of faults** *s.* l. of defects

localized örtlich begrenzt; ortsgebunden *(Vektor)*
l. **bearing** balligtragend
l. **corrosion** lokale (örtliche) Korrosion *f*
l. **tooth bearing** Balligtragen *n (Zähne)*
l. **vector** ortsgebundener Vektor *m*

localizer Ortsleitstrahlsender *m*; Ansteuerungsfunkfeuer *n*; Landekurssender *m*
l. **beam** Funkleitstrahl *m (Radar)*
l. **beam method** Leitstrahlverfahren *n*

locally busy *(Fmt)* ortsbesetzt
locate / to plazieren, anordnen, anbringen; orten *(Fehler)*; in Stellung bringen, einstellen; lagern
l. **a well** eine Bohrung ansetzen
l. **by pins** anstiften *(annageln)*
l. **centrally** auf Mitte stellen, zentrieren

locater beacon Nahentfernungsmeßbake *f (Radar)*
locating a hole Einstellung *f* der Bohrlage

l. defects Fehlerortsbestimmung *f*
l. pin Paßstift *m*; Fixierstift *m*, Anlegestift *m*
l. plunger Raststift *m* der Teilscheibe
l. ring *(Kfz)* Stellring *m*, Zentrierring *m*
l. surface Lagerfläche *f*, Auflagefläche *f*, Anlagefläche *f*
location 1. Stellung *f*, Lage *f*, Standort *m*; Einortung *f*; 2. *(Am)* Aufspannung *f* *(Werkzeuge)*; 3. *(Fs)* Speicherstelle *f*; 4. *(Brg)* Mutung *f*
l. of the defect Fehl[er]stelle *f*
l. shifting Verschiebung *f* der Informationsspeicherstelle
locator *(Flg)* Anflugfunkfeuer *n*
l. pin Führungsstift *m* *(Elektronenröhren)*
lock / to 1. *(Masch)* verriegeln *(Revolver)*; feststellen, festklemmen; sperren; sichern; [auf]spannen *(Werkzeug)*; 2. *(Schiff)* durchschleusen
l. in on einrasten auf
l. in place *s.* l. in position
l. in position festklemmen, feststellen, verriegeln, blockieren, arretieren; einrasten
l. the bolt Kolbendeckelschraube sichern
l. up verriegeln
lock 1. *(Masch)* Riegel *m*, Schloß *n*, Sperre *f*, Verschluß *m*; Verriegelung *f*; 2. *(Schiff)* Schleuse *f*, Kanalschleuse *f*, Kammerschleuse *f*, Dockschleuse *f*
l. angle *(Kfz)* Einschlagwinkel *m* *(Räder)*
l. bar Sperrschiene *f* *(Signal)*
l. bolt Sperriegel *m*, Sicherungsschraube *f*, Verriegelungsbolzen *m*
l. chamber Schleusenkammer *f*
l. die Gesenk *n* mit gebrochener Teilfuge
l. dowel Sicherungsstift *m*
l. gate Schleusentor *n*
l.-in amplifier *(BMSR)* Lock-in-Verstärker *m*
l.-in range *(El)* Mitnahmebereich *m*
l.-in relay selbsthaltendes Relais *n*
l.-joint dividers feststellbarer Meßzirkel (Spitzenzirkel) *m*
l. keeper Schleusenmeister *m*
l. knob Einrastknopf *m*, Verschlußknopf *m*
l. lever Arretierhebel *m*, Feststellhebel *m*; Riegelhebel *m* *(Signal)*
l. line Riegelleitung *f* *(Signal)*
l. nut Sicherungsmutter *f*, Gegenmutter *f*, Kontermutter *f*; Klemmutter *f* *(Schraublehre)*; Ringklemme *f*
l. out *(Fmt)* Leitungsblockierung *f*
l.-out valve Doppelschloßventil *n*
l. pawl Sicherungsklinke *f*, Sperrklinke *f*
l. pin *s.* locking pin
l. plate Sicherungsblech *n*
l. plunger Sperrklinke *f*, Sperrkolben *m*
l. ring Verschlußring *m*, Sicherungsring *m*
l.-up pawl Sperrklinke *f*
l. washer Federring *m*, Sicherheitsring *m*, Verschlußscheibe *f*; Sprengring *m*; Sicherungsscheibe *f*, Sicherungsblech *n*
l. wire Arretierungsring *m*; Arretierungsdraht *m*
locked coil rope patentverschlossenes Seil *n* *(Fördertechnik)*
l.-cover switch verschlossener Schalter *m*, Schalter unter Verschluß
l. octal tube Achtelektrodenröhre *f* mit Sockelverriegelung *f* *(Loktal-Röhre)*
l. rotor torque *(El)* Drehmoment *n* bei festgebremstem Läufer
l.-up stress Restspannung *f*
locker Schrank *m*, Kleiderschrank *m*, Spind *n*
lockhandle Spannhebel *m*
locking 1. Festklemmen *n*, Klemmen *n*, Klemmung *f*; Verriegeln *n*, Verriegelung *f*; Arretierung *f*; 2. *(Fs)* Netzsynchronisation *f*, Rasterverriegelung *f*
l. arrangement Festklemmeinrichtung *f*
l. arrangement securing sleeve to pillar Auslegerklemmung *f* *(Radialbohrmaschine)*
l. bolt Arretierung *f*; Haltebolzen *m*, Verriegelungsbolzen *m*, Verschlußbolzen *m*
l. brake Feststellbremse *f*
l. bushing Klemmhülse *f*, Klemmbuchse *f*
l. cabin Stellwerksbude *f*
l. cam Schließnocken *m*
l. cap Verschlußkappe *f*
l. clamp Verschlußklammer *f*
l. coupling Verschlußkupplung *f*
l. device Verriegelung *f*; Feststelleinrichtung *f*, Arretierung *f*, Arretiervorrichtung *f*
l. disk Arretierungsscheibe *f*, Rastscheibe *f*
l. fastener Sicherung *f* *(für Muttern)*
l. frame in signal tower Turmstellwerk *n*
l. gripping device Greifer *m* mit Sperrklinke, Schloßkralle *f*
l. handle Knebelgriff *m*
l.-in *(El)* Synchronisierung *f* *(durch Mitziehwirkung)*
l. key *(El)* Sperrtaste *f*, Taste *f* mit fester Stellung
l. lamp-holder *(El)* diebstahlsichere Fassung *f*
l. mechanism Verriegelung *f*
l. nut Sicherungsmutter *f*
l. pawl Sicherungsklinke *f*
l. pin Haltebolzen *m*, Riegelbolzen *m*, Sperrbolzen *m*, Verriegelungsstift *m*
l. plant Stellwerksanlage *f*
l. plate Sicherungsplatte *f*, Sicherungsscheibe *f*, Klemmplättchen *n*; Sicherungsblech *n*; *(Lw)* Scheuerplatte *f* *(Wagen)*
l. position *(Eis)* Sperrlage *f* *(Signal)*
l. ratchet Sperrklinke *f*
l. relay *(El)* Halterelais *n*

l. ring Klemmring *m*, Spannring *m*, Sperring *m*, Segerring *m*; Schloßring *m (Turbine)*
l. screw Spannschraube *f*, Klemmschraube *f*, Arretierschraube *f*
l. spring Haltering *m*, Sicherungsring *m*
l. switch Verriegelungsschalter *m*, Schalter *m* mit Verriegelung (Sperre); Verschlußschalter *m*
l. voltage *(El)* Mitnahmespannung *f*
l. washer Sicherungsscheibe *f*, Sicherungsring *m*
l. wedge Schloßkeil *m*
l. yoke Blockiergabel *f*
lockplate Sicherungsplatte *f*
locksmith Schlosser *m*
locksmith's shop Schlosserei *f*
l. trade Schlosserhandwerk *n*
lockwasher Unterlegscheibe *f*
loco garage *(Brg)* Lokomotivkammer *f*
l.-type boiler Lokomotivkessel *m*
locomobile Lokomobile *f*
locomotion Fortbewegung *f*; Zugförderung *f*
locomotive Lokomotive *f*
l. drive Lokomotivantrieb *m*
l. engine Lokomotive *f*
l. fire box Lokomotivfeuerkiste *f*, Lokomotivstehkessel *m*
l. furnace Feuerbüchse *f*
l. haulage Zugbetrieb *m*, Zugförderung *f*
l. propulsion Lokomotivantrieb *m*
l. reamer Nietlochreibahle *f*
l. shed Lokomotivschuppen *m*
l. steam engine Lokomotivdampfmaschine *f*
l. stock Lokomotivpark *m*
loctal tube *(El)* Allglasröhre *f* mit 8 Stiften, Loktalröhre *f*
locus 1. geometrischer Ort *m*; 2. *(BMSR)* Ortskurve *f*, Hodograf *m*, Trajektorie *f*
l.-dependency Ortsabhängigkeit *f*
l.-dependent ortsabhängig
l. of instantaneous centres *(Masch)* Polbahn *f*, Polkurve *f*
locust Heuschrecke *f*
lode *(Geol)* Erzgang *m*, Gang[zug] *m*; Deich *m*; Wasserlauf *m*
lodestone Magneteisenerz *n*, Magneteisenstein *m*, Gangerz *n*
lodge / to [sich] umlegen
lodged corn *(niedergeregnetes)* Lagerkorn *n*
loess Löß[boden] *m*
l. clay Lößlehm *m*
l. deposit Lößvorkommen *n*
l. formation Lößbildung *f*
l. soil Lößboden *m*
loft Boden[raum] *m*, Dachboden *m*, Trockenboden *m*; Speicher *m*
l. aerial Dachbodenantenne *f*
l. drying Trocknung *f* im Trockenhaus
lofting *(Schiff)* Schnürbodenarbeit *f*
loftsman *(Schiff)* Schnürbodenarbeiter *m*
log / to 1. Bäume fällen, durchforsten, entästen; 2. aufzeichnen, in eine Tabelle eintragen *(Daten)*
log 1. Logarithmus *m*; 2. Schiffstagebuch *n*, Logbuch *n*; 3. Holzklotz *m*, Block *m*, Stammholz *n*, Langholz *n*; 4. Log *n (Schiffsgeschwindigkeit)*
l. book Logbuch *n*, Schiffstagebuch *n*
l. cabin Blockhaus *n*
l. calipers *(Holz)* Gabelmaße *npl*
l. carrying bogie wagon Langholztransportwagen *m*
l. clean grown langschäftiges Holz *n*
l. deck Stammholzboden *m*
l. driver Flößer *m*
l. end Blockende *n*, Ausschnittende *n*, Stammabschnitt *m*
l. frame for deep cutting Hochgatter *n*
l. frame saw Sägegatter *n*
l. frame with rag carriage feed Schlittenvorschubgatter *n*
l. frame with roller feed Walzengatter *n*
l. frame with tension apparatus Spannwerkvollgatter *n*
l. handling Holztransport *m*
l. hauls Holzabfuhr *f*
l. hoist Stammwinde *f*
l. house Blockhaus *n*
l. line *(Schiff)* Logleine *f*
l.-log doppeltlogarithmisch
l.-log paper doppeltlogarithmisches Papier *n*
l.-log scale doppeltlogarithmischer Maßstab *m*
l.-magnitude and phase diagram *(BMSR)* logarithmischer Amplituden- und Phasengang *m*, Bode-Diagramm *n*
l.-magnitude-angle diagram *(BMSR)* Nichols-Diagramm *n*
l. measure Block[rund]maß *n*, Brettmaßstab *m (in Schneidemühlen)*
l. pond Holzteich *m*
l. raft Holzfloß *n*
l. reel *(Schiff)* Logrolle *f*
l. saw Blocksäge *f*
l. scale logarithmischer Maßstab *m*
l. scaling rules Langholzmeßstab *m*
l. timber Rundholz *n*
l. wall Blockwand *f*
logan *(Schiff)* Wrackgut *n*, versenktes Gut *n*
logarithm Logarithmus *m*
l. integral Integrallogarithmus *m*
l. to the base 10 Briggscher Logarithmus *m*
logarithmic logarithmisch
l. amplifier logarithmischer Verstärker *m (Analogrechner)*
l. capacitor *(Am)* logarithmischer Kondensator *m*, Mittellinienkondensator *m*

l. condenser *s.* l. capacitor
l. decrement logarithmisches Dekrement *n*
l. table Logarithmentafel *f*
l. tacheometric wedge *(Opt)* logarithmischer Tachymeterkeil *m*
logatom *(Ph)* Logatom *n*
logger 1. Heringslogger *m* *(Fischereifahrzeug)*; 2. *(Dat)* Mitschreiber *m*
logging 1. Holzfällen *n*; 2. *(Dat)* Registrieren *n*, Ausdrucken *n*, Mitschreiben *n*, Niederschreiben *n*, Notieren *n*
l. cycle *(Dat)* Ausdruckzyklus *m*, Ausdruckfolge *f*
l. of stations Feststellung *f* von Sendern *(z. B. durch Peilung)*
logic Logik *f*
logic *s.* logical
l. AND circuit logische UND-Schaltung *f*
l. circuit logische Schaltung *f*
l. module Logikmodul *m* *(Modulbauweise erleichtert das logische Programmieren)*
l. OR circuit logische ODER-Schaltung *f*
logical logisch
l. addition logische Addition *f*, Disjunktion *f* *(binäre Addition ohne Übertrag)*
l. circuit logischer Schaltkreis *m*, logische Schaltung *f*
l. comparison [logischer] Vergleich *m* *(programmgesteuerte Digitalrechner)*
l. conclusion logischer Schluß *m*
l. connective logische Verknüpfung *f*
l. constant logische Konstante *f* *(z. B. Konjunktion, Disjunktion, Verneinung)*
l. design logischer Entwurf *m*, Entwurf zum logischen Vergleich
l. machine logische Maschine *f*
l. majority function logische Majoritätsfunktion *f*
l. operation logische Operation *f*
l. piano *(Bezeichnung der ersten logischen Maschine)*
l. product logisches Produkt *n*, Konjunktion *f*
logistic logistisch
l. computer Rechner *m* für logistische Probleme *(Militärwissenschaft)*
logistics Logistik *f*
logs cut square-edged Kantholz *n*
l. sawn through and through unbesäumte Blockware *f*
long lang; Lang-
l. barrel staves langes Faßholz *n*
l. bearing langes Tragbild *n* *(Verzahnung)*
l.-boat Barkasse *f*, Großboot *n*
l.-boled langschäftig *(Holz)*
l. butt needle *(Text)* Hochfußnadel *f*
l.-chain molecule langkettiges Molekül *n*
l.-chip material langspanender Werkstoff *m*
l.-chord winding *(El)* übersehnte Wicklung *f*
l. continuous chip Langspan *m*
l. counter stern *(Schiff)* löffelförmiges Heck *n*
l.-cracking langbrechend
l.-delay echo Langzeitecho *n* *(Radar)*
l.-distance Fernverkehrs-, Langstrecken-, Weit-
l.-distance beam *(Kfz)* Weitstrahler *m*
l.-distance cable Fernkabel *n*
l.-distance call 1. Auslandsgespräch *n*; 2. *(Am)* Ferngespräch *n*
l.-distance circuit *(El)* Fernleitung *f*, Weitverkehrsleitung *f*
l.-distance communication *(Fmt)* Fernverkehr *m*
l.-distance connection *(El)* Weitverkehrsverbindung *f*
l.-distance control Fernsteuerung *f*, Fernlenkung *f*
l.-distance dialling *(Fmt)* Selbstwählfernverkehr *m*; Fernwahl *f*
l.-distance express train FD-Zug *m*, Fern-D-Zug *m*
l.-distance gas Ferngas *n*
l.-distance gas grid Ferngasnetz *n*
l.-distance gas main Ferngasleitung *f*
l.-distance gas pipeline Ferngasleitung *f*
l.-distance gas system Ferngasnetz *n*
l.-distance ignition Fernzündung *f*
l.-distance irradiation Fernbestrahlung *f*
l.-distance light *(Kfz)* Fernlicht *n*
l.-distance line *(Fmt)* Fernleitung *f*
l.-distance network *(El)* Weitverkehrsnetz *n*, Fernleitungsnetz *n*
l.-distance operation Fernbedienung *f*
l.-distance radio station Großfunkstelle *f*
l.-distance reception Fernempfang *m*
l.-distance reconnaissance *(Mil)* Fernaufklärung *f*
l.-distance road haulage Fernlastverkehr *m*, Überlandverkehr *m*
l.-distance scatter Entfernungsstreuung *f* *(elektromagnetische Wellen)*
l.-distance section *(Fmt)* Fernschrank *m*
l.-distance selection *(Fmt)* Fernwahl *f*
l.-distance station Großsender *m*
l.-distance telephone call Ferngespräch *n*
l.-distance telephony Fernsprechweitverkehr *m*
l.-distance thermometer Fernthermometer *n*
l.-distance trip *(Schiff)* Fernfahrt *f*
l.-duration echo Langzeitecho *n* *(Radar)*
l.-duration test Dauerversuch *m*
l.-fibred pulp Langschliff *m* *(Papierherstellung)*
l.-flame gas coal Gasflammkohle *f*
l.-flaming langflammig
l.-flaming coal langflammige Kohle *f*
l. focal langbrennweitig

l.-focus langbrennweitig
l.-focus lens Linse *f* mit großer Brennweite, Langfokuslinse *f*
l. grate Langrost *m (Feuerung)*
l.-handed tool Fernbedienungsgerät *n (Gerät mit verlängertem Griff)*
l.-horn *(Brg)* Abbau *m* mit weniger als 45° zu den Schlechten gestelltem Stoß
l.-legged vessel Schiff *n* mit großem Tiefgang (Aktionsradius)
l.-length twist drill langer Spiralbohrer *m*
l.-life 1. mit großer Standzeit; 2. *s.* l.-lived
l. life and ability to be resharpened many times lange Standzeit *f* und Lebensdauer *f*
l.-life tube Langlebensdauerröhre *f*
l. line *(Schiff)* Langleine *f*
l.-line method Spannbahnfertigung *f*, Spannbettverfahren *n (Spannbetonbau)*
l.-line system *s.* l.-line method
l.-lived *(Kern)* mit großer Halbwertszeit, langlebig
l.-lived material langlebige Substanz *f*
l.-lived radiation langlebige Strahlung *f*
l. oil-type varnish fetter Öllack *m*
l.-path gas cell Langweggasküvette *f*
l. persistence oscilloscope lang nachleuchtender Oszillograf *m*, Oszillograf mit langer Nachleuchtdauer
l.-pitch winding *(Elm)* Wicklung *f* mit verlängertem Schritt
l.-playing record Langspielplatte *f*
l.-playing tape Langspielband *n*
l. primer *(Typ) (etwa)* Borgis *f (Schriftgröße)*
l.-range 1. von großer Reichweite, weitreichend; 2. fernwirkend; Fern[wirkungs]-
l.-range accuracy radar system Langstreckenmeßverfahren *n*, Lorac-System *n (Radar)*
l.-range artillery Fernartillerie *f*
l.-range bomber Fern[strecken]bomber *m*, Fernkampfflugzeug *n*, strategisches Bombenflugzeug *n*
l.-range force Fernwirkungskraft *f*
l.-range navigation Langstreckennavigation *f*, Funkfernnavigationsverfahren *n*, Weitstreckennavigation *f*
l.-range particle *(Kern)* Teilchen *n* großer Reichweite, weitreichendes Teilchen
l.-range proving ground Versuchsgelände *n* für Langstreckenraketen
l.-range radar Langstreckenradar *n*
l.-range radar device Vorwarnfunkmeßgerät *n*, Langstreckenradargerät *n*
l.-range rocket Langstreckenrakete *f*
l.-range yacht Langstreckenjacht *f*
l. residue Toprückstand *m*
l.-run job *(Am) s.* l.-run repetition work
l.-run repetition work große Serie *f*, Serie mit großer Stückzahl
l. saw file Brettsägefeile *f*
l. screw double plane iron Doppelhobeleisen *n* mit langer Schraube
l. shot *(Fs)* Gesamtaufnahme *f*
l. shunt *(El)* langer Nebenschluß *m*
l.-sighted weitsichtig, übersichtig
l. slot burner Langlochbrenner *m*, Schlitzbrenner *m*
l.-span weitgespannt, mit großer Spannweite
l.-staple cotton langstaplige Baumwolle *f*
l.-staple flocks langstaplige Flockfasern *fpl*
l.-stem nozzle *(Kfz)* Langschaftdüse *f*
l. straw Schüttstroh *n*
l.-stroke langhubig
l.-stroke engine langhubiger Motor *m*, Langhuber *m*
l.-stroke steam engine langhubige Dampfmaschine *f*
l. taper flacher (schlanker) Kegel *m*
l.-term planning langfristige Planung *f*
l.-term trade agreement langfristiges Handelsabkommen *n*
l.-thread Langgewinde *n*
l.-thread peeling machine Langgewindeschälmaschine *f*
l. timber Langnutzholz *n*
l.-timber truck Langholztransportwagen *m*
l.-time behaviour Langzeitverhalten *n*
l.-time memory *(Dat)* Langzeitspeicher *m*
l.-time test Dauerversuch *m*, Langzeitversuch *m*
l. ton englische Tonne *f (1016 kg)*
l. travel motor Kranfahrmotor *m*
l. travelling Kranfahren *n*, Längsfahren *n*
l. tube vertical multiple effect distillation process Mehrstufenverfahren *n (Umwandlungsanlage von Seewasser in Süßwasser)*
l. wave Langwelle *f*
l.-wave coil Langwellenspule *f*
l.-wave transmitter Langwellensender *m*
l. wheel base *(Kfz)* großer Radstand *m*
longevity Dauerhaftigkeit *f*, lange Lebensdauer *f*, Langlebigkeit *f*
longimetry *(geografische)* Längenmessung *f*
longitude *(geografische)* Länge *f*, Längengrad *m*
l. effect *(Ph)* Längeneffekt *m*
longitudinal *(Schiff)* Längsspant *n*, Seitenträger *m*
longitudinal longitudinal, längs[laufend]; Längs-, Längen-
l. axis Längsachse *f*
l. bracing Längsverband *m (Brückenbau)*
l. bulkhead *(Schiff)* Längsschott *n*
l. carrier cable *(El)* Längstragseil *n*
l. centre of buoyancy *(Schiff)* Verdrängungsschwerpunkt *m* der Länge nach

l. chromatic aberration Farbenlängsabweichung *f*, chromatische Längsaberration *f*
l. clearance Durchmesserverjüngung *f (Spiralbohrer)*
l. coefficient *(Schiff)* Völligkeitsgrad *m* der Spantflächenskala, Zylinderkoeffizient *m*
l. conductivity longitudinale Leitfähigkeit *f*
l. control arm *(Kfz)* Längslenker *m*
l. copying attachment Längskopiereinrichtung *f*
l. crack Längsriß *m*
l. dead limit Längenbegrenzung *f*
l. deck beam Deckunterzug *m*
l. deformation Längenänderung *f*
l. diffusion Längsdiffusion *f*, Diffusion *f* in Längsrichtung
l. direction Längsrichtung *f*
l. elevation *(Schiff)* Längenplan *m*, Längsaufriß *m*
l. feed Längsvorschub *m*, Vorschub *m* in Längsrichtung, Längsverschiebung *f*
l. field *(Ph)* longitudinales Feld *n*, Längsfeld *n*
l.-finning längsgerippt
l. flow principle Längsflußprinzip *n (Mähdrescher)*
l. force 1. Längskraft *f*; 2. *(Masch)* Vorschubkraft *f*
l. framing system *(Schiff)* Längsspantenbauweise *f*
l. grinding Längsschliff *m*
l. impact Längsstoß *m*
l. laminated spring Längsblattfeder *f*
l. load Belastung *f* in Richtung der Längsachse
l. magnetization Längsmagnetisierung *f*
l. magnification *(Fs)* Längsvergrößerung *f*
l. member Längsträger *m*
l. motion *(Masch)* Längszug *m*
l. oscillation Längsschwingung *f*, longitudinale Schwingung *f*, Longitudinalschwingung *f*
l. passage *(Masch)* Längskanal *m*
l. raising *(Text)* Längsrauhen *n*
l. reinforcement Längsbewehrung *f (Stahlbetonbau)*
l. relief *(Am) s.* l. clearance
l. rib Längsrippe *f*
l. roll Horizontalwalze *f*
l. section Längsschnitt *m (technisches Zeichnen)*
l. shear Schubkraft *f*; Schubspannung *f*
l. shrinkage *(Holz)* Längsschwindung *f*
l. slide *(Masch)* Längsschlitten *m*
l. spherical aberration *(Opt)* sphärische Längsaberration (Längsabweichung) *f*
l. stop Längsanschlag *m*
l. stress Längsspannung *f*
l. table movement Tischlängsgang *m*
l. tricot *(Text)* Längstrikot *m*
l. turning Bearbeiten *n* von Längsdrehteilen
l. vial Längslibelle *f*
l. vibration longitudinale Schwingung *f*, Längsschwingung *f*, Longitudinalschwingung *f*
l. voltage *(El)* Längsspannung *f*
l. warping *(Holz)* Längskrümmung *f*
l. wave Längswelle *f*, Longitudinalwelle *f*, longitudinale Welle *f*
l. weld Längs[schweiß]naht *f*
longitudinally in Längsrichtung
l. adjustable längsverstellbar
l. corrugated grate bar *(Wärm)* Wellenroststab *m*
longshoreman Hafenarbeiter *m*, Kaiarbeiter *m* *(für Lade- und Löschbetrieb)*
longwall *(Brg)* Streb *m*; Strebbau *m*
l. caving Strebbruchbau *m*
l. coal cutter Schrämmaschine *f* für Strebbau
l. face Streb *m*
l. retreating Strebrückbau *m (zum Schacht hin)*
l. stoping Firstenstoßbau *m*
l. working Strebbau[verfahren *n*] *m*
longwalling *(Brg)* Strebbau *m*
look-out man Ausgucksposten *m*
looking aft *(Schiff)* von vorn gesehen
l. forward *(Schiff)* von hinten gesehen
l. outboard *(Schiff)* von innen gesehen, Ansicht *f* von innen
loom 1. Webstuhl *m*; 2. nichtmetallisches Isolierrohr *n*, Isolierschlauch *m*
l. beam *(Text)* Stuhlbaum *m*
l.-figured eingewebt, während des Webprozesses gemustert
l.-finished *(Text)* stuhlroh
l. fly *(Text)* Stuhlflug *m*
l. for weaving pile fabrics Florwebstuhl *m*
l. harness Webgeschirr *n*
l. shed Webfach *n*
l. slay Lade *f*, Weblade *f*
l.-state fabric *(Text)* Stuhlware *f*
l. waste *(Text)* Fadenabfall *m*, Stuhlabfall *m*
l. width Webbreite *f*
looming *(Text)* Einziehen *n* der Kettfäden in Geschirr und Blatt
loop / to 1. in Schleife legen, zusammenbinden; 2. *(Fmt)* zur Schleife schalten; 3. *(Met)* umführen *(Walzgut)*
l. a line *(El)* eine Leitung schleifen
loop 1. Masche *f*, Schleife *f*, Schlinge *f*, Schlaufe *f*; 2. *(Dat)* Schleife *(Art des Programmablaufs)*; 3. Wellenbauch *m*, Schwingungsbauch *m*; 4. Rahmen[antenne *f*] *m*; 5. *(Text)* ausgebildeter Polhenkel *m*; 6. *(Brg)* Schachtumtrieb *m*; 7. *(Kern)* Kreis[lauf] *m*; 8. *(Met)* Luppe *f*, Eisenluppe *f*
l. aerial Rahmenantenne *f*, Schleifenantenne *f*
l. attenuation *(Fmt)* Schleifendämpfung *f*

l. circuit *(El)* Doppelleitung *f*
l.-cut machine Polschneidemaschine *f*; Polnoppenschneidemaschine *f* *(Teppichherstellung)*
l. direction finder Drehrahmenpeiler *m*
l. drier Girlandentrockner *m*; Schleifentrockner *m*
l. expansion pipe *(Wärm)* Ausgleichschleife *f*
l. formation Schlingenbildung *f*, Durchhängung *f* *(beim Drahtwalzen)*
l. gain *(BMSR)* Kreisverstärkung *f*; *(El)* Rückkopplungsschleifenverstärkung *f*
l. galvanometer Schleifengalvanometer *n*
l. lifter *(Met)* Schlingenheber *m*
l. line Ringleitung *f*
l. of instructions *(Dat)* Befehlsschleife *f* *(im Programm)*
l. of the oscillation Schwingungsbauch *m*, Wellenberg *m*
l. oscillograph Schleifenoszillograf *m*
l.-pile fabric Schlingengewebe *n*, Schlingenware *f*
l. ration *s.* l. gain
l. resistance *(Fmt)* Schleifenwiderstand *m*
l. reversal of flow Schleifenumkehrung *f* *(Turbine)*
l. scavenging *(Kfz)* Umkehrspülung *f*
l. test bridge *(El)* Schleifendrahtbrücke *f*
l. thread *(Text)* Schlingenfaden *m*
l. winding Schleifenwicklung *f*, Scheitelwicklung *f*, überlappte Wicklung *f*
l. yarn *(Text)* Schleifengarn *n*, Schlingengarn *n*
looper 1. Schlingenkanal *m*, Schlingenspanner *m*, Umwalzer *m* *(Walzwerk)*; 2. *(Text)* Greifer *m* *(Nähmaschine)*; Kettelmaschine *f*
l. of wire rolling mill Schlingenspanner *m* im Drahtwalzwerk
loophole Schießscharte *f*
looping mill *(Met)* Umsteckwalzwerk *n*
l. roller *(Met)* Umwalzer *m*
l.-type rolling mill Umsteckwalzwerk *n*
loose 1. locker; abnehmbar; 2. lose; ungebunden *(Korn)*; 3. schlaff *(Seil)*; 4. mit totem Gang *(z. B. in der Lenkung)*
l. cargo lose (bewegliche) Ladung *f*
l.-ceiling *(Schiff)* lose Wegerung *f*
l. change gear Umsteckrad *n*
l. connection Wackelkontakt *m*
l. coupling *(El)* lose (unterkritische) Kopplung *f*; Schaltkupplung *f*
l. fit *(Masch)* Spielpassung *f*, Spielsitz *m*
l.-footed roadway arch *(Brg)* Streckenbogen *m* mit abnehmbarem Fußteil
l. gear Wechselrad *n*
l. ice Treibeis *n*
l. inserted peg *(Text)* Einsteckspindel *f*, Steckspindel *f*
l. knot herausgefallener Ast *m*, Durchfallast *m*
l.-leaf standard Normblatt *n*
l.-leaf system Loseblattsystem *n*
l. material Schüttgut *n*
l. place Kristallauflockerung *f*; Lockerstelle *f* *(Kristallgitterstörung)*
l. pulley Losscheibe *f*
l. reed *(Text)* fliegendes Blatt *n*, Losblatt *n*
l. reed loom Webstuhl *m* mit beweglichem Kamm, Webstuhl mit fliegendem Blatt
l.-roller needle bearing Nadellager *n* ohne Außen- und Innenring
l. soil lockerer Boden *m*
l. sticks Reisigholz *n*
l. stuffing box Stopfbüchseneinsatz *m*, angeschraubter Stopfbüchsenbehälter *m*; Einsatzstopfbüchse *f*
l. veneer in cold bonding *(Holz)* Fehlleimung *f* bei Kaltverleimung
loosely coupled *(Fmt)* lose gekoppelt
loosen / to 1. lösen, lockern; 2. auflockern; 3. abschrauben
loosening Lösen *n*, Lockern *n*; Auflockern *n*; Abschrauben *n*
l.-up Auflockern *n* *(Erde)*
lop / to *(El)* abschneiden, abrunden *(Amplituden)*
l. off *s.* lop / to
lopper *(El)* Abkapper *m*
lopping axe Astaxt *f*
loppings abgehauene Zweige *f*
lopsided *(Flg)* flügellastig, querlastig
l. vessel Schiff *n* mit Schlagseite
lopwood Astholz *n*
lorandite *(Min)* Lorandit *m*
Lorentz force Lorentz-Kraft *f*
Lorentz transformation Lorentz-Transformation *f*
lorry Last[kraft]wagen *m*; Lore *f*; *(Brg)* Hunt *m*
l. truck *(Eis)* Schemelwagen *m*
l. with platform body *(Kfz)* Pritschenwagen *m*
Loschmidt constant Avogadrosche Zahl *f* *(Moleküle/cm^3)*
Loschmidt number *s.* Loschmidt constant
lose / to verlieren
l. electrons Elektronen abgeben
l. the temper ausglühen
loss 1. Abfall *m*, Verlust *m*, Einbuße *f*; 2. *(Met)* Abbrand *m*; 3. *(El)* Dämpfung *f*; Übertragungsverlust *m*
l. angle Verlustwinkel *m*
l. by evaporation Verdunstungsverlust *m*
l. by radiation Strahlungsverlust *m*
l. due to condensation Niederschlagsverlust *m*, Kondensationsverlust *m*, Abkühlverlust *m*
l. due to evaporation Verdunstungsverlust *m*
l. due to friction Reibungsverlust *m*
l. due to leakage Undichtigkeitsverlust *m*, Undichtheitsverlust *m*

l. due to shocks Stoßverlust *m (Turbine)*
l. due to slippage Schlupfverlust *m*
l. factor Verlustfaktor *m*
l.-free dielectric verlustfreies (verlustloses) Dielektrikum *n*, ideales Dielektrikum, idealer Isolierstoff *m*
l. in efficiency Leistungsverlust *m*
l. in the noise Untergehen *n* im Rauschen *(Signale)*
l. in value Wertminderung *f*, Minderwert *m*
l. in weight Masseverlust *m*
l. indemnity Schadenersatz *m*
l. meter *(El)* Verlustzähler *m*
l. of charge *(El)* Ladeverlust *m*, Ladungsverlust *m*
l. of control *(BMSR)* Durchzünden *n (Stromrichter)*
l. of energy Energieverlust *m*
l. of gum in scouring Verlust *m* an Seidenleim, Entschälverlust *m*
l. of hardness Erweichen *n*
l. of heat Wärmeverlust *m*
l. of light Lichtverlust *m*
l. of performance Leistungsverlust *m*
l. of pressure Druckverlust *m*, Druckminderung *f*
l. of pressure head Druckhöhenverlust *m*
l. of velocity Geschwindigkeitsverlust *m*
l. of volume Mengenverlust *m*
l. of work Arbeitsverlust *m (Dampfdiagramm)*
l. resistance *(El)* Verlustwiderstand *m*
l.-summation method *(Elm)* Einzelverlustverfahren *n*
losser tube *(Am) (Fmt)* Absorptionsrohr *n*
lossless verlustfrei, verlustlos, verlustarm
lost motion toter Gang *m (Lenkung)*; Leerlauf *m*
l. time *(Ök)* bezahlte Ausfallzeit *f*
l. volts Spannungsabfall *m*
l.-wax [investment] casting Wachsausschmelz[gieß]verfahren *n*
l. work *(Ph)* Arbeit *f* der Bewegungswiderstände
lot Posten *m*; *(Masch)* Fertigungslos *n*, Serie *f*; Grundstück *n*, Parzelle *f*
l. number *(Text)* Partienummer *f*
l. plot Lot Plot, Standardstichprobenplan *m*
l. size Losumfang *m*, Losgröße *f*
l. tolerance percent defective Schlechtgrenze *f*, Schlechtlage *f*
lotion Gesichtswasser *n*, Hautwasser *n*, Lotion *f*
loud-speaking telephone Lautfernsprecher *m*
loudness Lautheit *f*
l. level Lautstärke *f*
l. level meter Lautstärkemesser *m*
l. scale Lautstärkeskala *f*
l. volume equivalent *(Fmt)* Bezugsdämpfung *f* eines Übertragungssystems
loudspeaker Lautsprecher *m*
l. baffle Sprechwand *f*; Schallwand *f*
l. capacitor Lautsprecherkondensator *m*
l. diaphragm Lautsprechermembrane *f*
l. horn Lautsprechertrichter *m*
l. telephone set Teilnehmerendverstärker *m*
l. voice coil Lautsprecherschwingspule *f*
loupe Lupe *f*
louver *s.* louvre
louvre 1. Jalousie *f*, Sonnenblende *f*; 2. Luftschlitz *m (Ventilator)*; 3. Schallöffnung *f (beim Rundfunkempfänger)*; 4. *s.* l. damper
l. damper *(BMSR)* Jalousiedrosselklappe *f*
l. deduster Jalousiesichter *m*
l. window Glasjalousie *f*
low niedrig; tief; gering; schwach
l. activity *(Kern)* geringe Aktivität *f*
l.-alloy niedriglegiert *(Stahl)*
l.-altitude passage *(Astr)* niedriger Durchgang *m*
l.-ash aschenarm
l.-background counter *(Kern)* Zähler *m* mit niedrigem Nulleffekt
l. beam *(Kfz)* Abblendlicht *n*
l. bearing tiefes Tragbild *n (Verzahnung)*
l.-bed trailer Tiefladeanhänger *m*, Tieflader *m*
l. binding schwache (lockere) Bindung *f*
l.-boiling leichtsiedend, tiefsiedend
l. boiling naphtha Leichtbenzin *n*
l.-built chassis Niederrahmenfahrgestell *n*
l.-built loom oberbauloser Webstuhl *m*
l.-built ship Niederbordschiff *n*
l. butt needle *(Text)* Tieffußnadel *f*
l.-capacity kapazitätsarm
l.-capacity cable kapazitätsarmes Kabel *n*
l.-capture *(Kern)* schlecht absorbierend, mit niedrigem Einfangquerschnitt
l.-capture material *(Kern)* Material *n* mit niedrigem Einfangquerschnitt, schlechter Absorber *m*
l.-carbon niedriggekohlt, kohlenstoffarm *(Stahl)*
l. carbon ferro-chrome *(Met)* weiches Ferrochrom *n*
l.-centre-of-gravity type niedrige Bauart *f*
l.-cetane gas oil zündträges Gasöl *n*
l. consistence bleaching Dünnstoffbleiche *f*
l. consumption cathode Sparkatode *f*
l. cooling Tiefkühlung *f*
l.-cost kostengünstig
l.-definition *(Fs)* niedrigzeilig
l.-definition television niedrigzeiliges Fernsehen *n (mit kleiner Bildpunktzahl)*
l. density particle board leichte Spanplatte *f*
l.-drag trailing aerial Schleppantenne *f (Flugzeugantenne)*
l.-drift crystal Schwing[quarz]kristall *m* mit kleinem Temperaturkoeffizienten
l.-energy niedrigenergetisch; energiearm *(Teilchen)*; weich *(Strahlung)*

l.-energy particle energiearmes Teilchen *n*
l.-energy physics Physik *f* niedriger Energien, Niederenergiephysik *f*
l.-enriched schwachangereichert, geringangereichert
l.-enrichment *s.* l.-enriched
l. excited state *(Kern)* schwach angeregter Zustand *m*
l.-expansion mit niedriger Wärmedehnung *(z. B. Legierung)*
l. explosive schiebend wirkender Sprengstoff *m*
l. field *(El)* schwaches Feld *n*
l.-flux reactor Reaktor *m* mit niedrigem Neutronenfluß
l.-forest Niederwald *m*
l.-frame chassis Niederrahmenfahrgestell *n*
l.-frequency Niederfrequenz-, NF-
l.-frequency amplifier Niederfrequenzverstärker *m*
l.-frequency choke Niederfrequenzdrossel *f*
l.-frequency compensation Niederfrequenzausgleich *m*
l.-frequency converter Niederfrequenzumsetzer *m*, NF-Umsetzer *m*
l.-frequency cutoff untere Grenzfrequenz *f*
l.-frequency dialling *(Am)* *(Fmt)* Unterlagerungsfernwahl *f*
l.-frequency disturbance Niederfrequenzstörung *f*
l.-frequency engineering Niederfrequenztechnik *f*
l.-frequency furnace Niederfrequenzofen *m*
l.-frequency generator Niederfrequenzgenerator *m*, NF-Generator *m*
l.-frequency omni range Langwellen *fpl* für alle Entfernungen *(Radar)*
l.-frequency response Niederfrequenzkurve *f*
l.-frequency speaker Tieftonlautsprecher *m*
l.-frequency transformer Niederfrequenztransformator *m*, Niederfrequenzübertrager *m*, Niederfrequenzüberträger *m*
l.-frequency valve Niederfrequenzröhre *f*
l. gear *(Kfz)* niedriger Gang *m*, Anfahrtsgang *m*
l.-grade arm, geringhaltig, geringwertig, minderwertig
l.-grade fuel Brennstoff *m* mit geringem Heizwert
l.-grade ore armes (geringhaltiges) Erz *n*
l.-grade varnish magerer Lack *m*
l. height geringe Höhe *f*
l. helix drill Bohrer *m* mit schlanker Spirale, Bohrer mit weitem Drall
l.-impedance niederohmig
l. in ore erzarm
l. in silicon siliziumarm
l.-induction induktionsarm
l. insulation schlechte Isolation *f*
l.-intensity intensitätsarm, von geringer Intensität
l.-intensity irradiation Schwachbestrahlung *f*
l.-intensity testing reactor Testreaktor *m* mit geringer Intensität
l.-level amplifier Eingangsverstärker *m*, Vorverstärker *m*
l.-level modulation Vorstufenmodulation *f*, Modulation *f* bei niedrigem Pegel
l.-level radiation schwache Strahlung *f*
l. lift pump Pumpe *f* für geringe Förderhöhe
l. limit of tolerance unteres Abmaß *n*
l.-load adjustment *(El)* Einstellvorrichtung *f* für den Kleinlastabgleich *(Zähler)*
l. loader Tiefladeanhänger *m*, Tieflader *m*
l.-loading trailer Tiefbettanhänger *m*
l.-loss verlustarm
l.-loss cable verlustarmes Kabel *n*
l.-melting mit niedrigem Schmelzpunkt, leichtschmelzend, niedrigschmelzend
l. note accentuation *(El)* Baßanhebung *f*
l. oil-content circuit breaker ölarmer Schalter *m*
l. pass Tiefpaß *m* *(Radio)*
l.-pass filter 1. *(Kfz)* Tiefpaßfilter *n* *(Schalldämpfer)*; 2. *(El)* Niederfrequenzfilter *n*; Drosselkette *f*, Spulenkette *f*, Tiefpaß *m*
l. pass selective circuit Hochfrequenzsperrkreis *m*
l. pitch niedrige Tonhöhe *f*, tiefer Ton *m*
l. power von geringer Leistung, Klein-
l.-power electrolytic capacitor Niederspannungselektrolytkondensator *m*
l.-power microphone Mikrofon *n* mit kleiner Ausgangsleistung
l.-power modulation Modulation *f* bei niedrigem Pegel, Vorstufenmodulation *f*
l.-power photomicrograph Lupenaufnahme *f*
l.-power reactor Niederleistungsreaktor *m*, Kleinreaktor *m*
l.-power stage *(Fmt, Fs)* Vorstufe *f*
l.-power water boiler Kochwasserreaktor *m* geringer Leistung
l. pressure Niederdruck *m*, Niederspannung *f*
l.-pressure air-blast Niederdruckluftgebläse *n*
l.-pressure air compressor Niederdruckluftverdichter *m*
l.-pressure blower Niederdruckgebläse *n*
l.-pressure boiler Niederdruck[dampf]kessel *m*
l.-pressure burner Niederdruckbrenner *m*
l.-pressure chamber Niederdruckkammer *f* *(Turbine)*
l.-pressure cylinder Niederdruckzylinder *m*
l.-pressure disk Niederdruckrad *n* *(Turbine)*
l.-pressure governor Niederdruckregler *m*, Unterdruckregler *m*

l.-pressure region Niederdruckzone *f*
l.-pressure stage Niederdruckstufe *f*
l.-pressure steam engine Niederdruckdampfmaschine *f*
l.-pressure test Niederdruckversuch *m*
l.-pressure turbine Niederdruckturbine *f*, ND-Turbine *f*
l.-pressure tyre *(Kfz)* Niederdruckreifen *m*
l.-priced billig, preiswert
l. range of gears *(Kfz)* niedrige Übersetzung *f*, Stadtgang *m*, Geländegang *m*
l. rank *(Brg)* niedriginkohlt
l. rate timber geringwertiges Holz *n*
l. resistive niederohmig
l. rudder *(Flg)* Tiefenruder *n*
l.-shaft furnace Niederschachtofen *m*
l.-sided [goods] wagon *(Eis)* Niederbordwagen *m*
l.-silicon pig iron siliziumarmes Roheisen *n*
l.-slump concrete erdfeuchter (schwachplastischer, steifer) Beton *m*
l.-speed niedrigtourig
l.-speed nozzle *(Kfz)* kleine Düse *f*
l.-speed unit Langsamläufer *m*
l. starting torque Leichtanlauf *m* *(Motor)*
l.-temperature Niedertemperatur-, Tieftemperatur-, mit niedriger Betriebstemperatur
l.-temperature carbonization Tieftemperaturverkokung *f*, Verschwelung *f*, Schwelung *f*
l.-temperature carbonization coal Schwelkohle *f*
l.-temperature carbonization furnace Schwelofen *m*
l.-temperature carbonizer Schwelofen *m*, Schweler *m*, Schwelgenerator *m*
l.-temperature characteristics Kälteverhalten *n*, Tieftemperaturverhalten *n*
l.-temperature coke Schwelkoks *m*
l.-temperature coking gasoline Schwelbenzin *n*
l.-temperature distillation Tieftemperaturdestillation *f*; Verschwelung *f*
l.-temperature furnace Schwelofen *m*
l.-temperature gas Schwelgas *n*, Urgas *n*
l.-temperature physics Tieftemperaturphysik *f*
l.-temperature plant Schwelanlage *f*
l.-temperature tar Schwelteer *m*, Urteer *m*
l.-tension Niederspannung *f*
l.-tension battery Heizbatterie *f*
l.-tension cable Niederspannungskabel *n*
l.-tension circuit Niederspannungsstromkreis *m*
l.-tension coil Niederspannungsspule *f*
l.-tension current niedriggespannter Strom *m*, Niederspannungsstrom *m*
l.-tension cut-out Niederspannungssicherung *f*
l.-tension dynamo Niederspannungsdynamo[maschine *f*] *m*, Niederspannungserzeuger *m*, Niederspannungsgenerator *m*
l.-tension electric detonator *(Brg)* Brückenzünder *m*, Glühzünder *m*
l.-tension fuse Niederspannungssicherung *f*
l.-tension generator Niederspannungsdynamo[maschine *f*] *m*, Niederspannungserzeuger *m*, Niederspannungsgenerator *m*
l.-tension insulator Niederspannungsisolator *m*
l.-tension line Niederspannungsleitung *f*
l.-tension magneto Niederspannungszündmaschine *f*, Niederspannungsmagnet[apparat] *m*
l.-tension motor Niederspannungsmotor *m*
l.-tension plant Niederspannungsanlage *f*
l.-tension supply Niederspannungsnetz *n*
l.-tension switch Niederspannungsschalter *m*
l.-tension switchboard Niederspannungsschalttafel *f*
l.-tension switchgear Schaltgerät *n* für Niederspannungsanlage
l.-tension transformer Niederspannungstransformator *m*
l.-tension voltmeter Niederspannungsvoltmeter *n*
l.-tension winding Niederspannungswicklung *f*
l.-twist viscose filament niedriggezwirnte (niedriggedrehte) Viskoseseide *f*
l.-viscosity oil dünnflüssiges Öl *n*
l. volatile coal gasarme Kohle *f*
l. voltage *(El)* Niederspannung *f*, niedrige Spannung *f*
l. voltage *(El)* Niederspannungs-
l. voltage apparatus Niederspannungsapparat *m*
l. voltage coil Niederspannungsspule *f*
l. voltage current niedergespannter Strom *m*
l. voltage dynamo *s.* l. voltage generator
l. voltage electrolytic capacitor Niedervoltelektrolytkondensator *m*
l. voltage frame Niederspannungsgehäuse *n*
l. voltage generator Niederspannungsdynamo[maschine *f*] *m*, Niederspannungsgenerator *m*, Niederspannungserzeuger *m*
l. voltage installation Niederspannungsanlage *f*
l. voltage insulator Niederspannungsisolator *m*
l. voltage lamp Niederspannungslampe *f*, Niedervoltlampe *f*
l. voltage machine *s.* l. voltage generator
l. voltage magneto Niederspannungsmagnet[apparat] *m*
l. voltage motor Niederspannungsmotor *m*
l. voltage plant Niederspannungsanlage *f*
l. voltage release Unterspannungsauslöser *m*
l. voltage switch Niederspannungsschalter *m*
l. voltage transformer Niederspannungstransformator *m*

l. **voltage voltmeter** Niederspannungsvoltmeter *n*
l. **voltage winding** Niederspannungswicklung *f*
l. **W/C mix** Betonmischung *f* mit niedrigem Wasserzementfaktor, erdfeuchter Beton *m*
l. **warp loom** Basselisse-Webstuhl *m*
l. **water** seichtes Wasser *n*; tiefster Wasserstand *m*
l.**-water alarm** *(Wärm)* Niedrigwasserstandrufer *m*, Speiserufer *m*
l.**-wing aircraft** *(Flg)* Tiefdecker *m*
lower / to 1. senken, abwärtsbewegen, herablassen, herunterfahren; 2. herunterfieren, aussetzen *(ein Boot)*; 3. *(Brg)* einhängen *(Seigerförderung)*
l. **a boat** ein Boot fieren
l. **the voltage** *(El)* die Spannung heruntersetzen
lower unter[er], niedriger; Unter-
l. **beam** Abblendlicht *n*
l. **beam of press brake** *(Masch)* Unterwange *f*
l. **bell** Unterglocke *f (Gichtverschluß)*
l. **bend of the characteristic** *(El)* unterer Kennlinienknick *m*
l. **blade** Untermesser *n (Schere)*
l. **boom** *(Bw)* Untergurt *m*
l. **bound** *(Math)* untere Schranke *f*
l. **chamber of the collector** Unterkammer *f* der Schleuse *(Turbine)*
l. **chord** *(Bw)* Untergurt *m*
l. **critical** Soliduslinie *f (Eisen-Kohlenstoff-Schaubild)*
l. **cutting blade** Untermesser *n*
l. **dead centre** unterer Totpunkt *m*
l. **deviation** unteres Abmaß *n*
l. **die** Gesenkunterteil *n*, Untergesenk *n*, Unterstempel *m*, Matrize *f*
l.**-grade fuel** minderwertiger Brennstoff *m*
l. **hole board** *(Text)* Chorbrett *n*, Harnischbrett *n*
l. **joist** Unterbalken *m*
l. **keel** Unterkiel *m (bei Holzschiffen)*
l. **launch slip** *(Schiff)* Vorhelling *f*, Unterwasserhelling *f*
l. **layer** Unterschicht *f*
l. **pair** niederes Elementenpaar *n (Getriebelehre)*
l. **part** Unterteil *n*, Unterstück *n*
l. **part of crankcase** *(Kfz)* Kurbelwanne *f*, Motorwanne *f*
l. **portion of bearing** Lagerunterteil *n*
l. **prop** *(Brg)* Außenstempel *m*, Unterstempel *m*
l. **pump box** *(Wärm)* Pumpenherz *n*
l. **roll** Unterwalze *f*
l. **shed** *(Text)* Unterfach *n*, Tieffach *n*
l. **shrouds** *(Schiff)* Unterwanten *pl*
l. **side** Unterseite *f*
l. **side-band** *(Fmt)* unteres Seitenband *n*
l. **slide** Unterschlitten *m*
l. **stage** untere Stufe *f (z. B. einer Rakete)*
l. **stern** *(Schiff)* Gillung *f (des Hecks)*
l. **twilling bar** *(Text)* Unterhebestab *m*
l. **work table** Werkstückuntertisch *m*, Untertisch *m*
lowerable senkbar
lowered herabgelassen, gesenkt, herabgesetzt
l. **subsoil water level** abgesenkter Grundwasserspiegel *m*
lowering 1. Herablassen *n*, Senken *n*, Niederlassen *n*; Herunterfahren *n*; 2. Aussetzen *n*, Fieren *n (Boot)*
l. **funnel** Klappschornstein *m*, einziehbarer Schornstein *m*
l. **harness** *(Text)* Zuggeschirr *n*
l. **heald** *(Text)* Drucklitze *f*
l. **mast** umlegbarer (einziehbarer) Mast *m*
l. **of the freezing point** Gefrierpunkterniedrigung *f*
l. **of the melting point** Schmelzpunkterniedrigung *f*
l. **of the supports** *(Bw)* Stützensenkung *f*
l. **of vapour pressure** Dampfdruckerniedrigung *f*
l. **shaft** *(Text)* Tiefschaft *m*
l. **the ground water** Entwässern *n (Tagebau)*
l. **the water table** Grundwassersenkung *f*
l. **winch** *(Brg)* Bremsberghaspel *f*
lowermost unterst
l. **deck** *(Schiff)* unterstes Deck *n*
lowest common multiple Hauptnenner *m*
l. **frequency** Grundfrequenz *f*
l. **percentage** Mindestgehalt *m*
lowland Niederung *f*; Tiefebene *f*
lozenge rautenförmig
lozenge Rhombus *m*, Raute *f*
L.P.G. tanker *s.* liquefied petroleum gas tanker
LS-coupling *(Kern)* Russell-Saunders-Kopplung *f*
L.T.V.-process *s.* long tube vertical multiple effect distillation process
lubber's line Steuerstrich *m (am Kompaßgehäuse)*
l. **mark** *s.* lubber's line
lube oil *(Am) s.* lubricating oil
luboil *(Am) s.* lubricating oil
lubricant 1. Schmiermittel *n*, Schmierflüssigkeit *f*, Schmierstoff *m*, Schmieröl *n*; Gleitmittel *n*; 2. *(Text)* Schmälze *f*, Spickmittel *n*; 3. *(Brg)* Gleitflüssigkeit *f (Senkschachtverfahren)*
l. **circuit** Schmiermittelkreis *m*
l. **container** Schmiermittelbehälter *m*
l. **cooler** Schmierölkühler *m*
l. **film** Schmierfilm *m*
l. **pump** Schmiermittelpumpe *f*
l. **return** Schmiermittelrückführung *f*

l. supply Schmierstoffversorgung *f*
lubricate / to 1. [ab]schmieren, fetten, ölen; 2. *(Text)* schmälzen, spicken
lubricated petrol Kraftstoff-Öl-Gemisch *n*, Zweitaktgemisch *n*
lubricating 1. Abschmieren *n*, Schmieren *n*, Schmierung *f*, Ölen *n*, Einfetten *n*; 2. *(Text)* Schmälzen *n*
l. action Schmierwirkung *f*
l. cup Schmiergefäß *n*
l. device Schmierapparat *m*
l. effect Schmierwirkung *f*
l. efficiency Schmierleistung *f*
l. equipment Abschmiergerät *n*, Abschmiervorrichtung *f*
l. film Schmierpolster *n*, Schmierfilm *m*
l. grease Abschmierfett *n*, Schmierfett *n*
l. groove Schmiernut *f*
l. hole Schmierloch *n*
l. nipple Schmiernippel *m*
l. oil 1. Schmieröl *n*; 2. *(Text)* Schmälze *f*, Spicköl *n*, Spinnöl *n*
l. oil drop test Tropfenprobe *f*, Löschblattprobe *f*
l. oil pump Schmierölpumpe *f*
l. oil pump intake strainer *(Kfz)* Schmierölpumpensaugfilter *n*
l. oil tank Schmieröltank *m*, Schmierölbehälter *m*
l. pad Schmierkissen *n*
l. piping Schmierleitung *f*
l. point Schmierstelle *f*
l. power Schmierleistung *f*
l. qualities Schmiereigenschaften *fpl*
l. schedule Schmierplan *m*, Schmierübersicht *f*
l. spot Schmierstelle *f*
l. wick Öldocht *m*
lubrication Schmierung *f*
l. by circulation of the oil Umlaufschmierung *f*
l. by gravity-feed Trockensumpfschmierung *f*
l. chart Schmierplan *m*
l. connection Schmierstutzen *m*
l. engineering Schmiertechnik *f*
l. groove Schmiernut *f*
l. hole Schmierbohrung *f*
l. nipple Schmiernippel *m*
l. of the pressure circulating type Druckumlaufschmierung *f*
l. point Schmierstelle *f*
l. schedule Schmierplan *m*
l. survey Schmierplan *m*
l. system Schmiersystem *n*
lubricator Öler *m*, Schmierapparat *m*, Schmiereinrichtung *f*, Schmierbüchse *f*
l. spring Schmierfeder *f*, Ölfeder *f*
l. working from a distance Fernschmierer *m*
lubricity Fettigkeit *f*, Schlüpfrigkeit *f*; Schmierergiebigkeit *f*, Schmierfähigkeit *f*; Filmfestigkeit *f*
l. of the lubricant Schlüpfrigkeit *f* des Schmiermittels
lubrify / to *s.* lubricate / to
Lüders lines Lüderssche Linien *fpl*
Ludlow typograph Ludlow-Typograf *m*
ludwigite *(Min)* Ludwigit *m*
luff / to *(Schiff)* [an]luven, aufluven
l. into the wind *(Schiff)* in den Wind luven, aufdrehen, aufschießen
luff Luvseite *f*, Luvkante *f* *(Segel)*; Vor[der]liek *n*, Mastliek *n* *(bei Gaffelsegeln)*; Stagliek *n*, Stagkante *f* *(bei Stagsegeln)*
luffing crane Einziehkran *m*, Kran *m* mit veränderlichem Radius; Hebekran *m*
l. davit *(Schiff)* Radialdavit *m (n)*, Klappdavit *m (n)*, Barkune *f*
lug 1. Henkel *m*, Ohr *n*, Anguß *m*, Ansatz *m*, Nase *f*, Öse *f*, Warze *f*, Zapfen *m*; 2. *(El)* Anschlußfahne *f*, Fahne *f*, Lageröse *f*, Lötfahne *f*, Lötöse *f*, Öse *f*; Polschuh *m*; Klemme *f*; Kabelschuh *m*
luggage boot *(Kfz)* Kofferraum *m*
l. car *(Eis)* Gepäckwagen *m*
l. compartment Gepäckraum *m*; Kofferraum *m*
l. dump Gepäckablage *f*
l. grid Gepäckrost *m*, Gepäckhalter *m*, Kofferbrücke *f*
l. locker *(Kfz)* Gepäckraumverschluß *m*
l. net Gepäcknetz *n*
l.-net bracket Gepäcknetzträger *m*
l. rack *(Eis)* Gepäckraufe *f*
l. van *(Eis)* Gepäckwagen *m*, Packwagen *m*
luke-warm handwarm, lauwarm
lumber / to Holz zuschneiden; versperren
lumber 1. Gerümpel *n*; 2. *(Am)* Bauholz *n*, Nutzholz *n*
l.-carrier Holztransportschiff *n*
l. jack Holzfäller *m*
l. man Holzfäller *m*
l. mill Sägemühle *f*
l. room Abstellraum *m*
l. trade *(Am)* Holzfahrt *f*
lumberer *(Am)* Holzfäller *m*
lumberman *(Am)* Holzfäller *m*
lumen Lumen *n*, lm *(Einheit des Lichtstroms)*
l.-hour Lumenstunde *f*, lmh *(Einheit der Lichtmenge)*
lumeter Lumenmesser *m*, Helligkeitsmesser *m*
luminaire *(Am)* Lampe *f*, Leuchte *f*, Leuchtkörper *m*
l. efficiency *(Am)* Leuchtenwirkungsgrad *m*
luminance 1. Leuchtdichte *f*, spezifische Lichtausstrahlung *f*; 2. *(Fs)* Helle *f*
l. factor *(Opt)* Remissionsgrad *m*

l. **signal** Helligkeitssignal *n (Steuerung des Fernsehbildes)*
luminator crystal Leuchtquarz *m*, Leuchtresonator *m*
luminesce / to lumineszieren
luminescence Lumineszenz *f*
l. **analysis** Lumineszenzanalyse *f*
l. **cell** Lumineszenzzelle *f*
l. **of crystallization** Kristallumineszenz *f*
l. **period** Nachleuchtdauer *f*
l. **quenching** Lumineszenzlöschung *f*
luminescent lumineszierend
l. **centre** Lumineszenzzentrum *n*, Leuchtzentrum *n*
l. **crystal** lumineszierender Kristall *m*
l. **emission** Lumineszenzemission *f*
l. **material** Leuchtstoff *m*, Luminophor *m*
l. **screen** Leuchtschirm *m*
l. **screen tube** Leuchtschirmröhre *f*
luminosity Helligkeit *f*, Leuchtstärke *f*, Leuchtkraft *f*
l. **curve** polare Lichtverteilungskurve *f*
l. **factor, of radiation** fotometrisches Strahlungsäquivalent *n*
l. **of stars** Helligkeit *f* der Sterne
luminous leuchtend; Leucht-, Licht-
l. **absorption** Lichtabsorption *f*
l. **advertisement** Lichtreklame *f*
l. **anode** Leuchtanode *f*
l. **buoy** *(Schiff)* Nachtrettungsboje *f*
l. **circuit diagram** Leuchtschaltbild *n*
l. **cone** Lichtkegel *m*
l. **coverage** Ausleuchtung *f (einer Fläche)*
l. **crystal** Leuchtresonator *m*, Leuchtquarz *m*
l. **current** Glimmstrom *m*
l. **dial** Leuchtzifferblatt *n*
l. **discharge** Glimmentladung *f*, Koronaentladung *f*
l. **effect** Leuchtwirkung *f*
l. **efficiency** Lichtausbeute *f (einer Lichtquelle)*; fotometrisches Strahlungsäquivalent *n (einer Strahlung)*
l. **emittance** spezifische Lichtausstrahlung *f*
l. **energy** Lichtenergie *f*, Lichtmenge *f*
l. **field** *(Opt)* Leuchtfeld *n*
l.**-field diaphragm** Leuchtfeldblende *f*
l. **flux** Lichtstrom *m*
l. **flux density** Lichtstromdichte *f*
l. **fore sight** *(Mil)* Leuchtkorn *n*
l. **gas lamp** Gasentladungslampe *f*
l. **hand** Leuchtzeiger *m*
l. **hand ophthalmoscope** Handaugenspiegel *m* mit elektrischer Beleuchtung
l. **intensity** Lichtstärke *f*
l. **intensity of a light source** Intensität *f* einer Lichtquelle, Lichtstärke *f*
l. **intensity reproduction** Helligkeitswiedergabe *f*
l. **mark** Lichtmarke *f*
l. **material** Leuchtmasse *f*, Leuchtstoff *m*
l. **matter** *s.* l. material
l. **numeral** Leuchtziffer *f*
l. **paint** Fluoreszenzfarbe *f*, Leuchtfarbe *f*
l. **pigment** Leuchtfarbe *f*; Leuchtpigment *n*
l. **pointer** Lichtzeiger *m*
l. **power** Leuchtkraft *f*
l. **quartz** Leuchtquarz *m*
l. **radiation** Lichtstrahlung *f*
l. **range** Sichtweite *f (eines Leuchtfeuers)*
l. **ray** Lichtstrahl *m*
l. **resonator** Leuchtresonator *m*
l. **sensitivity** Lichtempfindlichkeit *f*
l. **skiascope** *(Opt)* Skiaskop *n* mit elektrischer Beleuchtung
l. **source** Lichtquelle *f*
l. **spectrum** sichtbares Spektrum *n*
l. **spot** Leuchtfleck *m*
l. **standard** Normallichtquelle *f*
l. **stimulus** Lichtreiz *m*
l. **substance** Leuchtstoff *m*
l. **sulphide** Leuchtsulfid *n (Leuchtstoff)*
l. **track** Leuchtspur *f*
lump / to zusammenballen
lump Klumpen *m*, Stück *n*
l. **breaker** Gautschgegenwalze *f (Papierherstellung)*
l. **coal** Stückkohle *f*, Grobkohle *f*
l. **coke** Grobkoks *m*
l. **inductance** konzentrierte Induktivität *f*
l. **lime** Stückkalk *m*
l. **sum** Pauschalbetrag *m*, Pauschalsumme *f*
lumped punktförmig verteilt, konzentriert
l. **capacitance** punktförmig verteilte Kapazität *f*
l. **constant network** *(BMSR)* Schaltung *f* mit konzentrierten Parametern
l. **elements** konzentrierte Schaltelemente *npl*
l. **uranium lattice** *(Kern)* Uranblockgitter *n*
lumper Hafenarbeiter *m*, Kaiarbeiter *m (für Lade- und Löschbetrieb)*
lumpy [grob]stückig, großstückig *(z. B. Kohle)*; knollig *(Sand)*
l. **ore** stückiges Erz *n*
lunar base Mondbasis *f*
l. **caustic** Höllenstein *m*, Silbernitrat *n*
l. **crater** Mondkrater *m*
l. **eclipse** Mondfinsternis *f*
l. **probe** Mondsonde *f*
l. **satellite** Mondsatellit *m*
lune *(Math)* Möndchen *n*, Zweieck *n*
lunging *(Masch)* stotternde Bewegung *f*, Bewegung mit Geschwindigkeitsänderung
luster *s.* lustre
lustre Oberflächenglanz *m*, Schimmer *m*
l. **shrinking machine** Preßglanzdekatiermaschine *f*

l. terminal Lüsterklemme *f*
lustring *(Text)* Lüstrieren *n*
l. machine *(Text)* Lüstriermaschine *f*
lustrous carbon Glanzkohle *f*
lute / to dichten, [ver]kitten, verschmieren
lute 1. Dichtungsmasse *f*, Kitt *m*; 2. Laute *f* *(Musikinstrument)*
l. peg Lautenwirbel *m*
lutecia Lutetiumoxid *n*
lutecium Lutetium *n*
luteolin Luteolin *n* *(Farbstoff)*
luting agent Dichtungskitt *m*
lux Lux *n* *(Einheit der Beleuchtungsstärke)*
luxmeter *(Opt)* Luxmeter *n*, Helligkeitsmesser *m*
luxury liner Luxusschiff *n*
l. wood Edelholz *n*
luzonite *(Min)* Luzonit *m*
lycopene *(Ch)* Lykopin *n*
lycopodium Lykopodium *n* *(Formpuder)*
lye 1. Lauge *f*; Alkalilauge *f*; *(Text)* Beuche *f*; 2. *(Brg)* Weiche *f*, Wechsel *m*
l. bath Laugenbad *n*
l. preparation box Laugenansetzkasten *m*
lying aground *(Schiff)* Aufgrundliegen *n*
l.-in ward Wöchnerinnenstation *f*
Lyman series Lyman-Serie *f* *(Spektrum)*
lymegrass Strandhafer *m*
lyophilic lyophil
l. colloid lyophiles Kolloid *n*
lyophobic lyophob
l. colloid lyophobes Kolloid *n*
lyotropic *(Ch)* lyotrop
lyxonic acid Lyxonsäure *f*

M

M-band *(Ph)* M-Bande *f*
M-line M-Linie *f*
M-scope M-Schirm *m*
Mac'Intyre tank *(Schiff)* Mac-Intyre-Doppelboden *m*
macadam Makadam *m*
m. base Schottertragschicht *f* *(Straßenbau)*
m. foundation Schotterunterbau *m* *(Straßenbau)*
m. road Makadamstraße *f*, Schotterstraße *f*
m. surface *s.* m. surfacing
m. surfacing Makadamdecke *f*, Schotterdecke *f* *(Straßenbau)*
macadamize / to beschottern *(Straßenbau)*
macarthy gin *(Text)* Messerentkörnungsmaschine *f*, Plattenentkörnungsmaschine *f*, Macarthysche Maschine *f*
macerate / to ätzen, einweichen, mazerieren
macerate *(Am)* Füllstoff *m* für Preßmassen, Schnitzelmaterial *n*
m. molding *(Am)* Preßteil *n* mit Schnitzelfüllstoff, Schnitzelpreßstoffherstellung *f*, Verpressen *n* von Schnitzelmaterial
macerated fabric *(Am)* Gewebeschnitzel *n*
m. paper Papierschnitzel *n*
Mach front Machscher Kegel *m*
Mach number Mach-Zahl *f*, Machsche Zahl *f*
Mach stem *s.* Mach front
Mach wave Machsche Welle *f*
Mache unit *(Kern)* Mache-Einheit *f* *(1 ME = 3,64 Eman)*
machinability Spanbarkeit *f*, Zerspanbarkeit *f*, Bearbeitbarkeit *f*, Verarbeitbarkeit *f*, Verarbeitungsfähigkeit *f*
m. test Spanbarkeitsprüfung *f*
machinable bearbeitbar, [zer]spanbar; verarbeitbar
machine / to [spanend] bearbeiten, spanen; fertigen; verarbeiten
m. all over alle Oberflächen bearbeiten
m. from the solid aus dem Vollen herausarbeiten
m. in one setting of the work in einer Aufspannung bearbeiten
machine Maschine *f*, Arbeitsmaschine *f*
m. ancillary time *(Ök)* Maschinennebenzeit *f*
m. attendant Maschinenarbeiter *m*, Maschinenbedienungsmann *m*
m.-available time Maschinenwirkzeit *f*; Rechnerwirkzeit *f*; Wirkzeit *f*
m. bandsaw file Maschinenbandsägenfeile *f*
m. base Maschinenbett *n*, Maschinenunterteil *n*
m. base plate Maschinengrundplatte *f*
m. bed Maschinenbett *n*
m. bit Maschinenbohrer *m*
m. bit for wood Holzbohrer *m*
m. blacksmithing Maschinenschmieden *n*
m. bolt Maschinenschraube *f*
m. breakdown Maschinenschaden *m*
m. bridge reamer Maschinennietlochreibahle *f*
m. builder Maschinenbauer *m*
m. building Maschinenbau *m*
m. casting Maschinenguß *m*; Maschinengußstück *n*
m. chest Rührbütte *f*, Stoffkiste *f*, Stoffbütte *f* *(Papierherstellung)*
m. coating Maschinenauftrag *m*, Maschinenanstrich *m*
m. column Maschinenständer *m*
m. composition *(Typ)* Maschinensatz *m*
m. control unit Maschinensteuergerät *n*
m.-controlled time *(Ök)* Maschinenzeit *f*
m.-cut gear Zahnrad *n* mit geschnittenen Zähnen
m.-cut teeth geschnittene Zähne *mpl* *(Zahnrad)*

m. cutter Fräser *m*
m. cycle 1. *(Masch)* Bearbeitungszyklus *m*; 2. *(Dat)* Rechnerperiode *f*, Maschinenperiode *f*
m. direction Laufrichtung *f*, Maschinenrichtung *f (des Papiers)*
m. down time Maschinenausfallzeit *f*, Maschinenstillstandszeit *f*
m. drive Maschinenantrieb *m*
m. effective utilization index tatsächlicher Maschinenausnutzungsgrad *m*
m. element Maschinenelement *n*, Maschinenteil *n*
m. equation *(Dat)* Maschinengleichung *f*, Rechnergleichung *f*
m. filling Maschinenabfüllung *f*
m.-finished paper maschinenglattes Papier *n*
m. finishing Satinierung *f*, Satinage *f (Papier)*
m. fired mechanisch gefeuert
m. for parquetry work Parkettfügemaschine *f*
m. for roughing veneer Aufrauhmaschine *f (Furnier)*
m. for sheet metal working Blechbearbeitungsmaschine *f*
m. for stocking dyeing and boarding Damenstrumpffärbe- und -formmaschine *f*
m. for straightening spiral wood Aufbiegevorrichtung *f* für spiralige Holzplatten
m. forging Stauchschmiedestück *n*
m. foundation Maschinenfundament *n*
m. frame Maschinenrahmen *m*
m.-glazed einseitig geglättet *(Papier)*
m.-glazed paper einseitig glattes Papier *n*, maschinenglattes Papier
m. grinding Maschinenschliff *m*
m. gun squad Maschinengewehrbedienung *f*, MG-Bedienung *f*
m. hand *(Am)* Maschinenbedienungsmann *m*
m.-handling time *(Am)* Hilfszeit *f*, Maschinenhilfszeit *f*
m. idle time Maschinenausfallzeit *f*
m. key Längskeil *m*
m. key ringing *(Fmt)* halbselbsttätiger Ruf *m*
m. knitting Maschinenstricken *n (auf Handstrickmaschine)*; Wirken *n (auf mechanischem Webstuhl)*
m. language *(Dat)* Maschinensprache *f*, Rechnersprache *f*
m. lapping maschinelles Läppen *n*
m. learning *(Dat)* maschinelles Lernen *n*
m.-made maschinell hergestellt
m.-made brick Maschinenziegel *m*
m.-made index for technical literature *(Dat)* maschinell hergestellter Index *m* für technisches Schrifttum
m.-made lace *(Text)* Maschinenspitze *f*
m. maintenance Maschineninstandhaltung *f*
m. man Maschinist *m*, Maschinenwärter *m*
m.-man and man-machine communication system *(Dat)* Nachrichtensystem *n* Mensch-Maschine und Maschine-Mensch
m. moulding Maschinenformerei *f*
m. nut tap Maschinenmutterbohrer *m*
m. operator Maschinenarbeiter *m*, Maschinenbedienungsmann *m*
m. part Maschinenteil *n*; Maschinenelement *n*
m. pinking of cloth *(Text)* bogenförmiges Ausschneiden *n* von Musterabschnitten mit der Maschine
m. plane iron Maschinenhobelmesser *n*
m. proof *(Typ)* Maschinenabzug *m*
m. rating Leistung *f* der Maschine
m. reamer Maschinenreibahle *f*
m. reamer of the rose type Maschinengrundreibahle *f*
m. reamering Reiben *n* mit Maschinenwerkzeug
m. recognition of spoken words *(Dat)* maschinelles Wiedererkennen *n* gesprochener Worte
m.-relieved cutter hinterdrehter Fräser *m*
m. riveting Maschinennietung *f*
m. ruler Liniermaschine *f*, Linierwerk *n*
m. running time Maschinenlaufzeit *f*
m. saw Sägemaschine *f*
m. screw Maschinenschraube *f*
m. screw tap Maschinengewindebohrer *m*
m.-setter Einrichter *m*
m. setting Einrichten *n* einer Maschine
m. shop mechanische Werkstatt *f*, Maschinenwerkstatt *f*
m.-shop milling Produktionsfräsen *n*
m. slide Maschinenschlitten *m*
m. slug *(Typ)* Setzmaschinenzeile *f*, Satzzeile *f (Linotype)*
m. spindle Maschinenspindel *f*
m. steel Maschinenbaustahl *m*
m. table Maschinentisch *m*
m. tap Maschinengewindebohrer *m*
m. taper *(Masch)* Werkzeugkegel *m*, Aufnahmekegel *m*, Befestigungskegel *m*
m. taper reamer Maschinenkegelreibahle *f*
m. tool 1. Maschinenwerkzeug *n*; 2. Werkzeugmaschine *f*
m.-tool builder Werkzeugmaschinenbauer *m*
m.-tool centre Körnerspitze *f*
m.-tool circuit Flüssigkeitsgetriebe *n* einer Werkzeugmaschine, Werkzeugmaschinenhydraulik *f*
m.-tool component Baugruppe *f* einer Werkzeugmaschine
m.-tool designer Werkzeugmaschinenkonstrukteur *m*
m.-tool drive Werkzeugmaschinenantrieb *m*

m.-tool element Werkzeugmaschinenbaugruppe *f*
m.-tool exposition Werkzeugmaschinenausstellung *f*
m.-tool research and development institute Werkzeugmaschinenforschungs- und -entwicklungsinstitut *n*
m.-tool slideway Werkzeugmaschinenführung *f*
m.-tool table Werkzeugmaschinentisch *m*
m. translation maschinelles Übersetzen *n* *(von Sprachen)*
m. translator Übersetzungsmaschine *f* *(für Sprachen)*
m. transmitter Maschinensender *m*
m. turning on the spot *(Lw)* durchlenkbare Maschine *f*
m. upright Maschinenständer *m*
m. variable Rechnervariable *f* (*s. a.* computer variable)
m. vice Maschinenschraubstock *m*
m. welding maschinelles Schweißen *n*
m. with inherent self excitation *(El)* läufererregte Kommutatormaschine *f*
m. word *(Dat)* Rechnerwort *n*
machineable *s.* machinable
machined surface Arbeitsfläche *f* *(Spanen)*
machinery Maschinen[park *m*] *fpl*; Mechanismus *m*; Maschinenwesen *n*
m. casing *(Schiff)* Maschinenschacht *m*
m. casting Maschinenguß[stück *n*] *m*
m. catalogue Maschinenkatalog *m*
m. chock *(Schiff)* Holzbettung *f* unter Deckshilfsmaschinen, Holzunterklotzung *f*
m. layout Maschinenanlage *f*
m. space *(Schiff)* Maschinenraum *m*
m. steel Maschinenbaustahl *m*
machines arranged in flow lines in Fließreihe angeordnete Maschinen *fpl*
machine's hourly output Stundenleistung *f* einer Maschine in Stück
machining Fertigung *f*, Maschinenarbeit *f*; spanende Bearbeitung *f*, Verarbeitung *f*, Spanung *f*
m. accuracy Bearbeitungsgenauigkeit *f*
m. allowance Bearbeitungszugabe *f*
m. characteristic Spanungskennwert *m*; Spanbarkeit *f*
m. cost Bearbeitungskosten *pl*
m. cycle Arbeitsablauf *m*
m. instruction Bearbeitungsangabe *f*, Oberflächenangabe *f*
m. length Arbeitslänge *f*
m. method Bearbeitungsmethode *f*, Bearbeitungsverfahren *n*
m. operation Arbeitsvorgang *m*, Bearbeitungsvorgang *m*
m. plan Bearbeitungsplan *m*
m. rate Bearbeitungsgeschwindigkeit *f*, Spanungsgeschwindigkeit *f*
m. recommendation Richtwert *m* für die spanende Bearbeitung
m. sequence Folge *f* der Bearbeitungsgänge
m. steel Maschinenbaustahl *m*
m. step Bearbeitungsstufe *f*
m. technique Spanungsverfahren *n*, Zerspanungsverfahren *n*
m. technology Spanungslehre *f*, Zerspanungslehre *f*
m. time Hauptzeit *f*, Arbeitszeit *f*, Stückzeit *f*
m. time per component Stückzeit *f*
m. unit Bearbeitungseinheit *f*
machinist Maschinenbedienungsmann *m*; Spaner *m*; Maschinist *m*; Maschinenschlosser *m*; Feinmechaniker *m*
machinist's vice Parallelschraubstock *m*
machometer Machmeter *n* *(Fahrtmesser im Flugzeug)*
mackle / to *(Typ)* schmitzen, dublieren, doppeln
maco cotton Makobaumwolle *f*
macroanalysis *(Ch)* Grobanalyse *f*, Makroanalyse *f*
macroaxis Makroachse *f* *(Kristall)*
macrocrystalline makrokristallin
macroetching Grobätzung *f*, Makroätzung *f*
macrokinetics Makrokinetik *f*
macromolecule Makromolekül *n*
macroparticle Makroteilchen *n*
macrophysics Makrophysik *f*
macroscopic makroskopisch
m. scattering cross section *(Kern)* makroskopischer Streuquerschnitt *m*
macrosection *(Met)* Makroschliff *m*
macrosegregation *(Met)* Blockseigerung *f*
macrostate Makrozustand *m*
macrostructure Grobgefüge *n*, Grobstruktur *f*, Makrostruktur *f*
macroturbulence *(Strm)* Makroturbulenz *f*
madder Krapp *m*
m. red Krapprot *n*, Alizarin *n*
m. root Krappwurzel *f*
made to scale maßgerecht hergestellt
m.-up ground aufgeschütteter Boden *m*, Aufschüttung *f*
madia oil Sonnenblumenöl *n*
magamp *(sl)* *s.* magnetic amplifier
magazine 1. Warenlager *n*, Speicher *m*, Magazin *n*; 2. *(Masch)* Magazin
m. chute Magazinrutsche *f*
m.-fed mit Magazinzuführungseinrichtung
m. feed 1. Magazinzuführung *f*; 2. Zuführungsmagazin *n*
m. feed equipment Zuführungsmagazin *n*
m. feed press Presse *f* mit Magazinzuführungseinrichtung (Teilführungseinrichtung)
m. follower Zubringer *m* *(z. B. Gewehr)*

m. grinder *(Holz)* Magazinschleifer *m*
m. mining *(Brg)* Magazinbau *m*, Speicherbau *m*
m. spring Zubringerfeder *f (Gewehr)*
magic magisch
m. band valve magisches Band *n (Elektronenröhre)*
m. eye Abstimmanzeigeröhre *f*, magisches Auge *n (Elektronenröhre)*
m. lantern Laterna magica *f*
m. strip *(Fmt)* magischer Strich *m*
m. tee Wellenleiterkopplungseinrichtung *f (T-förmige Kopplung für gemeinsame Radarempfangs- und -sendeantenne)*
m. tube Abstimmanzeigeröhre *f*, magisches Auge *n (Elektronenröhre)*
magma Magma *n*, Schmelzfluß *m*
magmatic rock Eruptivgestein *n*
magnaflux testing for cracks *(Met)* Magnetpulverprüfung *f* auf Risse
m. testing method *(Met)* Magnetpulverprüfverfahren *n*
magnal base Elfstift-Sockel *m (für eine Katodenstrahlröhre)*
magnesia Magnesiumoxid *n*, Magnesia *f*, Bittererde *f*
m. cement Magnesiamörtel *m*
magnesian limestone Dolomit *m*
m. slate Talkschiefer *m*
magnesite Magnesit *m*, Talkspat *m*, Bitterspat *m*
m. brick Magnesitziegel *m*
m. flooring Magnesitfußboden *m*
m. lining Magnesitauskleidung *f*, Magnesitzustellung *f (des Ofens)*
magnesium Magnesium *n*
m. alloy Magnesiumlegierung *f*
m. chloride Magnesiumchlorid *n*, Chlormagnesium *n*
m. molybdate Magnesiummolybdat *n*
m. oxide Magnesiumoxid *n*, kalzinierte Magnesia *f*
m. sulphate Magnesiumsulfat *n*, schwefelsaures Magnesium *n*
magnestat *(El)* Spannungsregler *m*
magnesyn Magnesyn *(Firmenbezeichnung einer Drehmeldertype mit permanentmagnetischem Rotor)*
magnet Magnet *m*
m. carrier Magnethalter *m*
m. case Magnetgehäuse *n*
m. coil Magnetspule *f*, Feldspule *f*
m. core Magnetkern *m*
m.-core aerial Magnetstabantenne *f*, Magnetkernantenne *f*
m. frame Magnetgestell *n*; Feldgestell *n*, Joch *n*
m. gap Polschuhabstand *m*
m. housing Magnetgehäuse *n*
m. leg Magnetschenkel *m*
m. limb Magnetschenkel *m*
m. pole Magnetpol *m*, magnetischer Pol *m*
m. separator Magnetscheider *m*
m. steel Magnetstahl *m*
m. tester Magnetprüfer *m*
m. testing apparatus Magnetprüfer *m*
m.-type loudspeaker Magnetlautsprecher *m*
m. wheel Polrad *n*
m. winding Magnetwicklung *f*
m. wire Spulendraht *m*; Spulenwicklung *f*; Magnetspule *f*, Magnetwicklung *f*
m. with anisotropic properties Magnet *m* mit Vorzugrichtung
m. yoke Magnetjoch *n*, magnetisches Joch *n*
magnetic magnetisch; Magnet-; mißweisend *(Radar)*
m. after-effect magnetische Nachwirkung *f*
m. alloy magnetische Legierung *f*
m. amount printing *(Büro)* Drucken *n* des Betrages mit magnetischer Tinte
m. amplifier Magnetverstärker *m*, [magnetischer] Verstärker *m*, Transduktorverstärker *m*
m.-amplifier arithmetic unit Magnetverstärkerrechenwerk *n*
m. amplifier flip-flop Magnetverstärkerkippschaltung *f*
m. anomaly magnetische Anomalie *f*
m. antitank mine *(Mil)* Hafthohlladung *f*; Magnetpanzermine *f*
m.-armature loudspeaker magnetischer Lautsprecher *m*
m. attraction magnetische Anziehung[skraft] *f*
m. axis Magnetachse *f*, magnetische Achse *f*, Pol[ar]achse *f*
m. azimuth *(Flg)* Kurswinkel *m*
m. backing-pump valve with retarding relay magnetisch-automatisches Vorpumpenventil *n*
m. balance Feldwaage *f*, magnetische Waage *f*
m. bar Magnetstab *m*, Stabmagnet *m*
m. bearing magnetische (mißweisende) Peilung *f*, Magnetkompaßpeilung *f*
m. bias magnetische Vorspannung *f*
m. biasing Vormagnetisierung *f*
m. blow-out magnetische Bogenlöschung (Beblasung, Bogenbeeinflussung, Blaseinrichtung) *f*, Blasmagnet *m*
m. blow-out circuit-breaker Schalter *m* mit magnetischer Blasung, Magnetblasschalter *m*
m. blowing effect magnetische Blaswirkung *f*
m. brake Bremsmagnet *m*, Magnetbremse *f*
m. chuck Magnetspannfutter *n*
m. chuck of rectangular pattern Aufspannplatte *f*
m. circuit magnetischer Stromkreis *m*
m. clutch elektromagnetische Kupplung *f*,

Elektromagnetkupplung *f*, Magnet[pulver]-kupplung *f*
m. cobber Magnetscheider *m (Trockenscheider für Aufbereitung)*
m. cohesion magnetische Haftung *f*
m. conductance magnetischer Leitwert *m*
m. contactor elektromagnetisches Schütz *n*, Magnet[schalt]schütz *n*
m. core Ferritkern *m*, Magnetkern *m*, Speicherkern *m*
m. core coupling element Magnetkernkopplungselement *n (zur Verhinderung oder Zulassung der Übertragung eines bilateralen Stromes)*
m. core memory Magnetkernspeicher *m*
m. core memory matrix Magnetkernspeichermatrix *f*
m. counter Zählmagnet *m*, Zählrelais *n*
m. coupling Magnetkupplung *f*; magnetische Kopplung *f*
m. course mißweisender Kurs *m*
m. crack detection magnetische Rißprüfung *f*
m. creeping magnetische Nachwirkung *f*, kriechende Hysteresis *f*
m. cut-out Magnetschütz *n*
m. cycle Hysteresisschleife *f*; magnetischer Kreisprozeß *m*
m. damping Wirbelstromdämpfung *f*
m. declination Deklination *f*, Mißweisung *f* *(z. B. Radar)*
m. deflection magnetische Ablenkung *f*, Ablenkung *f* im Magnetfeld
m. deflector magnetischer Deflektor *m*
m. detector magnetischer Detektor (Gleichrichter) *m*, Magnetdetektor *m*
m. differential permeability Differentialpermeabilität *f*
m. dipole magnetischer Dipol *m*
m. dipole moment magnetisches Dipolmoment *n*
m. dipole radiation magnetische Dipolstrahlung *f*
m. disk memory Magnetplattenspeicher *m*
m. displacement magnetische Verschiebung *f*; [magnetische] Feldverschiebung *f*
m. disturbance Störung *f* des Erdmagnetfeldes, magnetische Störung
m. drum Magnettrommel *f*
m. drum memory Magnettrommelspeicher *m*
m. drum storage Magnettrommelspeicher *m*
m.-electric relay magnetodynamisches Relais *n*, Drehspulrelais *n*
m. electron microscope Elektronenmikroskop *n* mit magnetischen Linsen
m. false-twist spindle *(Text)* magnetische Falschdrahtspindel *f*
m. fatigue magnetische Nachwirkung *f*
m. field Magnetfeld *n*, magnetisches Feld *n*
m. field energy magnetische Feldenergie *f*
m. field intensity magnetische Feldstärke *f*
m. field line magnetische Feldlinie *f*
m. field strength magnetische Feldstärke *f*
m. figures magnetische Feilspanbilder *npl*
m. film memory magnetischer Dünnschichtspeicher *m*, Magnetfilmspeicher *m*
m. filter Magnetfilter *n*
m. flip-flop Magnetverstärkerkippschaltung *f*
m. flow magnetischer Induktionsfluß *m*
m. flowmeter *s.* electromagnetic flowmeter
m. flux magnetischer Fluß *m*, Magnetfeldwert *m*
m. flux density magnetische Flußdichte *f*
m. flux meter Flußmesser *m*
m. focusing magnetische Fokussierung (Strahlkonzentrierung, Bündelung) *f*
m. force magnetische Feldstärke *f*
m. forming magnetische Umformung *f*
m. gap *(Elm)* Polschuhabstand *m*, Luftspalt *m (im Magnetjoch)*
m. grader *(Met)* Magnetscheider *m*
m. head *(Dat)* Magnetkopf *m (zum Schreiben, Lesen und [oder] Löschen)*
m. hum magnetischer Brumm *m*
m. hysteresis magnetische Hysterese (Hysteresis) *f*
m. indicator for lightning currents Stahlstäbchen *n* zur Blitzstrommessung
m. induction magnetische Induktion (Flußdichte) *f*
m. initial permeability Anfangspermeabilität *f*
m. ink magnetische Tinte *f*; magnetische Druckfarbe *f*
m. ink character recognition Zeichenerkennung *f* mit magnetischer Tinte
m. ink imprinter Magnetschriftdrucker *m*
m. ink reader Magnetschriftleser *m*
m. ink ribbon Magnetschriftfarbband *n*
m. inspection *(Met)* Magnetdefektoskopie *f*, Magnetpulververfahren *n*
m. intensity magnetische Feldstärke *f*, Magnetfeldstärke *f*
m. interaction magnetische Wechselwirkung *f*
m. iron Magneteisenstein *m*
m. joint Stoßstelle *f* im Magnetkreis
m. lag magnetische Nachwirkung *f*
m. leakage magnetische Streuung *f*
m. leakage path magnetischer Streuweg (Streupfad) *m*
m. ledger card Magnetkontokarte *f*, magnetische Kontokarte *f*
m. lens magnetische Linse *f*
m. line of force magnetische Kraftlinie *f*
m. link *s.* m. indicator for lightning currents
m. lock Magnetschloß *n (z. B. Kühlschrank)*
m. matrix memory magnetischer Matrixspeicher *m*

m. memory Magnetspeicher *m*, magnetischer Speicher *m*; Magnetkern *m (elektronische Speichertechnik)*
m. memory matrix Magnet[kern]speichermatrix *f*
m. meridian magnetischer Meridian *m*
m. modulation magnetische Modulation *f*
m. modulator Tastdrossel *f*, Steuerdrossel *f*
m. moment magnetisches Moment *n*
m. multiplier magnetischer Vervielfacher *m*
m. needle Magnetnadel *f*, Kompaßnadel *f*
m. ore Magnetit *m*, Magneteisenerz *n*, Magneteisenstein *m*
m. ore concentration Magneterzanreicherung *f*
m. ore separator magnetischer Erzscheider *m*, Magneterzscheider *m*
m. particle coupling Magnetpulverkupplung *f*
m. particle method of crack detection Magnetpulverprüfung *f*
m. particles Magnetpulver *n*
m. permeability [magnetische] Permeabilität *f*
m. perturbation magnetische Störung *f*
m. plate memory Magnetplattenspeicher *m*
m. polarity magnetische Polarität *f*
m. polarization magnetische Polarisation *f*
m. pole Magnetpol *m*, magnetischer Pol *m*
m. potential magnetisches Potential *n*
m. potential difference magnetische Potentialdifferenz (Spannung) *f*
m. powder process Magnetpulververfahren *n*
m. powder testing Magnetpulverprüfung *f*
m. printing Kopiereffekt *m*
m. pull magnetische Anziehungskraft *f*, magnetischer Zug *m*
m. pyrites Magnetkies *m*
m. quantum number *(Kern)* magnetische Quantenzahl *f*
m. reactance induktiver Blindwiderstand *m*
m. recorder 1. magnetisches Registriergerät (Zählwerk) *n*; 2. Magnettongerät *n*
m. relaxation magnetische Nachwirkung (Relaxation) *f*
m. release magnetische Auslösung *f*
m. reluctance magnetischer Widerstand *m*, Reluktanz *f*
m. repulsion magnetische Abstoßung *f*
m. residual loss *(El)* Remanenzverlust *m*, Abfall *m* durch Remanenz
m. resistance magnetischer Widerstand *m*, Reluktanz *f*
m. resonance magnetische Resonanz *f*
m. retentivity Remanenz *f*, Remanenzinduktion *f*
m. return path magnetischer Rückschluß *m*
m. reversal Ummagnetisierung *f*
m. rigidity magnetische Steifigkeit *f*
m. roasting *(Met)* magnetisierende Röstung *f*
m.-rod memory Magnetstabspeicher *m*
m. saturation magnetische Sättigung *f*, Eisensättigung *f*
m. scattering Magnetstreuung *f*, magnetische Streuung *f*
m. scattering amplitude magnetische Streuamplitude *f*
m. separating belt Bandmagnetscheider *m*
m. separation Magnetscheiden *n*
m. separator *(Met)* Magnetscheider *m*, Eisenabscheider *m*
m. shield magnetische Abschirmung *f*
m. shunt magnetischer Nebenschluß *m*
m. slide coupling Magnetgleitkupplung *f*
m. sound Magnetton *m*
m. sound-playback Magnettonwiedergabe *f (Magnettontechnik)*
m. sound record copying machine Magnettonkopieranlage *f*
m. sound recording equipment (instrument) Magnettongerät *n*
m. spectrograph Magnetspektrograf *m*, magnetischer Spektrograf *m*, Geschwindigkeitsspektrograf *m*
m. spectrometer Magnetspektrometer *n*, magnetisches Spektrometer *n*
m. state magnetischer Zustand *m*
m. steel Magnetstahl *m*
m. stirrer Rührmagnet *m*
m. storm magnetische Störung *f*, magnetischer Sturm *m*, magnetisches Gewitter *n*
m. stray field Magnetstreufeld *n*
m. stray flux magnetischer Streufluß *m*
m. strip magnetischer Streifen *m (der auf Belegen vorgedruckt ist)*
m.-survey vessel Bereisungsschiff *n* für magnetische Messungen
m. susceptibility magnetische Aufnahmefähigkeit (Suszeptibilität) *f*
m. switching magnetisches Schalten *n*
m. tape Magnetband *n*, Magnetstreifen *m*; Magnettonband *n*, Tonband *n*
m. tape archive Bandarchiv *n*
m. tape control Magnetbandsteuerung *f*
m. tape memory magnetisches Speicherwerk *n*
m. tape recorder Magnettongerät *n*, Magnetbandaufnahmegerät *n*
m. track braking Schienenstrombremsung *f*
m. transfer Kopiereffekt *m*
m. transition point magnetischer Umwandlungspunkt *m*, Curie-Punkt *m*
m. trap magnetische Falle *f*
m. unit magnetische Einheit *f*
m. voltage stabilizer magnetischer Spannungskonstanthalter *m*
m. wire Magnetdraht *m (zur Datenspeicherung)*

m. wire recording Drahtaufnahme *f (Magnettontechnik)*
m. yoke magnetische Ablenkspule *f*, Magnetjoch *n*
magnetical *s.* magnetic
magnetically printed items *(Büro)* magnetisch gedruckte Posten *mpl*
magnetite Magnetit *m*, Magneteisenerz *n*, Magneteisenstein *m*
magnetizability Magnetisierbarkeit *f*
magnetizable magnetisierbar
magnetization Magnetisierung *f*
m. characteristic Magnetisierungskurve *f*
m. curve Magnetisierungskurve *f*
m. work Magnetisierungsarbeit *f*
magnetize / to magnetisieren
magnetizing Magnetisieren *n*, Aufmagnetisieren *n*
m. coil Magnetwicklung *f*, Magnetspule *f*; Feldspule *f*; Erregerwicklung *f*
m. current Magnetisierungsstrom *m*
m. field Erregerfeld *n*
m. force Magnetisierungsstärke *f*
magneto *(Kfz)* Magnet[zünder] *m*, Zündmagnet *m*
m. alternator Fahrraddynamo *m*; magnetelektrische Maschine *f*; *(Fmt)* Rufmaschine *f*
m. bearing Radialkugellager *n (einreihig, nicht selbsthaltend, einseitig wirkend)*
m. bell *(Fmt)* Wechselstromwecker *m*
m. board *(Fmt)* Klappenschrank *m (für Induktoranruf)*
m. booster *(El)* Magnetanlasser *m*, Anlaßmagnet *m*
m. call *(Fmt)* Induktor[an]ruf *m*
m.-electric relay magnetodynamisches Relais *n*, Drehspulrelais *n*
m. generator Magnetinduktor *m*; Lichtmagnetzünder *m*
m.-hydrostatic field magnetohydrostatisches Feld *n*
m. ignition Magnetzündung *f*
m. inductor Kurbelinduktor *m*
m.-motive *s.* magnetomotive
m.-ohmmeter Kurbelinduktor *m* für [Isolations-]Widerstandsmessung
m. ringer *(Fmt)* Rufinduktor *m*
m. ringing *(Fmt)* Induktor[an]ruf *m*
m. separator Magnetfilter *n*
m. speed indicator elektrischer Drehzahlmesser *m*, Tachogenerator *m*
m. switchboard *(Fmt)* Klappenschrank *m* für Induktoranruf
m. timing *(Kfz)* Magnetregelung *f*, Zündpunkteinstellung *f*
magnetoacoustic wave magnetoakustische Welle *f*
magnetochemistry Magnetochemie *f*

magnetohydrodynamics Magnetohydrodynamik *f*
magnetometer Magnetometer *n*, Neigungsmesser *m*
magnetomotive magnetomotorisch
m. force magnetomotorische Kraft *f*, MMK
magnetooptic rotation magnetische Drehung *f* der Polarisationsebene, Magnetorotation *f*, Faraday-Effekt *m*
magnetooptics Magnetooptik *f*
magnetor *s.* magnetic modulator
magnetostatic magnetostatisch *(z. B. Oszillator)*
magnetostriction Magnetostriktion *f*
m. oscillator Magnetostriktionsoszillator *m*, magnetostriktiver Schwinger *m*
magnetostrictive delay line magnetostriktives Verzögerungsglied *n*
magnetron Magnetron *n*, Magnetfeldröhre *f*
m. oscillator Magnetrongenerator *m*
magnification *(Opt)* Vergrößerung *f (s. a.* m. factor)
m. factor Gütezahl *f*, Gütefaktor *m*; Verstärkung *f (Funktechnik)*; Überspannungsfaktor *m (Oszillatorschaltung)*
m. in diameter lineare Vergrößerung *f*
m. of three diameters dreifache Vergrößerung *f*
magnifier Vergrößerungsglas *n*, Lupe *f*
m. spectacles Lupenbrille *f*
m. stand Lupenstativ *n*
magnify / to *(Opt)* vergrößern
magnifying *(Opt)* Vergrößern *n*
m. apparatus Vergrößerungsapparat *m*
m. glass Vergrößerungsglas *n*, Lupe *f*
m. lens Vergrößerungsglas *n*, Vergrößerungslinse *f*
m. power Vergrößerung[sfaktor *m*] *f (optisches System)*
magnitude Größe *f*, Wert *m*, Betrag *m*, Menge *f*
m. class Größenklasse *f (z. B. Sterne)*
m. of deflection Ablenkungsgröße *f (Turbine)*
m. of error Fehlergröße *f*
m. of oscillation Schwingungsgröße *f*
m. of the alternating current Wechselstromgröße *f*
m. of the command given *(Rak)* Aufschaltgröße *f*
magnolia metal Magnoliametall *n*, M-Metall *n (Bleilegierung)*
magslip Drehmelder *m*; Synchro-Servomechanismus *m*, Synchronantrieb *m (Radar)*
mahogany Mahagoni[holz] *n*
maiden trip *(Schiff)* Jungfernfahrt *f*, Jungfernreise *f*
m. voyage *s.* m. trip
Maier form Maier-Schiffsform *f*
mail 1. Post *f*; 2. *(Text)* Litzenauge *n (Jacquard)*

m. boat Postschiff *n*, Paketschiff *n*, Paketboot *n*; Linienfahrtschiff *n*
m. order catalogue Versandkatalog *m*
m. packet Postschiff *n*
m. steamer Postdampfer *m*
m. van *(Eis)* Postwagen *m*
m. wire *(Text)* Augendraht *m (Litze)*
mailed heald *(Text)* Helfenauge *n*, Maillon *n*, Litzenauge *n*
main 1. Hauptleitung *f*, Hauptrohr *n*, Hauptstrang *m (Rohrleitungsbau)*; 2. *(Brg)* Haupt[förder]strecke *f*
m. air gate *(Brg)* Hauptwettertür *f*
m. air reservoir Hauptluftbehälter *m*
m. airway *(Brg)* Hauptwetterstrecke *f*
m. anode Hauptanode *f*
m. assembly Haupt[bau]gruppe *f*
m. axis Hauptachse *f*
m. axis of inertia Hauptträgheitsachse *f*
m. bang Auslöseimpuls *m*, Steuerimpuls *m (Radar)*
m. bars Längsbewehrung *f (Stahlbeton)*
m. battery Hauptbatterie *f*
m. beam 1. *(Bw)* Unterzug *m*; Hauptträger *m*; 2. Hauptstrahl *m (Antennentechnik)*; 3. *(Kfz)* Fahrlicht *n*
m. bearing Grundlager *n*, Hauptlager *n*, Kurbelwellenlager *n*
m. bearing pedestal Kurbellagerbock *m*
m. bell *(Met)* Unterglocke *f (Gichtverschluß)*
m. belt Hauptförderband *n*
m. boiler *(Schiff)* Hauptkessel *m*
m. boom *(Schiff)* Großbaum *m*
m. bore Hauptbohrung *f*
m. brush *(El)* Hauptbürste *f*
m. bulk-head Hauptspant *n*
m. busbar *(El)* Hauptsammelschiene *f*
m. cable Hauptkabel *n*, Netzkabel *n*, Speisekabel *n*, Stammkabel *n*
m. carburetter Hauptvergaser *m*
m. carrier Hauptträger *m*
m. carrier cable Haupttragseil *n*
m. carrier wave Hauptträgerwelle *f*
m. centre office *(Fmt)* Knotenamt *n*
m. circuit Hauptstromkreis *m*; Primärstromkreis *m*
m. circuit connection Netzanschluß *m*
m. circuit of a switching device Hauptstrombahn *f* eines Schaltgerätes
m. circuit resistance Hauptstromwiderstand *m*
m. cock Haupthahn *m*; Hauptabsperrhahn *m (z. B. einer Gasleitung)*
m. column *(Masch)* Hauptständer *m*
m. condition Grundbedingung *f*
m. constituent Hauptbestandteil *m*, Hauptkomponente *f*
m. contact Hauptkontakt *m*
m. contractor *(Bw)* Hauptauftragnehmer *m*; Generalauftragnehmer *m*
m. control board Hauptschalttafel *f*
m. core 1. *(El)* Hauptkern *m*; 2. Kern *m* (Seele *f*) eines Seils
m. current 1. *(El)* Hauptstrom *m*, Netzstrom *m*; 2. Hauptströmung *f (z. B. der Luft)*
m. current-carrying contact Haupt[strom]kontakt *m*
m. current motor Hauptschlußmotor *m*, Hauptstrommotor *m*
m. current relay Primärrelais *n*
m. cut-out *(El)* Hauptsicherung *f*
m. cutoff signal *(Rak)* Hauptkommando *n*
m. cutting edge Hauptschneide *f*
m. cylinder roller Presseur *m*; Gegendruckwalze *f*
m. deck *(Schiff)* Hauptdeck *n*
m. detector Hauptdetektor *m*
m. dimensions Hauptabmessungen *fpl*
m. distribution centre *(El)* Hauptverteilung *f*
m. distribution frame *(El)* Hauptverteiler *m (Gestell)*
m. distributor Hauptverteiler *m*, Netzverteiler *m*
m. drain *(Schiff)* Hauptlenzleitung *f*
m. drive Hauptantrieb *m*
m. drive clutch Hauptantriebskupplung *f*
m. drive motor Hauptantriebsmotor *m*
m. drive shaft Hauptantriebswelle *f*
m. driving gear Kulissenrad *n*, Hubscheibe *f*
m. driving link Kulisse *f*, Kurbelschwinge *f*
m. electrode Hauptelektrode *f*
m. engine Hauptmaschine *f*; Hauptmotor *m*
m. entrance Haupteingang *m*
m. entrance signal Haupteinflugzeichen *n (Funknavigation)*
m. exchange *(Fmt)* Hauptamt *n*, Hauptvermittlungsstelle *f*
m. exciter *(El)* Haupterreger *m*, Haupterregermaschine *f*
m. exciter cable Haupterregerkabel *n*
m. factor Hauptfaktor *m*
m. fan Hauptlüfter *m*
m. feature Hauptmerkmal *n*
m. feeder *(El)* Hauptspeiseleitung *f*, Haupteinspeisung *f*
m. fleet Hochseeflotte *f*
m. floor *(Kfz)* Fondboden *m*
m. flue Sammelfuchs *m*, Hauptabzugskanal *m*
m. frame Grundspant *n (Holzschiffbau)*; Hauptrahmen *m*
m. frame of [a] machine Maschinengestell *n*
m. gallery *(Brg)* Hauptstollen *m*
m. gap Hauptentladungsstrecke *f (Röhren)*
m. gas pipe-line Hauptgasleitung *f*, Hauptgasrohr *n*
m. gas-tap Gashaupthahn *m*

m. gate *(Brg)* Grundstrecke *f*
m. girder Hauptträger *m*
m. hatchway *(Schiff)* große Ladeluke *f*
m. haulage drift *(Brg)* Hauptförderstrecke *f*
m. head lamp *(Kfz)* Hauptscheinwerfer *m*
m. hold *(Schiff)* Haupt[lade]raum *m*
m. hull *(Schiff)* Hauptrumpf *m (ohne Aufbauten)*
m. jet *(Kfz)* Hauptdüse *f*
m. keel *(Schiff)* Hauptkiel *m*
m. land Festland *n*
m. leakage 1. *(El)* Luftspaltstreuung *f*; 2. Leitungsverlust *m (Rohrleitung)*
m. light switch Hauptlichtschalter *m*
m. lighting Hauptbeleuchtung *f*
m. limit switch Grenzschalter *m*, Stromwächter *m*, Wächter *m*; Hauptendausschalter *m*
m. line 1. Sammelleitung *f*, Hauptleitung *f*, Hauptleitungsrohr *n*, Sammelkanal *m*; 2. *(Eis)* Normalbahn *f*, Hauptbahn *f*
m. line current *(El)* Hauptstrom *m*
m. line insulator Hauptleitungsisolator *m*
m. load-distributing centre Hauptlastverteilungszentrale *f (des Verbundnetzes)*
m. lobe Hauptkeule *f*, Hauptlappen *m (Antennentechnik)*
m. mast *(Schiff)* Großmast *m*, Hauptmast *m*
m. metering jet Hauptzumeßdüse *f (Vergaser)*
m. mirror Hauptspiegel *m (eines optischen Systems)*
m. motor *(Schiff)* [elektrischer] Hauptmotor *m*
m. panel Hauptschalttafel *f*
m. particulars Hauptdaten *pl*
m. phase Hauptphase *f*
m. piece of the rudder *(Schiff)* Ruderpfosten *m*, Ruderstock *m*
m. pipe Hauptleitung *f*
m. pipe-line *s.* m. line 1.
m. piping *s.* m. line 1.
m. potential *(El)* Hauptspannung *f*
m. propellant valve *(Rak)* Brennstoffhauptventil *n*
m. quantum number Hauptquantenzahl *f*
m. reinforcement Hauptbewehrung *f*
m. rescue centre *(Brg)* Hauptrettungsstelle *f*
m. reservoir Hauptbehälter *m*
m. resistance Hauptwiderstand *m*
m. ride *(Holz)* Hauptgestell *n*, Hauptschneise *f*
m. road 1. Hauptstraße *f*; Fernverkehrsstraße *f*; 2. *(Brg)* Haupt[förder]strecke *f*
m. roof *(Brg)* Hauptthangendes *n*
m. roof break Setzriß *m*, Hangendriß *m*
m. rope haulage *(Brg)* Einseilförderung *f*
m. rotor Hauptläufer *m*
m. rudder *(Schiff)* Hauptruder *n*
m. sail *(Schiff)* Großsegel *n*; Gieksegel *n*
m. sea Hochsee *f*
m. sewer Hauptkanal *m*, Hauptsammler *m (Entwässerung)*
m. shaft *(Masch)* Hauptwelle *f*
m. shaft bearing Hauptwellenlager *n*
m. shaft with helical splines Hauptwelle *f* mit Schraubenkeilen
m. shop Zentralwerkstatt *f*
m. sideband Hauptseitenband *n (Sendetechnik)*
m. silencer Auspuffhauptschalldämpfer *m*
m. sluice valve Hauptschieber *m*
m. span Mittelöffnung *f (Brückenbau)*
m. spindle Hauptspindel *f*
m. spindle bore Hauptspindelbohrung *f*
m. starting valve Hauptanfahrventil *n (Lokomotive)*
m. station Hauptbahnhof *m*
m. steam Frischdampf *m*
m. steam engine Hauptdampfmaschine *f*
m. steam-plant valve Hochdruck[dampf]-ventil *n*
m. stop Hauptregister *n (Orgel)*
m. stop valve Hauptabsperrventil *n*
m. stress Hauptbeanspruchung *f*, Hauptspannung *f*
m. switch Hauptschalter *m*, Netzschalter *m*
m. switch station Hauptschaltwarte *f*
m. switchboard Hauptschalttafel *f*
m. timber Grundspant *n (Holzschiffbau)*
m. topsail schooner Schonerbark *f*
m. transformer Haupttransformator *m*
m. transportation tunnel *(Brg)* Hauptförderstrecke *f*
m. turbine Hauptturbine *f*
m. undercarriage Hauptfahrwerk *n*
m. valve control unit *(Rak)* Schaltbatterie *f*
m. voltage Hauptspannung *f*
m. wale Plankengang *m (an der größten Schiffsbreite)*
m. watch spring Triebfeder *f* für Uhrwerk
m. winding *(El)* Hauptwicklung *f*
m. wire *(El)* Hauptleitung *f*
m. work-driving spindle Arbeitsspindel *f*
m. works Hauptwerk *n*
mainmast *(Schiff)* Großmast *m*, Hauptmast *m*
mains 1. Netz *n*, Lichtnetz *n*, Lichtleitungen *fpl*; 2. Hauptleitung *f*; Speiseleitung *f*; Rohrleitungsanlage *f*
m. aerial Lichtnetzantenne *f*
m. borne netzgebunden, über das Starkstromnetz kommend
m. connection Netzanschluß *m*
m. current Netzstrom *m*, Lichtstrom *m*
m. failure Netzausfall *m*
m.-failure set Notstromaggregat *n*
m.-fed netzgespeist
m. filter Netzsperre *f*, Störschutz *m*
m. fluctuation Netzschwankung *f*, Schwankung *f* der Netzspannung

m. frequency Netzfrequenz *f*
m. fuse Netzsicherung *f*, Starkstromsicherung *f*
m. hold *(Fs)* Netzsynchronisation *f*, Rasterverriegelung *f*
m. hum Netzbrumm *m*, Netzton *m*
m. lead Netzanschlußschnur *f*
m. leakage Leitungsverlust[e *pl*] *m*, Netz-[übertragungs]verlust[e *pl*] *m*
m. line hum Netzbrummen *n*, Wechselstromsummen *n*
m.-operated netzbetrieben, netzgespeist, für Netzanschluß
m. plug Netzstecker *m*
m. power unit Netzanschlußgerät *n*
m. receiver Netzanschlußempfänger *m (Radio)*
m. rectifier Netzgleichrichter *m*
m. set Netzanschlußgerät *n (Radio)*
m. supply Netzstromversorgung *f*, Netzanschluß *m*, Versorgung *f* aus dem Netz
m. switch Netzschalter *m*
m. transformer Netztransformator *m*, Netzwandler *m*
m. valve wechselstromgeheizte Röhre *f*
m. voltage Netzspannung *f*
m. voltage compensation Netzspannungskorrektion *f*
m. water Leitungswasser *n*
mainsail *(Schiff)* Großsegel *n*, Gieksegel *n*
maintain / to 1. *(El)* aufrechterhalten *(Strom)*; 2. pflegen, instandhalten, warten, unterhalten *(Gerät)*; 3. einhalten *(Toleranz)*
m. at a potential *(El)* auf einem Potential halten
maintained-vacuum rectifier *(Elm)* Gleichrichter *m* mit Vakuumhaltung, Pumpgleichrichter *m*
m. vibration Dauerschwingung *f*, erzwungene Schwingung *f*
maintaining Instandhaltung *f*
m. switch *(Am)* *s.* limit switch
maintenance Instandhaltung *f*, Pflege *f*, Unterhaltung *f*, Wartung *f (z. B. Maschinen, Geräte)*
m. chart Wartungsplan *m*
m. cost Instandhaltungskosten *pl*, Wartungskosten *pl*
m. free wartungsfrei
m. man Reparaturschlosser *m*
m. shop Reparaturwerkstatt *f*
m. testing schedule *(El)* Meßplan *m*, Plan *m* für regelmäßige Messungen
m. true bearing konstante wahre Peilung *f (Radar)*
m. works Wartungsarbeiten *fpl*, Pflegearbeiten *fpl*, Erhaltungsarbeiten *fpl*, Instandhaltungsarbeiten *fpl*
maisonette Maisonette-Wohnung *f (Zweietagenwohnung in einem mehrgeschossigen Wohnhaus)*
maize Mais *m*
m. sheller Maisentkerner *m*
major größere[r]; Haupt-
m. adjustment Grobeinstellung *f*
m. axis Hauptachse *f*, große Achse *f* der Ellipse
m. characteristic Haupteigenschaft *f*, typische Eigenschaft *f*
m. cycle Hauptperiode *f*, Hauptzyklus *m*
m. diameter Außendurchmesser *m (Außengewinde)*
m.-diameter fit Außenzentrierung *f (Keilprofil)*
m. diameter of a thread Gewindeaußendurchmesser *m*
m. diameter of external spline Keilwellenaußendurchmesser *m*
m. key *s.* m. mode
m. load Hauptlast *f (Rockwell-Härteprüfung)*
m. lobe Hauptkeule *f*, Hauptlappen *m (Antennentechnik)*
m. mode Dur *n (Tonart in der Musik)*
m. motor-vehicle component groups Kraftfahrzeughauptbaugruppen *fpl*
m. overhaul Grundüberholung *f*, Generalreparatur *f*
m. road Hauptverkehrsstraße *f*
m. scale Dur-Tonleiter *f (Musik)*
m. scale of equal intonation (temperament) temperierte Tonleiter *f*
m. segregation *(Met)* [normale] Blockseigerung *f*
m. triad Dur-Dreiklang *m (Musik)*
Majorana force *(Kern)* Majorana-Kraft *f*
Majorana particle *(Kern)* Majorana-Teilchen *n*
majorant *(Math)* Majorante *f*, obere Schranke *f (einer Punktmenge)*
majority *(Math)* Majorität *f*
m. carrier Majoritätsträger *m*
m. logic *(Dat)* Majoritätslogik *f*, Mehrheitslogik *f*
make / to 1. herstellen, erzeugen, fertigen, anfertigen, produzieren; 2. *(El)* einschalten *(Strom)*; aufbauen *(Feld)*; ansprechen *(Relais)*
m. a calculation eine Berechnung anstellen
m. a compromise einen Vergleich schließen *(Patentwesen)*
m. a connection *(El)* einen Anschluß herstellen
m. a constrained movement sich zwangsläufig bewegen *(Kinematik)*
m. a cosmic trip einen kosmischen Flug durchführen
m. a current einen Strom einschalten
m. a field *(El)* ein Feld aufbauen

m. a start in the work with *(Masch)* vorbohren mit
m. alive *(El)* Spannung legen
m. amends for Schadenersatz leisten für
m. and break the current den Strom ein- und ausschalten
m. blunt abstumpfen
m. close [ab]dichten
m. conducting leitend machen
m. dull abstumpfen
m. fast to *(Schiff)* festmachen an
m. for a port Hafen ansteuern (anlaufen)
m. interplanetary trips interplanetarische Flüge durchführen
m. lean magern *(Ton)*
m. level abgleichen
m. sail unter Segel gehen, Segel setzen, auslaufen
m. the space ship escape terrestrial gravity das Raumschiff aus dem Bereich der Erdanziehung bringen
m. up a quarrel eine Streitigkeit schlichten *(Patentwesen)*

make Bauart *f*, Fabrikat *n*, Fabrikmarke *f*
m.-and-brake Unterbrecher *m*
m.-and-brake ignition Abreißzündung *f*
m.-and-break contact Umschaltkontakt *m*
m.-and-break key Ein- und Ausschalter *m*
m.-and-break switching device Wechselkontaktschaltvorrichtung *f*
m. arc Einschaltlichtbogen *m*
m.-before-break Schließen *n* vor dem Öffnen *(eines Umschaltkontaktes)*
m.-before-break contact unterbrecherloser Umschaltkontakt *m*
m.-before-break contact unit Folgekontakt *m (Relais)*
m. contact Arbeitskontakt *m*, Schließkontakt *m*, Schließer *m*
m.-contact unit Arbeitskontakt *m (Relais)*
m. impulse *s.* m. pulse
m. of refuse *(Brg)* Bergeanfall *m*
m. pulse Schließungsimpuls *m*, Schließungsstromstoß *m*, Einschaltstoß *m*
m.-time Einschaltverzug *m*, Ansprechverzug *m*
m. transient *(El)* Einschwingung *f* beim Einschalten
m.-up Ausgleich *m*; Auffüllung *f*
m.-up hand *(Typ)* Metteur *m*
m.-up pan Verlängerungsblech *n (Förderer)*
m.-up water Zusatzwasser *n*, Frischwasser *n*

maker 1. Hersteller *m*; Herstellerfirma *f*; 2. *(sl, El)* Schließer *m*, Schließkontakt *m*, Arbeitskontakt *m*

makeready *(Typ)* Zurichtung *f*, Einrichten *n*

makeshift Notbehelf *m*
m. antenna Behelfsantenne *f*

making Herstellen *n*, Erzeugen *n*, Fertigen *n*
m. buoy *(Schiff)* Anseglungsboje *f*, Ansteuerungstonne *f*
m.-capacity Einschaltvermögen *n*, Einschaltleistung *f (Schalter)*
m. contact *(El)* Arbeitskontakt *m*, Schließer
m. current rating Nenneinschaltstrom *m*, Einschaltstrom *m*
m. light *(Schiff)* Landkennungsfeuer *n*, Anseglungsfeuer *n*, Ansteuerungsfeuer *n*
m. sketches Skizzieren *n*
m.-up 1. *(Schiff)* Stapelfüllholz *n*, Füllholz *n*; Aufpackung *f (der Ablaufschlitten)*; 2. *(Typ)* Umbruch *m*
m.-up piece Paßrohr *n*, Fassonrohr *n*, Formstück *n*

macro *s.* pseudo instruction

malachite *(Min)* Malachit *m*, Kupferspat *m*
m. green Malachitgrün *n*, Chinagrün *n*

maladjustment Fehljustierung *f*, Justierfehler *m*

malate Ester *m* (Salz *n*) der Äpfelsäure

malax / to weichmachen *(durch Kneten, Reiben, Rühren)*

malaxate / to verkneten, vermischen

male männlich *(Gewinde)*
m. coupling Außengewindekupplung *f*
m. die part Stempel *m*, Patrize *f*, Oberwerkzeug *n*
m. form Vollform *f*
m. gauge Lehrdorn *m*
m. part Innenteil *n*
m. plug *(El)* Steckerstift *m*
m. screw Schraubenbolzen *m*; Schraubenspindel *f*
m. spline Keil *m (Keilwelle)*
m. thread Außengewinde *n*, Bolzengewinde *n*
m. type of piston rod volle Kolbenstange *f*

maleate Ester *m* (Salz *n*) der Maleinsäure

maleic acid Maleinsäure *f*
m. anhydride Maleinsäureanhydrid *n*

malfunction *(Dat)* Störung *f*, Versagen *n*, Nichtfunktionieren *n*

malic acid Äpfelsäure *f*

mall *(Brg)* Schlägel *m*

malleability Verformbarkeit *f* unter Druckbeanspruchung *(Walzbarkeit, Hämmerbarkeit)*

malleable unter Druck verformbar *(schmiedbar, hämmerbar, walzbar)*
m. annealing furnace Temper[glüh]ofen *m*
m. cast iron Temperguß *m*
m. hard iron Temperrohguß *m*, Weißguß *m*
m. iron Schmiedeeisen *n*; Temperguß *m*
m.-iron foundry Tempergießerei *f*
m. pig iron Temperroheisen *n*

malleablize / to glühfrischen, tempern

malleablizing Tempern *n*
m. furnace Temperofen *m*

m. ore Tempererz *n*
mallet 1. Fäustel *m*, Holzhammer *m*, Schlägel *m*; 2. Poloschläger *m (Sport)*
malm *(Geol)* Malm *m*; künstlicher Mergel *m*
malonate Ester *m* (Salz *n*) der Malonsäure
malonic acid Malonsäure *f*
malt / to malzen, mälzen *(Brauerei)*
malt Malz *n*
m. beer Malzbier *n*
m. extract Malzextrakt *m*
m. grist Malzschrot *n (m)*
m. house Mälzerei *f*
m. kiln Malzdarre *f*
m. sugar Malzzucker *m*, Maltose *f*
maltese cross Malteserkreuz *n (Kinotechnik)*
maltha Erdteer *m*, Mineralteer *m*, Bergteer *m*
malting Herstellung *f* von Malz, Mälzen *n*
maltose Maltose *f*, Malzzucker *m*
mammoth aerial Mammutantenne *f*
m. pump Mammutpumpe *f*
man / to 1. bemannen *(Raumschiff)*; 2. besetzen *(Stellung; Arbeitsplatz)*
man 1. Mensch *m*; Mann *m*; Person *f*; Angestellter *m*; Arbeiter *m*; Soldat *m*; 2. Figur *f (Schach)*, Stein *m (Brettspiel)*
m. at the wheel *(Schiff)* Rudergänger *m*
m.-carrying rocket bemannte Rakete *f*
m.-created environmental activity *(Kern)* künstliche Untergrundaktivität *f*
m.-crimped yarn *(Text)* Garn *n* mit künstlicher Kräuselung
m.-hole *s.* manhole
m.-hour Arbeitsstunde *f*
m.-landing level *(Brg)* Seilfahrtbühne *f*
m.-made künstlich, synthetisch
m.-made earth satellite künstlicher Erdsatellit (Erdtrabant) *m*
m.-made fibre synthetische Faser *f*
m.-made isotope künstlich hergestelltes Isotop *n*
m.-made planet künstlicher Planet *m*
m.-made satellite künstlicher Satellit *m*, Sputnik *m*
m.-made static künstliche Funkstörung *f*
m.-of-war Kriegsschiff *n*
m.-riding *(Brg)* Personenbeförderung *f (söhlig)*; Seilfahrt *f (Schacht)*
m.-riding deck *(Brg)* Seilfahrtboden *m*
m.-riding platform *(Brg)* Seilfahrtbühne *f*
m.-riding station *(Brg)* Grubenpersonenbahnhof *m*
m. shaft *(Brg)* Seilfahrtschacht *m*
m. time Hilfszeit *f*
manageable manövrierfähig, handlich
management Betriebsleitung *f*, Direktion *f*; Handhabung *f*, Leitung *f*
m. of steam boilers Dampfkesselwartung *f*
m. of the port Hafenbetrieb *m*
m. of the ship Schiffsleitung *f*
m. personnel Verwaltungspersonal *n*
m. returns landwirtschaftlicher Reinertrag *m*
m. science Verwaltungswissenschaft *f*
mandelic acid Mandelsäure *f*
mandrel 1. Dorn *m*, Ziehdorn *m*; 2. Spanndorn *m*; Drehstift *m*, Lochdorn *m*, Spindel *f*; 3. Docke *f*, Dockenspindel *f*, Dockenstock *m (Drechslerei)*
m. drawing Stopfenzug *m*
m. for spring winding Wickeldorn *m*
m. press Dornpresse *f*
m. roll Dornwalze *f (Rohrwalzstraße)*
maneuver / to *s.* manoeuvre / to
manganate Manganat *n*, mangansaures Salz *n*
manganblende Manganblende *f*
manganese Mangan *n*
m. alloy Manganlegierung *f*
m. carbide Mangankarbid *n*
m. hard steel Manganhartstahl *m*
m. ore Braunstein *m*, Manganerz *n*
m. spar Manganspat *m*
m. steel Manganstahl *m*
manganesian manganhaltig
manganic acid Mangansäure *f*
m. compound Verbindung *f* des dreiwertigen Mangan
manganiferous manganhaltig
manganite Manganit *m*, Graumanganerz *n*
mangano-compound Verbindung *f* des zweiwertigen Mangan
manganous compound Verbindung *f* des zweiwertigen Mangan
manger 1. *(Schiff, Am)* Waschschott *n*, Waschbord *m*; 2. Futtertrog *m*, Krippe *f*
m. plate *(Am)* Wellenbrecher *m (auf dem Vorderdeck)*, Brechwasserplatte *f*
mangle / to mangeln
mangle Mangel *f*, Rolle *f*
manhole Mannloch *n (z. B. Kessel)*; Einsteigöffnung *f*, Einstieg *m*, Einsteigschacht *m*
m. chimney Arbeitsschacht *m*, Kabelschacht *m*
m. cover Mannlochdeckel *m*, Schachtdeckel *m*
m. dog Mannlochbügel *m*, Bügel *m* des Mannlochverschlusses
m. junction box *(begehbarer)* Schachtkabel[übergangs]kasten *m*
manifold 1. Rohrverzweigung *f*, Sammelleitung *f*, Sammelrohr *n*, Krümmer *m*; Verteilungsrohr *n*, Rohrverteiler *m*; Ladeleitung *f*; 2. *(Math)* Mannigfaltigkeit *f*
m. pressure Ladedruck *m*, Leitungsdruck *m*
m.-pressure control Ladedruckregler *m*
manila Manilahanf *m*
m. fibre Manilahanf *m*
m. hemp Manilahanf *m*

m. paper Manilapapier *n*, Hartpapier *n*
m. rope Hanfseil *n*
manipulate / to bedienen, betätigen, handhaben; umformen
manipulated variable Stellgröße *f*
manipulating valve Schaltventil *n*
manipulation Handhabung *f*, Bedienung *f*
manipulator 1. *(Met)* Kantvorrichtung *f*, Manipulator *m*; 2. *(Masch)* Wendeeinrichtung *f*; 3. *(Kern)* Manipulator, Ferngreifer *m*, Fernbedienungsgerät *n*
manless unbemannt
m. ship Schiff *n* ohne Besatzung
manned bemannt
m. balloon bemannter Ballon *m*
m. earth satellite bemannter Erdsatellit (Sputnik) *m*
m. face *(Brg)* belegter Abbau *m*
m. rocket bemannte Rakete *f*
m. space capsule bemannte Raumkapsel *f* *(für Flüge im kosmischen Raum)*
m. space flight bemannter Raumflug *m*
manner Art *f*, Weise *f*, Methode *f*
m. of representation Darstellungsweise *f*
m. of treatment Behandlungsweise *f*
Mannesmann process Schrägwalzverfahren *n* *(nach Mannesmann)*
Mannesmann tube Mannesmann-Rohr *n*
Mannesmann type rolling train Mannesmann-Walzstraße *f*
Mannheim gold Kupfergold *n*
manning Bemannung *f*, Besetzung *f*, Besatz *m* *(eines Schiffes)*
mannitol Mannit *m*
mannonic acid Mannonsäure *f*
manoeuvrability Lenkbarkeit *f*, Steuerbarkeit *f*, Wendigkeit *f*, Manövrierfähigkeit *f*
manoeuvre / to manövrieren
manoeuvring basin *(Schiff)* Wendebecken *n*
m. cam Steuernocken *m*
m. experiment *(Schiff)* Manövrierversuch *m*, Ruderversuch *m (mit einem Modell)*
m. station Steuerstand *m*, Manövrierstand *m* *(bei elektrischem Schiffsantrieb)*
m. test *(Schiff)* Drehkreisversuch *m*, Manövrierversuch *m*
m. trial *(Schiff)* Manövrierprüfung *f*, Steuerfähigkeitsprobe *f*
manograph Druckschreiber *m*
manometer Manometer *n*, Druckmesser *m*
manometric manometrisch
m. capsule Drucksonde *f*
m. lift manometrische Förderhöhe *f*, Druckhöhe *f*
m. pressure Förderdruck *m*
manostat Druckregler *m*, Manostat *m*, Pressostat *m*
manpower Arbeitskraft *f*
mansard roof gebrochenes Dach *n*, Mansardendach *n*
Mansfeld copper schist Mansfelder Kupferschiefer *m*
mantelpiece Kaminsims *m*
mantissa Mantisse *f (eines Logarithmus)*
mantle Glühkörper *m*; Überform *f (Gießerei)*; Wandverkleidung *f*
m. of chimney Rauchfang *m*
m. pillar Tragkranzsäule *f (Hochofen)*
m. ring Schachttragring *m (Hochofen)*
mantrip car *(Brg)* Personenwagen *m*
manual Handbuch *n*; Handakte *f*
manual manuell, von Hand
m.-automatic station Handautomatikumschalter *m*, Leitgerät *n*
m.-automatic transfer Handautomatikumschaltung *f*
m. blowpipe Handbrenner *m*
m. central office *(Fmt)* Handamt *n*, Handvermittlungsstelle *f*
m. control Handbetätigung *f*, Handsteuerung *f*, Handregelung *f*
m. controller handbetätigter (manueller) Steuerschalter *m*, Handregler *m*, Handsteuerschalter *m*
m. cranking Kurbeln *n* mit der Hand
m. element Griffelement *n*
m. exchange *(Fmt)* Handvermittlung *f*, Vermittlungsstelle *f* mit Handbetrieb
m. grind[ing] Handschliff *m*
m. high-pressure valve Hochdruckhandabsperrventil *n*
m. operation Handbetätigung *f*
m. release Handauslösung *f*
m. ringing *(Fmt)* Handruf *m*, Induktorruf *m*
m. start-up Handstart *m*
m. switch Handschalter *m*
m. switchboard *(Fmt)* Handschrank *m*, Handvermittlungsschrank *m*
m. tape relay Handübertragung *f* durch Lochstreifen
m. telephone system *(Fmt)* Handvermittlungssystem *n*
m. torch *s.* m. blowpipe
m. tuning Abstimmung *f* von Hand, Handabstimmung *f*
m. valve positioner Ventilhandsteuerungseinrichtung *f*
m. volume control handbetätigte Lautstärkeregelung *f*
m. welding Handschweißen *n*, Handschweißung *f*
manually von Hand
m. operated handbedient, handbetätigt, mit Handbetätigung
manufacture / to fertigen, herstellen, produzieren, erzeugen, fabrizieren, bauen

manufacture Fertigung *f*, Herstellung *f*, Erzeugung *f*, Produktion *f*, Bau *m*
manufactured head Setzkopf *m*
m. ice Kunsteis *n*
manufacturer Hersteller *m*, Produzent *m*
manufacturer's mark Fabrikzeichen *n*
manufacturing Fertigung *f*, Fertigen *n*, Herstellen *n*, Erzeugung *f*, Produktion *f*
m. branch Industriezweig *m*
m. cost Produktionskosten *pl*
m. efficiency Betriebsleistung *f*
m. engineer Betriebsingenieur *m*, Fertigungsingenieur *m*; Betriebswirtschaftler *m*
m. engineering Fertigungstechnik *f*
m. error Herstellungsfehler *m*
m. expenses Gestehungskosten *pl*
m. in series Serienfertigung *f*
m. lathe Produktionsdrehmaschine *f*
m. method Fertigungsmethode *f*, Produktionsmethode *f*, Bearbeitungsmethode *f*, Herstellungsgang *m*, Herstellungsmethode *f*
m. practice Betriebspraxis *f*, Fertigungspraxis *f*
m. process Fertigungsverfahren *n*, Arbeitsverfahren *n*, Herstellungsverfahren *n*, Produktionsverfahren *n*, Produktionsprozeß *m*, Produktionsgang *m*
m. program Fertigungsfolge *f*; Fabrikationsprogramm *n*, Fertigungsprogramm *n*
m. supervision Betriebsüberwachung *f*
m. technique Fertigungstechnik *f*
m. tolerance Herstellungstoleranz *f*
m. unit Fertigungseinheit *f*
manure Dünger *m*, Düngemittel *n*, Dung *m*, Mist *m*, Stalldung *m*, Jauche *f*
m. attachment Dungeinleger *m (Pflug)*
m. cart Dungkarre *f*
m. distributor Miststreuer *m*, Düngerstreumaschine *f*
m. drill Jauchedrillgerät *n*
m. dumping cart Mistkippwagen *m*
m. loader Mistlader *m*
m. pump Jauchepumpe *f*
m. shredder Mistzerreißer *m (Düngerstreuer)*
manurial value Düngewert *m*
manuring Düngen *n*
m. effect Düngewirkung *f*
m. lime Düngekalk *m*
m. salt Düngesalz *n*
manway compartment *(Brg)* Fahrtentrum *m*
many-angled polygonal
m.-body forces *(Ph)* Mehrkörperkräfte *fpl*
m.-body problem Mehrkörperproblem *n*
m.-coloured bunt
m.-electron spectrum Mehrelektronenspektrum *n*
m.-faced vielflächig
m.-line spectrum Viellinienspektrum *n*
m.-particle system *(Kern)* Mehrteilchensystem *n*
m.-to-many *(Math)* mehrmehrdeutig
m.-to-one *(Math)* mehreindeutig
m.-valued vielwertig, mehrwertig
manysided vielseitig
manystage mehrstufig, Mehrstufen-
map / to 1. *(Math)* abbilden *(Menge)*; einander zuordnen *(Elemente zweier Mengen)*; 2. kartieren; 3. aufnehmen, aufzeichnen
map 1. Karte *f*, Landkarte *f*; 2. *s.* mapping 2.
m. case Kartentasche *f*
m. drawer Kartenzeichner *m*, Kartograf *m*
m. location Kartenstandort *m* *(Navigation)*
m. of mine Grubenkarte *f*, Grubenriß *m*
m. of network *(El)* Netzplan *m*, Leitungsplan *m*
m. printing Kartografie *f*
m.-reader function generator kartenlesender Funktionsgenerator *m (Analogrechner)*
maple Ahornholz *n*
m. sugar Ahornzucker *m*
m. syrup Ahornsirup *m*
m. wood Ahornholz *n*
mapping 1. Kartierung *f (Radar)*; Landkartenherstellung *f*; 2. *(Math)* Abbildung *f*
m. instrument Kartiergerät *n*
mar / to beschädigen, zerkratzen
mar resistance Kratzfestigkeit *f*
marble / to marmorieren *(Papier)*
marble Marmor *m*
m. cutter Marmorschneider *m*
m. lime Marmorkalk *m*
m. polisher Marmorschleifer *m*
m. quarry Marmor[stein]bruch *m*
m. slab Marmorplatte *f*
marbled marmoriert, geadert, gemasert
m. edge *(Typ)* Marmorschnitt *m*, marmorierter Schnitt *m*
m. paper marmoriertes Papier *n*, Marmorpapier *n*
marblize / to marmorieren
marbling Marmorieren *n*
marcasite *(Min)* Markasit *m*, Graueisenerz *n*
Marconi antenna Marconi-Antenne *f*, Lambda-Viertel-Antenne *f*
Marconi transmitter Marconi-Sender *m*
mare Stute *f*
margaric acid Margarinsäure *f*
margarite *(Min)* Margarit *m*, Kalkglimmer *m*
margin / to *(El)* *(das Relais auf Ansprechwert)* einstellen
margin 1. Rand *m*, Grenze *f*; Spielraum *m*, Toleranz *f*; Steg *m*; 2. Fase *f*, Führungsfase *f*; Fasenbreite *f*
m. line *(Schiff)* Tauchgrenze *f*

m. of commutation Sicherheitswinkel *m* *(Stromrichter)*
m. of error Fehlerspielraum *m*, Streubereich *m*
m. of picture *(Fs)* Bildrand *m*
m. of safety Sicherheitsfaktor *m*, Sicherheitsspanne *f*
m. of stability *(Fmt)* Abstand *m* vom Pfeifpunkt
m. plank *(Schiff)* Leibholz *n*; Randplanke *f*, Begrenzungsplanke *f*
m. plate bracket *(Schiff)* Kimmstützplatte *f*
m. strip Randstreifen *m* *(Straßenbau)*
marginal am Rande befindlich, Rand-, Grenz-
m. checking *(Dat)* Grenzwertprüfung *f*, Randwertprüfung *f*
m. coefficient Randkoeffizient *m*, Leckfaktor *m*
m. condition Randbedingung *f*
m. current Grenzstrom *m*
m. discharge *(El)* Randentladung *f*
m. note Marginale *f*, Randbemerkung *f*
m. punched card Randlochkarte *f*
m. ray Randstrahl *m*
m. seam *(Brg)* bedingt bauwürdiges Flöz *n*
m. sharpness *(Opt)* Randschärfe *f*
m. strip Randstreifen *m* *(Straßenbau)*
m. value Randwert *m*
m. vortex Randwirbel *m*
m. width *(Am)* Fasenbreite *f*
marimeter Echolot *n*
marina *(Am)* Bootshafen *m*; Jachthafen *m*
marine Seewesen *n*, Marine *f*
m. architect Schiffbauingenieur *m*
m. boiler Schiffskessel *m*, Marinekessel *m*
m. cable Seekabel *n*
m. cargo handling equipment Schiffsumschlagseinrichtung *f*
m. coating Schiffsfarbanstrich *m*
m. construction Seebau *m*
m. correspondence system Schiffswechselsprechanlage *f*
m. crank *s.* m. crankshaft
m. crankshaft Schiffskurbelwelle *f*
m. Diesel engine Schiffsdieselmotor *m*
m. dredger Seebagger *m*
m. elevator Schiffsaufzug *m*, Bordaufzug *m*, Schiffslift *m*
m. engine Schiffsmaschine *f*, Schiffsmotor *m*
m. engineer Schiffsmaschinist *m*, Seemaschinist *m*, Schiffsingenieur *m*, Marineingenieur *m*
m. engineering Schiffsmaschinenbau *m*
m. fouling Anwuchs *m*, Schiffsbewuchs *m*
m. galvanometer Marinegalvanometer *n*
m. glue Marineleim *m*
m. insulation Schiffsisolierung *f*
m. insurance Seetransportversicherung *f*
m. motor *(elektrischer)* Schiffsmotor *m*
m. oil Marineöl *n*
m. paint Schiffs[anstrich]farbe *f*
m. primer Schiffsgrundierung *f*
m. propelling machinery Schiffsantriebsmaschinen *fpl*
m. propulsion Schiffsantrieb *m*
m. propulsion machinery Schiffsantriebsanlage *f*
m. radio communication Seefunk *m*
m. radio service Seefunkdienst *m*
m. radio station Bordfunkstelle *f*, Seefunkstelle *f*
m. railway Aufschlepphelling *f*
m. reduction gear Schiffsuntersetzungsgetriebe *n*
m. rope Schiffstau *n*
m. shaft Schiffswelle *f*
m. spar varnish Bootslack *m*
m. steam engine Schiffsdampfmaschine *f*
m. structure Küstenbauwerk *n*, Seebauwerk *n*
m. structures Küstenbauten *pl*, Seebauten *pl*
m. transport Seetransport *m*, Seeverkehr *m*
m. turbine Schiffsturbine *f*
m. welding Unterwasserschweißen *n*
m. works Küstenbau *m*, Seebau *m*; Küstenbauten *pl*, Seebauten *pl*
Mariotte's law Boyle-Mariottesches Gesetz *n* *(Gasgesetz)*
maritime See-; Schiffahrt[s]-; maritim
m. belt Hoheitsgewässer *n*
m. construction Seebau *m*
m. navigation Seeschiffahrt *f*
m. peril Seegefahren *fpl*, Seeunfall *m*
m. railway station Hafenbahnhof *m*
m. territory Hoheitsgebiet *n*
mark / to markieren, bezeichnen, anzeichnen, kennzeichnen, anreißen, signieren; vorzeichnen; stempeln; ankörnen
m. down aufzeichnen
m. off anreißen, abtragen *(Maß auf Zeichnung)*; abstecken *(Vermessung)*
m. out anreißen; auszeichnen
m. over übertippen
mark 1. Marke *f*; Riefe *f*; Eindruck *m*; 2. *(Dat)* Impuls *m*, Zeichen *n*
m. recess Beschriftungsstelle *f* [am Spiralbohrer], Einstich *m*
m. scraper Anreißnadel *f*
m. sensing Zeichenlochung *f* *(Lochkartenmaschinen)*
m.-sensing punch card Zeichenlochkarte *f*
m.-sensing reproducer (punch) Kartendoppler *m* mit Zeichenabfühlung
m.-space relation *(Dat)* Tastverhältnis *n*
marked riefig; bezeichnet
m. by letters buchstabengekennzeichnet
m. core Zählader *f* *(Kabel)*

m. pole Nordpol *m*

m. ratio *(El)* Nennübersetzungsverhältnis *n*

marker 1. *(Masch)* Anreißnadel *f*; 2. *(Lw)* Spurreißer *m*, Markör *m*; 3. *(Flg)* Einflugzeichen *n*, Signalbake *f*

m. beacon Funkfeuer *n*, Markierungsbake *f*; Einflugzeichensender *m*

m. frequency Kennfrequenz *f*

m. generator Frequenzmarkengenerator *m*, Markengenerator *m*

m. lamp Positionslampe *f*

m. light Positionslampe *f*

m. out Anreißer *m*

m. pip Markierungsimpuls *m (Impulstechnik)*

m. pulse Meßmarke *f*

m. range ring Abstandsring *m (Radarschirm)*

market research Marktforschung *f*

m. value Kaufwert *m*

marking 1. Anriß *m*, Anzeichnen *n*; Kennzeichnung *f*; 2. Arbeitszustand *m (Telegrafie)*

m. awl Reißspitze *f*

m. brush Signierpinsel *m*

m. chalk Signierkreide *f*

m. contact *(Fmt)* Zeichenstromkontakt *m*

m. current *(Fmt)* Zeichenstrom *m*

m. device Anreißgerät *n*

m. gauge Parallelreißer *m*

m. head Schreibkopf *m*

m. in degrees Gradeinteilung *f*

m. line Absteckleine *f (Vermessung)*

m. machine Signiermaschine *f*; Straßenmarkierungsmaschine *f*

m. microscope Punktiermikroskop *n*

m. off Anreißen *n*

m.-off board Reißboden *m*

m.-off operation Anreißarbeit *f*

m.-off table Anreißplatte *f (Vermessung)*

m.-off tool Reißzeug *n*, Anreißzeug *n*

m. out Anreißen *n*

m. punch Punktiereisen *n*

m. stencil Anreißschablone *f*

m. system Kennzeichnungssystem *n*

m. template Anreißschablone *f*

m. thread Kennfaden *m (des Kabels)*

m. tool Reißnadel *f*, Reißspitze *f*, Anreiß[werk]zeug *n*

m. wave Zeichenwelle *f (Telegrafie)*

markings Markierungszeichen *npl*; Straßenmarkierung *f*

Markov process *(Math)* Markow-Prozeß *m*

marks and deeps *(Schiff)* Lotleinenmarkierungen *fpl*

marl 1. Mergel *m*, Steinmergel *m*; 2. *(Text)* Mouliné *m*

m.-clay Mergelton *m*, Tonmergel *m*

m. slate Mergelschiefer *m*

m.-stone sandiger Kalkstein *m*

marling Mergeln *n (Acker)*

marly sandstone Mergelsandstein *m*

marmatite *(Min)* Marmatit *m*, Eisenzinkblende *f*

marmolite *(Min)* Marmolith *m*

marquenching Warmbadhärten *n*, Stufenhärtung *f*

marquetry Einlegearbeit *f*, Holzeinlegearbeit *f*

married fall system *(Schiff)* gekuppeltes Ladeseilsystem *n*

marsh Sumpf *m*; Marsch *f*

m. gas Grubengas *n*, Sumpfgas *n*

m. ore *(Met)* Raseneisenerz *n*

marshalling vehicle Rangierfahrzeug *n*

m. yard Rangierbahnhof *m*

Marsh's test Marshsche Probe *f (Arsennachweis)*

marshy sumpfig, morastig, versumpft

m. ground Moorboden *m*

m. meadow Moorwiese *f*

m. soil Moorboden *m*

martempering *(Met)* Warmbadhärten *n*, Stufenhärtung *f*

martensite *(Met)* Martensit *m*

m. formation Martensitbildung *f*

martensitic *(Met)* martensitisch

m. steel Martensitstahl *m*

Martin process *(Met)* Siemens-Martin-Verfahren *n*, Martinverfahren *n*

maser Maser *m*, Molekularverstärker *m*

m. amplifier Maser-Verstärker *m*

m. beam Maser-Strahl *m*

m. transition Maser-Übergang *m*

mash Maische *f*

m. tub Maischekochkessel *m*, Maischpfanne *f*

m. vat Braubottich *m*

m. wort Maischwürze *f*

masher Stampfer *m*

mask / to abdecken, verdecken, verblenden

mask 1. Maske *f*, Schablone *f*; Abdeckblende *f*; 2. *(Fs)* Bildröhrenmaske *f*, Strahlenmaske *f (Farbbildwiedergaberöhren)*

m. shield Bremsschild *m (Stapellauf)*

masked head lamp *(Kfz)* Tarnscheinwerfer *m*

masking 1. *(Fmt, Fs)* Verdeckung *f*, Verschleierung *f*; 2. *(Typ)* Maskieren *n*

m. colour Deckfarbe *f*

m. effect Verdeckungseffekt *m (z. B. bei akustischen Signalen, Impulsen usw.)*

m. ink *(Typ)* Abdecktusche *f*, Deckfarbe *f*

m. plate *(Fs)* Rasterblende *f*

mason Maurer *m*; Steinmetz *m*

masonry Mauerwerk *n*; Natursteinmauerwerk *n*, Werksteinmauerwerk *n*; Steinmauern *n*

m. anchor Maueranker *m*

m. cavity wall Hohlmauerwerk *n*

m. drill Mauerbohrer *m*, Steinbohrer *m*

m. lining *(Brg)* Mauerausbau *m*

m. mortar Baumörtel *m*

m. saw Steinsäge *f*
mason's hammer Steinspalthammer *m*
m. level Lot *n*
mass / to anhäufen, sammeln
m.-duplicate in Masse fertigen
m.-produce in Massenfertigung (Massenproduktion) herstellen
mass Masse *f*; Menge *f*
m. absorption coefficient Massenabsorptionskoeffizient *m*
m. action constant Massenwirkungskonstante *f*
m. analyzer Massenanalysator *m*
m. assignment Massenbestimmung *f*
m. centre Massenmittelpunkt *m*
m. coefficient of absorption Massenabsorptionskoeffizient *m*
m. colour absolut deckende Farbe *f*
m. concentration Massenkonzentration *f*, Konzentration *f* in Masseprozenten
m. concrete unbewehrter Beton *m*; Massenbeton *m*
m.-core Eisenpulverkern *m*, Massekern *m*, Pulverkern *m*
m. defect *(Kern)* Massendefekt *m*
m. density Massendichte *f*
m. destruction Massenvernichtung *f*
m. destruction weapon Massenvernichtungswaffe *f*
m. dispersion Massenstreuung *f*
m. effect *(Ph)* Packungseffekt *m*
m. energy equivalence Massen-Energie-Äquivalenz *f*
m. equation Massengleichung *f*
m. equivalent Masseäquivalent *n*
m. flow Massendurchsatz *m*, Massenstrom *m*
m. flowmeter Durchflußmesser *m*
m. formula Massenformel *f*
m. inspection Mengenkontrolle *f*
m. moment statisches Moment *n*
m. moment of inertia Massenträgheitsmoment *n*
m. motion Massenbewegung *f*
m. number Massenzahl *f*; Atomgewicht *n*
m. of the electron Elektronenmasse *f*
m. of the nucleus *(Ph)* Kernmasse *f*
m. polymerization Blockpolymerisation *f*
m.-produced in Massenfertigung hergestellt
m.-produced part Massenteil *n*
m. production Massenfertigung *f*, Massenproduktion *f*
m. production technique Massenfertigungsmethode *f*
m. radiator *(Ph)* Massenstrahler *m*
m. ratio Massenverhältnis *n*
m. segregation *(Met)* Blockseigerung *f*
m. separator Massentrenner *m*
m. spectrogram Massenspektrogramm *n*
m. spectrograph Massenspektrograf *m*
m. spectrometer Massenspektrometer *n*
m. spectrum Massenspektrum *n*
m. transfer Stoffaustausch *m*; Stoffübergang *m*, Massenübergang *m*
m. transfer coefficient *(Kern)* Massendurchgangszahl *f*, Stoffaustauschzahl *f*
m.-type plate *(El)* Masseplatte *f*
m. unit Masseeinheit *f*, Einheit *f* der Masse
massecuite Füllmasse *f (Zuckerherstellung)*
massicot Bleioxid *n*, Bleigelb *n*, Bleiglätte *f*
massif Gebirgsmassiv *n*
massive massiv, vollwandig
m. arch *(Bw)* Massivbogen *m*
m. tread *(Bw)* Blockstufe *f*
massless masselos
mast Mast *m*, Mastbaum *m*; Standmast *m*; Leitungsmast *m*, Gittermast *m*
m. ball *(Schiff)* Flaggenknopf *m*
m. base Mastfuß *m*
m. foundation Mastfundament *n*
m.-head lantern *(Schiff)* Topplaterne *f*, Topplicht *n*
m.-mooring gear *(El)* Mastgeschirr *n*
m. partners *(Schiff)* Mastfischung *f*, Mastpüsching *f (zwei Einspannhölzer für den Mast im Deck)*
m. rigging Mastverspannung *f*
m. rods Mastgestänge *n*
m. stack *(Schiff)* Macks *(Mast-Schornstein-Konstruktion)*
master 1. Kopierschablone *f*, Meisterstück *n (Kopieren)*, Musterstück *n*; Normale *f*, Einstellnormale *f*; Abnahmelehre *f*; 2. Hauptteil *n*; 3. Kapitän *m (eines Handelsschiffes)*, Schiffsführer *m*, Schiffer *m*; 4. Vaterplatte *f*, Vater *m (galvanische Kopie der Mutterschallplatte)*
m. aerial Gemeinschaftsantenne *f*
m. alloy *(Met)* Vorlegierung *f*
m.-and-slave control *(BMSR)* Kaskadenregelung *f*, Nachführregelung *f*
m.-and-slave mechanism *(BMSR)* Folgeregler *m*, Führungs- und Folgemechanismus *m*
m. bushing Grundbuchse *f*
m. cam Kopiernocken *m*
m. card *(Dat)* Hauptkarte *f*, Leitkarte *f*, Matrizenkarte *f*
m. carpenter selbständiger Zimmermeister *m*
m. chuck Zentrierknopf *m*
m. clock Taktgeber *m*, Zeitgeber *m*; Hauptuhr *f*
m. collet Patrone *f*
m. compass Mutterkompaß *m*, Kompaßmutter *f*, Hauptkompaß *m*
m. component Musterstück *n*, Bezugsstück *n*
m. connecting-rod assembly Pleuelstern *m*
m. control *(Fmt, Fs)* zentrales Mischpult *n*
m. control valve Hauptkontrollschieber *m*

m. controller *(El)* Meisterwalze *f*, Meisterschalter *m*; Führerschalter *m*; Hauptsteuerschalter *m*, Steuerwalze *f*; *(BMSR)* Führungsregler *m*, Primärregler *m (Kaskadenregelung)*
m. craftsman patent Meisterbrief *m*
m. crusher roll Meisterprofilrolle *f*
m. cylinder *(Kfz)* Hauptzylinder *m*
m. cylinder of a brake Hauptbremszylinder *m*
m. flask Musterformkasten *m*
m. form 1. Lehrenform *f*; Bezugsform *f*; Modell *n*; Kopiermodell *n*; Schablone *f*; 2. Urmodell *n (Elektrotypie)*
m. frequency Mutterfrequenz *f*
m. frequency meter integrierender Frequenzmesser *m*
m. gain control Hauptverstärkungsregler *m*
m. gauge Prüflehre *f*, Kontrollehre *f*
m. gear Meisterrad *n*, Urrad *n*, Lehrzahnrad *n* (*s. a.* m. gear tool)
m. gear of differential *(Kfz)* großes Differentialantriebskegelrad *n*
m. gear tool Zahnradwerkzeug *n (Schneidrad)*
m. gripping collet Spannpatrone *f*
m. gyro *(Schiff)* Kreiselmutter *f*, Mutterkompaß *m*
m. handle *(Kern)* Steuergriff *m (Fernbedienungsgerät)*
m. hole Bezugsbohrung *f*
m. jaw Grundbacke *f (Futter)*
m. key Hauptschlüssel *m*
m. lode *(Geol)* Hauptgang *m*, Hauptader *f*
m. mechanic Werkstattleiter *m*; Vorarbeiter *m*, Meister *m*
m. monitor Endkontrolleinrichtung *f (zur Fernsehbildkontrolle)*
m. number double contact *(Fmt)* Doppelsammelkontakt *m*
m. oscillator *(Fmt)* Hauptoszillator *m*, Röhrenoszillator *m*; Steuersender *m*, Steueroszillator *m*
m. panel Hauptschalttafel *f*
m. patent Grundpatent *n*
m. piece *(Masch)* Meisterstück *n*
m. pilot lamp Hauptüberwachungslampe *f*
m. plan Bebauungsplan *m*, Generalbebauungsplan *m*
m. plate Schablone *f*; Kopierlineal *n*
m. program *(BMSR)* Hauptprogramm *n*, Leitprogramm *n*
m. roll Meisterprofilrolle *f*
m. service *(Fmt)* Gemeinschaftsanschluß *m*
m. smelter Gießereimeister *m*
m. square Kontrollwinkel *m*, Prüfwinkel *m*
m. station Hauptsender *m (Decca)*
m. switch 1. Hauptschalter *m*; 2. *(Fmt)* Gruppenwähler *m*
m. tape *(Büro)* Hauptband *n*, Leitband *n*, Urband *n*
m. telephone Hauptapparat *m*
m. telephone transmission reference system Ureichkreis *m* für die Fernsprechübertragung
m. template *(Masch)* Kopierschablone *f*
m. transmitter Hauptsender *m*, Muttersender *m*
m. workman [ausgezeichneter] Facharbeiter *m*, Fachmann *m*; Meister *m*
master's certificate *s.* m. patent
m. patent Schiffskapitänsdiplom *n*, Kapitänspatent *n*
mastic Mastix *m*; Kitt *m*, Harz *n*; Gußasphalt *m*
m. asphalt Gußasphalt *m*
m. filler Mastixvergußmasse *f*
m. flooring Gußasphaltbelag *m*
m. pitch Pechharz *n*
masticate / to kneten, mastizieren *(Plaste)*
mastication Mastikation *f (Kautschuk)*
masticator Kneter *m*, Knetmaschine *f*, Mischkneter *m*
masting *(Schiff)* Bemastung *f*, Mastwerk *n*
m. sheers *(Schiff)* Kran *m* zum Masteinsetzen, Mastkran *m*
mastix *s.* mastic
masurium Technetium *n*
masut *s.* mazout
mat Faserfilz *m*, Filz *m*; Matte *f*; Teppichbelag *m (Straßenbau)*; *(Brg)* Abbaumatte *f*; *(Typ)* Mater *f*, Stereo *n*
mat matt, glanzlos, mattiert
m. finishing Mattierung *f*
m. foundation Plattengründung *f*; Fundamentplatte *f*
m. moulding Mattenpreßverfahren *n*
m. of fibres Faserfilz *m*, Matte *f*, Vlies *n*
m. paper Mattpapier *n*
m. revetment Mattenbekleidung *f*
m.-scoria Rohschlacke *f*
m. weave *(Text)* Mattenbindung *f*
match / to 1. aufeinanderpassen, zusammenpassen; 2. anpassen, angleichen *(z. B. Kabel)*; 3. *(Büro)* aussteuern *(Kartenmischer)*; 4. *(Text)* nach Muster färben
match 1. Zündholz *n*, Streichholz *n*; 2. *(Brg)* Zündlunte *f*
m. board Nut- und Federholz *n*, gespundetes Holz *n*, Spundbrett *n*
m.-boarding gespundete Bretter *npl*
m. box Streichholzschachtel *f*
m. cord Zündschnur *f*
m. joint Verzinkung *f*, Spundung *f (Holz)*
m. making machine Zündholzmaschine *f*
m.-merge *(Büro)* Mischen *n* mit Aussteuern
m. plane Nuthobel *m*, Spundhobel *m*
m. rope Zündschnur *f*

m. wood Holz *n* für Streichholzherstellung, Holzdraht *m*
matched boards gespundete Bretter *npl*, Spundbretter *npl*, Spundbohlen *fpl*
m. junction angepaßtes Verzweigungsglied *n*, reflexionsfreie Verzweigung *f (Wellenleiter)*
m. termination angepaßter Abschluß *m (Wellenleitertechnik)*; reflexionsfreier Abschluß
m. waveguide angepaßter Hohlleiter *m (Wellenleiter)*
matcher [Nut- und] Spundmaschine *f*
m. bit [Nut- und] Spundmesser *n*
matching 1. Anpassung *f*; 2. Zusammensetzen *n* von Furnieren; 3. Nut *f* und Feder *f*, Spundung *f*
matching aufeinanderpassend, zusammenpassend
m. attenuation *(El)* Anpassungsdämpfung *f*
m. diaphragm Blende *f (Wellenleiter)*
m. edge Anschlag *m*
m. for optimum power transfer *(El)* Leistungsanpassung *f*
m. impedance Anpassungsscheinwiderstand *m*
m. line Anpassungsleitung *f*, Leitungstransformator *m*
m. machine Nuthobelmaschine *f*
m. network Anpassungsnetzwerk *n*
m. part Anschlußteil *n*
m. pillar Anpassungsstift *m (Wellenleiter)*
m. plane [Nut- und] Spundhobel *m*
m. plate Impedanzanpassungsblende *f (Wellenleiter)*
m. section Anpassungsleitung *f*, Leitungstransformator *m*; Transformationsglied *n*; Anpassungsglied *n (Wellenleitertechnik)*
m. stub Anpaßblindleitung *f*, Anpaßstichleitung *f (Antennen)*
m. transformer Anpassungstransformator *m*, Anpaßtransformator *m*, Zwischentransformator *m*; Anpaßübertrager *m*
mate / to 1. paaren; ineinandergreifen, eingreifen, kämmen *(Zahnräder)*; 2. fügen
material materiell, stofflich; substantiell
material Material *n*, Werkstoff *m*; Stoff *m*, Substanz *f*; Stoff *m*, Zeug *n*, Gewebe *n*
m. allowance *(Masch)* Werkstoffzugabe *f*
m. buckling *(Kern)* Materialflußwölbung *f*
m. chain *(Ch)* Stoffkette *f (Reaktionskinetik)*
m. constant Werkstoffkonstante *f*
m. consumption Materialverbrauch *m*
m. damage Materialschaden *m*
m. damping Werkstoffdämpfung *f*
m. defect Werkstoffehler *m*, Materialfehler *m*
m.-dependent werkstoffabhängig
m. economy Materialausnutzung *f*
m. for investigation Untersuchungsmaterial *n*, Versuchsmaterial *n*
m. forced away *(Met)* verdrängter Werkstoff *m (beim Walzen)*
m. handling device *(Masch)* Werkstofffördereinrichtung *f*
m.-handling system *(Ök)* Transportsystem *n*
m.-independent werkstoffunabhängig
m. inducement *(Ök)* materieller Anreiz *m*
m. loss Werkstoffverlust *m*; Werkstoffverschwendung *f*
m. lowering *(Brg)* Materialeinhängen *n*
m. point Massenpunkt *m*
m. properties Werkstoffeigenschaften *fpl*
m. saving Werkstoffeinsparung *f*
m. shortage Werkstoffknappheit *f*
m. specification Gütevorschrift *f (für Werkstoff)*
m. stock Materiallager *n*
m. test specification Materialprüfvorschrift *f*
m. testing Werkstoffprüfung *f*; Defektoskopie *f*
m. testing laboratory Materialprüflaboratorium *n*, Werkstoffprüflaboratorium *n*
m. testing machine Werkstoffprüfmaschine *f*
m. testing reactor Materialprüfreaktor *m*
m. to be dried Trockengut *n*
m. to be ground Mahlgut *n*
m. to be mixed Mischgut *n*
m. to be roasted Röstgut *n*
m. under investigation Untersuchungsmaterial *n*, Versuchsmaterial *n*
m. well *(Plast)* Füllraum *m*
materialization *(Ph)* Materialisierung *f*, Paarbildung *f*
materials Baustoffe *mpl*; Werkstoffe *mpl*
m. engineering Werkstofftechnik *f*
m. handling Werkstoffhandhabung *f*, Förderung *f*
m. handling device Fördergerät *n*
m. handling engineering [in manufacturing] Fördertechnik *f*
m. removal technique spanendes Verfahren *n*
m. science Werkstoffkunde *f*
m. shaft *(Brg)* Hilfsschacht *m*, Schacht *m* für Materialförderung
m. testing Materialprüfung *f*, Werkstoffprüfung *f*
m. testing machine Werkstoffprüfmaschine *f*
m. testing reactor Materialprüfreaktor *m*
materiology [zerstörungsfreie] Werkstoffprüfung *f*, Defektoskopie *f*, Materialprüfung *f*
mathematical mathematisch; rechnerisch
m. analysis Analysis *f*
m. box Reißzeug *n*
m. check mathematische Kontrolle (Prüfung) *f*
m. composition *(Typ)* mathematischer Satz *m*
m. expression formelmäßiger (mathematischer) Ausdruck *m*
m. geography mathematische Erdkunde *f*

m. logic mathematische (formale, symbolische) Logik *f*
m. pendulum mathematisches Pendel *n*
m. proof mathematischer Beweis *m*
m. proposition mathematischer Satz *m*
m. table Rechentabelle *f*
mathematician Mathematiker *m*
mathematics Mathematik *f*
mating Paarung *f* *(Passung)*; formschlüssige Paarung *(Getriebe)*
mating kämmend *(Zahnrad)*; ineinandergreifend, fügend; aneinanderpassend; gepaart
m. gauge Gegenlehre *f*
m. gear Gegenrad *n*
m. gears miteinander im Eingriff stehendes Räderpaar *n*, gepaarte Räder *npl*
m. [of] materials Werkstoffpaarung *f* *(Lager)*
m. of [the] gears Paarung *f* der Zahnräder
m. part Paßstück *n*; fügendes Teil *n*
m. pulley Gegenscheibe *f* *(Riementrieb)*
m. surface Paßfläche *f*, Passungsfläche *f*; Anlagefläche *f*, Aufnahmefläche *f*
m. tooth flank Gegenflanke *f* *(Zahnrad)*
matrices *s.* matrix
matrix 1. *(Math)* Matrix *f*; 2. *(Masch)* Matrize *f*, Untergesenk *n*; Kalibrierung *f*; 3. *(Geol)* Matrix, Grundmasse *f*; Gangart *f*, Ganggestein *n*; 4. *(Min)* Muttergestein *n*; 5. *(Bw)* Gefüge *n*, Grundgefüge *n*, Grundmasse *f*; Bindemittel *n*, Binder *m*; 6. *(Typ)* Matrize, justierte Mater *f*; 7. Speichervorwärmer *m* *(Regenerativvorwärmer)*
m. algebra Matrizenrechnung *f*
m. board *(Typ)* Maternpappe *f*, Matrizenpappe *f*
m. calculus Matrizenkalkül *n*
m. case *(Typ)* Matrizenrahmen *m*
m. frame *(Typ)* Matrizenrahmen *m*
m. mechanics Matrizenmechanik *f*
m. memory *(Fs)* Matrixspeicher *m*
m. metal Grundmetall *n* *(Pulvermetallurgie)*
m. notation Matrizenschreibweise *f*, Matrixschreibweise *f*
m. product Matrixprodukt *n*
m. representation Matrizendarstellung *f*, Matrizenschreibweise *f*
m. unit *(Fs)* Matrizenschaltung *f*
matt *s.* matte
matte Rohstein *m*, Feinstein *m*
m. dip Mattbrenne *f* *(Galvanotechnik)*
m. of copper *(Met)* Kupfersau *f*
m. of lead Bleistein *m*, Hartblei *n*
matter 1. Materie *f*, Stoff *m*, Masse *f*; Bestandteil *m* (*s. a.* material); 2. *(Typ)* [gesetzter] Satz *m*; Manuskript *n*
m. annihilation Masseumwandlung *f* in Energie
m.-free materiefrei
m. in suspension Schwebstoffe *mpl*
m. ray Materiestrahl *m*
m. tensor *(Kern)* Materietensor *m*
m. wave Materiewelle *f*
matting Matte *f*
mattock Hacke *f*; Breithaue *f*
mattress Matratze *f*
mature / to reifen, ablagern, altern
mature ausgereift, abgelagert, gealtert
m. cotton pod ausgereifte Baumwollfrucht *f*
maturing Reifen *n*; Altern *n*, Alterung *f*
maturity Reife *f*
m. age Haubarkeitsalter *n* *(Holz)*
maul [schwerer] Holzhammer *m*
maxigraph Maxigraf *m*, Spitzenschreiber *m*
maximal *s.* maximum *(Adjektiv)*
maximum Maximum *n*, Höchstwert *m*, Scheitelwert *m*, Größtwert *m*
maximum maximal; höchst-, größt-; Maximal-, Höchst-, Größt-
m. admission Maximalfüllung *f*
m. amount removed in dressing Höchstabrichtbetrag *m* *(Schleifscheibe)*
m. amplitude Scheitelwert *m* *(Kurve)*; Maximalamplitude *f*
m.-and-minimum thermometer Maximum-Minimum-Thermometer *n*
m. anode dissipation größter Anodenverlust *m*
m. area of drilling größtes Bohrrechteck *n*
m. average anode current höchster mittlerer Anodenstrom *m* *(für Dauerbetrieb)*
m. average output power größte Durchschnittsausgangsleistung *f* *(eines Senders)*
m. beam *(Schiff)* größte Breite *f*
m. breadth größte Breite *f*
m. capacity 1. Höchstleistung *f*; 2. größtes Fassungsvermögen *n*
m. circuit-breaker Überstromschalter *m*
m. circuit breaking Hauptstromauslösung *f*
m. circuit control Hauptstromsteuerung *f*
m. clearance *(Masch)* Größtspiel *n*
m. clearance of cars Durchfahrtsprofil *n*
m. continuous output Dauerleistung *f*
m. current Höchststrom[stärke *f*] *m*, Maximalstrom *m*, Stromspitze *f*
m. current circuit breaker Höchststromschalter *m*, Maximalschalter *m*
m. current relay Maximalrelais *n*, Überstromrelais *n*
m. cut-out Maximalausschalter *m*, Überstromauslöser *m*, Überstromschütz *n*, Überstromschalter *m*
m. deflection 1. größte (maximale) Durchbiegung *f* *(z. B. eines Trägers)*; 2. maximale Ablenkung (Auslenkung) *f*, Ablenkweite *f*, Ablenkamplitude *f*; 3. größter Ausschlag *m*, Vollausschlag *m* *(eines Instrumentes)*

m. demand *(El)* Maximalbelastung *f*
m. demand indicator Höchstlastanzeiger *m*, Maximumzähler *m*, Spitzenzähler *m*
m. demand meter Maximumverbrauchszähler *m*
m. density größte (maximale) Dichte *f*
m. diameter größter Durchmesser *m*
m. diameter of work [that can be ground] größter zu schleifender Durchmesser *m*
m. diameter that can be turned größter Drehdurchmesser *m*
m. dose *(Kern)* Maximaldosis *f*, Grenzdosis *f*
m. economy cruising power *(Flg)* höchstzulässige Reisedauerleistung *f*
m. efficiency maximaler Wirkungsgrad *m*; Höchstleistung *f*
m. governed speed Höchstdrehzahl *f*
m. impulse indicator *(Fs)* Höchstwertanzeiger *m*
m. length that can be taken between centres größte Drehlänge *f*
m. limit Größtmaß *n*
m. load Höchstlast *f*, Höchstbelastung *f*, Höchstbeanspruchung *f*
m. load-bearing capacity *(Brg)* maximale Lastaufnahme *f (Ausbau)*
m.-minimum-thermometer Maximum-Minimum-Thermometer *n*
m. moment Größtmoment *n*
m. output Höchstleistung *f*, Spitzenleistung *f*
m. peak anode current maximaler Anodenstrom *m (für Kurzbetrieb)*
m. peak inverse voltage *(El)* Spitzensperrspannung *f*
m. peak plate current maximaler Spitzenanodenstrom *m*
m. peak-to-valley height größte Profilrauhtiefe *f*
m. permissible höchstzulässig, maximal zulässig
m. permissible concentration *(Kern)* höchstzulässige (maximal zulässige) Konzentration *f*
m. permissible deviation *(El)* zulässige Regelabweichung *f*
m. permissible exposure *(Kern)* höchstzulässige (maximal zulässige) Strahlenbelastung *f*
m. permissible level *(Kern)* höchstzulässige Konzentration (Dosis) *f*
m. permissible load Höchstlast *f*
m. permissible stress höchstzulässige Beanspruchung *f*
m. potential difference *(El)* höchste (höchstzulässige) Spannung *f*
m. power Höchstleistung *f*, Leistungsmaximum *n*
m. power demand *(El)* Leistungsbedarfsspitze *f*, Spitzenleistung *f*
m. power-demand time *(El)* Spitzenbelastungszeit *f*
m. power transfer größte Leistungsübertragung *f*, höchste Übertragungsleistung *f*
m. pressure Höchstdruck *m*
m. pressure governor Überdruckschalter *m*, Überdruckwächter *m*
m. pressure index Marke *f* für den höchsten Druck
m.-pressure lamp Höchstdrucklampe *f*
m. probability maximale Wahrscheinlichkeit *f*
m. range maximale Reichweite *f (z. B. der Strahlung)*
m. recording attachment Spitzenschreiber *m*
m. relay Maximalrelais *n*
m. retention time *(Fs)* Speicherdauer *f*
m. safety load zulässige Beanspruchung *f*
m. safety stress zulässige Spannung (Beanspruchung) *f*
m. size Größtmaß *n*
m. speed Höchstgeschwindigkeit *f*, Maximalgeschwindigkeit *f*
m. stress Höchstspannung *f*; Höchstlast *f*
m. surge current maximaler einmaliger Kurzschlußstromimpuls *m*
m. swing größter Drehdurchmesser *m*
m. system deviation *(Fmt)* höchstzulässige Frequenzabweichung *f*
m. telegraphy output *(Fmt)* Oberstrich *m (Leistung)*
m. tensile load Reißlast *f*
m. tension Höchst[wert]zug *m (elektrischer Leitungen)*
m. thermometer Maximumthermometer *n*
m. torque maximales Drehmoment *n*, Höchstdrehmoment *n*
m. transmission range *(Fmt)* Reichweite *f*
m. traverse *(Mil)* Bestreichungswinkel *m*
m. value Höchstwert *m*, Größtwert *m*, Maximalwert *m*, Scheitelwert *m*
m. variation Nennabmaß *n*
m. variation between basic dimension and maximum limit oberes Nennabmaß *n*
m. variation between basic dimension and minimum limit unteres Nennabmaß *n*
m. velocity Höchstgeschwindigkeit *f*, Maximalgeschwindigkeit *f*
m. voltage circuit-breaker Überspannungs[aus]schalter *m*
m. voltage relay Überspannungsrelais *n*
maxite *(Min)* Maxit *m*
maxterm Volldisjunktion *f (Logistik)*
maxwell Maxwell *n (nicht mehr zulässige Einheit des magnetischen Flusses)*
Maxwell bridge Maxwell-Brücke *f*
Maxwell distribution *s.* Maxwellian distribution

Maxwell field Maxwellsches Kräftefeld *n*
Maxwellian distribution *(Kern)* Maxwell-Verteilung *f*, Maxwellsche Verteilung *f*
Maxwellian distribution function Maxwellsche Verteilungsfunktion *f*
Maxwellian distribution of velocities Maxwellsche Geschwindigkeitsverteilung *f*
Maxwell's equation Maxwellsche Gleichung *f*
MAYDAY *(internationales Notsignal im Seefunksprechdienst)*
mazarine dunkelblau
mazout Heizöl *n*, Masut *n (Erdöl)*
mazun Joghurt *m*
meadow Wiese *f*, Grünland *n*
m. cultivation Wiesenbau *m*
m. fescue Wiesenschwingel *m*
m. oats Feldhafer *m*, Flughafer *m*, Windhafer *m*, Wildhafer *m*
m. ore Raseneisenstein *m*, Limonit *m*, Eisenerz *n*
m. plough Wiesenpflug *m*
meal Mehl *n*, Staub *m (Gestein)*
mean Mittel[wert *m*] *n*, Durchschnittswert *m*
m. axis of inertia mittlere Trägheitsachse *f*
m. camber *(Flg)* Durchschnittskrümmung *f*
m. carrier frequency mittlere Trägerfrequenz *f*
m. coefficient of absorption mittlerer Absorptionskoeffizient *m*
m. conical candle-power mittlere Lichtstärke *f* im Nutzkegel
m. dispersion Grunddispersion *f*
m. effective value Effektivwert *m*, quadratischer Mittelwert *m*
m. electrode Mittelelektrode *f*, Zentralelektrode *f*
m. error mittlerer Fehler *m*
m. free path *(Kern)* mittlere freie Weglänge *f*
m. frequency Mittelfrequenz *f*
m. holding time *(Fmt)* mittlere Belegungsdauer *f*
m. impulse indicator Mittelwertzeiger *m*
m. indicated pressure mittlerer indizierter Druck *m*
m. lethal time *(Kern)* mittlere Überlebenszeit *f*
m. life mittlere Lebensdauer *f*
m. output Ausgangsleistung *f*, mittlere Dauerleistung *f (Funktechnik)*
m. piston pressure mittlerer Kolbendruck *m*
m. piston speed mittlere Kolbengeschwindigkeit *f*
m. pitch mittlere Steigung *f (eines Gewindes)*
m. power mittlere Leistung *f*
m. power level mittlerer Leistungspegel *m*
m. pressure Mitteldruck *m*
m. proportional geometrisches Mittel *n*
m. pulse time mittlere Impulsdauer *f*
m. range mittlere Reichweite *f (z. B. der Strahlung)*
m. reference line mittlere Bezugslinie *f*
m. sample-tree Mittelstamm *m*
m. sea level mittlerer Wasserstand *m*, Seehöhe *f*, Normalnull
m. speed Durchschnittsgeschwindigkeit *f*
m. square deviation *(Math)* mittlere quadratische Streuung *f*, Standardabweichung *f*
m. square distance mittlerer quadratischer Abstand *m*
m. square value *s.* m. effective value
m. square velocity mittlere quadratische Geschwindigkeit *f*
m. temperature difference mittlere Temperaturdifferenz *f*, mittlerer Temperaturunterschied *m*
m. time between failures mittlere Lebensdauer *f (eines Bauteils)*
m. value Mittelwert *m*
m.-value brightness control *(Fs)* Mittelwertsteuerung *f*
m. value theorem *(Math)* Mittelwertsatz *m*
m. value with respect to time zeitlicher Mittelwert *m*
m. velocity Durchschnittsgeschwindigkeit *f*
m. water mittlerer Wasserstand *m*
meander wave *(Fs)* Mäanderwelle *f*
means 1. Mittel *npl*, Hilfsmittel *npl*; 2. mittlere Glieder *npl (einer Gleichung)*
m. of communication Nachrichtenmittel *npl*; Verkehrsmittel *npl*
m. of conveyance Transportmittel *npl*, Fördermittel *npl*, Beförderungsmittel *npl*
m. of dissolution *(Ch)* Aufschlußmittel *npl*
m. of instruction Lehrmittel *npl*
m. of preservation Konservierungsmittel *npl*
m. of production Produktionsmittel *npl*
m. of transport Transportmittel *npl*
m. of transportation *s.* m. of conveyance
measurable meßbar
measure / to [ab]messen, ausmessen, vermessen
m. again nachmessen
m. out dosieren
m. with a calorimeter kalorimetrieren
measure 1. Maßnahme *f*; 2. Maß *n*; 3. *(Typ)* Format *n*, Zeilenlänge *f*, Satzbreite *f*, Spaltenbreite *f*; 4. *(Geol)* Schicht *f*, Lager *n*; 5. *(Math)* Teiler *m*; 6. Maßstab *m*, Zollstock *m*; 7. Takt *m (in der Musik)*
m. of contraction Schwindmaß *n*
m. of deformation Deformationsmaß *n*
m. of depth Tiefenmessung *f*
m. of dispersion *(Math)* Streuungsmaß *n*, Schwankungsmaß *n*, Verteilungsmaß *n*
m. of gravity Schweremessung *f*
m. of safety Sicherheitsmaßnahme *f*
m. of shrinkage Schwindmaß *n*

measured gemessen
m. course *(Schiff)* Probefahrtsmeile *f*, gemessene Meile *f*
m. over the bark mit Rinde gemessen *(Holz)*
m. quantity Meßwert *m*
m. range Meßbereich *m*
m. under the bark ohne Rinde gemessen *(Holz)*
m. value Meßgröße *f*, Meßwert *m*, Prüfwert *m*
measurement 1. Messung *f*, Messen *n*, Vermessung *f*, Vermessen *n*; 2. *(Geol)* Markscheidung *f*
m. accuracy Meßgenauigkeit *f*
m. by planimeter Planimetrierung *f*, Flächenmessung *f*
m. capacity *(Schiff)* Ladefähigkeit *f* für Raumfracht
m. cargo *(Schiff)* Maßgüter[ladung *f*] *npl*, Maßgutladung *f*
m. equipment Meßeinrichtung *f*
m. error Meßfehler *m*
m. of angles Winkelmessung *f*
m. of cone[s] Kegelmessung *f*
m. of insulation Isolationsmessung *f*, Isolationsprüfung *f*
m. of ships Schiffsvermessung *f*
m. of standing timber Holzmassenaufnahme *f*
m. of threads Gewindemessung *f*
m. range Meßbereich *m*
m. result Meßergebnis *n*
m. technology Meßtechnik *f*
m. unit Maßeinheit *f*, Meßeinheit *f*
measurer Meßinstrument *n*, Dosierapparat *m*
measures of labour and health protection Arbeits- und Gesundheitsschutzmaßnahmen *fpl*
measuring Messen *n*; maßliches Prüfen *n*
m. accuracy Meßgenauigkeit *f*
m. amplifier Meßverstärker *m*
m. anvil Meßamboß *m*
m. apparatus Meßgerät *n*, Meßvorrichtung *f*
m. arrangement Meßanordnung *f*
m. bridge Meßbrücke *f*
m. cable Meßkabel *n*, Prüfkabel *n*
m. capacity Meßbereich *m*, Verstellbereich *m*, Verstellweite *f*
m. channel Meßkanal *m*
m. converter Meßwertgeber *m*
m. cylinder Meßzylinder *m*
m. device Meßeinrichtung *f*, Meßgerät *n*, Meßwerk *n*
m. dial Rundmeßskala *f*
m. diode Meßdiode *f*
m. edge Prüfkante *f*
m. element *s.* detecting element
m. equipment Meßgeräte *npl*
m. error Meßfehler *m*
m. face Prüffläche *f*
m. fault Meßfehler *m*
m. fixture Meßeinrichtung *f*
m. flask *(Ch)* Meßkolben *m*
m. glass *(Ch)* Meßzylinder *m*
m. head Meßkopf *m*
m. instrument Meßgerät *n*, Meßinstrument *n*, Meßmittel *n*, Meßzeug *n*
m. instrument for alternating currents Wechselstrominstrument *n*
m. instrument for continuous currents Gleichstrominstrument *n*
m. jaw Meßschnabel *m*
m. lag Verzögerung *f* des Meßglieds, Meßverzögerung *f*
m. loop Meßschleife *f*
m. machine 1. Meßmaschine *f*; 2. Ledermeßmaschine *f (Schuhproduktion)*
m. magnifier Meßlupe *f*
m. mark Meßmarke *f*, Meßstrich *m*
m. matters Meßwesen *n*
m. means *(Am) s.* m. unit
m. method Meßmethode *f*
m. microscope Meßmikroskop *n*
m. of the reverberation *(El)* Nachhallmessung *f*
m. orifice Stauscheibe *f*, Drosselscheibe *f*
m. outfit Meßeinrichtung *f*, Meßgerät *n*
m. pin Meßstift *m (zum Prüfen der Evolventenkeilverzahnung)*
m. plug Meßamboß *m*, Meßzapfen *m*
m. pocket *(Brg)* Fülltasche *f*, Meßtasche *f*
m. point 1. Prüfstelle *f*; 2. Meßanschlag *m*
m. pole Meßlatte *f*
m. probe Meßsonde *f*
m. range Meßbereich *m*; Skalenbereich *m (eines Instrumentes)*
m. rectifier Meßgleichrichter *m*
m. result Meßergebnis *n*
m. screw Meßspindel *f*
m. sensitivity Meßempfindlichkeit *f*, Ansprechempfindlichkeit *f (einer Meßeinrichtung)*
m. shade Meßblende *f*
m. spark-gap *(El)* Meßfunkenstrecke *f*
m. spindle Meßspindel *f*
m. system free from wear verschleißfreies Meßsystem *n*
m. tape Bandmaß *n*
m. transductor Meßtransduktor *m*, magnetischer Meßverstärker *m*
m. transformer Meßtransformator *m*, Meßwandler *m*
m. unit Meßeinrichtung *f*; Maßeinheit *f*, Meßeinheit *f*
m. voltage *(El)* Meßspannung *f*
meat / to *(Brg)* anfahren *(Seilfahrt)*
meat flour Fleischmehl *n*

m. ship Fleischtransportschiff *n*
mechanic gelernter Metallarbeiter *m*; Monteur *m (Maschinenmontage)*; Schlosser *m*; Installateur *m*
mechanical mechanisch; maschinell
m. adding device Summentrieb *m*, Summiertrieb *m (Analogrechner)*
m. assembly mechanisierte Montage *f*
m. assembly construction Baukastenkonstruktion *f*
m. assembly technique Baukastenprinzip *n*
m. axis mechanische Achse *f (des Kristalls)*
m. barking maschinelles Entrinden *n (von Bäumen)*
m. bench Werkstattbank *f*
m. bilge block *(Schiff)* Kimmschlitten *m*, Kimmpalle *f*
m. bur crusher *(Text)* Klettenwolf *m*
m. bur extracting Entfernung *f* von Kletten auf mechanischem Wege, maschinelle Klettenentfernung *f (Wolle)*
m. centring Bildzentrierung *f* mit mechanischen Mitteln, mechanische Zentrierung *f (des Fernsehbildes)*
m. composition *(Typ)* Maschinensatz *m*
m. core Installationskern *m*
m. cotton gin Baumwollegreniermaschine *f*, Baumwollentkörnungsmaschine *f*
m. cotton picking *(Am)* maschinelles Baumwollpflücken *n*
m. current *(El)* dynamischer Strom *m*
m. detent mechanische Auslösung *f*, mechanische Vorsteuerung *f*
m. dictionary *(Dat)* Übersetzungsmaschine *f*, mechanisches Wörterbuch *n*
m. digitizer mechanischer Analog-Digital-umsetzer *m*
m. dividing device *(Dat)* Dividiertrieb *m*, Divisionsgetriebe *n*
m. draught production *(Masch)* mechanische Zugerzeugung *f*
m. drawing Maschinenzeichnung *f*
m. efficiency mechanischer Wirkungsgrad *m*
m. electro-plating Plattierung *f* mit bewegten Katoden
m. endurance test mechanische Lebensdauerprüfung *f (Schaltgerät)*
m. engineer Maschinenbauer *m*; Metallarbeiter *m*; *(Am)* Ingenieur *m* für Haustechnik
m. engineering Maschinenbautechnik *f*, mechanische Technik *f*
m. equivalent of heat *(Masch)* mechanisches Wärmeäquivalent *n*, Wärmearbeitswert *m*
m. equivalent of light mechanisches Lichtäquivalent *n*
m. firing mechanische Feuerung *f*
m. floor *(Bw)* Installationsgeschoß *n*
m. governor mechanischer Regler *m*
m. hysteresis elastische Hysteresis *f*
m. hysteresis effect Dämpfung *f*
m. impedance mechanische Impedanz *f*, mechanischer Scheinwiderstand *m*
m. indicator mechanischer Fühlhebel *m*
m. integrating device *(Dat)* Integriertrieb *m*, Integriergetriebe *n*, mechanischer Integrator *m*, mechanische Integriereinrichtung *f*
m. knitting machine Motorflachstrickmaschine *f*
m. leather technisches Leder *n*
m. manipulation *(Plast)* Verarbeitung *f*
m. master oscillator *(El)* Stimmgabelsteuerstufe *f*
m. mixing maschinelles Mischen *n*
m. moulding Maschinenformerei *f*
m. multiplying device *(Dat)* Multipliziertrieb *m*, Multipliziergetriebe *n*, mechanischer Multiplikator *m*
m. ohm *(El)* akustisches (mechanisches) Ohm *n*
m. oscillation mechanische Schwingung *f*
m. output Leistungsmessung *f*
m. output of motor Motorenleistung *f*
m. pantograph manipulator *(Kern)* Scherenmanipulator *m*, Storchschnabelmanipulator *m*
m. picking *(Am)* *s.* m. cotton picking
m. pilot Autopilot *m*, Selbststeuergerät *n*, automatische Kurssteueranlage *f*
m. platform Bauaufzug *m*
m. plunger lubricator Schmierpresse *f*, Öldruckapparat *m*
m. press mechanische Presse *f*
m. pulp Zellstoff *m*; Holzschliff *m (Papierherstellung)*
m. pulp board weiße Holzpappe *f*
m. recording method *(El)* Nadeltonverfahren *n*
m. rectifier mechanischer Gleichrichter *m*
m. register mechanischer Zähler *m*
m. repeater automatische Umführung *f (Walzwerk)*
m. resistance mechanischer Widerstand *m*
m. resonance mechanische Resonanz *f*
m. shovel Bagger *m*
m. similitude mechanische Ähnlichkeit *f*
m. stage Kreuztisch *m (Mikroskop)*
m. stoking mechanische Rostbeschickung *f*
m. stowing *(Brg)* Schleuderversatz *m*
m. stowing with scraper-buckets Schrappversatz *m*
m. stowing with slingers Schleuderversatz *m*
m. subdivision Zerkleinerung *f*
m. testing mechanisches Prüfverfahren *n*
m. tissue Festigungsgewebe *n (Holz)*

m. transfer of direction *(Opt)* mechanische Richtungsübertragung *f*
m. tripping *(El)* mechanische Auslösung *f*
m. unit mechanische Einheit *f*
m. ventilation *(Brg)* künstliche Bewetterung *f*
m. vibration mechanische Schwingung *f*
m. woodpulp Holzstoff *m*
m. working Umformung *f*; Umformtechnik *f*
m. zero *(El)* mechanischer Nullpunkt *m*
mechanically-controlled mechanisch betätigt
m.-operated mechanisch betätigt
m.-operated brake mechanische Bremse *f*
m.-operated circuit breaker automatischer Schalter *m*
m.-operated switch Betätigungsschalter *m*, Schalter *m* mit mechanischem Antrieb
m.-operated valve gesteuertes Ventil *n*
mechanician Maschinenschlosser *m*, Mechaniker *m*
mechanics 1. Mechanik *f*; 2. Vorgang *m*
m. of cutting Spanungsvorgang *m*
m. of fluids Mechanik *f* der Flüssigkeiten und Gase
m. of liquids Mechanik *f* der Flüssigkeiten
m. of materials Festigkeitslehre *f*
m. of solid bodies Mechanik *f* der festen Körper, Festkörpermechanik *f*
mechanism 1. Mechanismus *m (Getriebelehre)*; 2. *(Masch)* Einrichtung *f*; Vorgang *m*; 3. Einsatzteil *n*
m. for movement in a curve Kurvenschubgetriebe *n*
m. of ag[e]ing Alterungsvorgang *m (Wärmebehandlung)*
m. of conduction Leitungsmechanismus *m*
mechanization Mechanisierung *f*
mechanize / to mechanisieren
mechanized firing mechanische Feuerung *f*
mechano-electronic transducer mechanisch-elektronischer Wandler *m (mechanische Schwingungen werden in elektrische Stromschwankungen umgesetzt)*
mecolite *(Met)* Kupfernickel *n*
median Median[wert] *m*, Zentralwert *m*; Mittellinie *f*; Mittelstreifen *m (Straßenbau)*
median mittlerer, Mittel-, Median-
m. control chart Mediankarte *f*
m. lethal dose mittlere Letaldosis *f*, LD_{50} *f*
m. plane Medianebene *f*
m. point Schnittpunkt *m* der Seitenhalbierenden
m. section Mittelschnitt *m*
m. strip Mittelstreifen *m (Straßenbau)*
medicated vinegar Arzneiessig *m*
medicinal flavours Geschmacksstoffe *mpl* für Arzneimittel
m. herb Arzneikraut *n*, Heilpflanze *f*
m. plant *s.* m. herb
m. substance Arzneistoff *m*
m. white oil Paraffinöl *n*, Paraffinum liquidum
medium Medium *n*, Mittel *n*
medium 1. mittlerer; Mittel-; 2. *(Typ)* halbfett *(Schrift)*
m. carbon steel Kohlenstoffstahl *m (0,20 bis 0,35% C)*
m. cut-off tube regelnde Empfängerröhre *f*
m. density fibreboard halbharte (mitteldichte) Faserplatte *f*
m. density particle board mittelschwere Spanplatte *f*
m.-duty mittelschwer; für mittelgroße Leistung
m. finger spacing cutter bar *(Lw)* Mittelschnittmähbalken *m*
m. fit *(Masch)* Feinpassung *f*; Schlichtpassung *f*
m. for refrigeration Kältemittel *n*
m. frequency Mittelfrequenz *f*
m. frequency band Mittelwellenbereich *m*
m. grained mittelkörnig
m. helix normaler Drall *m (20 bis 30°)*
m. high cut Mittelschnitt *m (Ernte)*
m. keying fit *(Masch)* mittlerer Keilsitz *m*
m. loading mittelschwere Bespulung (Pupinisierung) *f (Kabel)*
m. oil Mittelöl *n*
m. petrol Mittelbenzin *n*
m. plate mill Mittelstraße *f (Walzwerk)*
m.-pressure boiler Mitteldruck[dampf]kessel *m*
m. range missile Mittelstreckenrakete *f*
m. run *(Masch)* mittleres (mittelgroßes) Fertigungslos *n*
m. shot *(Fs)* Mittelaufnahme *f*, Porträtaufnahme *f*
m. sight *(Mil)* gestrichenes Korn *n (Anschlagart)*
m.-size *s.* m.-sized
m.-sized mittelgroß
m. sized sheet Mittelblech *n*
m. solids Beschwerungsstoff *m*
m.-temperature forming Halbwarmumformen *n*
m. voltage mittlere Spannung *f (Spannung zwischen 250 und 650 V)*
m. wave Mittelwelle *f (550 bis 1500 kHz)*
m.-wave receiver Mittelwellenempfänger *m*
m.-wave transmitter Mittelwellensender *m*
m.-weight mittelschwer
medullary ray *(Holz)* Markstrahl *m*, Spiegelfaser *f*
meet / to zusammentreffen, zusammenstoßen
m. demand den Bedarf decken
m. oneself's quota *(Am)* seine Arbeitsnorm erfüllen

m. requirements den Anforderungen entsprechen
m. the delivery date den Liefertermin einhalten
meet Durchschnitt *m (Mengen)*
meeting faces Berührungsflächen *fpl*
m. vessel *(Schiff)* Gegenkommer *m*, Gegensegler *m*
mega Mega *(1 M = 10⁶ Einheiten)*
m. electron volt Megaelektronenvolt *n*, *(Einheit der Arbeit und Energie in der Kernphysik: 1 MeV = 1,602 · 10⁻¹³ J)*
megabar Megabar *n*
megacycles per second Megahertz *n (Einheit der Frequenz; 1 Mc/s = 10⁶ Hz)*
megafarad Megafarad *n*
megaphone Megafon *n*, Sprachtrichter *m*, Sprachrohr *n*
megawatt Megawatt *n (Einheit der Leistung; 1 MW = 10³ kW)*
megger *s.* megohmmeter
megilp Leinöl-Terpentin-Farbverdünnungsmittel *n*
megohm Megaohm *n (Einheit des elektrischen Widerstandes; 1 MΩ = 10⁶ Ω)*
megohmmeter Megohmmesser *m*, Megohmmeter *n*, Megger *m*, Isolationsprüfapparat *m*, Isolationsprüfer *m*, Isolationsmeßapparat *m*, Isolationsmesser *m*, Isolationsmeßgerät *n*
Meissner-circuit *(Fmt)* induktive Röhrenrückkopplungsschaltung *f (nach Meißner)*
melamine *(Ch)* Melamin *n*
m.-formaldehyde-resin Melaminharz *n*
melanite *(Min)* Melanit *m*
melassigenic *(Ch)* melassebildend
melissic acid Melissinsäure *f*
melissyl alcohol Melissylalkohol *m*, Myrizylalkohol *m*
mellitic acid Mellitsäure *f*
mellow lime *(Led)* milder Äscher *m*
melt / to schmelzen
m. away wegschmelzen
m. on anschmelzen
melt Schmelze *f*, Schmelzung *f*; Charge *f*; Schmelzbad *n*
m.-down carbon Einschmelzkohlenstoffgehalt *m*, Anlaufkohlenstoff *m*
m.-down slag Einschmelzschlacke *f*
m.-down test Einschmelzprobe *f*
m. extrusion *(Plast)* Schmelzspinnen *n*
m. film coating *(Led)* Schmelzverfahren *n (Beschichtung)*
m. spinning Schmelzspinnverfahren *n*
m.-spun filament nach dem Schmelzspinnverfahren gesponnene Chemieseide *f*
meltable schmelzbar
meltableness Schmelzbarkeit *f*
melted verschmolzen
melting Schmelzen *n*; Erschmelzung *f*, Verschmelzung *f*
m. bath Schmelze *f*, Schmelzbad *n*
m. capacity Schmelzleistung *f*; Plastifizierleistung *f (Spritzgußmaschine)*
m. chamber Schmelzraum *m (des Ofens)*
m. cone *(Met)* Schmelzkegel *m*, Segerkegel *m*, Brennkegel *m*
m. crucible Schmelztiegel *m*
m. department Schmelzerei *f*, Schmelzbetrieb *m*
m. down Einschmelzen *n*
m.-down power Schmelzleistung *f (des Ofens)*
m. finery *(Met)* Frischhütte *f*
m. furnace Schmelzofen *m*
m. hearth Schmelzherd *m*
m. heat Schmelzwärme *f*
m.-hole furnace *(Met)* Unterflurofen *m*
m. in vacuo Schmelzen *n* im Vakuum
m. ladle Gießlöffel *m*
m. loss *(Met)* Abbrand *m*
m. point Schmelzpunkt *m*
m.-point curve Schmelzkurve *f*
m.-point diagram Schmelzdiagramm *n*
m. pot Schmelzkessel *m*
m. rate Schmelzgeschwindigkeit *f*
m. stock Schmelzgut *n*
m. temperature Schmelztemperatur *f*
m. time Schmelzzeit *f*, Schmelzdauer *f*
member Bauglied *n*, Bauteil *n*; Konstruktionsteil *n*; Stab *m*, Element *n (Fachwerk)*; Stabelement *n*; Stiel *m (Rahmen)*; Glied *n*; Organ *n*; [angetriebenes] Teil *n*
membrane Membran[e] *f*
m. pressure gauge Membranmanometer *n*
m. state of stress Membranspannungszustand *m*
m. theory Membrantheorie *f*
memo pad *(sl)* Notizblock *m*, Schmierblock *m*
memory Gedächtnis *n*; Informationsspeicher *m*, Speicher *m*, Datenspeicher *m*, elektronischer Speicher *m*
m. capacity Speicherkapazität *f (Rechenmaschine)*
m. cell Speicherzelle *f (Rechenmaschine)*
m. circuit Speicherschaltung *f*
m. clamp *(Fs)* Gleichspannungswiedereinführung *f (in Speicherschaltungen)*
m. core Speicherkern *m*
m. matrix Matrixspeicher *m*
m. register Speicherregister *n (Informations- und/oder Befehlsübertragung)*
m. tube Speicherröhre *f*
m. unit Speicher *m*
men-riding *(Brg)* Personenbeförderung *f (söhlig)*; Seilfahrt *f (Schacht)*
mend / to flicken, reparieren
m. broken threads *(Text)* Fadenbrüche heilen

Mendeleev's [periodic] law Periodengesetz *n* [nach Mendelejew]
mendelevium *(Ch)* Mendelevium *n*
mender *(Met)* Blech *n* zweiter Wahl
mending Ausbessern *n*, Stopfen *n*
meniscus Meniskus *m*
m. lens Meniskuslinse *f*
mensuration Flächenberechnung *f*, Planimetrie *f*; Körperberechnung *f*; Aufmaß *n*, Messung *f*; Vermessung *f*
mental arithmetic Kopfrechnen *n*
mercantile marine Handelsmarine *f*, Handelsflotte *f*
mercaptan *(Ch)* Merkaptan *n*
mercerization Merzerisierung *f*, Merzerisation *f*
mercerize / to merzerisieren
mercerized cotton merzerisierte Baumwolle *f*
mercerizer Merzerisieranlage *f*
mercerizing frame Merzerisierapparat *m*
m. mangle Merzerisierfoulard *m*
merchant bar Schweißeisenblock *m*
m.-fleet Handelsflotte *f*
m. iron Walzgrobeisen *n*
m. marine Handelsmarine *f*, Handelsflotte *f*
m. mill Feineisenstraße *f* *(Walzwerk)*
m. roll Reckwalzwerk *n*
m. service Handelsschiffahrt *f*
m. shipping Handelsschiffahrt *f*, Seeschiffahrt *f*, Linienschiffahrt *f*
m. tonnage Handelsschiffstonnage *f*
m. vessel Handelsschiff *n*
merchantman Handelsschiff *n*
mercuration *(Ch)* Merkurierung *f*
mercurial quecksilberähnlich, quecksilberhaltig
m. nitrogen thermometer Quecksilberthermometer *n* mit Stickstoffüllung
m. pendulum Quecksilberpendel *n*
mercurialism Quecksilbervergiftung *f*
mercuric compound Verbindung *f* des zweiwertigen Quecksilbers
m. oxide Merkurioxid *n*, Quecksilber(II)-oxid *n*
mercurous compound Verbindung *f* des einwertigen Quecksilbers
mercury Quecksilber *n*
m. air pump Quecksilberdiffusionspumpe *f*
m. arc Quecksilberlichtbogen *m*
m.-arc inverter Quecksilberdampfwechselrichter *m*, Quecksilberdampfumrichter *m*
m.-arc lamp Quecksilberdampflampe *f*
m.-arc rectifier Quecksilberdampfventil *n* *(Gleichrichter)*
m.-arc rectifier equipment Quecksilberdampfgleichrichter *m*
m. barometer Quecksilberbarometer *n*
m. break Quecksilberunterbrecher *m*, Quecksilberschalter *m*
m. cathode Quecksilberkatode *f*
m. collector Quecksilberfänger *m* *(Gleichrichter)*
m. column Quecksilbersäule *f*
m. contact Quecksilberkontakt *m*
m. delay line *(Fs)* Quecksilberverzögerungsleitung *f*
m. diffusion pump Quecksilberdiffusionspumpe *f*
m. fulminate *(Ch)* Knallquecksilber *n*, Quecksilberfulminat *n*
m. gauge Quecksilbermanometer *n*
m. halide Quecksilberhalogenid *n*
m.-in-glass switch Quecksilberschaltröhre *f*
m. jet rectifier Quecksilberstrahlgleichrichter *m*
m. jet rectifier equipment Quecksilberstrahlgleichrichter *m*
m. jet switch Quecksilberstrahlschalter *m*
m. memory *(Dat)* Quecksilberultraschallspeicher *m*
m.-plunger relay Statex-Relais *n*, Quecksilberverdrängerrelais *n*
m.-pool cathode Quecksilberkatode *f*
m.-pool rectifier Gleichrichter *m* mit flüssiger Quecksilberkatode
m. pressure gauge with reduced scale Quecksilbermanometer *n* mit verkürzter Skala
m. relay Quecksilberrelais *n*
m. ring balance Quecksilberringwaage *f*
m. rotating pump rotierende Quecksilberpumpe *f*, Kapselpumpe *f*
m. seal stopcock mit Quecksilber gedichteter Schliffhahn *m*
m. switch Quecksilberschalter *m*, Quecksilberschaltröhre *f*
m. tank *(Dat)* Quecksilberspeicher *m*, Quecksilberlaufzeitspeicher *m*
m. vacuum gauge Quecksilbervakuummeter *n*
m. vapour Quecksilberdampf *m*
m. vapour lamp Quecksilberdampflampe *f*
m. vapour rectifier Quecksilberdampfgleichrichter *m*
m. wetted contacts quecksilberbenetzte Kontakte *mpl*
merge / to *(Dat)* mischen *(höhere Stufe des Sortierens)*
meridian 1. Längenkreis *m*, Meridian *m*; 2. Mittelschnitt *m* *(Rotationsfläche)*
m. circle Meridiankreis *m*
m. finder Meridiansucher *m*
m. gyro Meridiankreisel *m*
m. line Mittagslinie *f*
meridional meridional, Meridian-
m. plane Meridionalebene *f*
m. section Meridionalschnitt *m*
merino sheep Merinoschaf *n*
merit Übertragungsgüte *f* *(im Funksprechbetrieb)*

merohedral meroedrisch
merohedrism Meroedrie *f*
merry-go-round Umspanngerät *n*, Wickelgerät *n (Spannbetonbau)*
mesa configuration Mesastruktur *f*
m. technique Mesatechnik *f (Transistorenfertigung)*
m. transistor Mesatransistor *m (Transistor mit tischförmiger Anordnung der Halbleiterschichten)*
mesh / to eingreifen, kämmen, ineinandergreifen *(Verzahnung)*
mesh 1. Geflecht *n*; Drahtgeflecht *n*; Masche *f*, Maschenweite *f*; Netzwerk *n*; Zelle *f*; 2. *(El)* Netz *n*, Vieleckschaltung *f*; 3. *(Fs)* Raster *m*; 4. Eingriff *m*, Ineinandergreifen *n (Zahnräder)*; 5. Messereingriff *m (Hackmaschine)*
m. analysis Siebanalyse *f*
m. aperture Maschenweite *f*
m. bottom Siebeinsatz *m*
m. connection *(El)* Maschenschaltung *f*, Deltaschaltung *f*, Dreieckschaltung *f*
m. current *(El)* Maschenstrom *m*
m. impedance *(El)* Maschenscheinwiderstand *m (Netzwerke)*
m. ionization chamber Ionisationskammer *f* mit Netzwandung
m. of an electric network Masche *f* eines elektrischen Netzes
m. pattern open work Knüpftrikot *m*
m. point Eingriffsstelle *f*
m. reinforcement Betonstahlgewebe *n*, Bewehrungsmatte *f (Stahlbetonbau)*
m. size Maschenweite *f (des Siebes)*
m. voltage *(El)* verkettete Spannung *f*
m. voltage of a polyphase system kleinste verkettete Spannung *f*
m. wire Maschendraht *m*
meshing Zahnradeingriff *m*, Kämmen *n*
m. gears miteinander im Eingriff stehendes Räderpaar *n*, gepaarte Räder *npl*
m. into each other im Eingriff befindlich
m. surface Überdeckung *f (Verzahnung)*
mesic *s.* mesonic
mesitylene *(Ch)* Mesitylen *n*
mesokurtic *(Math)* [von] mittlerer Krümmung *(Häufigkeitskurve)*
mesokurtosis *(Math)* mittlere Krümmung *f (Häufigkeitskurve)*
mesomerism *(Ch)* Mesomerie *f*
mesomorphous mesomorph
meson Meson *n (instabiles Elementarteilchen mittlerer Masse)*
m.-activ mesonaktiv
m. beam Mesonenstrahl *m*, Mesonenbündel *n*
m. decay Mesonenzerfall *m*
m. field Mesonenfeld *n*
m. mass Mesonenmasse *f*
m. physics Mesonenphysik *f*
m. production Mesonenerzeugung *f*, Mesonenbildung *f*
m. scattering Mesonenstreuung *f*
m. spectrum Mesonenspektrum *n*
m. theory Mesonentheorie *f*
m. theory of nuclear forces Mesonentheorie *f* der Kernkräfte
m. track Mesonenspur *f*
m. yield Mesonenausbeute *f*
mesonic Meso[nen]-, mesonisch
m. charge Mesonenladung *f*, Nukleonenladung *f*
mesonium Mesonium *n*
mesopic vision Dämmerungssehen *n*, Übergangssehen *n*, mesopisches Sehen *n*
mesotartaric acid Mesoweinsäure *f*
mesothorium Mesothor[ium] *n*
mesoton *s.* meson
mesotron *s.* meson
mess boy *(Schiff)* Kajütenjunge *m*
m. gear *(Schiff)* Eßbesteck *n*
m. trap *(Schiff)* Backsgeschirr *n*
message Information *f*; Nachricht *f*; Telegramm *n*
m. blank Telegrammformular *n*
m.-carrying rocket Melderakete *f*
m. form *s.* m. blank
m. minute *(Fmt)* Belegungsminute *f*
m. rate *(Fmt)* Einzelgebühr *f*
m. rate subscription Einzelgebührenvertrag *m*
m. register Registriergerät *n*, Zählwerk *n*; Gesprächszähler *m*
m. transition delay Telegrammlaufzeit *f*
messenger Telegrammzusteller *m*; Bote *m*
m. cable Tragseil *n*, Kabeltragseil *n*
m. clamp *(El)* Hängeschelle *f*
m. wire Luftkabeltragseil *n*; Aufhängedraht *m (Kabel)*
metabolic process Stoffwechselvorgang *m*
metabolism Metabolismus *m*, Stoffwechsel *m*
metaboric acid Metaborsäure *f*
metacentre Metazentrum *n*
metacentric height metazentrische Höhe *f*
m. radius *(Schiff)* Metazentrum *n* über Verdrängungsschwerpunkt
metachromatic metachromatisch
m. dye Metachromfarbstoff *m*
metachromotype process Abziehbilderverfahren *n*
metadynamo *(Elm)* Zwischenbürstenmaschine *f*
metadyne Metadyne *f (Elektromaschinenverstärker)*
m. converter Metadynumformer *m*
m. generator Metadyngenerator *m*
metal 1. Metall *n*; 2. Schotter *m (Straßenbau, Gleisbau)*

m. air-pump valve Luftpumpenmetallventil *n*
m. alloy Metallegierung *f*
m. amide *(Ch)* Metallamid *n*
m. arc Metallichtbogen *m*
m.-arc cutting elektrisches Brennschneiden *n* mit Metallelektrode
m.-arc electrode Metallelektrode *f* für elektrische Lichtbogenschweißung
m.-arc welding Lichtbogenschweißen *n* mit Schweißelektrode
m.-arc welding process Lichtbogenschweißverfahren *n* mit Schweißelektrode
m.-asbestos fabric-gasket Metallasbestgewebedichtung *f*
m.-backed screen metallhinterlegter Schirm *m (Bildwiedergaberöhre)*
m. backing Schirmverspiegelung *f (Fernsehbildröhre)*
m. bar extruding press Metallstrangpresse *f*
m. base 1. Metallsockel *m (Lampe)*; 2. Metallgrundplatte *f (Kristalle)*
m.-bonded metallgebunden
m. bonding Metallkleben *n*
m. borings Metallspäne *mpl*
m. braid Metallgeflecht *n*
m. brush Metallbürste *f*
m. cabinet Metallgehäuse *n*
m. carbide Metallkarbid *n*
m.-cased capacitor Becherkondensator *m*
m.-cased conductor Panzeraderleitung *f*, Stahlpanzerleitung *f*
m. casing Metallgehäuse *n*
m. casting Metallgußstück *n*; Metallguß *m*, Metallgießen *n*
m.-ceramic magnet Sintermagnet *m*, Metallkeramikmagnet *m*
m.-ceramic valve Metallkeramikröhre *f*
m. ceramics Metallkeramik *f*
m. charge *(Met)* Eisengicht *f*, Eiseneinsatz *m*
m.-clad metallplattiert; metallüberzogen; metallkaschiert
m.-clad cable Kabel *n* mit Metallmantel
m.-clad surface metallplattierte Oberfläche *f*
m.-clad switchgear gußgekapselte Schaltgeräte *npl*; gekapselte Schaltanlage *f*
m. cladding 1. Walzschweißplattierung *f*; 2. Metallplattierschicht *f*
m. coated bulb verspiegelter Kolben *m*
m. coating Metallüberzug *m*, metallischer Überzug *m*; Metallisieren *n (Tauchverfahren)*
m. cone kinescope Bildröhre *f* mit Metallhals
m. container Blechbehälter *m*, Blechkanister *m*
m. cooling Metallkühlung *f*
m. covering Metallmantel *m*
m. cutting Metallspanen *n*, spanende Fertigung *f*, Spanungsarbeit *f*
m. cutting spanend
m. cutting bandsaw Metallbandsäge *f*
m. cutting capacity Zerspanungsleistung *f*
m. cutting laboratory Zerspanungslabor *n*
m. cutting law Zerspanungsgesetz *n*
m. cutting material Schneidmetall *n*
m. cutting saw Metallsäge *f*
m. cutting science Zerspanungslehre *f*
m. cutting tool Schneidwerkzeug *n* für Metallbearbeitung
m. dark-slide Metallkassette *f*
m. degreasing Metallentfettung *f*
m. deposit Metallniederschlag *m*
m. detector Metallsuchgerät *n*
m. disk valve Metallplattenventil *n*
m. distribution ratio Niederschlagsverteilungsverhältnis *n (Galvanotechnik)*
m. drilling Metallbohren *n*
m. dropper *(Text)* Lamelle *f* aus Metall *(am Kettfadenwächter)*
m. electrode Metallelektrode *f*, Schweißelektrode *f*
m.-enclosed blechgekapselt, metallgekapselt (z. B. Schaltgeräte)
m. etching Metallätzen *n*
m. extruding Metallpressen *n*, Strangpressen *n*
m.-fabricating metallverarbeitend
m.-faced hardboard Hart[faser]platte *f* mit Metallüberzug
m. filament Metallfaden *m*
m. filament lamp Metallfadenlampe *f*
m. filings Feilspäne *mpl*, Metallspäne *mpl*
m. fittings Beschläge *mpl*
m. foil Metallfolie *f*, Blattmetall *n*
m. forming Umformung *f*
m.-forming process Umformverfahren *n*
m.-forming technique Umformtechnik *f*, Umformverfahren *n*
m.-founder Metallgießer *m*
m. foundry Metallgießerei *f*
m. gauze Metallgaze *f*
m. glu[e]ing Metallkleben *n*
m. grid Metallgitter *n*
m. grit Metallkies *m*
m. hose Metallschlauch *m*
m. in grains Metallgranalien *pl*, granuliertes Metall *n*
m. insert Metalleinlage *f*
m.-jointing method Metallfügemethode *f*
m. locator Metallsuchgerät *n*
m. mining Erzbergbau *m*
m. mould Kokille *f*
m. notch Abstichloch *n*, Stichloch *n*, Stich *m (des Schmelzofens)*
m. pattern Metallmodell *n*
m. physics Metallphysik *f*
m. plane Ganzmetallflugzeug *n*

m. planer Hobelmaschine *f* für Metallbearbeitung
m. plating *(galvanisches)* Plattieren *n* von Metallen
m. powder Metallpulver *n*
m. powder moulding press Metallpulverpresse *f*
m. processing Metallverarbeitung *f*
m. protector Metallschutzhülse *f*
m. radiography Röntgenmetallografie *f (zerstörungsfreie Werkstoffprüfung)*
m. rectifier Trockengleichrichter *m*, Metallgleichrichter *m*
m. reflector Metallspiegel *m*
m. removal Spanabnahme *f*; Zerspanung *f*
m. removal performance Zerspanungsleistung *f*
m. removal rate [spezifische] Spanmenge *f*, Spanvolumen *n*
m. removal research Zerspanungsforschung *f*
m. removing spanend
m. removing capacity Zerspanungsleistung *f*
m. rubbers *(Text)* Metallnitschelwerk *n*
m. saw Metallsäge *f*
m. screening Metallabschirmung *f*
m. section Metallschliff *m*, Schliff *m*
m. severing process Metalltrennverfahren *n*
m. sheet Blech *n*
m. sheeting Panzerung *f*
m. shot Schrot *m*; Metallkies *m*
m. slitting saw Metallschlitzsäge *f*
m. slotting saw Metallschlitzsäge *f*
m. slug Metallklumpen *m*
m. spinning Metalldrücken *n*
m.-spinning lathe Metalldrückbank *f*
m. splash Metallspritzer *m*
m.-sprayed glass envelope außenmetallisierter Glaskolben *m (Elektronenröhren)*
m. spraying Metallspritzen *n*; Spritzmetallisieren *n*
m. spraying pistol Metallspritzpistole *f*
m. spraying process Metallspritzverfahren *n*
m. spring Metallfeder *f*
m. steel wool Stahlwolle *f*
m. strip fuse *(El)* Streifensicherung *f*
m.-to-metal adhesive Metallkleber *m*
m.-to-metal bonding Metallkleben *n*
m.-to-metal contact metallische Berührung *f*
m.-to-wood laminating Metall-Holz-Laminierung *f*
m. tube Stahlrohr *n*, Metallrohr *n* (*s. a.* m. valve 2.)
m. tube making press Metallrohrpresse *f*, Rohrpresse *f*
m. turning 1. Drehen *n (Metall)*; 2. Metalldrehspan *m*
m. valve 1. *(Kfz)* Metallventil *n (Luftreifen)*; 2. Ganzmetallröhre *f*, Stahlröhre *f (Elektronenröhren)*
m.-vapour lamp Metalldampflampe *f*
m. varnish Metallack *m*
m. wind-instrument Metallblasinstrument *n*
m. wool Stahlwolle *f*
m. working Metallbearbeitung *f*, Umformung *f*
m. working metallbearbeitend; umformend
m. working factory Metallwarenfabrik *f*
m. working industry metallverarbeitende Industrie *f*
m. working machine Metallbearbeitungsmaschine *f*
m. working manufacturing Fertigungstechnik *f*

metaldehyde Metaldehyd *m*
metalled road Schotterstraße *f*
metallic metallisch, Metall-
m. antimony Antimonmetall *n*
m. arsenide Arsenmetall *n*
m. bed Erzlager *n*
m. bond *(Ch)* metallische Bindung *f*
m. brush Metallbürste *f*
m. button *(Met)* Metallkönig *m*
m. card clothing *(Text)* Ganzstahlgarnitur *f*
m. cloth metalldurchwirktes Gewebe *n*
m. coating Metallüberzug *m*
m. diaphragm indicator *(Wärm)* Plattenfederindikator *m*
m. drive screw Schneidschraube *f*, Blechschraube *f*
m. electricity metallische Elektrizität *f*
m. fibre Metallfaser *f*
m. lustre metallischer Glanz *m*, Metallglanz *m*
m. mixture *(Met)* Speise *f*; Metallmischung *f*
m. oxide Metalloxid *n*
m. packing Metallpackung *f*, Metalldichtung *f*
m. rectifier Metallgleichrichter *m*, Sperrschichtgleichrichter *m*
m. rectifier stack assembly Gleichrichtergruppe *f*
m. regulus *(Met)* Metallkönig *m*
m. resistance Metallwiderstand *m*
m. return circuit Kabelrückleitung *f*
m. selenide Selenmetall *n*
m. silicide Siliziummetall *n*
m. soap Schallplattenwachs *n*
m. sodium Natriummetall *n*
m. sponge Metallschwamm *m*
m. thermometer Metallthermometer *n*
m. tubing Metallschläuche *mpl*
m. yarn *(Text)* metallisiertes Garn *n*, Metallgarn *n*

metalliferous metallhaltig; *(Geol)* metallführend, erzführend
m. mine Erzgrube *f*, Erzbergwerk *n*, Mine *f*
m. mineral Erzmineral *n*

metalling Beschotterung *f (Straßenbau, Gleisbau)*
metallization Metallisierung *f*; Metallspritzverfahren *n*
metallize / to metallisieren, mit Metall überziehen
metallized dye *(Text)* Metallkomplexfarbstoff *m*
m. lamp verspiegelte Lampe *f*
m. paper mit Metall belegtes Papier *n*, kaschiertes Papier, Metallpapier *n*
m. paper capacitor Metallpapierkondensator *m*
m. screen Metallspiegel *m*, metallhinterlegter Schirm *m (Fernsehbildröhre)*
m. valve metallisierte Röhre *f (Elektronenröhre)*
metallizing Metallauftragung *f*, Metallisierung *f*
metallochromy galvanische Metallanfärbung *f*
metallographer Metallograf *m*
metallographic metallografisch
m. microscope Metallmikroskop *n*
metallography Metallografie *f*
metalloid Metalloid *n*, Nichtmetall *n*
metalloscope Metallmikroskop *n*
metallurgic *s.* metallurgical
metallurgical metallurgisch, hüttenmännisch
m. chemistry Eisenhüttenchemie *f*
m. coke Hüttenkoks *m*, Hochofenkoks *m*
m. engineer Hüttenmann *m*, Hütteningenieur *m*
m. engineer for iron and steel Eisenhütteningenieur *m*
m. engineering Hüttenwesen *n*, Hüttentechnik *f*
m. microscope Metallmikroskop *n*
m. process Hüttenprozeß *m*
m. treatment Verhüttung *f*
metallurgist Metallurge *m*
metallurgy Metallurgie *f*, Hüttenkunde *f*, Hüttenwesen *n*
m. of copper Kupferhüttenkunde *f*
m. of iron and steel Eisenhüttenwesen *n*, Eisenhüttenkunde *f*
metals Eisenbahnschienen *fpl*
metalworking *s.* metal working
metameric *(Ch)* metamer
metamorphic *(Geol)* metamorph
m. rock metamorphes Gestein *n*
metamorphosed *s.* metamorphic
metaphosphoric acid Metaphosphorsäure *f*
metasilicic acid Metakieselsäure *f*
metastable metastabil
m. atom metastabiles Atom *n*
m. diagram *(Met)* Eisen-Eisenkarbid-Diagramm *n*
m. molecular state metastabiler Molekülzustand *m*
m. nuclear state metastabiler Kernzustand *m*
m. nucleus metastabiler Kern *m*
m. state metastabiler Zustand *m*
m. state of atom metastabiler Atomzustand *m*
metathesis *(Ch)* Austausch[reaktion *f*] *m*, Umsetzung *f*
meteor Meteor *n (m)*
m. echo Meteorecho *n (Radar)*
m. shower Meteorenschwarm *m*
m. stream Meteorstrom *m*
meteoric iron Meteoreisen[stein *m*] *n*
meteorite Meteorit *m*
meteorograph Meteorograf *m*
meteorologic *s.* meteorological
meteorological meteorologisch
m. broadcasts Wetterfunk *m*
m. chart Wetterkarte *f*, meteorologische Karte *f*
m. conditions Wetterlage *f*
m. observatory Wetterwarte *f*; Seewarte *f*
m. rocket meteorologische Rakete *f*, Wetterrakete *f*
m. satellite meteorologischer Satellit *m*, Wettersatellit *m*
m. service Wetterdienst *m*
meteorology Meteorologie *f*, Wetterkunde *f*
meter / to messen; abgeben, zumessen *(Kraftstoff durch Pumpe)*
meter 1. [integrierendes] Meßgerät *n*, Meßuhr *f*, Uhr *f*, Zähler *m*, Zählwerk *n*; Anzeigeinstrument *n*; 2. *s.* metre
m. amplifier Meßgeräteverstärker *m*
m. armature Zähleranker *m*
m. base Zählergrundplatte *f*
m. board Zählertafel *f*
m. bottom bearing Unterlager *n* [eines Zählers]
m. calibration Zählereichung *f*, Zählerabgleich *m*
m. case Zählergehäuse *n*
m. change-over clock Zählerschaltuhr *f*
m. circuit Meßschaltung *f*
m. constant Zählerkonstante *f*; Instrumentenkonstante *f*
m. controller Meßwerkregler *m*
m. deflection Zeigerausschlag *m*, Instrumentenanzeige *f*
m. dial [runde] Zählerskala *f*
m. friction *(Am)* Zählerreibung *f*
m. key *(Fmt)* Gesprächszähltaste *f*, Zähltaste *f*
m. losses Verluste *mpl* im Zähler, Eigenverbrauch *m* des Zählers
m. movement Meßwerk *n*
m. needle Meßgerätezeiger *m*
m. panel Schalttafel *f*, Instrumententafel *f*, Zählertafel *f*
m. reading Zählerablesung *f*; Meßgeräteanzeige *f*

m. relay Kontaktgeberzähler *m*, Sendezähler *m*; Zählrelais *n*
m. resistance Innenwiderstand *m* des Zählers
m. switch Meßbereich[s][um]schalter *m*
m. tariff Zählertarif *m*
m. terminal cover Klemmendeckel *m (eines Zählers)*
m. unit Meßgerät *n*
m. winding Zählerwicklung *f*
m. with demand indicator Zähler *m* mit Maximumzeiger
metercandle Lux *n (Einheit der Beleuchtungsstärke; 1 metercandle = 1 lx)*
metered control system *(BMSR)* Kaskadenregelung *f*
metering Messen *n*, Dosieren *n*, Zuteilen *n*; Zählung *f*
m. clock Meßuhr *f*
m. clockwork Meßuhr *f*
m. element Meßglied *n*, Zählglied *n*
m. equipment Meßeinrichtungen *fpl*
m. presentation Meßanzeige *f*, Instrumentenanzeige *f*
m. relay *(Fmt)* Zählerrelais *n*
m. section *(Ch)* Homogenisierzone *f*
m. transfer Meßtransformator *m*, Meßwandler *m*
m. wheel Meßrädchen *n*, Meßrad *n*
methacrylate Methakrylat *n*, Methakrylsäureester *m*
methacrylic acid Methakrylsäure *f*
m. ester Methakrylsäureester *m*
methanal Methanal *n*, Formaldehyd *m*
methane Methan *n*
m.-burning ship Schiff *n* mit Methangas als Treibstoff
m. detector *(Brg)* Methanspürgerät *n*
m. drainage *(Brg)* Methanabsaugung *f*
m. emission *(Brg)* Methanausgasung *f*
methide *(Ch)* Methylverbindung *f* eines Metalls
method Methode *f*, Verfahren[sweise *f*] *n*, Arbeitsweise *f*
m. for drawing *(Met)* Ziehverfahren *n*
m. of actuating Betätigungsart *f*
m. of analysis Analysenverfahren *n*
m. of approximation *(Math)* Näherungsverfahren *n*, Approximationsverfahren *n*
m. of clearances Kahlschlagmethode *f (Forstwirtschaft)*
m. of combustion *(Kfz)* Verbrennungsart *f*
m. of control Steuerungsweise *f*
m. of disengagement Darstellungsverfahren *n*
m. of examination Prüfungsmethode *f*
m. of firing *(Wärm)* Beschickungsweise *f*, Feuerungsverfahren *n*
m. of fluctuation *(Math)* Differentialrechnung *f*
m. of forming Formgebungsverfahren *n*
m. of heating Beheizungsart *f*
m. of high forest with reserves Hochwaldbetrieb *m* mit Überhalt
m. of irrigation Berieselungsmethode *f*
m. of joints Rundschnittverfahren *n (Berechnung von Fachwerken)*
m. of least squares Methode *f* (Verfahren *n*) der kleinsten Quadrate
m. of measurement Meßmethode *f*, Meßverfahren *n*
m. of measuring Meßmethode *f*, Meßverfahren *n*
m. of modulation Modulationsverfahren *n*
m. of moment distribution Cross-Verfahren *n*, Methode *f* wiederholter Momentenverteilung, Momentenausgleichsverfahren *n (Statik)*
m. of production Fertigungsmethode *f*, Produktionsmethode *f*
m. of purification Reinigungsverfahren *n*
m. of sections [Rittersches] Schnittverfahren *n (Statik)*
m. of separation Trenn[ungs]verfahren *n*
m. of simple coppice Niederwaldbetrieb *m*
m. of successive approximations *(Math)* Iterationsverfahren *n*
m. of test Prüfverfahren *n*
m. of testing *s.* m. of test
m. study *(Ök)* Arbeitsstudie *f*
methods engineer Verfahrensingenieur *m*; *(Am)* Technologe *m*
methyl *(Ch)* Methyl *n*
m. acrylate Methylakrylat *n*, Akrylsäuremethylester *m*
m. alcohol Methylalkohol *m*
m. orange Methylorange *n*
m. violet Methylviolett *n*
methylate / to *(Ch)* methylieren
methylate *(Ch)* Methylat *n*
methylated spirit Brennspiritus *m*, denaturierter (vergällter) Alkohol *m*
methylene blue Methylenblau *n*
metol Metol *n (Entwickler)*
metre Meter *n*
m.-ampere *(El)* Meterampere *n*
m. bridge Schleifdrahtmeßbrücke *f*, Meterbrücke *f*
m.-kilogram-second unit MKS-Einheit *f*
m. rule Metermaß *n*
metric metrisch
m. carat Karat *n*
m. field metrisches Feld *n (Gravitationstheorie)*
m. horsepower Pferdestärke *f (CGS-System)*
m. involute gear cutter Modulfräser *m*
m. module Modul *m*
m. pitch Millimeterteilung *f (Verzahnung)*
m. screw thread metrisches Gewinde *n*

m. series metrische Reihe *f (Bohrer)*
m. taper metrischer Kegel *m*
m. thread metrisches Gewinde *n*
m. ton [metrische] Tonne *f (1000 kg)*
m. waves Meterwellen *fpl (1 bis 10 m)*
metrizability *(Math)* Metrisierbarkeit *f*, Abmeßbarkeit *f*, Meßbarkeit *f*
metrizable *(Math)* metrisierbar, [ab]meßbar; metrisch
metrology Metrologie *f*, Meßkunde *f*, Meßwesen *n*
metronom Metronom *n*, Taktmesser *m*, Taktzähler *m*
meyerhofferite *(Min)* Meyerhofferit *m*
mezzanine Halbgeschoß *n*, Mezzaningeschoß *n*, Zwischengeschoß *n*
mezzotint *(Typ)* Schabkunst *f*
MF *s.* medium frequency
M.F. paper *s.* machine-finished paper
M.G. paper *s.* machine-glazed paper
mho *(Am)* Siemens *n (Einheit des elektrischen Leitwertes; 1 mho = 1 S)*
mica *(Min)* Glimmer *m*
m. capacitor Glimmerkondensator *m*
m. condenser *s.* m. capacitor
m.-end-window counter *(Kern)* Fensterzählrohr *n*, Glockenzählrohr *n* [mit Glimmerfenster]
m. insulation Glimmerisolation *f*, Glimmerisolierung *f*
m. insulator Glimmerisolator *m*
m. punching machine Glimmerstanzmaschine *f (Elektronenröhrenherstellung)*
m. sandstone Mikasandstein *m*
m. schist Glimmerschiefer *m*
m.-sheet Glimmerplättchen *n*
m. slabs Blockglimmer *m*, Rohglimmer *m*
m. slate Glimmerschiefer *m*
m. splitting Glimmerfolie *f*
m. washer Unterlagsscheibe *f* aus Glimmer, Glimmerscheibe *f*
micaceous glimmerartig, glimmerhaltig
m. copper Kupferglimmer *m*
m. iron ore Eisenglimmer *m*
m. schist Glimmerschiefer *m*
micanite glimmerhaltiger Schichtstoff *m*
micatherm glimmerhaltiges Isoliermaterial *n*
micell *(Ch)* Mizelle *f*
Michell thrust bearing Michell-Drucklager *n*
Michigan-cut *(Brg)* Brennereinbruch *m (Schießarbeit)*
micro Mikro-
m.-alloy diffused transistor Mikrolegierungsdiffusionstransistor *m*
m.-analysis Feinanalyse *f*, Mikroanalyse *f*
m.-beam research equipment *(Opt)* Strahlensticheinrichtung *f*
m.-chemistry Mikrochemie *f*
m.-drilling Mikrobohren *n*
m.-etching Mikroätzen *n*
m.-fuse *(El)* Feinsicherung *f*
m.-hardness *(Met)* Mikrohärte *f*
m.-hardness tester Mikrohärteprüfer *m*
m.-inch Mikrozoll *m*, millionstel Zoll *m*
m.-jet roll coating *(Led)* Luftbürstverfahren *n (Beschichtung)*
m.-mesh stockings feinmaschige Damenstrümpfe *mpl*
m.-micro-farad *(El)* Picofarad *n*
m.-module Mikromodul *m*
m. pipe Mikrolunker *m*
m.-pressure-gauge Mikromanometer *n*
m.-second Mikrosekunde *f*
m.-speed motor Schleichfahrtmotor *m*
m.-wave *s.* microwave
microammeter Mikroamperemeter *n*
microampere Mikroampere *n*
microanalysis Feinanalyse *f*, Mikroanalyse *f*
microanalyzer Mikroanalysator *m*
microautoradiogram Mikroautoradiogramm *n*
microbalance Mikrowaage *f*
microbar Mikrobar *n*
mikrobarogram Mikrobarogramm *n*
microcalorimeter Mikrokalorimeter *n*
microchemical analysis *s.* microanalysis
microchemistry Mikrochemie *f*, Spurenchemie *f*
microcline *(Min)* Mikroklin *m*
microcomponent Mikrokomponente *f*
microcosmic mikrokosmisch
m. bead Phosphorsalzperle *f*
m. salt Natriumammoniumhydrogenphosphat *n*, Phosphorsalz *n*
microcrystalline feinkörnig, mikrokristallin
microcurie Mikrocurie *n (Einheit der Aktivität radioaktiver Strahlung)*
microelement Mikroelement *n*
microfilm Mikrofilm *m*
microfiltration Feinfiltrierung *f*
microfinishing Feinstbearbeitung *f*
microgalvanometer Galvanometer *n* mit fotoelektrischer Verstärkung
micrograined plates *(Typ)* mikrogekörnte Offsetplatte *f*
micrograph *(Met)* Mikrofotografie *f*, Mikro[schliff]bild *n*, Gefügebild *n*
microhardness *s.* micro-hardness
microhm Mikroohm *n (Einheit des elektrischen Widerstandes)*
microhone / to feinhonen
microindication Feinanzeige *f*
microlite *(Min)* Mikrolith *m*
micrologic element logisches Mikrobauelement *n*
micromagnetics Mikromagnetik *f*
micromanipulator Mikromanipulator *m*, Feinmanipulator *m*

micrometer Meßschraube *f*, Schraublehre *f*, Mikrometer *n*
m. adjustable feinsteinstellbar
m. adjusting screw Mikrometerfeinstellschraube *f*
m. caliper stand Schraublehrenhalter *m*
m. eyepiece Mikrometerokular *n*
m. for external measurements Meßschraube *f* für Außenmessungen
m. for internal measurements Meßschraube *f* für Innenmessungen
m. gap Feinmeßfunkenstrecke *f*, Funkenstrecke *f* mit Feineinstellung, Präzisionsfunkenstrecke *f*
m. ocular Okularmikrometer *n*
m. ratchet stop Meßschraubengefühlsschraube *f*; Meßschraubenratsche *f*
m. screw Mikrometerschraube *f*
m. setting Feinsteinstellung *f*
micromicrofarad *(El)* Picofarad *n*
microminiature electronic device elektronische Kleinstvorrichtung *f*
microminiaturization Kleinstausführungsart *f*, Mikrominiaturisierung *f*
micron Mikrometer *n*, μm
m.-sized vacuum tube Mikrominiaturröhre *f*
micronizer Feinstmahlvorrichtung *f*
micronutrient *(Ch)* Spurenelement *n*; Mikronährstoff *m*
microparticle Mikroteilchen *n*
microphone Mikrofon *n*
m. amplifier Mikrofonverstärker *m*
m. battery Mikrofonbatterie *f*, Mikrofonelement *n*
m. boom Mikrofongalgen *m*
m. button Sprechtaste *f*
m. cable Mikrofonkabel *n*
m. circuit Mikrofon[strom]kreis *m*
m. current Mikrofonstrom *m*
m. current feed Mikrofonspeisung *f*
m. current supply Mikrofonspeisung *f*
m. effect *s.* microphonic effect
m. gallows Mikrofongalgen *m*
m. hiss Mikrofonrauschen *n*
m. relay Mikrofonverstärker *m*, mikrofonischer Verstärker *m*, Mikrofonrelais *n*
m. transformer Mikrofontransformator *m*, Mikrofonübertrager *m*
microphonic mikrofonisch
m. carbon Mikrofonkohle *f*
m. effect Mikrofon[ie]effekt *m*
microphonics *s.* microphonic effect
microphonism *s.* microphonic effect
microphony *s.* microphonic effect
microphotograph mikrofotografische Aufnahme *f*, Mikrofotografie *f*, Mikrofotogramm *n*
microphotography Mikrofotografie *f*
microphotometer Mikrofotometer *n*
microphysics Mikrophysik *f*
microprogramming Mikroprogrammierung *f*
microradiogram Mikroradiogramm *n*
microradiography Mikroradiografie *f*
microradiometer Mikroradiometer *n*
microscope Mikroskop *n*
m. illuminant Mikroskopierleuchte *f*
microscopic mikroskopisch
m. cross section mikroskopischer Wirkungsquerschnitt *m*
m. depiction mikroskopische Abbildung *f*
m. hole Mikrobohrung *f*
m. metallography Metallmikroskopie *f*
m. sharpness gestochene Schärfe *f*
microscopy Mikroskopie *f*
microsecond Mikrosekunde *f*, μs
microsection Mikroschliff *m*
microsegregation Kristallseigerung *f*
microstate Mikrozustand *m*
microstrip Streifenleitung *f (Wellenleiter)*
microstructure Gefüge *n*, Feingefüge *n*, Mikrogefüge *n*, Feinstruktur *f*, Mikrostruktur *f*
microswitch Mikroschalter *m*
microsyn *(BMSR)* Mikrosyn *n (induktiver Geber für Drehbewegungen)*
microtelephone Handapparat *m*
microtome Mikrotom *n*
microtron *(Kern)* Mikrotron *n (Elektronenzyklotron)*
microturbulence *(Strm)* Mikroturbulenz *f*
microwatt *(El)* Mikrowatt *n*
microwave Mikrowelle *f*
m. breakdown Mikrowellendurchschlag *m*
m. cavity Mikrowellenresonator *m*, Mikrowellenhohlraum *m*
m. component Mikrowellenbauelement *n*
m. detector Mikrowellengleichrichter *m*
m. early warning Kurzwellenradaralarmsystem *n*
m. engineering Mikrowellentechnik *f*
m. fading Mikrowellenschwund *m*
m. interferometry *(Opt)* Mikrowelleninterferometrie *f*
m. length Mikrowellenlänge *f*
m. optics Mikrowellenoptik *f*
m. radio link Richtfunkstrecke *f (Radar)*
m.-relay system Richtfunkrelaiskette *f (UHF-Technik)*
m. resonator *s.* m. cavity
m. spectroscopy Mikrowellenspektroskopie *f*
m. spectrum Mikrowellenspektrum *n*
m. structure *s.* m. component
micrurgy *(Opt)* Mikrurgie *f*
mid-band *s.* midband
m.-beam Zwischenbalken *m*
m.-channel *(Schiff)* mitten im Fahrwasser
m.-channel buoy *(Schiff)* Mitte[l]fahrwassertonne *f*

m.-feather Mittelwand *f*, Scheidewand *f*, Zwischenwand *f (Holländer)*
m.-feather wall *(Wärm)* Trennungsmauer *f* der Längszüge
m.-frequency Mittelfrequenz *f*
m.-line Mittellinie *f*
m.-perpendicular Mittelsenkrechte *f*
m.-point *(El)* Mittelpunkt *m*, Sternpunkt *m*
m.-point tap Mittel[punkts]anzapfung *f*, Mittelabgriff *m*
m.-portion Mittelteil *n*; Mittelstellung *f*
m.-position Mittelstellung *f*
m.-position of the valve *(Masch)* Schiebermittellage *f*
m.-range mittlerer Bereich *m*
m. rib Mittelrippe *f*
m.-series termination *(El)* T-Kettenglied *n*
m.-shunt termination *(El)* L-Kettenglied *n*
m.-way in potential *(El)* auf Mittelpotential
m. wire Mittelleiter *m*
midband Bandmitte *f (Radio)*
m. frequency Bandmittenfrequenz *f*
middle Mitte *f*
middle Mittel-, mittlerer
m. body Mittelschiff *n*
m. colour Mittelfarbe *f*
m. conductor Mittelleiter *m (Drehstrom)*
m.-cutting *(Brg)* Schrämen *n* im Mittelteil des Flözes
m. edge Mittelgrat *m (Turbine)*
m. fraction Destillationsmittelschnitt *m*, Mittellauf *m*
m. grain Mittelkorn *n*
m. line bulkhead *(Schiff)* Mittellängsschott *n*
m. line keel-plate *(Schiff)* Mittelkielplatte *f*
m. marker Haupteinflugzeichen *n*, Halbwegmarkierungssender *m (Radar, Funknavigation)*
m.-mast *(Schiff)* Mittelmast *m*
m. of stroke Hubmitte *f*
m. oil Mittelöl *n*
m. plane Mittelebene *f*
m. position of eccentric Exzentermittellage *f*, Mittellage *f* des Exzenters
m. post Hängesäule *f*
m. pressure Mitteldruck *m*
m. roll Mittelwalze *f*
m. roving *(Text)* mittleres Vorgarn *n*
m. section of grate *(Wärm)* Mittelrostfeld *n*
m. shaft *(Masch)* Mittelstück *n* einer Welle
m.-shot mittelschlächtig *(Turbine)*
m. slide *(Masch)* Mittelschlitten *m*
m. spaces *(Typ)* Viertelspatien *npl*
m. strip Mittelstreifen *m (Straßenbau)*
m. third Kern *m*, mittleres Drittel *n (Statik)*
m. wire Nulleiter *m*, Mittelleiter *m*, Mitteldraht *m*
middlebreaker *(Am)* Häufelpflug *m*
middling von mittlerer Güte
middlings *(Brg)* Mittelgut *n*
midgap mittlere Fehlstelle (Störstelle) *f (Halbleitertechnik)*
midget miniaturisiert
m. base Mignonsockel *m*, Zwergsockel *m (Glühlampen)*
m. car Kleinstwagen *m*
m. excavator Klein[st]bagger *m*
m. receiver Kleinempfänger *m*, Zwergempfänger *m*
m. socket *s.* m. base
m. staff *(Opt)* Meßlättchen *n*
m. superhet Kleinsuper *m*, Zwergsuper *m*
m. tractor Kleinstschlepper *m*
m. valve Zwergröhre *f*
midposition *s.* mid-position
midsection coefficient *(Schiff)* Hauptspantvölligkeitsgrad *m*
midship mittschiffs
m. bend Hauptspant *n*, Hauptspantkurve *f*
m. coefficient Hauptspantvölligkeitsgrad *m*
m. frame Nullspant *n*, Hauptspant *n*
m. section *(Schiff)* Querschnitt *m*, Hauptspantquerschnitt *m*, Nullspant *n*, Hauptspant *n*
m. section coefficient Völligkeitsgrad *m* des Hauptspants
midships mittschiffs
midspan Feldmitte *f (Träger usw.)*
m. loading Einzellast *f* in Feldmitte *(Statik)*
midstrip Mittelstreifen *m (Straßenbau)*
midwing [mono]plane Mitteldecker *m*
migrate / to migrieren, wandern *(Ionen)*
migration 1. *(Text)* Farbwanderung *f*, Wanderung *f*; 2. *(Ph)* Ionenwanderung *f*; Migration *f*; 3. *(Geol)* Wanderung, Migration
m. area Wanderfläche *f*, Migrationsfläche *f (Reaktortheorie)*
m. length Wanderlänge *f*, Migrationslänge *f (Reaktortheorie)*
m. loss *(Ph)* Wanderungsverlust *m*
m. of dislocations *(Met)* Wanderung *f* von Versetzungen
m. of plasticizer *(Plast)* Weichmacherwanderung *f*
m. velocity of the ions Wanderungsgeschwindigkeit *f* der Ionen
mike *(sl) s.* micrometer
mil *Längeneinheit; 1 mil = 25,4 Mikrometer = 0,001 Zoll*
milanese warp knitting machine *(Text)* Milanesekettenwirkmaschine *f*
mild-carbon steel kohlenstoffarmer Stahl *m*
m.-centred steel Stahl *m* mit weichem Kern
m. steel Baustahl *m*; Betonstahl *m*; Flußstahl *m*; Schiffbaustahl *m*
mildew Brand *m (Pflanzenkrankheit)*, Mehltau *m*; Schimmel *m*, Moder *m*
m.-resistant schimmelbeständig

mile Meile *f* *(1 statute mile = 1,609 km)*
mileage Meilenzahl *f*; zurückgelegte Strecke *f* [in Meilen]
m. recorder *(Kfz)* Meilenzähler *m*
military port Kriegshafen *m*
milk / to melken
milk Milch *f*
m. churn Milchkanne *f*, Milchtank *m*
m. fat refractometer Milchfettrefraktometer *n*
m. flowmeter Milchdurchflußmeßgerät *n*
m. glass Milchglas *n*, Mattglas *n*
m. of lime Kalkmilch *f*
m. pail Milcheimer *m*
m. poise Galaktoskop *n*, Milchprüfer *m*
m. powder Trockenmilch *f*
m. pump Milchpumpe *f*, Melkpumpe *f*
m. sterilization Keimfreimachung *f* der Milch
m. tester Milchprüfgerät *n*
milker Melker *m*; Melkmaschine *f*
milking bucket Melkeimer *m*
m. cup Melkbecher *m (Melkmaschinen)*
m. generator *(El)* Einzellenlademaschine *f*
m. machine Melkmaschine *f*
m. parlour Melkstand *m*
m. plant Melkanlage *f*
m. teat cup Zitzensauggerät *n*, Zitzensaugschale *f*, Zitzenbecher *m*
mill / to 1. *(Masch)* fräsen; 2. *(Met)* walzen; 3. *(Text)* walken; 4. *(Ch)* rühren
m. away abfräsen
m. in an endwise direction in axialer Richtung fräsen
m. off abfräsen
m. out ausfräsen
mill 1. Mühle *f*; Sägemühle *f*; Mahlwerk *n*; 2. Hüttenwerk *n*; Walzwerk *n*; Straße *f*; Betrieb *m*; Fabrik *f*; 3. Spinnerei *f*, Weberei *f*; 4. *(Brg)* Aufbereitungsanlage *f*; Trichter *m (Trichterbau)*; 5. Fräser *m*; Fräsmaschine *f*
m. auxiliaries Walzwerkhilfsmaschinen *fpl*
m. bar Platine *f*, Rohschiene *f*, Schweißeisen *n*, Puddelluppe *f*
m. chips Walzspäne *mpl*
m. cinder Walzschlacke *f*, Walzzunder *m*
m. edge Walzkante *f*
m. extruder Maschine *f* zum kontinuierlichen Mischen und Strangpressen *(Plastverarbeitung)*
m. finished maschinenglatt *(Papier)*
m. grinder Mahlapparat *m*
m.-hole *(Brg)* Trichter *m (Trichterbau)*
m. housing Walzenständer *m*
m. limit Herstellungstoleranz *f*, Walztoleranz *f*
m. mixer Kneter *m*, Knetmischer *m*
m. mixing Mahlen *n* in der Kugelmühle, Kugeln *n*
m. race Mühlgraben *m*
m. run *(Holz)* unsortierte Platten *fpl*
m. scale Walzsinter *m*, Walzzunder *m*
m. scrap Rücklaufschrott *m*
m. stone Mahlstein *m*
m. term Betriebsausdruck *m*
m.-train arrangement Walzstraßenanordnung *f*
m. wheel Mühlrad *n*
millability Mahlbarkeit *f*
millboard Pappe *f*, Zellstoffpappe *f*; Handpappe *f*
milled [an]gefräst; gerändelt
m. cloth Filztuch *n*
m. file Fräserfeile *f*
m. goods *(Text)* Walkware *f*
m.-out portion Ausfräsung *f*
m. part Frästeil *n*
m. tooth cutter Fräser *m* mit gefrästen Zähnen
m. tooth file Fräserfeile *f*
m. tooth half round file Halbrundfräserfeile
miller 1. Müller *m*; 2. *(Am)* Fräsmaschine *f*
Miller bridge *(El)* Durchgriffsmeßbrücke *f*, Miller-Brücke *f*
Miller circuit Miller-Kreis *m (Impulstechnik)*
Miller indices Millersche Indizes *mpl*
Miller integrator *(Dat)* Miller-Integrator *m*, Miller-Schaltung *f*
Miller time-base Miller-Zeitbasis *f (Impulstechnik)*
millerite *(Min)* Millerit *m*, Nickelblende *f*
millet Hirse *f*
milli- Milli- *(10^{-3} Einheiten)*
m.-inch *s.* mil
milliammeter Milliamperemeter *n*
milliampere Milliampere *n*, mA
millibar Millibar *n*, mbar *(Einheit des Druckes)*
millibarn Millibarn *n*, mb *(Einheit des Wirkungsquerschnitts eines Atoms; 1 mb = 10^{-27} cm²)*
Milligoat counter *(Kern)* richtungsunabhängiger Zähler *m*, Milligoat-Zähler *m*
millihenry Millihenry *n*, mH *(Einheit der Induktivität)*
millilambert Millilambert *n*, mla *(Einheit der Leuchtdichte)*
millilitre Milliliter *n*, ml
millimetre Millimeter *n*, mm
m. pitch Millimeterteilung *f*
millimicrofarad Nanofarad *n*, nF *(Einheit der Kapazität; 1 nF = 10^{-9} C/V)*
millimicrohenry Nanohenry *n*
millimicron Nanometer *n*, nm *(Längeneinheit = 10^{-9} m)*
millimicrosecond Nanosekunde *f*, ns *(Zeiteinheit = 10^{-9} s)*
m. pulse *(El)* Nanosekundenimpuls *m*
m. pulse generator Nanosekundenimpulsgenerator *m*

millimu *s.* millimicron
milling 1. Mahlen *n*; Zerkleinern *n*; 2. Fräsen *n*; Fräsarbeit *f*; 3. *(Text)* Walken *n*; Walke *f*; 4. Trichterbau *m (Tagebau)*
m. and centring machine Abfläch- und Zentriermaschine *f*, Abläng- und Zentriermaschine *f*
m. attachment Fräsapparat *m*, Fräseinrichtung *f*
m. auxiliary *(Text)* Walkhilfsmittel *n*
m. azo red *(Text)* Azowalkrot *n*
m. capacity Fräsleistung *f*
m. chip Frässpan *m*
m. cutter Fräser *m*, Fräswerkzeug *n*
m. cutter for fluting taps Nutenfräser *m* für Gewindebohrer
m. cutter for wood Holzfräser *m*
m. cutter with inserted blades Fräser *m* mit Einsatzmessern, Fräskopf *m*
m. cycle Mahlgang *m*
m. dye *(Text)* walkfester Farbstoff *m*
m. fixture Fräsvorrichtung *f*
m. head Fräskopf *m*
m. in the grease *(Text)* Walken *n* im Schmutz
m. in transit Lohnschnitt *m (Sägewerk)*
m. length Fräslänge *f*, Fräsweg *m*
m. liquor *(Text)* Walkbrühe *f*, Walkflotte *f*
m. loss *(Text)* Walkverlust *m*
m. machine 1. Fräsmaschine *f*; 2. *(Text)* Walkmaschine *f*, Walke *f*
m. machine arbor Fräs[er]dorn *m*
m. machine arbor collar Frädornring *m*
m. machine template Frässchablone *f*, Nachformfrässchablone *f*
m. machine vice Maschinenparallelschraubstock *m*
m. off *(Led)* letztes Walken *n*
m. operation Fräsarbeit *f*
m. out Ausfräsung *f*, Ausfräsen *n*
m. pattern Frässchablone *f*, Nachformfrässchablone *f*
m. property *(Text)* Walkfähigkeit *f*, Walkvermögen *n*
m. quill Fräspinole *f*
m. test *(Text)* Walkversuch *m*
millings Frässpäne *mpl*
milliohm Milliohm *n (Einheit des elektrischen Widerstandes; 1 mΩ = 10^{-3} V/A)*
milliroentgen Milliröntgen *n (Einheit der Ionendosis; 1 mR = 2,58 · 10^{-7} As/kg)*
millisecond Millisekunde *f*, ms
millivolt Millivolt *n (Einheit der elektrischen Spannung; 1 mV = 10^{-3} W/A)*
millivoltage Millivoltspannung *f*
millivoltmeter Millivoltmeter *n*
milliwatt Milliwatt *n*, mW
millscale *s.* mill scale
millwork *(Bw)* Holzbauteile *npl*
millwright Schlosser *m*, Monteur *m*
milt Milch *f (Fisch)*
mimesis Mimese *f (Kristall)*
mimetic mimetisch *(Kristall)*
minable *(Brg)* abbauwürdig
mincing knife Hackmesser *n*, Wiegemesser *n*
mine / to graben, Bergbau treiben; abbauen, schürfen, gewinnen
m. by the open-cast method im Tagebau fördern
m. out abbauen
mine 1. Grube *f*, Bergwerk *n*, Zeche *f*; 2. *(Mil)* Mine *f*
m. air Grubenwetter *pl*, Wetter *pl*
m. air sample *(Brg)* Wetterprobe *f*
m. barrage *(Mil)* Minensperre *f*
m. cable Grubenkabel *n*
m. captain Obersteiger *m*
m. car Förderwagen *m*, Grubenwagen *m*, Hunt *m*
m. car axle catch Bremsvorrichtung *f*, die sich vor die Wagenachsen legt
m. cars circulation Grubenwagenumlauf *m*
m. climate Grubenklima *n*
m. coal Zechenkohle *f*
m. cover Abraum[schichten *fpl*] *m*, Deckgebirge *n*
m. crater Sprengtrichter *m*
m. damp Grubengas *n*
m. detector Minensuchgerät *n*
m. disaster Grubenunglück *n*
m. drainage Wasserhaltung *f*, Wasserhebung *f*
m. dump Grubenhalde *f*
m. explosion Schlagwetterexplosion *f*
m. fan Grubenlüfter *m*, Grubenventilator *m*
m. field Minenfeld *n*
m. filling Bergeversatz *m*, Grubenversatz *m*
m. fire Grubenbrand *m*
m. floor Grubensohle *f*
m. lamp Grubenlampe *f*, Grubenleuchte *f*
m. layer Minenleger *m*
m. level Grubensohle *f*
m. lighting Grubenbeleuchtung *f*
m. locomotive Grubenlokomotive *f*
m. office Bergamt *n*
m. openings Grubenräume *mpl*
m. prop Grubenstempel *m*
m. pumping steam engine Wasserhaltungsdampfmaschine *f*
m. railway Zechenbahn *f*
m. rescue brigade Grubenrettungsmannschaft *f*
m. rescue station Grubenrettungsstation *f*
m. resistance Grubenwiderstand *m*
m. run coal Förderkohle *f*
m. surveying Grubenvermessen *n*
m. surveying equipment Markscheiderausrüstung *f*
m. surveyor Markscheider *m*

m. surveyor's equipment Markscheiderausrüstung *f*
m.-sweeper Minenräumer *m*, Minensuchboot *n*
m. sweeper gear Minenräumgerät *n*
m. timber Grubenholz *n*
m. train dispatcher Grubenfahrdienstleiter *m*
m. truck Förderwagen *m*, Hunt *m*
m. tub kleiner Förderwagen *m*
m. walling Stoßverbauung *f*
m. worker Bergarbeiter *m*
m. working Grubenbau *m*
m. workings Grubengebäude *npl*
miner 1. Bergmann *m*, Bergarbeiter *m*; 2. *(selten)* Gewinnungsmaschine *f*
mineral Mineral *n*, Gestein *n*
mineral mineralisch, Mineral-
m. acid anorganische Säure *f*, Mineralsäure *f*
m. caoutchouc *(Min)* Elaterit *m*
m. carbon Graphit *m*
m. charcoal *(Geol)* Faserkohle *f*, Fusit *m*
m. coal Steinkohle *f*
m. dressing Aufbereitung *f*
m. fertilizer Mineraldünger *m*
m. filler Gesteinsmehl *n*, mineralischer Füllstoff *m*
m. green Scheelesches Grün *n*
m. jelly Petrolatum *n*, Vaseline *f*
m. lard oil Schneidöl *n*
m. lubricating oil Mineralschmieröl *n*
m. matter Mineralstoff *m*
m. oil Erdöl *n*, Mineralöl *n*, Petroleum *n*
m. oil blend Mineralölmischung *f*
m. oil product Mineralölprodukt *n*
m. pitch Asphalt *m*, Erdpech *n*, Mineralpech *n*
m. resources Bodenschätze *mpl*
m. salt Steinsalz *n*
m. spirit Lackbenzin *n*, Ligroin *n*, Lösungsbenzin *n*
m. spring Mineralquelle *f*
m. substance Mineralstoff *m*
m. surface roofing paper besandete Dachpappe *f*
m.-tanned mit Mineralsalz gegerbt, chromgar
m. tar Asphaltteer *m*, Asphaltpech *n*
m. tar oil Steinkohlenteeröl *n*
m. vein Mineralgang *m*, Erzgang *m*
m. water Mineralwasser *n*
m. wax Montanwachs *n*, Erdwachs *n*
m. white Gips *m*, Mineralweiß *n*
m. wool Mineralwolle *f*, Schlackenwolle *f*
mineralization Versteinerung *f*; Inkohlung *f*
mineralogist Mineraloge *m*
mineralogy Mineralogie *f*
miner's box *(Brg)* Gezähekiste *f*
m. lamp Grubenlampe *f*, Geleucht *n*
minette Minette *f*
miniature winzig [klein]
m. base Mignonsockel *m*, Zwergsockel *m (für Glühlampen)*
m. bulb Kleinlampe *f*; Zwerglampe *f*
m. camera Kleinbildkamera *f*
m. component Miniaturbaustein *m*
m. device Miniaturgerät *n*, Kleinstgerät *n*
m. Edison screwcap *s.* m. base
m. electronics Miniaturelektronik *f*
m. fuse Feinsicherung *f*
m. glow-discharge tube Zwergglimmlampe *f*
m. lamp Kleinstlampe *f*, Zwerglampe *f*
m. switch Kleinschalter *m*
m. transmitter Mikrosender *m*
m. triode Zwergtriode *f*
m. valve Kleinströhre *f*, Miniaturröhre *f*, Liliputröhre *f*
miniaturization Miniaturisierung *f (von Bauelementen)*
miniaturized unit miniaturisiertes Gerät *n*
minicar Kleinkraftwagen *m*, Kleinfahrzeug *n*
minim 1. halbe Note *f (Musik)*; 2. *(englische Volumeneinheit; 1 min = 59,19 mm³)*
minimal minimal, Kleinst-, Minimal-
m. curve Minimalkurve *f*
m.-state machine *(Dat)* Mindestzustandmaschine *f (die dasselbe leistet wie eine andere umfangreichere Maschine mit der gleichen Bestimmung)*
m. surface Minimalfläche *f*
m. value Mindestwert *m*
minimization of Boolean functions *(Dat)* Minimisierung *f* der Booleschen (logischen) Funktionen
minimize / to vermindern, auf das kleinste Maß herabsetzen; *(Dat)* minimalisieren, minimisieren
minimum Kleinstwert *m*, Minimum *n*
minimum minimal, kleinster; Kleinst-, Mindest-
m.-access programming Programmieren *n* mit Mindestzugriffszeit, schnelle Programmierung *f*, Minimalsuchzeitprogrammierung *f*
m. access time *(Dat)* Mindestsuchzeit *f*, Mindestzugriffszeit *f*
m. amount of clearance Mindestspiel *n*
m. audible field Hörschwelle *f* im freien Schallfeld
m. audible pressure Hörschwelle *f* des Schalldrucks im Ohr
m. circuit breaker Unterstromschalter *m*, Minimalschalter *m*, Schalter *m* mit Minimalauslösung
m. clearance Kleinstspiel *n*
m. crest clearance Mindestspitzenspiel *n*
m. deviation Minimalablenkung *f*
m. distance Mindestabstand *m*

m. flashover voltage Mindestüberschallspannung *f*; unterer Isolationspegel *m*
m. fusing-current Grenzstrom *m* *(einer Sicherung)*
m.-hardness Mindesthärte *f*
m. interference Kleinstübermaß *n* *(Passung)*
m.-ironing bügelarm
m. latency *(Dat)* Minimalsuchzeit *f*, Mindestlatenz *f*
m. limit Kleinstmaß *n*
m. limit dimension Kleinstmaß *n*
m. lobe Nebenkeule *f* *(Antenne)*
m. number of [gear] teeth Kleinstzähnezahl *f*
m. number of gear teeth which can be generated without undercut kleinste Grenzzähnezahl *f*
m. operating current Ansprechstrom *m* *(Relais)*
m. operating value Mindestansprechwert *m* *(Relais)*
m. orbital auf kleinster Umlaufbahn
m.-phase system *(BMSR)* Phasenminimumsystem *n*, System *n* minimaler Phase
m. potential *(El)* Potentialschwelle *f*
m. pressure governor Unterdruckschalter *m*, Unterdruckwächter *m*
m. size Kleinstmaß K *n*
m. strength Mindestfestigkeit *f*
m. value Kleinstwert *m*
m. voltage relay Unterspannungsrelais *n*
m. wavelength *(El)* Grenzwellenlänge *f*; kurzwellige Grenze *f*
m. working current Ansprechstrom *m*, Ansprechstromstärke *f*
mining Bergbau *m*, Abbau *m*, Ausbeutung *f*, Gewinnung *f*
m. cable Grubenkabel *n*
m. damage Bergschaden *m*
m. engineering Bergbautechnik *f*
m. hutch Grubenförderwagen *m*
m. locomotive Grubenlokomotive *f*
m. losses Abbauverluste *mpl*
m. machinery Bergbaumaschinen *fpl*
m. method Abbauverfahren *n*
m. property Grubenfeld *n*
m. subsidence Bergbausenkung *f*, Bergschaden *m*
minion *(Typ)* 7 britische Punkte *mpl* *(Schriftgröße)*
Minitrack localizing system Minitrack *(Empfangsortung zur Bahnverfolgung von Meßsatelliten)*
minium Mennige *f*
minor 1. *(Math)* Unterdeterminante *f*, Subdeterminante *f*; 2. Moll *n* *(Musik)*
m. axis kleinere Achse *f*, Nebenachse *f*
m. coinage Kleinmünzen *fpl* *(einer Währung)*
m. cycle *(Dat)* Nebenperiode *f*, Kleinperiode *f*, Kleinzyklus *m*
m. diameter Kerndurchmesser *m* *(Außengewinde)*
m.-diameter fit Innenzentrierung *f* *(Keilprofil)*
m. diameter of a thread Kerndurchmesser *m*, Gewindekerndurchmesser *m*
m. diameter of an internal thread Kernlochdurchmesser *m*, Innengewindekernlochdurchmesser *m*
m. diameter of external splines Keilwelleninnendurchmesser *m*
m. diameter of screw Gewindekerndurchmesser *m*
m. exchange *(Fmt)* Nebenamt *n*
m. load Vorlast *f* *(Rockwell-Härteprüfung)*
m. lobe Nebenzipfel *m*, Nebenkeule *f* *(Antennentechnik)*
m. road Nebenstraße *f*, Seitenstraße *f*
minorant *(Math)* Minorante *f*, untere Schranke *f* *(einer Punktmenge)*
minority carrier Minoritätsträger *m*
mint Münze *f* *(Münzstätte)*
m. mark Münzstättenangabe *f*
mintage Prägung *f* *(von Münzen)*
minterm Vollkonjunktion *f* *(Logistik)*
mintmark *s.* mint mark
mintmaster Münzmeister *m*
minuend *(Math)* Minuend *m*
minus minus, Minus-; negativ
m. allowance Untermaß *n*
m. charge negative Ladung *f*
m. gauge Minuslehre *f*
m. pole Minuspol *m*, negativer Pol *m*
m. sign Minuszeichen *n*
m. terminal *(El)* Minusklemme *f*, negative Klemme *f*
m. tolerance unteres Abmaß *n*, Minustoleranz *f*
m. wire *(El)* Minusader *f*
minute [winzig] klein, sehr gering
minute Minute *f*
m. hand Minutenzeiger *m*, großer Zeiger *m*
m. of [an] arc Bogenminute *f*
m. scale Minutenskale *f*
mirabilite Glaubersalz *n*
mirage Fata Morgana *f*, Luftspiegelung *f*
m. landing *(Flg)* Glattwasserlandung *f*
mirror / to [ab]spiegeln, widerspiegeln
mirror 1. Spiegel *m*; 2. *(Fmt)* unsymmetrischer Paraboloidreflektor *m*
m.-bright spiegelblank
m. coating Verspiegelung *f*
m. compass Spiegelkompaß *m*
m. condenser Spiegelkondensor *m*
m.-copying *(Masch)* spiegelbildliches Kopieren *n*

m. effect Spiegelwirkung *f (Elektronen)*
m. extensometer Spiegel[feinmeß]gerät *n*
m. finish spiegelnde Oberfläche *f*; Hochglanz *m*
m.-finished hochglanzpoliert
m. galvanometer Spiegelgalvanometer *n*, Reflex[ions]galvanometer *n*
m. image Spiegelbild *n*
m. image nuclei *(Kern)* Spiegelkerne *mpl*
m. instrument Spiegelinstrument *n*, Spiegelapparat *m*, Spiegeleinrichtung *f*
m.-inverted spiegelverkehrt
m.-like deposit Getterspiegel *m (Elektronenröhren)*
m. monochromator Spiegelmonochromator *m*
m. observation Spiegelbeobachtung *f*
m. oscillograph Spiegeloszillograf *m*
m. plane Spiegelebene *f*
m. reading Spiegelablesung *f (eines Meßgerätes)*
m. reflector Spiegelreflektor *m*
m. scanning wheel Spiegelrad *n*
m. sea spiegelglatte See *f*
m. surface Spiegeloberfläche *f*
m. symmetry Spiegelsymmetrie *f*
m. wheel Spiegelrad *n*, Spiegeltrommel *f*
m. writing Spiegelschrift *f*
mirrored verspiegelt
mis-taper nicht maßentsprechender Kegel *m*
misadjusted falsch eingestellt
misalign / to sich verlagern, sich versetzen; verschieben
misaligned außer Flucht, nichtfluchtend, versetzt, abweichend in der Richtung; falsch abgeglichen *(Radio)*
misalignment 1. Versatz *m*; Fluchtungsfehler *m*; Richtungsabweichung *f*; 2. *(BMSR)* Regelabweichung *f (Lageregler)*; 3. *(BMSR)* Fehlanpassung *f*, Verstimmung *f*; 4. schlechte Ausrichtung *f*, mangelnder Gleichlauf *m*, Verstimmung *f (Funktechnik)*
m. of shafts Verlagerung *f* der Wellen
miscalculate / to sich verrechnen, falsch errechnen
miscalculation falsche Rechnung *f*
miscibility Mischbarkeit *f*
m. gap Mischungslücke *f*
miscible [ver]mischbar, mischfähig
miscount / to falsch zählen
misfire Versager *m*; Zündausfall *m*; Fehlzündung *f*, Zündversager *m*
misfiring Fehlzündung *f*
misfit / to schlecht passen
misfit cast *(Met)* Fehlschmelze *f*
mismatch / to fehlanpassen *(Funktechnik)*
mismatch Fehlanpassung *f*, Verstimmung *f*
m. factor *(El)* Reflexionskoeffizient *m*
mismatching Nichtzusammenpassen *n*; Anpassungsfehler *m*, Fehlanpassung *f (Funktechnik)*
misnomer Fehlbezeichnung *f*
misphased *(El)* phasenvertauscht, falschphasig, phasenfalsch *(geschaltet)*
mispick *(Text)* fehlerhafter Schuß *m*
mispickel Arsenopyrit *m*, Giftkies *m*
misplaced core versetzter Kern *m (Formerei)*
m. hole *(Brg)* falsch angesetztes Bohrloch *n*
m. size Fehlkorn *n (Aufbereitung)*
misprint Druckfehler *m*; Fehldruck *m*
misrun casting Fehlguß *m*, nicht ausgelaufener Guß *m*
miss / to *(Brg)* versagen *(Schießarbeit)*; verpassen, verfehlen; fehlschlagen, mißglücken
m. an oil sand einen Ölsand überbohren
miss distance Fehlentfernung *f*
missed approach *(Flg)* Fehlanflug *m*
m. hole nicht gekommenes Bohrloch *n*
missile Rakete *f*, Geschoß *n*
m. angle Aggregatlagerwinkel *m*
m. axis Raketenachse *f*
m. base Raketenabschußbasis *f*
m. body Raketenkörper *m*
m. centre Raketenschwerpunkt *m*
m. centre line Raketenachse *f*
m. electronics Raketenelektronik *f*
m. equation Aggregatgleichung *f*
m. establishment wissenschaftliches Forschungsinstitut *n* für ferngelenkte Raketen
m. flight Raketenflug *m*
m. guide-beam receiver Leitstrahlempfänger *m*
m.-mounted recording instrument Bordregistriergerät *n*
m.-mounted terminal board Bordsammelschiene *f*
m. steering control Aggregatsteuerung *f*
missiles Raketenwaffen *fpl*
missing fehlend, verschollen
m. operation unterbliebenes Arbeiten *n (z. B. des Relais)*
m. word *(Typ)* Leiche *f*
mist / to *(Typ)* beschlagen, schwitzen *(Raster)*
m. up beschlagen *(optische Geräte)*
mist Nebel *m*, feiner Regen *m*; *(Masch)* Ölnebel *m*
m. blower Verneblungsgerät *n*
m. coolant Kühlnebel *m* für Ölnebelkühlung
m. cooling Düsenhochdruckkühlung *f*, Ölnebelkühlung *f*, Sprühnebelkühlung *f*
m. cooling nozzle Sprühkühldüse *f*
m.-lubricated düsenhochdruckgeschmiert
m. lubrication Ölnebelschmierung *f*
m. spray Sprühnebel *m (Schmierung)*
m. sprayer Nebelgerät *n*, Nebler *m*
mistake Fehler *m*; Irrtum *m*
m. in adding Additionsfehler *m*

m. in computation (calculation) Rechenfehler *m*, Berechnungsfehler *m*
mister Verneblungsgerät *n*
misting Beschlagen *n (Windschutzscheibe)*
mistune / to falsch abstimmen, verstimmen *(Radio)*
misty dunstig, neblig
Mitchell slicing system *(Brg)* Mitchell-Abbauverfahren *n*
mitre / to gehren
mitre Gehrung *f*; Winkel *m* von 45°
m.-angle Gehrungswinkel *m*
m.-box saw Gehrungssäge *f*
m. clamp Gehrungs[schraub]zwinge *f*
m. cut Gehrungsschnitt *m*
m. cutting machine Gehrungsschneidmaschine *f*, Gehrungsschneidgerät *n*
m. cutting shear Gehrungsschere *f*
m. gears Kegelradpaar *n* mit Übersetzungsverhältnis 1:1
m. rule Gehr[ungs]maß *n*
m. square Gehrungswinkel *m*, Gehrmaß *n*
mitred auf Gehrung geschnitten *(Holzbau)*
m. joint Gehrungsstoß *m*, Stoß *m* auf Gehrung
mitring Schrägschneiden *n*; Gehrung *f*
m. saw Gehrungssäge *f*
Mitrofanov group machining method Gruppenbearbeitung *f* [nach Mitrofanow], Mitrofanowmethode *f*
mitten Fausthandschuh *m*
m. money *(Schiff)* Zusatzlotsengebühr *f (im Winter)*
mix / to [ver]mischen, [ver]mengen; gattieren; anrühren, ansetzen; aufmaischen *(Zucker)*; verschneiden *(Wein)*
m. for average quality *(Text)* auf Durchschnittsgüte mischen
m. signals *(Rak)* aufschalten
mix Gemisch *n*, Mischung *f*; Mischgut *n*; Mischungsverhältnis *n*; Frischbeton *m*
mixability Mischbarkeit *f*
mixable mischbar
mixed gemischt
m. acid Mischsäure *f*, Nitriersäure *f*
m. acid tank *(Rak)* M-Tank *m*
m. admission gemischte Beaufschlagung *f (Turbine)*
m. admission turbine gemischt beaufschlagte Turbine *f*
m.-base notation *(Dat)* Gemischtbasisschreibweise *f*; Zifferndarstellung (Ziffernschreibweise) *f* mit gemischter Basis, Zifferndarstellung (Ziffernschreibweise) mit gemischter Grundzahl
m. base oil gemischtbasisches Öl *n*, Hybridrohöl *n*
m. batch Frischbeton *m*
m. batch capacity Mischerinhalt *m (Frischbetonmenge)*
m. card *(Text)* Halbdeckelkrempel *f*, gemischte Karde *f*
m. cargo *(Am) (Schiff)* gemischte Ladung *f*, Stückgut *n*
m. catalyst Mischkatalysator *m*
m. colour Mischfarbe *f*
m. construction Gemischtbauweise *f*
m. coupling gemischte Kopplung *f*
m. crop Mischbestand *m (Forstwirtschaft)*
m. crystal Mischkristall *m*
m. film lubrication Mischschmierung *f (Mischreibung)*
m. flow turbine vereinigte Axial- und Radialturbine *f*, Halbaxialturbine *f*
m. forest Mischwald *m*
m. highs Hochfrequenzkomponenten *fpl* des Farbfernsehbildsignales; Mischhöhen *fpl*
m. light lamp Mischlichtlampe *f*
m. melting point Mischschmelzpunkt *m*
m. polyelectrode potential Mischpotential *n (mehrerer Elektroden)*
m. retroaction magnetische und elektrische Rückkopplung *f*
m. strategy gemischte Strategie *f (Theorie der Spiele)*
m. straw board Strohmischpappe *f*
m. straw paper Strohmischpapier *n*
m. transmission gemischte Durchlassung *f (Licht)*
m. vessel Fracht- und Fahrgastschiff *n*
mixen *(sl)* Misthaufen *m*
mixer 1. Mischer *m*; Mischmaschine *f*, Betonmischer *m*; 2. Roheisenmischer *m*, Warmhalteofen *m*; 3. *(Fmt, Fs)* Mischstufe *f*, Überlagerungsstufe *f*; Mischschaltung *f*
m. circuit Mischschaltung *f*
m. driver *(Bw)* Mischerfahrer *m*
m. gain *(Fs)* Mischverstärkung *f*
m. paver Mischerfertiger *m (Straßenbau)*
m. platform Mischerbühne *f*
m.-preheater Mischvorwärmer *m*
m.-settler *(Kern)* Mixer-Settler *m*
m. stage *s.* mixer 3.
m. valve Mischröhre *f*, Modulatorröhre *f*, Überlagererröhre *f*
mixing 1. Mischen *n*, Vermischen *n*; Vermengen *n*; 2. *(Fmt, Fs)* Überlagerung *f*
m. and amplification by variable reactance Reaktanzverstärkung *f*
m. and fading facility Misch- und Überblendmöglichkeit *f*
m. arm Rührflügel *m*
m. chamber 1. *(Kfz)* Mischkammer *f*; 2. *(Rak)* Vorbrennkammer *f*
m. condenser *(Wärm)* Mischkondensator *m*

m. **efficiency** *(Kern)* Mischwirkungsgrad *m*, Homogenisierungsausbeute *f*
m. **feed table** *(Text)* Mischungszuführtuch *n*
m. **funnel** Mischtrichter *m*
m. **heat** Mischungswärme *f*
m. **hexode** Mischhexode *f*
m. **mill** Mischwalzwerk *n*, gemischtes Walzwerk *n*; Mischwalze *f*
m. **nozzle** *(Rak)* Mischdüse *f*
m. **ores and fluxes** Möllern *n*, Möllerung *f*
m. **picker** *(Text)* Mischmaschine *f*
m. **plant** Mischanlage *f*, Mischstation *f*
m. **platform** Mischbühne *f*
m. **point** *(BMSR)* Mischstelle *f*, Vergleichsstelle *f*
m. **ratio for angular acceleration signal** *(Rak)* Drehbeschleunigungsaufschaltung *f*
m. **ratio for angular velocity signal** *(Rak)* Geschwindigkeitsaufschaltung *f*
m. **ratio in the guide beam** *(Rak)* Aufschaltung *f* der Steuermaschine
m. **roll** Mischwalze *f*
m. **runner** Mischkollergang *m*
m. **stage** 1. Mischbühne *f*; 2. *(Fs)* Summierstufe *f*
m. **tank** Mischbehälter *m*
m. **time** Mischdauer *f*, Mischzeit *f*
m. **tower** Mischturm *m (Betontechnologie)*
m. **valve** Mischventil *n*; Mischbatterie *f (Rohrleitungsbau)*
m. **water** Anmachwasser *n (Betonherstellung)*
mixture Gemisch *n*, Mischung *f*, Gemenge *n*
m. **admission** *(Kfz)* Gemischzuführung *f*
m. **control** *(Kfz)* Gemischregelung *f*
m. **control valve** Gemischregelventil *n*
m.**-making** *(Met)* Gattieren *n*, Gattierung *f*
m. **preparation** Möllervorbereitung *f*
m. **strength** Mischungsverhältnis *n*, Stärke *f* des Gemisches
m. **turbine** Mischungsturbine *f*
m. **yarn** Melangegarn *n*
mizen-mast *(Schiff)* Achtermast *m*
mobile beweglich, fahrbar, ortsveränderlich, transportabel
m. **aerial** ortsveränderliche Antenne *f*
m. **crane** Mobil[dreh]kran *m*, Autokran *m*
m. **drilling platform** Ölbohrinsel *f*
m. **hearth** *(Met)* ausfahrbarer Herd *m*
m. **load** bewegliche Belastung *f*, wandernde Last *f*
m. **milking parlour** fahrbarer Melkstand *m*
m. **nuclear power unit** fahrbare (bewegliche) Kernenergieanlage *f*
m. **radio service** beweglicher Funkdienst *m*
m. **ramp** *(Rak)* fahrbare Startrampe *f*
m. **scaffolding** fahrbares Gerüst *n*
m. **searchlight** *(Mil)* Scheinwerferkraftwagen *m*, Scheinwerfer *m (auf Kfz)*
m. **testing unit** Kabelmeßwagen *m*
mobility Beweglichkeit *f*, Wendigkeit *f*; Elektronenbeweglichkeit *f*
m. **of ions** Ionenbeweglichkeit *f*
mock nachgemacht, unecht, Schein-
m. **attack** *(Mil)* Scheinangriff *m*
m. **seam** *(Text)* falsche (imitierte) Naht *f*, Scheinnaht *f*
m.**-up** Modell *n*, Muster *n*
m.**-up test** *(Kern)* Scheinversuch *m*
m.**-up tower** *(Flg)* Flugleitungsturm *m* für Ausbildungszwecke
mockup *s.* mock-up
modacrylic filament modifizierte Akrylseide *f*
mode 1. Art *f*, Weise *f*, Methode *f*, Form *f*; 2. *(Math)* Modus *m*, Scheitel[wert] *m (Häufigkeitskurve)*; 3. Welle[ntyp *m*] *f*, Schwingung[styp *m*] *f*, Wellenform *f*; 4. *(BMSR)* Eigenwert *m*; 5. Feldtyp *m (elektromagnetische Feldform, Wellenleiter)*
m. **changer** Wellentypwandler *m*, Wellenformwandler *m*, Wellentyptransformator *m*, Moduswandler *m (Wellenleiter)*
m. **coupler** Wellentypkoppler *m*
m. **filter** Wellentypfilter *n*, Modusfilter *n (Wellenleiter)*
m. **filter slot** Schlitzblende *f*; Schlitzwellenfilter *n (Wellenleiter)*
m. **of application** Anwendungsweise *f*
m. **of calculation** Berechnungsart *f*
m. **of control** Typ *m* der Regelung, Regelungsgesetz *n*
m. **of formation** Entstehungsart *f*
m. **of motion** Schwingungsform *f (Kristalle)*
m. **of operation** Arbeitsweise *f*, Betriebsweise *f*, Wirkungsweise *f*
m. **of oscillation** Schwingungstyp *m*, Schwingungsmodus *m*, Schwingungsart *f*, Eigenschwingung *f*; Schwingungsform *f*
m. **of propagation** Ausbreitungsform *f (elektromagnetische Wellen)*
m. **of vibration** Schwingungsart *f*
m. **separation** Frequenzabstand *m*
m. **shift** Frequenzinkonstanz *f (Magnetron)*
m. **transducer** *s.* m. changer
m. **transformer** *s.* m. changer
model / to gestalten, modellieren, formen
model Modell *n*, Bezugsstück *n*, Bezugsform *f (Nachformen)*; Ausführung *f*, Konstruktion *f (Maschine)*; Schiffsmodell *n*
m. **analysis** *(Bw)* Modellstatik *f*
m. **basin** *(Schiff)* Schleppversuchstank *m*, Modelltank *m*
m. **computation** Modellberechnung *f*
m. **experimental tank** *(Schiff)* Schleppversuchstank *m*, Modelltank *m*
m. **forest** Normalwald *m*

m. of nucleus Kernmodell *n*, Modell *n* des Atomkerns
m. plant *(Am)* Musterbetrieb *m*
m. study *s.* m. test
m. tank *(Schiff)* Schlepprinne *f*, Schleppkanal *m*
m.-tank establishment Schiffbauversuchsanstalt *f*, Schleppversuchsanstalt *f*
m.-tank testing *(Schiff)* Schleppversuch *m* mit Modellen
m. test Modellversuch *m*
modelling Modellierung *f*, Nachbildung *f*
moderate / to 1. mäßigen, mildern; 2. *(Kern)* [ab]bremsen, moderieren
moderated *(Kern)* [ab]gebremst, moderiert
m. neutron langsames (abgebremstes) Neutron *n*
m. reactor moderierter Reaktor *m*
moderately hard mittelhart, halbhart
moderating *(Kern)* Moderator-, Brems-
m. block Bremsblock *m*, Moderatorblock *m*
m. column Bremssäule *f*, Moderatorsäule *f*
m. power Bremsvermögen *n*, Bremskraft *f*
m. ratio Bremsverhältnis *n*
moderation *(Kern)* Bremsung *f*, Bremsen *n*, Abbremsung *f*, Abbremsen *n*
moderator *(Kern)* Moderator *m*, Bremsstoff *m*, Bremsmittel *n*
m. control Bremsmittelregelung *f*, Moderatorkontrolle *f (des Reaktors)*
m. effect Moderatoreffekt *m*
m. heat Moderatorwärme *f*
modification Modifikation *f*, Abart *f*, Abänderung *f*, Änderung *f*, Umwandlung *f*, Variante *f*
modified modifiziert, verändert, abgewandelt, variiert, umgebaut
m. cross-section fibre Profilfaser *f*
m. equation modifizierte Gleichung *f*
m. lip korrigierte Schneide *f (Spiralbohrer)*
m. wood Holzwerkstoff *m*
modifier *(Am) (Dat)* Umsteuergröße *f (d. Rechners)*
modify / to abändern, umgestalten, modifizieren; *(BMSR)* umsteuern
moding Frequenzspringen *n (Umspringen eines Oszillators)*
modular Modul-
m. co-ordination *(Bw)* Maßordnung *f*
m. construction Baukastenkonstruktion *f*
m. design Baukastensystem *n*
m. dimension modulare Abmessung *f*
m. furniture Baukastenmöbel *n*
m. grid Raster *m*, Systemnetz *n*
m. line Rasterlinie *f*, Systemlinie *f*
m. principle Baukastenprinzip *n*
m. products modulare Bauteile *npl*
m. ratio Verhältniszahl *f* der Elastizitätsmoduln von Stahl und Beton *(Stahlbetonberechnung)*
m. space grid dreidimensionaler Raster *m*, Raumraster *m*
m. unit Baustein *m*
modulate / to 1. modulgerecht bauen; 2. modeln, modulieren *(Funk)*
m. on aufmodulieren
m. upon aufmodulieren
modulated *(El)* moduliert, gemodelt
m. amplifier modulierter Verstärker *m*
m. at audible frequencies *(Fmt)* tonmoduliert
m.-carrier amplifier Zerhackerverstärker *m*, Trägerfrequenzverstärker *m*
m. continuous wave modulierter Dauerstrich *m*, modulierte ungedämpfte Welle *f*, unmodulierter Betrieb *m* mit Anodenwechselspannung
m. current gemodelter Strom *m*
m. keyed continuous waves *(Fmt)* A2-Wellen *fpl*, mit Tonfrequenztelegrafie modulierte ungedämpfte Wellen *fpl*
m. oscillation modulierte Schwingung *f*
m. stage Modulationsstufe *f*
m. wave modulierte Welle *f*
modulating *(El)* Modulations-
m. anode Modulationsanode *f*
m. choke *(Fmt)* Steuerdrossel *f*, Modulationsdrossel *f*
m. current Modulationsstrom *m*
m. frequency Modulationsfrequenz *f*
m. voltage Modulationsspannung *f*
modulation *(El)* Modulation *f*, Modelung *f*; Aussteuerung *f (Verstärker)*
m. bridge *(Fs)* Modulationsbrücke *f*
m. capability *(Fmt)* höchstzulässiger Modulationsgrad *m*, Modulationsfähigkeit *f*, Modulierbarkeit *f*
m. characteristic Modulationskennlinie *f*, Modulationskurve *f*
m. circuit Modulationskreis *m*
m. converter Modulationsumformer *m*
m. current Modulationsstrom *m*
m. depth Modulationsgrad *m*, Modulationstiefe *f*
m. distortion Modulationsverzerrung *f*
m. due to iron saturation *(Fmt)* Drosselmodulation *f*
m. electrode Steuer[ungs]elektrode *f*
m. envelope Modulationshüllkurve *f*, Modulationsverlauf *m*
m. factor Modulationsgrad *m*
m. frequency Modulationsfrequenz *f*
m.-frequency harmonic distortion Verzerrung *f* von Harmonischen, Oberwellenverzerrung *f*
m. index *(Fmt)* Phasenhub *m*, Modulationsindex *m*
m. keying Telegrafiemodulation *f*
m. measurement Modulationsmessung *f*

m. meter Modulationsmesser *m*
m. mirror *(Opt)* Modulationsspiegel *m*
m. monitor Modulationsüberwachungsgerät *n*
m. noise Modulationsrauschen *n*
m. stage Modulationsstufe *f*
m. suppression Modulationsunterdrückung *f*
m. transformer Modulationstransformator *m*
modulator *(El)* Modulator *m*, Zerhacker *m*; Kanalumsetzer *m*
m. electrode Steuerelektrode *f*, Wehnelt-Elektrode *f (Katodenstrahlröhren)*
module 1. *(Bw)* Grundmaß *n*, Modul *m*, Rastergrundmaß *n*; 2. Baustein *m*, Bauelement *n*, Element *n*, Baueinheit *f*
m. component Modulbaustein *m*
m. millimetres *(Masch)* Modul *m*
modulo n check *(Dat)* Prüfung *f* modulo, Quersummenrestkontrolle *f*
m.-2 counter *(Dat)* Binärzähler *m*
modulus 1. Modul *m*, Zahl *f*; 2. *(Text)* Verhältnis *n* von Gewichtszunahme zu Längenzunahme *(Strukturgarn)*
m. of elasticity Elastizitätsmodul *m*, Elastizitätszahl *f*
m. of elasticity for tension Dehnungskoeffizient *m*, Dehnungszahl *f*
m. of elasticity in shear Schub- oder Gleitmodul *m*
m. of resistance Festigkeitskennwert *m*, Festigkeitszahl *f*
m. of rigidity Schub[elastizitäts]modul *m*
m. of rupture statische Zerreißfestigkeit *f*, Bruchmodul *m*
m. of the foundation Flächenbettungszahl *f*
m. of torsion Torsionsmodul *m*, Schubmodul *m*, Gleitmodul *m*, Schermodul *m*
m. of transverse elasticity Schubelastizitätsmaß *n*, Gleitmaß *n*
m. of volume elasticity Kompressionsmodul *m*
modumite *(Min)* Hartkobalterz *n*
moellen Dégras *m*, Degras *n*, Moellon *n (Lederfett)*
mogul base *(El)* Goliath-Sockel *m*
mohair Angoraziegenhaar *n*, Mohär *m*
m. yarn Mohärgarn *n*
Mohr circle Mohrscher Kreis *m*
Mohr's salt Eisen(II)-ammoniumsulfat *n*, Mohrsches Salz *n*
Mohs' hardness scale Mohssche Härteskala *f*
moiré *(Typ)* Doppelraster *m*, Moiré *m*
moist feucht, naß
m. curing Feuchthaltung *f (Nachbehandlung von Beton)*
moisten / to anfeuchten, befeuchten, benetzen
moistening power Benetzungsfähigkeit *f*
moisture Feuchtigkeit *f*, Nässe *f*
m. content Feuchtigkeitsgehalt *m (in %)*, Wassergehalt *m*
m.-content meter Feuchtemesser *m*, Feuchtigkeitsmesser *m*, Materialfeuchtemesser *m*
m. equilibrium Feuchtigkeitsgleichgewicht *n*
m. extraction bin Trockenbunker *m (Belüftungstrockner)*
m. index Feuchtigkeitszahl *f (Bodenmechanik)*
m. loss Feuchtigkeitsverlust *m*
m. pick-up Feuchtigkeitsaufnahme *f*
m.-proof feuchtigkeitsgeschützt, feuchtigkeitsundurchlässig, wasserdampfundurchlässig, mit Feuchtigkeitsschutz
m. proofing Feuchtigkeitsschutz *m*, Imprägnierung *f*
m. regain Feuchtigkeitswiederaufnahme *f*
m.-repellent feuchtigkeitsabstoßend, hydrophob
m. separator Feuchtigkeitsabscheider *m*
m. test Feuchtigkeitsprüfung *f*
m. trap Trockenpatrone *f*
moiting Entfernung *f* von Verunreinigungen aus Wolle *(Blätter u. a.)*
moits Fremdbestandteile *mpl*, Fremdkörper *mpl (Wolle)*
moity durchsetzt mit Fremdbestandteilen *(Wolle)*
mol *s.* mole 1.
molal *s.* molar
molality *s.* molarity
molar *(Ch)* molar
m. boiling-point elevation molare Siedepunktserhöhung *f*
m. conductance molare Leitfähigkeit *f*
m. conductivity molare Leitfähigkeit *f*
m. heat Mol[ekular]wärme *f*
m. polarization Molpolarisation *f*
m. refraction Molrefraktion *f*
m. susceptibility Molsuszeptibilität *f*, molare Suszeptibilität *f*
m. volume Molvolumen *n*
molarity Molarität *f*, molekulare Gewichtskonzentration *f*
molasses Melasse *f*, Zuckersirup *m*
m. pulp Melasseschnitzel *npl*
m. pump Melassepumpe *f*
m. tank Melassebehälter *m*
mold *(Am) s.* mould
moldavite *(Min)* Moldavit *m*, Glasmeteorit *m*
mole 1. *(Ch)* Grammolekül *n*, Mol *n*; 2. Damm *m*, Mole *f*, Hafendamm *m*, Steindamm *m*, Wellenbrecher *m*; 3. Maulwurf[spelz] *m*
m. drainage Maulwurfsdränung *f*
m. fraction *(Ch)* Molenbruch *m*
m. hill Maulwurfshügel *m*
m. mining streifenweiser Verhieb *m* eines Kohlenflözes durch senkrecht zum Abbaustoß und nebeneinander liegende Bohrlöcher
m. plough Maulwurfspflug *m (für Drainagearbeiten)*
m. ratio *(Ch)* Mol[en]verhältnis *n*

molecular molekular, Molekül-, Molekular-
m. acoustic absorption molekulare Schallabsorption *f*
m. Ampere currents *(El)* Amperesche Molekularströme *mpl*
m. band *(Opt)* Molekülbande *f*, Bande *f*
m. beam Molekularstrahl *m*
m. cluster Molekülhaufen *m*, Molekülschwarm *m*
m. conductivity Molekularleitfähigkeit *f*
m. configuration Molekülanordnung *f*
m. crystal Molekularkristall *m*
m. diffusion Molekulardiffusion *f*
m. dipole moment Dipolmoment *n*
m. dissociation energy *(Kern)* Dissoziationsarbeit *f* der Moleküle
m. distillation Molekulardestillation *f*
m. electronics Molekularelektronik *f*; Festkörperkreistechnik *f*
m. flow molekulare Strömung *f*
m. formula Molekularformel *f*
m. heat Mol[ekular]wärme *f*
m. ion Molekülion *n*
m.-kinetic theory *(Ph)* molekularkinetische Theorie *f*
m. lattice Molekülgitter *n*
m. lowering of freezing point molekulare Gefrierpunktserniedrigung *f*
m. mass Molekülmasse *f*
m. motion Molekularbewegung *f*, Brownsche Bewegung *f*
m. orbit *(Kern)* molekulare Bahn *f*
m. polarization Molekularpolarisation *f*
m. pump *(Ph)* Molekular[luft]pumpe *f*
m. ray *(Kern)* Molekularstrahl *m*
m. reaction Molekülreaktion *f*
m. refraction Mol[ekular]refraktion *f*
m. rise of boiling point molekulare Siedepunktserhöhung *f*
m. rotation of the plane of polarization molekulare Drehung *f* der Polarisationsebene
m. scattering Molekularstreuung *f*, Molekülstreuung *f*
m. sound absorption molekulare Schallabsorption *f*
m. spectrum Molekularspektrum *n*
m. stopping *(Ph)* molekulares Bremsvermögen *n*
m. structure Molekularstruktur *f*, Molekülaufbau *m*, Molekülstruktur *f*
m. vacuum gauge Ionisationsvakuummeter *n*
m. velocity Molekulargeschwindigkeit *f*
molecularity Molekularzustand *m*
molecule Molekül *n*, Molekel *n*
m. lattice Molekülgitter *n*
molelectronics Molekularelektronik *f*; Festkörperkreistechnik *f*
Mollier diagram Mollier-Diagramm *n*, IS-Diagramm *n*
molten geschmolzen
m. bath Schmelzbad *n*, Schmelze *f*
m. glass Glasfluß *m*
m. metal Metallschmelze *f*, flüssiges Metall *n*
m. salt Salzschmelze *f*
m. solder Lötbad *n*
moly steel *(Am)* Molybdänstahl *m*
m. wire *(Am)* Molybdändraht *m*
molybdate Molybdat *n*
molybdenic compound Verbindung *f* des dreiwertigen Molybdän
molybdeniferous molybdänhaltig
molybdenite *(Min)* Molybdänglanz *m*, Molybdänit *m*
molybdenum 1. Molybdän *n*; 2. Molybdänmetall
m. glance *(Min)* Molybdänglanz *m*
m. steel Molybdänstahl *m*
molybdic acid Molybdänsäure *f*
m. anhydride Molybdänsäureanhydrid *n*
m. compound Verbindung *f* des sechswertigen Molybdän
molybdous compound Verbindung *f* des zweiwertigen Molybdän
moment Moment *n*; Drehmoment *n*
m. area Momentenfläche *f*
m. arm Kraftarm *m*, Dreharm *m*, Hebelarm *m*
m. axis Drehachse *f*
m. capacity aufnehmbares Moment *n* *(Statik)*
m. centre Drehpunkt *m*
m. compression zone Biegedruckzone *f*
m. curve of the shaft governor Momentenkurve *f* des Achsenreglers
m. diagram Momentendiagramm *n*, Momentenplan *m*
m. distribution Momentenausgleich *m*
m. distribution method *(Bw)* Cross-Verfahren *n*, Methode *f* wiederholter Momentenverteilung, Momentenausgleichsverfahren *n* *(Statik)*
m. limits curve Momentenhüllkurve *f* *(Stahlbetonbemessung)*
m. of a doublet *(Ph)* Moment *n* einer Doppelquelle
m. of a mass Massenmoment *n*
m. of an area Flächenmoment *n*
m. of bending Biegemoment *n*
m. of impulse Drall *m*
m. of inertia Trägheitsmoment *n*
m. of inertia of a body (solid) Massenträgheitsmoment *n*
m. of inertia of an area Flächenträgheitsmoment *n*
m. of momentum Impulsmoment *n*, Drehimpuls *m*
m. of momentum about an axis Impulsmoment *n* in bezug auf eine Achse

m. of resistance Widerstandsmoment *n*
m. of rotation Drehmoment *n*
m. of the couple of forces Moment *n* des Kräftepaares
m. of the mass Massenmoment *n*
m. zero point Momentennullpunkt *m* *(Statik)*
momentary momentan, kurzzeitig, Momentan-, Augenblicks-
m. current rating *(El)* Nenn-Kurzzeitstrom *m*
m. deviation vorübergehende Regelabweichung *f*
m. output augenblickliche Leistung *f*
m. overload kurzzeitige Überbelastung *f*
m. value Augenblickswert *m*
moments at supports Stützmomente *npl*
momentum Bewegungsgröße *f*, Impuls *m* (*s. a.* moment)
m. equation Momentengleichung *f*
m. space Impulsraum *m*
m. starter Schwungkraftanlasser *m*
monarda oil Pfefferminzöl *n*
monatomic einatomig; einwertig
m. gas einatomiges (nullatomares, nullwertiges) Gas *n*
monaural einohrig
monaxial einachsig
monazite *(Min)* Monazit *m*
m. sand Monazitsand *m*
Mond gas Mondgas *n*
Mond process Mondverfahren *n* *(Nickelgewinnung)*
Monel *s.* Monel metal
Monel metal Monelmetall *n* *(Ni-Cu-Legierung)*
monetite *(Min)* Monetit *m*
money-ore Münzerz *n*
m. standard gold Münzgold *n*, Dukatengold *n*
monitor / to kontrollieren, überwachen, steuern, vermitteln; mithören, abhören
monitor 1. *(El)* Monitor *m*, Kontrollbildgerät *n*, Kontrollempfänger *m*; Mithörgerät *n*; *(Kern)* Strahlenwarngerät *n*; 2. *(Brg)* Monitor, Wasserkanone *f*, Strahlrohr *n*
m. counter *(Kern)* Überwachungszählrohr *n*
m. desk *(Fmt)* Überwachungspult *n*
m. facility *(El)* Überwachungseinrichtung *f*
m. fan Monitorgebläse *n*
m. horn Kontrollautsprecher *m*
m. jack *(Fmt)* Mithörklinke *f*
m. loudspeaker Kontrollautsprecher *m*, Abhörlautsprecher *m*
m. man Tonmeister *m*
m. receiver Kontrollempfänger *m*
m. tube Bildkontrollröhre *f*
monitored control system *s.* closed-loop control system
monitoring Überwachung *f*; Grenzwertüberwachung *f*, Kontrolle *f*; *(Fmt)* Mithören *n*; *(Kern)* Strahlenwarnung *f*
m. amplifier Kontrollverstärker *m*, Mithörverstärker *m*
m. board *(Fmt)* Mithörschrank *m*
m. button *(Fmt)* Mithörtaste *f*
m. device Überwachungsanlage *f*
m. instrument Überwachungsinstrument *n*, Kontrollinstrument *n*
m. jack *(Fmt)* Mithörklinke *f*
m. key *(Fmt)* Mithörtaste *f*
m. loudspeaker Kontrollautsprecher *m*, Abhörlautsprecher *m*
m. radio receiver Kontrollempfänger *m*
m. station Überwachungsfunkstelle *f*
m. system Betriebsüberwachungsanlage *f*, Überwachungsanlage *f*, Überwachungseinrichtung *f*
m. tube Bildkontrollröhre *f*
monk bond *(Bw)* märkischer Verband *m*
monkey 1. Fallbär *m*, Fallhammer *m*, Rammbär *m* *(Pfahlgründung)*; 2. *(Am, Brg)* Wetterstrecke *f*; 3. Schlackenform *f* *(Hochofen)*
m. forecastle *(Schiff)* niedrige (halbe) Back *f*
m. island *(Schiff)* Peilkompaßstand *m*, Peildeck *n*
m. way *(Am)* *(Brg)* Wetterstrecke *f*
m. wrench Mutterschlüssel *m*, Schraubenschlüssel *m*, Universalschraubenschlüssel *m*; Engländer *m*, Franzose *m*
mono-atomic *s.* monatomic
m.-crystal Ein[zel]kristall *m*
m.-lever control *(Masch)* Einhebelschaltung *f*
m.-rail Einbahnschiene *f*, Einschienengleis *n*; Einschienenbahn *f*
m.-rail conveyor Einschienenbahn *f*
m.-rail tipping wagon Einschienenkippwagen *m*
m.-rail transporter Einschienenförderer *m*
m.-railway Einschienenbahn *f*
m.-tower crane Turmdrehkran *m*
monoacid einbasische Säure *f*
monoacid einsäurig
m. base einsäurige Base *f*
monoatomic *s.* monatomic
monobasic einbasisch
monobloc aus einem Stück gegossener Zylinderblock *m*
m. container Blockkasten *m* *(Batterie)*
monobrome-naphthalene Monobromnaphthalin *n*
monocentric ocular monozentrisches Okular *n*
monochloracetic acid Monochloressigsäure *f*
monochromat filter *(Opt)* Monochromatfilter *n*
monochromatic einfarbig, monochromatisch
m. disturbance monochromatische Störung *f*
m. light monochromatisches Licht *n*

m. radiation monochromatische Strahlung *f*
m. transmission Einfarbenübertragung *f*
monochromatism *(Opt)* Monochromasie *f*
monochromatization *(Opt)* Monochromatisieren *n*, Monochromatisierung *f*
monochromator *(Opt)* Monochromator *m*
monochrome einfarbig, monochromatisch
m. channel *(Fs)* Schwarz-Weiß-Kanal *m*
m. signal *(Fs)* Schwarz-Weiß-Signal *n*
monocle magnifier Monokellupe *f*
monoclinal *(Geol)* monoklin, in der gleichen Richtung geneigt
monoclinic monoklin[isch] *(Kristallsystem)*
monocoque construction Schalenbauweise *f*
m. fuselage *(Flg)* Schalenrumpf *m*
m. unit *(Kfz)* selbsttragende Karosserie *f*
monocrystal Einkristall *m*
monocular *(Opt)* einäugig, monokular
m. inclined tube monokularer Schrägtubus *m*
m. straight tube monokularer gerader Tubus *m*
m. terrestrial telescope monokulares Aussichtsfernrohr *n*
m. viewing monokulare Beobachtung *f*
monocyclic alternator Hilfsphasenwechselstromgenerator *m*
m. system *(El)* Einphasen-Dreiphasen-Verteilung *f*
monoenergetic monoenergetisch, monochromatisch, homogen
m. radiation monoenergetische Strahlung *f*
monofuel homogener Treibstoff *m*
monogyro system Einkreiselanordnung *f*
monohydric acid einbasische Säure *f*
monoisotopic *(Kern)* einisotopig, isotopenrein, ein Isotop besitzend
monolayer monomolekularer Film *m*
monolithic monolithisch, aus einem Stück gearbeitet
m. concrete Monolithbeton *m*
m. construction monolithische Bauweise *f*, Monolithbau *m (Betonbau)*
monomer Monomeres *n*
monomeric monomer
monomial eingliedrig
monomolecular monomolekular; einmolekülig
m. reaction monomolekulare Reaktion *f*
monomorphous monomorph *(Kristall)*
mononuclear einkernig
monophase einphasig
m. mains Einphasennetz *n*
monopolar *s.* unipolar
monopropellant Einzeltreibstoff *m*, homogener Treibstoff *m*
monopulse method Einpulsverfahren *n (Radar)*
monoreactant Einzeltreibstoff *m*, homogener Treibstoff *m*
monosaccharose *(Ch)* Mono[saccharo]se *f*
monoscope *(Fs)* Testbildröhre *f*
monostable monostabil
m. multivibrator monostabiler Multivibrator *m*, Univibrator *m*
monotectic monotektisch
monotooth alternator Wechselstromgenerator *m* mit konzentrierter Wicklung
monotron Bildgeberröhre *f* mit fester Testfigur, Schwarz-Weiß-Fernsehbildgeberröhre *f*, Monotron *n*
monotropic *(Ch)* monotrop
monotype composing machine *(Typ)* Monotype-Setzmaschine *f*, Einzelbuchstabensetzmaschine *f*
monovalence Einwertigkeit *f*
monovalent einwertig
montage Montage *f (Film)*
m. photograph Fotomontage *f*
montan-tar pitch Montanpech *n*
m. wax Montanwachs *n*
m.-wax tar Montanwachsteer *m*
montanic acid Montansäure *f*
montanite *(Min)* Montanit *m*
montant Senkrechtstab *m*
montejus Saftheber *m*, Saftpumpe *f*
montmorillonite *(Min)* Montmorillonit *m*
monzonite *(Min)* Monzonit *m*
moon rocket Mondrakete *f*
m.-sheered ship Schiff *n* mit sehr hohem Sprung vorn und achtern
moon's satellite Mondsatellit *m*
moonstone *(Min)* Mondstein *m*
moor / to anlegen, festmachen *(Schiff)*
moor Moor *n*
m. coal Moor[braun]kohle *f*
mooring anchor Vertäuanker *m*, Hafenanker *m*, Bojeanker *m*
m. bitts *(Schiff)* Poller *m*, Vertäupoller *m*
m. buoy Vertäuboje *f*, Ankerboje *f*, Muringboje *f*; Festmachetonne *f*
m. lighter *(Schiff)* Ankerleichter *m*
m. line Ankertau *n*
m. pipe *(Schiff)* Vertäuklüse[nrohr *n*] *f*, Seitenklüse *f*
m. post *(Schiff)* Festmachepfahl *m*, Vertäupfahl *m*, Haltepfahl *m*, Dückdalben *m*
Moorish maurisch *(Architektur)*
mop Schwabbel[scheibe] *f*, Wabbel[scheibe] *f*
m.-board Scheuerleiste *f*
mopping Schwabbeln *n*
m. grease Schwabbelfett *n*
moraine *(Geol)* Moräne *f*
morass Morast *m*
m. ore *(Min)* Sumpferz *n*
morbidity Krankenstand *m*, Erkrankungsziffer *f*
morbility *s.* morbidity
mordacity *(Ch)* Säurenschärfe *f*
mordant / to beizen, ätzen

mordant Beize *f*, Beizmittel *n*, Ätzmittel *n*; Grundiermittel *n*
mordant ätzend, beißend, beizend, scharf; fressend, zerstörend
m. dye *(Text)* Beizenfarbstoff *m*
mordanting Beizen *n*
more part flask mehrteiliger Formkasten *m*
morello Sauerkirsche *f*
Moresque Maureske *f*
morning shift Frühschicht *f*
morocco-binding *(Typ)* Maroquineinband *m*, Saffianeinband *m*
morphine *(Ch)* Morphin *n*
morse / to morsen, Morsezeichen geben (senden)
Morse alphabet Morsealphabet *n*
Morse code Morsezeichen *npl*, Morsealphabet *n*
Morse cone arbor Morsekegelaufspanndorn *m*
Morse dash Morsestrich *m*
Morse dot Morsepunkt *m*
Morse inkwriter *(Fmt)* Schreibempfänger *m*
Morse key Morsetaste *f*
Morse lamp Blinklampe *f*, Morselampe *f*
Morse receiver Morseempfänger *m*
Morse sender Morsegeber *m*
Morse signal Morsezeichen *n*
Morse tape Morsestreifen *m*
Morse tape perforator Morselocher *m*
Morse taper *(Masch)* Morsekegel *m*
Morse taper on the outer surface Außenmorsekegel *m*
Morse taper shaft Morsekegelschaft *m*
Morse telegraphy Morsetelegrafie *f*
Morse transmitter Morsesender *m*
mortar 1. *(Ch)* Mörser *m*, Reibschale *f*; 2. *(Mil)* Granatwerfer *m*, Minenwerfer *m*; 3. *(Bw)* Mörtel *m*
m. fabrication Mörtelaufbereitung *f*
m. floor Kalkestrich *m*
m. intrusion Mörtelinjektion *f*, Mörteleinpressung *f*
m. joint Mörtelfuge *f*
m. lining Mörtelverkleidung *f*
m. mill Mörtel[liefer]werk *n*; Mörtelmaschine *f*
m. mixer Mörtelmischer *m*
m. mixing Mörtelaufbereitung *f*
m. sand Mörtelsand *m*
m. squad *(Mil)* Granatwerfertrupp *m*
m. trowel Mörtelkelle *f*
m. walling Mörtelmauerwerk *n*
mortice / to *s.* mortise
mortise / to ausstemmen, einschlitzen, einstemmen, verzapfen *(Tischlerei)*
mortise Stemmloch *n*, Zapfenloch *n*; Nut *f*, Schlitz *m*
m. and tenon [joint] Zapfenverbindung *f*
m. axe Stichaxt *f*
m. chisel Stemmeisen *n*, Stemmeißel *m*
m. cleaner Bundaxt *f*, Queraxt *f*, Stichaxt *f*, Fugenputzer *m*
m. joint Zapfenverbindung *f*
m. knob latch Einsteckriegel *m*
mortiser *s.* mortising machine
mortising machine Stemmaschine *f*
mosaic Mosaik *n*
m. block Mosaikblock *m*
m. boundaries *(Met)* Mosaikgrenzen *fpl (Kristall)*
m. electrode Mosaikelektrode *f*
m. flooring Mosaikbelag *m*, Mosaikfußboden *m*
m. gold Musivgold *n*
m. paving Mosaikpflaster *n*
moses boat *(Am)* Ruderboot *n (zwischen Schiff und Land)*
moss Torfmoor *n*; Moos *n*
m. agate *(Min)* Moosachat *m*
m. cutter Moorzahn *m (Egge)*
mossy lead Bleischwamm *m*
m. litter Torfstreu *f*
most extreme fibre Randfaser *f*
m. probable velocity *(Ph)* wahrscheinlichste Geschwindigkeit *f*
m. remote fibre Randfaser *f*
m. significant digit *(Dat)* höchste Stelle *f*; bedeutsamste Ziffer *f*, gültige Ziffer der höchsten Ordnung
mote / to *(Text)* Knötchen entfernen
mote Knötchen *n (Gewebefehler)*
m. knife Abstreifmesser *n*
motel Motel *n*, Kraftfahrerhotel *n*
moth ball Mottenkugel *f*
m. repellent Mottenschutzmittel *n*
m. resistant *(Text)* mottenbeständig, mottensicher, mottenecht
mother Mutter-, Haupt-
m. belt Hauptförderband *n*
m. blank *s.* starting-sheet blank
m. crystal Mutterkristall *m*
m.-gate *(Brg)* Grundstrecke *f*
m. liquor Mutterlauge *f*, Stammlauge *f*
m. lye Mutterlauge *f*
m. missile Trägerrakete *f*; erste Raketenstufe *f*
m. of coal Faserkohle *f*, Fusit *m*
m.-of-pearl Perlmutt[er *f*] *n*
m. rock Urgestein *n*
m. ship Mutterschiff *n*
m. substance Mutterstoff *m*, Muttersubstanz *f*
mothproof / to mottenfest machen
motion Bewegung *f*, Lauf *m*, Gang *m*
m. about a fixed point *(Ph)* sphärische Bewegung *f*
m.-balance controller Wegvergleichsregler *m (pneumatischer Regler)*
m. bar Gleitstange *f*, Führungsstange *f*

m. box Triebkasten *m*
m. equation Bewegungsgleichung *f*
m. in a straight line geradlinige Bewegung *f*
m. of dislocations *(Met)* Versetzungsbewegung *f*
m. of molecules Molekularbewegung *f*
m. of the sea Seegang *m*
m. of translation Translation *f*, Verschiebung *f*
m. picture *(Am)* Filmbild *n*
m. picture camera *(Am)* Kinokamera *f*
m. picture projector *(Am)* Filmvorführgerät *n*, Kinoprojektor *m*
m. picture studio *(Am)* Filmstudio *n*
m. shaft Vorgelegewelle *f*
m. study *(Ök)* Bewegungsstudie *f*
m. unsharpness Bewegungsunschärfe *f*
motional displacement Bewegungsunschärfe *f*
m. impedance *(El)* Rückwirkungsimpedanz *f*
m. resistance *(El)* Ausgleichswiderstand *m* für mechanische Verluste, Rückwirkungswiderstand *m*
motionless bewegungslos, stillstehend
motive bewegend, [an]treibend
m. fluid *(flüssiger)* Treibstoff *m (z. B. Benzin)*
m. power Treibkraft *f*, Triebkraft *f*
motometer *(BMSR)* Umdrehungszähler *m*
motor 1. Kraftmaschine *f*; Motor *m*; Antriebsmaschine *f*; 2. Elektromotor *m*, Motor *m*
m. alternator Drehumformer *m (von Gleich- auf Wechselstrom)*
m. armature Motoranker *m*
m.-boat Motorboot *n*
m.-boating *(Fmt)* Motorbrummen *n (niederfrequente Störschwingungen)*
m.-car Kraftfahrzeug *n*, Pkw
m.-car engine Kraftfahrzeugmotor *m*
m.-car jack Wagenheber *m*
m.-car population Kraftfahrzeugbestand *m*
m.-car upholstery leather Autoleder *n*
m. casing Motorengehäuse *n*
m. characteristics Motorcharakteristik *f*, Kennlinie *f* des Motors
m. coach Reisebus *m*
m.-coach jack Automobilwinde *f*, Wagenheber *m*
m.-coaster Küstenmotorschiff *n*, Kümo *n*
m. combination *(Elm)* Gruppierung *f* von Motoren
m. control Motorsteuerung *f*
m. control switch *(Schiff)* Manövrierschalter *m*
m. converter *(El)* Kaskadenumformer *m*
m.-converter dynamotor Einankerumformer *m (von Dreh- auf Gleichstrom)*
m. cruiser Motorjacht *f*
m. current Motorstrom *m*
m.-cycle *s.* motorcycle
m. drifter Motorlogger *m*
m. drive Motorantrieb *m*
m. drive punch Motorlocher *m (Lochkartenmaschinen)*
m. drive verifier *(Büro)* Motorlochprüfer *m*
m.-driven mit Motorantrieb
m.-driven barrow Motorkarre *f*
m.-driven buzzer Motorsummer *m*
m.-driven combine Mähdrescher *m* mit Aufbaumotor
m.-driven interrupter Motorunterbrecher *m*
m.-driven pump Pumpe *f* mit elektrischem Antrieb
m.-driven ventilator Motorgebläse *n*
m.-field control Motorsteuerung *f* durch Feldschwächung, Feldregelung *f*
m. for build-in installation Einbaumotor *m*
m. frame Motorgehäuse *n*
m.-fuel gas storage Speichergasanlage *f*
m.-generator for welding Schweißumformer *m*
m.-generator set *(Elm)* Umformergruppe *f*, Motorgenerator *m*; *(Masch)* Motor-Generator-Aggregat *n*; Steuergeneratorensatz *m*
m. grader Motorstraßenhobel *m*
m. head *(Rak)* Motorkopf *m*, Kopf *m*
m. lorry Lastkraftwagen *m*, Lkw
m. man Lokomotivführer *m*; Wagenführer *m*
m. mine sweeper Minenräumboot *n*
m. of the reciprocating piston type Hubkolbenmotor *m*
m.-operated motorisch angetrieben
m. operator Stellmotor *m*
m. output Motorleistung *f*
m. power *(Rak)* Motorleistung *f*; Schub *m* des Triebwerkes
m. pulley Motorriemenscheibe *f*
m. racer Motorrennboot *n*
m. room Maschinenraum *m (Elektromotoren)*
m. sailer Motorsegler *m*, Segelschiff *n* mit Hilfsmotor
m. scooter Motorroller *m*
m. scraper *(Bw)* Motorschürfkübel *m*, Schürfkübelwagen *m*
m. scythe Rasenmotormäher *m*
m. shaft Motorwelle *f*
m. speed Motordrehzahl *f*
m. spirit Motortreibstoff *m*
m. sprinkler Sprengwagen *m*
m. starting switch Motoranlaßschalter *m*, Motoranlasser *m*
m. sweeper Straßenkehrmaschine *f*
m. testing plant Motorenprüfstand *m*
m. torque Motordrehmoment *n*
m. tractor Motorschlepper *m*
m. tree-sprayer Motorbaumspritze *f*
m. tricycle Kraftdreirad *n*
m. truck Motorkarren *m*; Lastkraftwagen *m*, Lkw

m.-truck body Lastkraftwagenaufbau *m*
m. uniselector *(Fmt)* Motorwähler *m*
m. vehicle Kraftfahrzeug *n*
m.-vehicle engine Kraftwagenmotor *m*
m.-vehicle fleet Kraftfahrzeugpark *m*, Kraftwagenpark *m*
m.-vehicle skidding Rücken *n* mittels Motorfahrzeuges *(Holztransport)*
m. vessel Motorschiff *n*
m. winch Motorwinde *f*
m. with combined ventilation *(Elm)* Motor *m* mit Eigen- und Fremdbelüftung
m. with compound characteristic *(Elm)* Motor *m* mit Doppelschlußverhalten
m. with reciprocating movement *(Elm)* Schwingmotor *m*
m. with series characteristic *(Elm)* Motor *m* mit Reihenschlußverhalten
m. with shunt characteristic *(Elm)* Motor *m* mit Nebenschlußverhalten
m. yacht Motorjacht *f*
m.-yacht with auxiliary sails Motorjacht *f* mit Hilfsbesegelung
motorboat / to *(Am, BMSR)* plubbern, pendeln; tiefstfrequent schwingen; tuckern *(Funktechnik)*
motorcycle Motorrad *n*, Kraftrad *n*, Krad *n*
m. messenger Kradmelder *m*
m. rifleman Kradschütze *m*
motoring Antreiben *n (mit Motor)*; Autotransport *m*; Motorsport *m*
m. offence Verstoß *m* gegen Verkehrsvorschriften
motorize / to motorisieren
motorized motorisiert
m. anti-aircraft artillery Kraftwagenflak *f*
m. cart Selbstfahrlafette *f*
motorship Motorschiff *n*
motorway Autobahn *f*
mottle / to marmorieren, sprenkeln
mottled marmoriert, geadert, gemasert; meliert *(Gußeisen)*
m. iron meliertes Gußeisen *n*
m. paper getupftes Papier *n*; marmoriertes Papier; Phantasiepapier *n*
mould / to formen, gestalten; gießen, pressen; schalen; auskehlen
mould 1. Form *f*, Kokille *f (Gießerei)*; *(Plast)* Werkzeug *n*, Form *f*, Modell *n*; Abdruck *m*; Schablone *f*; Preßwerkzeug *n*; Bütte *f*; 2. *(Bw)* Form, Schalung[sform] *f*; Matrize *f (Betontechnologie)*; 3. Humuserde *f*, Modererde *f*; 4. Mall *n (Schablone im Schiffbau)*; 5. Schimmel *m*
m. and die duplicator Formen- und Gesenknachformfräsmaschine *f*
m. carriage *(Met)* Formwagen *m*
m. core Formkern *m (Formerei)*
m. drying Trocknen *n* der Form
m. drying oven Form[en]trockenofen *m*
m. facing Schwärze *f (Gießerei)*
m. for casting steel Stahlgußform *f*
m. frame 1. *(Met)* Gießtisch *m*; 2. Schöpfrahmen *m (Papierherstellung)*
m. fungus Schimmelpilz *m*
m. grease Kokillenfett *n*, Kokillenschmiere *f*
m. locking pressure Formschließkraft *f*, Werkzeugschließkraft *f*
m.-loft [floor] *(Schiff)* Schnürboden *m*
m.-loft offsets *(Schiff)* Schnürbodenaufmaße *npl*
m. lubricant Formtrennmittel *n*, Entformungsmittel *n*, Gleitmittel *n*
m. made paper Schöpfpapier *n*; [unechtes] Büttenpapier *n*, Maschinenbütten *n*
m. oil Schalungsöl *n (Betontechnologie)*
m. paper *s.* m. made paper
m. release agent Entformungsmittel *n*, Formentrennmittel *n*, Trennmittel *n*
m. scrap Kokillenbruch *m*
m. wall Kokillenwand *f*
mouldability *(Plast)* Verpreßbarkeit *f*
mouldable formbar; verpreßbar
mouldboard Streichblech *n (Pflug)*
m. extension Streichschiene *f (Pflug)*
moulded geformt; gegossen, gepreßt
m. article Formkörper *m*, Formteil *n*, Preßteil *n*, Preßgut *n*
m. bakelite jacket Bakelitgehäuse *n*, Preßstoffgehäuse *n*
m. beam gemallte Breite *f*, Breite auf Spanten *(Stahlschiff)*; Breite über Planken *(Holzschiff)*
m. board gekehltes Brett *n*; Ziehpappe *f*, gezogene Pappe *f*
m. breadth *(Schiff)* Berechnungsbreite *f*, Breite *f* auf Spanten
m. depth *(Schiff)* Seitenhöhe *f (über Oberkante Kiel)*
m. displacement *(Schiff)* Form-Deplacement *n*
m. draught *(Schiff)* Tauchtiefe *f*, Konstruktionstiefe *f*
m. fibre board Hartfaserplatte *f*
m. form *(Schiff)* Form *f* auf Spanten
m. impregnated wood harzgetränktes gepreßtes Holz *n*
m.-in *(Plast)* eingepreßt, eingespritzt
m. insulating material Isolier[preß]stoff *m*
m. insulation Preßstoffisolierung *f*, Preßstoffisolation *f*
m. insulator Preßstoffisolator *m*
m. line *(Schiff)* gemallte Linie *f*
m. plywood Formpreßholz *n*
m. pulp geformtes Erzeugnis *n* aus Papierfaserstoff *(z. B. Pappenguß)*
moulder / to vermodern, faulen

moulder Former *m*
moulder's clamp Formkastenklammer *f*
m. pin Formstift *m*
m. rule Schwindmaßstab *m*
m. tool Formerwerkzeug *n*
moulding 1. Formen *n*, Formerei *f*; 2. *(Plast)* Formpressen; Preßverfahren *n*; Preßling *m*, Preßteil *n*, Formling *m*; 3. *(Holz)* Kehlung *f*; 4. Deckleiste *f*, Zierband *n*, Zierleiste *f*; Gesims *n*, Sims *m*
m. batch Formmasse *f*
m. blackening Formschwärze *f*
m. box Formkasten *m*
m. clay Formlehm *m*
m. compound *(Plast)* Formmasse *f*, Preßmasse *f*, Preßmischung *f*; Spritzgußmasse *f*
m. cutter Auskehlfräser *m*, Kannelierfräser *m (für Holzbearbeitung)*
m. cutter block Kehlkopf *m (für Holzbearbeitung)*
m. cycle *(Plast)* Preßdauer *f*, Preßvorgang *m*
m. flask Formkasten *m*
m.-generating *(Masch)* Wälzstoßen *n*
m.-generating process *(Masch)* Wälzstoßverfahren *n*
m. granulates *(Plast)* gekörnte Preßmasse *f*
m. in boxes Kastenformerei *f*
m. iron Kehlmesser *n*
m. loam *(Met)* Formlehm *m*
m. machine 1. *(Met)* Formmaschine *f (Gießerei)*; 2. *(Holz)* Leistenhobelmaschine *f*, Kehlmaschine *f*
m. machine for conical strips Kehlmaschine *f* für kegelige oder konische Leisten
m. material *(Plast)* Preßmasse *f*; Spritzgußmasse *f*; Strangpreßmasse *f*
m. of a beam *(Schiff)* Mallbreite *f eines Balkens*
m. operation Formarbeit *f*
m. pestle Formstampfer *m*
m. plane Fassonhobel *m*, Kehlhobel *m*, Leistenhobel *m*
m. powder 1. Formpuder *m (Gießerei)*; 2. *(Plast)* Preßpulver *n*
m. press 1. *(Met)* Formpresse *f*, Preßformmaschine *f*; 2. *(Typ)* Prägepresse *f (f. Matrizen)*
m. resin *(Plast)* Preßharz *n*
m. sand Formsand *m*
m. sand preparation Formsandaufbereitung *f*
m. sand preparation plant Formsandaufbereitungsanlage *f*
m. sand rammer Formsandstampfer *m*
m. shop Formerei *f*
m. time *(Plast)* Stehzeit *f*
mouldy schimmlig
m. soil Modererde *f*
mount / to aufstellen; montieren, zusammenbauen; aufspannen *(Werkzeug)*; anbringen
m. on cardboard auf Pappe aufziehen
m. the staves ein Faß binden (aufsetzen)
mount *(Opt)* Montierung *f*; Fassung *f (einer Linse)*; *(Masch)* Halterung *f* (*s. a.* mounting)
m. holding components Halterung *f*
m. holding lens Fassung *f* einer Linse
mountain Berg *m*; Gebirge *n*
m. ash Eberesche *f*
m. chain Gebirgskette *f*
m. chalk Bergkreide *f*
m. climber Bergsteiger *m*
m. cork *(Min)* Bergkork *m (leichter Asbest)*
m. crystal *(Min)* Bergkristall *m*
m. dew schottischer Whisky *m*
m. effect Abschattungsverlust *m (Ausbreitung elektromagnetischer Wellen)*
m. elm Bergulme *f*, Bergrüster *f*
m. flour Kieselgur *f*
m. fold Gebirgsfalte *f*
m. gorge Felsschlucht *f*
m. leather *(Min)* Bergleder *n (zäher Asbest)*
m. meal Kieselgur *f*
m. pine Bergkiefer *f*
m. pine oil Latschenkiefernöl *n*
m. railway Gebirgsbahn *f*
m. ridge Gebirgskamm *m*
m. wax *(Min)* Montanwachs *n*
m. wood *(Min)* Bergflachs *m (zäher Asbest)*
mounted attachment *(Lw)* Zusatzgerät *n*, Anhängegerät *n*
m. implement *(Lw)* Anbaugerät *n*
m. in line hintereinander angeordnet
m. mower Anbaumähbalken *m*
m.-on Anbau-, angebaut
m. on a truck in Rollbockausführung *(Schleifen)*
m. point Schleifstift *m*
m. spare valve *(El)* eingebaute Ersatzröhre *f*
mounting 1. Montage *f*; Anbringung *f*; Befestigung *f*; Halterung *f*; 2. Fassung *f (Linse)*
m. bracket Befestigungsschelle *f*, Befestigungsplatte *f*; Konsole *f*
m. cone Befestigungskegel *m*
m. flange Anbauflansch *m*, Befestigungsflansch *m*
m. mandrel Aufnahmedorn *m*; Aufspanndorn
m. nut Spannmutter *f*
m. of the grinding wheel Befestigung *f* der Schleifscheibe
m. of the Jacquard machine on a suspended frame Deckenaufhängung *f* der Jacquardmaschine
m. on centres Spannen *n* zwischen Spitzen *(Werkzeugmaschinen)*
m. plate Grundplatte *f*, Montageplatte *f*; Fixrosette *f*
m. position Aufstellung *f (z. B. einer Elektronenröhre)*

m. receptacles Steckdose *f* mit Grundplatte, Aufbausteckdose *f*
m. stand Montagebock *m*
m. surface Einbaufläche *f*, Auflagefläche *f*
m. taper Aufnahmekegel *m*
mountings Armaturen *fpl*, Zubehör *n*; Befestigungsteile *npl*
m. of a door Türbeschläge *mpl*
mourning Trauerkleidung *f*
mouth Öffnung *f*, Mündung *f*; Gicht[öffnung] *f* *(Hochofen)*; Maul *n* *(Brecher)*; Helm *m (Konverter)*; Stich *m*, Abstich[loch *n*] *m*, Schlackenloch *n*, Stichöffnung *f (SM-Ofen)*
m. of a blast furnace Gicht *f* eines Hochofens
m. of a harbour Hafenausfahrt *f*
m. of a river Flußmündung *f*
m. of breaker Brechermaul *n*
m. of the main flue *(Wärm)* Fuchsmündung *f*
m.-organ Mundharmonika *f*
mouthpiece Mundstück *n*, Ansatzrohr *n*
m. of microphone Sprechtrichter *m*, Mikrofontrichter *m*
m. of transmitter Sprechtrichter *m*, Mikrofontrichter *m*
mouthwash Mundwasser *n*
movability Beweglichkeit *f*, Verstellbarkeit *f*
movable beweglich, ortsveränderlich, verstellbar
m. blade Turbinenlaufradschaufel *f*; aufgesetzter Flügel *m (Schiffspropeller)*
m. brush bewegliche Bürste *f*
m. contact beweglicher Kontakt *m*
m. crane Laufkran *m*
m. grate beweglicher Rost *m*
m. link bewegliches Glied *n (Kinematik)*
m. member bewegliches Glied *n (Kinematik)*
m. path of instantaneous centre bewegliche Polbahn *f*
m. pulley lose Rolle *f*
m. scene Studiokulisse *f*
m.-sieve jig Stauchsetzmaschine *f (Aufbereitung)*
m. stop verschiebbarer Anschlag *m*
m. stud Scherenbolzen *m*
m. track *(Brg)* fliegendes Gestänge *n*
movableness Beweglichkeit *f*
move / to bewegen, [ver]schieben, laufen; verlegen *(Leitungen)*; wandern *(Feld, Welle)*; verholen *(Schiff)*
m. away fortbewegen
m. back zurückbewegen
m. by ship auf dem Wasserwege versenden
m. intermittently ruckweise [vorwärts]bewegen
m. over rücken *(Fördermittel)*
m. to and fro hin- und herbewegen
move Zug *m (z. B. beim Schachspiel)*
movement Bewegung *f*, Verschiebung *f*, Lauf *m*; Meßwerk *n (eines Meßinstrumentes)*; Uhrwerk *n*
m. area *(Flg)* Rollfeld *n*
m. blur Bewegungsunschärfe *f*
m. in a curve krummlinige Bewegung *f*
m. in endwise direction Axialbewegung *f*
m. in one plane ebene Bewegung *f*, Bewegung in einer Ebene
m. in space räumliche Bewegung *f*
m. joint *(Bw)* Bewegungsfuge *f*
m. of a watch Uhrwerk *n*
m. of stroke Hubbewegung *f (Verbrennungsmotor)*
m. of the ring rail *(Text)* Ringbankbewegung *f*
movie *(Am)* Film *m*
m. camera *(Am)* Filmkamera *f*
m. titler *(Am)* Titelzusatzgerät *n (Film)*
movies *(Am)* Kino *n*
moving Verstellung *f*, Verstellen *n*; Bewegen *n*, Treiben *n*
moving beweglich, bewegend; einstellbar, verstellbar
m. apparatus Triebwerk *n*
m. apron Rollboden *m (Stallmiststreuer)*
m. arm of the potentiometer Abgriff *m* des Potentiometers, Potentiometerarm *m*
m. bed *(Ch)* Wanderschicht *f*; *(Brg)* bewegtes Festbett *n*
m. bed process *(Ch)* Fließbettverfahren *n*
m.-belt production Fließbandproduktion *f*
m. blade Laufschaufel *f (Turbine)*
m. coil Drehspule *f (Meßinstrument)*; Schwingspule *f (Lautsprecher)*
m.-coil ammeter Drehspulstrommesser *m*, Drehspulamperemeter *n*
m.-coil galvanometer Drehspulgalvanometer *n*
m.-coil headphone dynamisches Telefon *n*
m.-coil headset elektrodynamischer Kopfhörer *m*
m.-coil high-speed recorder Drehspulschnellschreiber *m (Telegrafie)*
m.-coil instrument Drehspulmeßgerät *n*, Drehspulinstrument *n*
m.-coil loudspeaker [elektro]dynamischer Lautsprecher *m*
m.-coil microphone Tauchspulmikrofon *n*, [elektro]dynamisches Mikrofon *n*
m.-coil mirror galvanometer Drehspulspiegelgalvanometer *n*
m.-coil oscillograph Schleifenoszillograf *m*, Drehspuloszillograf *m*
m.-coil pick up [elektro]dynamischer Tonabnehmer *m*
m.-coil pointer galvanometer Drehspulzeigergalvanometer *n*
m.-coil receiver elektromagnetischer Fernhörer *m*

m.-coil relay Drehspulrelais *n*
m.-coil support Drehspultragkonstruktion *f*, Drehspulaufhängung *f*, Lagerung *f* der Drehspule
m.-coil transformer Transformator *m* mit beweglicher Sekundärwicklung, Drehspultransformator *m*, Drehspulwandler *m*
m.-coil vibration galvanometer Drehspulvibrationsgalvanometer *n*
m.-coil voltmeter Drehspulspannungsmesser *m*
m.-conductor loudspeaker [elektro]dynamischer Lautsprecher *m*
m.-conductor microphone [elektro]dynamisches Mikrofon *n*
m. contact *s.* m.-contact member
m.-contact member *(El)* beweglicher Kontakt *m*, bewegliches Schaltstück *n*; Anlaßstufe *f*
m.-contact spring bewegliche Kontaktfeder *f*
m.-dielectric voltmeter Strohhalmelektrometer *n*
m. field *(El)* Wanderfeld *n*
m. fillister stellbarer Falzhobel *m*
m. in the table *(Masch)* Einfahren *n* des Tisches
m. iron Dreheisen *n (Meßinstrument)*
m. iron ammeter Weicheisenstrommesser *m*, Dreheisenmeßwerk *n*
m. iron instrument Dreheiseninstrument *n*, Weicheiseninstrument *n*, Dreheisenmeßwerk *n*
m. iron loudspeaker [elektro]magnetischer Lautsprecher *m*
m. iron voltmeter Dreheisenspannungsmesser *m*, Weicheisenspannungsmesser *m*, Weicheisenvoltmeter *n*
m. load bewegliche Last *f*, Wanderlast *f*
m. magnet Drehmagnet *m*
m.-magnet galvanometer Nadelgalvanometer *n*, Drehmagnetgalvanometer *n*
m.-magnet instrument Drehmagnetinstrument *n*, Dreheiseninstrument *n*
m. mass *(Masch)* bewegliche Masse *f*
m. member bewegliches Glied *n (Kinematik)*
m.-mirror oscillograph Schleifenoszillograf *m*, Lichtstrahloszillograf *m*
m. out the table *(Masch)* Ausfahren *n* des Tisches
m. parts bewegliche Teile *npl*
m. picture Film *m*
m.-picture camera Filmaufnahmekamera *f*
m. power bewegende Kraft *f*
m. staircase Rolltreppe *f*
m. steady mitlaufender Setzstock *m*
m. tape transmitter *(Fmt)* Gleitbandgeber *m*, Gleitbandsender *m*
m. target indication *(Fs)* Anzeige *f* von beweglichen Zielen [mit Festzielunterdrückung]
m. target indicator *(Fs)* Radaranzeige *f* für bewegliche Ziele, Radaranzeige mit Festzielunterdrückung
m. the points *(Eis)* Auffahren *n* der Weichen
m. vanes *(Fmt)* bewegliche Platten *fpl*, Rotorplatten *fpl*
m. veneer Schälfurnier *n*
mow / to mähen
mow hay drying Heubelüftungsverfahren *n*
mower Mähmaschine *f*, Grasmäher *m*
m. blade grinder Mähmesserschleifmaschine *f*
m. knife grinder Mähmesserschleifmaschine *f*
mowing Mahd *f*
M.T. shaft Morsekegelschaft *m*
M.T. shank drill Bohrer *m* mit Morsekegelschaft
mu factor Verstärkungsfaktor *m (Röhren)*
mu meson My-Meson *n*, Myon *n*
mucic acid Schleimsäure *f*, Muzinsäure *f*
mucilage Pflanzenschleim *m*, Schleim[stoff] *m*; Büroleim *m*
mucinlike mukös, schleimstoffartig
muciparous schleimbildend, schleimformend
muck / to *(Brg)* wegladen, [weg]füllen
m. out 1. *(Brg)* wegladen, [weg]füllen; 2. ausmisten
muck 1. *(Brg)* Haufwerk *n*, hereingeschossenes Gestein *n*; 2. Stallmist *m*
m. bar *(Met)* Puddelluppe *f*, Rohschiene *f*
m. dozer Stallmistbulldozer *m*
m. fork Dunggabel *f*, Mistgabel *f*
m. heap Misthaufen *m*, Dunghaufen *m*
m. mill Puddelstahlwalzwerk *n*
m.-pile *(Brg)* Haufwerk[shaufen *m*] *n*
m. stack *(Brg)* Bergehalde *f*
mucker *(Brg)* Füller *m*
mucking *(Brg)* Wegfüllen *n*, Ladearbeit *f*, Wegfüllarbeit *f*
m. train *(Brg)* Bunkerzug *m (Streckenvortrieb)*
muconic acid Mukonsäure *f*
mucous [zäh]schleimig
m. fermentation Schleimgärung *f*
mud 1. Schlamm *m*, Schlick *m*, Moder *m*, Kot *m*; 2. *(Brg)* Schlamm, Trübe *f (Material unter 0,1 mm Korngröße)*
m. accumulation Schlammansammlung *f*
m. auger Schlammbohrer *m*, Bohrschappe *f*, Schappenbohrer *m (Tiefbohrtechnik)*
m. bath Schlammbad *n*, Moorbad *n*
m. broke *(Brg)* Schlammeinbruch *m*
m. coal Schlammkohle *f*
m. column Spül[ungs]säule *f (Tiefbohrtechnik)*
m. conditioning Bohrschlammaufbereitung *f*
m.-cracks *(Geol)* Trockenriß *m*, Schlammriß *m*
m. fluid Spültrübe *f*
m. guard Kotflügel *m*
m. inflow *(Brg)* Schlammeinbruch *m*

m. inrush *(Brg)* Schlammeinbruch *m*
m. lighter *(Schiff)* Baggerprahm *m*, Baggerschute *f*
m. pressure Spülungsdruck *m (Tiefbohrtechnik)*
m. sediment Schlammablagerung *f*
m. settling pond Schlammabsetzbecken *n*, Schlammteich *m*
m. tank Schlammbecken *n*
m. trap Schlammfang *m*
m. volcano *(Geol)* Schlammsprudel *m*, Schlammvulkan *m*
m. water Schlammwasser *n*
mudcapping *(Am) (Brg)* Auflegerschießen *n*, Schießen *n* mit aufgelegter Ladung
mudding up *(Lw)* Verschlammen *n* des Landes
muddy schlammig, schlammhaltig
m. ground Schlammboden *m*
m. structure *(Ch)* Schlammgefüge *n*
m. water schlammhaltiges Wasser *n*, Schmutzwasser *n*
mudrush *(Brg)* Schlammeinbruch *m*
muff Muffe *f*
m. coupling Muffenkupplung *f*
muffle / to dämpfen *(Schall)*
muffle 1. Muffel *f*, Schmelztiegel *m*; Schlackenscherbe *f*; *(Am) (El)* Blasrohr *n*, Löschrohr *n (des Ableiters)*
m. brazing Feuerlöten *n*
m. colour Muffelfarbe *f (Keramik)*
m. furnace Muffelofen *m*; Glasierofen *m*
m. kiln Muffel[brenn]ofen *m (Keramik)*
m. roaster Muffelröstofen *m*
muffler 1. Dämpfer *m*, Schalldämpfer *m*, Tondämpfer *m*; 2. Auspufftopf *m*
m. cut-out Auspuffklappe *f*
m. explosion Knallen *n* im Auspufftopf
muffling Dämpfung *f (von Geräuschen)*
mug Steinkrug *m*
muga moth Mugaspinner *m*
mulberry leaf Maulbeerblatt *n*
m. silk Maulbeerseide *f*
mulcher attachment *(Lw)* Dungeinleger *m*, Stoppeluntergraber *m*
mule 1. Maultier *n*; 2. *(Text)* Selfaktor *m*
m. carriage Selfaktorwagen *m*
m.-pulley Riemenspannrolle *f*
m. spindle Selfaktorspindel *f*
m. spinning Mulespinnerei *f*
m. spinning machine Selfaktor *m*, Mulemaschine *f*
m. yarn Selfaktorgarn *n*
mull / to pulverisieren; aufbereiten und mischen; kollern
mullanite *(Min)* Mullanit *m*
mullen tester *(Holz)* Berstfestigkeitsprüfgerät *n*, Mullenprüfer *m*
muller mixer Kollergang *m*
Müller's glass Hyalit *m*
mulling Kollern *n*, Mischen *n* im Kollergang
mullion senkrechter Mittelpfosten *m*
mullite *(Min)* Mullit *m*
mullock Abgang *m*
multi Viel-, Mehr-
m.-address instruction *(Dat)* Mehradressenbefehl *m*
m.-alloy Mehrstoff- *(Legierung)*
m.-anode rectifier *(Elm)* Mehranodenventil *n*
m.-arm signal *(Eis)* mehrflügeliges Signal *n*
m.-axle vehicle Mehrachsenfahrzeug *n*
m.-band aerial Allwellenantenne *f*, Mehrbandantenne *f*
m.-band drive Mehrbändertrieb *m*
m.-bank press Etagenpresse *f*
m.-bay *(Bw)* Mehrfeld-, mehrfeldrig
m.-bay biplane *(Flg)* mehrstieliger Doppeldecker *m*
m.-balded mit mehreren Messern *(Werkzeug)*
m.-balded fan mehrflügeliger Lüfter *m*
m. burner Mehrfachbrenner *m*
m.-channel *s.* multichannel
m.-chine boat Mehrknickboot *n*
m.-circuit Mehrfachstromkreis *m*
m.-circuit switch Serienschalter *m*
m.-circuit winding Schleifenwicklung *f*
m.-coin pay telephone Münzfernsprecher *m* für verschiedene Geldsorten
m.-colour[ed] *s.* multicolour
m.-component mixture Vielstoffgemisch *n*
m.-component signal *(El)* zusammengesetztes Signal *n*
m.-component system Vielstoffsystem *n*
m.-conductor cable Mehrleiterkabel *n*, mehradriges Kabel *n*
m.-cone friction clutch Mehrkegelkupplung *f*
m.-contact jack mehrpolige (vielpolige) Steckdose *f*, Vielfachklinke *f*
m.-contact plug Mehrfachstecker *m*; Stecker *m* mit mehreren Polen
m.-contact relay Vielkontaktrelais *n*
m.-contact switch Stufenschalter *m*
m.-core mehradrig, vieladrig *(Kabel)*
m.-core cable mehradriges (vieladriges) Kabel *n*, Mehraderkabel *n*, Mehrfach[leiter]kabel *n*, Vielfachkabel *n*
m.-core line mehradrige (vieladrige) Leitung *f*
m.-coupler Antennenverstärker *m*
m.-crank steam engine mehrkurblige Dampfmaschine *f*
m.-cut Vielschnitt-, Mehrfasen-
m.-cut drill Mehrfasenbohrer *m*; Programmwerkzeug *n*
m.-cut lathe Vielschnittdrehmaschine *f*
m.-cutting Vielmeißel[dreh]arbeit *f*
m.-cylinder Mehrzylinder-, mehrzylindrisch, Mehrgehäuse-

m.-cylinder compound steam engine mehrzylindrige Verbunddampfmaschine *f*
m.-cylinder engine Mehrzylindermotor *m*
m.-cylinder steam engine Mehrzylinderdampfmaschine *f*
m.-cylinder turbine Mehrgehäuseturbine *f*
m.-daylight press Etagenpresse *f*
m.-deck screen Mehrdecksiebmaschine *f*, Mehrdecker *m*
m.-diameter mehrfach abgesetzt, abgestuft
m.-diameter hole abgestufte (abgesetzte) Bohrung *f*
m.-digit number mehrstellige Zahl *f*
m.-dimensional *s.* multidimensional
m.-directional für mehrere Richtungen
m.-disk brake Lamellenbremse *f*
m.-disk clutch Mehrscheibenkupplung *f*
m.-disk friction clutch Mehrscheibenkupplung *f*
m.-drilling Mehrfachbohren *n*
m.-drilling operation Vielfachbohrarbeit *f*
m.-edge *s.* m.-edged
m.-edged mehrprofilig, mehrkantig, mehrschneidig
m.-electrode valve Mehrgitterröhre *f*, Mehrelektrodenröhre *f*
m.-electron system *(Kern)* Mehrelektronensystem *n*
m.-element array Strahlerwand *f (Antennentechnik)*
m.-engine *s.* m.-engined
m.-engined mehrmotorig
m.-feed circular knitting machine Vielsystemrundstrickmaschine *f*
m.-filament mehrfadig, multifil
m.-fingered mehrfingrig
m.-flute mehrnutig
m.-frequency generator frequenzumschaltbarer Generator *m*
m.-frequency heterodyne generator *(Fmt)* Schwebungssummer *m*, Überlagerungssummer *m*
m.-fuel burner Vielfachbrenner *m*
m.-furrow plow Mehrfurchenpflug *m*
m.-grid valve Mehrgitterröhre *f*
m.-group theory *(Kern)* Mehrgruppentheorie *f*
m.-head Mehrspindelkopf *m*
m.-hole Vielloch-, Mehrloch-
m.-hole brick Viellochstein *m*
m.-hole ferrite core *(BMSR)* Mehrlochferritkern *m*
m.-hole nozzle Mehrlochdüse *f*
m.-holed injector Mehrlocheinspritzdüse *f*
m.-image *(Fs)* Mehrfachbild *n (durch Echos)*
m.-impression mould *(Plast)* Mehrfachform *f*
m.-isotopic *(Kern)* mehrisotopig
m.-isotopic element Mischelement *n*
m.-jaw clutch Mehrklauenkupplung *f*
m.-jaw crusher Mehrbackenbrecher *m*
m.-lane highway mehrspurige (vielspurige) Straße *f*
m.-layer mehrschichtig
m.-layer board Mehrschichtenplatte *f*
m.-layer cathode Mehrschichtkatode *f*, L-Katode *f*
m.-layer coil Mehrlagenspule *f*, mehrlagige Spule *f*, Mehrschicht[en]spule *f*
m.-layer colour film Farbenschichtfilm *m*, Mehrschichtenfarbfilm *m*
m.-layer crystal Schichtkristall *m*
m.-layer fabric Mehrfachgewebe *n*
m.-layer film Mehrschicht[en]film *m*
m.-layer winding Mehrlagenwicklung *f*, mehrlagige Wicklung *f*, Mehrschichtenwicklung *f*
m.-lobe cam Mehrfachnocken *m*
m.-loom system Mehrwebstuhlsystem *n*
m.-metal plate *(Typ)* Mehrmetallplatte *f*
m.-metered call *(Fmt)* Gespräch *n* mit Mehrfachzählung
m.-metering Mehrfachablesung *f*
m.-motor drive Mehrmotorenantrieb *m*
m.-nave mehrschiffig
m.-needle machine *(Text)* Vielnadelmaschine *f*
m.-operator welding set Mehrstellenschweißgerät *n*
m.-pair vielpaarig *(Kabel)*
m.-pair cable mehrpaariges (vielpaariges) Kabel *n*
m.-party call *(Fmt)* Gruppenanruf *m*
m.-party line *(Fmt)* Gesellschaftsleitung *f*
m.-phase mehrphasig, vielphasig
m.-phase armature Mehrphasenanker *m*
m.-phase circuit Drehstromkreis *m*, Mehrphasenschaltung *f*
m.-phase current Mehrphasenstrom *m*, Drehstrom *m*
m.-phase generator Mehrphasengenerator *m*, Drehstromgenerator *m*
m.-phase meter Mehrphasenzähler *m*
m.-phase system Drehstromnetz *n*, Drehstromsystem *n*
m.-pin plug mehrpoliger Stecker *m*
m.-place vielstellig
m.-plate chamber *(Kern)* Kammer *f* mit mehreren Platten
m.-plate clutch Lamellenkupplung *f*, Vielscheibenkupplung *f*, Mehrscheibenkupplung *f*
m.-platen hot press Vieletagenheißpresse *f*
m.-platen press Etagenpresse *f*
m.-play board Multiplexpappe *f*, mehrlagige Pappe *f*
m.-polar mehrpolig, vielpolig
m.-pole mehrpolig, vielpolig
m.-position control *s.* m.-step control

m.-purpose Vielzweck-, Mehrzweck-
m.-purpose aircraft Mehrzweckflugzeug *n*
m.-purpose crop loader Feldlader *m*, Alleslader *m*
m.-purpose excavator Universalbagger *m*
m.-purpose gun Mehrzweckgeschütz *n*
m.-purpose kitchen Wohnküche *f*
m.-purpose machine Vielzweckmaschine *f*, Mehrzweckmaschine *f*
m.-purpose meter *(El)* Vielfachinstrument *n* *(für mehrere zu messende Größen)*
m.-purpose tool Vielfachgerät *n*; Mehrzweckwerkzeug *n*
m.-purpose vehicle Mehrzweckfahrzeug *n*
m.-range Mehrfach-, Mehrbereich-
m.-range instrument Vielfach[meß]instrument *n* *(für mehrere Meßbereiche)*
m.-range receiver Allwellenempfänger *m*
m.-rate meter *(El)* Mehrtarifzähler *m*
m.-region reactor Mehrzonenreaktor *m*
m.-rib *s.* m.-ribbed
m.-ribbed mehrprofilig; Mehrrillen-
m.-ribbed grinding wheel mehrprofilige Schleifscheibe *f*
m.-ribbed plunge-cut method of thread grinding Gewindeeinstechschleifverfahren *n* mit mehrprofiliger Scheibe
m.-ribbed thread cutter Mehrrillengewindefräser *m*
m.-ribbed traverse method of thread grinding Gewindelängsschleifverfahren *n* mit mehrprofiliger Scheibe
m.-rope grab bucket Mehrseilgreifer *m*
m.-rope hoisting Mehrseilförderung *f*
m.-rope winding Mehrseilförderung *f*
m.-screw extruder Mehrschneckenpresse *f*
m.-screw extrusion Mehrschneckenpreßverfahren *n*
m.-section filter Mehrkreisfilter *n*
m.-section valve *(El)* Mehrsystemröhre *f*, Verbundröhre *f*
m.-sectional mehrteilig, mehrfach unterteilt
m.-sector scanning *(Fs)* Mehrkanalmethode *f*
m.-segment magnetron *(El)* Vielschlitzmagnetron *n*
m.-segmented ship model Schiffsmodell *n* aus mehreren Segmenten bestehend
m.-shaft mehrwellig
m.-socket wrench Zehnlochschlüssel *m*
m.-span *(Bw)* Mehrfeld-, mehrfeldrig
m.-speed action *(BMSR)* Mehrlaufwirkung *f*
m.-speed floating control *(BMSR)* Mehrlaufregelung *f*
m.-speed motor *(Elm)* Motor *m* mit Drehzahlumschaltung; Motor mit Drehzahlregelung
m.-spindle mehrspindlig, vielspindlig, Mehrspindel (*s. a.* multiple-spindle)
m.-spindle arrangement Mehrspindelanordnung *f*
m.-spindle automatic Mehrspindelautomat *m*
m.-spindle automatic bar machine Mehrspindelstangenautomat *m*
m.-spindle bar automatic Mehrspindelstangenautomat *m*
m.-spindle construction Mehrspindelbauart *f*
m.-spindle design Mehrspindelbauart *f*
m.-spindle drilling machine Mehrspindelbohrmaschine *f*
m.-spindle drilling machine with circular or rectangular head Mehrspindelbohrmaschine *f* mit Anordnung der Spindeln auf einem Kreis oder rechtwinklig *(im Viereck)*
m.-spindle drilling machine with fixed spindles Mehrspindelbohrmaschine *f* mit fester Bohrspindelanordnung
m.-spindle drilling machine with universal-joint driven spindles Gelenkspindelbohrmaschine *f*
m.-spindle head Mehrspindelkopf *m*
m.-spindle model Mehrspindelbauart *f*
m.-spindle type Mehrspindelbauart *f*
m.-spiral *(Fs)* Mehrfachspirale *f*
m.-spline broaching Räumen *n* von Vielkeilwellen
m.-spline profile Mehrkeilprofil *n*
m.-spline shaft Vielkeilwelle *f*
m.-splined fit Mehrkeilsitz *m*
m.-stage vielstufig, mehrstufig
m.-stage action turbine mehrstufige Druckturbine *f*, Druckturbine mit Druckabstufung
m.-stage amplifier Mehrstufenverstärker *m*, mehrstufiger Verstärker *m*
m.-stage bleaching Mehrstufenbleiche *f*, Vielstufenbleiche *f*
m.-stage cluster rocket Mehrstufenrakete *f* mit paralleler Stufenanordnung
m.-stage compressor Mehrstufenverdichter *m*
m.-stage grate *(Wärm)* Staffelrost *m*, Etagenrost *m*
m.-stage impulse turbine *s.* m.-stage action turbine
m.-stage multiplier Mehrfachvervielfacher *m*
m.-stage process Mehrstufenverfahren *n*
m.-stage reaction turbine mehrstufige Überdruckturbine *f*
m.-stage rocket mehrstufige Rakete *f*, Mehrstufenrakete *f*, Stufenrakete *f*
m.-stage separation mehrstufige Trennung *f*
m.-stage tandem rocket Mehrstufenrakete *f* mit Hintereinanderanordnung der Stufen
m.-staged rocket Mehrstufenrakete *f*
m.-start mehrgängig *(Gewinde)*
m.-start thread mehrgängiges Gewinde *n*
m.-start worm mehrgängige Schnecke *f*
m.-step mehrstufig, gestuft, abgesetzt

m.-step control Mehrpunktregelung *f*
m.-step hydroplane Gleitboot *n* mit mehreren Stufen, Mehrstufen[gleit]boot *n*
m.-step reduction gear Mehrstufengetriebe *n*
m.-storey mehrgeschossig, mehrstöckig, Hochhaus-
m.-storey frame Stockwerkrahmen *m*
m.-storey parking garage Stockwerksgarage *f*, Hochhausgarage *f*
m.-storey switch mehretagiger Schalter *m*, Mehrebenenschalter *m*
m.-story *s.* m.-storey
m.-strand wire *(El)* feindrähtige Litze *f*
m.-strut biplane *(Flg)* mehrstieliger Doppeldecker *m*
m.-surface milling job Fräsarbeit *f* an mehreren Flächen
m.-tapping Mehrfachgewindebohrer *m*
m.-thread mehrgängiges Gewinde *n*
m.-thread milling cutter Gewinderillenfräser *m*, Mehrrillengewindefräser *m*
m.-thread worm mehrgängige Schnecke *f*
m.-threaded mehrgängig
m.-tool . . . *s.* multiple-tool . . .
m.-toothed mehrzahnig, mehrschneidig
m.-track pulse range Leitlinienimpulsverfahren *n*
m.-tube receiver Mehrröhrenempfänger *m*
m.-turn aus mehreren Wicklungen bestehend
m.-turn potentiometer mehrgängiges Potentiometer *n*, Wendelpotentiometer *n*
m.-unit drilling machine Reihenbohrmaschine *f*
m.-unit valve Mehrfachröhre *f*, Verbundröhre *f*
m.-valve Mehrfachröhre *f*, Mehrsystemröhre *f*, Verbundröhre *f*
m.-valve amplifier Mehrröhrenverstärker *m*
m.-valve receiver Mehrröhrenempfänger *m*
m.-vane condenser Mehrplattenkondensator *m*
m.-way mehrwegig
m.-way cable vielpaariges (mehrpaariges) Kabel *n*
m.-way drilling machine Mehrwegebohrmaschine *f*
m.-way machine Mehrwegemaschine *f*
m.-way switch Mehrfachumschalter *m*, Vielfachschalter *m*, Serienschalter *m*
m.-way valve Mehrwegeventil *n*, Mehrwegeschieber *m*
m.-wire mehradrig, vieladrig *(Kabel)*
m.-wire aerial Mehrleiterantenne *f*, Mehrdrahtantenne *f*, mehrdrähtige Antenne *f*
m.-wire cable mehradriges Kabel *n*, Mehraderkabel *n*, Mehrfach[leiter]kabel *n*, Mehrleiterkabel *n*, vieladriges Kabel, Vielfachkabel *n*
m.-work machine Universalmaschine *f*

multiagent control system Regelkreis *m* mit mehreren Stellgrößen
multibreak circuit-breaker Schalter *m* mit Vielfachunterbrechung (Mehrfachunterbrechung)
multicavity klystron *(El)* Mehrkammerklystron *n*
m. velocity modulated tube *(El)* Mehrkreistriftröhre *f*
multicellular mehrzellig, Mehrzellen-
m. counter *(Kern)* Mehrzellularzähler *m*
m. horn Facettentrichter *m*
m. instrument Multizellularinstrument *n*
m. voltmeter Multizellularvoltmeter *n*
multichannel Mehrkanal-, Vielkanal-
m. amplifier Vielbandverstärker *m*
m. analyzer Mehrkanalanalysator *m*, Vielkanalanalysator *m*
m. cable Vielbandkabel *n*
m. detector Mehrkanaldetektor *m*
m. loudspeaker *(El)* Lautsprecherkombination *f*; Mehrkanallautsprecheranlage *f*
m. pulse modulation Mehrkanalimpulsmodulation *f*
m. system Mehrkanalsystem *n*
m. telegraph Mehrfachtelegraf *m*
m. telegraphy Mehrfachtelegrafie *f*
m. television Mehrkanalfernsehen *n*
multiclone Multi[zy]klon *m*
multicolour mehrfarbig, Mehrfarben-
m. dyeing Batikfärberei *f*, Buntfärberei *f*
m. print[ing] Mehrfarbendruck *m*
m. printing process Vielfarbendruckverfahren *n*
m. recorder Mehrfarbenschreiber *m*
m. warp *(Text)* Buntkette *f*, mehrfarbige Kette *f*
multicoloured *s.* multicolour
multicore *s.* multi-core
multidimensional mehrdimensional
m. space *(Math)* mehrdimensionaler Raum *m*
multiemulsion film Mehrschichtenfilm *m*
multifilament Garn *n* aus Einzelfäden unbegrenzter Länge
multiform vielgestaltig
multiformity Vielgestaltigkeit *f*
multigap arrester Katodenfallableiter *m*, Vielfachfunkenstrecke *f*, Rollenblitzableiter *m*, Ableiter *m* mit Mehrfachfunkenstrecke
m. discharger Vielfachfunkenstrecke *f*
multigrade oil Mehrbereichsöl *n*, Multigrad-Öl *n*, Öl *n* für alle Jahreszeiten
multigrid valve Mehrgitterröhre *f*
multihearth roasting furnace mehrherdiger Röstofen *m*
multilayer *s.* multi-layer
multilead cable mehradriges Kabel *n*, Mehraderkabel *n*, Mehrfach[leiter]kabel *n*, Mehrleiterkabel *n*, vieladriges Kabel, Vielfachkabel *n*

multiloop system mehrschleifiger Regelkreis *m*, mehrschleifiges System *n*
multinuclear mehrkernig
multipart mehrteilig
multipartite vielteilig
multipass drier Mehrfachbahnentrockner *m*
m. heat exchanger Vielstufenwärme[aus]tauscher *m*
multipath mehrwegig
m. effect *(Fs)* Geisterbild *n*
m. magnetic core *(BMSR)* Mehrlochmagnetkern *m*, Mehrwegmagnetkern *m*
m. transmission *(El)* Mehrwegübertragung *f*
multiphase *s.* multi-phase
multiplace mehrstellig
multiplane *(Flg)* Mehrdecker *m*
multiple *(Math)* Vielfaches *n*; Mehrfachverbindungsstelle *f (z. B. bei Analogrechnern)*
multiple mehrfach, Mehrfach-, Vielfach-, Mehr-
m. acceleration Vielfachbeschleunigung *f*
m. address instruction *(Dat)* Mehradressenbefehl *m*
m. aerial Mehrfachantenne *f*, Strahlergruppe *f*
m. amplifier Mehrfachverstärker *m*
m. bead Mehrlagenraupe *f (Schweißtechnik)*
m.-beam interference Mehrstrahlinterferenz *f*
m.-belt entanglement *(Mil)* Flächendrahthindernis *n*
m. blade frame saw Bundgatter *n*, Vollgatter *n (Sägewerk)*
m. blade sawframe Vollgatter *n (Sägewerk)*
m. bond Mehrfachbindung *f*
m. box loom Wechselwebstuhl *m*, Karrierwebstuhl *m*
m. box slay *(Text)* Wechsellade *f*
m. cable mehradriges Kabel *n*, Mehraderkabel *n*, Mehrfach[leiter]kabel *n*, Mehrleiterkabel *n*, vieladriges Kabel, Vielfachkabel *n*
m. cable joint Kabelabzweigmuffe *f*
m. cable tile Kabelformstein *m (mit mindestens zwei Zügen)*
m. cabling Vielfachverkablung *f*
m.-capacity system *(Dat)* Mehrspeichersystem *n*
m.-cavity die *(Met)* Mehrfachwerkzeug *n (Spritzguß)*
m.-cavity mould *(Met)* Mehrfach[preß]form *f (Spritzguß)*
m. chain Mehrfachkette *f*
m. change-over switch Mehrfachumschalter *m*
m. circuit *s.* multi-circuit
m. coil Mehrfachspule *f*
m. coincidence *(Kern)* Vielfachkoinzidenz *f*
m. collision *(Kern)* Mehrfachstoß *m*; Vielfachstoß *m*
m. common-frame circuit-breaker gekuppelter mehrpoliger Schalter *m*
m.-conductor cable mehradriges Kabel *n*, Mehraderkabel *n*, Mehrfach[leiter]kabel *n*, Mehrleiterkabel *n*, vieladriges Kabel, Vielfachkabel *n*
m. conduit clip *(El)* Sammelschelle *f (für I-Rohre)*
m. connection Vielfachschaltung *f*
m. contact Vielfachkontakt *m*
m. contact plug Mehrfachstecker *m*
m. contact switch Zellenschalter *m*, Stufenschalter *m*; Mehrfachschalter *m*, Mehrwegschalter *m*; mehrpoliger Schalter *m*
m. converter Mehrfachstromrichter *m*
m.-core cable mehradriges Kabel *n*, Mehraderkabel *n*, Mehrfach[leiter]kabel *n*, Mehrleiterkabel *n*, vieladriges Kabel, Vielfachkabel *n*
m.-core meter Mehrleiterzähler *m*
m. cornered Mehrkant-
m. current generator Mehrstromgenerator *m*
m.-cutter boring head Bohrkopf *m* mit mehreren Meißeln
m.-cutter boring tool Bohrkopfstange *f*
m.-cutter method of thread milling Kurzgewindefräsen *n*
m.-cutting-edge mehrschneidig
m.-cutting-edge tool Mehrschneidenwerkzeug *n*
m. cylinder opener *(Text)* Trommelöffner *m* mit mehreren kleinen Trommeln
m. cylindrical boiler *(Wärm)* mehrfacher Walzenkessel *m*
m. decay *(Kern)* Mehrfachzerfall *m*
m. deflection Mehrfachablenkung *f*
m. diagram Vielfachdiagramm *n*
m. diameter turning Drehen *n* von Ansätzen
m.-die press Stufenpresse *f*
m. disintegration *(Kern)* Mehrfachzerfall *m*
m.-disk clutch Mehrscheibenkupplung *f*
m. drawing *(Met)* Vielfachzug *m*, Mehrfachzug *m*
m. drill head Mehrspindelbohrkopf *m (feste oder verstellbare Lochbilder)*
m. drilling machine Mehrspindelbohrmaschine *f*
m.-duct conduit Mehrlochkabelkanal *m*
m. edge *s.* m. edged
m. edged mehrschneidig
m. effect evaporator *(Kern)* Mehrfachverdampfer *m*, Gegenstromverdampfer *m*
m. electrode welding Bündelschweißung *f*
m. electrometer Multizellularvoltmeter *n*
m. excitation mehrfache Erregung *f*, Mehrfachanregung *f*
m.-expansion engine Mehrfachexpansionsmaschine *f*

m. exposure Mehrfachbelichtung *f*
m. farming machine landwirtschaftliches Vielfachgerät *n*
m. fibre material Mehrfachfasermaterial *n*
m. flight mehrarmig, mehrläufig *(Treppe)*
m.-flute mehrnutig
m. garages Reihengarage *f*
m.-grid valve Mehrgitterröhre *f*
m. head draw frame *(Text)* Mehrkopfstrecke *f*
m.-hearth roaster *(Met)* Plattenofen *m*, mehrherdiger Röstofen *m*
m. indexing Mehrfachteilen *n (Teilkopf)*
m. interference Mehrfachinterferenz *f*
m. interlacing *(Fs)* Mehrfachzeilensprung *m*
m. ionization Mehrfachionisation *f*
m. isomorphism *(Met)* Polymorphismus *m*, Polymorphie *f*
m. jack *(Fmt)* Multipelklinke *f*, Vielfachklinke *f*, Verbindungsklinke *f*
m.-jet carburetter *(Kfz)* Mehrdüsenvergaser *m*
m. keys Vielkeilprofil *n*
m. modulation Mehrfachmodulation *f*
m. multipole circuit-breaker mehrpoliger Vielfachumschalter *m*
m. needle flat bed sewing machine Mehrnadelflachmaschine *f (Schuhherstellung)*
m.-orifice injector Mehrloch[einspritz]düse *f*
m.-outlet television system Fernsehgemeinschaftsempfangsanlage *f*
m. parallel winding *(Elm)* mehrgängige Parallelwicklung *f*
m.-pin plug mehrpoliger Stecker *m*
m.-piston pump Mehrkolbenpumpe *f*
m.-plate clutch Mehrscheibenkupplung *f*
m. plunger press Stufenpresse *f*
m.-plunger pump Mehrkolbenpumpe *f*
m.-point *(Am)* mehrschneidig
m.-point recording potentiometer Mehrfachkompensations[punkt]schreiber *m*
m.-point switch mehrstufiger Schalter *m*, Stufenschalter *m*
m.-pole switch mehrpoliger Schalter *m*
m. position stop Anschlagwalze *f*
m. process Mehrfachprozeß *m*, pluraler Prozeß *m*
m. production Reihenfertigung *f*, Serienherstellung *f*
m. protective earthing *(El)* Nullung *f*
m.-purpose Mehrzweck- (*s. a.* multi-purpose)
m.-rate meter Mehrfach[tarif]zähler *m*
m. reception Vielfachempfang *m*
m. reflection Mehrfachreflexion *f*
m. repetition drilling Mehrfachserienbohren *n*
m.-riveted mehrreihig genietet
m. riveting Mehrreihennietung *f*
m. scattering *(Kern)* Vielfachstreuung *f*
m. set-up Mehrfachaufspannung *f (Werkstücke)*
m. ship production Serienschiffbauproduktion *f*, Schiffbau *m* in Serienproduktion
m. shot moulding *(Plast)* Mehrstufenpressen *n*
m. shuttle loom Wechselwebstuhl *m*, Karrierwebstuhl *m*
m. slide press Stufenpresse *f*
m. sound track *(Ak)* Vielfach[ton]spur *f*
m.-span frame *(Bw)* Durchlaufrahmen *m*, Mehrfeldrahmen *m*
m.-speed motor Motor *m* mit mehreren Drehzahlstufen
m.-spindle mehrspindlig, vielspindlig, Mehrspindel- (*s. a.* multi-spindle)
m. spindle automatic Mehrspindelautomat *m*
m. spindle automatic chucking machine Mehrspindelfutterautomat *m*
m.-spindle bar automatic Mehrspindelstangenautomat *m*
m.-spindle chucking machine Mehrspindelfutterautomat *m*
m. spindle drill head Mehrspindelbohrkopf *m*
m. spindle drill[ing machine] mehrspindlige Bohrmaschine *f*, Gelenkspindelbohrmaschine *f*
m. splined shaft Vielkeilwelle *f*
m. splining Vielkeilverzahnung *f*
m.-spot welding machine Vielfachpunktschweißmaschine *f*
m. stage *s.* multi-stage
m. stand rolling mill mehrgerüstiges Walzwerk *n*
m.-start *s.* multi-start
m. station machine Maschine *f* mit mehreren Arbeitsstellen
m.-step *s.* multi-step
m. stop device Anschlagkreuz *n*
m. storey *s.* multi-storey
m. story *s.* multi-storey
m. strand roller chain Mehrfachrollenkette *f*
m. switch Vielfachschalter *m*
m.-tariff meter *(Fmt)* Mehrfach[tarif]zähler *m*
m. thread *s.* multi-thread
m.-throw mehrfach gekröpft, mehrhubig
m. throw switch Stufenschalter *m*
m. tool block Mehrfachmeißelhalter *m*, Vielfachmeißelhalter *m*
m. tool holder Mehrfachmeißelhalter *m*, Vielfachmeißelhalter *m*
m.-tool lathe Vielschnittdrehmaschine *f*
m. tool post Mehrfachmeißelhalter *m*, Vielfachmeißelhalter *m*
m.-tool turning Vielmeißeldrehen *n*
m. tooling Vielfachmeißelanordnung *f*; Einspannen *n* mehrerer Werkzeuge

m. tooling saddle Werkzeugschlitten *m* mit Mehrfachmeißelhalter
m.-toothed mehrzahnig, mehrschneidig
m. track range radar *(El)* mehrspuriges Radarnavigationsverfahren *n*, Laufzeitdifferenzmeßverfahren *n*
m. tuned aerial Mehrfachantenne *f (mehrfach gespeist)*
m.-twin cable mehrpaariges (vielpaariges) Kabel *n*
m. unit condenser Mehrfachkondensator *m*
m. unit running Fahrt *f* in Vielfachsteuerung
m. unit semiconductor device Mehrfachhalbleitersystem *n*
m.-unit valve *(El)* Verbundröhre *f*
m.-valued mehrwertig, mehrdeutig
m. valve *s.* multi-valve
m. way *s.* multi-way
m. winding *(El)* Mehrfachwicklung *f*
m.-wire *s.* multi-wire
multiplet Multiplett *n*
multiplex connection *(Fmt)* Rundgesprächsanlage *f*, Sammelschaltung *f*
m. printing telegraphy Mehrfachdrucktelegrafie *f*
m. radio transmission Mehrfachübertragung *f*
m. switchboard Multiplexumschalter *m*
m. system Vielbandsystem *n*
m. telegraph Mehrfachtelegraf *m*
m. telegraphy Mehrfachtelegrafie *f*
m. telephony Mehrfachtelefonie *f*
m. wave winding Serienparallelwicklung *f*, Reihenparallelwicklung *f*
m. winding Reihenparallelwicklung *f*
multiplexer Mehrfachkoppler *m*
multiplexing Mehrfachausnutzung *f*, Schachtelung *f*
multiplicable multiplizierbar, vervielfachbar
multiplicand Multiplikand *m*
multiplication 1. *(Math)* Multiplikation *f*; 2. *(El)* Vervielfachung *f*, Verstärkung *f*; 3. *(Kern)* Vermehrung *f*, Multiplikation *(Neutronen)*
m. amplifier Vervielfacherverstärker *m*
m. bar Multiplikationstaste *f* *(Rechner)*
m. constant *(Kern)* Vermehrungsfaktor *m*, Multiplikationsfaktor *m*
m. factor 1. *(El)* Verstärkungsfaktor *m (eines Vervielfachers)*, Vervielfachungsfaktor *m*; 2. *s.* m. constant
m. length *(Kern)* Multiplikationslänge *f*
multiplicative conversion multiplikative Mischung *f (Funkempfangstechnik)*
multiplicity Multiplizität *f*
multiplier 1. Vervielfacher *m (z. B. von Neutronen)*; 2. *(Math)* Multiplikator *m*; *(Dat)* Multiplikator, Multiplizierer *m*; Multiplizierglied *n (Analogrechner)*; 3. *(El)* Vorsteckteiler *m*; Vor[schalt]widerstand *m*; Bereichsschalter *m*
m. chain Vervielfacherschaltung *f*
m. circuit Vervielfacherschaltung *f*
m. gain Verstärkungsfaktor *m (des Sekundärelektronenvervielfachers)*
m. photocell Fotosekundärelektronenvervielfacher *m*
m. phototube Fotovervielfacher *m*, Fotozelle *f* mit Elektronenvervielfacher
m. potentiometer Multiplikatorpotentiometer *n*, Multipliziererpotentiometer *n (Analogrechner)*
m. tube Fotovervielfacher *m*, Fotozelle *f* mit Elektronenvervielfacher
multiply / to 1. *(Math)* multiplizieren; 2. *(El)* vervielfachen, verstärken; 3. *(Kern)* vermehren *(Neutronen)*
multiply charged mehrfach geladen
m. connected region *(Math)* mehrfach zusammenhängender Bereich *m*
multiplying power Empfindlichkeitsfaktor *m (eines Galvanometernebenschlusses)*
multipoint recorder Punktschreiber *m*
m. switch *(El)* Stufenschalter *m*, Mehrstufenschalter *m*, vielstufiger (mehrstufiger) Schalter *m*
m. tap switch Schalter *m* mit mehreren Abgriffen
m. tool mehrschneidiges Werkzeug *n*
multipolar *(El)* mehrpolig, vielpolig
m. armature mehrpoliger Anker *m*; Sternanker *m*, vielpoliger Anker
m. winding mehrpolige Wicklung *f*, Mehrpolwicklung *f*, vielpolige Wicklung, Vielpolwicklung *f*
multipole Mehrpol *m*, Multipol *m*
multipole *s.* multipolar
multiprong plug socket case *(Rak)* Kastenhalterung *f*
multirate meter *(Fmt)* Mehrfach[tarif]zähler *m*
multiscaler Mehrfachuntersetzer *m*, Vielfachuntersetzer *m*, Universaluntersetzer *m*
multistage *s.* multi-stage
multistorey *s.* multi-storey
multistory *s.* multi-storey
multitubular vielröhrig, Vielröhren- *(z. B. Kessel)*
m. boiler Vielröhrenkessel *m*, vielröhriger Kessel *m*
m. boiler with withdrawable fire box and smoke tubes ausziehbarer Heizrohrkessel *m*, Heizrohrkessel mit Feuerbüchse und ausziehbarem Röhrenbündel
m. cooler Vielröhrenkühler *m*
multivalency Mehrwertigkeit *f*
multivalent mehrwertig

multivariable control system Mehrfachregelung *f*
multivibrator *(El)* Kippgenerator *m*, Multivibrator *m*
m. circuit Multivibratorschaltung *f*
multiwire *s.* multi-wire
mungo *(Text)* Mungo *m (geringwertige, kurzstapelige Reißwolle)*
m. from new rags Neumungo *m*
m. from old rags Altmungo *m*
municipal städtisch, Gemeinde-, Stadt-
m. airport städtischer Flughafen *m*
m. engineer Städteplaner *m*; Ingenieur *m* für Städtebau
m. engineering Städtebau *m*; städtischer Tiefbau *m*
m. planner Städteplaner *m*
m. planning Städtebau *m*, Städteplanung *f*
munnion *s.* mullion
Muntz metal Muntzmetall *n*
muon *(Kern)* My-Meson *n*, Myon *n*
murrain hide *(Led)* Sterbling *m*
musa aerial Mehrfachrautenantenne *f*
muscovado *(Ch)* Roh[rohr]zucker *m*
muscovite *(Min)* Muskowit *m*, Spaltglimmer *m*
mush Interferenz *f*, Störung *f (Radar)*
m. area Störgebiet *n*, Interferenzgebiet *(Radar)*
m.-wound coil *(El)* Träufelspule *f*
mushroom Pilz *m*; *(Brg)* Schrämpilz *m*
m. anchor *(Schiff)* Pilzanker *m*, Schirmanker
m. construction Pilzdecke *f (Stahlbetonbau)*
m. head bolt Pilzkopfschraube *f*
m. headed mit pilzförmigem Kopf
m. insulator *(El)* Pilzisolator *m*
m. jib *(Brg)* Schrämarm *m* mit Schrämpilz
m. mixer Pilzmischer *m*, Schrägkugelmischer *m*, schrägstehender Trommelmischer *m*
m. slab Pilzdecke *f (Stahlbetonbau)*
m. tappet *(Kfz)* Pilzstößel *m*
m.-type retainer *(Kfz)* Pilzsicherung *f*
m. valve Pilzventil *n*
m. ventilator Pilzlüfter *m*
music box Spieldose *f*
m. circuit Rundfunkleitung *f*
m. fidelity *(Ak)* Musiktreue *f*
m. line Notenlinie *f*
m. loading Musikpupinisierung *f*
m. rack Notenpult *n*
musical clock Spieluhr *f*
m. scale Tonsystem *n*
m. spark gap tönende Funkenstrecke *f*
musk ambrette Kresolmoschus *m*, synthetischer Moschus *m*
m. ketone Ketonmoschus *m*
m. rat Bisamratte *f*
m. xylene Xylolmoschus *m*
musquash Bisamrattenfell *n*
mustard gas Senfgas *n*, Dichlordiäthylsulfid *n*
musty schimmlig, dumpfig, muffig
mutation *(Kern)* Mutation *f*
m. constant Mutationskonstante *f*
mutator *(Elm)* Umrichter *m*
m. locomotive Umformerlokomotive *f*
mute / to [ab]dämpfen
mute Dämpfer *m*, Schalldämpfer *m*
m. aerial künstliche Antenne *f*
m. key-board stumme Klaviatur *f*
m. keys stumme Klaviatur *f*
mutilate / to *(Fmt)* verstümmeln
mutilated *(Fmt)* verstümmelt
m. selection unvollständige Wahl *f*
m. telegram verstümmeltes Telegramm *n*
mutilation *(Fmt)* Verstümmelung *f*, Entstellung *f*; Beschädigung *f*
muting Dämpfung *f*; Abschwächung *f*; Empfängersperre *f*, Rauschsperre *f*
m. switch Schalter *m* für Stummabstimmung; Dämpfungsschalter *m*
mutton tallow Unschlitt *n*, Hammeltalg *m*
mutual gegenseitig, Wechsel-
m. attraction gegenseitige Anziehung *f*
m. capacitance *(El)* Betriebskapazität *f (zwischen zwei Leitern bei Erdung aller übrigen)*, gegenseitige Kapazität *f*
m. conductance Steilheit *f (Elektronenröhren)*
m. conductance cutoff frequency Grenzfrequenz *f* der Steilheit
m. conductance slope Betriebssteilheit *f*
m. inductance Gegeninduktivität *f*, Koeffizient *m* der gegenseitigen Induktion, gegenseitige Induktivität *f*
m. induction gegenseitige Induktion *f*, Gegeninduktion *f*, wechselseitige Induktion
m. inductometer *(El)* regelbare Gegeninduktivität *f*, Variometer *n*
m. interaction gegenseitige Abstoßung *f*
m. reaction Wechselwirkung *f*
mutually perpendicular senkrecht aufeinander
muzzle 1. *(Mil)* Mündung *f*; 2. Maulkorb *m*
m. brake Mündungsbremse *f*
m. cover Mündungskappe *f*, Mündungsschoner *m*
m. energy Mündungsenergie *f*
m. flash Mündungsfeuer *n*
m.-loader Vorderlader *m*
m. velocity Anfangsgeschwindigkeit *f (des Geschosses)*
mycelium Myzel *n*, Pilzgewebe *n*
mycoderma Schimmelhaut *f*, Schimmelschicht *f*
myosin Muskeleiweiß *n*, Myosin *n*
myricinic acid Myrizinsäure *f*
myricyl alcohol Myrizylalkohol *m*, Melissylalkohol *m*
myristic acid Myristinsäure *f*
mysorine *(Min)* Mysorin *m*
mystery control *(sl)* drahtlose Fernsteuerung *f*

N

N-component nukleare Komponente *f* *(der kosmischen Strahlung)*
n-conducting n-leitend *(Halbleiter)*
N-display N-Darstellung *f* *(Radar)*
N-electron *(Kern)* N-Elektron *n*
n-fold n-zählig
N-level *(Kern)* N-Niveau *n*
n-person game *(Dat)* Spiel *n* mit n[-beteiligten] Personen
n-place shifter *(Dat)* n-stellige Schiebevorrichtung *f*
n-point switch *(Fmt)* Wähler *m* mit n-Kontakten
N-radiation *(Kern)* N-Strahlung *f*
N-shell *(Ph, Ch)* N-Schale *f*
n-step transformer n-stufiger Transformator *m*
N-truss Parallelfachwerk *n*
n-type n-leitend, n-Typ-, überschußleitend *(Halbleitertechnik)*
n-type conductivity N-Leitung *f*, Überschußleitung *f*, Elektronenleitung *f*
n-type semiconductor n-Halbleiter *m*, Überschußhalbleiter *m*
nabla Nabla *n*, Nabla-Operator *m*, Nabla-Vektor *m* *(Vektoranalysis)*
  n. operator *s.* nabla
nacelle Gondel *f*, Rumpf *m*
nacre Perlmutt[er *f*] *n*
nacreous mit Perlmuttglanz, perlmutt[er]artig
nacrite *(Min)* Nakrit *m*
nadir Nadir *m*
nail / to nageln
nail Nagel *m*
  n. extractor Nagelzieher *m*
  n. hammer Klauenhammer *m*
  n. head Nagelkopf *m*
  n. head pull-through test Nageldurchziehprüfung *f*
  n. holding Nagelhaltefestigkeit *f*
  n. lacquer Nagellack *m*
  n.-lifter Nagelzieher *m*
  n. making machine Drahtstiftmaschine *f*
  n. pass Nadelkaliber *n* *(Walzwerk)*
  n. polish Nagellack *m*
  n. polish remover Nagellackentferner *m*
  n. puller Nagelzange *f*; Nagelzieher *m*; Kistenöffner *m*
  n. set Nageltreiber *m*
  n. withdrawal test Nagelausziehwiderstandsprüfung *f*
nailability Nagelbarkeit *f*
nailable nagelbar
nailed truss *(Bw)* Nagelbinder *m*
nailing Nageln *n*, Annageln *n*; Benageln *n* *(Schuh)*
  n. concrete nagelbarer Beton *m*
  n. of climbing boots Bergschuhbenagelung *f*
  n. of the planks *(Schiff)* Verspiekerung *f* der Beplankung
naked blank *(Draht)*; offen *(Flamme)*; bloß *(Auge)*
  n. light *(Brg)* offenes Geleucht *n*
  n.-light mine *(Brg)* schlagwetterfreie Grube *f*
  n. model *(Schiff)* nacktes Modell *n* *(Schleppversuchsmodell ohne Anhänge)*
  n. wall *(Bw)* Kaldaune *f*
  n. wire *(El)* blanker (unisolierter) Draht *m*, Blankdraht *m*
name embroidery Namenstickerei *f*
naming of a ship Schiffstaufe *f*
nand *s.* NOT AND
nano Nano *n* *(10^{-9} Einheiten)*
nanosecond Nanosekunde *f* *(1 ns = 10^{-9} s)*
nap / to *(Text)* rauhen, aufkratzen, noppen
nap Noppe *f* *(im Gewebe)*; Gewebefehler *m*, Flaum *m*, Fluse *f*
  n. cloth weave Flockenstoffbindung *f*, Floconnébindung *f*
  n. fabric haariger Stoff *m*; Noppenstoff *m*
  n. finish Strichappretur *f*
  n. warp Florkette *f*, Samtkette *f*
  n. yarn Noppengarn *n*
napery Tischleinen *n*
naphtha Naphtha *n* *(f)*, Petroleum *n*, Erdöldestillat *n*, Leuchtpetroleum *n*
naphthalene Naphthalin *n*
  n. carboxylic acid Naphthoesäure *f*
  n. dicarboxylic acid Naphthal[indikarbon]-säure *f*
  n. dichloride Dichlornaphthalin *n*
  n. disulfonic acid Naphthalindisulfonsäure *f*
  n. sulfonic acid Naphthalinsulfonsäure *f*
naphthalic acid Naphthalsäure *f*
naphthenate Naphthenat *n*, Salz *n* der Naphthensäure
naphthene-base crude petroleum naphthenbasisches Erdöl *n*
naphthenes Naphthene *npl*
naphthenic acid Naphthensäure *f*
  n. oil naphthenbasisches Erdöl *n*
naphthionic acid Naphthionsäure *f*, α-Naphthylamin-4-sulfosäure *f*
naphthoate Ester *m* (Salz *n*) der Naphthoesäure
naphthoic acid Naphthoesäure *f*
naphthol Naphthol *n*
  n. dye Naphtholfarbstoff *m*
  n. sulfonic acid Naphtholsulfonsäure *f*
naphthology Erdölwissenschaft *f*
naphthoquinone Naphthochinon *n*
naphthylamine Naphthylamin *n*
Napierian logarithm natürlicher (hyperbolischer, Neperscher) Logarithmus *m*
napkin paper Papierserviette *f*
Naples yellow Neapelgelb *n*

nappe *(Math)* Schale *f* *(einer Fläche)*
napping *(Text)* Rauhen *n*, Strichappretieren *n*
narceine Narcein *n*
narcotic Betäubungsmittel *n*, Narkotikum *n*
narcotine Narkotin *n*
narrow / to sich verjüngen, enger werden; bündeln
narrow Flußenge *f*
narrow eng
n. angle lighting fitting Tiefstrahler *m*
n. band schmales Band *n* *(Funk)*
n.-band controller *(BMSR)* Regler *m* mit schmalem P-Bereich
n. beam enges Bündel *n*, dünner (schmaler) Strahl *m*, scharfe Keule *f*, schmale Strahlungskeule *f* *(Antennentechnik)*
n.-beam measurement Kleinfeldmessung *f*
n. bearing schmales Tragbild *n*
n. braid *(Text)* Schmalband *n*
n.-circled feinjährig *(Holz)*
n. edge Schmalkante *f*
n. frequency range schmales Frequenzband *n*
n. gauge *(Eis)* Schmalspur *f*
n. gauge railway Schmalspurbahn *f*, Kleinbahn *f*
n.-gauge traction motor Schmalspurbahnmotor *m*
n. guide Schmalführung *f*
n. joints between courses engfügig *(Holz)*
n. measure *(Typ)* Schmalsatz *m*
n.-mouth shovel schmale Schaufel *f*
n.-necked bottle Enghalsflasche *f*
n. resonance curve *(El)* schmale (scharfe) Resonanzkurve *f*
n.-ringed engringig *(Holz)*
n. side Schmalseite *f*
narrowing Bündelung *f*, Verengung *f*
n. and widening flat knitting machine Minder- und Zunahmeflachstrickmaschine *f*
n. of the steam passage *(Masch)* Verengung *f* des Dampfkanals
nascent hydrogen naszierender Wasserstoff *m*, Wasserstoff in statu nascendi
n. neutron naszierendes Neutron *n*, Neutron im Augenblick der Emission, Neutron im Entstehungszustand
natatorium Schwimmbad *n*
National Electric Code *(El)* *(US-Vorschriften, entsprechend den VDE-Bestimmungen)*
national markings *(Flg)* Hoheitszeichen *npl*
nationality plate Nationalitätszeichen *n*
nationally organized racer nationales Rennklassenboot *n*
n.-owned factory volkseigener Betrieb *m*
n.-owned undertaking volkseigener Betrieb *m*
native gediegen, natürlich vorkommend
n. asphalt Naturasphalt *m*, Erdasphalt *m*
n. copper gediegenes Kupfer *n*, Bergkupfer *n*
n. forge process Rennverfahren *n*, direktes Erz-Stahl-Verfahren *n*
n. gold gediegenes Gold *n*
n. hide Zahmhaut *f*, einheimische Haut *f*
n. metal gediegenes (reines) Metall *n*
n. occurence gediegenes Vorkommen *n*, natürliches Vorkommen im elementaren Zustand
n. ore bergfeines Erz *n*
n. rock gewachsener Fels *m*, anstehendes Gestein *n*, Muttergestein *n*
n. wood einheimisches Holz *n*
native's patent Inländerpatent *n*
natrocalcite *(Min)* Natrokalzit *m*
natrojarosite *(Min)* Natrojarosit *m*
natrolite *(Min)* Natrolith *m*, Faserzeolith *m*
natron Natriumkarbonat *n*, Soda *f*
natural natürlich, Natur-; Eigen-
n. abundance *(Kern)* natürliche Häufigkeit *f*
n. aging *(Met)* natürliches Altern *n*
n. asphalt Naturasphalt *m*
n. bevel Baumkante *f*, Wahnkante *f*, Waldkante *f*
n. building stone Naturbaustein *m*
n. capacity Eigenkapazität *f* *(von Antennen)*
n. cement hydraulischer Kalk *m*; Romanzement *m*
n. circulation natürlicher Umlauf *m*, Naturumlauf *m*
n. circulation water cooling Thermosiphonkühlung *f*
n. colour Eigenfarbe *f*, Naturfarbe *f*, natürliche Farbe *f*
n. cooling Selbstkühlung *f*
n. coordinates natürliche Koordinaten *fpl*
n. disintegration *(Kern)* natürlicher Zerfall *m*
n. divergence Eigenstreuung *f*
n. draught *(Wärm)* natürlicher Zug *m*
n. draught kiln Trockenkammer *f* mit natürlichem Zug
n. drying Trocknen *n* an der Luft
n. dye Naturfarbstoff *m*
n. element *(Ch)* natürliches Element *n*
n. event Naturereignis *n*
n. excitation natürliche Erregung *f*
n. fibre natürlicher Faserstoff *m*
n. fine grain Naturfeinkorn *n*
n. force Naturkraft *f*
n. frequency Eigenfrequenz *f*, natürliche Frequenz *f*
n. frequency of an antenna Antenneneigenfrequenz *f*
n. gas Erdgas *n*, Naturgas *n*
n. gas well Erdgasquelle *f*
n. gasoline Rohrkopfbenzin *n*
n. glass mineralisches Glas *n*
n. harbour natürlicher Hafen *m*

n. hydrogen natürlicher (gewöhnlicher) Wasserstoff *m*
n. joints Trenn[ungs]flächen *fpl (Kristall)*
n. law Naturgesetz *n*
n. leak Eigenverlust *m*
n. lighting Tageslicht[beleuchtung *f*] *n*, natürliche Beleuchtung *f*
n. load governing characteristic *(BMSR)* natürliche Netzkennlinie *f*
n. logarithm natürlicher (hyperbolischer, Neperscher) Logarithmus *m*
n. magnet natürlicher Magnet *m*
n. magnetization freie Magnetisierung *f*
n. mica Rohglimmer *m*
n. mode Eigenschwingung *f*
n. number natürliche Zahl *f*
n. occurence natürliches Vorkommen *n*, Auftreten *n* in der Natur
n. oscillation Eigenschwingung *f*; Grundschwingung *f*
n. period of vibration Eigenschwingungsdauer *f*, Eigenschwingungszeit *f*, Periode *f* der Eigenschwingung
n. phenomenon Naturerscheinung *f*
n. philosophy 1. Naturkunde *f*; 2. Naturwissenschaft *f*; 3. Naturlehre *f*, Physik *f*
n. radiation Eigenstrahlung *f*
n. radioactive natürlich radioaktiv
n. radioactive element natürliches radioaktives Element *n*
n. radioactivity natürliche Radioaktivität *f*
n. rectifier Kristalldetektor *m*
n. resin Naturharz *n*
n. resonance Eigenresonanz *f*, Eigenschwingung *f*
n. rotation Eigenrotation *f*
n. sag Durchbiegung *f* durch eigene Schwerkraft
n. science Naturwissenschaft[en *pl*] *f*, Naturkunde *f*
n. silk Naturseide *f*
n. speed Eigengeschwindigkeit *f*
n. stability limit of a transmission system *(El)* Grenze *f* der natürlichen Stabilität eines Übertragungssystems
n. steel Naturstahl *m*, selbsthärtender Stahl *m*
n. transient stability limit *(El)* Grenze *f* der natürlichen dynamischen Stabilität
n. ventilated motor Motor *m* mit Eigenbelüftung *(Kühlrippen und Kühlschlitze)*
n. ventilation natürliche Belüftung *f*; *(Brg)* natürliche Bewetterung *f*
n. vibration Eigenschwingung *f*
n. voltage *(El)* Eigenspannung *f*
n. water Bildungswasser *n*
n. water circulation natürlicher Wasserumlauf *m*
n. wave Eigenwelle *f*
n. wavelength Eigenwellenlänge *f*, natürliche Wellenlänge *f*; Grundwelle *f*
n. wax Naturwachs *n*
naturally aspirated ohne Aufladung *(Motor)*
nature Natur *f*; Eigenschaft *f*, Beschaffenheit *f*, Art *f*, Güte *f*, Wert *m*
n. of the ground (soil) Bodenbeschaffenheit *f*
nautical nautisch, See-
n. chart Seekarte *f*
n. mile Seemeile *f* *(1 imperial nautical mile = 1853,181 m; 1 US nautical mile = 1852 m = 1 sm)*
n. surveying nautische Vermessung *f*
n. term Schiffsausdruck *m*, Schifferausdruck
n. triangle *(Opt)* nautisches Dreieck *n*
nautics Nautik *f*
navaid *s.* navigational aid
naval Marine betreffend, zum Schiffswesen gehörend, naval
n. architect Schiffbauingenieur *m*
n. architecture Schiffbau *m*
n. brass Marinemessing *n (Cu-Zn-Sn-Legierung)*
n. constructor Marineingenieur *m*
n. forces Seestreitkräfte *fpl*
n. observatory Seewarte *f*
n. ordnance Schiffsartillerie *f*
n. port Kriegshafen *m*, Marinehafen *m*
n. propulsion Schiffsantrieb *m*
n. propulsion engine Schiffsantriebsmaschine *f*
n. propulsion gear Schiffsgetriebe *n*
n. radio station Seefunkstelle *f*
n. yard Marinewerft *f*
nave 1. Schiff *n (Kirche)*; 2. Nabe *f*
n. boring and mortising machine Radnabenbohr- und -stemmaschine
n. boring machine Nabenbohrmaschine *f*
n. mortising machine Maschine *f* zum Stemmen der Speichenlöcher
navigable befahrbar *(ein Gewässer)*, [be]schiffbar; seetüchtig *(Schiff)*
navigate / to befahren *(mit Schiff)*; lenken, steuern, navigieren
n. a vessel ein Schiff führen
navigating bridge Kommandobrücke *f (Handelsschiff)*
n. erections Kommandoaufbauten *pl*
n. lanterns Positionslichter *npl*, Positionslaternen *fpl*, Fahrtlaternen *fpl*
navigation Schiffahrt *f*, Navigation *f*, Schiffahrtskunde *f*; Fahrt *f*, Reise *f*
n. bridge Kommandobrücke *f*
n. by dead reckoning Koppelnavigation *f*
n. company Schiffahrtsgesellschaft *f*, Reederei *f*
n. lanterns Positionslaternen *fpl*, Fahrtlaternen *fpl*, Positionslichter *npl*

n. lights Befeuerung *f (Navigation)*
n. radio-light Navigationsfunkfeuer *n*
Navigation Board Seefahrtsamt *n*
navigational aid Navigationshilfe *f*
n. aids Betonnung *f* und Befeuerung *f*, Seezeichen *npl*; nautische Instrumente *npl*, Navigationsausrüstung *f (eines Schiffes)*
n. channel Fahrrinne *f*, Schiffahrtsrinne *f*
n. instruments nautische Instrumente *npl*
n. lighting Schiffahrtbefeuerung *f*
navigator Nautiker *m*, Navigateur *m*, Seefahrer *m*
navvy 1. Erdarbeiter *m*, Ausschachter *m*; 2. Bagger *m*, Löffel[hoch]bagger *m*
navy Kriegsmarine *f*, Marine *f*, Flotte *f*
n. aircraft Marineflugzeug *n*
n. transmitter Marinesender *m*
N/C machine *(Am)* numerisch gesteuerte Maschine *f*
neal / to *s.* anneal / to
neap-tide *(Schiff)* Nipptide *f*
near nahe, Nah-
n.-critical fastkritisch *(Reaktor)*
n. echo Nahecho *n (Radar)*
n.-end attenuation *(Fmt)* Nahnebensprechdämpfung *f*
n.-end crosstalk *(Fmt)* Nebensprechen *n*
n.-end crosstalk attenuation Nebensprechdämpfung *f*, Übersprechdämpfung *f*
n. fading Nahschwund *m (Funk)*
n. field Nahfeld *n*
n.-forward scattering *(Kern)* Streuung *f* überwiegend in Vorwärtsrichtung
n. infrared nahes Infrarot *n*
n.-mesh grain *(Met)* Grenzkorn *n (Kristalle)*
n. point Nahpunkt *m*
n.-sight tester *(Opt)* Nahsichtprüfgerät *n*
n. singing *(Fmt)* Pfeifneigung *f*
n.-sonic schallnahe
n. the earth in Erdnähe
n. to ground in Bodennähe
n. view Nahansicht *f*
nearest approach Mindestsicherheitsabstand *m*; sehr kurze Entfernung *f (Radar)*
nearly-parabolic circuit *(Astr)* parabelnahe Bahn *f*
nearness Nähe *f*
neat rein, unvermengt, unvermischt
n. casting sauberes Gußstück *n*
n. cement reiner Zementmörtel *m*, Zementpaste *f*, Zementschlämme *f*
n. house Kuhstall *m*
n. lime Kalkschlämme *f*, reiner Kalkmörtel *m*
neatly grouped übersichtlich angeordnet
neat's foot oil Klauenöl *n*
n. hide Rindshaut *f*
n. leather Rindsleder *n*
neb peg *(Text)* Einfederspindel *f* mit Nase, Nasenspindel *f*
nebula *(Astr)* Nebel *m*
nebular line Nebellinie *f*
nebulizer Zerstäuber *m*
necessary notwendig, zwangsläufig
n. and sufficient notwendig und hinreichend
n. bandwidth erforderliche Bandbreite *f (Funk)*
n. condition Vorbedingung *f*
neck / to 1. *(Masch)* einstechen; 2. *(Met)* einschnüren; 3. *(Led)* Nacken spalten
n. down einengen *(Strahlenbündel)*, bündeln
neck 1. Ansatz *m (einer Schraube)*; Zapfen *m*, Hals *m (Welle)*; Zapfen *(Walze)*; Mundstück *n (Turbine)*; Einstich *m*, Hals *m (Spiralbohrer)*; 2. *(Geol)* Neck *m*; Landenge *f*; 3. *(Brg)* Kammerhals *m*
n. bearing Halslager *n*
n. brick Spitzkeilstein *m*
n. bush Grundbuchse *f*
n. collar journal Halszapfen *m*
n. cord *(Text)* Schnurplatine *f*, Platinenschnur *f*
n. journal Halslager *n*
n. length Halslänge *f (Spiralbohrer)*
n. recess Halsausdrehung *f*
n. ring Grundring *m (Dichtung)*
necked-down eingeschnürt
n.-down portion eingeschnürter Teil *m (eines Maschinenteils)*
n. riser verengter Steiger *m (Gießerei)*
necking 1. *(Masch)* Einstechen *n*; 2. Einschnürung *f*, Verengung *f*, Verminderung *f* des Querschnitts
n. down Einschnürung *f*
n. operation *(Masch)* Aushalsungsarbeit *f*, Einstecharbeit *f*
n. tool Stechmeißel *m*
necklace microphone Kehlkopfmikrofon *n*
neddling of the comb strip *(Text)* Benadelung *f* des Nadelstabes *(Kämmaschine)*
need Bedarf *m*; Mangel *m*
needle / to nadelförmig kristallisieren
n.-punch *(Text)* nadeln *(Fasermasse)*
needle Nadel *f*; Zeiger *m*; Querbalken *m*; Kanüle *f*, Injektionsnadel *f*
n. antimony *(Min)* Antimonglanz *m*
n. beam Querbalken *m*
n. bearing Nadellager *n*
n. board *(Text)* Nadelbrett *n*
n. coal Nadelkohle *f*
n. control *(Kfz)* Nadelregelung *f*
n. counter *s.* point counter
n. coupling Nadelkupplung *f (Turbine)*
n. creel *(Text)* Nadelkette *f*
n. deflection Zeigerausschlag *m*
n. deviation Zeigerausschlag *m (Meßinstrumente)*; Nadelabweichung *f*

n. felting machine Nadelfilzmaschine *f*
n. file Nadelfeile *f*
n. lace *(Text)* Nadelspitze *f*
n. lead Nadelblei *n*
n. lubricator Nadelöler *m*
n. machine Stickmaschine *f*
n. ore Nadelerz *n*
n. plug Ventilnadel *f*
n. point Nadelspitze *f*
n.-point gap *(El)* Nadelfunkenstrecke *f*
n. point lace *(Text)* Nadelspitze *f*
n.-point spark gap *(El)* Nadelfunkenstrecke *f*
n.-point voltage *(El)* Überschlagspannung *f (der Nadelfunkenstrecke)*, Spitzenspannung *f*
n. pressure Nadeldruck *m*
n. protector *(Text)* Nadelbruchabsteller *m*
n. pulse Nadelimpuls *m*
n. punching machine Nadelfilzmaschine *f*
n. roller Nadel *f (Lager)*
n. roller bearing Nadelrollenlager *n*
n. scratch Nadelgeräusch *n*, Nadelrauschen *n (Schallplatten)*
n. segment *(Text)* Nadelsegment *n (Kämmmaschine)*
n.-shaped nadelförmig
n. tachometer Zeigertachometer *n*
n. throw Nadelausschlag *m*, Zeigerausschlag *m (Instrument)*
n. tip Nadelspitze *f*
n. travel *(Text)* Nadeldurchlauf *m* durch das Schloß
n. tying *(Text)* Nadelverschnürung *f*
n. valve Nadelventil *n*
n. vibrator *(Bw)* Flaschenrüttler *m*, Innenrüttler *m*, Tauchrüttler *m*
n. wire Nadeldraht *m*
n. work Näharbeit *f*; Stickarbeit *f*

negative Negativ *n (Foto)*

negative negativ, Negativ-
n. absorption negative Absorption *f*
n. after-potential *(El)* negatives Nachpotential *n*
n. area Stromeintrittszone *f (Galvanotechnik)*
n. back rake *(Am) (Masch)* positiver Neigungswinkel *m*
n. beta decay *(Kern)* Beta-minus-Zerfall *m*, β^--Zerfall *m*
n. bias *(El)* negative Vorspannung *f*
n. booster *(Elm)* Zusatzmaschine *f* in Gegenschaltung, Dévolteur *m*
n. boosting transformer *(Elm)* Saugtransformator *m*
n. brush *(Elm)* Minusbürste *f*, negative Bürste *f*
n. carrier Halteeinrichtung *f* für Negative
n. charge negative Ladung *f*
n. conductor *(El)* Minusleiter *m*
n. cyanotype Lichtpause *f*
n. distortion *(Fs)* kissenförmige Verzeichnung *f*
n. electricity negative Elektrizität *f*
n. electrode negative Elektrode *f*
n. entropy *(Dat)* Informationsentropie *f*, Nachrichtenentropie *f*
n. feedback Gegenkopplung *f*, negative Rückkopplung *f*
n.-feedback amplifier gegengekoppelter Verstärker *m*
n. feeder *(El)* Rückstromleitung *f*
n. film Negativfilm *m*
n. filtering infrared analyzer Absorptionsinfrarotanalysator *m*
n. frequency negative Frequenz *f*
n. ghost *(Fs)* negatives Geisterbild *n*
n. glow negatives Glimmlicht *n*, Katodenglimmlicht *n*
n. grid bias negative Gittervorspannung *f*
n. grid current umgekehrter Gitterstrom *m*
n. image negatives Bild *n*
n. impedance *(El)* negative Impedanz *f*
n. in sign mit negativem Vorzeichen
n. inside lap *(Wärm)* negative Auslaßüberdeckung *f*
n. ion Anion *n*
n. lens *(Opt)* Negativlinse *f*, Streulinse *f*
n. let-off motion *(Text)* negative Kettablaßvorrichtung *f*, negativ wirkender Kettbaumregler *m*
n. lever dobby Schaftwebstuhl *m* mit Hebel
n. loop negativer Teil *m* des Indikatordiagramms
n. main *(El)* Minusleiter *m* des Netzes
n. matrix Negativ *n*, Patrize *f (Galvanoplastik)*
n. meniscus lens Konvexkonkavlinse *f*
n. meson *(Kern)* negatives Meson *n*
n. modulation negative Modulation *f*, Negativmodulation *f*
n. muon *(Kern)* negatives My-Meson *n*
n. paper Negativpapier *n (Foto)*
n. picture 1. *(Fs)* Negativbild *n*; 2. Negativ *n*, negatives Bild *n (Foto)*
n. plate Minusplatte *f (Batterie)*
n. polarity negative Polarität *f*
n. pole Minuspol *m*, negativer Pol *m*
n. print negative Kopie *f*
n. proton *(Kern)* Antiproton *n*
n. quantity *(Math)* negative Größe *f*
n. rake negativer Spanwinkel *m*; *(Am)* positiver Spanwinkel
n. reaction Gegenkopplung *f*; negative Rückkopplung *f*
n. reactivity *(Ph)* negative Reaktivität *f*
n. reinforcement Bewehrung *f* im Bereich negativer Momente *(Stahlbetonbau)*

n. resistance negativer Widerstand *m*
n. resistance characteristic fallende (negative) Charakteristik *f*
n. resistance oscillator Oszillator *m* mit negativer Widerstandskennlinie; Dynatronoszillator *m*, Transitronoszillator *m*
n.-resistance repeater Zweidrahtverstärker *m* mit negativem Eingangswiderstand
n. resistor negativer Widerstand *m (Gerät)*; Negistor *m*
n. retroaction Gegenkopplung *f*, negative Rückkopplung *f*
n. segregation *(Met)* umgekehrte Blockseigerung *f*
n.-sequence active power *(El)* Wirkleistung *f* der Gegenkomponente
n.-sequence components of system of three-phase quantities Gegenkomponenten *fpl* eines Dreiphasensystems
n.-sequence coordinate of a three phase system Gegenkoordinate *f* eines Drehphasensystems
n.-sequence field impedance Impedanz *f* des gegenläufigen Feldes, Gegenimpedanz *f*
n.-sequence reactance Blindwiderstand *m* der Gegenkomponente
n.-sequence reactive power Blindleistung *f* der Gegenkomponente
n.-sequence resistance Widerstand *m* der Gegenkomponente
n.-sequence symmetrical component symmetrische Gegenkomponente *f*
n. sign Minuszeichen *n*
n. swing *(El)* negative Halbwelle *f*
n. temperature coefficient resistor *(El)* Heißleiter *m*, NTC-Widerstand[selement *n*] *m*
n. terminal negative Klemme *f*
n.-transconductance oscillator *(El)* Bremsfeldoszillator *m*, Dynatron *n*
n. valence negative Wertigkeit (Valenz) *f*
n. warp let-off motion *(Text)* Kettablaßvorrichtung *f*
n. work Gegendruckarbeit *f (Dampfdiagramm)*
negatively charged negativ geladen
n. oriented negativ orientiert *(Kristalle)*
negaton *s.* negatron 2.
negator Negator *m (elektron. Rechentechnik)*
negatron 1. *(El)* Negatron *n*, Doppelgitterröhre *f*; 2. *(Ph)* Negatron (e^-)
negentropy *(Dat)* Informationsentropie *f*, Nachrichtenentropie *f*
negistor *s.* negative resistor
neglected verwildert *(Land)*; vernachlässigt *(Berechnung)*
negligible vernachlässigbar
neighbourhood 1. Wohnkomplex *m*, Nachbarschaft *f (Siedlungsplanung)*; 2. *(Math)* Umgebung *f*, Nachbarschaft
n. unit Wohnkomplex *m*, Nachbarschaft *f (Siedlungsplanung)*
NEITHER-NOR circuit *(Dat)* WEDER-NOCH-Schaltung *f*
neo signal lamp Glimmlampe *f*
neoarsphenamine *(Ch)* Neosalvarsan *n*
neodymia *(Ch)* Neodymoxid *n*
neodymium Neodym *n*
neohexane *(Ch)* Trimethyläthylmethan *n*
neon Neon *n*
n. bulb 1. Neonlampe *f*, Neonröhre *f*; Glimmröhre *f*
n. charge Neonfüllung *f*
n. detector *(El)* Glimmgleichrichter *m*
n. discharge tube Neongasentladungsröhre *f*
n. filling Neonfüllung *f*
n. indicator Glimmlichtanzeigeröhre *f*, Glimmlichtindikator *m*, Kontrollglimmlampe *f*
n. induction lamp elektrodenlose Glimmlampe *f*
n. lamp Neonlampe *f*; Glimmlampe *f*
n. light Neonlicht *n*
n. luminescent tube Neonleuchtröhre *f*
n. oscillator Glimm[lampen]oszillator *m*
n. stabilizer *(El)* Glimmstabilisator *m*
n. tube Neonröhre *f*; Glimmröhre *f*
n. tube crest voltmeter Glimmröhrenscheitelspannungsmesser *m*
n. tube peak voltmeter Glimmlichtscheitelspannungsvoltmeter *n*
n. tube rectifier Glimm[licht]gleichrichter *m*
n. tube relaxation oscillator *(Fs)* Blinkschaltung *f (Erzeugung von Dreieckschwingungen)*
neopentane Tetramethylmethan *n*
neoprene *(Plast)* Neopren *n*, Dupren *n*
neozoic *(Geol)* Neozoikum *n*
neper Neper *n (1 Np = 8,686 dB)*
nepheline *(Min)* Nephelin *m*, Fettstein *m*, Sommit *m*
n. basalt Nephelinbasalt *m*
n. syenite Nephelinsyenit *m*
nephelitic rock *(Geol)* Nephelingestein *n*
nephelometer *(Opt)* Nephelometer *n*
nephelometric *(Opt)* nephelometrisch
n. analysis Nephelometrie *f*, Trübungsmessung *f*
nephelometry *(Opt)* Nephelometrie *f*
nephelynite *(Min)* Nephelinit *m*
nephoscope Nephoskop *n*, Bewölkungsmesser *m*, Wolkenzugmesser *m*
nephrite *(Min)* Nephrit *m*, Jadeit *m*
neptunite *(Min)* Neptunit *m*
neptunium *(Ch)* Neptunium *n*
n. family *(Kern)* Neptuniumreihe *f*
n. series *(Kern)* Neptunium[zerfalls]reihe *f*
neritic zone *(Geol)* neritische Zone *f*, Flachseezone *f*

Nernst bridge *(El)* Nernstbrücke *f* *(für Kapazitätsmessungen)*
Nernst burner Nernstbrenner *m*
Nernst distribution law Nernstscher Verteilungssatz *m*
Nernst effect Nernsteffekt *m*
Nernst filament Nernststift *m*
Nernst heat theorem Nernstsches Wärmetheorem *n*
Nernst lamp Nernstlampe *f*, Nernstbrenner *m*
nerve narcotic Nervenbetäubungsmittel *n*
n. structure *(Met)* adrige Struktur *f*
Nessler's solution *(Ch)* Neßlers Reagens *n*
nest / to [ineinander]schachteln
nest 1. Nest *n*; 2. *(Math)* Ineinanderschachtelung *f*, Schachtelung *f*; 3. Bootsgruppe *f* *(mehrere zusammen vertäute Boote)*
n. of intervals *(Math)* Intervallschachtelung *f*
n. of ore Erznest *n*
n. of screens Siebsatz *m*
n. of tubes Rohrbündel *n*
net 1. Netz *n*; 2. *(Math)* Feld *n*; 3. Mückenschleiertüll *m*, Netz- oder Filetarbeit *f*, Netzwerk *n*, Tüll *m*
net 1. Netto-, netto, rein, Gesamt-, Rein-, Wirk-; 2. *(Opt)* scharf
n. amplitude Gesamtamplitude *f*
n. attenuation *(El)* Restdämpfung *f*
n. balance counter positions *(Dat)* unmittelbar saldierende Zählerstellen *fpl*
n. belt pull Nutzkraft *f* *(Riementrieb)*
n. calorific value unterer Heizwert *m*
n. capacity *(Schiff)* Ladefähigkeit *f*, Tragfähigkeit *f* *(an reiner Ladung)*
n. chart *(Math)* Netztafel *f*
n. current Nettostrom *m*
n. effect Gesamtwirkung *f*
n. efficiency Nutzleistung *f*
n. gain 1. Nettogewinn *m*, Reinprofit *m*; 2. *(El)* Gesamtverstärkung *f*
n. gauze Kreuzgaze *f*, Netzgaze *f*
n. lace *(Text)* Netzspitze *f*
n. layer Netzleger *m* *(Schiff)*
n. lifter Netzheber *m* *(Schiff)*
n. loss Restdämpfung *f*
n. loss variation with amplitude *(Fmt)* amplitudenabhängige Dämpfung *f*
n. machine Tüllwebstuhl *m*
n. output *(Brg)* verwertbare Förderung *f*
n. point *(Math)* Gitterpunkt *m*, Netzpunkt *m*
n. proceeds Reinertrag *m*
n. production Nettoproduktion *f*
n. profits Reingewinn *m*
n. register ton *(Schiff)* Nettoregistertonne *f*
n. section Nettoquerschnitt *m*, Nutzquerschnitt *m*
n.-shaped electrode Netzelektrode *f*
n. ton *(Schiff)* Nettoregistertonne *f*
n. tonnage *(Schiff)* Nettoraumgehalt *m* *(in Registertonnen)*, Nettotonnage *f*, Nettotonnengehalt *m*
n. transmission equivalent *(Fmt)* Restdämpfung *f*
n. transport Nettodurchsatz *m*, Nettotransport *m*
n. volume Nettovolumen *n*
netter Netzleger *m*, Netzfischer *m* *(Schiff)*
netting *(Text)* Netz *n*, Tüll *m*; Filetarbeit *f*
n. needle Filetnadel *f*
nettle Nessel *f*
n. hemp Hanfnessel *f*
network *(El)* 1. Netz *n*, Leiternetz *n*, System *n*, Energienetz *n*; Netzwerk *n*, Sendernetz *n*, Nachrichtennetz *n*; 2. Stromverzweigung *f*; 3. *(Plast)* Vernetzung *f*
n. analog computor *(Dat)* Netzwerkanalog[ie]rechner *m*
n. analysis kritische Gesamtanalyse *f*
n. analyzer *(Dat)* Netzwerkgleichungslöser *m*; Netzwerkanalysator *m*
n. balancing rack *(El)* Nachbildungsgestell *n*
n. centre *(Fmt)* Netzgruppenmittelpunkt *m*, NGM
n. circuit *(El)* Maschenschaltung *f*
n. feeder *(El)* Netzspeiseleitung *f*
n.-group switch *(Fmt)* Netzgruppenschalter *m*
n.-junction exchange *(Fmt)* Netzgruppenknoten *m* *(Vermittlungsstelle in einer Netzgruppe)*
n. junction-point *(Fmt)* Netzknotenpunkt *m*
n. master relay Hauptanschlußrelais *n*
n. mesh *(El)* Kettenglied *n* *(Filter)*
n. of pipes Rohrnetz *n*
n. plane *(Fmt)* Netzebene *f*
n. primary distribution system Hochspannungsverteilungsnetz *n*
n. protector Netzschutzschalter *m*
n. region *(Math)* Netzgebiet *n*
n. secondary distribution [system] *(El)* Niederspannungsverteilungsnetz *n*
n. structure *(Met)* Netzstruktur *f*, netzförmige Ausbildung *f* des Gefüges
n. transformer Netztransformator *m*, Netzwandler *m*
Neumann lamellae Zwillingsstreifen *mpl* *(Kristalle)*
neutral Nulleiter *m*, Sternpunkt[leiter] *m*, Mittelleiter *m*; *(Am)* Indifferenzzone *f*
neutral 1. *(El)* neutral, ungeladen, Null-; 2. entspannt, entlastet *(Gebirgsmechanik)*; 3. indifferent *(Gleichgewicht)*; 4. flach *(Foto)*
n. armature *s.* non-polarized armature
n. atom *(Kern)* neutrales Atom *n*
n. auto-transformer Erdungsdrossel *f*
n. axis neutrale Achse *f*, Nullinie *f* *(Statik)*

n. bottom neutraler Boden *m*, neutrales Schiff *n*
n. compensator Löschtransformator *m*
n. conductor *(El)* Nulleiter *m*, Sternpunktleiter *m*
n. constitution of the lubricant neutrale Beschaffenheit *f* des Schmiermittels
n. direct-current telegraph system *(Fmt)* Morseeinstromsystem *n*
n. earthing Nullpunkterdung *f*, Sternpunkterdung *f*
n. electrode neutrale (indifferente) Elektrode *f*
n. equilibrium indifferentes Gleichgewicht *n*
n. fat Neutralfett *n*
n.-feeder *(El)* Nulleiter *m*
n. filter Neutralglas *n*
n. fit *(Masch)* Übergangspassung *f*, Übergangssitz *m*
n. ground Nullpunkterdung *f*, Sternpunkterdung *f*
n. line neutrale Achse *f*, Nullinie *f* *(Statik)*
n. meson neutrales (ungeladenes) Meson *n*
n. oil Neutralöl *n*
n. particle neutrales (ungeladenes) Teilchen *n*
n. pitch Grundsteigung *f* *(Luftschraube)*
n. plane Nullzone *f* *(Mehrphasenmaschine)*
n. point *(El)* 1. Sternpunkt *m*; 2. Nullpunkt *m*, Erdpunkt *m*
n. position Leerlaufstellung *f*; Nullstellung *f*, Ruhestellung *f*
n.-rake type cutter Fräser *m* ohne Spanwinkel
n. relay neutrales Relais *n*
n. solution *(Ch)* neutrale Lösung *f*
n. terminal *(Elm)* Sternpunktklemme *f*
n. water glass lösliches Natronwasserglas *n*, Natriumsilikat *n*
n. wedge Graukeil *m* *(Spektroskopie)*
n. wedge analysis Graukeilanalyse *f*
n. wedge method Graukeilverfahren *n*
n. wire *(El)* Nulleiter *m*, Mittelleiter *m*
n. zone 1. *(El)* Indifferenzzone *f*; neutrale Zone *f*; 2. *(Am)* *(BMSR)* Unempfindlichkeitsbereich *m*, tote Zone
neutralator Löschtransformator *m*
neutralization Neutralisation *f*, Neutralisierung *f*
n. number Säurezahl *f*, Neutralisationszahl *f*
neutralize / to aufheben; neutralisieren; kompensieren
n. the centrifugal force die Zentrifugalkraft ausgleichen
neutralized receiver *s.* neutrodyne receiver
n. series motor *(Elm)* Einphasenkommutatormotor *m* mit Kompensationswicklung
neutralizer Neutralisator *m*, Entsäurer *m*
neutralizing capacitor Neutralisationskondensator *m*, Neutrodynkondensator *m*, Neutralisierungskondensator *m*

n. transformer Kompensationstransformator *m*, Ausgleichstransformator *m*
neutretto Neutretto *n*, ungeladenes Elektron *n*
neutrino *(Kern)* Neutrino *n*
n. radiation Neutrinostrahlung *f*
neutrode circuit neutralisierte Triodenvorstufe *f* in Katodenbasisschaltung *(für Fernseh-VHF-Tuner)*
neutrodyne / to *(El)* neutrodyn schalten
neutrodyne capacitance *(El)* Neutralisationskapazität *f*
n. capacitor Entkopplungskondensator *m*
n. connection Neutrodynschaltung *f*
n. receiver Neutrodynempfänger *m*
n. reception neutralisierter Empfang *m*, Neutrodynempfang *m* *(Funk)*
neutron Neutron *n*
n. absorber Neutronenabsorber *m*, Neutronenempfänger *m*
n.-absorbing neutronenabsorbierend
n. absorption Neutronenabsorption *f*; Neutroneneinfang *m*
n. activation Neutronenaktivierung *f*, Aktivierung *f* durch Neutronen
n. activity Neutronenaktivität *f*
n. age Neutronenalter *n*
n. attenuation Neutronenschwächung *f*
n. beam Neutronenstrahl *m*, Neutronenbündel *n*
n. binding energy Neutronenbindungsenergie
n. bombardment Neutronenbeschuß *m*, Beschießung *f* mit Neutronen
n. burst Neutronenimpuls *m*
n. capture Neutroneneinfang *m*
n. capture cross section Einfangquerschnitt *m* für Neutronen, Neutroneneinfangquerschnitt *m*
n. chain reaction Neutronenkettenreaktion *f*
n. chamber Neutronenkammer *f*
n. concentration Neutronenkonzentration *f*
n. counter Neutronenzähler *m*, Neutronenzählrohr *n*
n. cross section Neutronenwirkungsquerschnitt *m*
n. current Neutronenstrom *m*
n. current density Neutronenstromdichte *f*
n. decay Neutronenzerfall *m*
n.-deficient isotope neutronenarmes Isotop *n*
n. density Neutronen[strom]dichte *f*
n. detection Neutronennachweis *m*
n. detector Neutronendetektor *m*, Neutronennachweisgerät *n*
n. diffraction Neutronenbeugung *f*
n. diffusion Neutronendiffusion *f*
n. distribution Neutronenverteilung *f*
n. dose Neutronendosis *f* *(Neutronen / cm^2)*
n. dosimeter Neutronendosismesser *m*, Neutronendosimeter *n*

n. emission Neutronenemission *f*
n. emitter Neutronenstrahler *m*
n. energy Neutronenenergie *f*
n. escape Neutronenausfluß *m*, Neutronenverlust *m*
n. evaporation Neutronenverdampfung *f*
n. excess Neutronenüberschuß *m*
n. excess number Neutronenüberschuß *m*
n. field Neutronenfeld *n*
n. flux Neutronenfluß *m*
n. hardening Zunahme (Erhöhung) *f* der mittleren Neutronenenergie
n.-induced neutronenausgelöst
n. interference Neutroneninterferenz *f*
n. irradiation Neutronenbestrahlung *f*
n. leakage Neutronenausfluß *m*, Neutronenverlust *m*
n. lifetime Neutronenlebensdauer *f*
n. mass Neutronenmasse *f*
n. moderator Neutronenmoderator *m*, Neutronenbremsmittel *n*
n. multiplication Neutronenvermehrung *f*, Neutronenmultiplikation *f*
n. optics Neutronenoptik *f*
n.-physical neutronenphysikalisch
n. physics Neutronenphysik *f*
n. poison effect Neutronenvergiftung *f (unerwünschte Neutronenabsorption des Reaktors)*
n. probe Neutronensonde *f*, Neutronenmeßkopf *m*
n. producer Neutronenerzeugungsreaktor *m*
n. production Neutronenerzeugung *f*
n.-proton force Neutron-Proton-Kraft *f*, Kernkraft *f*
n.-proton pair Neutron-Proton-Paar *n*
n. radiation Neutronenstrahlung *f*
n. ray Neutronenstrahl *m*
n. reaction Neutronenreaktion *f*
n. reflector Neutronenreflektor *m*, Reflektor *m (im Reaktor)*
n.-rich isotope neutronenreiches Isotop *n*
n. scattering Neutronenstreuung *f*
n. source Neutronenquelle *f*
n. spectrometer Neutronenspektrometer *n*
n. spectrometry Neutronenspektrometrie *f*
n. spectrum Neutronenspektrum *n*
n. transition Neutronenübergang *m*
n. transparency Neutronendurchlässigkeit *f*
n. trap Neutronenfalle *f*
n. vacancy Neutronenleerstelle *f*, Neutronenloch *n*
n. velocity Neutronengeschwindigkeit *f*
n. yield Neutronenausbeute *f (der Quelle)*
new neu, Neu-, unbekannt
n. building Neubau *m*
n. candle Candela *f (Grundeinheit der Lichtstärke)*
n.-cast / to umschmelzen, neu schmelzen
n.-dye / to *(Text)* auffärben
n. formation Neubildung *f*
n. lining *(Met)* Neuzustellung *f (des Ofens)*
n.-ploughed field Sturzacker *m*
n. sand Neusand *m (Gießerei)*
n.-type neuartig
newel *(Bw)* Treppenpfosten *m*; Antrittspfosten *m*; Austrittspfosten *m*
newjanskite *(Min)* Newjanskit *m*
newly discovered neuentdeckt
news felt Zeitungsdruckfilz *m*, Rotationsfilz *m*
n. print Zeitungs[druck]papier *n*
n. reader Nachrichtensprecher *m*
n. service Nachrichtendienst *m*
n. wrapper *(Typ)* Streifband *n*, Kreuzband *n*
newscast Nachrichtendienst *m*; Nachrichtensendung *f*
newscaster Nachrichtensprecher *m*
newspaper mailroom *(Typ)* Zeitungsexpedition *f*
newsprint Zeitungs[druck]papier *n*
newtape *(BMSR)* Frischband *n*, Frischstreifen *m*
newton Newton *n (Einheit der Kraft;* $1\,N = 1\,mkgs^{2}$*)*
Newtonian mechanics Newtonsche (klassische) Mechanik *f*
Newtonian telescope Newtons Spiegelteleskop *n*
Newtonian time Newtonscher Zeitbegriff *m*
newtonite *(Min)* Newtonit *m*
Newton's law of cooling Newtonsches Abkühlungsgesetz *n*
Newton's laws of motion Newtonsche Bewegungsgesetze *npl*
Newton's rings Newtonsche Ringe *mpl*
niacin *(Am) s.* nicotinic acid
nib Ende *n*, Meßende *n*; Spitze *f*; Stahlfeder *f*
nibble / to aushauen; ausschneiden *(Blech)*
nibbler *(Am)* Nibbelschere *f*
niccolic compound Verbindung *f* des dreiwertigen Nickel
niccolite Kupfernickel *n*, Rotnickelkies *m*
niccolous compound Verbindung *f* des zweiwertigen Nickel
niche Nische *f*
n. in a wall Mauernische *f*
Nichols' chart *(BMSR)* Nicholsdiagramm *n*
nichrome [steel] Chromnickelstahl *m*
n. wire Chromnickeldraht *m*
nichrotherm steel Nichrothermstahl *m*, wärmebeständiger Ni–Cr-Stahl *m*
nick / to [ein]kerben, schlitzen
nick Kerbe *f*, Einschnitt *m*, Schlitz *m*; Spanbrechernut *f*; Fallkerb *m*
n. bend test *(Met)* Kerbbiegeprobe *f*
nicked-cutting edge Schneide *f* mit Spanbrechernuten
nickel Nickel *n*
n. alloy Nickellegierung *f*

n. anode Nickelanode *f (Galvanotechnik)*
n. antimony glance Nickelantimonkies *m*
n. arsenic glance Arsennickelkies *m*
n. bath Nickelbad *n*, Vernickelungsbad *n (Galvanotechnik)*
n. bloom *(grüne)* Nickelblüte *f*, Nickelocker *m*
n. brass *(Am)* Neusilber *n*
n.-chrome steel Chromnickelstahl *m*
n.-clad nickelplattiert
n. coin Nickelmünze *f*
n. cube Würfelnickel *n*
n. glance Arsennickelkies *m*
n.-iron battery Nickel-Eisen-Batterie *f*
n. ore Nickelerz *n*
n. pellets *(Met)* Kugelnickel *n*
n. plated [galvanisch] vernickelt
n.-plating Vernickeln *n (galvanisch)*
n. pyrite Nickelblende *f*
n. sheet Nickelblech *n*
n. silver Neusilber *n*
n. speiss *(Met)* Nickelspeise *f*
n. steel nickellegierter Stahl *m*, Nickelstahl *m*
nickelic compound Verbindung *f* des dreiwertigen Nickel
nickeliferous nickelhaltig
nickelize / to vernickeln
nickelous compound Verbindung *f* des zweiwertigen Nickel
nicking Kerbung *f*, Einkerben *n*, Einarbeiten *n* von Kerben
niclad nickelplatiertes Stahlblech *n*
Niclausse boiler *(Wärm)* Wasserrohrkessel *m* mit geteilten Wasserkammern an einem Ende der Rohre
Nicol prism *(Opt)* Nicol[sches Prisma] *n*
nicollite *(Min)* Nickelin *m*, Rotnickelkies *m*
nicopyrite *(Min)* Pentlandit *m*
nicotine Nikotin *n*
nicotinic acid Nikotinsäure *f*
n. amide Nikotinsäureamid *n*
niello Niello *n (Metallverzierung)*
nigger heads *(Geol)* Kugelkohle *f*
nigging chisel Scharriereisen *n*
night Nacht *f*
n.-alarm switch Nacht[alarm]schalter *m*
n.-blind nachtblind
n.-blindness Nachtblindheit *f*
n. bomber Nachtbombenflugzeug *n*, Nachtbomber *m*
n. call *(Fmt)* Nachtruf *m*
n. current *(El)* Nachtstrom *m*
n. effect *s.* n. error
n. error Nachteffekt *m*, Nachtfehler *m*, Polarisationseffekt *m*, Polarisationsfehler *m (Radar)*
n. fighter Nachtjagdflugzeug *n*
n.-fighter detection apparatus Nachtjagdwarngerät *n (Radar)*
n. frequency *(El)* Nachtfrequenz *f*
n. glass *(Opt)* Nachtglas *n*
n. illumination Nachtbeleuchtung *f*
n. lamp Nachtlampe *f*
n. load *(El)* Nachtbelastung *f*
n. photograph Nachtaufnahme *f (Foto)*
n. position *(Fmt)* Nachtplatz *m*
n. practice *(Mil)* Nachtschießen *n*
n. range Nachtreichweite *f (Funk)*
n. school Abendschule *f*
n. service Nachtdienst *m*
n. shift Nachtschicht *f*
n. target practice *(Mil)* Nachtschießen *n*
n. tariff *(Fmt)* Nachttarif *m*
n. wave Nachtwelle *f (Funk)*
nightwear Nachtkleidung *f*
nigrosine Nigrosin *n*
nine-point circle Feuerbachscher Kreis *m*, Neunpunktekreis *m (eines Dreiecks)*
nine's complement *(BMSR)* Neunerkomplement
ninesided figure Neuneck *n*
niobate Niobat *n*, Salz *n* der Niobsäure
niobic acid Niobsäure *f*
niobium Niob *n*
nip / to *s.* n. off / to
n. off abkappen; abköpfen; abkneifen, abzwicken
nip 1. *(Text)* Kammspiel *n*; 2. *(Geol)* Verdrückung *f*
n. angle Einzugswinkel *m*
n. of the feed rollers *(Text)* Klemmstelle *f* der Speisewalzen
n. roll Haltewalze *f*
Nipkow disk *(Fs)* Nipkow-Scheibe *f*
nipped in Eisnot *(ein in Eis eingeschlossenes Schiff)*
nipper Kneifzange *f*
n. temple *(Text)* Klemmspannstock *m*
nippers Ziehzange *f*, Mitnehmer *m*; Kneifzange *f*, Beißzange *f*; Zange *f (Kämmaschine)*
nipping effect *(Text)* Klemmwirkung *f*
nipple Nippel *m*, Anschlußstück *n*; Gußwarze *f*; Ansatz *m*; Schraubenverbindung *f*
n. for pressure gauge Manometeranschlußstück *n*
n. plate *(Met)* Warzenblech *n*
niter *(Am) s.* nitre
niton *s.* radon
nitralloy Nitrierstahl *m*
nitramide *(Ch)* Nitramid *n*
nitratable nitrierbar
nitrate / to nitrieren
nitrate Nitrat *n*, Salz *n* der Salpetersäure
n. explosive Nitratsprengstoff *m*, Sprengsalpeter *n*
n. fertilizer Nitratdünger *m*
n. film Nitratfilm *m (Foto)*
nitratine Chilesalpeter *m*
nitrating acid Nitriersäure *f*
n. apparatus Nitrierapparat *m*

n. cotton Nitrozellulose *f*, Schießbaumwolle *f*
n. plant Nitrieranlage *f*
nitration *(Met)* Nitrierung *f*, Aufstickung *f*
n. furnace Nitrierofen *m*
n. paper Nitrierpapier *n*
n. process Nitrierhärteverfahren *n*
nitre Salpeter *m*, Kaliumnitrat *n*
n. in bars Stangensalpeter *m*
nitric stickstoffhaltig, Salpeter-, Stickstoff-
n. acid Salpetersäure *f*
n. anhydride Distickstoffpentoxid *n*, Salpetersäureanhydrid *n*
n. nitrogen Nitratstickstoff *m*
n. oxide Stick[stoff]oxid *n*
nitridation *(Ch)* Nitridbildung *f*
nitride / to *(Met)* nitrieren, aufsticken
nitride Nitrid *n*
n. hardening *(Met)* Nitrierhärtung *f*
nitrided steel Nitrierstahl *m*
nitriding *(Met)* Nitrierung *f*, Nitrierhärtung *f*, Aufstickung *f*
n. box Nitrierkasten *m*
n. furnace Nitrierofen *m*
n. process Nitrierhärteverfahren *n*
n. steel Nitrierstahl *m*
nitrifying bacteria Nitrifikationsbakterien *fpl*
nitrile *(Ch)* Nitril *n*, Blausäureester *m*
nitro acid sulfite Nitrosylschwefelsäure *f*
n. compound Nitroverbindung *f*
n. derivative *(Ch)* Nitroderivat *n*
n.-dye Nitrofarbstoff *m*
n. explosive Nitrosprengstoff *m*
n. gelatine Nitrogelatine *f*, Sprenggelatine *f*, Sprenggummi *m*
n. powder rauchschwaches Schießpulver *n*
n. starch Nitrostärke *f*
nitrobarbituric acid Nitrobarbitursäure *f*
nitrobarite Barytsalpeter *m*
nitrobenzaldehyde Nitrobenzaldehyd *m*
nitrobenzene Nitrobenzol *n*, Bittermandelöl *n*, Mirbanöl *n*
nitrobenzoic acid Nitrobenzoesäure *f*
nitrocalcite Nitrokalzit *m*, salpetersaurer Kalk *m*
nitrocarbol Nitromethan *n*
nitrocellulose *(Ch)* Nitrozellulose *f*, Schießbaumwolle *f*, Kollodiumwolle *f*
n. lacquer Nitrozelluloselack *m*
n. silk Nitrozelluloseseide *f*, Nitratseide *f*
nitrocopper Nitrokupfer *n*
nitrogen Stickstoff *m*
n. bridge Stickstoffbrücke *f*
n. case hardening *(Met)* Nitrierhärtung *f*
n. content Stickstoffgehalt *m*
n. cycle Stickstoffkreislauf *m*
n. estimation Stickstoffbestimmung *f*
n. fertilizer Stickstoffdünger *m*
n. harden / to *(Met)* nitrierhärten
n. hardening *(Met)* Nitrierhärtung *f*
n. monoxide Stickstoff[mon]oxid *n*, Stickoxid *n*, Lachgas *n*
n. oxide Stickoxid *n*
n. pentoxide Distickstoffpentoxid *n*, Salpetersäureanhydrid *n*
n. requirement of soil Stickstoffbedarf *m* des Bodens
n. trioxide Distickstofftrioxid *n*
nitrogenization Nitrierung *f*
nitrogenous fertilizer Stickstoffdünger *m*
nitroglycerine Nitroglyzerin *n*
nitroglycerol *(Am)* *s.* nitroglycerine
nitrogroup Nitrogruppe *f*
nitrolic acid Nitrolsäure *f*
nitrolime Kalkstickstoff *m*
nitromagnesite *(Min)* Nitromagnesit *m*
nitrometer Nitrometer *n*, Azotometer *n*
nitromuriatic acid Königswasser *n*
nitron *(Plast)* Nitron *n*
nitronic acid Nitronsäure *f*
nitrophenol Nitrophenol *n*
nitrosat *(Ch)* Nitrosat *n*
nitroso blue Nitrosoblau *n*
n. dyestuff Nitrosofarbstoff *m*
nitrotoluene Nitrotoluol *n*
nitrous Stickstoff-, salpetrig
n. acid salpetrige Säure *f*
n. anhydride Distickstofftrioxid *n*, Salpetrigsäureanhydrid *n*
n. compound Verbindung *f* des dreiwertigen Stickstoff
n. earth Salpetererde *f*
n. gases nitrose Gase *npl*
n. oxide Distickstoffoxid *n*, Lachgas *n*
nitroxanthic acid Pikrinsäure *f*
nitroxylene Nitroxylol *n*
nitryl chloride Nitrylchlorid *n*
nivellization Nivellierung *f*
n-n junction nn-Übergang *m* *(Halbleitertechnik)*
no-backlash *s.* no-clearance
no-bond tensioning Vorspannung *f* ohne Verbund *(Spannbetonbau)*
no-clearance *(Masch)* spielfrei; ohne Spiel
no-delay call *(Fmt)* Schnellgespräch *n*
no-delay circuit *(Fmt)* Schnellverkehrsleitung *f*
no-delay exchange *(Fmt)* Schnellamt *n*
no-delay operation *(Fmt)* Sofortverkehr *m*, Schnellverkehr *m*
no-delay service *(Fmt)* Sofortverkehr *m*, Schnelldienst *m*
no-delay working *(Fmt)* Sofortverkehr *m*, Schnellverkehr *m*
no-delivery position Nullhubstellung *f* *(Pumpe)*
no-field track *(Kern)* Nebelspur *f* in feldfreier Nebelkammer
no-fines concrete Einkornbeton *m*
no-go Ausschuß-
no-go limit Ausschußgrenze *f*

no-hang-up service *(Fmt)* Sofortverkehr *m*
no-load Leerlauf *m*; unbelasteter Zustand *m*
no-load adjustment Leerlaufeinstellung *f*
no-load characteristic *(Elm)* Leerlaufkennlinie *f*
no-load circuit-breaker Null[aus]schalter *m*, Nullstromschalter *m*
no-load condition Leerlaufzustand *m*, unbelasteter Zustand *m*
no-load connection Leerlaufschaltung *f*
no-load consumption Leerlaufverbrauch *m*
no-load current Leerlaufstrom *m*
no-load cut-out Leerlaufausschalter *m*, Nullausschalter *m*
no-load excitation Leerlauferregung *f*
no-load field Leerlauffeld *n*
no-load friction Leergangreibung *f*
no-load loss Leer[lauf]verlust *m*, Verlust *m* bei Leerlauf
no-load magnetizing current Leerlaufmagnetisierungsstrom *m*
no-load nozzle Leerlaufdüse *f (Turbine)*
no-load operation Leerlauf *m*
no-load point Leer[lauf]punkt *m*
no-load power Leerlaufleistung *f*
no-load resistance *(El)* Leerlaufwiderstand *m (Vierpol)*
no-load speed Leerlaufdrehzahl *f*
no-load stroke *(Kfz)* Leer[lauf]hub *m*
no-load voltage *(El)* Leerlaufspannung *f*
no-load work Leerlaufarbeit *f*
no-loss verlustlos
no-loss capacitor *(El)* verlustloser Kondensator
no-loss dielectric verlustloses (ideales) Dielektrikum *n*, idealer Isolierstoff *m*
no-loss insulant *(El)* verlustloses (ideales) Isoliermittel *n*
no-parallax parallaxenfrei *(Meßinstrument)*
no-pressure resin drucklos härtendes Harz *n*
no roll ohne Wälzung
no-rub wax gleitsicheres (rutschfestes) Wachs *n*
no-rust vapour wrapper imprägniertes Papier *n (als Korrosionsschutz)*
no-signal bias Gittergleichspannung *f*, Gittervorspannung *f*
no-signal current Ruhestrom *m*
no-stroke position Nullhubstellung *f (Pumpe)*
no-transmission interval Sendepause *f*
no-volt coil Unterspannungsspule *f*
no-volt release Nullspannungsauslösung *f*, Ruhestromauslöser *m*
no-voltage Nullspannung *f*
no-voltage circuit breaker Nullspannungsschalter *m*, Nullspannungsauslöser *m*
no-voltage relay Minimalspannungsrelais *n*, Nullspannungsrelais *n*
no-voltage release Nullspannungsauslöser *m*, Ruhestromauslöser *m*; Ruhestromauslösung *f*, Nullspannungsauslösung *f*

noble edel, kostbar
n. earths seltene Erden *pl*
n. gas Edelgas *n*
n. metal Edelmetall *n*
n. metal thermocouple edles Thermopaar *n*
Noble comb *(Text)* Noblesche Kämmaschine *f*
noctoviser Nachtsichtgerät *n*
noctovision Fernsehen *n* im Dunkeln mittels Infrarotstrahlung, Nachtfernsehen *n*
nodal *(Math)* Knoten-
n. cementite *(Met)* kugeliger Zementit *m*
n. diagram Knotendarstellung *f*
n. fitting Knotenpunktbeschlag *m (Flugzeug)*
n. line Knotenlinie *f*, Knotenachse *f*
n. plane Knotenfläche *f*
n. point Knotenpunkt *m*, Schwingungsknoten *m*
n. point of vibration Schwingungsknoten *m*
n. section Knotenschnitt *m*
nodding Nickbewegung *f (des Traktors)*
node Knoten[punkt] *m*, Schwingungsknoten *m*, Wellenknoten *m*; Nullpunkt *m*
nodistron *(El)* Nodistron *n (Glimmlampe mit 10 Katoden; Ziffernanzeige)*
nodular kugelig, kugelartig
n. cast iron Gußeisen *n* mit Kugelgraphit, Kugelgraphitguß *m*
n. graphite Kugelgraphit *m*
nodulation Knotenbildung *f*
nodule Granalie *f*, Kugel *f*; Knollen *m*, Knötchen *n*, Niere *f*; Druse *f*
n. of ore *(Geol)* Erzniere *f*
nodulized graphite cast iron Kugelgraphit-
nodulizing Agglomerieren *n*
n. plant Agglomerieranlage *f*
nog 1. Holzdübel *m*, Holzpflock *m*, Holzstift *m*; 2. *(Brg)* Keil *m*, Spreize *f (Schram)*
nogging Ausmauern *n* [zwischen Holzbalken]
noil *(Text)* Kämmling *m*
n. stripping comb Hacker *m*, Abstreifkamm *m*, Abnehmerkamm *m (Kämmaschine)*
noise Lärm *m*, Geräusch *n*; Rauschen *n*; regellose Störungen *fpl*, Störgeräusch *n*
n. abatement Geräuschunterdrückung *f*, Lärmbekämpfung *f*
n. background Rauschuntergrund *m*, Eigenrauschen *n*
n. behind the signal *(Fmt)* Modulationsrauschen *n*
n. cancellation *(El)* Rauschunterdrückung *f*; Störschallunterdrückung *f*
n. characteristics Geräuschverhalten *n*
n. component Rauschanteil *m*
n. conductance Rauschleitwert *m*
n. conduction Schalleitung *f*
n. control Lärmbekämpfung *f*
n. current Störstrom *m*, Rauschstrom *m*
n. diode Rauschdiode *f*
n. effect Rauscheffekt *m*

n. equivalent Rauschäquivalent *n*
n. equivalent power äquivalente Rauschleistung *f*
n. factor Rauschfaktor *m*
n. field *(El)* Störfeld *n*
n. figure Rauschfaktor *m*, Rauschzahl *f*; Rauschpegel *m*
n. filter *(El)* Störschutzfilter *n*; *(niederfrequente)* Geräuschsperre *f*
n. generator Rauschsender *m*
n. guards Lärmschutzmittel *npl*
n. input Eingangsrauschen *n*
n. killer Entstörnetzwerk *n*; Krachtöter *m*; Störschutzfilter *n*
n. level Störpegel *m*, Störspiegel *m*, Geräuschspiegel *m*, Störniveau *n*; Lautstärke *f* des Lärms
n. level measuring Fremdspannungsmessung *f*
n. limitation Störbegrenzung *f*
n. limiter Störbegrenzer *m*
n. matching Rauschanpassung *f*
n. meter Fremdspannungsmesser *m*, Geräuschmesser *m*; Lautstärkemesser *m*
n. number Rauschzahl *f*
n. output Rauschleistung *f (Antennen)*
n. pick-up Aufnahme *f* von Rauschstörungen
n. potential Rauschspannung *f*
n. reduction Lärmbekämpfung *f*, Lärmminderung *f*
n. silencer Krachtöter *m*, Störungsaustastung *f (bei Empfängern)*
n. source Lärmquelle *f*
n.-standard temperature Rauschbezugstemperatur *f*
n. suppression Entstörung *f*, Störschutz *m*; Geräuschbekämpfung *f*, Rauschunterdrückung *f*
n. suppression circuit *(El)* Geräuschunterdrücker *m*
n.-suppression device *(El)* Störschutz *m*
n. suppressor Geräuschfilter *n*, Krachtöter *m*, Rauschfilter *n*
n. thermometer Rauschthermometer *n (Gerät zur Messung des Verhältnisses absoluter Temperaturen)*
n. transmission impairment Minderung *f* der Übertragungsgüte durch Leitungsgeräusche
n. trap *(Fmt)* Störbeseitiger *m*
n. tuning Rauschabstimmung *f*
n. voltage Fremdspannung *f*; Störspannung *f*; Rauschspannung *f*
noiseless geräuschlos
noiselessness Rauscharmut *f*, Rauschfreiheit *f*
noisy geräuschvoll
n. valve *(El)* stark rauschende Röhre *f*
nomenclature Benennung *f*; Terminologie *f*; Nomenklatur *f*
n. of frequencies Benennung *f* der Frequenzen
nominal Nominal-, nominell; gering
n. allowance Nennabmaß *n*
n. black signal *(Fs)* Testsignal *n* für Schwarz
n. capacity Nennleistung *f*, Solleistung *f*
n. contents Sollgehalt *m*
n. cut-off frequency *(Fmt)* Nenngrenzfrequenz *f*
n. diameter Nenndurchmesser *m*, Nennweite *f*
n. diameter of thread Gewindenenndurchmesser *m*, Außendurchmesser *m*
n. dimension Nennmaß *n*, Richtmaß *n*, Sollmaß *n*, Normalmaß *n*
n. frequency *(El)* Nennfrequenz *f*
n. handling rate theoretische Geräteleistung *f (Bagger)*
n. line width *(Fs)* Zeilennennlänge *f*
n. load-bearing capacity *(Brg)* Nennlast *f (Ausbau)*
n. maximum circuit fiktiver Bezugskreis *m (Sendetechnik)*
n. measure Rauhmaß *n (für Holz)*
n. measurement *(Bw)* Nennmaß *n*, Richtmaß *n*
n. output Normalleistung *f*, Solleistung *f*, Nennleistung *f*
n. rating Nennleistung *f*
n. reinforcement konstruktive Bewehrung *f (Stahlbetonbau)*
n. size Nennmaß *n*, Sollmaß *n*, Normalmaß *n*, Nenndurchmesser *m*
n. speed Nenndrehzahl *f*
n. stress Nennspannung *f*, Nennbeanspruchung *f (Festigkeit)*
n. thread size Gewindenenndurchmesser *m*
n. value Nennwert *m*, Sollwert *m*, Nominalwert *m*
n. voltage *s.* rated voltage
n. voltage rating *(El)* Nennspannung *f*
n. white signal *(Fs)* Testsignal *n* für Weiß
nomogram *(Math)* Nomogramm *n*; Fluchtlinientafel *f*, Leitertafel *f*; Rechentafel *f*
nomograph Nomogramm *n*, Nomograf *m*
nomography Nomografie *f*
non-abrasive quality Abriebfestigkeit *f*
non-abrasiveness Abriebfestigkeit *f*
non-absorbing nichtabsorbierend
non-acid tannage *(Led)* nichtsaure Gerbung *f*
non-activated inaktiv, nichtaktiviert
non-additive type motor oil unlegiertes Motorenöl *n*
non-adiabatic nichtadiabatisch
non-adjustable nichtverstellbar
non-aging alterungsbeständig
non-aging property Alterungsbeständigkeit *f*
non-alloy steel nichtlegierter Stahl *m*
non-appealable judgment sofort rechtskräftiges Urteil *n*

non-arcing *(El)* funkenfrei
non-association cable nicht normgerechtes (den Standards entsprechendes) Kabel *n*
non-attenuated ungedämpft
non-baking coal Magerkohle *f*, Sandkohle *f*
non-bearing nichttragend
non-black nichtschwarz
non-bleed pneumatic relay pneumatischer Verstärker *m* ohne dauernden Luftverbrauch, nichtabblasender pneumatischer Verstärker
non-breakable unzerbrechlich
non-buckling *(Met)* knickfest
non-caking coal Magerkohle *f*, Sandkohle *f*
non-canned motor offener Motor *m*
non-capacitive *(El)* kapazitätsfrei
non-capsizable *(Schiff)* unkenterbar
non-carbonate hardness *(Bw)* bleibende Härte *f*, Mineralsäurehärte *f*, Nichtkarbonathärte *f*
non-causality *(Ph)* Akausalität *f*
non-cavitating flow *(Schiff)* nicht kavitierende Strömung *f*
non-charge-cooled engine Motor *m* ohne Ladeluftkühlung
non-chipping spanlos
non-circular nichtrund, nichtkreisförmig, unrund
non-circulating lubrication *(Kfz)* Durchlaufschmierung *f*, Verlustschmierung *f*
non-coherent scattering *(Ph)* inkohärente (nichtkohärente) Streuung *f*
non-cohesive *(Brg)* rollig
non-cohesive soil *(Bw)* nichtbindiger (rolliger) Boden *m*
non-combustible nicht brennbar, nicht entzündbar
non-compressed fibreboard nicht vorgepreßte Faserplatte *f*
non-condensing [steam] engine Auspuffdampfmaschine *f*, Dampfmaschine *f* ohne Kondensation
non-conducting nichtleitend
non-conducting material Isoliermittel *n*, Wärmeschutzmittel *n*, Wärmeschutzmasse *f*, Isoliermasse *f*
non-conducting material for covering the top of the boiler Kesselabdeckungsmaterial *n*, Deckenfüllung *f*, Kesselisolierung *f*
non-conductive *s.* non-conducting
non-conductor Nichtleiter *m*, Isolator *m*, Dielektrikum *n*
non-conformity *(Geol)* Diskordanz *f*
non-coniferous wood Hartholz *n*, Laubholz *n*
non-consumable nicht abschmelzbar
non-consumable [tungsten] electrode inert-gas welding Wolfram-Inertgasschweißen *n*
non-contact berührungsfrei
non-contact gauging berührungsloses Messen *n*
non-contact thickness gauge berührungslose Dickenmeßeinrichtung *f*
non-contacting berührungsfrei
non-continuous chip material kurzspanender Werkstoff *m*
non-controlled ungesteuert, nicht kontrolliert
non-cooled ungekühlt
non-corrective lens neutrales Brillenglas *n*
non-corrodibility Korrosionsbeständigkeit *f*
non-corrodible korrosionsbeständig, korrosionsfest, nichtkorrodierend
non-corroding *s.* non-corrodible
non-corrosive nicht ätzend; säurefrei; nicht korrodierend
non-corrosive grease säurefreies Fett *n*
non-corrugated riffelfrei
non-creasing knitterfest
non-critical nichtkritisch *(Reaktor)*
non-crucible furnace *(Met)* tiegelloser Ofen *m*
non-crucible tilting furnace tiegelloser Kippofen *m*
non-crush carpet nicht zerdrückbarer Teppich *m*
non-cutting spanlos
non-cutting shaping spanlose Formgebung *f*; Umformung *f*
non-cutting time *(Ök)* Totzeit *f*
non-cyclical field *(El)* wirbelfreies Feld *n*
non-cylindrical nicht zylindrisch, unrund
non-damped ungedämpft *(z. B. eine Welle)*
non-dazzling blendungsfrei
non-deformed fracture *(Met)* verformungsloser Bruch *m*
non-deforming steel verzugsfreier Stahl *m*
non-degenerated *(Math)* nichtausgeartet, nichtentartet
non-degenerated state nichtentarteter Zustand
non-delayed fuse *(El)* flinke (unverzögerte) Sicherung *f*
non-destructive 1. *(Met)* zerstörungsfrei; 2. *(Dat)* nicht löschend
non-destructive material testing zerstörungsfreie Werkstoffprüfung *f*
non-destructive reading nichtlöschendes (zerstörungsfreies) Lesen *n (Rechnerspeicher)*
non-destructive test[ing] zerstörungsfreie Prüfung *f*
non-detergent fatty matter Unverseifbares *n*
non-detonating quality Klopffestigkeit *f (Treibstoff)*
non-diathermic atherman
non-dimensional dimensionslos; *(Dat)* normiert
non-dimensional quantity dimensionslose Größe *f*, dimensionslose Zahl *f*
non-direction beacon *(El)* Funkfeuer *n* ohne Richtwirkung
non-directional richtungsunabhängig, ungerichtet *(Antenne)*

non-directional aerial Rundstrahlantenne *f*, ungerichtete Antenne *f*
non-directional counter *(Kern)* richtungsunabhängiger Zähler *m*
non-directional current protection ungerichteter (nicht gerichteter) Stromschutz *m*
non-directional microphone Kugelmikrofon *n*
non-directive ungerichtet
non-directive aerial Rundstrahlantenne *f*, ungerichtete Antenne *f*
non-dissipative stub *(El)* Blindleitung *f* zur Anpassung, Hohlleiteranpaßglied *n*
non-distorting verzugsfrei, verzerrungsfrei
non-draining cable Druckkabel *n*
non-durable finish *(Text)* nichtbeständige Ausrüstung *f*
non-eddying wirbelfrei
non-electric power mechanische Leistung *f*
non-elevating ohne Höhenverstellung *(Radialbohrmaschine)*
non-equilibrium instabil
non-equilibrium state instabiler Zustand *m*
non-erasable storage *(Dat)* unlöschbare (nicht löschbare) Speicherung *f*
non-Euclidian geometry nichteuklidische Geometrie *f*
non-exposed unbelichtet *(Film)*
non-feathering propeller fester (nicht verstellbarer) Propeller *m*
non-ferrous Nichteisen-, NE-
non-ferrous alloy NE-Legierung *f*, Metallegierung *f*
non-ferrous castings Metallguß *m*
non-ferrous foundry Metallgießerei *f*
non-ferrous metal NE-Metall *n*, Nichteisenmetall *n*
non-ferrous milling Buntmetallfräsen *n*, Fräsen *n* von Nichteisenmetallen
non-fiery *(Brg)* schlagwetterfrei
non-fissile *(Kern)* nichtspaltbar
non-fission capture *(Kern)* spaltungsloser Einfang *m*
non-fissionable *(Kern)* nichtspaltbar
non-foaming nichtschäumend, schaumlos
non-free point *(Ph)* unfreier Punkt *m*
non-freezing kältebeständig
non-fusing arc welding electrode nicht abschmelzende Elektrode *f*
non-fusion unvollständige Verschweißung *f*
non-gasifiable nichtvergasbar
non-gassing coal gasarme Kohle *f*
non-gassy mine schlagwetterfreie Grube *f*
non-grinding time in the grinding cycle Nebenzeit *f* beim Schleifen
non-guided satellite nichtgesteuerter Satellit *m*
non-halating *(Opt)* lichthoffrei
non-hardening steel nichthärtbarer (nichthärtender) Stahl *m*
non-harmonic unharmonisch
non-hydraulic mortar Luftmörtel *m*
non-hydrolyzable nichthydrolysierbar
non-ideal gas reales (nichtideales) Gas *n*
non-inductive induktionsfrei, induktionslos, induktivitätsfrei
non-inductive capacitor induktionsfreier Kondensator *m*
non-inductive resistance induktionsfreier (induktivitätsfreier, induktionsloser, reiner) Widerstand *m*, Bifilarwiderstand *m*
non-inductive shunt *(El)* induktionsfreier Nebenschluß *m*
non-inductive winding induktionsfreie (bifilare) Wicklung *f*
non-inflammable unentflammbar, flammsicher, unbrennbar
non-interchangeable fuse *(El)* unverwechselbare Sicherung *f*
non-interchangeable plug *(El)* unverwechselbarer Stecker *m*
non-ionizing nichtionisierend
non-ionizing radiation nichtionisierende Strahlung *f*
non-iron *(Text)* bügelfrei, bügelarm
non-irradiated nichtbestrahlt, unbestrahlt; nichtbelichtet, unbelichtet *(Film)*
non-irrigated soil nichtbewässertes Land *n*
non-isothermal nichtisotherm
non-isotopic *(Kern)* nichtisotop
non-isotropic anisotrop
non-kinkable flex *(Fmt)* Gummischlauchleitung *f*
non-lifting injector Injektor *m* mit zufließendem Wasser, nichtsaugender Injektor
non-linear nichtlinear
non-linear coil *(El)* Eisenspule *f*, nichtlineare Drossel *f*
non-linear distortion *(Fmt)* nichtlineare Verzerrung *f*, Klirrverzerrung *f*
non-linear distortion factor *(Fmt)* Klirrfaktor *m*
non-linear field theory nichtlineare Feldtheorie *f*
non-linear-resistance arrester *(El)* Ventilableiter *m*
non-linearity Nichtlinearität *f*
non-lined construction wickellose Bauweise *f* *(einer Batterie)*
non-linting nicht fasernd *(Baumwolle)*
non-load-bearing *(Bw)* nichttragend
non-loaded unbelastet; unbespult
non-loaded Q Leerlaufgütefaktor *m (einer elektrischen Impedanz)*
non-local nichtlokal
non-local field theory nichtlokale Feldtheorie *f*
non-locking key nicht einrastender Schalter *m*
non-machinable unzerspanbar
non-machining time Nebenzeit *f*

non-magnetic antimagnetisch, nichtmagnetisch, unmagnetisch
non-magnetic scattering nichtmagnetische Streuung *f*
non-magnetic vessel unmagnetisches Schiff *n*
non-metal Nichtmetall *n*
non-metallic nichtmetallisch
non-metallic element Nichtmetall *n*, Metalloid *n*
non-metallic inclusions nichtmetallische Einschlüsse *mpl*
non-metallic mineral industry Steine- und Erdenindustrie *f*
non-metallic minerals Steine *mpl* und Erden *fpl*; nichtmetallische Mineralien *npl*, mineralische Rohstoffe *mpl*
non-metallic sheathed cable Isoliermantelkabel *n*
non-metering *(Fmt)* Zählunterdrückung *f*
non-metering relay Zählverhinderungsrelais *n*
non-mobile stationär, ortsfest
non-orbital nichtkreisförmig
non-oscillating schwingungsfrei
non-oscillatory instability *(BMSR)* monotone Instabilität *f*
non-oxidizing annealing furnace *(Met)* Blankglühofen *m*
non-oxidizing annealing oil Blankhärteöl *n*
non-parallel nicht parallel
non-parallel equal crank Antiparallelkurbel *f*
non-penetrating nichtdurchdringend; energiearm *(Strahlung)*
non-penetrating particle *(Kern)* energiearmes Teilchen *n*
non-periodic aperiodisch, nichtperiodisch
non-permitted explosive Gesteinssprengstoff *m*
non-polar apolar, unpolar
non-polar bond unpolare Bindung *f*
non-polar generator Unipolargenerator *m*, Unipolarmaschine *f*
non-polarized armature *(El)* unpolarisierter Anker *m*
non-polarized electrolytic capacitor ungepolter Elektrolytkondensator *m*
non-polarized relay ungepoltes (unpolarisiertes, neutrales) Relais *n*
non-positive kraftschlüssig
non-premium motor oil unlegiertes Motoröl *n*
non-pressure welding Schmelzschweißung *f*
non-prestressed reinforcement schlaffe Bewehrung *f (Stahlbeton)*
non-productive unproduktiv
non-productive time Nebenzeit *f*
non-propelled vessel Schiff *n* ohne Eigenantrieb
non-proportional plastisch *(Verformung)*
non-quadded cable *(El)* mehrpaariges Kabel *n* ohne Vierer
non-quantized system *(El)* nicht quantisiertes System *n*, klassisches System
non-radiative strahlungslos *(Auger-Effekt)*
non-radioactive inaktiv, stabil, nichtradioaktiv
non-radioactive tracer stabiles Leitisotop *n*
non-ravel *(Am) (Text)* laufmaschensicher
non-reactive induktionsfrei; rückkopplungsfrei, ohne Rückkopplung; rückwirkungsfrei
non-reactive load ohmsche Belastung *f*; Belastung durch Wirkwiderstand
non-reactive power Wirkleistung *f*
non-reactive resistance induktionsfreier (reiner) Widerstand *m*
non-real *(Math)* nichtreell
non-recoverable rocket nicht rückgewinnbare Rakete *f*
non-recovery lubrication Durchlaufschmierung *f*, verlorene Schmierung *f*
non-recurrent triggering *(Fs)* einmalige Auslösung *f (der Zeitbasis)*
non-reflecting termination angepaßter Abschlußwiderstand *m*, reflexionsfreier Abschluß *m (Wellenleiter)*
non-reflection attenuation *(El)* Anpassungsdämpfung *f*
non-relativistic nichtrelativistisch *(z. B. Teilchen)*
non-removable discontinuity *(Math)* nicht hebbare Unstetigkeit *f*
non-renewable cartridge-fuse *(El)* Verbrauchssicherung[spatrone] *f*, Patronensicherung *f (für einmaligen Gebrauch)*
non-resinous harzfrei
non-resonance vibrating screen Resonanzfreischwingsieb *n*
non-resonant slot unabgestimmter Schlitzstrahler *m (Radar)*
non-resonating aerial verstimmte Antenne *f*
non-return ball valve Rückschlagkugelventil *n*
non-return-flow wind tunnel *(Flg)* offener Windkanal *m*
non-return-to-zero *(BMSR)* Nichtrückkehr *f* zu Null
non-return valve Rückschlagventil *n*
non-reversible nicht umsteuerbar
non-reversible mill nicht umkehrbare Walzstraße *f*, irreversible Walzstraße
non-reversing nicht umsteuerbar, ohne Umkehr der Drehrichtung
non-rigid weich[gestellt], nicht starr
non-rigid airship Pralluftschiff *n*
non-rigid plastic weiches Kunststoffmaterial *n*
non-rotating nicht umlaufend
non-rotating crank Schiefscheibe *f (Pumpe)*
non-rotating hoisting rope drallfreies Hubseil *n*
non-rotational field *(Ph)* wirbelfreies Feld *n*
non-rusting rostfrei, nichtrostend *(Stahl)*
non-salient pole rotor *(Elm)* Trommelläufer *m*, Trommelinduktor *m*

non-saturated ungesättigt
non scaling *(Met)* zunderbeständig, zunderfrei
non-scaling property *(Met)* Zunderbeständigkeit *f*
non-screwed ohne Gewinde
non-selective diffuser *(Ph)* nichtselektiv streuender Körper *m*
non-selective radiator nichtselektiver (grauer) Strahler *m*
non-self-maintained discharge *(El)* unselbständige Entladung *f*
non-separable untrennbar; nichtseparierbar, nichttrennbar
non-separator ohne Trennelement
non-shattering glass Sicherheitsglas *n*
non-shorting contact switch Stufenschalter *m* *(bei dem beim Umschalten die nebeneinanderstehenden Kontakte nicht kurzgeschlossen werden)*
non-shrinkable *(Text)* krumpfecht
non-silicious siliziumfrei
non-skid gleitsicher, griffig, rutschfest, mit Gleitschutz, profiliert *(Reifen)*
non-skid chain Gleitschutzkette *f*, Schneekette *f*, Spurkette *f*
non-skid device Gleitschutz *m*
non-skid flooring rutschfester Bodenbelag *m*
non-skid pattern rutschsicheres Reifenprofil *n*
non-skid surface rutschsichere Oberfläche *f*
non-skid tyre rutschsicherer Reifen *m*
non-slip gleitsicher, griffig, rutschfest; zwangsläufig, schlupffrei *(Antrieb)*
non-slip point *(Met)* Fließscheide *f (Walzen)*
non-solubility Nichtmischbarkeit *f*
non-solvent nicht lösendes Verdünnungsmittel *n*
non-sparking funkenfrei, funkenlos, funkensicher
non-spherical lens *(Opt)* asphärische Linse *f*
non-spill coupling selbstabdichtende Schlauchkupplung *f*
non-spinning 1. *(Flg)* trudelsicher; 2. drallfrei *(Seil)*
non-staining rostbeständig
non-starting motor *(Elm)* Anwurfmotor *m*
non-stationary flow *(Strm)* nichtstationäre Strömung *f*
non-stayed mast *(Schiff)* nicht abgestagter Mast *m*
non-steady nichtstationär
non-stepped ungestuft
non-stop switch *(El)* Durchfahrschalter *m*
non-strain ag[e]ing *(Met)* reckalterungsbeständig
non-structural *(Bw)* nichttragend
non-structural bulkhead *(Schiff)* Trennschott *n*
non-synchronous machine Asynchronmaschine *f*
non-synchronous motor Asynchronmotor *m*
non-tarnish[ing] metallic thread nicht blindwerdender Metallfaden *m*
non-volatile vehicle nichtflüchtiger Lacklösemittelanteil *m*, nichtflüchtiges Bindemittel *n*
non-vortical wirbelfrei
non-warping verzugsfrei
non-wearing verschleißfrei
non-wearing properties Verschleißfestigkeit *f*
non-wood verholztes Material *n*
non-workable unverarbeitbar
non-working of a patent Nichtausübung *f* eines Patentes
non-woven cotton Baumwollvliesstoff *m*
non-woven fabric Faservlies *n*
non-woven fabric industry Vliesstoffindustrie *f*
non-woven material Vliesstoff *m*
non-woven textile fabric textiles Flächengebilde *n*
non-yellowing nicht vergilbend
non-zero nicht verschwindend, ungleich Null
non-zero-sum game *(Dat)* Nichtnullsummenspiel *n*
non-zero term *(Kern)* nichtverschwindendes Glied *n*
nonanoic acid Pelargonsäure *f*
noncohesive *s.* non-cohesive
nonionics nichtionogene Verbindungen *fpl*
nonius Nonius *m*
nonmetal *s.* non-metal
nonoic acid Pelargonsäure *f*
non-tensioned reinforcement schlaffe Bewehrung *f (Stahlbeton)*
non-test chip board Schrenzkarton *m*
non-thermal nichtthermisch
non-throughed centreline bulkhead *(Schiff)* nichtdurchlaufendes Mittellängsschott *n*
non-torque *(Text)* verdrehungsfrei, ohne Verdrehungstendenz *(Texturgarn)*
non-tracking dielectric kriechstromfestes Dielektrikum *n*, kriechstromfester Isolator *m*
non-tracking quality Kriechstrombeständigkeit *f*
non-transparency Undurchsichtigkeit *f*
non-turbulent wirbelfrei
non-uniform ungleich[mäßig], ungleichförmig, nichtgleichförmig
non-uniform field *(Ph)* nichtgleichförmiges Feld *n*
non-uniform inflow *(Schiff)* ungleichförmige Anströmung *f*
non-uniform movement ungleichförmige Bewegung *f*
non-uniformity Ungleichmäßigkeit *f*, Ungleichförmigkeit *f*
non-valent *(Ch)* nullwertig
non-vanishing *s.* non-zero
non-vibrating schwingungsfrei
non-volatile nicht flüchtig
non-volatile memory Festwertspeicher *m*, permanenter Speicher *m*
non-volatile storage *(Dat)* leistungslose Steuerung *f*

nonpareil *(Typ)* 6 britische Punkte *mpl (Schriftgröße)*
nonreactive *s.* non-reactive
nonshatterable unbrechbar, splitterfrei
nonshrink *(Text)* krumpfecht
nonylic acid Pelargonsäure *f*
noose Schlinge *f*
NOR WEDER-NOCH *(logischer Operator)*
NOR-gate NOR-Glied *n*, WEDER-NOCH-Glied *n*, negiertes ODER-Glied *n*
Nordhausen [sulphuric] acid Nordhäuser Vitriolöl *n*, rauchende Schwefelsäure *f*
normal Normale *f*, Senkrechte *f*; Lot *n*
normal 1. normal, senkrecht; regelmäßig, regulär; 2. *(Ch)* normal, n, N
n. acceleration Normalbeschleunigung *f*
n. admission Normalfüllung *f (Dampfdiagramm)*
n. aspiration Selbstansaugen *n (Verbrennungsmotor)*
n. atmosphere physikalische Atmosphäre *f* *(1 atm = 101 325 N/m²)*
n. base pitch Teilung *f* im Normalschnitt des Grundzylinders
n. cathode fall normaler Katodenfall *m*
n. cell Normalelement *n*
n. circular pitch Teilung *f* im Normalschnitt *(Zahnrad)*
n. circular thickness Zahndicke *f* im Normalschnitt
n. concentration Normalkonzentration *f*
n. contact *(Am) (El)* Ruhekontakt *m*
n. curve Normalkurve *f*, Glockenkurve *f*, Gaußkurve *f*
n. distribution Normalverteilung *f*; Gaußsche Fehlerverteilung *f*
n. dose *(Kern)* Normaldosis *f*
n. duration of charge normale Ladedauer *f* *(einer Batterie)*
n. electrode Normalelektrode *f*
n. element Normalelement *n*
n. fault *(Geol)* Verwerfung *f*, Sprung *m*
n. force Abdrängkraft *f*, Normalkraft *f*, Normalkomponente *f* der Kraft
n. frequency Betriebsfrequenz *f*
n. hydrogen electrode Normalwasserstoffelektrode *f*
n. incidence senkrechter Einfall *m (z. B. einer Strahlung)*
n. loudness Normallautstärke *f*
n. magnetization curve Kommutierungskurve *f*
n. output Normalleistung *f*
n. pitch Normalteilung *f (Zahnrad)*
n. plane Normalebene *f*
n. position Grundstellung *f (Relais)*; Ruhelage *f*, Ruhestellung *f*; Normalstellung *f* *(Schalter)*
n. power *(Masch)* Normalleistung *f*
n. pressure Normaldruck *m*
n. radiation Hohlraumstrahlung *f*, schwarze Strahlung *f*, Normalstrahlung *f*, Temperaturstrahlung *f (Wärmestrahlung mit normaler spektraler Energieverteilung)*
n. reaction Normalkraft *f*
n. rock pressure *(Brg)* Überlagerungsdruck *m*, normaler Gebirgsdruck *m*
n. room temperature Zimmertemperatur *f*, Raumtemperatur *f*
n. rotative speed Betriebsdrehzahl *f*
n. salt *(Ch)* Neutralsalz *n*
n. solution *(Ch)* Normallösung *f*, n-Lösung *f*
n. speed Normalgeschwindigkeit *f*, Betriebsdrehzahl *f*
n. stress *(Mech)* Normalspannung *f*
n. time response *(BMSR)* normales Zeitverhalten *n (Übergangsfunktion bei verschwindender Anfangsenergie in den Energiespeichern)*
n. to the path *(Rak)* bahnsenkrecht
n. to the stratification *(Brg)* bankrecht
n. voltage *(El)* Betriebsspannung *f*, Normalspannung *f*
n. zero current *(El)* normaler Nulldurchgang *m* des Stromes
normality *(Ch)* Normalität *f (einer Lösung)*
normalization *(Math)* Normierung *f*
normalize / to 1. *(BMSR)* normalisieren; 2. *(Met)* normalisieren, normalglühen; 3. *(Math)* normieren
normalized admittance *(El)* relativer Leitwert *m*
n. impedance *(El)* relative Impedanz *f*, relativer komplexer Widerstand *m*, normalisierter Feldwellenwiderstand *m* einer Hohlleiterwelle *(Wellenleiter)*
normalizing *(Met)* Normalisieren *n*, Normalglühen *n*
normally *s.* normal
n. aspirated mit Selbstansaugung
n. bisecting senkrecht halbierend
n. closed [auxiliary] contact *(El)* Hilfsöffnungskontakt *m*, Öffner *m*, Ruhekontakt *m*
n. closed valve mit steigendem Steuerdruck öffnendes Ventil *n*
n. open *(El)* Arbeitskontakt *m*
n. open auxiliary contact *(El)* Hilfsschließkontakt *m*, Schließer *m*, Arbeitskontakt *m*
n. open valve mit steigendem Steuerdruck schließendes Ventil *n*
n. opened contact Arbeitskontakt *m*, Schließer *m (Relais)*
n. reinforced schlaff bewehrt *(Stahlbeton)*
Norman arch *(Bw)* Rundbogen *m*
Normand boiler Normand-Kessel *m (engrohriger Wasserrohrkessel mit einem Ober- und zwei Unterkesseln und gebogenen Rohren)*
norpinic acid Norpinsäure *f*

north Norden *m*, Nord

n.-magnetic nordmagnetisch

n. point Nordpunkt *m*

n. polar star Nordstern *m*

n. pole Nordpol *m*

n.-stabilized plan position indicator in Nordrichtung stabilisiertes Rundsichtgerät *n* *(Radar)*

northlight roof *(Bw)* S[c]heddach *n*

n. shell *(Bw)* S[c]hedschale *f*

Norton type gear Nortongetriebe *n*

Norton's theorem Zweipoltheorie *f*, Helmholtzscher Satz *m*, Satz von der Ersatzstromquelle

nose 1. Nase *f*; Ansatz *m*, Schnauze *f*; 2. *(Masch)* Spindelkegel *m*, Werkzeugspitze *f*; Schneidkopf *m*; 3. *(Schiff)* Bug *m*, Schiffsbug *m*, Schiffsnase *f*; 4. *(Rak)* Spitze *f*, Kopf *m*, Bugraum *m*, Bugzelle *f*

n. angle Spitzenwinkel *m (Drehmeißel)*

n.-bag Futterbeutel *m (für Pferde)*

n. cone *(Rak)* Schutzkegel *m*, Bugkegel *m*, Gefechtsteil *n*

n. dive / to *(Schiff)* stark kopflastig wegtauchen, mit der Nase wegtauchen

n. end of the nippers *(Text)* Zangenschnabel *m*, Zangenmaul *n*

n. fuse *(Mil)* Kopfzünder *m*

n. heaviness *(Flg)* Kopflastigkeit *f*

n. landing gear *(Flg)* Bugradfahrwerk *n*

n. of a Bessemer converter Hals *m* einer Bessemerbirne

n. of the rocket Kopf (Bug) *m* der Rakete

n. opening door *(Flg)* Kanzeleinstiegklappe *f*

n.-over vornüber

n. piece Revolver *m (des Mikroskops)*

n.-pipe Balgendüse *f (Orgel)*

n. radius Spitzenabrundung *f*

n. ring Nasenring *m*

n. suspension *(El)* Tatzlagerantrieb *m*

n. type collet chuck Spindelkopfspannzange *f*

n. wheel landing gear *(Flg)* Bugradfahrwerk *n*

nosean Nosean *m (Sodalithmineral)*

noselite *s.* nosean

nosian *s.* nosean

nosing 1. Ausladung *f*, Kante *f*; 2. Kantenschutzleiste *f*, Kantenschutzschiene *f (Treppen)*

NOT AND „NICHT UND", Unvereinbarkeit *f (logischer Operator)*

NOT-AND circuit Inversionsschaltung *f*, NICHT-UND-Schaltung *f (logischer Operator)*

NOT-AND connective NICHT-UND-Verknüpfung *f (logischer Operator)*

NOT BOTH NICHT BEIDE *(logischer Operator)*, Shefferscher Strich *m*

NOT-circuit Verneinungsschaltung *f*, NICHT-Schaltung *f (logischer Operator)*

NOT-gate Negator *m*, NICHT-Schaltung *f*, NICHT-Glied *n (logischer Operator)*

not-go end Ausschußseite *f (der Lehre)*

not-go gauge Ausschußlehre *f*

NOT OR NICHT ODER, WEDER-NOCH *(logischer Operator)*

not to be heard of verschollen *(Schiff)*

not to scale nicht maßstäblich

not-under-command manövrierunfähig

not-under-command light *(Schiff)* Fahrtstörungslaterne *f*, Notlaterne *f*, Manövrierunfähigkeitslaterne *f*

not-under-control manövrierunfähig, fahrtgestört *(Schiff)*

notable *(Ch)* wahrnehmbar, feststellbar

notation Bezeichnung *f*; Darstellung *f*, Schreibweise *f (Information)*; Setzung *f*

notch / to ausklinken; schlitzen, einschneiden; [ein]kerben, aussparen, einzapfen *(Zimmerei)*

notch Kerb *m*, Raste *f*; Einkerbung *f*, Schnitt *m*, Einschnitt *m*, Kerbe *f*; Fallkerb *m*, Falz *m*, Schnittfuge *f*; Schalterstellung *f*

n. aerial Tragflächenantenne *f (Flugnavigation)*

n.-bend impact test Kerbschlagbiegeversuch *m*

n.-bend test Kerbschlagbiegeversuch *m*

n. board Treppenwange *f (Holzbau)*

n. brittleness Kerbsprödigkeit *f*, Kerbzähigkeit *f*

n. diplexer Fernsehfilterweiche *f (Antennen)*

n. effect Kerbwirkung *f*

n. for tenon Schlitz *m (Holz)*

n. gun Stopfenmaschine *f (Hochofen)*

n. impact strength Kerbschlagzähigkeit *f*

n. sensitivity Kerbempfindlichkeit *f*

n. stress Kerbspannung *f*

n. test Kerb[schlag]versuch *m*

n. toughness *(Met)* Kerbzähigkeit *f*

notched geschlitzt; gekerbt

n. bar *(Met)* Kerbstab *m*

n. bar bending test *(Met)* Kerbbiegeprobe *f*

n.-bar impact strength *(Met)* Kerbschlagzähigkeit *f*

n.-bar impact test Kerbschlagversuch *m*

n.-bar test toughness *(Met)* Kerbzähigkeit *f*

n. chisel *(Masch)* Zahneisen *n*

n. disk Rastenscheibe *f*

n. impact strength Kerbzähigkeit *f*

n. pig gekerbte Massel *f*

n. pin Kerbstift *m*

n. plate Rastenscheibe *f*

n. susceptibility *(Met)* Kerbempfindlichkeit *f*

n. taper pin Kerbstift *m*

n. test specimen gekerbter Probestab *m*

n. wheel Klinkenrad *n*, Rastenrad *n*, Schaltrad *n*

notching Kerben *n*; Falzen *n*, Einzapfen *n*

n. cutter Falzfräser *m*
n. filter *(Fmt)* Saugfilter *n*
n. press Nutenstanzautomat *m*
n. ratio Ungleichförmigkeit *f*, Feinstufigkeit *f* *(einer Steuerung)*
n. relay Impulsspeicherrelais *n*, Fortschalterelais *n*
note 1. Note *f*; Notiz *f*; Merkmal *n*; 2. *(Ak)* Ton *m*
n. amplification Tonverstärkung *f*
n. amplifier Tonverstärker *m*
n. book Notizbuch *n*
n. of pitch Überlagerungston *m*
n. paper Briefpapier *n*
n. tuning Abstimmung *f* der Tonhöhe
notice 1. Nachricht *f*; Bericht *m*, Gedanke *m*, Vorstellung *f*; 2. *(Math)* Begriff *m*; 3. Meinung *f*, Ansicht *f*; 4. *(Am)* Galanteriewaren *fpl*
n. of abandonment *(Schiff)* Abandon-Erklärung *f*
n. of readiness *(Schiff)* Ladebereitschaftsnotiz *f*
notion Begriff *m*
notorious aktenkundig
noughts-complement representation *(BMSR)* Nullenkomplementdarstellung *f*
novarsenobenzene *(Ch)* Neosalvarsan *n*
novocaine *(Ch)* Novokain *n*
novolak Novolak *n*
nowel Unterkasten *m* *(Formerei)*
noxious surface *(Ph)* schädliche Fläche *f*
nozzle 1. Düse *f*, Gasdüse *f*, Ausflußdüse *f*; Mundstück *n*, Schlauchmundstück *n*; Stutzen *m*, Tülle *f*; 2. Brenner *m*; 3. Balgendüse *f* *(Orgel)*; 4. *(El)* Wellenleiteröffnung *f*; Düsenstrahler *m* *(Antenne)*
n. blade Leitschaufel *f* *(Turbine)*
n. block Düsenblock *m* *(Turbine)*
n. body Düsenkörper *m* *(Turbine)*
n. bore Düsenbohrung *f*
n. box *(Rak)* Düsenkammer *f*, Düsenkasten *m*
n. brick *(Met)* Lochstein *m* der Stopfenpfanne
n. burner Düsenbrenner *m*
n. channel Düsenkanal *m* *(Turbine)*
n. coefficient *(Kfz)* Düsenbeiwert *m*
n. consisting of several pieces mehrteilige Düse *f* *(Turbine)*
n. contraction Einschnürung (Verengung) *f* der Düse
n. diaphragm *(Rak)* Düsenzwischenboden *m*
n. efficiency Düsenwirkungsgrad *m*
n. efflux angle Leitschaufelaustrittswinkel *m* *(Turbine)*
n. exhaust velocity *(Rak)* Austrittsgeschwindigkeit *f*
n. exit diameter *(Rak)* Austrittsdurchmesser *m* der Düse
n. filter Düsenfilter *n*
n. fittings Düsengarnitur *f* *(Turbine)*
n. flap Düsenklappe *f*, Düsenzunge *f* *(Turbine)*
n.-flapper amplifier *(BMSR)* Düse-Prallplatte-Verstärker *m*
n. fouling Düsenverschmutzung *f*
n. gauge Düsenlehre *f*
n. head *(Kfz)* Düsenkörper *m*
n. holder Düsenhalter *m*; Strahlrohr *n* *(des Sandstrahlgebläses)*
n. inclination Anstellwinkel *m* der Düse, Düsenneigung *f* *(Turbine)*
n. injection Einspritzdüse *f*
n. jet Einspritzstrahl *m* *(Dieselmotor)*
n. pipe Strahlrohr *n*; Düse *f*
n. pipe of bellows Balgenrohr *n* *(Orgel)*
n. pressure Düsenspannung *f* *(Turbine)*
n. radiator Düsenkühler *m*
n. ring Düsenring *m* *(Turbine, Propeller)*
n. ring segment Düsensatz *m* *(Turbine)*
n.-shaped düsenförmig
n.-shaped hood Ringdüse *f*
n. spray Sprühdüse *f*
n. sprayer Streudüsenzerstäuber *m*, Düsenzerstäuber *m*
n. tip Düsenmündung *f*
n. trigger Zapfhahn *m*, Abfüllschlauchhahn *m*
n. with rectangular outlet Düse *f* mit rechteckiger Mündung *(Turbine)*
n. with smooth walls glatte Düse *f* *(Turbine)*
nozzlelike düsenähnlich
nozzles closely spaced eng aneinandergerückte Düsen *fpl* *(Turbine)*
n-p junction np-Übergang *m* *(Halbleitertechnik)*
nub kleiner weißer Fleck *m* *(Seidenfaden)*
n. yarn *s.* knop yarn
nubuck *(Led)* Narbenplüsch *m*
nuclear Kern-, nuklear
n. absorption Kerneinfang *m*
n. active kernaktiv
n. aero-engine Kernenergieflugmotor *m*, Atomflugmotor *m*
n. attraction Kernanziehung *f*
n. binding Kernbindung *f*
n. binding energy Kernbindungsenergie *f*
n. binding forces Kernbindungskräfte *fpl*
n. bomb Kernbombe *f*
n. bombardment Kernbeschuß *m*; Beschießen *n* von (mit) Kernen
n. building-up reactions [nukleare] Aufbauprozesse *mpl*
n. capture Kerneinfang *m*
n. cascade Kernkaskade *f*
n. central forces zentrale Kernkräfte *fpl*
n. chain reaction nukleare Kettenreaktion *f*, Kernkettenreaktion *f*
n. charge Kernladung *f*

n. charge distribution Ladungsverteilung *f* im Atomkern
n. charge number Kernladungszahl *f*
n. chemistry Kernchemie *f*
n. constant atomare Konstante *f*, Kernkonstante *f*
n. constitution Kernaufbau *m*, Kernstruktur *f*, Aufbau *m* des Atomkerns
n.-cooling-fluid rocket Flüssigkeitsatomrakete *f*
n. corrosion Kernkorrosion *f*
n. cross-section Wirkungsquerschnitt *m (Einheit: barn)*
n. decay Kernzerfall *m*, Kernumwandlung *f*
n. density Kerndichte *f*, Dichte *f* des Atomkerns
n. diameter Kerndurchmesser *m*
n. dimensions nukleare Dimensionen *fpl*, Kernabmessungen *fpl*
n. disintegration Kernzerfall *m*, Kernumwandlung *f*; Kernzertrümmerung *f*, Kernzersplitterung *f*
n. distance Kernabstand *m*
n. division Kernteilung *f*
n. electric moment elektrisches Moment *n* des Atomkerns
n. electron Kernelektron *n*
n. emulsion Kern[spuren]emulsion *f*
n. energetics Kernenergetik *f*, Kernenergietechnik *f*
n. energy Kernenergie *f*, Atomkernenergie *f*
n. energy law Kernenergiegesetz *n*
n. energy level Kernniveau *n*, Kernterm *m*
n. energy level spacing Kernniveauabstand *m*
n. energy rocket Kernenergierakete *f*, Atomrakete *f*
n. engine Kernenergietriebwerk *n*, Kernenergieantrieb *m*, Kernenergieanlage *f*
n. engineer Kerntechniker *m*, Kernverfahrensingenieur *m*
n. engineering Kern[verfahrens]technik *f*
n. evaporation Kernverdampfung *f*
n. excitation Kernanregung *f*
n. explosion 1. Kernbombenexplosion *f*; 2. Kernexplosion *f*
n. explosive Kernsprengstoff *m*
n. field Kernfeld *n*
n. fission Kernspaltung *f*
n. forces Kern[feld]kräfte *fpl*
n. fragment Spaltprodukt *n*; Kernbruchstück *n*
n. fuel Kernbrennstoff *m*; Spaltstoff *m*
n. fuel regeneration Spaltstoffrückgewinnung *f*
n. fusion Kernfusion *f*, Kernverschmelzung *f*
n. fusion bomb thermonukleare Bombe *f*, Fusionsbombe *f*
n. halogenation *(Ch)* Kernhalogenierung *f*
n. heat of fission Kernspaltungswärme *f*
n. impact Kernstoß *m*
n.-inactive kerninaktiv
n. increase Kernvermehrung *f*
n. interaction nukleare Wechselwirkung *f*, Kernwechselwirkung *f*
n. isobar Kernisobar *n*
n. isomer Kernisomer *n*
n. isomerism Kernisomerie *f*
n. level Kernniveau *n*, Kernterm *m*
n. linkage Kernbindung *f*
n. magnetic moment kernmagnetisches Moment *n*, magnetisches Kernmoment *n*
n. magnetic resonance magnetische Kernresonanz *f*
n. magnetic resonance spectroscopy Kernresonanzspektroskopie *f*
n. magneton Kernmagneton *n*
n. mass Kernmasse *f*
n. matter Kernmaterie *f*
n. medicine Kernmedizin *f*
n. membrane Kernwand *f*
n. model Kernmodell *n*
n. moment Kernmoment *n*
n. movement Kernbewegung *f*
n. particle Kernteilchen *n*, Nukleon *n*
n. photodisintegration Kernfotoprozeß *m*, fotonukleare Reaktion *f*
n. photoelectric effect Kernfotoeffekt *m*
n. physicist Kernphysiker *m*
n. physics Kernphysik *f*
n. plate Kern[foto]platte *f*, Kernspurplatte *f*
n. potential Kernpotential *n*
n. power Kernkraft *f*, Kernenergie *f*, Atomkraft *f*, Atomenergie *f*
n. power plant Atomkraftwerk *n*
n. power production Kernenergieerzeugung *f*, Kernenergiegewinnung *f*
n. power station Atomkraftwerk *n*, Kernkraftwerk *n*
n.-powered kernenergiebetrieben, mit Kernenergieantrieb
n.-powered icebreaker Eisbrecher *m* mit Kernenergieantrieb, Atomeisbrecher *m*
n.-powered rocket Rakete *f* mit Kernenergieantrieb, Kernenergierakete *f*, Atomrakete *f*
n.-powered ship Schiff *n* mit Kernenergieantrieb
n.-powered submarine Unterseeboot *n* mit Kernreaktorantrieb, Atom-U-Boot *n*, Atomunterseeboot *n*
n. propellant Kerntreibstoff *m*, nuklearer Treibstoff *m*
n.-propelled aeroplane Flugzeug *n* mit Kernenergieantrieb, Atomflugzeug *n*
n.-propelled submarine *s.* n.-powered submarine

n. propulsion Atomkraftantrieb *m*, Kernenergieantrieb *m*
n. proton Kernproton *n*
n. purification [nukleare] Feinreinigung *f*
n. radiation Kernstrahlung *f*, radioaktive Strahlung *f*
n. radius Kernradius *m*
n. reaction Kernreaktion *f*, Kernprozeß *m*
n. reactor Kernreaktor *m*, Reaktor *m*
n. reactor art Kernreaktorbau *m*
n. reactor research program Reaktorforschungsprogramm *n*
n. reactor safety Reaktorsicherheit *f*
n. recoil Kernrückstoß *m*
n. research Kernforschung *f*
n. resonance Kernresonanz *f*
n. resonance spectrograph Kernresonanzspektrograf *m*
n. resonance spectrometer Kernresonanzspektrometer *n*
n. retaliation *(Mil)* atomarer Gegenschlag *m*
n. rocket Rakete *f* mit nuklearem Treibstoff, Kernenergierakete *f*, Atomrakete *f*
n. saturation Kern[ab]sättigung *f*
n. scattering Kernstreuung *f*
n. shell 1. Kernschale *f*; 2. *(Mil)* nukleares Geschoß *n*, Granate *f* mit Kernsprengstoffladung
n. shielding Abschirmung *f*, Strahlenschutz *m*
n. ship *s.* n. powered ship
n. spectrometer Kernspektrometer *n*
n. spectroscopy Kernspektroskopie *f*
n. spin Kernspin *m*
n. stability Kernstabilität *f*
n. state Kernzustand *m*; Energiezustand *m* des Atomkerns
n. structure Kernstruktur *f*, Kernbau *m*
n. submarine *s.* n. powered submarine
n. substituted kernsubstituiert
n. surface Kernoberfläche *f*, Kernrand *m*
n. system of terms Termsystem *n* der Kerne
n. systematics Kernsystematik *f*
n. technique Kerntechnik *f*, angewandte Kernphysik *f*
n. technology Kern[verfahrens]technik *f*
n. temperature Kerntemperatur *f*
n. term Kernterm *m*
n. test Kernwaffentest *m*, Kernwaffenversuchsexplosion *f*
n. theory Kerntheorie *f*
n. track finding microscope Kernspursuchmikroskop *n*
n. track measuring microscope Kernspurmeßmikroskop *n*
n. transformation Kernumwandlung *f*
n. transition Kernübergang *m*
n. transmutation Kernumwandlung *f*
n. weapon Kernwaffe *f*, nukleare Waffe *f*

nucleation Kernbildung *f*, Kristallisationskeimbildung *f*
nuclei *s.* nucleus
nucleic acid Nukleinsäure *f*
nucleogenesis Atomkernbildung *f*, Bildung *f* der Atomkerne in der Natur, Elemententstehung *f*
nucleon Nukleon *n*, Kernbaustein *m*
n. binding Nukleonenbindung *f*
n. gas Nukleonengas *n*
n. isobar Nukleonenisobar *n*
nucleonic component Nukleonenkomponente *f*
nucleonics *s.* nuclear technique
nucleoprotein *(Ch)* Nukleoproteid *n*, Nukleoprotein *n*
nucleotide *(Ch)* Nukleotid *n*
nucleus Kern *m (Atom)*; Keim *m (Kristall)*; Zellkern *m*
n. bond energy Kernbindungsenergie *f*
n. crew *(Schiff)* Stammbesatzung *f*, Mindestbesatzung *f*
n. of an atom Atomkern *m*
n. of crystal Kristallkeim *m*; Kristallisationskeim *m*
n. of crystallization Kristallisationskeim *m*
nuclide *(Kern)* Nuklid *n*
nuclon *s.* nucleon
nugget Klumpen *m*, Nugget *n*
nuisance area *(El)* Störgebiet *n*; Interferenzzone *f*
n. raider Störflugzeug *n*
null Null *f*
n. balance *(El)* Nullabgleich *m*
n.-balance amplifier Nullabgleichverstärker *m*
n.-balance indicator Nullinstrument *n*
n. detector *(El)* Nullinstrument *n*
n. indicator *(El)* Nullanzeigegerät *n*, Nullindikator *m*
n.-setting device Nulleinstellgerät *n*
n. signal Nullzacke *f (Radar)*
n. valency *(Ch)* Nullvalenz *f*
n. vector Nullvektor *m*
nullify / to *(Dat)* annullieren, nullen, nullsetzen
number / to zählen, numerieren, beziffern
number Nummer *f*; Zahl *f*, Anzahl *f*; Ziffer *f*; Kernladungszahl *f*, Ordnungszahl *f*
n. couple Zahlenpaar *n*
n. detector *(Büro)* Kontonummernsucher *m*
n. field *(Math)* Zahlenkörper *m*
n. language Ziffernsprache *f*
n. module *(Math)* Zahlenmodul *m*
n. nail *(Fmt)* Bezeichnungsnagel *m*
n. of alterations Lastspielzahl *f (Werkstoffprüfung)*
n. of ampere turns Amperewindungszahl *f*
n. of atoms Atomzahl *f*
n. of calls *(Fmt)* Rufzahl *f*, Zahl *f* der Anrufe

n. of collisions *(Ph)* Stoßzahl *f*
n. of cycles 1. *(El)* Periodenzahl *f*, Wechselzahl *f*; 2. *(Met)* Lastspielzahl *f (Werkstoffprüfung)*
n. of electrons occupying a shell *(Kern)* Besetzungszahl *f*
n. of flutes Nutenzahl *f (Reibahle)*; Zähnezahl *f (Zahnrad)*
n. of impacts *(Ph)* Stoßzahl *f*
n. of lifestock Tierbestand *m*
n. of lines *(Fs)* Zeilenzahl *f*
n. of natural oscillations Eigenschwingungszahl *f*
n. of oscillations per minute Schwingzahl *f* je min
n. of picture elements Bildpunktzahl *f*, Rasterzahl *f*
n. of pieces per hour Stückzahl *f* [je Stunde], Stundenleistung *f* [in Stück]
n. of plants Pflanzenbestand *m*
n. of revolutions [per unit time] Drehzahl *f*, Tourenzahl *f*
n. of rotations Drehzahl *f*
n. of scanning lines *(Fs)* Anzahl *f* der Bildzeilen
n. of starts Gangzahl *f (Verzahnung)*
n. of steps Stufenzahl *f*
n. of strokes per minute Hubzahl *f* je min
n. of successive images Bildwechselzahl *f*
n. of teeth 1. *(Masch)* Zähne[an]zahl *f*; 2. *(Text)* Nasenzahl *f (Trommelöffner)*
n. of teeth to gauge Meßzähnezahl *f*
n. of teeth to measure over Meßzähnezahl *f*
n. of theoretical plates *(Kern)* Zahl *f* der theoretischen Böden
n. of threads per inch Gangzahl *f* auf 1 Zoll
n. of transfer units *(Kern)* [theoretische] Austauschzahl *f*
n. of turns *(El)* Windungszahl *f*
n. plane *(Math)* Zahlenebene *f*
n. plate *(Kfz)* Kennzeichenschild *n*, Nummernschild *n*
n. plate illumination lamp Kennzeichenlampe *f*, Nummernschildbeleuchtung *f*
n. punch Nummerneisen *n*
n. sequence Zahlenfolge *f*
n. series Zahlenreihe *f*
n. set Zahlenmenge *f*
n. theory Zahlentheorie *f*
n.-unobtainable tone *(Fmt)* Besetztzeichen *n*, Störungssignal *n*
n. wheel Zählrolle *f*
numbering Numerierung *f*, Bezifferung *f*
n. apparatus Paginierwerk *n*; Numerierwerk *n*
n. of cable cores *(El)* Adernzählfolge *f*
n. of silk Feinheitsbestimmung *f* der Seide
numeral reading machine *(Büro)* Zahlzeichenlesemaschine *f*
numerator Zähler *m (eines Bruches)*
numeric coding *(Dat)* numerische Kodierung (Verschlüsselung) *f*
numerical numerisch; zahlenmäßig; Zahlen-
n. aperture *(Opt)* numerische Apertur *f*
n. check Rechenprobe *f*
n. computation Zahlenrechnung *f*
n. control *(Masch)* numerische Steuerung *f*
n. digits *(Fmt)* Ziffernzeichen *n*
n. display Ziffernanzeige *f*
n. distance numerischer Abstand *m*
n. example Zahlenbeispiel *n*
n. indication numerische Anzeige *f*
n. quantity Zahlenwert *m*
n. ratio Zahlenverhältnis *n*
n. selector *(Fmt)* Nummernwähler *m*
n. sequence Ziffernfolge *f*
n. value Zahlenwert *m*
numerically controlled numerisch gesteuert
n. expressed zahlenmäßig ausgedrückt
numeroscope *(Dat)* Numeroskop *n (Ziffernanzeigeröhre)*
nun buoy *(Schiff)* Spitzboje *f*, Spitztonne *f*
nurl *s.* knurl
nursery 1. Kinderspielzimmer *n*; Kindergarten *m*; 2. Fischzuchtteich *m*; 3. Baumschule *f*, Pflanzschule *f*
n. for trees Forstgarten *m*
Nusselt number *(Ph)* Nußelt-Zahl *f*, Nußeltsche Kennzahl *f*
nut 1. Nuß *f*; 2. *(Masch)* Mutter *f*, Gewindering *m*
n. bevelling machine Mutternabkantmaschine *f*
n. brace Bohrwinde *f*
n. coal Nußkohle *f*
n. coke Perlkoks *m*, Nußkoks *m*
n. for removing Abdrückmutter *f*
n. guard Muttersicherung *f*
n. lock[ing] Muttersicherung *f*, Mutternverschluß *m*
n. milling machine Mutternfräsmaschine *f*
n. plier Mutterzange *f*
n. runner Schraubenanziehmaschine *f*
n. tap Mutter[gewinde]bohrer *m*
n. tapping Muttergewindebohren *n*
n. tapping attachment Muttergewindeschneideinrichtung *f*
n. tapping machine Muttergewindebohrmaschine *f*
nutating-piston meter *(BMSR)* Taumelscheibenzähler *m*
nutation *(Ph)* Nutation *f*
n. angle Nutationswinkel *m*
n. of the gyroaxis Nutation *f* der Kreiselachse

nutrient Nährflüssigkeit *f*, Nährstoff *m*
n. base Nährsubstrat *n*
n. solution Nährlösung *f*
nutritional quality Nährwert *m*
nutritious substances Nährstoffe *mpl*
nutritive value Nährwert *m*
nuts Nußkohlen *fpl*
nutty slack Nußgrus *m*, Nußgrießkohle *f*
nuvistor Nuvistor *m (besonders stoß- und temperaturunempfindliche Metallkeramikröhre)*
nylon Nylon *n*
n. corded tyre Autoreifen *m* mit Nylonkord
n. hawser Nylontrosse *f*
n. reinforced nylonverstärkt
nylons Damenstrümpfe *mpl* aus Nylon

O

O-band *(Ph)* O-Bande *f*
O-electron O-Elektron *n*
O-ring Dichtungsring *m* mit rundem Querschnitt
oak Eiche *f*; Eichenholz *n*
o. bark tan Eichenlohe *f*
o. bark tannage Eichenlohgerbung *f*
o.-tanned mit Eichenlohe gegerbt; lohgar
oakum Werg *n*, Kalfaterwerg *n*
oar Bootsriemen *m*, Skull *n*; Ruder *n*, Seitenruder *n*
oast Hopfendarre *f*
oat Hafer *m*
o. bin Haferkasten *m*
o. crusher Haferquetsche *f*
o. field Haferschlag *m*
oatmeal Hafermehl *n*
o. paper Holzmehlpapier *n (mit Holzmaterialfüllstoff)*
Obach cell *(El)* Obach-Element *n (Batterie)*
obedience to law Gesetzestreue *f*
obelisk Obelisk *m*, Spitzsäule *f*
obey / to *(Dat)* Befehl ausführen
object Ding *n*, Gegenstand *m*
o. air-lock Objektschleuse *f*
o. distance Gegenstandsweite *f*
o. holder Objektträger *m*
o. marker Objektmarkierer *m*
o. point Dingpunkt *m*
o. slide Objektträger *m*
o. space Dingraum *m*
o. stage Objekttisch *m*
o. traverser Objektführer *m*
objectionable beanstandenswert
objective *(Opt)* Objektiv *n*
o. function *(Dat)* Zielfunktion *f*
o. mount Objektivfassung *f (Mikroskop)*
o. prism Objektivprisma *n*
o. revolving nose-piece Objektivrevolver *m (Mikroskop)*

oblate abgeplattet, abgeflacht
o. spheroid abgeplattetes Rotationsellipsoid *n*, Rotationsellipsoid um die kleine Achse, Sphäroid *n*, an den Polen abgeflachte Kugel *f*
oblateness Abplattung *f*
oblique schief, schräg; schrägschenklig
o. angle schiefer Winkel *m*
o.-angled schiefwink[e]lig
o.-angled triangle schiefwink[e]liges Dreieck *n*
o. anode Schräganode *f (Katodenstrahlröhren)*
o. cone schiefer Kegel *m*
o. cut of a hip rafter Backenschmiege *f*, Klebeschmiege *f (Zimmerei)*
o. extruder head *(Plast)* Schrägspritzkopf *m*
o. frustum of cylinder schief abgeschnittener Zylinder *m*
o. incidence Schrägeinfall *m*
o. incidence transmission Übertragung *f* unter schrägem Einfall *(Nachrichtentechnik)*
o. photograph Schrägaufnahme *f*
o. position Schräglage *f*, Schrägstellung *f*
o. projection schiefe Parallelprojektion *f* auf lotrechte Bildebene
o. scarf with butt ends schräges Blatt *n (Zimmerei)*
o. seating valve Schrägsitzventil *n*
o. section Schrägschnitt *m*
o. split schräge Teilfuge *f*, schräger Schlitz *m*
o. triangle schiefwink[e]liges Dreieck *n*
o. type Kursiv[schrift] *f*, Schrägschrift *f*
o. view Schrägansicht *f*
o. winding *(El)* Schrägwicklung *f*
obliquely cut cylinder schief abgeschnittener Zylinder *m*
obliqueness Schräge *f*, Schrägstellung *f*
obliquity Schräge *f*, Schrägstellung *f*
obliterated *(Fs)* verwaschen *(Bild)*
obliterating pole piece Löschmagnet *m*
obliteration Auslöschung *f*, Tilgung *f*
o. magnet Löschmagnet *m*
oblong Rechteck *n*
oblong länglich, verlängert, gestreckt
o. aperture Langlochöffnung *f (Sieb)*
o. hole Langloch *n*
o. punching Längslochung *f*
oboe system Bumerangverfahren *n (Peilverfahren)*
obscuration Verdunk[e]lung *f*
obscure / to verdunkeln, verdecken
obscure radiation Dunkelstrahlung *f*
obscured glass Mattglas *n*, Milchglas *n*
observation Beobachtung *f*; Überwachung *f*
o. car Aussichtswagen *m*
o. circuit *(Fmt)* Mithöreinrichtung *f*
o. deck *(Schiff)* Beobachtungsdeck *n*
o. desk *(Fmt)* Überwachungsplatz *m*
o. grid Festpunktnetz *n (Vermessung)*

o. **hole** Schauloch *n*
o. **platform** Beobachtungsbühne *f*
o. **point** Festpunkt *m (Vermessung)*
o. **slit** Beobachtungsspalt *m*, Beobachtungsschlitz *m*, Sehschlitz *m*
o. **telescope** Beobachtungsfernrohr *n*
observational error Beobachtungsfehler *m*
observatory Sternwarte *f*
o. **dome** Sternwartenkuppel *f*
observe / to beobachten; messen
observe silent period! beachtet Funkstille!
observed altitude Beobachtungshöhe *f*
o. **output** tatsächlich gemessene Leistung *f*
o. **radio bearing** *(Fmt, Fs)* aufgenommene Funkpeilung *f*
observer Beobachter *m*
observing technique Beobachtungstechnik *f*
obsolete veraltet, ungebräuchlich
obstacle Hindernis *n*, Widerstand *m*; Widerstandskörper *m*; Ziel *n (Radar)*
obstruct / to verstopfen; [be]hindern, hemmen
obstruction Behinderung *f*; Blockierung *f*; Verstopfung *f*, Verschließung *f*, Verschmutzung *f*, Stauung *f* (*s. a.* obstacle)
o. **clearing** Hindernisbeseitigung *f*
o. **lights** Hindernisbefeuerung *f (Navigation)*
o. **marking** Hinderniskennzeichnung *f*
o. **to navigation** Schiffahrtshindernis *n*
obstructive hinderlich
obtain / to empfangen, erhalten, gewinnen, erreichen, erlangen
o. **a patent** ein Patent erhalten
o. **by scrubbing** herauswaschen
o. **the greatest economy** größte Einsparungen erzielen, größte Wirtschaftlichkeit erreichen
obtaining samples Probeentnahme *f*
obturator Verschluß *m (Geschütz)*
obtuse-angled stumpfwink[e]lig
obverse Vorderseite *f (einer Münze)*
occlude / to verschließen, einschließen, abschließen; absorbieren
occluded gas eingeschlossenes Gas *n (Elektronenröhren)*
o. **slag** *(Met)* Schlackeneinschluß *m*
occlusion 1. Absorption *f (Gas)*; 2. *(Met)* Einschluß *m*
occultation *(Astr)* Okkultation *f*, Verdeckung *f*, Verfinsterung *f*
occulting light unterbrochenes Leuchtfeuer *n (Navigation)*
o. **quick flashing light** unterbrochenes Blinkfeuer (Blitzfeuer) *n*
occupation 1. *(Ph)* Besetzung *f*; 2. Beruf *m*
o. **disease** Berufskrankheit *f*
o. **number** *(Ph)* Besetzungszahl *f*
occupational beruflich [bedingt]
o. **accident** Arbeitsunfall *m*
o. **disease** Berufskrankheit *f*
o. **exposure** berufliche Bestrahlung *f*, berufsbedingte Bestrahlung (Strahlenexponierung *f*)
o. **irradiation** *s.* o. exposure
occupied bandwidth belegte Bandbreite *f (Funk)*
o. **position** *(Fmt)* besetzter Platz *m*
o. **state** *(Ph)* besetzter Zustand *m*
occupy / to 1. beschäftigen; 2. *(Ph)* besetzen, belegen, einnehmen
occurence 1. Ereignis *n*; 2. *(Geol)* Auftreten *n*, Vorkommen *n*, Verbreitung *f*
ocean Meer *n*, Ozean *m*
o. **cable** Seekabel *n*, Unterseekabel *n*, Unterwasserkabel *n*, unterseeisches (unter Wasser verlegtes) Kabel *n*
o. **carrier** Prahmfähre *f*, Prahmmutterschiff *n*
o. **chart** Seekarte *f*
o. **cruiser** Seekreuzer *m (Jacht)*
o. **current** Meeresströmung *f*
o. **floor** Meeresboden *m*
o. **freight** *(Am)* Seefracht *f*
o.**-going** *(Schiff)* seegehend, Ozean-, Hochsee-
o.**-going ship** Hochseeschiff *n*, Ozeanschiff *n*
o.**-going tug** Hochseeschlepper *m*, Ozeanschlepper *m*
o. **greyhound** schnelles Ozeanschiff *n*
o. **highway** Schiffahrtsstraße *f*, Schiffahrtsweg *m*; Seewasserstraße *f*
o. **shipping** Handelsschiffahrt *f*; Seeschiffahrt *f*; Linienschiffahrt *f*
o. **steamer** Ozeandampfer *m*, Hochseedampfer *m*
o. **transport** Seeverkehr *m*
o. **water** Meerwasser *n*
o. **weather ship** Wetterbeobachtungsschiff *n*, Ozeanwetterschiff *n*
oceanic current Meeresströmung *f*
oceanography Meereskunde *f*, Ozeanografie *f*
ocher *s.* ochre
ochre Ocker *m*, Ockerfarbe *f*
ochreous ockerhaltig
o. **iron ore** Eisenocker *m*
octadecagon Achtzehneck *n*
octadecanedioic acid Hexadekandikarbonsäure *f*, Oktadekandisäure *f*
octadecanoic acid Oktadekansäure *f*
octagon Achteck *n*; Achtkant *m*
o. **bar steel** Achtkantstahl *m*
o. **pole** Achtkantbahn *f (Hammer)*
o. **stock** Achtkantmaterial *n*
octagonal achteckig, achtkantig
o. **coil** *(Fmt)* Achterspule *f*
octahedral oktaedrisch, achtflächig
octahedrite Oktaedrit *m*; Titanerz *n (Anatar)*, Rutil *m*
octahedron Oktaeder *n*
octal base *(El)* Oktalsockel *m*

o. digit *(Dat)* Oktalziffer *f*
o. notation *(Dat)* Oktalsetzung *f*, Oktalschreibung *f*, Oktaldarstellung *f*
octane Oktan *n*
o. booster Antiklopfmittel *n*, Oktanzahlverbesserer *m*
o. number Oktanzahl *f*
o. rating Klopffestigkeit *f*, Klopfwert *m*, Oktanzahl *f*
octanoic acid Oktansäure *f*, Kaprylsäure *f*
octant Oktant *m*
octantal error Oktantfehler *m (Radartechnik)*
octatomic ring *(Ch)* achtgliedriger Ring *m*
octavalence *(Ch)* Achtwertigkeit *f*
octavalent *(Ch)* achtwertig
octenic acid Oktenkarbonsäure *f*
octenoic acid Oktensäure *f*
octet shell *(Kern)* Achterschale *f*
octoalloy Achtfachlegierung *f*
octode Oktode *f*, Achtpolröhre *f*
octoid tooth Oktoidenzahn *m*
octuple / to verachtfachen
octuple achtfach
o. circuit Achterschaltung *f*
o.-phantom circuit *(Fmt)* Zweiunddreißigerleitung *f*
ocular Okular *n*
oculentum Augensalbe *f*
oculist's diagnostic set Augenarztbesteck *n*
odd ungerade; unpaarig
o. electron unpaariges Elektron *n*
o.-even ungerade-gerade
o.-even check *(BMSR)* Gerade-ungerade-Prüfung *f*
o.-even counter *(Dat)* Binärzähler *m*
o.-even nucleus Ungerade-gerade-Kern *m*, ug-Kern *m*
o.-fluted mit ungerader Nutenzahl
o. harmonics ungeradzahlige Oberschwingungen *fpl*
o. mass ungerade Massenzahl *f*
o. number ungerade Zahl *f*
o. number of flutes ungerade Nutenzahl *f*
o. numbered ungeradzahlig
o.-odd ungerade-ungerade
o.-odd nucleus Ungerade-ungerade-Kern *m*, uu-Kern *m*
o. parity ungerade Parität *f*
o.-parts einzelne Stücke *npl*
o.-shaped work unregelmäßig geformtes Werkstück *n*
oddleg caliper einseitiger Taster *m*
oddments *(Schiff)* Ladereste *mpl*, Einzelstücke *npl* der Ladung
odometer Kilometerzähler *m*
odorant Geruchsstoff *m*, Riechstoff *m*
odour control Geruchsbeseitigung *f (Gesundheitstechnik)*

oenanthic acid Önanthsäure *f*
o. aldehyde Önanthaldehyd *m*
oenanthylic acid Önanthsäure *f*
oenometer Aräometer *n* zur Weinprüfung
oersted Oerstedt *n (veraltete Einheit der magnetischen Feldstärke; 1 Oe = 79,58 A/m)*
of box-section construction kastenförmig ausgeführt
of different nature artfremd
of long focal distance *(Opt)* langbrennweitig
of opposite sign mit entgegengesetztem Vorzeichen
of regular pattern *(Led)* gut gestellt *(Narben)*
of short focal distance *(Opt)* kurzbrennweitig
of standardized building-block construction aus standardisierten Baueinheiten aufgebaut
of the same type typengleich
of the second order zweiten Grades *(z. B. Gleichung)*
of the slow-helix type schlankgenutet
off 1. *(El)* [aus]geschaltet, abgeschaltet; 2. geschlossen *(Hahn)*
off and on signal proving Signalmelder *m* für Halt- und Fahrtstellung
off-cast Fehlguß *m*
off-centre außermittig; seitlich versetzt; verlaufen *(Bohrer)*
off-centre hole verlaufene Bohrung *f*
off-centre position Unmittigkeit *f*
off-colour Fehlfarbe *f*
off-course correction *(Flg, Rak)* Kursverbesserung *f*
off-critical amount *(Kern)* nichtkritische Menge *f*
off-duty dienstfrei
off-gas pump Gasabsaugpumpe *f*
off-gas treatment Gasreinigung *f*
off-grade Abweichung *f (vom Gütegrad)*
off-grade iron *(Met)* Übergangseisen *n*
off-ground frei, vom Boden abgehoben
off-hand grinding Handschliff *m*; Schleifen *n* von Hand; Freihandschleifen *n*
off-hand grinding machine Freihandschleifmaschine *f*
off-heat *(Met)* Fehlschmelze *f*
off-iron *(Met)* Übergangsroheisen *n*, Ausfalleisen *n*
off-limit check *(BMSR)* Grenzwertüberwachung *f*
off-line *(Dat)* nicht eingegliedert, außerhalb, nebenher *(Kopplung Rechner-Prozeß)*
off-line computer selbständiger Rechner *m*
off-line operation unabhängig betriebene Datenverarbeitung *f*
off-load voltage elektromotorische Kraft *f*, EMK
off-normal contact *(Fmt)* Arbeitskontakt *m*
off-peak load *(El)* Normalbelastung *f*, Belastung *f* außerhalb der Spitzenzeit
off-period *(El)* Sperrzeit *f*

off-position Ruhestellung *f*, „Aus"-Stellung *f*, Ausschaltstellung *f (z. B. Relais)*
off-premises extension station *(El)* Außennebenstelle *f*
off-radial exzentrisch
off-resonance *(El)* vom Resonanzpunkt entfernt; verstimmt
off-resonance trip Verstimmungsschutz *m (Fernsprechtechnik)*
off-shore *(Schiff)* ablandig
off-shore cruiser See[fahrt]kreuzer *m (Jacht)*
off-shore dock L-Dock *n*
off-shore drill rig Ölbohrinsel *f*
off-shore drilling Unterwasserbohren *n*
off-shore drilling barge küstenferne Ölbohrinsel *f*, schwimmende Küstenölbohrinsel *f*
off-shore fisheries Hochseefischerei *f*, Tiefseefischerei *f*
off-shore navigation Fernfahrt *f*
off-shore oil rig Küstenölbohrinsel *f*
off-shore race Hochseeregatta *f*
off-shore wind Landwind *m*, ablandiger Wind *m*
off-shore yacht Hochseejacht *f*
off-side rechte Fahrbahnseite *f (in linksfahrenden Ländern)*, linke Fahrbahnseite *(in rechtsfahrenden Ländern)*
off-site außerhalb der Baustelle
off-size Maßabweichung *f*, Abmaß *n*, maßungenaues Werkstück *n*
off-sounding nicht mehr lotbar, nicht mehr lotbarer Grund *m*
off-spring *(Kern)* ausgelöstes Teilchen *n*, Sekundärteilchen *n*
off-street Nebenstraße *f*
off-take *s.* offtake
off-the-road geländegängig
off-the-road tyre Geländereifen *m*
off-the-shelf *(Am)* ab Lager
off-the-solid shotfiring *(Brg)* Schießen *n* aus dem Vollen
off-track equipment *(Brg)* Ausrüstung *f* für gleislosen Betrieb
off-watch *(Schiff)* Freiwache *f*
offal Abfall *m*
o. timber Abfallholz *n*
offals Abfälle *mpl*, Innereien *fpl*
offcut *(Typ)* Verschnitt *m*, Abfallpapier *n*, Papierabschnitt *m*
offer [up] / to probieren, einpassen
offer of proof Beweisangebot *n*
office *(Fmt)* Amt *n*
o. cable Amtskabel *n*, Zimmerkabel *n*, Innenkabel *n*
o. machine Büromaschine *f*
o. wiring Amtsverkabelung *f*
official logbook Schiffstagebuch *n*, Schiffsjournal *n*
o. meat inspector Fleischbeschauer *m*
o. telephone *(Am) (Fmt)* Dienstanschluß *m*
officinal bark Medizinalrinde *f*
o. weight Arzneigewicht *n*, Apothekergewicht *n*
offing Abstand *m* eines Schiffes vom Land; Höhe *f (der See)*; offene See *f*
offset / to versetzen, abbiegen, [ab]kröpfen, absetzen
offset Absatz *m*, Kröpfung *f (Werkzeug)*; Rücksprung *m (Mauerwerk)*; S-Rohr *n*; *(BMSR)* P-Abweichung *f*, bleibende Regelabweichung *f*; Ordinatenabstand *m*
offset versetzt, gekröpft, abgesetzt
o. angle Reibungswinkel *m*
o. carrier *(Fs)* versetzter Träger *m*
o. carrier system Sendersystem *n* mit versetzten Trägerwellen, versetztes Trägerwellensystem *n*
o. clamp *(Masch)* verstellbares Spanneisen *n*
o. computer Kursrechner *m*
o. connecting link gekröpftes Verbindungsglied *n (Rollenkette)*
o.-course computer automatischer Kursanzeiger *m (Radartechnik)*; Flugwegrechner *m*
o. funnel Trichter *m* mit Knierohr
o. in phase *(El)* phasenverschoben
o. lathe tool abgesetzter Drehmeißel *m*
o. line *(Brg)* Abhieblinie *f*
o. moulding *(Plast)* Spritzen *n* hitzehärtbarer Massen unter Verwendung von vorgeheizten Tabletten
o. operation *(El)* Offsetbetrieb *m*, Versatzbetrieb *m*
o. printing machine Offsetdruckmaschine *f*
o. printing press Offsetdruckmaschine *f*, Offsetpresse *f*
o. proof press Offsetandruckpresse *f*
o. rotary printing Offsetrotationsdruck *m*
o. screwdriver Winkelschraubenzieher *m*
o. tool abgesetzter (gekröpfter) Meißel *m*
offshore *s.* off-shore
offtake Abzugskanal *m*, Ableitungsrohr *n*, Abzugsröhre *f*
offtaker *(Brg)* Anschläger *m*, Abnehmer *m*
O.H.steel Siemens-Martin-Stahl *m*
ohm Ohm *n (Einheit des elektrischen Widerstandes; 1 Ω = 1 V/A)*
ohmage *(El)* Ohmwert *m*
ohmer *(Am) (El)* Widerstandsmesser *m*
ohmic contact ohmscher Kontakt *m*
o. drop ohmscher Spannungsabfall *m*
o. load ohmsche Belastung *f*
o. loss Stromwärmeverlust *m*, ohmscher Verlust *m*
o. overvoltage konzentrationsbedingter ohmscher Spannungsabfall *m*
o. resistance Gleichstromwiderstand *m*, ohmscher (phasenreiner) Widerstand *m*, Wirkwiderstand *m*

o. terminating-resistance ohmscher Abschlußwiderstand *m*
ohmmeter Widerstandsmesser *m*, Ohmmesser *m*, Ohmmeter *n*
Ohm's law Ohmsches Gesetz *n*
oil / to 1. ölen, schmieren; 2. *(Text)* schmälzen *(Garn)*
o.-impregnate mit Öl imprägnieren, ölen
o. toughen *(Met)* in Öl vergüten
oil Öl *n*
o. absorption Ölaufnahme *f*
o. additive Ölwirkstoff *m*, Ölzusatz *m*
o. areometer Ölwaage *f*
o. asphalt Erdölbitumen *n*
o. atomizer Ölzerstäuber *m*
o. baffle Ölabscheider *m*
o. barge *(Schiff)* Ölprahm *m*
o. barrel Ölfaß *n*
o. bath Ölbad *n*
o.-bath air cleaner Naßfilter *n*, Ölbadluftfilter *n*
o.-bath filter Ölbadfilter *n*
o.-bath lubrication Ölbadschmierung *f*, Tauch[bad]schmierung *f*
o.-bearing ölführend, ölhaltig
o.-blast circuit breaker Ölströmungsschalter *m*
o. bleeder Ölablaßvorrichtung *f*
o. boiling pot Öllacksiedekessel *m*
o.-bound mit Ölbinder
o.-bound distemper Temperafarbe *f* auf Ölgrundlage
o. bowl Gehäuse *n (des Ölluftfilters)*
o. box Schmierbüchse *f*
o. brake Öl[druck]bremse *f*
o.-break circuit breaker Ölschalter *m*
o.-break fuse *(El)* Ölsicherung *f*
o.-break switch Ölschalter *m*
o. brush Schmierbürste *f*, Ölbürste *f*
o. burner Ölbrenner *m*
o. burning installation Ölfeuerungsanlage *f*
o. by-pass valve Ölumleitventil *n*
o. cake Ölkuchen *m*
o. can Öl[vorrats]kanne *f*
o. capacitor Ölkondensator *m*
o. carbon Ölkohle *f*
o. case Ölwachstuch *n*, Ölwachsüberzug *m*
o. catch-ring Ölfangring *m (Dichtung)*
o. catcher Ölschöpfer *m*, Schöpffinger *m*
o. chamber Ölkammer *f (eines Lagers)*
o. change Ölwechsel *m*
o. circuit breaker Ölschalter *m*
o. circuit lubrication Ölumlaufschmierung *f*
o. circulating lubrication using force feed Öldruckumlaufschmierung *f*
o. circulation Ölumlauf *m*
o. circulation system Ölumlaufsystem *n*
o. cloth Wachstuch *n*; Öltuch *n*, Ölzeug *n*
o. coke Ölkoks *m*, Petrolkoks *m*
o. collector Ölfang *m (Radnabe)*
o.-conducting capacity Ölleitfähigkeit *f*
o. connection Ölzuführung *f*
o. consumption Ölverbrauch *m*
o. control ring Ölabstreifring *m*, Ölhalterring *m*
o. controlling valve Ölregelventil *n*
o. coolant Kühlöl *n*
o. coolant channel Kühlflüssigkeitskanal *m*
o.-cooled transformer Öltransformator *m*
o. core Ölkern *m*
o. country lathe Sonderdrehmaschine *f* zur Herstellung von Bohrturmeinrichtungen
o. cup Öler *m*, Schmierbüchse *f*, Ölbüchse *f*
o. cushion Öldruckpolster *n*
o. dashpot 1. *(Kfz)* ölgefüllter Stoßdämpfer *m*; 2. Ölbremse *f (Regler)*
o. deflector Ölablenkblech *n*
o. deposit Ölkruste *f*, Ölrückstand *m*
o.-diffusion pump Öldiffusionspumpe *f*
o. dilution Ölverdünnung *f*
o. dip-stick Ölmeßstab *m*
o. dipper Ölnase *f* an der Pleuelstange, Schöpffinger *m*
o. dipper rod Ölmeßstab *m*
o. dish Tropfschale *f*
o. distribution box Zentralschmiereinrichtung *f*
o. drag Reibungsverlust *m* durch Ölviskosität
o. drain Ölablaß[stutzen] *m*
o.-drain cock Ölablaßhahn *m*
o.-drain pan Ölablaufwanne *f*
o.-drain plug Ölablaßschraube *f*
o.-drainage container Ölfangbehälter *m*
o. drilling rig Ölbohrausrüstung *f*; Ölbohrinsel *f (schwimmender Bohrleichtertyp)*
o.-drilling vessel schwimmende Ölbohranlage *f*
o. dripping Öltropfenfall *m*
o. drop test Löschblattprobe *f*, Tropfenprobe *f*
o. duct Schmierkanal *m*
o.-electric drive dieselelektrischer Antrieb *m*
o. engine *s.* compression-ignition engine
o.-engined dieselgetrieben
o.-engined electric-transmission locomotive Diesellok *f* mit elektrischer Kraftübertragung
o. extracting plant Ölgewinnungsanlage *f*
o. feed Ölzufuhr *f*
o.-feed pump Öldruckpumpe *f*
o. feeder Ölkanne *f*
o. field Ölfeld *n*, Erdölfeld *n*
o.-filled cable Ölkabel *n*
o. filler Öleinfüllstutzen *m*
o. filler neck Öleinfüllstutzen *m*

o.-filler plug Öleinfüllstöpsel *m*, Öleinfüllverschraubung *f*
o.-filling pipe Öleinfüllstutzen *m*
o. film Ölschicht *f*, Ölfilm *m*
o.-film rupture strength Ölfilmzerreißfestigkeit *f*
o.-filter Ölreiniger *m*, Ölreinigungsvorrichtung *f*, Ölfilter *n*
o.-filter cartridge *(Kfz)* Ölfilterpatrone *f*, Ölfiltertopf *m*
o. filtering Ölreinigung *f*
o.-fired air-furnace *(Met)* Ölflammofen *m*, ölgefeuerter Flammofen *m*
o. firing Ölfeuerung *f*
o.-flinger ring *(Kfz)* Ölschleuderring *m*, Spritzring *m*
o. flotation Schwimmaufbereitung *f* durch Öl, Ölschwimmaufbereitung *f*
o.-flow indicator Schauglas *n*
o. flow metre Öldurchflußmesser *m*
o.-foam separator *(Rak)* Schmierstoffentschäumer *m*
o. foots Ölrückstand *m*
o. for drilling Bohröl *n*
o. fuel Treiböl *n*, flüssiger Kraftstoff *m*
o. fuel tank Treibstofftank *m*, Öltank *m*, Treiböltank *m*
o.-fuelled ölgeheizt
o. fuelling *(Schiff)* Ölübernahme *f*
o. furnace Ölfeuerung *f*, ölgeheizter Ofen *m*
o. gas Fettgas *n*
o.-gauge glass Ölschauglas *n*, Ölstandsglas *n*
o. groove Ölnut *f*, Schmiernut *f*
o.-groove cutter Ölnutenfräser *m*
o.-groove drill Spiralbohrer *m* mit äußerer Ölzuführung
o.-grooving lathe Schmiernutendrehmaschine *f*
o. gun Ölspritze *f*
o.-hardened ölgehärtet
o. hardening Ölhärtung *f*
o. hardening ölhärtend
o. hardening steel ölhärtender Stahl *m*, Ölhärter *m*
o. heating Ölheizung *f*
o. hole Schmierloch *n*, Schmierbohrung *f*, Ölbohrung *f*, Ölkanal *m*, Öltasche *f*
o.-hole drill Spiralbohrer *m* mit innerer Ölzuführung
o.-hydraulic ölhydraulisch
o. immersed im Ölbad, in Öl laufend
o.-immersed transformer Öltransformator *m*
o. immersion Ölimmersion *f*
o. in bulk Ladeöl *n*, Öl *n (als Ladung in Tankern)*
o.-insulated transformer Öltransformator *m*
o. insulation *(El)* Ölisolation *f*, Ölisolierung *f*
o. insulator *(El)* Ölisolator *m*
o. interceptor Abscheider *m*, Ölfänger *m*
o.-jacketed carburetter Vergaser *m* mit ölgespeistem Heizmantel
o. leak *s.* o. leakage
o. leakage Ölaustritt *m*; Schmiermittelverlust *m*
o. leakage pipe Leckölleitung *f*
o. leather Fettgarleder *n*
o. length Ölgehalt *m (Lack)*
o. level Ölstand *m*
o.-level checking Ölstandkontrolle *f*
o.-level dipstick *s.* oil dipper rod
o.-level gauge Ölstandanzeiger *m*
o. level glass Schauglas *n* für Ölstand
o. level indicator Ölstandzeiger *m*
o. lighter Ölprahm *m*
o. line Öl[tank]leitung *f*
o.-lubricated ölgeschmiert
o. lubrication Ölschmierung *f*
o. mill Ölmühle *f*, Ölschlägerei *f*
o.-minimum circuit breaker ölarmer Schalter *m*
o. mist Ölnebel *m*
o.-mist lubricating Ölnebelschmierung *f*
o.-mist lubricator Ölnebelschmiereinrichtung *f*
o. mixture Ölgemisch *n*
o. mud Ölschlamm *m*, Ölrückstand *m*
o. of resin Harzöl *n*
o. of vitriol Vitriolöl *n*
o. operation Dieselbetrieb *m*
o. outlet Ölablaß *m*
o. pad Schmierkissen *n*
o. paint Ölfarbe *f*
o. pan Kurbelwanne *f*, Ölfangschale *f*, Ölwanne *f*, Ölsumpf *m*
o. plant Erdölraffinerie *f*
o. pollution Verunreinigung *f* der Gewässer durch Öl, Ölpest *f*
o. pool Erdöllagerstätte *f*
o. pot Schmierkanne *f*
o.-powered unit ölhydraulischer Antrieb *m*
o. preservative Imprägnieröl *n*
o. presser Ölschläger *m*
o. pressing Ölschlagen *n*
o. pressure Öldruck *m*
o.-pressure cylinder Druckölzylinder *m*
o.-pressure gauge Ölmanometer *n*
o.-pressure governor Druckölregler *m*
o.-pressure hose Öldruckschlauch *m*
o.-pressure indicator lamp Öldruckkontrollleuchte *f*
o.-pressure relief valve Ölüberdruckventil *n*
o. proof öldicht, ölbeständig
o. pump Ölpumpe *f*
o.-pump carrier Ölpumpengehäuse *n*
o. putty Ölkitt *m*, Fensterkitt *m*, Glaserkitt *m*

o.-quenched cut-out *(El)* Sicherung *f* unter Öl, Ölsicherung *f*, Öllöschsicherung *f*
o.-quenched fuse *(El)* Ölsicherung *f*
o. quenching *(Met)* Ölabschreckung *f*, Ölhärtung *f*
o.-quenching bath Ölabschreckbad *n*
o. recovery barge Ölrückgewinnungsschute *f*
o. refinery Erdölraffinerie *f*
o. relay druckölbetriebener Servomotor *m*
o. relief valve Ölüberströmventil *n*
o. removal Entölung *f*
o. renewal Ölwechsel *m*
o. renewal period Ölwechselperiode *f*
o. reserve Ölvorrat *m*
o. reservoir Ölschleuse *f*, Ölbehälter *m*
o.-resistant ölbeständig, öldicht, ölfest
o. retainer Öldichtungsring *m*
o. return Ölrückführung *f*, Ölrücklauf *m*
o.-return pipe Ölrücklaufleitung *f*
o. rig vessel schwimmende Ölbohranlage *f*
o. ring Dampfdichtungsring *m*, Schmierring *m*, Ölring *m*
o.-ring bearing Ringschmierlager *n*
o. rubber 1. *(Plast)* Faktis *m*; 2. Ölstein *m*, Abziehstein *m*, Schleifstein *m*
o. run Ölnut *f*, Schmierloch *n*
o. sand Ölsand *m*
o. sandstone Ölsandstein *m*
o. sausage *(Schiff)* Ölwurst *f* *(flexibler Ladungsbehälter)*
o. save tank Ölsammeltank *m*, Ölauffangtank *m*
o.-scoop Ölschöpfer *m*, Ölschöpfnase *f* des Pleuelkopfes
o. scraper ring Ölabstreifring *m*
o. screen Ölsieb *n*
o. seal Öldichtung *f*, Simmerring *m*
o.-seal for rotating shafts Wellendichtung *f*
o.-seal ring Simmerring *m*
o. seed Ölsamen *m*, Ölsaat *f*
o. seep Erdölaustrittsstelle *f*
o. seepage Erdölaustritt *m*, Ölausbiß *m*
o.-separating barge *(Schiff)* Ölabscheiderleichter *m*
o. separator 1. Dampfentöler *m*, Entöler *m*; 2. *(Wärm)* Wasserentöler *m*; 3. *(Schiff)* Bilgewasserentöler *m*, Ölabscheider *m*
o. shale Ölschiefer *m*
o. shield Ölschutzring *m*
o. shock absorption Öldruckfederung *f*, Ölstoßdämpfung *f*
o.-shot system Eindruckschmierung *f*
o. showings Ölanzeichen *npl*, Ölspuren *fpl*
o. sight glass Ölschauglas *n*
o. skin Ölhaut *f*, Ölzeug *n* *(Schutz gegen Nässe)*
o. sleeve Schmierbüchse *f*
o. slinger Ölnase *f* an der Pleuelstange, Ölschleuderring *m*
o. sludge Ölschlamm *m*
o. sluice Ölschieber *m*
o.-soluble öllöslich
o. splash basin Öltrog *m*
o. splash guard Ölschirm *m*, Spritzblech *n*, Ölfang *m*
o. splasher Ölnase *f* an der Pleuelstange, Ölspritzblech *n*, Ölwerfer *m*
o. spray Ölnebel *m*
o. spring Mineralölquelle *f*
o. squirt Ölkännchen *n*
o. starvation mangelnde Schmierung *f*, Ölmangel *m*
o. stone Abziehstein *m*, Ölstein *m*
o. strainer Ölfilter *n*, Ölsieb *n*
o. strata *(Geol)* Ölschichten *fpl*
o. string Fördertour *f*, letzte Rohrfahrt *(Erdöl)*
o. sump Ölsumpf *m*, Ölwanne *f*
o. supply Ölzufuhr *f*, Ölversorgung *f*, Ölvorrat *m*
o.-supply vessel Ölversorgungsschiff *n*
o. switch Ölschalter *m* *(z. B. für Stopplicht)*
o. system Schmiersystem *n*
o. tank Ölbehälter *m*, Öltank *m*
o. tank car *(Eis)* Erdölwagen *m*
o.-tank fuse Ölkesselsicherung *f*
o.-tank vessel Öltankschiff *n*
o. tank wagon *(Eis)* Erdölwagen *m*
o.-tanker Öltanker *m*
o. tanning Fettgerbung *f*
o. temperature equalizer Ölwärmetauscher *m*
o.-tempered ölgehärtet
o.-tempered board ölvergütete Platte *f*
o. tempering 1. *(Met)* Anlassen *n* in Öl; 2. Ölvergütung *f* *(Holz)*
o. tester Ölprobiermaschine *f*, Ölprüfer *m*
o. thrower Spritzring *m*, Ölschleuderring *m*
o.-tight öldicht
o. tin Öl[vorrats]kanne *f*
o. transformer Öltransformator *m*, ölgekühlter Transformator *m*
o. trap Entöler *m*, Ölfalle *f*; Ölabscheider *m*
o.-treated wasserdicht *(Stoff)*
o.-treated board ölgetränkte Platte *f*
o. treating *(Met)* Ölvergütung *f*
o.-type clutch Naßkupplung *f*
o.-vapour baffle Öldampfsperre *f*
o. varnish Ölfirnis *m*, Öllack *m*
o. viscosity Ölzähigkeit *f*
o. well Erdölquelle *f*, Ölquelle *f*; Ölbohrung *f*
o. well derrick Ölbohrturm *m*
o. well drill Ölbohrgerät *n*
o.-wetted air cleaner *(Kfz)* Naßluftfilter *n*

o. **wick** Öldocht *m*
o. **wiper** Ölabstreicher *m*
oilcloth Wachstuch *n*
oildag Öldag *n*, Öl-Graphit-Emulsion *f*
oiled geölt; geschmiert
o. **cotton** Öltuch *n*
o. **linen** Ölleinen *n*, Ölleinwand *f*
o. **paper** Ölpapier *n*
o. **paper capacitor** *(El)* Ölpapierkondensator *m*
o. **paper photometer** *(Opt)* Fettfleckfotometer *n*
o.**-up sparking-plug** verölte Zündkerze *f*
oiler Öler *m*, Schmierkanne *f*
o. **level** Ölstand *m*
oiliness Schmierfähigkeit *f*, Schmierkraft *f*, Öligkeit *f*
oiling Ölen *n*, Fetten *n*, Schmieren *n*; *(Text)* Schmälzen *n*
o. **devil** *(Text)* Schmälzwolf *m*
o. **point** Ölstelle *f*, Schmierstelle *f*
o. **ring** Schmierring *m (Lager)*
o. **system** Schmiersystem *n*
o. **tube** Schmierrohr *n*
o. **willow** *(Text)* Schmälzwolf *m*
oilless öllos, selbstschmierend
o. **bearing** Dauerschmierlager *n*, Graphit-Bronze-Lager *n*
o. **transformer** Trockentransformator *m*, Lufttransformator *m*
oilskin Ölhaut *f*; Ölzeug *n*
oilstone / to [mit Ölstein] abziehen
oilstone Ölstein *m*, Abziehstein *m*, Wetzstein *m*
oiltight hatch *(Schiff)* öldichte Luke *f*
oilway Ölkanal *m*, Ölnut *f*
oily ölhaltig, fetthaltig, ölig
o. **top** *(Text)* fetthaltiger Kammzug *m*
o. **waste** *(Text)* fetthaltiger Abgang *m*
o. **water separator** Öl-Wasser-Separator *m*
o. **yolk** öliger Schweiß *m (Wolle)*
ointment-like salbenartig
O-L coded decimal notation *(Dat)* mit O und L verschlüsselte Dezimaldarstellung *f (L = binäre 1)*
old alt
o. **growth** Altwuchs *m*, Altholz *n*
o. **horse** *(Met)* Ofensau *f*
o. **man** 1. *(Brg)* alter Mann *m*; 2. *(Fmt) Bezeichnung für* Kurzwellenamateur *m*
o. **material** Altmaterial *n*
o. **sand** Altsand *m (Gießerei)*
o. **timber** Abbruchholz *n*, altes Bauholz *n*
Oldham coupling Oldham-Kreuzscheibenkupplung *f*
oleaginous ölartig, ölabgebend, ölig, fettig
olefiant gas Äthylen *n*
olefine Olefin *n*, Äthylenkohlenwasserstoff *m*
olefinic olefinisch
o. **bond** Äthylenbindung *f*, Doppelbindung *f*
oleic acid Ölsäure *f*, Olein *n*
oleiferous plants Ölpflanzen *fpl*
olein Olein *n*, Ölsäuretriglyzerid *n*
oleinic acid Ölsäure *f*, Olein *n*
oleo-pneumatic shock absorber Ölstoßdämpfer *m*
o.**-resinous varnish** Ölharzlack *m*
oleograph Öldruck *m*
oleography Öldruck *m*
oleomargarine Oleomargarin *n*
oleum Oleum *n*, rauchende Schwefelsäure *f*
oligiste Stahleisen *n*, Spiegeleisen *n*
o.**-iron** Eisenglanz *m*, Hämatit *m*
oligoclase *(Min)* Oligoklas *m*, Kalknatronfeldspat *m*
oligon spar *(Min)* Oligonspat *m*
olive oil Olivenöl *n*
olivenite *(Min)* Olivenit *m*
olivine *(Min)* Olivin *m*, Peridot *m*, Chrysolith *m*
Olympic dinghy Olympiajolle *f*
ombré *(Text)* schattierter Stoff *m*
o. **dyeing** Schattenfärbung *f*, schattierte Färbung *f*
ombrograph Regenschreiber *m*
ombrometer Regenmesser *m*
omission Wegfall *m*; Lücke *f*, Auslassung *f*
omit / to auslassen, weglassen
omitting device *(El)* Überbrückungseinrichtung *f (für Zeitschalter)*
omni-aerial Rundstrahler *m (Antenne)*
o.**-directional aerial** Rundstrahlantenne *f*, ungerichtete Antenne *f*
o.**-directional microphone** ungerichtetes Mikrofon *n*
omnibearing Allrichtungs-, Rund-
o. **indicator** Azimutanzeiger *m*
omnibus Omnibus *m*
o. **bill of lading** *(Schiff)* Sammelkonnossement *n*
omnidirectional Allrichtungs-; Rund-
o. **radio beacon** Allrichtungsfunkfeuer *n*
on a heavy cut bei großer Spantiefe
on a light cut bei geringer Spantiefe
on account of space limitation infolge Raummangels
on an even keel gleichlastig *(Schiff)*
on buttock *(Brg)* vor Kopf
on centres Achsabstand *m*, Achsmaß *n*
on-cost man Bergarbeiter *m* mit Schichtentlohnung
on-course zone *(Flg)* Dauertonsektor *m*
on edge hochkant
on end *(Brg)* auf Lagen *(Strebstellung)*
on face *(Brg)* senkrecht zu den Lagen *(Strebstellung)*
on level trim gleichlastig *(Schiff)*

on-line *(Dat)* eingegliedert, mitlaufend *(Kopplung Rechner-Prozeß)*
on-line computer *(Dat)* Prozeßrechner *m*
on-line data reduction *(Dat)* eingegliederte (parallele) Datenreduktion *f*, Parallelreduktion *f*
on-load switch *s.* on-load tap changer
on-load tap changer Stufenschalter *m*, Last[um]schalter *m*, Umschalter *m* für Lastregelung
on-load tap-changing equipment (switch) Last[um]schalter *m*, Umschalter *m* für Spannungsregelung unter Last
on-load voltage *(El)* Spannung *f* unter Last, Spannung bei Belastung
on-off action *s.* two-step action
on-off-control *s.* two-step action
on-off keying *(Fmt)* Ein-Aus-Tastung *f*
on open circuit *(El)* bei Leerlauf, unter Leerlaufbedingungen, im Leerlaufzustand
on period *(El)* Flußzeit *f*, Stromflußperiode *f*; Sperrzeit *f*
on phone auf Telefonie
on-position *(El)* Arbeitsstellung *f*, „Ein"-Stellung *f*
on-signal proving *(El)* Signalhaltmelder *m*
on-site Baustellen-
on-stream in Betrieb, in Gang
on-the-air monitor *(El)* Hauptmonitor *m*
on the fly *(Büro)* im Schwunge *(z. B. mit Druckhämmern im Schwunge drucken)*
on the port beam *(Schiff)* backbord querab
on the port bow *(Schiff)* backbord achteraus
on the port quarter *(Schiff)* backbord voraus
on-the-scene broadcast *(Fs)* Direktsendung *f*
once-through boiler Zwangsdurchlaufkessel *m*
o.-through cooling Kühlung *f* durch einmaligen Durchfluß des Kühlmittels
o.-through lubrication Frischölschmierung *f*, Durchlaufschmierung *f*
ondograph Ondograf *m*, Wellen[linien]schreiber *m*
ondometer Frequenzmesser *m*, Wellenmesser *m*
ondoscope Glimmlichtoszilloskop *n*, Ondoskop *n*
one-address instruction *(Dat)* Einadreßbefehl *m*
one-armed lever einarmiger Hebel *m*
one-bath dyeing *(Text)* Einbadfärbung *f*
one-bath process *(Led)* Einbadgerbung *f*
one-class vessel Einklassenschiff *n*
one-coil transformer Autotransformator *m*, Einspulentransformator *m*, Spartransformator *m*, Sparwandler *m*
one-compartment ship *(Schiff)* Einabteilungsschiff *n*, Einraumschiff *n*
one-compartment vessel *s.* one-compartment ship
one-conductor cable Einleiterkabel *n*, Einfachkabel *n*, einad[e]riges Kabel *n*
one-crop culture *(Lw)* Monokultur *f*
one-cut Brett *n (40 mm dick)*
one-cycle multivibrator *(Fs)* Einschlagmultivibrator *m*, monostabiler Multivibrator *m*, Start-Stop-Multivibrator *m*, Univibrator *m*
one-design clan Einheitsklasse *f (Segelboot)*
one-digit einstellig
one-dimensional eindimensional
one-direction ball thrust bearing einseitig wirkendes Axialrillenkugellager *n*
one-direction locating roller bearing Schulterrollenlager *n*
one-eighth bend *(Wärm)* Bogenrohr *n*
one-electron bond Einelektronenbindung *f*
one-eyed *(Opt)* einäugig
one-figure einstellig
one-fluid cell *(El)* Flüssigkeitselement *n*
one-handed action Einhandbedienung *f*
one-handle einschenklig, mit einem Griff
one-hop system Richtfunksystem *n* ohne Zwischenrelaisstationen
one-hour duty Stundenbetrieb *m*
one-hour rating Stundenleistung *f*
one-hour speed Stundendrehzahl *f*
one-line data processing schritthaltende Datenverarbeitung *f*
one-man control Einmannbedienung *f*
one-man cross-cut saw Einmannsäge *f*, Waldsteifsäge *f*
one-off part Einzelteil *n*
one-off production Einzelfertigung *f*
one-off work Bearbeitung *f* eines Einzelwerkstücks
one-pass *(Masch)* in einem Durchgang
one-pen recorder Einfachlinienschreiber *m*
one-phase einphasig
one-piece einteilig, ungeteilt, aus einem Stück gefertigt, Voll-, massiv
one-piece inductor Massivinduktor *m*
one-piece wing *(Flg)* durchgehende Tragfläche *f*
one-place einstellig
one-place number einstellige Zahl *f*
one-plane projection Eintafelprojektion *f*
one-quadrant multiplier Einquadrantmultiplikator *m*, Einquadrantmultiplizierer *m (Analogrechner)*
one-range winding *(El)* einetagiger Wickelkopf *m*, einetagige Wicklung *f*
one-shot lubrication Eindruck[zentral]schmierung *f*
one-shot multivibrator monostabiler Multivibrator *m (s. a. ~-cycle multivibrator)*
one-side box-tipping wagon *(Eis)* Einseitenkastenkipper *m*
one-sided einseitig

one-sided deflection of the steam jet einseitige Strahlablenkung *f (Turbine)*
one-sided paper einseitig gestrichenes Papier *n*
one-stage resin Resolharz *n*
one-step resin Resolharz *n*
one-story furnace Einetagenofen *m*
one-stroke bell Einschlagwecker *m*
one-termed *(Math)* eingliedrig
one throw crank einfach gekröpfte Kurbelwelle *f*
one-to-one *(Math)* eineindeutig, schlicht, umkehrbar eindeutig *(bei Zuordnungen, Abbildungen)*
one-to-one transformer *(El)* Isoliertransformator *m*, Transformator *m* 1:1
one volume edition *(Typ)* einbändige Ausgabe *f*
one-wave rectifier Halbwellengleichrichter *m*, Einweggleichrichter *m*
one-way automaton *(Dat)* Einwegautomat *m*
one-way clutch Freilaufkupplung *f*
one-way container Einwegbehälter *m (Verpackung, die nur einmalig benutzt wird)*
one-way packing verlorene Verpackung *f*, Einwegbehälter *m*
one-way rectifier *(El)* Einweggleichrichter *m*
one-way reinforced slab einsinnig bewehrte Platte *f (Stahlbetonbau)*
one-way road Einbahnstraße *f*
one-way slab einsinnig bewehrte Platte *f (Stahlbetonbau)*
one-way street Einbahnstraße *f*
one-way switch [einfacher] Schalter *m*, einpoliger Schalter, Einweg[um]schalter *m*
one-way traffic Einbahnverkehr *m*
one-wire line *(El)* eindrähtige Leitung *f*, Eindrahtleitung *f*
onker *(sl)* Holztransporter *m (Segelschiff für den Holztransport von der Ostsee nach England)*
onset time *(Ak)* Abklingzeit *f*
onsetting *(Brg)* Aufschieben *n*
onward transmission Weiterbeförderung *f*
onyx *(Min)* Onyx *m*
oolite *(Geol)* Oolithformation *f*
oolitic oolithisch, Oolith[en]-
  o. iron ore Eisenoolith *m*, Eisenrogenstein *m*, oolithisches Eisenerz *n*
  o. ore *(Min)* Bohnerz *n*
ooze / to durchsickern, langsam austreten (abscheiden)
oozing basin *(Bw)* Sickerbecken *n*
oozy schlickerig, schlammig
opacifier Opalisiermittel *n*, Trübungsmittel *n (Anstrichtechnik)*
opacity Opazität *f*, Trübung *f*, Undurchsichtigkeit *f*; Deckfähigkeit *f*, Deckkraft *f (Anstrichtechnik)*
opal *(Min)* Opal *m*
  o. glass Milchglas *n*, Mattglas *n*, Trübglas *n*, Opalglas *n*
  o. jasper *(Min)* Jaspopal *m*
  o. lighting fitting Opalglasleuchte *f*, Opalglasarmatur *f*
opalescence Opalisieren *n*, Opaleszenz *f*
opalescent opalisierend
  o. glass mitteltrübes Glas *n*, Opaleszenzglas *n*
opaline glass *s.* opalescent glass
opaque milchig, trüb, undurchsichtig; lichtundurchlässig
  o. colour Deckfarbe *f*
  o. enamel Opakemail *n*
  o. glass Opakglas *n*
  o. illuminator Opakilluminator *m (Mikroskopie)*
  o. meal Kontrastfüllung *f (Röntgendiagnostik)*
  o. object lichtundurchlässiges Objekt *n (Mikroskopie)*
  o. pattern lichtundurchlässiges Muster *n (Mikroskopie)*
  o. pigment Deckfarbe *f*
  o. to heat wärmeundurchlässig
  o. to infrared [rays] infrarotundurchlässig
  o. to ultrared [rays] ultrarotundurchlässig
opaqueness Lichtundurchlässigkeit *f*, Undurchsichtigkeit *f*, Trübung *f*
open / to 1. *(El)* öffnen, unterbrechen, trennen; 2. einmünden; 3. *(Ch)* aufschließen; 4. *(Met)* auflockern *(Formsand)*; 5. *(Brg)* anhauen *(Flöz)*
  o. a chamber *(Brg)* eine Kammer aufhauen
  o. a circuit ausschalten, einen Stromkreis öffnen (trennen)
  o.-anneal blauglühen *(Stahl)*
  o. out aufweiten
  o. the furnace anstechen, abstechen *(z. B. SM-Ofen)*
  o. the mould Gußform auswerfen (aufbrechen)
open 1. offen; 2. getrennt *(Stromkreis)*
  o. aerial Außenantenne *f*, offene Antenne *f*
  o.-air Außen-, Frei[luft]-
  o.-air exhibition space Freiausstellungsfläche *f*
  o.-air plant Freiluftanlage *f (Umspannwerk; Freibauweise)*
  o.-air space Freifläche *f (Messe)*
  o.-air transformer Freilufttransformator *m*, Freiluftumspanner *m*
  o. amortisseur winding *(Elm)* unterteilte Dämpferwicklung *f*
  o. and shut action *s.* two-step action
  o. annealing Blauglühen *n (Stahl)*
  o. anti-ballooning device *(Text)* halbringförmige Antiballonvorrichtung *f*
  o. assembly offene Verleimung *f (bei Plasterzeugnissen)*

o. assembly time Leimabbindezeit *f*, Wartezeit *f*
o. barge Schute *f*
o. basin *(Schiff)* Außendock *n*
o. beam offenes Strahlenbündel *n*
o. belt offener Riemen *m*, Parallelriemen *m*
o. belt-drive offener Riementrieb *m*
o. berth *(Schiff)* offener Ankerplatz *m*
o. blade seitlich nicht begrenzte Schaufel *f* *(Turbine)*
o. bore Durchgangsbohrung *f*
o. breast of a blast furnace *(Met)* offene Hochofenbrust *f*
o. cast *s.* opencast
o.-centre valve Steuerventil *n* mit Unterlappung in der Ruhelage
o. circuit offener Kreis (Stromkreis) *m*
o. circuit admittance Leerlaufscheinleitwert *m*, Scheinleitwert *m* einer offenen Leitung, Leerlaufadmittanz *f*
o. circuit alarm device Arbeitsstromalarmgerät *n*
o. circuit alarm system Alarmeinrichtung *f* mit Arbeitsstrom
o. circuit characteristic Leerlaufkennlinie *f*, Leerlaufverhalten *n*, Leerlaufcharakteristik *f*
o. circuit contact Ruhekontakt *m*, Arbeitskontakt *m*
o. circuit control Umschaltung *f* mit Unterbrechung, Schrittregler *m*, Stufenregler *m*
o. circuit impedance Leerlaufwiderstand *m*, Leerlaufimpedanz *f*
o. circuit loss Leerlaufverlust *m*
o. circuit operation Arbeitsstrombetrieb *m* *(Relais)*
o. circuit transformer Transformator *m* mit offenem Eisenkern (magnetischem Kreis)
o. circuit transition Reihen-Parallelumschaltung *f* mit Unterbrechung
o. circuit ventilated mit offenem Kühlkreislauf
o. circuit ventilation Frischwasserkühlung *f*, Frischluftkühlung *f*
o. circuit voltage Leerlaufspannung *f*
o. circuit winding offene Wicklung *f*
o.-circuited *(El)* leerlaufend, unter Leerlaufbedingungen
o. clinker double scull *(Schiff)* Gigdoppelzweier *m*
o.-closed loop control Regelung *f* mit Hilfsregelgröße
o.-closed shelter decker *(Schiff)* offener und geschlossener Schutzdecker *m*
o.-coil armature *(El)* Anker *m* mit nicht geschlossener Wicklung, offener Anker
o. contact Arbeitsstromkontakt *m*
o.-core transformer *s.* o. circuit transformer
o. cribbing *s.* cribbing
o. cup offener Tiegel *m* *(für Flammpunktbestimmungen)*
o. cup flash point Flammpunkt *m* *(bestimmt im offenen Tiegel)*
o. cup testing Flammpunktbestimmung *f* im offenen Tiegel
o. cut 1. offene Baugrube *f*; Graben *m*; 2. Tagebau *m*
o.-cut copper mine Kupfertagebau *m*
o. die offenes Gesenk *n*
o. discharge safety valve *(Wärm)* Sicherheitsventil *n* ohne Dampfabblasestutzen, offenes Sicherheitsventil
o. draw offene Bahnüberführung *f* *(Papierherstellung)*
o.-end forecastle *(Schiff)* offene Back *f*
o. ended clicking press Tellerstanze *f*
o.-ended line leerlaufende (offene) Leitung *f*
o.-ended well-type dredger Heckschlitzbagger *m*
o. eye offene Öse *f*
o. feeder Stichleitung *f*
o. feedheater *(Wärm)* offener Vorwärmer *m*
o. floor *(Schiff)* offene Bodenwrange *f*
o.-floor double bottom Doppelboden *m* mit offenen Bodenwrangen
o.-front cattle barn Rinderoffenstall *m*
o.-front design *(Masch)* Einständerbauart *f*
o.-front eccentric press Einständerexzenterpresse *f*
o.-front hog barn Schweineoffenstall *m*
o.-front press Einständerpresse *f*
o.-front shelter Offenstall *m*
o.-fronted ... *s.* o.-front ...
o. grain *(Met)* Grobkorn *n*
o. grain structure grobes Gefüge *n*, Grobgefüge *n*
o.-grained iron grobkörniges Eisen *n*
o. grid unterbrochener (defekter) Gitterkreis *m* *(bei Röhren)*
o. harbour offener Hafen *m* *(ohne Handelsbeschränkungen)*; ungeschützter Hafen
o. hearth *(Met)* Frischherd *m*, offener Herd *m*
o.-hearth cinder Martinofenschlacke *f*
o.-hearth furnace Siemens-Martin-Ofen *m*
o.-hearth furnace port end Siemens-Martin-Ofenkopf *m*
o.-hearth iron Siemens-Martin-Flußstahl *m*, Martinflußeisen *n*, Siemens-Martin-Material *n*
o.-hearth pig iron Stahleisen *n*
o.-hearth process Herdfrischverfahren *n*, Siemens-Martin-Verfahren *n*
o. hearth refining Herdfrischen *n*, Martinieren *n*
o.-hearth steel Siemens-Martin-Stahl *m*
o. jet of a liquid Freistrahl *m*
o. joint Schweißverbindung *f* ohne gegenseitige Berührung der zu verbindenden Teile

o. knot Astloch *n*
o.-line balancing network *(El)* Freileitungsnachbildung *f*
o. line between stations *(Eis)* freie Strecke *f*
o. loop offener Regelkreis *m*
o. loop control *(BMSR)* Steuerung *f*
o. loop control system Steuerkette *f*
o.-mouth buffer *(Eis)* Glockenpuffer *m*
o. pan mixer *(Bw)* Trogmischer *m*
o. pass offenes Kaliber *n (Walzwerk)*
o.-phase circuit breaker *(El)* Motorschutzschalter *m*, Netzschutzschalter *m*
o.-phase protection *(El)* Leiterbruchschutz *m*
o. piston *(Wärm)* einwandiger (offener) Kolben *m*
o. pit Tagebau *m*
o. pit rim Tagebaukante *f*
o. pitting Tagebau *m (Verfahren)*
o. plan office Großraumbüro *n*
o.-primary current transformer Schienenstromwandler *m*, Stabstromwandler *m*
o. road offene Reede *f*
o. roadstead offene Reede *f*
o. sand casting *(Met)* Herdguß *m*
o. sand mould Bodenform *f (Gießerei)*
o. screening area offene Siebfläche *f*
o. section of line *(Eis)* offene Strecke *f*
o. shed Stehfach *n*, Ruhefach *n (Jacquard)*
o.-shed Offenfach- *(Jacquard)*
o. shed cam dobby Offenfachschaftmaschine *f*
o. shed Jacquard machine Offenfachjacquardmaschine *f*
o. ship offenes Schiff *n (Schiffstyp)*
o.-side design Einständerbauart *f (Werkzeugmaschinen)*
o.-side jig boring machine Lehrenbohrwerk *n* in Einständerausführung
o.-side plate planing machine Einständerblechkantenhobelmaschine *f*
o.-site press Einständerpresse *f*
o.-side tool post Spannklaue *f*, Winkelklaue *f*, deutscher Meißelhalter *m*
o.-sided planer Einständerhobelmaschine *f*
o.-sided tool post *s.* o.-side tool post
o. socket Henkelmuffe *f*
o.-soil culture Freilandkultur *f*
o. square butt weld I-Naht *f*
o. stope *(Brg)* offene Abbaukammer *f (ohne Ausbau oder Versatz)*
o. subroutine *(Dat)* offenes Teilprogramm (Unterprogramm) *n*
o.-surface condenser *(Wärm)* Rieselkondensator *m*
o.-surface cooler Rieselkühler *m*
o.-surface evaporator Rieselverdampfer *m*
o. tail boom *(Flg)* Gitterschwanzträger *m*
o. tank method Trogtränkung *f*
o.-texture twill *(Text)* Durchbruchköper *m*
o. tie up *(Text)* offene (englische) Schnürung *f*
o.-top paper tub *(Text)* offene Papierhülse *f (Spinnerei)*
o.-type offen, gegen zufällige Berührung nicht geschützt
o.-type boiling plate elektrische Kochplatte *f*
o.-type cattle shelter Rinderoffenstall *m*
o. types *(Typ)* lichte Schrift *f*
o. water offenes Wasser *n (eisfrei)*; unbegrenztes Wasser *(Schiffahrt)*
o.-water efficiency of the propeller *(Schiff)* Propellerwirkungsgrad *m* des freifahrenden Propellers
o. water propeller test *(Schiff)* Propellerversuch *m (freifahrend)*
o. water test *(Schiff)* Freifahrtversuch *m*
o. well Treppenschacht *m*
o. wind tunnel *(Flg)* Freistrahlwindkanal *m*
o.-wire circuit *(El)* Freileitung *f*
o.-wire fuse offene Sicherung *f*
o.-wire pole line Freileitung *f* auf Holzmasten
o.-wire twin unabgeschirmte Doppelleitung *f*
opencast *(Brg)* Tagebau[betrieb] *m*
o. coal Tagebaukohle *f*
o. development Tagebauaufschluß *m*
o. lignite mine Braunkohlentagebau *m*
o. mining Tagebau *m (Verfahren)*, Tagebaubetrieb *m*, Tagebaugewinnung *f*
opener *(Text)* Öffner *m*, Ballenöffner *m*, Ballenbrecher *m*
o. for greasy wool Öffner (Reißwolf) *m* für Schweißwolle
o. waste Öffnerabfall *m*
opening 1. Öffnung *f*, Loch *n*, Kanal *m*, Bohrung *f*, Durchführung *f*; Weite *f*; Durchbruch *m*; Durchbrechung *f*; 2. *(El)* Öffnung, Trennung *f*, Öffnen *n*, Trennen *n*; *(Fmt)* Freigabe *f*; 3. Mündung *f*, Ausmündung *f (Turbine)*; 4. *(Ch)* Aufschluß *m*; 5. *(Brg)* Öffnung, Erschließung *f*; 6. *(Text)* Öffnen, Auflockern *n*; 7. Lichtung *f (Wald)*; 8. *(Geol)* Hohlraum *m*
o. capacity Spannweite *f (Schraubstock)*
o. cylinder *(Text)* Öffnertrommel *f*
o. drum *(Text)* Schlagtrommel *f*
o. for outflow Ausflußmündung *f*, Ausflußöffnung *f (Turbine)*
o. force Öffnungskraft *f*
o. in distance piece Laternenfenster *n*, Ausschnitt *m* des Mittelstücks, Schauloch *n*
o. in the plane face Hobelmaul *n*
o. light bewegliches Fenster *n*
o. material Magerungsmittel *n (Keramik)*
o. of a circuit breaker Öffnung *f* eines Schalters, Ausschaltung *f*, Abschaltung *f*

o. of groove Kaliberöffnung *f (Walzstraße)*
o. of ring Ringausschnitt *m*
o. out Ausfahrt *f*; Aufweitung *f*
o.-shot *(Brg)* Einbruchsschuß *m*
o. sidelight Drehfenster *n*, Pivotfenster *n*
o. size Maulweite *f*
o. spark *(El)* Öffnungsfunke *m*
o. time Ausschaltverzug *m (Kontakttrennung)*
o.-up *(Brg)* Aufschluß *m*, Ausrichtung *f*
openwork Durchbrucharbeit *f (Sticken)*
opera glass Opernglas *n*
o. house Opernhaus *n*
operable betriebsbereit
operand Rechengröße *f*
o. address Operandadresse *f*, Rechengrößenadresse *f*
operate / to bedienen, handhaben, betätigen, schalten, betreiben; arbeiten; speisen *(einen Elektromotor)*
o. a station eine Funkstelle betreiben
o. on frequencies auf Frequenzen arbeiten
operate margin *(Fmt)* Ansprechgrenze *f*
o. time Ansprechzeit *f (Relais)*
operating Betrieb *m*, Betätigung *f*
operating Betriebs-, betrieblich
o. angle *(El)* Stromflußwinkel *m*
o. by hand Handbedienung *f*, Handbetätigung *f*, Bedienung (Betätigung) *f* von Hand
o. characteristic Annahmekennlinie *f*, Operationscharakteristik *f*
o. characteristics Betriebskennzahlen *fpl*, Betriebsmerkmale *npl*, Betriebsdaten *pl*
o. condition Betriebsbedingung *f*, Betriebsverhältnis *n*, Arbeitsbedingung *f*; Betriebszustand *m*
o. contact *(El)* Arbeitskontakt *m*
o. control 1. *(Am) (Masch)* Betätigungselement *n*; 2. *s.* automatic controller
o. cost Betriebskosten *pl*
o. current *(El)* Arbeitsstrom *m*, Betriebsstrom *m*
o. current release Arbeitsstromauslöser *m*
o. curve Arbeitskennlinie *f*
o. cylinder *(Masch)* Arbeitszylinder *m*
o. data technische Daten *pl*; Betriebsdaten *pl*
o. desk Steuerpult *n*, Bedienpult *n*
o. device Bedienungseinrichtung *f*
o. efficiency Arbeitsleistung *f*
o. element Bedienelement *n*
o. engineer Betriebsingenieur *m*
o. expense[s] Betriebskosten *pl*
o. facilities Betriebseinrichtung *f*
o. frequency Betriebsfrequenz *f*; Häufigkeit *f* der Betätigung *(z. B. Schalter)*
o. friction Betriebsreibung *f (Lager)*
o. gear of interlocking plant *(Eis)* Antrieb *m* des Stellwerkes
o. handle Bediengriff *m*, Bedienungshebel *m*
o. handwheel Bedienungsrad *n*
o. head Nutzförderhöhe *f*
o. instruction[s] Bedienungsanleitung *f*, Bedienungsanweisung *f*
o. key Taste *f*
o. knob Bedienungsknopf *m*
o. lever Bedien[ungs]hebel *m*, Steuerhebel *m*
o. life Gesamtbetriebsdauer *f*, Lebensdauer *f (Festigkeit)*
o. line Arbeitskurve *f*
o. load Betriebsbelastung *f*
o. means Bedienelemente *npl*
o. mechanism Triebwerk *n*, Getriebe *n*
o. member Schaltelement *n*
o. number *(Dat)* Befehlsnummer *f*, Operationsordnungszahl *f*
o. overload *(El)* betriebsmäßige Überlastung *f*
o. pitch diameter *(Masch)* Betriebswälzkreisdurchmesser *m*
o. pole *(El)* Schaltstange *f*
o. position Schaltstufe *f (eines Stufenschalters)*
o. power Betriebsleistung *f*; Betätigungskraft *f*
o. pressure Arbeitsdruck *m*; Steuerdruck *m*; Betriebsdruck *m*
o. pressure angle Betriebseingriffswinkel *m*
o. principle Betriebsweise *f*, Arbeitsweise *f*
o. radius Ausladung *f (Ackergeräte)*
o. range Operationsbereich *m*, Wirkungsbereich *m*; Aktionsradius *m*, Fahrbereich *m (z. B. eines Schiffes)*
o. regulations Betriebsordnung *f*
o. reliability Betriebssicherheit *f*
o. requirements Betriebsanforderungen *fpl*
o. side *(Masch)* Bedienungsseite *f*
o. signal *(Eis)* Betriebszeichen *n*
o. speed Betriebsdrehzahl *f (z. B. beim Drehen)*; Arbeitsgeschwindigkeit *f*
o. stress Betriebsspannung *f*, Arbeitsbelastung
o. temperature Betriebstemperatur *f*
o. time 1. *(Masch)* Nutzungszeit *f*; Arbeitszeit *f*; 2. Ansprechzeit *f (z. B. Relais)*; 3. Ausschaltdauer *f*, Abschaltzeit *f (Sicherung)*
o. unit Bedienungsgerät *n*
o. voltage Betriebsspannung *f*; Brennspannung *f (der Entladung)*
o. wave *(El)* Arbeitswelle *f*, Betriebswelle *f*
o. wavelength *(El)* Betriebswellenlänge *f*
o. width Arbeitsbreite *f (Erntemaschine)*
operation 1. *(Masch)* Bedienung *f*, Handhabung *f*, Betriebsweise *f*; Arbeitsstufe *f*; Arbeitsvorgang *m*; Verfahren *n*; 2. Operation *f (Programmierung)*; 3. Rechnungsverfahren *n*, Rechnungsart *f*
o. code Operationsbefehl *m*, Operationskode *m (Programmierung)*
o. controls Einrichtungen *fpl* zur Prozeßführung

o. cycle Arbeitsperiode *f*
o. drawing Werkstattzeichnung *f (Einzelteil; Arbeitsgang)*
o. of a relay Ansprechen *n* eines Relais
o. part *(Dat)* Operationsteil *n*
o. plan Arbeitsplan *m*
o. producing fine finish Feinstbearbeitungsverfahren *n*
o. ratio *(Dat)* [effektiver] Betriebsfaktor *m*
o. sequence Folge *f* der Arbeitsstufen, Arbeitsfolge *f*
o. sheet Bearbeitungsplan *m*, Arbeitskarte *f*, Laufkarte *f*
operational betrieblich, betriebsmäßig
o. amplifier Rechenverstärker *m (Analogrechner)*
o. calculus Operatorenrechnung *f*
o. comfort Bedienungskomfort *m*
o. control Flugbetriebslenkung *f*
o. fluid Arbeitsflüssigkeit *f*, Druckflüssigkeit *f (Hydraulik)*
o. hazard Betriebsgefahr *f*
o. life Betriebsdauer *f*, Lebensdauer *f (in Betriebsstunden)*
o. planning operative Planung *f*
o. relay Rechenrelais *n (Analogrechner)*
o. sequence Arbeitsablauf *m*
operations area Bedienungsbühne *f*
o. control Betriebskontrolle *f*, Betriebsprüfung *f*
o. research Operations-Research, Ökonometrie *f*, Planungsforschung *f*, Unternehmensforschung *f*, Verfahrensforschung *f*
operative Arbeiter *m*; Maschinist *m*
o. attenuation *(Fmt)* Betriebsdämpfung *f*
o. condition Arbeitsbedingung *f*; betriebsfähiger Zustand *m*
o. position Betriebsstellung *f*; Arbeitslage *f*
o. speed Betriebsdrehzahl *f*
operator 1. Arbeiter *m*, Bedienungsmann *m*, Maschinenarbeiter *m*; 2. Vorführer *m (Filmprojektion)*; 3. Telefonistin *f*; 4. *(Math)* Operator *m*; 5. Reaktorfahrer *m*
o. attention Wartung *f*, Bedienung *f*
o. equation *(Math)* Operatorgleichung *f*
o. instruction sheet Arbeitsplan *m*
o. position Arbeitsplatz *m*
o. telephone jack *(Am)* *(Fmt)* Anschalteklinke *f*
operator's certificate Funkerzeugnis *n*
o. handbook Betriebshandbuch *n*
o. jack *(Fmt)* Anschaltklinke *f*
o. manual Bedienungshandbuch *n*, Betriebsanleitung *f*
o. plug *(Fmt)* Anschaltstöpsel *m*
o. set *(Fmt)* Sprechgarnitur *f*
ophthalmic ophthalmologisch
o. optician Augenoptiker *m*
o. optics Augenoptik *f*
ophthalmologist Ophthalmologe *m*
ophthalmology Ophthalmologie *f*
ophthalmoscope Ophthalmoskop *n*, Augenspiegel *m*
opiate Opiat *n*, Opiumpräparat *n*
oppose / to entgegenwirken
opposed entgegengesetzt, gegenläufig, Gegen-
o.-action screen Gegenschwingsieb *n*
o. current Gegenstrom *m*, Gegenströmung *f*
o. cylinder engine *s.* o. engine
o. engine Boxermotor *m*, Motor *m* mit gegenläufigen Kolben
o. force Gegenkraft *f*
o. piston engine Gegenkolbenmotor *m*
o. ram steering gear *(Schiff)* Gegenkolbenrudermaschine *f*
o. thrust cylinder Doppelkolbenzylinder *m*
opposing entgegengesetzt, Gegen-
o. currents entgegengesetzte Ströme *mpl*
o. field *(El)* entgegengesetztes Feld *n*, Gegenfeld *n*
o. magnetic field of armature *(Elm)* Ankergegenfeld *n*
o. reaction *(Ch)* Rückreaktion *f*, Gegenreaktion *f*
o. voltage *(El)* Gegenspannung *f*, entgegengesetzte Spannung *f*
o. winding *(El)* Gegenwicklung *f*
opposite Gegenteil *n*
opposite 1. gegenläufig, umgekehrt, entgegengesetzt; gegenüberliegend; 2. *(Math)* entgegengesetzten Vorzeichens; 3. ungleichnamig *(Pole)*
o. angle Gegenwinkel *m*
o. bank Gegenufer *n (eines Flusses)*
o. charges entgegengesetzte Ladungen *fpl*
o. contact Gegenkontakt *m*
o. course Gegenkurs *m*
o. direction entgegengesetzte Richtung *f*, Gegenrichtung *f*
o. electrode Gegenelektrode *f*
o. hand entgegengesetzter Steigungssinn *m*
o. in phase in Gegenphase, in Phasenopposition, gegenphasig
o. light Gegenlicht *n*
o. phase Gegenphase *f*
o. poles ungleichnamige Pole *mpl*
opposition *(Astr)* Opposition *f*
o. method Gegenschaltungsmethode *f*, halbpotentiometrische Schaltung *f*
optic optisch, Optik-, Seh-
o. angle Sehwinkel *m*
o. axis optische Achse *f*; Sehachse *f*
o. fiber [dünner] Lichtleiter *m*
o. spectrograph Lichtspektrograf *m*, optischer Spektrograf *m*
o. spectrum Lichtspektrum *n*, sichtbares Spektrum *n*

optical optisch, Optik-, Seh-, Licht-
o. activity optische Aktivität *f*
o. alignment testing telescope Fluchtungsfernrohr *n*
o. attenuator Kompensationsblende *f* *(Fotometrie)*
o. axis optische Achse *f*
o. balance optischer Abgleich *m*
o. bench optische Bank *f*
o. bleaching optische Bleiche *f*
o. cement optischer Kitt *m*
o. centre optischer Mittelpunkt *m*
o. comparator Optimeter *n*
o. correction optische Korrektion (Korrektur) *f*
o. crystallography Kristalloptik *f*
o. densitometer Schwärzungsmesser *m*
o. density optische Dichte *f*; Schwärzung *f*
o. density of smoke optische Rauchdichte *f*
o. depth optische Tiefe (Dichte) *f*
o. design optische Abbildung *f*
o. dividing head optischer Teilkopf *m*
o. dye *(Ch)* optischer Aufheller *m*
o. excitation optische Anregung *f*
o. fibre optische Faser *f*
o. flat Planglasplatte *f*
o. folding Knickung *f* des Strahlenganges
o. grating optisches Gitter *n*
o. isomerism optische Isomerie *f*
o. layout optische Anordnung *f*, optisches Schema *n*
o. limits *(Astr)* optische Schranken *fpl*
o. marking *(Schiff)* optisches Anzeichnen *n* *(Ersatz für Schnürbodenarbeit)*
o. maser *s.* laser
o.-mechanical system optisch-mechanisches System *n*
o. method of speed measurement optische Geschwindigkeitsmessung *f*
o. microscope Lichtmikroskop *n*
o. outlay optischer Aufwand *m*
o. parallel planparalleles Glasprüfmaß *n*
o. path length Lichtpfadlänge *f*
o. plummet optisches Lot *n*
o. profile grinding machine optische Profilschleifmaschine *f*
o. pumping optisches Pumpen *n*
o. pyrometer Teilstrahlungspyrometer *n*
o. rotary power optisches Drehvermögen *n*
o. rotary table Rundtisch *m* mit Teilkreis
o. sight Richtglas *n*
o. sound Lichtton *m*
o. square Winkelspiegel *m*
o. telephony Lichttelefonie *f*
o. tracing method optisches Anreißverfahren *n*
o. tracking optische Bahnverfolgung *f*; optische Ortung *f*
o. transition optischer Übergang *m*
o. transparency Lichtdurchlässigkeit *f*
o. vernier Nonius *m* mit Lupe
o. vibration optische Schwingung *f*
optically active optisch aktiv
o. excited state optisch angeregter Zustand *m*
o. generated durch Licht erzeugt
o. inactive optisch inaktiv
optician Optiker *m*
optics Optik *f*
optimal optimal, Best-
o. control optimale Regelung *f*
o. efficiency optimaler Wirkungsgrad *m*
o. strategy optimale Strategie *f* *(Theorie der Spiele)*
o. value Bestwert *m*
optimality criterion Optimalitätskriterium *n* *(lineares Programmieren)*
optimally coded program *(Dat)* Bestzeitprogramm *n*
optimization Optimierung *f*
optimize / to optimieren
optimizer Optimierungseinrichtung *f*, Optimierungsmaschine *f*; Optimisator *m*
optimizing control Bestwertregelung *f*
optimum Bestwert *m*, Optimum *n*, Optimalwert *m*
optimum optimal, Optimal-, Best-
o. amplitude optimale Amplitude *f*
o. control Optimalregelung *f*
o. coupling *s.* critical coupling
o. filter Optimalfilter *n*
o. programming optimales Programmieren *n*
o. source admittance *(El)* optimaler Quellenleitwert *m*
o. value Bestwert *m*, Optimalwert *m*
o. working frequency *(Fmt)* günstigste Betriebsfrequenz *f*
optional wahlweise
opto-constructional length optische Baulänge *f*
o.-electronic opto-elektronisch
o.-electronics Opto-Elektronik *f*
o.-mechanical opto-mechanisch
o.-physical opto-physikalisch
OR buffer ODER-Schaltung *f* *(ODER im Sinne von ODER AUCH)*
OR circuit ODER-Schaltung *f*
OR-element ODER-Glied *n*
OR-ELSE ausschließendes ODER *n*, ENTWEDER ODER *(logischer Operator)*
OR-ELSE circuit ENTWEDER-ODER-Schaltung *f*
OR-ELSE-gate ENTWEDER-ODER-Glied *n*, ENTWEDER-ODER-Schaltung *f*, exklusives ODER *n*
OR-gate ODER-Glied *n*, ODER-Schaltung *f*
oral lamp Mundleuchte *f*

orange chrome Chromrot *n*
o. oil Pomeranzenöl *n*
o.-peel bucket *(Brg)* Polypgreifer *m (Schachtabteufen)*
o.-peel effect *(Plast)* Apfelsinenschaleneffekt *m*, apfelsinenschalenartige Oberfläche *f*
orbit / to in Umlauf bringen; auf eine Umlaufbahn bringen
orbit Bahn *f*, Umlaufbahn *f*, Kreisbahn *f*, Flugbahn *f*
o. circumference Bahnlänge *f*
o. of electrons Elektronenbahn *f*
o. of the satellite Flugbahn *f* des Sputniks
o. plane Bahnebene *f*
o. radius Bahnradius *m*
o. shift coils *(Kern)* Ablenkspulen *fpl*
o. stability Bahnstabilität *f*
o. tuned-dipole belt künstlicher cm-Wellen-Reflektorgürtel *m (durch Ausstreuung von Metallnadeln mit Hilfe von Satelliten)*
orbital Bahn-; Hüllen-
o. angular momentum Bahndrehimpuls *m*
o. electron Bahnelektron *n*, Hüllenelektron *n*
o. elements *(Rak)* Bahnelemente *npl*
o. magnetic moment Bahnmagnetismus *m*, magnetisches Bahnmoment *n*
o. moment [of momentum] Bahndrehimpuls *m*
o. period Umlaufszeit *f*, Umlaufsperiode *f*
o. quantum number azimutale Quantenzahl *f*
o. refuelling Bahntanken *n*, Betankung *f* in der Luft
o. rocket Satellitenrakete *f*
o. speed Kreisbahngeschwindigkeit *f*
o. velocity Bahngeschwindigkeit *f*
orchard Obstgarten *m*, Obstpflanzung *f*, Fruchtgarten *m*; Baumschule *f*
order / to 1. bestellen; 2. befehlen, vorschreiben
order 1. Ordnung *f*, Größenordnung *f*, Grad *m*, Reihenfolge *f*, Aordnung *f*; 2. Bestellung *f*, Auftrag *m*; 3. Befehl *m*, Instruktion *f*; Befehlswort *n (Rechenautomat)*
o. code Befehlskode *m*, Befehlsliste *f*
o. form Bestellformular *n*, Bestellvordruck *m*
o. list *(Dat)* Befehlsliste *f*
o. of a root Wurzelexponent *m*
o. of accuracy Genauigkeitsgrad *m*
o. of connection *(El)* Schaltfolge *f*
o. of firing Zündfolge *f (beim Sprengen)*
o. of magnitude Größenordnung *f*
o. of movements Bewegungsfolge *f*
o. of sequence Reihenfolge *f*
o. of treadling *(Text)* Tretweise *f*, Trittfolge *f*
o. sorter *(Opt)* Vorzerleger *m*
o. wire circuit *(Fmt)* Dienstleitung *f*
ordering 1. Ordnen *n*; Bestellen *n*; 2. Befehlen *n*
orders Order *f (ein Schiff fährt für Order)*
ordinal number Ordnungszahl *f*
ordinary gewöhnlich; ordentlich; normal
o. concrete Beton *m*, Schwerbeton *m*
o. differential equation gewöhnliche Differentialgleichung *f*
o. dovetailing offene Zinken *fpl (Holzbau)*
o. key Längskeil *m*
o. lay rope Kreuzschlagseil *n (Förderung)*
o. milling Gegenlauffräsen *n*
o. ray ordentlicher Strahl *m*
o. seaman Leichtmatrose *f*
o. size Normalgröße *f*
o. wave Grundwelle *f*, ordentliche Welle *f*
o. wood Blindholz *n*
ordinate Ordinate *f*
o. axis Ordinatenachse *f*
ordnance Geschütze *npl*, Artilleriewaffen *fpl*; Arsenal *n*, Zeughaus *n*; Verwaltung *f* der Arsenale
o. rocket ungelenkte Rakete *f*, Artillerierakete *f*
o. survey amtliche Landesvermessung *f*
ore Erz *n*
o. agglomerating plant Erzsinteranlage *f*
o. and fluxes *(Met)* Möller *m*
o.-bearing erzhaltig, fündig *(Lagerstätte)*
o. bin Erzbunker *m*
o. bin gate Erzbunkerverschluß *m*
o. bloom *(Met)* Luppe *f*
o. body Erzlager[stätte *f*] *n*, Erzvorkommen *n*
o. boshes Erzansätze *mpl (im Hochofen)*
o. briquetting Erzbrikettierung *f*
o. bucket Erzkübel *m*
o. burden Erzgicht *f*, Erzmöller *m (Hochofen)*
o. calcining furnace Erzröstofen *m*
o. carrier Erzschiff *n*
o. charge Erzcharge *f*, Erzgicht *f (Hochofen)*
o. chute Erzrutsche *f*
o. crusher Erzbrecher *m*
o. crushing plant Erzzerkleinerungsanlage *f*
o. deposit Erzlagerstätte *f*, Erzvorkommen *n*
o. dressing Erzaufbereitung *f*
o. dressing plant Erzaufbereitungsanlage *f*
o. fines *(Met)* Erzklein *n*
o. funnel *(Met)* Erztrichter *m*
o.-grain carrier Erz-Getreideschiff *n*
o. hearth *(Met)* Schmelzherd *m*
o. in grains Graupenerz *n*
o. in pieces Stufenerz *n*
o. leaching *(Met)* Erzlaugung *f*
o. lode *(Geol)* Erzgang *m*
o. mining Erzbergbau *m*
o.-oil carrier Erz-Ölschiff *n*
o.-oil tanker Erz-Ölschiff *n*
o. pass Erzrolloch *n*, Erzrolle *f*, Schore *f*
o. picker Erzklauber *m*
o. picking belt Erzleseband *n*
o. pillar *(Brg)* Erzpfeiler *m*, Erzfeste *f*

o. **pocket** *(Geol)* Erznest *n*, Putzen *m*
o. **process** *(Met)* Erz[frisch]verfahren *n*
o. **puddling** *(Met)* Erzpuddeln *n*
o. **pulp** *(Met)* Erztrübe *f*
o. **reduction plant** *(Met)* Erzreduktionsanlage *f*
o. **reserves** Erzvorräte *mpl*
o. **roasting plant** Röstanlage *f* für Erze
o. **separator** *(Met)* Erzscheider *m*
o.**-shoot** *(Geol)* Erzfall *m*, Adelsvorschub *m*, Adelszone *f (reicher Lagerstättenteil)*
o. **sludge** *(Met)* Erzschlamm *m*
o. **smelting** Erzschmelzen *n*; Rohschmelzen *n*
o.**-smelting furnace** Rohofen *m*, Erzschmelzofen *m*
o. **stamp** Erzpochwerk *n*, Erzstampfwerk *n*
o. **steel** Rennstahl *m*
o. **vein** *(Geol)* Erzgang *m*
o. **washing** Erzschlämmen *n*
oreing down *(Met)* nachträgliches Herunterfrischen *n* durch Erzzugabe
orelline Orlean *n*, Bixin *n (Farbstoff)*
organ Orgel *f*
o. **bellows** Balgenwerk *n*, Orgelbalg *m*
o. **blowing gear** Orgelgebläse *n*
o. **key** Orgeltaste *f*
o. **pipes** Orgelpfeifen *fpl*
o. **stop** Registerstimme *f*, Registerzug *m*, Orgelregister *n*, Orgelzug *m*
organic organisch
o. **acid** organische Säure *f*
o. **analysis** organische Analyse *f*
o. **bond** organische Bindung *f*
o.**-bonded** mit organischer Bindung
o. **chemist** Organiker *m*
o. **chemistry** organische Chemie *f*
o. **coating** organischer Überzug *m*
o. **compound** organische Verbindung *f*
o. **crystal** organischer Kristall *m*
o. **dye** organischer Farbstoff *m*
o. **glass** organisches Glas *n (z. B. Plexiglas)*
o. **matter** organischer Stoff *m*
organo-metallic compound metallorganische Verbindung *f*
o. **moderated** *(Kern)* organisch moderiert
o.**-silicon** siliziumorganisch
organoleptic *(Ch)* organoleptisch, prüfbar durch Sinneswahrnehmung
o. **test** organoleptischer Test *m*
organzine *(Text)* Organzin[seide *f*] *n*, Ajouseide *f*, Kettseide *f*
oriental ruby Rubin *m*
orientation 1. Orientierung *f*, Ausrichtung *f*; 2. Ortung *f*, Peilung *f (Radar)*; 3. *(Kern)* Valenzrichtung *f*
o. **adjustment** Orientierungsverbesserung *f (Fotogrammetrie)*
o. **angle** Orientierungswinkel *m*
o. **effect** *(Kern)* Orientierungseffekt *m*
oriented orientiert, ausgerichtet, gerichtet
o. **surface** orientierte Fläche *f*
orienting *s.* orientation
orifice / to drosseln
orifice Öffnung *f*, Ausflußöffnung *f*; Mündung *f*; Austritt *m*, Düse *f*
o. **flowmeter** *s.* o. plate flowmeter
o. **in the spinneret** Düsenloch *n* (Düsenöffnung *f*) in der Spinndüse
o. **plate flowmeter** Blendendurchflußmengenmesser *m*
origin Anfangspunkt *m*, Ausgangspunkt *m*; Ursprung *m*; Ausgangsstelle *f*; Nullpunkt *m*
o. **distortion** *(El)* Nullpunktanomalie *f*, Nullpunktfehler *m (Gasentladungsröhren)*; Ursprungsverzerrung *f*
o. **of coordinates** Koordinatenursprung *m*
o. **of cosmic rays** Herkunft *f* (Ursprung *m*) der kosmischen Strahlung
original ursprünglich, original, Original-
o. **application** Erstanmeldung *f (Patent)*
o. **drawing** Urzeichnung *f*, Stammzeichnung *f*, Originalzeichnung *f*
o. **fission** *(Kern)* Primärspaltung *f*
o. **head** Setzkopf *m (Niet)*
o. **image** Original *n*, Urbild *n*
o. **ink drawing** Tuscheoriginal *n*
o. **nucleus** *(Kern)* Mutterkern *m*
o. **patent** Hauptpatent *n*
o. **pencilled drawing** Bleioriginal *n*
o. **position** Ausgangsstellung *f*, ursprüngliche Lage *f*
o. **sample weight** Einwaage *f (Probe)*
o. **stand** Urwaldbestand *m*
o. **state** Grundzustand *m*, Anfangszustand *m*
o. **tracing** Kopiervorlage *f (technisches Zeichnen)*
o. **value** Ausgangswert *m*
originating toll centre *(Fmt)* Abgangszentrale *f*
ornamental border Zierleiste *f*
o. **castings** *(Met)* Ornamentguß *m*
o. **concrete** Sichtbeton *m*
o. **disk** *(Kfz)* Zierscheibe *f*
o. **hood** Zierhaube *f*
o. **hub cap** *(Kfz)* Zierdeckel *m*
o. **steel** Zierstahl *m*
ornamented type *(Typ)* Zierschrift *f*
ornithine *(Ch)* Ornithin *n*
orogenesis *(Geol)* Gebirgsbildung *f*
orogenic pressure *(Geol)* Gebirgsdruck *m*
o. **thrust** *(Geol)* Gebirgsschub *m*
orographic orografisch *(Regen)*
orography Gebirgskunde *f*
orphan *(Am)* [alte] Maschine *f*, für die keine Ersatzteile erhältlich sind
orphitron Rückwärtswellenoszillatorröhre *f (Höchstfrequenztechnik)*

orpiment *(Min)* Auripigment *n*, Arsenblende *f*
Orsat apparatus Orsatapparat *m*, Rauchgasprüfer *m*
orsellinic acid Orsellinsäure *f*
orthicon *(Fs)* Orthikon *n*
orthite *(Min)* Orthit *m*
ortho-axis Orthoachse *f* *(Kristallografie)*
o.-orientating group *(Ch)* Substituent *m*, der in ortho-Stellung dirigiert
orthocentre *(Math)* Höhenschnittpunkt *m*
orthochromatic *(Opt)* orthochromatisch, farbrichtig
orthoclase *(Min)* Orthoklas *m*
orthogonal orthogonal, rechtwinklig
orthojector circuit-breaker *(El)* Ölstrahlschalter *m*, Querstrahlschalter *m*
orthonormal normiert orthogonal, orthonormal *(Funktionssysteme)*
orthophonic klangtreu, lautgetreu
orthophosphoric acid Orthophosphorsäure *f*
orthorhombic [ortho]rhombisch
orthoscopic *(Opt)* orthoskopisch, verzeichnungsfrei
o. eyepiece orthoskopisches Okular *n*
orthotropic plate orthotrope Platte *f* *(Brückenbau)*
oscillable schwingfähig
oscillate / to schwingen; oszillieren; hin- und herschwingen, pendeln; vibrieren
o. at a high frequency auf einer hohen Frequenz schwingen
oscillating Schwing[ungs]-
o. arm *(Masch)* Kurbelschwinge *f*, Schwingarm *m*, schwingende Kurbelschleife *f*
o. beacon Pendelfeuer *n*
o. charge schwingende (oszillierende) Ladung *f*
o. circuit Schwing[ungs]kreis *m*, Resonanzkreis *m*
o. component Wechselglied *n*, Wechselkomponente *f*, Wechselanteil *m*
o. contact Zitterkontakt *m*, Schwingkontakt *m*
o. crank gear *(Masch)* schwingende (oszillierende) Kurbelschleife *f*, schwingendes Schleifkurbelgetriebe *n*
o. crank mechanism *(Masch)* Bogenschubkurbel *f*
o. crystal Schwingkristall *m*
o. current oszillierender Strom *m*
o. detector Schwingaudion *n*
o. direction indicator *(Kfz)* Pendelwinker *m*
o. discharge oszillierende Entladung *f*
o. electrode Gitter *n* der Bremsfeldröhre
o. electron *(Kern)* Pendelelektron *n*
o. e.m.f. Wechsel-EMK *f*
o. klystron Oszillatorklystron *n*
o. meter oszillierender Zähler *m*, Schwingspulenzähler *m*, Schwingungszähler *m*; Pendelzähler *m*
o. motion Schwingbewegung *f*, schwingende Bewegung *f*, Schwingung *f*, Pendelbewegung *f*, Schaukelbewegung *f*
o. movement *s.* o. motion
o. of the steam engine Schwingen *n* der Dampfmaschine
o.-piston flowmeter Ringkolbenzähler *m*
o. potential difference Wechselspannung *f*
o. reflector *(Kfz)* Pendelrückstrahler *m*
o. riddle Rüttelsieb *n*
o. saw Kappsäge *f*, Pendelsäge *f*
o. screen Schwingsieb *n*
o. stress *(Mech)* schwingende Belastung *f*
o. trough Schwingrinne *f* *(Fördertechnik)*
oscillation Schwingung *f*, Schwingen *n*, Oszillation *f*; Schwankung *f*; Vibration *f*; Pendelung *f*, Schaukeln *n*
o. capability Schwingungsfähigkeit *f*
o. circuit *s.* oscillating circuit
o. energy Schwingungsenergie *f*
o. frequency Schwingungsfrequenz *f*, Frequenz *f*
o. galvanometer Schwingungsgalvanometer *n*, Vibrationsgalvanometer *n*
o. node Schwingungsknoten *m*
o. of the slay *(Text)* Ladenschwingung *f*
o. period Schwingungsdauer *f*
o. range Schwingungsbereich *m*
o. squeal *(Fmt)* Rückkopplungspfeifen *n*
o. start transaction *(Rak)* Einschwingvorgang *m*
o. starting process *(Rak)* Einschwingvorgang *m*
o. stop *(El)* Schwingloch *n*, Aussetzen *n* der Schwingung
o. transformer *(El)* HF-Übertrager *m*
oscillator Oszillator *m*, Schwinger *m*, Schwingungserzeuger *m*
o. alignment Oszillatorabgleich *m*
o. circuit Oszillatorkreis *m*, Schwingkreis *m*
o. coil Oszillatorspule *f*, Schwing[kreis]spule *f*
o. drift Abweichung *f* einer Oszillatorfrequenz
o. harmonic response Empfindlichkeit *f* für Oszillatorharmonische *(Funkempfangstechnik)*
o. padder Oszillatorserienkondensator *m*
o. strength Oszillatorstärke *f*
o. tank Schwingkreis *m*
o. valve Oszillatorröhre *f*, Generatorröhre *f*
o.-wavemeter Überlagerungswellenmesser *m*
oscillatory oszillierend, schwingend
o. circuit *s.* oscillating circuit
o. current Schwingstrom *m*, schneller Wechselstrom *m*
o. discharge *s.* oscillating discharge
o. moment Auftriebsmoment *n* *(eines Schiffes)*
o. motion Drehpendelbewegung *f*; oszillierende Bewegung *f*, Schwingung *f*
o. power Schwingleistung *f*

o. surge *(El)* HF-Stoß *m*
oscillogram Oszillogramm *n*
oscillograph Oszillograf *m*
o. loop Oszillografenschleife *f*
o. valve Oszillografenröhre *f*
oscillometer Dekameter *n*, DK-Meter *n* *(HF-Analysenmeßgerät)*
oscilloscope Oszillograf *m*, Oszilloskop *n*
o. display Oszillografenanzeige *f*
o. photograph Schirmbildaufnahme *f*
o. screen Betrachtungsschirm *m*, Oszillografenschirm *m*
oscilloscopic display oszillografische Darstellung *f*
osculating *(Math)* Krümmungs-, Schmiegungs-
o. circle *(Math)* Schmiegungskreis *m*
o. plane Schmieg[ungs]ebene *f*
osculation *(Math)* Berührung *f*
osmate Osmat *n*, Salz *n* der Osmiumsäure
osmic acid Osmiumsäure *f*
o. compound Verbindung *f* des vierwertigen Osmium
osmide Osmiumlegierung *f*
osmious osmiumhaltig
osmium Osmium *n*
o. alloy Osmiumlegierung *f*
o. lamp Osmiumlampe *f*
osmometer Osmometer *n*
osmosis Osmose *f*
osmotic pressure osmotischer Druck *m*
osteolite *(Min)* Osteolith *m*
ostrich farm Straußenfarm *f*
Ostwald's dilution law *(Ch)* Ostwaldsches Verdünnungsgesetz *n*
otavite *(Min)* Otavit *m*
otter Otter *m*; Otterpelz *m*
o. board *(Schiff)* Scherbrett *n* *(für Schleppnetz)*
o. trawl *(Schiff)* Scherbrettnetz *n*
Otto cycle Otto-Verpuffungsverfahren *n* *(Verbrennungsmotor)*
Otto cycle engine *s.* Otto engine
Otto engine Ottomotor *m*
ounce Unze *f (28,34953 g)*
o. system Garnnumerierung *f (Seide)*
out *(Typ)* Leiche *f*
out heraus, hinaus; aus[geschaltet], abgeschaltet *(elektrischer Strom)*
o.-cut method *(Masch)* Verfahren *n* des Gegenlauffräsens
o.-of-balance Unwucht *f*
o.-of-balance current unsymmetrischer Strom *m*
o.-of-balance force Unwucht *f*
o.-of-balance moment Schwerpunktmoment *n*
o. of cultivation verwildert *(Land)*
o.-of-date ship *(Schiff)* überaltertes Schiff *n*
o.-of-focus image unscharfes Bild *n*
o. of frame *(Fs)* Bildhälftenverschiebung *f*
o. of line außer Flucht
o. of mesh *(Masch)* ausgerückt, außer Eingriff
o. of operation außer Betrieb
o. of order außer Betrieb; gestört
o. of phase in der Phase verschoben, phasenverschoben
o.-of-phase component *(El)* falschphasige Komponente *f*
o.-of-phase switching *(El)* Schalten *n* in Phasenopposition
o. of plumb außer Lot, aus dem Lot
o. of print vergriffen *(beim Verlag)*
o.-of-round unrund
o. of roundness Unrundheit *f*
o.-of-section Querschnittsabweichung *f*, Querschnittsabmaß *n*
o. of square nicht senkrecht, nicht im Lot
o. of step *(Elm)* außer Tritt, unsynchron, asynchron
o.-of-step protection Pendelschutz *m* *(Relaistechnik)*
o. of stock vergriffen *(beim Buchhändler)*
o. of straight ungerade
o.-of-straightness Ungeradheit *f*
o. of trim *(Schiff)* schlecht getrimmt
o. of true Schlag aufweisend, nicht rundlaufend
o. of truth ungerade, unrund
o. of tune verstimmt
o. of wind windschief *(Holz)*
outboard bearing Nebenlager *n*, Gegenlager *n*, Außenlager *n*
o. brace Gegenhalterschere *f (Fräsmaschine)*
o. motor Außenbordmotor *m*
o. profile *(Schiff)* Längsansicht *f*, Außenansicht *f*
o. support Gegenhalterschere *f (Fräsmaschine)*
outbound bearing *(Flg)* Abflugpeilung *f*
outburst *(Geol)* Ausgehendes *n*; Ausbruch *m*
o. of carbon dioxide Kohlensäureausbruch *m*
o. of coal plötzlicher Kohlenausbruch *m*
o. of gas plötzlicher Gasausbruch *m*
outcoming electron herausfliegendes Elektron *n*
outcrop / to *(Geol)* zutage treten, ausgehen, ausstreichen
outcrop *(Geol)* Ausbiß *m*, Ausgehendes *n*, Ausstrich *m*
outdated veraltet
outdoor Freiluft-, Frei-; draußen, außen, Außen-
o. aerial Außenantenne *f*, Hochantenne *f*
o. circuit break Freiluft[leistungs]schalter *m*
o. equipment Freiluftanlage *f*
o. floodlight Außenscheinwerfer *m*
o. insulator Außenisolator *m*, Freiluftisolator *m*

o. lighting Außenbeleuchtung *f*, Beleuchtung *f* im Freien
o. substation *(El)* Freiluft[schalt]anlage *f*
o. temperature Außentemperatur *f*
o. transformer Freilufttransformator *m*, Freiluftumspanner *m*
outer 1. *(El)* Außenleiter *m*; 2. *(Met)* Gußhaut *f*
outer außenbefindlich, Außen-, äußerer; kernfrei *(Elektronen)*
o. air Außenluft *f*
o. arbor support äußeres Traglager *n*, Gegenlager *n (Fräsmaschine)*
o. barrel of pump Pumpengehäuse *n*
o. bearing Außenlager *n*; äußeres Traglager *n*, Gegenhalter *m (Fräsmaschine)*
o. body shell *(Kfz)* äußere Karosserieverkleidung *f*
o. brace Gegenhalterschere *f (Fräsmaschine)*
o. bremsstrahlung *(Kern)* äußere Bremsstrahlung *f*
o. cable Außenleiter *m*
o. casing Mantel *m (z. B. eines Hochofens)*
o. coating Außenbelag *m (Elektronenröhren)*
o. conductor Außenleiter *m*
o. conductor voltage Spannung *f* zwischen den Außenleitern (Phasen), Leiterspannung *f*
o. corner Schneid[en]ecke *f (Spiralbohrer)*
o. cover Laufdecke *f*, Mantel *m*, Reifen *m*; Außenhülle *f*
o. diameter Außendurchmesser *m*
o. edge corner Schneid[en]ecke *f (Spiralbohrer)*
o. electrons Elektronenhülle *f*
o. end-paper *(Typ)* äußeres Vorsatz *n*
o. flue *(Wärm)* Außenzug *m*
o. funnel Schornsteinmantel *m*
o. gates Unterhaupt *n*, Außenhaupt *n (einer Schiffsschleuse)*
o. harbour Außenhafen *m*, Vorhafen *m*
o. jacket Außenmantel *m*
o. marker *(Flg)* Voreinflugzeichen *n (ILS-Verfahren)*
o. member Außenring *m (Freilauf)*; Außenteil *n (Passung)*
o. piece Außenteil *n (Passung)*
o. pliers Außenfurnier *n (Holz)*
o. port Außenhafen *m*, Vorhafen *m*
o. product *(Math)* äußeres Produkt *n*, Vektorprodukt *n*
o. race Außenring *m (Kugellager)*
o. radiation-zone äußerer Strahlungsgürtel *m (der Erde)*
o. ring Begrenzungsring *m*; Außenring *m (Lager)*
o. screws *(Schiff)* Außenschrauben *fpl*
o. seal ring *(Masch)* Außensegerring *m*
o. shell *(Kern)* Außenschale *f*; Atomhülle *f*
o.-shell electron äußeres Elektron *n*; Valenzelektron *n*
o. skin Gußhaut *f*, Gußrinde *f*
o. sleeve Mantelrohr *n (Radialbohrmaschine)*
o. string Außenwange *f (Treppe)*
o. support äußeres Dorntraglager *n (Fräsmaschine)*
o. support bearing Gegenlager *n (Waagerechtbohrwerk)*
o. surface of cylinder Außenmantel *m*, Mantelfläche *f*
o. work function äußere Austrittsarbeit *f (Elektronenröhren)*
outermost äußerst, am weitesten außen
outfall Abflußleitung *f*
o. capacity *(Bw)* Vorflutleistung *f*
o. ditch *(Bw)* Vorfluter *m*
o. sewer *(Bw)* Sammelkanal *m*
o. structure Mündungsbauwerk *n*
o. works Auslaufbauwerk *n (Gesundheitstechnik)*
outfit 1. Ausrüstung *f*, Ausstattung *f*, Anlage *f*, Einrichtung *f*, Apparatur *f*, Geräte *npl*; 2. Belegschaft *f*, Arbeitsgruppe *f*
outfitter *(Schiff)* Bereederer *m*
outfitting pier Ausrüstungskai *m*
o. quay Ausrüstungskai *m*
outflow / to abfließen, ausfließen; ausströmen
outflow Ausfluß *m*; Abflußmenge *f*; Ausströmen *n*, Abströmung *f*; Auslaß *m*, Austritt *m*, Ausgang *m*
o. coefficient Ausflußkoeffizient *m (Turbine)*
o. period Ausflußzeit *f (Turbine)*
outgas / to entgasen
outgoing abgehend, ausströmend; austretend, abgegeben, emittiert
o. air Abluft *f*
o. cable Ausgangskabel *n*, Abgangskabel *n*
o. call *(Fmt)* abgehendes Gespräch *n*
o. current abgehender Strom *m*, Ausströmung *f*, Austrittsstrom *m*
o. feeder *(El)* abgehende Speiseleitung *f*
o. position *(Fmt)* Abgangplatz *m*, Ausgangsplatz *m*
o. repeater *(El)* Ausgangsverstärker *m*, Wiederholer *m*
o. selector *(Fmt)* Ausgangswähler *m*
o. side of rolls Auslaufseite *f* der Walzen *(Walzwerk)*
o. trunk call abgehendes Ferngespräch *n*
o. vessel auslaufendes Schiff *n*
o. wave abgehende (austretende) Welle *f*, Senderwelle *f*
outgrowth of crystals Kristallwachstum *n*
outlast / to überdauern, länger halten als
outlet Ablaß *m*, Ausfluß *m*, Austritt *m*; Auslaß *m (Sämaschine)*; Ausfahrt *f (z. B. eines Hafens)*; *(El)* Steckdose *f*

o. angle of the blade Endwinkel *m* der Schaufel *(Turbine)*
o. cable Anschlußkabel *n*
o. cam Auslaßnocken *m*
o. casing Austrittsgehäuse *n*
o. channel Auslaufkanal *m*
o. conduit Abflußleitung *f*
o. connection Auslaufstutzen *m*
o. edge Austrittskante *f (Turbine)*
o. flap Auslaßklappe *f*
o. from the nozzle Austritt *m* aus der Düse *(Turbine)*
o. funnel Auslauftrichter *m*
o. header Austrittsvorlage *f*
o. loss Austrittsverlust *m*, Auslaßverlust *m (Turbine)*
o. opening Ausflußmündung *f*, Ausflußöffnung *f*
o. orifice Auslaßöffnung *f*, Abführöffnung *f*
o. passage Auslaßkanal *m*
o. pipe Abflußrohr *n*, Ableitungsrohr *n*, Ablaßrohr *n*
o. pipe connection Austrittsstutzen *m*
o. port Austrittskanal *m*
o. sleeve Ablaßstutzen *m*
o. slide Auslaßschieber *m*
o. structure Mündungsbauwerk *n*
o. swirl Austrittsdrall *m*
o. temperature Auslaßtemperatur *f*, Abgastemperatur *f*
o. terminal Ausgangsklemme *f*, Anschlußklemme *f*
o. tube Ablaßrohr *n*, Ablaufrohr *n*
o. tunnel Ablaßstollen *m*
o. valve Ablaßventil *n*, Abzugsventil *n*, Ausblaseventil *n*, Auslaßventil *n*, Austrittsventil *n*, Druckventil *n*
outline / to umreißen, aufzeichnen
outline Umriß *m*, Zeichnung *f*, Profil *n*, Begrenzung *f*
o. drawing of the cross-sectional shape Querschnittsdarstellung *f*
o. of part Körperkante *f*
o. plan Perspektivplan *m*, Rahmenplan *m*
o. planning Perspektivplanung *f*, Vorplanung *f*
o. profile Profilansicht *f (eines Schiffes)*
o. type *(Typ)* Konturschrift *f*
outlined *(Typ)* umstochen *(Kapital)*, handgestochen
outperform / to in der Leistung übertreffen
outphase / to aus der Phase bringen, phasenverschieben
outphased phasenverschoben
outphasing modulation [system] *(Fmt)* Chireix-Modulation *f*
outport Außenhafen *m*, Vorhafen *m*
output Ausbringung *f*, Fertigungsmenge *f*, Arbeitsleistung *f*, Leistung *f*, Produktion *f*; Leistungsabgabe *f*, Leistungsfähigkeit *f*; *(Brg)* Fördermenge *f*, Förderung *f*; Ausstoß *m (Stahl)*; Fördermenge *(Pumpe)*; Abtrieb *m (Getriebe)*; *(Dat)* Ausgabe[information] *f*, Ausgangsinformation *f*, Ausgangsgröße *f*, Ausgangssignal *n*; Ausgabeausrüstung *f*
o. admittance *(El)* Ausgangsleitwert *m*
o. amplifier *(El)* Ausgangsverstärker *m*
o. at the draw-bar Leistung *f* am Zughaken *(elektrischer Zug)*
o. at the wheel rim Leistung *f* am Radumfang *(elektrischer Zug)*
o. attenuator *(El)* Ausgangsspannungsteiler *m*
o. block Ausgabespeicher *m*, Ausgangsspeicher *m*
o. capacitance Ausgangskapazität *f*
o. capacity *(Masch)* Leistungsvermögen *n*
o. curve Produktionskurve *f*
o. data Leistungsangaben *fpl (eines Motors)*
o. device Ausgabewerk *n (Rechner)*
o. electrode Ausgangselektrode *f*
o. element *s.* o. device
o. end Abtriebseite *f*
o. energy Nutzleistung *f*
o. equipment [außenliegendes] Ausgabegerät *n (bei Rechnern)*, Ausgabemaschine *f*
o. expected zu erwartende Produktion *f*
o. factor *(El)* Belastungsfaktor *m*
o. gap *(El)* Auskoppelraum *m*, Auskoppelspalt *m*
o. impedance *(El)* Ausgangsscheinwiderstand *m*, Ausgangsimpedanz *f*
o. indicator Leistungsmesser *m*
o. inductance coil *(El)* Ausgangsdrossel *f*
o. member Abtriebsglied *n*
o. meter *(El)* Ausgangsleistungsmesser *m*, Outputmeter *n*
o. of coal Kohleförderung *f*
o. of the sawing *(Holz)* Schnittergebnis *n*
o. pentode *(El)* Endpentode *f*
o. per annum Jahresproduktion *f*, Jahresförderung *f*
o. per hour Stundenleistung *f*
o. per litre Literleistung *f (eines Verbrennungsmotors)*
o. per man-shift *(Brg)* Förderung *f* je Mann und Schicht
o. per sharpening *(Am) (Masch)* Standmenge *f*, Stückzahl *f* je Nachschliff
o. per shift Schichtleistung *f*
o. per unit of displacement Hubraumleistung *f (eines Verbrennungsmotors)*
o. per wheel Einzelleistung *f (Turbine)*
o. potentiometer Ausgangspotentiometer *n (Analogrechner)*
o. power *(El)* Ausgangsleistung *f*, abgegebene Leistung *f*; Klemmenleistung *f*

o. raising Produktionssteigerung *f*, Erhöhung *f* der Fördermenge (Leistung)
o. regulator Leistungsregler *m*
o. resistance Ausgangswiderstand *m*
o. rise *s.* o. raising
o. shaft Abtriebswelle *f*
o. speed Ausgangsdrehzahl *f*
o. stage *(El)* Endstufe *f*
o. test Leistungsversuch *m*
o. torque Ausgangsmoment *n*
o. transmission lag *(BMSR)* Verzögerung *f* der Stellgröße
o. tube *s.* o. valve
o. turbine Nutz[leistungs]turbine *f*
o. unit *(Dat)* Ausgabeteil *n*, Ausgabe[gerät *n*] *f*
o. valve *(El)* Endröhre *f*
o. voltage *(El)* Ausgangsspannung *f*
o. voltage divider Ausgangsspannungsteiler *m*
outreach Ausladung *f*
outrigger 1. Ausleger *m*; ausfahrbarer Stützfuß *m (an Autokranen usw.)*; 2. Auslegerboot *n*; Maststütze *f*, Luvbaum *m*; Ruderstütze *f*
o. boat Auslegerboot *n (Ruderboot)*
o. clamp *(El)* zugentlastete Anschlußklemme *f*
outside Außenfläche *f*, Stirnseite *f*, Außenseite *f*
outside außen, außerhalb, Außen-
o. admission Außeneinströmung *f*
o. aerial Außenantenne *f*
o. air Außenluft *f*
o. ballast *(Schiff)* Außenballast *m*
o. beam Randträger *m*
o. caliper Außentaster *m*
o. chaser *(Masch)* Außenstrehler *m*
o. circle Kopfkreis *m (Zahnrad)*
o. coating Außenanstrich *m*
o. contractor Zulieferbetrieb *m*
o. crank Außenkurbel *f*
o. cylinder Kopfzylinder *m (Außenverzahnung)*
o. diameter Außendurchmesser *m*; Kopfkreisdurchmesser *m (Verzahnung)*
o. door handle Türaußengriff *m*
o. door panel Türaußenblech *n*
o. edge Außenkante *f*
o. firm-joint caliper Außentaster *m*
o.-flange tubbing ring englischer Tübbing *m (Schachtausbau)*
o. hand brake *(Kfz)* Außenhandbremse *f*
o. holding *(Masch)* Außenspannung *f*
o. lap Einlaßüberdeckung *f*, äußere Überdeckung *f*; Voreinströmung *f*, Voreintritt *m*
o. of the skin Blumenseite *f (Gerberei)*
o. panel Außenblech *n*
o. planking *(Schiff)* Außenbeplankung *f*
o. screw Außengewinde *n*
o. shoe Außenbackenbremse *f*
o. source Fremdquelle *f*
o. spring-joint caliper Außenfedertaster *m*
o. storage Freilagerung *f*
o. temperature Außentemperatur *f*
o. treading motion *(Text)* Außentrittvorrichtung *f*
o. valve regulator Außenregler *m (Ventil)*
outsides Decklage *f*, fehlerhafte äußere Schicht *f* eines Papierballens
outskirts Außenbezirk *m*, Randgebiet *n*; Stadtrandzone *f*, Vorort *m*, Vorstadt *f*
outsole Laufsohle *f (Schuh)*
o. stitcher Doppelmaschine *f (Schuhfertigung)*
outspeed / to an Schnelligkeit übertreffen
outturn sheet Ausfallmuster *n*, Vorlagemuster *n*
outward Außen-, äußerer, außen befindlich
o. batter Verbreiterung *f*
o.-bound road Ausfallstraße *f*
o. bulging Ausbeulung *f*, Ausbauchung *f*
o. double-flow turbine zweiflutige Turbine *f* mit Mitteleinströmung
o. flow turbine Turbine *f* mit innerer Beaufschlagung
o. run of the carriage *(Text)* Wagenausfahrt *f*
outwardly inclined bearing *(Masch)* nach außen geneigtes Lager *n*
outwash / to auswaschen, [aus]schlämmen
outwork Heimarbeit *f*
outworker Heimarbeiter *m*
oval oval, länglichrund
o. cable *(El)* Kabel *n* mit ovalem Mantel
o. countersunk head Linsensenkkopf *m*
o. groove Ovalkaliber *n (Walzwerk)*
o. head bolt Flachrundschraube *f*
o. head cap screw Linsenkopfschraube *f*
o. head elevator bolt Flachrundschraube *f* mit Vierkantansatz
o. head screw Linsenkopfschraube *f*
o. lathe Ovaldrehmaschine *f (für Holzbearbeitung)*
o. pane gewölbte Bahn *f (eines Hammers)*
o. pass Ovalkaliber *n (Walzwerk)*
o. point screw Schraube *f* mit Linsenkuppe
o. slide valve diagram *(Masch)* ovales Schieberdiagramm *n*, Schieberoval *n*
o. turning lathe Ovaldrehmaschine *f (Holzbearbeitung)*
oven Ofen *m (Backofen, Brennofen)*
o. basement raft Fundamentplatte *f (eines Ofens)*
o. battery Ofenbatterie *f (Koks)*
o. capacity Ofenfassungsvermögen *n*
o. case Thermostatgehäuse *n*
o. charging Ofenfüllung *f*
o. deck pad Ofentragplatte *f*, Düsendecke *f*
o. drying Ofentrocknung *f*
o. end Ofenkopf *m*
o. pad Ofenbetonsohle *f*
o. quoin Ofenkopf *m*

o.-to-table ware feuerfestes Geschirr *n*
o. with horizontal flues Horizontalofen *m*
over-all *s.* overall
o.-bias *(El)* zusätzliche *(zu hohe)* Vorspannung *f*
o.-compound excitation Überverbunderregung *f (eines Generators)*
o.-compounding *(El)* Überkompoundierung *f*, Überverbunderregung *f*
o.-compress / to überdrücken *(Briketts)*
o.-cut Oberhieb *m (Feile)*
o.-cutting *(Brg)* Schrämen *n* im oberen Teil des Flözes
o.-fuelling zu hohe Kraftstoffzufuhr *f*
o.-square überquadratisch
o.-the-top flight Flug *m* über den Wolken
o.-work / to *(Brg)* überbauen
overage Überschuß *m*
overall Gesamt-, Total-
o. amplification *(El)* Gesamtverstärkung *f*
o. attenuation *(Fmt)* Restdämpfung *f*
o. attenuation loss *(Fmt)* Betriebsdämpfung *f*
o.-brightness transfer characteristic *(Fs)* Kennlinie *f* der Gesamthelligkeitswiedergabe
o. characteristic Summencharakteristik *f*
o. coefficient of heat transfer *(Kern)* Wärmedurchgangszahl *f*
o.-contrast ratio *(Fs)* Gesamtgradationsverhältnis *n*
o. dimension Außenmaß *n*, Gesamtmaß *n*
o. dimensions gesamte Abmessung *f*; Raumbedarf *m*
o. efficiency Gesamtwirkungsgrad *m*
o. gain *(El)* Gesamtverstärkung *f*
o. heat-transfer coefficient *(Kern)* Wärmedurchgangszahl *f*
o. height Gesamthöhe *f*, Bauhöhe *f*
o. length Gesamtlänge *f*, Länge *f* über alles, Baulänge *f*
o. length of the drill Bohrerlänge *f*
o. logical plan of the system *(BMSR)* logischer Gesamtplan *m* des Systems
o. loss *(Fmt)* Betriebsdämpfung *f*, Gesamtverlust *m*
o. noise factor *(El)* mittlerer Rauschfaktor *m*
o. propulsive efficiency Gesamtwirkungsgrad *m* des Schiffsantriebes, Gesamtpropulsionswirkungsgrad *m*
o. pulse Gesamtimpuls *m*
o. stability Gesamtstabilität *f*
o. transfer function *(BMSR)* resultierende Übertragungsfunktion *f*, Übertragungsfunktion des gesamten Systems
o. transmission loss *(Fmt)* Restdämpfung *f*
o. voltage *(El)* Gesamtspannung *f*
o. width Gesamtbreite *f*
overamplification *(El)* Übersteuerung *f*
overarch / to *(Bw)* überwölben
overarm Gegenhalter *m (Fräsmaschine)*
o. braces Gegenhalterschere *f*
o. support Gegenhalterstütze *f*
overbleaching *(Text)* Überbleichen *n*
overblowing Überblasen *n*
overblown steel übergarer Stahl *m*
overburden *(Brg)* Abraum *m*, Deckgebirge *n*, Deckschicht[en *fpl*] *f*
o. disposal conveyor belt Abraumband *n*
o. dump Kippe *f*, Abraumhalde *f*, Abraumkippe *f*
o. moving Abraumbewegung *f*
o.-to-coal-ratio Verhältnis *n* Abraum zu Kohle
overburdening *(Met)* Übermöllerung *f*
overburnt totgebrannt *(Kalk)*
o. brittle iron sprödes, überhitztes Eisen *n*
overcast *(Brg)* Wetterkreuz *n*
overcast bedeckt, bewölkt
overcasting 1. *(Text)* Überwendlingsnähen *n*; 2. *(Brg)* Umsetzen *n* des Abraumes
o. machine Überwendlingsnähmaschine *f*
overcharge / to überladen
o. the furnace *(Met)* übersetzen
overcharge Überbeladung *f*, Überbelastung *f*; Überlast *f*; Überdruck *m*; Überladung *f (Akku)*
overcheck *(Text)* Überkaro *n*
overclimb / to *(ein Flugzeug)* überziehen
overcompensate / to überkompensieren
overcompensation Überkompensierung *f*
overcompressed überkomprimiert
overcurrent *(El)* Überstrom *m*
o. circuit breaker Überstrom[schutz]schalter *m*, Maximalschalter *m*
o. circuit breaking Maximalauslösung *f*, Überstromauslösung *f*
o. protection Überstromschutz *m*
o. relay Überstromrelais *n*, Maximal[strom]-relais *n*
o.-undervoltage circuit-breaker Maximalschalter *m* mit Nullspannungsauslösung, Schalter *m* mit Überstrom- und Unterspannungsauslösung
overcut Unterhieb *m*
overdamped system *(BMSR)* aperiodisches System *n*, System mit überkritischer Dämpfung
overdevelop / to überentwickeln *(Film)*
overdevelopment Überentwicklung *f (Film)*
overdimensioned überdimensioniert
overdrive / to übersteuern *(Elektronenröhren)*
overdrive *(Kfz)* Schnellgang *m*, Schongang *m*, Spargang *m*, Ferngang *m*
o. press Presse *f* mit obenliegendem Antrieb, Presse mit Oberantrieb
overdriven übersteuert *(Elektronenröhren)*
o. amplifier übersteuerter Verstärker *m*
o. press Presse *f* mit Oberantrieb

overdue überfällig *(Schiff, Flugzeug)*
overdye / to überfärben
overexcited übererregt
overexpose / to überbelichten *(Film)*
overexposure Überbelichtung *f (Film)*
overfall überbrechende See *f*
o. weir Überfallstauwehr *n*
overfault *(Geol)* Übersprung *m*
overfeeding 1. *(Ch)* Überdosierung *f*; 2. *(Masch)* Arbeiten *n* mit zu großem Vorschub
overflow / to überlaufen, überfließen, überströmen; überfluten *(Ufer)*
overflow 1. Überfall *m*; Überlauf *m*; 2. Überschreiten *n* einer Kapazität
o. buttress Überfallpfeiler *m*
o. channel Unterwassergraben *m*, Abflußkanal *m*
o. cone Überlaufkonus *m (Wasserbau)*
o. contact *(Fmt)* Durchdrehkontakt *m*, Überlaufkontakt *m*
o. crest Überlaufkrone *f (Wasserbau)*
o. dam Überfalldamm *m*, Überlaufdamm *m*
o. ejection Übertragvorschub *m (Lochkartenmaschinen)*
o. gutter Überlaufrinne *f*
o. indicator Überlaufmelder *m*
o. land Überschwemmungsgebiet *n*
o. oil line Leckölleitung *f*
o. oil line connection Leckölleitungsanschluß *m*
o. pipe Überlaufrohr *n*
o. position *(Fmt)* Durchdrehstellung *f*
o. regulator Abströmregler *m*
o. trap Überlaufventil *n*
o. valve Überlaufventil *n*
o. water Überfallwasser *n*
overfolded umgebördelt
overglaze Überglasur *f*
overgrate blast *(Wärm)* Oberluft *f*, Oberwind *m*
overgrind / to totmahlen *(Formsand)*
overground 1. über der Erde, über Tage; 2. totgemahlen *(Formsand)*
o. speed *(Schiff)* Totwassergeschwindigkeit *f*, Geschwindigkeit *f* über Grund
overgrown with grass vergrast
overhand *(Brg)* schwebend
o. stoping Firstenstoßbau *m*
o. stoping on ore Magazinbau *m*, Speicherbau *m*
overhang / to auskragen, ausladen, überstehen, vorstehen; überhängen
overhang Auskragung *f*, Ausladung *f*; Überhang *m*
overhanging überhängend, ausladend
o. arm Gegenhalter *m (Fräsmaschine)*
o. wall *(Brg)* überhängender Stoß *m*
overhaul / to überholen, durchsehen, überprüfen, reparieren
overhaul Überholung *f*, Wiederinstandsetzung *f*, Generalreparatur *f*, Generaldurchsicht *f*, Reparatur *f*, Überprüfung *f*
overhead darüber angebracht, obenliegend, Ober-, oberirdisch, Hoch-, Decken-
o. beam Deckenbalken *m*
o. bin Hochbehälter *m*
o. bucket Überkopflader *m*
o. cable Freileitungskabel *n*
o.-cable railway Drahtseilschwebebahn *f*
o. camshaft *(Kfz)* obenliegende Nockenwelle *f*
o. charges Gemeinkosten *pl*
o. cistern Hochbehälter *m*, Hochreservoir *n*
o. clearance lichte Höhe *f*
o. conductor rail Hängestromschiene *f*
o. construction Hochbau *m (von Leitungen)*
o. contact system Oberleitung[sanlage] *f*, Fahrleitung[sanlage] *f*
o. contact wire Oberleitung *f*
o. conveyor Hängebahn *f*
o. cost Gemeinkosten *pl*, Fertigungsgemeinkosten *pl*
o. countershaft Deckenvorgelege[welle *f*] *n*
o. crane Laufkran *m*; Brückenkran *m*
o. crossing Fahrdrahtkreuzung *f*
o. electric power line elektrische Freileitung *f*
o. expenses Gemeinkosten *pl*
o. handling Fördern *n* durch obenliegende Einrichtungen
o. hoist Hängekatze *f*
o. hopper Aufnahmebunker *m*, Baustellensilo *m*
o. insulator Freileitungsisolator *m*
o. junction crossing Fahrdrahtweiche *f* mit kreuzenden Fahrleitungen
o. junction knuckle Fahrdrahtweiche (Luftweiche) *f* mit Parallelführung der kreuzenden Fahrleitungen
o. line Freileitung *f*, Oberleitung *f*
o. line switch Streckenschalter *m*, Freileitungsschalter *m*
o. loader Überkopflader *m*
o. loading Überkopfladen *n*
o. pipe installation oberirdische Rohrleitung *f*
o. position Überkopflage *f (Schweißen)*
o. product Destillationsprodukt *n*
o. rail Hängeschiene *f*
o. railway Hängebahn *f*; Hochbahn *f*
o. slide Überhänger *m*, überhängender Schlitten *m*
o. switching Fahrdrahtweiche (Luftweiche) *f* mit angenäherten Fahrleitungen
o. tank Hochbehälter *m*
o. transmission Deckentransmission *f*
o. transmission gear Deckenvorgelege *n*
o. transmission line Überlandfreileitung *f*
o. traveller Laufkran *m*; Brückenkran *m*
o. travelling crane Laufkran *m*; Brückenkran

o. valve *(Kfz)* hängendes (obengesteuertes) Ventil *n*
o.-valve engine kopfgesteuerter (obengesteuerter) Motor *m*, Motor mit hängenden Ventilen
o. welding Überkopfschweißung *f*
overheat / to überhitzen, verbrennen
overheater Überhitzer *m*
overheating Überhitzung *f*; Warmlaufen *n* *(Lager)*
o. of the boiler plate Kesselblechüberhitzung *f*
overhung fliegend angeordnet, freihängend, überhängend, freitragend
o. beam Schwinghebel *m*, Schwingbalken *m*
o. casing freihängendes Gehäuse *n*
o. crank Stirnkurbel *f*, überhängende (außen aufgesteckte) Kurbel *f*
o. cylinder freihängender (schwebender, freitragender) Zylinder *m*
o. flywheel fliegendes Schwungrad *n*
o. governor fliegend gelagerter Regler *m*
o. mounting überhängende Lagerung *f*
o. pulley fliegende Riemenscheibe *f*
o. screw freitragende Schraube *f*
o.-type alternator Außenpolwechselstromgenerator *m*
o. wheel fliegendes Rad *n* *(Turbine)*
overirradiation *(Kern)* Überbestrahlung *f*, Überexponierung *f*
overlaid frequency *(Fmt)* Überlagerungsfrequenz *f*
overlap / to überlappen; überblatten *(Turbine)*; überblenden *(Film)*
overlap Überlappung *f*; Überblattung *f*, Überdeckung *f*
o. angle *(El)* Überlappungswinkel *m*, Überlappungszeit *f*
o. factor Überlappungsfaktor *m*
o. fault *(Geol)* Überschiebung *f*
o. of absorption lines Überlappen *n* (Überlagerung *f*) von Absorptionslinien
o. of lines *(Fs)* Zeilenüberlappung *f*
o. regulator Überlappungsregler *m*
overlapping Überlappung *f*, Überdeckung *f*, Übereinandergreifen *n*
overlapping überlappend; mit Überlappstaffelung *(Räumwerkzeug)*
o. length *(El)* Überlappungslänge *f*
o. of the belts *(Masch)* Übereinandergreifen *n* der Schüsse
o. the blade ends Überdecken *n* der Schaufelenden *(Turbine)*
overlay Belag *m*, Überzug *m*; Auflage *f*; Abraum *m*, Deckgebirge *n*, Deckschicht *f*
o. sheet *(Typ)* Zurichtebogen *m*
overleaf umseitig *(Rückseite im Buch)*
overload / to über[be]lasten; überbeanspruchen; überladen
overload Überlastung *f*; Übersteuerung *f* *(Analogrechner)*
o. capacity Überlastbarkeit *f*, Überlastungsfähigkeit *f*
o. capacity of a circuit Über[be]lastbarkeit *f* eines Stromkreises, Überlastfaktor *m*
o. circuit-breaker Maximalausschalter *m*, Überlastausschalter *m*, Überstromschalter *m*
o. clutch Überlastungskupplung *f*, Überlastsicherungskupplung *f*; Rutschkupplung *f*
o. coil Überstromspule *f*
o. cut-out Überlastschalter *m*, Überstromschalter *m*
o. indicator Überlastungsanzeiger *m*, Übersteuerungsanzeiger *m*
o. protection Überlastschutz *m*, Überstromschutz *m*
o. relay Überstromrelais *n*, Maximalrelais *n*, Höchststromrelais *n*
o. release Überstromauslöser *m*, Überstromwächter *m*
o. safety device Überlastsicherung *f*
o. switch Maximalschalter *m*, Überstromschalter *m*, Höchststromschalter *m*
o. temperature relay thermisches Überlastrelais *n*
o. trip Überstromautomat *m*, Überstromauslöser *m*, Überstromwächter *m*
o. tripping Überstromauslösung *f*, Maximalauslösung *f*
overloadable überlastbar
overloading Über[be]lastung *f*
overlying Überlagerung *f*, Superposition *f*
overlying rock *(Brg)* Deckgebirge *n*
o. seam *(Brg)* hangendes Flöz *n*
overman *(Brg)* Wettermann *m*, Wettergehilfe *m*
overmatch Überanpassung *f* *(Radio)*
overmeasure Zugabe *f*
overmodulation Übersteuerung *f*, Übermodulation *f*
overpass [bridge] Straßenüberführung *f*, Überführung *f*
overpath *(Bw)* Überweg *m*
overpick loom Oberschlagwebstuhl *m*
overplus Überschuß *m*
overpole / to überpolen *(Kupfer)*
overpoling Überpolen *n* *(Kupfer)*
overpotential Überpotential *n*
overpower protection *(El)* Leistungsbegrenzungsschutz *m*
o. scram Notstabschaltung *f* infolge Überlast *(Reaktortechnik)*
overpressure Überdruck *m*
overrange Überreichweite *f*
o. protection Überlastsicherung *f*, Übersteuerungsschutz *m*

overrefined *(Met)* übergar
overrich mixture zu fettes Gemisch *n*
override / to überschalten, übersteuern
override clutch gear change *(Kfz)* Überholklauenschaltung *f*
overrun Kolbenüberlauf *m*
o. brake Auflaufbremse *f*
o. control Nachlaufsteuerung *f*
o. speed Nachlaufgeschwindigkeit *f*
overrunning Überlauf[en *n*] *m*, Weiterlaufen *n*
o. brake Auflaufbremse *f*
o. clutch Freilaufkupplung *f*
o. of the piston rings Überlaufen *n* der Kolbenringe
overs 1. *(Typ)* Zuschuß *m (zum Einrichten)*; Ergänzungsbogen, Überdruckbogen *m*, Zuschußbogen *m*; 2. Siebüberlauf *m*, Siebrückstand *m*
oversailing course auskragende Ziegelschicht *f*
oversaturation *(Ch)* Übersättigung *f*
oversea trade seewärtiger Außenhandel *m*
o. vessel Überseeschiff *n*
overseas hide Wildhaut *f*
oversensitive überempfindlich
oversewn *(Typ)* geholländert, mit fliegendem Faden in der Falzmaschine geheftet
overshoot / to *(El)* überschwingen
overshoot *(El)* Durchschwingen *n*, Überschwingen *n (des Impulses)*; Überschwingung *f (Anzeigeinstrument)*
o. ratio Überschwingfaktor *m*
overshot Fanglocke *f (Kernen; Tiefbohrtechnik)*
overshot oberschlächtig *(Wasserrad)*
o. loader Überkopflader *m*
overside delivery Umladung *f (von Schiff auf Leichter)*
oversize 1. Übergröße *f*; Übermaß *n*; 2. Überkorn *n*; Siebrückstand *m*, Siebüberlauf *m*
o. material Siebrückstand *m*
o. particle Überkorn *n*
o. products Siebrückstand *m*
overspeed / to überdrehen *(Motor)*; mit zu hoher Drehzahl arbeiten
overspeed Schleuderdrehzahl *f (elektrischer Zug)*
o. device Sicherheitseinrichtung *f (Überdrehzahl)*
o. gear Schnellgang *m*
o. limiter Übergeschwindigkeitsbegrenzer *m*
o. monitor Drehzahlwächter *m*
overspun überdreht *(Garn)*
overstability Überstabilität *f*
oversteer / to übersteuern
overstitch fibre fleeces / to Watteschichten übernähen
overstrain / to 1. *(sl) s.* overstress / to; 2. plastisch verformen *(Festigkeit)*
overstress / to überbeanspruchen, überlasten
overswing *(Fs)* Überschwingen *n*
overtension Überbeanspruchung *f*
overthrow fault *(Geol)* Überschiebung *f*
overthrust fault *(Geol)* Überschiebung *f*
o. mountain Deckgebirge *n*
overtighten / to zu fest anziehen
overtime Mehrarbeit *f*; Überstunden *fpl*
o. premium Überstundenzuschlag *m*
o. rate Überstundenzuschlag *m*
overtone 1. *(Fmt)* harmonische Oberschwingung *f*, Oberton *m*; 2. absolut deckende Farbe *f*
o. band Oberschwingungsband *n (Spektroskopie)*
o. crystal *(Fmt)* Oberwellenquarz *m*
overtop flue *(Wärm)* Oberzug *m*
overtravel / to überregeln, überschwingen, überschreiten
overtravel *(Masch)* Überlauf *m*
overturn / to 1. umwerfen, umkippen; 2. *(Geol)* überkippen; 3. überdrehen *(Schraube)*
overturning discharge skip *(Brg)* kippbarer Skip *m*
o. moment Kippmoment *n (Statik)*
overtype armature *(Elm)* oberständiger Anker *m*
overvoltage *(El)* Überspannung *f*
o. arrester Überspannungsableiter *m*
o. circuit-breaker Überspannungsschalter *m*
o. device Überspannungsschutz *m*
o. of resonance Resonanzüberhöhung *f*
o. protection Überspannungsschutz *m*
overvoltages of atmospheric origin Gewitterüberspannungen *fpl*, atmosphärische Überspannungen *fpl*
overwind / to 1. überdrehen *(z. B. Uhrfeder)*; 2. überfahren *(Schachtförderung)*
overwind distance freie Höhe *f (Schachtförderung)*
overwinding Übertreiben *n (Schachtförderung)*
Ovitron *(Dat) Bezeichnung einer dem Neuron nachgebildeten Flüssigkeit enthaltenden Schaltzelle*
ovoid Eierbrikett *n*
owling Nachtfischen *n*, Fischen *n* bei Nacht
own eigen-, Eigen-
o. consumption Eigenverbrauch *m*
o. weight Eigenmasse *f*
owner Schiffseigner *m*, Reeder *m*
ownership Eigentum[srecht] *n*
oxgall Ochsengalle *f*
oxalate Oxalat *n*
oxalic acid Oxalsäure *f*
o. aldehyde Glyoxal *n*
o. ester Oxalsäureester *m*
oxalyl urea Oxalylharnstoff *m*, Parabansäure *f*
Oxford blue echtes Blau *n (Farbstoff)*
oxidant Oxydationsmittel *n*, Oxydierungsmittel *n*
o. tank *(Rak)* Sauerstoffträgerbehälter *m*, Oxydatorbehälter *m*

oxidate / to oxydieren
oxidate Oxydationsprodukt *n*
oxidation Oxydation *f*, Oxydierung *f*, Sauerstoffaufnahme *f*
o. inhibitor Antioxydans *n*, Antioxydationsmittel *n*
o. potential Oxydationspotential *n*
o.-resistant oxydationsbeständig
o. roasting *(Met)* oxydierendes Rösten *n*
o. stage Oxydationsstufe *f*
o. state Oxydationszustand *m*
o. style *(Text)* Oxydationsartikel *m*
oxide 1. *(Ch)* Oxid *n*; 2. *(Met)* Zunder *m*
o.-cathode *s.* o.-coated cathode
o. ceramic Oxidkeramik *f (Metallkeramik)*
o.-coated cathode Oxidkatode *f*
o. coating Oxidschicht *f (von Metallen)*
o. film Oxidschicht *f*, Rostbelag *m*
o. inclusion *(Met)* Oxideinschluß *m*
o. layer *(Met)* Zunderschicht *f*
o. of antimony Antimonoxid *n*
o. ore oxydisches Erz *n*
o.-proof oxidfest
o.-scaling Verzunderung *f*
o. skin *(Met)* Oxidhaut *f*
o. slagging Oxidausschlackung *f*
oxidiability Oxydierbarkeit *f*
oxidic oxidhaltig, oxydisch, oxydiert
oxidimetry Oxydimetrie *f*
oxidization Oxydation *f*
oxidize / to 1. oxydieren, Sauerstoff anlagern; 2. *(Met)* rosten, zundern; frischen; verschmoren
o. in the hanging room *(Text)* verhängen
o. off desoxydieren, entoxydieren, aboxydieren
oxidized oxydiert, sauerstoffhaltig
oxidizer Sauerstoffträger *m*, Oxydator *m*; Oxydationsmittel *n*
o. consumption Sauerstoffträgerverbrauch *m*, Oxydatorverbrauch *m*
o. jet Sauerstoffträgerstrahl *m*, Oxydatorstrahl *m*
o. tank Sauerstoffträgerbehälter *m*, Oxydatorbehälter *m*
oxidizing 1. Oxydieren *n*, Oxydationsvorgang *m*; 2. *(Met)* Frischen *n*; 3. Verkupferung *f (Elektrotypie)*
oxidizing oxydierend
o. agent Oxydationsmittel *n*, Oxydiermittel *n*
o. component Oxydator *m*, Sauerstoffträger *m*
o. constituent Oxydator *m*, Sauerstoffträger *m*
o. flame oxydierende Flamme *f*
o. loss *(Met)* Abbrand *m*
o. machine Oxydationsapparat *m*
o. period *(Met)* Oxydationsperiode *f*, Frischperiode *f*
o. process *(Met)* Frischverfahren *n*, Frischprozeß *m*
o. reaction *(Met)* Frischwirkung *f*
o. slag *(Met)* Frischschlacke *f*
oxidulated copper ore *(Min)* Rotkupfererz *n*
oxime Oxim *n (Isonitrosoverbindung)*
oxter plate *(Schiff)* Gillungsplatte *f*, Außenhautplatte *f* am Achtersteven
oxy-acetylene cutter Autogenschneidbrenner *m*
o.-acetylene cutting Autogenschneiden *n*, Schneidbrennen *n*
o.-acetylene cutting blowpipe Azetylensauerstoffschneidbrenner *m*
o.-acetylene flame Brenngas-Sauerstoff-Flamme *f*, Azetylen-Sauerstoff-Flamme *f*
o.-acetylene welding Autogenschweißen *n*, Azetylensauerstoffschweißung *f*
o.-arc cutting elektrisches Sauerstoffschneiden *n*
o.-arc cutting electrode Sauerstoffschneideelektrode *f*
oxycarbide of silicon *(Met)* Siliziumoxidkarbid *n*, Siliziumoxykarbid *n*
oxycellulose Oxyzellulose *f*
oxygen Sauerstoff *m*
o. apparatus Sauerstoffapparat *m*
o. breathing apparatus Sauerstoffschutzgerät *n*
o. carrier *(Rak)* Sauerstoffträger *m*, Oxydator *m*
o. compound Sauerstoffverbindung *f*
o. cylinder Sauerstoffflasche *f*
o. demand Sauerstoffbedarf *m*
o. developing agent sauerstoffabgebendes Mittel *n*
o.-enriched sauerstoffangereichert
o. filler valve *(Rak)* Betankungsventil *n*
o.-free sauerstofffrei
o. indicator gauge *(Rak)* Behälterstandsonde *f*
o. inhaling apparatus Sauerstoff[atmungs]apparat *m*, Sauerstoff[rettungs]gerät *n*
o. lance *(Met)* Sauerstofflanze *f*
o. mask Atemmaske *f*
o. point Sauerstoffanschluß *m (im Flugzeug)*
o.-replenisher coupling *(Rak)* Nachtankkupplung *f*
o.-replenisher valve *(Rak)* Nachtankventil *n*
o. tank pressure control line *(Rak)* Tankdruckregelleitung *f*
o. yielding Sauerstoff abgebend
oxygenated mit Sauerstoff gesättigt, sauerstoffangereichert
oxyhydrogen Knallgas *n*
o. blowpipe Knallgasgebläse *n*
o. welding Wasserstoff-Sauerstoff-Schweißung *f*
Ozalid print Ozalidlichtpause *f*
Ozalid printing process Ozalidlichtpausverfahren *n*
Ozalid white print process Ozalidlichtpausverfahren *n*

ozokerite Bergwachs *n*, Ozokerit *m*
ozone Ozon *n*
o. bleach *(Text)* Ozonbleiche *f*
ozoniferous ozonhaltig, ozonreich
ozonize / to ozonisieren
ozonizer Ozonisator *m*

P

P-band *(Fmt)* P-Band *n (225 bis 390 MHz)*
p-conducting p-leitend, defektleitend *(Halbleiter)*
P-controller *(BMSR)* P-Regler *m*
p-donor p-Donator *m (Halbleiter)*
P-electron P-Elektron *n*
p-hole Defektelektron *n (Halbleiter)*
p-material p-leitendes Material *n (Halbleiter)*
p-n boundary pn-Grenzschicht *f (Halbleiter)*
p-n junction pn-Übergang *m (Zone eines Halbleiters, in der sich der Dotierungscharakter ändert)*
p-n junction rectifier Sperrschichtgleichrichter *m*
p-p junction pp-Übergang *m (Halbleiter)*
p-type p-leitend, defektleitend *(Halbleiter)*
p-type conduction p-Leitung *f*, Löcherleitung *f*, Mangelleitung *f*
p-type conductivity p-Leitfähigkeit *f*, Löcherleitfähigkeit *f*, Defektelektronenleitfähigkeit *f*
p-type material p-leitendes Material *n*
p-type semiconductor p-[Typ-]Halbleiter *m*, Mangelhalbleiter *m*, Halbleiter *m* mit Defektelektronenleitung
P waves *(El)* P-Zacken *mpl (Ekg)*
pace a distance / to eine Strecke abschreiten
pace 1. Schritt *m*; Gangart *f*; 2. Podium *n*
p. setter Schrittmacher *m*
pachnolite *(Min)* Pachnolit *m*
pacing Abschreiten *n (von Entfernungen)*
pack / to [ab]dichten; unterbauen; paketieren; verpacken; stopfen *(Gleise)*; *(Brg)* versetzen *(insbesondere von Hand)*
p.-harden einsatzhärten
pack 1. Bündel *n*, Paket *n*; 2. Spiel *n (Karten)*; 3. *(Brg)* Bergedamm *m*, Bergerippe *f*; 4. *(Text)* 240 lb *(Wolle)*; 60000 yards *(Leinengarn)*
p.-annealing Kistenglühen *n*
p. dynamometer *(Brg)* Versatzdruckdose *f*
p. hardening *(Met)* Kisteneinsatzhärten *n*
p. ice Packeis *n*
p. line *(Brg)* Versatzkante *f*
p.-needle Packnadel *f*, Sacknadel *f*
p. of punched cards Stoß *m* Lochkarten
p. radio-set Tornisterfunkgerät *n*, tragbare Funkstation *f*
p. receiver Tornisterempfänger *m*
p. telephone Fernsprechtornister *m*
p.-type radio-set Tornisterfunkgerät *n*, tragbare Funkstation *f*
package 1. Paket *n*; Verpackungsbehälter *m*; 2. *(Dat)* kompakte Baugruppe *f*; 3. *(Text)* Garnkörper *m*
p. cargo Stückgut *n*
p. diameter Spulen[wicklungs]durchmesser *m*
p. drying machine *(Text)* Packtrockenmaschine *f*
p. dyeing *(Text)* Färben *n* im Wickelkörper, Apparatfärberei *f*, Färben auf Apparat, Packfärben *n*
p. freight *(Am)* Stückgut *n*
p. net Gepäcknetz *n*
p. plant [kleine] transportable Anlage *f*
p. tray Gepäckablage *f*
packaged timber Schnittholz *n* in Bündeln, gebündeltes Holz *n (Schiffsladung)*
packaging Verpackung *f*
p. industry Verpackungsindustrie *f*
p. material Verpackungsmaterial *n*
p. paper Verpackungspapier *n*
p. technique *(El)* Packungstechnik *f (Unterbringung auf engem Raum)*
packcloth Packleinwand *f*
packed gepackt; verpackt
p. cock Packhahn *m*
p. column Füllkörperkolonne *f*, Füllkörpersäule *f*, gefüllte Trennsäule *f (Gaschromatografie)*
p. height Schüttguthöhe *f*
p. petroleum Kanisteröl *n*, Faßöl *n (Schiffsladung)*
packer Packer *m*; Verpackungsmaschine *f*
packet Postschiff *n*, Paketboot *n*
packfilm Packfilm *m*
packing 1. Packen *n*, Verpacken *n*; 2. Abdichtung *f*, Dichtung *f*; Dichtungsmittel *n*, Dichtungsstoff *m*; Futter *n*; Füllung *f*, Packungsart *f*; Verdichtung *f (Straßenbau)*; Liderung *f (von Pumpen)*; 3. Schüttgut *n*; 4. Zusammendrükken *n (Fernsehbild)*; 5. *(Brg)* Versatz *m*
p. agent *(Met)* Packmittel *n*
p. bandage Dichtungsbinde *f (Kabel)*
p. block Unterlegeblock *m*
p. bolt Stopfbuchsenschraube *f*
p. box Stopfbüchse *f*, Stopfbuchse *f*
p. cloth Packleinwand *f*
p. coke Dichtungskegel *m*
p. collar Dichtungskragen *m*
p. cord Dichtungsschnur *f*, Dichtungszopf *m*
p. density *(Dat)* Informationsdichte *f*, Packungsdichte *f*
p. effect *(Kern)* Massendefekt *m*
p. fluid Sperrflüssigkeit *f*
p. gland Packungsstopfbuchse *f*
p. in belts *(El)* Streifenpacken *n (z. B. bei automatischer Bestückung mit Widerständen)*
p. liquid Sperrflüssigkeit *f*

p. machine 1. Verpackungsmaschine *f*; 2. *(Brg)* Versatzmaschine *f*
p. material 1. *(Masch)* Packungsmaterial *n*, Dichtungsmittel *n*; 2. Verpackungsmaterial *n*; 3. *(Met)* Packmittel *n*, Tempermittel *n (Temperei)*; 4. Füllkörper *mpl (Destillation)*; 5. *(Brg)* Versatzmaterial *n*
p. needle Packnadel *f*, Sacknadel *f*
p. of cable Kabelbett[ung *f*] *n*
p. of particles Gitteranordnung *f (Kristalle)*
p. of spheres Kugelpackung *f (Kristalle)*
p. of the gauge glass Glasrohrabdichtung *f*
p. paper Verpackungspapier *n*, Packpapier *n*
p. piece *(Bw)* Füllholz *n*, Futterholz *n*
p. press Ballenpresse *f*, Paketierpresse *f*
p. ring Dichtungsring *m*, Kolbenring *m*, Dichtungsscheibe *f*, Klemmring *m*, Gegenring *m (Turbine)*
p.-ring press Dichtungsfadenandrückmaschine *f*
p.-rod Stopfstange *f*
p. rubber Dichtungsgummi *m*
p. sand Füllsand *m (Formerei)*
p. sleeve Dichtungsbuchse *f*, Rillenbüchse *f*
p. space Packungsraum *m (Dichtung)*
p. stick *(Masch)* Packungstreiber *m*
p. support Schüttgutunterhalt *m*
p. surface *(Masch)* Dichtfläche *f*
p. washer Liderungsdeckel *m*, Stopfdichtung *f*
p. with copper Kupferabdichtung *f*, Abdichtung *f* durch Kupfereinlage
p.-worm *(Masch)* Packungszieher *m*
pad / to 1. wattieren; polstern; 2. klotzen
pad 1. Kissen *n*, Polster *n*; 2. Stoßdämpfer *m*, Puffer *m*; 3. Farbkissen *n*; 4. Block *m*, Papierblock *m*, Schreibblock *m*, Zeichenblock *m*; 5. *(El)* festes Dämpfungsglied *n*, festeingestellter Abschwächer *m (Wellenleitertechnik)*; 6. Eishockeybeinschiene *f*
p. dyeing *(Text)* Klotzfärben *n*, Klotzen *n*
p. lock Vorhängeschloß *n*
p. lubrication Polsterschmierung *f*, Kissenschmierung *f*
p. lubricator Polsterschmiereinrichtung *f*; Schmierfilz *m*, Schmierpolster *n*
p. piece Stoßkissen *n (seitlich an Rettungsbooten)*
p. roller Führungsrolle *f*, Druckrolle *f*
p. saw Fuchsschwanz[säge *f*] *m*
p.-steam vat-printing process *(Text)* Klotzdämpfküpendruckverfahren *n*
p. stone *(Bw)* Auflagerstein *m*
padded wulstig; gepolstert
p. resist style *(Text)* Klotzreserveartikel *m*
padder *(Text)* Foulard *m*, Klotzmaschine *f*, Paddingmaschine *f*
padding 1. Polster[ung *f*] *n*; Watteeinlage *f*; Wattierstoff *m*; 2. *(Text)* Pflatschen *n*, Klotzen *n*, Klotzverfahren *n*
p. ground shade Klotzgrundierfarbe *f*
p. liquor Foulardierlösung *f*, Klotz[flotte *f*] *m*
p. machine Foulard *m*, Klotzmaschine *f*
p. roller Klotzwalze *f*
paddle 1. Schaufel *f*, Rührarm *m (von Mischmaschinen)*; 2. Kratze *f*; 3. Ruder *n*
p.-box Radkasten *m*, Radgehäuse *n (bei Raddampfern)*
p. deck *(Schiff)* Radkastendeck *n*
p. dyeing machine *(Text)* Schaufelfärbemaschine *f*
p. dyer *(Text)* Schaufelfärbemaschine *f*
p. mixer Schaufelmischer *m*, Paddelrührer *m*
p. screw conveyor Rührschneckenförderer *m*
p. shaft Schaufelwelle *f (des Rührwerks)*
p. steamer Raddampfer *m*
p.-trough drying Trocknung *f* im Rührwerk *(Holzspäne)*
p.-trough type dryer Trockner *m* mit Rührwerk
p. tug Schaufelradschlepper *m*, Radschlepper *m*
p. wheel Schaufelrad *n (Raddampfer)*
p.-wheel steamer Raddampfer *m*
p. wheeler Raddampfer *m*
p. worm conveyor Misch[er]schnecke *f*
paddling pool Planschbecken *n*
paddock Gestüt *n*
padlock Vorhängeschloß *n*
padstone *(Bw)* Auflagerstein *m*
page / to *(Typ)* paginieren; umbrechen
page *(Typ)* Seite *f*
p. cord Kolumnenschnur *f*
p. depth Seitenhöhe *f*
p. gauge Kolumnenmaß *n*
p. printer *(Fmt)* Blattdrucker *m*, Blattschreiber *m*
paginate / to *(Typ)* paginieren; umbrechen
paginating *(Typ)* Umbruch *m*; Paginierung *f*
pail Eimer *m*
paint / to 1. malen, [an]streichen, tünchen; 2. *(Led)* schwöden
paint Farbe *f*, Anstrich[farbe *f*] *m*, Anstrichstoff *m*; Überzug *m*; Schminke *f*
p. and varnish remover Abbeizmittel *n* für Farbe und Lack
p. base Farbuntergrund *m*, Haftgrund *m*
p. box Tuschkasten *m*, Farbenkasten *m*
p. brush Malerpinsel *m*
p. coat Farbanstrich *m*, Anstrich *m*, Farbüberzug *m*, Lackierung *f*
p. for cocoonisation *(Plast)* Abdecklack *m*, Schutzhaut *f*
p. grinder mill Farbreibmühle *f*
p. locker Farbenraum *m*
p. practice Anstrichtechnik *f*

p. remover Abbeizmittel *n*, Farbabbeizmittel *n*, Farbabziehmittel *n*, Farbentferner *m*
p. resin Lackharz *n*
p. roller mill Farbwalzenstuhl *m*
p. spray gun Farbspritzpistole *f*
p. spraying Farbspritzen *n*, Spritzlackierung *f*
p. spraying apparatus Farbspritzpistole *f*
paintable streichbar
painter Maler *m*, Anstreicher *m*
painter's naphtha Testbenzin *n*
p. putty Glaserkitt *m*
painting Anstrich *m*, Farbauftrag *m*, Streichen *n*
pair Paar *n*; Elementenpaar *n (Getriebelehre)*
p. annihilation *(Kern)* Paarvernichtung *f*
p. conversion *(Kern)* Paarkonversion *f*
p. creation *(Kern)* Paarerzeugung *f*
p. forces *(Kern)* Paarkräfte *fpl*
p. furnace Platinenwärmofen *m (Walzwerk)*
p. generation *(Kern)* Paarbildung *f*
p. masts *(Schiff)* Doppelladepfosten *mpl*, Portalmasten *mpl*
p.-oar Zweier *m* ohne Steuermann *(Sportboot)*
p. of bevel gears Kegelradpaar *n*
p. of compasses Zirkel *m*
p. of cores *(El)* Adernpaar *n*, Doppelader *f*
p. of crossed helical gears *(Am)* Stirnradschraubgetriebe *n*
p. of drawing rollers *(Text)* Streckwalzenpaar *n*
p. of keys Keilpaar *n*
p. of lineman's climbers Steigeisen *npl (für Masten)*
p. of mating gears Radpaarung *f*
p. of pawls Doppelsperrklinke *f*
p. of pliers Zange *f*
p. of plugs *(El)* Zwillingsstecker *m*
p. of rolls Walzenpaar *n (Walzwerk)*
p. of spur gears Stirnräderpaar *n*
p. of toothed gears Zahnradpaar *n*
p. of wheels Räderpaar *n*; Radsatz *m*
p. of wheels on axle with grease box *(Eis)* Schmierbuchsenradsatz *m*
p.-oven *(Met)* Doppelofen *m*
p. production Paarerzeugung *f*, Elektronenpaarbildung *f*
p.-production absorption Absorption *f* von Gamma-Strahlen durch Paarerzeugung, Absorption durch Paarbildung *f*
p.-to-pair capacity *(Fmt)* Adernkapazität *f*, Schleifenkapazität *f*; Viererkapazität *f*
p. twisting Paarverseilung *f*
paired paarig
p. cable *(El)* paarverseiltes Kabel *n*
pairing 1. Paarung *f*; 2. *(Fs)* paarige Zeilenstruktur *f*; 3. *(El)* Paarverseilung *f (Kabel)*; 4. *(Led)* Paaren *n* in der Mulde

paktong *(Met)* Neusilber *n*, Alpaka *n*
palaite *(Min)* Palait *m*
pale Zaunlatte *f*; Holzpfahl *m*
p. fence Lattenzaun *m*, Staketenzaun *m*
paling Lattenzaun *m*, Staketenzaun *m*
p. fence Lattenzaun *m*, Staketenzaun *m*
palladic compound Verbindung *f* des vierwertigen Palladium
p. oxide Palladiumoxid *n*
palladinized asbestos Palladiumasbest *m*
palladium Palladium *n*
palladous compound Verbindung *f* des zweiwertigen Palladium
pallasite *(Geol)* Pallasit *m*
pallet 1. *(Masch)* Werkstückträger *m*; 2. Ladeplatte *f*, Transportpalette *f*, Palette *f*; 3. Dübel *m*; 4. Töpfer[dreh]scheibe *f*
p. tank *(Schiff)* auf Palette montierter Falttank *m*
palletization Palettieren *n* von Ladungen
palletized cargo ship Palettenfrachtschiff *n*
palm kernel oil Palmkernöl *n*
p. oil Palmöl *n*, Kokosöl *n*
palmic acid Palmitinsäure *f*
palmitic acid Palmitinsäure *f*
pan 1. *(Masch)*. Schale *f*, Wanne *f*; 2. Becken *n*, Napf *m*; Mulde *f*, Schüssel *f*, Trog *m*; Tiegel *m*; Toilettenbecken *n*; 3. Waagschale *f*; 4. *(Brg)* Räumpflug *m*; 5. *(Geol)* Sichertrog *m*
p. amalgamation Trogamalgamierung *f*, Pfannenamalgamierung *f*
p. arrest Waagschalenarretierung *f*
p. base Schalenfuß *m*
p. down *(Fs)* vertikale Kameraverschiebung *f* nach unten
p. furnace Pfannenofen *m*
p. glaze *(Plast)* Silikonharz *n (verhindert das Ankleben von Backwaren in der Form oder am Blech)*
p. grinder Kollergang *m*
p. grinding Kollern *n*
p. head Kegelkopf *m (Schraube)*
p. mixer Kollergang *m*; Zwangsmischer *m*, Trogmischer *m*
p. refractometer Einbaurefraktometer *n*
p.-tile Pfanne *f (Dachziegel)*
p.-type annealing furnace *(Met)* Topfglühofen *m*
p.-type mixer *(Bw)* Zwangsmischer *m*
p. up *(Fs)* vertikale Kameraverschiebung *f* nach oben
panary fermentation Brotgärung *f*
pancake / to durchsacken, absacken *(Flugzeug)*
pancake coil *(El)* Scheibenspule *f*, Flachspule *f*
panchromatic *(Opt)* panchromatisch
panchromatism *(Opt)* Panchromasie *f*

pancratic condenser *(Opt)* pankratischer Kondensor *m*
p. system *(Opt)* pankratisches System *n*
pandean pipe Pansflöte *f*
pane / to [aus]hämmern
pane Scheibe *f*, Tafel *f (Glas)*; Pinne *f (Hammer)*
p. of glass Glasscheibe *f*
panel 1. Platte *f*, Tafel *f*; Paneel *n*, Täfelung *f*; Türfüllung *f*; Feld *n (Fachwerk, Täfelung usw.)*; Beplankung *f*, Brett *n*, Füllung *f*; 2. *(El)* Schaltfeld *n*, Schalttafel *f*; 3. *(Brg)* Bauabteilung *f*, Abbau[feld *n*] *m*, Streb *m*
p. construction Plattenbauweise *f*, Tafelbauweise *f*
p. control Schalttafelsteuerung *f*
p. entry *(Brg)* Abbaustrecke *f*
p.-frame mounting Schalttafeleinbau *m*
p. heating Deckenstrahlungsheizung *f*
p. indicator Schalttafelinstrument *n*
p. instrument Einbauinstrument *n*, Schalttafelinstrument *n*
p. iron Hobelmesser *n (für Holzbearbeitung)*
p. mounting meter *(El)* Schalttafelinstrument *n*
p. raising cut Abplattung *f (Holzbearbeitung)*
p. raising cutterhead Abplattkopf *m (Holzbearbeitung)*
p. raising machine Türfüllungsplattenmaschine *f*
p. switch Schalttafelschalter *m*
panelescent lamp Leuchtkondensator *m (durch Elektrolumineszenz im elektrischen Feld)*
panelized house Großplattenhaus *n*, Großplattenbau *m*, Plattenhaus *n*
p. housing Wohnungsbau *m* in Großplattenbauweise
panelled getäfelt
p. ceiling Kassettendecke *f*; getäfelte Decke *f*
panelling Täfelung *f*; Wandbekleidung *f*; Beplankung *f*; *(Kfz)* Verkleidung *f*, Seitenverkleidung *f*
panicle Rispe *f*
panning 1. *(Geol)* Sichern *n*; 2. *(Fs)* Kameraschwenkung *f*
panorama 1. Panorama *n*; 2. Rundblickaufnahme *f (Film)*
p. scope Panoramasichtgerät *n*
p. sight Rundblickfernrohr *n (Geschütz)*
p. windscreen Rundsichtverglasung *f*
panoramic monitor *(El)* Frequenzspektrograf *m*
p. photograph Panoramaaufnahme *f*
p. view Panoramaansicht *f*; Rundblickaufnahme *f*
panscales Kesselstein *m*
pantelephone Pantelefon *n (verzerrungsfreier Fernhörer)*
pantile Dachziegel *m*
panting beam *(Schiff)* Piek[verstärkungs]balken *m*, Vorpiekbalken *m*
p. deck *(Schiff)* Piekdeck *n*
pantograph 1. *(Masch)* Pantograf *m*, Storchschnabel *m*; 2. *(El)* Scherenstromabnehmer *m*
pantographic die sinking and engraving machine Gesenkfräs- und Graviermaschine *f* mit Pantograf
pantograver Pantografgraviermaschine *f*
pantometer *(Opt)* Pantometer *n*
pantomiller Pantografnachformfräsmaschine *f*
pantry *(Schiff)* Anrichteraum *m*, Pantry *f*
papaverine *(Ch)* Papaverin *n*
paper / to einschlagen, einwickeln *(in Papier)*; tapezieren
paper 1. Papier *n*; 2. Zeitung *f*; 3. wissenschaftliche Abhandlung *f*; Beitrag *m* zu wissenschaftlicher Tagung
p. bag Papiertüte *f*, Papiersack *m*
p. band Papierband *n*, Papierstreifen *m*
p. board Karton *m*, Pappe *f*
p. board combination papierkaschierter Karton *m*
p.-bound book Broschüre *f*
p. cable Papierkabel *n*
p. capacitor Papierkondensator *m*
p. chromatography Papierchromatografie *f*
p. cloth Papiergewebe *n*
p. coating Papierbeschichtung *f*
p. conduit *(El)* Papierisolierrohr *n (gepreßt)*
p.-converting machine Papierverarbeitungsmaschine *f*
p. converting plant Papierverarbeitungsanlage *f*
p.-core cable Papierkabel *n*
p. covered wire *(El)* papierisolierter Draht *m*
p. cup Papierbecher *m*
p. dielectric capacitor *(El)* Papierkondensator *m*
p. drum Papiertrommel *f (Anzeigegerät)*
p. filter Papierfilter *n*
p. flong *(Typ)* Papiermater *f*
p. folder *(Typ)* Falzbein *n*; Falzeinrichtung *f*
p. gauge Papiernormalmaß *n*
p. goods manufacturer Papierwarenhersteller *m*
p. hanging *s.* paperhanging
p. in the flat Planopapier *n*
p. insulated enamelled wire Lackpapierdraht *m*
p. interleaf Papiereinschuß *m*, Papierfolie *f*, Papierzwischenschicht *f*
p. knit *(Text)* Papiergestrick *n*
p. lining Papierauskleidung *f*
p. machee (mâché) Papiermaché *n*, Pappmaché *n*
p. machine Papiermaschine *f*
p. maker's felt Papiermacherfilz *m*

p. making machine Papiermaschine *f*
p. mill Papiermühle *f*, Papierfabrik *f*
p. mounting machine Kaschiermaschine *f*
p. mulberry Papiermaulbeerbaum *m*
p. mulberry fibre Papiermaulbeerbaumfaser *f*
p. napkin Papierserviette *f*
p. number Quadratmetermasse *f*, Papiernummer *f*
p. pulley *s.* p. pulp pulley
p. pulp Papiermasse *f*
p. pulp pulley *(Am)* Faserstoffriemenscheibe *f*, Preßholzriemenscheibe *f*
p. shred Papierwolle *f*
p. shredder Papierzerfaserer *m*
p. shredding machine Papierwolf *m*
p. size Kleister *m*, Papierleim *m*
p. sizing Papier[ver]leimung *f*
p. skin Kunstdarm *m (für Wurst)*
p. stand Aktenständer *m*
p. stock Papierstoff *m*, Papierabfall *m*
p. strip chromatography Papierchromatografie *f*
p. tube Papphülse *f*
p. web Papierbahn *f*, Papiergewebe *n*
p. without cellulose holzfreies Papier *n*
p. without wood-pulp holzfreies Papier *n*
p. work Büroarbeit *f*
paperhanger Tapezierer *m*
paperhanging Tapezieren *n*, Tapezierarbeiten *fpl*, Tapezierung *f*
papier-mâché Papiermaché *n*, Pappmaché *n*
para-brake / to *(Flg)* durch Bremsfallschirm abbremsen
p.-directing group *(Ch)* Substituent *m*, der in Para-Stellung dirigiert
p. piassava Parapiassava *f (Blattfaser, besonders zur Herstellung von Besen und Bürsten)*
p. red Pararot *n*
p. rubber Paragummi *m*
p.-state *(Kern)* Parazustand *m*
parabanic acid Parabansäure *f*
parabola Parabel *f*
p. of higher order Parabel *f* höherer Ordnung
parabolic parabelförmig, Parabel-
p.-arc cambered aerofoil blade parabolisch gekrümmte Tragflügelschaufel *f (Turbine)*
p. girder *(Bw)* Parabelträger *m*
p. mirror Parabolspiegel *m*, parabolischer Spiegel *m*
p. orbit parabolische Bahn *f (Flugbahn)*
p. segment Parabelsegment *n*
p. truss Parabelfachwerk *n*
paraboloid Paraboloid *n*
p. of revolution Rotationsparaboloid *n*
parachor *(Ch)* Parachor *m*
parachute Fallschirm *m*
p. fabric Fallschirmseide *f*
p. flare Leuchtfallschirm *m*
p. jump Fallschirmabsprung *m*
p. raft Fallschirmschlauchboot *n*
p. recovery Fallschirmbergung *f*
p. silk Fallschirmseide *f*
parachutist Fallschirmjäger *m*
paracrystalline *(Ph)* parakristallin, pseudoamorph
paradox Paradoxon *n*
paraffin / to paraffinieren
paraffin Paraffin *n*, Erdwachs *n*, Mineralwachs *n*; Paraffinkohlenwasserstoff *m*
p.-base oil paraffinbasisches Erdöl *n*
p. oil Leuchtöl *n*, Petroleum *n*
p.-rash Paraffinkrätze *f*
p. slack wax Paraffingatsch *m*
p. wax Hartparaffin *n*
paraffinaceous *(Ch)* paraffinisch
paraffined wire *(El)* Wachsdraht *m*
paragon *(Typ) (etwa)* $1\frac{1}{2}$ Cicero *f (Schriftgröße)*
paragraph Absatz *m*, Alinea *n*
p.-vessel Paragraphschiff *n*
parallactic parallaktisch
p. equation Parallaxengleichung *f*
p. libration Parallaxschwingung *f*
p. movement Parallaxenbewegung *f*
parallax Parallaxe *f*
p. computer *(Dat)* Parallaxrechner *m*
p. refractometer Parallaxenrefraktometer *n*
parallel / to 1. *(El)* parallelschalten; 2. parallelisieren *(Fasern)*
parallel Parallele *f*; Richtschiene *f*; Parallelstück *n*, Paralleleisen *n*
parallel parallel; *(Masch)* zylindrisch
p. arrangement 1. *(El)* Parallelschaltung *f*; 2. Gleichtaktschaltung *f (von Röhren)*
p. attachment Reißschienenparallelführung *f*
p.-axes shaving Zahnradschaben *n* mit parallelen Achsen
p. bars Barren *m (Turngerät)*
p. beam Parallelstrahlenbündel *n*
p. capacitor Parallelkondensator *m*, Überbrückungskondensator *m*
p. chord truss Parallelfachwerkbinder *m*, parallelgurtiger Fachwerkbinder *m*
p. circuit *(El)* Parallelschaltung *f*
p. circular plate chamber Ionisationskammer *f* mit parallelen Kreisplatten
p. clamp Parallelschraubzwinge *f*
p. connected *(El)* parallelgeschaltet, nebeneinandergeschaltet
p. connection *(El)* Parallelschaltung *f*, Nebeneinanderschaltung *f*
p. crank Parallelkurbel *f*
p.-crank for bar linkage Parallelkurbelgetriebe *n*
p.-crank mechanism Parallelkurbel[getriebe *n*] *f*

p. current condenser *(Wärm)* Parallelstromkondensator *m*
p. current surface condenser *(Wärm)* Oberflächenparallelstromkondensator *m*
p. displacement Parallelverschiebung *f* *(Geometrie)*
p. equalizer *(El)* Querentzerrer *m*
p. fibre feed *(Text)* Längsfaserspeisung *f*
p. field of forces *(Ph)* Parallelfeld *n*
p. flat nose piece Planschnabel *m*
p.-flow condensation *(Wärm)* Gleichstromkondensation *f*, Parallelstromkondensation *f*
p. guide *(Masch)* Parallelführung *f*
p. guiding of the pencil Stiftgeradführung *f*
p. hand reamer Zylinderhandreibahle *f*
p. heading *(Brg)* Begleitstrecke *f*
p. helical gear Schrägzahnstirnrad *n*
p. hole zylindrische Bohrung *f*
p. jack *(Fmt)* Parallelklinke *f*
p. jet guiding parallele Strahlführung *f* *(Turbine)*
p. key Flachkeil *m*
p. mechanism Parallelführung *f* *(Zeichenmaschine)*
p. memory *(Dat)* Parallelspeicher *m*
p. operation Parallelbetrieb *m*
p. packings Schlittenspanner *m*
p. pin Zylinderstift *m*
p.-plate guide planparalleler Wellenleiter *m*
p.-plate spectrometer Parallelplattenspektrometer *n*
p. postulate *(Math)* Parallelenpostulat *n*
p. precission gauge Parallelendmaß *n*
p. projection Parallelprojektion *f*
p. reamer Zylinderreibahle *f*
p. resistance *(El)* Parallelwiderstand *m*, Nebenwiderstand *m*
p. resonance 1. *(El)* Stromresonanz *f*; 2. *(Ak)* Querresonanz *f*
p. resonance circuit *(Fmt)* Parallelschwingkreis *m*, Sperrkreis *m*
p. resonant circuit *s.* p. resonance circuit
p.-roller bearing Zylinderrollenlager *n*
p.-series connection *(El)* Reihenparallelschaltung *f*, gemischte Schaltung *f*
p. shank Zylinderschaft *m*
p. shank end mill Schaftfräser *m* mit zylindrischem Schaft
p. shank socket reamer for Morse tapers Kegelreibahle *f* für Morsekegel mit Zylinderschaft *(Werkzeugkegel)*
p. shift Parallelverschiebung *f* *(Geometrie)*
p. sliding jaw Parallelgleitbacken *m*
p. storage 1. Parallelspeicherung *f*; 2. Parallelspeicher *m*
p. strip Parallelstück *n* *(Maschinenelement)*
p. stroke milling Zeilenfräsen *n*

p.-T network *(Fmt)* Doppel-T-Netzwerk *n*, Doppel-T-Filter *n*, Doppel-T-Schaltung *f*
p.-to-grain cut Längsschnitt *m* *(Holzbearbeitung)*
p. to the axis achsparallel
p. to the path bahnparallel *(Flugbahn)*
p. transductor *(Elm)* Transduktor *m* in Parallelschaltung
p. vice Parallelschraubstock *m*
p. winding *(El)* Schleifenwicklung *f*
p. wire line *(El)* Lecherleitung *f*, Paralleldrahtleitung *f*
p. worm Zylinderschnecke *f*
paralleled *(El)* parallelgeschaltet
parallelepiped[on] Parallelepiped *n*
paralleling *(El)* Parallelschaltung *f*
p. reactor *(El)* Ausgleichsdrossel *f*
p. switch *(El)* Synchronisierungsschalter *m*
parallelism Gleichlauf *m*, Parallelität *f*
parallelogram Parallelogramm *n*
p. of forces Parallelogramm *n* der Kräfte, Kräfteparallelogramm *n*
p. of velocities *(Masch)* Geschwindigkeitsparallelogramm *n*, Parallelogramm *n* der Geschwindigkeiten
paramagnetic paramagnetisch
p. property paramagnetische Eigenschaft *f*
p. quantum-mechanical amplifier paramagnetischer quantenmechanischer Verstärker *m*
p. substance paramagnetischer Stoff *m*
p.-thermal oxygen analyzer paramagnetischer Sauerstoffanalysator *m*
paramagnetism Paramagnetismus *m*
parameter 1. *(Math)* Parameter *m* *(Hilfsvariable)*; 2. Kenngröße *f*, Kennwert *m*, Bestimmungsgröße *f*
parameters of state Zustandsgrößen *fpl*
p. of the orbit *(Rak)* Parameter *mpl* der Flugbahn, Flugbahnwerte *mpl*
parametric Parameter-, parametrisch
p. amplifier *(El)* parametrischer Verstärker *m*, Reaktanzverstärker *m*
p. representation Parameterdarstellung *f*
paramorphic paramorph *(Kristalle)*
paramorphism Paramorphose *f* *(Kristalle)*
paramorphous paramorph *(Kristalle)*
parapet Brüstung *f*; Brüstungsmauer *f*
paraphase amplifier *(El)* Symmetrieverstärker *m*
pararaft Fallschirmschlauchboot *n*
paraselene Nebenmond *m*
parasite Parasit *m*, Schmarotzer *m*
parasite *s.* parasitic
p. current Kriechstrom *m*
p. drag *(Flg)* schädlicher Widerstand *m*
parasitic unerwünscht, parasitär, Wirbel-, Kriech-
p. aerial mittelbar gespeiste Antenne *f*, strahlungsgekoppelte Antenne

p. capture parasitärer Neutroneneinfang *m* *(im Reaktor)*
p. current Fremdstrom *m*, parasitärer Strom *m*
p. element strahlungsgekoppelte Antenne *f*, strahlungsgekoppeltes Antennenelement *n*
p. lobe Nebenzipfel *m*, Störzipfel *m* *(Antennentechnik)*
p. noise *(El)* Interferenz *f*, Störung *f*, Störgeräusch *n*
p. oscillation *(El)* parasitäre (wilde) Schwingung *f*
p. reflector *(El)* ungespeister (strahlungsgekoppelter) Reflektor *m*, Störreflektor *m*
p. stopper *(El)* Sperre *f* für wilde Schwingungen
p. suppressor *(El)* Störungsunterdrücker *m*
parasitically excited mittelbar gespeist, strahlungsgekoppelt
parasitics *(El)* wilde Schwingungen *fpl*, Störgeräusch *n*; unerwünschte Nebenschwingungen *fpl*
parasol-type generator *(Elm)* Schirmgenerator *m*
paratrooper *(Mil)* Fallschirmjäger *m*
paraxial achsnahe; paraxial
parcel 1. Paket *n*; Stückgut *n*; 2. Parzelle *f*
p. drilling machine Parzellendrillmaschine *f*
p. net Gepäcknetz *n*
p. rack Gepäcknetz *n*
p. shelf Gepäckbrett *n*, Paketablage *f*
p. thresher Parzellendreschmaschine *f*
parcelled cargo Stückgut *n*
parchment Pergament *n*
p. ink Pergamentfarbe *f*
p. leather dressing Pergamentgerbung *f*
p. membrane Pergamentmembran *f*
p. paper Pergamentpapier *n*
parchmentize / to pergamentieren
parchmyn *(Ch)* Pergamin *n*
pare / to 1. verschneiden, schälen; 2. drechseln
parent Muttersubstanz *f*
p. atom Ausgangsatom *n*, Mutteratom *n*
p. distribution *(Math)* Verteilung *f* der Grundgesamtheit
p. element *(Kern)* Ausgangselement *n*, Ausgangssubstanz *f*, Muttersubstanz *f*, Stammelement *n*
p. field (El) Ausgangsfeld *n*, ursprüngliches Feld *n*
p. frequencies Primärfrequenzen *fpl*
p. hull Ausgangsschiffsform *f*, Grundschiffsform *f*
p. metal Grundmetall *n*, Mutterwerkstoff *m*
p. nucleus *(Kern)* Ausgangskern *m*, Mutterkern *m*
p.-plus-daughter fractions Mutter- und Tochteranteil *m* *(einer radioaktiven Substanz)*
p. population Grundgesamtheit *f* *(Statistik)*
p. stock Ausgangssubstanz *f*, Rohstoff *m*
p. substance 1. *(Kern)* Ausgangselement *n*, Ausgangssubstanz *f*, Muttersubstanz *f*, Stammelement *n*; 2. Grundverbindung *f*
parenthesis *(Math)* runde Klammer *f*
parfocalize / to abgleichen *(Mikroskopoptik)*
parfocalizing length *(Opt)* Abgleichlänge *f*
parget Putz *m*, Rapputz *m*
pargeting Putz *m*, Rapputz *m*; Berappen *n*, Putzen *n*, Verputzen *n*; Innenverputz *m* eines Schornsteins
parhelion Nebensonne *f*, Gegensonne *f*
paring axe Schälaxt *f*
p. chisel Breitstahl *m*, Flachstahl *m* *(Holzbearbeitung)*
p. plough Schälpflug *m*, Abstechpflug *m*
parings Hobelspäne *mpl*
Paris green Schweinfurter Grün *n*
Paris yellow Chromgelb *n*
parity 1. *(Math)* Restklassencharakter *m* mod 2; 2. *(Ph)* Parität *f*
p. check Paritätskontrolle *f*, Paritätsprüfung *f*
p. digit Paritätsziffer *f*
p. operator Paritätsoperator *m*
p. selection rule Paritätsauswahlregel *f*
park / to parken
parkerizing Parkerisieren *n*
Parke's process Parkes-Verfahren *n* *(Bleientsilberung durch Zink)*
parking Parken *n*
p. brake Feststellbremse *f*, Handbremse *f*
p. lamp Parkleuchte *f*, Standlicht *n*, Parklicht *n*
p. lane Parkspur *f*, Standspur *f*
p. light Parklicht *n* *(ausgestrahlt von einer Parkleuchte)*, Standlicht *n*
p. lot Parkplatz *m*
p. meter Parkzeitmesser *m*
p. place Parkplatz *m*
p. space Parkplatz *m*
p. strip Standspur *f* *(auf dem Parkplatz)*
parquet Parkett *n*
p. board Parkettbrett *n*
p. floor Stabparkettfußboden *m*
parquetry Parkett *n*; Stabparkettfußboden *m*, Parkettfußboden *m*
p.-fillet Parkettstab *m*
parrot tooth Kehlenzahn *m* *(Sägeblatt)*
parsec Parsek *n* *(Parallaxensekunde)*
parsley camphor *(Ch)* Apiol *n*
part / to trennen, teilen; scheiden, sich trennen
p. by melting abschmelzen
p. off [ab]trennen
p. off [to length] abstechen
part 1. Bauteil *n*, Baueinheit *f*, Organ *n*; Teil *n*, Werkstück *n*; 2. *(Typ)* Heft *n*, Lieferung *f*
p. chill roll Walze *f* mit gehärteter Oberfläche
p. drawing Einzelteilzeichnung *f*

p.-ejection device Teilauswerfeinrichtung *f*
p. handling Werkstücktransport *m*
p. list Stückliste *f*
p. load Teillast *f*
p.-load efficiency Teillastwirkungsgrad *m*
p.-load needle Teillastnadel *f (Vergaser)*
p. loads Stückgut *n*
p. of rotor carrying blades Schaufelträger *m (Turbine)*
p. of the boiler not supported *(Wärm)* freitragender Kesselteil *m*
p. pitch *(El)* Teilschritt *m (Wicklung)*
p. section Teilschnitt *m*
p.-time private wire circuit *(Fmt)* Privatleitung *f* mit begrenzter Mietdauer
p. to be ground Schleifteil *n*
p. to be hobbed Wälzfrästeil *n*
p. to be milled Frästeil *n*
parted geteilt, mehrteilig
p. hyperboloid *(Math)* zweischaliges Hyperboloid *n*
partial *(Math)* partiell; teilweise, unvollständig; Partial-, Teil-
p. admission teilweise (partielle) Beaufschlagung *f (Turbine)*
p. admission turbine Partialturbine *f*
p. automatic station *(Fmt)* halbautomatische Station *f*
p. awning deck vessel Schiff *n* mit teilweisem Sturmdeck
p. body irradiation *(Kern)* Teilkörperbestrahlung *f*
p. breakdown unvollständiger Durchbruch *m*, Teildurchbruch *m (der Isolierung)*
p. bulkhead *(Schiff)* Teilschott *n*, Rahmenschott *n*
p. carry *(Dat)* Teilübertrag *m*
p. cavitation teilweise Kavitation *f*
p. combustion Teilverbrennung *f*
p. cross section *(Kern)* partieller Wirkungsquerschnitt *m*
p. derivative partielle Ableitung *f*, partieller Differentialquotient *m*
p. differential equation partielle Differentialgleichung *f*
p. disintegration constant *(Kern)* partielle Zerfallskonstante *f*
p. dislocation partielle Versetzung *f*, Teilversetzung *f (von Kristallen)*
p. draft *(Text)* Einzelverzug *m*, Teilverzug *m*
p. drop of pressure Teildruckgefälle *n (Turbine)*
p. earth Erdschluß *m* über Widerstand, unvollkommener Erdschluß
p. fixation teilweise Einspannung *f (Statik)*
p.-flow filter *(Kfz)* Nebenstromfilter *n*
p. fraction *(Math)* Partialbruch *m*
p. integration partielle Integration *f*
p. ionization Teilionisation *f*
p. load Teillast *f*
p.-load region Teillastgebiet *n*
p.-load valve Teillastventil *n*
p. mode Teilwelle *f*
p. molar heat partielle Molwärme *f*
p. multiple *(El)* gestaffelte Vielfachschaltung *f*, Teilvielfachfeld *n*
p. payment Teilzahlung *f*
p. polarization partielle Polarisation *f*
p. pressure Partialdruck *m*
p.-radiation pyrometer Teilstrahlungspyrometer *n*
p. restoring time *(El)* Nachwirkzeit *f*, Teilnachwirkzeit *f*, Teilsperrzeit *f*
p. scanning *(Fs)* Teilabtastung *f*
p. sequence *(Math)* Teilfolge *f*
p. shadow Halbschatten *m*, Penumbra *f*
p. solubility *(Ch)* beschränkte Löslichkeit *f*
p.-span flaps *(Flg)* Klappen *fpl* über einem Teil der Tragfläche
p. stowing *(Brg)* Teilversatz *m*
p. subsidence *(Brg)* Teilsenkung *f*
p. superconstructure vessel Schiff *n* mit teilweisen (nicht durchlaufenden) Aufbauten
p. vacuum 1. *(Masch)* Unterdruck *m*; 2. Vorvakuum *n (Elektronenröhrenherstellung)*
p. valency *(Ch)* Nebenvalenz *f*, Partialvalenz *f*
p. vibration Partialschwingung *f*, Teilschwingung *f*
p. view Teilansicht *f*
p. wave Teilwelle *f*
partially occupied band teilbesetztes Energieband *n*
p. polarized teilweise polarisiert
p. prestressed teilweise vorgespannt, mit teilweiser (beschränkter) Vorspannung *(Spannbetonbau)*
particle Partikel *f*, Teilchen *n*; Korn *n*; Masseteilchen *n*
p. acceleration Teilchenbeschleunigung *f*
p. accelerator Teilchenbeschleuniger *m*
p. board Spanplatte *f*
p. board binder Bindemittel *n* für Plattenherstellung
p.-board core stock Spanplattenmittellagenmaterial *n*
p. board mat Spanplattenvlies *n*
p.-board panel stock Spanplattentäfelungsmaterial *n*
p. bombardement Teilchenbestrahlung *f*, Teilchenbeschuß *m*
p. counter Teilchenzähler *m*
p. emission Teilchenemission *f*, Korpuskularemission *f*
p. energy Teilchenenergie *f*
p. gradation *(Bw)* Kornabstufung *f*, Kornaufbau *m*

p. losses Teilchenverluste *mpl*
p. of mass Massenteilchen *n*, Punktmasse *f*
p. preparation Spanaufbereitung *f*
p. radiation Korpuskularstrahlung *f*, Teilchenstrahlung *f*
p. shape Kornform *f*, Korngestalt *f*
p. shape factor Kornformfaktor *m*
p. size Korngröße *f*
p. size distribution Korngrößenverteilung *f*
p. size range Korngrößenstufe *f*
p. spin *(Kern)* Teilchenspin *m*
p. trajectory *(Kern)* Teilchenbahn *f*
particular besonders; partikulär
p. average kleine Havarie *f*
p. solution *(Math)* partikuläre (spezielle) Lösung *f*
particulate aus Einzelheiten bestehend
p. fluidized bed homogene Wirbelschicht *f*
p. solid Feststoffpartikel *f*, Feststoffteilchen *n*
parting Lösen *n*, Trennung *f*; Scheidung *f* *(z. B. von Gold)*; *(Masch)* Trennen *n*, Abstechen *n*
p. agent *(Ch)* Scheidemittel *n*
p. compound Gleitmittel *n*, Trennmittel *n* *(für Formen)*
p. furnace *(Met)* Scheideofen *m*
p. line Trennfuge *f*, Teilfuge *f*, Trennlinie *f*
p. liquid *(Ch)* Scheideflüssigkeit *f*
p. medium *(Met)* Formpuder *m* *(Formerei)*
p.-off *(Masch)* Abstechen *n*, Trennen *n*
p.-off and recessing tool Abstech- und Einstechmeißel *m*
p.-off operation Trennarbeit *f*, Abstecharbeit *f*
p.-off tool Stechmeißel *m*
p.-off, turning and boring lathe Dreh-, Bohr- und Abstechmaschine *f*
p. plane Teilungsebene *f*
p. silver Scheidesilber *n*
p. surface Trennungsfläche *f*
p. tool Abstechdrehmeißel *m*, Einstechdrehmeißel *m*, Stechmeißel *m*
p. workshop Scheideanstalt *f*
partition leichte Trennwand *f*, Scheidewand *f*, Zwischenwand *f*, Membran *f*
p. boards Bretterverschlag *m*
p. bulkhead *(Schiff)* Teilschott *n*, Trennschott *n*, Kammerschott *n*
p. chromatography Verteilungschromatografie *f*
p. coefficient Verteilungskoeffizient *m*
p. function Zustandssumme *f*
p. line Trennfuge *f*
p. noise *(El)* Stromverteilungsrauschen *n*
p. panel Trennwand *f*
p. panel lining Trennwandverkleidung *f*
p. panel window Trennwandfenster *n*
p. rock *(Brg)* Nebengestein *n*
p. wall 1. Scheidewand *f*, Trennwand *f*, Zwischenwand *f*; 2. *(Brg)* Versatzmauer *f*
partly bleached halbgebleicht *(z. B. Papierfaserstoff)*
p. square wahnkantig, waldkantig
partner Teilhaber *m*
partnership Teilhaberschaft *f*
p. assets Gesellschaftsvermögen *n*
parts between regrinds *(Masch)* Standmenge *f*
p. inspection Werkstückgütekontrolle *f*
p. inventory Ersatzteilhaltung *f*
p. list Teileliste *f*, Stückliste *f*
party *(Fmt)* Teilnehmer *m*
p. called *(Fmt)* angerufener Teilnehmer *m*
p. calling *(Fmt)* anrufender Teilnehmer *m*
p.-coloured verschiedenfarbig
p. line *(Fmt)* Gemeinschaftsleitung *f*
p. reception Gemeinschaftsempfang *m*
p. wall Brandmauer *f*; Wohnungstrennwand *f*
Pascal's triangle Pascalsches Dreieck *n*
pass / to überholen, vorbeifahren; durchfließen *(Strom)*; durchfallen, durchstrahlen; durchlaufen, durchgehen, passieren; Durchgang haben *(Werkstück)*; strömen *(Flüssigkeit)*
p. a booking *(Fmt)* eine Gesprächsanmeldung übermitteln
p. along entlangstreichen *(Gase)*
p. an examination eine Prüfung bestehen
p. around umspülen *(Gase, Flüssigkeiten)*
p. back zurückführen *(Walzgut)*
p. into einleiten *(Gase)*
p.-off entweichen, abströmen
p. on weiterleiten
p. out of range *(El)* die Reichweite verlassen
p. over 1. überleiten; 2. *(Flg)* überfliegen; 3. überführen, kreuzen *(Straßenbau)*
p. through durchsickern *(Flüssigkeit)*; durchfallen *(durch ein Sieb)*, passieren; durchführen *(elektrische Leitungen durch Wände)*
p. through a point einen Punkt durchlaufen *(z. B. Bandfilterkurve)*
p. within range *(El)* in Reichweite kommen
pass Durchgang *m*, Arbeitsgang *m*; Zug *m*; Schweißlage *f*; Kaliber *n*, Stich *m* *(Walzen)*; Gebirgspaß *m*
p.-band *(El)* Durchlaßbereich *m*, Durchlaßband *n*
p.-band amplifier Bandpaßverstärker *m*
p.-band attenuation Durchlaßdämpfung *f*
p.-band width *s.* p.-band
p. cautiously signal *(Eis)* Langsamfahrt-Signal *n*
p. frequency Durchlaßfrequenz *f*
p.-key Hauptschlüssel *m*
p. lamp *(Kfz)* Abblendlicht *n*
p. line Walzbahn *f* *(Walzwerk)*
p.-over mill Übergabewalzwerk *n*, Überhebewalzwerk *n*; Profilwalze *f*

p. range Durchlaßbereich *m*
p. reduction Dickenabnahme *f* je Durchgang
p. round Umlauf *m (Wälzstoßen)*
p. sequence Kaliberfolge *f (Walzwerk)*
p. squeeze Kaliberfolge *f (Walzwerk)*
p. through capacitor Durchführungskondensator *m*
p. variation Kaliberwechsel *m (Walzwerk)*
passage 1. Durchgang *m*, Durchlauf *m*, Durchfluß *m*; Übergang *m*; 2. Kanal *m (Ventil)*; 3. *(Brg)* Durchörterung *f*
p. area Strömungsquerschnitt *m*, Kanalquerschnitt *m*
p. between blades Schaufelkanal *m (Turbine)*
p. boat Passagierschiff *n*, Fahrgastschiff *n*
p. of cards Kartendurchlauf *m*, Lochkartendurchlauf *m*
p. of current *(El)* Stromdurchgang *m*
p. of drawing *(Text)* Streckpassage *f*
p. of heat Wärmeübertragung *f*
p. of rays *(Opt)* Strahlendurchgang *m*
p. of the tool *(Masch)* Werkzeugdurchgang *m*
p. through the wheel Durchgang *m* durch das Rad *(Turbine)*
p.-way 1. *(Brg)* Verbindungsstrecke *f*; 2. Laufweg *m*
passageway *s.* passage-way
passband *s.* pass-band
passed for press *(Typ)* für den Druck freigegeben
passementerie *(Text)* Posamenten *npl*
passenger Fahrgast *m*
p. accomodation spaces Fahrgasträume *mpl*, Fahrgasteinrichtung *f*
p. automobile Personenkraftwagen *m*, Pkw
p. berth Fahrgasthafen *m*
p. boat Fahrgastboot *n*, Fahrgastschiff *n*
p. car 1. *(Eis)* Personenwagen *m*; 2. Personenkraftwagen *m*, Pkw
p.-car ferry *(Schiff)* Personen-Autofähre *f*
p. compartment Fahrgastraum *m*
p. counter Fahrgastzähler *m*
p. deck Fahrgastdeck *n*
p. elevator Personenaufzug *m*
p. handling Fahrgastabfertigung *f*
p.-hoist Personenaufzug *m*
p. launch Fahrgastboot *n*, Verkehrsboot *n*, Fahrgastbarkasse *f*
p. liner Fahrgastschiff *n (in der Linienfahrt)*, Linienpassagierschiff *n*
p. motor-ship Fahrgastmotorschiff *n*, Motorfahrgastschiff *n*
p. ship Fahrgastschiff *n*, Passagierschiff *n*
p. tonnage Fahrgastschiffsraum *m*
p. train Personenzug *m*
p.-vehicle ferry Personen-Fahrzeugfähre *f*
passing Durchgang *m*, Durchfluß *m*; Durchfahrt *f*; Siebdurchgang *m*, Siebdurchlauf *m*; Übergang *m*, Wechsel *m*
p. a vessel through a lock Durchschleusung *f (eines Schiffes)*
p. beam *(Kfz)* Abblendlicht *n*
p. contact Wischkontakt *m (eines Schalters)*
p. knot durchgehender Ast *m*
p. lane Überholspur *f*
p. light *(Kfz)* Abblendlicht *n*
p. of the slide beyond the port *(Wärm)* Überschleifen *n* des Kanals
p. place Ausweichstelle *f*, Überholstelle *f*
p. shaft durchgehende Welle *f*
p.-signal indicator *(Kfz)* Überholsignalgerät *n*
p. track Ausweichgleis *n*
passivate / to passivieren
passivating Passivierung *f*
passivation Passivierung *f*
passive passiv, inaktiv, reaktionsträge
p. aerial mittelbar gespeiste Antenne *f*, strahlungsgekoppelte Antenne
p. balance return loss Fehlerdämpfung *f* in Stromkreisen ohne Verstärker
p. earth pressure *(Bw)* Erdwiderstand *m*, passiver Erddruck *m*
p. electric network *(El)* passive Schaltung *f*
p. singing point Schwingungseinsatzpunkt *m* in Stromkreisen ohne Verstärker
p. state passiver Zustand *m*, Ruhezustand *m*; oberer Grenzzustand *m*
passivity Passivität *f*
past gegen, vorbei an
paste / to [be]kleben
p. on aufkleben
p. over bekleben
paste Paste *f*; Kleister *m*, Klebstoff *m*
p.-board *s.* pasteboard
p. of glass Glasfluß *m*
p.-up *(sl) (Typ) (geklebter)* Umbruch *m*
pasteboard geklebter Maschinenkarton *m*
p.-binding Papp[ein]band *m (Buchbinderei)*
p.-worker Kartonierer *m*
pasted board zusammengeklebte Pappe *f*, Mehrlagenkarton *m*
p. filament Pastekatode *f*
p. frame plate Rahmenplatte *f (Bleiakkumulator)*
p. plate pastierte Platte *f (der Batterie)*
pastel Pastellstift *m*
pasteurize / to pasteurisieren, keimfrei machen
pasteurized process cheese Schmelzkäse *m*
pasting of core parts Zusammenkleben *n* von Kernteilen *(Gießerei)*
pasture Viehweide *f*
p. milking plant Weidemelkanlage *f*
pasty pastös, teigig
p. iron *(Met)* teigiges Eisen *n*

pat Kuchen *m (Zementprüfung)*
patch / to 1. ausbessern, flicken; 2. *(El) (vorübergehend)* zusammenschalten
p. up aufflicken, zusammenflicken, ausbessern, reparieren
patch 1. *(Dat)* Korrekturbefehl *m*; 2. Flickstelle *f (Straßendecke)*; 3. *(Geol)* Erznest *n*; 4. Fleck *m*, Flecken *m*, Flickfleck *m*
p. bay *s.* patchcord bay
p. board *(BMSR)* Schalttafel *f*, Steckbrett *n*, Steckertafel *f*
p. effect *(El)* Fleckeneffekt *m (Halbleitertechnik)*
p. panel *(BMSR)* Steckbrett *n*; *(Fmt)* Klinkenfeld *n*
p. work Flickarbeit *f*, Stückwerk *n (schlechte Arbeit)*
patchcord Steckerschnur *f*; flexibles Verbindungskabel *n*; Überführungsschnur *f*, Umschaltungsschnur *f*; *(Fmt)* Vermittlungsschnur *f*
p. bay *(BMSR)* Schaltfeld *n*, Steckerfeld *n*, Steckverbindungsfeld *n*; Steckverbindungsgestell *n*, Steckverbindungsbucht *f*, Verbindungsschnurgestell *n*
patching cord *(Fmt)* Vermittlungsschnur *f*
p. mortar Flickmörtel *m*
p. of boiler plate Aufsetzen *n* eines Flickens auf das Kesselblech
p. of furnace Ausbessern (Flicken) *n* eines Ofens
patent / to patentieren *(Patent erteilen; Draht patentieren)*
patent Patent[schrift *f*] *n*
p. action Patentklage *f*, Patentstreit *m*
p. agent Patentanwalt *m*
p. allowance Patenterteilung *f*
p. anticipation Patentvorwegnahme *f*
p. anticipatory reference Patentvorwegnahme *f*
p. appeal Patenteinspruch *m*
p. application Patentanmeldung *f*, Patentgesuch *n*
p. attorney Patentanwalt *m*
p. claim Patentanspruch *m*
p. code Patentgesetz *n*
p. fee Patentgebühr *f*
p. fight Patentstreitigkeit *f*
p. for invention Erfindungspatent *n*
p. fuel Brikett *n (insbesondere aus Steinkohle)*
p. glass *(Schiff)* Patentglas *n*, Decksfenster *n*
p. grant Patenterteilung *f*
p. granting Patentgenehmigung *f*
p. holder Patentinhaber *m*
p. infringement Patentverletzung *f*
p. language Patentsprache *f*
p. law Patentgesetz *n*, Patentrecht *n*
p. leather Lackleder *n*, lackiertes Rindleder *n*
p. letter Patenturkunde *f*
p. licence Patentlizenz *f*
p. life Patentlaufzeit *f*
p. ligitation Patentstreit *m*
p. list Patentliste *f*
p. log Patentlog *n (Gerät zur Messung der Schiffsgeschwindigkeit mittels nachgeschleppten Propellers)*
p. of improvement Verbesserungspatent *n*
p. of utilization Gebrauchsmusterschutz *m*
p. office Patentamt *n*
p.-office action patentamtlicher Bescheid *m*
p.-office decision patentamtlicher Entscheid *m*
p.-office drawing Patentzeichnung *f*
p.-office notification patentamtlicher Bescheid *m*
p. renewal Patenterneuerung *f*
p.-renewal fee Patentjahresgebühr *f*, Patentverlängerungsgebühr *f*
p. right Patentrecht *n*
p. rights Schutzrechte *npl*
p. rolls Patentregister *n*
p. slip *(Schiff)* Aufschlepphelling *f*
p. specification Patentbeschreibung *f*, Patentschrift *f*
p. statutes Patentgesetz *n*
p. suit Patentklage *f*
p. survey Patentuntersuchung *f*
p. tax Patentgebühr *f*
patentability Patentierbarkeit *f*
patentable patentierbar
patented at home and abroad im In- und Ausland patentiert (patentgeschützt)
p. procedure gesetzlich geschütztes Verfahren *n*
p. steel patentierter Stahl *m*
p. wire patentierter Draht *m*
patentee Patentanmelder *m*, Patentinhaber *m*, Patentträger *m*
patenting furnace *(Met)* Patentierofen *m*
path 1. Bahn *f*; Flugbahn *f*; Weg *m*; Verlauf *m*; Weglänge *f*, Strecke *f*; 2. *(Elm)* Ankerzweig *m*; Pfad *m*, Strompfad *m*
p. acceleration Bahnbeschleunigung *f*
p. angle Bahnwinkel *m*, Bahnneigung *f*
p. control Bahnsteuerung *f (Werkzeugmaschinensteuerung)*
p. difference Wegdifferenz *f*, Wegunterschied *m*
p. difference of rays Gangunterschied *m* der Strahlen
p. element Bahnelement *n*, Kurvenelement *n*
p. force Bahnkraft *f*
p. of contact Eingriffskurve *f*, Eingriffsweg *m*
p. of integration Integrationsweg *m*

p. of lay *(Text)* Ladenweg *m*
p. of lines of force Kraftlinienverlauf *m*
p. of rays Strahlengang *m*
p. of the depression Depressionsbahn *f*
p. speed Bahngeschwindigkeit *f*
p.-time curve Weg-Zeit-Diagramm *n*
p. tracking Bahnvermessung *f*, Bahnverfolgung *f*
pathfinder *(Mil)* Gefechtsaufklärer *m*
patina Patina *f*, Edelrost *m*
patinate / to patinieren, mit Edelrost überziehen
patio Innenhof *m*
patrinite *(Min)* Patrinit *m*
patrol / to *(Fmt)* Leitungen begehen
patrol Polizeistreife *f*
p. car Streifenwagen *m*
p. vessel Vorpostenboot *n*, U-Bootjäger *m*
patronite *(Min)* Patronit *m*
pattering *(Ak)* Prasseln *n*
pattern 1. Muster *n*; Zeichnung *f*; Schablone *f*; Ausführung *f*; Schema *n*; 2. *(Met)* Modell *n* *(Formguß)*; 3. *(Fmt, Fs)* Strahlungscharakteristik *f*, Strahlungsdiagramm *n*, Oszillogramm *n*; 4. *(Text)* Schnittmuster *n*, Muster *n*
p. binding machine Modelleinfaßmaschine *f* *(Schuhherstellung)*
p. chain *(Text)* Musterkette *f*
p. changing *(Text)* Musterwechsel *m*
p. drum *(Text)* Mustertrommel *f*
p. grading machine Modellgradiermaschine *f* *(Schuhherstellung)*
p. lifter Modellheber *m* *(Gießerei)*
p. lumber *(Met)* Modellholz *n* *(für Modelltischlerei)*
p. of tyre tread Reifenprofil *n*
p. plate Modellplatte *f*
p. printing process Dessindruckverfahren *n*
p. recognition *(Dat)* Zeichenerkennung *f* *(im allgemeinen Sinne)*
p. repeat *(Text)* Bindungsrapport *m*
p. shop Modelltischlerei *f*
p. thread *(Text)* Musterfaden *m*
p. varnish *(Met)* Modellack *m* *(Gießerei)*
p. warp *(Text)* Figurkette *f*
p. weaving Musterweberei *f*
patternless moulding *(Met)* Formen *n* ohne Modell
patternmaker Modelltischler *m*
patternmaker's lathe Holzdrehmaschine *f*
p. rule Schwindmaßstab *m*
p. shop Modellmacherei *f*, Modelltischlerei *f*
Pauli['s] exclusion principle *(Kern)* Pauli-Prinzip *n*, Ausschließungsprinzip *n*
pave / to pflastern
pavement Pflaster[ung *f*] *n*; Bürgersteig *m*, Fußweg *m*, Gehweg *m*; Fahrbahn *f*; Straßenbefestigung *f*, Straßenbelag *m*, Straßendecke *f*, Straßenpflaster *n*
p. concrete Straßendeckenbeton *m*
paver *(Am)* Pflasterer *m*, Steinsetzer *m*
p.-finisher Schwarzdeckenfertiger *m*
p.-spreader Schwarzdeckeneinbaumaschine *f*
paver's dressing hammer Zurichtehammer *m*
pavilion Pavillon *m* *(für Messen)*
paving Deckeneinbau *m*; Pflastern *n* *(Straßenbau)*; Straßenbefestigung *f*, Straßenbelag *m*, Straßendecke *f*, Straßenpflaster *n*
p. block Pflasterstein *m*, Pflasterklotz *m*
p. brick Pflasterstein *m*, Straßenklinker *m*
p. finisher Deckenfertiger *m*
p. flag Gehwegplatte *f*
p. mixer Mischerfertiger *m* *(Straßenbau)*
p. sett Pflasterstein *m*
p. stone Pflasterstein *m*; Gehwegplatte *f*
p. train Deckeneinbauzug *m*
p. wood Pflasterholz *n*
paviour Pflasterer *m*, Steinsetzer *m*; Straßenklinker *m*
pawl / to 1. *(Masch)* einrasten, klinken; 2. *(Fmt)* einklinken
pawl 1. *(Masch)* Sperrklinke *f*, Schaltklinke *f*, Schnäpper *m*; Mitnehmerklinke *f*; Absperrklaue *f*; Sperrhebel *m*; Sperrhaken *m*; 2. *(Fmt)* Klinke *f*, Sperrklinke
p.-and-ratchet mechanism Sperrgetriebe *n*, Sperrwerk *n*; Rückschaltwerk *n*
p. arm Schwingarm *m*
p. coupling Klinkenkupplung *f*, Schnapperkupplung *f* *(z. B. Automatikempfänger)*
p. spring Sperrhakenfeder *f*
p. stud Sperrhakenstift *m*
p. wheel Sperrad *n*, Sperrglied *n* *(z. B. Automatikempfänger)*
pay / to [be]zahlen, entlohnen
p. duty verzollen
p. off 1. entlohnen; *(Schiff)* abmustern; 2. abspulen, abrollen *(Draht)*
p. out 1. auszahlen; 2. abspulen, abrollen *(Draht)*; auslegen *(Kabel)*
pay Bezahlung *f*, Gehalt *n*, Lohn *m*; *(Schiff)* Heuer *f*
p. as you see television Fernsehempfänger *m* mit Gebührenautomatik
p. button Zahlknopf *m*
p.-off machine Abspuleinrichtung *f*
p.-off matrix *(Dat)* Partieerfolgsmatrix *f*, Spielmatrix *f*
p. ore [ab]bauwürdiges Erz *n*
p.-out reel liegende Drahthaspel *f*
p. station Münzfernsprecher *m*
p.-streak *(Geol)* Erzfall *m*, Adelsvorschub *m*, Adelszone *f* *(reicher Lagerstättenteil)*
p. telephone Münzfernsprecher *m*
p. television Münzfernsehen *n*
p. vision Drahtfernsehen *n*

payability 1. Zahlbarkeit *f*; 2. *(Brg)* Ergiebigkeit *f*, Abbauwürdigkeit *f*
p. limit *(Brg)* Bauwürdigkeitsgrenze *f*
payable 1. zahlbar; 2. *(Brg)* ergiebig, [ab]bauwürdig
p. ore [ab]bauwürdiges Erz *n*
paying deadweight *(Schiff)* Ladungstragfähigkeit *f*
payload Nutzlast *f*, Zulast *f*
p. ratio Nutzlastverhältnis *n*
p.-structure-fuel weight ratio *(Rak)* Nutzlast-Aufbau-Kraftstoff-Verhältnis *n*
p. weight Nutzlastmasse *f*
payment Bezahlung *f*, Entlohnung *f*, Zahlung *f*, Auszahlung *f*
p. by instalments Ratenzahlung *f*
p. by results Leistungslohn *m*
p. by weight basis Bezahlung *f* nach Masse
p. in advance Vorauszahlung *f*, Vorschuß[zahlung *f*] *m*
p. in cash Barzahlung *f*
p. in kind Naturalleistung *f*
p. in part Teilzahlung *f*
paynize / to Holz imprägnieren
payroll Lohnliste *f*
p. deduction Lohnabzug *m*
pea Erbse *f*
p. cleaner and grader Erbsenreinigungs- und -sortiermaschine *f*
p. gravel Feinkies *m*, Perlkies *m*
p. harvester Erbsenerntemaschine *f*
p.-ore Bohnerz *n*
p. swather Erbsenschwader *m*
peacock ore *(Min)* Buntkupferkies *m*, Bornit *m*
peak 1. Spitze *f*, Gipfel *m*, Höhepunkt *m*, Maximum *n*, Scheitel[wert] *m*, Spitzenwert *m*; 2. *(Schiff)* Piek *f (Vorpiek, Achterpiek)*
p. achievement Spitzenleistung *f*
p. alternating gap voltage *(El)* Spaltspitzenspannung *f*
p. anode voltage Anodenspitzenspannung *f*
p. black *(Fs)* Schwarzspitze *f*, Maximum *n* an Schwarz
p. blocked voltage *(El)* Scheitelwert *m* der positiven Sperrspannung
p. bulkhead *(Schiff)* Piekschott *n*
p. cathode current Scheitelwert *m* des Katodenstromes; Katodenspitzenstrom *m*
p. cathode fault current maximaler Kurzschlußstrom *m*
p. chopper *(El)* Begrenzer *m*
p. chopper circuit *(El)* Schaltung *f* zur Beschneidung der Spitzen *(z. B. Spannungsspitzen)*
p. chopping *(El)* Beschneiden *n*, Begrenzen *n* *(der Amplitude)*
p. cross section maximaler Wirkungsquerschnitt *m*
p. current Spitzenstrom *m*; Scheitelwert *m* der Stromwelle, Scheitelstrom *m*
p. demand Spitzenbelastung *f*
p. energy Maximalenergie *f*, Höchstenergie *f*
p. factor 1. Extremwert *m*; 2. Scheitelfaktor *m (Strom)*
p. forward anode voltage Spitzenwert *m* der Anodenspannung in Durchlaßrichtung
p. forward current Durchlaßspitzenstrom *m*
p. frame *(Schiff)* Piekspant *n*
p. grid current Gitterspitzenstrom *m*
p.-holding optimalizing control Extremalwertregelung *f*, Optimalwertregelung *f* mit Maximumsuche
p. indicator *(El)* Höchstwertanzeiger *m*, Spitzenspannungszeiger *m*, Spitzen[wert]zeiger *m*
p. inverse anode voltage Spitzenwert *m* der Sperrspannung (Anodenspannung in der Sperrrichtung)
p. inverse voltage Sperrspannungsscheitelwert *m (eines Stromrichters)*, Spitzensperrspannung *f*
p. limiter Amplitudenbegrenzer *m*; Begrenzungsverstärker *m*
p. limiting amplifier Verstärker *m* mit automatischer Lautstärkebegrenzung
p. load Spitzenbelastung *f*, Spitzenlast *f*
p. load allowance *(El)* Überlastbarkeit *f*
p. load plant *s.* p.-load power plant
p. load power Spitzenlastenergie *f*
p.-load power plant Spitzenkraftwerk *n*
p. lopping Übernahme *f* der Spitzenlast
p. making current Scheitelwert *m* des Einschaltstromes *(für einen Pol eines Schalters)*
p. meter Spitzenzähler *m*
p. modulation indicator Höchstwertzeiger *m*
p. of oscillation Schwingungsmaximum *n*
p. of traffic Verkehrsspitze *f*
p. of wave Wellenberg *m*
p. output Spitzenleistung *f (eines Motors)*
p. plate current Scheitelwert *m* des Anodenstromes
p. plate voltage Anodenspitzenspannung *f*
p. power Spitzenleistung *f*, Höchstleistung *f*
p. power of the transmitter Oberstrichleistung *f* eines Senders
p. power output Oberstrichleistung *f*; Oberstrichwert *m*
p. programme meter *(El)* Pegelmesser *m*, Scheitelwertmesser *m*
p. pulse power Spitzen[im]pulsleistung *f*
p. response *(El)* Kurvenspitze *f*
p. reverse voltage rating *(El)* Spitzensperrspannung *f*
p.-signal level *(El)* Maximalsignalspannung *f*
p. stress *(Mech)* Spitzenspannung *f*, Spannungsspitze *f*

p.-to-peak Spitze-zu-Spitze *(größte Spannungsschwankung)*
p.-to-peak amplitude Spitze-zu-Spitze-Amplitude *f*, ss-Amplitude *f*
p.-to-peak rectification Spitze-zu-Spitze-Gleichrichtung *f*
p.-to-peak voltage *(El)* Differenz *f* zwischen Spannungsmaximum und Spannungsminimum
p.-to-valley height Rauhtiefe *f (Oberflächengüte)*
p.-to-valley ratio *(El)* Maximum-zu-Maximum-Verhältnis *n*
p.-to-zero *(El)* Spitze-zu-Null *(Bezeichnung für einen bestimmten Spannungsmeßbereich)*
p. value Höchstwert *m*, Scheitelwert *m*, Spitzenwert *m*
p. voltage *(El)* Scheitelspannung *f*
p. voltmeter Spitzenspannungsmesser *m*, Scheitelspannungsmesser *m*
p. white *(Fs)* Maximum *n* an Weiß, Weißspitze *f*
peakedness *(Math)* Kurvensteilheit *f*
peaker *(BMSR)* Differenzierschaltung *f*
peaking Durchlassen *n* einer Spannung ohne Amplitudenbegrenzung, Spitzenwertbildung *f*
p. capacity Spitzenleistung *f*
p. circuit Spitzenanhebungskreis *m (zur Anhebung der Amplitude bestimmter Frequenzen)*
p. coil Korrekturspule *f (für hohe Frequenzen)*
p. resistance *(Fs)* Linearisierungswiderstand *m (im Sägezahngenerator)*
p. resistor *(Fs)* Linearisierungswiderstand *m (Baueinheit)*
peanut Erdnuß *f*
p. oil Erdnußöl *n*
p. tube *(El)* Kleinströhre *f*
pear-shaped birnenförmig
p.-switch Schalterbirne *f*, Hängeschalter *m*, Schnurschalter *m* mit Druckknopf
p. tree Birnbaum *m*
pearl / to *(Text)* linksstricken
pearl 1. *(Typ)* 5 britische Punkte *mpl (Schriftgröße)*; 2. Perle *f*
p. lamp mattierte Lampe *f*, Opallampe *f*
p.-ore Perlerz *n*
pearlite *(Met)* Perlit *m*
pearlitic perlitisch
peat Torf *m*
p. bed *(Geol)* Torflager *n*, Torfschicht *f*
p. bog Torfmoor *n*
p. coal Torfkohle *f*
p. compressing machine Torfpreßmaschine *f*
p. cutter Torfgräber *m*, Torfpflug *m*
p. digger Torfgräber *m*, Torfstecher *m*
p.-drag Torfbagger *m*
p. dust Torfmull *m*
p. excavator Torfbagger *m*
p. litter Torfstreu *f*
p. moulding machine Torfpreßmaschine *f*, Torfpresse *f*
p. pit Torfgrube *f*
p. powder Torfmull *m*
p. press Torfpresse *f*
p. soil Torfboden *m*
p. tar Torfteer *m*
peatification Vertorfung *f*
peaty torfartig, torfhaltig, torfig, Torf-
p. fibrous coal Torffaserkohle *f*
p. flat *(Geol)* Torfniederung *f*
p. lay *(Geol)* Torfschicht *f*
p. water Moorwasser *n*
pebble / to *(Led)* chagrinieren
pebble Kiesel[stein] *m*, Flintstein *m*
p.-bed reactor *(Kern)* Kugelhaufenreaktor *m*
p. grain *(Led)* aufgepreßter Narben *m*, Chagrinnarben *m*
p. mill Kugelmühle *f*
p. reactor *(Kern)* Kugelhaufenreaktor *m*
pebbles körnige Trägersubstanz *f*; Körner *npl*; feines Geröll *n*
peck Peck *n (1. englische Volumeneinheit für Flüssigkeiten und trockene Substanzen: 1 pk = 9,092 dm³; 2. amerikanische Volumeneinheit für trockene Substanzen: 1 pk = 8,81 dm³)*
pectic acid Pektinsäure *f*
pectin sugar Arabinose *f*
pectinic acid Pektin *n (Gelierstoff)*
pectinose Arabinose *f*
pectolite *(Min)* Pektolith *m*
ped tooth saw Säge *f* mit stehendem (gleichschenklig-dreieckigem) Zahn *(für Holzbearbeitung)*
pedal 1. Fußauflage *f*, Hebel *m*, Tritt *m*; Kurbeltritt *m*, Pedal *n*; 2. Fußhebel *m*
p. bale breaker *(Text)* Muldenballenbrecher *m*, Ballenbrecher *m* mit Pedalvorrichtung
p. brake Fußbremse *f*, Fahrbremse *f*
p.-controlled fuß[hebel]betätigt
p. dynamo Tretdynamo *m*, Tretgenerator *m*
p. keyboard Tretklaviatur *f*
p. operated fuß[hebel]betätigt
p.-operated dip-switch *(Kfz)* Abblendfußschalter *m*
p.-pivot shaft Pedalachse *f*
p. stalk Fußstück *n* der Pedale
p. switch Fußschalter *m*
p. treadle Fußhebel *m*
pedalling Fußklaviatur *f (Orgel)*
pedals Fußklaviatur *f (Orgel)*
pedestal 1. Sockel *m*, Bock *m*, Auflager *n*, Fuß[stück *n*] *m*, Ständer *m*, Untersatz *m*; Lager[bock *m*] *n*, Postament *n*, Sockel; 2. Achslagerführung *f*; 3. *(Fs)* Basisimpuls *m (relativ langer Rechteckimpuls mit flachem Dach)*, Spannungspegel *m*
p. base *(Masch)* Lagerfuß *m*

p. bearing *(Masch)* Bocklager *n*, Deckellager *n*, Rumpflager *n*, Stehlager *n*
p. body *(Masch)* Lagerkörper *m*, Lagerrumpf *m*
p. control point *(Opt)* Sockelpunkt *m*
p. drilling machine Ständerbohrmaschine *f*
p. expansion Gehäusefußverschiebung *f* [infolge Wärmedehnung] *(Dampfturbine)*
p. forming a stuffing box *(Masch)* stopfbuchsenartiger Tragbock *m*
p. grinder Ständerschleifmaschine *f*
p. level *(Fs)* Austastpegel *m*
p. of a machine Maschinenfuß *m*
p. of the piston rod guide *(Masch)* Stützbock (Tragbock) *m* der Kolbenstange
p. signal *(Fs)* Austastsignal *n*
p.-type motor *(Elm)* Motor *m* mit Stehlager, Bocklagermotor *m*
pedestrian bridge Fußgängerbrücke *f*
p. crossing Fußgängerüberweg *m*
p. island Fußgängerschutzinsel *f*
p. mall Fußgängerstraße *f*
p. overpath Fußgängerüberweg *m*
p. precinct Fußgängerbereich *m*
p. subway Fußgängerunterführung *f*, Fußgängertunnel *m*
p. traffic Fußgängerverkehr *m*
pedion Pedion *n (Kristallografie)*
peel / to [ab]schälen, entrinden *(Baumstämme)*; abblättern *(metallische Überzüge)*
p. off abblättern, abspringen *(Farbe)*; [sich] abschälen
peelable protective coating *(Plast)* leicht abziehbarer, aufgetauchter Überzug *m*
peeled billet *(Holz)* Schälprügel *m*
p. billet-wood Schälholz *n*, entrindetes Holz *n*
peeler Schälblock *m*
peeling 1. *(Met)* Schälen *n (Blöcke)*; Schalenbildung *f (bei Gußstücken)*; 2. Rupfen *n (der Papieroberfläche)*
p. device *(Lw)* Schäler *m*, Schäleinleger *m*
p. machine Schälmaschine *f*
p. off Ausscheren *n* [aus dem Flugverband]
p. test *(Plast)* Schäl- und Abhebeprüfung *f*
peen / to [aus]hämmern, ausrichten
peening Hämmern *n*
peep hole Schauöffnung *f*, Schauloch *n*
peg / to verstiften, verdübeln
p. out abpflocken *(z. B. Kabeltrasse)*; abstecken *(z. B. eine Kurve)*
peg Stöpsel *m*; Drehstift *m*, Haken *m*; Dübel *m*
p. board Wirbelbalken *m*, Wirbelstock *m (Klavier)*; *(Masch)* Steckfeld *n (Steuerung)*
p. mill Schlagkorbmühle *f*, Schleudermühle *f*
p. of the violin Wirbel *m* der Violine
p. runner Dübelmaschine *f*
p. stake Markierstab *m (Vermessung)*
peganite *(Min)* Peganit *m*
pegged boots holzgenagelte Stiefel *mpl*
pegging rammer Spitzstampfer *m (Formerei)*
pein Pinne *f (Hammer)*
Peirce function *(BMSR)* Peircesche Funktion *f (logische Funktion)*
pelagic pelagisch, Tiefsee-
p. fishing Hochseefischfang *m*, Tiefseefischfang *m*
p. whaling Hochseewalfang *m*
p. zone Tiefseezone *f*
pelargonic acid Pelargonsäure *f*
pellet / to tablettieren
pellet 1. Kügelchen *n*; Pille *f*; Tablette *f*; Granalie *f*; Schrotkugel *f*, Kugel *f*; Schweißperle *f*, Perle *f*; 2. Pille *(Elektronenröhren)*
p. impact drilling Kugelschlagbohren *n*
p. primer Zündpille *f*
pelleter Tablettiermaschine *f*
pelleting property Tablettierfähigkeit *f*
pelletize / to pelletisieren *(Erz)*
pelletizing drum Granuliertrommel *f*
p. plant *(Met)* Pelletisierungsanlage *f*
p. rods *(Plast)* Strangpressen *n* von Stäben u. nachfolgendes Zerschneiden *n* zu Tabletten
pellicle of oxygen *(Met)* Oxidhäutchen *n*
pelmet board Gardinenleiste *f*
pelt *(Led)* Blöße *f*
p. wool abgebeizte Wolle *f*
Peltier cell *(El)* Peltier-Zelle *f*
Peltier effect *(El)* Peltier-Effekt *m*
Pelton turbine Peltonturbine *f*, Freistrahlturbine *f*
Pelton water wheel Peltonrad *n*
PEM-voltage *s.* photoelectromagnetic voltage
pen 1. Feder *f*; 2. *(Bw)* Box *f*, Verschlag *m*
p. compass Zirkel *m* mit Einsatz für Tusche
p.-knife Skalpell *n*; Federmesser *n*, Taschenmesser *n*
p. recorder Tintenschreiber *m*
pencil 1. *(Math)* Büschel *n*, einparametrige Schar *f*; 2. Bleistift *m*; 3. *(Opt)* Bündel *n*, Büschel
p. beam *(El)* Nadelstrahl *m*, Schreibstrahl *m*
p.-beam aerial Schmalbündelantenne *f*
p. cedar Zeder *f* für Bleistiftherstellung
p. compass Zirkel *m* mit Einsatz für Blei
p. drawing Bleistiftzeichnung *f*
p. electrode Stabelektrode *f*
p. grinder Schleifstift *m*, Spindelschleifkörper *m*
p. guide Schreibstiftführung *f*
p. lead Bleistiftgraphit *m*
p. of light *(Opt)* Lichtbündel *n*
p. of rays *(Opt)* Strahlenbündel *n*, Strahlenbüschel *n*
p. sharpener Bleistiftspitzer *m*
p. sharpening machine Bleistiftspitzmaschine *f*

p. sketch Blei[stift]skizze *f*
pencilled drawing Blei[stift]zeichnung *f*
pendant herabhängend, schwebend
pendant 1. Hängelampe *f*, Pendelleuchte *f*; 2. *(Masch)* Schwenkarm *m*; Hängedruckknopftafel *f*, Hängeschalter *m*
p.-actuated von Hängeschalter betätigt
p. control Steuerung *f* von der Hängedruckknopftafel aus
p. control panel Hängeschalttafel *f*
p. fitting Hängelampe *f*, Pendelleuchte *f*
p. light Hängeleuchte *f*, Pendelleuchte *f*
p. push button control unit Hängedruckknopfschalttafel *f*
p. push button panel Hängedruckknopf[schalt]tafel *f*
p. socket outlet Hänge[steck]dose *f*
p. station Hängedruckknopftafel *f*
p. switch Hängeschalter *m*, Schnurschalter *m*; Schaltpendel *n*
pending angemeldet, beantragt *(Patent)*
p. message unerledigter Funkspruch *m*
pendulous pendelartig, Pendel-
p. accelerometer Pendelbeschleunigungsmesser *m*
p. erection device *(BMSR)* Lotfühler *m* *(Kreiselgeräte)*
p. motion pendelnde Bewegung *f*
pendulum Pendel *n*
p. anemometer Pendelwindmesser *m*
p. bearings Pendellagerung *f*
p. bucket conveyor Pendelbecherwerk *n*
p. circular saw Pendelkreissäge *f*
p. clock Pendeluhr *f*
p. counterweight pendelndes Gegengewicht *n*
p. damper Pendel[schwingungs]dämpfer *m*
p. governor Pendelregler *m*
p. grinding machine Pendelschleifmaschine *f*
p. impact tester Pendelschlagwerk *n (Werkstoffprüfung)*
p. impact testing machine Pendelschlagwerk *n (Werkstoffprüfung)*
p. jigger *(Led)* Glanzmaschine *f*, Satiniermaschine *f*
p.-like vibration pendelartige Schwingung *f*
p. meter Pendelzähler *m*, Aronzähler *m*
p. mill Pendelmühle *f*
p. oscillation Pendelung *f*, Pendelschwingung *f*, Pendelbewegung *f*
p. rectifier Pendelgleichrichter *m*
p. roller Pendelwalze *f (Sohlleder)*
p. shears Pendelschere *f*
p. weighting arm *(Text)* Pendelträgerbelastungsarm *m*, Pendelträgerdruckarm *m (Streckwerk)*
peneplain *(Geol)* Peneplain *f*, Denudationsfläche *f*
penetrability Durchdringbarkeit *f*, Durchlässigkeit *f*
penetrant material Eindringmedium *n (Werkstoffprüfung)*
penetrate / to durchdringen, eindringen, einsickern; [durch]tränken, imprägnieren
p. deeply into tief eindringen in
penetrating hart, energiereich *(Strahlung)*
p. ability 1. Durchdringungsfähigkeit *f*, Eindringvermögen *n*; 2. Härte *f (Röntgenstrahlen)*; Durchstrahlungsfähigkeit *f*
p. component harte (durchdringende) Komponente *f (kosmische Strahlen)*
p. depth *(Met)* Eindringtiefe *f*
p. into space Eindringen *n* in den kosmischen Raum
p. particle energiereiches Teilchen *n*
p. power Durchdringungsfähigkeit *f*, Eindringvermögen *n (z. B. Hochfrequenz)*; Durchdringungsvermögen *n*, Härte *f (z. B. Röntgenstrahlen)*
p. radiation 1. durchdringende (harte) Strahlung *f*; 2. Durchstrahlung *f*
p. shower *(Kern)* durchdringender (harter, energiereicher) Schauer *m*
penetration 1. Durchdringung *f*, Eindringen *n*, Durchdringen *n*; 2. Durchgriff *m (der Anodenspannung)*; 3. *(Met)* Eindringtiefe *f*; 4. *(Mil)* Durchschlagskraft *f*; 5. Einbrand *m (Schweißen)*
p. advance Bohrfortschritt *m*
p. by rays Durchstrahlung *f*
p. depth *(Met)* Eindringtiefe *f*
p. factor Durchgriff *m*
p. hardness Eindruckhärte *f (Werkstoffprüfung)*
p. of arms *(Mil)* Durchschlagsstärke *f*
p. of current *(El)* Stromeindringtiefe *f*, Eindringtiefe *f* des Stromes
p. of heat Wärmedurchdringung *f*, Wärmedurchgang *m*
p. of rays Durchdringung *f* von Strahlen
p. potential Durchdringungspotential *n*
p. probability *(Kern)* Durchdringungswahrscheinlichkeit *f*
p. rate Bohrfortschritt *m*
p. speed Bohrgeschwindigkeit *f*
p. twins Durchdringungszwillinge *mpl (Kristalle)*
p. zone *(Met)* Einbrandzone *f*
penetrative durchdringend, Durchdringungs-, Durchschlags-
p. power Durchdringungsvermögen *n*
p. property Eindringungskraft *f*
penetrativeness Durchdringungsvermögen *n*, Durchschlagsvermögen *n*
penetrator Eindringkörper *m (Härteprüfung)*
penetrometer Eindruck[tiefen]messer *m*, Eindringgerät *n (Werkstoffprüfung)*
penicillin Penicillin *n*

penlite cell *(Am)* *(El)* Babyzelle *f*, Zwergstabbatterie *f*
penning Packlage *f*
pennite *(Min)* Pennin *m*
Pennsylvanian drilling system pennsylvanisches Bohren *n* *(Tiefbohrtechnik)*
penny-in-the-slot automat Münzautomat *m*
penstock 1. Düsenstock *m*; 2. Stauwehr *n*, Schleuse *f*; Druckleitung *f*; Rohrzuleitung *f*; Schleusentor *n*
pent roof V-förmig nach innen geneigtes Dach *n*; Pultdach *n*, Halbdach *n*
penta prism *(Opt)* Pentaprisma *n*
  p.-prism viewfinder *(Opt)* Prismenaufsatz *m*
pentacycle *(Ch)* Fünfring *m*
pentaerythritol *(Ch)* Pentaerythrit *n*
  p. tetranitrate Nitropentaerythrit *n*, Pentrit *n*
pentagon Fünfeck *n*, Fünfkant *n*
  p. turret *(Masch)* Fünfkantrevolverkopf *m*
pentagonal fünfeckig
  p. dodecahedron Pentagondodekaeder *n*
pentagrid Heptode *f*, Siebenpolröhre *f*, Mischheptode *f*
  p. converter Mischheptode *f*, Pentagrid-Mischröhre *f*
  p. valve Heptode *f*, Fünfgitterröhre *f*
pentahedral pentaedrisch, fünfflächig
pentahedron Pentaeder *n*, Fünfflächner *m*
pentahydric alcohol fünfwertiger Alkohol *m*
pentane *(Ch)* Pentan *n*
pentanoic acid Pentansäure *f*, Valeriansäure *f*
pentathionic acid Pentathionsäure *f*
pentatron Pentatron *n* *(Elektronenröhre)*
pentavalent *(Ch)* fünfwertig
penthouse Schutzdach *n*, Wetterdach *n*; Anbau *m* *(meistens mit Pultdach)*; Aufbau *m* auf einem Flachdach; offener Schuppen *m*
  p. roof Pultdach *n*
pentode Pentode *f*, Fünfpolröhre *f*
penumbra Halbschatten *m*, Penumbra *f*
peptizate / to peptisieren
peptize / to *(Ch)* in Lösung bringen, verteilen
peptogenous *(Ch)* peptonerzeugend
peptonization *(Ch)* Stärkeabbau *m*, Peptonisierung *f*
per capita production Pro-Kopf-Produktion *f*, Produktion *f* pro Kopf
per cent *s.* percent
per unit area je Flächeneinheit
peracetic acid Azetopersäure *f*, Peressigsäure *f*
perbenzoic acid Benzopersäure *f*
perborate *(Ch)* Perborat *n*
perbromo-hydrocarbon perbromierter Kohlenwasserstoff *m*
perceivable wahrnehmbar
perceiving computer erkennender (wahrnehmender) Automat *m*
percent Prozent *n*
  p. droop *(BMSR)* Ungleichförmigkeitsgrad *m*
  p. gain prozentueller Anstieg *m*
  p. ripple voltage *(El)* Brummspannungsverhältnis *n*
percentage Prozentsatz *m*
percentage prozentual
  p. by volume Volumenprozent *n*, Raumprozent *n*
  p. cost anteilige Kosten *pl*
  p. coupling *(El)* Kopplungsgrad *m*
  p. differential protection Prozentvergleichsschutz *m* *(Relais)*
  p. differential relay Prozent[differential]-relais *n*
  p. growth rate of industry Wachstumsrate *f* der Industrie
  p. harmonic content *(El)* Klirrfaktor *m*
  p. modulation prozentualer Modulationsgrad *m*
  p. of ashes Aschegehalt *m*
  p. of clinker [prozentualer] Schlackengehalt *m*
  p. of contact area Traganteil *m* *(Oberflächenmessung)*
  p. of moisture [prozentualer] Feuchtigkeitsgehalt *m*
  p. of reinforcement Bewehrungsanteil *m*, Bewehrungsprozentsatz *m* *(Stahlbetonbau)*
  p. purity Reinheitsgrad *m*
perceptibility Erkennbarkeit *f*
perceptible erkennbar
perception Wahrnehmung *f*, Empfindung *f*
  p. of loudness Lautstärkeempfindung *f*
Perceptron *(Dat)* Bezeichnung einer wahrnehmenden Maschine
perch / to 1. beschauen, absuchen; 2. *(Led)* schlichten, Narbenfehler ausschlichten
perch 1. Stange *f*; Sitzstange *f* *(für Geflügel)*; 2. *(Led)* Schlichtbaum *m*; 3. *(Längenmaß)* $5\frac{1}{2}$ Yard; *(Raummaß)* $\approx$ 25 Kubikfuß
percher Warenbeschauer *m*
perching Absuchen *n* auf Fehler
  p. knife *(Led)* Schlichtmond *m*
perchlorate Perchlorat *n*, Salz *n* der Perchlorsäure
perchloric acid Perchlorsäure *f*
perchloro-hydrocarbon perchlorierter Kohlenwasserstoff *m*
perchromate Perchromat *n*, Salz *n* der Perchromsäure
percolate / to *(Ch)* filtern durch Adsorptionsschicht, durchseihen, durchsickern [lassen], filtern
percolating filter Tropfkörper *m* *(Gesundheitstechnik)*
  p. leach *(Ch)* Sickerlaugung *f*
percolation Sickerung *f*; Versickerung *f*; Durchsickerung *f*

p. clay *(Ch)* Bleicherde *f*, Filtermittel *n*
p. filter Tropfkörper *m (Gesundheitstechnik)*
percolator Perkolator *m*, Filtrierbeutel *m*, Filtrierapparat *m*, Filtersack *m*
percuss / to abklopfen
percussion Schlag *m*, Stoß *m*, Erschütterung *f*, Perkussion *f*
p. boring schlagendes (stoßendes) Tiefbohren *n*
p. cap Zündhütchen *n*
p. deep boring schlagendes (stoßendes) Tiefbohren *n*
p. drilling Schlagbohrverfahren *n (Tiefbohrtechnik)*
p. drilling machine Schlagbohrmaschine *f*
p. effect Schlagwirkung *f*
p. feeder Stoßaufgabeeinrichtung *f*
p. frame Rüttelherd *m*, Stoßherd *m*
p. fuse Aufschlagzünder *m*, Schlagzünder *m*
p. hammer Perkussionshammer *m*, Schlaghammer *m*
p. instrument Schlaginstrument *n*
p. jig Stauchsetzmaschine *f*
p. lever Stoßhebel *m*
p. piston Schlagkolben *m*
p. power Durchschlagskraft *f*
p. press Spindelschlagpresse *f*, Spritzpresse *f*
p. priming Perkussionszündung *f*, Schlagzündung *f*
p. resistance Stoßfestigkeit *f*
p. riveting Schlagnietung *f*
p. riveting machine Schlagnietmaschine *f*
p. sieve Stoßsieb *n*
p. strength Stoßfestigkeit *f*
p. table Stoßherd *m*
p. wave Stoßwelle *f (bei Explosionen)*
p. welding Schlagschweißung *f*
percussive action Schlagwirkung *f*
p. drilling Schlagbohren *n*, schlagendes Bohren *n (Tiefbohrtechnik)*
p. fuse Aufschlagzünder *m*, Schlagzünder *m*
p. welding Schlagschweißung *f*
perfect / to 1. vervollkommnen, vervollständigen, überarbeiten, vollenden; 2. *(Typ)* schön- und widerdrucken, beidseitig bedrucken
perfect vollkommen, einwandfrei, fehlerfrei; ideal *(Flüssigkeit, Leiter)*; verlustlos *(Kondensator)*
p. absorber vollkommener Absorber *m*
p. alignment guter Gleichlauf *m (elektrischer Kreise)*
p. binding *(Typ)* Klebebindung *f*
p. combustion vollkommene Verbrennung *f*
p. condenser idealer (verlustloser) Kondensator *m*
p. condition fehlerfreier Zustand *m*
p. crystal Idealkristall *m*, idealer Kristall *m*
p. dielectric ideales (verlustfreies) Dielektrikum *n*, idealer Isolierstoff *m*
p. diffuser vollkommen mattweiße Fläche *f*; vollkommen streuender Körper *m*
p. elasticity vollkommene Elastizität *f*
p. fluid ideale (reibungsfreie) Flüssigkeit *f*
p. gas ideales Gas *n*
p. junction unlösbare Verbindung *f*
p. lattice ideales Gitter *n (Kristalle)*
p. liquid ideale Flüssigkeit *f*
p. lubrication vollkommene Schmierung *f*
p. plasticity vollkommene Plastizität *f*
p. rectifier idealer Gleichrichter *m*
perfecting cylinder *(Typ)* Widerdruckzylinder *m*
perfection Vervollkommnung *f*, Vollendung *f*
perfectly elastic ideal (vollkommen) elastisch
p. inductive *(El)* rein induktiv
p. inductive load rein induktive Belastung *f*
perfluoro-hydrocarbon perfluorierter Kohlenwasserstoff *m*
perforate / to durchbohren, [durch]lochen, durchstechen, perforieren, durchschlagen
perforated perforiert, gelocht
p. bottom Siebboden *m*
p. brick Lochziegel *m*, Lochstein *m*
p. card Lochkarte *f*
p. card system Lochkartensystem *n*
p. cathode Lochkatode *f*
p. circular saw hinterlochte Kreissäge *f*
p. code Lochschrift *f*
p. cone konischer Siebzylinder *m*
p. cylinder Siebzylinder *m*
p. diaphragm Siebboden *m*
p. disk Lochscheibe *f*
p. feed pipe Speiserohrverteiler *m*
p. filter plate Filtriersieb *n*
p. hardboard maschinell durchlochte Hartplatte *f*
p. metal screen gelochte Metallfolienabschirmung *f*
p. plate Siebboden *m*
p. plate column *(Ch)* Siebbodenkolonne *f*
p. seat *(Masch)* Lochsitz *m*
p.-strip-type transmitter Maschinentelegraf *m*
p. tape Lochstreifen *m*
p. tape recording Lochstreifenregistrierung *f*
p.-tape telegraph-recorder Lochschreiber *m*
p. tape transmission Lochstreifenübertragung *f*
p. tape transmitter *(Fmt)* Streifengeber *m*
p. tube perforierte Hülse *f*
perforating Lochen *n*, Durchlöchern *n*, Perforieren *n*
p. bar *(Typ)* Perforierleiste *f*
p. comb *(Typ)* Perforierkamm *m*
p. die Lochstempel *m*
p. machine *(Typ)* Perforiermaschine *f*
p. press Lochmaschine *f*, Lochstanze *f*, Perforiermaschine *f*, Perforierpresse *f*

p. rule *(Typ)* Perforierlinie *f*
perforation Lochung *f (Karten, Siebe)*; Perforation *f*; Loch *n*
p. dimensions Perforationsmaße *npl*
p. pitch Lochabstand *m (eines Films)*
perforator Locher *m*; Stanzer *m*
perform / to ausführen, leisten
performance *(Masch)* Leistungsfähigkeit *f*, Kenndaten *pl*, Betriebsverhalten *n*, Gebrauchsleistung *f*; Arbeitsweise *f (Gerät)*; Durchführung *f*; *(Dat)* Verhalten *n*, Reaktion *f*, Übertragungsverhalten *n*
p. characteristic Leistungskennwert *m*, Arbeitskenngröße *f*
p. check Funktionsprüfung *f*
p. criterion *(BMSR)* Gütekriterium *n*
p. curve Leistungskurve *f*, Leistungsdiagramm *n*
p. factor Gütefaktor *m (z. B. des Relais)*
p. index *(BMSR)* Gütekriterium *n*
p.-monitor transmitter Prüfechosender *m (Radar)*
p. of combustion unit *(Rak)* Brennkammerleistung *f*
p. operator *s.* transfer function
p. testing Leistungsprüfung *f*, Funktionsprüfung *f*
p. trial Leistungsprobe *f*
perfume Parfüm *n*, Duft[stoff] *m*, Riechstoff *m*
p. atomizer Parfümzerstäuber *m*
perfumery Parfümerie *f*
pergameneous pergamentartig, Pergament-
pergamyne Pergamin *n (Pergamentersatz)*
pergola Pergola *f*
perhalide *(Ch)* perhalogenierte (vollständig halogenierte) Verbindung *f*
perhalo-hydrocarbon perhalogenierter Kohlenwasserstoff *m*
perhalocarbon perhalogenierter Kohlenwasserstoff *m*
peri-cyclo compound *(Ch)* perizyklische Verbindung *f*, Verbindung mit kleinem innerem Ring
pericarp Fruchtgehäuse *n*
periclase *(Min)* Periklas *m*
pericline *(Min)* Periklin *m*
pericycloid Perizykloide *f*
peridot *(Min)* Peridot *m*, Olivin *m*
perigeal velocity Geschwindigkeit *f* im Perigäum, Geschwindigkeit in größter Erdnähe
perigee Erdnähe *f*, Perigeum *n*, Perigäum *n*
perigon Vollwinkel *m (360°)*
perimeter Umfang *m*; Umriß *m*
p. cable Umgehungskabel *n*
p. track *(Flg)* Rollfeldringstraße *f*
perimorphic perimorph *(Kristalle)*
period Dauer *f*; Frist *f*; Zeit[dauer] *f*; Periode *f*, Schwingungsdauer *f*
p. amplifier Periodenverstärker *m*
p. for entering opposition Einspruchsfrist *f (Patentwesen)*
p. in shipyard Werftliegezeit *f (eines Schiffes)*
p. indicator Periodenanzeiger *m*
p. limit Grenzperiode *f*
p. meter Periodenmesser *m*
p. of admission *(Wärm)* Einströmungsperiode *f*, Volldruckperiode *f*
p. of an instrument Schwingungsdauer *f (Meßtechnik)*
p. of beat Schwebungsdauer *f*
p. of combustion Verbrennungsperiode *f*
p. of contact *(El)* Eingriffszeit *f*, Kontaktdauer *f*
p. of distillation of gases Entgasungsperiode *f*
p. of dwell *(Masch)* Stillstandsdauer *f*, Bewegungspause *f*, Haltedauer *f*
p. of expansion Expansionsperiode *f*
p. of extension Stundungsfrist *f (Patentwesen)*
p. of half life Halbwertzeit *f*
p. of operation Grundzeit *f*, Arbeitszeit *f*
p. of oscillation Schwingungsdauer *f*
p. of revolution Umlaufzeit *f*
p. of roll *(Schiff)* Schlingerperiode *f*, Rollperiode *f*
p. of table dwell *(Masch)* Anhaltdauer *f* des Tisches
p. of training Lehrzeit *f*
p. of vibration Schwingungsdauer *f*
p. range Periodenbereich *m (eines Meßinstruments)*
p. resonance Resonanz *f* mit der Eigenschwingung
periodate Perjodat *n*, Salz *n* der Perjodsäure
periodic *(Ch)* Perjod-
p. acid Perjodsäure *f*, Überjodsäure *f*
periodic periodisch, Perioden-
p. aerial abgestimmte Antenne *f*
p. chart *(Ch)* periodisches System *n*
p. controller Schrittregler *m*, Abtastregler *m*
p. current periodischer Wechselstrom *m*
p. duty *(Elm)* Dauerbetrieb *m* mit periodisch veränderlicher Belastung; Aussetzbetrieb *m*
p. oil renewal regelmäßiger Ölwechsel *m*
p. potential difference *(El)* periodische Spannung *f*
p. state of resistance wiederkehrender Beharrungszustand *m*
p. table *(Ch)* periodisches System *n*
p. time Periodendauer *f*; Schwingdauer *f*
p. trigger-type receiver *(Fmt)* Pendelrückkopplungsempfänger *m*
p. weight Periodendruck *m (Gebirgsmechanik)*
periodical light Taktfeuer *n (Leuchtfeuer)*
p. literature Zeitschriftenliteratur *f*

periodically reversing magnetic field sich periodisch umkehrendes Magnetfeld *n*
periodicity Periodizität *f*, Frequenz *f*, regelmäßige Wiederholung *f*
periodo-hydrocarbon perjodierter Kohlenwasserstoff *m*
periodogram *(Math)* Periodogramm *n*
peripheral peripher, Umfangs-, Rand-
  p. blowhole *(Met)* Randlunker *m (im Gußblock)*
  p. component of the steam velocity Umfangskomponente *f* der Dampfgeschwindigkeit *(Turbine)*
  p. cutting edge Umfangsschneide *f*
  p. dressing of the grinding wheel Umfangsabrichten *n* der Schleifscheibe
  p. electron äußeres (kernfernes) Elektron *n*, Elektron der Außenschale
  p. land *(Masch)* Fase *f* an Umfangsschneiden
  p. layer Randschicht *f*, periphere Schicht *f*
  p. leakage *(El)* Luftspaltstreuung *f*
  p. milling Walzfräsen *n*
  p. nucleon Randnukleon *n*
  p. passage Umfangskanal *m*, Umführungskanal *m*
  p. pressure Umfangdruck *m*, Umfangkraft *f (Turbine)*
  p. ray Randstrahl *m*
  p. speed Peripheriegeschwindigkeit *f*, Randgeschwindigkeit *f*; *(Masch)* Umfangsgeschwindigkeit *f*; Schnittgeschwindigkeit *f (Kreissäge)*
  p. velocity *s.* p. speed
periphery Umfang *m*, Peripherie *f*; Umfläche *f*
  p. of the fundus *(Opt)* Funduschperipherie *f*
  p. of the grinding wheel Schleifscheibenumfläche *f*
periscope Periskop *n*, Sehrohr *n*, Scherenfernrohr *n*
periscopic-type telescope Periskop *n*, Sehrohr *n*, Scherenfernrohr *n*
perishable cargo verderbliche Ladung *f*
  p. food verderbliche Nahrungsmittel *npl*
perishables leicht verderbliche Güter *npl*
peristyle Peristyl *n*
peritectic *(Met)* peritektisch
  p. point Peritektikum *n*
perlite 1. *(Geol)* Perlit *m*, Perlstein *m*; 2. *(Met)* *s.* pearlite
permafrost Dauerfrostboden *m*
permanence Permanenz *f*, Dauer *f*, Beständigkeit *f*
  p. of size Maßbeständigkeit *f*, Maßhaltigkeit *f*
permanent permanent, [fort]dauernd, bleibend, ständig, stabil, Dauer-
  p.-arc producer Dauerbogenerzeuger *m*
  p. axis of rotation permanente Drehachse *f*
  p. ballast *(Schiff)* fester Ballast *m*
  p. blip feste Zacke *f (Radar)*
  p. bush fastened in the jig *(Masch)* Grundbuchse *f*
  p. chock *(Brg)* Standpfeiler *m*
  p. colour echte (beständige) Farbe *f*
  p. convergence *(Math)* unbedingte Konvergenz *f*
  p. coupling *(Masch)* nichtschaltbare *(starre und bewegliche)* Kupplung *f*
  p. crease *(Text)* Dauerfalte *f*
  p. crystal detector *(El)* fest eingestellter Detektor *m*
  p. current Dauerstrom *m*
  p. deformation bleibende Formänderung (Verformung) *f*
  p. dipole Dauerdipol *m*
  p. discharge *(El)* Dauerentladung *f*
  p. earth Dauererdschluß *m*
  p. echo Fest[zeichen]echo *n (Radar)*
  p. elongation bleibende Dehnung (Längenänderung) *f*
  p. fender *(Schiff)* Scheuerleiste *f*
  p. finish *(Text)* Dauerappretur *f*, Permanentappretur *f*
  p. fit *(Masch)* Ruhesitz *m*
  p. formwork verlorene Schalung *f (Betonbau)*
  p. glazed chintz *(Text)* Chintz *m* mit Dauerglanzausrüstung
  p. ground Dauererdschluß *m*
  p. hardness Permanenthärte *f (Wasser)*
  p. joint unlösbare Verbindung *f*
  p. linear deformation bleibende (plastische) Längenänderung *f*
  p. load ständige Last *f*; Dauerlast *f*; Eigenlast
  p. magnet Dauermagnet *m*, permanenter Magnet *m*
  p.-magnet chuck Dauermagnetfutter *n*
  p.-magnet dynamic loudspeaker permanentmagnetischer Lautsprecher *m*
  p.-magnet moving-coil instrument (El) Drehspulinstrument *n (mit Dauermagnet)*
  p.-magnet moving-iron instrument *(El)* Eisennadelinstrument *n*, Dreheiseninstrument *n* mit Magnet
  p. magnetic chuck *(Masch)* Dauermagnetspannfutter *n*
  p. magnetic field permanentes Magnetfeld *n*
  p. memory Dauerspeicher *m*, leistungsloser (permanenter) Speicher *m*, Permanentspeicher *m*, Festwertspeicher *m*
  p. metal mould Kokille *f*
  p. mould Kokille *f*
  p. mould casting Kokillenguß *m*
  p. note Dauerton *m*
  p. plant stationäre Anlage *f*
  p. ring *(Fmt)* Dauerruf *m*
  p. rope stretch *(Brg)* bleibende Seildehnung *f*
  p. set bleibende Verformung *f*, Dehnungsrest *m*

p. settlement bleibende Setzung *f (Grundbau)*
p. sheen finish *(Text)* Dauerglanzappretur *f*
p. shuttering verlorene Schalung *f (Betonbau)*
p. strain bleibende Dehnung *f*, Dehnungsrest *m*
p. support *(Brg)* endgültiger Ausbau *m*
p. way Gleisanlagen *fpl*; Gleisoberbau *m*, Oberbau *m (Eisenbahnbau)*
p. weight Eigenmasse *f*, Eigenlast *f*, ständige Last *f*
p. white Permanentweiß *n*
p. wiring *(El)* festverlegte Installation *f*; feste Verdrahtung *f*
permanganate *(Ch)* Permanganat *n*
p. of potassium Kaliumpermanganat *n*, übermangansaures Kali *n*
p. of sodium Natriumpermanganat *n*, übermangansaures Natrium *n*
permanganic acid Permangansäure *f*, Übermangansäure *f*
permeability Durchlässigkeit *f*; *(magnetische)* Permeabilität *f*; Wasserdurchlässigkeit *f*; Flutbarkeit *f (Schiffsräume)*
p. to air Luftdurchlässigkeit *f*
p. tuning *(El)* magnetische Abstimmung *f*, Permeabilitätsabstimmung *f*
permeable durchlässig, durchdringbar, permeabel
p. to air luftdurchlässig
p. to water wasserdurchlässig
permeameter Permeabilitätsmesser *m*, Permeameter *n*
permeance magnetischer Leitwert *m*, magnetische Leitfähigkeit *f*, Permeanz *f*
permeate / to durchdringen, diffundieren, sättigen; durchfluten *(Licht)*
permeation Durchdringung *f*; Sättigung *f*
permissibility Zulässigkeit *f*
permissible zulässig
p. admixture zulässige Beimengung *f*
p. amount of dimensional variation zulässige Maßabweichung *f*
p. body burden *(Kern)* zulässige Körperbelastung *f*
p. concentration zulässige (verträgliche) Konzentration *f*
p. deviation zulässige Abweichung *f*
p. dose *(Kern)* zulässige (verträgliche) Dosis *f*
p. draught zulässiger Tiefgang *m (Schiff)*
p. explosive *(Brg)* Wettersprengstoff *m*
p. impurity zulässige Verunreinigung *f*
p. limits Meßtoleranz *f*, Toleranz *f*
p. load zulässige Last (Belastung, Beanspruchung) *f*
p. parameter scattering *(Ph)* zulässige Parameterstreuung *f*
p. peak inverse voltage *(El)* maximal zulässiger Spitzenwert *m* der Sperrspannung
p. stress *(Mech)* zulässige Spannung (Beanspruchung) *f*
p. variation Toleranz *f*, Toleranzbereich *m*, zulässige Abweichung *f*
p. variation in diameter Durchmessertoleranz *f*
p. variation in length Längentoleranz *f*
p. variation in size Maßtoleranz *f*
permission to take off *(Flg)* Starterlaubnis *f*
permit / to erlauben, zulassen; *(Werkstück)* aufnehmen
permit for entry by air *(Flg)* Einflugerlaubnis *f*
permittance *s.* permittivity
permittivity [absolute] Dielektrizitätskonstante *f*, DK
permutability *(Math)* Permutabilität *f*, Austauschbarkeit *f*, Vertauschbarkeit *f*
permutation *(Math)* Permutation *f*, Vertauschung *f*
p. bar *(Fmt)* Wählerschiene *f*
p. code switching *(Fmt)* Wahl *f* mittels Tastatur
permutator Kommutatorgleichrichter *m*
permute / to *(Math)* permutieren, vertauschen
peroxidation *(Ch)* Überoxydation *f*, Überoxydierung *f*
peroxide *(Ch)* Peroxid *n*
perpendicular senkrecht, lotrecht, normal, rechtwinklig
perpendicular Senkrechte *f*, Normale *f*, Lot *n*
p. bisector Mittelsenkrechte *f (Geometrie)*
p. conveyance Senkrechtförderung *f*; *(Brg)* Seigerförderung *f*
p. cut Schnitt *m* senkrecht zur elektrischen Achse *f (Quarz)*
p. force Normalkraft *f*
p. line Vertikale *f*, Senkrechte *f*
p. magnetization Quermagnetisierung *f*, magnetische Queraufzeichnung *f*
p. sides Katheten *fpl (des Dreiecks)*
p. to the trajectory plane *(Rak)* Bahnnormale *f*
perpendicularity Rechtwinkligkeit *f*, senkrechte Stellung *f*, Lot *n*
perpetual beständig, ewig
perpetuum mobile Perpetuum mobile *n*
p. mobile first kind Perpetuum mobile *n* erster Art
perrhenate Perrhenat *n*, Salz *n* der Perrheniumsäure
perrhenic acid Perrheniumsäure *f*
perron Beischlag *m*, Treppenvorbau *m*
persalt Persalz *n*, Salz *n* einer Persäure
Persian leather Indisches Leder *n*
persist / to 1. beharren; bestehen bleiben; 2. nachleuchten *(Katodenstrahlröhre)*
persistence 1. Beständigkeit *f*, Persistenz *f*; Dauer *f*; 2. Nachleuchtdauer *f*, Nachleuchtzeit *f*; Nachleuchten *n (Katodenstrahlröhren)*

p. characteristic Nachleuchtcharakteristik *f* *(Katodenstrahlröhren)*
p. of light Nachleuchten *n*
persistent resonator *(Fmt)* schwach gedämpfter Resonator *m*
p. screen *(Fs)* Nachleuchtschirm *m*, Speicherschirm *m*
person-to-person call *(Fmt)* Gespräch *n* mit Voranmeldung, Voranmeldungsgespräch *n*
personal needs allowance *(Ök)* Zeitvorgabe *f* für persönliche Bedürfnisse
p. washing agent Hautreinigungsmittel *n*
personnel Mannschaft *f*, Belegschaft *f*; Begleitmannschaft *f* *(Zugpersonal)*
p. decontamination vehicle *(Mil)* Mannschaftsentgiftungsfahrzeug *n*
p. monitoring Strahlenüberwachung *f* des Betriebspersonals
p. protection Belegschaftsschutz *m*
p.-safety hazard Gesundheitsgefährdung *f* des Betriebspersonals
p. vehicle Personenfahrzeug *n*
perspective Perspektive *f*
perspective perspektivisch
p. distortion *(Opt)* perspektivische Verzeichnung *f*
p. drawing perspektivische Zeichnung *f*
p. plane *(Math)* Bildebene *f*
p. representation Sichtbild *n* *(Radar)*
perspex Plexiglas *n* *(Handelsname)*
perspiration corrosion Schwitzwasserkorrosion *f*
p. resistance *(Text)* Schweißbeständigkeit *f*
persuasive pill Entwöhnungsmittel *n*
pertinax Pertinax *n*
perturbance *s.* perturbation
perturbation Störung *f*, Störeffekt *m*
p. calculation Störungsrechnung *f*
p. energy Störungsenergie *f*
p. method Störungsmethode *f* *(Aberrationsmethode)*
p. theory *(Kern)* Störungstheorie *f*, Reaktorstörungsrechnung *f*
perturbed vision *(Astr)* Sehstörung *f*
perturbing body *(Astr)* störender Körper *m*
Peru[vian] balsam Perubalsam *m*
pervaporation Verdunstung *f* durch eine Membran
perveance *(El)* Raumladungskonstante *f*, Perveanz *f*
pervious durchlässig, permeabel; wasserdurchlässig
p. ground durchlässiger Boden *m*
p. to air luftdurchlässig
p. to water wasserdurchlässig
perviousity Durchlässigkeit *f*
perviousness Durchlässigkeit *f*
pest Ungeziefer *n*, Schädling *m*
p. destruction Ungezieferbekämpfung *f*, Ungeziefervernichtung *f*
pesthouse Isolierstation *f*
pesticide Pflanzenschutzmittel *npl*, Schädlingsbekämpfungsmittel *npl*, Schädlingsgifte *npl*
p. atomizer Zerstäubergerät *n*, Stäubegerät *n* *(für Pflanzenschutzmittel)*
p. dusting apparatus *s.* p. atomizer
pestle Pistill *n*, Mörserkeule *f*
pestology Schädlingsbekämpfung *f*
pet cock Kompressionshahn *m*, Kondenswasserhahn *m*, Probierhahn *m*
p. cock for pumps Pumpenprobehahn *m*
Petersburg standard *(Holz)* 165 Kubikfuß
Petersen coil *(El)* Petersen-Spule *f*
petrail Holzbalken *m*
Petri dish Petrischale *f*
petrification *(Geol)* Versteinerung *f*, Verhärtung *f*, Petrefakt *n*
petrified versteinert
petrify / to versteinern
petrochemical Erdölchemie-, Petrolchemie-, petrolchemisch
p. processing Erdölverarbeitung *f*
petrochemicals Produkte *npl* der Petrolchemie, Erdölderivate *npl*
petrochemistry Petrolchemie *f*, Erdölchemie *f*
petrographic microscopy Gesteinsmikroskopie *f*
petrography Petrografie *f*, Gesteinskunde *f*
petroil lubrication Gemischschmierung *f*, Zweitaktschmierung *f*
p. mixture Zweitaktmischung *f*, Zweitaktgemisch *n*
petrol Benzin *n*, Vergaserkraftstoff *m*, VK, Otto-Kraftstoff *m*
p. and lubricant bulk-storage plant Betriebsstofflager *n*
p.-assisted bicycle Moped *n*
p. barrel Benzinfaß *n*
p. blow torch Benzinlötlampe *f*
p. chemicals petrolchemische Produkte *npl*
p. chemistry Petrolchemie *f*
p. cock Benzinhahn *m*
p. combustion product Benzinrückstand *m*
p. consumption Benzinverbrauch *m*
p. container Benzinkanister *m*
p. engine Benzinmotor *m*, Vergasermotor *m*, Ottomotor *m*
p. filler Benzineinfüllstutzen *m*
p. furnace Benzinlötofen *m*
p. gauge Benzinstandanzeiger *m*, Kraftstoffanzeiger *m*, Benzinuhr *f*, Kraftstoffmesser *m*
p. injection Benzineinspritzung *f*
p.-injection engine Benzineinspritzmotor *m*
p.-injection system Benzineinspritzverfahren *n*; Benzineinspritzanlage *f*
p. jet Benzindüse *f*
p. level Benzinstand *m*

p. motor *s.* p. engine
p.-oil mixture Zweitaktgemisch *n*
p. pipe Benzinleitung *f*, Brennstoffleitung *f*
p. pressure indicator Benzindruckmesser *m*
p. pump 1. Benzinpumpe *f*; 2. Zapfsäule *f* *(Tankstelle)*
p. service truck Abfüllwagen *m*
p. station Tankstelle *f*
p. strainer Benzinfilter *n*
p. supply Benzinversorgung *f*; Benzinvorrat *m*
p. tank Benzintank *m*, Benzinbehälter *m*
p. tap Benzinhahn *m*
p. torch Benzinlötlampe *f*
p. trap Benzinabscheider *m*
p. used for blending (upgrading) Verschnittbenzin *n*
p. vapour Benzindampf *m*
petrolatum Vaseline *f*
p. oil Vaseline *f* *(für Kabel)*; Vaselinöl *n*, Petrolatum *n*
petroleum Erdöl *n*, Rohöl *n*; Petroleum *n*
p. coke Petrolkoks *m*
p. crude Rohöl *n*
p. distillate Erdöldestillat *n*
p. ether Gasolin *n*, Petroläther *m*
p. feeder Petroleuminjektor *m*
p. fraction Erdölfraktion *f*
p. jelly *s.* petrolatum
p. pitch Erdpech *n*, Petrolpech *n*
p.-refinery Petroleumraffinerie *f*
p. residues Petroleumrückstände *mpl*
p. soaps Benzinseifen *fpl (Erdöl)*
p. spirit Leichtbenzin *n*, Ligroin *n*
p. tanker Rohöltanker *m*
p. vapour Petroleumdampf *m*
p. wharf Petroleumkai *m*
petroliferous erdölführend, erdölhaltig
petrology Petrologie *f*, Gesteinskunde *f*
petroselinic acid Petroselinsäure *f*
petticoat *(El)* Isolatorglocke *f*
p. insulator Glockenisolator *m*
petty officer Maat *m*
pewter Weißmetall *n*, Kannenzinn *n*, Hartzinn *n*
pewtery Zinngießerei *f*
pH-acidometer pH-Messer *m*
pH indicator Anzeigegerät *n* für pH-Zahl
pH meter pH-Meßgerät *n*
phanotron Glühkatoden-Quecksilberdampfgleichrichter *m*; ungesteuerte Gleichrichterröhre *f*, Phanotron *n*
phant *s.* phantastron
phantastron *(Fs)* Phantastron *n*
phantom Phantom *n*; Phantomschaltung *f*, Viererschaltung *f*
p. aerial Ersatzantenne *f*, künstliche Antenne *f*
p. circuit Vierer[kreis] *m*, Viererschaltung *f*, Viererleitung *f*, Phantomschaltung *f*
p.-circuit loading coil Viererspule *f*, Phantomspule *f*
p.-circuit operation *(Fmt)* Viererbetrieb *m*
p. connection Phantomschaltung *f*, Vierer[schaltung *f*] *m*
p. frequency selector plug Tarnfrequenzwähler *m*
p.-loaded cable *(Fmt)* viererpupinisiertes Kabel *n*
p. loading *(Fmt)* Viererpupinisierung *f*
p. radar target Radartrugziel *n*
p. signal Falschzeichen *n*, Irrzeichen *n (Radar)*
p. telephony *(Fmt)* Vierersprechen *n*
phantoming *(Fmt)* Duplexierung *f* von Leitungen, Phantombildung *f*, Viererbildung *f*
pharmaceutical pharmazeutisch
p. chemist Pharmazeut *m*, Apotheker *m*
p. chemistry pharmazeutische Chemie *f*
pharmaceuticals Arzneimittel *npl*, Pharmazeutika *npl*
pharmacist Pharmazeut *m*, Apotheker *m*
pharmacolite *(Min)* Pharmakolith *m*
pharmacology Arznei[mittel]kunde *f*, Arzneimittellehre *f*, Pharmakologie *f*
pharmacopeia Arzneibuch *n*, Pharmakopöe *f*
pharmacy 1. Apotheke *f*; 2. Pharmazie *f*
phase / to in Phase bringen, gleichphasig machen, bündeln, phasieren
phase Phase *f*
p. advance Phasenvordrehung *f*, Phasenvoreilung *f*, Phasenvorhalt *m*
p. advancer *(voreilender)* Phasenschieber *m*
p. angle Phasenwinkel *m*
p. angle correction factor Phasenfehlerkorrektur *f*
p. angle error Winkelfehler *m*
p. angle of a transformer Fehlwinkel *m* eines Wandlers
p. balance Phasengleichheit *f*
p.-balance relay Phasenunterbrechungsrelais *n*
p. boundary Phasengrenzschicht *f*, Phasengrenzlinie *f*
p. break *(Elm)* Schutzstrecke *f*
p. centre Phasenzentrum *n*
p. change Phasenänderung *f*; Phasensprung *m*
p.-change coefficient Phasenkonstante *f*, Phasenmaß *n*, Winkelmaß *n*
p. changer Phasenumformer *m*, Phasenschieber *m*
p. coincidence Phasengleichheit *f*
p. comparison localizer *(Flg)* Landekurssender *m* mit Phasenvergleich
p.-comparison protection Phasenwinkelvergleichsschutz *m*
p. compensation Phasenkompensierung *f*, Phasenkompensation *f*, Phasenausgleich *m*, Phasenentzerrung *f*

p. compensator Phasenentzerrer *m*
p. condenser *(Opt)* Phasenkondensor *m*
p. condition Phasenbedingung *f*
p. conductor Phasenleiter *m*
p. constant Phasenkonstante *f*; Phasenmaß *n*, Wellen[längen]konstante *f*
p. contrast Phasenkontrast *m*
p.-contrast image Phasenkontrastbild *n*
p.-contrast method Phasenkontrastverfahren *n*
p. contrast microscope Phasenkontrastmikroskop *n*
p. contrast microscopy Phasenkontrastmikroskopie *f*
p. control 1. Phasenregelung *f*; 2. Phasenregler *m*; 3. Zündeinsatzsteuerung *f (Gleichrichter)*
p.-control factor Aussteuerung[sgrad *m*] *f (Stromrichter)*
p. converter Phasenschieber *m*
p.-corrected phasenkorrigiert
p. correction Phasenentzerrung *f*, Phasenausgleich *m*
p.-crossover frequency Phasenschnittfrequenz *f (Frequenz, bei der der Phasenwinkel 180° beträgt)*
p. delay Phasenverzögerung *f*, Phasenlaufzeit *f*
p. delay distortion Laufzeitfehler *m*
p. delay-time Phasenlaufzeit *f*
p. departure Phasenhub *m*
p. detector Phasendetektor *m*, Phasendiskriminator *m*
p. deviation Phasenhub *m*
p. diagram Zustandsdiagramm *n*
p. difference Phasendifferenz *f*, Phasenunterschied *m*, Phasenwinkel *m*
p. difference of the interfering beams *(Opt)* Gangunterschied *m* der interferierenden Strahlen
p. displacement Phasenverschiebung *f*
p. distortion Phasenverzerrung *f*
p. distortion index Phasenverzerrungsgrad *m*
p. equalizer Phasenentzerrer *m*
p. equation Phasengleichung *f*
p. error Phasenfehler *m*
p. factor Phasenfaktor *m*
p. fading Phasenschwund *m*
p. failer Phasenfehler *m*, Phasenausfall *m*
p. focusing Phasenfokussierung *f*
p.-frequency distortion index Phasenverzerrungsgrad *m*
p. improving Phasenverbesserung *f*
p. indicator Phasenanzeiger *m*
p. integral Phasenintegral *n*
p. inversion *(Fs)* Phasenumkehrung *f*
p. inverter *(Fmt)* Phasenumkehrschaltung *f*; Inverter *m*, Vorzeicheninverter *m*, Vorzeichenumkehrer *m*, Vorzeichenumkehrverstärker *m* *(Analogrechner)*
p. inverter amplifier Umkehrverstärker *m*
p. inverter stage Phasenumkehrstufe *f*
p. inverting Phasenumkehr *f*
p. jump Phasensprung *m*
p. lag nacheilende Phasenverschiebung *f*, Phasennacheilung *f*, Phasenverzögerung *f*, Phasenrückdrehung *f*
p. lead Vorhalt *m*, Phasenvordrehung *f*, Phasenvoreilung *f*
p. localizer Peilvorrichtung *f* mit Phasenverschiebung *(Radar)*
p. margin Phasenvorrat *m*, Phasenreserve *f*, Phasenspielraum *m*
p. meter Phasenmesser *m*
p. modulation Phasenmodulation *f*
p. modulator Phasenmodulator *m*
p. opposition Gegenphase *f*, Gegenphasigkeit *f*
p. plane Phasenebene *f*
p. plate Phasenplatte *f*
p. portrait Phasenebene *f*, Phasenebenendarstellung *f*, Phasenportrait *n*
p. position Phasenlage *f*
p. precorrection *(Fs)* Phasenvorentzerrung *f*
p. quadrature um 90° phasenverschoben
p. relation Phasenbeziehung *f*
p. response Phasengang *m*, Phasenverhalten *n*
p. retard Phasenrückdrehung *f*, Phasennacheilung *f*
p. reversal *(Fs)* Phasenumkehr[ung] *f*, Phasensprung *m*
p. reversal relay Phasenumkehrrelais *n*
p. rule Phasenregel *f*, Phasengesetz *n*
p. sensitive demodulator phasenabhängiger (phasenempfindlicher) Gleichrichter *m*
p. sensitive rectifier phasenabhängiger (phasenempfindlicher) Gleichrichter *m*
p. sequence Drehrichtung *f* des Feldes, Phasenfolge *f*
p.-sequence indicator Drehfeldrichtungsanzeiger *m*
p. shift Phasenverschiebung *f*; Gangunterschied *m (Phasenkontrast)*
p. shifter Phasenschieber *m*; Phasenschiebertransformator *m*, Phasenregler *m*
p.-shifting transformer Phasen[schieber]-transformator *m*
p. space Phasenraum *m*
p.-space cell Phasenraumelement *n*
p. spacing Phasenabstand *m*
p. splitter Phasentrenner *m*, Phasenspalter *m*
p. splitting Phasenteilung *f*, Phasenspaltung *f*
p. stability Phasenstabilität *f*
p. stabilization Phasenstabilisierung *f*
p. structure Phasenstruktur *f*
p. swinging *(Fs)* Außertrittfallen *n* des Bildgleichlaufes

p.-**synchronized** phasensynchronisiert
p. **trajectory** Trajektorie *f*, Zustandskurve *f*
p. **transformation** Phasenumwandlung *f*
p. **transformer** Phasen[schieber]transformator *m*
p. **turner** *(Am)* Phasenschieber *m*
p. **unbalance** Phasenverschiebung *f*, Phasenungleichheit *f*
p. **velocity** Phasengeschwindigkeit *f*
p. **voltage** Phasenspannung *f*
p. **wave** Materiewelle *f*
p. **white** Phasierung *f* auf Weiß *(Fernsehtechnik)*
p. **winding** Phasenwicklung *f*
phasing Phaseneinstellung *f*
p. **capacitor** Phasenabgleichkondensator *m*
p. **control** Phasenregelung *f*
p. **error in computer** Driftfehler *m* im Rechenautomat
p. **line** Phasenleitung *f (Antenne)*; *(Fs)* Phasenlinie *f*
p. **of a receiver crystal filter** Einstellung *f* des Antiresonanzpunktes bei Quarzfiltern in Funkempfängern
phasotron *(Kern)* Phasotron *n*, Synchrozyklotron *n*
phenanthraquinone *(Ch)* Phenanthrenchinon *n*
phenanthrene *(Ch)* Phenanthren *n*
phenate *(Ch)* Phenolat *n*
phenobarbital Luminal *n*, Phenyläthylbarbitursäure *f*
phenocryst *(Geol)* Einsprengling *m*
phenol Phenol *n*, Karbolsäure *f*
phenolic phenolisch
p. **adhesive** Phenolharzleim *m*, Phenolharzklebstoff *m*
p. **cement** Phenolharzkleber *m*
p. **ether** Phenoläther *m*
p. **foam** Phenolharzschaum *m*
p. **laminated sheet** Phenolharzschichtstoff *m*
p. **lined ways** *(Masch)* mit Phenolkunststoffleiste versehene Bahnen *fpl*
p. **lining** *(Masch)* Phenolkunststoffleiste *f*
p. **moulding compound** Phenolharzpreßmischung *f*
p. **resin glue** Phenolharzleim *m*
phenolphthalein Phenolphthalein *n*
phenomenon Phänomen *n*, Erscheinung *f*, Effekt *m*
p. **of diffraction** Beugungserscheinung *f*
phenoplast Phenolharz *n*, Phenoplast *m*
phenoxide einfacher Phenoläther *m*
phenyl *(Ch)* Phenyl *n*
p. **acetate** Phenylazetat *n*
p.-**acrylic acid** Phenylakrylsäure *f*, Zimtsäure *f*
p.-**benzoate** Benzoesäurephenylester *m*
p.-**chloride** Chlorbenzol *n*
p. **ether** Diphenyläther *m*
p. **ethyl alcohol** Phenyläthylalkohol *m*
p. **mustard oil** Phenylsenföl *n*
phenylate *(Ch)* Phenolat *n*
phenylation *(Ch)* Phenylierung *f*
phenylhydroxyacetic acid Mandelsäure *f*, Phenylglykolsäure *f*
phial Phiole *f*, Flakon *m*, gläsernes Fläschchen *n*
phloroglucine *(Ch)* Phlorogluzin *n*
phloroglucinol *(Ch)* Phlorogluzin *n*
pholerite *(Min)* Pholerit *m*
phon Phon *n (Kennzeichnung der Lautstärke)*
p.-**scale** Phonskala *f*
phone / to telefonieren
p. **up** anrufen
phone Telefon *n*, Fernsprecher *m* (*s. a.* telephone)
p. **adapter** Röhrenzwischensockel *m* für Grammofonanschluß
p. **operator** Telefonist *m*, Telefonistin *f*
p. **plug** *(Fmt)* Klinkenstecker *m*
phones Kopfhörer *mpl*, Hörer *mpl*
phonetic lettering *(Typ)* Lautumschrift *f*
phonevision system Fernsehsystem *n* mit teilweiser Übertragung über eine Telefonleitung
phonic phonisch
p. **drum** *s.* p. wheel
p. **image** Lautbild *n*
p. **wheel** *(Ak)* Tonrad *n*, phonisches Rad *n*
phonocardiogram Tonkardiogramm *n (Medizintechnik)*
phonogram Tonaufzeichnung *f*, Tonspur *f*; zugesprochenes Telegramm *n*
phonograph Grammofon *n*, Plattenspieler *m*
p. **amplifier** Grammofonverstärker *m*
p. **pick-up** Schallplattenabtaster *m*, Tonabnehmer *m*, Schalldose *f*, Aufnahmevorrichtung *f*
p. **recorder** Aufnahmevorrichtung *f*
phonographic recorder Diktafon *n*
phonometer Phonometer *n*, Lautstärkemesser *m*, Tonmeßgerät *n*, Klangmesser *m*
phonotelemetry akustische Entfernungsmessung *f*, Echolotung *f*
phonovision Abonnementsfernsehen *n*
phoronomics *s.* kinematics
phosgene *(Ch)* Phosgen *n*
phosgenite *(Min)* Phosgenit *m*, Bleihornspat *m*
phosphate / to phosphatieren
phosphate Phosphat *n*, Salz *n* der Phosphorsäure
p.-**coated** phosphatiert
p. **coating** Phosphatüberzug *m*
p. **fertilizer** Phosphatdünger *m*
p. **treatment** Phosphatierung *f*
phosphated metal phosphatiertes Metall *n*
phosphatic phosphatisch, phosphorsäurehaltig
p. **guano** phosphathaltiger Guano *m*

phosphatide Phosphatid *n*, Phospholipoid *n*
phosphatize / to phosphatieren
phosphatizing Phosphatieren *n*
phosphato-molybdic acid Phosphormolybdänsäure *f*
p.-tungstic acid Phosphorwolframsäure *f*
phosphide Phosphid *n*, Verbindung *f* eines Metalls mit Phosphor
p. of iron Phosphoreisen *n*
phosphine Phosphin *n*, Phosphorwasserstoff *m*
phosphinous acid hypophosphorige Säure *f*
phosphite Phosphit *n*, Salz *n* der phosphorigen Säure
phospho-guano phosphathaltiger Guano *m*
phosphomanganese Phosphormangan *n*
phosphor bronze Phosphorbronze *f*
p. removal Entphosphorung *f*
p. screen Leuchtschirm *m*
p. strip Leuchtstoffstreifen *m*
phosphoresce / to phosphoreszieren
phosphoric acid Phosphorsäure *f*
p. acid fertilizer Phosphorsäuredünger *m*
p. anhydride Phosphorsäureanhydrid *n*, Phosphorpentoxid *n*
p. chloride Phosphorpentachlorid *n*
p. compound Verbindung *f* des fünfwertigen Phosphor
p. ester Phosphorsäureester *m*
phosphorous acid phosphorige Säure *f*
p. chloride Phosphortrichlorid *n*
p. compound Verbindung *f* des dreiwertigen Phosphor
phosphorus Phosphor *m*
phosphuretted mit Phosphor verbunden
phot Phot *n (veraltete Einheit für die spezifische Lichtausstrahlung bzw. Beleuchtungsstärke; 1 ph = 1 lm/cm²)*
photistor lichtempfindlicher Transistor *m*
photo Fotografie *f*, Foto *n*
p.-cell *s.* photocell
p.-composition *(Typ)* Lichtsatz *m*
p.-conducting *s.* photoconducting
p.-conduction *s.* photoconduction
p.-conductive *s.* photoconductive
p.-decomposition fotochemische Zersetzung *f*
p.-detachment Elektronenablösung *f* durch Strahlung *(Fotoionisation)*
p.-dissociation Fotodissoziation *f*
p.-electric *s.* photoelectric
p.-emissive *s.* photoemissive
p.-multiplier *s.* photomultiplier
p.-plate Fotoplatte *f*
p.-potentiometer Fotopotentiometer *n*
p.-resistance *s.* photoresistance
p.-sensitive *s.* photosensitive
p.-telegraphic apparatus Bildtelegraf *m*
p.-telegraphy Bildtelegrafie *f*
p.-voltaic *s.* photovoltaic
photoactive lichtelektrisch aktiv, fotoaktiv
photocarrier Fototräger *m*
photocathode Fotokatode *f*
photocell 1. lichtelektrische Zelle *f*, Fotozelle *f*; 2. Fotoelement *n*, Halbleiterfotoelement *n*
p. amplifier Fotowiderstand *m*
p. current Fotostrom *m*, lichtelektrischer Strom *m*
p. noise Fotozellenrauschen *n*
p. receiver fotoelektrischer Strahlungsempfänger *m*
photochemical fotochemisch, lichtchemisch
p. activity fotochemische Wirkung *f*
p. colour printing *(Typ)* Lichtfarbendruck *m*
p. equivalence law Einsteinsches Äquivalenzgesetz *n*
p. memory *(Dat)* fotochemischer Speicher *m*
photochemicals Fotochemikalien *fpl*
photochemistry Fotochemie *f*
photochromatic farbfotografisch, fotochromatisch
photoconducting fotoleitend
p. cell *s.* photoresistance cell
photoconduction Fotoleitung *f*
p. response Fotoleitungsempfindlichkeit *f*
p. sensitivity fotoelektrische (lichtelektrische) Empfindlichkeit *f*
photoconductive fotoleitend, lichtelektrisch; Fotoleitungs-, Fotowiderstands-
p. cell Widerstandszelle *f*, Fotoleitzelle *f*, Fotowiderstand *m*
p. detector Detektorwiderstandszelle *f*, Detektorfotowiderstand *m*, Fotodetektor *m*
p. effect innerer lichtelektrischer Effekt *m*, innerer Fotoeffekt *m*, Halbleiterfotoeffekt *m*
p. ink lichtelektrische Tinte *f*
p. layer Fotowiderstandsschicht *f*
p. measurement Fotoleitfähigkeitsmessung *f*
p. property Fotoleitungseigenschaft *f*
p. sensitivity Fotoleitungsempfindlichkeit *f*
photocopy Fotokopie *f*
photodetector Fotodetektor *m*
photodielectric effect fotodielektrischer Effekt *m*
photodiffusion Fotodiffusion *f*
p. voltage Fotodiffusionsspannung *f*
photodiode Fotodiode *f*
photodisintegration Fotozerfall *m*, Zerfall *m* durch Lichteinwirkung
photoeffect Foto[leitungs]effekt *m*, fotoelektrischer (lichtelektrischer) Effekt *m*
photoelastic spannungsoptisch
p. analysis spannungsoptische Untersuchung *f*
p. optics Spannungsoptik *f*
p. stress analysis spannungsoptische Untersuchung *f*
photoelasticity Spannungsoptik *f*
p. test optische Spannungsprüfung *f*

photoelectric fotoelektrisch, lichtelektrisch
p. action fotoelektrische Wirkung *f*
p. cathode Fotokatode *f*
p. cell fotoelektrische (lichtelektrische) Zelle *f*, Fotozelle *f*
p. control lichtelektrische Steuerung *f*
p. current lichtelektrischer (fotoelektrischer) Strom *m*, Fotostrom *m*
p. detection fotoelektrischer Nachweis *m*
p. electron multiplier Fotoelektronenvervielfacher *m*
p. emission lichtelektrische Emission *f*, äußerer lichtelektrischer Effekt *m*
p. excitation lichtelektrische Anregung *f*
p. exposure meter fotoelektrischer Belichtungsmesser *m*
p. intrusion-detector fotoelektrische Einbruchmeldevorrichtung *f*
p. multiplier Fotovervielfacher *m*
p. relay Lichtrelais *n*
p. scanning lichtelektrische (fotoelektrische) Abtastung *f*
p. semiconduction lichtelektrische Halbleitung *f*
p. sensibility fotoelektrische (lichtelektrische) Empfindlichkeit *f*
p. sensibilization lichtelektrische Sensibilisierung *f*
p. sorter *(Büro)* fotoelektrische Sortiermaschine *f*
p. threshold lichtelektrische Grenze *f*, fotoelektrischer Schwellenwert *m*, Fotoschwelle *f*
p. tube *s.* phototube
p. valve *s.* phototube
p. yield fotoelektrische Ausbeute *f*
photoelectricity Fotoelektrizität *f*
photoelectromagnetic effect fotoelektromagnetischer Effekt *m*, PEM-Effekt *m*
p. voltage fotoelektromagnetische Spannung *f*
photoelectron Fotoelektron *n*
p. multiplier Fotoelektronenvervielfacher *m*
p. production Fotoelektronenerzeugung *f*
photoelectronic fotoelektronisch
photoemission Fotoemission *f*
photoemissive Fotoemissions-
p. cathode Fotokatode *f*
p. cell Emissionsfotozelle *f*, Fotoemissionszelle *f*
p. effect äußerer lichtelektrischer Effekt *m*, Fotoemissionseffekt *m*
p. tube *s.* phototube
p. yield Fotoemissionsausbeute *f*
photoemissivity Fotoemission *f*
photoemitter Fotokatode *f*, Fotoemitter *m*
photoexcitation Fotoanregung *f*, Strahlungsanregung *f*, lichtelektrische Anregung *f*
photoexcited lichtelektrisch angeregt
photofission Fotospaltung *f*
photoformer Fotoformer *m* *(Funktionsgenerator; Analogrechner)*
photogalvanomagnetic fotogalvanomagnetisch
photogeneration Fotoerzeugung *f*, lichtelektrische (fotoelektrische) Erzeugung *f*
photogrammetry Fotogrammetrie *f*, Bildmeßwesen *n*
photogramophone Schallplattentonfilm *m*
photograph Fotografie *f*, fotografisches Bild *n*
photographer Fotograf *m*
photographic fotografisch
p. apparatus Fotoapparat *m*
p. camera Fotoapparat *m*
p. detection fotografischer Nachweis *m*
p. developer Entwickler *m*
p. emulsion fotografische Schicht *f*
p. exposure Belichtung[szeit] *f*
p. exposure meter Fotobelichtungsmesser *m*
p. mission Bildflug *m*
p. paper Fotopapier *n*
p. printing Fotodruck *m*
p. recording fotografische Registrierung *f*
p. sensitivity fotografische Empfindlichkeit *f*
p. tracking *(Astr)* fotografische Bahnverfolgung *f*
photogravure Rastertiefdruck *m*, Kupfertiefdruck *m*
photohalide lichtempfindliches Halogenid *n*
photoionization Fotoionisation *f*
p. cross section Fotoionisationsquerschnitt *m*
photoionizing radiation fotoionisierende Strahlung *f*
photolayer Fotoschicht *f*
photolens Fotoobjektiv *n*
photoluminescence Fotolumineszenz *f*
photolysis Fotolyse *f*
photolyte Fotolyseprodukt *n*
photomacrography Makrofotografie *f*
photomagnetoelectric effect fotomagnetoelektrischer Effekt *m*, PME-Effekt *m*
photometer Fotometer *n*
p. bench Fotometerbank *f*
photometric sphere Ulbrichtsche Kugel *f*
p. unit fotometrische Einheit *f*
p. wedge Graukeil *m*
fotometry Fotometrie *f*
photomicrograph Mikrofotografie *f*
photomicrographic camera mikrofotografische Kamera *f*
p. change-over tube mikrofotografischer Wechseltubus *m*
photomicrography Mikrofotografie *f*
photomultiplier Foto[elektronen]vervielfacher *m*, Fotozelle *f* mit Elektronenvervielfacher, Vervielfacherfotozelle *f*, Sekundärelektronenvervielfacher *m*, SEV
p. cell Fotovervielfacherzelle *f*

p. counter Kristallzähler *m*, Szintillationszähler *m*
p. sensitivity Fotovervielfacherempfindlichkeit *f*
p. tube Fotovervielfacher *m*, Fotozelle *f* mit Elektronenvervielfacher; Sekundärelektronenvervielfacher *m*, SEV
photon Photon *n*, Lichtquant *n*, Strahlungsquant *n*
p. absorption Photonenabsorption *f*
p. annihilation Photonvernichtung *f*
p. bombardment Photonenbeschuß *m*, Gamma-Bestrahlung *f*
p. component Photonenkomponente *f*
p. counter Photonenzähler *m*
p. distribution Photonenverteilung *f*
p. energy Photonenenergie *f*
p. flux Photonenstrom *m*, Photonenfluß *m*
p. gain Photonenverstärkung *f*
p. gas Photonengas *n*
p. radiation Photonenstrahlung *f*
p. rocket Photonenrakete *f*
p. source Photonenquelle *f*
photoneutron Fotoneutron *n*
photonuclear fotonuklear, Kernfoto-, Fotokern-
p. interaction Kernfotowechselwirkung *f*
p. reaction Kernfotoreaktion *f*, fotonukleare Reaktion *f*, Kernfotoeffekt *m*
p. threshold Schwellenwert *m* für Kernfotoeffekt, fotonuklearer Schwellenwert *m*
photophone Lichttelefoniegerät *n*
photopic vision *(El)* Tagessehen *n*
photopolymer *(Typ)* durch [Chromat-]Sensibilisierung lichtempfindlich gemachte Plastfolie *f*
photopotential Fotospannung *f*
photoproduction Fotoerzeugung *f*, Erzeugung *f* durch Photonenabsorption
photoproton Fotoproton *n*
photoradio transmission Bildfunk *m*
photoresistance Fotowiderstand *m*
p. cell Fotowiderstand[szelle *f*] *m*, Lichtwiderstand *m*, Halbleiterfotowiderstand *m*, Halbleiterfotozelle *f*
photoresistor Fotowiderstand *m*
photoresponse Fotoempfindlichkeit *f*, lichtelektrische Empfindlichkeit *f*
photosensitive lichtempfindlich, fotoempfindlich
p. cathode Fotokatode *f*
p. layer lichtempfindliche (fotoempfindliche) Schicht *f*
p. surface lichtempfindliche Oberfläche *f*, Fotoschicht *f*
photosensitivity Fotoempfindlichkeit *f*, Lichtempfindlichkeit *f*
photosensitize / to lichtempfindlich (fotoempfindlich) machen
photostat Fotokopie *f*
p. copy Fotokopie *f*
photostimulated durch Licht angeregt
photosurface Fotoschicht *f*, strahlungsempfindliche (lichtempfindliche) Oberfläche *f*
photosyntesis Fotosynthese *f*
phototimer Bestrahlungsunterbrecher *m*, Fotoschaltuhr *f*
phototransistor Fototransistor *m*, lichtempfindlicher Transistor *m*
phototube lichtelektrische Zelle *f*, Fotozelle *f*
photovalve *s.* phototube
photovoltaic fotovoltaisch
p. cell Halbleiterfotoelement *n*, Halbleiterfotozelle *f*, Fotoelement *n*, Sperrschicht[foto]zelle *f*
p. detector Sperrschichtdetektor *m*, fotovoltaischer Strahlungsempfänger *m*
p. effect Sperrschichtfotoeffekt *m*, Fotovolteffekt *m*
p. junction fotovoltaischer Übergang *m*, pn-Übergang *m (Halbleiter)*
phthalate Phthalat *n*, Phthalsäureester *m*
phthalic acid Phthalsäure *f*
p. amide Phthalamid *n*
p. anhydride Phthalsäureanhydrid *n*
p. ester Phthalsäureester *m*
phthalocyanine Phthalozyanin *n (Farbstoff)*
physical physikalisch; materiell; körperlich, Körper-
p. adsorption physikalische (reversible) Adsorption *f*
p. analysis physikalische Analyse *f*
p. atomic weight physikalisches Atomgewicht *n*
p. chemistry physikalische Chemie *f*, chemische Physik *f*
p. interpretation physikalische Deutung *f*
p. metallurgy Metallkunde *f*
p. optics Wellenoptik *f*
p. photometry physikalische Fotometrie *f*
p. properties Festigkeitseigenschaften *fpl*, physikalische Eigenschaften *fpl (von Werkstoffen)*
p. quantity physikalische Größe *f*
p. scale of atomic weights physikalische Atomgewichtsskala *f*
p. unit physikalische Einheit *f*
physicist Physiker *m*
physico-chemical physikalisch-chemisch, physikochemisch
physics Physik *f*
p. of fission Spaltungsphysik *f*
p. of solids Festkörperphysik *f*
p. of X-rays Röntgenphysik *f*
phytochemistry Pflanzenchemie *f*
phytosterol *(Ch)* Phytosterin *n*
pi-electron Pi-Elektron *n*
pi-meson Pi-Meson *n*
piano Klavier *n*
p. head wood screw *(Am)* Halbrundholzschraube *f*

p. hinge Scharnier *n*
p. key Klaviertaste *f*
p. pedal Pianozug *m*
pianoforte Klavier *n*
p. wire Stahldraht *m*
p. wiring *(Fmt)* Blankverdrahtung *f*
piassava Piassava *f*, Piassavefaser *f*
pica *(etwa)* Cicero *f (12 britische Punkte; Schriftgröße)*
piccolo kleine Flöte *f*, Pikkoloflöte *f*
pick / to 1. [auf]hacken; pflügen; 2. lesen, ernten, sammeln *(Früchte)*; zupfen; 3. *(Brg)* [aus]klauben, lesen, sortieren; 4. [aus]wählen, sortieren
p. cotton Baumwolle pflücken
p. off abtasten *(Radar)*; abnehmen, ableiten, abzweigen
p. up 1. aufsammeln, lesen; aufnehmen; 2. ansprechen *(Relais)*; 3. aufreißen *(Holz)*
p. up moorings *(Schiff)* an die Boje gehen
p. with double weft *(Text)* den Einschuß zweifach eintragen
pick 1. Hacke *f*, Haue *f*; Kreuzhacke *f*, Pickhammer *m*, Pickhacke *f*; 2. *(Brg)* Keilhaue *f*; Schrämmeißel *m*; 3. *(Text)* [einzelner] Schußfaden *m*; 4. *(Text)* Schützenschlag *m*, Schußschlag *m*
p.-a-back exciter aufgesattelte Erregermaschine *f*
p.-and-pick bobbin changing *(Text)* Pick-à-pick-Spulenwechsel *m*
p.-and-shot *(Text)* Schuß *m*
p. axe Spitzhacke *f*, Kreuzhacke *f*
p. box *(Brg)* Meißelhalter *m*
p. breaker *(Brg)* Nadelbrecher *m*
p. clock Schußzähler *m (am Webstuhl)*
p. count *(Text)* Schußanzahl *f*, Feinheit *f* des Schusses
p. glass *(Text)* Weberglas *n*, Ausnehmerglas *n*
p. handle Hackenstiel *m*
p. mattock Maurerhammer *m* mit Spitze
p.-off Meßglied *n (Lageregelung)*; Abtastkopf *m*
p.-off attachment Abnehmevorrichtung *f*
p.-off change gear box Wechselradgetriebe *n*
p.-off gears Umsteckräder *npl*
p.-up 1. Herausreißen *n* eines Nukleons aus dem Kern; 2. Abtastgerät *n*; Tonabnehmer[kopf] *m*; 3. Ansprechen *n (Relais)*
p.-up amplifier Schallplattenverstärker *m*
p.-up arm Ton[abnehmer]arm *m*
p.-up bush Schalldosenbuchse *f*
p.-up capacitor Fühlerkondensator *m*
p.-up cartridge 1. *(El)* [auswechselbares] Abtastsystem *n*; 2. Tonabnehmereinsatz *m*
p.-up coil 1. Suchspule *f*, Sonde *f (Meßtechnik)*; 2. *(El)* Abnehmerspule *f*, Abgriffspule *f*
p.-up current Ansprechstrom *m (z. B. Relais)*; Anzugstrom *m*
p.-up drum Aufnehmertrommel *f*
p.-up grubber *(Lw)* Sammelroder *m*
p.-up hook Aufnahmehaken *m*
p.-up link transmitter *(Fs)* Kameraverbindungssender *m*
p.-up press *(Lw)* Sammelpresse *f*, Pick-up-Presse *f*
p.-up range *(El)* Auffaßbereich *m*
p.-up tube *(Fs)* Abtaströhre *f*; Aufnahmeröhre *f*
p.-up value Ansprechwert *m (Relais)*
p.-up velocity *(Fs)* Abtastgeschwindigkeit *f*
p.-up voltage Ansprechspannung *f (Relais)*
pickaback aircraft bordgestütztes Flugzeug *n*
pickage *(Text)* Schußanzahl *f* je Zoll, Schußdichte *f*
picker 1. Pflücker *m*; 2. *(Text)* Webervogel *m*; Schlagexzenter *m*; Entklettungsmaschine *f*; 3. *(Brg)* Klauber *m*, Klaubevorrichtung *f*; 4. Anfangsbohrer *m*
p.-lap *(Text)* Schlagmaschinenwickel *m*
picket Zaunbrett *n*; Pfosten *m*
picking 1. Pflücken *n*, Lesen *n*, Sortieren *n*; 2. *(Text)* Schützenschlag *m*; 3. Entkletten *n (Wolle)*; 4. *(Holz)* Rupfen *n (Herausreißen von Teilen aus der Papieroberfläche; Fehler)*
p. above *(Text)* Überschießen *n*
p. band Klaubeband *n*, Leseband *n*
p. bar Klauberost *m*, Leserost *m*
p. below *(Text)* Unterschießen *n*
p. belt Klaubeband *n*, Leseband *n*
p. by hand Handklauben *n*, Handlesen *n*
p. device mechanische Klaubevorrichtung *f*
p. dirt *(Brg)* Klaubeberge *pl*
p. drum *(Brg)* Läutertrommel *f*, Erzwäscher *m*
p. motion *(Text)* Eintragen *n (Weberei)*; Schußeintragevorrichtung *f*
p. plant Sortieranlage *f*
p. repeat *(Text)* Schlagrapport *m*
p. stick motion *(Text)* Schlagstockführung *f*
p. table *(Brg)* Klaubetisch *m*, Scheidetisch *m*
pickle / to 1. *(Met)* [ab]beizen, ätzen, dekapieren; gelbbrennen; 2. *(Led)* aufpöckeln, pickeln; 3. marinieren; pökeln
pickle 1. Beize *f*; 2. Säurelösung *f* zum Reinigen von Metallteilen; 3. Pökellauge *f*
p. bath Dekapierbad *n*
p. brittleness *(Met)* Beizsprödigkeit *f*
pickling elektrolytisches (elektrochemisches) Beizen *n*
p. agent Abbeizmittel *n*
p. bath Beize *f*, Beizbad *n*
p. compound Zusatz *m* zu Beizbädern, Beizzusatz *m*
p. crate Beizkorb *m*

p. fluid Beizflüssigkeit *f*
p. inhibitor Sparbeize *f*
p. plant Beizanlage *f*, Beizerei *f*, Gelbbrenne *f*
p. solution Beizflüssigkeit *f*
p. vat Beizkasten *m*
picklock Dietrich *m*
pickman *(Brg)* Häuer *m*
picks *(Text)* Schußzahl *f* je Zoll
picofarad Picofarad *n*
picramic acid Pikraminsäure *f*
picric acid Pikrinsäure *f*
pictogram method *(Am) s.* pictorial statistics
pictorial bildlich; illustriert
p. computer Bildrechner *m*, Rechengerät *n* mit Kartenbilddarstellung
p. representation bildliche Darstellung *f*
p. statistics Bildstatistik *f*
picture [projiziertes] Bild *n*
p. analyzer *(Fs)* Bildzerleger *m*
p. area Bildfeld *n*
p. black *(Fs)* Schwarzpegel *m*
p. border Bildkante *f*
p. carrier *(Fs)* Bildträger *m*
p. compression *(Fs)* Bildzusammendrückung *f*
p. content Bildinhalt *m*
p. control coil *(Fs)* Bildeinstellspule *f*
p. current Bildstrom *m*
p. definition *(Fs)* Bildschärfe *f*, Bildauflösung *f*
p. dissector Bildfeldzerleger *m*
p. duration *(Fs)* Bilddauer *f*
p. element *(Fs)* Bildpunkt *m*, Rasterpunkt *m*
p. feed roller Bildtransportrolle *f*
p. flyback *(Fs)* Bildrücklauf *m*
p. frequency *(Fs)* Bildfrequenz *f*, Bildwechselzahl *f*
p. halftone *(Fs)* Graukeil *m*
p. height Bildhöhe *f*
p. information Bildinhalt *m*
p. lock *(Fs)* [horizontale und vertikale] Bildverriegelung *f*
p. monitor Bildkontroll[empfangs]gerät *n*, Bildmonitor *m*
p.-phone Bildtelefon *n*
p. plane Bildebene *f (in der Perspektive)*
p. point *(Fs)* Bildpunkt *m*, Rasterpunkt *m*
p. quality Bildgüte *f*
p. ratio *(Fs)* Bildformat *n*, Seitenverhältnis *n* des Bildes
p. recreation Bildaufzeichnung *f*
p. resolution *(Fs)* Bildauflösung *f*
p. rich in contrast *(Fs)* kontrastreiches Bild *n*
p.-rotate control Regelung *f* der Bilddrehung *(Radartechnik)*
p. scan *(Fs)* Bildablenkung *f*
p. scanning *(Fs)* Bild[feld]zerleger *m*
p. sharpness Bildschärfe *f*
p. signal *(Fs)* Bildsignal *n*
p. size Bildgröße *f*
p. telegraph Bildtelegraf *m*
p. telegraph apparatus Bild[übertragungs]gerät *n*
p. telegraphy Bildtelegrafie *f*
p. time-base oscillation Bildkippschwingung
p. to synchronizing ratio *(Fs)* Verhältnis *n* des Helligkeitssignals zum Synchronisationssignal
p. tracing *(Fs)* Bildaufzeichnung *f*
p. transmission Bildtelegrafie *f*
p. transmission period *(Fs)* Bildübertragungsdauer *f*
p. transmitter Bildsender *m*
p. trap *(Fs)* Bildsperre *f*
p. tube Bildröhre *f*, Fernsehröhre *f*, Wiedergaberöhre *f*
p. voltage Bildspannung *f*
p. white *(Fs)* Weißpegel *m*
p. window großflächiges Fenster *n*
p. without contrast kontrastarmes Bild *n*
pie *(Typ)* Zwiebelfisch *m*, verfischter Satz *m*
piece 1. Stück *n*; Werkstück *n*; Bauteil *n*; 2. Geschütz *n*; 3. *(Text)* Warenlänge *f*
p. by piece stückweise
p. cost *(Ök)* Stückkosten *pl*
p. cut-off Abschnitt *m (Sägen)*
p. dyeing *(Text)* Stückfärben *n*; Färben *n* von Stückware
p.-glass *(Text)* Fadenzähler *m*
p. goods Stückgut *n*, Stückgüter *npl*; Stückware *f*
p. of scale Kesselsteinsplitter *m*
p. of work Werkstück *n*
p. part production Einzelteilfertigung *f*
p.-production cost *(Ök)* Stückkosten *pl*
p.-rate setting *(Ök)* Stückzeitermittlung *f*
p. time *(Ök)* Stückzeit *f*
p.-work grinding Schleifen *n* im Stücklohn
piecer Andreher[in *f*] *m (Spinnerei)*
pieces per hour Stundenleistung *f*, Stückzahl *f* je Stunde
piecing of single ends Andrehen *n* von Einzelfäden *(Spinnerei)*
p.-up leichtes Andrehen *n* des Fadens *(Spinnerei)*
pier Hafendamm *m*, Leitdamm *m*, Mole *f*, Pier *m (f)*, Anlegebrücke *f*, Landungsbrücke *f*; Pfeiler *m*
p. arch Pfeilerbogen *m*
pierce / to 1. [durch]lochen, durchbohren, durchstoßen, perforieren, stanzen; durchschlagen; 2. *(Brg)* durchörtern
pierced solder tag *(El)* Lötöse *f*
piercer 1. Locher *m*, Lochstanze *f*; 2. Durchschlag *m*; 3. Pilgerdorn *m (Rohrwalzwerk)*; 4. Luftspieß *m*; 5. Bohrer *m*, Faßbohrer *m (Küfnerei)*

piercing Lochen *n*; Perforieren *n*; Stanzen *n*; Aufdornen *n*
p. and drawing press Loch- und Ziehpresse *f*
p. die Lochmatrize *f*; Kaltschlagmatrize *f*
p. drill Lochbohrer *m*
p. mandrel Lochdorn *m*
p. mill Lochwalzstraße *f*; Lochwalzwerk *n*
p. point *(Math)* Durchstoßpunkt *m (einer Geraden mit einer Fläche)*
p. press *(Met)* Dornpresse *f*
p. punch Vorlochstempel *m*
p. rod *(Met)* Lochdornstange *f*
p. saw Lochsäge *f*, Stichsäge *f (Holzbearbeitung)*
p. tool Perforierwerkzeug *n*
pierman *(Am)* Kaiarbeiter *m*
piezo-electric *s.* piezoelectric
piezochemistry Piezochemie *f*
piezocrystallization Piezokristallisation *f*
piezoelectric piezoelektrisch
p. axis piezoelektrische Kristallachse *f*
p. coupler *(El)* Dreielektrodenquarz *m*, Koppelquarz *m*
p. crystal Piezokristall *m*
p. effect Piezoeffekt *m*
p. filter Quarzfilter *n*
p. frequency standard *(El)* Normalquarz *m*
p. loudspeaker Kristallautsprecher *m*, piezoelektrischer Lautsprecher *m*
p. microphone Kristallmikrofon *n*, piezoelektrisches Mikrofon *n*
p. pressure gauge *(El)* Druckmeßquarz *m*
p. quartz Piezoquarz *m*, piezoelektrischer Quarz *m*
p. resonator Quarzresonator *m*, Kristallresonator *m*, Quarzschwinger *m*
p. transducer piezoelektrischer Meßgrößenwandler *m*
piezoelectricity Piezoelektrizität *f*
piezomagnetic piezomagnetisch
piezometry Piezometrie *f*, Piezomessung *f*, Druckmessung *f*
piezotropic piezotrop
p. equation of state piezotrope Zustandsgleichung *f*
p. fluid piezotrope Flüssigkeit *f*
p. modulus of elasticity Piezotropiemodul *m*
pig / to *(Met)* Roheisen zugeben (zusetzen)
pig *(Met)* Massel *f*, Barren *m*, Block *m*
p.-and-ore process Roheisen-Erz-Verfahren *n*
p. bed Masselbett *n*, Gießbett *n*
p. boiling Puddeln *n*
p. breaker Masselbrecher *m*, Masselschläger *m*
p.-casting machine Masselgießmaschine *f*, Roheisengießmaschine *f*
p. copper Blockkupfer *n*
p. iron Roheisen *n*
p. iron charge Roheiseneinsatz *m*
p. iron for steel making Stahleisen *n*
p.-iron mixer Roheisenmischer *m*
p. iron-ore process Roheisen-Erz-Verfahren *n*
p. lead Rohblei *n*, Ofenblei *n*, Werkblei *n*; Silberblei *n*, Blockblei *n*
p. machine Roheisengießmaschine *f*
p. mould Masselform *f*
p. moulding machine Masselformmaschine *f*
p. process Roheisenverfahren *n*
pigeon hole *(Bw)* Fach *n*; Briefablage *f*
pigeonite *(Min)* Pigeonit *m*
piggery Schweinestall *m*
piggy-back control *s.* cascade control
pigment Pigment[farbstoff *m*] *n*, Farbe *f*, Farbkörper *m*, Farbstoff *m*, Körperfarbe *f*
p. dye Farbpigment *n*, Pigmentfarbstoff *m*
p. dyeing *(Text)* Pigmentfärbung *f (Oberflächenschattierung)*
p. gilding Beizvergoldung *f*
p. padding *(Text)* Pigmentklotzverfahren *n*
p. printing *(Text)* Pigmentdruck *m*
p. rayon pigmentierte Viskoseseide *f*
pigmented layer pigmentierte Oberfläche *f*
pigskin Schweinsleder *n*
pigsticker *(Am, sl) (Mil)* Seitengewehr *n*
pigsty *(Brg)* Holzkasten *m*, Holzpfeiler *m*
pigtaily *(El)* 1. Bürstenanschlußkabel *n*; Lötöse *f*; Lötfahne *f*; 2. geflochtene Litze *f*
p. lead Lötdrähtchen *n*, Anschlußdraht *m*
pilaster *(Bw)* Pilaster *m*, Halbsäule *f*, Wandpfeiler *m*
pile / to paketieren; Pfähle eintreiben, rammen; anhäufen; [auf]stapeln, schichten
p. afresh umschichten
p.-up anhäufen, aufstellen; sich stauen
p. up the waste *(Brg)* Berge versetzen
pile 1. Gründungspfahl *m*, Pfahl *m*, Rammpfahl *m*; 2. Stapel *m*; Hocke *f*; Haufen *m*; Satz *m*; 3. Kernreaktor *m*, Reaktor *m*; 4. *(Am)* Element *n*, Zelle *f (Batterie)*; 5. *(Text)* Flor *m (geschnittene Fäden)*; Pol *m (geschlossene Polhenkel)*
p. batter Pfahlneigung *f*
p. carpet Florteppich *m*
p. driver Pfahlramme *f*; Rammbär *m*
p.-driving barge Rammponton *m*
p.-driving formula Rammformel *f (Pfahlgründungen)*
p.-driving hoist Rammwinde *f*
p.-driving plant Ramme *f*, Rammanlage *f*
p. extractor Pfahlzieher *m*
p. fabric plain circular knitting machine einfonturige Plüschrundstrickmaschine *f*
p. fender *(Schiff)* Schutzpfahl *m*, Sturmpfahl *m*
p. foundation Pfahlgründung *f*
p. foundation structure Pfahlrostbauwerk *n*

p. frame Rammgerüst *n*
p. grating *(Wärm)* Pfahlrost *m*
p. hammer Pfahlramme *f*; Fallhammer *m*, Rammbär *m*, Rammhammer *m*
p. head Pfahlkopf *m*
p. loading test Probebelastung *f (Pfahlgründungen)*
p. moorings *(Schiff)* Dalben *fpl*, Dückdalben *fpl*
p. of plates Plattensatz *m*
p.-of-plates polarizer Plattensatzpolarisator *m*
p. of sheaves Garbenschober *m*
p. pick *(Text)* Florschuß *m*
p. plank Spundbohle *f*
p.-produced radioactive source im Reaktor hergestellte Strahlungsquelle *f*
p. shoe *(Bw)* Pfahlschuh *m*
p. sinker *(Text)* Polplatine *f*
p.-supported mit Pfahlgründung
p. thread *(Text)* Polfaden *m*
p. threshing Hockendrusch *m*
p.-up effect Aufeinandertürmen *n*, „pile-up"-Effekt *m*
p.-up of pulses *(Fs)* Aufeinandertürmen *n* der Impulse
p. warp *(Text)* Polkette *f*
p. weave Florbindung *f*, Plüschbindung *f*; Florgewebe *n*
p. weaving Samtweberei *f*
p. wire *(Text)* Samtnadel *f*, Samtrute *f*
p.-wound coil *(El)* Scheibenwicklung *f (Spule)*
piled cloud Kumuluswolke *f*
p. fabric *(Text)* Warenstapel *m*
p. foundation Pfahlgründung *f*
p. measure *(Holz)* Schichtmaß *n*
p. with sufficient air-space luftig gestapelt *(Holz)*
pilfer-proof abschraubsicher
pilger mandrel Pilgerdorn *m*, Pilgermandrill *m (Walzwerk)*
p. mill Pilgerstraße *f*; Pilgerwalzwerk *n*
p. roll Pilgerwalze *f*
p. rolling mill Walzwerk *n* mit Schrägwalzen, Pilgerwalzwerk *n*
p. step[-by-step-type seamless tube] rolling mill Pilgerschrittwalzwerk *n*; Pilgerschrittwalzstraße *f*
pilgrim Pilgerschiff *n*
piling 1. Stapeln *n*, Schütten *n*, Anhäufen *n*, Anhäufung *f*, Stapelung *f*, Schichtung *f*, Schüttung *f*; 2. Rammen *n (von Pfählen)*; Pfahlanordnung *f*; Haldenschüttung *f*; 3. *(Met)* Paketieren *n*, Paketierung *f*, Paketierverfahren *n*; 4. *(Brg)* Abtreibeverfahren *n*, Ansteckverfahren *n*, gewöhnliches Anstecken *n (Schachtabteufen)*
p. frame Pfahlramme *f*, Rammgerüst *n*
p. hammer Pfahlramme *f*; Rammbär *m*, Fallhammer *m*, Rammhammer *m*
p. in log form Stapeln *n* in Blockform *(Holz)*
p. ring *(Brg)* Pikotagering *m*
p. sticks Stapeln *n* mit Zwischenlage *(Holz)*
p. under cover Einschobern *n (Holz)*
p. underground Einmieten *n (Kartoffeln)*
p.-up Aufschobern *n*, Aufstellen *n* der Heuhaufen
pill resistant *(Text)* pillingfrei, pillfest, pillingfest
pillar / to *(Brg)* abpfeilern
pillar 1. Pfeiler *m*, Pfosten *m*, Säule *f*, Spindel *f*, Ständer *m*, Stütze *f*, Stiel *m*; 2. *(Brg)* Bergfeste *f*, Pfeiler *m*
p.-and-breast *(Brg)* Pfeilerbau[verfahren *n*] *m*
p. and chamber working Kammerbau[verfahren *n*] *m (Erz- und Salzbergbau)*; Pfeilerbau[verfahren *n*] *m (Kohlenbergbau)*
p. and stall method Kammerbau[verfahren *n*] *m (Erz- und Salzbergbau)*; Pfeilerbau[verfahren *n*] *m (Kohlenbergbau)*
p. burst *(Brg)* Pfeilerbruch *m*
p. caving *(Brg)* Abbau *m* von Erzschweben; *(Am)* kombinierter Magazin- und Blockbruchbau *m*
p. drawing *(Brg)* Pfeilerrückgewinnung *f*, Pfeilerabbau *m*
p. drill Säulenbohrmaschine *f*
p. drilling machine Säulenbohrmaschine *f*
p. extraction *(Brg)* Pfeilerrückgewinnung *f*
p. ladder *(Schiff)* Lukenleiter *f*, Stützenleiter *f*, Raumleiter *f*
p. mining method *(Brg)* Örterpfeilerbau *m*
p. of ore *(Brg)* Erzpfeiler *m*, Erzfeste *f*
p. recovery *(Brg)* Pfeilerrückgewinnung *f*, Pfeilerrauben *n*
p. robbing *(Brg)* Pfeilerrauben *n*, Pfeilerabbau *m*, Pfeilerrückgewinnung *f*
p. rolling mill Pilarwalzwerk *n*
p.-shaped säulenförmig
p. type switchgear *(El)* Schaltsäule *f*
p. working Pfeilerbau[verfahren *n*] *m (Kohlenbergbau)*; Kammerbau[verfahren *n*] *m (Erz- und Salzbergbau)*
pillaring *(Brg)* Abpfeilern *n*
pilling *(Text)* Knötchenbildung *f*, Pillchenbildung *f*
p. machine Tablettiermaschine *f*
pillion Soziussitz *m (Krad)*
p. passenger Soziusfahrer *m*
p. rider Soziusfahrer *m*
pillow Zapfenlager *n*
p. block Lagerbock *m*; Kurbel[wellen]lager *n*
p. distortion kissenförmige Verzeichnung *f*
p. lace Klöppelspitze *f*
p.-type lace making Spitzenklöppelei *f*
pilot / to führen; steuern, lenken

p. in einlotsen *(Schiff)*
p. the reamer in the hole die Reibahle in der Bohrung führen

pilot 1. Flugzeugführer *m*, Pilot *m*; 2. *(Schiff)* Lotse *m*; 3. Seehandbuch *n*; 4. Führungszapfen *m (Werkzeug)*; 5. Kraftglied *n*, Kraftschalter *m*; 6. *(El)* Meßader *f (eines Kabels)*; 7. *(Am) (BMSR)* Schnellverstärker *m*, pneumatischer Leistungsverstärker *m*, Schnellübertrager *m*
p. alarm *(Fmt)* Kontrollwecker *m*
p. balloon Pilotballon *m*
p. balloon observation Pilotballonbeobachtung *f*
p. bell *(Fmt)* Kontrollwecker *m*
p. boat Lotsen[versetz]boot *n*
p. bridge Lotsenbrücke *f*, Kommandobrücke *f*
p. brush *(Elm)* Prüfbürste *f*
p. bulb Anzeigelampe *f*
p. burner Zündbrenner *m*
p. bush Führungsbuchse *f*
p. bush for boring bar Bohrstangenführung *f*
p. cable Leitkabel *n*
p. carrier Pilotträger *m*
p. cell Prüfzelle *f (der Batterie)*
p. channel Steuerkanal *m*
p. circuit Steuerstromkreis *m*, Steuerleitung *f*
p. controller *(El)* indirekt angetriebener Steuerschalter *m*; Steuerschaltwalze *f*
p. drift *(Bw)* Richtstollen *m*
p.-ended grubscrew Gewindestift *m* mit Zapfen
p. flame Zündflamme *f*
p. form of countersink Spitzsenker *m* mit Führungszapfen
p. frequency Pilotfrequenz *f*, Steuerfrequenz *f*
p. handle Schaltgriff *m*
p. hole 1. *(Masch)* Führungsbohrung *f*, Vorbohrung *f* zur Führung des Werkzeugs; 2. *(Brg)* Vorbohrloch *n*
p.-in-command Flugzeugführer *m*
p. indicator 1. Empfänger *m* mit Lotsenflugzeug *(Radartechnik)*; 2. *(Fmt)* Platzlampe *f*
p. injection Voreinspritzung *f*
p. jack Lotsenflagge *f*
p. jet Vor[einspritz]düse *f*
p. lamp Kennlampe *f*, Kontrollampe *f*, Signallampe *f*; Notlampe *f*, Überwachungslampe *f*
p. light Kontrollampe *f*
p. lighting Notbeleuchtung *f*
p. loudspeaker Abhörlautsprecher *m*
p. model Versuchsmodell *n*
p. monitor Autopilot *m*, automatische Kurssteueranlage *f*
p. motor Kleinstmotor *m*, Servomotor *m*
p.-operated hydraulisch vorgesteuert
p. oscillator Steuersender *m*
p. pin Greiferstift *m*, Haltezahn *m*, Justierstift *m*, Sperrgreifer *m*; Führungszapfen *m (Senker)*
p. plant Versuchsanlage *f*, Versuchsbetrieb *m*, Modellanlage *f*, Prüfanlage *f*
p. point Anschnitt *m (Reibahle)*
p. protection Streckenschutz *m (Relaistechnik)*
p. pulse Auslöseimpuls *m*, Startimpuls *m*, Steuerimpuls *m*
p. radio service Lotsenfunk *m*
p. reactor Versuchsreaktor *m*
p. receiver Pegelempfänger *m*, Steuerfrequenzempfänger *m*
p. regulator Pilotregler *m*
p. relay Steuerrelais *n*; Melderelais *n*, Platzlampenrelais *n*
p. shaft Vorschacht *m (Schachtabteufen)*
p. signal Meldelampe *f*; Steuerzeichen *n*, Überwachungszeichen *n*
p. spark arc rectifier Zündfunkengleichrichter *m*, Lichtbogengleichrichter *m* mit Funkensteuerung
p. synchronizing *(Fmt)* Trägererzeugung *f* mit Hilfsfrequenz
p. tape *(Fmt)* Leitlochstreifen *m*
p. trainee Flugschüler *m*
p. transformer *(El)* Meßwandler *m*
p. tube *(Flg)* Staurohr *n*
p. tunnel *(Bw)* Richtstollen *m*
p. valve Schaltventil *n*, Steuerventil *n*, Steuerwelle *f*, Vorsteuerschieber *m*
p. valve governing mittelbare Regelung *f (Turbine)*
p. voltmeter Netzspannungskontrollvoltmeter *n*, Kontrollvoltmeter *n*
p. wire Hilfsleiter *m*, Prüfdraht *m*, Steuerleitung *f*
p. wire circuit Steuerleitung *f*
p. wire protection Streckenschutz *m* mit Hilfsleitung

pilotage *(Schiff)* 1. Lotsenkunde *f*, Lotsenwesen *n*; 2. Lotsengeld *n*, Lotsengebühr *f*
piloted bemannt *(z. B. eine Rakete)*
piloting Führen *n*, Führung *f*; Lotsen *n*
pilotless unbemannt *(z. B. eine Rakete)*
p. aircraft unbemanntes Flugzeug *n*
p. aircraft control Steuerung *f* unbemannter Flugzeuge
pilot's cockpit Flugzeugkabine *f*
pimelate *(Ch)* Ester *m* (Salz *n*) der Pimelinsäure
pimelic acid Pimelinsäure *f*
pimple *(Plast)* Preßfehler *m*, Pickel *m*
pin / to verstiften, verbinden, [an]heften; verbohren, befestigen
pin 1. Stift *m*, Bolzen *m*, Dorn *m*, Spurzapfen *m*, Zapfen *m*; Stecknadel *f*, Zwecke *f*; Stütze *f*; Warze *f*; Triebstock *m*; Klöppel *m*, Stütze

(Isolator); 2. *(El)* Durchführung *f (Elektronenröhren)*; Sockelstift *m*; Steckerstift *m*
p.-and-bushing coupling Kreuzgelenkkupplung *f*
p. bar Wagenrunge *f*
p. bar grid *(Text)* Nadelleistenrost *m*
p. barrel *(Masch)* Stiftbüchse *f*
p. base Stiftsockel *m*
p. bearing *(Masch)* Nadellager *n*, Stiftlager *n*
p. board *(El)* Steckerfeld *n*
p. bolt Federbolzen *m*
p. bushing *(El)* Steckerhülse *f*
p. carriage Stellstiftwagen *m*, Stellstückwagen *m*
p. chain Nietbolzenkette *f*
p. clutch Drehkeilkupplung *f*; Abscherkupplung *f*
p.-connected gelenkig angeschlossen (verbunden); mit Bolzen verbunden
p. cop *(Text)* Schußkötzer *m*
p. coupling Bolzenkupplung *f*
p.-cushion distortion kissenförmige Verzeichnung *f*
p. drill Zapfenbohrer *m*
p. feed platen device Stachelradführung *f*, Stachelwalzenführung *f (Büromaschinen)*
p. head Nietkopf *m*, Stecknadelkopf *m*
p. hole *s.* pinhole
p. insulator *(El)* Stützisolator *m*
p. joint Achsengelenk *n*, Bolzengelenk *n*
p.-jointed gelenkig angeschlossen (verbunden); mit Bolzen verbunden
p. key Rundkeil *m*
p. lifting Stiftabhebung *f*
p. link Gleitgelenk *n (einer Zahnkette)*, Außenglied *n*
p. link plate Außengliedplatte *f (Rollenkette)*
p. mark *(Typ)* Gießmarke *f*
p. movement Greiferantrieb *m (für Filmtransport)*
p. of bent crank *(Masch)* Krummzapfen *m*
p. of the driving plate *(Masch)* Mitnehmerbolzen *m*
p. oiler Stiftöler *m*
p. plug *(El)* Bananensteckerstift *m*, Steckstift *m*, Stiftstecker *m*
p. punch Durchtreiber *m*
p. rail Hakenbrett *n*
p. riveting Stiftnietung *f*
p. spanner Zapfenschlüssel *m*
p. support Stifthalterung *f (Halterungsart der Adern einer Koaxialleitung)*
p. tongs Werkzeughalter *m*
p.-type cap *(El)* Stiftsockel *m*, Stecksockel *m*
p.-type insulator *(El)* Stützisolator *m*, Stützer *m*
p.-type lubricator Stiftöler *m*
p.-type socket *(El)* Stiftsockel *m*
p.-type threshing machine Stiftendreschmaschine *f*
p. wire Nietstiftdraht *m*, Splintdraht *m*, Nadeldraht *m*
pinacoid Pinakoid *n (Kristallografie)*
pinacol *(Ch)* Pinakol *n*, Pinakon *n*
pinacolic rearrangement *(Ch)* Pinakolinumlagerung *f*
pinacoline *(Ch)* Pinakolin *n*
pincers Pinzette *f*; Beißzange *f*, [kleine] Kneifzange *f*
pincette Pinzette *f*
pinch / to zusammendrücken, kneifen, quetschen, klemmen
p. off abkneifen, abklemmen, abquetschen, abzwicken
pinch 1. *(El)* Quetschung *f (einer Röhre)*; 2. Verdrängung *f (einer Lagerstätte)*; 3. *s.* p. effect
p. bar Brechstange *f*
p. base *(El)* Preßfuß *m (Röhren)*
p.-cock Quetschhahn *m*
p. effect *(Ph)* Pincheffekt *m*, Einschnüreffekt *m*
p. off Abschnürung *f*
p.-off point Abschnürungsstelle *f*
p. roller Klemmrolle *f (Magnetband)*
p.-type lead-in Quetschfußdurchführung *f*
p. wheel Andruckrolle *f*
pinchbeck Rotguß *m*, Tombak *m*
pinched base *(El)* Quetschfuß *m (Röhren)*
pinching *(El)* Blockierung *f*, *(ungewollte)* Sperrung *f* von Röhren
pincushion distortion *(Fs)* kissenförmige Verzeichnung *f*
pine Kiefer *f*, Kiefernholz *n (bisweilen auch für andere Nadelbäume verwendet)*
p. cone Kiefernzapfen *m*
p. crystal Dendrit *m*, Tannenbaumkristall *m*
p. needle oil Kiefernnadelöl *n*, Fichtennadelöl *n*
p. resin Kiefernharz *n*, Fichtenharz *n*
p. tar Fichtenholzteer *m*
p. tree *(Masch)* Tannenbaumprofil *n*
p.-tree array Tannenbaumanordnung *f (Antennentechnik)*
p.-tree broach Tannenbaumprofilräumwerkzeug *n*
p.-tree form *(Masch)* Tannenbaumprofil *n*
p.-tree form broach Tannenbaumprofilräumwerkzeug *n*
p. trees *(Plast)* kalte Streifenbildung *f* bei Kalandermaterial
p. wood Kiefernholz *n*; Kiefernwald *m*
pineapple 1. Ananas *f*; 2. *(sl)* Handgranate *f*
p. cone Bikonus *m*, doppelkonischer Garnkörper *m*, doppelkonische Spule *f*

p. cone winder Doppelkonusspulmaschine *f*, Bikonusspulmaschine *f*
pinene *(Ch)* Pinen *n (bizyklisches Terpen)*
ping / to klingeln *(Motor)*
ping Sonarimpuls *m (Radartechnik)*
p.-pong Tischtennis *n*
pinhole 1. feines Loch *n*, Nadelloch *n*; 2. *(Met)* Gasblase *f*
p. camera Lochkamera *f*
p.-camera detector *(Met)* Feinlunkeranzeiger *m*
p.-diaphragm Lochblende *f*
p. grate *(Wärm)* perforierte Rostplatte *f*
p. plug *(Met)* Nadelboden *m (Konverter)*
p. porosity *(Met)* Mikrolunkerung *f*
p. sight *(Opt)* Lochvisier *n*
p. stop Lochblende *f*
pinion 1. *(Masch)* Ritzel *n*; Kammwalze *f*; 2. *(Typ)* Nuß *f*, Schließrolle *f*
p. cage Planetenträger *m*
p. drive Ritzelantrieb *m*
p. drive shaft Ritzelantriebswelle *f*
p. knob Triebknopf *m (für Meßinstrumente)*
p. leaf Ritzelzahn *m*
p. shaft Ritzelwelle *f*; Planetenradwelle *f*
p. spindle Zahnradwelle *f*
pinjoint ship Gelenkschiff *n*, Scharnierschiff *n*
pink-salt *(Ch)* Ammoniumhexachlorostannat *n*, Pinksalz *n*
p. stern *(Schiff)* Pinkenheck *n (hohes schmales Spiegelheck)*
pinking [leichtes] Klopfen *n (des Motors)*
pinnace Pinasse *f*, Beiboot *n*
pinned verstiftet
p. disk mill Stiftmühle *f*
p. fitting Verstiftung *f*
p. hinge Gelenk *n*; Bolzengelenk *n*
p. joint Gelenk *n*; Bolzengelenk *n*
pinpoint / to genau ausmachen *(Ziel)*
pinpoint accuracy *(Am)* höchste Genauigkeit *f*
p. bombing Bombenpunktwurf *m*
pint Pint *n (1. Flüssigkeitsmaß: 0,568 l in GB; 0,473 l in US; 2. dry pint: 0,55 l)*
pintle 1. Gelenkstift *m*; Achse *f*; Zapfen *m*; 2. Düsennadel *f*, Düsenzapfen *m*
p. chain Antriebskette *f*
p. nozzle Zapfendüse *f (Dieselmotor)*
p.-type injector Zapfen[einspritz]düse *f*
pion Pi-Meson *n*
pip *(El)* kurzer Impuls *m*, Pips *m*; Spannungsstoß *m*; Überlagerung *f* der Feinortung durch die Grobortung *(bei Decca)*
pipage Rohrleitungssystem *n*, Rohrnetz *n*; Rohrlegung *f*
pipe / to [in Rohrleitung] leiten
pipe 1. Rohr *n*, Leitungsrohr *n*; 2. *(Met)* Lunker *m*; 3. Pfeife *f (Orgel)*; 4. *(Geol)* Schlot *m*; 5. *(Brg)* Lutte *f*; 6. *(El)* Erd[er]platte *f*, Erdungsplatte *f*
p. assembly Leitungsanlage *f*
p. bell Rohrmuffe *f*
p. bend Knierohr *n*, Kniestück *n*, Rohrkrümmer *m*
p. bender Rohrbiegemaschine *f*
p. bracket Rohrschelle *f*, Rohrhalter *m*
p. branching Rohrverzweigung *f*
p. breakage Rohrbruch *m*
p. bridge Rohrbrücke *f*
p. burring reamer Rohrfräser *m*
p. casting *(Met)* Rohrguß *m (Verfahren)*
p. cavity *(Met)* Lunker[hohlraum] *m*
p. chase Rohrschlitz *m*
p. choking Rohrverstopfung *f*
p. clamp Rohrschelle *f*, Röhrenklammer *f*, Rohrsattel *m*
p. clamp fitting Rohrverbindungsschelle *f*
p. clip Rohrschelle *f*
p. clogging Rohrverstopfung *f*
p. coil Spiralrohr *n*
p. column Rohrtour *f*, Rohrsäule *f (Tiefbohrtechnik)*
p. compartment *(Brg)* Rohrtrum *m*
p. connection Anschlußstutzen *m*, Rohranschluß *m*, Rohrverbindung *f*
p. connection flange Rohrverbindungsflansch *m*
p. coupling Rohrverbindung *f*
p. covered with non-conducting material isoliertes Rohr *n*
p. covering Rohrumhüllung *f*, Rohrummantelung *f*, Rohrbekleidung *f*
p. cramp Rohrhaken *m*
p. crown Rohrscheitel *m*
p. culvert Rohrdurchlaß *m (im Erdboden)*
p. cutter Rohr[ab]schneider *m*
p. die Schneidbacke *f* für Rohrgewinde
p. dog Rohrfänger *m*
p. dredge Spülbagger *m*
p. elbow Kniestück *n*, Rohrknie *n*, Rohrkrümmer *m*
p. eliminator Lunkerverhütungsmittel *n (Gießerei)*
p. fitter Klempner *m*, Rohrleger *m*
p. fitting Fitting *n (m)*, Röhrenverbindung *f*, Rohrleitungsstück *n*
p. flange Rohrflansch *m*
p. foundry Röhrengießerei *f*
p. fracture Rohrbruch *m*
p. friction loss Rohrreibungsverlust *m*
p. hanger Rohrschelle *f*, Rohrsattel *m*
p. hoisting gear Saugrohrhebegeschirr *n (eines Saugbaggers)*
p. housing Rohrleitungsgehäuse *n*
p. installation Rohrverlegung *f*

p. joint Rohrmuffe *f*, Rohrverbindung *f*, Verbindungsrohr *n*
p. lagging Rohrumhüllung *f*, Rohrummantelung *f*, Rohrbekleidung *f*
p. laying Rohrverlegung *f*
p. layout Installationsplan *m*
p. leads Rohrleitungen *fpl*, Rohrleitungsplan *m (z.B. auf Schiffen)*
p. line Pipeline *f*, Rohr[fern]leitung *f*, Rohrstrang *m*
p. line milkers Melkrohrsystem *n*
p. line mixer Durchflußmischer *m*
p. line vessel Rohrlegerschiff *n*
p. manifold Rohrverteiler *m*
p. mill Rohrwalzwerk *n*
p. moulding *(Met)* Rohrformerei *f*
p. nipple Rohrnippel *m*
p. packing Rohrdichtung *f*, Rohrpackung *f*, Rohrliderung *f*
p. plug Rohrstopfen *m*
p. reducer Rohrreduktionsstück *n*
p. riser Standrohr *n*
p. run Leitungsführung *f*
p. saddle Rohrschelle *f*, Abzweigschelle *f*, Rohrsattel *m*
p. seal Rohrverschluß *m*
p. section 1. Rohrabschnitt *m*; 2. Rohrquerschnitt *m*
p. socket Rohrmuffe *f*, Rohrflansch *m*
p. spanner Rohrschraubenschlüssel *m*
p.-spinning machine Schleudergießmaschine *f* für Rohre
p. stock Kluppe *f*
p. stock and die Rohrkluppe *f*
p. strap Rohrschelle *f*
p. tap Schneidbohrer *m* für Rohrgewinde, Rohrgewindebohrer *m*
p. thread Rohrgewinde *n*
p. tongs Rohrzange *f*
p. trench Rohrkanal *m*, Rohrschacht *m*
p. tunnel *(Schiff)* Kieltunnel *m*
p. under pressure Druckleitung *f*
p. vice Rohrschraubstock *m*
p. way *(Brg)* Rohrtrum *m*
p.-way compartment *(Brg)* Rohrtrum *m*
p.-welding rolling mill Rohrschweißwalzwerk *n*
p. with fixed flange Rohr *n* mit festem Flansch
p. with flange brazed on Rohr *n* mit aufgelötetem Flansch
p. with flange riveted on Rohr *n* mit aufgenietetem Flansch
p. with flange screwed on Rohr mit aufgeschraubtem Flansch
p. wrench Rohr[schrauben]schlüssel *m*, Rohrzange *f*
pipeclay triangle Tondreieck *n (Laborgerät)*
pipeless 1. lunkerfrei, lunkerlos *(Gußblock)*; 2. ohne Pfeifen *(Elektronenorgel)*
p. organ Elektronenorgel *f*
piperazine *(Ch)* Piperazin *n*
piperic acid Piperinsäure *f*
piperidine *(Ch)* Piperidin *n*
piperine *(Ch)* Piperin *n*
pipette / to pipettieren
pipette Pipette *f*
p. stand Pipettengestell *n*, Pipettenständer *m*
piping 1. Rohr[anlage *f*] *n*, Rohrleitung *f*, Rohrnetz *n*, Leitungssystem *n*; 2. Rohrmaterial *n*; 3. Verlegen *n* von Rohren; 4. Leiten *n* [in Rohrleitungen]; 5. *(Met)* Lunkern *n*, Lunkerbildung *f*
p. connection Rohrstutzen *m*
p. effect *(Met)* Lunkerung *f*, Hohlraumbildung *f*
p. plan Rohrplan *m*
p. under the floor unterirdische Rohrleitung *f*
p.-up Verlegung *f* von Rohren
Pirani gauge Piranimanometer *n*, Wärmeleitungsvakuummeter *n*
pirate 1. Pirat *m*, Seeräuber *m*; 2. Piratenschiff *n*, Seeräuberschiff *n*
p. listener Schwarzhörer *m (Radio)*
p. power-consumer Stromdieb *m (auch im übertragenen Sinn von Geräten)*
pirn *(Text)* Schleifspule *f*; Schußspule *f*, Automatenspule *f*
p. changer Spulenwechseleinrichtung *f*, Spulenwechsler *m*
p. magazine Schußspulmagazin *n (Automatenstuhl)*
pirssonite *(Min)* Pirssonit *m*
pisciculture Fischzucht *f*
pisé de terre Piseebau *m*, Erdstampfbau *m*
pisiform erbsenförmig
p. iron-ore *(Min)* Bohnerz *n*
pisolite *(Min)* Pisolith *m*, Erbsenstein *m*
pisolitic erbsenförmig
p. ore *(Min)* Bohnerz *n*
pistacite *(Min)* Pistazit *m*
piston 1. Kolben *m (Verbrennungsmotor)*; 2. *(Masch)* Stempel *m*; 3. *(El)* Kolben *m*, Kurzschlußkolben *m (Wellenleiter)*
p.-actuator *(Wärm)* Kolbenstellantrieb *m*
p. area Kolbenfläche *f (Verbrennungsmotor)*
p. attenuator *(El)* Kolbenabschwächer *m*, Kolbenblindteiler *m*, [variables] Hohlleiterdämpfungsglied *n (Wellenleiter)*
p. barrel Kolbenkörper *m*
p. blower Kolbengebläse *n*, Kolbenverdichter *m*
p. boss Kolbenauge *n*, Kolbennabe *f*
p.-boss bushing Kolbenbüchse *f*
p. box Kolbenbüchse *f*
p. button Sicherungspilz *m* für Kolbenbolzen

p. clearance Kolbenspielraum *m*
p. compressor Kolbengebläse *n*, Kolben[vor]verdichter *m*
p. condition Kolbenlaufbild *n*; Zustand *m* des Kolbens
p. cooling Kolbenkühlung *f*
p. crown Kolbenboden *m*, Kolbenkrone *f*
p. cylinder side-thrust Kolbengleitbahnkraft *f*
p. dead centre Kolbentotlage *f*
p. displacement Hubraum *m*, Hubvolumen *n*, Kolbenverdrängung *f*, Zylinderinhalt *m*
p. drill Schlagbohrmaschine *f (Tiefbohrtechnik)*
p. engine Kolbenmotor *m*; Kolbenmaschine *f*
p. formed with rack teeth Zahnstangenkolben *m*
p. friction Kolbenreibung *f*
p. gap Ringspiel *n*, Kolbenringspiel *n*
p. guide Kolbenführung *f*, Stempelführung *f*
p. head Kolbenboden *m*
p. heating Kolbenheizung *f*
p. indicator Kolbenindikator *m*
p. jig *(Brg)* Kolbensetzmaschine *f*
p. knock Klopfen *n* der Kolben *(des Verbrennungsmotors)*
p. land Kolbensteg *m*, Nutensteg *m*, Ringsteg *m*
p. load Kolbenbeaufschlagung *f*
p. lubrication cock Kolbenschmierhahn *m*
p. motion Kolbenlauf *m*, Kolbenbewegung *f*
p. packing Kolbendichtung *f*, Kolbenliderung *f*
p. path Kolbenweg *m*
p. pin Kolbenbolzen *m*
p.-pin bushing Kolbenbolzenbuchse *f*
p. pin lock Kolbenbolzenbefestigung *f*, Kolbenbolzensicherung *f*
p. play Kolbenspiel *n*
p.-ported engine kolbengesteuerter Motor *m*, Motor mit Schlitzsteuerung (kolbengesteuerten Ein- und Auslaßschlitzen) *(Zweitaktmotor)*
p. position time diagram Kolbenweg[druck]-diagramm *n*, Raumdiagramm *n*, Volumendiagramm *n*
p.-pressure time diagram Kolbenkraftdiagramm *n*, Zeitdiagramm *n*
p. pulsator Kolbenpulsator *m (Melkmaschine)*
p. pump Kolben[luft]pumpe *f*
p. ring Kolbenring *m*, Kompressionsring *m*, Verdichtungsring *m*, Liderungsring *m*
p.-ring blow-by Durchblasen *n* am Kolben, Kolbengasdurchlaß *m*
p.-ring clamp Spannband *n* für Kolbenringeinbau
p.-ring compressor Kolbenringspanner *m*
p.-ring flutter Flattern *n* der Kolbenringe
p.-ring grinding machine Kolbenringschleifmaschine *f*
p.-ring groove Kolbennut *f*, Nut *f* des Kolbenkörpers
p.-ring joint Kolbenringfuge *f*
p.-ring lock Kolben[ring]schloß *n*
p.-ring side clearance Kolbenringspiel *n*
p.-ring tightener Kolbenringspanner *m*
p. rod Kolbenstange *f*, Pleuelstange *f*
p.-rod collar Kolbenstangenbund *m*
p.-rod guide Kolbenstangenführung *f*
p.-rod nut Kolbenstangenmutter *f*
p.-rod thread Kolbenstangengewinde *n*
p.-rod with tail rod durchgehende (durchlaufende, rückwärts verlängerte) Kolbenstange *f*
p. seal Kolbenmanschette *f*
p. shape Kolbenform *f*
p. shirt Kolbenhemd *n*
p. shoe Tragstück *n* (Stützvorrichtung *f*) des Kolbens
p. skirt Kolbenwand *f*, Kolbenschürze *f*, Kolbenmantel *m*, Kolbenschaft *m*
p. slack Kolbenschlackern *n*
p. slap seitliches Schlagen *n* des Kolbens, Kolbenschlackern *n*
p. speed Kolbengeschwindigkeit *f*
p. spring federnder Ring *m*, Kolbenfeder *f*
p. steam engine Kolbendampfmaschine *f*
p. stem Kolbenstange *f*
p. stroke Kolbenhub *m*, Kolbentakt *m*, Kolbenweg *m*
p. supported by piston getragener Kolben *m*
p. supported by the cylinder selbsttragender (von der Zylinderwand getragener) Kolben *m*
p.-testing rig Kolbenprüfeinrichtung *f*
p. top land *(Kfz)* oberster Kolbensteg *m*
p. travel Kolbenweg *m*
p.-type area flowmeter Schwebekolbendurchflußmesser *m*
p.-type flowmeter Kolbenzähler *m*
p.-type pressure gauge Kolbenmanometer *n*
p.-type scavenging pump Kolbenspülpumpe *f*
p.-type slide valve Kolbenschieber *m*
p.-type steam engine Kolbendampfmaschine *f*
p.-type viscosimeter Kolbenviskosimeter *n*
p. valve Kolbenschieber *m*
p. valve chest Kolbenschiebergehäuse *n*, Kolbenschieberkasten *m*
p. valve cylinder Steuerzylinder *m*
p. valve gear Kolbenschiebersteuerung *f*
p. valve liner eingesetzte Kolbenschieberbüchse *f*, Kolbenschiebereinsatz *m*
p. water-meter Kolbenwassermesser *m*
p. with slipper Schlittenkolben *m*
p. working alternately wechselläufiger Kolben *m*
p. working in opposite direction gegenläufiger Kolben *m*
p. working in the same direction gleichläufiger Kolben *m*

pit / to anfressen, angreifen, einfressen, zerfressen; auskolken *(Spanfläche)*
pit 1. Preßfehler *m*; Grube *f*; Kern *m*; Loch *n*; 2. *(Geol)* Schürfschacht *m*; 3. Box *f (Autorennen)*; 4. *(Kfz)* Reparaturgrube *f*
p. and quarry industry Industrie *f* für Steine und Erden
p. annealing *(Met)* Grubenglühen *n*
p.-bank *(Brg)* Hängebank *f*, Rasenhängebank *f*
p. bath *(Brg)* Waschkaue *f*
p. bin Tiefbunker *m*
p. bottom *(Brg)* Füllort *n*
p. bottom circuit Schachtumtrieb *m*
p. casting *(Met)* Grubengießen *n*
p. coal Steinkohle *f*
p. difference *(Brg)* Teufenunterschied *m*
p. fire Grubenbrand *m*
p. free from fire damp schlagwetterfreie Grube *f*
p. furnace *(Met)* Tiefofen *m*
p. gas Grubengas *n*
p. gate Schachttür *f*
p.-head *(Brg)* Hängebank *f*
p. heap *(Brg)* Bergehalde *f*
p. heating furnace *(Met)* Tiefofen *m*
p. hole *(Met)* Blasenloch *n (Block)*; Schlackenloch *n (Stahl)*
p. lands Grubengelände *n*
p. man Bergarbeiter *m*
p. mouth Schachtöffnung *f*, Schachtmündung *f*, Schachteinfahrt *f*
p. planer Grubenhobelmaschine *f*
p. prop Grubenstempel *m*; Strebe *f*
p. props Stempelholz *n*, Grubenholz *n*
p. rope *(Brg)* Förderseil *n*
p. sample *(Met)* Schmiedeprobe *f*, Schöpfprobe *f (Flußeisen)*
p. sand Grubensand *m*
p. saw Brettersäge *f*, Schrotsäge *f*, Blocksäge *f*, Langschnittsäge *f*, Kransäge *f*
p. saw tiller Kransägeangel *f*
p.-sleeper Grubenschwelle *f*
p. tannage *(Led)* Grubengerbung *f*
p.-top *(Brg)* Hängebank *f*, Rasenhängebank *f*
p.-type furnace *(Met)* Tiefofen *m*
p. worker Grubenarbeiter *m*
p. yard Zechenplatz *m*
pitch / to 1. errichten, aufstellen; 2. pflastern; 3. werfen, schleudern; fallen, sich senken, sich neigen; 4. bepichen *(Küfnerei)*; 5. verpichen, kleben, teeren; 6. stampfen und setzen *(Schiff)*; 7. stimmen *(Musikinstrument)*
p. and heave stampfen *(Schiff)*
p. at anchor vor Anker stampfen
pitch 1. Pech *n*, Teer *m*; Kunstasphalt *m*; 2. Ganghöhe *f*; Teilung *f (Zahnrad)*; Steigung *f (Gewinde)*; Nietabstand *m (Stahlbau)*; Lochabstand *m (Film)*; Abstand *m*; Flügelsteigung *f (Propeller)*; 3. *(Ak)* Tonhöhe *f*; 4. *(Brg)* Einfallen *n*, Neigung *f*; 5. *(Bw)* Gefälle *n*, Neigung *f*, Steigung; 6. Wurfhöhe *f (Häckselmaschine)*; 7. Stich *m (Vliesstoffherstellung)*; 8 *(El)* Wickelschritt *m*
p. angle Teilkegelwinkel *m (Zahnrad)*
p. attitude Inklinationswinkel *m (Radar)*
p. cam *(Masch)* Steigungskurve *f*
p.-changing range Steigungsverstellbereich *m*
p. circle Teilkreis *m*, Wälzkreis *m (Verzahnung)*; Schraubenlochkreis *m*
p. circle diameter Teilkreisdurchmesser *m*; Lochkreisdurchmesser *m*
p. circle of generation Erzeugungswälzkreis *m (Zahnrad)*
p. circumference Teilungsumfang *m (Zahnrad)*
p. coal Pechkohle *f*, Asphaltkohle *f*
p. cone Teilkegel *m (Zahnrad)*
p. cone apex Spitze *f* des Teilkreiskegels
p. control *(Fs)* Regelung *f* der Zeilendichte
p. cylinder Teilzylinder *m (Schnecke)*
p. diameter Flankendurchmesser *m (Gewinde)*; Teilkreisdurchmesser *m (Zahnrad)*, Wälzkreisdurchmesser *m*
p. diameter of tapped hole Flankendurchmesser *m* des Innengewindes
p. down *(Flg)* Abkippen *n*
p. error Steigungsfehler *m*; Teilfehler *m (Zahnrad)*
p. factor *(Elm)* Sehnungsfaktor *m (der Wicklung)*
p. gyro *(Rak)* Richtgeber *m* D
p. line Teilkreis *m*, Zahnteilbahn *f*
p. main *(Brg)* Pechvorlage *f*
p. measuring instrument Zahnteilungslehre *f*, Zahnteilungsmesser *m*
p. of a spiral Spiralsteigung *f*
p. of a thread Gewindesteigung *f*
p. of a tone *(Ak)* Tonhöhe *f*
p. of a winding *(Elm)* Wicklungsschritt *m*
p. of deflection sag *(Mech)* Biegungspfeil *m*
p. of dovetails Zinkenteilung *f*
p. of links Bügelabstand *m (Stahlbetonbau)*
p. of rivets Nietentfernung *f*, Nietabstand *m*
p. of spindles *(Text)* Spindelteilung *f*
p. of spiral Ganghöhe *f (Spiralbewehrung; Stahlbetonbau)*
p. of stay bolts *(Eis)* Stehbolzenteilung *f*
p. of the chain Kettenteilung *f*
p. of the creel *(Text)* Gatterteilung *f*
p. of the signal note Tonhöhe *f* des Zeichens, Zeichenhöhe *f*
p. of the teeth of a knurl Kordelteilung *f*
p. pipe Stimmpfeife *f (Orgel)*
p. pocket Harztasche *f (Holz)*
p. point Wälzpunkt *m (Zahnrad)*
p. range *(Flg)* Steigungsverstellbereich *m*

p. ratio Steigungsverhältnis *n (eines Propellers)*
pitchblende *(Min)* Pechblende *f*, Uranpecherz *n*, Nasturan *n*
pitched roof geneigtes Dach *n*, Schrägdach *n*, Steildach *n*
p. truss Dreieckbinder *m*, Dreieckfachwerk *n*; Trapezbinder *m*, Trapezfachwerk *n*
pitchfork 1. Forke *f*; Heugabel *f*; 2. Stimmgabel *f*
pitching 1. Aufstellen *n*, Errichten *n*; 2. Schleudern *n*; 3. Packlage *f (Straßenbau)*; Pflasterung *f (Böschungen)*; 4. Stampfen *n* und Setzen *n (eines Schiffes)*; Stampfschwingungen *fpl*
p. moment Längsmoment *n*
pitchpine Pechtanne *f*, Pechkiefer *f*
pitchy pechartig
pith Kern *m (von Bäumen)*; Mark *n (des Holzes)*
p. ball Holundermarkkügelchen *n*
p.-ball electroscope Holundermark[kugel]-elektroskop *n*
p. paper Reispapier *n*
pithead bank *(Brg)* Hängebank *f*, Rasen[hänge]-bank *f*
p. bath Waschkaue *f*
pitman 1. Verbindungsstange *f*; 2. Zugstange *f (Seilbohren)*; 3. Grubenarbeiter *m*, Bergmann *m*
p. arm *(Kfz)* Lenkstockhebel *m*, Lenkschnekkenarm *m*
Pitot tube Pitotsches Rohr *n*, Pitotrohr *n*, Staurohr *n*
Pitot-tube flowmeter Staudruckdurchflußmesser *m*, Durchflußmesser *m* mit Pitotrohr
pitted narbig *(Oberfläche)*
p. contact *(El)* [ausgefressener] Kontaktkrater *m*
p. structure *(Ch)* zellenartige Struktur *f*
p. surface *(Ch)* zerklüftete Oberfläche *f*
pitticite *s.* pittizite
pitting Grübchenbildung *f*; Korrosion *f*; Anpressung *f*; Lochfraß *m*; Auskolkung *f (Spanfläche)*
p. and corroding Kavitation *f*
p. corrosion Lochfraßkorrosion *f*
pittings Lochfraßkorrosion *f*
pittizite *(Min)* Pittizit *m*, Arseneisensinter *m*
pivot / to gelenkig verbinden (aufhängen), [sich] um einen Zapfen drehen; auf der Stelle drehen *(Raupenfahrzeuge)*
p. about gelenkig drehen um
p. around drehbar sein um
pivot Zapfen *m*, Tragzapfen *m*, Drehzapfen *m*, Spurzapfen *m*, Wellenzapfen *m*; Angel *f*, Türangel *f*, Angelzapfen *m*; Drehpunkt *m*
p. axis Drehachse *f*, Drehpunkt *m*
p. bearing Stützzapfenlager *n*
p. bolt Drehbolzen *m*, Lenkbolzen *m*
p. bridge Drehbrücke *f*
p. collar Zapfenkragen *m*
p. joint Zapfengelenk *n (Brückenbau)*
p.-jointed bar Zapfengelenklappe *f*
p. journal Spurzapfen *m*, Stützzapfen *m*
p. lever attachment bewegliche Hebelvorrichtung *f*
p. mounting *(Mil)* Drehzapfenlafette *f*
p. nozzle Zapfendüse *f*
p. pin Drehbolzen *m*, Kippzapfen *m*, Lagerzapfen *m*, Achsschenkelbolzen *m*
p. point Lagerspitze *f*; Drehpunkt *m*
p.-point setscrew Druckschraube *f* mit Zapfen, Stellschraube *f* mit abgerundetem Zapfen
p. pressure Zapfendruck *m*
p. suspension Zapfenaufhängung *f*
p. window Schwingflügelfenster *n*
pivotal centre Drehstern *m (Drehscheibe)*
p. fault *(Geol)* Drehverwerfung *f*
pivotally mounted gelenkig angebracht
pivoted [drehbar] gelagert, drehbar angelenkt (eingesetzt)
p. arm Schwenkarm *m*
p. bogie Drehschemel *m*
p. lever Schwenkhebel *m*, drehbar gelagerter Hebel *m*
p. link *(Masch)* Kurbelschwinge *f*, Kulisse *f*
p.-pad thrust bearing *(Masch)* Klotzlager *n*, Segmentlager *n*
p. sash Schwingflügel *m (Fenster)*
p. segmental thrust bearing Kippsegmentlager *n*
pivoting Drehung *f*, Zapfenlagerung *f*, drehbare Lagerung *f*
pivoting drehbar, schwenkbar
p. friction Bohrreibung *f*, bohrende Reibung *f*
place / to auflegen, anlegen, einlegen; anordnen; verlegen
p. a call *(Fmt)* ein Gespräch anmelden
p. a voltage across *(El)* eine Spannung anlegen an
p. ahead *(El)* vorschalten
p. an order with bestellen bei
p. concrete Beton einbringen, betonieren
p. in circuit einschalten *(in einen Stromkreis)*
p. in parallel *(El)* parallelschalten
p. in series *(El)* in Reihe schalten, hintereinanderschalten
p. on auflegen
p. out of service *(Fmt)* blockieren, sperren
p. the core den Kern einlegen *(Formerei)*
p. the satellite into its orbit den Satelliten auf seine Bahn bringen
place Ort *m*; *(Math)* Stelle *f*
p. gnawed by rats *(Led)* Rattenfraß *m (Fehler)*
p. value Stellenwert *m*

placement of concrete Einbringen *n* (Einbringung *f*) des Betons, Betonieren *n*, Betonierung *f*
placer [goldführende] Sandbank *f*
p. gold Seifengold *n*, Waschgold *n*
p. mining Seifenbergbau *m* *(Goldwäscherei)*
placing 1. Aufsetzen *n*, Aufstellen *n*, Auflegen *n*; Anordnen *n*; 2. Verlegen *n*, Einbau *m*
p. of concrete Einbringen *n* (Einbringung *f*) des Betons, Betonieren *n*, Betonierung *f*
p. of the rocket into its orbit Einsteuern *n* der Rakete auf ihre Flugbahn
plagioclase *(Min)* Plagioklas *m*
plagionite *(Min)* Plagionit *m*
plain *(Geol)* Ebene *f*, Flachland *n*
plain einfach, flach, eben; unlegiert *(Stahl)*; glatt *(Walze)*; blank *(Draht)*; einfarbig
p. aerial Einfachantenne *f*
p. annealed wire [blank]geglühter Draht *m*
p. back shovel flache Schaufel *f*
p. bar glatter Bewehrungsstab (Stabstahl) *m* *(Stahlbetonbau)*; Flachstab *m*
p. bearing Gleitlager *n*, Zapfenlager *n*
p. bearing plummer block Deckellager *n* als Gleitstehlager
p. bolt nicht abgesetzter Bolzen *m*
p. carbon steel unlegierter Stahl *m*
p. carbon structural steel unlegierter Baustahl *m*
p. carbon tool steel unlegierter Werkzeugstahl *m*
p. circular pattern wheel einfonturige Rundstrickmaschine *f* mit Ringeleinrichtung und Musterrädern
p. circular spring-beard needle knitting machine einfonturige Rundstrickmaschine *f* mit Hakennadeln
p. circular strong border knitting machine einfonturige Leistenrundstrickmaschine *f*
p. cloth glattes Gewebe *n*
p. coil chain steglose Rundgliederkette *f*
p. concrete unbewehrter Beton *m*
p. conductor blanker Leiter *m*; unverzinnter Leiter
p. contouring Langnachformdrehen *n*
p. cutter Walzenfräser *m*
p. cylindrical grinding machine Außenrundschleifmaschine *f*
p. detonator einfache Sprengkapsel *f*
p. die offenes Gesenk *n*
p. dividing apparatus einfacher Teilapparat *m*
p. draw wire gezogener ungeglühter Draht *m*
p. edged besäumt *(Holz)*
p.-ended pipe Glattrohr *n* *(Rohr ohne Muffe bzw. Flansch)*
p. feather geradstirnige Keilnutenfeder *f*
p. finish / to *(Led)* glatt zurichten
p. fishplate Flachlasche *f*
p. fixed-bed type milling machine einspindlige Bettfräsmaschine *f*
p. fuse *(El)* offene Sicherung *f*
p. grate Planrost *m*
p. grinding Rundschleifen *n*
p. head einfacher Spindelkasten *m* *(Werkzeugmaschinen)*
p. indexing *(Masch)* einfaches Teilen *n*
p. jolt-moulding machine Tischrüttelformmaschine *f*
p. key Keil *m* ohne Nase
p. knit Flachstrickerei *f*
p. language Klartext *m*
p. lead covered cable Kabel *n* mit blankem Bleimantel
p. loom Glattwebstuhl *m*
p. milling cutter Walzenfräser *m*
p. milling machine Einfachfräsmaschine *f*
p. nut Universalschraubenmutter *f*
p. parallel key Paßfeder *f*
p. plug gauge Normallehrdorn *m*, Lehrdorn *m* *(für Bohrungen)*
p. printing *(Typ)* Flachdruck *m*
p. protractor Transporteur *m*
p. radial drilling machine einfache Radialbohrmaschine *f*
p. ring gauge Normallehrring *m*, Lehrring *m* [für Wellen]
p. roll glatte Walze *f*, Glattwalze *f*, Walze mit glattem Ballen
p.-roll crusher Glattwalzwerk *n*
p. roller Glattwalze *f*
p. sawn im Hochschnitt *(Holz)*
p. screwdriver Werkstattschraubenzieher *m*
p. sectioning einfache Schraffur *f*
p. single-bit cutter Einzahn[schlag]fräser *m*
p. slide valve Flachschieber *m*
p. speech *(Fmt)* offene Sprache *f*
p. steel unlegierter Stahl *m*
p. steel bar glatter Bewehrungsstab (Stabstahl) *m* *(Stahlbetonbau)*
p. stitch Rechtsmaschine *f* *(Strickerei)*
p. stocking cutter Zahnformvorfräser *m*
p. superheating tube with flat diaphragm glattes Überhitzerdoppelrohr *n* mit gerader Trennungswand
p. superheating tube with twisted cross-shaped diaphragm glattes Überhitzerrohr *n* mit [eingewalzter] verdrehter Kreuzleiste
p. superheating tube with twisted diaphragm glattes Überhitzerdoppelrohr *n* mit verdrehter Trennungswand
p. surface machine eigenbelüftete Maschine *f* mit glatter Oberfläche
p. taper stock key flacher angeschrägter Keil *m*
p. tile Biberschwanz *m* *(Dachziegel)*

p. tuning error Abstimmpeilfehler *m (Radar)*
p. turning Langdrehen *n*
p. type combination centre (centring) drill Zentrierbohrer *m* ohne Schutzsenkung
p. washer [glatte] Unterlegscheibe *f*
p. weave Leinwandbindung *f*, Grundbindung *f*, Taftbindung *f*
plaining fin *(Flg)* Gleitflosse *f*
plait / to flechten; falten
p. round umflechten
plait Flechte *f*, Zopf *m*
p. point *(Kern)* kritischer Entmischungspunkt *m*
plaited filter Faltenfilter *n*
p. pattern Zopfmuster *n*
p. work Flechtwerk *n*, Geflecht *n*
plaiting Flechtwerk *n*, Geflecht *n*
p. machine Litzenmaschine *f*
plan / to planen; entwerfen, projektieren
plan Plan *m*; Grundriß *m*, Entwurf *m*; Lageplan *m*; Zeichnung *f*
p. angle Einstellwinkel *m (Meißel)*
p. chest Zeichnungsschrank *m*
p. elevation Draufsicht *f*
p. of cable layout Kabellageplan *m*
p.-position indicator Bildschirm *m*; Rundsichtanzeiger *m (Radartechnik)*
p. section Waagerechtschnitt *m*
p. target *(Ök)* Planziel *n*
p. view Ansicht *f* im Grundriß; Grundriß *m (technisches Zeichnen)*
planamilling *(Am)* Fräsen *n* mit Planetenspindelfräsmaschine
planar plan[ar], eben[flächig]
p. coordinates ebene Koordinaten *fpl*
p. diode planparallele Diode *f*, Planardiode *f*
p. junction planarer Übergang *m (Halbleitertechnik)*
p. network *(El)* ebenes Netzwerk *n*
p. transistor Planartransistor *m*
p. triode *(El)* Planartriode *f*
planathreading *(Am)* Gewindefräsen *n* mit Planetenspindelgewindefräsmaschine
planchet Platine *f*, Zuschnitt *m*, Rohling *m*
Planck's constant Plancksche Konstante *f*, [Plancksches] Wirkungsquantum *n*
Planck's function Plancksche Funktion *f*
Planck's law of radiation Plancksches Strahlungsgesetz *n*
plane / to [ab]hobeln, ebnen, glätten; gleiten
plane 1. Ebene *f*, ebene Fläche *f*; 2. Hobel *m*; 3. *(Brg)* Förderstrecke *f*
plane plan, eben, flach
p. aerial Flächenantenne *f*
p. annulus ebener Ring *m*
p. cathode Plankatode *f*, Flachkatode *f*
p.-centred lattice flächenzentriertes Gitter *n (Kristalle)*
p. coil *(El)* Scheibenspule *f*
p.-concave lens plankonkave Linse *f*, Plankonkavlinse *f*
p.-convex lens plankonvexe Linse *f*, Plankonvexlinse *f*
p. curve ebene Kurve *f*
p. cutter Hobeleisen *n*
p. diagonal Flächendiagonale *f*
p. electrode Planelektrode *f*
p. geometry Planimetrie *f*
p. grating *(Opt)* Plangitter *n*
p. hole *(Holz)* Spanloch *n*
p. iron Hobeleisen *n*, Hobelmesser *n*
p. knife Hobeleisen *n*, Hobelmesser *n*
p. lattice Flächengitter *n*, ebenes Gitter *n*
p. mirror ebener Spiegel *m*, Planspiegel *m*
p. of admission Beaufschlagungsebene *f (Turbine)*
p. of bedding *(Geol)* Schichtfläche *f*
p. of bending Biegeebene *f*
p. of comparison Vergleichsfläche *f*
p. of division Trennungsebene *f*, Teilungsebene *f (Turbine)*
p.-of-flight reference *(Flg)* Leitebene *f*
p. of flotation Schwimmebene *f*
p. of forces Kraftebene *f*
p. of fracture Bruchebene *f*
p. of grid Gitterebene *f*
p. of incidence Einfallsebene *f*
p. of joint Trennungsebene *f*, Teilungsebene *f (Turbine)*
p. of loading Belastungsebene *f*
p. of object Objektebene *f (Mikroskopie)*
p. of oscillation Schwingungsebene *f*
p. of polarization Polarisationsebene *f*
p. of projection Projektionsebene *f*
p. of reference *(Opt)* Einstellebene *f*, Bezugsebene *f*
p. of refraction Brechungsebene *f*
p. of rotation Drehungsebene *f*, Ebene *f* der Drehbewegung, Rotationsebene *f*
p. of rupture Bruchebene *f*
p. of shear Scherebene *f*, Scherfläche *f*
p. of sliding *(Geol)* Rutschebene *f*
p. of steering *(Kfz)* Steuerungsebene *f*
p. of stratification *(Geol)* Schichtungsebene *f*
p. of symmetry Symmetrieebene *f*
p. of the force system Zentralebene *f* des Kräftesystems
p. of the orbit Bahnebene *f*, Flugbahnfläche *f*, Bahnfläche *f*
p. of the outlet opening Mündungsebene *f (Turbine)*
p.-parallel planparallel
p. polarized linear polarisiert
p. sinusoidal wave ebene Sinuswelle *f*
p. spiral Planspirale *f*
p. stock Hobelgehäuse *n*, Hobelkasten *m*

p. stress ebener (zweiachsiger) Spannungszustand *m*
p. surface bearing Gleitlager *n*
p. surveying niedere Geodäsie *f*
p. table Meßtisch *m*
p. truss ebenes Fachwerk *n*
p. wave ebene Welle *f*
p. with single iron Schlichthobel *m (für Holzbearbeitung)*
planed measure *(Holz)* Spiegelmaß *n*, schmale Seite *f (unbesäumte Ware)*
p. sawn wood gehobeltes Schnittholz *n*
planer *(Masch)* Hobelmaschine *f*; *(Holz)* Abrichthobel *m*, Hobelmaschine
p. bed Hobelmaschinenbett *n*
p. cutting tool Hobelmeißel *m*
p. gauge Hobelmeißeleinstellehre *f*
p. head Hobelmaschinensupport *m*
p. platen Hobelmaschinentisch *m*
p. tool Langhobelmeißel *m*, Hobelmeißel *m*
p. tool holder Hobelmeißelhalter *m*
p.-type horizontal boring mill Waagerechtbohr- und -fräswerk *n* mit hobelmaschinenähnlichem Tisch
p.-type milling machine Bettfräsmaschine *f*, Langfräsmaschine *f*, Portalfräsmaschine *f*
p.-type table hobelmaschinenähnlicher Tisch *m*
planet Planet *m*
p. aberration planetarische Aberration *f*
p. carrier Planetenradträger *m*
p. pinion Planetenrad *n*
p. wheel umlaufendes Rad *n*, Planetenrad *n*
planetarium Planetarium *n*
planetary planetarisch, Planeten-
p. change-can mixer Planetenrührwerk *n* mit Wechselbehälter
p.-drive system Planetengetriebe *n (Verzahnung)*
p. electron Hüllenelektron *n*
p. gear Planetenzahnrad *n*
p. gear system Planetengetriebe *n*
p. gearing Planetengetriebe *n*
p. paddle mixer Planetenrührwerk *n*
p. transmission Planetengetriebe *n*
p.-type [internal] grinding machine Planetenspindelschleifmaschine *f*
p. wheel Planetenrad *n*
p. wheel spindle Planetenschleifspindel *f*
planimeter Planimeter *n*
planimetric distance Grundrißstrecke *f*
planimetry ebene Geometrie *f*, Planimetrie *f*
planing 1. Planieren *n*, Ebnen *n*, Glätten *n*, Einplanieren *n*, Einebnung *f*; 2. *(Masch)* Hobeln *n*; 3. *(Brg)* Hobeln, Schälen *n (Kohle)*
p. attachment Hobeleinrichtung *f*
p. chips Hobelspäne *mpl*
p. chisel Putzbeitel *m (Holzbearbeitung)*
p. craft Gleitboot *n*
p. crosswise Querhobeln *n*
p. fixture Hobelvorrichtung *f*
p. lengthwise Längshobeln *n*
p. machine Hobelmaschine *f*
p. machine for parquetry fillets Parkettstabhobelmaschine *f*
p. machine knife Hobelmesser *n (Holzbearbeitung)*
p. machine with cylindrical cutters Walzenhobelmaschine *f (Holzbearbeitung)*
p. machine with horizontal and vertical cutter blocks Walzenhobelmaschine *f* mit waagerechten und senkrechten Messerwellen
p. machine with independent cutter head Walzenhobelmaschine *f*
p.-mill products Sägewerkartikel *mpl*
p. off Abhobeln *n*
p. setup Aufspannung *f* auf dem Hobelmaschinentisch
p. speed Gleitgeschwindigkeit *f*
p. tool Hobelmeißel *m*
planish / to polierschlagen, glattwalzen
planished sheet poliertes Blech *n*
planisher Richtmaschine *f*; Vorschlichtkaliber *n (für Bleche)*
planishing Fertigpolieren *n (Walzgut)*; Planieren *n*, Flachstanzen *n*; Glätten *n*
p. hammer Schlichthammer *m*
p. mill Polierwalzwerk *n*
p. pass Schlichtkaliber *n*
p. roll Polierwalze *f*, Vorschlichtwalze *f*
p. stand Poliergerüst *n*
planispiral flachgewunden
plank / to dielen; verschalen; übertäfeln, verkleiden
plank 1. Planke *f*, Bohle *f*; Rohling *m*; 2. *(Brg)* Abtreibepfahl *m (Anstecken)*
p. lining Bretterverschalung *f (Betonbau)*
p. saw Bohlensäge *f*, Plankensäge *f*
p. sheer *(Schiff)* Schandeck *n*
p. strake *(Schiff)* Plankengang *m*
p. timbers Bohlenholz *n*, Plankenholz *n*
planking Dielung *f*; Bohlenbelag *m*; Verschalung *f*
plankway Bohlenweg *m*
planned economy Planwirtschaft *f*
p. figure Planzahl *f*
planner Planer *m*; Städteplaner *m*
planning Planung *f*; Projektierung *f*
p. building permission Baugenehmigung *f*
p. manager Planungsleiter *m*
p. technology Technologie *f*
plano-concave plankonkav
p.-convex plankonvex
p.-milling machine Bettfräsmaschine *f*, Langfräsmaschine *f*
p.-parallel planparallel
planographic printing Flachdruck *m*

planometer Richtplatte *f*
plant / to pflanzen; bepflanzen; ansiedeln; einsetzen; anreißen
plant 1. Anlage *f*, Betriebsanlage *f*; *(Am)* Betrieb *m*, Fabrik *f*, Werk *n*; 2. Pflanze *f*
 p. chemist *(Am)* Betriebschemiker *m*
 p. cultivation Pflanzenzucht *f*
 p. equipment Betriebseinrichtung *f*, Werkseinrichtung *f*
 p. fiber Pflanzenfaser *f*
 p. nursery Gewächshaus *n*; Pflanzenzucht *f*
 p. pigment Pflanzenfarbstoff *m*
 p. protective Pflanzenschutzmittel *n*
 p. security *(Am)* Betriebsschutz *m*
 p. units Betriebsaggregat *n*; einzelne Teile *npl* einer Anlage
 p. vermin Pflanzenschädlinge *mpl*
plantation Plantage *f*, Anpflanzung *f*, Pflanzung *f*
planted area Pflanzenbestand *m*, Nutzpflanzenbestand *m*
planter Züchter *m (Pflanzen)*; Pflanzmaschine *f*
planting drill Pflanzlocher *m*
 p. hoe Pflanzhacke *f*
 p. machine Pflanzmaschine *f*
plasma 1. *(Ph)* Plasma *n*; 2. *(Min)* Plasma *m*
 p. accelerator Plasmabeschleuniger *m*
 p. arc torch Plasmalichtbogenbrenner *m*
 p. balance Plasmagleichgewicht *n*
 p. beam Plasmastrahl *m*
 p. drive Plasmaantrieb *m*
 p. frequency Plasmafrequenz *f*
 p. ion source Plasmaionenquelle *f*
 p. of gas discharge Gasentladungsplasma *n*
 p. physics Plasmaphysik *f*
 p. propulsion Plasmaantrieb *m*
 p. sphere Plasmakugel *f*
 p. torch Plasmabrenner *m*
 p. wave Plasmawelle *f*
plasmolysis *(Ch)* Plasmolyse *f*
plastainer Kunststoffbehälter *m*
plaster / to [ver]putzen
 p. over übergipsen
plaster 1. Putz *m*; Putzmörtel *m*; Gips[mörtel] *m*; 2. Pflaster *n (Verbandsmaterial)*
 p. board Gipsbauplatte *f*, Gipsdiele *f*
 p. burning Gipsbrennen *n*, Gipsbrennerei *f*
 p. cast Gipsabdruck *m*
 p. coating Gipsüberzug *m*
 p. floor Gipsestrich *m*
 p. mortar Gipsmörtel *m*
 p. mould Gipsform *f (Formerei)*
 p. of Paris gebrannter Gips *m*, Stuckgips *m*
 p. pattern Gipsmodell *n*, Gipsschablone *f*
 p. powder Gipsmehl *n*
 p. priming solution Isoliervorstrich *m* für Zementuntergrund
 p.-shooting *(Brg)* Auflegerschießen *n*, Schießen *n* mit aufgelegter Ladung
 p. slab Gips[bau]platte *f*
 p. stone *(Min)* Gipsstein *m*
 p. work Stuckarbeit *f*
plasterboard *s.* plaster board
plasterer Putzer *m*; Stukkateur *m*
plastering Innenputz *m*; Putzen *n*, Verputzen *n*
 p. mix Putzmörtel *m*
 p. trowel Putzerkelle *f*
plasterwork Putz *m*, Putzarbeit[en *fpl*] *f*
plastic Kunststoff *m*, Plast *m*
plastic plastisch, bildsam, knetbar, [ver]formbar, geschmeidig, nachgiebig
 p. adhesive Kunststoffkleber *m*
 p. alloy Knetlegierung *f*
 p. binder *(Bw)* plastisches Bindemittel *n*
 p.-bodied mit Plastkarosserie
 p. cable Kunststoffkabel *n*
 p. cemented sleeve joint Einsteckklebeverbindung *f*
 p. clay Bildhauerton *m*, verformbarer Ton *m*
 p.-coated mit Plastüberzug
 p. deformation plastische Verformung (Deformation) *f*, bleibende Verformung
 p. effect *(Fs)* Verwaschung *f*, Bildplastik *f*
 p. engineering Plasttechnik *f*
 p. fabric Kunststoffgewebe *n*
 p. film Kunstharzfolie *f*
 p. flow *(Met)* Kriechen *n*; Fließen *n*
 p. foam Schaumstoff *m*
 p. foil Plastfolie *f*
 p. foil capacitor *(El)* Kunststoffolienkondensator *m*
 p. fracture *(Met)* Verformungsbruch *m*, Bruch *m* nach Verformung
 p. hearing *(Ak)* plastisches Hören *n*
 p.-hinge method *(Bw)* Fließgelenkverfahren *n*
 p. limit Plastizitätsgrenze *f*
 p. material Plast *m*, Kunststoff *m*
 p. metal joining process *(Am)* Metallklebeverfahren *n*
 p. mill *(Am)* Walze *f* für Kunststoff
 p. picture Raumbild *n*
 p. range unelastischer Bereich *m*, Fließbereich *m*, Plastizitätsbereich *m*
 p. recovery plastische Formänderung *f*
 p. resin Kunstharz *n*
 p. scintillator *(Kern)* Kunststoffszintillator *m*, Plastszintillator *m*
 p. slow persistence after flow *(Ph)* plastische Nachwirkung *f*
 p. stability theory *(Schiff)* Theorie *f* der plastischen Stabilität
 p. strain bleibende Dehnung *f*, Dehnungsrest *m*
 p. theory Plastizitätstheorie *f*, Traglastverfahren *n (Stahlbetonberechnung)*

p. way Führungsbahn *f* mit Kunststoffauflage
p. welding Feuerpreßschweißen *n*
p. wood Tischlerkitt *m*
p. workability Formänderungsvermögen *n*
plasticator *s.* plasticizer
plasticine Plastilin *n*, Knetmasse *f*
plasticity Verformbarkeit *f*, Formbarkeit *f*, Bildsamkeit *f*, Plastizität *f*, Geschmeidigkeit *f*
p. range unelastischer Bereich *m*, Bereich plastischer (bleibender) Formänderung *(Festigkeit)*
plasticize / to erweichen, plastifizieren, weich machen
plasticizer Weichmacher *m*, Plasti[fi]kator *m*, Betonverflüssiger *m*
p. migration Weichmacherwanderung *f*
plasticizing capacity Plastifizierleistung *f*, Verflüssigungsleistung *f*
p. efficiency weichmachende Wirkung *f*; Wirksamkeit *f* des Weichmachers
plasticodynamics Dynamik *f* plastischer Massen
plasticostatics Statik *f* plastischer Massen
plastics Plaste *mpl*, Kunststoff *m*, Preßstoff *m*
p. plating Metallüberzug *m* auf Kunststoff
p. press Kunststoffpresse *f*
p. sheet Kunststoffolie *f*
p. sheeting Kunststoffolie *f*
p. technology Plasttechnik *f*
plastify / to plastifizieren, weich machen, erweichen
plastifying agent Plastifikator *m*, Betonverflüssiger *m*
plastograph Plastograf *m*
plastomer Kunststoff *m*
plastometer Plastizitätsmesser *m*
plat *(Brg)* Anschlag *m*, Füllort *n*
plate / to plattieren, überziehen, mit Schutzschicht versehen; auskleiden; galvanisieren; überfangen *(Glas)*
plate Platte *f*; Blatt *n*; Blech *n*; Grobblech *n*, Mittelblech *n*; Scheibe *f*; Teller *m*; Schild *n*; *(Ch)* Boden *m (Destillation)*; Anode *f*; Kantholz *n*, Schwelle *f*; *(Typ)* Druckplatte *f*
p. bar Platine *f*
p. battery Anodenbatterie *f*
p. bending machine Blechbiegemaschine *f*
p. black *(Typ)* Kupferdruckschwarz *n*
p. bow *(Schiff)* Plattensteven *m*, Plattenbug *m*
p. cam *(Masch)* Scheibenkurve *f*
p. camera Plattenkamera *f*
p. capacitor *(El)* Plattenkondensator *m*
p. cathode Flächenkatode *f*
p. characteristic Anodenstrom-Anodenspannungs-Kennlinie *f*
p. circuit Anodenkreis *m*
p. clipping Anodenbeschneidung *f*
p. clutch Scheibenkupplung *f*
p. coil Anodenkreisspule *f*
p. column Bodenkolonne *f (Destillation)*
p. conductance Anodenleitwert *m*, Anodenableitung *f*
p. conveyor Plattenförderer *m*
p. coupling Scheibenkupplung *f*
p. culture *(Ch)* Plattenkultur *f*
p. current Anodenstrom *m*
p. current characteristic Anodenstromkennlinie *f*
p. current envelope Anodenstromverlauf *m*
p. current variation Anodenstromänderung *f*
p. cutters Blechschere *f*
p. cutting machine Blechschere *f*
p. decating *(Text)* Plattendekatur *f*
p. decoupling resistor Anodenentkoppelwiderstand *m*, Entkopplungswiderstand *m*
p. detection Anodengleichrichtung *f*
p. detector Anodengleichrichter *m*
p. die Gewindeschneidbacke *f*
p. distillation *(Ch)* Fraktionierung *f*, stufenweise Destillation *f*
p. doubler Blechdoppler *m*
p. drier Tellertrockner *m*
p. drop Anodenverlustleistung *f*, Anodenbelastung *f*
p.-dropping resistor *(Am)* Anodenschutzwiderstand *m*
p.-edge planing machine Blechkantenhobelmaschine *f*
p. edge profiling Blechkantennachformarbeit *f*
p. efficiency Anodenwirkungsgrad *m (von Röhren)*
p. efficiency factor Bodenwirkungsgrad *m (Destillation)*
p. electrode Plattenelektrode *f*
p. electrometer Plattenelektrometer *n*
p. feedback Anodenrückkopplung *f*
p. filter *(Ch)* Zellenfilter *n*
p. folding press Abkantpresse *f*
p. fuse *(El)* Streifensicherung *f*
p. gauge Grobblechlehre *f*, Blechlehre *f*
p. girder *(Bw)* Blechträger *m*, Vollwandbinder *m*
p. glass Tafelglas *n*, Flachglas *n*, Spiegelglas *n*
p.-graining *(Typ)* Plattenkörnen *n*
p. gravure printing *(Typ)* Plattentiefdruck *m*
p. grid Plattengitter *n (Batterie)*
p. heating furnace Blechglühofen *m*
p. holder Plattenhalter *m*, Kassette *f (Fotografie)*
p. impedance Gleichstromwiderstand *m (einer Röhre)*
p. insulator Tellerisolator *m*
p. jig Bohrplatte *f*
p. joggler Stauchverschränkungsmaschine *f*
p. keying *(Am)* Anoden[spannungs]tastung *f*, Tastung *f* im Anodenkreis

p. levelling machine Blechrichtmaschine *f*
p. lightning arrester Plattenblitzableiter *m*
p.-like plattenförmig
p. load resistance Anoden[belastungs]widerstand *m*
p. locking pin Haltestift *m* der Teilscheibe
p. mark Feingehaltsstempel *m (bei edelmetallhaltigen Gegenständen)*
p. mill Grobblechstraße *f*, schwere Blechstraße *f (Walzwerk)*
p. noise Anodenrauschen *n*
p. number Bodenzahl *f (Destillation)*
p. of a capacitor *(El)* Belegung *f* eines Kondensators; Kondensatorplatte *f*
p. orifice Meßblende *f*
p. planer Blech[kanten]hobelmaschine *f*
p. power detector Anodengleichrichter *m*
p. power input Anodeneingangsleistung *f (Produkt aus mittlerem Anodenstrom und Anodenspannung)*
p. power supply Anodenstromversorgung *f*
p. printing *(Typ)* Flachdruck *m*
p. protection anodischer Schutz *m (z. B. gegen Korrosion)*
p. roll Blechwalze *f*
p. rolling Blechwalzen *n*
p. rolling mill Grobblechwalzwerk *n*
p. scale Brückenwaage *f*
p. scrap Blechschrott *m*
p.-shaped tellerförmig
p.-shaped cutter Tellerfräser *m*
p.-shaped drill Tellerbohrer *m*
p. shear Blechtafelschere *f*, Tafelschere *f*
p. shearing machine *s.* p. shear
p. singer *(Text)* Plattensenge *f*
p. size Plattenformat *n (Fotografie)*
p. slab Bramme *f (Walzgut)*
p. spring Blattfeder *f*, Scheibenfeder *f*
p. stock Gewindeschneidkluppe *f*
p. straightener Blechrichtmaschine *f*
p. straightening press Blechrichtpresse *f*, Blechrichtmaschine *f*
p. strake *(Schiff)* Plattengang *m*
p. structure *(Bw)* Flächentragwerk *n*
p. supply Anodenstromversorgung *f*, Anodenquelle *f*
p. supply by-pass capacitor Anodenspannungssiebkondensator *m*
p. supply ripple kleine Schwankungen *fpl* der Anodenspannung, Anodenbrumm *m*
p. tank Anodenschwingkreis *m*
p. taper Plattenschnitt *m (Drehkondensator)*
p. thickness Blechdicke *f*
p. tuned circuit Anodenabstimmkreis *m*; abgestimmter Anodenkreis *m*
p.-type heat exchanger Plattenwärme[aus]tauscher *m*
p.-type lightning arrester Plattenableiter *m (z. B. für Gewitterüberspannungen)*
p.-type oil filter *(Kfz)* Spaltfilter *n*
p. valve Tellerventil *n*
p. vibrator Plattenrüttler *m*, Rüttelplatte *f*
p. voltage Anodenspannung *f*
p. voltage characteristic Anodenkennlinie *f*
p. voltage keying Anodentastung *f*
p. voltage swing Anodenspannungsamplitude *f*
p. welding Blechschweißen *n*
p. wheel Scheibenrad *n*
p. whirler *(Typ)* Plattenschleuder *f*

plateau 1. *(Geol)* Hochebene *f*, Plateau *n*; 2. *(Ph)* Plateau, Konstanzbereich *m*
p. characteristics *(Kern)* Zähl[rohr]charakteristik *f*
p. region *(Kern)* Plateau[bereich *m*] *n*

plated plattenförmig; plattiert, metallüberzogen
p. heel verstärkte Ferse *f (am Strumpf)*
p. mast *(Schiff)* Plattenmast *m*

platelayer's truck Bahnmeisterwagen *m*
p. wagon Oberbauwagen *m*

platelike plattenförmig

platen *(Am)* Platte *f* einer Etagenpresse, Pressentisch *m*; Maschinentisch *m*, Aufspanntisch *m*
p. press Plattenpresse *f*; *(Typ)* Tiegeldruckpresse *f*
p. release mechanism *(Büro)* Walzenfreilauf *m*
p. sliding surfaces Tischführungen *fpl*

plateworking machine Blechbearbeitungsmaschine *f*

platform 1. Bühne *f*, Plattform *f*, Bedienungsstand *m*, Laufbühne *f*, Podest *n (m)*, Rampe *f*, Podium *n*; Galerie *f*; Bahnsteig *m*
p. barrier Bahnsteigsperre *f*
p. conveyor Plattenförderer *m*
p. deck *(Schiff)* Plattformdeck *n*
p. lift Hebebühne *f*
p. outrigger *(Kfz)* Plattformträger *m*
p. roof Terrassendach *n*
p. scale Brückenwaage *f*
p. truck 1. Elektrokarren *m*; 2. *(Eis)* Rungenwagen *m*

platformate Produkt *n* des Platforming-Prozesses *(Erdöl)*

platforming Platforming *n*, Reformierungsprozeß *m* mit Platinkontakten *(Erdöl)*

platikurtic *(Math)* abgeflacht, abgeplattet *(bei der Häufigkeitskurve)*

platinate Platinat *n*, Salz *n* der Platinsäure

plating Galvanisieren *n*, galvanischer Überzug *m*
p. barrel Galvanisierfaß *n*
p. bath Galvanisierbad *n*
p. charge *(galvanischer)* Metallauftrag *m*
p. drum Galvanisiertrommel *f*
p. material Deckwerkstoff *m*, Deckmetall *n*

plumb lotrecht, im Lot
p. bob Lot *n*, Senkblei *n*
p. bow *(Schiff)* senkrechter Bug *m*
p. cord Maurerschnur *f*
p. line Schnurlot *n*
p. pile senkrechter Pfahl *m*
p. rod Stablot *n*
plumbago Graphit *m*
p. crucible Graphit[schmelz]tiegel *m*
plumbate Plumbat *n*
plumbeous bleiartig, bleiig
plumber 1. Klempner *m*, Rohrleger *m*, Installateur *m*; 2. *(Met)* Bleigießer *m*
p. block *s.* plummer block
plumber's solder Blei-Zinn-Weichlot *n*, Lötzinn *n*
p. tool Installationswerkzeug *n*, Klempnerwerkzeug *n*
plumbic compound Verbindung *f* des vierwertigen Blei
plumbiferous bleihaltig
plumbing 1. Einloten *n*, Loten *n*, Abloten *n*; 2. Installation *f*; Klempnerarbeit *f*; Rohrleitungen *fpl*; Rohrlegerarbeiten *fpl*, Rohrlegerhandwerk *n*; 3. Hohlleiterbau *m (Wellenleiter)*
p. core Installationskern *m*
p. unit Installationskern *m*, Installationszelle *f*
p. wall Installationswand *f*
plumbism Bleivergiftung *f*
plumbite Plumbit *n*
plumbous compound Verbindung *f* des zweiwertigen Blei
plume künstliches Echo *n (Radartechnik)*
plummer block Tunnelwellenlager *n*; Lauflager *n*, Traglager *n*, Lagerblock *m*; Bocklager *n*; Blocklager *n*
p. block bearing Rumpflager *n*
plummet 1. Lot *n*, Senkblei *n*, Senklot *n*; 2. Schwebekörper *m (Durchflußmesser)*
p. viscosimeter Schwebekörperviskosimeter *n*
plumping liquor *(Led)* Schwellfarbe *f*
plunge / to 1. [ein]tauchen, versenken; 2. [ab]-stürzen; 3. *(Masch)* einstechen
plunge battery *(El)* Tauchbatterie *f*
p.-cut grinding Einstechschleifen *n*
p.-cut grinding machine Einstechschleifmaschine *f*
p.-cut method Einstechverfahren *n (Schleifen)*
p.-cut motion *(Masch)* Einstechbewegung *f*
p. cutting Tauchfräsen *n*
p. grinding Einstechschleifen *n*
p. milling Tauchfräsen *n*
plunger 1. Plunger[kolben] *m*, Tauchkolben *m*, Druckkolben *m*, Kolben *m*; Riegelbolzen *m*; Stempel *m*, Stößel *m*; Ventilkolben *m*; Raststift *m*; 2. *(El)* Reglergestänge *n*; 3. *(El)* Kurzschlußkolben *m*, Kurzschlußschieber *m (Wellenleiter)*
p. block Sperriegel *m*
p. disk Kolbenscheibe *f*
p. door contact Haustür[öffnungs]kontakt *m*
p. electromagnet Tauchspulmagnet *m*
p. follower Kolbenscheibe *f*
p. moulding *(Plast)* Zweikolbenspritzpreßverfahren *n*
p. pin Raststift *m (Teilscheibe)*
p. piston Plungerkolben *m*, Tauchkolben *m*
p. poppet Kolbenkegel *m*
p. pump Tauchkolbenpumpe *f*, Plungerpumpe *f*
p. rod Druckstange *f (Bremse)*
p.-rod nut Kolbenmutter *f*
p. slide-type valve Kolbenschieber *m*
p. spring Kolbenfeder *f*
p. spring plate Kolbenfederteller *m*
p. stud Kolbenstutzen *m*
p. switch Schalter *m* mit Kontakttulpe
p.-type capacitor *(El)* Tauchkondensator *m*
p.-type coil Tauchspule *f*
p.-type pawl Sperrzahn *m*
p.-type relay Tauchankerrelais *n*, Tauchkolbenrelais *n*
p.-type slide valve Kolbenschieber *m*
p. valve Kolbenventil *n*, Pumpenkolbenventil *n*
p. working barrel Pumpenkolbenrohr *n*
plunging 1. Tauchen *n*; 2. *(Masch)* Einstechen *n*
p. battery Tauchbatterie *f*
p. siphon Stechheber *m*
plural Mehrfach-, mehr, mehrfach
p. formation Mehrfachbildung *f*
p. scattering Mehrfachstreuung *f*
plurivalent *(Ch)* mehrwertig
plus 1. *(Math)* plus; 2. *(El)* positiv
p. allowance [zulässiges] Übermaß *n*
p. electricity positive Elektrizität *f*
p. sign Pluszeichen *n*
plush Plüsch *m*
p. fabric Plüschgewebe *n*
p. loom Samtwebstuhl *m*
p. weaving Plüschweberei *f*
plutonium Plutonium *n*
p. bomb Plutoniumbombe *f*
p. breeder Plutoniumbrutreaktor *m*
p. enrichment Plutoniumanreicherung *f*
p. fission Plutonium[kern]spaltung *f*
p. pile Plutoniumreaktor *m*
p. poisoning Plutoniumvergiftung *f*
p. production Plutoniumerzeugung *f*
p. recycle Plutoniumrückführung *f*
p. regeneration Plutoniumrückgewinnung *f*
p. separation plant Plutoniumtrennanlage *f*
pluviometer Pluviometer *n*, Regenmesser *m*

ply / to 1. falten; 2. fachen, doppeln, dublieren *(Spinnerei)*; 3. *(Schiff)* regelmäßige Fahrten machen
ply 1. Furnierplatte *f*; 2. Zwirnung *f*; Zwischenlage *f*; Schicht *f*
p. metal Verbundmetall *n*
p.-rating *(Kfz)* Reifenunterbaubewertung *f (Kennziffer der Festigkeit und Tragfähigkeit eines Reifens)*
plywood Sperrholz *n*
p. pallet Palette *f*
p. standard normales Sperrholz *n*
p. three-ply dreifache Sperrholzplatte *f*
p. with equal plies Sperrholz *n* mit gleich dicken Lagen
PME effect *s.* photomagnetoelectric effect
pneumatic pneumatisch; Druckluft-, Preßluft-
p. boat Schlauchboot *n*
p. box opener *(Text)* kastenförmiger Saugöffner *m*
p. brake Druckluftbremse *f*
p. branch line *(BMSR)* pneumatische Signalleitung *f*
p. caisson Druckluftsenkkasten *m*
p. change Druckluftschaltung *f (Getriebe)*
p. chipping hammer Druckluftmeißel *m*
p. chucking *(Masch)* Druckluftspannung *f*
p. coal pick hammer *(Brg)* Preßluftabbauhammer *m*
p. collet Druckluftzange *f*
p. control Druckluftbetätigung *f*, Druckluftauslösung *f (des Leistungsschalters)*; Druckluftsteuerung *f*
p. conveying Druckluftförderung *f*, pneumatische Förderung *f*
p. conveyor pneumatische Förderanlage *f*, pneumatischer Förderer *m*
p. cushioning Luftfederung *f*, Luftpolsterung *f*
p. cylinder Preßluftzylinder *m*, Druckluftzylinder *m*
p. deduster Windsichter *m*
p. dedusting pneumatische Entstaubung *f*
p. dispatch Rohrpost *f*
p. dispatch carrier (container) Rohrpostbüchse *f*
p. drive Druckluftantrieb *m*, Preßluftantrieb *m*
p. expanding mandrel Druckluftspanndorn *m*
p. felt guide Druckluftfilzführung *f (Papierherstellung)*
p. flushing Druckluftspülung *f*
p. foundation *(Bw)* Druckluftgründung *f*
p. frogsuit Druckluft[schutz]anzug *m*
p. gauging *(Masch)* pneumatisches Prüfen *n*
p. governor Druckluftregler *m*
p. hammer Preßlufthammer *m*, Drucklufthammer *m*
p. hammer drill Druckluftbohrhammer *m*
p. jig *(Brg)* Luftsetzmaschine *f*
p. lifting device Drucklufthebezeug *n*
p. loader Druckluftlader *m*
p. loudspeaker Druckkammerlautsprecher *m*, pneumatischer Lautsprecher *m*
p. micrometer Druckluftfeinanzeiger *m*
p. mixing Mischen *n* durch Saugluft *(Baumwolle)*
p. nut-runner Preßluftmutteranziehmaschine *f*
p. pick Preßlufthammer *m*
p. piston Preßluftkolben *m*
p. pointed chisel *(Brg)* Abbauspieß *m*
p. pressure switch Luftdruckschalter *m*
p. raft aufblasbares Floß *n*
p. rammer Preßluftramme *f*, Druckluftramme *f*
p. riveter Preßluftnietmaschine *f*
p. separation Druckluftaufbereitung *f*
p. separator Windsichter *m*
p. sinking method *(Brg)* Caissonverfahren *n*, Senkschachtverfahren *n* mit Überdruckkammer
p. sizing Windsichtung *f*, Sichtung *f (Aufbereitung)*
p. spade Preßluftspaten *m*
p. stemmer *(Brg)* Druckluftbesatzgerät *n*
p. stowing *(Brg)* Blasversatz *m*
p. stowing machine Blasversatzmaschine *f*
p. tamper Druckluftstampfer *m*, Preßluftstampfer *m*
p. tool Preßluftwerkzeug *n*
p. tube blower Rohrpostgebläse *n*
p. tube central station Rohrpostzentrale *f*
p. tube conveyor Rohrpost *f*
p. tube plant (system) Rohrpostanlage *f*
p. tyres Gummibereifung *f* mit Luftfüllung, Luftbereifung *f*
p. vibrator Preßluftrüttler *m*, Druckluftrüttler *m*
pneumatically operated 1. *(BMSR)* pneumatisch betrieben, mit pneumatischer Hilfsenergie [betrieben]; 2. *(Masch)* druckluftbetätigt
p. operated drilling machine of the portable type Drucklufthandbohrmaschine *f*
p. operated switch Schalter *m* mit Druckluftbetätigung, Druckluftschalter *m*
pneumatics Pneumatik *f*, Mechanik *f* der Gase
pneutronic *(Am) (BMSR)* elektropneumatisch
pocket 1. Tasche *f*; Sack *m*; 2. Aussparung *f*; 3. *(Geol)* Druse *f*; 4. *(Brg)* Nest *n (z. B. Erznest)*; 5. *(Met)* Lunker *m*
p. accumulator Taschenakkumulator *m*
p. ammeter Taschenamperemeter *n*
p. blasting machine *(Brg)* Zündmaschine *f* zum gleichzeitigen Zünden von drei Schüssen
p. book leather Portefeuilleleder *n*, Täschnerleder *n*

p. counter Taschenumdrehungszähler *m*
p. dosimeter Taschendosisleistungsmesser *m*; *(Kern)* Ansteckdosimeter *n*
p. lamp Taschenlampe *f*
p. meter Taschenmeßinstrument *n*
p. of ore *(Geol)* Erznest *n*
p. rot Fäulnis *f (bei Holz)*
p. size Taschenformat *n*
p. tracer Taschenprüfgerät *n*
p. voltmeter Taschenvoltmeter *n*
pocketing *(Masch)* dreidimensionales Fräsen *n*, Gesenkfräsen *n*; Einschlagen *n (Ventil)*
pockets-type plate Taschenplatte *f (Batterie)*
pockety *(Geol)* drusig
pockwood-pillow Pockholzlager *n*
pod Sockel *m*, Halter *m*; Kokon *m*
p. auger Faßbohrer *m (Küfnerei)*
podium Podium *n*
podolite *(Min)* Podolit *m*
podsol Bleicherde *f*, Podsol *m*
point / to 1. spitzen, schärfen; ausspitzen *(Bohrer)*; 2. verfugen *(Mauerwerk)*; 3. punktieren
point 1. Punkt *m (Geometrie)*; Stelle *f*; Spitze *f*; Kuppe *f*; 2. Nadel *f (Reißnadel)*; 3. Komma *n (Dezimalbruch)*; 4. Schneidenkopf *m (Werkzeug)*; 5. *(Eis)* Weiche *f*; Zunge *f (einer Weiche)*; 6. Schneidzahn *m (Baggerlöffel)*; 7. handgearbeitete Spitze *f*; 8. Aufstoßnadel *f*; 9. Punkt *(englische und amerikanische Längeneinheit = typografischer Punkt = 0,351 mm)*
p. angle Spitzenwinkel *m (Spiralbohrer)*
p. approximation punktförmige Näherung *f*
p. at infinity *(Math)* unendlich ferner Punkt *m*
p.-bearing pile Spitzendruckpfahl *m (Grundbau)*
p. block Punkt[hoch]haus *n*
p. bridging *(Kfz)* Überbrückung *f* des Elektrodenabstandes *(durch Verbrennungsrückstände an Zündkerzen)*
p.-by-point evaluation punktweise Auswertung *f*
p. cathode Punktkatode *f*, Spitzenkatode *f*
p. charge *(Ph)* Punktladung *f*, punktförmige Ladung *f*
p. contact 1. Punktberührung *f (Zahnräder)*; 2. Spitzenkontakt *m (Halbleitertechnik)*
p.-contact transistor Spitzentransistor *m*
p. coordinate Punktkoordinate *f*
p. counter *(Kern)* Spitzenzähler *m*, Spitzenzählrohr *n*
p.-designation grid *(Mil)* Artilleriepunkttafel *f*
p. discharge *(El)* Spitzenentladung *f*
p.-focal *(Opt)* punktuell abbildend
p. focus *(Opt)* punktförmiger Brennpunkt *m*
p. geometry Schneidenausbildung *f (Bohrer)*
p. grinder Spiralbohrerschleifmaschine *f*
p. grinding Spitzenanschliff *m (Spiralbohrer)*
p. imperfection Punktfehlstelle *f*, punktförmige Störstelle *f (Halbleitertechnik)*
p. indicating signal *(Eis)* Weichensignal *n*
p. indicator *(Eis)* Weichensignal *n*
p. lace *(Text)* Nadelspitze *f*
p. lamp Punktlichtlampe *f*
p. lattice *(Ph)* Punktgitter *n*
p. lead Anschnitt *m (Werkzeug)*
p. lever Umleghebel *m (Stellwerk)*
p. load Punktlast *f*, konzentrierte Last *f*
p. locking *(Eis)* Weichensicherung *f*
p. locking frame brake Bremse *f (Stellwerk)*
p. mass Punktmasse *f*
p. of alternation *(Masch)* Wechselpunkt *m (Kurbelstellung)*
p. of application Angriffspunkt *m*
p. of application of a force Kraftangriffspunkt *m*, Angriffspunkt *m* einer Kraft
p. of application of a load Laststelle *f*
p. of connection *(El)* Anschlußort *m*, Anschlußstelle *f*, Anschlußpunkt *m*
p. of contact Berührungsstelle *f*, Berührungspunkt *m*; Eingriffspunkt *m (Zahnrad)*
p. of contraflexure Momentennullpunkt *m*
p. of cut-off Absperrpunkt *m (Dampfmaschine)*
p. of discharge Entladestelle *f*
p. of division *(Math)* Teil[ungs]punkt *m*
p. of engagement *(Masch)* Eingriffsstelle *f*
p. of entry Eintrittsstelle *f (z. B. von Elektronen)*
p. of ignition Zündzeitpunkt *m (Gasentladungsröhren)*
p. of impact Einschlag *m (Geschoß)*
p. of inflection of the curve *(Masch)* Wendepunkt *m* der Kurve
p. of initiation Ausgangspunkt *m*, Ausgangsstelle *f*
p. of intersection Schnittpunkt *m*
p. of leakage Leckstelle *f*
p. of lubrication Schmierstelle *f*
p. of pressure of the valve Druckpunkt *m* des Ventils
p. of reference Bezugspunkt *m*
p. of reflection *(El)* Stoßstelle *f*
p. of rest *(El)* Nullstellung *f*, Ruheausschlag *m*, stationärer Ausschlag *m*
p. of self-oscillation *(El)* Schwingungseinsatzpunkt *m*
p. of solidification Erstarrungspunkt *m*
p. of support Stützpunkt *m*; Auflagepunkt *m*; Auflagerungsstelle *f*; Lagerstelle *f*
p. of suspension Aufhängungspunkt *m*
p. of swell Staupunkt *m*
p. of tangency Berührungspunkt *m* von (an) Kreisen
p. of tooth engagement *(Masch)* Eingriffsstelle *f*

p. of voltage application *(El)* Spannungszuführungspunkt *m*
p. of weld Schweißstelle *f*
p. of zero moment Momentennullpunkt *m*
p. operating gear *(Eis)* Weichenantrieb *m*
p. particle 1. materieller Punkt *m*, Massenpunkt *m*; 2. *(Kern)* punktförmiges Teilchen *n*
p. probe Punktsonde *f*
p.-projection X-ray microscopy Röntgenschattenmikroskopie *f*
p. rail *(Eis)* Spitzenschiene *f*
p. recorder Punktschreiber *m*
p. shaped punktförmig
p. size *(Typ)* Kegelstärke *f (der Schrift)*
p. source 1. *(El)* punktartige Strahlungsquelle *f*; 2. *(Math)* Punktquelle *f*
p. suspension Spitzenaufhängung *f*
p. switch *(Eis)* Zungenweiche *f*
p.-thinned ausgespitzt *(Spiralbohrer)*
p. thinning Ausspitzen *n (des Spiralbohrers)*
p.-to-point communication feste Funklinie *f (Funkverkehr zwischen zwei festen Punkten)*
p.-to-point communication service fester Funkverkehr *m*
p.-to-ship communication Boden-Bord-Verbindung *f (Funkverkehr)*
p. welding Punktschweißen *n*
pointed spitz, Spitz-; angespitzt, zugespitzt; spitznasig; ausgespitzt *(Spiralbohrer)*
p. arch Spitzbogen *m*
p. bullet *(Mil)* Spitzgeschoß *n*
p. cathode *s.* point cathode
p. cone Spitzkegel *m*
p. copper bit Spitzlötkolben *m*
p.-end grubscrew Gewindestift *m* mit Spitze
p. lightning protector Spitzenblitzableiter *m*
p. resonance curve *(El)* spitze Resonanzkurve *f*
p. tie up *(Text)* Spitzenschnürung *f*
p. toe spitze Vorderkappe *f (Schuh)*
pointer 1. Zeiger *m*; Zeigermarke *f*; Zunge *f*; 2. Ausspitzmaschine *f (für Bohrer)*
p. deflection Zeigerausschlag *m*
p. eyepiece Zeigerokular *n*
p. instrument Zeigerinstrument *n*
p. knob Einstellknopf *m*, Zeigerdrehknopf *m*
p. reading Zeigerablesung *f*
p. setting Zeigerstellung *f*
pointing 1. Fugenverstrich *m*; Verfugen *n*, Verfugung *f (Mauerwerk)*; 2. Ausspitzen *n (Bohrer)*; Schärfen *n*, Anspitzen *n*
p. trowel Fugenkelle *f*; Spitzkelle *f*
pointlike punktförmig
points *(Eis)* Weiche *f*
p. and crossings *(Eis)* Weiche *f*
p. bar Aufstoßnadelbarre *f (Strickerei)*
pointsman Weichensteller *m*
poise / to balancieren, auswiegen, konstant halten
poise 1. Gleichgewicht *n*; Gewicht *n*; 2. Poise *n (Einheit der dynamischen Viskosität; 1 P = 10^{-1} Ns/m^2)*
Poiseuille formula Hagen-Poiseuillesches Gesetz *n*
poison / to vergiften
poison Gift *n*
p.-gas Giftgas *n*
poisoning 1. Vergiften *n*; 2. Einbrennen *n (Katodenstrahlröhre)*
p. effect Vergiftungseffekt *m*
p. override Vergiftungsausgleich *m (Reaktor)*
poisonous gas Giftgas *n*
Poisson's ratio *(Mech)* Querdehnungszahl *f*
poke / to schüren, stochern
poker Schürer *m*, Schüreisen *n*, Schürstange *f*, Stocheisen *n*, Löschspieß *m (Feuerungstechnik)*
p. iron *(Met)* Schlackenhaken *m*, Schlackenzieher *m*
p. vibrator Flaschenrüttler *m*, Innenrüttler *m*, Tauchrüttler *m*
poking Schüren *n (Feuer)*
polar *(Math)* Polare *f*
polar polar, polig, gerichtet
p. axis Polarachse *f*
p. coelostat Polarzoelostat *m*
p. coordinate Polarkoordinate *f*
p. coordinatograph Polarkoordinatograf *m*
p. crystal polarer Kristall *m*, Polarkristall *m*
p. curve Polare *f*
p. curve of light distribution polare Lichtverteilungskurve *f*
p. direct current telegraph system Morsedoppelstromsystem *n*
p. duplex *(Fmt)* Doppelstromgegensprechen *n*
p. form of a complex number Polarform *f* einer komplexen Zahl
p. head Polkopf *m*
p. liquid polare Flüssigkeit *f*
p. mode *(BMSR)* Polarkoordinatenmodus *m*
p. moment of inertia polares Trägheitsmoment *n*
p. planimeter Polarplanimeter *n*
p. radiation pattern Polarstrahlungsdiagramm *n (Antenne)*
p. reciprocals polar entsprechende Figuren *fpl (bezogen auf einen Kegelschnitt)*
p. section modulus polares Widerstandsmoment *n*, Widerstandsmoment gegen Torsion
p. tangent *(Math)* Polartangente *f*
p. vector polarer Vektor *m*
polarimeter *(Opt)* Polarimeter *n*
polariscope Polarisationsapparat *m*
polarity Polarität *f*
p. indicator *(El)* Polprüfer *m*
p. reverser *(El)* Polwechsler *m*
polarizability Polarisierbarkeit *f*
polarizable polarisierbar

polarization Polarisation *f*, Polarisierung *f*
p. capacitance *(El)* Polarisationskapazität *f*
p. charge Polarisationsladung *f*
p. coefficient Polarisationsgrad *m*
p. current *(El)* Polarisationsstrom *m*
p. direction Polarisationsrichtung *f*
p. effect Polarisationseffekt *m*, Einfluß *m* der Polarisation
p. energy Polarisationsenergie *f*
p. error Dämmerungseffekt *m*, Nachteffekt *m* *(Radar)*
p. interference filter Polarisationsinterferenzfilter *n*
p. microscope Polarisationsmikroskop *n*
p. of light Lichtpolarisation *f*
p. plane Polarisationsebene *f*
p. reactance Polarisationsblindwiderstand *m*, Polarisationsreaktanz *f*
p. rule Polarisationsregel *f*
p. sign Polarisationsvorzeichen *n*
p. voltage *(El)* Polarisationsspannung *f*
polarize / to polarisieren, polen
polarized polarisiert, gepolt
p. armature *(El)* gepolter (polarisierter) Anker *m*
p. bell *(Am) (Fmt)* Wechselstromwecker *m*
p. electrolytic capacitor gepolter Elektrolytkondensator *m*
p. induction Vormagnetisierungsinduktion *f*
p. light *(Ph)* polarisiertes Licht *n*
p. radiation polarisierte Strahlung *f*
p. relay polarisiertes Relais *n*
p. ringer *(Am) (Fmt)* Wechselstromwecker *m*
polarizer *(Opt)* Polarisator *m*, Polarisationsprisma *n*
polarizing Polarisations-
p. angle Polarisationswinkel *m*
p. current polarisierender Strom *m*, Polarisationsstrom *m*
p. field Vormagnetisierungsfeld *n*
p. magnetization Gleichstromvormagnetisierung *f*
p. magnetizing force Vormagnetisierungsfeldstärke *f*
p. photometer Polarisationsfotometer *n*
p. power Polarisationsvermögen *n*
p. source Polarisationsquelle *f*
p. spectacles Polarisationsbrille *f*
p. stress tester optischer Spannungsprüfer *m*
polarogram Stromspannungskurve *f*
polarographic wave *(Ph)* polarografische Welle (Stufe) *f*
polder *(Bw)* Polder *m*
Poldi hardness tester Poldihärteprüfer *m*, Poldihammer *m*
pole / to *(Met)* polen *(z. B. Kupfer)*
p.-dry *(Led)* auf der Stange trocknen
pole 1. *(El)* Pol *m*; 2. Mast *m*; Baustange *f*, Rundholz *n*; Deichsel *f*; 3. Telegrafenstange *f*; 4. *(Eis)* Schrankenbaum *m*
p. arc *(Elm)* Polbogen *m*
p. binder Polklemme *f (Batterie)*
p. chain Haltekette *f*, Schirrkette *f*, Steuerkette *f (Wagen)*
p.-change polumschaltbar
p.-change motor polumschaltbarer Motor *m*
p. changer Polwender *m*
p. changing Polumschaltung *f*, Umpolung *f*; Änderung *f* der Polzahl
p.-changing polumschaltbar
p.-changing control Regelung *f* durch Polumschaltung; Polumschaltung *f*
p.-changing motor polumschaltbarer Motor *m*
p.-changing switch Polwechselschalter *m*, Polumschalter *m*
p. clearance *(El)* Mastabstand *m*; Schlagweite *f (zwischen Polen)*
p. cleave Deichselschere *f*
p. climbers Steigeisen *n*, Steigestütze *f*
p. construction Mastenbauweise *f*
p.-core of a magnet Magnetkern *m*
p. cribbing *(Am) (El)* Querriegel *m (A-Mast)*
p. guy *(El)* Mastanker *m*, Abspannanker *m*, Spanndraht *m*
p. head *(El)* Mastkopf *m*
p. line *(Fmt)* Holzmastleitung *f*
p. mast *(Schiff)* Pfahlmast *m*
p.-mounted circuit-breaker *(El)* Mastschalter *m (an Holzmasten)*
p. operating frame *(Eis)* Drehgestell *n (Schranke)*
p. paper Polprüfpapier *n (Reagenzpapier)*
p. piece Polschenkel *m*, Pol[schuh] *m*
p. pitch *(Elm)* Polteilung *f*
p. retriever Stromabnehmerfänger *m*
p. reversal Polumkehr *f*, Umpolung *f*
p. reverser Polumschalter *m*
p. seasoning *(Holz)* Trocknen *n* von Stangen
p. shank Elektromagnetkern *m*; Polschenkel *m*
p. shoe Polschuh *m*
p. strength Polstärke *f*
p. switch Mastschalter *m (an Holzmasten)*
p. terminal Polklemme *f*
p. tester 1. Polprüfer *m*; 2. Mastsonde *f (kleiner Bohrer zur Prüfung auf Schädlingsbefall an Telegrafenstangen)*
p. timber Stangenholz *n*
p. transformer Masttransformator *m*, Mastumspanner *m*
p.-type capacitor Mastkondensator *m*
p.-type construction Mastenbauweise *f*
p.-type isolator Masttrennschalter *m*, Masttrenner *m (an Holzmasten)*
p.-type transformer Masttransformator *m (an Holzmasten)*

p. with cross arms *(Fmt)* Abzweiggestänge *n*, Verzweigungsstützpunkt *m*
p.-zero configuration *(BMSR)* Nullstellen-Pol-Verteilung *f*
poles and zeros Pole *mpl* und Nullstellen *fpl (bei Filtern)*
polhode cone Gangpolkegel *m*; körperfester Drehkegel *m*
poling *(Met)* Polen *n (z. B. von Kupfer)*
p. board *(Holz)* Absteifbrett *n*
p. planks Holzverschalung *f*
p. switch Polwechselschalter *m*
polish / to polieren, glätten; glanzschleifen; schlichten *(Walzgut)*
polish 1. Politur *f*, Glätte *f*, Glanz *m*; 2. Poliermittel *n*
p. attack Ätzpolieren *n*
polished sheet metal Glanzblech *n*, poliertes (geglänztes) Blech *n*
polisher Poliergerät *n*
polishing Polieren *n*, Glanzschleifen *n*
p. agent Poliermittel *n*
p. bit Schlichtbohrer *m*
p. cloth Polierfilz *m*
p. compound Polieremulsion *f*
p. disk Putzscheibe *f*, Polierscheibe *f*
p. felt Polierfilz *m*
p. mop Polierscheibe *f*
p. pad Polierscheibe *f*
p. powder Polierpulver *n*, Schleifpulver *n*
p. roll Polierwalze *f*
p. scraper Polierschaber *m*
p. slate Polierschiefer *m*
p. solution Polieremulsion *f*
p. tube Scheuertrommel *f*
p. wheel Polierscheibe *f*
p. with leather-covered wooden disk Pließ[t]en *n*, Pliesten *n*
p. wool Putzwolle *f*
poll / to beschneiden, kappen, köpfen *(Bäume)*
pollard 1. Kleie *f*; 2. gekappter (geköpfter) Baum *m*; 3. hornloses Tier *n*
pollen Blütenstaub *m*
pollute / to verunreinigen *(Gewässer, Luft)*
pollution Verunreinigung *f*, Verschmutzung *f*
poloidal magnetic field poloidales Magnetfeld *n*
polonium Polonium *n*
poly-paper mit Kunststoff beschichtetes Papier *n*
polyacidic mehrsäurig
polyacrylate Polyakrylat *n*, Polyakrylsäureester *m*
polyacrylonitrile fibre Polyakrylnitrilfaser *f*
polyamide fibre Polyamidfaser *f*
polyatomic vielatomig
p. alcohol mehrwertiger Alkohol *m*, Polyol *m*
polybasic mehrbasisch
polybasite Polybasit *m*
polyblend kautschukartiges Polymerengemisch *n*, Polyblend-Kautschuk *m*
polybromohydrocarbon mehrfach bromierter Kohlenwasserstoff *m*
polycarbonate Polykarbonat *n*
polychlorohydrocarbon mehrfach chlorierter Kohlenwasserstoff *m*
polychloroprene Polychloropren *n*; Chloroprenkautschuk *m*
polychromatic polychromatisch
polycondensation Polykondensation *f*
polyconic polykonisch
polycyclic mehrkernig; polyzyklisch
polydimensional mehrdimensional
polydirectional in mehreren Richtungen, Rund-
polydisperse polydispers
polydispersion Polydispersion *f*
polyelectrode Mehrfachelektrode *f*
polyester Polyester *m*
p. fibre Polyesterfaser *f*
p. putty Glasfaserkitt *m*
p. resin Polyesterharz *n*
polyethylene Polyäthylen *n*
polyfluohydrocarbon mehrfach fluorierter Kohlenwasserstoff *m*
polyfunctional für mehrere Zwecke verwendbar
polygenic mehr als eine Verbindung mit Wasserstoff bildend
polygon Vieleck *n*, Polygon *n*
p. method Polygonzugverfahren *n (bei Differentialgleichungen)*
p. of forces Kräftepolygon *n*, Kräftevieleck *n*
p. of instantaneous centre Polvieleck *n*
p. profile Mehrkantprofil *n*
polygonal vieleckig, vielkantig, polygonal, Polygon[al]-
p. connection *(El)* Polygonschaltung *f*, Polygonzug *m*
p. grate Polygonrost *m (Feuerungstechnik)*
p. number Polygonalzahl *f*
p. overhead contact system *(El)* polygonale Kettenfahrleitung *f*
p. rubble Zyklopenmauerwerk *n*
p. shape Mehreckprofil *n*
p. voltage *(El)* Polygonspannung *f*
polygraphic polygrafisch
polyhalide mehrfach halogenierte Verbindung *f*
polyhalohydrocarbon mehrfach halogenierter Kohlenwasserstoff *m*
polyhedral Vielkant *m*
polyhedral polyedrisch, vielflächig
polyhedron Polyeder *n*, Vielflach *n*
polyhydric alcohol mehrwertiger Alkohol *m*, Polyol *n*
polyisobutylene Polyisobutylen *n*
polymer Polymer[e] *n*, Polymerisat *n*
polymer polymer
p. gasoline Polymerbenzin *n*

polymeric polymer
polymerizable polymerisierbar
polymerization Polymerisation *f*
p. plant Polymerisationsanlage *f*
polymerize / to polymerisieren
polymerizing stove Polymerisierofen *m (zum Kunstharzhärten)*
polymerous polymer
polymethacrylate Polymethakrylat *n*, Polymethakrylsäureester *m*
polymorphic polymorph, vielgestaltig
p. inversion polymorphe Umwandlung *f*
polymorphism Polymorphie *f*, Vielgestaltigkeit *f*
polynomial *(Math)* Polynom *n*, mehrgliedriger Ausdruck *m*
p. computer *(Dat)* Polynomrechner *m*, Gerät *n* zur Berechnung von Polynomen
polynuclear mehrkernig, vielkernig
polypeptide *(Ch)* Polypeptid *n*
polyphase mehrphasig
p. alternating current Mehrphasen[wechsel]-strom *m*
p. alternator Mehrphasengenerator *m*, Drehstromgenerator *m*
p. circuit Drehstromkreis *m*, Mehrphasenkreis *m*
p. compound commutator motor Mehrphasenreihenschlußmotor *m* mit Begrenzung der Leerlaufdrehzahl
p. current Mehrphasen[wechsel]strom *m*
p. earth mehrphasiger Erdschluß *m*
p. equilibrium *(Ch)* heterogenes Gleichgewicht *n*
p. generator Drehstromgenerator *m*
p. induction motor Drehstromasynchronmotor *m*
p. meter Mehrphasenzähler *m*
p. motor Drehstrommotor *m*, Mehrphasenmotor *m*
p. power Drehstromleistung *f*
p. series commutator motor Mehrphasenreihenschlußmotor *m*
p. shunt commutator motor Mehrphasennebenschlußmotor *m*
p. shunt-conduction motor Drehstromnebenschlußkommutatormotor *m*
p. system Mehrphasensystem *n*, Drehstromnetz *n*, Drehstromsystem *n*
p. winding mehrphasige Wicklung *f*, Drehstromwicklung *f*
polyrod aerial *(Fmt)* dielektrische [Mehrfach]-Antenne *f*
polysaccharide *(Ch)* Polysa[c]charid *n*
polystage switch Stufenschalter *m*, Umschalter *m*
polystyrene Polystyrol *n*
p. capacitor *(El)* Polystyrolkondensator *m*
polysulphide Polysulfid *n*
polysynthetic polysynthetisch
polytainer Behälter *m* aus Polyäthylen
polytechnic[al] polytechnisch
polythene Polyäthylen *n*
polythionic acid Polythionsäure *f*
polytropic polytropisch
p. curve *(Ph)* Polytrope *f*, polytropische Kurve *f*
p. process *(Ph)* polytropische Zustandsänderung *f*
polyurethane backing Kaschieren *n* mit Polyurethan
p. lining Kaschieren *n* mit Polyurethan
polyvalency Mehrwertigkeit *f*
polyvalent mehrwertig
polyvinyl acetal Polyvinylazetal *n*
p. acetate Polyvinylazetat *n*, PVA
p. alcohol Polyvinylalkohol *m*
p. butyral Polyvinylbutyral *n*
p. carbazole Polyvinylkarbazol *n*
p. chloride Polyvinylchlorid *n*, PVC
p. chloride fibre *(Text)* Polyvinylchloridfaser *f*, PVC-Faser *f*
p. pyrrolidone Polyvinylpyrrolidon *n*
polyvinylidene chloride Polyvinylidenchlorid *n*
pome fruit Kernobst *n*
pomiculture Obstbau *m*
pomiform apfelförmig
pommel Krispelholz *n*
p. of cork Pantoffelholz *n*
pommeled horse Pauschenpferd *n (Turngerät)*
ponceau mohnrot
ponceau Ponceau *n (Azofarbstoff)*
pond Weiher *m*, Teich *m*; Wasserstauung *f*
p. retting Grubenröste *f (Flachs)*
pondage Teichfassungsvermögen *n*; Fassungsvermögen *n* der Stauwasserstelle
ponderable wägbar
ponderomotive force *(Ph)* ponderomotorische Kraft *f*
ponding Stauung *f* von Wasser auf einer frischen Betonfläche durch kleine Erddämme usw. *(Betonnachbehandlung)*
pontoon Ponton *m*; Prahm *m*
p. bridge Pontonbrücke *f*, Schwimmbrücke *f*
p. carrier *(Mil)* Pontonwagen *m*
p. dock Pontondock *n*, Schwimmdock *n*
p. lifeboat Halbklappboot *n*
pony motor Andrehmotor *m*, Anwurfmotor *m*, Anlassermotor *m*
p.-rougher *(Am)* Vorstreckgerüst *n (Walzwerk)*
pool / to ausmeißeln, ausstemmen
pool 1. Weiher *m*, Teich *m*; 2. Schmelzbad *n (Schweißen)*; 3. flüssige Katode *f*; 4. Erdölablagerung *f*, Erdöllagerstätte *f*
p. cathode flüssige Katode *f*
p. of oil Ölpfütze *f*

p. rectifier Ventil *n* mit flüssiger Katode
p.-type rectifier Quecksilberdampfgleichrichter *m* mit flüssiger Katode *f*
poop *(Schiff)* Poop *f*, Achterhütte *f*
p. deck Hüttendeck *n*, Kampanjedeck *n*
p.-lantern Achterlaterne *f*, Hecklaterne *f*
pooping *(Schiff)* achterliches Seeübernehmen *n*
poor arm, schlecht; gering[wertig]; mager
p. audibility *(Fmt)* schlechte Verständigung (Hörbarkeit) *f*
p. concrete Magerbeton *m*
p. conductor *(El)* schlechter Leiter *m*
p. contact schlechter Kontakt *m*
p. draught flauer Zug *m (Feuerung)*
p. finish geringe Oberflächengüte *f*
p. gas geringwertiges Gas *n*, Schwachgas *n*
p. ignition quality Zündträgheit *f*
p. in contrast *(Opt)* kontrastarm
p. lime Magerkalk *m*
p. ore mageres Erz *n*, Armerz *n*
poorly clamped *(Masch)* schlecht gespannt
p. conducting schlecht leitend
p. lit schlecht beleuchtet
poorness Armut *f*; Minderwertigkeit *f*
pop mark Meßmarke *f*
poplar 1. Pappel *f*; 2. Pappelholz *n*
p. tree rust Pappelrost *m (Holzkrankheit)*
poplin Popeline *f*
poppet 1. *(Masch)* Ventilkegel *m*; Gegenhalter *m*, Spannbock *m (Werkzeug)*; Spannschraube *f* mit Nutenstein; 2. Schlittenstütze *f*, Schlittendrehlager *n (eines Ablaufschlittens beim Stapellauf)*
p. head Docke *f*, Stock *m*; Überbau *m*, Fördergerüst *n*
p. valve pilzförmiges Ventil *n*, Ringventil *n*, Tellerventil *n*
p. valve gear Ventilsteuerung *f (Dampfmaschine)*
popping back Knallen *n* im Vergaser
p. rock *(Geol)* Steinschlag *m*
poppy Mohn *m*
p. head Mohnkapsel *f*
p. oil Mohnöl *n*
Poppy *s.* Poppymeter
Poppy meter *(Kern)* Poppymeter *n (amerikanisches Strahlenüberwachungsgerät)*
popular science publication populärwissenschaftliche Veröffentlichung *f*
populated 1. bevölkert; 2. besetzt *(z. B. mit Ladungsträgern)*
population 1. *(Math)* Gesamtheit *f*, Grundgesamtheit *f*, statistische Gesamtheit *f*; 2. Bahnzahl *f (Hochenergiephysik)*; 3. Bevölkerung *f*
p. mean *(Math)* Gesamtheitsmittel *n*
porcelain Porzellan *n*
porcelain porzellanartig
p. boat Porzellanschiffchen *n (Labortechnik)*
p. bond keramische Bindung *f*
p. clay Kaolin *n*, Porzellanton *m*, Porzellanerde *f*
p. enamel Email[le *f*] *n*
p. funnel Porzellantrichter *m*, Nutsche *f*
p. insulator Porzellanisolator *m*
p. lamp holder Porzellanfassung *f*
p. shelf Porzellaneinsatz *m (Exsikkator)*
porcelaneous porzellanartig
porch 1. Veranda *f*; Eingangshalle *f*, Vorhalle *f*, Windfang *m*; 2. *(Fs)* Schwarzschulter *f*
porcupine *(Text)* Igel *m*, Läufer *m*, Kammwalze *f*, Nadelwalze *f*, Stachelwalze *f*
p. cylinder Igeltrommel *f*, Zahntrommel *f*
p. drawing frame Nadelwalzenstrecke *f*, Igelstrecke *f*
p. opener Voröffner *m* mit Nasenwalze, Porcupine-Öffner *m*, Nasentrommelöffner *m*
pore Pore *f*, Öffnung *f*
p. damming Poren schließend
p. filler Porenfüller *m*, Vorlack *m*
p.-fluid anelasticity Porenflüssigkeitsanelastizität *f*
p. water Porenwasser *n (Bodenmechanik)*
porism Folgesatz *m*
pork pie furnace *(Met)* Maerz-Boelens-Ofen *m*
porosity Porosität *f*, Porigkeit *f*
p. determination Porenbestimmung *f*
p. of paper Porenvolumen *n* eines Papiers
porous porig, porös, löcherig, blasig, schwammig, locker
p. barrier *(Ch)* Trennwand *f*
p. concrete Schwammbeton *m*, Gasbeton *m*
p. plate Tonteller *m*
p. pot poröses Gefäß *n*, poröser Tonzylinder *m*, poröse Zelle *f*, Tonzelle *f*
p. spot Lunkerstelle *f*
porpezite *(Min)* Porpezit *m*, Palladiumgold[erz] *n*
porphyritic *(Min)* porphyrisch
porphyry *(Min)* Porphyr *m*
porpoising stampfend, wellenförmig
Porro prism Porro-Prisma *n*
Porro prism set Porroscher Prismensatz *m*
port / to *(Schiff)* [Ruder] nach Backbordseite umlegen
p. to leiten (führen) zu
port 1. Hafen *m*; 2. Backbord *n*; 3. Kanal *m*, Öffnung *f*, Schlitz *m*; 4. Kopf *m (eines SM-Ofens)*
p. charges Hafengelder *npl*, Hafenabgaben *fpl*, Hafenzoll *m*
p.-clearing Auslaufen *n (aus dem Hafen)*
p.-controlled scavenge *(Kfz)* Schlitzspülung *f*
p. dues Hafengebühr *f*, Hafengelder *npl*
p. engine Backbordmaschine *f*
p. facilities Hafeneinrichtungen *fpl*

p. hole *(Wärm)* Dampfweg *m* *(s. a.* porthole)
p. installations Hafenanlagen *fpl*
p. lantern Backbordlaterne *f*
p. light Seitenfenster *n*, Backbordlicht *n*; Hafenfeuer *n*, Molenfeuer *n*
p. management Hafenbetrieb *m*
p. nipple *(Kfz)* Schlitznippel *m*
p. of arrival Ankunftshafen *m*
p. of call Anlaufhafen *m*
p. of clearance *(Am)* Abfahrtshafen *m*, Abgangshafen *m*
p. of debarkation Landungshafen *m*, Ausschiffungshafen *m*
p. of delivery Ablieferungshafen *m*, Bestimmungshafen *m*, Lieferungshafen *m*
p. of departure Abgangshafen *m*, Abfahrtshafen *m*
p. of destination Bestimmungshafen *m*
p. of discharge Löschhafen *m*, Ausladungshafen *m*, Entladungshafen *m*
p. of distress Nothafen *m*, Schutzhafen *m*
p. of entry Ankunftshafen *m*, Eingangshafen *m*
p. of refuge Schutzhafen *m*, Zufluchtshafen *m*
p. of registry Heimathafen *m*, Registrierhafen *m*
p. of sailing Abgangshafen *m*, Abfahrtshafen *m*
p. of transit Durchgangshafen *m*, Transithafen *m*
p. of transshipment Umladehafen *m*, Umschlagshafen *m*
p.-opening diagram Steuerdiagramm *n* *(Schiebermotor)*
p. rudder Backbordruder *n*
p. side Backbord[seite *f*] *n*
p. timings Steuerzeiten *fpl* *(2-Takt-Verbrennungsmotor)*
portability Ortsbeweglichkeit *f* *(einer Maschine)*
portable *s.* portable receiver
portable tragbar, transportabel, ortsveränderlich, ortsbeweglich; fahrbar
p. aerial ortsveränderliche (tragbare) Antenne *f*
p. amplifier Kofferverstärker *m*
p. cable measuring set Kabelmeßkoffer *m*
p. drill Handbohrgerät *n*
p. drilling machine ortsveränderliche (tragbare, transportable) Bohrmaschine *f*
p. electric grinder elektrische Handschleifmaschine *f*
p. electro drill elektrische Handbohrmaschine *f*
p. fire extinguisher Handfeuerlöscher *m*
p. grinding outfit mounted on truck Elektroschleifmaschine *f* in Rollbockausführung
p. heating carpet [elektrischer] Heizteppich *m*
p. horizontal bars Spannreck *n* *(Turngerät)*
p. lamp Handlampe *f*
p. log frame *(Holz)* fahrbares Gatter *n*
p. measuring instrument tragbares Meßgerät *n*
p. measuring set *(Fmt)* Meßkoffer *m*
p. mixer *(Ch)* Anklemmrührer *m*, Ekatorührer *m*
p. plant bewegliche Anlage *f*
p. projector Kofferprojektor *m*
p. receiver Batterieempfänger *m*, Batterieempfangsgerät *n*, Reiseempfangsgerät *n*, Kofferempfänger *m*, Kofferempfangsgerät *n*, Kofferradio *n*
p. receiving set *s.* p. receiver
p. sawmill fliegendes Sägewerk *n*
p. scaffolding fahrbares Gerüst *n*
p. set *s.* p. receiver
p. spot welding gun Punktschweißzange *f*
p. sprayer Rückentragspritze *f*
p. staging fahrbares Gerüst *n*
p. steam engine bewegliche (transportable) Dampfmaschine *f*, Lokomobile *f*
p. telephone set Streckenfernsprecher *m*, tragbarer Fernsprechapparat *m*
p. transforming substation fahrbares Umspannunterwerk *n* *(für Baustellen usw.)*
p. voice frequency measuring set Fernmeldemeßkoffer *m*
portal Pforte *f*, Portal *n*, Tor *n*
p. crane Portalkran *m*, Torkran *m*
p. frame Portalrahmen *m*
portfire Lunte *f*
portfolio Aktentasche *f*
p. leather Aktentaschenleder *n*
porthole Bullauge *n*
portico Portikus *m*, Säulengang *m*, Säulenhalle *f*; Vordach *n*
portiere Portiere *f*, Türvorhang *m*
porting Schlitzanordnung *f*, Anordnung *f* der Schlitze *(2-Takt-Verbrennungsmotor)*
portion Anteil *m*, Teil *m*, Portion *f*, Quantum *n*, Menge *f*
p. of cylinder schief abgeschnittener Zylinder *m*, Zylinderhuf *m*
Portland blast-furnace cement Hochofenzement *m*, Schlackezement *m*, Eisenportlandzement *m*
Portland cement Portlandzement *m*
Portland clinker Portlandklinker *m*
portlandite *(Min)* Portlandit *m*
portlight Bullauge *n*
portrait size *(Typ)* Hochformat *n*
position / to 1. orten, einstellen, einorten; 2. positionieren; 3. zentrieren; in Stellung bringen

position 1. Lage *f*, Stelle *f*, Stellung *f*; Standort *m*; 2. Position *f* *(z. B. eines Schiffes)*; Ort *m*; 3. *(Fmt)* Platz *m*
p. at centre of gravity Schwerpunktlage *f*
p. by astro-calculation astronomischer Schiffsort *m*
p. control 1. Punktsteuerung *f*, Lagesteuerung *f*, Positionierung *f* *(Werkzeugmaschinensteuerung)*; 2. *(Fmt)* Einstellung *f* der Nulllinienhöhe
p. coupling key *(Fmt)* Verbindungsschalter *m*
p. error coefficient Lagefehler *m* *(Folgeregler)*
p. feedback *(BMSR)* Lagerückführung *f*
p. filar micrometer Positionsfadenmikrometer *n*
p. finding Ortung *f* *(Radar)*; Positionsbestimmung *f* *(eines Schiffes)*
p. fixing Ortung *f* *(Radar)*; Standortpeilung *f*
p. funktion *(Math)* Ortsfunktion *f*
p. grouping key *(Fmt)* Verbindungsschalter *m*
p. indicator 1. *(BMSR)* Stellungsgeber *m*, Stellungsmelder *m*; 2. Stabstandsanzeiger *m* *(Reaktor)*; 3. Bandzählwerk *n* *(Tonbandgeräte)*
p. light signal Lichtsignal *n*
p. message Positionsmeldung *f*
p. meter *(Fmt)* Belegungszähler *m*
p. of control surface *(Flg)* Ruderstellung *f*
p. of eccentric Exzenterstellung *f*
p. of equilibrium Gleichgewichtslage *f*
p. of installation Einbauort *m*
p. of maximum signal Peilmaximum *n* *(Radar)*
p. of minimum signal Peilminimum *n* *(Radar)*
p. of navigation lights *(Schiff)* Lichterführung *f*
p. of nozzle Düsenstellung *f* *(Turbine)*
p. of passage Durchschlaglage *f*, Durchganglage *f* *(Kurbelstellung)*
p. of rest Ruhelage *f*, Ruhestellung *f*
p. of the centre Mittelpunktslage *f*
p. of the load Laststellung *f*
p. of the valve at the dead point *(Masch)* Schiebertotlage *f*
p. pilot lamp *(Fmt)* Platzlampe *f*
p. report Standortmeldung *f*, Positionsmeldung *f*
p. vector Ortsvektor *m*
p. wiring *(Fmt)* Platzbelegung *f*
positionally accurate lagegenau
positioning Positionierung *f*, Lagesteuerung *f*, Punktsteuerung *f* *(Werkzeugmaschinensteuerung)*
p. accuracy Positionsgenauigkeit *f*
p. actuator Steuerelement *n*
p. controller Lageregler *m*, Folgeregler *m*
p. device Positioniereinrichtung *f*
p. error Lageeinstellungsfehler *m*, Einstellungsfehler *m*
positive *(Foto)* Abzug *m*, Positiv *n*
positive 1. *(Math)* positiv; fest[stehend]; 2. *(Masch)* formschlüssig, zwangsläufig
p. absorption positive Absorption *f*
p.-acting control *(Am)* *s.* two-step action
p. air pressure Luftüberdruck *m*
p. area *(El)* Stromaustrittszone *f*
p. beta particle positives Beta-Teilchen *n*
p. bias *(El)* positive Vorspannung *f*
p. blower *s.* p. displacement blower
p. booster *(Elm)* Zusatzmaschine *f*, Spannungserhöher *m*
p. brush *(Elm)* Plusbürste *f*
p. charge positive Ladung *f*
p. clutch formschlüssige Kupplung *f*
p. conductor *(El)* Plusleiter *m*
p. control zwangsläufige Regelung *f*
p.-direction component *(El)* mitläufige Komponente *f*, Mitkomponente *f*
p.-displacement blower Verdrängergebläse *n*, Kapselgebläse *n*, Drehkolbengebläse *n*
p.-displacement meter Verdrängungszähler *m*
p.-displacement supercharger *(Kfz)* Lader *m* der Verdrängerbauart
p. distortion tonnenförmige Verzeichnung *f*, Tonnenverzeichnung *f* *(Fernsehbild)*
p. drive zwangsläufiger Antrieb *m*, zwangsläufige Mitnahme *f*
p. electricity positive Elektrizität *f*
p. electron Positron *n*, positives Elektron *n*
p. excess Überschuß *m*
p. feedback *(El)* Mitkopplung *f*; Rückkopplung *f*
p. film stock Positivfilm *m*
p. gain factor positiver Vermehrungsfaktor *m* *(Reaktortheorie)*
p. glow Anodenglimmlicht *n*
p. going positiv [gerichtet] *(z. B. Impuls)*
p.-grid oscillator Bremsfeldoszillator *m*
p.-ion oscillation Schwingung *f* der positiven Ionen
p. lead *(El)* Plusleiter *m*
p. lens Sammellinse *f*
p. lift loom *(Text)* Gegenzugstuhl *m*, Kontermarschstuhl *m*
p. lock washer Federring *m* mit aufgebogenen Enden
p. lubrication Zwangsschmierung *f*
p. matrix positive Matrize *f*, Positiv *n*
p. meniscus Meniskus *m*
p. mixer Zwangsmischer *m*
p. modulation *(Fs)* Positivmodulation *f*
p. movement Zwangslauf *m*
p. plate Plusplatte *f*, positive Platte *f* *(Batterie)*
p. pole 1. Pluspol *m*, positiver Pol *m*; 2. Anode *f* *(Elektronenröhre)*
p. positioning relay *(Am)* *s.* valve positioner

p. print positive Kopie *f*
p. printing Positivdruck *m*
p. ray current Ionenstrahlstrom *m*, Kanalstrahlstrom *m*
p. rays positive Strahlen *mpl*, Kanalstrahlen *mpl*
p. reactance *(El)* Induktanz *f*
p. relay *(Am) (BMSR)* pneumatischer Verstärker *m* mit Zweipunktverhalten, pneumatisches Kontaktglied *n*
p. separation Positivfarbauszugkopie *f*
p.-sequence *(El)* mitläufig
p.-sequence active power Wirkleistung *f* des mitläufigen Systems
p.-sequence coordinate of a three phase system Mitkoordinate *f* eines Dreiphasensystems
p.-sequence polyphase system mehrphasiges Mitsystem *n*
p.-sequence reactive power Blindleistung *f* des mitläufigen Systems
p.-sequence symmetrical component mitläufige symmetrische Komponente *f*
p. sign positives Vorzeichen *n*
p. stop fester Anschlag *m*
p. sweepback *(Flg)* Rückenpfeilung *f*
p. terminal *(El)* Plusklemme *f*, positive Klemme *f*; Anodenklemme *f (Batterie)*
p. transparency Dia[positiv] *n*
p. type of clutch formschlüssige ausrückbare Kupplung *f*; Zahnkupplung *f*
p. valency positive Wertigkeit (Valenz) *f*
p. water-meter Kolbenwassermesser *m*
p. wire *(Fmt)* a-Ader *f*, positive Leitung *f*
positively actuated zwangsläufig betätigt
p. charged positiv geladen
p. connected zwangsläufig verbunden *(Signal)*
p. controlled valve gesteuertes Ventil *n*
positon *s.* positron
positron *(Kern)* Positron *n*
p. decay Positronenzerfall *m*
p.-electron annihilation Vernichtungsstrahlung *f*, Paarvernichtung *f*
p. emission Positronenemission *f*, Positronenstrahlung *f*
p. emitter Positronenstrahler *m*
p. spectrum Positronenspektrum *n*
positronium Positronium *n*
post / to absenden *(per Post)*
p.-accelerate nachbeschleunigen
p.-tension *(Bw)* mit nachträglichem Verbund vorspannen
post 1. Pfosten *m*, Säule *f*, Ständer[säule *f*] *m*, Stiel *m*, Stütze *f*, Pfeiler *m*, Strebe *f*; Mast *m*; 2. *(Brg)* Stempel *m*, Bein *n*; 3. Runge *f (eines Wagens)*; 4. *(El)* Klemme *f*; 5. Posten *m*; 6. Handelsstation *f*; 7. Post *f*
p. accelerating anode Nachbeschleunigungsanode *f*
p. acceleration Nachbeschleunigung *f*
p. accelerator Nachbeschleunigungselektrode *f*
p.-and-beam construction *(Bw)* Ständerbau *m*
p. and stall method *(Brg)* Pfeilerbau[verfahren *n*] *m*
p. arc current *(El)* Nachstrom *m*
p. bed sewing machine Säulenmaschine *f (Schuhfertigung)*
p.-defecation juice Nachscheidesaft *m (Zuckerindustrie)*
p.-deflection accelerating electrode Nachbeschleunigungselektrode *f (Katodenstrahlröhren)*
p.-deflection acceleration Nachbeschleunigung *f*
p. drill Säulenbohrmaschine *f*
p. exposure *(Foto)* Nachbelichtung *f*
p.-fermentation Nachgärung *f*
p. head Mastaufführung *f (Kabel)*
p. hole auger Tellerbohrer *m (Erdbohrer)*
p. hole digger Erweiterungsbohrer *m*
p.-ignition Glühzündung *f (nach Normalzündung)*
p.-irradiation day Tag *m* nach der Bestrahlung
p. jar squeezer *(Met)* Säulenrüttelpreßformmaschine *f*
p.-mark Poststempel *m*
p. mortem examination *(Dat)* Fehlersuche *f* im Fehlzustand
p. of a door Türstock *m*
p. office account Postscheckkonto *n*
p. office box Schließfach *n*
p. office exchange connection Amtsanschluß *m*
p. office pattern microphone Postmikrofon *n*
p.-process nach Beendigung des Arbeitsganges
p. puller Pfostenzieher *m*
p.-selection *(Fmt)* Nachwahl *f*
p. squeezer *(Met)* Säulenpreßformmaschine *f*
p.-synchronization Nachsynchronisierung *f*
p.-tensioned *(Bw)* mit nachträglichem Verbund vorgespannt
p.-tensioning *(Bw)* Vorspannung *f* mit nachträglichem Verbund
p.-tensioning jack Spannpresse *f (Spannbeton)*
p.-type mixer Wandkonsolrührer *m*, gegen die Wand gestellter Mischer *m* mit auswechselbarem Behälter *(chemische Technik)*
postage Porto *n*; Strafporto *n*
p. stamp Briefmarke *f*
postal Post-
p. service Postdienst *m*

postcard Postkarte *f*
postchlorinate / to *(Ch)* nachchlorieren
poster *(Typ)* Plakat *n*, Anschlag *m*
  p. lettering Plakatschrift *f*
  p. printing Plakatdruck *m*
  p. type Plakatschrift *f*
postforming laminates nachträgliche Formung *f* von Schichtstoffen
postqualize pulses Nachtrabanten *mpl (Fernsehnorm)*
postulate Postulat *n*
postweld treatment Schweißnachbehandlung *f*
pot / to *(Am)* vergießen, ausgießen
  p. galvanize feuerverzinken
pot Topf *m*, Tiegel *m*, Spritztopf *m*, Büchse *f*, Füllzylinder *m*, Gefäß *n*, Kübel *m*; *(Typ)* Gießkessel *m*, Gießtopf *m*
  p. annealing *(Met)* Kastenglühen *n*, Kastenglühung *f*
  p. annealing furnace Topfglühofen *m*
  p. circuit *(El)* Topfkreis *m*
  p. core *(El)* Topfkern *m*
  p. furnace Hafenofen *m*, Kesselofen *m*, Tiegelofen *m*, Büttenofen *m*, Gefäßofen *m*
  p. galvanizing Feuerverzinkung *f*
  p. head jointing sleeve Abschlußmuffe *f (für Kabel)*
  p. head terminal Kabelübergangskasten *m*
  p. herbs Gartenkräuter *npl*; Suppenkräuter *npl*, Würzkräuter *npl*
  p. hole 1. Schlagloch *n*; 2. Einsatzöffnung *f (Tiegelofen)*; 3. *(Brg)* Sargdeckel *m*; 4. *(Geol)* Gletschermühle *f (Störung im Kohlenflöz)*
  p. kiln Feldbrennofen *m*
  p. magnet Mantelmagnet *m*, Topfmagnet *m*
  p. melting furnace Tiegelschmelzofen *m*
  p.-moulding Preßspritzen *n (Plast)*
  p.-mouth *(Typ)* Gießmund *m*
  p. plant Topfpflanze *f*
  p. roasting Topfröstung *f*, Verblaseröstung *f*
  p. spinning Topfspinnen *n*
  p. spinning frame Zentrifugalspinnmaschine *f*
  p. steel Tiegel[guß]stahl *m*
  p. still Destillierblase *f*, Teerblase *f*
  p.-throat *(Typ)* Gießhals *m*
  p.-type breaker *(El)* Löschkammerschalter *m*
  p.-type coil *(El)* Topfkernspule *f*
  p.-type furnace Wannenofen *m*
  p.-type iron dust core *(El)* Eisenpulvertopfkern *m*
potable trinkbar
  p. water Trinkwasser *n*
potash Pottasche *f*, Kaliumkarbonat *n*
  p. alum Kalialaun *n*, Kaliumaluminiumsulfat *n*
  p. feldspar Kalifeldspat *m*
  p. fertilizer Kalidünger *m*
  p. from molasses Schlempekali *n*
  p. fusion Kalischmelze *f*
  p. manure Kalidünger *m*
  p. mine Kalibergwerk *n*
  p. nitre Kalisalpeter *m*
  p. salt Kalisalz *n*
potassium Kalium *n*
  p. antimony tartrate Kaliumantimonyltartrat *n*, Brechweinstein *m*
  p. ferricyanide Kaliumferrizyanid *n*, rotes Blutlaugensalz *n*
  p. ferrocyanide Kaliumferrozyanid *n*, gelbes Blutlaugensalz *n*
  p. fertilizer Kalidünger *m*
  p. hydroxide Kaliumhydroxid *n*, Ätzkali *n*
  p. iodide test paper Kaliumjodidstärkepapier *n*
  p. soap Kaliseife *f*
potato Kartoffel *f*
  p. crusher Kartoffelquetsche *f*
  p. cultivator Häufelpflug *m*
  p. digger Kartoffelroder *m*, Vorratsroder *m*
  p. elevator Kartoffelschrägförderer *m*, Kartoffelhöhenförderer *m*
  p. flour Kartoffelmehl *n*
  p. grader Kartoffelsortiermaschine *f*
  p. growing Kartoffelanbau *m*
  p. grubber Kartoffelroder *m*, Vorratsroder *m*
  p. harvester Kartoffelerntemaschine *f*
  p. masher Kartoffelstampfer *m*
  p. peeler Kartoffelschälmaschine *f*
  p. pit Kartoffelmiete *f*
  p. planter Kartoffellegemaschine *f*, Kartoffelsetzmaschine *f*, Kartoffelsteckmaschine *f*
  p. sorting machine Kartoffelsortieranlage *f*, Kartoffellesemaschine *f*
  p. spinner Kartoffelerntemaschine *f* zum Zeilenauffahren
  p. spirits Kartoffelbranntwein *m*
  p. steamer Kartoffeldämpfer *m*
  p.-stone *(Geol)* Druse *f*
  p. tops Kartoffelkraut *n*
potency *(Math)* Mächtigkeit *f (einer Menge)*
potential Potential *n*; *(El)* Spannung *f*; Potentialunterschied *m*, Potentialdifferenz *f*
potential potentiell; Potential-
  p. above ground *(Am)* Spannung *f* gegen Erde
  p. antinode *(El)* Spannungsbauch *m*
  p. attractive force *(Ph)* Anziehungspotential *n*
  p. barrier Potentialberg *m*, Potentialwall *m*
  p. capacity *(Masch)* theoretische Leistung *f*
  p. circuit *(El)* Spannungspfad *m*
  p. coil Spannungsspule *f*
  p. curve *(El)* Spannungskurve *f*
  p. diagram *(El)* Potential[schau]bild *n*
  p. difference Potentialunterschied *m*, Potentialdifferenz *f*; Spannungsunterschied *m*

p. divider *(El)* Spannungsteiler *m*, Potentiometer *n*
p. drop *(El)* Spannungsabfall *m*, Potentialgefälle *n*, Potentialabfall *m*; Potentialsprung *m*
p. energy potentielle (ruhende) Energie *f*, Energie der Lage, Lagenenergie *f*
p. energy curve *(El)* Potentialkurve *f*
p. energy of deformation Deformationsenergie *f*, potentielle Energie *f* der Formänderung
p. equalizer Spannungsregulator *m*
p. equation Potentialgleichung *f*
p. fall Anodenfall *m*, Katodenfall *m* (*s. a.* p. drop)
p. field Potentialfeld *n*
p. flow Potentialströmung *f*
p. function Potentialfunktion *f*
p. fuse *(El)* Spannungssicherung *f*; Sicherung *f* des Spannungspfades
p. gradient Potentialgradient *m*, Gradient *m* des Potentialwalls
p. head Geschwindigkeitshöhe *f*
p. hill *(El)* Potentialberg *m*
p. loop *(El)* Spannungsbauch *m*
p. motion Potentialströmung *f*
p. node *(El)* Spannungsknoten *m*
p. of excitation Anregungsspannung *f*
p. of stimulation Anregungsspannung *f*
p. of thermodiffusion Thermodiffusionspotential *n*
p. scattering Potentialstreuung *f*
p. slope *(El)* Feldstärke *f*, Potentialgradient *m*
p. theory Potentialtheorie *f*
p. threshold Potentialschwelle *f*
p. to breakdown *(El)* Durchbruchsspannung *f*
p. transformer Spannungswandler *m*, Spannungstransformator *m*
p. trough Potentialsenke *f*, Potentialtopf *m*
p. well Potentialkasten *m*, Potentialmulde *f*, Potentialtopf *m*
potentiograph *(El)* registrierender Spannungsteiler *m*
potentiometer 1. Potentiometer *n (einstellbarer Widerstand in Analogrechnern)*; Spannungsteiler *m*; 2. Kompensator *m*, Kompensationsschreiber *m (Meßtechnik)*; 3. *s. a.* self balancing potentiometer
p. band Potentiometerstreifen *m*
p. braking controller Bremsregler *m*
p. granularity *(BMSR)* Potentiometerkörnigkeit *f*
p. measuring circuit Potentiometermeßkreis *m*
p. recorder Kompensationsschreiber *m*
p. stud Stufe *f* des Spannungsteilers
p.-type resistor Potentiometerwiderstand *m*
p. wire *(El)* Schleifdraht *m*
potentiometric recorder *(El)* selbstabgleichender Widerstandsgeber *m*; Kompensationsschreiber *m*
p. titration *(Ch)* potentiometrische Titration *f*
pothead Kabelendmuffe *f*
Potier's coefficient of equivalence *(Elm)* Umrechnungsfaktor *m* nach Potier
Potier's electromotive force EMK *f* des Luftspaltfeldes, Potier-EMK *f*
potted capacitor *(El)* Becherkondensator *m*
potter Töpfer *m*
potter's clay Ton *m*
p. wheel Töpferscheibe *f*
pottery Steinzeug *n*; Töpferei *f*; Töpferware *f*
potting *(Text)* Potten *n*, Topffärben *n*
poultry Geflügel *n*
p. farmer Geflügelzüchter *m*
p. grit Futterkalk *m*
p. house Hühnerhaus *n*
pouncing machine Glättmaschine *f (Hutherst.)*
pound / to [zer]stoßen, schlagen, klopfen
pound Pound *n (Masseeinheit in England und USA; im Avoirdupois-System: 1 lb = 0,4535923 kg; ferner in USA: troy pound [für Edelmetalle] und apothecary pound = 1 lb.t. = 1 lb.ap. = 0,37324177 kg)*
Pound Centigrade Unit *englische Einheit der Wärmemenge: 1 PCU = 1,8 BTU = 4,536 · 10^{-1} kcal*
pound-foot *englische Einheit der Arbeit; 1 lb.ft. = 138,255 · 10^{-3} kpm*
pound-force *englische Krafteinheit; 1 Lb = 4,448 N = 0,454 kp*
pound-force per square foot *englische Druckeinheit; 1 Lb./sq.ft. = 4884 · 10^{-4} kp/cm²*
pound-force per square inch *englische Druckeinheit; 1 Lb./sq.in. [1 psi.] = 7,031 · 10^{-2} kp/cm²*
pound-force per square inch absolute *englische Einheit zur Angabe des absoluten Drucks; 1 psia = 7,031 · 10^{-2} at absoluter Druck*
pound-force per square inch gauge *englische Einheit zur Angabe von Überdruck; 1 psig = 7,031 · 10^{-2} at Überdruck*
pound-force per square yard *englische Druckeinheit; 1 Lb./sq.yd. = 0,543 · 10^{-4} kp/cm²*
pound-inch *englische Einheit der Arbeit; 1 lb.in. = 1,15 · 10^{-2} kpm*
pound per cubic foot *englische Einheit der Dichte; 1 lb./cu.ft. = 16,02 kg/m³*
pound per cubic inch *englische Einheit der Dichte; 1 lb./cu.in. = 27,68 g/dm³*
pound per cubic yard *englische Einheit der Dichte; 1 lb./cu.yd. = 0,593 kg/m³*
pound per gallon *englische und amerikanische Dichteeinheit: 1 lb./gal. = 9,98 · 10^{-2} kg/l in England; 11,98 · 10^{-2} kg/l in USA*
pound-weight *Krafteinheit in USA [engl. = pound-force]; 1 lb.wt. = 1 Lb = 4,448 N = 0,454 kp*

pound-yard *englische Einheit der Arbeit; 1 lb./yd. = 4,148 · 10^{-1} kpm*
poundal Poundal *n (englische Krafteinheit: 1 pdl = 0,138 N)*
pounder Stößel *m*, Mörserkeule *f*
pour / to gießen; vergießen, abgießen, kippen, schütten
p. again nachgießen
p. concrete Beton einbringen (schütten), betonieren
p. from the bottom *(Met)* steigend gießen
p. from the top *(Met)* fallend gießen
p. horizontally *(Met)* waagerecht gießen
pour point Stockpunkt *m (Schmiermittel)*
p. point depressant Stockpunkterniedriger *m*
pourability Gießbarkeit *f*, Vergießbarkeit *f*; Durchfließfähigkeit *f*
pourable [ver]gießbar
p. sealing compound *(Plast)* Vergußmasse *f*
poured-in-place *(Bw)* am Einbauort betoniert, monolithisch hergestellt
p.-in-place concrete Ortbeton *m*, monolithischer Beton *m*
pouring Vergießen *n*; Abguß *m*, Guß *m*
p. basin Gießtümpel *m*
p. car Gießwagen *m*
p. channel Gießrinne *f*
p. drum Trommelgießpfanne *f*
p. gate Anschnitt *m*, Gießtrichter *m*
p. hole Gießrinne *f*
p. ladle Gießpfanne *f*
p. lip Gießschnauze *f (der Gießpfanne)*
p. machine Gießmaschine *f*
p. of concrete Betonieren *n*, Betonierung *f*, Einbringen *n* (Einbringung *f*) des Betons
p. on end stehendes Gießen *n*
p. pit Gießgrube *f*
p. speed Gießgeschwindigkeit *f*
p. temperature Gießtemperatur *f*
p. test Fließprobe *f (Schmelzverfahren)*
powder / to pulverisieren, *(zu Pulver)* zerreiben, mahlen; einpudern
powder Pulver *n*, Mehl *n*, Preßpulver *n*, Puder *m*
p. charge Pulverladung *f*
p. duster Pulverzerstäubungsgerät *n*
p. man *(Brg)* Schießberechtigter *m*, Schießhauer *m*
p. metal Sintermetall *n*, Pulvermetall *n*, Metallpulver *n*
p. metallurgical pulvermetallurgisch
p. metallurgy Pulvermetallurgie *f*
p. particle Pulverteilchen *n*
p. rocket *s.* powdered-fuel rocket
p. sintering Pulversintern *n*, Pulversinterverfahren *n*
p. train Pulversatz *m*, Zündsatz *m*, Brandsatz *m*
p.-type fire extinguisher Trockenlöscher *m*
powdered pulverisiert
p. blast-furnace slag gemahlene Hochofenschlacke *f*
p. brown-coal Braunkohlenstaub *m*
p. charcoal Holzkohlenpulver *n*
p. coal Kohlenstaub *m*, Staubkohle *f*
p.-coal combustion Kohlenstaubverbrennung *f*
p. coke Koksmehl *n*
p. core Massekern *m (für elektrische Spulen)*
p. flux Schweißpulver *n*
p.-fuel rocket Feststoffrakete *f*
p. glue Leimpulver *n*
p. iron core *(El)* Eisenpulverkern *m*
p. metal part Sintermetallteil *m*
p. milk Trockenmilch *f*
p. mineral[s] Gesteinsmehl *n*
powdering 1. Feinstmahlung *f*; 2. Stäuben *n (einer Gießform)*
p. a mould Einpudern *n* einer Form
powderless etching *(Typ)* Einstufenätzung *f*
power 1. *(Math)* Potenz *f*; 2. Leistung *f*, nutzbare Energie *f*; Kraft *f*; 3. Leistungsvermögen *n*, Vermögen *n*, Fähigkeit *f*, Kapazität *f*; 4. *(Opt)* Stärke *f*; 5. *(Brg)* Mächtigkeit *f*
p. absorption *(El)* Leistungsaufnahme *f*, Leistungsverbrauch *m*
p. amplification *(El)* Leistungsverstärkung *f*
p. amplifier Leistungsverstärker *m*; Senderverstärker *m*; Endverstärker *m*
p.-amplifier stage Endverstärkerstufe *f*
p.-assisted control 1. *(Flg)* Stützmotorsteuerung *f*; 2. Boostersteuerung *f*
p.-assisted steering *(Kfz)* Kraftsteuerung *f*
p. attenuation Leistungsdämpfung *f*
p.-band merit *(Fmt)* Verstärkungsbandbreitezahl *f*
p. bracket Leistungsbereich *m*
p. breaker *(Am)* Leistungsschalter *m*
p. breeder Leistungsbrutreaktor *m*
p. brush finishing Bürsten *n (Polieren)*
p. bus-bar Starkstromsammelschiene *f*, Kraftschiene *f*
p. cable Starkstromkabel *n*
p. chain saw Kettensäge *f*
p. circuit Starkstromleitung *f*, Kraftstromleitung *f*; Hauptstromkreis *m*
p. coefficient Leistungskoeffizient *m (eines Kernreaktors)*
p. component Wirkkomponente *f*, Wattkomponente *f*
p. consuming point *(El)* Verbraucherort *m*, Verbraucherstelle *f*
p. consumption Strombedarf *m*, Kraftverbrauch *m*, Leistungsaufnahme *f*, Leistungsbedarf *m*, Energiebedarf *m*
p. control Leistungsregelung *f*, Leistungssteuerung *f*

p. control loop Leistungsregelkreis *m*
p. control rod Leistungsregelstab *m*, Steuerstab *m* für das Leistungsniveau
p. conversion Energieumwandlung *f*, Energieumformung *f*
p. converter Leistungswandler *m*, Energiewandler *m*
p. cross-feed *(Masch)* Planzug *m*
p. cruiser Motorkreuzer *m (Jacht)*
p. current Kraftstrom *m*, Starkstrom *m*
p. curve Leistungskurve *f*
p. cylinder *(Kfz)* Arbeitszylinder *m (Servolenkung)*
p. dam Kraftwerktalsperre *f*
p. delivery Leistungsabgabe *f*
p. density Leistungsdichte *f*
p. detection Gleichrichtung *f* mit hoher Leistung
p. detector Kraftdetektor *m*, Leistungsdetektor *m*
p. direction relay Leistungsrichtungsrelais *n*
p. dissipation Verlustleistung *f*; Energiezerstreuung *f*, Energiedissipation *f*, Energievernichtung *f*
p. distribution Stromverteilung *f*, Kraftverteilung *f*
p. divider Leistungsteiler *m (Wellenleitertechnik)*
p. drain *(El)* Leistungsaufnahme *f*
p. drive Kraftantrieb *m*
p.-driven durch Motor angetrieben, kraftbetrieben
p.-driven roller Motorwalze *f (Straßenbau)*
p. end of the engine Kraftabgabeseite *f* des Motors
p. engine Kraftmaschine *f*, Motor *m*
p. engineering Starkstromtechnik *f*
p. equalizer *(Am)* Echofalle *f (Radar)*
p. extraction Leistungsentzug *m*, Leistungsentnahme *f*
p. factor *(El)* Leistungsfaktor *m*
p.-factor adjustment Einstellvorrichtung *f* für den Phasenabgleich
p.-factor capacitor Phasenschieberkondensator *m*
p.-factor compensation Blindstromkompensation *f*
p.-factor improvement Phasenschiebung *f*, Leistungsfaktorverbesserung *f*
p.-factor indicator Phasenmesser *m*, Leistungsfaktormesser *m*
p. failure *(El)* Netzausfall *m*
p.-fed automatisch zugeführt, mit mechanisierter Zuführung
p. feed automatischer Vorschub *m*
p.-feeding maschineller Vorschub *m*
p. flash-up Leistungsspitze *f*
p. fluctuation Stromschwankung *f*, Netzschwankung *f*; Leistungsschwankung *f*
p. frequency Netzfrequenz *f*, Betriebsfrequenz *f*
p. frequency dew withstand voltage Stehwechselspannung *f* im betauten Zustand
p. frequency dry withstand voltage Stehwechselspannung *f* im trockenen Zustand
p. frequency wet withstand voltage Stehwechselspannung *f* im beregneten Zustand
p. frequency withstand voltage Stehwechselspannung *f*
p. fuse HL-Sicherung *f*, Hochleistungssicherung *f*, Grobsicherung *f*
p. gain Leistungsgewinn *m*, [praktischer] Antennengewinn *m*
p. gas Treibgas *n*; Generatorgas *n*
p. generating plant (station) Kraftwerk *n*, Stromversorgungsanlage *f*
p. given out abgegebene Leistung *f*
p. grid detector Gitterstromgleichrichter *m*
p. hacksaw Bügelsägemaschine *f*, Maschinenbügelsäge *f*, Hubsäge *f (für Metall)*
p. hacksaw blade Maschinenbügelsägeblatt *n*, Hubsägeblatt *n*
p. hammer Maschinenhammer *m*, Krafthammer *m*
p.-handling capacity Übertragungsleistung *f (z. B. des Netzes)*; Leistungsaufnahme *f (der Antenne)*; Belastbarkeit *f (Funk)*
p. house *(Am)* Kraftwerk *n*
p. impulse Stromstoß *m*, Laststoß *m*
p. input aufgenommene (zugeführte) Leistung *f*, Leistungsaufnahme *f*, Eingangsleistung *f*
p. kerosene Traktorenkraftstoff *m*
p. lathe Leit- und Zugspindeldrehmaschine *f*
p. laundry equipment mechanische Wäschereieinrichtung *f*
p. law Potenzgesetz *n*
p. lead Stromversorgung *f*, Stromzuführung *f*, Starkstromzuführung *f*
p. level Leistungsniveau *n*, Leistungspegel *m*, Energiepegel *m*
p. level recorder Leistungsschreiber *m*
p. limit Grenzleistung *f*, Leistungsgrenze *f*
p. line Starkstromleitung *f*
p.-line hum Netzbrumm *m*
p.-line ripple Netzflimmern *n*
p. load Strombelastung *f*
p.-loaded face *(Brg)* mit Schrämlader ausgerüsteter Streb *m*
p. loading mechanisiertes Laden *n*
p. locking frame *(Eis)* Kraftstellwerk *n*
p. longitudinal feed *(Masch)* Längszug *m*
p. longitudinal motion of carriage *(Masch)* Längszug *m*
p. longitudinal sliding motion *(Masch)* Längszug *m*

p. loom mechanischer Webstuhl *m*
p. loom weaver Maschinenweber *m*
p.-loomed carpet maschinell gewebter Teppich *m*
p. loss Leistungsverlust *m*
p. lost Verlustleistung *f*
p. mains Starkstromnetz *n*
p. margin Grenzleistung *f*
p. matching Leistungsanpassung *f*
p. meter *(Am)* Leistungsmesser *m*, Wattmesser *m*, Wattmeter *n*
p. navvy Bagger *m*; Hochlöffelbagger *m*, Löffelbagger *m*
p. of accommodation Akkommodationsvermögen *n*
p. of current *(Am)* Stromstärke *f*
p. of hydrolysis *(Ch)* Hydrolysierungsfähigkeit *f*
p. of the governor Arbeitsvermögen *n* des Reglers
p. operated mit Hilfsenergie, indirekt wirkend; kraftbetrieben, kraftbetätigt
p.-operated control Regelung *f* mit Hilfsenergie, indirekte Regelung
p.-operated hoist Kraftaufzug *m*
p.-operated roller conveyor Rollengang *m*
p. operation Betätigung *f* mit Kraftantrieb *(z. B. von Schaltern)*
p. oscillator Leistungsoszillator *m*, Leistungsgenerator *m*
p. outlet Netzsteckdose *f*, Steckdose *f*
p. output Leistungsabgabe *f*, abgegebene Leistung *f*; Ausgangsleistung *f*
p. overshoot Leistungsüberschlag *m*
p. pack 1. Netzanschlußgerät *n*, Netz[anschluß]teil *n*; 2. Hydraulikaggregat *n*
p. package Triebwerksblock *m*
p. per unit area Strahlungsdichte *f (Antennentechnik)*
p. per unit of displacement *(Kfz)* Hubraumleistung *f*, Literleistung *f*
p. per unit of surface *(Kfz)* Flächenleistung *f*
p. phase *(Rak)* Antriebsbahn *f*
p. piston Arbeitskolben *m*, Stellkolben *m*
p. plant 1. Motor *m*, Triebwerk *n*; 2. *(Am)* Kraftwerk *n*, Elektrizitätswerk *n*; 3. Energieerzeugungsanlage *f*
p. plant intake structure Einlaßbauwerk *n (zum Kraftwerk)*
p. plug Netzstecker *m*
p. press mechanische Presse *f*
p. production Energieerzeugung *f*, Erzeugung *f* nutzbarer Energie
p. rail Stromschiene *f*
p. range Leistungsbereich *m*
p. rapid traverse lever *(Masch)* Hebel *m* für maschinellen Eilgang
p. rating Belastbarkeit *f*, Nenn[eingangs]leistung *f*; Leistungsbestimmung *f*
p. ratio of a directional aerial Leistungsverhältnis *n* einer Richtantenne
p. reactor Leistungsreaktor *m*, Energiereaktor *m*, Kraftwerkreaktor *m*
p. recorder *(linearer)* Leistungsschreiber *m*
p. relay Leistungsrelais *n*
p. required Leistungsaufnahme *f*
p. reserve Kraftüberschuß *m*
p.-reversal relay Rückstromrelais *n*
p. room Maschinenraum *m*
p. saw Motorsäge *f*
p. screw Kraftschraube *f*
p. screw press mechanische Spindelpresse *f*
p. series Potenzreihe *f*
p. setting Leistungsregelung *f*, Leistungseinstellung *f*
p. shovel 1. Kraftschaufel *f*; 2. Bagger *m*; Hochlöffelbagger *m*, Löffelbagger *m*
p. source Stromquelle *f*; Netzgerät *n*
p. stage Leistungsstufe *f*, Kraftstufe *f*
p. standing wave ratio Leistungsreflektionsfaktor *m*, Stehwellenverhältnis *n*
p. station *(Am)* Kraftwerk *n*; Zentrale *f*
p.-steering mechanism *(Kfz)* Servolenkung *f*, Servolenkgetriebe *n*
p. storage Speicherung *f (von Wasser)* zur Energieerzeugung
p. stowing *(Brg)* mechanischer Versatz *m*
p. stroke Arbeitstakt *m*, Arbeitshub *m*, Expansionshub *m (Verbrennungsmotor)*
p. stub shaft Antriebswelle *f*
p. supply Netzanschluß *m*, Zuleitung *f*; Stromversorgung *f*; Energieversorgung *f*
p. supply cable Starkstromkabel *n*; Netz[anschluß]kabel *n*, Zuleitungskabel *n*
p. supply set Netzanschlußgerät *n*
p. supply system Stromversorgungsnetz *n*
p. supply unit Netzanschlußgerät *n*; Netzanschlußteil *n*
p. supply variation Stromversorgungsschwankung *f*
p. switch Leistungsschalter *m*, Netzschalter *m*
p. switchboard Kraftstromschalttafel *f*
p. take-off Zapfgetriebe *n*, Nebengetriebe *n*, Zapfwelle *f*; Antriebswelle *f*
p. take-off gear Zapfgetriebe *n*, Verteilergetriebe *n*
p. take-off shaft Zapfwelle *f*
p. test run Probelauf *m* bei Vollast
p.-to-weight ratio Leistungsgewicht *n*
p. train Kraftdurchgang *m*, Kraftübertragung[sweg *m*] *f*; Triebwerk *n*
p. transfer Energieübertragung *f*, Leistungsübertragung *f*
p. transfer relay Lastumschaltrelais *n*
p. transformer Leistungstransformator *m*,

Leistungsumspanner *m*, Krafttransformator *m*, Transformator *m* für Kraftzwecke
p. transmission Energieübertragung *f*; Leistungsübertragung *f*
p.-transmission chain Treibkette *f*
p.-transmission elements *(Masch)* Triebwerksteile *npl*
p.-transmission equipment *(Masch)* Triebwerk *n*
p.-transmission line Hochspannungsleitung *f*
p.-transmission screw thread Bewegungsgewinde *n*
p.-transmission shaft *(Masch)* Triebwerkswelle *f*
p. transmitting capacity Leistungsübertragung *f (Fähigkeit)*
p. traverse Selbstgang *m*
p. trowel Glättmaschine *f*, Glättscheibe *f*
p. turbine Arbeitsturbine *f*
p. unit 1. Leistungseinheit *f*; 2. Triebwerk *n*; 3. Leistungsverstärker *m*, Kraftverstärker *m*; Leistungsteil *n*, Netzteil *n*; 4. Antriebsgruppe *f*
p. valve Leistungsröhre *f*
p. water Druckwasser *n*, Preßwasser *n*
p.-weight ratio Leistungsgewicht *n*
p. winding *(El)* Leistungswicklung *f*
powered flight *(Rak)* Antriebsflug *m*
p.-flight phase Antriebsbahn *f*
p.-flight trajectory Antriebsbahn *f*
powerful leistungsfähig *(Maschine)*; *(Opt)* stark; stark vergrößernd
p. lens lichtstarke Linse *f*
Poynting vector Poyntingscher Vektor *m (Elektrodynamik)*
pozzolanic *s.* pozzuolanic
pozzuolane *(Geol)* Pozz[u]olanerde *f*
pozzuolanic puzzolanartig, puzzolanhaltig
p. cement Puzzolanzement *m*
PPI display Rundsichtschirmbild *n (Radar)*
PPI prediction theoretisches Radarbild *n*
PPI repeater Bildschirmreproduktionsgerät *n*, Radarbildfernübertragungsgerät *n*
praam Prahm *m*
practicability [praktische] Durchführbarkeit *f*, Anwendungsmöglichkeit *f*
practicable durchführbar, ausführbar, erreichbar, anwendbar, benutzbar
p. vacuum erreichbares Vakuum *n*
practical praktisch, angewandt, verwendbar, nützlich
p. application praktische Anwendung *f*, Nutzanwendung *f*
p. chemistry angewandte Chemie *f*
p. course Praktikum *n*
p. example Anwendungsbeispiel *n*
p. range praktische (tatsächliche) Reichweite *f*
p. unit praktische Einheit *f*
practice Praxis *f*, Übung *f*, Anwendung *f*, Arbeitsweise *f*, Betrieb *m (Verfahren)*, praktische Ausführung *f*
p. buzzer *(Fmt)* Übungssummer *m*
p. flight Übungsflug *m*
p. position *(Fmt)* Übungsplatz *m*
p. rocket Übungsrakete *f*
practise / to ausführen, durchführen, praktisch anwenden, praktizieren
pram Prahm *m*
Prandtl number Prandtlsche Zahl *f*
prang *(Flg)* Bruchlandung *f*
praseodymia Praseodymoxid *n*
praseodymium Praseodym *n*
pre-admission Voreinströmung *f*
pre-age / to künstlich altern, voraltern
pre-aging künstliche Alterung *f*, Voralterung *f*
pre-amplification *(Fmt)* Vorverstärkung *f*
pre-amplifier *(Fmt)* Vorverstärker *m*
pre-amplify / to vorverstärken
pre-arcing time *(El)* Ansprechdauer *f (einer Schmelzsicherung)*
pre-assemble / to *(Bw)* vormontieren
pre-bore / to vorbohren
pre-centred base Zentrierfassung *f*
pre-chamber Vorkammer *f*
pre-chamber engine Vorkammermotor *m*
pre-combustion Vorverbrennung *f*
pre-combustion chamber Vorkammer *f*
pre-combustion chamber engine Vorkammermotor *m*
pre-compounded vorbeharzt, vorimprägniert
pre-compress / to vordrücken; vorverdichten
pre-concentrate *(Brg)* Vorkonzentrat *n*
pre-consolidate / to *(Bw)* vorverdichten
pre-consolidation *(Bw)* Vorverdichtung *f*
pre-constructed section *(Schiff)* vorgearbeitete Sektion *f (für die Montage auf der Helling)*
pre-constructed unit vorgefertigtes Bauteil *n*
pre-cool / to vorkühlen
pre-cooler Vorkühler *m*
pre-crushing Vorzerkleinerung *f*
pre-cut / to vorschneiden
pre-cutter tooth Vorschneidezahn *m*
pre-cutting share of plough Vorschar *n (Pflug)*
pre-determination *s.* predetermination
pre-drill / to *s.* predrill / to
pre-dry / to vortrocknen
pre-drier Vortrockner *m*
pre-emphasis *(El)* Hervorhebung *f (diskreter Frequenzen)*, Preemphasis *f*, Vorbetonung *f*
pre-engineered vorgefertigt
pre-equalization *(El)* Aufnahmeentzerrung *f*, Wiedergabeentzerrung *f*
pre-equalize pulses *(Fs)* Vortrabanten *mpl (Halbzeilenimpulse)*
pre-escape Vorausströmung *f*

pre-evacuate / to vorevakuieren
pre-evaporate / to vorverdampfen
pre-evaporation Vorverdampfung *f*
pre-expansion chamber *(Kfz)* Auspuffvorschalldämpfer *m*
pre-fabricated unit vorgefertigtes Bauteil *n*
pre-fade listening *(Fmt)* Vorhören *n* eines laufenden Programms
pre-finish / to *(Bw)* vorfertigen
pre-finished *(Bw)* vorgefertigt
pre-finished board vorgefertigte Platte *f*
pre-finishing of plywood panel Vorfertigen *n* von Sperrholzplatten
pre-heat / to *s.* preheat / to
pre-heating *s.* preheating
pre-igniter Frühzünder *m*, Frühzündvorrichtung *f*
pre-ignition Frühzündung *f*, Glühzündung *f*
pre-ignition Diesel engine Vorkammerdieselmotor *m*
pre-ignition plug Frühzündkerze *f (Versuchsmotore)*
pre-launching Vorlauf *m (beim Stapellauf)*
pre-load / to vorbelasten, vorspannen, Vorspannung geben
pre-load Vorlast *f*
pre-loaded bearing vorgespanntes Lager *n*
pre-loading Vorspannung *f*
pre-mix / to vormischen
pre-mix vorgemischtes harzgetränktes Glasfasermaterial *n*
pre-mix[ed] carpet Teppichbelag *m (Straßenbau)*
pre-pierce / to vorlochen *(Stanzerei)*
pre-saturation *s.* presaturation
pre-screening Vorklassierung *f*
pre-select / to vorwählen
pre-selected heading *(Flg)* eingedrehter Kurs *m*
pre-selecting rotary line switch *(Am) (Fmt)* Drehwähler *m (als Vorwähler)*
pre-selection *s.* preselection
pre-sensitized [litho-]plate *(Typ)* vorbeschichtete Offsetplatte *f*
pre-set / to *s.* preset / to
pre-shave / to vorschaben *(Zahnrad)*
pre-shave lotion [vor der Rasur zu verwendendes] Rasierwasser *n*
pre-shrinking Vorschrumpfen *n*
pre-shrunk *(Text)* vorgeschrumpft, Vorschrumpf-
pre-slit vorgeschlitzt
pre-slung cargo method *(Schiff)* Pre-slung cargo-Methode *f (während des ganzen Transportzyklus wird das Ladegut in Schlingen oder Netzen befördert)*
pre-softened coal *(Brg)* aufgelockerte anstehende Kohle *f*
pre-stator stage Eintrittsleitvorrichtung *f*
pre-stress / to *s.* prestress / to
pre-superheater *(Wärm)* Vorüberhitzer *m*
pre-tension / to *(Bw)* mit sofortigem Verbund vorspannen
pre-tensioning *(Bw)* Vorspannung *f* mit sofortigem Verbund
pre-test / to vorprüfen
pre-tie / to *(Bw)* Bewehrung vorbereiten, Bewehrungskörbe vorfertigen
preact *(BMSR)* Vorhalt *m*
preadjusted *(Masch)* vorher eingestellt
preadjustment Voreinstellung *f*
  p. of homing device *(Rak)* Voreinweisung *f*
prealignment Vorabgleich *m (Funktechnik)*
prealloy Vorlegierung *f*
preamplification *s.* pre-amplification
préavis call *(Fmt)* Gespräch *n* mit Voranmeldung, V-Gespräch *n*
prebatching bin *(Bw)* Vormischsilo *m*
prebody / to vorverdicken *(Lackindustrie)*
precast vorgefertigt *(Betonfertigteile)*
  p. concrete Betonfertigteil[e *npl*] *n*, Betonwaren *fpl*; Stahlbetonfertigteil[e *npl*] *n*
  p. concrete construction Fertigteilbauweise *f*, Montagebauweise *f*
  p. concrete skeleton construction Stahlbetonskelettmontagebauweise *f*
  p. concrete unit Stahlbetonfertigteil *n*, Betonfertigteil *n*; Fertigteil *n*
  p. construction Fertigteilbauweise *f*, Montagebauweise *f*
  p. floor vorgefertigte Decke *f*, Fertigdecke *f*
  p. prestressed concrete unit Spannbetonfertigteil *n*
  p. reinforced concrete unit Stahlbetonfertigteil *n*
  p. slab Fertigbetonplatte *f*
  p. unit *s.* p. concrete unit
  p. unit construction Fertigteilbauweise *f*, Montagebauweise *f*
precasting Vorfertigung *f (Betonbau)*
  p. plant Betonwerk *n*
  p. yard offenes Betonwerk *n*; Vorfertigungsplatz *m*
precaution Vorsichtsmaßregel *f*, Arbeitsschutzmaßnahme *f*
preceding pass Vorkaliber *n (Walzwerk)*
  p. stage *(Fmt)* Vorstufe *f*
precess / to präzessieren
precessing track *(BMSR)* vorrückende Spur *f (z. B. auf der Magnettrommel)*
precession *(Ph)* Präzession *f*
  p. measurement Präzessionsmessung *f*
  p. of orbit Bahnpräzession *f*
  p. time Präzessionszeit *f*
precessional motion Präzessionsbewegung *f*
prechamber *s.* pre-chamber
precious edel, kostbar
  p. metal Edelmetall *n*

p. metal conductor Edelmetalleiter *m*
p. stone Edelstein *m*
precipitability *(Ch)* Fällbarkeit *f*
precipitable *(Ch)* fällbar, abscheidbar
precipitant Ausfällungsmittel *n*, Fäll[ungs]mittel *n*; Fällbad *n*
precipitate / to *(Ch)* niederschlagen, [sich] absetzen, [aus]fällen, ausscheiden, abscheiden, kondensieren
precipitate *(Ch)* Niederschlag *m*, Präzipitat *n*, Fällung *f*, Abscheidung *f*; abgeschiedener Stoff *m*, Fällungsprodukt *n*
p. of sulphur Schwefelniederschlag *m*
precipitated *(Ch)* niedergeschlagen, abgesetzt
p. chalk Schlämmkreide *f*
p. copper Zementkupfer *n*
p. sludge Klärschlamm *m*
precipitates atmosphärische Niederschläge *mpl*
precipitating *(Ch)* Fällen *n*, Niederschlagen *n*
p. ion Fällungsion *n*
precipitation 1. Ausfällung *f*, Ausscheidung *f*; Sedimentation *f*; 2. atmosphärischer Niederschlag *m*
p. analysis Fällungsanalyse *f*
p. apparatus Fäller *m*, Fällapparat *m*
p. brittleness *(Met)* Alterungssprödigkeit *f*
p. by electrolysis elektrolytische Fällung *f*
p. echo Niederschlagsecho *n (Radar)*
p. electrode Niederschlag[s]elektrode *f*
p.-harden / to *(Met)* aushärten
p. hardening *(Met)* Aus[scheidungs]härtung *f*
p. heat-treatment Vergütung *f (Leichtmetall)*
p. process Niederschlagsarbeit *f (Bleimetallurgie)*
p. tank Absetzbecken *n*
precipitator Abscheider *m*; Staubabscheider *m*; *(Ch)* Fällapparat *m*; Fällungsmittel *n*
precipitous steil abfallend
precise genau, präzis
p. adjustment Feineinstellung *f*
p. levelling Präzisionsnivellement *n*
p. measurement Feinmessung *f*
precision Genauigkeit *f*; Präzision *f*
precision Fein-; Feinst-, Präzisions-
p. altimeter Feinhöhenmesser *m*
p. balance Präzisionswaage *f*
p. boring Genaubohren *n*, Feinstbohren *n*
p. boring machine Feinstbohrmaschine *f*; Genauigkeitsbohrmaschine *f*
p. boring operation Genauigkeitsbohrarbeit *f*, Feinstbohrarbeit *f*
p. capacitor Präzisionskondensator *m*, Genauigkeitskondensator *m*
p. casting Genauguß *m*, Präzisionsguß *m*
p. component Präzisionsteil *n*
p. dial gauge Feintaster *m*
p. drilling Genaubohren *n*, Feinstbohren *n*
p. finishing process Feinstbearbeitungsverfahren *n*
p. forging 1. Formschmieden *n*, Fassonschmieden *n*; 2. Genauschmiedestück *n*
p. gauge Präzisionslehre *f*
p. gauge block Parallelendmaß *n*
p. glass scale Glasmaßstab *m*
p. grind[ing] Genauigkeitsschliff *m*, Präzisionsschliff *m*
p. instrument Präzisionsinstrument *n*
p. lathe Feindrehmaschine *f*, Mechanikerdrehmaschine *f* zum Feindrehen
p. levelling Feinnivellement *n*
p. master gear Lehrzahnrad *n*
p. measuring instrument Feinmeßgerät *n*, Präzisionsmeßinstrument *n*
p. measuring scale Präzisionsskale *f*
p. measuring set Feinmeßgerät *n*
p. mechanics Feinmechanik *f*
p. oscillograph Meßoszillograf *m*
p. resistance Präzisionswiderstand *m*
p. screw Präzisionsgewindespindel *f*
p. slit Präzisionsspalt *m*
p. stamping Genaupreßteil *n*
p. test Genauigkeitsprüfung *f*
p. timekeeper Präzisionsuhr *f*
p. toolmaker Feinmechaniker *m*, Feinmechanikerin *f*
p. tube Präzisionsrohr *n*
p. turning Feinstdreharbeit *f*
p.-type bearing Präzisionslager *n*
p.-type instrument Feinmeßinstrument *n*
p. voltmeter Präzisionsvoltmeter *n*
p. wavemeter Präzisionswellenmesser *m*
p. work Präzisionsarbeit *f*
precleaner Vorreiniger *m*
precleaning Vorreinigung *f*
p. process Vorreinigungsverfahren *n*
precoat / to grundieren, vorstreichen
precoat Grundanstrich *m*, Grundierung *f*, Grundlack *m*
precombustion *s.* pre-combustion
preconditioning Vorbehandlung *f*
precool / to *s.* pre-cool / to
precure *(Plast)* vorzeitige Härtung *f*
precursor 1. Vorläufer *m*; 2. *(Fs)* Unterschwingen *n*; 3. *(Kern)* Mutterkern *m*
predecessor *(Math)* Vorgänger *m*
predefecation Vorscheidung *f (Zuckerindustrie)*
predetermination Vorausberechnung *f*, Prädetermination *f*
predetermine / to vorausberechnen
predetermined vorausberechnet
p. breaking point *(El)* Sollbruchstelle *f*
p. flight path vorausberechnete Flugbahn *f*
p. number *(Fmt)* vorgewählte Zahl *f*
p. path vorausberechnete Flugbahn *f*
p. voltage *(El)* vorgegebene Spannung *f*

predetonation vorzeitige Explosion *f*
predicate calculus *(BMSR)* Prädikatenkalkül *m*
prediction *(BMSR)* Vorhersage *f*, Extrapolation *f*
predictor 1. Warngerät *n*; 2. Voraussager *m*, Zielrechenmaschine *f*; Rechenregler *m*, Rechengerät für Folgeregler *(Zweipunktfolgeregler)*
predissociation Prädissoziation *f*
predominance of one colour Farbstich *m*
predominant vorherrschend, dominierend, dominant
predrill / to vorbohren
predrilled hole Vorbohrung *f*
predrilling Vorbohren *n*
prefab 1. Fertighaus *n*; 2. Fertigbauteil *n*
prefab *s.* prefabricated
prefabricate / to vorbearbeiten, vorfertigen
prefabricated vorgefertigt; komplettiert
  p. construction Fertigteilbauweise *f*, Montagebauweise *f*
  p. unit Fertigteil *n*
prefabrication Vorfertigung *f*, Fertigteilbauweise *f*
preference Priorität *f*
preferential bevorzugt
  p. absorption bevorzugte Absorption *f*
  p. solvent *(Ch)* selektives Lösungsmittel *n*
preferred axis Hauptachse *f*, bevorzugte Achse *f*
  p. orientation vorzugsweise Orientierung *f*, Hauptrichtung *f* *(Kristalle)*
  p. series Auswahlreihe *f*
prefilling press *(Plast)* Presse *f* zur Verdichtung des Preßpulvers
prefiltration Vorfilterung *f*
prefine / to *(Met)* vorfrischen
prefining *(Met)* Vorfrischen *n*
  p. mixer Vorfrischmischer *m*
prefinish / to *s.* pre-finish / to
prefix [signal] *(Fmt)* Vorbereitungszeichen *n*; Vorimpuls *m*
preform / to vorformen, vorfertigen
preformed vorgeformt, vorgefertigt
  p. cable Kabelbaum *m*
  p. winding *(Elm)* Formspulenwicklung *f*, Schablonenwicklung *f*
preforming *(Ch)* Tablettierung *f*; Vorformverfahren *n*
prefractionator *(Ch)* Vorabscheider *m*
pregermination Vorkeimung *f*
preheat / to anwärmen, vorwärmen
preheater Vorerhitzer *m*, Vorwärmer *m*, Vorwärmgerät *n*
preheating Vorwärmen *n*, Vorwärmung *f*
  p. bench Vorwärmtisch *m* *(Elektronenröhrenherstellung)*
  p. by means of exhaust steam Vorwärmung *f* durch Abdampf
  p. chamber Vorwärmkammer *f*
  p. flame Vorwärmflamme *f*
  p. furnace Vorwärmofen *m*
  p. of air Luftvorwärmung *f*
  p. valve *(Kfz)* Vorwärmklappe *f*
prehnitic acid Prehnitsäure *f*
preignition *s.* pre-ignition
preliminary vorbereitend, einleitend; Vor-
  p. adjustment Vorabgleich *m* *(Funk)*
  p. air tank *(Kfz)* Vorluftbehälter *m*
  p. annealing Vorglühen *n*
  p. assembly drawing Entwurfszeichnung *f*
  p. bath Vorbad *n* *(Film)*
  p. breaker Vorbrecher *m*
  p. calculation Überschlagsrechnung *f*
  p. clarification tank Vorklärbecher *m* *(Gesundheitstechnik)*
  p. computation Vorberechnung *f*; Vorbemessung *f*
  p. condenser *(Wärm)* Vorkondensator *m*
  p. condition Vorbedingung *f*
  p. cooling Vorkühlung *f*
  p. cutoff signal *(Rak)* Vorkommando *n*
  p. design Vorentwurf *m*
  p. disintegration Vorzerkleinerung *f*
  p. draft *(Text)* Vorverzug *m*
  p. drilling Vorbohren *n*
  p. fire refining Vorraffination *f* im Schmelzfluß
  p. grinding Vorschliff *m*
  p. investigation Voruntersuchung *f*
  p. opener Voröffner *m*, Vorschläger *m* *(für Baumwolle)*
  p. sedimentation Vorklärung *f* *(Klärbecken)*
  p. selector *(Fmt)* Vorwähler *m*
  p. signal Vorsignal *n*
  p. signal-frequency stage Hochfrequenzvorstufe *f*
  p. stage [of propulsion] *(Rak)* Anfahrstufe *f*
  p. step Vorstufe *f*
  p. stress *(Mech)* Vorspannung *f*
  p. test Vorversuch *m*, Vorprüfung *f*
  p. trial Vorversuch *m*
  p. value vorläufiger Wert *m*
  p. work Vorarbeit *f*
preliming Vorscheidung *f* *(Zuckerindustrie)*
preloading *s.* pre-loading
premature vorzeitig
  p. burst *(Mil)* Frühzerspringer *m*
  p. ignition Frühzündung *f*, Glühzündung *f*
premelted iron Vorschmelzeisen *n*
premium oil mit Wirkstoffen versehenes Motorenschmieröl *n*, Premiumöl *n*
preoperation period Anfahrperiode *f*, Anfahrzeit *f*
preoptive *(Masch)* Vorwahl-, mit Vorwähleinrichtung
preoscillation current *(El)* Anschwingstrom *m*
preoxidize / to anoxydieren

prepacked aggregate concrete Prepaktbeton *m*
p. concrete Prepaktbeton *m*
preparating drawing *(Text)* Vorstrecken *n*
preparation 1. Vorbereitung *f*; Herstellung *f*; 2. *(Ch)* Darstellung *f*; 3. Aufbereitung *f*; 4. Präparat *n*; 5. *(Met)* Zustellung *f (eines Ofens)*
p. concentrate Aufbereitungskonzentrat *n*
p. for seasoning *(Led)* Glanzabkochung *f*
p. glass Präparateglas *n*
p. in a pure state Reindarstellung *f (Met)*
p. of logs Holzputzen *n*
p. of ore Erzaufbereitung *f*
p. of seed grain Saatgutbereitung *f*
p. plant Aufbereitungsanlage *f*
preparatory vorbereitend
p. drawing frame *(Text)* Vorstrecke *f*
p. process Vorbereitung *f*
p. shaping Vorformen *n*
p. treatment Vorbehandlung *f*
p. work *(Brg)* Vorrichtungsarbeiten *fpl*
prepare / to 1. vorrichten; vorbereiten, aufbereiten, präparieren; 2. *(Ch)* darstellen; 3. herstellen, erzeugen
p. mortar Mörtel anmachen
p. ores Erze aufbereiten
prepared chalk Schlämmkreide *f*
p. lapping abrasive Läpp-Pasta *f*
prepatch panel *s.* problem board
prepayment Voraus[be]zahlung *f*
p. coin box Münzfernsprecher *m*
p. meter Münzzähler *m*
p. telephone Münzfernsprecher *m*
p. telephone coin box Münzfernsprecher *m*
preplaced vorher aufgebracht
p. aggregate concrete Prepaktbeton *m*
preplanning Vorausplanung *f*; Vorwahl *f*
preponderate / to schwerer sein, Übergewicht haben; sich neigen *(Waage)*
p. over zahlreicher sein, an Zahl übertreffen
prepreg vorimprägniertes Glasfasermaterial *n*
p. process Vorimprägnierverfahren *n*
prepress lamination vorgepreßte Schichtung *f*
prepulse *(Fs)* Vorimpuls *m (z. B. zur Auslösung der Zeitbasis)*
prerequisite Voraussetzung *f*
preroast / to anrösten, vorrösten
prerun Vorlauf *m*
presaturate / to vorsättigen; vormagnetisieren
presaturation Vorsättigung *f*; Vormagnetisierung *f*
presbyopia *(Opt)* Alterssichtigkeit *f*
presbyopic *(Opt)* alterssichtig
prescribed trajectory *(Rak)* vorgeschriebene Flugbahn *f*
preselection Vorwahl *f*; Vorwählung *f*, Vorwählschaltung *f*
p. device *(Masch)* Vorwahleinrichtung *f*
preselective *(Masch)* vorwählbar, Vorwahl-
p. switching *(Masch)* Vorwahlschaltung *f*
preselector Vorwahlschalter *m*; Vorwähleinrichtung *f*; Vorwähler *m*
p. control *(Masch)* Vorwahlschalter *m*
p. dial *(Masch)* Vorwählerscheibe *f*
presence *(Geol)* Vorkommen *n*
p. of water Wasserführung *f*
presentation Darstellung[sweise] *f*, Form *f*
preservation Konservierung *f*, Erhaltung *f*, Imprägnierung *f*, Tränkung *f*, Haltbarmachung *f*
preservative Holzschutzmittel *n*; Konservierungsmittel *n*, Schutzmittel *n*; Schutzanstrich *m*
preserve / to konservieren, präparieren, haltbar machen, erhalten; imprägnieren *(Holz)*
preserve 1. Halbkonserve *f*, Konserve *f*, Präserve *f*; 2. Einbettungsmaterial *n*, Präparat *n*; 3. Wildbahn *f*
preserving agent Konservierungsmittel *n*
preset / to voreinstellen; *(Dat)* vorgeben
preset voreingestellt; vorgegeben
p. breaking point *(El)* Sollbruchstelle *f*
p. counter voreingestellter Zähler *m*
p. counts Impulsvorwahl *f*
p.-time counting Zählen *n* mit Zeitvorwahl
p. timer Zeitschaltgerät *n*
presetting Voreinstellung *f*
p. control gespeicherter Befehl *m (Eisenbahnsignalwesen)*; *(Masch)* Vorwählersteuerung *f*
p. ring *(Opt)* Vorwahlring *m*
p. tank Beruhigungstank *m*, Vorabscheider *m*
press / to 1. pressen, drücken; prägen; bügeln; quetschen; verdichten *(Formsand)*; 2. *(Typ)* drucken
p. down herunterdrücken, niederdrücken
p. flat plattdrücken
p. into place einpressen
press Presse *f*
p. bed Pressentisch *m*
p. brake *(Am)* Abkantpresse *f*, Abkantmaschine *f (schwere Bauart)*
p. button Druckknopf *m*, Drucktaste *f*, Taste *f*
p. button block Drucktastenaggregat *n*
p. button control Tastensteuerung *f*, Druckknopfsteuerung *f*
p. button locking with automatic release Taste *f* mit selbsttätiger Auslösung
p. button switch Druckknopfschalter *m*, Drucktaste *f*
p. cake Filterkuchen *m*; Preßpulverkuchen *m*
p. casting Druckguß *m*
p.-casting machine Druckgießmaschine *f*, Druckgußmaschine *f*
p. channel Preßkanal *m (Sammelpresse)*
p. charger in chaff cutter *(Lw)* Preßeinleger *m (Häcksler)*

p. column Pressenständer *m*
p.-drawing Preßziehen *n*
p. feeder in chaff cutter *(Lw)* Preßeinleger *m (Häcksler)*
p.-filling size voll ausgenutzte Preßfläche *f (Preßmaschine)*
p. filter Druckfilter *n*, Preßfilter *n*
p. fit *(Masch)* Preßpassung *f*, Preßsitz *m*
p.-fitting Aufpressen *n*; Fügen *n*; Einpressen *n*
p. frame Pressenkörper *m*
p. house *(Brg)* Pressendienst *m*, Pressenhaus *n*
p. jack *(Typ)* Preßbengel *m*
p. line Pressenfließreihe *f*
p. platen Pressentisch *m*
p. plug Verschlußdeckel *m* zum Eindrücken
p. plunger Preßkolben *m (Sammelpresse)*
p. polished hochglanzpoliert
p. ram Pressenstößel *m*
p. room Stanzerei *f*
p. spring *(Text)* Kartenpresse *f*
p. switch Druckschalter *m*
p. table Pressentisch *m*
p.-to-logs machine Holzbrikettmaschine *f*
p. tools Schnitt-, Stanz- und Ziehwerkzeuge *npl*
p. tub Kelterbütte *f*
p. upright Pressenständer *m*
p. vat Kelterbütte *f*
p.-welding Kaltpreßschweißen *n*
p. work Preßformen *n*; Stanzen *n*
p.-work pressure Stanzdruck *m*, Preßformdruck *m*
p. working Preßformen *n*; Stanzen *n*
p. working of metals Stanzereitechnik *f*
pressboard Preßpappe *f*; Preßspan *m*
pressed gepreßt
p. blank Preßling *m (Pulvermetallurgie)*
p. brass part Preßteil *n* aus Messing, Messingpreßling *m*
p. brick Preßziegel *m*, Preßstein *m*
p.-dust core *(El)* Massekern *m*
p. fit Preßpassung *f*, Preßsitz *m*
p.-glass base Preßfuß *m (Elektronenröhren)*
p. part Preßteil *n*
p. peat Preßtorf *m*
p. piece Preßling *m*, Preßteil *n*
p.-steel frame Preßstahlrahmen *m*
p.-steel part Preß[schmiede]teil *n*
pressel switch Schnurschalter *m* [mit Druckknopf], Hängeschalter *m*
presser Druckvorrichtung *f*; Presse *f*, Quetsche *f*
p. plate Druckplatte *f*
pressing 1. Pressen *n*, Drücken *n*, Quetschen *n*, Prägen *n*; 2. Bügeln *n*, Plätten *n*
p. bed Bügeltisch *m*
p. board Gautschbrett *n*
p. contact Druckkontakt *m*
p. die Preßwerkzeug *n*, Stanzwerkzeug *n*
p. lustre *(Text)* Preßglanz *m*
p. on Aufziehen *n (des Rades)*
p. plant Presserei *f*
p. ring Druckring *m*, Spannband *n*, Spannring *m*
p. roll Preßwalze *f*
p. screw Druckschraube *f*, Druckspindel *f*
pressman Pressenmann *m*, Pressenfahrer *m*
pressroom Drucksaal *m*
pressure 1. Druck *m*; 2. *(Am, El)* Spannung *f*
p. accumulator Druckakkumulator *m*
p.-actuated microphone auf Schalldruck ansprechendes Mikrofon *n*
p. adjustment Druckeinstellung *f*
p.-agent actuator *(Rak)* Fördermittel *n*
p. amplitude Druckamplitude *f*
p. angle Passungswinkel α *m*, Eingriffswinkel α *m (Verzahnung)*
p. antinode Druckbauch *m*
p. arch *(Brg)* Druckgewölbe *n*
p. at pitch line Zahndruck *m (Verzahnung)*
p. at rest Ruhedruck *m*
p. at the jet exit Strahlaustrittsdruck *m (Turboantrieb)*
p. at the outlet opening Mündungsdruck *m (Turbine)*
p. blower Druckgebläse *n*
p. boiler Druckkocher *m (Papierherstellung)*
p. burst *(Brg)* Gebirgsschlag *m*
p. cell Druckmeßdose *f*
p. circulating lubrication Druckumlaufschmierung *f*
p. circulating system of lubrication Druckumlaufschmierung *f*
p. coil *(Am) (El)* Spannungsspule *f*
p. compensation Druckausgleich *m*
p. compensation device Druckausgleicheinrichtung *f*
p. contact *s.* pressing contact
p. controller *(Am) (El)* Spannungskonstanthalter *m*, Spannungsregler *m*, Spannungsgleichhalter *m*
p. deep drawing Fließpressen *n*
p.-dependent druckabhängig
p. diagram Druckdiagramm *n*
p. die casting 1. Spritzgießen *n*, Spritzguß *m*, Druckgießen *n*; 2. Druckguß[stück *n*] *m*, Preßguß[stück *n*] *m*, Spritzguß[stück *n*] *m*
p. differential Druckgefälle *n*
p. digester Druckkocher *m*
p. distribution Druckverteilung *f*
p. drop Druckabfall *m*, Druckverlust *m*, Druckgefälle *n*; *(BMSR)* Wirkdruck *m*
p. due to head of water *(Wärm)* Gefälledruck *m*
p. due to shrinking Schrumpfspannung *f (Turbine)*

p. energy Druckenergie *f*
p. equalizing Druckausgleich *m*
p. equalizing reservoir Druckausgleichbehälter *m*
p. feed Druckförderung *f*
p.-feed lubrication Druckschmierung *f*
p.-feed oil lubrication Druckölschmierung *f*
p. field *(Ph)* Druckfeld *n*
p. filtration Druckfilterung *f*
p. float Druckschwimmer *m*
p. fluctuation Druckschwankung *f*
p. fluid Druckflüssigkeit *f (Hydraulik)*
p.-forged part Preßteil *n*, Gesenkpreßteil *n*
p. forging Gesenkpreßteil *n*
p. gas welding Gaspreßschweißen *n*
p. gauge Manometer *n*, Druckanzeiger *m*, Druckmesser *m*
p.-gauge test pump *(Wärm)* Manometerprobierpumpe *f*
p.-gauge with protecting screen Manometer *n* mit Sicherheitsgitter (Rostdeckel)
p. governor Druckregler *m*
p. gradient Druckgefälle *n*
p. grease cup Druckschmierkopf *m*
p. grease fitting Druckschmierkopf *m*
p. grease gun Hochdruckschmierpresse *f*, Fettpresse *f*
p. grease lubrication Preßfettschmierung *f*
p.-grout / to mit Mörtel auspressen, verpressen
p. grouting *(Bw)* Auspreßverfahren *n*, Einpreßverfahren *n*, Injektionsverfahren *n*
p. gun Druckschmierer *m*
p.-gun grease Hochdruckfett *n*
p. harness *(Text)* Vordergeschirr *n*, Vorderwerk *n*
p. head Druckhöhe *f*; Druckgefälle *n*
p. hull Druckkörper *m (U-Boot)*
p. hydraulics Hydraulik *f* der Druckflüssigkeit
p. increase Druckanstieg *m*
p.-independent druckunabhängig
p. indicating device Druckanzeigegerät *n*
p. indicator Druckmelder *m*; Druckmesser *m*, Manometer *n*
p. inflation Reifendruck *m*
p. injection Druckeinspülung *f (Dieselmotor)*
p. ionization Druckionisation *f*
p. law *(Wärm)* Spannungsgesetz *n*
p.-limiting valve Maximaldruckventil *n*, Höchstdruckventil *n*
p. line Druckleitung *f*; Drucklinie *f (Bogen)*
p. line of telescopic jack Hubraumdruckleitung *f*
p. liquid Druckflüssigkeit *f (Hydraulik)*
p. lock Druckluftverschluß *m*
p. loss Druckverlust *m*
p.-lubricated druckgeschmiert
p. lubrication Druck[umlauf]schmierung *f*
p. maintenance valve Druckregelventil *n*
p. measurement Druckmessung *f*
p. medium Druckmittel *n (Hydraulik)*
p. meter Druckmesser *m*, Manometer *n*
p. mixing Druckmischen *n*
p. node Druckknoten *m*
p. normal to blade surface Bahndruck *m*, Normaldruck *m (Turbine)*
p. of mixture *(Wärm)* Mischungsdruck *m*
p. of mountain mass Gebirgsdruck *m*
p. of rolling *(Met)* Walzdruck *m*
p. of steam Dampfspannung *f*
p. oil Drucköl *n*, Hydrauliköl *n*
p. on sleeve *(Masch)* Muffendruck *m*, Gleitkraft *f*
p.-operated druckbetätigt
p. partings *(Brg)* Abdrückungen *fpl*, Drucklagen *fpl (anstehende Kohle)*
p. pipe Druckrohr *n*, Druckleitung *f*
p. pipe tube Druckrohrstutzen *m*
p. plate Anpreßplatte *f (Kupplung)*
p.-plate lug Druckplattennase *f (Kupplung)*
p. produced by the fan *(Masch)* Ventilatorpressung *f*
p. propagation Druckfortpflanzung *f*
p. reducing transformer Abwärtstransformator *m*, Abwärtswandler *m*, spannungserniedrigender Transformator *m*, Reduktionstransformator *m*, Reduziertransformator *m*
p. reducing valve Druckreduzierventil *n*, Druckminderventil *n*; Druckminderer *m*, Reduzierstation *f*, Reduzierventil *n*
p. regulating valve Sicherheitsdruckregler *m*; Druckregelventil *n*
p. regulator Druckregler *m*
p. regulator valve Druckregelventil *n*
p. relay Druckrelais *n*
p. release valve Reduzierventil *n*
p. relief valve Druckreduzierventil *n*, Überdruckventil *n*, Sicherheitsventil *n*
p. resonance Spannungsresonanz *f*, Reihenresonanz *f*, Serienresonanz *f*
p.-responsive druckgesteuert
p. ring Verdichtungsring *m*
p. rise Druckanstieg *m*
p. roller *(Met)* Oberwalze *f*
p. seal Druckraumabschluß *m*
p.-sealed druckdicht
p. sensitive druckempfindlich; selbstklebend
p. sensitive adhesive Haftkleber *m*
p.-sensitive adhesive film klebefertige Kaschierfolie *f*
p.-sensitive switch Membranschalter *m*, druckempfindlicher Schalter *m*
p. sequence valve Ausgleichsventil *n*
p. servo oil Drucköl *n*
p.-set ink *(Typ)* Absorptionsdruckfarbe *f*

p. side *(Kfz)* Förderseite *f*; Druckseite *f*
p. sintering *(Met)* Drucksintern *n*
p. solenoid *(El)* Andruckmagnet *m*
p. spring Druckfeder *f*
p.-spring thermometer Federthermometer *n*
p. stage Druckstufe *f (Turbine)*
p. storage device Druckspeichereinrichtung *f*
p. stress Druckspannung *f*
p. system of lubrication Druckschmierung *f*
p. tank *s.* p. vessel
p. tap Druckentnahmebohrung *f (Durchflußmessung)*
p. thermit welding aluminothermisches Preßschweißen *n*
p. thermometer Tensionsthermometer *n*, Federthermometer *n*, manometrisches Thermometer *n*
p.-tight druckdicht
p. transformer *(Am)* Spannungswandler *m*, Spannungstransformator *m*
p. transmission Druckfortpflanzung *f*
p. transmitting seal Trennvorlage *f*, Druckmittler *m (Druckmesser)*
p. valve Druckventil *n*
p. valve cone Druckventilkegel *m*
p. valve spring Druckventilfeder *f*
p. variation Druckschwankung *f*
p. vessel Druckgefäß *n*; Druckbehälter *m*, Druckkessel *m*
p. welding Preßschweißen *n*
p. zone *(Mech)* Druckzone *f*
pressurize / to unter Druck setzen
pressurized druckdicht, mit Druckausgleich
p. air Druckluft *f*
p. blast furnace Hochdruckhochofen *m*
p. cable *(El)* Druckgaskabel *n*
p. water Druckwasser *n*
p. water reactor Druckwasserreaktor *m*
pressurizer Druckregler *m*, Druckbehälter *m*
prestige-type building repräsentativer Bau *m*, repräsentatives Gebäude *n*
prestore / to *(Dat)* vorspeichern
prestress / to vorspannen, Vorspannung aufbringen *(Spannbeton)*
prestress Vorspannung *f (Spannbeton)*
p. loss Spannungsabbau *m*, Spannungsverlust *m (Spannbeton)*
prestressed vorgespannt *(Spannbeton)*
p. concrete Spannbeton *m*
p. concrete construction Spannbetonbau *m*
prestressing *(Bw)* Aufbringen *n* der Vorspannung, Vorspannen *n (Spannbeton)*
p. bed Spannbett *n*
p. duct Spannkanal *m*
p. element Spannglied *n*
p. force Vorspannkraft *f*
p. jack Spannpresse *f*
p. lane Spannbahn *f*
p. steel Spannstahl *m*; Spannbewehrung *f*
p. wire Spanndraht *m*
presupposition Voraussetzung *f*
preventer deck *(Schiff)* loses Deck *n*, Schutzbelag *m*
p. stay *(Schiff)* Schlingerstag *n*
prevention device Sperre *f (z. B. optischer Geräte)*
p. of accidents Unfallverhütung *f*
p. of passage of noise Schalldämmung *f*
p. of scale formation Kesselsteinverhütung *f*
preventive Schutzmittel *n*, Vorbeugungsmittel *n*
p. against corrosion Korrosionsschutzmittel *n*
p. resistance *(El)* Überschaltwiderstand *m*
p. resistor *(El)* Schutzwiderstand *m*
previously bored vorgebohrt
p. drilled vorgebohrt
p. drilled hole Vorbohrung *f*
prevulcanization Vorvulkanisation *f*
prewired instruction *(Dat)* auswechselbares (vorbereitetes) Programm *n (Programmbrett)*
price Preis *m*
p. list Preisliste *f*
p. per unit Stückpreis *m*
p. vector Preisvektor *m (Ökonometrie)*
prick / to 1. stechen *(mit einer Nadel)*; 2. *(Schiff)* absetzen *(Kurs in die Karte eintragen)*, das Besteck machen
prick-out Pikieren *n* der Pflanzen
p. punch Anreißkörner *m*
pricked by worms wurmstichig *(Holz)*
pricker Luftspieß *m*, Pumpstock *m (Formerei)*
p. needle Pikiernadel *f*
pricking the chart *(Schiff)* Absetzen *n* des Bestecks
prickle Dorn *m*, Stachel *m*
prill *(Met)* Regulus *m*, König *m*
prilling Sprühkristallisation *f*
prillion Schlackenzinn *n*
primacord *(Brg)* detonierende Zündschnur *f*
primality *(Math)* Primzahlcharakter *m*, Primzahleigenschaft *f*
primary 1. Primärteilchen *n*; 2. *(sl) (Elm)* Ständer *m*, Stator *m*
primary primär, Primär-; primärseitig
p. alcohol primärer Alkohol *m*
p. aluminium pig Hüttenaluminium *n*
p. armament schwere Artillerie *f (an Bord von Kriegsschiffen)*
p. armature *(El)* Primäranker *m*
p. battery Primärbatterie *f*
p. benzol Benzolvorprodukt *n (Petrolchemie)*
p. cell Primärelement *n*, Batterie *f*
p. circuit 1. Primär[strom]kreis *m*, primärer (induzierender) Stromkreis *m*; 2. *(Kern)* Primärkreislauf *m*
p. clearance *(Masch)* gewetzte Freifläche *f*; Führungsfase *f*; Hinterschliff *m*; Freiwinkel *m*

p. clearance angle *(Masch)* Hinterwetzwinkel *m*, Wetzwinkel *m*
p. clock Hauptuhr *f*, Mutteruhr *f (zur elektrischen Steuerung der Nebenuhren)*
p. coating Grundanstrich *m*, Grundierung *f*
p. coil Primärwicklung *f*, primäre Spule *f*, Primärspule *f*
p. colour Grundfarbe *f*, Stammfarbe *f*
p. controls *(Flg)* Hauptsteuerorgane *npl (Höhen-, Quer- und Seitenruder)*
p. coolant *(Kern)* Kühlmittel *n* des Primärkreislaufs
p. cosmic radiation Primärstrahlung *f*, primäre kosmische Strahlung *f*
p. cosmic-radiation component Primärkomponente (primäre Komponente) *f* der kosmischen Strahlung
p. course Hauptrichtung *f*
p. coverage area Versorgungsgebiet *n (eines Senders)*
p. creep Anfangskriechen *n*
p. crusher Vorbrecher *m (Aufbereitung)*
p. crystal Primärkristall *m*
p. current *(El)* Primärstrom *m*, primärseitiger (induzierender) Strom *m*
p. cutting edge Hauptschneide *f*
p. deposit Lagerstätte *f* der magmatischen Folge, Primärlagerstätte *f*
p. desulphurization Vorentschwefelung *f*
p. distribution Hochspannungsverteilung *f*
p. effect Primäreffekt *m*
p. electron Primärelektron *n*
p. element *s.* detecting element
p. emission Primärstrahlung *f*, primäre Strahlung *f*; Primäremission *f*
p. emitter Primärstrahler *m*
p. feedback ratio *(BMSR)* Rückführverhältnis *n (Übertragungsfunktion, die Rückführgröße und Führungsgröße verknüpft)*
p. feedback variable *(BMSR)* [unmittelbare] Rückführgröße *f*, [unmittelbar] rückgeführte Größe *f*
p. field Primärfeld *n*
p. filter Primärfilter *n*
p. fission product primäres Spaltprodukt *n*
p. flame *(Brg)* Schußflamme *f*
p. furnace Vorfrischofen *m*
p. grinding Grobmahlen *n*, Vormahlen *n (Aufbereitung)*
p. harvest wilder Aufwuchs *m (Holz)*
p. ionization Primärionisation *f*
p. jig Vorsetzmaschine *f (Aufbereitung)*
p. keying Tastung *f* in der Primärwicklung der Anodenspeisung
p. layer Grundschicht *f*
p. leakage primäre Streuung *f*
p. line switch *(Am) (Fmt)* erster Anrufsucher *m*
p. luminous standard Normalkerze *f*, Lichteinheit *f*, fotometrische Einheit *f*
p. material Urstoff *m*, Ausgangsstoff *m*
p. matter Urstoff *m*
p. measuring element Meßfühler *m*, Meßwertgeber *m*
p. metal Primärmetall *n*
p. mole Grundmol[ekül] *n*
p. motion *(Masch)* Hauptbewegung *f*
p. nuclear charge primäre Kernladung *f*
p. particle Primärteilchen *n*
p. pipe primärer Lunker *m (im Gußblock)*
p. process Elementarprozeß *m*, Primärprozeß *m*
p. product Vorprodukt *n*
p. purification *(Met)* Vorfrischen *n*
p. radiation Primärstrahlung *f*
p. radiator Primärstrahler *m (Antennen)*
p. refining mixer *(Met)* Vorwindfrischer *m*
p. relay Primärrelais *n*, Hauptstromrelais *n*
p. skip zone *(Fmt)* primäre stille Zone *f*
p. spectrum chromatische Aberration *f*, Farbabweichung *f*, Farbenfehler *m*, chromatischer Fehler *m*
p. stage erste (untere) Stufe *f (z. B. einer Rakete)*
p. standard Urmaß *n*, Urnormal *n*
p. transmitting direction *(Fmt)* Hauptstrahlrichtung *f*
p. valence chain *(Ch)* Hauptvalenzkette *f*
p. voltage *(El)* Primärspannung *f*, Eingangsspannung *f*, angelegte (primärseitige) Spannung *f*, Erstspannung *f*
p. winding Primärwicklung *f*, Erstwicklung *f (eines Transformators)*

prime / to 1. grundieren *(Anstrichtechnik)*; 2. Anlaßkraftstoff einspritzen; zünden; 3. *(Math)* mit einem Strich versehen; 4. *(El)* vorspannen
prime 1. *(Math)* Strich *m*; 2. Qualitätsblech *n*
prime *(Math)* teilerfremd, Prim-
p. coat Grundanstrich *m*, Grundierung *f*
p.-coated board grundierte Platte *f*
p. cost Ankaufspreis *m*, Anschaffungskosten *pl*
p. focus Primärfokus *m*
p. mover Kraftmaschine *f*, Antriebsmotor *m*
p. number *(Math)* Primzahl *f*

primer 1. Anlaßkraftstoff *m*; Einspritzvorrichtung *f*; Zündvorrichtung *f*; 2. Zündhütchen *n*, Zündmittel *n*; Schlagpatrone *f*; 3. Grundanstrich *m*, Grundierung *f*; Grundiermittel *n*, Grundierlack *m*; Spachtelmasse *f*, Porenfüller *m*; 4. Vorionisator *m*
p. cap Zündkapsel *f*
p. cartridge Schlagpatrone *f*
p. coat Voranstrich *m*
p. electrode Starterelektrode *f*
p. pump Anlaßeinspritzpumpe *f*

primeval forest Urwald *m*

p. stand Urwaldbestand *m*

priming 1. Grundierung *f*, erster Anstrich *m*, Grundanstrich *m*; 2. Zündung *f*; Fertigmachen *n*, Vorbereiten *n* *(von Schlagpatronen zur Zündung)*; 3. Anlaßeinspritzung *f*; 4. Vorspannung *f* *(Speicherröhren)*

p. cap Zündkapsel *f*

p. coat Grundanstrich *m*, Grundierung *f*

p. fuel Anlaßkraftstoff *m*

p. fuel injector Anlaßeinspritzanlage *f*

p. handle Vorpumphebel *m*

p. of the water *(Wärm)* Überreißen (Mitreißen) *n* des Wassers

p. paint Grundierfarbe *f* *(Anstrichtechnik)*

p. potential *(El)* Vorspannung *f*

p. speed Vorspannungsgrad *m* *(von Speicherröhren)*

p. water *(Wärm)* übergerissenes (mitgerissenes) Wasser *n*

primitive colour *(Opt)* Grundfarbe *f*

primordial cell Urzelle *f*

principal Dachverband *m*; Dachstuhl *m*

principal Haupt-

p. angle of incidence Haupteinfallswinkel *m*

p. axis Hauptachse *f*

p. axis of inertia Hauptträgheitsachse *f*

p. channel filter *(Fmt)* Hauptbandfilter *n*

p. circuit *(El)* Grundschaltung *f*, Prinzipschaltbild *n*

p. clock Hauptuhr *f*, Mutteruhr *f* *(zur Steuerung von Nebenuhren)*

p. constituent Hauptbestandteil *m*

p. course of approach *(Flg)* Anfluggrundlinie

p. curvature Hauptkrümmung *f*

p. cutting edge Hauptschneide *f* *(Spiralbohrer)*

p. determinant Hauptdeterminante *f*

p. dimensions Hauptabmessungen *fpl*

p. dock *(Schiff)* Hauptdock *n*

p. element *(Masch)* Bauteil *n*

p. equation Hauptgleichung *f*

p. focus [point] Brennpunkt *m*

p. fusible cut-out Hauptsicherung *f*

p. meridian Hauptebene *f* *(bei Astigmatismus)*

p. mode Grundschwingung *f*

p. moment of inertia Hauptträgheitsmoment *n*

p. particulars Hauptabmessungen *fpl*

p. plane Hauptebene *f*

p. quantum number Hauptquantenzahl *f*

p. route Hauptstrecke *f*

p. strain Hauptdehnung *f*

p. stress Hauptspannung *f*

p. valency Hauptvalenz *f*

p. value Hauptwert *m*

p. wearing parts Hauptverschleißteile *npl*

p. wing rib *(Flg)* Hauptrippe *f*

principle Prinzip *n*, Grundsatz *m*; Theorem *n*; Gesetz *n*; Verfahren *n*

p. of action Aktionsprinzip *n*

p. of conservation of angular momentum Drehimpulssatz *m*, Erhaltungssatz *m* des Drehimpulses

p. of equipartition of energy Äquipartitionstheorem *n*, Gleichverteilungssatz *m* [der kinetischen Energie]

p. of exclusion *(Kern)* Ausschließungsprinzip *n*

p. of increase of entropy Prinzip *n* der Entropievermehrung, Satz *m* von der Vermehrung der Entropie

p. of least action Prinzip *n* der kleinsten Wirkung

p. of least constraint Prinzip *n* des kleinsten Zwanges

p. of mass-energy equivalence Prinzip *n* der Äquivalenz von Masse und Energie

p. of relativity Relativitätsprinzip *n*

p. of superposition Superpositionsprinzip *n*, Überlagerungsprinzip *n*

p. of the conservation of energy Prinzip *n* der Erhaltung der Energie, Energieerhaltungssatz *m*

p. of the lever Hebelgesetz *n*

p. of the parallelogram of forces Satz *m* vom Parallelogramm der Kräfte

print / to drucken, schreiben *(in Druckbuchstaben)*; kopieren, lichtpausen; abziehen *(Foto)*

print Druck *m*; Abdruck *m*; Kopie *f*; Abzug *m* *(Foto)*; Lichtpause *f* *(technisches Zeichnen)*

p.-roll coater Druckwalzenstreichmaschine *f*

p. selection common exit *(Dat)* Schreibsteuerungsausgang *m*

p. selection entry hubs *(Dat)* Schreibsteuerungseingangsbuchsen *fpl*

p.-through effect Kopiereffekt *m* *(Magnetband)*

printability *(Typ)* Bedruckbarkeit *f*; Pausfähigkeit *f*

printable bedruckbar; pausfähig; abziehbar

printed gedruckt

p. capacitor gedruckter Kondensator *m*

p. circuit *(El)* gedruckte Schaltung *f*

p. circuit board Leiterplatte *f*

p. circuit connectors Verbindungsklemmen *fpl* für gedruckte Schaltungen

p. circuit unit *(El)* gedruckter Baustein *m*

p. wiring gedruckte Verdrahtung *f*

p. wiring board Leiterplatte *f*

printer Drucker *m*, Druckwerk *n*, Meßwertdrucker *m*

p. control Druckwerksteuerung *f*

p. mechanism Druckwerk *n*

printer's ink Druckfarbe *f*, Druckerschwärze *f*

p. metal Schriftmetall *n*, Letternmetall *n*

p. proof Hauskorrektur *f*

p. reader Hauskorrektor *m*

printing 1. Drucken *n*; Bedrucken *n*; Lichtpausen *n*; Kopieren *n*; Abziehen *n*; 2. Auflage *f (von Druckerzeugnissen)*
p. apparatus Kopiergerät *n*
p. calculator schreibende Rechenmaschine *f*
p. frame Lichtpausrahmen *m*
p. ink Druckfarbe *f*
p. keyboard perforator druckender Tastaturlocher *m*
p. lamp Kopierlampe *f*
p. machine Druckmaschine *f*; polygrafische Maschine *f*
p. multiplex system *(Fmt)* Mehrfachtypendruckverfahren *n*
p.-out paper Auskopierpapier *n*
p. press Druckerpresse *f*
p. pressure *(Typ)* Druckspannung *f*
p. process Druckverfahren *n*; Lichtpausverfahren *n*; Kopierverfahren *n*; Abziehverfahren *n*
p. reception Druckempfang *m*
p. reperforator Drucker-Empfangslocher *m*
p. tabulator druckende (schreibende) Tabelliermaschine *f*
p. telegraphy Buchstabendrucktelegrafie *f*, Drucktelegrafie *f*, Fernschreibverfahren *n*
p.-type chronograph Druckchronograf *m*
prior machining *(Masch)* Vorbearbeitung *f*
p. processing Vorbehandlung *f*
priority call *(Fmt)* Vorranggespräch *n*
prism Prisma *n*
p. aerial Reusenantenne *f*
p. attachment Prismenvorsatz *m*
p. binoculars Prismenfeldstecher *m*
p. dioptre Prismendioptrie *f*
p. erecting set Prismenumkehrsatz *m*
p. power Prismenwirkung *f (Augenoptik)*
p.-shaped prismatisch
prismatic prismatisch
p. bearing surface *(Masch)* Prismenführung *f*
p. effect Prismenwirkung *f*
p. eyepiece Okularprisma *n*
p. grate bar prismatischer Roststab *m*
p. guideway *(Masch)* Prismenführung *f*, Prisma *n*
p. lens Prismenglas *n*
p. shell *(Bw)* [dünnwandiges] Faltwerk *n*
p. slab *(Bw)* Faltwerk *n*
p. spectrum prismatisches Spektrum *n*, Brechungsspektrum *n*
prismatical coefficient Zylindervölligkeitsgrad *m*, Völligkeitsgrad *m* der Spantflächenskala; Längenschärfegrad *m (einer Schiffsform)*
prismoid Prismatoid *n*
pritchel *(sl)* Lochdorn *m*
privacy system *(Fmt)* verschlüsseltes Sendeverfahren *n*
private armed vessel Hilfskreuzer *m*, bewaffnetes Handelsschiff *n*
p. automatic branch exchange *(Fmt)* private Nebenstellenanlage *f* mit Wählbetrieb
p. automatic exchange private Fernsprechzentrale *f* mit Wählbetrieb
p. branch exchange Nebenstellenanlage *f*
p. branch exchange switchboard *(Fmt)* Nebenstellenschrank *m*
p. branch extension *(Fmt)* Nebenanschluß *m*
p. connection *(Fmt)* Hausanschluß *m*
p. exchange *(Fmt)* Privatzentrale *f*, Hauszentrale *f*
p. manual branch exchange *(Fmt)* private Nebenstellenanlage *f* mit Handbetrieb
p. manual exchange *(Fmt)* handbediente Nebenstellenzentrale *f*
p. port Industriehafen *m*, Fabrikhafen *m*
p. telephone plant private Fernsprechanlage *f*
p. wire circuit *(Fmt)* Mietleitung *f*
privateer Kaperschiff *n*
privateering Kaperei *f*, Kaperfahrt *f*
privy Abort[anlage *f*] *m*
pro-knock klopfbegünstigend *(Verbrennungsmotor)*
pro-oxidant oxydationsbegünstigend
probability Wahrscheinlichkeit *f*
p. calculus Wahrscheinlichkeitsrechnung *f*
p. current Wahrscheinlichkeitsstrom *m*
p. curve Wahrscheinlichkeitskurve *f*, Häufigkeitskurve *f*, Fehlerkurve *f*
p. density Wahrscheinlichkeitsdichte *f*
p. law Wahrscheinlichkeitsgesetz *n*
p. of collision *(Kern)* Stoßwahrscheinlichkeit *f*
p. of ionization Ionisationswahrscheinlichkeit *f*
p. of the nuclear transformation *(Kern)* Umwandlungswahrscheinlichkeit *f*
p. paper *(Math)* Häufigkeitspapier *n (mit Wahrscheinlichkeitseinteilung)*, Wahrscheinlichkeitsnetz *n*
probable wahrscheinlich
p. error *(Math)* wahrscheinlicher Fehler *m*
p. value Wahrscheinlichkeitswert *m*
probe Sonde *f*; Prüfkopf *m*, Tastkopf *m*, Meßfühler *m*
p. amplifier Tast[kopf]verstärker *m (z. B. für Röhrenvoltmeter)*
p. coil Suchspule *f*
p. counter *(Kern)* Spürzähler *m*, Suchzähler *m*; Zählrohrsonde *f*
p. coupling Sondenkopplung *f*
p. scanning Sondenabtastung *f*
p. tip Sondenspitze *f*
p.-to-specimen contact Ankopplung *f (Werkstoffprüfung)*
p. transformer Stiftwandler *m (Wellenleitertechnik)*
probing plane Sondierungsebene *f*

problem board Aufgabentafel *f*, Problemtafel *f* *(Analogrechner)*
p. variable *(BMSR)* Aufgabenvariable *f*, Aufgabenveränderliche *f*, Problemveränderliche *f*
procedure Verfahren *n*, Arbeitsweise *f*, Arbeitsprozeß *m*, Arbeitsgang *m*, Behandlungsweise *f*, Prozedur *f*
p. track *(Flg)* Verfahrenskurs *m* über Grund
proceed / to 1. vorgehen; 2. promovieren
p. to sea in See gehen, auslaufen *(Schiff)*
proceed-to-dial signal *(Fmt)* Bereitschaftszeichen *n*, Wählbereitschaftszeichen *n*
p.-to-send signal *(Fmt)* Sendefreigabesignal *n*
proceeding Vorgehen *n*; Arbeitsgang *m*
proceeds Reinertrag *m*, Gewinn *m*
process / to 1. bearbeiten, behandeln, fertigen, herstellen; 2. verarbeiten *(Daten)*
p. data Daten verarbeiten
p. in groups gruppenweise bearbeiten
p. into rope verseilen
process 1. [technologischer] Prozeß *m*; Verfahren *n*; Methode *f*; Arbeits[vor]gang *m*; 2. *(BMSR)* Regelvorgang *m*, Regelstrecke *f*
p. annealing *(Met)* Zwischenglühen *n*, Glühen *n* bei der Weiterverarbeitung, Rekristallisationsglühen *n*
p. average mittlere Fertigungsgüte *f* *(Qualitätskontrolle)*
p. control Regelung *f* des Arbeitsvorganges, Verfahrensregelung *f*, Fertigungsregelung *f*
p. control engineer Regelungsingenieur *m*
p. controller Programmregler *m*, Programmschalteinrichtung *f*
p. engraving Chemigrafie *f*, Fotochemigrafie *f*
p. flow-sheet Verfahrensstammbaum *m*
p. industry verarbeitende Industrie *f*
p. line Fertigungsstraße *f*, Fertigungslinie *f*
p. monitoring Betriebsüberwachung *f*
p. of combustion Verbrennungsvorgang *m*, Verbrennungsprozeß *m*
p. of distribution Steuerungsvorgang *m* *(Dampf)*
p. of formation Entstehungsprozeß *m*
p. of manufacture Fertigungsverfahren *n*, Herstellungsverfahren *n*; [bestimmter] Arbeitsgang *m*, Herstellungsgang *m*
p. planning Fertigungsplanung *f*
p. reaction rate Ansprechgeschwindigkeit *f* der Regelstrecke
p. research Verfahrensforschung *f*
p. time Verfahrenszeit *f*
p. zinc *(Typ)* Klischeezink *n*
processed wood aufbereitetes Holz *n*
processibility Verarbeitbarkeit *f*
processible verarbeitbar
procession of wood verformende Holzbehandlung *f*
prochlorite *(Min)* Prochlorit *m*
Proctor test Proctor-Versuch *m* *(Bodenmechanik)*
produce / to herstellen, erzeugen, fertigen, produzieren
p. failure zum Bruch führen
p. in quantity *(Masch)* in Menge fertigen
p. mirror-image parts spiegelbildlich kopieren
produced quantity of steam Dampfleistung *f*
producer 1. Hersteller *m*, Produzent *m*; 2. Generator *m*
p. furnace Gasgenerator *m*
p. gas Generatorgas *n*
p. gas tar Gasteer *m*
p. gas wood Generatorholz *n*, Tankholz *n*
producing angular work by grinding Schrägschleifen *f*
p. horizon produzierender Horizont *m* *(Erdöl, Erdgas)*
p. well Förderbohrung *f*, Fördersonde *f*, Produktionssonde *f*
product 1. Produkt *n*, Erzeugnis *n*; 2. *(Math)* Produkt
p. of mass and velocity of the mixture Bewegungsgröße *f* der Mischung *(Turbine)*
p. of radioactive disintegration [radioaktives] Zerfallsprodukt *n*
production Produktion *f*, Herstellung *f*, Fertigung *f*; Bildung *f*, Entstehen *n*
p. allowance Herstellungsabmaß *n*
p. capacity Leistung *f*; Produktionskapazität *f*
p. cast *(Met)* Betriebsschmelze *f*
p. control Überwachung *f*, Regelung *f* und Steuerung *f* von Produktionsprozessen
p. control scheme Produktionssteuerungsplan *m*
p. cost Produktionskosten *pl*, Herstellungskosten *pl*
p. current Produktionsstrom *m*
p. department Betriebsabteilung *f*
p. efficiency Produktionsleistung *f*; Produktivität *f*
p. engineer Fertigungsingenieur *m*, Technologe *m*
p. engineering Fertigungstechnik *f*; Fertigungsplanung *f*
p. flow Fertigungsdurchlauf *m*
p. increase Produktionssteigerung *f*
p. life Lebensdauer *f*
p. line Fertigungsstraße *f*, Fließstraße *f*
p.-line method Fließbandfertigung *f*, Fließverfahren *n*
p.-line technique Fließbandfertigung *f*, Fließverfahren *n*
p. machine Fertigungsmaschine *f*
p. method Herstellungsverfahren *n*, Fertigungsverfahren *n*
p. of draught *(Wärm)* Zugerzeugung *f*

p. of metals Erzeugung (Darstellung) *f* von Metallen
p. of nuclear power Gewinnung (Freisetzung) *f* von Kernenergie
p. of oscillation Schwingungserzeugung *f*
p. of sparks Funkenbildung *f*
p. of velocity Geschwindigkeitserzeugung *f* *(Turbine)*
p. order Fertigungsauftrag *m*
p. planning Produktionsplanung *f*
p. process Herstellungsverfahren *n*, Fertigungsverfahren *n*
p. rate Produktionsziffer *f*; Ausbringung *f*; Mengenleistung *f*, Fertigungsstückzahl *f*
p. rate per hour Stückzahl *f* je Stunde, stündliche Produktionsleistung *f*
p. reactor Plutoniumerzeugungsreaktor *m*; Produktionsreaktor *m (zur Erzeugung von radioaktiven Isotopen)*
p. run Fertigungsablauf *m*; Fertigungsfolge *f*
p. scheduling Arbeitsplanung *f*
p. shift *(Brg)* Förderschicht *f*, Gewinnungsschicht *f*
p. shop Werkstatt *f*, Produktionsstätte *f*
p. specifications Herstellungsvorschriften *fpl*, Fertigungsrichtlinien *fpl*
p. spectrum Produktionsspektrum *n*, Erzeugungsspektrum *n*
p. supervisor Produktionsleiter *m*
p. technique Fertigungsmethode *f*
p. time Grundzeit *f*, Hauptzeit *f*, Stückzeit *f*
p. time per piece Stückzeit *f*
productive 1. produktiv, leistungsfähig; 2. *(Brg, Geol)* reich, ergiebig, mächtig
p. force Produktionskraft *f*
p. time Grundzeit *f*, Hauptzeit *f*
productivity Produktivität *f*
profession Beruf *m*
professional advice Berufsberatung *f*
p. desease Berufskrankheit *f*
p. education Berufsausbildung *f*
p. engineer Ingenieur *m*, Diplomingenieur *m*
p. school Berufsschule *f*
p. terminology Fachsprache *f*
professorship Professur *f*, Lehrstuhl *m*
proficiency flight Geschicklichkeitsflug *m*
profilated brick Profilstein *m*, Formstein *m*
profile / to fassondrehen; profilieren
profile Profil *n*, Umriß *m*
p. angle of the thread Flankenwinkel *m* des Gewindes
p. border Fugendeckleiste *f*
p. cutter Profilfräser *m*, Formfräser *m*
p. deviation Profilabweichung *f*
p. drag Profilwiderstand *m*
p. dressing diamond Formdiamant *m (Schleifscheibe)*
p. ease off Balligkeit *f (Zahnrad)*
p. error Profilfehler *m*
p. gauge Formlehre *f*
p. grinding Profilschleifen *n*; Profilhinterschleifen *n*
p.-ground profilhinterschliffen
p. milling Formfräsen *n*, Umrißfräsen *n*
p. milling machine Kopierfräsmaschine *f*, Nachformfräsmaschine *f*
p. of bobbin *(Text)* Spulenform *f*, Spulenprofil *n*
p. of gear tooth Zahnprofil *n*
p. of slope Böschungsprofil *n*
p.-relieved profilhinterschliffen
p. testing instrument [for gears] Zahn[rad]prüfgerät *n*
p. tool Formmeißel *m*, Profilwerkzeug *n*
p. tracer Fühler *m*, Fühlstift *m*
p. truing attachment Profilabrichteinrichtung *f (Schleifscheibe)*
p. washer *(Masch)* Profildichtung *f*
p. wire Profildraht *m*
profiled profiliert, geformt, Form-, Profil-
p. copper Fassonkupfer *n*, Formkupfer *n*, Profilkupfer *n*
p. fibre Profilfaser *f*
p. wire Profildraht *m*, Formdraht *m*
profiler Kopierfräsmaschine *f*
profiling Fassonieren *n*; zweidimensionales Nachformfräsen *n*; (360° p. Vollumrißkopieren *n*)
p. attachment Kopierfräseinrichtung *f*
p. lathe Formdrehmaschine *f*, Fassondrehmaschine *f (Holzbearbeitung)*
p. machine *s.* p. milling machine
p. milling machine Kopierfräsmaschine *f*
p. roller Formbacke *f (Holzbearbeitung)*
profilometer Oberflächenprüfgerät *n*
profit motive materieller Anreiz *m*
profitability Rentabilität *f*
profitable rentabel
program / to *s.* programme / to
program *s.* programme
programmable addressing machine programmierbare Adressiermaschine *f*
p. typing machine programmierbare Schreibmaschine *f*
programme / to programmieren
programme Programm *n*; Programmplan *m*
p. branch *(Rak)* Umlenkbogen *m*
p. circuit Rundfunk[übertragungs]leitung *f*
p. circuit loading Bespulung *f* für Musikübertragung
p. control Programmsteuerung *f*, Folgeregelung *f*
p. controlled mit Programmsteuerung, programmgesteuert
p. controlled computer programmgesteuerter Rechner *m*
p. controller Programmregler *m*

p. counter Befehlszähler *m*, Programmspeicher *m*, Steuer[befehls]speicher *m*
p. cycle control Programmsteuerung *f*
p. distribution system Drahtfunk *m*
p. evaluation and review technique *(Bw)* PERT-Verfahren *n (Netzwerkplanung)*
p. filler Beifilm *m (zum Hauptfilm im Kino)*
p. loop *(Dat)* Programmschleife *f*; *(Am)* Übertragungsleitung *f* zwischen zwei Sendestationen
p. of research Forschungsprogramm *n*
p. path *(Rak)* Umlenkbogen *m*
p. preselection Programmvorwahl *f*
p. register Steuer[befehls]speicher *m*, Programmspeicher *m*
p.-sensitive error *(Dat)* programmbedingter Versager (Fehler) *m*
p. step Programmschritt *m*
p. storage Programmspeicherung *f*
p. switch Programmschalter *m*
p. transmission Rundfunkübertragung *f*
programmed programmiert; mit Programmsteuerung
p. checking programmierte Prüfung *f*, Programmerprobung *f*
programmer Programmierer *m (Rechenmaschine)*; Programmplaner *m*, Programmbearbeiter *m*
programming Programmieren *n*, Programmierung *f*, Programmherstellung *f*, Aufgabe *f* der Arbeitsfolge
p. language Programmierungssprache *f*
p. problem Programmierungsaufgabe *f*
p. system Programmiersystem *n*
p. unit Programmiergerät *n*
progress 1. [technischer] Fortschritt *m*; 2. Vorwärtsbewegung *f*
p. chart Bauablaufplan *m*, Baufristenplan *m*
progression 1. Folge *f*; 2. *(Math)* Reihe *f*
progressive fortschreitend
p. assembly Fließbandmontage *f*, fließende Fertigung *f*
p. cut *(Holz)* Schälschnitt *m*
p. die Folgewerkzeug *n*
p. field tube *(El)* Lauffeldröhre *f*
p. kiln kontinuierlich arbeitende Trockenkammer *f*
p. motion fortschreitende Bewegung *f*
p. precession *(Ph)* progressive Präzession *f*
p. production work Fließarbeit *f*
p. ratio Stufensprung *m (von Drehzahlen)*
p. scanning *(Fs)* fortlaufende Abtastung *f*, Abtastung mit Zeilensprung
p. shrinkage *(Text)* Eingehen *n*, Schrumpfen *n*, Schrumpfung *f*
p. system of gear shifting *(Kfz)* Durchzugschaltung *f*
p. tool Folgewerkzeug *n*, Programmwerkzeug
p. total *(BMSR)* Staffelsumme *f*
p. wave fortschreitende Welle *f*, Wanderwelle *f*
progressively higher gestaffelt, ansteigend
project / to 1. *(Opt, Math)* projizieren; 2. projektieren, entwerfen, planen; 3. *(Bw)* auskragen, ausladen, überstehen, vorragen, vorspringen, vorstehen; schleudern, abschießen
project Plan *m*, Projekt *n*, Vorhaben *n*; Entwurf *m*; Bauvorhaben *n*
p. engineer Bauleiter *m*; Entwurfsingenieur *m*
p. of research Forschungsvorhaben *n*
p. superintendent Bauleiter *m*
projected area Projektionsfläche *f*, Preßfläche *f*
p.-scale instrument Projektionsskaleninstrument *n*
projectile Geschoß *n*, Projektil *n*
p. impact Geschoßaufschlag *m*
p. motion *(Mech)* Wurfbewegung *f*
projecting *(Bw)* vorspringend, vorstehend; überstehend, vorragend, auskragend, ausladend
p. apparatus *s.* projection apparatus
p. roof Vordach *n*
p. scaffold Auslegerrüstung *f*
projection 1. Projektion *f (technisches Zeichnen)*; 2. *(Bw)* Vorsprung *m*, Auskragung *f*, Ausladung *f*, Überstand *m*; 3. Warze *f (Buckelschweißung)*; Kante *f (Überlappung)*
p. aperture Fenster *n (zum Vorführraum)*
p. apparatus Bildwerfer *m*, Projektionsapparat *m*, Projektor *m*
p. attachment Projektionseinrichtung *f*
p. booth Projektorraum *m*; Vorführkabine *f*, Bildwerferraum *m*
p. box *s.* p. booth
p. distance Projektionsentfernung *f*
p. electrometer Projektionselektrometer *n*
p. lamp Projektionslampe *f*
p. lantern Bildwerfer *m*, Projektor *m*
p. microscope Projektionsmikroskop *n*
p. optics Projektoroptik *f*
p. pantometer Profilkontrollprojektor *m*
p. perimeter Projektionsperimeter *n*
p. plane Projektionsebene *f*
p. receiver Projektions[fernseh]empfänger *m*
p. room *s.* p. booth
p. screen Projektionsschirm *m*
p. thermography Strahlungsthermografie *f (Lumineszenzverfahren zur Temperaturmessung)*
p. tube *(Fs)* Projektionsröhre *f*
p. welding Buckelschweißung *f*
projectionist Filmvorführer *m*
projective projektiv *(Geometrie)*
projector 1. Projektor *m*, Bildwerfer *m*, Projektionsapparat *m*; 2. Scheinwerfer *m*; 3. *(Math)* projizierender Strahl *m*
p. head Projektorkopf *m*
p. lamp Projektionslampe *f*, Bildwerferlampe *f*
p.-type filament lamp Projektionsglühlampe *f*

prolamine Prolamin *n*
prolate verlängert; breit, ausgedehnt in der Breite; weit verbreitet; *(Math)* gestreckt, flach
p. cycloid geschwungene Zykloide *f*
p. spheroid gestrecktes Rotationsellipsoid *n*, Rotationsellipsoid um die große Achse
proliferation of plants Wuchern *n* der Pflanzen
p. of the nucleus *(Kern)* Kernwucherung *f*
prolin[e] *(Ch)* Prolin *n*
prolong / to verlängern, [aus]dehnen
prolongation Verlängerung *f*, Ausdehnung *f*
promenade deck Promenadendeck *n*
prometacentre *(Schiff)* falsches Metazentrum *n*
promethium Promethium *n*
prominence *(Astr)* Protuberanz *f*
p. observation attachment Protuberanzenbeobachtungsansatz *m*
p. spectroscope Protuberanzenspektroskop *n*
promoter *(Ch)* Beschleuniger *m*, Katalysebeschleuniger *m*
prompt-critical *(Kern)* promptkritisch
p.-critical reactor promptkritischer Reaktor *m*
p. gamma radiation prompte Gamma-Strahlung *f*
p. neutron promptes Neutron *n*
prong 1. Zacke[n *m*] *f*, Gabel *f*, Zinke[n *m*] *f*, Klaue *f*, Stachel *m*; 2. *(El)* Steckerstift *m*, Kontaktstift *m*; Röhrensockelstift *m*
p. terminal *(El)* Stiftklemme *f*
Prony brake Bremszaum *m*, Pronyscher Zaum *m*, Bremsdynamometer *n*
Prony motor Hilfsmotor *m*, Servomotor *m*
proof / to 1. *(Typ)* abziehen, andrucken; 2. undurchdringlich (dicht) machen
proof 1. Abzug *m*, Fahne *f* *(für Korrekturzwecke)*, Andruck *m*; 2. Beweis *m*; 3. Prüfung *f*; Prüffeld *n*; Versuchsgelände *n*; 4. Dichtigkeit *f*, Undurchdringlichkeit *f*; 5. polierte Platte *f* *(Erhaltungsgrad einer Münze)*; 6. *s.* proof-spirit
proof beständig, widerstandsfähig, dicht, undurchdringlich
p. correction Korrektur *f*
p. gold Feingold *n*
p. load Prüflast *f*; Versuchslast *f*
p. mark Korrekturzeichen *n*
p. of type Korrekturabzug *m*
p.-press *(Typ)* Abziehpresse *f*; Andruckpresse *f*
p.-reader Korrektor *m*
p.-reading 1. Korrekturlesen *n*; 2. *(Dat)* Programmüberprüfung *f*
p.-sheet Fahnenabzug *m*, Fahne *f*; Korrekturabzug *m*; Aushängebogen *m*
p. sign Korrekturzeichen *n*
p.-spirit Normalalkohol *m* *(mit 57,10 Vol.% reinem Alkohol)*
p. staff Normalmeßstab *m*
p. stress Dehngrenze *f*
proofing Durchlässigkeitsversuch *m*; Dichtprüfung *f*
prop / to absteifen, [ab]stützen; *(Brg)* versteifen *(durch Streben)*; abstempeln
p. up abstützen; absteifen
prop 1. Stütze *f*, Stempel *m*, Steife *f*; Strebe *f*; 2. Bolzen *m*; Pfahl *m*, Rohrstrebe *f*, Stützsäule *f*, Verstrebung *f*; 3. Baumpfahl *m*
p. density *(Brg)* Stempeldichte *f*
p. drawer *(Brg)* Stempelrauber *m*
p.-drawing shift *(Brg)* Raubschicht *f*
p.-free face *(Brg)* stempelfreie Abbaufront *f*
p.-jet [engine] Propellerstrahltriebwerk *n*
p. of a lever Hebelunterlage *f*
p. setting jack *(Brg)* Stempelsetzvorrichtung *f*
p.-slicing *(Brg)* Querbruchbau *m* mit Spreizenausbau in den Querbrechen
p. spacing *(Brg)* Stempelabstand *m*
p. stand Kippständer *m*
propagate / to vordringen, sich fortpflanzen, sich ausbreiten
propagating power Detonationsübertragungsfähigkeit *f* *(Sprengstoff)*
propagation Ausbreitung *f*, Fortpflanzung *f*
p. coefficient *s.* p. constant
p. conditions Ausbreitungsverhältnisse *npl*, Ausbreitungsbedingungen *fpl*
p. constant Fortpflanzungskonstante *f*, Ausbreitungsmaß *n*; *(Fmt)* Betriebsübertragungsmaß *n*
p. loss Ausbreitungsverlust *m*
p. of error Fehlerausbreitung *f*
p. of light Lichtausbreitung *f*
p. of waves Wellenfortpflanzung *f*, Wellenausbreitung *f*
p. speed Fortpflanzungsgeschwindigkeit *f*
p. velocity Fortpflanzungsgeschwindigkeit *f*
propane Propan *n*
propanoic acid Propansäure *f*, Propionsäure *f*
propel / to vorwärtstreiben, [vor]antreiben
propellant Treibstoff *m*, Kraftstoff *m*, Treibmittel *n*, Treibladung *f* *(Rakete)*; Pulverladung *f*, Sprengladung *f*
p. by-pass line Treibstoffumwälzleitung *f*, Kraftstoffumwälzleitung *f*
p. capacity Fassungsvermögen *n* des Treibstoffbehälters (Kraftstoffbehälters); Treibstoffvorrat *m*, Kraftstoffvorrat *m*
p. consumption Treibstoffverbrauch *m*
p. cutoff Brennschluß *m* *(einer Rakete)*
p. distributing ring *(Rak)* Brennstoffeinlaufsicke *f*
p. feed line Treibstoffdruckleitung *f*, Kraftstoffdruckleitung *f*
p. filler hose Kraftstoffbetankungsschlauch *m*, Treibstoffbetankungsschlauch *m*
p. flow rate Treibstoffverbrauch *m*

p. intake fold Kraftstoffeinlaufsicke *f*, Treibstoffeinlaufsicke *f*
p. mixture Treibstoffgemisch *n*, Treibstoffmischung *f*, Kraftstoffgemisch *n*
p. return line Kraftstoffumwälzleitung *f*, Treibstoffumwälzleitung *f*
p. spray nozzle Kraftstoffdüse *f*, Treibstoffdüse *f*
p. tank Treibstoffbehälter *m*, Kraftstoffbehälter *m*

propellent force schiebende Wirkung *f (Sprengstoff)*

propeller Antriebsschraube *f*, Flügelschraube *f*, Propeller *m*; *(Flg)* Luftschraube *f*, Propeller; Schiffsschraube *f*
p. aperture *(Schiff)* Propellerbrunnen *m*, Schraubenbrunnen *m*
p. arch *(Schiff)* Bogenstück *n* am Hintersteven *(über dem Propeller)*, Hinterstevenbogen *m*
p. area ratio Propellerflächenverhältnis *n*
p. blade Propellerflügel *m*
p. blower Wurfgebläse *n (zum Häckseln)*
p. boss *(Schiff)* Propellernuß *f*, Propellernabe *f*
p. boss plates *(Schiff)* Wellenhosenplatten *fpl*
p. bracket *(Schiff)* Wellenbock *m*
p. damping coefficient Propellerdämpfungskoeffizient *m*
p. eddy Propellerwirbelstrom *m*
p. efficiency Propellerwirkungsgrad *m*, Schraubenwirkungsgrad *m*
p. flows (flow fields) *(Schiff)* Propellerströmungsfelder *npl*
p. frame *(Schiff)* Propellersteven *m*, Schraubensteven *m*
p. guard Propellerschirm *m (Abweiser bei Zweischraubenschiffen)*
p. hub Propellernabe *f*, Schraubennabe *f*
p. inertia *(Schiff)* Propellermassenschwungmoment *n*
p. load coefficient Propellerbelastungsgrad *m*
p. nozzle ring Propellerdüsenring *m*
p. post Propellersteven *m*, Schraubensteven *m*
p. pump Kreiselpumpe *f*, Zentrifugalpumpe *f*
p. shaft Gelenkwelle *f*, Kardanwelle *f*; Schraubenwelle *f*, Schwanzwelle *f*, Propellerwelle *f*
p.-shaft stay *(Schiff)* Wellenbock *m*, Wellenträger *m*
p. shaft tunnel *(Schiff)* Propellerwellentunnel *m*
p. shaft with sliding joint Gelenkwelle *f* mit Gleitgelenk
p. speed Propellerdrehzahl *f*; Luftschraubendrehzahl *f*
p. strut *(Schiff)* Wellenbock *m*, Propellerbock *m*
p.-tailshaft Propellerschwanzwelle *f*
p. test in open water Versuch *m* mit freifahrendem Propeller
p. thrust Propellerschub *m*
p. torque Propellerdrehmoment *n*, Luftschraubendrehmoment *n*
p.-type anemometer Propelleranemometer *n*
p. wake *(Schiff)* Propellerstrom *m*, Schraubenstrom *m*
p. well *(Schiff)* Propellerbrunnen *m*, Schraubenbrunnen *m*

propelling machinery *(Schiff)* Maschinenanlage *f*
p. motor Schiffsantriebsmotor *m (E-Motor)*
p. movement Schiebefahrt *f (elektrischer Zug)*
p. pitch Wurfhöhe *f (Häckselmaschine)*
p. plant *(Schiff)* Antriebsanlage *f*

proper angemessen, echt, richtig, sachgemäß, eigen
p. cutting edge formrichtige Schneide *f*
p. field Eigenfeld *n*
p. fraction echter Bruch *m*
p. function Eigenfunktion *f*
p. motion [wahre] Eigenbewegung *f (Radartechnik)*
p. phase Eigenphase *f*
p. pilotage Hochseeschiffahrt *f*
p. speed mittlere Geschwindigkeit *f*, Normalgeschwindigkeit *f*; Eigengeschwindigkeit *f*
p. state of tillage *(Lw)* Bodengare *f*
p. time Eigenzeit *f*
p. value Eigenwert *m*
p. vector Eigenvektor *m*

properly seasoned verladetrocken *(Holz)*

properties *(Am) (Fs)* Dekors *npl*

property Eigenschaft *f*
p. of rock *(Brg)* Gesteinseigenschaft *f*

propine *(Ch)* Propin *n*, Allylen *n*

propiolic acid Azetylenmonokarbonsäure *f*
p. alcohol Propargylalkohol *m*

propionate Propionat *n*, Propionsäureester *m*

propionic acid Propionsäure *f*
p. aldehyde Propanal *n*, Propionaldehyd *m*

proportion / to bemessen, im [richtigen] Verhältnis zuteilen

proportion Verhältnis *n*, Proportion *f*; Anteil *m*, Prozentsatz *m*; Gleichheit *f*; Mischungsverhältnis *n (Betontechnologie)*
p. by volume Raumverhältnis *n*
p. of ingredients Mischungsverhältnis *n*
p. of mixture Mischungsverhältnis *n*
p. of water Wassergehalt *m*

proportional *(Math)* Proportionale *f*

proportional proportional, verhältnisgleich, Proportional-
p. acting controller Proportionalitätsregler *m*
p. band *(BMSR)* Proportionalbereich *m*, P-Bereich *m*
p. control Proportionalregelung *f*, P-Regelung *f*
p. control factor *(BMSR)* Proportionalverstärkung *f*
p. controller Proportionalitätsregler *m*

p. counter *(BMSR)* Proportionalzählrohr *n*, Proportionalzähler *m*
p. counter tube Proportionalzählrohr *n*
p. elastic limit Proportionalitätsgrenze *f*
p. feedback starre Rückführung *f (Regler)*
p. limit Proportionalitätsgrenze *f*
p. part Tafeldifferenz *f (in einer Logarithmentafel)*
p.-plus-derivative control Proportional-Differential-Regelung *f*, PD-Regelung *f*
p.-plus-integral control Proportional-Integral-Regelung *f*, PI-Regelung *f*
p.-plus-reset control *s.* p.-plus-integral control
p.-position action *s.* p. control
p. region Proportionalbereich *m (Zählrohr)*
p. sensitivity *(BMSR)* Proportionalverstärkung *f*, P-Verstärkung *f*
p.-speed floating action *s.* integral control
proportionality Proportionalität *f*
p. factor Proportionalitätsfaktor *m*; Verhältniszahl *f*, Proportionalität *f*
p. limit Proportionalitätsgrenze *f*
proportionately [gut] proportioniert, im rechten Verhältnis, angemessen
proportionating pump Dosierpumpe *f*
proportioned dimensioniert
proportioning Dimensionierung *f*; Dosierung *f*, Zumessung *f*
p. band *(Am) (BMSR)* Proportionalbereich *m*, P-Bereich *m*
p. control *(Am)* Proportionalregelung *f*, P-Regelung *f*
proposed cable scheme *(El)* Kabelprojekt *n*, Kabelplanung *f*
proposition Satz *m*, Lehrsatz *m*; Problem *n*
propositional calculus *(Dat)* Aussagenkalkül *m*
propping Absteifung *f*, Abstützung *f*; Grubenausbau *m*
proprietary gesetzlich geschützt, patentiert; Marken-; Eigentums-
propulsion Antrieb *m*, Schiffsantrieb *m*, Vortrieb *m*
p. branch *(Rak)* Antriebsbahn *f*
p. by recoil (reaction) Rückstoßantrieb *m*
p.-control desk Triebwerkspult *n*
p. motor Antriebsmotor *m*
p. period Antriebszeit *f*
p. portion Antriebsteil *m (der Flugbahn)*
p. reactor Antriebsreaktor *m*
p. stage Raketenstufe *f* mit Triebwerk
p. unit Triebwerk *n*
p.-unit contactor Getriebeschaltschütz *n*
p.-unit governor Triebwerksregler *m*
propulsive Vortriebs-, Antriebs-
p. jet Antriebsstrahl *m*
p. nozzle *(Rak)* Düse *f*
p. speed Antriebsgeschwindigkeit *f*
propylene *(Ch)* Propylen *n*
proscenium light Rampenlicht *n*
prospect / to *(Brg)* schürfen
prospect *(Brg)* Schurf[arbeit *f*] *m*
p. boring Untersuchungsbohrung *f*
p. shaft Schurfschacht *m*, Schürfschacht *m*
prospecting 1. Suche *f*, Aufsuchen *n*, Prospektieren *n*; 2. *(Brg)* Schürfen *n*, Schürfung *f*; Muten *n*, Mutung *f*
p. drill Schürfbohrmaschine *f*
prospective current *(El)* theoretisch zu erwartender Strom *m*
p. oil land Gebiet *n*, in dem Ölführung möglich ist; *(Geol)* ölhöffiges Gebiet *n*
prosthetic group *(Ch)* prosthetische Gruppe *f*
protactinium Protaktinium *n*
protected geschützt, Schutz-
p. by patent gesetzlich geschützt
p. motor *(Elm)* Motor *m* mit Berührungsschutz, gekapselter Motor
p. switch Schalter *m* mit Schutzkasten
protecting schützend, Schutz-
p. cap Schutzhaube *f*, Schutzkappe *f*
p. cone Schutzkegel *m*
p. gib *(Masch)* Deckleiste *f*
p. hose Schutzrohr *n*
p. sleeve Schutzmuffe *f*
p. tube Schutzrohr *n (Temperaturfühler)*; Mantelrohr *n*
p. zinc coating Zink[schutz]überzug *m*
protection Schutz[einrichtung *f*] *m*
p. against corrosion Korrosionsschutz *m*
p. against electric-shock hazard Berührungsspannungsschutz *m*
p. area Schutzzone *f*
p. box Schutzdose *f*
p. coating Schutzüberzug *m*
p. cover Schutzkappe *f*, Schutzhaube *f*
p. earth[ing] Schutzerde *f*, Schutzerdung *f*, Sicherungserdung *f*, Blitzschutzerde *f*
p. for interturn short-circuits *(El)* Windungsschlußschutz *m*
p. ground *s.* p. earthing
p. of inventors Erfinderschutz *m*
p. suit Schutzanzug *m*, Schutzkleidung *f*
p. survey Strahlenschutzüberwachung *f*
protective schützend, Schutz-
p. apron Bleischürze *f (Strahlenschutz)*
p. atmosphere Schutzgasatmosphäre *f*
p. capacitor Schutzkondensator *m*
p. casing Schutzgehäuse *n*
p. clothing Schutzkleidung *f*
p. coat[ing] Schutzüberzug *m (auf Metallen)*
p. colloid *(Ph)* Schutzkolloid *n*
p. cone Schutzkegel *m*
p. covering for cylinder cover Zylinderdeckelhaube *f*, Schutzhaube *f*, Verkleidungskappe *f*
p. deck *(Schiff)* Panzerdeck *n*

p. design Schutzmuster *n*
p. device Schutzeinrichtung *f*, Sicherheitseinrichtung *f*
p. earth[ing] *s.* protection earthing
p. film Schutzüberzug *m*
p. filter Schutzfilter *n*
p. furnace atmosphere Schutzgasatmosphäre *f*
p. gap Schutzfunkenstrecke *f*
p. glove Schutzhandschuh *m*
p. grating Schutzgitter *n*
p. hood Schutzhaube *f*, Schutzkappe *f*
p. iron lagging Schutzblechmantel *m*
p. lacquer Schutzlack *m*
p. layer Schutzschicht *f*
p. lead *(El)* Schutzdraht *m*, Schutzleitung *f*
p. lead glass Bleischutzglas *n*
p. lighting Schutzbeleuchtung *f*
p. motor switch Motorschutzschalter *m*
p. outfit Schutzkleidung *f*, Schutzausrüstung *f*
p. oxide film schützendes Oxidhäutchen *n*
p. plug reception Schutzkontakt *m (Steckdose)*
p. reactance coil Strombegrenzungsdrossel *f*
p. reactor *(El)* Schutzdrossel *f*
p. relay Schutzrelais *n*
p. resistance Schutzwiderstand *m*
p. resistance for rheostatic braking Bremsschutzwiderstand *m*
p. resistor Schutzwiderstand *m*
p. screen Schirm *m*
p. shield Schutzschild *m*
p. sleeve Schutzmuffe *f*; Isolierhülse *f*
p. spark gap Schutzfunkenstrecke *f*
p. suit Schutzkleidung *f*, Schutzanzug *m*
p. system Schutzart *f*
protector Schutz *m*, Schutzeinrichtung *f*, Schutzkappe *f*
p. ground *s.* protection earthing
protein Protein *n*, Eiweiß[körper *m*] *n*
p. base fibre Proteinfaser *f*, Eiweißfaser *f*
p. plastic Proteinplast *m*, Eiweißplast *m*
proteinaceous eiweißartig
proteolysis Eiweißabbau *m*, Eiweißspaltung *f*
protocatechuic acid Protokatechusäure *f*
proton Proton *n*
p. accelerator Protonenbeschleuniger *m*
p. beam Protonenstrahl *m*, Protonenbündel *n*
p. bombardment Protonenbeschuß *m*, Beschießung *f* mit Protonen
p. cloud Protonenwolke *f*
p. density Protonendichte *f*
p. emitter Protonenstrahler *m*
p. excess Protonenüberschuß *m*
p. mass Protonenmasse *f*, Masse *f* des Protons
p. microscope Protonen[über]mikroskop *n*
p. number Protonenzahl *f (Anzahl der Protonen im Kern)*
p.-proton scattering Proton-Proton-Streuung *f*
p. recoil counter Rückstoßprotonenzähler *m*
p. scattering Protonenstreuung *f*
p. source Protonenquelle *f*
p. spectrum Protonenenergiespektrum *n*
p. spin Protonenspin *m*
p.-synchrotron Protonensynchrotron *n*
p. transition Protonenübergang *m*
protoplasm Protoplasma *n*, Zellflüssigkeit *f*
prototype Musterinstrument *n*, Prototype *f*; Erstausführung *f*, Muster *n*, Prototyp *m*
p. house Musterhaus *n*
p. measure Urmaß *n*
p. of metre Meterprototyp *m*, Urmeter *n*
p. test Typenprüfung *f*
protract / to maßstäblich zeichnen; verlängern; verzögern
protracted protrahiert, Dauer-, Langzeit-
p. irradiation Dauerbestrahlung *f*, protrahierte Bestrahlung *f*, Langzeitbestrahlung *f*
protractor Schmiege *f*; Winkelmesser *m*, Winkelmeßgerät *n*, Transporteur *m*
protrude / to herausragen, [her]vorstehen
protruding knot vorstehender Ast *m*
proustite Rotgüldigerz *n*, Proustit *m*
provable beweisbar
prove / to beweisen, sich erweisen als; nachweisen; untersuchen, prüfen
p. a rule die Probe machen
proved oil land *(Geol)* ölführendes Gebiet *n*, Ölland *n*
provide / to vorsehen; liefern, bereitstellen, ergeben; lassen, sorgen für
p. more wrap den Umschlingungswinkel vergrößern *(Kettenrad)*
provided with versehen (ausgestattet, ausgerüstet) mit
proving Probe *f*, Erprobung *f*, Prüfung *f*; *(Ch)* Nachweis *m*
p. ground Prüffeld *n*; Prüfgelände *n*, Versuchsfeld *n*, Versuchsgelände *n*
p. of opening of switch blades Anzeige *f* für Mittelstellung (Halbstellung) von Weichenzungen
provisional vorläufig, behelfsmäßig, provisorisch
p. result Zwischenergebnis *n*
p. support *(Brg)* vorläufiger Ausbau *m*
proximate analysis *(Ch)* Immediatanalyse *f*, Kurzanalyse *f*
proximity Nähe *f*, Nachbarschaft *f*
p. effect 1. Stromverdrängungseffekt *m (durch benachbarte Leiter)*; 2. Nahewirkung *f*; Näherungseffekt *m*
p. fuse Abstandszünder *m*; Näherungszünder *m*; Richtungszünder *m*
p. theory *(Ph)* Nahewirkungstheorie *f*
p. zone Nahbereich *m (Antennentechnik)*
pruner Baumschere *f*, Gärtnersäge *f*
pruning Ästung *f*, Beschneiden *n (Bäume)*

p. **chisel** Baummeißel *m*
p. **hook** Baummesser *n*
p. **knife** Baummesser *n*
p. **saw** Astsäge *f*
p. **saw with teeth on both ends** zweischneidige Astsäge *f*
p. **shears** Baumschere *f*
p. **tree chisel** Baummeißel *m*
p. **trees** Stutzen *n* der Bäume (Sträucher)
Prussian blue Berliner Blau *n*, Preußischblau *n*
prussiate Salz *n* der Blausäure, Zyanid *n*
prussic acid Blausäure *f*, Zyanwasserstoffsäure *f*
pry Brecheisen *n*, Hebezeug *n*
pseudo-astatic governor pseudoastatischer Regler *m*
p.**-astaticity** Pseudoastasie *f*
p. **code** Pseudokode *m* *(muß für Maschine übersetzt werden)*
p.**-image** Pseudobild *n*
p. **instruction** Pseudobefehl *m*, symbolischer Befehl *m*
p.**-random result** *(Dat)* Pseudostreuergebnis *n*
p.**-urea** Isoharnstoff *m*
p.**-uric acid** Pseudoharnsäure *f*
pseudobinary quasibinär
pseudocolloid Pseudokolloid *n*
pseudocumene Pseudokumol *n*
pseudoharmonic pseudoharmonisch
pseudomoment Pseudomoment *n*
pseudomorph pseudomorpher Kristall *m*
pseudomorphism Pseudomorphose *f*
pseudomorphous pseudomorph
pseudoquadrupol Pseudoquadrupol *m*
pseudoscalar pseudoskalar
p. **effect** pseudoskalare Wechselwirkung *f*, Pseudoskalareffekt *m*
pseudoscopic viewing *(Opt)* pseudoskopisches Betrachten *n*
pseudovector Pseudovektor *m*
psilomelane Hartmanganerz *n*, Hartbraunstein *m*
psophometer Geräuschspannungsmesser *m*, Psophometer *n*
psophometric potential difference Geräuschabstand *m*; Geräuschspannung *f*
p. **voltage** Geräuschspannung *f*
psychrometer Psychrometer *n*, Luftfeuchtemesser *m*
psychrometric chart Kältediagramm *n*
Ptolemaic ptolemäisch
ptomaine Ptomatin *n*, Ptomain *n*, Leichengift *n*
public öffentlich
p. **address system** Lautsprecheranlage *f (z. B. Betriebsfunk)*
p. **buildings** öffentliche Gebäude *npl*, gesellschaftliche Bauten *pl*
p. **call office** öffentliche Fernsprechzelle *f*
p. **health engineering** Gesundheitstechnik *f*, Sanitärtechnik *f*
p. **mains** Verbrauchernetz *n*, Stromversorgungsnetz *n*
p.**-service vehicle** öffentliches Verkehrsfahrzeug *n*
p. **telephone-booth for local service** Ortsmünzfernsprecher *m*
p. **works** öffentliche Arbeiten *fpl*; staatlich geförderte Bauvorhaben *npl*
publication Veröffentlichung *f*
publish / to veröffentlichen
published work Veröffentlichung *f*
publisher Verleger *m*; Verlag *m*
publisher's colophon Verlagssignet *n*
p. **reader** Lektor *m*
puck Scheibe *f*, Puck *m (Eishockey)*
puddening *(Schiff)* Schamfielungsmatte *f*, Stoßwulst *m*
pudding *(Am) s.* puddening
puddle / to 1. *(Bw)* stochern, [von Hand] stampfen *(Betonverdichtung)*; 2. *(Met)* puddeln, rührfrischen
puddle ball Puddelluppe *f*, Rohluppe *f*, Luppe *f*
p. **bar** Luppenstab *m*
p. **cinder** Puddelschlacke *f*
p. **iron works** Puddelhütte *f*
p. **mill** Puddeleisenwalzwerk *n*
p. **train** Puddelstraße *f*
puddled mill Puddelstahlwalzwerk *n*
p. **steel** Puddelstahl *m*
puddler bar Puddelstab *m*
puddling 1. Stochern *n*, Stampfen *n* von Hand *(Betonverdichtung)*; 2. *(Met)* Rührfrischen *n*, Puddeln *n*
p. **furnace** Puddelofen *m*
p. **hearth** Puddelherd *m*
p. **rod** Rührstab *m*
puering paddle *(Led)* Beizhaspel *f*
puffing agent *(Plast)* Verdickungsmittel *n*
p. **tug** Schubboot *n*, Stoßboot *n*, Schubschlepper *m*
puffs and stiffeners Kappensteife *f (Schuh)*
pug Ziegellehm *m*
p. **mill** schwerer Mischer *m*, Knetfaß *n*, Schlägermühle *f*; Kollergang *m*
p. **mill mixer** Zwangsmischer *m*
pugging Schallisolation *f (Fußboden)*
pull / to 1. ziehen, zerren; 2. *(Typ)* abziehen, einen Abzug herstellen; 3. zerreißen *(Prüfstab)*; 4. rudern, holen, pullen, ziehen
p. **away** wegziehen
p. **in** einziehen *(Kabel)*
p. **off** abreißen, abziehen
p. **out** herausziehen, ausreißen; dehnen
p. **the cotton lumps to pieces** Baumwollklumpen zerrupfen
p. **the flax** Flachs raufen (ziehen)

p. to right nach rechts ziehen
p. up *(Schiff)* aufheißen
pull 1. *(Masch)* Zug[kraft *f*] *m*; Durchzugskraft *f*; 2. *(Schiff)* Zug *m*, Pull *m*, Schlag *m* *(mit dem Ruder)*; 3. *(Typ)* Abzug *m*; 4. *(Brg)* Abschlagslänge *f*, Abschlagstiefe *f*
p. box *(El)* Zwischendose *f*, Fädeldose *f*
p. broach Ziehnadel *f (Räumen)*
p. characteristic Anzugskennlinie *f (Relais)*
p. contact Zugkontakt *m*
p. cord Zugleine *f*, Zugschnur *f*
p. crack *(Met)* Haarriß *m*
p. fastener Reißverschluß *m*
p.-in torque *(El)* Intrittfallmoment *n*, synchronisierendes Moment *n*
p.-in-type synchronization *(Fs)* Mitnahmesynchronisierung *f*
p.-off spring Rückholfeder *f*, Zugfeder *f*
p.-on brake *(Kfz)* Zughandbremse *f*
p.-out radius *(Flg)* Abfangradius *m*
p.-out sheet Faltblatt *n*
p.-out test Ausziehversuch *m (Stahlbetonprüfung)*
p.-out torque Kippmoment *n*
p.-over hot mill Übergabewalzwerk *n*
p.-push rule Stahlbandmaß *n*
p. rod Zugstange *f*
p. rope Zugseil *n*, Vorderseil *n (Schrapper)*
p. switch *(El)* Zugschalter *m*, Zugkontakt *m*
p.-through winding *(Elm)* Fädelwicklung *f*, Durchzieherwicklung *f*
puller Abziehvorrichtung *f*
pullers Ziehkopf *m*, Spannkopf *m (Räumen)*
pulley Flasche *f*, Rolle *f*, Blockrolle *f*; Flaschenzug *m*; Scheibe *f*, Trommel *f*; *(Masch)* Scheibe für Riementrieb, Riemenscheibe *f*
p. bearing Rollager *n*
p. block Flaschenzug *m*, Seilblock *m*, Hakenflasche *f*
p. bushing Riemenscheibenbüchse *f*
p. chain Flaschenzugkette *f*
p. cord Antriebseil *n*, Skalenseil *n (für Skalenzeiger)*
p. crown Riemenscheibenwölbung *f*
p. drive Scheiben[an]trieb *m*, Rollenantrieb *m*
p. gear drive Bandgetriebe *n*
p. rim Riemenscheibenkranz *m*, Scheibenkranz *m*
p. tap Riemenscheibengewindebohrer *m*
pulling bolt Spannbolzen *m (Spannbetonbau)*
p. by a load impedance Lastverstimmung *f (Änderung der Arbeitsfrequenz eines Oszillators, bedingt durch eine angeschlossene Last, die sich ändert)*
p. factor Lastverstimmungsmaß *n (Oszillator)*
p. figure Lastverstimmungsmaß *n (Oszillator)*
p. flax Flachsraufen *n*
p. head Spannkopf *m (Spannbeton)*; Ziehkopf *m (Räumen)*
p.-in tape *(Am) (El)* Fädelband *n*
p. into tune *(Fmt)* Mitnahme[erscheinung] *f*
p. lifeboat Riemenrettungsboot *n*, Ruderrettungsboot *n*
p. machine Raufmaschine *f (Faserlein)*
p. mechanism Ziehwerk *n*, Zieheinrichtung *f (Räummaschine)*
p. on whites *(Fs)* Fahnenziehen *n*
pullscoop Tieflöffelbagger *m*
pulp / to einstampfen *(z. B. Makulatur)*; zu einem Brei anrühren
pulp 1. Brei *m*, Ganzzeug *n*, Masse *f*, Schnitzel *npl*; Pulpe *f*; 2. Papiermasse *f*, Papierfaserstoff *m*, Papierbrei *m*; Faserstoffaufschwemmung *f*; Ganzstoff *m*; 3. Trübe *f*, Schwertrübe *f*, Feinschlamm *m (Aufbereitung)*; 4. Fruchtmark *n*
p. bleaching Holzschliffbleiche *f*
p. board Zellstoffpappe *f*
p. cooking Holzaufschluß *m*, Zellstoffkochung *f*
p. drying Zellstofftrocknung *f*, Schlifftrocknung *f*
p. drying machine Schliffentwässerungsmaschine *f*, Zellstofftrockenmaschine *f*
p. engine Holländer *m*
p. felt Entwässerungsfilz *m*
p. hardness Aufschlußgrad *m* des Zellstoffs
p. in crumb form Zellstoffflocken *fpl*, Zellstoffknötchen *npl*
p. press Büttenpresse *f (Papierherstellung)*
p. pump Stoffpumpe *f*
p. saver Stoffänger *m*
p. stone Schleifstein *m (Holzschliffherstellung)*
p. strainer Zeugsichter *m*, Knotenfänger *m*
p.-vat Bütte *f*
pulper Holländer *m*; Breiapparat *m*, Zerfaserer *m*; Stofflöser *m*; Muser *m*
p. for feed preparation Muser *m* für Futterzubereitung
pulping Aufschlagen *n*, Aufschluß *m*, Kochung *f*
p. process Kochprozeß *m (Zellstoffherstellung)*
pulpstone Schleifstein *m (Holzschliffherstellung)*
pulpwood Papierholz *n*, Faserholz *n*
pulsatance *(El)* Kreisfrequenz *f*, Winkelgeschwindigkeit *f*
pulsate / to pulsieren; schwingen; intermittieren; schwanken; im Schwellbereich schwingen; schütteln, geschüttelt werden
pulsating pulsierend, wellig, stoßweise
p. bending fatigue limit *(Mech)* Biegeschwellfestigkeit *f*
p. component *(El)* Wechselkomponente *f*, Wechselglied *n*
p. current *(El)* pulsierender Strom *m*

p.-current-fed track circuit Impuls-Gleisstromkreis *m (elektrischer Zug)*
p. direct current pulsierender Gleichstrom *m*
p. fatigue strength *(Mech)* Schwellfestigkeit *f*
p. fatigue strength under bending stresses Biegeschwellfestigkeit *f*
p. field pulsierendes Feld *n*, Wechselfeld *n*
p. jig Pulsatorsetzmaschine *f*
p. magnetic field pulsierendes Magnetfeld *n*
p. potential difference Wechselspannung *f (mit überlagerter Gleichspannung)*
p. screen Schüttelsieb *n*
p. voltage Gleichspannung *f (mit überlagerter Wechselspannung)*; Wechselspannung *f (mit überlagerter Gleichspannung)*
p. wave pulsierende Welle *f*
pulsation Schwankung *f*; Stoßvorgang *m*, Schwingung *f*, Pulsation *f*
p.-free pulsationsfrei, stoßfrei, schwingungsfrei, gleichmäßig
p. welding Mehrimpulspunktschweißen *n*
pulsator Pulsator *m*
p. classifier Pulsatorsetzmaschine *f*
pulsatron *(El)* Pulsatron *n (Thyratron zur Impulserzeugung)*
pulse Impuls *m*, Stoß *m*, Spannungsstoß *m*, Stromstoß *m*
p. amplifier Impulsverstärker *m*
p. amplitude Impulshöhe *f*, Impulsamplitude *f*, Impulsgröße *f*
p. analyzer Impulshöhenanalysator *m*
p. bandwidth *(Fs)* Breite *f* des Impulsspektrums, Impulsbandbreite *f*
p. base *(Fs)* Impulsbasis *f*, Impulsfuß *m*
p.-bucking adder *(Dat)* Addierer *m* mit Impulskompensation
p. chamber Impulsionisationskammer *f*
p. circuit Impulsstromkreis *m*, Stoßkreis *m*
p. clipper *(Fs)* Impulsabtrenner *m*, Impulsabschneider *m*, Impulsbegrenzer *m*
p. code telemetering *(Fmt)* Impulskodefernübertragung *f*
p. coder Impulsmodulator *m*
p. coding Impulsverschlüsselung *f*
p. column *(Ch)* pulsierend arbeitende Kolonne *f*
p. control Impulsregelung *f*
p. correction Impulsentzerrung *f*, Impulsverbesserung *f*
p. counter Impulszähler *m*
p. counting circuit Impulszählschaltung *f*
p. counting methode Impulszählverfahren *n*
p. curve Impulskurve *f*
p. decay Impulsabfall *m*; Hinterflanke *f*
p. decay time Impulsabfallzeit *f*, Impulsabklingzeit *f*
p. depth Impulshöhe *f*
p. direction finder Impulspeiler *m*
p. discharge tube *(Fs)* Impulsentladungsröhre *f (Triode in der Bootstrapschaltung)*
p. divider Impulsteiler *m*
p. dividing circuit *(Fs)* Impulsuntersetzer[schaltung *f*] *m*
p. droop *(Fs)* negative Dachschräge *f*, Dachabfall *m* des Rechteckimpulses
p. duration Impulsdauer *f*, Impulsbreite *f*
p. duration modulation Impulsbreitenmodulation *f*, Impulslängenmodulation *f*, Pulsdauermodulation *f*, Pulsbreitenmodulation *f*
p. duration ratio *(Fs)* Impulstastverhältnis *n*
p. duty factor *(Fs)* Impulsleistungsverhältnis *n*
p. edge Impulsflanke *f*
p. excitation *(El)* Stoßerregung *f*
p.-forming line *(Fs)* Impulsformerleitung *f*, Impulserzeugerleitung *f*
p.-forming network *(Fs)* Impulskette *f*
p. frequency *(Fs)* Impulsfolgefrequenz *f*, Tastfrequenz *f*
p. front Impulsflanke *f*, Impulsfront *f*
p. gate Impulssperrschaltung *f*
p. generating device Impulsgeber *m*
p. generation Stoßerregung *f*; Impulserzeugung *f*
p. group Impulsreihe *f*
p. height Impulsamplitude *f*, Impulshöhe *f*
p. height analyzer Impulshöhenanalysator *m*
p. height meter Impulshöhenmesser *m*
p. intensity Impulsstärke *f*
p. interleaving *(Fmt)* Impulsverflechtung *f*; Pulsverflechtung *f*
p. interval Impulsabstand *m*, Impulsintervall *n*, Impulspause *f*
p. ionization chamber Impulsionisationskammer *f*
p. jet Pulso-Strahltriebwerk *n*
p. jitter *(Fs)* Impulsinstabilität *f*, Pulsinstabilität *f*
p. line Impulsleitung *f*
p. machine *(Am)* Maschinenzahlengeber *m*
p. measuring Impulsmessung *f*
p. meter Impulsmesser *m*
p. mixing *(Fs)* Impulsmischung *f*
p. mode Impulsart *f*, Pulsart *f*
p.-mode multiplex Kanalwahl *f* durch Impulsschwingungsarten
p.-modulated amplifier *(Fs)* Leistungsverstärker *m* für Impulsbetrieb, Tastverstärker *m (Impulstechnik)*
p. modulation Impulsmodulation *f*, Pulsmodulation *f*
p. multiplex Impulsverschlüsselung *f*
p. operation Impulsbetrieb *m*
p. overlap Impulsüberlagerung *f*
p.-position modulation Impulsphasenmodulation *f*

p. power Impulsleistung *f*
p.-power amplifier Leistungsverstärker *m* für Impulsbetrieb, Tastverstärker *m (Impulstechnik)*
p.-power calibrator Impulsleistungsmesser *m*
p. radar Pulsradar *n*
p. rate Impulsdichte *f*
p. recorder Impulsschreiber *m*
p. recurrence frequency Impulsfolgefrequenz *f*
p.-reflection flaw detector Echo-Impulsfehlerprüfgerät *n*
p. regeneration *(Fs)* Impulsentzerrung *f*
p. repetition Impulsfolge *f*
p. repetition frequency Impulsfolgefrequenz *f*, Tastfrequenz *f*
p. repetition rate Impuls[folge]frequenz *f*, Impulshäufigkeit *f*, Wiederholungsfrequenz *f*
p. reply Impulsecho *n*
p. responder Impulswiederholer *m*
p. response Impulscharakteristik *f*; Impulswiedergabe *f*
p. restoration Impulsentzerrung *f*, Impulsverbesserung *f*, Impulserneuerung *f*
p. rise time Impulsanstiegszeit *f*
p. selection Impulsauswahl *f*
p. separation 1. Impulsabstand *m*, Impulsintervall *n*, Impulspause *f*; 2. Impulstrennung *f*; Impulsabtrennung *f*
p. separator Impulstrennstufe *f*
p. shape Impulsform *f*
p. shaper Impulsformer[stufe *f*] *m*
p.-shaping circuit impulsformender Kreis *m*
p. slope Impulsflankensteilheit *f*
p. slope modulation Impulssteilheitsmodulation *f*
p. spacing Impulsabstand *m*
p. spectrum Impulsspektrum *n*
p. spike Impulsspitze *f*
p. steepness Impulsflankensteilheit *f*
p. stretcher *(Am)* Impulskorrektor *m*
p. stretching Impulsverbreiterung *f*, Nachleuchtschleppe *f*
p. stretching circuit Impulsverbreiterungsschaltung *f*
p. switch Impulsschalter *m*
p. technique Impulstechnik *f*
p. testing method Impulsverfahren *n* der Werkstoffprüfung
p. tilt positive Dachschräge *f*, Dachanstieg *m* eines Rechteckimpulses
p.-time delay Impulslaufzeit *f*
p. top Impulsdach *n (z. B. beim Rechteckimpuls)*
p. train Impulsreihe *f*, Impulsserie *f*
p.-train spectrum Frequenzspektrum *n* einer Impulsreihe
p. transformer Impulstransformator *m*, Impulsübertrager *m*
p. transistor Impulstransistor *m*
p. transmission 1. Impulsübertragung *f*; 2. Impulsdurchgang *m*
p. triggering Impulsanregung *f*, Impulsauslösung *f*
p. tube Pulsatron *n*, Impulsröhre *f*
p. voltage Stoßspannung *f*
p. width Impulsdauer *f*, Impulsbreite *f*, Impulslänge *f*
p. width modulation Impulsbreitenmodulation *f*
p. work Impulstechnik *f*
pulsed impulsgesteuert
p. discharge Stoßentladung *f*, Impulsentladung *f*
p. frequency modulation Impulsfrequenzmodulation *f*, Impulstempomodulation *f*
p.-guide path *(Flg)* Impulslandeverfahren *n*
p. infusion shotfiring *(Brg)* Tränkungsschießen *n*
p. ion source Impulsionenquelle *f*
p. magnetron Impulsmagnetron *n*
p. neutron source Impulsneutronenquelle *f*
p. operation Impulsbetrieb *m*
p. oscillator [im]pulsgesteuerter Oszillator *m*
p.-oscillator starting time Oszillatoransprechzeit *f* bei Tastung
p. radar Impulsradar *n*
p. technique *s.* pulse technique
pulser Impulsgenerator *m*, Impulsgeber *m*; Impulszentrale *f*
p. unit Impulsgerät *n*
pulsing *(Fmt)* Impulsgabe *f*, Stromstoßgabe *f*
p. circuit *s.* pulse circuit
p. signal *(Fmt)* Zifferzeichen *n*
pulsometer Pulsometer *n*, Dampfwasserheber *m*, Dampfdruckpumpe *f*
pulverizable pulverisierbar
pulverization Pulverisierung *f*; Zerstäubung *f*; Fein[st]mahlung *f*
pulverize / to pulverisieren, fein[st]mahlen, [fein] zerkleinern, zerreiben, zerstäuben, zerstoßen
pulverized pulverisiert, pulverförmig
p. coal Kohlenstaub *m*, Staubkohle *f*
p. coal burner Kohlenstaubbrenner *m*
p. coal firing Kohlenstaubfeuerung *f*
p. coal plant Kohlenstaubmahlanlage *f*
pulverizer Pulverisierapparat *m*, Zerkleinerer *m*, Zerstäuber *m*, Fein[st]mühle *f*
p. mill Fein[st]zerkleinerungsanlage *f*
pulverizing Pulverisieren *n*, Pulverisierung *f*, Fein[st]mahlung *f*
p. equipment Fein[st]mahlanlage *f*
p. mill Staubmühle *f*
p. roll Mahlwalze *f*
pulverulent pulverig, staubig, staubbedeckt; zerkrümelnd
pulvinated *(konvex)* gewölbt

pumice Bimsstein *m*
p. concrete Bimsbeton *m*
p. powder gemahlener Bimsstein *m*
p. slag *(Met)* Schaumschlacke *f*; Hüttenbims *m*
p. stone Bimsstein *m*
pumiceous bimssteinähnlich, bimssteinartig
pummel Stampfer *m*, Ramme *f*
pump / to pumpen
p. out auspumpen
p. up aufpumpen *(Luftreifen)*
pump Pumpe *f*
p. action of piston rings Pumpen *n* der Kolbenringe
p. and motor unit Hydraulikaggregat *n*
p. barrel Pumpenzylinder *m*, Pumpenstiefel *m*
p. body Pumpenkörper *m*
p. borer Löffelbohrer *m (Brunnenbau)*
p. bucket Pumpenkolben *m* mit Klappe
p. casing Pumpengehäuse *n*
p. chamber Pumpenkammer *f*, Pumpenrohr *n*
p. circulated cooling Pumpenumlaufkühlung *f*
p. cistern Pumpenkasten *m*
p. cylinder Pumpenzylinder *m*
p. delivery Fördermenge *f* (Fördergut *n*) der Pumpe
p. discharge Verdrängung *f* (Fördergut *n*, Fördermenge *f*) der Pumpe
p. dredge Pumpenbagger *m*, Saugbagger *m*
p. drive gear treibendes Pumpenrad *n*
p. driven by power Transmissionspumpe *f*
p. duty Pumpenförderleistung *f*
p. edge Schmalschnitt *m (Schuh)*
p. efficiency Pumpennutzleistung *f*
p.-fed basin Rückpumpbecken *n*
p.-fed power station Pumpspeicherkraftwerk *n*
p. for water of condensation Kondensatpumpe *f*
p. gear Pumpenbeschlag *m*
p. governor Pumpenregler *m*
p. handle Pumpenschwengel *m*
p. head Druckhöhe *f*, Förderhöhe *f (Pumpe)*; *(Wärm)* Kesselboden *m*
p. idler gear getriebenes Pumpenrad *n*
p. impeller 1. Kreiselpumpe *f*; 2. Pumpenrad *n*
p. inlet valve Pumpeneinlaßventil *n*
p. jack Pumpvorgelege *n*, Pumpvorrichtung *f*; Pumpengestell *n*
p. jet Pumpendüse *f*
p. kettle Pumpenkorb *m*
p. lever Pumpenschwengel *m*
p. lift Pumpenhub *m*
p. lubrication Druckschmierung *f*
p. outlet pressure Förderdruck *m* der Pumpe
p. outlet valve Pumpenauslaßventil *n*
p. packing Pumpenliderung *f*
p. piston lever Pumpenhebel *m*
p. plunger Pumpenkolben *m*
p. plunger drive Pumpenstempelantrieb *m*
p. rodding Pumpengestänge *n*
p. rods Pumpengestänge *n*
p. shaft Pumpenschacht *m*, Wasserhaltungsschacht *m*
p. spears Pumpengestänge *n*
p. staff Pumpenstock *m*
p. station Pumpwerk *n*
p. stock Pumpenkasten *m*
p. strainer Pumpenansaugkorb *m*
p. stroke Pumpenhub *m*
p. sucker Pumpenschuh *m*
p. valve Pumpenklappe *f*
p. well 1. Pumpensod *m*, Pumpensumpf *m*; 2. Pumpenbrunnen *m*
p. well cover Pumpen[brunnen]deckel *m*
p. wheel Pumprad *n*
pumpability Pumpfähigkeit *f*
pumpable pumpbar, förderbar [durch Pumpen]
pumpage Pumpwirkung *f*
pumpcrete Pumpbeton *m*
pumped concrete Pumpbeton *m*
p. rectifier [elektrisches] Ventil *n* mit Vakuumhaltung
p. storage Pumpspeicher *m (in einer Wasserkraftanlage)*
pumping capacity Pumpleistung *f*
p. engine Pumpwerkdampfmaschine *f*
p. gear Pumpenwerk *n*, Pumpgetriebe *n*
p. head Förderhöhe *f (Pumpe)*
p. jack Pumpenbock *m*, Pumpenwinde *f*
p. lead Pumpstengel *m*
p. plan *(Schiff)* Lenz- und Ballasteinrichtungsplan *m*
p. set Pumpensatz *m*
p. station Pumpwerk *n*, Pumpstation *f*, Hebewerk *n*
p. stem Pumpstengel *m (Lampe)*
pumpman Pumpenwärter *m*
pun / to rammen, stampfen
punch / to ankörnen; lochen *(Band)*; durchstoßen; ausschneiden *(stanzen)*; durchtreiben
p. holes lochen, [Löcher] ausstanzen
p. out aushauen, ausschneiden
punch Stempel *m*, Preßstempel *m*, Schnittstempel *m*, Oberwerkzeug *n*; Patrize *f*; Körner *m*, Lochstempel *m*, Locheisen *n*; Ziehstempel *m*; Schnitt *m*; Ausschneidwerkzeug *n*; *(sl)* Stanzloch *n*
p. and form shaper Form- und Stempelhobelmaschine *f*
p. card Lochkarte *f*
p. clutch *(Büro)* Locherkupplung *f*
p. cutter Stempelmacher *m*, Stempelschneider *m*

p.-die assembly Lochermatrix *f*, Lochermatrizenvorrichtung *f*
p. mark Ankörnung *f*, Körnermarke *f*
p. position *(Büro)* Lochstellung *f*
p. press Ausschneidpresse *f (Stanzmaschine)*
p. reader Leselocher *m*, Leser *m* und Locher *m*
p. shaper Stempelhobler *m*
p. shaping machine Stempelhobelmaschine *f*
p.-tape control Lochbandsteuerung *f*, Lochstreifensteuerung *f*
p.-tape controlled lochbandgesteuert, lochstreifengesteuert
punched gestanzt, gelocht
p. card *(Dat)* Lochkarte *f*
p. card control Lochkartensteuerung *f*
p. card controlled mit Lochkartensteuerung, lochkartengesteuert
p. card department Lochkartenabteilung *f*
p. card file Lochkartenablage *f*
p. card machine accounting Buchführung *f* mit Lochkartenmaschinen
p. galosh *s.* p. golosh
p. golosh ausgelochtes Oberteil *n (Schuh)*
p. hole vorgeschmiedete Bohrung *f*; ausgeschnittenes (gestanztes) Loch *n*
p. paper tape Lochpapierstreifen *m*
p. tape Lochband *n*, Lochstreifen *m*
p.-tape control Lochbandsteuerung *f*, Lochstreifensteuerung *f*
p.-tape editing equipment Lochstreifenredigierausrüstung *f*
p.-tape recorder Lochstreifenschreiber *m*
puncheon 1. Pfosten *m*, Ständer *m*; Steife *f*, Stempel *m*; Zwischenpfosten *m*; 2. *(Typ)* Patrize *f*
puncher Lochstanze *f*; Ausschneider *m*; Locher *m*
punching Ankörnen *n*; Lochen *n*, Ausschneiden *n (Stanzen)*; Durchstoßen *n*, Durchtreiben *n*
p. bag Birne *f*, Plattformball *m (für Boxtraining)*
p. ball Doppelendball *m (für Boxtraining)*
p. die Aushaustempel *m*, Ausschneidstempel *m*
p. machine Ausschneidpresse *f*, Stanze *f*, Stanzmaschine *f*; Perforiermaschine *f*
p. out Aushauen *n*, Ausschneiden *n*
p. press Stanzpresse *f*; Ausschneidpresse *f*
p. recorder Lochstreifen *m*
p. test Stanzprüfung *f*, Lochprüfung *f*
p. tool Lochwerkzeug *n*, Ausschneidwerkzeug *n*
punchings Stanzabfälle *mpl*
punctiform punktförmig
punctuation mark Satzzeichen *n*, Interpunktionszeichen *n*
puncture / to 1. punktieren; 2. *(El)* durchschlagen; 3. ein Loch bekommen *(Reifen)*; 4. aufstechen; durchstechen
puncture 1. *(El)* Durchschlag *m*, Spannungsdurchschlag *m*; 2. Einstich *m*; 3. Reifenpanne *f*
p. cut-out *(El)* Durchschlagsicherung *f (mit Membrane)*
p. of the insulation Durchschlag *m* des Isolierstoffs
p.-proof insulator durchschlagsicherer (durchschlagsfester) Isolator *m*
p. strength Durchschlagfestigkeit *f*
p. voltage *(El)* Durchschlagsspannung *f*
pungent scharf *(schmeckend)*; stechend *(riechend)*
punk Zunder *m*; faules Holz *n*
punner Stampfer *m*, Ramme *f*
punning Stampfen *n (Bodenverdichtung)*
punt *(Schiff)* Prahm *m*, Scheuerprahm *m*, Pünte *f*
pup Junghund *m*
pupa Puppe *f (Seidenraupe)*
pupal stage Puppenstadium *n (Seidenraupe)*
pupil Pupille *f*
pupillary dilatation Pupillenerweiterung *f*
Pupin coil *(El)* Pupinspule *f*
Pupin loading *(El)* Pupinisierung *f*
pupinization *(El)* Pupinisierung *f*
pupinized cable pupinisiertes (bespultes) Kabel *n*
purchase 1. Handgriff *m*; Hebezeug *n*; 2. *(Schiff)* Takel *n*, Hebevorrichtung *f*; 3. Kauf *m*
p. price Kaufpreis *m*
purchasing cost Anlagekosten *pl*
p. department Bestellungsabteilung *f*, Einkaufsabteilung *f*, Einkaufsdienst *m*
p. power Kaufkraft *f*
pure *(Ch)* rein; *(Met)* gediegen
p. bending reine Biegung *f*
p. carbon arc lamp Reinkohlenbogenlampe *f*
p. coal Reinkohle *f*
p. continuous wave ungedämpfte Welle *f*
p. element Reinelement *n*
p. gold Feingold *n*
p. mathematics reine Mathematik *f*
p. metal Reinstmetall *n*
p. rag paper Reinhadernpapier *n*
p. resistance Wirkwiderstand *m*
p. science reine Wissenschaft *f*
p. scrap Edelschrott *m*
p. sound *(Ak)* reiner Ton *m*
p. strategy *(Dat)* originäre (reine) Strategie *f (Theorie der Spiele)*
p. substance Reinstoff *m*
p. time delay *(BMSR)* reine Laufzeit (Totzeit) *f*
purely resistive rein ohmisch, winkelfrei
purgative Abführmittel *n*
purification *(Ch, Met)* Reindarstellung *f*; Raffinierung *f*; Reinigung *f*; Aufbereitung *f*
p. of electrolyte Elektrolytaufbereitung *f*
p. of gas Gasreinigung *f*

p. tub Waschbütte *f*, Wässerungsbütte *f*
purified steel Frischstahl *m*
purifier Reinigungsmittel *n*
purify / to *(Ch)* reinigen, klären, läutern, waschen, abschlämmen; raffinieren *(Metall)*
purifying Reinigen *n*
p. furnace Frischofen *m*
p. mass Reinigungsmasse *f*
p. reaction *(Met)* Frischwirkung *f*
purine *(Ch)* Purin *n*
purity Reinheit *f*; Feingehalt *m*
p. adjustment *(Fs)* Farbreinheitseinstellung *f*
p. control *(Fs)* Farbreinheitsregelung *f*
p. test Reinheitsprüfung *f*
Purkinje effect *(Opt)* Purkinjesches Phänomen *n*
purl / to rieseln; murmeln; wirbeln
purl Drahtspirale *f*
p.-edge abgekettelter Rand *m (Stricken)*
p. knitting Linksstricken *n*
purlin *(Bw)* Dachpfette *f*, Pfette *f*
purple Purpur *m*
p. of Cassius Cassiusscher Goldpurpur *m*
purpose Zweck *m*; Aufgabe *f*
p.-made brick Formziegel *m*
purposely beabsichtigterweise, mit Absicht; willkürlich
purse seine Beutelnetz *n (Fischerei)*
pursuit curve (line, path) Verfolgungskurve *f*, Hundekurve *f*, Anflugkurve *f*
purview Wirkungsbereich *m (eines Gesetzes)*
push / to stoßen, drücken, schieben; treiben; *(Brg)* rücken *(Fördermittel)*; *(Schiff)* schieben *(einen Verband von Prahmen)*
p. broach *(Masch)* stoßräumen
push Stoß *m*, Schub *m*; Pressung *f (Strangpresse)*; *(Schiff)* Schubverband *m*
p. and pull rod *(Flg)* Steuerstange *f*
p. bench *(Met)* Stoßbank *f*; Warmziehbank *f (für Hohlkörper)*
p. bench bloom *(Met)* Stoßbankluppe *f*
p.-boat Schubboot *n*, Stoßboot *n*, Schubschlepper *m*
p.-boating *(Schiff)* Schubbetrieb *m*, Schubverkehr *m*
p.-button Druckknopf *m*, Drucktaste *f*
p.-button control Druckknopfsteuerung *f*
p.-button controlled druckknopfgesteuert
p.-button dialling *(Fmt)* Tastenwahl *f*
p.-button key Drucktaste *f*, Druckknopfschalter *m*
p.-button mine *(Brg)* Grube *f* mit hohem Automatisierungsgrad
p.-button operation Druckknopfbetätigung *f*
p.-button panel Druckknopfschalttafel *f*
p.-button pendant am Schwenkarm hängender Druckknopfschalter *m*
p.-button printing *(Text)* druckknopfgesteuertes Drucken *n*
p.-button radio Druckknopfradio *n*, Rundfunkapparat *m* mit Stationswählertasten
p.-button switch Druck[knopf]schalter *m*
p.-button tuner Druckknopfabstimmvorrichtung *f*
p.-button-type radio set Drucktastenempfänger *m*, Drucktastenempfangsgerät *n*
p. contact Druckkontakt *m*
p. down store listenförmiger Datenspeicher *m (Zugriff nur von oben)*
p. finger Stoßfühler *m (Webstuhl)*
p. loading zusätzliches Schieben *n* eines Schürfkübelwagens durch einen oder mehrere Traktoren während des Ladevorgangs
p.-off pin Abdrückbolzen *m*
p.-on filter *(Opt)* Aufsteckfilter *n*
p.-pull *(El)* Gegentakt *m*
p.-pull action *(BMSR)* Zweiseitenbetrieb *m*, Zweirichtungsbetrieb *m*; Gegentaktbetrieb *m*
p.-pull amplifier Gegentaktverstärker *m*
p.-pull arrangement Gegentaktschaltung *f*, Gegentaktanordnung *f*
p.-pull carbon microphone Doppelkohlemikrofon *n*
p.-pull cascade *(El)* Gegentaktstufe *f*
p.-pull connection *s.* p.-pull arrangement
p.-pull coupling Gegentaktkopplung *f*
p.-pull cylinder Gegentaktzylinder *m*
p.-pull detector Gegentaktgleichrichter *m*
p.-pull frequency doubler Gegentaktfrequenzverdoppler *m*
p.-pull microphone Gegentaktmikrofon *n*
p.-pull oscillator Gegentaktgenerator *m*, Gegentaktoszillator *m*
p.-pull output stage Gegentaktendstufe *f*
p.-pull power amplifier Gegentaktverstärker *m*
p.-pull power stage Gegentaktendstufe *f*
p.-pull rectifier Gegentaktgleichrichter *m*
p.-pull stage Gegentaktstufe *f*
p.-pull switch Druck-Zug-Schalter *m*
p.-pull transformer Gegentakttransformator *m*, symmetrischer Übertrager *m*
p.-push circuit *s.* p.-pull arrangement
p. rod Stößelstange *f*; Schubstange *f*, Stoßstange *f*
p. shovel Löffelbagger *m* mit Löffelstiel
p.-to-talk button *(Fmt)* Sprechtaste *f*
p. tow *(Schiff)* Schubverband *m*, Schubeinheit *f*; Schubschleppzug *m*
p.-tow-boat Schubschlepper *m*
p.-tow-tug Schubschlepper *m*
p.-towing trial Probefahrt *f* mit Schubschlepper
p.-tug unit Schubschlepper[verband] *m*
p.-type broaching machine Räumpresse *f*
p.-type slotter Nutenstoßmaschine *f (Holzbearbeitung)*

p.-type towboat Schubschlepper *m*
pusher 1. Schubboot *n*, Stoßboot *n*; 2. Ausdrückmaschine *f (Kokerei)*
p. beam Druckstange *f*, Zahnstange *f*
p.-conveyor furnace Schubofen *m*, Stoßofen *m*
p. furnace Schubofen *m*, Stoßofen *m*
p. navigation Schubschiffahrt *f*
p. operation Nachschiebefahrt *f (elektrischer Zug)*
p. rack Ausdrückstange *f*
p. side Ausdrückseite *f*, Maschinenseite *f*
p. tug Schubschlepper *m*, Stoßschlepper *m*; Schubboot *n*, Stoßboot *n*; Schubbugsierer *m*
p.-type furnace Stoßofen *m*, Durchstoßofen *m (Walzwerk)*
pushing figure Stromverstimmungsmaß *n (der Oszillatorfrequenz)*
p. machine Ausdrückmaschine *f*
p. motorbarge *(Schiff)* schiebender Selbstfahrer *m*
put / to setzen; stellen; legen
p. a gloss on *(Led)* glanzstoßen, Stoßglanz geben
p. a transmitter on the air einen Sender ausstrahlen lassen, senden
p. a voltage across a resistance Spannung an einen Widerstand anlegen
p. about Kurs ändern
p. back 1. zurückfahren *(Schiff)*; 2. zurückstellen *(Uhr)*
p. back the receiver *(Fmt)* den Hörer anhängen
p. forth auslaufen *(Schiff)*
p. in circuit *(El)* einschalten
p. in mesh *(Masch)* in Eingriff bringen
p. in motion in Bewegung setzen
p. in parallel with parallelschalten mit
p. into a port einen Hafen anlaufen; ansegeln
p. into commission in Dienst stellen *(Schiff)*
p. into gear *(Masch)* [ein]schalten
p. into operation in Betrieb setzen
p. on aufsetzen
p. on a disk eine Schallplatte auflegen
p. out of commission außer Dienst stellen; abrüsten *(Schiff)*
p. out of tune verstimmen *(Radio)*
p. the wool together by fleeces Wolle vliesweise zusammenlegen
p. through *(Fmt)* durchverbinden
p. to sea in See stechen *(Schiff)*
put-in energy aufgewandte Energie *f*
putlog *(Bw)* Rüststange *f*
p. hole Rüstloch *n*
putrefaction Fäulnis *f*; Fäulnisgärung *f*; Verwesung *f*
putrefactive fermentation Fäulnisgärung *f*
putrescence Fäulnis *f*
putrid faul, verfault
p. soak *(Led)* faule Weiche *f*
putting out of action Außerbetriebsetzen *n*
p. the shot Kugelstoßen *n (Sport)*
p. the yarn out of action *(Text)* Fadenauslegen *n*
putty / to spachteln; auskitten
p. up verkitten *(z.B. Fenster)*
putty Spachtelmasse *f*, Kitt *m*, Glaserkitt *m*
p. chaser Kollergang *m*
p. knife Spachtel *m*; Spatel *m*
p. powder Zinnasche *f*
pycnometer Dichtemesser *m*, Pyknometer *n*
pylon 1. Pylone *f (Brückenbau)*; 2. Antennenturm *m*, Gittermast *m*, Mast *m*
p. aerial Schlitzrohrantenne *f*
pyramid Pyramide *f*
p. cut *(Brg)* Pyramideneinbruch *m*
p. diamond hardness Vickershärte *f (Werkstoffprüfung)*
p.-type plate bending rolls Dreiwalzenblechbiegemaschine *f*
pyramidal pyramidenförmig, pyramidal
p. horn Reusenantenne *f*, Reusenstrahler *m*, Pyramidenhorn *n*
p. point Pyramidenspitze *f*
pyrargyrite *(Min)* Pyrargyrit *m*, Antimonsilberblende *f*
pyrene *(Ch)* Pyren *n*
pyrheliometer Pyrheliometer *n*, Sonnenstrahlenmesser *m*
pyridine *(Ch)* Pyridin *n*
p. carboxylic acid Pyridinkarbonsäure *f*
p. sulfonic acid Pyridinsulfonsäure *f*
pyrimidine *(Ch)* Pyrimidin *n*
pyrite Pyrit *m*, Schwefelkies *m*
p. cinder Kiesabbrand *m*, Pyritabbrand *m*
p. furnace *(Met)* Pyritofen *m*
pyritic pyritisch; kiesähnlich, kieshaltig
p. ore Pyriterz *n*
p. shales Pyritschiefer *m*
pyritiferous pyritisch, kieshaltig
pyritohedron Pyritoeder *n*, Pentagondodekaeder *n (Kristall)*
pyroacid Brenzsäure *f*
pyrocatechin Brenzkatechin *n*
pyrocatechol Brenzkatechin *n*
pyrochemistry Pyrochemie *f*
pyrochlore *(Min)* Pyrochlor *m*
pyrochroite *(Min)* Pyrochroit *m*
pyrocollodion cotton Schieß[baum]wolle *f*
pyrocrystalline pyrokristallin
pyroelectricity Pyroelektrizität *f*
pyrogallic acid *(Ch)* Pyrogallol *n*
pyrogenic *(Ch)* pyrogen
pyroligneous acid Holzessig *m*
pyrolusite *(Min)* Pyrolusit *m*, Braunstein *m*, Mangandioxid *n*

pyrolysis *(Ch)* Pyrolyse *f*, Zersetzung *f* unter Hitzeeinwirkung
pyrometallurgy Pyrometallurgie *f*
pyrometre Pyrometer *n*, Strahlungsthermometer
p. couple Thermoelement *n*
p. fire-end Pyrometereintauchende *n*
p. probe Temperaturfühler *m*
p. tube Temperaturfühler *m* mit Schutzrohr
pyrometric pyrometrisch
p. cone Brennkegel *m*, Schmelzkegel *m*, Segerkegel *m*, Pyrometerkegel *m*
p. effect Heizwert *m*
p. scale Temperaturskala *f*
pyrometry Pyrometrie *f*
pyromucic acid Brenzschleimsäure *f*
pyrophoric pyrophor, selbstentzündlich
pyrophosphoric acid Diphosphorsäure *f*, Pyrophosphorsäure *f*
pyrophosphorous acid diphosphorige (pyrophosphorige) Säure *f*
pyroracemic acid Brenztraubensäure *f*
pyrosulphuric acid Dischwefelsäure *f*, Pyroschwefelsäure *f*
pyrosulphurous acid dischweflige (pyroschweflige) Säure *f*
pyrotartaric acid Brenztraubensäure *f*
p. aldehyde Brenztraubensäurealdehyd *m*, Methylglyoxal *n*
pyrotechnic composition Feuerwerkskörper *m*, Leuchtsatz *m*
pyrotechnics Feuerwerkerei *f*
pyroxene *(Min)* Pyroxen *m*
pyroxylin Kollodiumwolle *f*, Nitrozellulose *f*
p. lacquer Nitrozelluloselack *m*
pyrrhotite *(Min)* Pyrrhotin *m*, Magnetkies *m*
pyrrole *(Ch)* Pyrrol *n*
pyrrolidine *(Ch)* Pyrrolidin *n*
pyruvaldehyde Brenztraubensäurealdehyd *m*, Methylglyoxal *n*
pyruvate *(Ch)* Pyruvat *n*, Ester *m* (Salz *n*) der Brenztraubensäure
pyruvic acid Brenztraubensäure *f*, Pyruvinsäure *f*
Pythagoras's theorem Lehrsatz *m* des Pythagoras, pythagoreischer Lehrsatz
Pythagorean pythagoreisch

Q

Q-aerial mit Stichleitung angepaßter Dipol *m*, Q-Antenne *f*
Q-electron Q-Elektron *n*
Q.F. gun Schnellfeuergeschütz *n*
Q-factor Güte *f*, Gütefaktor *m*; Kreisgüte *f (Parallelschwingkreis oder Spule)*
Q-shell *(Kern)* Q-Schale *f*
Q-ship U-Bootfalle *f*
Q-waves Q-Zacken *mpl (Ekg)*
quad / to *(El)* zum Vierer verseilen
quad cable *(El)* Viererkabel *n*
q. pair formation Doppelsternverseilung *f (Kabel)*
quadded viererverseilt *(Kabel)*
q. cable Viererkabel *n*
quadding machine Verseilmaschine *f (für Kabel)*
quadrangle Viereck *n*
quadrangular viereckig, vierseitig
quadrant 1. Quadrant *m*; Winkelmesser *m*; Viertelkreis *m*; 2. Segment *n (Gangschaltung)*; Schaltbogen *m*; 3. *(Masch)* Wechselräderschere *f*; Wendeherz *n*
q. angle of departure *(Mil)* Abgangswinkel *m*
q. electrometer Quadrant[en]elektrometer *n*
q. lever Zahnsegmenthebel *m*
q.-pipe Rohrkrümmer *m*
q. steering gear Quadrantenrudermaschine *f*
quadrate / to quadrieren, zu einem Quadrat umbilden
quadrate quadratisch *(nicht mathematisch)*
quadratic quadratisch *(Geometrie)*; zweiten Grades *(z. B. eine Gleichung)*
q. equation quadratische Gleichung *f*
q. factor *(BMSR)* Verzögerung *f* zweiter Ordnung
q. response *(BMSR)* Zeitverhalten *n* zweiter Ordnung, Schwingungsverhalten *n*
q. system tetragonales System *n (Kristall)*
quadraton Hochvakuumvierelektrodenröhre *f*
quadrats *(Typ)* Quadraten *mpl*; Ausschlußstücke *npl*
quadrature 1. *(Math)* Quadratur *f*; 2. *(BMSR)* Phasenverschiebung *f* einer Schwingung um 90° gegenüber einer Bezugsschwingung
q.-axis component *(El)* Querfeldkomponente *f*
q.-axis component of the electromotive force Quer-EMK *f*
q.-axis component of the voltage Querspannung *f*
q.-axis subtransient electromotive force subtransitorische Quer-EMK *f*
q.-axis subtransient impedance subtransitorische Querimpedanz *f*
q.-axis transient electromotive force transitorische Quer-EMK *f*
q.-axis transient impedance transitorische Querimpedanz *f*
q. coil Phasenschieberspule *f*
q. component *(El)* Blindkomponente *f*
q. rejection *(BMSR)* Unterdrückung *f* um 90° versetzter Störungen
quadrennial vierjährig
quadribasic *(Ch)* vierbasisch
quadric *s.* quadratic
quadrilateral Viereck *n*

quadrilateral vierseitig; viereckig
quadrinomial viergliedrig
quadripartite vierteilig
quadripolar *(El)* vierpolig
quadripole *(El)* Vierpol *m*
q. attenuation factor Vierpoldämpfungsfaktor *m*
q. propagation factor Übertragungsfaktor *m* eines Vierpols
quadrivalence Vierwertigkeit *f*
quadrivalent vierwertig
quadruplane Vierdecker *m*
quadruple / to vervierfachen
quadruple vierfach
q. bond *(Ch)* Vierfachbindung *f*
q. pair cable Vierfachzwillingskabel *n*, achterverseiltes Kabel *n*
q. phantom circuit *(Fmt)* Sechzehnerleitung *f*
q. point *(Ch)* Quadrupelpunkt *m*
q. riveted vierreihig genietet
q. screw ship Vierschraubenschiff *n*
q.-strand chain Vierfachkette *f*
q. telegraph Vierfachtelegraf *m*
q. thread viergängiges Gewinde *n*
q.-threaded viergängig *(Gewinde)*
q. turret Vierlingsturm *m (Kriegsschiff)*
quadrupled scanning *(Fs)* vierfaches Zwischenzeilenverfahren *n*
quadrupler Frequenzvervierfacher *m*
quadruplex operation *(Fmt)* Doppelgegensprechbetrieb *m*
q. system *(Fmt)* Doppelgegensprechsatz *m*, Quadruplexsatz *m*
q. telegraphy Quadruplextelegrafie *f*, Vierfachtelegrafie *f*, Doppelgegenschreiben *n*
quadrupole Quadrupol *m*
q. coupling Quadrupolkopplung *f*
q. interaction Quadrupolwechselwirkung *f*
q. lens Quadrupollinse *f*
q. moment Quadrupolmoment *n*
q. objective turret vierfacher Objektivrevolver *m*
q. radiation Quadrupolstrahlung *f*
q. spectrum Quadrupolspektrum *n*
q. transition Quadrupolübergang *m*
quads *(Typ)* Quadraten *mpl*, Ausschlußstücke *npl*
quaggy kernrissig *(Holz)*
quagmire Sumpfboden *m*
qualification[s] Kenntnisse *fpl*, Ausbildung *f*, Qualifikation *f*
qualify / to [sich] qualifizieren
qualifying examination Facharbeiterprüfung *f*
qualimeter Qualimeter *n (Meßgerät für die Strahlenhärte)*
qualitative qualitativ
q. analysis qualitative Analyse *f*
q. determination qualitative Bestimmung *f*
q. routine analysis qualitative Routineanalyse *f*
q. spectral analysis qualitative Spektralanalyse *f*
quality Güte *f*, Qualität *f*
q. characteristic Qualitätsmerkmal *n*
q. complaint Qualitätsbeanstandung *f*
q. control Gütekontrolle *f (besonders statistische)*
q. factor *(El)* Gütefaktor *m*
q. inspection Qualitätskontrolle *f*, Gütekontrolle *f*
q. level Gütegrad *m*, Qualitätslage *f*
q. loss Güteminderung *f*
q. mark Gütezeichen *n*
q. monitoring *(selbsttätige)* Gütekontrolle *f*
q. of concrete Betongüte *f*
q. of finish Oberflächengüte *f*
q. of fit Passungsgüte *f*
q. of sound Klangfarbe *f*
q. of speech *(Fmt)* Sprachgüte *f*
q. of surface finish Oberflächengüte *f*
q. of transmission *(Fmt)* Übertragungsgüte *f*
q. sheet Qualitätsblech *n*
q. specification Gütevorschrift *f*; Abnahmevorschrift *f*
quantifiable quantitativ bestimmbar
quantify / to quantitativ bestimmen
quantitative quantitativ
q. analysis quantitative Analyse *f*
q. change quantitative Umsetzung *f*
q. measurement quantitative Messung *f*
q. spectrum analysis quantitative Spektralanalyse *f*
quantities by volume Raumteile *mpl*
q. by weight Masseteile *mpl*
quantity Größe *f*; Menge *f*, Masse *f*, Betrag *m*, Quantität *f*
q. description *(Bw)* Leistungsbeschreibung *f*
q. discharged Ausflußmenge *f (Turbine)*
q. efficiency Mengenwirkungsgrad *m*
q. governing Mengenregelung *f*, Quantitätsregelung *f (Turbine)*
q. of cooling water required Kühlwasserbedarf *m*
q. of electricity Elektrizitätsmenge *f*
q. of heat Wärmemenge *f*, Wärmequantum *n*
q. of heat passed through a partition durch eine Scheidewand hindurchgehende Wärmemenge *f*
q. of light Lichtmenge *f*
q. of radiation Strahlungsmenge *f*
q. of steam to be condensed zu kondensierende Dampfmenge *f*
q. of X-rays Röntgendosis *f*
q. passed Durchflußmenge *f (Turbine)*
q. passed per second sekundliche Durchflußmenge *f*

q.-produced in Mengenfertigung hergestellt
q. production Massenerzeugung *f*, Massenherstellung *f*, Serienfertigung *f*
q. surveyor Kalkulator *m*, Kostenplaner *m*
q. to be measured Meßgröße *f*, zu messende Größe *f*
quantization *(Ph)* Quantelung *f*; Quantisierung *f*
quantize / to *(Ph)* quanteln; quantisieren
quantized *(Ph)* gequantelt
q. field gequanteltes Feld *n*
q. system gequanteltes System *n*
quantizer *(Dat)* Quantisierer *m*, Umsetzer *m* in Digitalschreibweise; Zuordner *m*
quantizing *s.* quantization
quantum Quant *n*, Teilchen *n*, Strahl *m*
q. chemistry Quantenchemie *f*
q. condition Quantenbedingung *f*
q. efficiency Quantenausbeute *f*
q. electrodynamics Quantenelektrodynamik *f*
q. emission Quantenemission *f*
q. energy Quantenenergie *f*
q. field theory Quantenfeldtheorie *f*
q. jump Quantensprung *m*
q. limit kurzwellige Grenze *f*, Grenzwellenlänge *f*
q. mechanical quantenmechanisch
q. number Quantenzahl *f*
q. of action Plancksches Wirkungsquantum *n*
q. of energie Energiequant *n*, Energiequantum *n*
q. optics Quantenoptik *f*
q. orbit Quantenbahn *f*
q. physics Quantenphysik *f*
q. radiation Quantenstrahlung *f*
q. state Quantenzustand *m*
q. statistics Quantenstatistik *f*
q. theory Quantentheorie *f*
q. theory of light Lichtquantentheorie *f*, Quantentheorie *f* des Lichtes
q. transition Quantensprung *m*, Quantenübergang *m*
q. yield Quantenertrag *m*, Quantenausbeute *f*
quarantine anchorage gekennzeichneter Platz *m* im Hafen *(gelbe Bojen)* für Schiffe zur Quarantäneinspektion
q. harbour Quarantänehafen *m*
quarry / to brechen *(Steine)*
quarry 1. Steinbruch *m*; 2. *s.* q. tile
q. rocks Bruchgestein *n*
q. stone Bruchstein *m*, Naturstein *m*, Werkstein *m*
q. tile Natursteinplatte *f*; Fußbodenplatte *f*; Gehwegplatte *f*
quarrying 1. Gewinnung *f* von Steinen und Erden; 2. Brechen *n*; 3. Steinbruch *m*; Steinbruchbetrieb *m*; Steinbrucharbeit *f*
quarryman Steinbrucharbeiter *m*
quart *englische und amerikanische Volumeneinheit: 1. 1 qt = 1,136 l (England); 2. 1 dry qt = 1,102 l (in USA für trockene Substanzen); 3. 1 liq.qt = 0,946 l (in USA für Flüssigkeiten)*
quarter *1. englische Masseeinheit: 1 qr. = 12,701 kg; 2. englische Längeneinheit: 1 qr. = 1 span = 22,860 cm; 3. englische Volumeneinheit: 1 qr. = 289,5 l; 4. amerikanische Volumeneinheit: 1 qr. = 281,92 l*
quarter 1. Viertel *n*; 2. Revier *n*; 3. Quartier *n*
q. adder *(Dat)* Vierteladdierer *m (binär, ohne Übertrag)*
q. bend Knie[rohr] *n*, Kniestück *n*, 90°-Krümmer *m*
q. boat Seitenboot *n (an Davits);* Barkasse *f*, Schaluppe *f*
q. cloth Broschur *f (mit Geweberücken)*
q. cloth binding Halbgewebeband *m (ohne Gewebeecken)*
q. davit *(Schiff)* Seitendavit *m (n)*
q. deck *(Schiff)* Quarterdeck *n*, Achterdeck *n*, Hinterdeck *n*
q.-elliptic spring *(Kfz)* Viertelfeder *f*
q. leather binding Halblederband *m (ohne Lederecken)*
q. of a ship Achterschiff *n*, Hinterschiff *n*
q.-phase circuit *(El)* Zweiphasenschaltung *f*
q. pillar *(Schiff)* seitliche Deckstütze *f*, Seitenstütze *f*
q. point *(Bw)* Viertelpunkt *m*
q. rudder Seitenruder *n*
q. sawing Radialschnitt *m (Holzbearbeitung)*
q. sawn radial geschnitten *(Holz)*
q.-square multiplier *(Dat)* Viertelquadratmultiplizierer *m (Analogrechner)*
q. stuff vierzölliges Brett *n*
q. timber Kreuzholz *n*
q.-turn Vierteldrehung *f*
q.-turn arrangement [of belts] Halbkreuzriementrieb *m*
q.-wave 1. *(Schiff)* Hecksee *f*; 2. *(El)* Viertelwelle *f*
q.-wave aerial Viertelwellenantenne *f*
q.-wave length Viertelwellenlänge *f*
q.-wave plate Viertelwellenblättchen *n*, Viertelwellen[längen]platte *f*, $\lambda/4$-Blättchen *n*
q.-wave transformer *(Fs)* $\lambda/4$-Transformator *m*
q.-wavelength plate *s.* q.-wave plate
q.-wheel pusher tug Heckseitenradschubboot *n*
q. wheeler Heckseitenradschiff *n*
quarterdecker Quarterdecker *m (Schiffstyp)*
quartic Gleichung *f* vierten Grades; Kurve *f* vierten Grades
quartic vierten Grades
q. equation Gleichung *f* vierten Grades
quartz Quarz *m*, Bergkristall *m*

q.-bearing quarzhaltig
q.-controlled quarzgesteuert
q.-crystal chronometer Quarzuhr *f*
q. crystal-controlled oscillator Quarzoszillator *m*, quarzgesteuerter Oszillator *m*
q.-excited control stage Kristallsteuerstufe *f*
q. fibre Quarzfaden *m*, Quarzfaser *f*
q.-fibre dosimeter Quarzfadendosimeter *n*
q. filter Quarzfilter *n*, Kristallfilter *n*
q. glass Quarzglas *n*
q. lamp Quarzlampe *f*
q. oscillator Quarzoszillator *m*, quarzgesteuerter Oszillator *m*
q. resonator Schwingquarz *m*
q. spectrograph Quarzspektrograf *m*
q. thread Quarzfaden *m*
q. wavemeter Quarzwellenmesser *m*
quartziferous quarzhaltig, quarzig
quartzite *(Min)* Quarzit *m*; *(Geol)* Quarzfels *m*
q. brick Quarzitstein *m*
q. rock Quarzitgestein *n*
quartzous quarzähnlich, quarzartig
quasi quasi
q.-binary quasibinär
q.-binary alloy quasibinäre Legierung *f*
q.-classic quasiklassisch
q.-elastic quasielastisch
q.-harmonic quasiharmonisch *(Schwingungen)*
q.-multiple Quasivielfaches *n*
q.-optical quasioptisch
q.-optical frequency quasioptische Frequenz *f*
q.-rigid quasistarr
q.-stable quasistabil
q.-stable state quasistabiler Zustand *m*
q.-static process quasistatischer Vorgang *m*
q.-statical quasistationär, quasistatisch
q.-stationary quasistationär
q.-stationary discharge quasistationäre Entladung *f*
q.-stationary field quasistationäres Feld *n*
q.-stationary state quasistationärer Zustand *m*
quassia Quassiaholz *n*, Bitterholz *n*, Fliegenholz *n*
quaternary quaternär
quaternions *(Math)* Quaternionen *fpl*
quay Kai *m*
q. accommodations Kaianlagen *fpl*
q. wall Kaimauer *f*, Ufermauer *f*
quayage Kailänge *f*; Kaigeld *n*
quebracho Quebracho *n* *(Quebrachoholz, Quebrachogerbstoffe)*
q.-extract Quebracho *n* *(Gerbstoff)*
queen bee ferngesteuertes, unbemanntes Übungszielflugzeug *n*
q. closer Meisterquartier *n*, Riemchen *n* *(Mauerwerk)*
q. post Hängesäule *f* *(Holzbau)*
quench / to abschrecken *(Stahl)*; abkühlen, [ab]löschen *(Lichtbogen, Feuer)*; lösen
q. age *(Met)* aushärten, vergüten *(für Leichtmetalle)*; abschreckaltern
quench Abschreckmittel *n*
q. aging *(Met)* Abschreckalterung *f*
q. and hot bend test Abschreckbiegeversuch
q. frequency Pendelfrequenz *f* *(Nachrichtentechnik)*
q. hardening *(Met)* Abschreckhärten *n*
q. tank *(Met)* Abschreckbad *n*, Abschreckbehälter *m*
quenchant Härteflüssigkeit *f*, Abschreckflüssigkeit *f*, Abschreckmittel *n*
quenched abgeschreckt *(Stahl)*; gelöscht *(Funke, Kalk)*
q. and subsequently tempered steel Vergütungsstahl *m*
q. arc circuit breaker Deion-Schalter *m*, Leistungsschalter *m* mit Lichtbogenlöschung
q. lime gelöschter Kalk *m*
q. spark Löschfunke *m*
q. spark gap Löschfunkenstrecke *f*
quencher 1. *(Ch)* Lösekessel *m*; 2. *(El)* Lichtbogenlöscheinrichtung *f*
q. side Löschseite *f* *(Kokerei)*
quenching 1. Abschrecken *n*, Härten *n* *(Stahl)*; 2. Löschen *n* *(Koks)*; 3. *(Plast)* Spritzen *n* aus flacher Düse, Flachspritzen *n*
q. and tempering furnace Vergüteofen *m*
q. bath *(Met)* Abschreckbad *n*
q. car Löschwagen *m* *(Kokerei)*
q. chamber *(El)* Löschkammer *f*
q. coil *(El)* Sperrdrossel *f*
q. crack *(Met)* Härteriß *m*
q. deformation *(Met)* Härteverzug *m*
q. frequency *(El)* Pendelfrequenz *f*
q. liquid Abkühlflüssigkeit *f*, Abschreckflüssigkeit *f*
q. medium 1. *(El)* [Lichtbogen-]Löschmittel *n*; 2. Abschreckmittel *n* *(Stahl)*
q. of discharge *(El)* Löschung *f* der Entladung
q. oil Härteöl *n*, Kühlöl *n*
q. plant Löschanlage *f* *(Kokerei)*
q. stress Härtespannung *f*, Abschreckspannung *f* *(Stahl)*
q. tower Löschturm *m*
quern Handmühle *f*
queueing theory *(Dat)* Wartezeittheorie *f*, Warteschlangentheorie *f*
quibinary coded decimal notation quibinär kodierte (verschlüsselte) Dezimaldarstellung *f*, quibinär kodierte (verschlüsselte) Dezimalschreibweise *f*
quick schnell, geschwind; kräftig *(Feuer)*; ungelöscht *(Kalk)*
q. access store Schnellzugriffsspeicher *m*

q.-acting chuck *(Am) (Masch)* Schnellwechselfutter *n*
q.-acting clamping *(Masch)* Schnellspannung *f*
q.-acting disengagement Schnellausrückung *f (Kupplung)*
q.-acting engagement Schnelleinrückung *f (Kupplung)*
q.-acting regulator Eilregler *m*
q.-acting relay Schnellrelais *n*
q.-acting valve Schnellschlußventil *n*
q. action clamp *(Masch)* Schnell[spann]-schraubzwinge *f*
q. action draw-in arbor *(Masch)* Schnellspanndorn *m*
q. action fuse flinke (unverzögerte) Sicherung *f*
q. action vice *(Masch)* Schnellspannzeug *n*
q.-ash Flugasche *f*
q. break Momentunterbrecher *m*
q. break cut-out Momentschalter *m*
q.-break fuse flinke Sicherung *f*, Schnellsicherung *f*
q.-break switch Schnellschalter *m*, Sprungschalter *m*, Momentschalter *m*
q.-change chuck *(Masch)* Schnellwechselfutter *n*
q.-change drill chuck Schnellwechselbohrfutter *n*
q.-change gear mechanism Schnellwechselgetriebe *n*
q. charge Schnelladung *f (der Batterie)*
q. clamp *(Masch)* Momentschraubknecht *m*, Schnell[spann]schraubzwinge *f*
q. clamp jig Schnellspanneinrichtung *f*
q.-connector parachute *(Flg)* Einklinkfallschirm *m*
q. curing schnellhärtend *(Plast)*
q. detachable connector Schnellkupplung *f*
q. drying Schnelltrocknung *f*
q. fermentation Schnellgärung *f*
q.-firing *(Mil)* Schnellfeuer-
q. flashing light Funkelfeuer *n*, Blitzfeuer *n (Navigation)*
q. freezing Schnellkühlung *f*, Tiefkühlung *f*
q.-frozen pack Tiefgefrierkonserve *f*
q. fuse detonierende Zündschnur *f*
q. gripping *(Masch)* Schnellspannung *f*
q. helix *(30° to 40°)* kurzer (enger, flacher) Drall *m (35° bis 45°)*
q. hitch Schnellkupplungseinrichtung *f*
q.-lime *s.* quicklime
q.-make-and-break switch Momentschalter
q.-motion camera Zeitraffer[kamera *f*] *m*
q.-motion screw Grobeinstellschraube *f*
q.-opening valve Tellerventil *n*
q.-operating relay schnell ansprechendes Relais *n*
q. operation schnelle Betätigung *f*
q.-release Schnellauslösung *f*; schnelle Trennung (Unterbrechung) *f*
q.-release fastener Schnellverschluß *m*
q.-release valve Schnellöseventil *n (einer Druckluftbremse)*, Schnellablaßventil *n*
q. return (reverse) *(Masch)* Eilrücklauf *m*
q.-setting Schnellverstellung *f*
q.-setting cement schnellabbindender Zement *m*, Schnellbinderzement *m*
q. solder Schnellot *n*
q.-starting motor Motor *m* mit Schnellanlauf
q. test Vorprobe *f (Stahlwerk)*
q. traverse *(Masch)* Schnellverstellung *f*, Eilgang *m*
q. vinegar process *(Ch)* Schnellessigverfahren
quicklime Branntkalk *m*, gebrannter Kalk *m*, Ätzkalk *m*
q. mill Kalkmühle *f*
quickness Schnelligkeit *f*; Detonationsgeschwindigkeit *f (Sprengstoff)*
quicksand Treibsand *m*, Mahlsand *m*; Schwimmsand *m*
quicksilver Quecksilber *n*
q. mine Quecksilberbergwerk *n*
quiescent ruhend, ruhig
q. anode current Anodenruhestrom *m*
q. carrier unterdrückter Träger *m*
q. condition Ruhezustand *m*
q. period Ausgarzeit *f (Stahlschmelze)*
q. point statischer Arbeitspunkt *m (der Elektroden)*
q. push-pull amplifier *(Fmt)* Gegentaktschaltung *f* ohne Ruhestrom, B-Verstärker *m* in Gegentaktschaltung, Gegentakt-B-Betrieb *m*, Gegentaktverstärker *m* Klasse B
q. transmitter system Sender *m* mit unterdrückter Trägerwelle
q. value Leerlaufwert *m*, Ruhewert *m (Strom oder Spannung einer Röhrenelektrode)*
quiet / to beruhigen
q. steel Stahl beruhigen
quiet ruhig, geräuscharm
q. automatic grid-control *(Fmt)* geräuschfreie automatische Verstärkungsregelung *f*
q. interval *(Fmt)* Modulationspause *f*
q. melt ruhiges Schmelzbad *n*
q. steel beruhigter Stahl *m*
q. tuning *(Fmt)* Stummabstimmung *f*
quieting 1. Beruhigung *f (Schmelzbad)*; 2. Schallisolierung *f*; 3. *(Fmt)* Krachtötung *f*
q. sensitivity Empfindlichkeitsschwelle *f (Funkempfangstechnik)*
quietness Geräuschlosigkeit *f*
q. in operation ruhiger Lauf *m*
q. of running Laufruhe *f*
quill / to falten, kräuseln, knittern
quill 1. *(Masch)* Hülse *f*, Pinole *f*; Hohlwelle *f*; 2. *(Text)* Schützenspule *f*

q. **shaft** Hohlwelle *f*
quiller *(Text)* Schußspulmaschine *f*
quilling machine *(Text)* Schußspulmaschine *f*
quilt / to steppen, polstern, füttern
quilt Steppdecke *f*; Bettdecke *f*
quinaldine *(Ch)* Chinaldin *n*
quinary *(Dat)* quinär, zur Basis 5
quinhydrone *(Ch)* Chinhydron *n*
q. electrode Chinhydronelektrode *f*
quinic acid Chininsäure *f*
quinidine *(Ch)* Chinidin *n*
quinine Chinin *n*
quinitol *(Ch)* Chinit *n*
quinizarine *(Ch)* Chinizarin *n*
quinoid *(Ch)* chinoid
quinol *(Ch)* Chinol *n*
quinoline *(Ch)* Chinolin *n*
quinolinic acid Chinolinsäure *f*
quinone *(Ch)* Chinon *n*
quinonoid *(Ch)* chinoid
quinoxaline *(Ch)* Chinoxalin *n*
quinquevalence Fünfwertigkeit *f*
quinquevalent fünfwertig
quintic fünften Grades
quintuple fünffach
q. objective turret fünffacher Objektivrevolver *m*
q.-strand chain Fünffachkette *f*
quire folding *(Typ)* Falzung *f* in Lagen, Lagenfalzung *f*
quirk Nut *f*, Hohlkehle *f*
quoin / to *(Typ)* schließen, einkeilen
quoin Eckstein *m*; Gebäudeecke *f*; Mauerecke *f*
q. spaces *(Typ)* Keilspatien *npl*, Schließspatien *npl*
quoins *(Typ)* Schließzeug *n*
quota restriction on the import Einfuhrbeschränkung *f (durch Festlegen von Quoten)*
q. sample *(Math)* feste Teilstichprobe *f*, Quotenstichprobe *f*
quotient Quotient *m*, Teilzahl *f*
q. relay Quotientenrelais *n*

R

R-display R-Darstellung *f (Radar)*
r-meter chamber Röntgenmeter *n*, Röntgenmeßgerät *n*, r-Meter *n*
R-scope R-Schirm *m (Radar)*
R waves *(El)* R-Zacken *mpl (Ekg)*
rabbet / to [ein]falzen; [ein]fugen
rabbet Falz *m*, Nut *f*
r. joint Einfalzung *f*, Einfügung *f*, Fuge *f (Holz)*
r. plane Falzhobel *m*, Nuthobel *m*, Grathobel
rabbeting Falzen *n (Holz)*
r. cutter Falzfräser *m*
rabbit hair Kaninchenhaar *n*
r. skin Kaninchenfell *n*
r. tube *(Kern)* Rohrpostbüchse *f (zur Aufnahme von Bestrahlungsproben)*
rabble / to [durch]rühren
rabble Rührhaken *m*, Rührkrücke *f*, Puddelkrücke *f*, Krähler *m*
r. arm Krählarm *m*, Krählhalter *m*
r. furnace Krählofen *m*, Rührofen *m*
rabbling Krählen *n*, Kratzen *n*, Rühren *n*
r. hole *(Met)* Arbeitsöffnung *f*, Schummelloch *n (des Puddelofens)*
r. operation Krählarbeit *f*
race / to durchgehen *(Motor)*
race 1. Strom *m*, Strömung *f*; 2. Einlaufkanal *m*; 3. Laufring *m*, Laufbahn *f (Lager)*; 4. Regatta *f*, Wettfahrt *f*; 5. Rasse *f*, Zucht *f*
r. board *(Text)* Schützenbahn *f*
r. boat Rennboot *n*
r. of yachts Jachtwettrennen *n*
r. saddle and head gear Rennsattel *m* und Zaumzeug *n*
r.-track *(Kern)* Bahn *f*, Rennbahn *f (Beschleuniger)*
r. way Laufbahn *f (Lager)*
racemate *(Ch)* Razemat *n*
racemic *(Ch)* razemisch, optisch inaktiv
r. [tartaric] acid razemische Weinsäure *f*, Traubensäure *f*
racemism *(Ch)* Razemie *f*, optische Inaktivität *f*
racemization *(Ch)* Razemisierung *f*
racemize / to *(Ch)* razemisieren
racer 1. Laufring *m (Lager)*; 2. Rennboot *n*, Rennjacht *f*
raceway 1. Kabelkanal *m*, Leitungskanal *m*, Leitungsrohr *n*; 2. Laufbahn *f (Lager)*
racing Blindlaufen *n (der Schiffsschraube beim Stampfen)*
r. boat Rennboot *n*
r. car Rennwagen *m*
r. fuel Rennkraftstoff *m*
r. of the steam engine Durchgehen *n* der Dampfmaschine
r. skates Rennschlittschuhe *mpl.*
r. skiff Einer *m*, Einskuller *m (Sportboot)*
rack / to stapeln, einlegen, Futter (Heu) aufstecken
rack 1. *(Masch)* Zahnstange *f*; 2. Gestell *n*, Gerüst *n*, Traggerüst *n*, Regal *n*, Rahmen *m*; 3. Gepäcknetz *n*; 4. *(Flg)* Steuergestell *n*; 5. *(Lw)* Raufe *f*
r.-and-pinion [gear] drive Zahnstangentrieb *m*
r.-and-pinion jack Zahnstangenheber *m*, Zahnstangenwinde *f*
r.-and-pinion steering Zahnstangenlenkung *f*
r.-and-pinion steering gear Zahnstangenlenkgetriebe *n*

r. cutter Zahnstangenfräser *m* (*s. a.* r.-form cutter)
r. cutting machine Zahnstangenfräsmaschine *f*
r.-form cutter Kammeißel *m*, Zahnstangenwerkzeug *n (Schneidkamm)*
r. indexing attachment Zahnstangenteileinrichtung *f*
r. jack Zahnstangenwinde *f*
r. locomotive Zahnradlokomotive *f*
r. method of gear shaving Zahnradschaben *n* mit Zahnstange
r. milling attachment Zahnstangenfräseinrichtung *f*
r. mode Gestellbauweise *f (von Geräten)*
r. pinion Zahnstangenritzel *n*
r.-railway Zahnradbahn *f*
r. spacing attachment Zahnstangenteileinrichtung *f*
r. steering Zahnstangenlenkung *f*
r. teeth Zahnstangenverzahnung *f*
r.-type gear cutter Kammeißel *m (Verzahnung)*
r. with helical teeth schrägverzahnte Zahnstange *f*
racked geschichtet, gestapelt *(Holz in Klaftern)*
r. drill pipe im Turm stehendes Bohrgestänge *n (Tiefbohrtechnik)*
racket Tennisschläger *m*
racking back Abtreppung *f (Mauerwerk)*
racon Antwortbake *f*, Bodenantwortstation *f*, Wiederholerbake *f (Radar)*
radar (radio detection and ranging) Radar *n*, Funkmeßtechnik *f*
r.-aid Funkmeßgerät *n*
r.-aid to navigation Funkmeßnavigationsgerät *n*
r. band Radarband *n*
r. beacon Antwortgerät *n (Funknavigation)*
r. beam Radarstrahl *m*
r. camouflage Tarnung *f*
r. chain Radarkette *f*
r. contact Radarerfassung *f*
r. countermeasure Radargegenmaßnahme *f*, Gegenmaßnahme *f* gegen Radar
r. deception Radartäuschung *f*
r. detector Radarortungsgerät *n*
r. device Funkmeßeinrichtung *f*
r. display Radarschirmbild *n*
r. engineering Radartechnik *f*, Funkmeßtechnik *f*
r. equation Radargleichung *f*
r. equipment Radargerät *n*
r.-equipped mit Radar ausgerüstet
r.-fitted mit Radar ausgerüstet
r. flip [Radar-]Echozacken *m*
r. frequency Ortungsfrequenz *f*
r. gun-laying radargesteuerte Fernlenkung *f*
r. homing bomb Radarzielbombe *f*
r. homing guidance gelenkter Radarzielflug *m*
r. identification unit Radarkenngerät *n*
r. image Radarbild *n*
r. interference Radarstörung *f*
r. island Radarinsel *f*
r. jammer Radarstörsender *m*
r. jamming Radarstörung *f*
r. landscape scanner Panoramagerät *n*, Rundsuchradar *n*
r. link Radarnetz *n*
r. net Radarnetz *n*
r. network Radarnetz *n*
r. on Doppler effect principle Dopplereffektradar *n*
r. picket-vessel Radarsicherungsfahrzeug *n*
r. plot Radarschaubild *n*
r. plotter system Radarsystem *n* zum Kursabsetzen
r. plotting computer Radarschaubildrechner *m (Umwandlung des Polarkoordinatensystems in rechtwinkliges Koordinatensystem)*
r. presentation Radarschirmbilddarstellung *f*
r. presentation restitutor Radarbildentzerrer *m*
r. range Radarreichweite *f*
r. receiver Radarempfänger *m*
r. response Radarecho *n*
r. screen Radarschirm *m*
r. screen picture Radarschirmbild *n*
r. screen watching Radarschirmbeobachtung *f*
r. set Radargerät *n*
r. spike Impulsspitze *f*
r. station Radarstation *f*, Radaranlage *f*
r.-swamping coverage Radartarnüberzug *m (z. B. für Flugzeuge)*
r. target Radarziel *n*
r. technique Funkmeßverfahren *n*, Funkmeßtechnik *f*
r. tower Funkmeßturm *m*
r. transmitter Radarsender *m*, Funkmeßsender *m*
r. watch Radarüberwachung *f*
r. weather report Radarwetterbericht *m*
raddle *(Text)* Reihkamm *m*
radial Radialbohrmaschine *f*
radial radial, sternförmig
r. acceleration Radialbeschleunigung *f*
r. admission radiale Beaufschlagung *f (Turbine)*
r. arm Auslegerarm *m (Radialbohrmaschine)*
r. arm switch *(Elm)* Kurbelschalter *m*
r. armature *(Elm)* Radialanker *m*
r. arrangement Sternanordnung *f*
r. bearing Radiallager *n*
r. blower Radialgebläse *n*
r. brush *(Elm)* Radialbürste *f*

r. clearance Ringspalt *m (Turbine)*
r. clutch bearing Kupplungsdrucklager *n*
r. commutator *(Elm)* Flachkommutator *m*, Scheibenkollektor *m*
r. component Radialkomponente *f*
r. compressor Radialgebläse *n*
r. cylinder engine Sternmotor *m*
r. davit *(Schiff)* Klappdavit *m (n)*, Radialdavit *m (n)*, Barkune *f*
r. distribution feeder *(El)* Speiseleitung *f* im Radialnetz
r. drill Radialbohrmaschine *f*
r. drilling Radialbohren *n*
r. drilling machine Radialbohrmaschine *f*
r. drilling machine with fixed height arm Radialbohrmaschine *f* ohne Höhenverstellung
r. dyeing apparatus *(Text)* Radialfärbeapparat *m*
r. engine Sternmotor *m*, sternförmiger Motor *m*
r. extent radiale Ausdehnung *f*
r. feed *(Masch)* radiale Zustellung *f*
r. flow impeller Turbinenmischer *m*
r. flow turbine Radialturbine *f*
r. flux radialer Fluß *m*
r. focusing Radialfokussierung *f*
r. force Radialkraft *f*; Abdrängkraft *f*; radiale Komponente *f*
r. forcing out of the honing sticks Zustellung *f* der Honsteine
r. grating Radialfilter *n (Wellenleiter)*
r. height of blade radiale Schaufelhöhe *f (Turbine)*
r. inwards-flow turbine Radialturbine *f* mit Strömung von außen nach innen
r. joint radialer Fugenschnitt *m*
r. line Mittelpunktstrahl *m*
r. loading Radialbelastung *f*
r. piston pump Radialkolbenpumpe *f*
r. planing attachment Rundhobeleinrichtung *f*
r. play Radialspiel *n*
r. pressure [between hub and shaft] Fugenpressung *f*
r. relief Hinterwetzwinkel *m*
r. rim strain radiale Randspannung *f*
r. runout Radialschlag *m (Wälzlager)*
r. shrinkage Radialschwindung *f (Trocknen von Holz)*
r. stretching device Radialspannvorrichtung *f*, Radialstellvorrichtung *f*
r. thrust Radialdruck *m*, Radialbelastung *f*
r. tooth Stirnzahn *m (Fräser)*
r. transmission feeder *(Fmt)* Radialleitung *f*
r. triangulation *(Opt)* Radialtriangulierung *f*
r. velocity Radialgeschwindigkeit *f*
r. wall drilling machine Wandradialbohrmaschine *f*
r. widening of the passage between the blades radiale Erweiterung *f* der Schaufelkanäle *(Turbine)*
radially cut teeth achsparallele Zähne *mpl*
r. expanding clutch Spreizringkupplung *f*
radian *(Math)* Radiant *m (Einheit im Bogenmaß)*
r. frequency *(El)* Kreisfrequenz *f*, Winkelfrequenz *f*
r. measure *(Math)* Bogenmaß *n*
radiance of a surface Strahlungsintensität *f*
radiant Strahler *m*, Strahlungsquelle *f*
radiant strahlend, Strahlung abgebend
r. black body Schwarzstrahler *m*
r. boiler Strahlungskessel *m*
r. ceiling heating Deckenstrahlungsheizung *f*
r. cooling Strahlungskühlung *f*
r. density Strahlungsdichte *f*
r. emittance *(El)* spezifische Ausstrahlung *f (in einem Punkt einer Fläche)*
r. energy Strahlungsenergie *f*
r. energy density Strahlungsdichte *f*
r. field *(Opt)* Strahlungsfeld *n*
r. flux Strahlungsstrom *m*, Strahlungsfluß *m*
r. heat Wärmestrahlung *f*; Strahlungswärme *f*
r. heating Strahlungsheizung *f*
r. intensity *(El)* Strahlstärke f *(einer Strahlungsquelle in einer Richtung)*, Strahlungsintensität *f*
r. power Strahlungsfluß *m*, Energiefluß *m*; Strahlungsleistung *f*
r. tube Strahlrohr *n (Härteofen)*
radiate / to strahlen; ausstrahlen, abstrahlen; [aus]senden
r. back zurückstrahlen
r. heat Wärme abstrahlen
r. through durchstrahlen
radiated ausgestrahlt; strahlenförmig
r. beam Richtstrahl *m*
r.-crystalline *(Min)* strahlig
r. energy Strahlungsenergie *f*
r. field Strahlungsfeld *n*
r. power ausgestrahlte Leistung *f*, Strahlungsleistung *f*
r. wave ausgestrahlte Welle *f*
radiating Ausstrahlung *f*
radiating strahlend, emittierend
r. capacity Strahlungsvermögen *n*, Ausstrahlungsvermögen *n*, Emissionsvermögen *n*
r. electricity strahlende Elektrizität *f*
r. element Strahlungselement *n (Antennentechnik)*
r. heat Strahlungswärme *f*
r. power *s.* radiated power
radiation Strahlung *f*
r. absorption Strahlungsabsorption *f*
r. analysis Strahlungsanalyse *f*
r. attenuation Strahlenschwächung *f*, Strahlungsschwächung *f*

r. background Strahlungsuntergrund *m*, Umgebungsstrahlung *f*
r. barrier Strahlenschranke *f*, Strahlenabschirmung *f*
r. burn Strahlungsverbrennung *f*
r. chamber Ionisationskammer *f*; Bestrahlungsraum *m*, Bestrahlungskammer *f*, Bestrahlungszelle *f*
r. chemistry Strahlenchemie *f*, Strahlungschemie *f*, Radiationschemie *f*
r. constant Strahlungskonstante *f*
r. content Strahlungsinhalt *m*
r. contrast Strahlungskontrast *m*
r. counter Strahlenzähler *m*, Strahlenzählrohr *n*
r. coupling Strahlungskopplung *f*
r. damage Strahlenschaden *m*
r. damping Strahlungsdämpfung *f*
r. danger Strahlengefahr *f*
r. danger zone strahlengefährdete Zone *f*
r. death Strahlentod *m*
r. decomposition Strahlenzersetzung *f*, Radiolyse *f*
r. density Strahlungsdichte *f*
r. detection instrument Strahlennachweisgerät *n*, Strahlungsdetektor *m*, Strahlungsempfänger *m*
r. detector *s.* r. detection instrument
r. dosage Strahlungsdosis *f*, Strahlendosis *f*; Bestrahlungsdosis *f*
r. dose *s.* r. dosage
r. dose-meter Dosismesser *m*, Dosimeter *n*
r. dosimetry Strahlendosimetrie *f*, Dosismessung *f*
r. effect Strahlenwirkung *f*
r. efficiency Strahlungswirkungsgrad *m (Antennen)*
r. energy Strahlungsenergie *f*; ausgestrahlte Energie *f*, strahlende Energie
r. equilibrium Strahlungsgleichgewicht *n*
r. excitation Anregung *f (Gas)* durch elektromagnetische Strahlen; Strahlenerzeugung *f*
r. exposed strahlenexponiert, bestrahlt
r. field Strahlungsfeld *n*; Bestrahlungsfeld *n*
r. furnace *(Met)* Strahlungsofen *m*
r. hazard Strahlengefahr *f*
r. heating surface *(Wärm)* Strahlungsheizfläche *f*
r.-induced strahleninduziert, durch Strahlung entstanden
r.-initiated polymerization *(Plast)* strahlungsinduzierte Polymerisation *f*
r. injury Strahlungsschaden *m*, Strahlenschädigung *f*, Bestrahlungsschaden *m*
r. intensity Strahlungsintensität *f*, Strahlenintensität *f*; Strahl[ungs]dichte *f*
r. ionization Strahlungsionisation *f (Gasentladungsröhren)*
r. laboratory Strahlenlaboratorium *n*
r. leakage Leckstrahlung *f*
r. length Strahlungslänge *f*
r. lethality Sterblichkeit *f* infolge Bestrahlung, Strahlenletalität *f*
r. level Strahlungsniveau *n*
r. lobe Strahlungskeule *f (Antennentechnik)*
r. loss Strahlungsverlust *m*
r. meter Strahlungsmeßgerät *n*, Strahlungsmesser *m*, Radiameter *n*
r. monitor Strahlenüberwachungsgerät *n*, Strahlenwarngerät *n*
r. monitoring Strahlenüberwachung *f*
r. of heat Wärme[aus]strahlung *f*, Temperaturstrahlung *f*
r. of light Licht[aus]strahlung *f*
r. of waves Wellenabstrahlung *f*
r. pattern Strahlungsdiagramm *n*, Strahlungscharakteristik *f*
r. pattern of an aerial array Richtkennlinie *f (Antennentechnik)*
r. physics Strahlenphysik *f*
r. potential Strahlungspotential *n*
r. pressure Strahlungsdruck *m*
r. processing Strahlenbehandlung *f*
r. proof strahlensicher
r. protection Strahlungsschutz *m*; Strahlenabschirmung *f*
r. pyrometer Strahlungspyrometer *n*, Strahlungshitzemesser *m*
r. quality Röntgenstrahlenhärte *f*
r. resistance Strahlungsdämpfung *f*, Strahlungswiderstand *m*
r. safety Strahlensicherheit *f*, Strahlungssicherheit *f*
r. shield Strahlenschild *m*, Strahlenschutz *m*
r. source Strahlenquelle *f*, Strahlungsquelle *f*, Strahler *m*
r. survey Strahlenüberwachung *f*
r. survey meter Strahlenüberwachungsgerät *n*
r. temperature Strahlungstemperatur *f*
r. therapy Strahlenbehandlung *f*
r. tolerance zulässige Strahlenbelastung *f*
r. trap Strahlenschleuse *f*
r. treatment Strahlenbehandlung *f*
r. tube continuous bright annealing furnace *(Met)* Strahlrohrdurchlaufblankglühofen *m*
r. unit 1. Strahlungseinheit *f*; 2. Meßteil *n* eines Pyrometers
r. zone Strahlungsgürtel *m (der Erde)*; Einstrahlungsgebiet *n*, Bestrahlungszone *f*
radiative strahlend
r. capture Strahlungseinfang *m*
r. equilibrium Strahlungsgleichgewicht *n*
r. loss Strahlungsverlust *m*
r. recombination Strahlungsrekombination *f*
r. transition Strahlungsübergang *m*
radiator 1. Radiator *m*, Rippenheizkörper *m*,

Gliederheizkörper *m*, Heizkörper *m*; 2. *(Kfz)* Kühler *m*; 3. Erzeugerplatte *f (Spektrometer)*
r. baffle plate Kühlerspritzblech *n*
r. block Kühlerblock *m*
r. bonnet Kühlerhaube *f*
r. cooling Umlaufkühlung *f*
r. core Kühlerblock *m*
r. cowl Kühlerverkleidung *f*
r. element Kühlerteilblock *m*
r. fin Kühlrippe *f*
r. frame Kühlergehäuse *n*
r. grille Kühlerschutzgitter *n*
r. guard Kühlerschutzbügel *m*
r. inlet connection Kühlereinlaufstutzen *m*
r. lamination Kühlerlamelle *f*
r. mounting Kühlerfuß *m*
r. muff Kühler[frostschutz]haube *f*
r. outlet connection Kühlerauslaufstutzen *m*
r. rib Kühlrippe *f*
r. safety ring Kühlerschutzring *m*
r. segment Kühlerlamelle *f*
r. shutter Kühlerabdeckung *f*
r. strut Kühlerstrebe *f*
r. support Kühlerfuß *m*, Kühlerstütze *f*
r. tank Wasserkasten *m*
radical 1. *(Ch)* Radikal *n*; 2. *(Math)* Wurzel [ausdruck *m*] *f*
r. quantity Radikand *m*
r. sign Wurzelzeichen *n*
radicand *(Math)* Radikand *m*
radio / to 1. funken, drahtlos senden; 2. mit Röntgenstrahlen durchleuchten
radio Radio *n*, Rundfunk *m*; Rundfunkempfänger *m*
r. abbreviation Funkabkürzung *f*
r. accessories Radiozubehör *n*
r.-active *s.* radioactive
r.-activity *s.* radioactivity
r. advertizing Rundfunkwerbung *f*
r. aerial Autoantenne *f*
r. aids to navigation Funknavigationshilfen *fpl*, Radionavigationshilfsmittel *npl*
r. alert Funkbereitschaft *f*
r. altimeter Hochfrequenzhöhenmesser *m*, Funkhöhenmesser *m*
r. amateur Funkamateur *m*, Kurzwellenamateur *m*
r. analyzer Funkprüfgerät *n*
r. announcer Rundfunksprecher *m*, Rundfunkansager *m*
r. approach system Blindlandeeinrichtung *f*
r. astronomy Radioastronomie *f*
r. attenuation Funkdämpfung *f*
r. autocontrol Funkselbststeuerung *f*
r. autopilot coupler radiogesteuerte automatische Lenkeinrichtung *f (Funknavigation)*
r. beacon Funkbake *f*, Peilfunksender *m*, Funkfeuer *n*
r. beacon identification *(Flg)* Funkfeuerkennung *f*
r. beacon receiver Bakenempfänger *m*
r. beam Richtstrahl *m*; Funkleitstrahl *m*
r. bearing Azimut *m (n)*, Radiopeilwinkel *m*, Peilwinkel *m*, Funkpeilung *f*
r. bearing installation Peilvorrichtung *f*
r. bearing instrument Funkpeiler *m*
r. buoy Funkboje *f*
r. cabin Funkerkabine *f*
r. call Funkruf *m*
r. car Funkkraftwagen *m*
r.-carrier frequency Funkträgerfrequenz *f*
r. channel Funkkanal *m*, Funkweg *m*
r. circuit Funknachrichtenkanal *m*, Funkverbindung *f*
r. coil Hochfrequenzspule *f*
r. communication Funkverkehr *m*, Funkverbindung *f*, drahtlose Verbindung *f*
r. communication centre Funkempfangszentrale *f*
r. communication service Funkdienst *m*
r. communications Funkwesen *n*
r. communications between two stations Funklinienverkehr *m*
r. compass Radiokompaß *m*
r. compass station Funkpeilstation *f*, Funkpeilstelle *f*
r. constructor Radiobastler *m*
r.-control Funksteuerung *f (von Motoren usw.)*
r.-control cage Radiokäfig *m*
r.-control desk Fernlenkpult *n*
r.-controlled drahtlos gesteuert
r.-controlled ship model drahtlos [fern]-gesteuertes Schiffsmodell *n*
r. conversation Funkgespräch *n*
r. deception Funktäuschung *f*
r. detecting and ranging Radar *n*
r. diffusion Radiorundspruch *m*
r. direction finder Funkpeilgerät *n*, Funkpeiler *m*, Peilempfänger *m*, Radiopeilgerät *n*, Bordpeiler *m*
r. direction finding Funkpeilung *f*
r. direction finding service Funkpeildienst *m*
r. directive device Leitstrahlanlage *f*
r. distribution system [hochfrequenter] Drahtfunk *m*
r. disturbance Funkstörung *f*
r. drama Hörspiel *n*
r. engineer Funktechniker *m*, Rundfunktechniker *m*
r. engineering Funktechnik *f*, Radiotechnik *f*, Rundfunktechnik *f*
r. equipment Funkgerät *n*, Funkausrüstung *f*
r. equipment store organization Funkzeugamt *n*

r. examination of materials Röntgen[strahlen]materialprüfung *f*
r. exchange installation Rundfunkvermittlungsanlage *f*
r. facsimile Bildfunk *m*, drahtlose Bildtelegrafie *f*
r. facsimile service Bildfunkdienst *m*
r. frequency Hochfrequenz *f* (*s. a.* high frequency)
r. frequency accelerator Hochfrequenzbeschleuniger *m*, HF-Beschleuniger *m*
r. frequency alternator Hochfrequenzgenerator *m*, HF-Generator *m*
r. frequency amplifier Hochfrequenzverstärker *m*, HF-Verstärker *m*
r. frequency bridge Hochfrequenz[meß]brücke *f*
r. frequency by-pass capacitor HF-Ableitkondensator *m*
r. frequency cable Hochfrequenzkabel *n*, HF-Kabel *n*
r. frequency carrier hochfrequenter Träger *m*
r. frequency ceramic Hochfrequenzkeramik *f*
r. frequency choke Hochfrequenzdrossel *f*, HF-Drossel *f*
r. frequency circuit Hochfrequenzkreis *m*, HF-Kreis *m*
r. frequency coil Hochfrequenzspule *f*, HF-Spule *f*
r. frequency current Hochfrequenzstrom *m*, HF-Strom *m*
r. frequency diode Hochfrequenzdiode *f*
r. frequency direction finding Funkpeilung *f*
r. frequency discharge Hochfrequenzentladung *f*
r. frequency engineering Hochfrequenztechnik *f*, HF-Technik *f*
r. frequency generator Hochfrequenzgenerator *m*, HF-Generator *m*
r. frequency heater Hochfrequenzheizung *f*, HF-Heizung *f*
r. frequency interference Hochfrequenzstörung *f*
r. frequency noise Hochfrequenzstörung *f*
r. frequency oscillator Hochfrequenzoszillator *m*, HF-Oszillator *m*, HF-Generator *m*
r. frequency powdered iron HF-Eisen *n*
r. frequency power Hochfrequenzleistung *f*, HF-Leistung *f*
r. frequency pulse impulsmodulierter Träger
r. frequency radiation Hochfrequenzstrahlung *f*, hochfrequente Strahlung *f*
r. frequency resistance Hochfrequenzwiderstand *m*
r. frequency signal generator Hochfrequenzmeßsender *m*, HF-Meßsender *m*, Service-Meßsender *m*, Service-Oszillator *m*
r. frequency stage Hochfrequenzstufe *f*, HF-Stufe *f*
r. frequency telegraphy Hochfrequenztelegrafie *f*
r. frequency transformer Hochfrequenztransformator *m*
r. frequency unit Hochfrequenzteil *m*, HF-Teil *m*
r. frequency vacuum furnace Hochfrequenzvakuumofen *m*
r. frequency voltage hochfrequente Spannung *f*
r. fuse Rundfunkapparatesicherung *f*
r. grammophone Radioempfänger *m* für Schallplattenwiedergabe, Rundfunkgerät *n* mit Plattenspieler, Musikschrank *m*, Musiktruhe *f*
r. intelligence Funkaufklärung *f*
r. intercept service Funkhorchdienst *m*
r. intercept truck Funkhorchkraftwagen *m*, Funkabhörfahrzeug *n*
r. interception Funkaufklärung *f*
r. interference Funkstörung *f*, Funküberlagerung *f*
r. interference abatement (control) Bekämpfung *f* von Funkstörungen
r. interference elimination service Entstörungsdienst *m*
r. interference filter Hochfrequenzstörschutzfilter *n*, Störschutzfilter *n*
r. interference rejection Entstörung *f*
r. interference suppression Rundfunkstörschutz *m*; Entstörung *f*
r. interference suppression resistor Störschutzwiderstand *m*
r. interference suppression service Rundfunkentstörungsdienst *m*
r.-jamming service Funkstördienst *m*
r. leg Funkleitstrahl *m*
r. letter telegram Funkbrief *m*
r. line of sight Funksichtlinie *f*, quasioptische Reichweite *f*
r. link Richtfunksystem *n*; Richtfunkstrecke *f*, Richtfunklinie *f*
r. location *s.* radiolocation
r. log Funktagebuch *n*
r. magnetic indicator Funk[peiltochter]kompaß *m*
r. marker beacon Funkfeuer *n* mit Kennung, Markierungsfunkbake *f*
r. mast Antennenmast *m*; Sendemast *m*
r. materiology Röntgen[strahlen]materialprüfung *f*
r. message Funkmeldung *f*, Funkspruch *m*
r. metal locator HF-Metallsuchgerät *n*
r. navigation *s.* radionavigation
r. network Funknetz *n*
r. network chart Funkverkehrskarte *f*
r. noise Funkstörung *f*

r. noise field intensity Rauschfeldstärke *f*
r. noise suppression Funkentstörung *f*
r. noise-suppression device Störschutz *m*
r. observation Funkbeobachtung *f*
r. operator Funker *m*
r.-optical range optische Reichweite *f* eines Senders, radiooptische Reichweite
r. patrol car Funkstreifenwagen *m*
r. photo Bildfunktelegramm *n*
r. picture Funkbild *n*
r. picture transmission drahtlose Bildtelegrafie *f*
r. position line determination Funkkurspeilung *f*
r. procedure Funkbetriebsverfahren *n*
r. range Kursfunkfeuer *n*
r. range beacon Richtfunkbake *f*
r. range finding Funkmeßverfahren *n*
r. range receiver Vierkursfunkfeuerempfänger
r. receiver Radio[apparat *m*] *n*, Rundfunkempfänger *m*
r. reception Rundfunkempfang *m*
r. relay exchange Rundfunkvermittlungsanlage *f*
r. relay transmission system Richtfunkstrecke *f*
r. repairman Rundfunkmechaniker *m*
r. repeating station Ballsender *m*
r. room Funkraum *m*
r. screening Funkentstörung *f*
r. service 1. Funkbetrieb *m*, Funkdienst *m*; 2. Rundfunkservice *m (Reparatur von Rundfunkempfängern)*
r. set Rundfunkgerät *n*, Empfänger *m*, Radio[gerät] *n*
r. shielding Abschirmung *f*, Störschutz *m*, Entstörung *f*
r. shielding cap Entstörkappe *f*
r. shielding unit Entstörer *m*
r. signal Funksignal *n*
r. spectrum Funk[wellen]spektrum *n*
r.-star *(Astr)* Radiostern *m (Radioquellen)*
r. station Funkstation *f*
r. system Funknetz *n*
r. technician Rundfunkmechaniker *m*
r. technology Funktechnik *f*, Rundfunktechnik *f*
r. telecontrol Funkfernsteuerung *f*
r. telegraphy *s.* radiotelegraphy
r. telephone *s.* radiotelephone
r. telephony *s.* radiotelephony
r. teleprinter Funkfernschreiber *m*
r. teletypewriter Funkfernschreiber *m*
r. tower Sendeturm *m*
r. traffic Funkverkehr *m*
r. traffic operation Funkbetrieb *m*
r. transmission Radioübertragung *f*, Radiosendung *f*
r. transmission formula Ausbreitungsformel *f*, Übertragungsformel *f*
r. transmission of pictures Bildfunk *m*, drahtlose Bildübertragung *f*
r. transmitter Rundfunksender *m*
r. transmitting apparatus Funksendegerät *n*
r. valve Radioröhre *f*
r. wave Radiowelle *f*, Funkwelle *f*
r. wave propagation Radiowellenausbreitung *f*
r. weather broadcast station Wetterfunkstelle *f*
radioacoustics Radioakustik *f*
radioactinium Radioaktinium *n*
radioactivate / to radioaktiv machen (verseuchen), aktivieren
radioactive [radio]aktiv
r. ash radioaktive Asche *f*
r. background radioaktive Untergrundstrahlung *f*
r. burn radioaktive Verbrennung *f*
r. by-product radioaktives Nebenprodukt *n*
r. chain Zerfallsreihe *f*
r. clock Atomuhr *f*
r. constant Zerfallskonstante *f*
r. contamination radioaktive Verseuchung *f*
r. decay radioaktiver Zerfall *m*, radioaktive Umwandlung *f*
r. decay law Zerfallsgesetz *n*
r. deposit 1. radioaktiver Niederschlag *m*; 2. radioaktive Erzlagerstätte *f*
r. disintegration *s.* r. decay
r. dust radioaktiver Staub *m*
r. effluents radioaktive Abwässer *npl*
r. element radioaktives Element *n*, Radioelement *n*
r. equilibrium radioaktives Gleichgewicht *n*
r. fall-out radioaktiver *(atmosphärischer)* Niederschlag *m*
r. fission product radioaktives Spaltprodukt *n*
r. fragment radioaktives Spaltbruchstück *n*
r. irradiation radioaktive Bestrahlung *f*
r. material aktives Material *n*, aktiver Stoff *m*
r. nucleus radioaktiver Kern *m*
r. product radioaktives Zerfallsprodukt *n*
r. radiation radioaktive Strahlung *f*
r. radiation counter Zählrohr *n*
r. rain radioaktiver Regen *m*
r. recoil radioaktiver Rückstoß *m*
r. relationship radioaktive Verwandtschaft *f*
r. relay Strahlenschranke *f*
r. resistance radioaktiver Widerstand *m*
r. series radioaktive Zerfallsreihe *f*
r. standard Bezugspräparat *n*, radioaktives Standardpräparat *n*
r. tracer radioaktives Leitisotop *n*, radioaktiver Indikator *m*

r. tracer method Indikatorverfahren *n (Markierung durch radioaktive Isotope)*
r. transformation radioaktive Umwandlung *f*
r. transmutation radioaktive Umwandlung *f*
r. waste radioaktiver Abfall *m*
r. waste disposal Beseitigung *f* radioaktiver Abfälle
radioactivity Radioaktivität *f*
r. detection Radioaktivitätsnachweis *m*
radioautograph Autoradiografie *f*, Autoradiogramm *n*
radiobiologic[al] strahlenbiologisch
radiobiology Strahlenbiologie *f*
radiocarbon radioaktiver Kohlenstoff *m*
r. age Kohlenstoffalter *n*
radiochemical radiochemisch
r. analysis radiochemische Analyse *f*
r. purity radiochemische Reinheit *f*
radiochemist Radiochemiker *m*
radiochemistry Radiochemie *f*
radiochromatography Radiochromatografie *f*
radiocobalt radioaktives Kobalt *n*; Radiokobalt *n*
radiocontamination radioaktive Verseuchung *f*
radiode Radiumkapsel *f*, Radiumbehälter *m*
radiodetermination Funkortung *f*
r. service Ortungsfunkdienst *m*
radioelement radioaktives Element *n*, Radioelement *n*
radiogenic durch radioaktiven Zerfall entstanden (erzeugt)
r. heat radiogene Wärme *f*
radiogold radioaktives Gold *n*; Radiogold *n*
radiogoniometer Funkpeiler *m*, Radiogoniometer *n*
radiogoniometry Radiogoniometrie *f*
radiogram 1. Funktelegramm *n*, Radiotelegramm *n*; 2. Musiktruhe *f*, Musikschrank *m*; 3. Röntgenaufnahme *f*
radiograph / to durchstrahlen mit Röntgenstrahlen oder Gamma-Strahlen; röntgen
radiograph Röntgenaufnahme *f*, Röntgenbild *n*; Aufnahme *f* mit Gamma-Strahlen
radiographic radiografisch; Röntgen-
r. film Röntgenfilm *m*
radiography 1. Radiografie *f*; Gammagrafie *f*; 2. Röntgenografie *f*, Röntgenaufnahme *f (Verfahren)*
radioiodine radioaktives Jod *n*; Radiojod *n*
radioiron radioaktives Eisen *n*; Radioeisen *n*
radioisotope radioaktives Isotop *n*, Radioisotop *n*
r. concentration Radioisotopenkonzentration *f*
r. production Erzeugung *f* radioaktiver Isotope, Radioisotopengewinnung *f*
radiolead radioaktives Blei *n*; Radioblei *n*
radiolocation Funkortung *f*
radiologic *s.* radiological
radiological radiologisch, röntgenologisch, Röntgen-, Strahlen-
r. bomb radiologische (schmutzige) Bombe *f*
r. consultation report Röntgenbefund *m*
r. physics Strahlenphysik *f*
r. safety Strahlungsschutz *m*; Strahlensicherheit *f*
r. unit 1. radiologische Einheit *f*; 2. Strahlentherapiegerät *n*, Durchstrahlungsanlage *f*
radiologist 1. Radiologe *m*, Strahlungsforscher *m*; 2. Röntgenfacharzt *m*, Röntgenologe *m*; 3. Strahlenschutzarzt *m*
radiology Radiologie *f*
radiolucent röntgenstrahlendurchlässig
radiolysis Radiolyse *f*, Strahlendissoziation *f*
radiomateriology [zerstörungsfreie] Werkstoffprüfung *f* mittels [radioaktiver] Strahlung, Gamma-Defektoskopie *f*
radiometallography Röntgenmetallografie *f*
radiometeorograph Funksonde *f*, Radiometeorograf *m*
radiometer Radiometer *n*
radiometric radiometrisch
r. quantity Strahlungsgröße *f*
r. titration radiometrische Titration *f*
radiometry Radiometrie *f*, Strahlungsmessung *f*
radionavigation Funknavigation *f*
radionuclide Radionuklid *n*
radioopacity Strahlenundurchlässigkeit *f*
radioopaque strahlenundurchlässig
radiophone distress signal SOS-Notrufsignal *n*
radioplay Hörspiel *n*
radioresistance Strahlenresistanz *f*, Strahlenbeständigkeit *f*, Strahlenfestigkeit *f*
radioscope Radioskop *n*
radioscopy Radioskopie *f*, Röntgenoskopie *f*, Durchleuchtung *f*, Durchstrahlung *f*
radiosensitivity Strahlenempfindlichkeit *f*, Strahlungsempfindlichkeit *f*
radiosodium radioaktives Natrium *n*; Radionatrium *n*
radiosonde [system] Radiosonde *f*
radiosulphur radioaktiver Schwefel *m*; Radioschwefel *m*
radiotelegraph / to funken, drahtlos telegrafieren
radiotelegraph Funktelegraf *m*
r. service Funktelegrafendienst *m*, Funktelegrafiedienst *m*
radiotelegraphy drahtlose Telegrafie *f*, Funktelegrafie *f*, Hochfrequenztelegrafie *f*
radiotelemetering Funkfernmessen *n*
radiotelephone / to funksprechen
radiotelephone Funkfernsprecher *m*, Funktelefon *n*
r. call Funkgespräch *n*
r. circuit Funksprechverbindung *f*
r. communication Funksprechverbindung *f*

r. operation Funkfernsprechbetrieb *m*
r. transmitter Telefoniesender *m*
radiotelephony Funktelefonie *f*
r. certificate Funksprechzeugnis *n*
r. code Funksprechkode *m*
r. exchange Funksprechzentrale *f*
r. network Sprechfunknetz *n*
r. service Funksprechdienst *m*
radiotelescope Radioteleskop *n*
radiotherapist 1. Strahlentherapeut *m*; 2. Röntgenfacharzt *m*, Röntgenologe *m*
radiothorium Radiothorium *n*
radiotron Summerröhre *f*
radium Radium *n*
r. age Radiumalter *n*
r. drug Radiumpräparat *n*
r. irradiation Radiumbestrahlung *f*
r. series Radiumreihe *f*
r. source Radiumpräparat *n*
radius / to *(Masch)* ausrunden
radius Radius *m*, Halbmesser *m*; Rundung *f*, Ausrundung *f*
r. at the root [of a tooth] *(Masch)* Zahngrundrundung *f*
r. bar Leitstange *f* für das Rundhobeln
r. cutter Radiusfräser *m*, Halbkreisfräser *m*
r. dressing Radienabrichten *n (Schleifscheibe)*
r. dressing fixture Radienabrichteinrichtung *f (Schleifscheibe)*
r. gauge Radienschablone *f*, Halbmesserlehre *f*
r. of action Aktionsradius *m*, Fahrbereich *m*
r. of bend Biegungshalbmesser *m*, Krümmungsradius *m*
r. of centre of gravity Schwerpunktradius *m*
r. of convergence Konvergenzradius *m*
r. of curvature Krümmungsradius *m*, Krümmungshalbmesser *m*
r. of gyration Trägheitshalbmesser *m*, Trägheitsradius *m*
r. of inertia Trägheitsmomentradius *m*
r. of oil drainage Aktionsradius *m* einer Erdölbohrung, Entölungshalbmesser *m*
r. of operation Aktionsradius *m*
r. of oscillation Schwingungsradius *m*, Amplitude *f*
r. of outside circle Kopfkreishalbmesser *m (Zahnrad)*
r. of the nucleus Kernradius *m*
r. of web Stegrundung *f*
r. rod *(Flg)* Einziehstrebe *f (Fahrwerk)*
r. unit Schwingarm *m*
r. vector Leitstrahl *m*, Fahrstrahl *m*, Radiusvektor *m*, Ortsvektor *m*
radiused gerundet, halbrund, mit Radius
radiusing Abrunden *n*
radix *(Math)* Wurzel *f*; Basis *f*, Grundzahl *f (einer Potenz, eines Logarithmus)*
r. complement code Basiskomplementkode *m*, Grundzahlkomplementkode *m*
r. converter Basiswandler *m (Kodewandler, der Zahlen in ein Zahlensystem mit einer anderen Basis umsetzt)*
r. 2 counter Binärzähler *m*
r. notation *(Dat)* Radixschreibweise *f*, Grundzahlschreibweise *f*
r. point *(Dat)* Komma *n*
radome Antennenabdeckung *f*, Radarhaube *f*, Radarkuppel *f*, Wetterschutzhaube *f (für Antennen)*
radon Radon *n*, Radiumemanation *f*
raffia Raffiabast *m*
raffinate Raffinat *n*
raffinose Raffinose *f*
raft Floß *n*
r. bridge Floßbrücke *f*
r. foundation *(Bw)* Plattenfundament *n*, Plattengründung *f*
r. port Ladepforte *f*, Bugpforte *f (Ladeöffnung im Bug oder Heck eines Schiffes)*
r. wood Floßholz *n*
rafted timber Floßholz *n*
rafter Dachsparren *m*, Sparren *m*
rag 1. Lappen *m*, Lumpen *pl* (*s. a.* rags); 2. *(Geol)* Kalksandstein *m*; 3. Dachschieferplatte *f (36″ × 24″)*
r. boiler Hadernkocher *m (Papierherstellung)*
r. bolt Hakenbolzen *m*, Steinschraube *f*
r. buffing wheel Schwabbelscheibe *f*
r. chopper Hadernschneider *m*, Lumpenschneider *m (Papierherstellung)*
r. cleaning machine Lumpenreinigungsmaschine *f*
r. cutter Hadernschneider *m*, Lumpenschneider *m (Papierherstellung)*
r. cutting machine *s.* r. cutter
r. engine Hadernholländer *m (Papierherstellung)*
r. oiler *(Text)* Lumpenschmälzmaschine *f*
r. paper Hadernpapier *n*
r. picker *(Text)* Lumpenklauber *m*
r. pulp Hadernstoff *m*, Lumpenhalbstoff *m (Papierherstellung)*
r. shaker *(Text)* Lumpenentstaubungstrommel *f*
r. stone grobkörniger Sandstein *m*
r. tearing machine Lumpenwolf *m (Papierherstellung)*
ragged zackig
r. note *(El)* unreiner Ton *m*
r. picture *(Fs)* welliges Bild *n*
r. roll aufgerauhte (geriffelte) Walze *f*
ragger Ragger *m*, Zopfwinde *f (Aufarbeitung von Altpapier)*
ragging grobe Anreicherung *f*, Vorsortieren *n (von Erzen)*

rags Hadern *pl*, Papierhadern *pl*, Lumpen *pl* für Papierherstellung
r. for filtering paper Filtrierhadern *pl* *(Papierherstellung)*
rail 1. Schiene *f (Eisenbahn, Straßenbahn)*; Fahrschiene *f (Drehscheibe)*; 2. Gleitschiene *f*; Leiste *f*; 3. *(Masch)* Führung *f*, Querführung *f*, Querträger *m*; 4. Geländer *n*
r. base Schienenfuß *m*
r. bending machine Schienenbiegemaschine *f*
r. bond Schienen[strom]verbinder *m (elektrische Eisenbahn)*
r.-borne gleisgebunden, schienengebunden; schienenverfahrbar
r.-borne crane Schienen[dreh]kran *m*
r. breakage Schienenbruch *m*
r. car Schienenwagen *m*
r.-car barge *(Schiff)* Leichter *m* für Eisenbahnwaggon
r.-car underfloor engine Triebwagenunterflurmotor *m*
r. clamping Tischträgerfestklemmung *f* *(Waagerechtstoßmaschine)*; Querbalkenfestklemmung *f (Hobelmaschine)*
r. elevating screw Spindel *f* zur Senkrechtverstellung des Querbalkens *(Hobelmaschine)*
r. flange Schienenfuß *m*
r. gauge Spurweite *f*
r. grate Schienenrost *m*
r. guard Bahnräumer *m*, Schienenräumer *m*
r. head 1. Support *m (Hobelmaschine)*; 2. Schienenkopf *m*
r. jack Schienenhebewinde *f*
r. joint Schienenstoß *m*; Fußlasche *f*
r. layer Schienenleger *m*
r. mill Schienenwalzwerk *n*
r. motor coach Triebwagen *m*
r.-mounted gleisgebunden, schienengebunden; schienenverfahrbar
r.-mounted crane Schienen[dreh]kran *m*
r. nail Schienennagel *m*
r. oil engine Diesellokmotor *m*, Dieselmotor *m* für Schienenfahrzeuge
r. plane Schienenhobel *m*
r. port Eisenbahnhafen *m*
r. rocket launcher *(Rak)* Startschiene *f*, Startgestell *n*
r.-rolling mill Schienenwalzwerk *n*; Schienen[walz]straße *f*
r. stanchion Relingstütze *f*, Geländerstütze *f*
r. steel Schienenstahl *m*
r. straightener Schienenrichtmaschine *f*
r. switch Weiche *f*
r. switchblade planer Weichenzungenhobelmaschine *f*
r. track Gleis *n*, Schienenstrang *m*, Schienenweg *m*
r. traction unit Bahnmotor *m*
r. vehicle Schienenfahrzeug *n*
r. with butt joint Schiene *f* mit Stumpfstoß
r. with continuous fishplates Schiene *f* mit Fußlaschen
r. with wide base Breitfußschiene *f*, Schiene *f* mit breitem Fuß
r. with Z-joint Schiene *f* mit Halbstoß
railbus Schienenbus *m*
railing Maschinengeländer *n*, Schutzgeländer *n*; *(Schiff)* Reling *f*, Geländer *n*
railless schienenlos
railroad *(Am) s.* railway
r. ballast Gleisbettungsmaterial *n*, Schotter *m*
r. bedding Gleisbettung *f*
railway Eisenbahn *f*, Bahn *f*
r. anti-aircraft artillery Eisenbahnflak *f*
r. bridge crossing Eisenbahnüberführung *f*
r. carriage Eisenbahnwagen *m*
r. conductor Eisenbahnschaffner *m*
r. construction Eisenbahnbau *m*, Bahnbau *m*
r. cradle Slipwagen *m*, Aufschleppwagen *m* *(für Schiffe)*
r. dry-dock *(Schiff)* Aufschlepphelling *f*, Aufschleppe *f*, Slip[anlage *f*] *m*
r. embankment Eisenbahndamm *m*
r. engineering Eisenbahnbau *m*
r. ferry Eisenbahnfähre *f*
r. gate Bahnschranke *f*
r. gauge Spurweite *f (Eisenbahn)*
r. grade crossing Bahnübergang *m*, Schienenübergang *m*
r. guide Kursbuch *n*
r. junction Eisenbahnknotenpunkt *m*
r. level crossing Bahnübergang *m*, Schienenübergang *m*
r. line Eisenbahnlinie *f*
r. man Eisenbahner *m*
r. motor Bahnmotor *m*
r. mount *(Mil)* Eisenbahnlafette *f*
r. protective pillar Bahnschutzpfeiler *m*
r. repair shop Eisenbahnreparaturwerkstatt *f*
r. sleeper Eisenbahnschwelle *f*
r. slip *(Schiff)* Aufschlepphelling *f*, Aufschleppe *f*, Slip *m*
r. switch Weiche *f*
r. system Eisenbahnnetz *n*
r. terminal Endstation *f*
r. time-table Eisenbahnfahrplan *m*
r. track gauge Spurweite *f*
r.-track scale Gleiswaage *f (für Waggons)*
r. trackage Gleisanlage *f*
r. traffic Eisenbahnverkehr *m*
r. train Eisenbahnzug *m*
r. yard Abstellbahnhof *m*
rain 1. Regen *m (atmosphärisch)*; 2. Regen *(Bildstörung durch unregelmäßigen Einfall von Sekundärelektronen auf das Mosaik der Bildaufnahmeröhre)*

r. cap Regenkrone *f*, Regenkappe *f (Schornstein)*
r. gauge Regenmesser *m*
r.-like distribution *(Wärm)* regenförmige Verteilung *f (Kondensator)*
r. test Beregnungsversuch *m*
r. wear Regenbekleidung *f*
rainbow Regenbogen *m*
raincoat Regenmantel *m*
rainfall Niederschlag *m*; Niederschlagsmenge *f*
r. recorder Regenschreiber *m*
rainglass Barometer *n*
rainproof impregnation Regendichtimprägnierung *f*
rainwater Regenwasser *n*
r. gutter Regenrinne *f*
r. head Dachrinnenkessel *m*, Rinnenkessel *m*
raise / to 1. [hoch]heben; abheben *(Werkzeug)*; erhöhen; aufstellen; 2. *(El)* hochfahren *(Generator usw.)*; 3. *(Brg)* fördern, [zutage] heben *(Seigerförderung)*; aufhauen, überhauen, hochbrechen *(Schacht)*; aufkimmen *(Schiffsboden)*
r. a selector *(Fmt)* einen Wähler heben
r. on edge hochkanten
r. steam Dampf aufmachen
r. the anchor Anker lichten
r. to a power *(Math)* potenzieren
raise Erhöhung *f*, Steigerung *f*
r. cam Heber *m (Strickmaschine)*
r. climber *(Brg)* Kletterbühne *f*
r.-lower push-button Druckknopf *m* für Höhenverstellung
r. of the needle *(Text)* Nadelaustrieb *m*
r. platform *(Brg)* Kletterbühne *f*
raised erhaben
r.-and-weighed tonnage *(Brg)* Rohförderung *f*
r. cabin *(Schiff)* Kajütsaufbau *m*
r. cords *(Typ)* echte Bünde *pl*
r. deck cruiser Backdeckkreuzer *m (Motorboot)*
r.-grain *(Holz)* maserig
r. head gewölbter Kopf *m (einer Schraube)*
r. head screw Linsensenkholzschraube *f*
r. quarter deck *(Schiff)* erhöhtes Quarterdeck *n*
r. stern *(Schiff)* Entenheck *n*
r. tuck stitch *(Text)* erhabene Fangmasche *f*
raiser 1. Züchter *m (Pflanzen)*; 2. *(Text)* gezeichneter Bindepunkt *m*
raising 1. Heben *n*, Anheben *n*; Abhebung *f (Werkzeug)*; Erhöhung *f*; 2. Aufzucht *f (Tiere)*; 3. *(Flg)* Einfahren *n (Fahrwerk)*; 4. *(Text)* Rauhen *n*; Bürsten *n*, Kratzen *n*
r. card *(Text)* Rauhkarde *f*
r. of water Wasserförderung *f*
r. screw Hebeschraube *f*, Schraubwinde *f*
rake / to rechen; krählen; schüren *(Feuer)*; kratzen
r. out auskratzen
r. together zusammenkratzen
rake 1. Rechen *m*; Harke *f*; Kratze *f*, Rühreisen *n*, Rührstange *f*; Krammeisen *n*, Schürkelle *f*; 2. *(Schiff)* Fall *m*, Neigung *f (z. B. des Schornsteins)*; 3. *s.* r. angle; 4. *s.* r. face
r. angle Spanwinkel *m (Werkzeug)*; Schneidwinkel *m (Presse)*; Steigungswinkel *m (Bohrer)*
r. classifier Rechenklassierer *m (Aufbereitung)*
r.-ended barge *(Schiff)* Schute *f* mit aufgezogenen Enden
r. face Spanfläche *f (Werkzeug)*
r. of pole *(Fmt)* Neigung *f* des auf Zug gestellten Gestänges
r. of propeller blade Propellerflügelneigung *f*
r. tedder *(Lw)* Rechenwender *m*
raked geneigt, schräg; abfallend, ansteigend
r. stem *(Schiff)* überhängender Steven *m*
raker 1. Diagonalstab *m (Fachwerk)*; 2. Schüreisen *n*
r. beam schräger Balken (Träger) *m*
r. tooth Räumzahn *m*, Räumer *m (Säge)*
raking geneigt, schräg; abfallend; ansteigend
r. bond Diagonalverband *m*, Fischgrätenverband *m*, Schrägverband *m (Mauerwerk)*
r. mechanism Krählwerk *n (Tellertrockner)*
r. operation *(Met)* Krählarbeit *f*
r.-out *(Bw)* Auskratzen *n* von Mörtelfugen
r. pile Schrägpfahl *m*
r. stem *s.* raked stem
rakish boat schnittiges Boot *n*
ram / to stampfen; treiben *(Pfahl)*; rammen; besetzen *(Bohrloch)*
r. up aufstampfen *(Form)*
ram 1. Ramme *f*; Stoßheber *m*, [hydraulischer] Widder *m*; 2. Bär *m*, Rammbär *m*, Rammblock *m*; 3. *(Masch)* Stempel *m*, Druckkolben *m (einer Presse)*; Stoßkolben *m*, Plunger[kolben] *m*, Tauchkolben *m*; Schieber *m*, Stößel *m*; 4. Stempel *m (Kokerei)*; 5. *(Am)* Arbeitszylinder *m*; Pumpe *f (Hydraulik)*; 6. *s.* random access memory
r. balance weight Stößelausgleichsgewicht *n*
r. bar Ausdrückstange *f*
r. bench Maschinenseite *f*
r. bow *(Schiff)* Rammsteven *m*, Rammbug *m*, Rammsporn *m*
r. charger *(Rak)* Belüftungsrohr *n*
r. drive Stößelantrieb *m*
r. face Stößelkopf *m*
r. guide Stößelführung *f*
r. head slide Hobelkopfschlitten *m (Schnellhobler)*
r. jet [engine] Staustrahltriebwerk *n*
r. rack Ausdrückstange *f (Kokerei)*
r. rails Druckmaschinengleis *n (Kokerei)*
r. side Ausdrückseite *f*, Maschinenseite *f (Kokerei)*

r. slide Stößelführung *f*
r. socket Rammwelle *f*
r. stem *(Schiff)* Rammsteven *m*
r. stroke Stößelhub *m*
r. track Ausdrückgleis *n (Kokerei)*
r. way Stößelführung *f*
Raman effect Raman-Effekt *m*
Raman spectrum Raman-Spektrum *n*
ramie Ramie *f*, Chinagras *n*
r. decocticator *(Text)* Entholzungsmaschine *f* für Ramie, Ramiestengelbrechmaschine *f*
ramification 1. *(Math)* Verzweigung *f*; 2. Aufspaltung *f*, Gabelung *f*, Verästelung *f*
ramify / to [sich] verzweigen, gabeln, verästeln, [sich] aufspalten
rammable stampfbar, verdichtbar
rammed bottom lining *(Met)* Stampfherd *m*
r. clay Stampflehm *m*
r. lining *(Met)* Stampfauskleidung *f*
rammer 1. Ramme *f*, Stampfer *m*; 2. *(Mil)* Ansetzer *m*, Geschoßansetzer *m*
ramming Rammen *n*, Stampfen *n*, Feststampfen *n*
r. head *(Met)* Schleuderkopf *m (des Sandslingers)*
r. mass Stampfmasse *f*
r. rate *(Met)* Schleudergeschwindigkeit *f (des Sandslingers)*
ramp / to geneigt sein; steigen; fallen
ramp 1. Rampe *f*; 2. Krokodilkontakt *m (elektrische Eisenbahn)*
r. function *(BMSR)* Rampenfunktion *f*, zeitlineare Anstiegsfunktion *f*
r. lane Auffahrtrampe *f*, Auffahrtstraße *f*; Auffahrt *f (Autobahn, Brücke usw.)*
r. of frog *(Eis)* Auflaufherzstück *n*
r. response *(BMSR)* Reaktion *f* auf Rampenfunktion
rampant arch steigender Bogen *m (Widerlager nicht horizontal)*
ram's wool Widderwolle *f*
rand cutting machine Lederschneidmaschine *f (Schuhproduktion)*
random Zufall *m*
random regellos, wahllos; wahlweise; zufällig; Zufall[s]-; ungeordnet *(Kristall)*; *(Dat)* aleatorisch, stochastisch
r. access wahlfreier (beliebiger, stochastischer, wahlweiser) Zugriff *m (bei Rechenspeichern)*
r. access memory (store) Speicher *m* mit beliebigem (wahlfreiem) Zugriff
r. access storage Speicher *m* mit beliebigem Zugriff
r. coincidence *(Kern)* Zufallskoinzidenz *f*, zufällige Koinzidenz *f*
r. distribution regellose Verteilung *f*
r. error Zufallsfehler *m*
r. fluctuation Zufallsschwankung *f*
r.-incidence response *(El)* Empfindlichkeit *f* im diffusen Schallfeld
r. inspection Stichprobe *f*
r. noise Grundgeräusch *n*; statistisches Rauschen *n*
r. operation zufälliges Ansprechen *n*
r. orientation regellose Orientierung *f*
r.-position scattering *(Kern)* regellose Streuung *f*
r. rubble Bruchsteinmauerwerk *n*; Schichtenmauerwerk *n*
r. rubble in courses [hammerrechtes] Schichtenmauerwerk *n*
r. sample Stichprobe *f*
r. stoping *(Brg)* Weitungsbau *m*
r. variable Zufallsvariable *f*
r. voltage *(El)* Rauschspannung *f*
r. walk 1. unregelmäßiger Gang *m*; 2. ungeordnete (chaotische) Bewegung *f*
randomization zufällige Anordnung *f*
randomize / to zufällig zuordnen
range / to einordnen, einreihen; sich erstrecken, reichen
r. out *(Bw)* einfluchten
range 1. Strecke *f*, Abstand *m*; Bereich *m*; Skalenbereich *m*; Meßbereich *m*; Stufe *f*, Reihe *f*; Anwendungsbereich *m*; Satz *m*; 2. *(Kern)* Reichweite *f (Teilchen)*; Materie[schicht]-dicke *f*; 3. *(El)* Band *n*, Durchlässigkeitsbereich *m*; Reichweite; Bereich, Frequenzbereich *m*, Wellenbereich *m (Radio)*; Auffaßbereich *m (Radar)*; 4. *(Opt)* Entfernung *f*; 5. Schießplatz *m*, Schießstand *m*; 6. Herd *m*, Kochherd *m*; 7. *(Geol)* Hügelkette *f*
r. beacon *(Schiff)* Leitbake *f*
r. boiler Gliederkessel *m (Dampfheizung)*
r. circle Entfernungskreis *m (Radar)*
r. control Entfernungsregelung *f (Radar)*
r. discrimination Abstandsunterscheidung *f (Radar)*
r. distribution *(Ph)* Reichweitenverteilung *f*
r.-energy relation *(Kern)* Reichweite-Energie-Beziehung *f*
r. extension Meßbereichserweiterung *f*
r. finder Entfernungsmeßgerät *n*; Entfernungsmesser *m (an der Kamera)*
r. for alternating stresses Wechselbereich *m (Werkstoffprüfung)*
r. in density Rohdichte *f*
r. light *(Schiff)* Leitfeuer *n*, Richtungsfeuer *n*
r. mark[er] Entfernungsmeßmarke *f (Radarschirm)*
r. measurement Abstandsmessung *f*
r. multiplier Vorschaltwiderstand *m* eines Voltmeters
r. of a transmitter Senderreichweite *f*
r. of action Aktionsradius *m*
r. of adaptation Adaptationsbreite *f*

r. of adjustment Verstellbereich *m*; Regelbereich *m*
r. of application Anwendungsbereich *m*
r. of audibility Hörbereich *m*
r. of boring feed Bohrvorschubbereich *m*
r. of composition Konzentrationsbereich *m* *(Metallografie)*
r. of fatigue under pulsating tensile stresses Zug-Schwell-Bereich *m (Werkstoffprüfung)*
r. of feeds Vorschubbereich *m*
r. of elasticity elastischer Bereich *m*
r. of magnification Vergrößerungsbereich *m*
r. of measurement Meßbereich *m*
r. of sensitivity Ansprechbereich *m*; empfindlicher Bereich *m*
r. of setting Einstellbereich *m*
r. of speed regulation Drehzahlregelbereich *m*
r. of speeds Drehzahlbereich *m*
r. of stability Stabilitätsgrenze *f*
r. of transmission *(Fmt, Fs)* Reichweite *f*
r. of uniform control *(El)* Aussteuer[ungs]-bereich *m*
r. of validity *(Math)* Gültigkeitsbereich *m*
r. of values Wertevorrat *m (einer Funktion)*
r. of variation Variationsbereich *m (Statistik)*
r. of vision Blickfeld *n*, Gesichtsfeld *n*, Sichtbereich *m*; Sichtweite *f*
r.-plotting truck *(Mil)* Vermessungsauswertekraftwagen *m*
r. region Reichweitebereich *m*
r. resolution Entfernungsauflösungsvermögen *n*
r. scale Entfernungsskala *f (Radar)*
r. selector *s.* r. switch
r. suppression *(BMSR)* Nullpunktunterdrükkung *f*, Bereichseinengung *f*
r. switch Meßbereichsumschalter *m*, Bereichs[um]schalter *m*
r. transmission unit Abstandgeber *m (Radartechnik)*
rangeability *(BMSR)* Stellverhältnis *n (Ventil)*
ranger vessel Bereisungsschiff *n*
ranging rod *(Bw)* Fluchtstab *m*
rank Rang *m (einer Matrix)*; Reihe *f*
r. correlation *(Math)* Rangkorrelation *f*
r. of a matrix *(Math)* Rang *m* einer Matrix
r. of selectors *(Fmt)* Wählerstufe *f*
r. weed Wucherpflanze *f*
Rankine cycle Clausius-Rankine-Prozeß *m*
Rankine diagram *(Wärm)* rankinisiertes Diagramm *n*
Rankine's combined vortex *(Strm)* trichterförmiger Wirbel *m*
ranking of plants Wuchern *n* der Pflanzen
r. weed Wucherpflanze *f*
rap / to *(Met)* losklopfen, abklopfen *(Form)*
rape Raps *m*
r. oil Rüböl *n*
rapeseed Rübsaat *f*
r. oil Rüböl *n*
rapic acid Oleinsäure *f*
rapid schnell, Schnell-, Eil-
r. advance *(Masch)* Eilvorlauf *m*
r.-advance / to *(Masch)* schnell heranführen; schnell herangeführt werden
r. approach [motion] *(Masch)* Eilvorlauf *m*, schnelles Heranführen *n*
r. charging Schnelladung *f (Batterie)*
r. closing valve Schnellschlußventil *n*
r. combustion lebhafte Verbrennung *f*
r. control Schnellregelung *f*
r. curing schnellabbindend
r. determination Schnellanalyse *f*, Schnellbestimmung *f*
r. discharger Schnellentlader *m*
r. drivage *(Brg)* Schnellstreckenvortrieb *m*
r.-fire weapon Schnellfeuerwaffe *f*
r. hardening cement frühhochfester (schnellhärtender) Zement *m*
r. idle movement *(Masch)* Eilrücklauf *m*
r. index plate Direktteilscheibe *f*
r. levelling instrument *(Opt)* Schnellhorizontiergerät *n*
r. memory *(Dat)* Schnellspeicher *m*
r. method Schnellverfahren *n*
r. motion *(Masch)* Eilbewegung *f*
r. photometer *(Opt)* Schnellfotometer *n*
r. power traverse *(Masch)* Eilgang *m*, Schnellverstellung *f*
r. return *(Masch)* Eilrücklauf *m*
r. reverse traverse *(Masch)* Eilrücklauf *m*
r. scan *(Fs)* Schnellabtastung *f*
r. shut-down Notabschaltung *f*, Schnellabschaltung *f*, Sicherheitsabschaltung *f*
r. starting *(Fmt)* guter Schwingeinsatz *m*, Schnellanlauf *m*, schneller Anlauf *m*
r. test Kurzzeitversuch *m*
r. transfer rabbit *(Kern)* Eilrohrpost *f*, Blitzrohrpost *f*
r. travel Schnellgang *m*
r.-traverse / to *(Masch)* schnell verstellen, im Eilgang bewegen
r. traverse *(Masch)* Eilgang *m*
r. traverse control Eilgangschaltung *f*
r. traverse rate Eilganggeschwindigkeit *f*
rapidity Schnelligkeit *f*, Geschwindigkeit *f*
r. of vaporization Verdampfungsgeschwindigkeit *f*
rapidly melting schnellschmelzend, leichtschmelzend *(Metalle)*
rapier loom Lanzenwebstuhl *m*
rapper Abklopfapparat *m (Formerei)*
rare 1. selten, rar; 2. verdünnt
r. earth elements seltene Erdmetalle *npl* (Erden *fpl*)
r. earths seltene Erden *fpl*

r. gas Edelgas *n*; seltenes Gas *n*
r.-gas rectifier Edelgasgleichrichter *m*
r.-gas valve Edelgasröhre *f*
r. metal Edelmetall *n*, seltenes Metall *n*
r.-metal couple edles Thermoelement *n*
r. mixture armes Gemisch *n* *(Treibstoff für Verbrennungsmotor)*
rarefaction Verdünnung *f*
r. of the air Luftverdünnung *f*
rarefiability Verdünnbarkeit *f*
rarefiable verdünnbar
rarefied evakuiert, verdünnt
r. air verdünnte Luft *f*
r. air space luftverdünnter Raum *m*
r. strata of the earth's atmosphere verdünnte Schichten *fpl* der Atmosphäre
raschel knitting *(Text)* Raschelwirken *n*
r. warp knitting machine *(Text)* Raschelmaschine *f*
Raschig rings Raschigringe *mpl*
rasp / to raspeln
rasp Raspel *f*
r. file Raspelfeile *f*
rasping drum Reibtrommel *f*
r. file Holzfeile *f*
r. machine *(Text)* Entfaserungsmaschine *f* *(Sisal)*
r. strip Reibleiste *f*
raster *(Fs)* Raster *m*
rat destroyer Rattenvertilgungsmittel *n*
r. killer Rattenvertilgungsmittel *n*
r. race Differentialringübertrager *m* *(Wellenleitertechnik)*
r. tail *(Fmt)* Antennenzuführungsleitung *f*
ratch 1. Streck[feld]weite *f* *(Spinnen)*; 2. Sperrstange *f*
r. of the fallers *(Text)* Nadelstababstand *m*
ratchet Sperrad *n*, Klinkenrad *n*; Gesperr *n*, Knarre *f*, Ratsche *f*; Sperre *f*, Sperrhaken *m*, Sperrklinke *f*
r. and pawl Zahngesperre *n*
r.-and-pawl mechanism Ruckschaltwerk *n*, Sperrgetriebe *n*, Sperrwerk *n*
r. bit brace Bohrratsche *f*, Bohrknarre *f*
r. box Knarrengehäuse *n*
r. brace *s.* r. bit brace
r. brace bit Bohrknarrenbohrer *m*
r. circuit Blinkschaltung *f*
r. counter Hubzähler *m*
r. die stock Ratschengewindeschneidkluppe *f*
r. dog *(Masch)* Ratschenhebel *m*
r. drill Bohrratsche *f*, Knarrenbohrer *m*
r. drum *(Fmt)* Sperrzahnkranz *m*
r. gear Gesperre *n*
r. gearing Sperrgetriebe *n*
r. handle Klinkengriff *m*, Knarrenhebel *m*
r. lever 1. *(Masch)* Klinkhebel *m*; 2. *(Fmt)* Sperrhebel *m*
r. mechanism Sperrwerk *n*, Sperrgetriebe *n*
r. nut Sperradschraube *f*
r. pawl Sperrklinke *f*
r. pipe stock Knarrenkluppe *f*
r. rod Klinkenstange *f*
r. spanner Knarrenschlüssel *m*
r. spring Klinkenfeder *f*
r. stop Gefühlsschraube *f*, Ratsche *f*
r. thread Sägegewinde *n*
r. threading device Ratschenkluppe *f*
r. time base *(Fs)* Sperrzeitbasis *f*
r. tooth 1. Sperrzahn *m*; 2. Schaltzahn *m*
r. wheel *(Fmt)* Sperrad *n*, Sperrklinkenrad *n*; Sperrglied *n* *(z. B. Automatikempfänger)*
r. wrench Ratschenhebel *m*, Radschlüssel *m*
rate / to bemessen; schätzen, veranschlagen; taxieren
rate Größe *f*, Menge *f*, Wert *m*, Rate *f*; Geschwindigkeit *f*; Zahl *f*, Stückzahl *f*; Fahrtgeschwindigkeit *f* *(eines Schiffes)*
r. action Differentialquotienteneinfluß *m*, D-Einfluß *m* *(Regler)*; PID-Regelung *f*, PI-Regelung *f* mit Vorhalt
r. gyro *(Dat)* Differenzierkreisel *m*
r. of air flow Luftdurchflußmenge *f*
r. of amplitude change Steilheit *f* *(Elektronenröhren)*
r. of burning Brennbarkeit *f*
r. of carbon drop *(Met)* Frischgeschwindigkeit *f*
r. of change Markierungsdifferenz *f* *(Radar)*
r.-of-change indicator *(Flg)* Sink- oder Steiggeschwindigkeitsmesser *m*
r.-of-change relay *(El)* Gradientenrelais *n*, Stoßrelais *n*
r. of circulation Umlaufgeschwindigkeit *f*
r. of climb *(Flg)* Steiggeschwindigkeit *f*
r.-of-climb indicator Steiggeschwindigkeitsmesser *m*
r. of combustion *(Wärm)* Leistung *f* der Feuerung
r. of combustion per hour stündliche Brenngeschwindigkeit *f*
r. of cooling [kalorimetrische] Abkühlungsgeschwindigkeit *f*
r. of corrosion Korrosionsgeschwindigkeit *f*
r. of creep Kriechgeschwindigkeit *f*
r. of crystalline growth Kristallisationsgeschwindigkeit *f*
r. of current change Stromänderungsgeschwindigkeit *f*
r. of cut *(Masch)* Spantiefe *f*
r. of delivery Fördermenge *f* *(Pumpe)*
r. of deposition Abschmelzgeschwindigkeit *f* *(Schweißen)*
r. of development Entwicklungstempo *n*
r. of drill[ing] feed Bohrvorschub[größe *f*] *m*

r. of energy loss *(El)* spezifischer Energieverlust *m*, spezifische Ionisation *f*
r. of escape Austrittsgeschwindigkeit *f*
r. of evacuation Sauggeschwindigkeit *f*
r. of feed Vorschubgröße *f*, Vorschubstufe *f*, Vorschubgeschwindigkeit *f*
r. of fire *(Mil)* Feuergeschwindigkeit *f*
r. of flame propagation Zündgeschwindigkeit *f*
r. of flow Fließgeschwindigkeit *f*, Strömungsgeschwindigkeit *f*; Strömungsmenge *f*, Durchflußmenge *f*
r. of growth Wachstumstempo *n*, Wachstumsrate *f*; Zuwachsprozent *n*, Zuwachsrate *f*
r. of heat transfer Wärmedurchsatz *m*
r. of metal removal Zerspanungsleistung *f*
r. of power input Leistungsaufnahme *f*
r. of precipitation Niederschlagsmenge *f*
r. of production Mengenleistung *f*; Fertigungsmenge *f*; Ausstoß *m*; Stückzahl *f*
r. of rise *(El)* Anstiegssteilheit *f*
r. of sending Gebetempo *n (Funktechnik)*
r. of stock removal Spanungsleistung *f*
r. of subsidence *(Brg)* Senkungsgeschwindigkeit *f*
r. of tool wear Verschleiß *m* des Werkzeuges, Verschleißfortschritt *m* beim Werkzeug
r. of travel *(Masch)* Geschwindigkeit *f*; Fahrgeschwindigkeit *f*
r. of travel of sound Schallausbreitungsgeschwindigkeit *f*
r.-of-turn gyroscope *(BMSR)* Wendezeiger *m*
r. of vibration Schwingungszahl *f*
r. of voltage change Spannungsänderungsgeschwindigkeit *f*
r. of wear Abnutzungsbetrag *m*, Abnutzungsgrad *m*, Abnutzungsmaß *n*
r. response *(Am) s.* derivative action
r. time *(Am)* Vorhaltezeit *f (Regler)* (*s. a.* derivative action time)
rated [ab]geschätzt; Nenn-, Nominal-
r. breaking current Nennabschaltstrom *m*, Nennausschaltstrom *m*
r. burden Nennleistung *f (Transformator)*
r. capacity Nennleistung *f*, Solleistung *f*; Belastungsfähigkeit *f*
r. circuit voltage *(El)* Nennspannung *f*
r. continuous current Nenndauerstrom *m*; Nenngleichstrom *m*
r. cross section Nennquerschnitt *m (des Leiters)*
r. current Nennstrom[stärke *f*] *m*
r. current load Nennstrombelastung *f*
r. duty Nennbetriebsart *f*
r. efficiency Nennwirkungsgrad *m*
r. frequency Nennfrequenz *f*
r. interrupting capacity Nennabschaltleistung *f*, Nennausschaltvermögen *n*
r. load Nennlast *f*, Tragzahl *f (Lager)*
r. mechanical terminal load *(El) (mechanischer)* Nennklemmenzug *m*
r. output Nennleistung *f*
r. power Nennleistung *f*
r. power supply Nennetzspeisung *f*, Netzspeisung *f* mit vorgegebenen Daten
r. primary voltage primäre Nennspannung *f*
r. short-circuit current thermischer Grenzstrom *m*; Nennkurzschlußstrom *m*
r. short-time current Nennkurzzeitstrom *m*
r. speed Nenndrehzahl *f*
r. temperature-rise current *(El)* höchster zulässiger Betriebsstrom *m (eines Meßgerätes)*
r. value Nennwert *m*, Sollwert *m*
r. voltage Nennspannung *f*
r. yield load *(Brg)* Nennlast *f (Ausbau)*
rater Klassenboot *n*
rating Leistung *f*, Auslegung *f*; Belastungsziffer *f*; Nennbetrieb *m*, Nennbereich *m*; Nenndaten *pl*; genaue Bezeichnung *f*, Pflichtenblatt *n*
r. certificate *(Schiff)* Meßbrief *m*
r. of an instrument Nennmeßbereich *m* eines Instrumentes
r. plate Leistungsschild *n (an Maschinen)*
ratio 1. Verhältnis *n*; Verhältniszahl *f*; Maßstab *m*; Übersetzung *f*; 2. Übersetzung[sverhältnis *n*] *f (Transformator)*
r. amplifier Quotientenverstärker *m*
r.-arm box *s.* resistance box
r. between the channels *(El)* Stärkeverhältnis *n* der Kanäle
r. control Verhältnisregelung *f*
r. correction factor Korrekturfaktor *m* für das Übersetzungsverhältnis *(Wandler)*
r. detector *(Fmt)* Verhältnisdetektor *m*
r. error of a transformer *(El)* [prozentualer] Übersetzungsfehler *m* eines Wandlers
r. of a transformer *(El)* Windungsverhältnis *n*, Übersetzungsverhältnis *n*
r. of axle drive Übersetzung *f* des Achsantriebs
r. of bearing contact area to total area (surface) Traganteil *m*
r. of compression Kompressionsverhältnis *n*, Verdichtungsverhältnis *n*
r. of contraction of volume relative Volumenkontraktion *f*, kubische Zusammenziehung *f*
r. of cylinder volumes Zylindervolumenverhältnis *n*
r. of driving and driven wheels Übersetzungsverhältnis *n* bei Zahnrädern
r. of eccentricity Ausschlag *m (Statik)*
r. of expansion Ausdehnungsverhältnis *n*
r. of gearing (gears) *(Masch)* Übersetzungsverhältnis *n*
r. of net weight of missile to thrust Schubbelastung *f (von Raketen)*

r. of power to weight Leistungsgewicht *n*
r. of slenderness Schlankheitsgrad *m*
r. of the number of teeth Zähnezahlverhältnis *n*
r. of transformation *(El)* Übersetzungsverhältnis *n*, Windungsverhältnis *n*
r. of transmission *(Masch)* Übersetzungsverhältnis *n*
r. paper halblogarithmisches Papier *n*
rational rational *(Zahl)*
r. horizon *(Schiff)* wahre Kimm *f*, wahrer Horizont *m*
r. number rationale Zahl *f*
rationalize / to rational machen *(Zahl)*; rationalisieren
rattan rope *(Schiff)* Rotangtauwerk *n*
ratter Rätter *m (Sieb)*
rattle barrel Gußputztrommel *f*, Scheuerfaß *n*, Rommelfaß *n*
r. flax *(Text)* Klanglein *m*, Springlein *m*
rattler star Putzstern *m*, Stern *m (Gußputzerei)*
r. test Abnutzungsversuch *m*, Trommelversuch *m*, Verschleißversuch *m (mittels Stahlkugeln)*
rattles Rattergeräusch *n*, Rattern *n*
rattling noise Rasseln *n (Elektronenröhren)*
raw 1. roh; 2. unbearbeitet *(Kristalle)*; ungeschliffen *(Diamanten)*; 3. roh, ungebleicht, unverarbeitet; 4. nichtstabilisiert *(Spannung)*; 4. unbelichtet *(Film)*
r. brown coal Rohbraunkohle *f*
r.-cast roh gegossen
r. clay Rohton *m*
r. coal Rohkohle *f*, Förderkohle *f*
r. coal bunker Rohkohlenturm *m*
r. coal output Rohkohlenförderung *f*
r. grinding Vormahlen *n*
r. hide Rohhaut *f*
r.-hide mallet Rohhauthammer *m*, Rohhautschlegel *m*
r.-hide store *(Led)* Rohhäutelager *n*
r. ingot Rohblock *m*
r. lead Werkblei *n*
r. lignite Rohbraunkohle *f*
r. material Rohmaterial *n*, Einsatzmaterial *n*, Ausgangswerkstoff *m*
r. ore Frischerz *n*, Roherz *n*
r. piece *(Masch)* Rohling *m*, unbearbeitetes Teil *n*
r. silk Rohseide *f*
r. silk cleaning machine Rohseidenputzmaschine *f*
r. steel Rohstahl *m*
r. wool scouring Rohwollwäsche *f*
r. zinc Werkzink *n*
rawness Rohzustand *m*
ray Strahl *m*
r. control electrode Strahlsteuerelektrode *f*
r.-locking device Strahl[en]falle *f (Blockierungsvorrichtung für Katodenstrahlen)*
r. of light Lichtstrahl *m*
r. of sparks Funkengarbe *f*
r. path Strahlenbahn *f*, Strahlengang *m*, Strahlenverlauf *m*
r. tube Katodenstrahlröhre *f*
raying Bestrahlung *f*
Rayleigh disk Rayleighsche Scheibe *f*
Rayleigh scattering Rayleighsche Streuung *f*, Rayleigh-Streuung *f*
rayon Chemieseide *f*, Viskoseseide *f*, Kupferseide *f*
r. staple fibre Viskosefaser *f*
raze / to abbrechen *(Gebäude)*
razor steel Rasiermesserstahl *m*
RC constant *(El)* RC-Konstante *f*, Zeitkonstante *f*
RC coupling *(El)* RC-Kopplung *f*, Kondensatorwiderstandskopplung *f*
RC network *(El)* RC-Schaltung *f*, RC-Glied *n*
re-building Umbau *m*, Wiederaufbau *m*
re-coil / to neu wickeln *(Spule)*, wieder aufspulen (auftrommeln) *(Kabel)*
re-edify / to wieder aufbauen
re-edit / to neu (nochmals) auflegen
re-edition Neuauflage *f*
re-egg / to *(Led)* nachgerben mit Eigelb
re-emission *(Kern)* Reemission *f*
re-employ / to wiederverwenden
re-engage / to wieder in Eingriff kommen (bringen), wieder einrücken (einstellen, einschalten)
re-enlarger *(Opt)* Rückvergrößerungsgerät *n*
re-enrichment *(Kern)* Neuanreicherung *f*
re-enter / to wieder einführen (ansetzen)
re-fuse / to wieder[zusammen]schmelzen
re-lay / to wieder verlegen *(Kabel)*
re-record / to umspielen
re-tannage *(Led)* Nachgerbung *f*
re-tanning of chrome-leather Nachgerben *n* von Chromleder
reabsorb / to resorbieren
reabsorber Resorber *m*
reach / to erreichen
r. the apogee die größte Erdferne erreichen
r. the perigee die größte Erdnähe erreichen
reach 1. Weite *f*, Strecke *f*; Reichweite *f*, Aktionsradius *m*; 2. *(Masch)* Spannbereich *m*; 3. *(El)* Stufenreichweite *f (Relais)*; 4. Stromstrecke *f (eine gerade Strecke eines schiffbaren Flusses)*
r. of call *(Fmt)* Rufweite *f*
r. of trade *(Schiff)* Fahrtbereich *m*
react / to reagieren, einwirken, zurückwirken
reactance Blindwiderstand *m*, Reaktanz *f*, induktiver Widerstand *m*, Scheinwiderstand X *m*

(Blindanteil der Impedanz, bestehend aus induktivem und kapazitivem Widerstand)
r. circuit Drosselkreis *m*
r. coil Drossel[spule] *f*
r. dimmer Verdunkelungsdrossel *f*, Verdunkelungswiderstand *m*
r. drop induktiver Spannungsabfall *m*
r. earthed induktiv geerdet
r. fed track circuit induktiv gespeister Gleisstromkreis *m*
r. grounded *(Am)* induktiv geerdet
r. oscillator Blindleistungsgenerator *m*
r. transformer Reaktanztransformator *m*
r. tube Blindröhre *f*, Reaktanzröhre *f*
r. tube for frequency control Nachstimmröhre *f*
r. voltage Reaktanzspannung *f*, Streuspannung *f*, induktiver Spannungsabfall *m*, Blindspannung *f*
reactant Reaktant *m*, reagierender Stoff *m*, Reaktionspartner *m*, Reaktionsteilnehmer *m*, Reagenz *f*, Reaktionsmittel *n*
reacting Reagieren *n*
reacting rückwirkend
r. duration Ansprechdauer *f (Relais)*
r. weight der in Reaktion tretende Masseanteil *m* einer Substanz
reaction 1. *(Ch)* Reaktion *f*, Umsetzung *f*; Verhalten *n*; 2. *(Mech)* Gegenwirkung *f*; Reaktionskraft *f*; 3. *(El)* Rückwirkung *f*, Rückkopplung *f*
r. amplifier Rückkopplungsverstärker *m*
r. apparatus *(Ch)* Reaktionsgefäß *n*, Reaktionsapparat *m*
r. brush holder *(Elm)* schräger Bürstenhalter *m*
r. capacitor Rückkopplungskondensator *m*
r. chain Reaktionskette *f (Kettenreaktion)*
r. coefficient Rückkopplungsgrad *m*
r. coil Rückkopplungsspule *f*
r. conductance *(El)* Rückwirkungsleitwert *m*
r. coupling Rückkopplung *f*
r. cross section *(Kern)* Reaktionsquerschnitt *m*, Wirkungsquerschnitt *m* der Reaktion
r. curve Übergangsfunktion *f* der Regelstrecke
r. energy Energietönung *f* [der Reaktion], Reaktionsenergie *f*
r. engine Strahltriebwerk *n*
r. force Reaktionskraft *f*; Auflagerreaktion *f*
r. inhibition *(Ch)* Reaktionshemmung *f*
r. material *(Rak)* Rauchstoff *m*
r. mixture Reaktionsgemisch *n*
r. of formation Bildungsreaktion *f*
r. of the valve gear *(Masch)* Rückwirkung *f* der Steuerung
r. power Rückstoßkraft *f*
r. principle Reaktionsprinzip *n*
r. probability Reaktionswahrscheinlichkeit *f*
r. propelled nach Rückstoßprinzip angetrieben
r. propulsion Strahltriebwerk *n*, Strahlantrieb *m*, Rückstoßantrieb *m*
r. propulsion engine Flugtriebwerk *n*, Reaktionstriebwerk *n*
r. rate Reaktionsgeschwindigkeit *f*, Umwandlungsgeschwindigkeit *f*
r. ring Druckring *m*
r. stage Reaktionsstufe *f*
r. suppressor Rückkopplungssperre *f*
r. time Reaktionszeit *f*
r. torque Gegendrehmoment *n*
r. turbine Überdruckturbine *f*, Rückdruckturbine *f*, Gegendruckturbine *f*, Reaktionsturbine *f*
r. turbine with pressure stages mehrstufige Überdruckturbine mit Druckabstufung
r.-type blading Überdruckbeschaufelung *f*
r. upon the nozzle Rückdruck *m* auf die Düse *(Turbine)*
r. vessel Reaktionsschiff *n (durch Wasserrückstoß angetriebenes Schiff)*
r. wheel Reaktionsrad *n (Turbine)*
reactivate / to reaktivieren, regenerieren; wiedereinsetzen
reactivation *(Ch)* Reaktivierung *f*, Regeneration *f*
reactive reagierend, reaktionsfähig, reaktiv; rückwirkend
r. attenuator *(El)* nicht absorbierender Abschwächer *m*; verlustfreier Abschwächer
r. component *(El)* Blindkomponente *f*
r. current Blindstrom *m*
r. current meter Blindstromzähler *m*
r. drop induktiver Spannungsabfall *m*
r. dye[stuff] *(Text)* Reaktivfarbstoff *m*
r. effect Rückwirkung *f*
r.-energy meter *s.* var-hour meter
r. force Stützkraft *f*, Zwangskraft *f*, Reaktionskraft *f*; Gegenwirkung *f*
r. impedance Blindwiderstand *m*
r. load induktive Last (Belastung) *f*, Blindlast *f*; Belastung durch Wechselstromwiderstand, Wechselstrombelastung *f* durch komplexen Widerstand
r. load regulator Blindlastregler *m*
r. meter Blindlastzähler *m*, Blindleistungszähler *m*
r. mixture Treibstoffgemisch *n*
r. permeability Blindpermeabilität *f*
r. power Blindleistung *f*
r. power factor nacheilender Leistungsfaktor *m*
r. power generator Blindleistungsmaschine *f*
r. power relay Blindleistungsrelais *n*
r. typewriter rückkoppelnde Schreibmaschine *f (mit dem Rechner gekoppelte Schreibmaschine zur Informationserschließung)*

r. volt-amperes Blindleistung *f* [in VA]
r. voltage Blindspannung[sabfall *m*] *f*
reactivity Reaktivität *f*; Reaktionsfähigkeit *f*, Reaktionsfreudigkeit *f*
r. change Reaktivitätsänderung *f*
r. coefficient Reaktivitätskoeffizient *m*
r. excess Reaktivitätsüberschuß *m*
reactor 1. *(El)* Drosselspule *f*; Drossel *f*; 2. *(Ch)* Reaktionsgefäß *n*, Reaktionsapparat *m*; 3. Kernreaktor *m*, Reaktor *m*
r. behaviour Reaktorverhalten *n*
r. campaign Reaktorkampagne *f*
r. cell Zelle *f* im Reaktor, Reaktorzelle *f*
r. chemistry Reaktorchemie *f*
r. control Reaktorsteuerung *f*; Reaktorregelung *f*
r. core Reaktorzone *f*, Spaltzone *f*
r. engineering Reaktorbau *m*, Reaktortechnik *f*
r. envelope Reaktormantel *m*
r. equation Reaktorgleichung *f*
r. feed Reaktorbeschickung *f*
r. for irradiation Bestrahlungsreaktor *m*
r. fuel Kernbrennstoff *m*
r. instrumentation Reaktorinstrumentierung *f*
r. loading Reaktorbeschickung *f*
r. material Reaktorwerkstoff *m*
r. metallurgy Reaktormetallurgie *f*
r. period Reaktorperiode *f*, Periode *f* [des Reaktors]
r. physics Reaktorphysik *f*
r. plant Kernenergieanlage *f*
r. power Reaktorleistung *f*
r. process Reaktorbetrieb *m*
r. purity Reaktorreinheit *f*
r. safety Reaktorsicherheit *f*
r. shell Reaktormantel *m*
r. shielding Reaktorabschirmung *f*
r. sphere Reaktorkuppel *f*
r.-start motor *(Elm)* Motor *m* mit Drosselanlasser
r.-start split-phase motor Einphasenwechselstrommotor *m* mit Anlaßdrossel
r. technology Reaktortechnik *f*
r. theory Reaktortheorie *f*
r. vessel Reaktor[druck]gefäß *n*, Reaktormantel *m*
r. waste Reaktormüll *m*
read / to *(Dat)* abtasten, lesen; ablesen
r. in einlesen, schreiben *(in einen Datenspeicher)*
r. off ablesen
r. out auslesen *(mittels Maschinen)*
r. proofs Korrektur lesen
read amplifier Leseverstärker *m*
r.-around number *(Dat)* Zahl *f* der aufeinanderfolgenden Ablesungen; Zugriffzahl *f*
r.-disturbed core lesegestörter Magnetkern *m*
r. number Zahl *f* der wiederholten Ablesungen *(Speichertechnik)*
r.-only-memory Auslesespeicher *m*, Totspeicher *m (z. B. für automatische Sprachübersetzung)*
r.-out device *(Dat)* Ausgabeeinrichtung *f*, Ableseeinrichtung *f*, Anzeigeeinrichtung *f*
r.-out memory Auslesespeicher *m (Ziffernrechner)*
r.-out timer Schaltuhr *f* mit Zeitdrucker
r.-record head kombinierter Lese-Schreibkopf *m (Magnetschichtspeicher)*
readability Lesbarkeit *f*; Ablesemöglichkeit *f*, Ablesbarkeit; Verständlichkeit *f*
readable lesbar; ablesbar
reader 1. Korrektor *m*; 2. Lektor *m*; 3. Lesegerät *n*
reader's mark Korrekturzeichen *n*
r. proof Fahnenkorrektur *f*, Hauskorrektur *f*
readily levelling *(Text)* gut egalisierend
r. machined leicht zerspanbar
reading Lesen *n*; Ablesung *f*, Ablesen *n*; Ablesemöglichkeit *f*
r. accuracy Ablesegenauigkeit *f*
r. brushes *(Büro)* Lesebürsten *fpl*
r. error Ablesefehler *m*, Ablesungsfehler *m*
r. glass Leseglas *n*
r. head Lesekopf *m (Magnetschichtspeicher)*
r. height abgelesene Höhe *f*, Stand *m*
r. instrument Anzeigeinstrument *n*
r. lamp Leseleuchte *f*
r. mechanism Ableseeinrichtung *f*
r. portion *(Opt)* Nahteil *m (im Zweistärkenglas)*
r. precision Ablesegenauigkeit *f*
r.-type encoder *s.* spatial encoder
readjust / to verstellen, nachrichten, nachstellen
readjustable nachstellbar
readjusting device for bearing bolts Nachstellvorrichtung *f* für Lagerbolzen
readjustment Nachregelung *f*; Verstellung *f*, Nachstellung *f*
ready bunk festeingebaute Koje *f (auf Schiffen)*
r. for operation (service, use) betriebsbereit, betriebsfertig
r. for take-off *(Flg)* startklar
r. lifeboat *s.* emergency boat
r.-mix concrete Fertigbeton *m*, Frischbeton *m*, Transportbeton *m*, Lieferbeton *m*
r.-mix plant Fertigbetonwerk *n*, Frischbetonwerk *n*, Transportbetonwerk *n*
r. mixed streichfertig *(Farbe)*
r.-mixed concrete Fertigbeton *m*, Frischbeton *m*, Transportbeton *m*, Lieferbeton *m*
r.-to-fit einbaufertig
r. to sail fahrbereit *(Schiff)*
r. to speak *(Fmt)* sprechbereit
r.-to-wear clothing Konfektionskleidung *f*

r.-use tank Verbrauchsbehälter *m*
reagency Reaktionsvermögen *n*
reagent Reagens *n*
r. bottle Chemikalienflasche *f*, Reagenzflasche *f*
real reell *(Zahl)*
r. circuit *(Fmt)* Stammleitung *f*
r. circular pitch Stirnteilung *f (Verzahnung)*
r. cost Grundkosten *pl*
r. course *(Schiff)* beschickter Kurs *m*, Kartenkurs *m*
r. crystal Realkristall *m*
r. cut-off wirksame Füllung *f (Dampfdiagramm)*
r. estate survey Grundstücksmessung *f*
r. fluid wirkliche Flüssigkeit *f*
r. image reelles Bild *n*
r. measure Sollmaß *n*
r. module Stirnmodul *m (Verzahnung)*
r. number reelle Zahl *f*
r. output wirkliche Leistung *f*
r. power *(El)* Wattleistung *f*, Wirkleistung *f*
r. pressure angle Stirneingriffswinkel *m (Verzahnung)*
r. time *(Dat)* Echtzeit *f*, Realzeit *f*
r.-time computer Echtzeitrechner *m*, Realzeitrechner *m*
r.-time simulation Echtzeitbetrieb *m*
r.-to-random ratio *(Fs)* Verhältnis *n* der echten Impulse zu den zufälligen
realgar *(Min)* Realgar *m*, Rauschrot *n*
realign / to nachrichten
realignment Nachrichten *n*
ream / to reiben *(Reibahle)*, aufreiben, ausreiben
r. the hole out to gauge das Bohrloch auf vollen Durchmesser aufreiben
r. to size auf das Endmaß [auf]reiben
reamed bolt Paßschraube *f*, eingepaßte Schraube *f*
reamer 1. Reibahle *f (spanende Fertigung)*; 2. Nachschneider *m*, Räumer *m (Tiefbohrtechnik)*
r. arbor Reibahlendorn *m*
r. blade Reib[ahlen]messer *n*, Reibahlenschneide *f*
r. body Reibahlenschneidenteil *n*
r. cutter Reibahlennutenfräser *m*
r. diameter Reibahlendurchmesser *m*
r. fluting cutter Reibahlennutenfräser *m*
r. holder Reibahlendorn *m*
r. indexing attachment (head) Reibahlenteilgerät *n*
r. shank Reibahlenschaft *m*
r. with pilot guide Führungsreibahle *f*
r. with unequally spaced teeth Reibahle *f* mit ungleicher Zahnteilung
reamering *(sl)* Ausreiben *n*
reaming Reiben *n*, Reibarbeit *f*, Aufreiben *n*, Aufreibarbeit *f*
r. allowance Reibzugabe *f*, Untermaß *n* zum Reiben
r. bar Reibestange *f*
r. out Ausreiben *n*
r. to the final size Reiben *n* auf Fertigmaß
reanneal / to *(Met)* nachglühen
reap / to mähen, schneiden; ernten *(Getreide)*
reaper Mähmaschine *f*; Getreidemäher *m*
reaping beam Mähbalken *m*
r. machine Erntemaschine *f*, Mähbinder *m*
rear / to 1. aufrichten, errichten; 2. züchten
rear Hintergrund *m*, Rückseite *f*
rear rückwärtig, hintere, Hinter-
r. axle Hinterachse *f*
r. axle assembly Hinterachskörper *m*
r. axle casing Hinterachsbrücke *f*
r. axle casing cover Deckel *m* zur Hinterachsbrücke
r. axle drive Hinterradantrieb *m*
r. axle flared tube Hinterachstrichter *m*
r. axle housing Hinterachsgehäuse *n*
r. axle housing section Hinterachsgehäusehälfte *f*
r. axle radius rod Hinterachsschubstange *f*
r. axle shaft Hinterachswelle *f*
r. axle strut Hinterachsstrebe *f*
r. axle tie rod Hinterachsstrebe *f*, Hinterachszugstange *f*
r. axle tube Hinterachsrohr *n*
r. brake Hinterradbremse *f*
r. bumper *(Kfz)* Heckstoßstange *f*, hintere Stoßstange *f*
r. connection *(El)* Anschluß *m* von hinten, rückseitiger Anschluß
r. curtain hintere Antennenwand *f*, Reflektorwand *f*
r. door Rückwandtür *f*
r. drive Heckantrieb *m*
r. driving mirror Rück[blick]spiegel *m*
r.-dump truck Hinterkipper[-LKW] *m*, Muldenhinterkipper *m*
r. dumper Hinterkipper *m*, Muldenhinterkipper *m*
r. engine Heckmotor *m*
r.-engined mit Heckmotor
r. flap *(Kfz)* Heckklappe *f*
r. focal plane *(Opt)* hintere Brennebene *f*
r. gunner *(Mil)* Heckschütze *m*
r. gunner's station *(Mil)* Heckstand *m*
r. lamp *(Kfz)* Schlußlampe *f*, Schlußlicht *n*, Hecklampe *f*
r. lamps of train *(Eis)* Zugschlußsignal *n*
r. light *(Kfz)* Schlußlicht *n*, Rücklicht *n*
r. number plate light *(Kfz)* Kennzeichenbeleuchtung *f*
r. panel *(Kfz)* Rückwand *f*; Heck *n* der Karosserie
r. parcel shelf *(Kfz)* hinteres Ablagebrett *n*

r. pilot hintere Führung *f (Werkzeug)*
r. plough Hinterpflug *m*
r. projection *(Fs)* Durchprojektion *f*
r. red reflex reflector Rückstrahler *m*, Katzenauge *n*
r. reflector Rückstrahler *m*
r. seat Hintersitz *m*
r. section *(Rak)* Heckzelle *f*
r. shock absorber hinterer Stoßdämpfer *m*
r. side Rückseite *f*
r.-side view Ansicht *f* von hinten
r. sight *(Mil)* Kimme *f*
r. sight slide Visierschieber *m*
r. spring Hinterfeder *f*
r. spring hanger Hinterfederbock *m*
r. spring support Hinterfederstütze *f*
r. suspension Hinterradaufhängung *f*
r. terminal block *(El)* rückwärtige Anschlußleiste *f*
r.-to-front ratio Vorwärts-zu-rückwärts-Verhältnis *n*, Vor-rück-Verhältnis *n (Antennen)*
r. valve rod guide hinteres Lager *n (Schieberstangenführung)*
r. view Rückansicht *f*
r. view mirror Rück[blick]spiegel *m*
r. vision mirror Rück[blick]spiegel *m*
r. wheel Hinterrad *n*
r. wheel brake Hinterradbremse *f*
r. wheel drive Hinterradantrieb *m*
r. wheel set *(Eis)* Hinterradsatz *m*
r. wheel springing Hinterradfederung *f*
r. window Rückwandfenster *n*
r. wing Hinterkotflügel *m*, hinterer Kotflügel *m*
rearrange / to 1. umordnen, umgruppieren; 2. *(El)* umschalten, neu schalten; 3. *(Math)* ordnen
rearrangement *(Ch)* Umlagerung *f*, Umgruppierung *f*, Umstellung *f*, Neuordnung *f*
r. of crystals Neuordnung (Umlagerung) *f* von Kristallen
rearward action Rückwärtswirkung *f*
reasoning computer *(Dat)* schließender (urteilender) Automat *m*
reassemble / to wieder zusammenbauen
Réaumur scale Reaumurskale *f*
rebacking off *(Am)* Nachschliff *m* des Freiwinkels
rebalance / to nachwuchten
rebalancing Nachwuchten *n*
rebar *s.* reinforcing bar
rebate 1. Falz *m*; 2. Kittfalz *m (am Fenster)*; 3. *(Ök)* Vergütung *f*; Rabatt *m*
r. meter *(El)* Vergütungszähler *m*
rebated gefalzt, genutet
rebating cutter Falzfräser *m*
r. machine Latten- und Leistenschlitzmaschine *f*
reblending *(Text)* zweites Mischen *n*
reboil / to wiederaufkochen, wiederholt kochen, nachkochen
rebore / to nachbohren; ausschleifen *(z. B. einen Zylinder)*
reboring Nachbohren *n*
rebound / to abspringen, zurückprallen
rebound 1. Rückprall *m*, Rückstoß *m*, Rückschlag *m*; 2. Rückprallhöhe *f (Härteprüfung)*; 3. Verdichtungsstoß *m (Turbine)*
r. crusher *(Met)* Prallmühle *f*
r. hardness *(Met)* Rückprallhärte *f*
r. strap Fangband *n*
r. tester Rückprallhärteprüfer *m*
r. valve Rückschlagventil *n (z. B. im Stoßdämpfer)*
rebricking *(Brg)* Neuausmauerung *f*
rebroadcast / to *(El) (Sendung)* wiederholen; übertragen *(durch Relaisstationen)*
rebroadcast receiver *(El)* Ballempfänger *m*
r. transmitter *(El)* Ballsender *m*
rebuild / to 1. wiederaufbauen; umbauen, umkonstruieren; 2. runderneuern *(Reifen)*
rebuilding 1. Wiederaufbau *m*; Umbau *m*; Aufarbeitung *f*; 2. Runderneuerung *f (Reifen)*
recalesce / to wieder warm werden
recalescence *(Met)* Wärmeentwicklung *f*; Haltepunkt Ar *m*
r. curve Abkühlungskurve *f*
r. point Haltepunkt Ar *m*
recalibrate / to nacheichen
recalling key *(Fmt)* Rufschalter *m*
recalming zone Beruhigungszone *f (Gebirgsmechanik)*
recapture constant Einfangkonstante *f*
recarburization *(Met)* Rückkohlung *f*
recarburize / to *(Met)* rückkohlen, [wieder] aufkohlen
recarburizing agent *(Met)* Aufkohlungsmittel *n*
recast / to umschmelzen, wieder einschmelzen
receding side Ablaufseite *f (Treibriemen)*
r. table *(Masch)* Schiebetisch *m*
receipt of goods Wareneingang *m*
receive / to 1. erhalten, empfangen; 2. empfangen *(Wellen)*; 3. aufnehmen *(Werkstück)*
received band Empfangsband[breite *f*] *n*
r. frequency *(El)* Empfangsfrequenz *f*
r. voltage Spannung *f* am Leitungsende
r. wave *(El)* Empfangswelle *f*
receiver 1. Empfänger *m*; 2. *(El)* Empfänger, Rundfunkempfänger *m*, Empfangsapparat *m*; 3. *(Fmt)* Hörer *m*, Fernhörer; 4. Behälter *m*, Sammelbehälter *m*, Auffänger *m*, Auffangschale *f*; 5. Vorlage *f (Destillation)*; 6. *(Wärm)* Aufnehmer *m*, Zwischenkammer *f (Dampfleitung)*; Dampfsammler *m (Niclausse-Kessel)*
r. alignment *(El)* Empfängerabgleich *m*
r. amplifier Empfangsverstärker *m*

r. cabinet Empfängergehäuse *n*, Radiogehäuse *n*
r. cap *(Fmt)* Hörmuschel *f*, Schallöffnung *f*
r. circuitry Empfängerschaltung *f*
r. cord Fernhörerschnur *f*
r. cradle *(Fmt)* Hörergabel *f*
r. earpiece Hörermuschel *f*
r. earth *(Fmt)* Empfängererdanschluß *m*
r. earth terminal *(Fmt)* Empfängererdklemme *f*
r. failure Empfängerausfall *m*
r. for condensed water Dampfniederschlaggefäß *n*
r. for local reception *(Fmt)* Ortsempfänger *m*
r. output Empfängerausgangsleistung *f*
r. output voltage Empfängerausgangsspannung *f*
r. output volume Empfängerlautstärke *f*
r. potentiometer Empfängerpotentiometer *n*
r. radiation Empfängereigenstrahlung *f (Funkempfangstechnik)*
r. recovery time Erholungszeit *f* eines Empfängers *(Funkempfangstechnik)*
r. rectifier Empfangsgleichrichter *m*
r. response time Trägheit *f* eines Empfängers *(Funkempfangstechnik)*
r. rest *(Fmt)* Apparategabel *f*, Gabel *f*
r. shell *(Am)* Hörergehäuse *n*
r. steam Aufnehmerdampf *m*
r. tuning Empfängerabstimmung *f*
r. watch Prüfhörer *m*

receiving Empfangen *n*
r. aerial Empfangsantenne *f*
r. amplifier Empfangsverstärker *m*
r. capacity Aufnahmefähigkeit *f*
r. end *(El)* Ausgang *m*; Ausgang[sseite *f*], Empfangsseite *f*, Ende *n*, Verbraucherseite *f*
r. end crosstalk *(Fmt)* Gegennebensprechen *n*, Nebensprechen *n*
r. frequency Empfangsfrequenz *f*
r. hopper *(Met)* Schüttrumpf *m*, Fülltrichter *m*
r. inspection *(Masch)* Abnahmeprüfung *f*
r. installation *(El)* Empfangsanlage *f*
r. perforator *(Fmt)* Empfangslocher *m*
r. port Empfangshafen *m*
r. recess Aussparung *f* zur Aufnahme
r. repeater *s.* r. amplifier
r. set 1. Empfänger *m*, Empfangsgerät *n*; 2. *(Fmt)* Hörer *m*; Empfangssatz *m*
r. slit *(El)* Eintrittsspalt *m*
r. tank Aufnahmebehälter *m*; Windkessel *m*
r. valve Empfängerröhre *f*
r. vessel Sammelgefäß *n*

recellulose / to nachlackieren, neulackieren

receptacle 1. Sammelbecken *n*, Aufnahmegefäß *n*, Behälter *m*; 2. *(El)* Steckdose *f*, Steckbuchse *f*; Steckerloch *n*; Fassung *f*
r. outlet Steckdose *f*, Steckeranschlußstelle *f*

reception 1. Annahme *f*, Empfang *m*; 2. *(El)* Empfang *m*; 3. Aufnahme *f (Werkstück, Werkzeug)*
r. aerial *s.* receiving aerial
r. amplifier *s.* receiving amplifier
r. by buzzer *(Fmt)* Summerempfang *m*
r. diagram Empfangsdiagramm *n (Antennentechnik)*
r. taper *(Masch)* Befestigungskegel *m*
r. zone Empfangszone *f*

receptivity Aufnahmevermögen *n*, Aufnahmefähigkeit *f*

recess / to einstechen, aussparen

recess 1. *(Masch)* Einstich *m*; Vertiefung *f*; Auskehlung *f*; 2. *(Bw)* Aussparung *f*; Nische *f*; Rücksprung *m*
r. length Einstichlänge *f (Spiralbohrer)*

recessed ausgespart; mit Schlitz *(Schraube)*
r. bulkhead *(Schiff)* Knickschott *n*, Stufenschott *n*
r. on one side einseitig ausgespart
r. on two sides beiderseitig ausgespart
r. portion *(Masch)* Ausdrehung *f*, Aussparung *f*; Einstich *m*
r. switch *(El)* Einbauschalter *m*; Unterputzschalter *m*, Dosenschalter *m* für Montage unter Putz, Einlaß[dosen]schalter *m*, Einsatz[dosen]schalter *m*

recesses *(Flg)* Aussparungen *fpl (für das Einziehfahrwerk)*

recessing Einstechen *n*; Auskehlen *n*
r. and shaping machine Oberfräse *f (Holzbearbeitung)*
r. feed Einstechvorschub *m*
r. slide Einstechschlitten *m*
r. tool Hakenmeißel *m*; Einstechmeißel *m*

recharge / to 1. nachfüllen; umladen; 2. *(El)* nachladen *(z. B. Sammler)*; 3. *(Met)* nachsetzen

recharging 1. Auffüllung *f*; 2. *(El)* Aufladung *f*; 3. *(Met)* Nachsetzen *n*
r. trailer for flame throwers *(Mil)* Flammenwerferfüllwagen *m*

rechuck / to umspannen, wieder einspannen *(Werkstück)*

rechucking Umspannen *n*, Umspannung *f*, Wiedereinspannung *f (Werkstück)*

recipe Rezept *n*, Vorschrift *f*

recipient *(Met)* Sammelherd *m*; Rezipient *m*

reciprocal Reziprokwert *m*, reziproke (inverse) Größe *f*

reciprocal reziprok, invers, umgekehrt
r. action Wechselwirkung *f*
r. bearings gegenseitige Peilungen *fpl*
r. course Gegenkurs *m*
r. cross section *(Kern)* reziproker Wirkungsquerschnitt *m*

r. effect Wechselwirkung *f*
r. excitation wechselseitige Erregung *f*
r. interlocking Wechselsperre *f (Relais)*
r. lattice reziprokes Gitter *n*
r. length reziproke Länge *f*
r. milling Pendelfräsen *n*
r. network *(El)* duales Netzwerk *n*, Dualnetzwerk *n*
r. ohm *(El)* Siemens *n (Leitwerteinheit)*
r. of amplification factor *(El)* Durchgriff *m*
r. of screen amplification factor *(El)* Schirmgitterdurchgriff *m*
r. polar Gegenpolare *f*
r. ratio Kehrwert *m*, Reziprokwert *m*
r. relation Wechselverhältnis *n*
r. relationship Wechselbeziehung *f*
r. theorem *(Math)* Reziprozitätssatz *m*
r. value Kehrwert *m*, reziproker Wert *m*
r. value of the coefficient of rigidity Schubzahl *f*
r. velocity reziproke Geschwindigkeit *f*
reciprocate / to hin- und hergehen, hin- und hergehende Bewegung ausführen
reciprocate-bending test Hin- und Herbiegeversuch *m (Werkstoffprüfung)*
reciprocating hin- und hergehend, abwechselnd wirkend
r. blade Wendeschaufel *f*
r. compressor Kolbenverdichter *m*
r. elevator Wurfelevator *m (Dreschmaschine)*
r. engine Kolbenmaschine *f*, Kolbenmotor *m (Verbrennungs- und Gasmotor)*, Kolbentriebwerk *n*
r. grate mechanisch bewegter Rost *m*
r. machinery Maschinen *fpl* mit hin- und hergehender Hauptbewegung
r. motion in vertical direction Auf- und Abbewegung *f*
r. moulding sand riddle *(Met)* Formsandschüttelsiebmaschine *f*
r. movement hin- und hergehende Bewegung *f*, Pendelbewegung *f*
r. movement in vertical direction auf- und abwärtsgehende Bewegung *f*
r.-piston blower Kolbengebläse *n*
r.-plate feeder *(Brg)* Schubwagenspeiser *m*
r. plate feeding Wurfförderung *f (Dreschmaschine)*
r. pump Kolbenpumpe *f*
r. rolling process Pilgerschrittverfahren *n (Walzverfahren)*
r. saw Säge *f* mit hin- und hergehender Schnittbewegung
r. screen Schwingsieb *n (Aufbereitung)*
r. steam engine Kolbendampfmaschine *f*
r. stroke Gegenzug *m (Säge)*
r. table Längstisch *m*
reciprocation Hin- und Herbewegung *f*
reciprocity *(Math)* umgekehrtes Verhältnis *n*; Gegenseitigkeit *f*
recirculating ball steering *(Kfz)* Kugelumlauflenkung *f*
r. heating Umluftheizung *f*
r. memory *(Dat)* Umlaufspeicher *m*
r. pump Umlaufpumpe *f*
r. register *(Dat)* Umlaufspeicher *m*
recirculation Umlauf *m*; Wiederaufnahme *f* in den Kreislauf
r. system Umlaufsystem *n*
recirculatory heater *(Kfz)* Umluftheizer *m*
reckon / to rechnen, berechnen, errechnen
reckoning *(Schiff)* [gegißtes] Besteck *n*, Koppelbesteck *n*; Besteckrechnung *f*
reclaim / to wiedergewinnen, [zu]rückgewinnen
reclaim Regeneratgummi *m*
reclaimed rubber Regeneratgummi *m*
r. soil Neuland *n*, Schwemmland *n*
r. wool Reißwolle *f*, Lumpenwolle *f*
r. wool yarn Reißwollgarn *n*
reclaiming tank Rücklaufbehälter *m*
reclamation Rückgewinnung *f (Formsand)*
r. dredger Spülbagger *m*
r. of land Landgewinnung *f*, Neulandgewinnung *f*
r. vessel Spülschiff *n*
reclamp / to umspannen *(Werkstück)*
reclamping Neuaufspannung *f (Werkstück)*
reclean / to nachsortieren, nachwaschen *(Aufbereitung)*
recleaning Nachwäsche *f (Aufbereitung)*
reclining seat *(Eis)* Liegesitz *m*
reclose / to *(El)* wiedereinschalten
reclosing fuse Sicherung *f* mit selbsttätiger Wiedereinschaltung
r. relay Wiedereinschaltrelais *n*
r. time Wiedereinschaltzeit *f*
recognition circuit *(Dat)* Erkennungsschaltung *f*
r. letter *(Flg)* Kennbuchstabe *f*
recoil / to zurückprallen, zurückschnellen
recoil Rückstoß *m*; Rückprall *m*, Rückschlag *m*; Zurücklaufen *n*
r. atom Rückstoßatom *n*
r. buffer *(Mil)* Rohrrücklaufbremse *f*
r. counter *(Kern)* Rückstoßzähler *m*
r. effect *(Kern)* Rückstoßeffekt *m*
r. electron Rückstoßelektron *n*
r. energy *(Kern)* Rückstoßenergie *f*
r. ion Rückstoßion *n*
r. neutron Rückstoßneutron *n*
r. nucleus Rückstoßkern *m*
r. particle Rückstoßteilchen *n*
r. range *(Kern)* Rückstoßreichweite *f*
r. track Rückstoßteilchenspur *f*
recoiler Aufwickeleinrichtung *f*
recoilless rückstoßfrei

recombination Rekombination *f*, Wiedervereinigung *f*, Wiederverbindung *f*
r. centre Rekombinationszentrum *n*
r. coefficient Rekombinationskoeffizient *m*
r. light Rekombinationsleuchten *n*
r. of ions Rekombination (Wiedervereinigung) *f* von Ionen
r. process Rekombinationsprozeß *m*
r. radiation Rekombinationsstrahlung *f*
r. rate Rekombinationskoeffizient *m*, Zahl *f* der Rekombinationen je Zeit- und Volumeneinheit
r. velocity Rekombinationsgeschwindigkeit *f*
recommended feed Vorschubrichtwert *m*
r. value Richtwert *m*
recondition / to instandsetzen, ausbessern, erneuern, überholen, reparieren
reconditioned carrier *(Fmt)* angehobene Trägerwelle *f*, aufgefrischter Träger *m*
r. carrier reception Empfangsart *f* mit Wiedereinführung des entstörten Trägers
reconditioning Instandsetzung *f*, Überholung *f*
reconduct / to [zu]rückleiten
reconnaissance Erkundung *f*, Aufklärung *f*
r. tank Aufklärungspanzerkampfwagen *m*
reconnection *(El)* Umschaltung *f*, Wiedereinschaltung *f*
reconstruct / to umarbeiten, umbauen, umgestalten
reconstruction Rekonstruktion *f*, Umbau *m*, Wiederherstellung *f*
reconvert / to rückverwandeln
recool / to rückkühlen
recooling Rückkühlung *f*
r. plant Rückkühlanlage *f*, Rückkühlwerk *n*
r. tower Rückkühler *m*, Rückkühlturm *m*
record / to aufzeichnen, eintragen, registrieren, schreiben, speichern; einprägen; aufschreiben, aufzeichnen; aufnehmen
record 1. Aufzeichnung *f*, gespeicherte Information *f*; Eintragung *f*, Registrierung *f*; 2. Schallplatte *f*
r. changer Plattenwechsler *m*
r. circuit *(Fmt)* Meldeleitung *f*
r. head Aufnahmekopf *m*
r. ink Dokumententinte *f*
r. line *(Fmt)* Meldeleitung *f*
r. player Plattenspieler *m*
r. relay Melderelais *n*
r.-repeat head Hör-Sprechkopf *m*
recorder 1. Aufzeichnungsgerät *n*, Registriergerät *n*, Schreiber *m*; Zähler *m*; 2. *(Fmt)* Gesprächskontrolluhr *f*; Registrierapparat *m*; Schreiber *(Telegrafieempfang)*; Schreibtelegraf *m*; Zeichenschreiber *m*
r. chart Registrierstreifen *m*
r.-controller schreibender Regler *m*
recording Aufzeichnung *f*, Eintragung *f*; Speicherung *f*; Aufnahme *f*
r. accelerometer Beschleunigungsschreiber *m*
r. adaptometer *(Opt)* Registrieradaptometer *n*
r. anemometer Windschreiber *m*
r. apparatus Aufzeichnungsgerät *n*
r. balloon Registrierballon *m*, Ballonsonde *f*
r. barometer Barograf *m*
r. chart Registrierpapier *n*
r. clockwork Registrierwerk *n*
r. controller schreibender Regler *m*
r. cutter Schneidnadel *f*, Schneidstichel *m*
r. density Aufzeichnungsdichte *f*
r. device Registrierapparat *m*
r. draught gauge *(Wärm)* registrierender Zugmesser *m*
r. drum Registriertrommel *f*, Schreibtrommel *f*
r. element Registrierteil *n*
r. furnace filling counter *(Met)* registrierender Gichtenzähler *m*
r. gas balance registrierende Gaswaage *f*
r. head Aufzeichnungskopf *m*, Schreibkopf *m* *(magnetischer Wandler)*
r. instrument Aufnahmeapparat *m*; Registriergerät *n*, Registrierinstrument *n*, registrierendes Meßgerät *n*, Schreibinstrument *n*, Schreibwerk *n*, Schreiber *m*
r. level indication Aussteuerungskontrolle *f* *(Tonbandgerät)*
r. manometer Druckschreiber *m*
r. meter schreibendes Meßinstrument *n*
r. oscillometer Oszillograf *m*, Schwingungsschreiber *m*
r. paper Registrierpapier *n*
r. photometer Registrierfotometer *n*
r.-playback head Hör-Sprechkopf *m*
r. potentiometer Kompensations[linien]schreiber *m*
r. precision indicator Schreibfeinzeiger *m*
r. pressure gauge Registriermanometer *n*, Manometer *n* mit selbsttätiger Druckaufzeichnung
r. principle Registrierverfahren *n*
r. pyrometer Registrierpyrometer *n*
r. speed Schreibgeschwindigkeit *f*, Registriergeschwindigkeit *f*
r. stylus Schreibstift *m*, Schreibfeder *f* *(Selbstschreiber)*
r. tachometer Tachograf *m*, registrierendes Tachometer *n*
r. telegraph Schreibtelegraf *m*
r. transmission measuring set Pegelschreiber *m*
r. trunk *(Fmt)* Überweisungsmeldeleitung *f*
r. unit Registrierwerk *n*; Aufnahmegerät *n*
r. voltmeter Registriervoltmeter *n*, registrierender (schreibender) Spannungsmesser *m*

r. watt and varmeter Wirk- und Blindleistungsschreiber *m*
recordist Tonmeister *m*
recover / to wiedergewinnen; sich erholen; rückverformen; zurückgewinnen; *(Schiff)* bergen
r. damages Schadenersatz erlangen
r. the properties by tempering anlassen, vergüten *(Stahl)*
recover Erholung *f (Kristall)*
recoverability Rückverformbarkeit *f*; Rückgewinnbarkeit *f*
recoverable oil rückgewinnbares Öl *n*
recovered waste paper aufgearbeitetes Papier *n*
recovering of tailings *(Met)* Schlammgewinnung *f*
recovery Rückgewinnung *f*, Wiedergewinnung *f*; Aufbereitung *f*; Erholung *f (Kristalle)*; Ausbringen *n*, Mengenausbringen *n*, Ausbringung *f*, Gewinn *m*; Bergung *f (Raketen, Schiffe, Flugzeuge usw.)*
r. curve 1. Störübergangsfunktion *f* der Regelstrecke; 2. Erholungskurve *f*
r. cycle Restitutionszyklus *m*, Erholungszyklus *m*
r. of by-metals Gewinnung *f* von Begleitmetallen (Nebenmetallen)
r. of the digestion liquor Regeneration *f* der Ablauge
r. parachute Bergungsfallschirm *m*
r. period Erholungspause *f (z. B. des Oszillators)*
r. plant Regenerierungsanlage *f*
r. service Abschleppdienst *m*, Bergungsdienst *m*
r. time *(El)* Erholzeit *f*, Abklingzeit *f*, Nachklingzeit *f*; innere Totzeit *f*; Nachholzeit *f*, Sperrträgheit *f (Diode)*; Entionisierungszeit *f (Gasentladungsröhre)*
r. vehicle Bergungsfahrzeug *n*; Abschleppwagen *m*
r. vessel Bergungsschiff *n*
r. voltage *(El)* wiederkehrende Spannung *f*, Wiederkehrspannung *f*
r. well Gewinnungsbohrung *f*
recreation period Erholungspause *f (z. B. des Oszillators)*
recrushing Nachzerkleinerung *f*
recrystallization Rekristallisation *f*; Umkristallisation *f*, Umkristallisierung *f*
r. welding Feuer[preß]schweißen *n*
recrystallize / to rekristallisieren; auskristallisieren, umkristallisieren
rectangle Rechteck *n*
rectangular rechteckig, rechteckförmig, rechtwinklig
r. cathode Rechteckkatode *f*
r. function Rechteckfunktion *f*
r. hyperbola gleichseitige Hyperbel *f*
r. parallelepiped Rechtflach *n*, Quader *m*
r. plate spring Rechteckfeder *f*
r. pulse Rechteckimpuls *m*
r. section Rechteckquerschnitt *m*
r. shaft *(Brg)* Schacht *m* mit rechteckiger Schachtscheibe
r. table Langtisch *m*
r. tank kastenförmiger Behälter *m*
r. timber Kantholz *n*
r. triangle rechtwinkliges Dreieck *n*
r. wave Rechteckwelle *f*
r. waveguide Rechteckhohlleiter *m*
r. wire spring Vierkantstahlfeder *f*
rectangularity Rechtwinkligkeit *f*
rectifiable *(Math)* rektifizierbar
rectification 1. *(Fmt)* Demodulation *f*; Gleichrichtung *f*; 2. *(Math, Ch)* Rektifikation *f*, Rektifizierung *f*; 3. Richten *n*, Ausrichten *n*
r. characteristic Gleichrichtungscharakteristik *f*, Gleichrichterkennlinie *f*
r. column *(Ch)* Fraktionierkolonne *f*; Rektifiziersäule *f*
r. curve *s.* r. characteristic
r. factor Gleichrichtungsfaktor *m*, Richtfaktor *m*
r. of river Flußbegradigung *f*
r. valve Detektorröhre *f*
rectified [gleich]gerichtet
r. current gleichgerichteter Strom *m*; Richtstrom *m (Elektronenröhren)*
r. voltage gleichgerichtete Spannung *f*, Richtspannung *f*
rectifier 1. *(Ch)* Rektifizierapparat *m*, Rektifikator *m*, Rektifizierer *m*; 2. Gleichrichter *m*; 3. *(Opt)* Entzerrungsgerät *n*
r. anode Gleichrichteranode *f*
r. bulb Gleichrichterkolben *m*
r. cell Sperrschichtzelle *f*, Gleichrichterelement *n*, Gleichrichterzelle *f*
r. connection Gleichrichterschaltung *f*
r. device Richtleiter *m (elektrischer Leiter, dessen Widerstand sich mit der Polarität der angelegten Spannung ändert)*
r. disk Gleichrichterscheibe *f*
r. element *s.* r. cell
r. equipment Gleichrichter[gruppe *f*] *m*
r. exciter Erregergleichrichter *m*
r. for battery-charging Akkumulatorladegleichrichter *m*
r. in bridge connection Brückengleichrichter *m*
r. instrument Meßgerät *n* mit Gleichrichter
r. load resistance Richtwiderstand *m (Belastungswiderstand eines Gleichrichters)*
r. photocell Sperrschichtfotozelle *f*
r. photoelectric cell Gleichrichterfotozelle *f*
r. plant Gleichrichteranlage *f*
r. stack Gleichrichterblock *m*, Gleichrichtersatz *m*

r. substation Gleichrichterunterwerk *n*, Gleichrichterstation *f*
r.-type instrument Gleichrichter[meß]instrument *n*
r. unit Gleichrichtersatz *m*
r. valve Gleichrichterröhre *f*
rectify / to 1. *(Ch)* läutern, reinigen, rektifizieren; 2. *(Math)* rektifizieren *(Kurve)*; 3. *(Opt)* entzerren; 4. *(El)* gleichrichten
rectifying 1. *(Ch)* Läutern *n*; Reinigen *n (Alkohol)*; 2. *(El)* Gleichrichten *n*
r. action Gleichrichterwirkung *f*
r. effect Gleichrichterwirkung *f*, Richtwirkung *f*
r. film Gleichrichterschicht *f*, Sperrschicht *f*
r. section Verstärkersäule *f (Destillation)*
r. tray *(Ch)* Boden *m* einer Rektifizierkolonne, Rektifizierboden *m*
r. valve Gleichrichterröhre *f*
r. valve and voltage doubler Einweggleichrichterröhre *f* und Spannungsverdoppler *m*
r. washer Gleichrichterscheibe *f*
rectilinear geradlinig, zeilenförmig
r. comber *(Text)* Mehrkopfflachkämmer *m*
r. combing machine Flachkämmaschine *f*
r. manipulator *(El)* Koordinatenmanipulator *m*
r. motion geradlinige Bewegung *f*
r. movement geradlinige Parallelbewegung *f*, Parallelgleitung *f*
r. moving mechanism geradlinig bewegter Maschinenteil *m*
r. scanning *(Fs)* Streifenabtastung *f*
rectilinearity Geradlinigkeit *f*
recto side *(Typ)* Schöndruckseite *f*
recuperate / to zurückgewinnen *n*, wiedergewinnen
recuperation *(Met)* Vorwärmung *f (in Rekuperatoren)*; Wärmeaustausch *m*
recuperative capacity Erholungsvermögen *n (z. B. der Batterie)*
r. furnace *(Met)* Rekuperativofen *m*
recuperator Rekuperator *m*, Lufterhitzer *m*, Winderhitzer *m*, Wärme[aus]tauscher *m*, Abwärmeverwerter *m*
recurrence *(Math)* Rekursion *f*
r. formula Rekursionsformel *f*
r. frequency Impuls[folge]frequenz *f*
r. network Kettenleiter *m*, Kettenschaltung *f (Vierpole)*
r. rate Impulsfrequenz *f*, Wiederholungsfolge *f*
recurrent *(Math)* rekursiv
r. circuit *(El)* Kettenleiter *m*
r. network *(El)* Kettenleiter *m*, Vierpolkette *f*
recursion *(Math)* Rekursion *f*
r. formula Rekursionsformel *f*
recut / to nachhauen *(Feile)*
recut Nachhieb *m*
recycle / to umpumpen, zurückführen; wiederholt umlaufen; wiederholt in Umlauf bringen
recycling Wiedereinsetzen *n*, Wiederverwendung *f*, Rückführung *f*
r. detector trägergesteuerter Schalterdetektor *m*
r. system Kreislaufsystem *n*, Kreislaufverfahren *n*
red rot
r.-abstracting filter *(Opt)* rotarmes Filter *n*
r. adder Rotbeimischer *m (Farbfernsehen)*
r. alder Erle *f*
r. antimony Antimonblende *f*, Kermesit *m*
r. arsenic glass *(Min)* Realgar *m*, Rauschrot *n*, rote Arsenblende *f*
r. beech Rotbuche *f*
r. birch Rotbirke *f*
r. brass Rotguß *m*, Rotmessing *n*, Tombak *m*
r. bronze metal Rotguß *m*
r. cinchona bark Chinarinde *f*
r. cinnabar Zinnober *m*
r. copper foundry Rotgießerei *f*
r. copper ore Rotkupfererz *n*
r. corpuscle Erythrozyt *m*, rotes Blutkörperchen *n*
r. haematite Roteisenerz *n*
r.-hardness Rotgluthärte *f*, Rotglühhärte *f*, Rotwarmhärte *f*
r.-hearted rotkernig *(Holz)*
r. heat Rotglut *f*
r.-hot rotglühend
r. iron ore Roteisenstein *m*
r. lane Rot-Streifen *m (Decca-Navigationssystem)*
r. lead [rote] Mennige *f*, Bleimennige *f*
r.-lead cement Mennigekitt *m*
r. lead ore Rotbleierz *n*
r. liquor *(Text)* Rotbeize *f*
r. oxide of zinc Rotzinkerz *n*
r. pencil Rotstift *m*
r. phosphorus roter Phosphor *m*
r. pine Rotkiefer *f*
r. print Rotpause *f*
r. shift Rotverschiebung *f (Spektrallinien)*
r. shortness Rotbruch *m*; Warmbrüchigkeit *f*
r. silver ore Rotgiltigerz *n*
r. tape Bürokratie *f*
r.-tape operation *(Am)* *(Dat)* Routineoperation *f*
r. tinged rotstichig
r. turmaline roter Turmalin *m*
reddle 1. Mennige *f (Öl)*; 2. Roteisenerz *n*; 3. roter Ocker *m*, Rötel *m*
redesign / to umkonstruieren, umarbeiten, umgestalten
redevelopment Neuplanung *f (Siedlungsplanung)*

redistill / to umdestillieren, wiederholt destillieren
redistilled bidestilliert
redox potential Redox-Potential *n*, Reduktions-Oxydations-Potential *n*
redraw / to nachzeichnen; weiterziehen
redrawing Nachzeichnen *n*
redress / to 1. gleichrichten; 2. nachpolieren, weiteraufarbeiten
redruthite Kupferglanz *m*
redrying Nachtrocknung *f*
r. face veneer Rücktrocknung *f* von Oberflächenfurnier
reduce / to reduzieren, herabsetzen, kürzen, vermindern, verkleinern; ins Langsame übersetzen; abmindern; verengen
r. by boiling eindicken, konzentrieren
r. by liquidation *(Met)* ausseigern
r. charge [elektrische] Ladung abführen
r. gain *(El)* die Verstärkung herabsetzen
r. to chip form zerspanen
r. to swarf zerspanen
reduced 1. reduziert, vermindert; 2. verschnitten, verdünnt
r. admittance *(El)* relativer Leitwert *m*
r. draught *(Schiff)* beschränkter Tiefgang *m*
r. draught vessel Schiff *n* mit Freibord
r. duty Teillast *f*
r. length [of column] Knicklänge *f*
r. moment of inertia reduziertes Trägheitsmoment *n*
r. pigment *(Plast)* verschnittenes Pigment *n*
r. pressure verminderter Druck *m*
r. scale verkleinerter Maßstab *m*
r. stress reduzierte Spannung *f*, Ersatzspannung *f*
r. to dust zerstäubt
reducer 1. *(Bw)* Reduzierstück *n*, Übergangsstück *n*; Verengung *f*; 2. Verdünnungsmittel *n*; 3. Abschwächer *m (Foto)*
reducibility Reduzierbarkeit *f*
reducible reduzierbar
reducing Reduzieren *n*, Vermindern *n*
r. adapter *(El)* Zwischenstecker *m*, Zwischenstück *n*
r. agent *(Ch)* Reduktionsmittel *n*, reduzierendes Mittel *n*
r. bath *(Led)* Reduktionsbad *n (Zweibadverfahren)*
r. coupling Reduktionskupplung *f*
r. die Reduziergesenk *n*, Reduziermatrize *f*
r. electrode Reduktionselektrode *f*
r. fitting Reduzierstück *n*
r. flame *(Met)* reduzierende Flamme *f*, Reduktionsflamme *f*
r. gear Reduziervorgelege *n*
r. gears Untersetzungsgetriebe *n*
r. mill Reduzierwalzwerk *n*
r. pass Reduzierung *f (Walzprozeß)*
r. pipe Reduzierrohr *n*, Übergangsrohr *n*
r. press Ziehpresse *f*
r. pulley Hubverminderungsrolle *f*, Reduktionsrolle *f*
r. resistor Abschwächungswiderstand *m*
r. roasting reduzierendes Rösten *n*
r. roll stand Reduziergerüst *n (Walzstraße)*
r. slag Reduktionsschlacke *f*
r. sleeve Reduziermuffe *f*, Reduktionsmuffe *f*, Übergangsmuffe *f*
r. smelting process reduzierender Schachtofenprozeß *m*
r. socket Reduziermuffe *f*, Übergangsmuffe *f*
r. tacheometer (tachymeter) *(Opt)* Reduktionstachymeter *n*
r. telemeter *(Opt)* Reduktionsentfernungsmesser *m*, Basisreduktionstachymeter *n*
r. transformer Abwärtstransformator *m*, Abwärtswandler *m*, Reduktionstransformator *m*, Reduziertransformator *m*, spannungserniedrigender Transformator *m*
r. turbine Zwischenanzapfturbine *f*, Reduktionsturbine *f*
r. valve Reduzierventil *n*, Druckminderer *m*
reduction Reduktion *f*; Verkleinerung *f (Maßstab)*; Verengung *f (Rohr)*; Dickenabnahme *f (Walzen)*; Übersetzung *f* ins Langsame
r. discharge *(Text)* Reduktionsätze *f*
r. discharge agent Reduktionsätzmittel *n*
r. gear Reduktionsgetriebe *n*, Getriebe *n* mit Übersetzung ins Langsame
r. gear ratio Übersetzungsverhältnis *n* bei Übersetzung ins Langsame
r. of admission Füllungsverkleinerung *f*
r. of area Einschnürung *f (Zugversuch)*; Querschnittverringerung *f*
r. of cross-section due to contraction Querschnittverengung *f* durch Kontraktion
r. of detection efficiency Modulationsverflachung *f (bei Einseitenbandempfang)*
r. of harmonic content *(Ak)* Schwächung *f* der Obertöne
r. of harmonic content in sound analysis Abbau *m* von Schallvorgängen
r. of roll passes Kaliberanstellung *f*, Kaliberdruck *m*; Kaliberabnahme *f*, Stichabnahme *f (Walzwerkstechnik)*
r. of sectional area Querschnittverminderung *f*
r. of speed Übersetzung *f* ins Langsame
r. point Reduktionspunkt *m*, Bezugspunkt *m*
r. ratio Verkleinerungsverhältnis *n*; Übersetzungsverhältnis *n* bei Übersetzung ins Langsame
r. scale Reduktionsmaßstab *m*, Verjüngungsmaßstab *m*
r. spur pinion Antriebsstirnrad *n*

r. zone Reduktionszone *f (Hochofen)*
reductive *(Ch)* Reduktionsmittel *n*
reductor Abschwächer *m*; Druckminderer *m*
redundancy *(Dat)* Redundanz *f*, Überflüssigkeit *f*, mehr als notwendiger Aufwand *m*; Überzähligkeit *f*
redundant überzählig *(Variable, Gleichungen, Statik)*; *(Dat)* redundant, mehr als notwendig
r. bar überzähliger Stab *m (Statik)*
r. [digit] check *(Dat)* Redundanzprüfung *f (mit Hilfe von Prüfziffern)*, redundante Prüfung *f*
r. member *s.* r. bar
r. terms überflüssige Glieder *npl (in Gleichungen)*
redwood Rotholz *n*
Redwood second Redwood-Sekunde *f*
redye / to auffärben, umfärben
redyeing process Umfärbeverfahren *n*
reed 1. Ried *n*, Schilf *n*; Schilfstengel *m*; 2. Zunge *f (Orgel und Blasinstrumente)*; 3. *(Text)* Schärblatt *n*, Riet *n*, Blatt *n*, Webblatt *n*
r. bank Rohrgebüsch *n*
r. binding Blattbinden *n*
r. blade Blattstab *m*
r. contact luftdicht verschlossener Federkontakt *m (für Relais)*
r. counting *(Text)* Blattnumerierung *f*
r.-flute Rohrpfeife *f*
r. frequency meter Zungenfrequenzmesser *m*
r. hook Blattstecher *m*, Einziehmesser *n*
r. horn Nebelhorn *n*
r. instrument Zungeninstrument *n*
r. loudspeaker Zungenlautsprecher *m*
r. mark Blattstreifen *m*, Blattmarkierung *f*
r. pipe Zungenpfeife *f*; Rohrflöte *f*
r. relay Schutzgasrelais *n*, Reedrelais *n*, Trockenzungenrelais *n*, Zungenrelais *n*
r. space Blattbreite *f*
r. width Blattbreite *f*
r. wire Webeblattzahn *m*, Rietzahn *m*
Reed roller bit Zylinderrollenmeißel *m*, Vierrollenmeißel *m*
Reed taper Reed-Kegel *m (Drehmaschinenspindel)*
reeded edge gerippter Rand *m (einer Münze)*
reediness 1. *(Text)* Rietstreifigkeit *f*; Blattmarkierung *f*; 2. *(Bw)* Rundstabverzierung *f*
reeding *(Text)* Blattstechen *n*
reedit *s.* re-edit
reedy *(Text)* kettstreifig
r. fabric kettstreifige Ware *f*
reef / to reffen
reef 1. Erzgang *m (australisch)*; Goldgang *m*; 2. Riff *n*
reefer *(Am)* Kühlschiff *n*
r. ship *(Am)* Kühlschiff *n*
reefing a sail Reffen *n* eines Segels
reel / to spulen, [auf]wickeln; abhaspeln *(Seide)*
r. in *(El)* aufhaspeln
r. off abrollen, abwickeln, abziehen, abspulen
r. out 1. *(El)* abspulen, herausspulen; 2. auskurbeln *(Antenne)*
r. up aufhaspeln, aufspulen, auftrommeln
reel 1. Aufspuler *m*; Bobine *f*; Drahthaspel *f*, Rolle *f*, Scheibe *f*, Spule *f*, Trommel *f*; 2. *(Text)* Haspel, Weife *f*, Winde *f*; 3. 1000 Fuß Film *(Standardlänge)*
r. antenna *(Flg)* Schleppantenne *f*
r. cable trailer *(Am)* Anhänger *m* für Kabeltransport
r.-drum hoist Bobinenfördermaschine *f*
r.-end signal Aktend-Alarm *m (Kino)*
r.-fed gravure rotary [printing press] Rollenrotationstiefdruckmaschine *f*
r. of wire Kabeltrommel *f*
r. paster *(Typ)* Rollenanklebevorrichtung *f*
reeler *(Text)* Abspulmaschine *f*
reeling Abhaspeln *n*, Abspulen *n*
r. device Haspelvorrichtung *f*
r. end Aufwickelseite *f* der Papiermaschine
r. frame Abspulmaschine *f*
r. machine Abhaspelmaschine *f*
r. mandrel rod Glättdornstange *f*
r. mill Glättwalzwerk *n*
r. plant Haspelanlage *f*
r. water Haspelwasser *n (Seide)*
reentrant angle einspringender Winkel *m*
r. oscillator Serienoszillator *m* mit Hohlraumresonatoren
r. winding *(El)* wiedereintretende (mehrgängige) Wicklung *f*
reentry velocity *(Astr)* Wiedereintrittsgeschwindigkeit *f*
reevaporate / to nachverdampfen
reevaporation Nachverdampfung *f*
ref *s.* reference
reface / to 1. erneuern *(Fassade usw.)*; 2. nachschleifen *(Ventilsitz)*
refacing 1. Fassadenerneuerung *f*; 2. *(Masch)* Nachschleifen *n*, Ausdrehen *n*
reference Verweis *m*, Quellennachweis *m*
reference Bezugs-; Vergleichs-; Nachschlage-
r. axis Bezugsachse *f*
r. black level *(Fs)* Schwarzwertbezugspegel *m*, Vergleichsschwarzpegel *m*
r. book Nachschlagewerk *n*
r. capacitor Normalkondensator *m*, Vergleichskondensator *m*
r. circuit Bezugs[strom]kreis *m*
r. data Vergleichswerte *mpl*, Richtwerte *mpl*
r. distortion *(Fmt)* Bezugsverzerrung *f*
r. edge Bezugskante *f*
r. electrode Bezugselektrode *f*, Vergleichselektrode *f*; indifferente Elektrode *f*
r. element Bezugsnormal *n*, Sollwertgeber *m*, Sollwertquelle *f*

r. equivalent Bezugsdämpfung *f* eines Übertragungssystems, Nutzdämpfung *f*
r. face Bezugsfläche *f*
r. frequency Bezugsfrequenz *f*, Vergleichsfrequenz *f*
r. fuel Bezugskraftstoff *m*, Eichkraftstoff *m*
r. gauge Normallehre *f*, Prüflehre *f*, Revisionslehre *f*; Einstellmaß *n*, Einstellnormale *f*, Vergleichslehre *f*, Vergleichsmaß *n*
r. grid Bezugsraster *m*
r. height Festpunkthöhe *f*
r. hole Bezugsbohrung *f*
r. impedance Bezugsscheinwiderstand *m*
r. input *(Am)* *(BMSR)* Führungsgröße *f*; Sollwert *m*; umgeformter Aufgabenwert *m*
r. instrument Vergleichsinstrument *n*, Bezugsinstrument *n*, Normalinstrument *n*
r. junction Vergleichsstelle *f*, kalte Lötstelle *f* *(Thermoelement)*
r. level Bezugspegel *m*
r. line Bezugslinie *f* *(Anreißen)*; Basislinie *f* *(Radar)*
r. magnitude Bezugsgröße *f*
r. mark Bezugspunkt *m*
r. model Bezugsstück *n* *(Nachformen)*
r. noise *s.* r. telephonic noise
r. plane Bezugsebene *f*
r. point Bezugspunkt *m*
r. position Bezugsstellung *f*
r. record Bezugsdaten *pl*
r. resistance Vergleichswiderstand *m*
r. source Bezugspräparat *n*, radioaktives Standardpräparat *n*
r. source of supply Bezugsspannungsquelle *f*
r. standard Bezugsnormale *f*
r. standard for recording sound head Bezugssprechkopf *m* *(Magnettontechnik)*
r. surface Bezugsfläche *f*
r. system Bezugssystem *n*
r. telephonic noise *(Fs)* Bezugspegel *m*, Fernsprechbezugsleistung *f*
r. telephonic power Fernsprechbezugsleistung *f*
r. temperature Bezugstemperatur *f*, Justiertemperatur *f*
r. tone Eichton *m*, Normalton *m*
r. value Bezugswert *m*
r. volume *(Fmt)* Bezugspegel *m*, Fernsprechbezugsleistung *f*; Bezugslautstärke *f*
r. white level *(Fs)* Weißbezugspegel *m*
reference Literaturverzeichnis *n*, Literaturnachweis *m*, Quellenangabe *f*
refill / to *(Kfz)* betanken, nachfüllen; nachgießen; [Ölfüllung] erneuern
refill unit Patrone *f*, Schmelzeinsatz *m* *(Sicherung)*
refilling Nachfüllen *n*, Nachfüllung *f*; Betanken *n*; Erneuern *n* der Füllung
r. deck seams *(Schiff)* Nachdichten *n* von Decksfugen
refine / to raffinieren, reinigen, verfeinern; garfrischen, läutern; treiben *(Silber)*
r. by hardening vergüten *(Stahl)*
r. tin Zinn gattern
refined raffiniert, gereinigt
r. abrading process Feinbearbeitungsverfahren *n*
r. benzol Reinbenzol *n*
r. coke Edelkoks *m*
r. copper Feinkupfer *n*, Garkupfer *n*, Raffinatkupfer *n*
r. gold Brandgold *n*
r. iron froth *(Met)* Garschaum *m*
r. lead Raffinatblei *n*
r. pulp Edelzellstoff *m* *(Papierherstellung)*
r. state *(Met)* Gare *f*
r. sugar Zuckerraffinade *f*
r. zone Übergangszone *f* *(Schweißnaht)*
refinement Raffinieren *n*, Raffination *f*, Reinigung *f*, Veredelung *f*
refiner 1. Reibmaschine *f*, Mühle *f*; 2. Stoffausschläger *m* *(Papierherstellung)*
r. stock Raffinierstoff *m*
r. stock-chest Raffinierstoffbütte *f*
refinery Raffinerie *f*, Raffinationsanlage *f*; Erdölraffinerie *f*; Zuckerraffinerie *f*
r. coke Petrolkoks *m*
r. hearth *(Met)* Frischherd *m*, Frischfeuer *n*
r. operation *(Met)* Herdfrischarbeit *f*
r. process *(Met)* Herdfrischprozeß *m*
r. slag *(Met)* Raffinationsschlacke *f*
refining 1. Raffination *f*, Raffinierung *f*, Raffinieren *n*; 2. Veredlung *f* *(Zellstoff)*; 3. Verfeinern *n* *(Korn)*
r. assay Garprobe *f* *(Kupfergewinnung)*
r. cinder *(Met)* Frischschlacke *f*
r. furnace Raffinierofen *m*; Treibofen *m*
r. of the hackled flax Veredeln *n* des gehechelten Flachses
r. of waste *(Met)* Krätzfrischen *n*
r. period *(Met)* Feinungsperiode *f*
r. plant Raffinerie *f*, Raffinationsanlage *f*
r. procedure *(Met)* Veredelungsverfahren *n*
r. process Herdfrischen *n*; Treibverfahren *n*
r. slag *(Met)* Garschlacke *f*, Fertigschlacke *f*
refitting Wiedereinpassen *n*, Wiedereinbauen *n*
reflect / to reflektieren, zurückwerfen, zurückstrahlen, spiegeln
reflectance 1. *(Ph)* Reflexionskoeffizient *m*; 2. *(Opt)* Reflexionsvermögen *n*, Remission *f*
reflected gespiegelt, reflektiert
r. capacity *(El)* Rückwirkungskapazität *f*
r. code *(Dat)* reflektierter Kode *m*
r. control circuit resistance transformierter Widerstand *m* eines Steuerkreises
r. image Spiegelbild *n*

r. light Auflicht *n*; reflektiertes Licht *n*
r. pulse Echoimpuls *m*
r. radiation reflektierte Strahlung *f*
r. ray reflektierter Strahl *m*
r. reactor reflektierter Reaktor *m*, Reaktor mit Reflektor
r. resistance Rückwirkungswiderstand *m*, übertragener Widerstand *m*
r. wave Echowelle *f*, reflektierte (zurückgeworfene, rücklaufende) Welle *f*
reflecting zurückstrahlend, zurückwerfend
r. capacity Reflexionsvermögen *n*, Rückstrahlvermögen *n*
r. cross section Rückstrahlquerschnitt *m*
r. disk Reflektorscheibe *f*
r. exophthalmometer *(Opt)* Spiegelexophthalmometer *n*
r. factor Reflexionsgrad *m*
r. galvanometer Spiegelgalvanometer *n*
r. glass Reflexionsglas *n*
r. layer Reflexionsschicht *f*, reflektierende Schicht *f*
r. material Reflektormaterial *n*, Reflexionsmaterial *n*
r. measuring set Rückflußdämpfungsmesser *m*
r. medium Reflexionsmittel *n*, reflektierendes Medium *n*
r. microscope Spiegelmikroskop *n*
r. object Rückstrahler *m (Radar)*
r. plane Reflektorebene *f (Antenne)*
r. power Reflexionsvermögen *n*, Rückstrahlvermögen *n*
r. state Rückstrahlzustand *m*
r. surface Spiegelungsfläche *f*; Rückstrahlungsfläche *f*
r. telescope Spiegelteleskop *n*
reflection Reflexion *f*, Spiegelung *f*
r. angle Reflexionswinkel *m*
r. coefficient Reflexionsfaktor *m (Hohlleiter)*; Reflexionskoeffizient *m (Neutronen)*
r. effect Reflexionseffekt *m*
r. factor Reflexionsgrad *m (Licht)* (*s. a.* r. coefficient)
r. grating Reflexionsgitter *n*
r. loss *(Opt)* Reflexionsverlust *m*; *(El)* Rückflußdämpfung *f*; Stoßdämpfung *f*; Echoverlust *m*
r. measuring set *(El)* Rückflußdämpfungsmesser *m*
r. microscope Reflexionsmikroskop *n*
r. microscopy Reflexionsmikroskopie *f*
r. mode filter Reflexionswellentypfilter *n (Wellenleitertechnik)*
r. sounding Echolotung *f*
r. spectrum Reflexionsspektrum *n*
reflective reflektierend, spiegelnd; *(selten)* reflektiert, gespiegelt
r. power Reflexionsvermögen *n*
r. surface spiegelnde Oberfläche *f*
reflectivity Reflexionsvermögen *n*, Reflexionsfähigkeit *f*
reflector Reflektor *m (Antenne)*; Katzenauge *n*, Rückstrahler *m*
r. aerial Spiegelantenne *f*
r. electrode Reflektorelektrode *f*
r. lamp verspiegelte Lampe *f*
r. telescope Spiegelteleskop *n*
r. voltage Reflektorspannung *f*
reflectoscope Episkop *n*; Reflektoskop *n*
reflex Reflex *m*; Spiegelbild *n*
reflex reflektiert, gespiegelt
r. amplification Reflexverstärkung *f (Radio)*
r. angle überstumpfer Winkel *m*
r. camera Spiegelreflexkamera *f*
r. circuit Reflexschaltung *f (Radio)*
r. klystron Reflex[ions]klystron *n*
r. receiver Reflexempfänger *m*
r. reception Reflexempfang *m*
r. sight *(Mil)* Reflexvisier *n*
reflexible reflektierbar
reflexion *s.* reflection
refloat / to wieder flottmachen, abarbeiten, abbringen *(ein gestrandetes Schiff)*
refluent zurückfließend
reflux 1. Rückfluß *m*, Rücklauf *m*, Rückstrom *m*; 2. Rückstau *m (Turbine)*
r. condenser Rückflußkühler *m*, Dephlegmator *m*
r. ratio Rücklaufverhältnis *n*
reforest / to aufforsten
reforestation Aufforstung *f*
reforge / to umschmieden
reforming Reformieren *n*; Reforming-Verfahren *n*
refound / to umgießen, umschmelzen
refract / to brechen *(Licht, Schall)*
refracted wave gebrochene Welle *f*
refracting [strahlen]brechend
r. angle brechender Winkel *m*, Brechungswinkel *m*
r. edge brechende Kante *f*
r. surface brechende Fläche *f*
refraction Brechung *f*, Refraktion *f*, Strahlenbrechung *f*
r. angle Brechungswinkel *m*
r. balance test *(Opt)* Prüfung *f* auf Refraktionsgleichheit
r. index Brechungszahl *f*, Brechungsindex *m*
r. law Brechungsgesetz *n*
r. meter Brechungsmesser *m*
r. of light Lichtbrechung *f*
r. of lines of forces Brechung *f* von Kraftlinien
r. of sound Brechung *f* des Schalls

r. spectrum Brechungsspektrum *n*
refractive brechend, lichtstrahlenbrechend
r. angle Brechungswinkel *m*
r. index Brechungszahl *f*, Brechungsexponent *m*, Brechungsindex *m*, Brechungskoeffizient *m*
r. power Brechkraft *f*
refractometer Refraktometer *n* *(optisches Analysenmeßgerät)*
refractometry Refraktometrie *f*
refractor Linsenfernrohr *n*, Refraktor *m*
refractories feuerfeste Steine *mpl*, Schamottesteine *mpl*
refractoriness Feuerfestigkeit *f*, Feuerbeständigkeit *f*
refractory feuerfester Baustoff *m*
refractory feuerfest; wärmebeständig; schwerschmelzbar; widerspenstig *(Erze)*
r. alloy schwerschmelzbare Legierung *f*
r. arch feuerfestes Gewölbe *n*
r. brick feuerfester Mauerstein *m*; Schamottestein *m*, Schamotteziegel *m*
r. clay feuerfester Ton *m*, Feuerton *m*, Schamotte[erde] *f*
r. concrete feuerfester (hitzebeständiger) Beton *m*
r. lined feuerfest ausgemauert (ausgekleidet)
r. lining feuerfeste Ausmauerung *f*
refrangibility Brechungsvermögen *n*, Brechkraft *f*
refrigerant Kältemittel *n*, Kälteträger *m*, Kühlmittel *n*
refrigerate / to kühlen
refrigerated gekühlt, Kühl-
r. cargo *(Schiff)* Kühlladung *f*
r. cargo-motorship Motorkühlschiff *n*
r. cargo space *(Schiff)* Ladekühlraum *m*
r. cargo vessel Kühlschiff *n*
r. carrier Kühlschiff *n*
r. chamber Kältekammer *f*, Kühlraum *m*
refrigerating Kühl-, Kälte-
r. agent Kältemittel *n*, Kälteträger *m*
r. brine Kälteflüssigkeit *f*
r. capacity Kälteleistung *f*
r. coil Kühlschlange *f*
r. cylinder Kühlzylinder *m*
r. effect Kühlwirkung *f*
r. hold *(Schiff)* Kühlraum *m*
r. machine Kältemaschine *f*, Kühlanlage *f*
r. plant Kälteanlage *f*, Kühlanlage *f*
refrigeration Kältetechnik *f*, Kälteerzeugung *f*; Kühlung *f*
r. compressor Kälteverdichter *m*
r. engineering Kältetechnik *f*
r. plant Kühlanlage *f*, Gefrieranlage *f*
r. technology Kältetechnik *f*
r. vessel Kühlschiff *n*
refrigerator Kälteerzeugungsmaschine *f*, Kälteanlage *f*; Kühlschrank *m*
r. capacity Kühlvermögen *n* der Kühlanlage; Aufnahmevermögen *n* der Kühlanlage
r. train Kühlzug *m*
refringent lichtbrechend
reftone *(Ak)* Bezugston *m*, Normalton *m*, Eichton *m*
refuel / to nachbunkern *(Treibstoff)*, tanken
refuelling Tanken *n*, Treibstoffaufnahme *f*; Nachbunkern *n*
r. and maintenance duty squad technische Bereitschaft *f* *(auf dem Flugplatz)*
r. point Zapfstelle *f*; Tankstelle *f*
refuge Fußgängerschutzinsel *f*, Verkehrsinsel *f*
refuse 1. Müll *m*, Abfall[stoff] *m*, Abschaum *m*, Ausschuß *m*, Rückstand *m*, Kehricht *m*; Altpapier *n*; 2. *(Brg)* Berge *pl*, Abgänge *mpl*
r. chute Müllabwurfschacht *m*, Müllschlucker *m*
r.-collecting vehicle Müllabfuhrwagen *m*
r.-collector vehicle Müllabfuhrwagen *m*
r. destructor furnace Müllfeuerung *f*
r. disposal Abfallbeseitigung *f*, Müllbeseitigung *f*; Abfallverwertung *f*, Müllverwertung *f*
r. disposal drum Mülltonne *f*
r. duct Müll[schlucker]schacht *m*
r. extractor *(Brg)* Bergeaustrag *m*
r. lighter *(Schiff)* Abfallschute *f*, Abfallprahm *m*
r. matter Abfallstoffe *mpl*
r. tip *(Brg)* Bergehalde *f*
refusion *(Met)* Schlackeumschmelzen *n*
regain Zunahme *f*; Wiedergewinnung *f*
regenerate / to regenerieren; auffrischen, wiederaufbereiten, wiedergewinnen
regenerated cellulose Regeneratzellulose *f*, regenerierte Zellulose *f*
r. cellulose fibre Zelluloseregeneratfaser *f*
r. dense-medium Dicktrübe *f* *(Aufbereitung)*
regenerating receiver *(Fmt)* Rückkopplungsempfänger *m*
regeneration 1. Regenerieren *n*, Regeneration *f*; 2. Rückkopplung *f*
r. boiler Regenerierungskessel *m*
r. felling Verjüngungshieb *m* *(Forst)*
regenerative 1. regenerierend, wiedererzeugend; 2. selbstansaugend *(Pumpe)*; 3. *(El)* Rückkopplungs-, rückkoppelnd
r. amplification Rückkopplungsverstärkung *f*
r. amplifier Rückkopplungsverstärker *m*
r. braking elektrische Nutzbremsung *f* *(elektrischer Zug)*
r. chamber Regenerativkammer *f*
r. circuit Rückkopplungskreis *m*
r. control Nutzbremsschaltung *f* *(elektrischer Zug)*
r. detection Rückkopplungsgleichrichtung *f*
r. detector Rückkopplungsaudion *n*, rückgekoppelter Detektor *m*

r. feedback Mitkopplung *f*, Rückkopplung *f*
r. furnace Regenerativfeuerung *f*, Regenerativofen *m*
r. gas firing Regenerativgasfeuerung *f*
r. receiver Rückkopplungsempfänger *m*
r. reception Rückkopplungsempfang *m*
r. temperature booster Brennkammer *f* mit Regenerativvorwärmer
r. valve detector Rückkopplungsaudion *n*
regenerator 1. Regenerator *m*, Wärmespeicher *m*, Wärmeakkumulator *m*; 2. *(Met)* Winderhitzer *m*, Regenerator *m*; 3. *(El)* rückgekoppelter Verstärker *m*
r.-combustor Brennkammer *f* mit Regenerativvorwärmer
regeneratory cutting Verjüngungshieb *m* *(Forst)*
regime of cavitation Kavitationsart *f*
region 1. Bereich *m*, Bezirk *m*; Zone *f*, Fläche *f*; 2. *(Brg)* Revier *n*
r. of deformation Formänderungsbereich *m*
r. of initial current flow *(El)* Anlaufgebiet *n*
r. of non-operation Ruhebereich *m* *(Relais)*
r. of phase Phasenbereich *m*
r. of stability Stabilitätsbereich *m*
regional broadcasting station *(Fmt)* Bezirkssender *m*, Leitsender *m*
register / to aufzeichnen, eintragen, aufnehmen; zählen
r. zero auf Null zeigen
register 1. Register *n*, Verzeichnis *n*; Tabelle *f*; 2. Zählwerk *n*; Register *(meist Maschinenwortspeicher)*, Registrierwerk *n*, Addierwerk *n*; 3. Speicherzelle *f*; 4. Register *(Orgel)*
r. book Schiffsregister *n*
r. frame Anlegerahmen *m* *(Fotogrammetrie)*
r. key Zähltaste *f*
r. mark *(Typ)* Paßkreuz *n*
r. mark aligning tube Richtrohr *n* für Paßmarken
r. mark templet Paßkreuzscheibe *f*
r. of a meter Zählwerk *n* eines Zählers
r. of organ Orgelregister *n*, Registerstimme *f*, Registerzug *m*
r. of shipping Schiffsregister *n*
r. pilot lamp *(Am)* *(Fmt)* Zählüberwachungslampe *f*
r. sheet *(Typ)* Zurichtebogen *m*, Einrichtebogen *m*
r. slide Registerschieber *m* *(Orgel)*
r. ton Registertonne *f*
r. tonnage Registertonnengehalt *m*
r. translator *(El)* Nummernumsetzer *m*
r. wheel Rosettenrädchen *n*
registered breadth Vermessungsbreite *f* *(z. B. eines Schiffes)*
r. depth Raumtiefe *f*, Vermessungstiefe *f* *(z. B. eines Schiffes)*
r. design [eingetragenes] Gebrauchsmuster *n*
r. tonnage Nettoregistertonnengehalt *m*
registering 1. Zählung *f*; 2. Meßwertaufzeichnung *f*
r. galvanometer Registriergalvanometer *n*
r. instrument zählendes Meßinstrument *n*
r. meter Schreiber *m*, schreibendes (zählendes) Meßgerät *n*
r. screw Paßschraube *f*
r. surface Bezugsfläche *f*
registration 1. Registrierung *f*, Eintragung *f*; 2. genaue Einstellung *f*; 3. Registrieren *n* *(Orgel)*
r. arm *(El)* Seitenhalter *m* *(Fahrdraht)*
r. book Fahrzeugpapiere *npl*
r. number *(Kfz)* Kennzeichennummer *f*, Zulassungsnummer *f*
r. of designs Gebrauchsmusterschutz *m*
r. of the yield Schlagaufnahme *f* *(Holz)*
r. of trade marks Gebrauchsmusterschutz *m*
reglet *(Typ)* Reglette *f* *(Durchschuß)*
regression Regression *f*
r. coefficient Regressionskoeffizient *m*
r. equation Regressionsgleichung *f*
r. line Regressionsgerade *f*
regrind / to 1. *(Masch)* nachschärfen; nachschleifen *(z. B. Ventile)*; 2. wieder mahlen, wieder aufarbeiten
regrind *(Masch)* Nachschliff *m*
r. life Standzeit *f*
regrindable *(Masch)* nachschleifbar
regrinding *(Masch)* Nachschleifen *n*, Nachschliff *m*
reground material *(Plast)* wieder aufgearbeitete (vermahlene) Abfälle *mpl*
regular regulär, regelmäßig; normal; *(Opt)* isometrisch
r. component reguläre Komponente *f*, regulärer Teil *m*
r. countersink *(Masch)* Senker *m* für Körnerloch ohne Schutzsenkung
r. coursed ashlar regelmäßiges Schichtenmauerwerk *n*; Quadermauerwerk *n*
r. coursed rubble regelmäßiges Schichtenmauerwerk *n*
r. equipment Normalausstattung *f*
r. field theory *(Ph)* reguläre Feldtheorie *f*
r. grade of gasoline Normalbenzin *n* *(im Gegensatz zum Superbenzin)*
r. lay Kreuzschlag *m* *(Drahtseil)*
r. octagon regelmäßiges Achteck *n*
r. planer Zweiständerhobelmaschine *f*
r. polygon regelmäßiges Vieleck *n*
r. precession *(Ph)* regelmäßige Präzession *f*
r. reflection regelmäßige Reflexion *f*, Spiegelung *f*; gerichtete Reflexion
r. rubble in courses regelmäßiges Schichtenmauerwerk *n*

r. spacing of teeth gleichmäßige Zahnteilung *f (Reibahle)*
r. trader Tourendampfer *m*
regularity Regelmäßigkeit *f*, Stetigkeit *f*, Gleichmäßigkeit *f*
r. of the lap *(Text)* Gleichmäßigkeit *f* des Wickels
regulate / to [nach]stellen, regeln, regulieren; stabilisieren
regulated condition Stellgröße *f*
r. quantity *(BMSR)* Stellgröße *f*
r. voltage source stabilisierte Stromquelle *f*
r. voltage supply unit Spannungskonstanthalter *m*
regulating regelnd, Steuer[ungs]-, Regulier-
r. amplifier Regelverstärker *m*
r. cell Gegenzelle *f*; Regelzelle *f*, Schaltzelle *f*
r. contact Reglerkontakt *m*
r. flap valve Regulierdrehklappe *f*
r. inductor Regeldrossel *f*
r. lever Reglerarm *m*, Reglerhebel *m*
r. power of the governor Stellkraft *f* des Reglers
r. range Regelbereich *m*
r. relay Regelrelais *n*
r. resistance Regelwiderstand *m*, Regulierwiderstand *m*
r. resistor Regelwiderstand *m*
r. rod Regelstab *m* für selbsttätige Regelung *(Kernreaktor)*
r. screw Stellschraube *f*
r. starter Regelanlasser *m*
r. switch Regelschalter *m*
r. system Regelkreis *m*
r. transformer Regeltransformator *m*, Stufentransformator *m*, Reguliertransformator *m*
r. unit Stelleinrichtung *f*, Stellglied *n*, Stellmotor *m*
r. valve Regelröhre *f*; Regelventil *n*
r. wheel Regelscheibe *f (spitzenloses Schleifen)*
r. wheel spindle Regelspindel *f*
r. wheel spindle bearing Regelspindellagerung *f*
r. wheel tru[e]ing attachment Regelscheibenabrichteinrichtung *f*
r. winding Regulierwicklung *f*
regulation 1. Regelung *f*, Steuerung *f*; Stabilisierung *f*; 2. Vorschrift *f*, Verordnung *f*; 3. Regulierung *f (Orgel)*
r. by throttling Regelung *f* durch Dampfdrosselung
r. in steps stufenweise Regelung *f*
r. lights *(Schiff)* Positionslichter *npl*, Positionslaternen *fpl*
r. of output *(Fmt)* Ausgangssteuerung *f*
r. of wires *(Fmt)* Durchhangsregelung *f*
regulator 1. *(El)* Regler *m*, Regelschalter *m*; Konstanthalter *m*; 2. chemische Substanz *f*, die den Reaktionsablauf beeinflußt
r. and cut-out relay Reglerschalter *m*
r. characteristic Reglerkennlinie *f*
r. cut-out Reglerschalter *m*
regulex Regulex *(Verstärkermaschine)*
reguline *(El)* regulinisch
r. deposit gut galvanisierte Schicht *f*
regulus 1. *(Math)* Erzeugendenschar *f*, Regelschar *f (einer Regelfläche)*; 2. Metallkönig *m*, Metallkorn *n*, Regulus *m*
r. furnace Regulusofen *m*
r. of antimony Antimonregulus *m*, handelsübliches Antimonmetall *n*
r. of gold Goldkönig *m*
reheat / to anlassen, aufwärmen, nachhitzen, tempern, wiedererwärmen
reheat Zwischenüberhitzung *f (Dampfanlage)*; Nachverbrennung *f (Turbostrahltriebwerk)*; Wiedererwärmen *n*, Zwischenüberhitzen *n*
r. engine Dampfmaschine *f* mit Wärmeaustausch
reheater steam engine Dampfmaschine *f* mit Wärmeaustausch
reheating furnace Nachwärmofen *m*
r. furnace slag Wärmofenschlacke *f*, Schweißofenschlacke *f*
reignition voltage *(El)* Wiederzündspannung *f*
reimbursement of the outlay Deckung (Rückerstattung) *f* der Auslagen
reinforce / to verstärken, versteifen; abstützen; armieren, bewehren; kräftigen
reinforced cement-mortar construction Armozementbauweise *f*, Ferrozementbauweise *f*
r. concrete Stahlbeton *m*
r. concrete beam Stahlbetonträger *m*
r. concrete construction Stahlbetonbau *m*
r. concrete engineering Stahlbetonbau *m*
r. concrete floor Stahlbetondecke *f*
r. concrete floor slab Stahlbetondeckenplatte *f*
r. concrete frame Stahlbetonrahmen *m*; Stahlbetonskelett *n*
r. concrete vessel Betonschiff *n*, Stahlbetonschiff *n*
r. fillet weld Wölbkehlnaht *f (Schweißen)*
r. hosiery Strümpfe *mpl* mit verstärkten Fersen und Fußspitzen
r. paper textilverstärktes Papier *n*
r. plastics verstärkte Plaste *mpl*, Schichtstoff *m*
r. seam Wulstnaht *f*
r. union paper textilverstärktes Teer- und Bitumenpapier *n*
r. weld überwölbte Kehlnaht *f (Schweißen)*
reinforcement Bewehrung *f*, Armierung *f (Stahlbetonbau)*; Verstärkung *f*
r. bar Betonstahl *m*, Bewehrungsstahl *m*
r. cage Bewehrungskorb *m (Stahlbetonbau)*

r. mat Bewehrungsmatte *f (Stahlbetonbau)*
r. of pole butt *(El)* Mastschuh *m*, Mastfußverstärkung *f*
r. of the blades Versteifung *f* der Schaufel *(Turbine)*
r. of the weld Überhöhung *f* der Schweißnaht
reinforcing bar Betonstahl *m*, Bewehrungsstahl *m*
r. bead Verstärkungs[schweiß]raupe *f*
r. cage Bewehrungskorb *m (Stahlbetonbau)*
r. crease *(Masch)* Sicke *f*
r. fin *(Masch)* Sicke *f*
r. material *(Plast)* Füllstoff *m*, Verstärkungsmaterial *n*
r. mesh Betonstahlgewebe *n*, Bewehrungsmatte *f*
r. plate Verstärkungsblech *n*
r. rib Verstärkungsrippe *f*
r. ring Verstärkungsring *m*
r. rod Betonstahl *m*, Bewehrungsstab *m*
r. seam *(Masch)* Sicke *f*
r. steel Betonstahl *m*, Bewehrungsstahl *m*
reintroduce / to *(Math)* wiedereinführen, wiedereinsetzen
reirradiation *(Kern)* wiederholte Bestrahlung *f*
reject / to zurückweisen
r. a frequency *(El)* eine Frequenz nicht durchlassen; eine Frequenz sperren
reject Ausschußteil *n*
r. backs ausgemusterte Rückseiten *fpl (Furnier)*
r. rate Ausschußquote *f*
rejected material Siebrückstand *m*
r. part Ausschußteil *n*
rejection Beanstandung *f*, Zurückweisung *f*
r. criterion Rückweisekriterium *n*
r. limit Ausschußgrenze *f*
r. of the accompanying sound *(Fs)* Tonunterdrückung *f*
rejector Ausstoßer *m*, Rückwerfer *m*, Sperre *f*; Stromresonanzkreis *m*
r. circuit Drosselkreis *m*, Sperrkreis *m*
r. impedance Impedanz *f* bei Parallelresonanz
rejects 1. Ausschuß *m*; Abfall *m*; 2. *(Brg)* Abgänge *mpl*
rejuvenate / to [sich] verjüngen
r. a producer eine Fördersonde wieder beleben
relation[ship] Beziehung *f*, Verhältnis *n*
relative relativ, verhältnismäßig, Relativ-
r. absorption relative Absorption *f*
r. abundance relative Häufigkeit *f (Isotope)*
r. acceleration Relativbeschleunigung *f*
r. activity *(Kern)* relative Aktivität *f*
r. address *(Dat)* relative Adresse *f*
r. aperture *(Opt)* Öffnungsverhältnis *n*; relative Apertur *f*
r. bearing Seitenpeilung *f*; Peilwinkel *m (Radar)*
r. biological effectiveness relative biologische Wirksamkeit *f (der Strahlung)*
r. coding *(Dat)* relative Befehlsfolge *f*
r. delay relative Verzögerung *f*
r. density Dichtezahl *f*, Dichteverhältnis *n*, relative Dichte *f*
r. dielectric constant relative Dielektrizitätskonstante *f*
r. displacement gegenseitige Verschiebung *f*
r. efficiency relativer Wirkungsgrad *m*
r. entrance velocity relative Eintrittsgeschwindigkeit *f (Turbine)*
r. harmonic content Klirrfaktor *m*, Oberschwingungsgehalt *m*
r. humidity relative Feuchtigkeit *f (z. B. der Luft)*
r. isotopic abundance relative Isotopenhäufigkeit *f*
r. level Bezugspegel *m*, relativer Pegel *m*
r. movement gegenseitige Bewegung *f*, Relativbewegung *f*, scheinbare Bewegung
r. permittivity relative Dielektrizitätskonstante *f*
r. programming relatives Programmieren *n*
r. response Relativempfindlichkeit *f*, relative Empfindlichkeit *f*; relativer Übertragungsfaktor *m*
r. rolling motion between the work and the cutter Abwälzbewegung *f (Zahnradherstellung)*
r. rotative efficiency of the propeller *(Schiff)* Anordnungsgütegrad *m* des Propellers
r. spectral energy distribution Strahlungsfunktion *f*
r. stopping power relatives Bremsvermögen *n*
r. velocity Relativgeschwindigkeit *f*
r. voltage drop relativer Spannungsabfall *m*
r. voltage rise relativer Spannungsanstieg *m*
relatively prime *(Math)* teilerfremd
relativistic mechanics Relativitätsmechanik *f*
relativity Relativität *f*
r. theory Relativitätstheorie *f*
relaxation 1. Entspannung *f*, Erschlaffung *f*, Relaxation *f*; 2. *(El)* Kippung *f*, Kippen *n*
r. amplitude Kippamplitude *f*
r. circuit Kippkreis *m*
r. diagram Kippdiagramm *n*
r. effect Relaxationseffekt *m*
r. frequency Kippfrequenz *f*; Relaxationsfrequenz *f*
r. generator Kipp[schwingungs]generator *m*
r. oscillation Kippschwingung *f*; Relaxationsschwingung *f*
r. oscillator Kippschwingschaltung *f*, Kippschwingungsoszillator *m*
r. scanning Kippablenkung *f*

r. time Halbwertzeit *f*, Relaxationszeit *f*
r.-type oscillator Kippschwingungsgenerator *m*
r. vibration Kippschwingung *f*, Relaxationsschwingung *f*
relaxed 1. gelockert, entspannt *(Feder)*; 2. entspannt, entlastet *(Gebirgsmechanik)*
relay Relais *n*
r. amplifier Relaisverstärker *m*
r. armature Relaisanker *m*
r. box *(Fmt)* Relaisschrank *m*
r. broadcast station Relaissender *m*, Ballsender *m*
r. calculating machine *(Dat)* Relaisrechenmaschine *f*, Relaisrechner *m*
r. chain circuit Relaiskette *f*
r. chatter Relaiskontaktprellungen *fpl*
r. computer *(Dat)* Relaisrechenmaschine *f*, Relaisrechner *m*
r. contact Relaiskontakt *m*
r. control Relaissteuerung *f*
r. core Relaiskern *m*
r. frame *(Fmt)* Relaisschrank *m*
r. group Relaiswähler *m*; Relaissatz *m*
r. magnet Relaismagnet *m*
r.-operated relaisgesteuert
r.-operated controller *(Am)* Regler *m* mit Hilfsenergie *(indirekter Regler)*
r. point Relaiskontakt *m*
r. protection Relaisschutz *m*
r. reception Ballempfang *m (Rundfunk)*
r. releasing time Relaisabfallzeit *f*
r. station Neben[rundfunk]sender *m*; Relaisstation *f*
r. switch Schaltschütz *n*, Schütz *n*
r. television Fernsehübertragung *f* durch Relaisstrecke
r. transmitter Relaissender *m*
r.-type recording instrument Fallbügelinstrument *n*
r. valve Überströmventil *n*
r. with holding winding Halterelais *n*, Relais *n* mit Haltewicklung
r. with make and break contacts Relais *n* mit Wechselkontakten
release / to 1. *(El)* freigeben, auslösen, loslassen; 2. *(Masch)* entriegeln, ausspannen *(Werkstück)*
r. the clutch pedal *(Kfz)* einkuppeln
r. the line *(Fmt)* die Leitung freigeben
release 1. Freisetzung *f*, Freigabe *f*, Entbindung *f (z. B. von Elektronen)*, Abgabe *f*, Auslösung *f*; 2. Abfall *m (des Ankers)*; 3. Auslöser *m (Fotoapparat)*; 4. Entlastung *f (z. B. einer Gleiswaage)*
r. agent *(Ch)* Gleitmittel *n*, Trennmittel *n*
r. button Auslöseknopf *m*
r. catch Auslöseknagge *f*
r. clutch Rutschkupplung *f*
r. contact Auslösekontakt *m*
r. cord Reißleine *(Fallschirm)*
r. current Auslösestrom *m*; Abfallstrom *m*; Ansprechstrom *m (Relais)*
r. cycle Entlastungstakt *m (Melkmaschine)*
r. disk Auslösescheibe *f*
r. gear Ausklinkvorrichtung *f*, Auslöseeinrichtung *f*, Entlastungsgetriebe *n*, Klinkwerk *n*
r. guard signal *(Fmt)* Auslösungszeichen *n*
r. handle Auslösegriff *m*
r. key *(Dat)* Lösungstaste *f*; *(Fmt)* Auslösetaste *f*
r. lever Auslösehebel *m*
r. of abrasive grains Ausbrechen *n* des Kornes *(Schleifscheibe)*
r. of electrons Elektronenbefreiung *f*, Elektronenloslösung *f*
r. of energy Energiefreisetzung *f*, Entbindung *f* von Energie, Energieabgabe *f*
r. of heat Wärmeentbindung *f*
r. pawl Auslöseklinke *f*
r. pressure of the nozzle Düsenöffnungsdruck *m*
r. pulse Auslöseimpuls *m*, Startimpuls *m*
r. retardation Auslöseverzögerung *f*
r. sleeve Auslösemuffe *f*
r. spring Auslösefeder *f*, Auslösungsfeder *f*, Rückzugsfeder *f*
r. switch Auslöseschalter *m*, Auslösungsschalter *m*, Freigabeschalter *m*
r. time Abfallzeit *f*, Abklingzeit *f (Relais)*; Auslösezeit *f*
r. trigger Auslöser *m*
r. valve Entlastungsventil *n*, Sicherheitsventil *n*, Verminderungsventil *n*
r. wire *(Fmt)* Auslöseader *f*
releasing Auslösen *n (Vorgang)*; Abfallen *n (Relais)*; Freigabe *f*; Ausklinken *n*; Lockern *n*
r. delay Abfallverzögerung *f*, Ansprechverzögerung *f*, Auslösungsverzögerung *f*
r. device Auslösevorrichtung *f*
r. time *s.* release time
reliability Zuverlässigkeit *f*
r. in (of) service Betriebszuverlässigkeit *f*, Betriebssicherheit *f*
r. test Zuverlässigkeitsprüfung *f*
relief 1. Relief *n*, erhöhte Form *f*; 2. Beseitigung *f*; Abhebung *f*; Entlastung *f*; 3. Hinterschliff *m (Werkzeug)*; Freiwinkel *m*
r. angle Hinterschleifwinkel *m*; Freiwinkel *m*
r. block *(Typ)* Reliefklischee *n*
r. bookprinting Hochdruck *m*; Buchdruck *m*
r. cam *(Masch)* Hinterdrehkurve *f*
r. cock Entlüftungsventil *n*
r. embossing *(Typ)* Blindprägung *f*; Blinddruck *m*
r. etch litho plate hochgeätzte Offsetplatte *f*

r. face Freifläche *f (Werkzeug)*
r.-grind / to hinterschleifen
r. grinding Hinterschliff *m*
r.-grinding attachment Hinterschleifeinrichtung *f*
r. halftone *(Typ)* Reliefautotypie *f*
r.-mill / to hinterfräsen
r. motion Abhebebewegung *f*
r. of tool Meißelabhebung *f*, Werkzeugabhebung *f*
r. period Erholungspause *f (z. B. Oszillator)*
r. pilot vessel Lotsenversetzboot *n*
r.-polish / to reliefpolieren
r.-polishing Reliefpolieren *n (Metallografie)*
r. printing *(Typ)* Hochdruck *m*
r. roadway *(Brg)* Entspannungsort *n*
r. sprue Drucksteiger *m (Gießerei)*
r.-turn / to hinterdrehen
r. valve Ablaufventil *n*, Entlüftungshahn *m*, Entspannungsventil *n*, Rückflußventil *n*, Rückschlagklappe *f*; Überströmventil *n*
r. vent Explosionsdeckel *m (z. B. am Kurbelgehäuse)*
r. well Entlastungsbohrung *f (Tiefbohrtechnik)*
relieve / to 1. *(Masch)* hinterdrehen, hinterschleifen; abheben *(Meißel)*; 2. entlasten
r. the watch *(Schiff)* ablösen *(Wache)*
relieved formed cutter hinterdrehter Formfräser *m*
relieving 1. *(Schiff)* Ablösung *f (Wache)*; 2. *(Masch)* Hinterdrehen *n*, Hinterschleifen *n*; Abhebung *f (Meißel)*; 3. Entlasten *n*, Entlastung *f*
r. anode Hilfsanode *f*
r. arch Entlastungsbogen *m (Mauerwerk)*
r. attachment (device) Hinterdreheinrichtung *f*
r. groove Entlastungsrille *f*
r. lathe Hinterdrehmaschine *f*
r. operation Hinterdreharbeit *f*
r. shot *(Brg)* Hilfsschuß *m*
r. tool box Meißelklappe *f*; Meißelträger *m* mit Abhebung *(Hobelmaschine)*
r. work Hinterdreharbeit *f*
relifting table Überhebetisch *m*, Wippe *f (Walzwerk)*
relining 1. *(Met)* Neuzustellung *f*; 2. *(Brg)* Erneuerung *f* des Ausbaus
r. a shell with white metal Wiederausgießen *n* einer Schale mit Weißmetall
reload / to 1. umladen; 2. neu einlegen *(Film)*
reloading Umladen *n*
r. run *(Kern)* Vorlauf *m*
relocate / to verlagern; verrücken; verstellen, umsetzen, versetzen; umstellen; umspannen *(Werkstücke)*
relubrication interval Schmierhäufigkeit *f*
reluctance *(El)* Reluktanz *f*, magnetischer Widerstand *m*
r. generator Reaktionsgenerator *m*
r. motor Reluktanzmotor *m*
reluctivity *(El)* spezifische Reluktanz *f*, spezifischer magnetischer Widerstand *m*
remachine / to nacharbeiten
remain aloft / to in der Luft bleiben *(Flugzeug)*
r. set in Einstellung bleiben
remainder 1. *(Math)* Rest *m*; 2. *(Ch)* Rest, Rückstand *m*
remanence Remanenz *f*, remanenter Magnetismus *m*
remanent remanent, zurückbleibend
r. flux remanenter Fluß *m*, Remanenzfluß *m*
r. flux density remanente [magnetische] Induktion *f*, remanente [magnetische] Flußdichte *f*
r. magnetism Remanenz *f*, remanenter Magnetismus *m*
remedy Gegenmittel *n*; Arzneimittel *n*, Heilmittel *n*
remelt / to umschmelzen, wieder[ein]schmelzen
remelt heat *(Met)* Umschmelze *f*
remelted metal Umschmelzmetall *n*
remelting furnace *(Typ)* Umschmelzofen *m*
r. process Umschmelzverfahren *n*
remetal / to neu ausgießen *(Lager mit Lagerwerkstoff)*
remission of the fee Gebührenerlaß *m (Patentwesen)*
remnant 1. *(Brg)* Restpfeiler *m*; 2. Reststück *n (Gewebe)*
remodel / to umgestalten, umarbeiten, umändern
remoistening adhesive beim Anfeuchten haftender Klebstoff *m*
remote Außenaufnahme *f*; Außenreportage *f*
remote 1. entfernt, abseits, abgelegen, weit weg; 2. fernbedient, ferngesteuert, fernbetätigt, Fern[wirk]-
r. action equipment Fernwirkeinrichtung *f*
r. aiming Ferneinstellung *f*, Ferneinrichtung *f*
r. area monitoring system Fernüberwachungssystem *n*
r.-bulb thermostat Temperaturregler *m* mit Federthermometer
r. control Fernbedienung *f*, Fernlenkung *f*, Fernsteuerung *f*, Fernbetätigung *f*, Fernüberwachung *f*
r. control apparatus Fernsteuerapparat *m*
r. control change Fernschaltung *f*
r. control facility Fernsteuereinrichtung *f*
r. control steering gear elektrische Rudersteuerung *f (von Schiffen)*
r. control switch Fern[steuer]schalter *m*
r.-controlled ferngesteuert
r.-controlled vessel Fernlenkschiff *n*, ferngelenktes Schiff *n*

r. cut-off tube Exponentialröhre *f*
r. desk Fernbedienpult *n*
r. drilling site weit abgelegene Bohrstelle *f*
r. feeder Fernaufgabe *f*, Ferneinleger *m*
r. feeding device Fernaufgabe *f*, Ferneinleger *m*
r. handling equipment Fernbedienungsgeräte *npl* [des Reaktors]
r. ignition *(Mil)* Fernzündung *f*
r. indicating instrument fernanzeigendes Gerät *n*, Instrument *n* mit Fernanzeige
r. indication Fernanzeige *f*, Fernmessung *f*
r. loading Fernbeschickung *f (z. B. des Reaktors)*
r. manipulating equipment Fernbedienungsgerät *n* [des Reaktors]
r. manipulator Fernmanipulator *m*
r. meter Fernmeßgerät *n*
r. metering Fernmessung *f*
r.-operated switch Fernschalter *m*
r. position indicator Stellungsgeber *m*
r.-reading tank gauge Tankfüllstandsmesser *m* mit Fernanzeige
r.-reading thermometer Fernthermometer *n*
r. recorder Fernregistrierinstrument *n*
r. recording instrument Fernregistrierinstrument *n*
r. registration Fernregistrierung *f*
r. tuning device Fernabstimmgerät *n*
r. valve fernbedientes Ventil *n*, Fernsteuerungsventil *n*
r. viewing Fernbeobachtung *f*
remotely controlled ferngesteuert, ferngeregelt, fernbetätigt, fernbedient, indirekt betätigt
r. controlled instrument ferngesteuertes Gerät *n*
r. operated *s.* r. controlled
remoulded sample *(Bw)* gestörte Bodenprobe *f*
removable abnehmbar, auswechselbar, herausnehmbar
r. bottom *(Met)* Losboden *m (des Konverters)*
r. discontinuity *(Math)* behebbare Unstetigkeit *f*
r. stanchion *(Eis)* Einsteckrunge *f*
r. truck *(El)* Schaltwagen *m*
removal 1. Beseitigung *f*, Entfernung *f*; 2. *(Masch)* Abnahme *f (Span)*; Abfuhr *f (Späne)*; Ausspannen *n (Werkstück)*; 3. Abfluß *m*, Abführung *f*, Entzug *m*
r. of after-heat Restwärmeabfuhr *f*
r. of carbon *(Met)* Entkohlen *n*, Entkohlung *f*
r. of chips *(Masch)* 1. Spanabhebung *f*, Spanabnahme *f*; 2. Spanabfuhr *f*
r. of colour *(Text)* Abziehen *n* der Farbe
r. of copper Entkupfern *n*
r. of faults Entstörung *f*, Fehlerbeseitigung *f*
r. of iron Enteisenung *f*
r. of metal Zerspanung *f*, Zerspanen *n*
r. of nitrogen *(Met)* Entstickung *f*
r. of overburden *(Brg)* Abraumbewegung *f*
r. of slag Entschlackung *f*, Abschlackung *f*
r. of stains Fleckenentfernung *f*, Detachieren *n*
r. of stress *(Met)* Entspannen *n*, Entspannung *f*
r. rate abgenommene Spanmenge *f*, Zerspanungsleistung *f*, gespantes Volumen *n*
remove / to 1. entfernen, beseitigen; 2. ausbauen, demontieren; 3. abheben *(Span)*; abführen *(Späne)*; ausspannen *(Werkstück)*; 4. abführen, ableiten
r. a short-circuit einen Kurzschluß beseitigen
r. by suction absaugen
r. cinder *(Met)* entschlacken, Schlacke abziehen
r. fins entgraten
r. lead entbleien
r. the burr entgraten
r. the gloss *(Met)* dekatieren, Preßglanz entfernen
r. the lustre *(Text)* abglänzen
r. the receiver den Hörer abnehmen
r. the rost entrosten
r. the slag entschlacken
r. the sludge entschlammen
r. water entwässern
removing of kinks Recken *n* des Drahtes
r. of shavings by trough conveyor Spänebeförderung *f* durch Kette in offener Rinne
r. of the drill Austreiben *n* des Bohrers
render / to machen; reproduzieren; ausschießen; putzen *(Mauer)*; ausschmelzen
r. visible on a screen auf einem Schirm sichtbar machen, vorführen
rendered gespalten *(Holz)*
renderer *(Led)* Abdeckerhaut *f*
rendering 1. Ausschießen *n*, Anwerfen *n* des Unterputzes; 2. Unterputz *m*
renew / to erneuern, auswechseln
renewable erneuerbar, auswechselbar
r. bushing *(Masch)* Einsatzbuchse *f*
r. fuse unit offene Sicherung *f*, Streifensicherung *f*
renewal Erneuerung *f*
r. of the grease Fettwechsel *m*
r. parts Ersatzteile *npl*
reniform nierenförmig
rennet casein *(Ch)* Labkasein *n*
rennin *(Ch)* Labferment *n*
rental unit Mietwohnung *f*, Wohnungseinheit *f*; Wohnung *f*
reopen / to *(Brg)* aufwältigen, sümpfen
reopening *(Brg)* Aufwältigung *f*, Sümpfen *n*
reorder / to nachbestellen
repair / to 1. instandsetzen, reparieren; ausbessern; flicken *(z. B. Ofenfutter)*; 2. *(Brg)* nachbauen, durchbauen *(Strecke)*

repair Instandsetzung *f*, Reparatur *f*
r. and maintenance Instandsetzung *f* und Unterhaltung *f*
r. berth *(Schiff)* Absetzplatz *m*
r. dock Ausbesserungsdock *n*
r. kit Werkzeugkasten *m* für Reparatur
r. part Ersatzteil *n*
r. pit Reparaturgrube *f*
r. service *(Am) (El)* Instandsetzungsdienst *m*
r. ship Werkstattschiff *n*, Reparaturschiff *n*
r. shop Instandsetzungswerkstatt *f*, Reparaturwerkstatt *f*
r. welding Reparaturschweißen *n*
r. work Instandsetzungsarbeit *f*, Reparaturarbeit *f*
r. workshop Reparaturwerkstatt *f*
repairing Reparatur *f*, Instandsetzung *f*
r. operations Instandsetzungsarbeiten *fpl*
repeat / to wiederholen; weitergeben *(mit Relais)*
repeat *(Text)* Rapport *m*, sich wiederholendes Muster *n*
r. mechanism automatische Rapporteinstellung *f*
r. of pattern Fadenrapport *m*
r. of warp threads Kettenrapport *m*, Rapportbreite *f*
r. of weft thread Schußrapport *m*, Rapporthöhe *f*
r. order Nachbestellung *f*
r. point *(Fmt)* Spiegelabstimmpunkt *m*
r. station Relaisstation *f*
repeatability Wiederholbarkeit *f (eines Arbeitsganges)*; Nachformbarkeit *f*, Reproduzierbarkeit *f*
repeatable wiederholbar, reproduzierbar
repeated bending stress strength Biegeschwingungs[schwell]festigkeit *f*
r. direct stress test Dauerzugversuch *m*
r. flexural stress Dauerbiegebeanspruchung *f*
r. impact bending strength Dauerschlagbiegefestigkeit *f*
r. impact bending test Dauerschlagbiegeversuch *m*
r. impact tensile test Dauerschlagzugversuch *m*
r. impact test Dauerschlagversuch *m*
r. solidification wiederholte Erstarrung *f (Halbleitertechnik)*
r. stress test Schwingungsversuch *m*
r. torsion test Dauerverdrehversuch *m*
r.-until-acknowledged signal *(Fmt)* Zeichen *n* mit Wiederholung bis zur Bestätigung
repeater 1. *(El)* Verstärker *m*, Übertrager *m*, Folge[stufe] *f*; 2. periodischer Dezimalbruch *m*; 3. Umführung *f*, Schlingenumführung *f (Walzstraße)*; 4. Repetiergewehr *n*; 5. *s.* r. compass
r. bay *(Fmt)* Verstärkergestell *n*
r. circuit Verstärkerschaltung *f*
r. compass Tochterkompaß *m*, Kreiseltochter *f*, Kompaßtochter *f*
r. gain *(Fmt)* Verstärkungsziffer *f*
r. rack *(Fmt)* Verstärkergestell *n*
r. room *(Fmt)* Verstärkerraum *m*
r. selector *(Fmt)* Mitlaufwähler *m*
r. station Relaissender *m*, Relaisstelle *f*
r. test rack Verstärkerprüfgestell *n*, Meßgestell *n (für Verstärkerämter)*
r. transmitter Nebensender *m*, Tochtersender *m*, Wiederholungssender *m*
repeating coil *(Fmt)* Leitungsübertrager *m*, Ringübertrager *m*, Übertrager[spule *f*] *m*
r. relay Übertragungsrelais *n*
r. rolling mill halbkontinuierliches Walzwerk *n*
r. station Zwischenstation *f*, Zwischensender *m*; Übertrageramt *n*
r. telegraph Antworttelegraf *m*, Rückmeldetelegraf *m*
r. the signal Rückmeldung *f (Signal)*
r. tie up in bands *(Text)* bandförmige Schnürung *f*
r. transformer Kopplungstransformator *m*
repel / to *(Ph)* zurückstoßen, abstoßen
repellent Insektenschutzmittel *n*
repeller Reflexionselektrode *f (Klystron)*
r. force Abstoßungskraft *f*, Repulsionskraft *f*
reperforation *(Fmt)* Empfangslochung *f*
reperforator *(Fmt)* Empfangslocher *m*, Lochstreifenempfänger *m*
repetend Periode *f (eines periodischen Dezimalbruches)*
repetition Wiederholung *f*
r. angular measurement repetitionsweises Winkelmessen *n*
r. equivalent *(Fmt)* Verkehrsgüte *f*
r. frequency Impuls[folge]frequenz *f*, Folgefrequenz *f*
r. job *(Masch)* Serienarbeit *f*
r. of pictures *(Fs)* Bildwechsel *m*
r. part *(Masch)* Wiederholteil *n*
r. production *(Masch)* Reihenfertigung *f*
r. rate *(Fs)* Impulsfrequenz *f*, Wiederholungsfrequenz *f*
r. rate of pulse Impulsfolgefrequenz *f*
r. theodolite *(Opt)* Repetitionstheodolit *m*
r. turning of curved profiles Nachformdrehen *n*
r. work Serienfertigung *f*
repetitive *(periodisch)* wiederkehrend, sich wiederholend
r. accuracy Wiederholgenauigkeit *f*, Reproduktionsgenauigkeit *f*
r. analog[ue] computer repetierender Analogrechner *m'*
r. error Regelabweichung *f*

r. housing Serienwohnungsbau *m*
r. part Serienstück *n*
replace / to auswechseln, ersetzen, erneuern; zurücklegen
r. a fuse eine Sicherung auswechseln
r. the oil die Ölfüllung erneuern
r. the receiver den Hörer auflegen
replaceability Austauschbarkeit *f*, Ersetzbarkeit *f*
replaceable auswechselbar, austauschbar
replacement 1. Ersatz *m*, Auswechslung *f*; 2. Ersatzteil *n*; 3. *(Geol)* Metasomatose *f*, Verdrängung *f*
r. bobbin Ersatzspule *f*
r. collision *(Kern)* Austauschstoß *m*
r. deposit *(Geol)* metasomatische Lagerstätte *f*
r. part Ersatzteil *n*, Auswechselteil *n*
r. reaction Austauschreaktion *f*
r. valve Ersatzröhre *f*
replant / to umpflanzen
replenish / to auffüllen, nachfüllen, [nach]tanken
replenishing device Nachtankvorrichtung *f*
r. line Nachtankleitung *f*
replenishment Auffüllung *f*, Nachtankung *f*
replica Abdruck *m*, Nachbildung *f*; Reproduktion *f*
r. of the cam Leitnocken *m*
r. technique Abdrucktechnik *f*
replication 1. Replikation *f*, Verdopplung *f*; Wiederholung[sprüfung] *f*
reply / to antworten
reply Antwort *f*; *(Fmt)* Rückmeldung *f*
r. bell *(Fmt)* Rückmeldeläutewerk *n*
r. frequency *(Fmt)* Antwortfrequenz *f*
r. pulse generator *(Fmt)* Antwortimpulsgenerator *m*
repolish / to nachpolieren
reporting service for distress at sea Seenotmeldedienst *m*
repost / to umbuchen
repour / to *(Met)* umgießen
repower / to *(sl)* wiedereinschalten
reprecipitation *(Ch)* Umfällung *f*
represent / to 1. darstellen; 2. verkörpern
r. diagrammatically zeichnerisch (schematisch) darstellen
representation *(Math)* Darstellung *f (Funktionen, Gruppen)*; Darstellungsweise *f (technisches Zeichnen)*
r. transmission Schirmbildübertragung *f (Radar)*
repress / to nachpressen; zurückdrängen
repressuring [process] Gaseinpreßverfahren *n* zur Wiederbelebung bzw. zur Erhaltung des Lagerstättendruckes *(Erdöl)*
reprint Nachdruck *m*; Wiederabdruck *m*, Neudruck *m*; Sonderdruck *m*
reprocess / to wiederaufarbeiten; wieder verarbeiten
reprocessed wool wiederverarbeitete Wolle *f*
reprocessing *(Kern)* Aufarbeitung *f* [von verbrauchten Brennstoffelementen]
reproduce / to nachformen, kopieren, reproduzieren; wiedergeben
reproducer 1. *(Dat)* Kartendoppler *m*; 2. *(Ak)* Geber *m*
reproducibility Reproduzierbarkeit *f*; *(Masch)* Wiederholbarkeit *f*, Nachformbarkeit *f*
reproducible reproduzierbar; nachformbar; wiederholbar
reproducing Wiedergabe *f*; Vervielfältigung *f*
r. amplifier *(Dat)* Wiedergabeverstärker *m*
r. head Tonabnehmer *m*, Wiedergabekopf *m*
r. lathe Schablonendrehmaschine *f (Holzbearbeitung)*
r. puncher *(Dat)* Kartendoppler *m*
reproduction 1. Reproduktion *f*, fotografische Wiedergabe *f*; 2. *(Masch)* Wiederholung *f*, Nachformung *f*, Kopierarbeit *f*; 3. *(Ak, Fs)* Wiedergabe; 4. Kopie *f*
r. felling Verjüngungshieb *m (Forst)*
r. of sound Tonwiedergabe *f*
r. speed Wiedergabegeschwindigkeit *f*
reproductive fortpflanzungsfähig
reproof / to wieder wasserdicht machen
repulp / to einstampfen *(z. B. Makulatur)*; aufbereiten *(Papier)*
repulse / to abstoßen
repulse Rückstoß *m*, Repulsion *f*; Abstoßung *f (Magnet)*
repulsion Abstoßung *f*, Repulsion *f*
r. ammeter Rundspulamperemeter *n*
r. induction motor Repulsionsmotor *m* mit Käfiganker
r. motor Repulsionsmotor *m*
r. potential Abstoßungspotential *n*
r.-start induction motor Asynchronmotor *m* mit Repulsionsanlauf
r.-start single-phase motor Einphasenrepulsionsmotor *m*
repulsive abstoßend
r. centre *(Ph)* abstoßendes Zentrum *n*
r. force abstoßende Kraft *f*, Abstoßungskraft *f*
r. potential Abstoßungspotential *n*
require / to erfordern, verlangen
required erforderlich; Soll-
r. accuracy verlangte (geforderte) Genauigkeit *f*
r. power Leistungsaufnahme *f*, Leistungsbedarf *m*
r. subscriber *(Fmt)* verlangter Teilnehmer *m*
r. value Sollwert *m*
r. vane angle *(Rak)* Rudersollwinkel *m*
reradiate / to wiederausstrahlen, [zu]rückstrahlen

reradiating transmitter Ballsender *m*, Relaissender *m*
reradiation 1. Rückstrahlung *f*, Zurückstrahlung *f*; 2. Ballsenden *n*, Relaissenden *n*
rereceiver Ballempfänger *m*
rerecording Kopie *f*, Umschnitt *m*
r. of tapes Überspielen *n* von Bändern
r. room Mischraum *m*
rereel / to umhaspeln
rereeling Umhaspeln *n (Seide)*
rerip *(Brg)* Nachbauen *n*, Durchbauen *n*
reroll / to umwalzen
reroute / to umleiten
rerun / to *(Am) (Dat)* wiederholen
rerun point *(Am) (Dat)* Wiederholpunkt *m*
r. program *(Am) (Dat)* Wiederholprogramm *n*
rerunning still Anlage *f* zum Feinfraktionieren
resaw / to *(Holz)* [auf]trennen
resaws *(Holz)* Trenngatter *n*
reschedule / to umplanen
rescrape / to *(Masch)* nachschaben
rescud / to *(Led)* nachstreichen
rescue Rettung *f*
r. apparatus *(Brg)* Rettungsapparat *m*, Gasschutzgerät *n*
r. boat Seenotfahrzeug *n*, Seenotboot *n*
r. centre *(Brg)* Rettungsstelle *f*
r. craft Seenotfahrzeug *n*, Seenotboot *n*
r. cruiser Seenotkreuzer *m*
r. ship Rettungsschiff *n (im Konvoi)*
r. tug Rettungsschlepper *m*
research Forschung *f*
r. activities Forschungstätigkeit *f*, Forschungsarbeit *f*
r. and testing station Versuchsanstalt *f*
r. council Forschungsrat *m*
r. engine Versuchsmaschine *f*
r. engineer Forschungsingenieur *m*
r. funds Forschungsmittel *pl*
r. institute Forschungsinstitut *n*
r. laboratory Forschungslabor *n*
r. microscope Forschungsmikroskop *n*
r. program[me] Forschungsprogramm *n*
r. project Forschungsvorhaben *n*
r. report Forschungsbericht *m*
r. result Forschungsergebnis *n*
r. rocket Forschungsrakete *f*
r. technique Untersuchungsverfahren *n*
r. vessel Forschungsschiff *n*
reseat / to einschleifen, nachschleifen *(Ventile)*
reservation Schutzwald *m*; Reservat *n*, Reservation *f*
reserve Reserve *f*
r. buoyancy *(Schiff)* Reserveschwimmfähigkeit *f*, Reserveverdrängung *f*, Restauftrieb *m*
r. buoyancy tank *(Schiff)* Adjustiertank *m*
r. position *(Fmt)* freier Platz *m*
r. power Kraftreserve *f*
r. wire Reserveader *f*
reservice / to instandsetzen, reparieren, aufarbeiten
reservicing Reparatur *f*, Wiederinstandsetzung *f*; Aufarbeitung *f*
reservoir 1. Talsperre *f*, Wasserstaubecken *n*, Sammelbecken *n*, Speicherbecken *n*, Stausee *m*; 2. Behälter *m*, Vorratsbehälter *m*, Bassin *n*, Becken *n*, Tank *m*; Klärbottich *m*; 3. Lagerstätte *f*
r. capacitor *(El)* Speicherkondensator *m*
r. capacity Fassungsvermögen *n* des Bassins (Staubeckens)
r. ladle *(Met)* Mischpfanne *f (Kupolofen)*
r. lining Beckenauskleidung *f*
r. outflow Speicherablaß *m*
r. pressure Druck *m* an der Bohrlochsohle, Lagerstättendruck *m (Erdöl, Erdgas)*
reset / to 1. *(Masch)* umrüsten, umrichten, umspannen, neu einstellen; 2. zurücksetzen, löschen, nullen, nullstellen; zurückstellen *(elektronische Schaltung)*; umklappen *(Multivibrator)*; zurückgehen *(Relais)*
reset Rückstellung *f*, Rückstellen *n*, Rücksetzen *n*
r. action integrierende Regelung *f*, Integralregelung *f*
r. constant *s.* integral action time
r. contact Löschkontakt *m*
r. control *(Am) s.* r. action
r. control circuit Rückstellregelungskreis *m (Magnetverstärker)*
r. counter Tourenzähler *m*, Umdrehungszähler *m*
r. key *(Fmt)* Berichtigungstaste *f*, Korrekturtaste *f*
r. magnet Richtmagnet *m*, Rückstellmagnet *m*
r. pulse Rückstellimpuls *m*; Löschimpuls *m*
r. rate *s.* integral action time
r. time Nachstellzeit *f*, Rückführzeit *f*, Isodromzeit *f*, Rückstellzeit *f*
resetting 1. *(Masch)* Umrüsten *n*; Umrichten *n (Schneidwerkzeug)*; Umspannen *n (Werkstück)*; 2. Neueinstellung *f*; Nullstellung *f*, Zurückstellung *f*, Rückstellung *f*, Nachgleich *m*
r. device Rückstellvorrichtung *f*
r. of the cutting tool Umrichten *n* des Schneidwerkzeugs
r. of the work Umspannen *n* des Werkstücks
r. ratio Rückgangsverhältnis *n (Relais)*
r. time 1. Rücklaufzeit *f*, Rückgangszeit *f (z. B. des Relais)*; 2. Umrichtezeit *f (Maschine)*
resharpen / to nachschleifen, nachschärfen
resharpenable nachschleifbar, nachschärfbar
resharpening Nachschliff *m*, Nachschärfung *f*, Anschliff *m*, Scharfschleifen *n*
reshuttling Schützenwechsel *m (Weberei)*
residence telephone Wohnungsanschluß *m*
r. time Verweilzeit *f*, Standzeit *f*

residential area Wohngebiet *n*
r. construction Wohnungsbau *m*
r. house Wohnhaus *n*; Mietshaus *n*
r. neighbourhood unit Wohnkomplex *m* *(Siedlungsplanung)*
r. population Einwohnerschaft *f*, Einwohnerzahl *f*
r. street Wohnstraße *f*
r. unit Wohnung[seinheit] *f*
residual übrigbleibend, zurückbleibend, remanent; Rest-, Eigen-
r. aberration *(Opt)* Restfehler *m*
r. activity Restaktivität *f*
r. aerial effect Antenneneffekt *m* *(des Rahmens)*
r. affinity Restaffinität *f*
r. air Luftrückstand *m*
r. atom Restatom *n*
r. central intensity *(Ph)* Restintensität *f* in der Linienmitte
r. charge Restladung *f*, Rückstand[s]ladung *f*, Ladungsrückstand *m*, remanente Ladung *f*
r. classes *(Math)* Restklassen *fpl*
r. comatic aberration *(Opt)* Komarestfehler *m*
r. cover solid ore *(Brg)* Erzschwebe *f*
r. current *(El)* Reststrom *m*; Anlaufstrom *m*
r. current state *(El)* Anlaufzustand *m*, Anlaufstromgebiet *n*
r. deflection Nullpunktfehler *m* *(Elektronenstrahlröhre)*
r. deviation Restablenkung *f*
r. elasticity elastische Nachwirkung *f*
r. excitation remanente Erregung *f* *(eines Magneten)*
r. exhaust gas Abgasrückstand *m*
r. exhaust product Auspuffrückstand *m*, Verbrennungsrückstand *m*
r. fat content Restfettgehalt *m*
r. field *(El)* Remanenzfeld *n*, Restfeld *n*
r. flash *(Masch)* Gratansatz *m*
r. flux density *(El)* zurückbleibende Induktion *f*, Remanenz *f*
r. gas Gasrückstand *m*, Restgas *n* *(Verbrennungsmotor)*
r. heat Resthitze *f*
r. hum voltage *(El)* Restbrummspannung *f*
r. impulse Restimpuls *m*, Reststromstoß *m*
r. induction *(El)* Remanenz[induktion] *f*, remanente Induktion *f*
r. intensity Restintensität *f*
r. loss magnetische Nachwirkung *f*, Nachwirkungsverlust *m*
r. magnetism remanenter Magnetismus *m*, Remanenz *f*
r. magnetization Remanenz *f*, zurückbleibende Magnetisierung *f*
r. matter Residuum *n*, Rückstand *m*
r. moisture Restfeuchtigkeit *f*
r. oil Rückstandsöl *n*
r. oil cooler Ölrückstandskühler *m*
r. phase difference Phasenrestfehler *m*
r. pillar *(Brg)* Restpfeiler *m*
r. plate *s.* r. stud
r. pulse Restimpuls *m*, Reststromstoß *m*
r. radiation Reststrahlung *f*
r. radioactivity Restradioaktivität *f*
r. rays Reststrahlen *mpl*
r. relay Erdschlußrelais *n*, Erdstromrelais *n*
r. shrinkage *(Text)* Restkrumpfung *f*, Restschrumpfung *f*
r. solvent Lösungsmittelrest *m*
r. stress Eigenspannung *f*, Restspannung *f*
r. stretch *(Text)* Restkräuselung *f*
r. stud Klebestift *m*, Klebeblech *n* *(z. B. des Relais)*
r. time constant Eigenzeitkonstante *f*
r. valence *(Ch)* Restvalenz *f*
r. voltage *(El)* Remanenzspannung *f*
residuary resistance Formwiderstand *m*; Normalwiderstand *m* *(z. B. eines Schiffes)*
residue 1. *(Ch)* Rückstand *m*, Überrest *m*, Rest *m*; 2. *(Math)* Residuum *n* *(einer analytischen Funktion)*
r. class *(Math)* Restklasse *f*
r. code *(Dat)* Rest[klassen]kode *m*
r. curve Kurvenform *f* des Reststromes, Oberwellenstromkurve *f*
r. from combustion *(Kfz)* Verbrennungsrückstand *m*
r. from roasting *(Met)* Röstrückstand *m*
r. gas Armgas *n*, von Gasolin befreites Erdgas *n*
r. number system *(BMSR)* Restklassenzahlensystem *n*
r. oil Heizöl *n*
r. on ignition *(Met)* Glührückstand *m*
residuum Rest *m*, Rückstand *m*, Residuum *n*
resile / to zurückfedern; elastisch sein
resilience 1. Zurückspringen *n*, Rückprall *m*, Zurückprallen *n*; 2. elastische Energie *f*, Formänderungsenergie *f*; 3. Federung *f*; Federungsvermögen *n*, Federwirkung *f*; 4. Elastizität *f* *(textiler Flächengebilde)*
r. bending test Biegeschlagversuch *m*
resiliency Elastizität *f*; Steifigkeit *f*; Prallkraft *f*
resilient elastisch, federnd
r. contact *(El)* Federkontakt *m*
resin / to harzen, mit Harz tränken
resin Harz *n*
r. acid Harzsäure *f*
r. adhesive Harzkleber *m*, Klebharz *n*
r. bearing *(Masch)* Kunstharzlager *n*
r. binder Harzträger *m*
r. blender Harzmischer *m*
r. board Hartpappe *f*, Kunstharzpappe *f*

r.-bonded 1. kunstharzverleimt; 2. harzgebunden, mit Harzbindung
r.-bonded chip board harzgebundene Faserplatte *f*
r. bonding 1. Kleben *n* mit Kunstharz; 2. naturharzhaltige Bindung *f*
r. canal Harzgalle *f*, Harzstreifen *m (Holz)*
r. cement Harzkleber *m*
r. coating Kunstharzüberzug *m*
r.-core[d] solder Kolozinn *n*, Kolophoniumlötdraht *m*, Tinoldraht *m*, Lötzinndraht *m* mit Harzeinlage
r. duct Harzkanal *m (Holz)*
r. ester Harzester *m*, Lackester *m*
r. finish[ing] *(Text)* Kunstharzausrüstung *f*
r. for paints and varnishes Lackharz *n*
r. glass boat Boot *n* aus glasfaserverstärktem Harz
r. glue Kunstharzleim *m*
r. milk Harzlösung *f*
r. oil varnish Harzölfirnis *m*
r. paper harzgeleimtes Papier *n*
r. resist Harzreserve *f*
r. size Harzleim *m (Papierherstellung)*
r. sizing Massenleimung *f*, Stoffleimung *f*
r. soap Harzseife *f*
r. tapping Harznutzung *f*
r.-treated fabric *(Text)* mit Kunstharz ausgerüstete Ware *f*
r. varnish Tränkharz *n*
resinate / to harzen, mit Harz tränken
resinate Resinat *n*
resiniferous harzgebend, harzhaltig
resinification Verharzung *f*
r. of the lubricating oil Verharzung *f* des Schmieröls
resinify / to verharzen
resinking *(Brg)* Sümpfen *n*
resinoid Resinoid *n*, harzförmiger Plast *m*; Kunstharz *n*
resinoid harzähnlich, Kunstharz-
r. bonded wheel kunstharzgebundene Schleifscheibe *f*
r. bonding kunstharzhaltige Bindung *f*
resinol Resinol *n*, Harzalkohol *m*
resinophore zur Harzbildung neigend
resinous harzartig, harzhaltig, harzig, harzreich
resist / to 1. widerstehen, aushalten; *(Kräfte)* aufnehmen; widerstandsfähig sein gegen; 2. *(Text)* reservieren
r. in colours *(Text)* bunt reservieren
r. in white *(Text)* weiß reservieren
resist *(Text)* 1. Reserve *f*; 2. Schutzpräparat *n*; 3. Reservedruck *m*, Schutzdruck *m*
r. paste Ätzpaste *f (Textildruck)*
r. printing *(Text)* Reservedruck *m*, Schutzdruck *m*
resistance 1. Widerstand *m*; Beständigkeit *f*, Festigkeit *f (Werkstoff)*; 2. *(El)* [ohmscher] Widerstand
r. acceleration Widerstandsbeschleunigung *f*
r. alloy Widerstandslegierung *f*
r. attenuation Widerstandsdämpfung *f*
r. box Widerstandskasten *m*, Widerstandsgehäuse *n*; Stöpselkasten *m*, Stöpselwiderstand *m*
r. braking Widerstandsbremsung *f*
r. brazing Hartlöten *n* mittels elektrischer Widerstandserwärmung
r. bridge Wheatstonesche Brücke *f*, Widerstands[meß]brücke *f*
r. butt welding Widerstandsstumpfschweißen *n*
r.-capacitance coupled amplifier RC-Verstärker *m*, widerstandsgekoppelter Verstärker *m*, Widerstandsverstärker *m*
r.-capacity coupling RC-Kopplung *f*, Widerstands-Kapazitäts-Kopplung *f*
r. coefficient Widerstandskoeffizient *m*, Widerstandsbeiwert *m*, Widerstandszahl *f*
r. coil Widerstandsspule *f*
r. control Widerstandsregelung *f*
r. coupled widerstandsgekoppelt
r. coupling Widerstandskopplung *f*
r. drop *(sl)* ohmscher Spannungsabfall *m*, Widerstandsabnahme *f*
r.-earthing Widerstandserdung *f*, Erdung *f* über Widerstand
r. element Widerstandselement *n*; Heizwiderstand *m*, Heizelement *n*
r. flash welding Widerstandsabschmelzschweißung *f*
r. furnace [elektrischer] Widerstandsofen *m*
r. gauge Widerstandsmanometer *n*
r. grounded über einen Widerstand geerdet
r. heating Widerstandserwärmung *f*, Widerstandsheizung *f*
r. inductance coupling induktive Kopplung *f*, Drosselkopplung *f*
r. loss ohmscher Verlust *m*, Joulesche Wärme *f*; Widerstandsdämpfung *f*
r. mat Widerstandsgewebe *n*, Schniewindtband *n*
r. of charge Ladewiderstand *m*
r. of friction Widerstandsreibung *f*
r. of motion Bewegungswiderstand *m*
r. of ship Schiffswiderstand *m*
r. of the adjusting gear Stellzeugwiderstand
r. of valve Innenwiderstand *m* einer Röhre
r. oven Widerstandsofen *m*
r. projection welding Widerstandsbuckelschweißung *f*
r. protection Resistanzschutz *m (Relais)*
r. pyrometer Widerstandsthermometer *n*
r. reduced to the crank circle auf den Kurbelkreis bezogener Widerstand *m*

r. relay Resistanzrelais *n*
r. spot welding Widerstandspunktschweißung *f*
r. standard Normalwiderstand *m*
r.-start motor Motor *m* mit Widerstandsanlasser
r. step Widerstandsstufe *f*
r. strain gauge Dehnmeßstreifen *m (Werkstoffprüfung)*
r. test *(Schiff)* Schleppversuch *m*
r. thermometer Widerstandsthermometer *n*
r. to abrasion Abnutzungswiderstand *m*
r. to aging Alterungsbeständigkeit *f*
r. to alkalis *(Text)* Alkalibeständigkeit *f*
r. to atmospheric conditions Witterungsbeständigkeit *f*
r. to buckling Knickwiderstand *m*
r. to cold Kältebeständigkeit *f*
r. to compression Druckfestigkeit *f*
r. to conduction Fortleitungswiderstand *m*
r. to cracking Rißfestigkeit *f*
r. to creep Kriechfestigkeit *f*
r. to creep stress Dauerstandfestigkeit *f*
r. to deformation Formbeständigkeit *f*, Verformungswiderstand *m*
r. to fatigue Ermüdungsfestigkeit *f*
r. to heat Hitzebeständigkeit *f*, Wärmebeständigkeit *f*, Warmfestigkeit *f*
r. to impact Schlagfestigkeit *f*; Stoßfestigkeit *f*
r. to lateral bending Wechselbiegefestigkeit *f*
r. to pressure Druckfestigkeit *f*
r. to propulsion *(Schiff)* Propulsionswiderstand *m*, Vortriebswiderstand *m*
r. to repeated impact Dauerschlagfestigkeit *f*
r. to shearing Scherwiderstand *m*, Scherfestigkeit *f*
r. to shock Stoßfestigkeit *f*, Schlagfestigkeit *f*
r. to stretching Streckfestigkeit *f*
r. to tearing Zerreißfestigkeit *f*
r. to thermal shocks Temperaturwechselbeständigkeit *f*
r. to torsion Drehfestigkeit *f*, Torsionswiderstand *m*
r. to tracking Kriechstromfestigkeit *f*
r. to twisting Verdrehungsfestigkeit *f*
r. to wear [and tear] Abnutzungsbeständigkeit *f*, Verschleißfestigkeit *f*
r. to yield *(Brg)* Einsinkwiderstand *m (Stempel)*
r.-tuned oscillator *(El)* RC-Generator *m*, RC-Summer *m*
r. tuning *(El)* Abstimmung *f* durch veränderlichen Widerstand
r. voltage ohmscher Spannungsabfall *m*
r. voltage divider ohmscher Spannungsteiler *m*; Widerstandsspannungsteiler *m*
r. welding Widerstandsschweißen *n*
r. welding electrode Widerstandsschweißelektrode *f*
r. wire Widerstandsdraht *m*; Heizdraht *m*
resistant widerstandsfähig, beständig, fest
r. to acid säurebeständig
r. to bending biegungsfest
r. to boiling kochbeständig
r. to corrosion korrosionsbeständig
r. to fracture bruchsicher
r. to rotting verrottungsbeständig
r. to rusting rostbeständig
r. to scaling zunderbeständig
r. to wear verschleißfest
resistibility Widerstandsfähigkeit *f*, Beständigkeit *f*
resisting beständig
r. force Widerstandskraft *f*
r. moment Widerstandsmoment *n (Festigkeit)*
r. power Widerstandskraft *f*
resistive 1. widerstandsfähig; 2. *(El)* widerstandsbehaftet, mit Widerstand; Widerstands-, Wirk-
r. attenuator *s.* absorptive attenuator
r. branch ohmscher Zweig *m*, Widerstandszweig *m*
r. coil Widerstandsspule *f*, Spule *f* mit Widerstand, widerstandsbehaftete Spule
r. component Wirkkomponente *f*
r. conductor widerstandsbehafteter Leiter *m*, Widerstandsschnur *f*
r. film Widerstandsfolie *f*, Widerstandsschicht *f*
r. inductor Spule *f* mit Widerstand, widerstandsbehaftete Spule
r. load ohmsche Belastung *f*, Wirklast *f*
r. voltage drop ohmscher Spannungsabfall *m*
resistivity 1. *(El)* spezifischer Widerstand *m*, Eigenwiderstand *m*, Widerstandsbeiwert *m*, Widerstandseigenschaft *f*; 2. Widerstandsfähigkeit *f*, Beständigkeit *f (Werkstoffe)*
r. change Widerstandsänderung *f*
r. to heat Wärmedurchgangswiderstand *m*
resistor Widerstand *m (technische Ausführung)*, Widerstandselement *n*, Widerstandsgerät *n*, Rheostat *m*; Vorschaltwiderstand *m*
r. core Widerstandskörper *m*
r. material Widerstandswerkstoff *m*
r.-quenched counter circuit *(Kern)* Zählrohrlöschschaltung *f* mit hohem Außenwiderstand
r.-transistor logic circuit logische Schaltung *f* mit Widerstand und Transistor, logische Widerstandstransistorschaltung *f*
resistron *(Fs)* Resistron *n*
resite Resit[harz] *n*
resnatron Resnatronröhre *f (Strahlbündelungshochleistungsröhre der UKW-Technik)*

resol Resol[harz] *n*
resolder / to nachlöten
resoluble lösbar
resolution 1. *(Math)* Auflösung *f*, Zerlegung *f*; 2. *(Ch)* Auflösung; 3. *(Opt)* Auflösung; Auflösungsvermögen *n*; 4. Auflösungsvermögen *(Radar)*; 5. *(Fs)* Bildauflösung *f*; 6. Trennung *f*; Trennschärfe *f*
r. limit *(Opt)* Auflösungsgrenze *f*
r. of forces *(Mech)* Zerlegung *f* von Kräften, Kräftezerlegung *f*
r. of velocities *(Mech)* Zerlegung *f* von Geschwindigkeiten
r. pattern *(Am) (Fs)* Probebild *n*, Testbild *n* (*s. a.* definition chart)
r. power *(Dat)* Auflösungsvermögen *n* *(z. B. bei der Speicherung und Wiedergabe von Daten)*
resolutive [auf]lösend
resolvability *(Ch)* Lösbarkeit *f*
resolve / to 1. *(Ch)* auflösen; 2. zerlegen *(Kräfte)*
resolvent *(Math)* Resolvente *f*
r. kernel lösender Kern *m*
resolver *(BMSR)* Koordinatenwandler *m*, Vektorzerleger *m*
resolving Lösen *n*, Auflösen *n*; Zerlegen *n*; Trennen *n*
r. power *(Opt)* Auflösungsvermögen *n* *(Mikroskop)* (*s. a.* resolution power)
r. time *(Dat)* Auflösungszeit *f* *(Mindestimpulsabstand, z. B. zwischen zwei Zählungen)*
resonance Mitschwingen *n* *(von Tönen usw.)*; Resonanz *f*
r. absorber Resonanzabsorber *m*
r. absorption Resonanzabsorption *f*
r. acceleration Resonanzbeschleunigung *f*
r. accelerator Resonanzbeschleuniger *m*
r. activation Resonanzaktivierung *f*
r. amplification Resonanzverstärkung *f*
r. amplifier Resonanzverstärker *m*, abgestimmter Verstärker *m*
r. bridge Resonanzbrücke *f*
r. capacitor transformer kapazitiver Spannungswandler *m* in Resonanzschaltung
r. capture Resonanzeinfang *m*
r. chamber Hohlraumresonator *m*
r. circuit Resonanzkreis *m*
r. curve Resonanzkurve *f*
r. detector Resonanzdetektor *m*
r. effect Resonanzerscheinung *f*
r. energy Resonanzenergie *f*
r. escape probability *(Kern)* Resonanzfluchtfaktor *m*, Resonanzentkommwahrscheinlichkeit *f*
r. excitation Resonanzanregung *f*
r. fission Resonanzspaltung *f*
r. frequency Resonanzfrequenz *f*
r.-frequency meter Resonanzfrequenzmesser *m*
r. hump Resonanzbuckel *m*
r. indicator Resonanzanzeiger *m*, Resonanzindikator *m*
r. induction Resonanzinduktion *f*
r. level Resonanzniveau *n*
r. line Resonanzlinie *f*
r. loss Resonanzverlust *m*
r. neutron Resonanzneutron *n*
r. peak Resonanzspitze *f*
r. point Resonanzpunkt *m*, Resonanzstelle *f*
r. potential Resonanzpotential *n*
r. proton Resonanzproton *n*
r. radiation Resonanzstrahlung *f*
r. ratio *(BMSR)* Resonanzüberhöhung *f*
r. relay Resonanzrelais *n*
r. resistance Resonanzwiderstand *m*
r. scattering Resonanzstreuung *f*
r. screen Resonanz[schwing]sieb *n*
r. series Resonanzserie *f*
r. shape Resonanzkurvenform *f*
r. state Resonanzzustand *m*
r. transformer Resonanztransformator *m*, abgestimmter Transformator *m*
r. velocity Resonanzgeschwindigkeit *f*
r. vibration Resonanzschwingung *f*
r. vibro-screen Resonanzschwingsieb *n*
r. voltage Resonanzspannung *f*
r. wavemeter Resonanzwellenmesser *m*
resonant mitschwingend, resonant
r.-absorption silencer *(Kfz)* Reflexionsschalldämpfer *m*, Schalldämpfer *m* mit Resonatoren
r. cavity Hohlraumresonator *m* *(Wellenleitertechnik)*; Resonanzhohlraum *m*
r. chamber *(Kfz)* Resonatorraum *m*
r. choke Resonanzdrosselspule *f*
r. circuit Resonanz[strom]kreis *m*; Schwing[ungs]kreis *m*, Spannungsresonanzkreis *m*
r. circuit excitation Schwingkreisanregung *f*
r. conductor mitschwingender Leiter *m*
r. diaphragm Anpassungsblende *f* *(Wellenleiter)*
r. frequency Resonanzfrequenz *f*, Eigenfrequenz *f*
r. line Leitungsschwingkreis *m*, schwingende Leitung *f*
r.-line oscillator Oszillator *m* mit abgestimmten Resonanzleitungen
r. machine Resonanzmaschine *f* *(Dauerschwingprüfung)*
r. mode Resonanzmodus *m* *(Wellenleiter)*
r. mode filter Resonanz[wellen]typfilter *n* *(Wellenleitertechnik)*
r. relay Frequenzrelais *n*
r. rise Resonanzanstieg *m*, Resonanzüberhöhung *f*; Aufschaukeln *n* *(einer Schwingung)*

r. rise of signals Aufschaukeln *n* von Signalen
r. shunt Resonanznebenschluß *m*; Überbrükkungsschwingkreis *m*
r. transformer Resonanztransformator *m*
r. vibration Resonanzerschütterung *f*
r. width Resonanzbreite *f*
resonate / to mitschwingen; auf Resonanz bringen; mitklingen
resonator 1. Resonator *m*; Schwinger *m*; Schwingkreis *m*; 2. Hohlraumresonator *m* *(Mikrowellentechnik)*
r. grid Hohlraumgitter *n*
r. intake silencer *(Kfz)* Reflexionsansauggeräuschdämpfer *m*
resonoscope Resonoskop *n*
resorb / to resorbieren, aufsaugen, einsaugen
resorcinol *(Ch)* Resorzin *n*
resound / to widerhallen; schallen; ertönen
respirable dust *(Brg)* lungengängiger Staub *m*
respiratory protection apparatus Atemschutzgerät *n*
respond to / to ansprechen auf; reagieren auf
respondentia Schiffsladungsverpfändung *f*
responder Antwortsender *m*, Empfänger *m (Radartechnik)*
r. beacon Antwortbake *f*, Bodenantwortstation *f*, Wiederholerbake *f (Radartechnik)*
responding speed Ansprechgeschwindigkeit *f (Relais)*; Einstellgeschwindigkeit *f*, Einstelldauer *f (Meßinstrument)*
response 1. Ansprechen *n*; Anziehen *n (Relais)*; 2. Ausschlag *m*, Anzeige *f (Meßinstrument)*; 3. Frequenzgang *m*, Frequenzkurve *f*, Empfindlichkeit *f*, Verhalten *n*, Reaktion *f*; Frequenzbereich *m*; 4. Kennlinie *f*, Kurve *f*; 5. Ansprechschwelle *f*
r. characteristic Frequenzgang *m*; Ansprechcharakteristik *f*; Empfangscharakteristik *f (Antennen)*; Filterkurve *f*; Empfindlichkeitskurve *f*
r. curve *s.* r. characteristic
r.-frequency diagram *(BMSR)* Frequenzgang *m (getrennte Auftragung von Amplitude und Phase)*
r. latency Ansprechverzögerung *f*; Schrecksekunde *f*
r. time Ansprechzeit *f (z. B. des Relais)*, Einschwingzeit *f*, Beruhigungszeit *f*; Einstellzeit *f (elektrische Messung)*
r. to current Stromübertragungsfaktor *m*
r. to power Leistungsübertragungsgrad *m*
r. to voltage Spannungsübertragungsfaktor *m*
responsive to ansprechend auf, ansprechfähig auf
responsor Abfrageempfänger *m*
rest / to in Ruhelage sein, ruhen
r. on ruhen auf, lagern auf, aufgelagert sein auf, Auflager haben auf
rest 1. *(Masch)* Auflage *f*, Stütze *f*; Schlitten *m*, Support *m*; 2. Stillstand *m*, Ruhe *f*; 3. Rest *m*
r. action integrierende Regelung *f*
r. contact Ruhekontakt *m*
r. current Reststrom *m*; Ruhestrom *m*
r. mass of the electron Ruhemasse *f* des Elektrons
r. period Ruhezeit *f*, Stillstandszeit *f*
r. potential Ruhepotential *n*
r. room Aufenthaltsraum *m*
r. system *(Ph)* ruhendes System *n*, Ruhesystem *n*
restake / to *(Led)* nachstollen
restarting injector *(Wärm)* Sicherheitsinjektor *m*, selbstanziehender (wieder ansaugender) Injektor *m*
restiform seilförmig
resting carrier frequency feste Trägerfrequenz *f*
r. potential *(El)* Ruhepotential *n*
restitution delay Wiedergabeverzögerung *f*
r. of protective rights Freigabe *f* von Schutzrechten *(Patentwesen)*
restock / to 1. *(Led)* nachwalken; 2. aufforsten
restoration Instandsetzung *f*, Wiederherstellung *f*; Rückstellung *f*; Rekultivierung *f (Landschaftspflege)*
r. of voltage *(El)* Wiederhochfahren *n* der Spannung
restore / to 1. instandsetzen, wiederherstellen, restaurieren; 2. rückstellen; 3. abfallen *(Relais)*; 4. *(Dat)* nullen, zurückführen; 5. rekultivieren *(Landschaftspflege)*
r. the receiver *(Am) (Fmt)* den Hörer anhängen
r. to zero auf Null abgleichen (zurückbringen), wieder auf Null bringen
restoring force Rückstellkraft *f*; Standmoment *n*
r. key *(El)* Rückstelltaste *f (Tippbetrieb)*, Rückstellknopf *m*
r. moment *(El)* Rückstellmoment *n*, Richtmoment *n*
r. moment due to vanes *(Schiff)* Ruder[rückstell]moment *n*
r. time Rückstellzeit *f*
r. torque Richtmoment *n*, Rückstellmoment *n (eines Instrumentes)*
restowing *(Schiff)* Umstauung *f*
restraighten / to wieder ausrichten, nachrichten
restrained eingespannt *(Statik)*
restrainer 1. *(Met)* Sparbeize *f*; 2. Verzögerer *m (Fotografie)*
restraining end moment Einspannmoment *n (Statik)*
r. flange Bord *m (Rollenlager)*
restraint Beschränkung *f*, Hinderung *f*; Bewe-

gungsbegrenzung *f*, Beschränkung; Zwanglauf *m*; Zwang *m*; Einspannung *f (Statik)*
restrict / to beschränken; drosseln, einschnüren, einschränken, verengen
restricted beschränkt, gedrosselt; zwangsläufig
r.-hour maximum demand indicator Maximumanzeiger *m* mit aussetzender Registrierung
r. scale gedrängte Skala *f*
r. theory of relativity spezielle Relativitätstheorie *f*
r. waters *(Schiff)* Gewässer *n* beschränkten Querschnitts
restriction Beschränkung *f*, Einschränkung *f*, Begrenzung *f*; Verengung *f (Rohr)*; Drossel[ung] *f (Pneumatik)*
r. valve Drosselventil *n*
restrictive circuit *(El)* Sparschaltung *f*
restrictor Drossel *f*
r. valve Drosselventil *n*
restriking *(El)* Wiederzündung *f*
r. voltage Einschwingspannung *f*; Nachzündspannung *f*; Rückzündspannung *f*; Wiederzündspannung *f*
resublime / to doppelt sublimieren, resublimieren
resuing *(Brg)* Abbau *m* eines sehr dünnen Gangs durch Mitnehmen des Nebengesteins
resulphurization *(Met)* Rückschwefelung *f*
resultant Resultierende *f*; Resultante *f*
resultant resultierend
r. acceleration resultierende Beschleunigung *f*
r. motion resultierende Bewegung *f*
r. radial force radiale Mittelkraft *f (Turbine)*
r. velocity resultierende Geschwindigkeit *f*
r. voltage resultierende (verkettete) Spannung *f (Drehstrom)*
resume control / to *(El)* Steuerfunktion wiedererhalten
resupply the carrier / to *(El)* den Träger wieder zusetzen
resurvey *(Opt)* Neuaufnahme *f*
ret / to rösten *(Flachs)*
retain / to hemmen, festhalten, zurückhalten; stauen *(Wasser)*; behalten
r. a keen cutting edge schneidhaltig sein
r. an image *(Fs)* ein Bild festhalten
r. the cutting ability schneidhaltig sein
r. the lubricant den Schmiermittelaustritt verhindern
retained austenite *(Met)* Abschreckaustenit *m*
r. by friction mit Selbsthemmung, durch Reibschluß
r. charge *(El)* zurückbleibende Ladung *f*
r. on sieve Siebrückstand *m*, Überkorn *n*
retainer *(Masch)* 1. Käfig *m (Lager)*; 2. Feststellvorrichtung *f*, Anschlag *m*, Raste *f*, Klinke *f*, Arretierung *f*
r. pin Haltestift *m*
r. ring 1. Sperring *m (Fernschreiber)*; 2. Sprengring *m (Lager)*
r. ring gauge Sprengringlehre *f*
retaining Hemmung *f*; Festhalten *n*; Unterbindung *f* des Austretens
r. band Kappe *f (Axiallager)*
r. basin Rückhaltebecken *n*
r. bush Spannhülse *f*
r. clip Halteklammer *f (Fernschreiber)*
r. dam Staudamm *m*
r. device Halteeinrichtung *f*
r. member Führungskäfig *m (Nadellager)*
r. nut Befestigungsmutter *f*, Haltemutter *f*, Sicherungsring *m*, Überwurfmutter *f*; Feststellschraube *f*, Klemmschraube *f*
r. plate Rückhalteblech *n*
r. ring Klammerring *m*, Sicherungsring *m*; Federring *m*; Sprengring *m*; Klemmring *m*, Haltering *m*
r. spring Haltefeder *f*; Sperrfeder *f*
r. strip *(Schiff)* Stevenrohrlängsschiene *f*
r. wall Stützmauer *f*
r. washer Sicherungsscheibe *f*
r. zone *(Fs)* Haltebereich *m*, Synchronisationsbereich *m*
retake *(Opt)* Neuaufnahme *f*, Wiederholungsaufnahme *f*
retan / to nachgerben
retard / to verzögern, verlangsamen
r. the ignition *(Kfz)* Spätzündung einstellen
retardation 1. Bremsung *f*, Bremswirkung *f*; Verzögerung *f*; Hemmung *f*; 2. Gangunterschied *m (Polarisation)*
r. area *(Wärm)* Verzögerungsfläche *f*
r. coil Phasenabgleichspule *f*, Siebspule *f*; Drossel[spule] *f*
r. method *(Elm)* Auslaufverfahren *n*
r. network *(El)* Laufzeitkette *f*, Verzögerungsschaltung *f*
r. of boiling Siedeverzug *m*
r. of ebullition Siedeverzug *m*
r. of neutrons Neutronenbremsung *f*, Neutronenverzögerung *f*
r. rocket Bremsrakete *f*
r. spectrum Bremsspektrum *n (Röntgenstrahlung)*
r. time Verzögerungszeit *f*
retarded verzögert; nacheilend
r. action *(Opt)* Zeitdehnung *f*
r. coking time Nachhinken *n* der Garung
r. ignition *(Kfz)* Spätzündung *f*
r. motion verzögerte Bewegung *f*
r. phase nacheilende Phase *f*
r. potential retardiertes Potential *n*
r. spark *(Kfz)* Spätzündung *f*
retarder Verzögerer *m*, Verzögerungsmittel *n*; Abbindeverzögerer *m (Beton)*; Hemmwerk *n (Telegrafie)*

retarding Verzögern *n*, Verzögerung *f*, Verlangsamung *f*; Nacheilen *n*
retarding verzögernd; Brems-, Verzögerungs-
r. agent Abbindeverzögerer *m (Betontechnologie)*
r. conveyor Bremsförderer *m*
r. disk conveyor Stauscheibenförderer *m*
r. electrode Bremselektrode *f (Fernsehaufnahmeröhre)*
r. field Bremsfeld *n*
r. field detector Brems[feld]audion *n*
r. field oscillator Bremsfeldgenerator *m*
r. field tube Bremsfeldröhre *f*
r. force Verzögerungskraft *f*
r. lens Verzögerungslinse *f*
r. potential Bremspotential *n*
r. power Verzögerungskraft *f (Bremse)*
r. pressure Verzögerungsdruck *m*
r. torque Bremswiderstand *m*; Bremsmoment *n*
retention Beibehaltung *f*; Retention *f*, Zurückhaltung *f*; Rückstand *m*; Restaktivität *f*
r. analysis Retentionsanalyse *f (Chromatografie)*
r. coefficient Retentionsgrad *m*, *(Ch)* Retentionsfaktor *m*
r. mixer Speichertankmischer *m*
r. time Retentionszeit *f*, Rückhaltezeit *f*, Verweilzeit *f*, Haltezeit *f (z. B. der Temperatur)*
retentive zurückhaltend, festhaltend
r. field *(El)* Remanenzfeld *n*, remanentes Feld *n*
retentivity Remanenz *f*
reticle Fadennetz *n*
reticular netzartig, Netz-
reticulate *(Plast)* netzartig verworren *(Faserstruktur)*
reticulated vault[ing] *(Bw)* Netzgewölbe *n*
reticule *(Opt)* Fadennetz *n*
retile / to neu eindecken, umdecken *(Dach)*
retimbering *(Brg)* Erneuerung *f* des Ausbaus
retinal camera *(Opt)* Netzhautkamera *f*
r. illumination *(Opt)* Netzhautbeleuchtung *f*
retool / to umrichten, umrüsten *(Werkzeugmaschinen)*
retort Retorte *f*, Destillierblase *f*, Kolben *m*
r. bench Koksofenbatterie *f*, Koksofenanlage *f*
r. carbonization *(Met)* Retortenverkohlung *f*
r. coal Gaskohle *f*, Retortenkohle *f*
r. coke Gaskoks *m*, Retortenkoks *m*
r. coking Retortenverkokung *f*
r. furnace *(Met)* Retortenofen *m*, Muffelofen *m*, Gaseinsatzofen *m*
r. neck Retortenhals *m*
retorting Destillieren (Kracken) *n* in der Retorte
retouch / to retuschieren
retouching Retusche *f*, Retuschieren *n*
r. device Retuschiergerät *n*
retour trace blanking *(Fs)* Rücklauf[strahl]-austastung *f*
retrace / to *(Fmt)* noch einmal überschreiben, rückwärts (wieder) durchlaufen
r. the course *(Schiff)* Kurs wieder aufnehmen
retrace Rücklauf[spur *f*] *m (des Katodenstrahles)*
retract / to zurückziehen, einziehen; einfahren, einziehen *(Fahrwerk)*; zusammenfalten *(Tragfläche)*
r. landing gear *(Flg)* Fahrwerk einfahren
retractable rückziehbar; einfahrbar, einziehbar *(Fahrwerk)*
r. landing gear *(Flg)* Einziehfahrwerk *n*
r. radiator ausziehbarer Kühler *m*
r. roller bridge Rollbrücke *f*
retraction Einfahren *n (Fahrwerk)*; Zurückziehung *f*
r. device of chaff cutter Einzugsvorrichtung *f (Häcksler)*
r. of the drill Herausziehen *n* des Bohrers
r. of the milling head Fräserabhebung *f*
retransfer Rückübertragung *f*
retransformation Rücktransformation *f*, Rücktransformierung *f*, Rückumspannung *f*
retranslation Rückübersetzung *f*
retransmission *(Fmt)* Vermittlung *f*, Weitergabe *f*, Weiterübermittlung *f*; Ballsenden *n*
retransmit / to *(Fmt)* weitergeben, weitersenden
r. a booking *(Fmt)* eine Gesprächsanmeldung weitermelden
retransmitter Zwischensender *m*, Ballsender *m*
retransmitting installation *(Fmt)* speichernde Übertragung *f*
retread / to runderneuern *(Reifen)*
retreat / to 1. nachbehandeln, wiederholt behandeln; 2. *(Brg)* zurückbauen
retreat *(Brg)* Rück[wärts]bau *m*, Heimwärtsbau *m*
r. mining *s.* retreat
r. working *s.* retreat
retreating *s.* retreat
r. face *(Brg)* Streb *m* im Rückbau
r. working *s.* retreat
retreatment wiederholte Behandlung *f*
r. concentration stufenweise Anreicherung *f (Flotationsaufbereitung)*
r. plant *(Brg)* Nachbehandlungsanlage *f*
retrieval of digital information *(Dat)* Wiederauffinden *n* digitaler Information
retriever head Zubringerkopf *m (Räummaschine)*
r. unit Zubringeeinrichtung *f (Räummaschine)*
retro- rückwärts, Rück-; hinter
r.-rocket Bremsrakete *f*
retroact / to [zu]rückwirken
retroaction Rückwirkung *f*; Rückkopplung *f*
retroactive rückwirkend; rückgekoppelt

r. rocket Bremsrakete *f*
r. valve detector Rückkopplungsaudion *n*
retrogradation *s.* retrogression
retrograde / to zurücklaufen, sich rückwärts bewegen, sich zurückziehen, rückläufig sein
retrograde rückläufig
r. motion rückläufige Bewegung *f*, Rückwärtsbewegung *f*
r. precession retrograde Präzession *f*
r. rocket Bremsrakete *f*
retrogression Rücklauf *m*, Rückwärtsbewegung *f*; Rückentwicklung *f*
retrogressive rückläufig, rückschreitend
r. movement rückläufige Bewegung *f*, Kehrbewegung *f*
rettery Flachsröste[rei] *f*
retting Flachsröste *f*, Röste *f*; Rösten *n*
r. ripeness Röstreife *f (Flachs)*
return / to zurückführen, zurückstellen, zurücklaufen; zurückkehren; zurücksenden; *(El, Fmt)* ablaufen *(Nummernscheibe)*
r. to *(El)* rückverbinden mit, zurückschalten auf
return Rückführung *f*, Wiederkehr *f*; Rückkehr *f*; Rückgang *m*; Rückleitung *f*; Rücklauf *m (z. B. Katodenstrahl)*
r. activated sludge Rücklaufbelebtschlamm *m (Abwasserreinigung)*
r. basin Rücklaufölbehälter *m*
r. beam Rückstrahl *m*
r. belt Rückführband *n (Fördertechnik)*
r. bend Doppelbogen *m*, Doppelkrümmer *m (Rohrkrümmer um 180°)*, Doppelkniestück *n*
r. block Umlenkblock *m*
r. cable Stromrückleitungskabel *n*, Rückstromkabel *n*
r. circuit Fahrstromrückleitung *f*; Rücklaufschaltung *f*
r. condenser Rückflußkühler *m*
r. connection Rückleitungsstutzen *m*
r. current Rückstrom *m*; Erdstrom *m*
r. current coefficient *(Fmt)* Reflexionsfaktor *m*
r. difference *(BMSR)* Rückführdifferenz *f (Signalflußdiagramm)*
r. drum Umkehrtrommel *f*, Umlenktrommel *f*
r. echo pulse Echoimpuls *m*, Rückimpuls *m*
r. feeder Rückleitung *f (mit besonderem Leiter)*
r. flow Rückfluß *m*, Rücklauf *m*
r.-flow system Gegenflußsystem *n*
r. gate-road *(Brg)* ausziehende Wetterstrecke *f*
r. interval *(Fs)* Rücklaufzeit *f*
r. line Rücklauflinie *f*, Rücklaufstrahl *m (Katodenstrahlröhre)*; Rückleitung *f (Hydraulik)*
r. loss Fehlerdämpfung *f*; Echodämpfung *f*, Rückflußdämpfung *f*
r. measuring set *(Am) (El)* Nachbildmesser *m*
r. movement *(Masch)* Rücklauf *m*
r. of dial *(Fmt)* Rücklauf *m* der Nummernscheibe
r. of pallet to loading station Werkstückträgerrücklauf *m (Taktstraße)*
r. of the spindle Spindelrücklauf *m*
r. pass Rückstich *m (Walzwerk)*
r. path Rückweg *m*; Rücklauf *m*; Rückwärtskaliber *n (Walzwerk)*
r. pipe *(Bw)* Rücklaufleitung *f*, Rückflußrohr *n*
r. pipeline Rücklaufleitung *f*
r. piston stroke Kolbenrückgang *m*
r. question frequency *(Fmt)* Rückfragehäufigkeit *f*
r. screen Rücklaufsieb *n*
r. speed Rücklaufgeschwindigkeit *f (Hobelmaschinentisch)*
r. spiral Kehrwendel *f*
r. spring Gegenfeder *f*, Rückholfeder *f*, Rückstellfeder *f*
r. stop Rücklaufsicherung *f*, Rücklaufsperre *f*
r. stroke *(Masch)* Kolbenrückgang *m*; Einwärtshub *m*; Rücklauf *m*, Leerhub *m*
r. tank Rücklauftank *m*
r. ticket Hin- und Rückflugschein *m*
r. to standard path *(Rak)* Bahneinlenkung *f*
r. trace *(El)* Rücklaufweg *m*
r. trip Rückfahrt *f*
r.-type steam trap Kondenswasserrückleiter *m*
r. valve Rückflußventil *n*
r. wire *(El)* Nulleiter *m*, Rückleitung[sdraht *m*] *f*
returning Rücklaufen *n*, Rückkehr *f*
r. time Rücklaufzeit *f (Presse)*
Reuleaux-Müller valve diagram *(Wärm)* Reuleaux-Müllersches Schieberdiagramm *n*
reunion Wiedervereinigung *f*
reusable wiederverwendbar
reused wool Altwolle *f*
reveal Leibung *f*; Fensterleibung *f*; Türleibung *f*
revenue passenger zahlender Fluggast *m*
r. vessel Zollfahrzeug *n (Schiff)*
reverberant *(Ak)* nachhallend
r. sound Nachhall *m*
reverberate / to zurückwerfen, reflektieren; *(Ak)* widerhallen, nachhallen
reverberating zurückstrahlend; *(Ak)* nachhallend
r. roof *(Met)* Flammofenkuppel *f*
reverberation *(Ak)* Nachhall *m*, Widerhall *m*
r. chamber Nachhallraum *m*, Nachhallkammer *f*
r.-controlled nachhallgesteuert
r. period (time) Nachhallzeit *f*
reverberator Reflektor *m*

reverberatory hallend, reflektierend
r. calciner Fortschaufelungsofen *m*, Krählofen *m (Erzröstung)*
r. furnace refinery Flammofenfrischen *n*
r. hearth furnace Herdflammofen *m*
r. tin furnace Zinnflammofen *m*
revers Revers *n*
reversal 1. Umkehr *f*, Umkehrung *f*; 2. Umsteuern *n*, Umsteuerung *f (Motor)*; 3. Lastspiel *n*; 4. Umpolung *f (Batterie)*
r. development Umkehrentwicklung *f (Fotografie)*
r. dwell Bewegungspause *f* bei der Umsteuerung
r. film Umkehrfilm *m*
r. in the direction Richtungsumkehr *f*; Umsteuern *n* [der Bewegung]
r. of flow Umkehr *f* der Strömungsrichtung
r. of magnetization Ummagnetisierung *f*
r. of motion Bewegungsumkehr *f*
r. of movement Bewegungsumkehr *f*
r. of phase *(El)* Phasenumkehr *f*
r. of polarity *(Ph)* Umkehr *f* der Polarität, Umpolen *n*
r. of poles *(El)* Umpolung *f*, Polwendung *f*; Polumschaltung *f*
r. of stress Lastwechsel *m (Festigkeit)*
r. of stroke *(Wärm)* Hubwechsel *m*
r. of the current *(El)* Stromrichtungsumkehr *f*, Stromwendung *f*
r. of the direction of rotation Umsteuerung *f* der Drehrichtung
r. of the direction of the flow of the steam Umkehrung *f* der Dampfstromrichtung *(Turbine)*
r. of the eccentric motion Umkehrung *f* der Exzenterbewegung
r. of the piston Kolbenumsteuerung *f*
r. of the table *(Masch)* Tischumsteuerung *f*
r. of [the] travel direction Umkehr (Umsteuerung) *f* der Bewegungsrichtung
r. point Wendepunkt *m*
r. test Umkehrprobe *f*
reversals *(Fmt)* Stromwechsel *m*
reverse / to umdrehen, umlegen, umschalten, umsteuern; rückwärts fahren, rückwärts laufen lassen
r. connections Anschlüsse umstecken
r. direction die Drehrichtung umkehren
reverse 1. Umsteuerung *f*; Umkehr *f*; Rücklauf *m*; 2. Gegenstück *n*; 3. Gegenseite *f*; Rückseite *f*; Gegenteil *n*
reverse rückwärts, umgekehrt, Umkehr-, Gegen-
r. acting rückwärts wirkend; reversierend *(Pneumatik; steigender Eingangsgröße entspricht fallende Ausgangsgröße)*
r. acting valve reversierendes Ventil *n*, mit steigendem Steuerdruck öffnendes Ventil
r. action Gegenwirkung *f*
r. bending strength Biegewechselfestigkeit *f*
r. bending test Hin- und Herbiegeversuch *m*
r. bias *(El)* Sperrvorspannung *f*
r. circulation drilling *(Brg)* Saugspülbohrverfahren *n*
r. clutch Rückwärtskupplung *f*
r. compound winding *(El)* Gegenkompoundwicklung *f*
r. current Sperrstrom *m*; Rückstrom *m*
r. current circuit breaker Rückstrom[selbst]schalter *m*, Rückstromautomat *m*
r.-current cleaning anodische Reinigung *f (Galvanotechnik)*
r. current cut-out Rückstrom[schalt]schütz *n*
r. current relay Rückstromrelais *n*
r. current switch Rückstromschalter *m*
r. curve S-Kurve *f*
r. danger light *(Eis)* Ersatzhaltlampe *f*
r. direction *(Am) (El)* Sperrichtung *f*
r. dog *(Masch)* Umsteuerknagge *f*, Anschlag *m* zum Umsteuern
r. drum Umkehrtrommel *f*
r. duplicating *(Masch)* spiegelbildliches Nachformen (Kopieren) *n*
r. feedback *(Fmt)* Gegenkopplung *f*, Regeneration *f*
r.-field detector Bremsfeldaudion *n*
r.-flow combustion chamber Brennkammer *f* mit Umkehrströmung
r.-flow cooling Kühlung *f* durch Rückbelüftung
r.-flow scavenging Umkehrspülung *f*
r. fluctuation Rückwärtsschwankung *f*
r. frame Gegenspant *n*
r. gear Umsteuergetriebe *n*, Wendegetriebe *n*; Rückwärtsgang *m*
r. gear box Wendegetriebe *n*
r. gear shaft *(Kfz)* Vorgelegewelle *f*
r. gear stop *(Kfz)* Rückwärtsganganschlag *m*
r. grid current Gitterrückstrom *m*
r. idler gear *(Kfz)* Rücklaufrad *n*
r. idler gear bushing Rücklaufbuchse *f*
r. idler shaft Rücklaufachse *f*
r. image attachment Spiegelbildkopiereinrichtung *f*
r. lamp Rückfahrscheinwerfer *m*
r. leakage current *(Am) (El)* Sperrstrom *m*
r. lever Umsteuerhebel *m*
r. lever ratchet Steuerbock *m*
r. locking relay Rückstromsperrelais *n*
r. mirror duplicating *(Masch)* spiegelbildliches Nachformen (Kopieren) *n*
r. order umgekehrte Reihenfolge *f*
r. phase *(El)* Gegenphase *f*
r. pitch negative Steigung *f (Gewinde)*
r. power *(El)* Rückleistung *f*

r. power circuit breaker Rückleistungsschalter *m*
r. power circuit-breaking Rückleistungsauslösung *f*
r. power protection Rückleistungsschutz *m*
r. power relay Rückleistungsrelais *n*
r. power tripping Rückleistungsauslösung *f*
r. press Steigpresse *f*, Wendepresse *f*
r. resistance Sperrwiderstand *m*
r. roll coating Umkehrwalzenbeschichtung *f*
r. rolling mill Umkehrwalzwerk *n*; Umkehr[walz]straße *f*
r. scavenging *(Kfz)* Umkehrspülung *f*
r. shaft Rücklaufwelle *f*
r. suction-press Wendesaugpresse *f (Papierherstellung)*
r.-thrust rocket Bremsrakete *f*
r. traverse Rücklauf *m*
r. twill *(Text)* Gegenköper *m*
r. twin gear Rücklaufdoppelrad *n*
r. valve *(Kfz)* Umsteuerventil *n*, Umsteuerschieber *m*
r. valve plunger Umsteuerkolben *m*
r. voltage *(El)* Sperrspannung *f*; Gegenspannung *f*
r. wheel Rücklaufrad *n*

reversed umgekehrt, verkehrt
r. bending Hin- und Herbiegung *f*
r. block *(Typ)* Negativklischee *n*
r. fatigue strength Wechselfestigkeit *f*
r. fault *(Geol)* Überschiebung *f*
r. feedback *(Fmt)* Gegenkopplung *f*
r. image *(Fs)* negatives Bild *n*; umgekehrtes Bild
r. right-to-left *(Opt)* seitenvertauscht
r. segregation *(Met)* umgekehrte Seigerung *f*
r. strength Wechselfestigkeit *f*
r. top to bottom *(Opt)* höhenverkehrt

reverser *(El)* Fahrtrichtungsschalter *m*, Fahrtwender *m*, Richtungswender *m*

reversibility Umkehrbarkeit *f*; Umsteuerbarkeit *f*

reversible umkehrbar; umsteuerbar; umdrehbar; umschaltbar; reversierbar; umsteckbar; umpolbar *(Polarisation)*
r. back of seat *(Eis)* klappbare Rückenlehne *f*
r. booster *(Elm)* Zusatzmaschine *f* für Zu- und Gegenschaltung, Survolteur-Dévolteur *m*
r. cell umkehrbares Element *n (Akkumulator)*
r. chain umkehrbare Kette *f*
r. colour film Farbumkehrfilm *m*
r. current Nachwirkungsstrom *m*
r. cycle umkehrbarer Kreisprozeß *m*
r. film Umkehrfilm *m*
r.-globe control valve reversierbares Regelventil *n*
r. in direction umsteuerbar
r. level *(Opt)* Reversionslibelle *f*
r. motor Umkehrmotor *m*, umsteuerbarer Motor *m*
r. motor drive Antrieb *m* durch umsteuerbaren Motor
r. plough Reversierpflug *m*, Umkehrpflug *m*, reversibler Pflug *m*
r. pressure curve *(Wärm)* Gleichgewichtsdruckkurve *f*
r. process reversibler Vorgang (Prozeß) *m*
r. propeller *(Schiff)* Drehflügelschraube *f*, Umsteuerpropeller *m*, Wendepropeller *m*, Verstellpropeller *m*
r. tape drive *(Text)* Bandumkehrantrieb *m*
r. winch Winde *f* mit Umsteuergetriebe

reversing umsteuerbar, mit Umkehr der Drehrichtung; umsteuernd, Umsteuer-
r. air valve Luftwechselklappe *f*
r. attachment for tapping Einrichtung *f* zur Drehrichtungsumkehr für Gewindebohrungen
r. blooming mill train Reversierblock[walz]straße *f*
r. bucket Umkehrschaufel *f (Turbine)*
r. cam Umschaltkurve *f*
r. clutch Umsteuerkupplung *f*
r. commutator Umkehrschalter *m*, Polwechselschalter *m*
r. controller Drehrichtungsumschalter *m*, Umlaufrichtungswendeschalter *m*, Fahrtwendeschalter *m*, Fahrtrichtungsschalter *m*
r. conveyor belt reversierbares Förderband *n*
r. dog *(Masch)* Umsteuerknagge *f*, Anschlag *m* zum Umsteuern
r. electric motor Umkehrmotor *m*
r. field *(El)* Kommutierungsfeld *n*
r. friction clutch Umsteuerreibkupplung *f*
r. gear Wendegetriebe *n*
r. handle Richtungshebel *m*, Umsteuerhebel *m*
r. lever *(Masch)* Umsteuerhebel *m*
r. mechanism Wendegetriebe *n*
r. mill Umkehrwalzwerk *n*, Reversierwalzwerk *n*
r. mirror Umkehrspiegel *m*
r. motor Umkehrmotor *m*
r. nozzle Umlenkungsdüse *f*, Umkehrkanal *m* *(Turbine)*
r. of the spindles Rückdrehung *f* der Spindeln
r. piston valve Kolbenumsteuerventil *n*
r. pole Wendepol *m*
r. rolling mill Umkehrwalzwerk *n*
r. starter Umkehranlasser *m*
r. steam engine Umkehrdampfmaschine *f*, Dampfmaschine *f* mit Umsteuerung
r. switch Wendeschalter *m*, Umschalter *m*, Polwendeschalter *m*, Polwechselschalter *m*

r. turbine Umkehrturbine *f*, Umsteuerungsturbine *f*, Reversierturbine *f*
r. valve Umsteuerventil *n*
r. valve of the sliding piston type Umsteuerlängsschieber *m*
reversion Rückkehr *f*; Umkehr *f*; Umkehrung *f*
reverting call switch *(Fmt)* Rückrufwähler *m*
revertive communication apparatus *(Fmt)* Rückmeldeanlage *f*
revet / to verkleiden *(Böschung)*
revetment Stützmauer *f*; Verkleidung *f*, Böschungsverkleidung *f*; Uferschutz *m*
r. wall Futtermauer *f*, Verkleidungsmauer *f*
review [of book] Buchbesprechung *f*, Rezension *f*
revise Zweitkorrektur *f*, dritte Korrektur *f*; Revision *f*
revive / to auffrischen
revivification Auffrischung *f*; Reaktivierung *f*
revivify / to auffrischen *(Formsand)*; reaktivieren
revolution 1. Zentralbewegung *f*; 2. *(Masch)* Umdrehung *f*; Umlauf *m*
r. counter *(Masch)* Drehzahlmesser *m*; Umdrehungszähler *m*
r. ellipsoid Rotationsellipsoid *n*
r. of table Tischumdrehung *f*
r. of the drill Bohrerumdrehung *f*
r. press *(Led)* Brückenstanze *f*
revolutions per minute *(Masch)* Umdrehungen *fpl* je Minute, U/min; Drehzahl *f*
revolve / to [sich] drehen, umdrehen; umlaufen
revolver Trommelrevolver *m*, Revolver *m*
revolving drehbar, kreisend, rotierend; Dreh-
r. armature *(Elm)* Drehanker *m*
r. bobbin *(Text)* Laufspule *f*
r. box loom Revolverwechselwebstuhl *m*
r. cage *(Text)* umlaufender Korb *m*
r. commutator rotierender Umschalter *m*, Drehumschalter *m*
r. contact-maker Typenrad *n (Telegrafie)*
r. crane Drehkran *m*, Schwenkkran *m*
r. cutter head dredger Kugelschaufler *m*
r. cylindrical furnace Drehrohrofen *m*
r. damper Drehrauchschieber *m*, Drehklappe *f*, Rauchklappe *f*
r. disk planing machine Scheibenhobelmaschine *f*
r. door Drehtür *f*
r. drum mixer Freifallmischer *m*
r. endless wire umlaufendes Langsieb *n (Papierherstellung)*
r. field Drehfeld *n*
r. field motor Asynchronmotor *m*
r. filter Zellenfilter *n*
r. flat card *(Text)* Wanderdeckelkrempel *f*
r. furnace Dreh[rohr]ofen *m*
r. grate *(Wärm)* Drehrost *m*
r. kiln Drehofen *m*
r. light Drehfeuer *n*
r. mechanism *(Masch)* Umschalteinrichtung *f*
r. of the flyer *(Text)* Flügeldrehung *f*
r. reverberatory furnace Drehflammofen *m*
r. screen Drehsieb *n*, Sortiertrommel *f*, Trommelsieb *n*
r. shutter Umlaufblende *f*
r. stage Drehbühne *f*
r. switch Drehschalter *m*, Knebelschalter *m*
r. table *(Masch)* Drehtisch *m*
r. tippler Kreiselwipper *m*
r. tower crane Turmdrehkran *m*
r. tubular kiln Drehrohrofen *m*
r. vane anemometer Flügelradanemometer *n*
rewash / to nachsortieren, nachwaschen *(Aufbereitung)*
rewash box Nachwaschsetzmaschine *f*, Nachsetzkasten *m (Aufbereitung)*
r. jig Nachwaschsetzmaschine *f (Aufbereitung)*
rewasher Nachwaschmaschine *f (Aufbereitung)*
reweigh / to [zu]rückwägen
rewelding of root Wurzelnachschweißung *f*
rewind / to rückspulen, umspulen
rewind Rückwicklung *f*, Umwicklung *f*; Rücklauf *m*
rewinder Umroller *m*; Umspuler *m*, Umwickler *m*
rewinding Rückspulen *n*, Umspulen *n*, Umwickeln *n*
r. key *(El)* Rücklauftaste *f*
rewire / to Leitungen neu legen, neuinstallieren
rework / to aufarbeiten, aufbereiten, umarbeiten, wiederaufarbeiten, nacharbeiten
reworking Umarbeitung *f*, Nacharbeit *f*, Nachbearbeitung *f*
Reynolds number Reynoldssche Zahl *f*
RF . . ., r.f. . . ., r-f . . . *s.* radio-frequency . . .
R.H. rail head rechter Querbalkensupport *m (Hobelmaschine)*
rhabdite *(Min)* Rhabdit *m*, Schreibersit *m*
rhamnazin Rhamnazin *n (Farbstoff)*
rhamnitol *(Ch)* Rhamnit *n*
rhea fibre Ramiefaser *f*
rhenate Rhenat *n*
rhenite Rhenit *n*
rhenium Rhenium *n*
rheo-box Rheoapparat *m (Aufbereitung)*
rheodyne circuit widerstandsgekoppelter Sender *m*
rheolaveur washery Rheowäsche *f (Aufbereitung)*
rheological rheologisch
rheology Rheologie *f*
r. properties Fließeigenschaften *fpl*
rheopexy Rheopexie *f (Kolloidchemie)*
rheostat Rheostat *m*, Widerstandsregler *m*, Regelwiderstand *m*, Kurbelwiderstand *m*, Vor[schalt]widerstand *m*
r. dimmer Verdunk[e]lungswiderstand *m*

r. slider *(El)* Widerstandsabgriff *m*
rheostatic braking Widerstandsbremsung *f (elektrischer Zug)*
r. braking controller Widerstandsbremsschalter *m*
r. control Widerstandsregelung *f*
r. controller Feldregelanlasser *m*
r. loss Joulesche Wärme *f*, Widerstandsverlust *m*
r. starter Regelanlasser *m*
r. starting Anlassen *n* mit Widerständen
rheostriction Rheostriktion *f*, Pincheffekt *m*
rheotron *(Kern)* Rheotron *n*
rho meson *(Kern)* Rho-Meson *n*
r. ratio Querzahl *f*
rhodanic acid Rhodanwasserstoffsäure *f*, Thiozyansäure *f*
rhodanide Rhodanid *n*, Ester *m* (Salz *n*) der Rhodanwasserstoffsäure
rhodanizing [galvanisches] Rhodinieren *n*
rhodium Rhodium *n*
r.-plated rhodiniert
rhodochrosite *(Min)* Rhodochrosit *m*, Manganspat *m*
rhodonite *(Min)* Rhodonit *m*, Kieselmanganerz *n*
rhomb Rhombus *m*
rhombic rhombisch
r. aerial Rhombusantenne *f*
r. dodecahedron Rhombendodekaeder *n*
rhombododecahedron Rhombendodekaeder *n*
rhombohedral rhomboedrisch
rhombohedron Rhomboeder *n*
rhomboid Rhomboid *n*
rhombus Rhombus *m*
rhumb *(Math)* Loxodrome *f (s. a.* r. line)
r. line Kurs *m*; Kompaßstrich[abstand] *m*
rhumbatron Hohlraumresonator *m*, Rhumbatron *n (Mikrowellentechnik)*
rhythmic light Taktfeuer *n (Navigation)*
rib / to mit Rippen versehen, rippen
rib 1. *(Masch)* Rippe *f*; Steife *f (Platte)*; *(Am)* Bohrerlippe *f*; 2. *(Brg)* anstehende Kohle *f (senkrecht zur Strebfront)*; 3. *(Geol)* Erzader *f*; 4. Furche *f*; 5. Spant *n (Boot)*
r. clearance Fasenhöhe *f*
r.-knitting machine Ripp[strick]maschine *f*
r. of the [grinding] wheel Schleifrippe *f*
r. stitching *(Flg)* Rippennaht *f (Tragfläche)*
r. velvet Rippensamt *m*
r. weave *(Text)* Ripsbindung *f*
ribband Kantholz *n*, Riegel *m*, Verbindungsstück *n*
ribbed *(Masch)* verrippt, durch Rippen versteift; gerippt, geriffelt; Riffel-, Rippen-
r. cover gerippter Deckel *m*, Rippengußdeckel *m*
r. fire bar Rippenroststab *m*
r. floor Rippendecke *f (Stahlbetonbau)*
r. glass Riffelglas *n*
r. radiator Rippenheizkörper *m*
r. slab Rippenplatte *f*, Rippendecke *f (Stahlbetonbau)*
r. slab filler Deckenfüllkörper *m (Stahlbetonbau)*
r. steel bar Rippenstahl *m*
r. surface machine *(Elm) (selbstgekühlte)* Maschine *f* mit Rippengehäuse
r. tube Lamellenrohr *n*
r.-type tread Rippenprofil *n (Reifen)*
r. upholstery Möbelrips *m*
ribbing *(Masch)* Verrippung *f*; Rippenversteifung *f*; Rippung *f*
ribbon Band *n*, Streifen *m*
r. agate *(Min)* Bandachat *m*
r. cable *(Fmt)* Bandkabel *n*, Flachkabel *n*
r. clip Bandschelle *f*
r. coil Bandspule *f*
r. conductor *(Fmt)* Bandleitung *f*, Rippenleiter *m*
r. development Bandbebauung *f (Siedlungsplanung)*
r. grained wood gestreiftes Holz *n*
r. guide Bandführung *f*
r. lap machine *(Text)* Wickelstrecke *f*, Kehrstrecke *f*
r. loom Bandwebstuhl *m*
r. mixer Band[schnecken]mischer *m*, Gegenstromdoppelschneckenmischwerk *n*
r. screw Bandschnecke *f*
r. steel armouring Bandstahlbewehrung *f*, Bandstahlarmierung *f (Kabel)*
r.-type dynamic speaker Bandlautsprecher *m*
r. weaving Bandweberei *f*
ribbonlike chip Bandspan *m*
ribbons Zügel *mpl*
ribonic acid Ribonsäure *f*
ribonucleic acid Ribonukleinsäure *f*
ribose Ribose *f*
ribside *(Brg)* anstehende Kohle *f (senkrecht zur Strebfront)*
r. road Abbaustrecke *f* mit einseitig anstehender Kohle
rice flour Reismehl *n*
r. paper Reispapier *n*, chinesisches Markpapier *n*
r. starch Reisstärke *f*
r. straw Reisstroh *n*
rich 1. reich, reichlich; 2. *(Brg)* ergiebig, mächtig, reich, fündig *(Lagerstätte)*; 3. satt *(Farbe)*; 4. fett *(Beton)*; 5. fruchtbar *(Boden)*
r. air-fuel mixture *(Kfz)* fettes Kraftstoff-Luft-Gemisch *n*
r. concrete fetter Beton *m*, zementreiche Betonmischung *f*

r. gas Starkgas *n*, Reichgas *n*
r. gas collecting main Starkgasvorlage *f*
r. in carbon kohlenstoffreich
r. in contrast kontrastreich *(z. B. Fernsehbild)*
r. in harmonics oberwellenreich
r. in oxygen sauerstoffreich
r. in precipitate niederschlagsreich
r. in tone *(Ak)* volltönend
r. lead Reichblei *n*
r. milk Vollmilch *f*
r. mix *s.* r. concrete
r. mixture fettes Gemisch *n*
r. ore hochprozentiges Erz *n*, Edelerz *n*
r. source ergiebige Quelle *f*
richness in colour Farbfreudigkeit *f*
r. of tone Klangfülle *f*
richterite *(Min)* Richterit *m*
ricinoleate Ester *m* (Salz *n*) der Rizinolsäure, Rizinoleat *n*
ricinoleic acid Rizinolsäure *f*, Rizinussäure *f*
rick Schober *m (Heu)*
r.-cloth Schoberdecktuch *n*
rickers Rundhölzer *npl (Durchmesser* $< 2\frac{1}{2}''$*)*
ricking Schoberbau *m*
ricochet / to abprallen
ricochet 1. Abprall *m*; 2. *(Mil)* Abpraller *m*
riddle / to 1. [ab]sieben, durchsieben; 2. durchlöchern
riddle Schüttelsieb *n*, Rätter *m*; Schüttler *m*
r. sieve Rätter *m*
riddling Grobsieben *n*
r. screen Schüttelsieb *n*
ride / to 1. reiten; 2. fahren *(Motorrad)*; 3. [vor Anker] liegen; 4. aufsitzen auf; sich erstrecken über; 5. *(Brg)* fahren
r. at anchor vor Anker liegen
r. in[bye] *(Brg)* einfahren
r. on the teeth aufklettern *(Kette)*
r. out[bye] *(Brg)* ausfahren
rider 1. Reiter *m*; 2. Fahrer *m (Motorrad)*; 3. problematische Mathematikaufgabe *f*; 4. Aufbau *m*, überbrückendes Teil *n*; 5. Aufsetzgewicht *n*; Laufgewicht *n*
r. keelson *(Schiff)* Oberkielschwein *n*, Kielschweinsohle *f*
riderless weaving harness *(Text)* reiterloses Webgeschirr *n*
ridge / to riefen; furchen
ridge 1. Gewinderippe *f*, Gang *m*; angeschliffener Grat *m*; Wulst *m*, 2. Dachfirst *m*, First *m*; Firstbohle *f*, Firstbalken *m*; 3. Furche *f*; Riefe *f*; Kamm *m*, Rücken *m*; 4. *(Geol)* Höhenrücken *m*
r. bar *(Text)* Helfenschiene *f*, Litzenschiene *f*
r. capping Firsthaube *f*, Firstabdeckung *f*
r. course Firstschar *f (Dachdeckung)*
r. covering *s.* r. capping
r. drill Reihensämaschine *f*
r. left by the tool Bearbeitungsriefe *f*
r. of a wave Wellenkamm *m*
r. plough Häufelpflug *m*
r. prism Dachkantprisma *n*
r. purlin *(Bw)* Firstpfette *f*
r. soaring Hangsegelflug *m*
r. tile Firstziegel *m*
r. waveguide Steghohlleiter *m*, Hohlleiter *m* mit Längssteg *(Wellenleitertechnik)*
ridgeless gratlos, riefenlos, wulstlos
ridger Häufelpflug *m*
ridging share *(Lw)* Häufelschar *n*
riding anchor *(Schiff)* tragender Anker *m*
r. boom *(Schiff)* Backspiere *f*
r. boots Reitstiefel *mpl*
r. breeches Reithose *f*
r. chock Ankerkettenstopper *m*
r. light Ankerlaterne *f*, Ankerlicht *n*
r. of the links on the teeth Aufklettern *n* der Kette auf Zahnflanken
r. out too near the ends of the teeth Aufklettern *n* der Kette auf Zahnflanken
r. stopper Anker[ketten]stopper *m*
riebeckite *(Min)* Riebeckit *m*
riffle / to furchen, riffeln
riffle Rinne *f*, wellenartige Riffelung *f*
riffled table Rillenherd *m (Aufbereitung)*
riffler Lochfeile *f*, Raspel *f*; Riffelblech *n*; Sandfang *m (Zellstoffherstellung)*
rifle / to 1. mit Gewehr schießen; 2. Züge einarbeiten; 3. riffeln, furchen
rifle 1. Gewehr *n*; 2. Zug *m (gezogener Lauf)*; 3. wellenartige Riffelung *f*
r. butt Gewehrkolben *m*
r. grenade Gewehrgranate *f*
r. range Schießstand *m*
r. sling Gewehrriemen *m*
rifled barrel Lauf *m* mit Zügen
rifling Zug *m (im gezogenen Lauf einer Schußwaffe)*
r. groove Gewehrlaufzug *m*
rift / to aufreißen, springen
rift Schlucht *f*, Kluft *f*; Riß *m*, Spalte *f*
r.-sawn zu Hobelbrettern gesägt
rig / to *(Schiff)* auftakeln; ausrüsten, ausstatten
r. the derrick *(Schiff)* Ladebaum klarmachen
r. up aufbauen *(Gerüste)* aufrüsten, auftakeln
rig 1. Vorrichtung *f*, Anlage *f*, Aufbau *m*, Ausrüstung *f*, Gerät *n*, Maschine *f*; 2. Bohranlage *f*, Bohrgestell *n*; 3. *(Schiff)* Takelung *f*, Takelage *f*, Takelwerk *n*
r. irons Eisenteile *npl (Seilbohren)*
rigged up *(Schiff)* aufgetakelt
rigger Transportarbeiter *m*
rigging 1. Abspannung *f*, Verspannung *f*; 2. *(Schiff)* Takelage *f*, Takelung *f*, Betakelung *f*; Gut *n (Tauwerk)*; Wanten *pl*

r. angle of incidence *(Flg)* Einstellwinkel *m*
r. plan *(Schiff)* Takelriß *m*
right / to aufrichten, ins Gleichgewicht bringen; korrigieren
r. the aircraft *(Flg)* Flugzeug abfangen
right rechts; rechtwinklig; gerade; richtig; direkt; aufrecht
r.-and-left-hand helix cutter Fräser *m* mit Pfeilverzahnung
r. angle rechter Winkel *m*
r.-angle drive Winkelantrieb *m*
r.-angle fold[ing] *(Typ)* Kreuzbruchfalz[ung *f*] *m*
r.-angle head [of planer] Querhobelsupport *m*
r. angled rechtwinklig
r. angularity Rechtwinkligkeit *f*
r. ascension *(Ph, Astr)* Rektaszension *f*
r. bent lathe tool gebogener rechter Drehmeißel *m*
r.-cut[ting] rechtsschneidend
r.-hand recht; Rechts-, rechts-
r.-hand cut[ting] rechtsschneidend
r.-hand helical cutting edges Schneiden *fpl* mit Rechtsdrall
r. hand propeller rechtsdrehender Propeller *m*, Rechtspropeller *m*
r.-hand rotation Rechtsdrehung *f*
r.-hand rule *(Ph)* Rechte-Hand-Regel *f*
r.-hand side rechte Seite *f*
r.-hand spindle rechter Wickelteller *m (Tonbandgerät)*
r.-hand spiral Rechtsspirale *f*, Rechtsdrall *m*
r.-hand thread Rechtsgewinde *n*
r.-hand twill *(Text)* Rechtsgratköper *m*
r.-hand winding *(El)* Rechtswicklung *f*, rechtsläufige (nach rechts laufende) Wicklung *f*
r.-handed rechtsgängig, rechtsläufig, rechtsdrehend; Rechts-
r.-handed helix Rechtsdrall *m*, Rechtsspirale *f*
r.-handed motor rechtsläufiger Motor *m*
r.-handed quartz rechtsdrehender Quarz *m*
r.-handed system [of coordinates] Rechtssystem *n*
r. line Gerade *f*
r.-lined geradlinig
r. prisma gerades Prisma *n*
r. rudder Steuerbordruder *n*
r. section Normalschnitt *m*
r. side view rechte Seitenansicht *f*
r. triangle rechtwinkliges Dreieck *n*
right Recht *n*
r. of co-use Mitbenutzungsrecht *n*
r. of exploitation Nutzungsrecht *n*
r. of recovery Regreßanspruch *m*, Regreßrecht *n*
r. of way Vorfahrt *f*
r. to recall a contract Kündigungsrecht *n* für einen Vertrag
righting Aufrichten *n*
r. capacity *(Schiff)* Aufrichtungsvermögen *n*
r. moment Standmoment *n (Statik)*
rightward welding Nachrechts-Schweißen *n*
rigid starr, unbiegsam, steif, biegesteif; formhaltig *(Keramik)*
r. adjustable die Schneidkluppe *f* mit verstellbarem Schneideisen
r. airship Starrluftschiff *n*
r. arch *(Bw)* eingespannter Bogen *m*
r. body starrer Körper *m (Statik)*
r.-body displacement Starrkörperverschiebung *f*
r. bond *(Ch)* starre Bindung *f*
r. coupling starre Kupplung *f*
r. double-tube core-barrel Doppelkernrohr *n* mit feststehendem Innenrohr *(Tiefbohrtechnik)*
r. driving mechanism starr gelagerter Antrieb *m*
r.-extensible prop *(Brg)* verstellbar-starrer Stempel *m*
r. frame *(Bw)* Rahmen *m* [mit biegesteifen Knoten]
r. in phase phasenstarr
r. insulation board steife Isolierplatte *f*
r.-jointed biegesteif verbunden
r. linkage *(Ch)* starre Bindung *f*
r. mould biegesteife Form *f* *(Spannbetonbau)*
r.-plastic starrplastisch
r. prop *(Brg)* starrer Stempel *m*
r. sheet harte Platte *f*, Hartplatte *f*, Hartfolie *f*
r. steel conduit Stahlpanzerrohr *n (Kabelschutz)*
r. suspension *(El)* starre Aufhängung *f* *(Leitung)*
r. tower *(Fmt)* Abspannmast *m*
r. voltage *(El)* steife Spannung *f*
rigidity Starrheit *f*, Steifigkeit *f*
rigidly attached to starr befestigt an
r. coupled *(El)* starr gekoppelt
rill stope *(Brg)* Firstenstoß *m* mit schräggestellter Firste
rim / to beranden *(Münzen)*
rim 1. Felge *f*, Kranz *m (Zahnrad)*; 2. Rand *m*, Bord *m*; 3. Wasseroberfläche *f*; 4. Siebkörper *m*
r. band Felgenband *n*
r. beam Randbalken *m*, Randträger *m*
r. brake Felgenbremse *f*
r. clutch Bandkupplung *f*
r. flange Felgenflansch *m*, Felgenrand *m*
r. lock Kastenschloß *n*
r. of disk Scheibenrand *m (Turbine)*
r. of guide blade disk (diaphragm) Leitkranz *m (Turbine)*

r. of the flywheel Schwungring *m*, Schwungradkranz *m*
r. of the gear Zahnkranz *m*
r. socket wrench Felgenwinde *f*
r. speed Umfangsgeschwindigkeit *f*, Radkranzgeschwindigkeit *f*
r. vortex Randwirbel *m (Aerodynamik)*
r. well Felgenbett *n*
r. with taper bead-seats Schrägschulterfelge *f*
r. wrench Radmuttersteckschlüssel *m*
rima Kluft *f*, Riß *m*
rime 1. Rauhreif *m*; 2. Leitersprosse *f*
rimmed unberuhigt vergossen *(Stahl)*
rimming unberuhigtes Vergießen *n (Stahl)*
r. ingot unberuhigt vergossener Block *m*, unberuhigter Block
r. steel unberuhigter (unberuhigt vergossener) Stahl *m*
rimose rissig
rimous *s.* rimose
rind Rinde *f*, Schale *f*, Hülse *f*; Schwarte *f*
ring / to klingen, läuten; klingeln; [an]rufen; *(hochfrequent)* überschwingen *(Impulstechnik)*
r. back *(Fmt)* zurückrufen
r. off *(Fmt)* abläuten, abrufen
r. up *(Fmt)* anrufen
ring 1. Ring *m*; Kreisring *m*; Torus *m*; Öse *f*, Wulst *m*; Arena *f*; Hülse *f*; 2. Klang *m*; 3. *(Fmt)* Anruf *m*, Ruf *m*
r. armature *(Elm)* Ringanker *m*
r.-back key *(Fmt)* Rückruftaste *f*
r. balance manometer Ringwaage *f*
r. beam Ringbalken *m*, Ringträger *m*; Ringanker *m*
r. burner Ringbrenner *m*, Kronenbrenner *m*
r. bushing Büchsenführung *f*
r. carrier Ringträger *m*, Tragring *m (Kolben)*
r. circuit Ringkreis *m (Mikrowellentechnik)*
r. cleavage Ringspaltung *f*
r. clogging *(Kfz)* Verkleben *n* der Kolbenringe
r. compound *(Ch)* zyklische Verbindung *f*, Ringverbindung *f*
r. connection *(Fmt)* Ruf[strom]schaltung *f*
r. core Ringkern *m (Spule)*
r. crush Ringstauchprüfung *f (Papier)*
r. current Kreisstrom *m*, Ringstrom *m*
r. demodulator Ringdemodulator *m*
r. discharge Ringentladung *f*
r. doubling frame Ringzwirnmaschine *f*
r. down connection *(Fmt)* Abfragebetrieb *m*
r. expander Ringdehner *m (für Kolbenringe)*
r. extractor Ringzieher *m (für Kolbenringe)*
r. filament ringförmig angeordneter Leuchtdraht *m*
r. filter *s.* r. mode filter
r. for the sealing liquid Fangschale *f*
r. formation *(Ch)* Ringbildung *f*
r.-forward signal *(Fmt)* Vorwärtsruf *m*
r. frame *s.* r. spinning frame
r. gasket Dichtungsring *m*
r. gate Ringverschluß *m*
r. gauge Hülsenlehre *f*, Lehrring *m*
r. gear Glockenrad *n*, Ringrad *n (Tellerrad)*, Zahnkranz *m*
r. groove Kolbennut *f*, Ringnut *f*
r. groove bearing Ringrillenlager *n*
r. header ringförmige Sammelleitung *f*
r. kiln Ringofen *m*
r. ladle carrier Gabelpfanne *f*
r. lamps Tiefstrahler *mpl (Boxring)*
r. land Ringsteg *m (Kolben)*
r. lock Ringsicherung *f (von Kolbenringen)*
r.-lubricated bearing Ringschmierlager *n*
r. lubrication Ringschmiervorrichtung *f*
r. mains *(El)* Ringnetz *n*; Sammelleitung *f*
r. mode filter Ringmodusfilter *n*, ringförmiges Wellenfilter *n*, Wellentypringfilter *n (Wellenleitertechnik)*
r. modulator Ringmodulator *m*
r. nozzle *(Masch)* ringförmige Düse *f*, Ringdüse *f*
r. nut Gewindering *m* [mit Zahnkranz]; Ringmutter *f*
r. of axle *(Kfz)* Achsenring *m*
r. of starter *(Kfz)* Anlasserzahnkranz *m*
r.-off indicator *(Fmt)* Schlußklappe *f*
r.-oil bearing Lager *n* mit Schmierring
r.-oiled mit Ringschmierung
r.-oiled bearing Ringschmierlager *n*
r. oiler Ölring *m*
r. oiling Ringschmierung *f*
r.-oiling bearing Ringschmierlager *n*
r. pendulum Drehpendel *n*
r. pivot Ringzapfen *m*
r. quartz Ringquarz *m*
r. rail Ringbank *f (Ringspinnen)*
r. rail motion Ringbankschubvorrichtung *f*
r. retainer Ringsicherung *f (für Kolbenringe)*
r. roll press Ringwalzenpresse *f*
r. scaler *(Fmt)* rückgekoppelte Untersetzerschaltung *f*, Untersetzer *m* in Ringschaltung
r. screw gauge Gewindelehrring *m*
r. seal 1. Ringverschmelzung *f (Elektronenröhren)*; 2. Ringdichtung *f*
r. seam Ringnaht *f*
r. segment Ringsegment *n*, Ringstück *n*, Kreisringsegment *n*
r. shake Ringriß *m (Holzfehler)*
r.-shaky kernschälig *(Holz)*
r.-shaped ringförmig
r. sight Kreiskornvisier *n*
r. spanner for square nuts Vierkantschlüssel *m*

r. spinning Ringspinnen *n*
r. spinning frame Ringspinnmaschine *f*
r. spinning frame for pin cops Ringspinnmaschine *f* für Schußkötzer
r. spinning frame for warp cops Ringspinnmaschine *f* für Zettelkötzer
r. spinning frame with twisting head Ringspinnmaschine *f* mit Drehkopf
r. spring Ringfeder *f*
r.-spun yarn *(Text)* Watergarn *n*
r. sticking *(Kfz)* Festkleben (Hängenbleiben) *n* der Kolbenringe, Ringstecken *n*, Ringverkleben *n*
r. stop peg Kolbenringsicherungsstift *m*
r. support Tragring *m*, Ringträger *m (Kolben)*
r. switch Resonanz[ring]schalter *m (Wellenleitertechnik)*
r. system Ringleitung *f*
r. temple *(Text)* Stachelscheibenbreithalter *m*
r. thread gauge Gewindelehrring *m*
r. transformer Ringtransformator *m*
r. traveller *(Text)* Ringläufer *m*
r. twister Ringzwirnmaschine *f*
r. twisting frame Ringzwirnmaschine *f*
r.-type modulator Doppelgegentaktgleichrichter *m (Trägerfrequenztechnik)*
r.-type recording head ringförmiger Aufzeichnungskopf (Schreibkopf) *m (für magnetische Datenaufzeichnung)*
r. wheel Schleifring *m*
r. winding *(El)* Ringwicklung *f*
r. yarn Ringgarn *n*
ringer *(Fmt)* Ruf[strom]maschine *f*, Rufstromgeber *m*
ringing 1. *(Fmt)* Anruf *m*, Ruf *m*; 2. *(Fs)* Bildverdopplung *f*; 3. Klingen *n (Elektronenröhren)*; 4. abklingende Schwingung *f (Funktechnik)*; 5. *(Fs)* gedämpfte (gedämpft verlaufende) Schwingung
r. amplifier *(Fmt)* Rufverstärker *m*
r. battery *(Fmt)* Klingelbatterie *f*
r. change-over switch *(Fmt)* Weckerumschalter *m*
r. circuit getastete Oszillatorstufe *f*, Klingelschaltung *f*
r. current *(Fmt)* Rufstrom *m*
r. dynamo *(Fmt)* Rufstrommaschine *f*
r. frequency *(Fmt)* Rufstromfrequenz *f*
r. interrupter *(Fmt)* Rufzeichenerzeuger *m*
r. periodicity *(Fmt)* Rufphase *f*, Wiederkehr *f* der Rufstromsendung
r. relay *(Fmt)* Ruf[anschalte]relais *n*
r.-trip relay *(Fmt)* Rufabschalterelais *n*; Trennrelais *n*
r. wire Klingeldraht *m*
Rinmann's green Rinmanns Grün *n*
rinse / to [ab]schwemmen, [ab]spülen, abwaschen, ausspülen, auswaschen, wässern; abbrausen
rinsing agent Spülmittel *n*
r. machine *(Text)* Spülmaschine *f*
r. machine with propeller Spülmaschine *f* mit Treiber
r. screen Brausesieb *n (Aufbereitung)*
r. solution *(Led)* Pickelbad *n*
rip / to reißen; auftrennen; längssägen *(mit der Faser)*; schlitzen
r. out herausreißen, heraustrennen
r. up aufschlitzen
rip Riß *m*
r. cord Reißschnur *f*; Handabzug *m*
r.-rap 1. *(Bw)* Steinschüttung *f*; Steinpackung *f*; 2. hartes Gestein *n (Tagebau)*
r.-rapping *(Bw)* Steinschüttung *f*; Steinpakkung *f*
r.-saw Säge *f* für Längsschnitt *(mit der Faser)*
r.-saw teeth Kernsägezähne *mpl*
ripper *(Bw)* Straßenaufreißer *m*; *(Brg)* Aufreißer *m*
ripping *(Brg)* Nachreißstelle *f*; Nachreißen *n* des Hangenden in Abbaustrecken
r. bar Kistenöffner *m*
r. circular saw Trennkreissäge *f*
r. knife Trennmesser *n*
r. lip *(Brg)* Nachreißstelle *f*
ripple / to brummen; riffeln; [sich] kräuseln *(Wasser)*
r. the flax Flachs riffeln
ripple 1. *(El)* Brummen *n*, Netzbrummen *n*, Welligkeit *f (eines Gleichstroms)*; Rauhigkeit *f*, Störgeräusch *n*, überlagerte Störungen *fpl*; 2. *(Text)* Riffelkamm *m*
r. component Welligkeitsanteil *m* des Stromes
r. current Kräuselstrom *m*, Brummstrom *m*
r. effect Brummeffekt *m*
r. factor Brummfaktor *m*, Welligkeitsfaktor *m*
r. filter Siebschaltung *f*
r. filter choke Glättungsdrossel *f*
r. frequency Welligkeitsfrequenz *f*, Brummfrequenz *f*
r. percentage *(El)* Welligkeitsgrad *m*
r. ratio Brummspannungsverhältnis *n*
r. voltage Welligkeitsspannung *f*; Brummspannung *f*
rippling comb *(Text)* Riffelkamm *m*
riprap *s.* rip-rap
rise / to steigen, ansteigen, zunehmen, sich steigern, sich heben, sich erhöhen, sich erheben, aufstehen; aufgehen *(Sonne)*; aufsteigen *(Gase)*; heißen *(Flagge)*; abheben *(Werkzeug)*; aufquellen, anschwellen; gären; aufsprudeln
rise 1. Steig[er]ung *f*, Anstieg *m*, Erhöhung *f*, Aufstieg *m*; Aufgang *m (Sonne)*; Erhebung *f*, Hebung *f*, Zunahme *f*, Wachstum *n*; Abhebung *f (Werkzeug)*; 2. *(Bw)* Pfeil[höhe *f*] *m*,

Stich[höhe *f*] *m*; Steigungshöhe *f*; 3. *(Brg)* Überhauen *n*, Aufhauen *n*

r.-and-fall mechanism *(Masch)* Hebe- und Senkeinrichtung *f*

r. face *(Brg)* schwebender Streb *m*

r. heading *(Brg)* Aufhauen *n*

r. in pressure Druckanstieg *m*

r. of bottom Aufkimmung *f* *(Schiffsboden)*

r. of current Stromanstieg *m*; Stromerhöhung *f*, Stromzunahme *f*

r. of floor Aufkimmung *f* *(Schiffsboden)*

r. of temperature Temperaturanstieg *m*, Temperaturerhöhung *f*, Temperaturzunahme *f*

r. of water nach oben verlaufender Wasserstau *m* *(Kondensator)*

r. per tooth Überhöhung *f* von Zahn zu Zahn *(Räumwerkzeug)*

r. time Anstiegszeit *f*

r. time correction *(Fs)* Laufzeitentzerrung *f*

r. time distortion *(Fs)* Laufzeitverzerrung *f*

riser 1. *(El)* Steigleitung *f*; 2. Steiger *m* *(Gießform)*; 3. Aufstiegsöffnung *f* *(von den Glockenböden)*; 4. Futterbrett *n*, Setzstufe *f* *(Holztreppe)*; 5. *(Schiff)* Decksunterzug *m*; 6. *(Geol)* Verwerfungskluft *f*

rising Anwachsen *n*, Schwellen *n*, Ansteigen *n*

r. box *(Text)* Steigkasten *m*, Fallkasten *m*, Hubkasten *m*

r. casting steigender Guß *m*

r. floor hohe Bodenwrange *f* *(an den Schiffsenden)*

r. gas aufsteigendes Gas *n* *(im flüssigen Metall)*

r. gate Steiger *m* *(am Gußstück)*

r. head *(Met)* verlorener Kopf *m*

r. main[s] Steigleitung *f*

r. platform Hebebühne *f*

r. sea aufkommende See *f*

r. space *(Typ)* Spieß *m*

r. sun-type magnetron *(Fmt)* Magnetron *n* mit Schlitzresonatoren ungleicher Tiefe

r. wood *(Schiff)* Totholz *n*

risk function Risikofunktion *f*

Ritter method Rittersches Schnittverfahren *n* *(Statik)*

riven timber Nutzspaltholz *n*

river Fluß *m*, Strom *m*

r. and sea-going vessel Fluß-Seeschiff *n*

r. arm Flußarm *m*

r. bank Flußufer *n*

r. barge Flußkahn *m*

r. boat Flußfahrzeug *n*, Flußschiff *n*

r. cable Flußkabel *n*

r. construction Flußbau *m*

r. craft Flußfahrzeug *n*

r. crossing Flußübergang *m*

r. engineering Flußbau *m*

r. floor Flußsohle *f*

r. gravel Flußkies *m*

r. harbour *s.* r. port

r. launch Flußbarkasse *f*

r. mud Flußschlamm *m*

r. navigation Flußschiffahrt *f*

r. navigation lock Flußschleuse *f*

r. paddle-wheel steamer Flußraddampfer *m*

r. pier Strompfeiler *m* *(Brückenbau)*

r. pollution Flußverunreinigung *f*

r. port Flußhafen *m*, Stromhafen *m*

r. sand Flußsand *m*

r. service Flußfahrt *f*

r. steamer Flußdampfer *m*

r. survey Flußvermessung *f*

r. vessel Flußschiff *n*

r. weir Flußwehr *n*

rivet / to nieten

rivet Niet *m*

r. arrangement Nietbild *n*

r. body Nietschaft *m*

r. buster Nietenquetscher *m*

r. chisel Nietmeißel *m*

r. driving Nietenschlagen *n*

r. fastening Nietverbindung *f*

r. gun pneumatische Nietmaschine *f*

r. hammer Niethammer *m*

r. head Nietkopf *m*; Setzkopf *m*; Schließkopf *m*

r. header Nietkopfsetzer *m*, Nietkopfanstauchmaschine *f*, Kopfsetzer *m*, Nietendöpper *m*

r. hole Nietloch *n*

r. joint Nietverbindung *f*

r. line Nietrißlinie *f*

r. pitch Nietabstand *m*

r. pliers Nietzange *f*

r. set Döpper *m*, Kopfmacher *m*, Schellhammer *m*, Nietkopfsetzer *m*

r. shank Nietschaft *m*

r. snap *s.* r. set

r. stock Nieteisen *n*

r. tongs Nietenhalter *m*, Nietzange *f*

rivetability Nietbarkeit *f*

riveted genietet

r. butt joint with cover plate[s] Laschennietung *f*

r. butt joint with single cover plate einseitige Laschennietung *f*

r. combination lap-joint durch Lasche verstärkte Überlappungsnietung *f*

r. connection Nietverbindung *f*

r. girder Nietträger *m*

r. joint Nietverbindung *f*

r. joint with double row of rivets zweireihige Nietverbindung *f*

r. joint with single row of rivets einreihige Nietverbindung *f*

r. lap-joint Überlappungsnietung *f*

r. section Nietfeld *n*

r. to aufgenietet auf
riveting Nieten *n*, Nietung *f*
r.-clamp Nietzwinge *f*
r. machine Nietmaschine *f*
r. press Nietpresse *f*
r. rig Ausnietgroßvorrichtung *f*
r. set Kopfmacher *m*, Schellhammer *m*, Schellkopf *m*, Döpper *m*, Nietdöpper *m*
r. snap *s.* r. set
r. tongs Nietzange *f*
rms . . . *s.* root mean square . . .
road 1. Straße *f*, Landstraße *f*, Weg *m*; Strecke *f*; 2. *(Schiff)* Reede *f*; 3. *(Brg)* Pfeilerstrecke *f* *(Pfeilerbau)*; 4. *s. a.* roadway
r. bed *(Eis)* Unterbau *m*, Schienenkörper *m*
r. bifurcation *(Brg)* Streckenabzweigung *f*
r. breaker Straßenaufbruchhammer *m*, Straßenaufreißer *m* *(Straßenbau)*
r. bridge Straßenbrücke *f*
r. builder Straßenbauer *m*
r. building Straßenbau *m*
r. building machinery Straßenbaumaschinen *fpl*
r. burner Asphaltdeckenerhitzer *m*
r. carpet Teppichbelag *m* *(Straßenbau)*
r. channel Rinnstein *m*; Straßenrinne *f*
r. clearance Bodenabstand *m*, Bodenfreiheit *f* *(eines Fahrzeugs)*
r. construction Straßenbau *m*, Wegebau *m*
r. ditch Straßengraben *m*
r. drag *(Kfz)* Rollwiderstand *m*
r. drill Straßenaufbruchhammer *m*
r. engineering Straßenbau *m*
r. finisher Deckenfertiger *m*, Straßenfertiger *m*
r. finishing machine Deckenfertiger *m*, Straßenfertiger *m*
r. flusher Sprengwagen *m*, Straßensprengwagen *m*
r. foundation Straßenunterbau *m*
r. friction Fahrbahnreibung *f*
r. furniture Straßenausrüstung *f*
r. grader Straßenhobel *m*, Straßenplanierer *m*
r. gravelling Straßenbeschotterung *f*
r. grinder Straßenschleifmaschine *f*
r. grip *(Kfz)* Kraftschluß *m* zur Straße *(des Laufstreifens eines Reifens)*
r. heater *s.* r. burner
r. holding *(Kfz)* Straßenlage *f*
r. joint Straßenfuge *f*
r. kettle Straßenkocher *m* *(Straßenbau)*
r. leveller Straßenwalze *f*
r. life Straßennutzungsdauer *f*
r. making *s.* roadmaking
r. marking Fahrbahnmarkierung *f*
r.-marking machine Straßenmarkierungsmaschine *f*, Strichziehgerät *n* *(zur Anzeichnung von Verkehrslinien)*
r. mat Teppichbelag *m* *(Straßenbau)*
r. metal Schotter *m*, Straßenschotter *m*
r. mix Straßendeckenmischung *f*; Bodenvermörtelung *f* *(Straßenbau)*
r. overseer Straßenmeister *m*
r.-paving block[s] Holzpflaster *n*
r. repair Straßenausbesserung *f*
r. ripper Straßenaufreißer *m* *(Straßenbau)*
r. roller Straßenwalze *f*
r. shelf Straßenanschnitt *m* *(Straßenbau)*
r. sign Verkehrszeichen *n*
r. stud Markierungsknopf *m*
r. surface *s.* r. surfacing
r. surfacing Straßenbelag *m*, Straßendecke *f*, Verschleißschicht *f*
r. tar Straßenteer *m*
r. timbering *(Brg)* Streckenausbau *m*
r. traffic Straßenverkehr *m*
r. transport[ation] Straßentransport *m*
roadability Fahreigenschaft *f*, Straßenlage *f*, Straßenverhalten *n* *(eines Fahrzeugs)*
roadhead *(Brg)* Streckenort *n*
roadholding straßengängig *(Fahrzeug)*
roadmaking Straßenbau *m*, Wegebau *m*
roadman Straßenbauarbeiter *m*; Straßenwärter *m*
roadmixer Bodenvermörtelungsmaschine *f* *(Straßenbau)*
roads *(Schiff)* Reede *f*
roadside ditch Straßengraben *m*
r. filling station Tankstelle *f*, Zapfstelle *f*
r. pack *(Brg)* Streckendamm *m*, Bergemauer *f*
r. planting Straßen[seiten]bepflanzung *f*
r. restaurant Raststätte *f*
roadstead *(Schiff)* Reede *f*, Ankergrund *m*
roadster 1. auf Reede liegendes Schiff *n*; 2. Tourenrad *n* *(Fahrrad)*; Straßenfahrzeug *n*; 3. Zugtier *n*
roadway 1. Straße *f*; Fahrbahn *f*, Straßendamm *m*; 2. *(Brg)* Abbaustrecke *f*; 3. *s. a.* road
r. arch Streckenbogen *m* *(Ausbau)*, Streckengewölbe *n* *(Gebirgsmechanik)*
r. excavation *(Brg)* Streckenraum *m*
r. flooring Fahrbahndecke *f*
r. flusher Straßensprengwagen *m*
r. junction *(Brg)* Streckenabzweigung *f*
r. maintenance *(Brg)* Streckenunterhaltung *f*
r. ripping machine *(Brg)* Streckenvortriebsmaschine *f*
r. support *(Brg)* Streckenausbau *m*
roadworks Straßenbauarbeiten *fpl*
roadworthiness Verkehrstüchtigkeit *f*
roadworthy straßentüchtig
roaring flame nichtleuchtende Flamme *f*
r. of the engine *(Kfz)* Brüllen *n* des Motors
roast / to [ab]rösten, abschwelen, brennen
r. and sinter sintern *(Erze)*
roast *s.* roasting

r. gas Röstgas *n*
r. heap Rösthaufen *m*
r.-reaction process Röstreaktionsarbeit *f*, Röstreaktionsprozeß *m* *(Bleidarstellung)*
r.-reduction process Röstreduktionsarbeit *f*
roasted geröstet
r. material Röstgut *n*
r. pyrites Kiesabbrand *m*
roaster Röstofen *m*
r. gas Röstgas *n*
roasting Rösten *n*, Abrösten *n*, Brennen *n*
r. and smelting Röstschmelzen *n*
r. bed Röstbett *n*
r. blast-furnace Röstschachtofen *m*
r. chamber Röstkammer *f*
r. furnace Röstofen *m*, Brennofen *m*
r. hearth Röstherd *m*
r. in heaps Haufenrösten *n*
r. kiln Röstofen *m*
r. plant Röstanlage *f*
r. reaction method Röstreaktionsverfahren *n*
r. reduction method Röstreduktionsverfahren *n*
r. residue Abbrand *m*, Verlust *m*
r.-smelting Röstschmelzen *n*
r. temperature Rösttemperatur *f*
rob / to *(Brg)* rauben
robot *(Am)* Automat *m*, automatisches Gerät *n*, Roboter *m*
r. pilot Autopilot *m*, automatische Kurssteueranlage *f*
robotize / to *(Am)* automatisieren
rock / to wiegen, schwanken, schaukeln, beben, erschüttern
rock Fels *m*, Gestein *n*
r. asphalt Asphaltgestein *n*, asphaltdurchtränktes Gestein *n*
r. beam *(Brg)* Gesteinsplatte *f*, Gesteinsbalken *m*
r. bed *(Geol)* Gesteinsschicht *f*
r. blasting Felssprengung *f*
r.-blasting explosive Gesteinssprengmittel *n*, Gesteinssprengstoff *m*
r. bolt *(Brg)* Anker[bolzen] *m*, Ausbauanker *m*, Gebirgsanker *m*, Gesteinsanker *m*
r. breaker Steinbrecher *m*
r. burst *(Brg)* Gebirgsschlag *m*
r. capping *(Am)* *(Brg)* Deckgebirge *n*
r. cork *(Min)* Korkstein *m*, Bergkork *m*
r. crusher Steinbrecher *m*
r. crystal Bergkristall *m*, kristallisierter Quarz *m*
r. cutter Felsbrecher *m*
r. dating method geologische Altersbestimmung *f*
r. de-stressing *(Brg)* Gebirgsentspannung *f*
r. decay Verwitterung *f* *(von Gestein)*
r. dressing Gesteinsaufbereitung *f*
r. drill Bohrhammer *m*, Gesteinsbohrer *m*, Felsbohrer *m*
r. driller 1. Felsenbohrschiff *n*; 2. Gesteinshauer *m*
r. drilling Gesteinsbohren *n*, Felsbohren *n*
r. fall Steinschlag *m*
r.-fill dam Steinschüttdamm *m*
r. flint Bergkiesel *m*
r. flour Gesteinsmehl *n*
r. formation Gesteinsschicht *f*
r.-forming gesteinsbildend, gebirgsbildend
r. foundation Gründungsfels *m*
r. gas Erdgas *n*, Naturgas *n*
r. industry Steinindustrie *f*
r. mass *(Brg)* Gebirgskörper *m*
r. mechanics Gebirgsmechanik *f*
r. minerals gesteinsbildende Mineralien *npl*
r. movement *(Brg)* Gebirgsbewegungen *fpl*
r. oil Erdöl *n*, Steinöl *n*
r. pentice Bergfeste *f* *(Schachtabteufen)*
r. plant Gesteinsaufbereitungsanlage *f*, Schotteraufbereitungsanlage *f*
r. pressure Gebirgsdruck *m*
r. salt Steinsalz *n*, Kochsalz *n*
r.-salt mine Salzbergwerk *n*
r.-salt prism Steinsalzprisma *n*
r. temperature gradient *(Geol)* geothermische Tiefenstufe *f*
rockair Raketenstart *m* vom Flugzeug aus
rocked tube kaltgepilgertes Rohr *n*
rocker 1. Schwinge *f*, Kulisse *f*, Kurbelschwinge *f*; 2. Schwinghebel *m*, Kipphebel *m*; 3. Wiege *f*, Wippe *f*; Wiegekufe *f*, Rollkufe *f*; 4. *(Elm)* Bürstengestell *n*
r. arm Kulisse *f*, Schwinge *f*, Kurbelschwinge *f*; Kulissenhebel *m*, Kipphebel *m*, Schwinghebel *m*; Unterbrecherhammer *m*, Unterbrecherhebel *m*; Gleitstein *m*; Steuerhebel *m*
r. arm shaft Kipphebelwelle *f*
r. bar Kippriegel *m*
r.-bar type furnace *(Met)* Balkenherdofen *m*, Schrittmacherofen *m*, Hubbalkenofen *m*
r. bearing Kipplager *n* *(Brückenbau)*
r. box Kipphebelgehäuse *n*
r. clearance Ventilspiel *n*
r. gear Kipphebelsteuerung *f* *(Flugzeugmotor)*
r. pivot Schwenklager *n*
r. shaft Schaukelwelle *f*, Kipphebelwelle *f*
r. shovel [loader] Überkopflader *m*, Schaufellader *m* *(Gesteinsstrecken)*
rockered keel *(Schiff)* gekrümmter Kiel *m*
rocket Rakete *f*; Raketentriebwerk *n*
r. aircraft Raketenflugzeug *n*
r. airframe Raketenkörper *m*
r.-assisted take-off Starthilfstriebwerk *n*, Starthilfe *f*, Startbeschleuniger *m*
r. base Raketenversuchsgelände *n*, Raketenstützpunkt *m*; Raketenstartbasis *f*

r. battery *(Mil)* Raketenbatterie *f*
r. blast Flamm[en]strahl *m*
r. body Raketenkörper *m*
r.-borne Raketen-
r. carrier Raketenträger *m*
r. combustion chamber Raketenbrennkammer *f*, Raketenofen *m*
r. composition Raketentreibsatz *m*
r. control Raketensteuerung *f*
r. drive Raketenantrieb *m*
r.-driven aircraft Raketenflugzeug *n*
r. engine Raketentriebwerk *n*
r. engineering Raketentechnik *f*
r. escaping *(Rak)* Verlassen *n* des Bereiches der Erdanziehung
r. exhaust drilling Bohren *n* mit heißen Abgasen *(Tiefbohrtechnik)*
r. expert Raketenfachmann *m*
r. firing duration Brennzeit *f*
r. flight Raketenflug *m*
r. fuel Raketenbrennstoff *m*, Raketentreibstoff *m*
r.-propelling charge Raketentreibsatz *m*
r. propulsion Raketenantrieb *m*
r. sounding Raketensondierung *f*, Raketenaufstieg *m*
r. stabilization Raketenstabilisierung *f*
r. stage Raketenstufe *f*
r. technique Raketentechnik *f*
r. track Startgestell *n*, Startschiene *f*
rocketeer Raketenfachmann *m*
rocketry Raketentechnik *f*
rocking schwingend, oszillierend
r. arc furnace Lichtbogenschaukelofen *m*
r. arm *(Masch)* Schwingarm *m (Shaper)*
r. lever Kipphebel *m*, Schwenkhebel *m*, Schwinghebel *m*, Schwunghebel *m*, Unterbrecherhebel *m*
r. motion Schaukelbewegung *f*, Drehschwingung *f*
r. pier Pendelpfeiler *m (Brückenbau)*
r. plate micrometer Planplattenmikrometer *n*
r. runner *(Met)* Schaukelrinne *f*
r. guidance Raketenlenkung *f*
r. igniter Zünder *m* für Raketentriebwerk
r. instrumentation Raketeninstrumentierung *f*
r. launching Raketenstart *m*
r. missile Raketengeschoß *n*
r. model Modellrakete *f*
r. motor Raketentriebwerk *n*
r. motor assembly Raketentriebwerk *n*
r. nozzle Raketendüse *f*
r. power Schub *m* des Raketentriebwerkes
r.-powered speedboat Schnellboot *n* mit Raketenantrieb
r. projectile Raketengeschoß *n*
r.-propelled raketengetrieben, mit Raketenantrieb
r.-type furnace *(Met)* Schaukelofen *m*, Kippofen *m*
r. vibration Schaukelschwingung *f*
rockoon [rocket] Raketenstart *m* vom Ballon aus
Rockwell hardness Rockwellhärte *f (Werkstoffprüfung)*
Rockwell hardness test Härteprüfung *f* nach Rockwell, Rockwell-Härteprüfung *f*
rocky felsig, steinig
r. ledge Felsenriff *n*
r. matter *(Brg)* Gangart *f*
rod / to stochern *(Betonverdichtung)*
rod 1. Stab *m*; Rundstab *m*, Rundstahl *m*; Stange *f*; Strebe *f*; Bügel *m*; Querhaupt *n*; Kolbenstangenschaft *m*; Draht *m*; 2. 5½ Yards *(Längenmaß)*; 3. 306 Kubikfuß Ziegelmauerwerk *(Raummaß)*
r. aerial Stabantenne *f*
r.-and-tube thermometer Stabausdehnungsthermometer *n*
r. bender Biegemaschine *f*, Biegetisch *m (Bewehrungsbau, Stahlbeton)*
r. bore *(Masch)* Schaftausbohrung *f (Schubstange)*
r. chuck Stangenspannfutter *n*
r. collector Stangenstromabnehmer *m*
r. cutter Stangenschneidmaschine *f*
r. drive *(Kern, El)* Stabantrieb *m*
r. feed Stangenvorschub *m*
r. insertion *(Kern)* Stabeinführung *f*
r. lattice *(Kern)* Stabgitter *n*
r. mill 1. Rohrmühle *f*; Stabmühle *f*; 2. Feineisenstraße *f*; Stabstraße *f (Walzwerk)*
r. milling and fluting machine Rundstabfräs- und -furchmaschine *f (Holzbearbeitung)*
r. mirror *(Fs)* Stabreflektor[wand *f*] *m*
r.-operated switch Stangenschalter *m*, Schalter *m* mit Betätigungsstange
r. puller *(Brg)* Gestängeziehvorrichtung *f*
r. reel Drahthaspel *f*
r.-rolling mill Drahtwalzwerk *n*; Draht-[walz]straße *f*
r.-shaped stabförmig
r. sulphur Stangenschwefel *m*
r. support *(Fmt)* Hängedraht *m*, Hängeseil *n*
r.-type carbon resistor *(El)* Karbowid-Widerstand *m*
r. wire Walzdraht *m*
rodding Stochern *n*, Stampfen *n* von Hand *(Betonverdichtung)*
rodenticide Rattenvertilgungsmittel *n*, Rodentizid *n*
rodent Nagetier *n*
rods Gestänge *n*; Stabstahl *m*; Walzdraht *m*
roentgen Röntgen *n (Einheit der Ionendosis einer ionisierenden Strahlung; 1 R = 2,58 · 10^{-4} C/kg)*
r. equivalent Röntgenäquivalent *n*

r. meter Röntgenmeter *n*, Röntgenmeßgerät *n*
r. meter chamber Röntgenmeterkammer *f*
r. radiation Röntgenstrahlung *f*
r. rays Röntgenstrahlen *mpl*
r. spectrum Röntgenspektrum *n*
Roentgen apparatus Röntgengerät *n*, Röntgenapparat *m*
Roentgen tube Röntgenröhre *f*
roentgenofluorescence Röntgenfluoreszenz *f*
roentgenogram Röntgenbild *n*, Röntgenaufnahme *f*, Röntgenogramm *n*
roentgenography Röntgenografie *f*, Röntgenaufnahme *f (Verfahren)*
roentgenology Röntgenologie *f*, Röntgenstrahlenkunde *f*
roentgenoluminescence Röntgenlumineszenz *f*
roentgenomateriology Röntgen[werkstoff]prüfung *f*, zerstörungsfreie Werkstoffprüfung *f* mit Röntgenstrahlen
roentgenoscope Durchleuchtungsgerät *n*
roentgenospectroscopic röntgenspektroskopisch
rogersite Rogersit *m (Uranmineral)*
rogue rocket vom Ziel abweichende Rakete *f*, ihr Ziel verfehlende Rakete
roll / to 1. *(Masch)* rollen *(Gewinde)*; walzen; druckumformen; 2. *(Schiff)* schlingern
r. along abrollen auf
r. away wegrollen, abrollen
r. down herunterwalzen *(Kaliber)*
r. in einwalzen
r. off abwälzen
r. out *(Dat, Met)* auswalzen
r. thread Gewinde walzen
r. up aufrollen, zusammenrollen
roll 1. Rolle *f*; Walze *f*; Ablaufhaspel *f*, Auflaufhaspel *f*; Welle *f*; Wulst *m*; 2. *(Geol)* Flözfaltung *f*
r. body Walz[en]ballen *m*
r. coater Aufwalzvorrichtung *f*
r. coating Walzzunder *m*
r. contouring lathe Walzennachformdrehmaschine *f*
r. copper Walzkupfer *n*
r. core Walzenkern *m*
r. crusher Walzenbrecher *m*
r. design Walzkalibrierung *f*
r. designing Kalibrierung *f (von Walzen)*
r. drawing *(Text)* Walzenverzug *m*
r.-end milling machine Abflächmaschine *f*, Fräsmaschine *f* zur Bearbeitung der Stirnseiten von Walzen
r. feed Walzenvorschub *m*
r. film Rollfilm *m*
r. for cold-rolling metals Kaltwalze *f (Walzwerk)*
r. foundry Walzengießerei *f*
r. gap Walzspalt *m (am Walzgerüst)*
r. grinder *s.* r. grinding machine
r. grinding machine Walzenschleifmaschine *f*; Spinnwalzenschleifmaschine *f*
r. groove Walz[en]kaliber *n*
r. heart Walzenkern *m*
r. heat-treating furnace Walzenvergüteofen *m*
r. housing Walzenständer *m (Walzwerk)*
r. in steps Staffelwalze *f*, Stufenwalze *f (Walzwerk)*
r.-jaw crusher Walzenbackenbrecher *m*
r. laminating *(Plast)* Kaschieren *n* auf der Walze
r. line Walzstraße *f*
r. mandrel Walzdorn *m*
r. mill *(Plast)* Mischwalze *f*, Walzenstuhl *m*, Walzwerk *n*
r. neck Walzenzapfen *m*
r.-neck collar *(Text)* Rollkragen *m*
r.-on-roll-off ship Roll-on-roll-off-Schiff *n (beladene Kraftfahrzeuge oder Hänger fahren mit Eigenantrieb oder Trecker über Auffahrtsrampen auf das Frachtschiff)*
r.-over fixture Wendevorrichtung *f*, schwenkbare Werkstückspannvorrichtung *f*
r.-over skip Kippkübel *m*
r.-over station Wendestation *f (Taktstraße)*
r.-over-type furnace *(Met)* Rollofen *m*
r. pass 1. Stich *m*, Durchgang *m* des Walzgutes; 2. Kaliber *n*
r. pass design Kalibrierung[sschema *n*] *f*
r. piercing mill Schrägwalzwerk *n (Rohrherstellung)*
r. pin Rundbolzen *m*, Walzenzapfen *m*
r. press Walzenpresse *f*
r. pressure Walzdruck *m*
r. quenching *(Schiff)* Schlingerdämpfung *f*
r. rate indicator *(Flg)* Rollgeschwindigkeitsmesser *m*
r. scale Walzzunder *m*
r. screen Rollsichter *m (Papierherstellung)*
r. shaft Walzenspindel *f*
r. slag Walzzunder *m*, Walzschlacke *f*
r. slippage Walzenschlupf *m*
r. spot weld Rollnaht *f (Schweißen)*
r. spot welding Rollnahtschweißung *f*
r. spot welding machine Rollenschweißgerät *n*
r. stand Walzgerüst *n*
r. standard Walzenständer *m*
r. straightener *s.* r.-straightening machine
r.-straightening machine Blechrichtmaschine *f*
r. stud Rollstößel *m*
r. surface Walz[en]ballen *m*
r. surface velocity Walzenumfangsgeschwindigkeit *f*
r. table Rollgang *m (zur Zuführung des Walzgutes)*

r. train Walzstraße *f*
r. turning lathe Walzendrehmaschine *f*
r. wabbler *s.* r. wobbler
r. welding Walzschweißen *n*
r. wobbler Walzenkleeblattzapfen *m*
r. wound on a core Rollenpapier *n* auf Kern
rollability Walzbarkeit *f*
rollback / to *(Dat)* nullen
rolled gewalzt, Walz-
r. and moulded laminated section *(Plast)* gewickelter Schichtkörper *m*
r. bellies *(Led)* Stoßseiten *fpl*
r. concrete Walzbeton *m*
r. edge Walzkante *f*
r. girder Walzträger *m*
r. lead Walzblei *n*
r. leather gewalztes Leder *n*
r. part Walzteil *n*
r. plate Walzblech *n*
r. product Walz[werks]erzeugnis *n*
r. rod Walzdraht *m*
r. section Walzprofil *n*
r. section brass Profilmessing *n*
r. selvedge eingerollte (gefaltete) Leiste *f* *(Webfehler)*
r. shape Walzprofil *n*
r. sheet metal Walzblech *n*
r. steel Walzstahl *m*
r. steel joist Walz[stahl]träger *m*
r. steel section Walz[stahl]profil *n*
r. steel single beam *(El)* Querjochtragwerk *n* *(aus Walzeisen oder Stahlgitter)*
r. timber Windbruch *m* *(Forst)*
r. wire Walzdraht *m*, gewalzter Draht *m*
roller Läufer *m*, Rädchen *n*, Röllchen *n*; Rolle *f*, Walze *f*, Welle *f*, Zylinder *m*; Laufrolle *f*, Bandrolle *f*; Kopierrolle *f*
r. beam *(Text)* Zylinderbaum *m*, Zylindertank *m* *(Aufsteckgatter)*
r. bearing Rollenlager *n*, Walzenlager *n*, Zylinderrollenlager *n*, Rollenlagerung *f*
r. bearing axle box Rollenachslager *n*
r. bearing drop hanger Rollenhängelager *n*
r. bearing pillow block Rollenstehlager *n*, Stehrollenlager *n*
r. bearing spindle *(Text)* Rollenlagerspindel *f*, auf Rollenlager gelagerte (laufende) Spindel *f*
r. bearing with shoulder Schulterrollenlager *n*
r. blind Rolladen *m*, Rollvorhang *m*, Rouleau *n*
r. blind shutter *(Opt)* Schlitzverschluß *m*
r. bow Rollenspriegel *m*
r. bridge Rollbrücke *f*
r. cage Rollring *m*, Walzenkäfig *m*
r. card *(Text)* Walzenkrempel *f*
r. card with sliver delivery Walzenkrempel *f* mit Bandablieferung
r. centre Rollenstern *m*
r. chain Rollenkette *f*
r. chain drive Rollenkettentrieb *m*
r. chain sprocket Rollenkettenrad *n*
r. chain sprocket hob Kettenradwälzfräser *m*, Rollenkettenradwälzfräser *m*
r. coating Walzenbezug *m*
r. conveyor Rollenförderer *m*, Rollenförderband *n*, Rollenförderstrecke *f*
r.-conveyor furnace Rollenherdofen *m*
r. engraver *(Text)* Walzengravierer *m*
r. engraving *(Text)* Walzengravur *f*
r. face Laufkranz *m*
r. fading Seegangschwundeffekt *m* *(Radartechnik)*
r. feeder Walzenspeiser *m*
r. follower Rollenstößel *m*; Fühlerwalze *f*
r. gear Rollenlaufwerk *n*
r. gear bed Rollgang *m*
r. gin *(Text)* Walzenegreniermaschine *f*
r. grate Walzenrost *m*
r. idler Bandrolle *f*, Laufrolle *f*, Tragrolle *f*
r. leveller Rollenrichtmaschine *f*
r. levelling Richtwalzen *n*
r. levelling machine Rollenrichtmaschine *f*
r.-like rollenförmig, walzenförmig
r. link Innenglied *n* *(Rollenkette)*
r. mill Läufermühle *f*, Walzenmühle *f*
r. path Rollbahn *f* *(Rollenlager)*
r. press Walzenpresse *f*; *(Text)* Walzenquetschwerk *n*, Quetschpresse *f*, Muldenpresse *f*
r. press washing machine *(Text)* Preßwalzenwaschmaschine *f*
r. pressure Walzdruck *m*
r. printing *(Text)* Walzendruck *m*, Rouleauxdruck *m*
r. race Lagerring *m*, Rollbahnring *m*, Rollenlaufkranz *m*
r. rail Rollengang *m*; Laufschiene *f*
r. ring bearing Rollenkranzlagerung *f*
r. seating Rollenlagerung *f*
r. squeezer *(Opt)* Rollenquetscher *m*
r. straightening Richtwalzen *n*
r.-straightening machine Rollenrichtmaschine
r. table Rollenhebevorrichtung *f*
r. tappet Rollenstößel *m*, Gleitstößel *m*
r. track Rollenbahn *f*
r.-type capacitor *(El)* Rollblock *m*, Rollkondensator *m*; Wickelkondensator *m*
r.-type pipe cutter Rohrabschneider *m* mit Schneiderädchen
r. way Rollenbahn *f*
rolling 1. Rollen *n*; Einwalzen *n*, Walzen *n* *(Straßenbau; Stahl)*; Wenden *n* *(Werkstück)*; 2. *(Fs)* Transversaldrehung *f* *(der Kamera)*
r. barrel Putztrommel *f* *(Gußputzen)*
r. billet Walzknüppel *m*, Knüppel *m* *(vorgewalzter Block)*

r. blind Rolladen *m*
r. burr Walznaht *f*
r. by hand Freihandwalzen *n*
r. calender Rollkalander *m*
r. chock *(Schiff)* Schlingerkiel *m*, Kimmkiel *m*
r. circle Rollkreis *m*, Wälzkreis *m (Zahnrad)*
r. contact Wälzkontakt *m*
r.-contact bearing Wälzlager *n*
r. crack *(Met)* Walzriß *m*
r. crusher Walzenbrecher *m*
r. defect *(Met)* Walzfehler *m*
r. die Walzbacke *f*
r. edge Walzkante *f*
r. element Wälzkörper *m*
r. error Wälzfehler *m (Zahnrad)*
r. face Wälzfläche *f (Zahnrad)*
r. fin Walznaht *f*
r. friction Rollreibung *f*; Wälzwiderstand *m*
r. frictional resistance *s.* r. friction
r. harrow *(Lw)* Rollegge *f*
r. heat *(Met)* Walzhitze *f*
r. ingot Walzbarren *m*
r. load Verkehrslast *f*, bewegliche Last *f*
r. machine Kalander *m*
r. member Wälzkörper *m*
r. mill Walzwerk *n*; Walzstraße *f*
r. mill engineer Walzwerker *m*
r. mill for circular shapes Rund[material]-walzwerk *n*
r. mill gauge Bandeisenlehre *f*
r. mill shears Walzwerksschere *f*
r. mill steam engine Walzenzugdampfmaschine *f*
r. mill train Walzstraße *f*
r. motion Rollbewegung *f*; Wälzbewegung *f*
r. movement Wälzbewegung *f*
r. plane Walzebene *f*
r. platform Walztisch *m*
r. pressure Walzdruck *m*
r. property Walzbarkeit *f*
r. resistance Rollwiderstand *m*, Laufwiderstand *m*
r. scale Sinter *m*, Walzsinter *m*, Walzzunder *m*, Walzhaut *f*
r. schedule Walzprogramm *n*
r. skin Walzhaut *f*
r. slab Walzbarren *m*
r. slide Walzenschlupf *m*
r. speed Walzgeschwindigkeit *f*
r. stand Walzgerüst *n*; Walzenständer *m*
r. stock 1. *(Eis)* rollendes Material *n*, Wagenpark *m*; 2. Walzgut *n*
r. tester Wälzprüfgerät *n (Zahnrad)*
r. together abwälzend aufeinander
r. tolerance Walztoleranz *f*
r. train Walzstraße *f*
r. train for heavy products Grobstraße *f*
rolls Walzenmühle *f (Aufbereitung)*
Roman type Antiqua *f (Schrift)*
Roman cement Romankalk *m*
Roman steelyard Laufgewichtswaage *f*
rongalite Rongalit *n (Natriumformaldehydsulfoxalat)*
röntgen *s.* roentgen
rood *englische Flächeneinheit; 1 rood = 1011,71 m²*
roof / to überdachen; eindecken
roof 1. Dach *n*; 2. *(Brg)* Hangendes *n*, Dach *n*, First *m*; 3. Deckel *m (Lichtbogenofen)*
r. arch *(Brg)* Firstengewölbe *n*
r. bar *(Brg)* Kappe *f (Ausbau)*
r. batten Dachlatte *f*
r. boarding Dachschalung *f*
r. boards Dachschalung *f*
r. bolt *(Brg)* Anker[bolzen] *m*, Ausbauanker *m*, Gebirgsanker *m*, Gesteinsanker *m*
r.-bolting *(Brg)* Ankerausbau *m*, Verankerung *f* der Firste, Verankerung des Hangenden
r. bow Dachspriegel *m*
r. control *(Brg)* Hangendpflege *f*, Beherrschung (Lenkung, Pflege, Beeinflussung) *f* des Hangenden
r. covering Dach[ein]deckung *f*, Dachhaut *f*
r. decking Dach[ein]deckung *f*, Dachhaut *f*
r. dome Dachkuppel *f*
r. fall *(Brg)* Hereinbrechen *n* des Hangenden, Strebbruch *m*
r. frame Dachrahmen *m*
r. garden Dachgarten *m*
r. guard Schneefang *m (Dach)*
r. ice refrigerator wagon Deckeneiskühlwagen *m*
r. insulator spindle Dachstütze *f (für Antennen)*
r. lamp Deckenleuchte *f*, Innenlampe *f*
r. lantern Dachlaterne *f*; Lichtkuppel *f*
r. lath Dachlatte *f*
r.-light Dachfenster *n*; Dachlaterne *f*; Oberlicht *n*
r.-lighted mit Oberlicht
r. membrane Dachhaut *f*
r. observation point *(Brg)* Firstenfestpunkt *m*
r. of furnace Ofengewölbe *n*, Ofendecke *f*
r. panel Dachbeplankung *f*
r.-panel assembly Dachbaugruppe *f*
r. pitch Dachneigung *f*
r. pressure *(Brg)* Hangenddruck *m*, Überlagerungsdruck *m*
r. prism *(Opt)* Dachkantprisma *n*
r. profile Dachform *f*
r. rack *(Kfz)* Gepäcknetz *n (Dach)*
r. rafters Dachsparren *mpl*
r. rail Regenrinne *f*
r. rib Dachsparren *m*
r. ridge Dachfirst *m*, First *m*
r.-shaped grate Dachrost *m (Feuerung)*

r. shell Dachschale *f*
r. slate Dachschiefer *m*
r. standard *(El)* Dachabspannmast *m*, Dachständer *m*, Dachgestänge *n*
r. stick Dachspriegel *m*
r. structure Dachkonstruktion *f*, Dachtragwerk *n*
r. tile Dachziegel *m*, Dachstein *m*
r. tiling Ziegeleindeckung *f*
r. truss Dachbinder *m*
r. type Dachform *f*
r.-type dachförmig
r.-type supporting insulator *(El)* Dachstütze *f (Isolator)*
r. ventilation *(Eis)* Fahrwindbelüftung *f*
r. ventilator *(Eis)* Fahrwindlüfter *m*
roofbolter *(Am) (Brg)* 1. Ankerlochbohrmaschine *f*; 2. Bohrhauer *m* für das Herstellen der Ankerbohrlöcher
roofed-over berth *(Schiff)* überdachte Helling *f*
roofer 1. Brett *n* für Dachhaut; 2. Dachdecker *m*
roofing Überdachung *f*; Dach[ein]deckung *f*, Dachhaut *f*, Dachschalung *f*
r.-base-paper Rohdachpappe *f*
r. felt Dachpappe *f*
r. material Eindeckungsmaterial *n*
r. sheet Dachblech *n*
r. shingle Schindel *f*
r. slate Dachschieferplatte *f*
r. tile Dachziegel *m*, Dachstein *m*
rooflight *s.* roof-light
room Raum *m*; Zimmer *n*; Platz *m*
r. and pillar method Kammerbau[verfahren *n*] *m (insbesondere Erz- und Salzbergbau)*; Pfeilerbau[verfahren *n*] *m (insbesondere Kohlenbergbau)*
r. paper Tapete *f*
r.-size [wall] panel Wand[groß]platte *f*, Großplatte *f*
r.-sized raumgroß, zimmergroß
r.-sized [wall] panel Wand[groß]platte *f*, Großplatte *f*
r. temperature Zimmertemperatur *f*
rooming *s.* room and pillar method
roose / to einsalzen
root 1. *(Math, Lw)* Wurzel *f*; 2. Fuß *m*, Grund *m (Zahn, Gewinde)*; 3. *(Ak)* Grundton *m*
r. angle Grundwinkel *m (Verzahnung)*
r. area Kernfläche *f (Gewinde)*
r. breaker *(Lw)* Wurzelschneidemaschine *f*
r. bruiser *(Lw)* Wurzelschneidemaschine *f*
r. chopper *(Lw)* Krautschläger *m*
r. circle Fußkreis *m (Zahnrad)*
r. clearance Grundspiel *n*, Kopfspiel *n (Gewinde, Verzahnung)*
r. collar Wurzelhals *m (Baumstamm)*
r. cone Fußkegel *m (Zahnrad)*
r. cutter *(Lw)* Wurzelmesser *n*
r. diameter Fußkreisdurchmesser *m (Verzahnung)*; Kerndurchmesser *m (am Gewinde)*
r. diameter of screw Kerndurchmesser *m* des Schraubengewindes
r. lifter *(Lw)* Wurzelroder *m*, Entwurzelungsmaschine *f*
r. locus *(BMSR)* Wurzelort[skurve *f*] *m*
r. mean square quadratischer Mittelwert *m*, Effektivwert *m*
r.-mean-square deviation mittlerer quadratischer Fehler *m*
r.-mean-square error criterion Kriterium *n* des mittleren quadratischen Fehlers
r.-mean-square pulse amplitude *(Fs)* effektive Impulsamplitude *f*
r.-mean-square value *s.* r. mean square
r.-mean-square voltage *(El)* effektive Spannung *f*, Effektivspannung *f*
r. neck Wurzelhals *m (Baumstamm)*
r. of blade Schaufelfuß *m*, Schaufelwurzel *f (Turbine)*
r. of notch Kerbgrund *m*
r. of seam Nahtwurzel *f*
r. of thread Gewindegrund *m*, Gewindefuß *m*
r. of tooth Zahnfuß *m (Verzahnung)*
r. radius Fuß[ab]rundung *f (Verzahnung)*
r. shoot *(Lw)* Wurzelableger *m*
r. stock *(Lw)* Wurzelstock *m*
r.-sum-square value quadratischer Mittelwert *m*, Effektivwert *m*
r. swelling Wurzelanlauf *m (Baumstamm)*
r. topper *(Lw)* Krautschläger *m*
r. truncation Gewindefußabrundung *f*
r. vegetable harvester Hackfruchterntemaschine *f*
r. vortex Nabenwirbel *m (Kavitation)*
r. welding Wurzellagenschweißung *f*
rooter Tiefaufreißer *m (Straßenbau)*
rooting-out a forest Abholzen *n (Wald)*
Roots blower Rootslader *m*, Rootsgebläse *n*
rope / to anseilen; mit einem Seil ziehen; seilförmig verdrallen
rope Seil *n*, Tau *n*
r. barrel Seiltrommel *f*
r. bending ratio Seilbiegungsfaktor *m*
r. bleaching Strangbleiche *f*
r. block Seilkloben *m*
r. breaking load Seilbruchlast *f*
r. capping *(Brg)* Seileinband *m*
r. carrier Aufführseil *n (Papiermaschine)*
r. changing Seilwechsel *m*
r. clamp Seilklemme *f*, Seilschloß *n*
r. clip Seilklemme *f*, Seilschloß *n*
r. core Seele *f* eines Seiles
r. coupling Seilkupplung *f*
r. cramp Seilklemme *f*, Seilschloß *n*

r. creep correction device Seilkriechausgleichvorrichtung *f*
r. drive Seiltrieb *m*
r. drum Seiltrommel *f*
r. expander *(Text)* Strangausbreiter *m*
r. ferry Seilfähre *f*
r. fibre Seilfaser *f*
r. fleeting *(Brg)* Seilablenkung *f*
r. framed conveyor structure Seilfördebandgerüst *n*
r. groove Seilrille *f (Schachtförderung)*
r. guide Seilführung *f*
r. haulage Seilförderung *f*
r. haulage aggregate Seilzugaggregat *n*
r. installation Seilauflegen *n*
r.-lay cable verseiltes Kabel *n*
r. life Seilaufliegezeit *f*, Seillebensdauer *f*
r. loop Seilbucht *f*
r. maker Seiler *m*
r. making machine *(Text)* Seilschlagmaschine *f*
r. packing *(Masch)* Kabelpackung *f*
r. piling device *(Text)* Rüsselvorrichtung *f*
r. polygon Seilpolygon *n (Statik)*
r. pulley Seilscheibe *f*, Seilrolle *f*
r. pulley block Seilflaschenzug *m*
r.-pulley flywheel Seilscheibenschwungrad *n*
r. retainer Seilspanner *m*
r. sheave Seilscheibe *f*, Seilleitrolle *f*
r. sideframe belt Förderband *n* mit Seiltraggerüst
r. sling *(Schiff)* Taustropp *m*
r. slip Seilrutsch *m*
r. socket Seilflasche *f*
r. speed Fördergeschwindigkeit *f*, Seilgeschwindigkeit *f*
r. strand Rundseil *n*
r. tension compensating equipment Seilausgleichsvorrichtung *f (Schachtförderung)*
r. tread Seilrillenfutter *n (Schachtförderung)*
r. trolley Seillaufkatze *f*
r. twister Seildreher *m*
r.-ware Seilerware *f*
r. winch Seilwinde *f*
ropebelt conveyor Bandförderer *m* mit Seiltraggerüst, Seilbandförderer *m*
ropemaking Seilerei *f*
ropeway Drahtseilbahn *f*, Luftseilbahn *f*, Seilbahn *f*
roping [together] Anseilen *n*
r. up Seilauflegen *n*
ropy fadenziehend, klebrig, zäh[flüssig]
rosaniline Rosanilin *n (Farbstoff)*
roscoelite *(Min)* Roskoelith *m*
rose chucking reamer Grundreibahle *f*
r. copper Scheibenkupfer *n*
r. counter sink Bohrfräse *f (Drechslerei)*
r.-quartz Rosenquarz *m*
r. reamer Stirnreibahle *f*, Grundreibahle *f*
r. shell reamer Aufsteckgrundreibahle *f*
r. wood Rosenholz *n*
rosette Rosette *f*
rosin Kolophonium *n*, harzförmiger Rückstand *m* aus der Rohterpentindestillation
r. soap Harzseife *f*
Rosin-Rammler-Bennett size distribution curve Kornverteilungskurve *f* nach Rosin-Rammler-Bennett *(Aufbereitung)*
resolic acid Resolsäure *f*
ross / to schälen *(Holz)*
ross Borke *f*
rossite *(Min)* Rossit *m*
roster Dienstplan *m*
rot / to vermodern, [ver]faulen, verrotten
rot Fäulnis *f*, Fäule *f*
r.-proof fäulnisbeständig, fäulnissicher, nichtfaulend
r.-proof finish fäulniswidrige Ausrüstung *f*
rotameter Rotamesser *m*, Rotameter *n*, Schwebekörperdurchflußmesser *m*
rotary rotierend, drehend, drehbar, umlaufend; Dreh-
r. actuator Drehantrieb *m*, Hebelantrieb *m*, Stellantrieb *m* mit Winkelbewegung
r. agitator Rührwerk *n*
r. amplifier *s.* rotating amplifier
r. armature magnet *(Elm)* Drehankermagnet *m*
r. attachment feed gear Rundtischantrieb *m* *(Fräsmaschine)*
r. attenuator Drehscheibenabschwächer *m* *(Wellenleitertechnik)*
r. automatic telephone system Drehwählersystem *n*
r. beam aerial rotierende (drehbare) Richtstrahlantenne *f*
r. bit Drehbohrmeißel *m*, Rotarybohrmeißel *m*, Spülbohrmeißel *m (Tiefbohrtechnik)*
r. blower Kreiselgebläse *n*
r. boring tool Drehbohrmeißel *m*
r. breaker Kreiselbrecher *m*, Diskusbrecher *m*
r. broaching machine Rundtischräummaschine *f*
r. brush Bürstenwalze *f*
r. bur Umlauffeile *f*, rotierende Fräserfeile *f*
r. calcining kiln Drehtrommelröstofen *m*
r. capacitor Drehkondensator *m*; Synchronphasenschieber *m*, Drehphasenschieber *m*
r. casting Schleuderguß *m*
r. cement kiln Zementdrehofen *m*
r. converter *(Elm)* Einankerumformer *m (mit einer einzigen Wicklung)*
r. crane Drehkran *m*
r. crossed-axis gear shaving Zahnradschaben *n* mit gekreuzten Achsen

r. crossed-axis shaving Achskreuzschaben *n*, Schaben *n* mit gekreuzten Achsen
r. crusher Kreiselbrecher *m*
r. current Drehstrom *m*
r. cutter Schneidmühle *f*, Pralltellermühle *f* mit Raspelfläche; Fräser *m*
r. deep drilling [process] drehendes Tiefbohren *n*, Rotaryverfahren *n*
r. dial system Drehwählersystem *n*
r. discharger Drehfunkenstrecke *f*, rotierende Funkenstrecke *f*
r.-disk grizzly Scheibenrost *m*
r. disk loader Frässcheibenlader *m*
r. disk valve Kreisschieberventil *n*
r. drier Trockentrommel *f*, Drehrohrofen *m*
r. drilling Dreh[spül]bohren *n*, Rotarybohren *n*, Rotarybohrverfahren *n* *(Tiefbohrtechnik)*
r.-drum broaching machine Trommelräummaschine *f*
r. drum filter Trommeldrehfilter *n*
r. drum fixture milling machine Trommelfräsmaschine *f*
r. drum mixer Freifallmischer *m*
r. drying Rotationstrocknung *f*
r. dump Kreiselwipper *m*
r. engine Umlaufmotor *m*; Kreiskolbenmotor *m* *(Wankel)*
r. eyepiece selector Okularrevolverkopf *m*
r. feed Rundvorschub *m*
r. field Drehfeld *n*
r. file Umlauffeile *f*
r. fine crusher Glockenmühle *f*
r. flange Drehflansch *m*
r. fluid motor Flüssigkeitsmotor *m* [mit drehender Abtriebsbewegung]
r. furnace Drehofen *m*
r. gear shave (shaving) cutter Schabrad *n*, Zahnschabrad *n*, Werkzeugrad *n*, Schneidrad *n* zum Zahnradschaben
r. generator *s.* rotating amplifier
r. grate Drehrost *m*
r. harrow *(Lw)* Rollegge *f*
r. hearth-furnace Drehherdofen *m*
r. hoe *(Lw)* Bodenfräse *f*
r. hysteresis *(El)* Drehfeldhysterese *f*, Drehfeldhysteresis *f*
r. indexing automatic Rundtaktmaschine *f*
r. indexing machine Rundtischschaltmaschine *f*
r. indexing unit *(Am)* Rundtischschaltmaschine *f*
r. interruptor *(umlaufender)* Unterbrecher *m* *(z. B. Kfz-Zündung)*
r. jack Drehkolben *m*, Motor *m* für pendelnde Drehbewegung
r. kiln Dreh[rohr]ofen *m*
r. knife cutter Schneidmühle *f*, Pralltellermühle *f* mit Raspelfläche
r. lathe Schälmaschine *f* *(für Furniere)*
r. letterpress Buchdruckrotationsmaschine *f*
r. lift Paternoster *m*
r. lime kiln Kalkdrehofen *m*
r. line switch *(Fmt)* Drehwähler *m* *(als Vorwähler)*
r. method of gear shaving Zahnradschaben *n* mit umlaufenden Rädern
r. milling Rundfräsen *n*
r. milling machine Rundtischfräsmaschine *f*
r. mixer Trommelmischer *m*
r. motion Kreisbewegung *f*, Drehbewegung *f*
r. motor of the gear type Zahnradflüssigkeitsmotor *m*
r. motor of the vane type Flügelzellenmotor *m*
r. movement *s.* r. motion
r. mud Bohrschlamm *m*, Spülschlamm *m* *(Tiefbohrtechnik)*
r. multipoint cutter Fräser *m*
r. oscillation Drehschwingung *f* *(Festigkeit)*
r. out-trunk switch *(Fmt)* Ausgangsdrehwähler *m*
r. pelleting machine rundlaufende Tablettiermaschine *f*
r.-percussive drilling *(Brg)* Drehschlagbohren *n*, drehend-schlagendes Bohren *n*, Vibrobohren *n*
r. phase advancer *(El)* rotierender Phasenschieber *m*, Drehphasenschieber *m*
r. phase changer Drehphasenschieber *m* *(Wellenleitertechnik)*
r. phase converter *s.* r. phase changer
r. phase shifter *s.* r. phase changer
r. piercing mill Hohlwalzwerk *n*
r. piston compressor *(Kfz)* Drehkolbenverdichter *m*
r. piston-type of selector valve Umsteuerdrehschieber *m*
r. plough Drehpflug *m*, Wendepflug *m*
r. plug valve Drehschieber *m*
r. power Drehvermögen *n*
r. press *(Typ)* Rotationspresse *f*
r. printing *(Typ)* Rotationsdruck *m*
r. puddling furnace *(Met)* Drehpuddelofen *m*, Puddeldrehofen *m*
r. pump Drehkolbenpumpe *f*, Rotationspumpe *f*; Kreiselpumpe *f*, Zentrifugalpumpe *f*
r. ram *(Am)* hydraulischer Drehantrieb *m*; Drehflügelmotor *m*
r. rheostat Drehwiderstand *m*, Kurbelwiderstand *m*
r. scale Drehwaage *f*
r. screen Trommelsieb *n*; Siebtrommel *f*; Zentrifugalsortierer *m*, Rundsieb *n*
r. selector Umsteuerdrehschieber *m*
r. shaft seal Wellendichtung *f*, Lippendichtung *f*, Simmerring *m* *(Handelsname)*

r. shear Roll[en]schere *f*, Kreisschere *f*
r. sheeter *(Am) (Typ)* Rotationsquerschneider *m*
r. sintering kiln Trommelsinterofen *m*, Drehrohrsinterofen *m*
r. slide valve Drehschieber *m*
r. sliver screen rotierender Astfänger *m (Zelluloseherstellung)*
r. slotting Rundstoßnachformen *n*
r. stage Drehtisch *m (für optische Geräte)*
r. steam engine Rotationsdampfmaschine *f*, rotierende Dampfmaschine *f*
r. step Drehschritt *m (Fernsprechtechnik)*
r. suction dredger Saugbagger *m* mit drehbarem Saugkopf
r. switch Drehschalter *m*, Knebelschalter *m*
r. swivel Bohrwirbel *m*, Drehkopf *m*, Spülkopf *m (Tiefbohrtechnik)*
r. table Drehtisch *m*, Rotarytisch *m (Rotarybohren)*; Rundtisch *m (Werkzeugmaschine)*
r. table machine with the machine heads inside the table Ringtischmaschine *f*
r. table machine with the machine heads outside the table Rundtischmaschine *f*
r. table surface grinding machine Rundtischflächenschleifmaschine *f*
r. tracing Rundstoßnachformen *n*
r. transfer machine Rundtaktmaschine *f*
r. transformer *(Elm)* Dynamotor *m*
r. tubular kiln Drehrohrofen *m*
r. turret Drehturm *m (z. B. eines Panzers)*
r.-type regulator Drehregler *m*
r. veneer Schälfurnier *n*
r. veneer lathe rotierende Furnierschneidemaschine *f*
r. veneer quality Schälqualität *f (Furnier)*
r. watermeter Flügelradwassermesser *m*
r.-wing aircraft Hubschrauber *m*, Drehflügelflugzeug *n*
rotatable drehbar, schwenkbar
rotate / to rotieren, [sich] drehen, umlaufen; die Umlaufbewegung haben (ausführen); umlaufen lassen; kreisen
r. freely [on a spindle] lose laufen [auf einer Spindel]
r. in truth genau rundlaufen
r. the plane of polarization die Polarisationsebene drehen
rotating Drehen *n*, Kreisen *n*; Umlaufen *n*, Mitlaufen *n*
rotating rotierend, kreisend, umlaufend; mitlaufend *(Spitze)*
r. amplifier Elektromaschinenverstärker *m*, rotierender Verstärker *m*
r.-anode tube Drehanodenröhre *f (Röntgenröhren)*
r. armature *(Elm)* Drehanker *m*
r. axis of the drill Bohrerdrehachse *f*
r. beacon Dreh[funk]feuer *n*
r. beacon transmitter Drehbakensender *m*
r.-beam machine Umlaufbiegemaschine *f*
r. centre mitlaufende Spitze (Reitstockspitze) *f*
r. coil variometer Dreh[spul]variometer *n*
r. crank umlaufende (rotierende) Geradschubkurbel *f*, umlaufende (rotierende) Schubkurbel *f*; Taumelscheibe *f*
r. crystal method Drehkristallmethode *f*
r. cylindrical kiln Drehrohrofen *m*
r.-disk deduster Kreiselsichter *m (Aufbereitung)*
r. disk shutter Drehscheibenverschluß *m (bei optischen Geräten)*
r. distributing box *(Lw)* rotierender Streutopf *m*
r. double crank *(Masch)* durchschlagendes Doppelkurbelgetriebe *n*, umlaufende (rotierende) Doppelkurbel *f*, Kniekupplung *f*
r. drum drier Rotationstrommeltrockner *m (Papierherstellung)*
r. field Drehfeld *n*
r. field aerial Drehfeldantenne *f*
r. field instrument Drehfeldinstrument *n*, Ferraris-Instrument *n*
r. field-type alternator Innenpolwechselstromgenerator *m*
r. harrow *(Lw)* Rotoregge *f*
r. hearth furnace Drehherdofen *m*
r. hoe *(Lw)* Rotorhacke *f*
r. insulator switch Schalter *m* mit Drehstützer
r. joint Drehkupplung *f (Wellenleiter)*
r. loop aerial Drehrahmenantenne *f*
r. loop radio beacon Drehfunkfeuer *n*
r. mechanism Schwenkeinrichtung *f*
r. mixer Freifallmischer *m*
r. plate capacitor Drehkondensator *m*
r. soldering machine Lötmaschine *f (Elektronenröhrenherstellung)*
r. strip column Drehbandkolonne *f*
r. tippler Kreiselwipper *m*
r. tower crane Turmdrehkran *m*
r. valve Umlaufventil *n*, Walzenschieber *m*
r. vane anemometer Flügelradanemometer *n*
r. vane [flow]meter Drehflügelzähler *m*
rotation Rotation *f*; Drehung *f*; Umlaufbewegung *f*
r. angle Dreh[ungs]winkel *m*
r. axis Drehachse *f*
r. centre Drehzentrum *n*
r. ellipsoid Rotationsellipsoid *n*
r. energy Rotationsenergie *f*
r. frame aerial drehbare Rahmenantenne *f*
r. interrupter Rotationsunterbrecher *m*
r.-invariant rotationsinvariant
r. irradiation Rotationsbestrahlung *f*

r. matrix Drehmatrix *f*
r. method Drehkristallverfahren *n*
r. moulding *(Plast)* Rotationspressen *n*
r. of crops Fruchtfolge *f*, Fruchtwechsel *m*, Saatwechsel *m*
r. of cutting Schnittrichtung *f (des Spiralbohrers)*
r. of the flyer *(Text)* Flügeldrehung *f*
r. of the plane of polarization Drehung *f* der Polarisationsebene
r. of work kreisförmige Arbeitsbewegung *f*; Umlaufbewegung *f* des Werkstücks
r. spectrum Rotationsspektrum *n*
r. speed Umlaufgeschwindigkeit *f*
r. symmetry Drehsymmetrie *f*
r. time Umlaufzeit *f*
r. to the left Linksdrehung *f*
r. to the right Rechtsdrehung *f*
r. twin Rotationszwilling *m (Kristall)*
r.-vibration band Rotationsschwingungsbande *f*
r.-vibration energy Rotationsschwingungsenergie *f*
r.-vibration spectrum Rotationsschwingungsspektrum *n*
rotational Umlauf-, Dreh-, Rotations-
r. band Rotationsbande *f*
r. broadening Rotationsverbreitung *f (Spektrum)*
r. constant Rotationskonstante *f*
r. distortion Drehungsverzeichnung *f*
r. electromotive force Rotations-EMK *f*
r. excited level *(Kern)* Rotationsanregungsniveau *n*
r. field Wirbelfeld *n*
r. fine structure Rotationsfeinstruktur *f*
r. frequency Umlauffrequenz *f*
r. level *(Opt, Kern)* Rotationsterm *m*, Rotationsniveau *n*
r. line Rotationslinie *f*
r. quantum number Rotationsquantenzahl *f*
r. resistance Drehungswiderstand *m (Turbine)*
r. shell *(Bw)* Rotationsschale *f*
r. slide Grundbruch *m (Grundbau)*
r. spectrum Rotationsspektrum *n*
r. speed Umlaufgeschwindigkeit *f*; Drehzahl *f*
r. state Rotationszustand *m*
r. state excitation Rotationsanregung *f*
r. structure Rotationsstruktur *f*
r.-type viscosimeter Rotationsviskosimeter *n*
rotative umlaufend, drehend; Dreh-
r. and percussive drilling machine *(Brg)* Drehschlagbohrgerät *n*
rotator *s.* rotor
rotatory rotierend, drehbar
r. dispersion Drehungsstreuung *f*
r. motion Rotationsbewegung *f*
r. polarization Rotationspolarisation *f*
r. power Drehungsvermögen *n*
rotogravure *(Typ)* Rollentiefdruck *m*, Rotationstiefdruck *m*
rotor 1. Läufer *m*, Rotor *m (E-Maschine)*; 2. Laufrad *n (Turbine)*; 3. Drehkörper *m*, Schaufelrad *n*; Unterbrecher *m*; Trommelkonverter *m*
r. baler *(Lw)* Heuroller *m*
r. barrel Läufertrommel *f*
r. blade Laufschaufel *f*, Rotorflügel *m*
r. circuit *(El)* Läuferkreis *m*
r. core *(Elm)* Ankerkern *m*
r. current Läuferstrom *m*; Induktorstrom *m*
r. disk Laufradscheibe *f*
r. fed commutator motor läufergespeister Kommutatormotor *m*
r.-resistance starter *(Elm)* Läuferanlasser *m*
r. rheostat *(Elm)* Läuferanlasser *m*, Regelanlasser *m*
r. shaft Läuferwelle *f*, Rotorwelle *f*
r. ship Rotorschiff *n*
r. slot milling machine Rotornutenfräsmaschine *f*
r. starter *(Elm)* Läuferanlasser *m*
r. voltage Läuferspannung *f*
r. wheel Laufrad *n (Turbine)*
r. winding Läuferwicklung *f*; Induktorwicklung *f*
rototrol *(Wärm)* Rototrol *n (Verstärkermaschine)*
rotten 1. *(Met)* rotbrüchig, warmbrüchig; 2. verfault, vermodert, verrottet; stockig, schwammig *(Holz)*
r. knot Faulast *m (Zelluloseherstellung)*
r. spot *(Led)* verstunkene Stelle *f*; *(Text)* verrottete Stelle *f*; Verrottungsfleck *m*
rottenness Verfall *m*; Fäulnis *f*; Sprödigkeit *f*, Brüchigkeit *f (Stahl)*
rotting Verfaulen *n*, Verrotten *n*
r. pond Rottgrube *f*
r. vat Faulbütte *f*, Lumpenfäule *f (Papierherstellung)*
rotunda Rotunde *f*, Rundbau *m*
rouge Polierrot *n (Metalle)*
rough / to 1. aufrauhen, rauh werden; 2. vorbearbeiten; vorwalzen; schruppen
r. bore vorbohren
r. broach vorräumen, schruppräumen
r. cut vorschneiden; vorbearbeiten, schruppen; vorfräsen
r. down *(Met)* vorwalzen, herunterwalzen, [vor]strecken
r. down blooms vorwalzen
r. drill vorbohren
r. face planschruppen
r. forge vorschmieden
r.-grind 1. *(Masch)* schruppschleifen, vorschleifen; 2. zerschroten *(Getreide)*

r.-machine vorbearbeiten, schruppen
r.-mill vorfräsen, schruppfräsen
r.-plane vorhobeln, schrupphobeln
r.-polish vorpolieren
r.-position grob einstellen (positionieren)
r.-ream vorreiben
r. up aufrauhen
rough 1. rauh *(Oberfläche)*; 2. ungeschält *(Reis)*; 3. grob, uneben
r. adjustment Grobeinstellung *f*
r. broaching Vorräumen *n*, Schruppräumen *n*
r. cast *(Bw)* Rapputz *m*, Rauhputz *m*
r.-casting *(Met)* Rohguß *m*; Rohgußstück *n*
r. coal Rohkohle *f*, Förderkohle *f*
r.-coat *(Bw)* Unterputz *m*
r. concentrate Vorkonzentrat *n (Aufbereitung)*
r. contour planing Nachformvorhobeln *n*
r.-cut file Schruppfeile *f*
r.-cutter Vorschneider *m*; Vorfräser *m*
r. diamond ungeschliffener Diamant *m*
r. edge Baumkante *f (eines Brettes)*
r. edged baumkantig *(Brett)*
r.-edged beam baumkantiger Balken *m*
r. estimate roher Überschlag *m*
r. file Grobfeile *f*, Schruppfeile *f*
r. finishing *(Masch)* Grobschlichten *n*
r. forging Schmiederohling *m*
r. gas main Gichtgasleitung *f (Hochofen)*
r. grained grobnarbig *(Leder)*
r.-grinding 1. Vorschliff *m*; Schruppschleifen *n*; 2. Schrotmahlen *n*, Zerschroten *n (Getreide)*
r.-grinding mill Schrotmühle *f*
r. lumber *(Am)* unbehobeltes Schnittholz *n*
r.-machining Vorbearbeitung *f*; Schruppen *n*
r.-machining operation Schrupparbeit *f*
r.-milling Vorfräsen *n*, Schruppfräsen *n*
r.-pierced tube blank *(Met)* Rohrluppe *f*
r.-planing Vorhobeln *n*, Schrupphobeln *n*
r. positioning Grobeinstellung *f*, Grobpositionierung *f*
r. pressing Vorpressen *n*
r. reaming Vorreiben *n*
r. rolled ingot Rohblock *m*, vorgewalzter Block *m*
r. service lamp stoßfeste Lampe *f*
r.-shaping Vorhobeln *n*, Schrupphobeln *n* *(Waagerechtstoßmaschine)*
r. size Rohmaß *n*
r. steel Frischstahl *m*
r.-surfaced mit rauher Oberfläche
r.-tanned leather lohgares Leder *n*
r. timber unbehobeltes Schnittholz *n*
r.-turning tool Schruppdrehmeißel *m*
r. vacuum Vorvakuum *n*
roughage Rauhfutter *n*
roughed 1. aufgerauht; 2. geschruppt, vorbearbeitet
r. slab Vorbramme *f (Walzwerk)*
roughen / to aufrauhen *(Oberfläche)*
rougher 1. *(Met)* Vorgerüst *n*; Vorwalzer *m*; Vorschmiedegesenk *n*; Vorstraße *f*; 2. *(Brg)* Vorsetzmaschine *f*; 3. *(Masch)* Schruppwerkzeug *n*; Schruppzahn *m*
roughing 1. Schruppen *n*, Vorbearbeitung *f* *(spanend)*; 2. Grobwalzen *n*, Grobzug *m*; 3. *(Text)* Hecheln *n*
r. block Grobzug *m (Draht)*
r. cut Schruppschnitt *m*; Schruppspan *m*
r. cutter Schruppfräser *m*, Vorfräser *m*
r. depth *(Masch)* Schruppspantiefe *f*
r. down mill Grobwalzstraße *f*; Grobwalzwerk *n*
r. gear milling cutter Zahnformvorfräser *m*
r. hand taper reamer Handkegelreibahle *f* zum Vorreiben
r. hob Schruppwälzfräser *m*, Vorwälzfräser *m*
r. knife Vorschneidmesser *n (Holzbearbeitung)*
r. lathe Schruppdrehmaschine *f*
r. lathe tool Schruppdrehmeißel *m*
r. mill Vor[walz]straße *f*
r. operation *(Masch)* Schrupparbeit *f*; Schruppschnitt *m*
r. pass *(Met)* Blockkaliber *n*, Vorstich *m*; *(Masch)* Schruppdurchgang *m*
r. reamer Schruppreibahle *f*, Vorreibahle *f*
r. roll *(Met)* Vorwalze *f*, Streckwalze *f*
r. stand Vorgerüst *n*; Vorwalzwerk *n*
r. taper reamer Schruppreibahle *f* für kegelige Bohrungen
r. tool Schruppmeißel *m*, Schruppwerkzeug *n*
r. tooth Schruppzahn *m*, Vorschneidezahn *m*
r. train Vor[walz]straße *f*
r.-up machine *(Led)* Aufrauhmaschine *f*
roughly grob, näherungsweise, ungefähr
roughness Rauheit *f (Oberflächengüte)*
r. width Rillenabstand *m*
roulette Roulette *f (Kupferstechen)*
round / to abrunden; umkreisen
r. a curve *(Kfz)* eine Kurve ausfahren
r. off *(Math)* abrunden, [auf]runden
round 1. Triebstock *m*; 2. Beschickung *f (Hochofen)*; 3. *(Brg)* Satz *m (Schießarbeit)*; *(Mil)* Patrone *f*; 4. Leitersprosse *f*
round rund, mit kreisförmigem Querschnitt
r. bar Rundstab *m*, Rundstahl *m*
r.-bar davit *(Schiff)* Drehdavit *m (n)*
r.-bar steel Rundstahl *m*
r. bastard file Bastard *m*, Rundfeile *f*
r. billet Rundblock *m*, Rundknüppel *m*
r. billet wood Knüppelholz *n*
r. blank *(Met)* Ronde *f*
r. bloom Rundblock *m (Walzwerk)*
r.-bottom flask *(Ch)* Rundkolben *m*
r.-bottomed mit Nutgrundausrundung
r. brackets runde Klammern *fpl*

r.-can capacitor runder Becherkondensator *m*
r. carry Rückübertrag *m*
r. coal Stückkohle *f*
r. column Rundsäule *f*
r.-column [upright] drilling machine Säulenbohrmaschine *f*
r. conical spring Kegelfeder *f* mit rundem Querschnitt
r. corner *(Masch)* Hohlkehle *f*, Ausrundung *f*
r.-cornered mit ausgerundeten Ecken
r.-edged unbesäumt *(Brett)*
r.-ended rundstirnig, mit abgerundeter Stirnfläche
r. escapement file Rundnadelfeile *f*
r. file Rundfeile *f*
r. flask *(Ch)* Rundkolben *m*
r. hatch *(Schiff)* Kugelschott *n*
r. head Halbrundkopf *m (Niet)*
r.-head bolt Halbrundschraube *f*
r.-head bolt with square shank Halbrundschraube *f* mit Vierkantschaft
r.-head screw Halbrundschraube *f*
r. headed bolt Halbrundschraube *f*
r. headed screw Halbrundschraube *f*
r.-hole card *(Büro)* Rundlochkarte *f*
r. kiln Ringofen *m*
r. knitting frame *(Text)* Rundrändermaschine *f*
r. needle file Rundnadelfeile *f*
r. nose iron plane Schrupphobel *m*
r.-nose tool Hohlkehlwerkzeug *n*
r. of beam *(Schiff)* Balkenbucht *f*
r. of shots *(Brg)* Abschlag *m*
r. pass Rundkaliber *n (Walzwerk)*
r. pin runder Dübel *m*
r. plane Stabhobel *m*
r.-point setscrew Druckschraube *f* mit Linsenkuppe
r. punch Runddurchschlag *m*, Runddurchschläger *m*, Durchschläger *m*, Durchtreiber *m*
r.-robin test *(Am) (Ch)* Ringanalyse *f*, Analyse *f* von verschiedenen unabhängigen Bearbeitern
r. rod Rundstab *m*
r. screen *(Typ)* Rundraster *m*
r. shank Zylinderschaft *m*
r. specimen Rundstab *m (Probe)*
r. steel Rundstahl *m*
r. strand rope Rundlitzenseil *n*
r. table Rundtisch *m*
r. timber Rundholz *n*
r. top countersunk rivet Halbversenkniet *m*
r. wire spring Rundstahlfeder *f*
r. work table Rundtisch *m (Werkzeugmaschinen)*
roundabout Kreisverkehr *m*
rounded abgerundet, ausgerundet
r. bilge *(Schiff)* runde Kimm *f*
rounding error Rundungsfehler *m*, Abrundungsfehler *m*
r. off 1. *(Masch)* Ausrundung *f*; 2. *(Math)* Abrundung *f*
r. off the inlet Abrundung (Ausrundung) *f* der Eintrittskante *(Turbinendüse)*
r. off the top Kopf[ab]rundung *f (Verzahnung)*
roundness Rundheit *f*
roundwood Rundholz *n*
rouse / to einsalzen
roustabout Hafenarbeiter *m*
rout / to ausschneiden; ausarbeiten; den Grund hobeln; mit Handvorschub nachfräsen; abschneiden, verschneiden
route 1. Trasse *f*; Linienführung *f*; 2. *(El)* Leitungsführung *f*, Leitungsweg *m*
r. indicating signal *(Eis)* Wegesignal *n*
r. indication *(El)* Leitwegangabe *f*
r. locking Fahrstraßenfestlegung *f (Eisenbahnsignalwesen)*
r. mapping Linienführung *f*, Trassierung *f (Eisenbahnbau, Straßenbau)*
r. section *(Eis)* Teilstrecke *f*
r. surveying Trassenvermessung *f*
router Grundhobel *m*; Langlochfräser *m*; Nachformfräsmaschine *f* für Handvorschub
r. plane Grundhobel *m*
routine Programm *n (das für einen Rechner entsprechend angefertigt ist)*
r. maintenance regelmäßige Unterhaltung (Wartung) *f*
r. radio message Betriebsfunkspruch *m*
r. repair work laufende Instandsetzungsarbeit *f*, Pflege *f*
r.-work microscope Gebrauchsmikroskop *n*
routing 1. Ausfräsen (Gesenkfräsen, Umrißfräsen) *n* mit Handvorschub; Aushobeln; 2. Arbeitsvorbereitung *f*
r. cutter Gesenkfräser *m*; Langlochvorfräser *m*
r. indicator Leitwegkenngruppe *f (Fernschreibdienst)*
r. of a ship Seeverhalten *n* eines Schiffes
rove Strähne *f (s. a.* roving)
r.-wind system *(Text)* Vorgarnaufwickelsystem *n*
roving 1. Glasseidenstrang *m*; Vorgarn *n*; [feines] Vorgespinst *n*; 2. [feines] Vorspinnen *n*
r. bobbin Vorgarnspule *f*
r. build Vorgarnwindung *f*
r. frame Feinflyer *m*, Vorspinnmaschine *f*
r. indicator Vorgarnzähler *m*
row / to rudern
row 1. *(Math)* Zeile *f (Matrix)*; 2. Reihe *f (Häuser; Sitze)*; Flucht *f*; Schicht *f*
r. average *(Math)* Zeilenmittel *n*
r. crop tractor Reihensaatbehandlungstraktor *m*

r. of boreholes *(Brg)* Gang *m*
r. of needles Nadelreihe *f*, Kammreihe *f* *(Kämmaschine)*
r. of radiators Strahlerzeile *f* *(Antennen)*
r. rivets Nietnaht *f*, Nietreihe *f*
rowed barley Futtergerste *f*
rower's seat Ruderbank *f* *(Sportboot)*
rowing boat Ruderboot *n*, Riemenboot *n*
r. match Ruder[boot]rennen *n*
royal 1. Royalrah *f*, Royal[segel] *n*; 2. *(Papierformat; Druckpapier: 20×25"; Schreib- und Zeichenpapier; 19×24")*
r. sail Royal[segel] *n*
royalty Lizenzabgabe *f*
rub / to [ab]reiben; abziehen; [ab]wischen; schleifen, polieren
r. down *(Bw)* abreiben
r. with wax bohnern, wachsen *(Fußboden)*
rub resistance Scheuerfestigkeit *f* *(von Drucken)*
rubber 1. Gummi *m*, Kautschuk *m*, Weichgummi *m*; 2. Reibzeug *n*; 3. *(Bw)* leichter, einfach zu bearbeitender Ziegelstein *m*; 4. Ablenkseil *n* *(Schachtförderung mit Führungsseilen)*
r. apron *(Text)* Nitschelhose *f*
r. bag moulding *(Plast)* Gummisackverfahren *n*
r. belt Gummiriemen *m*
r. belt conveyor Gummigurtförderband *n*, Gummi[förder]gurt *m*
r. blanket Gummiplane *f*
r. blanket coating *(Led)* Gummituchstreichverfahren *n*
r. block 1. Gummiunterlage *f*, Gummipolster *n*; 2. *(Typ)* Gummiklischee *n*
r. boat Schlauchboot *n*, Gummiboot *n*
r. bonding Gummibindung *f*; naturharzhaltige Bindung *f* *(Schleifscheibe)*
r. boot Gummistiefel *m*
r. box Gipsmolle *f*, Gipsbecher *m*
r. buffer Gummipuffer *m*
r. buffer coupling elastische Kupplung *f* mit Gummibuchsen
r. bung Gummistopfen *m*, Gummistöpsel *m*
r. bush Gummimuffe *f*, Gummibuchse *f*, Gummihülse *f*
r. cable Gummi[ader]kabel *n*
r. cement Gummilösung *f*, Kautschukkitt *m*
r. cloth Gummituch *n*
r.-coated gummiert
r. coating Gummibelag *m*
r. cord Gummischnur *f*
r. core Gummieinlage *f* *(Dichtung)*
r.-covered braided wire Gummilitze *f*
r.-covered cable Gummikabel *n*
r.-covered wire Gummiader *f*
r. cushion Gummipuffer *m*, Gummipolster *n*
r.-cushioned mit Gummipuffer, gummigelagert
r.-cushioned spring hanger Gummifederlager *n* *(Fahrzeugaufhängung)*
r. cylinder Druckzylinder *m* *(Offsetdruck)*
r. die process Gummipufferziehverfahren *n*, Gummiformpressen *n*
r. dinghy Gummiboot *n*, Schlauchboot *n*
r. drawing *(Text)* Frottierstrecke *f*, Nitschelstrecke *f*
r. dressing *(Typ)* Gummiaufzug *m* *(z. B. Walze)*
r. gaiter Gummimanschette *f*
r. gasket Gummidichtung *f*, Gummilasche *f*
r. gear *(Text)* Nitschelzeug *n*, Nitschelwerk *n*, Nitschler *m*
r. glove Gummihandschuh *m*
r. heel Gummiabsatz *m*
r. hose Gummimuffe *f*, Gummischlauch *m*
r.-hose connection Gummischlauchmuffe *f*
r.-impregnated gummiimprägniert
r.-insulated gummiisoliert
r.-insulated cable Gummikabel *n*
r.-insulated core Gummiaderschnur *f*
r.-insulated lead-covered cable Gummibleikabel *n*
r. joint Gummi[ab]dichtung *f*
r.-lead-covered cable Gummibleikabel *n*
r. mat Gummimatte *f*
r. moulding Gummiformteil *n*
r.-mounted gummibereift
r. packing Gummizwischenlage *f*, Gummidichtung *f*
r. pad Gummikissen *n*, Gummiklotz *m*
r. plug Gummistopfen *m*
r.-ring packing Gummiringdichtung *f*
r. sac Gummispeicherblase *f*, Gummisack *m*
r. sealing ring Gummidichtring *m*
r. shock absorber Gummistoßdämpfer *m*
r. solution Gummilösung *f*
r. sponge Gummischwamm *m*
r. spreader *(Plast)* Streichmaschine *f*
r. spring Gummifeder *f*
r. stamp Gummistempel *m*
r. stopper Gummipfropfen *m*, Gummistopfen *m*
r. substitute Gummiaustauschstoff *m*
r. tape Isolierband *n*
r. tube Gummirohr *n*, Gummischlauch *m*
r. tyred gummibereift
r. universal joint Gummikreuzgelenk *n*
r. valve Gummiventil *n*
r. washer Gummizwischenlage *f*
rubberite Vulkanit *m*
rubberize / to gummieren
rubberized gummiert
r. fabric gummiertes Gewebe *n*
rubberizing Gummierung *f*
rubbery gummiartig, gummielastisch

rubbing 1. *(Text)* Nitscheln *n*, Würgeln *n*; 2. Reiben *n*; Einreiben *n*; Gleiten *n*; Abwischen *n*
r. agent Einreib[ungs]mittel *n*
r. apron Nitschelhose *f*
r. condenser Nitschelwerk *n*
r. contact *(Am)* Gleitkontakt *m*; Wälzkontakt *m*
r.-fastness tester *(Text)* Reibechtheitsprüfer *m*
r. leather Nitschelhose *f*
r. leather with cut geschnittene Nitschelhose *f*
r. out Ausradieren *n*
r. pad Führungsleiste *f*
r. pads Reibbacken *fpl*
r. roller Nitschelwalze *f*
r. rope Ablenkseil *n* *(Schachtförderung mit Führungsseilen)*
r. surface 1. Reibfläche *f (Bremse)*; Gleitfläche *f (Reibung)*; 2. *(Text)* Nitschelfläche *f*
r. varnish Schleiflack *m*
r. velocity Gleitgeschwindigkeit *f*
rubbish 1. Müll *m*; Abfall *m*; 2. *(Brg)* Berge *pl*, Bergeklein *n*
r. chute Müllschlucker *m*, Müllabwurfschacht *m*
r. dump Schuttkippe *f*, Schuttabladeplatz *m*
r. heap Schutthalde *f*
rubble 1. Schutt *m*, Trümmer *pl*; Grobkies *m*; 2. *s.* r. masonry
r. concrete Bruchsteinbeton *m*
r. masonry Bruchsteinmauerwerk *n*, Natursteinmauerwerk *n*
r. work *s.* r. masonry
rubbles *(Brg)* Bruchmassen *fpl*
rubeanic acid Rubeanwasserstoff[säure *f*] *m*
rubellite *(Min)* Rubellit *m*
rubidium Rubidium *n*
rubiginous rostfarben
rubious rubinrot
ruby 1. Rubin *m*, edler Korund *m*; 2. Rubinrot *n*; 3. *(Typ) (etwa)* Perl *(Schriftgröße)*
ruby rubinrot, dunkelrot
r.-coloured rubinrot, dunkelrot
r. glass Rubinglas *n*
r. laser Rubin-Laser *m*
r. light Rotlicht *n (Fotografie)*
r. maser Rubin-Maser *m*
r. optical maser optischer Rubin-Maser *m*
r. silver ore Rotgiltigerz *n*
r. spinel edler Spinell *m*
rudder *(Schiff)* Steuerruder *n*, Ruder *n*
r.-angle indicator Ruder[lagen]zeiger *m*
r. blade Blatt *n*, Ruderblatt *n*
r. heel Ruderhacke *f*
r. post Rudersteven *m*
r. tiller Ruderpinne *f*
r. yoke Ruderjoch *n*
ruddle / to markieren [mit rotem Ocker]
ruddle roter Ocker *m*
rude rauh, unpoliert
ruffle / to kräuseln
rug 1. *(kleiner)* Teppich *m*; 2. Reisedecke *f*
rugged kräftig, stabil, nicht störanfällig
r. construction robuste Konstruktion *f*
r. unit widerstandsfähiges Bauelement *n*
ruggedized stoßfest, *(mechanisch)* widerstandsfähig
ruggedness geringe Störanfälligkeit *f*, Unempfindlichkeit *f*
Ruhmkorff [induction] coil Ruhmkorffscher Funkeninduktor *m*
ruin / to zerstören; beschädigen
ruin Verfall *m*, Baufälligkeit *f*
rule / to 1. lini[i]eren; 2. anordnen
rule 1. Regel *f*; Anordnung *f*; 2. Lineal *n*, Maßstab *m*, Schiene *f*
r. bender and cutter *(Typ)* Linienbiege- und -schneidemaschine *f*
r. of operation Bedienungsvorschrift *f*
r. of three *(Math)* Dreisatz *m*, Regeldetri *f*
r. of thumb Faustregel *f*
ruled paper lini[i]ertes Papier *n*
r. surface *(Math)* Regelfläche *f*
ruler Lineal *n*, Linienreißer *m*
rules of navigation Seestraßenordnung *f*, Seeschiffahrtsstraßenordnung *f*
Rules of the Road at Sea Seestraßenordnung *f*; Seeschiffahrtsstraßenordnung *f*
ruling 1. *(Math)* Erzeugende *f (einer Regelfläche)*; 2. Lineatur *f*, Linierung *f*; Linierungsstrich *m*
r. device Linienziehgerät *n*
r. machine Liniermaschine *f*
r. of a lattice Gitterstrich *m*
r. pen Reißfeder *f*, Ziehfeder *f*, Zeichenfeder *f*
r. scale Zeichenmaßstab *m*
rum ether Rumessenz *f*
rumble / to trommeln, rommeln, fertigputzen
rumble Gußputztrommel *f*, Scheuerfaß *n*
run / to 1. laufen [lassen]; fahren *(Maschine)*; leiten *(Betrieb)*; betreiben, betätigen, antreiben; funktionieren; arbeiten *(Motor)*; 2. verlegen *(Kabel)*; führen *(Leitungen)*; 3. zerfließen, zerlaufen *(Farbe)*; fließen *(Flüssigkeiten)*; 4. auftauen *(Eis)*; 5. auslaufen *(Gefäße)*
r. a cable eine Leitung [ver]legen, ein Kabel [ver]legen
r. a casing into the hole Bohrloch verrohren, Futterrohre ins Bohrloch einbringen *(Tiefbohrtechnik)*
r. aground *(Schiff)* auf Grund laufen, festkommen
r. ashore *s.* r. aground / to
r. concentric rundlaufen
r. copal Kopal schmelzen *(Lackindustrie)*

r. down 1. kollidieren; 2. stehenbleiben, auslaufen *(Motor)*; 3. überentladen *(Batterie)*
r. hot warmlaufen, heißlaufen *(Motor)*
r. idle leer laufen *(Motor)*
r. in einlaufen; einströmen; einfahren
r. into *(Schiff)* kollidieren
r. into the grid-current *(El)* ins Gitterstromgebiet kommen
r. off 1. ablaufen, abfließen; abstechen; 2. abspulen *(Draht)*; abtrommeln *(Kabel)*
r. off centre verlaufen *(Bohrer)*
r. on petrol mit Benzin fahren
r. out of centre verlaufen *(Bohrer)*
r. out of true (truth) *(Kfz)* schlagen, Schlag aufweisen; verlaufen *(Bohrer)*
r. the engine all out Motor voll beanspruchen
r. the engine on the chocks Flugzeugmotor abbremsen
r. together einfluchten; einlaufen; einspielen lassen
r. true rundlaufen
r. untrue schlagen, taumeln *(Schwungrad)*; verlaufen *(Bohrer)*
r.-up 1. hochfahren *(Gleichrichter)*; 2. einlaufen *(Gewebe)*
r. wires overhead Leitungen oberirdisch verlegen
r. Z twist in Z-Richtung zwirnen (drehen)
run 1. Lauf *m*, Verlauf *m*, Gang *m*, Weg *m*; Durchlauf *m*; Anlauf *m*; Ansatz *m*; 2. Fabrikation *f*; Fertigungslos *n*; 3. Meßreihe *f*; 4. Versuch[sreihe *f*] *m*; 5. *(Math)* ununterbrochene Folge *f*; 6. *(El)* Führung *f*, Verlauf *(Kabel)*; 7. *(Schiff)* Schiffsform *f* unter Wasser, Formverlauf *m*, Linienverlauf *m*, Schiffsbodenverlauf *m*; Hinterschiffsform *f*; Sog *m*
r. aft Schiffsbodenverlauf *m* hinten
r.-down battery entladene (erschöpfte) Batterie *f*
r. downstreams *(Schiff)* Talfahrt *f*
r. forward Schiffsbodenverlauf *m* vorn
r. grain *(Led)* losnarbig
r.-in of the carriage *(Text)* Wageneinfahrt *f*
r.-in period Einlaufzeit *f*
r. of cable Kabelweg *m*
r.-of-mine coal Förderkohle *f*, Roh[förder]-kohle *f*
r.-of-mine ore Roherz *n*
r.-of-oven coke Mischkoks *m*
r. of rays Strahlengang *m*
r. of the bubble *(Opt)* Libellenweg *m*
r. of work 1. Arbeitsfolge *f*; Arbeitsstufen *fpl*; 2. Serienarbeit *f*
r.-off 1. *(Met)* Abstich *m*; Abschlackung *f*; Abfluß *m*, Ablauf *m*; Ablaufmenge *f*; Ablaß *m*; 2. Verlaufen *n (Bohrer)*
r.-out Verlaufen *n (Bohrer)*; Auslauf *m*
r.-out time Auslaufzeit *f*
r.-over frog *(Eis)* Anlaufherzstück *n* der Weiche
r.-proof *s.* runproof
r.-resist[ant] *s.* runproof
r. test *(Math)* Prüfverfahren *n* der ununterbrochenen Folgen
runabout Verkehrsboot *n*, Fahrgastboot *n*, Wassertaxi *n*
runaway 1. *(Flg)* Start- und Landebahn *f*, Rollbahn *f*; 2. *(El, Fs)* Abrutschen *n*, Durchstoßen *n*, Instabilwerden *n*, Umkippen *n*; 3. Durchgehen *n (Reaktor)*
r. flags *s.* flags of convenience
r. speed Durchgangsdrehzahl *f*
rung [step] Leitersprosse *f*
runner 1. Läufer *m (Ausbau, Bergwerk; Rechenschieber)*; 2. Einguß *m*; Anschnitt *m*, Gießlauf *m (Gießen)*; 3. Laufrad *n (Turbine)*; Schaufelrad *n*, Laufrolle *f*; Ventilationsrad *n*; 4. Hebeklaue *f*; 5. *(Schiff)* Decksunterzug *m*, Unterzug *m*; Schlittenläufer *m (Stapellaufschlitten)*; 6. Schlittenkufe *f*; 7. Ausläufer *m*, Wurzelableger *m*, Wurzelschößling *m*
r. and tackle *(Schiff)* Mantel *m (mit Talje)*
r. basin *(Met)* Eingußsumpf *m*, Gießtümpel *m*
r. head *(Met)* Gießkopf *m*, Gießtrichter *m*
r. of sleigh Schlittenkufe *f*
r. stick Gußlochzapfen *m*
runnerless angußlos
running Lauf *m*, Fahren *n (Maschine)*
r. accuracy Laufgenauigkeit *f*
r. aground *(Schiff)* Aufgrundlaufen *n*
r. balance dynamische Auswuchtung *f*, dynamisches Gleichgewicht *n*
r. board *(Kfz)* Trittbrett *n*; *(Eis)* Laufsteg *m (zwischen Wagen)*
r.-board aerial Trittbrettantenne *f*
r.-board support Trittbretthalter *m*
r. bowsprit *(Schiff)* einzuziehendes Bugspriet *n*
r. condition Laufbedingung *f (Motor)*
r. cost laufende Unkosten *pl*, Betriebskosten *pl*
r. current Laststrom *m*, Fahrstrom *m*, Betriebsstrom *m*
r. fit Laufsitz *m*, Spielpassung *f*
r. fix *(Schiff)* Besteck *n*
r. gear Fahrgestell *n*, Chassis *n*
r. hot Warmlaufen *n (Lager)*
r. in Einlaufen *n*, Einfahren *n*
r. in bearings *(Kfz)* gelagert
r. knot Schifferknoten *m*, Laufknoten *m*
r. lights Positionslichter *npl*, Positionslaternen *fpl*
r. of cables Kabelverlegung *f*
r. of wires Leitungsführung *f*

r.-off reel Ablaufhaspel *f*
r. on position *(Text)* Aufstoßstellung *f*
r. out Verlaufen *n (Bohrer)*
r. packing bewegliche Dichtung *f*
r. plate *(Kfz)* Trittplatte *f*
r. points *(Ch)* Destillationspunkte *mpl*
r. position Betriebsstellung *f*, Laufstellung *f*
r. powers Recht *n* zur Benutzung einer Eisenbahnstrecke
r. rail Laufschiene *f*
r. resistance Rollwiderstand *m*
r. rigging *(Schiff)* laufendes Gut *n (Tauwerk)*
r. sand Schwimmsand *m*, Treibsand *m*, Triebsand *m*
r. ship Schiff *n (in Fahrt, in Betrieb)*
r. stitch Steppstich *m*
r. stopper undichter Pfannenstopfen *m*
r. surface Lauffläche *f*
r. true schlagfrei laufend
r. tubing Einbau *m* von Steigrohren *(Tiefbohrtechnik)*
r.-up Abbremsen *n (Flugzeugtriebwerk)*; Hochfahren *n (Motor)*
r. voltage *(El)* Vollastspannung *f*, Lastspannung *f*; Brennspannung *f (Lichtbogen)*
r. ways *(Schiff)* Ablaufbahn *f (beweglicher Teil)*; Schlitten *m*, Ablaufschlitten *m*, Schlittenläufer *m (Stapellauf)*
r. with no load Leerlauf *m*
runnings *(Ch)* Füllmittel *npl*, Streckmittel *npl*
runout Schlag *m*, Unrundheit *f*
runproof *(Text)* [lauf]maschenfest, laufmaschensicher
runway Start- und Landebahn *f*, Rollbahn *f*, Piste *f*
r. lighting Pistenbefeuerung *f*
r. lights Pistenbefeuerung *f*
r.-localizing beacon Anflugrichtungsfunkbake *f*
rupture / to [zer]reißen, zu Bruch gehen, zerbrechen
rupture 1. Bruch *m*, Reißen *n*, Zerreißen *n*; Trennbruch *m*; Bersten *n*, Zerbrechen *n*; 2. Durchschlag *m (Isolation)*
r. curve Mohrsche Umhüllungskurve *f (Gebirgsmechanik)*
r. load Bruchlast *f*
r. of the grinding wheel Bersten *n* der Schleifscheibe
r. of wire Drahtbruch *m*
r. strength Trennfestigkeit *f*
rupturing Bruchbildung *f*
r. capacity Unterbrecherleistung *f*, Schaltleistung *f*, Ausschaltleistung *f*
rural ländlich, Land-
r. construction ländliches Bauwesen *n*
r. district *(Fmt)* Landnetzgruppe *f*
r. telephone system Landfernsprechnetz *n*, ländliches Fernsprechnetz *n*
rush 1. Ausbruch *m*; Andrang *m*; Ansturm *m*; Übereilung *f*; Herabstürzen *n*; 2. Binse *f*
r. hours Hauptverkehrszeit *f*
r. job 1. eilige Arbeit *f*; 2. *(Typ)* Schnellschuß *m*
r. order *(Typ)* Schnellschuß *m*
russet upper leather Fahlleder *n*
Russian leather Juchtenleder *n*
rust / to rosten
r. in einrosten, festrosten
rust Rost *m*
r. cement *(Masch)* Rostkitt *m*
r. coating Rostschicht *f*, Rostbelag *m*
r. damage Rostschaden *m*
r. formation Rostbildung *f*, Verrosten *n*
r. inhibiting paint Rostschutzfarbe *f*
r. layer Rostschicht *f*
r. prevention Rostschutz *m*
r. preventive Rostschutzmittel *n*
r. preventive paint Rostschutzfarbe *f*
r.-proof nichtrostend
r.-proof quality Korrosionsbeständigkeit *f*
r. proofing grease Rost- und Korrosionsschutzfett *n*
r. remover Entrostungsmittel *n*
r. removing agent Rostentfernungsmittel *n*, Rostlösemittel *n*, Entroster *m*
r.-resisting rostbeständig, rostfrei
r.-resisting paint Rostschutzfarbe *f*
r.-resisting property Rostbeständigkeit *f*
rusting Rosten *n*, Verrosten *n*
rustless rostfrei, nichtrostend, rostbeständig
rustlessness Rostbeständigkeit *f*, Rostfreiheit *f*
rusty 1. rostig, verrostet; 2. ranzig *(Speck)*
ruthenate *(Ch)* Ruthenat *n*
ruthenic compound Verbindung *f* des vierwertigen Ruthenium
ruthenium Ruthenium *n*
ruthenous compound Verbindung *f* des zweiwertigen Ruthenium
Rutherford atom Rutherfordsches Atom *n*
Rutherford-Bohr atom model Rutherford-Bohrsches Atommodell *n*
Rutherford scattering Rutherfordsche Streuung *f*
rutherfordine Rutherfordin *m (Uranmineral)*
Ruths's accumulator Ruthsspeicher *m*
rutilant rotglänzend, rotstrahlend
rutile Rutil *m*, Titandioxid *n*
rydberg Rydberg *n (Energiemaß der Spektroskopie)*
Rydberg constant Rydberg-Konstante *f*
Rydberg frequency Rydberg-Frequenz *f*
rye Korn *n*, Roggen *m*
r. gluten Roggenkleber *m*

S

S band *(Fmt)* S-Band *n (1550 bis 5200 MHz)*
S bend gekröpftes Rohr *n*
s-conductivity s-Leitfähigkeit *f (Halbleiter)*
S-distortion *(El)* S-Verzeichnung *f*
S-shaped hook S-Haken *m*
S value Streufaktor *m*
S waves S-Zacken *mpl (Ekg)*
sable Zobel[pelz] *m*
sac Sack *m*; Speicherblase *f*
saccharase Saccharase *f*, Invertase *f*
saccharic acid Zuckersäure *f*
sacchariferous zuckerhaltig
saccharification Verzuckerung *f*, Zuckerbildung *f*
saccharimeter Saccharimeter *n*
saccharimetry Saccharimetrie *f (Bestimmung der Konzentration von Zuckerlösungen)*
saccharin[e] Sa[c]charin *n*
saccharine zuckerartig, zuckerhaltig, zuck[e]rig
saccharoid glukoseartig, zuckerähnlich, zuckerartig
saccharomyces Saccharomyces *f (Hefe)*
saccharose Rohrzucker *m*, Saccharose *f*
saccharous zuckerähnlich, zuckerartig
sack / to absacken, in Säcke abfüllen
sack Sack *m*, Beutel *m*
  s. cell Beutelelement *n (Batterie)*
  s. cloth Sackleinwand *f*, Sacktuch *n*
  s. element Beutelelement *n (Batterie)*
  s. hoist Sackaufzug *m*
  s. shooter Sackrutsche *f*
  s. truck Sackkarren *m*
sacking machine Sackpackmaschine *f*, Absackmaschine *f*
sad dunkel *(Farbe)*
sadden / to nachdunkeln *(Farbe)*
  s. down glanzlos werden
saddening *(Text)* Regenbogendruck *m*, Irisdruck *m*
saddle / to satteln; aufbürden, aufladen *(Last)*
saddle 1. Sattel *m (Pferd, Fahrrad)*; 2. *(Masch)* Bettschlitten *m*, Längsschlitten *m (Drehmaschine)*; Werkzeugschlitten *m*; Querschieber *m*; Stößelschlitten *m*; 3. Auflager *n*; 4. *(Math)* Sattelpunkt *m*
  s. and harness Sattelzeug *n*
  s. apron Schloßplatte *f*, Räderplatte *f (Drehmaschine)*
  s. clamp *(Masch)* Schlittenklemmung *f*
  s. clamp bolt Schlittenfeststellschraube *f*
  s. clamping lever Feststellhebel *m* für Support
  s. girth Sattelgurt *m*
  s. key Hohlkeil *m*
  s. of a planer Grundschieber *m* einer Langhobelmaschine
  s. packing Sattelfüllkörper *m*
  s. point *(Math)* Sattelpunkt *m* (*s. a.* minimax)
s.-shaped sattelförmig
  s. stitching *(Typ)* Heften *n* im Falz
  s. surface Sattelfläche *f*
saddler Sattler *m*
saddler's awl Sattlerahle *f*
safe sicher; zulässig; ungefährlich
  s. carrying capacity *(El)* zulässige Belastung (Strombelastung) *f*
  s. clearance path *(Flg)* hindernisfreie Flugbahn *f*
  s. filament current zulässiger Heizstrom *m (Elektronenröhren)*
  s. health level *(Kern)* ungefährliches Strahlenniveau *n*
  s. load zulässige Last (Belastung) *f*
  s. operating temperature betriebszulässige Temperatur *f (Verbrennungsmotor)*
  s. strain *(sl) s.* s. stress
  s. stress zulässige Spannung (Beanspruchung) *f*
safeguard Schutz *m*, Sicherung *f*
  s. against overload Überlastschutz *m*
safely handled handhabungssicher
safety Sicherheit *f*, Ungefährlichkeit *f*, Unschädlichkeit *f*
safety Sicherheits-, Schutz-
  s. against fracture Bruchsicherheit *f*
  s. appliance Sicherheitsvorrichtung *f*
  s. belt Anschnallgurt *m*, Rettungsgürtel *m*, Sicherheitsgürtel *m*, Sicherheitsgurt *m*
  s. bolt Sicherheitsbolzen *m*, Sicherungsbolzen *m*
  s. boot Arbeitsschutzschuh *m*
  s. brake Notbremse *f*
  s. buoy Rettungsboje *f*; Rettungsring *m*
  s. button Aufnahmesperre *f (Tonbandgerät)*
  s. cage ladder Korbleiter *f*, Sicherheitsleiter *f*
  s. cap Sicherungskappe *f*
  s. catch Abzugssicherung *f*, Arretierklinke *f*, Sicherungshebel *m*, Sicherungsklinke *f*, Sicherungsknagge *f*, Sicherungsriegel *m*; Gesperre *n*, Sperreinrichtung *f*, Sperrhebel *m*; Fangvorrichtung *f*
  s. chain Sicherheitskette *f*; *(Eis)* Notkette *f (Kupplung)*
  s. clip Sicherungsvorstecker *m*
  s. clutch Sicherheitskupplung *f*, Überlastungskupplung *f*
s.-cock Sicherheitshahn *m*
  s. contactor Sicherheitsschalter *m*
  s. container Sicherheitsbehälter *m*
  s. control Sicherheitsüberwachung *f*
  s. controls Sicherheitseinrichtungen *fpl*, Schutzeinrichtungen *fpl*
  s. coupling Notkupplung *f*; Sicherheitskupplung *f*

s. cut-out Abschmelzsicherung *f*, Schmelzsicherung *f*, Sicherheitsausschalter *m*
s. deck Sicherheitsdeck *n (beim Schwimmdock)*
s. device Sicherheitseinrichtung *f*, Sicherung *f*, Schutzeinrichtung *f*
s. explosive Sicherheitssprengstoff *m*; *(Brg)* Wettersprengstoff *m*
s. factor Sicherheitsfaktor *m*, Sicherheit[sgrad *m*] *f*
s. factor for drop-out Abfallsicherheitsfaktor *m (Relais)*
s. factor for holding Haltesicherheitsfaktor *m (Relais)*
s. factor for pick-up Ansprechsicherheitsfaktor *m (Relais)*
s. film Sicherheitsfilm *m*
s.-first engineering Arbeitsschutztechnik *f*; Sicherheitsmaßnahmen *fpl*
s.-friction clutch Sicherheitsreibkupplung *f*
s. from interception *(Fmt)* Abhörsicherheit *f*
s. fuse 1. *(El)* Schmelzsicherung *f*, Sicherung *f*; 2. Zündschnur *f*
s. gap *(El)* Sicherheitsfunkenstrecke *f*
s. gear Fangvorrichtung *f*
s. glass Sicherheitsglas *n*, splittersicheres (splitterfreies) Glas *n*; Verbundglas *n*
s. glasses Schutzbrille *f*
s. gloves Sicherheitshandschuhe *mpl*, Unfallverhütungshandschuhe *mpl*
s. goggles Schutzbrille *f*
s. governor Sicherheitsregler *m*, Hilfsregler *m*, Schnellschlußregler *m (Turbine)*
s. guard Schutzeinrichtung *f*
s. handle *(Mil)* Sicherungsflügel *m*
s.-in-mines Grubensicherheit *f*
s. instructions Sicherheitsvorschriften *fpl*
s. interlock *(BMSR)* Sicherheitsverblockung *f*
s. interlock device *(Masch)* Sicherheitsverriegelungseinrichtung *f*, Abhängigkeitsschaltung *f*
s. island Fußgängerverkehrsinsel *f*
s. joint Sicherheitsmuffe *f*
s. keel *(Schiff)* Sicherheitskiel *m*
s. lamp Sicherheitslampe *f*
s.-lamp gasoline *(Brg)* Wetterlampenbenzin *n*
s.-lamp mine Schlagwettergrube *f*, schlagwettergefährdete Grube *f*
s. lap *(Wärm)* Sicherheitsdeckung *f*
s. level *(Wärm)* Sicherheitswasserstand *m*
s. lever Sicherheitshebel *m*, Sicherungshebel *m*
s. lock Sicherheitsschloß *n*
s. measure Sicherheitsmaßnahme *f*
s. nut Sicherungsmutter *f*
s. ordinances Sicherheitsvorschriften *fpl*, Schutzbestimmungen *fpl*
s. pawl Sicherungsklinke *f*
s. pillar *(Brg)* Sicherheitspfeiler *m*
s. pilot *(Am)* Zündflammenüberwachungsanlage *f (Brenner)*
s. pin Arretierstift *m*, Sicherheitsnadel *f*
s. plug [with earthing contact] *(El)* Schukostecker *m*, Schutzkontaktstecker *m*
s. precautions Arbeitsschutzmaßnahmen *fpl*
s. regulations Sicherheitsvorschriften *fpl*, Schutzbestimmungen *fpl*
s. rod Schnellschlußstab *m*, Sicherheitsstab *m*, Abschaltstab *m (Reaktorregelung)*
s. screw Sicherungsschraube *f*
s. shut-down Abschaltung *f* aus Sicherheitsgründen, Sicherheitsabschaltung *f (Reaktor)*
s. shut-down scram Notabschaltung *f*, Schnellabschaltung *f*, Sicherheitsabschaltung *f*
s. stirrup Sicherungsbügel *m*
s. switch Sicherheitsschalter *m*
s. tap chuck Sicherheitsfutter *n* für Gewindebohrer
s. valve Sicherheitsventil *n*, Ablaßventil *n*
s. valve loaded with weight Sicherheitsventil *n* mit Gewichtbelastung
s. valve with high lift Hochhubsicherheitsventil *n*
s. washer aufgebogene Unterlegscheibe *f*
s. wire Plombendraht *m*
saffian Saffian *m*
saffron Safran *m*
safranine Safranin *n (Farbstoff)*
sag / to 1. durchhängen *(Leitungen)*; 2. absinken *(Kurve)*; 3. abtreiben *(Schiff durch Wind)*; durchsacken
sag 1. Durchhang *m*, Schnellhöhe *f* von Freileitungen; Durchbiegung *f*, Senkung *f*; Pfeilhöhe *f*; 2. Scheitel *m (Sagitta)*
s. of piston rod Durchbiegung *f* der Kolbenstange
s. test Durchhangsprüfung *f (Freileitung)*
s. tie Hängesäule *f*, Hängestab *m (Fahrdrahtaufhängung)*
sage Salbei *f*
s. oil Salbeiöl *n*
sagenite *(Min)* Sagenit *m*
saggar 1. Brennkapsel *f*; 2. Kerntrockenkasten *m (Formerei)*
s. clay Kapselton *m*
sagging 1. Durchbiegung *f*, Durchhang *m*; 2. *(Schiff)* Durchsackung *f*, Kielsenkung *f*; Kielbucht *f*; 3. außergewöhnliche Abdrift *f (durch Wind)*
s. point Erweichungspunkt *m (Keramik)*
sagittal Sagittal-
s. focal line *(Opt)* sagittale Brennlinie *f*
sago 1. Sago *m*, Palmenmehl *n*; 2. Sagopalme *f*
s. palm 1. Sagopalme *f*; 2. Sagopalmenfaser *f*
s. powder Sagomehl *n*
sail / to segeln; abgehen
s. close-hauled am Wind segeln

s. in line in Kiellinie laufen
s. line ahead in Kiellinie laufen
s.-over hervorstehen, überhängen
sail 1. Segel *n*; 2. Flügel *m (Windmühle)*
s. area Segelfläche *f*, Besegelung *f*
s. boat *(Am)* Segelboot *n*
s. cloth Segeltuch *n*
s. plan Segelriß *m*
s. ship *(Am)* Segelschiff *n*
s. sprit Sprietbaum *m*
sailer Segler *m (Schiff)*
sailing Segelsport *m*; Abfahrt *f (eines Schiffes)*
s. barge Ewer *m*, Frachtewer *m*
s. boat Segelboot *n*
s. day Abfahrtstag *m (eines Schiffes)*
s. dinghy Segeljolle *f*, Segelding[h]i *n*
s. directions Segelhandbuch *n*, Seehandbuch *n*
s. drifter Segellogger *m*
s. navigation Segelschiffahrt *f*
s. ship with auxiliary motor Motorsegler *m*, Segelschiff *n* mit Hilfsmotor
s. training ship Segelschulschiff *n*
s. trim Segelfähigkeit *f*
sailor Matrose *m*, Seefahrer *m*
sailor's knot Schifferknoten *m*
sails Segelwerk *n*, Besegelung *f*
sal Salz *n*
s. ammoniac Ammoniumchlorid *n*, Salmiak[salz *n*] *m*
s.-ammoniac cell *(El)* Leclanché-Element *n*
salamander *(Met)* Bodensau *f*, Ofensau *f*, Bär *m*
salaried staff Gehaltsempfänger *mpl*
salary earner Gehaltsempfänger *m*
sale by agent Kommissionsverkauf *m*, Verkauf *m* durch Kommissionär
s. by area Festmeterkauf *m*, Flächenkauf *m (Holz)*
saleable coal verkaufsfähige Kohle *f*
s. output *(Brg)* absatzfähige Förderung *f*
sales agreement Verkaufsvertrag *m*
s. analysis Umsatzstatistik *f*
s. executive Verkäufer *m*
s. expense Vertriebskosten *pl*
s. slip Kassenabschnitt *m*, Kassenzettel *m*
salicin[e] *(Ch)* Salizin *n*
salicylate Ester *m* (Salz *n*) der Salizylsäure, Salizylat *n*
salicylic acid Salizylsäure *f*
salient angle vorspringender Winkel *m*
s. pole *(Elm)* ausgeprägter Pol *m*
saliferous salzführend, salzhaltig
salifiable zur Salzbildung fähig
salification Salzbildung *f*
saline Saline *f*; Salzbergwerk *n*; Salzquelle *f*
saline salzartig, salzig
s. deposit Abraumsalz *n*
s. lake Salzsee *m*
s. manure Düngesalz *n*
s. water salinisches Wasser *n*, Salinenwasser *n*
salinity Salzgehalt *m*, Salzhaltigkeit *f*
salinous salzartig, salzhaltig
sallow Weide *f*
sally *(Am)* Bootshafen *m*
salmon boat *(Schiff)* Lachsfänger *m*
saloon Kajüte *f*
s. deck Salondeck *n*
salt / to [ein]salzen
s. a well Fündigkeit einer Bohrung vortäuschen *(Tiefbohrtechnik)*
salt Salz *n*; Siedesalz *n*, Steinsalz *n*
s. bath Salzbad *n*
s.-bath case hardening Salzbadhärtung *f (Stahl)*
s.-bath furnace Salzbadofen *m*
s.-bath patenting Salzbadpatentierung *f (Stahl)*
s. brine Salzlake *f (Gerberei)*
s. cake Salzkuchen *m*
s. deposit Salzlagerstätte *f*
s. displacement *(Ch)* Aussalzen *n*
s. dome Salzkuppel *f*, Salzdom *m*
s. dried salztrocken
s. glaze Salzglasur *f*
s.-gun *(Brg)* Salzspritzmaschine *f*
s. lake Salzsee *m*
s.-lick Salzlecke *f*
s. mine Salzbergwerk *n*
s. pen Salzlast *f (Fischereischiff)*
s. pits Salzgrübchen *npl (Fehler im Leder)*
s. plug Salzdurchbruch *m*, Salzhorst *m*, Salzstock *m*
s. trip Fangreise *f (Salzfisch)*
s. water Salzwasser *n*
s.-water bath Solbad *n*
s.-water fish Seefisch *m*
salting-out *(Ch)* Aussalzen *n*
saltpetre Salpeter *m*
s. lye Salpeterlauge *f*
s. paper Salpeterpapier *n*
saltwater-proof seewasserbeständig
salvage / to 1. bergen; 2. wiedergewinnen, rückgewinnen; 3. retten
salvage 1. *(Schiff)* geborgenes Gut *n*; Bergung *f*; Bergegeld *n*; 2. Wiedergewinnung *f*, Reparatur *f*
s. dock Hebeschiff *n*
s. service Schiffsbergung *f*
s. sheet Sprungtuch *n (z. B. für die Feuerwehr)*
s. ship Bergungsschiff *n*
s. tug Bergungsschlepper *m*
s. vessel Bergungsschiff *n*, Bergungsfahrzeug *n*
salve / to bergen, retten
s. a ship ein Schiff heben
salve Salbe *f*

salvor Berger *m (Schiff)*
sam / to *(Led)* abwelken
samaric compound Verbindung *f* des dreiwertigen Samarium
samarium Samarium *n*
samarous compound Verbindung *f* des zweiwertigen Samarium
samarskite *(Min)* Samarskit *m*
same-type parts typengleiche Teile *npl*
samel brick zu wenig gebrannter Ziegel *m*
sameness Gleichheit *f*, Identität *f*
s. of colours Farbgleichheit *f*
samming machine *(Led)* Abwelkmaschine *f*
sample / to Proben entnehmen, [be]mustern; probieren
sample Probe *f*, Muster *n*, Versuchsstück *n*; Stichprobe *f*
s. changer Probenwechsler *m*
s. fair Mustermesse *f*
s. mean *(Math)* Stichprobenmittel *n*
s. port Öffnung *f* zum Einführen von Proben, Probeöffnung *f (Reaktor)*
s. range Spannweite *f* der Stichprobe
s. taken at random Stichprobe *f*
s. workpiece Musterstück *n*, Bezugsstück *n*
sampled-data control system *(BMSR)* Abtastregelung *f*
s.-data system *(BMSR)* Abtastsystem *n*
sampler 1. Probeentnahmeeinrichtung *f*, Probenahmegerät *n*; 2. Probenehmer *m (Person)*; 3. *(BMSR)* Abtastglied *n*, Abtaster *m*; 4. *(Fs)* Farbschalter *m*; 5. elektronischer Kommutator *m*
sampling 1. Probe[n]entnahme *f*, Probenahme *f*; 2. *(Math)* zufällige Auswahl *f*, Stichprobenerhebung *f*; Stichprobenprüfung *f*, Stichprobenentnahme *f*; 3. Abtastung *f (mittels Impulsen)*; 4. *(Text)* Bemustern *n*
s. apparatus Probennehmer *m*
s. device Probenahmegerät *n*
s. instant *(Dat)* Abtastzeitpunkt *m*, Abtastmoment *m*
s. interval *(Dat)* Abtastperiode *f*
s. pipe Entnahmepipette *f*, Entnahmerohr *n*
s. plan Stichprobenplan *m*
s. procedure elektronische Abtastung *f*
s. rate Abtastfrequenz *f*
s. theorem Abtasttheorem *n*, Probensatz *m*
Sampson post Schwengelbock *m*, Schwengelständer *m (Seilbohren)*
samsonite *(Min)* Samsonit *m*
sanadine *(Min)* Sanadin *m*
sand / to schmirgeln, mit Sand scheuern; mit Sand bestreuen
s. blast *s.* sandblast / to
sand Sand *m*
s. aerator Sandauflockerungsgerät *n*
s. bag Sandsack *m*
s. bank *s.* sandbank
s. bath Sandbad *n (Warmbehandlung)*
s. bed Sandbett *n (Gießerei)*
s. belt Schleifband *n*
s. blast *s.* sandblast
s.-blasting *s.* sandblasting
s. bottom Sandherd *m*
s. box 1. Sandform *f (Formerei)*; 2. Sandkasten *m (Streumaterial)*
s.-cast pig im Masselbett vergossenes Roheisen *n*
s. casting Kastenguß *m*; Sandgußstück *n*
s. castings Sandguß *m*; Sandgußstücke *npl*
s. compacting Sandverdichtung *f (Formerei)*
s. core Sandkern *m*
s. deposit *(Geol)* Sandvorkommen *n*
s. disintegrator Sandschleuder[maschine] *f*
s. drum Schleiftrommel *f (zum Polieren von Holz)*
d. dust Sandstaub *m*
s. expansion Wachsen *n* des Formsandes
s. feed Sandzuführung *f*
s. filling socket Sandeinfüllstutzen *m*
s. filter Sandfilter *n*
s. flotation process Sandschwimmverfahren *n (Aufbereitung)*
s. frame Füllrahmen *m (Kernpreßmaschine)*
s. grain Sandkorn *n*
s. gutter Sandrinne *f*
s. heap Sandhaufen *m*
s. jet Sandstrahl *m*
s.-lime brick Kalksandstein *m*
s. line Schmantseil *n (Seilbohren)*
s. mould Sandform *f*
s. muller Formsandaufbereiter *m*
s. nozzle Sanddüse *f*
s. packing Sandverdichtung *f (Formerei)*
s. paper Sandpapier *n*, Schleifpapier *n*, Schmirgelpapier *n*
s. pit Sandgrube *f*
s. preparing Sandaufbereitung *f (Formerei)*
s. quarry Sandgrube *f*
s. rammer Formsandstampfer *m*
s. reclamation Sandaufbereitung *f*
s. reel Schmanttrommel *f (Seilbohren)*
s. roll *s.* s. roller
s. roller *(Text)* Sandbaum *m*, Warenabzugsbaum *m*
s. screw conveyor Sandförderschnecke *f*
s. skin Sandkruste *f*, sandbehaftete Oberfläche *f (Gußstück)*
s. slinger *(Met)* Sandslinger *m*, Sandschleuderformmaschine *f*
s. stopping Schlackenwehr *n (Roheisenabstich)*
s. thrower Sandschleuder *f*
s. trap Sandfang *m*
s. working Sandgrube *f*

sandal brick *s.* samel brick
s. wood Sandelholz *n*
sandbank Untiefe *f*, Watt *n*, Sandbank *f*
s. buoy Untiefentonne *f*
sandblast / to sandstrahlen
sandblast Sand[gebläse]strahl *m*
s. action Sandblaswirkung *f*
s. cleaning Sandstrahlputzen *n*
s. machine Sandstrahlgebläse *n*; Gußputzmaschine *f*
sandblasting Sandstrahlen *n*, Sandstrahlreinigung *f*
s. machine Sandstrahlgebläse *n*
s. sand Gebläsesand *m*
sander Schmirgelschleifmaschine *f (Holzbearbeitung)*
sanding Sanden *n*, Schleifen *n (Holzbearbeitung)*
s. device Sandstreuvorrichtung *f (für Schienenfahrzeuge)*
s. disk Sandpapierscheibe *f*, Schmirgelscheibe *f*
s. sealer Schleifgrund *m*, Schnellschliffgrund *m*
sandpaper / to abschleifen *(Holz)*
sands Sandbank
sandstone Sandstein *m*
s. masonry facing Sandsteinverkleidung *f*
sandwich / to aufeinanderschichten, schichtweise anordnen
sandwich Schichtelement *n*
sandwich dreischichtig; mehrschichtig; zwischen zwei Schichten
s. arrangement *(Text)* Zwischenlage *f*
s. blending *(Text)* Mischen *n* im Bett
s. blending in bins Mischen *n* von Fasern durch Auflegen von Faserschichten in einem Fach
s. board Verbundplatte *f*
s. construction Verbundbauweise *f*
s. detector Schichtendetektor *m*
s. metal doppelseitig plattiertes Blech *n*
s. moulding Pressen *n* von Verbundbauteilen
s. panel Dreischichtenplatte *f*, Mehrschichtenplatte *f*, Sandwichplatte *f*
s. photocell Sperrschichtfotozelle *f*
s. structure *(Plast)* Verbundbauteil *n* aus fester Außenhaut und lichtem Stützstoff im Innern
s. technique *(Kern)* Schichtemulsionstechnik *f*
s.-type panel Verbundplatte *f*
s. windings Scheibenwicklungen *fpl (z. B. des Transformators)*
sandy shale sandiger Schieferton *m*, Sandschiefer *m*
sanforizing *(Text)* Sanforisieren *n*
sanidine *(Min)* Sanidin *m*
sanitary sanitär
s. engineering Gesundheitstechnik *f*, Sanitärtechnik *f*
s. sewage Abwässer *npl*
s. tank Sanitärtank *m*, Spülwassertank *m*, Hochbehälter *m*, Klosettank *m*
s. wallpaper abwaschbare Tapete *f*
sans-serif, sanserif Grotesk *f (Schrift)*
santicizer *(Am) (Plast)* Weichmacher *m*
sap 1. *(Mil)* Sappe *f*; 2. Saft *m (Baum)*
s.-stain *(Holz)* Splintbläue *f*, Splintfleck *m*
sapling junger Baum *m*
saponaceous seifenartig, seifig
saponifiable verseifbar
s. fat verseifbares Fett *n*
saponification Seifenbildung *f*, Verseifung *f*
s. number Verseifungszahl *f*
s. value Verseifungszahl *f*
saponify / to verseifen
saponite *(Min)* Saponit *m*, Seifenstein *m*
sapphire *(Min)* Saphir *m (Edelkorund)*
sappiness Saftgehalt *m*
sappy fracture *(Met)* feinkörniger Bruch *m*
s. structure *(Met)* feinkörniges Gefüge *n*
sapropel Faulschlamm *m*, Sapropel *n*
sapropelic sapropelitisch
sapwood Splintholz *n*
sarcolactic acid Fleischmilchsäure *f*, rechtsdrehende Milchsäure *f*
sardonyx *(Min)* Sardonyx *m*
sash schiebbarer Fensterrahmen *m*
s. cord Schnur *f* zwischen Fensterrahmen und Gegengewicht
s. door Schiebetür *f*
s. door furniture Schiebetürbeschläge *mpl*
s. fillister Nutenhobel *m*
s. line *s.* s. cord
s. pocket Laufbahn *f* für Gewicht *(Fenster)*
s. saw Schließsäge *f*, Schlitzsäge *f* *(Holzbearbeitung)*
s. sticker Fensterkehlmaschine *f*
s. stuff Fensterrahmenholz *n*
s. weight Schiebefenstergewicht *n*
s. window Schiebefenster *n*
sate Schrotmeißel *m*
sateen Baumwollsatin *m*, Baumwollatlas *m*
s. finish Satinappretur *f*
s. weave *(Text)* Atlasbindung *f*
satellite Satellit *m*
s. electron Hüllenelektron *n*
s. exchange Hilfsamt *n*, Fernsprechnebenamt *n*, Teilamt *n*
s. instrumentation Meßgeräteausrüstung *f* eines Satelliten
s.-launching rocket Trägerrakete *f* für einen Satelliten
s. orbit Flugbahn *f* des Satelliten
s. period [of revolution] Umlaufzeit *f* des Satelliten [um die Erde]
s. rocket Satellitenrakete *f*

s. separation Trennung *f* des Satelliten von der Trägerrakete
s. studio Hilfsstudio *n*
s. town Satellitenstadt *f*, Trabantenstadt *f*
s. tracking Satellitenverfolgung *f*, Satellitenortung *f*
s. transmitter 1. Hilfssender *m*, Tochtersender *m*; 2. Satellitensender *m*
s. vehicle Satellitenrakete *f*, Trägerrakete *f*; künstlicher Satellit *m*, Sputnik *m*
s. velocity Bewegungsgeschwindigkeit (Bahngeschwindigkeit) *f* des Satelliten
satiate / to sättigen, saturieren
satin / to satinieren *(Papier)*
satin-etched bulb innenmattierter Kolben *m* *(einer Lampe)*
s.-frosted seidenmattiert
s. glaze *(Led)* Satinglanz *m*
s. leather satiniertes Leder *n*
s. paper satiniertes Papier *n*
s. rib *(Text)* Atlasgrat *m*
s. spar *(Min)* Atlasspat *m*, Faserkalk *m*
s. twill *(Text)* Köperatlas *m*
s. weave *(Text)* Atlasbindung *f*
satinize / to satinieren, glätten *(Papier)*
satisfy / to 1. zufriedenstellen; 2. genügen *(Bedingung; Statik)*
saturable sättigungsfähig
s. inductor *(El)* Sättigungsspule *f*
s. reactor *(El)* Sättigungsdrossel *f*, Eisendrossel *f*
saturant Imprägniermittel *n*, Sättigungsmittel *n*
saturate / to absättigen, sättigen; [durch]tränken, saturieren, schlämmen
saturated gesättigt
s. activity Sättigungsaktivität *f*
s. compound *(Ch)* gesättigte Verbindung *f*
s. hydrocarbon gesättigter Kohlenwasserstoff *m*, Grenzkohlenwasserstoff *m*
s. iron core modulation *(Fmt)* Drosselmodulation *f*
s. solution gesättigte Lösung *f*
s. steam gesättigter Dampf *m*, Sattdampf *m*
s. vapour *s.* s. steam
saturation Absättigung *f*, Sättigung[szustand *m*] *f*
s. activation Sättigungsaktivierung *f*
s. backscattering *(Kern)* Rückstreusättigung *f*
s. concentration Sättigungskonzentration *f*
s. current Sättigungsstrom *m*
s. curve *(El)* Sättigungskennlinie *f*, Sättigungscharakteristik *f*; Magnetisierungskurve *f*
s. deficit Untersättigung *f*
s. degree Sättigungsstufe *f*
s. density Sättigungsdichte *f*
s. inductance Sättigungsinduktivität *f*
s. limit Sättigungsgrenze *f*
s. line Sättigungslinie *f*
s. magnetization Sättigungsmagnetisierung *f*
s. of the nuclear forces Absättigung *f* der Kernkräfte
s. point Sättigungspunkt *m*
s. pressure Sättigungsdruck *m*
s. reactance *(El)* Sättigungsreaktanz *f*
s. state Sättigungszustand *m*
s. temperature Sättigungstemperatur *f*
s. voltage Sättigungsspannung *f*
saturator Sättigungsapparat *m*, Saturateur *m*
sauce boat Soßenschüssel *f*, Sauciere *f*
sausage aerial *(Fmt)* Reusenantenne *f*
s. stuffer Wurststopfmaschine *f*
save / to 1. einsparen; 2. retten, erhalten, bergen
save-all Siebtisch *m*, Stoffänger *m*, Abwasserklärtrichter *m*, Rückgewinnungsanlage *f* *(Papierherstellung)*
saving in cost (economies) Einsparung *f*
s. in material Materialeinsparung *f*
s. in time Zeiteinsparung *f*
s. of heat Wärmeersparnis *f*
s. of steam due to condensation Dampfersparnis *f* durch Kondensation
s.-oil Ölwanne *f*, Öltrog *m*, Ölsammler *m*
saw / to sägen
s. off to dead length accuracy maßgenau absägen
s.-off to length ablängen
s. untrue verlaufen *(Säge)*
saw Säge *f*
s. band Sägeblatt *n* für Bandsäge *(endlos)*, Bandsägeblatt *n*
s. blade Sägeblatt *n*
s.-blade body Stammblatt *n*
s.-blade carriage Sägeschlitten *m*
s.-blade for cutting in both directions vor- und rückwärtslaufendes Sägeblatt *n*
s.-blade guidance Sägeblattführung *f*
s.-blade holder Sägeangel *f*
s. buck Sägebock *m*
s. carriage Sägeschlitten *m*
s. clamp Sägefeilkluppe *f*, Sägespannkluppe *f*
s. cut Sägeschnitt *m*
s. cutting on the draw stroke Gegenzugsäge *f*
s. dust *s.* sawdust
s. file Sägefeile *f*, Sägeschärffeile *f*
s. filer Sägefeilmaschine *f*
s. filing machine Sägefeilmaschine *f*
s. for mitre cutting Gehrungssäge *f*
s. for parallel sawing Besäumsäge *f*
s.-formed boat Stufenboot *n*
s. frame Sägebügel *m*
s. gate Sägegatter *n*, Gatter *n*
s. gin *(Text)* Sägeegreniermaschine *f*, Sägeentkörnungsmaschine *f*
s. grinding wheel Sägeschärfschleifscheibe *f*
s. groove Sägenut *f*, Sägeschnitt *m*
s. guide Gatterführung *f*

s. **gummer** Sägeschärfmaschine *f*
s. **horse** Sägebock *m*
s. **kerf** Sägeschnitt *m*, Schnittfuge *f*
s. **kerf gauge** Schnittfugenlehre *f*
s. **log** Sägeblock *m*
s. **log size** abgelängt *(Rundholz)*
s. **mill** 1. Sägewerk *n*; 2. Sägegatter *n*
s. **mill with one saw** Blockgatter *n*
s. **pitch** Zahnteilung *f (von Sägen)*
s. **segment** Sägezahnsegment *n*
s. **set** Schränkeisen *n (Säge)*
s. **set pliers** Schränkzange *f* für Sägezähne
s. **setting machine** Schränkmaschine *f*
s.**-shaped tooth** Sägezahn *m*
s. **sharpening machine** Sägeblattschärfmaschine *f*, Sägeblattschleifmaschine *f*
s.**-timber** Schneideholz *n*, Sägeblock *m*
s. **tooth** Sägezahn *m*
s.**-tooth current** *(El)* Sägezahnstrom *m*
s.**-tooth generator** Sägezahngenerator *m*
s.**-tooth system** *(Brg)* Sägeblattanordnung *f (Ausbau)*
s.**-tooth voltage** *(El)* Sägezahnspannung *f*
s.**-tooth wave** *(El)* Sägezahnschwingung *f*
s.**-tooth wire filleting** Sägezahndrahtbeschlag *m*
s. **vice** Sägefeilkluppe *f*
s. **wood** Schnittholz *n*
sawdust Sägemehl *n*, Sägespäne *mpl*
s. **concrete** Sägemehlbeton *m*
s. **room** Sägemehllager *n*
sawing Sägearbeit *f*; Sägen *n*, Zersägen *n*, Absägen *n*
s. **machine** Sägemaschine *f*
s.**-off to dead length accuracy** [maßgenaues] Ablängen *n*
s. **practice** Sägetechnik *f*
sawn timber Schnittholz *n*
s. **veneer** Sägefurnier *n*
s. **wood** Schnittholz *n*
sawyer 1. Sägewerker *m*, Sägearbeiter *m*; 2. Treibholz *n*
sax Dachhammer *m*
scab 1. *(Met)* Schorf *m*, Schuppe *f*, Schale *f*; Schülpe *f*; 2. *(Brg)* Unterzug *m (Ausbau)*
scaffold / to Gerüst bauen, einrüsten
scaffold 1. Baugerüst *n*, Gerüst *n*, Rüstung *f*; 2. Schmelzsäule *f (Hochofen)*
s. **board** Rüstbrett *n*
s. **pole** Gerüststange *f*; Rundholz *n*
scaffolding 1. Baugerüst *n*, Gerüst *n*, Rüstung *f*; 2. Absteifung *f*, Schalung *f*; 3. Gerüstbau *m*, Einschalung *f*; 4. Hängen *n (der Gicht)*
s. **plank** Rüstbohle *f*
s. **standard** Gerüstbaum *m*
scalar *(Math)* Skalar *m*, ungerichtete Größe *f*
s. **coupling** skalare Kopplung *f*
s. **field** Skalarfeld *n*
s. **flux** skalarer Fluß *m*
s. **meson field** skalares Mesonenfeld *n*
s. **potential** skalares Potential *n*
s. **product** skalares Produkt *n*, Arbeitsprodukt *n*
scald / to verbrühen, abbrühen, abkochen, [aus-]brühen
scalding by escaping steam Verbrühen *n* durch ausströmenden Dampf
scale / to 1. abblättern, abbröckeln, abschuppen, Kesselstein abklopfen; 2. Schuppen bilden, verzundern; 3. wiegen, messen; maßstäblich ändern; hinsichtlich Größenordnung ändern; 4. Abschätzen *(z. B. Holzbestand auf Stamm)*
s. **a boiler** Kessel abpicken (ausklopfen)
s. **down** [im Maßstab] verkleinern, *maßstabsgetreu* verkleinern
s. **off** abblättern, entzundern, schälen
s. **up** [im Maßstab] vergrößern, *maßstabsgetreu* vergrößern
scale 1. Schuppe *f*; 2. Gußhaut *f*; Zunder *m*; Sinterschlacke *f*; Kesselstein *m*; Hammerschlag *m*; Brandkruste *f*; Abblätterung *f*; Glühspan *m*; Sinter *m*; 3. Skala *f*; Maßeinteilung *f*, Maßstab *m*; 4. Waage *f*
s. **beam** Waagebalken *m*
s. **breaker** Zunderbrecher *m*
s.**-collecting cup** Abbrandschale *f (Funkenstativ)*
s. **crust** Kesselsteinschale *f*
s. **deflection** Skalenausschlag *m*, Zeigerausschlag *m*
s. **dial** Skalenblatt *n*, Skalenscheibe *f*
s. **division** Skalenteil *n*, Teilstrich[breite *f*] *m*
s. **effect** Maßstabseffekt *m*
s. **factor** Skalenfaktor *m*, Maßstabbeiwert *m*
s. **formation** Kesselsteinbildung *f*; Zunderbildung *f*
s.**-forming** kesselsteinbildend
s.**-free** zunderfrei
s. **galvanometer** Skalengalvanometer *n*
s. **hopper** Dosiertrichter *m*, Meßtrichter *m*
s. **interval** Skalenintervall *n*, Teilstrichabstand *m*
s. **model** maßstabgerechtes (maßstabgetreues, maßstäbliches) Modell *n*, Maßstabsmodell *n*
s. **model test** Modellversuch *m*
s. **of breadths** Breitenmaßstab *m*
s. **of hardness** [Mohssche] Härteskala *f*
s. **of length** Längenmaßstab *m*
s.**-of-n circuit** *(BMSR)* Untersetzungsschaltung *f* 1 : n
s. **of pressure** Druckmaßstab *m*
s.**-of-ten circuit** Dekadenzähler *m*, dekadischer Untersetzer *m*, Zehnfachuntersetzer *m*
s.**-of-two circuit** Binärzählstufe *f*, einstufiger Impulsuntersetzer *m*; Flip-Flop-Schal-

tung *f*; bistabile Schaltung *f*; Dualzähler *m*, binärer Zähler *m*, Zweifachuntersetzer *m*
s.-of-two counter Binäruntersetzer *m*, Binärzähler *m*
s. of water hardness Wasserhärteskala *f*
s. of waterplane areas Schwimmflächenskala *f*
s. of work Arbeitsmaßstab *m*
s. pan Waagschale *f*
s. pit Sintergrube *f*, Schlackengrube *f*
s.-producing salt Kesselsteinbildner *m*, kesselsteinbildendes Salz *n*
s. ratio of a model *(Schiff)* linearer Modellmaßstab *m*
s. resisting steel zunderbeständiger Stahl *m*
s. with omitted (suppressed) zero Skala *f* mit unterdrücktem Nullpunkt
scaled maßstabsgerecht, maßstäblich
scalene ungleichseitig *(Dreieck)*
scaler Impulszähler *m*; Untersetzerschaltung *f* *(für Impulse)*
scales Waage *f*
scaling 1. Verzunderung *f*, Zundern *n*, Abblättern *n*, Splittern *n*; Schuppenbildung *f*; 2. Entzundern *n*; 3. *(Brg)* Bereißen *n*, Beräumen *n*; 4. *(BMSR)* Ändern *n* der Größenordnung
s. circuit Untersetzer *m*, Zähler *m*, Zählgerät *n*; Untersetzerschaltung *f*, Zählschaltung *f*
s. furnace Blankglühofen *m*
s. hammer Kesselsteinabklopfer *m*
s.-off Entzunderung *f*
s. resistance Zunderfestigkeit *f*
scallop 1. Zipfel *m (Tiefziehen)*; 2. Muschel *f*
scalloped gezackt; gekerbt
scalloping machine *(Led)* Auszackmaschine *f*
scalp / to vorsieben, grobsieben
scalpel Skalpell *n*
scalper Stückgutscheider *m*, Grobsieb *n (Aufbereitung)*
scalping 1. mechanisches Entfernen *n* der Oberfläche; Schälen *n (Blöcke)*; 2. Vorabsiebung *f*, Grobsiebung *f (Aufbereitung)*
s. machine Schälmaschine *f (für Holz)*
scaly schuppig, geschuppt, zundrig
scan / to [periodisch] abtasten, zerlegen, auflösen; bestreichen *(Radar)*
scan *(Fs)* Abtastung *f*, Auflösung *f* in Bildelemente
s. current generator Kippstromgenerator *m*
s. generator Kippschwingungsgenerator *m*
scandium Skandium *n*
scanned area abgetastetes Gebiet *n (Radar)*
scanner Abtastgerät *n*, Abtaster *m*, Bildabtaster *m*; Drehantenne *f*
s. turning motor Drehantennenmotor *m*
s. unit Drehantenne *f*, Radarabtaster *m*
scanning Abtastung *f*, Ablenkung *f*, Zerlegung *f*; Bildabtastung *f*, Bildzerlegung *f*; Antennenbewegung *f*
s. aerial Drehantenne *f*, Rundsuchantenne *f* *(Radar)*
s. amplifier Abtastverstärker *m*
s. aperture Abtastöffnung *f*, Strahlöffnung *f*; Blende *f*
s. beam Abtaststrahl *m*
s. cycle *(Dat)* Abtastzyklus *m*, Abtastfolge *f*
s. density Abtastdichte *f*
s. device Bildabtaster *m*, Bildabtastvorrichtung *f*
s. diaphragm Abtastblende *f*, Rasterblende *f*
s. disk Rasterscheibe *f (Oszillografenröhre)*
s. electron beam abtastender Elektronenstrahl *m*
s. electron microscope Rastermikroskop *n*
s. generator Ablenkgenerator *m*
s. head Abtastkopf *m*
s. interference Rasterstörung *f*
s. line Bildzeile *f*, Abtastzeile *f*
s. pitch *(Fs)* Zeilenabstand *m*
s. raster Abtastraster *m*
s. sequence Abtastfolge *f*
s. slit Abtastschlitz *m*
s. spot Elektronenfleck *m*; Abtastlichtfleck *m*
s. stage *(Fs)* Bildablenkstufe *f*
s. traverse *(Fmt)* Abtastvorschub *m*
s. voltage Abtastspannung *f*
s. yoke Ablenkspule *f*
scansion *s.* scanning
scantling Halbholz *n*, Kantholz *n*
s. gauge Holzmaß *n*
s. sawn for specification Listenkantholz *n*
scapement Hemmung *f*
scapolite *(Min)* Skapolith *m*, Wernerit *m*
scar of peeling *(Holz)* Schälwunde *f*
scare-crow Vogelscheuche *f*
scarf / to anschärfen, abschrägen; kerben; mit schrägem Stoß verbinden; zusammenblatten
s. by the square anblatten
scarf Laschenverbindung *f*, Verblattung *f*, Laschung *f*, Verlaschen *n*, Verbindung *f* zweier Holzteile in Längsrichtung
s. and key gerades doppeltes Blatt *n*, Hakenblatt *n*, Hakenkamm *m*
s. brazed joint schräger Lötverbindungsstumpfstoß *m*
s. cut Fallkerb *m*
s. joint Stumpfstoß *m* unter 45°
s. tenon Blattzapfen *m (Tischlerei)*
s. with oblique cut ends schräg eingeschnittenes Blatt *n*
scarfed joint Balkenstoß *m*
s. lengths Stoßlängen *fpl*
scarfer Zuschärfmaschine *f*
scarfing Ausschärfung *f (Blechkante)*; Abschrägung *f*; Verzahnung *f*

scarifier Aufreißkamm *m (Straßenbau)*; *(Brg)* Aufreißer *m*; *(Lw)* Messeregge *f*
scarify / to *(Bw)* aufreißen; *(Lw)* kultivieren
scarifying factor *(Brg)* Auflockerungsfaktor *m*
scarlet Scharlach *m*, Scharlachfarbe *f*
scarp / to senkrecht (steil) abböschen
scarp Steilböschung *f*
scatter / to *(Ph)* [zer]streuen
s. back zurückstreuen
scatter Streuung *f*; Streuecho *n (durch Reflexionen an ionisierten Schichten; Radar)*
s. of indicator Streuung *f (eines Feinzeigers)*
s. propagation method Streuausbreitungsverfahren *n*
s. transmission Streustrahlübertragung *f*
scattered gestreut, zerstreut, diffus
s. electron Streuelektron *n*, gestreutes Elektron *n*
s. heat radiation Wärmestreustrahlung *f*
s. light Streulicht *n*, zerstreutes Licht *n*
s. light method Streulichtmethode *f*
s. neutron Streuneutron *n*, gestreutes Neutron *n*
s. radiation Streustrahlung *f*, gestreute Strahlung *f*
s. reflections zerstreute Reflexionen *fpl*
s. wave Streuwelle *f*, gestreute Welle *f*
s. X-rays Röntgenstreustrahlung *f*
scatterer Streusubstanz *f*
scattering *(Ph)* Streuung *f*
s. amplitude Streuamplitude *f*
s. angle Streuwinkel *m*
s. area Streufläche *f*
s. capacity Streuvermögen *n*; Streufähigkeit *f*
s. coefficient Streukoeffizient *m*
s. cross-section Streu[ungs]querschnitt *m*
s. error Streufehler *m*
s. factor Streufaktor *m*
s. frequency Streufrequenz *f*
s. intensity Streuintensität *f*
s. medium Streusubstanz *f*, streuendes Medium *n*
s. of electromagnetic waves Streuung *f* der elektromagnetischen Wellen
s. of radio waves Streuung *f* von Funkwellen
s. power Streuvermögen *n*
s. probability Streuwahrscheinlichkeit *f*
s. transmission Streustrahlübertragung *f*
s. volume Streuvolumen *n*
scavenge / to durchspülen, reinigen, [aus]spülen; entionisieren
scavenge air Spülluft *f (Verbrennungsmotor)*
s. port Spülschlitz *m (Schieber)*
s. pump *(Kfz)* Absaugpumpe *f*, Spülpumpe *f*
s. stroke *(Kfz)* Auspuffhub *m*, Spültakt *m*
scavenger piston *(Kfz)* Spülkolben *m*
s. pump Rundförderpumpe *f (Turboantrieb)*
scavenging Reinigung *f*, Säuberung *f*; Reinigungsfällung *f*; Ausspülung *f*, Spülung *f (Verbrennungsmotor)*
s. action *(Kfz)* Spülwirkung *f*
s. air *(Kfz)* Spülluft *f*
s. loop *(Kfz)* Umkehrspülung *f*
sceleton Schiffsgerippe *n (Rumpf eines Schiffes ohne Außenhaut und Wegerung)*
scenario Drehbuch *n (Film)*
scend Tauchschwingungsamplitude *f*, Aufwärtsbewegung *f (eines Schiffes beim Stampfen)*
scene Szene *f*, Bühne *f*; Kulissenstück *n*
s. painter Kulissenmaler *m*
s. shifter Bühnenarbeiter *m*
scenograph perspektivische Gesamtansicht *f* eines Gebäudes
scent compound Duftstoff *m*
scented parfümiert
scentless geruchlos
schappe Schappeseide *f*
s. silk Schappeseide *f*
s. silk spinning Schappespinnerei *f*
schedule / to ansetzen; planen *(Elektronenrechner)*
schedule Liste *f*; Plan *m*, Zeitplan *m*; Fahrplan *m*; Programm *n*
s. and trim *(Am) (BMSR)* Regelung *f* mit Störgrößenaufschaltung (Hilfsregelgröße)
s. of periodic tests *(Am)* Meßplan *m*, Plan *m* für regelmäßige Messungen
s. speed Reisegeschwindigkeit *f*
s. time fahrplanmäßige Zeit *f*
s. timer *(BMSR)* Zeitplangeber *m*, Programmgeber *m*
scheduled air service planmäßiger Luftverkehr *m*
scheelite Scheelit *m*, Tungstein *m*
schematic schematische Zeichnung *f*
schematic schematisch
s. circuit diagram *(El)* Prinzipschaltbild *n*
s. diagram Schemazeichnung *f*, schematische Darstellung *f*; *(El)* Schaltbild *n*, Schaltschema *n*, Schemaschaltbild *n*
s. representation schematische Darstellung *f*
scheme Schema *n*, Plan *m*, Anordnung *f*
s. of connections *(El)* Schaltanordnung *f*
Schering bridge Schering-Brücke *f (Brückenschaltung)*
schist Schiefer *m*
schistose *s.* schistous
schistous schieferartig, schieferig
s. lignite Braunkohlenschiefer *m*
schliere Schliere *f*
schlieren method Schlierenmethode *f*
s. photograph Schlierenaufnahme *f*
s. picture Schlierenbild *n*
Schlippe's salt Schlippesches Salz *n*
school Schule *f*
s. of fish Fischschwarm *m*

s. **of mines** *(Am)* Bergakademie *f*
s. **of navigation** Seefahrtsschule *f*
s. **telescope** Schulfernrohr *n*
s. **television** Schulfernsehen *n*
s. **textbook** Lehrbuch *n*
schoolship Schulschiff *n (Handelsmarine)*
schooner *(Schiff)* Schoner *m*, Rahschoner *m*, Gaffelschoner *m*
schorl Schörl *m (Turmalin)*
Schottky defect Schottkysche Fehlstelle (Fehlordnung, Störstelle) *f (Kristalle)*
Schottky disorder Schottkysche Fehlordnung *f (Kristalle)*
Schootky effect Schottky-Effekt *m*
Schrage motor *(El)* Schrage-Motor *m*
Schreiner calender *(Text)* Kalander *m* mit gravierten Walzen, Seidenglanzkalander *m*, Schreiner-Kalander *m*
Schuchoff's vertical water tube boiler *(Wärm)* stehender Wasserrohrkessel *m* von Schuchoff
Schulz boiler *(Wärm)* Schulz-Kessel *m*
Schweinfurt green Schweinfurter Grün *n*
Schwörer superheater *(Wärm)* Schwörer-Überhitzer *m*
science Wissenschaft *f*; Naturwissenschaft *f*; Lehre *f*; Kunde *f*
s. **of colours** Farbenlehre *f*
s. **of electricity** Elektrizitätslehre *f*
s. **of forestry** Forstkunde *f*
s. **of materials** Werkstoffkunde *f*
s. **of the strength of materials** Festigkeitslehre *f*
scientific wissenschaftlich
s. **article** Fachaufsatz *m*, Fachartikel *m*
s. **worker** Wissenschaftler *m*; wissenschaftlicher Mitarbeiter *m*
scientifically educated mit wissenschaftlicher Ausbildung
scientist Wissenschaftler *m*
scintillate / to flimmern, szintillieren, funkeln; funken
scintillating light Funkelfeuer *n*, Leuchtfeuer *n*
scintillation Szintillation *f*, Flimmern *n*, Funkeln *n*; Funken *n*, Sprühfunkenbildung *f*
s. **chamber** *(Kern)* Szintillationskammer *f*
s. **counter** *(Kern)* Szintillationszähler *m*
s. **crystal** Szintillatorkristall *m*
s. **detector** Szintillationsdetektor *m*
s. **head** Szintillationsmeßkopf *m*
s. **logging** Bohrlochuntersuchung *f* mit Hilfe von Szintillationszählern
s. **method** Szintillationsverfahren *n*
s. **spectrometer** Szintillationsspektrometer *n*
s. **well counter** *(Kern)* Szintillationszähler *m* mit Bohrlochkristall
scintillator Szintillator *m*, szintillierendes Material *n*
scintillometer Szintillometer *n*
scintilloscope Szintilloskop *n*
scintiscanner Gerät *n* zur Aufnahme von Szintigrammen, Szintiscanner *m*
scion *(Lw)* Reis *n*, Pfropfreis *n*
scissel Blechabfall *m*
scissile spaltbar
scissor head Greiferkopf *m*
s. **vibration** Scherenschwingung *f*
scissors Schere *f*
s. **telescope** Scherenfernrohr *n*
sclera Sklera *f*, Lederhaut *f*
scleral lamp Skleraleuchte *f*
sclerometer *(Met)* Ritzhärteprüfer *m*
scleroscope Rücksprunghärteprüfer *m*, Skleroskop *n*
s. **hardness** *(Met)* Kugelfallhärte *f*, Rücksprunghärte *f*
scobs Späne *mpl*
scoop / to ausschaufeln; schöpfen
scoop Rührflügel *m*, Schaufel *f*, Schöpfgefäß *n*, Schöpfkelle *f*; Löffelschaufel *f*; Aufnehmer *m*; Schaufler *m*; Greifer *m*; Eimer *m*
s. **all-round tipping wagon** *(Eis)* Schnabelrundkipper *m*
s. **chain dredge** Schaufelkettenbagger *m*
s. **end tipping wagon** *(Eis)* Schnabelvorderkipper *m*
s. **plough** Schaufelpflug *m*
s. **tipping wagon** *(Eis)* Schnabelkipper *m*
s.**-tyre air intake** *(Flg)* Luftfänger *m*
s. **wheel** Schöpfrad *n*, Wurfrad *n*
s.**-wheel lubrication** Schöpfradschmierung *f*
scooped *(Led)* stark ausgefleischt
scope 1. Bereich *m (n)*; Länge *f*; 2. Oszilloskop *n*
s. **of measurement** Meßbereich *m*
s. **of work** Arbeitsbereich *m*
scorch / to anschmoren, dörren, rösten, [ver-]brennen, [ver]sengen; durchschmoren *(Kabel)*
scorching *(Plast)* Anvulkanisation *f*, Halbvulkanisation *f*
score / to [ein]kerben; zerkratzen, beschädigen; einritzen *(Linien)*; furchen; ausstreichen
s. **a direct hit** einen Volltreffer erzielen
score 1. Punktzahl *f (bei Spielen, Wettbewerben u. ä.)*; 2. Kerb *m*, Kratzer *m*, eingeritzte Linie *f*
scored skin verschnittenes Fell *n*
scoria Schlacke *f*
s. **of raw matte** Röstschlacke *f*
scoriaceous schlackenähnlich, schlackenartig, schlackig
scorification Verschlackung *f*; Schlackenbildung *f*
scorifier Ansiedescherben *m*, Scherbe *f*
scorify / to [ver]schlacken
scoring hardness Ritzhärte *f*
s. **knife** Ritz[el]messer *n*
s. **machine** *(Typ)* Rillmaschine *f*; Ritzmaschine *f*, Nutmaschine *f*

Scotch boiler Schottischer Kessel *m (Heizrohrkessel mit Flammrohr, rückkehrenden Heizröhren und innerer Wendekammer)*; Zylinderkessel *m*
Scotch crank (yoke) Kreuzschleife *f*
scotoma *(Opt)* Skotom *n*
scotopic vision Nachtsehen *n*
scototron Scototron *n (Speicherröhre; Ladungen werden in einem Festkörper gespeichert)*
Scott connection *(Elm)* Scottsche Schaltung *f*, Scott-Schaltung *f*
scour / to 1. abbeizen, abputzen, abscheuern, ausschwemmen, beizen, entzundern, reinigen, scheuern, schrubben, blank reiben; putzen *(Gußstücke)*, säubern; 2. ausspülen, auswaschen, unterspülen; 3. entfetten, entschweißen *(Wolle)*; 4. *(Met)* ausfressen *(Ofenfutter)*
scour Ausspülung *f*, Auswaschung *f*, Kolkung *f*, Unterspülung *f*
scouring 1. Beizung *f*; Abputzen *n*, Abscheuern *n*; Säubern *n*; Entzundern *n*, Putzen *n (Guß)*; Reinigen *n*; 2. Ausspülung *f*, Auswaschung *f*, Kolkung *f*, Unterspülung *f*; 3. Entfetten *n*, Entschweißen *n (Wolle)*; 4. *(Brg)* durch den Versatz getriebene Strecke *f*; 5. *(Met)* Ausfressung *f (Ofenfutter)*
 s. addition Waschzusatzmittel *n*
 s. barrel Scheuertrommel *f*, Scheuerfaß *n (Draht)*; Scheuertonne *f*
 s. basin Spülbecken *n*, Hafenreinigungsbecken *n*, Schleusenbecken *n*
 s. machine Ausreckmaschine *f*, Auswaschmaschine *f*
 s. of wool Wollwäsche *f*, Wollentschweißung *f*
 s. section Entbastungsteil *n*, Waschteil *n*
 s. slag aggressive Schlacke *f*
 s. stone *(Led)* Ausreckstein *m*, Auswaschstein *m*
 s. table *(Led)* Ausrecktafel *f*, Auswaschtafel *f*
scout Spähschiff *n*, Aufklärer *m*
scow Prahm *m*, Schute *f*, Punt *n*, Leichterschiff *n*
scragging Hochtrainieren *n (Feder)*
scram Havarieabschaltung *f*, Schnellabschaltung *f*, Schnellschluß *m (Reaktorregelung)*
 s. button Notschalter *m*, Schnellschlußschalter *m*
 s. clutch Abschaltkupplung *f*
 s. delay Abschaltverzögerung *f*
 s. drift *(Am) (Brg)* Verbindungsstrecke *f* zwischen Überhauen
 s. power Abschaltleistung *f*
 s. rod Abschaltstab *m*, Schnellschlußstab *m*, Sicherheitsstab *m*
 s. rod transport mechanism Rückholmechanismus *m*
scrambled speech *(Fmt)* verschlüsselte Sprache *f*
scrambler apparatus *(El)* Sprachverzerrer *m*
 s. circuit *(El, Fmt)* Bandspalter *m*
scramming *(Brg)* Querbruchbau *m*, Scheibenbruchbau *m*
 s. rod Havariestab *m*, Schnellschlußstab *m (Reaktorregelung)*
scrap / to verschrotten
scrap Schrott *m*; Ausschuß *m*; Abfall *m*
 s. baling press Schrottpaketierpresse *f*
 s. briquetting press Schrottpaketierpresse *f*
 s. castings Gußbruch *m*, Gußschrott *m*
 s. chopper Abfallschere *f*
 s. coke Abfallkoks *m*
 s. component Ausschußteil *n*
 s. copper Altkupfer *n*
 s. cutting attachment Abfallschneider *m*
 s. fagotting Schrottpaketieren *n*
 s. heat Schrottschmelze *f*
 s. iron Alteisen *n*, Schrott *m*
 s. lead Altblei *n*
 s. metal Schrott *m*, Metallschrott *m*
 s. paper Altpapier *n*, Makulatur *f*
 s. parts Ausschuß *m*
 s. percentage Ausschußquote *f*
 s. pile Schrotthaufen *m*
 s. piling machine Paketierpresse *f*
 s. rubber Abfallgummi *m*, Altgummi *m*
 s. shear[s] Schrottschere *f*
 s. side Ausschußseite *f*; Seitenabfall *m*; Seitenabschnitt *m*
 s. smelting Schrottverhüttung *f*
 s. stock yard Schrottplatz *m*
 s. wire Drahtabfall *m*
 s. work Ausschuß *m*
scrape / to schaben; kratzen
 s. in einschaben, einkratzen
 s. off abkratzen; *(Led)* abschaben *(Fleischreste von der Fleischseite)*
scraper 1. Schaber *m*, Schabeisen *n*; Ziehklinge *f*; 2. Abstreifer *m*, Kratzer *m*; Abstreicher *m*; Kesselsteinabkratzer *m*, Kratzeisen *n*; 3. Bodenhobel *m*, Flachbagger *m*; Schürfwagen *m*; Schrapper; *(Brg)* Krätzer *m (Schießarbeit)*
 s. bit Blattmeißel *m (Tiefbohrtechnik)*
 s. box Schrapperkasten *m*
 s. bucket Schürfkübel *m*, Schleppschaufel *f*
 s.-chain conveyor Kratzkettenförderer *m*
 s. conveyor Kratzband[förderer *m*] *n*
 s.-dozer Schürfkübelraupe *f*
 s. loader Schrapplader *m*
 s. rake Abstreicher *m*, Abstreifer *m*
 s. ramp Schrappbühne *f*
 s. ring Abstreifring *m*, Ölabstreifring *m*
 s. slide Schrappbühne *f*
 s.-type classifier Kratzbandklassierer *m*
scraping tool Schaber *m*, Schabeisen *n*
scrapless ohne Ausschuß, ausschußfrei
scrapped rolls Walzenbruch *m*, gebrochene Walzen *fpl*

scrapping Verschrottung *f*, Verschrotten *n*; Abwracken *n (Schiff)*
s. works *(Schiff)* Abwrackbetrieb *m*
scripyard Autofriedhof *m*; Schrottplatz *m*
scratch / to zerkratzen, ritzen
scratch Kratzer *m*, Schramme *f*
s. brush Drahtbürste *f*
s. coat Unterputz *m*
s. depth Rillentiefe *f*
s. filter Nadelgeräuschfilter *n*
s. gauge Streichmaß *n (zum Anreißen)*
s. hardness *(Met)* Ritzhärte *f*
s. hardness number Ritzhärtezahl *f*
s. resistance Kratzfestigkeit *f*; Ritzhärte *f*
s. resistant kratzfest
s. test *(Met)* Ritzversuch *m*
scratching noise Kratzgeräusch *n*
screech / to flattern *(Sägeblätter)*
screed board Abziehbohle *f*, Glättbohle *f*
s. plate Abziehbohle *f*, Glättbohle *f*
screeding board Abziehbohle *f*, Glättbohle *f*
s. plate Abziehbohle *f*, Glättbohle *f*
s. template Abziehlehre *f*
screen / to 1. [aus]sieben, durchsieben, klassieren, sichten, filtern; 2. vernebeln, verschleiern; 3. abschirmen; abblenden; schützen; 4. projizieren
s. in sizes der Größe nach aussieben, klassieren
s. off abschirmen
screen 1. Wand *f*; Schirm *m*, Schutz *m*; Drahtgitter *n*; Rost *m*; Ofenschirm *m*, Schutzwand *f*; Filmleinwand *f*, Projektionsfläche *f*; 2. Sieb *n*, Filter *n*; Rundlochsieb *n*; 3. Bildschirm *m*; 4. Raster *m (Druck, Foto)*; 5. Brennrahmen *m (Elektronenröhrenherstellung)*; 6. *(Schiff)* Geleitschutz *m*; 7. Siebpartie *f (Papierherstellung)*
s. analysis Siebanalyse *f*
s. aperture Maschenweite *f*
s. area Siebfläche *f*
s.-back board Platte *f* mit Siebabdruck
s. body Siebkasten *m*
s. bulkhead *(Schiff)* Staubschott *n*, Schutzschott *n*, Schirmschott *n*
s. burning *(El)* Schirmeinbrennen *n*
s. by-pass capacitor *(El, Fmt)* Schirmgitterblock *m*
s. cap *(El)* Gitterkappe *f*
s. chain conveyor Siebkettenförderer *m*
s. classification Siebfraktionierung *f*; Siebsortierung *f*
s. classifying Siebklassierung *f*
s. cloth Siebbelag *m*, Siebgewebe *n*
s. cover 1. Schutzdecke *f*; 2. Abschirmgehäuse *n*
s. current Schirmgitterstrom *m*
s. deck Siebboden *m*
s. dissipation Schirmgitterverlustleistung *f*
s. dropping resistor Schirmgitter[vor]widerstand *m*
s. drum Siebtrommel *f*
s. fabric Siebgewebe *n*
s. factor Schirmfaktor *m (eines Gitters)*
s. frame Siebkasten *m*
s. grading Siebklassierung *f*
s. grid Schirmgitter *n*
s.-grid dissipation Schirmgitterverlustleistung *f*
s.-grid feedback Schirmgitterrückkopplung *f*
s.-grid valve Schirmgitterröhre *f*
s.-grid voltage supply Schirmgitterspannungsquelle *f*
s. hammer mill Siebhammermühle *f*
s. hole Sieböffnung *f*
s. image Schirmbild *n*
s. magnification Direktvergrößerung *f (Elektronenmikroskop)*
s. opening Sieböffnung *f*, Siebweite *f*
s. overflow Siebrückstand *m*, Siebüberlauf *m*
s. oversize Siebgrobes *n*
s. photography Schirmbildfotografie *f*
s. pipe Filterrohr *n*
s. plate Sortierblech *n*, Sortierplatte *f*; Siebblech *n*
s. printing Siebdruck *m*, Durchdruck *m*; *(Text)* Schablonendruck *m*, Filmdruck *m*
s.-protected mit Berührungsschutz *(Maschine)*
s. riddle Vibrationssieb *n*, Rüttelsieb *n*
s. room Sortieranlage *f*
s. separation Siebklassierung *f*, Siebung *f*
s.-side deflection plates schirmnahes Ablenkplattenpaar *n (Katodenstrahlröhren)*
s. sizing Kornklassierung *f*, Siebklassierung *f*
s. stencil printing *(Text)* Siebschablonendruck *m*
s. surface Siebfläche *f*
s. underflow Siebdurchgang *m*, Siebdurchlauf *m*
s. underframe Siebboden *m*
s. undersize Siebfeines *n*
s. varnish Schablonenlack *m*
s. voltage Schirmgitterspannung *f*
s. wire Maschendraht *m*
s. with harmonics Sieb *n* mit Oberschwingungen *n*, Siemob *n*
s. with twin crank-shaft drive Kurbelschwingsieb *n*
s. with two counterbalanced frames Gegenschwingsieb *n*
screenage Abschirmung *f*, Abschirmen *n*
screened 1. abgeschirmt; abgeblendet; 2. gesiebt; klassiert; gefiltert; 3. gegen zufällige Berührung geschützt *(elektrische Geräte)*
s. aerial abgeschirmte Antenne *f*
s. cable Schirmkabel *n*, abgeschirmtes Kabel *n*, abgeschirmte Leitung *f*

s. colour separation *(Typ)* Rasterfarbauszug *m*
s. conductor cable Kabel *n* mit abgeschirmten Leitern
s. down-lead abgeschirmte Niederführung *f* *(Antenne)*
s. headlight Tarnscheinwerfer *m*
s. lead abgeschirmte Zuleitung *f*
s. pair abgeschirmte Doppelader *f*
s. pentode abgeschirmte Pentode *f*
s. plate valve Schirmgitterröhre *f*
s. valve Schirmgitterröhre *f*
screener Knotenfang *m* *(Papierindustrie)*; Sieb[vorrichtung *f*] *n*
screening 1. Durchsiebung *f*, Sieben *n*; Klassierung *f*, Siebklassierung *f*; Sortieren *n*; 2. Vernebelung *f*, Verschleierung *f*; 3. Abschirmen *n*, Abschirmung *f*; Abblendung *f*; Schutz *m*; 4. Projektion *f*; 5. 100%-Prüfung *f* *(Qualitätskontrolle)*
s. action Schirmwirkung *f*
s. ball mill Siebkugelmühle *f*
s. cage *(El)* Faradaykäfig *m*
s. can *(El)* Abschirmbecher *m*
s. capacity Siebleistung *f*
s. cloth Siebtuch *n*
s. constant Abschirmkonstante *f*
s. current Abschirmstrom *m*
s. drum Siebtrommel *f*
s. effect Abschirmwirkung *f*
s. efficiency Siebgütegrad *m*, Siebtrennungsgrad *m*; Siebleistung *f*
s. enclosure *(El)* abgeschirmtes Gehäuse *m*, Gehäuse mit Berührungsschutz
s. factor Abschirmfaktor *m*
s. harness Kabelgeschirr *n* für die Abschirmung
s. machine Siebmaschine *f*
s. material Siebgut *n*
s. medium Siebboden *m*
s. plate Siebblech *n*
s. refuse Siebrückstand *m*
s. surface Siebfläche *f*
s. table Plansichter *m*
s. tubing Abschirmschlauch *m*
s. unit *(El)* Entstörer *m*
s. wire Abschirmdraht *m*, Schutzdraht *m*
screenings 1. Ausgesiebtes *n*, Siebdurchfall *m*; 2. Brechsand *m*; Splitt *m*, Steinsplitt *m*; 3. Koksgrus *m*, Feinkoks *m*
screenless sieblos
s. hammer mill sieblose Hammermühle *f*
screens Siebrückstand *m*; Schrenz *m*
screeve mark Ritzzeichen *n* *(Forst)*
screw / to festschrauben, verschrauben; schneiden *(Außengewinde)*
s. back zurückschrauben
s. into einschrauben in
s. off abschrauben, losschrauben
s. on aufschrauben
screw 1. Schraube *f*; Bolzen *m*; 2. Gewindespindel *f*, Schraubspindel *f*; 3. Schnecke *f*; 4. Schiffsschraube *f*
s. action Schraubwirkung *f*
s. anchor Schraubenklemme *f*
s.-and-nut steering gear Schraubenlenkung *f*, Spindellenkgetriebe *n*
s.-and-nut table feeding Tischvorschub *m* durch Spindel und Mutter
s. aperture *(Schiff)* Schraubenbrunnen *m*, Propellerbrunnen *m*; Schraubenöffnung *f* *(im Hintersteven)*
s. axis Schraubenachse *f*
s. base Schraubsockel *m*, Schraubenfuß *m*
s. blade *(Schiff)* Schraubenflügel *m*, Propellerflügel *m*
s. blank Schraubenrohling *m*
s. bolt Schraubbolzen *m*
s. brake Spindelbremse *f*
s. bushing Schraubhülse *f*
s. caliper Mikrometerschraube *f*
s. cap 1. Gewindestopfen *m*, Schraubdeckel *m*, Überwurfmutter *f*, Verschlußkappe *f*, Verschlußstück *n*, Verschraubung *f*; 2. Gewindesockel *m*, Edisonsockel *m* *(einer Lampe)*
s. chasing Gewindestrehlen *n*
s. clamp Schraubknecht *m*, Schraubzwinge *f*
s. clip Schraubklemme *f*
s. collar ring Überwurfring *m*
s. connector *(El)* Schraubverbinder *m*, Schraubstecker *m*
s. conveyor Förderschnecke *f*; Schneckenförderer *m*
s. coupling Verschraubung *f*, Schraubenkupplung *f*
s. cover Schraubdeckel *m*
s.-cutting Gewindeschneiden *n*
s.-cutting lathe Zug- und Leitspindeldrehmaschine *f*
s. cutting operation Gewindeschneidarbeit *f*
s. die Backe *f*, Gewindebacke *f*, Schneidbacke *f*
s.-down movement Anstellung *f* *(Walzwerk)*
s.-down speed Anstellgeschwindigkeit *f* *(Walzwerk)*
s. driver *s.* screwdriver
s. extractor Schraubenauszieher *m*
s. extruder Schnecken[strang]presse *f*
s. extrusion press Schnecken[strang]presse *f*
s. fastening Schraubverbindung *f*
s. feeder Schneckenförderer *m*
s. gauge Schraublehre *f*, Mikrometer *n*
s. gear Zahnrad *n* auf der Leitspindel
s. gill drawing frame *(Text)* Schraubenstrecke *f*
s. head Schraubenkopf *m*

s.-head file Schraubenkopffeile *f*, Einstreichfeile *f*
s.-head slotter Schraubenschlitzautomat *m*
s.-in eyepiece *(Opt)* Schraubokular *n*
s.-in-type cartridge fuse *(El)* Patronensicherung *f*
s. jack Schraubspindel *f*, Schraubwinde *f*
s.-jack prop *(Brg)* Schraubstempel *m*
s. joint Schraub[en]verbindung *f*
s. lifting jack Wagenheber *m*, Schraubenheber *m*
s. log *(Schiff)* Patentlog *n*
s. machine tap Automatengewindebohrer *m*
s. micrometer Schraubenmikrometer *n*
s. mixer Schneckenmischer *m*
s. neck Schraubstutzen *m*
s. nut Mutterschraube *f*
s. pair Schraubenpaar *n (Getriebelehre)*
s. pitch Gewindehöhe *f*
s.-pitch gauge Gewindekamm *m*; Gewindeschablone *f*
s. plasticizing Schneckenplastifizierung *f*
s. plate Gewindeschneidkluppe *f*
s. plate die Kluppenbacke *f*
s. plate stock Gewindeschneidkluppe *f*, Schneideisenhalter *m*
s. plug 1. Gewindestopfen *m*, Gewindestöpsel *m*, Verschlußstopfen *m*, Verschlußschraube *f*; 2. *(El)* Schraubstecker *m*
s.-plug cartridge fuse *(El)* Schraubstöpselsicherung *f*, Stöpselkopfsicherung *f*
s.-plug gauge Gewindelehrdorn *m*
s.-plug washer Verschlußschraubendichtung *f*
s. press 1. *(Typ)* Stockpresse *f*; 2. *(Masch)* Spindelpresse *f*
s. propeller Schiffsschraube *f*, Schraubenpropeller *m*, Schiffspropeller *m*
s. releasing wedge clamping Schraubkeilklemmung *f*
s. retainer Schraubsicherung *f*
s. ring Augenbolzen *m*, Gewindeöse *f*
s. ring gauge Gewindelehrring *m*, Gewindelehrmutter *f*
s. shackle Spannschloß *n*
s.-shaft *(Masch)* Schraubenwelle *f*, Schwanzwelle *f*, Propellerwelle *f*
s.-shaft liner *(Schiff)* Schraubenwellenbezug *m*
s.-shaft tunnel *(Schiff)* Schraubenwellentunnel *m*
s. ship Schraubenschiff *n*
s. slotter Schraubenschlitzmaschine *f*
s. slotting Schraubenschlitzen *n*
s. slotting attachment Schraubenschlitzeinrichtung *f*
s. slotting cutter Schraubenschlitzfräser *m*
s. slotting machine Schraubenschlitzmaschine
s. slotting saw Schraubenschlitzfräser *m*, Schraubenschlitzsäge *f*
s. socket Gewindebuchse *f*
s. stay bolt Stehbolzenschraube *f*
s. steamer Schraubendampfer *m*
s. steering gear *(Schiff)* Schraubenspindelsteuervorrichtung *f*
s. stock Gewindeschneidkluppe *f*
s.-tensioning Spannen *n* mittels Schraubspindeln *(Spannbeton)*
s. terminal Schraubklemme *f*
s. thread Schraubengewinde *n*
s. thread caliper Gewindeschraubmeßlehre *f*
s. thread cutting tool Gewindeschneidwerkzeug *n*
s. thread gauge Gewindelehre *f*
s.-thread limit gauge Gewindegrenzlehre *f*
s. thread micrometer Feinmeßschraublehre *f* für Gewindemessungen
s. thread micrometer caliper Gewindeschraublehre *f*
s. thread standards Gewindenormen *fpl*
s. thread tolerance Gewindetoleranz *f*
s. threading die Gewindeschneideisen *n*
s. tug *(Schiff)* Schraubenschlepper *m*
s.-type lamp Lampe *f* mit Edison-Sockel
s. union Rohrverschraubung *f*
s. wake *(Schiff)* Schraubenstrom *m*; Propellerstrom *m*
s. with tapered head Kegelschraube *f*
s. within a telescopic cover Teleskopspindel *f*
s. wrench Schraubenschlüssel *m*
screwable verschraubbar
screwdriver Schraubenzieher *m*
s. blade Schraubenzieherklinge *f*
s. tip Schraubenzieherschneide *f*
screwed angeschraubt, verschraubt
s. bush Gewindebuchse *f*
s. cap Überwurfmutter *f*; Schraubkappe *f*, Stöpselkopf *m (Sicherung)*
s. connection Schraubverbindung *f*
s. contact *(El)* Schraubkontakt *m*
s. coupler Schraubmuffe *f*, Gewindemuffe *f*, Muffenverbinder *m*
s. flange joint Flanschgewindeverbindung *f*
s. holder Schraubfassung *f*
s.-in extension Einschraubverlängerung *f*
s. internally mit Innengewinde
s. lamp-holder Schraubfassung *f*
s.-on cover Schraubdeckel *m*
s.-on extension Aufschraubverlängerung *f*
s. peg-type counterboring cutter Zapfensenker *m* mit eingeschraubtem Führungszapfen
s. pin Gewindestift *m*
s. pipe joint Rohrverschraubung *f*
s. plug 1. Verschlußpfropfen *m*, Verschlußstutzen *m*; 2. *(El)* Schraubstecker *m*
s. shank Gewindeschaft *m*

s. **sleeve joint** Muffenverbindung *f*
s. **socket** Gewindemuffe *f*, Schraubmuffe *f*, Schraubfassung *f*
s. **spigot** Führungszapfen *m* mit Gewindeansatz
s. **spindle** Gewindespindel *f*
s. **spindle hose** Spindelkopf *m* mit Gewinde
s.**-type socket** Edison-Fassung *f*
s. **with spring action** elastisch verschraubt
screwing 1. Verschraubung *f*; 2. Gewindeschneiden *n*
scribble / to *(Text)* krempeln
scribe / to anreißen
scribe awl Reißnadel *f*
scribed circle Lochkreis *m (Bohren)*
s. **line** Anrißlinie *f*, Rißlinie *f*
scriber Reißnadel *f*
s. **point** Reißnadelspitze *f*
s. **with knurled sleeve** Reißnadel *f* mit kordierter Hülse
scribing Anreißen *n*
s. **block** Parallelreißer *m*
s. **point** Reißnadel *f*
scrieve board *(Schiff)* Spantenplan *m*, Holzboden *m (Spantenplan wird 1:1 aufgerissen)*
s. **hook** *(Schiff)* Reißhaken *m*, Anreißhaken *m*, Reißahle *f (Schnürbodenwerkzeug)*
script Schreibschrift *f*, Handschrift *f*
scrive board *s.* scrieve board
scroll 1. Schnecke *f*; Spirale *f (Turbine)*; Spindel *f*; Triebkranz *m*; 2. Rolle *f (Pergament)*
s. **case** Spirale *f (Turbine)*
s. **chuck** selbstzentrierendes Spannfutter *n*, Universalspannfutter *n*
s. **gear** Triebkranz *m*
s. **of violin** Schnecke *f (Violine)*
s. **saw** Dekupiersäge *f*
s. **shear** Kurvenschere *f*
scroop / to *(Text)* knirschen, krachen *(z. B. Seide)*
scroop *(Text)* Knirschen *n*, Krachen *n*, Seidenschrei *m*
scroopy feel *(Text)* Knirschgriff *m*, Seidengriff *m*
scrub / to scheuern, schrubben; [aus]waschen; berieseln
scrub Waschlösung *f*
s. **section** Waschabteilung *f*, Waschanlage *f (Extraktionskolonne)*
scrubber 1. Schrubbürste *f*, Schrubber *m*; 2. Skrubber *m*, Rieselturm *m*, Berieselungsturm *m*; Absorptionsturm *m*; 3. Wäscher *m*, Abscheider *m*; 4. Ackerschleppe *f*
s. **with hoods** Glockenwäscher *m*
s. **with stages** Stufenwäscher *m*
scrubbing action Scheuerwirkung *f*
s. **surface** Waschfläche *f*

scruff Anwuchs *m (am Schiffskörper)*
scruple 1. *englische und amerikanische Maßeinheit für Apothekerwaren: 1 s. (England) = 1 s.ap. (USA) = 1,296 g;* 2. *englische Volumeneinheit für flüssige Apothekerwaren: 1 s. = 1 sc. = 1 fl.s. = 1,184 cm³*
scrutineer Gleichlaufkontrolleinrichtung *f (Telegrafie)*
scrutinize / to genau untersuchen (prüfen)
scud / to lenzen, mit dem Wind segeln
scud *(Led)* Schleim *m*, Schmutz *m*
scudding beam *(Led)* Streichbaum *m*
s. **knife** *(Led)* Streicheisen *n*
scuff / to abnutzen, verschleißen; fressen; abschurren
scuff-resistant *(Plast)* beständig gegen rauhe Behandlung durch Abschurren
scuffing Verschleiß *m*; Fressen *n (z. B. am Ventilschaft)*; Abschurren *n*
scuffle hoe Schaufelhacke *f*
scuffler Messeregge *f*, Pflugscharegge *f*, Räumpflug *m*
scuffling with disk plough *(Lw)* Schälpflügen *n*
scull 1. Ruder *n*; 2. Skullboot *n*; Skull *n*, Riemen *m*
s. **and sweep** Skull *n* und Riemen *m*
s. **blades** Ruderblatt *n*
s. **handle** Rudergriff *m*
sculptor Bildhauer *m*; Schnitzer *m*
sculptress Bildhauerin *f*
sculpture / to gestalten *(in Holz, Stein, Ton)*
sculpture 1. Bildhauerei *f*, Bildhauerkunst *f*; Plastik *f*; Skulptur *f*; 2. Bildwerk *n (Vollplastik, Rundplastik, Relief)*
scum / to entschäumen, abschäumen, abstreifen, schlacken, abschöpfen
scum Schaum *m*, Abschaum *m*, Abstrich *m*, Abzug *m*; Schlacke *f*
s. **guard** Spritzschutz *m*
s. **riser** Schaumkopf *m*, Schlackenkopf *m (Gießtechnik)*
s. **washing** Schlammabsüßen *n*
scumble / to erweichen, aufweichen
scummer Schaumlöffel *m*, Abstreichlöffel *m*
scumming Abschäumen *n*, Abstreifen *n*, Entschäumen *n*, Schlacken *n*; Abschöpfen *n*
scummy schaumbedeckt, schaumig
scupper *(Schiff)* Speigatt *n*
s. **hole** Speigattöffnung *f*
scutch / to *(Text)* schwingen; schlagen
scutch Maurerhammer *m*
s. **blade** *(Text)* Schwinge *f*, Schwingbeil *n*, Schwingmesser *n*
scutched flax gepochter Flachs *m*, Schwingflachs *m*
scutcher Schläger *m*, Schlagmaschine *f (Flachs)*
s. **lap** Schlagmaschinenwickel *m*
s. **waste** Schlägerabfall *m*

scutching Schwingen *n (Flachs)*
s. roller Schlagwalze *f (Flachs)*
scutchion 1. Namenschild *n*; 2. Schlüssellochverdeckung *f*
scuttle / to anbohren *(ein Schiff)*; Seeventil öffnen *(um ein Schiff zu versenken)*
scuttle 1. *(Schiff)* Springluke *f*, kleine Luke *f*; 2. *(Kfz)* Windlaufquerteil *n*; Querstrebe *f* über der Frontscheibe
s. hatch *(Schiff)* Deckloch *n*, Bunkerluke *f*
scythe / to mähen [mit der Sense]
scythe Sense *f*
s. hammer Dengelhammer *m*
s. sharpener Sensenstreicher *m*
s. stone Wetzstein *m*
s. whetting hammer Blankhammer *m (Schärfer)*
sea 1. Meer *n*, See *f*; 2. Seegang *m*
s. anchor Treibanker *m*, Seeanker *m*
s.-barge clipper Prahmfähre *f*; Prahmmutterschiff *n*
s. bed Meeresgrund *m*
s. boat Seeboot *n*, Seeschiff *n*; Unfallboot *n (an Bord stets ausgeschwungen gefahren)*
s.-borne commerce Seehandel *m*
s.-borne goods traffic Seegüterverkehr *m (seewärtiger Güterverkehr)*
s.-borne trade Seehandel *m*
s. bottom Meeresgrund *m*
s. buoy Anseglungstonne *f*
s. canal Seekanal *m*
s. chart Seekarte *f*
s. clutter Seegangsreflex *m*, Wellenreflexion *f (Radartechnik)*
s. cock Seeventil *n*, Seehahn *m*; Flutventil *n*, Bodenventil *n*
s. connections Außenbordanschluß *m (z. B. für Bodenventile, Seeventile, Seehähne usw.)*
s.-damaged terms Seebeschädigungsbedingungen *fpl*
s. dog 1. Seehund *m*; 2. *(sl)* Seebär *m (erfahrener Seemann)*
s. dyke Seedeich *m*, Meeresdeich *m*
s. fisheries Seefischerei *f*
s. fishing Seefischerei *f*
s. fog Küstennebel *m*
s. gate Seeschleuse *f*
s.-going seegehend, Hochsee-
s.-going barge Seeleichter *m*
s.-going ship Seeschiff *n*
s.-going steamer Seedampfer *m*
s.-going trawler Hochseefischdampfer *m*, Hochseetrawler *m*
s.-going tug Hochseeschlepper *m*, Seeschlepper *m*
s.-going vessel Hochseeschiff *n*, Seeschiff *n*
s. horizon Gesichtskreis *m*, Kimm *f*
s.-keeping basin Seegangsbecken *n (Modellversuchswesen)*
s.-keeping qualities Seeeigenschaften *fpl*
s.-keeping tank Modellversuchstank *m* für Seegangsversuche
s.-keeping trial Erprobung *f* des Schiffsverhaltens im Seegang, Seeverhaltenversuch *m*
s. ladder Seefallreep *n*, Steigeisen *n* an der Außenhaut
s.-lane Fahrwasser *n*, Seeweg *m*, Seestraße *f*
s. law Seerecht *n*
s. level Meeresspiegel *m*, Wasserstand *m*
s. log Seelogbuch *n*
s. mark Seezeichen *n*, Seemarke *f*
s. mile Seemeile *f (1 sm = 1 852 m)*
s. mule Seebulldog *m*, Schubschlepper *m (Hafenbetrieb)*
s. otter 1. Seeotter *m*, Meerotter *m*; 2. Seeotterfell *n*
s.-protest Seeverklarung *f*
s.-quake Seebeben *n*
s.-rescue aircraft Seenotflugzeug *n*
s. return See-Echo *n (Radar)*
s. returns Reflexion *f* durch die Seeoberfläche *(Radar)*
s. route Seeweg *m*, Track *m*
s. routine Seebetrieb *m*
s. side Küste *f*
s.-side excursion ship Seebäderschiff *n*
s. sled Seeschlitten *m (Gleitboot)*
s. stores Schiffsvorräte *mpl*
s. traffic Seeverkehr *m*
s. train Eisenbahnfähre *f*, Trajektschiff *n*
s. trial Probefahrt *f*
s. turn Seetörn *m*, Seefahrzeit *f (eines Schiffes)*
s.-unworthiness Fahrunfähigkeit *f (Schiff)*
s. valve Seeventil *n*, Bodenventil *n*; Außenbordventil *n*
s. water Seewasser *n*, Meereswasser *n*
s.-water evaporator Seewasserverdampfer *m*
s. wave Meereswelle *f*
s.-worthy packed seemäßig verpackt
seabord Seeküste *f*, Wasserkante *f*
seafarer Seefahrer *m*
seafaring seefahrend, zur See fahrend
seal / to [ab]dichten, absperren, einschmelzen, isolieren, verkitten, verkleben, [ver]schließen, versiegeln, zuschmelzen, plombieren; abschmelzen *(Vakuumtechnik)*
s. hermetically luftdicht verschließen
s. in einschmelzen
s. off absperren
s. through glass in eine Glaswand einschmelzen
seal 1. Abdichtung *f*, Dichtung *f*, Einschmelzstelle *f (Elektrodenanschluß)*, Kapsel *f*, Abschlußvorrichtung *f*, Plombe *f*, Siegel *n*, Verschluß *m*; Geruchverschluß *m*; 2. Dichtungsfenster *n (in einem Wellenleiter)*, Wellenleitersiegel *n*; 3. Seehund *m*

s. casing Dichtungsgehäuse *n*
s. coat Verschlußschicht *f (Straßenbau)*
s. face Dichtungsfläche *f*
s. ring Dichtungsring *m*
s. wire Dichtungsdraht *m*
sealant Isoliergrund *m*, Isoliermittel *n*; Abdichtungsmittel *n*
sealed abgedichtet, abgesperrt, abgeschlossen, versiegelt, dicht; überkappt
s. against dust staubdicht
s. cabin *(Flg)* Höhenkabine *f*, Überdruckkabine *f*
s. cable end verbleites (zugebleites, überkapptes) Kabelende *n*, Kabelende mit Bleischutzkappe
s. chamber terminal *(Am)* [vergossener] Kabelendverschluß *m*
s.-in battery vergossene Batterie *f*
s.-in fuel unit *(Kern)* eingehülstes Brennstoffelement *n*
s. radioactive material geschlossenes radioaktives Präparat *n*
s. rectifier *(Elm)* pumpenloses (abgeschmolzenes) Ventil *n*
s. tube Einschmelzrohr *n*
sealer 1. *(Plast)* Porenfüller *m*, Porenschließer *m*; 2. Grundierung *f (Anstrichtechnik)*; 3. Robbenfänger *m (Schiffstyp)*
s. coat Versiegelungsüberzug *m*
sealing Abdichtung *f*, Dichtung *f*; Versiegeln *n*, Plombieren *n*; Verschluß *m*, Verschließen *n*; Vergießen *n*, Ausgießen *n (Muffen)*
s. assembly rolling process Walzschweißverfahren *n*
s. box *(vergießbare)* Muffe *f*
s. compound 1. Dichtungsmasse *f*; Isoliermasse *f*; Verschlußmasse *f*; 2. Oberflächenschutzmittel *n* für Frischbeton; 3. Kabel[verguß]masse *f*
s. current Ansprechstrom *m*
s. device Abdichtung[svorrichtung] *f*
s. diaphragm Abschlußmembran *f*
s. end Endverschluß *m*, Kabelendverschluß *m*
s. face Dichtfläche *f*
s. flange Verschlußflansch *m*
s.-in joint Einschmelzung *f (Elektronenröhrenherstellung)*
s.-in machine Einschmelzmaschine *f (Elektronenröhrenherstellung)*
s. liquid Absperrflüssigkeit *f*, Sperrflüssigkeit *f*, Sperrwasser *n*
s. material Dichtungsmaterial *n*; Verschlußmaterial *n*; Vergußmaterial *n*
s. off 1. Abdämmen *n*, Abdichten *n*, Eindämmen *n*; 2. Abschmelzen *n*, Abziehen *n*
s.-off burner Abschmelzbrenner *m (Elektronenröhrenherstellung)*
s. ring Dicht[ungs]ring *m*
s. screw Dichtungsschraube *f*
s. surface Dicht[ungs]fläche *f*
s. thread Dicht[ungs]gewinde *n*
s. voltage Ansprechspannung *f (Relais)*
s. wax Siegellack *m*; Abdichtwachs *n*
s. wire Plombendraht *m*
seam / to säumen; falzen
seam 1. Saum *m*; Naht *f*; Lötstelle *f*; Ader *f*; Grat *m*; Bruchstelle *f*, Riß *m*; Falz *m*, Fuge *f*; 2. *(Geol)* Flöz *n*, geschichtete Lagerstätte *f*
s. batten *(Schiff)* Nahtspant *n*
s. joint Falzstoß *m*
s. memoirs Flözarchiv *n*
s. of casting Gußnaht *f*, Gußgrat *m*
s. output Flözausbringen *n*
s. rip Riß *m* zwischen den Nietlöchern
s. rubbing machine Nahtausreibmaschine *f (Schuhfertigung)*
s. section Flözprofil *n*
s. thickness Flözmächtigkeit *f*
s. welding Nahtschweißen *n*
seamer Besäumvorrichtung *f*
seaming Säumen *n*, Falzen *n*
s. machine Verschließmaschine *f (Konservendosenherstellung)*
seamless nahtlos
s. boat nahtloses Stahlboot *n*
s. drawn nahtlos gezogen
s. hose nahtloser Strumpf *m*
s. tube nahtloses Rohr *n*
seamstress Näherin *f*
seaplane Wasserflugzeug *n*
s. hull Flugbootrumpf *m*
seaport Seehafen *m*, Hafen *m*
seaquake Seebeben *n*
sear / to welken; sengen, brennen
search / to suchen; untersuchen, durchsuchen, durchforschen; abtasten
search and rescue and homing equipment Seenotrettungsgerät *n*
s. coil Prüfspule *f*
s. radar Aufklärungsradar *n*
s. radar set Funkmeßbeobachtungsgerät *n*, Funkmeßsuchgerät *n*
s.-type reflector Parabolspiegel *m*
s. unit Tastkopf *m*
s. winding Meßwicklung *f*, Prüfwicklung *f*, Prüfspule *f*, Meßspule *f*
searching-fire device Tiefenfeuereinrichtung *f*
searchlight Scheinwerfer *m*, Suchscheinwerfer *m*
searchlighting Absuchen *n* mittels Radar
searchtracking radar Aufklärungsradar *n*
season / to 1. ablagern, altern; [aus]trocknen; 2. *(Text)* appretieren; 3. *(Lw)* reifen; 4. würzen
season check *(Holz)* Trockenriß *m*
s. crack Alterungsriß *m*
s. wanting in light lichtarme Jahreszeit *f*

seasonal oil renewal saisonbedingter Ölwechsel *m*
seasoning 1. Altern *n*, Alterung *f* *(Metalle)*; 2. Trocknung *f*, Ablagerung *f* *(Holz)*; 3. *(Text)* Appretieren *n*, Appretur *f*; 4. *(Lw)* Reifen *n*; 5. Würzen *n*
s. kiln Trockenofen *m*
s. machine Appretiermaschine *f*
seat / to 1. einpassen, einsetzen; 2. aufnehmen; 3. aufliegen, eingepaßt sein
seat 1. Sitz *m*; 2. Auflagefläche *f*, Anlagefläche *f*; 3. Herd *m* *(Erdbeben)*
s. adjuster Sitzverstellung *f*
s. back Sitzlehne *f*; Rückenlehne *f*
s. cover Sitzbezug *m*
s. cushion Sitzkissen *n*
s. dresser Ventilsitzfräsapparat *m*
s. fastening machine Fersenheftmaschine *f* *(Schuhfertigung)*
s. frame Sitzgestell *n*
s. frame bending machine Sitzreifenbiegemaschine *f*, Sitzringbiegemaschine *f*
s. of valve Ventilsitz *m*
s. spring Sitzfeder *f*
s. upholstery Sitzpolster *n*
seated on aufgesteckt auf
seating 1. Einpassen *n*, Einsetzen *n*; 2. Aufnahme *f*, Aufnehmen *n*; 3. Sitz *m*, Auflagefläche *f*, Anlagefläche *f*
s. face Sitzfläche *f*, Auflagefläche *f*, Anlagefläche *f*
s. surface *s.* s. face
s. valves Ventileinschleifen *n*, Feinschleifen *n* von Ventilen
seatrain *s.* sea train
seaward seewärts
seaway Seegang *m*
seaweed Meeresalgen *fpl*, Seegras *n*, Seetang *m*
seaworthiness certificate Seetüchtigkeitsattest *n*
seaworthy seetüchtig, seefähig
sebacamid Sebazinsäureamid *n*
sebacate Ester *m* (Salz *n*) der Sebazinsäure, Sebazat *n*
sebaceous talgähnlich, talgartig, talgig
sebacic acid Sebazinsäure *f*
secant Sekante *f*
secant schneidend, Schnitt-
s. cut *(Holz)* Flachschnitt *m*, Fladerschnitt *m*
s. line Schnittlinie *f*; Sekante *f*
s. plane Schnittebene *f*
secohm *s.* henry
second Sekunde *f*
s. anode Beschleunigungsanode *f*
s. bath Reduktionsbad *n* *(Gerberei)*
s. breaker *(Text)* Feinkrempel *f*, Pelzkrempel *f*, Vlieskrempel *f*; Mittelkrempel *f*
s.-channel frequency Spiegelfrequenz *f*
s.-channel interference Spiegelfrequenzstörung *f*
s.-channel rejection Nebenwellenunterdrückung *f*
s. charge Nachladung *f* *(Batterie)*
s.-class conductor *(El)* Leiter *m* zweiter Klasse
s. crusher Nachbrecher *m*
s. cut Oberhieb *m* *(Feile)*
s.-cut file Halbschlichtfeile *f*
s. detector ZF-Gleichrichter *m*; zweiter Gleichrichter *m* *(Funktechnik)*
s. drying Nachtrocknung *f*
s. growth Folgewald *m*
s. gun *s.* s. anode
s. hay Spätheu *n*, Grünmahd *f*, Grummet *n*
s. high water niedrigeres Hochwasser *n*
s. ionization potential zweites Ionisationspotential *n*
s. law of thermodynamics zweiter Hauptsatz *m* der Thermodynamik
s. lengths Mittelware *f* *(Holz)*
s. light indirektes Licht *n*
s. mate Steuermannsmaat *m*
s. moment of area Propellerschwungmoment *n*
s. of an arc Bogensekunde *f*
s. operation Nacharbeit *f*
s.-order activity *(Kern)* Sekundäraktivität *f*
s.-order reaction Reaktion *f* zweiter Ordnung
s. order terms *(Math)* Glieder *npl* zweiter Ordnung
s. order transition Phasenumwandlung *f* zweiter Ordnung
s. order transition temperature *(Plast)* Einfrierpunkt *m*, Einfriertemperatur *f*
s. power *(Math)* zweite Potenz *f*, Quadrat *n*
s. product coal Nachwaschkohle *f*
s. quantization Supraquantelung *f*, zweite Quantelung *f*
s. rolls Fertigwalzwerk *n*
s.-setting *(Masch)* in zweiter Aufspannung
s. stage zweite Raketenstufe (Stufe) *f*
s.-stage burnout Brennschluß *m* der zweiten Raketenstufe
s.-stage engine Triebwerk *n* der zweiten Raketenstufe
s.-stage rocket zweite Raketenstufe *f*
s.-trace echo Sekundärecho *n* *(Radartechnik)*
s. washing Nachwaschen *n*
s. wire *(El)* Meßader *f*
secondary *(El)* Sekundärwicklung *f*
secondary sekundär; *(El)* sekundärseitig, niederspannungsseitig
s. air Oberluft *f*, Sekundärluft *f*; Zweitluft *f*
s. alcohol sekundärer Alkohol *m*
s. amine sekundäres Amin *n*
s. axis Nebenachse *f*

s. base Hilfsbasis *f (Vermessung)*
s. battery Sekundärbatterie *f*
s. beam Querträger *m*, Zwischenträger *m*
s. blast Zweitluft *f*
s. blasting Nachzerkleinerung *f* durch Schießarbeit
s. bond *(Ch)* Nebenbindung *f*
s. cell Sekundärelement *n*
s. chamfer angle *(Am)* Übergangswinkel *m (Reibahle)*
s. circuit Sekundärkreis[lauf] *m*, zweiter Kreislauf *m*
s. clearance Freifläche *f (an einer Reibahle)*
s. clock Nebenuhr *f*
s. collision *(Kern)* Sekundärstoß *m*
s. component Sekundärkomponente *f (kosmische Strahlung)*
s. condenser *(Wärm)* Nachkondensator *m*
s. condition Nebenbedingung *f*
s. contact Hilfskontakt *m*, Kontakt *m* an Sekundärseite
s. controlled variable *(BMSR)* Aufgabengröße *f*
s. controller geführter Regler *m*, Sekundärregler *m (Kaskadenregelung)*
s. cost Nebenkosten *pl*
s. coverage Fernempfangszone *f (Funktechnik)*
s. crop Nebenbestand *m*, Zwischenbestand *m (Forst)*
s. current Sekundärstrom *m*, sekundärseitiger Strom *m*
s. cutting edge *(Masch)* Nebenschneide *f*
s. deposit *(Geol)* Lagerstätte *f* der metamorphen Folge, Sekundärlagerstätte *f*
s. distribution bus-bar Niederspannungsverteilungssammelschiene *f*
s. drift *(Brg)* Auslenkungsstrecke *f*
s. drive Nebenantrieb *m*
s. drying Nachtrocknung *f*
s. electrode Hilfselektrode *f*
s. electron Sekundärelektron *n*
s. electron emission Sekundärelektronenemission *f*
s. electron multiplier Sekundärelektronenvervielfacher *m*, SEV
s. electron spectrum Sekundärelektronenspektrum *n*
s. emission *(Kern)* Sekundäremission *f*
s. emission multiplier Sekundärelektronenvervielfacher *m*, SEV
s. emission rate Sekundäremissionsfaktor *m*, Sekundäremissionsausbeute *f*
s. felling Nachhieb *m (Forst)*
s. fermentation Nachgärung *f*
s. filter Sekundärfilter *n*
s. gluing Montageleimung *f*
s. harvest grain Zweitfrucht *f*
s. ion Sekundärion *n*
s. ionization Sekundärionisation *f*
s. leakage Sekundärstreuung *f*
s. lobe Nebenzipfel *m (Antennen)*
s. luminous standard Lampen[hilfs]normal *n*
s. metal Sekundärmetall *n*, aus Schrott erschmolzenes Metall *n*
s. neutron Sekundärneutron *n*, sekundäres Neutron *n*
s. nuclear fuel sekundärer Kernbrennstoff *m*; sekundärer Spaltstoff *m*
s. particle Sekundärteilchen *n*
s. phenomenon Nebenerscheinung *f*, Sekundärerscheinung *f*
s. pipe *(Met)* sekundärer Lunker *m*
s. piping Nebenrohrleitung *f*
s. port Anschlußort *m (Hafen, dessen Tideunterschiede nach einem Standardhafen berechnet wurden)*
s. power *s.* second power
s. process Sekundärprozeß *m*
s. quantum number Nebenquantenzahl *f*
s. radiation Sekundärstrahlung *f*
s. reaction Nebenreaktion *f*
s. reactor Sekundärreaktor *m*
s. reinforcement Querbewehrung *f*, Verteilerstäbe *mpl (Stahlbetonbau)*
s. resistance Hilfswiderstand *m*, Sekundärwiderstand *m*
s. road Landstraße *f*, Straße *f* II. Ordnung; Nebenstraße *f*
s. roof-break *(Brg)* Setzriß *m* in den Dachschichten
s. sedimentation basin Nachklärbecken *n (Gesundheitstechnik)*
s. service area Fernempfangszone *f (Funktechnik)*
s. settler Nachklärbecken *n (Abwasserreinigung)*
s. shaft Sekundärwelle *f*, Vorgelegewelle *f*
s. spectrum sekundäres Spektrum *n*, Sekundärspektrum *n*, Spektrum zweiter Ordnung
s. standard abgeleitetes Normal *n*, Unternormal *n*; Hilfsnormal *n*
s. steel Querbewehrung *f*, Verteilerstäbe *mpl (Stahlbetonbau)*
s. storage *(Dat)* Fremdspeicherung *f*
s. store *(Dat)* Hilfsspeicher *m*
s. stress Nebenspannung *f*, Zusatzspannung *f*, Sekundärspannung *f*
s. terminal Sekundärklemme *f*
s. treatment Nachbehandlung *f*
s. valence bond *(Ch)* Nebenvalenzbindung *f*
s. valency *(Ch)* Nebenvalenz *f*, Partialvalenz *f*
s. voltage Sekundärspannung *f*
s. winding Sekundärwicklung *f*
s. yield Sekundärelektronenausbeute *f*
seconds Waren *fpl* zweiter Wahl, Artikel *mpl* zweiter Qualität

secrecy 1. Geheimhaltung *f*; 2. *(Fmt)* Sprachverschlüsselungsgerät *n*
s. of telecommunication Fernmeldegeheimnis *n*
s. of telephone service Fernsprechgeheimnis *n*
secret geheim
s. door Tapetentür *f*
s. dovetailing ganz verdeckte Gehrungszinken *mpl*, geschlossene Gehrungszinken *(Holzbau)*
s. language Geheimsprache *f*
s.-nailed verdeckt genagelt
s. nailing verdeckte Nagelung *f*
s. switch Verschlußschalter *m*
s. transmission Geheimübertragung *f*
sectile teilbar, abtrennbar
section / to profilieren; abteilen, unterteilen
section Absatz *m*, Abschnitt *m*, Abteilung *f*; Profil *n*; Querschnitt *m*; Schnitt *m (technisches Zeichnen)*; Sektion *f*, Zarge *f (Destillationskolonne)*; Strecke *f*; Glied *n*; Segment *n (Zahnrad, Kommutator)*; *(Typ)* Lage *f*, Heftlage *f*; *(El)* Leitungsglied *n*, Kettenglied *n*, Wicklungsabschnitt *m*, Wicklungsteil *m*; *(Schiff)* Konstruktionsspant *n*
s. blocking Streckenverblockung *f*, Streckenblock *m*, Fahrstreckenverriegelung *f*
s. box Trennmuffe *f*
s. building *(Schiff)* Sektionsbauweise *f*
s. circuit-breaker Streckenschalter *m*; Zweigschalter *m*
s. clearing Streckenentblockung *f*, Freigabe *f* einer Strecke
s. fuse board Sicherungsverteilungstafel *f*
s. groove Walzkaliber *n*
s. insulator Streckentrenner *m*
s. line Schraffurlinie *f (technisches Zeichnen)*
s. lineman Leitungsaufseher *m*
s. lining Schnittflächenschraffur *f*, Schraffur *f* von Schnittflächen
s. mill frame Profilgerüst *n (Walzwerk)*
s. modulus Widerstandsmoment *n* [des Querschnitts]
s. modulus of bending Biegungswiderstandsmoment *n*
s. modulus of torsion polares Widerstandsmoment *n*, Widerstandsmoment gegen Torsion
s. of core Kernquerschnitt *m*
s. of line 1. *(Eis)* Streckenabschnitt *m*; 2. *(Fmt)* Leitungsabschnitt *m*
s. of passage Durchgangsquerschnitt *m*, Durchtrittsquerschnitt *m*
s. plane Schnittebene *f (Schnitt)*
s. road *(Brg)* Abteilungsstrecke *f*
s. roll Profilwalze *f (Walzwerk)*
s. steel Formstahl *m*
s. switch Zweigschalter *m*, Verteilungsschalter
s. tapering Querschnittsverjüngung *f*
s. telephone Streckenfernsprecher *m*
s. warping machine *(Text)* Sektionalschärmaschine *f*
s. wire Profildraht *m*, profilierter Draht *m*
sectional gebietsweise; Querschnitts-; Schnitt-; Teil-; Profil-; Sektions-; mehrteilig; geteilt; teilbar
s. area Spantfläche *f*, Querschnittsfläche *f*; Schnittfläche *f*
s. area of chip Spanquerschnitt *m*
s. assembly view Montagezeichnung *f* im Schnitt
s. beam *(Text)* Teilbaum *m (für Kettenstuhl)*
s.-belt conveyor Teilbandförderer *m*
s. boiler Sektionalkessel *m*, Gliederkessel *m*
s. bullet Teilkerngeschoß *n*
s. coil Scheibenspule *f*
s. dies Schneideisen *n* mit zweiteiligen Backen
s. dock Sektionsdock *n (Schwimmdock)*
s. drawing Schnitt[zeichnung *f*] *m*
s. drill rod aus Einzelstangen bestehendes Bohrgestänge *n*
s. economizer *(Wärm)* Gliedervorwärmer *m*
s. flywheel geteiltes (mehrteiliges) Schwungrad *n*
s. grate geteilter Rost *m*
s. header boiler Teilkammerkessel *m*, Sektional[kammer]kessel *m*
s. iron Profileisen *n*, Formeisen *n*
s. plan [section] Waagerechtschnitt *m (technisches Zeichnen)*
s. pontoon dock Sektionspontondock *n (Schwimmdock mit mehreren Bodenpontons)*
s. shape Profil *n*, Querschnittsform *f*
s. steel Formstahl *m*, Profilstahl *m*
s. toll switching *(Fmt)* Auftrenntechnik *f (Verfahren des Selbstwählfernverkehrs)*
s. view Schnitt[darstellung *f*] *m (technisches Zeichnen)*
s. warp machine *(Text)* Scheibenschärmaschine *f*
s. warping *(Text)* Schären *n* in Abschnitten
sectionalization Eingrenzung *f*
sectionalize / to in Abschnitte zerlegen
sectionalizing switch Streckenschalter *m*, Streckentrenner *m*, Zellenschalter *m (Batterie)*
sectioned 1. im Schnitt dargestellt; 2. schraffiert; mit schraffierten Schnittflächen
sectioning 1. Schnittdarstellung *f*; 2. Schraffieren *n* von Schnittflächen
sections *(Met)* Profile *npl*, Formstahl *m*
sector 1. Sektor *m (Kreis)*; 2. *(Masch)* Zeigerpaar *n (Teilscheibe)*
s. arm *(Masch)* Zeiger *m*
s. arms *(Masch)* Zeigerpaar *n*
s. of circle Kreissektor *m*
s. pattern instrument *(El)* Sektorinstrument *n*

sectoring Sektorbildung *f (Radartechnik)*
secular equation *(Math)* Säkulargleichung *f*
s. radioactive equilibrium dauerndes radioaktives Gleichgewicht *n*, radioaktives Dauergleichgewicht *n*
secure / to sichern; befestigen; *(Masch)* spannen; *(Schiff)* festzurren; abläuten *(mit Maschinentelegraf)*
s. by pins verstiften
s. the steam engine to the foundation die Dampfmaschine verankern
s. the weight against shifting das Gewicht gegen Verschieben sichern
s. under terminals *(El)* anklemmen, unterklemmen
securing Befestigung *f*
s. line Festmacheleine *f*, Verbindungsleine *f (vom Schiff zum Land)*
s. of hatch *(Schiff)* Lukensicherung *f*
s. pin Sicherungsstift *m*
s. wedge *(Schiff)* Lukenkeil *m*
security Sicherheit *f*
s. anchor Notanker *m*
s. signal Sicherheitszeichen *n (auf der Notfrequenz 500 kHz gesendet)*
sedan geschlossener, viertüriger Pkw *m (auch: Karosserie dieses Wagens)*
sedative Beruhigungsmittel *n*, Schlafmittel *n*
sediment Sediment *n*, Sinkstoff *m*, Ablagerung *f*, abgesetzter Feststoff *m*, Bodensatz *m*, Niederschlag *m*, Schlamm *m*
s. in boilers Kesselstein *m*
s. vessel Absetzgefäß *n*
sedimentary sedimentär, Sediment-
s. rock Sedimentgestein *n*
s. solids Sinkstoffe *mpl*
s. water schlammhaltiges Wasser *n*
sedimentate / to ablagern, absetzen, niederschlagen; sedimentieren
sedimentation Absetzen *n*; Ablagerung *f*; Sedimentation *f*
s. analysis Absetzprobe *f*, Schlämmprobe *f*
s. method Absetzverfahren *n (Gesundheitstechnik)*
s. potential Sedimentationspotential *n*
s. process Absetzverfahren *n (Gesundheitstechnik)*
s. tank Absetzbecken *n*
s. test 1. Absetzversuch *m*; 2. Blutsenkung *f*
Seebeck effect Seebeck-Effekt *m*, thermoelektrischer Effekt *m*
seed Saat *f*, Samen *m*; Keim *m*
s. bag Sätuch *n*
s. bed cultivation Saatbettkultivierung *f*
s. box Saatkasten *m*
s. broadcaster Breitsämaschine *f*
s.-case Samenhülle *f*
s. cleaning plant Saatreinigungsanlage *f*
s. cloth Sätuch *n*
s. corn Saatgetreide *n*, Saatgut *n*
s. crystal Impfkristall *m*, Kristallkeim *m*
s. cultivating station Saatzuchtanstalt *f*
s. drilling Saatdrillen *n*
s. fibre Samenfaser *f*
s. grain Samenkorn *n*
s.-harrow Saategge *f*
s.-hull Samenhülse *f*
s.-husk Samenhülse *f*; Samenkapsel *f (Baumwolle)*
s.-pod *s.* s.-husk
s. potatoes Saatkartoffeln *fpl*
s. tree Samenbaum *m*
seeder Säapparat *m*
seeding box Säkasten *m*, Drillkasten *m*
s. grain Saatgetreide *n*, Saatgut *n*
s. grain rasper Saatgutreiber *m*
s. of crystallization Animpfen *n*, Impfen *n (Kristallbildung)*
s. plough Säpflug *m*
seedling Samenpflanze *f*, Pflanzling *m*
s. enclosure Forstgarten *m*
s. nursery Forstgarten *m*
seedy körnig
seep / to [ver]sickern, durchsickern
seepage 1. Sickerung *f*, Versickerung *f*, Durchsickern *n*; Einsickern *n*; Lecken *n*; Durchdringen *n*; Eindringen *n*; Durchströmung *f*; 2. Leckflüssigkeit *f*; Ölsumpf *m*; Ölausbiß *m*
s. face Sickerfläche *f*
s. flow Sickerströmung *f*
s. force Sickerdruck *m*
s. loss Sickerverlust *m*
s. pressure Sickerdruck *m*
s. water Sickerwasser *n*
seesaw / to schaukeln, wippen, hin- und hergehen
seethe / to kochen, sieden
Seger cone Segerkegel *m*, Brennkegel *m*, Schmelzkegel *m*
seggar Brennkapsel *f (Keramik)*
s.-clay Kapselton *m (Keramik)*
segment Segment *n*, Abschnitt *m*; Teilstück *n*
s.-core armature *(Elm)* Segmentanker *m*
s.-fed loop Endkoppelschleife *f (Mehrkammermagnetron)*
s. gear Zahnradsegment *n*
s. of a circle Kreisabschnitt *m*, Kreissegment *n*
s. of a line Strecke *f*
s. of a sphere Kugelabschnitt *m*, Kugelsegment *n*
s. of the eye *(Opt)* Augenabschnitt *m*
s. pitch Lamellenteilung *f*
segmental abrasive block Segmentschleifkörper *m*
s. arch Flachbogen *m*, Segmentbogen *m*, Stichbogen *m*

s. bearing Segmentlager *n*, Klotzlager *n* *(Mitchell)*
s. chip Bruchspan *m*, Reißspan *m*
s. circular saw Segmentkreissägeblatt *n*
s. grinding wheel Segmentschleifscheibe *f*
s.-type blade Sägeblatt *n* mit Zahnsegmenten
segregate / to absondern, [aus]seigern, entmischen, trennen
segregated band *(Met)* Seigerungsstreifen *m*
s. spot *(Met)* Seiger[ungs]stelle *f*
segregation Seigerung *f*, Segregation *f*; Entmischung *f*; Sortierung *f*; Einschluß *m*; Preßfehler *m*, Streifen *m*
s. constant Segregationskonstante *f*, Verteilungskoeffizient *m* *(Halbleitertechnik)*
s. of circuits Abspalten *n* von Stromkreisen *(Zweipole)*
s. streaks *(Met)* Seigerungszellen *fpl*
s. zone *(Met)* Seigerungszone *f*
Seidel sums (terms) Seidelsche Summen *fpl*
Seignette electric crystal Seignettesalzkristall *m*
Seignette salt Seignettesalz *n*
seine stehendes Fischnetz *n*, Wade *f* *(Netz)*
seiner Waden[fischer]boot *n*, Seiner *m*
seismic seismisch
s. disturbance seismische Erschütterung *f*
s. observatory Erdbebenwarte *f*
s. resistance Widerstandsfähigkeit *f* gegen Erd[beben]stöße
s. shock Erd[beben]stoß *m*
s. wave Erdbebenwelle *f*
seismicrophone Erdmikrofon *n*
seismogram Seismogramm *n*
seismograph Seismograf *m*
s. station Erdbebenwarte *f*
seismology Seismik *f*, Erdbebenkunde *f*, Seismologie *f*
seismometer Seismometer *n*
seismometry Seismometrie *f*
seismoscope Seismoskop *n*
seize / to 1. [sich] festklemmen, [sich] festsetzen; hängenbleiben; 2. fressen; 3. *(Schiff)* festmachen, zurren
seizing 1. Festklemmen *n*, Hängenbleiben *n*; 2. Fressen *n*; Anfressen *n* *(Fasen des Bohrers)*; 3. *(Schiff)* Festmachen *n*, Zurren *n*
s. of the piston Kolbenfressen *n*
seizure Fressen *n* *(Lager)*
s. resistance Widerstand *m* gegen Fressen *(Lagerwerkstoff)*
select / to 1. wählen, auswählen, aussuchen; auslesen, ausklauben; 2. *(Dat)* wählen; abgreifen; 3. *(El)* aussieben *(Frequenz)*
s. in advance *(Masch)* vorwählen
selectable frequency wählbare Frequenz *f*
selectance *s.* selectivity
selected copper Raffinatkupfer *n*
s. course track vorgewählte Kurslinie *f*
selecting bar Auswahlschiene *f* *(bei Matrixschaltern)*
s. comb *(Fmt)* Wählkamm *m*
s. mechanism *(Fmt)* Wählorgan *n*
s. pin Fühlnadel *f*
s. switch Wählschalter *m*
selection 1. Auswahl *f*, Wahl *f*; 2. Klauben *n*; 3. *(Fmt)* Wählen *n*
s. of fits Passungsauswahl *f*
s. pulse Auswahlimpuls *m*
s. rule Auswahlregel *f*
s. series Auswahlreihe *f*
selective auswählend, selektiv, trennscharf; wahlweise; lokal, örtlich, Teil-, Zonen-
s. absorber selektiver Absorber *m*
s. absorption selektive Absorption *f*, Selektivabsorption *f*
s. acting selektiv wirkend (ansprechend) *(Relais)*
s. amplifier selektiver Verstärker *m*
s. calling *(Fmt)* Selektivruf *m*
s. circuit *(El)* Siebkreis *m*
s. coefficient of absorption wahrer Absorptionskoeffizient *m*
s. corrosion selektive (örtliche) Korrosion *f*
s. fading *(Fmt)* selektive Schwunderscheinung
s. hardening Teilhärtung *f*, Zonenhärtung *f*, schichtweises Härten *n*
s. heating Teilerhitzung *f*
s. impulse *(Fmt)* Wählimpuls *m*
s. list wahlweise Schreibung *f* *(Lochkartenmaschinen)*
s. logging Durchforstung *f*
s. network *(El)* Siebschaltung *f*
s.-ordinates method *(Opt)* Auswahlordinatenverfahren *n*
s. protection Selektivschutz *m*
s. radiator Selektivstrahler *m*
s. receiver selektives (trennscharfes) Empfangsgerät *n*
s. reflection selektive Reflexion *f*
selectivity Trennvermögen *n*, Trennschärfe *f*, Selektivität *f*
s. characteristic Schutzkennlinie *f* *(Relais)*; Abstimmkurve *f*
s. control Selektivitätsregelung *f*, Trennschärferegelung *f*
selectode Selektode *f* *(Elektronenröhre)*
selector 1. *(Kfz)* Gangwähler *m*, Schaltgriff *m*; 2. *(Fmt)* Wähler *m*; 3. *(Dat)* Wähler[mechanismus] *m*; 4. Auswähler *m* *(Strickmaschine)*
s. bank *(Fmt)* Wählerkontaktfeld *n*
s. carrying capacity *(Fmt)* Belastbarkeit *f* der Wähler
s. fork *(Kfz)* Gangwählerschaltgabel *f*, Schaltgabel *f*
s. installation *(Fmt)* Wähler[anlage *f*] *m*

s. lubricating oil *(Fmt)* Wähleröl *n*
s. of the slide type Umsteuer[kolbenlängs]-schieber *m*
s. position-marker *(Fmt)* Markierer *m*
s. rack *(Fmt)* Wählergestell *n*
s. repeater *(Am) (Fmt)* Mitlaufwähler *m*
s. rod *(Fmt)* Bürstenträger *m*
s. shaft *(Fmt)* Schaltwelle *f*
s. shaft guide Führungsschlitz *m*
s. shelf *(Fmt)* Wählerrahmen *m*
s. switch Wähl[er]schalter *m*, Wahlschalter *m*, Betriebsartenschalter *m*
s. valve Umsteuerschieber *m*
selectron Selektron *n*, Selektronröhre *f (Speicherröhre für Elektronenrechner)*
selenate Salz *n* der Selensäure, Selenat *n*
selenic acid Selensäure *f*
selenide Selenid *n*
selenious acid selenige Säure *f*
selenite 1. Salz *n* der selenigen Säure, Selenit *n*; 2. *(Min)* Selenit *m*
selenium Selen *n*
s. barrier cell Selen[sperrschicht]zelle *f*
s. bridge Selenwiderstand *m*, Selenzelle *f*
s. conductive cell Selenwiderstandszelle *f*
s. film Selenschicht *f*
s. layer Selenschicht *f*
s. magnifier fotoelektrischer Kompensator *m*, Fotozellengalvanometerverstärker *m*
s. metal Selenmetall *n*
s. photocell Selensperrschichtzelle *f*
s. photoelement Selenfotoelement *n*
s. rectifier Selengleichrichter *m*
s. rectifier cell Selengleichrichterzelle *f*
self *s.* s.-acting
s.-absorption Selbstabsorption *f*
s.-absorption coefficient Selbstabsorptionskoeffizient *m*
s.-absorption effect Selbstabsorption *f*
s.-acting selbsttätig, automatisch
s.-acting brake selbsttätige Bremse *f*
s.-acting controller *s.* s.-actuated controller
s.-acting cutout automatisches Schaltschütz (Schütz) *n*, Selbstschütz *n*
s.-acting mule *(Text)* Wagenspinner *m*, Selfaktor *m*
s.-acting plane *(Brg)* Bremsberg *m*
s.-activated selbstaktiviert
s.-activated conductivity selbstaktivierte Leitfähigkeit *f*
s.-actor *(Text)* Selfaktor *m*, Wagenspinner *m*
s.-actuated controller Regler *m* ohne Hilfsenergie, direkt[wirkend]er Regler
s.-actuating clutch selbsteinrückende Kupp-
s.-adjoint *(Math)* selbstadjungiert
s.-adjoint operator selbstadjungierter (Hermitescher) Operator *m*
s.-adjusting selbsteinstellend *(Lager)*; selbstoptimierend
s.-adjusting bearing selbsteinstellendes Lager *n*
s.-adjusting control system selbstanpassende (selbsteinstellende) Regelung *f*; selbstoptimierende Regelung
s.-adjusting [hydraulic] tappet Ventilstößel *m* mit automatischer Spieleinstellung *(Verbrennungsmotor)*
s.-adjustment Selbsteinstellung *f*, automatische Einstellung *f*; Selbstabgleich *m*
s.-advancing [roof-]support *(Brg)* schreitender Ausbau *m*, Wanderausbau *m*
s.-aligning selbsteinstellend, [sich] selbst ausrichtend
s.-aligning ball bearing Pendelkugellager *n*
s.-aligning bearing Pendellager *n*
s.-aligning level Kompensatornivellierinstrument *n*
s.-aligning radial roller bearing Pendelrollenlager *n*
s.-aligning roller bearing Tonnenlager *n*
s.-aligning system selbstsynchronisierendes System *n*
s.-aligning torque Rückstellmoment *n (Lenkung)*
s.-alignment Selbsteinstellung *f*
s.-anchored *(Bw)* selbstverankert
s.-aspirating engine Ansaugmotor *m*, kompressorloser Motor *m*
s.-bailing boat selbstlenzendes Rettungsboot *n*
s.-baking electrode kontinuierlich arbeitende Elektrode *f*, selbstbrennende Elektrode, Dauerelektrode *f*
s. balancing selbstabgleichend, Kompensations-
s.-balancing bridge circuit selbstabgleichende Brückenschaltung *f*
s.-balancing instrument Kompensationsinstrument *n*
s.-balancing potentiometer selbsttätiger (selbstabgleichender) Kompensator *m*; selbstabgleichendes Potentiometer *n*, Nachlaufwiderstand *m*, selbstabgleichender Spannungsteiler *m*
s.-balancing system Selbstabgleichsystem *n*
s.-baling boat *(Am) s.* s.-bailing boat
s.-bearing *(Schiff)* Eigenpeilung *f*
s.-bias [voll]automatische Gittervorspannung *f*, Vorspannung *f* durch Audionschaltung
s.-bias arrangement automatische Gittervorspannungserzeugung *f*

s.-biased valve Röhre *f* mit automatischer Gittervorspannungserzeugung
s.-biasing automatische Gittervorspannungserzeugung *f*
s.-blocking oscillator Sperrschwinger *m*
s.-capacitance Eigenkapazität *f*; Windungskapazität *f*
s.-centring selbstzentrierend
s.-centring chuck Zentrierfutter *n*
s.-centring punch Zentrierglocke *f*
s.-centring steering *(Kfz)* selbstzentrierende Lenkung *f*
s. charge Eigenladung *f*
s.-checking *(Dat)* selbstprüfend
s.-checking code selbstprüfender Kode *m*
s.-clamping selbstspannend
s.-cleaning selbstreinigend
s.-cleaning contact Kontakt *m* mit Selbstreinigung, Schleifkontakt *m*, Wischkontakt *m*
s.-closing selbstschließend, Selbstverschluß-; selbstzuschaltend
s.-closing circuit breaker Leistungsschalter *m* mit automatischer Wiedereinschaltung, Selbsteinschalter *m*
s.-coiling hose *(Text)* sich selbsttätig zusammenringelnder Schlauch *m*
s.-collimating telescope Autokollimationsfernrohr *n*
s.-collision time Stoßzeit *f* bei Stößen von Partikeln gleicher Art
s.-compensating motor läufergespeister (rotorgespeister) kompensierter Induktionsmotor *m*
s.-consistent field Eigenfeld *n*
s.-contained Einbau-; für sich zusammengebaut, in sich abgeschlossen, selbständig; mit eigenem Antrieb
s.-contained battery eingebaute Batterie *f*
s.-contained construction geschlossener Aufbau *m*
s.-contained controller *(Am)* *s.* s.-actuated controller
s.-contained dock Dock *n* mit eigener Kraftzentrale
s.-contained instrument Multiinstrument *n*, Vielfachmeßgerät *n*
s.-contained navigation system bodenunabhängiges Navigationssystem *n*
s.-contained radio set Kofferempfänger *m*
s.-contained steam engine freistehende Dampfmaschine *f*
s.-containing turn-table *(Eis)* Tellerdrehscheibe *f*
s.-cooled selbstgekühlt, eigenbelüftet, eigenventiliert *(Maschine)*
s.-cooling Selbstkühlung *f (Transformator)*
s.-damping Eigendämpfung *f*
s.-decay Eigenabfall *m*
s. diffusion Selbstdiffusion *f*, Eigendiffusion *f*
s.-discharge *(Ph)* Selbstentladung *f*
s.-discharge wagon *(Eis)* Selbstentlader *m* *(z. B. Sattelwagen)*
s.-discharging facility Selbstentladeeinrichtung *f*
s.-discharging vessel Selbstentladeschiff *n*
s.-docking selbstdockend
s.-docking dock selbstdockendes Dock *n* *(Schwimmdock)*
s.-draining cockpit *(Schiff)* selbstlenzende Plicht *f*
s.-dumping barge *s.* s.-dumping scow
s.-dumping scow *(Schiff)* Klappschute *f*, selbstklappende Schute *f*
s.-dumping skip *(Brg)* kippbarer Skip *m*
s.-energizing brake selbstverstärkende Bremse *f*
s. energy Eigenenergie *f*, Selbstenergie *f*
s.-exchange Selbstaustausch *m*
s.-excitation Selbsterregung *f*, Eigenerregung *f*
s.-excitation capability Selbsterregungsfähigkeit *f*
s.-excitation winding Rückkopplungswicklung *f*
s.-excited selbsterregt *(elektrische Maschine)*
s.-excited a.c. generator with revolving armature selbsterregter Wechselstromgenerator *m* mit umlaufender Kommutatorwicklung
s.-excited valve selbsterregte Röhre *f*
s.-exciter Selbsterreger *m*
s.-exciting selbsterregend
s.-feeder Selbsteinleger *m*
s.-feeding mit eigenem Vorschub
s.-feeding furnace Schüttrostfeuerung *f*, Füllschachtfeuerung *f*
s.-feeding grate Schüttrost *m*
s.-field *(Ph)* Eigenfeld *n*
s.-fluxing ore *(Met)* selbstgehendes Erz *n*
s.-focusing selbstfokussierend
s.-gating selbststeuernd
s.-generating dynamo selbsterregter Dynamo *m*
s.-gripping selbstspannend
s.-hardening steel Selbsthärtestahl *m*, selbsthärtender Stahl *m*, Selbsthärter *m*
s.-heating Selbsterhitzung *f*
s.-heterodyne receiver Rückkopplungsempfänger *m*
s.-heterodyne reception Schwingaudionempfang *m*
s.-heterodyning mixer selbstschwingende Mischröhre *f*
s.-holding selbsthemmend
s.-ignitible selbstzündend
s.-ignition Selbstzündung *f*
s.-ignition temperature Selbstzündungstemperatur *f*

s.-impedance Eigenimpedanz *f*
s.-indexing selbstschaltend
s.-indicating mit automatischer Anzeige
s.-induced selbstinduziert
s.-inductance Selbstinduktivität *f*, Selbstinduktionskoeffizient *m*
s.-induction Selbstinduktion *f*
s.-inductive selbstinduktiv
s.-inductor Drosselspule *f*; Selbstinduktivität *f* *(Schaltelement)*
s.-instructed carry *(Dat)* selbsttätiger Übertrag *m*
s. interference audion Rückkopplungsaudion *n*, selbstschwingendes Audion *n*
s.-interruption Selbstunterbrechung *f*
s.-leading die Gewindeschneidkluppe *f* mit Führung
s.-loading ship Selbstladeschiff *n*
s.-locking 1. Selbstsperrung *f*; 2. *(Brg)* Selbsthemmung *f* *(Stempel)*
s.-locking selbstsperrend, selbsthemmend, selbstsichernd; selbstsynchronisierend
s.-locking gear selbstsperrendes Getriebe *n*
s.-locking nut selbstsichernde Mutter *f*
s.-lubricating selbstschmierend
s.-lubricating bearing Dauerschmierlager *n*; Ringschmierlager *n*
s.-luminous selbstleuchtend, Leucht-
s.-magnetic eigenmagnetisch
s.-maintained discharge selbständige Entladung *f*
s.-modulation Eigenmodulation *f*
s.-moving roof-support *(Brg)* schreitender Ausbau *m*, Wanderausbau *m*
s.-noise Eigenrauschen *n*
s.-opening selbstöffnend
s.-opening die head selbstöffnender Gewindeschneidkopf *m*
s.-opening stake pocket selbstauslösende Rungenöse *f* *(Eisenbahnwaggon)*
s. operated ohne Hilfsenergie, direktwirkend
s.-operated controller *(Am)* direkter Regler *m*, Regler ohne Hilfsenergie
s.-operating oscillator selbsterregter Oszillator *m*
s.-optimization Selbstoptimierung *f* *(Kybernetik)*
s.-organizing learning machine selbstorganisierende Lernmaschine *f*
s.-organizing system selbstorganisierendes System *n*
s.-orientating selbstorientierend, selbstausgleichend
s.-oscillate / to in Schwingung geraten
s.-oscillating selbstschwingend
s.-oscillation selbsterregte Schwingung *f*
s.-oscillator transmitter eigenerregter Sender *m*
s.-oscillatory *(El)* eigenerregt
s.-pinched plasma *(Ph)* durch den eigenen Strom zusammengeschnürtes Plasma *n*, gepinchtes Plasma
s.-polymerization Autopolymerisation *f*
s.-priming selbstansaugend *(Pumpe)*
s.-propelled mit Eigenantrieb
s.-propelled excavator Mobilbagger *m*
s.-propelled experiment Modellschleppversuch *m* mit selbstangetriebenem Modell
s.-propelled gun mounting *(Mil)* Selbstfahrlafette *f*
s.-propelled vessel Schiff *n* mit Eigenantrieb, Selbstfahrer *m*
s.-pulse modulation Modulation *f* durch selbsterregte Impulse
s.-purification Selbstreinigung *f* *(Gesundheitstechnik)*
s.-quenched detector Pendelrückkopplungsaudion *n* mit selbsterzeugter Pendelfrequenz
s.-quenching Selbst[aus]löschung *f*
s.-quenching oscillator *(Fmt)* Sperrschwinger *m*
s.-quenching super-regenerative detector Pendelrückkopplungsaudion *n*
s.-radiation Eigenstrahlung *f*
s.-reading instrument registrierendes Gerät *n*
s.-recording potentiometer selbstregistrierendes Potentiometer *n*
s.-recording unit Selbstschreiber *m*
s.-registering selbstaufzeichnend
s.-regulating selbstregelnd, selbstregulierend
s.-regulating arc-welding transformer selbstregelnder Lichtbogenschweißumspanner *m*
s.-regulating belt-brake selbstregulierende Riemenbremse *f*
s.-regulating d.c. welding generator selbstregelnder Gleichstromschweißgenerator *m*
s.-regulation Selbstregelung *f*
s.-rendering winch *(Schiff)* selbstaufspulende Winde *f*
s.-repulsion Selbstabstoßung *f*
s. reset rückfallendes Relais *n*
s.-resetting maßgesteuert
s.-resonance Eigenschwingung *f*, Eigenresonanz *f*
s.-resonance point Resonanzpunkt *m*
s.-restoring drop *(Fmt)* Rückstellklappe *f*
s. restoring indicator *(Fmt)* Rückstellklappe *f*
s.-retaining selbsthemmend
s.-retention Selbsthemmung *f*
s.-reversal Selbstumkehr *f*, Selbstumkehrung *f*
s.-reversal of spectral lines Selbstumkehr *f* von Spektrallinien
s.-righting boat sich selbst aufrichtendes Boot *n*
s.-rotation Eigenrotation *f*

s.-saturation direkte Selbsterregung *f*, innere Mitkopplung *f*
s.-scattering Selbststreuung *f*, Eigenstreuung *f*
s.-sealed bearing selbstabdichtendes Lager *n*
s.-selecting selbstwählend
s.-service lift Fahrstuhl *m* mit Selbststeuerung
s.-service shop Selbstbedienungsladen *m*
s.-sharpening action [of the grinding wheel] Selbstschärfung *f* [der Schleifscheibe]
s.-shielding Selbstabschirmung *f*
s.-shifting gear Selbstschaltgetriebe *n*
s.-stabilization Eigenstabilisierung *f*, Selbststabilisierung *f*
s.-stabilizing dynamo Querfelddynamo *m*
s.-starter automatischer Anlasser *m*
s.-starting oscillator selbsteinsetzender Sender *m*, Sender mit weichem Schwingungseinsatz
s.-steering selbststeuernd
s.-stressed concrete Eigenspannbeton *m*
s.-stressing concrete Eigenspannbeton *m*
s.-supporting *(Bw)* freitragend, selbsttragend; *(Masch)* freistehend
s.-supporting aerial mast freistehender (freitragender) Antennenmast *m*
s.-supporting piston selbsttragender (von der Zylinderwand getragener) Kolben *m*
s. sustained discharge Selbstentladung *f*
s.-sustaining discharge selbständige Entladung *f*
s.-sustaining pile Reaktor *m* mit selbständiger Kettenreaktion
s.-sustaining speed Selbstlaufdrehzahl *f*
s.-switching amplifier selbstschaltender Verstärker *m*
s.-tannage Alleingerbung *f*
s.-tapping screw selbstschneidende Schraube *f*, Schneidschraube *f*
s.-tipping log barge *(Schiff)* selbstkippende Holzschute *f*
s.-transformer *(El)* Autotransformator *m*
s.-trapping Selbstanlagerung *f (Elektronen)*
s.-trimming vessel Selbsttrimmerschiff *n*
s.-unloader Selbstentladeschiff *n*, Selbstentlader *m*, selbstentladendes (selbstlöschendes) Schiff *n*
s.-unloading vessel selbstlöschendes Schiff *n*, Selbstentlader *m*
s. whistles *(Fmt)* Eigenpfeifen *n*, Überlagerungspfeifen *n*
s.-wiping contact selbstreinigender Kontakt *m*, Wischkontakt *m*

Sellers screw thread Sellersgewinde *n*
Sellers taper Sellerskegel *m*
selling expenses Vertriebskosten *pl*
selsyn Gleichlaufanlage *f*, Gleichlaufgerät *n*; Synchronisieranlage *f*; Ringfeldgeber *m*; Drehmelder *m*, Selsyn *n*
s. generator *(El)* Drehfeldgeber *m*, Synchronisierspannungsgenerator *m*
s. motor Drehmelder *m*
s. system Drehfeldsystem *n (s. a.* selsyn*)*

selvage 1. *(Geol)* Salband *n*, Lettenbesteg *m*; 2. *(Text)* Webkante *f*, Salleiste *f*, Webleiste *f*, Leiste *f*, Gewebeleiste *f*
s. loop Randmasche *f*
s. printing device Randdruckeinrichtung *f*
s. steaming apparatus Leistendämpfapparat *m*
s. uncurler Leistenaufroller *m*
s. uncurling device Kantenausroller *m*

selvagee strop *(Schiff)* Garnstropp *m*, Wantstropp *m*
selvedge *s.* selvage
semantic semantisch
semaphore direction indicator Winker *m*, mechanischer Richtungsanzeiger *m*, Pendelwinker *m*
s. flag-signalling *(Schiff)* Flaggensignalisieren *n* mit zwei Flaggen
s. signal *(Eis)* Armsignal *n*, Flügelsignal *n*

semi Halb-
s.-acetal *(Ch)* Halbazetal *n*, Semiazetal *n*
s.-adder Halbadder *m*
s.-analytical halbanalytisch
s.-angle Halbwinkel *m*
s.-anthracite Magerkohle *f*; *(Am)* Steinkohle *f (mit flüchtigen Bestandteilen von 8 bis 12%)*
s.-automatic Halbautomat *m*
s.-automatic halbautomatisch, halbselbsttätig
s.-automatic bar machine Stangenhalbautomat *m*
s.-automatic chucking machine Futterhalbautomat *m*
s.-automatic cycle halbautomatischer Arbeitsablauf *m*
s.-automatic exchange halbselbsttätiges Fernsprechamt *n*
s.-automatic key *(Fmt)* halbautomatische Taste *f*
s.-automatic lathe Drehhalbautomat *m*, halbautomatische Drehmaschine *f*
s.-automatic machine Halbautomat *m*
s.-automatic ringing *(Fmt)* halbselbsttätiger Ruf *m*
s.-automatic screw machine Drehhalbautomat *m*
s.-automatic tandem working *(Fmt)* halbselbsttätiger Durchgangsdienst *m*
s.-axis Halbachse *f*
s.-balanced rudder *(Schiff)* Halbschweberuder *n*, Halbbalanceruder *n*
s.-basement Souterrain *n*
s.-beam Kragträger *m*, Freiträger *m*, Kragbalken *m*

s.-bituminous coal Magerkohle *f*; *(Am)* Steinkohle *f (mit flüchtigen Bestandteilen von 12 bis 18%)*
s.-cantilever *(Flg)* halbfreitragend *(Tragfläche)*
s.-carbazide *(Ch)* Semikarbazid *n*
s.-chemical pulp Halbzellstoff *m*
s.-chemical straw board gelbe Strohpappe *f*, Strohstoffpappe *f*
s.-chemical straw pulp Strohstoff *m*
s.-circle *s.* semicircle
s.-circular *s.* semicircular
s.-classical method halbklassische Methode *f*
s.-coated einseitig gestrichen *(Papier)*
s.-coke Halbkoks *m*
s.-coke oven Halbkoksofen *m*
s.-conductor *s.* semiconductor
s.-continuous mill halbkontinuierliche Walzstraße *f*
s.-crystalline halbkristallin[isch]
s.-cylinder Halbzylinder *m*
s.-cylindrical halbzylindrisch
s.-diameter Halbdurchmesser *m*
s.-Diesel engine Halbdieselmotor *m*
s.-direct lighting halbdirekte Beleuchtung *f*; vorwiegend direkte Beleuchtung
s.-drop-centre rim Halbtiefbettfelge *f (Reifen)*
s.-dry concrete erdfeuchter (schwachplastischer, steifer) Beton *m*
s.-dry electrolytic capacitor feuchter Elektrolytkondensator *m*
s.-elliptic leaf spring halbelliptisches Blattfederpaket *n*
s.-elliptic spring Halbfeder *f*
s.-enclosed fuse halbgeschlossene Sicherung *f*
s.-enclosed motor Motor *m* mit Berührungsschutz
s.-finish grind / to vorfertigschleifen
s.-finish ream / to halbfertig reiben
s.-finished vorgearbeitet
s.-finished article Halbfabrikat *n*
s.-finished flat Breiteisen *n (Walzwerk)*
s.-finished goods (products) Halbfertigwaren *fpl*, Halbzeug *n*
s.-finishing reamer Vorreibahle *f*
s.-flat-base rim Halbflachbettfelge *f (Reifen)*
s.-floating axle halbtragende Achse *f*
s.-fluid halbflüssig, viskos
s.-fluid friction Mischreibung *f*
s.-fluid lubrication halbflüssige Schmierung *f*, Grenzschmierung *f*
s.-flush switch halbversenkter Schalter *m*, Impu-Schalter *m*
s.-fused halbgeschmolzen, halbgesintert; teigig
s.-girder Konsolbalken *m*, Kragbalken *m*
s. gutter cutter Halbhohlkehlfräser *m*
s.-hardboard halbharte Platte *f*
s.-hot press Halbwarmpresse *f*
s.-image Halbbild *n*
s.-indirect lighting halbindirekte Beleuchtung *f*; vorwiegend indirekte Beleuchtung
s.-infinite body Halbkörper *m*
s.-infinite space unendlicher Halbraum *m*
s.-killed halbberuhigt *(Stahl)*
s.-logarithmic einfachlogarithmisch, halblogarithmisch
s.-logarithmic scale halblogarithmischer Maßstab *m*
s.-lunar halbmondförmig
s.-metal Halbmetall *n*
s.-open shed Halboffenfach- *(Jacquardmaschine)*
s.-permanent memory *(Dat)* Halbfestwertspeicher *m*
s.-permanent parallel form *(Dat)* halbpermanente Parallelform *f*
s.-permeable halbdurchlässig, semipermeabel
s.-permeable membrane halbdurchlässige (semipermeable) Membran *f*
s.-plant Versuchsanlage *f*
s.-polar halbpolar *(Bindung)*
s.-polar bond *(Ch)* Koordinationsbindung *f*
s.-porcelain Halbporzellan *n*
s.-portable boiler stationärer (feststehender) Lokomobilkessel *m*
s.-portable Cornish multitubular boiler Lokomobilkessel *m* [mit vorgehenden Heizröhren]
s.-precious gem[stone] Halbedelstein *m*
s.-producer type furnace Halbgasfeuerung *f*
s.-quantitative halbquantitativ
s.-quantitative method halbquantitatives Verfahren *n*
s.-recessed switch halbversenkter Schalter *m*
s.-remote control Halbfernsteuerung *f*
s.-remote manipulation Halbfernsteuerung *f*
s.-rigid halbstarr, halbsteif
s.-rigid insulation board halbsteife Isolierplatte *f*
s.-rigid suspension halbstarre Aufhängung *f* *(Kabel)*
s.-scale plant Anlage *f* im halbtechnischen Maßstab, Versuchsanlage *f*
s.-selective ringing *(Am)* *(Fmt)* Einzelanruf *m* mit verabredeten Zeichen
s -self-maintained discharge halbselbständige Entladung *f*
s.-skilled worker angelernter Arbeiter *m*
s.-solid lubricant Starrschmiere *f*
s.-spherical halbkugelförmig, Halbkugel-
s.-steel Gußeisen *n* mit Stahlschrottzusatz *(hochwertiger Grauguß)*
s.-steep *(Geol)* halbsteil
s.-trailer *(Kfz)* Sattelanhänger *m*, Sattelschlepper *m*; *(Eis)* Sattelwagen *m*
s.-trailer unit Sattelschleppzug *m*

s.-transparent halbdurchlässig
s.-tubular boiler Halbröhrenkessel *m*, zusammengesetzter Kessel *m* mit Heizröhren
s.-universal radial drilling machine Halbuniversalradialbohrmaschine *f*
s.-vector halbvektoriell
s.-vulcanized board Vulkanfiberimitation *f*
s.-water gas Halbwassergas *n*
s.-wildcat well Suchbohrung *f* in einiger Entfernung von einer bereits bekannten Erdöllagerstätte
semicircle Halbkreis *m*
semicircular halbrund, halbkreisförmig, Halbkreis-
s. arc Halbkreisbogen *m*
s. area Halbkreisfläche *f*
s. focusing Halbkreisfokussierung *f*
s. groove Halbkreiseinschnitt *m*
s. path halbkreisförmige Bahn *f*
s. shape Halbkreisform *f*, Halbrundform *f*
s. spectrometer Halbkreisspektrometer *n*
s. spectroscope Halbkreisspektroskop *n*
semiconductor Halbleiter *m*
s. barrier layer Halbleitersperrschicht *f*
s. block Halbleiterblock *m*
s. capacitor Halbleiterkondensator *m*
s. detector Halbleiterdetektor *m*
s. device Halbleiteranordnung *f*, Halbleiterbauelement *n*
s. diode Halbleiterdiode *f*
s. electronics Halbleiterelektronik *f*
s. element Halbleiterbauelement *n*
s. junction Halbleiterübergang *m*, Halbleitersperrschicht *f*
s. junction diode Halbleiterflächendiode *f*
s. laser Halbleiter-Laser *m*
s. manufacturing Halbleiterfertigung *f*
s. material Halbleiterwerkstoff *m*
s. performance Halbleiterleistung *f*
s. physicist Halbleiterphysiker *m*
s. physics Halbleiterphysik *f*
s. property Halbleitereigenschaft *f*
s. rectifier Halbleitergleichrichter *m*
s. switching device Halbleiterschaltelement *n*
s. technique Halbleitertechnik *f*
semiconductored mit Halbleitern bestückt
semiflat halbmatt
seminal capsule Samenkapsel *f*
s. leaf Samenblatt *n*
send / to [ver]senden; aussenden; [ver]schicken; geben
s. by water verschiffen, auf dem Wasserwege versenden
s. out aussenden *(Funktechnik)*
send distortion *(El)* Sendeverzerrung *f*
s.-receive contact *(Fmt)* Sende-Empfang-Kontakt *m*
sending Sendung *f*, Übermittlung *f*, Geben *n* *(Funktechnik)*
s.-end crosstalk *(Fmt)* Nebensprechen *n*
s.-end impedance Eingangswiderstand *m*
s.-end voltage Eingangsspannung *f*, Spannung *f* am Leitungsanfang
s. key Morsetaste *f*; Sendetaste *f*
senegal gum Gummiarabikum *n*
sensation Empfindung *f*
s. of heat Wärmeempfindung *f*
s. of light Lichtempfindung *f*
sense / to abtasten, erfassen *(Lochstreifen, Lochkarten)*
sense *(Math)* Richtung[ssinn *m*] *f*
s. adjustment Seitenbestimmung *f* *(Peilung)*
s. aerial Seitenbestimmungsantenne *f*
s. ambiguity Richtungsdoppeldeutigkeit *f*
s. finder Schalter *m* für Seitenbestimmung, Seitenschalter *m* *(Peilung)*
s. finding Zweideutigkeitsaufhebung *f* *(Radartechnik)*
s. of a bearing Peilsinn *m*
s. of current Stromrichtung *f*
s. of rotation Drehsinn *m*, Umlaufrichtung *f*, Drehrichtung *f*
s. research Zweideutigkeitsaufhebung *f* *(Radartechnik)*
sensed *(Math)* gerichtet, orientiert
sensibility Empfindlichkeit *f* (*s. a.* sensitivity)
s. to heat Hitzeempfindlichkeit *f*, Temperaturempfindlichkeit *f*
sensible heat fühlbare Wärme *f*
s. horizon *(Schiff)* scheinbare Kimm *f*, scheinbarer (terrestrischer) Horizont *m*
s. to daylight tageslichtempfindlich
sensing element Meßfühler *m*; Meldeorgan *n* *(Regler)*
sensitive empfindlich; feinstufig; feinfühlig; feinempfindlich
s. adjustment Feineinstellung *f*, Feinverstellung *f*
s. drill [press] *(kleine, feinfühlige)* Tischbohrmaschine *f*
s. element *s.* detecting element
s. feed *s.* s. hand feed
s. flame sensitive Flamme *f*
s. hand feed Gefühlshebelvorschub *m*
s. heat fühlbare Wärme *f*
s. position Stellung *f*, bei der ein Gerät anspricht
s. tapping machine Gefühlsgewindebohrmaschine *f*
s. time Ansprechzeit *f*
s. to corrosion korrosionsanfällig, rostempfindlich
s. to heat wärmeempfindlich
s. to light lichtempfindlich

sensitiveness Empfindlichkeit *f (Film)*; Ansprechempfindlichkeit *f (Regler)*
s. to direction Richtungsempfindlichkeit *f*
s. to percussion Schlagempfindlichkeit *f*
s. to shock Erschütterungsempfindlichkeit *f*
sensitivity *(BMSR)* Empfindlichkeit *f*; Ansprechwahrscheinlichkeit *f*; Feinfühligkeit *f*; Genauigkeit *f (Meßgerät)*
s. control Empfindlichkeitsregelung *f*
s. decrease Empfindlichkeitsabnahme *f*
s. drift Empfindlichkeitsänderung *f*
s. of deflection Ablenkempfindlichkeit *f*
s. of gasoline Bleiempfindlichkeit *f* von Benzin
s. test Empfindlichkeitsprüfung *f*
s. to hardening Härteempfindlichkeit *f*
s. to overheating Überhitzungsempfindlichkeit *f*
sensitization Aktivierung *f*; Sensibilisierung *f*
sensitize / to lichtempfindlich machen, sensibilisieren
sensitized sensibilisiert, lichtempfindlich
s. discomposition sensibilisierte Zersetzung *f*
s. paper lichtempfindliches (sensibilisiertes) Papier *n*
sensitizer Sensibilisator *m*; Aktivator *m*
sensitizing agent Sensibilisierungsmittel *n*
sensitometer Sensitometer *n*
sensitometry Sensitometrie *f*
sensor Meßfühler *m*, Fühler *m*
sentential calculus *(Dat)* Aussagenkalkül *m*
sentinel Marke *f (bei Rechnern)*
separable abscheidbar, trennbar, separierbar
separate / to [ab]scheiden, absondern, abspalten, [ab]trennen, auslesen, ausscheiden, klassieren, klauben, separieren, sichten, sortieren, teilen, zerlegen; ablösen *(Strömung)*; zentrifugieren
s. as a flocculent ausflocken
s. into parts entmischen
s. out entmischen
s. the dross *(Met)* abschlacken, Schlacke abziehen
s. to the sizes klassieren *(Erz)*
separate Separatdruck *m*
separate getrennt, abgesondert, isoliert, separat
s. application *(Plast)* Getrenntauftragen *n*, Vorstreichverfahren *n*
s. application adhesive Zweikomponentenklebstoff *m*
s. combustion chamber Teilkammer *f*
s. coupling aufgesetzte Kupplung *f*
s. drive Einzelantrieb *m*
s. excitation *(Elm)* Fremderregung *f*
s. gear-change *(Kfz)* getrennte Schaltung *f*
s. pot Druckkammer *f*, Füllraum *m*
s. pot mould *(Plast)* Spritzkopf *m*, Preßspritzwerkzeug *n* mit separater Druckkammer
s. self-excitation *(El)* getrennte Selbsterregung *f*, äußere Mitkopplung *f*
s. sound *(Fs)* Parallelton *m*
s. ventilation Fremdbelüftung *f*
separated cavity abgelöster Hohlraum *m (Kavitation)*
s. phase switchgear Schaltanlage *f* mit Phasentrennung; Phasentrennschalter *m*; Schalter *m* für Einzelphasenschaltung
separately-cooled mit Fremdkühlung *(durch Gase oder Flüssigkeiten)*
s.-excited fremderregt
s.-excited dynamo fremderregter Dynamo *m*
s.-instructed carry *(Dat)* befohlener Übertrag *m*
s. lead-covered cable Dreibleimantelkabel *n*, gürtelloses Kabel *n*
separating Lösen *n*, Trennen *n*, Trennung *f (Raketenstufe)*; Scheidung *f (Metalle)*; Sortieren *n*, Klassieren *n (Erz)*; Teilen *n*; Lösen *n (Strömung)*
s. agent Scheidemittel *n*
s. allowance *(Ök)* Trennungszulage *f*
s. bowl *(Lw)* Separatortrommel *f*
s. drum Dampfsammler *m (Etagenkessel)*; *(Lw)* Separatortrommel *f*
s. factor Trennfaktor *m*
s. filter Frequenzfilter *n*; elektrische Weiche *f*
s. funnel Scheidetrichter *m*
s. layer Trennschicht *f*
s. line Trennungslinie *f*
s. lug Trennungszunge *f*
s. nozzle *(Kern)* Trenndüse *f*
s. partition Scheidewand *f*
s. plant Sicht[ungs]anlage *f*, Sortieranlage *f*
s. screen Sortiersieb *n*
s. stage *(El)* Trennstufe *f*
s. wall Scheidewand *f (Kreuzkopf)*
separation Lösen *n*, Trennen *n*, Trennung *f (z. B. Raketenstufe)*; Impulsabtrennung *f*; Entmischung *f*, Scheidung *f (Metalle)*; Scheidung, Sortierung *f*, Klassierung *f (Erze)*; Teilung *f*, Abtrennung *f*; Abstand *m*; Ablösung[svermögen *n*] *f (Strömung)*
s. between cut-off frequencies Bandbreite *f*
s. between frequencies Frequenzabstand *m*
s. chamber *(Kern)* Trennkammer *f*
s. column Trennsäule *f*, Trennkolonne *f*
s. factor Trennfaktor *m*
s. failure (fracture) Trennbruch *m*
s. method Trennverfahren *n*
s. nozzle Trenndüse *f*
s. of elements Elementtrennung *f*
s. of isotopes Isotopentrennung *f*
s. of property Gütertrennung *f*
s. of the boundary layer Grenzschichtablösung *f*

s. of variables *(Math)* Trennung *f* der Variablen
s. of water from oil Wasserentölung *f*
s. plant Trennanlage *f*
s. point Ablösungspunkt *m (Strömung)*, Grenzschichtablösungspunkt *m*
s. potential Trennpotential *n*
s. process Abtrennprozeß *m*; Trennprozeß *m*, Trennvorgang *m*; Trennverfahren *n*
s. surface Trennungsfläche *f*
separator 1. Trennanlage *f*, Trennschleuder *f*, Separator *m*; Abscheider *m*, Scheider *m*; 2. Separator, Trennelement *n (Batterie)*; 3. *(El)* Trennstufe *f*, Pufferstufe *f*; 4. *(Text)* Teilplatte *f*; 5. *(Masch)* Abstandshalter *m*, Trennstück *n*, Distanzhalter *m*; 6. Trennzeichen *n (Rechner)*
s. spring Trennfeder *f*
separatory funnel Scheidetrichter *m*, Schütteltrichter *m*
sepia printing process Sepialichtpausverfahren *n*
sepiolite *(Min)* Sepiolith *m*, Meerschaum *m*
septangle Siebeneck *n*
septangular siebeneckig
septate cable radial unterteiltes konzentrisches Kabel *n*
s. waveguide Hohlleiter *m* mit axialer Scheidewand *(Wellenleiter)*
septic Fäulniserreger *m*
septic faulig, fäulniserregend, septisch
s. tank Faulbecken *n*, Faulbehälter *m*, Klärbehälter *m*
septivalence Siebenwertigkeit *f*
septivalent siebenwertig
septuble / to versiebenfachen
septuble siebenfach
septum Scheidewand *f*, Diaphragma *n*, Membran *f*
sequence / to *(Dat)* anreihen, aufreihen, einreihen
sequence Reihenfolge *f*, Folge *f*; *(Masch)* Bearbeitungsfolge *f*; Arbeitsablauf *m*
s. calling *(Fmt)* Anmeldung *f* von Reihengesprächen, Reihenanmeldung *f*
s. change-over contact *(Fmt)* Folgeumschaltkontakt *m*
s. checking routine Folgeprüfprogramm *n*
s. contact *(Am)* Folgekontakt *m*
s. control Ablaufsteuerung *f*, Folgesteuerung *f*
s.-control register Befehlszähler *m*, Befehlsfolgeregister *n (Rechner)*
s.-controlled computer Rechner *m* mit Folgesteuerung
s.-controlled contact Folgekontakt *m*
s. interdependence Folgeabhängigkeit *f (Signalwesen)*
s. of motions Bewegungsablauf *m*
s. of numbers Zahlenfolge *f*
s. of operation Schaltfolge *f*
s. of operations Arbeitsfolge *f*, Programmablauf *m*
s. signal *(Fmt)* pausenfreies Signal *n*, Folgesignal *n*
s. switch Programmschalter *m*, Folgeschalter *m*, Steuerschalter *m*
s. switch cam Folgeschalterkamm *m*, Steuernocken *m*
s. zones Zonenfolge *f*
sequencing *(Masch)* Folgesteuerung *f*
sequential aufeinanderfolgend
s. analysis *(Math)* Sequenzanalyse *f*
s. circuit Schaltwerk *n*
s. colour system Rasterfolgeverfahren *n*, Rasterwechselverfahren *n (Farbfernsehtechnik)*
s. computer programmgesteuerte Rechenanlage *f*
s. control Folgesteuerung *f (Rechner)*
s. operator sequentieller Operator *m (Rechner)*
s. sampling Sequentialstichprobenverfahren *n*
s. scanning *(Fs)* Zeilenfolgeabtastung *f*, Folgeabtastung *f*, fortlaufende Bildabtastung *f*
s. searching sequentielles Suchen *n (Rechner)*
s. switching Folgeschalten *n*
s. test Folgeprüfung *f*
sequi-side-band system *(Fs)* Anderthalbseitenbandverfahren *n*
sequoia Mammutbaum *m*
serge weave *(Text)* Köperbindung *f*
serial Fortsetzungsfilm *m*; Hörspielfolge *f (im Radio)*
serial Reihen-, Serien-, aufeinanderfolgend, laufend, serienmäßig
s. computing machine Serienrechner *m*
s. connection Reihenschaltung *f*, Serienschaltung *f*, Hintereinanderschaltung *f*
s. construction Serienbau *m*
s. development *(Math)* Reihenentwicklung *f*
s. entry printer Drucker *m* mit Serieneingang
s. film Fortsetzungsfilm *m*
s. intercommunication set *(Fmt)* Reihenanlage *f*
s. number Fabrikationsnummer *f*, Seriennummer *f*
s. operation Serienbetrieb *m (Rechner)*
s. storage Serienspeicherung *f (Datenspeicherung)*; Serienspeicher *m*
s. symbol Aktenzeichen *n*
s. tap Gewindebohrersatz *m*
s. transfer *(Dat)* Serienübertragung *f*, Serientransport *m*
serially connected *(El)* hintereinandergeschaltet, in Reihe geschaltet
seriate / to in Reihe anordnen
seriate in Reihe [angeordnet]

seriation Anordnung *f* in Reihe
sericeous seidig, seidenartig
sericite *(Min)* Serizit *m (Kaliglimmer)*
s. slate Serizitschiefer *m*
sericulture Seidenraupenzucht *f*
sericulturist Seidenraupenzüchter *m*
series *(Math)* Reihe *f*; Serie *f*; Baureihe *f*
series *(El)* Reihen[schluß]-, Serien-
s. arrangement Reihenanordnung *f*
s. arrangement of impulses and velocity stages Hintereinanderschaltung *f* von Druck- und Geschwindigkeitsstufen *(Turbine)*
s. capacitor reihengeschalteter Kondensator *m*, Vorschaltkondensator *m*
s. characteristic Reihenschlußverhalten *n (eines Motors)*
s. circuit in Reihe (Serie) geschalteter Stromkreis *m*
s. coil Hauptschluß[feld]spule *f*; Reihenschlußspule *f*
s. commutator motor Reihenschlußkommutatormotor *m*
s. connected hintereinander geschaltet, in Reihe (Serie) geschaltet
s. connected station *(Fmt)* Reihenstelle *f*
s. connection Reihenschaltung *f*, Serienschaltung *f*, Hintereinanderschaltung *f*
s. control unterlagerte Regelung *f*
s. diode *(Fs)* reihengeschaltete Diode *f*, Vorschaltdiode *f*
s.-efficiency diode *(Fs)* Schalterdiode *f*
s. element Serien[bau]element *n*
s. equalizer *(El)* Längsentzerrer *m*
s. excitation *(Elm)* Reihenschlußerregung *f*
s. expansion *(Math)* Reihenentwicklung *f*
s. feedback *(El)* Reihenrückkopplung *f*, Serienrückkopplung *f*
s. grid resistor Gitterdämpfungswiderstand *m*, Gittervorwiderstand *m*
s. inductance Serieninduktivität *f*, Reiheninduktivität *f*, Längsinduktivität *f*
s. inductance input filter Tiefpaß *m* in T-Schaltung
s. loading Pupinisierung *f*, Bespulung *f (von Kabeln)*
s. modulation Anodenmodulation *f*, Vorröhrenmodulation *f*
s. motor Reihenschlußmotor *m*, Hauptstrommotor *m*, Hauptschlußmotor *m*
s. mounting Serienschaltung *f*
s. multiple jack *(Fmt)* Vielfachunterbrechungsklinke *f*
s. notch Reihenschaltstufe *f*
s. of diagrams *(Wärm)* Diagrammsatz *m*
s. of diameter ranges Reihe *f* der Durchmesserbereiche
s. of formations *(Geol)* Schichtenpaket *n*
s. of measurements Meßreihe *f*, Meßserie *f*
s. of preferred tolerances *(Masch)* Auswahlreihe *f*
s. of strata *(Geol)* Schichtfolge *f*
s. of tests Versuchsreihe *f*
s. of tolerances *(Masch)* Toleranzreihe *f*
s. parallel Gruppenschaltung *f (Batterie)*
s.-parallel circuit Reihenparallelschaltung *f*, Serienparallelkreis *m*
s.-parallel connection Reihenparallelschaltung *f*, gemischte Schaltung *f*
s.-parallel control Regelung *f* durch Gruppierung der Motoren
s.-parallel starter Reihenparallelanlasser *m*
s.-parallel winding Reihenparallelwicklung *f*, Serienparallelwicklung *f*
s. peaking Erweitern *n* der Frequenzkurve durch Serieninduktivität im Koppelglied
s. phase resonance Reihenresonanz *f*, Phasenresonanz *f*
s. production Reihenfertigung *f*, Reihenherstellung *f*, Serienherstellung *f*, Serienfertigung *f*
s. production of small batches Serienfertigung *f* kleiner Stückzahlen
s. reactor Vorschaltdrossel *f*; *(Am)* Saugspule *f*, Wellensauger *m*
s. representation Reihenersatzschaltbild *n*
s. resistance Vor[schalt]widerstand *m*; Reihenwiderstand *m*, Serienwiderstand *m*
s. resistor Vorwiderstand *m*; Serienwiderstand *m*
s. resonance Phasenresonanz *f*, Reihenresonanz *f*, Spannungsresonanz *f*, Serienresonanz *f*
s.-shunt network symmetrisches Netzwerk *n*
s. starter Hauptstromanlasser *m*
s. stator *(Elm)* Stromkern *m*
s. T Serienverzweiger *m*, Impedanz-T-Glied *n (Wellenleitertechnik)*
s. transductor Transduktor *m* in Reihenschaltung, Reihentransduktor *m*
s. transformer Reihentransformator *m*, Vorschalttransformator *m*, Stromwandler *m*
s. undercurrent release direkte Unterstromauslösung *f*
s. undercurrent tripping direkte Nullstromauslösung *f*
s. ventilation Reihenbelüftung *f*
s. voltage Reihenspannung *f*
s. winding Reihenwicklung *f*
s.-wound Hauptschluß-, Reihenschluß-
s.-wound dynamo Reihenschlußdynamo *m*
s.-wound generator Reihenschlußgenerator *m*
s.-wound motor Hauptschlußmotor *m*, Reihenschlußmotor *m*
serif *(Typ)* Haarstrich *m*, Serife *f*
serigraphy Siebdruck *m*, Serigrafie *f*

serpentine 1. Schlangenlinie *f*, Schlangenkurve *f*; 2. *(Min)* Serpentin *m*; 3. *(Geol)* Serpentinit *m*, Serpentinschiefer *m*
s. asbestos Serpentinasbest *m*
s. cooler Schlangenkühler *m*
s. grate bar Schlangenroststab *m*
serrate / to einkerben, kerbverzahnen; riefen, riffeln
serrated geriffelt, gerippt, gezackt, sägeartig; kerbverzahnt, mit Kerbzahnprofil
s. blade geriffeltes Messer *n (Fräser)*; Schneidbacke *f (Gewindeschneidkopf)*
s. hub Kerbzahnnabe *f*
s. profile Sägezahnprofil *n*, Kerbzahnprofil *n*
s. shaft Kerbzahnwelle *f*
s. vertical pulse gezahnter vertikaler Synchronisationsimpuls *m*
serration Kerbverzahnung *f*, Verzahnung *f*; Zacken *m*; Einkerbung *f*
s. hob Kerbverzahnungswälzfräser *m*
serve / to [be]dienen; versorgen; absolvieren
s. a district ein Gebiet versorgen
served cable umhülltes Kabel *n*
s. lead-covered cable Kabel *n* mit Bleimantel und Umhüllung, Bleikabel *n* mit Korrosionsschutz
service / to bedienen, warten, pflegen, unterhalten
service Dienst *m*; Betrieb *m*, Gebrauch *m*, Verwendung *f*, Nutzung *f*; Bedienung *f (Maschine)*
s. aboard *s.* s. afloat
s. afloat Borddienst *m*, Schiffsdienst *m*
s. area Empfangsbereich *m*; Sendebereich *m*
s. band *(Fmt)* zugeteiltes Frequenzband *n*, Kabelanschluß *m*, Versorgungskabel *n*, Zuleitungskabel *n*
s. cartridge scharfe Patrone *f*
s. ceiling *(Flg)* Dienstgipfelhöhe *f*
s. channel Leitungskanal *m*; Rohrkanal *m*
s. circuit 1. Betriebsstromkreis *m*; 2. *(Fmt)* Dienstleitung *f*
s. condition Betriebsbedingung *f*
s. condition on board Bordbetriebsbedingungen *fpl*
s. conditions Betriebsbedingungen *fpl*; Betriebsbeanspruchung *f*; Betriebsverhältnisse *npl*
s. conductor Hausanschlußleitung *f*
s. connection Betriebsschaltung *f*; Kabelanschluß *m*
s. cord *(El)* Netzlitze *f*
s. department Kundendienstabteilung *f*
s. diagnosis Fehlersuche *f*
s. door Arbeitstür *f*, Beschickungstür *f*
s. draught Betriebstiefgang *m (eines Schiffes)*
s. drop Hausanschlußleitung *f*
s. duct Leitungskanal *m*; Rohrkanal *m*
s. engineer Reparaturtechniker *m*
s. entrance Hausanschluß *m*
s. entrance conductor [äußere] Hausanschlußleitung *f*
s. fuel tank Brennstofftagestank *m*, Brennstoffverbrauchsbehälter *m*
s. ground *(El)* Betriebserde *f*
s. hatch Ausgabefenster *n*, Durchgabe *f*, Durchreiche[fenster *n*] *f*, Durchreicheöffnung *f*
s. instruction Bedien[ungs]anweisung *f*, Bedien[ungs]vorschrift *f*
s. instrument Betriebsmeßinstrument *n*
s. kit Werkzeugkasten *m*; Ersatzteilausstattung *f*
s. life Lebensdauer *f*; Nutz[ungs]dauer *f*; Betriebsfähigkeit *f*
s. line Hausanschlußleitung *f*
s. men Wartungspersonal *n*, Pflegepersonal *n*; Bedienungspersonal *n*
s. meter *(Fmt)* Gesprächszähler *m*
s. output Nutzleistung *f (Batterie)*
s. personnel Pflegepersonal *n*, Wartungspersonal *n*; Bedienungspersonal *n*
s. pipe Anschlußrohr *n*
s. platform Bedienungsbühne *f*, Laufbühne *f*
s. radio message Betriebsfunkspruch *m*
s. radio station Betriebsfunkstelle *f*
s. raise *(Brg)* Fahrüberhauen *n*, Fahraufhauen *n*
s. regulation *(Eis)* Fahrdienstvorschrift *f*
s. reservoir Trinkwasserspeicher *m*, Brandwasserspeicher *m*
s. road Anliegerstraße *f*
s. routine Wartungsprogramm *n (Elektronenrechner)*
s. runs Versorgungsleitungen *fpl*
s. speed Dienstgeschwindigkeit *f*, Fahrgeschwindigkeit *f*
s. stress Betriebsspannung *f*
s. tank Tagestank *m*, Verbrauchsbehälter *m*
s. telephone *(Fmt)* Dienstanschluß *m*
s. test Arbeitsprüfung *f*
s. tool-kit *(Flg)* Bordwerkzeug *n*
s. trench Leitungsgraben *m*
s.-value Gebrauchswert *m*
s. vehicle Dienstfahrzeug *n*
s. voltage Gebrauchsspannung *f*, Betriebsspannung *f*
s. water Brauchwasser *n*, Leitungswasser *n*
serviceability 1. Funktionstüchtigkeit *f*; Verwendbarkeit *f*; Tauglichkeit *f*; 2. *(Am, BMSR)* Wartbarkeit *f*, Wartungsmöglichkeit *f*
serviceable tauglich, funktionstüchtig, wartbar
servicing Wartung *f (Motor)*
s. facilities Tankgelegenheit *f*, Tankstelle *f* mit Wartung
s. schedule Wartungsvorschrift *f*
s. tower Startturm *m (für Raketen)*
serving hatch Ausgabefenster *n*, Durchgabe *f*, Durchreiche[fenster *n*] *f*, Durchreicheöffnung *f*

servo-action Servowirkung *f*, verstärkende Wirkung *f*
s.-actuator Stellantrieb *m*, Stellwerk *n*
s.-amplifier Reglerverstärker *m*; Hilfsverstärker *m*, Servoverstärker *m*
s.-assisted mit Servoeinrichtung
s. brake Servobremse *f*
s.-control surface *(Flg)* Hilfsruder *n*
s.-controlled mit Hilfssteuerung, mit Vorsteuerung
s.-drive Hilfsantrieb *m*, Servoantrieb *m*
s.-integrator *(Dat)* Servointegrator *m*, Geschwindigkeitsfolgeregler *m*
s. loop Folgeregelkreis *m*
s.-mechanism Folgeregler *m*, Nachlaufregler *m*
s.-motor Servomotor *m*, Hilfsmotor *m*; Stellmotor *m*, Nachlaufmotor *m*
s.-motor operated controller Steuerschaltung *f* mit Servomotor
s. multiplier *(Dat)* Servomultiplikator *m*, Servomultiplizierer *m*
s.-operated control *s.* power-operated control
s.-operated instrument *(El)* Nachlaufinstrument *n*
s. piston Stellkolben *m*, Arbeitskolben *m*
s. prop *(Brg)* Servostempel *m*
s.-system *(El)* Servoregelsystem *n*
s. unit *(Flg)* Rudermaschine *f*
sesqui- anderthalb
sesquialteral im Verhältnis 3:2 *(stehend)*
sesquinonal im Verhältnis 10:9 *(stehend)*
sesquioctaval im Verhältnis 9:8 *(stehend)*
sesquiquartal im Verhältnis 5:4 *(stehend)*
sesquiquintal im Verhältnis 6:5 *(stehend)*
sesquiseptimal im Verhältnis 8:7 *(stehend)*
sesquisextal im Verhältnis 7:6 *(stehend)*
sesquitertial im Verhältnis 4:3 *(stehend)*
set / to 1. *(Ch)* [ab]binden, aushärten, erhärten, erstarren, fest werden, fixieren, gerinnen, hart werden, [ver]härten, verfestigen, anziehen; 2. [ein]stellen; setzen; aufspannen *(Werkzeug)*; rücken; einrichten *(Maschine)*; 3. hissen *(Flagge)*
s. at an angle schrägstellen
s. at zero auf Null einstellen
s. beacons *(Schiff)* abbaken
s. casing Futterrohre im Bohrloch absetzen *(Tiefbohrtechnik)*
s. colours Farben fixieren
s. cores *(Fmt)* Kerne einlegen
s. for clearness *(Opt)* scharf einstellen
s. free *(Ch)* entwickeln, freisetzen
s. going in Betrieb (Gang) setzen
s. in air an der Luft härten *(Stahl)*
s. out anreißen
s.-over umsetzen, umstellen, verstellen
s. sail abfahren, auslaufen *(Schiff)*
s. the saw teeth schränken *(Säge)*
s. to work in Betrieb setzen
s. to zero auf Null einstellen
s. up 1. aufstellen *(Gleichung)*; errichten; aufspannen *(Werkstück)*; einrichten *(Maschine)*; 2. aushärten *(Stahl)*; erstarren; abbinden *(Beton)*
s. up a call *(Fmt)* eine Verbindung herstellen
s. up a job in a machine eine Maschine einrichten
s. up a number *(Fmt)* wählen
set 1. *(Typ)* Satz *m*; Set *n (Maß beim Monotypesatz)*; 2. Aggregat *n*; Gerät *n*; Batterie *f*; Garnitur *f*; 3. Schränkung *f (Säge)*; Haltung *f*, Lage *f*; Sitz *m*; 4. bleibende Verformung *f*, Dehnungsrest *m*; 5. Erhärtung *f*; Erstarrung *f*; Abbinden *n*; 6. *(Text)* Fadendichte *f*; 7. *(Math)* Menge *f*; 8. Überputz *m*; 9. Strömungsrichtung *f*; Neigungsrichtung *f*
s.-back *(Kern)* Rückstellen *n*, Tiefstellen *n* *(Regulierung der Überschußreaktivität)*; Rückschlag *m*
s.-bolt Stellschraube *f*, Klemmschraube *f*
s. collar Einstellbüchse *f*, Stellring *m*
s. in air Lufthärtung *f (Stahl)*
s. noise Eigenrauschen *n*
s. of blades Schaufelsatz *m (Turbine)*
s. of cards *(Text)* Krempelsatz *m*
s. of change gears Wechselrädersatz *m*
s. of compasses Reißzeug *n*
s. of curves Kurvenschar *f*
s. of dies Backenpaar *n*
s. of drawing (drafting) instruments Reißzeug *n*
s. of fallers *(Text)* Nadelstabfeld *n*, Gillfeld *n*
s. of gauge blocks Endmaßsatz *m*
s. of meters Zählersatz *m*
s. of nozzles Düseneinsatz *m (Turbine)*
s. of packing rings Dichtungsringlage *f*, Dichtungsringsatz *m*
s. of pulleys Flaschenzug *m*
s. of pumps Pumpensatz *m*
s. of the reed *(Text)* Blattdichte *f*
s. of three pairs of drawing rollers *(Text)* Dreiwalzenstreckwerk *n*
s. of unipolar accumulator plates Akkumulatorplattensatz *m*
s. of valves Röhrensatz *m*
s. of warp in the reed *(Text)* Blatteinstellung *f*
s.-off sheet *(Typ)* Schmutzbogen *m*, Einschießbogen *m*
s. pin Stellstift *m (für elektromechanische Büromaschinen)*; Paßstift *m*
s. point Sollwert *m*
s. quoin Setkeil *m (Monotype-Satz)*

s.-screw Klemmschraube *f*, Befestigungsschraube *f*, Stiftschraube *f*
s.-screw for adjusting the bearing Nachstelldruckschraube *f*
s. solid *(Typ)* kompreß, ungesperrt, ohne Durchschuß
s. square Winkel[schiene *f*] *m*, Zeichendreieck *n*
s. squares for chart work *(Schiff)* Kursdreieck *n*
s. theory *(Math)* Mengenlehre *f*
s. union *(Math)* Vereinigungsmenge *f*
s.-up 1. Aufspannung *f (Werkstück)*; Einrichten *n (Maschine)*; 2. Anordnung *f*, Gruppe *f (z. B. in Gruppen zusammengefaßte Pressen, Mehrfachpressen)*; 3. Aushärtung *f*, Erstarrung *f*, Abbinden *n*; 4. *(Kern)* Hochstellen *n (Regulierung der Überschußreaktivität)*; 5. *(Math)* Ansatz *m*
s.-up fixture *(Masch)* Aufspannvorrichtung *f*
s.-up man *(Am)* Einrichter *m*
s.-up of the work Werkstückaufspannung *f*
s.-up procedure Programmierung *f*, Programmvorbereitung *f (Analogrechner)*
s.-up scale Skala *f* ohne Nullpunkt, Skala mit unterdrücktem Nullpunkt
s.-up time Rüstzeit *f*, Vorbereitungszeit *f*, Umrichtezeit *f*
s. value Sollwert *m*
sett 1. Pflasterstein *m (Quader)*; 2. Meißel *m*; 3. *(Text)* Fadendichte *f*
setter Einrichter *m (Maschine)*
setting 1. Abbinden *n*, Abbindung *f*; Aushärten *n*; Erstarren *n*; Gerinnen *n*; Anziehen *n*; 2. Einstellung *f*; Einstellwert *m*; Einrichten *n (Maschine)*; Fixieren *n*; Richten *n*; Justierung *f*; Aufspannen *n*, Aufspannung *f (Werkzeug, Werkstück)*; 3. Ofenblock *m*, Ofengruppe *f*; 4. Fassung *f*, Einfassung *f*; Fassen *n*, Einsetzen *n*
s. accuracy Einstellgenauigkeit *f*
s. angle Aufspannwinkel *m*; Einstellwinkel *m*
s. clamp *(Brg)* Setzklaue *f (Ausbau)*
s. coat Überputz *m*
s. device *(Brg)* Setzvorrichtung *f (Stempel)*; Einstellehre *f*
s. dial Stellscheibe *f*
s. gauge Einstellehre *f*
s. heat Abbindewärme *f*
s.-in oscillation Schwingungseinsatz *m*
s. in width Breiteneinstellung *f*
s. load *(Brg)* Klemmlast *f*, Anfangstraglast *f (Ausbau)*, Setzlast *f (Ausbau)*
s. magnet Richtmagnet *m*
s. mark Einstellmarke *f*
s. of flags Flaggensetzen *n*, Flaggenführung *f*
s. of zero reference Nullpunkteinstellung *f*
s. out *(Masch)* Anreißen *n*
s. out machine *(Led)* Ausreckmaschine *f*
s.-out process *(Masch)* Anreißverfahren *n*
s. over the tailstock centre *(Masch)* Verstellen *n* der Reitstockspitze
s. point Erstarrungspunkt *m*, Stockpunkt *m (Flüssigkeit)*; Erstarrungspunkt *(Metall)*
s. quality Abbindefähigkeit *f*, Abbindeeigenschaft *f*
s. range Einstellbereich *m*
s. ring Stellring *m*
s. screw Stellschraube *f*
s. slicker *(Led)* Ausreckeisen *n*, Schlicker *m*
s. spindle Einstellspindel *f*
s. stick *(Typ)* Winkelhaken *m*
s. the tension of the [saw] blade Spannen *n* des Sägeblattes
s. time 1. Abbindedauer *f*, Abbindezeit *f*; Erstarrungszeit *f*; 2. *(Masch)* Einrichtezeit *f*; 3. Beruhigungszeit *f (des Zeigers)*
s. to work Betätigung *f*
s.-up 1. *(Plast)* Aushärtung *f*; 2. *(Masch)* Einrichten *n*; 3. Errichten *n*, Aufrichten *n*, Montage *f*; 4. Aufspannen *n (Werkstück)*
s. up of a connection *(Fmt)* Herstellung *f* einer Verbindung
s.-up time 1. Einrichtezeit *f*, Einstellzeit *f*; Umrichtezeit *f*; 2. Aufspannzeit *f (Werkzeuge)*; Montagezeit *f*
s.-up work Spannen *n* der Werkstücke
settle / to [an]siedeln; absetzen, klären; sich setzen, sich senken; absenken; lagern *(Bier)*; begleichen
s. an account eine Rechnung bezahlen
s. down to abklingen auf, absinken auf
settle rate Absetzgeschwindigkeit *f (Aufbereitung)*
settled ground *(Brg)* beruhigtes Gebirge *n*
settlement 1. Setzung *f*; Senkung *f*; Absacken *n*; 2. Siedlung *f*
s. of accounts Begleichung *f* der Rechnungen
s. stress *(Bw)* Setzungsspannung *f*
settler Absetzgefäß *n*, Trenngefäß *n*
settling Senkung *f*; Bodensatz *m*, Rückstand *m*, Sedimentation *f*
s. apparatus Klärapparat *m*, Dekantierapparat *m*
s. chamber Staubabscheider *m*
s. cone Absetztrichter *m*
s. glass Absetzglas *n (Baustoffprüfung)*
s. of the grinding sludge Absetzen *n* des Schleifschlammes
s. pond Klärteich *m*
s. rate Absetzgeschwindigkeit *f*
s. speed Absenkungsgeschwindigkeit *f (Gebirgsmechanik)*
s. tank Absetzbecken *n*, Absetzkasten *m*, Absetzbehälter *m*, Klärbecken *n*, Klärgefäß *n*; Ölabscheider *m*, Setzöltank *m*
s. vessel Absetzgefäß *n*

s. water Klärwasser *n*
s. well Scheidegrube *f (Aufbereitung)*
settlings Sinkstoffe *mpl*
seven can slasher *(Text)* Siebenzylinderschlichtmaschine *f*
s. line pica *(Typ)* 84 britische Punkte *mpl (Schriftgröße)*
s.-membered ring *(Ch)* siebengliedriger Ring *m*
s.-wire three-phase system Drehstromsiebenleiternetz *n*
Seven Seas Weltmeere *npl*
sever / to trennen; brechen; reißen
Sèvres Sèvresporzellan *n*
sew / to 1. *(Typ)* heften; 2. nähen
s. on aufnähen
s. up zunähen, einnähen
sewage Abwasser *n*, Abwässer *npl*
s. chlorination Abwasserchlorung *f*
s. clarification plant Abwasserreinigungsanlage *f*, Kläranlage *f*
s. disposal Abwasserbehandlung *f*, Abwasserbeseitigung *f*
s. disposal works Abwasserbeseitigungsanlage *f*
s. engineering Abwasserbehandlung *f*, Abwassertechnik *f*
s. farming Abwasserverrieselung *f*, Rieselfelderwirtschaft *f*
s. gas Faulgas *n*, Klärgas *n*, Schlammfaulgas *n*
s. irrigation Abwasserbewässerung *f*, Bewässerung *f* mit Abwasser
s. pipe Kanalisationsrohr *n*
s. pump Abwasserpumpe *f*
s. purification Abwasserklärung *f*, Abwasserreinigung *f*
s. sedimentation plant Abwasserkläranlage *f*
s. sludge Abwasserschlamm *m*
s. tank Abwasserbehälter *m*
s. treatment Abwasserbehandlung *f*, Abwasserklärung *f*
s. treatment plant Kläranlage *f*
sewer / to kanalisieren
sewer Kanalisationsrohr *n*, Abwasserkanal *m*, Abwasserleitung *f*, Abflußkanal *m* [für Abwässer]; Abzugsrohr *n*
s. bottom Abwasserkanalsohle *f*
s. brick Kanalstein *m*
s. castings *(Met)* Kanalguß *m*
s. cleaning Kanalreinigung *f*
s. flushing Kanalspülung *f*
s. gas Klärgas *n*, Sumpfgas *n*; Kanalgas *n*
s. pipe Abwasserrohr *n*
sewerage Abwasserbeseitigung *f*, Kanalisation *f*, Siedlungsentwässerung *f*, Stadtentwässerung *f*
s. pram Prahm *m* für Abortabwässer, Kloakenprahm *m*
s. system Kanalisation *f*
sewing awl Nähahle *f*
s. cotton Nähzwirn *m*
s. machine Nähmaschine *f*
s. maschine needle Nähmaschinennadel *f*
s. machine oil Nähmaschinenöl *n*, Weißöl *n*
s. silk Nähseide *f*
sewn on tapes *(Typ)* auf Bänder geheftet
s. through gauze *(Typ)* auf Heftgaze geheftet
sewrounds rundum genähte Schuhe *mpl*
sexagesimal division Sexagesimalteilung *f*
sexangle Sechseck *n*
sexangular sechseckig
sexivalence Sechswertigkeit *f*
sexivalent sechswertig
sexless connector *(El)* Zwitterstecker *m*
sexpartite sechsteilig
sextant Sextant *m (Winkelmeßgerät)*
s. telescope Sextantenfernrohr *n*
s. with artificial horizon Kreiselsextant *m*
sextic sechsten Grades
sextuple / to [sich] versechsfachen
sextuple sechsfach
s.-strand chain Sechsfachkette *f*
sgraffito Sgraffito *n*
shabrack Satteldecke *f*
shack *(Am)* Kate *f*, Bude *f*
shackle Schäkel *m*; Kettenglied *n*; Feder *f*, Einspannbund *m*, Einspannlasche *f*, Lasche *f*
s. bolt Schäkelbolzen *m*
s. insulator Schäkelisolator *m*, Abspannisolator *m*, Schnallenisolator *m*
s. joint Laschengelenk *n*
s. pin Federlaschenbolzen *m*
shade / to schraffieren *(Zeichnung)*; abschirmen, abdecken; abtönen, abstufen *(Farbton)*
shade 1. Schatten *m*, Schattierung *f*; Farbton *m*; 2. Leuchtenschirm *m*, Lampenschirm *m*
s.-deck vessel Schattendeckschiff *n*
s. line Schraffurlinie *f*, Schattenlinie *f*, Schraffierlinie *f*
s. of gray *(Fs)* Graustufe *f*
shaded schraffiert; [dunkel] getönt
s. goods *(Text)* geflammte Färbung *f (Fehler)*
s. pole *(El)* Spaltpol *m*
s.-pole instrument Spaltpolinstrument *n*, Ferraris-Instrument *n*
s.-pole motor Spaltpolmotor *m*
s. twill *(Text)* Köperschatten *m*
shading 1. *(El)* Dunkelsteuerung *f*; Rauschpegeleinregelung *f*; 2. Schraffur *f*; Schraffieren *n*; Schattieren *n*
s.-compensation signal Schattenkompensationssignal *n (Fernsehkamera)*
s. dyestuff *(Text)* Abtönungsfarbstoff *m*
s. generator Schattengenerator *m (Fernsehbildaufnahmeröhre)*

s. ring Kurzschlußring *m*
s. signal *(Fs)* Schattensignal *n*, Störsignal *n*
shadow 1. Schatten *m*; 2. *(Fmt)* Empfangsloch *n*; 3. *(Fs)* dunkle Bildstelle *f*
s. angle Schattenwinkel *m*
s. area Schattengebiet *n (Radar)*
s. column instrument Schattenzeigerinstrument *n*
s. cone Schattenkegel *m*
s. effect Abschattungsverlust *m (Nachrichtentechnik)*; Schatteneffekt *m (Dosimetrie)*
s. figure Schattenbild *n*
s. microscope Schattenmikroskop *n*
s. outline projector Profilprojektor *m*
s. photometer Schattenfotometer *n*
s. print *(Text)* Schattendruck *m*
s. region Schattenzone *f*, Schweigezone *f (Radio)*; Schattengebiet *n (Radar)*
s. scattering *(Kern)* Schattenstreuung *f*
s.-transit Vorübergang *m* des Schattens des Trabanten
shadowgraph system Schattenbildsystem *n*
shadowing Schattenbildung *f*; Abschirmung *f*; Bedampfung *f*
shady schattig, Schatten-
s. kind Schattenholzart *f*
shaft 1. *(Masch)* Welle *f*; 2. *(Text)* Webschaft *m*; 3. *(Brg)* Schacht *m*; 4. Griff *m*; Deichsel *f*; 5. Schacht *(Hochofen)*
s. air-lock *(Brg)* Schachtschleuse *f*
s. alignment Wellenausrichtung *f (z. B. von Schiffswellen)*
s. alley *(Schiff)* Wellentunnel *m*
s. angle Achsenwinkel *m (Kegelrad)*
s. bearing Wellenlager *n*
s. boring *(Brg)* Schachtbohrverfahren *n*
s. bossing *(Schiff)* Wellennuß *f*
s. bossings *(Schiff)* Wellenhosen *fpl*
s. bracket *(Schiff)* Wellenbock *m*
s. butt *(Masch)* Wellenstumpf *m*
s. by-pass *(Brg)* Schachtumtrieb *m*, Schachtumfahrungsstrecke *f*
s. cable 1. Schachtkabel *n*; 2. Förderschachtseil *n*
s. collar 1. Wellenbund *m*; 2. Schachtmündung *f*
s. construction Schachtbau *m*
s. contact *(Fmt)* Wellenkontakt *m (Kontaktgeber auf Welle)*
s. core Wellenseele *f*
s. counter Stichdrehzahlmesser *m*, Umdrehungszähler *m*
s. coupling Kurbelwellenkupplung *f*, Wellenkupplung *f*
s. coupling flange Wellenkupplungsflansch *m*
s. cover *(Brg)* Schachtdeckel *m*
s. cross-section *(Brg)* Schachtscheibe *f*, Schachtquerschnitt *m*
s. deflection Wellendurchbiegung *f*
s. digitalizer *s.* s. position-to-digital converter
s. drilling technique Schachtbohrverfahren *n*
s. driving Wellenantrieb *m*
s. end *(Masch)* Wellenende *n*, Wellenstumpf *m*
s. extension Achsverlängerung *f*
s. fire *(Brg)* Schachtbrand *m*
s. flange Wellenflansch *m*
s. frame work *(Brg)* Fördergerüst *n*
s. furnace Schachtofen *m*
s. gate *(Brg)* Schachttor *n*
s. generator Wellengenerator *m*
s. governor Achsenregler *m*
s. hall *(Brg)* Schachthalle *f*
s. hoist Schachtwinde *f*
s. hoisting equipment Schachtfördereinrichtung *f*
s. horsepower Wellenleistung *f*; Wellenpferdestärken *fpl*, WPS
s. house *(Brg)* Schachtkaue *f*
s. inset *(Brg)* Schachtanschlag *m*
s. journal Wellenzapfen *m*
s. keyseat Wellennut *f*
s. keyslot Wellen[keil]nut *f*
s. lime kiln Kalkschachtofen *m*
s. liner *(Schiff)* Propellerwellenüberzug *m*
s. lining 1. *(Brg)* Schachtausbau *m*; 2. *(Met)* Schacht[ofen]ausmauerung *f*
s. loom Ruderschaft *m*
s. malalignment *s.* s. misalignment
s. misalignment Verachsung *f*
s. packing Wellendichtung *f*
s. pillar *(Brg)* Schachtsicherheitspfeiler *m*
s. pivot *(Masch)* Wellenende *n*
s. placing point *(Brg)* Schachtansatzpunkt *m*
s. position-to-digital converter *(Dat)* Lageverschlüßler *m*, Lageumsetzer *m*, Winkelumsetzer *m*
s. raising *(Brg)* Hochbrechen *n* eines Schachts
s. ring Wellenscheibe *f (Axialkugellager)*
s. roasting *(Met)* Schachtofenrösten *n*
s. safety pillar *(Brg)* Schacht[sicherheits]pfeiler *m*
s.-sealing ring Simmerring *m (Dichtungsring)*
s. section Schafttrommel *f*
s. signal recorder Schachtsignalschreiber *m*
s. signalling device Schachtsignalanlage *f*
s. sinking *(Brg)* Schachtabteufen *n*, Niederbringen *n* eines Schachts
s. sinking pump Abteufpumpe *f*
s. speed Wellendrehzahl *f*
s. speed transducer Drehzahlgeber *m*
s. structure *(Brg)* Schachtsäule *f*
s. strut *(Schiff)* Wellenbock *m*
s. stuffing box Wellenstopfbuchse *f*
s. timbering *(Brg)* Schachtzimmerung *f*

s. tower *(Brg)* Schachtgerüst *n*
s. twist Wellenverdrehung *f*
s. with one crank einfach gekröpfte Kurbelwelle *f*
s. with three cranks dreifach gekröpfte Kurbelwelle *f*
s. with two cranks doppelt gekröpfte Kurbelwelle *f*
shafting 1. *(Brg)* Schachtabteufen *n*, Niederbringen *n* eines Schachtes; 2. Wellenleitung *f*; 3. Transmission *f*
shag Gerstenstroh *n*
shake / to schütteln, rütteln; vibrieren
s. out entleeren *(Formkasten)*
shake 1. Kernriß *m*, Riß *m*, Sprung *m* *(Holz)*; 2. *(Am)* handgefertigte Schindel *f*
s. culture *(Ch)* Schüttelkultur *f*
shaker Rütteltisch *m*; Rüttelmaschine *f*; Schüttelmaschine *f*; Schüttelsieb *n*; *(Text)* Kopfwolf *m*
s. chute Schurre *f*, Rüttelschurre *f*
s. conveyor Schüttelrutsche *f*, Schüttelrinne *f*, Förderschwinge *f*
s. conveyor trough Schüttelrinne *f*
s. feeder Schüttelspeiser *m*, Schüttelaufgabeeinrichtung *f*
s. loader Schwingförderrinne *f*
s. screen Schüttelsieb *n*, Schwingsieb *n*
s.-type chip screen Schwingsichter *m* für Hackschnitzel *(Zelluloseherstellung)*
shakes Rissigkeit *f* *(von Holz)*
shakiness Schälrissigkeit *f* des Holzes, Kernschäle *f*, Ringschäle *f*
shaking Rütteln *n*, Schütteln *n*; Erschütterung *f*
s. barrel Gußputztrommel *f*, Scheuerfaß *n*, Rommelfaß *n*
s. chute *(Brg)* Schüttelrinne *f* *(mit Handbetrieb)*
s. grate Schüttelrost[feuerung *f*] *m*; beweglicher Rost *m*
s. grizzly Schüttelrost *m*, Schüttelsieb *n*
s. motion Schüttelbewegung *f*, Rüttelbewegung *f*
s. screen Rüttelsieb *n*, Schüttelsieb *n*
s. sieve Schüttelsieb *n*, Rüttelsieb *n*
s. sliver screen Schüttelsortierer *m*, schüttelnder Astfänger *m* *(Zelluloseherstellung)*
s. table Schüttelherd *m*, Schwingherd *m*; Rütteltisch *m*
shale 1. Kohlenschiefer *m*, Ölschiefer *m*, Schiefer[ton] *m*; 2. *(Brg)* Berge *pl*, Abgänge *mpl*
s. break Schiefertonzwischenlage *f*
s. oil Schieferöl *n*
s. partings dünne Schiefertoneinlagen *fpl* in Sandsteinen
s. tar Schieferteer *m*
shallow Untiefe *f* *(flaches Wasser)*
shallow flach; seicht; muldenförmig
s. draught vessel flachgehendes Schiff *n*
s. floor *(Schiff)* niedrige Bodenwrange *f*
s. forming *(Met)* Napfziehen *n*, Tiefziehen *n*
s. modulation *(Fmt)* Modulation *f* geringer Tiefe
s. pitched roof flach geneigtes Dach *n*
s. water beschränkte Wassertiefe *f*, Flachwasser *n*, flaches (beschränktes) Wasser *n*
s.-water cable Küstenkabel *n*
shaly schieferartig, schiefertonig
sham unecht, vorgetäuscht, Blend-
s. title-page *(Typ)* Schmutztitel *m*
shank 1. Schaft *m*; Werkzeugschaft *m*; Ankerschaft *m*; Griff *m*; 2. Beinlänge *f* *(Strumpf)*; 3. Gelenkstück *n* *(Schuh)*
s. diameter Schaftdurchmesser *m* *(Werkzeug)*
s. end Einsteckende *n* *(Bohren)*
s. ladle *(Met)* Gabelpfanne *f*
s. of connecting rod Schubstangenschaft *m*
s. painter *(Schiff)* Rüstleine *f*, Kippleine *f*
s. taper Schaftkegel *m*, Konus *m* *(Werkzeug)*
s. with a square cross section Vierkantschaft *m*
shape / to [ver]formen, bilden, gestalten, profilieren, vorwalzen; kurzhobeln, stoßen *(Metall)*; fräsen *(Holz)*
s. by the generating method wälzstoßen
s. course for *(Schiff)* Kurs absetzen (festlegen)
shape Gestalt *f*, Form *f*, Fasson *f*; Umriß *m*, Profil *n*; Verfassung *f*
s. above *(Schiff)* Überwasserform *f*
s. cutter Formfräser *m*
s. factor *(El)* Formfaktor *m*
s. mill *s.* s. rolling mill
s. of a curve Kurvenverlauf *m*
s. of blade Schaufelform *f* *(Turbine)*; Flügelform *f* *(der Schiffsschraube)*
s. of outlet Mündungsform *f* *(Turbinendüse)*
s. of submerged part of hull *(Schiff)* Unterwasserform *f*
s. of the cross section Profilquerschnitt *m*
s. of tooth Zahnform *f*
s. of tube Hülsenform *f*
s. retention Formbeständigkeit *f*
s. roll Profilwalze *f*
s. rolling mill Profilstahlwalzwerk *n*
shaped geformt, profiliert, Form-
s. base Topfsockel *m* *(Elektronenröhren)*
s. bloom Vorblock *m* *(Walzwerk)*
s. brick Formstein *m*, Formziegel *m*
s.-casting *(Plast)* Formstück *n*
s. charge *(Brg)* Hohlladung *f*
s. conductor *(El)* Sektorleiter *m*, Profilleiter *m*
s. copper Fassonkupfer *n*, Formkupfer *n*, Profilkupfer *n*
s. die Formgesenk *n*
s. ingot Profilwalzblock *m*

s. profile cutter profilhinterschliffener (spitzgezahnter) Formfräser *m*
shapeless formlos, gestaltlos
shapelessness Formlosigkeit *f*, Gestaltlosigkeit *f*
shapely gut proportioniert
shaper 1. Waagerechtstoßmaschine *f*; 2. Holzfräser *m*
shapes 1. Fliegersichtzeichen *npl*; 2. *(Met)* Walzprofile *npl*
shaping 1. Formen *n*, Gestalten *n*; Formgebung *f*; 2. Kurzhobeln *n (Metallbearbeitung)*; 3. Fräsen *n (Holzbearbeitung)*
s. attachment Kurzhobeleinrichtung *f (Metallbearbeitung)*
s. groove Profilkaliber *n (Walzwerk)*
s. machine Waagerechtstoßmaschine *f*, Kurzhobelmaschine *f*
s. mill Vorwalzgerüst *n*
s. network *(El)* Korrektionsschaltung *f*
s. pass Fertigstich *m*, Formstich *m (Walzwerk)*
share 1. Anteil *m*; Kontingent *n*, Quote *f*; 2. Schar *n (Pflug)*
s. grinder Pflugscharschleifmaschine *f*
s. point Scharspitze *f*
s. tip Scharspitze *f*
shared channel broadcasting Gleichwellenrundfunk *m*, Gemeinschaftswellenbetrieb *m*
sharp 1. scharf[kantig], spitz; 2. griffig *(Schleifscheibe)*; 3. mit großem Terpentinanteil *(Ölfarbe)*; 4. *(Ak)* deutlich; 5. *(Opt)* scharf, deutlich
s. bilge *(Schiff)* scharfe Kimmrundung *f*
s. break scharfer Knick *m (einer Kurve)*
s.-cornered scharfkantig
s.-crested weir scharfkantiges Wehr *n*
s. cut-off scharfer unterer Knick *m* der Kennlinie *(einer Elektronenröhre)*
s.-edged scharfkantig
s. end *(Schiff)* Spitzgatt *n*
s. increase steiler Anstieg *m*
s. inlet edge scharfe Einlaufkante *f*, scharfkantige Einströmstelle *f (Turbinendüse)*
s. kink Knick *m (z. B. in einem Schlauch)*
s. maximum steiles Maximum *n*
s. picture scharfes Bild *n*
s.-pointed mit scharfer Spitze
s. rise of floor *(Schiff)* starke Aufkimmung *f*
s. stern *(Schiff)* Spitzgatt *n*
s. thread Spitzgewinde *n*
s. tone *(El)* scharfer (zu hoher) Ton *m*
s. tuning Feinabstimmung *f*, Scharfabstimmung *f*, scharfe Abstimmung *f*
s. turn *(Flg)* enge Kurve *f*, Steilkurve *f*
s. V-thread Spitzgewinde *n*, scharfgängiges Gewinde *n*
s. wavefront *(El)* steile Wellenfront *f*
sharpen / to schärfen, scharfschleifen, nachschleifen; wetzen; spitzen; abziehen
s. edge tools wetzen
s. scythes dengeln *(Sense)*
s. the grinding wheel die Schleifscheibe griffig machen, die Schleifscheibe aufrauhen
s. the pencil den Bleistift spitzen
sharpening 1. Schärfen *n*, Scharfschleifen *n*, Nachschleifen *n (Werkzeug)*; Abziehen *n*; Wetzen *n*; Spitzen *n (Bleistift)*; 2. Verschärfung *f*, Versteilerung *f (Funktechnik)*
s. allowance Schleifzugabe *f*
s. anvil for scythes Dengelamboß *m*, Dengelstock *m*
s. device Schärfapparat *m*
s. equipment (fixture) Scharfschleifgerät *n*, Scharfschleifeinrichtung *f*
s. machine Schärfmaschine *f*
s. stone Abziehstein *m*
sharpie Knickspantboot *n*
sharply resonant *(El)* resonanzscharf
s. selective *(El)* hochselektiv, scharf trennend
sharpness Schärfe *f*
s. of definition Bildschärfe *f*; Zeichenschärfe *f*
s. of directivity *(Fmt)* Bündelungsschärfe *f*, Richtschärfe *f*
s. of the image Bildschärfe *f*
s. of tuning Abstimmschärfe *f (Radio)*
shatter / to zerschmettern, zersplittern, zertrümmern
shatter Splitter *mpl*, Scherben *mpl*, Trümmer *pl*
s. crack Innenriß *m (Schmiedestück)*
s. cut *(Brg)* Brennereinbruch *m*, Kanonenschießen *n*, Röhreneinbruch *m*
s. test Rüttelprobe *f*, Fallprobe *f* von Koks
shattering force Brisanz[kraft] *f*, zertrümmernde Wirkung *f (Sprengstoff)*
s. power Brisanz[kraft] *f (Sprengstoff)*
shatterproof bruchsicher, splitterfrei
s. glass Sicherheitsglas *n*
shave / to 1. schaben *(Zahnräder)*; *(Led)* falzen
s. on the flesh side *(Led)* scheren *(Häute entfleischen)*
s. out of the limes *(Led)* aus dem Kalk falzen
shave knife Schälmesser *n (Furnier)*
s. milling Fräsen *n* mit geringer Spanabnahme
shaving Hobelspan *m*, Span *m*
s. beam Falzbock *m*
s. board Hobelspanplatte *f*
s. knife Falzeisen *n*, Falzmesser *n*
s. machine 1. *(Led)* Falzmaschine *f*; 2. *(Masch)* Schabmaschine *f (für Zahnräder)*
s. of gears Schaben *n* von Zahnrädern
s. shoulder gears Schulterradschaben *n*
s. slicker *(Led)* Falzeisen *n* für schwere Häute
shavings bin Spänefangkasten *m (an der Pappschere)*
s. separator Spänefänger *m*, Späneabscheider *m*

sheaf / to bündeln; Garben binden
sheaf Garbe *f*; Bündel *n (Strahlen usw.)*
s. binder Garbenbinder *m*
s. binding machine Garbenbinder *m*
sheafing Garbenbinden *n*
shear / to [ab]scheren, [ab]schneiden, beschneiden, besäumen
s. a gate *(Plast)* Anguß abscheren
s. off abscheren
shear 1. Schere *f*; Wange *f*; 2. Schub *m*; Scherung *f*; Gleitung *f*; Schiebung *f (s. a.* shearing*)*
s. action Scherkraft *f*, Schubkraft *f*
s. angle Abscherwinkel *m*, Scherwinkel *m*
s. blade Scher[en]messer *n*, Scher[en]blatt *n*
s. chip Scherspan *m*
s. cut 1. *(Masch)* Scherschnitt *m*, scherender Schnitt *m*; 2. *(Brg)* senkrechter Schram *m*
s. diagram Querkraftdiagramm *n*
s. edge *(Plast)* Abquetschfläche *f*, Abquetschrand *m*
s. failure *s.* s. fracture
s. force Schubkraft *f*
s. fracture Schubbruch *m*, Scherbruch *m*
s. hulk Bockprahm *m*, Bockschiff *n*, Scherenkranhulk *m*
s. joint Überlapp[t]stoß *m (Löten)*
s. leg derrick Dreibeinkran *m*
s. legs Dreibeinkran *m*
s. modulus Schubmodul *m*, Gleitmodul *m*, Torsionsmodul *m*
s. modulus of elasticity Scherelastizitätsmodul *m*
s. pin Scherstift *m*; Abscherbolzen *m*
s. plane Scherebene *f*
s. plate Bruchplatte *f*
s. reinforcement Schubbewehrung *f (Stahlbetonbau)*
s. strength Scherfestigkeit *f*; Schubfestigkeit *f*
s. stress Scherspannung *f*; Schubspannung *f*
s. tool Schermeißel *m*
s. viscosity Scherungsviskosität *f*
s. zone *(Geol)* Zertrümmerungszone *f*
shearer-loader *(Brg)* Schrämwalzenlader *m*
shearing 1. Scheren *n*; 2. Abscheren *n*, Schubumformen *n*; 3. *(Brg)* Schlitzen *n*; Schrämen *n* mit einer Schrämwalze oder einer vertikalen Schrämscheibe; 4. *s. a.* shear
s. action Abscherwirkung *f*, Scherwirkung *f*
s. blade Schermesser *n*
s. bolt Abscherbolzen *m*
s. crack Scherriß *m*
s. cut schälender Schnitt *m*, Schälschnitt *m (Reibahle)*; Scherschnitt *m*
s. effect Scherwirkung *f*
s. force Schubkraft *f*, Scherkraft *f*
s. frame *(Text)* Schermaschine *f*
s. jib *(Brg)* Senkrechtschrämarm *m*
s. load Scherbelastung *f*
s. machine Schermaschine *f*
s. resistance Scherfestigkeit *f*, Schubfestigkeit *f*
s. strain Gleitung *f*, Schiebung *f*
s. strength Scherfestigkeit *f*; Schubfestigkeit *f*
s. stress Scherspannung *f*, Scherbeanspruchung *f*; Schubspannung *f*
s. stress line Schublinie *f*
s. test Scherversuch *m*, Schubversuch *m*
s. yield strength Fließgrenze *f* beim Abscheren
shearings *(Text)* Scherflocke *f*; Scherwolle *f*
shearling geschorenes Lammfell *n*, Scherling *m*
shears Schere *f*; Wangen *fpl (Drehmaschine)*
sheath / to armieren; bekleiden *(Kabel)*
sheath Hülle *f*, Hülse *f*, Ummantelung *f*, Umhüllung *f*, Umkleidung *f*, Mantel *m*, Verkleidung *f*; Schutzhülle *f*, Futteral *n*
s. current Mantelstrom *m (eines Kabels)*
s. eddies Mantelwirbelstrom *m (eines Kabels)*
s. electron Hüllenelektron *n*
s. of electrons Elektronenhülle *f*
sheathe / to verbrettern, verkleiden, verschalen
sheathed ummantelt, umhüllt
s. cable Mantelkabel *n*, bewehrtes Kabel *n*
s. electrode umhüllte Elektrode *f*, Mantelelektrode *f*
s. explosive umhüllter (ummantelter) Sprengstoff *m*
s. wire umhüllte Leitung *f*, Manteldraht *m*, bewehrter Draht *m*
sheathing 1. Umhüllung *f*, Ummantelung *f (Kabel)*; 2. Verschalung *f*; 3. Bodenbeschlag *m*, Metallhaut *f*, Verkleidung *f (Schiffsboden)*; 4. *(Am)* Dachschalung *f*
s. compound Kabelmasse *f*
s. material Hüllenmaterial *n*
sheave Scheibe *f (mit Rille)*, Rillenscheibe *f*; Seilscheibe *f*; Rolle *f (Flaschenzug)*
shed / to abwerfen; fallen lassen, verschütten; abfallen [lassen] *(Belastung)*
shed 1. Schuppen *m*; Schutzdach *n*; 2. Fach *n*, Webfach *n*
s. for wire Nadelfach *n*
s. of fixed length gleichbleibendes Fach *n*
s. of the loom Webfach *n*
s. roof Pultdach *n*
s. stick *(Text)* Einführstab *m*
shedding Fachbildung *f (Weberei)*
s. by treadles Fachtreten *n*
s. motion *(Text)* 1. Fachbildevorrichtung *f*; 2. Fachen *n (Weben)*
s. tappet *(Text)* Daumenrad *n*, Fachexzenter *m*
s. wiper *(Text)* Fachexzenter *m*
sheen Glanz *m*, Schein *m*

sheep-clip Schafschur *f*
s. cropping Schafschur *f*
s. fleece Wollvlies *n*; Schaffell *n*
s. pelt Schaffell *n*
s.-shearing Schafschur *f*
sheer / to vom Kurs abweichen
sheer rein; einfach; unverdünnt; ungemischt, nicht in Verbindung; direkt, absolut; feinmaschig
sheer *(Schiff)* Sprung *m*, Decksprung *m*, Strak *n*
s. draught Längsriß *m*, Aufriß *m* *(Zeichnung)*
s. draught plan Linienriß *m*
s. fabric sehr dünnes (feines) Gewebe *n*
s. line Decksprunglinie *f*
s. mast Zweibeinmast *m*
s. plane Längsriß *m*, Aufriß *m*
s. rail *(Schiff)* Schandeckleiste *f*
s. strake *(Schiff)* Schergang *m*
sheerleg Mastenkran *m*
sheet / to *(Plast)* ausrollen, auswalzen
sheet 1. Bogen *m*; Tafel *f*; Blatt *n* *(Papier)*; Platte *f*; Scheibe *f* *(Glas)*; Fläche *f*; Folie *f*, Grobfolie *f*, Blech *n*, Platte; 2. *(Plast)* Folie; Folienzuschnitt *m*; 3. Feinblech *n*, Blech
s. anchor Reserveanker *m*, Notanker *m*
s. bar Platine *f*, Vorblech *n* *(Walzwerk)*
s. bar reheating furnace Platinenwärmofen *m*
s. bar rolling mill Platinenwalzwerk *n*
s. bending machine Blechbiegemaschine *f*, Blechbiegewalze *f*
s. billet Platine *f* *(Walzwerk)*
s. blank Blechzuschnitt *m*
s. board Bogenpappe *f*
s.-bordering machine Blechbördelmaschine *f*
s. brass Messingblech *n*
s.-by-sheet feeder schuppenförmiger Bogenvorschub *m*
s. cavitation Schichtkavitation *f*
s. coater Bogenstreichmaschine *f* *(Papierherstellung)*
s. copper Kupferblech *n*
s. copper in rolls Rollenkupfer *n*
s. cutting shears Blechschere *f*
s. drill Blechbohrer *m*
s.-fed gravure [printing] press Bogentiefdruckmaschine *f*
s.-fed rotary [printing] press Bogenrotationsdruckmaschine *f*
s. gasket Dichtungsscheibe *f*
s. gauge Blechlehre *f*
s. glass Tafelglas *n*
s. heating furnace *(Met)* Paketwärmofen *m*, Sturzenwärmofen *m*
s.-iron casing Blechhaube *f* *(Schornstein)*
s.-iron cover Eisenverkleidungsblech *n*, Eisenschutzblech *n*
s.-iron foundation Blechfundament *n*
s.-iron gauge Blechdickenmesser *m*
s.-iron lining Blechauskleidung *f*
s. lead Bleiblech *n*
s.-lead in rolls Rollenblei *n*
s. leveller Blechrichtmaschine *f*
s. lightning Flächenblitz *m*
s. material Plattenmaterial *n*
s. metal Feinblech *n*; Blech *n*
s. metal casing Blechgehäuse *n*
s. metal container Blechbehälter *m*
s. metal coolant tray Blechfangschale *f* für Tropföl
s. metal folding machine Blechbiegemaschine *f*
s. metal gauge Blechlehre *f*
s. metal housing *s.* s. metal casing
s. metal lagging Blechverkleidung *f*
s. metal part Blechteil *n*
s. metal screw Blechschraube *f*
s. metal templet Blechschablone *f*
s. metal tube *(Text)* Blechhülse *f*
s. metal working Blechbearbeitung *f*
s. metal working machine Blechbearbeitungsmaschine *f*
s. mica Spaltglimmer *m*
s. mill Blechwalzwerk *n*
s. of foam Schaumstoffolie *f*
s. of veneer Furnierblatt *n*
s. of warp yarns Kettband *n* *(Kettfadenbreite bei Eintritt in den Webstuhl)*
s. pack[et] *(Met)* Blechpaket *n*
s. pile *(Bw)* Spundbohle *f*
s.-pile wall Spundwand *f*
s. piling 1. Spundwand *f*; 2. *(Brg)* Spundwandverfahren *n* *(Schachtabteufen)*
s. plastic Plastfolie *f*
s. polarizer Flächenpolarisator *m*
s. reinforcements for laminates *(Plast)* Furnierbahn *f*, geharzte Bahn *f*, Lackgewebebahn *f*, Lackpapierbahn *f*, Rohbahn *f*
s. rolling Blechwalzen *n*
s. rolling mill Blech[walz]straße *f*
s. rubber Plattengummi *m*
s. scrap Blechschrott *m*
s. signature *(Typ)* Bogensignatur *f*
s. size Blattgröße *f* *(technisches Zeichnen)*
s. stamping press Blechstanzpresse *f*
s. steel Stahlblech *n*
s.-steel disk wheel Stahlblechscheibenrad *n*
s.-steel lagging Blechverkleidung *f*
s.-steel piling Spundwand *f*
s.-steel template Stahlblechschablone *f*
s. straightening machine Blechrichtmaschine *f*
s. zinc Zinkblech *n*
sheeted coating compound *(Led)* zur Folie ausgewalzte Beschichtungsmasse *f*
sheeter *(Am)* *(Typ)* Querschneider *m*
s. lines *(Plast)* Schneidriefen *fpl*

sheeting 1. Bleche *npl*; Blechverkleidung *f*; Bahn[enmaterial *n*] *f*; Folie[nmaterial *n*] *f*; 2. Absteifung *f*; Schalwand *f*; 3. Bettuchleinen *n*; 4. Straßendecke *f*
s. calender Folienkalander *m (Herstellung von Folien)*
Sheffer stroke Shefferscher Strich *m (logische Funktion)*
shelf 1. Gestell *n*, Regal *n*; Schrankbrett *n*, Bücherbrett *n*; Trockenblech *n*; 2. Anschnitt *m (Straßenbau, Eisenbahnbau)*
s. drier Trockenschrank *m*
s. furnace Schüttofen *m*
s.-life Lagerbeständigkeit *f*
s. piece *(Schiff)* Balkenweger *m*
shell / to 1. enthülsen; [sich] abschälen *(Holz)*; 2. beschießen *(Granatfeuer)*; 3. umhüllen
shell 1. Schale *f (Elektronen; Mechanik; Frucht)*; Außenhaut *f*; Hülle *f*, Kapsel *f*; 2. Lagerstützschale *f*; Schild *m*; Mantel *m*; Gehäuse *n*, Scheinwerfergehäuse *n*; 3. *(Led)* Spiegel *m*; 4. Galvano *n*, Shell *(Elektrotypie)*; 5. Geschoß *n*, Projektil *n*; Geschoßmantel *m*; 6. Bohrlöffel *m (Tiefbohrtechnik)*
s. attachment 1. *(Schiff)* Außenhautwinkel *m (Rahmenspant)*; 2. Granate *f*
s. bearing Schalenlager *n (im Pleuel)*
s. bossing *(Schiff)* Wellenhose *f*
s. breaker card *(Text)* Muldenvorkrempel *f*
s. chock *(Schiff)* Aufklotzung *f*, Außenhautfüllstück *n (Spantnische)*
s. circuit *(El)* Topfkreis *m*
s. conversion *(Ph)* Schalenumwandlung *f*
s. core transformer Manteltransformator *m*, Panzertransformator *m*, gepanzerter Transformator *m*
s. crane Munitionsaufzug *m*
s. crater Granattrichter *m*
s. door *(Schiff)* Außenhautpforte *f*
s. end mill Walzenstirnfräser *m*
s. filler *(Plast)* Schalen *fpl* als Kunststofffüller
s. insulator Sattelisolator *m*
s. longitudinal *(Schiff)* Außenhautlängsspant *n*
s. model *(Kern)* Schalenmodell *n*
s. moulding [process] Formmaskenverfahren *n*, Schalengußverfahren *n*
s. of bosh Rastmantel *m (Hochofen)*
s. of revolution *(Bw)* Rotationsschale *f*
s. of the feed heater Vorwärmermantel *m*
s. plating Außen[haut]beplattung *f*
s. reamer Aufsteckreibahle *f*
s. ring *(Wärm)* Kesselschuß *m*, Mantelschuß *m*
s. rock *(Geol)* Muschelgestein *n*
s. roof Schalendach *n*
s. section in the broaching tool Schneideneinsatz *m* im Räumwerkzeug
s.-still Destillierblase *f*
s. structure 1. *(Bw)* Schale[nkonstruktion] *f*; 2. Schalenstruktur *f*, Schalenaufbau *m (Atomkern)*
s. tap Aufsteckgewindebohrer *m*
s. tool Aufsteckwerkzeug *n*
s. transformer *s.* s.-type transformer
s.-type magnet Schalenmagnet *m*
s.-type motor Anbaumotor *m*, Flanschmotor *m*
s.-type reamer Aufsteckgrundreibahle *f*
s.-type transformer Manteltransformator *m*, gepanzerter Transformator *m*, Panzertransformator *m*
s. vacancy *(Kern)* Leerstelle *f* in der Schale
shellac Schellack *m*
s. bond Schellackbindung *f*
s.-bonded schellackgebunden
s. varnish Schellackfirnis *m*
shelling 1. Schälen *n*, Enthülsen *n*; 2. Artilleriebeschuß *m*; 3. Umhüllung *f*
s. feeder Schäleinleger *m (Dreschmaschine)*
s. of legumes Schälen *n* der Hülsenfrüchte
shelly schalig
s. limestone Muschelkalkstein *m*
shelter / to schützen
shelter Schutzraum *m*; Schutz *m*, Dach *n*; Wartehalle *f*
s. deck *(Schiff)* Schutzdeck *n*
s.-deck vessel Schutzdeckschiff *n*, Shelterdeckschiff *n*, Schutzdecker *m*, Shelterdecker *m*
shelve / to 1. ausfachen, mit Fächern versehen; einstellen; 2. sanft abfallen
sherardize / to *(Met)* sherardisieren, mit Zink überziehen
shide *s.* shingle
shield / to abschirmen, schützen
shield 1. Abschirmung *f*, Schild *m*; Schutzschirm *m*; 2. *(Brg)* Abbauschild *m*
s. base *(El)* Abschirmungsträger *m*
s. braid *(El)* [geklöppelter] Abschirmschlauch *m*, Abschirmlitze *f*
s. grid Schirmelektrode *f*
s. layer *(El)* Abschirmlage *f*, Abschirmschicht *f*; Störschutzschicht *f*
s. terminal Abschirmanschluß *m*, Abschirmungsklemme *f*
s. winding Abschirmwicklung *f*
shielded abgeschirmt
s. aerial abgeschirmte Antenne *f*
s. arc abgeschirmter Lichtbogen *m*, Arkatombogen *m*
s. arc welding verdecktes Lichtbogenschweißen *n*
s. cable Panzerkabel *n*, Abschirmkabel *n*
s. camera abgeschirmte Kamera *f*
s.-conductor cable Höchstädter Kabel *n*
s. counter *(Kern)* abgeschirmtes Zählrohr *n*

s. electromagnet *(Am)* Topfmagnet *m*
s. galvanometer Panzergalvanometer *n*
s. grid valve Schirmgitterröhre *f*
s. joint abgeschirmte Muffe *f*
s. pair *(El)* abgeschirmte Doppelleitung *f*
s.-pole instrument *(El)* Ferraris-Instrument *n*
shielding Abschirmung *f*; Strahlenschutz *m*
s. effect Abschirmeffekt *m*
s. enclosure Abschirmung *f*
s. factor Schirmfaktor *m*, Abschirmfaktor *m*
s. harness Abschirmgeflecht *n*
s. material Abschirmmaterial *n*
s. medium Schutzmittel *n* *(Lichtbogenschweißen)*
s. width Abschirmdicke *f*
s. window Schutzfenster *n*
shift / to 1. verschieben, verstellen, schalten, [ein]rücken, ausrücken, umlegen, umstellen; wechseln; umsetzen; 2. schieben, drehen *(die Phase)*
s. a berth den Ankerplatz wechseln
s. the lobe die Keule schwenken *(Antennen)*
s. to another wave *(Fmt)* auf eine andere Welle übergehen
s. tracks Gleise rücken
shift 1. Schicht *f* *(Arbeitsschicht)*; 2. *(Dat)* Kommaverschiebung *f*, Verschiebung *f* *(z. B. Linksverschiebung oder Rechtsverschiebung der Ziffern um eine oder mehr Stellen)*; 3. *(Text)* Äschergang *m*; 4. Verrückung *f*, Verschiebung *f*; Schaltung *f*, Verstellung *f*; Umstellung *f*
s. angle Verschiebungswinkel *m*
s. boss *(Am) (Brg)* Schichtsteiger *m*, Steiger *m*
s. cable Behelfskabel *n*; Rangierkabel *n*, Schaltkabel *n*
s. claw Schaltgabel *f*
s. control Zentrierung *f* *(Katodenstrahlröhren)*
s. dog Schaltklaue *f*
s. foreman *(Am)* Schichtmeister *m*
s. fork Schaltgabel *f*
s. in frequency Frequenzänderung *f*, Frequenzverwerfung *f*, Frequenzwanderung *f* *(Röhrenoszillator usw.)*
s. in phase Phasensprung *f*, Phasenverschiebung *f*
s. key Wechseltaste *f* *(Fernsprechtechnik)*; Umschalttaste *f* *(Schreibmaschine)*
s. lever Riemenrückhebel *m*, Schalthebel *m*, Umlenkhebel *m*
s.-lever dust cover Schaltdom *m*
s. lock Feststeller *m*, Feststelltaste *f*, Umschaltefeststeller *m*
s.-lock keyboard *(Fmt)* Tastatur *f* mit Verriegelung
s. of spectral lines Verschiebung *f* der Spektrallinien, Spektrallinienverschiebung *f*
s. order *(Dat)* Verschiebebefehl *m*
s. plate Schaltblech *n*
s. register Schieberegister *n*, Verschieberegister *n*
s. signal *(Fmt)* Wechselzeichen *n*
s.-worker Schichtarbeiter *m*
shiftability Verrückbarkeit *f*, Verschiebbarkeit *f*, Verstellbarkeit *f*, Umstellbarkeit *f*
shiftable verrückbar, verschiebbar; umstellbar
shifter 1. *(Brg)* Anschläger *m*; 2. Steuerschieber *m*; Schalthebel *m*; Einrückhebel *m*, Ausrückhebel *m*
shifting 1. Schalten *n*, Verschieben *n*, Verstellen *n*; Einrücken *n*, Ausrücken *n*, Umlegen *n*, Umstellen *n*; 2. Schieben *n*, Drehen *n* *(Phase)*
s. ballast *(Schiff)* fliegender Ballast *m*
s. beam *(Schiff)* Schiebebalken *m* *(Luke)*
s. belt Umlegeriemen *m*
s. board stanchion *(Schiff)* Getreideschottstütze *f*
s. boards *(Schiff)* Schlingerschott *n*, losnehmbares Getreideschott *n*
s. camshaft verschiebbare Nockenwelle *f*
s. chock klappbare Bootsklampe *f*
s. coupling ausrückbare (lösbare) Kupplung *f*, Schaltkupplung *f*
s. device Ausrückvorrichtung *f*
s. gears Schieberäder *npl*
s. iron *(Schiff)* Getreideschotthalter *m* *(Laderaum)*
s. lever Umstellhebel *m*
s. of cargo *(Schiff)* Übergehen *n* der Ladung
s. sand wandernde Sandbank *f*, Flugsand *m*, Treibsand *m*
s. spanner Franzose *m*, Stellschraubenschlüssel *m*
shiftless transmission automatisches Getriebe *n*
shim / to unterlegen, unterbauen
shim 1. Beilage *f*, Zwischenlage *f*, Unterlage *f*, Unterlegscheibe *f*, Beilegescheibe *f*; Blechbeilage *f*; Beilageblech *n*; Klemmstück *n*; Shim *m*; Lamelle *f*; 2. Einstellring *m*, Stellring *m*; 3. Flachpflug *m*, Unkrautpflug *m*
s. rod Anpassungsstab *m*; Grobregelstab *m*, Regelstab *m*, Trimmstab *m* *(Reaktorregelung)*
s. safety signal arm Trimmabschaltarm *m* *(Reaktortechnik)*
s. velocity Trimmgeschwindigkeit *f* *(Reaktortechnik)*
shimmer / to schimmern
shimming 1. Unterlegen *n*, Unterbauen *n*; 2. Einlegen *n* dünner Bleche zur Feldkorrektion, Verwendung *f* von Shims, Feldfeinkorrektion *f* *(Reaktortechnik)*
s. plough Flachpflug *m*, Unkrautpflug *m*
shimmy / to flattern, vibrieren
shimmying *(Flg)* Flattern *n*, Schwänzeln *n*
shin Lasche *f*
shine / to scheinen; glänzen

shingle / to mit Schindeln eindecken
shingle 1. Schindel *f*, Dachschindel *f*; 2. Kieselstein *m*
ship / to versenden, aufgeben; einschiffen, verladen
ship Schiff *n*, Wasserfahrzeug *n*
s. account *(Schiff)* Besteckrechnung *f*
s. at anchor vor Anker liegendes Schiff *n*
s.-based antiaircraft artillery Bordflak *f*
s.-breaker Abwracker *m*
s. broker Schiffsmakler *m*
s. canal Seekanal *m*
s. canal lift Schiffshebewerk *n*
s. channel Fahrwasser *n*, Fahrrinne *f*
s. designing department Schiffbaukonstruktionsbüro *n*
s. direction finding station Bordpeilstelle *f*
s. handling tug Verholschlepper *m*
s. keepers Instandhaltungspersonal *n*
s. labourer Kaiarbeiter *m*
s. landing Anlandeplatz *m*
s. lift Schiffshebewerk *n*
s. lofting Aufschnüren *n* des Schiffskörpers
s. model testing tank Schiffsmodellversuchstank *m*
s. not under control manövrierunfähiges Schiff *n*
s. oscillation Schiffsschwingung *f*
s. out of commission außer Dienst gestelltes Schiff *n*
s.-propeller Schiffsschraube *f*
s. propulsion Schiffsantrieb *m*
s.-propulsion reactor Schiffs[antriebs]reaktor *m*
s. propulsion trial *(Schiff)* Meilenfahrt *f*
s. repair Schiffsreparatur *f*
s. resistance Schiffswiderstand *m*
s. slamming Stampfstöße *mpl (Schlagen des Vorschiffes in schwerer See)*
s. steel Schiffbaustahl *m*
s. stores Schiffsvorräte *mpl*, Schiffsbedarf *m*
s.-supplier Schiffsausrüster *m*
s. test tank Schleppversuchstank *m*
s. timber Schiffbauholz *n*
s.-to-shore traffic *(Fmt)* Bord-Land-Verkehr *m*
s.-transmitter Schiffssender *m*
s. trials Probefahrt *f*, Schiffserprobungen *fpl*
s. under construction im Bau befindliches Schiff *n*
s. with flag of truce Kartellschiff *n*
shipboard fighter Deck[s]jäger *m*, bordgestütztes Jagdflugzeug *n*, Bordjagdflugzeug *n*
shipbreaking Abwracken *n*, Abwrackung *f*
shipbuilder Schiffbauer *m*
shipbuilding Schiffbau[wesen *n*] *m*
s. research establishment Schiffbauversuchsanstalt *f*
s. yard Werft *f*, Schiffbauplatz *m*, Schiffswerft *f*
shipkeeper Wachmann *m (auf einem Schiff ohne Besatzung an Bord)*
shiplap Spundbretter *npl*
s. edge Überplattung *f (Holzbau)*
shipload Schiffslast *f*, Tragfähigkeit *f*
shipman Hafenarbeiter *m (an Bord arbeitend)*
shipment Verschiffung *f*, Verladung *f*
s. on deck Decksverladung *f*
shipmotion Schiffsbewegung *f*
shipowner Schiffsreeder *m*, Schiffseigner *m*
shipper 1. Belader *m*, Ablader *m*, Verlader *m*; 2. verformter Ziegel *m*
shipping 1. Versand *m*; 2. Schiffahrt *f*
s. agent Seehafenspediteur *m*
s. authority Schiffahrtsbehörde *f*
s. bill Zollfreischein *m*
s. company Reederei *f*, Schiffahrtsgesellschaft *f*
s. container Transportbehälter *m (für radioaktive Stoffe)*
s. lane Fahrtroute *f*
s. master Heuerbaas *m*, Heuerbeamter *m*
s.-ton *englische Einheit für den Laderaum eines Schiffes = 1,1327 m^3*
s. trade Reedereigewerbe *n*
shiproom Frachtraum *m*, Schiffsraum *m*; Tonnage *f*
ship's boat Beiboot *n*, Schiffsboot *n*
s. clock Schiffsuhr *f*
s. crew Schiffsmannschaft *f*
s. direction finder *(Fmt)* Bordpeiler *m*
s. framing Bespantung *f*, Spantwerk *n*
s. funnel Schiffsschornstein *m*
s. hull Schiffsrumpf *m*
s. husband Schiffsdisponent *m*, Korrespondentreeder *m*
s. management Schiffsführung *f*
s. measurement office Schiffsvermessungsamt *n*
s. position Schiffsort *m*, Schiffsposition *f*, Standort *m*, Besteck *n*
s. propeller Schiffsschraube *f*
s. propulsion Schiffsantrieb *m*
s. reckoning Besteckrechnung *f*
s. shape Schiffsform *f*
s. side Bordwand *f*
s. skeleton Schiffsgerippe *n*
s. staff Schiffsstab *m (Personal)*
s. telegraph Schiffstelegraf *m*, Maschinentelegraf *m*
s. time Schiffszeit *f*
s. time [piece] Schiffsuhr *f*
s. wall Bordwand *f*
shipway Schiffahrtsstraße *f*, Schiffahrtsweg *m*
shipways Helling *f*
shipwreck Schiffbruch *m*
shipwrecked schiffbrüchig
shipwright Schiffbauer *m*, Schiffszimmermann *m*
shipwright's yard Schiffswerft *f (Holzschiffbau)*

shipyard Schiffswerft *f*, Werft *f*
s. facilities Werfteinrichtungen *fpl*
shiring Schußstreifen *m (Gewebefehler)*
shirt 1. Hemd *n*; 2. *(Am, sl)* Ausmauerung *f* eines Hochofens
shirting [fabric] Hemdenstoff *m*
shive *(Text)* Schäbe *f*
shiver / to zersplittern, aufspalten
shiver Splitter *m*; Bruchstein *m*; Scherbe *f*
shoal / to seichter werden
shoal 1. Untiefe *f*, flache Stelle *f*; 2. Fischschwarm *m*
shoal seicht, flach
s. buoy Untiefentonne *f*
s.-draft boat flachgehendes Boot *n*
s. water Flachwasser *n*
shock / to 1. [an]stoßen; 2. Puppen aufstellen *(Getreide)*
s. into oscillation *(El)* zum Schwingen anstoßen
shock 1. Erschütterung *f*, Stoß *m*, Prall *m*; *(El)* Schlag *m*; 2. Getreidepuppe *f*
s. absorber Stoßdämpfer *m*
s. absorber bracket Stoßdämpferbock *m*
s. absorber mounting Stoßdämpferhalter *m*
s.-absorbing stoßdämpfend
s.-absorbing mounting elastische Aufhängung *f*
s. absorption Stoßdämpfung *f*
s. and vibration pick-up Stoß- und Schwingungsaufnehmer *m*
s. at entry Kantenstoß *m* des Dampfes *(Turbine)*
s. bending test *(Mech)* Schlagbiegeversuch *m*
s. component Stoßkomponente *f*
s. damper Stoßdämpfer *m*
s. damping stoßdämpfend
s. deflection test *(Mech)* Schlagbiegeversuch *m*
s. excitation *(Fs)* Stoßerregung *f (Sägezahngenerator)*
s.-excitation-type transmitter stoßerregter Sender *m*
s.-excited oscillator Blockschwinger *m*, stoßerregter Generator *m*
s. factor Stoßzahl *f (Festigkeit)*
s.-free stoßfrei
s. frequency Stoßzahl *f*
s. hazard *(El)* Berührungsgefahr *f*
s.-hazard voltage Berührungsspannung *f*
s. load Stoßbelastung *f (Festigkeit)*
s. mounting stoßsichere Befestigung (Montage) *f*
s. of steam striking edge Kantenstoß *m* des Dampfes *(Turbine)*
s. potential Stoßspannung *f*
s.-proof berührungsschutzsicher, mit Berührungsschutz; stoßfest
s.-proof lamp socket Berührungsschutzfassung *f*
s.-proof switch *(El)* berührungsschutzsicherer (geerdeter, stoßfester) Schalter *m*, gegen elektrischen Schlag geschützter Schalter, Gußschalter *m* mit Gehäuseerdung
s. propagation Stoßausbreitung *f*
s.-protection capacitor Berührungsschutzkondensator *m*
s.-resistant stoßfest, schlagfest
s. stress *(Met)* Schlagbeanspruchung *f*
s. surface Stoßfläche *f (Turbine)*
s.-wave Schockwelle *f*, Schockfront *f*; Druckwelle *f*; *(Ak)* Stoßwelle *f*; Verdichtungswelle *f*
s. wave drilling Bohren *n* mittels Stoßwellen
shocker Elektrisierapparat *m*
shockless stoßfrei
shoddy Lumpenwolle *f*, Reißwolle *f*, Mischwolle *f*
shoe / to 1. beschlagen *(Pferd)*; 2. mit Schuh versehen *(Mast usw.)*
shoe 1. Schuh *m*; 2. Hufeisen *n*; 3. Mastschuh *m*; 4. Tragklotz *m*, Klotz *m*, Segment *n (Segmentlager)*; 5. Gleitschuh *m*, Gleitplatte *f*, Gleitstück *n*; 6. Backen *m (Bremse)*; Klotz; 7. Schleifstück *n (Schienenstromabnehmerkontaktschuh)*
s. area Bremsfläche *f*
s. black Schuhcreme *f*
s. board Schuhpappe *f*
s. braid Flechtschnur *f*, Schlinge *f*
s. brake Backenbremse *f*; Klotzbremse *f*
s. gear Schienenstromabnehmerkontaktschuh *m*
s.-lace Schnürsenkel *m*
s. of a keel *(Schiff)* Loskiel *m*, loser Kiel *m*, Kielschuh *m*
s. plough Stelzpflug *m*
s. thread Schuhdraht *m*
shoed pole *(El)* angeschuhte Stange *f*, angeschuhter Mast *m*
shoelace Schnürsenkel *m*
shoemaker Schuhmacher *m*
shoestring *s.* shoelace
shook Kistenbretter *npl*; Faßstäbe *mpl*
shoot / to 1. [ab]schießen, sprengen; 2. aufnehmen *(Foto)*; 3. anpeilen; 4. Kanten genau hobeln *(Holz)*; 5. changieren
shoot 1. Reis *n*, Schoß *m*, Schößling *m*, Trieb *m (Pflanzen)*; 2. Auslaufrinne *f (Walzwerk)*; 3. Schurre *f*, Rutsche *f*; Schräge *f*
shooting Schießen *n*, Sprengen *n*; Schütten *n*
s. and blasting *(Brg)* Schießarbeit *f*
s. board Stoßlade *f (Hobeln)*
s. plane Abkanthobel *m*
s. rifle Jagdgewehr *n*
shop 1. Laden *m*, Verkaufsstelle *f*; 2. Werkstatt *f*; Werkhalle *f*; Produktionsabteilung *f*, Fertigungsabteilung *f*; Betrieb *m*

s. accident Betriebsunfall *m*
s.-built selbstgebaut, selbstgefertigt *(im Betrieb)*
s. drawing Werkstattzeichnung *f*
s. gauge Arbeitslehre *f*
s. practice Werkstattpraxis *f*, Werkstattbetrieb *m*, Betriebspraxis *f*; Produktion *f*
s. rivet Werkstattniet *m*
s. shaft Transmissionswelle *f*
s. term Werkstattausdruck *m*
s. traveller Laufkran *m*
s. weld Werkstattschweißung *f*
s. worker Produktionsarbeiter *m*; Fabrikarbeiter *m*
shopping knife Hackmesser *n*
shopwindow Schaufenster *n*
SHORAN *s.* short range navigation system
shore / to [ab]stützen, absteifen
s. up abbäumen *(Schiff seitlich abstützen)*
shore 1. Küste *f*, Ufer *n*; 2. Steife *f*, Strebe *f*, Stütze *f*
s.-anchored suspension bridge erdverankerte Hängebrücke *f*
s.-based radar Küstenradar *n*
s. delivery pipe Landrohr *n (Spülbagger)*
s.-end cable Küstenkabel *n*
Shore hardness *(Mech)* Shore-Härte *f*
shoring Abstützung *f (z. B. eines beladenen Decks)*; Absteifung *f*; Verschalung *f*
shorn wool Schurwolle *f*
short / to *(sl)* kurzschließen
s.-circuit kurzschließen
s. out a resistor einen Widerstand kurzschließen
s.-ship nicht zur Verschiffung kommen lassen
s. to chassis gegen Masse kurzschließen
s. to earth erden, an Erde legen, gegen Erde kurzschließen
s. to frame gegen Masse kurzschließen
s. to ground erden, an Erde legen, gegen Erde kurzschließen
short *(sl)* Kurzschluß *m*
short kurz; niedrig *(z. B. Schornstein)*; knapp; stramm, streng *(Druckfarbe)*; in engem Bereich; spröde *(Werkstoff)*, brüchig, krümelnd; plötzlich, abrupt; nicht ausgepreßt, nicht voll gespritzt *(Formteil)*; kleinstückig
s. annealing *(Met)* Kurztemperung *f*
s. barrel staves kurzes Faßholz *n*
s.-brittle *(Met)* faulbrüchig
s. chip Kurzspan *m*
s.-chip kurzspanend
s. circuit Kurzschluß *m*
s.-circuit armature Kurzschlußanker *m*
s. circuit between plates Plattenkurzschluß *m (Batterie)*
s.-circuit characteristic Kurzschlußkennlinie *f*
s.-circuit component Kurzschlußglied *n*, Kurzschlußkomponente *f*
s.-circuit contact Kurzschlußkontakt *m*
s.-circuit current Kurzschlußstrom *m*
s.-circuit current to earth Erdschlußstrom *m*
s.-circuit frequency Kurzschlußfrequenz *f*
s.-circuit impedance Kurzschlußwiderstand *m*, Kurzschlußimpedanz *f*; Kurzschlußscheinwiderstand *m*
s.-circuit key Kurzschlußtaste *f*
s.-circuit plug Kurzschlußstöpsel *m*
s.-circuit proof kurzschlußsicher
s.-circuit protection Kurzschlußschutz *m (Relais)*
s.-circuit ratio Leerlauf-Kurzschluß-Verhältnis *n*
s.-circuit rotor Kurzschlußläufer *m*, Käfigläufer *m*
s.-circuit strength Kurzschlußfestigkeit *f*
s.-circuit transition *s.* shunt transition
s.-circuit voltage Kurzschlußspannung *f*
s.-circuit winding Kurzschlußwicklung *f*, kurzgeschlossene Wicklung *f*
s.-circuited kurzgeschlossen, Kurzschluß-
s.-circuited armature Kurzschlußanker *m*
s.-circuited line Kurzschlußleitung *f*
s.-circuited loop Kurzschlußschleife *f*
s.-circuited rotor Kurzschlußläufer *m*
s.-circuited turns test Prüfung *f* auf Windungsschluß
s.-circuited winding Kurzschlußwicklung *f*, kurzgeschlossene Wicklung *f*
s.-circuiting Kurzschließen *n*
s.-circuiting plug Kurzschlußstecker *m*
s.-circuiting winding Kurzschlußwicklung *f*
s.-code signal Kurzsignal *n (Funktechnik)*
s. cut Kurzverfahren *n*, verkürztes Verfahren *n (Spinnerei)*
s.-delay electric detonator Schnellzeitzünder *m*, Kurzzeitzünder *m (für Sprengungen)*
s. delivery *(Schiff)* Fehlladung *f*
s.-distance control Nahsteuerung *f*
s.-distance irradiation Nahbestrahlung *f*
s.-distance navigation Nahschiffahrt *f*; Kurzstreckennavigation *f*
s.-distance receiving Nahempfang *m (Funktechnik)*
s.-distance scatter *(Fmt)* Nahstreuung *f*
s.-distance trunk equipment *(Fmt)* Nahverkehrseinrichtung *f*
s.-duration pulse Kurz[zeit]impuls *m*
s.-fibred pulp Kurzfaserzellstoff *m (Papierherstellung)*
s.-flame explosive *(Brg)* Wettersprengstoff *m*
s.-focus kurzbrennweitig
s.-focus lens kurzbrennweitige Linse *f*, Kurzfokuslinse *f*

s. hair knife *(Led)* Putzmesser *n* zum Entfernen der Grundhaare
s.-haul Kurzstrecken-, Nahverkehrs- *(Transport)*
s.-haul service Luftverkehr *m* auf kurzen Strecken
s. iron brüchiges, sprödes Eisen *n*
s. irradiation Kurzzeitbestrahlung *f*
s.-lead mit kleiner Steigung
s.-length bore Bohrung *f* geringer Tiefe
s.-length twist drill kurzer Spiralbohrer *m*
s. life Kurzlebigkeit *f*
s.-line fault Abstandskurzschluß *m*
s.-lived kurzlebig
s.-lived activity kurzlebige Aktivität *f*
s. log Reststück *n (Baumabfall)*
s. measure Mindermaß *n*
s. oil Lack *m* mit geringerem Ölgehalt
s.-on *(Brg)* Abbau *m* zwischen 45° und 90° zu den Schlechten gestelltem Stoß
s. period Vorperiode *f (periodisches System)*
s. persistence kurze Nachleuchtdauer *f*
s.-pitch winding *(Elm)* Wicklung *f* mit verkürztem Schritt
s. range von geringer Reichweite; nahwirkend
s.-range fading Nahschwund *m (Radio)*
s.-range field of the aerial Antennennahfeld *n*
s.-range force Nahwirkungskraft *f*
s. range navigation Kurzstreckennavigation *f*
s.-range particle *(Kern)* Teilchen *n* geringer Reichweite
s.-range rocket Kurzstreckenrakete *f*
s. run kleine Losgröße *f*; Kleinreihe *f*
s. run production Kleinreihenfertigung *f*
s.-sea coaster Küstenfrachter *m*
s.-sea trade große Küstenfahrt *f*
s. shot *(Plast)* ungenügende Füllung *f*, unvollständiges Spritzteil *n*
s.-sightedness Kurzsichtigkeit *f*
s.-staple cotton kurzstapelige Baumwolle *f*
s.-stop bath Unterbrechungsbad *n*
s.-stopping of chain *(Ch)* Kettenabbruch *m*
s. stowage Kurzholzstauung *f*
s. straw Streu *f*
s.-stroke kurzhubig
s.-stroke steam engine kurzhubige Dampfmaschine *f*
s. taper Steilkegel *m*
s.-term kurzfristig, kurzzeitig
s.-term current Kurzzeitstrom *m*
s.-term memory Kurzzeitspeicher *m*
s.-term operation kurzzeitiger Betrieb *m*
s. thread Kurzgewinde *n*
s. thread milling Kurzgewindefräsen *n*
s. thread milling machine Kurzgewindefräsmaschine *f*
s. throat press Presse *f* mit kurzer Ausladung
s.-time kurzzeitig, Kurzzeit-
s.-time behaviour Kurzzeitverhalten *n*
s.-time current Kurzzeitstrom *m*
s.-time duty kurzzeitiger Betrieb *m (mit gleichbleibender Belastung)*, Kurzzeitbetrieb *m*
s.-time stability Frequenzkonstanz *f* über kurze Zeit, Kurzzeitkonstanz *f*
s.-time test Kurzzeitversuch *m (Materialprüfung)*
s. ton *Masseeinheit in den USA; 1 sh.tn. = 2000 lb. = 907,185 kg*
s. trade *(Schiff)* kleine Fahrt *f*
s. wave *(El)* Kurzwelle *f*
s.-wave adapter Kurzwellenvorsatz *m*
s.-wave aerial Kurzwellenantenne *f*
s.-wave coil Kurzwellenspule *f*
s.-wave communication Kurzwellenverbindung *f*
s.-wave converter Kurzwellenvorsatz *m*
s.-wave interference Kurzwellenstörung *f*
s.-wave radiation *(Ph)* kurzwellige Strahlung *f*
s.-wave receiver Kurzwellenempfänger *m*
s.-wave reception Kurzwellenempfang *m*
s.-wave telegraphy Kurzwellentelegrafie *f*
s.-wave transmission Kurzwellenübertragung *f*
s.-wave transmitter Kurzwellensender *m*
shortage Knappheit *f*, Mangel *m*
s. of skilled labour (men, workers) Facharbeitermangel *m*
shorted *(sl)* kurzgeschlossen, Kurzschluß-
s. turn Kurzschlußwindung *f*, kurzgeschlossene Windung *f*
s. wires kurzgeschlossene Drähte *mpl*
shorten / to [ver]kürzen; elektrisch verkürzen
shortening 1. Kürzung *f*, Kürzen *n*; Verkürzung *f*; 2. *(Am)* Pflanzenfett *n*
s. capacitor Verkürzungskondensator *m*
s. capacity Verkürzungskapazität *f*
shorthand Stenografie *f*, Kurzschrift *f*
shorthanded *(Schiff)* unterbemannt, unterbesetzt, ungenügend bemannt
shorting *(sl)* Kurzschließen *n*
s. bar Kurzschlußbügel *m*
s. bridge Kurzschlußbrücke *f*
s. link Kurzschlußlasche *f*
s. plunger Kurzschlußschieber *m (Mikrowellentechnik)*
shortness 1. Kürze *f*; Knappheit *f*; 2. Brüchigkeit *f*, Sprödigkeit *f*; 3. *(Plast)* geringe Einreißfestigkeit *f*
shortwall coal cutter Schrämmaschine *f* für kurzen Abbaustoß *(bis 10 m)*
s. working *(Brg)* Auffahren *n* im breiten Blick, Breitauffahren *n*
shot 1. Schrot *m (n)*; Geschoß *n*; Abschuß *m*;

Schütze *m*; 2. *(Plast)* Füllgewicht *n*, Pressung *f*, Schußgewicht *n*, Spritzung *f*; 3. Stoßkugel *f*, Kugel *f (Sportgerät)*; 4. Schuß *m (Teppichweben)*

shot changierend, schillernd
s. aluminium Aluminiumgranalien *fpl*
s. bag moulding *(Plast)* Schrotpressen *n*
s. bit Schrotkrone *f (Tiefbohrtechnik)*
s. blasting 1. Strahlreinigen *n*; 2. Kugelstrahlen *n*
s.-blasting medium Strahlmittel *n*
s. boring Bohren *n* mit Schrotkrone, Schrotbohren *n*, Schrotbohrverfahren *n (Tiefbohrtechnik)*
s. capacity Spritzvolumen *n*, Schußvolumen *n (Umformtechnik)*
s. drilling *s.* s. boring
s. effect 1. *(El)* Schroteffekt *m*; 2. *(Text)* Changeanteffekt *m*
s.-fire cable *(Brg)* Schießkabel *n*
s.-firer *(Brg)* Schießmeister *m*, Schießhauer *m*; Sprengmeister *m*
s.-firing *(Brg)* Schießarbeit *f*, Schießen *n*
s.-firing cable Schießkabel *n*
s.-gun Jagdgewehr *n*
s.-hole Sprengbohrloch *n*, Bohrloch *n*
s. moulding *(Plast)* Schrotpressen *n*
s.-noise reduction factor Schwächungsfaktor *m (Röhrenrauschen)*
s. peening Kugelstrahlen *n (Oberflächenhärteverfahren)*
s. pellet Schrotkugel *f*
s.-proof kugelsicher, schußfest
s. silk schillernde Seide *f*
s. welding Schußschweißen *n*, kurzzeitiges Punktschweißen *n*

shotting Granulieren *n*

shoulder / to schultern; absetzen *(Wellen)*

shoulder 1. Absatz *m*, Ansatz *m*, Bund *m*, Kröpfung *f*, Schulter *f*, Vorsprung *m (Spannungsschulter)*; angeschliffener Grat *m*; 2. Zentrierwulst *m (Granate)*; 3. Sicherheitsstreifen *m (beiderseits der Start- und Landebahn)*
s. cutting Ansatzfräsen *n*, Stufenfräsen *n*
s. milling *s.* s. cutting
s. of piston rod Kolbenstangenbund *m*
s. of rolls Walzenschulter *f*
s. of the cut Schnittfläche *f (Spanen)*
s. of tyre *(Eis)* Sprengring *m* der Bandage
s. on crank shaft Kurbelwellenbund *m*
s. screw Ansatzschraube *f*
s. tool Seitenmeißel *m*

shouldered abgesetzt
s. work grinding Ansatzschleifen *n*

shovel / to schaufeln, schippen
s. out herausschaufeln

shovel 1. Schaufel *f*; 2. Löffelbagger *m (insbesondere Hochlöffelbagger)*
s. blade Schaufelblatt *n*
s. bucket Löffel *m (Löffelbagger)*
s. cable Baggerseil *n*
s. dipper Löffel *m* eines Hochlöffelbaggers
s. engine Baggermotor *m*
s. excavator Löffelbagger *m*
s. handle Schaufelstiel *m*
s. loader Überkopflader *m*, Schaufellader *m*, Wurfschaufellader *m*
s. plough Schaufelpflug *m*

shovelling Schaufeln *n*, Schaufelarbeit *f (Handarbeit)*; Baggern *n*, Laden *n (Löffelbagger)*

show / to 1. zeigen; darstellen; beweisen; sichtbar machen; ausstellen; 2. sichtbar sein (werden)

show-case Schaukasten *m*, Vitrine *f*

shower / to 1. stark regnen, gießen; 2. abbrausen; bespritzen
s.-proof regendicht (wasserabstoßend) imprägnieren *(textile Stoffe)*

shower 1. Regenschauer *m*, Schauer *m*; 2. *(Kern)* Teilchenschauer *m*, Schauer; 3. Schwarm *m (Meteoriten)*; 4. Dusche *f*, Brause *f*
s. bath Brause *f*, Dusche *f*; Brauseraum *m*, Duschraum *m*
s. cooler Rieselkühler *m*
s. head Brausekopf *m*
s. of cosmic radiation Schauer *m* der kosmischen Strahlung
s.-proof regendicht
s.-proof impregnation *(Text)* Regendichtimprägnierung *f*
s. stall Brausenische *f*, Duschnische *f*; Brausekabine *f*, Duschkabine *f*

showroom Ausstellungsraum *m*, Ausstellungshalle *f*

shrapnel 1. Schrapnell *n*; 2. Schrapnelladung *f*; 3. Granatsplitter *m*
s. bomb Splitterbombe *f*
s. shell Schrapnell *n*

shred / to zerfasern, zerkleinern, zerschnitzeln; schroten; zerreißen

shredder Schnitzelmühle *f*, Zerfaserer *m*; Reißwolf *m*; Lumpenreißer *m*; Schrotmühle *f*; Zerreißer *m*, Reißer *m*, Futterzerreißer *m*, Futterschneider *m*

shredding drum Zerreißertrommel *f*
s. machine Zerreißer *m*

shrink / to schrumpfen; schwinden; krumpfen; einlaufen, eingehen; stauchen; sich verwerfen *(Holz)*
s. on [hot] warm aufziehen (einziehen), aufschrumpfen *(Reifen)*

shrink band size Schrumpfmaß[stab *m*] *n (Modelltischlerei)*
s. fit Schrumpfverbindung *f*; Schrumpfpreßpassung *f*; Aufschrumpfsitz *m*

s. head Saugmassel *f*, verlorener Kopf *m* *(Gießtechnik)*
s. hole *(Met)* Lunker *m*, Schwindungshohlraum *m*
s. joint Schrumpfverbindung *f*
s. point *(Met)* Abhebepunkt *m* *(Stranggießen)*
s. press fit *s.* s. fit
s. resistant *(Text)* krumpfbeständig, schrumpfecht
s. ring Schrumpfring *m*
s. rule Schwindmaß[stab *m*] *n*
shrinkability Schwindfähigkeit *f*, Schwindvermögen *n*
shrinkable schwindfähig, schrumpfbar
shrinkage Schrumpfung *f*; Schwinden *n*, Schwund *m*; Schwindmaß *n*
s. allowance *(Met)* Schwind[maß]zugabe *f*
s. cavity *(Met)* Lunker *m*, Schwindungshohlraum *m*
s. control *(Text)* Einlaufkontrolle *f*
s. crack *(Met)* Schrumpfriß *m*, Schwind[ungs]riß *m*
s. fit *s.* shrink fit
s. from mould dimensions Schwindmaß *n*
s. in width *(Text)* Breiteneingang *m*, Einlaufen *n*
s. method *s.* s. stoping
s. piping Schrumpflunker *m*, Schwindungslunker *m* *(im Gußblock)*
s. reinforcement Schwindbewehrung *f* *(Stahlbetonbau)*
s. rule *s.* shrink rule
s. stope and pillar-caving *(Brg)* kombinierter Magazin- und Blockbruchbau *m*
s. stoping *(Brg)* Magazinbau *m*, Speicherbau *m*
s. stress Schwindspannung *f*, Schrumpfspannung *f*
shrinking 1. Schrumpfen *n*, Schwinden *n*; 2. Lunkerbildung *f*, Lunkern *n*; 3. *(Text)* Krumpfen *n*; Schrumpfen *n*; Einlaufen *n*
s. on Aufschrumpfen *n*
s. process *(Text)* Krumpfen *n*
s. resistant schrumpfungsfrei, schrumpffest; krumpffrei
s. stress *(Met)* Schrumpfspannung *f*
s. test *(Met)* Schrumpfungsversuch *m*
shroud / to einhüllen
shroud Verstärkungsflansch *m*; kreisförmiger Steg *m*
s. line Fangleine *f* *(Fallschirm)*
s. plate Deckscheibe *f*
s. ring Abschlußring *m*, Bandage *f* *(Turbine)*
shrouded propeller *(Schiff)* Düsenpropeller *m*
shrouds *(Schiff)* Wanten *pl*
shrub Staude *f*
s. planter Staudensetzer *m*, Staudensetzmaschine *f*
shrubbery Gesträuch *n*
shrubs Gesträuch *n*
shrunk gekrumpft; geschrumpft; eingelaufen
s. dowel Schrumpfklammer *f*
s. on aufgeschrumpft [auf]
s. ring Schrumpfring *m*, Schrumpfband *n* *(Turbine)*
shunt / to 1. rangieren, auf Seitengleis stellen; 2. *(El)* in Nebenschluß schalten; parallelschließen
s. off *(El)* abzweigen
shunt 1. *(El)* Neben[schluß]widerstand *m*; Shunt *m*, Nebenschluß *m*; 2. *(Eis)* Nebengleis *n*, Abzweigung *f*
s. admittance komplexer Querleitwert *m*
s. admittance equalization Querentzerrung *f*
s. arc lamp Nebenschlußbogenlampe *f*
s. arm Querglied *n* *(Vierpol)*
s.-arm electric wave filter Abzweigfilter *n*
s. brake Nebenschlußbremse *f*
s. capacitance Parallelkapazität *f*
s. capacitor Parallelkondensator *m*, Nebenschlußkondensator *m*, Überbrückungskondensator *m*
s. characteristics Nebenschlußverhalten *n*
s. circuit Parallelkreis *m*, parallelgeschalteter Stromkreis *m*; Nebenschluß[strom]kreis *m*; Feldstromkreis *m*
s. coil Nebenschlußfeld *n*, Nebenschlußwicklung *f*, Nebenschlußspule *f*
s. component Querkomponente *f*, Querglied *n*
s. conductance Querleitwert *m*
s.-conduction motor Motor *m* mit Nebenschlußverhalten
s. connection Parallelschaltung *f*, Nebeneinanderschaltung *f*
s. current Nebenschlußstrom *m*
s. damping resistor Nebenschlußdämpfungswiderstand *m* *(Schwingkreis)*
s. dynamo Nebenschlußdynamo *m*
s. element Nebenschlußglied *n*, Querglied *n* *(Filter usw.)*
s. excitation Nebenschlußerregung *f*
s.-excited dynamo Nebenschlußdynamo *m*
s.-fed parallelgespeist
s.-fed aerial parallelgespeiste Antenne *f*
s.-fed vertical aerial Vertikalantenne *f* mit Anzapfspeisung
s. feed Parallelspeisung *f*
s.-field relay Relais *n* mit magnetischem Nebenschluß
s.-field rheostat Nebenschlußfelderreger *m*
s. impedance Querimpedanz *f*; Querableitung *f*
s. inductance Querinduktivität *f*
s. induction motor *(läufergespeister)* Nebenschlußkommutatormotor *m*, Nebenschlußkollektormotor *m*

s. **inductor** Paralleldrossel *f*, parallelgeschaltete Induktivität *f*
s. **lead** Nebenschlußzuleitung *f*
s. **line** Abzweigleitung *f*, Rückleitung *f (Hydraulik)*
s. **loading** Querbelastung *f*
s. **loss** Ableitungsdämpfung *f*
s. **magnet** Nebenschlußmagnet *m*
s. **motor** Nebenschlußmotor *m*
s. **neutralization** induktive Neutralisation *f*
s. **rectifier** Nullanode *f (Stromrichter)*
s. **regulation** Nebenschlußregelung *f*, Feldregelung *f*
s. **regulator** Nebenschlußregelwerk *n*, Nebenschlußregler *m*, Nebenschlußregulator *m*
s. **release** Sekundärauslösung *f*
s. **resistance** Nebenschlußwiderstand *m*
s. **resistor** parallelgeschalteter Widerstand *m*, Nebenschlußwiderstand *m*
s. **T** Parallelverzweiger *m*, Admittanz-T-Glied *n (Wellenleitertechnik)*
s. **terminal** Nebenschlußklemme *f*
s. **transformer** Nebenschlußtransformator *m*
s. **transition** Nebenschlußübergangsschaltung *f*
s. **tripping** Spannungsauslösung *f (Relais)*
s. **winding** Nebenschlußwicklung *f*, Parallelwicklung *f*
s.**-wound motor** Nebenschlußmotor *m*
shunted buzzer Nebenschlußsummer *m*
s. **capacitor** Kondensator *m* mit Parallelwiderstand
s. **field control** Regelung *f* durch Feldschwächung
s. **instrument** Instrument *n* mit Nebenwiderstand
shunter *(Eis)* Rangierer *m*
shunting 1. *(Eis)* Rangieren *n*, Verschieben *n*; 2. *(El)* Nebenschlußbildung *f*, Parallelschalten *n*, Parallelschaltung *f*
s. **capacitor** *(El)* Querkondensator *m*
s. **crane** Verschiebekran *m*
s. **effect** Nebenschlußwirkung *f*
s. **engine** Schiebelokomotive *f*, Rangierlokomotive *f*
s. **frame** *(Brg)* Umgleiser *m*
s. **locomotive** Verschiebelokomotive *f*, Rangierlokomotive *f*
s. **resistance** *(Elm)* Widerstandnebenschluß *m*, ohmscher Shunt *m*
s. **siding** Abstellgleis *n*
s. **slope** Ablaufberg *m (zum Rangieren)*
s. **station** Verschiebebahnhof *m*, Rangierbahnhof *m*, Güterbahnhof *m*
s. **tower** Rangierstellwerk *n*
s. **track** Rangiergleis *n*
s. **yard** Rangierbahnhof *m*
shut / to [ver]schließen; absperren, [ver]sperren
s.**-down** 1. herunterfahren *(Gleichrichter)*; 2. zum Stillstand bringen, außer Betrieb setzen
s. **hermetically** hermetisch (luftdicht) abschließen
s.**-off** abschalten, abdrehen; absperren *(Dampf)*
shut *(Met)* Kaltschweißstelle *f (Guß)*
s.**-down** Stillegung *f (Betrieb)*; Abschaltung *f (Reaktor)*
s.**-down delay** Abschaltverzögerung *f*
s.**-down development** Abschaltverlauf *m*
s.**-down switch** Abschaltvorrichtung *f*
s.**-off rod** Abschaltstab *m*, Schnellschlußstab *m*, Sicherheitsstab *m*
s.**-off valve** Absperrventil *n*, Absperrschieber *m*
shutter / to einschalen *(Betonbau)*
s. **up** einschalen *(Betonbau)*
shutter Riegel *m*, Sicherheitsvorrichtung *f*, Verschluß *m*; Kameraverschluß *m*; Blende *f*; Fallklappe *f*, Klappe *f*
s. **bow** *(Kfz)* Verschlußspriegel *m*
s. **release** Verschlußauslöser *m*
s. **speeds** Verschlußzeiten *fpl*
shuttering Schalung[sform] *f (Betonbau)*
shutting Verschließen *n*
s. **off** Verschluß *m*, Abschluß *m*
shuttle / to pendeln
shuttle 1. Schützen *m (Webstuhl)*; Schiffchen *n (Nähmaschine)*; 2. *(Kern)* Rohrpostbüchse *f*; Rohrpost *f*
s. **armature** Doppel-T-Anker *m*
s. **box** *(Text)* Schützenkasten *m*
s. **braking** *(Text)* Schützenbremsung *f*, Schützenabbremsen *n*
s. **car** Pendelwagen *m (Fördertechnik)*
s. **carriage** Springwagen *m (z. B. an Buchungsautomaten)*
s. **catcher** *(Text)* Schützenfänger *m*
s. **changing automatic loom** *(Text)* Schützenwechselautomat *m*
s. **course** *(Text)* Lauf *m* des Webschützens, Schützenlauf *m*
s. **guard** *(Text)* Schützenfänger *m*
s. **guard wire** *(Text)* Schützenfangdraht *m*
s. **mark** *(Text)* Markierung *f* durch unsauberen Schützen
s. **movement** *(Text)* Hin- und Herbewegung *f* des Schützens
s. **race** *(Text)* Schützenbahn *f*
s. **service** Pendelbetrieb *m*, Pendelverkehr *m*
s. **tongue** *(Text)* Schützenzunge *f*
s. **truck** Pendelwagen *m (Fördertechnik)*
s. **with several bobbins** *(Text)* Melierschützen *m*
shuttlecock Federball *m*
shuttleless *(Text)* schützenlos
s. **loom** schützenloser Webstuhl *m*, schützenlose Webmaschine *f*

siccative Sikkativ *n*, Trockenmittel *n*, Trockenstoff *m*
sick bay Schiffslazarett *n*
sickle Sichel *f*
s.-shaped sichelförmig
side 1. Seite *f*; Seitenfläche *f*; Schenkel *m* *(Winkel)*; 2. *(Brg)* Stoß *m* *(einer Strecke usw.)*
s. aisle *(Bw)* Seitengang *m*; Seitenschiff *n*
s.-arm 1. Seitengewehr *n*; 2. *(El)* Mastausleger *m*
s. armature relay Seitenankerrelais *n*, Flachrelais *n*
s. band *s.* sideband
s. bearing Längslager *n*
s.-bearing-type fit Flankenzentrierung *f*
s.-blower converter *(Met)* seitlich blasender Konverter *m*
s.-bracket bearing Flanschlager *n*
s. bulkhead *(Schiff)* Seitenschott *n*
s. by side nebeneinander, Bord an Bord, längsseits, Seite an Seite *(bei zwei Schiffen)*
s. car Beiwagen *m*
s. chain *(Ch)* Seitenkette *f*; Nebenkette *f*
s.-chain conformation Seitenkettenkonformation *f* *(Proteinstruktur)*
s. charging seitliches Beschicken *n* *(eines Schmelzofens)*
s. circuit *(Fmt)* Stammkreis *m*, Stammleitung *f*
s.-circuit loading coil Stammspule *f*
s. clearance Flankenspiel *n*
s. coaming *(Schiff)* Längssüll *m (n)*
s. conductor rail seitliche Stromschiene *f*
s. contact Seitenkontakt *m* *(Elektronenröhren)*
s. counter timbers *(Schiff)* Heckstütze *f*
s. cut Seitenschnitt *m* *(Destillation)*
s. cutting pliers Seitenschneider *m*
s. cutting turning tool Seitendrehmeißel *m*
s. discharge wagon Seitenentleerer *m* *(Eisenbahnwaggon)*
s.-draw cut *(Brg)* Stoßlöseneinbruch *m* *(Schießarbeit)*
s. drift horizontaler Tunnel (Schacht) *m*
s. dump car Seitenentleerer *m* *(Eisenbahnwaggon)*
s. dump loader Seitenkipplader *m*
s. edge Seitenkante *f*
s. elevation Seitenriß *m* *(technisches Zeichnen)*
s. emission *(Fmt)* Streuemission *f*
s. face Seitenfläche *f*
s. flap Seitenklappe *f* *(LKW)*
s. flue *(Wärm)* Seitenzug *m*
s. frequency Seiten[band]frequenz *f*
s. friction brake Backenbremse *f*
s. guide Außenführung *f* bei Zahnkette
s. head Seitensupport *m*
s.-lamp *(Kfz)* Positionslampe *f*, Seitenlampe *f*
s. launching Querablauf *m*, Querstapellauf *m*
s. light 1. *(Kfz)* Begrenzungslicht *n*; 2. *(Schiff)* Bullauge *n*, Seitenfenster *n*
s.-lights *(Schiff)* Positionslichter *npl*, Seitenlichter *npl*
s. line 1. *(El)* Nebenleitung *f*, Abzweigleitung *f*; 2. Nebenlinie *f* *(Bahn)*
s. lobe Nebenzipfel *m*, Nebenkeule *f*, Seitenzipfel *m* *(Antenne)*
s.-lobe echo Nebenzipfelecho *n*, Seitenzipfelecho *n*; Seitenecho *n*
s.-lobe sector unsicherer Sektor *m* *(Funknavigation)*
s. marker Peilstange *f*
s.-marker lamp Peilstableuchte *f*
s. member Seitenholm *m*, Seitenträger *m*; Längsträger *m*
s. mill *s.* s. milling cutter
s. milling Walzenstirnfräsen *n*
s. milling cutter *(Am)* Scheibenfräser *m*
s. of the pulse Impulsflanke *f*
s. of tooth Zahnflanke *f*
s.-paddle-wheel steamer Seitenraddampfer *m*
s. panel Seitenwand *f* *(LKW)*
s.-panel frame Seitenwandrahmen *m* *(LKW)*
s. planking *(Schiff)* Seitenbeplankung *f*
s. plate seitliche Gleitschiene *f*, Seitenleiste *f*
s.-plate rudder Zweiplattenruder *n*, Doppelplattenruder *n*
s. pressure 1. *(Brg)* Stoßdruck *m*; 2. Seitendruck *m*
s. punt *(Schiff)* Scheuerprahm *m*
s. reaction *(Ch)* Nebenreaktion *f*
s. ring Seitenring *m*
s.-rope Fallreep *n*
s. roughing tool Seitenschruppmeißel *m*
s. saw blade frame Besäumgatter *n*
s. scattering *(Ph)* Seitenstreuung *f*, seitliche Streuung *f*
s. section *(Brg)* Stoßsegment *n* *(Ausbau)*
s. section of grate *(Wärm)* Seitenrost *m*
s. shore *(Schiff)* Deckseitenstütze *f*, Seitenstütze *f*
s. slip *(Schiff)* Querslip *m*, Querhelling *f*
s. spiling *(Brg)* Streckengetriebe *n*
s. stitching *(Typ)* Querheftung *f*
s. stop Seitenausschlag *m*
s. tank *(Schiff)* hoher Wasserballastkasten *m*, Seitentank *m*
s. thrust seitlicher Schub *m*, Querdruck *m*
s.-thrust effect *(Kern)* Lateralverschiebung *f*, seitliche Ablenkung *f*
s. tilt stability *(Kfz)* Neigungsstabilität *f*
s. tilting shovel loader Seitenkipplader *m*
s.-to-far-end crosstalk *(Fmt)* Gegenmitsprechen *n*
s.-to-phantom crosstalk *(Fmt)* Mitsprechen *n*

s.-to-side capacity *(El)* Vierer[schleifen]-kapazität *f*
s.-tone *(Fmt)* Seitengeräusch *n*, Rückhören *n*
s.-tone attenuation Mitsprechdämpfung *f*
s.-tone circuit Mithörschaltung *f*
s.-tone level Kontrollautstärke *f*, Mithörlautstärke *f*
s.-tone reduction wiring mitsprechfreie Schaltung *f*
s.-tone reference equivalent Rückhörbezugsdämpfung *f*
s.-tone telephone set Fernsprechapparat *m* ohne Rückhördämpfung, Mithörtelefon *n*
s. tool Seitenmeißel *m*
s. track Abstellgleis *n*; Nebengleis *n*
s. tracking Ablenken *n* von Bohrlöchern *(Tiefbohrtechnik)*
s. trawler Seitentrawler *m*
s. trimmer Seitenbesäumer *m (Holzbearbeitung)*
s. valve seitliches Ventil *n*
s.-valve engine seitengesteuerter Motor *m*
s. view Seitenriß *m*, Seitenansicht *f (technisches Zeichnen)*
s.-wall hovercraft *(Schiff)* Seitenwandschwebefahrzeug *n*
s. weft-fork motion *(Text)* seitlicher Gabelschußwächter *m*
s.-wheel cylinder sewing machine Zylindermaschine *f (Schuhfertigung)*
s.-wheel steamer Seitenraddampfer *m*, Raddampfer *m*
s. wheeler Seitenraddampfer *m*, Raddampfer *m*
s. window Seitenfenster *n*
s. wing *(Bw)* Nebengebäude *n*, Seitenflügel *m*
sideband Seitenband *n*; Nebenbande *f*
s. analyzer Seitenbandanalysator *m*
s. attenuation Seitenbanddämpfung *f*
s. echo Seitenbandecho *n*
s. frequency Seitenbandfrequenz *f*
s. interference Seitenbandinterferenz *f*
s. spectrum Seitenbandspektrum *n*
s. splash Störung *f* durch einen Nachbarkanal
sidelong seitlich geneigt, schräg
sidereal siderisch, Sideral-, Stern-
s. time Sternzeit *f*, Sideralzeit *f*
s. year Sternjahr *n*, siderisches Jahr *n*
siderite *(Min)* Siderit *m*, Eisenspat *m*, Spateisen[stein *m*] *n*
siderotilate *(Min)* Siderotilat *m*
sideslip / to seitlich abgleiten
sideswiping *(Brg)* Abbau *m* in Form schmaler senkrechter Streifen von der Ortsbrust
sidewalk *(Am)* Fußweg *m*
sideways engine zur Horizontalen geneigt eingebauter Motor *m*
sidewise seitlich, in seitlicher Richtung, seitwärts
s. movement Querverschiebung *f*
siding 1. Anschlußgleis *n*, Gleisanschluß *m*, Abstellgleis *n*, Nebengleis *n*; 2. *(Schiff)* Höhe *f (eines Spants)*
s. machine Besäummaschine *f*, Besäumsäge *f*
s. track *s.* siding 1.
sidings Abstellbahnhof *m*
Siemens Siemens *n (Einheit des elektrischen Leitwertes; 1 S = 1 A/V)*
Siemens-Martin furnace Siemens-Martin-Ofen *m*
Siemens-Martin process SM-Verfahren *n*
Siemens moving coil-type telewriter Siemens-Drehspulfernschreiber *m*
Siemens-multiplex printing telegraph Siemens-Rekorder *m*, Siemens-Schnellschreiber *m*
Siemens ozone tube Siemensscher Ozonisator *m*
siemensite Siemensit *m (feuerfester Stoff)*
sienna Siena[erde] *f*
sieve / to [ab]sieben, durchseihen
sieve Sieb[kasten *m*] *n*, Prüfsieb *n*; Durchschlag *m*
s. analysis Siebanalyse *f*
s. grate *(Wärm)* Siebrost *m*
s.-plate column *(Ch)* Siebbodenkolonne *f*
s. series Siebfolge *f*
s.-tray column *(Ch)* Siebbodenkolonne *f*
sieving Durchsieben *n*, Durchseihen *n*
s. filter Siebfilter *n*
s. residue Siebaustrag *m*
sift / to [ab]sieben, sichten
sifted coal Siebkohle *f*
s. coke Siebkoks *m*
sifter Sieb *n*, Schüttelsieb *n*, Siebmaschine *f*; Siebkette *f*; Sichter *m*; Plansichter *m*, Ausleser *m*
siftings Siebdurchgang *m*, Siebfeines *n (Aufbereitung)*
sight / to 1. anvisieren; 2. sichten, beobachten; 3. Visiereinrichtung anbringen
s. at (on) anvisieren
sight 1. Sicht *f*; Sehfähigkeit *f*; Ansicht *f*; 2. Visiereinrichtung *f*
s. axis Sehachse *f*
s. distance Sichtbereich *m*, Sichtweite *f*
s. disturbance Sichtbehinderung *f*
s. feed gravity oiler Öltropfapparat *m* mit Schauglas
s.-feed lubrication Schmierung *f* durch Tropföler mit Schauglas
s.-feed oiler Tropföler *m*, Tropfschmiergefäß *n*
s. glass Schauglas *n*
s. hole Schauöffnung *f*, Schauloch *n*
s.-testing chart Sehprobentafel *f*
sighting device Visiereinrichtung *f*
s. distance Zielweite *f*

s. platform *(Rak)* Richtstand *m*
s. shot Zielübung *f*
s. telescope Visierfernrohr *n*, Zielfernrohr *n*
s. tube Beobachtungsrohr *n*
sights Visiereinrichtung *f*, Zieleinrichtung *f*
sign / to markieren, anzeichnen
s. on *(Schiff)* anmustern, anheuern
sign Zeichen *n*; Vorzeichen *n*; Marke *f*
s. convention Vorzeichenregel *f*
s. digit *(Fs)* Vorzeichenbit *n*, Vorzeichen *n*, Vorzeichenziffer *f (binär)*
s. of current Vorzeichen *n* des Stromes, Stromrichtung *f*
s. of equality Gleichheitszeichen *n*
s. of integration Integrationszeichen *n*
s.-off signal *(Fmt)* Schlußzeichen *n*
s. post Wegweiser *m*
s. reversal Vorzeichenumkehrung *f*
signal / to signalisieren, melden
signal Signal *n*; Meldung *f*
s. apparatus Nachrichtengerät *n*
s. attenuation Signalabschwächung *f*
s. backlight Signalrücklicht *n*
s. bat *(Flg)* Signalkelle *f*
s. board Signaltafel *f*
s. box *(Eis)* Stellwerksanlage *f*, Stellwerksbude *f*
s. centre Nachrichtenzentrum *n*
s. connection Signalverbindung *f*
s. converter Zeichenwandler *m*
s. current Meßstrom *m*
s. delay Signalverzögerung *f*
s. device Signaleinrichtung *f*
s. discrimination Signalunterscheidung *f*
s. distortion Signalverzerrung *f*
s. element *(Fmt)* Stromschritt *m*
s. engineering Fernmeldetechnik *f*
s. equipment Nachrichtengerät *n*
s.-flow diagram *s.* s.-flow graph
s.-flow graph *(BMSR)* Signalflußdiagramm *n*, grafische Signalflußdarstellung *f*
s. fluctuation Signalschwankung *f*; Signalabweichung *f*
s. frequency Signalfrequenz *f*, Zeichenfrequenz *f*
s. gear Läutewerk *n (Signalwesen)*
s. generator *(El)* Signalgenerator *m*, Meßsender *m*, Prüfsender *m*, Service-Oszillator *m*
s. glass Blendenglas *n (Signal)*
s. grid *(El)* Steuergitter *n*; Zeichenfeld *n*
s. handling Signalverarbeitung *f*
s. indicator Signalisator *m*
s. injection Signaleinspeisung *f*
s. input Signaleingabe *f*
s. intensity Signalstärke *f*, Zeichenstärke *f*
s. junction working *(Fmt)* Anrufbetrieb *m*
s. lamp Blinkgerät *n*; Leuchtzeichen *n*; Signallicht *n*
s. level Signalpegel *m*
s. light proving *(Eis)* Signallichtüberwachung *f*
s. lock *(Eis)* Signalstellwerk *n*
s. machine Signalantrieb *m*
s. mixer *(Fs)* Signalmischer *m*
s. mixer unit Signalverteilungskasten *m (Radartechnik)*
s. mixing Kursaufschaltung *f*, Aufschaltung *f*
s. mixing for guide beam Leitstrahlaufschaltung *f*
s. mixing motor Aufschaltmotor *m*
s.-noise ratio Rauschabstand *m*, Störabstand *m*
s. operating cable Signalzug *m*
s. output Signalausgabe *f*
s. output level Ausgangssignalpegel *m*
s. plate Signalanode *f (Elektronenröhren)*
s. propagation Signalausbreitung *f*
s. proving *(Eis)* Signalmelder *m*
s. recording Signalaufzeichnung *f*
s. recording telegraphy Telegrafie *f* mit Aufzeichnung von kodierten Zeichen
s. relay Melderelais *n*, Signalrelais *n*
s. repeating Signalwiedergabe *f*
s. representation Signaldarstellung *f*
s. resolution Signalauflösung *f*
s. return Signalrückführung *f*
s. separation Signaltrennung *f*
s.-shaping network Telegrafiezeichenentzerrer *m*
s.-summation point *(Dat)* Signalsummierpunkt *m (automatische analoge Zeichenerkennung)*
s.-to-crosstalk-ratio *(Fmt)* Signalnebensprechverhältnis *n*
s.-to-noise level Rauschabstand *m (Verhältnis von Nutz- zu Störpegel)*
s.-to-noise ratio Störabstand *m*, Verhältnis *n* Signal/Rauschen, Signal-Rauschspannungsverhältnis *n*
s. tower Stellwerksturm *m*
s. tracer Signalverfolger *m (Meßtechnik)*
s. tracing Signalverfolgung *f (Meßtechnik)*
s. transfer Signalübertragung *f*
s. velocity Signalgeschwindigkeit *f*
s. wave *(Fmt)* Betriebswelle *f*, Zeichenwelle *f*
signalize / to signalisieren, melden
signaller Signalgeber *m*
signalling Signalisierung *f*, Signalisieren *n*; Zeichengebung *f*, Meldung *f*
s. battery Signalbatterie *f*
s. bell *(Fmt)* Signalglocke *f*, Warnglocke *f*
s. capacitor Sendekondensator *m (Telegrafie)*
s. current Telegrafierstrom *m*
s. faults *(Fmt)* Rufstörungen *fpl*
s. frequency Rufstromfrequenz *f*, Telegrafierfrequenz *f*

s. installation Signalanlage *f*
s. lamp Signalgrubenlampe *f*; Signallampe *f*
s. relay *(Fmt)* Rufrelais *n*
s. set *(Fmt)* Niederfrequenzrufsatz *m*
s. troubles Rufstörungen *fpl*
s. voltage *(Fmt)* Rufspannung *f*
s. voltmeter Signalvoltmeter *n*
signature tune *(Fmt)* Stationszeichen *n*
signet *(Typ)* Verlagszeichen *n*, Verlegerzeichen *n*, Signet *n*
significance Bedeutung *f*
s. point *(Math)* Sicherheitspunkt *m*
significant *(Math)* mit [bestimmtem] Zahlenwert
s. digit *(Dat)* bedeutsame (geltende, Zahlenwert besitzende) Ziffer *f*
s. overtone charakteristischer Oberton *m*
silage / to silieren
silage Silofutter *n*
silence Ruhe *f*, Stille *f*
s. in operation Geräuscharmut *f*; Laufruhe *f*
s. period Funkstille *f (für SOS-Zeichen)*
silencer Schalldämpfer *m*
silencing geräuschdämpfend
s. device Krachtöter *m (Funktechnik)*
s. switch Schalter *m* für Stummabstimmung
silent ruhig; tonlos; still; *(Masch)* geräuscharm
s. arc ruhiger Lichtbogen *m*
s. area *(Fmt)* tote (empfangslose) Zone *f*, Schweigezone *f*
s. balance *(Fmt)* Abgleich *m* auf Tonlosigkeit
s. block Gummilager *n*
s. chain [geräuscharme] Zahnkette *f*
s.-chain drive Zahnkettentrieb *m*
s. discharge *(El)* dunkle Entladung *f*
s. film Stummfilm *m*
s. period Funkstille *f*
s. projektor Stummfilmprojektor *m*
s. running geräuscharmer Lauf *m*
s. space Nullzone *f*, Schwebungslücke *f*
s. tooth chain [geräuschlose] Zahnkette *f*
s. zone *(Fmt)* tote (empfangslose) Zone *f*, Schweigezone *f*
silhouetting method Schattenbildverfahren *n*
silica Kieselerde *f*, Siliziumdioxid *n*
s. brick Silikastein *m*, Silikaziegel *m*
s. brick with clay bond Dinasstein *m*
s. gel Kieselgel *n*, Silikagel *n*
s. gel grease Siliziumgelfett *n*, Kieselgelfett *n*
s. glass Silikatglas *n*
s. lamp Quarzlampe *f*
s. sand Quarzsand *m*
silicate Salz *n* der Kieselsäure, Silikat *n*
s. bond Silikatbindung *f*
s. chemistry Silikatchemie *f*
s. clinker Silikatschlacke *f*
s. enamel Silikatemaille *f*
siliceous kieselartig, kieselig; kieselhaltig, siliziumhaltig, kieselsauer
s. sinter Kieselsinter *m*, Quarzsinter *m*
silicic acid Kieselsäure *f*
silicide Silizid *n*, Silizium[-Metall]verbindung *f*
silicification Verkieselung *f*
silicify / to verkieseln
silico-manganese steel siliziumhaltiger Manganstahl *m*
silicochloroform Trichlorsilan *n*
silicoethane Disilan *n*
silicofluoric acid Hexafluorokieselsäure *f*
silicoiodoform Trijodosilan *n*
silicomethane Monosilan *n*, Silan *n*
silicon Silizium *n*
s. alloy transistor Legierungssiliziumtransistor *m*
s. block Siliziumkristall *m*
s. carbide Karborund *n*, Siliziumkarbid *n*
s. cell Siliziumzelle *f*
s. compound Siliziumverbindung *f*
s. controlled rectifier steuerbare Siliziumzelle *f*, steuerbarer Siliziumgleichrichter *m*
s. crystal rectifier Siliziumkristallgleichrichter *m*
s. diffused transistor Diffusionssiliziumtransistor *m*
s. ethyl Siliziumtetraäthyl *n*, Tetraäthylmonosilan *n*
s. ingot Siliziumblock *m*
s. junction element Siliziumgleichrichter *m*
s. methyl Siliziumtetramethyl *n*, Tetramethylmonosilan *n*
s. rectifier Siliziumgleichrichter *m*
s. resistor Siliziumwiderstand *m*
s.-rich casting siliziumreicher Guß *m*
s. solar cell Siliziumsonnenzelle *f*
s. transistor Siliziumtransistor *m*
silicone *(Ch)* Silikon *n*
s. fluid Silikonflüssigkeit *f*
s. grease Silikonfett *n*
s. oil Silikonöl *n*
s. resin Silikonharz *n*
s. rubber Silikonkautschuk *m*
s. varnish Silikonlack *m*
siliconize / to silizieren
silicopropane *(Ch)* Trisilan *n*
silicosis Silikose *f*
silit Silit *n (Werkstoff für elektrische Widerstände)*
s. resistor Silitwiderstand *m*
silk 1. Seide *f*; 2. *(Min)* Seidenglanz *m*
s. and cotton covered cable Baumwollseidenkabel *n*
s. covered seidenumsponnen *(elektrischer Leiter)*
s.-covered wire Seidendraht *m*, seidenumsponnener Draht *m*, Draht mit Seidenumflechtung
s. degumming Entbasten *n* der Seide
s. finish effect Seidenfinisheffekt *m*

s. **gum** Seidenbast *m*
s. **loom** Seiden[web]stuhl *m*
s. **manufacture** Seidenverarbeitung *f*
s. **moth** Seidenspinner *m*
s. **reeling** Abhaspeln *n* der Seide
s. **scarf** Seidentuch *n*
s. **screen printing** Siebdruck *m*, Filmdruck *m*, Seidenrasterdruck *m*, Serigrafie *f*
s.**-screen printing process** Siebdruckverfahren *n*
s. **screening** Siebdruck *m (z. B. für Leiterplatten geeignet)*
s. **throwster** Seidenzwirner *m*
s. **waste** entbasteter Seidenabfall *m*
s. **weaving** Seidenweberei *f*
s. **winding frame** Seidenspulmaschine *f*
silken in lustre seidig glänzend
silkiness Seidenglanz *m*
silkworm Seidenraupe *f*
s. **farm** Seidenraupenzucht *f*
silky 1. seidig, seidenartig; 2. *(Led)* zart *(Narben)*
s. **lustre** Seidenglanz *m*
sill 1. *(Geol)* Intrusivlager *n*, Lagergang *m*; 2. *(Brg)* Liegendes *n*; Grundschwelle *f*, Sohlenholz *n (Ausbau)*; 3. *(Schiff)* Süll *m (n)*, Dockdrempel *m*; 4. Schwelle *f*; Fensterbrett *n*; Unterzug *m*
s.**-high** in Fensterhöhe
s. **plate** Schwelle *f (Holzbau)*
sillimanite *(Min)* Sillimanit *m (Tonerdesilikat)*
silo Bunker *m*, Großspeicher *m*, Silo *m*, Speicher *m*
siloing Einmieten *n*, Silierung *f*
siloxan Siloxan *n*
siloxen Siloxen *n*
silt 1. Schluff *m*; Schlamm *m*; Silt, Schwemmsand *m*, Triebsand *m*, Feinstsand *m*, Mehlsand *m*; 2. Feinkohle *f*, Kohlenschlamm *m*; 3. Schwebstoff *m*, Schwemmstoff *m*
s. **box** Schlammfang *m (Sinkkasten)*
s. **content** Schwebstofführung *f*, Schwemmstofführung *f*
s. **loam** Schlufflehm *m*
silting 1. Anschwemmung *f*, Sedimentation *f*, Schlammablagerung *f*; 2. *(Brg)* Spülversatz *m*
s. **up of soil** Verschlammen *n* des Landes
silty schluffartig, schluffhaltig; schlammig
silumin alloy Silumin *n (Handelsbezeichnung für Silizium-Aluminium-Legierung)*
silver / to versilbern
s.**-plate** versilbern
silver Silber *n*
silver silbern, silbrig, Silber-
s. **alloy** Silberlegierung *f*
s. **amalgam** Silberamalgam *n*
s. **amalgamation** Silberamalgamation *f*
s. **bark** *(Holz)* Glanzrinde *f*, Spiegelrinde *f*
s.**-bearing** silberhaltig
s. **bromate** Silberbromat *n*
s. **bromide** Silberbromid *n*, Bromsilber *n*
s. **bullion** Rohsilber *n*, Blocksilber *n*
s. **chloride** Silberchlorid *n*, Chlorsilber *n*
s. **chloride battery** Chlorsilberelement *n*
s. **chromate** Silberchromat *n*
s.**-clad** silberplattiert
s.**-coated** versilbert
s. **coin** Silbermünze *f*
s. **content** Silbergehalt *m*
s. **cyanide** Silberzyanid *n*
s. **deposit** Silberbelag *m*
s. **detector** Silberindikator *m*
s. **electrode** Silberelektrode *f*
s.**-embroidery** Silberstickerei *f*
s. **extraction** Silbergewinnung *f*
s. **fir** Edeltanne *f*
s. **fluoride** Silberfluorid *n*
s. **fog** dichroitischer Schleier *m (Fotografie)*
s. **foil** Blattsilber *n*, Silberfolie *f*
s. **glance** *(Min)* Silberglanz *m*, Argentit *m*
s. **globule** Rasterkorn *n (Fotografie)*
s. **grain** Silberkorn *n*
s.**-grey** silbergrau
s. **halide** Halogensilber *n*, Silberhalogenid *n*
s. **impurity** Silberverunreinigung *f*
s. **ingot** Silberbarren *m*
s. **iodide** Silberjodid *n*, Jodsilber *n*
s. **lamina** Silberblech *n*
s. **lattice** Silbergitter *n*
s. **leaching plant** Silberlaugerei *f*
s. **lead ore** silberhaltiger Bleiglanz *m*
s.**-leafed fir-tree** Edeltanne *f*
s. **lining** Silberbeschlag *m*
s. **mine** Silbergrube *f*, Silberbergwerk *n*
s.**-mounted** silberbeschlagen
s. **nitrate** Silbernitrat *n*, Höllenstein *m*
s. **nitrate stick** Höllenstift *m*, Ätzstift *m*
s. **nucleus** Silberkeim *m (Fotografie)*
s. **ore** Silbererz *n*
s. **plated** versilbert, silberplattiert *(auf trockenem Wege)*
s. **plug** Rasterkorn *n (Fotografie)*
s. **point** Erstarrungspunkt *m* des Silbers, Silberpunkt *m (Temperaturskala)*
s.**-pointed style** Silberstift *m*
s. **refinery** Silberscheideanstalt *f*
s. **refining** Silberscheidung *f*, Silberbrennen *n*
s. **solder** Silberlot *n*
s. **steel** Silberstahl *m*, Präzisionsstahl *m*
s. **sulphur** Schwefelsilber *n*
s. **telluride** Tellursilber *n*
s. **voltameter** Silbervoltameter *n*
s. **white** Bleiweiß *n*, Silberweiß *n*
s. **wire** Silberdraht *m*
silvered mirror oberflächenversilberter Spiegel *m*

silvering Versilberung *f (Elektrotypie)*; Spiegelbelegung *f*
s. shop Spiegelbelegerei *f*
silvery silberartig, silberfarbig, silbern
s. getter film Getterspiegel *m (Elektronenröhren)*
s. luster Silberglanz *m*
silviculture Forstkultur *f*
similar ähnlich; gleichartig; gleichnamig
s. charges gleichnamige Ladungen *fpl*
s. terms gleichnamige Glieder *npl*
s. triangles ähnliche Dreiecke *npl*
similarity Ähnlichkeit *f*
s. law *(Ph)* Ähnlichkeitsgesetz *n*
s. principle *(Ph)* Ähnlichkeitsgesetz *n*
s. transformation *(Math)* Ähnlichkeitstransformation *f*
similitude Ähnlichkeit *f*; Ähnlichkeitstheorie *f*
simmer / to sieden, wallen
simple einfach; einteilig; ungeteilt; nicht verzweigt; einseitig; einadrig; rein; einwellig; eingliedrig
s. beam *(Bw)* Einfeldträger *m*, Träger *m* auf zwei Stützen
s. bending Biegung *f* ohne Längskraft, reine Biegung
s. bridge 1. *(Bw)* Einfeldbrücke *f*; 2. *(El)* Einfach[meß]brücke *f*
s. cable Einfachkabel *n*, Einleiterkabel *n*, einadriges Kabel *n*
s. D-bit Kanonenbohrer *m*
s. detached layer *(Ph)* einfache Ablösung *f*
s. distillation einfache Destillation *f*
s. electrode Einfachelektrode *f*
s. fraction *(Math)* gemeiner Bruch *m*
s. gear train einfache Wechselräderübersetzung *f*
s. harmonic current Sinusstrom *m*, Einwellenstrom *m*, einwelliger (sinusförmiger) Strom *m*
s. indexing *(Masch)* einfache Teilung *f*
s. lattice primitives Gitter *n (Kristalle)*
s. line Einfachleiter *m*, Einfachleitung *f*, Einzelleiter *m*, Einzelleitung *f*
s. load einfache Beanspruchung *f*
s. machine einfache Maschine *f*
s. mechanism einfaches Getriebe *n*
s. microscope Lupe *f*
s. parallel winding *(Elm)* eingängige Parallelwicklung *f*
s. pendulum mathematisches Pendel *n*
s. piston rod einseitige Kolbenstange *f*
s. relay einfaches Relais *n*
s. sinusoidal potential difference *(El)* rein sinusförmige Spannung *f*
s. steel unlegierter Stahl *m*
simplex Simplex-
s. algorithm Simplexverfahren *n* des linearen Programmierens
s. brake Simplexbremse *f*
s. circuit Einfachleitung *f*, Simultanleitung *f*
s. communication einfache Verbindung *f*
s. criterion Simplexkriterium *n (lineares Programmieren)*
s. line Einfachleiter *m*, Einfachleitung *f*, Einzelleitung *f*
s. machine Einspindelmaschine *f*
s. method of linear programming *s.* s. algorithm
s. operation *(Fmt)* Einfachbetrieb *m*, Simplexbetrieb *m*, Simplexverkehr *m*, Wechselverkehr *m*
s. steam pump Simplexpumpe *f*
s. telegraphy Einfachtelegrafie *f*
s. winding eingängige Wicklung *f*
s. working *s.* s. operation
simplification Vereinfachung *f*
simplified analysis Näherungsverfahren *n (Statik)*
s. ophthalmoscope einfacher Augenspiegel *m*
s. schematic diagram *(El)* Prinzipschaltbild *n*
s. thread counter Fadenzähllupe *f*
simplify / to vereinfachen
simply connected region *(Math)* einfach zusammenhängender Bereich *m*
s. mesh-connected circuit *(Fmt)* Polygonschaltung *f*
Simpson's rule Simpsonsche Regel *f*
simulate / to [analog] nachbilden, nachahmen, reproduzieren
simulation *(Fs)* Simulation *f*, analoge Nachbildung *f (Echtzeitbetrieb)*; Nachbildung *f (eines physikalischen Vorganges)*
simulator Simulator *m*, Nachbildner *m (Analogrechner)*
simultaneity Gleichzeitigkeit *f*
s. factor Gleichzeitigkeitsfaktor *m*
simultaneous gleichzeitig, simultan
s. broadcasting Gleichwellen[rund]funk *m*
s. carry *(Dat)* Parallelübertrag *m*
s. contrast *(Ph)* Simultankontrast *m*
s. equations simultane Gleichungen *fpl*
s. reception *(Fmt)* Simultanempfang *m*
sine Sinus *m*
s. bar Sinuslineal *n*
s. condition Sinusbedingung *f*
s. current Sinusstrom *m*, sinusförmiger Strom *m*
s. curve Sinuskurve *f*
s. function Sinusfunktion *f*
s. galvanometer Sinusbussole *f*
s. integral Integralsinus *m*
s. law Sinussatz *m*
s. law variation sinusförmige Änderung *f*, sinusförmiger Verlauf *m*
s. meter Blindverbrauchszähler *m*, Blindstromzähler *m*, Blindlastzähler *m*

s. series Sinusreihe *f*
s.-shaped sinusförmig
s. table Sinustisch *m*
s. term Sinusglied *n*
s. wave Sinuswelle *f*, sinusförmige Welle *f*; Sinusschwingung *f*
s.-wave current Sinusstrom *m*
s.-wave generator Sinuswellenerzeuger *m*, Sinuswellengeber *m*
s.-wave response Kennlinie *f* für Sinusspannungen, Sinuswellenempfindlichkeit *f*
s.-wave voltage Sinusspannung *f*
sing / to heulen, pfeifen, singen *(Verstärker)*
singe / to brennen, flammen, [ver]sengen; gasieren; *(Text)* absengen
singeing *(Text)* Sengen *n*, Abflammen *n*, Gasieren *n*
singing Pfeifen *n (Verstärker)*
s. arc singender (tönender, musikalischer, sprechender) Lichtbogen *m*
s. margin Abstand *m* vom Pfeifpunkt
s. path *(Fmt)* Rückkopplungsweg *m*
s. point *(Fmt)* Pfeifpunkt *m*, Schwingungseinsatzpunkt *m*
s. propeller singender Propeller *m*
s. spark tönender Funken (Lichtbogen) *m*
s. suppressor *(Fmt)* Rückkopplungssperre *f*
single / to aussondern; vereinzeln
single Bohrgestängeeinheit *f*, Bohrstange *f*
single einzeln, Einzel-; Ein-; einfach *(Kristall)*; einteilig; einzig
s. acceleration Einfachbeschleunigung *f*
s.-acting einfachwirkend
s.-acting barring gear einfachwirkendes Schaltwerk *n (Schwungrad)*
s.-acting cylinder *(Wärm)* einfachwirkender (einseitig beaufschlagter) Zylinder *m*
s.-acting steam engine einfachwirkende Dampfmaschine *f*
s.-action procedure *(Kern)* Einfachprozeß *m*
s.-address system *(Dat)* Einadreßsystem *n*
s. angle [milling] cutter Winkelstirnfräser *m*
s.-anode rectifier *(Elm)* Einanodenventil *n*
s. arm anchor Hafenanker *m*
s. arm signal einflügeliges Signal *n*
s. armature converter *(Elm)* Einankerumformer *m*
s.-axle einachsig
s.-banked boat Einmannruderboot *n*
s. battery switch Einzelzellenschalter *m*
s. beam Einzelstrahl *m*
s.-beam equipment *(Opt)* Einstrahleinrichtung *f*
s.-beam oscilloscope Einstrahloszillograf *m*
s.-berth cabin *(Schiff)* Einzelkabine *f*, Einmannkammer *f*, Einbettkammer *f*
s.-bit double cutter zweischneidiges Bohrmesser *n*
s.-blade[d] tool Einzahnwerkzeug *n*
s.-block brake Einbackenbremse *f*
s. boiler Einfachkessel *m*
s. bottom einfacher Boden *m*
s.-bottom vessel Schiff *n* ohne Doppelboden
s. braid einmal umsponnen *(Leiter)*
s.-break switch Schalter *m* mit Einfachunterbrechung
s. bridge Einfach[meß]brücke *f*
s. cabin *s.* s.-berth cabin
s. cable Einleiterkabel *n*, Einfachkabel *n*, eindriges Kabel *n*
s. cavity mould *(Plast)* Einfachform *f*
s.-cell battery Monozelle *f (Batterie)*
s. chain Einfachkette *f*
s.-chamber brake cylinder Einkammerbremszylinder *m*
s.-chamber trailer control valve Einkammeranhängersteuerventil *n*
s.-channel Einkanal-
s.-channel analyzer Einkanalanalysator *m*
s.-circuit receiver Einkreisempfänger *m*, Einkreisempfangsgerät *n*
s.-circuit reception Einkreisempfang *m*
s. clip wool Einschurwolle *f*
s.-coated polyurethane foam tape *(Text)* einseitig mit Polyurethanschaumstoff kaschiertes Band *n*
s.-coil einspulig
s.-coil lamp Einfachwendellampe *f*
s.-coil ringer *(Fmt)* Einspulwecker *m*
s.-coil spring einfach gewundene Indikatorfeder *f*
s.-coloured einfarbig
s.-column construction Einständerbauart *f*
s. column press Einständerpresse *f*
s.-column vertical boring [and turning] mill Einständerkarusselldrehmaschine *f*
s.-column vertical turret lathe Einständerkarusselldrehmaschine *f*
s. commutation direct current signalling Gleichstromsignalisierung *f* mit Stromrichtungsregelung *(Nachrichtentechnik)*
s.-component signal Einfachsignal *n*, Einfachzeichen *n*
s. conductor Einzelleiter *m*
s.-conductor cable Einleiterkabel *n*, Einfachkabel *n*, einadriges Kabel *n*
s. conductor cord einadrige Schnur *f*, Einfachschnur *f*, Einleiterschnur *f*
s. conical cover *(Wärm)* einwandiger konischer Boden *m*
s. contact Einfachkontakt *m (Relais)*
s.-contact einpolig
s.-contact relay einpoliges Relais *n*
s.-contact switch einpoliger Schalter *m*
s.-contact system einfacher Fahrdraht *m*
s. control *(Fmt)* Einknopfbedienung *f*

s.-control tuning Einknopfabstimmung *f*
s. cord *s.* s. conductor cord
s.-core einadrig *(Kabel)*
s.-core barrel Einfachkernrohr *n (Tiefbohrtechnik)*
s.-core cable einadriges Kabel *n*, Einleiterkabel *n*
s.-corner rounding cutter einseitiger Radiusfräser *m*
s.-cotton-covered mit Baumwolle einfach umsponnen *(Leiter)*
s.-cotton enamelled wire *(El)* einmal mit Baumwolle umsponnener Lackdraht *m*
s. counter *(Kern)* Einzelzähler *m*
s. coverage einfache Deckung *f (Dach)*
s.-covered wire einfach umsponnener (umklöppelter) Draht *m*
s. crank control Einhebelsteuerung *f (Kurbelgriff)*
s. crank press Kurbelpresse *f* mit einfach gekröpfter Welle
s. crank selection Einhebelwahl *f* [der Vorschübe]
s. crank steam engine einkurblige Dampfmaschine *f*
s.-crank toggle drawing press Kniehebelziehpresse *f* mit einfach gekröpfter Welle
s. cross over einfache Gleisverbindung *f*
s. cruising vessel allein fahrendes Schiff *n*
s. crystal Einkristall *m*
s.-crystal ferrite Ferriteinkristall *m*
s.-crystal growth Einkristallzüchtung *f*
s.-crystal spectrometer Einkristallspektrometer *n*
s. cubic crystal kubischer Einkristall *m*
s.-cup insulator Einfachglockenisolator *m*
s. current key *(Fmt)* Einstromtaste *f*
s. current telegraph system *(Fmt)* Einstromsystem *n*
s.-current working *(Fmt)* Einfachstrombetrieb *m*
s.-cut einschnittig; einhiebig
s. cutter Einformfräser *m*, scheibenförmiger Gewindefräser *m*
s.-cutter method of thread milling Langgewindefräsen *n (Einformfräser)*
s. cutter-type of thread milling machine Langgewindefräsmaschine *f*
s.-cycle [formate cutting] method Formwälzmethode *f* mit Fertigstellung einer Zahnlücke je Räumkopfumdrehung
s.-cylinder Einzylinder-, einzylindrig
s. cylinder diagram *(Wärm)* Einzylinderdiagramm *n*
s. cylinder multitubular boiler einfacher Heizrohrkessel *m*
s. cylinder steam engine einzylindrige Dampfmaschine *f*, Einzylinderdampfmaschine *f*
s. cylindrical boiler einfacher Walzenkessel *m*, Zylinderkessel *m*
s.-deck screen Eindeckersieb *n*
s.-deck vessel Eindeckschiff *n*, Eindecker *m*
s.-decker *s.* s.-deck vessel
s.-dial control *(Fmt)* Einknopfabstimmung *f*
s. diode Einfachdiode *f*
s. dislocation Einzelversetzung *f (Kristallgitter)*
s. dividing error Einzelteilfehler *m (Teilkopf)*
s. dry plate clutch Einscheibentrockenkupplung *f*
s.-duct einzügig *(Kabel)*
s. earth Einfacherdschluß *m*, Einzelerdschluß *m*
s.-edge wheel Einprofilschleifscheibe *f (Gewindeschleifen)*
s.-edged tool einschneidiges Werkzeug *n*
s. electron Einzelelektron *n*
s.-electron capture Einelektroneneinfang *m*
s.-end control Führerstandssteuerung *f*
s.-end[ed] einseitig; einmäulig *(Lehre)*
s.-ended action *(BMSR)* Einseitenbetrieb *m*, Ein-Richtungsbetrieb *m*
s.-ended amplifier *(Fmt)* Seriengegentaktverstärker *m*
s.-ended boiler Einenderkessel *m*
s. ended circuit *(Fmt)* Eintaktendverstärker *m*
s.-ended push-pull transformatorlose Gegentaktendstufe *f*
s. ended Scotch marine boiler Einenderschiffskessel *m*
s. ended spanner einmäuliger Schraubenschlüssel *m*
s.-engined einmotorig
s.-face tool einbahniges Werkzeug *n*
s.-figure einstellig
s.-filament bulb Eindrahtlampe *f*
s. fillet weld Einfachkehlnaht *f*
s. flat cover einwandiger ebener Boden *m (Kessel)*
s.-flute einlippig
s.-fluted drill for deep-hole drilling Einlippentieflochbohrer *m*
s. footing *(Bw)* Einzelfundament *n*, Einzelgründung *f*
s.-formed cutter Einformfräser *m*
s.-frame exposure Einzelbildaufnahme *f*
s. frequency line trap einwellige Trägerfrequenzsperre *f*
s.-grid valve Eingitterröhre *f*
s. guide einseitige Schlittenführung *f*
s.-gyro system Einkreiselanordnung *f*
s.-hand drilling *(Brg)* einmännisches Bohren *n*
s.-handle einarmig, mit einem Griff
s. head rectilinear comber *Text)* Einkopfflachkämmer *m*

s. head wrench einmäuliger Schraubenschlüssel *m*
s. header boiler with circulating water *(Wärm)* Einkammerkessel *m* mit Doppelrohren
s. header boiler with plain water tubes *(Wärm)* Einkammerkessel *m* mit einfachen Wasserrohren
s. headphone Einzelkopfhörer *m*
s.-hole Einloch-
s. impression mould *(Plast)* Einfachform *f*
s. inductive shunt einfacher induktiver Nebenschluß (Shunt) *m*
s. ionization Einfachionisation *f*
s.-item ejection *(Büro)* Einzelpostenausstoß *m*
s.-J butt weld Jotnaht *f*, J-Naht *f*
s.-jet eindüsig, einflammig
s.-knob Einknopf-
s.-lane traffic einspuriger Verkehr *m*
s.-lap machine for flat work Einscheibenläppmaschine *f* für Flachläppen
s.-layer einlagig
s.-layer coil einlagige Spule *f*
s.-layer winding einlagige (einschichtige) Wicklung *f (Zylinderspule)*
s.-leaf electrometer Einblattelektrometer *n*
s.-leaf spring einfache Blattfeder *f*
s.-leaf swing door einflügelige Drehtür *f*
s.-lens reflex camera einäugige Spiegelreflexkamera *f*
s. lever Einhebel *m*
s.-lever control Einhebelbedienung *f*, Eingriffbedienung *f*
s. lift Jacquard machine for open shedding Einhubjacquardmaschine *f* für Offenfach
s.-line eingleisig
s.-lined board einseitig gedeckter Karton *m*
s.-lined spectrum Einlinienspektrum *n*
s.-lip einlippig
s.-lipped rifle drill Gewehrlaufbohrer *m*, Spindelbohrer *m*
s. lock *(Schiff)* einfache Kammerschleuse *f*
s. lubricator Einzelöler *m*
s. machine Einständermaschine *f*
s.-masted vessel einmastiges Schiff *n*, Einmaster *m*
s. molecule Einzelmolekül *n*
s. needle flat bed sewing machine Einnadelflachmaschine *f (Schuhfertigung)*
s.-note distortion Eintonverzerrung *f*
s. nozzle Einzeldüse *f*
s. open-end wrench Einfachschraubenschlüssel *m*
s. operation *(Fmt)* Simplexbetrieb *m*
s. outrigger shell Einer-Auslegerboot *n*
s. part production Einzelteilfertigung *f*
s. particle Einzelteilchen *n*
s.-particle equation Einteilchengleichung *f*
s.-particle model Einteilchenmodell *n*
s.-particle state Einteilchenzustand *m*
s.-particle transition Einteilchenübergang *m*
s.-particle wave-function Einteilchenwellenfunktion *f*
s. pass tap Einzelschneider *m*
s. path *(Kern)* Einzelbahn *f*
s.-phase einphasig, Einphasen-
s.-phase armature Einphasenanker *m*
s.-phase arrangement Einphasenanordnung *f*
s.-phase bridge circuit Einphasenbrückenschaltung *f*
s.-phase cable Einphasenkabel *n*
s.-phase circuit Einphasen[strom]kreis *m*
s.-phase commutator motor with self-excitation kompensierter Repulsionsmotor *m*
s.-phase commutator motor with series compensating winding Einphasenreihenschlußmotor *m* mit Kompensationswicklung
s.-phase current Einphasen[wechsel]strom *m*
s.-phase generator Einphasengenerator *m*, Einphasendynamo[maschine *f*] *m*, Einphasen[strom]erzeuger *m*
s.-phase induction motor Einphaseninduktionsmotor *m*
s.-phase inverter einphasiger Wechselrichter *m*
s.-phase line Einphasenleitung *f*
s.-phase meter Einphasenzähler *m*
s.-phase motor Einphasen[wechselstrom]-motor *m*
s.-phase reactor Einphasenreaktor *m*
s.-phase rectifier Einphasengleichrichter *m*
s.-phase starter Einphasenanlasser *m*
s.-phase system Einphasensystem *n*
s.-phase transformer Einphasentransformator *m*
s.-phase winding Einphasenwicklung *f*
s.-phase wiring *s.* s.-phase line
s. phenomenon Einzelerscheinung *f*
s.-phonon process Einphononenprozeß *m*
s. photograph Einzelaufnahme *f*
s. piece Einzelstück *n*, Einzelteil *n*
s. piece work Einzelfertigung *f*
s. pitch error Einzelteilfehler *m* *(Zahnrad)*
s. pivot instrument Meßinstrument *n* mit Hängelager
s.-plane image Einebenenbild *n*
s. plate Einzelblech *n*
s.-plate [disk-type friction] clutch Einscheibenkupplung *f*
s.-plate dry-type clutch Einscheibentrockenkupplung *f*
s.-plate rudder Ein[zel]plattenruder *n*, Einblattruder *n*
s.-ply belt Einfachriemen *m*
s.-point einschneidig, einschnittig

s.-point application of pressure Einpunktsystem *n (Umformtechnik)*
s.-point boring Aufbohren *n* mit einschneidigem Werkzeug
s.-point boring tool Bohrmeißel *m*
s.-point cutter Einzahnschlagfräser *m*
s.-point recorder Einpunktschreiber *m*
s.-point threading tool Gewindedrehmeißel *m*
s.-point tool einschneidiges Werkzeug *n*
s.-pole einpolig
s.-pole circuit breaker einpoliger Schalter *m*
s.-pole cut-out einpolige Sicherung *f*; einpoliges Schaltschütz *n*
s.-pole disconnection einpolige Abschaltung *f*
s.-pole double-throw switch einpoliger Umschalter *m*
s.-pole plug einpoliger Stecker *m*
s.-pole valve Einpolröhre *f*
s. potential Einzelpotential *n*
s. prism Einzelprisma *n*
s. production Einzelfertigung *f*
s.-pulley drive Einscheibenantrieb *m*
s.-pulley lathe Einscheibendrehmaschine *f*
s. pulse Einzelimpuls *m*
s.-purpose machine Einzweckmaschine *f*
s.-purpose reactor Einzweckreaktor *m*
s.-purpose tool Einzweckwerkzeug *n*
s.-ram *(Masch)* mit einem Räumschlitten
s. resonance einfache Resonanz *f*
s.-rib einprofilig
s.-rib grinding wheel einprofilige Schleifscheibe *f*, Einprofilschleifscheibe *f*
s.-ribbed einprofilig
s.-ribbed wheel Einprofilschleifscheibe *f (Gewindeschleifen)*
s. ring coil *(El)* Kreisringspule *f*, Toroidspule *f*
s.-riveted einreihig genietet
s. riveted joint einreihige Nietverbindung *f*
s.-roller chain Einfachrollenkette *f*
s. rolling mill eingerüstiges Walzwerk *n*
s.-rope grap bucket Einseilgreifer *m*
s. row einreihig *(Lager; Nietung)*
s. row construction einreihige Konstruktion *f*
s. sampling einfache Stichprobenprüfung *f*
s. scan Einzelabtastung *f*
s. scattering einfache Streuung *f*, Einzelstreuung *f (Elektronen)*
s.-screw extruder Einschneckenpresse *f*
s.-screw propulsion Einschraubenantrieb *m*
s.-screw stern frame *(Schiff)* Einschraubensteven *m*
s.-screw vessel Einschraubenschiff *n*
s. screw vice einfacher Schraubstock *m*
s. scull Einer *m (Sportboot)*
s. scutcher *(Text)* einfache Schlagmaschine *f*
s.-seat fighter Jagdeinsitzer *m*, einsitziges Jagdflugzeug *n*
s. semaphore einflügeliges Signal *n*
s. set-up Einzelaufspannung *f*
s. shading einfache Schraffierung *f*
s.-shear einschnittig
s. shell *(Am) s.* s. scull
s.-shot blasting machine *(Brg)* Zündmaschine *f* zum gleichzeitigen Zünden von max. 3 Schüssen
s.-shot blocking oscillator *(Fs)* monostabiler Sperrschwinger *m*
s.-shot firing (fire) *(Mil)* Einzelfeuer *n*
s.-shot instrument Gerät *n* für Einzelaufnahmen
s.-shot survey Einpunktvermessung *f*
s.-shot trigger circuit *(Fs)* monostabile Triggerschaltung *f*, Univibratorkippschaltung *f*
s. sideband *(Fmt)* Einseitenband *n*
s. sideband amplitude modulation Einseitenbandamplitudenmodulation *f*
s. sideband receiver Einseitenbandempfänger *m*
s.-sideband transmission Einseitenbandübertragung *f*
s.-sided amplifier Einseitenbandverstärker *m*
s. silk covered mit Seide einfach umsponnen *(Leiter)*
s. solder joint Einzellötstelle *f*
s. solution etching *(Typ)* Einbadätzung *f*, Einstufenätzung *f*
s.-speed floating control *(BMSR)* Zweilaufregelung *f*
s.-spindle einspindlig, Einspindel-
s.-spindle automatic bar machine Einspindelautomat *m* für Stangenarbeit
s.-spindle automatic [screw machine] Einspindelautomat *m*
s.-spindle construction (design) Einspindelbauart *f*
s.-spindle drilling machine Einspindelbohrmaschine *f*
s.-spindle drive *(Text)* Ein[zel]spindelantrieb *m*
s.-spindle model Einspindelbauart *f*
s.-spindle sliding-head type automatic screw machine Einspindelautomat *m* mit verschiebbarem Spindelstock
s.-spindle type Einspindelbauart *f*
s. split mould *(Plast)* zweiteilige Preßform *f*
s.-spool engine Einkreistriebwerk *n*
s.-spot welder Einfachpunktschweißgerät *n*
s. spread einseitig aufgetragen
s.-stage einstufig
s.-stage amplifier einstufiger Verstärker *m*
s.-stage grinding mill Einstufenmahlwerk *n*
s.-stage impulse turbine einstufige Druckturbine *f*
s.-stage recycle *(Kern)* Einstufenrückführung *f*, Stufe *f* mit Rückführung

s.-stage resin Resolharz *n*
s.-stage rocket einstufige Rakete *f*
s.-start eingängig *(Schnecke usw.)*
s.-start thread eingängiges Gewinde *n*
s.-station fixture Einstückspannvorrichtung *f*
s.-step hydroplane Einstufengleitboot *n*
s.-step resin Resolharz *n*
s.-step rocket einstufige Rakete *f*
s. sticker einmastiges Schiff *n*, Einmaster *m*
s.-storey einstöckig
s.-strand chain Einfachkette *f*
s. strand process *(Met)* Einstrangverfahren *n*
s.-strand roller chain Einfachrollenkette *f*
s.-strand roller-chain drive Einfachrollenkettentrieb *m*
s.-strap [riveted] butt joint einseitige Laschennietung *f*
s. stroke *(Wärm)* einfacher Hub *m*
s.-stroke bell Einschlagwecker *m*
s.-term eingliedrig
s.-thread[ed] eingängig
s.-throw circuit-breaker Ein-Aus-Schalter *m*
s.-throw crankshaft einfach gekröpfte Kurbelwelle *f*
s.-throw switch Kippschalter *m*, Hebelausschalter *m*
s.-tone keying *(Fmt)* Eintonverfahren *n*
s.-tool post (holder) Einfachmeißelhalter *m*
s.-tooth[ed] einzahnig, Einzahn-
s.-tower Einmast-
s.-tracked eingleisig, einbahnig
s. transformer Einzeltransformator *m*
s.-transistor oscillator Einzeltransistoroszillator *m*
s.-transistor stage Eintransistorstufe *f*
s.-trip trigger circuit *(Fs)* monostabile Triggerschaltung *f*
s.-trunnion tipping wagon Einzapfenkipper *m*
s.-tube core-barrel Einfachkernrohr *n (Tiefbohrtechnik)*
s.-turn current transformer Stabstromwandler *m*; Durchsteckwandler *m*
s.-turn transformer Einleiterwandler *m*
s. U butt weld U-Naht *f*
s. unit face *(Brg)* einflügliger Streb *m*
s.-unit frequency converter Einankerfrequenzwandler *m*
s. V butt weld V-Naht *f*
s.-valued einwertig, eindeutig
s.-valuedness Einwertigkeit *f*, Eindeutigkeit *f*
s.-valve amplifier Einröhrenverstärker *m*
s.-valve arrangement Einröhrenanordnung *f*
s.-valve receiver Einröhrenempfangsgerät *n*
s.-vision lens *(Opt)* Einstärkenglas *n*
s. warp *(Text)* Einfachkette *f*
s. water gauge einfacher Wasserstandanzeiger *m*
s. wave Einzelwelle *f*
s.-wave current Einwellenstrom *m*, einwelliger Strom *m*
s. wave-length working *(Fmt)* Einwellenbetrieb *m*
s.-wave operation Einwellenbetrieb *m*
s.-wave rectification Einweggleichrichtung *f*
s.-wave rectifier Einweggleichrichter *m*
s.-way furnace Einwegofen *m*
s.-way launching Kielablauf *m*, Stapellauf *m* mit Mittelschlitten
s.-way machine Einwegmaschine *f*
s.-way switch Einwegschalter *m*
s. wheel einfaches Rad *n*
s.-wheel grinding machine einscheibige Schleifmaschine *f*
s.-wheel knurl holder Rändelhalter *m*
s.-wheel knurling tool Rändelwerkzeug *n*
s.-wheel load Einzelradbelastung *f*
s.-wing door einflügelige Tür *f*
s.-wire eindrähtig, Eindraht-, Einleiter-
s.-wire aerial Eindrahtantenne *f*, Einleiterantenne *f*
s.-wire cable einadriges Kabel *n*, Einleiterkabel *n*, Einfachkabel *n*
s.-wire circuit Eindrahtleitung *f*
s.-wire feeder Eindrahtspeiseleitung *f*
s.-wire memory *(Dat)* Einzeldrahtspeicher *m (z. B. sogenannter Twistor)*
s. wire transmission line *(Fmt)* Eindrahtwellenleiter *m*
s. yarn backing *(Text)* Bindefadenfutter *n*
s. yarn warp *(Text)* einfädige Garnkette *f*
s. zone double apron arrangement *(Text)* Einzonendoppelriemchenvorrichtung *f (Streckwerk)*
singlet *(Opt)* Singulett *n*
singling-out of beets Vereinzeln *n* der Rüben
singly reinforced einfach bewehrt *(Stahlbetonbau)*
singular singulär
s. component *(Math)* singuläre Komponente *f*, singulärer Teil *m*
s. lattice singuläres Gitter *n*
s. point *(BMSR)* singulärer (ausgezeichneter) Punkt *m*
s. points of the trajectory of the principal stresses *(Ph)* singuläre Punkte *mpl* des Spannungsfeldes
s. proton-track Protoneneinzelbahn *f*
singularity Singularität *f*, singuläre Stelle *f*
sinistrorse [von rechts] nach links gewunden; linksgängig
sinistrorsum linksgewundene Kurve *f*
sink / to 1. [ver]sinken, [sich] setzen, [her]untergehen; eindringen; absacken; 2. [ab]senken, [ab]sinken, [ab]teufen, einsinken, niedersinken, versenken; 3. einarbeiten, hohlformen; einfräsen

s. a cut ausfräsen
s. a hole ein Bohrloch herunterbringen
s. a mine eine Mine abschießen
s. a mould *(Plast)* abwälzen, prägen
s. a pole einen Mast setzen
s. a well eine Bohrung abteufen (herunterbringen)
sink 1. Abwaschbecken *n*, Abzugsschleuse *f*; Ausguß *m*, Senkgrube *f*, Spülbecken *n*; 2. *(Math)* Senke *f*
s. basin Spülbecken *n*; Abwaschbecken *n*, Ausgußbecken *n*
s. hole 1. Abzugsgrube *f*; Senkgrube *f*; 2. *(Met)* Lunker *m*
s. mark *(Plast)* Einfallstelle *f*, Einsackstelle *f*, Mulde *f* *(Preßfehler)*
s. trap Abzugsgrube *f*
sinker 1. *(Text)* Platine *f*; gezeichneter Bindepunkt *m*; 2. *(Brg)* Bohrhammer *m* *(Abteufen)*
s. wheel machine Rundwirkmaschine *f*
sinking 1. Absenken *n*, Absenkung *f*, Abteufen *n*, Abteufung *f*, Senken *n*, Senkung *f*, Niederbringen *n*; 2. Einarbeitung *f*, Einfräsung *f*; Aussparung *f*, Vertiefung *f*
s. advance *(Brg)* Abteuffortschritt *m*
s. compartment *(Brg)* Abteuftrum *m* *(Weiterteufen)*
s. engine Abteuffördermaschine *f*
s. head frame Abteufgerüst *n*
s. head gear Abteufförderger üst *n*
s. hoist Abteuffördermaschine *f*
s. mill Reduzierwalzwerk *n*
s. of a support *(Bw)* Stützsenkung *f*
s. plant Abteufausrüstung *f*
s. scaffold winch Abteufbühnenwinde *f*
sinter / to sintern, fritten; [fest]brennen; verschlacken
s. fuse *(Plast)* aufsintern
sinter cake Sinterkuchen *m*, Agglomeratkuchen *m* *(Sinterröstung)*
sintered gesintert
s. alloy Sinterlegierung *f*
s. cake Agglomerat *n*, Anhäufung *f*, Ballung *f*, Sintererzeugnis *n*, Sinterkuchen *m*
s. carbide Sinterhartstoff *m*, Sinterkarbid *n*, Hartmetall *n*
s. carbide tip Sinterhartmetallschneidplättchen *n*
s. carbide tipped tool hartmetallbestücktes Werkzeug *n*
s. iron Sintereisen *n*
s. metal Sintermetall *n*
s. product Sinterwerkstoff *m*, Sinterteil *n*
s. steel Sinterstahl *m*
sintering 1. Sintern *n*; Fritten *n*; Verschlacken *n*; 2. Sinterteil *n*
s. furnace Sinterofen *m*
s. grate *(Met)* Verblaserost *m*
s. heat Sinterhitze *f*, Sintertemperatur *f*
s. pallet Sinterpfanne *f*
s. point Erweichungspunkt *m*
s. press Sinterpresse *f*
sinuosity Turbulenz *f*; Krümmung *f*
sinuous turbulent
sinus Höhlung *f*
sinusoid *(Math)* Sinuslinie *f*, Sinusoide *f*
sinusoidal sinusförmig, Sinus-
s. current sinusförmiger (einwelliger) Strom *m*, Sinusstrom *m*
s. deflection voltage sinusförmige Ablenkspannung *f*
s. input voltage sinusförmige Eingangsspannung *f*
s. motion sinusförmige Bewegung *f*
s. quantity sinusförmige Wechselgröße *f*, Sinusgröße *f*
s. signal sinusförmiges Signal *n*
s. sweep *(El)* sinusförmige Zeitablenkung *f*
siphon / to [ab]hebern *(Entleerung mit Hilfe eines Siphons)*
siphon Siphon *m*, Heber *m*, Saugheber *m*
s. barometer Heberbarometer *n*
s. pipe Heberleitung *f*
s. recorder Heberschreiber *m*
s. trap Geruchverschluß *m*; Siphon *m*
siren *(Ak)* Sirene *f*
sisal grass Sisal *m*, Sisalhanf *m*
sister keelson *(Schiff)* Seitenkielschwein *n*
s. ship Schwesterschiff *n*
site / to anbringen, aufstellen *(Geräte)*; unterbringen *(Einzelteile im Gehäuse)*
site Platz *m*; Baustelle *f*; Aufstellungsort *m*; Platz *(im Kristallgitter)*; Lage *f*; Gebäude *n*
s. arrangement Baustelleneinrichtung *f*
s. assembly Baustellenmontage *f*
s. concrete Ortbeton *m*
s. engineer Bauführer *m*, Bauleiter *m*
s. error Lagemißweisung *f* *(Peilfehler)*
s. hopper Baustellensilo *m*
s. office Bau[stellen]büro *n*
s. organization Baustelleneinrichtung *f*; Baustellenorganisation *f*
s. rivet Baustellenniet *m*; Montageniet *m*
s. survey Baustellenvermessung *f*
s. test Baustellenprüfung *f*
siting *(Bw)* Standortbestimmung *f*, Standortwahl *f*
sitting room Wohnzimmer *n*
six-angle[d] sechseckig
s.-colour recorder Sechsfarbenschreiber *m*
s.-cylinder engine Sechszylindermotor *m*
s.-cylinder [vertical] in-line engine stehender Sechszylinderreihenmotor *m*
s.-edged sechskantig
s.-figure sechsstellig
s.-fold sechsfach; sechszählig

s.-head comber Nähmaschine *f* mit 6 Köpfen
s. line pica *(etwa)* sechs Cicero *(Schriftgröße)*
s.-membered ring *(Ch)* Sechsring *m*
s.-phase current Sechsphasenstrom *m*
s.-phase rectifier Sechsphasengleichrichter *m*
s.-point bit Kronenschneide *f*, Rosettenschneide *f*
s.-point socket Sechskanteinsatz *m (für Sechskantschlüssel)*
s.-speed gear box sechsstufiger Räderkasten *m*
s.-spindle sechsspind[e]lig
s.-spindle automatic machine Sechsspindelautomat *m*
s.-spindle [horizontal] automatic bar machine Sechsspindelstangenautomat *m*
s.-spline shaft Sechskeilwelle *f*
s.-stage sechsstufig
s.-step filter *(Opt)* Sechsstufenfilter *n*
s. to pica *(etwa)* Viertelpetit *f (Schriftgröße)*
s.-wire three-phase system offenes Drehstromsystem (Dreiphasensystem) *n*
size / to 1. auf Fertigmaß bringen, kalibrieren; klassieren; sortieren; zurichten; 2. leimen, mit Leim überstreichen; schlichten
size 1. Abmessung *f*; Größe *f*; Maßgenauigkeit *f*; Dimension *f*; Baumaß *n*, Baugröße *f*; Format *n*; 2. Durchmesser *m (Bohrung)*; 3. Appreturmasse *f*; Grundiermasse *f*; Isoliergrund *m*; Leim *m*; Schlichte[mittel *n*] *f*; Untergrund *m*, Vorlack *m*
s. block Endmaßblock *m*
s. box *(Text)* Schlichtetrog *m*
s. consistency Maßhaltigkeit *f*
s.-control Meßsteuerung *f*
s.-control device Meßsteuereinrichtung *f*
s. cutting Harzverseifung *f*
s. degradation Korngrößenverminderung *f*
s. difference Größenunterschied *m*
s.-distribution curve Körnungskennlinie *f*, Kornverteilungskurve *f (Aufbereitung)*
s. drill Vollmaßbohrer *m*
s. grading Klassierung *f*, Sortierung *f*
s. limit Größenbereich *m*
s. loss Maßverlust *m*
s. mixing apparatus *(Text)* Schlichtekocher *m*
s. of cavity Größe *f* des Hohlraumes *(Kavitation)*
s. of commutator Kommutatorgröße *f*
s. of drawing Zeichnungsformat *n*
s. of drawing paper Blattgröße *f (technisches Zeichnen)*
s. of grain Korngröße *f*
s. of land *(Masch)* Fasenbreite *f*
s. of lump Stückgröße *f*
s. of meshes Maschenweite *f (Sieb)*
s. of sheet Bogengröße *f*, Format *n (Papier)*
s. range Korngrößenbereich *m*, Kornklasse *f*, Körnungsband *n*, Körnungsbereich *m (Aufbereitung)*
s. reduction Zerkleinerung *f*
s. stain *(Text)* Schlichtefleck *m*
s. vat Leimbütte *f*, Leimtrog *m*
sized tool Formmeißel *m*
sizer 1. Siebklassierer *m*; 2. Kalibrierwerkzeug *n*; Abrichtmaschine *f*
sizing 1. Kalibrieren *n*; Dimensionieren *n*; Bearbeiten *n* auf Maß; Klassieren *n*, Sortieren *n*; Zurichten *n*; Sieben *n*; 2. Leimen *n*; Grundieren *n*; Schlichten *n*; 3. Appreturmasse *f*; Grundiermasse *f*; Schlichte *f*; 4. Meßsteuerung *f*
s. agent *(Text)* Schlichtemittel *n*
s. bath *(Text)* Schlichtebad *n*, Schlichteflotte *f*
s. broach Kalibrierräumwerkzeug *n*
s. device Meßsteuereinrichtung *f*
s. drum Klassiertrommel *f*, Siebtrommel *f*
s. machine *(Text)* Schlichtmaschine *f*
s. material *(Text)* Schlichte *f*
s. mix *(Text)* Schlichtegemisch *n*
s. of ores Klassieren *n* der Erze
s. paste *(Text)* Schlichtemasse *f*
s. pin Kaliberdorn *m*, Sortierstift *m*
s. preparation Appreturmasse *f*
s. press Kaliberpresse *f*
s. reamer Kaliberreibahle *f*
s. roll Kalibrierwalze *f*
s. rolls Maßwalzwerk *n*
s. screen Klassiersieb *n*
s. test Siebanalyse *f*
s. the warp threads Schlichten *n* der Kettfäden
s. tool Kalibrierwerkzeug *n*
s. tooth Kalibrierzahn *m (Räumwerkzeug)*
sizzle / to zischen; knistern
sizzle Knattern *n*, Knistern *n*, Rasseln *n (Funkempfangstechnik)*
skarn *(Geol)* Skarn *m*
skate 1. Schlittschuh *m*; 2. Gleitkontakt *m*
skater Eisläufer *m*
skating ground Eisbahn *f*
skatole *(Ch)* Skatol *n*
skeg *(Schiff)* Totholz *n*
s. of keel Kielhacke *f (Holzschiff)*
s. shore Kielstütze *f*
skein *(Text)* Strang *m*, Docke *f*
s. drier Strangtrockenapparat *m*
s.-dyed stranggefärbt
s. dyeing Strangfärben *n*
skeined hank *(Text)* abgebundener Strähn *m*
skeleton Gerippe *n*
s. construction Skelettbauweise *f*
s. diagram Prinzipschaltung *f*
s. floor *(Schiff)* offene Bodenwrange *f*
s. key Dietrich *m*

s. sketch Strichzeichnung *f*, schematische Zeichnung *f*
s. structure Skelettbau *m*
s.-type switchboard Gestellschaltpult *n*
s.-type tube mount Rohrmontierung *f* mit offener Gitterkonstruktion
skelp *(Met)* Röhrenstreifen *m*
s. rolling mill Rohrstreifenwalzwerk *n*
skep Weidenkorb *m*, Korb *m*; Bienenkorb *m*
sketch / to skizzieren
sketch Skizze *f (technisches Zeichnen)*
s. of interlacing *(Text)* bildliche Darstellung *f* der Bindungen
s. plate Skizzenblech *n*
sketching Skizzieren *n*
sketchmaster Luftbildumzeichner *m*
skew Schrägverzerrung *f (Bildfunk)*
skew schräg *(Verzahnung, Kegelrad)*; [schief-]winklig, verzerrt; geschweift *(Säge)*; [wind-]schief
s. arch *(Bw)* schief[winklig]er Bogen *m*
s.-arched vault *(Bw)* schief[winklig]es Gewölbe *n*
s.-back blade geschweifter Blattrücken *m* einer Säge
s.-back saw Säge *f* mit geschweiftem Rücken
s. bevel gear Schrägzahnkegelrad *n*
s. bridge schief[winklig]e Brücke *f*
s. factor Schrägungsfaktor *m (der Wicklung)*
s. gear Schrägzahnrad *n*
s. plane iron Plattenbankhobeleisen *n*
s. roller table Schrägrollengang *m (Walzwerk)*
s.-rolling mill Schrägwalzwerk *n*
s.-symmetric *(Math)* antisymmetrisch, schiefsymmetrisch
s. teeth Schrägzähne *mpl (Kegelrad)*
s. wheel Hyperboloidenrad *n*, Hyperbelrad *n (Verzahnung)*
s.-whiff schief, gewunden, krumm *(Verzahnung)*
skewer *(Text)* Aufsteckspindel *f*
skewness Schräge *f*; Schiefheit *f*; Einseitigkeit *f*, Asymmetrie *f*
ski Schi *m*, Ski *m*
s.-hoist Schilift *m*
s. press Schispannzeug *n*
s.-wear Schibekleidung *f*
skiagram Röntgenaufnahme *f*
skiagraph Röntgenaufnahme *f*
skiascope *(Opt)* Skiaskop *n*
skiatron Skiatron *n*, Schwärzungsröhre *f*, Dunkelschriftröhre *f*
skid / to gleiten, rutschen
skid 1. Gleitbahn *f (für Bootsaussetzvorrichtung)*; Ladeschlitten *m*; Ladebaum *m*; 2. Hemmschuh *m*; Kufe *f*; Gleitschiene *f*; Tragbalken *m*; 3. *(Plast)* Markierungen *fpl* durch überfließendes Material; 4. *(Flg)* Sporn *m*
s. beam Bootsgalgen *m*, Gleitbahn *f (für Bootsaussetzvorrichtung)*
s. for paper Ablagebock *m* für Papier
s. plate Gleitblech *n*
s. resistance *(Kfz)* Rutschsicherheit *f*
s. resistant gleitsicher; rutschsicher
s. undercarriage Kufenfahrgestell *n*
skidder Holzschleifer *m*, Holzrücker *m*
skidding Gleiten *n*, Rutschen *n*; Rücken *n (Holztransport)*; Schleudern *n (Wagen)*
skidless decking *(Schiff)* Gleitschutzdecksbelag *m*
skids Unterleghölzer *npl*
skidway Schleifweg *m (Holztransport)*
skill Geschick *n*, Übung *f*, Geschicklichkeit *f*, Handfertigkeit *f*
skilled-labour shortage Facharbeitermangel *m*
s. lathe operator Dreher *m*
s. rate of pay Facharbeiterlohn *m*
s. worker Facharbeiter *m*
skillet Gußstahltiegel *m*
skim / to abheben, abrahmen, abschöpfen, entrahmen, entschäumen; abschäumen *(Beize)*; abheben *(Gleitboot)*
s. off abziehen; abschöpfen; abschäumen
skim coulter Schäleinleger *m*, Vorschäler *m*, Vorschneider *m*, Hobelpflug *m*
skim[med] milk Magermilch *f*
skimmer 1. Abheber *m*, Abschäumlöffel *m*, Abstreicher *m*, Kratze *f*, Krampstock *m*, Schaumlöffel *m*; 2. Flachbagger *m*; Flachlöffel *m*, Planierlöffel *m*; Flachlöffelbagger *m*, Planierlöffelbagger *m*
s. gate Schaumtrichter *m (Gießtechnik)*
skimming Abheben *n*; Abrahmen *n*; Abschöpfen *n*; Entschäumen *n*; Abschäumen *n (Beize)*; Abkrammen *n*
s. bottom Gleitboden *m (eines Gleitbootes)*
s. head Schruppschälkopf *m*
s. screw Außenhautbefestigungsschraube *f*
s. speed *(Schiff)* Gleitgeschwindigkeit *f*
skimmings 1. Schlacke *f*, Schaum *m*, Abgeschäumtes *n*; 2. Krätze *f*
skin / to 1. [ent]häuten, abbalgen, abziehen; 2. abisolieren, Isolierung entfernen; 3. *(Bw)* abputzen, tünchen
skin 1. Fell *n (von kleinen Tieren)*; Haut *f*, Balg *m*, Pelz *m*; 2. *(Met)* Gußhaut *f*, Walzhaut *f*; Randzone *f*; 3. *(Lw)* Hülse *f (Getreide)*; 4. *(Schiff)* Außenhaut *f*
s. boat Fellboot *n*
s. covering Außenhaut *f*
s. decarburization *(Met)* Randentkohlung *f*
s. depth Eindringtiefe *f*
s. effect Hautwirkung *f*, Skineffekt *m*, Stromverdrängung *f*
s. fault Oberflächenfehler *m*

s. friction Mantelreibung *f (Pfahlgründung)*; Oberflächenreibung *f*, Außenhautreibung *f*
s. friction drag Oberflächenwiderstand *m (Strömung)*
s. hardness Oberflächenhärte *f*
s.-irritating hautreizend
s. of casting Gußhaut *f*
s. of the satellite Hülle *f* des Satelliten
s. pass Dressierwalzwerk *n*, Dressiergerüst *n*
s. plate kunststoffplattiertes Blech *n*
s. protective Hautschutzmittel *n*
s. resistance Reibungswiderstand *m (Strömung)*
s. wool Gerberwolle *f*, Hautwolle *f*; Schlachtwolle *f*
skinner 1. isoliertes Drahtende *n (einer Kabelform)*; 2. *(Am)* leere Spule *f (Betriebsausdruck)*
skinner's wool Blötwolle *f (Kadaverwolle)*
skip / to überspringen, auslassen; springen
skip 1. Förderkübel *m*, Transportkübel *m*, Fördergefäß *n*, Skip *m*; Aufzugskübel *m (Mischmaschine)*; Gichtkübel *m*; 2. *(Dat)* Auslassung *f*, Aussetzer *m*; 3. *(Text)* Webnest *n*
s. bars Sprungschienen *fpl (Lochkartenmaschinen)*
s. bridge Schrägaufzug *m (Hochofen)*
s. charging Kippgefäßbegichtung *f*
s. compartment *(Brg)* Fördertrum *m*
s. distance *(Fmt)* Sprungentfernung *f (Ausbreitung)*; *(Ak)* Zone *f* des Schweigens, tote Zone
s.-feeding for rapid traverse between gaps in the work *(Masch)* Sprungvorschub *m* für Leerwege im Eilgang
s. filling Kippkübelbegichtung *f*
s. frequency *(Fmt)* Grenzfrequenz *f*
s. hoist Schrägaufzug *m*, Kübelaufzug *m*, Kippaufzug *m*, Muldenheber *m (Betonmischer)*
s. operation *(Fmt)* Verschränkung *f* [von Wählergruppen]
skipjack Knickspantboot *n (mehrfach geknickte Spanten)*
skipped distance *(Fmt)* Sprungentfernung *f*, tote Zone *f*
skipper Schiffer *m*, Schiffsführer *m*, Kapitän *m (eines kleinen Handelsschiffes)*
skipping off Springen *n (eines Flugzeugs bei der Landung)*
s. rope Sprungseil *n*
skirt / to ausspritzen [lassen]
skirt 1. Rock *m*; 2. Rand *m*; Saum *m*; Grenze *f*; 3. Wellenfallenrand *m (Wellenleitertechnik)*; 4. Strahl *m (Flüssigkeit)*
skirted valve-plug *(BMSR)* Schlitzkegel *m*
skirting board Fußleiste *f*, Scheuerleiste *f*, Sockelleiste *f*
skive / to *(Plast)* aufspalten, Kanten glätten (schärfen)
skiving machine Schärfmaschine *f (Schuhfertigung)*
skot Skot *n (Einheit der Dunkelleuchtdichte bei einer Farbtemperatur von 2360° K)*
skotograph Dunkelleuchtbild *n*
skull *(Met)* Pfannenbär *m*
sky Firmament *n*, Himmel *m*
s. wave *(Fmt)* Raumwelle *f*, reflektierte Welle *f*
skylight Oberlicht *n*; Deckfenster *n*
s. cover Oberlichtbezug *m*
skyscraper Wolkenkratzer *m*
slab / to 1. *(Brg)* abblättern, ablösen, abplatzen *(Gestein aus dem Streckenstoß)*; 2. *(Masch)* Flächen bearbeiten; 3. *(Met)* platinieren, flachwalzen *(Blöcke)*
slab 1. *(Met)* Bramme *f*, Platine *f*; Platte *f*; Tafel *f*; 2. *(Brg)* Gesteinsablösen *n*, Gesteinsplatte *f*; 3. *(Holz)* Schwarte *f*
s. and girder floor *(Bw)* Plattenbalkendecke *f*
s. billet *(Met)* Flachknüppel *m*
s. block Scheibenhochhaus *n*
s. bridge Plattenbrücke *f*
s. cogging mill Brammenwalzwerk *n*
s. coil Flachspule *f*
s. cut Besäumschnitt *m*
s. foundation Plattenfundament *n*
s. heating furnace Brammentiefofen *m*
s. ingot Brammenblock *m*, Rohbramme *f*
s. keel *(Schiff)* Schutzkiel *m*
s. mill Feinstraße *f*
s. of stock Rohstoff *m* in Tafeln
s.-on-grade nichtunterkellert *(Gebäude)*
s. on grade nichtunterkellerte Fußbodenplatte *f*
s. shears Brammenschere *f*; Schlagschere *f*
s. turning device Brammendrehvorrichtung *f*
s. yard crane Grobblechlagerkran *m*
slabbing 1. *(Brg)* Abbau *m* in Form schmaler senkrechter Streifen von der Ortsbrust; 2. *(Met)* Flachwalzen *n*
s.-cut *(Brg)* Fächereinbruch *m (Schießarbeit)*
s. mill Brammenwalzwerk *n*
s. mill roll Flachwalze *f*
s. mill train Platinen[walz]straße *f*
s. pass Platinenkaliber *n*, Brammenkaliber *n*
slabstock *(Text)* Streichbaum *m*
slack / to nachlassen, lockerlassen
s. up [sich] lockern *(Schraube)*
slack 1. Gruskohle *f*, Klarkohle *f*, Feinkohle *f*; 2. *(El)* loser Durchhang *m* der Leitung; Zuschlag *m* zur Leitungsdrahtlänge *(infolge Durchhangs der Leitung)*
slack locker, schlaff, lose
s. cable switch Schlaffseilschalter *m*
s. coal Gruskohle *f*, Kohlengrus *m*
s. hours *(Fmt)* verkehrsschwache Zeit *f*
s. joint Wackelkontakt *m*

s. **lime** *(Led)* Ascherkalk *m*
s. **period** *(Fmt)* verkehrsschwache Zeit *f*
s. **rope** Hängeseil *n (Schachtförderung)*
s. **selvage** *(Text)* wellige Leiste *f*
s. **side** Leertrum *m (Riemen)*
s. **water** *(Schiff)* Stauwasser *n*, Totwasser *n*, Stillwasser *n*
s. **wax** Rohparaffin *n*, Paraffingatsch *m*
slacken / to entspannen, [sich] lockern, zurückdrehen, lösen *(Schraube)*
s. **in speed** mit der Fahrt heruntergehen
slackening off Lockern *n*, Lösen *n*, Zurückdrehen *n*
slackline cable plant Kabelbagger *m*
slackness Spiel *n*; Schlaffheit *f*; Durchhängen *n*
slag / to schlacken; ausschlacken; sintern; verschlacken
s. **out** abschlacken
slag Schlacke *f*, Gekrätz *n*; Asche *f*
s. **action** *(Met)* Schlackenangriff *m*
s. **addition** *(Met)* Schlackenzuschlag *m*
s. **brick** Schlackenziegel *m*, Schlackenstein *m*
s. **bucket** Schlackenkübel *m*
s. **cement** Hüttenzement *m*, Hochofenzement *m*
s. **channel** Schlackenrinne *f*
s. **concrete** Hochofenschlackenbeton *m*
s. **control** *(Met)* Schlackenführung *f*
s. **conveying truck** Schlackentransportwagen *m*
s. **dump** Schlackenhalde *f*
s. **entrapment** Schlackeneinschluß *m*
s. **former** Schlackebildner *m*, schlackenbildendes Mittel *n*
s.**-forming** schlackenbildend
s.**-forming period** Verschlackungsperiode *f*
s. **furnace** Schlackenofen *m*
s. **hole** Schlackenabstichloch *n*
s. **hook** Schlackenhaken *m*
s. **ladle** Schlackenpfanne *f*
s. **ladle car** Schlackenwagen *m*
s. **lead** Krätzblei *n*
s. **notch** Schlackenabstichloch *n*
s. **occlusion** Schlackeneinschluß *m*
s.**-off** Schlackenabzug *m*
s. **penetration** Verschlackung *f*
s. **pile** Schlackenhalde *f*
s.**-pot truck** Transportkübelwagen *m*
s. **ratio** *(Met)* Basizitätsverhältnis *n* $(CaO:SiO_2)$
s. **regulation** Schlackenführung *f*
s. **removal** Entschlackung *f*
s. **skimmer** Krampstock *m*, Schlackenabscheider *m*
s. **spout** Schlacken[abstich]rinne *f*
s. **streak** Schlackenzeile *f (im Gefüge)*
s.**-sulphate cement** Gipsschlackenzement *m*
s. **tap** Schlackenabstich *m*
s. **transporter** Schlackentransportwagen *m*
s. **truck** Schlackentransportwagen *m*
s. **tuyere** Schlackenform *f*
s. **wool** Schlackenwolle *f*
slaggability Verschlackungsfähigkeit *f*, Verschlackungsvermögen *n*
slagging 1. Schlackenbildung *f*; Entschlackung *f*; 2. *(Kern)* Spaltproduktausschlackung *f*
slaggy schlackenähnlich, schlackenartig; schlakkenreich, schlackig
slake / to [Kalk] löschen
slake-lime pit Äscher *m*
slaked lime gelöschter Kalk *m*, Löschkalk *m*
slaking Ablöschen *n*, Kalklöschen *n*
s. **slag** Zerfallschlacke *f*
slamming loads *(Schiff)* Stampfstöße *mpl*, Seeschlagbelastung *f*
slant / to schräg liegen (sitzen); abböschen, [sich] neigen
slant Neigung *f*, Gefälle *n*, Schräge *f*, Schrägung *f*
s. **drilling** Schrägbohren *n*
slanting Schrägung *f*
slanting schief, schräg, seitlich abfallend
s. **bottom** Schrägboden *m (eines Bunkers)*
s. **cut** *(Holz)* Krummschnitt *m*
s. **drill hole** Bohrloch *n* mit schiefer Richtung *(Tiefbohrtechnik)*
s. **lift** Schrägaufzug *m*
slash / to schlitzen; fällen
slash grain *(Holz)* Tangentialschnitt *m*
slasher [sizing machine] *(Text)* Trommeltrokkenschlichtmaschine *f*
slashes gefällte Bäume *mpl*
slashing *(Brg)* Abbau *m* in Form schmaler senkrechter Streifen von der Ortsbrust
slashings Schlagabraum *m*, abgehauene Äste *mpl*, gefällte Bäume *mpl*
slat 1. Latte *f*, Steg *m*; 2. *(Flg)* Vorflügel *m*
s. **conveyor** Taschenbandelevator *m*
s. **grating** Lattenrost *m*
s. **machine** Holzstabmaschine *f*
s.**-packing tower** Füllkörpersäule *f* mit bandförmiger Füllung
slate / to [mit Schiefer] decken
slate Schiefer *m*, Tonschiefer *m*, Schieferplatte *f*, Dachschiefer *m*
s. **axe** Dachhammer *m*
s.**-black** schieferschwarz
s. **black** Zeichenschiefer *m*, Schieferschwarz *n*
s.**-blue** schieferblau
s. **board** Schiefertafel *f*
s.**-clay** Schieferton *m*
s. **coal** Blattkohle *f*, Schieferkohle *f*, Splitterkohle *f*
s.**-grey** schiefergrau
s. **in leaves** Tafelschiefer *m*
s. **quarry** Schieferbruch *m*
s. **quarrying** Schieferbergbau *m*

s. roof Schieferdach *n*
slater 1. Dachdecker *m*, Schieferdecker *m*; 2. *(Led)* Glätteisen *n*
slater's hammer Dachhammer *m*
slating Schieferdeckung *f*
s. machine *(Led)* Glättmaschine *f*
s. stone *(Led)* Glättstein *m*
slatted deck *(Schiff)* Lattendeck *n*
slaty schieferähnlich, schieferartig, schieferhaltig, schief[e]rig
slaughtered hide Schlachthaushaut *f*
slaughterhouse Schlachthaus *n*
slave arm Arbeitsarm *m (Fernbedienungsgerät)*
s. clock *s.* secondary clock
s. end Arbeitsseite *f (Fernbedienungsgerät)*
s. station Nebensender *m (Decca)*
s. transmitter Tochtersender *m*, Nebensender *m*
s. wrist joint Arbeitsgelenk *n (Fernbedienungsgerät)*
Slavianoff arc welding process Slawjanow-Lichtbogenschweißverfahren *n*
slay cap *(Text)* Ladendeckel *m*
s. check *(Text)* Ladenbaum *m*, Ladenklotz *m*
s. movement *(Text)* Ladenbewegung *f*
s. top *(Text)* Ladendeckel *m*
sled / to schlitteln *(Holztransport)*
sled Schlitten *m*
s. plane Kufenflugzeug *n*
s.-road Schlittenweg *m (Holztransport)*
s. runner Schlitten *m*
sledge / to schlitteln *(Holztransport)*
sledge 1. *(Brg)* Vor[schlag]hammer *m*; 2. Schlitten *m*
s. hammer schwerer Schlegel (Treibfäustel) *m*
sledger-runner Schlittenkufe *f*
sledging Zerkleinern *n* (Nachzerkleinerung *f*) mit dem Fäustel
sleek / to glätten; *(Led)* reinmachen, mit dem Schlicker bearbeiten
s. off *(Led)* Fett abstoßen
sleeker Dämmbrett *n (Formerei)*
sleeper 1. Schlafwagen *m*; 2. Eisenbahnschwelle *f*; 3. *(Bw)* Unterzug *m*
s. block Doppelschwelle *f*
s. bolt Schwellenschraube *f*
s. laying machine Schwellenverlegemaschine *f*
s. pass Schwellenkaliber *n*
s. plate Schwelle *f (Holzbau)*
sleeping car Schlafwagen *m*
sleeve Muffe *f*; Büchse *f*; Buchse *f*, Laufbuchse *f*, Zylinderbuchse *f*; Hülse *f*, Schafthülse *f*, Bohrhülse *f*; Pinole *f*; Führungskörper *m*; Manschette *f*; Muffe, Gleitmuffe *f*; Verbindungsmuffe *f*; Lagerschale *f*; Mantelrohr *n (Radialbohrmaschine)*
s. bearing Gleitlager *n*
s. brace Bohrwinde *f*
s. buffer *(Eis)* Hülsenpuffer *m*
s. compound Muffenvergußmasse *f*
s. coupling Muffenkupplung *f*, Hülsenkupplung *f*, Schalenkupplung *f*, Kupplungsmuffe *f*, Durchgangsmuffe *f*, Verbindungsmuffe *f*
s. dipole Hülsendipol *m*, Rohrdipol *m (Antennen)*
s. hem Ärmelsaum *m*
s. joint Muffenverbindung *f*, Vermuffung *f*, Hülsenbund *m*
s.-lock Pinolenfestklemmung *f*
s. nut Hülsenmutter *f*, Überwurfmutter *f*
s. packing Ringbüchse *f*
s. port Schieberschlitz *m*, Schieberöffnung *f*
s.-pressure curve *(Wärm)* Muffendruckkurve *f*, Kurve *f* des Muffenreglers
s. protector Ärmelschoner *m*
s. screw Hohlspindel *f*
s. socket Paßeinsatz *m*
s. splice *(Bw)* Hülsenstoß *m*
s.-type chain Hülsenkette *f*
s. valve Buchsenventil *n*, Kolbenschieber *m*, Rohrschieber *m*, Schiebeventil *n*, Zylinderlaufbuchse *f*
s. valve engine schiebergesteuerter Motor *m*, Schiebermotor *m*
s. wire *(Fmt)* Stöpselringanschluß *m*
sleeving Buchse *f*; Muffe *f*
s. machine Ärmeleinnähmaschine *f*
sleigh Schlitten *m*
slender schlank
slenderness ratio Schlankheitsgrad *m*
slew / to schwenken, drehen
slewability Schwenkbarkeit *f*
slewable schwenkbar, drehbar
slewing Schwenken *n*, Drehen *n*
s. angle Schwenkwinkel *m (Kran)*
s. crane Drehkran *m*, Schwenkkran *m*
s. drive Schwenkantrieb *m (Kran)*
s. journal Drehzapfen *m*
s. motion Drehbewegung *f*, Schwenkbewegung *f*
s. ring Drehkranz *m (Drehkran)*
sley *(Text)* 1. Weblade *f*; 2. *(Am)* Kettendichte *f*
sleying *(Text)* Blattstechen *n*
slice / to in Platten (Scheiben) schneiden; in Scheiben abbauen
slice 1. Scheibe *f*, Keil *m*; 2. *(Brg)* Scheibe, Abbauscheibe *f*; 3. Staulatte *f*, Stauvorrichtung *f (Papierherstellung)*
s.-drift *(Brg)* Querbrechen *n (Teilsohlen- und Querbruchbau)*
s.-galley *(Typ)* Zungen[setz]schiff *n*
sliced sheet vom Block geschnittene Folie *f*
s. veneer Messerfurnier *n*

slicer 1. Schneidmaschine *f*, Schnitzelmaschine *f*, Stauvorrichtung *f (Papierherstellung)*; 2. *(El)* doppeltwirkender Begrenzer *m*
slicing 1. Schneiden *n (Platten)*; Schneiden vom Block *(Zelluloid)*; 2. Messern *n*; 3. *(Brg)* Scheibenbau *m*
s.-and-caving [method] *(Brg)* Querbruchbau *m*, Scheibenbruchbau *m*
s. machine *(Led)* Abschärfmaschine *f*
s. method *(Brg)* Scheibenbau *m*
slickenside *(Geol)* Gleitfläche *f*, Harnisch *m*, Rutschfläche *f*
slickensiding *(Geol)* Gleitschicht *f*
slicking Glätten *n* von Gießformflächen
slidable verschiebbar
s. axially verschiebbar (verschieblich) in Achsrichtung
slide / to gleiten, rutschen
slide 1. Gleitbacke *f*; Gleitbahn *f*, Gleitfläche *f*, Führung *f*; Rutsche *f*; 2. Support *m*, Führungsschlitten *m*, Schlitten *m*; Stößel *m*; Schieber *m*; Zunge *f*, Läufer *m (Rechenschieber)*; 3. Rollsitz *m (Ruderboot)*; 4. Gleitstein *m (Getriebelehre)*; 5. Lichtbild *n*, Diapositiv *n*
s.-back valve voltmeter Kompensationsröhrenvoltmeter *n*
s.-back voltmeter Kompensationsvoltmeter *n*
s. bar 1. Gleitschiene *f*; Geradführung *f (Kreuzkopf)*; Lineal *n*, Schieber *m*; 2. *(Brg)* Schiebekappe *f*, Gleitkappe *f*
s.-bar bearer Gleitschienenbock *m*
s.-bar head *(Brg)* Stempelkopf *m* für Schiebekappen
s. bearing Gleitlager *n*
s. block *(Masch)* Kreuzkopfschleifer *m*, Gleitklotz *m*, Schlitten *m*
s. bow *(Kfz)* Gleitspriegel *m*
s. box Gleitbüchse *f*, Schieberkasten *m*, Ventilkasten *m*
s. bridge Brücke *f* mit Gleitkontakt *(Meßtechnik)*
s. caliper Schieblehre *f*, Schublehre *f*
s. catch Bajonettverschluß *m*
s. changer Bildwechsler *m*
s. contact Gleitkontakt *m*, Schleifkontakt *m*, Schiebekontakt *m*
s. control stoßfreie (stetige) Regelung *f*
s. damper Schieber *m*; *(Wärm)* Rauchschieber *m*
s.-damper opening Rauchschieberöffnung *f*
s.-damper regulation Schieberregelung *f*, Schieberregulierung *f*
s. division Schiebereinteilung *f*
s. dog Anschlagknagge *f*
s.-fastener link Reißverschlußglied *n*
s. fit *(Masch)* Gleitsitz *m*
s. for cylinder base Schlittenführung *f* für den Zylinderfuß
s. gate Schleusentor *n*
s. gate valve Schieberventil *n*; Schleusenschütz *n*, Schleusenschieber *m*
s. gauge Schieblehre *f*, Schublehre *f*
s. head Schlitten *m (Räummaschine)*
s.-in chassis Einschubchassis *n*
s. lecture Lichtbildervortrag *m*
s. lock Bajonettverschluß *m*, Gleitriegel *m*
s. needle Schiebernadel *f*
s. plate Schieber *m*, Zungenplatte *f*
s.-protective spring mount Präparateschutz *m*
s. rail 1. Stellschiene *f*, Gleitschiene *f*, Laufschiene *f*; 2. *(Brg)* fliegendes Gestänge *n*
s. resistance Schiebewiderstand *m*
s. resistor Schiebewiderstand *m*
s. rest Support *m*, Werkzeugschlitten *m*
s. ring Gleitstein *m*, Gleitring *m*
s. rod Schieberstange *f*
s. rule Rechenschieber *m*, Rechenstab *m*
s.-rule caliper Schieblehre *f*, Schublehre *f*
s. selector Umsteuer[kolben]schieber *m*
s.-spindle grinding attachment Verschiebespindelschleifeinrichtung *f*
s. tappet Stößel *m*
s. valve Flachschieber *m*; Absperrschieber *m*; Schieberventil *n*; Schieber *m*
s. valve chest Schieberkasten *m*
s. valve circle Schieberkreis *m*
s. valve diagram Schieberdiagramm *n*
s. valve ellipse Schieberellipse *f*
s. valve pump Schieberluftpumpe *f*
s. way Gleitbahn *f*
s. window Fallfenster *n*
s. wire Widerstandsschleifdraht *m*, Schleifdraht *m*; Meßdraht *m*
s.-wire attenuator Schleifdrahtabschwächer *m*
s.-wire bridge Schleifdrahtmeßbrücke *f*
s.-wire brush Schleifdrahtbürste *f*
s.-wire circuit Schleifdrahtschaltung *f*
s.-wire potentiometer Schleifdrahtkompensator *m*
s.-wire resistor Schleifdrahtwiderstand *m*
slider 1. Gleitkontakt *m*; Schleifer *m (eines Potentiometers)*; 2. *(Text)* Kulisse *f*, Lauffläche *f*; 3. Schieber *m*, Schiebeeinrichtung *f*; 4. Laufgewichtsstück *n*
s. coil *(El)* Spule *f* mit Gleitkontakt
slideway Führung *f*, Gleitbahn *f*
s. grinder Führungsbahnenschleifmaschine *f*
s. grinding machine Führungsbahnenschleifmaschine *f*
sliding 1. Gleiten *n*, Rutschen *n*, Schlupf *m*; 2. Langdrehen *n*
sliding gleitend; verschiebbar
s. and screwcutting lathe Leit- und Zugspindeldrehmaschine *f*

s. and surfacing lathe Zugspindeldrehmaschine *f*
s. bar 1. Führungsstange *f*, Geradführung *f*, Gleitstange *f*; 2. Meßschenkel *m*
s. bearing Gleitlager *n*, Lagerung *f* für hin- und hergehende Bewegung
s. block 1. Kulissenstein *m*, Führungsstein *m*, Schwingenstein *m*; Gleitstück *n*, Gleitblock *m*; 2. *(Brg)* Gleitfuß *m* *(Ausbau)*
s. block linkage Schubkurbel *f*
s. bolt Gleitriegel *m*, Riegel *m*
s. bow Stromabnehmerbügel *m*, Kontaktbügel *m*
s.-bow carbon brush Bügelschleifkohle *f*
s. bridge Meßdrahtbrücke *f*
s. bulkhead door Schottschiebetür *f*
s. caliper Schieblehre *f*, Schublehre *f*
s. cam Gleitnocken *m*
s. cam plate Kurvenschieber *m*
s. cockpit enclosure *(Flg)* Cockpitschiebehaube *f*
s. collar Schiebehülse *f*
s. contact Gleitkontakt *m*, Schleifkontakt *m*, Schiebekontakt *m*
s.-contact bearing Gleitlager *n*
s.-contact maker Schleifkontakt *m*
s. coupling Schiebekupplung *f*
s. cover Schiebedeckel *m*
s. curve *(Masch, Wärm)* Gleitkurve *f*
s. damper *(Masch, Wärm)* Senkschieber *m*
s. door Schiebetor *n*, Schiebetür *f*, Rolltür *f*
s. door for bunker Bunkerschieber *m*, Bunkerschiebetür *f*
s. double gear Zweiräderblock *m* *(Zahnrad)*
s. electrode verstellbare Elektrode *f*
s. feather [key] Gleitfeder *f*
s. feed Langvorschub *m* *(Drehmaschine)*
s. fit *(Masch)* Gleitsitz *m*
s. flue damper Rauchschieber *m*
s. form[work] Gleitschalung *f* *(Betonbau)*
s. friction gleitende Reibung *f*, Gleitreibung *f*
s. gear Schieberad *n*, verschiebbares Zahnrad *n* *(Getriebe)*
s. gears Schieberäder[block *m*] *npl*
s. grate Schieberost *m*
s. hatch Schiebefenster *n*, Schiebeluke *f*
s. head Schieber *m*, Schlitten *m*, Höhenschieber *m*
s. head-type automatic screw machine Langdrehautomat *m*
s. inductance Schiebespule *f*
s. jaw Gleitbacke *f*, verschiebbare Backe *f*
s. keel *(Schiff)* Steckschwert *n*, Hängekiel *m*
s. key Gleitfeder *f*; Ziehkeil *m*
s. lathe Langdrehmaschine *f*
s. lever Gleithebel *m*
s. lock Verschlußschieber *m*
s. member Gleitstein *m* *(Getriebelehre)*; Schieber *m* *(Meßschieber)*; bewegliches Teil *n*
s. mesh gear Schieberad *n*
s.-mesh gear train Schieberädergetriebe *n*
s. micromanipulator Gleitmikromanipulator *m*
s. motion Gleitbewegung *f*
s. mould Gleitschalung *f* *(Betonbau)*
s. movement Gleitbewegung *f*; Schubbewegung *f*
s. nut Gleitmutter *f*, Schraubenmutter *f*
s. panel Schiebewand *f*
s. part Gleitstück *n*
s. pin drive Kurbelschwinge *f*
s. pin handle Knebel *m* *(Schraubzwinge)*
s. piston-type of selector valve Umsteuerkolbenschieber *m*
s. plane Gleitebene *f*
s. plug Schieber *m*
s. resistance Gleitwiderstand *m*
s. resistor Schiebewiderstand *m*, Stellwiderstand *m*
s. ring Schieberring *m*
s. roadway arch *(Brg)* Gleitbogen *m* *(Ausbau)*
s. roof Schiebedach *n*
s.-roof fastener Schiebedachverschluß *m*
s. rotor Verschiebeanker *m*
s. rupture Verschiebungsbruch *m*
s. sash Schiebefenster *n*
s. scale Gleitskala *f*
s. screen-grid voltage gleitende Schirmgitterspannung *f*
s. seat Rollsitz *m*
s. selector shaft *(Kfz)* Schaltstange *f*
s. shackle bearing Abwälzlager *n*
s. shaft Schiebewelle *f*
s. shutter Längsschieber *m*, Schieberverschluß *m*
s. shuttering Gleitschalung *f* *(Betonbau)*
s. shuttle *(Text)* Schleifschützen *m*, Gleitschützen *m*, Rutscherschützen *m*
s. slay *(Text)* Gleitlade *f*
s. sleeve Abhebezylinder *m*, Ausrückmuffe *f*, Gleitbüchse *f*, Schiebehülse *f*, Schiebemuffe *f* *(Getriebe)*
s. speed Gleitgeschwindigkeit *f*
s. spool-type valve Kolbenschieber *m*
s. stage Gleittisch *m* *(Mikroskop)*
s. support Gleitfuß *m*
s. surface Gleitfläche *f*; Lauffläche *f*
s. T-bevel Stellschmiege *f*
s. table Schiebetisch *m*
s. tappet Gleitstößel *m*
s. ticker Schleifdetektor *m*
s. triple gear Dreiräderblock *m* *(Zahnrad)*
s. vane Schieber *m* *(Zellenverdichter)*
s.-vane compressor Zellenverdichter *m*
s. velocity Gleitgeschwindigkeit *f* *(Zahnrad)*

s. vernier caliper Schieblehre *f*
s. ways 1. *(Schiff)* Ablaufschlitten *m*, Stapellaufschlitten *m*; 2. *(Masch)* Führungen *fpl*, Gleitbahnen *fpl*
s. wedge [roof-]bolt *(Brg)* Doppelkeilanker *m*
s. window Schiebefenster *n*
slight gering, schwach, leicht, wenig
s. bevel kurzer Anschnitt *m (Reibahle)*
s. dip *(Geol)* schwaches Einfallen *n*
s. sea leicht bewegte See *f*
s. taper schlanker Kegel *m*
slightly curved crimp *(Text)* flachbogige Kräuselung *f*
s.-inclined *(Geol)* schwach einfallend
s. rusty angerostet, mit Flugrost versehen
slim hole enges Bohrloch *n*
s. taper file dünne Feile *f*
slime / to 1. ausstoßen *(von Schmutz und Fett bei Schaf- und Lammfellen)*; 2. schlämmen *(Erze)*
slime Schlamm *m*; Schmant *m*, Bohrschlamm *m*, Bohrschmant *m*
s. concentrating table *(Met)* Plan[stoß]herd *m*
s. fermentation Schleimgärung *f*
s.-jig Schlammsetzmaschine *f*
s. layer Schlammschicht *f*
s. pit Schlammgrube *f*, Schlammsumpf *m*
s. pulp Schlammtrübe *f*
s. screen Schlammsieb *n*
sliming Schlammbildung *f*
slimy schlammig, schlammhaltig, schlammartig; schleimig, schmierig, glitschig
sling / to schleudern, werfen
sling *(Schiff)* Schlinge *f*, Tauschlinge *f*, Heißstropp *m*, Hängestropp *m*, Stropp *m*
s. bolt Bügelschraube *f*
s. chain Anschlagkette *f*; Schlingkette *f*; Kettenstropp *m*
s. load *(Schiff)* Hieve *f*, Hiev *m*
s. psychrometer Schleuderpsychrometer *n*
slinger moulding machine Schleuderformmaschine *f*
s. ring Schleuderring *m*
slinging wire Einhängedraht *m (Galvanotechnik)*
slip / to 1. *(Met)* gleiten *(an den Korngrenzen)*; 2. [aus]rutschen; schlüpfen; 3. durchblättern *(Papiersortierung)*
s. a plug into the jack den Stecker in die Klinke stecken
s. in place einschieben
s. off herunterrutschen, heruntergleiten
s. over aufstecken auf, überstreifen, überstülpen
slip 1. Gleiten *n*, Gleitbewegung *f*; Rutschen *n*; Schlupf *m*; Falte *f*; 2. Zettel *m*, Notizzettel *m*; Fahnenabzug *m*; Papierstreifen *m*; Einschießbogen *m*; 3. *(Schiff)* Slip *m*, Helling *f*, Helgen *m*; 4. Schlicker *m (Keramik)*; Emailmischung *f*, Auftragmasse *f*; 5. Dübel *m*
s. band Gleitband *n*, Gleitlinie *f*; Fließfigur *f*
s. bolt Riegel *m*, Schubriegel *m*
s. bushing Schiebebuchse *f*
s. clutch Schlupfkupplung *f*
s. conduit *(El)* Falzrohr *n*, I-Rohr *n*
s. dock *(Schiff)* Slipdock *n*, Aufschleppdock *n*
s. fold *(Typ)* Fälzel; Falzstreifen *m*
s. form Gleitschalung *f (Betonbau)*
s.-form paver Gleitfertiger *m (Straßenbau)*
s. forming *(Plast)* Streckformverfahren *n*
s. fracture winkliger Bruch *m (Zerreißstab)*
s. frequency Schlupffrequenz *f*
s. galley *(Typ)* Spaltenschiff *n*
s. gauge Meßblock *m*; Endmaß *n*, Parallelendmaß *n*; Einstellmaß *n*
s. jaw Gleitbacke *f*
s. joint 1. biegsame Verbindung *f*, Dehnungsschlitz *m*, Gleitfuge *f*; 2. *(Am)* I-Rohrmuffe *f*
s. loss Schlupfverlust *m*
s. of belt Riemenschlupf *m*
s. of magnetic field lines Schlupf *m* der magnetischen Feldlinien
s.-over current transformer Kabelstromwandler *m*
s. plane Gleitebene *f*, Gleitfläche *f*
s. point Kreuzungsweiche *f*
s. power *(Elm)* Schleifringleistung *f*
s.-proof *(Typ)* Fahne *f*, Korrekturfahne *f*
s. regulator *(Elm)* Schlupfregler *m*
s. renewable bushing Einsteck[bohr]buchse *f*
s. ring Gleitring *m*, Schleifer *m*, Schleifring *m*
s.-ring brush Schleifringbürste *f*
s.-ring bush Schleifringhülse *f (aus Isolierstoff)*
s.-ring clutch Fliehkraftbandkupplung *f*
s.-ring [induction] motor Schleifringmotor *m*, Asynchronmotor *m* mit Schleifringläufer
s.-ring rotor *(Elm)* Schleifringläufer *m*
s.-ring starter Schleifringanlasser *m*
s. roll *(Text)* Kettbaum *m*
s. sheet *(Typ)* Einschießbogen *m*, Einschußbogen *m*
s.-sheet board Zwischenlagekarton *m (gegen Rollen)*
s. socket Hornbüchse *f (Fangwerkzeug; Tiefbohrtechnik)*
s. stopper Slipstopper *m (Ankerkettenstopper)*
s.-type drill bushing Einsteckbohrbuchse *f*
s. universal joint Gleitkreuzgelenk *n*
slippage Rutschen *n (Kupplung)*; Verschiebung *f*; *(Text)* geringe Fadenverschiebefestigkeit *f*; Neigung *f* zum Verschieben der Fäden *(Nähen)*, Fadenschieben *n*
slipper Gleitschuh *m*, Gleitplatte *f*, Gleitstück *n*, Bremskeil *m*
s. pin Gleitschuhzapfen *m*

s. piston Gleitschuhkolben *m*
slippers Hausschuhe *mpl*, Reiseschuhe *mpl*
slippery schlüpfrig
slipping Gleiten *n*; Rutschen *n*, Nachrutschen *n*; Schlupf *m*
s. clutch Rutschkupplung *f*
slipstream Propellerwind *m*
slipway Helling *f*, Helgen *m*, Bauhelling *f*; Aufschlepphelling *f*
s. crane Hellingkran *m*
slit / to schlitzen, aufschneiden, spalten
slit Schlitz *m*; Spalt *m*; Einschnitt *m*; Blende *f*, Spaltblende *f*
s. burner Schlitzbrenner *m*, Schnittbrenner *m*, Schwalbenschwanzbrenner *m*
s. coupling of waveguides Kopplung *f* zweier Hohlleiter durch Schlitze
s. diaphragm Schlitzblende *f*
s. drum Schlitztrommel *f*
s. irradiator *(Kern)* Spaltstrahler *m*, Schlitzstrahler *m*
s.-lamp apparatus Spaltleuchtengerät *n*
s. ring Schlitzring *m (Turbine)*
s. source *(Kern)* spaltförmige Quelle *f*
s. spectrograph Spaltspektrograf *m*
s.-stop *(Opt)* Schlitzblende *f*, Spaltblende *f*
s. width Spaltquelle *f*
slither Stapellaufkeil *m*
slitted geschlitzt, gespalten
slitter 1. Keilhaue *f*; 2. Längsschneider *m*, Streifenschneider *m*, Rollenschneidemaschine *f (Papierindustrie)*
slitting and rewinding machine Längsschneider *m* und Umroller *m (Papierherstellung)*
s. cutter Schlitzfräser *m (für Schrauben)*
s. file Schwertfeile *f*
s. knife Trennmesser *n*
s. plane Schneidhobel *m*
s. saw Furniersäge *f (Holz)*; Schlitzsäge *f (Metall)*
s. shear Hebelblechschere *f*
sliver 1. Holzsplitter *m*, Schiefer *m*, Holzspan *m*; Walzsplitter *m*; 2. Stapellaufkeil *m*; 3. *(Text)* Faserband *n*
s. board Feinspanplatte *f*
s. bobbin Vorgarnspule *f*
s. can Spinnkanne *f*
s.-dyeing machine Spinnbandfärbemaschine *f*
s. lap machine Bandwickelmaschine *f*, Bandwickler *m*
s. screen Astfänger *m (Papierindustrie)*
slog 1. schwere Arbeit *f*; 2. großer Span *m*
sloop Schaluppe *f*
slop pad printing *(Text)* Pflatschdruck *m*
s.-padded black Klotzschwarz *n (Textildruck)*
s. sink Sinkkasten *m*
s. tank Nachlauftank *m*
s. treatment Schlempenbehandlung *f*
slope / to *(Lw)* abböschen; abfallen [lassen]; ansteigen [lassen]
s. down sich neigen, abfallen
s. steeply steil abfallen
slope Gefälle *n*, Neigung *f*, Steigung *f*, Schrägfläche *f*; Böschung[swinkel *m*] *f*; Abhang *m*, Hang *m*; Quergefälle *n*, Querneigung *f (Straßenbau)*; *(El)* Anstieg *m (Kurve)*; Steilheit *f (Elektronenröhren)*
s. angle Böschungswinkel *m*; Richtungswinkel *m*; Inklinationswinkel *m*, Neigungswinkel *m*
s. change Gefälleänderung *f*
s. of amplifier valve Verstärkerröhrensteilheit *f*
s. of bench *(Brg)* Strossenböschungswinkel *m*
s. of density Dichteabfall *m*
s. of repose Böschungsneigung *f*
s. of the slipway Helling[bahn]neigung *f*
s. resistance Steilheitswiderstand *m*
sloped geneigt, schräggestellt; geschweift *(Kurve)*
sloping geneigt, schief, schräg [abfallend]
s. furnace ansteigender Feuerungsraum *m*
s. grate Schrägrost *m*
s. outward frame *(Schiff)* ausfallendes Spant *n*
s. soil Hanglage *f*
slot / to schlitzen, einschneiden, nuten; stoßen *(Senkrechtstoßmaschine)*; *(Brg)* durchörtern
slot Kerbe *f*, Nut *f*, Schlitz *m*, Spalte *f*
s. aerial Schlitzantenne *f*
s. and key cutter Nut- und Spundfräser *m*
s. and key iron Nut- und Spundmesser *n*
s. and keyway milling machine Langloch- und Keilnutenfräsmaschine *f*
s.-and-wedge [roof-]bolt *(Brg)* Schlitzkeilanker *m*
s. array Schlitz[strahler]kombination *f (Antennen)*
s. boring machine Langlochbohrmaschine *f*
s. bunker Schlitzbunker *m*
s. covering *(Rak)* Spaltverkleidung *f*
s. depth Nutentiefe *f*
s. die method *(Plast)* Schlitzpressen *n*, Strangpressen *n (Bänder, Folien)*
s. drill Langlochfräser *m*, Fingerfräser *m*, Nutenfräser *m*
s. factor *(Elm)* Wicklungsfaktor *m*
s. for cotter *(Wärm)* Keilloch *n*
s. for holding work Aufnahmenut *f (Aufspanntisch)*
s. for work *s.* s. for holding work
s. grinder Nutenschleifmaschine *f*
s. insulator Nutisolation *f*
s. leakage *(El)* Nutenstreuung *f*
s. liner Nuthülse *f*, Nutisolation *f*, Isolierhülse *f*
s. meter *(Fmt)* Münzzähler *m*

s. milling Nutenfräsen *n*, Schlitzfräsen *n*, Schlitzen *n*
s. mortising machine Langlochbohrmaschine *f*, Schlitzmaschine *f*
s. perforation Strichperforierung *f*, Schlitzperforierung *f*
s. pitch *(Elm)* Nutteilung *f*, Nutenschritt *m* einer Wicklung
s. radiator Schlitzstrahler *m*
s. sampling *(Geol)* Schlitzbemusterung *f*
s.-siphoning Spaltabsaugung *f (Turboantrieb)*
s. space factor *(Elm)* Nutenfüllfaktor *m (Anker)*
s. sprayer Schlitzzerstäuber *m (Düse)*
s.-type magnetron Schlitzmagnetron *n*, Spaltmagnetron *n (Typ eines Vielkammermagnetrons)*
s. wedge *(Elm)* Nutenkeil *m (des Ankers)*
s. width Nutenbreite *f*
s. winding *(Elm)* Nutenwicklung *f*
slotted geschlitzt, genutet
s. arm *(Masch)* Kulisse *f*, Kurbelschwinge *f*
s. armature *(Elm)* genuteter Anker *m*, Nut[en]anker *m*
s. bolt geschlitzter Bolzen *m*
s. cheesehead metal screw Zylinderschraube *f*
s. conduit Schlitzstromzuführung *f*
s. core armature *(Elm)* genuteter Anker *m*
s. countersunk head screw Senkschraube *f*
s. cup-head screw Halbrundschraube *f*
s. cylinder aerial Schlitzrohrstrahler *m*
s. disk Rastenscheibe *f*, Schlitzscheibe *f*
s. drill driver geschlitzte Kegelklemmhülse *f*
s. flap Spaltklappe *f*
s. hole Langloch *n*
s.-hole screen Feinspaltrost *m*
s. lever Langlochhebel *m*
s. line Meßleitung *f*
s. link *(Masch)* Kulisse *f*, Kurbelschwinge *f*
s. measuring section [geschlitzte] Meßleitung *f (Wellenleiter)*
s. nut flache Kronenmutter *f*
s. piece Nutkörper *m*
s. pin Kerbstift *m*
s. piston Schlitzmantelkolben *m*
s. rocket jet *(Rak)* Ringspaltdüse *f*
s. scraper ring geschlitzter Ölabstreifring *m*
s. screw Schlitz[kopf]schraube *f*
s. set screw geschlitzte Stiftschraube *f*
s. shutter Schlitzverschluß *m (einer Kamera)*
s. waveguide geschlitzter Wellenleiter *m*
slotter Senkrechtstoßmaschine *f*
s. tool Stoßmeißel *m*
slotting Schlitzen *n*; Senkrechtstoßen *n*; Nutenfräsen *n*, Nutenstoßen *n*; Langlochfräsen *n*
s. attachment Senkrechtstoßeinrichtung *f*
s. cutter Nutenfräser *m*
s. file Nutenfeile *f*
s. machine Senkrechtstoßmaschine *f*
s. saw Schraubenschlitzfräser *m*
s. tool Stoßmeißel *m*
slough-off abgeschlagener Schuß *m*
slow / to verlangsamen, verzögern; nachgehen *(Uhr)*
s. down [ab]bremsen, verlangsamen
slow langsam, mit geringer Geschwindigkeit
s.-acting mit Ansprechverzögerung, langsam ansprechend (wirkend)
s.-acting relay Verzögerungsrelais *n*, Langzeitrelais *n*
s.-burning iron stove eiserner Dauerbrandofen *m*
s. combustion langsame Verbrennung *f*
s.-down path Verzögerungsweg *m*
s.-draining stuff schmieriger Stoff *m (Papierherstellung)*
s. feed motion Langsamgang *m*
s. frequency drift Frequenzverwerfung *f*
s.-grown langsam gewachsen *(Holz)*
s.-helix schlankgenutet
s. helix langer (steiler) Drall *m*
s. match Lunte *f*, Zündschnur *f*
s. memory *(Dat)* langsamer Speicher *m*
s. motion langsame Bewegung *f*
s. motion camera Zeitlupenkamera *f*
s. motion control Feinregelung *f (z. B. Funkgerät)*
s. motion control knob Feinabstimmknopf *m*
s. motion dial Feinstellskala *f*
s. motion drive Feinstelltrieb *m*
s. motion head Feinstellschraube *f*
s. motion link Langsambewegungsgetriebe *n*
s. motion photography Zeitlupenaufnahme *f*
s. movement Nachgehen *n (von Uhren)*
s. neutron langsames Neutron *n*; thermisches Neutron
s.-neutron bombardment Beschießung *f* mit langsamen Neutronen
s.-neutron fission Spaltung *f* durch langsame Neutronen
s.-operating relay Relais *n* mit Ansprechverzögerung (Anzugsverzögerung), Verzögerungsrelais *n*, Langzeitrelais *n*
s. ray *(Opt)* außerordentlicher Strahl *m*
s. reactor thermischer Reaktor *m*
s. release verzögerte Auslösung *f*; verzögerte Ausschaltung (Trennung, Unterbrechung) *f*
s.-release relay Relais *n* mit verzögerter Auslösung
s.-running engine langsam laufender Motor *m*
s.-running jet Leerlaufdüse *f*
s.-running position Leerlaufstellung *f*
s.-running screw Leerlaufeinstellschraube *f*
s.-setting cement langsam abbindender Zement *m*

s.-speed mit geringer Drehzahl, langsam laufend
s.-speed nozzle Leerlaufdüse *f*
s.-speed steam engine langsam laufende Dampfmaschine *f*
s. tuning Feinabstimmung *f*
slowed-down neutron gebremstes Neutron *n*
slowing Verzögerung *f*
s. controller Fahrtminderungsschalter *m*
s. down 1. Bremsung *f*, Abbremsung *f*, Verlangsamung *f*; 2. *(Masch)* Drehzahlverminderung *f*; Herabsetzung *f* der Geschwindigkeit
s.-down area Bremsfläche *f*
s.-down cross section Bremsquerschnitt *m*, Wirkungsquerschnitt *m* der Bremsung
s.-down efficiency Bremswirkung *f*
s.-down length Bremslänge *f*
s.-down power *(Kern)* Bremsvermögen *n*
s.-down radiation Bremsstrahlung *f*
s.-down time Bremszeit *f (für Neutronen)*
slowness Mahlgrad *m (Zellstoff und Papier)*
slub Wulst *m (Garn)*
s. twist Knotenzwirn *m*
s. yarn doubling frame Knotengarnzwirnmaschine *f*
slubber *(Text)* Lunte *f (Spinnen)*
slubbing Vorgespinst *n*, Lunte *f*, grobes Vorgarn *n*
s. frame Grobbank *f*, Grobflyer *m*, Vorflyer *m*
s. machine Vorspinnmaschine *f*
sludge Schlamm *m*; Faulschlamm *m (Gesundheitstechnik)*; Ölschlamm *m (Flugzeugmotor)*; Bohrschmant *m*, Schmant *m (Bohren)*
s. activation Schlammbelebung *f (Abwasserreinigung)*
s. bottom Schlammsumpf *m*
s. cake Schlammkuchen *m*
s. centrifuge Schlammschleuder *f*
s. decomposition Schlammzersetzung *f*
s. densifier Schlammeindicker *m*
s. deposition Schlammablagerung *f*
s. dewatering Schlammentwässerung *f*
s. digestion Schlammfaulung *f*
s. drain cock *(Kfz)* Abschlämmhahn *m*
s. draining Schlammentwässerung *f*
s. drying ground Schlammtrockenplatz *m*
s. extractor Schlammfänger *m*
s. formation Schlammbildung *f (Hydraulik)*
s. gas Klärgas *n*
s. gate Schlammschleuse *f*
s. incineration Schlammverbrennung *f*
s. incinerator Schlammverbrenner *m*
s. outlet Schlammauslaß *m*
s. pump Schlammpumpe *f*
s. pumping system Jauchenpumpsystem *n*
s. removal Schlammbeseitigung *f*
s. return Schlammrückführung *f*
s. sample Spülprobe *f (Tiefbohrtechnik)*
s. sump Schlammsumpf *m*
s. tanker Klärschlammtanker *m*
s. thickener Schlammeindickbehälter *m*
s. thickening Schlammeindickung *f*
s. utilization Schlammverwertung *f*
s. vessel Klärprahm *m*, Kloakenprahm *m*
s. water Trübe *f*; Schlammtrübe *f*
sludger 1. Schlammlöffel *m*, Bohrlöffel *m (Tiefbohrtechnik)*; 2. Schlammpumpe *f*
sludging Entschlammung *f*
slug 1. Block *m*, Rohling *m*; Preßbarren *m*; Metallklumpen *m*; 2. [kurzer dicker] Stab *m (aus Kernbrennstoff)*; 3. *(Typ)* Setzmaschinenzeile *f*, Satzzeile *f*; gegossene Zeile *f*; 4. Transformationshülse *f (Hohlleiter)*; Anpaßkörper *m (Wellenleiter)*; 5. weiche Stelle *f* im Seidenfaden
s. channel Brennstoffkanal *m* des Graphit-Natururan-Reaktors
s. composing machine Zeilensetzmaschine *f (z. B. Linotype)*
s. tuner Abstimmkörper *m (Wellenleiter)*
s. tuning Eisenkernabstimmung *f*
slugging fluidized bed stoßende Wirbelschicht *f*
sluggish träge *(Sicherung)*; strengflüssig, zäh[flüssig]
s. measuring instrument träges Meßinstrument *n*
sluggishness Trägheit *f*, Unempfindlichkeit *f (Regler)*
slugwise continuous flow mixer kontinuierlicher Mischer *m* mit portionsweiser Materialzuführung
sluice / to mit Schleuse versehen; spülen
sluice Schleuse *f*; Schütz *n (Schleusenventil)*
s. gate Schottschleuse *f*
s. keel Schleusenkiel *m*
s. pump Spülpumpe *f (Entaschung)*
s. valve Schieber *m*
sluicing dump *(Brg)* Spülkippe *f*
s. system Spülanlage *f (Entaschung)*
slum 1. Elendsviertel *n*; 2. Harzrückstand *m (Öl)*
s. clearance Slumsanierung *f*, Stadtsanierung *f (Räumung und Abbruch von Slums)*
slump 1. Abfall *m*; Senkung *f*; 2. Ausbreitmaß *n (Betonprüfung)*
s. test Ausbreitversuch *m (Betonprüfung)*
slur Schlichte *f (Formerei)*
slurry Brei *m*, Kohlenschlamm *m*, Schlamm *m*; Aufschlämmung *f*; nasse Suspension *f*; Rohschlamm *m (Zementherstellung)*
s. fuel *(Kern)* Brennstoffschlamm *m*
s. preforming *(Plast)* Saugling-Vorformverfahren *n*
s. reactor Suspensionsreaktor *m*, Reaktor *m* mit nasser Suspension

s. tank attachment Jauchetankanhänger *m*
s. tank car Jauchetankwagen *m*
s. tank trailer Jauchetankanhänger *m*
slush Schmiere *f*, Schmutz *m*
s. casting *(Plast)* Gießen *n* von Hohlkörpern
s. pit Absetzbecken *n* *(Aufbereitung)*
s. pump Spülpumpe *f*; Dickstoffpumpe *f*
s. stock eingedickter Stoff (Zellstoff) *m*
slusher *(Brg)* Schrapper *m*
s. hoist Schrapperhaspel *f*
s. ramp Schrapperbühne *f*
s. train *(Brg)* Bunkerzug *m*
slushing oil Rostschutzöl *n*
small klein, schwach, dünn; schmal, verengt
s. accumulator Kleinakku[mulator] *m*, Kleinsammler *m*
s.-angle grain boundary Kleinwinkelkorngrenze *f*
s.-angle scattering Kleinwinkelstreuung *f*
s. area kleinflächig
s. arm Handfeuerwaffe *f*
s. batch production Kleinreihenfertigung *f*
s. bell Oberglocke *f* *(Gichtverschluß)*
s. bend *(Text)* Bögchen *n* *(Spitzennadel)*
s.-bore gauge Kleinbohrungsmeßgerät *n*
s.-bore glass tube Kapillarrohr *n*, Haarröhrchen *n*
s.-bore measuring microscope Kleinbohrungsmeßmikroskop *n*
s.-bore pipe Kleinrohr *n* *(Heizungstechnik)*
s.-bore system Kleinrohrheizung *f*
s. calorie Grammkalorie *f*
s.-capacity cable Kabel *n* mit geringer Adernzahl
s. cap *(Typ)* Kapitälchen *n*
s. centre cap Kleinlampenfassung *f*
s. coasting trade kleine Küstenfahrt *f*
s. craft kleines Schiff *n* *(für Fischfang und Handel)*
s. deep-sea fishing kleine Hochseefischerei *f*
s.-delivery pump Pumpe *f* mit geringer Fördermenge
s.-diameter drilling *(Brg)* Kleinkaliberbohren *n*
s. end Pleuelkopf *m*, Pleuelauge *n*; Kreuzkopfende *n* *(Schubstange)*
s. end bearing Kolbenbolzenlager *n* der Pleuelstange
s. end bush[ing] Pleuelbuchse *f*
s.-end hole kleines Pleuelauge (Kolbenbolzenauge) *n*
s. generator Kleindynamo[maschine *f*] *m*, Kleingenerator *m*
s. hand-vice Stielfeilkloben *m*
s. hardware Kleineisenzeug *n*
s.-hole drill Kleinstbohrer *m*
s.-lot manufacture (production) Kleinlosfertigung *f*
s.-lot run *(Am)* Kleinserie *f*, Kleinreihe *f*; Kleinreihenfertigung *f*
s. motor Klein[kraft]motor *m*
s. pica *(etwa)* Korpus *f* *(Schriftgröße)*
s. pit-prop *(Holz)* Grubenspitzenknüppel *m*
s. plane Fausthobel *m*
s. power electromotor Kleinelektromotor *m*
s. quantity batch production Kleinreihenfertigung *f*
s.-scale series production Kleinserienbau *m*, Kleinserienfertigung *f*
s. section rolling mill Feineisenwalzwerk *n*, Feinstahlwalzwerk *n*
s. ship navigation Kleinschiffahrt *f*
s. signal Kleinsignal *n*
s.-signal behaviour Kleinsignalverhalten *n*
s.-signal forward transadmittance Leitwert *m* bei kleinem Signal
s.-signal gain Kleinsignalverstärkung *f*
s.-size assembly kleine Baugruppe *f*
s.-size component part Miniaturbauteil *n*
s.-size drill Bohrer *m* mit kleinem Durchmesser
s.-sized klein *(von kleinen Dimensionen)*
s.-sized cable Kabel *n* mit geringer Adernzahl
s. strand Bändchen *n* *(Spinnen)*
s. transformer Kleintransformator *m*
s.-type locomotive Kleinlokomotive *f*
s.-upright Pianino *n*
s. wiring Signal- und Steuerleitungen *fpl*
smallest cross-section of nozzle engster Düsenquerschnitt *m* *(Turbine)*
smalls Feinkohle *f*
s. jig Feinkornsetzmaschine *f* *(Aufbereitung)*
smallware loom Bändchenwebstuhl *m*, Schmalwebstuhl *m*
smallwares Kurzwaren *fpl*
smalt Smalte *f*, Kobaltblau *n*
smaltine *s.* smaltite
smaltite Glanzkobalt *m*, Speiskobalt *m*
smaragdite *(Min)* Smaragdit *m*
smash / to [zer]brechen; zerschmettern, zertrümmern; zusammenstoßen; pressen *(Buch)*
smash *(Text)* Schützenschlag *m*
smear / to [ver]schmieren, streichen; plastisch werden
smear 1. Ausstrich *m* *(Mikroskoppräparat)*; 2. wachsartige (elfenbeinartige) Glasur *f*
smeary schmierig
smell of gas Gasgeruch *m*
s.-proof geruchsicher
smelt / to [ab]schmelzen, erschmelzen, einschmelzen, blasen, verhütten
smeltable verhüttungsfähig, verhüttbar *(Erze)*
smelted copper Schmelzkupfer *n*
smelter 1. Schmelzer *m*; 2. Schmelztiegel *m*
smeltery *s.* smelting plant

smelting Schmelzen *n*, Verhütten *n*, Verhüttung *f*
s. furnace Schmelzofen *m*
s. hearth Schmelzherd *m*
s. loss Abbrand *m*
s. plant Hütte[nwerk *n*] *f*, Schmelzhütte *f*
s. point Schmelzpunkt *m*
s. pot Schmelztiegel *m*
s. work Hüttenwerk *n*
s. zone Schmelzzone *f (des Ofens)*
smith forging 1. Freiformschmieden *n*; 2. Freiformschmiedeteil *n*
s. welding Hammerschweißen *n*
smithery Schmiede *f*
smith's bellows Blasebalg *m*, Schmiedegebläse *n*
s. hearth Schmiedeherd *m*
s. shop Schmiede *f*
smithsonite *(Min)* Smithsonit *m*, Zinkspat *m*
smithy Schmiede *f*
smoke / to rauchen; ausräuchern; verrußen; vernebeln; räuchern *(Gerberei)*
smoke Rauch *m*; Nebel *m*
s. box Rauchkammer *f (Lokomotive)*
s. box superheater Rauchkammerüberhitzer *m*
s.-consuming furnace rauchverzehrende Feuerung *f*
s. damper *(Wärm)* Rauchschieber *m*
s. deflector Windleitblech *n (Lokomotive)*
s. density indicator Rauchdichtemesser *m*
s. detector fire alarm Rauchmeldeanlage *f*
s. dust Flugstaub *m*
s. gas Rauchgas *n*
s.-laying apparatus *(Mil)* Nebelgerät *n*
s. mortar *(Mil)* Nebelwerfer *m*
s. nuisance Rauchbelästigung *f*
s.-preventing furnace rauchverhütende Feuerung *f*
s. projectile Nebelgeschoß *n*
s. scale Rauchskala *f*
s. shell Nebelgranate *f*
s.-shell mortar Nebelwerfer *m*
s. stack Schornstein *m*, Kamin *m*, Esse *f*, Schlot *m*
s. tanning *(Led)* Rauchgerbung *f*
s. tube Rauchrohr *n*
s.-tube boiler Feuerrohrkessel *m*
s. tube superheater Rauchrohrüberhitzer *m*
smoked 1. mit Ruß geschwärzt, verräuchert; 2. geräuchert; 3. vernebelt
s. glass Rauchglas *n*, Rußglas *n*
smokeless rauchlos, rauchfrei
s. furnace rauchverhütende Feuerung *f*
smoker *(Eis)* Raucherabteil *n*
smokestack *(Schiff)* Schornstein *m*
smokestone Rauchtopas *m*
smoking compartment *(Eis)* Raucherabteil *n*
s. paper Zigarettenpapier *n*
s. tobacco Pfeifentabak *m*
smoky rauchend, rauchig
s. flame rußende Flamme *f*
s. quartz Rauchquarz *m*
s. topaz Rauchtopas *m*
smooth / to glätten; abschleifen; abziehen; blank schleifen, ebnen; schlichten *(Formen)*; ausspachteln; *(Lw)* entfurchen
smooth glatt *(Kurve)*; ruhig, stoßfrei; stufenlos
s. armature *(El)* glatter Anker *m*
s. bore Schlichtbohrung *f*
s.-core armature *(Elm)* glatter Anker *m*
s.-core motor Volltrommelmotor *m*, Motor *m* ohne ausgeprägte Pole
s.-cut file Schlichtfeile *f*
s. cutter combination saw Feinkreissäge *f*
s. drying cotton fabric glatt trocknendes Baumwollgewebe *n (bügelfrei)*
s. file Schlichtfeile *f*
s. flow Laminarströmung *f*
s. fracture feinkörnige Bruchfläche *f*; glatter Bruch *m*
s. hide-leather glattes Rindleder *n*
s. oats Glatthafer *m*
s. passage yarn glattes Ablaufen *n* des Garnes *(Spinnerei)*
s. plane Schlichthobel *m*
s. planer Abrichte[r *m*] *f*, Abrichtmaschine *f*
s. roller Glattwalze *f*
s. roofing paper unbesandete Dachpappe *f*
s. running Laufruhe *f*
s. surface Glanzfläche *f (Briketts)*
s. winding glatte Wicklung *f*
smoothed current ausgeglichener (geglätteter) Strom *m*
s. voltage beruhigte (geglättete) Spannung *f*
smoothing Beruhigen *n*, Glätten *n*; Abziehen *n*; Einebnung *f*, Einebnen *n*; Schlichten *n*; *(Lw)* Entfurchen *n*
s. capacitance *(El)* Beruhigungskapazität *f*
s. capacitor Abflachkondensator *m*, Beruhigungskondensator *m*
s. chisel Schlichtmeißel *m*
s. choke *(El)* Siebdrossel *f*, Glättungsdrossel[spule] *f*
s. circuit *(El)* Abflachschaltung *f*, Glättungskreis *m*, Siebkreis *m*
s. factor Glättungsfaktor *m*
s. file Polierfeile *f*
s. network *(El)* Glättungskreis *m*, Glättungsschaltung *f*
s. plane Schlichthobel *m*
s. reactor *s.* s. choke
s. resistance *(El)* Beruhigungswiderstand *m*
s. rolls Glättkalander *m (Beschichtung)*; Glättwalzen *fpl*
s. tool Schlichtmeißel *m*
smoothness Glattheit *f*; Sauberkeit *f*

smoulder / to schwelen, glimmen, langsam verbrennen
smudge / to beschmutzen, verschmieren
smudge Schmutz *m*; Schmutzfleck *m*; *(Plast)* Schmierstelle *f* am Spritzling
smudging *(Fs)* Fahnenziehen *n (Bildorthikon)*
smut / to berußen; vom Brand befallen werden *(Getreide)*
smut Ruß[fleck] *m*, Rußflocke *f*; Brand *m (Getreidekrankheit)*
Smyrna rug Smyrnateppich *m*
S/N curve Wöhlerkurve *f*
snaffle Trense *f (Rennpferd)*
 s. bit Knebelgebiß *n*
snag / to abgraten, putzen; auf Felsen laufen; Felsen (Baumstümpfe, Hindernisse) entfernen
snag Baumstumpf *m*; Felszacke *f*; unerwartetes Hindernis *n*; *(Text)* Fadenzieher *m (Damenstrumpf)*
snagging *(Masch)* Abgraten *n (Schleifarbeit)*, Putzen *n*
 s. machine Gußputzmaschine *f*
snake over / to umlegen *(Fördermittel)*
snake poison Schlangengift *n*
snaked wire verdrillter Draht *m*
snakeholing *(Brg)* Schießen *n* mit untergelegter Ladung
snap / to [ab]schnappen; abfedern; nieten *(Köpfe machen)*; abreißen; knipsen *(Foto)*
 s. in einschnappen
 s. in place einschnappen *(Schloß)*
 s. on anknipsen, einschalten
 s. the line anreißen *(Linie mit Schnur)*
snap 1. Döpper *m*, Kopfmacher *m*, Schellhammer *m (Nieten)*; 2. Verschluß[vorrichtung *f*] *m*; 3. *(Text)* 320 yards *(Längeneinheit von Wollgarn)*
 s.-acting controller Regler *m* mit Sprungschaltung
 s. action Ausgangsstromsprung *m (Magnetverstärker)*
 s.-action contact Sprungkontakt *m*, Schnappkontakt *m*
 s.-back forming *(Plast)* Aufschrumpfen *n* unter Vakuum
 s. bench Formerbank *f*
 s. catch Schnapper *m*
 s. cover type pump Klappdeckelpumpe *f*
 s. die Döpper *m*, Nietkopfsetzer *m*
 s.-fastener Druckknopf *m*
 s. flask Abziehformkasten *m (Gießerei)*
 s. gauge Rachenlehre *f*; Taster[lehre *f*] *m*
 s. hammer Döpper *m (Nieten)*
 s. head Halbrundkopf *m*
 s. hook Karabinerhaken *m*; Bergsteigerkarabiner *m*
 s.-in contact Einschnappkontakt *m*
 s.-lock Einrastung *f*
 s. magnet Schnappmagnet *m*
 s. piston ring federnder Ring *m*, Kolbenfeder *f*; Überstreifring *m*
 s. ring Sprengring *m*, Sicherungsring *m*; Federring *m*; Schnappring *m*; Bergsteigerkarabiner *m*
 s. riveting Schellkopfnietung *f*
 s. switch Schnappschalter *m*, Sprungschalter *m*, Drehschalter *m*
 s. tool Döpper *m*, Kopfmacher *m*, Schellhammer *m (Nieten)*
snapped header Halbziegel *m (Strecker)*
snapping line Anreißschnur *f*
snapshot Momentaufnahme *f (Foto)*
snare-head Saitenfell *n (Trommel)*
snarl / to verwickeln, überdrehen; verzieren *(mit Relief)*
snarl Schleife *f*, Kräuselung *f*
 s. catcher Schlaufenfänger *m (Ringspinnmaschine)*
 s. yarn *(Text)* Schlingengarn *n*
sneak current Kriechstrom *m (Starkstromanlagen)*, Irrstrom *m*, Fremdstrom *m*
snick Einschnitt *m*
snifter valve *s.* snifting valve
snifting valve Schnarchventil *n*, Schnüffelventil *n*
sniper Scharfschütze *m*
snipper *(Text)* Schnippmaschine *f*
snips Handschere *f*, Blechschere *f*
snout Düse *f*, Mundstück *n*
snow 1. Schnee *m*; 2. *(Fs)* Grieß *m*, Schnee *(Bildstörungen)*
 s.-blocked road verschneite Straße *f*
 s.-blowing machine Schneeschleuder *f*
 s. cement Schneezement *m*
 s. clearing Schneeräumung *f*
 s. drift Schneeverwehung *f*
 s. fence Schneezaun *m*
 s. goggles Schneebrille *f*
 s. load Schneelast *f*
 s. loader Schnee[auf]lader *m*
 s. plough Schneepflug *m*
 s. track Schneekette *f (Reifen)*
 s. tread Schneeprofil *n (Reifen)*
Snowcrete weißer eisenoxidfreier Portlandzement *m (Handelsname)*
snub / to zum Halten bringen
snubber *(Am) (BMSR)* pneumatischer Pulsationsdämpfer *m*
snug / to anziehen *(Mutter)*
snug bolt Nasenschraube *f*
 s. fit enge Passung *f (engste, von Hand zu erreichende Form)*
 s. harbour geschützter Hafen *m*
soak / to 1. aufsaugen, [auf]weichen, durchfeuchten, durchnässen, einquellen, einweichen, tränken, quellen [lassen]; 2. imprägnieren,

tränken *(z. B. Transformatoren)*; 3. *(Met)* durchwärmen, ausgleichen, durchweichen, durchglühen, weichglühen; vorwärmen
s. in lime *(Led)* äschern
soak liquor *(Led)* Weichwasser *n*
s. pit *(Led)* Weichkasten *m*
soaker 1. Reaktionskammer *f*; 2. *(Met)* Tiefofen *m*
soaking *(Led)* Weiche *f*
s. furnace *(Met)* Wärmofen *m*
s. hearth Schweißherd *m*
s. pit *(Met)* Ausgleichsgrube *f*, Durchweichungsgrube *f*, Tiefofen *m*
s. pit crane Tiefofenkran *m*
s. pit furnace *(Met)* Herdtiefofen *m*, Zellentiefofen *m*
s. time *(Met)* Ausgleichzeit *f (Tieföfen)*
s. weight *(Led)* Einweichmasse *f*
s. wheel *(Led)* Weichhaspel *f*
soap / to abseifen, [ein]seifen
s. in full width *(Text)* breitseifen
soap Seife *f*
s. base Grünseife *f*
s. bath Seifenbad *n*
s. boiler Seifensieder *m*
s. bubble Seifenblase *f*
s. builders Seifenzusatzstoffe *mpl*
s. chippings Seifenflocken *fpl*, Seifenschnitzel *npl*
s. dispenser Seifenbehälter *m (für flüssige Seife)*
s. factory Seifensiederei *f*
s. film Seifenhaut *f*
s. in bars Riegelseife *f*
s.-making (manufacture) Seifenherstellung *f*
s. mill Piliermaschine *f*
s.-milling *(Text)* Seifenwalke *f*
s. paste Seifenleim *m*
s. powder Seifenpulver *n*
s.-refining outfit Pilieranlage *f*
s.-resistant seifenbeständig
s.-slab cooling plant Seifenkühlpresse *f*
s. solution Seifenwasser *n*
s. suds Seifenlauge *f*
s. tincture Seifenspiritus *m*
s. water Seifenwasser *n*
s. works Seifenfabrik *f*, Seifensiederei *f*
soaping aftertreatment *(Text)* Seifennachbehandlung *f*
soapless soap Einweichmittel *n*
soapstone Speckstein *m*, Steatit *n*, Seifenstein *m*
soapy seifenartig, seifig
soar / to 1. in die Höhe schnellen; 2. segeln *(Flugzeug)*; Höhe gewinnen
soaring *(sl)* Segelfliegerei *f*
social services Sozialleistungen *fpl*
sock lining Deckbrandsohle *f (Schuh)*
socket Hülse *f*; Muffe *f*, Schliffhülse *f*, Steckhülse *f*, Steckdose *f*, Fassung *f (Lampe)*; Flansch *m*; Rohransatz *m*; Rohrstutzen *m*, Sockel *m*; Sitz *m*; Aufnahmebohrung *f*; Einsatz *m*
s. adapter *(El)* Zwischensockel *m*
s. base *(Bw)* Hülsenfundament *n*
s.-head cap screw Zylinderkopfschraube *f*
s.-head screw Zylinderschraube *f* mit Innenaussparung *(Innensechskant usw.)*
s. joint 1. Steckverbindung *f*; 2. Muffenverbindung *f*
s. of the spindle *(Masch)* Aufnahmebohrung *f* der Spindel
s.-outlet Steckdose *f*, Wandsteckdose *f*, Steckdosenanschluß *m*
s.-outlet adapter Abzweigstecker *m*, Doppelstecker *m*, Mehrfachstecker *m*
s.-outlet and plug Steckvorrichtung *f*
s.-power unit *(El)* Netzanschlußgerät *n*
s. reamer for Morse taper Kegelreibahle *f* für Morsekegel *(Werkzeugkegel)*
s. ring Taumelscheibe *f*
s. screw Innensechskantschraube *f*
s. spanner Steckschlüssel *m*
s. switch Sockelschalter *m*, Fassungsschalter *m*, Hahnschalter *m*; Hahnfassung *f*
s. terminal Röhrensockel *m*, Sockelanschluß *m*
s. wrench Steckschlüssel *m*, Aufsatzschlüssel *m*, Aufsteckschlüssel *m*, Stangensteckschlüssel *m*
s. wrench for square nuts Vierkantsteckschlüssel *m*
socketed pipe Muffenrohr *n*
socle Sockel *m*, Fundament *n*
sod Rasenstück *n*, Sode *f*
s.-lifter Rasenheber *m*
s.-mattock Rasenhaue *f*
s.-plough Schälpflug *m*
soda Natriumkarbonat *n*, Soda *f*; *(Am)* Natron *n*
s. ash kalzinierte Soda *f*
s. aspirator *(Wärm)* Sodainspirator *m*
s. cellulose *(Am)* Natronzellstoff *m*, Natronzellulose *f*
s. crystals Kristallsoda *f*; kristallisierte Soda *f*; Sodakristalle *mpl*
s. extraction Sodaauszug *m (Analytik)*
s. feldspar Natronfeldspat *m*
s. furnace Soda[schmelz]ofen *m*
s. glass Sodaglas *n*, Natronglas *n*
s. lime Natronkalk *m*
s.-lime glass Sodakalkglas *n*, Kalknatronglas *n*
s. lye Natronlauge *f*
s. nitre Natriumnitrat *n*, Chilesalpeter *m*, Natronsalpeter *m*
s. pulp Natron[zell]stoff *m (Papierherstellung)*

s. recovery Sodarückgewinnung *f* *(Papierherstellung)*
s. soap Natronseife *f*
s. works Sodafabrik *f*
sodalite *(Min)* Sodalith *m*
sodamide Natriumamid *n*
sodammonium Natriumammoniakat *n*
sodium Natrium *n*
s. acetate Natriumazetat *n*
s. bicarbonate Natriumbikarbonat *n*, Natriumhydrogenkarbonat *n*
s. bichromate Natriumbichromat *n*, Natriumpyrochromat *n*
s. bromide Natriumbromid *n*
s. carbonate Natriumkarbonat *n*
s. cellulose Natronzellulose *f*
s. chlorate Natriumchlorat *n*
s. chloride Natriumchlorid *n*, Kochsalz *n*, Steinsalz *n*
s. chloride lattice Natriumchloridgitter *n*
s.-cooled natriumgekühlt
s.-cooled reactor natriumgekühlter Reaktor *m*
s. cyanide Natriumzyanid *n*
s. discharge lamp Natriumdampflampe *f*
s. excess Natriumüberschuß *m*
s. fluoride Natriumfluorid *n*
s.-graphite reactor Natrium-Graphit-Reaktor *m*
s. hydroxide Natriumhydroxid *n*, Ätznatron *n*, kaustische Soda *f*
s. iodide Natriumjodid *n*
s. metal metallisches Natrium *n*, Natriummetall *n*
s. nitrate Natriumnitrat *n*, Chilesalpeter *m*
s. paper Natronpapier *n*
s. peroxide Natriumperoxid *n*
s. phosphate Natriumphosphat *n*
s. phosphide Natriumphosphid *n*
s. position Natriumplatz *m*, Natriumstelle *f* *(im Kristallgitter)*
s. pump Natriumpumpe *f*
s. silicate Natriumsilikat *n*
s. site *s.* s. position
s. spectral lamp Natriumspektralleuchte *f*
s. sulphate Natriumsulfat *n*, Glaubersalz *n*
s. tetraborate Natriumtetraborat *n*, Borax *m*
s. thiosulphate Natriumthiosulfat *n*, Fixiersalz *n*
s.-vapour lamp Natrium[dampf]lampe *f*
soffit Tunnelscheitel *m*
s. arch Gewölbebogen *m*
s. boards Schalboden *m*
soft weich, mild, sanft; geschmeidig; schmiedbar, dehnbar; alkoholfrei *(Getränk)*; weich, ungehärtet *(Stahl)*; *(Opt)* weich; kontrastarm; locker *(Boden)*; energiearm
s.-annealed *(Met)* weichgeglüht
s.-annealed wire weichgeglühter Draht *m*
s. annealing *(Met)* Weichglühung *f*, Weichglühen *n*
s. bow *(Schiff)* Plattenbug *m*
s. carbon steel kohlenstoffarmer weicher Stahl *m*
s. coal bituminöse Kohle *f*, Fettkohle *f*, Weichkohle *f*
s. component *(Kern)* weiche Komponente *f*
s. copper Weichkupfer *n*
s.-drawn wire weichgezogener Draht *m*
s. drink alkoholfreies Getränk *n*
s.-ended vessel Schiff *n* mit Plattensteven
s.-focus lens *(Opt)* Weichzeichner *m*
s. ground lockerer Boden *m*
s. iron Weicheisen *n*
s.-iron ammeter Weicheisenstrommesser *m*
s.-iron armature Weicheisenanker *m*
s.-iron core Weicheisenkern *m*
s.-iron instrument Weicheiseninstrument *n*
s.-iron vane instrument Weicheiseninstrument *n*
s. lead Weichblei *n*
s. packing Weichpackung *f*, Weichdichtung *f*
s. paraffin Weichparaffin *n*
s. picture *(Fs)* weiches Bild *n*
s. pulp leicht bleichbarer Stoff *m* *(Papierherstellung)*
s. radiation weiche Strahlung *f*, Weichstrahlung *f*
s. ray weicher Strahl *m*
s. resin [natürliches] Weichharz *n*
s. soap Schmierseife *f*
s.-solder / to weichlöten
s. solder Schnellot *n*, Weichlot *n*
s. soldering Weichlöten *n*
s. spongy weichfaul *(Holz)*
s. steel Flußstahl *m*, weicher Stahl *m*
s. vacuum tube gasgefüllte Röhre *f* *(nicht ganz evakuiert)*
s. valve Niedervakuumröhre *f*, weiche Röhre *f*
s. water weiches Wasser *n*
s. wood Nadelholz *n*, Weichholz *n*
s. wood lumber *(Am)* Nadelschnittholz *n*
s. wood pulp Papierfaserstoff *m* aus Nadelholz
s. wood sulphate pulp Sulfatzellstoff *m* aus Nadelholz
s. X-rays weiche Röntgenstrahlen *mpl*
soften / to erweichen, weich machen, aufweichen, plastifizieren, geschmeidig machen; abschwächen, abtönen; enthärten; schmälzen; quetschen *(Jute)*; dämpfen *(Farbton; Stimme)*
s. down *(Ak)* abschwächen
s. water Wasser enthärten
softened lead bullion Werkblei *n*
softener *(Plast)* Weichmacher *m*; Enthärtungsmittel *n* *(Wasser)*

softening Weichmachen *n*, Erweichen *n*, Weichwerden *n*, Aufweichen *n*; Abschwächen *n*, Dämpfen *n*; Erweichung *f (Schneide)*; *(Plast)* Plastifizierung *f*; Enthärtung *f (Wasser)*
s. agent Weichmacher *m*, Weichmachungsmittel *n*
s. of the feed water Enthärtung *f* des Speisewassers
s. point Erweichungspunkt *m*, Erweichungstemperatur *f*; Kegelfallpunkt *m*
s. range Erweichungsbereich *m*, Erweichungszone *f*
soil / to verschmutzen
soil Boden *m*, Erde *f*, Erdreich *n*, Grund *m*; Ackerboden *m*
s. analysis Bodenanalyse *f*
s. characteristics Bodenbeschaffenheit *f*
s. chisel Bodenmeißel *m*
s. classification Bodenklassifizierung *f*
s. clod Bodenscholle *f*
s. compression Bodenverdichtung *f*
s. condition Bodenbeschaffenheit *f*
s. crumb Bodenkrümel *m*, Erdkrümel *m*
s. cutting Bodeneinschnitt *m*
s. drainage Trocknen *n* des Bodens
s. engineering Erdbau *m*; Grundbau *m*
s. grain Bodenteilchen *n*
s. injection Poreninjektion *f*
s. investigation Bodenuntersuchung *f*; Baugrunduntersuchung *f*
s. mechanics Baugrundmechanik *f*, Bodenmechanik *f*
s. moisture Bodenfeuchtigkeit *f*
s. nutrient Bodennährstoff *m*
s. packing harrow Krumenpacker *m*
s. particle Bodenteilchen *n*
s. physics Bodenphysik *f*
s. pipe Fallrohr *n (Installation)*
s. pressure Bodendruck *m*, Erddruck *m*; Bodenpressung *f*, Sohlpressung *f*
s. produce Bodenertrag *m*
s. property Bodeneigenschaft *f*
s. pulverizer Schollenbrechmaschine *f*
s. removing capacity Schmutzlösevermögen *n*
s. sample Bodenprobe *f*
s. science Bodenkunde *f*
s. stabilization Baugrundstabilisierung *f*, Baugrundverbesserung *f*, Baugrundverfestigung *f*, Bodenstabilisierung *f*, Bodenverbesserung *f*, Bodenverfestigung *f*, Bodenvermörtelung *f*
s. stabilizer Bodenstabilisator *m*
s. state of tillage Bodengare *f*
s. stratum Bodenschicht *f*
s. suction Bodensaugvermögen *n*
s. survey Bodenerkundung *f*
s. suspending power Schmutztragevermögen *n*
s. test Bodenprüfung *f*
s. texture Bodenstruktur *f*
s. tormentor Bodenlockerungsgerät *n*
sojourn time Verweilzeit *f*
sol *(Ch)* kolloidale Lösung *f*, Sol *n*
solar solar, Solar-, Sonnen-
s. activity Sonnentätigkeit *f*
s. activity period Sonnentätigkeitsperiode *f*
s. atmosphere Sonnenatmosphäre *f*
s. battery *s.* s. cell
s. cell Solarzelle *f*, Sonnenbatterie *f*, Sonnenzelle *f*
s. cell power supply Stromversorgung *f* mit Sonnenzellen
s. component of cosmic rays Sonnenkomponente *f* der kosmischen Strahlung
s. constant Solarkonstante *f*
s. corona Sonnenkorona *f*
s. day Sonnentag *m (Zeit)*
s. eclipse Sonnenfinsternis *f*
s. electricity Sonnenelektrizität *f*
s. energy Sonnenenergie *f*
s. energy conversion Sonnenenergieumwandlung *f*
s. engine Solarmaschine *f*, Sonnenkraftmaschine *f*
s. eruption Sonneneruption *f*
s. evaporation natürliche Verdampfung *f* unter Sonneneinwirkung
s. filter Sonnenfilter *n*
s. flare Sonneneruption *f*
s. generator Sonnenkraftmaschine *f*
s. heat Sonnenwärme *f*
s. heating *(Bw)* Sonnenheizung *f*
s. irradiation Sonnenbestrahlung *f*
s. limb Sonnenrand *m*
s. magnetic moment solares magnetisches Moment *n*, magnetisches Moment der Sonne
s. magnetism Sonnenmagnetismus *m*
s. noise Sonnenrauschen *n (Funk)*
s. observation Sonnenbeobachtung *f*
s. oil Solaröl *n*, Gasöl *n*
s. parallax Sonnenparallaxe *f*
s. photovoltaic cell Sperrschichtsonnenzelle *f*
s. power Sonnenenergie *f*
s. power plant Sonnenkraftwerk *n*
s. power station Sonnenkraftwerk *n*
s. prominence Sonnenprotuberanz *f*
s. radiation Sonnenstrahlung *f*
s. second Sonnenzeitsekunde *f*
s. spectrograph Sonnenspektrograf *m*
s. spectrum Sonnenspektrum *n*
s. spectrum analyzer automatisches Sonnenspektrometer *n*
s. stream solarer Teilchenstrom *m*
s. system Sonnensystem *n*
s. telescope Sonnenteleskop *n*
s. time Sonnenzeit *f*

solarization Solarisation *f*, Bildumkehrung *f* *(Fotografie)*; Einstrahlung *f*
solder / to löten; weichlöten
solder Lot *n*, Lötmittel *n*, Weichlot *n*
s. bath Lötbad *n*
s. tin Lötzinn *n*
s. wire Lötdraht *m*
solderability Lötbarkeit *f*
solderable lötbar
s. finish *(Plast)* lötbare Lackierung *f*
soldered connection piece Lötanschlußstück *n*
s. joint Lötverbindung *f*; Lötstelle *f*
soldering Löten *n*; Weichlöten *n*
s. agent Lötmittel *n*
s. box Lötmuffe *f* *(Kabel)*
s. compound Lötpulver *n*
s. copper Lötkolben *m*
s. ear Lötfahne *f*, Lötöse *f*
s. fluid Lötwasser *n*
s. flux Lötflußmittel *n*
s. furnace Lötofen *m*
s. iron Lötkolben *m*
s. lamp Lötlampe *f*
s. lug *(Am)* Lötfahne *f*, Lötöse *f*
s. paste Lötpaste *f*
s. pewter Lötzinn *n*
s. pin Lötstift *m*
s. post Stützpunkt *m* *(für Verdrahtung, Bauelemente usw.)*
s. process Lötverfahren *n*
s. sleeve Lötmuffe *f*
s. solution Lötwasser *n*
s. surface Lötfläche *f*
s. tag Lötfahne *f*, Lötanschluß *m*, Lötstift *m*
s. terminal Lötöse *f*
s. tongs Lötzange *f*
s. wire Lötdraht *m*
solderless unverlötet; lötlos
s. joint lötlose Verbindung *f* *(Rohr)*
soldier 1. *(Flg)* Pfosten *m* *(Tragflächengerippe)*; 2. aufrechtstehende Ziegelschicht *f*
sole Sohle *f*; Unterseite *f*; Bahn *f*; Bettplatte *f*; Holztraggerüst *n*
sole alleinig, Allein-
s. cutting machine Sohlenschneidemaschine *f*
s. inventor Alleinerfinder *m*
s. laying machine Sohlenauflegemaschine *f*
s. leather Sohlleder *n*
s. owner Alleininhaber *m*
s. piece of the stern-frame *(Schiff)* Hinterstevensohle *f*
s. plate frame of box section Kastenträger *m*
s. runner *(Bw)* Schwelle *f*
s. sewing machine Durchnähmaschine *f*
s.-shoe of plough Pflugsohle *f*
s. tacking machine Sohlenheftmaschine *f*
solenoid *(El)* Solenoid *n*, Zylinderspule *f*
solenoid [elektro]magnetisch
s. actuator *(BMSR)* Elektromagnetstellantrieb *m*
s. brake Magnetbremse *f*
s. core Solenoidkern *m*
s. operation magnetische Betätigung *f*
s. starter Magnetanlasser *m*
s. switch Magnetschalter *m*, magnetisch betätigter Schalter *m*
s. valve Magnetventil *n*
solenoidal field quellenfreies Feld *n*, Solenoidalfeld *n*
Solex pneumatic micrometer Solexgerät *n*, Druckluftfeinzeiger *m*
solid fester Körper *m*, Festkörper *m*
solid fest; massiv, stabil; aus einem Stück gefertigt, ungeteilt *(Rad)*; voll, einteilig; anstehend *(Kohle)*; formschlüssig; räumlich; beruhigt vergossen *(Stahl)*
s. angle Raumwinkel *m*
s. anode massive Anode *f*, Vollanode *f*
s. ballast fester Ballast *m*
s. bearing Vollager *n*
s. bed-type milling machine Planfräsmaschine *f*
s. bit Vollbohrer *m*
s. blank Rohling *m*
s. body fester Körper *m*, Festkörper *m*
s. bottom fester Boden *m*
s. brick Vollziegel *m*
s. cable Kabel *n* mit massiver Isolierung, massiv isoliertes Kabel, Massivkabel *n*
s. carbide drill Vollhartmetallbohrer *m*
s.-carbide special-purpose reamer Sonderreibahle *f* in Vollhartmetall
s. carbide tip Vollhartmetallschneidenkopf *m*
s. carbide tool Vollhartmetallwerkzeug *n*
s.-carbide twist drill Vollhartmetallspiralbohrer *m*
s. casting massiver Guß *m*
s. circuit Festkörperschaltung *f* *(aus einem einzigen Halbleiterblock bestehend)*
s. conductor massiver Leiter *m*, Massivleiter *m*
s. connection *(El)* massiver Anschluß *m*
s. construction Vollbauweise *f*
s. contact *(El)* fester Kontakt *m*
s. contents of stacked wood Derbgehalt *m*, Derbmasse *f* *(Holz)*
s. core removed ausgebohrter Kern *m* *(Hohlbohren)*
s. curve ausgezogene Kurve *f*
s. cutter einteiliger (massiver) Fräser *m*
s. dielectric Festdielektrikum *n*
s. earthing starre Erdung *f*
s. electrolyte Festelektrolyt *m*
s. engine Feststoffraketentriebwerk *n*
s. expansion reamer Spreizreibahle *f*

s. felt wheel for polishing Filzscheibe *f*
s. fibreboard Starkpappe *f*
s. firing Verfeuerung *f* fester Brennstoffe
s. forged coupling aufgeschmiedete Kupplung *f*
s. friction Trockenreibung *f*
s. fuel fester Brennstoff *m*, Festtreibstoff *m*
s.-fuel rocket Feststoffrakete *f*; Feststoffraketentriebwerk *n*
s.-fuel rocket engine Feststoffraketentriebwerk *n*
s. geometry Stereometrie *f*
s. gob *(Brg)* Vollversatz *m*
s. hand stowing *(Brg)* Handvollversatz *m*
s. height Länge *f* einer Feder in zusammengedrücktem Zustand
s.-height load erforderliche Last *f* zum völligen Zusammendrücken einer Feder
s.-hub pulley Riemenscheibe *f* mit ungeteilter Nabe
s. injection direkte (kompressorlose) Einspritzung *f*, Strahleinspritzung *f* *(Dieselmotor)*
s. insulation Massivisolation *f*
s. jaw feste (ungeteilte) Spannbacke *f* *(Futter)*
s. line 1. ausgezogene Linie *f*, Vollinie *f*; 2. Soliduslinie *f*
s. log bulkhead *(Schiff)* Holzschott *n*
s. lower portion of bearing ungeteiltes Lagerunterteil *n*
s. lubricant Starrschmiere *f*
s. matter 1. Feststoff *m*; Materie *f* in festem Zustand; 2. *(Typ)* kompresser Satz *m*
s. measure Festmaß *n*, Festgehalt *m* *(Holz)*
s. medium festes Medium *n*
s. motor Feststoffraketentriebwerk *n*
s. nickel Hartnickel *n*
s. non-metallic inclusion fester, nichtmetallischer Einschluß *m*
s. object räumliches Objekt *n*
s. of revolution Rotationskörper *m*, Umdrehungskörper *m*
s. paraffin Hartparaffin *n*
s. phase feste Phase *f*
s.-phase welding Feuer[preß]schweißen *n*
s. piston Scheibenkolben *m*
s. propellant fester Treibstoff *m*, Festtreibstoff *m*
s. propellant engine Feststoffraketentriebwerk *n*
s.-propellant motor Feststoffraketentriebwerk *n*
s.-propellant rocket Feststoffrakete *f*; Feststoffraketentriebwerk *n*
s.-propellant rocket motor Feststoffraketentriebwerk *n*
s. propeller Propeller *m* *(Nabe und Flügel aus einem Stück gegossen)*
s. punch Durchschläger *m*, Durchtreiber *m*
s. reamer feste (nicht verstellbare) Reibahle *f*
s. rim ring ungeteilter Felgenring *m*
s. rocket unit Feststoffraketentriebwerk *n*
s. rubber tyre Vollgummireifen *m*
s. scrap Kernschrott *m*, stückiger Schrott *m*
s. shaft Vollwelle *f*
s. shrinkage strain *(sl)* Schrumpfspannung *f*
s. solution feste Lösung *f*; Mischkristall *m*, Einlagerungsmischkristall *m*
s. specimen Festprobe *f*
s. spirit Hartspiritus *m*
s. state fester Zustand *m*, Festkörperzustand *m*
s. state amplifier Festkörperverstärker *m*
s. state analogue memory analoger Festkörperspeicher *m*
s. state circuit Festkörperschaltung *f*
s. state component Festkörperbauelement *n* *(Halbleiter)*
s. state detector Festkörperdetektor *m*
s. state device Festkörperbaustein *m*
s. state digital computer Festkörperdigitalrechner *m*
s. state electronics Festkörperelektronik *f*
s. state element Festkörperbauelement *n*
s. state maser Festkörper-Maser *m*, Festkörpermolekularverstärker *m*
s. state memory Festkörperspeicher *m*
s. state microwave amplifier Festkörperverstärker *m* für Mikrowellen
s. state optical maser optischer Festkörper-Maser *m*
s. state photodetector Festkörperfotodetektor *m*
s. state physics Festkörperphysik *f*
s. state printed circuit Festkörperdruckschaltung *f*
s. state switch Festkörperschalter *m*
s. steel beruhigt vergossener Stahl *m*
s. stowing *(Brg)* Vollversatz *m*
s. stress [of a spring] Spannung *f* einer völlig zusammengedrückten Feder
s. timber Massivholz *n*
s. turbine *(Am)* Einwellenturbine *f*
s.-type cable Massekabel *n*
s. tyre Vollgummireifen *m*
s. web girder *(Bw)* Vollwandbinder *m*, Vollwandträger *m*
s. web section *(Bw)* Vollwandträger *m*
s. wood Vollholz *n*
s.-wood pulley Vollholzscheibe *f*

solidification Erstarrung *f*, Verfestigung *f*
s. point Erstarrungspunkt *m*, Verfestigungspunkt *m*
s. range Erstarrungsbereich *m*, Erstarrungsintervall *n*
s. shrinkage *(Met)* Schrumpfung *f*, Schwindung *f*

solidify / to erstarren, fest werden, [sich] verfestigen
solidifying point Erstarrungspunkt *m*
s. point of oil Kältepunkt *m* des Öls
solidly earthed system starr geerdetes Netz *n*, starre Erdung *f*
solidus *s.* s. line
s. curve *s.* s. line
s. line Soliduslinie *f*
solim inclusion nichtmetallischer Einschluß *m*
solion Solion *n*
Solnhofen stone Solnhofer Lithografenkalk *m*
solo 1. Solomaschine *f*; 2. Soloflug *m*
solodyne receiver Solodynempfänger *m*
solubility Löslichkeit *f*; Lösbarkeit *f*
s. curve Lösungskurve *f*
s. limit Löslichkeitsgrenze *f*
solubilize / to löslich machen
solubilizer Lösungsvermittler *m*
soluble lösbar *(Problem)*; löslich *(Stoff)*
s. glass Wasserglas *n*
s. in fat fettlöslich
s. in water wasserlöslich
s. oil wasserlösliches Öl *n*
solute [auf]gelöster Stoff *m*
solution *(Ch, Math)* Lösung *f*; Auflösung *f*
s.-annealed lösungsgeglüht
s. dyed spinngefärbt, in der Spinnlösung (Spinnmasse) gefärbt
s. heat-treatment Lösungsglühen *n* *(von Leichtmetallen)*
s. of caustic soda Natronlauge *f*, Ätznatronlösung *f*
s. of tannic acid Gerbsäurelösung *f*
s.-polymerization Lösungspolymerisation *f*, Polymerisation *f* in Lösung
s. pressure Lösungsdruck *m*
solvability Lösbarkeit *f*
solvable löslich *(Stoff)*; *(Math)* lösbar
solvation Solvatation *f*, Solvatisierung *f*
solve / to *(Ch, Math)* [auf]lösen; losbinden
solvent lösend, löslich
solvent Lösungsmittel *n*, Auflösungsmittel *n*, Lösemittel *n*
s. laminating Schichtstoffherstellung *f* unter Anwendung von Lösungsmitteln
s. method Lösungsmittelverfahren *n* *(Chemiefaserherstellung)*
s. moulding *(Plast)* Tauchen *n*, Tauchformen *n*
s. naphtha leichtes Teeröldestillat *n*, Schwerbenzin *n*, Solventnaphtha *n* *(f)*
s. power Lösungsvermögen *n*
s. preservative lösliches Schutzmittel *n* *(Holztränkung)*
s. rubber cement Gummilösung *f*
s. tolerance Lösemittelaufnahmefähigkeit *f* *(Lack)*
solvolysis *(Ch)* Solvolyse *f*

somersault / to überschlagen *(Flugzeug)*
sonde Registrierballon *m*, Ballonsonde *f*
sone Sone *n*
(Kennwort zur Angabe der Lautheit)
s. scale Sone-Skala *f*
sonic Ton-, Schall-
s. analyzer Ultraschallprüfgerät *n*, Defektoskop *n*
s. barrier Schallmauer *f*
s. buoy Geräuschboje *f*
s. delay line (memory) *(Dat)* Ultraschallspeicher *m*, akustischer Laufzeitspeicher *m*, akustische Verzögerungsleitung *f*, Ultraschallverzögerungsleitung *f*
s. depth-finder Schallwellenecholot *n*, Echolot *n*
s. depth gauge Ultraschallot *n*, Ultraschalltiefenmesser *m*
s. flowmeter Ultraschalldurchflußmesser *m*
s. frequency Tonfrequenz *f*
s. pressure Schalldruck *m*
s. probe akustische Sonde *f*
s. speed Schallgeschwindigkeit *f*
s.-type screening machine Schallsiebmaschine *f* *(Aufbereitung)*
s. value Schallgrenze *f*
s. velocity Schallgeschwindigkeit *f*
sonometer Sonometer *n*, Schallmesser *m*
soot / to berußen, verrußen; verschmutzen *(Kontakte)*
soot Ruß *m*
s. and white wash Schatteneffekt *m* *(Ikonoskop)*
s. black Lampenruß *m*, Rußschwarz *n*
s. black niker Rußschreiber *m*
s. blower Rußbläser *m*
s. blowing Rußblasen *n*
s. catcher Rußfänger *m*
s. coal Rußkohle *f*
s. collector Rußvorlage *f*
s.-coloured rußig, rußfarbig
s. door Schornsteinreinigungsklappe *f*
s. filter Rußvorlage *f*
s. pit Rußsammelkasten *m*
s. receiver Rußvorlage *f*
sophisticated verfeinert, kompliziert, hochgezüchtet *(Gerät)*
soporific Schlafmittel *n*
sorbate adsorbierter Stoff *m*
sorbent Adsorbens *n*, adsorbierender Stoff *m*, Sorbens *n*
sorbic acid Sorbinsäure *f*
sorbite *(Met)* Anlaßgefüge *n* nach Zerlegung des Martensits
sorbitol Sorbit *n* *(Kohlehydrat)*
sordine Dämpfer *m*, Schalldämpfer *m*, Tondämpfer *m* *(Geige)*
Sorel's cement Sorelzement *m*

Soret effect *(Kern)* Soret-Effekt *m*, Thermodiffusion *f*
sorption Sorption *f*, Aufnahme *f*
sort / to [aus]sortieren, aussondern, auslesen, klassieren, scheiden; *(Lw)* auslesen
s. out *(Brg)* ausklauben, [aus]sortieren
sorter 1. *(Brg)* Klauber *m*, Sortierer *m*; 2. Sortiermaschine *f*, Lochkartensortierer *m*, Sortierer; 3. *(Lw)* Sichter *m*
s. reader Sortierer *m* und Leser *m*, Sortierleser *m*
sorting Sortieren *n*, Sortierung *f*
s. aggregate Sichter *m*
s. arrangement Sortiergerät *n*, Sortieranlage *f*
s. belt Leseband *n*, Sortierband *n*
s. cylinder Sortierzylinder *m*
s. gates Sortierweiche *f*
s. room Sortierraum *m*
s. screen Sortiersieb *n*
s. table Sortiertisch *m*, Lesetisch *m*; *(Brg)* Klaubetisch *m*
sough Abflußgraben *m*
sound / to 1. schallen, tönen; 2. [aus]loten, peilen
sound lunkerfrei, fehlerfrei *(Guß)*
sound 1. Klang *m*, Schall *m*, Ton *m*; 2. Meerenge *f*, Sund *m* *(schiffbare Wasserstraße)*; 3. Sonde *f*
s. absorbant Schalldämmstoff *m*, Schluckstoff *m*
s.-absorbent geräuschdämpfend, schalldämmend, schalldämpfend, schallschluckend, schallabsorbierend
s. absorber Schallschlucker *m*, Schluckstoff *m*
s. absorbing geräuschdämpfend, schalldämmend, schalldämpfend, schallschluckend, schallabsorbierend
s. absorbing material Dämpfungsmaterial *n*, Schallschluckstoff *m*
s. absorption Schallabsorption *f*, Schallabsorbierung *f*
s. absorption coefficient Schallschluckgrad *m*, Schallabsorptionskoeffizient *m*
s. absorptivity Schluckgrad *m*
s. amplification Tonverstärkung *f*
s. amplifier Tonverstärker *m*
s. analyzer Klanganalysator *m*
s. and wave tuning Abstimmung *f* der Tonhöhe und Welle
s. antenna Tonsender *m*
s. aperture Schallöffnung *f*
s. attenuation Schalldämpfung *f*, Schalldämmung *f*
s. barrier Schallmauer *f*
s. beam Schallstrahl *m*
s. broadcasting Hörfunk *m*
s. camera Tonkamera *f*
s. carrier *(Fs)* Tonträger *m*
s. change-over Tonüberblendung *f*
s. coefficient Schluckgrad *m*
s. conductivity Schalleitfähigkeit *f*
s. conductor Schalleiter *m*
s. corrector Schallregler *m*, Tonregler *m*, Klangregler *m*
s. damper Dämpfer *m*, Schalldämpfer *m* *(Klavier)*
s. damping Schalldämmung *f*, Schallisolierung *f*
s. deadening schalldämpfend
s. deadening material Schalldämmstoff *m*
s. detector Horchgerät *n*
s. diffraction Schallbeugung *f*
s. discriminator *(Fs)* Tonfrequenzgleichrichter *m*
s. dissociation Klangzerfall *m*
s. distortion Klangverzerrung *f*
s. distribution Tonverteilung *f*
s. effect Lautwirkung *f*, Tonwirkung *f*, Klangwirkung *f*
s. energy Schallenergie *f*
s. energy density Schall[energie]dichte *f*
s. energy flux Schallfluß *m* *(Produkt aus Schallschnelle und Schallquerschnitt)*
s. energy flux density objektive Schallstärke *f*
s. engineer Tonmeister *m*
s. excitation Schallerregung *f*
s. fading Tonüberblendung *f*
s. field Schallfeld *n*
s. film Tonfilm *m*
s. film installation Tonfilmanlage *f*
s. filter Klangfilter *n*
s. focusing Schallbündelung *f*
s. frequency Tonfrequenz *f*
s. funnel Schalltrichter *m*
s. generation Schallerzeugung *f*
s. generator Schallquelle *f*
s. groove Schallrille *f* *(Schallplatte)*
s. hardness Schallhärte *f*
s. head Tonkopf *m*
s. in solids Körperschall *m*
s. incidence Schalleinfall *m*
s.-insulated schallisoliert
s. insulation Schalldämmung *f*, Schallisolation *f*, Schallisolierung *f*
s. intensity Schallintensität *f*, Schallstärke *f*
s. level Lautstärke *f*, Schallpegel *m*
s. level indicator Lautstärkeanzeiger *m*
s. level meter Lautstärkemesser *m*, Schalldruckmesser *m*
s. localization Lokalisierung *f* des Schalls
s. locating akustische Ortung *f*
s. locator Horchgerät *n*
s. measurement Schallmessung *f*
s. mirror Schallspiegel *m*
s. mixer Tonmischpult *n*

s. mixing Tonmischung *f*
s.-modulated wave tonmodulierte Welle *f*
s. motion picture Tonfilm *m*
s. motion picture print Tonfilmkopie *f*
s. on disk system Nadeltonverfahren *n*
s.-on-film system Lichttonverfahren *n*
s. on vision *(Fs)* Ton *m* im Bild
s. oscillation Schallschwingung *f*, Tonschwingung *f*, akustische Schwingung *f*
s. overshooting Mikrofonübersteuerung *f*
s. path length Schallweglänge *f*
s. pattern Chladnische Klangfiguren *fpl*
s. permeability Schalldurchlässigkeit *f*
s. phase fehlerfreie Phase *f*
s. pick-up Tonabnehmer *m*
s. picture projector Tonfilmprojektor *m*
s. plate Stimmplatte *f*
s. post Stimmholz *n (Geige)*
s. pressure Schalldruck *m*
s. pressure level Schalldruckpegel *m*
s. pressure microphone Schalldruckmikrofon *n*
s. projector Tonfilmprojektor *m*
s.-proof schalldicht, schallsicher
s.-proof chamber schallsicherer (schalltoter) Raum *m*
s.-proofing Schalldämmung *f*, Schallschutz *m*; Schalldichtmachung *f*
s. propagation Schallausbreitung *f*, Schallfortpflanzung *f*
s. pulse Schallimpuls *m*
s. quality Tonqualität *f*
s. radiation Schall[ab]strahlung *f*
s. radiation pressure Schallstrahlungsdruck *m*
s. radiation reactance Schallstrahlungsreaktanz *f*
s. radiation resistance Schallstrahlungswiderstand *m*
s. range Schallumfang *m*
s.-range altimeter elektroakustisches Echolot *n*
s. ranging Schallortung *f*, Schallradar *n*, Schallmeßaufklärung *f*
s. ranging microphone Schallmeßmikrofon *n*
s. receiver Schallempfänger *m (Mikrofon)*; *(Fs)* Tonempfänger *m*
s. reception Schallempfang *m*
s. recorder Tonaufzeichnungsgerät *n*, Tonaufnahmegerät *n*
s. recording Tonaufnahme *f*, Tonaufzeichnung *f*, Schallaufzeichnung *f*
s. recording amplifier Tonaufzeichnungsverstärker *m*
s. recording device *s.* s. recorder
s. reduction factor Schalldämpfungsfaktor *m*
s.-reflecting schallreflektierend
s. reflection coefficient Schallreflexionskoeffizient *m*
s. rejection Tonunterdrückung *f*
s. reproduction Tonwiedergabe *f*
s.-scanning slit Tonabtastspalt *m*
s. shadow Schallschatten *m*
s. signal akustisches Signal *n*
s. source Schallquelle *f*, Schallgeber *m*, Schallsender *m*
s. spectroscopy Schallspektroskopie *f*
s. steel lunkerfreier Stahl *m*
s. suppression Schalltilgung *f*
s. telegraphy Wechselstromtelegrafie *f*
s. track Tonspur *f (Tonfilm)*
s. transmission Schallübertragung *f*
s. transmitter *(Fs)* Tonsender *m*
s. tuning Abstimmung *f* der Tonhöhe
s. velocity Schallgeschwindigkeit *f*
s. vibration Schallschwingung *f*, Tonschwingung *f*, akustische Schwingung *f*
s. volume Klangfülle *f*, Tonumfang *m*, Lautstärke *f*
s. wave Schallwelle *f*, Welle *f* im Hörbereich
s. waveform Tonwellenform *f*
sounder 1. Lotgerät *n*, Tiefenmeßgerät *n*; 2. Klopfer *m (Telegrafie)*
sounding 1. Lotung *f*; 2. Tönen *n*
sounding tönend, schallend
s. arc tönender Lichtbogen *m*
s. balloon Registrierballon *m*, Ballonsonde *f*
s. board Klangboden *m*, Decke *f*, Schallboden *m (Saiteninstrumente)*
s. boat Peilboot *n*
s. buoy Heultonne *f*, Heulboje *f*
s. device Echolot *n*
s. out the grinding wheel Klangprobe *f* an Schleifscheibe
s. pipe Peilrohr *n*
s. pole Lotstock *m*, Lotstange *f*
s. post Stimmholz *n (Geige)*
s. rocket Registrierrakete *f*
s. sextant Lotsextant *m*, Peilsextant *m*
soundings *(Schiff)* lotbarer Grund *m*
soundness Fehlerlosigkeit *f*, Lunkerfreiheit *f (Guß)*; Raumbeständigkeit *f (Beton)*
soundproofing *s.* sound-proofing
soup *(sl)* Entwickler *m (Foto)*
sour / to [an]säuern; sauer werden
sour sauer
s. fermentation saure Gärung *f*
s. milk acid Gärungsmilchsäure *f*
s. natural gas schwefelwasserstoffhaltiges Erdgas *n*
source Quelle *f*; Stromquelle *f*, Spannungsquelle *f*; Strahlenquelle *f*, Strahlungsquelle *f*, *(radioaktives)* Präparat *n*
s. and load resistances Betriebswiderstände *mpl (Vierpol)*

s. container Behälter *m* für Strahlenquellen
s. flux Quellfluß *m*
s.-free quellenfrei
s. function *(Math)* Quellenfunktion *f*
s. mover *(Kern)* Präparateschieber *m*, Quellenschieber *m*
s. of energy Energiequelle *f*, Kraftquelle *f*
s. of errors Fehlerquelle *f*
s. of heat Wärmequelle *f*
s. of light Lichtquelle *f*
s. of radiation Strahlungsquelle *f*, Strahlenquelle *f*
s. of sound Schallquelle *f*
s. of supply Stromquelle *f*, Spannungsquelle *f*
s. point Quellpunkt *m*
s. resistance Generatorwiderstand *m*; Quellenwiderstand *m*
s. rock Muttergestein *n*; gesunder Fels *m*, gesundes Gestein *n*
s.-sink method *(Math)* Quellsenkenverfahren *n*
s. strength Stärke (Aktivität) *f* der Strahlenquelle, Quellenaktivität *f*
s.-to-skin distance *(Kern)* Strahlenquelle-Haut-Abstand *m*, Fokus-Haut-Abstand *m*
sourdene Schwingungsdämpfer *m (Freileitungen)*
souring bath Säurebad *n*
sourness Säuregehalt *m*, Säuregrad *m*
south Süden *m*
south südlich
s. latitude südliche Breite *f*
s. pole Südpol *m*
southeasterly südöstlich
southern südlich
southwesterly südwestlich
sow / to säen
sow *(Met)* Ofensau *f*
s. channel Masselgraben *m*
sowing Aussaat *f*, Saat *f*
s. area Aussaatfläche *f*
s. machine Sämaschine *f*, Drillmaschine *f*, Reihensämaschine *f*
s. period Säzeit *f*
s. plough Säpflug *m*
s. quantities Aussaatmenge *f*
s. time Säzeit *f*
soya bean Sojabohne *f*
s. bean oil Soja[bohnen]öl *n*
space / to 1. in Abstand bringen, teilen; 2. *(Typ)* sperren, spationieren, ausschließen
s. apart im Abstand halten
s. irregularly ungleich teilen
space 1. Raum *m*; Abstand *m*, Lücke *f*; Platz *m*; Weg *m*; 2. *(Typ)* Spatium *n*, Durchschuß *m*; 3. *(Dat)* Impulspause *f*, Impulslücke *f*
s. above grate Feuerstätte *f*
s. age Weltraumzeitalter *n*
s. angle 1. Raumwinkel *m*; 2. Zahnlückenwinkel *m (Verzahnung)*
s. aviation Weltraumschiffahrt *f*, Kosmonautik *f*
s. bar Leertaste *f*
s. bulkhead *(Schiff)* Gitterschott *n*, Lattenschott *n*
s. capsula Raumkapsel *f*
s. charge *(El)* Raumladung *f*
s. charge accumulation Raumladungsanhäufung *f*
s. charge boundary Raumladungsgrenzschicht *f*
s. charge capacity Raumladungskapazität *f*, Sperrschichtkapazität *f*
s. charge current Raumladungsstrom *m*
s. charge density Raumladungsdichte *f*
s. charge detector Raumladungsdetektor *m*
s. charge distortion Raumladungsverzerrung *f*
s. charge effect Raumladungseffekt *m*
s. charge formation Raumladungsbildung *f*
s. charge grid Raumladungsgitter *n*
s. charge layer Raumladungsrandschicht *f*, Raumladegebiet *n*; Raumladungszone *f*, Sperrschicht *f (Halbleitertechnik)*
s. charge limitation Raumladungsbegrenzung *f*
s. charge modulation Raumladungsmodulation *f*
s. charge polarization Raumladungspolarisation *f*
s. charge region Raumladungsgebiet *n*, Raumladungszone *f*, Raumladungsbereich *m*, Raumladungsrandschicht *f*
s. charge wave Raumladungswelle *f*
s. charge wave tube Elektronenwellenröhre *f (Lauffeldröhre ohne statisches Querfeld)*
s. code Abstandkode *m*, Pausenkode *m*, Zwischenraumkode *m*
s. communications Weltraumnachrichtenverbindung *f*
s. coordinates Raumkoordinaten *fpl*
s. craft Raumschiff *n*, Raumfahrzeug *n*
s. current Raumladestrom *m*, raumladungsbegrenzter Strom *m*; Arbeitsstrom *m*, Strom im Belastungswiderstand *(Elektronenröhren)*
s. curvature Raumkrümmung *f*
s. curve Raumkurve *f*
s.-cutter Zahnlückenfräser *m*
s. dependence Raumabhängigkeit *f*
s. depth *(Opt)* Tiefenraum *m*
s. distribution Raumverteilung *f*, räumliche Verteilung *f*
s. distribution of mass räumliche Massenverteilung *f*
s. electronics Raumfahrtelektronik *f*
s. expert Fachmann *m* für Raumschiffahrt
s. factor *(Fmt)* Gruppencharakteristik *f*, Gruppenfaktor *m (Antennen)*

s. factor of winding Ausnutzungsfaktor *m* des Wicklungsraumes *(Transformator)*
s. filter Spaltfilter *n (Hydraulik)*
s. flight Raumflug *m*, Raumfahrt *f*, Raumschiffahrt *f*, Weltraumfahrt *f*; Raumfahrt eines bemannten Flugkörpers
s.-flight physiology Raumfahrtphysiologie *f*
s. frame *(Bw)* räumliches Tragwerk *n*; Raumfachwerk *n*, räumliches Fachwerk (Gittertragwerk) *n*
s. frequency Raumfrequenz *f*
s. girl *(sl)* Kosmonautin *f*
s.-group *(Ph)* Raumgruppe *f*
s. heating *(Bw)* Raumheizung *f*
s. lattice Raumgitter *n*, Kristallgitter *n*; räumliches Gitter[trag]werk *n*
s.-lattice cell Gitterzelle *f*
s.-lattice distance Netzebenenabstand *m*
s. law Raumrecht *n*, Weltraumrecht *n*
s. limitation Raumbeschränkung *f*
s. man Kosmonaut *m*
s. medicine Raumfahrtmedizin *f*
s. model Raummodell *n*
s. nautics Raumschiffahrt *f*
s. permeability *(El)* Induktionskonstante *f*, Permeabilität *f* des leeren Raumes
s. phasing Phasendifferenz *f (Antennentechnik)*
s. probe Raumsonde *f*
s. quantization *(Ph)* Raumquantelung *f*
s. radiation Raumstrahlung *f*
s. ray Raumstrahl *m*
s. reflection Raumreflexion *f*
s. requirement Raumbedarf *m*, Platzbedarf *m*
s. research Weltraumforschung *f*
s. rocket kosmische Rakete *f*
s. saving raumsparend, platzsparend
s. saving construction raumsparende Konstruktion *f*
s. ship Raumschiff *n*, Raumfahrzeug *n*
s. signal *(Fmt)* Zwischenraumzeichen *n*
s. station Weltraumstation *f*, Außenstation *f*, Raumstation *f*
s. structure *(Bw)* räumliches Tragwerk *n*; Raumfachwerk *n*, räumliches Fachwerk (Gittertragwerk) *n*
s. suit Raumanzug *m*
s. telecommunications Weltraumnachrichtenverkehr *m*
s. telephony drahtlose Telefonie *f*, Funktelefonie *f*
s.-time Raumzeit *f*
s.-time conception Raum-Zeit-Vorstellung *f*
s. travel Raumflug *m*, Raumfahrt *f*, Raumschiffahrt *f*, Weltraumfahrt *f*; Raumfahrt eines bemannten Flugkörpers
s. traveller Astronaut *m*, Raumschiffer *m*
s. truss Raumfachwerk *n*, räumliches Fachwerk *n*
s. vehicle Raumschiff *n*, Raumfahrzeug *n*
s. wave Raumwelle *f*
s.-wound *(El)* auf (mit) Abstand gewickelt
spaced unterteilt
s. aerial reception Empfang *m* mit Mehrfachantenne
s. frame aerial nachteffektfreie Antenne *f*
spacer Abstandshalter *m*, Abstandshülse *f*; Abstandsring *m*; Beilage *f*, Zwischenlage *f*; Distanzscheibe *f*; Distanzstück *n*, Zwischenstück *n*; Distanzbüchse *f*
s. sleeve Abstandsrohr *n*
s.-wave *(Fmt)* Pausenwelle *f*
spacing 1. Abstand *m*, Zwischenraum *m*; Teilung *f*; Lücke *f*; Entfernung *f*; 2. *(Typ)* Ausschließen *n*; Sperren *n*; 3. *(Fmt)* Unterbrechung *f*; 4. Abstandsbestimmung *f*
s. accuracy Teilgenauigkeit *f*
s. accuracy of holes (bores) Abstandsgenauigkeit *f* von Bohrungen
s. attachment Teileinrichtung *f*
s. block Abstandshalter *m*
s. collar Spindelring *m*; Distanzring *m*, Abstandsring *m (Fräsdorn)*
s. contact Trennkontakt *m*
s. current Zwischenzeichenstrom *m*, Pausenstrom *m (Telegrafie)*
s. error Teilungsfehler *m*
s. jig Abstandslehre *f*; Stichmaß *n (Elektronenröhrenherstellung)*
s. key *(Typ)* Ausschlußtaste *f*
s. lug Trennungszunge *f*
s. of abrasive grains Gefüge *n* der Schleifscheibe
s. of file teeth Feilenhiebteilung *f*
s. of teeth Zahnteilung *f*
s. of the cutting edges Schneidenteilung *f*
s. of the flutes Zahnteilung *f (Reibahle)*
s. of the humps *(El)* Abstand *m* der Buckel (Resonanzspitzen)
s. piece Abstandsstück *n*
s. piled luftig gestapelt *(Holz)*
s. pulse *(Fs)* Trennimpuls *m*
s. ring Abstandsring *m*
s. signal *(Fmt)* Abstandszeichen *n*, Ruhezeichen *n*
s. stay Abstandshalter *m*
s. washer Abstandsscheibe *f*
s. wave *(Fmt)* negative Tastwelle *f*, Pausenwelle *f*
s. wedge Abstandskeil *m*
spacious geräumig, weiträumig
spacistor *(El)* Spacistor *m (Verstärkerelement)*
spade / to umgraben
spade Spaten *m*
s. drill Spitzbohrer *m*
s. end offene Lötöse *f*; offener Kabelschuh *m*

s. rudder *(Schiff)* Steckruder *n*, Balanceschweberuder *n*, Vollschweberuder *n*
spading 1. Umgraben *n*; 2. Mischen *n* von Hand *(Beton)*
spaghetti 1. Isolierschlauch *m*; 2. Filmsalat *m*
spall / to abplatzen, abblättern *(Gestein)*; abbröckeln *(Schneide)*
s. off *(Brg)* abplatzen *(Nebengestein)*
spall Splitter *m*; Span *m*; Ziegelbruch *m*
spallation Kernzersplitterung *f*, Kernzertrümmerung *f*, Spallation *f*
s. cross section Spallationsquerschnitt *m*
s. fragment Spallationsbruchstück *n*
s. reaction Spallation *f*, Spallationsreaktion *f*
spalling Abblättern *n*; Schülpen *n*; Abplatzen *n*; Abspalten *n*, Abmeißeln *n*
s. resistance *(Met)* Temperaturwechselbeständigkeit *f*
s. test *(Met)* Hitzebeständigkeitsversuch *m* *(gegen Abblättern)*
span / to überspannen; überbrücken; umspannen; einfassen
span 1. *(Bw)* Spannweite *f*, Stützweite *f*, Stützlänge *f*; 2. Meßbereich *m*; 3. *(Schiff)* Hanger *m* *(Ladebaum)*
s. length Stützweite *f*, Stützlänge *f*
s. of pressure arch *(Brg)* Weite *f* des Druckgewölbes
s. pole Abspannpfahl *m*
s. rope Abspannseil *n*
s. web blade with riveted tangs Spannsägeblatt *n* mit Nietangeln
s. web blade with solid tangs Spannsägeblatt *n* mit festen Angeln
s. wire Abspanndraht *m*, Spanndraht *m*, Tragdraht *m*
s.-wire suspension Spanndrahtaufhängung *f*
spandex yarn Spandexgarn *n* *(hochelastisches Polyurethangarn)*
spandrel Bogenzwickel *m* *(Brückenbau)*
s. steps Dreieckstufen *fpl*
spangle 1. Metallfolie *f*; 2. Zinkblume *f*
Spanish guitar Schlaggitarre *f*
spanner 1. Bolzenschlüssel *m*, Mutterschlüssel *m*, Schlüssel *m*; 2. Pleuelstange *f*
spar 1. *(Min)* Spat *m*; 2. Spiere *f*, Rundholz *n*; Sparren *m*, Längsholm *m*, Holm *m*
s. and boom gear *(Schiff)* Ladegeschirr *n* *(für gekuppelte Ladebäume)*
s. and skin miller Flügelholm- und Außenhautfräsmaschine *f*
s. and stringer milling machine Flügelholm- und Stützenfräsmaschine *f*
s. buoy *(Schiff)* Spierentonne *f*
s. ceiling *(Schiff)* Lattenwegerung *f*
s. cutter Holmfräser *m*
s. fender *(Schiff)* Holzfender *m*, Reibholz *n*
s. varnish Außenlack *m*, Luftlack *m*; Bootslack *m*
sparable Stift *m* *(kopfloser Stift zum Besohlen)*
spardeck ship Spardeckschiff *n*
spare / to [ein]sparen; schonen
spare Ersatzteil *n*
spare Reserve-, Ersatz-
s. anchor *(Schiff)* Notanker *m*, Reserveanker *m*
s. assembly Ersatzbaustein *m* *(Elektronik)*
s. battery Reservebatterie *f*
s. boiler Reservekessel *m*
s. brush *(El)* Ersatzbürste *f*
s. cable Ersatzkabel *n*, Reservekabel *n*, Vorratskabel *n*
s. carrier Reserveradhalter *m*
s. conductor Schutzleiter *m*, Reserveleiter *m*
s. contact freier Kontaktstift *m*, Leerkontakt *m*
s. current source Netzersatzanlage *f*
s. engine Ersatzmotor *m*
s. fuel capacity Kraftstoffreserve *f*, Treibstoffreserve *f*
s. fuse Ersatzsicherung *f*
s. insulator Ersatzisolator *m*, Reserveisolator *m*
s. jaw Ersatzbacke *f* *(Futter)*
s. level *(Fmt)* unbesetzter Höhenschritt *m*
s. part Ersatzteil *n*, Reserveteil *n*
s. part catalogue Ersatzteilliste *f*
s. part depot Ersatzteillager *n*
s. piston Ersatzkolben *m*
s. repeater *(Fmt)* Reserveverstärker *m*, Vorratsverstärker *m*
s. shaft Reservewelle *f*
s. termination set *(Fmt)* Vorratsgabelschaltung *f*
s. wheel Reserverad *n*
s. wheel carrier Reserveradhalter *m*
s. wire *(Fmt)* Vorratsader *f*
sparging Einmaischen *n*; Hindurchperlen *n*; Durchblasen *n* *(Gase durch Flüssigkeiten)*
spark / to funken, feuern
s. out ausfeuern
s. over überschlagen, überspringen *(Funken)*
s. through *(Am)* durchschlagen *(Funken)*
spark Funke *m*
s. adjustment Zündverstellung *f*
s. advance Frühzündung *f*, Vorzündung *f* *(Verbrennungsmotor)*
s. advance and retard mechanism Zündzeitpunktverstellvorrichtung *f*
s. arrester *(El)* Funkenableiter *m*
s. arrester funnel Funkenfängerschornstein *m* *(Lokomotive)*
s. at break *(El)* Öffnungsfunke *m*
s. blow-out *(El)* Funkenlöscher *m*
s. breakdown Funkenüberschlag *m*

s. capacitor Störschutzkondensator *m*
s. catcher Funkenfänger *m*
s. coil Funkeninduktor *m*; Zündspule *f*
s. cutting Elektrofunkenbearbeitung *f*
s. discharge Funkenentladung *f*, Funkenübergang *m*; Funkenstrecke *f*
s. eroder Abfunkmaschine *f*; Funkenerosionsmaschine *f*
s. erosion 1. elektroerosive Metallbearbeitung *f*, Funkenerosion *f*; 2. Zündkerzenabbrand *m*
s. erosion machine Funkenerosionsmaschine *f*
s. finned plug Zündkerze *f* mit Kühlrippen
s. forming elektrohydraulische Umformung *f*
s. gap 1. Funkenstrecke *f*; 2. Elektrodenabstand *m (z. B. einer Zündkerze)*
s.-gap flashover Funkenstreckenüberschlag *m*
s.-gap modulation *(Fmt)* Modulation *f* durch Funkenstrecke
s.-gap oscillator Funkenstreckensender *m*
s. generator Funkenerzeuger *m*
s. igniter elektrischer Zünder *m*
s. ignition Fremdzündung *f*, Kerzenzündung *f*, Funkenzündung *f*
s.-ignition engine Ottomotor *m*, Vergasermotor *m*, Motor *m* mit elektrischer Zündung
s. killer *(sl)* Störschutzkondensator *m*
s. lighter Anzünder *m*, Feueranzünder *m*
s.-like discharge funkenähnliche Entladung *f*
s. line Funkenlinie *f*
s. machining Ausfunken *n*, elektroerosive Metallbearbeitung *f* nach dem Elektrofunkenprinzip
s. metal working process Ausfunkverfahren *n*
s.-over Funkenüberschlag *m*, Überspringen *n* eines Funkens
s.-over test Überschlagsprüfung *f*
s.-over voltage Überschlagsspannung *f*
s. plug Zündkerze *f*
s. plug electrode Zündkerzenelektrode *f*
s. plug gap Elektrodenabstand *m (Zündkerze)*
s. plug interference Zündkerzenstörung *f (z. B. im Fernsehbild)*
s. plug position Zündkerzenanordnung *f*
s. plug tap Zündkerzengewindebohrer *m*
s. plug terminal Zündkerzenstecker *m*
s. plug wrench Zündkerzenschlüssel *m*
s. quench *(El)* Funkenlöschkreis *m*
s. quenching *(El)* Funkenlöschung *f*
s.-quenching capacitor Funkenlöschkondensator *m (Kfz.-Zündung)*
s. retard Nachzündung *f*
s. setting Zündeinstellung *f*
s. spectrum Funkenspektrum *n*
s. stand Funkenstativ *n*
s. timing Zündmomenteinstellung *f*, Zündzeitpunkt *m*
s.-timing range Zündverstellbereich *m*
s. transmitter Funkensender *m*
sparking Funken *n*; Funkenbildung *f*
s. alloy pyrophore Legierung *f*
s. coil Zündspule *f*
s. distance Funkenstrecke *f*, Entladeweite *f*
s. out Ausfunken *n*
s.-out equipment Ausfeuereinrichtung *f*
s.-out period Ausfeuerzeit *f (Schleifen)*
s. plug Zündkerze *f*
s.-plug brush Zündkerzenbürste *f*
s.-plug fouling Zündkerzenverschmutzung *f*
s.-plug shield Funkenabschirmung *f*
s.-plug spanner Zündkerzenschlüssel *m*
s.-plug wire Zündkabel *n*
s. potential Funkenpotential *n*
s. voltage Überschlagsspannung *f*, Funken[zünd]spannung *f*
sparkle / to funkeln, glitzern, glänzen
sparkless funkenfrei
s. breaking funkenlose Unterbrechung *f (Schalter)*
sparkling Funkeln *n*, Geflimmer *n*, Flimmern *n*
sparkover *s.* spark-over
sparry spatig, spatartig
sparry iron ore Spateisenstein *m*
spartalite Zinkit *m*, Rotzinkerz *n*
spasmodic krampferzeugend, krampfhaft, spasmotisch
spathic ore Eisenspat *m*, Siderit *m*, Spateisenstein *m*
spatial räumlich, Raum-
s. arrangement räumliche Anordnung *f*
s. charge Raumladung *f*
s. encoder *(Dat)* Lageverschlüßler *m*, Lage-A-D-Umsetzer *m*, Lageumsetzer *m*
s. landing path *(Flg)* dreidimensionaler Landeleitstrahl *m*
s. model Raummodell *n*
spatter / to herausschleudern, [heraus]spritzen, spratzen, verspritzen
spatter cloth Spritzleinewand *f (Dreschmaschine)*
spatula Spachtel *m*, Spatel *m*
speak / to sprechen
speaker 1. Sprecher *m (Person)*; 2. Lautsprecher *m*
s. cloth Lautsprecherseide *f*
s. cone Lautsprecherkegel *m*
s. housing Lautsprechergehäuse *n*
s. wire Sprecherleitung *f*
speaking Sprechen *n*
s. battery Sprechbatterie *f*
s. circuit Sprechstromkreis *m*
s. current Sprechstrom *m*
s. current supply Mikrofonspeisung *f*
s. key *(Fmt)* Sprechtaste *f*, Abfragetaste *f*

s. pair *(Fmt)* Sprechadern *fpl*
spear Stange *f (eines Gestänges)*; Seilspeer *m*, Seilfanghaken *m (Fangwerkzeug)*
s. joint Gestängekupplung *f*
special Spezial-, Sonder-, extra, speziell, besonders
s. attachment Zusatzeinrichtung *f*, Sondereinrichtung *f*
s. attachments Sonderzubehör *n*
s. broadcasting messages Rundfunksonderdienste *mpl*
s. case Sonderfall *m*
s. casting Sonderguß *m*
s. circuit Sonderleitung *f*
s. cutter Sonderfräser *m*
s. densified hardboard extraverdichtete Platte *f*
s. design Sonderbauart *f*, Sonderausführung *f*, Sonderkonstruktion *f*
s. device Sondervorrichtung *f*; Sondereinrichtung *f*
s. drilling machine Sonderbohrmaschine *f*
s. duty Sondereinsatz *m*
s. equipment Sonderausrüstung *f*
s. equipment for hobbing crowned gears Einrichtung *f* für Balligverzahnung
s. field Fachrichtung *f*
s. fit Sonderpassung *f*
s. form Sonderform *f*
s. holder Sonderfassung *f (für Lampen)*
s. holding fixture Sonderspannzeug *n*
s. knowledge Sachkenntnis *f*
s. lathe tool Spezialdrehmeißel *m*
s. observation post *(Fmt)* Mithör- und Sprechapparat *m (in Nebenstellenanlagen)*
s. optical glass optisches Spezialglas *n*
s. partnership Kommanditgesellschaft *f*
s. pig iron Sonderroheisen *n*
s.-purpose computer Spezialrechner *m*
s.-purpose grinding machine Sonderschleifmaschine *f*
s.-purpose lathe Sonderdrehmaschine *f*
s.-purpose machine Sondermaschine *f*, Spezialmaschine *f*
s.-purpose reamer Sonderreibahle *f*
s.-purpose tool Sonderwerkzeug *n*
s.-purpose vehicle Sonderfahrzeug *n*
s. quality Sondergüte *f*, Sonderqualität *f*
s. radio service Funksonderdienst *m*
s. section Sonderprofil *n*
s. shape Sonderform *f*
s. ship Sonderschiff *n*
s. steel Sonderstahl *m*, Edelstahl *m*
s. theory of relativity spezielle Relativitätstheorie *f*
s. thick-web drill Sonderbohrer *m* mit verstärktem Kern
s. thread Sondergewinde *n*
s. twist drill Sonderspiralbohrer *m*
s. type *s.* s. design
s. work-holding fixture Sonderspannvorrichtung *f*
specialization Spezialisierung *f*
specialized staff Kräfte *fpl* mit Sonderausbildung
s. training [in the field of] Sonderausbildung *f* [auf dem Gebiet]
specific spezifisch
s. absorption selektive Absorption *f*
s. activity spezifische Aktivität *f*
s. binding energy spezifische Bindungsenergie *f*
s. capacity dynamische Tragzahl *f*
s. charge spezifische Ladung *f*
s. coding *(Dat)* Speicherfolgekodierung *f*
s. concentration spezifische Konzentration *f*
s. conductivity spezifische Leitfähigkeit *f*
s. consumption spezifischer Verbrauch *m*
s. dielectric strength *(El)* spezifische Durchschlagsfestigkeit *f*
s. dose *(Kern)* spezifische Dosis *f*
s. furnace capacity Herdflächenleistung *f*
s. gravity Dichte *f*
s. heat spezifische Wärme *f*
s. heat at constant volume spezifische Wärme *f* bei konstantem Volumen
s. heat of the combustible mixture spezifische Wärme *f* des brennbaren Gemisches
s. impulse *(Rak)* spezifischer Schub *m*
s. inductive capacity Dielektrizitätskonstante *f*
s. ionization coefficient differentieller Ionisationskoeffizient *m*
s. ionization loss spezifischer Ionisationsverlust *m*
s. ionization power spezifisches Ionisationsvermögen *n*
s. line capacity Leitungskapazität *f* je Längeneinheit
s. mass spezifische Masse *f*
s. materials Sonderwerkstoffe *mpl*
s. power output Literleistung *f (Flugzeugmotor)*
s. propellant consumption Einheitsbrennstoffverbrauch *m*
s. refraction spezifische Refraktion *f*
s. reluctance Kehrwert *m* der Permeabilität
s. resistance *s.* resistivity
s. thrust *(Rak)* spezifischer Schub *m*
s. train resistance *(El)* [spezifischer] Fahrwiderstand *m*
s. volume spezifisches Volumen *n*
s. volumetric dilatation spezifische räumliche Ausdehnung (Dilatation) *f*
s. weight Dichte *f*
specification Bezeichnung *f*; genaue Angabe *f*,

Detaillierung *f*; Nachweis *m (Verkehr)*; Richtlinie *f*; Pflichtenheft *n*; Vorschrift *f*; Festlegung *f*; Bedingung *f*; Beschreibung *f*
s. of sale Lieferungsbedingung *f*
s. of size Kennziffer *f*; Größenangabe *f*
s. test Materialprüfung *f*, Werkstoffprüfung *f*
specifications Hauptabmessungen *fpl (z. B. einer Werkzeugmaschine)*; technische Daten *pl*
specified diameter Solldurchmesser *m*
s. dimension Sollmaß *n*
s. output Garantieleistung *f*, garantierte Leistung *f*
s. size Sollmaß *n*
specify / to spezifizieren, vorschreiben; detailliert angeben
specimen Exemplar *n*; Formstück *n*, Muster *n*; Probe[stück *n*] *f*, Probestab *m*, Prüfling *m*; Objekt *n*; Beispiel *n*, Musterstück *n*
s. chamber Objektkammer *f*
s. copy Probeexemplar *n*
s. holder Objekthalter *m*, Objektträger *m*
s. of type Schriftmuster *n*
s. sheet *(Typ)* Aushängebogen *m*
s. stage Objekttisch *m (Mikroskop)*
speck *(Plast)* Oberflächenfehler *m*, Flecken *m*; Teilchen *n*
s. of dust Staubpartikel *f*, Staubteilchen *n*
specking Fleckenbildung *f (Fehler)*
speckle / to tüpfeln, sprenkeln
speckled gesprenkelt, geadert, gemasert, gefleckt
s. background getüpfelter Hintergrund *m*; verrauschte Wiedergabe *f (Radartechnik)*
specs 1. *(specifications)* Vorschriften *fpl*; 2. *(pair of spectacles)* Brille *f*
spectacle frame 1. Brillengestell *n*; 2. *(Schiff)* Wellenbock *m*
s. frame gauge Brillengestellmaß *n*
s. furnace zweiherdiger Spurofen *m*, Brillenofen *m*
s. lens Brillenglas *n*
s.-lens condenser Brillenglaskondensor *m*
s. toric lens torisches Brillenglas *n*
spectacles Brille *f*
spectral spektral, Spektral-
s. absorption spektrale Absorption *f*
s. analysis Spektralanalyse *f*
s. band Spektralbande *f*
s. bandwidth spektrale Bandbreite *f*
s. centroid Schwerpunktswellenlänge *f*
s. colour Spektralfarbe *f*
s. composition spektrale Zusammensetzung *f*
s. concentration spektrale Dichte *f*
s. density Spektraldichte *f*
s. dispersion spektrale Zerlegung *f (des Lichtes)*
s. distribution spektrale Verteilung *f*, spektraler Verlauf *m*
s. distribution of energy *s.* s. energy distribution
s. distribution of sensitivity spektrale Empfindlichkeitsverteilung *f*
s. energy distribution spektrale Energieverteilung *f*, Energieverteilung im Spektrum, Spektralverteilung *f*
s. filter Spektralfilter *n*, Farbfilter *n*
s. frequency Spektralfrequenz *f*
s. function Spektralfunktion *f*
s. intensity spektrale Intensität *f*
s. limit spektrale Empfindlichkeitsgrenze *f*
s. line Spektrallinie *f*
s.-line displacement Verschiebung *f* der Spektrallinien, Spektrallinienverschiebung *f*
s. range Spektralbereich *m*
s. region Spektralgebiet *n*
s. resolution spektrale Auflösung *f*
s. response Spektralempfindlichkeit *f*
s.-response characteristic spektrale Verteilungscharakteristik *f*
s. selectivity spektrale Empfindlichkeit *f*
s. sensitivity spektrale Empfindlichkeit *f*
s. series Spektralserie *f*
s. term Spektralterm *m*
s. transmission curve spektrale Durchlässigkeitskurve *f*, Kurve *f* der spektralen Durchlässigkeit
s. transmittance spektraler Durchlaßgrad *m*
spectre of the Brocken Brockengespenst *n*
spectrochemical spektrochemisch
s. analysis spektrochemische Analyse *f*
spectrogram Spektrogramm *n*
spectrograph Spektrograf *m*
s. tube Spektrografenröhre *f*
spectrographic analysis Spektralanalyse *f*
spectroheliogram Spektroheliogramm *n*
spectroheliograph Spektroheliograf *m*
spectrohelioscope Spektrohelioskop *n*
spectrometer Spektrometer *n*
spectrometry Spektrometrie *f*
spectrophotoelectrical sensitivity spektrale Empfindlichkeit *f (Fotozelle)*
spectrophotometer Spektralfotometer *n*
spectrophotometry Spektrofotometrie *f*
spectroscope Spektroskop *n*
spectroscopic analysis Spektralanalyse *f*
s. unit spektroskopische Einheit *f*
spectroscopy Spektroskopie *f*
spectrum 1. Spektrum *n*; 2. Frequenzband *n*
s. analysis Spektralanalyse *f*
s. analyzer Spektralanalysator *m*
s. line Spektrallinie *f*
s. map Spektralatlas *m*
s. of turbulence Turbulenzspektrum *n*
s. of violin Spielblättchen *n (Violine)*
s. pressure level Schalldruckpegel *m* je Hz Bandbreite

s. projector Spektrumprojektor *m*
s. recorder Spektrumschreiber *m*
s. recording Spektralaufzeichnung *f*
s. scanner Spektrumabtaster *m*
specular spiegelnd, reflektierend
s. angle Reflexionswinkel *m*
s. effect Spiegelung *f*
s. galena Bleispiegel *m*
s. haematite *(Min)* Eisenglanz *m*
s. iron Eisenspiegel *m*, Eisenglanz *m*
s. iron ore *(Min)* Eisenglanz *m*, Hämatit *m*, Glanzeisenerz *n*
s. reflection regelmäßige Reflexion *f*; Spiegelung *f*, Spiegelreflexion *f*, Normalreflexion *f*, gerichtete Reflexion (Rückstrahlung) *f*
s. symmetry Spiegelsymmetrie *f*
speculum [metal] Spiegelmetall *n*
speech Sprache *f*
s. amplifier Niederfrequenzverstärker *m*, Modulationsvorverstärker *m*
s. analysis Sprechanalyse *f*
s. analyzer Sprechanalysator *m*
s. articulation *(Fmt)* Silbenverständlichkeit *f*
s. band Sprechfrequenzband *n*
s. channel *(Fmt)* Sprechkanal *m*, Sprechweg *m*
s. coil Lautsprecherspule *f*
s. communication Sprechverbindung *f*
s. current Sprechstrom *m*
s. defect Sprachfehler *m*
s.-excited *(Fmt)* durch Sprache angesteuert (ausgelöst)
s. input amplifier Mikrofonverstärker *m*
s. input equipment *(Am)* Mikrofonsender *m*
s. inverter *(Fmt)* Banddehner *m*
s. level Sprechpegel *m*
s.-modulated sprachmoduliert
s.-modulated wave sprachmodulierte Welle *f*
s. recognition *(Dat)* Spracherkennung *f*, Wiedererkennen *n* des Gesprochenen
s. recording Sprechaufnahme *f*
s. restorer *(Fmt)* Trägerzusatzeinrichtung *f*
s. signal *(Fmt)* Modulation[sspannung] *f*, Sprechstrom *m*
s. trainer Sprachschulgerät *n*
s. transmission Sprachübertragung *f*
s. voltmeter Aussteuerungsmeßgerät *n* für Sprache
speed / to [zu] schnell fahren; die zugelassene Geschwindigkeit überschreiten
s. up beschleunigen
speed 1. Geschwindigkeit *f*; 2. Drehzahl *f*; 3. Gang *m*; Fahrt *f*; 4. *(Foto)* Empfindlichkeit *f*; Helligkeit *f*
s. adjustment 1. Einstellung *f* der Geschwindigkeit; 2. Drehzahleinstellung *f*
s. at continuous rating *(Elm)* Motordrehzahl (Geschwindigkeit) *f* bei Dauerleistung
s. at one-hour rating *(Elm)* Motordrehzahl (Geschwindigkeit) *f* bei Stundenleistung
s. at trial Probefahrtgeschwindigkeit *f (z. B. eines Schiffes)*
s. boat Schnellboot *n*
s. change Drehzahlwechsel *m*; Änderung *f* der Geschwindigkeit
s. change back gear lever Vorgelegehebel *m*
s. change gear Gangschaltung *f*, Schaltgetriebe *n*
s. changer Drehzahlverstellmotor *m (Kraftmaschinenregelung)*
s.-changing mechanism Schaltgetriebe *n*
s. cone Stufenscheibe *f*
s. control Drehzahlregelung *f*
s. controller Drehzahlregler *m*
s. counter Tachometer *n*, Drehzahlmesser *m*
s. drop Drehzahlabfall *m*
s. frame Vorspinnmaschine *f*; Spindelbank *f*, Flyer *m*
s. governing Drehzahlregelung *f*
s. governor Geschwindigkeitsregler *m*; Fahrtregler *m (Fördermaschine)*
s. graduation *(Elm)* Drehzahlstufung *f*
s. increase Übersetzung *f* ins Schnelle
s. increaser Übersetzungsgetriebe *n*
s.-increasing drive Übersetzungsgetriebe *n*
s. indicator Geschwindigkeitsmesser *m*; Drehzahlmesser *m*, Umlaufzähler *m*, Umdrehungsanzeiger *m*, Tachometer *n*
s. lathe Schnelldrehmaschine *f*
s. lever Reglerhebel *m*
s. limit Geschwindigkeitsbeschränkung *f*, *(polizeilich zugelassene)* Höchstgeschwindigkeit *f*
s.-limiting device *(El)* Geschwindigkeitsbegrenzer *m*
s. of accomodation Anpassungsgeschwindigkeit *f (z. B. des Auges)*
s. of advance *(Schiff)* Anströmgeschwindigkeit *f*; Fortschrittsgeschwindigkeit *f*
s. of answer *(Fmt)* Wartezeit *f*
s. of circulation Umlaufgeschwindigkeit *f*
s. of combustion Verbrennungsgeschwindigkeit *f*
s. of design Konstruktionsgeschwindigkeit *f*
s. of diffusion Diffusionsgeschwindigkeit *f*
s. of discharge Auspreßgeschwindigkeit *f (Chemiefaserherstellung)*
s. of electrons Elektronengeschwindigkeit *f*
s. of fall Fallgeschwindigkeit *f*
s. of flow Durchflußgeschwindigkeit *f*
s. of light Lichtgeschwindigkeit *f*
s. of propagation Fortpflanzungsgeschwindigkeit *f*, Ausbreitungsgeschwindigkeit *f*
s. of resonant rise *(El)* Aufschaukelungsgeschwindigkeit *f*
s. of response Ansprechgeschwindigkeit *f (z. B. Relais)*; Reaktionsgeschwindigkeit *f*

s. of rotation Drehgeschwindigkeit *f*, Drehzahl *f*
s. of signalling Telegrafiergeschwindigkeit *f*
s. of sound Schallgeschwindigkeit *f*
s. of wear Abnutzungsgeschwindigkeit *f*
s. preselection Drehzahlvorwahl *f*
s. range Drehzahlbereich *m*, Drehzahlreihe *f*; Geschwindigkeitsbereich *m*
s. ratio Übersetzungsverhältnis *n (Getriebe)*
s. ratio of a fuse *(El)* Geschwindigkeitsverhältnis *n* einer Sicherung, Ansprechzeitverhältnis *n*
s. ratio of a pair of gears Übersetzung *f* einer Radpaarung
s.-reducing gears Übersetzungsgetriebe *n* ins Langsame
s. reduction Übersetzung *f* ins Langsame
s. regulating rheostat Drehzahlregelwiderstand *m*
s. regulation Drehzahlregelung *f*, Geschwindigkeitsregelung *f*
s. rise Drehzahlanstieg *m*
s. selector Drehzahlwähler *m*
s. setting Drehzahleinstellung *f*
s. skates Rennschlittschuhe *mpl*
s. stage Fahrstufe *f*
s. trial trip Geschwindigkeitsprobefahrt *f (z. B. eines Schiffes)*
speeding up lamp *(Fmt)* Drängelampe *f*
speedometer Tachometer *n*, Wirbelstromtachometer *n*; Drehzahlmesser *m*; Geschwindigkeitsmesser *m*
s. drive Tachometerantrieb *m*
s. drive cover Tachometerantriebsdeckel *m*
s. drive gear Tachometerantriebsrad *n*
s. drive housing Tachometerantriebsgehäuse *n*
s. drive pinion Tachometerantriebsritzel *n*
s. reading Kilometerstand *m*, Meilenstand *m*
speedy advancing *(Brg)* Schnellstreckenauffahren *n*, Schnellstreckenvortrieb *m*
s. tunnelling *(Brg)* Schnellstreckenauffahren *n*
speise *s.* speiss
speiss *(Met)* Speise *f*
s. cobalt Speiskobalt *n*
spell / to 1. buchstabieren, schreiben; 2. ablösen *(Schichtwechsel)*
s. below *(Schiff)* keine Wache haben
spell Törn *m*, Schicht *f (bestimmter Zeitabschnitt einer Arbeit bis zur Ablösung)*
spelter Rohzink *n*
s. solder [Kupfer-Zink-]Hartlot *n*, Schlaglot *n*
spend / to 1. ausgeben; verbringen; 2. verbrauchen, aufbrauchen, aufgebraucht werden; 3. *(Brg)* ergiebig sein
s. power Leistung verbrauchen
spending on research Forschungsmittel *pl*
spent erschöpft, verbraucht
s. electrolyte Elektrolysenendlauge *f*, verbrauchter Elektrolyt *m*
s. fuel *(Kern)* erschöpfter (ausgebrannter) Brennstoff *m*, erschöpftes Spaltmaterial *n*
s. fuel element ausgebranntes (verbrauchtes) Brennstoffelement *n*
s. gas Abgas *n*
s. grain Treber *pl (Brauerei)*
s.-grains remover Austrebermaschine *f (Brauerei)*
s. nuclear fuel ausgebrannter Kernbrennstoff *m*, ausgebranntes Spaltmaterial *n*
s. oxide verbrauchte Gasreinigungsmasse *f*
s. reaction beendete (erschöpfte) Reaktion *f*
s. reactor fuel *s.* s. nuclear fuel
s. tan press *(Led)* Lohpresse *f*
sperrylite *(Min)* Sperrylith *m*
spessartine *s.* spessartite
spessartite *(Min)* Spessartin *m*
sphalerite *(Min)* Sphalerit *m*, Zinkblende *f*
sphene *(Min)* Sphen *m*, Titanit *m*
sphenoid keilförmig
sphenoid Sphenoid *n (Kristallform)*
spheradian Steradiant *m*
sphere Kugel *f*; Globus *m*; Sphäre *f*; Bereich *m*, Kreis *m*
s. gap Kugelfunkenstrecke *f*
s. of action Wirkungssphäre *f*
s. of research Forschungsgebiet *n*
s. packing Kugelpackung *f (Kristallgitter)*
spherical sphärisch [gekrümmt]; kugelförmig, Kugel-
s. aberration sphärische Aberration (Abweichung) *f*, Öffnungsfehler *m*
s. buoy Kugelboje *f*, Kugeltonne *f*
s. cap Kugelkalotte *f*
s. coil-type variometer Kugelvariometer *n*
s. counter *(Kern)* Kugelzählrohr *n*
s. cup Kugelkalotte *f*, Kugeleindruck *m (Härteprüfung)*
s. electrode Kugelelektrode *f*
s.-end measuring gauge Kugelendmaß *n*
s. fender *(Schiff)* Kugelfender *m*
s. function Kugelfunktion *f*
s. geometry sphärische Geometrie *f*; Kugelgeometrie *f*
s. harmonic Kugelfunktion *f*
s. head Halbrundkopf *m (Niet)*
s. head rivet Halbrundniet *m*
s. indentation Kugeleindruck *m (Härteprüfung)*
s. ionization chamber Kugel[ionisations]kammer *f*
s. joint Kugelgelenk *n*
s. lens Kugellinse *f*
s. lune sphärisches Zweieck *n*
s. magnet Kugelmagnet *m*

s. milling Fräsen *n* von Halbkugelformen (Kugelformen)
s. mirror sphärischer Spiegel *m*, Kugelspiegel *m*
s. nucleus kugelförmiger Kern *m*
s. packing Kugelpackung *f*
s. pendulum sphärisches Pendel *n*
s. reactor kugelförmiger Reaktor *m*
s. resolver Kugelkoordinatenwandler *m*
s. resonator Hohlkugelresonator *m*
s. roller bearing Pendelrollenlager *n*
s. rotary digester Kugelkocher *m (Papierherstellung)*
s. segment Kugelabschnitt *m*, Kugelkalotte *f*
s. shell *(Bw)* Kugelschale *f*
s. spectacle lens achsensymmetrisches Brillenglas *n*
s. surface Kugelfläche *f*
s. symmetry Kugelsymmetrie *f*
s. triangle sphärisches Dreieck *n*
s. trigonometry sphärische Trigonometrie *f*
s. turning attachment Kugeldreheinrichtung *f*
s. washer kugelballige Unterlegscheibe *f*
s. wave Kugelwelle *f*
spherically seated kugelig gelagert
sphericity Kugelförmigkeit *f*; Genauigkeit *f* der Kugelform (Kugelgestalt)
sphero-symmetrical kugelsymmetrisch
spherobacterium Kugelbakterie *f*
spheroid Sphäroid *n*, Rotationsellipsoid *n*
spheroidal cementite kugliger (globularer) Zementit *m (Metallgefüge)*
s. graphite cast iron Kugelgraphitguß *m*
s. harmonic Sphäroidfunktion *f*
spheroidize annealing *(Met)* Weichglühen *n*
spherolite Sphärolith *m*
s. formation Sphärolithbildung *f*
spherometer Sphärometer *n*, Wölbungsmesser *m*
sphygmograph Sphygmograf *m*, Pulsschreiber *m (Elektromedizin)*
spicule crystal Nadelkristall *m*, nadelförmiger Kristall *m*
spider Drehkreuz *n*, Spinne *f*, Speichenstern *m*; Radstern *m*, Armstern *m*; Arm *m (Zahnrad)*; Verteilerkasten *m*
s. differential pinion Lagerkreuz *n* für Trabantenräder
s. line Fadenstrich *m*
s. threads Spinnfäden *mpl*
s.-type work-holder for lapping *(Masch)* Läppkäfig *m*
s.-web aerial Fächerantenne *f*
s.-web winding Korbbodenwicklung *f*
spiegel iron *(Met)* Spiegeleisen *n*
spigot Drucklager *n*; Zapfen *m*, Drehzapfen *m*; Zentrieransatz *m*
s. joint Rohrsteckverbindung *f*; Muffenrohrverbindung *f*
s. nut Überwurfmutter *f*
spike 1. Zimmermannsnagel *m*; Schienennagel *m*; Stift *m*; 2. *(Lw)* Eggenzahn *m*; 3. *(Fs)* Impulszacke *f*; 4. *(Kern)* Erhitzungsbereich *m*
s. anchor *(Brg)* Schienenfederklammer *f*
s. driver Schienennagelhammer *m*
s. hammer Schienennagelhammer *m*
s. knot Flügelast *m*, Hornast *m*
s. maul Schienennagelhammer *m*
s. of harrow *(Lw)* Eggenzinke *f*
s. puller Schienennagelzange *f*
s. roller *(Lw)* Stachelwalze *f*, Zahnwalze *f*
s. tooth harrow Netzegge *f*, Unkrautstriegler *m*
s. train Impulsserie *f*
spiked chain temple *(Text)* Stachelkettenbreithalter *m*
s. lattice *(Text)* Spitzenlattentuch *n*, Nadellattentuch *n*
spile 1. Holzstapel *m*; 2. *(Brg)* Getriebepfahl *m*; 3. Faßspund *m*
spiling *(Brg)* Anstecken *n (Schachtabteufen)*; Getriebeverfahren *n (Vorrichtung)*; Firstengetriebe *n*
spill / to vergießen, verschütten; überströmen [lassen]
spill 1. Rückstreuungsverlust *m (Speichertechnik)*; 2. Getriebepfahl *m*
s.-control Überströmregelung *f*
s. current *(El)* Fehlerstrom *m*
s. line Überströmleitung *f*
s. port Überströmschlitz *m (Zweitaktverbrennungsmotor)*
s. shield Raster *m (Leuchten)*
s. valve Überströmventil *n*
spillage Ladungsverlust *m*, Schüttverlust *m*, Fegsel *n*; Rieselgut *n*, verschüttetes Gut *n*
spilling over *(El)* Schwingungsneigung *f (zur Selbsterregung)*
spillway Überlauf *m*, Hochwasserentlastungsanlage *f*
s. bucket Überfallsturzbecken *n*
s. channel Hochwasserentlastungskanal *m*, Überlaufkanal *m*
s. dam Stauwehr *n*, Überlaufdamm *m*
spin / to 1. *(Text)* spinnen; 2. drücken *(Metallblech)*; 3. *(Flg)* trudeln, sich abtrudeln lassen
spin *(Ph, Kern)* innerer Drehimpuls *m*, Spin *m*, Eigendrehimpuls *m*
s. counter Umdrehungszähler *m*, Drehzahlmesser *m*
s. dependence Spinabhängigkeit *f*
s. doublet Spindublett *n*
s. echo Spinecho *n*
s. echo memory Spinechospeicher *m (Datenspeicher)*

s. effect Spineffekt *m*
s. forming Streckdrücken *n*
s. function Spinfunktion *f*
s. independent spinunabhängig
s. interaction Spinwechselwirkung *f*
s.-magnetic moment magnetisches Spinmoment *n*
s. moment Spinmoment *n*, Spin *m*
s.-orbit interaction Spin-Bahn-Wechselwirkung *f*
s. orientation Spinorientierung *f*, Spinrichtung *f*
s. quantum number Spinquantenzahl *f*
s. saturation Spinabsättigung *f*
s. space Spinraum *m*
s.-spin interaction Spin-Spin-Wechselwirkung *f*
s. state Spinzustand *m*
s. unit Spineinheit *f*
s. wave Spinwelle *f*
spindle Spindel *f*; Laufspindel *f (Turbine)*; Achsschenkel *m*
s. band *(Text)* Spindelschnur *f*
s. bearing Spindellager *n*, Zapfenlager *n*
s.-bolt-bearing box Achsschenkellagerbüchse *f*
s.-bolt nut Achsschenkelmutter *f*
s. bore Spindelbohrung *f*
s. bored to pass bars Stangendurchlaß *m* durch Spindel
s. brake Spindelbremse *f*
s. carrier drum Spindeltrommel *f*
s. collar *(Text)* Spindelhalslager *n*; Spindelbund *m*
s. diameter Spindeldurchmesser *m*
s. drive Spindelantrieb *m*
s. drive motor Spindelantriebsmotor *m*, Hauptantriebsmotor *m*
s.-end nut Befestigungsmutter *f* der Schleifscheibe
s. extension Spindelverlängerung *f*
s. footstep *(Text)* Spindelfußlager *n*
s. guiding *(Text)* Spindelführung *f*
s. head Spindelkopf *m*, Spindelstock *m*
s. hole Spindelbohrung *f*
s. moulder Fräsmaschine *f (Holzbearbeitung)*
s. moulder for dovetailing Zinkenfräsmaschine *f*
s. nose Spindelkegel *m*
s. oil Spindelöl *n*
s. point Spindelspitze *f*
s. quill Spindelhülse *f*, Spindelpinole *f*
s. r.p.m. Spindeldrehzahl *f*
s. rail *(Text)* Spindelbank *f*
s. reversal Umkehr *f* der Spindeldrehrichtung
s. reverse [gears] Wendegetriebe *n (Gewindebohren)*
s. sleeve Spindelhülse *f*, Spindelpinole *f*, Bohrhülse *f (Bohrmaschine)*
s. slide Spindelschlitten *m*
s. speed Spindeldrehzahl *f*
s. speed change Spindeldrehzahländerung *f*
s. stop Spindelstillsetzung *f*
s. tape *(Text)* Spindelband *n*
s. top *(Text)* Spindelkopf *m*, Spindelspitze *f*
s. travel Spindelhub *m*
s. turn during outward running of carriage *(Text)* Drehung *f* der Spindeln während der Wagenausfahrt
s. yoke Spindelgabel *f*
spindleless pirn winder *(Text)* spindellose Schußspulmaschine *f*
spindrift Schaum *m*, Gischt *f*
spine 1. Rücken *m (des Buches)*; 2. Gitterstab *m (des Antennenreflektors)*
s. beam *(Bw)* Mittelträger *m*
s. wall *(Bw)* Mittellängswand *f*
spinel *(Min)* Spinell *m*
spinner 1. Spinnmaschine *f*; 2. Spinner[in *f*] *m*; 3. *(mitlaufende)* Propellernabe *f*; 4. Schleuder *f (Schleuderguß)*
spinneret Spinndüse *f*, Mehrlochdüse *f*
spinning 1. Spinnen *n*; 2. Drücken *n (Metall)*; 3. Schleudern *n*, Schleuderverfahren *n (Herstellung von Betonrohren)*; 4. Spritzen *n* von Fäden durch Düsen; 5. schraubenartige Drehung *f*; *(Flg)* Abtrudeln *n*
s. casting process Druckgußverfahren *n*
s. dope Spinnflüssigkeit *f*
s. electron Spinelektron *n*; Kreiselektron *n*
s. jenny *(Text)* Jennymaschine *f*
s. lathe Metalldrückbank *f*, Drückbank *f*
s. material *(Text)* Spinngut *n*
s. mill Spinnerei *f*
s. nose dive *(Flg)* Sturzspirale *f*
s. of coloured cotton yarn Baumwollbuntspinnerei *f*
s. of imitation yarn Imitatspinnerei *f*
s. of waste Abfallspinnen *n*
s. out wrinkles Drücken *n*, Planieren *n*
s. regulator *(Text)* Spinnregler *m*
s. roller Drückrolle *f*
s. tool Drückwerkzeug *n*
s. waste Spinnereiabfall *m*; Abfallspinnen *n*
s. with twisting at the head Spinnen *n* mit Nachdraht
spinode *(Math)* Spitze *f (einer Kurve)*
spinor Spinor *m*, Spinvektor *m*
s. component Spinorkomponente *f*
spinthariscope Spinthariskop *n*
spintharometer *(El)* Funkenmikrometer *n*
spiral Spirale *f*; Drall *m*
spiral spiralförmig, spiralig, Spiral-
s. aerial Spiralantenne *f*
s. angle Schrägungswinkel *m*, Spiralwinkel *m (Zahnrad)*; Steigungswinkel *m*
s. bevel gear Spiral[zahn]kegelrad *n*

s. binding *(Typ)* Spiralbindung *f*, Spiralheftung *f*
s. chamber *(Kern)* Spiralbahnkammer *f*
s. chute Wendelrutsche *f*
s. cleaner Wendelscheider *m*, Spiralscheider *m (Aufbereitung)*
s. coil Spiralspule *f*
s. coiled spring Spiralfeder *f*
s. column spiralbewehrte Säule *f (Stahlbetonbau)*
s. condenser Schlangen[rohr]kühler *m*
s. conveyor Schneckenförderer *m*, Förderschnecke *f*
s. curve Spirallinie *f*
s. cutter Spiralfräser *m*
s. drawing frame *(Text)* Spiralstrecke *f*
s. drill Schneckenbohrer *m*
s. drive gear Tellerrad *n*
s. drive pinion Antriebsritzel *n*, Kegelrad *n*
s. feeder Spiralaufgabevorrichtung *f*
s. flute Spiralnut *f*
s. flute milling Spiralnutenfräsen *n*
s. fluted spiralgenutet, drallgenutet
s. fluted reamer spiralverzahnte Reibahle *f*
s. gashed spiralgenutet
s. gear schrägverzahntes Zahnrad *n*, Schraubenrad *n*
s. gearing Schrägverzahnung *f*
s. gears Stirnradschraubgetriebe *n*
s. grain *(Holz)* Krummfaser *f*; Drehwuchs *m*
s.-grain[ed] *(Holz)* drehwüchsig
s. groove Spiralnut *f*
s. housing Spiralgehäuse *n*
s. key spring Keildruckfeder *f*
s. line Schraubenlinie *f*
s. microscope Spiralmikroskop *n*
s. milling attachment Spiralfräseinrichtung *f*
s. nebula Spiralnebel *m*
s. of Archimedes archimedische Spirale *f*
s. orbit spiralförmige Bahn *f*, Spiralbahn *f*
s. point drill Spiralbohrer *m* mit schraubenförmiger Verschiebung der Freifläche
s. quad formation Sternverseilung *f*, Sternvierer *m (Kabel)*
s. reinforcement Spiralbewehrung *f (Stahlbetonbau)*
s. scanning Spiralabtastung *f*
s. seed line Spiralsaatleitung *f*
s.-shaped duct Spiralgehäuse *n*
s. shell end mill Walzenstirnfräser *m* mit Spiralverzahnung
s. spring Spiralfeder *f*
s. stair[case] Wendeltreppe *f*
s. superheater Spiralrohrüberhitzer *m*
s. superheating tube Überhitzerschraubenrohr *n*
s. teeth Spiralverzahnung *f*; Schrägverzahnung *f*
s.-tooth[ed] spiralverzahnt; schrägverzahnt
s. trajectory Spiralbahn *f*
s. tube seed line Spiralsaatleitung *f (Sämaschine)*
s. turn Schraubenwindung *f*
s. winding Spiralwicklung *f*, Wellenwicklung *f*, Schraubenwicklung *f*, spiralförmige Wicklung *f*
s. worm Spiralschnecke *f*
spirality 1. Spiralform *f*; 2. *(Text)* Spiralneigung *f*
spiralling Verdrehungsneigung *f (von Schmaltextilien)*
spirally reinforced column spiralbewehrte Säule *f (Stahlbetonbau)*
spire 1. spitzer Körper *m*, Spitze *f*; Turmhelm *m*, Turmspitze *f*; 2. *(El)* Windung *f (einer Wicklung)*; 3. *(Brg)* Zündlunte *f*
spiring *(Brg)* Eindrücken *n (des Bogens in die Firste)*
spirit Spiritus *m*, Sprit *m*
s.-based mordant Spiritusbeize *f*
s. lacquer Spirituslack *m*
s. lamp Spirituslampe *f*
s. level 1. Ätherlibelle *f*; 2. Richtwaage *f*, Wasserwaage *f*
s.-level sextant Libellensextant *m*
s. mordant Spiritusbeize *f*
s. varnish Spirituslack *m*, Spritlack *m*
spirits 1. Auszug *m*, Extrakt *m*; 2. Spirituosen *pl*; Branntwein *m*
s. of wine denaturierter Alkohol *m*, Spiritus *m*
spirituous alkoholisch
s. liquors Spirituosen *pl*
spiro compound *(Ch)* Spiranverbindung *f*
spirule *(BMSR)* Spirule *(Zeichengerät für Wurzelortskurven)*
spit 1. Spatenstich *m*; Spatentiefe *f*; 2. Landzunge *f*, Nehrung *f*
spitter fuse *(Brg)* gekerbte Zündschnur *f* zum Zündschnuranzünden
spittings *(Met)* Auswurf *m (eines Konverters)*
spitzkasten Spitzkasten *m (Aufbereitung)*
splash / to spritzen
splash 1. Spritzen *n*; 2. *(Plast)* spritzerförmige Markierung *f* am Spritzling; 3. Spiegel *m (Holz)*
s. baffle Ablenkplatte *f (Quecksilberdampfröhren)*
s. board Spritzbrett *n*
s.-board leather Spritzleder *n*
s. cup Spritzteller *m*
s. guard Spritzblech *n*, Spritz[wasser]schutz *m*
s. hood Spritzhaube *f*
s. lubrication Tauchschmierung *f*, Spritzschmierung *f*, Schleuderschmierung *f*
s.-proof spritzwassergeschützt
s. ring Schirmring *m (Quecksilberdampfröhren)*

splasher 1. Spritzschutz *m*; Spritzblech *n*; 2. Nase *f* des Pleuelkopfes
splat Deckstreifen *m*
splatter *(Fmt, Fs)* Störung *f* durch einen Nachbarkanal
splay / to abschrägen
splay-legged roadway arch *(Brg)* breitbeiniger Streckenbogen *m (Ausbau)*
s. of the wheel *(Kfz)* Radsturz *m*
splayed abgekantet, abgeschrägt
s. jamp schräge Leibung *f (Holzbau)*
splendent mit Metallglanz
s. fracture *(Met)* glänzender Bruch *m*
splice / to 1. pfropfen *(in den Spalt)*; 2. spleißen *(Seil)*; überlaschen, verblatten *(Holzbau)*; kleben *(Film)*
splice Stoß *m*; Spleiße *f*, Spleißung *f (Seilverbindung)*
s.-grafting *(Lw)* Pfropfen *n*
s. piece Lasche *f*, Stoßlasche *f*
s. plate Lasche *f*, Stoßlasche *f*; Stoßblech *n*
spliced 1. *(Text)* verstärkt; 2. gespleißt *(Seil)*
s. connection Spleißverbindung *f*
s. joint Spleißstelle *f*
splicer 1. Spleißer *m*; 2. Filmkittlehre *f*
splicing 1. Spleißen *n*, Spleißung *f*; 2. Kleben *n (Film)*
s. cement Filmkleister *m*
s. ear *(Fmt)* Verbindungsklemme *f*, Verbindungsöse *f*
s. fitting *(El)* Fahrdrahtstoßklemme *f*
s. gauge Kittlehre *f*
s. of the cable Kabelspleißung *f*
s. sleeve Lötmuffe *f*, Lötverbinder *m*
s. tape Klebeband *n*
s. unit Klebepresse *f (für Filme)*
s. wire Bindedraht *m (Stahlbetonbau)*
spline / to längsnuten, Keilprofile fräsen
spline 1. Paßfeder *f*; Gleitfeder *f*; Schiebekeil *m*; 2. Keil[wellen]nut *f*; 3. Kurvenlineal *n*
s. bar Keilnabe *f*
s. broach Innenräumwerkzeug *n* für Vielkeilprofil
s. cutter Keilwellenfräser *m*
s. gauge Keilnutenlehre *f*
s. hob Keilwellenwälzfräser *m*
s. hole Keilnabe *f*
s. mill Keilwellenfräsmaschine *f*
s. milling Keilwellenfräsen *n*
s. profile Keilprofil *n*
s. running lengthwise keyway Längsnut *f*
s. shaft Schiebewelle *f*; Keilnutwelle *f*, Federwelle *f*
s. shaft hob Keilwellenwälzfräser *m*
s.-shaft milling cutter Keilwellenfräser *m*
splined mit Keilprofil
s. arbor Keilwelle *f*
s. bolt Keilschraube *f*
s. hub Keilnabe *f*
s. nut kerbverzahnte Mutter *f*
s. pin Kerbstift *m*
s. shaft Keilwelle *f*
splines Keilprofil *n*; Kerbverzahnung *f*
splining Keilnutenfräsen *n*
splint / to schienen
splint Splint *m*, Vorsteckkeil *m*; Schiene *f*
s. chopping machine Holzdrahtabschlagmaschine *f (Zündholzherstellung)*
s. coal *(Am)* Splintkohle *f*, Splitterkohle *f*
s. drying drum Holzdrahttrockentrommel *f (Zündholzherstellung)*
s. levelling machine Gleichlegemaschine *f (Zündholzherstellung)*
s. pin Vorsteckkeil *m*
s. shaping machine Holzdrahthobelmaschine *f (Zündholzherstellung)*
splinter / to aufsplittern
splinter Splitter *m*, Span *m (Holz)*; Schiefer *m*
s. of glass Glassplitter *m*
splinterable splitternd, splitterbar
splintery splittrig
s. fracture *(Met)* schieferartiger Bruch *m*
split / to aufspalten, aufreißen; schlitzen
s. up the chips Kurzspäne erzeugen
split 1. Schlitz *m*; Spaltfuge *f*; 2. falscher Zwischenraum *m (Telegrafie)*
split geschlitzt *(Reibahle)*; geteilt *(Lager)*; zweiteilig; gespalten; getrennt
s. anode Schlitzanode *f*
s.-anode magnetron Schlitzanodenmagnetron *n*, Zweischlitzmagnetron *n*
s.-anode magnetron tube Habannröhre *f*
s. bearing geteiltes Lager *n*
s. board Sägeblock *m*
s. bridge Windfeuerbrücke *f*, Feuerbrücke *f* mit Luftzuführung
s. cable grid Kabelnachziehschlauch *m*, Kabelziehstrumpf *m*
s. chuck Spannpatrone *f*
s. coil relay Spaltspulenrelais *n*
s. collar Klemmbüchse *f*
s.-conductor protective system *(El)* Parallelzweigschutz *m*, Querdifferentialschutz *m*
s. core geschlitzter Kern *m*
s.-core type transformer *(El)* Anlegewandler *m*, Zangenwandler *m*
s. cotter pin Splint *m*
s. coupling Schalenkupplung *f*
s. course Schicht *f* aus gespaltenen Ziegelsteinen
s. cylinder test Spaltzylinderprüfung *f (Betonprüfung)*
s. die zweiteilige Backe *f*, geschlitztes (offenes) Schneideisen *n*
s. drum *(Text)* Schlitztrommel *f*

s. drum winder for cheeses Kreuzspulmaschine *f* mit Schlitztrommel
s. elbow *(El) (auseinandernehmbare)* Winkelmuffe *f* für I-Rohr
s. end vane geschlitzte Platte *f (Drehkondensator)*
s. fire wood Scheitholz *n*
s. flap Spreizklappe *f*
s.-flow reactor Reaktor *m* mit geteiltem Kühlmittelfluß
s. flywheel gesprengtes Schwungrad *n*
s. frequency Spaltfrequenz *f*
s. hides *(Led)* Vachetten *pl*
s. hub getrennte Nabe *f*
s. hub pulley geteilte Riemenscheibe *f*
s. image 1. *(Fs)* Bildvervielfachung *f*, geteiltes Bild *n (durch fehlerhafte Synchronisation)*; 2. Detailbild *n (Film)*
s.-image range finder Schnittbildmeßsucher *m*
s. ingot mould aufklappbare Kokille *f*
s. knob insulator Klemmisolator *m*
s. lower portion of bearing geteiltes Lagerunterteil *n*
s. nuts [on lead screw] Mutterschloß *n*
s. of the reed *(Text)* Zahn *m* des Webblatts, Rietstab *m*, Rohr *n*
s. pattern geteiltes Modell *n (Formguß)*
s.-phase motor Einphasenmotor *m*, Wechselstrommotor *m* mit Spaltphase
s.-phase starting Anlassen *n* mit Hilfsphase
s.-phase winding gespaltene Phasenwicklung *f*
s. picture Detailbild *n (Film)*
s. pin Splint *m*
s.-pin hole Splintloch *n*
s. plug Bananenstecker *m*
s. pole Spaltpol *m*
s.-pole motor Spaltpolmotor *m*
s. pulley geteilte Riemenscheibe *f*
s. reel zweiteilige Spule *f*
s. response Doppelzeichen *n (Radar)*
s. rim ring geteilter Felgenring *m*
s. rudder *(Schiff)* Spaltruder *n (Ruder wirkt in geöffnetem Zustand als Bremse)*
s. second detonator *(Brg)* Millisekundenzünder *m*
s. skirt piston geschlitzter Kolben *m*
s.-stator variable capacitor Doppelstatordrehkondensator *m (für Gegentaktschaltungen)*
s. tank circuit Gegentaktschwingkreis *m (mit geerdetem elektrischem Mittelpunkt)*
s. taper pin Paßkerbstift *m*, Kerbstift *m*
s. taper sleeve Kegelklemmhülse *f (Bohrer)*
s. test Spaltprüfung *f (Materialprüfung)*
s. timber Spaltholz *n*, Scheitholz *n*
s.-timber truck Scheitholzwagen *m*
s.-trail carriage *(Mil)* Spreizlafette *f*
s. transformer Anzapftransformator *m*
s. weld Klauenschweißung *f*
splitting Aufspaltung *f*, Spaltung *f*
s. key Abtrennschalter *m*
s. machine *(Led)* Spaltmaschine *f*
s. of chain *(Ch)* Kettenaufspaltung *f*
s. of printer Druckwerksplittung *f (Tabelliermaschine)*
s. time Trennzeit *f (Nachrichtentechnik)*
s. tool Spaltgeschirr *n*
s. up Aufspaltung *f*; Aufspalten *n*; Zersetzung *f*
s. wedge Spaltkeil *m (Holzbearbeitung)*
splutter print *(Plast)* Spritzdruck *m*
spluttering *(El)* Knackgeräusche *npl*, Kratzgeräusche *npl*, Nebengeräusche *npl*, Prasseln *n*
spodumene *(Min)* Spodumen *m*
spoil / to 1. verderben; 2. *(Brg)* absetzen, Abraum verstürzen, verkippen
spoil 1. Abraum *m*, Aushub[boden] *m*, Aushubmasse *f*; 2. *(Typ)* Ausschußbogen *m*, Fehlbogen *m*; Makulatur *f*
s. bank Abraumkippe *f*
s. bank wall Kippenstoß *m*
s. capacity Kabeltankfassungsvermögen *n*
s. ground Abladeplatz *m (Baggergut)*
s. moving Abraumbewegung *f*
s. pile toe Kippenfuß *m*
spoilage Ausschuß *m*
spoiled casting *(Met)* Wrackguß *m*
s. meat verdorbenes Fleisch *n*
spoke / to Speichen einziehen
spoke Speiche *f*; Sprosse *f*
s. planing machine Radspeichenhobelmaschine *f*
s. shave Speichenhobel *m*, Schabhobel *m*
s. wheel Speichenrad *n*
s. wheel centre Radstern *m*
spoked wheel Speichenrad *n*
spokes mast Kreuzantenne *f*
spoking Radeffekt *m* [durch schlechte Kontaktgebung] *(Radar)*
sponge Schwamm *m*, schwammförmiger Schaumstoff *m*
s. cloth Putztuch *n*
s. grease Schwammfett *n*
s. iron Schwammeisen *n*, Eisenschwamm *m*
s. rubber Schaumgummi *m*
spongeous schwammig
spongin Schwammskelett *n*, Schwammstoff *m*
sponginess Porosität *f*, Porigkeit *f*, Durchlässigkeit *f*, Schwammigkeit *f*
sponging *(Text)* Krumpen *n*, Krimpen *n*
spongy schwammig; blasig *(Guß)*
s. casting blasiger Guß *m*
s. coke schaumiger (blasiger) Koks *m*
s. platinum Platinschwamm *m*
s. zinc Zinkschwamm *m*

sponson 1. Ausleger *m*, ausladende Plattform *f*; 2. Radkasten *m (Dampfer)*
s.-beam äußerer Radkastenbalken *m (Raddampfer)*
sponsored [industrial] research Auftragsforschung *f*
s. television Werbefernsehen *n*
spontaneous spontan
s. combustibility Selbstentzündlichkeit *f*
s. combustion Selbstentzündung *f*
s. decay *(Kern)* spontaner Zerfall *m*
s. discharge selbständige (spontane) Entladung *f*
s. emission Spontanemission *f*, spontane Emission *f*
s. excitation Selbsterregung *f*
s. fission Spontanspaltung *f*, spontane Spaltung *f*
s. glowing Spontanleuchten *n*
s. ignition Selbst[ent]zündung *f*
s. ionization spontane Ionisation *f*
s. magnetization spontane Magnetisierung *f*
s. nuclear reaction spontane Kernreaktion *f*
s. radiation charakteristische Strahlung *f (Gasentladungsröhren)*
s. radioactivity spontane Radioaktivität *f*
s. transmutation natürlicher radioaktiver Zerfall *m*, spontane Kernumwandlung *f*
spontaneously inflammable selbstentzündlich
spool / to spulen; aufwinden, aufwickeln
spool 1. Haspel *f*, Wickelspule *f*, Spule *f*; Rolle *f*; 2. Steuerkolben *m (Hydraulik)*; 3. ringförmige Auskehlung *f*
s. flange Spulenflansch *m*
s.-type throttling valve Drosselkolbenschieber *m*
s. valve Steuerventil *n*, Kolbenschieberventil *n*
spooler Spulmaschine *f*
spooling Spulen *n*, Wickeln *n*, Wicklung *f*
spoon Löffel *m*, Kelle *f*
s. bit Löffelbohrer *m*
s. bow *(Schiff)* Löffelbug *m*
s. sample Löffelprobe *f*, Sondenprobe *f*
s.-shaped blades Löffelschaufelung *f (Turbine)*
sporadic sporadisch, vereinzelt
s. reflections abnormale Reflexionen *fpl (Nachrichtentechnik)*
sports clothing Sportbekleidung *f*
sportsview *(Fs)* Sportschau *f*
spot / to 1. Position feststellen, orten; 2. masern *(Holz)*; fleckig machen; 3. *(Masch)* anschneiden, aussenken
s.-drill anbohren
s.-face ansenken; anschneiden, [aus]senken; [ab]flächen
spot Fleck *m*; Stelle *f*; Leuchtfleck *m*; Bildelement *n*, Bildpunkt *m*; Lichtmarke *f*
s. analysis *(Ch)* Tüpfelanalyse *f*
s. and stain removing agent Fleckenentfernungsmittel *n*
s. contact Punktkontakt *m*
s. cotton *sofort lieferbare* Baumwolle *f*
s. deflection Punktablenkung *f*
s. diameter Punktdurchmesser *m*; Leuchtfleckdurchmesser *m*
s. distortion Fleckverzerrung *f*
s. drilling Anbohren *n*
s. facing Abflächen *n*, Flächen *n*; Ansenken *n*; Plansenken *n*
s.-facing cutter Abflächmesser *n*
s. glueing *(Plast)* Punktleimverfahren *n*
s. lamp Punktstrahllampe *f*, Punktstrahler *m*; Sucher *m*, Suchscheinwerfer *m*
s. lamp bulb Sucherlampe *f*
s. light Suchscheinwerfer *m*
s. light scanning Lichtstrahlabtastung *f*
s. lighting Punktscheinwerferbeleuchtung *f*, Punktanstrahlung *f*
s. noise factor *(El)* Punktrauschzahl *f*
s. noise figure *(El)* Punktrauschzahl *f*
s. of light Lichtpunkt *m*, Lichtfleck *m*; Lichtzeiger *m*
s. position Stellung *f* des Leuchtfleckes *(Katodenstrahlröhren)*
s. reaction *(Ch)* Tüpfelreaktion *f*
s. remover Fleckenreiniger *m*, Fleckenwasser *n*
s. removing agent *(Text)* Detachiermittel *n*
s. resistant fleckenbeständig, fleckabweisend
s. size Punktgröße *f*
s. test 1. Stichprobe[nversuch *m*] *f*; 2. *(Ch)* Tüpfelreaktion *f*, Tropfenreaktion *f*
s. twist Knotenzwirn *m*
s.-type scanning punktförmige Abtastung *f (Fernschreiber)*
s. weld Punkt[schweiß]naht *f*; Punktschweißung *f*; Schweißpunkt *m*
s.-welded punktgeschweißt
s. welding Punktschweißen *n*
s. welding equipment Punktschweißzeuge *npl*
s. welding machine Punktschweißmaschine *f*
s. wobble *(Fs)* Strahlwobblung *f*
spotfacer Stirnsenker *m*, Plansenker *m*
spotted gesprenkelt, gefleckt, fleckig
s. wood maseriges Holz *n*
spotter Artilleriebeobachter *m*
spottiness Fleckigkeit *f (Fernsehbild)*
spotting Fleckigwerden *n (Galvanotechnik)*
s. drill Anbohrer *m*
s. plane Artillerieflugzeug *n*
s. scope Sucherfernrohr *n*
spout 1. Ausgußrinne *f*; Auslauf *m*; Schnauze *f*; Strahl *m*, Fontäne *f*; 2. Düse *f (Wellenleiter)*, Wellenleiteröffnung *f*

s. angle Einlaufwinkel *m*
s. nozzle Rinnenschnauze *f (Gießrinne)*
s. plane Rund[kehl]hobel *m*
sprag / to *(Brg)* abstempeln, abspreizen, verspreizen
sprag 1. *(Brg)* Strebe *f*, Bolzen *m*, Spreizstempel *m*; 2. Bremsknüppel *m*; 3. Bergstütze *f*
spragger Zugbegleiter *m*, Bremser *m*
spray / to [auf]spritzen, [auf]sprühen, berieseln, besprengen, besprühen, zerstäuben
s. apply aufspritzen
spray Zerstäubung *f*; [komplette] Spritzung *f*; Spritzer *m*; Spritzgußteile *npl* mit Angußspinne; Ölnebel *m*; Sprühnebel *m*; Spritzwasser *n*; Spritzer *m*, Spray *m*
s. application Spritzauftrag *m*
s. booth Spritzkabine *f*
s. carburetter Düsenvergaser *m*
s. coater Aufsprühstreichmaschine *f*
s. column Rieselturm *m*
s. cone Spritzkegel *m (Kondensator)*
s. cooler Berieselungskühler *m*; Rieselkühler *m*
s. cooling Kühlung *f* durch Zerstäubung; Ölnebelkühlung *f*
s. deposit Spritzbelag *m*
s. discharge *(El)* Sprühentladung *f*
s. drier Zerstäubungstrockner *m*
s. guard Spritzschutz *m*
s. gun Spritzpistole *f*
s. hole Spritzloch *n*
s. hood Bootsbezug *m*
s. nozzle Spritzdüse *f*; Streudüse *f*; *(Kfz)* Vergaserdüse *f*
s. oils Sprühöle *npl*
s. painting Farbspritzen *n*
s.-painting plant Farbspritzgerät *n*
s. pak Siebbodenkolonne *f* mit schräggestellten Siebböden
s. scrubber Spritzwascher *m (Aufbereitung)*
s. test Spritzprüfung *f (Prüfung der Gewebe auf wasserabstoßende Eigenschaften)*
s. tower Berieselungsturm *m*, Rieselturm *m*
s. valve Einspritzventil *n*
s. washer Berieselungsanlage *f*
s. washing machine *(Text)* Sprühwaschmaschine *f*
s. webbing Kokonisierung *f (auf Gitterwerk aufgespritzter Kunststoffüberzug)*
sprayability Spritzfähigkeit *f*
sprayed cathode bespritzte Katode *f (Elektronenröhren)*
s. metal coating Spritzmetallisierung *f*
sprayer Sprenkler *m*, Spritzapparat *m*; Sprühgerät *n*; Spritze *f*
spraying aggregate Beregnungsanlage *f*
s. apparatus Spritzapparat *m*
s. furnace Zerstäubungsfeuerung *f*
s. grease Spritzfett *n*
s. lacquer Spritzlack *m*
s. liquor Spritzbrühe *f (Pflanzenschutz)*
s. screen Brausesieb *n (Aufbereitung)*
s. varnish Spritzlack *m*
spread / to 1. auftragen, ausbreiten, bestreichen, spreiten; verteilen *(Last)*; 2. streuen *(Düngemittel)*; 3. spreizen *(Werkzeug)*; 4. *(Text)* anlegen; 5. verlaufen *(Farbe)*
s. foam direct to fabric *(Text)* Schaum direkt auf Textilien aufschäumen
s. out sich ausbreiten
spread 1. Auftrag *m*, Klebstoffauftrag *m*; Aufstrich *m*; spezifische Auftragsmenge *f*; 2. Streuung *f*; Fortpflanzung *f*; 3. Keulenbreite *f (Antennen)*; Strahlbreite *f*; 4. Spreizfähigkeit *f*; 5. Verdienstspanne *f*; 6. *(Am) (Bw)* Maschinen- und Gerätepark *m*
s. beam Breitstrahler *m*
s. coater Streichmaschine *f*
s. coating Leimauftragen *n*, Überzugaufbringen *n* mit Rakel; Streichen *n (Beschichtung)*
s. factor Zonenfaktor *m (der Wicklung)*
s. footing *(Bw)* Flachgründung *f*
s. foundation *(Bw)* Flachgründung *f*
s. of the wing *(Flg)* Spannweite *f* [der Tragfläche]
spreadable life *(Plast)* Gebrauchsdauer *f*, Lebensdauer *f*
spreader 1. Leimauftragmaschine *f*; Verteiler *m*; Verteilerstück *n*, Torpedo *m*; 2. Anlege[maschine] *f (Flachs)*
s. beam Traverse *f*
s. of clutch *(Kfz)* Kupplungsexpansionsring *m*
s. roll Auftragswalze *f*
s.-type insulator Stützer *m*, Stützisolator *m*
spreading 1. Auftragen *n*, Ausbreiten *n*; Bestreichen *n*; Verteilen *n (Last)*; 2. Streuen *n*; 3. Spreizen *n*; 4. *(Text)* Anlegen *n*
s. machine 1. Verteiler *m (Straßenbau)*; 2. Anlegemaschine *f (Flachs)*
s. of manure Düngerstreuen *n*
s. of the steam jet Streuung (Zersplitterung) *f* des Dampfstrahls *(Turbine)*
s. rate Ergiebigkeit *f*
s. the load Verteilung *f* der Belastung
sprig Drahtstift *m*; Kernnagel *m*
spring / to 1. springen; splittern, zerbrechen, reißen; 2. [ab]federn; schnellen; 3. zuschnappen lassen; 4. zur Explosion bringen; 5. entspringen; quellen
s. back zurückfedern, zurückschnellen
s. off abspringen
s. out ausspringen
s. up aufspringen
spring 1. *(Masch)* Feder *f*; 2. Federung *f*; Elastizität *f*; 3. Quelle *f*; Brunnen *m*; Becken *n*;

4. Rückbewegung *f*, Rückfederung *f*; 5. Frühling *m*
s. action Federantrieb *m*, Federwirkung *f*, Federung *f*
s. assembly Federsatz *m*, Federpaket *n*
s. balance Federwaage *f*
s.-balanced federentlastet
s. balancing device Federausgleichvorrichtung *f*
s. barley Sommergerste *f*
s. beard needle *(Text)* Hakennadel *f*, Spitzennadel *f*
s. bearing Federlager *n*
s. board Federsprungbrett *n*, Sprungbrett *n*, Trampoline *f*
s. bolt Federbolzen *m*
s. bow compass Nullenzirkel *m*
s. bracket Federbock *m*; Federträger *m*
s.-brake cylinder *(Kfz)* Federspeicherbremszylinder *m*
s. bumper pad Federpuffer *m*
s. bushing Federbuchse *f*
s. cage Federhülse *f*
s. caliper Federtaster *m*
s. cap Federteller *m*
s. catch Federklinke *f*; Schnappschloß *n*, Schnappriegel *m*, Schnapper *m*
s. centre federnde Spitze *f*
s. centre bolt Federbefestigungsstift *m*
s. clamp Federbüchse *f*, Spannblech *n*
s. clip Federbügel *m*, Federbund *m*, Federklammer *f*, Quetschhahn *m*; Schlauchklemme *f*
s. coil Federwindung *f*
s. coiling Federwickeln *n*
s.-coiling machine Federwickelmaschine *f*
s. collar Federteller *m*; Spannbüchse *f*, Federspannfutter *n*
s. constant Federkonstante *f*
s. contact Federkontakt *m*
s. contact plug Bananenstecker *m*
s. deflection Federdurchbiegung *f*, Federweg *m*, Durchfederung *f*
s. divider Federzirkel *m*
s.-driven clockwork Federuhrwerk *n*
s. end Federende *n*
s. eye eingerolltes Federende *n*
s. force Federkraft *f*
s. galvanometer Federgalvanometer *n*
s. governor Federregler *m*
s. grain Sommergetreide *n*
s. grease cup Federdruckbüchse *f*
s. guide Gleitfeder *f*
s. hammer Federhammer *m*, Blattfederhammer *m*
s. hanger Federlager *n*; Federgehänge *n*
s. inside caliper Innentaster *m* mit Spannfeder
s. jack *(Fmt)* Klinke *f*
s. jaw Stahlfederbacke *f (Bohrwinde)*
s.-joint inside caliper Innentaster *m* mit Spannfeder
s.-joint outside caliper Außentaster *m* mit Spannfeder
s. latch Schnappschloß *n*; Federklinke *f*
s. leaf Federblatt *n*
s. leaf sight Federvisier *n*
s. load-and-fire mechanism federgespannte Auslöseeinrichtung *f*
s.-loaded federnd, federbelastet, gefedert, federgespannt
s.-loaded diaphragm motor federbelasteter Membranmotor *m*
s.-loaded lever safety valve Sicherheitsventil *n* mit Federwaage (Hebelfederbelastung), Sicherheitsventil mit mittelbarer Federbelastung
s.-loaded type of accumulator Federakkumulator *m*
s. lock washer Federscheibe *f (Unterlegscheibe)*
s. loop Federöse *f*
s. material Federwerkstoff *m*
s. needle *(Text)* Hakennadel *f*, Spitzennadel *f*
s. needle machine Hakennadelstrickmaschine *f*
s. nut Spannmutter *f*
s.-operated federbetätigt
s.-opposed federbelastet
s. outside caliper Außentaster *m* mit Spannfeder
s. packing ring elastischer Dichtungsring *m*
s. pad Federunterlage *f*
s. pick loom *(Text)* Federschlagstuhl *m*
s. pile Federsatz *m (Relais)*
s. pin Federstift *m*
s. plate Federteller *m*
s. plunger 1. Federkolben *m*; 2. Federstift *m*
s.-pole drilling stoßendes Bohren *n* mit Schlagbaum
s. power Federung[svermögen *n*] *f*
s. pressure Federdruck *m*
s.-protected objective Objektiv *n* mit Präparateschutz
s. rebound Rückfederung *f*
s. ring Federring *m*, Sprengring *m*; selbstspannender Liderungsring *m (Dichtungsring)*; Selbstspanner *m*
s. roller bearing Federrollenlager *n*
s. saddle Federschuh *m*
s. scale Federmaßstab *m*
s. screw Federschraube *f*
s.-screw thread die Gewindeschneidpatrone *f*
s. seating pan Federauflage *f*, Federteller *m*
s. seed Frühsaat *f*
s. shackle Federlasche *f*, Federklammer *f*; Federgehänge *n*
s. sleeve Federbuchse *f*

s. sowing Frühsaat *f*
s. steel Federstahl *m*
s. steel tooth insert *(Text)* einsetzbarer Abschlagzahn *m* aus Federstahl
s. stiffness Federsteifigkeit *f*
s. stop nut Federring *m*
s. suspension Federaufhängung *f*
s. switch *(Fmt)* Federklinke *f*
s. temper Federhärte *f*
s.-tempered federhart
s. tension Federspannung *f*
s. tension plate Federspannplatte *f*
s. terminal *(El)* Federklemme *f*
s. testing Federprüfung *f*
s. tide Springflut *f*, Springtide *f*
s. tine cultivator *(Lw)* Federzinkengrubber *m*
s. tooth federnder Zinken *m*
s.-tube manometer Bourdonmanometer *n*, Röhrenfedermanometer *n*
s. U-bolt Federbügel *m*
s. valve federbelastetes Rückschlagventil *n*
s. washer Federring *m*, Federteller *m*, federnde Unterlegscheibe *f*
s. water Quellwasser *n*
s. winding Federwickeln *n*
s. wire Federdraht *m*
springback Rückfederung *f*
springiness Rückfederung *f*; Elastizität *f*
springing 1. Springen *n*; Splittern *n*; Brechen *n*; 2. Abfederung *f*; Durchfedern *n*, Federn *n*; Schnellen *n*; 3. Schnappen *n*; 4. Entspringen *n*, Quellen *n*; 5. *(Brg)* Auskesseln *n* eines Bohrlochs
s. out Spreizen *n*
springwood Frühholz *n*
springy rückfedernd, elastisch
sprinkle / to berieseln, anfeuchten, bespritzen
sprinkler Berieselungsapparat *m*, Berieselungsanlage *f*, Brause *f*, Feuerlöschbrause *f*, Sprenger *m*, Sprinkler *m*
s. installation Beregnungsanlage *f (z. B. Feuerlöschanlage eines Schiffes)*
s. nozzle Brause *f*, sprühende Düse *f*
sprinkling apparatus *(Lw)* Sprengapparat *m*
s. device *(Lw)* Sprengvorrichtung *f*
s. installation Beregnungsanlage *f*
s. machine Einsprengmaschine *f*
s. stoker Wurffeuerung *f*
sprocket Ketten[zahn]rad *n*; Transportrolle *f*; Zahntrommel *f (Film)*; Transportrolle *f (Stachelwalze)*
s. chain Kettenradkette *f*, Zahnkette *f*
s. cutter Kettenradfräser *m*
s. diameter Kettenraddurchmesser *m*
s. hob Kettenradwälzfräser *m*
s. holes Transportlochung *f (von Lochstreifen)*
s. shaft Kettenwelle *f*
s. tooth Kettenradzahn *m*
s. wheel Kettenrad *n*, Kettenscheibe *f*, Zahnkettenrad *n*
s. wheel [milling] cutter Kettenradfräser *m*
sprout / to 1. *(Lw)* [ent]sprießen; keimen; 2. *(Met)* spratzen
sprout Keimspitze *f*; Schoß *m*, Schößling *m*; Trieb *m (Pflanze)*; Ansatz *m (Blüte)*
s. land Ausschlagwald *m*, Niederwald *m*
sprouting 1. Knospen *n*, Sprossen *n*; Keimen *n*; 2. *(Met)* Spratzen *n (Silber)*
spruce Fichte *f*; Fichtenholz *n*
s. rosin Fichtenharz *n*
sprue *(Met)* Eingußtrichteransatz *m*; Gießtrichter *m*; Anschnitt *m*
s. cutter Eingußabschneider *m*
s. hole Trichterloch *n*
s. opening Trichteröffnung *f*
spru[e]ing Enttrichtern *n* von Gußstücken
sprung abgefedert
s. hole ausgekesseltes Bohrloch *n (Tiefbohrtechnik)*
spud / to 1. eine Bohrung beginnen; 2. Unkraut jäten *(mit kleinem Spaten)*
spudder Stoßbohrvorrichtung *f (Tiefbohrtechnik)*
spudding bit Flachmeißel *m*
s. shoe Stoßbohrschuh *m*
s. the well Beginn *m* der Bohrung, Bohrbeginn *m*
spun 1. gesponnen; 2. gedrückt *(Metall)*
s. concrete Schleuderbeton *m*
s.-dyed *(Text)* spinngefärbt
s.-dyed colour Spinnfärbung *f*
s.-dyeing *(Text)* Düsenfärbung *f*, Färben *n* in der Spinnlösung *(Chemiefaser)*
s. ferrule Kappenzwinge *f*
s. glass Glaswatte *f*, Glaswolle *f*
s. pipe Schleuderbetonrohr *n*
s. rayon Gespinst *n* aus Viskose- oder Kupferfasern
spur Sporn *m*; Strebe *f*
s. cable *(El) (kurze)* Abzweigleitung *f*
s. differential Stirnraddifferential *n*
s. friction wheel Zylinderreibrad *n*
s. gear Stirnrad *n*
s.-gear cutter Stirnradfräser *m*
s.-gear drive Stirnrad[an]trieb *m*
s.-gear generation grinding Abwälzschleifen *n* von Stirnrädern
s.-gear hob Stirnradwälzfräser *m*
s.-gear train Stirnradgetriebe *n*
s. gearing Stirnradverzahnung *f*; Stirnradgetriebe *n*
s. insulator Nasenisolator *m*, Krückenisolator *m*
s. pinion Geradzahnritzel *n*
s. point Führungsspitze *f (Bohrer)*
s. rack Zahnstange *f* mit Geradverzahnung
s. shore *(Schiff)* Dockseitenstütze *f*, Abhalter *m*

s. wheel Stirnrad *n*
s. wheel drive Stirnradantrieb *m*
s. wheel gearing Stirnradgetriebe *n*
spurious künstlich; falsch, fehlerhaft; ungewollt, unerwünscht
s. bearing *(Fmt)* fehlerhafte Peilung *f*, Fehlpeilung *f*
s. capacitance (capacity) Streukapazität *f*
s. coincidence *(Ph)* zufällige Koinzidenz *f*
s. count *(Kern)* unechter Zählstoß *m*
s. coupling Streukopplung *f*
s. emission Störstrahlung *f*
s. frequency Störfrequenz *f*
s. modulation Störmodulation *f*
s. noise Störrauschen *n*
s. oscillation parasitäre (wilde) Schwingung *f*
s. pulse mode ungewollter Impulsmodus *m*
s. radiation Nebenausstrahlung *f*, Störstrahlung *f*, ungewollte Aussendung *f*
s. resonance Nebenresonanz *f*, unerwünschte Resonanz *f*
s. response Unselektivität *f* *(Nachrichtentechnik)*
s. sidebands *(Fmt)* störende Seitenbänder *npl*
s. signal Störsignal *n*
s. voltage Nebenspannung *f*
sputter / to sprühen, spritzen; zerstäuben
sputtered material zerstäubtes Material *n*
sputtering 1. Sprühen *n* *(Gasentladungsröhren; Schweißen)*; 2. Bestäuben *n*; Zerstäuben *n*; 3. Aufspritzen *n*, Spritzmetallisieren *n*
s. of cathode Katodenzerstäubung *f*
s. voltage Zerstäubungsspannung *f*
spy hole Schauloch *n*
spyndle *(Text)* 14400 yards *(Längenmaß)*
squab Rückenlehne *f*, Sitz *m*; Stützkissen *n*
squabble / to *(Typ)* quirlen *(Satz)*
squabble *(Typ)* gequirlter (eingeworfener) Satz *m*
squalene *(Ch)* Squalen *n*
squall *(Schiff)* Bö *f*
square / to 1. ins Quadrat erheben; 2. rechtwinklig schneiden
s. off rechtwinklig abschneiden
s. up rechtwinklig beschneiden; besäumen
square 1. Quadrat *n*; 2. *(rechteckiger)* Platz *m*; 3. Vierkant *m*; 4. Anschlagwinkel *m* *(Zeichengerät)*
square quadratisch; kantig, viereckig, senkrecht, rechtwinklig, gerade
s. back *(Typ)* nichtgerundeter (gerader, glatter) Rücken *m*
s. bar Quadratstahl *m*, Vierkantstahl *m*
s. bars Vierkantmaterial *n*
s. base quadratische Grundfläche *f*
s. bastard cut file Vierkantfeile *f*
s. bit Kronenbohrer *m*, Kreuzbohrer *m*
s. blade Vierkantmesser *n* *(Fräser)*
s. bloom vorgewalzter Vierkantblock *m* *(Walzwerk)*
s. body paralleles Mittelschiff *n*
s. bolt Vierkantschraube *f*
s. box-section column Kastenständer *m*
s. box wrench Vierkantschraubenschlüssel *m*, Vierkantsteckschlüssel *m*
s. bracket eckige Klammer *f*
s. butt joint I-Stoß *m* *(bei Nietungen oder Schweißungen)*
s. cascade *(Kern)* rechteckige Kaskade *f*
s. chain *amerikanische Flächeneinheit; 1 sq.ch. = 404,687* m^2
s. corner halving Ecküberblattung *f* mit geradem Schnitt *(Holzbau)*
s. die Vierkantschneidmutter *f*
s. dimension Flächendimension *f*
s. edge Breitschnitt *m* *(Schuh)*
s. edge weld Stirnflachnaht *f*
s. egg Druckbirne *f*
s. file Vierkantfeile *f*
s. finish tooth Nachschneidezahn *m*
s. frame *(Schiff)* Richtspant *n*, Querspant *n*, Mittelspant *n*
s. groove Quadratkaliber *n* *(Walzwerk)*
s. grooving and tonguing *(Holz)* Quadratspundung *f*
s. guide *(Masch)* Flachführung *f*
s. head Vierkantkopf *m*
s.-head bolt Vierkantschraube *f*
s.-head bolt with square shank Vierkantschraube *f* mit Vierkantschaft
s.-head screw Vierkantschraube *f*
s.-headed plug Vierkantstöpsel *m*
s. hole Innenvierkant *m*, Vierkantloch *n*
s. hole broach Vierkanträumnadel *f*
s. inch Quadratzoll *m* *(englische Flächeneinheit: 1 sq.in. = 6,452* cm^2*)*
s. joint Stumpfstoß *m*
s. key Quadratkeil *m*, Vierkantkeil *m*
s. law Entfernungsgesetz *n*
s.-law capacitor Nierenplattenkondensator *m*, wellengerader Kondensator *m*
s.-law dependence quadratische Abhängigkeit
s.-law detection quadratische Gleichrichtung *f*
s.-law detector quadratischer Detektor *m*
s.-law function generator *(Dat)* Quadrierglied *n*
s.-law multiplier *(Dat)* Viertelquadratmultiplikator *m*
s.-law rectification quadratische Gleichrichtung *f*
s. link *amerikanische Flächeneinheit: 1 sq.li. = 40,469* cm^2
s. marking gauge Winkelstreichmaß *n*
s. mile Quadratmeile *f* *(englische und amerikanische Flächeneinheit: 1 sq.mi. = 2,59* km^2*)*

s.-mile foot *27,88 Millionen Kubikfuß, 174 Millionen Gallonen (Wassermenge, die eine Fläche von einer Quadratmeile einen Fuß hoch bedeckt)*
s. nut Vierkantmutter *f*
s. paper Millimeterpapier *n*
s. pattern *(Text)* Karomuster *n*
s. plain taper key Treibkeil *m*
s. pulse Rechteckimpuls *m*
s. pulse generator Rechteckimpulsgeber *m*
s. punch Vierkantdurchschlag *m*, Vierkantdurchschläger *m*
s. rod *englische Flächeneinheit: 1 sq.rd. = 25,293 m²*
s. root Quadratwurzel *f*
s. scraper Flachschaber *m*
s. set *(Brg)* geschlossener Türstock *m*, Ausbaurahmen *m*, Rahmenausbau *m*, Rahmenzimmerung *f*
s.-set-and-fill Abbau *m* mit Rahmenzimmerung und Bergeversatz
s.-set method Abbau *m* mit Rahmenzimmerung
s.-set slicing Querbruchbau *m* mit geschlossenem Türstockausbau in den Querbrechen
s.-set stoping Abbau *m* mit Rahmenzimmerung
s.-set timbering Rahmenzimmerung *f*, Rahmenausbau *m*
s.-setting *s.* s.-set stoping
s. shoulder rechtwinkliger Absatz *m*
s. sight *(Mil)* Rechteckkorn *n*
s. socket screw wrench Vierkantstiftschlüssel *m*
s. stern *(Schiff)* Spiegelheck *n*; Plattheck *n*, Plattgatt *n*; Barkassenheck *n*
s. stern boat Plattgattboot *n*
s. thread flachgängiges Gewinde *n*, Flachgewinde *n*
s.-thread screw flachgängige Schraube *f*
s.-threaded tap Flachgewindebohrer *m*
s. tool post Vierkantrevolverkopf *m*
s. tooth clutch Klauenkupplung *f*
s.-tuck stern *(Schiff)* Plattgattheck *n*
s. turret Vierkantrevolverkopf *m*
s. washer Vierkantscheibe *f*
s. washer with corner bent-up [to secure the bolt] aufgebogene Unterlegscheibe *f*
s. wave *(Fs)* Quadratwelle *f*, Rechteckwelle *f*
s.-wave current pulse Rechteckstromimpuls *m*
s.-wave frequency Eckfrequenz *f*
s.-wave generator *s.* s.-wave oscillator
s.-wave modulated rechteckwellenmoduliert
s.-wave modulation Rechteckwellenmodulation *f*
s.-wave oscillator Rechteck[wellen]generator *m*
s.-wave pulse Rechteck[wellen]impuls *m*, Rechteckschwingung *f*
s.-wave response Sprungcharakteristik *f (Verhalten eines Verstärkers gegenüber sehr steiler Spannungsänderung)*
s.-wave signal Rechteckwellensignal *n*
s.-wave voltage Rechteckspannung *f*
s. with in rechtem Winkel auf, senkrecht zu
s. wrench Vierkantsteckschlüssel *m*
s. yard *englische Flächeneinheit: 1 sq.yd. = 0,836 m²*
squared *(Math)* [ins] Quadrat, hoch zwei
s. shank Vierkantschaft *m*
s. taper shank verjüngter Vierkantschaft *m*
s. timber Balken *m*, Kantholz *n*
squareness Rechtwinkligkeit *f*; rechtwinklige Lage *f*
s. ratio *(Dat)* Rechteckigkeitsverhältnis *n (Magnetwerkstoffe)*
squaring Bearbeiten *n* auf rechten Winkel; Beschneiden *n*, Bestoßen *n*; *(El)* Rechteckformung *f*
s. circuit *(Dat)* Quadrierschaltung *f*; *(Fs)* Kreis *m* zur Umformung von Sinus in Rechteck, Rechteckkreis *m*
s. up machine Querhobelmaschine *f*
squash / to [zer]quetschen, zerdrücken, zu Brei machen
squash Quetschfuß *m (Lampe)*
squat *(Schiff)* Trimmänderung *f*, Fahrttrimm *m*
squeak Quieken *n*, Quietschen *n*, Piepsen *n (Funkempfangstechnik)*
squeal Heulen *n*, Pfeifen *n*, Quieken *n (Funkempfangstechnik)*
squealing Selbsttönen *n (von Verstärkern)*
squeegee 1. Quetscher *m*; Rollenquetscher *m*, Gummitrockner *m (Foto)*; 2. *(Typ)* Siebdruckrakel *f*, Schaber *m*; 3. Straßenreinigungsbesen *m*; Räumbalken *m (zur Schlammbeseitigung)*
squeeze / to 1. [aus]drücken, [aus]pressen, [aus-]quetschen; 2. *(Brg)* verdrücken; 3. anstauchen; druckumformen
s. out abquetschen, herausquetschen
squeeze Druck *m*, Preßdruck *m*; Quetschung *f*; Quetschfuß *m (Elektronenröhren)*
s. block *(Brg)* Quetschholz *n*
s. bottle Plastikspritzflasche *f*
s.-cementing mixture Zementmischung *f* für Druckzementierung
s. riveting Quetschnietung *f*
s. riveting machine Nietpresse *f*
s. roll Vorgautschwalze *f*, Abquetschwalze *f*
s. section Quetschleitung *f (Hohlleiter)*
squeezer Quetschwalzwerk *n*; Preßformmaschine *f*
squeezing 1. Ausdrücken *n*, Drücken *n*, Ausquetschen *n*, Quetschen *n*, Auspressen *n*, Pressen *n*; 2. *(Brg)* Verdrücken *n*; 3. Anstauchen *n*, Druckumformen *n*
s. effect *(Text)* Abquetscheffekt *m*

s. machine *(Text)* Abquetschmaschine *f*
s. mangle *(Text)* Abquetschfoulard *m*
s. press Abquetschpresse *f* *(Wollwaschmaschine)*
s. roller *(Text)* Abquetschwalze *f*
s. time Preßzeit *f* *(Punktschweißen)*
squegger *(Fs)* Sperrschwinger *m*, Pendeloszillator *m*
s. super-regenerative receiver Pendelrückkopplungsempfänger *m*
squeggering *(El)* Instabilität *f*, Unstabilität *f*
squegging Pendelung *f* *(Oszillator)*
s. oscillator Stoßoszillator *m*, Pendeloszillator *m*
squelch circuit *(Fmt)* Krachtöterschaltung *f*; Stummabstimmschaltung *f*
squelching *(Fmt)* stille Regelung *f*, Stummabstimmung *f*
squib Zündladung *f*, Zünder *m*; einfache Sprengkapsel *f*
s.-initiated igniter Zünder *m* mit Zündkerze
squibbing *(Brg)* Auskesseln *n* eines Bohrlochs
squint / to schielen
squint Winkelfehler *m* *(Antenne)*
squinting Schielen *n*
s. lobe schielende Charakteristik *f* *(Antennen)*
squirrel cage 1. Kurzschlußwicklung *f*, Käfiganker *m*; 2. Siebtrommel *f*
s. cage aerial Reusenantenne *f*
s. cage grubber Siebtrommelroder *m*
s.-cage induction motor Käfigankerasynchronmotor *m*; Motor *m* mit Kurzschlußläufer; Kurzschlußmotor *m*
s.-cage magnetron Käfigmagnetron *n*
s.-cage motor Käfig[anker]motor *m*, Kurzschlußläufermotor *m*, Asynchronmotor *m* *(ohne Schleifringe)*
s.-cage rotor Kurzschlußläufer *m*, Käfigläufer *m*
s.-cage winding Kurzschlußkäfig *m*, Käfigwicklung *f*
squirt / to [heraus]spritzen
s. in einspritzen
squirt-gun welding halbautomatisches Unterpulverschweißen *n*
s. oiler Ölspritzkanne *f*
squirting machine Spritzmaschine *f*
St. Elmo's fire Sankt-Elms-Feuer *n*
stab / to abteufen, in den Boden bringen, niederbringen *(Bohrung)*; aufrauhen; [durch]stechen, steppen
stab Anstich[bohrloch *n*] *m*, Entspannungsbohrloch *n* *(Tiefbohrtechnik)*
stabbing *(Typ)* Querheftung *f*; Klammern *n*, Drahtheftung *f*
stabilitron Stabilitron *n* *(cm-Wellenröhre für Radarzwecke)*
stability Stabilität *f*; Standfestigkeit *f*, Standsicherheit *f*; Haltbarkeit *f*; Dauerhaftigkeit *f*; Beständigkeit *f*, Widerstandsfähigkeit *f*
s. condition Stabilitätsbedingung *f*
s. constant Stabilitätskonstante *f*
s. criterion Stabilitätskriterium *n*
s. in probability Stabilität *f* in Wahrscheinlichkeit
s. in the mean *(Math)* Stabilität *f* im Mittel
s. in volume Volumenbeständigkeit *f*
s. limit Stabilitätsgrenze *f*
s. of atom nucleus Kernstabilität *f*
s. of capacitance Kapazitätskonstanz *f*
s. of cavity Stabilität *f* der Kavitation
s. of frequency Frequenzkonstanz *f*
s. of shape Formbeständigkeit *f*, Formfestigkeit *f*
s. test *(Schiff)* Stabilitätsnachweis *m* *(Krängungsversuch)*
stabilization Stabilisierung *f*
s. network *(BMSR)* Stabilisierungsnetz *n*
s. resistance *(El)* Stabilisierungswiderstand *m*, Beruhigungswiderstand *m*
stabilizator *(Ch)* Stabilisator *m*, Antikoagulationsmittel *n*
stabilize / to stabilisieren, haltbar machen, konstant halten
stabilized stabilisiert
s. fabric fixiertes (formfest gemachtes) Gewebe *n* *(durch Vorschrumpfen oder Veredlung)*
s. feedback *(El)* Gegenkopplung *f*
s. feedback amplifier gegengekoppelter Verstärker *m*
s. rectifier Konstantgleichrichter *m*
s. shunt-wound motor kompoundierter Nebenschlußmotor *m*
stabilizer 1. Stabilisator *m*, abbauverhinderndes Mittel *n*, Stabilisier[ungs]mittel *n*, Stahlveredler *m*; Stützstoff *m*; 2. *(Schiff)* Stabilisator; 3. *(Rak)* Flosse *f*; 4. *(El)* Glättungsschaltung *f*; Stabilisator; Spannungsgleichhalter *m*, Spannungskonstanthalter *m*
stabilizing 1. Stabilisieren *n*; 2. *(Text)* Fixieren *n*, Stabilisieren, Formbeständigmachen *n*
s. fin *(Schiff)* Stabilisierungsflosse *f*
s. network Stabilisierungsschaltung *f*, Korrekturnetzwerk *n*
s. transformer Stabilisierungstransformator *m*, Vorhalttransformator *m*
s. winding *(Elm)* Tertiärwicklung *f*
stable 1. Stall *m*; 2. *(Brg)* Maschinenstall *m*
stable beständig, stabil; haltbar; standfest; stetig, beständig
s. at red heat hitzebeständig, wärmebeständig
s. circuit *(El)* stabiler Kreis *m*
s. diagram *(Met)* Eisen-Graphit-Diagramm *n*
s. element stabiles Element *n*
s. equilibrium stabiles Gleichgewicht *n*
s. gain konstante Verstärkung *f*

s. **governor** stabiler Regler *m*
s. **implements** Stallgeschirr *n*
s. **in frequency** frequenzkonstant, frequenzstabil
s. **in water** wasserbeständig
s. **isotope** stabiles Isotop *n*
s. **manure** Stalldünger *m*
s. **nucleus** stabiler Kern *m*
s. **particle** stabiles Teilchen *n*
s. **phase** beständige Phase *f*
s. **range** stabiler Bereich *m*
s. **ship** kursstabiles Schiff *n*
s. **source** konstante Quelle *f*
stack / to stapeln, lagern, aufschichten
stack 1. Schornstein *m*, Kamin *m*; 2. Schober *m* *(Heu)*; Feime *f*; 3. Schacht *m* *(Hochofen)*; 4. Glättwerk *n*; 5. Stapel *m*; Stoß *m*; Paket *n*; 6. 108 Kubikfuß *(Holzmaß)*
s. **band** Schachtband *n* *(Schachtofen)*
s. **brickwork** Schachtmauerwerk *n* *(Hochofen)*
s. **casing** Schachtpanzer *m* *(Hochofen)*
s. **gas** Gichtgas *n*, Rauchgas *n*
s. **guard** Schoberdach *n*
s. **lining** Schachtfutter *n*, Schachtausmauerung *f* *(Hochofen)*
s. **moulding** *(Plast)* Übereinanderpressen *n*
s. **of bales** *(Text)* Ballenstapel *m*
s. **of boards** Bretterstoß *m*
s. **of portable engine** Lokomobilschornstein *m*
s. **of sheaves** Garbenschober *m*, Miete *f* *(Getreide)*
stackable stapelbar
stacked übereinandergestapelt
s. **aerial** gebündelte Antennenanordnung *f*
s.**-diaphragm controller** Stapelregler *m*
s. **dipol array** gebündelte Antennenanordnung *f*
s. **timber** Schichtnutzholz *n*
s. **wood** Schichtholz *n*, Klafterholz *n*
stacker *(Am)* Absetzer *m*, Stapler *m*
s. **belt** Absetzer[förder]band *n*
s. **dump** *(Am)* Absetzerkippe *f*
s. **truck** Hubstapler *m*
stacking 1. Stapeln *n*; Lagern *n*; Aufschichten *n*; Aufschobern *n*, Schoberbau *m*; 2. Rieselanlage *f* *(Kühlturm)*
s. **ground** Stapelplatz *m*; Lager[fläche *f*] *n*, Lagerplatz *m*
s. **of hay** Aufstellen *n* der Heuhaufen
s. **of sheaves** Aufstellen *n* der Garben
s. **yard** Stapelplatz *m*; Lager[hof *m*] *n*, Lagerplatz *m*
stactometer *s.* stalagmometer
stadia lines ratio Multiplikationskonstante *f* *(Vermessung)*
staff 1. Stab *m*; Stock *m*; Meßlatte *f*; Runge *f*; Dorn *m*; Achse *f*; 2. Personal *n*; Belegschaft *f*; Personalbestand *m*; Stab; 3. Gipsmörtel *m*
s. **intercept** Lattenabschnitt *m* *(Vermessungswesen)*
s. **shortage** Personalmangel *m*
s. **target point** Lattenzielpunkt *m* *(Vermessungswesen)*
s. **wood** Daubholz *n*, Faßdauben *fpl*
staffman Vermessungsgehilfe *m*
stag headedness Gipfeldürre *f* *(Forst)*
stage 1. Stufe *f*; Raketenstufe *f*; Arbeitsstufe *f* *(Turbine)*; *(Am)* *(El)* Senderstufe *f*; Stadium *n*; Grad *m*; 2. Plattform *f*; Bühne *f*; Absatz *m*; Gerüstboden *m*
s. **filter** Etagenfilter *n*
s. **hoisting** *(Brg)* gebrochene Förderung *f*
s. **lighting** Bühnenbeleuchtung *f*
s. **micrometer** Objektmikrometer *n*
s. **of a rocket** Raketenstufe *f*
s. **of manufacture** Verarbeitungsstufe *f*, Verarbeitungsgrad *m*
s. **of operation** Arbeitsstufe *f*
s. **rocket** Stufenrakete *f*, mehrstufige Rakete *f*
s. **separation** *(Rak)* Stufentrennung *f*
s. **turbine** Stufenturbine *f*
staged rocket Stufenrakete *f*, mehrstufige Rakete *f*
stagger / to auf Lücke setzen (stehen), versetzen, versetzt anordnen, staffeln; schwanken, taumeln
stagger Zickzack *m* *(der Fahrleitung)*; Zickzackanordnung *f*, versetzte (gestaffelte) Anordnung *f*
s.**-tuned** *(Fs, Fmt)* gegeneinander verstimmt, gestaffelt abgestimmt
s.**-tuned amplifier** Breitbandresonanzverstärker *m*, Verstärker *m* mit versetzter Abstimmung
staggered gestaffelt, versetzt, auf Lücke
s. **arrangement of teeth** Kreuzverzahnung *f*
s. **brush** *(Elm)* versetzt angeordnete Bürste *f*
s. **circuits** *(El, Fs)* verstimmte Kreise *mpl*
s. **feed** Zickzackbandzuführung *f*
s. **joints** versetzt liegende Stoßstellen *fpl*
s. **mill** Zickzackduo *n* *(Walzwerk)*
s. **poles** *(El)* versetzte Pole *mpl*
s. **riveting** Zickzacknietung *f*
s. **rows of rivets** Zickzacknietung *f*
s. **scanning** *(Fs)* Zeilensprungabtastung *f*
s. **spot weld** Zickzackpunktschweißung *f*
s. **teeth** Kreuzverzahnung *f*
s. **tooth cutter** Fräser *m* mit Kreuzverzahnung
s. **tooth side milling cutter** Kreuzzahnscheibenfräser *m*, Scheibenfräser *m* mit Kreuzverzahnung
s. **tooth system** Kreuzverzahnung *f*
staggering *(Fmt)* Kanalversetzung *f*
s. **advantage** *(Fmt)* Versetzungsgewinn *m*

staging 1. Anordnung *f* der Raketenstufen; 2. *(Schiff)* Hellinggerüst *n*, Stellage *f*; 3. Gerüstbau *m*
s. altitude *(Rak)* Stufentrennungshöhe *f*
stagnancy *s.* stagnation
stagnant stillstehend, unbewegt *(Flüssigkeit)*
stagnation Stagnation *f*, Stillstand *m*; Stockung *f*; Aufstau *m (Strömung)*
s. point Staupunkt *m*
s. pressure Staudruck *m*
stain / to 1. beflecken; anlaufen; verfärben, abfärben, abflecken; 2. anfressen, korrodieren, rosten; 3. beizen, farbig polieren *(Holz)*; bedrucken
stain Fleck[en] *m*; Verfärbung *f*; Farbe *f*, Farbstoff *m*, Färbungsmittel *n*; Beize *f (für Holz)*; Rost *m*
s. removal Fleckentfernung *f*
s. remover Fleckentferner *m*, Fleckenreiniger *m*
s.-repellent finish fleckabweisende Ausrüstung *f*
s. resistance Korrosionsbeständigkeit *f*, Korrosionswiderstand *m*; Rostbeständigkeit *f*; Fleckenbeständigkeit *f*
s.-resistant finish *s.*-repellent finish
stained cotton fleckige Baumwolle *f*
s. specimen gefärbtes Präparat *n (Mikroskopie)*
staining 1. Flecken *n*; Verfärbung *f*; *(Led)* Bürstfärbung *f*; Färben *n*; Beizen *n (Holz)*; 2. Korrodieren *n*, Korrosion *f*, Rostbildung *f*, Rosten *n*
s. of specimens Einfärben *n* von Präparaten
stainless 1. nichtrostend, rostbeständig, rostfrei, rostsicher; 2. fleckenfrei, fleckenlos; farblos
s. property Korrosionsbeständigkeit *f*, Rostbeständigkeit *f*
s. steel nichtrostender (rostfreier) Stahl *m*
stair Treppe *f*
s. builder's saw Gratsäge *f*
s. horse Treppenwange *f*
s. landing Podest *n (m)*, Treppenabsatz *m*, Treppenpodest *n (m)*
s. lighting Treppenbeleuchtung *f*
s.-step generator Kipptreppengenerator *m*
s. well Treppenhaus *n*
staircase 1. Treppe *f*; Treppenhaus *n*; 2. *(El)* Treppe *(Treppenspannung)*
s. curve *(Math)* Treppenkurve *f*, Treppenpolygon *n*
s. generator Treppen[spannungs]generator *m*
s. pulse Treppenimpuls *m*
s. voltage *(El)* Treppenspannung *f*
s. waveform *(El)* Treppenwellenform *f*
stairway Treppe *f*; Treppenlauf *m*; Treppenhaus *n*
stake / to anpflocken, abstecken
stake Pflock *m*, Pfahl *m*; Runge *f*
staking machine *(Led)* Stollmaschine *f*
s. out Abpflocken *n*; Abstecken *n* der Ausschachtung
stalagmometer *(Ch)* Stalagmometer *n*, Tropfenzähler *m*
stalagmometry *(Ch)* Stalagmometrie *f (Analyse von Lösungen mittels Oberflächenspannungsmessung)*
stale flesh *(Led)* alte (matte) Fleischseite *f*
stalk Einspannzapfen *m*; Stengel *m*
s. fibre Stengelfaser *f*
stall / to bis zum Stillstand abgebremst werden, stillstehen; zum Stehen kommen (bringen); abwürgen
s. the engine den Motor abwürgen
stall 1. Stillstand *m*; 2. Stand *m*; Stall *m*
s. feeding Trockenfütterung *f*
s. road *(Brg)* Pfeilerstrecke *f*, Abbaustrecke *f*
stalling flight *(Flg)* Langsamstflug *m*
s. of the engine Spucken *n* des Motors
s. of the tool when cutting Stehenbleiben *n* des Werkzeugs im Schnitt
stamina Widerstandskraft *f*; Ausdauer *f*
stamp / to 1. stampfen; 2. stanzen; pressen; prägen *(stempeln)*; 3. aufdrucken
stamp 1. Stempel *m*, Stempelung *f*; 2. Marke *f*
s. dedusting Pressenmaulentstaubung *f*
s. mill Pochmühle *f*, Pochwerk *n*, Stampfwerk *n*
s. pad Stempelkissen *n*
s. pad ink Stempelfarbe *f*
s. pulp Pochtrübe *f*
s. rock Pochgestein *n*
stamped poles *(Fmt)* lamellierte Pole *mpl*
stamper Schlagmaschine *f*; Stampfer *m*
stamping 1. Stampfbetrieb *m (Aufbereitung)*; 2. Stanzen *n*; Pressen *n*; Prägen *n (Stempeln)*; 3. Preßteil *n*; Stanzteil *n*; Gesenkschmiedestück *n*; 4. *(El)* Trafoblech *n*
s. automatic Stanzautomat *m*
s. calender *(Plast)* Prägekalander *m*
s. die Stanzmatrize *f*, Stanzwerkzeug *n*
s. mill Pochwerk *n*
s. press Gesenkpresse *f*; Stanzpresse *f*, Stanze *f*
s. tool Stanzwerkzeug *n*
stanchion 1. *(Schiff)* Decksstütze *f*, Pfosten *m*; 2. *(Eis)* Runge *f*; 3. Strebe *f*, Stütze *f*
s. car Rungenwagen *m*
s. pocket Rungenöse *f*
s. socket Rungenöse *f*
s. wagon Rungenwagen *m*
stand / to 1. [aufrecht] stehen; unbewegt sein; 2. hoch (groß) sein; eine Höhe haben von; 3. bestehen bleiben; gültig bleiben; standhalten, aushalten; 4. stellen; 5. Kurs halten

s. by 1. bereit stehen; 2. auf Empfang bleiben *(Funktechnik)*
s. out hervorstehen, hervorragen
s. rough work *(Fmt)* rauhen Betrieb aushalten
stand 1. Gestell *n*, Objektträger *m*, Ständer *m*, Stativ *n*, Untergestell *n*; 2. Walzenständer *m*
s.-by Bereitschaft *f*; *(Fmt)* Betriebsbereitschaft *f*; Anheizstellung *f*
s.-by Reserve-, Hilfs-, Ersatz-
s.-by battery Ersatzbatterie *f*
s.-by boiler Reservekessel *m*
s.-by current Anheizstrom *m*
s.-by generator Notstromgenerator *m*, Notstromerzeuger *m*; Reservegenerator *m*, Ersatzgenerator *m*
s.-by generator set Notstromaggregat *n*
s.-by loss Leerlaufverlust *m*
s.-by machine Reservemaschine *f*
s.-by plant Reserveanlage *f*
s.-by steam engine Reservedampfmaschine *f*
s.-by supply relay Notstromrelais *n*
s.-by switch *(Fmt)* Bandbreitenregler *m*, Bandbreitenschalter *m*; Bereitschaftsschalter *m*
s.-by tank Reservebehälter *m*, Hilfstank *m*
s. of rolls Walzgerüst *n*
s.-pipe *(Bw)* Standrohr *n*
s.-pipe for air-pump suction Luftsaugstutzen *m*
s.-pipe for condenser suction Kondensatorsaugstutzen *m*
s. sheet *(Bw)* feststehendes Fenster *n*
standard 1. Standard *m*, Norm *f*; 2. Normal *n*, Normale *f (Meßtechnik)*; Eichnormal *n*; Einheit *f*; 3. *(Masch)* Ständer *m*; 4. Mast *m*; Pfahl *m*, Pfosten *m*, Stütze *f*; Tragbalken *m*
standard handelsüblich, normal; standardisiert, genormt; geeicht
s. absorber Standardabsorber *m*
s. assembly Standardbaugruppe *f*
s. atmosphere Normatmosphäre *f*
s. aural-transmitter input signal *(Fs)* Standardtonsendereingangssignal *n*
s. base Einheitsröhrenfassung *f*, Einheitsfassung *f*
s. basic dimension Normalmaß *n*
s. beam approach *(Flg)* Standardleitstrahlanflug *m*
s. block Endmaß *n*
s. burden *(Met)* Normalgattierung *f*
s. cable Normalkabel *n*
s. caliper gauge Grenzrachenlehre *f*
s. cap *(El)* Normalsockel *m*
s. capacitor Normalkondensator *m*, Eichkondensator *m*
s. capacity Normalfassungsvermögen *n*; Normalleistung *f*
s. casing Standardgehäuse *n*
s. cell Normalelement *n*
s. charge *(El)* normale Ladung *f*
s. clock Normaluhr *f*
s. component 1. Normteil *n*, standardisiertes Teil *n*; 2. Baueinheit *f*
s. composite picture signal *(Fs)* vollständiges Standardbildsignal *n*
s. counter *(Kern)* Einheitszählrohr *n*
s. cutting speed Normalschnittgeschwindigkeit *f*
s. design Normalausführung *f*, Normalbauweise *f*, Einheitsbauweise *f*
s. deviation Standardabweichung *f*; normale Abweichung *f (Frequenzmodulation)*
s. dimensioning einheitliche Bemaßung *f*
s. electrode potential Normalpotential *n*
s. element component Baueinheit *f*
s. equipment Normalzubehör *n*
s. error mittlerer Fehler *m*, Standardfehler *m*
s. field *(El)* Eichfeld *n*, Standardfeld *n*
s. film Normalfilm *m*
s. fit *(Masch)* Normalsitz *m*
s. fit table Passungstafel *f*
s. fitment Standardausrüstung *f*
s. for handwheel Handradständer *m*, Handradsäule *f (Turbine)*
s. for setting Einstellmaß *n*, Kontrollmaß *n*
s. frequency Normalfrequenz *f*, Eichfrequenz *f*, Einheitsfrequenz *f*
s. frequency generator Eichsender *m*, Normalfrequenzgenerator *m*
s. frequency signal Normalfrequenzsignal *n*
s. frequency station Normalfrequenzsender *m*
s. fuel Eichkraftstoff *m*
s. gap gauge Normalrachenlehre *f*
s. gauge 1. Normallehre *f*, Normalmaß *n*; 2. *(Eis)* Normalspur *f*, Regelspur *f*
s. gauge block *(BMSR)* Eichsatz *m*, Normal *n*
s. geometry Standardgeometrie *f*, Normalgeometrie *f*
s. instrument Standardgerät *n*
s. lamp 1. Lampennormal *n*, Normallampe *f*; 2. Ständerlampe *f*, Ständerleuchte *f*, Stehlampe *f*
s. lamp holder Edisonsockel *m*, Normalfassung *f*
s. length Handelslänge *f*
s. lens Normalobjektiv *n*, Standardobjektiv *n*
s.-lens camera Normalwinkelkamera *f*
s. level Normalpegel *m*
s.-level generator Normalgenerator *m*
s. line Eichleitung *f*
s. load progression *(Opt)* Normlastreihe *f*
s. loudspeaker Normallautsprecher *m*
s. measure 1. Normale *f*, Normal *n*; 2. Normalmaß *n*, Eichmaß *n*
s. metal Regelmetall *n*

s. microphone Eichmikrofon *n*, Normalmikrofon *n*
s. mix Beton *m (Mischungsverhältnis: 1 Teil Zement, 2 Teile Kies, 4 Teile Splitt)*
s. module *(El)* Standardbaustein *m*
s. molar volume Molnormalvolumen *n*
s. Morse taper Normalmorsekegel *m*
s. mount Normalfassung *f*
s. musical pitch Kammerton *m*, Normalton *m*
s. neutron source geeichte Neutronenquelle *f*
s. noise generator Normalrauschgenerator *m*, Rauschsummer *m*
s. of comparison Vergleichsmaßstab *m*
s. of length Längennormale *f*
s. of reference Vergleichsmaßstab *m*
s. ohm Normal-Ohm *n*
s. part Normteil *n*
s. picture-line amplifier output signal *(Fs)* Standardausgangssignal *n* des Bildkanalverstärkers
s. pitch Normalteilung *f (Zahnräder)*
s. planer Doppelständerhobelmaschine *f*
s. plough *(Brg)* Einheitshobel *m*
s. port Basishafen *m*, Bezugshafen *m*
s. position Normallage *f*, Normalstellung *f*
s. price Richtpreis *m*
s. quartz plate Normalquarzplatte *f*
s. railway track *(Eis)* Normalgleis *n*
s. rating Normleistung *f*
s. receiver Normal[fern]hörer *m*
s. reference frequency Vergleichsfrequenz *f*
s. resistance Eichwiderstand *m*, Normalwiderstand *m*, Widerstandsnormale *f*
s. rules for dimensioning drawings Bemaßungsregeln *fpl*
s. scale Normalmaßstab *m*
s. screw driver Normenschraubenzieher *m*
s. screw socket Lampensockel *m* mit Edisongewinde, Normalsockel *m*
s. set Seriengerät *n*
s. sheet Normblatt *n*
s. sheet stock Standardfurniere *npl*
s. signal generator Meßsender *m*
s. signal lamp Normallampe *f (Signal)*
s.-signal oscillator Meßsender *m*
s. size Normalgröße *f*; Standardgröße *f*; Nennweite *f*, Nennmaß *n*; Normalformat *n (Papier)*; Regelgröße *f*
s. soiled fabric *(Ch)* angeschmutzter standardisierter Teststoff *m*
s. solution *(Ch)* Maßlösung *f*, Normallösung *f*, Standardlösung *f*; Titrationslösung *f*
s. sowing table Einheitssätabelle *f*
s. specification Ausführungsbestimmung *f*, Norm[vorschrift] *f*, Standard[isierungsvorschrift *f*] *m*
s. teeth Normalverzahnung *f*
s. television signal Fernsehnormsignal *n*, Standardfernsehsignal *n*
s. test bar Normalprobestab *m*
s. thread Normalgewinde *n*, Einheitsgewinde *n*
s. time Normzeit *f*
s. titrimetric substance *(Ch)* Urtitersubstanz *f*
s. train loading Normallastenzug *m (Brückenbau)*
s. transformer Einheitstransformator *m*
s. transmitter Eichmikrofon *n*, Normalmikrofon *n*
s. twist *(Text)* Normaldrehung *f*
s. type Regelausführung *f*, lagermäßige Ausführung *f*
s. unit Aufbaueinheit *f*
s. value Eichwert *m*
s. version Standardausführung *f*
s. visual-transmitter input signal *(Fs)* Standardbildsendereingangssignal *n*
s. voltage Norm[al]spannung *f*
s. wavemeter Normalwellenmesser *m*
s. weave *(Text)* Grundbindung *f*
s. wheel faces Randformen *fpl* der Schleifscheibe
standardization Standardisierung *f*
standardize / to standardisieren
standardized unit Baueinheit *f*
s. unit construction Aufbau *m* nach Baukastensystem
standardizing laboratory Eichamt *n*; Prüfamt *n*, Zählerprüfamt *n*
s. room Eichraum *m*; Prüfraum *m*, Zählereichwerkstatt *f*
standards proposals Normenvorschläge *mpl*
standing stehend; stationär
s. bath *(Text)* altes Bad *n*
s. bowsprit *(Schiff)* festes Bugspriet *n*
s. fillister einfacher Falzhobel *m*
s. matter *(Typ)* Stehsatz *m*
s. pump feste Pumpe *f*
s. rope Halteseil *n*, Tragseil *n*
s. timber Stammholz *n*, stehendes Holz *n*, Holz auf dem Stamm; Nutzholz *n*
s. wave stehende Welle *f*
s. wave indicator Stehwellenmeßgerät *n*, Welligkeitsmesser *m*
s. wave loss factor Verlust *m* durch Fehlanpassung
s. wave meter Welligkeitsmesser *m*
s. wave ratio Stehwellenverhältnis *n*, Wellenziffer *f*
s. ways Ablaufbahn *f*, Stapellaufbahn *f (Helling)*
standpipe Standrohr *n*; Hydrant *m*
standstill Stillstand *m*
stannate Salz *n* der Zinn(IV)-säure, Stannat *n*

stannic compound Verbindung *f* des vierwertigen Zinns
s. salt Zinnsalz *n*
stanniferous zinnführend, zinnhaltig
stannite Salz *n* der Zinn(II)-säure, Stannit *n*
stannous compound Verbindung *f* des zweiwertigen Zinns
stannum Zinn *n*
staple / to 1. sortieren; klassifizieren; stapeln, aufschichten; 2. *(Text)* gleich lang schneiden
staple 1. Rohstoff *m*; Hauptartikel *m*; 2. Krampe *f*; Haspe *f*; Öse *f*; Klammer *f*; 3. *(Typ)* Heftklammer *f*; 4. *(Text)* Stapel *m*, Faser *f*
s. class *(Text)* Stapelsorte *f*
s. fibre Stapelfaser *f*, Chemiefaser *f*
s. fibre content Chemiefaseranteil *m*
s. fibre fabric Gewebe *n* aus Chemiefasern
s. fibre waste Chemiefaserabgang *m*, Chemiefaserabfall *m*
s. hole *(Brg)* Überhauen-Bohrloch *n*
s. length *(Text)* Stapellänge *f*
s. of the fibre Stapellänge *f* der Faser
s. shaft *(Brg)* Blindschacht *m*, Stapelschacht *m*
stapling 1. *(Text)* Klassifizierung *f* nach Stapellänge; 2. *(Typ)* Drahtheftung *f*
s. machine *(Typ)* Heftmaschine *f*
s. press Klammerheftmaschine *f*
s. test Stapelprüfung *f*
s. wire Heftdraht *m*
star Stern *m*
s. aerial Sternantenne *f*
s. bit Kreuzbohrmeißel *m*
s. chain Sternkette *f (Radar)*
s. chart Sternkarte *f*
s. circuit Sternschaltung *f*
s. cloud Sternwolke *f*
s. cluster Sternhaufen *m*
s.-connected sterngeschaltet, im Stern geschaltet
s.-connected circuit Sternschaltung *f*
s. connection Sternschaltung *f*
s. connection equivalent to delta connection Ersatzsternschaltung *f* einer Dreieckschaltung
s.-delta connection Sterndreieckschaltung *f*
s.-delta starter Sterndreieckschalter *m (Anlasser)*, Sterndreieckanlasser *m*
s.-delta starting Sterndreieckanlassen *n*, Sterndreieckanlauf *m*
s.-delta switch Sterndreieckschalter *m*
s.-delta throw-over switch Sterndreieckumschalter *m*
s. feeder Zellenrad *n*, Kammerschleuse *f*
s. handwheel Sterngriff *m*, Kreuzgriff *m*
s. knob Griffkreuz *n*, Kreuzgriff *m*, Sterngriff *m*
s. magnitude Sterngröße[nklasse] *f*
s. map Sternkarte *f*
s. navigation Astronavigation *f*
s. of first magnitude Stern *m* erster Größe
s.-plate comparator Sternplattenkomparator *m*
s.-plate photometer Sternplattenfotometer *n*
s. point Sternpunkt *m*
s. polygon Sternvieleck *n*
s. production *(Kern)* Sternbildung *f*, Sternerzeugung *f*
s. quad [cable] Sternvierer[kabel *n*] *m*
s. reel *(Text)* Sternhaspel *f*
s. roller *(Lw)* Sternwalze *f*
s. shake *(Holz)* Sternriß *m*, Strahlenriß *m*
s. shell Leuchtgranate *f*
s. spectrum Sternspektrum *n*
s.-star-connection Sternsternschaltung *f*
s. steamer *(Text)* Sterndämpfer *m*
s.-tracking guidance *(Rak)* Astrolenkung *f*, Lenkung *f* mit Hilfe astronomischer Ortung
s. twisting Sternverseilung *f (eines Kabels)*
s. voltage Sternspannung *f*, Phasenspannung *f*
s. wheel 1. Handkreuz *n*, Griffkreuz *n*; 2. Flatterrädchen *n* zum Abrichten *(Schleifscheibe)*
s.-wheel dresser Abrichter *m* mit Flatterrädchen
starboard Steuerbord *m*
starch Stärke[mehl *n*] *f*
s. glue Stärkeleim *m*
s. gum Stärkegummi *n*, Dextrin *n*
s. paste Stärkekleister *m*
s. spirits Kartoffelspiritus *m*
s. sugar Glukose *f*, Stärkezucker *m*, Traubenzucker *m*
starching Stärken *n*
s. clay Stärkeglanz *m (Baumwolle)*
Stark effect Stark-Effekt *m*
starling Eisbrecher *m (am Brückenpfeiler)*; Pfeilerschutz *m*
start / to anfangen, beginnen, starten; anfahren, in Gang setzen, einrücken, einschalten *(Maschinen)*, anlassen *(Motoren)*; anblasen *(Hochofen)*; anschneiden *(Spanen)*; anlaufen; einsetzen *(Schwingung)*; locker werden
s. a blast furnace Hochofen anblasen
s. a hole anbohren, anschneiden
s. the drill into the workpiece anschneiden
s.-up anfahren, in Betrieb nehmen
start Anlauf *m*, Beginn *m*, Start *m*; Ingangsetzung *f*; Anschnitt *m*, Anschneiden *n (Spanen)*
s. button Startknopf *m*
s. of the cut Anschnitt *m*
s. point Ausgangspunkt *m*
s. signal *(Fmt)* Anlaufschritt *m*; Öffnungsimpuls *m*; Beginnzeichen *n*
s.-stop apparatus Springschreiber *m*

s.-stop multivibrator Einschlagmultivibrator *m*, monostabiler Multivibrator *m*, Start-Stop-Multivibrator *m*
s.-stop printer Geh-Steh-Schreiber *m*, Start-Stop-Schreiber *m*
s.-stop telegraphy Geh-Steh-Telegrafie *f*, Springschreibertelegrafie *f*
s.-up Anlauf[vorgang] *m*, Anfahren *n*, Hochfahren *n*, Start *m*
s.-up accident Startzwischenfall *m*, Unfall *m* beim Anfahren des Reaktors
s.-up time Hochfahrzeit *f*; Startzeit *f*; Vorbereitungszeit *f*

starter 1. *(Kfz)* Anlasser *m*; 2. *(Brg)* Anfänger *m (Bohrersatz)*
s. anode Starteranode *f*
s. battery Anlaßbatterie *f*, Starterbatterie *f*
s. bit Anbohrmeißel *m*
s. button *(Kfz)* Anlaßdruckknopf *m*; *(Masch)* Einschaltknopf *m*
s. cable Anlasserleitung *f*
s. contact Anlaßstufe *f*
s. drum Anlasserwalze *f*
s. gap Zündstrecke *f (Gasentladungsröhren)*
s. pinion Anlaßritzel *n*
s. push-button Anlasser[druck]knopf *m*
s. rheostat Regelanlasser *m*
s. ring Anlasserzahnkranz *m*
s. step Anlaßstufe *f*

starting Anfang *m*; Starten *n*; Einschaltung *f*, Inbetriebnahme *f*, Ingangsetzen *n*, Einrücken *n*, Anfahren *n*; Anlassen *n (Motor)*; Anblasen *n (Hochofen)*; Anschneiden *n (Spanung)*; Einsetzen *n (Schwingung)*
s. ability Starteigenschaft *f*, Startvermögen *n*
s.-air bottle *(Kfz)* Anlaßluftflasche *f*
s.-air cylinder *(Kfz)* Anlaßluftflasche *f*
s. anode Zündanode *f*
s. bath Ansatzbad *n*
s. butterfly valve Starterdrosselklappe *f*
s. capacitor Anlaßkondensator *m*
s. capacity Anlaßleistung *f*
s. carburetter Startvergaser *m*
s. circuit Anlasser[strom]kreis *m*, Anlasserschaltung *f*, Anlauf[strom]kreis *m*
s. coil Anlaufspule *f*; Reibungskompensationsspule *f (elektrische Meßtechnik)*
s. connection Anlaßschaltung *f*
s. course Anfangsreihe *f*
s. crank Andrehkurbel *f*
s. crank arm Andrehkurbelarm *m*
s. crank dog Andrehkurbelklaue *f*
s. crank handle Andrehkurbelgriff *m*
s. crank jaw Andrehkurbelklaue *f*
s. crankshaft Andrehkurbelwelle *f*
s. current Einschaltstrom *m*, Anfahrstrom *m*, Anlaßstrom *m*, Anlaufstrom *m*; Anschwingstrom *m*
s. cut Anschnitt *m*
s. delay Einschaltzeit *f*
s. diameter Anfangsdurchmesser *m (Bohrloch)*
s. distance Anlaufstrecke *f*, Anfahrstrecke *f*; Anfahrweg *m (Lokomotive)*
s. dog Andrehklaue *f*
s. electrode Zündelektrode *f*
s. force Anfahrzugkraft *f (Lokomotive)*
s. friction Anlaufreibung *f (Lager)*
s. gate valve *(Wärm)* Anlaßschieber *m*
s. gear Anlaßvorrichtung *f*
s. hand reamer *(Am) s.* s. taper
s. handle Andrehkurbel *f*
s. hole Führungsbohrung *f*, Vorbohrung *f*
s. machine reamer *(Am) s.* s. taper
s. material Ausgangsprodukt *n*, Ausgangsstoff *m*
s. mixture Anlaßgemisch *n*
s. motor Anlasser *m*, Anlaßmotor *m*, Anwurfmotor *m*, Anwerfmotor *m (für Verbrennungsmotoren)*
s. of oscillations Schwingungseinsatz *m*
s. position Anfahrstellung *f*; Ausgangsstellung
s. ram Anlaßhebel *m (Stapellauf)*
s. relay Anlaßrelais *n*, Einschaltrelais *n*
s. resistance Widerstand *m* beim Anlassen, Anlasserwiderstand *m*, Anlaufwiderstand *m*
s. resistance of a valve Anlaufwiderstand *m* einer Röhre
s. resistor Anlaßwiderstand *m*
s. section Ausgangsquerschnitt *m*
s. sheet Mutterblech *n*
s. signal Ausfahrtsignal *n*
s. surface Ausgangsfläche *f*
s. switch Anlaßschalter *m*, Anlasser *m*
s. taper *(Masch)* Anschnitt *m*; Zwischenabschrägung *f*
s. temperature Anlaßtemperatur *f*
s. the hole Anbohren *n*
s. tool Zentrierbohrer *m*
s. torque Anzugsmoment *n*, Anlaufdrehmoment *n*, Anlaufmoment *n*; Anfahrmoment *n*
s. tractive effort Anfahrzugkraft *f*
s. transformer Anlaßtransformator *m*
s.-up Anlassen *n*, Anwerfen *n*; Ingangsetzen *n*
s. valve Anlaßventil *n*
s. velocity Anfahrgeschwindigkeit *f*, Startgeschwindigkeit *f*; Anfangsgeschwindigkeit *f (Geschoß)*
s. voltage Anlaßspannung *f*
s. vortex *(Rak)* Anfahrwirbel *m*, Anfangswirbel *m*
s. winding Anlaßwicklung *f*

starve / to ungenügend Flüssigkeit erhalten; ungenügende Füllung *(Leim usw.)* haben

state Zustand *m*, Beschaffenheit *f*, Erscheinungsform *f*

s. **change** Zustandsänderung *f*
s. **density** Zustandsdichte *f*
s. **diagram** Zustandsdiagramm *n*
s. **formula** Zustandsformel *f*
s. **function** Zustandsfunktion *f*
s. **of aggregation** Aggregatzustand *m*, Formart *f*
s. **of charge** Ladezustand *m*
s. **of combustion** Verbrennungszustand *m*
s. **of conservation** Erhaltungszustand *m*
s. **of deformation** Deformationszustand *m*; Verformungszustand *m*
s. **of disrepair** nicht mehr reparaturfähiger Zustand *m*
s. **of equilibrium** Gleichgewichtszustand *m*
s. **of inertia** Beharrungszustand *m*
s. **of insulation** Isolationszustand *m*
s. **of matter** Aggregatzustand *m*
s. **of resistance** Beharrungszustand *m*
s. **of rest** Ruhezustand *m*
s. **of saturation** Sättigungszustand *m*
s. **of strain** Formänderungszustand *m*, Verzerrungszustand *m*
s. **of stress** Spannungszustand *m*
s. **of the art** Stand *m* der Technik (Forschung)
s. **of weightlessness** Zustand *m* der Schwerelosigkeit
s. **room** Kabine *f*, Kajüte *f*
s. **value** Zustandsgröße *f*

stated point *(Dat)* Festkomma *n*

statement Angabe *f*, Darlegung *f*, Feststellung *f*; Anweisung *f (Rechner)*; Information *f (Kybernetik)*
s. **of account** Kontoauszug *m*

static atmosphärische Störungen *fpl*

static statisch, ruhend
s. **airscrew** *(Flg)* Bremsluftschraube *f*
s. **atom** ruhendes Atom *n*
s. **balance** statisches Gleichgewicht *n*
s. **balancer** Mittelpunkttransformator *m*, Phasenschieber *m*; Ausgleichstransformator *m*
s. **balancing** statische Ausgleichung *f*
s. **balancing apparatus** statischer Ausgleichsapparat *m*
s. **bollard pull** *(Schiff)* Trossenzug *m* am Pfahl, Pfahltrossenzug *m*
s. **casting** Standguß *m*
s. **characteristic** statische Kennlinie *f*
s. **charge** *(El)* statische (ruhende) Ladung *f*
s. **coefficient of friction** Haftreibungskoeffizient *m*
s. **converter** Stromrichter[gruppe *f*] *m*
s. **current** Ruhestrom *m*
s. **discharge head** statische Förderhöhe *f*
s. **disturbance** atmosphärische Störungen *fpl*
s. **electricity** statische Elektrizität *f*
s. **electrification** statische Aufladung *f*
s. **elimination** Störbefreiung *f (Funkempfangstechnik)*
s. **eliminator** Krachtöter *m*; Störbefreiungsschaltung *f (Funkempfangstechnik)*; Antistatikgerät *n*
s. **error** *(BMSR)* Lagefehler *m*, Regelabweichung *f* bei Sollwertänderung, statischer Folgefehler *m*
s. **field** statisches Feld *n*
s. **friction** Haftreibung *f*, Ruhereibung *f*, statische Reibung *f*
s. **head** Druckhöhe *f (als Flüssigkeitssäule ausgedrückter statischer Druck)*
s. **hysteresis** *(Elm)* Wechselfeldhysterese *f*, Wechselfeldhysteresis *f*
s. **load** ruhende (statische) Last *f*
s. **load capacity** statische Tragfähigkeit (Tragzahl) *f*
s. **loading** ruhende (statische) Belastung *f*
s. **luminous sensitivity** statische Lichtempfindlichkeit *f*
s. **moment** statisches Moment *n*
s. **noise** atmosphärischer Störpegel *m (Funkempfangstechnik)*
s. **potential** statisches Potential *n*
s. **pressure** statischer (ruhender) Druck *m*
s. **seal** Abdichtung *f* ortsfester Teile
s. **sensibility** statische Empfindlichkeit *f*
s. **stability** Standsicherheit *f*
s. **steam pressure** statischer Dampfdruck *m*
s. **storage** statischer Speicher *m*
s. **subroutine** statisches Teilprogramm (Unterprogramm) *n (Elektronenrechner)*
s. **suction lift** statische Saughöhe *f*
s. **switching system** *(BMSR)* kontaktlose (ruhende) Steuerung *f*
s. **tensile test** Zerreißprüfung *f* mit ruhender Belastung
s. **tension endurance test** Dauerzugversuch *m* bei ruhender Belastung
s. **test** *(Schiff)* Pfahlprobe *f (eines Schleppers)*; statische Berechnung *f*
s. **test stand** *(Rak)* Brennstand *m*
s. **thrust** *(Schiff)* statischer Schub *m*, Standschub *m (eines Propellers)*
s. **torque** *(El)* Drehmoment *n* bei festgebremstem Läufer
s. **transformer** ruhender (statischer) Transformator *m*
s. **trial** Standprobe *f*

statical analysis statische Berechnung (Untersuchung) *f*
s. **deflection** statische Durchsenkung *f*
s. **induction** Influenz *f*, elektrostatische Induktion *f*

statically balanced system statisch ausgeglichenes System *n*
s. **determinable** statisch bestimmbar

s. determinate statisch bestimmt
s. indeterminable statisch unbestimmbar
s. indeterminate statisch unbestimmt
s. indeterminate to a large degree hochgradig statisch unbestimmt
s. undeterminated statisch unbestimmt
staticizor *(Dat)* Serienparallelwandler und -speicher *m (Flip-Flop-Schaltung)*
statics 1. Statik *f*, Mechanik *f* ruhender Körper; 2. atmosphärische (elektrische) Störungen *fpl*, Luftstörungen *fpl*
station 1. Station *f*; Bahnhof *m*; Stelle *f*; Depot *n*; Aufbewahrungsplatz *m*; 2. *(Masch)* Arbeitsstelle *f*, Spannstelle *f*; 3. trigonometrischer Punkt *m*; Ausgangspunkt *m* für Vermessung; 4. Normalabstand *m (100 Fuß oder 66 Fuß)*
s. bill *(Schiff)* Musterrolle *f*, Mannschaftsverzeichnis *n*
s. calender Tafel *f* der Zugabfahrtszeiten
s. master Stationsvorsteher *m*
s. on the air sendende Funkstation *f*, Funkstation im Sendebetrieb
s. point Standort *m*
s. ringer *(Fmt)* Apparatwecker *m*
s.-to-station call Gespräch *n* zwischen zwei Sprechstellen
stationary ortsfest, unbeweglich, ruhend, [fest-] stehend, Stand-
s.-anode tube Festanodenröhre *f*, Röhre *f* mit fester Anode *(Röntgenröhre)*
s. armature *(El)* ruhender (stationärer) Anker *m*
s. ballast *(Schiff)* fester Ballast *m*
s. battery ortsfeste (stationäre, fixe) Batterie *f*
s. blade Leitschaufel *f (Turbine)*
s. boiler feststehender (stationärer, ortsfester) Dampfkessel *m*, Landkessel *m*
s. centre *(Masch)* feste Körnerspitze *f*
s. coil stehende Spule *f*, Feldspule *f*, Statorspule *f*
s. contact member stehendes (stationäres) Schaltstück *n*, fest[stehend]er Kontakt *m*
s. discharge stationäre Entladung *f*
s. face *(Brg)* gestundeter Streb *m*
s. field *(Ph)* Stehfeld *n*
s. flow stationäre Strömung *f*
s. furnace stationärer Ofen *m*
s. grate *(Wärm)* fester Rost *m*, Festrost *m*
s. housing nicht verschiebbares Gehäuse *n*
s. jaw feststehende Backe *f (Schraubstock)*
s. load Standlast *f*
s. method of flame hardening Aufsatzhärteverfahren *n*, Stillstandshärtung *f*
s. motion stationäre Bewegung *f*
s. net *(Schiff)* Stellnetz *n*, stehendes Netz *n*
s. pattern ruhendes (stillstehendes) Bild *n*
s. picture stehendes Bild *n*
s. point of a curve Rückkehrpunkt *m* einer Kurve
s. position Ruhelage *f*, Ruhestellung *f*
s. screen Festsieb *n*
s. shaft *(Text)* Stehschaft *m*
s. state stationärer Zustand *m*
s. steam engine ortsfeste (stationäre) Dampfmaschine *f*
s. wave stehende Welle *f*
s. winding stillstehende (stehende, stationäre) Wicklung *f*
stationery Schreibmaterialien *npl*
statistical statistisch
s. accuracy statistische Genauigkeit *f*
s. analysis statistische Analyse *f*
s. distribution statistische Verteilung *f*
s. error statistischer Fehler *m*
s. fluctuation statistische Schwankung *f*
s. interpretation statistische Deutung *f*
s. mechanics statistische Mechanik *f*
s. quality control statistische Qualitätskontrolle *f*
s. separation statistische Trennung *f*
s. straggling *(Kern)* statistische Streuung *f*
s. theory of heat statistische Theorie *f* der Wärme
s. thermodynamics statistische Thermodynamik *f*
s. uncertainty statistische Unsicherheit *f*
s. weight *(El)* thermodynamische Wahrscheinlichkeit *f*, statistisches Gewicht *n*
statistician Statistiker[in *f*] *m*
statistics Statistik *f*; statistische Maßzahlen *fpl*
stator 1. *(Elm)* Ständer *m*, Stator *m*; 2. *(Kfz)* Leitrad *n*
s. blade Leitschaufel *f (Turbine)*
s. conductor *(Elm)* Ständerstab *m*
s. current *(Elm)* Ständerstrom *m*, Statorstrom *m*
s. e.m.f. Stator-EMK, induzierte Ständerspannung *f*
s. frame Ständergehäuse *n*
s. lamination Ständerblech[paket] *n*
s.-resistance starter Ständeranlasser *m*
s. slot Ständernut *f*
s. vane Statorplatte *f (Drehkondensator)*
s. winding Ständerwicklung *f*, Statorwicklung *f*, Gehäusewicklung *f*
statoscope Statoskop *n*, Feinhöhenmesser *m*
statute mile *amerikanische Längeneinheit; 1 mi = 1,609 km*
statutory deck-line gesetzliche Deckslinie *f (Freibord wird hiervon abgesetzt)*
s. load-line gesetzliche Tiefladelinie *f (eines Schiffes)*
Stauffer lubricator Staufferbüchse *f*
Stauffer screw-down greaser Staufferbüchse *f*
staunch wasserdicht, luftdicht

staunchness Dichtheit *f*
staurolite *(Min)* Staurolith *m*
stave 1. Daube *f*, Faßdaube *f*; Runge *f*; Sprosse *f*; 2. Triebstock *m*
s. backing and hollowing machine Hobel- und Aussparmaschine *f (Küfnerei)*
s. bending machine Faßdaubenbiegemaschine *f*
s. cross-cut saw Daubenabkürzsäge *f*
s. dressing machine Daubenabrichtmaschine *f*
s. iron Daubenhobelmesser *n*
s. jointing machine Faßdaubenfügemaschine *f*
stay / to stützen, verankern; bleiben; aussetzen; *(Schiff)* wenden
s. put in Stellung bleiben
stay Anker *m*, Bolzen *m*, Knagge *f*, Stempel *m*, Strebe *f*, Stütze *f*, Träger *m*, Zuganker *m*; Verankerung *f (Mast)*; Abspannseil *n*; Ständer *m*; Baumpfahl *m*; *(Schiff)* Stag *n*, Leiter *f*
s. bearing Setzstocklager *n*
s. block Ankerklotz *m*, Ankerpfahl *m (Leitungsbau)*
s. bolt Stehbolzen *m*
s. bolt driver Stehbolzenzieher *m*
s. bolt tap Stehbolzengewindebohrer *m*
s. clamp Kreuzungsklemme *f* für Luftkabeltragseile
s. plate Versteifungsplatte *f*
s.-put switch *(Am)* Schalter *m* ohne Ruhestellung
s. rod Stehbolzen *m*
s. rods terminal pole *(Fmt)* Abspanngestänge *n*
s. wire Ankerdraht *m (vom Mast zum Anker)*, Abspanndraht *m*
stayed mast 1. *(El)* abgespannter Mast *m*; 2. *(Schiff)* abgestagter Mast
s. pole *(Fmt)* Abspannmast *m*; Linienfestpunkt *m*
s. terminal pole *(Fmt) (verankertes)* Endgestänge *n*
staying Abspannung *f*, Verspannung *f*
steadiness Stetigkeit *f*, Gleichmäßigkeit *f*; Standfestigkeit *f*; Zügigkeit *f*; Stabilität *f*, Konstanz *f*
s. of cavity Stetigkeit *f* der Kavitation
s. of the image Bildstabilität *f*
steady / to stabilisieren; in Richtung bringen; zügig (stetig) machen
steady *(Masch)* Setzstock *m*; Lünette *f*
steady gleichmäßig, stetig; ständig; stationär *(Strömung)*; zügig *(Beanspruchung)*; stabil, konstant; fest, standfest; in Richtung
s. anode current Anodenruhestrom *m*
s. arm Seitenhalter *m*
s. beam *(Ph)* konstantes (ständiges) Bündel *n*
s. bearing *(Schiff)* Wellenlager *n (im Tunnel)*
s. cavity stetiger Hohlraum *m (Kavitation)*
s. component Gleichstromkomponente *f*, Gleichstromanteil *m*
s. condition Beharrungszustand *m*
s. current stationärer Strom *m*, Dauerstrom *m*; stationäre Strömung *f*
s. current impulses nicht abklingende Stromimpulse *mpl*
s. current resistance Gleichstromwiderstand *m*
s. dowel Stellstift *m*
s. flow stationäre Strömung *f*
s. hum Dauerton *m (Funknavigation)*
s. load Dauerbelastung *f*, Dauerlast *f*
s. motion stationäre Strömung *f*
s. pin Stellstift *m*, Paßstift *m*
s. plastic flow stationäres plastisches Fließen *n*
s. position Ruhelage *f*, Gleichgewichtslage *f*
s. rest feststehender Setzstock *m*; Lünette *f*
s. source konstante Strahlenquelle *f*
s. span [wire] *(Fmt)* Spanndraht *m* für seitliche Festlegung
s. state eingeschwungener (stationärer) Zustand *m*; Dauerzustand *m*, Beharrungszustand *m*
s.-state condition Beharrungszustand *m*
s.-state current *(El)* eingeschwungener (stationärer) Strom *m*
s.-state error Fehler *m* im eingeschwungenen Zustand, stationäre (bleibende) Regelabweichung *f*
s.-state stability limit Grenze *f* der statischen Stabilität
s. value Ruhewert *m*, stationärer Wert *m*
s. voltage Gleichspannung *f*; konstante Betriebsspannung *f*
steadying circuit *(El)* Beruhigungsglied *n*
s. resistance Beruhigungswiderstand *m*
steam / to dampfen; dämpfen, kochen *(Holz)*; beschlagen
steam Dampf *m*, Wasserdampf *m*
s. accumulator Dampfsammler *m*, Dampfspeicher *m*
s. acting on one side *(Wärm)* einseitige Dampfwirkung *f*
s. admission Dampfzutritt *m*, Dampfeintritt *m*, Dampfeinströmung *f*
s. admission and exhaust port Dampfeintritts- und -austrittskanal *m*
s. admission line Dampfeinströmlinie *f*, Volldrucklinie *f*
s. admission pipe Dampfeinströmungsrohr *n*, Dampfeintrittsrohr *n*
s. admission side Dampfeinlaßseite *f*
s. ager *(Text)* Dämpfer *m*, Dämpfkasten *m*
s. air heater Dampfluftvorwärmer *m*

s. airpump Dampfluftpumpe *f*
s. ash-hoist *(Wärm)* Dampfaschwinde *f*
s. baking oven Dampfbackofen *m*
s. bath Dampfbad *n*
s. bent timber in Dampf gebogenes Spant *n*, eingebogenes Spant
s. blow-off pipe Dampfabblasrohr *n*
s. blower Dampfstrahlgebläse *n*
s. boiler Dampfkessel *m*
s. boiler inspection Dampfkesselrevision *f*, Dampfkesseluntersuchung *f*
s. boiler machine Kesseldampfmaschine *f*
s. boiler working Dampfkesselbetrieb *m*
s. box Dampfkasten *m (zum Holzdämpfen)*
s. bubble Dampfblase *f*
s. calorimeter Kondensationskalorimeter *n*
s. capstan *(Schiff)* Dampfspill *n*
s. chamber Dampfkammer *f*
s. chemicking *(Text)* Dampfbehandlung *f* mit Chemikalien
s. chest Schieberkasten *m*, Dampfkasten *m*, Düsenkasten *m*; Dampfeinlaßbüchse *f (Turbine)*
s.-chest cover *(Schiff)* Schieberkastendeckel *m*
s. cleaner Dampfstrahlreiniger *m*
s. coal Bunkerkohle *f*; Kesselkohle *f*; Eßkohle *f*; kurzflammige Kohle *f*
s. cock Dampfhahn *m*
s. coil Dampfschlange *f*
s. collector Dampfsammler *m*
s. condenser Dampfkondensator *m*
s. conduit Dampfleitung *f*
s. conduit pipe Dampfleitungsrohr *n*
s. connection Dampfstutzen *m*
s. consumption Dampfverbrauch *m*
s. converter Dampfumformer *m*
s. crane Dampfkran *m*
s. cured bedampft, dampferhärtet, mit Dampf nachbehandelt *(Betonbau)*
s. curing Bedämpfung *f*, Dampfbehandlung *f*, Dampferhärtung *f (Betonbau)*
s. cylinder Dampfzylinder *m*
s.-cylinder lubricator Dampfzylinderschmierapparat *m*
s.-cylinder oil Dampfzylinderöl *n*
s. density Dampfdichte *f*
s.-developing process *(Text)* Dämpfentwicklungsverfahren *n*
s. distillation Wasserdampfdestillation *f*
s.-distributing pipe Dampfverteilungsrohr *n*
s. distribution Dampfverteilung *f*
s. distribution passage Dampfverteilungskanal *m (Turbine)*
s. dome Dampfdom *m*, Dampfsammler *m*
s. drier Dampftrockner *m*
s. drifter Dampflogger *m*
s. drive Dampfantrieb *m*
s.-driven dampfangetrieben, mit Dampfantrieb
s.-driven dynamo Dampfdynamo *m (direkt gekuppelter Dynamo)*
s. drum Obertrommel *f*, Dampfsammler *m*
s. drying apparatus Dampftrockner *m*
s. dynamo Dampfdynamo *m (direkt gekuppelter Dynamo)*
s. economy Dampfersparnis *f*
s. ejector Dampfbläser *m*
s. end Dampfteil *m (Pumpe, Turbine usw.)*
s. engine Dampfmaschine *f*
s. engine control gear Dampfmaschinensteuerung *f*
s. engine driving shafting Transmissionsdampfmaschine *f*
s. engine part Dampfmaschinenteil *n*
s. engine plant Dampfmaschinenanlage *f*
s. engine with drop valve gear Ventildampfmaschine *f*, Dampfmaschine *f* mit Ventilsteuerung
s. engine with fixed expansion Dampfmaschine *f* mit fester Expansion
s. engine with oscillating cylinders schwingende Dampfmaschine *f*, Dampfmaschine mit schwingenden Kolben
s. engine with receiver Aufnehmerdampfmaschine *f*, Receiverdampfmaschine *f*
s. engine with slide valve Schieberdampfmaschine *f*, Dampfmaschine *f* mit Schiebersteuerung
s. engine working expansively Expansionsdampfmaschine *f*
s. engine working without expansion Volldruckdampfmaschine *f*, Dampfmaschine *f* ohne Expansion
s. exhaust Dampfaustritt *m*, Dampfausströmung *f*
s. exhaust pipe Dampfausströmungsrohr *n*
s. exhaust side Dampfauslaßseite *f*
s. expansion Dampfdehnung *f*
s.-fan furnace *(Wärm)* Dampfschleierfeuerung *f*
s. feedpump Dampfspeisepumpe *f*
s. fire-extinguishing plant Dampffeuerlöschanlage *f*
s. for heating purposes Heizdampf *m*
s. friction loss Dampfreibungsverlust *m (Turbine)*
s. friction work Dampfreibungsarbeit *f (Turbine)*
s. funnel dampfbeheizter Trichter *m*
s. gauge Dampfkesselmanometer *n*
s.-generating plant Dampferzeugungsanlage *f*
s. generator Dampferzeuger *m*
s. hailer Dampfheuler *m*
s. hammer Dampfhammer *m*; Dampframme *f (Pfahlgründungen)*; Dampfhammerdampfmaschine *f*

s.-heated can *(Text)* Dampftrockenzylinder *m*, Dampftrockentrommel *f*
s.-heated cylinder *(Text)* Trockentrommel *f*, Trockenzylinder *m*
s.-heated jacket Dampfheizmantel *m*
s.-heated press plate mit Dampf heizbare Preßplatte *f (Sperrholz, Furniere)*
s. heater Dampfüberhitzer *m*
s. heating Dampfheizung *f*
s. heating coil Dampfheizschlange *f*
s. hose Dampfschlauch *m*
s. hydraulic dampfhydraulisch
s.-hydraulic steering gear dampfhydraulische Ruderanlage *f*
s. indicator diagram Dampfdruckdiagramm *n*, Dampfdruckschaubild *n*
s. inlet valve Dampfeinlaßventil *n*
s. jacket Dampfmantel *m*, Dampfhemd *n*, Zylinderheizraum *m*
s. jet Dampfstrahl *m (Turbine)*
s.-jet air ejector Dampfstrahlluftsauger *m*
s.-jet atomizer Dampfstrahlzerstäuber *m*
s.-jet blower Dampfstrahlgebläse *n*
s.-jet burner Dampfstrahlbrenner *m*
s.-jet-operated air ejector Dampfstrahler *m*, Dampfstrahlluftpumpe *f*
s.-jet pump Dampfstrahlpumpe *f*
s.-jet sprayer Dampfstrahlzerstäuber *m*
s. jib crane Dampfauslegerkran *m*
s. lap *(Wärm)* Einlaßüberdeckung *f*, äußere Überdeckung *f*
s. leaking Dampfentweichung *f*
s. lighter Dampfprahm *m (Baggerprahm)*
s.-limit curve *(Wärm)* Dampfgrenzkurve *f*
s. line Dampfleitung *f*
s. locomotion Dampfzugförderung *f*
s. locomotive Dampflokomotive *f*
s. lubrication Dampfschmierung *f*
s. lubrication apparatus Dampfschmierapparat *m*
s. navigation Dampfschiffahrt *f*
s. navvy Trockenbagger *m*
s. nozzle Dampfdüse *f*
s. packer Dampffüllapparat *m (Hackmaschine)*
s. passage Dampfkanal *m*, Dampfweg *m*
s. pile-driver Dampframme *f*
s. pinace Dampfpinasse *f*
s. pipe Dampf[leitungs]rohr *n*, Dampfleitung *f*
s. piston Dampfkolben *m*
s. piston valve Dampfkolbenventil *n*
s. platen drier Dampfplattentrockner *m*
s. plough Dampfpflug *m*
s. point Siedepunkt *m* des Wassers
s. port Dampf[austritts]öffnung *f*, Dampfweg *m*, Dampf[eintritts]kanal *m*
s. port side Dampfeinlaßseite *f*
s. power Dampfkraft *f*
s. power plant Dampfkraftwerk *n*; Dampfkraftanlage *f*
s. power station Wärmekraftwerk *n*
s. pressure Dampfdruck *m*, Dampfspannung *f*
s. pressure above atmospheric Dampfüberdruck *m*
s. pressure area Dampfdruckfläche *f*
s. pressure diagram Dampfdruckdiagramm *n*
s. pressure reducer Dampfdruckminderer *m*
s. pressure reducing valve Druckminderungsventil *n*, Reduktionsventil *n*, Reduzierventil *n*
s. pressure regulator Dampfdruckregler *m*
s. purifier Dampfreiniger *m*
s. purple Dampfpurpur *m (Indigofarbstoff)*
s. raising Dampfaufmachen *n*; Dampferzeugung *f*
s.-raising plant Dampferzeugungsanlage *f*
s. rate erzeugte Dampfmenge *f*
s.-reducing valve Dampfdruckminderventil *n*
s. reheat[ing] Dampfzwischenüberhitzung *f*
s. retting *(Text)* Wasserdampfröste *f*
s. reversing gear *(Wärm)* Dampfumsteuerung *f*
s. roller Dampfwalze *f*
s. saving Dampfersparnis *f*
s. saw mill Dampfsägemühle *f*
s. separator Dampfabscheider *m*
s. service Dampfbetrieb *m*
s.-set ink *(Typ)* Dampftrockenfarbe *f*
s. shovel *s.* s. navvy
s. shutting-off device Dampfabsperrorgan *n*
s. signal gear *(Eis)* Dampfläutewerk *n*
s. siren Dampfsirene *f*
s. slide valve Dampfschieber *m*
s. smothering Dampffeuerlöschung *f*
s. smothering line Dampffeuerlöschleitung *f*
s. space Arbeitsraum *m (Zylinder)*
s. steering gear Dampfsteuerapparat *m*, Dampfrudermaschine *f*
s. stop valve Dampfabsperrventil *n*; Dampfabsperrung *f*
s. strainer Dampfreinigungssieb *n (Turbine)*
s. superheater Dampfüberhitzer *m*
s. superheating Dampfüberhitzung *f*
s. supply Dampfzufluß *m*, Dampfzuführung *f*, Dampfzuleitung *f*, Dampfzutritt *m*
s. supply pipe Dampfzuleitungsrohr *n*
s. supply valve Dampfeinlaßventil *n*; Dampfentnahmeventil *n*
s. sweat *(Led)* Dampfschwitze *f*
s. syringe *(Wärm)* Dampf[feuer]spritze *f*
s. throttling Dampfdrosselung *f*
s.-tight *(Wärm)* dampfdicht
s. tiller Dampfpinne *f (Ruderpinne mit darauf montierter Antriebsmaschine)*
s. transformer Dampfumformer *m*
s. trap Kondenstopf *m*, Wasserabscheider *m*, Kondenswasserabscheider *m*

s. **trawler** Fischdampfer *m*
s. **tug** Dampfschlepper *m*, Schleppdampfer *m*, Bugsierdampfer *m*
s. **turbine** Dampfturbine *f*
s. **valve** Dampfabsperrventil *n*
s. **vessel** Dampfschiff *n*
s. **vulcanizer** Dampfvulkanisator *m*
s. **wharf** Dampferkai *m*
s. **whistle** Dampfpfeife *f*
s. **winch** Dampfwinde *f*
s. **windlass** Dampfwinde *f*
steamer Dampfer *m*
s. **column** *(Lw)* Dämpfkolonne *f*
steaming Dampfen *n*; Dämpfen *n*, Dämpfung *f*, Kochen *n (Holz)*
s. **boiler** Dampfkessel *m*
s. **duty** Verdampfungsvermögen *n*
s. **light** Dampferlicht *n*, Topplicht *n*, Topplaterne *f*
s. **pressing machine** *(Text)* Dekatiermaschine *f*
s. **radius (range)** Aktionsbereich *m*, Fahrbereich *m (eines Schiffes)*
s. **rate** Dampferzeugung *f*, Dampfleistung *f*, erzeugte Dampfmenge *f*
steamproof dampfdicht
steamship Dampfschiff *n*
steamy dampfig, dunstig
stearate Ester *m* (Salz *n*) der Stearinsäure, Stearat *n*
stearic acid Stearinsäure *f*
stearin Stearin *n*
s. **candle** Stearinkerze *f*
stearylic alcohol Stearylalkohol *m*
steatite *(Min)* Steatit *m*, Speckstein *m*, Talk *m*
Steckel mill Steckelwalzwerk *n*
steel / to verstählen
s. **face** verstählen
steel 1. Stahl *m*; 2. Gesteinsbohrer *m*
s. **alloy** legierter Stahl *m*, Sonderstahl *m*
s. **and iron industry** Eisenhüttenindustrie *f*
s. **angle** *(Bw)* L-Stahl *m*, Winkelstahl *m*
s. **armour** Stahlpanzer *m*
s.**-backed bearing** Stahlstützschalenlager *n*
s. **ball** 1. *(Met)* Stahlluppe *f*; 2. Stahlkugel *f (Werkstoffprüfung)*
s. **band** Stahlband *n*
s. **band armouring** Stahlbandbewehrung *f*
s. **band conveyor** Stahlbandförderer *m*
s. **band governing** Stahlbandregelung *f (Turbine)*
s. **bar** Stahlschiene *f*
s. **beam** Stahlträger *m*
s. **belt** Stahlband *n*
s. **belt conveyor** Stahlbandförderer *m*
s. **belt drive** Stahlbandantrieb *m*
s. **blade** Stahlklinge *f*, Stahlmesser *n*
s. **body** Eisenkasten *m (Eisenbahnwaggon)*
s.**-body tipping wagon** Stahlmuldenkipper *m*
s. **brush** Drahtbürste *f*
s. **casting** Stahlgußstück *n*; Gußstahl *m*
s.**-casting foundry** Stahlgießerei *f*
s. **casting with milled nozzles** Stahlgußstück *n* mit eingefrästen Düsen *(Turbine)*
s.**-clad** gepanzert
s. **conduit** Stahlpanzerrohr *n*
s. **construction** Stahlbau *m*
s. **container** Stahlblechzellenkasten *m (der Batterie)*; Stahlbehälter *m*
s. **converting furnace** *(Met)* Zementierofen *m*
s.**-core bullet** Stahlkerngeschoß *n*
s. **cored aluminium conductor** Stahlaluminiumseil *n*, Staluseil *n (Freileitung)*
s. **cotter** Stahlkeil *m*
s. **cutting wheel** Stahlrädchen *n (Glasschneider)*
s. **cylinder** Stahlflasche *f*
s. **disk** Stahlscheibe *f*; Reibtrennsägeblatt *n*
s. **electrode** Stahlelektrode *f*
s. **engraving** Stahlgravierung *f*
s. **facing** galvanische Verstählung *f*
s. **finery** Stahlfrischherd *m*
s. **flooring** *(Eis)* Stahlabdeckung *f*
s. **foundry** Stahlgießerei *f*
s. **frame** Stahlskelett *n*
s. **framing** Stahlskelett *n*; Stahlskelettbau *m*
s. **friction disk metal sawing machine** Reib- und Schnelltrennsäge *f*
s. **furnace** Stahlofen *m*
s. **furniture** 1. *(Typ)* Stahlstege *mpl*, Formatstege *mpl*; 2. Stahlmöbel *pl*
s. **girder** [zusammengesetzter] Stahlträger *m*
s. **grade** Stahlsorte *f*
s. **grit** Stahlkies *m*
s. **ingot** Rohstahlblock *m*
s. **ladle** Stahlgießpfanne *f*
s. **lamination coupling** Stahllamellenkupplung *f*
s. **lattice mast** Stahlgittermast *m*
s. **letter** Buchstabenstempel *m*
s. **liner** Stahlbüchse *f*
s. **magnet** Stahlmagnet *m*
s. **maker** Stahlwerker *m*
s. **making** Stahlgewinnung *f*
s.**-making furnace** Stahlschmelzofen *m*
s.**-making process** Stahlgewinnungsverfahren *n*
s. **measuring tape** Stahlbandmaß *n*
s. **melting furnace** Stahlschmelzofen *m*
s. **mesh fabric** Bewehrungsmatte *f (Stahlbetonbau)*
s. **metallurgist** Eisenhüttenmann *m*, Stahlwerker *m*
s. **mill** Stahlwerk *n*, Eisenhüttenwerk *n*
s. **mould** Stahlgußform *f*
s. **pellets** Stahlkies *m*

s. pin Stahlbolzen *m*
s. plate Stahl[grob]blech *n*, Stahlmittelblech *n*
s.-plate conveyor Plattenband *n*, Stahlglieder[förder]band *n*
s. plating 1. Verstählung *f*; 2. Panzerung *f*
s. punch *(Typ)* Stahlpunze *f*
s. refining ore Stahlwerksfrischerz *n*
s. reinforcing worker Eisenbieger *m*, Eisenflechter *m* *(Stahlbetonbau)*
s. rods Stabstahl *m*
s. rope Stahlseil *n*
s.-rope drive Drahtseiltrieb *m*
s. rule Stahllineal *n*
s. scrap Stahlschrott *m*
s. scriber Stahlreißnadel *f*
s. section Profilstahl *m*; Stahlprofil *n*
s. set *(Brg)* Stahltürstock *m*
s. sheet Stahlblech *n* *(Feinblech, Mittelblech)*
s. sheet piling Spundwand *f*
s. shelf Stahlregal *n*
s. shipbuilding Stahlschiffbau *m*
s. shot Stahlkies *m*, Stahlsand *m*
s. sleeper *(Eis)* Stahlschwelle *f*
s. slitting shear Hebelblechschere *f*
s. socket Stahlmuffe *f*; Stahlbuchse *f*, Stahlhülse *f*
s.-spoked wheel Stahlspeichenrad *n*
s. spring Stahlfeder *f*
s. square Anschlag[stahl]winkel *m*, Stahlwinkel *m*
s. straight edge Stahllineal *n*
s. string Stahlsaite *f*
s. strip Bandstahl *m*
s. tank rectifier *(Elm)* Eisengleichrichter *m*
s. tape dünnes Stahlband *n*
s. tape [for measuring] Stahlbandmaß *n*
s. tape measure Stahlbandmaß *n*
s. tape rule Stahlbandmaß *n*
s. torsion bar Stahldrehstabfeder *f*
s. tower Gittermast *m*
s. trellis-work Stahlgitter *n*
s. try square Anschlag[stahl]winkel *m*
s.-tube mercury thermometer Stahlquecksilberthermometer *n*
s.-using industry eisenverarbeitende Industrie *f*
s. wire Stahldraht *m*
s. wire armoured cable stahldrahtbewehrtes Kabel *n*
s. wire gauge Blechlehre *f*
s. wool Stahlwolle *f*
s. works Stahlwerk *n*
steelify / to verstählen
steeliness Stahlhärte *f* *(Stahlähnlichkeit)*
steely stählern, stahlartig, stahlhart
s. iron niedrig gekohltes Eisen *n*
steelyard Laufgewichtswaage *f*
steening *s.* steining
steep / to einquellen; [ein]tauchen; einweichen; [be]netzen; einspritzen, bespritzen, besprengen; quellen; tränken, imprägnieren; *(Led)* äschern
s. and full einweichen und walken *(Leder)*
s. in lime water einkalken *(Gerberei)*
s. in mordant beizen
s. in tan *(Led)* lohen, lohgar machen
s. the hides in a week ooze Leder in Treiblauge legen
steep 1. steiler Abhang *m*; 2. Flachsröste *f*
steep steil, schroff
s. angle taper Steilkegel *m*
s.-bank flight *(Flg)* steiler Kurvenflug *m*
s. edge steile Flanke *f* *(Impuls)*
s.-fronted steil ansteigend *(Impuls)*
s. side steile Flanke *f* *(Impuls)*
s.-sight prism *(Opt)* Steilsichtprisma *n*
s. slope steiler Abfall *m*; steile Signalflanke *f*
s.-spiral steilgängig; schäl-spiralgenutet *(Reibahle)*, mit steiler Spirale
s.-spiral reamer Schälreibahle *f* [mit großem Drall]
s.-spiral taper reamer Kegelschälreibahle *f*
s. taper steiler Kegel *m*, Steilkegel *m*
steepen / to versteilern *(z. B. Signalflanke)*
steeper Gefäß *n* zum Einquellen (Imprägnieren, Einweichen, Netzen)
steeping 1. Imprägnierung *f*, Tränkung *f*; Einweichen *n*; Eintauchen *n*; Netzen *n*; Quellen *n*; 2. *(Text)* Rösten *n*, Flachsröste *f*
s. bath *(Text)* Abkochbad *n*
s. bowl Einweichbottich *m* *(für Wolle)*
s. of flax Flachsrösten *n*, Flachsrotten *n*
s. pond Röstgrube *f*, Rottgrube *f*
s. press Tauchpresse *f* *(Chemiefaserherstellung)*
steeple Spitzturm *m*
s. head spitzer Nietkopf *m*
s. roof Turmdach *n*
steeply-falling side *(Fmt)* steile Signalflanke *f*
s. sloping [downward] steil abfallend
steepness Steilheit *f*
steer / to steuern, lenken
s. for Kurs nehmen auf
s. towards ansteuern
steerability Lenkbarkeit *f*; Steuerbarkeit *f*
steerable lenkbar, steuerbar
s. aerial einstellbare Antenne *f*
steerage *(Schiff)* 1. Steuervermögen *n*, Steuerfähigkeit *f*; 2. Zwischendeck *n*
s.-way Kielwasser *n*
steering Lenkung *f*
s. arm Lenkhebel *m*, Lenkschenkel *m*
s.-arm stop Lenkhebelanschlag *m*
s. axle Lenkachse *f*
s. bolster *(Lw)* Lenkschemel *m*
s. box Lenkgehäuse *n*
s. cam Lenkschraube *f*

s. column Lenksäule *f*, Lenkspindel *f*
s.-column assembly Lenkstock *m*
s.-column bearing Lenkrohrlager *n*
s.-column bracket Lenksäulenhalter *m*
s.-column gear-change Lenkstockschaltung *f*
s.-column lock Lenkschloß *n*
s.-column tube Lenksäulenrohr *n*
s. control arm Lenkhebel *m*
s. control shaft Lenkachse *f*
s. drag link Lenkschubstange *f*
s. drop arm Lenkhebel *m*
s. finger Lenkfinger *m*
s. fork Lenkgabel *f*
s. gear 1. Lenkgetriebe *n*; 2. *(Schiff)* Rudermaschine *f*, Rudergeschirr *n*
s. gear arm Lenkstockhebel *m*
s. gear box Lenkgehäuse *n*
s. gear case Steuergehäuse *n*
s. gear compartment (flat) *(Schiff)* Rudermaschinenraum *m*
s. gear housing Lenkgehäuse *n*
s. gear mounting Lenkungsbock *m*
s. idler Ablenkrolle *f*, Lenkrolle *f*
s. knuckle Vorderachsschenkel *m*, Lenkhebel *m*
s.-knuckle pillar Achsschenkellenker *m*
s.-knuckle pin Lenkzapfen *m*, Vorderradlenkzapfen *m*, Steuerzapfen *m*
s.-knuckle pivot Vorderachszapfen *m*
s.-knuckle shaft Fingerhebelwelle *f*
s. lever Lenkhebel *m*
s. linkage Lenkgestänge *n*
s. lock Lenkungsausschlag *m*, Lenkungseinschlag *m*, Volleinschlag *m* *(der Vorderräder)*
s. lock angle Einschlagwinkel *m* der Vorderräder
s. manoeuvre *(Schiff)* Rudermanöver *n*
s. mechanism Lenkeinrichtung *f*; Steuermechanismus *m*
s. nut Lenkmutter *f*
s. pillar Lenksäule *f*
s. pilot Schaltkreis *m* einer Selbststeueranlage
s. pivot Lenkzapfen *m*
s. pivot pin Achsbolzen *m*
s. play toter Gang *m* der Lenkung
s. qualities *(Schiff)* Steuerfähigkeit *f*, Kursbeständigkeit *f*
s. rods Lenkorgane *npl*
s. roller Lenkrolle *f*
s. screw Lenkspindel *f*
s. segment Lenksegment *n*
s. shaft sector Lenkstockspindelsegment *n*
s. stability Spurhaltigkeit *f*
s. stop Lenk[ungs]anschlag *m*
s. swivel Achsschenkel *m*
s. swivel bolt Achsschenkelbolzen *m*
s. swivel bush Achsschenkelbuchse *f*
s. swivel pin Achsschenkelbolzen *m*
s. track rod Lenkspurstange *f*, Lenkschubstange *f*
s. tube Lenkrohr *n*
s. tube extension Lenkrohrstummel *m*
s. valve Lenkungsventil *n*
s. wheel Steuerrad *n*, Lenkrad *n*
s.-wheel lock Lenkungsschloß *n*, Lenkradschloß *n*
s. wobble Schlagen *n* der Lenkung
s. worm Lenkschnecke *f*
s. worm wheel Lenkschneckenrad *n*
Stefan-Boltzmann constant Stefan[-Boltzmann]-Konstante *f*
Stefan-Boltzmann law Stefan-Boltzmannsches Gesetz *n (der Strahlung)*
steining Ausmauern (Verschalen) *n* eines Brunnens; Brunneneinfassung *f*
stellar stellar, Stern-
s. body Himmelskörper *m*
s. constellation Sternbild *n*
s. guidance *(Rak)* Astrolenkung *f*
s. magnetic field stellares Magnetfeld *n*
s. matter Sternmaterie *f*
s. parallax Sternparallaxe *f*
s. spectrum Sternspektrum *n*
s. time Sternzeit *f*
stellerite *(Min)* Stellerit *m*
stellite Stellit *n (Hartmetallegierung)*
s. tipping Hartmetallbestückung *f*
stellited tool mit Stellit bestücktes Werkzeug *n*
stelznerite *(Min)* Stelznerit *m*
stem / to entstielen, entrippen; anstauen; dichten; verstopfen, besetzen *(Bohrloch)*
stem 1. Stengel *m*; Stamm *m*; Stiel *m*; Rippe *f (Tabak)*; 2. Steg *m (Träger, Plattenbalken)*; Rippe *(Balken)*; Schaft *m*; Stift *m*; Stutzen *m*; Griff *m*; Dorn *m*; 3. *(BMSR)* Stengel, Schiebestange *f*, Stange *f (einer Taste)*; 4. *(Text)* oberer Teil *m* des Nadelschaftes; 5. *(El)* Elektrodenfuß *m*, Fuß *m*; Röhrenfuß *m (Elektronenröhre)*; Lampenfuß *m*; Querglied *n*; Querwiderstand *m*; 6. *(Schiff)* Vorschiff *n*, Vor[der]steven *m*; 7. Beladung *f* eines Schiffes *(zu einer bestimmten Zeit)*; 8. Kolbenstange *f (Ventil)*; 9. Röhre *f (Thermometer)*
s. correction Fadenkorrektur *f (Thermometer)*
s. fibre Stengelfaser *f*
s. foot Vorstevenschuh *m*, Stevenfuß *m*
s. guide Schaftführung *f (Ventil)*
s. plate Vorstevenplatte *f*
s. post Vor[der]steven *m*
s. radiation *(Kern)* Gehäusestrahlung *f*, Stielstrahlung *f*
s. rot *(Lw)* Stengelfäule *f*
s. timber Langnutzholz *n*, Stammholz *n*
stemming Besatz[material *n*] *m* *(für Bohrlöcher)*

stempost Vor[der]steven *m*
stench trap Geruchverschluß *m* *(Sanitärtechnik)*
stencil / to schablonieren
stencil *(Typ)* Schablone *f*; Matrize *f*, Wachsmatrize *f*, Vervielfältigungsmatrize *f*
  s. paint Schablonenfarbe *f*
  s. printing *(Text)* Drucken *n* mit gravierten Walzen
stenter [frame] *(Text)* Spannrahmen *m*
  s. hook *(Text)* Kluppe *f*
  s. with lateral ventilation *(Text)* Querluftspannrahmen *m*
stentering *(Text)* Aufrahmen *n*, Aufspannen *n*
step / to [ab]stufen; aufstufen, stufenweise aufsetzen; absetzen; überhöhen; stufenweise regeln, in Stufen einstellen
  s. down herabsetzen; *(El)* erniedrigen *(Spannung)*, heruntertransformieren
  s. drill stufenweise vorbohren
  s. off abstufen
  s. on the clutch pedal auskuppeln, auf die Kupplung treten
  s. on the gas *(sl)* den Gashebel durchtreten
  s. up erhöhen *(Produktion, Zuwachsrate, Spannung)*
  s. up the voltage hochtransformieren
step Schritt *m*; Stufe *f*; Absatz *m*; Sprosse *f*; Tritt[brett *n*] *m*; Setztreppe *f*; Schwelle *f* *(Impulstechnik)*
  s.-and-repeat machine *(Typ)* Kopiermaschine *f*
  s. attenuator *(El)* Stufenteiler *m*
  s.-back Abtreppung *f*; Absatz *m*
  s.-back relay Rückschaltrelais *n*
  s.-back welding Pilgerschrittschweißung *f*
  s. bearing Spurlager *n*, Axiallager *n*; Stützlager *n*; Lagerpfanne *f*
  s. board Fußbrett *n*
  s. bolt Flachrundschraube *f* mit Vierkantansatz
  s.-bottomed boat Stufenboot *n*
  s.-by-step stufenweise, schrittweise
  s.-by-step action schrittweise Wirkung *f*, Weiterschaltung *f*; Schrittschaltwerk *n*
  s.-by-step automatic system *(Fmt)* Schrittwählersystem *n*
  s.-by-step control Schrittregelung *f*
  s.-by-step dial system *(Fmt)* Schrittwählersystem *n*
  s.-by-step excitation stufenweise Anregung *f*
  s.-by-step regulation stufenweise Regelung *f*
  s.-by-step selection *(Fmt)* Schrittwahl *f*
  s.-by-step switch Schrittschalter *m*
  s.-by-step system Folgeregler *m* mit Schrittmotoren
  s. chuck Stufenfutter *n* *(Spannfutter)*
  s. cone [pulley] Stufenscheibe *f* *(Antrieb)*
  s.-cone pulley-driven mit Stufenscheibenantrieb
  s. control Schrittregelung *f*
  s. counterbore Profilstufensenker *m*, mehrstufiges Senkwerkzeug *n*
  s.-down Herabsetzung *f*; Übersetzung *f* ins Langsame
  s.-down [spannungs]erniedrigend
  s.-down amplifier Schwellenverstärker *m* mit negativer Anode
  s.-down gear Reduktionsgetriebe *n*, Übersetzungsgetriebe *n* ins Langsame
  s.-down ratio Untersetzungsverhältnis *n*
  s.-down transformation Abwärtstransformierung *f*
  s.-down transformer Abwärtstransformator *m*, Abspanner *m*, Reduziertransformator *m*, spannungserniedrigender Transformator *m*
  s. drilling stufenweises Vorbohren *n*
  s. filter *(Opt)* Stufenfilter *n*
  s. function Sprungfunktion *f*, Treppenfunktion *f*
  s.-function response Übergangsfunktion *f*, Sprungantwort *f*; Einschwingverhalten *n* bei Sprungfunktionen
  s.-function response characteristic Sprungcharakteristik *f*
  s.-function signal *(Fs)* Einheitsnadelimpuls *m*, Einheitsstoß *m*
  s. grate *(Wärm)* Treppenrost *m*, Stufenrost *m*, Etagenrost *m*
  s. grate furnace Treppenrostfeuerung *f*
  s. grate side bearers Treppenwange *f*
  s. hardening *(Met)* Stufenhärtung *f*
  s. jaw Stufenspannbacke *f* *(Spannfutter)*
  s. lens *(Opt)* Stufenlinse *f*
  s. photometer *(Opt)* Stufenfotometer *n*
  s. pulley Stufenscheibe *f*
  s. rocket Stufenrakete *f*, mehrstufige Rakete *f*
  s. switch Stufenschalter *m*
  s. transformer Kaskadentransformator *m*, Kaskadenwandler *m*, Stufentransformator *m*
  s. twist *(Am)* *s.* binomial twist
  s.-up [spannungs]erhöhend
  s.-up ratio Übersetzungsverhältnis *n*
  s.-up transformer Aufwärtstransformator *m*, Aufspanner *m*, Aufwärtsübertrager *m*, spannungserhöhender Transformator *m*
  s. voltage regulator Spannungsstufenregler *m*
  s. wedge *(Fs)* Stufenkeil *m*
stephanite *(Min)* Stephanit *m*
stepless stufenlos
  s. speed changing stufenlose Geschwindigkeitsänderung (Drehzahländerung) *f*
steplessly adjustable stufenlos einstellbar
  s. variable stufenlos veränderlich
stepney [speichenloses] Ersatzrad *n*

stepped abgetreppt, treppenförmig; [ab]gestuft; [mehrfach] abgesetzt *(im Durchmesser) (Werkzeuge, Bohrung)*
s. bore abgesetzte Bohrung *f*
s. bulkhead *(Schiff)* Stufenschott *n*, versetztes Schott *n*
s. cavity mould *(Plast)* Form *f* zum Pressen tiefer Teile mit dünnen Wänden
s. cone Stufenscheibe *f*
s. curve distance-time protection Distanzschutz *m* mit Stufenkennlinie
s. cutting edge Schneide *f* mit Spanbrecherstufen
s. diameter mit abgestuftem Durchmesser
s. drill Stufenbohrer *m*
s. drum Stufentrommel *f*, Stufenwalze *f* *(Turbine)*
s.-face overhand stoping *(Brg)* Firstenstoßbau *m* mit sägeblattartigem Verhieb
s. gear cutter gestufter Zahnformfräser *m*
s. gear set Stufenrädersatz *m (Verzahnung)*
s. grate *s.* step grate
s. hardening Warmbadhärtung *f*
s. hole Ansatzbohrung *f*
s. pulley Stufenscheibe *f*
s. quenching gebrochenes Härten *n*
s. roll Staffelwalze *f*, Stufenwalze *f*
s. sector Stufensektor *m*
s. through hole Ansatzdurchgangsbohrung *f*
s. wedge Stufenkeil *m*
s. winding *(Elm)* Treppenwicklung *f*
stepping 1. *(Bw)* Abtreppung *f*; 2. *(Masch)* treppenförmiger Anschliff *m (Bohrer)*; [mehrfache] Abstufung *f* [im Durchmesser]; Überhöhung *f*
s. motor Schrittmotor *m*
s. of the drill point treppenförmiger Anschliff *m* des Schneidkopfes
s. operation Schrittschaltoperation *f*
s. pulse Fortschaltimpuls *m*
s. relay Schrittschaltrelais *n*, Schrittschaltwerk *n*
s. switch Schrittwähler *m*
stepwise schrittweise; stufenweise
sterad *s.* steradian
steradian *(Math)* Steradiant *m*, Sterad *m (Einheit des Raumwinkels)*
s. measure Raumwinkelmaß *n*
steradiancy Richtstrahlungsdichte *f (Beleuchtungstechnik)*
stere *Volumeneinheit der Forstwirtschaft; 1 st = 1 m^3*
stereo 1. *(Typ)* Stereo[platte *f*] *n*, Klischee *n*; 2. Stereo *(Rundfunk)*
s. amplifier Stereoverstärker *m*
s. attachment Stereovorsatz *m*
s.-mate Raumklangbox *f*
s.-telescope Scherenfernrohr *n*
stereobate *(Bw)* Stereobat *m*
stereochemical stereochemisch
stereochemistry Stereochemie *f*
stereogram, stereograph Stereobilder *npl*, Bildpaar *n*, Raumbild *n*
stereomicroscope Stereomikroskop *n*
stereopair *(Opt)* Bildpaar *n*
stereophonic method of sound reproduction stereofonische Schallwiedergabe *f*
s. sound Raumton *m*
s. sound system stereofone Übertragungsanlage *f*
stereoprojector Stereoprojektor *m*
stereoscope Stereoskop *n*
stereoscopic camera Stereokamera *f*
s. mapping Stereokartierung *f*
s. mapping instrument Zweibildkartiergerät *n*
s. rangefinder Raumbildentfernungsmesser *m*
s. television dreidimensionales (plastisches, stereoskopisches) Fernsehen *n*
stereotype / to *(Typ)* klischieren
stereotyped command *(Dat)* Standardbefehl *m*
steric hindrance *(Ch)* sterische Hinderung *f*
sterilize / to entkeimen, sterilisieren
sterilizer Sterilisator *m*, Sterilisierapparat *m*
sterlingite *(Min)* Zinkit *m*, Rotzinkerz *n*
stern *(Schiff)* Heck *n*, Gatt *n*
s. anchor Heckanker *m*
s. bearing Stevenlager *n*
s. board Rückwärtsbewegung *f* des Schiffes
s. boat Heckboot *n*
s. boss Schraubenöffnung *f*, Hinterstevennuß *f*
s. bush Stevenrohrbuchse *f*, Stevenrohrlager *n*
s. chute Heckaufschleppe *f*
s. davit Heckdavit *m (n)*
s. flag Heckflagge *f*
s. frame Achtersteven *m*, Hintersteven *m*, [stählernes] Heckspant *n*
s. gland Sternbuchse *f*, Stevenrohrbuchse *f*
s. hawse Heckklüse *f*
s. hawse hole Heckankerklüse *f*
s. light Hecklaterne *f*
s. port Hinter[lade]pforte *f (Ladeöffnung im Heck eines Schiffes)*
s. post *s.* sternpost
s. recess Heckschlitz *m (Eisbrecher)*
s. rudder Heckruder *n*
s. sheave Heckscheibe *f*
s. tank Hecktank *m*
s. timber [hölzernes] Heckspant *n*
s. transom Heckspiegel *m*
s. trawler Hecktrawler *m*
s. tube Wellenaustrittsrohr *n*

s. tube bearing Stevenrohrlager *n*
s. tube bulkhead Stopfbuchsenschott *n*
s. tube bushing Stevenrohrlager *n*
s. wave Heckwelle *f*, Hecksee *f*
s.-well dredger Bagger *m* mit Heckschlitz
s. wheeler Heckradschiff *n*
s. wheeler steamer Heckraddampfer *m*
sternpost Achtersteven *m*, Hintersteven *m*, Rudersteven *m* *(ohne Propellerbrunnen)*
sternutator Nasenreizstoff *m*, Rachenreizstoff *m* *(chemischer Kampfstoff)*
sternway *(Schiff)* Rückwärtsfahrt *f*
sterol *(Ch)* Sterin *n*
stevedore *(Schiff)* Stauer *m*, Schauermann *m*
stew / to schmoren, kochen, dünsten
s. down eindünsten, eindicken
stew 1. Fischteich *m*; Fischtank *m*; Austerbett *n*; 2. unerwünschter Ton *m* *(Tonfilm)*
stibial antimonhaltig, antimonisch
stibialism Antimonvergiftung *f*
stibiate Antimonat *n*
stibic compound Verbindung *f* des fünfwertigen Antimons
stibine Antimonwasserstoff *m*, Stibin *n*
stibious compound Verbindung *f* des dreiwertigen Antimons
stibium Antimon *n*
stibnide Antimonid *n*
stibnite *(Min)* Stibnit *m*, Antimonglanz *m*
stibnous compound Verbindung *f* des dreiwertigen Antimons
stick / to haften; schmieren; stehen; stecken; steckenbleiben; festsitzen; festfahren; verklemmen; hängenbleiben; kleben *(Farbe; Relaisanker usw.)*; anpflocken
s. to haften an
s. together zusammenkleben *(Kontakte)*
stick 1. Stock *m*; Stange *f*; Stiel *m*; 2. *(El)* Schaltstange *f*; 3. *(Typ)* Winkelhaken *m*; Steg *m*
s.-bombing Bombenreihenabwurf *m*
s. grenade Stielhandgranate *f*
s. of solder Lötdraht *m*, Lötstreifen *m*
s.-type incendiary bomb Stabbrandbombe *f*
stickiness Klebrigkeit *f* *(z. B. des Schmiermittels)*
sticking 1. Kleben *n*; Haften *n*; 2. *(Masch)* Festklemmen *n*, Festfahren *n*, Hängenbleiben *n* *(Ventil)*; 3. *(Brg)* Feststecken *n* *(Bohrarbeit)*; 4. *(Fs)* Bildkonservierung *f*
s. of piston rings Verkleben *n* der Kolbenringe
s. of the cover [due to overheating] Festbrennen *n* des Deckels
s. plaster Heftpflaster *n*
s. temperature *(Plast)* Erweichungspunkt *m*
s. voltage Sperrspannung *f* *(Sekundärelektronenvervielfacher)*
sticky klebrig
s. slag zähe Schlacke *f*
Stiefel piercing mill Stiefelsches Lochwalzenwerk *n*
Stiefel tube-rolling process Stiefelrohrwalzverfahren *n*
stiff steif, starr; stramm, streng *(Farbe)*
s. against torsion verdrehungssteif, torsionssteif, verdrehungsfest
s. frame starrer Rahmen *m*
s.-leg derrick Derrickkran *m*, Scherenkran *m*
stiffen / to versteifen, verstärken
stiffened cable suspension bridge Hängebrücke *f* mit steifer Fahrbahntafel
stiffener Versteifung *f*, Aussteifung *f*, Steife *f*
stiffening Absteifung *f*, Versteifung *f*, Verankerung *f*
s. agent Imprägnierungsmittel *n*, Versteifungsmittel *n*
s. booms *(Schiff)* Ballastbäume *mpl*
s. of the corner Eckversteifung *f*
s. plate Verstärkungsplatte *f*, Versteifungsplatte *f*
s. rib Versteifungsrippe *f*, Versteifungsstab *m*
s. ring Versteifungsring *m*, Eckring *m*
s. varnish Spannlack *m*
stiffness Steifigkeit *f* *(z. B. einer Welle)*; Starrheit *f*
s. coefficient Momentensteifheit *f* *(Folgeregler)*
s. module Steifigkeitsmodul *m*
stifle / to ersticken *(Brand)*
stigmatic *(Opt)* stigmatisch
s. depiction stigmatische Abbildung *f*
stigmatism *(Opt)* Stigmatismus *m*
stigmator *(Opt)* Stigmator *m*
stilb Stilb *n* *(Einheit der Leuchtdichte; 1 sb = 1 cd/cm²)*
stile 1. Senkrechtstreifen *m*; Senkrechtstab *m*; 2. Übergang *m* *(Zaun)*; Durchgang *m*
still / to destillieren
still Abtreiber *m*, Destillierapparat *m*, Destillierblase *f*; Destillationsanlage *f*
still ruhend, unbewegt, stehend; lautlos; windstill
s. head Destillierkolben *m*
s. house Branntweinbrennerei *f*
s. image *(Fs)* stehendes Bild *n*
s. picture Stehlichtbild *n*
s. projector Stehbildprojektor *m*
s. water *(Schiff)* Stauwasser *n*, Totwasser *n*
stillage Plattform *f*, Pritsche *f*, Ladeplatte *f*, Gestell *n*
s. truck Hubwagen *m*
stilling basin *(Bw)* Tosbecken *n*, Toskammer *f*
stillpot Absetzbecken *n*, Klärbecken *n*
Stillson wrench [Stillson-]Rohrzange *f*
stilt Stelze *f* *(Regler)*; *(Brg)* Fußteil *m* *(Ausbau)*
stimulant Anregungsmittel *n*, Reizmittel *n*
stimulating corrosion korrosionsfördernd

stimulation energy Anregungsenergie *f*
stimuli 1. Lichtreize *mpl*, Farbreize *mpl*; 2. materielle Anreize *mpl*
stimulus Wirkungsgröße *f (jedes Signal, das einen Einfluß auf die Regelgröße ausübt)*; Anreiz *m*
stink damp *(Brg)* stinkende Wetter *pl (mit H_2S angereichert)*
s. trap Geruchverschluß *m*
stipple / to tupfen
stippling *(Met)* Stippenbildung *f*
stir / to rühren
stirrer Rührer *m*, Rührwerk *n*, Rührapparat *m*
s. driving gear Rührwerkantrieb *m*
stirring Rühren *n*
s. apparatus Rührwerk *n*
s. arm Rührarm *m*
s. machine Rührwerk *n*
s. paddle Rührschaufel *f*
s. tub Rührbütte *f*
stirrup 1. Bügel *m (Stahlbetonbau)*; 2. Steigeisen *n*, Steigbügel *m*
s. reinforcement Bügelbewehrung *f (Beton)*
stitch / to *(Typ)* heften; *(Text)* steppen
s.-bond *(Text)* übernähen
stitch Stich *m*, Masche *f*, Nadelstich *m*
s.-bonding element *(Text)* Nähwirkstelle *f*
s.-bonding method Nähwirkverfahren *n*
s. cam Kulierexzenter *m*
s.-knitting Nähwirken *n (z. B. Malimo)*
s. regulator *(Text)* Stichsteller *m*
s. resistance Stichausreißfestigkeit *f*
s. tear resistance Weiterreißfestigkeit *f*
s. transfer design *(Text)* Deckmuster *n*
s. wheeling machine Ränderiermaschine *f (Schuhfertigung)*
stitched twill *(Text)* Mehrgratköper *m*
stitcher *(Typ)* Broschürendrahtheftmaschine *f*
stitching gauze *(Typ)* Heftgaze *f*
s. machine Heftmaschine *f*
stoat / to zusammennähen
stochastic stochastisch, aleatorisch, zufällig
s. linear programming stochastisches lineares Programmieren *n*
s. variable stochastische Variable *f*, Zufallsgröße *f*, Zufallsvariable *f*
stochasticity Stochastizität *f*, zufälliger Charakter *m*
stock / to 1. aufbewahren; vorsorgen; 2. vorfräsen
stock 1. Baumstumpf *m*, Stamm *m*; 2. Werkstoff *m*, Material *n*; 3. Hauptteil *n*; Halter *m*; Griff *m*; 4. Grundkapital *n*; Inventar *n*; Viehbestand *m*; Lagerbestand *m*, Vorrat *m*; Lager[vorrat *m*] *n*; 5. Vorratshalde *f*, Halde *f*; 6. Papierstoff *m*
s. allowance Materialzugabe *f*, Werkstoffzugabe *f (Spanen)*
s. and die Schneidkluppe *f*, Außengewindeschneidwerkzeug *n*
s. bin Vorratsbehälter *m*
s. block *(Typ)* Vignette *f*
s. boat Serienboot *n (in Serie gebautes Schiff)*
s. brick Hintermauerungsstein *m*
s. brush Weißbürste *f*
s. car Viehwagen *m*
s. card Artikelkarte *f*
s. chest Vorratsbütte *f*
s. coal Haldenkohle *f*
s. coke Haldenkoks *m*
s.-dyed flockegefärbt, in der Flocke gefärbt, in der Wolle gefärbt
s.-dyeing Färben *n* in der Flocke
s. farmer Viehzüchter *m*
s. farming Viehzucht *f*
s. food Viehfutter *n*
s. furnace Schmiedeofen *m*
s. inlet Stoffauflauf *m (Papierherstellung)*
s. inventory Bestandsnachweis *m*
s. keeper Lagerverwalter *m*
s. left Werkstoffzugabe *f*
s. level indicator Gichtsonde *f*, Möllersonde *f*
s. line diameter Gichtdurchmesser *m*
s. line indicator Teufenanzeiger *m (Hochofen)*
s. market Viehmarkt *m*
s. number Fabrikationsnummer *f*
s. of machinery Maschinenpark *m*
s. of ores Erzmöller *m*
s. on hand Lagervorrat *m*
s. pile *(Masch)* Werkstückvorrat *m*
s.-pond Setzteich *m (für Fische)*
s. preparation Stoff[auf]bereitung *f*, Stoffherstellung *f (Papierherstellung)*
s. rail Backenschiene *f*; Mutterschiene *f*, Stammschiene *f*, Anschlagschiene *f*
s. raiser Viehzüchter *m*
s. removal Spanabnahme *f*, Werkstoffabnahme *f*; Spanen *n*
s. removing capacity Zerspanungsleistung *f*
s. solution *(Ch)* Stammlösung *f*
s. taking Bestandsaufnahme *f*, Inventaraufnahme *f*, Inventur *f*, Lageraufnahme *f*
s. waste Werkstoffverschwendung *f*
stocked anchor *(Schiff)* Stockanker *m*, Normalanker *m*
stockhead *(Opt)* Stativaufsatz *m*
stocking 1. Lagerung *f*, Lagern *n*; 2. Vorfräsen *n*, Vorschneiden *n*; 3. Strumpf *m*
s. cutter Zahnformvorfräser *m*
stockpile Kohlenhalde *f*, Kohlenlagerplatz *m*
stockpiling Stapeln *n*, Lagern *n*; Aufspeichern *n*
stocks 1. *(Schiff)* Helling *f* mit Stapelklötzen, Stapelplatz *m*, Kielstapel *m*; 2. Haldenbestände *mpl*; Lagervorrat *m*; 3. Qualitätsziegel *mpl (mit Kohle- oder Kalkzusatz)*

s. **for shipbuilding** Stapel *m*, Schiffbauplatz *m*, Helling *f*
stockwork *(Geol)* Stockwerk *n*
stockyard 1. Viehhof *m*; 2. Kohlenhalde *f*, Kohlenlagerplatz *m*
stoichiometry Stöchiometrie *f*
stoke / to beschicken, schüren, heizen *(Heizung bedienen)*
s. **by hand** mit der Hand beschicken
stoke Stokes *n* *(Einheit der kinematischen Viskosität; 1 St = 1 cm²/s)*
s. **hole** Heizerstand *m*, Schürraum *m*, Stockraum *m*
s.**-hole control** Überwachung *f* der Feuerung
stokehold Heizraum *m*, Kesselraum *m* *(Schiff)*
s. **bulkhead** *(Schiff)* Kesselraumschott *n*
stoker 1. Heizer *m*; 2. Schürvorrichtung *f*; Befeuerungsvorrichtung *f*
s. **arch** Feuerungsgewölbe *n*
stoking diagram *(Wärm)* Beschickungsdiagramm *n*
s. **test** Heizversuch *m*
s. **the fire** Bedienung *f* der Feuerung
stolzite *(Min)* Stolzit *m*
stone / to 1. abziehen [mit Ölstein]; 2. entkernen, entsteinen, auskernen *(Früchte)*
stone 1. Stein *m*; Gestein *n*; 2. *englische Masseeinheit; 1 stone = 6,350 kg*
s. **age** *(Geol)* Steinzeit *f*
s. **bit** Gesteinsbohrer *m*, Steinbohrer *m*
s. **bolt** Steinschraube *f*
s. **breaker** Bergebrecher *m*, Backenbrecher *m*
s. **breaker's hammer** Stein[schlag]hammer *m*
s. **chisel** Steinmeißel *m*
s. **crusher** Steinbrecher *m*
s. **cutter** 1. Steinbohrer *m*; 2. Steinmetz *m*
s.**-cutting saw** Steinsäge *f*
s. **drift** *(Brg)* Gesteinsstrecke *f*
s. **drill** Gesteinsbohrer *m*
s. **dust** *(Brg)* Gesteinsstaub *m*, Inertstaub *m*
s. **dusting** *(Brg)* Gesteinsstaubverfahren *n*, Einstäuben *n* mit Gesteinsstaub
s. **expansion** *(Masch)* Zustellung *f* der Honsteine
s. **fork** Steinschlaggabel *f*
s. **frame saw** Steingattersäge *f*
s. **mason** Steinmetz *m*
s. **pit** Steinbruch *m*
s. **rest** *(Met)* Gichtstein *m*, Windstein *m*
s. **sledge** Steinspalthammer *m*
s. **trap** Steinfang *m* *(in der Erntemaschine)*
s. **ware** Steinzeug *n*
stoning *(Masch)* Abziehen *n*
stool Bock *m*; Schemel *m*; Fensterbrett *n*
stoop and room method *(Brg)* Pfeilerbau[verfahren *n*] *m*
stop / to 1. anhalten, stillsetzen, arretieren, sperren; ausschalten; 2. stopfen *(Stichloch)*; verstöpseln; dichten
s. **a leak** *(Schiff)* ein Leck dichten
s. **down** abblenden *(Fotoapparat)*
stop 1. Absatz *m*, Anschlag *m*, Auflager *n*, Feststellknagge *f*, Hubbegrenzer *m*, Sperre *f*; 2. Blende *f*; 3. Halt *m*; Stillstand *m*; Aus-Stellung *f*
s. **and tail lamp** Bremsschlußleuchte *f*
s. **band** Sperrbereich *m* *(eines Filters)*
s. **bar** 1. Anschlagleiste *f*, Arretierleiste *f*; 2. *(El)* Halttaste *f*
s. **bath** Unterbrechungsbad *n* *(Fotografie)*
s. **blade** Paßschaufel *f* *(Turbine)*
s. **block** Prellblock *m*, Anschlagblock *m*
s. **bolster** Sperrpolster *n*
s. **button** Ausschaltknopf *m*
s. **cam** Sperrnocken *m*
s. **collar** Anschlagbund *m*
s. **disk** Festhaltescheibe *f* *(Signal)*
s. **dog** Schaltknagge *f*, Schaltnocken *m*, Anschlagknagge *f*
s. **for adjusting length of stroke** *(Masch)* Endanschlag *m*
s. **frequency** Sperrfrequenz *f*
s. **image** Blendenbild *n*
s. **impulse** *(Fmt)* Stoppstromstoß *m*
s. **lamp** Bremsleuchte *f*
s. **lever** Begrenzungshebel *m*, Bremshebel *m*; Auslösehebel *m*; Blockhebel *m*
s. **light** Bremslicht *n*, Stopplicht *n*, Bremsschlußleuchte *f*
s.**-light pull switch** Bremslichtzugschalter *m*
s.**-light rotating switch** Bremslichtdrehschalter *m*
s. **lock** Haltsperre *f* *(Signal)*
s. **motion device** Abstelleinrichtung *f*
s. **motion for yarn breakages and yarn drag** *(Text)* Fadenbruch- und -zupfabsteller *m*
s. **motion lever** Abstellhebel *m*
s. **number** Blendenzahl *f* *(Fotografie)*
s. **of bellows** Balgenregister *n* *(Orgel)*
s. **pawl** Sperrklinke *f*
s. **pin** Anschlagstift *m*, Arretierschraube *f*, einsteckbarer Bolzen *m*, Vorsteckstift *m*
s. **position** Haltlage *f* *(Signal)*
s. **riveting** Abdichten *n* durch Einziehen eines Niets
s. **screw** Anschlagschraube *f*, Arretierschraube *f*
s. **signal** 1. *(Fmt)* Sperrschritt *m*; 2. Haltsignal *n*
s. **spring** Rastfeder *f*
s.**-tail lamp** Bremsschlußleuchte *f*
s. **valve spindle** Abschlußspindel *f*, Absperrspindel *f* *(Turbine)*
s. **watch** Stoppuhr *f*

stopcock Absperrhahn *m*, Hahn *m*, Küken *n*
s. grease Hahnfett *n*
stope / to *(Brg)* abbauen
s.-out stoßweise abbauen
stope *(Brg)* Abbau *m (Ort)*
s. cut stufenförmiger Einbruch *m (Schachtabteufen)*
s. drift Strecke *f* unter der Schwebe zur Einleitung des Strossenbaus; Strecke über der Schwebe zur Einleitung des Firstenstoßbaus
s. drive *s.* s. drift
s. hammer Abbauhammer *m*
stoper *(Brg)* Abbauhammer *m*, Bohrhammer *m*
stoping *(Brg)* Abbau *m (Vorgang, insbesondere Firstenbau)*
s. back Abbau *m* der Teilsohle über und neben dem Querbrechen *(Teilsohlenbruchbau)*
stoppage 1. *(Masch)* Stillsetzung *f*; Stillstand *m*; 2. *(Mil)* Ladehemmung *f*
s. time *(Masch)* Stillstandszeit *f*
stopped furnace kaltliegender Hochofen *m*
stopper / to verstöpseln
stopper Pfropfen *m*, Stopfen *m*, Stöpsel *m*
s. circuit *(El, Fmt)* Entkopplungsglied *n*; Sperrkreis *m*
stoppered bottle Stöpselflasche *f*
stopping 1. Stillsetzen *n*, Anhalten *n*; 2. Verstopfen *n*; Verstöpseln *n*; 3. Spachtelmasse *f*, Füllkitt *m*; 4. *(Brg)* Damm *m*, Sperre *f*
s. agent *(Ch)* Inhibitor *m*
s. brake *(El)* Anhaltebremse *f (elektrischer Zug)*
s. cross section atomares Bremsvermögen *n*
s. device Anhaltevorrichtung *f*
s. distance Bremsweg *m*, Bremsstrecke *f*
s. lever Arretierhebel *m*
s. of chain *(Ch)* Kettenabbruch *m*
s. of the drawing frame *(Text)* Streckenabstellung *f*
s. potential Bremspotential *n*
stopple / to verstöpseln
stopple Stöpsel *m*
storage 1. Lagerung *f*; Lagerraum *m*; Lagerungskosten *pl*; 2. *(Dat)* Speicher *m*; Speicherung *f*; 3. *(Masch)* Vorratsbildung *f*; Magazinieren *n*
s. bank Zwischenpuffer *m (Taktstraße)*
s. battery Akkumulator[enbatterie *f*] *m*
s. battery acid Akkumulatorsäure *f*
s. battery cell Akkumulatorzelle *f*
s. battery locomotive Akkulok *f*, Akkumulatorlokomotive *f*
s. battery of iron-nickel type Nickel-Eisen-Akkumulator *m*, Nife-Akkumulator *m*
s. battery truck Elektrokarren *m*
s. bin *(Bw)* Vorratsbehälter *m*, Vorratssilo *m*; Lagerbehälter *m (Getreide)*
s. camera tube Bildspeicherröhre *f*
s. capacitor Speicherkondensator *m*
s. capacity Speichervermögen *n*; Lagerungsmöglichkeit *f*
s. cell Akkumulatorenbatterie *f*, Sammlerbatterie *f*; *(Dat)* Speicherzelle *f*, Speicherelement *n*
s. characteristic Alterungscharakteristik *f*
s. depreciation Leistungsverminderung *f* durch Lagerung *(der Batterie)*
s. drum Magnettrommel *f*, Speichertrommel *f*
s. expenses Lagerkosten *pl*
s. integrator Speicherintegrator *m*
s. keyboard Speichertastatur *f*
s. life Lagerfähigkeit *f (z. B. der Batterie)*
s. location Speicherplatz *m*, Speicherzelle *f (für Daten)*
s. magazine Speichermagazin *n*
s. power station Speicherkraftwerk *n*
s. property Lagerfähigkeit *f*
s. register *(Dat)* Speicherzelle *f*
s. section Speicherwerk *n (Rechner)*
s. space Lagerraum *m*; Aufstellungsfläche *f*
s. space-heater elektrischer Speicherofen *m*
s. stability Lagerfähigkeit *f*
s. tank Lagertank *m*, Reservoir *n*, Sammelbehälter *m*, Vorratsbehälter *m*
s. test Lagerfähigkeitsprüfung *f*, Alterungsprüfung *f*
s. tube *(Dat)* Speicherröhre *f*
s. unit *(Dat)* Speicherwerk *n*
s. wall *(Bw)* Schrankwand *f*
s. yard Lager[fläche *f*] *n*, Lagerhof *m*, Lagerplatz *m*
store / to speichern; einlagern
s. energy Energie speichern
store 1. Abstellraum *m*, Lager *n*, Magazin *n*, Vorratsraum *m*; 2. *(Dat)* Speicher *m*
s. boat Vorratsschiff *n*, Lagerkahn *m*
s.-carrier Versorgungsschiff *n*
s. room Vorratsraum *m*; Speicher *m*
s. ship Versorgungsschiff *n*
stored coal Lagerkohle *f*
s.-energy welding Speicherenergieschweißen *n*
s.-up aufgespeichert
storehouse Lagerhaus *n*, Magazin *n*, Speicher *m*
storey Stock[werk *n*] *m*; Geschoß *n*; Etage *f*; Terrasse *f*
s. culture Terrassenanbau *m (Reis, Wein usw.)*
s. drying machine *(Text)* Etagentrockenmaschine *f*
s. farming Terrassenanbau *m (Reis, Wein usw.)*
s. furnace *(Met)* Etagenofen *m*
storeyed mit Stockwerken (Geschossen)
storm guyed pole *(Fmt)* Abspannmast *m*
s. oil tank *(Schiff)* Wellenölbehälter *m*
s. signal Sturmsignal *n*

story *(Am) s.* storey

stove / to einbrennen *(Farbe)*; hart werden, härten *(Plaste)*; bleichen *(textile Gewebe)*

stove *(kleiner)* Ofen *m*, Zimmerofen *m*; Einbrennofen *m (für Lack)*; *(Met)* Winderhitzer *m*; Darre *f*

s. damper Ofenklappe *f*

s. fitter Ofensetzer *m*

s. heating Ofenheizung *f*

s. pipe Ofenrohr *n*

s. plate Ofenplatte *f*

s. room Heizungsraum *m*

s. tile Ofenkachel *f*

s. wood Brennholz *n*

stoving Einbrennen *n (Lack)*

s. varnish Einbrennlack *m*, ofenhärtender Lack *m*

stow / to 1. *(Schiff)* stauen *(Güter)*; festmachen *(Segel)*; 2. *(Brg)* versetzen *(Berge)*

s. away [ver]stauen

stowage 1. *(Schiff)* Stauen *n*; Stauraum *m*; 2. *(Brg)* Bergeversatz *m*

s. factor Raumbedarf *m*; Staukoeffizient *m*

s. plan *(Schiff)* Ladeplan *m*, Stauungsplan *m*

s. space Stauraum *m*

stowaway 1. blinder Passagier *m*; 2. abgestellte Güter *pl*; 3. Abstellraum *m*

stowed goaf *(Brg)* Versatzfeld *n*

stower *(Brg)* Versatzmaschine *f*

stowing *(Brg)* Versetzen *n*, Versatzarbeit *f*

s. material Versatz[gut *n*] *m*, Versatzberge *pl*

strabismus *(Opt)* Strabismus *m*

straddle / to spreizen

straddle carrier Torladewagen *m*, Torlader *m*

s. cutter (mill) Scheibenfräserpaar *n* zum Planparallelfräsen

s. milling planparalleles Fräsen *n* mit einem Satz Scheibenfräser

s. mounting Gabellagerung *f*

s. truck *(Am) s.* straddle carrier

s. wrench Gabelschlüssel *m*

straggling Schwankung *f*, Streuung *f*, Streuen *n*, Verteilung *f*, Straggling *n*

straight gerade, geradlinig; zylindrisch *(Bohrung, Schaft)*; unverdünnt

s. acceleration of mass reine Massenbeschleunigung *f*

s. amplifier Geradeausverstärker *m*

s. angle gestreckter Winkel *m*

s. approaching line Näherungsgerade *f*

s. arm Schreibarm *m (Ellipsenzeichner)*

s.-back saw Säge *f* mit geradem Rücken *(Holzbearbeitung)*

s.-backed blade Schaufel *f* mit geradem Rücken *(Turbine)*

s. bevel gear Geradezahnkegelrad *n (Verzahnung)*

s. binary *(Dat)* direkt-binär

s.-boled geradstämmig

s. bore zylindrische Bohrung *f*

s. bow *(Schiff)* gerader (senkrechter) Vorsteven *m*

s. carbon steel reiner Kohlenstoffstahl *m*

s.-chain *(Ch)* geradkettig, normalkettig

s. chrome stainless steel nichtrostender reiner Chromstahl *m*

s. circuit Geradeausschaltung *f (Empfänger)*

s. conductor *(El)* gestreckter Leiter *m*

s. cup wheel gerade Topfschleifscheibe *f*

s. cylinder Stangenzylinder *m (Hydraulik)*

s. cylindrical nozzle zylindrische Düse *f*

s. edge Richtlatte *f*, Richtscheit *n*; Lineal *n*, Maßstab *m*

s. edge of steel Stahllineal *n*

s.-face tooth Zahn *m* mit gerader Flanke

s. filament glatter Leuchtdraht *m*

s. flute gerade Spannut *f*

s. fluted geradnutig, geradgenutet

s. fluted reamer geradzahnige Reibahle *f*

s. forward trunking method *(Am) (Fmt)* Anrufbetrieb *m*

s. framed ship Geradspantenschiff *n*

s. grinding wheel gerader Schleifkörper *m*

s. hole zylindrische Bohrung *f*

s.-in approach *(Flg)* Geradeausanflug *m*

s. joint 1. Durchgangsmuffe *f*, Muffe *f*, Verbindungsmuffe *f*; 2. Fuge *f* auf Fuge *(Mauerwerk)*; 3. I-Stoß *m*

s. knurl / to rändeln

s. knurl Rändelwerkzeug *n*, Rändelrädchen *n*

s.-knurling pitch Rändelteilung *f*

s. lathe tool gerader Drehmeißel *m*

s.-legged roadway arch *(Brg)* geradbeiniger Streckenbogen *m (Ausbau)*

s. length gerades Rohrstück *n*

s. line Gerade *f*

s.-line geradlinig

s.-line capacitance capacitor kapazitätsgerader Drehkondensator *m*, kapazitätsproportionaler Kondensator *m*

s.-line curve Gerade *f*

s.-line detector linearer Gleichrichter *m*

s.-line distance Luftlinienentfernung *f*

s.-line fit geradlinige Anordnung *f*

s.-line frequency capacitor frequenzgerader Drehkondensator *m*, frequenzgerader Kondensator *m*

s.-line mechanism Geradführungsmechanismus *m*

s.-line motion Geradführung[sbewegung] *f*

s.-line nature Geradlinigkeit *f*

s.-line path *(Ph)* geradlinige Bahn *f*

s.-line rectification lineare Gleichrichtung *f*

s.-line trajectory *(Ph)* geradlinige Bahn *f*

s.-line type of drilling machine Reihenbohrmaschine *f*
s.-line wavelength capacitor wellengerader Drehkondensator (Kondensator) *m*
s.-lined geradlinig
s. magnet Stabmagnet *m*, Magnetstab *m*
s. manganese steel Manganhartstahl *m*
s. mineral oil Blankwalzöl *n*
s. motor oil unlegiertes Motorenöl *n*
s. multiple *(Fmt)* gerade Vielfachschaltung *f*
s. pin Zylinderstift *m*
s.-profile geradflankig
s. rack geradverzahnte Zahnstange *f*
s. reamer Zylinderreibahle *f*
s. receiver Geradeausempfänger *m*
s. recessing tool Kopfdrehmeißel *m*
s. ribbon feed *(Text)* Bandübertragung *f* ohne Kreuzung
s. rod aerial *(Fmt)* gestreckter Strahler *m*
s. run *(El)* gerades Leitungsstück *n*
s.-run gasoline Straight-run-Benzin *n*, Destillationsbenzin *n*
s. scanning *(Fs)* fortlaufende Bildabtastung *f*
s. scarf with saddle-backed ends gerades Blatt *n* mit Grat *(Zimmerei)*
s.-set saw blade geschränktes Sägeblatt *n*
s. shank *(Am)* Zylinderschaft *m*
s.-side eccentric press Zweiständerexzenterpresse *f*
s.-side tyre wulstloser Reifen *m*
s.-sided gear Zahnrad *n* mit geraden Zahnflanken
s.-sided trimming press Zweiständerabgratpresse *f*
s. sliding link Geradschubkurbel *f*
s. slot burner Rundlochbrenner *m*
s. spar *(Flg)* Balkenholm *m*
s. spur gear Geradzahnstirnrad *n*
s. spur gear teeth Stirnradverzahnung *f*
s. stem *(Schiff)* Stampfsteven *m*, gerader Vorsteven *m*
s. storage system *(Fmt)* reines Batteriesystem *n*
s. thread zylindrisches Gewinde *n*
s.-through cooling Durchflußkühlung *f*
s.-through ray direkter Strahl *m (Antennenemission)*
s.-through reactor Durchlaufreaktor *m*
s. tie up *(Text)* gerade Schnürung *f*, Harnischstich *m*
s. tooth Geradzahn *m (Zahnrad)*; Schabezahn *m (Räumwerkzeug)*
s.-tooth geradverzahnt, mit Geradverzahnung
s.-tooth bevel gear Geradzahnkegelrad *n (Verzahnung)*
s.-tooth blade Sägeblatt *n* mit gerader Zahnspitzenlinie *(Holzbearbeitung)*
s.-tooth clutch Geradzahnkupplung *f*, Klauenkupplung *f*
s.-toothed *s.* s.-tooth
s. tracing Zeilentasten *n (Kopieren)*
s. turning Langdrehen *n*
s.-up-and-down filament *(El)* Langdrahtleuchtdraht *m*
s.-vacuum forming *(Plast)* Tiefziehen *n*, Vakuumsaugverfahren *n*, Vakuumtiefziehverfahren *n*
s.-way cock Zweiwegehahn *m*
s.-way valve Durchgangsventil *n*
s.-wedge mechanism Keilschubgetriebe *n*
s. wheel *s.* s. grinding wheel
s. worm hob zylindrischer Schneckenfräser *m*
straightedge *s.* straight edge
straighten / to begradigen; [aus]richten, geradrichten
straightener Blechrichtmaschine *f*
straightening Begradigung *f*; Ausrichten *n*, Richten *n*
s. anvil Richtamboß *m*
s. machine Richtmaschine *f*
s. plant Zurichterei *f*
s. plate Richtplatte *f*
s. press Flachstanze *f*, Richtpresse *f*
s. roll Richtwalze *f*
s. the top *(Text)* Entkräuseln *n* des Kammzugs
s. tongs Richtzange *f*
s. tool Richtwerkzeug *n*
straightness Geradheit *f*, Geradlinigkeit *f*; Zylindrizität *f (Bohrung, Schaft)*
strain / to [an]spannen, beanspruchen; deformieren, recken, dehnen; durchgießen, durchpressen, [durch]seihen, filtrieren, sieben; spannen *(die Haut zum Trocknen)*
s. harden verfestigen
strain 1. Dehnung *f*; Formänderung *f*; Deformation *f*; Verzerrung *f*; 2. *(nichttechnisch und Betriebsausdruck)* Beanspruchung *f*; Spannung *f*; 3. Rasse *f*, Stamm *m*
s. age hardening *(Met)* Reckalterung *f*, Streckalterung *f*
s. ag[e]ing *(Met)* Reckalterung *f*, Streckalterung *f*
s. ear *(El)* Abspannöse *f*
s. element *(El)* Zugentlastung *f*
s. energy Formänderungsenergie *f*, Formänderungsarbeit *f*
s. gauge Dehnungsmeßstreifen *m*, Dehnungsmeßgerät *n*
s. hardening Kaltverfestigung *f*
s. insulator Abspannisolator *m*; Verankerungsisolator *m*
s. measurement Dehnungsmessung *f*
s. rate Dehnungsgeschwindigkeit *f*, Verformungsgeschwindigkeit *f*

s. velocity *s.* s. rate
strained honey Schleuderhonig *m*
s. wire Streckdraht *m*
strainer Filter[einsatz *m*] *n*, Durchschlag *m*, Sieb[einsatz *m*] *n*, Seihetuch *n*; Knotenfänger *m*, Splitterfänger *m (Papierherstellung)*
s. core Siebkern *m*
s. gate Siebzulauf *m*
strainfree 1. *(Brg)* entspannt, entlastet *(Gebirgsmechanik)*; 2. dehnungsfrei; 3. spannungsfrei *(Betriebsausdruck)*
straining 1. *(Betriebsausdruck)* Beanspruchung *f*; Spannen *n (Haut)*; 2. Filtrieren *n*; Sieben *n*; 3. Dehnen *n*; Verzerren *n*
s. box Siebkasten *m*
s. clamp *(El)* Abspannklemme *f*
s. cloth Seihtuch *n*
s. drum Siebtrommel *f*
s. ring Spannring *m*
s. screw Spannschraube *f*
s. tower *(El)* Abspannmast *m*
strake *(Schiff)* Plattengang *m*
stramineous strohartig, strohfarben
strand / to 1. verlitzen *(Draht)*; verseilen *(Kabel)*; 2. stranden, auf Strand setzen, auflaufen; 3. zurückgelassen werden
strand 1. *(El)* Litze *f*; 2. Faserbündel *n*; 3. Fadenspan *m*, Splitter *m*; 3. Trum *m (Förderband)*; 4. Strand *m*, Ufer *n*
s. rolls *(Met)* Zwischengerüst *n*, Strangwalzen *fpl*
stranded aerial wire Antennenlitze *f*
s. cable verseiltes Kabel *n*
s. conductor Litzenleiter *m*, Seilleiter *m*
s. copper wire Kupferlitze *f*
s. rope Litzenseil *n*
s. two-wire lead *(El)* Doppellitze *f*
s. wire verseilter Draht *m*, Litze[ndraht *m*] *f*, Drahtlitze *f*
s. wire coil Litzenspule *f*
s. wire line verkabelte Leitung *f*
stranding Schiffbruch *m*
strangler 1. Drossel *f (Vergaser)*; 2. *(Kfz)* Bowdenzug *m* zur Lüfterklappe
strap / to 1. festbinden, anschnallen; verlaschen; 2. abziehen *(Messer)*; 3. *(Schiff)* einstroppen
s. in festschnallen, anschnallen
s. together verbinden *(durch Laschen)*
strap 1. Riemen *m*; Gurt *m*; Schnürsenkel *m*; Festhaltegriff *m (Straßenbahn)*; 2. Band *n*, Spannband *n*, Steg *m*; Metallband *n*; 3. Lasche *f*; 4. *(Schiff)* Stropp *m*, Ladestropp *m*
s. bolt Bügelschraube *f*
s. brake Bandbremse *f*
s. coil Bandspule *f*
s. connection Laschenverbindung *f*
s. drive Riemen[an]trieb *m*
s. end *(Masch)* Kappenkopf *m (Schubstange)*
s. fork mechanism *(Text)* Riemchensteuerung *f*
s. hinge Bandscharnier *n*
s. iron *s.* s. steel
s. joint Laschenstoß *m*
s. polisher Polierleder *n*
s. steel Bandstahl *m*
s. weaving Gurtweberei *f*
strapped joint Laschenstoß *m*
strapping 1. Anschnallen *n*; Verlaschen *n*; 2. *(El)* Koppelleitung *f (Magnetron)*; *(Fmt)* Blankverdrahtung *f*
strass Straß *m (Edelsteinimitation)*
strata Gesteinsschichten *fpl*
s. adjacent Nebengestein *n*
s. bolting Ankerausbau *m*
s. control Hangendpflege *f*, Beherrschung (Lenkung, Pflege) *f* des Hangenden
s. movement Gebirgsbewegung *f*
s. section Schichtenschnitt *m*
s. sequence Schichtenfolge *f*
s. series Schichtenfolge *f*
stratification *(Geol)* Schichtung *f*
stratified geschichtet
s. lattice Schichtengitter *n*
s. rock Schichtgestein *n*
stratigraphy *(Geol)* Stratigrafie *f*, Formationskunde *f*
stratonaut *(Astr)* Stratonaut *m*
stratosphere Stratosphäre *f*
s. plane Stratosphärenflugzeug *n*
stratum Schicht *f*; Ablagerung *f*; Bodenschicht *f*, Gesteinsschicht *f*; Bank *f*, Lage *f*, Flözschicht *f*
s. thickness Schichtdicke *f*
straw 1. Stroh *n*; 2. *(Opt)* Gelbpunkt *m (weniger als ein Newtonscher Ring)*
s. baler Strohpresse *f*
s. baling press Strohballenpresse *f*
s. carrier Strohelevator *m (Dreschmaschine)*
s. charger Stroheinleger *m (Dreschmaschine)*
s. charging device Stroheinleger *m (Dreschmaschine)*
s. chopper Strohschneider *m*
s. compost Strohkompost *m*
s. crusher Strohquetscher *m*
s.-cutting bench Häckselbank *f*
s.-feeding rake Zuführrechen *m*
s. lifter Strohelevator *m (Dreschmaschine)*
s. paper Strohpapier *n*
s.-piled der Länge nach gestapelt *(Rundholz)*
s. pulp Stroh[faser]stoff *m*, Strohhalbstoff *m (Papierherstellung)*
s. rope spinner Strohseilspinnmaschine *f*
s. shredder Strohzerreißer *m*
s. stack Strohfeime *f*
strawboard 1. Stroh[bau]platte *f*; 2. starke Pappe *f*; Packpappe *f*, Strohpappe *f*

stray / to streuen; abweichen; verirren; vagabundieren
stray verirrt; vagabundierend; parasitär; Streu-; Irr-; Stör-
s. capacitance *(El)* Streukapazität *f*
s. coupling *(El)* Streukopplung *f*, ungewollte (wilde) Kopplung *f*
s. current Erdstrom *m*, Irrstrom *m*, vagabundierender Strom *m*
s. electron Fremdelektron *n*, vagabundierendes Elektron *n*
s. electrostatic field elektrostatisches Streufeld *n*
s. emission Streuemission *f*
s. field *(El)* Streufeld *n*
s. flux Streufluß *m*
s. induction Streuinduktion *f*
s. light Streulicht *n*
s. line *(El)* Streulinie *f*
s. magnetic field magnetisches Streufeld *n*
s. parameters parasitäre (zusätzliche) Parameter *mpl*
s. pick-up *(El)* Einwirkung *f* von Streufeldern; Störaufnahme *f*
s. radiation Streustrahlung *f*; Störstrahlung *f*; gestreute Strahlung *f*
s. reactance Streureaktanz *f*
s. voltage Streuspannung *f*
s. wiring capacitance Schaltungskapazität *f*
straying *(El)* Streuung *f*; Streuverlust *m*
s. from great-circle path *(Fmt)* Ausbrechen *n* vom Großkreis *(Abweichung von der Großkreisausbreitung bei Langwellen)*
strays atmosphärische Störungen *fpl*
streak Schliere *f*, Streifen *m*; Maserung *f*, Ader *f (Holz)*; *(Min)* Strich *m*
s. cavitation Streifenkavitation *f*
s. formation Schlierenbildung *f*
s. test *(Min)* Strichprüfung *f*
streaking Nachziehen *n*, Fahnenziehen *n (Fernsehbildwiedergabe)*
stream / to strömen
stream 1. Strom *m*; Strömung *f*; 2. fließendes Gewässer *n*
s. feeder *(Typ)* Staffelanleger *m*
s. jig washer Stromsetzmaschine *f (Aufbereitung)*
s. line Stromfaden *m*
s. line axle *(Flg)* Profilachse *f*
s. line continuous flow mixer kontinuierlicher Mischer *m* mit kontinuierlicher Materialzuführung
s. line flow laminare Strömung *f*
s.-lined stromlinienförmig; schnittig
s. of exhaust gases ausströmende Gase *npl*
s. of ions Ionenstrom *m*
s. tin *(Met)* Seifenzinn *n*, Zinnseife *f*
s.-tube Stromröhre *f*
s. velocity Strömungsgeschwindigkeit *f*
streamer Schliere *f*; Schwaden *m*; Streifen *m*; Wimpel *m*
streaming potential Strömungspotential *n*
streamline / to stromlinienförmig gestalten
streamline *s.* stream line
streamlined *s.* stream-lined
street Straße *f (innerhalb geschlossener Ortschaften)*
s. car *(Am)* Straßenbahnwagen *m*
s. cleaning Straßenreinigung *f*
s. flusher Straßensprengwagen *m*
s. inlet Straßenabfluß *m*, Straßenablauf *m*
s. lighting Straßenbeleuchtung *f*
s. manhole *(Am) (Fmt)* Kabelschacht *m (Fahrbahn)*
s. marker Spurnagel *m (zur Fahrbahnkennzeichnung)*
s. organ Drehorgel *f*, Leierkasten *m*
s. refuge Verkehrsinsel *f*, Fußgängerschutzinsel *f*
s. roller Straßenwalze *f*
s. sprinkler Straßensprengwagen *m*
s. sweeper Straßenkehrmaschine *f*
strengite *(Min)* Strengit *m*
strength 1. Konzentration *f (Säure)*; 2. Festigkeit *f (Festigkeitslehre)*; 3. Kraft *f*; Intensität *f*; Stärke *f*; Härte *f*
s. at high (increased) temperatures Warmfestigkeit *f*
s. at low temperatures Kältefestigkeit *f*
s. bulkhead *(Schiff)* Festigkeitsschott *n*
s. calculation Festigkeitsberechnung *f*
s. deck *(Schiff)* Festigkeitsdeck *n*
s. depending on shape Gestaltsfestigkeit *f*
s.-duration curve *(El)* Reizzeitstromkurve *f*, I-t-Kurve *f*, Reizzeitspannungskurve *f*
s. in compression Druckfestigkeit *f*
s. in shear Scherfestigkeit *f*
s. in tension Zugfestigkeit *f*
s. limit Festigkeitsgrenze *f*, Bruchgrenze *f*
s. of a magnetic shell *(Ph)* Stärke *f* der magnetischen Doppelfläche, Stärke des magnetischen Blattes
s. of compression Druckfestigkeit *f*
s. of current Stromstärke *f*
s. of draught *(Wärm)* Zugstärke *f*
s. of materials Werkstoffestigkeit *f*; Festigkeitslehre *f*
s. of reception Empfangslautstärke *f*
s. of rupture Bruchfestigkeit *f*
s. of shearing Scherfestigkeit *f*
s. of signals Signalstärke *f*, Zeichenstärke *f*
s. of [the] bond of the [grinding] wheel Bindungshärte *f* der Schleifscheibe
s. of the combustion unit *(Rak)* Brennkammerfestigkeit *f*

s. property Festigkeitseigenschaft *f*
s. test Festigkeitsprüfung *f*
s. theory Festigkeitslehre *f*
s. value Festigkeitswert *m*
strengthen / to verstärken, versteifen; die Festigkeit vergrößern
strengthening *(mechanische)* Verstärkung *f*
s. by cold working Kaltverfestigung *f*
s. for navigation in ice Eisverstärkung *f* *(von Schiffen)*
s. rib Verstärkungsrippe *f*
s. tissue Festigungsgewebe *n*
stress / to spannen; beanspruchen, belasten
stress Spannung *f*; Beanspruchung *f*
s. analysis Spannungsberechnung *f*, Spannungsermittlung *f*, Spannungsnachweis *m*
s. concentration Spannungsspitze *f*; Beanspruchungsspitze *f*; Spannungskonzentration *f*
s. corrosion Spannungskorrosion *f*
s. crack Spannungsriß *m*
s. cycle Lastspiel *n*; Lastwechsel *m*
s.-cycle diagram Wöhlerkurve *f*
s. determination Spannungsermittlung *f*
s. diagram Spannungsdiagramm *n*; Kräfteplan *m*
s. distribution Spannungsverlauf *m*, Spannungsverteilung *f*
s.-equalizing Spannungsausgleich *m*
s.-equalizing spannungsausgleichend
s. fluctuation Spannungsveränderung *f*, Spannungsschwankung *f*
s.-free spannungsfrei
s. modulus E-Modul *m*
s.-number curve Dauerfestigkeitsschaubild *n*, Wöhlerkurve *f*
s.-optical constant spannungsoptische Konstante *f*
s.-raising spannungserhöhend
s. redistribution Spannungsumlagerung *f*
s. relief Entspannung *f*; Spannungsentlastung *f*
s. relief annealing Spannungsfreiglühen *n*
s.-relieved spannungsentlastet
s. relieving 1. Entspannen *n*, Spannungsfreiglühen *n*; 2. Entspannung *f* *(Gebirgsmechanik)*
s. removal Entspannung *f*
s. reversal Lastwechsel *m*
s. state Spannungszustand *m*
s.-strain curve (diagram) Spannungs-Dehnungs-Kurve *f*; Spannungs-Dehnungs-Diagramm *n*, Zerreißschaubild *n*
s.-strain relation Spannungs-Dehnungs-Beziehung *f*
stressability Belastbarkeit *f*, Beanspruchbarkeit *f*
stressbed Spannbett *n* *(Spannbeton)*
stresscoat Reißlackschicht *f*
stressed unter Spannung, unter Beanspruchung; beansprucht
s.-skin construction *(Flg)* Schalenbauweise *f*
stressing Beanspruchung *f*; Spannen *n*; Anziehen *n*
s. bed Spannbett *n* *(Spannbeton)*
s. head Spannkopf *m* *(Spannbeton)*
s. jack Spannpresse *f* *(Spannbeton)*
s. tendon Spannglied *n* *(Spannbeton)*
s. unit Spannglied *n* *(Spannbeton)*
stretch / to recken, längen, strecken, dehnen; verstrecken; ziehen; überdehnt werden; straffen; sich erstrecken
s.-form streckformen
s. the fibres longitudinally *(Text)* die Fasern in Längsrichtung strecken
stretch 1. Länge *f*, Strecke *f*; Abschnitt *m*; 2. Ausdehnung *f*; Weitung *f*; elastische Dehnung *f*
s. cotton fabric hochelastisches Gewebe *n* aus Baumwolle
s. forming Streckformen *n*
s. press Reckziehpresse *f*
s. resistant streckfest
s.-spinning process Streckspinnverfahren *n*
s.-strain limit *(Plast)* Fließgrenze *f*
s. test *(Met)* Reckprüfung *f*
s. yarn Kräuselgarn *n*
stretchability *(Plast)* Reckfähigkeit *f*, Verstreckbarkeit *f*
stretchable [ver]streckbar, reckbar; spannbar
stretched steel Streckstahl *m*
s. wire Streckdraht *m*
stretcher 1. Läufer *m* *(Ziegelmauerwerk)*; 2. Stemmbrett *n* *(Boot)*; 3. Trage *f*; 4. Strecker *m*; Leisten *m*; Spannkopf *m*; Steg *m*
s. bond Läuferverband *m* *(Mauerwerk)*
s. course Läuferschicht *f* *(Mauerwerk)*
s. lines Fließfiguren *fpl*, Kraftwirkungsfiguren *fpl*
s. strain *(Met)* Reckspannung *f*
s. strains Ziehriefen *fpl*; Fließfiguren *fpl*
stretchforming Streckformen *n*
stretching Einspannen *n*, Festspannen *n*, Spannen *n*; *(Text)* Breithalten *n*
s. ring Spannring *m* *(beim Kondensatormikrofon)*
s. strain *(Betriebsausdruck)* Zugbeanspruchung *f*; Dehnung *f* durch Zugbeanspruchung; Zugverformung *f*
s. vibration *(Ch)* Valenzschwingung *f*
s. wire Spanndraht *m*; *(El)* Abspanndraht *m*
strew / to [be]streuen
stria Schliere *f*; Riefe *f*, Furche *f*
striated geriffelt *(Kavitation)*
striation Schichtung *f*, Streifenbildung *f*; Strieme *f*

strickle / to schablonieren, mit Ziehbrett formen
strickle 1. Schablone *f*, Lehre *f*, Ziehlehre *f*, Ziehbrett *n (Formen)*; 2. Meßstab *m*; 3. Wetzstein *m*
s. board Ziehbrett *n*
strickling Lehrenformerei *f*
striction 1. Einengung *f*, Striktion *f*, Verengung *f*; 2. *(Ph)* Pincheffekt *m*
stricture *s.* striction
strident schrill, schneidend *(Ton)*
striding level Reiterlibelle *f*
strike / to schlagen; prägen *(Münzen)*; [auf-]treffen; stoßen auf, prallen auf; streichen; [an]zünden; abbauen, niederholen; ausschalen; Arbeit niederlegen
s. a contact auf einen Kontakt auftreffen, einen Kontakt berühren
s. an arc einen Lichtbogen ausbilden, überschlagen, zünden
s. formwork ausschalen, entschalen *(Betonbau)*
s. ground *(Flg)* aufschlagen
s. the ship's bell *(Schiff)* glasen
strike 1. *(Geol)* Streichen *n*; 2. Abstreichlineal *n (Formerei)*
s. bath Vorgalvanisierbad *n*
s. deposit Vorgalvanisierung *f*, Untergalvanisierung *f*
s. face *(Brg)* streichender Streb *m*
striker Mitnehmer *m*, Stößel *m*; *(Mil)* Schlagbolzen *m*
s. bar Anschlagschiene *f*; Mitnehmerschiene *f*
s. clutch Schaltgabel *f*
s. fork Schaltgabel *f*, Verschiebegabel *f*
s. magnet Druckmagnet *m*
s. nut Schlagbolzenmutter *f*
s. spring *(Mil)* Schlagbolzenfeder *f*
striking 1. Auftreffen *n*, Aufprallen *n*; 2. Zündung *f (Lichtbogen)*; Überschlag *m*; 3. *(Bw)* Ausschalen *n*; 4. Formen *n* mit Ziehbrett
s. by spark Funkenentzündung *f*
s. distance Schlagweite *f*
s. face of a hammer Hammerpinne *f*
s. fork Schaltgabel *f*
s. hammer Schlegel *m*
s. of sparks *(El)* Funkenziehen *n*
s.-out table *(Led)* Ausrecktisch *m*
s. potential Zündspannung *f*
s. power Schlagkraft *f*
s. rod Schaltstange *f*
s.-up Schablonenformen *n*
s. voltage Zündspannung *f (Lichtbogen, Stromrichter)*
s. work Schlagwerk *n (Uhr)*
string / to binden, schnüren, knüpfen; mit Saiten beziehen
string 1. Schnur *f*, Bindfaden *m*, Faden *m*; Saite *f*; Seil *n*; 2. *(Bw)* Treppenwange *f*, Wange *f*; 3. Kette *f*, Bündel *n*
s. board Treppenwange *f*, Wange *f*
s. construction Fadenkonstruktion *f (Ellipse)*
s. efficiency *(El)* Kettenwirkungsgrad *m*
s. electrometer Saitenelektrometer *n*, Fadenelektrometer *n*
s. galvanometer Saitengalvanometer *n*, Fadengalvanometer *n*
s. holder Saitenfessel *f*, Saitenhalter *m (der Musikinstrumente)*
s. instrument Saiteninstrument *n*, Streichinstrument *n*; Bogeninstrument *n*
s. of insulators Isolator[en]kette *f*
s. oscillograph Saitenoszillograf *m*
s. polygon Seileck *n*
stringed instrument Saiteninstrument *n*, Streichinstrument *n*
stringer 1. Längsbalken *m*, Längsträger *m*, Stützbalken *m*; 2. *(Flg)* Holm *m*, Längsversteifung *f*; 3. *(Schiff)* Stringer *m*; 4. *(Brg)* Unterzug *m (Ausbau)*; 5. *(Kern)* Verschlußstück *n (für Strahlenkanal)*
stringy langspanend; faserig
strip / to abstreifen *(Kokillen)*; abisolieren; reinigen; abkratzen *(Farbe)*; abschälen, ablösen; entbasten; entrippen *(Tabak)*; überdrehen *(Gewinde)*; lockern *(Schraube)*; ausschalen, entschalen *(Betonbau)*; abwickeln *(Draht)*; ausstoßen *(Karde)*; *(Schiff)* abtakeln
s. a wire einen Draht abisolieren
s. off abschälen *(Hanf)*
s. the bark entrinden, entborken, abschälen
s. the hemp Hanf schleißen
strip 1. Band *n*; Streifen *m*; Leiste *f*; 2. *(Kern)* Waschlösung *f (zur Rückextraktion)*
s. attenuator *(El)* Streifenabschwächer *m*
s. chart 1. Registrierstreifen *m*; Meßstreifen *m*; 2. Streifendiagramm *n*
s.-chart recorder Bandschreiber *m*, Linienschreiber *m*, Streifenschreiber *m*
s. coating abstreifbarer Überzug *m*
s. conductor *(El)* Bandleiter *m*
s. cutter Streifenschneider *m*
s. feed Werkstoffzubringung *f (bei Stanz- und Schnittarbeiten)*
s.-feed press Presse *f* mit Streifenzuführung
s. footing Streifenfundament *n*
s. foundation Streifenfundament *n*
s. fuse Streifensicherung *f*, Lamellensicherung *f*
s. heater Heizband *n*
s. light *(Flg)* Landestreifenfeuer *n*
s.-light lamp Soffittenlampe *f*
s. line Bandleiter *m*, Bandleitung *f (Wellenleitertechnik)*
s. load Streifenlast *f*
s. mill Brammenstraße *f (Walzwerk)*
s. mining Tagebauverfahren *n (Kohlenbergbau)*

s.-mounted set *(Fmt)* auswechselbarer Gerätesatz *m*
s. of ground Schlag *m (Feld)*
s. of indicators *(Fmt)* Klappenstreifen *m*
s. of keys *(Fmt)* Tastenstreifen *m*
s. of web *(Text)* Florbändchen *n*
s. packing *(Brg)* Rippenversatz *m*
s.-planking Leistenbauweise *f (Bootsbau)*
s. printer Streifendrucker *m (z. B. Addiermaschine)*
s. rolling mill Bandagenwalzwerk *n*, Bandwalzwerk *n*, Streifenwalzwerk *n*
s. solder Lötstange *f*
s. steel Bandstahl *m*
s. stock Bandmaterial *n*
s.-type earthing conductor *(El)* Banderder *m*
stripe Streifen *m*
s. iron Streifenhobelmesser *n*
striped gestreift
strippable abstreifbar, ablösbar, abschälbar
stripped nackt *(Draht)*
s. atom hochionisiertes (nacktes) Atom *n*
s. thread beschädigtes (abgenutztes) Gewinde *n*
s. wash oil abgetriebenes Waschöl *n (Aufbereitung)*
stripper 1. *(Met)* Abstreifer *m*, Blockabstreifer *m*; 2. *(Kern)* Abscheider *m*, Abtriebssäule *f*; 3. Abtreiber *m (Aufbereitung)*; 4. *(Text)* Wendewalze *f*, Wender *m*; 5. *(Masch)* Führungsplatte *f*
s. anode Mutterblechanode *f*
s. bar *(Text)* Abnehmerleiste *f*
s. finger Abstreifer *m*
s. plate mould *(Plast)* Abstreiferform *f*
s. pressure Niederhaltedruck *m (Tiefziehen)*
s. roller *(Text)* Abnehmerwalze *f*
s. tank Mutterblechbad *n*
stripping 1. Ablösen *n*, Abziehen *n (Farbe)*; 2. Ausschalen *n*, Entschalen *n (Betonbau)*; 3. *(Brg)* Tagebauverfahren *n (Kohlenbergbau)*; Abbau *m* eines sehr dünnen Gangs durch Mitnehmen des Nebengesteins *(Abbauverfahren)*; Abbau der Teilsohle über und neben dem Querbrechen *(Teilsohlenbruchbau)*; 4. *(elektrolytische)* Entplattierung *f*; 5. Abstreifen *n*; 6. Abgraten *n*; 7. Abziehen *n* der Kokille, Strippen *n (Gießerei)*
s. comb Abschlagkamm *m (Spinnerei)*
s. compound *(Plast)* Masse *f* für abstreifbaren Überzug
s. device Abschiebevorrichtung *f*
s. installation *(Text)* Ringelapparat *m*
s. knife Küferzugmesser *n*
s. locomotive Abraumlokomotive *f*
s. plate Abstreifplatte *f (Gießerei)*
s. process Bandangießverfahren *n (Lager)*
s. pumping system *(Schiff)* Nachlenzpumpensystem *n*
s. roller *(Text)* Wender *m*, Wendewalze *f*
s. section Abtriebsäule *f (Destillation)*
s. still *(Brg)* Abtreiber *m*
strobile Zapfen *m (Nadelbaum)*; Dolde *f (Hopfen)*
strobing Signalauswertung *f*
s. pulse Auswertimpuls *m*
s. pulse generator Auswerteimpulsgenerator *m*
stroboglow Stroboskop *n* mit Neonröhre
stroboscope Stroboskop *n*
stroboscopic stroboskopisch
s. calibrating of a meter stroboskopische Zählereichung *f*
s. checking of a meter *s.* s. calibrating of a meter
s. direction finder stroboskopischer Peilempfänger *m*
s. disk stroboskopische Scheibe *f*, Stroboskopscheibe *f*
s. effect stroboskopischer Effekt *m*
s. meter disk stroboskopische Läuferscheibe *f* eines Zählers
s. method stroboskopische Methode *f*, stroboskopisches Verfahren *n*
s. tube Stroboskopröhre *f*
strobotac stroboskopisches Tachometer *n*
strobotron Strobotron *n (Kaltkatodenröhre mit Edelgasfüllung)*
stroke 1. Hub *m*, Kolbenhub *m*; Takt *m*; Schlag *m (Uhr)*; 2. Strich *m (Feile)*; 3. *(Met)* Abstrich *m*; 4. *(Schiff)* Ruderschlag *m*; 5. Federzug *m*
s. adjustment Hubverstellung *f*
s.-bore ratio Hub-Bohrungs-Verhältnis *n*
s. indicator dial Hublängenskala *f*
s. length Hublänge *f*
s. length adjustment Hublängenverstellung *f*
s. of admission Füllungsstrecke *f*
s. of compression Verdichtungshub *m*
s. of exhaust Auspuffhub *m*, Auslaßhub *m*
s. of explosion Arbeitstakt *m*
s. of lathe *(Text)* Ladenhub *m*
s. of piston Kolbenhub *m*
s. of straight sliding link Geradführungslänge *f (Getriebelehre)*
s. positioning shaft Welle *f* für Verlegung des Hubs
s. setting shaft Welle *f* für Hublängenverstellung
s. test *(Plast)* Härtezeittest *m*
s. volume Hubraum *m*
s. wheel Hubrad *n*
stroking speed Hubgeschwindigkeit *f (Honen)*
strong fest *(Festigkeitslehre)*; widerstandsfähig; stark
s.-back *(Bw)* Traverse *f*
s. field *(El)* starkes Feld *n*

s. lining feste Deckschicht *f (Papier)*
s. liquor *(Brg)* Starkwasser *n*
s. signal attenuator Antennenabschwächer *m*
s. wind steife Brise *f*
strontia Strontiumoxid *n*
strontium Strontium *n*
strop / to abziehen, schärfen *(Messer)*
strop Abziehriemen *m*; Abziehapparat *m (für Messer)*
stropping Abziehen *n (Messer)*
struck abgebaut *(Gerüst)*; abmontiert
s. through *(Led)* durchgegerbt
structural baulich, konstruktiv; Gefüge-, Struktur-; Bau-, Konstruktions-
s. adhesive *(Plast)* Montageleim *m*
s. alloy steel legierter Baustahl *m*
s. analysis Baustatik *f*
s. bulkhead *(Schiff)* Verbandsschott *n*, Festigkeitsschott *n*
s. calculation baustatische Berechnung *f*
s. change *(Met)* Strukturänderung *f*, Gefügeumwandlung *f*
s. chemistry Strukturchemie *f*
s. condition *(Met)* Gefügezustand *m*
s. constituent Gefügebestandteil *m*
s. constitution Gefügeaufbau *m*
s. defect Gitterstörstelle *f*
s. design feature Konstruktionsmerkmal *n*
s. drawing Konstruktionszeichnung *f*
s. element Bauteil *n*, Konstruktionselement *n*, Bauelement *n*, Bauglied *n*; Maschinenteil *n*
s. engineer Hochbauingenieur *m*; Hochbaustatiker *m*
s. engineering Ingenieurhochbau *m*
s. fault Strukturfehler *m*, Gefügefehler *m*
s. fibreboard Faserbauplatte *f*
s. form Bauform *f*
s. formula *(Ch)* Strukturformel *f*
s. geology tektonische Geologie *f*
s. insulation board Isolierbauplatte *f*
s. irregularity Gitterstörstelle *f*
s. isomerism *(Ch)* Strukturisomerie *f*
s. lightweight concrete konstruktiver (tragender) Leichtbeton *m*
s. lumber Bauholz *n*
s. member *s.* s. element
s. module Bauraster *m*, Raster *m*; Rastermaß *n*
s. properties *(Brg)* Verbandseigenschaften *fpl*
s. reamer Nietlochreibahle *f*
s. return loss *(El)* Rückflußdämpfung *f*
s. rolling mill Fassonwalzwerk *n*, Formstahlwalzwerk *n*
s. statics Baustatik *f*
s. steel 1. Baustahl *m*; 2. Feinkornstahl *m*
s. steel erection Stahlbau *m*
s. technique *(Text)* Verbindungstechnik *f*
s. theory Baukonstruktionslehre *f*
s. timber Bauholz *n*
s. transformation Gefügeumwandlung *f*, Gefügeneubildung *f*
s. unit *s.* s. element
structurally strukturell, konstruktionsmäßig; gefügemäßig
structure 1. Struktur *f*, Gefüge[aufbau *m*] *n*; 2. Konstruktion *f*; Bauwerk *n*; Tragwerk *n*
s. analysis Strukturuntersuchung *f*
s. anomaly Strukturanomalie *f*
s. of alloys Legierungsaufbau *m*
s. of the grinding wheel Gefüge *n* der Schleifscheibe
s. sensitive strukturempfindlich, strukturabhängig
s. theory Strukturtheorie *f*
structureless gefügelos, amorph
structures Bauten *pl*
strummed instrument Zupfinstrument *n*
strut / to versteifen, verstreben; abstützen
strut 1. Strebe *f*; Druckglied *n*; Druckstab *m*; Verstrebung *f*; Stützsäule *f*, Tragsäule *f*; Stempel *m*; 2. *(Schiff)* Wellenbock *m*; 3. *(Fmt)* Gestänge *n*; Mast *m*; Stange *f*; 4. *(Brg)* Bolzen *m*; 5. Druckfinger *m (am Kupplungshebel)*
s. frame Strebenfachwerk *n*
s. of hornplate *(Eis)* Achsgabelstrebe *f*
struts *(Schiff)* Wellenbockarme *mpl*
strutted pole *(El)* verstrebter Mast *m*; verstrebte Stange *f*
s. terminal pole *(Fmt) (verstrebtes)* Endgestänge *n*
strutting Absteifung *f (Mast)*; Verstrebung *f*; *(Brg)* Verbolzung *f*; Versteifung *f*
stub / to roden; ausreißen
stub 1. Stumpf *m*; Baumstumpf *m*; 2. Stummel *m*; 3. Blindschwanz *m*; Stichleitung *f (Antennen)*
s. aerial Stichleitungsantenne *f*
s. arbor fliegend angeordneter Dorn *m*, freitragender (kurzer) Dorn, Aufsteckdorn *m*
s. axle Achsschenkel *m*, Vorderachsschenkel *m*; Achsenansatz *m*
s. beam kurzer Balken *m*
s. cable *(Fmt)* Abzweigkabel *n*
s. card Abreißkarte *f*, Abschnittkarte *f (gelocht)*
s. drilling Bohren *n* mit kurzem Bohrer
s. line *(El)* Leitungsabzweig *m*; Leitungsansatz *m*
s. matching *(El)* Anpassen *n* mittels Stichleitung
s. reinforced pole *(Fmt)* angeschuhte Stange *f*
s. shaft Zapfwelle *f*; Wellenstumpf *m*
s.-shaft harvester Nockenwellenernter *m*
s. shaft harvesting machine Nockenwellenernter *m*

s. support Stichleitungsträger *m (Wellenleiter)*
s.-supported line Leitung *f* auf Metallisolatoren
s. teeth Stumpfverzahnung *f*
s. tooth Stumpfzahn *m*
s.-tooth gear Rad *n* mit Stumpfverzahnung
s. tuner Abstimmstichleitung *f (Wellenleiter)*
stubble Stoppel *f*; Stoppeln *pl*; Stoppelfeld *n*
s. clearing Stoppelräumung *f*
s. crop Stoppelfrucht *f*
s. cutting adjustment Stoppelverstellung *f (Schneidwerk)*
s. field Stoppelfeld *n*
s. fruit Stoppelfrucht *f*
s. plough Schollenpflug *m*, Schollenwender *m*; Stoppelpflug *m*
stubbling Stoppelsturz *m*
stubby gedrungen
stucco moulding Stuckplastik *f*
stuck-on sole geklebte Schuhsohle *f*
stud Anschlag *m*; Stehbolzen *m*; Wechselbolzen *m*; Stift[schraube *f*] *m*; Warze *f*; Steg *m*; Runge *f*; Stütze *f*
s. bolt Stiftbolzen *m*, Stiftschraube *f*; Gewindestift *m*
s. farm Gestüt *n*
s. gear Drehspindelzahnrad *n (treibendes Rad an Drehmaschine)*
s. gun Bolzenschießgerät *n*
s. inserting machine Bolzenlochbohrmaschine *f*
s. nut Bolzenmutter *f*
s. pin Kettensteg *m*
studded tyre Stahlseilreifen *m*
studdle *(Brg)* Bolzen *m (Bolzenschrotausbau)*
student apprentice Praktikant *m*
studio Studio *n*, Aufnahmeraum *m*; Senderaum *m*, Sendesaal *m*
s. broadcast Studioübertragung *f*, Studiosendung *f*
s. camera Atelierkamera *f*
s. equipment Studioausrüstung *f*
s. light boards Beleuchtungsbühne *f*, Studiolampenbrett *n*
s. pick-up Studioaufnahme *f*
study / to untersuchen
study Untersuchung *f*; Arbeitszimmer *n*
stuff / to polstern; vollstopfen
stuff Grundstoff *m*, Zeug *n*, Materie *f*; Stoff *m*, Gewebe *n*; Mörtel *m*, Speise *f*; Papierzeug *n*, Papierstoff *m*, Papierbrei *m*; taubes Gestein *n*
s. box Stoffauflauf *m (Papierherstellung)*
s. grinder Zerfaserer *m*, Holzschleifmaschine *f (Papierherstellung)*
stuffer *(Plast)* hydraulische Strangpresse *f*
stuffing 1. Polstern *n*; Vollstopfen *n*; Ausstopfen *n*; Füllen *n*; 2. Lederschmiere *f*; Einfetten *n* von Leder
s. aggregate of straw press Stopfer *m* der Strohpresse
s. box Packungsstopfbüchse *f*, Stopfbüchse *f*
s. box bearing Dichtungslager *n*
s. box with separate rings Stopfbuchse *f* aus Einzelringen
s. drum *(Led)* Schmierwalkfaß *n*
stull *(Brg)* Spreize *f*, Strebe *f (Holzausbau)*
stulled *(Brg)* mit Spreizen ausgebaut
s. overhand stoping Firstenstoßbau *m* mit Spreizenausbau
stump 1. Baumstumpf *m*, Stubben *m*; 2. *(Brg)* kleiner Pfeiler *m*
s. puller Wurzelzieher *m*, Baumstumpfzieher *m*, Wurzelroder *m*
s. wood Stockholz *n*
stumpage *(Am) (Holz)* Schnittverlust *m*
stumpiness Völligkeit *f (eines Schiffes)*
stunt / to reißen beim Abkühlen
stunt flying *(Flg)* Kunstflug *m*
sturdiness Stabilität *f*, Festigkeit *f*, Widerstandsfähigkeit *f*
sturdy stabil, robust, gedrungen, kräftig, massiv
sty Koben *m*
style Stil *m*; Art *f*, Bauart *f*, Modell *n*, Ausführung *f*; Fasson *f*, Form *f*; Fall *m*; Sorte *f*; Säule *f*
s. of technical writing technische Ausdrucksweise *f*
s. of tooth Zahnform *f (Verzahnung)*
stylus Fühler[stift] *m*, Taster *m (Nachformverfahren)*; Griffel *m*; Nadel *f (Grammofon)*
s. excursion Nadelauslenkung *f*
s. force Nadeldruck *m*, Auflagedruck *m*
s. mounting Nadelbefestigung *f*
s. tip Nadelspitze *f*
s. trajectory *(El)* Nadelbewegung *f*
s. wear Nadelverschleiß *m*
styphnic acid Styphninsäure *f*, Trinitroresorzin *n*
styptic blutstillend
styptic blutstillendes Mittel *n*
styrene *(Ch)* Styrol *n*
s. plastic Styrolplast *m*
s. resin Styrolharz *n*
styrol *s.* styrene
sub *(Am) s.* subdrift *und* sublevel
subacoustic Unterschall-
subaqueous Unterwasser-
s. cable Unterwasserkabel *n*, Unterseekabel *n*, Seekabel *n*, Flußkabel *n*, unter Wasser verlegtes Kabel *n*
s. loudspeaker Unterwasserlautsprecher *m*
subassembling Teilmontage *f*
subassembly 1. Bauteil *n*, Baueinheit *f*; 2. Teilmontage *f*
s. of positive and negative accumulator plates Akkumulatorplattenblock *m*

subatmospheric unteratmosphärisch
subatomic subatomar
s. particle Elementarteilchen *n*, subatomares Teilchen *n*
subaudio infraakustisch
s. frequency Infraschallfrequenz *f*, unhörbar tiefe Frequenz *f (unter 30 Hz)*
s. telegraphy *(Fmt)* Unterlagerungstelegrafie *f*
subbase Frostschutzschicht *f*, Sauberkeitsschicht *f*, Unterbau *m (Straßenbau)*; Fundamentplatte *f*
subbasement zweites Kellergeschoß *n*
subbing *(Am) (Brg)* Teilsohlenbruchbau *m*
subbituminous coal Moorkohle *f*
subcarrier *(Fs)* Zwischenträger *m*, Unterträger *m*, Sekundärträger *m*
s. amplitude modulation *(El)* Zwischenträgeramplitudenmodulation *f*
s. frequency modulation Zwischenträgerfrequenzmodulation *f*
s. generator Nebenträgergenerator *m*
s. modulation Hilfsträgermodulation *f*, Unterlagerungsmodelung *f*
subcavity mould *(Plast)* Abquetschform *f* mit tiefliegender Matrize
subcircuit Abzweigstromkreis *m*, Nebenstromkreis *m*; Abzweigung *f*
subconcrete Unterbeton *m*
subcontract work Zulieferarbeit *f*
subcontracting factory Zulieferbetrieb *m*
subcontractor *(Bw)* Nachauftragnehmer *m*, Subunternehmer *m*
subcooled unterkühlt
subcooling Unterkühlung *f*
subcritical unterkritisch
s. annealing *(Met)* Rekristallisationsglühen *n*
s. assembly unterkritische Anordnung *f*
s. mass unterkritische Masse *f*
s. reactor unterkritischer Reaktor *m*
subcutaneous blowholes Randblasen *fpl (im Gußblock)*
subdeterminant Unterdeterminante *f*
subdivide / to unterteilen
subdivided shaft *(Masch)* unterbrochene Welle *f*
s. winding *(El)* Doppelwicklung *f*
subdivision 1. Unterteilung *f*; 2. *(Schiff)* Schotteneinteilung *f*, wasserdichte Unterteilung
s. draught Schottentiefgang *m*
s. load line Schottenladelinie *f*
s. of the drop of pressure Druckaufteilung *f*, Unterteilung *f* des Druckgefälles *(Turbine)*
subdomain *(Math)* Teilbereich *m*, Unterbereich *m*
subdrift *(Brg)* Strecke *f* unter der Schwebe zur Einleitung des Strossenbaus
s. caving Teilsohlenbruchbau *m*
subdue / to dämpfen; unterdrücken
subdued light gedämpftes Licht *n*
suberic acid Korksäure *f*, Suberinsäure *f*
suberification Verkorkung *f*
suberination Korkbildung *f*, Verkorkung *f*
subexchange *(Fmt)* Teilamt *n*, Hilfsamt *n*, Unteramt *n*, Unterzentrale *f*
subfeeder Nebenspeiseleitung *f*
subfield Teilkörper *m*
subfluvial cable Flußkabel *n*
subframe *(Kfz)* Unterbau *m*; Untergestell *n*; Nebenrahmen *m*
subfusion temperature Temperatur *f* unter dem Schmelzpunkt
subgrade Unterbau *m (Eisenbahn)*
subgrain Unterkorn *n*, Subkorn *n*
subgroup Untergruppe *f*, Unterabteilung *f*
subharmonic Subharmonische *f*, Unterharmonische *f*, Unterfrequenz *f*
subhead[ing] Untertitel *m*
subincline shaft *(Brg)* tonnlägiger Blindschacht *m*
subinterval Teilintervall *n*
subject to / to beanspruchen auf
subject Fach[gebiet] *n*; Thema *n*
s. contrast Objektkontrast *m*
s. field Fachgebiet *n*
s. index Sachwörterverzeichnis *n*
s. matter Fachgebiet *n*
s. of examination Prüfungsfach *n*
s. to ausgesetzt, unterworfen, abhängig von
s. to authorization genehmigungspflichtig
subjected to beansprucht auf
s. to bending biegebeansprucht
s. to compression druckbeansprucht
s. to tension zugbeansprucht
s. to torsion verdrehbeansprucht
subland drill Programmwerkzeug *n*
s. drill-reamer Folgewerkzeug *n* mit Bohr- und Reibwerkzeug
sublet / to *(Arbeit an Zulieferbetrieb)* vergeben
sublethal *(Kern)* subletal, nichttödlich
s. dose nichttödliche (subletale) Dosis *f*
sublevel 1. *(Ph)* Unterniveau *n*; 2. *(Brg)* Querbrechen *n (Querbruchbau)*; Teilsohle *f*
s. caving Teilsohlenbruchbau *m*
s. crosscut Ortsquerschlag *m*, Teilsohlenquerschlag *m*
s. slicing [with ore caving] *(Am)* Teilsohlenbruchbau *m*
s. stoping Teilsohlenkammerbau *m*, Teilsohlenweitungsbau *m*
sublimability Sublimierbarkeit *f*
sublimable sublimierbar
sublimate / to sublimieren
sublimate Sublimat *n*, Quecksilber(I)-chlorid *n*
sublimation Sublimation *f*
s. furnace *(Met)* Kapellenofen *m*
s. heat Sublimationswärme *f*
sublime / to sublimieren

sublimed sublimiert
s. sulphur Schwefelblüte *f*
subliminal advertizing unterschwellige Reklame *f (im Fernsehen der USA)*
subliming furnace Sublimierofen *m*
submains *(Am) (El)* Verteilersammelschienen *fpl*
submarginal unrentabel, unwirtschaftlich
submarine Unterseeboot *n*, U-Boot *n*
submarine Unterwasser-, Untersee-, unterseeisch
s. cable Seekabel *n*, Unterseekabel *n*, unter Wasser verlegtes Kabel *n*
s. cable of shallow water type Flachseekabel *n*, Küstenkabel *n*
s. camera Unterwasserkamera *f*
s. cargo-carrier Handels-U-Boot *n*, Unterseefrachtboot *n*
s. chaser *(Am)* U-Bootjäger *m*
s. current Unterwasserströmung *f*
s. depot ship U-Bootmutterschiff *n*, Begleitschiff *n* für U-Boote
s. electric detonator Unterwasserzünder *m*
s. hunter U-Bootjäger *m*
s. intermediate reactor mittelschneller Reaktor *m* für Unterseebootantrieb
s.-oil formation ölführendes Gebirge *n* unter dem Meeresspiegel
s. ore-carrier Unterwassererztransporter *m*, Unterwassererzfrachter *m*
s. oscillator Unterwasserschallsender *m*
s. photograph Unterwasseraufnahme *f*
s. sentry Tiefenmelder *m*
s. signal Unterwassersignal *n*
s. telegraph system Seekabeltelegrafie *f*
s. tender *(Am)* U-Bootmutterschiff *n*
s. thermal reactor thermischer Reaktor *m* für Unterseeboote
s. vessel Unterwasserschiff *n*
submaster controller geführter Regler *m*, Sekundärregler *m*
submerge / to überfluten; [unter]tauchen, eintauchen
submerged eingetaucht, untergetaucht
s. aerial Unterwasserantenne *f*
s. arc welding Unterpulverschweißen *n*, UP-Schweißen *n*
s. cable Unterwasserkabel *n*; Seekabel *n*
s. coil condenser Eintauchschlangenkühler *m*, Schlangenkühler *m*
s. condenser *(Wärm)* Unterwasserkondensator *m*, Bassinkondensator *m*
s. propulsion motor Elektromotor *m* für Unterwasserfahrt
s. speed Unterwassergeschwindigkeit *f (U-Boot)*
submersible eintauchbar, tauchfähig; wasserdicht *(Maschine)*
s. boat Tauchboot *n*
s. motor Unterwassermotor *m*
s. pump Tauchpumpe *f*, Unterwasser[motor]-pumpe *f*; *(Schiff)* Notlenzpumpe *f*
submicron *Teilchen n mit Durchmesser von* 5×10^{-7} *bis* 2×10^{-5} *cm*
submicroscopic submikroskopisch
submicrostructure Submikrostruktur *f*, Submikrogefüge *n*
subminiature component Kleinstbauelement *n*
s. engineering Subminiaturtechnik *f*
submodulator *(Fmt)* Modulationsvorverstärker *m*
submultiple resonance untersynchrone Resonanz *f*
subnormal Subnormale *f*
subnormality [elektrische] Untererregbarkeit *f*
subordinate line Nebenlinie *f*
subpanel *(Fmt)* Brett *n* für Schaltelemente, Zwischenbrett *n*
subproduct Teilprodukt *n*
subprogramme Teilprogramm[plan *m*] *n*, Unterprogramm *n*
subrepeater *(Fmt)* Hilfsverstärker *m*
subroutine Nebenrechnung *f*; Teilprogramm *n*, Unterprogramm *n (Rechner)*
subsalt basisches Salz *n*
subscriber *(Fmt)* Teilnehmer *m*
s. dial-pulse repeater Namentaster *m*
s. exchange Teilnehmerzentrale *f*
s. trunk dialling Selbstwählferndienst *m*
s.-vision Abonnementsfernsehen *n*
s. with several lines Sammelanschlußteilnehmer *m*
subscriber's automatic telephone Fernsprechapparat *m* für Wählbetrieb
s. cable Anschlußkabel *n*, Teilnehmerkabel *n*, Ortskabel *n*
s. extension station Nebenstelle *f* mit freier Amtswahl, Privatnebenstelle *f*, Teilnehmeranschluß *m*
s. jack Teilnehmerklinke *f*
s. line Teilnehmer[anschluß]leitung *f*
s. main station Teilnehmerhauptanschluß *m*
s. meter Gesprächszähler *m*
s. number Rufnummer *f*
s. register Gesprächszähler *m*
s. rotary line switch Vorwähler *m*
s. set Teilnehmerapparat *m*, Teilnehmersprechstelle *f*
subscript *(Math)* unterer Index *m*
subscription Abonnement *n*
s. television Abonnementsfernsehen *n*
s. television [set] Fernsehempfänger *m* mit Gebührenautomat
subsequence *(Math)* Teilfolge *f*
subsequent nachfolgend

s. **compensation** Nachkompensierung *f*
s. **enlargement** *(Opt)* Nachvergrößerung *f*
s. **machining** Nacharbeit *f*, Nachbearbeitung *f*
s. **treatment** Nachbehandlung *f*; Weiterverarbeitung *f*
subset 1. *(Math)* Teilmenge *f*; 2. *s.* subscriber's set
subshell *(Ph)* Zwischenschale *f*, Unterschale *f*
subside / to [sich] senken, sich setzen
subsidence Sinken *n*, Einsinken *n*, Senkung *f*, Setzung *f*; Bodensenkung *f*
s. **damage** *(Brg)* Bergschaden *m*
s. **development curve** Verlaufskurve *f*, Entwicklungskurve *f (Aufbereitung)*
s. **ratio** *(BMSR)* Abklingverhältnis *n*
s. **slope** Schieflage *f (Aufbereitung)*
s. **trough** Senkungstrog *m (Aufbereitung)*
subsider Klärbottich *m*
subsidiary Hilfs-, Neben-, mitwirkend
s. **battery** Zusatzbatterie *f*
s. **circuit** *(El)* Hilfskreis *m*
s. **company** Tochtergesellschaft *f*
s. **drift** *(Brg)* Auslenkungsstrecke *f*
s. **lamp** Zusatzlampe *f*
s. **line** Hilfslinie *f (Geometrie)*
s. **lobe** Nebenkeule *f (Antennentechnik)*
s. **patent** Hilfspatent *n*
s. **quantity** Hilfsgröße *f*
s. **resonances** *(El)* Nebenresonanzen *fpl*; Mehrwelligkeit *f*
s. **road** Nebenstraße *f*, Seitenstraße *f*
s. **triangulation** Kleintriangulierung *f*; Hilfstriangulation *f*
s. **unit** Scheidemünze *f*
s. **vibration** *(El)* Hilfsschwingung *f*, Nebenresonanz *f*
subslicing *(Am) (Brg)* Teilsohlenbruchbau *m*
subsoil Baugrund *m*, Boden *m*, Untergrund *m*
s. **chisel** *(Lw)* Untergrundmeißel *m*
s. **cultivator** Tieflockerer *m*, Wühlgrubber *m*
s. **drain** Dränung *f*
s. **packer** *(Lw)* Untergrundpacker *m*
s. **plough** Untergrundpflug *m*
s. **water** Grundwasser *n*
s. **water level** Grundwasserspiegel *m*
s. **water packing** Grundwasserabdichtung *f*
subsonic Unterschall-
s. **power amplifier** Tieftonleistungsverstärker *m*
s. **speed** Unterschallgeschwindigkeit *f*
s. **velocity** Unterschallgeschwindigkeit *f*
substage Beleuchtungsapparat *m (für Mikroskope)*
s. **control** Kondensorbewegung *f (des Mikroskops)*
substance 1. Stoff *m*, Substanz *f*; Material *n*; 2. Wanddicke *f*; 3. *(Brg)* Mächtigkeit *f*; 4. Quadratmetergewicht *n (von Papier)*
s. **density** Substanzdichte *f*
substandard Substandard *m*
s. **film** Schmalfilm *m*
s. **film stock** Schmalfilmmaterial *n*
s. **meter** *(El)* Vergleichszähler *m*, Hilfsnormal *n (für Zählereichung)*
s. **stock** Schmalfilmmaterial *n*
substantive dye direktziehender (substantiver) Farbstoff *m*
s. **dyeing** *(Text)* Direktfärbung *f*
substation *(El)* Unterwerk *n*; *(Fmt)* Teilnehmersprechstelle *f*
substituent Austauschstoff *m*, Substituent *m*
substitute / to austauschen, ersetzen, substituieren; einsetzen
substitute Austauschstoff *m*, Ersatz[stoff] *m*, Surrogat *n*
s. **material** Austauschwerkstoff *m*
substitution Substitution *f*, Ersetzen *n*, Einsetzung *f*; Ersatz *m*
s. **conduction** Substitutionsleitung *f*
s. **isomerism** Substitutionsisomerie *f*
s. **kernel procedure** Ersatzkernverfahren *n (bei Integralgleichungen)*
s. **method** Substitutionsmethode *f*
substitutional resistance Ersatzwiderstand *m*
s. **solid solution** Substitutionsmischkristalle *mpl*, Austauschmischkristalle *mpl*
substoping Teilsohlenkammerbau *m*, Teilsohlenweitungsbau *m*
substrate Schichtträger *m*, Substrat *n*, Trägersubstanz *f*
substratoplane Höhenflugzeug *n*
substratosphere Substratosphäre *f*
substratum *(Geol)* unterlagernde Schicht *f*
substructure Unterbau *m*
subsurface irrigation Untergrundberieselung *f*
s. **pipe drain** Dränstrang *m*
s. **stress** Spannung *f* unter der Oberfläche
subsynchronous untersynchron
subtangent Subtangente *f*
subtend / to entgegengesetzt liegen zu, sich erstrecken unter
subtense bar Basislatte *f (Vermessungswesen)*
subterranean unterirdisch
s. **cable** Erdkabel *n*, Untergrundkabel *n*, unterirdisch verlegtes Kabel *n*
subthreshold dose *(Kern)* unterschwellige Dosis *f*
subtitle Untertitel *m*, Nebentitel *m*
subtolerance *(Kern)* Dosis *f* unter der Toleranzgrenze, verträgliche Dosis
subtract / to abziehen, subtrahieren
subtraction Subtraktion *f*
s. **circuit** Subtraktionsschaltung *f*
subtractive *(Opt)* subtraktiv
s. **colour** subtraktive Farbe *f*, Subtraktionsfarbe *f*

s. **colour process** subtraktives Dreifarbenverfahren *n*
s. **primary colour** subtraktive Grundfarbe *f*
subtractor *(Opt)* Subtraktivfilter *n*
subtrahend Subtrahend *m*
subtropics Subtropen *pl*
subunit array *(Ch)* Untereinheitenanordnung *f* *(Proteinstruktur)*
suburb Vorort *m*, Vorstadt *f*, Stadtrandsiedlung *f*
suburban vorstädtisch, Vorstadt-, Vorort-
s. **bus** Vorortomnibus *m*
s. **call** *(Fmt)* Vorortsgespräch *n*
s. **connection** *(Fmt)* Vorortsverbindung *f*
s. **traffic** *(Eis, Fmt)* Vorortverkehr *m*
s. **train** Vorortbahn *f*, Vorortzug *m*
subway 1. Untergrundbahn *f*, U-Bahn *f*; 2. Unterführung *f*; 3. unterirdischer Kanal *m*
s. **line** Untergrundbahnlinie *f*, Untergrundbahnstrecke *f*
subzero cooling Unterkühlen *n*
s. **treatment** Tieftemperaturbehandlung *f*, Kältebehandlung *f* *(Vergüten)*
succession 1. Reihenfolge *f*; 2. *(Geol)* Abfolge *f*, Ausscheidungsfolge *f*, Folge *f*
s. **of beds** *(Geol)* Schichtenfolge *f*
s. **of cranks** *(Masch)* Kurbelfolge *f*
s. **of crops** *(Lw)* Fruchtfolge *f*
s. **of pass** Kaliberfolge *f* *(Walzwerk)*
s. **of phase** *(El)* Phasenfolge *f*
s. **of pulses** *(Fs)* Impulsfolge *f*, Impulsreihe *f*
s. **of strata** *(Geol)* Schichtfolge *f*
successive aufeinanderfolgend
s. **carry** *(Dat)* Serienübertrag *m*
s. **detection** *(El)* Kettengleichrichtung *f*
succinate Succinat *n*, Ester *m* (Salz *n*) der Bernsteinsäure
succindialdehyde Succindialdehyd *m*, Bernsteinsäuredialdehyd *m*
succinic acid Bernsteinsäure *f*
s. **anhydride** Bernsteinsäureanhydrid *n*
succinite Bernstein *m*
suck / to saugen
s. **away** absaugen
s. **off** absaugen
s. **up** aufsaugen
sucker 1. Wurzelableger *m*, Wurzelschößling *m*; 2. Saugrohr *n*; Saugnapf *m*; Gummisauger *m*; Kolben *m*
s. **rod** Pumpenstange *f*
s. **rods** Pumpengestänge *n*
suckering Ausgeizen *n* *(Weinbau)*
sucking Ansaugen *n*, Einsaugen *n*
s. **coil** *(El)* Tauchkernspule *f*
s. **transformer** Saugtransformator *m*
sucrose Rohrzucker *m*, Sa[c]charose *f*
suction Saugen *n*, Saugwirkung *f*, Ansaugung *f*; Saugmenge *f*; Sog *m*; Unterdruck *m*; Zug *m*
s. **aerofoil** *(Flg)* Unterdrucktragfläche *f*
s. **air drier** Sauglufttrockner *m*
s. **ball** Gummisauger *m*, Saugball *m*
s. **basket** Saugkopf *m*
s. **blank feed attachment** Zuführeinrichtung *f* mit Saugnapf *(Presse)*
s. **boost** Verstärkung *f* der Ansaugmenge durch Hilfspumpe
s. **bottle** Saugflasche *f*, Vakuumflasche *f*
s. **box** Absauger *m*, Saugkasten *m*
s. **capacity** Ansaugleistung *f*
s. **chamber** Saugraum *m*
s. **cleaner** Sauger *m*, Aufsauggerät *n*
s. **connection** Ansaugstutzen *m*
s. **couch roll** Zellensaugwalze *f* *(Papierherstellung)*
s. **cutter** Saugkopf *m* mit Schneidwerk *(beim Bagger)*
s.**-cutter dredger** Saugbagger *m* mit Schneidkopf, Schneidkopfbagger *m*
s. **cycle of milker** Saugtakt *m* der Melkmaschine
s. **drain** Saugablaß *m*
s. **draught** Saugzug *m*
s. **dredger** Saugbagger *m*, Pumpenbagger *m*
s. **drier** Saugtrockner *m*
s. **dust remover** *(Masch)* Staubabsauger *m*
s. **effect** Sogwirkung *f*
s. **filter** Saugfilter *n*, Nutsche *f*
s. **flange** Saugflansch *m*
s. **flask** Saugflasche *f*, Saugkolben *m*
s. **funnel** Nutschentrichter *m*
s. **gas** Sauggas *n*
s. **gas main** Saugleitung *f*
s. **head** Saughöhe *f*
s. **head of the pump** Saughöhe *f* der Pumpe
s. **height** Saughöhe *f*
s. **hopper-dredger** Saughopperbagger *m*
s. **intake** Ansaugöffnung *f*
s. **lift** Saughöhe *f*
s. **line** Ansaugleitung *f*
s. **main** Ansaugleitung *f*
s. **pipe** Saugleitung *f*; Ansaugrohr *n*, Saugkopf *m*
s. **pipe connection** Saugstutzen *m*
s. **pipe discharge** Saugrüssel *m* *(Getreideentladung)*
s. **pipe friction** Saugrohrreibung *f*
s. **pipe line** Saugrüssel *m* *(Getreideentladung)*
s. **port** Ansaugkanal *m*; Ansaugschlitz *m* *(Zweitaktverbrennungsmotor)*
s. **pump** Saugpumpe *f*
s. **resistance** Saugwiderstand *m*
s. **speed** Saugleistung *f*
s. **strainer** Ansaugkorb *m*
s. **stroke** Ansaugtakt *m*
s. **valve** Saugventil *n* *(Pumpe)*; Einlaßventil *n* *(Motor)*

s. valve bushing Saugventilbuchse *f*
s. valve cone Saugventilkegel *m*
s. valve spring Saugventilfeder *f*
s. ventilator Sauglüfter *m*, Luftabsauger *m*
sudden plötzlich, unvermittelt, schroff, schlagartig
s. break in the weather Witterungsumschlag *m*
s. change in frequency Frequenzsprung *m*
s. change in programme velocity *(Rak)* Programmsprung *m*
s. fall of temperature Temperatursturz *m*
s. interruption plötzliche Unterbrechung *f*
s. load stoßartige Beanspruchung *f*; Stoßbelastung *f*
s. pull-up hartes Abfangen *n (des Flugzeuges)*
sudorific schweißtreibendes Mittel *n*
sudorific schweißtreibend
suds Seifenschaum *m*; Schmiermittel *n* auf Wasserbasis; Kühlmittel *n*
s. booster *(Am)* Schaumverbesserer *m*
s. pump Kühlmittelpumpe *f*
s. tank Kühlmittelbehälter *m*
sue for damages / to auf Schadenersatz verklagen
suede leather Plüschleder *n*, Sämischleder *n*
sugar / to [über]zuckern; auskristallisieren
sugar Zucker *m*
s. alum Alaunzucker *m*
s. beet Zuckerrübe *f*
s.-beet cossette Zuckerrübenschnitzel *n*
s. beet harvester Zuckerrübenerntemaschine *f*
s.-beet juice Zuckerrübensaft *m*
s. candy Zuckerkand *m*
s. cane Zuckerrohr *n*
s.-cane crusher Zuckerrohrquetsche *f*
s. cube press Würfelzuckerpresse *f*
s. dye Karamel *n*, Zuckerfarbe *f*
s. factory Zuckerfabrik *f*
s. industry Zuckerindustrie *f*
s. manufacture Zuckerherstellung *f*
s. meter Zuckergehaltsmesser *m*
s. mill Zuckerfabrik *f*
s. of lead Bleizucker *m*, Bleiazetat *n*
s. polarimeter Zuckerpolarimeter *n*
s. powder Puderzucker *m*
s. scale *(Opt)* Zuckerskale *f*
sugaring Überzuckern *n*, Verzuckern *n*
sugary zuckerhaltig, zuckerig; süß
suggestion box Einwurfkasten *m* für Verbesserungsvorschläge *(in USA und Großbritannien)*
suggestions for improvements *(Ök)* Verbesserungsvorschläge *mpl*
suit / to [an]passen, sich eignen, taugen
suitable brauchbar, geeignet, zweckmäßig, sachgemäß
suite of cores *(Brg)* Kernfolge *f*, Kernsatz *m*
sulf . . . *s.* sulph . . .
sull coat / to *(Met)* anlaufen lassen *(Draht)*
sullage Schlammablagerung *f*; Abwässer *npl*; Abfall *m*
s. lighter Aschenprahm *m*
sulphamic acid Amidosulfosäure *f*
sulphanilic acid Sulfanilsäure *f*
sulphate / to sulfat[is]ieren, schwefeln
sulphate Sulfat *n*, Salz *n* der Schwefelsäure
s. cellulose Sulfatzellulose *f*
s. of lead Bleisulfat *n*
s. paper Natronpapier *n*
s. pulp Sulfatzellstoff *m*
sulphating roasting *(Met)* sulfatisierende Röstung *f*
s. roasting furnace *(Met)* Sulfatisierofen *m*
sulphide / to sulfidieren, mit Schwefelwasserstoff umsetzen
sulphide sulfidisch, schwefelhaltig
sulphide Sulfid *n*
s. inclusion *(Met)* Sulfideinschluß *m*
s. ore sulfidisches Erz *n*, Sulfiderz *n*, Schwefelerz *n*
sulphidic schwefelhaltig, sulfidisch
sulphitation Sulfitierung *f*, Schwefelung *f*
sulphite Sulfit *n*, Salz *n* der schwefligen Säure
s. liquor Sulfit[ab]lauge *f*
s. pulp Sulfitzellstoff *m*
sulphoacid Sulfonsäure *f*, Sulfosäure *f*
sulphocyanate Rhodanid *n*, Sulfozyanat *n*, Thiozyanat *n*
sulphocyanic acid Rhodanwasserstoffsäure *f*, Thiozyansäure *f*
sulphonate / to sulfonieren, sulfurieren
sulphonated castor oil sulfoniertes Rizinusöl *n*, Türkischrotöl *n*
s. oil geschwefeltes (sulfoniertes) Öl *n*
sulphonic acid Sulfonsäure *f*, Sulfosäure *f*
s. amide Sulfonamid *n*
s. chloride Sulfochlorid *n*
sulphoxylate Sulfoxylat *n*, Salz *n* der Sulfoxylsäure
sulphoxylic acid Sulfoxylsäure *f*, unterschweflige Säure *f*
sulphur / to schwefeln, ausräuchern
sulphur Schwefel *m*
s. black Schwefelschwarz *n*
s. cement Schwefelzement *m*
s. compound Schwefelverbindung *f*
s. determination Schwefelbestimmung *f*
s. dye[stuff] Schwefelfarbstoff *m*
s. halide Schwefelhalogen *n*
s. point Schwefelpunkt *m (Temperaturskala)*, Siedepunkt *m* des Schwefels
s. printing Schwefelabdruck *m*
s. removal Entschwefelung *f*, Schwefelentfernung *f*

s. segregation Schwefelseigerung *f*
sulphurate / to schwefeln
sulphureous schweflig, schwefelartig, schwefelhaltig
sulphuretted hydrogen Schwefelwasserstoff *m*
sulphuric acid Schwefelsäure *f*
s. anhydride Schwefeltrioxid *n*, Schwefelsäureanhydrid *n*
s. ether Diäthyläther *m*, Schwefeläther *m*
sulphurize / to schwefeln, sulfieren
sulphurous schweflig, schwefelartig, schwefelhaltig
s. acid schweflige Säure *f*
s. anhydride Schwefeldioxid *n*, Anhydrid *n* der schwefligen Säure
s. pyrites Schwefelkies *m*
sulphuryl chloride Sulfurylchlorid *n*
sum / to addieren; summieren
s. up zusammenzählen, addieren
sum Summe *f*
s. frequency Summenfrequenz *f*
s. of digits Quersumme *f*
s. rule Summenregel *f*
s. voltage Summenspannung *f*
sumach tanned sumachgar *(Holz)*
Sumatra camphor Sumatrakampfer *m*, d-Borneol *n*
summarize / to zusammenfassen
summary Zusammenfassung *f*, Kurzreferat *n*
s. card Summenkarte *f*
s. punch Summenlocher *m*, Summenstanzer *m*
s. recorder Summenschreiber *m*
summation Summierung *f*, Zusammenzählung *f*
s. check Summenkontrolle *f*, Summenprobe *f*, Summenprüfung *f*
s. equation Summengleichung *f*
s. index Summationsindex *m*
s. instrument summierendes Meßgerät *n*
s. tone Summationston *m*, Summenton *m*
s. transformer Summenstromwandler *m*
summator amplifier *(El)* Summationsverstärker *m*
summer 1. Sommer *m*; 2. Balken *m* *(für ruhende Last)*; 3. *(Dat)* Summator *m*, Summierverstärker *m*
s. draught *(Schiff)* Sommerfreibordtiefgang *m*
s. freebord *(Schiff)* Sommerfreibord *m*
s. grain Sommergetreide *n*
s. tree Balken *m* *(für ruhende Last)*
summerwood Spätholz *n*
summing amplifier *(Dat)* Summierverstärker *m*, Summator *m*
s. circuit Additionsschaltung *f*, Summierwerk *n*, Zählwerk *n*
s. integrator *(Dat)* Summierintegrator *m*
s. junction Summierverbindungsstelle *f* *(Rechenverstärker)*
s. point *(Dat)* Mischstelle *f*, Summierstelle *f*
s.-up Zusammenrechnen *n*
s.-up Addier-
summit Gipfel *m*; Krone *f*; Kappe *f*; Spitze *f*
s. of trajectory *(Mil)* Gipfelpunkt *m* der Flugbahn
sump 1. *(Brg)* Sumpf *m*, Einbruch *m*; 2. Pumpensumpf *m*; 3. *(Masch)* Ölfangschale *f*, Ölwanne *f*, Ölsumpf *m*; 4. *(Met)* Ofensumpf *m*
s.-hole Sickeranlage *f*
s. shaft Pumpenschacht *m*
sumper *(Brg)* Einbruchschuß *m*
s. drill Einbruchbohrmaschine *f*
sun Sonne *f*
s.-and-planet gearing Planetengetriebe *n* *(Verzahnung)*
s. bleaching Sonnenbleiche *f*
s. blind Markise *f*
s. compass Sonnenkompaß *m*
s. deck *(Schiff)* Sonnendeck *n*
s. dial Sonnenuhr *f*
s. energy Sonnenenergie *f*
s. gear Sonnenrad *n* *(Getriebe)*
s. irradiation Sonnenbestrahlung *f*
s. lamp Höhensonne *f*
s. prism Sonnenprisma *n*
s. roof Schiebedach *n*
s. shade *(Opt)* Sonnenblende *f*
s. visor *(Kfz)* Sonnenblende *f*
s. wheel Sonnenrad *n* *(Verzahnung)*
sunfast sonnebeständig, sonnenlichtecht
sunflower oil Sonnenblumenöl *n*
sunk eingefallen; versenkt; graviert; eingearbeitet
s. forecastle versenkte Back *f* *(teilweise unter und teilweise über dem Oberdeck eines Schiffes)*
s. key Einlegekeil *m*
s. spot *(Plast)* Einfallstelle *f*, Einsackstelle *f*
s. switch versenkter Schalter *m*, Unterputzschalter *m*, Dosenschalter *m* für Montage unter Putz, Einlaß[dosen]schalter *m*
s. wall switch Unterputz[dosen]schalter *m*, Einsatz[dosen]schalter *m*
s. wreck gesunkenes Wrack *n*
sunken torpedo Grundgänger *m* *(Torpedo)*
sunrise Sonnenaufgang *m*
sun's corona Sonnenkorona *f*
s. corpuscular radiation Korpuskularstrahlung *f* der Sonne
s. disk Sonnenscheibe *f*
s. interior Sonneninneres *n*
s. magnetism Sonnenmagnetismus *m*
s. motion Sonnenbewegung *f*
s. parallax Sonnenparallaxe *f*
sunscreens Sonnenschutzmittel *npl*
sunseeker *(Astr)* Sonnensucher *m*
sunset Sonnenuntergang *m*
sunspot Sonnenfleck *m*

s. activity Sonnenfleckenaktivität *f*
s. period Sonnenfleckenperiode *f*
superacidity Superazidität *f*
superacidulation Übersäuerung *f*
superacoustic telegraphy Überlagerungstelegrafie *f*
superaerodynamics *(Rak)* Superaerodynamik *f*
superallowed *(Ph)* übererlaubt
s. transition übererlaubter Übergang *m*
superaudic telegraphy Überlagerungstelegrafie *f*
superballoon tyre *(Kfz)* Superballonreifen *m*
supercalender Satinierkalander *m (Papierherstellung)*
supercalendered paper hochsatiniertes Papier *n*, Illustrations[druck]papier *n*
supercarburization *(Met)* Überkohlung *f*
supercavitation Überkavitation *f*, Superkavitation *f*
supercharge / to aufladen *(Verbrennungsmotor)*
supercharge Aufladung *f (Verbrennungsmotor)*
s. pressure Aufladedruck *m*
supercharged engine aufgeladener (vorverdichteter) Motor *m*; Kompressormotor *m*
supercharger Aufladegebläse *n*, Vorverdichter *m*, Kompressor *m*, Auflader *m*, Lader *m*
s. power absorption Leistungsaufnahme *f* des Laders
s. pressure Ladedruck *m*
supercharging Aufladung *f*; Vorverdichtung *f*
supercompressed überkomprimiert, überverdichtet
superconducting *(El)* supraleitend, supraleitfähig
s. magnet supraleitender Magnet *m*
s. memory supraleitfähiger Speicher *m (für Daten)*
superconductive *(El)* supraleitend, supraleitfähig
s. state supraleitender Zustand *m*
s. thin film memory supraleitender Dünnschichtspeicher *m*
superconductivity *(El)* Supraleitfähigkeit *f*
superconductor Supraleiter *m*
supercontrol tube *(El)* Exponentialröhre *f*, Regelröhre *f*
supercooling Unterkühlen *n*, Unterkühlung *f*
supercosmic rays ultraharte kosmische Strahlen *mpl*
supercritical überkritisch
s. mass überkritische Masse *f*
s. pile überkritischer Reaktor *m*
superdrill Hochleistungsbohrer *m*
superelastic collision *(Ph)* Stoß *m* zweiter Art
s. stress überelastische Beanspruchung *f*
superelevation 1. Überhöhung *f (Straßenbau, Eisenbahnbau)*; 2. *(Fs)* Abhebung *f*
s. of outer rail Schienenüberhöhung *f*
superemitron *(Fs)* Superikonoskop *n*
superexcitation *(El)* Übererregung *f*
superfatting agent Überfettungsmittel *n*
superficial an der Oberfläche; Oberflächen-
s. current Oberflächenstrom *m*
s. friction Oberflächenreibung *f*
s. hardening Oberflächenhärtung *f*
s. hardness Oberflächenhärte *f*
s. velocity Oberflächengeschwindigkeit *f*
superfinish / to *(Masch)* feinstbearbeiten, feinstziehschleifen, schwingschleifen
superfinishing Feinstziehschleifen *n*; Kurzhubhonen *n*
s. process Feinstbearbeitungsverfahren *n*; Superfinish-Verfahren *n*
superfluidity *(Ph)* Supraflüssigkeit *f (Eigenschaft)*
superfrequency Hochfrequenz *f (3000 bis 30000 MHz)*
supergrid Hochspannungsnetz *n*
supergroup Übergruppe *f (Trägerfrequenztechnik)*
s. distribution frame Übergruppenverteiler *m*
s. translating equipment Übergruppenmodulationseinrichtung *f*
superheat *s.* superheating
superheated überhitzt
s. steam überhitzter Dampf *m*
superheater Überhitzer *m*
s. chamber Überhitzerkammer *f*
s. chamber in smoke box Überhitzerfeuerkiste *f*
s. for horizontal boiler Langkesselüberhitzer *m*
s. for marine boiler Schiffskesselüberhitzer *m*
s. for semi-portable boiler Lokomobilüberhitzer *m*
s. not placed in a chamber kammerloser Überhitzer *m*
s. placed in a superposed chamber Überhitzer *m* in aufgesetzter Kammer
s. placed in the flue Kesselzugüberhitzer *m*, Feuerzugüberhitzer *m*
s. regulator Überhitzerregler *m*
s. with by-pass flue ausschaltbarer Überhitzer *m*
s. with sliding damper Überhitzer *m* mit versenkbarem Schieber
superheating Überhitzen *n*, Überhitzung *f*; Überheizung *f*; Wärmestau *m*
s. between expansion stages Zwischenüberhitzung *f*
s. plant Überhitzeranlage *f*
s. surface Überhitzerheizfläche *f*
s. tube Überhitzerrohr *n*
superhet *s.* superheterodyne

superheterodyne Überlagerungsempfänger *m*, Super[het] *m*
s. converter Überlagerungsvorsatzgerät *n*
s. receiver *s.* superheterodyne
s. reception Überlagerungsempfang *m*
s. tracking Gleichlauf *m* zwischen Eingangs- und Oszillatorkreis
superheterodyning Überlagerung *f*
superhigh frequencies superhohe Frequenzen *fpl (Zentimeterwellen von 3000 bis 30000 MHz)*
s. speed steel Hochleistungsschnellstahl *m*
supericonoscope *(Fs)* Superikonoskop *n*
superimpose / to überlagern
superimposed circuit *(Fmt)* Phantomkreis *m*
s. circuit coil *(Fmt)* Viererspule *f*
s. code Überlagerungskode *m (z. B. auf Lochkarten)*
s. current überlagerter Strom *m*
s. decks übereinanderliegende Decks *npl*
s. telegraphy Überlagerungstelegrafie *f*
s. time markers *(Fmt, Fs)* eingetastete Zeitmarken *fpl*
superimposing *(Fmt)* Superponierung *f*, Überlagerung *f*
s. the web in layers *(Text)* schichtweises Aufeinanderlegen *n* des Flors
superionoscope *(Fs)* Bildwandlerionoskop *n*, Superionoskop *n*
superior überlegen
s. stand Hauptbestand *m (Forst)*
superlattice structure Überstruktur *f*
superliner Riesenschiff *n*
supermarket Einkaufszentrum *n*
supernatant layer *(Ch)* obenschwimmende (überlagernde) Schicht *f*
supernumerary überzählig
superorthicon *(Fs)* Superorthikon *n*
superphantom circuit *(Fmt)* Achterleitung *f*
s. telegraphy Achtertelegrafie *f*
superphosphate of guano Guanosuperphosphat *n (Düngemittel)*
superposable überlagerbar
superpose / to überlagern, überschichten
superposed circuit Überlagerungskreis *m*, Simultanleitung *f*
s. current überlagerter Strom *m*
s. field *(El)* überlagertes Feld *n*
s. magnetization Vormagnetisierung *f*
s. ringing *(Fmt)* überlagerter Gleichstromruf *m*
s. superheater chamber with swivel damper *(Wärm)* aufgesetzte Überhitzerkammer *f* mit Regulierdrehklappe
s. telegraph Simultantelegraf *m*
superposition Superposition *f*, Überlagerung *f*
s. principle Superpositionsprinzip *n*
superpower Hochleistungs-
s. distribution Großstromversorgung *f*
s. valve Kraftverstärkerröhre *f*, Leistungsröhre *f*
superpressure Höchstdruck *m*; Überdruck *m*
superradio waveband Grenzwellenband *n* zwischen Infrarot und Millimeterwellen
superrefined steel Edelstahl *m*
superregeneration *(Fmt)* Pendelrückkopplung *f*
superregenerative detector Pendelrückkopplungsaudion *n*
s. reception Pendelrückkopplungsempfang *m*, Überrückkopplungsempfang *m*, Superregenerativempfang *m*
superregenerator *(Fmt)* Pendelrückkopplungsempfänger *m*
superroad *(Am)* Autobahn *f*
superroyal *Druckpapierformat* $20^1/_2 \times 27^1/_2''$
supersaturate / to übersättigen
supersaturated übersättigt
s. solution übersättigte Lösung *f*
s. steam übersättigter Dampf *m*
supersaturation Übersättigung *f*
superscaler *(Kern)* 1000-fach-Untersetzer *m*, 1000-fach-Zähler *m*
superscript *(Math)* oberer Index *m*
supersonic ultraschallfrequent, Ultraschall-
s. cavitation Überschallkavitation *f*
s. communication Ultraschallverkehr *m*, Ultraschallverbindung *f*
s. echo sounder Ultraschallecholot *n*
s. flight Flug *m* mit Überschallgeschwindigkeit
s. generator Ultraschallgenerator *m*
s. soldering Ultraschallöten *n*, Löten *n* mit Ultraschall
s. sound wave Ultraschallwelle *f*
s. speed Überschallgeschwindigkeit *f*
s. testing Überschallprüfung *f*, Ultraschallprüfung *f (Materialprüfung)*
s. transport aircraft Überschallverkehrsflugzeug *n*
s. velocity Überschallgeschwindigkeit *f*
superstructure *(Bw)* Überbau *m*; Aufbau *m*; Schiffsaufbau *m (Decksaufbau)*
s. deck Aufbaudeck *n*
s. frame Aufbauspant *n*
supertanker Supertanker *m*, Großtanker *m*
supertension cable Höchstspannungskabel *n*
superturnstile aerial Mehrfachschmetterlingsantenne *f*
supervise / to überwachen
supervision of air navigation Flugüberwachung *f*
supervisory control Fernsteuerung *f (Energiesysteme)*
s. equipment Überwachungseinrichtung *f*
s. magnet switch Überwachungsschalter *m (Signalwesen)*
s. rack *(Fmt)* Überwachungsgestell *n*

s. relay *(El, Fmt)* Kontrollrelais *n*, Überwachungsrelais *n*
s. signal *(Am) (Fmt)* Schlußzeichen *n*
superwide angle objective *(Opt)* Überweitwinkelobjektiv *n*
s. angle projector *(Opt)* Überweitwinkelprojektor *m*
superwood harzgetränktes Holz *n*
supplant / to ersetzen, verdrängen
supple biegsam, schmiegsam
supplement 1. Ergänzung *f*, Nachtrag *m*; 2. Supplementwinkel *m*; Nebenwinkel *m*
supplementary supplementär, zusätzlich
s. air Nachluft *f*
s. angle Supplementwinkel *m*; Nebenwinkel *m*
s. anode Hilfsanode *f*
s. bridge *(Masch)* Einsatzbrücke *f*
s. condition Nebenbedingung *f*
s. drawing *(Text)* Nachstrecken *n*
s. equipment Zusatzeinrichtung *f*; Ergänzungsausrüstung *f*
s. lens Vorsatzlinse *f*, Vorsatzobjektiv *n*
s. loss Zusatzverluste *mpl*
s. part Zusatzteil *n*
s. relay Hilfsrelais *n*, Zwischenrelais *n*
s. resistance *(El)* Ergänzungswiderstand *m*
s. surgery lamp Operationshilfsleuchte *f*
s. work Nachtragearbeit *f*
supplies Zubehör *n*; Zulieferungen *fpl*
s. road *(Brg)* Strecke *f* für Materialtransporte
supply / to versorgen; zuführen; zuleiten; liefern
supply 1. Versorgung *f*, Belieferung *f*, Lieferung *f*, Zufuhr *f*, Zuführung *f*; Vorrat *m*; 2. *(El)* Zuleitung *f*, Stromzuführung *f*; Speisung *f*; Stromversorgung[snetz *n*] *f*
s. area Versorgungsgebiet *n*
s. bin Vorratssilo *m*
s. cable Zuführungskabel *n*
s. circuit Speisestromkreis *m*
s. cord Netzanschlußschnur *f*
s. creel *(Text)* Magazingatter *n*
s. current Netzstrom *m*
s. district Versorgungsgebiet *n*
s. fluctuations *(El)* Netzschwankungen *fpl*
s. frequency *(El)* Netzfrequenz *f*
s. hoisting *(Brg)* Materialeinhängen *n*
s. lead Zuführung[sleitung] *f*, Speiseleitung *f*, Stromzuführung *f*
s. line Zuleitung *f*; Netz *n*, Hilfsenergieversorgung *f (elektrisch oder pneumatisch)*
s. mains *(Fmt)* Hausanschluß[leitung *f*] *m*
s. metre Stromverbrauchszähler *m*
s. network Stromversorgungsnetz *n*
s. of circulating water Kühlwasserzuleitung *f*
s. of current Stromversorgung *f*
s. of energy Energiezufuhr *f*
s. of power 1. *(Masch)* Kraftabgabe *f*; 2. *(El)* Stromversorgung *f*
s. point Stromanschluß[stelle *f*] *m*
s. raise *(Brg)* Materialüberhauen *n*, Versorgungsüberhauen *n*
s. reel Transporttrommel *f*
s. relay *(Fmt)* Speiserelais *n*
s. rocket Versorgungsrakete *f*
s. ship Proviantschiff *n*, Versorgungsschiff *n*
s. spool Transporttrommel *f*
s. station Kraftwerk *n*
s. steam pipe Dampfeintrittsrohr *n*
s. system Versorgungsnetz *n*
s. transformer Speisetransformator *m*
s. voltage Netzspannung *f*, Versorgungsspannung *f*, Ladespannung *f*
support / to [ab]stützen, tragen; aufgelagert haben, aufliegen; *(Brg)* ausbauen
support Auflage *f*; Auflager *n*, Lagerung *f*, Stütze *f*; Absteifung *f*; Gestell *n*, Halter *m*, Lager *n*, Säule *f*, Ständer *m*, Sockel *m*; Stativ *n*; *(Plast)* Träger[material *n*] *m*; *(Brg)* Ausbau *m*
s. arm Tragarm *m*
s. box *(El)* Befestigungsdose *f*
s. bracket Stützwinkel *m*
s. conditions Auflagerbedingungen *fpl*
s. for bearing *(Masch)* Lagerträger *m*
s. movement *(Bw)* Stützensenkung *f*
s. of the upper drum Oberkesselstütze *f*
s. pattern *(Brg)* Ausbauschema *n*
s. plate Zwischenplatte *f (Preßform)*
s. rule *(Brg)* Ausbauregel *f*
s. strand Entlastungsseil *n*, Tragseil *n (für Luftkabel)*
s. tube Stützrohr *n*
supported 1. gestützt, getragen, aufgelagert; 2. mit Füllstoff (Trägerstoff) versehen
s. piston getragener Kolben *m*
s. wire aerial aufgehängte Drahtantenne *f*
supporting tragend, Stütz-, Trag-
s. base Träger *m*
s. bracket Tragknagge *f*, Stützkonsole *f*
s. cable Tragseil *n*
s. force Auflagerkraft *f*, Stützkraft *f (Statik)*
s. frame tragendes Gestell *n (Getriebelehre)*
s. insulator Stützisolator *m*, Stützer *m*
s. long thin work Spannen *n* langer zylindrischer Werkstücke
s. loop *(Kfz)* Handschlaufe *f*
s. of the boiler Stützung (Auflagerung) *f* des Kessels
s. post Aufschlagständer *m (Eisenbahnschranke)*
s. reaction Auflagerkraft *f*, Auflagerreaktion *f*, Stützkraft *f (Statik)*
s. rod Trägerstab *m (Elektronenröhrenherstellung)*
s. strap Haltegriff *m*

s. strength of filling *(Brg)* Tragfähigkeit *f* des Versatzes
s. structure tragende Konstruktion *f*, Tragwerk *n*
s. surface Auflagefläche *f*; *(Flg)* tragende Fläche *f*, Tragfläche *f*
s. tower *(El)* Abspannmast *m*; Brückenturm *m*
suppress / to unterdrücken; verzögern; dämpfen; sperren
suppressed automatic volume control *(Fmt)* verzögerte Schwundregelung *f*
s. balloon spinning Spinnen *n* bei begrenztem Fadenballon, Spinnen mit unterdrücktem Ballon
s. carrier transmission Senden *n* mit unterdrückter Trägerwelle
s. frame method *(Fs)* Halbbildverfahren *n*
s. frequency band *(Fmt)* Sperrbereich *m* *(eines Filters)*
s. scale Skala *f* mit unterdrücktem Nullpunkt
s. zero instrument Meßinstrument *n* mit unterdrücktem Nullpunkt
suppression Unterdrückung *f*
s. capacitor *(El)* Störschutzkondensator *m*
s. device Entstörvorrichtung *f*
s. filter *(Fmt)* Sperrfilter *n*, Sperrkreis *m*
s. of noise Schalldämmung *f*, Schallabwehr *f*, Geräuschbeseitigung *f*
suppressor Begrenzerschaltung *f*, Bremsgitter *n* *(einer Röhre)*, Drosselschaltung *f*; Nullode *f*, Nahechodämpfungsschaltung *f* *(Radar)*
s. choke *(Fmt)* Hochfrequenzentstördrossel *f*, Entstördrossel *f*
s. circuit Begrenzerschaltung *f* *(Funkempfangstechnik)*
s. grid Bremsgitter *n*, Fanggitter *n* *(Elektronenröhren)*
s. grid-mesh Bremsgitterwindungen *fpl*
s. modulation Bremsgittermodulation *f*
s. resistance Entstörungswiderstand *m*
supra-acoustic frequency *(Fmt)* Ultraschallfrequenz *f*
s.-conductor Supraleiter *m*
supradyne reception *(Fmt)* Überlagerungsempfang *m*
Supreme Mining Authority Oberste Bergbehörde *f*
surcharge / to überlasten; überspannen *(Dampf)*
surcharge Überbeanspruchung *f*
surd irrationale Zahl *f*
surface / to 1. abflächen; abrichten; glatthobeln; plandrehen; 2. auftauchen *(U-Boot)*
surface Fläche *f*, Oberfläche *f*, Außenfläche *f*, Außenseite *f*; Oberseite *f*; Belag *m* *(z. B. einer Straße)*; *(Brg)* Tagesoberfläche *f*, Tag *m*; Erdoberfläche *f*
s. abrasion Abriebfestigkeit *f*
s. acceleration Flächenbeschleunigung *f*
s. active oberflächenaktiv
s. adhesion Flächenanziehung *f*
s.-air missile Boden-Luft-Rakete *f*
s. analyzer Oberflächenabtaster *m*
s. appearance Tragbild *n* *(Kolben usw.)*
s. appearance of cavity Erscheinung *f* der Oberfläche *(Kavitation)*
s. area Oberfläche *f*; wirksame Oberfläche *f*; Flächenausdehnung *f*
s. attraction Flächenanziehung *f*
s. blowhole *(Met)* Außenlunker *m*
s. brightness Flächenhelligkeit *f*
s. broach Außenräumwerkzeug *n*
s. broaching Außenräumen *n*
s.-broaching machine Außenräummaschine *f*
s. carburizing *(Met)* Oberflächenkohlung *f*
s. charge Oberflächenladung *f*; Flächenladung *f*
s. coating Oberflächenüberzug *m*, Oberflächenauflage *f*
s. combustion Glühkopfzündung *f*; Vorverbrennung *f*
s. concentration Oberflächenkonzentration *f*
s. condenser *(Wärm)* Oberflächenkondensator *m*
s. conditions Oberflächenbeschaffenheit *f*
s. conductance Oberflächenleitfähigkeit *f*
s. conduction Oberflächenleitung *f*
s. contact Flächenberührung *f*
s.-contact rectifier Flächengleichrichter *m* *(Halbleitertechnik)*
s. contamination *(Kern)* Oberflächenverseuchung *f*
s. content Flächeninhalt *m*
s. cooling Oberflächenkühlung *f*
s. crack Oberflächenriß *m*
s. craft Überwasserfahrzeug *n*
s. cratering *(Masch)* Spanflächenauskolkung *f*
s. current Oberflächenstrom *m*
s. curvature Flächenkrümmung *f*
s. decarburization *(Met)* Oberflächenentkohlung *f*
s. defect Oberflächenfehler *m*
s. density Flächendichte *f*
s. density of electric charge elektrische Flächendichte *f*
s. density of luminous flux Flächenlichtstromdichte *f*
s. Diesel-engine Dieselmotor *m* für Überwasserfahrt *(U-Boot)*
s. diffusion Oberflächendiffusion *f*
s. discharge Oberflächenentladung *f*
s. disintegration Verwitterung *f*
s. distribution Flächenverteilung *f*
s. distribution of curl *(Ph)* Flächenwirbel *m*
s. dose *(Kern)* Oberflächendosis *f*
s. drift *(Brg)* tonnlägiger Tagesschacht *m*

s. duct *(Fmt)* bodennaher Kanal *m*
s. effect ship Luftkissenfahrzeug *n*
s. element Flächenelement *n*
s. energy Oberflächenenergie *f*, Grenzflächenenergie *f*
s. expansion Flächenausdehnung *f*
s. feed heater Oberflächenvorwärmer *m*
s. finish 1. Oberflächengüte *f*; Oberflächenausführung *f*, Oberflächenbeschaffenheit *f*
s. finish tester Oberflächenprüfgerät *n*
s. flaw Oberflächenfehler *m*
s. force Oberflächenkraft *f*, Flächenkraft *f*
s. forces Überwasserstreitkräfte *fpl*
s. friction Oberflächenreibung *f*
s. gauge Parallelreißer *m*, Parallelmaß *n*, Reißstock *m*
s. grinder *s.* s. grinding machine
s. grinding Flächenschliff *m*
s. grinding machine Flächenschleifmaschine *f*
s.-hardened oberflächengehärtet
s. hardening Oberflächenhärtung *f*
s. hardness Oberflächenhärte *f*
s. heating Oberflächenvorwärmung *f*
s. ignition Glühzündung *f*
s. imperfection Oberflächenfehler *m*
s. integral Flächenintegral *n*
s. ionization Oberflächenionisierung *f*
s. irradiation Oberflächenbestrahlung *f*
s. irregularities Oberflächenfehler *mpl*
s. jig *(Text)* Oberflottenjigger *m*
s. layer Oberflächenschicht *f*; Randschicht *f*
s. leakage Oberflächenableitung *f*, Kriechen *n* *(von Strömen)*
s. leakage current Kriechstrom *m*
s. length of roll Walzenballenlänge *f*, Ballenlänge *f*
s. machine Hobelmaschine *f*
s. magnetization Oberflächenmagnetisierung *f*
s. meter Oberflächenmeßgerät *n*
s. microphone Körperschallmikrofon *n*
s. migration Oberflächenwanderung *f*
s. milling Planfräsen *n*, Flächenfräsen *n*
s. moisture Oberflächenwasser *n*
s. noise Nadelgeräusch *n* *(Plattenspieler)*
s. of a liquid Flüssigkeitsspiegel *m*
s. of absorption Absorptionsfläche *f*
s. of action Eingriffsfläche *f*
s. of bore Bohrungswandung *f*
s. of contact Kontaktfläche *f*; Berührungsfläche *f*
s. of discontinuity Diskontinuitätsfläche *f*, Unstetigkeitsfläche *f*
s. of equal phase Fläche *f* gleicher Phase
s. of revolution Rotationsfläche *f*
s. of rotation Rotationsfläche *f*
s. of separation Grenzfläche *f*
s. of the earth Erdoberfläche *f*
s. of the piston Stirnfläche *f* des Kolbens
s. of the thread *(Text)* Fadenmantel *m*
s. oil indications Oberflächenanzeichen *npl* von Erdöllagerstätten
s. orientation Oberflächenorientierung *f*
s. photoelectric effect lichtelektrischer Oberflächeneffekt *m*
s. pipe Leitrohrtour *f*, Standrohr *n*
s. planing machine Planhobelmaschine *f*
s. plate Abrichtplatte *f*, Anreißplatte *f*, Richtplatte *f*, Tuschierplatte *f*
s. preparation Oberflächen[vor]behandlung *f*
s. pressure Flächendruck *m*, Flächenpressung *f*, Anpreßdruck *m*
s. printing Flachdruck *m*
s. protection Oberflächenschutz *m*
s. quality Oberflächengüte *f*
s. refinement Oberflächenveredlung *f*, Oberflächenvergütung *f*
s. resistance Oberflächenwiderstand *m*
s. rough grinding Flächenschruppschleifen *n*
s. roughness Oberflächenrauheit *f*
s. scratching test *(Met)* Ritzhärteprüfung *f*
s. search radar Radargerät *n* zur Abtastung der Erdoberfläche
s. sheet Deckblatt *n*, Deckbogen *m*, Oberflächenbogen *m*
s. smoothness Oberflächenglätte *f*
s. socket Aufputzsteckdose *f*
s. source Flächenquelle *f*
s. specimen block Oberflächennormale *f*
s. speed 1. Überwassergeschwindigkeit *f* *(U-Boot)*; 2. *(Masch)* Umfangsgeschwindigkeit
s. stability *(Schiff)* Formstabilität *f*
s. stress *(Met)* Oberflächenspannung *f*
s. string Leitrohrtour *f*, Standrohr *n*
s. structure *(Met)* Oberflächengefüge *n*
s.-surface missile Boden-Boden-Rakete *f*
s. switch Aufputzschalter *m*
s. table Abrichtplatte *f*
s. tension Oberflächenspannung *f*
s. tester Flächenprüfer *m*, Oberflächenprüfgerät *n*
s. thermometer Oberflächenthermometer *n*, Außenthermometer *n*
s.-to-air-missile Boden-Luft-Rakete *f*
s.-to-surface missile Boden-Boden-Rakete *f*
s. treatment Oberflächenbearbeitung *f*, Oberflächenbehandlung *f*
s.-underwater missile Boden-Unterwasser-Rakete *f*
s. unit Oberflächeneinheit *f*, Flächeneinheit *f*
s. vessel Überwasserschiff *n*
s. vibration Oberflächenrüttlung *f*
s. vibrator Oberflächenrüttler *m*
s. voltage gradient *(El)* Schrittspannung *f*
s. water inflow *(Brg)* Oberflächenwasserzufluß *m*

s. wave Oberflächenwelle *f*
s. wear *(Text)* Oberflächenverschleiß *m*, Abtragen *n*
s. winding *(El)* Oberflächenwicklung *f*
s. wiring Leitungsverlegung *f* auf Putz, Aufputzinstallation *f*
s. working Oberflächenbearbeitung *f*
surfacer 1. *(Plast)* Oberflächenkitt *m*, stark pigmentiertes Grundmittel *n*; 2. Planbearbeitungsmaschine *f*
surfacing 1. Flächenbearbeiten *n*; Planen *n* *(z. B. Plandrehen)*; 2. *(Plast)* Deckbogen *m*, Deckplatte *f*; 3. Decken[ein]bau *m* *(Straßenbau)*; Straßenbelag *m*, Straßendecke *f*, Verschleißschicht *f*
s. end Kalanderpartie *f* *(Papiermaschine)*
s. feed Planvorschub *m* *(Drehmaschine)*
s. lathe Plandrehmaschine *f*
s. motion Planzug *m*
s. of cores Oberflächenbehandlung *f* von Mittellagen *(z. B. bei Furnieren)*
s. operation Plan[dreh]arbeit *f*
s. slide Plandrehschlitten *m*
surfboat Brandungsboot *n*
surge / to wogen; pumpen
s. back and forth *(El)* hin- und herschwingen
surge Spannungsstoß *m*, Wanderwelle *f* [mit steiler Front]; Überspannung *f*, Überstrom *m* *(Halbleitertechnik)*
s. absorber Wellenschlucker *m*
s. amplitude Stoßamplitude *f*
s. arrester Überspannungsableiter *m*
s. bin Ausgleichsbunker *m*
s. characteristic Sprungkennlinie *f*
s. dissipator Überspannungsableiter *m*
s. diverter Überspannungsableiter *m*
s. forward current Durchlaßstromstoß *m*
s. front Wellenfront *f*
s. guard Pendelsperre *f* *(Schutzrelais)*
s. hopper Ausgleichsbunker *m*
s. impedance Wellenwiderstand *m*
s. limit Pumpgrenze *f*
s. line Pumpgrenze *f*
s. load Stoßlast *f*
s. protection Wanderwellenschutz *m*; Übergangsschutz *m*
s. relay Stromanstiegsrelais *n*
s. tank Beruhigungsbehälter *m*, Windkessel *m*, Zwischenbehälter *m*
s. voltage Stoßspannung *f*, Wellenspannung *f*
surgery Operationssaal *m*
s. hand lamp Operationshandleuchte *f*
s. lamp Operationsleuchte *f*
s. microscope Operationsmikroskop *n*
surgical instrument chirurgisches Instrument *n*
s. wool Verbandwatte *f*
surging Pumpen *n*
s. shock Verdichtungsstoß *m* *(Turbine)*
surglace hochsatiniertes Papier *n*; scharfsatiniertes Papier
surmount / to bedecken, überheben, überhöhen, überragen
surplus Überschuß *m*; Mehrwert *m*
s. gas Überschußgas *n*
s. gas burner Abgasfackel *f*
s. gas screw Gasballastschraube *f*
s. load Mehrbelastung *f*
s. production Mehrerzeugung *f*, Überproduktion *f*
surrosion Massezunahme *f* infolge Korrosion
surrounding rock *(Brg)* Nebengestein *n*
surveillance Überwachung *f*
s. radar Rundsuchradar *n*, Rundsichtradargerät *n*
s. radar equipment Rundsuchradaranlage *f*
survey / to vermessen
survey 1. Übersicht *f*, Überblick *m*; 2. Überwachung *f*; 3. Vermessung *f* *(s. a.* surveying)
s. aircraft Vermessungsflugzeug *n*
s. measurement markscheiderische Messung *f*
s. of the trajectory *(Rak)* Bahnvermessung *f*
s. ship Vermessungsschiff *n*
surveying Vermessung *f*; Feldmessung *f*, Landmessung *f*; Vermessungskunde *f* *(s. a.* survey 3.)
s. boat Peilboot *n*, Vermessungsboot *n*
s. of boreholes Bohrlochneigungsmessungen *fpl*
s. ship Vermessungsschiff *n*, Bereisungsschiff *n*
s. vessel Vermessungsschiff *n*
surveyor Landmesser *m*; Überprüfer *m* für Meßeinrichtungen und Gewichte; *(Schiff)* Schiffsbesichtiger *m*, Vermessungsingenieur *m*
surveyor's chain Absteckkette *f* *(Vermessungswesen)*
s. level Nivelliergerät *n*
s. rod Meßlatte *f*
susceptance Blindleitwert *m*, Suszeptanz *f*
susceptibility 1. Empfindlichkeit *f*, Anfälligkeit *f*, Neigung *f*; 2. *(El)* Suszeptibilität *f*
s. of magnet core *(Ph)* Kernsuszeptibilität *f*
s. to aging Alterungsempfindlichkeit *f*
s. to corrosion Korrosionsanfälligkeit *f*
s. to cracking Rißempfindlichkeit *f*
s. to glare Blendungsempfindlichkeit *f*
s. to shocks Stoßempfindlichkeit *f*
s. to welding cracks Schweiß[rissigkeits]-empfindlichkeit *f*
susceptible empfindlich, anfällig
s. to aging alterungsempfindlich
suspend / to 1. aufhängen; frei tragen; einseitig unterstützen; 2. schweben *(Teilchen in Flüssigkeit)*; 3. einstellen, aufheben
s. traffic den Funkverkehr einstellen

suspended aufgehängt, Hänge-, Schwebe-, mit Aufhängung
s. by aufgehängt an
s. coil type galvanometer *(El)* Drehspulgalvanometer *n* mit Fadenaufhängung
s. furnace front aufgehängtes Geschränk *n*
s. lamp Hängeleuchte *f*
s. loop hängende Schleife *f (Schleifengalvanometer)*
s. matters Schwebestoffe *mpl*
s. platform Schwebebühne *f*
s. railway Schwebebahn *f*
s. roof Hängedach *n*
s. rudder Schweberuder *n*
s. slay *(Text)* Hängelade *f*
s. span Einhängeträger *m (Brückenbau)*
s. table *(Schiff)* Hängetisch *m*, Schlingertisch *m*
suspension 1. Aufhängung *f*; 2. Suspension *f*; Aufschlämmung *f*
s. arm Schwingarm *m*; Tragarm *m*
s. bracket Traverse *f*
s. bridge Hängebrücke *f*
s. by means of rods Aufhängung *f* durch Hängestangen
s. cable Tragseil *n*
s. chain Hängekette *f*, Tragkette *f*
s. clamp Hängeklemme *f*
s. crane Hängekran *m*
s. device Aufhängevorrichtung *f*
s. in water Aufschlämmung *f* in Wasser
s. insulator Hängekette *f*, Hängeisolator *m*, Kettenisolator *m*
s.-polymerization Perlpolymerisation *f*, Suspensionspolymerisation *f*
s. railway Hängebahn *f*
s. rod Hängestange *f (Regler)*
s. segment Tragkranz *m*
s. switch Hängeschalter *m*
s. tower Tragmast *m*
s. tube Tragrohr *n*
s. type drier Suspensionstrockner *m*
s. wire Tragdraht *m*, Spanndraht *m*
suspensoid Schwebeteilchen *n*, Suspensoid *n*, Suspensionskolloid *n*
sustained ungedämpft; Dauer-, kontinuierlich, ununterbrochen
s. audio signal Dauerton *m*
s. flight Dauerflug *m*
s. operation *(Fmt)* Dauerbetrieb *m*
s. oscillation ungedämpfte (kontinuierliche) Schwingung *f*
s. reaction ununterbrochene Reaktion *f*; Kettenreaktion *f*
s. short-circuit Dauerkurzschluß *m*
s. short-circuit current Dauerkurzschlußstrom *m*
s. wave ungedämpfte Welle *f*
sustenance Nährwert *m*
suture / to vernähen
swab Wattebausch *m*; Tampon *m*, Tupfer *m*; Anfeuchtschwamm *m*; Bürste *f* zum Benetzen
s. pot Wassertopf *m*
swabbing of oil wells Pistonieren *n* von Ölsonden
swage / to tiefziehen *(Plaste)*; fassonschmieden, gesenkdrücken *(Metall)*; stauchen *(Sägezähne)*
swage Matrize *f*; Gesenk *n*
s. block Lochplatte *f*, Gesenkplatte *f*
swaged tooth point gestauchte Zahnspitze *f (Sägeblatt)*
swaging Gesenkschmieden *n*; Gesenkdrücken *n*; Stauchen *n*; Tiefziehen *n (Plaste)*
s. die Zahnstaucher *m (Sägeblatt)*
swamp cypress Sumpfzypresse *f*; Sumpfzypressenholz *n*
swamping resistance Spulenvorwiderstand *m (elektrische Meßtechnik)*
swampy soil Moorboden *m*, Sumpfboden *m*
swan-neck ventilator Schwanenhalslüfter *m*
s.-necked gekröpft *(Meißel)*
s.-necked insulator Isolator *m* mit Schwanenhalsstütze
Swan cap *(El)* Swan-Sockel *m*
swansdown *(Text)* Barchent *m*
swarf Späne *mpl*; Schleifschlamm *m*
s. clearance Spanabführung *f*; Spanauswurf *m*, Entfernen *n* der Späne
s. congestion Spanstauung *f*
s. conveyor Späneförderer *m*
s. disposal Spanabfuhr *f*
s. pan Spanfangschale *f*
s. relieving stroke Rücklauf *m* zum Ausspänen *(Gewindebohren)*
s. removal Spanabfuhr *f*, Ausspänen *n*
swash bulkhead *(Schiff)* Schlagschott *n*, Schlingerschott *n*
s. plate Taumelscheibe *f*
s.-plate engine Taumelscheibenmotor *m*
s.-plate pump Axialkolbenpumpe *f* mit Taumelscheibe
swath clearing Schwadenaufnahme *f (Erntemaschine)*
s. clearing machine Schwadenräumer *m*
s. lifting Schwadenaufnahme *f*
s. press Schwadenpresse *f*
s. tedder Schwadenwender *m*
s. turning arm Schwadenwendearm *m*
swather *(Lw)* Schwader *m*
swathing press Schwadenpresse *f*
s. rake Schwadenrechen *m*
sway / to schwingen
sway Pendelausschlag *m*; Schwung *m*
s. bracing Schlingerverband *m (Brückenbau)*
swaying pendelartige Schwingung *f*
sweal / to abbrennen *(Schiffsboden)*

sweat / to 1. schwitzen; fermentieren lassen *(Tabak)*; 2. löten; 3. *(Met)* seigern
sweat Schwitzwasser *n*
s. band Schweißband *n (Hut)*
s. board Schweißlatte *f (Verkleidung gegen Schwitzwasser im Schiff)*
sweated-on lead caps Kabelendverschluß *m*
sweating *(Met)* Seigern *n (Zinn)*; Löten *n (durch Erwärmen des Grundwerkstoffes)*
s. thimble Kabelschuh *m*, Löt[kabel]schuh *m*
swedge *(Am) s.* swage
Swedish turnip Steckrübe *f*
S. vaulting box schwedischer Sprungkasten *m (Sport)*
S. wall bars schwedisches Wandreck *n (Sportgerät)*
sweep / to 1. fegen, kehren *(Schornstein)*; 2. ablenken *(Oszillograf)*; 3. schweifen; überstreichen, bestreichen; abtasten
s.-back pfeilen *(Tragfläche)*
s. clear off abstreifen [von]
s. out überstreichen
sweep 1. Schwung *m*; überstreichende Bewegung *f*; Abtastung *f*; 2. Drehbrett *n*; 3. langes Ruder *n*; 4. Bereich *m*
s. amplifier Ablenkverstärker *m*
s. amplitude Kippamplitude *f*
s.-back Pfeilung *f (Tragfläche)*
s. band-width Kippbandbreite *f*
s. circuit Zeitablenkschaltung *f (Katodenstrahloszillografie)*
s. cutting saw Schweifsäge *f*
s. deflection Kippablenkung *f*
s. diagram Kippdiagramm *n*
s. frequency Ablenkfrequenz *f*, Kippfrequenz *f*, Wobblerfrequenz *f*
s. frequency generator Kippschwingungsgenerator *m*
s. gas *(Kern)* Spülgas *n*
s. generator Kippgenerator *m*, Kippschaltung *f*; Frequenzwobbler *m*, Wobbelgenerator *m*
s. hand Sekundenzeiger *m*
s. initiating pulse Auslöseimpuls *m* für Ablenkspannung
s. linearity *(Fs)* Ablenkgeradlinigkeit *f*
s. magnification Zeitdehnung *f*
s. oscillator Kipp[schwingungs]generator *m*, Kipp[schwingungs]oszillator *m*
s. range Wobbelbereich *m*
s. rate Wobbelgeschwindigkeit *f*; Zeitablenkgeschwindigkeit *f*
s. reversion Schleifenumkehrung *f (Turbine)*
s. saw Schweifsäge *f*
s. table Stoßherd *m*, Rüttelherd *m*
s. unit *(Fs)* Zeilenablenkgerät *n*; Zeitablenkeinheit *f*
s. voltage Kippspannung *f*, Zeitablenkspannung *f*
sweeper Kehrmaschine *f*, Kehrer *m*
sweeping 1. Wobbeln *n*, Durchlauf *m*; 2. Schablonieren *n*
s. beam Schreibstrahl *m*
s. coil Ablenkspule *f*
s. oscillator Wobbeloszillator *m*
s. system *(Fs)* Ablenksystem *n*
s. voltage *(Fs)* Ablenkspannung *f*
sweepings Kehricht *m*
sweeps 1. *(Fs)* Ablenkspannungen *fpl*; 2. lange Ruder *npl*
sweet 1. süß; 2. gasfrei
s. coal schwefelfreie Kohle *f*
s. gasoline geruchfreies Benzin *n*
s. oil Olivenöl *n (dickflüssig, minderer Qualität)*
s. pea Zuckererbse *f*
s. tester *(Wärm)* Kondenswasserprüfer *m*
s. wort Würze *f (Brauerei)*
sweetie barrel Pilzmischer *m*, schrägstehender Trommelmischer *m*
swell / to 1. [auf]quellen; schwellen, ausbauchen, bombieren; verdicken; 2. aufgehen *(Kalk)*; 3. *(Brg)* quellen *(Gebirge)*; 4. blähen *(Kohle)*
swell 1. Schwellung *f*; Wölbung *f*; 2. Ausbuchtung *f*, Ausbauchung *f*, Bombage *f*; Verdickung *f*; Wulst *m*; Verstärkung *f*; 3. Dünung *f*; Stauung *f*, Gegenströmung *f*
s.-proof finish *(Text)* Quellfestappretur *f*
swelling 1. Quellen *n*, Quellung *f*, Aufquellung *f*; Quellvermögen *n*; 2. Verstärkung *f*, Wulst *m*; Verdickung *f*; 3. Anwachsen *n*, Anschwellen *n*; 4. *(Brg)* Treiben *n*; Blähen *n*
s. agent Quellmittel *n*
s. capacity Blähvermögen *n*
s. coal treibende Kohle *f*
s. pressure Treibdruck *m*
s. resistant finish *(Text)* quellfeste Ausrüstung *f*
swept-back wing *(Flg)* Pfeilflügel *m*
s. capacity *s.* s. volume
s. volume Hubraum *m*, Hubvolumen *n*, Zylinderinhalt *m*
swerve / to abweichen *(vom Kurs)*
swerve Kursabweichung *f*
swift / to *(Schiff)* banken
swift *(Text)* Seidenhaspel *f*
swifter *(Schiff)* Stoßkissenfender *m*
swill / to spülen, waschen
swim barge *(Schiff)* Zille *f*, Schute *f* mit hochgezogenem Boden
s.-fin Schwimmflosse *f*
swimming-pool reactor Schwimmbadreaktor *m*, Swimming-Pool-Reaktor *m*
s. suit Badebekleidung *f*, Badeanzug *m*
swing / to schwenken, schwingen; pendeln; ausschlagen
s. about schwenken um
s. back zurückschwingen

s. off *(Flg)* [von der Startbahn] abkommen
s. out ausschwenken, ausschwingen
s. the plane *(Flg)* Flugzeug herumziehen
swing 1. *(Masch)* Drehdurchmesser *m*, Umlaufdurchmesser *m* *(Drehmaschine)*; 2. Ausschlag *m*; Schwingung *f*; 3. Schaukel *f* *(Turngerät)*
s. angle Drehwinkel *m*, Schwenkbereich *m* *(Kran)*
s.-away wegdrehbar
s. axle Pendelachse *f*, Schwingachse *f*
s.-back zurückklappbar
s. bearing Pendellager *n*
s. berth Schlingerkoje *f*
s. bolster Pendelwiege *f*; Wiege *f* *(Drehgestell)*
s. bolt Gelenkschraube *f*, Klappschraube *f*
s. boom *(Schiff)* Backspiere *f*
s. bridge Drehbrücke *f*
s. conveyor Schwingförderer *m*
s. cut *(Brg)* Fächereinbruch *m* *(Schießarbeit)*
s. diameter *(Masch)* Umlaufdurchmesser *m*, Schwingdurchmesser *m*
s. feeder Schwingzubringer *m*; Schüttelspeiser *m*
s. frame Schere *f* *(Drehmaschine)*
s.-frame grinder Pendelschleifmaschine *f*
s. grinding machine Pendelschleifmaschine *f*
s. gripper *(Typ)* Schwinggreifer *m*
s.-hammer crusher Hammermühle *f* *(Aufbereitung)*
s.-jib crane Kran *m* mit Drehausleger
s. joint Drehanschluß *m*, Gelenkverbindung *f*, Kugelgelenk *n*
s. loader Schwenk[schaufel]lader *m*, Schwenkschaufler *m*
s. needle box *(Text)* schwingendes Nadelgehäuse *n*
s. of frequency Frequenzabweichung *f*
s. on the dial Ausschlag *m* auf der Skala, Skalenausschlag *m*
s.-out ausschwenkbar, ausschwingbar
s.-out condenser Gelenkkondensator *m* *(Mikroskop)*
s. pipe Schwenkrohr *n*, Schwingrohr *n*
s. plate Schere *f* *(Drehmaschine)*
s. plough Schwingpflug *m*
s. rest schwenkbarer Support *m*
s.-round method Umschlagverfahren *n* *(Prüfen)*
s. saw Pendelsäge *f*
s. screen Schwingsieb *n*
s. sieve Schwingsieb *n*; Rätter *m* *(f)*
s. sledge mill schwere Hammermühle *f*, Schwingmühle *f*
swinging schwingend; schwenkbar
swinging Schwenkbewegung *f* *(Bagger)*; Schwenken *n*; Schwingen *n*
s. arm Schwenkarm *m*; schwingende Kurbelschleife *f*
s.-block mechanism Kurbelschleifenmechanismus *m* *(Getriebelehre)*
s.-buoy *(Schiff)* Deviationsboje *f*
s. choke *(El)* Schwingdrossel *f*, Siebdrossel *f*
s. lever Schwenkarm *m*
s. radius Ausladung *f* eines Kranes
s. rings Aufzugringe *mpl* *(Turnen)*
s. screen Schwingsieb *n*
s. shovel Wurfwidder *m*, schwingende Wurfschaufel *f*
s. table Schwenktisch *m*
s. vane Schwingflügel *m*
swingle / to schwingen, brechen *(Flachs, Hanf)*
swirl / to wirbeln, strudeln
swirl Drall *m*, Wirbel *m*
s. chamber Wirbelkammer *f*
s.-chamber engine Wirbelkammermotor *m*
s. combustion chamber Drallbrennkammer *f*
s. nozzle *(Rak)* Dralldüse *f*
s.-type combustion chamber Wirbelkammer *f*
swirler Drallvorrichtung *f*, Drallkörper *m*; Wirbelkammer *f*
swirling Wirbelung *f*; Bewegung *f* *(Schmelzbad)*
swishtail / to *(Flg)* abbremsen *(durch Seitenruderbetätigung)*
Swiss file Nadelfeile *f*
S. milk kondensierte Milch *f*
S.-type automatic screw machine Schweizer Langdrehautomat *m*
swissing calender *(Text)* Rollkalander *m*, Druckkalander *m*
switch / to 1. [um]schalten; 2. rangieren
s. back zurückschalten
s. into circuit einschalten
s. off ausschalten, abschalten
s. on einschalten, zuschalten
s. over [to] umschalten [auf]
s. through durchschalten
s. to low beam abblenden *(Scheinwerfer)*
switch 1. *(El)* Schalter *m*; Lastschalter *m* *(bei Hochspannungsgeräten)*; 2. *(Eis)* Weiche *f*
s. assembly Schalttafel *f*
s. base Schaltersockel *m*, Schaltergestell *n*, Schaltergrundplatte *f*
s. blade Weichenzunge *f*
s. box 1. Schaltkasten *m*; 2. Fernsprechvermittlungsschrank *m*
s. cabinet *(Fmt)* Schaltschrank *m*
s. circuit Schalterstromkreis *m*
s. column Schaltsäule *f*
s. connection Schalteranschluß *m*
s. contact Schalterkontakt *m*
s. cover Schalterkappe *f*, Schalterdeckel *m*
s. cradle *(Fmt)* Apparategabel *f*, Gabelschalter *m*
s. cubicle Schaltschrank *m* *(Werkzeugmaschine)*

s. cycle Schaltspiel *n*
s. desk Schaltpult *n*
s. for surface mounting Aufputzschalter *m*
s. frame *(Fmt)* Wählergestell *n*
s. frog Weichenherzstück *n*
s. fuse Einsatzsicherung *f*
s. gallery Schaltwarte *f*
s. gear *s.* switchgear
s. hook Schaltstange *f*, Schalthaken *m*; Hakenumschalter *m*
s. isolator Lasttrennschalter *m*
s. knob Schalterknopf *m*, Schalterknebel *m*
s. lamp-holder Hahnfassung *f*
s. lamp-socket *s.* s. lamp-holder
s. lever 1. Schalthebel *m*; 2. Weichenstellhebel *m*
s.-over plug umschaltbarer Stecker *m*
s. pillar Schaltsäule *f*
s. plug Steckkontakt *m*
s.-plug box Steckdose *f*
s. point Schaltort *m*, Schaltstelle *f*
s. position Schalterstellung *f*
s. rail Zunge *f (Weiche)*
s. relay Umschaltrelais *n*
s. section Schaltfeld *n*
s. stand Stellbock *m*, Weichenstellbock *m*
s. starter Anlaßschalter *m*
s. tank Schalterkessel *m*
s. unit Schaltpaket *n*; Schaltereinheit *f*
s. wafer Schaltscheibe *f*
switchable [um]schaltbar
s. television set Fernsehempfänger *m* mit umschaltbarer Zeilennorm
switchboard Schalttafel *f*, Schaltbrett *n*; *(Fmt)* Vermittlungsschrank *m*
s. connection Schalttafelanschluß *m*
s. cord Vermittlungsschnur *f*
s. drop *(Fmt)* Anrufklappe *f*
s. fuse unit Schalttafelsicherungselement *n*, Einbauelement *n*
s. gallery Schaltwarte *f*
s. instrument Schalttafelinstrument *n*
s. mounting Schalttafelmontage *f*
s. position Vermittlungsplatz *m*
switchgear Schaltgerät[e *pl*] *n*; Schaltungsteile *npl*
s. cubicle Schaltzelle *f*, Hochspannungsschaltzelle *f*
switching 1. Schalten *n*, Schaltung *f (mit Schaltgerät)*; 2. Rangieren *n*
s. action Schaltvorgang *m*
s. algebra Schaltalgebra *f*
s. arrangement Schaltanordnung *f*
s. circuit Hilfskreis *m*, Schalteinrichtung *f*; Impulsschaltung *f*; Schaltstromkreis *m*, Schaltung *f* für logische Verknüpfung
s. current Schaltstrom *m*
s. delay Schaltverzögerung *f*
s. desk Schaltertisch *m*, Schaltpult *n*
s. device Schaltgerät *n*
s. diode Schaltdiode *f*
s. equipment Schalteinrichtung *f*
s. frequency Schalthäufigkeit *f*
s. function *(Dat)* Schaltfunktion *f*
s. key Kippschalter *m*, Schalttaste *f*
s. line Schaltgrenze *f*, Schaltkurve *f (Zweipunktregler)*
s. locomotive Rangierlok[omotive] *f*, Verschiebelokomotive *f*
s. member Schaltorgan *n*
s. mode Schalt[ungs]art *f*
s.-off Abschalten *n*, Ausschalten *n*
s.-on Einschalten *n*, Zuschalten *n*
s. operation Schaltvorgang *m*
s.-out Ausschalten *n*
s.-over contact Umschaltekontakt *m*
s. period Umschaltperiode *f*
s. point *(Fmt)* Verteilfernamt *n*
s. process Schaltvorgang *m*
s. programme Schaltprogramm *n*
s. pulse Schaltimpuls *m*
s. rate Schalthäufigkeit *f*
s. relay Schaltrelais *n*
s. sequence Schaltfolge *f*
s. speed Schaltgeschwindigkeit *f*
s. stage Schaltstufe *f*
s. starter Anlaßschalter *m*
s. surge Schaltüberspannung *f*
s. system Wählersystem *n*
s. time Schaltzeit *f*
s. transient Schalteinschwingvorgang *m*
s. transistor Schalttransistor *m*
s. unit Schaltgerät *n*
switchman Weichenwärter *m*
switchyard Hochspannungsschaltanlage *f*
swivel / to schwenken, drehen; schrägstellen
swivel 1. Bohrwirbel *m*, Drehkopf *m*, Drehwirbel *m*, Rotationskopf *m*; Spülkopf *m (Tiefbohrtechnik)*; 2. Drehteil *n*; Drehscheibe *f (Shaper)*; Ankerspannschraube *f*; Kardandrehzapfen *m*; Spannschloß *n*
s. angle Schwenkwinkel *m (Verzahnung)*
s. arm Schwenkarm *m*
s. base Drehteil *n (einer Maschine)*; Drehunterteil *n (Schraubstock)*
s. damper Drehrauchschieber *m*, Drehklappe *f*, Rauchklappe *f*
s. double-tube core-barrel Doppelkernrohr *n* mit mitdrehendem Innenrohr *(Tiefbohrtechnik)*
s. flow head drehbarer Spülkopf *m*, Drehspülkopf *m (Tiefbohrtechnik)*
s. head Drehsupport *m*; schwenkbarer Spindelkopf *m*
s. hook Wirbelhaken *m (Tiefbohrtechnik)*
s. joint Drehgelenk *n* , Kreuzgelenk *n*
s. lock Drehdolle *f (Rudern)*

s. loom *(Text)* Broschierstuhl *m*
s. pin Achsschenkelbolzen *m*, Achszapfen *m*, Drehzapfen *m*, Lenkzapfen *m*; Drehachse *f*, Schwingbolzen *m*
s. shackle *(Schiff)* Patentschäkel *m*, Wirbelschäkel *m*
s. tap Schwenkhahn *m*
s. tool holder Schwenkmeißelhalter *m*
s. union Gelenkverbindung *f*
s. vice drehbarer Parallelschraubstock *m*, Drehschraubstock *m*
s. weave *(Text)* Bindung *f* in Broché
s. window Drehfenster *n*
swivelling Schwenken *n*, Drehen *n*; Schrägverstellen *n*
swivelling schwenkbar; schräg verstellbar
s. arrangement schwenkbare Anordnung *f*
s. axis Schwenkachse *f*
s. block schwenkbares Gehäuse *n (Pumpe)*
s. link Drehgelenk *n*
s. piece Drehgelenk *n*
s. position Drehlage *f*
syenite *(Min)* Syenit *m*
syllabic articulation *(Fmt)* Silbenverständlichkeit *f*
syllable Silbe *f*
s. articulation (intelligibility) *(Fmt)* Silbenverständlichkeit *f*
syllogism logischer Schluß *m*, Syllogismus *m*
sylvanite *(Min)* Sylvanit *m*
sylvic acid Abietinsäure *f*
sylvine *(Min)* Sylvin *m*
sylvinite *(Min)* Sylvinit *m*
symbol Sinnbild *n*, Symbol *n*; Kurzzeichen *n*
s. designating fits Passungskurzzeichen *n*
s. of the crystal face Symbol *n* der Kristallfläche
s. printer Zeichendrucker *m*
s. weight Atomgewicht *n*
symbolic symbolisch, sinnbildlich
s. address *(Dat)* symbolische Adresse *f*
s. circuit Funktionsschema *n*, Blockschaltbild *n*
s. language symbolische Sprache *f (des Rechners)*
symmetrical symmetrisch
s. about an axis achsensymmetrisch
s. about the zero axis höhensymmetrisch *(Kristalle)*
s. angle cutter *(Am)* Prismenfräser *m*
s. conductivity richtungsabhängige Leitfähigkeit *f*
s. double bucket symmetrische Doppelschaufel *f (Turbine)*
s. electrometer symmetrisches Elektrometer *n*
s. heterostatic circuit symmetrisch-heterostatische Schaltung *f*; Nadelschaltung *f* [nach Mascart]
s. three-phase system symmetrisches Drehstromnetz *n*
s. winding symmetrische Wicklung *f*
symmetrization Symmetrierung *f*
symmetry Symmetrie *f*
s. axis Symmetrieachse *f*
s. operation *(Math)* Symmetrieoperation *f*, Deckoperation *f*, Deckbewegung *f*
sympathetic mitschwingend; Resonanz-
sync discriminator *(Fs)* Impulssieb *n*, Synchronisierungsgenerator *m*
s. pulse *(Fs)* Synchronisier[ungs]impuls *m*
s. separator Amplitudensieb *n*
s. signal *(Fs)* Synchronsignal[gemisch] *n*
synchro Drehmelder *m*, Fernübertragungssystem *n*, FÜ-System *n*, Selsyn *n*, Synchro *n*
s. control transformer Drehmelderrückmelder *m*, Drehmeldersteuerempfänger *m*
s. differential receiver Drehmelderdifferentialempfänger *m*
s. differential transmitter Drehmelderdifferentialgeber *m*
s. indicator Drehmelderanzeigeempfänger *m*
s. receiver Drehmelderempfänger *m*, Momentenempfänger *m*
s. repeater Drehmelderempfänger *m*
s. torque receiver Synchrondrehmomentempfänger *m*
s. torque transmitter Synchrondrehmomentsender *m*
s. transmitter Drehmeldergeber *m*, Drehfeldgeber *m*
s. two speed system Drehmelder-grob-fein-Übertragung *f*
synchrocyclotron *(Kern)* Phasotron *n*, Synchrozyklotron *n*
synchromesh gear Synchrongetriebe *n*
s. gearbox Synchrongetriebe *n*, synchronisiertes Getriebe *n*
s. sleeve Synchronisiermuffe *f*, Synchronisierhülse *f*
s. unit Synchroneinheit *f*, Synchronkupplung *f*
synchronal *s.* synchronous
synchronism Synchronismus *m*, Gleichgang *m*; Gleichheit *f*
s. of phases Phasengleichheit *f*
synchronization Synchronisierung *f*, Parallelschaltung *f*
s. control Synchronsteuerung *f*, Gleichlaufregelung *f*
s. stretcher *(Fs)* Pegelstabilisierung *f*
synchronize / to synchronisieren, gleichschalten
synchronizer Synchronisiervorrichtung *f*, Synchronisator *m*; Synchronimpulsgeber *m*
synchronizing ball *(Kfz)* Synchronkugel *f*
s. block *(Kfz)* Synchronriegel *m*

s. bus-bar *(El)* Synchronisierungsschiene *f*
s. clamp *(Fs)* Synchronisationsabtrennung *f*
s. cone *(Kfz)* Synchronkegel *m*
s. contact Synchronkontakt *m*
s. current synchronisierender Strom *m*
s. discriminator Gleichlauffrequenzgleichrichter *m*
s. disk *(Kfz)* Synchronscheibe *f*
s. generator *(Fs)* Synchronisierungsgenerator *m*
s. grid Synchronisiergitter *n*
s. impulse Synchronisationsimpuls *m*, Synchronisier[ungs]impuls *m*, Gleichlaufimpuls *m*
s. mechanism *(Kfz)* Synchronisiereinrichtung *f*
s. of images *(Fs)* Bildsynchronisation *f*
s. pulse *(Fs)* Synchronisier[ungs]impuls *m*
s.-pulse separation Impuls[ab]trennung *f*
s. reactor Synchronisierungsdrossel *f*
s. signal *(Fs)* Gleichlaufzeichen *n*, Synchronisationssignal *n*
s. slide collar *(Kfz)* Synchronschiebehülse *f*
s. spring *(Kfz)* Synchronfeder *f*
synchronoscope Synchronoskop *n*
synchronous synchron, gleichlaufend, gleichgehend; gleich
s. amplifier Synchronverstärker *m*
s. capacitor Phasenschieber *m*
s. computer Synchronrechenautomat *m*
s. converter Einankerumformer *m*
s. detector Synchrodetektor *m (UKW- und FM-Technik)*; Synchrongleichrichter *m*
s. drive Synchronantrieb *m*
s. electric clock Synchronuhr *f*
s. electromotive force *(Elm)* fiktive synchrone EMK *(des Polradfeldes)*
s. field *(El)* synchrones Feld *n*
s. gate *(Fs)* Synchrontor *n*
s. generator Synchrongenerator *m*
s. induction motor synchronisierter Induktionsmotor *m*
s. mixer Synchronmischstufe *f*
s. modifier synchroner Phasenschieber *m*, Synchronphasenschieber *m*
s. motor Synchronmotor *m*
s. phase advancer Synchronphasenschieber *m*
s. rectifier Synchrongleichrichter *m*
s. reed *(Fmt)* Zerhackerfeder *f*
s. rotary converter Einankerumformer *m*
s. scanning Synchronabtastung *f*
s. self-rectifying vibrator *(Fmt)* Pendelumformer *m*, Wechselgleichrichter *m*
s. timer Synchronuhr *f*
s. vibrator Synchronzerhacker *m*
synchrotron Synchrotron *n*
s. radiation Synchrotronstrahlung *f*
synclinal fold *(Geol)* Mulde *f*
syncline *(Geol)* Mulde *f*
syndet *(Am) s.* synthetic detergent
synoptic [weather] chart Wetterkarte *f*
synthesis Synthese *f*, Aufbau *m*, Darstellung *f*
s. gas Synthesegas *n*
s. of oscillations Schwingungssynthese *f*
synthesize / to synthetisieren, aufbauen, darstellen
synthetic künstlich, synthetisch
s. detergent synthetisches Waschmittel *n*
s. enamel Kunstharzlack *m*
s. fertilizer Kunstdünger *m*
s. fibre synthetische Faser *f*
s. manure Kunstdünger *m*
s. reagent *(Ch)* Ausgangsstoff *m* für Synthesen
s. resin adhesive Kunstharzklebstoff *m*
s. resin bonded laminate *(Plast)* Schichtstoff *m*
s. resin bonding kunstharzhaltige Bindung *f (Schleifscheibe)*
s. resin finish Kunstharzüberzugslack *m*
s. resin varnish Kunstharzlack *m*
s. resinoid-bonded grinding wheel kunstharzgebundene Schleifscheibe *f*
s. rubber synthetischer Kautschuk *m*
s.-speech machine Sprechmaschine *f (zur synthetischen Erzeugung von Sprechlauten)*
s. trainer Flugübungsgerät *n*
syntonic wireless telegraphy abgestimmte drahtlose Telegrafie *f*
syntonizing coil Abstimmspule *f*
syntony *(Fmt)* Abstimmung *f*, Resonanz *f*
syphon *s.* siphon
syringe Injektionsspritze *f*, Spritze *f*
syst *s.* system
system System *n*; Anordnung *f*; Plan *m*; Schema *n*; Verfahren *n*; *(Math)* Schar *f*; Formation *f*; Netz *n*
s. engineering Untersuchung *f* und Bearbeitung *f* komplexer Systeme
s. of balancing Massenausgleich *m*
s. of buffer battery *(Fmt)* Pufferbetrieb *m*
s. of coordinates Koordinatensystem *n*
s. of coplanar forces ebenes Kräftesystem *n*
s. of fits Passungssystem *n*
s. of forces Kräftesystem *n*, Kräftebild *n*
s. of lubrication Schmieranordnung *f*
s. of particles Punkthaufen *m*
s. of powering Antriebsart *f (eines Schiffes)*
s. of the meter Zählersystem *n*
s. pressure Betriebsdruck *m*
systems engineer Systemingenieur *m (für höhere Organisationsaufgaben)*
s. engineering *(Dat)* Systemtechnik *f*

T

T-B cell *s.* Transmitter-blocker cell
T-bar clamp Spannknecht *m*, Spanner *m*
T-beam T-Träger *m*; Plattenbalken *m (Stahlbetonbau)*
T-bevel Doppelschmiege *f*
T-bolt T-Bolzen *m*
T-girder T-Träger *m*
T-joint 1. T-Stoß *m*; 2. *(El)* Abzweigmuffe *f*, T-Muffe *f*; Abzweigklemme *f*
T-junction T-Verzweigung *f*; T-Glied *n (Wellenleiter)*
T-network T-Schaltung *f*
T-nut T-Nutenstein *m*
T-pad *(El)* T-Dämpfungsglied *n*
T-parameter equivalent circuit *(El)* Ersatzschaltung *f* in T-Form
T-R cell Empfangssperrzelle *f (Wellenleitertechnik)*
T-R switch *s.* transmit-receive switch
T-section T-Profil *n*
T-shaped aerial T-Antenne *f*
T-slot T-Nut *f*, Aufspannut *f*
T-slot cutter T-Nutenfräser *m*
T-slot planing tool T-Nuten-Hobelmeißel *m*, Hakenmeißel *m*
T-splice *(El)* Leitungsabzweigung *f*
T-square Reißschiene *f*, Anschlaglineal *n*
tab 1. Aufreißband *n*, Aufreißstreifen *m*; Lasche *f*; Öse *f*; Leiste *f*; 2. *(El)* Drucktaste[nauslöser *m*] *f*; 3. Beschriftungsschild *n*, Bezeichnungsschild *n*, Kennschild *n*; 4. Hilfsruder *n*; 5. [leichter] Vorhang *m*
t. speed *(Büro)* Sammelganggeschwindigkeit *f*
tabaret 1. Stickrahmen *m*; 2. gestreifter seidener Stoff *m*
tabasheer Bambuszucker *m*
tabby / to *(Text)* moirieren
tabby weaving Leinwandbindung *f*
table / to tabellarisch zusammenstellen
table 1. Tabelle *f*, Tafel *f*; 2. *(Masch)* Tisch *m*; Planscheibe *f (Karusselldrehmaschine)*
t. area Tischfläche *f*
t. column Tabellenspalte *f*
t. dog Tischanschlag *m*
t. drive Tischantrieb *m*
t. dwell control Tischhaltesteuerung *f*
t. elevating screw Tischhebespindel *f*
t. feed Tischvorschub *m*
t. feed screw Tisch[vorschub]spindel *f*
t. for buddling Stoßherd *m*, Rüttelherd *m*
t. gear rim Planscheibenzahnkranz *m (Karusselldrehmaschine)*
t. lamp Tischlampe *f*
t. lead screw Tischleitspindel *f*
t. lever device Tischstellwerk *n (Signal)*
t. look up logic *(Dat)* Logik *f* mit Tabellen[nachschlagen]
t. of logarithms Logarithmentafel *f*
t. of the vertical boring and turning mill Planscheibe *f* der Karusselldrehmaschine
t. positioning Tischeinstellung *f*, Einortung (Positionierung) *f* des Tisches
t. positioning servo motor Servomotor *m* zur Bohrtisch[lage]einstellung, Servomotor zur Tisch[lage]einstellung
t. power traverse Tischselbstgang *m*
t. pull Durchzugskraft *f (Hobelmaschine)*
t. rack Tischzahnstange *f*
t. rapid traverse Tischeilgang *m*
t. relief Tischentlastung *f*
t. return Tischrücklauf *m*
t. reversal Tischumsteuerung *f*, Tischumkehr *f*
t. reversal dwell Verzögerung *f* bei Tischumkehr
t. reverse clutch Tischumsteuerkupplung *f*
t. revolution Tischumdrehung *f*
t. roller Rollgang *m*
t. side surface Tischseitenfläche *f*
t. speed Planscheibendrehzahl *f (Karusselldrehmaschine)*; Tischgeschwindigkeit *f*
t. standard Tischlampe *f*
t. stop Tischanschlag *m*
t. strip Tischleiste *f*
t. support Tischstütze *f*
t. support arm Tischträger *m (Bohrmaschine)*
t. support clamp *(Masch)* Tischstützenklemmung *f*
t. telephone set Tischfernsprecher *m*
t. top surface Tischoberfläche *f*
t. track Tischführung *f*; Führungsbahn *f* der Planscheibe
t. travel Tischweg *m*
t. travel per tooth Vorschub *m* je Zahn *(Fräsen)*
t. traverse Tischweg *m*
t. trip Anschlag *m* für die Tischumsteuerung, Tischknagge *f*
t.-type horizontal boring and facing machine (mill) Tischbohrwerk *n*, Waagerechtbohr- und -fräswerk *n* in Tischausführung mit beweglichem Ständer
t. way Tischführung *f*
t. working surface Aufspannfläche *f* des Tisches
tablecloth Tischdecke *f*
tablet / to 1. tablettieren, zu Pillen pressen; 2. vorverdichten
tablet Pastille *f*, Pille *f*, Tablette *f*, Täfelchen *n*
t. compressing machine Tabletten[form]-maschine *f*
t. of slate Schieferplatte *f*
t. paper Papier *n* für Notizbücher
tabletting machine Tablettiermaschine *f*

tabloid Tablette *f*
taboret *s.* tabaret
tabular tabellarisch; tafelförmig
  t. matter *(Typ)* Tabellensatz *m*
  t. value Tabellenwert *m*
tabulate / to tabell[aris]ieren, in Tabellenform zusammenstellen
tabulating machine *(Büro)* Tabelliermaschine *f*, Tabulatormaschine *f*
tabulator *s.* tabulating machine
tac *s.* tachometer
tacheometer Tachymeter *n*
  t. theodolite Tachymetertheodolit *m*
tacheometric tachymetrisch
  t. land survey tachymetrische Landaufnahme *f*
tacheometry Tachymetrie *f*
tachograph Tachograf *m*, registrierendes Tachometer *n*
tachometer Geschwindigkeitsmesser *m*, Tachometer *n*, Umdrehungszähler *m*, Drehzahlmesser *m*
tachymeter *s.* tacheometer
tacitron Tacitron *n (Typ einer Gastriode)*
tack / to 1. nageln; heften; befestigen *(mit Reißzwecken, Dachpappennägeln usw.)*; 2. beidrehen *(Flugzeug)*; 3. kreuzen *(Segler)*
tack 1. Stift *m*; Zwecke *f*; Nagel *m (mit breitem Kopf)*, Dachpappennagel *m*; Täcks *m*; 2. *(Typ)* Zügigkeit *f (von Druckfarben)*; 3. *(Plast)* Klebrigkeit *f*
  t. claw Nagelklaue *f*
  t. range Dauer *f* der Trockenklebrigkeit
  t. riveting Heftnietung *f*
  t. welding Heftschweißung *f*
tackiness *(Plast)* Haftfähigkeit *f*, Klebrigkeit *f*
tacking Heften *n*
  t. machine Lederheftmaschine *f*, Täcksheftmaschine *f (Schuhfertigung)*
  t. rivet Heftniet *m*
tackle 1. *(El)* Seilgeschirr *n*, Spanngeschirr *n*, Spannzeug *n (Freileitungsbau)*; 2. Talje *f*, Flaschenzug *m*; 3. Kloben *m*; 4. Gerät *n*; Ausrüstung *f*
  t. block Flaschenzug *m*, Hebezeug *n*
  t. block wire stretcher Drahtspanner *m*
tacky *(Plast)* klebrig
tacnode Selbstberührungspunkt *m*, Doppelspitze *f (einer Kurve)*
tactical nuclear weapons taktische Kernwaffen *fpl*
tactile taktil, den Tastsinn betreffend
tactual [ab]tastbar, durch Abtastung
tafferel *s.* taffrail
taffeta weave Taftbindung *f*, Leinwandbindung *f*
taffrail Heckgeländer *n*, Heckreling *f*, Heckbord *m*
  t. log Decklog *n*, Patentlog *n*, Heckrelingslog *n*
taft joint Lötverbindung *f* von Bleirohren
tag 1. Anhänger *m*, Anhängezettel *m*, Anhängeetikett *n*; Aufklebeadresse *f*; Markierung *f*; 2. *(El)* Kontaktstift *m*, Sockelstift *m*, Verbindungsstift *m*; Anschlußfahne *f*, Lötfahne *f*, Lötöse *f*; 3. Schlaufe *f*
tagged 1. mit Lötösen versehen; 2. *(Kern)* markiert *(Atom)*
  t. atom Indikatoratom *n*, markiertes Atom *n*
  t. molecule markiertes Molekül *n*
tagger dünnes Feinblech *n*
tagging [radioaktive] Markierung *f*
tail off / to allmählich abklingen *(Schwingungen)*
tail 1. Schwanz *m*, Stert *m*, Heck *n (eines Schiffes)*; *(Flg)* Rumpfende *n*; 2. Impulsschwanz *m*; Nachleuchtschleppe *f*; Signalschwanz *m (Impulstechnik)*; 3. Schwanz *(der Kennlinie)*
  t. absorption Ausläuferabsorption *f*
  t. assembly *(Flg)* Leitwerk *n*
  t. bearing *(Masch)* Nebenlager *n*, Außenlager *n*
  t. board Ladeklappe *f (LKW)*
  t. boom *(Flg)* Leitwerksträger *m*
  t. centre *(Masch)* [feste] Körnerspitze *f* am Reitstock
  t. cone *(Rak)* Heckkonus *m*, Schwanzkonus *m*; *(Flg)* hintere Halbsphäre *f*
  t. control surface *(Flg)* Leitwerksfläche *f*
  t. drum Umkehrtrommel *f*, Umlenktrommel *f*
  t. fin Heckflosse *f*, Schwanzflosse *f*
  t. fraction *(Ch)* Nachlauf[fraktion *f*] *m*
  t.-gate *(Brg)* Kopfstrecke *f*
  t. group *(Flg)* Leitwerk *n*
  t.-gun position *(Mil)* Heckstand *m*
  t. gunner *(Mil)* Heckschütze *m*
  t. hull *(Rak)* Heckzelle *f* ohne Einbauten
  t. lamp Schlußlicht *n (eines Zuges)*; Nummernschildlicht *n*; Schlußleuchte *f*; Rücklicht *n*
  t. lamps of train Zugschlußlicht *n*
  t. machine gun Heckmaschinengewehr *n*, Heck-MG *n*
  t. of the dog Mitnehmerarm *m* am Drehherz
  t. parachute Brems[fall]schirm *m*
  t.-piece of violin Saitenbrett *n*
  t. pipe Schubdüse *f*, Strahlrohr *n*, Saugrohr *n*
  t. rod Führungsstange *f (eines Kolbens)*
  t. rod guide hintere Kolbenstangenführung *f*
  t. rope Hinterseil *n (Schrapper)*; Unterseil *n*
  t.-rope haulage *(Brg)* Förderung *f* mit Vorder- und Hinterseil *(Streckenförderung)*
  t. run *(Ch)* Destillationsnachlauf *m*
  t. shaft 1. *(Schiff)* Schwanzwelle *f*, Propellerwelle *f*; 2. Kurbelwellenstumpf *m*
  t. sheave Umlenkrolle *f (Schrapper)*

t. skid *(Flg)* Schwanzsporn *m*, Schwanzkufe *f*
t. turret *(Mil)* Heckkanzel *f*
t. unit *(Flg)* Leitwerk *n*, Heckleitwerk *n*
tailed characteristic langsam auslaufende Kennlinie *f*
tailer Rübenroder *m*, Rübenzieher *m*
tailheavy schwanzlastig
tailing Schwanzbildung *f*, Schweifbildung *f* *(Aufzeichnung von Chromatogrammen und Spektrogrammen)*; *(Fs)* Fahnenbildung *f*
tailings 1. Abfall *m*, Nachlauf *m*, Rückstände *mpl*; Destillationsrückstand *m*; Bodenkörper *m*; 2. Rückstände, Berge *pl* *(Erzaufbereitung)*, Abgänge *mpl*
tailor-made maßgeschneidert
tailor's chalk Schneiderkreide *f*
t. dummy Schneiderbüste *f*
tails *(Plast)* Grat *m*, Schwänzchen *n*, überschüssiges Material *n* *(Spritzfehler)*
tailspin Autorotation *f*
tailstock *(Masch)* Reitstock *m*
t. barrel Reitstockoberteil *n*
t. base Reitstockunterteil *n*
t. body Reitstockkörper *m*
t. centre Reitstockspitze *f*
t. quill Reitstockpinole *f*
t. set-over Reitstockversetzung *f* *(beim Kegeldrehen)*
t. sleeve Reitstockpinole *f*
tain feines Weißblech *n*; Spiegelbelag *m*
take / to nehmen; aufnehmen *(Werkstück)*; fassen, annehmen *(Farbe)*; abnehmen *(Span)*; ansaugen, einsaugen
t. a bearing peilen
t. a curve eine Kurve nehmen
t. a cut einen Span abnehmen
t. a photograph eine Aufnahme machen
t. a polish polierfähig sein
t. aim anvisieren
t. an angle einen Winkel messen
t. an edge eine Schneide bilden
t. apart auseinandernehmen
t. away entfernen, fortbringen
t. down auseinandernehmen
t. dyes Farbstoffe aufnehmen
t. hold of a piece angreifen, fassen *(ein Walzstück)*
t. in laden *(Schiffsladung)*
t. in tow *(Schiff)* ins Schlepptau nehmen
t. little floor space geringen Platzbedarf haben, wenig Bodenfläche einnehmen
t. off abführen, ableiten; abheben *(Span)*; starten; abzapfen *(elektrischen Strom)*
t. off the handset (receiver) den Hörer abnehmen
t. out entnehmen
t. out a licence under a patent eine Lizenz auf ein Patent erwerben
t. out of service außer Betrieb setzen; blokkieren; sperren
t. readings Meßwerte ablesen
t. the average den Mittelwert bilden
t. the feed from den Vorschub ableiten von
t. to chassis (frame) *(El)* an Masse legen
t. to pieces zerlegen *(in einzelne Teile)*
t. to the air *(Flg)* aufsteigen
t. up abnehmen *(den Hörer)*; hochführen; auffangen
t. up backlash Spiel beseitigen
t. up slack spannen, Durchhang ausgleichen
t. up wear Verschleiß ausgleichen
t. watts *(El)* Leistung aufnehmen
take-apart model zerlegbares Modell *n*
t.-down angle Abzugswinkel *m*
t.-down mechanism Abzug *m*
t.-down roller Abzugswalze *f*
t.-off 1. *(Kfz)* Anfahren *n*, Start *m*; 2. *(sl)* *(Fs)* Tonunterdrückung *f* *(im Bildteil)*
t.-off arm Ablaufarm *m*
t.-off clearance Startfreigabe *f*
t.-off distance Startstrecke *f*
t.-off point *(BMSR)* Verzweigungsstelle *f* *(Blockschaltbild)*
t.-off rocket Startrakete *f*
t.-off weight Startmasse *f*
t.-up cassette Aufwickelkassette *f*
t.-up friction Aufwickelfriktion *f*
t.-up machine Rohrabzieher *m* *(Strangpresse)*
t.-up roll Warenabzugbaum *m* *(Webstuhl)*
t.-up roller Sandbaum *m*, Abzugbaum *m*, Warenbaum *m*
t.-up sand *s.* t.-up roll
t.-up spool Aufwickelspule *f*
t.-up strip Nachstelleiste *f*
taken from abgeleitet von
taking fine sight Feinkornnehmen *n* *(Schießen)*
t. full sight Vollkornnehmen *n* *(Schießen)*
t.-in band *(Text)* Einzugsseil *n* *(Selfaktor)*
t. lens Aufnahmeobjektiv *n*
t. of current Stromaufnahme *f*
t. of power *(El)* Leistungsentnahme *f*
t. of watts *(El)* Leistungsaufnahme *f*
t.-off 1. Starten *n*; 2. Veranschlagung *f*
talc Talk *m*
t. coiling Talkschnur *f* *(Dichtungsmittel)*
t. schist (slate) Talkschiefer *m*
talcous talkartig
talcum powder Talkum *n*
talk / to *(Fmt)* sprechen, Gespräche führen
talk-back circuit *(Fmt)* Gegensprechschaltung *f*; Rundfunkmeldeleitung *f*
t.-listen switch *(Fmt)* Hör-Sprech-Schalter *m*
talker echo *(Fmt)* Sprecherecho *n*, Gegenecho *n*, gegenläufiges Echo *n*
talkies *(sl)* Tonfilm *m*
talking *(Fmt)* Sprechen *n*

t. beacon akustische Bake *f*, akustischer Leitstrahlsender *m (Radar)*
t. circuit Sprechkreis *m*
t. current Sprechstrom *m*
t. echo Sprecherecho *n*
t. film Tonfilm *m*
t. key *(Am)* Sprechschalter *m*
t. movies Tonfilm *m*
tall oil *(Ch)* Tallöl *n*
tallate Ester *m* (Salz *n*) des Tallöls, Tallat *n*
tallboy cupboard Besenschrank *m*
tallow / to schmieren
tallow Talg *m*
tally / to nachprüfen; stimmen; registrieren; zählen
t. to übereinstimmen mit
tally 1. *(Math)* Kontrollrechnung *f*, Probe *f*; 2. *(Brg)* Fahrmarke *f*, Kontrollmarke *f*
tallyman Ladungskontrolleur *m*, Ladungsanschreiber *m*
tallymeter *(Am)* Meßinstrument *n* mit Fernanzeige, Fernmeßinstrument *n*
tamp / to stampfen, feststampfen; [unter]stopfen *(Gleisbau)*; besetzen, Besatz einbringen, verdämmen *(Bohrlöcher)*
tamp Ramme *f*, Handstampfer *m*
t.-concrete Stampfbeton *m*
tamped bottom Stampfboden *m (SM-Ofen)*
t. clay Stampfmasse *f*
t. concrete Stampfbeton *m*
t. well verdämmte Bohrung *f (Tiefbohrtechnik)*
tamper 1. Stampfer *m*; Stampfbohle *f*, Stampfbalken *m*; Schwellenstopfer *m*; 2. Reflektor *m (Reaktortechnik)*
Tampico brushing Bürsten *n (Polieren)*
tampin konischer Holzstopfen *m*
tamping Stampfen *n*; Unterstopfen *n (Gleisbau)*; *(Brg)* Eindrücken *n* der Ladung mit dem Ladestock
t. bag Besatzpatrone *f*
t. bar Ladestock *m*
t. board Stampfbohle *f*
t. compound Stampfmasse *f*
t. iron Stampfer *m*
t. pick Stopfhacke *f*
t. rod Stocherstange *f*; Ladestock *m*
t. stick Ladestock *m*
tampon Mullbausch *m*, Pfropfen *m*, Tampon *m*, Wattebausch *m*
t. memory Pufferzwischenspeicher *m*, Pufferspeicher *m*
tan / to gerben, beizen, lohen, lohgar machen
tan Lohe *f*, Gerberlohe *f*
t. liquor Gerbebrühe *f*
t. net Lohnetz *n*
t. pit Lohgrube *f*
tandem hintereinander [an]geordnet
tandem *(El)* Kaskade *f*; Tandem *n*
t. area *(Am) (Fmt)* Knotenamtsbereich *m*
t. arrangement Tandemanordnung *f*, Tandemaufstellung *f (Zylinderanordnung)*
t. blower engine Tandemgebläse *n*
t. central office *(Am) (Fmt)* Knotenamt *n*
t. compound engine Tandemverbunddampfmaschine *f*; *(El)* Reihenverbundmaschine *f*
t. connection *(El)* Kaskadenschaltung *f*
t. dialling *(Fmt)* Durchgangswahl *f*
t. drive Tandemantrieb *m*
t. engine Tandem[dampf]maschine *f*
t. exchange *(Fmt)* Knotenamt *n*
t. head flat knitting links-and-links machine zweiköpfige Links-links-Flachstrickmaschine *f*
t. hoisting Tandemförderung *f*
t. knife switch mehrpoliger Messerschalter *m*, Messerschalter mit gleichzeitig wirkenden Schaltstücken
t. mill Tandemwalzwerk *n*
t. network *(Fmt)* Sternnetz *n*
t. operation *(Fmt)* Knotenamtsbetrieb *m*
t. position *(Am) (Fmt)* Durchgangsplatz *m*
t. roller Tandemwalze *f*
t. selector *(Fmt)* Durchgangswähler *m*
t. toll circuit dialling *(Am) (Fmt)* Schnellverkehr *m* über Knotenämter
t. tractor Tandemtraktor *m*
t. trunk *(Fmt)* Durchgangsleitung *f*
t. winding Tandemförderung *f*
t. working *(Fmt)* Knotenamtsbetrieb *m*
tang / to lappen, mit Zapfen versehen *(Werkzeuge)*
tang Lappen *m*; Austreiblappen *m*; Keilzapfen *m*; Angel *f (Feile)*; Schaft *m (Schraubenzieher)*
t. drive collet Einsatzfutter *n (Fräsmaschine)*
t. for mill saw Gattersägeangel *f*
t. for web blade Sägeangel *f*
t. for web blade with strengthening Sägeangel *f* mit Verstärkung
t. slot Austreiberschlitz *m*; Angelnut *f*
tanged end shank Schaft *m* mit Mitnehmerlappen
t. file Feile *f* mit Angel
tangency *(Math)* Berührung *f*
tangent Tangens *m*; Tangente *f*
tangent tangential, tangierend
t. arc Tangentenbogen *m*
t. bar Kegellineal *n*, Leitlineal *n (Kegeldrehen)*
t. galvanometer *(El)* Tangentenbussole *f*
t. image Tangentenbild *n*
t. of motion *(Rak)* Bahntangente *f*
t. plane Tangentialebene *f*
t. point Tangentialpunkt *m (einer Kurve)*; Krümmungsendpunkt *m (der Straße)*
t. to the trajectory *(Rak)* Bahntangente *f*
tangential tangential
t. acceleration Tangentialbeschleunigung *f*

t. admission tangentiale Beaufschlagung *f* *(Turbine)*
t. casting Tangentialguß *m*
t. channel Tangentialkanal *m*
t. component Tangentialkomponente *f*
t. feed Tangentialvorschub *m*
t. force Tangentialkraft *f*
t. key Tangent[ial]keil *m*
t. law Tangentensatz *m*
t. load Tangentiallast *f*, Umfangslast *f*
t. plane Berührungsebene *f*
t. point Berührungspunkt *m*
t. pressure Drehkraft *f*, Tangentialkraft *f*, Tangentialdruck *m*, drehender Druck *m*
t. pressure area Tangentialdruckfläche *f*
t. pressure diagram Drehkraftdiagrammlinie *f*, Tangentialdruckdiagramm *n*
t. resistance Tangentialwiderstand *m*
t. speed Tangentialgeschwindigkeit *f*
t. stress Tangentialspannung *f*
t. tensile stress Tangentialzugspannung *f*
t. vector Tangentenvektor *m*
t. velocity Tangentialgeschwindigkeit *f*
t. wave-path *(Fmt)* Ausbreitungsrichtung *f* der Bodenwelle
tangible tastbar, fühlbar
tangle / to verwirren, verfitzen
tangle *(Text)* Webnest *n*, Nest *n*
tangled growth Gestrüpp *n*
t. roots Wurzelwerk *n*
tangling Verwirren *n* *(Faden)*
t. of magnetic field lines Verwirrung *f* von magnetischen Feldlinien
tank 1. Becken *n*, Behälter *m*, Bottich *m*, Kessel *m*, Tank *m*, Wanne *f*; 2. Windkessel *m* *(Hydraulik, Pneumatik)*; 3. *(Dat)* Schleifenspeicher *m*; 4. Panzer[kampfwagen] *m*
t. air charging line *(Rak)* Tankbelüftungsleitung *f*
t. air pressurizing line *(Rak)* Tankbelüftungsleitung *f*
t. barge *(Schiff)* Tankleichter *m*
t. capacity 1. *(Fmt)* Kapazität *f* eines Schwingkreises; 2. Behälterinhalt *m*, Fassungsvermögen *n* des Behälters
t. car Kesselwagen *m*, Zisternenwagen *m*
t. carrier Panzertransportfahrzeug *n*
t. circuit *(Fmt)* Leistungsschwingkreis *m*
t. construction Behälterbau *m*
t. cooling Verdampfungskühlung *f*
t. destroyer *(Am)* Selbstfahrlafette *f* für Panzerbekämpfung
t. ditch Panzergraben *m*
t. dome Kesseldom *m* *(Kesselwagen)*
t. farm Tanklager *n*
t. furnace Wannenofen *m*, Tankofen *m*
t. gauge Tankfüllstandmesser *m*
t. gun Kampfwagenkanone *f*, Panzerkanone *f*
t. identification service Panzererkennungsdienst *m*
t. level Flüssigkeitsspiegel *m* im Behälter
t. locomotive Tenderlokomotive *f*
t. pressurizing *(Rak)* Tankbelüftung *f*
t. resistance Badwiderstand *m* *(Elektrolyse)*
t. road Panzerstraße *f*
t. screw-cap *(Kfz)* Tankverschluß *m*
t.-side bracket Kimmstützplatte *f* *(Schiffsdoppelboden)*
t. skin Behälterhaut *f*
t. standpipe Tankbelüftungsrohr *n*
t. top *(Schiff)* Doppelbodendecke *f*; Tankdecke *f*
t.-top plating Innenboden *m*, Doppelbodenbeplattung *f*
t. trap Panzerfalle *f*
t. vessel Tankschiff *n*, Tanker *m*
t. voltage Badspannung *f* *(Elektrolyse)*
t. wagon Kesselwagen *m*
tankage Tankaufnahmekapazität *f*, Tankfassungsvermögen *n*
t. capacity *s.* tankage
tanker Tanker *m*, Tankschiff *n*
tankship *s.* tanker
tannage Gerberei *f*, Gerbung *f*
tanned lohgar
t. hides Garleder *n*
t. leather gegerbtes Fell *n*
tanner Gerber *m*
tanner's wool Gerberwolle *f*
tannery Gerberei *f*
t. liquor Lohbrühe *f*, Lohe *f*
tannic gerbsauer
t. acid Gerbsäure *f*, Tannin *n*
tanniferous gerbstoffhaltig
tannin Gerbstoff *m*, Tannin *n*
tanning Gerben *n*, Gerberei *f*
t. agent Gerbstoff *m*
t. pit Gerbgrube *f*
t. process Gerberei *f*
t. stain Angerbfleck *m*
t. tumbler Gerbfaß *n*, Gerbtrommel *f*
t. value Gerbwert *m*
tannometer Gerbsäuremesser *m*
tantalate Salz *n* der Tantalsäure, Tantalat *n*
tantalic compound Verbindung *f* des fünfwertigen Tantal
tantalite Tantalit *m*, Tantalerz *n*
tantalous compound Verbindung *f* des dreiwertigen Tantal
tantalum Tantal *n*
t. anode Tantalanode *f*
t. capacitor Tantalkondensator *m*
t. carbide Tantalkarbid *n*
t. filament lamp Tantallampe *f*
t. film Tantalschicht *f*
t. lamp Tantallampe *f*

t. ore Tantalerz *n*
t. rectifier Tantalgleichrichter *m*
tanyard Gerberei *f*, Gerbanlage *f*
tap / to 1. anzapfen, anbohren; anstechen; entnehmen; 2. *(El)* abgreifen, anschließen; *(Fmt)* abhören; 3. Gewinde bohren; 4. anschlagen, abklopfen
t. a telephone line [sich] in eine Telefonleitung einschalten
t. into *(El)* einschalten in; zwischenschalten
t.-off 1. *(El)* abgreifen; abzweigen; 2. abstechen *(Hochofen)*
t. off from *(El)* abgreifen an (von)
t. the slag Schlacke abstechen
tap 1. *(El)* Anzapfung *f*, Abzweigung *f*, Abgriff *m*; 2. Hahn *m*, Zapfen *m*, Zapfhahn *m*, Zapfstelle *f*; 3. Gewinde[schneid]bohrer *m*; 4. *(Met)* Abstich *m*; 5. leichter Schlag *m*
t. and reamer fluting cutter Fräser *m* zum Nuten von Gewindebohrern und Reibahlen
t. and reamer wrench Windeisen *n* für Gewindebohrer und Reibahlen
t. bolt Kopfschraube *f*, Stiftschraube *f*; Kolbendeckelschraube *f*
t. bolt secured by drilling verbohrte Deckelschraube *f*
t. breakage Gewindebohrerbruch *m*
t. changer *(El)* Anzapfumschalter *m*
t. changing *(El)* Anzapfungsumschaltung *f*
t. cinder *(Met)* Puddelschlacke *f*, Rohschlacke *f*
t. connection *(El)* Anzapfstelle *f*
t. cutter Gewindebohrernutenfräser *m*
t. depth Gewindebohrtiefe *f*
t. drill Gewindekernlochbohrer *m*, Kernlochbohrer *m*
t. end Einschraubende *n*
t. extractor Gewindebohrerauszieher *m*
t. failure Gewindebohrerbruch *m*
t. furnace Stichofen *m*
t. holder Gewindebohrerhalter *m*
t. hole 1. *(Met)* Abstichloch *n*, Stichloch *n*; 2. Gewindebohrung *f*
t. hole gun *(Met)* Stichlochstopfmaschine *f*
t. ladle *(Met)* Abstichpfanne *f*
t. life Gewindebohrerstandzeit *f*
t. sharpening machine Gewindebohrerschleifmaschine *f*
t. water Leitungswasser *n*
t. with cut thread Gewindebohrer *m* mit geschnittenem Gewinde
t. with ground thread Gewindebohrer *m* mit geschliffenem Gewinde
t. wrench Windeisen *n*
tape / to umwickeln
t.-record eine Bandaufnahme machen
tape Band *n*, Streifen *m*; Riemchen *n*; Gurt *m*; Bandmaß *n*; Isolierband *n*; Lackband *n*
t. adhesive Klebeband *n*, Klebestreifen *m*
t. aerial Bandantenne *f*
t. armouring Band[eisen]bewehrung *f (von Kabeln)*
t. background noise *(El)* Bandrauschen *n*
t. breakage Bandriß *m*
t. cable Bandkabel *n*
t. capacitor Bandkondensator *m*
t. chronograph Bandchronograf *m*, Bandzeitschreiber *m*
t. control Lochstreifensteuerung *f*
t. control system Bandsteuersystem *n*
t.-controlled bandgesteuert, lochstreifengesteuert
t.-controlled typesetting *(Typ)* lochbandgesteuerte Satzherstellung *f*
t. counter Bandzählwerk *n*
t. covering Isolierband *n*, Wärmeschutzband *n*
t. data preserver *(Dat)* Magnetstreifendatenschützer *m (vor magnetischen Feldern, die die Daten zerstören)*
t. dial Bandskala *f*
t. differential *(Dat)* Seilzugdifferential *n*
t. divider *(Text)* Riemchenflorteiler *m*
t. drive Bandantrieb *m*
t. drive motor Tonbandmotor *m*
t. eraser Tonbandlöschgerät *n*
t. feed Bandvorschub *m*, Streifenvorschub *m*
t. file Bandablage *f*
t. guidance Bandführung *f*
t. guide Bandführung *f*
t. level gauge Pegelmesser *m*
t. loop endloses Band *n*
t. measure Bandmaß *n*
t. merging *(Typ)* Lochbandverschmelzung *f (Vereinigung von korrigiertem und unkorrigiertem Lochband)*
t. microphone Bändchenmikrofon *n*
t. perforator Bandlocher *m*, Streifenlocher *m*
t. playback Wiedergabe *f (vom Tonband)*
t. printer Streifendrucker *m*
t. puller Streifenzieher *m*
t. punch Bandlocher *m*, Streifenlocher *m*
t. reader Bandleser *m*, Streifenleser *m*, Streifenlesekopf *m*
t. reading-head Streifenlesekopf *m*
t. recorder Bandaufnahmegerät *n*, Tonbandgerät *n*
t. recording Bandaufnahme *f*
t. retransmission Lochstreifenübertragung *f*
t.-rule Bandmaß *n*
t. sensing Bandabtastung *f*
t. sizer *(Text)* Schlichtmaschine *f* für Riemchen
t. speed Bandgeschwindigkeit *f*
t. spool Bandspule *f*
t. store Bandspeicher *m (für Daten)*
t. tension Bandzug *m*

t.-to-card conversion *(Dat)* Lochstreifen-Lochkarten-Übersetzung *f*
t.-to-card converter *(Dat)* Lochstreifen-Lochkarten-Umsetzer *m*, Magnetband-Lochkarten-Umsetzer *m*; lochstreifengesteuerter Kartenlocher *m*
t.-to-printer converter *(Dat)* lochstreifengesteuerter Drucker *m*
t. transmitter Lochstreifengeber *m*, Streifengeber *m*
t. velocity Banddurchlaufgeschwindigkeit *f*
t. weaving Gurtweberei *f*
taped wire *(El) (mit Band)* umwickelter Draht (Leiter) *m*
taper / to sich verjüngen; kegelig machen; zuspitzen; spitz auslaufen
t.-bore Kegel bohren, kegelig bohren
t. off schräg abfallen
t.-ream kegelig aufreiben
taper 1. Kegel *m*; Verjüngung *f*, Abschrägung *f*, Konizität *f*; Keil *m*; 2. *(El)* Pyramidenrohr *n*, Trichterhorn *n (Wellenleiter)*, keilförmiger Wellenleiter *m*
taper kegelig, Kegel-, konisch, verjüngt
t. adaptor Kegeleinsatz *m*
t. angle Kegelwinkel *m*
t. arbor Bohrfutterkegel *m*
t. attachment Kegeldreheinrichtung *f*
t. bar Konuslineal *n*
t. bore Kegelbohrung *f*
t. boring Kegeligbohren *n*
t. bridge reamer Nietlochreibahle *f*
t. drift [pin] *(Masch)* Austreiber *m*
t. drilling Kegeligbohren *n*
t. expanding plug Spreizkegel *m*
t. file Spitzfeile *f*; Dreikantsägefeile *f*
t.-flute die sinking cutter kegeliger Gesenkfräser *m*
t. gauge Kegellehre *f*
t. gib *(Masch)* Nachstelleiste *f*, Keilleiste *f*
t. grinding Kegelschleifen *n*
t. grinding wheel kegelige Schleifscheibe *f*
t. guide bar Kegellineal *n* *(Kegeldreheinrichtung)*
t. hole Kegelbohrung *f*; Innenkegel *m*
t. key *(Am) (Masch)* Austreiber *m*
t. lead Zwischenabschrägung *f*
t.-loaded cable Kabel *n* mit längsveränderlicher Bespulung
t. measurement Kegelmessung *f*
t. milling Kegelfräsen *n*
t. of groove Kaliberanzug *m*, Kaliberverjüngung *f (Walzwerk)*
t. of surfaces Kantenabfall *m*
t. parallel Keilstück *n*
t. per unit length Steigung *f* eines Kegels, Kegelsteigung *f*
t. pin Kegelstift *m*
t.-pin hole Kegelstiftbohrung *f*
t.-pin reamer Stiftlochreibahle *f*
t. plug Hahnküken *n*
t. plug [gauge] Kegellehrdorn *m*
t. pulley Stufenscheibe *f*
t. reamer Nietlochreibahle *f*, Kegelreibahle *f*, konische Reibahle *f*, Stiftlochreibahle *f*
t. reamer with steep spiral angle[s] Kegelschälreibahle *f*
t. ring gauge Kegellehrhülse *f*
t.-roller bearing Kegelrollenlager *n*, Schrägrollenlager *n*
t. shank Kegelschaft *m*
t. shank arbor *(Masch)* Aufsteckhalter *m*
t. shank cutter Schaftfräser *m* mit [Morse-]-Kegel
t. shank drill Kegelschaftbohrer *m*
t. shank reamer Kegelschaftreibahle *f*
t. shank socket reamer Kegelreibahle *f* [für Werkzeugkegel] mit Kegelschaft
t. shank with a square cross section verjüngter Vierkantschaft *m*
t.-sided piston-ring Trapez[kolben]ring *m*
t. sleeve Kegelhülse *f*, Reduziereinsatz *m*
t. sleeve for drill chuck Bohrfutterkegel *m*
t. socket kegelige Einsteckhülse (Einsatzhülse) *f (Bohrer)*
t. tap Vorschneider *m (Gewindebohrer)*
t. thread Kegelgewinde *n*, Rohrgewinde *n*
t. turning Kegeldrehen *n*
t. turning attachment Kegeldreheinrichtung *f*
t. wheel kegelige Schleifscheibe *f*
tapered sich verjüngend, zugespitzt, spitz auslaufend; kegelförmig, konisch; keilförmig
t. collar Ringkeil *m*
t. end Konus *m* der Kolbenstange, Ansatzkegel *m*
t. end mit kegeligem Anschnitt
t. gib *(Masch)* Keilleiste *f*, Nachstelleiste *f*
t. hand reamer Kegelhandreibahle *f*
t. hole Innenkegel *m*, kegelige Bohrung *f*
t. key *(Am) (Masch)* Austreiber *m*
t. on both sides doppeltkonisch
t. overlap abgeschrägte Überlappung *f*, Schäftung *f*
t. plug 1. *(Masch)* Spreizkegel *m*; 2. konischer Ventilkegel *m*
t. pole piece konischer Polschuh *m*
t. potentiometer konusförmiges Potentiometer *n (zum analogen Berechnen bestimmter Funktionen)*
t. recess kegelige Aussparung *f*
t.-roller bearing Kegelrollenlager *n*
t. section verjüngter Querschnitt *m*
t.-sleeve wedge bearing Spannhülsenlager *n*
t. thread konisches Gewinde *n*
t. transition Kegelübergang *m (Mikrowellentechnik)*

t. waveguide sich erweiternder Hohlleiter *m* *(Wellenleiter)*
t. wing *(Flg)* Trapezflügel *m*
tapering kegelige Verjüngung *f*
tapering verjüngt; kegelig; spitz zulaufend; *(Holz)* abförmig, abholzig
t. ball-end die-sinking cutter *(Am)* kegeliger Gesenkfräser *m* mit runder Stirn
t. bore Kegelbohrung *f*
t. off spitz auslaufend
t. pass abgesetztes Kaliber *n* *(einer Walze)*
t. scale verkleinerter Maßstab *m*
tapestry Gobelin *m*; Polstermöbelstoff *m*; Tapisseriegewebe *n*
t. braid Tapezierborte *f*
t. brick [mehrfarbiger] Ziegel *m* mit rauher Oberfläche
taping 1. Steuerung *f* durch Lochstreifen; 2. *(El)* Bandbewicklung *f*
tapioca Maniokstärke *f*, Tapioka *f*
tapped 1. *(El)* angezapft, abgegriffen; mit Anzapfung, mit Abgriff; 2. mit Innengewinde, mit Anzugsgewinde
t. air Anzapfluft *f*
t. back-pressure turbine Gegendruckentnahmeturbine *f*, Gegendruckanzapfturbine *f*
t. capacitor Mehrfachkondensator *m*
t. coil *(El)* Abzweigspule *f*
t. end shank Schaft *m* mit Anzugsgewinde
t. hole Gewindebohrung *f*
t. potentiometer Potentiometer *n* mit Anzapfungen
t. resistor *(El)* Anzapfwiderstand *m*
t. shank end mill Schaftfräser *m* mit Anzugsgewinde
t. transformer Anzapftransformator *m*, Transformator *m* mit Anzapfungen
t. variable inductor *(El)* Anzapfdrossel *f*, Drossel *f* mit stellbarer Anzapfung
t. voltage transformer Anzapftransformator *m*
tapper 1. Abgreifer *m* *(Fernsprechtechnik)*; 2. Gewindebohrmaschine *f*
t. tap Maschinenmuttergewindebohrer *m* mit langem Schaft
tappet Greifer *m*, Knagge *f*, Mitnehmer *m*; Nocken *m*; Stößel *m*, Ventilstößel *m* *(Verbrennungsmotor)*; Exzenter *m* *(Webstuhl)*
t. adjustment Einstellung *f* des Stößelspiels
t. block Nockenwellengehäuse *n*
t. clearance Stößelspiel *n*, Ventilspiel *n*
t. face Stößelfläche *f*
t. guide Stößelführung *f*
t. guiding screw Stößelführungsschraube *f*
t. head Stößelteller *m*
t. loom Exzenterwebstuhl *m*, Trittwebstuhl *m*
t. rod Riegelschieber *m* *(Signal)*
t. roller Stößelrolle *f*, Anschlagrolle *f*
t. wheel Daumenrad *n* *(Webstuhl)*
t. wrench Ventileinstellschlüssel *m*
tapping 1. Abstich *m*, Anzapfen *n*; 2. *(Met)* Abstich, Abstechen *n*; 3. *(El)* Anzapfung *f*, Abzweigung *f* *(Leitung)*; 4. Innengewindebohren *n*; 5. *(Fmt)* Abhören *n*
t. attachment Gewindebohreinrichtung *f*; Wendegetriebe *n* zum Gewindebohren
t. contact intermittierender Kontakt *m*
t. contactor *(El)* Anzapfschütz *n*, Stufenschütz *n*
t. cutter Gewindefräser *m*
t. depth Gewindebohrtiefe *f*
t. door *(Met)* Abstichöffnung *f*
t. hole 1. Stichloch *n*, Eisenabstich *m*; 2. Gewindekernloch *n*
t. jig Gewindebohreinrichtung *f*
t. job Gewindebohrarbeit *f*
t. machine Gewindebohrmaschine *f*
t. of nuts Muttergewindebohren *n*
t. platform *(Met)* Abstichbühne *f*
t. point Anzapfung *f*, Anzapfort *m*; Abzweigungspunkt *m*
t. reversal Bewegungsumkehr *f* der Spindel beim Gewindebohren
t. service *(Fmt)* Abhördienst *m*
t.-size hole Gewindekernloch *n*
t. splice *(El)* Leitungsabzweigung *f*
t. spout *(Met)* Abstichrinne *f*
t. switch *(El)* Stufenschalter *m*
t. temperature Abstichtemperatur *f*
tappings *(Met)* Puddelschlacke *f*
tar / to teeren, mit Teer bestreichen
t. over beteeren
tar Teer *m*
t. acid Teersäure *f*, saure Teerfraktion *f*
t. barrel Teerfaß *n*
t. board Teerpappe *f*
t. boiler Teerkessel *m*, Asphaltkessel *m*
t. burner Teerbrenner *m*
t. circulation Teerspülung *f*
t. coke Pechkoks *m*
t. collecting main Teervorlage *f*
t. concrete Teerbeton *m* *(Straßenbau)*
t. condensate Teerkondensat *n*
t. cooler box Teerkühler *m*
t. decantation under pressure Teerdruckscheidung *f*
t. distillery Teerschwelerei *f*
t. dyes Teerfarbstoffe *mpl*
t. extraction Teerabscheidung *f*, Teerwäsche *f*
t. extraction plant Teergewinnungsanlage *f*
t. extractor Teer[ab]scheider *m*
t. felt Teerpappe *f*
t. main Teervorlage *f*, Teerleitung *f*
t. oil Teeröl *n*
t. paper Teerpapier *n*
t. residue Teerrückstand *m*

t.-saturated felt Teerpappe *f*
t. separation Teerscheidung *f*
t. separator Teerscheider *m*
t. settler Teerscheider *m*
t.-storage tank Teergrube *f*
t. tank car Teer[kessel]wagen *m*
t. tanker Teerwagen *m*
t. value Verteerungszahl *f*
t. vapour Teerdampf *m*
tare / to tarieren
tare Tara *f*, Masse *f* der Verpackung
target 1. Ziel *n*; Zielscheibe *f (Sportschießen)*; 2. Auffänger *m*, Auffangplatte *f*; Prallplatte *f*; Auffangschirm *m*; Target *n*; Fangelektrode *f*; Antikatode *f*; 3. Radarziel *n*; Signalplatte *f (Elektrode einer Bildaufnahmeröhre)*
t. angle Anodenwinkel *m (Röntgenröhren)*
t. blip Zielzeichen *n (Radar)*
t. date Endtermin *m*, Fertigstellungstermin *m*, Termin *m*
t. electrode Auffängerelektrode *f*
t. element Speicherplattenelement *n*; Targetelement *n*
t.-homing Zielsuche *f (Flugkörper durch Radar)*
t. nucleus Zielkern *m*, Targetkern *m*
t. practice Schießübung *f*, Übungsschießen *n*, Scheibenschießen *n*
t. range Schießstand *m*
t. shooting Scheibenschießen *n*
t. signal Zieltafel *f*
t. signal equipment Tafelsignalausrüstung *f*
t. surface Speicherplattenoberfläche *f*
t.-towing aircraft Scheibenschlepper *m (Flugzeug)*
tariff Tarif *m*; Zolltarif *m*
tarmacadam Asphaltteermakadam *m*
tarnish / to anlaufen, glanzlos werden (machen), [sich] trüben
tarpaulin Persenning *f*; Plane *f*; Verdeckleinwand *f*; Zeltplane *f*; Abdeckplane *f*
t. cover Planenüberzug *m*
t. frame Planengestell *n*, Spriegel *m*
tarragon oil Estragonöl *n*
tarred geteert
t. board Teerpappe *f*
t. brown paper Teerpapier *n*, Bitumenpapier *n*
tarring Teeren *n*, Teerung *f*
tarry teerartig, teerig
tarsia Holzeinlegearbeit *f*, Intarsia *f*
tartar Weinstein *m*, Kaliumhydrogentartrat *n*
t. emetic Brechweinstein *m*, Kaliumantimonyltartrat *n*
tartaric acid Weinsäure *f*
tartarous weinsteinhaltig
tartrate Ester *m* (Salz *n*) der Weinsäure, Tartrat *n*
tartrated antimony Brechweinstein *m*, Kaliumantimonyltartrat *n*
tartronic acid Hydroxymalonsäure *f*, Tartronsäure *f*
tarviated mit Teerbindung
task Aufgabe *f*
tasteless geschmackfrei
tau meson *(Kern)* Tau-Meson *n*
taurine *(Ch)* Taurin *n*
taurocholic acid Taurocholsäure *f*
taut straff, gespannt
t. band suspension Spannbandhalterung *f (bei Meßinstrumenten)*
t. measuring wire *(Masch)* Meßdraht *m*
t. steel wire gespannter Stahldraht *m*
t.-wire suspension Spanndrahtaufhängung *f (Meßinstrumente)*
tautomeric *(Ch)* tautomer
t. transformation tautomerer Übergang *m*
tautomerism *(Ch)* Tautomerie *f*
tautomerize / to *(Ch)* tautomerisieren
taw / to *(Led)* weißgerben, weißgar machen, abbeizen
tawer Weißgerber *m*
tawing Weißgerben *n*, Alaungerben *n*
t. matter Weißgerbstoff *m*
tax / to taxieren
tax indicator *(Fmt)* Gebührenmelder *m*, Gebührenanzeiger *m*
taxi / to rollen *(Flugzeug)*
t. off abrollen *(Flugzeug vom Boden)*
taxi-holding position Haltepunkt *m* [auf dem Rollfeld]
t. way light *(Flg)* Rollbahnfeuer *n*, Rollbahnbefeuerung *f*
taxiing trial *(Flg)* Rollversuch *m*
taximeter Taxameter *n (m)*
teaching aid Lehrmittel *n*
t. staff Lehrkörper *m*
teak Tiekholz *n*, Teakholz *n*
tear / to [ein]reißen; zerreißen; verschleißen
tear chip Reißspan *m*
t. drops *(Plast)* Tropfenbildung *f* an Spritzlingen *(Spritzfehler)*
t. gas *(Ch)* Tränengas *n*
t. of glass Glasträne *f*
tearing 1. Zerreißen *n*, Reißen *n*; 2. *(Fs)* Bildzerreißung *f*
t. loose Ausbrechen *n* [des Kornes] *(Schleifscheibe)*
t. machine Reißwolf *m*
t. strength Zerreißfestigkeit *f*
teasel *(Text)* Karde *f*
t. frame Kardenrahmen *m*
teaseling *(Text)* Rauhen *n*
teaser Reißwolf *m*
teat cup Zitzenbecher *m (Melkstand)*
t. rubber cup Zitzengummi *m* im Zitzenbecher *(Melkstand)*
teazle *(Text)* Rauhkarde *f*

t. raising machine Kardenrauhmaschine *f*
technetium Technetium *n*
technical technisch
t. advisory service technischer Beratungsdienst *m*
t. aid agreement Abkommen *n* über technische Hilfe
t. aptitude test technische Eignungsprüfung *f*
t. association Fachverband *m*
t. data technische Daten *pl* (Information *f*)
t. delivery conditions technische Lieferbedingungen *fpl*
t. education Fachausbildung *f*
t. journal Fachzeitschrift *f*
t. literature Fachliteratur *f*
t. school Fachschule *f*, Technikum *n*; Berufsschule *f*, Gewerbeschule *f*
t. standard Fachnorm *f*, Standard *m*
t. term Fachausdruck *m*, Fachwort *n*
t. training Fachausbildung *f*
technically pure technisch rein
technician Techniker *m*; Arbeiter *m* im technischen Beruf
technicist *s.* technician
technique Technik *f*, Methode *f*, Verfahren *n*, Praxis *f*
t. of welding Schweißtechnik *f (Methode)*
techniques of measurement Meßtechnik *f*
technologist Techniker *m*
technology Technik *f*
t. of textiles Faserstofftechnik *f*
tecnetron Tecnetron *n (n-Germaniumtransistor, der nach dem Feldeffektprinzip arbeitet)*
tectonic *(Geol)* tektonisch
tectonics *(Geol)* Tektonik *f*
ted / to wenden, ausbreiten
tedder Heuwendemaschine *f*, Heuwender *m*
tee / to *(El)* abzweigen
tee T-Stück *n*
t. beam Plattenbalken *m (Stahlbetonbau)*
t.-conduit box *(El)* T-Dose *f (für I-Rohr)*
t. head bolt Hammerschraube *f*
t.-joint 1. *(El)* Abzweigklemme *f*, T-Klemme *f*; 2. T-Stoß *m (Schweißen)*
t.-piece T-Stück *n*
t. slot T-Nut *f*
t.-slotted mit T-Nuten
t. square Reißschiene *f*
t. steel T-Stahl *m*
teed feeder *(El)* Gabelleitung *f*, Stichspeiseleitung *f*
teem / to *(Met)* in Kokillen abgießen *(aus einer Stopfenpfanne)*
teeming *(Met)* Vergießen *n* [in Kokillen] *(aus einer Stopfenpfanne)*
t. ladle Stopfenpfanne *f*
t. platform Gießbühne *f*
t. rate Gießgeschwindigkeit *f*
teeth Verzahnung *f*; Zähne *mpl*
t. arranged for depth cutting Zähne *mpl* mit Tiefenstaffelung *(Räumwerkzeug)*
t. having circular recess unechte Pfeilverzahnung *f*
t. lock washer federnde Zahnscheibe *f*
t. on the end Stirnzähne *mpl (Fräser)*
t. on the periphery Umfangszähne *mpl (Fräser)*
t. with side set geschränkte Sägezähne *mpl*
teleammeter Stromfernmeßgerät *n*
teleautography Teleautografie *f*
telecast / to durch Fernsehen übertragen
telecast Fernsehsendung *f*
telecentric path of rays *(Opt)* telezentrischer Strahlengang *m*
t. pencil *(Opt)* telezentrischer Strahlengang *m*
telecine projector Fernsehgroßbildprojektor *m*
telecommunication Fernmeldetechnik *f*, Fernmeldewesen *n*
t. engineer Fernmeldetechniker *m*
t. network Fernmeldenetz *n*, Nachrichtennetz *n*
t. service Fernmeldedienst *m*
t. system Fernmeldenetz *n*, Nachrichtennetz *n*
telecommunications engineering Nachrichtentechnik *f*
t. service Fernmeldedienst *m*
telecontrol Fernbedienung *f*; Fernsteuerung *f*
t. engineering Fernwirktechnik *f*
telefilm Fernsehfilm *m*
telegram Telegramm *n*
t. address Kurzanschrift *f*
telegraph / to telegrafieren
telegraph Telegraf *m*
t. blank Telegrammformular *n*
t. cable Telegrafenkabel *n*
t. channel Telegrafenkanal *m*
t. character Telegrafiezeichen *n*
t. code Telegrafenkode *m*, Telegrafenalphabet *n*
t. construction Telegrafenbau *m*
t. construction office Telegrafenbauamt *n*
t. cross-talk Telegrafiernebensprechen *n*
t. current Telegrafierstrom *m*
t. distributor Kanalumschalter *m*, Verteilerschalter *m (Telegrafie)*
t. equipment stores Telegrafenzeugamt *n*
t. frequency Telegrafierfrequenz *f*
t. insulator Telegrafenisolator *m*
t. key Morsetaste *f*
t. line Telegrafenleitung *f*
t. modulated wave getastete ungedämpfte Welle *f*
t. network Telegrafennetz *n*
t. pole Telegrafenmast *m*
t.-pulse distortion Telegrafierzeichenunschärfe *f*

t. **reception** Telegrafieempfang *m*
t. **regenerative repeater** entzerrender Telegrafieübertrager *m*
t. **regulations** Telegrafenvollzugsordnung *f*
t. **repeater** Telegrafieverstärker *m*; Telegrafenübertrager *m*
t. **signal distortion** Zeichenverzerrung *f (Telegrafie)*
t. **sounder** Klopfer *m (Telegrafie)*
t. **station** Telegrafenamt *n*
t. **transmission speed** Telegrafiergeschwindigkeit *f*
t. **wire** Telegrafendraht *m*
telegrapher Telegrafist *m*
telegraphic telegrafisch
t. **address** Telegrammadresse *f*
t. **distortion** Telegrafierverzerrung *f*
t. **frequency** Telegrafierfrequenz *f*
t. **noise** Telegrafiergeräusch *n*
t. **restitution** telegrafische Wiedergabe *f*
t. **speed** Telegrafiergeschwindigkeit *f*
telegraphy Telegrafie *f*
teleguided ferngelenkt, ferngesteuert
teleindicator of level elektrischer Fernmelder *m* für Flüssigkeitsstände
telelens Fernobjektiv *n*, Teleobjektiv *n*
telemeter 1. Entfernungsmesser *m*; 2. Fernmesser *m*, Fernmeßinstrument *n*
telemetered data Funkfernmeßdaten *pl*
telemetering 1. Fernmessen *n*, Fernmessung *f (elektrischer Größen)*; 2. Entfernungsmessung *f*
t. **data** Funkfernmeßdaten *pl*
t. **device** Fernmeßeinrichtung *f*
telemetry Fernmessung *f*, Telemetrie *f*
telemotor steering-gear control Fernsteuerapparat *m (ferngesteuerte Rudermaschine)*
telephone Telefon *n*, Fernsprecher *m*
t. **adaptor** Telefonadapter *m*
t. **apparatus** Fernsprechapparat *m*
t. **apparatus with restore-to-normal switch** Fernsprechapparat *m* mit Ruhestellung
t. **area** Anschlußbereich *m*, Sprechbereich *m*
t. **booth** Fernsprechzelle *f*
t. **box** Fernsprechzelle *f*
t. **breakdown** Fernsprechstörung *f*
t. **cabin** Fernsprechzelle *f*
t. **cable** Fernsprechkabel *n*, Telefonkabel *n*
t. **call** Telefonanruf *m*
t. **carrier current** Telefonieträgerstrom *m*
t. **central office** *(Am)* Fernsprechamt *n*, Fernsprechvermittlungsstelle *f*
t. **channel** Fernsprechkanal *m*, Fernsprechweg *m*
t. **charge** Fernsprechgebühr *f*, Telefongebühr *f*
t. **circuit** Fernsprechleitung *f*, Sprechstromkreis *m*
t. **circuit with telegraph circuit superposed** Simultanleitung *f*
t. **communications** Fernsprechwesen *n*
t. **conversation** Telefongespräch *n*
t. **counter** Gesprächszähler *m*
t. **current** Telefoniestrom *m*
t. **dial** Telefonwählscheibe *f*
t. **directory** *(Am)* Telefonbuch *n*, Fernsprechverzeichnis *n*, Teilnehmerverzeichnis *n*
t. **drop** Telefonhausanschlußleitung *f*
t. **engineering** Fernsprechtechnik *f*
t. **exchange** Fernsprechamt *n*, Telefonamt *n*
t. **fee control section** Gebührenüberwachungsstelle *f*
t. **flex** Telefonschnur *f*
t. **frequency** Fernsprechfrequenz *f*
t. **head-gear receiver** Kopf[fern]hörer *m*
t. **hook switch** Hakenumschalter *m*
t. **influence factor** Fernsprechformfaktor *m* der Spannung
t. **installation** Fernsprechanlage *f*, Telefonanlage *f*
t. **insulator** Fernsprechisolator *m*
t. **interference factor** Fernsprechstörfaktor *m*; Rauschfaktor *m*
t. **line** Fernmeldeleitung *f*, Telefonleitung *f*, Fernsprechkreis *m*, Fernsprechlinie *f*
t. **message** fernmündliche Mitteilung *f*
t. **network** Fernsprechnetz *n*, Telefonnetz *n*
t. **office** Fernsprechamt *n*
t. **operation** Fernsprechbetrieb *m*
t. **pair** Aderpaar *n*
t. **plant** Telefonanlage *f*
t. **plug** Telefonstecker *m*
t. **receiver** Fernhörer *m*, Telefonhörer *m*
t. **regulations** Fernsprechordnung *f*
t. **relay** Fernsprechrelais *n*, Telefonrelais *n*
t. **repeater** Telefonverstärker *m*, Leitungsverstärker *m*, Fernsprechverstärker *m*
t. **service** Fernsprechdienst *m*
t. **set** Handapparat *m*, Fernsprechapparat *m*, Fernsprecher *m*
t. **station** Fernsprechstelle *f*
t. **subscriber** Fernsprechteilnehmer *m*
t. **switchboard** Vermittlungsschrank *m*
t. **switchhook** Aufhängebügel *m*, Telefonhaken *m*
t. **traffic** Fernsprechverkehr *m*, Telefonverkehr *m*
t. **traffic unit** Belegungszahl *f*
t. **transmission reference system** Fernsprecheichkreis *m*
t. **truck** Fernsprechwagen *m*
t. **trunk zone** Fernsprechverkehrsbereich *m*
t. **voice recorder** Gerät *n* zum Aufzeichnen von Gesprächen
telephonic telefonisch
t. **connection** Telefonverbindung *f*
telephonograph Telefonograf *m*, Fernsprechschreiber *m*

telephony Telefonie *f*
t. and telegraphy Drahtnachrichtentechnik *f*, Fernmeldewesen *n*
telephoto Fernaufnahme *f*
t. attachment Vorsatzlinse *f*
t. lens Teleobjektiv *n*
t. transmission Bildübertragung *f*
telephotography *(El, Fs)* Fernbildübertragung *f*
teleprint exchange Fernschreibvermittlung *f*
teleprinter Fernschreibmaschine *f*, Springschreiber *m*
t. exchange Fernschreibvermittlung *f*
t. network Fernschreibnetz *n*
Teleran Teleran *n (Blindflugfunkverfahren mit Fernsehkursanweisung vom Boden aus)*
telerecorder Fernregistrierinstrument *n*
telerecording Fernregistrierung *f*
telerepeating device Fernsteuervorrichtung *f (mit Rückmeldung)*
telescope / to ineinanderschieben; ausziehen
telescope Fernrohr *n*
t. drive Fernrohrantrieb *m*
t. level Nivellierlibelle *f*
t. support Fernrohrstütze *f*
telescopic ausziehbar, teleskopisch
t. alidade *(Opt)* Kippregel *f*
t. berth *(Schiff)* Ausziehkoje *f*
t. cylinder Teleskopzylinder *m*
t. damper Teleskopstoßdämpfer *m*
t. flow *(Plast)* laminares Fließen *n*
t. funnel Teleskopschornstein *m*
t. gauge Innentaster *m*, verstellbare Bohrungslehre *f*
t. joint Teleskopverbindung *f*
t. magnifier Fernrohrlupe *f*
t. rifle sight Gewehrzielfernrohr *n*
t. screw Teleskopspindel *f*
t. shaft Teleskopwelle *f*, längsveränderliche Welle *f*
t. shock absorber Teleskopstoßdämpfer *m*
t. sight Visierfernrohr *n*
t. spectacles Fernrohrbrille *f*
t. tube Teleskoprohr *n*
telescoping coil transformer Tauchkerntransformator *m*
teletranscriptions Fernsehaufzeichnungen *fpl*
teletype / to fernschreiben
teletype Fernschreiber *m*
t. apparatus Fernschreiber *m*
t. circuit Fernschreibleitung *f*
t. communication Fernschreibverbindung *f*
t. exchange Fernschreibvermittlung *f*
t. network Fernschreibnetz *n*
t. service Fernschreibdienst *m*
t. setter *(Typ)* Setzmaschine *f* mit Fernsteuerung durch Lochband, Teletypesetter *m*
t. setting *(Typ)* Fernsatz *m*
teletypewriter Fernschreiber *m*
t. exchange Fernschreibamt *n*, Fernschreibstelle *f*
t. service Teilnehmerfernschreibverkehr *m*
teleview / to fernsehen
televiewer Fernsehteilnehmer *m*
televise / to durch Fernsehen übertragen
television Fernsehen *n*
t. aerial Fernsehantenne *f*
t. announcer Fernsehansager *m*
t. band Fernsehband *n*
t. broadcast Fernsehsendung *f*
t. broadcast band Fernsehband *n*
t. broadcast station Fernsehsender *m*
t. broadcasting Fernsehrundfunk *m*
t. by electronic methods Katodenstrahlfernsehen *n*
t. by mechanical methods mechanisches Fernsehen *n*
t. cable Fernsehkabel *n*
t. camera Fernseh[aufnahme]kamera *f*
t. camera tube Fernsehaufnahmeröhre *f*
t. car Aufnahmewagen *m*, Fernsehwagen *m*
t. channel Fernsehkanal *m*, Bildkanal *m*
t. cinema Fernsehkino *n*
t. engineering Fernsehtechnik *f*
t. film scanner Filmabtaster *m*
t. image Fernsehbild *n*
t. in natural colours Farb[en]fernsehen *n*
t. information storage tube Fernsehbildspeicherröhre *f*
t. microscope Fernsehmikroskop *n*
t. microscopy Fernsehmikroskopie *f*
t. picture Fernsehbild *n*
t. picture tube Bildröhre *f*
t. receiver Fernseh[empfangs]gerät *n*
t. reception Fernsehempfang *m*
t. receptor *(Am)* Fernsehnetzantenne *f*
t. recording Fernsehbildaufzeichnung *f*
t. recording unit Bildaufzeichnungsgerät *n*
t. set Fernseh[empfangs]gerät *n*
t. signal Bildsignal *n*, Fernsehsignal *n*
t. standard Fernsehnorm *f*
t.-standard converter Normwandler *m*
t. station link Fernsehrelaisstrecke *f*
t. studio Fernsehaufnahmeraum *m*, Fernsehstudio *n*
t. studio technique Bildaufnahmetechnik *f*
t. test pattern generator Fernsehbildmustergenerator *m*
t. tower Fernsehturm *m*
t. transmission Fernsehübertragung *f*
t. transmission standards Fernsehübertragungsnormen *fpl*
t. transmitter Fernsehsender *m*
televisor Fernsehempfangsgerät *n*
televoltmeter Spannungsfernmeßgerät *n*
telewattmeter Leistungsfernmeßgerät *n*

telex Telex, Fernschreibdienst *m*; Fernschreiber *m*; Eintontelegrafie *f*
t. line Fernschreibleitung *f*
t. on multiplex Übertragung *f* von Fernschreibzeichen über Funk
t. service Teilnehmerfernschreiben *n* auf Fernsprechleitungen, Teilnehmerfernschreibdienst *m*
telltale Anzeiger *m*
t. board Anzeigetafel *f*
t. lamp Kontrollampe *f*, Merklampe *f*, Überwachungslampe *f*, Signallicht *n*
t. level plug Flüssigkeitsmeßstab *m*
t. light Kontrollampe *f*, Signallicht *n*, Leuchtzeichen *n*, Anzeigelampe *f*
t. transmitter Kontrollsender *m*
t. watch Wächterkontrolluhr *f*
tellurate Tellurat *n*, Salz *n* der Tellursäure
telluric acid Tellursäure *f*
t. compound Verbindung *f* des vierwertigen Tellur; Verbindung des sechswertigen Tellur
telluride Tellurid *n*, Salz *n* der Tellurwasserstoffsäure
t. ore Tellurerz *n*
telluriferous tellurhaltig
tellurite Tellurit *n*, Salz *n* der tellurigen Säure
tellurium Tellur *n*
tellurous acid tellurige Säure *f*
t. compound Verbindung *f* des zweiwertigen Tellur; Verbindung des vierwertigen Tellur
telly *(sl)* Fernsehempfangsgerät *n*
telomer *(Plast)* Telomer[es] *n*
telpher elektrisch betriebenes Fahrzeug *n*; Einschienenhängebahn *f*
t. line Einschiene *f (für elektrisch betriebenes Fahrzeug)*
telpherage Kabelkrananlage *f*
TEM mode L-Typ *m*, TEM-Modus *m*, TEM-Typ *m (Wellenleiter)*
temper / to 1. anlassen *(Metall)*; 2. kneten *(Ton)*; 3. stimmen
temper 1. Härtegrad *m (nach Anlassen)*; 2. Feuchtigkeitsgehalt *m (Sand)*; 3. Mischung *f*; Zustand *m (Mörtel)*
t. brittleness Anlaßsprödigkeit *f (Stahl)*
t. carbon Temperkohle *f*
t. colour Anlaßfarbe *f*
t. hardening Anlaßhärtung *f (Stahl)*
t. mill Dressiergerüst *n*, Dressierwalzwerk *n*
t. screw Nachlaßvorrichtung *f (Seilbohren)*
temperability Härtbarkeit *f (Stahl)*
temperate gemäßigt *(Klima)*
temperature Temperatur *f*
t. balance Temperaturausgleich *m*
t. change Temperaturschwankung *f*, Temperaturänderung *f*
t. coefficient Temperaturkoeffizient *m*, Temperaturbeiwert *m*, Temperaturfaktor *m*, Temperaturzahl *f*; Wärmekoeffizient *m*, Wärmezahl *f*
t. coefficient of resistance Temperaturkoeffizient *m* des Widerstandes
t. conductivity Temperaturleitfähigkeit *f*
t. control equipment Temperaturregler *m*
t. curve *(Wärm)* Temperaturkurve *f*, Dörflische Charakteristik *f*
t. dependent temperaturabhängig
t. differential Temperaturdifferenz *f*, Temperaturgefälle *n*, Temperaturunterschied *m*
t. drop Temperaturabfall *m*
t.-entropy diagram Wärmebild *n*
t. field *(Ph)* Temperaturfeld *n*
t. fuse *(El)* thermische Sicherung *f*
t. gradient Temperaturgefälle *n*, Temperaturgradient *m*
t. increase Temperaturzunahme *f*, Temperaturanstieg *m*
t. independent temperaturunabhängig
t. influence Temperatureinfluß *m*
t. interval Temperaturintervall *n*
t. ionization Thermoionisation *f*, Wärmeionisation *f*
t. measurement Temperaturmessung *f*
t. measuring device Temperaturmesser *m*, Temperaturmeßgerät *n*
t. noise *(Ph)* Temperaturrauschen *n*
t. of combustion Verbrennungstemperatur *f*
t. of evaporation Verdampfungstemperatur *f*
t. of reference Bezugstemperatur *f*
t. radiation Wärme[aus]strahlung *f*, Temperaturstrahlung *f*
t. range Temperaturbereich *m*
t. recorder Temperaturschreiber *m*
t.-regulating device Temperiereinrichtung *f*
t. regulator Temperaturregler *m*
t. relay thermisches Relais *n*, Thermorelais *n*, Temperaturrelais *n*
t. response Temperaturabhängigkeit *f*
t. rise Temperaturanstieg *m*
t.-rise computation Wärmeberechnung *f*
t. scale Temperaturskale *f*
t. sensor Temperaturfühler *m*
t. stability Temperaturbeständigkeit *f*
t. stress Wärmespannung *f*
t. switch Temperaturschalter *m*, Thermoschalter *m*
t. variation Temperaturschwankung *f*
tempered [hard]board gehärtete (vergütete) Faserplatte *f*
tempering 1. Anlassen *n (Metall)*; 2. Anfeuchten *n (Sand)*; 3. Kneten *n (Ton)*
t. flame furnace Flammhärteofen *m*, Flammofen *m* zum Härten
t. furnace Anlaßofen *m*
t. medium Anlaßmittel *n*
t. oil Anlaßöl *n*

t. powder Härtepulver *n*
t. steel Vergütungsstahl *m*
tempet *s.* temple
template Lehre[nform] *f*, Schablone *f*; Kopierlineal *n*, Kopierschiene *f*
t. bolt *(Masch)* Paßschraube *f*, eingepaßte Schraube *f*
t. copying attachment Schablonennachformeinrichtung *f*
t. maker 1. Modelltischler *m*, Modellbauer *m*; 2. Schablonenfräsmaschine *f*
t. milling machine Schablonenfräsmaschine *f*
t. moulding Schablonenformerei *f*
t. pipe Paßrohr *n*, Fassonrohr *n*, Formstück *n*
t. tracer technique *(Masch)* Nachformung *f* mit Schablone und Fühler
temple Breithalter *m (Webstuhl)*
t. blade Breithalterschere *f*
t. mark Breithaltermarkierung *f*, Breithalterstreifen *m (Webfehler)*
t. roller Stachelwalze *f*
templet *s.* template
temporary zeitweise, temporär, vorübergehend; behelfsmäßig, provisorisch
t. chock *(Brg)* Wanderkasten *m*, Wanderpfeiler *m*
t. connection *(Fmt)* zeitweilige Verbindung *f*
t. earth vorübergehender (zeitweiser) Erdschluß *m*
t. hardness vorübergehende (temporäre) Härte *f*, Bikarbonathärte *f (Wasser)*
t. installation provisorische Anlage *f*
t. magnet temporärer (zeitlicher) Magnet *m*
t. repair Notausbesserung *f*, vorläufige Reparatur *f*
t. stop key Schnellstopptaste *f (Tonbandgerät)*
t. storage *(Dat)* Zwischenspeicher *m*; Zwischenspeicherung *f*
t. support *(Brg)* vorläufiger Ausbau *m*
t. switch *(Eis)* Notweiche *f*
t. timbering *(Brg)* verlorene Zimmerung *f*
t. wire fliegende (provisorisch verlegte) Leitung *f*
temse Sieb *n*
ten Zehn *f*; Zehner *m*; Zehnerstelle *f*
t.-angled zehneckig
t.-digit number zehnstellige Zahl *f*
t.-part zehnteilig
t.-sided zehnflächig, zehnseitig
t.-to-one synchronization *(El)* Zehn-zu-eins-Frequenzteilung *f*
t.-way sorting zehnwegige Sortierung *f*
tenacious zäh, widerstandsfähig; beharrlich
tenacity Zähigkeit *f*; Zugfestigkeit *f*
t. test Festigkeitsuntersuchung *f*
Tenbrink flue furnace Flammrohr-Tenbrinkfeuerung *f*
tend / to *(Masch)* warten, bedienen
tendency Neigung *f*, Tendenz *f*
t. of solution *(Ch)* Lösungstendenz *f*
t. to oscillation *(El)* Schwingneigung *f*
t. to sing *(Fmt)* Pfeifneigung *f*
tender / to 1. bewerben, Kostenanschlag einreichen; 2. brüchig machen, schädigen, schwächen; morsch werden
tender 1. Kostenanschlag *m*; 2. Wärter *m*; 3. Tender *m*; 4. Beischiff *n*, Beiboot *n*, Tender; Versorgungsschiff *n*, Zubringerschiff *n*
tending Pflege *f*; *(Masch)* Wartung *f*
tendon Spannglied *n (Spannbeton)*
tennantite Tennantit *m*, Arsenfahlerz *n*, lichtes Fahlerz *n*
tennis court Tennisplatz *m*
tenon / to Zapfen schneiden *(Holzbau)*
t. and mortise verzapfen
tenon Mitnehmerzapfen *m*; Vorsprung *m*, Zinken *m*; Zapfen *m*
t. head Zapfenschneidkopf *m*
t. saw Fuchsschwanz *m (Säge)*
t. saw blade Absatzsägeblatt *n*
tenoner Zapfenschneidmaschine *f (Holzbearbeitung)*
t. plane Zahnhobel *m*
tenoning apparatus Zapfenschneidgerät *n (Holzbearbeitung)*
t. cutter Zapfenschneidmesser *n*
tenorite Tenorit *m*, Kupferoxid *n*, Kupferschwarz *n*
tens carry pulse *(Dat)* Zehnerübertragimpuls *m*
t.-carrying mechanism *(Dat)* Zehnerübertragmechanismus *m*
t. complement form *(Dat)* Zehnerkomplementform *f*
t. digit *(Fmt)* Zehnerstufe *f*
tensibility Zugbelastbarkeit *f*, Streckbarkeit *f*
tensible zugbelastbar, streckbar
tensile and compression test *(Met)* Zugdruckversuch *m*
t. bar Zugstab *m*
t. dynamometer Zugkraftdose *f*
t. fatigue test *(Met)* Zugdauerschwingversuch *m*
t. force Zugkraft *f*
t. impact stress *(Met)* Schlagzerreißbeanspruchung *f*
t. load Zugbeanspruchung *f*, Zuglast *f*
t. reinforcement Zugbewehrung *f (Stahlbetonbau)*
t. strength Zugfestigkeit *f*, Zerreißfestigkeit *f*
t. stress Zugspannung *f*
t. test Zugversuch *m*, Zerreißversuch *m*
t. testing machine Zerreißmaschine *f*
tension / to auf Zug beanspruchen; strecken
tension 1. Zug *m (Festigkeitslehre)*; 2. Spannung *f (z. B. Oberflächenspannung; auch Elektrotechnik)*; 3. Streckung *f (Garn)*

t. **bar** Zugstab *m*
t. **between electrodes** Elektrodenspannung *f*
t. **cable** Spanndraht *m*
t. **crack** Spannungsriß *m*
t. **dynamometer** Zugkraftmesser *m*
t. **insulator** *(El)* Abspannisolator *m*
t. **load[ing]** Zugbelastung *f*
t. **member** Zuganker *m*, Zugband *n*, Zugglied *n*, Zugstab *m*
t. **of belt** Riemenspannung *f*
t. **of spring** Federspannung *f*
t. **of vapour** *(Wärm)* Gasdruck *m (bei Kavitation)*; Dampfspannung *f*
t. **pulley** Spannrolle *f*
t. **reinforcement** Zugbewehrung *f (Beton)*
t. **resonance** Spannungsresonanz *f*, Reihenresonanz *f*, Serienresonanz *f*
t. **rod** Zugstab *m*; Spannkabel *n*
t. **screw** Federspannschraube *f*, Zugfederschraube *f*
t. **specimen** Zerreißstab *m*
t. **spring** Spannfeder *f*, Zugfeder *f*
t. **strength** Zugfestigkeit *f*
t. **stress** Zugspannung *f*
t. **test** Zerreißversuch *m*, Zugversuch *m*
t. **test specimen** Probekörper *m* für Zugversuch
t. **transformer** Spannungstransformator *m*, Spannungswandler *m*
t. **wire** Spanndraht *m*
tensional strain Längsdehnung *f*
tensioned zugbeansprucht; straff [gespannt]
tensioning 1. Zugbeanspruchung *f*, Beanspruchen *n* auf Zug; 2. Spannen *n*; Vorspannen *n*, Aufbringen *n* von Vorspannung *(Spannbeton)*
t. **device** Nachspannvorrichtung *f (Freileitung)*; Spannvorrichtung *f*
t. **equipment** Nachspannvorrichtung *f (Freileitung)*
t. **the [saw] blade** Spannen *n* des Sägeblattes
tensor *(Math)* Tensor *m*
t. **condition** Tensorbedingung *f*
t. **coupling** Tensorkopplung *f*, tensorielle Kopplung *f*
t. **field** Tensorfeld *n*
t. **force** Tensorkraft *f*
t. **interaction** Tensorwechselwirkung *f*
t. **of the second rank** Tensor *m* zweiter Stufe
t. **operator** Tensoroperator *m*
tensorial tensoriell
tentative standard Vornorm *f*, Versuchsnorm *f*, vorläufiger Standard *m*
tenter 1. Wärter *m*; 2. Spannrahmen *m*
t. **drier** Bahnentrockner *m*, Rahmentrockner *m*
tenth Zehntel *n*
t.**-metre** *10^{-10} m; Ångströmeinheit*
t. **of a degree** Zehntelgrad *m*
t.**-value** Zehntelwerts-
tenth's scale Zehntelskale *f*
tephigram Temperatur-Entropie-Diagramm *n*
tepid lauwarm
teracidic base *(Ch)* dreisäurige Base *f*
terbia Terbiumoxid *n*
terbium Terbium *n*
terephthalate Terephthalat *n*, Ester *m* (Salz *n*) der Terephthalsäure
terephthalic acid Terephthalsäure *f*
terhalide Trihalid *n*, Trihalogenid *n*, Trihalogenverbindung *f*
term / to benennen
term 1. Term *m*, Energieterm *m*, Energiezustand *m*, Energieniveau *n*; 2. Begriff *m*, Fachausdruck *m*; 3. Zeitraum *m*; Frist *f*; 4. *(Math)* Glied *n*; Seite *f (Gleichung)*
t. **diagram** Termschema *n*, Energieschema *n*, Niveauschema *n*
t. **displacement** Termverschiebung *f*
t. **of a protective right** Laufzeit *f* eines Schutzrechtes
t. **of limitation** Verjährungsfrist *f (Patentwesen)*
t. **of notice** Kündigungsfrist *f*
t. **of patent** Dauer *f* des Patentschutzes
t. **of payment** Zahlungsfrist *f*
t. **of prescription** Verjährungsfrist *f*
t. **of the apprenticeship** Lehrdauer *f*, Lehrzeit *f*
t. **scheme** Termschema *n*, Energieschema *n*, Niveauschema *n*
terminable begrenzbar, festlegbar
terminal 1. *(El)* Klemme *f*, Anschluß[klemme *f*] *m*; Pol *m*; Kabelschuh *m*; 2. *(Fmt)* Endamt *n*; 3. Endhafen *m*, Zielhafen *m*; Endstation *f*, Abfahrtsstelle *f*, Bahnhof *m (Bus)*
terminal Grenz-, End-, endständig
t. **amplifier** Endverstärker *m*
t. **assembly** Klemmenstreifen *m*, Rahmenverteiler *m*
t. **at the input side** Eingangsklemme *f*
t. **at the output side** Ausgangsklemme *f*
t. **block** Klemmenblock *m*, Blockklemme *f*; Klemmbrett *n*
t. **board** Klemmbrett *n*, Klemmleiste *f*; Lötösenleiste *f*
t. **box** Abschlußmuffe *f*, Kabelendverschluß *m*; Kabelübergangskasten *m*, Klemmenkasten *m*, Anschlußkasten *m*
t. **bracket** *(Fmt)* Abspannstütze *f*
t. **building** Empfangsgebäude *n (Flughafen)*
t. **clamp** *(Fmt)* Abspannklemme *f*
t. **condition** Abschlußbedingung *f*, Bedingung *f* am Leitungsende
t. **connection piece** Klemmenanschlußstück *n*
t. **connector** Klemmenverbinder *m*

t. contact Endkontakt *m*, Schlußkontakt *m*
t. control area Nahverkehrsbereich *m (Flugsicherung)*
t. echo suppressor *(Fmt)* Gabelechosperre *f*
t. edge Polkante *f*
t. fishplate Anschlußlasche *f (Eisenbahnschiene)*
t. forecast Wettervorhersage *f* für den Zielflughafen
t. impedance Anschlußimpedanz *f*
t. insulator Abspannisolator *m*
t. lead äußere Zuführung *f*, Klemmenzuleitung *f*
t. limit switch Endschalter *m*
t. lug *(Am)* Anschlußöse *f*
t. marking Klemmenbezeichnung *f*
t. pillar Polbolzen *m*
t. plate Klemmenbrett *n*, Klemmenblock *m*
t. plug Anschlußklemme *f*
t. pole Endmast *m*, Abspannmast *m (am Leitungsende)*
t. port Endhafen *m*, Zielhafen *m*
t. post Batterie[pol]klemme *f*
t. pressure *(Wärm)* Expansionsendspannung *f*
t. punching *(Am)* Kontaktnietöse *f*
t. repeater Endverstärker *m (von Fernleitungen)*
t. resistance Abschlußwiderstand *m*, Endwiderstand *m*
t. return loss *(El)* Rückflußdämpfung *f*
t. screw Klemm[en]schraube *f*
t. socket *(Fmt)* Endtülle *f*, Überführungsdose *f*
t. spindle Abspannstütze *f*
t. state Endzustand *m*
t. station 1. *(Fmt)* Endamt *n*; 2. Zielbahnhof *m*
t. strain insulator Abspannisolator *m*
t. strip Anschlußleiste *f*, Klemm[en]leiste *f*, Klemmstreifen *m*
t. switch End[aus]schalter *m*
t. trunk exchange *(Fmt)* Endfernamt *n*
t. velocity Endgeschwindigkeit *f*
t. voltage Klemmenspannung *f*, Endspannung *f*; Arbeitsspannung *f*
t. wire Anschlußdraht *m*, Zuleitung *f*
terminate / to *(El)* abschließen, begrenzen
terminated level *(El)* Pegel *m (bei abgeschlossener Leitung)*
terminating capacitance (capacity) *(El)* Endkapazität *f*
t. impedance *(El)* Abschlußwiderstand *m*
t. resistance ohmscher Abschlußwiderstand *m*, Endwiderstand *m*
termination 1. Beendigung *f*; 2. *(El)* Abschluß *m (Widerstand)*, Begrenzung *f*; Endverschluß *m (Kabel)*
t. cable Abschlußkabel *n*
t. of combustion *(Rak)* Brennschluß *m*
t. of wire Abspannbindung *f* der Leitung
t. of wires on terminal poles Abspannung *f* der Leitungen
terminology Terminologie *f*
termolecular *(Ch)* trimolekular
terms of delivery Lieferbedingungen *fpl*
t. of sale Verkaufsbedingungen *fpl*
tern schooner Dreimastgaffelschoner *m*
ternary *(BMSR)* aus drei Einheiten bestehend, dreigliedrig, ternär; Dreistoff-, ternär
t. alloy *(Met)* Dreistofflegierung *f*, ternäre Legierung *f*
t. fission *(Kern)* Dreifachspaltung *f*, ternäre Spaltung *f*, Dreierspaltung *f*
t. system Dreistoffsystem *n*, ternäres System *n*
terne plate Mattblech *n (mattes Weißblech)*
ternitrate Trinitrat *n*
teroxide Trioxid *n*
terpene *(Ch)* Terpen *n*
terpenic acid Terpensäure *f*
terra-cotta Terrakotta *f*, Terrakotte *f*
terrace Terrasse *f*
t. house Reihenhaus *n*, Zeilenhaus *n*
terraced culture Terrassenanbau *m (Reis, Wein usw.)*
t. housing Reihenbebauung *f*, Zeilenbebauung *f*; Reihenhäuser *npl*, Zeilenhäuser *npl*
terrain clearance indicator *(Flg)* Bodenhindernisanzeiger *m*, Funkfeinhöhenanzeiger *m*
terrazzo Terrazzo *m*
terrestrial *(Geol)* terrestrisch
t. electricity Geoelektrizität *f*, Erdelektrizität *f*
t. magnetic field erdmagnetisches Feld *n*
t. magnetic poles magnetische Pole *mpl* der Erde
t. magnetism Erdmagnetismus *m*
t. navigation terrestrische Navigation *f*
t. orbit Erdbahn *f*
t. photogrammetry terrestrische Fotogrammetrie *f*
t. telescope Aussichtsfernrohr *n*
territorial waters Hoheitsgewässer *npl*, Küstenmeer *n*
terry fabric Frottégewebe *n*, Frottiergewebe *n*
t. towel Frottétuch *n*, Frottiertuch *n*
t. towelling Frottégewebe *n*, Frottiergewebe *n*
t. velvet Halbsamt *m*
tersulphate Trisulfat *n*
tertiary *(sl) (El)* Tertiärwicklung *f*
t. air Drittluft *f*, Tertiärluft *f*
t. compound tertiäre Verbindung *f*
t. overburden Tertiärgebirge *n*
t. spectrum tertiäres Spektrum *n*
t. winding Tertiärwicklung *f*
tervalence *(Ch)* Dreiwertigkeit *f*
tervalent *(Ch)* dreiwertig

Tesla coil Tesla-Transformator *m*
tessela Mosaikstein *m*
tesselated schachbrettförmig, schachbrettartig, gewürfelt
t. mirror *(Opt)* Facettenspiegel *m*
t. pavement Mosaikfußweg *m*
tessera Mosaikstein *m*
tesseral tesseral *(Kristall)*
test / to prüfen, erforschen, erproben, untersuchen
t. a tube *(El)* eine Röhre [auf Emission] prüfen
t. again nachprüfen
t. for prüfen auf
t. for balance *(Masch)* [den] Wuchtzustand prüfen
test Prüfung *f*; Versuch *m*
t. adapter *(El)* Prüfzwischensockel *m*
t. aerial Meßantenne *f*
t. balloon Ballonsonde *f*, Registrierballon *m*
t. bar 1. *(Met)* Zugstab *m*, Zerreißprobestab *m*; 2. *(El)* Prüfspitze *f*
t. battery Meßbatterie *f*
t. bay Prüfstand *m*
t. board *(Am)* *(Fmt)* Prüfgestell *n*, Prüfschrank *m*
t. brush *(Fmt)* C-Kontaktarm *m*, Prüfkontaktarm *m*
t. burner number Prüfbrennerzahl *f*
t. call *(Fmt)* Leitungsprobe *f*, Probeanruf *m*
t. cell *(El)* Ölprüfeinrichtung *f* *(Gerät für Durchschlagprüfungen an Flüssigkeiten)*
t. certificate Abnahmeprotokoll *n*, Prüfschein *m*
t. chart 1. Prüfanleitung *f*; 2. *(Fs)* Testbild *n*
t. circuit Meßkreis *m*
t. cock Probierhahn *m*, Wasserstandhahn *m*
t. cord *(Fmt)* Prüfschnur *f*
t. coupon Probe *f*, Prüfkörper *m*
t. department Prüffeld *n*
t. drive Probefahrt *f*
t. engineer Prüffeldingenieur *m*
t.-figures projector Sehzeichenprojektor *m*
t. final selector *(Fmt)* Prüfwähler *m*
t. finger *(El)* Prüffinger *m* *(Gerät zur Prüfung des Berührungsschutzes)*
t. floor Prüfstand *m*
t. for resolving power *(Opt)* Auflösungstest *m*
t. frequency *(Fmt)* Meßfrequenz *f*, Prüffrequenz *f*
t. hut *(Fmt)* Untersuchungshäuschen *n*
t. indicator Meßzeigergerät *n*
t. instruction Prüfvorschrift *f*
t. insulator Trennisolator *m* *(für Prüfzwecke)*
t. jack *(Fmt)* Prüfklinke *f*
t. jack panel Prüfklinkenfeld *n*
t. lamp Prüflampe *f*
t. launching *(Rak)* Versuchsstart *m*
t. level Meßpegel *m*
t. line Prüfleitung *f*
t. load Prüflast *f*, Prüfbelastung *f*
t. loop Meßschleife *f*
t. mandrel Prüfdorn *m*; Meßdorn *m* *(Abnahme)*
t. mark Prüfzeichen *n*
t. method Prüfverfahren *n*
t. model Versuchsmodell *n*
t. of long duration Langzeitversuch *m*
t. of short duration Kurzzeitversuch *m*
t. of significance Signifikanztest *m*
t. outfit Prüfeinrichtung *f*
t. output *(El)* Meßausgang *m*
t. paper Reagenzpapier *n*
t. pattern *(Fs)* Testbild *n*
t. pick *(Am)* *(El)* Prüfspitze *f*, Prüfdraht *m*
t. piece Prüfling *m*, Prüfstück *n*, Probestück *n*
t. pit *(Geol)* Schürfschacht *m*, Schurfschacht *m*
t. plug *(Fmt)* Prüfstöpsel *m*
t. point Meßpunkt *m*
t. probe Meßsonde *f*
t. procedure Prüfverfahren *n*
t. reading Meßablesung *f*, Meßergebnis *n*
t. record Prüfungsprotokoll *n* [für Maschinenabnahme]
t. relais Prüfrelais *n*
t. result Versuchsergebnis *n*, Prüfbefund *m*, Meßergebnis *n*
t. rig Prüfeinrichtung *f*; Versuchseinrichtung *f*; Versuchsaufbau *m*; Versuchsanlage *f*; Prüfstand *m*
t. ring *(Met)* Herdrahmen *m* *(des Treibofens)*
t. room Prüfraum *m*
t. routine *(Dat)* Prüfprogramm *n*
t. run 1. Versuchslauf *m*; 2. *(Rak)* Durchschaltversuch *m*
t. sample Prüfstück *n*; Probe *f*
t. satellite Versuchssatellit *m*
t. selector *(Am)* *(Fmt)* Prüfwähler *m*
t. set Meßgerät *n*
t. setup Versuchsanordnung *f*; Meßanordnung *f*
t. shop *(sl)* *(El)* Prüffeld *n*
t. shot *(Fs)* Probeaufnahme *f*
t.-signal generator *(Fs)* Prüfsignalgeber *m*
t. specification Prüfvorschrift *f*
t. specimen Probekörper *m*, Prüfkörper *m*, Versuchskörper *m*; Probestab *m*
t. stand Prüfstand *m*
t.-strip attachment *(Opt)* Richtreihenansatz *m*
t. terminal Prüfklemme *f*, Versuchsklemme *f*
t. terminal box Prüfklemmenblock *m* für Zähler
t. tone Prüfton *m*
t. transformer Prüftransformator *m*
t. tube Reagenzglas *n*, Probeglas *n*, Probierglas *n*, Reagenzröhrchen *n*

t. tube stand Reagenzglasgestell *n*
t. voltage *(El)* Prüfspannung *f*
tester Prüf[meß]gerät *n*
testing Prüfen *n*, Prüfung *f*, Testen *n*
t. apparatus Prüfgerät *n*; Prüfvorrichtung *f*
t. battery Meßbatterie *f*, Prüfbatterie *f*
t. circuit Meßkreis *m*
t. cord Prüfschnur *f*, Prüfleitung *f*
t. crew Meßtrupp *m*
t. device Prüfgerät *n*
t. edge Prüfschneide *f*
t. engineer Prüfingenieur *m*
t. equipment Prüfmittel *npl*
t. fixture Prüfeinrichtung *f*
t. for true running Schlagfehlerprüfung *f*
t. gallery *(Brg)* Versuchsstrecke *f*
t. ground Versuchsgelände *n*
t. instrument Prüfgerät *n*
t. laboratory Prüflabor *n*
t. lamp Prüflampe *f*
t. machine Prüfmaschine *f*
t. materials Werkstoffprüfung *f*, Prüfen *n* von Werkstoffen
t. of insulation Isolationsmessung *f*
t. of materials Werkstoffprüfung *f*, Materialprüfung *f*
t. of the efficiency of a line balance *(Fmt)* Fehlerdämpfungsmessung *f*
t. outfit Prüfeinrichtung *f*
t. potential *(El)* Prüfspannung *f*
t. room Prüffeld *n*; Prüfraum *m*
t. screen Brennrahmen *m* *(Elektronenröhrenherstellung)*
t. sieve Prüfsieb *n*
t. sizes Prüfen *n* von Größen
t. spike *(El)* Prüfspitze *f*, Prüftaster *m*
t. surface Meßfläche *f*
t. switch Prüfschalter *m*
t. technique *(El)* Meßpraxis *f*, Meßverfahren *n*
t. techniques Meßtechnik *f*
t. van Kabelmeßwagen *m*, Meßwagen *m*
t. wire Prüfdraht *m*
tetartohedral tetartoedrisch *(Kristall)*
tetartohedron Tetartoeder *n* *(Kristall)*
tetra-atomic alcohol vierwertiger Alkohol *m*
t.-polar *(El)* vierpolig
tetrabasic acid vierbasische Säure *f*
tetraborate Tetraborat *n*
tetraboric acid Tetraborsäure *f*
tetracid base viersäurige Base *f*
tetracosanic acid Tetrakosansäure *f*
tetracuspid *(Math)* Astroide *f*, vierspitzige Hypozykloide *f*
tetracuspid vierspitzig
tetracyclic compound *(Ch)* Verbindung *f* mit einem viergliedrigen Ring, Vierkernverbindung *f*
tetrad 1. *(BMSR)* Tetrade *f*; 2. vierwertiges Atom *n*
tetradecanoic acid Myristinsäure *f*
tetraethyl lead Bleitetraäthyl *n*
tetragon Viereck *n*
t. connection *(El)* Trapezschaltung *f*
tetragonal viereckig; tetragonal *(Kristall)*
tetrahalogenated vierfach halogeniert
tetrahedral tetraedrisch
tetrahedrite Tetraedrit *m*; Antimonfahlerz *n*
tetrahedron Tetraeder *n*
tetrahydric alcohol vierwertiger Alkohol *m*
tetrakisazo-dyestuff Tetrakisazofarbstoff *m*
tetramorphous tetramorph
tetrasubstituted *(Ch)* vierfach substituiert
tetrathionic acid Tetrathionsäure *f*
tetravalence *(Ch)* Vierwertigkeit *f*
tetravalent *(Ch)* vierwertig
tetrode *(Fs)* Tetrode *f*, Vierpolröhre *f*
tewing Schlagen *n* des Hanfes
tex-system *(Text)* Tex-System *n*
t.-system of yarn counts Tex-Numerierung *f*
text Text *m*
t. book *s.* textbook
t. in clear *(Fmt)* Klartext *m*
t. matter *(Typ)* glatter Satz *m*
textbook Leitfaden *m*; Lehrbuch *n*
textile Gewebe *n*, textiles Flächengebilde *n*
textile gewebt; textil, Textil-
t. belting Textilriemen *m*, Faserstoffriemen *m*
t. finishing Textilveredlung *f*
t. industry Textilindustrie *f*
t. machinery Textilmaschinen *fpl*
t. oil Textilöl *n*, Appreturöl *n*
t. paper tube Papierhülse *f* *(Spinnerei)*
t. printing Zeugdruck *m*
t. reinforced plastics mit Textilien verstärkte Plaste *mpl*
t. soap Textilseife *f*
textiles Textilien *pl*
texture 1. Gewebe *n*, Textur *f*; 2. *(Brg)* Schicht *f*, Gefüge *n*; 3. *(Met)* Gefüge, Bruchbild *n*, Gefügebild *n*; 4. *(Geol)* Textur; 5. Struktur *f*
t. of wood Holzgewebe *n*
textured yarn Texturgarn *n*
textureless amorph, strukturlos
texturing finishing Appretur *f*
t. printing Zeugdruck *m*
thallic compound Verbindung *f* des dreiwertigen Thallium
thalliferous thalliumhaltig
thallium Thallium *n*
thallous compound Verbindung *f* des einwertigen Thalliums
thatch / to mit Schilf (Stroh) eindecken *(Dach)*
thatchboard Strohbauplatte *f*
thatched roof Rieddach *n*, Rohrdach *n*, Schilfdach *n*, Strohdach *n*, weich eingedecktes Dach *n*

thaw point Taupunkt *m*
thawing Auftauen *n*
theatre organ Kinoorgel *f*
Thenard's blue Thenards Blau *n*
theobromine Theobromin *n*
theodolite Theodolit *m*
theorem Satz *m*, Lehrsatz *m*
  t. of centre of mass Schwerpunktsatz *m*
  t. of corresponding states Theorem *n* der übereinstimmenden Zustände
  t. of three moments Dreimomentengleichung *f (Statik)*
  t.-proving machine *(Dat)* lehrsatzbeweisende Maschine *f*
theoretical theoretisch
  t. cutoff angle *(Rak)* Programmendwinkel *m*
  t. plate Trennstufe *f*, theoretischer Boden *m* *(Destillation)*
  t. value errechneter Wert *m*; Soll-Wert *m*
theory Theorie *f*, Lehre *f*
  t. of alternating currents Wechselstromlehre *f*
  t. of circuits Stromkreistheorie *f*
  t. of dispersion Dispersionstheorie *f*
  t. of failure Bruchhypothese *f*
  t. of functions Funktionentheorie *f*
  t. of games *(Dat)* Spieltheorie *f*; Theorie *f* der Spiele
  t. of heat Wärmelehre *f*; Wärmetheorie *f*
  t. of information Informationstheorie *f*
  t. of interlacing *(Text)* Bindungslehre *f*
  t. of numbers Zahlentheorie *f*
  t. of probability Wahrscheinlichkeitslehre *f*, Wahrscheinlichkeitsrechnung *f*
  t. of radiation Strahlungstheorie *f*
  t. of relativity Relativitätstheorie *f*
  t. of specific heat Theorie *f* der spezifischen Wärme
  t. of stops *(Opt)* Blendentheorie *f*
  t. of strain Spannungstheorie *f*
  t. of strength of materials Festigkeitslehre *f*
  t. of vibrations Schwingungslehre *f*
  t. of wave motion Wellentheorie *f*, Undulationstheorie *f*
theralite *(Geol)* Theralith *m*
therm *Einheit der Wärmemenge des englischen Einheitensystems; 1 therm = 10^5 British Thermal Units = $1{,}056 \cdot 10^8$ J*
thermactinic radiation *(Ph)* Temperaturstrahlung *f*
thermal thermisch, kalorisch, Wärme-
  t. absorption Wärmeabsorption *f*
  t. action Wärmewirkung *f*
  t. agitation thermische Anregung *f*
  t. agitation noise Wärmerauschen *n*
  t. agitation voltage Spannung *f* infolge Wärmerauschens
  t. ammeter Hitzdrahtamperemeter *n*
  t. analysis thermische Analyse *f*
  t. atomic rocket thermische Kernrakete *f*
  t. beam thermisches Neutronenbündel *n*
  t. breakdown Wärmedurchschlag *m*
  t. breeder *(Kern)* thermischer Brutreaktor *m*
  t. bridge *(Bw)* Kältebrücke *f*; Wärmebrücke *f*
  t. capacity Wärmekapazität *f*
  t. change point *(Met)* Umwandlungspunkt *m*
  t. circuit-breaker Schalter *m* mit thermischer Auslösung, Thermoschalter *m*
  t. circuit-breaking *(El)* thermische Auslösung *f*
  t. coefficient of expansion Wärmeausdehnungszahl *f*
  t. conduction Wärmeleitung *f*
  t. conductivity Wärmeleitfähigkeit *f*, thermische Leitfähigkeit *f*
  t. contact Thermokontakt *m*
  t. contraction Wärmeschrumpfung *f*
  t. convection Wärmekonvektion *f*, konvektive Wärmeübertragung *f*
  t. converter *(El)* Thermoumformer *m*, Thermokreuz *n*
  t. crack *(Met)* Warmriß *m*
  t. cross section *(Kern)* Wirkungsquerschnitt *m* für thermische Neutronen
  t. current rating *(Am) (El)* thermischer Nennstrom (Kurzschlußstrom) *m*, Nennkurzzeitstrom *m*
  t. curve *(Met)* Haltepunktkurve *f*
  t. cycle Wärmekreislauf *m*, Kühlkreislauf *m*
  t. cycling *(Met)* Temperaturwechselbeanspruchung *f*
  t. detector Detektor *m* für thermische Neutronen
  t. diffusion Thermodiffusion *f*
  t. diffusion method Thermodiffusionsverfahren *n*
  t. diffusion plant Thermodiffusionsanlage *f*
  t. diffusivity Wärmeausbreitungsvermögen *n*; Temperaturleitstrahl *m*, Temperaturleitfähigkeit *f*
  t. E.M.F. Thermo-EMK *f*, thermoelektrische Kraft *f*
  t. effect *(El)* Thermoeffekt *m*, Wärmewirkung *f*
  t. effectiveness Wärmewirkungsgrad *m*
  t. efficiency thermischer Wirkungsgrad *m*, Wärmewirkungsgrad *m*
  t. efficiency of the boiler thermischer Wirkungsgrad *m* des Kessels
  t. effusion Wärmediffusion *f*
  t. emissivity Wärmeabgabevermögen *n*
  t. energy Wärmeenergie *f*, thermische Energie *f*
  t. energy yield Wärmeenergieausbeute *f*
  t. engine Wärmekraftmaschine *f*
  t. equilibrium thermisches Gleichgewicht *n*

t. expansion Wärme[aus]dehnung *f*
t. expansivity Wärmeausdehnungsvermögen *n*
t. explosion Wärmeexplosion *f*
t. extension Wärme[aus]dehnung *f*
t. fission *(Kern)* thermische Spaltung *f*, Spaltung durch thermische Neutronen
t. flowmeter thermischer Durchflußmesser *m*
t. fluctuation Wärmeschwankung *f*
t. gradient Temperaturgefälle *n*, Temperaturabfall *m*, Temperaturgradient *m*
t. grid emission thermische Gitteremission *f*
t. impulse heat sealing *(Plast)* Wärmeimpulsschweißen *n*
t. inertia Wärmeträgheit *f*
t. instability thermische Instabilität *f*
t. instrument [elektro]thermisches Instrument *n*; Thermomeßinstrument *n*
t. insulation Wärmedämmung *f*, Wärmeschutz *m*
t. insulation material Wärmeisolierstoff *m*, Wärmedämmstoff *m*
t. ionization thermische Ionisation (Ionisierung) *f*, Temperaturionisation *f*, Thermoionisation *f*, Wärmeionisation *f*
t. jet engine Heißstrahltriebwerk *n (Turboantrieb)*
t. jet-propulsion unit Heißstrahltriebwerk *n*
t. lag Wärmeverzug *m*, thermische Verzögerung *f*
t. level thermisches Niveau *n*, Wärmeniveau *n*
t. life-time Lebensdauer *f* thermischer Neutronen
t. load Wärmebelastung *f*
t. microphone Thermomikrofon *n*, thermisches Mikrofon *n*
t. molecular flow thermische Molekularströmung *f (Knudsen-Effekt)*
t. motion Wärmebewegung *f*
t. neutron thermisches Neutron *n*
t. neutron absorption thermische Neutronenabsorption *f*, Absorption *f* thermischer Neutronen
t. neutron cross section Wirkungsquerschnitt *m* für thermische Neutronen
t. noise Wärmerauschen *n*
t. noise generator Wärmerauschgenerator *m*
t. power Wärmeleistung *f*
t. power station Wärmekraftwerk *n*
t. precipitator *(Brg)* Thermalpräzipitator *m*
t. radiation thermische Strahlung *f*, Temperaturstrahlung *f*, Wärmestrahlung *f*
t. radiator Temperaturstrahler *m*, Wärmestrahler *m*
t. rating Wärmeleistung *f*
t. reactor thermischer Reaktor *m*; langsamer Reaktor
t. relay Thermorelais *n*
t. resistance Wärmefortleitungswiderstand *m*; Wärmedurchgangswiderstand *m*
t. resistivity Wärmebeständigkeit *f*, Wärmewiderstandsfähigkeit *f*
t. shield Wärme[schutz]schild *m*, Wärmeabschirmung *f*
t. shock test Wärmestoßprüfung *f*
t. siphon Thermosiphon *m*
t. spectrum thermisches Spektrum *n*
t. spike *(Kern)* thermischer Störungsbereich *m*
t. spring Thermalquelle *f*
t. stability Wärmebeständigkeit *f*, Hitzebeständigkeit *f*
t. state Wärmezustand *m*
t. station Wärmekraftwerk *n*
t. storage water heater [elektrischer] Heißwasserspeicher *m*
t. strain 1. thermale Verformung *f (unzulässige Erwärmung des Halbleiterkristalls)*; 2. *(sl)* Wärmebeanspruchung *f*
t. stress Temperaturspannung *f*, Wärmespannung *f*; Wärmebeanspruchung *f*
t. superconductivity Suprawärmeleitfähigkeit *f*
t. treatment *(Met)* Wärmebehandlung *f*
t. tuning sensitivity thermische Abstimmempfindlichkeit *f*
t. unit Wärmeeinheit *f*
t. upcurrent thermischer Aufwind *m*
t. utilization *(Kern)* thermische Nutzung *f*
t. value Heizwert *m*; Wärmewert *m (Zündkerze)*
t. velocity thermische Geschwindigkeit *f*
t.-velocity neutron thermisches Neutron *n*
t. voltmeter thermischer Spannungsmesser *m*
t. well Schutzrohr *n (Temperaturfühler)*
t. working Warmbearbeitung *f*
thermally compensated temperaturkompensiert
t. insulated mit Wärmedämmung, wärmeisoliert
t. stable temperaturkonstant
t. unstable thermisch instabil
thermel Thermoelement *n*
thermic thermisch, Wärme-; kalorisch
t. equivalent kalorisches Äquivalent *n*, Wärmewert *m* der mechanischen Arbeit
t. siphon Wasserkammer *f (Feuerbüchse)*
t. unit Wärmeeinheit *f*
thermion Thermion *n*, Glühelektron *n*
thermionic glühkatodisch, glühelektrisch, thermionisch; Elektronen-; Röhren-
t. cathode Glühkatode *f*
t. constant Glühemissionskonstante *f*
t. current thermischer Elektronenstrom *m*, Glühelektronenstrom *m*
t. dark current thermischer Dunkelstrom *m*

t. **diode** Röhrendiode *f*
t. **discharge** Glühelektronenentladung *f*
t. **emission** Glüh[elektronen]emission *f*, thermische Elektronenemission *f*, thermionische Emission *f*
t. **generator** Röhrengenerator *m*
t. **ion source** Thermionenquelle *f*
t. **rectifier** Glühkatoden[-Gasentladungs]ventil *n*, Glühkatodengleichrichter *m*
t. **valve** Elektronenröhre *f*
t. **valve detector** Röhrendetektor *m*, Lampendetektor *m*
t. **valve receiver** Röhrenempfänger *m*
t. **valve transmitter** Röhrensender *m*
t. **voltmeter** Röhrenvoltmeter *n*
thermionics *(El)* Theorie *f* der Elektronenröhrentechnik
thermistor Thermistor *m*, Heißleiter *m*, temperaturabhängiger Widerstand *m*
t. **calorimeter** Thermistorkalorimeter *n (für Impulsleistungsmessungen)*
t. **mount** Bolometerabschluß *m (Wellenleiter)*
thermit[e] Thermit *n*
t. **welding** Thermitschweißung *f*
thermo-ammeter Thermoamperemeter *n*
t.**-bimetallic instrument** *(El)* Bimetallmeßinstrument *n*
t.**-E.M.F.** Thermo-EMK *f*, thermoelektrische Kraft *f*, Thermokraft *f*, Thermospannung *f*
t.**-electric** *s.* thermoelectric
t.**-sensitive element** Temperaturfühlglied *n*
thermoanalysis thermische Analyse *f*
thermochemistry Thermochemie *f*
thermoconverter Wärmeumformer *m*
thermocouple Thermoelement *n*, Thermopaar *n*
t. **ammeter** Amperemeter *n* mit Thermoumformer
t. **amplifier** Thermoelementverstärker *m*
t. **element** Thermoelement *n*
t. **junction** Lötstelle *f* des Thermoelements
t. **thermometer** Temperaturmeßgerät *n* mit Berührungsthermoelement
thermodiffusion Thermodiffusion *f*
thermodynamic thermodynamisch
t. **potential** thermodynamisches Potential *n*
t. **probability** thermodynamische Wahrscheinlichkeit *f*
t. **similitude** thermodynamische Ähnlichkeit *f*
t. **state** thermodynamischer Zustand *m*
t. **transformation** thermodynamische Änderung *f*, thermodynamischer Prozeß *m*
thermodynamically unstable thermodynamisch instabil
thermodynamics Thermodynamik *f*
thermoelastic wärmeelastisch, thermoelastisch
thermoelasticity Thermoelastizität *f*
thermoelectric wärmeelektrisch, thermoelektrisch
t. **cell** Thermoelement *n*, thermoelektrisches Element *n*
t. **cooler module** thermoelektrischer Kühlmodul *m*
t. **couple** Thermoelement *n*
t. **current** Thermostrom *m*
t. **effect** thermoelektrische Wirkung *f*; Seebeck-Effekt *m*
t. **force** Thermokraft *f*
t. **generator** Thermosäule *f*
t. **inversion** Polaritätsumschlag *m* der Thermospannung
t. **junction** Thermoelementkontaktstelle *f*
t. **pile** Thermoelement *n*, Thermosäule *f*
t. **series** thermoelektrische Spannungsreihe *f*
t. **traction** thermoelektrische Zugförderung *f*
t. **voltage** Thermospannung *f*
thermoelectricity Wärmeelektrizität *f*, Thermoelektrizität *f*
thermoelectron Glühelektron *n*, Thermion *n*
thermoelement Thermoumformer *m*, Thermoelement *n*
thermoexpansion instrument Hitzdrahtinstrument *n*
thermofixation Heißfixierung *f*
thermograph Thermoelement *n*, Temperaturschreiber *m*
thermography 1. Thermografie *f (Lumineszenzverfahren zur Temperaturmessung)*; 2. *(Typ)* Stahlstichimitationsdruck *m*
thermohardening *s.* thermosetting
thermohydrometer Thermoaräometer *n*
thermoionic pressure gauge Ionisationsvakuummeter *n*, Ionisationsmanometer *n*
t. **vacuum gauge** Ionisationsvakuummesser *m*
thermojunction Lötstelle *f (eines Thermoelements)*; Thermoübergang *m (Halbleitertechnik)*
thermokinetic fuel equivalent *(Rak)* thermokinetisches Brennstoffäquivalent *n*
thermoluminescence Thermolumineszenz *f*
thermolysis thermische Zersetzung *f*, Thermolyse *f*
thermomagnetic effect thermomagnetischer Effekt *m*
thermometer Thermometer *n*
t. **controller** Thermometerregler *m*
t. **probe** Thermometersonde *f*, Temperaturfühler *m*
t. **resistor** Widerstandsthermometer *n*
t. **tube** Temperaturfühler *m* mit Schutzrohr
thermometric scale Temperaturskale *f*
thermometrical rate of cooling thermometrische Abkühlungsgeschwindigkeit *f*
thermometry Temperaturmessung *f*
thermomotor Heißluftmaschine *f (Motor)*
thermonuclear thermonuklear, Fusions-
t. **bomb** thermonukleare Bombe *f*, Fusionsbombe *f*

t. reaction thermonukleare Reaktion *f*
t. reactor thermonuklearer Reaktor *m*
t. transmutation thermonukleare Umwandlung *f*
thermopile Thermosäule *f*, Thermoelement *n*
thermoplastic Thermoplast *m*
thermoplastic thermoplastisch
t. adhesive Schmelzklebstoff *m*, thermoplastischer Harzklebstoff *m*
t. fibre wärmeverformbare Faser *f*; thermoplastische Faser
t. recording *(BMSR)* thermoplastisches Aufzeichnen *n*
t. tape *(BMSR)* thermoplastisches Band *n*
t. tape recorder Thermoplastbandaufzeichner *m*
thermoregulator Temperaturregler *m*
thermorelay Thermorelais *n*, thermoelektrischer Verstärker *m*
thermoset duroplastisch, hitzehärtbar
thermoset hitzehärtbares Material *n*, Duroplast *m*
thermosetting Thermofixierung *f*
thermosetting hitzehärtbar, in der Wärme aushärtend; duroplastisch
t. adhesive hitzehärtbarer Harzklebstoff *m*, Schmelzklebstoff *m*
t. plastic Duroplast *m*, hitzehärtbarer Plast *m*
thermosiphon action Thermosiphonwirkung *f* *(Heizungs- und Kühlsystem)*
t. cooling Wärmeumlaufkühlung *f*
thermostable hitzebeständig, thermostabil, wärmebeständig
thermostat Temperaturschalter *m*, Thermostat *m*, Signalthermometer *n*
t. relay Thermostatrelais *n*, thermisches Relais *n*
thermostate / to in einem Thermostaten unterbringen, thermostatisieren
thermostatic control Thermostatsteuerung *f*, Thermostatregelung *f*
t. controller Thermostatregler *m*
t. valve direktwirkender Temperaturregler *m*, Temperaturregler ohne Hilfsenergie
thermotape *s.* thermoplastic tape
thermoweld / to *(Plast)* unter Einwirkung von Wärme verschweißen
thermowell Schutzrohr *n* *(Temperaturfühler)*
thesis These *f*; Dissertation *f*; Diplomarbeit *f*
Thevenin's theorem *(Dat)* Zweipoltheorie *f*, Helmholtzscher Satz *m*, Satz von der Ersatzspannungsquelle
thick dick; stark; *(Geol)* mächtig
t.-film lubrication Vollschmierung *f*, flüssige Reibung *f*
t. full line dicke Vollinie *f* *(technisches Zeichnen)*
t. juice blowup Dicksaftkocher *m* *(Zuckerindustrie)*
t. mud Dickschlamm *m*
t. sheet iron Grobblech *n*
t.-walled dickwandig
t. wire drawing bench Grobdrahtziehbank *f*
thicken / to eindicken; einkochen; gelieren
thickener 1. Verdickungsmittel *n*; 2. Eindickzylinder *m*; 3. Eindicker *m* *(Aufbereitung)*
thickening Dickwerden *n*; Eindicken *n*, Eindickung *f*; Verdickung *f*
t. chest Eindickbütte *f* *(Papierherstellung)*
t. of the blade root Verdickung *f* der Schaufelplatte *(Turbine)*
thicket Dickicht *n*
thickness / to auf bestimmte Dicke bringen (hobeln)
thickness Dicke *f*
t. at point of tooth Randdicke *f* *(Sägeblatt)*
t. gauge Dickenmesser *m*; Dickenmeßgerät *n*; Fühlerlehre *f*
t. gauging Dickenmessung *f*
t. measurement Dickenmessung *f*
t. of cavity Dicke *f* des Hohlraumes *(Kavitation)*
t. of fire Schichthöhe *f*, Schütthöhe *f*, Höhe *f* der Schüttung *(Feuerungstechnik)*
t. of pipe Rohrwanddicke *f*
t. of tooth Zahndicke *f*
t. vibrations *(Fmt)* Dickenschwingungen *fpl*
thicknessing machine Dickenhobelmaschine *f*
t. machine for boards Bretterhobelmaschine *f*
thicktar Dickteer *m*
t. decanter Dickteerabscheider *m*
thill wagon Gabelwagen *m*
thimble Fingerhut *m*; Hülse *f*; Halter *m*; Mantelhülse *f*; Muffe *f*; Zwinge *f*; Kabelschuh *m*; Seilkausche *f*; Meßtrommel *f*
t.-tube waste-heat boiler Clarkson-Abgaskessel *m*
thin / to 1. verdünnen; 2. ausspitzen *(Spiralbohrer)*; 3. ausschmieden, ausbreiten; 4. verziehen *(Rüben)*; 5. durchforsten
t. out strecken, recken
thin dünn; dünnflüssig; *(Brg)* geringmächtig
t.-bodied dünnflüssig
t. casting dünnwandiger Guß *m*
t. cut section Dünnschnitt *m*
t. film Dünnfilm *m*, Dünnschicht *f*
t.-film capacitor Dünnfilmkondensator *m*
t.-film cold cathode Dünnschichtkaltkatode *f*
t.-film lubrication Teilschmierung *f*; Mischreibung *f*
t.-film magnetic storage magnetisches Speichern *n* auf dünnen Schichten
t.-film microminiaturization Dünnfilmmikrominiaturisierung *f*
t.-film resistor Dünnfilmwiderstand *m*

t.-film semiconductor Dünnschichthalbleiter *m*
t.-film technique Dünnfilmtechnik *f (z. B. zur Herstellung elektrischer Widerstände)*
t.-film transistor Dünnfilmtransistor *m*
t. flame Stichflamme *f*
t. full line dünne Vollinie *f (technisches Zeichnen)*
t. ground section Dünnschliff *m*
t.-layer chromatography Dünnschichtchromatografie *f*
t.-layer evaporator Dünnschichtverdampfer *m*
t. magnetic film dünne magnetische Schicht *f (Datenspeicherung)*
t. polished section Dünnschliff *m*
t.-section mit geringem Querschnitt; dünnwandig *(Guß)*
t. section Dünnschnitt *m*; Dünnschliff *m*
t.-sheet plastic Plastfolie *f*
t.-shell dünnschalig
t.-wall dünnwandig
t. walled casting dünnwandiger Guß *m*
thinner Verdünner *m*, Verdünnungsmittel *n (Anstrichtechnik)*
thinning 1. Verdünnen *n*; 2. Ausspitzen *n (Spiralbohrer)*; 3. Durchforsten *n*; 4. Ausschmieden *n*, Ausbreiten *n*; 5. Verziehen *n (Rüben)*
t. in the suppressed wood Niederdurchforstung *f*
t. operation Ausspitzarbeit *f (Spiralbohrer)*
t. out of beets Verziehen *n* der Rüben
t. the point of the drill Ausspitzen *n* des Bohrers
thinnings Durchforstungsholz *n*
thioacetic acid Thioessigsäure *f*
thioalcohol Merkaptan *n*, Thioalkohol *m*
thiocarbonic acid Thiokohlensäure *f*
thiocyanate Rhodanid *n*, Thiozyanat *n*
thiocyanic acid Rhodanwasserstoffsäure *f*, Thiozyansäure *f*
thiocyanogen value Rhodanzahl *f*
thioic acid Thiosäure *f*, Thiolsäure *f*, Thionsäure
thiolactic acid Thiomilchsäure *f*
thiolcarbonic acid Thiolkohlensäure *f*
thiolic acid Thiolsäure *f*
thionamic acid Sulfaminsäure *f*
thioncarbonic acid Thionkohlensäure *f*
thionic acid Thionsäure *f*
thionyl chloride Thionylchlorid *n*
thiophene Tiophen *n*
thioplast Thioplast *m*
thiosulphuric acid Thioschwefelsäure *f*
thiourea Thioharnstoff *m*
third-angle orthographic representation amerikanische Darstellungsweise *f (technisches Zeichnen)*
t. anode dritte Anode *f*, Nachbeschleunigungselektrode *f*
t.-brush control Stromregelung *f (durch eine dritte Bürste)*
t.-brush dynamo stromregelnde Lichtmaschine *f*
t. order reaction *(Ch)* Reaktion *f* dritter Ordnung
t. party liability Haftpflicht *f*
t. power of dritte Potenz *f* von
t. stage dritte Raketenstufe (Stufe) *f*
t. tap Fertigschneider *m*
t. wire Mittelleiter *m (im Gleichstromnetz)*; *(Fmt)* Prüfleiter *m*
t. wire system Dreileitersystem *n*, Mittelleitersystem *n*
thistle funnel Tropftrichter *m*
t. head *(Text)* Kardendistel *f*
thixotropic thixotrop *(umkehrbare Eigenschaft einiger Gele, beim Rühren flüssig, beim Stehen wieder fest zu werden)*
thixotropy Thixotropie *f*
thole pin Ruderdolle *f*, Riemendolle *f*, Bootsdolle *f*
Thomas-Gilchrist process *(Met)* Thomasverfahren *n*
Thomas-Gilchrist steel Thomasstahl *m*
Thomas low carbon steel kohlenstoffarmer Thomasstahl *m*
Thomas pig iron Thomasroheisen *n*
Thomson bridge *(El)* Thomsonbrücke *f*, Doppelmeßbrücke *f*
Thomson cable Thomsonkabel *n*
Thomson effect *(El)* Kreisrauschen *n*, Thomsoneffekt *m*
Thomson meter elektrodynamischer Zähler *m*
thoria Thoriumoxid *n*
thorianite *(Min)* Thorianit *m*
thoriate / to thorieren
thoriated tungsten cathode thorierte Wolframkatode *f*
t. tungsten filament thorierter Wolframfaden *m*
thorite *(Min)* Thorit *m*
thorium Thorium *n*
t. breeder Thoriumbrutreaktor *m*
t. fission *(Kern)* Thoriumspaltung *f*
t.-free pitchblende thoriumfreie Uranpechblende *f*, Nasturan *n*
t. ore Thoriumerz *n*
t. poisoning Thoriumvergiftung *f*
t. reactor Thoriumreaktor *m*
t. series Thorium[zerfalls]reihe *f*
thoron Thoron *n*, Thorium-Emanation *f*
thorough dry kerntrocken
thoroughfare Durchfahrt *f*
thortveitite *(Min)* Thortveitit *m*
thou *0,001 Zoll*

thrash / to *s.* thresh / to
thrasher *s.* thresher
thread / to 1. einfädeln; durchführen, durchstecken; 2. Gewinde schneiden
t. the broach through the work das Räumwerkzeug durch das Werkstück hindurchführen
t. through a circuit durch einen Stromkreis hindurchtreten lassen, mit einem Stromkreis verketten
thread 1. Faden *m*; 2. Gewinde[gang *m*] *n*; Windung *f (Schraube)*; 3. *(Geol)* dünne Erzader *f*
t. angle Flankenwinkel *m*
t. axis Gewindeachse *f*
t. board Faserbrett *n*
t. bottom Gewindefuß *m*
t. breakage *(Text)* Fadenbruch *m*
t. caliper Flankentaster *m*, Kugeltaster *m*, Rundtaster *m*
t. caliper gauge Gewinderachenlehre *f*
t. chaser Gewindesträhler *m*
t. chasing Gewindestrählen *n*
t.-chasing attachment Gewindesträhleinrichtung *f*
t. comparator Gewindekomparator *m*
t. count *(Text)* Feinheitsnummer *f*
t. counter *(Text)* Fadenzähler *m*
t. counting glass *(Text)* Fadenzähler *m*
t. cutter Gewindefräser *m*
t. cutting Gewindeschneiden *n*
t.-cutting attachment Gewindeschneideinrichtung *f*
t.-cutting die Gewindeschneideisen *n*
t.-cutting indicator Gewindeganganzeiger *m*
t.-cutting operation Gewindeschneidarbeit *f*
t.-cutting tool Schneidzeug *n* [für Gewinde]
t. flank Gewindeflanke *f*
t. for transmitting motions Bewegungsgewinde *n*
t. form Gewindeform *f*
t. gauge Gewindelehre *f*, Gewindering *m*
t. grinding Gewindeschleifen *n*
t.-grinding machine Gewindeschleifmaschine *f*
t.-grinding wheel Gewindeschleifscheibe *f*
t. groove Gewindelücke *f*, Gewinderille *f*
t. guidance *(Text)* Fadenführung *f*
t. guide *(Text)* Fadenauge *n*, Fadenöse *f*, Fadenführer *m*
t.-guiding elements *(Text)* Fadenleitorgane *npl*
t. hob *s.* t.-milling hob
t. hobbing Gewindewälzfräsen *n*
t. holder Fadenhalter *m (Sammelpresse)*
t. in a nut Muttergewinde *n*
t. insertion apparatus *(Text)* Fadeneinlegeapparat *m*
t. limit plug gauge Gewindegrenzdorn *m*
t.-measuring wire Meßdraht *m*
t. miller Gewindefräsmaschine *f*
t. milling Gewindefräsen *n*
t.-milling cutter Gewindefräser *m*
t.-milling hob Mehrrillengewindefräser *m*
t. of quartz Quarzfaden *m*
t. pitch Gewindesteigung *f*
t. plug gauge Gewindelehrdorn *m*
t. profile Gewindeform *f*, Gewindeprofil *n*
t. profile angle Gewindeflankenwinkel *m*
t. reach Länge *f* der Schraubfassung *(Zündkerzenmantel)*
t. recorder *(Fmt)* Fallbügelschreiber *m*
t. ring gauge Gewindemeßschraube *f*, Lehrmutter *f (Gewindelehrring)*
t. roll Rolle *f* zum Gewinderollen
t.-roll-snap gauge Gewinderollenlehre *f*
t. rolling Gewindewalzen *n*, Gewinderollen *n*
t. rolling die Gewindewalzbacke *f*
t. rolling machine Gewindewalzmaschine *f*
t. series Gewindereihe *f*
t. setting gauge Einstellgewindelehre *f*
t. setting plug Gewindelehrdorn *m*
t. sewing *(Typ)* Fadenheftung *f*
t.-shaped 1. fadenförmig; 2. gewindeförmig
t. specification Gewindebezeichnung *f*
t. suction plant *(Text)* Fadenabsauganlage *f*
t. symbol Gewindekurzzeichen *n*
t. system Gewindesystem *n*
t. take-up spring *(Text)* Fadenanzugsfeder *f*
t. tap Gewindebohrer *m*
t. tapping Gewindebohren *n*
t.-tapping operation Gewindebohrarbeit *f*
t. template Gewindeschablone *f*
t. tool Gewindemeißel *m*
t. traverse grinding Gewindelängsschleifen *n*
t. whirling Gewindewirbeln *n*
threaded mit Gewinde
t. arbor Gewindedorn *m*
t. bushing Gewindebuchse *f*
t. collar Gewindering *m*
t. connection Schraubenverbindung *f*
t. coupling Gewindestück *n*
t.-end hand reamer Handreibahle *f* mit Führungsgewinde
t. fastener Befestigungsschraube *f*
t. journal Gewindezapfen *m*
t. knuckle Gewindegelenk *n*
t. pin Gewindestift *m*
t. pipe connection Gewinderohrverbindung *f*
t. ring Gewindebuchse *f*
t.-shank drill Bohrer *m* mit Innenanzugsgewinde
t. sleeve Gewindemuffe *f*; Gewindebuchse *f*
t. winding *(El)* gefädelte Wicklung *f*
threading Gewindeherstellung *f*, Gewindeschneiden *n*

t. die Gewindeschneideisen *n*, Gewinde[schneid]backe *f*, Gewindeschneidpatrone *f*
t. externally Außengewindeschneiden *n*
t. machine 1. Gewindeschneidmaschine *f*; 2. *(Text)* Fädelmaschine *f*
t. mechanism Gewindeschneideinrichtung *f*
t. needle *(Text)* Einziehnadel *f*
t. operation Gewindeschneidarbeit *f*
t. roll Gewindewalze *f*
t. the tape Einlegen *n* des Tonbandes
t. tool Gewindeschneidwerkzeug *n*, Gewindemeißel *m*
threads per inch Gänge *mpl* je Zoll
three-address system *(Dat)* Dreiadressensystem *n*
t.-armed flange Dreiarmflansch *m*
t.-bay Dreifeld-, dreifeldrig; dreischiffig
t.-bay frame Dreifeldrahmen *m*
t.-bed hand-controlled flat knitting machine dreibettige Handflachstrickmaschine *f*
t.-blade rudder Dreiblattruder *n*, Dreiflächenruder *n*
t.-body decay *(Kern)* Dreikörperzerfall *m*
t.-body problem Dreikörperproblem *n*
t.-body rudder *s.* t.-blade rudder
t.-body system Dreikörpersystem *n*
t. card set *(Text)* Dreikrempelsatz *m*
t.-circuit receiver Dreikreisempfänger *m*, Dreikreiser *m*
t.-coat work on walls dreilagiger Putz *m*
t.-colour method *(Fs)* Dreifarbenverfahren *n*
t.-colour photography Dreifarbenfotografie *f*
t.-colour process of printing Dreifarbendruck *m*
t. column transformer dreisäuliger Transformator *m*
t.-component system Dreistoffsystem *n*
t.-component wave Dreikomponentenwelle *f*
t.-conductor cable Dreileiterkabel *n*, Dreifachkabel *n*, dreiadriges Kabel *n*
t.-contact plug dreistiftiger (dreipoliger) Stecker *m*
t.-core cable dreiadriges Kabel *n*
t.-core flex dreiadriges Kabel *n*
t.-cornered dreieckig, dreikantig
t.-cornered drill Dreikantbohrer *m*
t.-cornered file Dreikantfeile *f*
t.-cutter boring head Bohrkopf *m* mit drei Messern
t.-cylinder Dreizylinder-
t.-cylinder steam engine Drillingsdampfmaschine *f*
t.-deck[ed] vessel *s.* t.-decker
t.-decker Dreidecker *m*, Dreideckschiff *n*
t.-digit display dreiziffrige Anzeige *f*
t.-digit group *(Dat)* Trigramm *n*
t.-digit number dreiziffrige Zahl *f*
t.-dimensional dreidimensional, in den drei Ebenen; räumlich, Raum-
t.-dimensional block *(Bw)* Raumzelle *f*
t.-dimensional curve Raumkurve *f*
t.-dimensional direction finding Raumpeilung *f (Radar)*
t.-dimensional display dreidimensionale Wiedergabe *f*
t.-dimensional element *(Bw)* Raumzelle *f*
t.-dimensional lattice Raumgitter *n*
t.-dimensional milling Raumnachformen *n*
t.-dimensional milling device Raumnachformfräseinrichtung *f*
t.-dimensional reproduction Raumnachformen *n*
t.-dimensional stress räumlicher Spannungszustand *m*
t.-edged dreikantig
t.-electrode cell Dreielektrodenzelle *f*
t.-electrode photocell Steuergitterfotozelle *f*
t.-electrode valve Triode *f*
t. end twill *(Text)* dreibindiger Köper *m*
t.-engine dreimotorig; Dreimaschinen-
t.-figure dreistellig
t.-filament lamp Dreifadenlampe *f*
t.-finger rule *(El)* Dreifingerregel *f*
t.-flute drill Dreischneidenbohrer *m*, Kaliberbohrer *m*
t.-fold dreifach; dreizählig
t.-fold collision *(Kern)* Dreierstoß *m*
t.-furrow plough Dreifurchenpflug *m*
t.-gang capacitor Dreigangdrehkondensator *m*, Dreifachdrehkondensator *m*
t.-grid valve Dreigitterröhre *f*
t.-groove dreinutig, dreischneidig
t.-halves-power characteristic Raumladungskennlinie *f*
t.-handle dreiarmig, dreigriffig
t.-heat switch Dreistufenschalter *m*, Kochplattenumschalter *m*
t.-high mill Triowalzwerk *n*
t.-high rolling stand Dreiwalzengerüst *n*
t.-high train Triostraße *f*
t.-hinged Dreigelenk-
t.-hinged arch Dreigelenkbogen *m*
t.-hinged arch slab Dreigelenkbogenscheibe *f*
t.-hinged frame Dreigelenkrahmen *m*
t.-hole memory core Dreilochspeicherkern *m* *(magnetische Datenspeicherung)*
t.-in-one transport ship Mehrzwecktransportschiff *n*
t.-in-one-valve Dreifachröhre *f*
t.-island vessel Dreiinselschiff *n (Schiff mit drei Aufbauten: Poop, Brücke, Back)*
t.-jaw chuck Dreibackenfutter *n*
t.-jaw gear scroll chuck Universaldreibackenfutter *n*
t.-jaw universal Universaldreibackenfutter *n*

t.-layer board Dreischichtenverbundplatte *f*
t.-layer transistor dreischichtiger Transistor *m*
t.-lead cord Dreifachschnur *f*, Dreileiterschnur *f*, dreiadrige Schnur *f*
t.-leaf dreibindig
t.-legged Dreifuß-
t.-legged transformer dreisäuliger Transformator *m*
t.-lip[ped] dreilippig, dreischneidig, dreinutig
t.-masted fore and aft schooner Dreimastgaffelschoner *m*
t.-masted topsail schooner Dreimastschoner *m (Fock-, Groß- und Besanmast)*
t.-membered ring *(Ch)* dreigliedriger Ring *m*, Dreiring *m*
t.-miles limit Dreimeilengrenze *f*
t.-mirror contact lens *(Opt)* Dreispiegelhaftglas *n*
t.-moment theorem Dreimomentengleichung *f (Statik)*
t.-necked bottle *(Ch)* Dreihalskolben *m*
t.-out-of-seven-code *(Dat)* Drei-aus-sieben-Kode *m*
t.-part alloy ternäre Legierung *f*, Dreistofflegierung *f*
t.-part bearing *(Masch)* dreiteiliges Lager *n*
t.-particle problem Dreikörperproblem *n*
t.-phase alternating current Dreiphasenwechselstrom *m*, Drehstrom *m*
t.-phase alternator Drehstromgenerator *m*, Drehstromerzeuger *m*, Dreiphasengenerator *m*, Dreiphasenstromerzeuger *m*
t.-phase amplifier Dreiphasenverstärker *m*
t.-phase arc lamp Drehstrombogenlampe *f*
t.-phase armature Drehstromanker *m*, Dreiphasenanker *m*
t.-phase asynchronous motor Drehstromasynchronmotor *m*, Drehstrominduktionsmotor *m*, Dreiphasenasynchronmotor *m*, Dreiphaseninduktionsmotor *m*
t.-phase cable Drehstromkabel *n*, Dreiphasenkabel *n*
t.-phase circuit Drehstromkreis *m*, dreiphasiger Stromkreis *m*
t.-phase commutator motor Drehstromkollektormotor *m*, Drehstromkommutatormotor *m*
t.-phase connection Dreiphasenschaltung *f*, Dreiphasenstromkreis *m*, dreiphasiger Stromkreis *m*
t.-phase current Drehstrom *m*, Dreiphasen-[wechsel]strom *m*
t.-phase current furnace Drehstromofen *m*, Dreiphasenofen *m*
t.-phase current installation Drehstromanlage *f*
t.-phase current oil transformer Drehstromöltransformator *m*
t.-phase current output Drehstromleistung *f*
t.-phase exciter Dreiphasenerregermaschine *f*
t.-phase four-core cable Drehstrom-Vierleiterkabel *n*
t.-phase four-wire system Drehstrom-Vierleiteranlage *f*, Drehstromnetz *n* mit Nulleiter, Drehstrom-Vierleitersystem *n*
t.-phase induction motor *s.* t.-phase asynchronous motor
t.-phase line Drehstromleitung *f*
t.-phase magnet Drehstrommagnet *m*
t.-phase mains Drehstromnetz *n*, Dreiphasennetz *n*
t.-phase meter Drehstromzähler *m*
t.-phase output Drehstromleistung *f*
t.-phase power Dreiphasenleistung *f*
t.-phase rectifier Dreiphasengleichrichter *m*
t.-phase reversing starter Drehstromumkehranlasser *m*, Drehstromreversieranlasser *m*
t.-phase series wound motor Drehstromreihenschlußmotor *m*
t.-phase six-wire system offenes Drehstromsystem *n*, Drehstromsystem in offener Schaltung
t.-phase starter Drehstromanlasser *m*, Dreiphasenanlasser *m*
t.-phase switch Drehstromschalter *m*, Dreiphasenschalter *m*
t.-phase synchronous motor Drehstromsynchronmotor *m*
t.-phase synchroscope Dreiphasensynchroskop *n*
t.-phase system Drehstromsystem *n*
t.-phase three-core cable Drehstrom-Dreileiterkabel *n*
t.-phase three-wire circuit Drehstrom-Dreileiternetz *n*, Drehstromnetz *n* ohne Nulleiter
t.-phase turbogenerator Drehstromturbogenerator *m*
t.-phase voltage Drehstromspannung *f*, Dreiphasenspannung *f*
t.-phase wattmeter Dreiphasenwattmeter *n*
t.-phase winding Drehstromwicklung *f*, Dreiphasenwicklung *f*
t.-piece timber set *(Brg)* dreiteiliger Türstock *m*
t.-pin dreipolig *(Stecker)*
t.-pin plug dreipoliger Stecker *m*, Dreistiftstecker *m*, Dreifachstecker *m*, Dreifachstöpsel *m*
t.-pin plug socket dreipolige Steckdose *f*
t.-pinned arch Dreigelenkbogen *m*
t. pit system *(Led)* Dreiäscherverfahren *n*
t.-place dreistellig
t.-plane projection *(Math)* Dreitafelprojektion *f*

t.-plate rudder *(Schiff)* Dreiblattruder *n*, Dreiflächenruder *n*
t.-ply dreilagige Furnierplatte *f*
t.-point bearing Dreipunktlagerung *f*
t.-point connection Dreipunktschaltung *f*
t.-point fixing Dreipunktbefestigung *f*
t.-point landing Dreipunktlandung *f*
t.-point linkage Dreipunktaufhängung *f* *(Ackergeräte)*
t.-point location Dreipunktlagerung *f*
t.-point mounting Dreipunktaufhängung *f*
t.-point suspension Dreipunktaufhängung *f*
t.-point switch Dreifach[um]schalter *m*, Dreiwegschalter *m*
t.-pole circuit breaker dreipoliger Schalter (Ausschalter, Leistungsschalter) *m*
t.-pole jack dreipolige Klinke *f*; dreipolige Steckdose *f*
t.-pole socket dreipolige Steckdose *f*
t.-pole switch dreipoliger Schalter *m*
t.-position action *s.* t.-step action
t.-position controller Dreipunktregler *m*
t.-position fixture Vorrichtung *f* mit drei Spannstellen
t.-prism spectrograph Dreiprismenspektrograf *m*
t.-product separation Dreigutscheidung *f* *(Aufbereitung)*
t.-purpose valve Dreifachröhre *f*
t.-quarter bat Dreiviertelziegel *m*
t.-quarter set *(Brg)* normaler Türstock *m*
t.-range winding *(El)* dreietagige Wicklung *f*
t.-response controller *(BMSR)* PID-Regler *m*
t.-ribbed thread grinding wheel dreiprofilige Schleifscheibe *f*
t.-roll mill Dreiwalze *f*
t. roller calender *(Text)* Dreiwalzenkalander *m*
t.-row dreireihig *(Lager)*
t.-screw dreispindelig
t.-section band-pass filter *(Fmt)* Dreifachbandfilter *n*
t.-shift working Dreischichtenbetrieb *m*
t.-slot winding *(El)* Dreinutwicklung *f*, Dreilochwicklung *f*
t.-span *(Bw)* Dreifeld-, dreifeldrig; dreischiffig
t.-speed Dreigang-; Dreistufen-; mit drei Geschwindigkeiten (Drehzahlen)
t.-square dreikantig, Dreikant- *(Schaber, Feile)*
t.-square file Dreikantfeile *f*
t.-stage dreistufig
t.-stage rocket Dreistufenrakete *f*
t.-start dreigängig *(Gewinde)*
t.-step dreifache Stufe *f*
t.-step dreistufig, Dreistufen-
t.-step action *(BMSR)* Dreipunktregelung *f*, Dreipunktwirkung *f*
t.-step cone pulley Dreistufenscheibe *f*
t.-step filter *(Opt)* Dreistufenfilter *n*
t.-step pulley Dreistufenscheibe *f*
t.-step rocket Dreistufenrakete *f*
t.-stepped drill Stufenbohrer *m* für dreifach abgesetzte Bohrung
t.-storeyed furnace *(Met)* Dreietagenofen *m*
t.-storeyed switch dreietagiger Schalter *m*, Dreiebenenschalter *m*
t.-term controller *(BMSR)* PID-Regler *m*
t.-terminal dreipolig
t.-terminal capacitor dreipoliger Kondensator *m*
t.-terminal device Dreipolgerät *n*
t.-throw crank dreifach gekröpfte Kurbel *f*
t.-throw crankshaft dreifach gekröpfte Welle *f*
t.-track reaction *(Kern)* Dreispurenreaktion *f*
t.-valued logic *(Dat)* dreiwertige Logik *f*
t.-valve amplifier Dreiröhrenverstärker *m*
t.-view drawing Zeichnung *f* mit drei Ansichten
t.-voltmeter method Dreivoltmetermethode *f*, Dreivoltmeterverfahren *n*
t.-way dreibahnig; Dreiwege-; Dreileiter-
t.-way cock Dreiwegehahn *m*
t.-way cord *(El)* Dreileiterschnur *f*, dreiadrige Schnur *f*, Dreifachschnur *f*
t.-way drilling machine Dreiwegebohrmaschine *f*
t.-way pilot valve *(BMSR)* Ventil *n* mit kombinierter Einlaß-Auslaß-Düsensteuerung *(pneumatischer Regler)*
t.-way pipe Dreiwegestück *n*
t.-way switch Dreifachumschalter *m*, Dreifachserienschalter *m*
t.-way tap Dreiwegehahn *m*
t.-way valve Dreiwegeschieber *m*
t.-weft binding *(Text)* Dreischußbindung *f*
t.-wire meter Dreileiterzähler *m*
t.-wire method of thread measurement Dreidrahtmethode *f* des Gewindemessens
t.-wire system Dreileitersystem *n*, Dreileiternetz *n*, Dreidrahtsystem *n*
threephaser Drehstromgenerator *m*, Drehstromdynamomaschine *f*, Drehstromerzeuger *m*, Dreiphasengenerator *m*, Dreiphasenstromerzeuger *m*
thresh / to dreschen
thresher Dreschmaschine *f*
threshing Dreschen *n*, Drusch *m*
t. aggregate Druschvorrichtung *f*
t. and shredding pin Quetschstift *m* *(Stiftenmaschine)*
t. basket Dreschkorb *m*
t. cylinder Dreschzylinder *m* *(Erntemaschine)*
t. drum Dreschtrommel *f*
t. drum basket Dreschtrommelkorb *m*
t. flail Dreschflegel *m*

t. floor Dreschtenne *f*, Scheunentenne *f*, Tenne *f*
t. gear Dreschwerk *n*
t. gin Göpeldreschmaschine *f*
t. machine Dreschmaschine *f*
t. machinery Dreschwerk *n*
t. mill Dreschmühle *f*
t. operation Ausdrusch *m*
t. roller Dreschwalze *f*
t. strip Schlagleiste *f (Dreschmaschine)*
threshold 1. *(Bw)* Schwelle *f*, Türschwelle *f*; 2. *(Ph)* Schwelle, Schwellenwert *m*; Schwellenenergie *f*, Schwellenwert *m* der Energie; Energieschwelle *f*
t. bias Schwellenvorspannung *f*
t. condition Schwellenbedingung *f*
t. detector Schwell[en]wertdetektor *m*
t. dose Schwellendosis *f*, Schwell[en]wertdosis *f*
t. energy Schwell[en]wert *m* der Energie, Schwellenenergie *f*
t. frequency [fotoelektrische] Grenzfrequenz *f*
t. indicator Schwell[en]wertindikator *m*
t. light Schwellenfeuer *n (Luftfahrt)*
t. of audibility *(Ak)* Hörschwelle *f*
t. of feeling Schmerzgrenze *f*, Schmerzschwelle *f*
t. of hearing Hörschwelle *f*
t. of luminescence Lumineszenzschwelle *f*
t. of response Ansprechschwelle *f*, Reizschwelle *f*
t. of sensation Reizschwelle *f*
t. of sensitiveness Schwellenwert *m* der Empfindlichkeit
t. of time Zeitschwelle *f*
t. of visibility Sehschwelle *f*; Sichtbarkeitsschwelle *f*
t. reaction Schwellenreaktion *f*
t. value Schwell[en]wert *m*
t. voltage Schwellenspannung *f*
t. wavelength *(El)* [obere] Grenzwellenlänge *f*
thrive / to gedeihen, gut wachsen
throat 1. Kehle *f*; *(Masch)* Ausladung *f*; Lücke *f*; Maul *n*; Rachen *m (Lehre)*; Hals *m*; 2. Nahthöhe *f (Schweißen)*; 3. Verengung *f (Venturirohr)*; Einschnürung *f*; 4. Gicht *f (Hochofen)*; 5. Ansatzstück *n (Wellenleiter)*; Halsstück *n* des Hornstrahlers *(Wellenleiter)*; 6. *(Büro)* Kartenhals *m*; 7. Anschnitt *m (Werkzeug)*
t. angle Anschnittwinkel *m (Gewindeschneidbacke)*
t. clearance *(Masch)* Ausladung *f*
t. depth *(Masch)* Ausladung *f*
t. microphone Kehlkopfmikrofon *n*
t. of a furnace Feuerluke *f (des Kessels)*
t. of a hook Hakenkehle *f*
t. of the shuttle *(Text)* Schützenkehle *f*
t. stopper Gichtverschluß *m (Hochofen)*
throated gekehlt
throstle Flügelspinnmaschine *f*
throttle / to drosseln
t. down abdrosseln
throttle Drossel[klappe] *f*
t. control lever Drosselklappenhebel *m*
t. governing Drosselregelung *f*, Regelung *f* durch Dampfdrosselung *(Turbine)*
t. governor Kraftmaschinenregler *m*
t. lever Drosselklappenhebel *m*
t. pedal Gaspedal *n*
t. response Beschleunigungsvermögen *n*
t. ring Drosselring *m*
t. switch Endschalter *m*
t. twist-grip Gasdrehgriff *m (Motorrad)*
t. valve Drosselventil *n*; Drosselklappe *f*
t. valve shaft Drosselklappenwelle *f*
throttling Drosselung *f*
t. band *s.* proportional band
t. control *s.* proportional control
t. curve Drosselkurve *f (Turbine)*
t. path Drosselweg *m (Turbine)*
t. range *s.* proportional band
t. turbine Drosselturbine *f*
through-and-through [coal] *(Am)* Rohkohle *f*, Förderkohle *f*, Rohförderkohle *f*
t. bolt 1. *(Rak)* Zuganker *m*; 2. Durchsteckschraube *f*
t. bore Durchgangsbohrung *f*
t. circuit *(Fmt)* Durchgangsleitung *f*
t.-circulation drier Umlufttrockner *m*
t. clearing *(Fmt)* Fernamtstrennung *f*, Trennung *f* zugunsten des Fernverkehrs
t. connection *(Fmt)* Durchgangsverbindung *f*
t. dialling *(Fmt)* Durchwahl *f*
t.-drill / to durchbohren
t.-feed process Durchgangsverfahren *n*
t.-feed rate 1. Durchlaufvorschub *m*; 2. Ausbringung *f*, Durchgang *m*, Durchsatz *m*
t. freight *s.* transit freight
t.-going shaft durchgehende Welle *f (Turbine)*
t. grinding Durchgangsschleifen *n*
t. hardening *(Met)* Durchhärtung *f*
t. hole 1. Durchgangsbohrung *f*; 2. Lochstein *m (Uhr)*
t. hole jewel Lochstein *m (Uhr)*
t. hole tap Gewindebohrer *m* für Durchgangsbohrungen
t. hole with shoulder Ansatzdurchgangsbohrung *f*
t.-joint *(El)* Durchgangsmuffe *f*
t. magnetization Durchmagnetisierung *f*
t.-plate centre keelson *(Schiff)* durchlaufendes Mittelkielschwein *n*
t.-put Ausstoß *m*, Durchsatz *m*
t.-put rate Durchsatzgeschwindigkeit *f*
t. scavenging Querspülung *f*
t. signal *(Eis)* Liniensignal *n*

t. switching board *(Fmt)* Durchgangsfernschrank *m*
t. switching position *(Fmt)* Durchgangsfernplatz *m*
t. traffic Durchgangsverkehr *m*
t.-transmission ultrasonic testing Durchschallungsverfahren *n* der Ultraschallprüfung
t. trunk exchange *(Fmt)* Durchgangsfernamt *n*
t. ventilation Querlüftung *f*
throughout durchgehend
throughs Siebdurchgang *m*, Siebfeines *n*
throw / to werfen; schleudern; rücken; umstellen; kippen; umlegen *(Schalter)*; *(Geol)* verwerfen; [ver]zwirnen; moulinieren *(Seide)*; formen *(Ton)*
t. a beam einen Strahl aussenden
t. in einrücken, einschalten *(Kupplungen)*
t. into gear einrücken, einkuppeln
t. into mesh einrücken *(Getriebe)*
t. into vibrations in Schwingung versetzen
t. out ausrücken, ausschalten *(Kupplungen)*
t. out of gear ausrücken, auskuppeln
t. out of mesh ausrücken *(Getriebe)*
t. shuttle *(Text)* Schützen werfen
throw 1. Wurf *m*; 2. *(El)* ballistischer Ausschlag *m*, Stoßausschlag *m*; 3. Hub *m (Kurbelwelle)*; Weg *m*; Kröpfung *f*
t.-away tips Wegwerfplatten *fpl (Hartmetallbestückung)*
t. of crankshaft Kurbelkröpfung *f*
t.-out bearing of the clutch Kupplungsausrücklager *n*
t.-over contact Umschaltkontakt *m*
t.-over relay Kipprelais *n*, Umschlagrelais *n*
t.-over switch Umschalter *m*; Meßbereichsumschalter *m*
thrower Schutzring *m*
thrower's engine Töpferscheibe *f*
throwing over the switch Umstellen *n* der Weiche
t. power Streuvermögen *n (Galvanotechnik)*
t. the discus Diskuswerfen *n*
t. the hammer Hammerwerfen *n*
thru-and-thru *(Am)* *s.* through-and-through [coal]
thrus *(Am)* *s.* throughs
thrust 1. Druck[kraft *f*] *m*, Axialdruck *m*; Schub[kraft *f*] *m*; 2. *(Geol)* Überschiebung *f*, Aufschiebung *f*
t. augmentation *(El)* Schubverstärkung *f*
t. ball bearing Druckkugellager *n*
t. bearing Längslager *n*, Axiallager *n*, Schublager *n*, Spurzapfenlager *n*, Kammlager *n*
t. block Axiallager *n*, Drucklager *n*, Kammlager *n*
t. cleavage *(Geol)* Druckschlechten *pl*
t. coefficient *(Rak)* Schubbeiwert *m*
t. collar *(Masch)* Stoßring *m*; Tragring *m* *(Welle)*; Druckring *m*
t. cone Schubreglerkegel *m*
t. cutoff *(Rak)* Schubabschaltung *f*
t. deduction Sogziffer *f (Strömung)*; Schubverlust *m*
t. deduction coefficient Schubbeiwert *m*
t. developed by the rocket motor Schubkraft *f* des Raketentriebwerkes
t. die Druckstempel *m (Werkzeug)*
t. disk Druckscheibe *f*
t. fault *(Geol)* Aufschiebung *f*, inverse Verwerfung *f*
t. force Schubkraft *f*; Abdrängkraft *f*; Radialkraft *f*
t. journal Kammzapfen *m*
t. load Axialbeanspruchung *f*, Längsbeanspruchung *f*
t. measurement Schubmessung *f*
t. nozzle Schubdüse *f*
t. rating Belastungsfähigkeit *f* in Achsrichtung *(Lager)*
t. ring Stützring *m*, Druckring *m*
t. shaft Drucklagerwelle *f*
t. spring Kupplungsfeder *f*
t. strip Druckleiste *f*
t. washer Druckscheibe *f*, Stoßscheibe *f*, Druckring *m*
thulia Thuliumoxid *n*
thulium Thulium *n*
thumb nut Flügelmutter *f*
t. screw Knebelschraube *f*, Druckschraube *f*, Flügelschraube *f*, Klemmschraube *f*, Kordelschraube *f*, Ohrenschraube *f*
t. tack Reißzwecke *f*, Heftzwecke *f*, Reißbrettstift *m*
thump *(Am)* Telegrafiergeräusch *n (Knacken)*
thwart *(Schiff)* Ruderducht *f*, Ruderbank *f*, Bootsducht *f*
thwartship plank *(Schiff)* Leibholz *n*
thwartships querschiffs, dwars
thyme camphor *(Ch)* Thymol *n*
thymonucleic acid Thymonukleinsäure *f*
thyratron *(El)* Thyratron *n*, Stromtor *n*
t. relay Thyratronrelais *n*
t. valve Thyratronröhre *f*
thyristor Thyristor *m (Transistor mit negativer Widerstandskennlinie)*, steuerbare Siliziumzelle *f*, steuerbarer Gleichrichter *m*, Steuerschalter *m*, Vierschichtentriode *f*
thyrite Thyrit *m (spannungsabhängiges Widerstandsmaterial)*
thyroxine *(Ch)* Thyroxin *n*
tick / to ticken *(z. B. Geigerzähler)*
tick mark Zeckenschaden *m (Lederfehler)*
t. rate Tickfolge *f*
ticker *(El)* Schnellunterbrecher *m*, Ticker *m*; Zerhacker *m*

ticketing building Schalterhalle *f*
tickler *(El)* Rückkopplungsspule *f*
t. machine Deckmaschine *f* *(Strumpfherstellung)*
tidal basin Flutbecken *n*, Tidebecken *n*
t. harbour Fluthafen *m*, Tidehafen *m*
t. light *(Schiff)* Gezeitenfeuer *n*
t. quay Flutkai *m*
t. stream Gezeitenstrom *m*
t. waters Flutgebiet *n*
t. wave Flutwelle *f*, Gezeitenwelle *f*
tide Gezeiten *pl*
t.-dial *s.* t.-gauge
t. gate Flutschleuse *f*
t.-gauge Gezeitenpegel *m*; Tidenmesser *m*, Flutmesser *m*, Flutuhr *f*
t. lock Kammerschleuse *f*, Dockschleuse *f*, Flutschleuse *f*
t. range Tidenhub *m*
t.-tables Flutkarte *f*, Tidenatlas *m*, Gezeitentafeln *fpl*
tie / to verankern, verbinden, befestigen
t. back *(Bw)* rückwärtig verankern
t. double weaver's knots doppelte Weberknoten machen
t. ends anknüpfen
t. off a cable ein Kabel abbinden
t. piece ends anknüpfen
t. together zusammenschnüren
t. up *(Text)* anschnüren
t. up the harness *(Text)* Harnisch vorrichten
t. up the treadles *(Text)* Tritte anschnüren
tie 1. Krawatte *f*, Schlips *m*, Selbstbinder *m*; 2. Zuganker *m*, Zugband *n*, Zugglied *n*, Zugstab *m*; 3. *(Am)* Eisenbahnschwelle *f*, Schienenschwelle *f*; 4. Verbindungsstück *n*; Drahtbund *m*; Abbindung *f*; 5. Bügel *m* *(Stahlbeton)*; 6. *(El)* Anker *m*, Mastanker *m*; Strebe *f* *(des Mastes)*
t. bar Zuganker *m*, Zugband *n*, Zugglied *n*, Zugstab *m*; Ankerbolzen *m*, Spurstange *f*; Ankerstab *m*, Fugenbewehrung *f* *(Stahlbetonbau)*
t. beam 1. Spannbalken *m*; 2. *(Schiff)* Lukenverschlußbügel *m*
t. block Lasche *f* *(Holzbau)*
t. bolt Ankerbolzen *m*, Verbindungsbolzen *m*; Ankerschraube *f*
t. line *(Fmt)* Querverbindung *f*; *(El)* Verbundleitung *f*
t.-plate *(Schiff)* Längsband *n* *(Deck)*; Lukenstringer *m*
t. rod Zuganker *m*, Zugband *n*, Zugglied *n*, Zugstab *m*; Ankerbolzen *m*; Stahlanker *m*; Kuppelstange *f*, Riegel *m*, Verbindungsstange *f*, Zugstange *f*; Spurstange *f* *(Lenkung)*
t.-rod end Spurstangenkopf *m*
t.-rod yoke Spurstangengabel *f*
t. trunk *(Fmt)* Querverbindung *f*
t.-up *(Text)* Stuhlzettel *m*, Schnürung *f*
t.-up Jacquard *(Text)* Schaftmaschine *f*, Trittmaschine *f*
t. wall *(Bw)* aussteifende Trennwand *f*
t.-wire *(El)* Bindedraht *m*, Abbindedraht *m* *(für Kabel)*; *(Bw)* Rödeldraht *m*
tied arch *(Bw)* Bogen *m* mit Zugband
t.-arch bridge Bogenbrücke *f* mit Zugband
t.-arch truss Bogenfachwerk *n*
t. column bügelbewehrte Säule *f* *(Stahlbetonbau)*
tieing of the brickwork Mauerwerkverankerung *f*
tier / to [auf]stapeln
tier Stapel *m*; Etage *f*, Ebene *f* *(Wicklung)*
t. drier Etagentrockner *m*
tiering machine Stapelgerät *n*
tige Säulenschaft *m*
tight Strumpfhose *f*
tight dicht, fest; straff, stramm; undurchlässig, impermeabel
t. alignment *(Fs)* Abgleich *m* der Durchlaßkurve *(von versetzten Kreisen)*
t. beam dicht gebündelter Strahl *m*
t. bond feste (überkritische) Kopplung *f*
t. contact *(El)* guter Kontakt *m*
t. core cable *(Am)* Kabel *n* mit Massivisolierung
t. coupling *(El)* feste (enge) Kopplung *f*
t. end Spannfaden *m* *(Weberei)*
t. fit *(Am)* Preßpassung *f*
t.-fitting genau passend, dicht abschließend, dicht anliegend
t. pick Spannfaden *m* *(Weberei)*
t. side *(Brg)* Arbeitstrum *m*, Zugtrum *m*
t. weld joint dichte Schweißverbindung *f*
tighten / to anziehen *(Mutter)*; spannen; bündeln *(Strahl)*; festklemmen
tightener pulley Riemenspannrolle *f*
tightening Anziehen *n*, Festziehen *n*, Festdrehen *n* *(Mutter)*; Spannen *n*; Bündeln *n* *(Strahlen)*
t. device Anzieheinrichtung *f*
t. nut Spannmutter *f*
t. ring Spannring *m*
t. screw Befestigungsschraube *f*, Spannschraube *f*, Zugspannschraube *f*
t. wedge Stellkeil *m*, Nachstellkeil *m*
tights Trikot *m (n)* *(Sport)*
tiglic acid Tiglinsäure *f*
tile / to fliesen, kacheln; decken *(Dach)*; belegen
tile 1. Fliese *f*, Kachel *f*, Platte *f*; Formstück *n* *(Steingut, Ton)*; Dachstein *m*, Dachziegel *m*; 2. *(Met)* Schmelz[tiegel]decke *f*
t. batten Dachlatte *f*

t. lining Kachelauskleidung *f*, Kachelverkleidung *f*
t. pipe Tonrohr *n*
t. stove Kachelofen *m*
tiler Fliesenleger *m*; Dachdecker *m*
tiling Fliesenbelag *m*, Plattenbelag *m*; Kachelung *f*, Verkachelung *f*; Kacheln *n*, Verkacheln *n*, Verlegen *n* von Fliesen (Platten)
till / to bebauen, pflügen, bestellen
t. the soil ackern
tillage of soil Bestellung *f* des Bodens
tiller *(Schiff)* Pinne *f*, Ruderpinne *f*
tilling Beackerung *f*; Bestellung *f* des Bodens (Ackers)
t. beam Ackerschiene *f*
t. harrow Ackeregge *f*
t. of soil *(Lw)* Bodenbearbeitung *f*
t. of vines Weinbau *m*
t. rail of tractor Ackerschiene *f*
t. the ground Landbebauung *f*
tilt / to kippen, schrägstellen; umlegen *(Schalter)*
tilt 1. *(Masch)* Schrägstellung *f*, Neigung *f*; 2. *(Brg)* Schieflage *f*; 3. Dachschräge *f (Impulstechnik)*; *(Fs)* Rasterverzerrung *f*; 4. Nadirdistanz *f (Fotogrammetrie)*; 5. Plane *f (Wagen)*
t. angle Kippwinkel *m*, Neigungswinkel *m*, Schrägstellungswinkel *m*
t. boat Klappboot *n*
t. bow Planbügel *m*
t. mill Hammerwerk *n*, Hammerschmiede *f*
t. mixer *(Fs)* Rasterverzerrungskompensator *m*
t.-type mixer Kipptrommelmischer *m*
t.-up method Tilt-up-Bauweise *f (am Einbauort liegend vorgefertigte Wandplatten werden anschließend hochgekippt und montiert)*
tiltable kippbar, schräg[ver]stellbar
t. shoe Kippsegment *n (Segmentlager)*
tilted drum mixer Schrägtrommelmischer *m*
t. turret Schrägrevolverkopf *m*
tilting Kippen *n*, Schrägstellen *n*; Umlegen *n (Schalter)*
tilting schrägstellbar, Kipp-
t. arrangement Kippeinrichtung *f*
t. body schrägstellbare Stößelführung *f (Senkrechtstoßmaschine)*
t. box Schwenkscheibe *f (Pumpe)*
t. car Kippwagen *m*; Kipper *m*
t. cart Schüttkarren *m*
t. coil *(El)* Kippspule *f*
t. digester Sturzkocher *m (Papierherstellung)*
t.-drum mixer Kipptrommelmischer *m*
t. force Kippkraft *f*
t. furnace Kippofen *m*, Schaukelofen *m*
t. head press Kipppresse *f*
t.-insulator switch Schalter *m* mit Kippstützer
t. manometer Schrägrohrmanometer *n*
t. mixer Kipp[trommel]mischer *m*
t. moment Kippmoment *n*
t. motion Kippbewegung *f*
t. oscillation *(Fs)* Kippschwingung *f*, Relaxationsschwingung *f*
t. ram guide schrägstellbare Stößelführung *f (Senkrechtstoßmaschine)*
t. system Kippsystem *n*
t. table Kipptisch *m*
t.-turning tripod head Kinostativkopf *m*
timber / to *(Brg)* ausbauen, Ausbau einbringen, zimmern; verschalen, einschalen
timber Bauholz *n*, Nutzholz *n*, Schnittholz *n*; Ganzholz *n*, Bäume *mpl*; Baumstämme *mpl*; Holzspant *n*; Balken *m*
t. bogie Forstwagen *m*, Waldbahntruck *m*
t. bogie car Langholztransportwagen *m*
t. brick Holzziegel *m*
t. capacity Holzladefähigkeit *f*
t. carriage Bretterwagen *m*
t. carrier Holztransportschiff *n*
t. connector Holzdübel *m*; Industriedübel *m*
t. construction Holzbau *m*; Holzkonstruktion *f*
t. deck cargo *(Schiff)* Holzdecksladung *f*, Holzdeckslast *f*
t. dolphin Holzdalbe *f*
t. dresser Dickten-, Flächen- und Seitenabrichtmaschine *f*
t. end Balkenende *n*
t. for sounding boards Resonanzbodenholz *n*
t. forest hochstämmiger Wald *m*, Hochwald *m*
t. haulage vehicle Langholzwagen *m*
t. head *(Schiff)* Poller[kopf] *m*
t. joint Holzverbindung *f*
t. lagging Holzverkleidung *f*
t. line *(Geol)* Baumgrenze *f*
t.-lined shed Bretterverschlag *m*
t. load line *(Schiff)* Holz[last]ladelinie *f*
t. pentice *(Brg)* Holzschutzbühne *f*, Sicherheitsbühne *f*
t. preservative Holzschutzmittel *n*
t. set *(Brg)* Türstock *m*, Ausbaurahmen *m*
t. set without posts *(Brg)* Kappenausbau *m*
t. sizer Dickten- und Abrichthobelmaschine *f*
t. slabs Schwarten *fpl (Holzabfälle)*
t. trade *(Schiff)* Holzfahrt *f*
t. tree Nutzholzbaum *m*
t. trestle *(Am)* Holzeisenbahnbrücke[ngerüst *n*] *f*
t. truck Forstwagen *m*
t. used by the cartwright Stellmacherholz *n*
t. wharf Holzlager *n*, Holzhof *m*, Zimmerplatz *m*
t. yard Bauhof *m*, Holzlager *n*, Zimmerplatz *m*, Holzplatz *m*
timbered bestockt; eingeschalt
t. building Fachwerk *n*

timberer *(Brg)* Zimmerling *m*, Zimmerhauer *m*
timbering 1. Einschalen *n*, Einschalung *f*; 2. *(Brg)* Ausbauen *n*, Zimmern *n*; 3. Zimmerholz *n*
t. of a mine Grubenschwarten *fpl*
timberman *(Brg)* Zimmerling *m*, Zimmerhauer *m*
timbre [of sound] Klangfarbe *f*
time / to Zeit messen; [zeitlich] einteilen, zeitlich bemessen (abstimmen); zeitliche Lage bestimmen; [Zeit] stoppen; regulieren *(Uhr)*
time Zeit[dauer] *f*; Zeitpunkt *m*; Uhrzeit *f*
time zeitlich, Zeit-
t. analyzer Zeitanalysator *m*
t. and zone metering circuit *(Fmt)* Zählstromkreis *m*, Zeitstromkreis *m*, Zonenstromkreis *m*
t. assignment speech interpolation zeitabhängiges Sprechinterpolierverfahren *n*
t. axis Zeitachse *f*
t. ball Zeitball *m*
t. base 1. Zeitbasis *f*; Zeitlinie *f*; 2. Zeitablenkgerät *n*
t.-base amplifier Ablenkungsverstärker *m*
t.-base calibration Eichung *f* der Zeitbasis *(Elektronenstrahloszillografie)*
t.-base encoder *(Dat)* Zeitbasisverschlüßler *m*, Zeitbasis-A-D-Umsetzer *m*
t.-base extension Zeitbasisdehnung *f*
t.-base range Zeitablenkungsbereich *m*
t.-base voltage Ablenkspannung *f*
t. behaviour Zeitverhalten *n*, zeitliches Verhalten *n*, Zeitverlauf *m*
t. chart *(Masch)* Laufzeitdiagramm *n*
t. charter index *(Schiff)* Zeitcharterindex *m*
t. check Zeitkontrolle *f*
t. clock *(Am)* Stempeluhr *f*, Stechuhr *f*
t. compression Zeitraffung *f*
t. constant 1. Zeitkonstante *f*; 2. Trägheit *f* *(eines Meßgerätes)*
t. constant of fall *(El)* Abfallzeit *f*
t. constant of rise *(El)* Einschwingzeit *f*
t. constant of the aperiodic component Zeitkonstante *f* des Gleichstromgliedes
t.-consuming zeitraubend
t. control Programmregler *m*
t. control pulse Zeitkontrollimpuls *m* *(Radartechnik)*
t. creep curve Zeitdehnungskurve *f*
t.-current characteristic Strom-Zeit-Kennlinie *f* *(Sicherung)*
t. decrease of permeability *(El)* Permeabilitätsabnahme *f* durch Alterung
t. delay Laufzeit *f*, Verzögerungszeit *f*; Schaltverzögerung *f*, Zeitverzögerung *f*
t. delay circuit Verzögerungskreis *m*, Verzögerungsschaltung *f*
t. delay circuit breaker Schalter *m* mit verzögerter Auslösung
t. delay fuse Zeitsicherung *f*
t. delay relay Verzögerungsrelais *n*
t. delay switch Verzögerungsschalter *m*
t. delay unit Zeitverzögerungsgerät *n*
t. delay valve Verzögerungsventil *n*
t. dependence Zeitabhängigkeit *f*
t.-dependent zeitabhängig
t. determination Zeitbestimmung *f*
t. difference Zeitunterschied *m*, Zeitdifferenz *f*
t. dilatation Zeitdehnung *f*, Zeitdilatation *f*
t. displacement Zeitverschiebung *f*
t. division multiplier *(Dat)* Time-division-Multiplikator *m*, Pulshöhen-Pulsbreiten-Multiplikator *m*
t. division telegraph system Zeitmultiplextelegrafie *f*
t. domain *(BMSR)* Zeitbereich *m*
t.-edge effect in photoelastic measurements Randveränderung *f* während spannungsoptischer Messungen
t. electrical distribution system Zentraluhrenanlage *f*
t. element Zeitglied *n*
t. evaluation zeitliche Auswertung *f*
t. exposure Zeitbelichtung *f*
t. for running out *(El)* Ablaufzeit *f*, Auslaufzeit *f*
t. for unloading Entladefrist *f*
t. fuse Zeitzünder *m*
t. gate *(Dat)* Zeitselektor *m*; Tor *n* mit spezifischem Zeitintervall
t.-independence Zeitunabhängigkeit *f*
t.-independent zeitunabhängig
t.-intensity curve *s.* strength-duration curve
t. interval Zeitraum *m*, Zeitabstand *m*, Zeitabschnitt *m*, Zeitintervall *n*
t. interval meter Zeitintervallmesser *m*
t.-invariant zeitunabhängig
t. jitter *(El)* Flackern *n*, Zittern *n*, störende Zeitmodulation *f*
t. lag Verzögerungszeit *f* *(z. B. des Relais)*; Schaltverzug *m*; Nacheilung *f*
t.-lag circuit breaking verzögerte Auslösung *f*
t.-lag relay Zeitschalter *m*, Zeitrelais *n*
t. lapse motion Zeitraffer *m*
t.-lead zeitliche Voreilung *f*
t. limit Grenzzeit *f* *(z. B. Relais)*
t.-limit circuit breaker Zeitsicherung *f*
t.-limit switch Zeitschalter *m*
t. limiter Zeitablaufwerk *n*
t. loss Zeitverlust *m*; tote Zeit *f*
t. mark Zeitmarke *f*
t. measuring equipment Zeitmeßgerät *n*
t. modulation Impulszeitmodulation *f*
t. multiplex Zeitmultiplexsystem *n* *(Farbfernsehen)*

t. multiplexing *(Dat)* Zeitschachtelung *f*, Zeitteilung *f*
t. of charge *(El)* Ladedauer *f*, Ladezeit *f*
t. of climb *(Flg)* Steigzeit *f*
t. of collision *(Ph)* Stoßzeit *f*
t. of discharge *(El)* Entladedauer *f*
t. of dwell *(Masch)* Stillstandszeit *f*, Bewegungspause *f*
t. of efflux Ausflußzeit *f*
t. of energy exchange Zeitdauer *f* für den Energieaustausch, Austauschzeit *f* der Energie
t. of exposure Belichtungszeit *f*, Bestrahlungszeit *f*
t. of flight Laufzeit *f*, Flugzeit *f (z. B. von Elektronen)*
t.-of-flight method *(Kern)* Laufzeitmethode *f*, Flugzeitmethode *f*
t.-of-flight spectrometer *(Kern)* Flugzeitspektrometer *n*, Laufzeitspektrometer *n*
t. of liberation *(El)* Auslösezeit *f*
t. of loading *(El)* Ladedauer *f*, Ladezeit *f*
t. of oscillation Schwingungsdauer *f*
t. of passage *(Met)* Durchlaufzeit *f*, Durchsatzzeit *f*
t. of relaxation *(El)* Relaxationszeit *f*, Kippzeit *f*
t. of reverberation *(El)* Nachhallzeit *f*
t. of rise Anstiegsdauer *f (von Impulsen)*
t. of swing Schwingungsdauer *f*
t. of transmission Sendezeit *f*
t. of vibration Schwingungsdauer *f*
t. parity *(Ph)* Parität *f* der Zeit, Zeitparität *f*
t.-path curve Zeit-Weg-Kurve *f*
t. per piece Stückzeit *f*
t. printer Zeitdrucker *m*
t. pulse Zeitimpuls *m*
t. pulse relay Impulsrelais *n*
t. quench hardening *(Met)* Zeithärten *n*, Auslagerungshärtung *f*
t. range Zeitbereich *m*
t. recorder Zeitschreiber *m*, Zeitdrucker *m*; Kontrolluhr *f*
t. recorder for a watchman's round Wächterkontrolluhr *f*
t. relay Zeitrelais *n*
t. required for roll changing *(Met)* Walzenumrüstzeit *f*
t. resolving power zeitliches Auflösungsvermögen *n*
t. response Zeitverhalten *n*; Zeitkonstante *f*; Anstiegszeit *f*
t. reversal Zeitumkehr *f*
t.-saving zeitsparend
t.-saving Zeitersparnis *f*
t. scale Zeitmaßstab *m*
t. scale factor *(Dat)* Zeitmaß[stabs]faktor *m*
t. schedule Terminplan *m*
t.-selector transducer *(El)* Zeitdiskriminatorkreis *m*
t. setting Zeiteinstellung *f*
t.-shared amplifier *(El)* Zeitmultiplexverstärker *m*
t.-shared computer *(Dat)* Simultanrechner *m*
t.-sharing *(BMSR)* Zeitschachtelung *f*, Zeitteilung *f*
t. sharing multiplex Zeitmultiplex *m*
t. shutter *(Opt)* Zeitverschluß *m*
t. signal Zeitzeichen *n (Radio)*
t. signal service Zeitzeichendienst *m*
t. signalling system Zeitzeichenanlage *f*
t. stability Zeitkonstanz *f*
t. study Arbeitszeitstudie *f*
t.-switch Schaltuhr *f*, Zählerschaltuhr *f*, Zeitschalter *m*
t. table Zeitplan *m*; Fahrplan *m*; Stundenplan *m*
t. tapper Taktgeber *m*
t. to puncture *(El)* Durchschlagsverzögerung *f*
t. unit Zeiteinheit *f*
t. variant zeitvariabel
t.-varied gain circuit zeitabhängige Empfindlichkeitsregelung *f (Radar)*
timed control Programmregler *m (Elektronenrechner)*
t. pulse zeitlich definierter Impuls *m*
t. sequence Takt *m*
t. spark gap gesteuerte Funkenstrecke *f*
timekeeper Uhr *f (jeden Typs)*
timepiece Uhr *f (die nicht schlägt)*
timer 1. Stoppuhr *f*; Schaltuhr *f*; 2. Zeitrelais *n*; Schalter *m* für Zeiteinstellung; Zeitplangeber *m*; Programmgeber *m*; Zeitmarkengeber *m*
t. switch Programmzeitschalter *m*
timetable *s.* time table
timing 1. Stellen *n (Uhr)*; 2. Einstellung *f* [des Zeitpunktes]; Einregeln *n*; Zeiten *n*; zeitliche Bemessung *f*, Abstimmen *n*; Zündzeiteinstellung *f*; Regulierung *f*; 3. *(Am) (Fmt)* Zählung *f* nach der Zeitdauer, Zeitzählung *f*; 4. Zeitnahme *f (Arbeitszeitstudie)*
t. adjustment Einstellung *f* der Steuerung
t. capacitor Ladekondensator *m* für Zeitsteuerung, Verzögerungskondensator *m*
t. chain Steuerkette *f*
t. constant Zeitkonstante *f*
t. diagram Steuerschaubild *n*
t. electrode Zeitablenkungselektrode *f*
t. element Zeitglied *n (des Relais)*; Taktgeber *m*, Zeitgeber *m*
t. gear Steuerung *f*
t. housing Steuergehäuse *n*
t. interrupter Zeitkontakteinrichtung *f*, Zeitrelais *n*
t. lever Verstellhebel *m*

t. **mark** Einstellmarke *f*, Totpunktsmarke *f*
t. **mechanism** Steuermechanismus *m*
t. **pulse** Synchronisierungsimpuls *m*; Einstellimpuls *m*, Taktimpuls *m*
t. **pulse generator** *(Dat)* Taktgeber *m*, Unabhängige-Variable-Geber *m*, UV-Geber *m*, Zeitgeber *m*
t. **range** *(Kfz)* Verstellbereich *m* der Zündung
t. **register** *(Am) (Fmt)* Gesprächszeitmesser *m*, Zeitzähler *m*
t. **relay** Zeitrelais *n*, Verzögerungsrelais *n*, Relais *n* mit Ansprechverzögerung
t. **resistance** Verzögerungswiderstand *m*
tin / to [feuer]verzinnen
t.-**coat** [feuer]verzinnen
t.-**plate** [feuer]verzinnen
tin 1. Zinn *n*; 2. verzinnte Blechdose *f*, Blechbüchse *f*
t. **ashes** Zinnkrätze *f*, Zinnasche *f*
t. **bar** Zinnbarren *m*
t.-**base alloy** zinnhaltige Legierung *f*
t.-**base solder** Lötzinn *n*
t. **bronze** *(Am)* Zinnbronze *f*
t. **coating** Zinnüberzug *m*; Feuerverzinnung *f*
t. **cry** Zinngeschrei *n* *(Knirschen der Zinnkristalle)*
t. **deposit** Zinnlagerstätte *f*
t. **dross** Zinnabstrich *m*, Zinnkrätze *f*
t. **foil** Zinnfolie *f*, Stanniol *n*
t.-**foil capacitor** Stanniolkondensator *m*
t.-**foil electrode** Stanniolelektrode *f*
t. **plague** Zinnpest *f*
t. **plate** Weißblech *n*
t. **refuse** Zinngekrätz *n*, Zinnrückstand *m*
t. **solder** Lötzinn *n*, Weichlot *n*
t. **tetrachloride** Zinnbutter *f*, Zinn(IV)-chlorid *n*
t. **works** Zinnhütte *f*
t. **yield** Zinnausbringen *n*
tinctorial power Färbevermögen *n*
tine cage Zinkenkorb *m*
t. **of harrow** Eggenzinke *f*
tinge Färbung *f*, Schattierung *f*
tinman Zinngießer *m*; Klempner *m*
tinman's snips Blechschere *f*
t. **solder** Zinn-Blei-Lot *n*
tinned [feuer]verzinnt
t. **conductor** *(El)* verzinnter Leiter *m*
t. **copper wire** verzinnter Kupferdraht *m*
t. **steel sheet** Weißblech *n*
tinner's snip Handblechschere *f*
tinning Verzinnen *n*, Feuerverzinnen *n*; Reiblöten *n*
tinsel cord *(El)* Litzenleitung *f*, Litzenschnur *f*
t. **fabric** *(Text)* Lamé *m*
t. **yarn** Metallfäden *mpl*
tinstone Zinnstein *m*, Zinnerz *n*, Kassiterit *m*
tint / to abtönen; aufhellen
tint Abtönfarbe *f*; Farbaufhellung *f*, hellere Farbnuance *f*, Helltönung *f*; Farbton *m*, Tönung *f*, Schattierung *f*; Farbstoff *m*, Farbbeimengung *f*
tinted paper Buntpapier *n*, Bezugspapier *n*
t. **rinses** Haarfärbemittel *npl*
tinting 1. *(Text)* Nuancierung *f*, Nuancieren *n*, Abtönen *n*; 2. *(Typ)* Tonen *n*
t. **colour** Abtönfarbe *f*
t. **material** Abtönfarbstoff *m*
t. **power** Abtönvermögen *n*; Aufhellungsvermögen *n*; Färbevermögen *n*
t. **value** Aufhellungswert *m*, Farbzahl *f*
tiny crack *(Met)* Haarriß *m*
tip / to [ab]kippen; bestücken, Plättchen auflöten *(Werkzeug)*
t. **over** auf den Kopf gehen *(Flugzeuglandung)*
t. **over the lip** *(Met)* über die Schnauze gießen
tip Spitze *f*; Pumpspitze *f (Elektronenröhrenherstellung)*; Ende *n (Meßspitze)*; Kontaktspitze *f*; Schneide *f*; Schneidplatte *f*; Kopf *m*; Bestückung *f*
t. **cart** Japaner[karren] *m*, Kippkarren *m (Betontransport)*
t. **cavitation** Spitzenwirbelkavitation *f*
t. **clearance** 1. Kopfspiel *n (Gewinde, Verzahnung)*; 2. *(Schiff)* Propellerspiel *n (kleinster Abstand zwischen Propellerflügelspitzendrehkreis und Außenhaut)*
t. **cutting edge** Schneidplattenschneide *f*
t. **flank** Schneidplattenfreifläche *f*
t. **formation** *(Met)* Zipfelbildung *f*
t. **frame** Kipprahmen *m*
t. **jack** Steckdose *f (für Anschlußstifte von Fernmeldeschnüren)*
t. **material** Schneidenwerkstoff *m*
t. **of share** Scharspitze *f (Pflug)*
t. **platform** Kippbühne *f*
t. **radius** Kopfabrundung *f (Werkzeuge)*
t. **relief** Kopfkürzung *f (Zahnrad)*
t. **vortex** Spitzenwirbel *m (Kavitation)*
t. **wagon** Kippwagen *m*
t. **wire** *(Fmt)* Ader *f* zur Stöpselspitze
tipless bulb spitzenloser Kolben *m (Elektronenröhren)*
tipped bestückt *(Werkzeug)*
t. **contact** *(El)* Kontakt *m* mit Spitzenüberzug *(z. B. versilberter Kontakt)*
t. **solid cutter** massiver Fräser *m* mit bestückter Schneide
t. **with cemented carbide** hartmetallbestückt
tipper Kipp-LKW *m*, Kipper *m*
t. **barge** Kippschute *f*
tipping Bestückung *f (Werkzeug)*
t. **barrow** Kippkarren *m*
t. **cradle** *(Eis)* Abrollwiege *f*
t. **device** Kippvorrichtung *f*
t. **drum** Kipptrommel *f*

t. **ear** *(Eis)* Abrollohr *n (Kipper)*
t. **grate** Klapprost *m*, Kipprost *m*
t.-**lorry** Kipper *m*
t. **rail** *(Eis)* Abrollbahn *f (Kipper)*
t. **sill** *(Eis)* Abrollträger *m*
t. **skip** Kippkübel *m*
t. **torch** Abschmelzbrenner *m (Elektronenröhrenherstellung)*
t. **trailer** Kippanhänger *m*
t. **trough** Kippmulde *f*
tipple box *(Text)* Flachsstuhl *m*
tippler Kipper *m*; Wipper *m*
t. **floor** Wipperboden *m*
tire *(Am) s.* tyre
tissue Gewebe *n*, Zellgewebe *n*; Stoff *m (Papierherstellung)*
t. **paper** Seidenpapier *n*
titanate Titanat *n*, Salz *n* der Titansäure
titania Titandioxid *n*
titanic acid Titansäure *f*
t. **iron ore** Titaneisenerz *n*
titaniferous titanführend, titanhaltig
t. **iron** Titaneisen *n*
titanite *(Min)* Titanit *m*
titanium Titan *n*
t. **carbide** Titankarbid *n*
t. **chloride** Titanchlorid *n*
t. **white** Titanweiß *n*
titanometry Titanometrie *f*
titanous titanhaltig
title block Schriftfeld *n (Zeichnung)*
t. **of a drawing** Zeichnungsschriftfeld *n*
t. **page** Titelseite *f (eines Buches)*
titling type *(Typ)* Versalschrift *f*
titlings *(Typ)* Versalien *mpl*
titrant Titrationslösung *f*
titrate / to titrieren
t. **back** rücktitrieren
titrate zu titrierender Stoff *m*
titrating beaker Titrierbecher *m*
titration Maßanalyse *f*, Titration *f*; Titrieren *n*
t. **voltmeter** Titrationsvoltameter *n*
titre Titer *m*
titrimetric titrimetrisch
t. **standard** Urtitersubstanz *f*
to-and-fro bend test *(Met)* Hin- und Herbiegeversuch *m*
to-and-fro motion Hin- und Hergang *m*
to customer requirements nach Anforderung des Kunden; entsprechend dem Wunsch des Käufers
to scale maßstabsgetreu, maßstäblich
toad's eye [tin] Holzzinn *n*
toast / to rösten *(Brot)*
toaster *(elektrischer)* Brotröster *m*
tobacco-grater Tabakmühle *f*
t.-**juice** Tabakbeize *f*, Tabakbrühe *f*
toe 1. Fuß *m*; 2. *(Brg)* Rippenbildung *f* am Bohrlochtiefsten *(Sprengarbeit)*
t.-**cap perforating machine** Kappenlochmaschine *f (Schuhfertigung)*
t.-**cut** Sohllöseneinbruch *m (Schießarbeit)*
t. **dog** Niederhalter *m*, Spannfinger *m*
t. **hole** *(Brg)* waagerechtes Bohrloch *n* beim Wandsprengen
t.-**in** Vorspur *f (Auto)*
t. **irons of ski** Schibacke *f*
t. **of bead** Wulstspitze *f (Reifen)*
t. **of stope** *(Brg)* oberer Stoß *m (Strossenbau)*; unterer Stoß *(Firstenstoßbau)*
t. **switch** Fußstarterknopf *m*
t. **wing** Zunge *f (Schuh)*
toeboard *(Kfz)* geneigtes Fußbrett *n*
toeing *s.* toenailing
toenailing Schrägnageln *n*
toggle Gelenk *n*; Kniehebel *m*; *(El)* Froschklemme *f*
t. **block** Gleitstück *n (eines Backenbrechers)*
t. **drawing press** Kniehebelziehpresse *f*
t. **joint** Winkelverbindung *f*
t. **lever** Kniehebel *m*
t. **lever press** Kniehebelpresse *f*
t. **link** Gelenkstange *f*
t. **linkage** Kniegelenkgetriebe *n*
t. **mechanism** Kniehebelsystem *n*
t. **motion** Kniehebelbewegung *f*
t. **press** Kniehebelpresse *f*
t. **switch** Kipp[hebel]schalter *m*
t. **wedge** *(Brg)* Schwenkkeil *m*
toilet bowl Toilettenbecken *n*
tolerable zulässig
tolerance Toleranz *f*
t. **envelope** Toleranzbereich *m*
t. **gauge** Toleranztasterlehre *f*, Grenzlehre *f*
t. **given for running true** Rundlauftoleranz *f*
t. **in size** Maßtoleranz *f*
t. **rate** *(Kern)* zulässige (verträgliche) Dosisleistung *f*
t. **zone** Toleranzfeld *n*
tolerant tree Schattholz *n*
toll *(Fmt)* Läuten *n*
t. **area** *(Fmt)* Vorortverkehrsbereich *m*
t. **board** Schnellverkehrsschrank *m*
t. **cable** Fernkabel *n* für Schnellverkehr
t. **call** Anruf *m* im Vorortsverkehr, Fernverbindung *f*, Schnellgespräch *n*
t. **district** Fernnetzgruppe *f*
t. **exchange** Fernamt *n*
t. **final selector** Fernleitungswähler *m*
t. **line** Fernleitung *f*
t. **line dialling** Fernwahl *f (Schnellverkehr)*; Selbstwählfernverkehr *m (Netzgruppenbetrieb)*
t. **network** Schnellverkehrsleitungsnetz *n*
t. **office** Fernamt *n*, Schnellamt *n*

t. rate table Gebührentafel *f* für Schnellverkehr
t. recording Gesprächsanmeldung *f*
t. switching plan Fernleitungsschaltplan *m*; Bauschaltplan *m*
t. switching trunk Überweisungsleitung *f*
t. terminal Teilnehmerleitung *f* für Schnellverkehr
t. test panel Fernleitungsprüfschrank *m*

Tolu balsam Tolubalsam *m*

toluene *(Ch)* Toluol *n*
t. sulfonyl chloride Toluolsulfochlorid *n*

toluic acid Toluylsäure *f*

tolylation *(Ch)* Tolylierung *f*, Einführung *f* der Tolylgruppe

tomato-pulp Tomatenmark *n*

tombac *s.* tombak

tombak Tombak *m (Kupfer-Zink-Legierung)*

tommy [bar] Einsetzstab *m*

tomography Tomografie *f*, Schichtaufnahmeverfahren *n*

ton I 1. *Masseeinheit in England: 1 ton = 2240 lb. = 1016,047 kg;* 2. *Masseeinheiten in den USA: 1 short ton (sh.tn.) = 2000 lb. = 907,185 kg; 1 long ton (tn.) = 2240 lb. = 1016,047 kg.; 1 metric ton = 1000 kg = 1 Tonne;*
II *(Raummaß)* Registertonne *f (100 cubic feet)*

ton-foot *englische Einheit der Energie: 1 ton ft. = 309,7 kpm*

ton-force *englische Einheit der Kraft: 1 ton-force (Ton) = 9 964,015 N = 1016,047 kp*

ton-force per square foot *englische Druckeinheit: 1 Ton/sq.ft. = 1,094 kp/cm²*

ton-force per square inch *englische Druckeinheit: 1 Ton/sq.in. = 157,48 kp/cm²*

ton-force per square yard *englische Druckeinheit: 1 Ton/sq.yd. = 1,215 Mp/m³*

ton-kilometre Tonnenkilometer *n*

ton of shipping Frachttonne *f*, Schiffstonne *f*

ton per cubic foot *Dichteeinheit* 1. *in England: 1 ton/cu.ft. = 35,881 kg/dm³;* 2. *in USA: 1 sh.tn./cu.ft. = 32,037 kg/dm³*

ton per cubic yard *Dichteeinheit* 1. *in England: 1 ton/cu.yd. = 1,329 t/m³;* 2. *in USA: 1 sh.tn./cu.yd. = 1,187 t/m³*

ton-weight *s.* ton-force

tonal quality Tonqualität *f*
t. value Helligkeitsstufe *f*

tonalite *(Geol)* Tonalit *m*

tonality 1. Tonart *f*; 2. Färbung *f*

tone / to tonen, abstufen *(Farbe)*
t. down abschwächen *(Farbe)*

tone Ton *m*, Klang *m*; Färbung *f*, Farbton *m*
t. arm Tonarm *m (Plattenspieler)*
t. arm bearing Tonarmlager *n*
t. channel Tonkanal *m*
t. colour *(Ak)* Klangfarbe *f*
t.-compensated gehörrichtig, tonrichtig
t.-compensated volume control gehörrichtiger Lautstärkenregler *m*
t. control Aussteuerung *f (Rundfunksender)*; Klangfarbenregler *m*; Klangfarbenregulierung *f*; Tonblende *f*
t. control for bass Tiefenregler *m*
t. control for treble Höhenregler *m*
t. control measuring set Aussteuerungsmesser *m*
t. correction Klangfarbenkorrektur *f*
t. filter Tonsieb *n*, Tonfilter *n*
t. modulated tonfrequent moduliert
t. oscillator Tongenerator *m*
t. print Tonschrift *f*
t. quality Klangfarbe *f*
t. regulator Klang[farben]regler *m*
t.-reproduction *(Typ)* Tonwertwiedergabe *f*
t. scale *(Ak)* Tonskala *f*
t. source Ton[frequenzspannungs]quelle *f*
t. wedge *(Fs)* Gradationskeil *m*, Graukeil *m*
t. wheel *(Ak)* Tonrad *n*

toner *(Plast)* klarer organischer Farbstoff *m* hoher Färbekraft, organischer Pigmentstoff *m*

tongs Zange *f*

tongue / to 1. federn *(Holz)*; 2. vernuten; lappen

tongue Dorn *m*, Lasche *f*; Zunge *f (z. B. Rechenstab)*; [prismatische] Leiste *f*; Ansatz *m*
t. and groove Feder *f* und Nut *f (Holzbearbeitung)*
t. and groove cutter Nut- und Spundfräser *m*
t.-shaped inclined bar Zungenschrägroststab *m*

tongued gefedert *(Holzbau)*
t. and grooved gespundet *(Holzbau)*
t. and grooved rough boarding Rauhspund *m*

tonguing Spunden *n*; Federn *n (Holz)*
t. cutter Spundfräser *m*

tonic Tonikum *n*, Stärkungsmittel *n*
t. train signalling *(Fmt)* Tonfrequenzzeichengebung *f*

toning bath Tonfixierbad *n (Fotografie)*

tonnage 1. *(Schiff)* Tonnage *f*, Raumgehalt *m (in Registertonnen)*; Vermessung *f (nach Registertonnen)*; 2. Leistung *f* in Tonnen; 3. *(Masch)* Druckleistung (Zugleistung) *f* in Tonnen
t. breadth Vermessungsbreite *f (eines Schiffes)*
t. certificate Schiffsmeßbrief *m*
t. deck Vermessungsdeck *n*
t. depth Vermessungstiefe *f (eines Schiffes)*
t. formula *(Schiff)* Meßformel *f*, Vermessungsformel *f*
t. law Schiffsvermessungsordnung *f*
t. measurement Vermessung *f*, Schiffsvermessung *f*

t. measures Vermessungsmaße *npl*, Registermaße *npl (eines Schiffes)*
t. scale Vermessungsfrachttarif *m*
tonne Tonne *f (1000 kg)*
tonne-force *abgeleitete Einheit der Kraft: 1 tf = 1 Mp*
tonneau *(Kfz)* Fahrgastraum *m*
tonometer *(Opt)* Tonometer *n*
tool / to 1. mit Werkzeug bearbeiten; steinmetzmäßig bearbeiten; 2. einrichten; [Werkzeuge] aufspannen
tool 1. Werkzeug *n*; Meißel *m*; Gerät *n*; Instrument *n*; 2. Werkzeugmaschine *f*
t. allowance Zeitvorgabe *f* für Werkzeugeinstellung und -wartung
t. and cutter grinding machine Scharfschleifmaschine *f*
t. and die milling machine Werkzeug- und Gesenkfräsmaschine *f*
t. and die shop Werkzeug- und Gesenkmacherei *f*
t. angle Winkel *m* an der Schneide
t. apron Meißelhalterklappe *f (Hobelmaschine)*
t. bag Werkzeugtasche *f*
t. bar Stoßstange *f (Senkrechtstoßmaschine)*
t. bit Bohrkrone *f*; *(Am)* Bohrmeißel *m*; Meißeleinsatz *m*; Schneidplatte *f*
t. bit for very large holes in thin metal Ausstechmeißel *m*
t. block Meißelklappe *f (Hobelmaschine)*
t. box 1. Werkzeugkasten *m*; 2. Support *m*, Werkzeugträger *m*; Quersupport *m*
t. cabinet Werkzeugschrank *m*
t. carrier Werkzeugträger *m*
t. changing Werkzeugwechsel *m*
t. chest Werkzeugkasten *m*; Werkzeugschrank *m*
t. chuck Werkzeugfutter *n*
t. clamping Werkzeugspannung *f*
t. cradle Werkzeugwiege *f (Wälzfräsen)*
t. crib *(Am)* Werkzeugausgabe *f*
t. cupboard Werkzeugwandschrank *m*
t. cutting edge Werkzeugschneide *f*, Meißelschneide *f*
t. edge *s.* t. cutting edge
t. ejector Werkzeugauswerfer *m*
t. engineering Werkzeugbau *m*
t. face Spanfläche *f*
t. feed movement Vorschubbewegung *f*
t. flank wear Freiflächenverschleiß *m*
t. frame Haltearm *m (Feil- und Sägemaschine)*
t. geometry Schneidengeometrie *f*
t. grind Werkzeugnachschliff *m*
t. grinder *s.* t. grinding machine
t. grinding Scharfschleifen *n*
t. grinding machine Werkzeugschleifmaschine *f*, Scharfschleifmaschine *f*
t. gripping plate Spannplatte *f*
t. handle Werkzeuggriff *m*, Werkzeugheft *n*
t. handling Werkzeugzubringung *f (Räumen)*
t. handling mechanism Räumwerkzeugzubringer *m*
t. head Support *m*, Werkzeugträger *m (Hobelmaschine)*; Stößelkopf *m (Senkrechtstoßmaschine)*
t. head slide Hobelkopfschlitten *m (Waagerechtstoßmaschine)*; Schieber *m (Hobelmaschine)*
t. holder Meißelhalter *m*, Werkzeughalter *m*
t. holder bit Einsteckmeißel *m*
t. holder flap Meißelklappe *f*
t. holder slide Schieber *m*, Meißelschlitten *m (Hobelmaschine)*
t. holding device Werkzeugspannvorrichtung *f*
t. holding fixture Werkzeugspannvorrichtung *f*, Spannkopf *m*
t.-issuing department Werkzeugausgabe *f*
t. kit Werkzeugtasche *f*, Werkzeugkasten *m*, Werkzeugkoffer *m*
t. lay-out Werkzeuganordnung *f*
t. life 1. Lebensdauer *f* des Werkzeugs; 2. [Werkzeug-]Standzeit *f*
t. life curve Standlängenkurve *f (Taylor)*
t.-lifter Meißelabheber *m*
t. lifting gear Meißelabhebeeinrichtung *f*
t. maintenance Werkzeuginstandhaltung *f*
t. manufacture Werkzeugbau *m*
t. mark Drehriefe *f*, Rattermarke *f*
t. point Schneidenkopf *m*
t. post Werkzeughalter *m*, Meißelhalter *m*
t. post slide Oberschlitten *m (Drehmaschine)*
t.-rake angle Spanwinkel *m*
t. regrind[ing] Werkzeugnachschliff *m*, Meißelnachschliff *m*
t. releasing Ausspannen *n* des Werkzeuges
t. relief *s.* t. relieving
t. relieving Werkzeugabhebung *f*, Meißelabhebung *f*
t. relieving unit Abhebeeinrichtung *f*
t. rest Oberschlitten *m (Drehmaschine)*
t. setter Einrichter *m*, Werkzeugeinrichter *m*
t. setting gauge Werkzeugeinstellehre *f*
t. setting time Einrichtezeit *f*, Einstellzeit *f*
t. shank Werkzeugschaft *m*
t. shop Werkzeugmacherei *f*
t. slide Werkzeugschlitten *m*, Meißelschlitten *m*, Werkzeugsupport *m*; Vertikalschieber *m*
t. slide of vertical turning and boring machine Werkzeugschieber *m*
t. steel Werkzeugstahl *m*
t.-storage unit Werkzeugspeicher *m*
t. storeroom Werkzeugmagazin *n (Betrieb)*
t. support Werkzeugauflage *f*
t. tip Werkzeugschneide *f*; Schneidplatte *f*
t. tip material Schneidenwerkstoff *m*
t. turret Revolverkopf *m*, Mehrfachmeißelhalter *m*

t. wear Werkzeugabnutzung *f*, Werkzeugverschleiß *m*
tooling 1. Bearbeitung *f* mit Werkzeugen; steinmetzmäßige Bearbeitung; 2. Aufspannen *n* von Werkzeugen; Einrichten *n*
t. change-over *(Masch)* Umrüsten *n*
t. time *(Masch)* Rüstzeit *f*
toolmaker Werkzeugmacher *m*
toolmaker's divider Spitzenzirkel *m*
t. knife-edge Haarlineal *n*
t. lathe Werkzeugmacherdrehmaschine *f*
t. microscope Werkstattmikroskop *n*
t. universal vice Universalschraubstock *m*
toolmaking department Werkzeugmacherei *f*
toolroom Werkzeugmacherei *f* und Vorrichtungsbau *m*
t. lathe Werkzeugmacherdrehmaschine *f*
tooth / to verzahnen; zacken
tooth Zahn *m*; Messer *n (Messerkopf)*
t. bearing Zahntragen *n*, Zahntragbild *n*
t. bearing area Zahntragfläche *f*
t. blade Messer *n (Messerkopf)*
t. bottom Zahngrund *m*, Zahnfuß *m*
t. breakage Zahnbruch *m*
t. chamfer Zahnabrundung *f*
t. chamfering machine Zahnabrundfräsmaschine *f*
t. clearance angle Zahnfreiwinkel *m*
t. clutch schaltbare Zahnkupplung *f*
t. contact Zahneingriff *m*
t. crest track Kopfbahn *f (Verzahnung)*
t. curve Zahnkurve *f*
t. depth Zahnhöhe *f*
t. engagement Zahneingriff *m*
t. face Spanfläche *f (Fräserzahn)*
t. form correction Zahnformkorrektur *f*
t. front Spanfläche *f (Fräserzahn)*
t. gullet Zahngrund *m*
t. height Zahnhöhe *f*
t. increment Überhöhung *f* je Zahn, Zahnabstufung *f (Räumwerkzeug)*
t. lock washer Zahnscheibe *f (Unterlegscheibe)*
t. number Zähnezahl *f (Verzahnung)*
t. of harrow Eggenzahn *m*
t. of the fluted type gefräster Zahn *m*
t. outline Zahnprofil *n (Verzahnung)*
t. pitch Zahnteilung *f*
t. profile Flankenprofil *n (Zahnrad)*
t. profiles Flanken *fpl*
t. pulsation frequency *(Elm)* Nutfrequenz *f*, Zahnfrequenz *f*
t. rack Zahnstange *f*
t. rest Zahnstütze *f*
t.-roller *(Led)* Karrierrolle *f*, Zahnholz *n*
t. saturation *(Elm)* Nutsättigung *f*, Zahnsättigung *f*
t. sector Zahnbogen *m*
t. shape correction Zahnformkorrektur *f*, Korrektur *f* des Zahnprofils
t. space Zahnlücke *f*
t. spacing Hiebteilung *f (Feilenzähne je Zoll)*, Zahnteilung *f*
t. system Verzahnung *f*, Zahnung *f*
t. thickness 1. Zahndicke *f*; 2. Rückenbreite *f (Räumwerkzeug)*
t. thickness along pitch circle Zahndicke *f* am Teilkreis (Erzeugungswälzkreis)
t. to tooth composite error Wälzsprung *m*
t. width Zahnbreite *f*
toothed gezahnt, verzahnt
t. chain Zahnkette *f*
t. clutch Zahnkupplung *f*
t. flywheel rim Schwungradzahnkranz *m*
t. gear Zahnrad *n*
t.-gear drive Zahnradtrieb *m*
t. lockwasher Zahnscheibe *f*
t. rack Zahnstange *f*
t. rim Zahnkranz *m*
t. roll crusher Stachelwalzenbrecher *m*, Stachelwalzwerk *n*
t. roller *(Text)* Reißwalze *f*, Brechwalze *f*, Stachelwalze *f (Ballenbrecher)*
t. segment Zahnsegment *n*
t. wheel Zahnrad *n*
t. wheel dynamometer Zahnraddynamometer *n*
t. wheel interrupter *(El)* Zahnradunterbrecher *m*
toothing Zahnung *f*; Verzahnung *f*
top / to toppen *(Rohöl)*; herausdestillieren *(von flüchtigen Bestandteilen)*; köpfen *(Rüben)*, beschneiden; auftoppen *(Schiffe)*; überdecken *(mit Farbe)*, überfärben
t.-pour *(Met)* fallend gießen
top 1. Oberteil *n*; Oberfläche *f*; Oberseite *f*; Spitze *f*; Kopf *m*; Deckel *m*; Kreisel *m*; Scheitel *m*; Krone *f*; First *m*; 2. *(Schiff)* Topp *m*, Mars *m*, Spitze *f (Mast)*; 3. Gicht *f*, oberer Teil *m (Hochofen)*; 4. *(Text)* Kammzug *m*; 5. *(Holz)* Zopf *m*
t. and bottom shedding machine Jacquardmaschine *f* für Tief- und Hochstellung
t. band Oberteil *n (Quartier-Schuh)*
t. beam Traverse *f (Mast)*
t. blade Obermesser *n (Schere)*
t. boots Schaftstiefel *mpl*
t. cage *(Text)* obere Siebtrommel *f (Trommelöffner)*
t. cap Anodenkappe *f*, Gitterkappe *f*, Kappenanschluß *m*, Kopfanschlußkappe *f (Elektronenröhre)*
t. cap connector Anodenkappenanschluß *m*
t.-capacitor aerial kapazitiv beschwerte Antenne *f*
t. capacity Endkapazität *f (Antenne)*

t. capacity coupling Kopplung *f* zweier Schwingkreise durch eine Serienkapazität
t. casting fallender Guß *m*; Gießen *n* von oben
t. centre *s.* t. dead centre
t. charging Füllen *n* von oben
t. circle Kopfkreis *m*
t. cloth weave *(Text)* Bindung *f* der Oberware
t. coal *(Brg)* hangende Kohle[nbank] *f*
t. coat Oberschicht *f*; Deckanstrich *m*; Überzugslack *m*
t. coat paste Oberschichtpaste *f*, Oberflächenbeschichtungsmasse *f*
t. coming cutter rechtsdrehender Fräser *m*
t. cover Stirnwand *f* des Zylinders, Zylinderstirnwand *f*
t. cut *(El)* Höhenbeschneidung *f (Beschneidung des Anteiles von hohen Frequenzen)*
t. dead centre oberer Totpunkt *m*
t. die Obergesenk *n*
t. digit *(Dat)* bedeutsame Ziffer *f* der höchsten Ordnung
t. drafting roller obere Streckwerkswalze *f*, Oberwalze *f*
t. dressing *(Lw)* Kopfdüngung *f*
t.-dyed *(Text)* im Kammzug gefärbt, überfärbt
t. edge 1. Zahnspitzenlinie *f (Sägeblatt)*; 2. Oberkante *f*
t. end Zopfende *n (eines Baumes)*
t. face 1. Oberseite *f*; 2. Spanfläche *f*
t.-fed aerial obengespeiste Antenne *f*
t. feeding *(Wärm)* Firstbeschickung *f*
t. fermentation *(Ch)* Obergärung *f*
t.-fermented *(Ch)* obergärig
t. fertilization *(Lw)* Kopfdüngung *f*
t. fittings Kronenarmatur *f (Hochofen)*
t. flame Gichtflamme *f (Schachtofen)*
t. flange Obergurt *m*
t. flask obere Formhälfte *f*, Oberkasten *m (Formerei)*
t. flue *(Wärm)* Oberzug *m*
t. fuller Ballhammeroberteil *n*
t. gas Gichtgas *n*
t. gate 1. *(Brg)* Kopfstrecke *f*; 2. Obertor *n (Schleuse)*
t. gear *(Kfz)* schnellster Gang *m*
t. gig *(Text)* Barchentrauhmaschine *f*
t. going cutter linksdrehender Fräser *m*
t.-grade coal hochwertige Kohle *f*
t.-hat section Hutprofil *n*
t.-heavy *(Schiff)* topplastig, oberlastig
t. hide *(Led)* oberste Haut *f (des Stapels)*
t.-hinged flap door *(Eis)* hochschwingende Klappe *f*
t. housing *(Masch)* Querhaupt *n*
t.-hung oben befestigt, aufgehängt
t. keel *(Schiff)* Gegenkiel *m*
t. land 1. *(Kfz)* oberster Kolbensteg *m*; 2. Fase *f* an Umfangsschneide *(Fräser)*
t. limit obere Grenze *f*
t.-loaded aerial kapazitiv beschwerte Antenne
t.-loading capacity Dachkapazität *f (Antennentechnik)*
t. logs Zopfware *f*
t. longeron *(Flg)* Rumpfoberholm *m*
t. measure Zopfmaß *n*
t.-mounted magnetic separator Überbandmagnetscheider *m*
t. of blade Kopfplatte *f (Turbine)*
t. of desk calculator Tischrechenmaschine *f*
t. of key Keilrücken *m*
t. of rail Schienenoberkante *f*
t. of table *(Masch)* Tischoberfläche *f*
t. of tank *(Schiff)* Wasserballasttankdeck *n*
t. of tooth Zahnkopf *m (Verzahnung)*
t. opening Gichtöffnung *f (Schachtofen)*
t. overhaul Teilüberholung *f*, Zwischenausbesserung *f*
t. panel Frontplatte *f*
t. part Oberteil *n*, oberes Formkastenteil *n (Formerei)*
t. position oberste Stellung *f*
t. pressure *(Brg)* Firstendruck *m*
t. quality Spitzenqualität *f*
t. radius Kopfabrundung *f*
t. rail *(Masch)* Oberhaupt *n*, Traverse *f*
t. rake Spitzenspanwinkel *m*; Spanfläche *f*
t. ring Topring *m*, Minutenring *m*
t. road *(Brg)* Kopfstrecke *f*
t. roll Oberwalze *f (Walzgerüst)*
t. roller Würgewalze *f (Walzgerüst)*
t. season *(Led)* letzter Glanz *m*
t. section Haubenkörper *m (Konverter)*; *(Brg)* Firstensegment *n (Ausbau)*
t. shed Oberfach *n (Weben)*
t. side Oberseite *f*
t.-side scantlings *(Schiff)* Deckstringer *m* mit Schergang
t.-slicing [and cover-caving] *(Brg)* Querbruchbau *m*, Scheibenbruchbau *m*
t.-slicing with partial ore caving *(Am)* Teilsohlenbruchbau *m*
t. slide *(Masch)* Oberschlitten *m*
t. soil Mutterboden *m*
t. speed Höchstgeschwindigkeit *f*, Höchstdrehzahl *f*
t. spindle moulder Oberfräse *f*, Versenkmaschine *f*
t. swage Obergesenk *n*, Gesenkoberteil *n*
t. temperature Gichttemperatur *f (Schachtofen)*
t. view Draufsicht *f*, Grundriß *m (technisches Zeichnen)*
t. voltage *(Am) (El)* Höchstspannung *f*

t. wood Draufholz *n*
t. yeast *(Ch)* obergärige Hefe *f*
topaz *(Min)* Topas *m*
t. rock *(Geol)* Topasfels *m*
topazolite *(Min)* Topazolith *m*
topgallant bulwark *(Schiff)* Oberschanzkleid *n*, hohes Schanzkleid *n*
t. forecastle *(Schiff)* hohe (feste) Back *f*
topless loom oberbauloser Webstuhl *m*
topmaking *(Text)* Vorstrecken *n* des Kammzugs
topman *(Brg)* beim Abteufen über Tage beschäftigter Arbeiter *m*
topmark buoy Bakentonne *f*
topochemistry Topochemie *f*
topographer Topograf *m*
topographical topografisch
t. map topografische Karte *f*
topography Topografie *f*
topology Topologie *f*, Geometrie *f* der Lage
topped crude Topprückstand *m* *(durch Toppen von Benzin, Petroleum und Gasöl befreites Rohöl)*
t. gasoline Toppbenzin *n*
t. out *(Bw)* rohbaufertig
topper Rübenköpfer *m*
topping 1. Toppen *n*, Abtrennen *n* leichter Fraktionen aus dem Rohöl, Destillieren *n* unter Atmosphärendruck; 2. *(Bw)* Decke *f*, Deckschicht *f*, Verschleißschicht *f*; Straßendecke *f*; 3. Überfärben *n*, Überdrucken *n*
t. of beets Köpfen *n* der Rüben
t.-out ceremony Richt[e]fest *n*
t. up Auffüllen *n*, Nachfüllen *n*
tops 1. *(Ch)* Kopfprodukte *npl*; 2. Kraut *n* *(Kartoffeln, Rüben)*; 3. Spitzenleistung *f*
topside *(Schiff)* Deck *n*
t. deck Gurtungsdeck *n*
t. tank seitlicher Deckballasttank *m*, Toppseitentank *m*, Flügeltank *m*
topsides *(Schiff)* Überwasserteile *mpl* eines Schiffes
topsoil *s.* top soil
torbernite *(Min)* Torbernit *m*
torch *(Am)* Brenner *m*; Flamme *f*; Fackel *f*
t. battery Stabbatterie *f* *(für Taschenlampen)*
t. cell Taschenlampenbatterie *f*, Element *n* für Taschenlampe
t. cut / to brennschneiden
t. deseaming *(Met)* Brennputzen *n*
t. handle Griffrohr *n* *(Schweißbrenner)*
t. head Brennereinsatz *m*, Brennerkopf *m*
t. lamp Taschenlampe *f*
t. light *s.* t. lamp
t. tube Brenner *m*
torching Mörtelverstrich *m* *(Dachdeckung)*
toric torisch
tormenting of soil Bodenlockerung *f*
tormentor Lockerpflug *m*
torn selvedge [in Kettrichtung] abgerissene Leiste *f* *(Webfehler)*
t. tape Bandriß *m* *(Tonbandgerät)*
t.-up cloth Lumpen *pl*
toroid *(Kern)* Ringkammer *f*, Ringröhre *f*, [ringförmige] Vakuumkammer *f*
toroidal ringförmig
t. coil *(El)* Ringspule *f*
t. core Ringkern *m*
t. magnetic field toroidales Magnetfeld *n*
t. repeater coil *s.* t. repeating coil
t. repeating coil Ringübertrager *m* *(zur Anpassung der Wellenwiderstände der zusammengeschalteten Leitungen; Fernsprechtechnik)*
t. transformer Ring[kern]transformator *m*, Ringübertrager *m*, Toroid[al]transformator *m*
t. winding Ringwicklung *f*
torpedo aircraft Torpedoflugzeug *n*
torpedoboat stern Torpedobootsheck *n*
torque Drehmoment *n*
t. amplifier Momentenverstärker *m*, Drehmomentenverstärker *m*
t. at rated load Nenn[dreh]moment *n*
t. ball Schubkugel *f*
t. control Drehmomentenkontrolle *f*
t. conversion Drehmomentenumwandlung *f*
t. converter Drehmomentwandler *m*
t. flux Drehmomentfluß *m*
t. gradient *(BMSR)* [Drehmelder-]Momentengradient *m*, Richtmomentzunahme *f*
t. limiting spanner Drehmomentenschlüssel *m*
t. motor *(BMSR)* Momentengeber *m*, Momentenmotor *m*, Motor *m* mit [lotrecht-] stehender Welle
t. spanner Drehmomentschlüssel *m*
t.-synchro receiver *(El)* Drehfeldempfänger *m*
t.-synchro transmitter *(El)* Drehfeldgeber *m*
t. tube ball joint Schubkugelgelenk *n*
t.-weight ratio bezogenes Drehmoment *n* *(eines Zählers)*
t. wrench Drehmomentenschlüssel *m*
torqueless yarn Garn *n* ohne Verdrehung
torr Torr *n* *(Einheit des Drucks; 1 Torr* $= 1{,}316 \cdot 10^{-3}$ *atm; in England: 1 Torr* $= 10^3\ \mu$*)*
torrefy / to darren *(Kupfer)*
torrent Bergstrom *m*, Gebirgsfluß *m*
Torricellian vacuum Torricellische Leere *f*
torsion Torsion *f*, Verdrehung *f*, Drillung *f*
t. angle Torsionswinkel *m*, Verdrehungswinkel *m*
t. balance Drehwaage *f*
t. bar Drehstab[feder *f*] *m*
t.-bar spring Drehstabfeder *f*
t.-bar springing Drehstabfederung *f*, Torsionsstabfederung *f*

t.-bar stabilizer Drehstabstabilisator *m*
t.-bar suspension Drehstabaufhängung *f*
t. dynamometer Torsionsdynamometer *n*
t. elasticity Torsionselastizität *f*, Verdrehungselastizität *f*
t. electrometer Torsionselektrometer *n*
t. failure Verdrehungsbruch *m*
t. fracture Verdrehungsbruch *m*
t. galvanometer Torsionsgalvanometer *n*
t. indicator Torsionsindikator *m*
t. modulus Torsionsmodul *m*
t. moment Torsionsmoment *n*
t. pendulum Drehpendel *n*, Torsionspendel *n*
t. rod Drehstab *m*, Torsionsstab *m*, Verdrehungsstab *m*
t. rod spring Drehstabfeder *f*
t. spring Drehstabfeder *f*
t. test Torsionsversuch *m*, Verdrehungsversuch *m*
t. testing machine Verdrehungsprüfmaschine *f*
t. vector Drallvektor *m*
torsional Torsions-, Drehungs-, Verdreh[ungs]-
t. damper Drehschwingungsdämpfer *m*
t. frequency Drehschwingungszahl *f*
t. hydromagnetic wave hydromagnetische Torsionswelle *f*
t. moment *(Am)* Drehmoment *n*, Verdrehmoment *n*
t. oscillation Verdrehungsschwingung *f*
t. resistance Drillwiderstand *m*, Torsionswiderstand *m*, Verdreh[ungs]festigkeit *f*
t. rigidity Torsionssteifigkeit *f*, Verdreh[ungs]steifigkeit *f*
t. stiffness Torsionssteifigkeit *f*, Verdreh[ungs]steifigkeit *f*
t. strain *(sl)* Torsionsbeanspruchung *f*, Verdrehungsbelastung *f*
t. strength Verdreh[ungs]festigkeit *f*, Torsionsfestigkeit *f*, Drillfestigkeit *f*
t. stress Torsionsspannung *f*
t. test Torsionsversuch *m*, Verdrehungsversuch *m*
t. vibration Drehschwingung *f*
t. vibration damper Drehschwingungsdämpfer *m*
t. wave Torsionswelle *f*
torsionally stiff drehsteif
t. stressed verdrehbeansprucht
tortoise [beleuchteter] Verkehrsteiler *m*, „Schildkröte" *f*
t. shell Schildpatt *n*
tortuous [mehrfach] gewunden, [mehrfach] gekrümmt, verdreht; in verschiedenen Ebenen
torus *(Math)* Torus *m*, Ringfläche *f*, Ringkörper *m*
tosylation *(Ch)* Tosylierung *f*, Einführung *f* der Toluolsulfonylgruppe

total Gesamtbetrag *m*, Gesamtzahl *f*, Gesamtsumme *f*
total total, gesamt, Gesamt-
t. absorption Gesamtabsorption *f*
t. accumulative dividing error Gesamtfehler *m (Zahnrad)*
t. activity Gesamtaktivität *f*
t. amplification Gesamtverstärkung *f*
t. amplitude doppelte Schwingungsweite (Amplitude) *f*
t. angular momentum Gesamtdrehimpuls *m*
t. body irradiation Ganzkörperbestrahlung *f*
t. boiler efficiency Gesamtwirkungsgrad *m* des Kessels
t. carbon content Gesamtkohlenstoffgehalt *m*
t. caving [method] *(Brg)* Totalbruchbau *m*
t. composite error Wälzfehler *m (Zahnrad)*
t. cross section gesamter (totaler) Wirkungsquerschnitt *m*
t. current Gesamtstrom *m*
t. cycle *(Büro)* Summengang *m*
t. depth Zahnhöhe *f (Frästiefe)*; Zahntiefe *f (Verzahnung)*
t. derivative *(Math)* totale Ableitung *f*
t. differential *(Math)* totales Differential *n*
t. distortion Gesamtverzerrung *f*; Summenklirrfaktor *m*
t. dose *(Kern)* Gesamtdosis *f*
t. draft *(Text)* Gesamtverzug *m*
t. effective collision cross-section totaler Wirkungsquerschnitt *m (Gasentladungsröhren)*
t. efficiency Gesamtwirkungsgrad *m*
t. error Gesamtfehler *m*
t. field *(El)* Gesamtfeld *n*
t. flight time Gesamtflugzeit *f*, Gesamtflugdauer *f*
t. growth of plants Pflanzenbestand *m*
t. hardness of water Gesamtwasserhärte *f*
t. harmonic content *(El)* Gesamtgehalt *m* an Oberwellen
t. head Förderhöhe *f*; Gesamthöhe *f*; gesamte Energie *f*
t. head pumped against Nutzförderhöhe *f (Pumpe)*
t. internal reflection *(Fmt)* Totalreflexion *f*
t. ionic concentration Gesamtionenkonzentration *f*
t. ionization Gesamtionisation *f*
t. labour force Gesamtbelegschaft *f*
t. laid-up tonnage Gesamttonnage *f*
t. length Gesamtlänge *f*
t. load volle Beladung (Belastung) *f*
t.-loss lubrication Durchlaufschmierung *f*, verlorene Schmierung *f*
t. magnification Gesamtvergrößerung *f*
t. moisture Gesamtfeuchtigkeit *f*
t. operating speed Löffelspiel *n*, Arbeitsspiel *n (Bagger)*

t. overshoot *(BMSR)* maximale Überschwingweite *f*
t. pressure Gesamtdruck *m*
t. printing Drucken *n* der Summe, Summenschreibung *f*
t. production time for one part Stückzeit *f*
t. quantum number Gesamtquantenzahl *f*, Hauptquantenzahl *f*
t. radiation Gesamtstrahlung *f*
t. radiation pyrometer Gesamtstrahlungspyrometer *n*
t. reduction gesamte Dickenabnahme *f (Walzen)*
t. reflectance Reflexionsgrad *m*
t.-reflecting prism totalreflektierendes Prisma *n*
t. reflection Totalreflexion *f*
t. resistance Gesamtwiderstand *m*; Gesamtfahrwiderstand *m (einschließlich Kurven- und Steigungswiderstand)*
t. scattering amplitude Gesamtstreuamplitude *f*
t. sensitivity Gesamtempfindlichkeit *f*
t. telegraph distortion *(Fmt)* Gesamtverzerrung *f*
t. time constant Gesamtzeitkonstante *f*
t. transfer Summenübertragung *f*
t. voltage Gesamtspannung *f*
t. water hardness Gesamtwasserhärte *f*
t. weight Gesamtmasse
totalizator Summenzählwerk *n*
totalizing counter summierender Zähler *m*
totally enclosed *(El)* geschlossen, gekapselt *(Gerät)*
t.-enclosed motor ungelüfteter (geschlossener) Motor *m*
tote box *(Am)* Transportbehälter *m*
tottering contact *(El)* Wackelkontakt *m*
touch / to berühren
t. a port einen Hafen anlaufen *(zum Löschen oder Laden)*
t. down *(Flg)* aufsetzen *(auf der Rollbahn)*
t. ground *(Schiff)* Grund berühren
t. up ausbessern
touch 1. Berührung *f*; Strich *m*; Meßgefühl *n*, Tastgefühl *n*; 2. Talg *m (für Klempnerarbeiten)*
t.-control *(Am) (El)* Druckknopfsteuerung *f*, Taststeuerung *f*, Tippsteuerung *f*
t. spark Abreißfunken *m*
t.-tuning Druckknopfabstimmung *f (Funkempfangstechnik)*
touching of the blades Anstreifen *n* der Schaufeln *(Turbine)*
t. the ground *(Schiff)* Grundberührung *f*
touchstone *(Min)* Lydit *m*, Basanit *m*; *(Ch)* Probierstein *m*, Prüfstein *m*
tough zäh
t.-pitch hammergar *(Kupfer)*
t. texture zähes Gefüge *n*
toughen / to zäh machen; zäh werden
t. by poling zähpolen *(Kupfer)*
t. copper Kupfer polen
toughened glass Hartglas *n*
toughness Zähigkeit *f*, Zähfestigkeit *f*
touring aircraft Reiseflugzeug *n*
tourist class Touristenklasse *f (Fahrgastschiff)*
tourmalin[e] *(Min)* Turmalin *m*
tow / to 1. *(Schiff)* schleppen, bugsieren, treideln; 2. entgraten, glätten
t. along treideln
t. astern achteraus schleppen
t. in einschleppen
t. into sluices einschleusen *(Schiff)*
t. off abschleppen
t. out ausschleppen, ausschleusen
t. through sluices durchschleusen *(Schiff)*
tow 1. *(Schiff)* Schleppzug *m*, Stoßzug *m*, Schubverband *m*; Schleppanhang *m*, geschleppte Schiffe *npl*, Schlepper *m*; 2. Schlepptau *n*; 3. Hede *f*, Werg *n*; 4. Spinnkabel *n*
t.-bar coupling *(Kfz)* Abschleppkupplung *f*
t. boat Schubboot *n*, Stoßboot *n*, Schubschlepper *m*; Schlepper *m*, Schleppboot *n*
t. card *(Text)* Wergkratzmaschine *f*, Wergkrempel *f*
t. combing *(Text)* Kämmen *n* des Wergs
t. hook Schlepphaken *m*; Zughaken *m*
t.-hook traveller Schleppbügel *m (für Schlepphaken)*
t. rail Schleppbogen *m*, Schleppgalgen *m (am Schlepper)*
t. rope Schlepptau *n*, Schlepptrosse *f*; *(Kfz)* Abschleppseil *n*
t.-rope pull Trossenzug *m*, Schlepptrossenzug *m*
t.-rope resistance Schleppwiderstand *m*, Formwiderstand *m*
t. roving frame Wergvorspinnmaschine *f*
t. spinning frame Wergspinnmaschine *f*
t. start Anschleppen *n (Kraftwagen)*; Schleppstart *m*
t. target *(Mil)* Schleppscheibe *f*
t. yarn Werggarn *n*
towage Schleppen *n*, Bugsieren *n*, Schleppschifffahrt *f*; Schleppgebühr *f*, Bugsierlohn *m*
t. service Schlepperhilfe *f*
towards the weather side *(Schiff)* leewärts
towed barge Schleppkahn *m*, Leichter *m*
t. load Anhängelast *f*
t. target *(Mil)* Schleppscheibe *f*
t. vibratory roller *(Bw)* Anhängevibrationswalze *f*, Vibrationsanhängewalze *f*
towel rack Handtuchhalter *m*
towelling Handtuchpapier *n*, Handtuchkrepppapier *n*

tower 1. Turm *m*; 2. Gittermast *m*; 3. *(Schiff)* Bugsierer *m*
t. acid Glover-Säure *f*, Turmsäure *f*
t. aerial Mastantenne *f*
t. base Mastfuß *m*; Mastfundament *n*
t. clock Turmuhr *f*
t. crane Turmdrehkran *m*
t. drier Schachttrockner *m*
t. earthing resistance *(Fmt)* Masterdungswiderstand *m*
t. lighting Mastbeleuchtung *f*
t. line Gittermastleitung *f*
t.-mounted winder *(Brg)* Turmfördermaschine *f*
t. packed column *(Ch)* Füllkörperkolonne *f*
t. packing *(Ch)* Füllkörper *m*
t. pond Turmsumpf *m (Kühlturm)*
t. silo Turmsilo *m*
t.-type crane Turmdrehkran *m*
towing Schleppen *n*, Bugsieren *n*; Schleppschifffahrt *f*
t. bitt *(Schiff)* Schleppoller *m*, Schleppbeting *m*
t. bollard *(Schiff)* Schleppoller *m*, Schleppbeting *m*
t. carriage Schleppwagen *m*, Schleppeinrichtung *f (Modellversuchswesen)*
t. efficiency Schleppwirkungsgrad *m*
t. engine Schleppwinde *f*, Trossenwinde *f*
t. experiment Schleppversuch *m (Großversuch)*
t. gear *(Schiff)* Schleppgeschirr *n*
t. hitch Abschleppzugvorrichtung *f*
t. hook Schlepphaken *m*
t. launch Schleppbarkasse *f*
t. log Schlepplog *n*
t. power Schleppleistung *f*
t. pull test Schlepptrossenzugmessung *f*
t. rod *(Kfz)* Abschleppstange *f*
t. shackle *(Schiff)* Schlepptauschäkel *m*
t. tank Schlepptank *m (Modellversuchstank)*
t. tractor *(El)* Treidellokomotive *f*
t. vehicle *(Kfz)* Zugmaschine *f*
t. vessel Schleppschiff *n*, Schlepper *m*
t. winch Schleppwinde *f*, Trossenwinde *f*
town Stadt *f*
t. gas Stadtgas *n*, Leuchtgas *n*
t. mains *(El)* Stadtleitung *f*
t. planning Städtebau *m*; Stadtplanung *f*
townscape Stadtbild *n*, Städtebild *n*, Stadtlandschaft *f*
toxic toxisch, giftig, gifthaltig
t. agent 1. Toxikum *n*, Giftstoff *m*; 2. *(Mil)* Kampfstoff *m*
t. war gas Kampfstoff *m*
toxicant Toxin *n*, Gift[stoff *m*] *n*
toxicity Toxizität *f*, Giftigkeit *f*
toxicological toxikologisch
toxin Toxin *n*, Gift[stoff *m*] *n*
toxoid giftähnlich
toy industry Spielwarenindustrie *f*
t. lacquer Spielzeuglack *m*
trace / to 1. [nach]zeichnen, nachziehen, ausziehen *(Zeichnungen)*; [durch]pausen; 2. abfahren *(kopieren)*; 3. suchen; verfolgen, abtasten *(Signal)*; 4. aufzeichnen, schreiben
t. a fault einen Fehler suchen
t. out umreißen
trace 1. *(Math)* Spur[punkt *m*] *f*; Durchstoßpunkt *m*; 2. Leuchtfleck *m*, Leuchtspur *f* *(Katodenstrahlröhren)*; 3. *(El)* Strahlhinlauf *m*
t. element Spurenelement *n*
t. hook Stranghaken *m (Fuhrwerk)*
t. interval *(Fs)* Ablenkweite *f*, Zeilenabtastdauer *f*
t. leather Strangleder *n*, Strangriemen *m*
t. of the rays Strahlenbahn *f*, Strahlengang *m*, Strahlenverlauf *m*
t. of trajectory Bahnspur *f*
t. rotation system Zeitachsendrehung *f (Radar)*
t. strap Strangleder *n*, Strangriemen *m*
tracer 1. Fühler *m*, Fühlfinger *m*, Fühlstift *m*, Taster *m*, Taststift *m*, Tastnase *f*; 2. *(Kern)* Indikator *m*, Tracer *m*; 3. Leuchtspurgeschoß *n*
t. ammunition Leuchtspurmunition *f*
t. atom *(Kern)* Indikatoratom *n*
t. compound *(Kern)* Indikatorverbindung *f*
t. control *(Masch)* Fühlersteuerung *f*
t.-controlled *(Masch)* fühlergesteuert
t. detector Spurenfinder *m*
t. element Indikatorelement *n*, Markierungselement *n*
t. finger Taster *m*, Fühlfinger *m*, Taststift *m*
t. isotope Leitisotop *n*, Indikatorisotop *n*, isotoper Indikator *m*, Markierungsisotop *n*
t. method 1. *(Kern)* Indikatormethode *f*, Leitisotopenmethode *f*; 2. *(Masch)* Abtastverfahren *n*
t. pin *(Masch)* Taststift *m*
t. stylus Fühler *m*, Fühlfinger *m*, Taststift *m* *(Kopierfräsmaschine)*
tracery Maßwerk *n*
tracheid Tracheide *f (Holz)*
tracing 1. Nachziehen *n*; 2. Pause *f*, durchgepaustes Bild *n*; Pausen *n*; 3. Abtastung *f*
t. attachment *(Masch)* Kopiereinrichtung *f*, Fühlernachformeinrichtung *f*
t. cloth Pausleinen *n*
t. cord Absteckleine *f*
t. paper Pauspapier *n*; Transparentpapier *n*
t. point *(Masch)* Fahrstift *m*; Taststift *m*
t. routine *(Dat)* Programm *n* zum Überwachen, Programmprüfprogramm *n*, Folgeprüfprogramm *n*

t. stereometer Zeichenstereometer *n*
t. work fühlergesteuerte Nachformarbeit *f*
track / to 1. *(Schiff)* schleppen, bugsieren, treideln; 2. spuren; verfolgen
track 1. Bahn *f*, Fahrbahn *f*; Laufbahn *f*; *(Masch)* Gleitbahn *f*; Spur *f*; 2. Laufring *m* *(Lager)*; 3. Schiffahrtsweg *m*, Route *f*; 4. Schienenstrang *m*, Schienen *fpl*, Gleis *n*; 5. Spurweite *f*; Radabstand *m*; 6. Kursprojektion *f (Radar)*; 7. *(Kfz)* Gleiskette *f*, Laufkette *f*, Raupenkette *f*; 8. Spur *(z. B. auf Magnetband, Magnettrommel)*; 9. *(Brg)* Gestänge *n*, Gleis; Ausbaufeld *n*, Feld *n (im Streb)*
t. accessories Gleiszubehör *n*
t. alignment *(Kfz)* Radausfluchtung *f*
t. alignment gauge Radspurmesser *m*
t. arm *(Kfz)* Spur[stangen]hebel *m*
t. beacon Leitstrahl *m (Radar)*
t. bearing rechtweisende Richtung *f (Flugnavigation)*
t. bed Gleisbett *n*
t. bolt Schloßschraube *f*
t. brake Schienenbremse *f*
t. circuit Gleisstromkreis *m (elektrische Eisenbahn)*
t. condensation *(Kern)* Spurenbildung *f (Kondensation längs der Teilchenbahn in Nebelkammern)*
t. construction car Gleisbauwagen *m*
t. control arm *(Kfz)* Spurstangenhebel *m*
t. gauge Schienenspurweite *f*
t. guard Kettenschutz *m*
t. idler Kettenleitrad *n*
t. jack Gleishebewinde *f*
t. keeping *(Kfz)* Spurhaltigkeit *f*, Spuren *n*
t.-laying tractor Raupenschlepper *m*
t. length *(Kern)* Spurenlänge *f*
t. link Kettenglied *n*
t. lock Gleissperre *f*
t. lock signal Gleissperrsignal *n*
t. loosening device Spurlockerer *m (an landwirtschaftlichen Geräten)*
t. lug Kettenauge *n*
t. magnet Gleismagnet *m*
t. rail bond Schienenverbinder *m*
t. relay Gleisrelais *n*
t. resistance Kriechpfadwiderstand *m*
t. return system Schienenrückleitung *f*
t. ring Laufring *m*
t. rod *(Kfz)* Spurstange *f*
t. rod arm Spurstangenhebel *m*
t. rod end Spurstangenkopf *m*
t. scales Gleiswaage *f*
t. section *(Eis)* Teilstrecke *f*
t.-sectioning cabin Schaltstation *f*
t.-sensitive volume *(Kern)* empfindliches Volumen *n (Nebelkammer)*
t. shifting machine Gleisrückmaschine *f*
t. shoes Rennschuhe *mpl*
t. spike Schienennagel *m*
t. sprocket *(Kfz)* Raupenrad *n*
t. stop Gleissperre *f*
t. switch Streckenschalter *m*, Schienenschalter *m*
t. tamper Gleisstopfer *m*
t. tamping machine Gleisstopfmaschine *f*
t. tensioner Kettenspanner *m*
t.-type vehicle Kettenfahrzeug *n*, Raupenfahrzeug *n*
t. warping Gleisverwerfung *f*
t.-width *(Brg)* Feldesbreite *f*
t. wrench Laschenschraubenschlüssel *m*
trackage Gleisstrecke *f*
tracked vehicle Raupenfahrzeug *n*
tracker wire Schaltdraht *m*, Rangierdraht *m*
tracking 1. *(Schiff)* Schleppen *n*, Bugsieren *n*, Treideln *n*; 2. Parallellauf *m* gemeinsam abgestimmter Schwingungskreise; 3. Spuren *n*; 4. Folgen *n (Radar)*; 5. *(El)* Kriechspurbildung *f*
t. beam *(Fs)* Spurstrahl *m*
t. capacitor Gleichlauftrimmer *m*, Padding-Kondensator *m*
t. circuit Nachlaufschaltung *f (Radar)*
t. disk Abgleichscheibe *f (Drehkondensator)*
t. error *(El)* Gleichlaufmangel *m*; Kursfehler *m (Radar)*
t. radar Zielverfolgungsradar *n*
t. shot *(Fs)* Folgebild *n*
t. system Zielverfolgungsverfahren *n*
t. the satellite Suchen *n* des Satelliten
trackless vehicle schienenloses Fahrzeug *n*
trackmaster Trackmaster *m (zur Elektronenstrahlsteuerung in Radarsichtgeräten; Decca-Radarverfahren)*
tractility Dehnbarkeit *f*, Streckbarkeit *f*
traction Zug *m*
t. battery Traktionsbatterie *f*, Fahrzeug[antriebs]batterie *f*
t. current Fahrstrom *m*
t. gas Treibgas *n*
t. generator Fahrstromerzeuger *m*
t. motor *(Elm)* Fahrmotor *m*, Bahnmotor *m*
t. resistance Zugwiderstand *m*
t. rope Zugseil *n*
t. sheave Treibscheibe *f*
t. sheave elevator Treibscheibenaufzug *m*
t. switch Fahrschalter *m*
t. transformer Fahrzeugtransformator *m*, Traktionstransformator *m*
t.-type hoist Treibscheibenaufzug *m*
tractive effort Zugkraft *f* am Radumfang *(elektrischer Zug)*
t. effort at continuous rating Dauerzugkraft *f*, Zugkraft *f* bei der Dauerleistung

t. effort at hourly rating Stundenzugkraft *f*, Zugkraft *f* bei der Stundenleistung
t. force Zugkraft *f*; Zugleistung *f (Traktor usw.)*
t. power Zugleistung *f*; Haltekraft *f*, Zugkraft *f*, Tragkraft *f (Magnet)*
t. resistance Fahrwiderstand *m*
tractor 1. Traktor *m*, Schlepper *m*; 2. Luftschraube *f*; Zugschraube *f*
t.-brake valve Zugwagenbremsventil *n*
t. crane Drehkran *m* mit Raupenfahrwerk
t. cultivator Schlepperpflug *m*
t. driver Traktorist *m*, Schlepperfahrer *m*
t. haulage Traktorzug *m*
t. hitch Traktorenanhängevorrichtung *f*
t.-locomotive Kleinlokomotive *f*
t.-mounted implement *(Lw)* Anbaugerät *n*
t. mowing machine Schleppermähwerk *n*
t. plough Schlepperpflug *m*
t.-scraper *(Brg)* Bodenhobel *m* mit Eigenantrieb
t.-trailer brake valve Lastzugbremsventil *n*
t. vaporizing oil Traktorenkraftstoff *m*
tractrix curve Traktrix *f*, Schleppkurve *f*
trade Handel *m*; Gewerbe *n*; Beruf *m*
t. designation Handelsbezeichnung *f*
t. fleet Handelsflotte *f*
t. journal Fachzeitschrift *f*
t. mark Warenzeichen *n*
t.-mark protection Musterschutz *m*
t.-marked camera Markenkamera *f*
t. partner Handelspartner *m*
t. union Gewerkschaft *f*
trader Handelsschiff *n (in Linienfahrt)*, Tourendampfer *m*
trading area Verkehrsgebiet *n*
t. port Handelshafen *m*, Seehafen *m*
t. vessel Handelsschiff *n*
traffic Verkehr *m*; Handel *m*; Gewerbe *n*
t. accident Verkehrsunfall *m*
t. analysis Verkehrsanalyse *f*
t. artery Hauptverkehrsstraße *f*
t. beam *(Kfz)* Abblendlicht *n*
t. capacity *(Fmt)* Verkehrsleistung *f (eines Leitungsbündels)*
t. concentration Verkehrsdichte *f*
t. congestion Verkehrsstauung *f*
t. control Verkehrsregelung *f*
t. control clearance Einfluggenehmigung *f (Flugsicherung)*
t. control sign Straßenverkehrszeichen *n*
t. control signal Verkehrsampel *f*
t. counting Verkehrszählung *f*
t. disturbance *(El, Fmt)* Betriebsstörung *f*
t. fluctuation Verkehrsschwankung *f*
t. handling Verkehrsabwicklung *f*
t. island Verkehrsinsel *f*
t. lane Fahrspur *f*, Spur *f (der Straße)*
t. light Verkehrsampel *f*
t. line Verkehrsstrich *m (Straßenmarkierung)*
t. list Sammelanrufliste *f (Seefunk)*
t. load bewegliche Belastung *f*, wandernde Last *f*; Verkehrslast *f*
t. meter Verkehrszähler *m*
t. offender Verkehrssünder *m*
t. paint Straßenmarkierungsfarbe *f*
t. planning Verkehrsplanung *f*
t. recorder *(Fmt)* Gesamtbelegungszähler *m*
t. stream Verkehrsstrom *m*
t. structures Verkehrsbauten *pl*
t. unit *(Fmt)* Belegungsstunde *f*, Verkehrseinheit *f*
t. wave *(Fmt)* Verkehrswelle *f*
trafficator *(Kfz)* Fahrtrichtungsanzeiger *m*, Winker *m*
tragacanth gum Tragant *m*
trail / to schleppen
trail *(Kfz)* Nachlauf *m (Vorderräder)*
t. bridge Fährbrücke *f*
t. edge Nebenschneide *f*
t. eye Protzöse *f (Geschütz)*
trailer 1. Anhänger *m*, Anhängewagen *m*; 2. Voranzeige *f (Film)*; 3. *(Fmt)* Bürstenarm *m*
t.-brake pressure regulator *(Kfz)* Anhängerbremskraftregler *m*
t.-brake valve *(Kfz)* Anhängerbremsventil *n*
t. brush Anhängebürstenwalze *f*, Schleppwalze *f*, Anhängebesenwalze *f*
t. coach Wohnwagenanhänger *m*
t. coupling Anhängekupplung *f*
t. door Bodenklappe *f (für Bodenentleerung)*
t. hopper Anhängebunker *m*, Anhängesilo *m*
t. plow Anhängepflug *m*
t. ship Lastfahrzeuganhängertransportschiff *n*, Trailerschiff *n (die LKW-Anhänger werden mit Traktorenhilfe über eine Auffahrtsrampe an und von Bord gebracht)*
t. spreader Düngerstreuanhänger *m*
trailing nacheilend
t. aerial freihängende Antenne *f*, Schleppantenne *f*
t. arm Trapezlenker *m (Radaufhängung)*
t. attachment *(Lw)* Anhängegerät *n*
t. axle *(Kfz)* Hinterachse *f*
t. beam Anhängebalken *m*, Anhängeschiene *f*
t. blade *(Typ)* Schlepprakel *f*
t. blade coater Schleppklingenstreichmaschine *f (zur Papierherstellung)*
t. brush holder schrägstehender Bürstenhalter *m*
t. cable Schleppkabel *n*
t. cavity Kavitation *f*, die sich vom Körper in den Nachlauf erstreckt
t. contact Folgekontakt *m*, Schleppkontakt *m*
t. device Anhängevorrichtung *f*

t. edge 1. *(Fs)* Rückflanke *f*; 2. *(Flg)* Tragflächenhinterkante *f*; 3. *(Schiff)* Propellerflügelhinterkante *f*
t. edge rise time *(Fs)* Anstiegzeit *f* der Rückflanke
t. edge synchronization *(Fs)* Rückfrontsynchronisierung *f*
t. link *(Kfz)* Längslenker *m*
t. load Schlepplast *f*, Anhängelast *f*
t. pole horn *(Elm)* ablaufende Polkante *f*
t. rail Anhängeschiene *f*
t. shoe Ablaufbacke *f (Bremse)*
t. spring *(Kfz)* Hinterachsfeder *f*
t. suction dredger Schlepprohrsaugbagger *m*, Schleppkopfsaugbagger *m*
t. vortex Nachlaufwirbel *m (Kavitation)*
t. wire aerial Hängedrahtantenne *f*, Schleppdrahtantenne *f*
train / to 1. ausbilden; 2. zielen, richten
train 1. Reihe *f*; Zug *m*; 2. Straße *f (Walzen)*; 3. Getriebe *n (Räder usw.)*; Räderwerk *n (Uhr)*
t. announcement Vormeldung *f (Eisenbahnsignalwesen)*
t. control Zugsteuerung *f*, Fahrsteuerung *f*
t. covering signal Zugdeckungssignal *n*
t. crew Zugbegleitpersonal *n*
t. crew compartment Dienstabteil *n*
t.-ferry Eisenbahnfähre *f*
t. guard Zugschaffner *m*
t. of barges Schleppzug *m*, Schleppanhang *m*
t. of friction gears Reibrädergetriebe *n*
t. of gears Rädergetriebe *n*, Stufenrädergetriebe *n*
t. of harmonic waves *(Ph)* harmonischer Wellenzug *m*
t. of mechanism Getriebe *n*
t. of pulleys Riemen[scheiben]getriebe *n*
t. of pulses Impulskette *f*, Impulsreihe *f*
t. of rope sheaves Seilscheibengetriebe *n*
t. of sprockets Kettengetriebe *n*
t. of toothed gears Zahnrädergetriebe *n*
t. of waves Wellenzug *m*
t. of wheels Räderwerk *n*
t. oil Tran *m*
t. personnel Zugbegleitpersonal *n*
t. reverser *(Eis)* Richtungswender *m*
t. staff Zugbegleitpersonal *n*
t. starting signal *(Eis)* Abmeldesignal *n*
t. with mail service Zug *m* mit Postdienst
trainee Anlernling *m*
trainer Ausbilder *m*; Schulschiff *n*
training 1. Ausbilden *n*; 2. Richten *n*, Zielen *n*
t. allowance Lehrlingslohn *m*
t. director *(Am)* Ausbildungsleiter *m*
t. in firing Schießausbildung *f*
t. methods Ausbildungsmethoden *fpl*
t. scheme Ausbildungsplan *m (für Lehrlinge)*
t. ship Schulschiff *n*
t. staff Ausbildungspersonal *n*
t. workshop Lehrwerkstatt *f*, Ausbildungswerkstätte *f*
trainloader *(Brg)* Bunkerzug *m (Streckenvortrieb)*
trajectories of principal extensions *(Ph)* Hauptdehnungslinien *fpl*, Hauptdehnungstrajektorien *fpl*
t. of principal stresses *(Ph)* Hauptspannungslinien *fpl*, Hauptspannungstrajektorien *fpl*
trajectory 1. *(Math)* Trajektorie *f (Kurve, die alle Kurven oder Flächen einer Schar unter dem gleichen Winkel schneidet)*; 2. Flugbahn *f*; Bahn *f (Elektron)*; ballistische Kurve *f*
t. chart Flugbahnbild *n*
t. error Bahnfehler *m*
t. of climb Aufstiegsbahn *f*
t. of electrons Elektronenbahn *f*
t. oscillation Bahnschwingung *f*
t. peak maximale Flugbahnhöhe *f*
t. plane Bahnebene *f*
tram 1. Hunt *m*, Förderwagen *m*; Straßenbahnwagen *m*; 2. Trameseide *f*, Schußseide *f*
t. depot Straßenbahndepot *n*, Straßenbahnhof *m*
t. mains Fahr[draht]leitung *f*, Straßenbahnoberleitung *f*
t. tower Fahrleitungsmast *m*
tramcar Straßenbahntriebwagen *m*
trammel *(Math)* Ellipsenzirkel *m*, Ellipsenzeichner *m*
t. method of drawing an ellipse Zeichnen *n* der Ellipse mit Ellipsenzeichner
trammels Ellipsenzirkel *m*
tramming *(Brg)* Förderung *f (insbesondere Schlepperförderung)*
t. level Fördersohle *f*
tramp Trampschiff *n*, Tramper *m*
t. iron Eisenteile *npl (im Fördergut)*; Fremdeisen *n*
t. material Fremdkörper *m*
t. ship Trampschiff *n*
tramper Trampschiff *n*, Tramper *m*
tramping Trampschiffahrt *f*
tramway Straßenbahn *f*
t. depot Straßenbahnhof *m*
t. pole Fahrleitungsmast *m*
t. rail Straßenbahnschiene *f*
tranquillizer *(Ch)* Beruhigungsmittel *n*
trans-isomer *(Ch)* Transisomeres *n*
transaction [wissenschaftliche] Abhandlung *f*
t. counter Postenzähler *m (z. B. einer Kontrollkasse)*
t. posting unit Einheit *f* zum Buchen der Kontenbewegungen
transadmittance Gegenscheinleitwert *m*, Transadmittanz *f (Elektronenröhren)*
transatlantic cable Überseekabel *n*
t. steamer Transatlantikdampfer *m*

transauroral line *(Ph)* Transauroralinie *f*
transceiver Sende-Empfangsgerät *n*
t. filter Simultanweiche *f (Sende- und Empfangsfunktion mit nur einer Antenne)*
transcendental *(Math)* transzendent
t. equation transzendente Gleichung *f*
t. function transzendente Funktion *f*
transconductance Gegenwirkleitwert *m*; Mischsteilheit *f (Elektronenvolumen)*
t. required to start oscillation Anschwingsteilheit *f*
transcribe / to *(Dat)* umschreiben, abgreifen
transcriber *(Dat)* Übersetzer *m (Rechner)*
transcrystalline intrakristallin
transcrystallization Transkristallisation *f*
transducer Wandler *m (z. B. Magnetkopf)*, Energiewandler *m*, Leistungswandler *m*, Umformer *m*, Meßgrößenwandler *m*; *(BMSR)* Geber *m*, Fühler *m*, Meßfühler *m*, Meßglied *n*, Meßwertgeber *m*; Übertrager *m*, Übertragungssystem *n*
t. bridge Geberbrücke *f*
t. displacement Meßwandlerverschiebung *f*
t. gain Übertragungsgewinn *m*
t. loss Übertragungsverlust *m*
transductor Transduktor *m*, magnetischer Verstärker *m*, Magnetverstärker *m*
t. amplifier magnetischer Verstärker *m*, Transduktorverstärker *m*
t. reactor Blindlasttransduktor *m*
transfer / to 1. übertragen; übersetzen *(Fähre)*, hinüberschaffen; *(Masch)* transportieren; 2. *(Typ)* umdrucken; abziehen *(Abziehbild)*; 3. umschlagen, kippen *(Relais)*
t. a patent to somebody ein Patent an jemanden abtreten
t. back *(El)* rückübertragen, rückwirken
t. money by telegraph Geld telegrafisch überweisen
transfer 1. Übertragung *f*; Hinüberbringen *n*; 2. Umdruck *m*; Abziehen *n*; 3. *(Masch)* Transport *m*; 4. Umschlagen *n*, Kippen *n (Relais)*; 5. Querschlepper *m (Walzwerk)*
t. admittance *(El)* Übertragungsleitwert *m*
t. area Berührungsfläche *f (Destillation)*
t. bar *(Masch)* Transportstange *f (Taktstraße)*
t. box *(El)* Hausanschlußkasten *m*
t. butt *(Text)* Übergabefuß *m*
t. cam *(Text)* Übertragheber *m*
t. case *(Kfz)* Verteilergetriebe *n*
t.-case differential Ausgleich *m* im Verteilergetriebe
t. chamber Druckkammer *f*; Füllraum *m*; Heizkammer *f*
t. chamber retainer plate *(Plast)* Füllplatte *f (Preßspritzverfahren)*
t. characteristic *(El)* Übertragungscharakteristik *f*, Übertragungskennlinie *f*, Übergangskennlinie *f (Vierpol)*; *(Am)* Gitterspannung-Anodenstrom-Kennlinie *f*
t. check Transportkontrolle *f*
t. circuit Anschaltleitung *f*, Verbindungsleitung *f*
t. constant Übertragungskonstante *f*
t. contact *(El)* Umschaltkontakt *m*
t. container Transportbehälter *m*
t. curve statische Kennlinie *f*, Übergangskennlinie *f (Vierpol)*
t. cylinder *(Plast)* Preßspritzzylinder *m*
t. element *(Dat)* Transportglied *n*
t. equation Übertragungsgleichung *f*
t. exchange *(Fmt)* Überleitungsamt *n*, Überweisungsamt *n*
t. factor *(Dat)* Übertragungsfaktor *m*
t. ferry Übersetzfähre *f*
t. function Übertragungsfunktion *f*
t. function analyzer Frequenzgangmeßplatz *m*, Übertragungsfunktionsanalysator *m*
t. impedance *(El)* Kopplungswiderstand *m*, Übertragungsscheinwiderstand *m*
t. ink *(Typ)* Umdruckfarbe *f*
t. instruction *(Dat)* Transportbefehl *m*
t. jack *(Fmt)* Umschaltklinke *f*, Überführungsklinke *f*, Verlegeklinke *f*
t. key *(Fmt)* Zuteilschalter *m*
t. ladle *(Met)* Transportpfanne *f*
t. lag Verzögerung *f*, Zeitverzögerung *f*
t. line 1. *(Fmt)* Überweisungsleitung *f*; 2. *(Masch)* Fertigungsstraße *f*, Transferstraße *f*
t. line control Transferstraßensteuerung *f*
t. line for drilling Bohrstraße *f*
t. link *(Masch)* Verkettungseinrichtung *f*
t. locus *(BMSR)* Ortskurve *f* [des Frequenzganges]
t. loss *(Fmt)* Übergangsdämpfung *f (Kopplung)*
t. machine Transfermaschine *f*
t. mould Preßspritzform *f*
t. moulding [process] Preßspritzverfahren *n*, Spritzpreßverfahren *n*
t. of balance Saldoübertrag *m*
t. of direction Richtungsübertragung *f*
t. of energy Energieübertragung *f*
t. of heat Wärmeübertragung *f*
t. of heat by convection Wärmeübertragung *f* durch Konvektion
t. of heat by radiation Wärmeübertragung *f* durch Strahlung
t. of momentum Impulsübertrag *m*
t. of needles *(Text)* Nadelübergabe *f*
t. order *(Dat)* Sprungbefehl *m*
t. picture Abziehbild *n*
t. pipette *(Ch)* Vollpipette *f*
t. plunger *(Plast)* Preßspritzkolben *m*
t. point Übergabestation *f*; Übergabestelle *f*
t. port Überströmkanal *m*
t. pot *(Plast)* Druckkammer *f*, Spritztopf *m*

t. press 1. *(Masch)* Stufenumformautomat *m*, Stufen[umform]presse *f*; 2. *(Typ)* Umdruckpresse *f*
t. ratio *(El)* Ausbreitungsfaktor *m*; *(Dat)* Übertragungsfaktor *m*
t. shed Umschlaghalle *f*
t. signal *(Fmt)* Umlegungszeichen *n*
t. switch unterbrechungsfreier (verzögerungsfreier) Umschalter *m*, Momentumschalter *m*
t. table *(Bw)* Schiebebühne *f*
t. tool *(Plast)* Preßspritzform *f*
t. track Transportband *n*, Förderband *n*
t. trunk exchange *(Fmt)* Überweisungsfernamt *n*
t. unit *(Fmt)* Umschaltgerät *n*
t. well *(Plast)* Füllzylinder *m*, Spritztopf *m*
transferability Übertragbarkeit *f* *(von Größen)*
transferee of a protective right Erwerber *m* eines Schutzrechtes
transference Übertragung *f*
t. number Überführungszahl *f*
t. of heat Wärmedurchgang *m*
t. of ions Wanderung *f* der Ionen, Ionenwanderung *f*
transferred charge call *(Fmt)* R-Gespräch *n*
t. cylinder needle loop *(Text)* überhängte Zylindermasche *f*
Transfluxor Transfluxor *m* *(Warenzeichen für Mehrlochmagnetkern)*
transform / to umsetzen, umformen, umwandeln; *(El)* umspannen, transformieren
transform *(Math)* Transformierte *f*
transformability Umsetzbarkeit *f*, Umwandelbarkeit *f*, Transformierbarkeit *f*
transformable umsetzbar, umwandelbar, transformierbar
transformation 1. *(Math)* Transformation *f*; 2. *(Ch)* Umformung *f*; Umwandlung *f*; Umsetzung *f*; 3. *(El)* Umwandlung, Phasenumwandlung *f*; Umspannung *f*, Transformierung *f*; 4. *(Ph)* Zustandsänderung *f*
t. equation Transformationsgleichung *f*
t. of coordinates Koordinatentransformation *f*
t. of energy Energieumwandlung *f*
t. of heat into work Umwandlung *f* von Wärme in Arbeit
t. of mechanical work Umwandlung *f* mechanischer Arbeit
t. period Umwandlungszeit *f*
t. point *(Met)* Haltepunkt *m*, Umwandlungspunkt *m*
t. range Umwandlungsbereich *m*
t. ratio of a transformer Übersetzung *f* eines Wandlers
t. temperature Umwandlungstemperatur *f*, Umwandlungspunkt *m*
t. theory Transformationstheorie *f*
t. theory of quantum mechanics Transformationstheorie *f* der Quantenmechanik
transformer *(Elm)* Transformator *m*, Umspanner *m*, Wandler *m*; Übertrager *m*
t. amplifier Transformatorverstärker *m*
t. bank Transformatorengruppe *f*, Transformatorensatz *m*, Umspannergruppe *f*, Transformatorenbank *f*
t. carriage Transformator[en]wagen *m*
t. coil Transformatorspule *f*
t.-connected meter Wandlerzähler *m*
t. core Transformatorkern *m*
t. coupling Transformator[en]kopplung *f*
t. diagram Transformatordiagramm *n*, Betriebsdiagramm *n* eines Transformators
t. electromotive force Transformations-EMK *f*
t. for alternating current Wechselstromtransformator *m*
t. for continuous current Gleichstromtransformator *m*
t. fuse Wandlersicherung *f*
t. generator Drehfeldumformer *m*
t. grade sheet Transformatorenblech *n*
t. grid control ruhende Gittersteuerung *f* *(von Stromtoren)*
t. iron *s.* t. sheet
t. oil Transformator[en]öl *n*
t. plate Transformatorenblech *n*
t. primary primäre Transformatorwicklung *f*, Primärseite *f* des Transformators
t. principle Transformatorenprinzip *n*
t. secondary sekundäre Transformatorwicklung *f*, Sekundärseite *f* des Transformators
t. sheet Transformatorenblech *n*
t. shell Transformatormantel *m*
t. starter Anlaßtransformator *m*
t. substation Umspann[unter]werk *n*
t. switch Transformatorenschalter *m*, Wandlerschalter *m*
t. tap Transformatoranzapfung *f*
t. tripping Auslösen *n* über Wandler, indirekte Auslösung *f*, Wandlerauslösung *f*
t. truck Transformator[en]wagen *m*
t.-type instrument Wandlerinstrument *n*
t. winding Transformatorwicklung *f*
t. with closed magnetic circuit Transformator *m* mit geschlossenem Eisenkern (magnetischem Kreis)
t. with iron core Eisen[kern]transformator *m*
t. with open magnetic circuit Transformator *m* mit offenem Eisenkern (magnetischem Kreis)
t. yoke Transformatorjoch *n*
transformerless receiver Allstromempfangsgerät *n*
transforming Transformierung *f*, Umspannung *f*
t. station Umspannstation *f*, Umspannwerk *n*

transgression of a law Gesetzesübertretung *f*
tranship / to *s.* transship / to
transhipment *s.* transshipment
transhorizon propagation Überreichweiteausbreitung *f*, transhorizontale Ausbreitung *f (von Wellen)*
transient Ausgleichsvorgang *m*, Einschwingvorgang *m*; Einschaltstoß *m*, Schaltvorgang *m*; vorübergehende Überspannung *f*; Übergangsvorgang *m*
transient einschwingend, kurzzeitig, vorübergehend, momentan; nichtstationär
t. analysis Einschwinganalyse *f*
t. bearing Kurzzeitpeilung *f (Radar)*
t. behaviour of a machine Anlaufverhalten *n* einer Maschine
t. cavity abschwimmender Hohlraum *m (Kavitation)*
t. characteristic Übergangsverhalten *n*
t. condition Einschwingvorgang *m*
t. creep Übergangskriechen *n*
t. current Einschwingstrom *m*, Ausgleichsstrom *m*, Stromstoß *m*
t.-decay current Nachwirkungsstrom *m*
t. distortion Verzerrung *f* durch Ein- und Ausschwingen
t. effect Laufzeitverzerrung *f*; Einschwingvorgang *m*
t. error vorübergehende Regelabweichung *f*
t. facilities Umschlagsanlagen *fpl (Hafen)*
t. imperfection vorübergehende Störung *f*
t. oscillation Ausgleichsschwingung *f*
t. period Einschwingzeit *f*
t. phenomenon Einschwingvorgang *m*, Ausgleichsvorgang *m*
t. potential difference Einschwingspannung *f*, Augenblicksspannung *f*
t. problem Einschwingproblem *n*
t. protection Wanderwellenschutz *m*
t. protective device Wanderwellenschutz *m*
t. reactance Einschwingreaktanz *f*, Stoßkurzschlußreaktanz *f*
t. response Einschwingkurve *f*, Ausgleichkurve *f*; Sprungcharakteristik *f*; Beruhigungszeit *f*, Schaltzeit *f*; Übergangsfunktion *f*, Übergangsverhalten *n*
t. rise kurzzeitiger Anstieg *m*, vorübergehende Verstärkung *f*
t. short-circuit current transitorischer Kurzschlußwechselstrom *m*
t. signal Einschwingsignal *n*
t. stability dynamische Stabilität *f*
t. stability limit Grenze *f* der dynamischen Stabilität
t. state Einschwingzustand *m*, Einschwingvorgang *m*; Übergangszustand *m*
t. stress condition veränderliche Druckbeanspruchung *f*
t. surge Einschwingstoß *m*
t. time Einschwingzeit *f*
t. time difference Laufzeitunterschied *m*
t. voltage Einschwingspannung *f*, Schaltüberspannung *f*, Überspannung *f*, Ausgleichsspannung *f*; Stoßspannung *f*
t. wave Wanderwelle *f*
transients Einschwingvorgänge *mpl*, Einschwingerscheinungen *fpl (beim Schalten)*
transistor Transistor *m (aktives Halbleiterbauelement mit drei oder mehr Elektroden)*
t. action Transistorwirkung *f*
t. aging Transistoralterung *f*
t. amplifier Transistorverstärker *m*
t. assembly Transistorbaugruppe *f*
t. checker Transistorenprüfgerät *n*
t. chopper Transistorzerhacker *m*
t. circuit Transistorschaltung *f*
t. electronics Transistorelektronik *f*
t. enclosure Transistorumhüllung *f*
t. engineering Transistortechnik *f*
t. fabrication Transistorherstellung *f*
t. failure Transistorausfall *m*
t. gain Transistorverstärkung *f*
t. inverter Transistorwechselrichter *m*
t. life Lebensdauer *f* von Transistoren
t. maintained transistorisiert
t. manufacture Transistorenherstellung *f*
t. multiplier Transistorvervielfacher *m*
t. performance Transistorleistung *f*
t. process technology Transistorverfahrenstechnik *f*
t. radio Transistorradio *n*, Transistorempfangsgerät *n*
t. radio receiver *s.* t. radio
t. reliability Betriebssicherheit *f* von Transistoren
t. stabilization Transistorstabilisierung *f*
t. stage Transistorstufe *f*
t. structure Transistoraufbau *m*
t. switch Schalttransistor *m*, Transistorschalter *m*
t. tester Transistorprüfgerät *n*
t. theory Transistortheorie *f*
t. timer transistorisierter Zeitschalter *m*
t. transmitter Transistorsender *m*
t. triode Transistortriode *f*
transistorization Transistorisierung *f*, Transistorbestückung *f*
transistorized transistorisiert, mit Transistoren bestückt
t. amplifier Transistorverstärker *m*
transit Durchgang *m*, Durchlauf *m*; Drift *f*, Abtrieb *m*, Abweichung *f*
t. angle *(El)* Laufwinkel *m*
t. bearing *(Schiff)* Deckpeilung *f*
t. cargo *(Schiff)* Transitladung *f*
t. exchange *(Fmt)* Durchgangsamt *n*

t. **freight** *(Schiff)* Durchfracht *f*
t. **goods** Durchfuhrgut *n (Schiffsfracht)*
t. **line** *(El, Fmt)* durchgehende Leitung *f*
t.-mix concrete Transport[misch]beton *m*
t.-mix truck Transportmischer *m (Transportbeton)*
t. **mixer** Transportmischer *m (Transportbeton)*
t. **phase angle** *(El)* Lauf[zeit]winkel *m*
t. **port** Transithafen *m (Übergangshafen)*
t. **shed** Kaischuppen *m*, Transitlagerschuppen
t. **state current** *(El)* Anschwingstrom *m*
t. **station** Übergangsbahnhof *m*
t. **time** Umschlagzeit *f (z. B. Relais)*, Kippzeit *f*; Laufzeit *f (in Elektronenröhren)*; Übertragungszeit *f*
t. **time compensation** *(El)* Laufzeitausgleich *m*, Laufzeitkompensation *f*
t. **time correction** *(Fs)* Laufzeitentzerrung *f*
t. **time delay** Laufzeitverzögerung *f*
t. **time tube** *(El)* Laufzeitröhre *f*
t. **truck mixer** Transportmischer *m (Transportbeton)*
transition Übergang *m*; Übergangsschaltung *f*
t. **capacitance** Übergangskapazität *f*
t. **coil** *(Elm, El)* Überschaltdrossel *f*
t. **crossing** *(Eis)* Übergangskreuzung *f*
t. **curve** Übergangsbogen *m*, Übergangskurve *f*
t. **effect** Übergangseffekt *m*
t. **element** *(Ch)* Übergangselement *n*
t. **fit** Übergangspassung *f*, Übergangssitz *m*
t. **frequency** Übergangsfrequenz *f*, Dämmerungsfrequenz *f*
t. **interval of indicator** Farbumschlagsbereich *m* eines Indikators
t. **loss** Übergangsverlust *m*
t. **of heat** Wärmedurchgang *m*
t. **point** *(Met)* Umwandlungspunkt *m*, Haltepunkt *m*, Übergangspunkt *m*
t. **probability** Übergangswahrscheinlichkeit *f*
t. **resistance** Überschaltwiderstand *m*
t. **temperature** Umwandlungstemperatur *f*, Sprungtemperatur *f*
t. **time** Laufzeit *f*; Sprungzeit *f*
t. **zone** Übergangszone *f*
transitron Transitron *n*, Bremsfeldröhre *f*
t. **oscillator** Bremsfeldröhrenoszillator *m*
translate / to 1. *(Math)* parallel verschieben; translatorisch bewegen; 2. *(Dat)* übersetzen, umwandeln, zuordnen
translating field *(Fmt)* Umrechnerfeld *n*
t. **relay** *(Fmt)* Zeichenumsetzrelais *n*
translation *(Math)* Translation *f*
t. **lattice** Translationsgitter *n*
translational shell *(Bw)* Translationsschale *f*
translator *(Dat)* Zuordner *m*, Übersetzer *m*; Zeichenumsetzer *m (Telegrafie)*; Frequenzbandumsetzer *m*
translatory translatorisch, fortschreitend
translucent halbdurchsichtig; durchscheinend, transparent, lichtdurchlässig
t. **body** *(El)* durchscheinender Körper *m*
t. **light** *(Opt)* Durchlicht *n*
t. **plastic** durchscheinender Plast *m*
translumination base *(Opt)* Leuchtuntersatz *m*
transmissibility Übertragbarkeit *f*
transmission 1. Übertragung *f*, Transmission *f*; 2. Durchgang *m*, Durchlassung *f*, Transmission; 3. Durchlässigkeit *f*, Durchlaßgrad *m*; 4. Durchstrahlung *f (Elektronenmikroskopie)*; 5. Sendung *f (Rundfunkprogramm)*; 6. *(Ak)* Durchschallung *f*; 7. *(Masch)* Übersetzung *f (Getriebe)*; Trieb *m*
t. **brake** Triebwerkbremse *f*, Getriebebremse *f*
t. **by double current** *(Fmt)* Doppelstrombetrieb *m*
t. **by simplex current** *(Fmt)* Einfachstrombetrieb *m*
t. **cable** Fernleitungskabel *n*
t. **channel** *(El)* Übertragungsweg *m*
t. **constant** Übertragungskonstante *f*
t. **dynamometer** Transmissionsdynamometer *n*
t. **efficiency [in percent]** 1. Wirkungsgrad *m* der Kraftübertragung; 2. Durchlässigkeit *f*; 3. Übertragungswirkungsgrad *m (von Antennen)*
t. **efficiency test** *(Fmt)* Streckendämpfungsmessung *f*
t. **equivalent** *(El)* Bezugsdämpfung *f*
t. **factor** Durchlassungsvermögen *n*, Durchlässigkeit *f*
t. **frequency meter** *(El)* Durchgangswellenmesser *m*, Frequenzmesser *m (im Durchgangsverfahren)*
t. **grating** *(Opt)* Transmissionsgitter *n*
t. **impairment** *(Fmt)* Minderung *f* der Übertragungsgüte
t. **lag** Übertragungsverzögerung *f (Signalübertragung)*
t. **lay-out** *(Fmt)* Pegelplan *m* einer Übertragung
t. **level meter** Pegelmesser *m*
t. **line** 1. Hochspannungsleitung *f*; Fernleitung *f*, Überlandleitung *f*; 2. Transmission *f*, Wellenstrang *m*
t.-line amplifier Kettenverstärker *m*
t. **line tower** Überlandleitungsmast *m*
t. **loss** 1. Durchgangsverlust *m*; 2. *(Fmt)* Übertragungsverlust *m*, Dämpfung *f*
t. **measuring set** Dämpfungsmesser *m*, Pegelmesser *m*, Pegelschreiber *m*, Pegelzeiger *m*; *(El)* Streckendämpfungsmesser *m*, Übertragungsmeßgerät *n*
t. **method** Durchstrahlungsverfahren *n*, Durchschallungsverfahren *n (Werkstoffprüfung)*

t. of bearing information Peilwertübertragung *f*
t. of energy Energieübertragung *f*
t. of heat Wärmeübergang *m*
t. of motion Bewegungsübertragung *f*
t. of pictures Bildübertragung *f*
t. of power Kraftübertragung *f*, Leistungsübertragung *f*
t. of sound Schalldurchgang *m*, Schallübertragung *f*
t. of waves Wellenfortpflanzung *f*
t. performance *(Fmt, El)* Übertragungsgüte *f*
t. performance rating *(Fmt)* Wert *m* der Übertragungsgüte
t. range *(El)* Durchlaßbereich *m*, Reichweite *f*, Übertragungsbereich *m*
t. ratio Übersetzungsverhältnis *n*
t. rope Transmissionsseil *n*, Förderseil *n*
t. shaft Transmissionswelle *f*
t. shafting Transmission *f*
t. speed *(Fmt)* Übermittlungsgeschwindigkeit *f*
t. standards *(Am)* Übertragungsnormen *fpl*
t. time Echolaufzeit *f (Radar)*
t. tower Hochspannungsmast *m*
t. trouble *(Fmt)* schlechte Verständigung *f*
t. tunnel *(Kfz)* Getriebetunnel *m*
t. unit Übertragungseinheit *f*
transmissivity Durchlässigkeit *f*; Durchlässigkeitsgrad *m*
transmit / to 1. übertragen, übermitteln; leiten *(Strom)*; 2. *(Fmt)* [aus]senden; 3. befördern; [durch]geben; 4. durchschicken, durchlassen; 5. fortpflanzen, ausbreiten
t. heat Wärme durchlassen
transmit-receive switch Sende-Empfangsschalter *m (Wellenleitertechnik)*
t.-receive-tube Empfängerspeicherröhre *f*
transmittance Transparenz *f*; Durchlässigkeit *f*, Durchlassungsvermögen *n*, Durchlässigkeitsgrad *m*; Durchsichtigkeit *f*
transmitted-carrier operation *(Fmt)* Betrieb *m* mit Trägerübertragung
t. light *(Opt)* Durchlicht *n*
transmitter Sender *m*; Schallkopf *m*; Mikrofon *n*; *(Fmt)* Geber *m*; Meßumformer *m*, Meß[wert]wandler *m*, Transmitter *m*
t. amplifier Sendeverstärker *m*
t. arrangement *(Fmt)* Gebevorrichtung *f*
t. attenuation *(Fs)* Senderdämpfung *f*
t. battery Mikrofonbatterie *f*
t. beacon Richtungssender *m*
t.-blocker cell Sendesperrzelle *f (Wellenleitertechnik)*
t. button *(Fmt)* Sprechtaste *f*
t. capsule Mikrofonkapsel *f*
t. capsule protector Schutzgehäuse *n* für die Mikrofonkapsel
t. circuit Senderschaltung *f*
t. control station Sendezentrale *f*
t. crystal Senderkristall *m*
t. frequency Senderfrequenz *f*
t. housing Mikrofongehäuse *n*
t. inset Sprechkapsel *f*
t. noise Mikrofongeräusch *n*, Mikrofonrauschen *n*
t. outage Senderausfall *m*
t. output valve Senderendröhre *f*
t. potentiometer Geberpotentiometer *n*
t. power Senderleistung *f*
t. preheating Sendervorheizung *f*
t. receiver cell Sende-Empfangs-Weiche *f*
t. storage *(Fmt)* Sendespeicher *m*
t. supply 1. Mikrofonspeisung *f*; 2. Senderspeisung *f*
t. tuning Senderabstimmung *f*
t. unit 1. Sprechkapsel *f*; 2. Sender *m*
t. valve Senderöhre *f*
transmitting Übertragen *n*, Senden *n*
t. aerial Sendeantenne *f*
t. factor Durchlässigkeitsgrad *m*
t. frequency Sendefrequenz *f*
t. laser Sende-Laser *m*
t. medium Übertragungsmedium *n*
t. power Sendeleistung *f*
t. pulse Übertragungsimpuls *m*
t. set Sendegerät *n*, Sender *m*
t. station Sendestation *f*
t. stylus Abtaststift *m*
t. tape Sendestreifen *m*
t. terminal Senderendstufe *f*
t. valve Senderöhre *f*
transmutation Umwandlung *f*, Transmutation *f* *(Atomkerne)*
t. of elements Umwandlung (Transmutation) *f* von Elementen
t. product Umwandlungsprodukt *n*
t. reaction Umwandlungsreaktion *f*
transoceanic communication *(Fmt)* Überseeverkehr *m*
t. radio service drahtloser Überseeverkehr *m*
transom 1. *(Schiff)* Heck *n*, Spiegel *m*; Heckkonstruktion *f*; 2. waagerechte Fenstersprosse *f*
t. beam Heckbalken *m*, Worpbalken *m*
t. drive Heckantrieb *m*
t. floor Heckbodenwrange *f*, Transomwrange *f*, Transomplatte *f*
t. frame Heckspant *n*
t. stern Spiegelheck *n*, Transomheck *n*
transonic speed (velocity) *(Rak)* schallnahe Geschwindigkeit *f*
transparency Transparenz *f*, Durchlässigkeit *f*; Strahlendurchlässigkeit *f*; Diapositiv *n*
t. of the potential barrier Durchlässigkeit *f* der Potentialschwelle
t. tester Lichtdurchlässigkeitsprüfer *m*

transparent durchsichtig, transparent
t. cut Dünnschliff *m*
t. front cathode Durchsichtsfrontkatode *f*
t. paper Transparentpapier *n*
t. slide Diapositiv *n*
transpierce / to durchstechen
transpiration cooling Schwitzkühlung *f*
t. method *(Ph)* Durchflußmethode *f*
transplanting Umpflanzen *n*, Auspflanzen *n*
transponder Antwortbake *f*, Antwortsender *m*, Antwortgerät *n*, Bodenantwortstation *f*, Wiederholerbake *f (Radartechnik)*
transpondor *s.* transponder
transport / to transportieren, [be]fördern
transport 1. Transport *m*, Beförderung *f*, Güterverkehr *m*; Transportschiff *n*
t. bridge Schiebebühne *f*
t. equation *(Ph)* Transportgleichung *f*, Boltzmann-Gleichung *f*
t. factor Transportfaktor *m*
t. kernel *(Kern)* Transportkern *m*
t. number *(Ch)* Überführungszahl *f*
t. phenomenon in gases *(Ph)* Transporterscheinung *f* in Gasen
t. ship Lastschiff *n*, Transportschiff *n*
transportable ortsveränderlich, tragbar, [be-]förderbar
t. lamp Handlampe *f*
transportation Transport *m*, Beförderung *f*
t. damage Transportschaden *m*
t. engineering Verkehrstechnik *f*
t. lag *s.* distance velocity lag
transporter Transporter *m*, Transportwagen *m*; Transportschiff *n*
t. bridge Schiebebühne *f*
transpose / to 1. *(Math)* transponieren, versetzen, vertauschen *(Elemente)*; stürzen *(eine Matrix)*; hinüberbringen *(ein Glied einer Gleichung auf die andere Seite)*; 2. *(Fmt)* kreuzen *(Drähte zum Induktionsschutz)*; vertauschen *(Leitungen)*; 3. *(Mech)* verschieben
transposed transmission line *(Fmt)* Leitung *f* mit Induktionsschutzkreuzungen
t. twill *(Text)* versetzter Köper *m*
transposition Transposition *f*; Verdrillung *f*, Kreuzung *f (Freileitung)*; Vertauschung *f (Leitungen)*
t. bracket *(Fmt)* Einschiebestütze *f*
t. insulator Verdrillungsisolator *m (Freileitungen)*
t. of wires Drahtkreuzung *f*
t. pole Verdrillungsmast *m*; Kreuzungsgestänge *n*
t. receiver Zwischenfrequenzempfänger *m*
transrectifier *(El)* Gleichrichter *m* mit Gittersteuerung
transrector *(Elm)* Transformator *m* mit Gleichrichter

transship / to umladen *(Ladung)*
transshipment Umschlag *m (Ladung)*, Umladung *f*
t. harbour Umschlagshafen *m*
t. port Umschlagshafen *m*
transuranic elements *(Ch)* Transurane *npl*
transvaal tipping wagon *(Eis)* Transvaalkipper *m*
transversal *(Math)* Transversale *f*
transversal transversal, quer, Quer-
t. modulation Quermodulation *f*
t. strength Biegefestigkeit *f*
t. turning Plandrehen *n*
t. vibration Querschwingung *f*, transversale Schwingung *f*
transverse quer; transversal
t. acceleration Transversalbeschleunigung *f*
t. adjustment Querverstellung *f*
t. axis *(Math)* Hauptachse *f (Hyperbel)*
t. beam 1. Querträger *m*; 2. Querstrahl *m*
t. beam travelling-wave tube Querstrahlwanderfeldröhre *f*
t. bending test *(Mech)* Biegeprüfung *f*
t. bracing Querverband *m*, Windverband *m (Brückenbau)*
t. bulkhead Querschott *n*
t. coaming *(Schiff)* Quersüll *m (n)*
t. conductance Querleitwert *m*
t. control arm Querlenker *m*
t. copying attachment Plankopiereinrichtung *f*
t. crack *(Met)* Querriß *m*
t. crosstalk coupling *(Fmt)* Nebensprechkopplung *f*, Übersprechkopplung *f*
t. current Querstrom *m*, Transversalstrom *m*
t. deck beam Deck[s]balken *m*
t. deflection Durchbiegung *f (Festigkeit)*
t. differential protection *(El)* Quervergleichsschutz *m*
t. direction Querrichtung *f*
t. electric querelektrisch
t. electric and magnetic mode TEM-Typ *m*, TEM-Modus *m*, L-Typ *m*, L-Modus *m (Wellenleiter)*
t. electric field elektrisches Querfeld *n*
t. electric wave elektrische Transversalwelle *f*, H-Welle *f*, TE-Welle *f*
t. electromagnetic elektromagnetisch quer, querelektromagnetisch
t. extension Querdehnung *f*
t. fatigue test *(Mech)* Biegedauerversuch *m*
t. feed *(Masch)* Planzug *m*, Quervorschub *m*
t. fibre Querfaser *f*
t. fibre feed *(Text)* Querfaserspeisung *f*
t. field Transversalfeld *n*, Querfeld *n*
t. field loss Querfeldverlust *m*
t. field travelling-wave tube Querfeldwanderfeldröhre *f*, Lauffeldröhre *f* mit statischem Querfeld

t. **finishing** Planschlichten *n*
t. **force** Querkraft *f*
t. **frame** Querspant *n*
t. **framing** Querspanten *npl*, Querbespantung *f*
t. **leaf spring** Querblattfeder *f*
t. **level** Querlibelle *f*
t. **link** Querlenker *m*
t. **load** Querbelastung *f*
t. **magnetic wave** magnetische Transversalwelle *f*, E-Welle *f*, TM-Welle *f*
t. **magnetization** Quermagnetisierung *f*
t. **module** Stirnmodul *m*
t. **motion** *(Masch)* Planzug *m*
t. **movement** Querbewegung *f*
t. **pillar extension** *(Brg)* Pfeilerquerdehnung *f*
t. **pitch** Stirnteilung *f (Zahnrad)*
t. **plane** Stirnebene *f (Zahnrad)*
t. **planing** Querhobeln *n*
t. **propeller** Querpropeller *m*
t. **reinforcement** Querbewehrung *f (Stahlbetonbau)*
t. **sailing** *(Schiff)* Koppelnavigation *f*
t. **section** Querschnitt *m*
t. **slope** Quergefälle *n (Straßenbau)*
t. **slot** Quernut *f*
t. **spherical aberration** *(Opt)* sphärische Seitenaberration (Queraberration) *f*
t. **spring** Querfeder *f*
t. **stiffening** Querversteifung *f*
t. **strain** Querdehnung *f*, Einschnürung *f*
t. **strength** Biegefestigkeit *f*; Querbruchfestigkeit *f*
t. **stress** Biegespannung *f*
t. **test** *(Mech)* Biegeversuch *m*
t. **traverse** *(Masch)* Planzug *m*; Planbewegung *f*
t. **tricot** *(Text)* Quertrikot *m (n)*
t. **turning** Plandrehen *n*
t. **vibration** Transversalschwingung *f*
t. **voltage** *(El)* Transversalspannung *f*
t. **warping** *(Holz)* Breitenkrümmung *f*
t. **wave** Transversalwelle *f*, Querwelle *f*, transversale Welle *f*

transversely in Querrichtung, plan
t. **corrugated grate bar** Schlangenroststab *m*
t. **loaded** senkrecht zur Achse belastet

transverter *(El)* 1. Hochspannungskommutatorgleichrichter *m*; 2.Umrichter *m*

trap / to 1. einfangen, einschließen; 2. Quetschflüssigkeit bilden; 3. mit Entwässerungsstutzen versehen

trap 1. Falle *f*; Abscheider *m*; Ausscheider *m*; Auffangvorrichtung *f*, Auffanggefäß *n*, Sammelgefäß *n*; Geruchverschluß *m*; Klappe *f*, Rückstauklappe *f*; Kühlfalle *f*, Siphon *m*; 2. *(El)* Fangstelle *f*; *(Kern)* Abscheider, Falle; Auffänger *m*, Fänger *m*; 3. *(Fmt)* Sperre *f (Funkempfangstechnik)*; Sperrkreis *m*; 4. Entwässerungsstutzen *m*
t. **amplifier** *(El)* Trennverstärker *m*
t. **circuit** Haltestromkreis *m*
t. **door** 1. Falltür *f*; Fallklappe *f*; 2. Bodenklappe *f (im Flugzeug)*; 3. Schachtklappe *f (Schachtabteufen)*
t. **tree** Fangbaum *m*
t. **trench** Fanggraben *m*
t. **valve amplifier** *(Fmt)* Leitungsverstärker *m*, Trennverstärker *m*

trapeze Trapez *n*, Schaukelreck *n*

trapezium 1. Trapez *n*; 2. *(Am)* beliebiges Viereck *n*
t. **distortion** *(Fs)* Trapezfehler *m*; Trapezverzerrung *f*, Schlußsteinverzerrung *f*

trapezohedron *(Min)* Trapezoeder *n*, Ikositetraeder *n*

trapezoid 1. beliebiges Viereck *n*; 2. *(Am)* Trapez *n*

trapezoidal field Trapezfeld *n*
t. **linkage** Trapezgelenkeck *n*
t. **profile** Trapezprofil *n*
t. **thread** Trapezgewinde *n*

trapped mit Geruchverschluß
t. **air-bubble** *(Plast)* eingeschlossene Luftblase *f*
t. **fluid** Quetschflüssigkeit *f (Pumpe)*
t. **oil** Quetschöl *n (Pumpe)*
t.**-radiation belt** Strahlungsgürtel *m (Erde)*

trapping Einfang *m*; Einschluß *m*; Bildung *f* von Quetschflüssigkeit *(Pumpe)*
t. **the chip in the tooth gullet** Spanstauung *f* im Zahngrund

trappings Pferdegeschirr *n*

trash Abfall *m (im Fasergut)*; Ausschuß *m*; Reisig *n*
t. **removal** *(Text)* Entfernung *f* des Ausschusses

trass *(Geol)* Traß *m*; Tuffstein *m*
t. **cement** Traßzement *m*

travel / to 1. *(Masch)* sich bewegen, sich verschieben; 2. wandern *(Feld, Welle)*

travel 1. *(Masch)* Bewegung *f*, Verschiebung *f*; Verschiebbarkeit *f*; Arbeitshub *m*, Hub *m*; Weg *m*; 2. Wanderung *f (Elektronen)*
t. **indicator** Hubindikator *m*
t. **of radio waves** *(El)* Wellenausbreitung *f*
t. **of the slide valve** Schieberhub *m*, Schieberweg *m*
t. **shot** *(Fs)* Folgebild *n*

traveller Läufer *m (Spinnen)*
t. **burn** Läufererhitzung *f*
t. **tension** Läuferzug *m*
t. **weight** Läufermasse *f*

travelling beweglich, verschiebbar
t. **bar-grizzly** Wanderrost *m (Aufbereitung)*
t. **belt** Transportband *n*
t. **cavity** abschwimmender Hohlraum *m (Kavitation)*

t. chain block Laufkatzenflaschenzug *m*
t. cleaner Wanderreiniger *m (Spinnerei)*
t. column *(Masch)* waagerecht verschiebbarer Ständer *m*
t. contact *(El)* Wanderkontakt *m*
t. crane Laufkran *m*
t. gear Fahrwerk *n*
t. gib crane Portalkran *m*
t. grate Wanderrost *m*
t. grate firing Wanderrostfeuerung *f*
t. head planer Hobelmaschine *f* mit beweglichem Hobelwerkzeug
t. hopper Zubringerwagen *m*
t. limit angle dynamischer Grenzwinkel *m*
t. load Verkehrslast *f*
t. loading hopper *(Met)* Trichterwagen *m*
t. microscope Reisemikroskop *n*
t. piston rod wandernde Kolbenstange *f*
t. plane wave fortschreitende ebene Welle *f (Wellenleiter)*
t. platform Schiebebühne *f*
t. shaft *(Brg)* Fahrschacht *m*
t. track *(Brg)* Fahrgasse *f*, Fahrpaß *m*, Fahrweg *m*
t. wave fortschreitende Welle *f*, Wanderwelle *f*
t.-wave aerial Antenne *f* mit fortschreitenden Wellen
t.-wave valve Wanderfeldröhre *f*, Lauffeldröhre *f*
t.-wave valve interaction circuit Verzögerungsleitung *f* in einer Lauffeldröhre
t. way *(Brg)* Fahrpaß *m*, Fahrfeld *n*, Fahrstrecke *f*
travelway *(Brg)* Fahrpaß *m*, Fahrfeld *n*, Fahrstrecke *f*
traverse / to überqueren; *(Masch)* fahren, verschieben, verstellen; [durch]laufen; *(Opt)* polygonieren
traverse *(Masch)* Bewegung *f*, Gang *m*, Lauf *m*, Verschiebung *f*, Verstellung *f*; *(Opt)* Polygonzug *m*
t. guide *(Text)* Luntenführer *m*
t. leg Polygonseite *f*
t. limiting switch Endausschalter *m*
t. of the ring rail *(Text)* Ringbankbewegung *f*
t. oscillation Querschwingung *f*
t.-spindle grinding attachment Verschiebespindelschleifeinrichtung *f*
traverser Schiebebühne *f*
traversing *(Opt)* Polygonierung *f*
traversing *(Masch)* verschiebbar, verstellbar, traversierend
t. condenser roller *(Text)* Nitschelwalze *f*
t. equipment Polygonausrüstung *f*
t. force Vorschubkraft *f*
t. gear Fahrwerk *n (Kran)*
t. lever Richtbaum *m (Geschütz)*
t. thread guide *(Text)* schwingender Fadenführer *m*
t.-type radial drilling machine Radialbohrmaschine *f* auf Fahrbett
travertine Kalktravertin *m*, Kalktuff *m*, Travertin *m*
travolator *(Am)* rollender Fußgängerweg *m*
trawl Grundschleppnetz *n*, Schleppnetz *n*, Fischnetz *n*
t. board Scherbrett *n*
t. gallow Fischgalgen *m (zum Einholen des Fischnetzes)*
t. gear Schleppnetzgeschirr *n*
t. net Fischnetz *n*, Grundschleppnetz *n*
t. smack Schleppnetzfischkutter *m*
t. warp gallow *s.* t. gallow
t. winch Fischnetzwinde *f*, Netzwinde *f*
trawler Trawler *m (Schleppnetzfischereifahrzeug)*
t.-factory ship Fischfabrikschiff *n*, Fischereiverarbeitungsschiff *n*
trawling Schleppnetzfischerei *f*
trawlline gallow Fischgalgen *m*
tray Mulde *f*, Trog *m*; Schale *f*, Wanne *f (Kühlmittel)*; *(Masch)* Beschickungsbehälter *m*; Batterietrog *m*; Kolonnenboden *m*, Trockenblech *n*
t. drier 1. *(Ch)* Trockenschrank *m*; 2. *(Lw)* Hordentrockner *m*
treacle *(Am)* Melasse *f*; Sirup *m*
t.-stage *(Plast)* viskoser Zustand *m* eines Gießharzes
tread 1. Treppenstufe *f (obere Fläche)*, Trittstufe *f*; Stufenschutz *m*; Stufenbreite *f*; 2. Lauffläche *f (Schiene)*; Reifenprofil *n*; Laufkranz *m (Rad)*; 3. Sohle *f*
t. loom Pedalwebstuhl *m*, Tretwebstuhl *m*
t. pattern Laufflächenprofilierung *f (Reifen)*
t. roll *(Met)* Laufflächenwalze *f*
treadle Fußhebel *m*, Pedal *n*; Durchbiegungskontakt *m (elektrische Eisenbahn)*
t. loom Trittwebstuhl *m*, Exzenterwebstuhl *m*
t. motion *(Text)* Trittvorrichtung *f*
treat / to 1. behandeln; vergüten; bearbeiten, verarbeiten; 2. beschichten; präparieren; tränken; imprägnieren; 3. aufbereiten
treated ore aufbereitetes Erz *n*
t.-paper filter getränktes Papierfilter *n*
t. pole getränkter (imprägnierter) Mast *m*
treating characteristic Imprägnierfähigkeit *f*
treatment 1. Behandlung *f*; Vergütung *f*; Bearbeitung *f*; Verarbeitung *f*; 2. Beschichtung *f*; Präparierung *f*; Tränkung *f*, Imprägnierung *f*; 3. Aufbereitung *f*
t. cone *(Kern)* Bestrahlungstubus *m*, Lokalisator *m*
t. of water Wasseraufbereitung *f*
treaty port Vertragshafen *m*, Bestimmungshafen *m*

treble / to verdreifachen
treble dreifach
t. band-pass filter *(Fmt)* Dreifachbandfilter *n*
t. control Hochtonblende *f*
t. corrector Höhenentzerrer *m*
t.-cut Höhenbeschneidung *f*, Beschneidung *f* des Anteiles von hohen Frequenzen
t. figure dreistellige Zahl *f*
trebles Nußkohle *f*
trebling 1. Verdreifachung *f*; 2. Eisverstärkung *f (eines Schiffes)*
tree Baum *m*; Stamm *m*
t.-agate *(Min)* Baumachat *m*
t. felling Einschlag *m (Wald)*
t.-felling axe Baumaxt *f*
t.-like crystal *(Geol)* Dendrit *m*
t. limit Baumgrenze *f*
t.-nursery Baumschule *f*
t. planter Baumsetzmaschine *f*
t. puller Baumumleger *m*
t. scraper Rindenreiniger *m*
treenail Dübel *m*
treillage Spalier *n*
trellis / to mit Spalier versehen
trellis Gitter *n*; Spalier *n*
t. fence Lattenzaun *m*; Gitterzaun *m*
t. fruit Spalierobst *n*
t. mast Gittermast *m*
t. work 1. *(Bw)* Gitterwerk *n*; 2. Spalierwerk *n*
trellised Gitter-, vergittert
t. mast Gittermast *m*
trembler *(El)* Hammerunterbrecher *m*, Kontakthammer *m*, Wagnerscher Hammer *m*, Zungenunterbrecher *m*
t. bell *(Fmt)* Gleichstromwecker *m*, Wecker *m* mit Selbstunterbrechung
t coil Funkeninduktor *m*, Spule *f* mit Unterbrecher
tremie Betontrichter *m*
tremolite *(Masch)* Tremolit *m*
tremor Beben *n*
trenail *s.* treenail
trench / to 1. umgraben; 2. *(Geol)* schürfen, graben; 3. *(Masch)* nuten
trench Graben *m*, Rinne *f*; *(Geol)* Schürfgraben *m*
t. excavator Grabenbagger *m*
t. hoe Tieflöffelbagger *m*
t.-mortar Granatwerfer *m*
trencher 1. Grabenziehmaschine *f*, Grabenziehgerät *n*; 2. Grabenfräse *f*, Grabenbagger *m*
trenching Grabenaushub *m*, Herstellen *n* eines Grabens
t. attachment Grabenzieh-Anbaugerät *n*
trend Tendenz *f*
trepan / to hohlbohren, kernbohren, ringbohren
trepan *(Brg)* Schachtbohrer *m (Kind-Chaudron-Verfahren)*
trepanning Kernbohren *n*, Ringbohren *n*, Hohlbohren *n*
t. bar Bohrrohr *n*
t. bar cutter head Kernbohrkopf *m*, Anbohrkopf *m*, Hohlbohrkopf *m*
t. cutter Hohlbohrer *m*
t. head *s.* t. bar cutter head
t. head cutting bit Kernbohrkopfmesser *n*
t. lathe Hohlbohrbank *f*
trestle Bock *m*, Gerüst *n*
t. bridge *(Bw)* Bockbrücke *f*, Jochbrücke *f*
tri-lobe dreilappig *(Antenne)*
tri-maran Dreirumpfschiff *n*, Dreirumpfboot *n*
tri-pack film Dreischichtenfilm *m*
triacetin Triazetin *n*, Triazetylglyzerin *n*
triacid base dreisäurige Base *f*
triacontahedral dreißigflächig
triad *(Ch)* Triade *f*
trial Versuch *m*; Abnahmeprüfung *f*; Überschlag *m*
t.-and-error *(Dat)* Probierverfahren *n*
t.-and-error method Annäherungsverfahren *n*, empirisches Ermittlungsverfahren *n*, Rechnungsverfahren *n* mit fortschreitenden Näherungswerten
t. boring Probebohrung *f*, Versuchsbohrung *f*
t. calculation Überschlagsrechnung *f*
t. dive Tauchversuch *m (U-Boot)*
t. frame *(Opt)* Probierbrille *f*
t. launching *(Rak)* Versuchsstart *m*
t. lens *(Opt)* Probierglas *n*
t. order Probeauftrag *m*
t. period Probezeit *f*
t. record Prüfbericht *m*, Prüfprotokoll *n*
t. result Versuchsergebnis *n*
t. run Probefahrt *f*, Probelauf *m*, Probebetrieb *m*; Einrichtlauf *m*
t. tank *(Schiff)* Modelltank *m*, Schleppversuchstank *m*, Schlepprinne *f*
t. test Probefahrt *f*
t. trip Probefahrt *f*
trials upon the measured mile *(Schiff)* Meilenfahrt *f*
triangle Dreieck *n*; Zeichendreieck *n*, Winkel *m*; Triangel *m*
t. of forces Kräftedreieck *n (Statik)*
t. pulse *(Fs)* Dreieckimpuls *m*
triangular dreieckig, dreikantig; scharfgängig *(Gewinde)*; dreiseitig *(Pyramide)*, dreischenklig *(Zirkel)*; Dreigurt-
t. classification chart Dreistoffsystem *n*
t. cross section Dreiecksquerschnitt *m*
t. file Dreikantfeile *f*
t. insert *(Masch)* Dreieckplatte *f*
t. load Dreiecklast *f*
t. motion Dreiecklenker *m*

t. numbers Dreieckszahlen *fpl*
t. oscillation Dreieckschwingung *f*
t. plate spring Dreieckfeder *f*
t. pyramid dreiseitige Pyramide *f*
t. reflector Dreieckrückstrahler *m*
t. strand winding rope Dreikantlitzenförderseil *n*
t. system *(Brg)* Dreiecksanordnung *f (Ausbau)*
t. thread scharfgängiges Gewinde *n*
t. transient pulse *(Fs)* Dreieckimpuls *m*
t. truss Dreiecksbinder *m*
triangulate / to triangulieren; in Dreiecke unterteilen; dreieckig machen
triangulated engine Dreiecks-Gegenkolbenmotor *m*
triangulation Triangulierung *f*, Dreiecksversteifung *f*
t. from photographs Bildtriangulation *f*
t. point Triangulationspunkt *m*, TP
t. system Dreiecksnetz *n*
triatomic *(Ch)* dreiatomig
triaxial dreiachsig; räumlich *(Spannungszustand)*
tribasic acid dreibasische Säure *f*
triblet Dorn *m*
triboelectric reibungselektrisch
triboelectricity Reibungselektrizität *f*
tributary canal *(Bw)* Nebenkanal *m*
trichromatic Dreifarben-
trick *(Text)* Nadelkanal *m*
t. switching arrangement Trickaufnahmevorrichtung *f*
trickle / to tropfen
trickle charge *(El)* Pufferladung *f*
t. cooler Berieselungskühler *m*
t.-furnace Sickerfeuerung *f*
trickling cooling plant Rieselkühler *m*
t. water Rieselwasser *n*, herabrieselndes Wasser *n*
triclinic *(Min)* triklin
tricolour filter Dreifarbenfilter *n*
t. tube *(Fs)* Dreifarbenröhre *f*; Farbbildröhre *f* mit drei Komponenten
tricosoic acid Trikosansäure *f*
tricot machine *(Text)* Kettenstuhl *m*
t. underwear Trikotagen *fpl*
tricuspid dreispitzig
tricuspid dreispitzige Hypozykloide *f*, Steinersche Kurve *f*
tricyclic compound *(Ch)* Dreikernverbindung *f*, Verbindung *f* mit einem dreigliedrigen Ring
tridecagon Dreizehneck *n*
tridecoic acid Tridekansäure *f*
tridimensional dreidimensional
t. plotting *(Opt)* räumliche Auswertung *f*
tridymite Siliziumdioxid *n*, Tridymit *m*
triethyl aluminium Aluminiumtriäthyl *n*, Triäthylaluminium *n*
triethylamine Triäthylamin *n*
trieur Trieur *m*
trifurcating box *(El)* Gabelmuffe *f*, Dreifachverzweigungsmuffe *f*, Dreileiterendverschluß *m*
trigatron Trigatron *n*, gesteuerte Funkenschaltröhre *f (Impulsmodulation)*
trigger / to auslösen, triggern *(Kippschwingungen)*; einleiten *(Vorgang)*; hervorrufen *(Reaktion)*
trigger 1. Ausrücker *m*, Auslöser *m*, Drücker *m*; Schwingauslöser *m*; 2. *(Dat)* Trigger[schaltung *f*] *m*, Kippschaltung *f*; *(Mil)* Abzug *m*
t. circuit *(Fs)* Triggerkreis *m*, Triggerschaltung *f*, Impulsgeberschaltung *f*; einmalig kippende Schaltung *f*, Univibrator *m*; Kippschwingkreis *m*
t. current Triggerstrom *m*
t. electrode Triggerelektrode *f*
t. guard *(Mil)* Abzugsbügel *m*
t. level Triggerpegel *m*
t. pulse Triggerimpuls *m*, Auslöseimpuls *m*
t. relay Fallklappenrelais *n*, durch Stromstöße weiterschaltendes Relais *n*
t. sharpener *(Fs)* Schaltung *f* zur Versteilerung eines Auslöseimpulses, Triggerimpuls-Versteilerungskreis *m*
t. spark Zündfunke *m*
t. switch Kippschalter *m (für einmalige Auslösung)*, Tumblerschalter *m*, Triggerschalter *m*
t. timing pulse Zeitkontrollimpuls *m (Radar)*
t. triode Triggertriode *f*, Auslösetriode *f*
t. tube Triggerröhre *f*
triggered start-stop sweep circuit getriggerter monostabiler Ablenkkreis *m (Katodenstrahloszillografie)*
triggering Triggerung *f*, Auslösung *f*
trigistor Trigistor *m (pnpn-Halbleiter, der als Flip-Flop arbeitet)*
trigonal dreieckig; dreikantig; dreizählig; trigonal *(Kristall)*
trigonometric trigonometrisch
t. function trigonometrische Funktion *f*
t. height determination trigonometrische Höhenmessung *f*
t. interpolation trigonometrische Interpolation *f*
trigonometrical survey Vermessung *f* durch Triangulation
trigonometry Trigonometrie *f*, Dreiecksmessung *f*
trihalogenated *(Ch)* dreifach halogeniert
trihedral dreiflächig
trihydric alcohol dreiwertiger Alkohol *m*
trilateral dreiseitig
trilling Drilling *m (Kristall)*
trillion Trillion *f* ($\hat{=} 10^{18}$); *(Am)* Billion *f* ($\hat{=} 10^{12}$)
trim / to 1. *(Masch)* abgraten; entgraten, bestoßen; putzen *(Gußstücke)*; besäumen; 2. *(Typ)* beschneiden; zurichten; 3. trimmen *(richtige*

Verteilung der Fracht im Schiff); 4. senken *(Unkosten)*
t. aback backtrimmen
trim by the bow *(Schiff)* Kopflastigkeit *f*, Buglastigkeit *f*
t. by the head *(Schiff)* Vorlastigkeit *f*, Kopflastigkeit *f*
t. by the stern *(Schiff)* Achterlastigkeit *f*, Steuerlastigkeit *f*, Hecklastigkeit *f*
t. saw Schrotsäge *f*, Kerbsäge *f*, Quersäge *f*
t. tab *(Rak)* Trimmruder *n*
t. tab motor Trimmotor *m*
trimer *(Ch)* Trimer[es] *n*
trimethide Trimethylverbindung *f* eines Metalls
trimetric trimetrisch
trimmed size Endformat *n*, beschnittenes Format *n (Papier)*
trimmer 1. *(Typ)* Beschneider *m*; Beschneidemaschine *f*, Papierbeschneidemaschine *f*; 2. Wechsel[balken] *m (Holzbau)*; 3. *(El)* Abgleichkondensator *m*, Trimmer *m*; 4. *(Brg)* Profilschuß *m*; 5. *(Schiff)* Kohlentrimmer *m*
t. beam Wechselbalken *m (Holzbau)*
t. joist Wechselbalken *m (Holzbau)*
trimming 1. *(Masch)* Entgraten *n*, Abgraten *n*; Bestoßen *n*, Besäumen *n*; 2. *(Typ)* Beschnitt *m*; 3. Durchhieb *m (Forst)*; 4. Trimmen *n*; Ausrüsten *n*; 5. Auswechselung *f*, Wechsel *m (Holzbau)*
t. castings Putzen *n (Gußstücke)*; Gußputzerei *f*
t. cutter Abgratfräser *m*
t. die Beschneidestempel *m (Stanze)*, Beschneidewerkzeug *n*; Abgratmatrize *f*
t. edge Beschnittkante *f*, Beschnittrand *m*
t. hatch Getreideluke *f*, Trimmluke *f*
t. hole *(Brg)* Kranzbohrloch *n (Schießarbeit)*
t. machine Beschneidmaschine *f*
t. moment Trimmoment *n (Flugzeug)*
t. press Abgratpresse *f*
t. punch Seitenschneider *m (Stanze)*
t. resistor Justierwiderstand *m*
t. shear Abgratschere *f*
t. tab Trimmklappe *f*
t. tank Trimmtank *m (U-Boot)*
t. to size Zuschneiden *n* auf Maß *(Furnier)*
t. tool Abgratwerkzeug *n*
trimmings *(Text)* Posamenten *npl*; Verzierungsleisten *fpl*
t. and edgings Besatz *m*
t. industry Posamentenindustrie *f*
trimorphic trimorph, dreigestaltig
trimorphism Trimorphie *f*
trinistor *(El)* Trinistor *m*
trinitrate Trinitrat *n*
trinitrobenzene Trinitrobenzol *n*
trinitrotoluene Trinitrotoluol *n*
trinity-house tender *(Schiff)* Leuchtturmtender *m*

trinomial dreigliedrig
trinuclear dreikernig
trio mill Drillingswalzwerk *n*
triode Triode *f*, Dreipolröhre *f*
t. amplifier Triodenverstärker *m*, Verstärkertriode *f*
t. connection Triodenschaltung *f*
t. hexode frequency changer Triode-Hexode-Mischröhre *f*
t. valve Triodenröhre *f*
trioxide Trioxid *n*
trip / to auslösen *(z. B. Relais)*; ausrücken, schalten
t. a gas tube ein Gasentladungsrohr zünden
t. out auslösen
trip 1. *(Masch)* Auslösung *f*; Anschlag *m*, Knagge *f*; 2. *(Brg)* Fahrspiel *n*, Förderspiel *n (söhlige Förderung)*; Förderzug *m (Schachtförderung)*; 3. kurze Seereise *f (einmalige Fahrt)*
t. cam Schaltnocken *m*
t. charter index *(Schiff)* Reisecharterindex *m*
t. circuit Auslösestromkreis *m*
t. coil Auslösespule *f*, Erregerspule *f*, Relaisspule *f*
t. cycle Fahrspiel *n*
t. device Auslösungseinrichtung *f*
t. dog Anschlag[bolzen] *m*, Sperrklinke *f*
t.-free mit Freiauslösung *(Leistungsschalter)*
t. hammer Stampfe *f*, Hammer *m*
t. lever Schalthebel *m*
t. magnet Auslösemagnet *m*
t.-out *(Masch)* Auslösung *f*
t. pawl Auslöseklinke *f*
t. plunger Auslösekolben *m*
t. rod *(Fmt)* Bürstenwähler *m*
t. switch Auslöseschalter *m*, Betätigungsschalter *m*
t. valve Schnellschlußventil *n*
tripartite dreiteilig
tripeptide *(Ch)* Tripeptid *n*
triphase dreiphasig
triphenylmethane dye Triphenylmethanfarbstoff *m*
triplane 1. Spodumen *m*; 2. Dreidecker *m*
triple / to verdreifachen, dreifach sein
triple dreifach
t. arm signal dreiflügeliges Signal *n*
t.-articulated locomotive dreiteilige Lokomotive *f*
t. bollard *(Schiff)* Dreifachpoller *m*
t. bond *(Ch)* Dreifachbindung *f*
t. coincidence *(Kern)* Dreifachkoinzidenz *f*
t. coincidence counter Dreifachkoinzidenzzähler *m*
t. core cable dreiadriges Kabel *n*, Dreileiterkabel *n*
t. cylinder steam engine Drillingsdampfmaschine *f*

t.-detector reception *(Fmt)* Doppelsuperempfang *m*
t. diffused transistor dreifachdiffundierter Transistor *m*, Dreifachdiffusionstransistor *m*
t.-edge thread-grinding wheel dreiprofilige Schleifscheibe *f*
t. engine Drillingsmaschine *f*
t. expansion [steam] engine Dreifachexpansionsdampfmaschine *f*, dreizylindrige Verbunddampfmaschine *f* mit dreifacher Expansion
t. flue boiler Dreiflammrohrkessel *m*
t. gear chain block Stirnradflaschenzug *m*
t. grate bar dreifacher Roststab *m*, Trioroststab *m*
t. grid amplifier *(Am) (El)* Pentode *f*
t. helical gear Stirnrad *n* mit Doppelpfeilverzahnung
t. helical teeth Doppelpfeilverzahnung *f*
t. interlaced scanning *(Fs)* Dreifachzeilensprungabtastung *f*
t. jack *(Fmt)* Drillingsklinke *f*
t. modulation Dreifachmodulation *f*
t.-ply belt Dreifachriemen *m*
t. point Tripelpunkt *m* *(Zustandsdiagramm)*
t.-pole circuit breaker dreipoliger Schalter *m*
t.-pole single-throw switch dreipoliger Ausschalter *m*
t. riveted dreireihig genietet
t. riveting Dreireihennietung *f*
t. roller chain Dreifachrollenkette *f*
t. screw vessel Dreischraubendampfer *m*
t.-sector clutch hub Dreiarmnabe *f*
t. semaphore dreiflügeliges Signal *n*
t.-strand chain Dreifachkette *f*
t.-substituted *(Ch)* dreifach substituiert
t. tackle block Flaschenzug *m* mit drei Rollen
t. thread dreigängiges Gewinde *n*
t.-threaded dreigängig
t.-throw switch Dreifachumschalter *m*
t.-track bed Dreibahnenbett *n*
t.turret Drillingsturm *m* *(Kriegsschiff)*
t. valve Luftdrucksteuerventil *n*
tripler Frequenzverdreifacher *m*
triplet Triplett *n*
t. crystal Drillingskristall *m*
t. state Triplettzustand *m*
triplex board Triplexpappe *f* *(dreilagige Pappe)*
t. cable Dreileiterkabel *n*, dreipoliges Kabel *n*
t. chain Dreifachkette *f*
t. jib coal cutter Schrämmaschine *f* mit dreifachem Schrämarm
t. winding Dreischleifenwicklung *f*, Dreifachwicklung *f*, Dreilagenwicklung *f*
triplicate / to verdreifachen
triplicate dreifach
triplicity verdreifachter Zustand *m*, Verdreifachung *f*
triplite Triplit *m*, Manganeisenstein *m*, Manganpecherz *n*
tripod Dreibein *n*, Dreifuß *m*; Stativ *n*
t. head Stativkopf *m*
t. head plate Kopfplatte *f* des Stativs
t. leg Stativbein *n*
t. mast Dreibeinmast *m*
t. mounting Dreibeinlafette *f*
t. screw Stativschraube *f*
tripolar *(El)* dreipolig
t. aerial Tripolantenne *f*
Tripoli powder Tripel *m*; Tripelerde *f*
tripolite *s.* Tripoli powder
tripper 1. Auslöser *m*; 2. *(Brg)* Bandschleifenwagen *m*
tripping Auslösung *f*
t. battery Auslösebatterie *f*
t. circuit Auslösestromkreis *m*
t. device Auslöser *m*
t. fuse Sicherung *f* mit Auslösemechanismus, Auslösesicherung *f*, Sicherungsautomat *m*
t. latch Schaltklinke *f*
t. line *(Schiff)* Aufholleine *f*, Tripleine *f*
t. link *(Schiff)* Sliphaken *m*
t. plate Schaltplatte *f*
t. relay *(Am) (Fmt)* Rufabschaltrelais *n*
trirectangular dreifach rechtwinklig *(Geometrie)*
trisect / to dreiteilen
trisection Dreiteilung *f*
t. of an angle Winkeldreiteilung *f*
trisectrix Trisektrix *f*
tristimulus values Farbwerte *mpl (eines Lichtes)*
tritane *(Ch)* Triphenylmethan *n*, Tritan *n*
trithiocarbonic acid Trithiokohlensäure *f*
trithionic acid Trithionsäure *f*
tritium Tritium *n*, überschwerer Wasserstoff *m*
t. source *(Kern)* Tritiumquelle *f*
triton *(Kern)* Tritiumkern *m*, Triton *n*
triturate / to aufschlämmen, fein mahlen, verkneten, verreiben, zerreiben
trituration Verreibung *f*, Zerreibung *f*, Mahlung *f*
trityl Triphenylmethyl-
trivalence *(Ch)* Dreiwertigkeit *f*
trivalent *(Ch)* dreiwertig
trochoid *(Math)* Zykloide *f*
trochoidal analyzer Zykloidenanalysator *m*, Trochoidenanalysator *m*
trochotron *s.* t. tube
t. tube Trochotron *n* *(Magnetfeldröhre mit Schalt- bzw. Zählfunktion)*
trolley 1. Stangenstromabnehmer *m*; 2. Katze *f*, Laufkatze *f*; 3. Hunt *m*, Förderwagen *m*
t. base Stromabnehmerbock *m*
t. car *(Am)* Straßenbahntriebwagen *m*
t. frog Fahrdrahtweiche *f*
t. head Stromabnehmerkopf *m*
t. locomotive Fahrdrahtlokomotive *f*

t. pivot Drehachse *f* des Stangenstromabnehmers, Königszapfen *m*
t. pole Stromabnehmerstange *f*
t. shield Rollengabel *f (Stromabnehmer)*
t. tower Fahrdrahtmast *m (für Zweileiterbetrieb)*
t. wheel Stromabnehmerrolle *f*
t. wire Fahrdraht *m*, Oberleitung *f*
t. wire strainer Fahrdrahtspanner *m*
trolleybus Obus *m*, Trolleybus *m*
trombone 1. Posaune *f (Hohlleiterkreis, Wellenleiter)*; 2. Baßposaune *f*
trommel [screen] Siebtrommel *f*
trompe Gebläse *n*
tronics *(sl)* elektronische Buchungsmaschinen *fpl*
troop carrier Truppentransporter *m*, Truppentransportschiff *n*
t.-carrying glider Lastensegler *m*
t.-ship Truppentransporter *m*, Truppentransportschiff *n*
trooper *s.* troop-ship
troostite *(Min, Met)* Troostit *m*
tropic acid Tropasäure *f*
tropicalized tropengeschützt
tropics-proof tropensicher
troposphere Troposphäre *f*
t. wave troposphärische Welle *f*
tropospheric forward scatter troposphärische Streustrahlverbindung *f*
trot boat Routineboot *n*
trouble Störung *f*
t. bell Störungsglocke *f*
t.-free service störungsfreier Betrieb *m*
t.-location problem *(Dat)* Stör[ungs]suchaufgabe *f*
t. shooter Störungssucher *m (Person)*
t. shooting Fehlersuche *f*, Störungssuche *f*; Fehlereingrenzung *f*, Ortung *f* des Fehlers, Ortung der Störung
t. tone *(Fmt)* Störungszeichen *n*
t. tracking Störungssuche *f*, Fehlersuche *f*
trough 1. Trog *m*, Spänetrog *m*; Wanne *f*, Schale *f*, Behälter *m*; 2. Wellental *n*; Tiefpunkt *m (einer Zeitreihe)*
t. bridge Brücke *f* mit untenliegender Fahrbahn, Trogbrücke *f*
t. furnace Schalenfeuerung *f*
t. grate Muldenrost *m*
t. gutter Kastenrinne *f (Dachrinne)*
t. mixer Trogmischer *m*
t. shaped idler Muldenrolle *f*
t. share Muldenschar *n (Pflug)*
t. tipper Muldenkipper *m*
t. washer Muldenwäscher *m*, Rinnenwäscher *m*
troughed belt conveyor Muldenband *n*
t. chain conveyor Trogkettenförderer *m*
troughing idler Muldenrolle *f*

trowel / to [auf]spachteln, dick auftragen; glätten
trowel 1. Kelle *f*; Maurerkelle *f*; Glättkelle *f*; Spachtel *m (f)*; 2. Streichblech *n (Formerei)*
trowelling Glätten *n (mit Kelle)*; Aufspachteln *n*
troy pound Troy-Pound *n (= 12 troy ounces = 373,2418 g für Gold und Silber)*
t. weight Troysystem *n*, Massemaße *npl* für Gold und Silber
TR-tube Empfängersperröhre *f*
truck 1. *(Am)* Lastkraftwagen *m*, LKW; 2. offener Güterwagen *m*; 3. Untergestell *n*, Fahrgestell *n*; Drehgestell *n*; Rollbock *m*; 4. Karre *f*; Förderwagen *m*; 5. *(Schiff)* Flaggenknopf *m*, Mastknopf *m*, Klotje *f*, Wantklotje *f*
t. body LKW-Aufbau *m*
t. concrete mixer Mischerfahrzeug *n*, Transportmischer *m (Transportbeton)*
t. crane Autokran *m*, Mobilkran *m*
t. drier Hordentrockner *m*
t. farming Gemüsebau *m*
t. haulage Kraftwagenförderung *f*
t. ladle Wagengießpfanne *f*
t. lifter Wagenheber *m*
t. mixer Mischerfahrzeug *n*, Transportmischer *m (Transportbeton)*
t.-mounted crane Autokran *m*
t.-mounted workshop Werkstattwagen *m*
t. tractor Sattelschlepper *m*
t.-type switchgear Schaltwagen *m*
t. tyre LKW-Reifen *m*
trucking Flurförderung *f*, Lastwagenverkehr *m*
true / to ablehren; abdrehen, abrichten *(Schleifscheiben)*; auf richtiges Maß bringen; zentrieren; begradigen; auswuchten
t. up ausrichten; formgenau machen
true richtig, genau, maßgerecht, maßgenau, formgerecht, maßhaltig, paßgenau, richtungsgenau; unwuchtfrei; genau rund; rechtweisend; wahr; absolut; echt
t. airspeed Eigengeschwindigkeit *f (Flugzeug)*
t. [compass] bearing rechtweisende Peilung *f*
t. course rechtweisender Kurs *m*; Grundkurs *m (Radar)*
t. current Wirkstrom *m*
t. motion true motion *(Darstellung der wahren Schiffsbewegung; Radar)*
t. motion radar Radar *n* für absolute Bewegungsanzeige
t. ohm absolutes Ohm *n*
t. resistance echter Widerstand *m*, Echtwiderstand *m*
t. running [genauer] Rundlauf *m*
t.-running mit genauem Rundlauf
t.-running spindle accuracy Spindelrundlaufgenauigkeit *f*
t. solution *(Ch)* echte (molekulare) Lösung *f*

t. spinel Edelspinell *m*
t. to gauge lehrenhaltig
t. to nature naturgetreu
t. to shape formgerecht
t. to side *(Opt)* seitenrichtig
t. to size maßgerecht, maßhaltig, maßgenau
t. track *(Flg)* rechtweisender Kurs *m*
t. undersize recovery Feinkornausbringen *n*
t. watts *(El)* Wirkleistung *f*
t. within genau innerhalb *(Toleranz)*
trueness of the way Geradheit *f* der Führungsbahn
truing device Abdreheinrichtung *f*
t. the grinding wheel [in the manufacturing process] Abdrehen *n* der Schleifscheibe
trumpet Trompete *f*; Eingußtrichter *m (Gußblock)*; Schalltrichter *m (Lautsprecher)*; *(Kfz)* Signalhupe *f*
t. without piston Fanfare *f*
truncate / to abstechen; abflachen *(Gewinde)*; *(Math)* abstumpfen; abstützen; *(Dat)* abrunden
truncated cone Kegelstumpf *m*
t. picture *(Fs)* abgeschnittenes Bild *n*
t. right circular cylinder schief abgeschnittener gerader Kreiszylinder *m*
truncating Abflachung *f*; Abstumpfung *f*; Abrundung *f*
truncation Abrundung *f (von Gewinden an Spitze und Fuß)*; Abflachung *f*; Stutzen *n*, Abstechen *n*
t. error 1. *(Math)* Abbruchfehler *m*; 2. Fehler *m* durch Abbrechen einer Ziffernstellenreihe, Verfahrensfehler *m*
trundle / to rollen
trundle Rädchen *n*; Walze *f*; Karren *m*; Triebstockzahnrad *n*
t. head *(Schiff)* Spillkopf *m*
t. wheel Triebstockzahnrad *n*
trunk 1. Stamm *m (Baum)*; Schaft *m (Säule)*; 2. Hauptlinie *f*; Hauptleitung *f*; Fernleitung *f*; Verbindungsleitung *f*; 3. Koffer *m (Schiffsaufbau)*; 4. Kofferraum *m*; 5. Rüssel *m*; 6. *(Am) (Dat)* Kanal *m*, Sammelschiene *f*; Weg *m*; 7. *(Am)* Badehose *f*
t. buoy Vertäuboje *f*
t. cabin Kofferkajüte *f*
t. call Ferngespräch *n*
t. circuit with dialling facilities Fernwahlleitung *f*
t. congestion lamp Drängelampe *f* im Fernverkehr
t. congestion signal Anhäufungszeichen *n*
t. conveyor Sammelband *n*
t. deck Kofferdeck *n (Schiffsaufbau)*
t. deck vessel Kofferdeckschiff *n*, Trunkdeckschiff *n*
t. dialling Fernwahl *f (Fernleitung)*
t. directory inquiry Fernamtsauskunftstelle *f*
t. distribution frame Fernleitungshauptverteiler *m*
t. exchange Fernamt *n*
t. hunting Leitungsuchen *n*
t. junction jack Fernvermittlungsklinke *f*
t. lid Kofferraumdeckel *m*
t. line *(Dat)* Hauptlinie *f*, Sammelschiene *f*; *(Fmt)* Fernvermittlungsleitung *f*
t. line conduit Hauptkabelkanal *m*
t. loss Nutzdämpfung *f* der Fernleitung
t. operating supervision Fernbetriebsüberwachung *f*
t. pipe line Hauptrohrleitung *f*
t. piston Gleitschuhkolben *m*; Hohlkolben *m*; Tauchkolben *m*
t. road Fernverkehrsstraße *f*
t. service observation Fernbetriebsüberwachung *f*
t. switching scheme Schaltplan *m (Fernleitungen)*
t.-type engine Tauchkolbenmotor *m*
t. vessel Kofferdecker *m*, Trunkdecker *m*
t. zone cable Bezirkskabel *n (Fernsprechtechnik)*
trunked casing *(Schiff)* Schacht *m (über das Deck hinausgehend)*
t. hatchway *(Schiff)* Schachtluke *f*
trunking diagram *(Fmt)* Gruppenverbindungsplan *m*
trunnel Dübel *m*
trunnion Achse *f*, Auflager *n*, Drehzapfen *m*, Kurbelzapfen *m*, Lagerzapfen *m*, Stirnzapfen *m*, Tragzapfen *m*; Kupplungsbetätigungshebel *m*
t. axis height Kippachsenhöhe *f*
t. bearing Drehzapfenlager *n*
t.-mounted link Kulisse *f*, Kurbelschwinge *f*
t. screw Schildzapfenschraube *f*
t. tipping truck Zapfenkipper *m*
truss / to unterstützen, halten, befestigen; versteifen
truss 1. Binder *m*; Fachwerk *n (Flugzeug)*; Hängewerk *n*; 2. Bruchband *n*; 3. Heubündel *n (altes: 56 lbs oder frisches: 60 lbs)*; Strohbündel *n (36 lbs)*
t. bridge Fachwerkbrücke *f*
t. girder Fachwerkträger *m*
t. guyed pole *(Am) (Fmt)* Stange *f* mit Verspannung
t. head rivet Flachrundkopfniet *m*
t. head screw Halbrundschraube *f* mit flachem Kopf
t. joint Fachwerkknoten *m*, Knotenpunkt *m (eines Fachwerks)*
t. member Fachwerkstab *m*
trussed arch Fachwerkbogen *m*
t. brake *(Eis)* Dreieckbremse *f*
t. girder Fachwerkbinder *m*

t. girder bridge Fachwerkbalkenbrücke *f*
t. pole *(Fmt)* Stange *f* mit Verspannung
t. with sag rods *(Bw)* unterspannt
trussing of the frame *(Kfz)* Rahmenversteifung *f*
truth 1. Richtigkeit *f (Berechnung usw.)*; Wahrheit *f*; 2. Genauigkeit *f* der Form (Lage, Richtung)
t. table Wahrheitstafel *f (Tafel oder Tabelle für logische Funktionen)*
t.-functional expression Formel *f* für eine logische Funktion
try / to versuchen, experimentieren
try Versuch *m*, Experiment *n*
t. and mitre square Stahlwinkel *m* mit Gehrung
t. square Anschlagwinkel *m*
trying plane Langhobel *m*
trypaflavine Trypaflavin *n*
Tshebycheff's inequality Tschebyscheffsche Ungleichung *f*
TTT-diagram ZTU-Schaubild *n (Zeit-Temperatur-Umwandlungsschaubild)*
tub 1. Bottich *m*, Bütte *f*, Faß *n*; Klärbottich *m*; Kübel *m*; Küpe *f*; Wanne *f*; Badewanne *f*; 2. Förderwagen *m*, Hunt *m*; Fördereimer *m*; 3. *(sl)* Übungsboot *n (Ruderboot)*
t. cleaning *(Brg)* Wagenreinigung *f*
t. pusher *(Brg)* Vorziehvorrichtung *f*
t. sizing 1. Büttenleimung *f*; 2. Oberflächenleimung *f*
t. washed nach dem Scheren gewaschen *(Wolle)*
tuba Tuba *f*
tubbing *(Brg)* Tübbing *m*; Tübbingausbau *m*, Stahlringausbau *m*
tube 1. Rohr *n*; Tube *f*; 2. *(Am)* Elektronenröhre *f (s. a.* valve); 3. *(Opt)* Tubus *m*; 4. Untergrundbahn *f*; 5. Schlauch *m (Bereifung)*
t. bender Rohrbieger *m*
t. bending machine Rohrbiegemaschine *f*
t. bending press Rohrbiegepresse *f*
t. blank *(Met)* Rohrluppe *f*, Rohling *m*
t. boiler Röhrenkessel *m*
t. brush Rohrreinigungsbürste *f*
t. cleaner Kesselsteinabklopfer *m*
t. closing Rohrverschluß *m*
t. cold bending machine Rohrkaltbiegemaschine *f*
t. connection Rohrverbindung *f*
t. cooler Röhrenkühler *m*
t. cutter Rohr[ab]schneider *m*
t. drawing Rohrabziehen *n*
t. end plug Rohrverschluß *m*
t. expander Aufweitezange *f*
t. expanding press Rohraufweitpresse *f*
t. extension *(Opt)* Tubusverlängerung *f*
t. extruding press Rohr[strang]presse *f*
t. filter Pumpenkorb *m*
t. furnace Rohrofen *m*
t. fuse *(El)* Rohrsicherung *f*, Röhrchensicherung *f*, Hülsensicherung *f*
t. iron Hohleisen *n*
t. joint Rohrverbindung *f*
t. lamp Röhrenlampe *f*, Soffittenlampe *f*
t. lathe Rohrdrehmaschine *f*
t. loom Röhrchenwebstuhl *m*
t. mill Rohrmühle *f*
t. nest Rohrbündel *n*
t. plate *(Wärm)* Rohrboden *m*
t.-plate heating surface Rohrplattenheizfläche *f*
t. preheater Röhrenvorwärmer *m*
t. protector *(Wärm)* Schutzrohr *n*, Schutzhülse *f*
t. push-bench Rohrstoßbank *f*
t. railway Untergrundbahn *f*
t. rolling mill Rohrwalzwerk *n*, Röhrenwalzwerk *n*
t. rounds Halbzeug *n* für Röhren, Röhrenrundstahl *m*
t. scraper Rohrreiniger *m*, Rohrauskratzer *m*, Rohrschaber *m*
t. sheet Rohrplatte *f (Dampfkessel)*
t. shutter Verschluß *m* für Röntgenröhre[n]
t.-spring manometer *(BMSR)* Bourdonmanometer *n*, Röhrenfedermanometer *n*
t. stopper *(Wärm)* Rohrstopfer *m*
t.-works Berohrung *f*
tubeless tyre schlauchloser Reifen *m*
tubes strip winding machine Rohrhaspelmaschine *f*
t.-type plate Röhrchenplatte *f (Batterie)*
tubing 1. Berohrung *f*; Rohrleitung *f*; Rohrmaterial *n*; Schlauch *m*; Steigrohr *n*, Pumprohr *n*; Pumpenrohr *n*; 2. *(Met)* Rohrziehen *n*
t. catcher Pumpenrohrkrebs *m*
t. connection Rohrverbindung *f*
t. cutter Rohr[ab]schneider *m*
t. string Förderrohr *n*, Steigrohr *n (Tiefbohrtechnik)*
tubular Rohr-, röhrenförmig, rohrförmig; in Rohrkonstruktion
t. aerial Rohrantenne *f*
t. anode Hohlanode *f*
t. axle Rohrachse *f*, Hohlachse *f*
t. blade aus Rohr gepreßte Schaufel *f (Turbine)*
t. body Hohlkörper *m*
t. boiler Röhrenkessel *m*, Flammrohrkessel *m*
t. box spanner Vierkantsteckschlüssel *m*
t. bulb Röhrenlampe *f*
t. bulk cargo container *(Schiff)* Lastrohr *n (mehrere zusammengekoppelt ergeben ein Lastfloß)*
t. capacitor *(El)* Rohrkondensator *m*; Papierkondensator *m*

t. cell Stabbatterie *f*
t. compass *(Opt)* Röhrenbussole *f*
t. construction Rohr[rahmen]konstruktion *f (Kfz.-Chassis)*
t. cop spinning frame Schlauchkopsspinnmaschine *f*
t. core Hohlkern *m (einer Spule)*
t. cross member Rohrquerträger *m (Kfz.-Chassis)*
t. drier Röhrentrockner *m*
t. fabric Hohlgewebe *n*
t. feed heater Röhrenvorwärmer *m*
t. felt Rundfilz *m*
t. fluorescent lamp Leuchtstoffröhre[nlampe] *f*
t. frame Rohrrahmen *m (Kfz-Chassis)*
t. fuse Röhrensicherung *f*
t. inside micrometer verschiebbare Innenschraublehre *f*
t. insulation Rohrisolierung *f*
t. insulator Röhrenisolator *m*
t. knitted fabric Rundstrickware *f*, Schlauchgestrick *n*
t. knitted goods Rundstuhlware *f*
t. lamp Röhrenlampe *f*, Soffitte[nlampe] *f*
t. lamp holder Fassung *f* für Leuchtröhren, Soffittenfassung *f*
t. lamp wire Soffittenleitung *f*
t. level *(Opt)* Röhrenlibelle *f*
t. line *(Am) (El)* konzentrische Leitung *f*
t. magnet Röhrenmagnet *m*
t. mast Rohrmast *m*
t. planing machine Fallhobelmaschine *f (Zündholzfabrikation)*
t. prop *(Brg)* Rohrstempel *m*
t. radiator Röhrenheizkörper *m*; *(Kfz)* Röhrenkühler *m*
t. recuperative air heater Röhrenwärme[aus]tauscher *m*
t. rivet Hohlniet *m*
t. rotary kiln Drehrohrofen *m*
t. scaffolding Stahlrohrgerüst *n*
t. shaft Hohlwelle *f*; Gelenkrohrwelle *f*
t. shank Rohrschaft *m*
t. slide valve Rundschieber *m*; Kolbenschieber *m*
t. spirit level Röhrenlibelle *f*
t. sprayer Rohrzerstäuber *m*
t. steel Stahlrohr *n*, Röhrenstahl *m*
t. steel scaffolding Stahlrohrgerüst *n*
t. structure Rohrkonstruktion *f*
t.-type resistor *(El)* Rohrwiderstand *m*
t. valve Rundschieber *m*; Kolbenschieber *m*
t. vaporizer Röhrenverdampfer *m*
t. woven fabrics Schlauchgewebe *n*
tubulating cardboard for electric accounting machines Lochkartenkarton *m*
t. machine Stengelansatzmaschine *f (Elektronenröhrenherstellung)*
tubule Tubulus *m*; Röhrchen *n*, Kanüle *f*
tuck / to falten
tuck cement pointer Fugenkelle *f*
tucking fit Haftsitz *m*
t. position *(Text)* Fangstellung *f*
tufa Tuff[stein] *m*; Kalktuff *m*, Süßwassertuff *m (in Quellen)*
tufaceous earth Tufferde *f*
tuff Tuff *m (vulkanischen Ursprungs)*
tuft Faserbart *m*, Bart *m (Baumwolle)*; Haarbüschel *n*
tufting *(Text)* Noppensetzverfahren *n*
tug / to schleppen
tug Schlepper *m*, Schleppschiff *n*
t. and tow Schleppzug *m*, Schlepper *m* mit Schleppanhang
t.-barge combination Schlepper-Leichter-Verband *m*
t. chain Steuerkette *f (Wagen)*
t. charge Schlepplohn *m*, Bugsierlohn *m*
t. service Schleppdienst *m*, Bugsierdienst *m*; Hafenbugsierdienst *m*; Schleppschiffahrt *f*
tugboat Schleppboot *n*, Schlepper *m*
tugger *(Brg)* Preßlufthaspel *f*
tuition and research Lehre *f* und Forschung *f*
tulle Tüll *m*
tumble / to 1. trommeln, rommeln *(Guß)*; 2. *(Led)* walken; 3. stürzen, einfallen
t. home nach innen neigen
t. in einpassen; nach innen neigen
tumble *(Met)* Putztrommel *f*
t. polishing Polieren *n* in Poliertrommel
t. shore Hauptstopper *m (Stapellauf)*
tumbler 1. Gußputztrommel *f*, Scheuerfaß *n*, Rommelfaß *n*; 2. beschwerter Hebel *m*; 3. Trinkglas *n*; 4. *(Schiff)* Slipvorrichtung *f (Anker)*, Ankerschlipper *m*
t. bearing Pendellager *n*
t. gear Wendeherz *n*; Schwenkrad *n*
t. lock Zuhaltung *f*, Verschluß *m*
t. shaft Schwingenwelle *f (Nortongetriebekasten)*
t. switch Kipp[hebel]schalter *m*, Tumblerschalter *m*
t. yoke Hebelschwinge *f*, Nortonschwinge *f*
tumbling 1. Gußputzen *n* in Trommeln; 2. Aufbringen *n* von Überzügen in Lackiertrommeln
t. barrel 1. Gußputztrommel *f*; 2. Lackiertrommel *f*, Poliertrommel *f*, Trommelmühle *f*
t. bob beschwerter Hebel *m*
t. mixer Mischtrommel *f*, rotierender Trommelmischer *m*
tun 1. Bottich *m*; 2. 252 Gallonen *(Wein)*
t. fermentation Bottichgärung *f*
tundish *(Met)* Gießwanne *f*
tune / to 1. *(Ak)* stimmen; 2. *(El)* abstimmen
tuned 1. *(Ak)* gestimmt; 2. *(El)* abgestimmt
t. aerial abgestimmte Antenne *f*

t. amplifier abgestimmter Verstärker *m*, Resonanzverstärker *m*
t. anode circuit abgestimmter Anodenkreis *m*
t. anode coupling Schwingkreiskopplung *f*
t. buzzer abgestimmter Summer *m*
t. cavity abgestimmter Hohlraumresonator *m*
t. circuit abgestimmter Kreis (Stromkreis) *m*, Abstimmkreis *m*, Resonanzkreis *m*
t. circuit element Abstimmkreisglied *n*
t. grid abgestimmter Gitterkreis *m*
t.-grid tuned-anode oscillator Oszillator *m* mit abgestimmten Gitter- und Anodenkreisen; Huth-Kühn-Schaltung *f*
t.-grid tuned-plate oscillator Huth-Kühn-Schaltung *f*
t. intermediate circuit abgestimmter Zwischenkreis *m*
t.-not-tuned oscillator Oszillator *m* mit abgestimmtem Anodenkreis und unabgestimmtem Gitterkreis
t.-plate circuit abgestimmter Anodenkreis *m*
t.-reed indicator Vibrationstachometer *n*
t.-reed rectifier *(El)* Pendelgleichrichter *m*
t.-reed relay Resonanzrelais *n*
t. ringing *(Fmt)* abgestimmter Anruf *m*
t. to resonance auf Resonanz abgestimmt
t. transformer Resonanztransformator *m*

tuner 1. *(Fs)* Tuner *m*, Kanalwähler *m*; 2. Abstimmeinrichtung *f (für Antennen)*, Antennenabstimmgerät *n*; 3. Abstimmspule *f*

tung oil Tungöl *n*, chinesisches Holzöl *n*

tungstate Wolframat *n*, Salz *n* der Wolframsäure

tungsten Wolfram *n*
t. alloyed wolframlegiert *(Stahl)*
t. anode Wolframanode *f*
t. arc lamp Wolfram[licht]bogenlampe *f*
t. carbide Wolframkarbid *n*
t. carbide ball Wolframkarbidkugel *f*
t. carbide tipped wolframkarbidbestückt
t. carbide tool tip Wolframkarbidplättchen *n*, Wolframkarbidschneide *f*
t. chromium steel Chromwolframstahl *m*
t. concentrate Wolframkonzentrat *n*
t. contact Wolframkontakt *m*
t. disk Wolframscheibe *f (Röntgenröhren)*
t. electrode Wolframelektrode *f*
t.-electrode inert-gas-shielded arc welding Wolfram-Inertgasschweißen *n*
t. filament Wolframheizdraht *m*, Wolfram[glüh]faden *m*
t. filament lamp Wolframfadenlampe *f*
t. high speed steel Wolframschnellstahl *m*
t. lamp Wolframlampe *f*
t. rectifier Wolframgleichrichter *m*
t. steel Wolframstahl *m*

tungstic acid Wolframsäure *f*
t. anhydride Wolframsäureanhydrid *n*

tungstide Wolframid *n*

tuning 1. *(Ak)* Stimmen *n*; 2. *(El)* Abstimmung *f*, Abstimmen *n*; 3. Einstellen *n*
t. capacitor Abstimmkondensator *m*
t. circuit Abstimmkreis *m*
t. coil Abstimmspule *f*
t. control Abstimmknopf *m*
t. core Abstimmkern *m*
t. curve Durchstimmkurve *f*, Resonanzkurve *f*
t. dial Abstimmskala *f*
t. drift Abstimmänderung *f*
t. eye Abstimmauge *n*, Abstimmanzeigeröhre *f*
t. fork Stimmgabel *f*
t.-fork chronoscope schreibende Stimmgabel *f*
t.-fork clock stimmgabelgesteuerte Uhr *f*
t. hammer Stimmhammer *m*, Stimmschlüssel *m*
t. indicator Abstimmanzeiger *m*, magisches Auge *n*
t. inductance *(El)* Abstimmspule *f*
t. key Stimmhammer *m*, Stimmschlüssel *m*
t. knob Abstimmknopf *m*
t. means Abstimmittel *npl*
t. meter Meßinstrument *n* für Abstimmanzeige
t. method Abstimmverfahren *n*
t. note Abstimmton *m*
t. of carburetter Vergasereinstellung *f*
t. pipe Stimmpfeife *f (Orgel)*
t. plate Stimmplatte *f (Akkordeon)*
t. precision Abstimmschärfe *f*
t. range Abstimmbereich *m*
t. scale Abstimmskala *f*
t. switch Wellenschalter *m*
t. unit Abstimmittel *n*

tunnel 1. Tunnel *m*; Gang *m*; Kanal *m*; 2. *(Brg)* Stoll[e]n *m*, Strecke *f*; 3. Wellentunnel *m (für Schiffswellen)*
t. current Durchtunnelungsstrom *m*
t. diode *(El)* Tunneldiode *f*, Eskasi-Diode *f*
t. diode amplifier Tunneldiodenverstärker *m*
t. diode circuit Tunneldiodenschaltung *f*
t. drier Kanaltrockner *m*
t. driving *(Brg)* Streckenvortrieb *m*, Streckenauffahren *n*
t. drying oven Kanalofen *m*
t. effect Tunneleffekt *m*
t.-form crankcase Tunnelgehäuse *n*
t. invert *(Bw)* Tunnelsohle *f*
t. junction Tunnelübergang *m (Elektronik)*
t. kiln Kanalofen *m*, Tunnelofen *m*
t. kiln with cars Kanaltrockner *m* mit Wagen
t. kiln with racks Kanaltrockner *m* mit Gestellen
t. screw Tunnelpropeller *m*
t. shaft bearing Tunnelwellenlager *n*
t. stern Tunnelheck *n*

t.-stern tug *(Schiff)* Tunnelschlepper *m*
t. support *(Brg)* Streckenausbau *m*
t.-type radiator in Luftführung eingeschlossener Kühler *m*
t. vault *(Bw)* Tunnelgewölbe *n*
t. well *(Schiff)* Tunnelbrunnen *m*
t.-wound armature *(Elm)* gefädelter Anker *m*
tunnelling 1. *(Kern)* Tunneleffekt *m*, Durchtunnelung *f (Durchdringung des Potentialwalles)*; 2. *(Brg)* Streckenvortrieb *m*, Streckenauffahren *n*
t. current Durchtunnelungsstrom *m*
t. machine Streckenvortriebsmaschine *f*, Streckenauffahrmaschine *f*
t. resistance Durchtunnelungswiderstand *m*
tunneltron Tunneltron *n (elektronisches Bauelement)*
tup Bär *m*, Fallbär *m*; Fallgewicht *n*, Fallhammer *m (Materialprüfmaschinen, Rammgeräte)*
turbid trübe; schlammig; dickflüssig
t. glass prism Trübglasprisma *n*
turbidimeter Trübungsmesser *m*
turbidimetric analysis *s.* turbidimetry
turbidimetry Nephelometrie *f*, Trübungsmessung *f*
turbidity Trübe *f*, Trübung *f*, Schleier *m*, Streulicht *n*
turbine Turbine *f*
t. blade Laufschaufel *f*, Turbinenschaufel *f*
t. blower Turbinengebläse *n*
t. casing Schaufelzylinder *m*, Turbinengehäuse *n*
t. drier Turbinentrockner *m*
t. drilling Turbinenbohren *n*
t.-driven pump Pumpe *f* mit Turbinenantrieb
t.-driven set Turbogeneratorsatz *m*
t.-flowmeter Turbinen[durchfluß]zähler *m*; Pottermeter *n*, Woltman-Zähler *m*
t. foundation Turbinenfundament *n*
t. mixer Schaufelradmischer *m*, Turbinenmischer *m*, Turbomischer *m*
t. nozzle blade Turbinendüsenschaufel *f*
t. rotor Turbinenrotor *m*, Laufrad *n*
t. steamer Turbinendampfer *m*
t. torpedo boat Turbinentorpedoboot *n*
t. unit Turbinensatz *m*
t. wheel Laufrad *n*
t. with horizontal shaft Horizontalturbine *f*
t. with nozzles and blades rotating in opposite directions gegenläufige Turbine *f*, Gegenlaufturbine *f*
t. with rimstraddling buckets Reiterschaufelturbine *f*
t. with vertical shaft Vertikalturbine *f*
t. working condensing mit Kondensation arbeitende Turbine *f*
t. working non-condensing mit Auspuff arbeitende Turbine *f*
turbo-alternator Turboalternator *m*
t.-blower Kreiselgebläse *n*, Turbogebläse *n*
t.-charged Diesel engine Dieselmotor *m* mit Turboaufladung, turboaufgeladener Dieselmotor
t.-dynamo Turbodynamo *m*, Turbinendynamo *m*
t.-electric turboelektrisch
t.-electric drive turboelektrischer Antrieb *m*
t.-expander Entspannungsturbine *f*
t.-gas absorber Turbomischer *m* mit Gasabsorption
t.-generator Turbogenerator *m*
t.-generator set Turbogeneratorsatz *m*, Turbogeneratorgruppe *f*
t.-jet Strahlturbine *f*
t.-jet engine Turbostrahltriebwerk *n*
t.-jet fan Doppelstromturbotriebwerk *n*
t.-liner Strahlverkehrsflugzeug *n*, Verkehrsflugzeug *n*
t.-oil Turbinenöl *n*
t.-prop engine Turboproptriebwerk *n*
t.-supercharger Abgasturbolader *m*, Turbokompressor *m*
turboblown engine Motor *m* mit Abgasturbolader
turbocharge / to mit Abgasturbolader aufladen
turbocharger Turbolader *m*
turbocharging Turboaufladung *f*
turbojet *s.* turbo-jet
turbomachinery Strömungsmaschine *f*
turboprop engine Propellerturbine *f*, PTL
turbulator *(Am) (El)* Querstromlöschkammer *f*
turbulence Turbulenz *f*, Wirbelung *f*
t. chamber Wirbelkammer *f*
t. head Wirbelkopf *m*, Brennraum *m* mit besonders guter Wirbelung *(Ottomotor)*
t. rod *(Schiff)* Stolperdraht *m (bei Modellschleppversuchen)*
turbulent wirbelnd, turbulent
t. boundary layer *(Strm)* turbulente Grenzschicht *f*
t. burner Wirbelbrenner *m*
t. flow turbulente Strömung *f*
t. fluidized bed turbulente Wirbelschicht *f*
t. friction *(Strm)* turbulente Reibung *f*
t. motion Turbulenzbewegung *f*
turf Muttererde *f*; Rasensode *f*; Torf *m (Irland)*
t. cutter Rasenpflug *m*
t. drain Abzugskanal *m (Torfmoor)*
t. spade Abstechschaufel *f*, Torfstechschaufel *f*
t. spud Abstechschaufel *f*, Torfstechschaufel *f*
Turkey-blue Türkischblau *n*
T.-brown Türkischbraun *n*
T.-red Türkischrot *n*, Alizarinrot *n*
T.-red oil Türkischrotöl *n*
turmalin[e] *s.* tourmalin[e]

turmeric test paper Kurkumapapier *n*
turn / to 1. wenden; kreisen; 2. *(Masch)* drehen; sich drehen; 3. gerinnen, sauer werden; 4. umschlagen *(Beize)*; 5. *(Flg)* kurven
t. a shoulder einen Absatz drehen
t. a stepped shaft Absätze an einer Welle drehen
t. bottom up kentern *(Schiff)*
t. cylindrically along the outside überdrehen
t. dark schwarz werden
t. first diameter andrehen
t. inside diameters ausdrehen
t. off abdrehen *(vom Kurs)*; *(Flg)* abkurven; abdrehen *(Drehmaschine)*; ausschalten *(Licht)*
t. on andrehen, einschalten *(Licht)*
t. on the chuck lathe aus freier Hand drechseln
t. out abzweigen *(Gleise)*
t. outside diameter[s] überdrehen
t. over umwenden; umkippen; kentern; umlegen *(Fördermittel)*
t. round umdrehen, umlaufen
t. the crank kurbeln
t. turtle kentern *(ein Schiff)*
t. up a converter einen Konverter aufrichten
t. wood drechseln
turn Windung *f (einer Wicklung)*; Drehung *f*, Wendung *f*; Umdrehung *f*
t.-about 1. Umdrehung *f*, Wendung *f*; 2. Hafentörn *m (Hafenaufenthalt zwischen Ankunft und Abfahrt z. B. einer Fähre)*; Gegenkurs *m*
t.-around Schiffsumlauf *m*, Hafenabfertigung *f*
t. bridge Drehbrücke *f*
t.-buckle *s.* turnbuckle
t. control Richtungsgeber *m (Kurssteuerung)*
t. of bilge *(Schiff)* Kimm[rundung] *f*
t. of the propeller *(Schiff)* Propellerschlag *m (beim Manövrieren)*
t.-out Abzweigung *f (Schienen)*; Ausstoß *m*
t.-out track *(Eis)* Stichgleis *n*
t.-over Umführung *f (Walzwerk)*
t.-over fixture *(Masch)* Wendevorrichtung *f (Taktstraße)*
t.-over plough Unterwendepflug *m*
t. ratio *(El)* Windungsverhältnis *n*
t.-round Umschlag *m*, Hafenumschlag *m*
t. signals Fahrtrichtungsanzeiger *m*
t. switch Drehschalter *m*, Knebelschalter *m*
t.-table 1. Drehscheibe *f*, Drehtisch *m*; 2. Plattenspieler *m*
t.-table on ball bearings *(Eis)* Kugeldrehscheibe *f*
t.-table on roller bearings *(Eis)* Rollendrehscheibe *f*
t.-table press Revolverpresse *f*
t.-table with centre bearing *(Eis)* Zapfendrehscheibe *f*
t.-table with four-way-crossing *(Eis)* Kletterdrehscheibe *f*
t.-wrest plough Unterwendepflug *m*
turnback strap Rückenriemen *m* des Pferdes
turnbarrel Spannband *n*
turnbuckle Spannschloß *n*
turndown *(BMSR)* Betriebsstellbereich *m (Ventil)*
turned *(Schiff)* kieloben
t. bate *(Led)* Schwarzbeize *f*
t. letter *(Typ)* Blockade *f*
t. part Drehteil *n*
t.-vertical shaft *(Brg)* Seigerschacht *m*, dessen unteres Ende tonnlägig fortgesetzt ist *(mit einer gemeinsamen Fördereinrichtung)*
turner Dreher *m*
t. apprentice Dreherlehrling *m*
t. bar *(Typ)* Wendestange *f*
turnery Drechslerarbeit *f*; Drechslerei *f*
turning 1. Drehen *n*; Drehspan *m*; 2. Bogenmauern *n*; 3. Wendung *f*
t. accuracy Drehgenauigkeit *f*
t. attachment Dreheinrichtung *f*
t. bar *(Typ)* Wendestange *f*
t. basin *(Schiff)* Wendebecken *n*
t. blank gebogener Schichtholzformling *m*
t.-block linkage umlaufende Kurbelschleife *f*
t. bridge *(Schiff)* Drehbrücke *f*
t. circle Wendekreis *m*; Drehkreis *m (beim Schiffswendemanöver)*
t. down a converter Kippen *n* eines Konverters
t. external diameters Außendrehen *n*
t. gouge Drechslerrohr *n*
t. inside diameters Ausdrehen *n*
t. internal surfaces Ausdrehen *n*, Innendrehen *n*
t. jaw Drehkloben *m*, Spannkloben *m*
t. job Dreharbeit *f*
t. lathe Drehmaschine *f*
t. length Drehlänge *f*
t. machine 1. Wendemaschine *f*; 2. Drehmaschine *f*
t. moment Drehmoment *n*
t. off centre Außermittedrehen *n*
t. operation Dreharbeit *f*
t.-over station Wendestation *f*
t. platform mounting Drehscheibenlafette *f (Geschütz)*
t. point of a curve Rückkehrpunkt *m* einer Kurve
t. radius Wenderadius *m (Fahrzeuge)*
t. saw for unhinging Aushängesäge *f*
t. speed Drehzahl *f*
t. station Wendestation *f (Fließstraße)*
t. tool Drehmeißel *m*

t.-tool holder Drehmeißelhalter *m*
t. trial *(Schiff)* Rudermanövrierprobe *f*
t. vane Umlenkschaufel *f*, Leitschaufel *f*
turnings Drehspäne *mpl*
turnip cutter Rübenschneidemaschine *f*; Rübenstecher *m*
t. cutter blade Rübenmesser *n*
t. harvester Rübenroder *m*, Rübenstecher *m*, Rübenerntegerät *n*
t. pulper Rübenstampfmaschine *f*
turnover 1. *(Ök)* Umsatz *m*, Umschlag *m*; 2. *(Schiff)* eingezogene Bordwand *f*; 3. *(Fs)* Umpoleffekt *m*; Umpolung *f*
turnpike *(Am)* gebührenpflichtige Autobahn *f*
turnround Umlauf *m*, Umsatz *m*
turns of spindle per minute *(Text)* Anzahl *f* der Spindeldrehungen je Minute
t. per inch *(El)* Windungen *fpl* je Zoll *(Spulenlänge)*
t. ratio *(El)* Windungsverhältnis *n* *(Spulen)*
turnstile *(Bw)* Drehkreuz *n*
t. aerial Schmetterlingsantenne *f*
t. dipol *(El)* Kreuzdipol *m*
turntable *s.* turn-table
turpentine Terpentinöl *n*
turquoise *(Min)* Türkis *m*
turret 1. *(Masch)* Revolver[kopf] *m*; 2. Turm *m*; Drehkuppel *f*
t. automatic Revolverdrehautomat *m*
t. block Revolverkopf *m*
t. bushing Spannhülse *f* *(Revolverkopf)*
t. carriage Revolverkopfschlitten *m*
t. coal-cutter *(Brg)* Turmschrämmaschine *f*
t. crane Turmdrehkran *m*
t.-deck vessel Turmdeckschiff *n*, Turmdecker *m*
t. drilling machine Revolverkopfbohrmaschine *f*
t. gun Turmgeschütz *n*
t. hatch Turmluke *f*
t. head Revolverkopf *m*
t. jib *(Brg)* Hangendschrämarm *m*
t. lathe Revolverdrehmaschine *f* [mit auf Bett verschiebbarem Revolverschlitten]
t. overcutter *(Brg)* Turmschrämmaschine *f*
t. position stop Anschlagwalze *f*
t. punch press Revolverpresse *f*
t. slide Revolverwerkzeugschlitten *m*
t. switch Trommelschalter *m*
t. tool Revolverkopfwerkzeug *n*
t. tool post Revolverkopf *m*
t.-type drill head Revolverbohrkopf *m*
turtle-back deck *s.* t. deck
t.-back forecastle *(Schiff)* gewölbte Back *f*
t.-backed stern *(Schiff)* Biberschwanzheck *n*
t. deck *(Schiff)* Schildkrötendeck *n*; Walrükkendeck *n*; rundes Deck *n*
tussah silk wilde Seide *f*, Tussahseide *f*
t. silkworm Tussahspinner *m*
tussur silk *s.* tussah silk
tutty Zinkschwamm *m*, Gichtschwamm *m*
tuyere *(Met)* Düse *f*; Blasform *f*, Windform *f*
t. belt Formenzonenring *m* *(Hochofen)*
t. cooling plate Windformkühlkasten *m* *(Hochofen)*
t. level Blasformebene *f*
t. zone Verbrennungszone *f* *(Kupolofen)*
TV demodulator Nyquist-Demodulator *m*
TV picture Fernsehbild *n*
TV picture monitor Bildkontrollempfänger *m*
'tween deck *(Schiff)* Zwischendeck *n*
't.-deck hatchcover Zwischendecklukendeckel *m*
't.-deck tonnage Zwischendeckvermessung *f*
tweeter Hochtonlautsprecher *m*
t. loudspeaker Hochtonlautsprecher *m*
tweezers kleine Kneifzange *f*
twelve-point box wrench Zwölfkantringschlüssel *m*
t.-point socket Zwölfkanteinsatz *m* *(Steckschlüssel)*
t. to pica *(Typ)* ein Punkt *(etwa ein Achtelpetit; Schriftgröße)*
twenty-four hour service *(Am)* *(Fmt)* durchgehender Dienst *m*
twice-normal solution *(Ch)* zweinormale Lösung *f*
t.-substituted *(Ch)* zweifach substituiert
twilight zone Mischzone *f* *(Flugfunknavigation)*
twill *(Text)* geköperter Stoff *m*, Köper *m*
t. fabric weave Köperbindung *f*
t. interlacing gratartige Bindung *f*
t. line Köpergrat *m*
t. weave *(Text)* Köperbindung *f*
twilling bars under the comber board *(Text)* Hebestäbe *mpl* unter dem Schnürbrett
twin / to verzwillingen
twin Zwilling[skristall] *m*, Kristallzwilling *m*
t. aerial Zwillingsantenne *f*, Doppelantenne *f*, Antennenpaar *n*
t.-bank engine Doppelreihenmotor *m*
t.-barreled anti-aircraft gun Zwillingsflak *f*
t.-barreled machine gun Zwillingsmaschinengewehr *n*, Zwillings-MG *n*
t.-boom aircraft Doppelrumpfflugzeug *n*
t. boundary Zwillingsgrenze *f*
t.-briquetting press Zwillingsbrikettpresse *f*
t.-bulkhead tanker Dreitankschiff *n*, Zweischottentanker *m* *(Tanker mit zwei Längsschotten)*
t. cable Zweileiterkabel *n*, Zweifachkabel *n*, Zwillingskabel *n*, doppeladriges (paar-verseiltes) Kabel *n*
t.-cam flat knitting machine Doppelschloßflachstrickmaschine *f*

t. carbon arc lamp Doppel[kohlen]bogenlampe *f*
t. carburetter Doppelvergaser *m*; Zwillingsvergaser *m*
t. chain Zwillingskette *f*, Zweifachkette *f*
t. chamber furnace Zweikammerofen *m*
t. check *(Dat)* Dup-Kontrolle *f*; Zwillingsprüfung *f (Verdopplung der Bauelemente und Vergleich der Ergebnisse)*
t. column Doppelsäule *f*, Doppelstütze *f (z. B. an Dehnungsfugen)*
t. conductor Doppelleiter *m*
t. cone ring Doppelkegelring *m*
t. contact wires Doppelfahrdraht *m*, Trolleyleitung *f*
t.-core cable Zweileiterkabel *n*, zweiadriges Kabel *n*
t. Cornish boiler Flammrohrdoppelkessel *m*, Zwillingskessel *m*
t.-coupled capacitor *(El)* Zweifachkondensator *m*
t. crystal Zwillingskristall *m*, Kristallzwilling *m*
t.-cylinder mixer Zwillingstrommelmischer *m*
t. digital store Doppeldigitalspeicher *m*
t. diode Duodiode *f* mit gemeinsamer Katode
t.-disk clutch Zweischeibenkupplung *f*
t. downdraught carburetter Doppelfallstromvergaser *m*
t. electron Elektronenzwilling *m*
t. engine Zweizylindermotor *m*
t.-engined zweimotorig
t.-feed circular hosiery machine Zweisystemstrumpfrundstrickmaschine *f (für nahtlose Damenstrümpfe)*
t. feeder Paralleldrahtleitung *f*
t. flex *(Am)* doppeladrige Litze *f*
t. flexible cord doppeladrige Litze *f*, Doppellitze *f*
t.-flued oven Zwillingsofen *m*
t. jib Doppelschrämarm *m*, Doppelschrämausleger *m*
t.-lens reflex camera zweiäugige Spiegelreflexkamera *f*
t. machine gun Zwillingsmaschinengewehr *n*, Zwillings-MG *n*
t. mushroom *(Brg)* Zwillingsschrämpilz *m*
t.-panel tube *(Fs)* Bildröhre *f* mit direkt auflamierter Schutzscheibe
t. plane Zwillingsebene *f*
t.-plate clutch Zweischeibenkupplung *f*
t.-pointer fuel meter Doppelzeigerbrennstoffmeßuhr *f*
t.-pump with belt drive *(Wärm)* Zwillingsriemenpumpe *f*
t.-row potato harvester Zweireihenkartoffelerntemaschine *f*
t. rudder *(Schiff)* Doppelruder *n*
t. saddle Wechselschlitten *m (Stauchautomat)*
t.-screw Doppelschrauben-; Doppelschnecken-
t.-screw passenger ship Doppelschraubenfahrgastschiff *n*, Zweischraubenfahrgastschiff *n*
t. screw propulsion Doppelschraubenantrieb *m*, Zweischraubenantrieb *m*
t.-screw vessel Doppelschraubenschiff *n*, Zweischraubenschiff *n*
t. screws *(Schiff)* Doppelschrauben *fpl*
t.-sector clutch hub Zweiarmnabe *f*
t. shaft *(Brg)* Doppelschacht *m*
t. ships Schwesterschiffe *npl*
t. simplex pump Zwillingssimplexpumpe *f*
t. steam engine Zwillingsdampfmaschine *f*
t. strainer Doppelfilter *n*
t. tandem engine Zwillingstandemdampfmaschine *f*
t. tyre Zwillingsbereifung *f*
t. wheel Zwillingsrad *n*, Doppelrad *n*
t. work-holding fixture Doppelwerkstückträger *m*
t.-worm mixer Doppelschneckenmischer *m*
twinax cable *s.* twin cable
twine / to [ver]zwirnen; zusammendrehen
twine Zwirn *m*; Kordel *f*; Bindfaden *m*; Abbindegarn *n*; Kettenzug *m*
t. holder Bindfadenbrücke *f (Mähbinder)*
twiner mule *(Text)* Zwirnselfaktor *m*
twines and ropes Seile *npl* und Taue *npl*
twinned verzwillingt *(Kristall)*
twinning 1. *(Fs)* Paarigkeit *f*; Paarigstehen *n* der Zeilen; 2. Zwillingsbildung *f (Kristall)*
twist / to verdrillen, verdrallen, [sich] verdrehen, verwinden; auf Verdrehung beanspruchen; *(Text)* [ver]zwirnen, flechten
t. on andrehen *(Faden)*
t. together verdrillen, zusammendrehen, verseilen
twist 1. Verdrehung *f*, Torsion *f*; Verdrillung *f*, Verwindung *f*; Drall *m*; 2. *(El)* tordierter (geschränkter) Hohlleiter *m*; Hohlleiterwendelstück *n (Wellenleiter)*; 3. *(Text)* Drehung *f*; Zwirn *m*
t. bit Holzbohrer *m*
t. change wheel Drehungswechselrad *n (Spinnerei)*
t. drill Spiralbohrer *m*, Wendelbohrer *m*
t. drill centre grinding attachment Spiralbohrerspitzenschleifapparat *m*
t. drill cutter [Spiralbohrer-]Nutenfräser *m*
t. drill grinding attachment Spiralbohrerschleifeinrichtung *f*
t. drill grinding machine Spiralbohrerschleifmaschine *f*
t. drill milling machine Spiralbohrerfräsmaschine *f*
t. drill nomenclature Spiralbohrerbenennungen *fpl*

t. drill point grinding attachment Schleifeinrichtung *f* für Spiralbohrer, Spiralbohrerspitzenschleifeinrichtung *f*
t. drill point grinding machine Spiralbohrerspitzenschleifmaschine *f*
t. drill point sharpening machine Spiralbohrerspitzenschleifmaschine *f*
t. drill relief grinding machine Spiralbohrerhinterschleifmaschine *f*
t. drill shank Spiralbohrerschaft *m*
t. drill sharpening machine Spiralbohrerschleifmaschine *f*
t. drill with exterior oil channel Spiralbohrer *m* mit äußerer Ölzuführung
t. factor Drehungsfaktor *m*, Zwirnfaktor *m* *(Zwirnerei)*
t. fluted drallgenutet
t. gimlet Nagelbohrer *m* mit Kegelschaft *(Holz)*
t.-hand *(Am) (Text)* Spitzenhersteller *m*
t. joint *(El)* Würgeverbindung *f*, Würgebund *m*
t. lock connection Bajonettverbindung *f*
t. motion *(Text)* Drahtzähler *m*, Drehungszähler *m*
t. multiplier Drehungsfaktor *m*, Zwirnfaktor *m*
t. nozzle *(Rak)* Dralldüse *f*
twisted 1. verdreht, verseilt, verdrallt; 2. krumm, gekrümmt; schraubenförmig, gewunden, verwunden; 3. auf Torsion beansprucht
t. cable verseiltes Kabel *n*
t. drill gewundener Bohrer *m*
t. growth *(Holz)* Drehwuchs *m*
t. hemp Hanfzopf *m*
t. jute strick zopfförmig zusammengedrehte Juteriste *f*
t. magnetic field verdrilltes Magnetfeld *n*
t. pair verdrillte Doppellitze *f* *(Kabel)*
t. silk kardonnierte Seide *f*
t. strap gedrehter Riemen *m*
t. wire verseilter Draht *m*
t. yarn Zwirn *m*
twister Zwirnmaschine *f*
t. ring Zwirnring *m*
t. spindle Zwirnspindel *f*
twisting 1. Verwinden *n*, Verdrehen *n*, Verdrallen *n*; 2. *(Text)* Verzwirnen *n*, Zwirnen *n*; Verdrehungsneigung *f* *(Schmaltextilien)*; 3. Biegung *f*, Krümmung *f* *(des Holzes durch Trocknen)*
t. angle Verdrehungswinkel *m* *(Festigkeit)*
t. bobbin Zwirnspule *f*
t. flyer Zwirnflügel *m*
t. force Drehkraft *f*; Verdreh[ungs]kraft *f*
t. frame with several tiers Etagenzwirnmaschine *f*
t. hook *(Text)* Anschirrhaken *m*
t.-in *(Text)* Andrehen *n*
t.-in frame Andrehapparat *m*
t. machine Zwirnmaschine *f*
t. moment Drillmoment *n*, Torsionsmoment *n*, Verdrehungsmoment *n* *(Festigkeit)*
t. paper Wachsrohpapier *n*
t. resistance Torsionswiderstand *m*
t. spindle Zwirnspindel *f*
t. strength Verdrehungsfestigkeit *f*, Torsionsfestigkeit *f*
t. stress Verdrehungsbeanspruchung *f*, Torsionsbeanspruchung *f*
t. test *(Met)* Torsionsversuch *m*, Drehversuch *m*, Verdrehungsversuch *m*
t. vibration Torsionsschwingung *f* *(Spektroskopie)*
twistor Twistor *m* *(Einzeldrahtdatenspeicher)*
two-address system *(Dat)* Zweiadressensystem *n*
two-armed zweiarmig, doppelarmig
two-armed flange *(Kfz)* Zweiarmflansch *m*
two-armed lever zweiarmiger Hebel *m*
two-bay *(Bw)* Zweifeld-, zweifeldrig; zweischiffig
two-bay aerial Zweiebenenantenne *f*
two-bay frame Zweifeldrahmen *m*
two-beam interferences Zweistrahlinterferenzen *fpl*
two-belt drive Doppelriemenantrieb *m*
two berth cabin *(Schiff)* Doppelkabine *f*
two-bladed floating reamer Pendelmesserreibahle *f*
two-body decay *(Kern)* Zweiteilchenzerfall *m*
two-body forces Zweikörperkräfte *fpl*
two-body problem Zweikörperproblem *n*
two-card set *(Text)* Zweikrempelsatz *m*
two-chamber brake cylinder *(Kfz)* Zweikammerbremszylinder *m*
two-channel amplifier Zweikanalverstärker *m*
two-circuit receiver Zweikreisempfänger *m*, Zweikreiser *m*
two-circuit tuner *(Fmt)* Gleichlaufabstimmelement *n* von zwei Kreisen
two-circuit tuning Zweikreisabstimmung *f*
two-circuit turbine jet unit Doppelstromturbostrahltriebwerk *n*
two-coat work on walls zweilagiger Putz *m*
two-colour zweifarbig, Zweifarben-
two-column machine Zweiständermaschine *f*, Doppelständermaschine *f*
two-compartment jig Zweibettsetzmaschine *f* *(Aufbereitung)*
two-compartment ship Zweiraumschiff *n* *(durch Schotte so unterteilt, daß es bei Leck nicht kentert)*
two-contact plug Zweipolstecker *m*
two-core zweiadrig
two-core cable Zweileiterkabel *n*
two-cycle engine Zweitaktmotor *m*

two-cylinder steam engine Zweizylinderdampfmaschine *f*
two-decked vessel Zweideckschiff *n*
two-decker Zweidecker *m*
two-digit group *(Dat)* Bigramm *n*
two-dimensional zweidimensional, eben
two-dimensional lattice Flächengitter *n (Kristalle)*
two-direction ball thrust bearing zweiseitig wirkendes Axialrillenkugellager *n*
two-direction roller bearing Führungsrollenlager *n*
two-direction thrust bearing zweiseitig wirkendes Axiallager *n*
two-doffer card *(Text)* Zweisammlerkrempel *f*
two-doffer two-card set Doppelflorzweikrempelsatz *m*
two-edged zweischneidig
two-electrode rectification Diodengleichrichtung *f*
two-electrode [vacuum] valve Zweipolröhre *f*, Diode *f*, Richtröhre *f*, Ventilröhre *f*
two-electron problem Zweielektronenproblem *n*
two-electron shell *(Ph)* K-Schale *f*, Zweierschale *f*
two-eyed *(Opt)* beidäugig
two-family *(Math)* zweischarig
two-figure number zweistellige Zahl *f*
two-filament bulb Doppelfadenlampe *f*, Zweifadenlampe *f*
two-fluid cell *(El)* Voltaelement *n*
two-flute zweinutig
two-flute cutter Zweischneider *m*
two-flute end mill Langlochfräser *m (Zweischneider)*
two-flute slot drill Finger- oder Nutenfräser *m*, Langlochfräser *m (Zweischneider)*
two-for-one twister DD-Zwirnmaschine *f*, Doppeldrahtzwirnmaschine *f*
two-frequency signalling *(Fmt)* Tonfrequenzzeichengabe *f* unter Anwendung zweier Frequenzen
two-furrow plough Zweifurchenpflug *m*
two-gang capacitor *(El)* Zweifachkondensator *m*
two-high reversing mill Reversierduostraße *f*, Reversierduowalzwerk *n*
two-high rolling mill Duowalzwerk *n*
two-hinged arch *(Bw)* Zweigelenkbogen *m*
two-hinged frame *(Bw)* Zweigelenkrahmen *m*
two-housing *(Masch)* Zweiständer-
two-image Zweibild-
two-jack chuck Zweibackenfutter *n*
two-lane zweispurig *(Straßenbau)*
two-leaf zweiflügelig *(Tür)*
two-leaf cavity wall zweischalige Hohlmauer *f*, zweischaliges Hohlmauerwerk *n*
two-leg propeller strut *(Schiff)* zweiarmiger Wellenbock *m*
two-line brevier *(Typ)* 16 britische Punkte *(Schriftgröße)*
two-line great primer *(etwa)* Doppeltertia *(Schriftgröße)*
two-line paragon *(etwa)* Kanon *(Schriftgröße)*
two-lip countersink zweischneidiges Senkwerkzeug *n*
two-lippe *s.* two-lipped
two-lipped zweischneidig
two-lipped end mill Langlochfräser *m*, Finger- oder Nutenfräser *m (Zweischneider)*
two-lipped slot drill Langlochfräser *m*, Finger- oder Nutenfräser *m (Zweischneider)*
two-lipped twist drill Spiralbohrer *m*
two-mandrel method Zweidornverfahren *n (Abnahmeprüfung)*
two-masted ship Zweimastschiff *n*, Zweimaster *m*
two-master *s.* two-masted ship
two-men saw Zugsäge *f*
two-motion selector *(Fmt)* Hebdrehwähler *m*
two-needle vamping flatbed sewing machine Doppelnahtmaschine *f (Schuhfertigung)*
two-node vibration Zweiknotenschwingung *f*
two-oared shell with coxswain Rennzweier *m* mit Steuermann
two-parameter family *(Math)* Zweiparameterschar *f*
two-part bearing *(Masch)* zweiteiliges Lager *n*
two-part fastening ring *(Masch)* zweiteiliger Befestigungsring *m*
two-part flask zweiteiliger Formkasten *m (Gießerei)*
two-part packing cone *(Masch)* zweiteiliger Dichtungskegel *m*
two-pass broaching operation Räumarbeit *f* in zwei Durchgängen
two-phase zweiphasig
two-phase alternating current Zweiphasen[wechsel]strom *m*
two-phase alternator Zweiphasengenerator *m*
two-phase armature Zweiphasenanker *m*
two-phase cable Zweiphasenkabel *n*
two-phase circuit Zweiphasenschaltung *f*
two-phase connection Zweiphasenschaltung *f*
two-phase current Zweiphasen[wechsel]strom *m*
two-phase dosimeter *(Kern)* Zweiphasendosimeter *n*
two-phase equilibrium Zweiphasengleichgewicht *n*
two-phase four-wire system unverkettetes Zweiphasensystem *n*, Zweiphasen-Vierleitersystem *n*
two-phase induction motor Zweiphaseninduktionsmotor *m*
two-phase line Zweiphasenleitung *f*
two-phase meter Zweiphasenzähler *m*
two-phase modulation *(Fs)* Zweiphasenmodulation *f*

two-phase reactor Zweiphasenreaktor *m*
two-phase three-wire connection Zweiphasenschaltung *f (im Drehstromsystem)*, Zweiphasen-Dreileiterschaltung *f*
two-phase transformer Zweiphasentransformator *m*
two-phase winding Zweiphasenwicklung *f*
two-piece zweiteilig, geteilt
two-piece ring geteilter Ring *m (Lager)*
two-pin plug zweipoliger Stecker *m*
two-pinned arch *(Bw)* Zweigelenkbogen *m*
two-place zweistellig
two-plane winding *(El)* zweietagige Wicklung *f*
two-ply belt Doppeltreibriemen *m*
two-ply board Duplexpappe *f (zweilagige Pappe)*
two-ply yarn *(Am)* Zweifachgarn *n*
two-point attachment Zweipunktaufhängung *f*
two-point contact bearing Zweipunktlager *n*
two-pole armature *(Elm)* zweipoliger Anker *m*, Doppel-T-Anker *m*
two-pole switch zweipoliger Schalter *m*
two-position action *s.* two-step action
two-position action controller *s.* two-step action controller
two-position differential gap action *s.* two-step action with overlap
two-position range selector *(El)* Zweibereich[wahl]schalter *m*
two-product separation Zweigutscheidung *f*
two-propellants tank Doppelkraftstoffbehälter *m*
two-quadrant multiplier *(Dat)* Zweiquadrantenmultiplikator *m*, Zweiquadrantenmultiplizierer *m*
two-rate meter Doppeltarifzähler *m*, Zweitarifzähler *m*
two-region reactor Zweizonenreaktor *m*
two-resonator klystron Zweikammerklystron *n*
two-revolution machine *(Typ)* Zweitourenmaschine *f*
two-room[ed] apartment Zweizimmerwohnung *f*
two-row tedder zweireihiger Heuwender *m*
two-section filter Zweikreisfilter *n*
two-segment magnetron Zweischlitzmagnetron *n*
two-shear zweischnittig
two-shoe brake Zweibackenbremse *f*
two-side box tipping wagon Zweiseitenkastenkipper *m*
two-side planer zweiseitige Walzenhobelmaschine *f*
two-source frequency keying *(Fmt)* Tastung *f* durch Frequenzwechsel
two-span *(Bw)* Zweifeld-, zweifeldrig; zweischiffig
two-speed action *(BMSR)* Zweilaufwirkung *f*
two-spindle automatic Zweispindelautomat *m*
two-spindle machine Doppelspindelmaschine *f*
two-split magnetron Zweischlitzmagnetron *n*
two-stage amplifier zweistufiger Verstärker *m*
two-stage centrifugal pump Verbundkreiselpumpe *f*, Verbundzentrifugalpumpe *f*
two-stage compressor Zweistufenverdichter *m*
two-stage rocket Zweistufenrakete *f*
two-stage sulphate digestion Zweistufensulfataufschluß *m*
two-start zweigängig
two-state device *(Dat)* binäres Bauelement *n*, Binärspeicherelement *n*
two-station fixture *(Masch)* Zweistückspannvorrichtung *f*
two-step zweistufig
two-step action *(BMSR)* Zweipunktregelung *f*, Zweipunktwirkung *f*, Zweipunktverhalten *n*
two-step action controller *(BMSR)* Zweipunktregler *m*
two-step action with overlap *(BMSR)* Zweipunktregelung *f* mit Hysterese
two-step rocket Zweistufenrakete *f*
two-stepped zweistufig
two-stepped inserted-tooth reamer zweistufige Reibahle *f* mit eingesetzten Messern
two-storey zweistöckig
two-strand roller chain Doppelrollenkette *f*, Zweifachrollenkette *f*
two-stranded zweiadrig *(Kabel)*
two-stroke cycle Zweitaktverfahren *n*, Zweitaktlauf *m*
two-stroke Diesel engine Zweitaktdieselmotor *m*
two-stroke engine Zweitaktmotor *m*
two-stroker Zweitakter *m*
two-subcarrier system *(Fs)* Zweihilfsträgersystem *n*
two-term[ed] *(Math)* zweigliedrig
two-terminal capacitor *(El)* Zweipolkondensator *m*
two-terminal measurement Zweipolmessung *f*
two-terminal network *(El)* Zweipol *m*
two-throw crank shaft doppelt gekröpfte Kurbelwelle *f*
two-tier winding *(El)* zweietagige Wicklung *f*
two-tone *(Plast)* Preßfehler *m*
two-valued *(Math)* zweiwertig
two-valve receiver Zweiröhrenempfänger *m*
two-volume edition zweibändige Ausgabe *f*
two-volume pump zweistufige Zwillingspumpe *f*
two-wave rectification *(Elm)* Vollwellengleichrichtung *f*, Zweiweggleichrichtung *f*
two-way automaton *(Dat)* Zweiwegeautomaton *n*
two-way boring Zweiwegebohren *n*
two-way boring and drilling machine Zweiwegebohrmaschine *f*
two-way break-before-make contact Umschaltekontakt *m* mit Unterbrechung *(Relais)*

two-way circuit *(Am) (Fmt)* Leitung *f* für wechselseitigen Verkehr
two-way communication *(Fmt)* Doppelverkehr *m*, Duplexverkehr *m*, Gegensprechverkehr *m*, Wechselverkehr *m*
two-way connection *(El)* Wechselschaltung *f*
two-way contact Umschaltekontakt *m*
two-way drilling machine Zweiwegebohrmaschine *f*
two-way flat slab Pilzdecke *f (Stahlbetonbau)*
two-way launching *(Schiff)* Stapellauf *m* auf zwei Schlitten, Doppelbahnablauf *m*
two-way make-before-break contact unterbrechungsloser Umschaltekontakt *m (Relais)*
two-way pilot valve Ventil *n* mit Alternativsteuerung *(pneumatischer Regler)*
two-way radio *(Fmt)* Sende- und Empfangsgerät *n*; Gegensprechanlage *f*
two-way reinforced [concrete] slab kreuzweise bewehrte Stahlbetonplatte *f*
two-way reinforcement kreuzweise Bewehrung *f (Stahlbetonbau)*
two-way slab kreuzweise bewehrte Platte *f (Stahlbetonbau)*
two-way switch Umschalter *m*, Wechselschalter *m*, Zweiweg[um]schalter *m*, Umschalter *m* für zwei Stromkreise
two-way tap Zweiwegehahn *m*
two-way telephone conversation Gegensprechverkehr *m*, Doppelsprechen *n*
two-way valve Wechselventil *n*
two-wheel[ed] zweirädrig
two-wire aerial Bifilarantenne *f*, Zweidrahtantenne *f*, zweidrähtige Antenne *f*
two-wire automatic telephone system *(Fmt)* Schleifensystem *n*
two-wire line *(El)* Doppelleitung *f*
two-wire meter Zweileiterzähler *m*
two-wire system Zweileitersystem *n*, Zweidrahtsystem *n*
two-wire winding bifilare Wicklung *f*, Doppelfadenwicklung *f*, Zweidrahtwicklung *f*
twofold zweifach; zweizählig
t. yarn Zweifachgarn *n*
TWT phototube Fotozelle *f* mit Wanderwellenstruktur
twyer *s.* tuyere
tying *(Fmt)* Bindung *f (Leitungsbau)*
t.-in machine *(Text)* Knüpfmaschine *f*, Anknüpfmaschine *f*
t. machine *(Typ)* Verschnürmaschine *f*
t.-up berth Anlegestelle *f*, Ankerplatz *m*
tymbal Kesselpauke *f*, Pauke *f*
tympan Handpauke *f*
tympano Pauke *f*
tympanum *(Bw)* Giebelfeld *n*, Tympanon *n*, Tympanum *n*
Tyndall cone Tyndallkegel *m*
Tyndall effect Tyndalleffekt *m*
type / to [mit Maschine] schreiben, tippen
type 1. *(Masch)* Bauart *f*, Modell *n*, Type *f*, Ausführung *f*; 2. Form *f*, Muster *n*, Sorte *f*; 3. Drucktype *f*, Type *f*, Schrift *f*, Letter *f*
t. A0 wave *(Fmt)* ungedämpfte Welle *f*
t. A1 wave *(Fmt)* getastete ungedämpfte Welle *f*
t. A2 wave *(Fmt)* getastete tonüberlagerte ungedämpfte Welle *f*
t. A3 wave *(Fmt)* sprachmodulierte ungedämpfte Welle *f*
t. A4 wave Rundfunkwelle *f*
t. A5 wave Fernsehwelle *f*
t. area *(Typ)* Satzspiegel *m*
t. B wave *(Fmt)* gedämpfte Welle *f*
t. body *(Typ)* Kegel *m*, Schriftkegel *m*
t. family *(Typ)* Schriftfamilie *f*
t. fount *(Typ)* Schriftart *f*, Schrifttyp *m*
t. gauge *(Typ)* Typometer *n*, Schrifthöhenmesser *m*
t. locality Originalfundort *m*, Originalvorkommen *n*
t. matter *(Typ)* Schriftsatz *m*
t. metal Schriftmetall *n*, Letternmetall *n*
t. of boiler Kesseltyp *m*
t. of cable Kabeltyp *m*
t. of equilibrium Gleichgewichtsform *f*
t. of radiation Strahlenart *f*
t. of reaction Reaktionstyp *m*
t. of vessel Schiffstyp *m*
t. of vibration Schwingungsform *f*
t. of winding *(Text)* Spulart *f*
t. page *(Typ)* Satzspiegel *m*
t. planer *(Typ)* Schrifthobel *m*
t. pool Schreibmaschinenzimmer *n*
t. printing *(Fmt)* Typendruck *m*
t. printing telegraph Typendrucktelegraf *m*
t. 0 servomechanism *(BMSR)* Folgeregler *m* ohne Integration, P-Folgeregler *m*
t. 1 servomechanism *(BMSR)* Folgeregler *m* mit einer Integration, I-Folgeregler *m*
t. 2 servomechanism *(BMSR)* Folgeregler *m* mit zwei Integrationen, I_2-Folgeregler *m*
t.-setter *(Typ)* Setzer *m*
t.-setting *(Typ)* Schriftsetzen *n*
t. ship Schiffsprototyp *m*, erstes Schiff *n* einer Bauserie
t. size *(Typ)* Schriftgrad *m*, Kegelstärke *f*
t. wheel *(Fmt)* Typenrad *n*
t. wheel translator Typenradübersetzer *m*
types of radio communication Funkverkehrsformen *fpl*
typewriter Schreibmaschine *f*
t. component Schreibmaschinenteil *n*
t. composition *(Typ)* Schreibmaschinensatz *m*
t. eraser Schreibmaschinenradiergummi *m*
t. finish Schreibmaschinenlack *m*

t. **keyboard** Schreibmaschinentastatur *f*
t. **paper** Schreibmaschinenpapier *n*
t. **ribbon** Farbband *n*, Schreibmaschinen[farb]band *n*
t. **roller** Schreib[maschinen]walze *f*
typewriting machine *s.* typewriter
t. **telegraph** Fernschreiber *m*
typing Schreiben *n (mit Maschine)*
t. **[re]perforator** *(Dat)* Schreiblocher *m*
typographer Typograf *m*
typographic typografisch; Druck-
typography Typografie *f*; Schriftsatz *m*
tyre 1. Reifen *m*; 2. *(Eis)* Bandage *f*
t. **boring and turning mill** Bandagenbohr- und -drehmaschine *f*
t. **casing** Reifenlaufdecke *f*, Reifenlaufmantel *m*, Reifenhülle *f*
t. **casing material** Reifeneinlagematerial *n (Kord)*
t. **cord** Reifenkord *m*
t. **cord fabric** Reifenkord *m*
t. **cover** Decke *f*, Reifenmantel *m*
t. **deflection** Reifeneinsenkung *f*
t. **fabric weaving machine** Reifenkordwebeanlage *f*
t. **fitting** Reifenmontage *f*, Reifenreparatur *f*
t. **gauge** Luftdruckmesser *m*, Reifenfülldruckmesser *m*
t. **heating furnace** Band[agen]glühofen *m (Walzwerk)*
t. **inflating cock** Reifenfüllhahn *m*
t. **inflator** Reifenfüller *m*
t.**-inflator cylinder** Reifenfüllflasche *f*
t. **ingot** Bandagenblock *m (Walzwerk)*
t. **iron** Reifenaufzieheisen *n*, Reifenheber *m*
t. **lever** Montierhebel *m*
t. **marks** Bremsspur *f*, Reifenspur *f*
t. **mileage** Reifenlaufzeit *f*
t. **pump** Luftpumpe *f*
t. **remover** Radabzieher *m*
t. **shoulder** Reifenschulter *f*
t. **tape** Isolierband *n*
t. **with lip** *(Eis)* Bandage *f* mit Ansatz *(Sprengring)*
tyred luftbereift
tyres Bereifung *f*
tyro *(sl)* Flugschüler-Anfänger *m*
tyrosine *(Ch)* Tyrosin *n*

U

U-bend Doppelkrümmer *m*, U-Rohr *n*
U-bolt Bügelbolzen *m*, Bügelschraube *f*
U-channel U-Profil *n*
U-clamp Gabelspanneisen *n*, Bogenaufspannklaue *f*
U-frame U-Rahmen *m*
U-free *(Kern)* uranfrei
U-groove weld U-Naht *f*, Tulpennaht *f*
u.h.f. ultrahohe Frequenz *f*, UHF
u.h.f. band Ultrahochfrequenzband *n*, UHF-Band *n*
u.h.f. engineering Hochfrequenztechnik *f*
u.h.f. mixer Ultrahochfrequenzmischer *m*, UHF-Mischer *m*
u.h.f. technique Dezimeterwellentechnik *f*
u.h.f. transmitter frequency UHF-Senderfrequenz *f*
U-leather Ledermanschette *f*
U-link *(Fmt)* Kurzschlußbügel *m*
U-plug *(El)* Bananendoppelstecker *m*
U-section U-Profil *n (Stahl)*
U-shaped expansion pipe *(Wärm)* U-förmiger Federrohrbogen *m*
U-shaped frame *(Schiff)* Sackspant *n*, U-Spant *n*
U-shaped vessel Schiff *n* mit U-Spantform
U-troughing *(Fmt)* Kabelkanal *m*, Kabelrohr *n (zum Schutz)*, Leitungskanal *m*
U-type furnace *(Met)* Umkehrofen *m*, U-Ofen *m*
udometer Regenmesser *m*
Ulbricht globe *(Opt)* Ulbrichtsche Kugel *f*
Ulbricht globe photometer Ulbricht-Kugelfotometer *n*
ullmanite Ullmannit *m*, Nickelantimonkies *m*, Nickelspießglanz *m*
ulmous earth Humuserde *f*
ultimate äußerst, höchstmöglich, elementar
u. **analysis** *(Ch)* Elementaranalyse *f*
u. **bearing capacity** Grenztragfähigkeit *f*, Tragfähigkeit *f*
u. **consumer** Endverbraucher *m*
u. **creep value** Kriechendwert *m*
u. **elongation** Bruchdehnung *f (Festigkeit)*
u. **limit switch** Endausschalter *m*; Grenzschalter *m*, Maximalschalter *m*; Höchststromschalter *m*
u. **load** Bruchlast *f*, Traglast *f*
u. **load design** Traglastverfahren *n (Stahl- und Stahlbetonberechnung)*, n-freies Berechnungsverfahren *n (Stahlbetonberechnung)*
u. **recovery** Gesamtausbeute *f* [einer Bohrung]
u. **ripening** *(Ch)* Nachreife *f*
u. **sensitivity** Grenzempfindlichkeit *f (Funkempfangstechnik)*
u. **stability** Endstabilität *f*
u. **state** Grenzzustand *m*
u. **strength** Bruchfestigkeit *f*, Zerreißfestigkeit *f*
u. **strength design** *s.* u. strength theory
u. **strength theory** Traglastverfahren *n (Stahl- und Stahlbetonberechnung)*, n-freies Berechnungsverfahren *n (Stahlbetonberechnung)*
u. **stress** Bruchbelastung *f*, Bruchspannung *f*

u. structure of matter Elementaraufbau *m* (Elementarstruktur *f*) der Materie
u. user Endverbraucher *m*
u. value *(Math)* Endwert *m*
ultimately controlled variable *(BMSR)* Aufgabengröße *f*
ultra-acoustic teleprinter *(Fmt)* Überlagerungsfernschreiber *m*
u.-audible frequency Überhörfrequenz *f*
u.-audion Ultraaudion *n*, Rückkopplungsaudion *n*
u.-centrifuge Ultrazentrifuge *f*
u.-fine structure Ultrafeinstruktur *f*
u.-high frequency ultrahohe Frequenz *f* (*s. a.* u.h.f.)
u.-high vacuum höchstes Vakuum *n*, Ultrahochvakuum *n*
u.-microscope Ultramikroskop *n*
u.-red infrarot, ultrarot (*s. a.* ultrared)
u.-short wave Ultrakurzwelle *f*, UKW
u.-short wave transmission Ultrakurzwellenübertragung *f*
u.-sonic Ultraschall- (*s. a.* ultrasonic)
u.-violet ultraviolett, Ultraviolett- (*s. a.* ultraviolet)
ultradyne receiver Ultradynempfänger *m*
ultramarine blue Lasurblau *n*, Ultramarin *n*
ultrared infrarot, ultrarot
u. analyzer Infrarotanalysator *m*
u. emission Ultrarotemission *f*
u. radiation Ultrarotstrahlung *f*
u. rays Ultrarotstrahlen *fpl*, Ultrarotstrahlung *f*
u. transmittancy Ultrarotdurchlässigkeit *f*
ultrasoft ultraweich *(Röntgenstrahlen)*
ultrasonic Ultraschall-
u. amplification Ultraschallverstärkung *f*
u. attenuation Ultraschalldämpfung *f*
u. beam Ultraschallstrahl *m*
u. cleaning Ultraschallreinigung *f*, Reinigen *n* mit Ultraschall
u. device Ultraschallbaustein *m*
u. drilling Ultraschallbohren *n*
u. drilling machine Ultraschallbohrmaschine *f*
u. flaw detector Ultraschallprüfgerät *n*, Ultraschalldefektoskop *n (für zerstörungsfreie Werkstoffprüfung)*
u. intensity Ultraschallintensität *f*
u. intensity meter Ultraschallintensitätsmesser *m*
u. material tester Ultraschallwerkstoffprüfgerät *n*
u. power Ultraschalleistung *f*
u. power generator Ultraschallgenerator *m*
u. probe Ultraschallsonde *f*
u. signal Ultraschallsignal *n*
u. sounding Ultraschallotung *f*
u. thickness gauge Ultraschalldickenmeßgerät *n*, Ultraschalldickenmesser *m*
u. transmitter Ultraschallgeber *m*, Ultraschallsender *m*
u. vessel Ultraschalltopf *m*
u. viscosimeter Schwingungsviskosimeter *n*, Ultraschallviskosimeter *n*
ultrasonics Ultraschall *m*
ultraviolet ultraviolett, Ultraviolett-
u. band Ultraviolettbande *f*
u. engineering Ultraviolettechnik *f*
u. lamp Ultraviolettlampe *f*, Ultraviolettstrahler *m*
u. light ultraviolettes Licht *n*, UV-Licht *n*
u. rays Ultraviolettstrahlung *f*, ultraviolette Strahlen *mpl*
u. region Ultraviolettbereich *m*
u. response Ultraviolettempfindlichkeit *f*
ultrawhite region Ultraweißgebiet *n*
umber *(Ch)* Umbra *f*
umbra Kernschatten *m*
umbrella 1. Schirm *m*; Regenschirm *m*; Sonnenschirm *m*; 2. Schornstein[regen]kragen *m*
u. aerial Schirmantenne *f*
u. cloth Schirmstoff *m*
u. cover Schirmhülle *f*
u. insulator Pilzisolator *m*
u.-shaped aerial Schirmantenne *f*
u. shell *(Bw)* Pilzschale *f*
u. stand Schirmständer *m*
u.-type generator *(Elm)* Schirmgenerator *m*
u.-type insulator Schirmisolator *m*
unaccelerated unbeschleunigt
unactivated unaktiviert
unadulterated unverschnitten, unverfälscht
unaffected by light lichtecht
u. by wear verschleißfrei
unaided eye unbewaffnetes Auge *n*
unalloyed unlegiert
unaltered unverändert
unambiguous *(Math)* eindeutig
u. reading eindeutige Anzeige *f*
unamplified *(El)* unverstärkt
unannealed ungeglüht, nicht wärmebehandelt *(Stahl)*
u. malleable iron Temperrohguß *m*, Weißguß *m*, Hartguß *m*
unarmoured cable unbewehrtes Kabel *n*
unassorted scrap unsortierter Schrott *m*
unate truth function *(BMSR) Wahrheitsfunktion, in der keine Variable sowohl bejaht als auch verneint ist*
unattended 1. *(Masch)* wartungsfrei; 2. unbemannt; 3. unbewacht, unbeaufsichtigt
u. beacon unbemannte Bake *f (Navigation)*
u. light unbewachtes Feuer *n (Navigation)*
u. operation unbeaufsichtigter Betrieb *m*; vollautomatischer Betrieb

u. **power-station** vollautomatisches Kraftwerk *n*
u. **ship** unbemanntes Schiff *n*
u. **station** unbemannte Station *f*
unbalance Unausgeglichenheit *f*; Unwucht *f*; Wuchtfehler *m*
u. **attenuation measuring set** *(El)* Fehlerdämpfungsmesser *m*
u. **spiral** asymmetrische Spiralantenne *f* *(Flugfunk)*
unbalanced unsymmetrisch; unausgewuchtet, mit Unwucht; unausgeglichen; ohne Druckausgleich, unkompensiert
u. **component** *(El)* unsymmetrische Komponente *f*, unsymmetrisches Glied *n*
u. **lacing pattern** *(Brg)* unsymmetrische Anordnung *f* der Schrämmeißel in der Schrämkette
u. **load** unsymmetrische Belastung *f*
u. **rudder** nichtausbalanciertes (nichtausgeglichenes) Ruder *n*
u. **signal** unsymmetrisches Signal *n*
u. **voltage** unsymmetrische Spannung *f*
unballast / to *(Schiff)* Ballast löschen (ausschießen), entballasten
unbedded schichtungslos, ungeschichtet
unbiased nichtvormagnetisiert; nichtvorgespannt
unbleached linen Rohleinen *n*
u. **pulp** ungebleichter Papierfaserstoff *m*
u. **straw pulp** Gelbstrohstoff *m*
unblock / to entriegeln
unblocking *(Fmt)* Deblockierung *f*, Freigabe *f*
unbolt / to losschrauben, abschrauben
unbonded prestressing Vorspannen *n* ohne Verbund *(Spannbetonbau)*
unbound nichtgebunden, frei
u. **electron** freies (nichtgebundenes) Elektron *n*
u. **state** ungebundener Zustand *m*
unbounded *(Math)* unbegrenzt
unbracketed ohne Klammern; außerhalb der Klammern
unbreakable bruchfest, unzerbrechlich
unbroken line Vollinie *f*
unbuffered ungepuffert
unbuttoning Abbau *m* von Stahlgerüsten
uncaging *(Brg)* Abziehen *n*
uncalibrated ungeeicht
uncanned fuel element *(Kern)* nacktes Brennstoffelement *n*
uncased part of the hole offenes Loch *n*, unverrohrter Teil *m* des Bohrlochs *(Tiefbohrtechnik)*
uncertainty Unbestimmtheit *f*, Unsicherheit *f*
u. **of measurement** Meßunsicherheit *f*
u. **principle** Unbestimmtheitsprinzip *n* *(Quantentheorie)*
unchanged unverändert
uncharged ohne Ladung, ladungslos, ladungsfrei, ungeladen, neutral
u. **particle** ungeladenes (neutrales) Teilchen *n*
unchopped beam ununterbrochener Strahl *m*
uncial *(Typ)* Unziale *f*
uncirculated mit Stempelglanz *(Münzen)*
unclad unplattiert
unclamp / to ausspannen; entriegeln *(Revolver)*; lösen *(Schlitten)*
unclamping Ausspannen *n* *(Werkstück)*; Entriegeln *n* *(Revolver)*; Lösen *n* *(Schlitten)*
uncleaned coal Rohkohle *f*, Förderkohle *f*
unclosed pair *(Mech)* kraftschlüssige Paarung *f*
unclutch / to auslösen, entkuppeln
uncoagulable ungerinnbar
uncoated paper ungestrichenes Papier *n*
u. **wire** Blankdraht *m*
uncoil / to abspulen, abwickeln *(Draht)*
uncollimated ungebündelt
uncoloured unverfärbt *(z. B. ein Kristall)*
uncombined carbon *(Met)* freier (graphitischer, ungebundener) Kohlenstoff *m*
u. **heat** freie Wärme *f*
uncompensated attractions *(Ch)* Nebenvalenzkräfte *fpl*
unconditional jump *s.* u. transfer
u. **transfer** *(Dat)* unbedingter (nichtbedingter) Sprung *m*, Einschleusung *f*
unconformity *(Geol)* Diskordanz *f*, ungleichförmige Überlagerung *f*
unconsolidated *(Geol)* rollig, unverfestigt
uncontaminated unverseucht; verunreinigungsfrei
uncontrollable motor Motor *m* ohne Drehzahlbeeinflussung, nicht steuerbarer Motor
uncontrolled nichtgesteuert, ungesteuert, unkontrollierbar
uncooled ungekühlt
uncountable unzählbar; unzählig
uncouple / to auslösen, ausrücken, entkuppeln; entkoppeln
uncover / to *(Brg)* abräumen, abtragen, freilegen
uncover *(Brg)* Beseitigung *f* des Abraums, Freilegen *n* der Lagerstätte
uncovered wire blanker Draht *m*
uncovering *(Brg)* Beseitigung *f* des Abraums, Freilegen *n* der Lagerstätte
uncrystallizable unkristallisierbar
unctuous fettig, ölig, salbenartig
u. **clay** fetter Ton *m*
unctuousness Fettigkeit *f*, Öligkeit *f*, Schmierigkeit *f*
uncut ungeschliffen *(Stein)*
u. **pile** *(Text)* Schlingen *fpl*
u. **pile fabrics** gezogene Florgewebe *npl*
u. **plane iron** einfaches Hobeleisen *n*, Schlichthobeleisen *n*

u. plush gezogener Plüsch *m*
u. velvet gezogener Samt *m*
undamped ungedämpft
u. oscillation ungedämpfte Schwingung *f*
u. wave ungedämpfte Welle *f*
undecagon Elfeck *n*
undecked offen, ungedeckt *(Schiff)*
u. ship Schiff *n* ohne Deck, ungedecktes (offenes) Schiff
undecomposable unzerlegbar
undecylenic acid Undezylensäure *f*
undecylic acid Undezylsäure *f*
undecynic acid Undezinsäure *f*
undeformed unverformt
under-car aerial Chassisantenne *f*
u.-compound excitation Unterverbunderregung *f*
u.-exposure Unterbelichtung *f*
u. weigh mit gelichtetem Anker
underbase Untergestell *n*, Fundament *n*; Unterbau *m*, Grundplatte *f*
underbed Fundament *n*, Bett *n*, Untergestell *n*
underbody Unterwasserteil *m (des Schiffskörpers)*
underbridge Unterführung *f*
underbunching unterkritische Ballung (Paketbildung) *f (Elektronik)*
underburdening *(Met)* Untermöllerung *f*
undercarriage Fahrgestell *n*, Fahrwerk *n*
u. wheel brake *(Flg)* Laufradbremse *f*
undercoat Grundierung *f*
undercooling Unterkühlung *f*
undercorrection Unterkorrektion *f*
undercure *(Plast)* nicht ausgehärtet, unterhärtet *(Preßfehler)*
undercurrent circuit-breaker Minimal[strom]-schalter *m*; Unterstromwächter *m*
u. relay Minimalrelais *n*
undercut / to 1. *(Masch)* unterschneiden; hinterstechen; 2. *(Brg)* unterbauen, unterfahren, unterschrämen
undercut 1. Unterschneidung *f*, Unterschnitt *m*; 2. Grundhieb *m (Feile)*, Unterhieb *m*; 3. *(Holz)* Fallkerb *m*
u. caving *(Brg)* Blockbruchbau *m*
undercutting Schrämen *n* im unteren Teil des Flözes
underdamped system *(BMSR)* System *n* mit unterkritischer Dämpfung
underdeck space *(Schiff)* Unterdeckraum *m*
u. tonnage *(Schiff)* Unterdeckraumgehalt *m*, Unterdecktonnage *f*
underdrive press Presse *f* mit untenliegendem Antrieb, Presse mit Unterantrieb
underexpose / to unterbelichten
underexposure Unterbelichtung *f*
u. region Schwellenwert *m*, unterbelichteter Bereich *m (Fotografie)*

underfeed furnace Unterschubfeuerung *f*, Feuerung *f* mit Beschickung von unten
underfired gas furnace Gasunterfeuerung *f*
u. oven Unterbrennerofen *m*
underfiring from external producers Fremdgasbeheizung *f*
underfloor *(Bw)* Decke *f*
u. engine Unterflurmotor *m*
u. extraction Unterflurstaubabführung *f (Spinnerei)*
underframe Fahrgestell *n (Wagen)*; Untergestell *n*; Unterbau *m*
undergrate blast *(Wärm)* Unterwind *m*
u. blower Unterwindgebläse *n*
u. firing *(Met)* Unterfeuerung *f*
underground unterirdisch, unterhalb der Erdoberfläche; erdverlegt; *(Brg)* unter Tage, untertägig
u. aerial Untergrundantenne *f*
u. burst unterirdische Explosion *f*
u. cable Erdkabel *n*, Untergrundkabel *n*, unterirdisches (unterirdisch verlegtes) Kabel *n*
u.-cable railway Standseilbahn *f*
u. cooler Unterflurkühler *m*
u. distribution chamber *(Fmt)* Kabelkeller *m*
u. excavation *(Brg)* Grubenbau *m*
u. explosion unterirdische Explosion *f*
u. fertilization Bodendüngung *f*
u. fire Grubenbrand *m*
u. furnace *(Met)* Unterflurofen *m*
u. gasification Untertagevergasung *f (UTV)*
u. laying *(El)* Erdverlegung *f*
u. main flue *(Wärm)* unterirdischer Fuchs *m*
u. milling *(Brg)* untertägiger Trichterbau *m*
u. mine levelling Nivellement *n* unter Tage
u. silo Miete *f (z. B. für Kartoffeln)*
u. store *s.* u. silo
u. tank Erdbehälter *m*, unterirdischer Behälter *m*
u. water Grundwasser *n*
underhand *(Brg)* fallend, in fallender Reihenfolge
u. stoping Strossenbau *m*
underhung rudder *(Schiff)* Schweberuder *n*, Hängeruder *n*
underjet oven Unterbrennerofen *m*
underload circuit-breaker Minimal[last]schalter *m*, Unterstromschalter *m*
u. relay Minimalrelais *n*
underlying seam *(Brg)* liegendes Flöz *n*
undermine / to untergraben, unterhöhlen; auskolken
undermining *(Brg)* Schießen *n* mit untergelegter Ladung
undermodulation unvollständige Modulation *f*
underoxidized unteroxydiert
underpass Unterführung *f*
underpick loom Unterschlagwebstuhl *m*

underpin / to unterfangen, abfangen; unterbauen *(Bauwerke)*
underpinning Unterfangen *n*, Unterfangung *f* *(von Bauwerken)*
underpoled copper kaltbrüchiges (unterpoltes) Kupfer *n*
underpowered mit zu geringer Leistung
underriver cable Flußkabel *n*
undersea level Tauchtiefe *f (Taucher)*
undershoot *(El)* kriechendes Einschwingen *n*, Unterschwingen *n*
undershot unterschlächtig *(Wasserrad)*
underside Unterseite *f*
u. elevation *(Bw)* Untersicht *f*
u. view *(Bw)* Untersicht *f*
undersize 1. Untermaß *n*, Untergröße *f*; 2. Unterkorn *n*, Siebdurchgang *m*
u. drill Untermaßbohrer *m*, Minusbohrer *m*
u. for reaming *(Masch)* Unterrreibmaß *n*, Untermaß *n* für das Aufreiben
u. reamer Untermaßbohrer *m*, Minusbohrer *m*
underslung hängend *(z. B. Flugzeugmotor)*
u. crankshaft hängende Kurbelwelle *f*
undersoil packer Untergrundpacker *m*
u. plough Untergrundpflug *m*
understood sign *(Fmt)* Verstandenzeichen *n*
understructure Sockel *m (Hochofen)*
undertighten / to zu locker anziehen *(Mutter)*
undertone *(Plast)* dünne Farbschicht *f* auf weißem Untergrund
undertype armature *(Elm)* unterständiger Anker *m*
undervoltage Unterspannung *f*
u. circuit-breaker Unterspannungsschalter *m*
u. protection Unterspannungsschutz *m (Relais)*
u. relay Unterspannungsrelais *n*, Nullspannungsrelais *n*
u. release *s.* u. tripping
u. tripping Unterspannungsauslösung *f*, Nullspannungsauslösung *f*
u. winding Unterspannungswicklung *f (Relais)*
underwater aerial Unterwasserantenne *f*
u. body Unterwasserschiff *n*
u. burst Unterwasserexplosion *f*
u. camera Unterwasserkamera *f*
u. concrete Unterwasserbeton *m*
u. cutting Unterwasserbrennschneiden *n*
u. exhaust Unterwasserauspuff *m*
u. explosion Unterwasserexplosion *f*
u. level *(Schiff)* Tauchtiefe *f*, Unterwassertiefe *f*
u. lightning Unterwasserblitz *m*
u. protection *(Schiff)* Unterwasserschutz *m*
u. reactor Unterwasserreaktor *m*
u. spark Unterwasserblitz *m*
u. vehicle Unterwasserfahrzeug *n*
u. welding Unterwasserschweißen *n*
underwood Buschholz *n*
underwork / to *(Brg)* unterbauen
undesired mode unerwünschter Wellentyp *m*
u. noise *(El)* Nebengeräusch *n*, Störgeräusch *n*
u. reflection Störreflexion *f*
undeterminable unbestimmbar
undetermined coefficients *(Math)* unbestimmte Koeffizienten *mpl*
undeveloped unentwickelt *(Film)*
undiffused nichtdiffundiert
undiluted unverdünnt
undirected ungerichtet
undirectional ungerichtet
undissociated *(Ch)* undissoziiert
undissolved un[auf]gelöst
undistorted unverzerrt, ungestört
u. power output *(El)* unverzerrte Ausgangsleistung *f*
undisturbed störfrei
u. beds ungestörte Schichten *fpl*
u. orbit ungestörte Bahn *f*
u. soil sample ungestörte Bodenprobe *f*
undo / to aufmachen, lockern, lösen, losschrauben; demontieren
undock / to ausdocken *(ein Schiff)*
undoped ungedopt, undotiert, rein, frei von Zusätzen (Beimischungen)
u. crystal undotierter (ungedopter) Kristall *m*
u. fuel Kraftstoff *m* ohne Zusätze
undressed castings Rohguß *m*
undue unangemessen, unzulässig
undulating light-signal *(El)* Flackersignal *n*
undulation Wellenbewegung *f*
undulator Undulator *m*
undulatory wellenförmig, wellenartig
u. current Wellenstrom *m*
u. theory of light Wellentheorie *f* des Lichtes
undyed ungefärbt
unearthed *(El)* ungeerdet
uneconomical unwirtschaftlich
unenriched nichtangereichert, unangereichert
u. uranium Natururan *n*
unequal ungleich[schenklig]
u. spacing of the flutes ungleiche Schneidenteilung *f*
uneven ungleichmäßig; uneben, ungerade
unevenly spaced ungleich geteilt
unevenness Unebenheit *f*, Welligkeit *f*
u. of the wool fibre Untreue *f* der Wollfaser
unexcitable unerregbar
unexcited unerregt
unexposed unbelichtet *(Film)*
unfading nicht nachbleichend; echt *(Farbe)*
unfermentable unvergärbar
unfilled unbesetzt *(Elektronenschalen)*

unfiltered ungefiltert
unfinished paper ungeglättetes Papier *n*
unfinned rocket Rakete *f* ohne Stabilisierungsflächen
unfissionable unspaltbar
unfluted ungenutet
ungalvanized unverzinkt
ungeared räderlos
ungerade-gerade ... *s.* odd-even ...
ungerade-ungerade ... *s.* odd-odd ...
unglazed unglasiert; maschinenglatt, unsatiniert *(Papier)*
unground ungeschliffen; nicht gemahlen
  u. resin undispergiertes Harz *n*
ungrounded *(El)* ungeerdet
unguent Schmiere *f*; Einreibung *f*; Schmiermittel *n*
unguided rocket ungesteuerte (ungelenkte) Rakete *f*
ungula schiefer Stumpf *m (von Zylinder oder Kegel mit gegeneinander geneigter Grund- und Deckfläche)*
unhairing beam *(Led)* Haarbaum *m*
  u. knife *(Led)* Haareisen *n*, Haarmesser *n*, Putzmesser *n*
  u. machine *(Led)* Enthaarmaschine *f*
unhardened ungehärtet *(Stahl)*
unharmonious unharmonisch
unhewn unbehauen
  u. timber Rundholz *n*
unhindered-settling classifier Freifallklassierer *m*
unhinge / to ausheben *(aus Angeln)*
unhinging tang Aushängehaken *m*
uniacidic base einsäurige Base *f*
uniaxial einachsig
  u. crystal einachsiger Kristall *m*
  u. stress einachsige Spannung *f*, einachsiger Spannungszustand *m*
unicellular einzellig
unicolour einfarbig, monochromatisch
unidentified flying object unidentifiziertes fliegendes Objekt *n*
unidimensional eindimensional
unidirectional einseitig gerichtet, gleichgerichtet; *(Masch)* einseitig wirkend; *(BMSR)* gerichtet, mit gerichteter Wirkung; rückwirkungsfrei
  u. action einseitige Richtwirkung *f* *(Antennen)*
  u. characteristic Gleichrichtungscharakteristik *f*
  u. circuit einseitig betriebene Leitung *f*
  u. current Strom *m* gleichbleibender Richtung, gleichgerichteter Strom
  u. electrode Sperrelektrode *f*
  u. pulse einseitig gerichteter Impuls *m*
  u. spectra Spektren *npl* einheitlicher Richtung *(zur Bestimmung des Seegangsverhaltens)*
  u. transducer *(El)* Einrichtungswandler *m*
unificated vereinheitlicht
unification Vereinheitlichung *f*
unified field theory *(Ph)* einheitliche Feldtheorie *f*
unifilar electrometer Einfadenelektrometer *n*
  u. projection electrometer Einfadenprojektionselektrometer *n*
  u. suspension *(El)* Einfadenaufhängung *f*
uniflow Gleichstrom *m*
  u. drier Gleichstromtrockner *m*
  u. furnace *(Met)* Einwegofen *m*
  u. heating Gleichstromheizung *f*
  u. preheater Gleichstromvorwärmer *m*
  u. scavenging Gleichstromspülung *f*
  u. steam engine Gleichstromdampfmaschine *f*
uniform gleichförmig, gleichmäßig, gleichbleibend, homogen
  u. continuity of a function gleichmäßige Stetigkeit *f* einer Funktion
  u. diffuser *(El)* vollkommen matte Fläche *f*, vollkommen streuender Körper *m*, Lambert-Fläche *f*
  u. distribution homogene Verteilung *f*
  u. field *(El)* homogenes Feld *n*; *(Ph)* uniformes Feld
  u. magnetic field gleichförmiges Magnetfeld *n*
  u. mode gleichförmige Welle *f*
  u. motion gleichförmige Bewegung *f*
  u. pitch konstante Steigung *f (eines Propellerblattes)*
  u. pitch of teeth gleiche Zahnteilung *f*
  u. radiant flux gleichmäßiger Strahlenfluß *m*
  u.[-spectrum] random noise *(El)* weißes Rauschen *n*
  u. static field homogenes statisches Feld *n*
  u. waveguide *(El)* homogener Hohlleiter *m*
uniformity Einheitlichkeit *f*; Gleichförmigkeit *f*
  u. of running *(Masch)* Gleichförmigkeit *f* des Ganges
uniformly accelerated motion gleichförmig beschleunigte Bewegung *f*
  u. retarded motion gleichförmig verzögerte Bewegung *f*
  u. variable motion gleichförmig veränderliche Bewegung *f*
unilateral einseitig
  u. deflection einseitiger Ausschlag *m (eines Meßinstrumentes)*
  u. transducer *(El)* Einrichtungswandler *m*
unimolecular *(Ch)* monomolekular
uninflammable unverbrennbar, unentzündlich
  u. finish flammfeste Ausrüstung *f (von Geweben)*
uninsulated *(El)* unisoliert, blank
  u. wire blanker Draht *m*
unintelligibility *(Fmt)* Unverständlichkeit *f*

unintelligible crosstalk *(Fmt)* unverständliches Nebensprechen *n*
uninterrupted ununterbrochen
unintersected kreuzungsfrei *(Straßenbau)*
union Stutzen *m*, Verschraubung *f*, Überwurfmutter *f*, Rohrverschraubung *f*
u. dye Halbwollfarbstoff *m*
u. dyeing Halbwollfärberei *f*, Mischwollfärberei *f*
u. flange Ansatzflansch *m*
u. joint hose Schlauchverbindung *f*
u. nut Anschlußmutter *f*, Verschraubungsmutter *f*
u. socket Verbindungsmuffe *f*
unionized nichtionisiert
uniphase einphasig
uniplanar motion ebene Bewegung *f*
unipolar einpolig
u. armature einpoliger Anker *m*
u. electrode system unipolare Elektrodenanordnung *f*, Unipolarsystem *n*
u. induction Unipolarinduktion *f*
unipotential cathode Äquipotentialkatode *f*
unique *(Math)* eindeutig; einzig
uniqueness *(Math)* Eindeutigkeit *f*
unirefringent *(Opt)* einfachbrechend
uniselector *(Fmt)* Drehwähler *m* mit einer Ebene, Wähler *m* mit nur einer Bewegungsrichtung; Vorwähler *m*
unit 1. Einheit *f*, Maßeinheit *f*; 2. Bauglied *n*, Bauteil *n*; Fertigteil *n (Stahlbetonbau)*; 3. *(Masch)* Aufbaueinheit *f*; Aggregat *n*; 4. *(El)* Apparat *m*, Gerät *n*; 5. Schritt *m (sprunghafte Zunahme einer Größe)*; 6. *(Math)* Einer *m*
u. area Flächeneinheit *f*
u. area capacitance Kapazität *f* je Flächeneinheit; Kapazitätsbelag *m (Elektrolytkondensator)*
u.-assembly system *(Masch)* Baukastensystem *n*
u. automatic exchange *(Fmt)* Kleinwählerzentrale *f*, Landzentrale *f*
u.-block system *(Masch)* Baukastensystem *n*
u. call *(Fmt)* Gesprächseinheit *f*
u. cell Elementarzelle *f (Kristallgitter)*
u. charge *(El)* Ladungseinheit *f*, Einheitsladung *f*
u. circle Einheitskreis *m*
u. coil *(El)* auswechselbare Spule *f*
u.-constructed aus Baueinheiten zusammengesetzt, nach dem Baukastenprinzip konstruiert
u. construction Baukastenbauweise *f*; Regelbauweise *f*, Einheitsbauweise *f*, Blockbauweise *f*, Blockkastenkonstruktion *f*
u. construction all-metal body *(Kfz)* selbsttragende Ganzstahlkarosserie *f*
u. construction on the building-block principle Aufbau *m* nach Baukastensystem
u. construction principle Baukastensystem *n*
u. cost Kosten *pl* je Einheit
u.-counting code *(Dat)* Pulszahlkode *m*
u. current Stromeinheit *f*
u. duration of signal *(Fmt)* Schrittlänge *f*
u. electrostatic charge elektrische Einheitsladung *f*
u. function response *(El)* Sprungkennlinie *f*
u. head Aufbaueinheit *f*
u. head machine Aufbaumaschine *f*
u. heater Lufterhitzer *m*
u. interval [of a winding] *(Elm)* Spulen[seiten]teilung *f*
u. lattice Einheitsgitter *n*
u. length Längeneinheit *f*
u. load *(Flg)* Flächeneinheitsbelastung *f*
u. magnetic mass elektromagnetische Polstärkeneinheit *f*, Polstärkeneinheit im elektromagnetischen System
u. magnetic pole magnetischer Einheitspol *m*
u. matrix Einheitsmatrix *f*
u. of absorption Absorptionseinheit *f*
u. of accommodation Wohnungseinheit *f*
u. of angle Winkeleinheit *f*
u. of electric charge Einheit *f* der elektrischen Ladung
u. of electric conductance Einheit *f* der elektrischen Konduktanz
u. of electric current Einheit *f* der Stromstärke
u. of electric displacement Einheit *f* des elektrischen Verschiebungsstromes
u. of electric field strength Einheit *f* der elektrischen Feldstärke
u. of electric power Einheit *f* der elektrischen Leistung
u. of electric resistance Einheit *f* des elektrischen Widerstandes
u. of electricity Einheit *f* der Elektrizitätsmenge
u. of electromotive force Einheit *f* der elektromotorischen Kraft
u. of energy Energieeinheit *f*, Einheit *f* der Arbeit
u. of force Krafteinheit *f*
u. of frequency Frequenzeinheit *f*
u. of heat Wärmeeinheit *f*
u. of illumination Einheit *f* der Beleuchtungsstärke
u. of inductance Einheit *f* der Induktivität
u. of length Längeneinheit *f*
u. of light Lichteinheit *f*, fotometrische Einheit *f*
u. of mass Masseeinheit *f*
u. of measurement Maßeinheit *f*
u. of notation Maßeinheit *f*, Meßeinheit *f*
u. of output Leistungseinheit *f*

u. of potential Potentialeinheit *f*
u. of power Leistungseinheit *f*
u. of pressure Druckeinheit *f*
u. of quantity of electricity Einheit *f* der Elektrizitätsmenge
u. of radioactivity radioaktive Einheit *f*
u. of resistance Widerstandseinheit *f*
u. of resistivity Einheit *f* des spezifischen Widerstandes
u. of space Volum[en]einheit *f*, Raumeinheit *f*
u. of specific resistance Einheit *f* des spezifischen Widerstandes
u. of speed Geschwindigkeitseinheit *f*, Einheit *f* der Geschwindigkeit
u. of tension Spannungseinheit *f*
u. of time Zeiteinheit *f*
u. of velocity Einheit *f* der Geschwindigkeit
u. of voltage Spannungseinheit *f*
u. of volume Volum[en]einheit *f*, Raumeinheit *f*
u. of weight Masseeinheit *f*
u. of work Arbeitseinheit *f*, Energieeinheit *f*
u. operation Grundoperation *f*, Grundverfahren *n*
u. potential Einheit *f* des Potentials, Potentialeinheit *f*
u. power factor *(El)* Leistungsfaktor 1 *m*, Phasenwinkel 0 *m*
u. pressure spezifischer Preßdruck *m*
u. process Grundprozeß *m*
u. production *(Am)* Einzelfertigung *f*
u. quantity of electricity elektrostatische Ladungseinheit *f*, Ladungseinheit im elektrostatischen System
u. radius Einheitsradius *m*
u. resistance *(El)* Widerstandseinheit *f*, Widerstand 1 *m*
u. solid angle Raumwinkeleinheit *f*
u. sphere Einheitskugel *f*
u. step *(El)* Einheitssprung *m*
u. strain *(Am)* Dehnung *f*
u. stress *(Am)* Spannung *f*
u. surface Einheitsfläche *f*
u. tube *(El)* Einheitsröhre *f*
u.-type machine Aufbaumaschine *f*
u. volume Einheitsvolumen *n*
unitary *(Math)* unitär
unitized construction Aufbau *m* nach Baukastensystem
u. machine Aufbaumaschine *f*
u. machine tool Werkzeugmaschine *f* nach Baukastensystem
unity Einheit[lichkeit] *f*; Eins *f*
u. power factor *(El)* Leistungsfaktor 1 *m*, Phasenwinkel 0 *m*
UNIVAC *s.* universal automatic computer
univalence *(Ch)* Einwertigkeit *f*
univalent *(Ch)* einwertig
univariant *(Ph)* monovariant, univariant
universal universal
u. amplifier *(Fmt)* Allverstärker *m*
u. angle plate dreh- und kippbarer Aufspannwinkel *m*
u. applicability universelle Anwendbarkeit *f*
u. automatic computer universeller automatischer Rechner *m*
u. ball joint *(Masch)* Kugelkardan *n (m)*
u. bevel protractor Universalwinkelmesser *m*
u. bridge *(El)* Universal[meß]brücke *f*
u. chuck *(Masch)* Universalfutter *n*
u. condenser *(Wärm)* Universalkondensator *m*
u. constant universelle Konstante *f*
u. cord circuit *(Fmt)* Einheitsschnurpaar *n*
u. coupling *(Masch)* Kardangelenkkupplung *f*, kardanisches Gelenk *n*, Kreuzgelenk[kupplung *f*] *n*
u. cutter and tool grinder Universalwerkzeugschleifmaschine *f*
u. cylindrical grinding machine Universalrundschleifmaschine *f*
u. dividing head Universalteilkopf *m*
u. equivalent circuit *(El)* Ersatzschaltung *f* mit Universalparametern
u. face plate Universalplanscheibe *f*
u. fuse link *(El)* Universalschmelzeinsatz *m*
u. galvanometer Universalgalvanometer *n*
u. gas constant universelle Gaskonstante *f*
u. grinder Universalschleifmaschine *f*
u. grinding machine Universalschleifmaschine *f*
u.-head *(Masch)* mit schwenkbarem Spindelkopf
u. index centre Universalteilgerät *n*
u. instrument Mehrzweckinstrument *n*
u. joint Kardangelenk *n*, Kniegelenk *n*, Kreuzgelenk *n*, Kugelgelenk *n*
u.-joint coupling Kreuzgelenkkupplung *f*
u.-joint housing Antriebsgelenkgehäuse *n*
u.-joint multiple drilling machine Gelenkspindelbohrmaschine *f*
u.-joint shaft Gelenkwelle *f*
u.-joint spindle Gelenkspindel *f*
u.-joint spindle holder Gelenkspindelhalter *m*
u.-joint spline Nutenstück *n* des Kardangelenks
u.-joint type drilling machine Gelenkbohrmaschine *f* mit einstellbaren Spindeln
u.-joint yoke *(Kfz)* Kreuzgelenkgabel *f*
u. knee and column type milling machine Universal[konsol]fräsmaschine *f*
u. lathe Universaldrehmaschine *f*
u. lathe chuck Universal[drehmaschinen]futter *n*
u.-mains radio set Allstromradiogerät *n*

u. match machine Universalzündholzmaschine *f*
u. miller Universalfräsmaschine *f*
u. milling attachment Universalfräskopf *m*; Universalfräseinrichtung *f*
u. milling machine Universalfräsmaschine *f*
u. motor Universalmotor *m*
u. network *(El)* Allpaß[netzwerk *n*] *m*
u. pressure boiler Vieldruckkessel *m*
u. rolling mill Universalwalzwerk *n*, Mehrzweckwalzwerk *n*
u. shaft Gelenkwelle *f*
u. shaping machine Universalwaagerechtstoßmaschine *f*
u. shunt *(El)* Mehrfachnebenwiderstand *m*, kombinierter Nebenwiderstand *m*
u. shunt box *(El)* Universalnebenschluß *m*
u. space Weltraum *m*
u. spanner Universalschraubenschlüssel *m*
u. spark plug wrench Universalzündkerzenschlüssel *m*
u. suspension Kardanaufhängung *f*
u. time Weltzeit *f*
u. time clock Weltzeituhr *f*
u. tool and cutter grinder Universalwerkzeugschleifmaschine *f*
u. turret lathe Universalrevolverdrehmaschine *f*
u. valve Universalröhre *f*
u. varnish Universallack *m*
u. welding machine Universalschweißmaschine *f*
universe Weltraum *m*, Universum *n*
university education Hochschulbildung *f*
u. graduate engineer Hochschulingenieur *m*
univibrator Univibrator *m*
univocal *(Math)* eindeutig
unkilled steel unberuhigter Stahl *m*; unberuhigt vergossener Stahl
unknown *(Math)* unbekannt
unknown *(Math)* Unbekannte *f*
unlace / to aufschnüren
unlaid rope aufgedrehtes Seil *n*
unlatch / to ausklinken, lösen
unleaded unverbleit, bleifrei *(Kraftstoff)*
unlike 1. ungleich; 2. unähnlich
u. currents *(El)* Ströme *mpl* entgegengesetzter Richtung
u. poles *(El)* ungleichnamige Pole *mpl*
unlimber / to *(Mil)* abprotzen
unlimited unbegrenzt
unlined 1. nichtausgemauert; *(Brg)* ohne Ausbau, unausgebaut; 2. ohne Belag *(Futter, Besatz)*
unload / to 1. entladen, abladen; löschen *(Schiffe)*; 2. abnehmen, abspannen, abführen *(Werkstücke)*; ausspannen *(Werkzeuge)*; 3. entlasten, entspannen; 4. herausnehmen *(Film)*
unloaded cable nichtbespultes Kabel *n*, unbespultes Kabel
unloading 1. Abladen *n*; Löschen *n (Schiffe)*; Entladung *f (Wagen)*; 2. Entlasten *n*, Entlastung *f*, Entspannen *n*; 3. Ausspannen *n*; Abführung *f*, Abführen *n (Werkstück)*; 4. Herausnehmen *n (Film)*
u. chute Abführrinne *f*
u. components Abnehmen *n* von Werkstücken
u. device *(Masch)* Entladeeinrichtung *f*, Abführungseinrichtung *f*
u. point Entladestelle *f*
u. station *(Masch)* Entladestation *f (Taktstraße)*
u. time Entladefrist *f (eines Schiffes)*
unlock / to entriegeln, lösen
unlocking Entriegelung *f*, Lösen *n*
unlubricated ungeschmiert
unmachinable unzerspanbar
unmanageable nicht manövrierfähig *(z. B. ein Schiff)*
unmanned rocket unbemannte Rakete *f*
unmeasurable unmeßbar
unmeltable unschmelzbar
unmesh / to ausrücken
unmixed batch capacity Füllmenge *f (Betonmischer)*
unmodulated unmoduliert
u. signal *(Fmt)* Dauerstrich *m*, unmoduliertes Zeichen *n*
unmoor / to abankern, losmachen *(Schiff)*
unnavigable nicht schiffbar, unbefahrbar für Schiffe
unnecessary operation unnötiges Arbeiten *n (z. B. des Relais)*
unobstructed approach hindernisfreier Anflug *m*
u. chip flow ungehinderter Spänefluß *m*
u. view ungehinderte Sicht *f*
unoccupied time *(Fmt)* Leer[lauf]zeit *f*
unoil / to entölen
unoperated nicht eingeschaltet
u. position Ruhestellung *f*
unoriented ungerichtet, nichtorientiert
unpack / to *(Dat)* entpaketieren
unpack ball hardness test *(Met)* Fallhärteprüfung *f*
unpaired unpaar
unperforated unperforiert
unplasticized *(Plast)* ohne Weichmacher, unplastifiziert
unpolarized unpolarisiert
unproductive unproduktiv
unquenched coke ungelöschter Koks *m*
unravel / to zerfasern
unrefined antimony Rohantimon *n*
unreinforced unbewehrt
unrelated beziehungslos, ohne Beziehung

unresisted flow *(Wärm)* verlustlose Strömung *f*
unrestricted ungehindert
u. plastic flow uneingeschränktes plastisches Fließen *n*
unrig / to abtakeln *(Schiff)*
unriveting hammer Entnietungshammer *m*
unroof / to abdecken *(Dach)*
unrust / to entrosten
unsaponifiable unverseifbar
u. matter Unverseifbares *n*
unsaturated *(Ch)* ungesättigt, unsaturiert
unscoured silk Bastseide *f*
unscreened 1. *(El)* unabgeschirmt; 2. ungesiebt
unscrew / to abschrauben, lösen, losschrauben
unsealed radioactive material offenes radioaktives Präparat *n*
unseaworthiness Seeuntüchtigkeit *f*, Fahruntüchtigkeit *f*
unserviceable unbrauchbar; betriebsunfähig, untauglich
unsettled ground *(Brg)* noch nicht zur Ruhe gekommenes Gebirge *n*
unshaded unschraffiert
unsheathed explosive *(Brg)* nichtummantelter Sprengstoff *m*
unshielded ohne Strahlenschutz; unabgeschirmt
u. [arc] welding offenes Lichtbogenschweißen *n*
unshift-on-space selbsttätige Zwischenraumbildung *f (Telegrafie)*
unship / to löschen, entladen *(Schiffe)*
unshrinkable *(Text)* nicht einlaufend
unshunted *(El)* ohne Nebenschluß, ohne Nebenwiderstand; unüberbrückt, ohne Brücke
unsimilar charges *(El)* ungleichnamige Ladungen *fpl*
unsinkable unsinkbar *(Schiff)*
unsized paper ungeleimtes Papier *n*
unskilled female labour ungelernte weibliche Arbeitskräfte *fpl*
u. male labour ungelernte männliche Arbeitskräfte *fpl*
unslaked lime ungelöschter Kalk *m*
unsmeltable unverhüttbar
unsolder / to ablöten, loslöten
unsoling pincers Hufabreißzange *f*
unsolvable unlösbar
unsolved ungelöst
unsound castings fehlerhafter Guß *m*
u. steel unberuhigter (blasiger) Stahl *m*
unspillable accumulator Akkumulator *m* mit gelatiniertem Elektrolyt
unsplittable unspaltbar
unspool / to abspulen
unsprung ungefedert
unstable unstabil, instabil; radioaktiv
u. atom instabiles (angeregtes) Atom *n*
u. equilibrium instabiles Gleichgewicht *n*
u. governor labiler Regler *m*
u. ground *(Brg)* nichtstandfestes Gebirge *n*
u. nucleus instabiler Kern *m*
u. oscillation labile Schwingung *f*
u. ship kursunstabiles Schiff *n*
unsteady unstet[ig], unbeständig; nichtstationär *(Strömung)*
u. cavity unstetiger Hohlraum *m (Kavitation)*
unstick / to abheben *(Flugzeug vom Boden)*
unstitch / to *(Text)* auftrennen
unstrained member Nullstab *m*, 0-Stab *m (Fachwerk)*
unstratified *(Geol)* schichtungslos, ungeschichtet
unstress / to entspannen
unstressed spannungsfrei
u. reinforcement schlaffe Bewehrung *f (Betonbau)*
unsupercharged ohne Aufladung *(Motor)*
unsupported trägerlos; nichtaufliegend, nichtgelagert
u. length of column freie Knicklänge *f*
u. sheeting Folie *f* ohne Füller, trägerlose Folie
unsymmetrical grading *(Fmt)* unsymmetrische Staffelung *f*
untearable unzerreißbar
untempered unangelassen *(Stahl)*
untensioned reinforcement schlaffe Bewehrung *f (Betonbau)*
unthread / to *(Text)* den Faden ausziehen
untie / to *(Text)* aufdrehen, abknüpfen
untile / to abdecken *(Dach)*
untilled land (soil) Brachland *n*, Brache *f*
untreated pole rohe Stange *f*, roher Holzmast *m*, nichtimprägnierter Mast *m*
untrimmed size Rohformat *n (Papier)*
untrue ungenau; unrund; versetzt
untuned aerial aperiodische Antenne *f*
u. secondary circuit *(El)* aperiodischer Sekundärkreis *m*
untuning *(Fmt)* Verstimmen *n*
untwist / to *(El)* entdrillen, aufdrehen *(Litze)*
untwisting machine Abwickelmaschine *f*
unused unbenutzt
unvisible radiation unsichtbare Strahlung *f*
unwashed coal Förderkohle *f*, ungewaschene Kohle *f*
unwatched unbewacht *(Feuer)*; unbemannt *(Schiff)*
unwieldy sperrig
unwind / to abwickeln *(Geometrie)*; abspulen
unworkable seam nicht [ab]bauwürdiges Flöz *n*, unbauwürdiges Flöz
unworked *(Brg)* unverritzt
up / to erhöhen

u.-date modernisieren, auf den neuesten Stand *(der Technik)* bringen
u.-end das Ende hochhalten
up-and-down line Hin- und Rückleitung *f*
up-and-down movement auf- und abgehende Bewegung *f*
up-and-over door Kipptor *n*
up-cut milling Gegenlauffräsen *n*
up-cutting Gegenlauffräsen *n*
up-draught carburetter Steigstromvergaser *m*
up-dredge Hochschnittbagger *m*
up-feed method of milling Gegenlauffräsverfahren *n*
up-grading Veredelung *f*
up hole *(Brg)* aufwärts gerichtetes Bohrloch *n*
up-milling Gegenlauffräsen *n*
up-movement Aufwärtsbewegung *f*
up-stream *s.* upstream
uparching of strata *(Geol)* Aufwölbung *f* der Schichten
upcast shaft *(Brg)* Ausziehschacht *m*, ausziehender Wetterschacht *m*
upcut *(Masch)* Oberhieb *m*
updraft carburetter Steigstromvergaser *m*
u. fire *(Met)* Oberfeuer *n*
u. type furnace *(Met)* Ofen *m* mit steigender Flamme
upgraded verbessert
uphill cast / to steigend gießen
upholster / to polstern
upholstered seat Polstersitz *m*
upholstery fabric Möbelstoff *m*
upkeep Instandhaltung *f*, Wartung *f*, Unterhaltung *f (von Anlagen)*
uplift Auftrieb *m*; Aufrichtung *f* von Bodenschichten, Bodenhebung *f*
upper 1. Schaft *m (Oberteil des Schuhes)*; 2. aufwärts gerichtetes Bohrloch *n*
u.-air research rocket Rakete *f* zur Erforschung der oberen Schichten der Atmosphäre
u. atmosphere instrumentation Apparatur *f* zur Erforschung der oberen Schichten der Atmosphäre
u. beam *(Kfz)* Fernlicht *n*
u. beam of press brake *(Masch)* Oberwange *f*
u. blade Obermesser *n*
u. bosh line Kohlensack *m (Hochofen)*
u. bound *(Math)* obere Schranke *f*
u. box *(Met)* Oberform *f*
u. calorific value oberer Heizwert *m*
u. chamber of the collector Oberkammer *f* der Schleuse *(Turbine)*
u. cutting blade Obermesser *n*
u. cylinder Oberkessel *m*
u. cylinder lubricant *(Kfz)* Oberschmiermittel *n*, Obenschmieröl *n*
u. dead centre oberer Totpunkt *m*
u. deck *(Schiff)* Oberdeck *n*, Hauptdeck *n*
u. deviation oberes Abmaß *n*
u. die Obergesenk *n*
u. edge of keel *(Schiff)* Oberkante Kiel *(Vermessungslinie)*
u. flue *(Wärm)* Oberzug *m*
u. gates Oberhaupt *n*, Binnenhaupt *n (einer Schleuse)*
u. harmonic *(El)* Oberwelle *f*
u. heat Oberhitze *f*
u. keel *(Schiff)* Oberkiel *m*
u. knife Obermesser *n*
u. limiting frequency obere Grenzfrequenz *f*
u. longeron *(Flg)* Rumpfoberholm *m*
u. loop *(Text)* Oberschlinge *f*
u. platen Pressentisch *m*
u. port sill Oberdrempel *m (Dock)*
u. press knives Oberlederstanze *f (Schuhfertigung)*
u. prop *(Brg)* Innenstempel *m*, Oberstempel *m (Schachtausbau)*
u. rim Oberfelge *f (eines Rades)*
u. roll Oberwalze *f*
u. shed *(Text)* Oberfach *n*, Hochfach *n*
u. slide Oberschlitten *m*
u. thread Oberfaden *m*
u. work Oberschiff *n*, Oberwerk *n*, Überwasserschiff *n*
u. work table Werkstückobertisch *m*
uppermost position oberste Stellung *f*
upright *(Masch)* Ständer *m*
upright aufrecht, stehend
u. carrying the boring spindle Bohrständer *m*
u. converter *(Met)* stehender Konverter *m*
u. drilling machine Säulenbohrmaschine *f*
u. format Hochformat *n*
u. housing *(Masch)* Ständer *m*
u. lattice *(El)* Steiglattentuch *n*, Steiggitter *n*
u. loop stehende Schleife *f*
u. spindle Stange *f (Parallelreißer)*
u. steam engine stehende Dampfmaschine *f*
u. twill *(Text)* Diagonalköper *m*, steiler Köper *m*
u.-type engine Standmotor *m*
uprights *(Schiff)* Hellingmast *m*; *(Masch)* Ständer *mpl*
uproot / to entwurzeln, roden
uprooting Entwurzeln *n*, Roden *n*
upset / to stauchen, anstauchen *(Umformung)*
upset butt welding [Widerstands-]Preßstumpfschweißung *f*
u. forging Stauchen *n*
u. pass Stauchkaliber *n (Walzwerk)*
u. point Schließkopf *m*
u. welding Stumpfschweißen *n*; Preßschweißung *f*

upsetting Stauchen *n*; Kaltstauchen *n*
u. die Stauchmatrize *f*
u. moment *(Schiff)* Kentermoment *n*, Kippmoment *n*
upstand beam *(Bw)* Überzug *m*
upstream flußaufwärts, oberwasserseitig, stromauf; wasserseitig
u. navigation *(Schiff)* Bergfahrt *f*
u. pressure Druck *m* hinter dem Ventil
u. traffic *(Schiff)* Bergverkehr *m (auf Flüssen)*
upstructure deflection of the bore hole Bohrlochabweichung *f* gegen das Schichtenfallen *(Tiefbohrtechnik)*
uptake Schornsteinschacht *m*
u. ventilator Entlüfter *m*, Saug[kopf]lüfter *m*
upthrown fault *(Geol)* Sprung *m* ins Hangende
upthrust *(Bw)* nach oben gerichtete Auflagerkraft *f*; Auftrieb *m*
uptwister Etagenzwirnmaschine *f*
upward-current classifier Aufstromklassierer *m (Aufbereitung)*
u. growth Längenwachstum *n*
u. modulation *(Fmt)* additive Modulation *f*
u. pull Saugzug *m (Schornstein)*
uranate Uranat *n*, Salz *n* der Uransäure
uranic acid Uransäure *f*
u. compound Verbindung *f* des sechswertigen Uran
uraniferous uranhaltig
uranite Uranit *n*, Uranglimmer *m*
uranium Uran *n*
u. bomb Uran[spalt]bombe *f*
u. cartridge Uranstab *m*; Uranstabhülse *f*
u. compound Uranverbindung *f*
u. concentrate Urankonzentrat *n*
u. content Urangehalt *m*
u. deposit Uranlagerstätte *f*, Uranerzvorkommen *n*
u. fission Uran[kern]spaltung *f*
u. free uranfrei
u. fuel element Uranbrennstoffelement *n*
u. nucleus Urankern *m*
u. ore Uranerz *n*
u. ore detector Uransuchgerät *n*
u. prospecting Uranprospektion *f*
u. radioactive series *s.* u. series
u. reactor Uranreaktor *m*
u. salt Uransalz *n*
u. series Uran[zerfalls]reihe *f*
uranous compound Verbindung *f* des vierwertigen Urans
uranyl acetate Uranylazetat *n*
urate Urat *n*, Ester *m* (Salz *n*) der Harnsäure
urea Harnstoff *m*
u.-formaldehyde resin Formaldehydharz *n*, Karbamidharz *n*
u. laminated sheet Harnstoffharzschichtstoff *m*
u. resin Harnstoffharz *n*
urena Urenastengelfaser *f*
urgency frequency Dringlichkeitsfrequenz *f*
urgent call *(Fmt)* dringendes Gespräch *n*
u. message dringendes Telegramm *n*
usability Brauchbarkeit *f*, Gebrauchseignung *f*, Verwendbarkeit *f*
usable nutzbar, verwendbar
usage Ausnutzung *f*
use Anwendung *f*, Verwendung *f*; Nutzen *m*; Verwendungszweck *m*
used fuel *(Kern)* abgebrannter Brennstoff *m*
u. oil Altöl *n*
u. paper Altpapier *n*
u. sand Altsand *m*, gebrauchter Sand *m (Gießerei)*
useful cross-section *(El)* Nutzquerschnitt *m*, Wirkquerschnitt *m (des Leiters)*
u. current Wirkstrom *m*, Nutzstrom *m*
u. damping Nutzdämpfung *f*
u. deadweight Ladefähigkeit *f*, Tragfähigkeit *f* an reiner Ladung
u. effect Nutzeffekt *m*
u. life Nutzlebensdauer *f*, Betriebslebensdauer *f*
u. load Nutzladefähigkeit *f*
u. magnification *(Opt)* förderliche Vergrößerung *f (nach Abbe)*
u. output Nutzleistung *f*
u. power Nutzleistung *f*
u. power output effektive Leistung *f*, Nutzleistung *f*
u. resistance Nutzwiderstand *m*
u. screen area *(Fs)* Nutzbildfläche *f*
u. voltage Nutzspannung *f*
user Benutzer *m*; Verbraucher *m*
utensil socket Gerätesteckdose *f*
utensils Geräte *npl*
utility Installation *f*; Versorgungseinrichtung *f*
u. aircraft Mehrzweckflugzeug *n*
u. core *(Bw)* Installationskern *m*, Installationszelle *f*
u. line Versorgungsleitung *f*
u. operating method *(Am)* *(Fmt)* ständige Bereitschaft *f*
u. run *(Bw)* Leitungskanal *m*
u. trench *(Bw)* Leitungsgraben *m*
utilization coefficient Ausnutzungskoeffizient *m*
u. factor Ausnutzungsfaktor *m*, Wirkungsgrad *m*; *(Fmt)* Belegungswahrscheinlichkeit *f*
u. of heat Wärmeausnutzung *f*
u. of steam Dampfausnutzung *f*
u. of waste heat Abdampfverwertung *f*
u. time Nutzzeit *f*, Hauptnutzzeit *f*
utilize / to nutzbar machen, [aus]nutzen, verwerten

V

V- *(Am)* V-förmig ausgebildet
V-aerial V-Antenne *f*
V band V-Band *n (46000 bis 56000 MHz)*
V-belt *(Am)* Keilriemen *m*
V-belt pulley Keilriemenscheibe *f*
V-block Anreißprisma *n*, Anreißblock *m*
V-bottom boat V-Bodenboot *n*, Knickspantboot *n*
V-bottom form *(Schiff)* V-Bodenform *f*, Wellenbinderform *f*
V-cut *(Brg)* Keileinbruch *m (Schießarbeit)*
V-cylinder block *(Kfz)* Zylinderanordnung *f* in V-Form
V-drive Keilriemenantrieb *m*
V-groove Kerbnute *f*; Gewinderille *f*
V-notch Spitzkerbe *f*
V-port plug Schlitzkegel *m*
V-shaped V-förmig ausgebildet
V-shaped body V-förmige Mulde *f*
V-shaped milling cutter Prismenfräser *m*
V-shaped thread Spitzgewinde *n*
V-shaped vessel Schiff *n* mit V-Spantform
V-shed *(Text)* reines (schräges) Fach *n*, Schrägfach *n*
V-thread Spitzgewinde *n*
V-type V-förmig ausgebildet
V-type contact *(El)* Messerkontakt *m*
V-type engine V-Motor *m*
V-unit *(Bw)* Falte *f*, Faltwerk *n*
VV-drive Keilriementrieb *m (beide Räder mit Keilnuten)*
V-way *(Masch)* V-Bahn *f*, prismatische Führung *f*
V-weld V-Schweißnaht *f*
vacancy Leerstelle *f*, Gitterlücke *f*, Gitterleerplatz *m (von Kristallen)*
  v. concentration Leerstellenkonzentration *f*
  v. migration Leerstellenwanderung *f*
vacant unbesetzt, leer, frei
  v. contact Leerkontakt *m*, freier Kontakt *m*
  v. electron site Elektronenfehlstelle *f*, Elektronenloch *n*, Loch *n*, Defektelektron *n (im Kristallgitter)*
  v. lattice site *s.* v. site in the lattice
  v. line *(Fmt)* freie Leitung *f*
  v. site in the lattice Gitterlücke *f*, Gitterfehlstelle *f*, Leerstelle *f* im Gitter
  v. terminal *(Am) (El)* Leerklemme *f*, Leerkontakt *m*
vaccin Vakzin *n*, Impfserum *n*, Impfstoff *m*
vacuum [away] / to absaugen, trocknen
vacuum Vakuum *n*; Leere *f*; Unterdruck *m*
vacuum evakuiert
  v. advance mechanism Unterdruckzündversteller *m*
  v. augmenter *(Wärm)* Vakuumvermehrer *m*
  v. back *(Typ)* Vakuumfilmhalter *m*, Filmsaugplatte *f (einer Kamera)*
  v. brake Saugluftbremse *f*
  v. brake system Unterdruckbremsanlage *f*
  v. breaker *(Wärm)* Vakuumreglerventil *n*, Vakuumleck *n*
  v. casting Vakuumguß *m*
  v. chamber Vakuumkammer *f*
  v. chuck *(Masch)* Saugluftspannfutter *n*
  v. cleaner Staubsauger *m*
  v. concrete Vakuumbeton *m*
  v.-controlled carburetter *(Kfz)* Vergaser *m* mit Unterdrucksteuerung
  v. conveyor tube Saugluftförderer *m*
  v. curve *(Wärm)* Vakuumlinie *f*
  v. cylinder former Vakuumrundsiebmaschine *f*
  v. deposition Vakuumbedampfung *f*
  v. distillation Vakuumdestillation *f*, Destillation *f* im Vakuum
  v. distributor Saugluftverteiler *m*
  v. doughnut *(Kern)* ringförmige Vakuumkammer *f*
  v. drier Vakuumeindampfer *m*
  v. effect Sogeinwirkung *f*
  v. filter Vakuumfilter *n*
  v. forming *(Plast)* Vakuumverformung *f*
  v. frame *(Typ)* Vakuumkopierrahmen *m*, pneumatischer Kopierrahmen *m*
  v. freeze drier Vakuumgefriertrockner *m*
  v. fuel pump Unterdruckförderpumpe *f*
  v. furnace Vakuumofen *m*
  v. fusion Vakuumschmelze *f*
  v. gauge Vakuummeter *n*, Unterdruckmesser *m*
  v. governor Unterdruckregler *m*
  v. ignition adjustment Unterdruckzündverstellung *f*
  v. ignition advance Unterdruckzündverstellung *f*
  v. impregnation Vakuumimprägnierung *f*
  v. ionization gauge Ionisationsvakuummesser *m*
  v. line Saugluftleitung *f*
  v. lock Vakuumverschluß *m*; Vakuumschleuse *f*
  v. melting plant Vakuumschmelzanlage *f*
  v. metalling Aufdampfen *n*
  v. metallizing Vakuumbedampfung *f (zur Schaffung von Metallüberzügen)*
  v. metallurgy Vakuummetallurgie *f*
  v.-operated hydraulic brake saugluftbetätigte hydraulische Bremse *f*
  v. pan Vakuum[gieß]pfanne *f*
  v. polarization Vakuumpolarisation *f*, Polarisation *f* des Vakuums
  v.-power change Saugluftschaltung *f*
  v. preparative process Vakuumpräparationsverfahren *n*

v. pump Saugluftpumpe *f*, Vakuumpumpe *f*
v. pump element Unterdruckförderer *m*
v. reservoir Saugluftbehälter *m*
v. seal Vakuumdichtung *f*
v. servo brake Saugluftbremse *f*
v. shelf drier Vakuumtrockenschrank *m*
v. shift cylinder Saugluftschaltzylinder *m*
v. snap-back Aufschrumpfen *n* unter Vakuum
v. spectrograph Vakuumspektrograf *m*
v. steaming equipment Vakuumdämpfeinrichtung *f*
v. switch Vakuumschalter *m*
v. system of broken end collection Fadenabsaugsystem *n (Spinnerei)*
v. technique Vakuumtechnik *f*
v.-tight vakuumdicht
v. timing control Unterdruckzündverstellung *f*
v. tube *s.* v. valve 2.
v.-type induction furnace Vakuuminduktionsofen *m*
v. unit Unterdruckregler *m*
v. valve 1. Sicherheitsluftventil *n*; 2. Vakuumröhre *f*
v. valve amplifier Röhrenverstärker *m*
v. valve detector Röhrendetektor *m*, Lampendetektor *m*
v. valve noise *(Fmt)* Verstärkerrauschen *n*
v. valve oscillator Röhrengenerator *m*
v. valve oscillator set Röhrensender *m*
v. valve rectifier Hochvakuumgleichrichter *m*
v. valve switch Elektronen[röhren]schalter *m*
v. valve voltmeter *(El)* Röhrenvoltmeter *n*
vagabond current Erdstrom *m*, vagabundierender Strom *m*
vagrant vagabundierend
valence Valenz *f*, Wertigkeit *f*
v. angle Valenzwinkel *m*
v. band Valenzband *n*
v. band distortion Valenzbandverzerrung *f*
v. band level Valenzbandniveau *n*
v. band state Valenzbandzustand *m*
v. bond Valenzbindung *f*
v. electron Valenzelektron *n*, Wertigkeitselektron *n*, äußeres Elektron *n*
v. force Valenzkraft *f*
v. lattice Valenzgitter *n*
v. linkage Valenzbindung *f*
v. orbital Valenzbahn *f*, Valenzzustand *m*
v. shell Valenzschale *f*
v. vibration Valenzschwingung *f*
Valenciennes bobbin lace *(Text)* Valenciennesspitze *f*
valency Valenz *f*, Wertigkeit *f*; Ladungszahl *f*
v. force Valenzkraft *f*
valent wertig
valeraldehyde Valeraldehyd *m*
valerianic acid Valeriansäure *f*
valeric acid Valeriansäure *f*
v. aldehyde Valeraldehyd *m*
valid up to gültig bis zu
validation Gültigkeitserklärung *f (Patentwesen)*
validity limit Gültigkeitsgrenze *f*, Gültigkeitsbereich *m*
valley *(Bw)* Kehle *f*, Tal *n (Dach)*
value / to bewerten, abschätzen
value Wert *m*; Größe *f*, Betrag *m*, Menge *f*
v. of game *(Dat)* Spielwert *m*
valueless 1. wertlos; 2. taub *(Gestein)*
valve 1. Ventil *n*; Klappe *f*; Schieber *m*; Absperrorgan *n*, Regelorgan *n*; 2. *(El)* Röhre *f*, Elektronenröhre *f*
v. action *(El)* Ventilwirkung *f*, Gleichrichterwirkung *f*
v. actuating gear Stellgetriebe *n*, Ventilstellgetriebe *n*
v. adapter Zwischensockel *m*, Röhrenzwischensockel *m*
v.-adjusting stud Ventil[ein]stellschraube *f*
v. adjustment Ventileinstellung *f*
v. aging Röhrenalterung *f*
v. alternating current resistance Innenwiderstand *m (von Röhren)*
v. amplifier Röhrenverstärker *m*
v. anode characteristic resistance Innenwiderstand *m (von Röhren)*
v. arm Ventilhebel *m*
v. base Röhrensockel *m*
v. body Ventilgehäuse *n*, Ventilkörper *m*
v. box Ventilgehäuse *n*
v. cage Ventilkäfig *m*, Ventilkorb *m*
v. cap Ventilaufsatz *m*, Ventildeckel *m*, Ventilkappe *f*
v. casing Ventilgehäuse *n*
v. chamber Ventilgehäuse *n*, Ventilkammer *f*
v. characteristic Röhrenkennlinie *f*
v. checker unit Röhrenprüfgerät *n*
v. checking Röhrenprüfung *f*
v. chest Ventilkasten *m*; Ventilgehäuse *n*
v. clamp Röhrenhalter[ung *f*] *m*, Röhrenklemme *f*
v. clearance Ventilspiel *n*
v. cock Ventilhahn *m*
v. collar Ventilklemmkegel *m*
v. complement Röhrenbestückung *f*, Röhrensatz *m*
v. cone Ventilkegel *m*
v. control Ventilsteuerung *f*
v. control linkage Ventilsteuergestänge *n*
v. control motor Ventilgetriebe *n*, Stellantrieb *m*, Stellmotor *m*
v. control rod Ventilregelstange *f*
v. cotter Ventilkeil *m*
v. cover Ventildeckel *m*
v. cover plate Ventilgehäusedeckel *m*

v. detector Röhrendetektor *m*, Lampendetektor *m*
v. diagram *(Wärm)* Steuerungsdiagramm *n*
v. disk Ventilteller *m*
v. drive elektronisch gesteuerter Oszillator *m*
v. drop Anodenspannungsabfall *m* in der Röhre
v. dropping Schließen *n* des Ventils
v. electrode Röhrenelektrode *f*
v. electrometer Röhrenelektrometer *n*
v. extension Ventilverlängerung *f (Luftreifen)*
v. extractor Röhrenauszieher *m*
v. face Ventilsitzfläche *f*
v. facing Hartmetallauflage *f (eines Ventils)*
v. failure Röhrenausfall *m*
v. fitting Ventileinsatz *m*
v. flap Ventilkappe *f*
v. for superheated steam Heißdampfventil *n*
v. gear Ventil[an]trieb *m*, Ventilgestänge *n*, Ventilmechanismus *m*, Ventilsteuerung[sorgan *n*] *f*
v.-gear arrangement Ventilanordnung *f*
v.-gear housing Ventilgehäuse *n*, Ventilverkleidung *f*
v. grinding Ventileinschleifen *n*
v. guide Ventilführung *f*
v. head Ventilteller *m*, Ventilkegel *m*
v. holder Röhrenfassung *f*, Röhrensockel *m*
v. hood Ventilhaube *f*, Ventilkappe *f*
v.-in-head engine kopfgesteuerter (obengesteuerter) Motor *m*
v. insulator Röhrenisolator *m*
v. lag Ventilverzögerung *f*
v. lead Ventilvor[er]öffnung *f*
v. leather Leder *n* für Dichtungen
v. lift Hubhöhe *f*, Ventilerhöhung *f*, Ventilhub *m*
v.-lift period Ventilöffnungsdauer *f*, Ventilöffnungszeit *f*
v.-lift stop Ventilhubbegrenzung *f*
v. lifter Ventilmitnehmer *m*, Ventilstößel *m*
v.-lifting velocity Ventilöffnungsgeschwindigkeit *f*
v. line-up Röhrenbestückung *f*
v. location Ventilanordnung *f*, Ventillage *f*
v. manual Röhren[hand]buch *n*
v. of bellows Balgenklappe *f (Orgel)*
v. opening duration Ventilöffnungsdauer *f*
v. opening pressure Ventilöffnungsdruck *m*
v. operating mechanism Ventilantrieb *m*, Ventilbetätigungsmechanismus *m*, Ventilsteuerungsvorrichtung *f*
v. oscillator Röhrengenerator *m*, Röhrensummer *m*
v. outlet Ventilauslaß *m*
v. output Röhrenleistung *f*
v. overlap Überschneidung *f* [der Ventilöffnungszeiten]
v. packing disk Ventildichtungsscheibe *f*
v. petticoat Ventilschutzhaube *f*
v. pin Röhrensockelstift *m*
v. plate Ventilplatte *f*
v. plug Ventilgehäuseverschluß *m*; Ventilkegel *m*
v. poppet Ventilkegel *m*
v. port Ventilkanal *m*, Ventilöffnung *f*
v.-position indicator Ventilstellungsmelder *m*
v.-position potentiometer Hubpotentiometer *n*
v. positioner Positionär *m*, Stell[ungs]macher *m*; Ventilstellwerk *n*
v. power Röhrenleistung *f*
v. prong Röhrensockelstift *m*
v. push rod Ventilanhubstange *f*, Ventilschubstange *f*, Ventilstößelstange *f*, Ventilstoßstange *f*
v. push rod ball Ventilstoßstangenkugelpfanne *f*
v. reactor Reaktanzröhre *f*, Hubröhre *f*
v.-reactor frequency modulator Frequenzmodulator *m* mit Induktivitätsröhre *(Blindstromröhre)*
v.-reactor modulator Röhrenschaltung *f* für Frequenzmodulation
v.-reactor phase modulator Phasenmodulator *m* mit Reaktanzröhre
v. receiver Röhrenempfänger *m*
v. rectifier Röhrengleichrichter *m*
v. refacer Ventilnachschleifmaschine *f*
v. rejection test Röhrenprüfung *f (im Betrieb)*
v. remover Ventilhebezange *f*
v. reseater Ventilsitznachschleifmaschine *f*, Ventilsitzfräser *m*
v. rocker Kipphebel *m*, Schwinghebel *m*, Ventilsteuerhebel *m*, Ventilstößel *m*
v.-rocker arm Ventilschwinghebel *m*, Ventilkipphebel *m*
v.-rocker bracket Kipphebelblock *m*, Schwinghebelblock *m*, Ventilsteuerhebelstütze *f*
v.-rocker fulcrum pin Kipphebelzapfen *m*, Schwinghebeldrehbolzen *m*, Ventilsteuerhebelzapfen *m*
v.-rocker pedestal Kipphebelblock *m*, Schwinghebelblock *m*
v. rod Ventilschieberstange *f*, Ventilspindel *f*
v. rod packing Schieberstangendichtung *f*, Schieberstangenliderung *f*, Schieberstangenpackung *f*
v. roller Kipphebelrolle *f*, Schwinghebelrolle *f*
v. rotator Ventildreheinrichtung *f*
v. scavenging Ventilspülung *f*
v.-sealing surface Ventildichtungsfläche *f*
v. seasoning Röhrenalterung *f*
v. seat Ventilsitz *m*, Aufsitzfläche *f*, Sitzfläche *f* des Ventiltellers

v.-seat contact Ventilschluß *m*
v.-seat grinding machine Ventilsitzschleifmaschine *f*
v.-seat insert Ventileinsatz[ring] *m*
v.-seat ring Ventilsitzring *m*
v. seating *(Kfz)* Aufsitzen *n* des Ventils; *(Wärm)* Klappensitz *m*
v.-seating machine Ventilsitzfräsmaschine *f*
v. setting gauge Ventileinstellehre *f*
v. shaft Ventilschaft *m*
v. shield Röhrenabschirmung *f*
v. shutter Ventilverschluß *m*
v. spindle Ventilspindel *f*
v. spring Ventilfeder *f*
v.-spring cap Ventilfedergehäuse *n*, Ventilfederhaube *f*, Ventilfederteller *m*
v.-spring cotter Ventilkeil *m*
v.-spring lifter Ventilfedermitnehmer *m*
v.-spring retainer Ventilfederteller *m*
v.-spring retainer lock Ventilkeil *m*
v.-spring surge Ventilfederschwingung *f*
v.-spring washer Ventilfederteller *m*
v. stabilizer Röhrenspannungsgleichhalter *m*
v. stem Ventilschaft *m*, Ventilspindel *f*, Ventilstößel *m*
v. stem guide Ventilführung *f*
v. sticking Verklemmung *f* des Ventils
v. stroke Ventilhub *m*
v. support Ventilträger *m*
v. tap Druckspüler *m (WC)*
v. tappet Ventilstößel *m*
v.-tappet guide Ventilstößelführung *f*
v.-tappet roller pin Ventilstößelrollenbolzen *m*
v. tester Röhrenprüfer *m*, Röhrenprüfgerät *n*
v. throat Ventilhals *m*
v. timing Ventileinstellung *f*, Ventilsteuerung *f*, Ventilsteuerzeit *f*
v. timing gear Ventilsteuerungsvorrichtung *f*
v. top Membrankammer *f (Ventil)*
v. train Ventilantrieb *m*, Ventilmechanismus *m*, Ventilsteuergestänge *n*
v. travel Hubweg *m* eines Ventils, Ventilhub *m*
v. voltage drop Spannungsabfall *m (in einer Röhre)*
v. voltmeter Röhrenvoltmeter *n*, Röhrenspannungsmesser *m*
v. wattmeter Röhrenwattmeter *n*
v. with variable slope Röhre *f* mit veränderlichem Durchgriff, Exponentialröhre *f*, Regelröhre *f*
valved röhrenbestückt
valveless compressor Kapselgebläse *n*
vamp *(Led)* Blatt *n*, Vorderblatt *n (Schuh)*
van de Graaff accelerator (generator) *(Kern)* Van-de-Graaff-Beschleuniger *m*, Van-de-Graaff-Generator *m*, Bandgenerator *m*
van der Waals crystals Van-der-Waals-Kristalle *mpl*
van der Waals equation van der Waalssche Gleichung *f*
vanadate Vanadat *n*, Salz *n* der Vanadinsäure
vanadic acid Vanadinsäure *f*
v. compound Verbindung *f* des fünfwertigen Vanadins
vanadium Vanadium *n*
v.-alloyed steel vanadiumlegierter Stahl *m*
vanadous compound Verbindung *f* des dreiwertigen Vanadiums
vane 1. Blatt *n*, Flügel *m*; Schaufel *f (Turbine)*; Rippe *f*; 2. *(El)* Rotorplatte *f (eines Drehkondensators)*; 3. Wetterhahn *m*
v. alignment box *(Rak)* Ruderabgleichkasten *m*
v. anemometer *(BMSR)* Klappenströmungsmesser *m*
v. attenuator Streifenabschwächer *m*, Streifenabsorptionsteiler *m (Wellenleitertechnik)*
v. flowmeter Klappendurchflußmesser *m*, Strömungswächter *m*
v. magnetron *(Fmt)* Fahnenmagnetron *n*
v. motor *(Rak)* Ruderantrieb *m*
v. of vacuum pump Pumpenlamelle *f*
v. position *(Rak)* Ruderstellung *f*
v. pump Flügel[zellen]pumpe *f*
v. ring Leitschaufelkranz *m (Turbine)*
v.-screw propeller Flügelradpropeller *m*
v. shape *(El)* Plattenschnitt *m (eines Drehkondensators)*
v. spindle *(Schiff)* Flügelspill *n*
v.-type hydraulic actuator hydraulischer Drehflügelmotor *m*
v.-type piston Flügelkolben *m*
v.-type relay Stellungszeiger *m*, Stellungsgeber *m*, Fähnchenrelais *n*
v. water meter Flügelradwassermesser *m*
v. wattmeter Fähnchenwattmeter *n (Wellenleitertechnik)*
v. wheel Flügelradpropeller *m*
v.-wheel ship Flügelradschiff *n*
v. wheeler Flügelradschiff *n*
vaned mine *(Mil)* Flügelmine *f*
vanillic acid Vanillinsäure *f*
vanish / to verschwinden; Null werden
vanishing cream Tagescreme *f*
v. line Fluchtlinie *f (Vermessung)*
v. point Fluchtpunkt *m (Vermessung)*
vapor *(Am) s.* vapour
vaporizability of the lubricant Verdampfbarkeit *f* des Schmiermittels
vaporization Verdampfung *f*
vaporize / to verdampfen, verdunsten
vaporizer Verdampfer *m*, Zerstäuber *m*
v. holder Verdampferhalter *m*
vaporizing carburetter Verdampfungsvergaser *m*

v. deposition Aufdampfen *n*, Aufdampfung *f*
v. oil Traktorenkraftstoff *m*
v. oil engine Traktorenmotor *m*
vaporous dampfförmig
vapour-coat / to bedampfen, aufdampfen
v.-deposit aufdampfen, bedampfen
vapour Dampf *m*, Dunst *m*, Brodem *m*, Brüden *pl*
v.-actuated thermometer Dampfspannungsthermometer *n*, Tensionsthermometer *n*
v. barrier Dampfsperre *f*
v. blasting Strahlläppen *n*
v. density Dampfdichte *f*
v. escape Brüdenabzug *m*, Dunstabzug *m*
v. hood Brüdenabzug *m*, Wrasenhaube *f*
v.-phase chromatography Gaschromatografie *f*, Dampfphasenchromatografie *f*
v. plating Aufdampfverfahren *n*
v. pressure Sättigungsdruck *m*; Dampfdruck *m*, Gasdruck *m*, Verdunstungsdruck *m (Benzin)*
v.-proof gasdicht
v.-proof installation *(El)* Feuchtraumanlage *f*
v. trail Kondensstreifen *m*
var-hour meter *(El)* Blindverbrauchszähler *m*
variability Veränderlichkeit *f*
v. of mass Massenveränderlichkeit *f*
v. of the steam distribution Verstellbarkeit *f* der Dampfverteilung
variable Veränderliche *f*, Variable *f*
variable veränderlich, variabel
v. air capacitor *(El)* Luftdrehkondensator *m*
v.-area flowmeter Schwebekörperdurchflußmesser *m*, Rotamesser *m*
v.-area nozzle Einstelldrossel *f*, Regeldrossel *f*, veränderliche Drossel *f*
v.-area recording Amplitudenschrift *f (Lichtton, Tonfilm)*
v.-area sound tracks Amplitudenschrift *f (Lichtton, Tonfilm)*
v. attenuator Dämpfungsregler *m*
v. base length veränderliche Base *f*
v. capacitance standard *(El)* Normaldrehkondensator *m*
v. capacitor *(El)* Drehkondensator *m*
v. capacitor with linear capacity increase kapazitätslinearer Drehkondensator *m*
v.-capacity pump Verstellpumpe *f*
v. compensator Drehkompensator *m*
v.-compression engine Motor *m* mit veränderlicher Verdichtung
v.-control switch *(El)* Serienschalter *m*
v.-correction unit *(El)* Tonblende *f (Entzerrung)*
v. cross section with constant ratio of divergence veränderlicher Querschnitt *m* bei beständigem Erweiterungsverhältnis *(Turbine)*
v.-cycle operation *(Dat)* Asynchronbetrieb *m*
v.-delivery pump Verstellpumpe *f*
v.-density recording Sprossenschrift *f*, Intensitätsschrift[methode] *f (Tonfilm)*
v.-density tracks Sprossenschrift *f*, Intensitätsschrift *f (Lichtton)*
v.-displacement pump Verstellpumpe *f*
v.-filling coupling füllungsgesteuerte Strömungskupplung *f*
v. frequency abstimmbare Frequenz *f*
v.-gain amplifier Regelverstärker *m*
v. ignition Zündpunktverstellung *f (Verbrennungsmotor)*
v.-incidence wing *(Flg)* Verstelltragfläche *f*
v. inductance *(El)* veränderliche Induktivität *f*
v. intermittent duty *(Elm)* aussetzender Betrieb *m* mit während eines Spiels veränderlicher Belastung
v.-length line Posaune *f (Wellenleiter)*
v.-mu veränderliche Steilheit *f (einer Elektronenröhre)*
v.-mu action *(El)* Exponentialwirkung *f*
v.-mu R.F. pentode HF-Regelpentode *f*
v.-mu tetrode Exponentialtetrode *f*
v. mutual-conductance valve Regelröhre *f*
v. pad *(El)* Kreuzgliedregler *m*
v.-pitch propeller (screw) Verstellpropeller *m*, Drehflügelschraube *f*
v.-pitch tooth surface broach Außenräumwerkzeug *n* mit ungleichmäßiger Zahnteilung
v. ratio of divergence veränderliches Erweiterungsverhältnis *n (Turbine)*
v.-ratio transformer Regeltransformator *m*, Regelumspanner *m*, Reguliertransformator *m*
v.-reactance line Blindleitung *f (Wellenleiter)*
v. resistance Regelwiderstand *m*
v. resistor Regelwiderstand *m*, Stellwiderstand *m*
v. slider *(El)* Abgriffschelle *f*
v.-slope valve Exponentialröhre *f*
v.-speed d.c. motor regelbarer Gleichstrommotor *m*
v.-speed drive Antrieb *m* mit gestuften Drehzahlen, Regelgetriebe *n*
v.-speed gear Verstellgetriebe *n*, Drehzahlverstellgetriebe *n*, Regelgetriebe *n*
v.-speed motor Motor *m* mit veränderlicher Drehzahl
v.-speed scanning *(Fs)* Abtastung *f* mit Geschwindigkeitsmodulation
v. star *(Astr)* veränderlicher Stern *m*
v. temporary duty *(Elm)* kurzzeitiger Betrieb *m* mit veränderlicher Belastung
v. transformer Regeltransformator *m*, Reguliertransformator *m*
v.-velocity scanning generator Zeilengenerator *m* für veränderliche Abtastgeschwindigkeit
v.-voltage control Regelung *f* durch Änderung der Spannung

v. with frequency frequenzabhängig
v. without steps stufenlos veränderlich
variance *(Math)* Abweichung *f (Mittelwert der Quadrate der Abweichungen)*, Streuung *f*; Varianz *f*
variation 1. Änderung *f*; Abweichung *f*; Ablenkung *f*; Schwankung *f*; 2. Abmaß *n (Passungslehre)*; 3. Mißweisung *f (magnetische Abweichung)*
v. between basic dimension and limit Nennabmaß *n*
v. in apparent bearing Peilstrahlschwankung *f*, Peilstrahlwanderung *f*
v. in combustion cut-off point *(Rak)* Brennschlußsteuerung *f*
v. in load Lastschwankung *f*
v. in path angle *(Rak)* Bahnwinkeländerung *f*
v. in signal strength *(Fmt)* Lautstärkeschwankung *f* im Empfang
v. in size Maßabweichung *f*; Größenunterschied *m*
v. in spindle speed Spindeldrehzahländerung *f*
v. in temperature Temperaturschwankung *f*
v. in voltage *(El)* Spannungsschwankung *f*
v. needle Kompaßnadel *f*
v. of cut-off Füllungsänderung *f (Dampfdiagramm)*
v. of frequency Frequenzschwankung *f*
v. of inverse amplification factor Durchgriffsänderung *f (Elektronenröhren)*
v. of load Lastspiel *n*
v. of quantity of water required Wasserbedarfsschwankung *f*
v. of stroke *(Masch)* Hubänderung *f*
variational calculus Variationsrechnung *f*
v. equation Variationsgleichung *f*
v. method Variationsrechnung *f*
v. principle Variationsprinzip *n*
v. resistance *(El)* Wechselstromwiderstand *m*
varicap *(BMSR)* Varicap, spannungsabhängiger Kondensator *m*
varicoupler veränderlicher Hochfrequenztransformator *m*
variegated verschiedenfarbig, bunt
v. copper ore Buntkupferkies *m*
variety 1. Vielzahl *f*; 2. Abart *f*; Sorte *f (z. B. Stahl)*
variocap spannungsabhängige Kapazität *f*
variolosser regelbares Dämpfungsglied *n*
variometer Drehdrossel *f*, Schiebedrossel *f*, Variometer *n*
v. coil Variometerspule *f*
varistor Varistor *m*, spannungsabhängiger Widerstand *m (Halbleiterwiderstand)*
varitron *(Kern)* Varitron *n*
varmeter *(El)* Varmeter *n*, Blindleistungsmesser *m*
varnish / to lackieren; tränken, imprägnieren
varnish Lack *m*, Firnis *m*; Lackfarbe *f*
v. coat Lacküberzug *m*
v. paint Lackfarbe *f*
v. remover Lackentferner *m*; Abbeizmittel *n*
varnished cambric Isolierband *n*
v. cambric cable *(El)* Lackbaumwollkabel *n*
v. fabric Lackgewebe[bahn *f*] *n*, geharzte Bahn *f*
v. hose Lackschlauch *m*
v. insulating tape Isolierlackband *n*
v. paper Lackpapier *n*; imprägniertes Papier *n*; Lackpapierbahn *f*
v. paper-insulated wire Lackpapierdraht *m*
v. silk-braided wire Seidenlackdraht *m*
v. sleeving Lackschlauch *m*
v. tape Lackband *n*
v. wire Lackdraht *m*
varnishing machine Lackiermaschine *f*, Beharzungs- und Imprägniermaschine *f*
varved clay *(Geol)* Bänderton *m*
vary / to variieren, abändern, sich ändern; schwellen; wechseln; schwanken; im Bereiche liegen; verstellen
v. steplessly stufenlos ändern
varying component Wechselstromkomponente *f*, Wechselstromglied *n*
v. output *(Wärm)* schwankende Leistung *f*
v. speed motor Motor *m* mit lastabhängiger Drehzahl
v. the point of cut-off Verstellen *n* der Füllung *(Dampfdiagramm)*
v. voltage control *(El)* Hauptstromregelung *f*
vat / to verküpen
vat Bottich *m*, Faß *n*, Kübel *m*, Küpe *f*, Trog *m*, Wanne *f*; Bütte *f (Papierherstellung)*
v. dye[stuff] Küpenfarbstoff *m*
v. dyeing Küpenfärberei *f*
v. press Büttenpresse *f (Papierherstellung)*
vatting Verküpen *n*
vault / to wölben *(Mauerwerk)*
vault Gewölbe *n*; Bogen *m*; Kappe *f*; Wölbung *f*; Keller *m (des Flammofens)*
vaulted gewölbt *(Mauerwerk)*
vaulting Gewölbe *n*
v. box Sprungkasten *m (Sport)*
v. buck Sprungbock *m*, Bock *m (Turngerät)*
v. horse Sprungpferd *n*, Turnpferd *n*
veal Kalbfleisch *n*
v. skin Mastfell *n*
vector Vektor *m*
v. addition Vektoraddition *f*
v. admittance *(El)* Leitwertoperator *m*, komplexer Leitwert *m*
v. algebra Vektoralgebra *f*
v. analysis Vektoranalyse *f*, Vektoranalysis *f*
v. calculus Vektorrechnung *f*
v. component Vektorkomponente *f*

v. coupling vektorielle Kopplung *f*, Vektorkopplung *f*
v. diagram Vektordiagramm *n*, Vektorzeichnung *f*, Zeigerdiagramm *n*
v. division Vektordivision *f*
v. electrocardiogram Vektor[elektro]kardiogramm *n*
v. equation Vektorgleichung *f*
v. field Vektorfeld *n*
v. function Vektorfunktion *f*
v. impedance *(El)* Widerstandsoperator *m*, komplexer Widerstand *m*
v. interaction Vektorwechselwirkung *f*
v. meson *(Kern)* Vektormeson *n*
v. model Vektormodell *n (eines Atoms)*
v. potential Vektorpotential *n*, vektorielles Potential *n*
v. power factor *(El)* Leistungsfaktor *m*
v. product Vektorprodukt *n*, vektorielles (äußeres) Produkt *n*
v. quantity Vektor[größe *f*] *m*, vektorielle Größe *f*
v. representation Vektordarstellung *f*
v. space Vektorraum *m*
vectorcardiogram *s.* vector electrocardiogram
vectorial vektoriell
v. angle Vektorwinkel *m*
v. display Vektordarstellung *f*
v. field Vektorfeld *n*
vectoring Einweisung *f (Flugzeuge)*
vee 1. V-; V-förmig; 2. Kimme *f (Meßtechnik)*; Keil *m*; Prisma *n*
vee-aerial V-Antenne *f*
vee belt Keilriemen *m*
vee-belt drive Keilriemenantrieb *m*
vee-belt pulley Keilriemenscheibe *f*
vee block Prismenstück *n*
vee engine V-Motor *m*
vee-groove *(Opt)* prismatische Nut *f*
vee grooving and tonguing Keilspundung *f*
vee-guide Prismenführung *f*
vee-jaw gauge block *(Opt)* Kimmenendmaß *n*
vee thread Spitzgewinde *n*
vee-type V-förmig ausgebildet
vee-way *(Masch)* prismatische Führungsfläche *f*, Prismenführung *f*
veering 1. Abdrehen *n*; 2. Lockern *n*, Entstraffen *n*
vees *(Masch)* prismatische Führungsflächen *fpl*, Prismen *npl*
vegetable adhesive Pflanzenleim *m*, vegetabilischer Klebstoff *m*
v. dye pflanzlicher Naturfarbstoff *m*, Pflanzenfarbstoff *m*
v. earth Humuserde *f*
v. farming Gemüseanbau *m*
v. fibre Pflanzenfaser *f*; pflanzliche Naturfaser *f*, pflanzlicher Faserstoff *m*
v. glue Pflanzenleim *m*
v. gum Pflanzengummi *m*, Pflanzenschleim *m*
v. oil Pflanzenöl *n*, pflanzliches Öl *n*
v. parchment echtes Pergamentpapier *n*
v. planter Gemüsepflanzgerät *n*; Pflanzensetzmaschine *f*
v. resin Naturharz *n*
v. shortening Pflanzenfett *n*
v. soil Ackerkrume *f*
v. tannage vegetabilische (pflanzliche) Gerbung *f*
v. tanned lohgar *(Leder)*
v. tanning material vegetabilischer (pflanzlicher) Gerbstoff *m*
vehicle 1. Fahrzeug *n*; 2. *(Plast)* Bindemittel *n*, flüssiger Lackanteil *m*, Lösungsmittel *n*, Trägersubstanz *f*; Trägerflüssigkeit *f*; Trägermaterial *n*
v. battery Fahrzeugbatterie *f*
v. deck Fahrzeugdeck *n*
v. engine Fahrzeugmotor *m*
v. industry Fahrzeugindustrie *f*
v. winch Fahrzeugwinde *f*
vehicular ferry Wagenfähre *f*
v. interference *(El)* Störung *f* durch Kraftfahrzeuge
v. radar Verkehrsradargerät *n*
veil Schleier *m*; [fotografische] Unschärfe *f*
veiling glare *(Opt)* Verschleierung *f*
vein / to masern
vein 1. *(Geol)* Gang *m*, Ader *f*; 2. Maserung *f (Holz, Marmor)*
v. fissure Gangspalte *f*
v. of silver Silberader *f*
v. ore Gangerz *n*
v. structure Ganggefüge *n*
veined growth *(Holz)* Maserwuchs *m*
veins on a casting Adern *fpl* auf einem Gußstück
vellum Pergamentpapier *n*
velocity Geschwindigkeit *f*
v. amplitude Geschwindigkeitsamplitude *f*
v. analyzer Geschwindigkeitsanalysator *m*
v. at main cut-off signal *(Rak)* Hauptkommandogeschwindigkeit *f*
v. at preliminary cut-off signal *(Rak)* Vorkommandogeschwindigkeit *f*
v. band Geschwindigkeitsbereich *m*
v. coefficient Geschwindigkeitsziffer *f*; Ausflußziffer *f (Strömung)*
v. dependent geschwindigkeitsabhängig
v. diagram Geschwindigkeitsdreieck *n*, Geschwindigkeitsplan *m*, Geschwindigkeitsdiagramm *n (Turbine)*
v. distance lag *s.* distance velocity lag
v. distribution Geschwindigkeitsverteilung *f*
v. error constant Geschwindigkeitsfehler *m (Folgeregler)*
v. filter Geschwindigkeitsfilter *n*

v. fluctuation Geschwindigkeitsschwankung *f*
v. focusing Geschwindigkeitsfokussierung *f*
v. head Geschwindigkeitshöhe *f (als Flüssigkeitssäule ausgedrückte kinetische Strömungsenergie)*
v.-head speed indicator *(Flg)* Staudruckfahrtmesser *m*
v. in trajectory Bahngeschwindigkeit *f*
v. microphone Druckgradientenmikrofon *n*, Geschwindigkeitsempfänger *m*, Geschwindigkeitsmikrofon *n*
v. misalignment Geschwindigkeitsfehler *m (Folgeregler)*
v.-modulated geschwindigkeitsmoduliert
v.-modulated electron beam geschwindigkeitsgesteuerter Elektronenstrahl *m*
v.-modulated valve Röhre *f* mit Geschwindigkeitssteuerung, geschwindigkeitsmodulierte Röhre
v. modulation Geschwindigkeitsmodulation *f*, Laufzeitmodulation *f*
v. of a periodic wave Phasengeschwindigkeit *f* einer Welle
v. of crystallization Kristallisationsgeschwindigkeit *f*
v. of detonation Detonationsgeschwindigkeit *f (Sprengstoff)*
v. of diffusion Diffusionsgeschwindigkeit *f*
v. of discharge Ausflußgeschwindigkeit *f*
v. of dissolution Auflösungsgeschwindigkeit *f*
v. of efflux Abströmgeschwindigkeit *f*
v. of electrons Elektronengeschwindigkeit *f*
v. of emission Emissionsgeschwindigkeit *f*
v. of energy transmission Geschwindigkeit *f* der Energiefortpflanzung
v. of fall Fallgeschwindigkeit *f*
v. of falling water Fallwassergeschwindigkeit *f*
v. of flow Strömungsgeschwindigkeit *f*, Stromgeschwindigkeit *f*; Durchflußgeschwindigkeit *f (Turbine)*
v. of ionization Ionenbildungsgeschwindigkeit *f*, Bildungsgeschwindigkeit *f* der Ionen
v. of light Lichtgeschwindigkeit *f*, Fortpflanzungsgeschwindigkeit *f* des Lichtes
v. of light in vacuum Vakuumlichtgeschwindigkeit *f*
v. of precession Präzessionsgeschwindigkeit *f*
v. of propagation Fortpflanzungsgeschwindigkeit *f*
v. of pure rolling Rollgeschwindigkeit *f*
v. of rotation Dreh[ungs]geschwindigkeit *f*, Rotationsgeschwindigkeit *f*, Winkelgeschwindigkeit *f*
v. of sound Schallgeschwindigkeit *f*
v. of the moving space Führungsgeschwindigkeit *f*
v. of transportation Führungsgeschwindigkeit *f*
v. potential Geschwindigkeitspotential *n*
v. ratio Übersetzungsverhältnis *n*
v. reduced by friction durch Reibung verminderte Geschwindigkeit *f*
v. resonance *(El)* Phasenresonanz *f*
v. space Geschwindigkeitsraum *m*
v. stage Geschwindigkeitsstufe *f*
v. staging Geschwindigkeitsabstufung *f*
v. through ports Durchgangsgeschwindigkeit *f*
velodyne integrator *(Dat)* Velodyne, Geschwindigkeitsfolgeregler *m*
velour finish / to *(Led)* buffen, veloutieren
velvet Samt *m*
v. copper ore Kupfersamterz *n*
v. cutting machine Samtschneidemaschine *f*
v. leather Plüschleder *n*
v. loom Samtwebstuhl *m*
v. weaving Samtweberei *f*
veneer / to furnieren, kaschieren
veneer Furnier *n*; Straßendecke *f*
v. clippers Furnierschere *f*
v. cutter Furnierschneidemaschine *f*
v. cutting knife Furniermesser *n*
v. cutting machine Furnierhobelmaschine *f*
v. face Furnierdecke *f*
v. lathe Furnierschälmaschine *f*
v. log Furnierblock *m*
v. panelling Furnierverkleidung *f*
v. peeling machine Furnierschälmaschine *f*
v. saw Klobsäge *f*
v. spoke shave Furnierschabhobel *m*
v. tape Umleimer *m*
v. taping Umleimung *f*
veneering Furnierschneiden *n*; Furnierverarbeitung *f*
v. press Furnierbock *m*, Furnierpresse *f*
v. rolls Furnierpreßwalzwerk *n*
v. saw Feinsäge *f*, Furniersäge *f*
Venetian blind Jalousie *f*
V. mosaic Terrazzo *m*
Venn diagram *(Dat)* Venn-Diagramm *n (zum Veranschaulichen aussagenlogischer Gesetze)*
venom Gift *n (Schlange, Skorpion, Spinne)*
vent / to entlüften; ablassen; luftstechen *(Form)*
vent 1. Entlüftungsöffnung *f*; Luftkanal *m (Gießform)*; Entlüftung *f*, Lüftung *f*, Abzug *m*; 2. *(Mil)* Zündloch *n*
v. hole Entlüftungsloch *n*; Lüftungsloch *n*, Lüftungsöffnung *f*, Lüftungsschlitz *m*; Abblasöffnung *f (Pneumatik)*
v. line *(Am) (Schiff)* Entgasungsleitung *f*
v. pipe *(Met)* Windpfeife *f (Formerei)*; Entlüftungsrohr *n*; Abzugsrohr *n*; Abdampfleitung *f*
v. plug Verschlußstopfen *m (Batterie)*
v. screw Entlüftungsschraube *f*
v. tube *(Am) (Brg)* Wetterlutte *f*

v. **valve** Lüftungsventil *n*, Entlüfter *m*, Atmungsventil *n*
v. **wire** Luftspieß *m (Formen)*
ventilate / to [ent]lüften, belüften; luftkühlen; bewettern; ventilieren
ventilated frame machine geschlossene Maschine *f* mit äußerer Eigenbelüftung (Eigenventilation)
v. **kiln** Trockenraum *m* mit natürlicher Luftbewegung
v. **motor** belüfteter Motor *m*
v. **radiator machine** Maschine *f* mit Radiatorkühlung
v. **ribbed surface machine** Maschine *f* mit belüftetem (ventiliertem) Rippengehäuse
v. **totally-enclosed motor** Motor *m* mit Mantelkühlung
v. **transformer** luftgekühlter Transformator *m*
ventilating district *(Brg)* Wetterabteilung *f*
v. **duct** 1. Kühlschlitz *m*; 2. Lüftungskanal *m*
v. **fan** Lüfter *m*, Ventilator *m*
v. **flap** Belüftungsklappe *f*
v. **grille** *(Eis)* Rosette *f (Wagenlüftung)*
v. **opening** Belüftungsöffnung *f*
v. **panel** Luftklappe *f*
v. **pipe** *(Brg)* Wetterlutte *f*, Lutte *f*
ventilation Lüftung *f*; Belüftung *f*; Entlüftung *f*; Bewetterung *f*; Ventilation *f*
v. **conduit** Lüftungskanal *m*
v. **flap** Luftklappe *f*, Ventilationsklappe *f*
v. **flue** Entlüftungsrohr *n*, Abzugsrohr *n*, Wrasenabzug *m*, Wrasenrohr *n*
v. **funnel** *(Brg)* Wetteraufhauen *n*
v. **lock** *(Brg)* Wetterschleuse *f*
v. **rib** *(Brg)* Wettermauer *f*
v. **road** *(Brg)* Wetterstrecke *f*
v. **shaft** *(Brg)* Wetterschacht *m*
v. **system** Belüftungsanlage *f*
v. **trunk** Lüftungsschacht *m*
v. **velocity** *(Brg)* Wettergeschwindigkeit *f*
ventilator Lüfter *m*, Ventilator *m*; Gebläse *n*
v. **head** *(Schiff)* Lüfterkopf *m*
v. **window** *(Kfz)* Ausstellfenster *n*
venting 1. Entlüften *n*, Lüftung *f*; 2. *(Met)* Luftstechen *n*, Luftspießen *n (Gießerei)*
v. **channel** Pumpsteiger *m (Formerei)*
v. **of moulds** Entlüften *n* von Formen
v. **property** Luftdurchlässigkeit *f (Formsand)*
v. **quality** Gasdurchlässigkeit *f (Gießform)*
v. **slot** Kühlschlitz *m*, Luftschlitz *m*, Lüftungsschlitz *m*
ventside of a furnace Stichseite *f* eines Ofens
Venturi [pipe] Lufttrichter *m*, Venturirohr *n*
Venturi tube *s.* Venturi [pipe]
venturine Goldstaub *m*
veranda[h] Veranda *f*
v. **deck** Verandadeck *n*
veratric acid Veratrumsäure *f*
Verdan system automatische Signalwiederholung *f*, Verdan-Verfahren *n (Nachrichtentechnik)*
verde antico Patina *f (Grünspan)*
verdigris Grünspan *m*
verge 1. Rand *m*, Außenkante *f*; Giebelkante *f*; Randstreifen *m*, Seitenstreifen *m (Straßenbau)*; 2. Welle *f*; Spindel *f*; Schaft *m*; Achse *f*
v. **bit** *(Text)* Abschlagzahn *m*
verification Bestätigung *f*; Beweis *m*, Kontrolle *f*, Prüfung *f*
verifier *(Dat)* Lochprüfgerät *n*, Verifizierer *m*, Verifiziergerät *n*
verify / to nachprüfen; *(Dat)* verifizieren
v. **the fuses** *(El)* die Sicherungen prüfen
v. **whether a line is engaged** *(Fmt)* auf Besetztsein prüfen *(Leitungen)*
verifying attachment *(BMSR, Dat)* Überwachungseinrichtung *f*, Kontrolleinrichtung *f*
vermilion roter Zinnober *m*
vermin Ungeziefer *n*
v. **destruction** Schädlingsbekämpfung *f*
v.**-infested soil** verseuchtes Land *n*
vernalization Vorkeimen *n*, Vorkeimung *f*; Jarowisation *f*
vernier *(Math)* Nonius *m*
v. **adjustment** Feineinstellung *f*
v. **caliper** Meßschieber *m*, Schieblehre *f*
v. **capacitor** *(El)* Feinabstimmkondensator *m*
v. **control capacitor** *(El)* Feineinstellkondensator *m*
v. **division** Noniusteilung *f*
v. **drive** Feinantrieb *m*
v. **height gauge** Höhenschieblehre *f*; Höhenreißer *m*, Höhenreißlehre *f*
v. **height scriber** Höhenreißer *m*
v. **mark** Noniusteilstrich *m*
v. **reading** Noniusablesung *f*
v. **scale** Nonienteilung *f*; Feinabstimmungsskala *f*
v. **tuning** *(El)* Feinabstimmung *f*
v. **zero** Noniusnullstrich *m*
versatile anpassungsfähig, vielseitig einsetzbar
versatility Vielseitigkeit *f*, vielseitige Verwendbarkeit *f*
version Ausführung[sart] *f*
versorium Elektrometernadel *f*
versus in Abhängigkeit von, als Funktion von
vertex 1. Scheitel[punkt] *m*, Spitze *f*, Ecke *f*; Eckpunkt *m*; Gipfelpunkt *m (Flugbahn)*; 2. Mittelpunkt *m*, Träger *m (Bündel)*
v. **feed** Antennenspeisung *f* an der Konusspitze
v. **of lens** Linsenscheitel *m*
v. **plate** Scheitelplatte *f (Wellenleiter)*
vertical Senkrechte *f*
vertical senkrecht, vertikal

v. adjustment Höhenverstellung *f*, Höheneinstellung *f*, Vertikaleinstellung *f*
v. aerial Vertikalantenne *f*, senkrechte Antenne *f*
v. air photograph Senkrechtluftbild *n*
v. and rotary selector *(Fmt)* Hebdrehwähler *m*
v. angle Vertikalwinkel *m*, Scheitelwinkel *m*
v. ascent *(Rak)* Vertikalaufstieg *m*
v. axis Vertikalachse *f*
v. band saw Vertikalbandsäge *f*, stehende (senkrechtlaufende) Bandsäge *f*
v. blanking *(Fs)* Zeilenaustastung *f*
v. boiler Vertikalkessel *m*, stehender Dampfkessel *m*
v. boring and turning mill Karusselldrehmaschine *f*
v. boring and turning mill of the fixed table type Zweiständerkarusselldrehmaschine *f* mit fester Planscheibe
v. boring and turning mill of the sliding table type Zweiständerkarusselldrehmaschine *f* mit verschiebbarer Planscheibe
v. boring and turning mill with side head Karusselldrehmaschine *f* mit Seitensupport
v. boring mill Senkrechtbohrwerk *n*
v. boring mill table Planscheibe *f (Karusselldrehmaschine)*
v. box column drill Kastenständerbohrmaschine *f*
v. box-type upright drilling machine Kastenständerbohrmaschine *f*
v. break switch Schalter *m* mit senkrechter Schaltstrecke
v. broaching machine Senkrechträummaschine *f*
v. camera Vertikalkamera *f*
v. centring *(Fs)* Bildmitteneinstellung *f*
v. coelostat *(Opt)* Vertikalzölostat *m*
v. construction Senkrechtbauart *f*
v. cross tube boiler stehender [Flammrohr-]Kessel *m* mit waagerechten Quersiedern, Lachapellekessel *m*
v. cylinder boring mill Vertikalzylinderausbohrmaschine *f*
v. damper *(Wärm)* Senkschieber *m*
v. definition *(Fs)* vertikale Bildauflösung *f*
v. deflecting electrodes Vertikalablenkplatten *fpl (Katodenstrahlröhren)*
v. deflection *(Fs)* vertikale Bildablenkung *f*
v. directive pattern senkrechte Richtcharakteristik *f (Antennentechnik)*
v. drilling Senkrechtbohren *n*
v. drilling machine Senkrechtbohrmaschine *f*, Säulenbohrmaschine *f*; Ständerbohrmaschine *f*; freistehende Bohrmaschine *f*
v. drilling machine with multiple spindle head Mehrspindelbohrmaschine *f* mit einstellbaren Spindeln
v. effect Hochantenneneffekt *m (Funkempfangstechnik)*
v. electric wave senkrechtstehende elektrische Welle *f*
v. engine stehender Motor *m*
v. feed Tiefenvorschub *m*
v. flue boiler stehender Flammrohrkessel *m*
v. force Senkrechtkraft *f*; Hauptschnittkraft *f*
v. frequency *(Fs)* Rasterfrequenz *f*
v. grain *(Holz)* Spiegelschnitt *m*
v. hold control *(Fs)* Regelung *f* der Vertikalablenkung
v. honing machine Senkrechthonmaschine *f*
v. hunting *(Fs)* senkrechte Lageschwankung *f*
v. illumination Auflicht *n (Mikroskopie)*
v. interrupter contact *(Fmt)* Ruhekontakt *m* des Hebmagneten
v. iron-plate keel *(Schiff)* Balkenkiel *m*
v. joint Stoßfuge *f*
v. launching *(Rak)* Vertikalstart *m*, Senkrechtstart *m*
v. lock spring *(Fmt)* Hebsperrfeder *f*
v. log frame Vertikalgatter *n* mit Oberantrieb *(Sägewerk)*
v. log frame saw Vollvertikalgatter *n (Sägewerk)*
v. log frame with roller feed Walzengatter *n (Sägewerk)*
v. log frame with side driving gear Vertikalgatter *n* mit Seitenantrieb
v. loom Senkrechtwebstuhl *m*
v. magnet *(Fmt)* Hebmagnet *m*
v. member Vertikalstab *m*, Senkrechtstab *m*; Ständer *m (Fachwerk)*
v. member of strutted pole *(Fmt)* senkrechte Stange *f (von einem verstrebten Gestänge)*
v. milling attachment Senkrechtfräseinrichtung *f*
v. milling machine Senkrechtfräsmaschine *f*
v. motion *(Fmt)* Hebbewegung *f*, Heben *n (des Wählers)*; Senkrechtbewegung *f*
v. multitubular boiler stehender Heizröhrenkessel *m*
v. multitubular boiler with fire box stehender Feuerbüchsenkessel *m*
v. off-normal contacts *(Fmt)* Kopfkontakte *mpl*
v. opener combined with porcupine opener *(Text)* Kegelöffner *m* mit Voröffner
v. photograph Senkrechtaufnahme *f*
v. piling *(Brg)* senkrechtes Anstecken *n (Schachtabteufen)*
v. polarization Vertikalpolarisation *f*
v. radiation senkrechte Abstrahlung *f*
v. radiation pattern vertikales Strahlungsdiagramm *n*, Vertikaldiagramm *n (Strahlung)*
v. relief movement *(Masch)* Senkrechtabhebung *f*

v. **roll** Kopfwalze *f*
v. **round-column upright drilling machine** Säulenbohrmaschine *f*
v. **screw** Senkrechtspindel *f*
v. **seam** *(Geol)* stehendes Flöz *n*
v. **section** Senkrechtschnitt *m* *(technisches Zeichnen)*
v. **section of the chimney** Aufrißfläche *f* des Schornsteins
v. **shaft** 1. *(Kfz)* Königswelle *f*; 2. *(Brg)* Seigerschacht *m*, seigerer Schacht *m*
v. **shaper** 1. *(Fs)* Bildbegrenzer *m*; 2. *(Am)* *(Masch)* Senkrechtstoßmaschine *f*
v. **shear** Querkraft *f*
v. **sheet stripper bar hackling machine** stehende Jutehechelmaschine *f* mit Abnehmerleisten
v. **side of intermediate distributing frame** *(Fmt)* Zwischenverteiler *m*
v. **side of main distributing frame** *(Fmt)* Hauptverteiler *m*
v. **size control** *(Fs)* Rasterhöhenregler *m*
v. **slide** *(Masch)* Senkrechtschlitten *m*; Schieber *m* *(Hobelmaschine)*
v.**-spindle grinding machine with twin rotary tables** Doppelrundtischflächenschleifmaschine *f* senkrechter Bauart
v.**-spindle machine** Senkrecht[spindel]maschine *f*
v.**-spindle pump** Pumpe *f* in Senkrechtanordnung
v. **stabilizer** *(Flg)* Seitenflosse *f*
v. **steam engine** stehende Dampfmaschine *f*
v. **step** *(Fmt)* Hebschritt *m*, Höhenschritt *m*
v. **surface grinding machine** Senkrechtflachschleifmaschine *f*
v. **synchronization employing power-line frequency** *(Fs)* Netzsynchronisierung *f*
v. **take-off** Senkrechtstart *m*
v. **throw** lotrechter (senkrechter) Wurf *m*
v. **tool clearance** *(Masch)* Senkrechtabhebung *f* des Meißels
v. **turbine** Vertikalturbine *f*
v. **turning and boring mill** Karusselldrehmaschine *f*
v. **turret head** Trommelrevolverkopf *m*
v. **turret lathe** 1. Trommelrevolverdrehmaschine *f*; 2. Karussellrevolverdrehmaschine *f*
v. **turret lathe with side head** Drehwerk *n* mit Seitensupport
v. **valve** *(Kfz)* stehendes Ventil *n*
v. **wall-duct** *(Fmt)* Kabelhochführungsschacht *m*
v. **wire aerial** Vertikaldrahtantenne *f*
vertically adjustable höhenverstellbar
v. **movable** höhenbeweglich
very high frequency Ultrakurzwellenfrequenz *f*, Ultrakurzwelle *f*, UKW (*s. a.* v.h.f. ...)
v. **high tension generator** *(Elm)* Höchstspannungsgenerator *m*
Very light Leuchtpatrone *f*, Signalpatrone *f*
vesicant blasenbildend, hautätzend
vesicant ätzender Kampfstoff *m*
vesicatory Zugpflaster *n*
vessel 1. Hohlkörper *m*; 2. Behälter *m*, Birne *f*, Bottich *m*, Gefäß *n*, Kessel *m*; 3. Schiff *n*, Fahrzeug *n*
v. **carrying petroleum in bulk** Rohöltanker *m*, Öltankschiff *n*
v. **of full scantlings** Vollbordschiff *n*
v. **sanitation** Schiffshygiene *f*
vestibule 1. Diele *f*, Vestibül *n*, Vorhalle *f*; 2. *(Eis)* Einsteigraum *m*
vestigial restlich
v. **sideband** *(Fmt)* Restseitenband *n*
vesuvian[ite] *(Min)* Vesuvian *m*
veterinary surgeon Tierarzt *m*
v.f.t. Tonfrequenztelegrafiesystem *n*
v.h.f. carrier UKW-Träger *m*
v.h.f. coverage UKW-Bedeckung *f*
v.h.f. direction-finding UKW-Peilung *f*
v.h.f. frequency meter UKW-Frequenzmesser *m*
v.h.f. omnirange UKW-Drehfunkfeuer *n*
v.h.f. range UKW-Bereich *m*
v.h.f. receiver UKW-Empfänger *m*
v.h.f. region UKW-Bereich *m*
v.h.f. transmitter UKW-Sender *m*
v.h.f. transmitter frequency UKW-Senderfrequenz *f*
v.h.f. tuner unit UKW-Baustein *m*
v.h.f. waves Ultrakurzwellen *fpl*
via center *(Am)* *(Fmt)* Verteilfernamt *n*
v. **circuit** *(Am)* *(Fmt)* Durchgangsleitung *f*, Transitstromkreis *m*
v. **net loss** *(Fmt)* Durchgangsdämpfung *f*
v. **office** *(Am)* *(Fmt)* Durchgangsamt *n*
v. **traffic** *(Am)* *(Fmt)* Durchgangsverkehr *m*
viaduct Viadukt *m*
vial 1. Ampulle *f*, Fläschchen *n*, Phiole *f*; 2. Röhrenlibelle *f*, Libelle *f*
vibrate / to 1. vibrieren, schwingen; zittern; erschüttert werden; 2. rütteln, vibrieren *(Betontechnologie)*
vibrated concrete Rüttelbeton *m*
vibratile schwingbar, schwingfähig
vibrating ball mill Kugelschwingmühle *f*
v. **beam finisher** *(Bw)* Rüttelbohlenfertiger *m*, Vibrierbohlenfertiger *m*
v. **bell** *(Fmt)* Wecker *m* mit Selbstunterbrechung
v. **break** *(El)* Hammerunterbrecher *m*
v.**-diaphragm electrometer** Schwingkondensatorelektrometer *n*
v. **galvanometer** Schwingungsgalvanometer *n*, Vibrationsgalvanometer *n*
v. **grate** Schüttelrost *m*

v. of the turbine Zittern *n* der Turbine
v. pin Membranstift *m*
v. reed break *(El)* Zerhacker *m*, Zungenunterbrecher *m*
v. reed electrometer Gleichstromverstärker *m* mit Schwingkondensator
v.-reed frequency meter Zungenfrequenzmesser *m*
v. reed instrument Zungenfrequenzmesser *m*
v. reed rectifier Schwing[kontakt]gleichrichter *m*
v. relay Vibrationsrelais *n*
v. roller *(Bw)* Rüttelwalze *f*, Vibrationswalze *f*
v. screed *(Bw)* Rüttelbohle *f*, Vibrationsglättbohle *f*, Vibrierbohle *f*
v. screen Schüttelsieb *n*, Vibrationssieb *n*
v.-screen type deduster Vibrosichter *m*
v. shuttle Schwingschiff *n (Nähmaschine)*
v. table 1. *(Bw)* Rütteltisch *m*, Vibriertisch *m*; 2. Schwingherd *m (Aufbereitung)*
v. tamper *(Bw)* Rüttelstampfer *m*, Vibrationsstampfer *m*
v.-wire strain gauge Schwingsaitendehnungsmesser *m*
vibration Schwingung *f*, Vibration *f*; Erschütterung *f*
v. absorber Schwingungsdämpfer *m*
v. absorbing schwingungsdämpfend
v. acceleration Schwingbeschleunigung *f*
v. amplifier Schwingungsverstärker *m*
v. amplitude Schwingungsamplitude *f*
v. antinode Schwingungsbauch *m*
v. damper Schwingungsdämpfer *m*
v. detuner Schwingungsdämpfer *m*
v. engineering Schwingungstechnik *f*
v.-free mounting schwingungsfreie Montage *f*
v. galvanometer Vibrationsgalvanometer *n*
v. generator Schwingungserzeuger *m*, Schwingungstisch *m*
v. hoe *(Lw)* Vibrationshacke *f*
v. loss Schwingungsverlust *m*
v. meter Schwingungsmesser *m*
v. mill Schwingmühle *f*
v. node Schwingungsknoten *m*
v. of concrete Betonrüttelverdichtung *f*
v. of the piston Brummen *n* des Kolbens
v. pick-up Schwingungsgeber *m*
v.-producer Schwingungserzeuger *m*, Schwingungserreger *m*
v. resistance Schwingungsfestigkeit *f*; Erschütterungsfestigkeit *f*, Rüttelfestigkeit *f*
v. state Schwingungszustand *m*
v. table Rütteltisch *m*, Vibriertisch *m*
vibrational level Schwingungsterm *m*, Schwingungsniveau *n*
v. loading Schwingbeanspruchung *f*
v. quantum number Schwingungsquantenzahl *f*
v. spectrum Schwingungsspektrum *n*
v. stress Schwingbeanspruchung *f*
vibrationless schwingungsfrei, erschütterungsfrei
vibrations from outside sources von außen her verursachte Schwingungen *fpl*
vibrator 1. Rüttler *m*, Verdichter *m*, Vibrator *m (Betontechnologie)*; 2. *(El)* Unterbrecher *m*, Wechselrichter *m*, Zerhacker *m*, elektromagnetischer Modulator *m*
v. converter *(El)* Zerhackerumformer *m*
v. conveyor Schwingförderer *m*
v. moulding machine *(Met)* Rüttelformmaschine *f*
v. table Schütteltisch *m*
v. unit *(El)* Zerhackerpatrone *f*
vibratory schwingend, Schwing-
v. converter *(El)* Zerhacker *m*
v. feeder (hopper) *(Masch)* Schwingförderer *m*
v. roller Rüttelwalze *f*, Vibrationswalze *f*
v. screen Schüttelsieb *n (Altpapieraufbereitung)*
v. strength Schwingungsfestigkeit *f*
v. table Rütteltisch *m*, Vibriertisch *m*
v. test Dauer[schwing]versuch *m*
vibro-deduster Vibrosichter *m (Aufbereitung)*
v.-massage machine Vibromassagegerät *n*
vibrograph Vibrograf *m*, Schwingungsschreiber *m*, Schwingungsaufzeichner *m*
vibromotive schwingungserregend
vibroplex *(Am) (El)* halbautomatische Schnellsendetaste *f*
Vicat apparatus *s.* Vicat needle
Vicat needle Vicatapparat *m (Prüfgerät)*
vice Schraubstock *m*, Spanner *m*
v. body Schraubstockkörper *m*, Schraubstockunterteil *n*
v. dog Schraubstockklaue *f*
v. jaws Schraubstockbacken *mpl*, Schraubstockfuttereinsatz *m*
v. screw Schraubstockspindel *f*
v. versa umgekehrt
v. with replaceable jaws Schraubstock *m* mit eingesetzten Backen
Vickers [diamond] hardness Vickershärte *f*
Vickers hardness test Härteprüfung *f* nach Vickers
Vickers tester Vickershärteprüfer *m*
Victory ship *(Am)* Frachtschiff *n* der Victory-Klasse *(10700 tdw)*
victualling vessel Proviantschiff *n*
video amplifier Bildsignalverstärker *m*, Videoverstärker *m*
v. broadcasting Fernsehrundfunk *m*
v. carrier Bildträger *m*
v. detection *(Fs)* Bildgleichrichtung *f*

v. frequency Videofrequenz *f*, Bildpunktfrequenz *f*, Fernsehfrequenz *f*
v. frequency amplifier Videoverstärker *m*
v.-gain control *(Fs)* Intensitätsbegrenzer *m*
v. map Videokarte *f* *(Radar)*
v. mapping unit Videokartendarstellungsgerät *n* *(Radar)*
v. mixer *(Fs)* Bildmischpult *n*
v. monitoring and mixing desk *(Fs)* Bildmischpult *n*
v. signal Videosignal *n*, Bildsignal *n*
v. technique Videofrequenztechnik *f*
v. telephony Fernsehtelefonie *f*
v. transmitter *(Fs)* Bildsender *m*
v. trigger *(Fs)* Bildauslösung *f*
vidicon Vidikon *n* *(Bildaufnahmeröhre)*
v. camera tube Vidikonaufnahmeröhre *f*
Vierendeel girder *(Bw)* Vierendeelträger *m*
Vierendeel truss Vierendeelfachwerk *n*, Vierendeelfachwerkbinder *m*
view / to beobachten, betrachten
view Bild *n*; Betrachtung *f*; Riß *m*; Ansicht *f* *(technisches Zeichnen)*
v. finder Sucher *m* *(Foto)*, Suchereinrichtung *f*
v. port Beobachtungsfenster *n*
viewed object area Betrachtungsausschnitt *m*
viewer Fernsehteilnehmer *m*, Fernsehzuschauer *m*
viewing angle Gesichtswinkel *m*, Sehwinkel *m*
v. distance Betrachtungsabstand *m*
v. mirror Spiegelreflektor *m*
v. screen Bildschirm *m*
v. storage tube Sichtspeicherröhre *f*
vignetting *(Opt)* Vignettierung *f*
v. effect Vignettierungseffekt *m*
vigorous hart *(Fotopapier)*
v. emulsion film Dokumentenfilm *m*
Vigoureux printing Vigoureuxdruck *m*, Kammzugdruck *m*
vinaconic acid Äthylenmalonsäure *f*
vinculum *(Math)* Verbindungsstrich *m* über einem Ausdruck *(dient als Ersatz von Klammern)*
vine Weinstock *m*, Rebe *f*
v. branch Rebschoß *m*
v. culture Weinbau *m*
v. grower Winzer *m*
v. sprayer Rebenspritze *f*
v. twig Weinrebe *f*
vinegar Essig *m*
vineyard Weinberg *m*
vinic acid Äthylschwefelsäure *f*
v. ether Diäthyläther *m*
vintage Weinlese *f*, Weinernte *f*
vintager Winzer *m*
vinyl compound Vinylverbindung *f*
v. elastomeric compounds weichgestellte Vinylmassen *fpl*
v. foam laminate Polyvinylschaumstofflaminat *n*
v. plastic Vinylplast *m*
v. resin Vinylharz *n*
vinylation *(Ch)* Vinylierung *f*, Einführung *f* der Vinylgruppe
vinylidene chloride resin Vinylidenchloridharz *n*
v. compound Vinylidenverbindung *f*
violarite *(Min)* Violarit *m*
violent heftig, stark; brisant *(Sprengstoffe)*
v. bubbling *(Met)* Schäumen *n*, Aufschäumen *n* *(Schmelzbad)*
v. fermentation Schnellgärung *f*
v. formation of steam stürmische Dampfbildung *f*
violet acid Sulfonitronsäure *f*
v. shift Violettverschiebung *f* *(Spektrallinien)*
violin Geige *f*, Violine *f*
v. bow Bogen *m*, Violinbogen *m*, Geigenbogen *m*
v. bridge Geigensattel *m*, Geigensteg *m*, Sattel *m*
v. peg Geigenwirbel *m*
v. rosin Geigenharz *n*
v. stop Geigenregister *n*, Geigenwerk *n* *(Orgel)*
v. stop of organ Geigenwerk *n*, Geigenzug *m* *(Orgel)*
violuric acid Violursäure *f*
virgin *(Brg)* unverritzt
v. forest Urwald *m*
v. growth Urwaldbestand *m*
v. material frisch hergestellter (zum ersten Male verarbeiteter) Stoff *(Thermoplaste)*
v. metal gediegenes (reines) Metall *n*
v. neutron *(Kern)* jungfräuliches Neutron *n*
v. neutron flux jungfräulicher Neutronenfluß *m*
v. oil Erstöl *n*, Frischöl *n*, hochwertigstes Öl *n*
v. pulp system Zellstoffaufbereitungsanlage *f*
v. rock temperature Temperatur *f* des anstehenden Gebirges
v. wool Originalwolle *f*, Neuwolle *f*
virial coefficient *(Kern)* Virialkoeffizient *m*
v. law *(Kern)* Virialsatz *m*
virtual scheinbar, virtuell
v. cathode virtuelle (scheinbare) Katode *f*
v. displacement virtuelle Verschiebung (Verrückung) *f*
v. image virtuelles Bild *n*
v. value Effektivwert *m*, quadratischer Zeitmittelwert *m*
v. voltage *(El)* Effektivspannung *f*
v. work *(Ph)* virtuelle Arbeit *f*
vis-à-vis engine Boxermotor *m*
viscid klebrig, haftend, haftfähig
visco-elastic viskoelastisch
v.-plastic viskoplastisch
viscometer *(Opt)* Viskometer *n*
viscose *(Ch)* Viskose[lösung] *f*

v. rayon Viskoseseide *f*
v. staple fibre Viskosefaser *f*, Zellwolle *f*
viscosimeter Konsistenzmesser *m*, Viskosimeter *n*, Viskositätsmesser *m*, Flüssigkeitsgradmesser *m*
viscosity Viskosität *f*, Zähigkeit *f*
v. chart Viskogramm *n*
v. index Viskositätsindex *m*
viscous viskos, dickflüssig, zähflüssig; laminar
v. fermentation Schleimgärung *f*
v. flow Reibungsströmung *f*
v. fluid viskose (zähe) Flüssigkeit *f*
v. friction Flüssigkeitsreibung *f*
v. friction damper Reibungs[schwingungs]-dämpfer *m*
viscousness Viskosität *f*, Zähflüssigkeit *f*
vise *(Am) s.* vice
visibility Sicht *f*; Sichtbereich *m*, Sichtmöglichkeit *f*
v. factor Sichtfaktor *m (Radar)*
v. meter Sichtmesser *m*
visible sichtbar
v. horizon Seehorizont *m*
v. line sichtbare Linie (Körperkante) *f (technische Zeichnung)*
v. range sichtbarer Bereich *m (des Spektrums)*
v. region *s.* v. range
v. trajectory sichtbare Flugbahn *f*
vision 1. Sehen *n*; 2. Sehvermögen *n*
v. amplifier Bildverstärker *m*
v. bandwidth Bildbandbreite *f*
v. carrier *(Fs)* Bildträger *m*
v. frequency Bildträgerfrequenz *f*, Fernsehfrequenz *f*
v. modulation *(Fs)* Bildmodulation *f*
v. on sound *(Fs)* Bild *n* im Ton
v. pick-up Bildaufnahme *f*
v. signal Bildsignal *n*
v. transmitter Bildsender *m*
v. transmitter output (power) Bildsendeleistung *f*
vista shot Fernaufnahme *f*, Gesamtbild *n (Fernsehbildaufnahme)*
visual visuell, optisch
v. acuity Sehschärfe *f*, Sehleistung *f*
v. angle *(Ph)* Sehwinkel *m*, Gesichtswinkel *m*
v.-aural radio range Vierkursfunkfeuer *n*, Leitstrahlbakensystem *n* VAR *(akustisch und optisch anzeigendes Vierkursfunkfeuer)*
v. busy lamp *(Fmt)* Besetztlampe *f*
v. capacity in twilight *(Opt)* Dämmerungssehleistung *f*
v. deficiency *(Opt)* Fehlsichtigkeit *f*
v. direction finder Sichtpeilgerät *n*
v. display Sichtanzeige *f (von Zahlen)*
v. distance Sehweite *f*
v. engaged lamp *(Fmt)* Besetztlampe *f*
v. engaged signal *(Fmt)* [optisches] Besetztzeichen *n*
v. engaged test *(Fmt)* optische Besetztprüfung *f*
v. flight rules Sichtflugregeln *fpl*
v. frequency *(Fs)* Bild[träger]frequenz *f*
v. ground aid *(Flg)* optische Bodeneinrichtung *f*
v. indication Sichtanzeige *f*
v. indication equipment Einrichtung *f* für Sichtanzeige
v. indicator *(Fmt)* Schauzeichen *n*
v. indicator valve Anzeigeröhre *f*
v. inspection Sichtprüfung *f*
v. observation visuelle Beobachtung *f*
v. omnirange UKW-Richtfunkfeuer *n*
v. photometry visuelle (subjektive) Fotometrie *f*
v. purple Sehpurpur *m*
v. reception Schreibempfang *m*
v. sensation Seheindruck *m*
v. signal optisches Zeichen *n*, Schauzeichen *n*
v. transmitter *(Fs)* Bildsender *m*
v. tuning *(Fmt)* optische Abstimmung *f*
v.-tuning indicator Abstimmanzeiger *m*
visualization Sichtbarmachung *f*
Vita glass Vitaglas *n (UV-durchlässiges Glas)*
vital staining Vitalfärbung *f*
vitaminize / to Vitamine zusetzen
vitiated air verbrauchte Luft *f (Lüftungstechnik)*; *(Brg)* schlechte Wetter *pl*
vitrain *(Min)* Vitrit *m*, Glanzkohle *f*
vitreous glashart, gläsern
v. body Glaskörper *m*
v. clinker glasige Schlacke *f*
v. enamelling Emaillierung *f*
v. humour Glaskörper *m*
v. state Glaszustand *m*
vitrification Verglasung *f*, Übergang *m* in den glasartigen Zustand
vitrified bond anorganische (keramische) Bindung *f*
v.-bonded anorganisch gebunden
v. brick Klinker *m*
v. [grinding] wheel Schleifkörper *m* mit keramischer Bindung
vitrify / to glasieren, sintern, überglasen, verglasen
vitriol Vitriol *m (n)*
vitriolic acid Schwefelsäure *f*, Vitriolöl *n*, Vitriolsäure *f*
vivianite *(Min)* Vivianit *m*, Blaueisenerz *n*
v.l.f. oscillator Oszillator *m* für sehr niedrige Frequenzen
v.l.f. phase meter Phasenmesser *m* für sehr niedrige Frequenzen
v.l.f. transmission Längstwellenübertragung *f*
vocational guidance service Berufsberatung *f*

v. **school** Berufsschule *f*
v. **training** Berufsausbildung *f*
voice Stimme *f*
v. **coil** Schwingspule *f (Lautsprecher)*
v. **communication channel** Funksprechkanal *m*
v. **controlled** *(Fmt)* sprachgesteuert
v.-**ear test** Sprech-Hör-Versuch *m*, Prüfung *f* durch Hörvergleich mit Sprache
v. **frequency** Sprechfrequenz *f*, Sprachfrequenz *f*
v. **frequency carrier telegraphy** Tonfrequenzträgertelegrafie *f*, Wechselstromtelegrafie *f*
v. **frequency dialling** *(Fmt)* Tonfrequenzfernwahl *f*
v. **frequency generator** Tonfrequenzgenerator *m*
v. **frequency multichannel** WT-System *n (Telegrafie)*
v. **frequency ringing** *(Fmt)* Tonfrequenzruf *m*
v. **frequency selecting system** Einrichtung *f* für Tonfrequenzfernwahl
v. **frequency telephony** Niederfrequenzfernsprechen *n*
v.-**modulated** sprachmoduliert
v.-**operated** *(Fmt)* sprachgesteuert
v. **power** Sprachenergie *f*
v. **recording** Sprachaufzeichnung *f*
v. **recording equipment** *(El)* Gerät *n* zur Aufzeichnung von Gesprächen
v. **transmission** Sprachübertragung *f*
voicing Stimmen *n (Orgel)*
void Leere *f*; Blase *f*; Hohlraum *m*, Pore *f*
v. **ratio** Porenziffer *f (Bodenmechanik)*
v. **spaces between grains [in wheel]** Porenraum *m* [in der Schleifscheibe]
v. **water** *(Brg)* Porenwasser *n*
Voith-Schneider propeller Voith-Schneider-Propeller *m*
volatile flüchtig, leicht verdampfbar
v. **content** Gehalt *m* an flüchtigen Bestandteilen
v. **fission product** *(Kern)* flüchtiges Spaltprodukt *n*
v. **loss** Verdampfungsverlust *m*
v. **matter** flüchtige Bestandteile *mpl*
v. **storage** *(Dat)* leistungsabhängige Speicherung *f*
v. **vehicle** flüchtiger Anteil *m* der Trägerflüssigkeit
volatility Flüchtigkeit *f*, Verdampfbarkeit *f*
volatilization Verflüchtigung *f*; Verdunstung *f*
v. **roasting** *(Met)* Verdampfungsröstung *f*
volatilize / to abdampfen, verdampfen, verdunsten, verflüchtigen
volley fire *(Mil)* Gruppenfeuer *n*
volt Volt *n*, V
v.-**ampere** Voltampere *n*
v.-**ampere-hour meter** Scheinverbrauchszähler *m*
v.-**ampere meter** Scheinleistungsmesser *m*
v.-**hour meter** Voltstundenzähler *m*
v.-**second** Voltsekunde *f*
Volta cell Voltasches Element *n*, Voltaelement *n*
Volta effect Voltaeffekt *m*
voltage [elektrische] Spannung *f*
v. **above earth (ground)** hochliegende Spannung *f*
v. **across the circuit** Kreisspannung *f*
v. **amplification** Spannungsverstärkung *f*
v. **amplification factor** Spannungsverstärkungsfaktor *m*
v. **amplifier** Spannungsverstärker *m*
v. **antinode** Spannungsbauch *m*
v. **balance** Spannungsgleichheit *f*
v. **between lines** verkettete Spannung *f*, Leiterspannung *f*
v.-**carrying** spannungführend, unter Spannung stehend
v. **change** Spannungsänderung *f*
v. **characteristic** Spannungscharakteristik *f*
v. **circuit** Spannungspfad *m*
v. **coil** Spannungsspule *f*
v. **component** Spannungskomponente *f*
v. **control** Spannungssteuerung *f*; Spannungsregelung *f*; Spannungsregler *m*
v. **curve** Spannungskurve *f*
v. **cut-out** Überspannungssicherung *f*
v. **detector** Spannungsprüfer *m*, Spannungsanzeiger *m*
v. **difference** Spannungsdifferenz *f*
v. **discharge gap** *(Am)* Schutzfunkenstrecke *f*
v. **distribution** Spannungsverteilung *f*
v. **divider** Spannungsteiler *m*
v. **doubler** Spannungsverdoppler *m*, spannungsverdoppelnde Gleichrichterschaltung *f*
v. **drop** Spannungsabfall *m*
v. **dropping resistor** Vor[schalt]widerstand *m*
v. **equivalent of thermal energy** Temperaturspannung *f*, Temperaturäquivalent *n*
v. **factor** Verstärkungsfaktor *m (zwischen zwei Elektroden)*
v.-**fed** spannungsgespeist
v.-**feed** Spannungsspeisung *f (Antennentechnik)*
v. **fluctuation** Spannungsschwankung *f*
v. **gain** Spannungsverstärkung *f*
v. **gradient** Spannungsgefälle *n*, Potentialgradient *m*
v. **impulse** Spannungsstoß *m*
v. **in non-active position** Ruhespannung *f*
v. **in normal position** Ruhespannung *f*
v. **increase** Spannungserhöhung *f*
v. **indicator** Spannungsanzeiger *m*
v. **jump** Spannungssprung *m*

v. level Spannungspegel *m*
v. limiter Spannungsbegrenzer *m*
v. loop Spannungsbauch *m*
v. loss Spannungsabfall *m*, Spannungsabnahme *f*, Spannungsverlust *m*
v. maximum Spannungsmaximum *n*
v. multiplier *(El)* Vor[schalt]widerstand *m*
v. node Spannungsknoten *m*
v. of a system Leiterspannung *f*; verkettete Spannung *f*
v. path Spannungspfad *m*
v. peak Spannungsspitze *f*
v.-proof spannungsfest, durchschlagsfest
v. protection Spannungsschutz *m (Relais)*
v. range Spannungsmeßbereich *m*
v. range switch Spannungsmeßbereichumschalter *m*
v. ratio Spannungsverstärkungsfaktor *m*; Spannungsübersetzung *f (Transformator)*
v. reaction Spannungsrückwirkung *f*
v. recorder Spannungsschreiber *m*
v. reference valve Stabilisator *m*
v. reflection factor Spannungsreflexionsfaktor *m*
v. regulation Spannungsregelung *f*, Spannungsgleichhaltung *f*
v. regulation system Spannungsregelungsanlage *f*
v. regulator Spannungskonstanthalter *m*, Spannungsgleichhalter *m*
v. regulator diode Spannungsreglerdiode *f*; Zenerdiode *f*
v. relay Spannungsrelais *n*
v. resonance Spannungsresonanz *f*, Reihenresonanz *f*, Serienresonanz *f*
v. ripple Spannungswelligkeit *f*, Kräuselspannung *f*; Spannungsschwankung *f*
v. rise Spannungsanstieg *m*
v. sensitive oscillator spannungsgesteuerter Oszillator *m*
v. sensitivity Spannungsempfindlichkeit *f*
v. stabilizer Spannungskonstanthalter *m*, Spannungsstabilisator *m*
v. stabilizing valve Glimmstreckenstabilisator *m*
v. standing wave ratio Stehwellenverhältnis *n*, Spannungsstehwellenverhältnis *n*, Wellenziffer *f*
v. step Spannungsstufe *f*
v. surge Spannungs[stoß]welle *f*, Spannungsstoß *m*, Stoßüberspannung *f*
v. surge indicator Stoßspannungsanzeiger *m*
v. tapping switch Spannungskarussell *n*, Spannungsumschalter *m*
v. test Spannungsprüfung *f*
v. time conversion Spannungszeitumwandlung *f*
v. to earth (ground) Spannung *f* gegen Erde
v. to neutral Sternspannung *f*, Leitererdspannung *f*, Phasenspannung *f*
v. transfer Spannungsübersetzung *f*
v. transformation Spannungsübersetzung *f*
v. transformer Spannungstransformator *m*, Spannungswandler *m*, Transformator *m*
v.-tunable spannungsabgestimmt
v.-tunable oscillator Oszillator *m* mit abstimmbarer Spannung
v.-type telemeter *(El)* Fernmeßgerät *n* mit Spannungsübertragung
v.-variable capacitor durch Spannung veränderlicher Kondensator *m*
v. variation Spannungsschwankung *f*
v. vector Spannungsvektor *m*
v. wave Spannungswelle *f*

voltaic galvanisch
v. cell galvanisches Element *n*, Voltaelement *n*
v. current galvanischer Strom *m*
v. electricity Berührungselektrizität *f*, Kontaktelektrizität *f*

voltmeter Spannungsmesser *m*, Voltmeter *n*
v. for high tension Hochspannungsmesser *m*, Hochspannungsvoltmeter *n*
v. for low tension Niederspannungsvoltmeter *n*

volume Volumen *n*, Rauminhalt *m*, Inhalt *m*; Lautstärke *f*
v. adjustment Lautstärkeregelung *f*
v.-batching Dosierung *f* nach Raumteilen *(Betontechnologie)*
v.-built serienmäßig gebaut (gefertigt)
v. calliper Kubikgabelmaß *n*
v. change Volumenänderung *f*, Raumänderung *f*
v. charge Raumladung *f*
v. control 1. Mengenregelung *f*, Füllstandsregelung *f*; 2. Pegelregelung *f*, Lautstärkeregelung *f (Elektroakustik)*; Verstärkungsregelung *f*; 3. Lautstärkeregler *m*
v. displaced verdrängtes Volumen *n*, Flüssigkeitsverdrängung *f*, Wasserverdrängung *f*
v. elasticity Volumenelastizität *f*
v. equivalent *(El)* Bezugsdämpfung *f* eines Übertragungssystems, Nutzdämpfung *f*
v. expansion 1. räumliche (kubische) Ausdehnung *f*; 2. *(Fmt)* Dynamiksteigerung *f*, Dynamikdehnung *f*
v. flow Volumendurchsatz *m*
v. governor Mengenregler *m*
v. indicator *(El)* Aussteuerungsmesser *m*, Aussteuerungsanzeiger *m*, UV-Meter *n (effektiv anzeigender Aussteuerungsmesser)*
v. integral Raumintegral *n*, Volum[en]integral *n*
v.-limiting amplifier dynamikbegrenzender (lautstärkebegrenzender) Verstärker *m*

v. loss per unit length *(Fmt)* Bezugsdämpfung *f* je Längeneinheit
v. of chips removed *(Masch)* Spanvolumen *n*, Volumen *n* des zerspanten Werkstoffs
v. of compression space Kompressionsraum *m*; Kompressionsvolumen *n*
v. of displacement *(Schiff)* Rauminhalt *m* der Verdrängung, Verdrängung *f*
v. of fluid Flüssigkeitsvolumen *n*
v. of metal removed Spanvolumen *n*, zerspantes Metallvolumen *n*
v. of sound Lautstärke *f*
v. of speech Lautstärke *f*
v. of the displaced liquid Volumen *n* der verdrängten Flüssigkeit
v.-pressure diagram p,V-Diagramm *n*
v. rate of flow Strömungsmenge *f*
v. regulator Lautstärkeregler *m*
v. resistivity spezifischer Widerstand *m*, Würfelwiderstand *m*
v. unit Volum[en]einheit *f*, Raumeinheit *f*; Lautstärkeeinheit *f*, Pegeleinheit *f*
v. voltameter Volum[en]voltameter *n*
volumetric volumetrisch, im Volumen gemessen; je Volumeneinheit
v. analysis Maßanalyse *f*
v. capacity Fassungsvermögen *n*, Rauminhalt *m*, Kubikinhalt *m*
v. delivery Fördervolumen K *n*, Fördermenge *f*
v. displacement flowmeter Verdrängungszähler *m*, [unmittelbarer] Volumenzähler *m*
v. efficiency volumetrischer Wirkungsgrad *m*; Füllungsgrad *m*; Liefergrad *m*
v. flask Meßkolben *m*
v. flowmeter Durchflußzähler *m*, Volumenzähler *m*
v. force Volumkraft *f*
v. measure Raummaß *n*
v. pipette Vollpipette *f*, Meßpipette *f*
v. solution Maßlösung *f*, Titrationslösung *f*
v. voltameter Volumenvoltameter *n*
volute spring Kegelfeder *f*
volution Schneckenwindung *f* *(Verzahnung)*
vortex Wirbel *m*, Strudel *m*
v. centre Wirbelkern *m*
v. field Drehfeld *n*, Wirbelfeld *n*
v. filament Wirbelfaden *m*
v. flow Wirbelströmung *f*
v. formation Wirbelbildung *f*
v. line Wirbellinie *f*
v. motion Wirbelbewegung *f*
v. region Wirbelgebiet *n*
v. ring Wirbelring *m*
v. row Wirbelreihe *f*
v. street Wirbelstraße *f*
v. surface Wirbelfläche *f*
v. tube Wirbelröhre *f*
vortical wirbelnd
vorticity Wirbelstärke *f*, Wirbligkeit *f*
v. zone Wirbelbereich *m*
voucher copy *(Typ)* Belegexemplar *n*
voussoir Keilstein *m*
v. arch *(Brg)* Druckgewölbe *n*
voyage Reise *f*; Schiffsreise *f*
vug *(Geol)* Druse *f*
vulcanite Ebonit *n*, Hartgummi *m*
vulcanizable vulkanisierbar
vulcanization of rubber Vulkanisieren *n* von Gummi
vulcanize / to vulkanisieren
vulcanized fibre Vulkanfiber *f*
v. fibre valve Vulkanfiberventil *n*
v. paper Vulkanfiber *f*
vulgar fraction *(Math)* gemeiner Bruch *m*

W

wabble / to schwanken, taumeln
wabbler *(Met)* Kleeblattzapfen *m* *(Walze)*
wabbling disk Taumelscheibe *f*
wad / to füllen, ausstopfen; wattieren
wad 1. Erdkobalt *m*; Asbolan *m*; 2. Filzscheibe *f*; Watte *f*
wadding Watte *f*
w. thread *(Text)* Füllfaden *m*
w. warp *(Text)* Füllkette *f*
w. weft *(Text)* Füllschuß *m*
wading pool Planschbecken *n*
wafer *(El)* Scheibe *f* *(Flachbahnschalter)*, Schalterscheibe *f*
w. socket Plättchenfassung *f*
w. switch Flachbahnschalter *m*, Scheibenschalter *m*
w.-type switch *s.* w. switch
waffle floor Waffelboden *m*
w. form backing for carpets waffelförmige Teppichrückseitenbeschichtung *f*
w. plate Kassettenplatte *f*
w. slab Kassettenplatte *f*
w. weave *(Text)* Waffelbindung *f*
wage earner Lohnempfänger *m*
wages Arbeitslohn *m*; *(Schiff)* Heuer *f*
wagging vibration Nickschwingung *f* *(Spektroskopie)*
waggon *s.* wagon
Wagner earth Wagnersche Erde *f* *(Meßtechnik)*
wagon Eisenbahnwagen *m*, Waggon *m*; Lore *f*
w. body Wagengestell *n*
w. box Korntankwagen *m*
w. circulation Wagenumlauf *m*
w. drill *(Brg)* Bohrwagen *m*
w. load Waggonladung *f*
w. shed Wagenschuppen *m*
w. tippler Waggonkipper *m*

w. trace Zugriemen *m*
w. unloader Waggonentlader *m*
wagonroad *(Brg)* Förderstrecke *f*
wainscot / to täfeln
wainscot Holztäfelung *f*; *(Schiff)* hölzerne Schottwand *f*
wainscot[t]ing Täfelung *f*, Wandtäfelung *f*; Innenverkleidung *f*
waist 1. Einbuchtung *f*; Einschnürung *f*; Sattel *m*; 2. Mitteldeck *n (auf Kriegsschiffen)*
w. belt *(Flg)* Bauchgurt *m*
w.-level finder *(Opt)* Sucherschacht *m*
w. pad Gelenkeinlage *f (Schuhfertigung)*
waisting Einschnürung *f*
waistline dart Taillenabnäher *m*
waiting-passenger indicator Stockwerksruf *m (Fahrstuhl)*
w. room Wartezimmer *n*
w. time Wartezeit *f*
waive a right / to Verzicht leisten auf ein Recht
wake *(Schiff)* Kielwasser *n*, Nachstrom *m*, Vorstrom *m*, Mitstrom *m*, Sog *m*; Totwasser *n*
w.-adapted propeller Nachstrompropeller *m*
w. cavitation Kavitation *f* im Nachlauf
w. coefficient Nachstromziffer *f*, Nachstromkoeffizient *m*
w. fraction Nachstromziffer *f*, Sogziffer *f*
w. of the torpedo Blasenbahn *f (Torpedo)*
w. space Totraum *m*
wale 1. Bergholz *n*, Krummholz *n*; Bordleiste *f*; Riegel *m*, Verbindungsstück *n*; 2. *(Text)* Maschenstäbchen *n*; Rippe *f*, Strieme *f*
waling *(Bw)* Brustholz *n*, Gurt[holz *n*] *m*, Holm *m*, Riegel *m (Grabenverbau)*
walk off center / to *(Am, shop term)* verlaufen *(Bohrer)*
w. out *(Brg)* ausfahren *(Seilfahrt)*
walk way Laufsteg *m*
walkie-lookie tragbares Fernsehaufnahmegerät *n*
w.-talkie tragbares Gegensprechgerät (Wechselsprechgerät) *n*, tragbare Sender-Empfänger-Kombination *f*, Tornisterfunkgerät *n*
walking beam Bohrschwengel *m (Seilbohren)*
w.-beam annealing furnace *(Met)* Balkenherdglühofen *m*
w.-beam furnace *(Met)* Schwingbalkenofen *m*
w. dragline Schreitbagger *m*
w. excavator Schreit[kufen]bagger *m*
w. plough Schwingpflug *m*
w. support *(Brg)* schreitender Ausbau *m*
wall 1. Wand[ung] *f*; Mauer *f*; 2. *(Brg)* Stoß *m (einer Strecke)*
w. absorption Wandabsorption *f*
w. arch Wandbogen *m*
w. bars Sprossenwand *f*, Wandreck *n (Turnen)*
w. bed Klappbett *n*
w. bolt Steinschraube *f*
w. bracket Mauerbügel *m*; Wandkonsole *f*
w. bushing Mauerdurchführung *f*, Wanddurchführung *f (I-Rohr usw.)*
w. clamp Rohrschelle *f (für Wandbefestigung)*
w. crack Wandriß *m*
w. drilling machine Mauerbohrmaschine *f*
w. effect Wand[ungs]effekt *m*
w. facing Wandverkleidung *f*
w. fan Wandlüfter *m*
w. fastening Wandbefestigung *f*
w. finish Lochwandungsoberflächengüte *f*
w. fitting Wandleuchte *f*
w. footing Mauergründung *f*; Streifenfundament *n*
w. fruit Spalierobst *n*
w. fruit tree Spalierbaum *m*
w. hoist Wandwinde *f*
w. hook Rohrhaken *m*, Wandhaken *m*
w. line Mauerflucht *f*
w. lining *(Met)* Mauerfutter *n*; Wandbelag *m*
w. losses Wandverluste *mpl*
w. of casing Gehäusewand *f*
w. of nozzle Düsenwandung *f (Turbine)*
w. of the hole Bohrungswandung *f*
w. outlet box *(Am)* Wandsteckdose *f*
w. panel Wandtafel *f*, Wandplatte *f*
w. panel heating Wandflächenheizung *f*
w. paper Tapete[npapier *n*] *f*
w. paper machine Tapetendruckmaschine *f*
w.-pattern switchboard *(El)* Wandschalttafel *f*
w. plate 1. Fußpfette *f*; Mauerlatte *f*; 2. *(Brg)* langes Joch *n (Bolzenschrotausbau)*
w. plug Stecker *m* für Wandsteckdose
w. radial drill Wandradialbohrmaschine *f*
w. raw paper Tapetenrohpapier *n*
w. recombination Wandrekombination *f*
w. rocks *(Brg)* Nebengestein *n*
w. rubber Seitengummi *m (Reifen)*
w. rug Wandbehang *m*
w. scraper Spachtel *m*
w. screw Steinschraube *f*
w. set *s.* w. telephone
w.-sided vessel Schiff *n* mit geraden Seitenwänden
w.-siphoning *(Rak)* Wandabsaugung *f*
w. socket Anschlußdose *f*, Steckdose *f*
w. steam engine Wanddampfmaschine *f*
w. telephone [set] Wandfernsprecher *m*
w. thickness Wanddicke *f*
w. thickness gauging Wanddickenmessung *f*
w. tie Maueranker *m*
w.-tube insulator Durchführungsisolator *m*
w.-type telephone Wandfernsprecher *m*
wallboard Leichtbauplatte *f (für Wände)*, Wand[faser]platte *f*
wallet Brieftasche *f*; Werkzeugtasche *f*; Satteltasche *f (Fahrrad)*

walling Mauerwerk *n*, Steinmauern *fpl*
w. crib *(Brg)* Mauerfuß *m*, Mauerkranz *m*
walls Bordwand *f (Güterwagen)*
walnut Walnuß *f*; Walnußholz *n*
wander from the centre / to verlaufen *(Bohrer)*
wanted subscriber *(Fmt)* verlangter Teilnehmer *m*
warble / to *(El)* wobbeln
warble frequency *(El)* Wobblerfrequenz *f*
w. lump *(Led)* Dasselbeule *f (Fehler)*
w. tone Wobbelton *m (Meßtechnik)*
warbled hide mit Dassellarven behaftete Haut *f*
warbler oscillator *(El)* Schwebungsoszillator *m*, Schwebungssummer *m*, Überlagerungssummer *m*
Ward-Leonard set [Ward-]Leonard-Satz *m*
Ward-Leonard-type drive [Ward-]Leonard-Antrieb *m*
warding file Schlüsselfeile *f*
ware potato Speisekartoffel *f*
warehouse Lagerhaus *n*, Magazin *n*, Speicher *m*
warehousing Lagerung *f*; Unterbringung *f*
warhead Gefechtskopf *m (einer Granate)*
warm / to [er]wärmen, erhitzen
w.-press warmpressen
w. up warm werden; anwärmen, aufwärmen
w. up a kiln *(Holz)* eine Trockenkammer anwärmen
w. up the cylinder den Zylinder vorwärmen
warm warm
w. air Warmluft *f*
w.-air heating Warmluftheizung *f*
w. bath Warmbad *n*
w. brittlement *(Met)* Warmversprödung *f*, Warmbrüchigkeit *f*
w. extrusion Strangpressen *n*
w.-formable warm[um]formbar, warmverformbar
w. front Warmfront *f (Meteorologie)*
w. store warmgehaltener Lagerraum *m*
w.-up drift *(Fmt)* Frequenzabweichung *f* beim Anheizen
warmed-up warmgelaufen; betriebswarm
warming-up period Anwärmzeit *f*
warmwater-shrinkage Heißwasserkrumpfung *f*
warning 1. Alarm *m*; 2. Warnung *f*, Warnmeldung *f*; 3. Warnsignal *n*
w. circuit *(Fmt)* Vormeldestromkreis *m*
w. device Warnanlage *f*
w. lamp Anzeigelampe *f*, Kontrollampe *f*, Kontrollicht *n*, Warnlampe *f*
w. light Signalfeuer *n*
w. signal *(Eis)* Ankündigungssignal *n*; Warn[ungs]signal *n*
warp / to 1. verwölben, sich krümmen, sich verziehen, sich werfen; 2. *(Text)* schären, kettschären, zetteln, anschären, anschirren; 3. *(Schiff)* verholen, bugsieren
warp 1. *(Text)* Kette *f*, Zettel *m*; 2. *(Schiff)* Verholleine *f*, Warpleine *f*; 3. Schlick *m*, Modder *m*
w.-backed fabric Kettendouble *n*
w. beam Kettbaum *m*
w. beam brake Kettbaumbremse *f*
w. beam creel Kettbaum[ablauf]gestell *n*
w. beam flange Kettbaumscheibe *f*
w. beam truck Kettbaum[transport]wagen *m*
w. bobbin Kettspule *f*, Zettelspule *f*
w.-bowed cloth Gewebe *n* mit kurvenförmiger Verzerrung der Kette
w. breakage Kettfadenbruch *m*
w. changer Kettenwechsler *m*
w. cop Kettgarnkötzer *m*
w. cords Kettenrippen *fpl*
w. count Kettnummer *f*
w. cylinder Schärtrommel *f*
w. drop wire Kettfadenreiter *m*
w. dyeing Kettfärberei *f*
w. dyeing machine Kettfärbemaschine *f*
w. float Kettflottierung *f*
w. frame Kettrahmen *m*
w.-knit fabric Kettwirkware *f*
w.-knitted kettengewirkt
w. knitting loom Kettenwirkmaschine *f*, Kettenstuhl *m*
w. loom fabric Kettgewebe *n*
w. pattern Schärmuster *n*
w. pile carpet Kettflorteppich *m*
w. pile fabric Kettflorgewebe *n*, Kettsamt *m*
w. plush Kettplüsch *m*
w. preparation Kettvorbereitung *f*
w. printing Kettdruck *m*
w. printing machine Kettdruckmaschine *f*
w.-proof verzugsfrei
w. rib Kettrips *m*, Querrips *m*
w. roller Kettzylinder *m*
w. rope *(Schiff)* Verholleine *f*
w.-saving kettschonend
w. setting Kettfadeneinstellung *f*; Kettendichte *f*
w. sizing Kettschlichten *n*
w. stop motion Kettfadenwächter *m*
w. streak Schußstreifen *m (Fehler)*
w. strip Kettstreifen *m*
w. take-up Einarbeitung *f* der Kette
w. tension Kettspannung *f*, Längsspannung *f*
w. thread Kettfaden *m*
w. twill Kettköper *m*
w. tying machine Kettenanknüpfmaschine *f*
w. winder Kettspulmaschine *f*
w. winding frame Kettgarnspulmaschine *f*
w. yarn Kettgarn *n*
warpage Verzug *m*, Verziehen *n*, Werfen *n*
warped verwölbt, gekrümmt, geworfen, verworfen, verzogen; doppelt gekrümmt *(Schalen)*
w. ends gezettelte Webketten *fpl*

warper *(Text)* Zettelmaschine *f*, Schärmaschine *f*
warper's knot Weberknoten *m*, Kettfadenknoten *m*
warping 1. Krümmen *n*, Werfen *n*, Ziehen *n*; Verwerfung *f*; 2. Verwölbung *f*; Werfung *f* *(Torsion)*; 3. *(Text)* Schären *n*; 4. Schlammdüngung *f*
w. beam *(Text)* Zettelbaum *m*
w. bridge *(Schiff)* Achterbrücke *f*, Dockbrücke *f*
w. buoy *(Schiff)* Verholboje *f*
w. capstan *(Schiff)* Verholspill *n*
w. chock *(Schiff)* Verholklüse *f*
w. cone *(Text)* Schärkonus *m*
w. creel *(Text)* Aufsteckzeug *n* für Kettschären; Schärgatter *n*
w. drum 1. *(Text)* Schärtrommel *f*; 2. Spillkopf *m*
w. frame *(Text)* Schärmaschine *f* *(kurze Kette)*; Schärrahmen *m*
w. guide *(Schiff)* Verholpoller *m*
w. mill *(Text)* Schärmaschine *f* *(lange Ketten)*
w. mill for ribbons Schärmaschine *f* für Bänder
w. of the grate *(Wärm)* Werfen *n* des Rostes
w. pipe *(Schiff)* Verholklüse *f*
w. plant *(Text)* Schäranlage *f*
w. reed *(Text)* Schärriet *n*
w. roller *(Text)* Zettelwalze *f*
w. winch *(Schiff)* Verholwinde *f*, Verholspill *n*
Warren girder Strebenfachwerkträger *m*
Warren truss Warren-Fachwerk *n* *(parallelgurtiges Fachwerk ohne Vertikalstäbe)*
warship Kriegsschiff *n*
warted plate Warzenblech *n*
wash / to 1. waschen, spülen; wässern; schwemmen; 2. löschen *(Tonband)*; 3. läutern *(Aufbereitung)*
w. ashore anspülen, anschwemmen *(an das Ufer)*
w. away wegspülen
w. out löschen *(Tonband)*
wash 1. Aufschlämmung *f*; 2. Wäsche *f* *(Aufbereitung)*; 3. *(Schiff)* Ruderblatt *n*; 4. Schlichte *f* *(Formerei)*; 5. Luftstrudel *m*; Wellenschlag *m*, Kielwasser *n*, 6. Schwemmland *n*
w. and wear finish *(Text)* Wasch- und Trageausrüstung *f*
w. bottle Spritzflasche *f*; Waschflasche *f*
w. box Setzmaschine *f* *(Aufbereitung)*
w. bulkhead *(Schiff)* Schlingerschott *n*, Schlagschott *n*
w. column Berieselungsturm *m*, Rieselturm *m*, Skrubber *m*, Wäscher *m*
w.-down toilet Spülklosett *n*
w.-fast waschecht, waschfest
w. heat *(Met)* Entzunderungswärme *f*
w. heating furnace *(Met)* Entzunderungsofen *m*, Blankglühofen *m*
w. leather Waschleder *n*
w.-out heat Spülcharge *f*, Vorcharge *f* *(SM-Prozeß)*
w. plate *(Schiff)* Waschplatte *f*, Schlag[wasser]platte *f*, Schlingerplatte *f* *(zwischen den Bodenwrangen)*
w. primer Grundiermittel *n* [mit Rostschutz]; Haftgrund *m*
washability 1. Waschbarkeit *f*, Auswaschbarkeit *f*; 2. Aufbereitbarkeit *f*, Sortierbarkeit *f*
washable 1. [auf]waschbar; waschfest; 2. aufbereitbar, sortierbar
washboard 1. Scheuerleiste *f*; 2. Waschbrett *n*; 3. Dollbord *m* *(Boot)*
washed fines gewaschene Feinstkohle *f*
washer 1. Wäscher *m* *(Gas)*; Waschapparat *m*; Reinigungsapparat *m*; 2. Naßsortiermaschine *f* *(Aufbereitung)*; 3. Berieselungsturm *m*, Rieselturm *m*; 4. Unterlegscheibe *f*, Beilage *f*, Zwischenlage *f*; Abdichtungsring *m*
w. box Setzkasten *m* *(Aufbereitung)*
w. of compressible material elastische Zwischenlage *f*
washery Wäsche *f* *(Anlage zur Aufbereitung)*
w. dirt Waschberge *pl* *(Aufbereitung)*
w. refuse Waschberge *pl* *(Aufbereitung)*
w. superintendent Wäschesteiger *m* *(Aufbereitung)*
washes *(Am)* Waschwasser *n*
washing 1. Waschen *n*, Spülen *n*; Naßreinigung *f*; Wässerung *f*, Wässern *n*; 2. Löschen *n* *(Tonband)*
w. agent Waschmittel *n*
w. away Unterspülen *n*
w. bath Wässerungsbad *n* *(Foto)*
w. blue Waschblau *n*
w. bottle Waschflasche *f*, Spritzflasche *f*
w. cylinder 1. Läutertrommel *f* *(Aufbereitung)*; 2. Waschtrommel *f* *(Papierherstellung)*
w. department Entlaugerei *f*
w. drum Waschtrommel *f* *(Papierherstellung)*
w. engine Halbzeugholländer *m* *(Papierherstellung)*
w. liquor Waschlauge *f*, Flotte *f*, Waschflotte *f* *(Spinnerei)*
w. machine Waschmaschine *f*
w. oil Waschöl *n*
w. on sheep's back *(Text)* Rückenwäsche *f*
w. out Löschen *n* *(Tonband)*
w.-out loss Auswaschverlust *m*
w. powder Waschpulver *n*, pulverförmiges Waschmittel *n*
w. power Waschvermögen *n*, Waschwirkung *f*
w. process Waschprozeß *m*, Waschvorgang *m*; Flotationsverfahren *n*, Schlämmverfahren *n*

w. **rack** Wässerungsgestell *n*
washings 1. Waschwasser *n*; 2. Waschverlust *m*, Waschabgänge *mpl*
Washita stone Washitastein *m*
washy wäßrig, dünn; verblaßt
waste / to abreiben, abnützen; vergeuden, verschwenden
w. **watts** Energie vergeuden, Strom verschwenden
waste Abfall *m*, Abgang *m*, Abfallprodukt *n*, Ausschuß *m*; Müll *m*; Makulatur *f*; Vergeudung *f*, Verschwendung *f*
w. **area** *(Brg)* Alter Mann *m*
w. **beating willow** *(Text)* Abfallklopfwolf *m*
w. **bin** Abfallbehälter *m*
w. **breaker** *(Text)* Abfallreißer *m*
w. **casting** Fehlguß *m*
w. **chamber** *(Text)* Abfallkammer *f*
w. **chute** *(Brg)* Bergerolle *f*, Versatzrolloch *n*
w. **cleaner** Abfallreinigungsmaschine *f*
w. **cleaning** Abfallreinigung *f*
w. **coal** Abfallkohle *f*
w. **coke** Abfallkoks *m*
w. **collecting tank** Abfallspeicher *m*, Abfallsammelbehälter *m*
w.-**filled area** *(Brg)* Versatzfeld *n*
w. **gas** Abgas *n*, Rauchgas *n*
w.-**gas flue** Rauchgaskanal *m*, Abgaskanal *m*
w. **head** verlorener Kopf *m (Gießtechnik)*
w. **heap** Grubenhalde *f*, Bergehalde *f*
w. **heap ore** Haldenerz *n*
w. **heat** Abhitze *f*, abgehende Wärme *f*, Abwärme *f*, Wärmeabfuhr *f*
w.-**heat boiler** Abhitzekessel *m*
w.-**heat engine** Abwärmekraftmaschine *f*
w.-**heat loss** Abwärmeverlust *m*
w.-**heat recovery** Abhitzeverwertung *f*
w.-**heat utilization** Abhitzeverwertung *f*
w. **liquor** Ablauge *f*, Sulfitablauge *f*
w. **metal** Metallkrätze *f*; Abfallmetall *n*, Metallabfall *m*
w. **motion** *(Masch)* Leerlauf *m*
w. **of power** Leistungsverlust *m*
w. **oil** Altöl *n*
w. **paper** Papierabfall *m*; Altpapier *n*
w. **paper dressing plant** Altpapieraufbereitungsanlage *f*
w. **picker** *(Text)* Abfallklauber *m*
w. **processing** Abfallverarbeitung *f*
w. **pulp preparation** Abfallzellstoffaufbereitung *f*
w. **raise** *(Brg)* Versatzüberhauen *n*
w. **recovery** Abfallaufbereitung *f*, Abfallverwertung *f*
w. **removal** Abraumgewinnung *f*
w. **rock** taubes Gestein *n*, Gangart *f*
w. **rubber** Altgummi *m*
w. **shaker** *(Text)* Abfallreinigungstrommel *f*, Abfallklopfwolf *m*
w. **shaker with dust hood** *(Text)* Abfallreinigungstrommel *f* mit Staubhaube
w. **sheet** *(Typ)* Fehlbogen *m*
w. **sorter** Abfallsichter *m*
w. **spinning** Abfallspinnen *n*
w. **steam** Abdampf *m*
w.-**steam pipe** Dampfauslaßrohr *n*, Dampfabblasrohr *n*
w.-**steam turbine** Abdampfturbine *f*
w. **stripping** *(Brg)* Abraumgewinnung *f*
w. **stuff** Ausschuß *m*
w. **treatment** Abfallbehandlung *f*, Abfallkonzentrierung *f*
w. **veneer core** Schälrestrollen *fpl (Furnier)*
w. **water** Abwasser *n*; Kondenswasser *n*; Schmutzwasser *n*
w. **weir** Überlauf *m*
w. **well** Senkgrube *f*
w. **wood** Abfallholz *n*
w.-**wood chipper** Hackmaschine *f* für Abfallholz
watch / to beobachten; überwachen
watch 1. Armbanduhr *f*, Taschenuhr *f*; 2. Schiffswache *f*; 3. Wache *f*; Hörbereitschaft *f (Funk)*
w. **and clock industry** Uhrenindustrie *f*
w. **below** *(Schiff)* Freiwache *f*
w. **boat** Wachboot *n*
w. **buoy** Markierungsboje *f*
w. **case** Uhrengehäuse *n*
w. **part** Uhrenteil *n*
w. **spring** Uhrfeder *f*
watchboat Wachboot *n*
watchmaker's file Präzisionsfeile *f*
w. **screwdriver** Uhrmacherschraubenzieher *m*
watchman Wächter *m*, Wachmann *m*
watchman's clock Wächterkontrolluhr *f*
watchwork Uhrwerk *n*
water / to gießen *(Pflanzen)*; sprengen *(Straßen)*; tränken *(Vieh)*; Wasser aufnehmen (fassen) *(Lokomotive)*
water Wasser *n*
w.-**absorbing capacity** Wasseraufnahmefähigkeit *f*, Wasseraufsaugfähigkeit *f*
w. **absorption** Wasseraufnahme *f*; Quellung *f*
w. **balance** Wasserhaushalt *m (z. B. des Bodens)*
w. **ballast tank** Wasserballasttank *m*
w. **base ink** Druckfarbe *f* auf Wasserbasis
w.-**bearing** wasserführend
w. **boiler** Wasserkessel *m*
w. **bottom** *(Am) (Schiff)* Doppelboden[tank] *m*
w. **bottoms** *(Schiff)* Bodentanks *mpl*
w. **brake** *(Schiff)* Wasserbremse *f*
w. **buoyancy** Wasserauftrieb *m*
w. **carriage** Wassertransport *m*, Wasserfracht *f*

w.-cement ratio Wasser-Zement-Faktor *m*, WZ-Faktor *m (Betontechnologie)*
w. channel Wasserkanal *m (Strömung)*
w.-checked regendicht
w. circulating pump Wasserumlaufpumpe *f*, Kühlwasserpumpe *f*
w. clerk Schiffsmakler *m* im Außendienst
w. closet Wasserklosett *n*
w.-closet pan WC-Becken *n*
w. collecting area Wassereinzugsgebiet *n*
w. colour Wasserfarbe *f*
w. column Wassersäule *f*
w. column arrester *(El)* Überspannungsableiter *m* mit Wasserwiderstand
w. condenser Frischwassererzeuger *m*
w. contact *(El)* Wasserkontakt *m*
w. container Wasserbehälter *m*
w. content Wassergehalt *m*
w. coolant pump Kühlwasserpumpe *f*
w.-cooled wassergekühlt
w. cooling Wasserkühlung *f*
w. cooling and circulating system Kühlwasserkreislauf *m*
w. course Wasserlauf *m*
w. crack *(Met)* Wasserriß *m*, Kühlriß *m* durch Wasser
w. craft Wasserfahrzeug *n*
w. cross-section Kanalquerschnitt *m*
w. cross-section of a river Flußquerschnitt *m*
w. curing Feuchthalten *n* des Betons, Nachbehandlung *f* mit Wasser *(Betontechnologie)*
w. decomposition Wasserzersetzung *f*, Wasserradiolyse *f*
w. deflector Siebabstreicher *m (Papierherstellung)*
w. displacement Wasserverdrängung *f*
w. diviner Wünschelrutengänger *m*
w. drain Wasserablaß *m*, Wasserabfluß *m*
w. draught gauge Wassersäulenzugmesser *m*, Wassermanometer *n*
w. drip *(Bw)* Tropfnase *f*, Wassernase *f*
w. entrained with the steam *(Wärm)* übergerissenes (mitgerissenes) Wasser *n*
w. equivalent Wasserwert *m (Kalorimetrie)*
w. equivalent of the calorimeter Wasserwert *m* des Kalorimeters
w.-equivalent of the condenser body Wasserwert *m* des Kondensatorkörpers
w. factor Wasserwert *m (Viskosimetrie)*
w. film *(Wärm)* Wasserschleier *m (Kondensator)*
w. flow relay Wasserströmungswächter *m*
w. flushing Wasserspülung *f*
w. garland *(Brg)* Wasserring *m*, Träufelrinne *f*
w. gas Wassergas *n*
w. gauge Wasseruhr *f*; Wasserstandsanzeiger *m*, Wasserstandsglas *n*
w. gauge arm with cock Wasserstandshahnkopf *m*
w. gauge arm with valves Wasserstandsventilkopf *m*
w. gauge protector of armoured glass Drahtglasschutzhülse *f*, Drahtglasschutzmantel *m*
w. gauge with protector (guard) Wasserstandsapparat *m* mit Schutzmantel
w. gauge with test cocks Wasserstandsglas *n* mit Probierhahnreihe
w. gland Flüssigkeitsabschluß *m*
w. glass Wasserglas *n*
w. hammer *(Wärm)* Wasserschlag *m*
w. hardening steel wassergehärteter Stahl *m*, Wasserhärter *m*
w. head Druckhöhe *f*, Wassersäule *f*; Wasserstand *m*
w. header Wasservorlage *f*
w. heater Heißwasserspeicher *m*
w. highway Schiffahrtsstraße *f*, Schiffahrtsweg *m*; Seewasserstraße *f*
w. hoisting Wasserhebung *f*
w. hole *(Brg)* abwärts gerichtetes Bohrloch *n*
w. imbibition value *(Text)* Quellwert *m*
w. in the calorimeter Kalorimeterwasser *n*
w. incrustation Kesselstein *m*
w. inflow *(Brg)* Wassereinbruch *m*
w. influx *(Brg)* Wasserzufluß *m*
w. infusion *(Brg)* Stoßtränkung *f*
w. inlet connection Kühlwassereinlaufstutzen *m*
w. inrush *(Brg)* Wassereinbruch *m*
w.-insoluble dye[stuff] wasserunlöslicher Farbstoff *m*
w. irrigation Wasserberieselung *f*, Bewässerung *f*
w. jacket Wassermantel *m*, Kühl[wasser]mantel *m*, Kühljackett *n*
w.-jacketed wassergekühlt, mit Wasserkühlmantel; mit Wasserheizmantel
w.-jacketed furnace *(Met)* Wassermantelofen *m*
w. jet heater *(El)* Wasserstrahlelektrodenkessel *m*
w. jet pump Wasserstrahlpumpe *f*
w. leaf nasse Papierbahn *f (Preßwalze)*
w.-leaf paper ungeleimtes Papier *n*, Wasserpapier *n*
w. leaf-type of paper wasserglasverleimtes Papier *n*
w. level Wasserwaage *f*; Wasserstand *m*, Wasserspiegel *m*
w. level indicator Wasserstandsanzeiger *m*, Füllstandsanzeiger *m*
w. level pointer Wasserstandsmarke *f*
w. level recorder Wasserpegelschreiber *m*
w. level scale Wasserstandsskala *f*

w. line Wasserlinie *f (Schiff)*
w. line coefficient Völligkeitsgrad *m* der Wasserlinie
w. line length Wasserlinienlänge *f (Länge in der Wasserlinie)*
w. lines Konstruktionswasserlinien *fpl*
w.-load method Wasserbelastungsverfahren *n (cm-Wellenenergiemeßverfahren)*
w. main Wasserrohr *n*
w. mark 1. Pegel *m*, Wasserstandsmarke *f*; 2. *(Typ)* Wasserzeichen *n*
w. mark apparatus Wasserzeichengerät *n (Papierherstellung)*
w. meter Wasseruhr *f*
w. metering Wassermessung *f*
w. mill Wassermühle *f*
w. miscibility Wasserentmischfähigkeit *f*, Emulgierbarkeit *f*
w.-moderated wassermoderiert, wassergebremst
w.-moderated reactor wassermoderierter Reaktor *m*
w. of capillarity Kapillarwasser *n*
w. of condensation Kondenswasser *n*, Niederschlagswasser *n*, Kondensat *n*
w. of constitution *(Ch)* Konstitutionswasser *n*
w. of crystallization Kristallwasser *n*
w. of hydration Hydrat[ations]wasser *n*
w. of open capillaries Grobkapillarwasser *n*
w. outlet connection Kühlwasserauslaufstutzen *m*
w. permeability Wasserdurchlässigkeit *f*
w. pier Strompfeiler *m (Brückenbau)*
w. pipe Wasserleitung[srohr *n*] *f*
w. plane Wasserlinienebene *f (Schiff)*
w. plane coefficient Wasserlinienvölligkeitsgrad *m*, Völligkeitsgrad *m* der Konstruktionswasserlinie
w. pollution Wasserverunreinigung *f*
w. polo Wasserball *m (Sportdisziplin)*
w. power station Wasserkraftwerk *n*
w.-proof wasserfest, wasserdicht, wasserbeständig
w.-proof finish *(Text)* wasserbeständige Appretur *f*
w.-proof property Wasserbeständigkeit *f*
w. pump Wasserpumpe *f*, Kühlwasserpumpe *f*
w. pump body Wasserpumpengehäuse *n*
w. pump gland Wasserpumpenstopfbuchse *f*, Wasserpumpendichtung *f*
w. purification Wasseraufbereitung *f*, Wasserreinigung *f*
w. recooling *(Wärm)* Wasserrückkühlung *f*
w. regain Feuchtigkeitsgehalt *m*
w. repellency *(Text)* Abperleffekt *m*, wasserabweisende Eigenschaft *f*
w.-repellent hydrophob, wasserabstoßend, wasserabweisend
w. reserve Wasservorrat *m*, Wasserreserve *f*
w. reservoir Wasserbehälter *m*, Wassertank *m*
w. resistance *(El)* Wasserwiderstand *m*
w.-resisting wasserbeständig, wasserecht
w. resource Wasserhaushalt *m (z. B. des Bodens)*
w. ret[ting] Wasserröste *f (Flachs)*
w. ring *(Brg)* Wasserrinne *f*, Wasserring *m (Schachtausbau)*
w. route Wasserstraße *f*, Wasserweg *m*
w. seal Wasserverschluß *m (Geruchverschluß)*
w.-sealed gland Sperrwasserdichtung *f*
w. separator Wasserabscheider *m*
w. sluice Spülrinne *f (Entaschung)*
w. slurry wäßrige Suspension *f (Aufschlämmung)*
w.-smoking Trocknen *n* von Schamottesteinen vor dem Brennen
w. softener Wasserenthärtungsmittel *n*
w. softening Wasserenthärtung *f*
w. softening plant Wasserreinigungsanlage *f*
w.-soluble wasserlöslich
w.-soluble size wasserlösliche Schlichte *f (Weberei)*
w. space Wasserraum *m*
w. spindle *(Text)* Ringspindel *f*
w.-spout Wasserspeier *m*
w. sprinkling Wasserberieselung *f*, Beregnung *f*
w. standage *(Brg)* Wassersumpf *m*
w. standages *(Brg)* Sumpfstrecken *fpl*
w. stemming Wasserbesatz *m*
w. storage Wasserspeicherung *f*
w. string wasserabsperrende Rohrtour *f (Tiefbohrtechnik)*
w. supply Wasserversorgung *f*
w. supply line Wasserversorgungsleitung *f*
w. suspension *s.* w. slurry
w. table Grundwasserspiegel *m*
w. tank waggon Wasserwagen *m*
w. tap Wasserhahn *m*
w. tapping *(Brg)* Lösen *n* von Wasser
w. tower Wasserturm *m*
w. tractor *(Schiff)* Wassertrecker *m*, Schlepper *m*
w. treating equipment Wasseraufbereitungsanlage *f*
w. treatment Wasseraufbereitung *f*; Klärung *f*
w. treatment plant *(Am)* Kläranlage *f*
w. tube boiler Wasserrohr[dampf]kessel *m*, Siederohrkessel *m*
w. tube boiler with divided header at both ends Wasserrohrkessel *m* [nach Babcock und Wilcox] mit geteilten Wasserkammern an beiden Enden der Röhren
w. tube boiler with header Wasserrohrkessel *m* mit Wasserkammer, Wasserkammerkessel *m*

w. tube boiler with junction boxes at the tube ends Wasserrohrkessel *m* mit Kapselverbindungen an den Rohrenden
w. tube boiler with large tubes weitrohriger Wasserrohrkessel *m*
w. tube boiler with two headers Wasserrohrkessel *m* mit zwei Wasserkammern, Zweikammerkessel *m*
w. tunnel Wasserstollen *m*
w. turbine Wasserturbine *f*
w. value Wasserwert *m*
w. void Wasserpore *f (im Beton)*
w. voltameter Wasservoltameter *n*
w. well Wasserbrunnen *m*
w. wheel Wasserrad *n*
watering Gießen *n (Pflanzen)*; Sprengen *n (Straße)*; Tränken *n (Vieh)*; Wasseraufnahme *f (Lokomotive)*
w. basin system *(Lw)* Tränkbeckensystem *n*
w. cart Sprengwagen *m*
w. hose Sprengschlauch *m*
w. place Tränke *f*
w. pot Gießkanne *f*
w. trough *(Lw)* Tränktrog *m*
waterlevel *s.* water level
waterline *s.* water line
w. plan *(Schiff)* Wasserlinienriß *m*
waterlogged wassergesättigt
waterman Flußschiffer *m*, Binnenschiffer *m*
watermark *s.* water mark
waterport Wasserpforte *f*
waterproof / to wasserdicht machen, wasserundurchlässig machen, imprägnieren
waterproof wasserdicht, wasserundurchlässig
w. paper wasserdichtes Papier *n*
w. tissue imprägniertes Gewebe *n*
waterproofing *(Bw)* Grundwasserabdichtung *f*, Abdichtung *f*, Dichtung *f*, Feuchtigkeitssperre *f*; baulicher Feuchtigkeitsschutz *m*, Bauwerksabdichtung *f*; *(Text)* Wasserbeständigmachen *n*, Imprägnieren *n*
w. agent *(Text)* Mittel *n* zur wasserbeständigen Ausrüstung
watershed Wasserscheide *f*
watertight wasserdicht, wasserundurchlässig
w. flat wasserdichtes Plattformdeck *n*; Plattformzwischendeck *n*
w. frame Vollspant *n*
w. gland wasserdichte Kabelstopfbuchse *f*
w. integrity *(Schiff)* Schiffssicherheit *f* durch wasserdichte Unterteilung
w. line wasserdichte Leitung *f*
w. machine *(Elm)* [schwall]wassergeschützte Maschine *f*
w. stopping *(Brg)* Wasserdamm *m*
waterway Wasserstraße *f*
waterwhite wasserhell, wasserklar
waterworks Wasserwerk *n*

watery fusion *(Ch)* Schmelzen *n* im eigenen Kristallwasser, Zerfließen *n* der Kristalle
watt Watt *n (Einheit der Leistung: 1 W = 1 J/s)*
w. component Wattkomponente *f*
w. consumption Wirkleistungsverbrauch *m*
w.-hour Wattstunde *f*
w.-hour capacity Kapazität *f* in Wattstunden *(Batterie)*
w.-hour meter Wattstundenzähler *m*, Wirkverbrauchszähler *m*
w.-second Wattsekunde *f*
wattage *(El)* Leistung *f (in W)*, Wattzahl *f*
w. rating Nennleistung *f*
wattful current Wirkstrom *m*
w. power Wirkleistung *f*
wattless wattlos, leistungslos
w. component Blindkomponente *f*
w. component meter Blindverbrauchszähler *m*, Blindstromzähler *m*
w. current *(El)* Blindstrom *m*
w. output Blindleistung *f*
w. power Blindleistung *f*
wattmeter Wattmeter *n*, Leistungsmesser *m*
watts input *(Masch)* Leistungsaufnahme *f*
wave / to wogen, hin- und herbewegen
wave Welle *f*
w. absorption Wellenabsorption *f*
w. aerial Wellenantenne *f*
w. amplification Wellenverstärkung *f*
w. amplitude Wellenamplitude *f*
w. band *(Fmt)* Wellenband *n*, Wellenbereich *m*
w.-band filter Bandfilter *n*
w. beam Wellenbündel *n*
w. calming oil Wellenberuhigungsöl *n*
w. changing switch Wellen[um]schalter *m*
w. clutter Seegangsreflex *m*, Wellenreflexion *f (Radartechnik)*
w. component Wellenanteil *m*
w. concentration Wellenbündelung *f (Antennentechnik)*
w. converter Wellenumformer *m (Wellenleiter)*
w. crest Wellenberg *m*, Wellenkamm *m*
w. curve Wellenlinie *f*
w. detector *(Fmt)* Wellenanzeiger *m*
w. diffraction Wellenbeugung *f*
w. equation *(Ph)* Wellengleichung *f*
w. field Wellenfeld *n*
w. filter Siebkette *f*, Wellenfilter *n*
w.-form *s.* waveform
w. front Wellenfront *f*
w.-front plotter Wellenfrontschreiber *m (Meßtechnik)*
w. function Wellenfunktion *f*
w. generator *(Schiff)* Wellenerzeuger *m*, Wellenerreger *m (Modellversuch)*
w. image Wellenbild *n*

w. impedance [komplexer] Feldwiderstand *m* *(Hohlleitertechnik)*; Wellenwiderstand *m*
w. interference Welleninterferenz *f*
w.-length *s.* wavelength
w. load Belastung *f* im Seegang
w. loop Wellenbauch *m*
w. mechanics Wellenmechanik *f*
w. mode Wellentyp *m* *(Mikrowellen)*
w. motion Wellenbewegung *f*
w. nature of light Wellennatur *f* des Lichtes
w. node Wellenknoten *m*
w. number Wellenzahl *f*
w. of replacement Heckwelle *f*
w. of translation Fortpflanzungswelle *f*
w. optics Wellenoptik *f*
w. packet Wellenpaket *n*, Wellengruppe *f*
w. path Wellenbahn *f*
w. pattern Wellenbild *n*
w. polarization Wellenpolarisation *f*
w. propagation Wellenfortpflanzung *f*, Wellenausbreitung *f*
w. propagation over built-up areas Wellenausbreitung *f* über Stadtgebieten
w. propagation over open country Wellenausbreitung *f* über freiem Gelände
w. quantization Wellenquantisierung *f*
w. radiation Wellenstrahlung *f*
w. range switching Wellenbereichsumschaltung *f*
w.-set [saw] blade gewelltes Sägeblatt *n*
w. shape Wellenform *f*, Wellengestalt *f*
w. subduer Wellenberuhiger *m*
w. surface Wellenfläche *f*
w. tail Wellenrücken *m*
w. theory Wellenlehre *f*, Wellentheorie *f*
w. tilt Stirnsteilheit *f*, Wellenfrontwinkel *m*
w. trace Wellenlinie *f*
w. train *(El)* Wellenzug *m*
w. trap Sperrkreis *m*; Wellenfalle *f*; Wellensaugkreis *m*
w. trough Wellental *n*
w. velocity Wellengeschwindigkeit *f*
w. winding Wellenwicklung *f*
waved twill *(Text)* Wellenköper *m*
w. twist *(Text)* Kräuselzwirn *m*
waveform Wellenform *f*, Kurvenform *f*
w. analyzer Wellenformanalysator *m*
w. converter Wellenformwandler *m*
w. distortion Wellenformverzerrung *f*
w. generator Wellenformgeber *m*
waveguide Wellenleiter *m*; Hohlleiter *m*
w. attenuation Wellenleiterdämpfung *f*
w. circuit Wellenleiterschaltung *f*
w. construction Wellenleiteraufbau *m*
w. cross-section Hohlleiterquerschnitt *m*
w. elbow Wellenleiterknie *n*
w. field Hohlleiterfeld *n*
w. gasket Wellenleiterdichtung *f*
w. mount Hohlleiterfassung *f*
w. shim Wellenleiterfolie *f*; Flanschzwischenlage *f* *(Wellenleiter)*
w. shutter Hohlleiterblende *f*
wavelength *(El)* Wellenlänge *f*
w. constant Phasenbelag *m* *(des Kondensators)*
w. coverage Wellenlängenbereich *m*
w. dependence Wellenlängenabhängigkeit *f*
w. limit Grenzwellenlänge *f*
w. of the spectral centroid Schwerpunktwellenlänge *f*
w. scale Wellenlängenskale *f*
w. sensitivity Wellenlängenempfindlichkeit *f*
w. switch Wellenbereichsumschalter *m*
w. threshold Wellenlängengrenze *f*
wavemaker *(Schiff)* Wellenerzeuger *m* *(Modellversuch)*
wavemeter Wellenmesser *m*
waviness Welligkeit *f*
wavy wellenartig, wellenförmig; wellig *(Oberfläche)*; wimmerig *(Holz)*
w.-grained krummfaserig *(Holz)*
w. set gewellte Zahnung *f* *(Säge)*
wax / to 1. wachsen, größer werden, zunehmen; 2. wachsen, bohnern
wax Wachs *n*, Paraffin *n*, Hartparaffin *n*
w. cloth Wachstuch *n*
w.-coated wire *(El)* Wachsdraht *m*
w.-like wachsartig
w. size Paraffinleim *m*, Wachsleim *m*
w. tailings paraffinhaltige Rückstände *mpl*
waxed cotton-covered wire *(El)* Wachsbaumwolldraht *m*
w. paper Paraffinpapier *n*, Wachspapier *n*
waxing Wachsen *n*, Einwachsen *n*
waxy wachsähnlich, wachsartig, wachsig, weich
way 1. Weg *m*; Richtung *f*; Fahrt *f*; Strecke *f*; 2. Bahn *f*; *(Masch)* Führung[sbahn] *f*; 3. Art *f*, Weise *f*; Branche *f*
w.-type drilling machine Mehrwegebohrmaschine *f*
w.-type machine Mehrwegemaschine *f*
wayleave charges *(Fmt)* Anerkennungsgebühr *f*, Benutzungsgebühr *f*, Konzessionsgebühr *f*
ways 1. *(Schiff)* Ablaufbahn *f*, Stapellaufbahn *f*, Slip *m*; 2. *(Masch)* Führungen *fpl*
w.-end Vorhelling *f*, Ende *n* der Ablaufbahn
weak schwach, undeutlich; dünn *(Flüssigkeiten)*; flau *(Fotografie)*
w. concrete Magerbeton *m*
w. coupling *(El)* lose Kopplung *f*
w. current Schwachstrom *m*; schwacher Strom *m*
w.-current cable Schwachstromkabel *n*
w.-current engineering Schwachstromtechnik *f*
w. draught flauer Zug *m* *(Feuerungstechnik)*

w. explosive *(Brg)* schiebend wirkender Sprengstoff *m*
w.-mixture cruising *(Flg)* Sparflug *m*
w. roof *(Brg)* gebräches (unsicheres) Hangendes *n*
w. sand magerer Sand *m*
weaken / to [ab]schwächen; verdünnen *(Flüssigkeiten)*
weakening Schwächung *f*, Abschwächung *f*
w. of a permanent magnet Alterung *f* eines Dauermagneten, Nachlassen *n* eines Permanentmagneten
w. of cross section Querschnittsschwächung *f*
w. ratio Feldschwächungsverhältnis *n*
wear / to 1. tragen; 2. abtragen, abnutzen, verschleißen, verschleißend wirken
w. in einlaufen *(Lager)*
w. out 1. auslaufen *(Lager)*; 2. abnutzen
wear Abnutzung *f*, Verschleiß *m*
w. and tear Abnutzung *f*; Abschreibung *f*
w. condition of liner *(Kfz)* Zylinderlaufbild *n*, Zylindertragbild *n*
w.-free verschleißfrei
w. land Verschleißfase *f*, Abstumpfungsfase *f* *(Werkzeug)*
w. land width Abstumpfungsfasenbreite *f*, Verschleißfasenbreite *f*
w. life Standzeit *f* *(Werkzeug)*
w. mark Verschleißmarke *f* *(Meißel)*
w. of brakes Bremsenabnutzung *f*
w. of grooves Verschleiß *m* der Kaliber *(Walzwerk)*
w.-reducing verschleißmindernd
w. resistance Scheuerfestigkeit *f*, Verschleißfestigkeit *f*, Abriebfestigkeit *f*
w. resisting verschleißfest, abriebfest, scheuerfest
w. testing thread plug gauge Abnutzungsgewindeprüfdorn *m*
wearability 1. *(Text)* Trageeigenschaft *f*, Gebrauchstüchtigkeit *f*; 2. Abnutzwiderstand *m*
wearer's spectacles Gebrauchsbrille *f*
wearing capacity Abnutzbarkeit *f*
w. course Verschleißschicht *f*, Straßendecke *f*
w. out-of-truth of the cylinder Unrundwerden *n* des Zylinders
w. parts dem Verschleiß unterworfene Teile *npl*, Verschleißteile *npl*
w. plate Reibungsplatte *f*, Verschleißplatte *f*; Schleißblech *n*, Schleißplatte *f*
w. property Abnutzbarkeit *f*, Verschleißfestigkeit *f*
w. quality Gebrauchstüchtigkeit *f*
w. service Verschleißverhütung *f*
w. strips *(Elm)* Schleifleisten *fpl*
w. surface *s.* w. course
w. test Verschleißprüfung *f*
wearless verschleißfrei
weather / to 1. auswittern, verwittern; 2. lagern, altern *(Metalle)*; abwettern *(ein Schiff)*
weather anchor *(Schiff)* Luvanker *m*
w. conditions Witterungsbedingungen *fpl*, Witterungsverhältnisse *npl*
w. deck Wetterdeck *n*, Freideck *n*
w. drip Tropfnase *f*, Wassernase *f*
w. forecast Wettervorhersage *f*
w. guard Wetterschutz *m*
w. indicator *(sl)* Wetterglas *n*, Barometer *n*
w. map Wetterkarte *f*
w. observation station Wetterbeobachtungsstation *f*
w. observatory Wetterwarte *f*
w. performance *(Am)* Witterungsbeständigkeit *f*
w. radar Wetterradar *n*
w. rail *(Schiff)* Luvreling *f*
w. resistance Witterungsbeständigkeit *f*
w. service meteorologischer Dienst *m*
w. ship Wetterschiff *n*
w. side Luv[seite] *f*; Wetterseite *f*
w. station Wetterwarte *f*
weathering 1. Bewitterung[sversuch *m*] *f*; 2. Verwitterung *f*, Auswitterung *f*; 3. Altern *n*, Lagern *n* *(Metalle)*
w. test Naturversuch *m*
weatherly qualities Seefähigkeit *f* *(eines Schiffes)*
weatherproof / to wetterfest machen
weatherproof wetterbeständig, wetterfest
w. box wetterfestes Gehäuse *n*
w. installation *(El)* Freiluftanlage *f*, wetterfeste (wettersichere) Anlage *f*
w. line *(El)* wetterfeste Leitung *f*
weatherstrip Dichtungsstreifen *m*, Regenleiste *f*
weathertight wetterdicht, wetterfest
weave / to 1. weben; 2. *(sl)* ausweichen
w. the heading Kopf (Marke, Schlag) weben
w. white goods *(Text)* weißweben
weave *(Text)* Bindung *f*
w. analysis Bindungsanalyse *f*, Musterausnehmen *n*
weaver Weber *m*
weaver's knot Weberknoten *m*
w. nippers Weberzange *f*, Putzeisen *n*
weaving Weben *n*, Weberei *f*
w. of trimmings Bortenweberei *f*
w. repeat Bindungsrapport *m*
web 1. Schenkel *m*, Steg *m* *(Träger)*; Rippe *f*; Aussteifung *f*, Versteifung *f*; 2. Kern *m* *(Bohrer)*, Seele *f*; 3. *(Text)* Gewebe *n*; Gurtband *n* Flor *m*; 4. Blatt *n* *(der Spannsäge)*; 5. Papierbahn *f*; Rolle *f*
w. divider Florteiler *m*
w. frame Rahmenspant *n*; Gestell *n*
w. guide *(Typ)* Seitenregistersteuerung *f*
w. member Fachwerkstab *m*, Füllungsstab *m*

w. of rail Schienensteg *m*
w. of the drill Bohrerkern *m*
w. plate Stegblech *n*
w. plate stiffener Stegblechaussteifung *f*
w. printing press Rollendruckpresse *f*, Rollendruckmaschine *f*
w. punching Steglochung *f*
w. reinforcement Stegbewehrung *f*; Schubbewehrung *f (Stahlbeton)*
w. rib *(Flg)* Stegrippe *f*
w. saw Bogensäge *f*, Bügelsäge *f*, Spannsäge *f*
w. saw file Schwertfeile *f*
w. section Kernquerschnitt *m (Bohrer)*
w. sling Gurtstropp *m*, Gurtschlinge *f (zum Laden und Löschen)*
w. taper Kernsteigung *f*, Verstärkung *f* der Kerndicke *(Bohrer)*
w. tension *(Typ)* Papierbahnspannung *f*, Bahnspannung *f*
w. thickness Kerndicke *f (Spiralbohrer)*
w. thinning Ausspitzen *n (Spiralbohrer)*
w. thinning attachment Ausspitzvorrichtung *f (Spiralbohrer)*
w. to be coated Trägermaterial *n (Beschichtung)*
weber Weber *n (Einheit des magnetischen Flusses; 1 Wb = 1 Vs)*
Weber-Fechner law Webersches Gesetz *n*, Weber-Fechnersches Gesetz *n*
Weber's cavities Webersche Hohlräume *mpl (Gebirgsmechanik)*
weddings Anzeigenpapier *n*, Reklamepapier *n*
wedge / to verkeilen; mit Keil spalten
w. in the work im Werkstück steckenbleiben *(Bohrer)*
wedge 1. Keil *m*; Backe *f*; Spitzkeil *m*; Kegelstück *n*; 2. keilförmiger Absorber *m (Wellenleiter)*; *(Fs)* Keil *(Testbild)*; 3. Zwickel *m*; 4. Keilstein *m (Gewölbemauerung)*
w.-actuated collet chuck *(Masch)* Keilspannfutter *n*
w. angle Keilwinkel *m*
w. block *(Schiff)* Stapelklotz *m*
w. bolt Keilbolzen *m*, Keilschraube *f*
w. breech mechanism Keilverschluß *m (Geschütz)*
w. breechblock Keilverschluß *m (Geschütz)*
w. contact *(Fmt)* Zungenkontakt *m*
w. cut *(Brg)* Keileinbruch *m*
w.-draw test Zipfelzugversuch *m*; Keilzugversuch *m*
w. effecting rectilinear motion Keilschubgetriebe *n*
w. grip Beißkeil *m*
w. joint Keilverbindung *f*
w.-like fold Abnäher *m (Konfektion)*
w. photometer *(Opt)* Keilfotometer *n*
w. prop *(Brg)* Stempel *m* mit Keilschloß
w. ram Keilramme *f (Stapelung)*
w.-shaped kegelförmig, keilförmig; kommaförmig *(Span)*
w.-shaped gate Keiltrichter *m*
w.-shaped grate bar keilförmiger Roststab *m*
w.-shaped oil film Schmierkeil *m*
w.-shaped piece Keilstück *n*
w.-shaped [piston-]ring Trapez[kolben]ring *m*
w.-type collet chuck Keilspannfutter *n*
w.-type transistor Grattransistor *m*, Keiltransistor *m (Halbleitertechnik)*
wedges Bugfender *mpl (kleine Fender bei Ruderbooten)*
wedging 1. Verkeilen *n*; 2. Abkeilen *n (Gestein)*; 3. Aufbereiten *n* von Ton
w. crib *(Brg)* Keilkranz *m*, Tragkranz *m (Schachtausbau)*
weed / to jäten
weed Unkraut *n*
w. control Unkrautbekämpfung *f*
w. control aggregates Entkrautungsgeräte *npl*
w. extermination Unkrautbekämpfung *f*
w. exterminator Unkrautvertilgungsmittel *n*
w. [grass] killer Unkrautvertilgungsmittel *n*
w. seed Unkrautsamen *m*
w. sprayer Unkrautspritze *f*
weeded ver[un]krautet
weeder *(Lw) (Am)* Federzahngrubber *m*; Netzegge *f*, Unkrautegge *f*; Entkrauter *m*
weedicide Unkrautvertilgungsmittel *n*
weeding comb Netzegge *f*, Unkrautstriegel *m*
w. harrow Unkrautegge *f*
w. hoe Unkrauthaue *f*, Jätehacke *f*
w. screen Unkrautsieb *n (Dreschmaschine)*
weedy ver[un]krautet
weekly earning Wochenlohn *m*
w. season ticket Wochenkarte *f*
weep hole Entwässerungsloch *n*, Sickerloch *n*, Tropfloch *n*; Entwässerungsschlitz *m*, Sickerschlitz *m*; Leckloch *n*
weeping willow Trauerweide *f*
weft *(Text)* Schuß *m*, Einschuß *m*, Eintrag *m*
w.-bowed cloth Gewebe *n* mit kurvenförmiger Verzerrung des Schusses
w. change Schußwechsel *m*
w. float Schußflottung *f*
w. fork Schußwächtergabel *f*
w. fork motion Gabelschußwächter *m*
w. knit [fabrics] Schußwirkware *f*
w. pile tube Florschußschlauch *m*
w. rib Langrips *m*, Schußrips *m*
w. straightener Schußfadenrichtgerät *n*, Schußausrichter *m*, Schußregler *m*
w. streak Kettstreifen *m (Webfehler)*
w. thread Schußfaden *m*
w. winder Schußspulmaschine *f*

w. winding machine Schußspulmaschine *f*
weigh / to wiegen, die Masse feststellen
w. anchor Anker lichten
weigh-batcher Dosierwaage *f (Betontechnologie)*
w.-batching Massedosierung *f*
w. beam Wiegebalken *m*
w. beam scale Balkenwaage *f*
w.-bridge Gleiswaage *f*, Brückenwaage *f*
weighable wiegbar, mit feststellbarer Masse
weigher Dosiervorrichtung *f*, Wiegevorrichtung *f*
weighing Wiegen *n*; Feststellen *n* der Masse; Kraftmessung *f*
w. bridge Brückenwaage *f*, Fuhrwerkswaage *f*, Wiegehäuschen *n*; Gleiswaage *f*
w. device Wägevorrichtung *f*
weight / to beschweren, belasten
w.-stretch mittels Gewichten spannen *(Spannbeton)*
weight 1. Masse *f*; 2. Wägestück *n*, Gewichtstück *n*
w. arm Lastarm *m*
w. at combustion termination *(Rak)* Masse *f* bei Brennschluß, Brennschlußmasse *f*
w.-batching Dosieren *n* nach Masseteilen *(Betontechnologie)*
w. flow Massedurchsatz *m*, Durchflußmenge *f*
w.-giving substance Belastungskörper *m* *(Turbine)*
w. limitation Massebeschränkung *f*
w.-loaded massebelastet
w.-loading Massebelastung *f*
w. loss Masseverlust *m*
w. of armature *(El)* Masse *f* des Ankers
w. of metal removed Spanmasse *f*
w. of rotor *s.* w. of armature
w. of ship Masse *f* des Schiffs; Eigenmasse *f* des Schiffs
w. voltameter Massenvoltameter *n*
weighted flow-sheet Mengenstrombild *n*
weighting Beschwerung *f*, Belastung *f*; *(Text)* Erschweren *n*
w. agent Beschwerungsmittel *n*
w. network *(Ak)* Bewertungsfilter *n*
w. of the cloth Erschweren *n* des Gewebes
weightless masselos, schwerelos
weightlessness Schwerelosigkeit *f*
weir Stauwehr *n*, Wehr *n*
w. crest Wehrrücken *m*
w. flowmeter Überfalldurchflußmesser *m*
w. pier Wehrpfeiler *m*
w. sill Wehrschwelle *f*
weld / to [ver]schweißen
w. down-hand abwärtsschweißen
w. from the bottom up an der stehenden Wand schweißen
w. from the top down an der stehenden Wand schweißen
w. harden schweißhärten
w. in overhead position überkopfschweißen
w. on aufschweißen, anschweißen
w. up-hand aufwärtsschweißen
weld Schweißung *f*; Schweißstelle *f*; Naht *f*, Schweißnaht *f*
w. all around ringsumlaufende Naht *f*
w. bead Schweißraupe *f*, Schweißlage *f*
w. crack Schweißnahtriß *m*
w. cracking Schweißrissigkeit *f*
w. face Schweißnahtoberfläche *f*
w. failure Schweißnahtbruch *m*
w. groove Schweißfuge *f*
w. material eingeschweißtes Material *n*
w. puddle Schweißbad *n*, Schweiße *f*
w. reinforcement Schweißnahtüberhöhung *f*, Nahtüberhöhung *f*
w. size Nahtschenkellänge *f (Schweißen)*
w. throat Schweißnahthöhe *f*
weldability Schweißbarkeit *f*
weldable schweißbar
welded geschweißt
w. butt joint geschweißte Stumpfnaht *f*
w. connection Schweißverbindung *f*
w. construction Schweißkonstruktion *f*
w. girder Schweißträger *m*
w. joint Schweißverbindung *f*, Schweißstoß *m*, Schweißnaht *f*
welder Schweißer *m*
welding Schweißen *n*, Verschweißung *f*
w. additive Schweißzusatzwerkstoff *m*
w. alternator Wechselstromschweißgenerator *m*
w. and cutting torch Schweiß- und Schneidbrenner *m*
w. apparatus Schweißapparat *m*, Schweißmaschine *f*
w. arc Schweißlichtbogen *m*
w. arc voltage *(El)* Schweißspannung *f*
w. bead Schweißperle *f*
w. bench Schweißtisch *m*
w. burner Schweißbrenner *m*
w. cable Schweißkabel *n*
w. capacity Schweißleistung *f*
w. cinder Schweißschlacke *f*
w. compound Schweißpulver *n*
w. contraction Schweißschrumpfung *f*
w. converter Schweißumformer *m*
w. crack Schweißriß *m*
w. current *(El)* Schweißstrom *m*
w. defect Schweißfehler *m*
w. dynamo Schweißgenerator *m*
w. electrode Schweißelektrode *f*
w. engineering Schweißtechnik *f*
w. equipment Schweißausrüstung *f*
w. flame Schweißflamme *f*
w. flux Schweißflußmittel *n*
w. goggles Schweißbrille *f*

w. groove Schweißfuge *f*
w. head Schweißkopf *m*
w. head tip Schweißbrennerdüse *f*
w. heat Schweißhitze *f*
w. joint Schweißverbindung *f*, Schweißnaht *f*
w. mill Schweißwalzwerk *n*
w. operator Schweißer[in *f*] *m*
w. point Schweißpunkt *m*
w. powder Schweißpulver *n*
w. process Schweißverfahren *n*
w. quality Schweißgüte *f*
w. rod Schweißstab *m*, Zusatzwerkstoff *m*
w. seam Schweißnaht *f*
w. set Schweißaggregat *n*, Elektroschweißgerät *n*
w. shop Schweißerei *f*
w. speed Schweißgeschwindigkeit *f*
w. spot Schweißpunkt *m*
w. steel schweißbarer Stahl *m*
w. stick Schweißdraht *m*
w. strain *(sl)* Schweißspannung *f*
w. tongs Schweißzange *f*
w. torch Schweißbrenner *m*
w. transformer Schweißtransformator *m*, Schweißumspanner *m*
w. upset Schweißdruck *m*
w. wire Schweißdraht *m*, Zusatzdraht *m*
weldless tube nahtloses Rohr *n*
weldment Schweißteil *n*
welfare building Sozialgebäude *n*
well 1. Bohrung *f*, Bohrloch *n*; Brunnen *m*, Quelle *f*; Schacht *m*; 2. Gestell *n (Hochofen)*; 3. *(Kfz)* Felgenbett *n*; 4. Cockpit *n*
w.-air intake jet *(Kfz)* Luftdüse *f*
w.-base rim Tiefbettfelge *f*
w. boat Bünnschiff *n (Fischereifahrzeug)*
w. deck Brunnendeck *n*, Welldeck *n*
w.-deck vessel Welldeckschiff *n*
w.-equipped gut ausgerüstet (ausgestattet)
w. foundation *(Bw)* Brunnengründung *f*, Senkbrunnengründung *f*
w. lining segment Brunnenring *m*
w. location Bohrstelle *f (Tiefbohrtechnik)*
w.-maintained gut erhalten (gepflegt)
w. smack *s.* w. boat
w.-type counter *(Kern)* Zähler *m* mit Probenkanal
w.-type dredger Heckschlitzbagger *m*
welled smack *s.* well boat
welt / to falzen; säumen
welt 1. Falz *m (Blechbearbeitung)*; 2. *(Text)* Rand *m*; Einfassung *f*; Saum *m*; 3. Rahmen *m (Schuhfertigung)*
w. sewing machine Einstechmaschine *f (Schuhfertigung)*
welting Biesenleder *n*
w. leather Leder *n* für Einstechrahmen
wernerite *(Min)* Skapolith *m*
Weston [normal] cell *(El)* Westonelement *n*
wet / to benetzen, anfeuchten
wet feucht, naß
w. abrasion effect Naßscheuerwirkung *f*
w. arcing *(El)* Regenüberschlag *m*
w. ball milling Naßmahlung *f* in Kugelmühlen
w. basin *s.* w. dock
w. bleaching Naßbleiche *f*
w.-brined in Salzlake gelaugt *(Leder)*
w. bulb temperature Feuchttemperatur *f*
w. bulb thermometer Feuchtstrumpfthermometer *n*
w. cell *(El)* Naßelement *n*, Füllelement *n*
w. cleaning Naßaufbereitung *f*
w. concrete Frischbeton *m*
w. cutting *(Brg)* Naßschrämen *n*
w. decating *(Text)* Naßdekatur *f*
w. dock Dock *n*, Schleusendock *n*, Hafenbecken *n*, Flutbecken *n*; Fluthafen *m*
w. drawing Naßzug *m*
w. dredger Naßbagger *m*
w. dressing 1. *(Text)* Naßappretur *f*; 2. Naßaufbereitung *f*
w. elongation *(Text)* Naßdehnung *f*
w. end Naßpartie *f*, Siebpartie *f (Papiermaschine)*
w. fastness to light *(Text)* Naßlichtechtheit *f*
w.-felted fibreboard naßverfilzte Faserplatte *f*
w. finishing Naßappretur *f*
w. galvanizing Naßverzinkung *f*
w.-glover Weißgerber *m*
w. grinding Naßschliff *m*
w. grinding attachment Naßschleifeinrichtung *f*
w. liner *(Kfz)* nasse Laufbuchse *f (unmittelbar vom Kühlwasser umspült)*
w. mechanical naßmechanisch
w. method Naßverfahren *n (Chemiefaserherstellung)*
w.-on-wet printing *(Typ)* Naß-in-Naß-Druck *m*, Naßdruck *m*
w. pick *(Brg)* Naßabbauhammer *m*
w.-pressed board gegautschter Karton *m*
w. process *(Text)* Klebebeschichtungsverfahren *n*; Naßverfahren *n*, Naß- und Schlämmverfahren *n*
w. processing *(Text)* Naßveredlung *f*
w. processing method *(Text)* Naßdekatur *f*
w. purification *(Met)* Naßscheidung *f*
w. quenching Naßlöschen *n*, Naßlöschung *f*
w. raising *(Text)* Naßrauhen *n*
w. ring spinning frame Naßringspinnmaschine *f*

w. screening Naßklassierung *f*, Naßsiebung *f* *(Aufbereitung)*
w. season hide Haut *f* der nassen Jahreszeit
w. section *(Brg)* Naßdienst *m*
w. sheet formation Naßvliesbildung *f*
w. sheet forming machine Naßvliesformmaschine *f*
w. slag dünnflüssige Schlacke *f*
w. spinning Naßspinnen *n*, Lösungsspinnen *n*
w.-spun naß gesponnen
w.-spun yarn Naßgespinst *n*
w. stamp Naßpochwerk *n*
w. steam nasser Dampf *m*, Naßdampf *m*
w. strength relative Naßfestigkeit *f (Papier)*
w. strength paper naßfestes Papier *n*
w.-test meter *(Am)* Gasuhr *f*, nasser Gaszähler *m*
w. test run *(Rak)* nasser Durchschaltversuch *m*
w. tumbler *(Met)* Naß[putz]trommel *f*
w. tumbling *(Met)* Naßtrommeln *n*, Naßscheuern *n*
w. twisting *(Text)* Naßzwirnen *n*
w.-washing table Naßherd *m (Aufbereitung)*
wetness of steam Dampffeuchtigkeit *f*, Dampfnässe *f*
wettability Benetzbarkeit *f*
w. agent Netzmittel *n*, Entspannungsmittel *n*
w. test Prüfung *f* auf Benetzbarkeit
wetted midship section *(Schiff)* benetzte Hauptspantfläche *f*
w. perimeter benetzter Umfang *m*
w. surface Unterwasserfläche *f*, benetzte Oberfläche *f*
wetter explosive schlagwettersicherer Sprengstoff *m*
wetting 1. Benetzung *f*; 2. Verzinnen *n*
w. agent Netzmittel *n*, Benetzungsmittel *n*
w. bath *(Text)* Netzbad *n*
w. deterioration Schädigung *f* durch Nässe
w. machine Anfeuchtmaschine *f*
w. power Benetzungsvermögen *n*
Weymouth pine Weymouthskiefer *f*
whale catcher Walfangboot *n*, Walfänger *m*
w.-factory ship Walfangmutterschiff *n*
w. jib *(Brg)* Schrämarm *m* für das Naßschrämen
w.-lance Harpune *f*
w. oil Waltran *m*
w.-oil factory vessel Walverarbeitungsschiff *n*
whaleback steamer Runddeckdampfer *m*
whaler Walfänger *m*
whaling gun Harpunengeschütz *n*
w. vessel Walfangschiff *n*, Walfänger *m*
wharf / to Ladung löschen *(am Kai)*; anlegen
wharf Kai *m*, Dock *n*; Landungssteg *m*
w. crane Hafenkran *m*, Hellingkran *m*
w. of fascinage Faschinendamm *m*
wharfage Kaigebühren *fpl*, Löschgeld *n*, Umschlagsgebühren *fpl*
wharfboat *(Am)* schwimmendes Lagerhaus *n* *(kein Schiff im Sinne des Seerechts)*
wharfinger *(Am)* Kaibesitzer *m*, Kaimeister *m* *(Kaivertreter des Schiffseigners)*
wheal Bergwerk *n (Cornwall)*
wheat flour Weizenmehl *n*
Wheatstone bridge *(El)* Wheatstonebrücke *f*, Wheatstonesche Brücke *f*
Wheatstone telegraph Wheatstonetelegraf *m*
wheel Rad *n*; Schleifscheibe *f*; Ruder *n*
w.-and-disk device Reibradtrieb *m (Rechner)*
w.-and-disk integrator Reibradintegrator *m*, Gonella-Trieb *m*
w. arrangement Achsanordnung *f*, Achsfolge *f*
w. axle Radachse *f*
w. balance Wuchtzustand *m* der Schleifscheibe
w. base *(Kfz)* Radstand *m*, Achs[ab]stand *m*
w. body Radkörper *m*
w. bolt Radbolzen *m*
w. bond Schleifscheibenbindung *f*
w. cage-type Gitterrad *n (Traktor)*
w. centre Radkörper *m*
w. centre distance Rad[ab]stand *m*
w. chuck Schleifscheibenfutter *n*
w. circumference 1. Radumfang *m*; 2. Schleifkörperumfang *m*
w. crushing Einrollverfahren *n (Schleifscheibe)*
w. crushing attachment Abroll-Abrichtgerät *n*
w. cylinder *(Kfz)* Radzylinder *m*
w. disk stamped out of a plate Blechradscheibe *f*, Radscheibe *f* aus Blech *(Turbine)*
w. dresser Abrichtwerkzeug *n (Schleifscheibe)*
w. dresser for shaping angles Winkelabrichteinrichtung *f*
w. dresser for shaping radii Radienabrichteinrichtung *f*
w.-dressing operation Abrichtarbeit *f*, Abrichtvorgang *m*
w. excavator Schaufelradbagger *m*
w. feed Beistellung *f* der Schleifscheibe
w. flange 1. Felgenhorn *n*, Radflansch *m*; Spurkranz *m*; 2. Schleifscheibenflansch *m*
w. friction Radreibung *f (Turbine)*
w. guard Schleifscheibenschutzhaube *f*
w. head Schleifkopf *m*
w.-hop Springen *n* der Räder *(Kraftwagen)*
w. house Ruderhaus *n*, Steuerhaus *n*
w. house top Peildeck *n*
w. hub Radkörper *m*, Radnabe *f*
w.-hub lubrication Radnabenschmierung *f*
w.-hub mortiser Nabenstemmaschine *f*

w.-hub screw stock Nabenschneidkluppe *f*
w. lathe Radsatzdrehmaschine *f*
w. mount Schleifscheibenbefestigung *f*, Scheibenaufnahme *f*; Radbefestigung *f*
w. mounting bolt Radbefestigungsbolzen *m*
w. mounting nut Radbefestigungsmutter *f*
w.-nut spanner Radmutternschlüssel *m*
w. ore *(Min)* Rädelerz *n*
w. package Radverpackung *f* *(Verpackung elektronischer Geräte in Raketenflugkörpern)*
w. port *(Am)* *(Schiff)* Schraubenbrunnen *m*, Schraubenöffnung *f*, Propellerbrunnen *m*
w.-position indicator *(Kfz)* Radstellungsanzeiger *m*
w. puller *(Kfz)* Radabzieher *m*
w. shimmy *(Kfz)* Radflattern *n*
w. slide Schleif[spindel]schlitten *m*, Schleifscheibensupport *m*
w. slip Radschlupf *m*
w. speed Scheibendrehzahl *f* *(Schleifen)*
w. spindle Schleifspindel *f*
w. spindle axis Schleifspindelachse *f*
w. spindle bearing Schleifspindellagerung *f*
w. spindle head Schleifspindelstock *m*
w. spindle infeed Zustellung *f* der Schleifscheibe
w. spindle speed Schleifspindeldrehzahl *f*
w. spoke Radspeiche *f*
w.-spoke planing machine Radspeichenhobelmaschine *f*
w. stand Schleifbock *m*
w. stud Radstift *m*
w. suspension Radaufhängung *f*
w. track Radspur *f*, Wagenspur *f*
w. truing attachment Schleifscheibenabrichteinrichtung *f*
w.-type armature *(El)* Radanker *m*
w.-type pipe cutter Rohrabschneider *m* mit Schneidrädchen
w.-type tractor Radschlepper *m*
w.-type trenching machine Grabenfräse *f*
w. tyre Radkranz *m*
w. tyre vertical boring and turning mill Radreifenbohrwerk *n*
w. wobble Radflattern *n*
wheelbarrow Schubkarre[n *m*] *f*, Handkarre[n *m*] *f*
w. sprayer Karrenspritze *f* *(Pflanzenschutz)*
wheeled cable drum carriage Kabelwagen *m*
wheeler Stellmacher *m*
Wheeler hopper-barge *(Schiff)* Wheeler-Tankreinigungsprahm *m*
Wheeler-system of tank cleaning *(Schiff)* Wheeler-Tankreinigungssystem *n*
wheelman *(Am)* *(Schiff)* Rudergänger *m*
wheels down *(Flg)* mit ausgefahrenem Fahrwerk
wheelwright Stellmacher *m*
w. machine Wagenradmaschine *f*
wheelwright's shop Stellmacherei *f*
whet / to abziehen, wetzen
whetstone Schleifstein *m*, Wetzstein *m*
whey Molke *f*
whim Göpel *m*
whip / to 1. schlagen *(Bohrer, Welle)*; Schaum schlagen; 2. peitschen; 3. überwendlich nähen; abbinden, umwickeln
w.-lash upwards nach oben schnellen
whip 1. Peitsche *f*; 2. Schlag *m* *(Welle)*
w. aerial Peitschenantenne *f*, Stabantenne *f*
w. roll *(Text)* Kettbaum *m*; Musterfadenspule *f*
whipping Bewicklung *f*, Umwicklung *f* *(Kabel)*
whippletree *(Lw)* Ortscheit *n*
whipstock Richtkeil *m*, Ablenkkeil *m*
whirl Wirbel *m*
w. combustion chamber Wirbelkammer *f*
w. tube Wirbelröhre *f*
whirling of the heating gases Durchwirbelung *f* der Heizgase
w. speed Schwirrgeschwindigkeit *f* *(Gelenkwelle)*
w. test stand *(Flg)* Schleuderprüfstand *m*
whirlpool combustion chamber Wirbelkammer *f*
whisker Spiralfeder *f* *(eines Kristalldetektors)*
whistle buoy Heultonne *f*, Heulboje *f*
whistling tone Pfeifton *m* *(Funkempfangstechnik)*
white annealing Weißglühen *n*
w. arsenic Arsenik *n*, Arsentrioxid *n*
w. cast iron Vollhartguß *m*, Hartguß *m*
w. damp *(Brg)* giftige (böse) Wetter *pl*
w. German sole leather altgegerbtes Sohlleder *n*
w. heart malleable cast iron Weiß[kerntemper]guß *m*
w. heat Weißhitze *f*, Weißglut *f*
w.-hot weißglühend
w. iron casting Vollhartgußstück *n*, Hartgußstück *n*
w. iron pyrite *(Min)* Markasit *m*
w. lacquer weißer Schellack *m*
w. lead Bleiweiß *n*
w. leather Alaunleder *n*, Weißleder *n*
w. line *(Typ)* Blindzeile *f*, blinde Zeile *f*
w. liquor Abfallauge *f* *(Natronzellstoffprozeß)*, Weißlauge *f*, Frischlauge *f*
w. metal Weißmetall *n*
w.-metal lined bearing Lager *n* mit Weißmetallausguß
w.-metal lining Weißmetallausfütterung *f*, Weißmetallgarnierung *f*, Weißmetalleinlage *f*, Weißmetallausguß *m*
w.-metal strip Weißmetallstreifen *m*
w. noise weißes Rauschen *n*, Rauschen mit Gaußscher Verteilung
w. oil Weißöl *n*, Nähmaschinenöl *n*

w. peak *(Am) (Fs)* Maximum *n* an Weiß, Weißspitze *f*
w. pig iron weißes Roheisen *n*
w. pine Edeltanne *f*
w. rot Weißfäule *f (Holz)*
w. saturation *(Fs)* Weißpegelsättigung *f*
w. slag Fertigschlacke *f*, Reduktionsschlacke *f*
w. star indicator *(Fmt)* Sternschauzeichen *n*
w.-to-black amplitude range *(Fs)* Weiß-Schwarz-Amplitudenbereich *m*
w. water Siebwasser *n*, Abwasser *n (Papierherstellung)*
w. wire *(El)* Nulleiter *m*
whiten / to 1. kalken, tünchen, weißen; 2. *(Led)* blanchieren
whitening Schlämmkreide *f* (*s. a.* whiting)
w. machine *(Led)* Blanchiermaschine *f*
w. slicker *(Led)* Blanchiereisen *n*
w. steel *(Led)* Blanchierstahl *m*
whitenings *(Led)* Blanchierspäne *mpl*
whitewash / to kalken, tünchen, weißen
whitewash Tünche *f*; Kalkmilch *f*
whiting Kreidegrund *m*, Schlämmkreide *f*; Perlweiß *n*, Kreideweiß *n*
Whiting system *(Brg)* Whiting-Verfahren *n (Schachtförderung)*
whitleather *s.* white leather
whittle / to schnitzen
whittle Schnitzmesser *n*, Schnitzer *m*
Whitworth pipe thread Whitworth-Rohrgewinde *n*
Whitworth [screw] thread Whitworth-Gewinde *n*
Whitworth's variable speed gear umlaufende (rotierende) Kurbelschleife *f*
whiz[z] / to schleudern, zentrifugieren
w. out ausschleudern *(Wasser)*
whiz[z] out-effluent Schleuderwasser *n (Aufbereitung)*
whizzer Trockner *m*, Windsichter *m*; Zentrifuge *f*
whole depth of tooth Zahnhöhe *f*
w. log Ganzholz *n*
w. number ganze Zahl *f*
wholeness *(Math)* Ganzheit *f*
wholesome seetüchtig *(Schiff)*
whorl Spiralenwindung *f*; Scheibe *f (Spindel)*
Wichert truss Fachwerkgelenkträger *m*, Gerber-Fachwerkbinder *m*
wick Docht *m*
w.[-feed] lubrication Dochtschmierung *f*
w.-oiler Dochtöler *m*
w. oiling Dochtschmierung *f*
wicker basket Weidenkorb *m*
w. work Flechtarbeit *f (Korb)*
wicks Dochtgewebe *n*
wide weit; breit; fern
w.-angle lens *(Opt)* Weitwinkelobjektiv *n*
w.-angle lighting fitting Breitstrahler *m (Lichttechnik)*
w.-angle projector Weitwinkelprojektor *m*
w.-angle viewing Weitwinkelbeobachtung *f*
w. band 1. Breitband *n (Walzprodukt)*; 2. Breitband, breites Frequenzband *n*
w.-band aerial Breitbandantenne *f*
w.-band amplifier *(Fmt)* Breitbandverstärker *m*
w.-band cable Breitbandkabel *n*
w.-band controller Regler *m* mit breitem P-Bereich
w.-band dipole Breitbanddipol *m*
w.-band filter breitbandiges Filter *n*
w.-band microphone Breitbandmikrofon *n*
w.-band oscilloscope Breitbandoszillograf *m*
w.-band ratio Frequenzbandausbeute *f*
w.-band reception Breitbandempfang *m*
w.-band telegraphy Breitbandtelegrafie *f*
w. bearing breites Tragbild *n*
w. face square nose finishing tool Breitschlichtmeißel *m*
w. finishing Breitschlichten *n*
w. finishing tool Breitschlichtmeißel *m*
w. flange Breitflansch *m*, weit auskragender Flansch *m*
w.-flange beam Breitflanschträger *m*
w. frequency band breites Frequenzband *n*
w. gauge Breitspur *f*
w.-ringed grobjährig *(Holz)*
w.-screen film Breitwandfilm *m*
w. spacing großporiges (offenes) Gefüge *n*
w. spectral range breiter Spektralbereich *m*
w. strip mill Breitbandstraße *f*
w. wheel rim Breitbettfelge *f*
widen / to verbreitern, [auf]weiten
widia *(gesintertes)* Hartmetall *n*, Widia *n*
Widmanstätten structure Widmanstättensches Gefüge *n*
width Breite *f*, Dicke *f*, Weite *f*
w. between jaws Rachenweite *f*
w. control *(Fs)* Bildbreitenregelung *f*
w.-milling operation Breitfräsarbeit *f*
w.-modulated längenmoduliert *(Impuls)*
w. of a reel Rollenbreite *f (Papier)*
w. of blade axiale Schaufellänge (Schaufelbreite) *f (Turbine)*; *(Flg)* Flügelbreite *f*
w. of clearance space Spaltweite *f*, Spaltbreite *f (Turbine)*
w. of cut Schnittbreite *f (Mähbalken)*; *(Masch)* Spanbreite *f*, Schnittbreite *f*
w. of fabric *(Text)* Spannbreite *f*
w. of jaws Spannbreite *f (Schraubstock)*
w. of joint Fugenweite *f*
w. of kerf Schnittfugenbreite *f (Sägen)*
w. of land Zahnstollenbreite *f*, Stegbreite *f (Gewindebohrer)*; Fasenbreite *f*; Zahnrückenbreite *f (Fräser)*

w. of line 1. *(Fs)* Zeilenbreite *f*; 2. Strichdicke *f (technisches Zeichnen)*
w. of opening Eröffnungsweite *f*; Öffnungsweite *f*
w. of passage Kanalbreite *f (Turbine)*
w. of port Spaltweite *f*, Kanalweite *f*
w. of rim Felgenbreite *f*
w. of scale division Skalenteilgröße *f*
w. of shock wave Breite *f* der Stoßwelle
w. of shock zone Breite *f* der Stoßschicht
w. of space Zahnlückenweite *f*
w. of the grate openings Schlitzweite *f* der Rostspalten
w. of the knitted fabric tube Schlauchbreite *f (Rundstrickware)*
w. of the land Zahnstollenbreite *f*; Stegbreite *f (Gewindebohrer)*; Fasenbreite *f*; Zahnrückenbreite *f (Fräser)*
w. of the steam port *(Wärm)* Kanalweite *f*
w. of the tooth Zahnbreite *f*
w. of track Spurbreite *f*, Spurweite *f*
w. of transition Impulsflankensteilheit *f*
w. of wear land Verschleißfasenbreite *f (Werkzeug)*
w. series Breitenreihe *f (Lager)*
Wien bridge Wien-Brücke *f (Meßtechnik)*
Wien's displacement law Wiensches Verschiebungsgesetz *n (Wärmestrahlung)*
Wien's radiation law Wiensches Strahlungsgesetz *n*
wild oats Flughafer *m*, Windhafer *m*, Glatthafer *m*
w. steel unberuhigter (unberuhigt vergossener) Stahl *m*
wildcat well Schürfbohrung *f*, Prospektionsbohrung *f*, Suchbohrung *f*, Aufschlußbohrung *f*, Bohrung *f* auf unerforschtem Gelände
willemite *(Min)* Willemit *m*
Williams tube *(Dat)* Williams-Speicherröhre *f*
Williot diagram Verschiebungsplan *m* nach Williot
willow / to klopfen, reinigen, putzen
willow 1. Weide *f*; Weidenholz *n*; 2. *(Text)* Klopfwolf *m*, Droussierwolf *m*
w. pin *(Text)* Wolfstift *m*
w. rust Weidenrost *m (Holzkrankheit)*
Wilson chamber [Wilsonsche] Nebelkammer *f*, Wilson-Kammer *f*
wilting Heuwerbung *f*, Heugewinnung *f*
Wilton carpet Wiltonteppich *m*
winch Haspel *f*, Kurbel *f*; Winde *f*
w. below the surface of the liquor *(Text)* Unterflottenhaspel *f*
w. dyeing machine *(Text)* Haspelfärbeapparat *m*
w. vat *(Text)* Haspelkufe *f*
winchreel Haspel *f*
wind / to 1. sich winden; wickeln; umwickeln; *(El)* [be]wickeln *(Spule)*; 2. *(Brg)* fördern, treiben, zutage heben
w. off abspulen; abwickeln *(Kabel)*
w. on reels aufspulen, auf Spulen wickeln
w. up aufwickeln; aufziehen *(Uhr)*
wind 1. Wind *m*; 2. Verdrehung *f (Holz)*; 3. Windung *f*, Biegung *f*; 4. Winde *f (Fördermittel)*
w. belt Windmantel *m*, Windkasten *m*, Windring *m (Kupolofen)*
w. box Windkasten *m*
w. bracing *(Bw)* Windverband *m*
w. channel Windkanal *m*, Windtunnel *m*
w. direction indicator Windrichtungsanzeiger *m*
w.-driven windgetrieben
w.-driven generator Windkraftmaschine *f*
w.-driven power station Windkraftwerk *n*
w. erosion Winderosion *f*
w. force Windkraft *f*
w. furnace Blasofen *m*
w. gauge *s.* w.-speed indicator
w. indicator Wetterfahne *f*
w. instrument Blasinstrument *n*
w. lantern Sturmlaterne *f*
w. motor Windkraftmaschine *f*
w.-off spool *(Fmt)* Ablaufspule *f*
w. power Windkraft *f*
w.-power engine Windkraftmaschine *f*
w. pressure Winddruck *m*
w. road Wetterschacht *m*, Ventilationskanal *m*
w. rose Windrose *f*
w.-speed indicator Anemometer *n*, Windstärkemesser *m*
w. tunnel Windkanal *m*, Windtunnel *m*
w. way *s.* w. road
windage loss Ventilationsverlust *m (Turbine)*
w. resistance Ventilationswiderstand *m (Turbine)*
w. work Ventilationsarbeit *f (Turbine)*
winder Haspel *f*, Kurbelvorrichtung *f*, Aufwinder *m*
w. operator Fördermaschinist *m*
w. tower Förderturm *m*
windfall 1. Fallobst *n*; 2. Windbruch *m (Forst)*
winding 1. Haspeln *n*; Winden *n*; Wickeln *n*; Umwickeln *n*; Spulen *n*; 2. *(Elm)* Wicklung *f*; 3. spiralförmige Verdrehung *f (Holzfehler)*; 4. Kurve *f*
w. bobbin *(Text)* Aufwickelspule *f*
w. by hand *(El)* Handwicklung *f*
w. capacitance *(El)* Wicklungskapazität *f*
w. capacity *(Brg)* Schachtförderleistung *f*
w. coefficient *(El)* Wicklungsfaktor *m*
w. compartment Fördertrum *m*
w. drum elevator Trommelaufzug *m*
w. drum score Seilrille *f (Winde)*

w. element *(El)* Wicklungselement *n*, Wicklungsglied *n*
w. engine *(Brg)* Fördermaschine *f*
w. factor *(El)* Wicklungsfaktor *m*
w. faller *(Text)* Aufwinder *m (Selfaktor)*
w. faller wire Aufwinderdraht *m (Selfaktor)*
w. frame *(Text)* Spulmaschine *f*
w. handle Kurbel *f*
w. inductance *(El)* Wicklungsinduktivität *f*
w. machine 1. *(El)* Wickelmaschine *f*, Spulenwickelmaschine *f*; 2. *(Text)* Spulmaschine *f*
w. mandrel Wickeldorn *m*
w. motion *(Text)* Windevorrichtung *f*, Windezeug *n*
w. on Aufwickeln *n*
w. pitch *(El)* Wicklungsteilung *f*, Wicklungsschritt *m*
w. rope Förderseil *n*
w. section *(El)* Wicklungselement *n*, Wicklungsglied *n*
w. shaft Förderschacht *m*
w. space *(El)* Wickelraum *m*
w. speed regulator Fahrtregler *m (Fördermaschine)*
w. stair Wendeltreppe *f*
w. stand *(El)* Wickelbock *m*
w. steam engine Förderdampfmaschine *f*
w. surface of the bobbin *(Text)* Fadenmantelfläche *f* der Spule
w. tension of the yarn *(Text)* Aufwindespannung *f* des Garns
w. tower Förderturm *m*
w.-type current transformer Durchsteckwandler *m* ohne Primärleiter
w. wire *(El)* Spulendraht *m*, Wicklungsdraht *m*, Wickeldraht *m*
windlass Kreuzhaspel *f*, Haspel *f*; Räderwinde *f*, Spill *n*, Zahnradwinde *f*; Pumpspill *n*; Ankerspill *n*, Ankerwinde *f*
windmill Windmühle *f*
w. on trestles Bockmühle *f*
window 1. Fenster *n*; Austrittsfenster *n (Röntgenröhre)*; 2. Koppelschlitz *m*, Schlitz *m (Wellenleiter)*; 3. Düppel *m (Radarstörung)*; 4. *(Fs)* Ausblendungsschaltkreis *m*
w. amplifier *(Kern)* Fensterverstärker *m* Kanalverstärker *m*
w. board Fensterbrett *n*
w. crank *(Kfz)* Fensterkurbel *f*
w. curtain Gardine *f*
w. fastener Fensterfeststeller *m*, Fensterhaken *m*, Sturmhaken *m*
w. handle Fenstergriff *m*
w. head Fenstersturz *m*
w. latch Fensterriegel *m*
w. lift Fenstergriff *m*
w. moulding Fensterleiste *f*
w. pane Fensterscheibe *f*
w. sash bar making machine Fenstersprossenmaschine *f*, Kreuzsprossenmaschine *f*
w. shutter Fensterladen *m*
w. sill Fensterbrüstung *f*, Sohlbank *f*
w. stay Fensterhaken *m*, Sturmhaken *m*
w. sticker Klebeplakette *f (z. B. für Kfz.-Scheiben)*
w. strap Fenstergurt *m*
w. surrounds Fenstergewände *n*
w. transom Fensterriegel *m*
w. wedge Fensterriegel *m*
windrow Schwaden *m*, Getreidemandel *f*
w. loader Schwadenaufnehmer *m*
windrower Garbenableger *m*, Getreidemähableger *m*
windrowing cornmower Getreideablegemäher *m*
windscreen Windschutzscheibe *f*
w. frame Windschutzscheibenrahmen *m*
w. opener Windschutzscheibenaussteller *m*
w. washer Scheibenwäscher *m*
w. wiper Scheibenwischer *m*
windshield Windschutz[scheibe *f*] *m*
w. header Windschutzscheibenkopfleiste *f*
w. vizor Blendschutz *m*
w. wiper Scheibenwischer *m*
w. wiper motor Scheibenwischermotor *m*
wine cask Weinfaß *n*
w. press Weinkelter *f (Presse)*
w. vault Weinkeller *m*, Weinlager *n*
wing 1. Flügel *m*; Tragfläche *f*; 2. Kotflügel *m*; 3. Stellbogen *m (Zirkel)*
w. assembly *(Flg)* Tragwerk *n*
w. bulkhead *(Schiff)* Flügelschott *n*, Außenschott *n*, Seitenschott *n*
w. cannon Flügelkanone *f*, Tragflächenkanone *f*
w. deck Seitenkastendeck *n (Schwimmdock)*
w. disk Flügelscheibe *f*
w. divider Bogen[spitz]zirkel *m*
w. dropping plötzliches Durchsacken *n* einer Tragfläche
w. gun Tragflächenmaschinengewehr *n*, Flügelmaschinengewehr *n*, Tragflächen-MG *n*, Flügel-MG *n*
w. lamp Kotflügelleuchte *f*
w. nut Flügelmutter *f*
w. of a line Linienflügel *m*
w. of the [navigating] bridge *(Schiff)* Brükkennock *n*
w.-over *(Flg)* hochgezogene Kehrtkurve *f*
w. propeller *(Schiff)* Außenpropeller *m*
w. rail Flügelschiene *f*, Hornschiene *f (Drehscheibe)*
w. shelter *(Schiff)* Nockhaus *n*
w. span *(Flg)* Spannweite *f*
w. stay Kotflügelstütze *f*
w. structure Dockseitenkasten *m*, Seitenkasten *m (Schwimmdock)*

w. **strut** *(Flg)* Tragflächenstrebe *f*
w. **tank** *(Schiff)* Seitentank *m*
w. **taper** *(Flg)* Tragflächenzuspitzung *f*
w. **tip** *(Flg)* Tragflächenende *n*
w. **tip flare** *(Flg)* Landefackel *f*
w. **tip rake** *(Flg)* Tragflächenüberhang *m*
w. **transom** Heckbalken *m (Holzschiff)*
w. **wall** Flügelmauer *f*
winnow *(Lw)* Schwinge *f*
w.-shaker Schwingschüttler *m*
winnower Trieur *m*, Gebläserohr *n*; Kornschwinge *f*, Windfege *f*, Putzmühle *f (Getreidereinigung)*
winnowing Windfegen *n*, Schwingen *n*, Getreidereinigen *n*
w. **basket** Futterschwinge *f*
w. **blower** Getreidereinigungsgebläse *n*
w. **conveyor** Schwingförderer *m*
w. **fan** Futterschwinge *f*
w. **machine** Getreidereinigungsmaschine *f*, Kornschwinge *f*, Staubmühle *f*, Unkrautsamenentferner *m*; Vorreinigungsmaschine *f*, Aspirateur *m*
w. **screen** Unkrautsieb *n*, Schwingsieb *n*
w. **trough** Futterschwinge *f*
winter barley Futtergerste *f*, Wintergerste *f*
w. **freebord** Winterfreibord *m*
wintering Auswintern *n*; Überwintern *n*
w. **harbour** Winter[schutz]hafen *m*
winze *(Brg)* Gesenk *n*
wipe / to abwischen, abtrocknen; abstreifen [von]
w.-out löschen *(Tonband)*
wipe *(Fs)* Tricküberblendung *f*, Überblendung *f*
w. **contact** Schleifkontakt *m*, Gleitkontakt *m*; Wischkontakt *m*, Wischer *m*
w.-on *(Typ)* Offsetplatte *f* für Wischbeschichtung
w.-out Gitterblockierung *f*, Zustopfung *f (Funkempfangstechnik)*
w.-out effect Zustopfen *n* des Empfängers durch übermäßig hohe Feldstärke
wiped joint Rohrlötfuge *f*
wiper 1. Streifblech *n*, Schmierkamm *m*, Schmierbügel *m (Schmiervorrichtung)*; 2. Bürstenträger *m*, Kontaktbürste *f*, Kontaktarm *m*; 3. *(Masch)* Abstreifer *m (für Späne)*
w. **arm** *(Kfz)* Wischerarm *m*
w. **blade** *(Kfz)* Wischerblatt *n*
w. **contact** *(Kfz)* Wischerkontakt *m*
w. **set** *(Fmt)* Kontaktarmsatz *m*
w. **shaft** Bürstenwelle *f*, Bürstenträger *m*, Kontaktarmträger *m*
wiping-proof wischfest
w. **solder** Blei-Zinn-Weichlot *n*
w. **spring contact** *(Fmt)* Zungenkontakt *m*
wire / to *(El)* verdrahten, [be]schalten; Leitungen verlegen, installieren; telegrafieren
w. **buried** unter Putz verlegen
w. **for** *(Fmt)* beschalten mit
w. **on the surface** auf Putz verlegen
w. **up** *(Fmt)* verdrahten
wire Draht *m*; Leitung *f*; Leiter *m*, Ader *f*; Telegramm *n*
w. **annealing** *(Met)* Drahtglühen *n*
w.-armoured cable drahtbewehrtes Kabel *n*, Drahtpanzerkabel *n*, Kabel mit Stahldrahtarmierung
w. **armouring** Drahtbewehrung *f*
w. **bar** *(Met)* Drahtbarren *m*
w. **bobbin** Drahtspule *f*
w. **braiding machine** Draht[um]flechtmaschine *f*
w. **break[age]** Drahtbruch *m*
w. **broadcasting** Draht[rund]funk *m*
w. **brush** Drahtbürste *f*
w. **clamp** Drahtklammer *f*, Drahtklemme *f*
w. **cloth** Drahtgeflecht *n*, Drahtgewebe *n*, Maschendraht *m*, Metallgewebe *n*
w. **cloth weaving** Drahtweberei *f*
w. **coil spring** Schraubenfeder *f*
w. **comb** Kratzer *m (Außenputzbearbeitung)*
w. **compensating device** Drahtzugspannwerk *n (Signal)*
w. **core** Drahteinlage *f (Dichtungsmaterial)*; Drahtseele *f*, Drahtkern *m*
w. **core coil** Drahtkernspule *f*
w. **cross section** Drahtquerschnitt *m*
w. **crossing** *(Fmt)* Drahtkreuzung *f*
w. **cutter** Drahtschere *f*, Drahtschneider *m*, Drahtzange *f*
w. **drawing** Drahtziehen *n*, Drahtzug *m*
w. **drawing bench** Drahtziehbank *f*
w. **drawing die** Ziehstein *m (Drahtziehmaschine)*
w. **drawing machine** Drahtziehmaschine *f*
w. **drawing mill** Drahtzieherei *f*, Drahtzug *m*
w. **drawing passage** Drosselweg *m (Turbine)*
w. **drawing plate** Zieheisen *n*, Drahtzieheisen *n*, Ziehstein *m*
w. **enamel** Drahtemaille *f*, Drahtlack *m*
w. **end** Siebpartie *f (Papierherstellung)*
w. **entanglement** *(Mil)* Drahthindernis *n*, Drahtverhau *m*
w. **eye** Drahtauge *n (Litze)*
w. **fabric** Drahtgeflecht *n*, Drahtgewebe *n*, Maschendraht *m*, Metallgewebe *n*
w. **fence** Drahtzaun *m*
w. **for indoor installation** Zimmerleitungsdraht *m*
w. **frame** Siebpartie *f (Papierherstellung)*
w. **fuse** Drahtsicherung *f*, Schmelzdraht *m*
w. **gauge** Drahtlehre *f*
w. **gauze** Drahtgaze *f*, Drahtgeflecht *n*
w. **glass** Drahtglas *n*
w. **grating** Drahtgitter *n (Wellenleiter)*

w. grip Froschklemme *f*, Drahtklemme *f*; *(Am)* Kniehebelklemme *f*
w. guide *(Fmt)* Leitungshalterung *f*
w. heald shaft Drahtlitzenschaft *m*
w. helix Drahtspirale *f*
w. hook lacing Drahthakenverbindung *f (Riemen)*
w. lashing Drahtbund *n*
w. manifold Kabelgeschirr *n*
w. mesh Drahtgeflecht *n*, Drahtgewebe *n*, Maschendraht *m*, Metallgewebe *n*
w.-mesh guard Drahtmaschenschutz *m*
w.-mesh reinforcement Bewehrungsmatte *f*, Mattenbewehrung *f (Stahlbetonbau)*
w. mill Drahtstraße *f (Walzwerk)*
w. nail Drahtstift *m*, Drahtnagel *m*
w. nail machine Drahtstiftmaschine *f*
w. netting 1. Drahtgeflecht *n*, Drahtgewebe *n*, Drahtnetz *n*, Maschendraht *m*, Metallgewebe *n*; 2. *(Text)* Wollauflegeboden *m* aus Drahtgeflecht
w.-netting tapping *(Fmt)* Abzweigung *f* einer Leitung
w. nippers Drahtschneidezange *f*
w. part Siebpartie *f (Papierherstellung)*
w. particulars Drahtangaben *fpl*, Drahtdaten *pl*
w. patenting furnace Drahtpatentierofen *m*
w. pin Drahtstift *m*
w. pliers Drahtschneidezange *f*
w. plug Kabelstecker *m*
w. printer Drahtdrucker *m*, Nadeldrucker *m*
w. program distribution *(Am)* Drahtfunk *m*, Rundfunkvermittlung *f*
w. recorder Drahtspeichergerät *n*
w. reel Drahthaspel *f*; *(Am)* Drahtspule *f*; Rolle *f*, Spule *f* für Draht
w. rod Walzdraht *m*
w. rope Drahtseil *n*
w.-rope clip Drahtseilklemme *f*, Seilklemme *f*
w.-rope core Drahtseele *f*, Drahtkern *m*
w.-rope drum Seiltrommel *f*
w.-rope tackle block Kabelflaschenzug *m*, Seilflaschenzug *m*
w. screen Drahtsieb *n*; Drahtschirm *m*
w. section Siebpartie *f (Papierherstellung)*
w.-sewn drahtgenäht *(Schuhfertigung)*
w. side Siebseite *f (Papierherstellung)*
w. sieve Drahtsieb *n*
w. sling Drahtschlinge *f*, Drahtstropp *m*
w. socket Seilbuchse *f*
w. solder Lötdraht *m*
w.-spoked wheel Drahtspeichenrad *n*
w. stock Drahtmaterial *n*
w. straightening and cutting-off machine Drahtricht- und -abschneidemaschine *f*
w. stretcher *(Fmt)* Drahtspanner *m*; Froschklemme *f*
w. stripper Abisoliermaschine *f*
w. tapping *(Fmt)* Abhören *n*
w. tester Kabelmeßgerät *n*
w.-to-wire capacity *(El)* Schleifenkapazität *f*
w. wedge cramp Drahtseilklemme *f*
w. wheel Speichenrad *n*
w. winding Drahtwicklung *f*
w.-wound armature *(Elm)* gewickelter Anker *m*
w.-wound heating element Drahtheizkörper *m*
w.-wound potentiometer Drahtpotentiometer *n*, Drahtregler *m*
w.-wound resistance drahtgewickelter Widerstand *m*
w.-wound resistor Drahtwiderstand *m*
wired verdrahtet
w. broadcasting Draht[rund]funk *m*
w. glass Drahtglas *n*
w. radio Draht[rund]funk *m*
w. wireless Drahtfunk *m*
w. wireless amplifier Drahtfunkverstärker *m*
w. wireless transmitter Drahtfunksender *m*
wireless Radio *n*, Funk *m*
wireless drahtlos; Rundfunk-
w. aircraft service Flugfunkdienst *m*
w. bearing Funkpeilung *f*
w. call drahtloser Anruf *m*
w. communication Funkverkehr *m*
w. compass Funkkompaß *m*, Funkpeiler *m*
w. control drahtlose Fernsteuerung *f*
w. direction-finder Funkpeiler *m*, Bordpeiler *m*
w. engineering Funktechnik *f*, Radiotechnik *f*
w. exhibition Funkausstellung *f*
w. installation Funkanlage *f*
w. message Funkmeldung *f*
w. navigation Funknavigation *f*, Funkortung *f*
w. officer Funker *m*, Bordfunker *m*
w. operator Funker *m*, Funktelegrafist *m*; Bordfunker *m*
w. picture telegraph Bildfunkgerät *n*
w. plant Funkanlage *f*
w. range finder Funkmeßgerät *n*
w. receiver Rundfunkempfänger *m*
w. receiving set Funkempfänger *m*, Rundfunkgerät *n*
w. reception Radioempfang *m*, Funkempfang *m*
w. record player Plattenspieler *m* mit kleinem Lokalsender
w. room Funkraum *m*
w. set Rundfunkempfänger *m*
w. station Funkstelle *f*, Funkstation *f*
w. telegram Funktelegramm *n*
w. telegraphist Funker *m*

w. telegraphy drahtlose Telegrafie *f*, Funktelegrafie *f*
w. telephone traffic Funksprechverkehr *m*
w. telephony drahtlose Telefonie *f*, Funkfernsprechen *n*
w. transmitting and receiving apparatus Sende-Empfangs-Funkgerät *n*
wiremark Siebmarkierung *f (Papierherstellung)*
wirework 1. Drahtgeflecht *n*; 2. Drahtherstellung *f*
wiring *(El)* Verdrahtung *f*; Drahteinlegen *n*; Installation *f*, Leitungsverlegung *f*; Schaltung *f*; Verdrahtung *f*; Drahteinrollen *n*
w. capacity Kapazität *f* der Leitungen, Schaltkapazität *f*
w. connection Drahtverbindung *f*
w. diagram Verdrahtungsplan *m*, Schaltbild *n*, Schaltplan *m*, Schaltschema *n*, Leitungsplan *m*, Stromlaufplan *m*
w. in diagonal pairs Diagonalgruppierung *f*
w. material Installationsmaterial *n*
w. plan Schaltschema *n*
w. point Anschlußort *m*
w. resistance Zuleitungswiderstand *m*
w. scheme Leitungsplan *m*, Schaltplan *m*, Schaltschema *n*
w. symbol Schaltsymbol *n*, Schaltzeichen *n*
with automatic size control and compensation *(Masch)* meßgesteuert
w. equal legs gleichschenklig
w. helical flutes drallgenutet, gedrallt
w. large annual rings grobringig, grobjährig *(Holz)*
w. localized tooth bearing balligtragend
w. small annual rings engringig, feinjährig *(Holz)*
w. straight cutting edges geradgenutet
w. straight flutes geradgenutet
w. unequal legs ungleichschenklig
w. wide annual rings grobjährig *(Holz)*
withdraw / to 1. zurückziehen, ausheben *(Bohrer)*; 2. *(Kern)* ausfahren *(Stab)*; 3. *(Brg)* rauben *(Ausbau)*
withdrawable mast *(Schiff)* Steckmast *m*
w. superheater ausziehbarer Überhitzer *m*
withdrawal 1. Zurückziehen *n*, Zurücknahme *f*, Rücknahme *f*; Rücklauf *m*; 2. *(Kern)* Ausfahren *n (Stab)*
w. collar Ausrückmuffe *f*
w. lever Ausrückhebel *m*
w. of the quill *(Masch)* Pinolenrückzug *m*
w. sleeve Abziehhülse *f*
w. winch *(Brg)* Raubwinde *f (Ausbau)*
withdrawing Zurückziehen *n*
wither / to verwelken; trocknen, darren
within a tolerance innerhalb einer Toleranz
without attention by the operator wartungsfrei
w. changing the position of the part (workpiece) *(Masch)* ohne Umspannung *(des Werkstückes)*
w. dead centre totpunktfrei
w. [tooling] marks riefenfrei
w. undercut unterschnittfrei
withstand / to aushalten, widerstehen
withstand voltage *(El)* Stehspannung *f*
witness line Maßhilfslinie *f*, Bezugslinie *f*
wittichenite *(Min)* Wittichenit *m*
wobble / to flattern *(Kfz.-Räder)*; taumeln
w. from side to side hin- und hertanzen *(Bohrer)*
wobble engine Taumelscheibenmotor *m*
w. modulation Wobbeln *n (Funktechnik)*
w. plate Taumelscheibe *f*
w. plate engine Taumelscheibenmotor *m*
w. plate pump Taumelscheibenpumpe *f*
wobbler *(El)* Wobbelgenerator *m*
wobbling Flattern *n (Räder)*; Taumeln *n*; *(El)* Wobbeln *n*
w. frequency Wobbelfrequenz *f*
w. of the drill from side to side Schlagen *n* des Bohrers
wobbulator *(El)* Wobbelgenerator *m*
wolfram s. tungsten
wolframite *(Min)* Wolframit *m*
Wollaston wire Wollastondraht *m*
womp *(Am)* intensiver Lichtfleck *m (Katodenstrahlröhre)*
wood 1. Holz *n*; Gehölz *n*, Wald *m*; 2. Holzblasinstrument *n*
w. auger Holzbohrer *m*
w. axe Holzhacke *f*, Holzbeil *n*
w. bending-template Holzbiegeform *f*
w. bit Holzbohrer *m*
w. blind factory Jalousiefabrik *f*
w. board Holzschliffpappe *f*
w. boiler Dampfkasten *m (zum Holzdämpfen)*
w. boring Holzbohrarbeiten *fpl*
w.-boring bit Holzbohrer *m*
w. cellulose Holzzellstoff *m*
w. chips Splitterholz *n*
w. chisel Beitel *m*, Holzmeißel *m*
w. craft vice Spannkluppe *f (Holzbearbeitung)*
w. dowel Holzdübel *m*
w. drilling Holzbohren *n*
w.-engraving Holzschnitt[kunst *f*] *m*
w. felling Holzung *f*, Holzfällerei *f*
w. felloe Holzfelge *f*
w. fibre building board Holzfaserbauplatte *f*
w. fired furnace Holzfeuerung *f*
w. flour Holzmehl *n*
w.-free holzfrei *(Papier)*
w. hydrolysis Holzverzuckerung *f*
w. insulator Holzisolator *m*
w. joint Holzverbindung *f*

w. lagging Holzverschalung *f*, Holzverkleidung *f*, Holzschutzmantel *m*
w. marking gauge Streichmaß *n*
w. moulding factory Kehlleistenfabrik *f*
w. moulding machine Kehlhobelmaschine *f*
w. oil Holzöl *n*, Tungöl *n*
w. pattern shop Modelltischlerei *f*
w. pegging machine Holznagelmaschine *f*
w. piling Holzverschalung *f* zum Absteifen eines Kabelgrabens
w. planing machine for cord fibres Hobelmaschine *f* für Holzgeflechtspäne
w. polish Holzpolitur *f*
w. preservative Holzkonservierungsmittel *n*
w. pulp Holzschliff *m*, Holzzellstoff *m*, Pulpe *f*
w.-pulp paper Holzschliffpapier *n*
w. rasp Holzraspel *f*
w. refuse Holzabfälle *mpl*
w. rim Holzreifen *m (Küfnerei)*
w. rim bending machine Holzreifenbiegemaschine *f*
w. room Holzputzerei *f*
w. saw Gestellsäge *f*, Tischlergestellsäge *f*
w. screw Holzschraube *f*
w. shavings Hobelspäne *mpl*, Holzspäne *mpl*
w. shed Holzschuppen *m*
w. ship Holzschiff *n*
w. shoring Holzverschalung *f (zum Absteifen eines Kabelgrabens)*
w. soot Holzruß *m*
w. spirit Methylalkohol *m*, Holzalkohol *m*, Holzspiritus *m*
w.-stack Holzschober *m*
w. tar Holzteer *m*
w. trimmer Holzkantenbestoßmaschine *f*
w. turner Drechsler *m*
w. turning lathe Holzdrehmaschine *f*, Drehbank *f*, Drechselbank *f*, Drechslerbank *f*, Holzdrehbank *f*
w. waste board Holzabfallplatte *f*
w. wind instrument Holzblasinstrument *n*
w. wool Holzwolle *f*
w.-wool board Holzwollplatte *f*
w.-wool making machine Holzwollmaschine *f*
w.-wool slab Holzwollplatte *f*
w. working lathe Holzbearbeitungs[dreh]maschine *f*, Drechselbank *f*
woodcut Holzschnitt *m*
wooden bar bending machine Stangenbiegemaschine *f*
w. bobbin Holzspule *f*
w. bulkhead Holzschott *n*
w. grid Holzhorde *f*
w. guide Holzspurlatte *f*
w. hurdle Holzhorde *f*
w. lagging Holzmantel *m*
w. pavement Holzpflaster *n*
w. reinforcement *(Am)* Stützpfahl *m*
w. ship Holzschiff *n*
w. shuttle Holzschützen *m*, Webschützen *m* aus Holz
w. sleeper Holzschwelle *f*
w. template for the bolt hole Holzlehre *f* für das Ankerloch
w. toy factory Holzspielwarenfabrik *f*
woodland Forst *m*
Woodruff key Scheibenfeder *f*
Woodruff keyway cutter Schlitzfräser *m* für Scheibenfedernuten
Wood's alloy Woodsches Metall *n*
woodworking Holzbearbeitung *f*
w. machine Holzbearbeitungsmaschine *f*
w. planer Hobelmaschine *f* für Holzbearbeitung
woof *(Text)* Schuß *m*, Einschuß *m*, Eintrag *m*
woofer Tieftonlautsprecher *m*, Baßlautsprecher *m*
wool Wolle *f*
w. blending Mischen *n* der Wolle
w. card[er] Wollkratze *f*, Wollkrempel *f*, Wollkämmgerät *n*
w. drying apparatus with staggered lattices Wolltrockenvorrichtung *f* mit versetzt übereinanderliegenden Fördertüchern
w.-dyed in der Wolle gefärbt
w. fat Wollfett *n*
w. felt Wollfilz *m*
w. fleece Wollvlies *n*
w. grading Wollklassierung *f*
w. oiling Schmälzen *n* der Wolle
w. scale Wollschuppenschicht *f*
w. scouring Wollwäsche *f*
w. sorting Wollsortierung *f*
w. top Wollkammzug *m*
w. washing machine Wollwaschmaschine *f*
w. weaving Wollweberei *f*, Tuchweberei *f*
woollen count Feinheit *f* des Wollgarns
w. fabric Streichgarngewebe *n*
wools Wollkleidung *f*, Wollsachen *pl*
word *(Dat)* Wort *n (eine Gruppe von Zeichen, gewöhnlich von einer bestimmten Länge)*, Buchstabengruppe *f*, Pulsgruppe *f*, Zahlengruppe *f*
w. time Wortlaufzeit *f*
work / to 1. arbeiten; funktionieren; 2. bearbeiten; *(Am)* umformen; 3. abbauen, gewinnen *(Steine, Erden, Kohle usw.)*
w. a machine eine Maschine bedienen
w. a ship manövrieren *(Schiff)*
w. a single shift in einer Schicht arbeiten
w. condensing mit Kondensation arbeiten
w.-harden verfestigen
w. into a load auf eine Belastung arbeiten
w. loose locker werden *(Mutter)*
w. overtime Überstunden machen
w. with counterflow mit Gegenstrom arbeiten *(Dampfmaschine)*

w. with parallel flow mit Gleichstrom arbeiten *(Dampfmaschine)*

work 1. Arbeit *f*; 2. Werkstück *n*

w. accuracy Arbeitsgenauigkeit *f*

w. arbor *(Masch)* Aufspanndorn *m*

w. axis Werkstückachse *f*

w. bench Installationstisch *m*; Werk[statt]-bank *f*, Werktisch *m*

w. blank Rohling *m*, unbearbeitetes Werkstück *n*

w. boat Arbeitsboot *n*

w. capacity Arbeitsbereich *m*

w. chuck Werkstückspannfutter *n*

w. clamping Werkstückaufspannung *f*

w. cycle Arbeitsspiel *n*

w. done by stroke Arbeit *f* je Hub

w. done during admission Admissionsarbeit *f* *(Dampfdiagramm)*; Eintrittsarbeit *f* *(Turbine)*

w. done during change of state Disgregationsarbeit *f* *(Dampfdiagramm)*

w. done during compression Kompressionsarbeit *f* *(Dampfdiagramm)*

w. done during exhaust Ausström[ungs]-arbeit *f* *(Dampfdiagramm)*

w. done during expansion Expansionsarbeit *f* *(Dampfdiagramm)*; Dehnungsarbeit *f* *(Turbine)*

w. drive *(Masch)* Mitnahme *f*

w. driver *(Masch)* Mitnehmer *m*

w. due to back pressure Gegendruckarbeit *f* *(Dampfdiagramm)*

w. ejector mechanism Werkstückauswerfeinrichtung *f*

w.-equivalent of one kilogramme of steam Arbeitswert *m* eines Kilogrammes Dampf

w.-fixture Spannvorrichtung *f*

w. flow Werkstückfluß *m*

w. function Austrittsarbeit *f*

w. function difference Differenz *f* in der Austrittsarbeit *(Kontaktpotentialdifferenz)*

w. function of electrons Elektronenaustrittsarbeit *f*

w. gear Werkrad *n*

w. guide [in centreless grinding] Führungsleiste *f* [beim spitzenlosen Schleifen]

w. handling Werkstückhandhabung *f*, Werkstückförderung *f*, Werkstücktransport *m*

w. handling equipment Werkstückfördereinrichtung *f*, Werkstücktransporteinrichtung *f*

w. hardening Kaltverfestigung *f*

w. head Werkstückspindelstock *m*

w. head transformer Anpaßtransformator *m*

w.-holding attachment Spannvorrichtung *f*

w.-holding centre Aufnahmespitze *f*

w.-holding chuck Werkstückspannfutter *n*

w.-holding equipment Spannvorrichtung *f*

w.-holding fixture Werkstückspannvorrichtung *f*

w.-holding slot Aufspann-Nut *f*

w.-holding surface Aufspannfläche *f*

w.-holding table Spannplatte *f*

w. layout *(Am)* Anreißen *n*

w. length *(Masch)* Arbeitsweg *m*

w. lighting Arbeitsbeleuchtung *f*

w. loading device Werkstückzuführeinrichtung *f*

w. locating fixture Spannvorrichtung *f*

w. material Werkstückwerkstoff *m*

w. mounting mandrel Werkstückaufspanndorn *m*

w. of acceleration Beschleunigungsarbeit *f*

w. of deformation Deformationsarbeit *f*, Formänderungsarbeit *f*

w. of excitation *(El)* Erregerarbeit *f*

w. of friction Reibungsarbeit *f*

w. of polarization Polarisationsarbeit *f*

w. of reaction Reaktionsarbeit *f*

w. of resistance Widerstandsarbeit *f*

w. on hand vorliegende Arbeit *f*

w. order Arbeitsauftrag *m*

w. repositioning Werkstückumspannung *f*

w. rest Werkstückbock *m*

w. rest blade Werkstückauflageschiene *f* *(für spitzenloses Schleifen)*

w. revolution Werkstückumdrehung *f*

w. roll Arbeitswalze *f*

w. rotation Umlaufbewegung *f* des Werkstücks

w. sample Werkstückmuster *n*

w. seat Werkstückauflage *f*, Werkstückanlage *f*

w. setup Werkstückaufspannung *f*

w. site Baustelle *f*; Arbeitsstelle *f*

w. specification Arbeitsauftrag *m*

w. spindle Arbeitsspindel *f*

w. spoilage Ausschuß *m*

w. study Arbeitsstudie *f*

w.-support[ing] blade Werkstückauflageschiene *f* *(für spitzenloses Schleifen)*

w. surface finish Oberflächengüte *f* des Werkstücks

w. table Aufspanntisch *m*, Maschinentisch *m*, Werkstücktisch *m*

w. table in grinding Schleiftisch *m*

w. transfer Werkstücktransport *m* *(Taktstraße)*

w. unit Arbeitseinheit *f*, Energieeinheit *f*

workability Umformbarkeit *f* *(Metall)*; Bearbeitbarkeit *f* *(Stein usw.)*; Verarbeitbarkeit *f* *(Beton)*; *(Brg)* Gängigkeit *f*, Gewinnbarkeit *f* *(Kohle)*

workable umformbar *(Metall)*; bearbeitbar *(Stein usw.)*; verarbeitbar; gewinnbar; abbauwürdig *(Kohle)*

w. control range Aussteuer[ungs]bereich *m*

w. hatch *(Schiff)* arbeitsfähige Luke *f* *(Luke mit bestimmten Abmessungen und einer bestimmten Umschlagkapazität)*

w. in cold state kalt umformbar

w. seam abbauwürdiges Flöz *n*
worked hardboard bearbeitete Hartplatte *f*
worker 1. Arbeiter; 2. Arbeitswalze *f*
working Umformen *n*, Umformung *f (Metall)*; Bearbeitung *f (Stein)*; Verarbeitung *f*; *(Brg)* Verhieb *m*, Abbau *m*, Verhau *m (Vorgang)*; Abbau[raum] *m (Ort)*
w. accuracy Arbeitsgenauigkeit *f*
w. allowance Bearbeitungszugabe *f*
w. area of table nutzbare Tischfläche *f*
w. boiler Betriebskessel *m*
w. clearance Arbeitsspiel *n*
w. condensing Kondensationsbetrieb *m (Dampfmaschine)*
w. condition Betriebszustand *m*
w. connections Betriebsschaltung *f*
w. current *(El)* Betriebsstrom *m*
w. cycle time Arbeitstakt *m*
w. cylinder Arbeitszylinder *m (Servolenkung)*
w. depth Eingriffstiefe *f*
w. distance 1. Reichweite *f*; 2. Objektabstand *m*; 3. Schutzabstand *m*
w. drawing Ausführungszeichnung *f*, Werk[statt]zeichnung *f*
w. edge 1. Anlegekante *f (Winkel)*; 2. *(Am)* Hauptschneide *f*
w. face 1. *(Wärm)* Auflagefläche *f*, Druckfläche *f*; 2. Bezugsfläche *f*; 3. Druckseite *f (Luftschraube)*; 4. *(Brg)* in Betrieb befindlicher Streb *m*
w. face locomotive Abräumlokomotive *f (elektrischer Zug)*
w. fit Bewegungssitz *m*
w. fluid Arbeitsflüssigkeit *f (Hydraulik)*
w. foreman Vorarbeiter *m*; Brigadier *m*
w. frequency Arbeitsfrequenz *f*; Verkehrsfrequenz *f*
w. gauge 1. Spurbreite *f*; 2. Arbeitslehre *f*, Arbeitsmaß *n*
w. gear *(Brg)* Gezähe *n*
w. gear economy Gezähewirtschaft *f*
w. hatch *s.* workable hatch
w. home *(Brg)* Rückbau *m*, Heimwärtsbau *m*
w. hours Arbeitszeit *f*, Arbeitsstunden *fpl*, Betriebszeit *f*, Betriebsstunden *fpl*
w. instructions Bedienungsvorschrift *f*, Bedienungsanleitung *f*, Bedienungsanweisung *f*
w. joint Arbeitsfuge *f (Betonbau)*
w. length Lauflänge *f (Zylinder)*
w. life Nutzungsdauer *f*; Dauer *f* der Brauchbarkeit
w. line Belastungskurve *f (Funktechnik)*
w. lines Festmacheleinen *fpl (für Schiffe)*
w. load Gebrauchslast *f*, Nutzlast *f*; zulässige Last *f*; Betriebsbeanspruchung *f*
w. method 1. Umformverfahren *n*, Bearbeitungsverfahren *n*; 2. Abbauverfahren *n*, Arbeitsverfahren *n*

w. non-condensing Auspuffbetrieb *m (Dampfmaschine)*
w. of furnace Ofengang *m*
w. panel *(Brg)* Gewinnungsabteilung *f*, Abbauabteilung *f*, Abbaurevier *n*, Baufeld *n*
w. part Verschleißteil *n*
w. partnership Zusammenarbeit *f*
w. platform Arbeitsbühne *f*
w. point [dynamischer] Arbeitspunkt *m*
w. position Arbeitsstellung *f*
w. pressure Betriebsdruck *m*, Kesseldruck *m*
w. pull Durchzugskraft *f*
w. radius Ausladung *f (Ackergeräte)*
w. reference system *(Fmt)* Arbeitseichkreis *m*
w. scheme 1. Arbeitsplan *m*; 2. *(Brg)* Abbauschema *n*
w. ship Arbeitsboot *n*
w. side of furnace Bedienungsseite *f* eines Ofens
w. sketch Werkskizze *f*
w. speed Arbeitsgeschwindigkeit *f*
w. station Bearbeitungsstation *f*
w. steam *(Wärm)* Betriebsdampf *m*
w. storage Arbeitsspeicherung *f*, Zwischen[ergebnis]speicherung *f*
w. store Arbeitsspeicher *m*, Zwischen[ergebnis]speicher *m*
w. stress zulässige Spannung *f*, Betriebsspannung *f*
w. stroke Arbeitshub *m*
w. surface 1. Angriffsfläche *f*, Arbeitsfläche *f*; Gleitfläche *f (Zylinder)*; 2. *(Masch)* Aufspannfläche *f*
w. surface of cylinder Zylinderlauffläche *f*
w. surface of piston Kolbenlaufbahn *f*, Kolbenlauffläche *f*
w. tempo Arbeitstempo *n*
w. the heat Führung *f* der Schmelze
w. time Bearbeitungszeit *f*
w. unit stress *(Am)* höchstzulässige Spannung
w.-up of ammonia Ammoniakverarbeitung *f*
w. voltage Arbeitsspannung *f*, Betriebsspannung *f*
w. with acid slag *(Met)* mit saurer Schlackenführung
w. with basic slag *(Met)* mit basischer Schlackenführung
w. with saturated steam *(Wärm)* Sattdampfbetrieb *m*
w. with superheated steam *(Wärm)* Heißdampfbetrieb *m*
workmanlike fachgemäß, fachkundig
workmanship 1. Werkstattarbeit *f*; 2. Fachkönnen *n*
workpiece Werkstück *n*, Teil *n*
w. cam contour Werkstücknocken *m*
w. transfer equipment Werkstücktransporteinrichtung *f*

works Fabrik *f*, Werk *n*
w. area Werksgelände *n*
w. drawing Werkstattzeichnung *f*
w. engineer Betriebsingenieur *m*
w. inspector Gütekontrolleur *m*
w. manager Werkleiter *m*
w. school Betriebsschule *f*
w. specification Werksvorschrift *f*
w. standard specification Werksnorm *f*, Betriebsstandard *m*
w. training Berufspraktikum *n* *(für Studenten)*
w. visit Betriebsbesichtigung *f*, Exkursion *f*
workshop 1. Werkstatt *f*, Werkstätte *f*; 2. Produktionsabteilung *f (Betrieb)*; Werkhalle *f*
w. drawing Konstruktionszeichnung *f*; Werkstattzeichnung *f*
w. equipment Werkstatteinrichtung *f*
w. erection Werkhallenmontage *f*
w. line shafting Transmission *f*
w. measuring gauge Betriebsmeßgerät *n*
worm Schnecke *f*
w. agitator Rührwelle *f (Saatkasten, Düngerstreuer)*
w.-and-sector steering gear *(Kfz)* Schneckenlenkung *f*
w.-and-wheel steering gear *(Kfz)* Schneckenlenkung *f*
w.-and-worm wheel drive Schneckengetriebe *n*; Schneckentrieb *m*
w. barring gear *(Wärm)* Schaltwerk *n* mit Schnecke *(Schwungrad)*
w. blank Schneckenkörper *m*
w. condenser Kühlschlange *f*
w. conveyor Schneckenförderer *m*
w. cutter Schneckenfräser *m*
w. drive Schnecken[rad]getriebe *n*
w.-eaten wurmstichig *(Holz)*
w. gear Schneckenrad *n*
w. gear cutter Schneckenradfräser *m*
w. gear drive Schnecken[rad]antrieb *m*
w. gear hob Schneckenradabwälzfräser *m*
w. gear hob for tangential feed Schneckenradabwälzfräser *m* für tangentiale Zustellung
w. gear stud Schneckenwelle *f* *(Verzahnung)*
w. gearing Schnecken[rad]getriebe *n*
w. generating hob Schneckenwälzfräser *m*
w. hob Schneckenfräser *m*
w. milling Schneckenfräsen *n*
w. reduction gear Schneckenreduziergetriebe *n*, Schneckenradgetriebe *n* für Übersetzung ins Langsame
w. shaft Schneckenwelle *f*
w. thread Schneckengewinde *n*
w. thread cutter Schneckengewindefräser *m*
w. thread grinding machine Schneckenschleifmaschine *f*
w. thread milling cutter Schneckenfräser *m*
w. wheel Schneckenrad *n*
w. wheel cutter Schneckenradfräser *m*
w. wheel generating machine Schneckenradwälzfräsmaschine *f*
w. wheel generation Schneckenradherstellung *f* nach dem Wälzverfahren
w. wheel hobbing by the radial-feed method Radialverfahren *n* zur Schneckenradherstellung
w. wheel hobbing by the tangential-feed method Axialverfahren *n* zur Schneckenradherstellung
w. wheel hobbing process Fräsverfahren *n* zur Schneckenradherstellung
w. wheel throat Zahnwurzelbogen *m*
w. wheel tooth Schneckenradzahn *m*
worn abgetragen, abgenutzt
w. contact verschmorter Kontakt *m*
w. cylinder unrunder Zylinder *m*
w.-out abgenutzt, verbraucht
w.-out bearing ausgelaufenes Gleitlager *n*
worsted Kammgarn[stoff *m*] *n*; Kammwolle *f*
w. count Feinheit *f* des Kammgarns
w. drawing Verstrecken *n* des Kammzugs
w. fabric Kammgarngewebe *n*
w. milling Kammgarnwalke *f*
w. spinning Kammgarnspinnverfahren *n*
w. stocking loom Wollstrumpfwirkstuhl *m*
w. warp Kettkammgarn *n*
w. wool Kammwolle *f*
wort Maische *f*, Würze *f*
w. masher Maischapparat *m*
Woulfe's bottle Woulfesche Flasche *f*
wound gewickelt; gespult; gewunden
w. rotor *(El)* gewickelter Läufer *m*
w.-rotor induction motor Schleifringmotor *m*
w. torsion spring gewundene Drehfeder *f*
woven gewebt, Web-
w. label Webetikett *n*
w. material Webware *f*, Gewebe *n*
w. resistors elektrisches Heizgewebe *n*
w.-screen memory (store) *(BMSR)* Schirmgewebespeicher *m* *(Datenspeicher aus nichtleitendem Gewebe, durch das leitende Drähte [kreuzweise] gezogen sind)*
wow [langsame] Tonhöhenschwankungen *fpl*
wrack Seetang *m*
wraith *(Text)* Schärblatt *n*
w. holder Schärblatthalter *m*
wrap / to einwickeln, einschlagen, einpacken
wrap-around plate *(Typ)* Wickelplatte *f*
w.-around windshield *(Kfz)* Rundsichtverglasung *f*, Vollsichtverglasung *f*
wrapper Hülle *f*; Kreuzband *n*
wrapping Bewehrung *f*, Umhüllung *f*, Umwicklung *f*; Bewicklung *f*
w. machine Verpackungsmaschine *f*

w. material Verpackungsmaterial *n*
w. paper Einwickelpapier *n*, Packpapier *n*
w. test *(Met)* Wickelversuch *m*
wreck Wrack *n*, Schiffstrümmer *pl*; Strandgut *n*
w. buoy Wracktonne *f*
wreckage Wrackgut *n*, Wrackteile *npl*, Schiffstrümmer *pl*
wrecker Berger *m (Wrackbergung)*; Strandräuber *m*
wrecking bar Brechstange *f*
w. bar with gooseneck claw Nageleisen *n*
w. car Abschleppwagen *m*
w. crane Abschleppkran *m*
w. truck Abschleppwagen *m*
w. tug Bergungsschlepper *m*
wrench Mutternschlüssel *m*, Schraubenschlüssel *m*
w. cutter Schraubenschlüsselfräser *m*
wrenchless schlüssellos
wrest plate Wirbelbalken *m*, Wirbelstock *m (Klavier)*
wring / to 1. ansprengen *(Endmaße)*; 2. [aus]drücken, [aus]pressen, [aus]wringen
w. together aneinanderschieben, ansprengen, zusammensetzen *(Endmaße)*
wrinkle finish Schrumpflack *m*, Kräusellack *m*, Runzellack *m*
w.-free runzelfrei *(Farbe)*
w. recovery *(Text)* Knittererholung *f*
w. resistance *(Text)* Knitterwiderstand[svermögen *n*] *m*, Knitterarmut *f*
w. resistant *(Text)* knitterarm
w.-resistant finish *(Text)* knitterarme Ausrüstung *f*
w. washer federnde (gewellte) Unterlegscheibe *f*
wrinkling 1. Knittern *n*; Kräuseln *n*; Wellen *n*; 2. Faltenbildung *f* an den Tragflächen
wrist pin Gelenkstift *m*; Kolbenbolzen *m*
wristlet Armband *n*; Armbanduhr *f*
write / to [ein]schreiben
write amplifier Schreibverstärker *m*
writing Schreiben *n*; Schrift *f*
w. bar Schreibstange *f*
w. edge Schreibstange *f*
w. projector Schreibprojektor *m*
w. speed Schreibgeschwindigkeit *f*
w. surface Schreibfläche *f*
wrong checking pattern Karierfehler *m*
w. colour of weft Farbstreifen *m*, Farbband *n*
w. connection Fehlanruf *m*, Fehlverbindung *f*
w. draft *(Text)* falscher Geschirreinzug *m*, Einzugfehler *m*
w. ducting *(Text)* falscher Blatteinzug *m*; Blatteinzugfehler *m*
w. fount *(Typ)* Fisch *m*, verfischter Buchstabe *m*
w. lift *(Text)* Bindungsfehler *m*
w. number call *(Am) s.* w. connection
w. reading *(Typ)* seitenverkehrt
w. take-off *(Flg)* Fehlstart *m*
wrought bearbeitet; zugerichtet; geschmiedet; geknetet
w. alloy Knetlegierung *f*
w. iron Puddelstahl *m*, Schweißstahl *m*, Schmiedestahl *m*
w. steel Schweißstahl *m*
wulfenite Wulfenit *m*, Gelbbleierz *n*
wurtzite Wurtzit *m*, Zinkblende *f*
wye Hosenrohr *n*, Rohrverzweigung *f*
w.-delta connection Sterndreieckschaltung *f*
w. junction Y-Kopplung *f (Wellenleiter)*

X

X-amplifier X-Verstärker *m*, Verstärker *m* für horizontale Ablenkung
X-amplitude X-Amplitude *f*
X-arm Querträger *m*
X-axis x-Achse *f*
X-band X-Band *n (Radar)*
X-bit X-Schneide *f*
X-contact Blitzlichtkontakt *m*
X-coordinate x-Koordinate *f*
X-cut crystal Schwingkristall *m* mit X-Schnitt
X-irradiation Röntgenbestrahlung *f*
X-plates Horizontalablenkplatten *fpl*
X-radiation Röntgenstrahlung *f*
X-ray / to röntgen, durchleuchten
X-ray absorption edge Röntgenabsorptionskante *f*
X-ray absorption spectrum Röntgenabsorptionsspektrum *n*
X-ray apparatus Röntgenapparat *m*
X-ray crystal analysis Röntgenfeinstrukturanalyse *f*
X-ray crystal density method Röntgenstrahlen-Kristalldichtemethode *f*
X-ray diascope Röntgendiaskop *n*
X-ray diffraction Beugung *f* von Röntgenstrahlen
X-ray diffraction camera Röntgenbeugungskamera *f*
X-ray dose Röntgenstrahlendosis *f*
X-ray effect Röntgenstrahlenwirkung *f*
X-ray emission Röntgenstrahlenemission *f*
X-ray examination Röntgenuntersuchung *f (Personen)*; Röntgenprüfung *f (Metalle usw.)*
X-ray film Röntgenfilm *m*
X-ray fluorescence radiation Röntgenfluoreszenzstrahlung *f*
X-ray interference Röntgeninterferenz *f*
X-ray level Röntgenterm *m*
X-ray line Röntgenlinie *f*

X-ray luminescence Röntgenlumineszenz *f*
X-ray materiology zerstörungsfreie Werkstoffprüfung *f* mit Röntgenstrahlen, Röntgen[werkstoff]prüfung *f*
X-ray measurement Röntgenmessung *f*; Röntgenmeßkunde *f*
X-ray optics Röntgenoptik *f*
X-ray output Strahlenausbeute *f (Röntgenröhren)*
X-ray photograph Röntgenaufnahme *f*, Röntgenbild *n*
X-ray picture Röntgenogramm *n*, Röntgenbild *n*, Röntgenaufnahme *f*
X-ray quantum Röntgen[strahlen]quant *n*
X-ray radiation Röntgenstrahlung *f*
X-ray scattering Streuung *f* von Röntgenstrahlen, Röntgenstreuung *f*
X-ray spectrography Röntgenspektrografie *f*
X-ray spectrometer Röntgenspektrometer *n*
X-ray spectroscopy Röntgenspektroskopie *f*
X-ray spectrum Röntgenspektrum *n*
X-ray structure Röntgenfeinstruktur *f*
X-ray transformer Röntgentransformator *m*
X-ray tube Röntgenröhre *f*
X-ray unit Röntgenanlage *f*, Röntgeneinrichtung *f*
X-ray vacuum Röntgenröhrenvakuum *n*, Vakuum *n* unter 0,01 mm Hg
X-rays Röntgenstrahlung *f*, Röntgenstrahlen *mpl*
X-stopper Krachtöter *m (Funkempfangstechnik)*
X-tube *(Am) s.* X-ray tube
X-unit X-Einheit *f (Längeneinheit in der Röntgenspektroskopie; 1 XE = 10^{-13} m)*
xanthate Xanthogenat *n*, Xanthat *n*, Ester *m* (Salz *n*) der Xanthogensäure
xanthating churn Sulfidiertrommel *f*
xanthene Xanthen *n*
xanthic acid Xanthogensäure *f*
  x. disulphide Dixanthogen *n*
  x. oxide Xanthogenoxid *n*
xanthogenate Xanthogenat *n*, Ester *m* (Salz *n*) der Xanthogensäure
xanthogenic acid Xanthogensäure *f*
xanthone Xanthon *n*
xanthophyll Xanthophyll *n*, Blattgelb *n*
xanthoproteic reaction Xanthoproteinreaktion *f*
xanthosiderite Gelbeisenerz *n*, Xanthosiderit *m*
xenon Xenon *n*
  x. discharge lamp Xenonlampe *f*
  x. poisoning *(Kern)* Xenonvergiftung *f*
xenotime *(Min)* Xenotim *m*, Ytterspat *m*
Xmit / to senden, geben *(Telegrafie)*
Xmitter Sender *m*
Xmitting Senden *n*, Geben *n (Telegrafie)*
Xs atmosphärische Störungen *fpl (beim Funkempfangsbetrieb)*
XS-3 code *(Dat)* Dreier-Exzeß-Kode *m*
Xtal Kristall *m*, Steuerquarz *m*
XY-plotter *s.*XY-recorder
XY-recorder XY-Schreiber *m*, XY-Kompensationsschreiber *m*
xylene Xylol *n*
  x. musk Xylolmoschus *m*
xylenol resin Xylenolharz *n*
xylic acid Xylylsäure *f*
xylitol Xylit *n*
xyloid holzähnlich, holzartig
xylolite Steinholz *n*
xylonic acid Xylonsäure *f*
xylonite Zelluloid *n*

Y

Y-aerial Y-Antenne *f*, Gabelantenne *f*
Y-alloy Y-Legierung *f*
Y-amplifier Y-Verstärker *m*, Verstärker *m* für vertikale Ablenkung
Y-axis y-Achse *f*
Y-connected *(El)* sterngeschaltet
Y-connection *(El)* Sternschaltung *f*
Y-contact unit Y-Kontaktsatz *m (spät arbeitender Kontakt)*
Y-joint Gabelmuffe *f*, Abzweigmuffe *f*
Y-matrix Leitwertmatrix *f*
Y-parameter equivalent circuit Ersatzschaltung *f* mit Y-Parametern
Y-pipe Hosenrohr *n*
Y-valve Schrägsitzventil *n*
Y-voltage *(El)* Sternspannung *f*; Phasenspannung *f*
Y-Y-connection *s.* star-star-connection
Y-zigzag-connection *s.* delta-zigzag-connection
yacht Jacht *f*
  y.-building yard Jachtwerft *f*
  y. racing Jachtwettrennen *n*
  y. stern Jachtheck *n*
yachting Jachtsegeln *n*, Jachtwesen *n*, Segelsport *m*
Yagi aerial Yagiantenne *f*
Yankee machine Selbstabnahmepapiermaschine *f*
yard 1. Hof *m*; Lagerplatz *m*; 2. *(Schiff)* Stapelplatz *m*, Werft *f*; Rah[e] *f*; 3. Yard *n (Längenmaß; 1 yd = 0,91440·m)*
  y. number Baunummer *f (eines Schiffes)*
  y.-pound *englische Einheit der Arbeit; 1 yd.lb. = 4,148 · 10^{-1} kpm*
  y.-service lighter Werftleichter *m*
  y. stick Zollstab *m*
  y. trial trip Werftprobefahrt *f*
yardage *(Brg)* Leistung *f* in Yard (Kubikyard)
  y. goods *(Text)* Schnittware *f*
yarn Garn *n*; Faden *m*
  y. carrier Garnträger *m*
  y. clearer Fadenreiniger *m*, Garnreiniger *m*

y. conditioning Garnkonditionierung *f*
y. count Garnnummer *f*
y.-dyed im Garn gefärbt, garngefärbt
y. dyeing Garnfärben *n*
y. evenness tester Garngleichheitsprüfer *m*
y. feed installation Fadenzubringer *m*
y. irregularity Garnungleichmäßigkeit *f*
y. package Garnkörper *m*
y. ply Garnfachung *f*
y. shading Flammgarnfärberei *f*
y. singer Garnsengmaschine *f*
y. size *s.* y. count
y. stop motion detection installation Fadenüberwachungsgerät *n*
y. tension Fadenbremse *f*; Fadenspannung *f*
y. testing Garnprüfung *f*
Yarrow boiler *(Wärm)* Yarrow-Kessel *m (engrohriger Wasserrohrkessel mit einem Ober- und zwei Unterkesseln und geraden Rohren)*
yaw / to *(Schiff)* gieren
yaw angle Gierwinkel *m*
yawing Gieren *n*, Gierbewegung *f (Abweichen eines Schiffes vom Kurs)*
y. moment Giermoment *n*
yawl Ruderboot *n*, Segeljolle *f*, Jolle *f*
yawmeter Gierwinkelmeßgerät *n*
year ring Jahresring *m*
yearling's wool Jährlingswolle *f*
yearly dose *(Kern)* Jahresdosis *f*
y. make Jahresleistung *f*
yeast Hefe *f*
y. germ Hefekeim *m*
y. nutrient Nährhefe *f*
yeasty hefeartig
yellow / to vergilben
yellow Gelb *n*
y. brass Messing *n*, Gelbguß *m*
y. discharge *(Text)* Gelbätze *f*
y. heat *(Met)* Gelbglut *f*
y. jack *(Schiff)* Quarantäneflagge *f*
y. metal Neumessing *n*
y. mica Goldglimmer *m*
y. ochre *(Min)* Berggelb *n*
y. resin Kolophonium *n*
y. screen Gelbfilter *n (Fotografie)*
y. spot gelber Fleck *m (Retina)*
y. strawboard Strohpappe *f*
y. strawpaper Strohpapier *n*
y. strawpulp Gelbrohstoff *m*
y. tinged gelbstichig
y. uranium oxide Urangelb *n*
yellowroot Gelbkraut *n*, Färberwau *m*, Wau *m*
yield / to 1. ergeben, liefern, hergeben, freigeben; 2. nachgeben; 3. fließen, sich strecken *(Werkstoffe)*
y. neutrons Neutronen liefern
yield 1. Anfall *m*, Ausbeute *f*, Ergebnis *n*, Ertrag *m*, Fertigung *f*, Förderung *f*, Produktion *f*, Ausstoß *m*, Ausbringen *n*; 2. *(Brg)* Einsinkweg *m*, Zusammenschub *m (Ausbau)*; 3. *(Bw)* Fließen *n*, Strecken *n*; plastische Verformung *f*
y. condition Fließbedingung *f*
y. in secondary electrons Sekundärelektronenausbeute *f*
y. limit Fließgrenze *f*
y. line theory Bruchlinientheorie *f*
y. load *(Brg)* Knickpunkt *m (Ausbau)*
y.-mass distribution *(Kern)* Ausbeute-Masse-Verteilung *f*
y. of radiation Strahlungsausbeute *f*
y. of soil Bodenertrag *m*
y. point Fließgrenze *f*; Streckgrenze *f*, Quetschgrenze *f*
y. point at elevated temperatures Warmfließgrenze *f*
y. resistance Fließwiderstand *m*
y. stress Fließspannung *f*
y. value *(Plast)* Ausgiebigkeitsfaktor *m*; Fließwert *m*, unterste Fließgrenze *f*
yielding fließend, nachgiebig
y. phenomenon Fließvorgang *m*
y. roadway arch *(Brg)* Gleitbogen *m (Ausbau)*
yoke 1. Joch *n (Relais)*; Polgehäuse *n*; 2. Bügel *m*, Gabelkopf *m*, Joch; 3. Traglager *n (Fräsdorn)*
y. adjuster Jochverstellhebel *m (Katodenstrahlröhren)*
y. end Gabelkopf *m*
y. hoop Ziehband *n (Wagen)*
y. of [a] magnet Magnetjoch *n*, magnetisches Joch *n*
Young's modulus [of elasticity] Elastizitätsmodul *m*, E-Modul *m*
yperite Yperit *n*, Dichlordiäthylsulfid *n (Kampfstoff)*
ytterbia Ytterbiumoxid *n*
ytterbium Ytterbium *n*
yttria Yttriumoxid *n*
yttrium Yttrium *n*
yucca fibre Yuccafaser *f*
Yukawa field *(Kern)* Yukawa-Feld *n*
Yukawa particle *(Kern)* Yukawa-Teilchen *n*
Yukawa potential *(Kern)* Yukawa-Potential *n*

Z

Z-axis z-Achse *f*; Applikatenachse *f*
Z-axis modulation Aufhellungssteuerung *f (Katodenstrahlröhren)*
Z-bit Z-Schneide *f (Bohrmeißel)*
Z-crank Taumelscheibe *f (fest auf Antriebswelle)*
Z-crank pump Taumelscheibenpumpe *f*
Z-joint Halbstoß *m (Schiene)*
Z-joint plate Halbstoßlasche *f*

Z-twist *(Text)* Z-Drehung *f*
zaffer *s.* zaffre
zaffre Glasurblau *n*; Kobaltoxid *n (für Kobaltblau)*
Zanzibar copal Sansibarkopal *m*
Zapon lacquer Zaponlack *m*
zax Dachhammer *m*
zed Z-Profil *n (eines Trägers)*
zee *s.* zed
Zeeman effect Zeeman-Effekt *m*
zein Maismehlkleber *m*
Zener current *(El)* Zenerstrom *m*
Zener diode Zenerdiode *f*
Zener effect Zenereffekt *m*
Zener formula Zenersche Formel *f*
Zener voltage Zenerspannung *f*
zenith Zenit *m*; Scheitelpunkt *m*
z. distance Zenitdistanz *f*
z. mirror Zenitspiegel *m*
z. telescope Zenitteleskop *n*
zeolite *(Min)* Zeolith *m*
Zerener process Zerener-Verfahren *n*
zero / to auf Null einstellen, nullen
zero Null *f*; Nullpunkt *m*; Nullstelle *f*
z.-access storage zugriffszeitfreie Speicherung *f*
z.-access store zugriffszeitfreier Speicher *m*, Speicher ohne Suchzeit, Schnellspeicher *m*
z.-address instruction Nulladreßbefehl *m*
z.-address system Nulladreßsystem *n*
z. adjuster Nullsteller *m*, Rückstelleinrichtung *f*
z. adjustment Null[punkt]einstellung *f*, Nulljustierung *f*
z. adjustment screw Nulleinstellschraube *f*
z. admission *(Wärm)* Nullfüllung *f*
z. anode Nullpotentialanode *f*
z. axis Nullinie *f*
z. balancing Nullabgleich *m*; Nullkontrolle *f*
z. beat Schwebungsnull *f*, verschwindende Schwebung *f*
z.-beat method Nullschwebungsmethode *f*
z.-beat reception Homodynempfang *m (Empfang mit schwingendem Audion im Schwebungsnull)*
z. bias Gittervorspannung *f* Null
z. blocking-contact *(Fmt)* Nullkontakt *m*
z. burden äußerer Kurzschluß *m (des Meßwandlers)*
z. carrier Trägernull *n*, Nullamplitude *f* eines Trägers
z. ceiling Wolkenhöhe *f* unter 100 Fuß *(zum Fliegen ungeeignetes Wetter)*
z. centre instrument Meßinstrument *n* mit Nullpunkt in der Skalenmitte
z.-compensating method *(Opt)* Nullmethode *f*
z. compensation Nullstellenkompensation *f*
z. conductor *(El)* Nulleiter *m*
z. count rate Nullzählrate *f*
z. current Nullstrom *m*
z. cut-out *(El)* Nullausschalter *m*
z.-delivery position Nullhubstellung *f*
z. deviation Nullpunktabweichung *f*
z. disk Meßscheibe *f (Lehrenbohrtisch)*
z. displacement Nullpunktverschiebung *f*
z. drift Nullpunktverschiebung *f*, Nullpunktwanderung *f*
z. energy Nullenergie *f*
z.-energy level Nullenergieniveau *n*
z.-energy reactor Null[energie]reaktor *m*, Nullexperiment *n*
z. error Abweichung *f* des Nullpunkts, Nullpunktfehler *m*
z.-field feldfrei, feldlos
z. frequency Nullfrequenz *f*
z.-frequency component Gleichstromkomponente *f*, Nullkomponente *f*
z. graduation Nullmarke *f*
z.-gravity Schwerelosigkeit *f*
z. incidence 1. *(Flg)* Nullanstellung *f (Tragfläche)*; 2. stoßfreier Eintritt *m*
z. indication Nullanzeige *f*
z. indicator Nullanzeiger *m*
z. jitter Nullpunktunruhe *f*
z. level Nullpegel *m*, Nullniveau *n*
z. line Nullinie *f*; Nullriß *m*
z. line of vernier Noniusnullstrich *m*
z. loss circuit *(El)* verlustfreie Leitung *f*
z.-magnetic moment magnetisches Moment *n* Null, verschwindendes magnetisches Moment
z. mark Nullmarke *f*; Nullstrich *m*
z. mark of the vernier scale Noniusnullstrich *m*
z. modulation verschwindende Modulation *f*; unmodulierter Zustand *m*
z. moment Nullmoment *n*
z. moment method Nullmomentmethode *f*
z. neutron-absorption cross section verschwindender Neutronenabsorptionsquerschnitt *m*
z. offset Nullpunktunterdrückung *f*
z.-order ray Strahl *m* nullter Ordnung *(Beugung)*
z. output Nulleistung *f*
z. passage Nulldurchgang *m*
z.-phase modulation Nullphasenmodulation *f*
z.-phase-sequence protection Nullsystemschutz *m (Relais)*
z.-phase-sequence relay Erdschlußrelais *n*
z. point Nullpunkt *m*; Sternpunkt *m*
z.-point anode Sternpunktanode *f*
z.-point energy Nullpunktenergie *f*
z.-point entropy Nullpunktentropie *f*
z.-point field Nullpunktsfeld *n*
z.-point stability Nullpunktstabilität *f*

z. position Nullage *f*, Nullstellung *f*
z. potential *(El)* Nullpotential *n*
z. power Nullenergie *f*; Nulleistung *f*
z.-power range Leerlaufbereich *m*
z.-power reactor Nullenergiereaktor *m*, Nullleistungsreaktor *m*
z.-power system Nullenergiesystem *n*
z. range Nullreichweite *f*
z. reaction Gleichdruckwirkung *f*
z.-reader *(Flg)* Kreuzzeigergerät *n*
z. reading Nullablesung *f*
z.-recorder Kompensationsschreiber *m*
z.-reference level *(El)* Bezugspegel *m* Null
z.-sequence component Nullkomponente *f*
z.-sequence field impedance Nullimpedanz *f*
z.-sequence reactance Blindwiderstand *m* für das Nullsystem
z.-sequence symmetrical component symmetrische Nullkomponente *f*
z. setting 1. Null[ein]stellung *f*, Stellung *f* auf Nullwert; 2. Nullstellknopf *m*
z. speed Stillstand *m*
z.-sum game Nullsummenspiel *n*
z. suppression Entnullen *n*, Null[wert]unterdrückung *f*, Nullpunktunterdrückung *f*; Beseitigung *f* führender Nullen *(links vor einer Zahl)*
z.-time activity Scheinaktivität *f (Programmierung)*
z. valence Nullwertigkeit *f*
z. valent nullwertig
z. variation Nullpunktabweichung *f (eines Instruments)*
z. visibility Fehlen *n* jeglicher Sicht
z. voltage *(El)* Nullspannung *f*, Nulldurchgang *m* der Spannung, Spannungsnulldurchgang *m*, Nullung *f*
zeroing Einstellung *f* auf Null
zeta (*z*ero *e*nergy *t*hermonuclear *a*ssembly) Nullenergie-Kernfusionsgerät *n*
z. potential Zetapotential *n (Elektrobiologie)*
Zeuner valve diagram Zeunersches Schieberdiagramm *n*
zigzag / to Zickzackkurs fahren
zigzag aerial Zickzackantenne *f*, Sägezahnantenne *f*
z. bulkhead Zickzackschott *n*
z. connection Zickzackschaltung *f*
z. feed Zuführung *f* im Zickzackschritt
z. folding *(Typ)* Zickzackfalzung *f*
z. grate bar Schmiegeroststab *m (Feuerungstechnik)*
z. reflection Mehrfachreflexion *f (elektromagnetischer Wellen)*
z. rule Gliedermaßstab *m*
z. superheater *(Wärm)* Schlangenrohrüberhitzer *m*
z. superheating tube *(Wärm)* schlangenförmiges Überhitzerrohr *n*, Schlangenüberhitzerrohr *n*, Überhitzerheizschlange *f*
zinc / to verzinken
z.-plate *(galvanisch)* verzinken
zinc Zink *n*
z. ash Zinkasche *f*
z. battery Zinkbatterie *f*
z. blende Zinkblende *f*, Sphalerit *m*
z. box *(Met)* Fällkasten *m (Zinkfällung)*
z. coat Zinkauflage *f*
z.-coated verzinkt
z.-coated sheet verzinktes Blech *n*
z.-coating furnace Verzinkungsofen *m*
z. contact Zinkkontakt *m*
z. crust Zinkschaum *m*
z. desilverization plant Zinkentsilberungsanlage *f*
z. die-casting Zinkdruckguß *m*; Zinkspritzguß *m*
z. distillation Zinkdestillation *f*
z. distilling furnace Zinkdestillierofen *m*, Zinkmuffelofen *m*
z. distilling process Zinkmuffelverfahren *n*
z. dross Zinkschlicker *m*, Zinkgekrätz *n*
z. dust Zinkdampf *m*, Zinkstaub *m*
z. dust discharge *(Text)* Zinkstaubätze *f*
z. dust distillation Zinkstaubdestillation *f*
z. electrode Zinkelektrode *f*
z. engraving *s.* z. etching
z. etching Zinkätzung *f*
z. extraction Zinkgewinnung *f*
z. finish *(Text)* Kalandern *n* mit Zinkwalzen
z.-iron cell Zink-Eisen-Element *n*, Hawkins-Element *n*
z.-lime vat *(Text)* Zinkstaubkalkküpe *f*
z. lining Zinkeinlage *f*, Zinkausguß *m*
z. metal sheet Zinkblech *n*
z. ore Zinkerz *n*
z. ore roasting plant Zinkerzrösthütte *f*
z.-plating Zinkplattierung *f*, Verzinken *n*, Verzinkung *f*
z. pole Zinkpol *m (Batterie)*
z. refinery Zinkraffinerie *f*
z. scum Zinkschaum *m*
z. smelting plant Zinkhütte *f*
z. spar Zinkspat *m*, Galmei *m*
z. vapour Zinkdampf *m*
z. voltameter Zinkvoltameter *n*
zincate Zinkat *n*
zinciferous zinkhaltig
zincify / to verzinken
zincite Rotzinkerz *n*, Zinkit *m*
zinco *s.* zincograph
zincograph Zinkplatte *f (Chemigrafie)*
zincography Zinkografie *f*
zinnwaldite *(Min)* Zinnwaldit *m*
zip / to mit Reißverschluß schließen

zip-fastener Reißverschluß *m*
zipper-closed belt Reißverschlußband *n*
zippertubing Reißverschlußverkabelungsrohr *n*
zircon *(Min)* Zirkon *m*
zirconate Zirkonat *n*
zirconia Zirkondioxid *n*, Zirkonerde *f*
zirconic Zirkon-, zirkonhaltig
zirconium Zirkonium *n*, Zirkon *n*
z. lamp Zirkonlampe *f*
zoisite *(Min)* Zoisit *m*
zonal zonal, Zonen-
zone Zone *f* *(Kristall)*; Bereich *m*; Gürtel *m*; *(Dat)* *(Am)* Sonderspeicher *m*; Zonenteil *m* *(oberste drei Reihen einer Lochkarte)*
z. ahead the face *(Brg)* Strebvorfeld *n*
z. centre *(Fmt)* Durchgangsfernamt *n*, Nahverkehrsamt *n*, Verteilfernamt *n*
z.-law Zonengesetz *n* *(Kristallografie)*
z. marker Zonenbake *f* mit vertikalem Leitstrahl *(Radar)*
z. melting Zonenschmelzen *n*, Zonenschmelzverfahren *n* *(Halbleitertechnik)*
z.-melting furnace Zonenschmelzofen *m*
z.-melting process Zonenschmelzverfahren *n*
z. metering *(Fmt)* Zonenzählung *f*
z. of admission Beaufschlagungsgebiet *n*, Beaufschlagungszone *f* *(Turbine)*
z. of aeration Grundluftzone *f*
z. of audibility Hörbarkeitszone *f*
z. of combustion Verbrennungszone *f*
z. of concentration *(Geol)* B-Horizont *m*, Anreicherungshorizont *m*
z. of contact *(El)* Berührungsfläche *f*
z. of fusion *(Met)* Schmelzzone *f*
z. of high temperature Hochtemperaturzone *f*
z. of incandescence *(Met)* Glühzone *f* *(Generatorenbetrieb)*
z. of lost returns Spülverlustzone *f* *(Tiefbohrtechnik)*
z. of negative pressure Unterdruckzone *f*
z. of removal *(Geol)* A-Horizont *m*, Auslaugungshorizont *m*
z. of saturation *(Wärm)* Sättigungsgebiet *n*
z. of silence Zone *f* des Schweigens *(Schall)*; tote Zone *(Funktechnik)*
z. position indicator Einweisungsradar *n*, Zielverfolgungsradar *n*
z. punching *(Büro)* Zonenlochung *f* *(Form der Überlochung)*
z. purification Zonenreinigung[sverfahren *n*] *f* *(Halbleitertechnik)*
z. refining Zonenreinigung *f*; Zonenraffinationsverfahren *n*
z. refining process Zonenschmelzverfahren *n*
z. registration *(Fmt)* Zonenzählung *f*
z. television *(Fs)* Streifenzerlegung *f*, partielle Bildaufnahme *f*
zoning 1. *(Met)* Zonenbildung *f*; 2. *(Bw)* Zonierung *f*; Aufstellen *n* des Flächennutzungsplanes; 3. Zonung *f* *(Lochkarten)*; 4. Zonarbau *m* *(Kristalle)*
z. map *(Bw)* Flächennutzungsplan *m*
zoom 1. *(Fs)* Wegreißen (Schnellschwenken) *n* der Kamera; 2. Hochreißen *n*; Hochziehen *n* *(Flugzeug)*
z. lens *(Fs)* Gummilinse *f*, Linse *f* veränderlicher Brennweite
zwitterion Zwitterion *n*
zymogenic gärungserregend
zymology Gärungschemie *f*
zymo[si]meter Gärungsmesser *m*
zymosis Gärung *f*
zymotechnic gärungstechnisch

VERZEICHNIS WICHTIGER ENGLISCHER ABKÜRZUNGEN

A

A academician; academy; arts; associate *(of society)*; admiral; aircraft; ambulance; amphibian; army; attack; artillery; aviation; aeronautics; aspect ratio; ammeter; ampere; amperemeter; amplification; amplifier; anode; antenna; armature; acceleration; arrest; atomic weight; acre; annealed; airborne; area of immersed midship section; reciprocal of mechanical equivalent of heat

Å Ångstrom unit

A° absolute temperature

a ante; at; acceleration *(linear)*; ampere; amplitude; anode; atomic weight; attenuation [coefficient]

a. accepted; annual; axial; acre

a′ anode of section 1

a″ anode of section 2

AA arithmetical average; anti-aircraft; approach aid *(navigation)*; American Airlines Inc.

A.A., a.a. always afloat *(chartering)*

AAA anti-aircraft artillery

A.A.A. American Automobile Association

AAAS American Association for the Advancement of Science

AAC automatic amplitude control

AACR All America Cables and Radio

AAE American Association of Engineers

AAF Allied Air Forces; army air field

A.A. gun anti-aircraft gun

AAM air-to-air missile

A. & C.P. anchors & chains proved

A & R assembly and repair

AAODC American Association of Oilwell Drilling Contractors

AAP army air personnel

A.A.P.G. American Association of Petroleum Geologists

AAR Association of American Railroads

A.A.R., a.a.r. against all risks *(marine insurance)*

AAS American Astronautical Society

AASM Association of American Steel Manufacturers

AASMTC Association of American Steel Manufacturers' Technical Committee

AASR airport and surveillance radar

AAT anti-aircraft tank

AATCC American Association of Textile Chemists and Colorists

AATT American Association for Textile Technology

AB air base; airborne

AB, A.B. able-bodied; able-bodied seaman

A / B, ab airborne

ab. above

a.b. as before

A.B.B.F. Association of Bronze and Brass Founders

abbr., abbrev. abbreviation

ABC atomic, biological, and chemical; automatic binary computer; American Broadcasting Company

A.B.C. alum, blood, clay

A.B.C.C. Association of British Chambers of Commerce

abc automatic bass compensation; automatic bias control; automatic brightness control

abd abandoned

ABMA American Boiler Manufacturers' Association & Affiliated

ABn aerodrome beacon

Abn airborne

Abn Div airborne division

ABNMA Aircraft Bolt and Nut Manufacturers' Association

above s.l. above sea-level

ABS American Bureau of Shipping

A.B.S. American Bureau of Shipping

Abs. abstracts

abs, abs. absolute; absolute value

ABSIE American Broadcasting Station in Europe

abt. about

a.b.t. air-blast transformer

abv. above

AC absolute ceiling; account current; air corps; Army corps; alternating current; Automobile Club; aviation cadet

A.C. alternating current; air-cooled; air-craft carrier

A / C account current

Ac acetyl; *(symbol)* actinium

ac aerial current; anti-clutter

ac. accumulator; acid

a.c. *(Am)* aerodynamic center; alternating current

a / c account; alter course

a–c alternating current

Ac.A. acetic acid

ACAN Army Command and Administrative Network

A.C.A.R. Australian Coal Association Research

ACC area control centre; automatic colour control

Acc. accessories

acc account

acc. acceleration; according to

accel. accelerate; accelerator

acct. account

A.C. dial alternating current dialling

ACE automatic computing equipment; altimeter control equipment; Assistant Chief Engineer

A.C.E. alcohol, chloroform, ether; Association of Consulting Engineers; automatic computing equipment

ace absorption of conversion electrons
Ac Em actinium emanation, actinon
ACESA Australian Commonwealth Engineering Standards Association
acet. acetone
acet.ac. acetic acid
AcH. acetaldehyde
A.C.I. Alloy Castings Institute; American Concrete Institute; Assistant Chief Inspector
acid. acidified
ack. acknowledged
ACM Association for Computing Machinery
A.C.M.E. Association of Consulting Management Engineers
ACN automatic celestial navigation
ACNA analogue computer for net adjustment
ACO Air Congo
acous. acoustics; acoustic; acoustical
ACR aerodrome control radar; approach control radar
A.C. rel. alternating-current relay
ACRMA Air Conditioning & Refrigerating Machinery Association of Engineers
ACS Alaska Communications System
ACS, A.C.S. American Chemical Society
ACSM American Congress on Surveying and Mapping
ACSR aluminium cable, steel reinforced
ACT air-cooled triode
act., act.- active
ACTH, Acth adrenocorticotropic hormone
ACV air cushion vehicle
AC VTVM alternating current vacuum tube voltmeter
ACW alternating continuous wave
AD amplifier detector
AD, Ad aerodrome
A.D. air dried
a.d. average depth
a / d after date
ADA Aluminium Development Association
A.D.C. Ardeer Double Cartridge
adc aerodrome control
addl. additional
addn. addition
ADE audible Doppler enhancer
ADF, A.D.F. automatic direction finder
ADM, adm aerodrome
ADPT adapter
ADR advisory route
A.D.T.S. air-dried and tub-sized
ADW air defence warning
AE assistant engineer; aviation engineer; aviation engineering
A.E. assistant engineer
ae aerial
A.E.A. Atomic Energy Authority
AEC American Engineering Council
A.E.C. Atomic Energy Commission
AEI Association of Electrical Industries
aer. aerodrome; aerodynamics; aeronautics; aeroplane; aerial
AERE, A.E.R.E. Atomic Energy Research Establishment
aernl. aeronautical
AERO aeronautics
aerp. aeroplane
AES American Electrochemical Society; American Electroplaters' Society; Audio Engineering Society
AEW airborne early warning
AF Admiral of the Fleet; Air Force
A.F. audio frequency; advanced freight
af audio frequency; aureal frequency
a.f. ash-free
AFA American Foundrymen's Association; Air Force Academy
A.F.A. audio-frequency amplifier
A.F.A.S. Associate of the Faculty of Architects & Surveyors
AFB Air Force Base
AFBMA Anti-friction Bearings Manufacturers' Association
A.F.C. automatic frequency control; armoured flexible conveyor
A.F.C.E. automatic flight control equipment
Afld airfield
A.F.M. amplitude-frequency modulation
AFMTC Air Force Missile Test Center
AFN American Forces Network
AFS aeronautical fixed service
AFT aeronautical fixed telecommunication; audio-frequency transformer
AFTN aeronautical fixed telecommunication network
AG American gage; air gunner; assault gun; automatic gun
A.G. American gallon
Ag *(symbol)* silver
ag agent
ag. agglomerating
a-g air to ground
AGA American Gas Association
AGARD Advisory Group for Aeronautical Research and Development
A.G.C. automatic gain control
AGCA automatic ground control approach
AGCL automatic ground controlled landing
agg. aggregate
AGI American Geological Institute
agl airborne gun-laying radar
a.g.l. above ground level
aglt airborne gun-laying turret
AGMA American Gear Manufacturers' Association
AGS automatic gain stabilization

ags airborne radar gunsight
agt agent
AGU American Geophysical Union
A.H. air hardened; ampere hour; after hatch; access hole
Ah, ah ampere hour
ah'd ahead
a.h.m. ampere-hour meter
AHq Army Headquarters
AI airborne interceptor; aircraft interception
A.I. Air India; all-iron; artificial insemination
AIA airborne interception of aircraft; American Institute of Aeronautics
AIA, A.I.A. Aircraft Industries Association
AICBM anti-intercontinental ballistic missile
AICE, A.I.Ch.E. American Institute of Chemical Engineers
A.I.D. Aeronautical Inspection Directorate
AIEE American Institute of Electrical Engineers
AIHA American Industrial Hygiene Association
AIL Airborne Instruments Laboratory; amber indicating lamp
A.I.M. aerotriangulation by the observation of independent models
AIME American Institute of Mining Engineers
AIMME American Institute of Mining & Metallurgical Engineers
A.I.O.A. American Iron Ore Association
AIP Aeronautical Information Publications
A.I.P. American Institute of Physics
AIR All India Radio
AIRE American Institute of Radio Engineers
Airtps airborne troops
AIS Aeronautical Information Service
A.I.S.C. American Institute of Steel Construction
AISE Association of Iron & Steel Engineers
AISI American Iron and Steel Institute
AJ, aj anti-jamming
AL amber glow lamp
Al *(symbol)* aluminium
A. lab analytical laboratory
ALC automatic locking circuit
alc. alcohol
alc., alcoh. alcoholic
ald. aldehyde
ALGOL algorithmic language
Alk alkyl
alk. alkali[ne]
ALM absorption limiting frequency
A.L.S. Associate of the Linnean Society
ALT alternator
Alt., alt. altitude
alt. alteration
Alt.Lt. alternating light
AM amplitude; amplitude modulation; ammeter
A.M. Associate Member; amplitude modulation
Am *(Symbol)* americium; amyl
a.m. ante meridiem; amplitude modulated; amplitude modulation; above mentioned
a-m amplitude modulation; amplitude modulated
AMA automatic message accounting; Automobile Manufacturers' Association
amb ambient
AMC, A.M.C. American Mining Congress
AMC, AmC armoured car
A.M.C.C.W. Association of Manufacturers of Chilled Car Wheels
Am. Chem. Soc. American Chemical Society
A.M.E.M.E. Association of Mining Electrical and Mechanical Engineers
A.M.I.A.E. Associate Member of the Institution of Automobile Engineers
A.M.I.C.E. Associate Member of the Institution of Civil Engineers
A.M.I.E.E. Associate Member of the Institution of Electrical Engineers
A.M.I.M.E. Associate Member of the Institution of Mining Engineers
A.M.I.Mech.E. Associate Member of the Institution of Mechanical Engineers
AMM anti-missile missile
A.M.M. American Metal Market
ammon. ammonia; ammoniacal
amorph. amorphous
amp ampere[s]
Am. Perfumer Ess. Oil Rev. American Perfumer and Essential Oil Review
amp.hr., amp-hr ampere-hour
AMPL, amplr., AMPR amplifier
AMS Aeronautical Materials Specification
A.M.S.E. Associate Member of the Society of Engineers
amsl above mean sea level
amt. amount
amtank amphibious tank
Amtec Ampex-time-element-compensator
AMTI airborne moving-target indicator
A.M.T.P.I. Associate Member of the Town Planning Institute
amu atomic mass unit
AN air navigation; Army-Navy specification
An normal atmosphere
anal. analysis
Analytic.Chem. analytical chemistry
ANB Air Navigation Bureau
ANC American National Coarse *(thread)*; Air Navigation Commission
ANF American National Fine *(thread)*; American National Form
ANFO, AN/FO ammonium nitrate/fuel oil
anhyd., anhydr. anhydrous
A.N.L. automatic noise limiter
Anl. analyst
anmt announcement
ann. annealed

Ann. Conf. annual conference
Ann. Conv. annual convention
Ann. Meet. annual meeting
ANRAC aids to navigation radio control
ant. antenna
antilog antilogarithm
A. / O. account of
a.o. and others
AOB aircraft observer
A.O.I.V. automatically operated inlet value
A.O.P. artillery observations plåne
AOQ average outgoing quality
AOQL average outgoing quality limit
AOSO advanced orbiting solar observatory
A.O.V. automatically operated valve
AP analytical plotter; axial (linear) pitch; access panel; armour-piercing; automatic pistol; aviation pilot; aircraft park; airplane; airport
A.P. author's proof; armour piercing; Amsterdam level; arithmetical progression; American patent; aft peak; aft perpendicular; average payable; additional premium
A / P account paid
ap approach [lights]
ap. apothecaries
APA American Pharmaceutical Association; axial pressure angle
A.P.B.X. automatic private branch exchange
APC analytical plotter civilian model; approach control
A.P.C. automatic phase control; automatic picture control
APCA Air Pollution Control Association
APEC all purpose electronic computer
a.ph.c. automatic phase control
API, A.P.I. American Petroleum Institute
A.P.I. air position indicator
APIC American Petroleum Industries Committee
APL average picture level; airplane
A.P.L.E. Association of Public Lighting Engineers
A.P,M.A. Annealing Potmakers' Association
APO Army Post Office
APOTA automatic positioning of telemetering antenna
App. appendix
app. apparatus
Appar. apparent, apparently
APPITA Australian Pulp and Paper Industries Technical Association
appl. applied
approx. approximate[ly]
APR airborne profile recorder
APS American Peat Society; armour-piercing shell; accessory power supply
APT system automatic picture transmission system
A.P.U. auxiliary power unit
APV armoured personnel vehicle
A.Q. air quenched
aq. aqueous
AQL acceptable quality level
aqu. aqueous
AR arrester; air reconnaissance; aspect ratio
A.R. analytical reagent; all risks *(marine insurance)*
Ar *(symbol)* Argon; aryl
ar amplifier; effective area of antenna; arrester
a / r at the rate of
A.R.A. Associate of the Royal Academy; Aerial Ropeways Association; airborne radar attachment
ARB Air Research Bureau
ARBA American Railway Bridge & Building Association
ARC Aeronautical Research Council *(GB)*
ARC, A.R.C. automatic remote control; automatic range control
ArC armoured car
Arc automatic relay computer
Archip. archipelago
A.R.C.S. Associate of the Royal College of Science
arcsin a arc the sine of which is a
A.R.D. Armament Research Department *(Woolwich)*
A.R.E. Aeronautical Research Establishment
AREA American Railroad Engineers' Association, American Railway Engineering Association
ARI airborne radio installation; airborne radio instrument
ARIA Apollo range instrumented aircraft
A.R.I.B.A. Associate of the Royal Institute of British Architects
ARL Aircraft Radio Laboratory
ARM armature
ARO airborne range only
arom. aromatic
ARR aircraft radio regulations
ARRL American Radio Relay League
ARS American Rocket Society
A.R.S. aircraft radiosight
ARSB anchored radio sound buoy
ART automatic range tracking; additional reference transmission
ARTA American Radio Telegraphist Association
ARTCC Air Route Traffic Control Center
ARU automatic range unit
AS armature shunt; air speed; automatic sprinkler
A.S. air-seasoned *(timber)*; Academy of Sciences
A / S antisubmarine; alongside
A-S altostratus
As *(symbol)* arsenic

as armature shunt
as. asymmetric
as- asymmetrically substituted
ASA, A.S.A. American Standards Association; American Statistical Association
ASAC automatic selection of any channel
ASAE American Society of Aeronautical Engineers
ASB American Society of Biochemists; asymmetric side-band
asb. asbestos
ASC automatic selectivity control
A.S.C. automatic sensitivity control; armoured snaking conveyor; American Society of Cinematographers
ASCC automatic sequence controlled calculator; American small centre cap
ASCE American Society of Civil Engineers
ASDE airport surface detection equipment
A.S.D.I.C. anti-submarine depth indicating control
ASE, A.S.E. Amalgamated Society of Engineers; Associate of the Society of Engineers; American Society of Automotive Engineers
a.s.e. air standard efficiency
A.S.E.E. Association of Supervising Electrical Engineers
ASFA American Steel Foundrymen's Association
ASG aeronautical standards group
ASH & VE American Society of Heating & Ventilating Engineers
ASI air-speed indicator
ASL atomic safety line
A.S.L. Anderton shearer-loader
ASLE American Society of Lubricating Engineers
ASLEF, A.S.L.E. & F. Associated Society of Locomotive Engineers and Firemen
A.S.L.I.B. Association of Special Libraries & Information Bureaus
ASM air[-to-]surface missile; airfield surface movement indicator; American Society for Metals
ASME, A.S.M.E. American Society of Mechanical Engineers
ASNE American Society of Naval Engineers
ASP American Society of Photogrammetry
asph. asphalt
ASQC American Society for Quality Control
ASR airborne search radar; airport surveillance radar; automatic strength regulation
ASRE American Society of Refrigerating Engineers
A.S.R.S. Amalgamated Society of Railway Servants
Ass. association
ass. assistant
ASSE American Society of Safety Engineers; American Society of Sanitary Engineering
Ass'n, assocn. association
ASST American Society for Steel Treating
Asst assistant
assy assembly
AST anti-sidetone; Atlantic Standard Time
ASTE American Society of Tool Engineers
ASTM American Society for Testing Materials
ASTRO artificial satellite time and radio orbit
A-sub. atomic submarine
ASV air-to-surface vessel radar
ASW anti submarine warfare
A.S.W.D.K.W. Amalgamated Society of Wire Drawers and Kindred Workers
asym. asymmetric
AT attenuator; ampere turn[s]; air transport; antitank; army training; army troops
A.T. aerial triangulation; ampere turn[s]; antitank; air transport; automatic telephone; automatic transmitter
At *(symbol)* astatine
at. atomic
a.t. ampere turn[s]
ATA American Transit Association
ATB all trunks busy; antitank battery
A.T.B. aeration test burner
ATC air traffic control
ATC, A.T.C. automatic tone correction; automatic tuning control
a.t.c. aerial tuning capacitor
AT gn antitank gun
athw. athwarthship
a.t.i. aerial tuning inductance
ATK aviation turbine kerosene
a / tk r. antitank rifle
ATM aerial turning motor
atm, atm. atmosphere
atm.abs. atmosphere absolute
At.No., at.no. atomic number
ATO, A.T.O. assisted take-off
A.T.P. anode tapping point
ATR anti-transmit-receive
a.tr. auto-transformer
ATS aided tracking system; air traffic services; American Television Society; administrative terminal system; applications technology satellites
A.T.S. automatic telephone set; animal tub sized
att. attenuator; attention [of]
ATTC American Telephone and Telegraph Company
ATTEN attenuator; attenuation
ATU aerial tuning unit
atv air-to-vessel radar
at.vol. atomic volume
ATW aircraft tail-warning radar

At.Wt., at.wt. atomic weight
AU, Å.U. Ångstrom unit
Au *(symbol)* gold
AUD audible
aud.equip. audio equipment
aud.snl. audio signal
AUM air-to-underwater missile
AUT, auto automatic
Autoscan automatic stereo broadcast scanner
AUW all-up weight
AUX auxiliary
A.V. acid value
Av., av. average; avoirdupois
A.V.C. automatic volume control
avdp avoirdupois
AVE automatic volume expansion
av.eff. average efficiency
avg. average
AVI, AVN aviation
Av.No. average number
avoir. avoirdupois
AVR automatic volume regulation; autovoltage regulator
av.wt. avoirdupois weight
AW articles of war; atomic weapon; aircraft warning
A.W. atomic weight; all widths *(timber)*
a.w. above water
a-w atomic weight
AWA American Wire Association
AWG American Wire Gage
AWPA American Wood Preservation Association
AWS American Welding Society
awu, a.w.u. atomic weight unit
AWWA American Waterworks Association
AWY airway
AX aftercastle; quarterdeck
AZ, Az., az azimuth
azon azimuth only

B

B black *(pencil lead)*; best; British thermal unit; ballast; base; breadth; brightness; *(symbol)* boron; break; break-contact; baud; resinoid; rough; buoyancy; centre of buoyancy; breadth; beam; breadth moulded; bearing; boatswain; bale
B. breadth
b. boiling at; boiling point; barn; breadth; box
BA bending allowance
B.A. British Association; Bachelor of Arts; Buenos Aires
Ba *(symbol)* barium
BABS beam (blind) approach beacon system
B.A.C.M. British Association of Colliery Management
B. Ag[ric]. Bachelor of Agriculture
B.A.L., bal British anti-lewisite
ball. ballast
bal.net balancing network
Baltcon Baltic and International Maritime Conference
bal.tr. balancing transformer
balun balanced-to-unbalanced
B. and S., B & S Brown and Sharpe *(taper, wire gauge)*
B & W TV black-and-white television
BAR butylacetyl ricinoleate
BAS beacon airborne S-band
B.A.S. British Association Standard
BATCHG battery charging
BAU, B.A.U. British Association Unit
Baud Baudot
BAX beacon airborne X-band
BAY bayonet
BB double-black *(pencil-lead)*; best best; double rough
B.B. below bridges
bb bomb
b.b. ball bearing
BBB treble-black *(pencil lead)*
BBC British Broadcasting Corporation
B.B.D.C., b.b.d.c. before bottom dead centre
BBL beacons and blind landing
bbl barrel[s] *(31½ gallons)*
B.B.no. bromide-bromate number
BC broadcast; breaking capacity; between centres; bayonet catch; battle cruiser; bonded single cotton
B.C. bayonet cap; British Corporation
bc bare copper; base connection
b.c. between centres; broadcast
BCB broadcast band
B.C.C. body-centred cubic
b.c.c. bi-cylindro-conical *(drum)*
BCD binary coded decimal
BCG Bacillus-Calmette-Guerin
B / Ch. Bristol Channel
b.c.i. broadcast interference
B.C.I.R.A. British Cast Iron Research Association
b.c.l. broadcast listener
BCN beacon
B.coeff block coefficient
B.C.P.A. British Commonwealth Pacific Airlines
B.C.P.M.A. British Chemical Plant Manufacturers Association
B.C.S.O. British Commonwealth Scientific Office
bcstn broadcasting station
B.C.U.R.A. British Coal Utilization Research Association
BD board; base diameter

B.D. barrels daily; below day
B / D bulldozer; bar draft
Bd boundary
bd board
bd. bundle; board
b/d barrels per day
BDC bonded double cotton
BDC, B.D.C., b.d.c. bottom dead centre
bdct broadcast
BDG bridging
bdg building
BDI, B.D.I. bearing deviation indicator
b.d.i. both days inclusive
BDM bomber defence missile
BDP bonded double paper
Bds. boards *(timber)*
BDSA Business and Defence Services Administration
B.D.S.T. British double summer time
BDV breakdown voltage
B.E. binding energy
B/E bill of exchange
Be *(symbol)* beryllium
Bé Baumé [scale]
be band elimination
BEA, B.E.A. British Electricity Authority; British European Airways [Corporation]
B.E.A.I.R.A. British Electrical & Allied Industries Research Association
BEAMA, B.E.A.M.A. British Electrical and Allied Manufacturers' Association
B.E.D.A. British Electrical Development Association
BEF band-elimination filter
Bell Bell Telephone Laboratories
bemf back electromotive force
B.Eng. Bachelor of Engineering
BEPO British Experimental Pile at the Atomic Energy Research Establishment at Harwell
BeV billion electron volts [of energy]
Bev. bevel; bevelled
B extreme extreme beam; extreme breadth
BF barrage fire; battle fleet; British Forces
BF, bf beat frequency
b./f. brought forward
BFN British Forces Network
BFO, B.F.O., bfo beat frequency oscillator
BG background; Birmingham Gauge
B.G. bevel gear; blasting gelatine; Birmingham gauge
B/G bonded goods
bg bag[s]; bearing
bg. bearing; bevel gear
BGS beacon ground S-band
B.H. Brinell hardness
B/H Bordeaux-Hamburg range of ports; bill of health
BHA base helix angle
BHC benzene hexachloride
BHD. bulkhead
BHN, B.H.N., Bhn Brinell hardness number
BHP, B.H.P., bhp, b.hp., b.h.p. brake horsepower; boiler horsepower
bhp-hr brake horsepower-hour
B.I. bodily injury
Bi *(symbol)* bismuth
B.I.C.E.R.A. British Internal Combustion Engine Research Association
B.I.F. British Industries Fair
bi-m bi-monthly
BINAC binary automatic computer
biol. biological
B.I.O.S. British Intelligence Objectives Sub-Committee
BIRE British Institution of Radio Engineers
BIS, B.I.S. British Interplanetary Society
B.I.S.R.A. British Iron and Steel Research Association
bit binary digit
bitn. bitumen
BK break-in keying *(communication)*
B.-K. book-keeping
Bk *(symbol)* berkelium; bank
bk. black
Bkpr book-keeper
BKT. bracket
bkt basket[s]
BL blue lamp
B.L. boundary line; breech-loading
B/L bill of lading
bl bale[s]; black
bl. barrel; blue
BLC balance; boundary layer control
blk black
bls bales; barrels
BM beam; height of metacentre above centre of buoyancy
B.M. bending moment; ball mill; board measure; bench mark *(surveying)*; metacentric radius
B.m. board measure *(timber)*, board measurement *(timber)*
B.M.C. British Metals Corp. Ltd.
B.M.E.P., bmep, b.m.e.p. brake mean effective power (pressure)
B.mld. breadth moulded
B.M.N. British Merchant Navy
BMO Ballistic Missile Office
BN, Bn. beacon
BNEC British National Export Council
B.N.F.M.R.A. British Non-Ferrous Metals Research Association
BNL Brookhaven National Laboratory
BNRC Baltic and North Sea Radiotelephone Conference
BNTM Baltic and North Sea Telecommunication Meeting

BO blocking oscillator
B.O. B-operator; back-out; beat frequency oscillator
bo, b/o beat frequency oscillator
B.O.A. British Optical Association
BOAC British Overseas Airways Corporation
Boats. boatswain
BOB bobbin
B.O.D. biochemical oxygen demand
B.O.E. brick on edge
B of M Bureau of Mines
B.O.M.A. British Overseas Mining Association
bomst bombsight
B.O.T., B.o.T. Board of Trade
bot bought
bot. bottle
botts bottles
B.O.V. brown oil of vitriol *(77% H_2SO_4)*
BP by-pass; base pitch; battle practice; bonded single paper; British Petroleum Co.
B.P. between perpendiculars; British Pharmacopoeia; back-pressure; boiling point; British patent; band pass
B/P blue print; bills payable
bp, b.p. boiler pressure; boiling point; beam plates
b.p.d. barrels per day
B.P.F. band-pass filter; British Pacific Fleet
b.p.h. barrels per hour
bpi bits per inch
BPIT basic parameter input tape
BPO British Post Office
bpt. boiling point
BPV back-pressure valve
B.P.V. by-pass valve
bq. barque
B.R. British Railways; basic requirements; boiler room
B/R. bills receivable
Br *(symbol)* bromine
Br. brass; bridge
b.r. boiling range
BRC British Radio Communication
B.R.C. British reinforced concrete
BREMA British Radio Equipment Manufacturers' Association
B.R.F. British Road Federation
Brg., brg, brg. bearing
Brl. barrel
brn brown
bro bronze
brom. bromide
Bros, bros. brothers
B.R.R.A. British Refractories Research Association
Br.sh. brown shale
brw. brown
BS broadcast station; band setting; bonded single silk
B.S. boiler survey; balance sheet; British Standard
B/S. bill of sale; bill of store
b.s. basic sediment; bottom settlings
B.S.A.A.C. British South American Airways Corporation
BS & W basic sludge and water
B.S.B. British Standard beam
BSC, B.S.C. British Standard cycle *(thread)*; British Standard channel
B.Sc. Bachelor of Science
B.S.F. British Standard fine [screw thread]
B.S.F.A. British Steel Founders Association
BSFC brake specific fuel consumption
B.S.G. British Standard gauge
bsh. bushel
B.S.I. British Standards Institution
BSL Biochemical Society of London
B.S.M. Bureau of Steel Manufacturers
BSP, B.S.P. British Standard pipe screw thread
BSRA British Sound Recording Association
B.S.R.A. British Shipbuilding Research Association
BSS, B.S.S. British Standard specification
BST booster
B.S.T. British Summer Time
B/St bill at sight
b'st ballast
B.S.W. British Standard Whithworth screw thread
BSWG British Standard wire gauge
B.S.Whit. British Standard Whitworth screw thread
bt bitumen
b.t., b/t. berth terms
B.T.D.C., b.t.d.c. before top dead centre
B.Th.U. British thermal unit
Btm. bottom
BTO blocking tube oscillator
Btry battery
BTU, B.T.U. Board of Trade unit *(kilowatt-hour)*; British thermal unit
Btu British thermal unit; Board of Trade Unit *(kilowatt-hour)*
B.t.u. British thermal unit
Btu p.lb. °F British thermal unit per pound per degree Fahrenheit
Btu p.sq.ft.hr. °F British thermal unit per square foot per hour per degree Fahrenheit
Btu./sq.ft./min. British thermal unit per square foot per minute
Btu./sq.ft./°MTD/HR British thermal unit per square foot per degree temperature difference per hour
Bty battery
Bu butyl

Bull. Bulletin
bus. bushel; business
Bus.Mgr Business Manager
BUT button
buzz buzzer
BV balanced voltage
B.V. Bureau Veritas
B/V book value
bv balanced voltage
BVA British Radio Valve Manufacturers' Association
BW biological weapon
B.W. butt welded
B/W black and white *(television)*
bw bandwidth
b.w. butt welded; bell wire
B.W. & S. basic water and sediments
BWG Birmingham wire gauge
bwk. brickwork; bulwark
b.w.o. backward wave oscillator
BWR boiling-water reactor
B.W.R.A. British Welding Research Association
bx box
Bz benzoyl
bz buzzer
bz. benzene

C

C candle; capacitance; capacity; centrum; centre; collector; concentration; cotton; coulomb; cycle, cyclo; Celsius; *(symbol)* carbon; centigrade
C. captain; case; centigrade; cruiser
°C centigrade degree; degrees centigrade
c candle; carat; cent[i]; chapter; cis-; concentration; cubic; curie; cycle, cyclic, cyclo-; specific heat
$\mu\mu$c micromicrocurie
CA cathode; capacitor; Chemical Abstracts; coast artillery; civil aviation; controlled approach
C.A. chief accountant; Chemical Abstracts; close annealed
C/A capital account; credit account; current account
Ca *(symbol)* calcium
ca candle
CAA Central African Airways Corporation; Civil Aeronautics Administration
CAAC Civil Aviation Administration of China
CAB cellulose acetate butyrate; cabinet; captured air bubble; Civil Aeronautics Board
C.A.B. critical air blast
CABOWV captured air bubble over water vehicle
CABRA Copper and Brass Research Association
CAC current account
c.a.d. cash against documents
c.a.f. cost and freight
CAI Compressed Air Institute
c.a.i. cost and insurance
CAL calibration; Cyprus Airways Ltd.
C.A.L. concentrated ammonia liquor
cal calory
calc[d] calculated
CAMA centralized automatic message accounting
can. cancelled; cancelling date
Canad. J. Chem. Canadian Journal of Chemistry
canc. cancelled; cancelling date
cancl, canclg. cancelling
c & c coal and coke
C & D collected and delivered; communication and data
C and E commission and exchange
c. & f. cost and freight
c. & i. cost and insurance
C. & M. care and maintenance
C. & W. coal and waggon
Can. G.P. Canadian Government Purchasing Standards Committee
CAP capacitor
cap. capillary
caps capital letters
CAR Civil Air Regulations
car carrier
carb. carburetter
carr.pd carriage paid
cart. cartridge
CAS calibrated air speed
CAT College of Advanced Technology; cooled-anode transmitting valve; centralized automatic testing; clear air turbulence
Cat caterpillar
cat. catalytic
CATO catapult-assisted take-off
CATV cooled-anode transmitting valve; Community Antenna Television
caust. caustic
CAV cavity
CAX community automatic exchange
CB circuit breaker
C.B. central battery, common battery; control button; centre board; centre of buoyancy
C/B cash book
Cb *(symbol)* columbium; coulomb
Cb, C_b block coefficient
cb centibel
CBC Canadian Broadcasting Corporation
c.b.d. cash before delivery
CBE central battery exchange
CBM constant boiling mixture
CBRM cash by return mail
CBS Columbia Broadcasting System

C.B.S. central (common) battery signalling
CC combined carbon; crystal controlled; continuous current; conductive channel; carrier current; ceramic capacitor; communications centre
C.C. combined carbon; crystal controlled; continuous current; conductive channel; carrier current; ceramic capacitor; communications centre; continuation clause; chief constructor
cc cubic centimetre; continuous current; closed circuit; cotton-covered *(cable)*
cc. cubic capacity
c.c. continuous current; closed circuit; cubic centimetre; current cost; cotton-covered *(cable)*; closing coil
c-c centre to centre
CCB Combined Communications Board
ccb close control bombing
CCBP Combined Communications Board Publications
CCC carrier-current communication
CCH. industrial calcium hypochlorite
CCKW, cckw counter clockwise *(rotation)*
C.C.P. critical compression pressure; central control position
ccpm cubic centimetre per minute
C.C.R. critical compression ratio
CCS continuous commercial service; hundred calls/second
ccsr copper cable steel reinforced
ccsw copper clad steel wire
CCT, cct circuit
CCTV closed-circuit television
CCW, C.C.W. counterclockwise *(rotation)*
CD cathode of diodes; continuous duty; current density; *(Am)* center distance; civil defence; coast defence
C.D. corrected depth; current density
C/D commercial dock
Cd *(symbol)* cadmium
Cd. cored
cd candela, new candle; cord
c.d. current density
CDA Copper Development Association
CDC course and distance calculator
C.D.F. combined distribution frame
C.D.H. cable distribution head
CDI course deviation indicator
C.D.M. Consolidated Diamond Mines of South West Africa
CDO community dial office
CE combustion engine
C.E. civil engineer; chief engineer; compass error; centre of effort *(naval architecture)*
Ce *(symbol)* cerium
C.E.A. Central Electrical Authority
CEB Central Electricity Board
CEC Commonwealth Exports Council
C.E.E. Central Engineering Establishment
C.E.G.B. Central Electricity Generating Board
C.E.I. Cycle Engineers' Institute screw thread
cem. cement
C.E.M.F., cemf, c.e.m.f. counter-electromotive force
cent centigrade; century
CER ceramics
CERCA Commonwealth and Empire Conference on Radio for Civil Aviation
cert. certificate
Cert.Inv. certified invoice
CF current forcing (force); cathode follower; cost and freight
C.F. centre of flotation; carrier frequency; centrifugal force
C/F carried forward
Cf performance coefficient for fuel consumption; *(symbol)* californium
cf centrifugal force; carrier frequency; cubic foot
cf. confer, compare
c.f. cost and freight; centrifugal force; carrier frequency; cubic foot
cfh cubic feet per hour
c.f.i. cost, freight and insurance
cfm, c.f.m. cubic feet per minute
CFN Canadian Forces Network
C.F.O. calling for orders
CFR co-operative fuel research; Co-operative Fuel Research Committee
cfs, c.f.s. cubic feet per second
CG *(Am)* commercial-ground *(tap)*; centre of gravity; control grid; commanding general
C.G. centre of gravity; Consul General
Cg capacitance of grid; grid capacitance
cg centigram; centre of gravity
cg. coring; centigram
c.g. centre of gravity
Cgf grid-filament capacitance
CGH commercial-ground high *(tap)*
cgo. cargo
Cgp grid-plate capacitance
CGS, C.G.S. centimetre-gramme-second
C.G.S. coast guard ship
c.g.s. centimetre-gramme-second
cgse electrostatic system
cgsm electromagnetic system
CH choke; compass heading
C.H. case hardened; custom house
Ch, Ch. chamfer[ed]
ch candle hour; chapter
ch. chain
ch., CHAN, Chan. channel
chap chapter
charac. characteristic
CHC choke coil; cyclohexylamine carbonate
chem. chemical

Chem. Engng. Chemical Engineering *(New York)*
Cheq. chequered
chl. chloroform
ch.opt. charterer's option
CHR condenser heat rejection
CHU, c.h.u. centigrade heat unit *(pound calorie, 453.6 cal)*
CI cold-iron soldered joint; crystal impedance; chief inspector
C.I. cast iron; compression ignition; consular invoice
Ci Curie; admiralty performance coefficient; cirrus
c.i. cost and insurance
c/i., c./i. certificate of insurance
Ci-Cu cirro-cumulus
cim *(Am)* cubic inches per minute
C.I.N.A. Coal Industry Nationalization Act *(of 1946)*
CIP cast-iron pipe
ciphony enciphered telephony
CIPRA Cast Iron Pipe Research Association
cir.bkr circuit breaker
circ. circumference
cir mil circular mils
Ci-S cirro-stratus
CIT coordinates input tape
CIU computer interface unit
c.j. cold junction
CKW clockwise
CL centre line
C.L. car-load lots; centre line
C/L craft loss
C_L centre line
Cl *(symbol)* chlorine
cl, cl. centilitre; claim; class; clause; clearance; cylinder
c.l. car load
CLA centre line average
class. classification
cld. cleared
clg. cancelling
clkw. clockwise
CLR Co-ordinating Lubricant and Equipment Research
C.L.R. centre of lateral resistance
C.L.T. code language telegram
Cm *(symbol)* curium; midship area coefficient
cm, cm. centimetre
CMA contact making ammeter
C.M.B. coastal motor-boat
CMC carboxymethylcellulose; contact-making clock
cmn communication
CMP combination multiclone-precipitator [unit]; compression amplifier
C.M.P. controlled materials plan; central monitoring position
CMR common mode rejection; continuous maximum rating
cmtd.csg. cemented casing
CMV contact making voltmeter
CN, C.N. cetane number; co-ordination number
C/N, c/n consignment note; cover note; credit note
Cn *(symbol)* centurium
C.N.B. Dutch Standardization Association
cnd conduit
CNL cancellation message
C.N.S.L. cashew nutshell liquid
cntr. container
CO classifier overflow
CO, C.O. cut-out; crystal oscillator; cleaning out
C/O certificate of origin; cash order
Co *(symbol)* cobalt
Co. company
co. course
c/o care of; certificate of origin
COAX coaxial cable
C.O.B. cleaning, opening, blowing *(textiles)*
C.O.C., C.o.C. Chamber of Commerce
COD, C.O.D., c.o.d. cash on delivery; collection on delivery; collect on delivery
CODAN carrier operated device, antinoise
coef, coef., coeff. coefficient
C. of G., COG centre of gravity
COHO coherent oscillator
col. column; colour
co-L. co-latitude *(navigation)*
coll. colloidal; collective
COLLR collector
coln. coloration
colog *(Am)* cologarithm of
COM communications
Com. commission
com. commercial
comb. combine
combd. combined
combn. combination
comd command
COMECON Council for Mutual Economic Aid
COMM communication
Comm. committee
comm. commercial
compd. compound
comp.g. compressed gas
compn. composition
compt. compartment
Comsat communication satellite
CON connecting
conc. concentration; concentrate
concd. concentrated
concn. concentration

cond conductivity
cond. condensance; condenser; conductivity; conductor; condition[s]
condens. condensation
CONDX conditions for DX communication
Conf. conference
CONN connect[ing]
Cons.M.S. Consolidated Mines Selection Comp.
const[.] constant
cons't consignment
constit. constituent
Constr. construction
CONT, cont. contact
cont. control; contain; containing; continent [of Europe]
CONT. B.H., Cont. B.H. Bordeaux-Hamburg range of continental ports
contd. continued
contg. containing
CONT.H.H., Cont. H.H. Le Havre-Hamburg range of continental ports
cont hp continental horsepower
CONTL control
Contr contractor
contr. control
conts contains
CONV converter
Conv. convention; conversion
conv. convert; conveyor
convn. conversion
Co-op. Co-operative Society
C.O.P. coefficient of performance
cop. copper
c.o.p. customs of port
cor. corrected
Corp. corporation
corr. corrected; corrugated
corresp. corresponding
C.O.S. cash on shipment
cos cosine
cosec cosecant
cosh hyperbolic cosine
cot cotangent
coul coulomb
C.O.V. concentrated oil of vitriol
covers coversed sine [of]
cox. coxswain
Coy Company
CP computer; coupling; *(Am)* circular pitch
C.P. constant potential; candle power; chemically pure; circular pitch; change point; casing pressure; cement through perforations; conveying panel; charter party
C/P custom of the port; charter party
Cp centipoise; [at] constant pressure; plate capacitance; constant potential; *(symbol)* cassiopeium
C_p prismatic coefficient
cp coupling; constant potential; candle power; *(Am)* chemically pure
cp. counterpoise; compare
c.p. chemically pure; candle power, centre of pressure; circular pitch
CPA colour phase alternation
CPAL Canadian Pacific Airlines Ltd.
CPC card-programmed calculator; computer process control
Cpc plate-cathode capacitance
C.p.cycle constant pressure cycle
CPD contact potential difference
C.P.D. charterer pays dues
cpd. compound
c.p.d. charterer pays dues
Cpf plate-filament capacity
c.p.f.f. cost plus fixed fee
CPI carbon preference index
CPLG coupling
CPM critical path method
cpm, c.p.m. cycles per minute; calls per minute
c.p.m. cathode pulse modulation
C.P.P. coal preparation plant
c.p.p., c.p.propeller controllable pitch propeller
C.P.R. Canadian Pacific Railway
Cpr. copper
CPS cathode-potential stabilized; cycles per second
cps, c.p.s. cycles per second
Cpt., cpt. captain
CQ commercial quality; general inquiry call
C.Q. call to quarters
CQD, c.q.d. come quick, danger!
CR control relay; Columbia resin
C.R. carrier's risk; company's risk; current rate; cold rolled; cathode ray
Cr crystal; *(symbol)* chromium
Cr. credit; creditor
cr carrier's risk; cathode ray
cr. cruiser; cutter
c.r. current rate; compression ratio
CRA Canadian Research Association
CRC Coordinating Research Council, Inc.
C.R.C.A. cold rolled, close annealed
CRcn counterreconnaissance
Crd cored
CRDF cathode-ray direction finding
C.R.E. Coal Research Establishment
CREI Capitol Radio Engineering Institute
crit critical mass [of untampered reactor]
crit. critical
C.R.O. cathode-ray oscilloscope
C Rock Rockwell C hardness
C.R.P. Central Recovery Plant
CRPL Central Radio Propagation Laboratory
c.r.p.m. crankshaft revolutions per minute
CRS cold rolled steel; chain-radar system
Crs creditors

crs coarse
CRST cold rolled steel
CRT, C.R. tube, crt, c.r.t. cathode-ray tube
cryst. crystalline; crystallized
crystd. crystallized
crystn. crystallization
CS constant stress
C.S. *(Am)* commercial standard; cast steel
C/S colliery screened; cycles per second; chief of staff
Cs communications; performance coefficient for shaft horsepower; *(symbol)* cesium
C's hydrocarbons
cS centistoke
cs cotton/silk
c.s. cross section
c/s cycles per second; case
CSA Canadian Standards Association
csc *(Am)* cosecant
csch hyperbolic cosecant
Csg, csg casing
C'sink countersink
C.S.I.R. Council for Scientific and Industrial Research
C.S.I.R.O. Commonwealth Scientific and Industrial Research Organisation *(Australia)*
Csk countersink
CSL. Chemical Society of London
C.S.M. Camborne School of Mines
CSP control switching point
CSp chief specialist
CSR constant stress rate
CSSB compatible single-sideband
cSt, c.st. centistoke
CT centre-tap; current transformer; control transformer; carrier telephony; conductivity transmitter; conduit
C.T. cable transfer; chronometer time; cooling transformation
C/T conference terms
Ct total capacitance
ct, ct. carat; circuit; current; centre tap
c/t connecting tube
C.T.C. centralized traffic control
C.T. diagram cooling transformation diagram
CTG cartridge
ctg cotangent
CTI *(Am)* Color Television Incorporated
C.T.L., c.t.l. constructive total loss
C.T.L.O. constructive total loss only
ctn cotangent
C.T.O. Central Telegraph Office
C. to C., c to c centre to centre
CTR counter
C.T.R.A. Coal Tar Research Association
ctrs centres
C.T.S. controlled thermal severity *(welding test)*; Consolidated Tin Smelters
c.t.s. cable cable with cabtyre-sheathing
CTU, C.T.U., c.t.u. centigrade thermal unit
CTV command (control) test vehicle; colour television
ctwt counterweight
cty capacity
CU crystal unit
C.U. Cambridge University
Cu *(symbol)* copper; cumulus
cu crystal unit
cu, cu., cub. cubic
C.U.C. Coal Utilization Council
cu.cm. cubic centimetre
cu.dm. cubic decimetre
CuFt., cu.ft. cubic foot, cubic feet
cu.in. cubic inch
cu.in./cutting hr cubic inches per cutting hour
cu.in./min.h.p., cu.in. per min. per h.p. cubic inches per minute per horse power
cu.in. per rev. cubic inches per revolution
cu m, cu.m. cubic metre
cum. cumulative
C.U.M.M. Council of Underground Mining Machinery Manufacturers
Cu mm, cu.mm. cubic millimetre
cu mu, cu.mu. cubic micron
Cu-N cumulo-nimbus
CUR., cur. current
cur. currency
CURT current
cu rt cubic root
cusec, cu.-sec. cubic feet per second
cu yd, cu.yd. cubic yard
CV coulomb volt; constant volume; single cotton varnished
C.V. calorific value; cheval vapeur *(metric hp)*
c.v. constant volume
c.v.cycle constant volume cycle
CW carrier wave; code word; continuous wave
C/W cement-water ratio
C-W. chronometer watch
cw continuous wave
cw. clockwise
c.w. continuous wave
CWO carrier wave oscillator
c.w.o. cash with order
CWP circulating water pump
CWS centre wireless station
CWT carrier-wave telegraphy; carrier-wave transmission
cwt. hundredweight *(50.8 kg)*
CX control transmitter
cx convex
CXR carrier
cy cycle; capacity
cy. copy; cycle
c.y. cubic yard
cyl, cyl. cylinder; cylindrical

D

D deca; density; distortion; displacement; diameter; dissipation factor; dimensional; *(symbol)* deuterium; deflection plate; pitch diameter; declination; deadweight; destroyer; difference; depth moulded
D- dextro *(in configurational sense)*
d deci-; density; depth; diameter; differential; deuteron, deuton; duty cycle; delivered; depth of hold; displacement; draught moulded
d. diameter; penny, pence; day
d- dextro[rotatory]
DA double armoured; dummy aerial
D.A. delayed action; depth average
D/A documents against acceptance
D/A, D/a days after acceptance
d.a. de-asphalt
d/a days after acceptance
DAB, D.A.D. documents against discretion of collecting bank
d.a.f. dry ash-free
dag. decagram
D.A.L. Diffusion Alloys Ltd.
dal decalitre
dam decametre
D. & A. dry and abandoned
D and P [mass-production] development and printing
D. and T. double and twist
DAP. diffused-alloy power *(Transistor)*
D.A.S. delivered alongside ship
das decastere
DAVC, D.A.V.C. delayed automatic volume control
D.B. dry bulb; double bottom; deals and battens
db. decibel
d.b. double bottom; deals and battens
DBA dense blasting agent
D.B. and B. deals, battens, and boards
DBB detector balanced bias
D.B.B., d.b.b. deals, battens, and boards
DBD double base diode
DBLR doubler
DBP dibutylphthalate
D.B. pull draw bar pull
dbRN decibel referred to the reference noise
D.B.S.T. double British summer time
DBWP double-braided weatherproof
DC direct current; diode cathode; double contact; directional coupler; double cotton
D.C. deviation clause; direct current; downcast
D/C deviation clause
d.c. direct current; double current; dot cycle; double column; double crown; double cap; direct coupled
dc/ac direct current / alternating current
d.c.c. double cotton covered
DCCC double current cable code
DCDS double-cotton double-silk covered
D.C.F. deal-cased frame
DCG Doppler-controlled gain
DCLSP digging cellar and slush pits
DCPV direct current peak voltage
DCSS double-cotton single-silk
DCU decade counter unit
DCV double cotton varnished
DCWV direct current working voltage
DD destroyer; duplex-drive
D.D. diamond-drill; drilling deeper
D/D. delivered at docks; demand draft
D./D. delivered at destination
D/d delivered
D./d. days after date
dd. delivered
d/d delivered dock
DDA, d.d.a. digital differential analyzer
D.D. & Shpg. dock dues and shipping
DDC direct digital control
DDM difference depth of modulation
d.d.o. dispatch discharging only
DDP double diode-pentode
D.D.T. dichlorodiphenyltrichloroethane *(insecticide)*
ddt. deduct
DE deemphasis
D.E. Diesel-electric; degree of elasticity
Dec. declination
dec., decim. decimetre
decomp. decompose; decomposition
decompd. decomposed
decompn. decomposition
DED Diesel engine driven
def *(Am)* definition
DEFL deflecting
deg, deg. degree
deg.cent. degree centigrade
DEL'D delayed
dely and r-dely delivery and re-delivery
DEM demodulator
den. density
Dep. departure
depolym. depolymerize
depolymn. depolymerization
Dept., dept. Department
deriv. derivative
derivd. derived
desic. desiccator
Det., det. detector
detectn. detection
DEUCE digital electronic universal computing engine
DEV, Dev. deviation
DEW distant early warning

DF direction finding; double feeder; distortion factor; distribution frame; day of fire; defence; defensive fire
DF, D/F., D.F. direction finder
D.F., d.f. dead freight
d.f. diversity factor
D factor dissipation factor
DFC double frequency changing
df/dt derivation of frequency with respect to time
DFM distortion factor meter
DFO decade frequency oscillator
DFq day frequency
DF Stn direction-finding station
DG double glass insulated
dg, d.g. degree; decigram
DGC double glass insulated
dgm. decigram
DGS double green silk
DH directly heated
D.H. double-hung
dh, d.h. directly heated
DHD down-the-hole drill; double-heatsink-diode
DHP designed horse-power
DI demand indicator
D.I. Diesel index; direct ignition; District Inspector
dia, dia. diameter
DIAD drum information assembler and dispatcher
diam, diam. diameter; diametral
DIAPH diaphragm
DICBM detection intercontinental ballistic missile system
d.i.d. dielectric inner diameter
DIEL dielectric
DIFF differential
Diff. difference
diff. differential
dil. dilute
diltd. diluted
diltg. diluting
diltn. dilution
din dinuclear
Dina direct noise amplifier
DIO diode
DIR directional; director
dis. dissolve
DISC disconnect
Disc. discontinued
DISCH discharge
disct. discount
dislvd. dissolved
displ. displacement
dissocd. dissociated
dissocg. dissociating
dissocn. dissociation
DIST, Dist. distance
dist. distillable; distillate
dist., distd. distilled
distg. distilling
distn. distillation
DISTR distribute
distribn. distribution
Dk derrick
dk deca, deka
dk. deck; dock; dark
dkg dekagramme
dkl dekalitre
dkm dekametre
DL dead load; racemic
D.L. dead load; difference of latitude; difference of longitude
dl, dl. decilitre
d.l. dead load
dl- racemic
DLCO decade LC oscillator
D.L.O., d.l.o. dispatch loading only
dlt deck landing trials
D.L.W.L. designed load waterline
D.M. disconnecting manhole
dm. decimetre
DME, D.M.E. distance measuring equipment
DMH drop manhole
d.mld. depth moulded
D.M.O. Division Medical Officer
D.M.P. dimethyl phthalate
DMPR damper
d.m.r. digital meter reader
DN deci-neper
D/N debit note
Dn. dolphin
DNA deoxyribonucleic acid
DNF Dominion Naval Forces
DNL do not load
DNT dinitrotoluene
D.O. dissolved oxygen; duty officer; Diesel oil
D/O delivery order
Do Diesel oil
D.o. drawing office
d.o.d. dielectric outer diameter
Dodar determination of direction and range
D.O.F. direct-on-face *(machine)*
DON distribution octane number
DOP dioctyl phthalate
DOR daily operation report
DOS dioctyl sebacate
doub. doubler
doz. dozen
D.P. diametral pitch; double-pole; double paper; diode plate; dial pulse; direct port
D/P documents against payment
dP differential pressure
dp double pole
d.p. direct port; duty paid; diametral pitch
d/p documents against payment

DPC double paper single cotton; double paper covered
DPDC double paper double cotton
D.P.H. diamond pyramid hardness
dplx duplex
D.P.N. diamond pyramid hardness number
DPR double pure rubber
DPS, DPSS double-pole snap switch
d.p.s.t. switch double-pole single-throw switch
DPSW double pole switch
D.P.T. directed practical training
DR dead reckoning; drill rod
D.R. differential relay; distant reading; dead reckoning
Dr dispatch rider; drum
dr debtor; dram
DRAD drill adapter
DRBG drill bushing
DRC distant reading compass
DRFX drill fixture
drg. drawing
d.r.g. double-reduction gear
Drlct. derelict
DRSH drilling shell
DS directing station; double-silk
D.S. directing station; distant surveillance; data sheet
D/S double screened
D/S, d/s days [after] sight
DSB, D.S.B. double sideband
D.Sc. Doctor of Science
d.s.c. double-silk-covered *(cable)*
DSCC double-silk-cotton covered *(cable)*
D.S.I.R. Department of Scientific and Industrial Research
DSL deep scattering layer
DSRC double sideband reduced carrier
DSSC double sideband suppressed carrier; double-silk single-cotton covered *(cable)*
DSV double-silk varnish *(cable)*
DT discharge tube; double-throw switch
D.T. dialling tone
Dt. draught
DTA differential thermal analysis
D.T.B. delayed time base
D.T.D. Directorate of Technical Development
D.T.I. dial test indicator; distortion transmission impairment
DT-logic diode-transistor-logic
DTM duration time modulation
DTMB David Taylor model basin
Dt.m[ld]. draught moulded
dtng detuning
D.T. reaction deuterium-tritium reaction
dup. duplicate
Dura. duration
DV double vibration
D.V. direct voltage
dv double vibration
DVM digital voltmeter
D.W. deadweight
D.W., D/W dock warrant
d.w. deadweight
DWA double wire amoured *(cable)*
d.w. all told deadweight all told
d.w.c. deadweight capacity
D.W.F. daily water flow
d.w.f. dry-weather-flow
dwg. drawing
D.W.L. designed waterline
DWS double white silk *(cable)*
d.w.s. double wound silk
dwt deadweight
dwt. pennyweight
DX duplex; distance
DY deflection yoke
Dy *(symbol)* dysprosium
dy delivery
dy. dynode
dyn[am] dynamics
DYNM dynamotor

E

E east; Egyptian; earth; elasticity; shellac bond; electromotive force; electric potential; electric voltage; engineer; [Young's] modulus of elasticity
e east; electric field strength; electric intensity; electric potential; electromotive force; electron charge; emitter; [electric] voltage; efficiency
EA enemy area; electrostatic analyzer
Ea accelerating voltage; earth
EAAC East African Airways Corporation
EAL Eastern Air Lines Inc.
EAN. effective atomic number
E & OE, E. & O.E., e.a.o.e. errors and omissions excepted
E.A.O.N. except as otherwise noted
EAP *(Am)* equivalent air pressure
EAR electronic aural responder
EARC Extraordinary Administration Radio Conference
EAS extended area service; electronic automatic switch
E.A.S. equivalent air speed
eas. easily
EAT expected approach time
EAX electronic automatic exchange
E.B. Encyclopaedia Britannica
EBC enamel bonded single cotton
EBDC enamel bonded double cotton
EBDS enamel bonded double silk
E-B junction emitter-base junction
E.b.N. east by north

EBP enamel bonded single paper
EBS enamel bonded single silk
E.b.S. east by south
EBU European Broadcasting Union
EC emergency call; enamelled copper; enamel single cotton
E.C. east coast; emergency call
Ec electronic computer
ec electron-coupled; enamel covered
ECAFE [United Nations] Economic Commission for Asia and Far East
ECC electron-coupled control
E.C.C.P. east coast coal port
ECE Export Council for Europe
E.C.E. [United Nations] Economic Commission for Europe
E.C.G. electrocardiogram
ECGD Export Credits Guarantee Department
Ecl. eclipse
ECME electronic circuit making equipment
E.C.O. electron-coupled oscillator
econ economics
E.C.P.D. Engineers Council for Professional Development
ECSA European Communications Security Agency
E.C.S.A. East Coast South America
E.C.S.C. European Coal and Steel Community
ED electronic device; editor
E.D. existence doubtful
Ed, ed editor
EDA Electrical Development Association
EDC enamelled double-cotton
EDCD ethylene dichloride derivative
EDCV enamel double-cotton varnished
EDM electric discharge machining
EDP, edp electronic data processing
EDR equivalent direct radiation
EDS enamelled double-silk covered; every day stress
EDSAC electronic delay storage automatic calculator
EDSV enamel double-silk varnished
EDT ethylene diamine tartrate
EDVAC electronic discrete variable automatic calculator
EE electrical engineer
E.E. electrical engineer
E.E., e.e. errors excepted
EEA Electronic Engineering Association
EEG electro-encephalogram
EEI Edison Electric Institute
EF effective fire; expeditionary force
eff, eff. efficiency
effl. efflorescent
EG enamel single glass
e.g. exempli gratia, for instance, for example
EG & FA Eastern Gas and Fuel Association
EH extra high; extremely high
EHF electrohydraulic forming
E.H.F., ehf extremely high frequency
EHP, E.H.P., ehp, e.h.p. effective horsepower
E.H.P.h. effective horse-power hour
EHT, eht extremely (extra) high tension
EHV, ehv extra high voltage
E_i exponential integral
EID electronics intelligence digest
E.I.S. electric induction steel
el. elevation; elongation
ELBUG electronic key
Elec, elec electric
ELECOM electronic computer
ELECT electrician; electrolyte
elect., electr. electricity
elem.ch. elementary charge
Elev. elevation
elev, elev. elevated; elevation; elevator
E.long. east longitude
elong *(Am)* elongation
Ely. easterly
EM electromagnetic; emanation
E.M. electromagnetic; mining engineer
em, e.m. mechanical efficiency
em. emanation
e/m ratio of the electron charge to the mass of an electron
emb. embargo
EMC eight-meter-community
EMF, E.M.F., emf, e.m.f. electromotive force
emf-series electromotive series
EMN electromagnetic moving coil and neutralized winding
E.M.T. electric metallic tubing
EMU, emu electromagnetic unit; electromagnetic system of cgs units
EMW, emw electromagnetic wave
en ethylenediamine
ENAM enamelled
encl. enclosed; enclosure
end. endurance
ENE east-northeast
eng, eng. engine
engg. engng engineering
engr, engr. engineer
ENI equivalent noise input
Eniac electronic numeral integrator and computer
E.N.S.I. equivalent noise-sideband input
Entr. entrance
EO education officer; engineer officer
E.O. equivalent orifice; extra over price
E.O.M. end of month following date of sale
EOTS electronic optical tracking system
EP estimated position; extreme pressure; extended play *(record)*
E.P. electroplate; English Patent; estimated position

EPC easy processing channel
EPD earliest possible date
EPI expanded position indicator; electronic position indicator
E.P.L., e.p.l. extreme pressure lubricant
E.P.N.S. electroplated nickel silver
EPR equivalent parallel resistance
e.p.s. electrical power storage
EPUT events per unit time
EQ equalizer
Eq., eq, eq., eqn. equation
Eq.s. equivalent to sheathed [explosive]
Eq.T. equation of time
equil. equilibrium
equiv. equivalent
ER engine relay
E.R. engine room
Er *(symbol)* erbium
E.R.A. engine room artificer; Electrical and Allied Industries Research Association
Erd. eastward
ERMI electronic-radio manufacturing industry
Ern. eastern
ERP effective radiated power
E.R.W. electric resistance welding
ES enamel single silk; Electrochemical Society
E.S. echo sounding
Es *(symbol)* einsteinium
e.s. earth switch
ESE east-southeast
esp. especial[ly]
Esq. esquire
ESR equivalent series resistance; effective signal radiated
Est. established
estn. estimation
ESU, E.S.U., esu, e.s.u. electrostatic unit
ESV enamel single silk varnished *(cable)*; earth satellite vehicle
ET electric telegraph
Et, et ethyl [radical]
ETA., E.T.A. expected to arrive; estimated time of arrival
e.t.a. expected to arrive
et.ac. ethyl acetate
e.t.d. embedded temperature detector
ETE External Telecommunications Executive
ETF engine test facility
eth. ether
E. to E. end to end
EtOH ethyl alcohol
E.T.S. Electrodepositors' Technical Society; expected to sail; estimated time of sailing
e.t.s. expected to sail
ETV educational television
E.U. energy unit
Eu *(symbol)* europium
E.U.E. external upset ends
EV error voltage; electron volt; escort vessel
ev, e.v. electron volt[s]
evap *(Am)* evaporate; evaporation
evapn. evaporation
evoln. evolution
EW early warning; electronic war
E.W.T. elsewhere taken
Ex. excluding; examined; exchange; executed
ex. excluding; excess
Exam. examined
examn. examination
ex. & ct. excavate and cart
EXC excitation
excl. exclusive, excluding
Exd. examined
Exh. exhibited; exhibiting
EXHY extra heavy
exp *(Am)* exponential function
exp. export; express; explode
Exper. experimental
expos. exposure
expt. experiment; experimental
exptl. experimental
exsec *(Am)* exterior secant
ex.sur.tr. & ct. excavate surface trenches and cart away
ext *(Am)* external
ext. extract
ext.bat. extension battery
extd. extracted
ext.freq. extension frequency
extn. extraction
Extr. extreme
ext.T-phone extension telephone

F

F Fahrenheit; farad; faraday [constant]; free energy; magnetomotive force; face; flat; fuse; filament; fine *(pencil-lead)*; luminous flux; fighter; fire; flagship; force[s]; *(symbol)* fluorine; centre of flotation
F. fuel; flagship; fellow
°F degree Fahrenheit
μμF micro-micro-farad
f foot; farad; frequency; filament; fugacity; *(Am)* finish drawing; free
f. modulus of rupture
FA field artillery
f.a. free alongside
F.A.A., f.a.a. free of all average
fab. fabricate
FAC Frequency Allocation Centre
f.a.c. fast as can *(loading or discharging)*
F.A.C.E.M. Federation of Associations of Colliery Equipment Manufacturers
facs[im]. facsimile

Fahr Fahrenheit
F.A.I. fresh air inlet
F. and D., F. & D. freight and demurrage
F.A.O. *(Am)* finish all over
f.a.q. fair average quality; free alongside quay
F.A.S., f.a.s. free alongside ship
FAST fully automatic sorting and testing
fath. fathom
FAX facsimile
F.B. freight bill
F/B fighter-bomber
FBA, F.B.A. Fellow of the British Academy
fbd. freeboard
F.B.I. Federation of British Industries
FBL friction braked landing
FBM fleet ballistic missile
fbm *(Am)* feet board measure
F.B.O.A. Fellow of the British Optical Association
FBP, F.B.P., f.b.p. final boiling point
fbr. fibre
F.B.T. fuel ballast tank
FC fire control; furnace control; frequency changer
F.C. furnace cooled
F/C flight control
f.c. frequency changer; fixed carbon; foot candle
FCB Frequency Control Board
FCC [United States Government's] Federal Communications Commission
F.C.C. face-centred cubic
f'cle forecastle
FCS flight control system
FCT filament centre tap
FD frequency divider
F.D. forced draught; frequency doubler
f.D. free [delivery to the] docks
fd farad
fd. framed
f.d. free discharge, free dispatch; free dock; free delivery
Fd Ldg forced landing
F.D.M. frequency-division modulation
f.d.p. free dispatch
FDS Fermi-Dirac-Sommerfeld
FDU, f.d.u. frequency divider unit
Fe *(symbol)* iron
feet/min feet per minute
F.E.F. Foundry Educational Foundation
F.E.S. feather-edged springer
FET field effect transistor
F.f.a., f.f.a. free from alongside
F.F. & W. furnish fuel and water
FFAR folding fin aircraft rocket
F.F.A.S. Fellow of the Faculty of Architects & Surveyors
FFC free from chlorine
f.f.i. free from infection
FFLT familiarization flight
FFRR full frequency range record; full frequency range records system
FG field gun
F.G. fine grain
F.G.A. foreign general average; free of general average
F.G.S. Fellow of the Geological Society
F.H. fire hydrant
f.h. fore hatch
F.H.P., fhp, f.h.p. friction horsepower; fractional horsepower
F.I.A., f.i.a. full interest admitted
FIAT Field Information Agency, Technical Report
f.i.b. free into bunker, free into barge
FIC flight information centre
fig. figure
figs. figures
FIL filament
Fil. filter
fil. filament
Fil. Hd. *(Am)* fillister head
filn. filtration
fil.sup. filament supply
filt. filtrate
F.I.O., f.i.o. free in and out
f.i.o.s. free in and out and stowed
FIR flight information region
FIS flight information service
F.I.U. frequency identification unit
f.i.w. free in wag[g]on
FL filter
fl, fl. fluid; flake
fl. florin
FLBE filter for band elimination
FLBII high pass filter
FLBP band pass filter
FLHP high-pass filter
FLLP low pass filter
Fl.Lt. flashing light
floc. flocculent
flot flo[a]tation
FLOX fluorine plus liquid oxygen
fl.ozs. fluid ounces
fl.pt. flash point
F.Lt. fixed light
F/Lt flight lieutenant
fluores. fluorescent
FM fulminate of mercury
FM, F.M. frequency modulation
Fm *(symbol)* fermium
fm. fathom; fireman
f.m., f-m *(Am)* frequency modulation
F.M.E.P., f.m.e.p. friction mean effective pressure
f./min. feet per minute
FMQ frequency-modulated quartz oscillator

fms. fathoms *(timber)*
FNA final approach
fnp *(Am)* fusion point
FO, F.O. fuel oil
F.O. for orders
F/O flying officer; fuel-to-oxidizer
f.o. for orders; firm offer; firm order
F.O.B., f.o.b. free on board
f.o.c. free of charge[s]; free on car
fo'c'sle, foc's'le forecastle
f.o.d. free of damage
Fog Sig. fog signal
Fog W./T. fog wireless telegraph
FONE telephony
F.o.q., f.o.q. free on quay
f.o.r. free on rail
forams foraminifera
f.o.r.t. full out rye terms
FORTRAN formula translation
F.O.S. factor of safety
f.o.s. free on ship
FOT, F.O.T. frequency of optimum traffic
f.o.t. free on truck
F.O.W., f.o.w. first open water
f.o.w. free on wag[g]on
FP flight position
FP, F.P. fission product; forward perpendicular; freezing point
F.P. floating policy; fore perpendicular
fp, f.p. freezing point; fire-proof; freight paid (prepaid)
F.P.A., f.p.a. free of particular average
F.P.A.A.C. free of particular average, American conditions
F.P.A.E.C. free of particular average, English conditions
F.P.I.L. full premium if lost
fpm, f.p.m. feet per minute
fps *(Am)* feet per second; foot-pound-second
f.p.s. foot-pound-second
f.p.sec. feet per second
F.P.S. system foot-pound-second system
F.P.T. fore peak tank
Fq-cy frequency
FR frequency meter; fast-release relay
F/R. freight release
fr. frame
fract. fractionate
fractn. fractionation
fractnl. fractional
F.R.A.S. Fellow of the Royal Astronomical Society
FRED figure reading electronic device
FREQ frequency
FREQM, freq.m. frequency meter
F.R.I.B.A. Fellow of the Royal Institute of British Architects
f.r.o.f. fire risk on freight
Fr.P. French patent
F.R.S. Fellow of the Royal Society
F.R.S.S. Fellow of the Royal Statistical Society
Frt., frt. freight
Frt.fwd. freight forward
Frt.ppd. freight prepaid
frz. freezes
FS frequency shift[ing]
F.S. *(Am)* Federal Specifications; factor of safety; field switch; fog signal; forged steel
fs factor of safety, safety factor
fs. facsimile
f/s, f/s. factor of safety, safety factor
FSB *(Am)* Federal Specifications Board
FSD, f.s.d. full scale deflection
F.S.E. Fellow of the Society of Engineers
FSG frequency of signal generator
FSI free-swelling indices
FSK frequency shift keying
FSR flight safety research
FSS flying spot scanner; field sequential system
FT flight test; flame thrower
Ft. fittings
F.t., f.t. full terms
ft foot; feet
FTC fast time control; Federal Trade Commission
ft-c foot-candle
Ft.Hd. feet head *(of liqui[illegible]*
ft-L foot-Lambert
ft-lb foot-pound
FTM frequency time modulation
Ft./min. feet per minute
FU fuse
fumg. fuming
fund.freq. fundamental frequency
fur, fur. furlong
fus. fuse; melt; fuselage
fuslg. fuselage
fusn. fusion
FV fighting vehicle
FW *(Am)* face width
F.W. full-wave rectifier; fresh water
f.w.b. four-wheel brake
FWD forward
F.W.D., f.w.d. fresh water damage; front-wheel drive
F.W.T. fresh water tank
F.X. foreign exchange

G

G gravitational constant; [centre of] gravity; giga; generator; grid; gain; general staff; general staff officer; gun
g gram[me]; gaseous; acceleration due to gravity
g. gram[me]

g′ grid of section 1
g″ grid of section 2
g_1 control grid
g_2 screen grid
g_3 suppressor grid
G.A. grate area
G.A., G/A general average
Ga *(symbol)* gallium
g.a. general average
g/a gliding angle
G.A. and S. general average and salvage
G/A Con. general average contribution
G/A Dep. general average deposit
gal gallon
galls./min. gallons per minute
GALV galvanometer
galv. galvanized
GAM guided aircraft missile
GAMA Gas Appliances Manufacturers' Association
GAPA ground-to-air pilotless aircraft
GAR guided aircraft rocket
G.A.T. Greenwich apparent time
GB grid bias; gun boat
G.B. Great Britain
gb gilbert
g.b. grid bias; gun boat
Gbo gumbo
GC gain control; grounded collector; gun cotton
G.C. general cargo
gc grounded collector
GCA ground controlled approach
g.cal., g-cal gram[me]-calorie
GCD, gcd greatest common divisor
G.C.E. General Certificate of Education
GCE(A) General Certificate of Education (Advanced Level)
GCE(O) General Certificate of Education (Ordinary Level)
G.C.F., gcf greatest common factor
GCI ground-controlled interception
GCL ground-controlled landing
G.C.M. greatest common measure
GCP gain control pulse
Gc/s gigacycles/second
G.C.T. Greenwich civil time
Gd *(symbol)* gadolinium
g.d. gravimetric density
Gd.A. gauged arch
GDES Government Department Electrical Specification
GDF gas dynamic facility; group distribution frame
gds. goods
GE grounded emitter
G.E. geological engineer
Ge *(symbol)* germanium
G.E.C. General Electric Company
gel. gelatinous
GEM ground effect machine
GEN generator
Ger.pat. German patent
G.F. government form
g-factor Landé g-factor, interval factor
GGS gyro gun sight
G.H.A. Greenwich hour angle
GI ground interception
G.I. gas injection
gi gill *($\frac{1}{4}$ pint)*
Gibfo Gibraltar for orders
G.L. German Lloyd; ground level
gl. glass
glac. glacial
GLEEP Graphite Low Energy Experimental Pile *(at the Atomic Energy Research Establishment at Harwell)*
glob. globular
glyc. glycerin[e]
GM metacentric height; mutual conductance; General Manager; guided missile
G.M. good middling; General Motors; transverse metacentric height; gold medal
gm. gram[me]
g_m mutual conductance
G.M.counter Geiger-Muller counter
GMT, G.M.T. Greenwich mean time
G.M.V. gram[me]-molecular volume
G.M.W. gram[me]-molecular weight
GND, gnd ground
gns. guineas
GO gas-oil
G.O good ordinary
gold. golden
GOR gas-oil ratio
Govt. government
govtel government telegram
GP gopher protection
G.P. German patent; Grand Prix
Gp gas pressure
g.p. gauge pressure
g.p.d. gallons per day
gpf grain per foot
Gp.Fl.L. group flashing light
g.p.h. gallons per hour
G.P.I. ground position indicator
g.p.l. grams per litre
GPM groups per minute
G.P.M., g.p.m. gallons per minute; gallons per mile
G.P.O. General Post Office
Gp.Occ.L. group occulting light
GPR general purpose radar
gps., g.p.s. gallons per second
g.p.t. gallons per ton
GPV general-purpose vehicle
GR gas ratio [of ionization]

Gr. *(Am)* grind
gr. grain; grey
grad. gradient; gradually; graduate; graduated
gran. granular; granulated
GRD ground
GRE ground radar equipment
grm. gram[me]
grn. green
GR-S *(Am)* styrene-type government rubber
G.R.T. gross register tonnage
g.r.t. gross register tons
g.r.v. graphic recording voltmeter
g.r.w. graphic recording wattmeter
Gr.wt., gr.wt. gross weight
GS ground surface; general staff
G.S. grain size; geological survey
gs guineas
G.S.A. Geological Society of America; General Services Administration
G.S.C. Geological Survey of Canada
G.S.T. Greenwich sidereal time
g.s.w. galvanized steel wire ropes
GT glass tube
g.t. gross tonnage
GTC gas turbine compressor
GTPU gas turbine power unit
GTV, Gt.V. gate valve
guar. guarantee; guaranteed
G.V.W. gross vehicle weight
GWL ground water level
GZ ground zero

H

H hour; henry; *(symbol)* hydrogen; horizontal force; hardness; heat content; heater; hoy *(small vessel)*; headquarters; helicopter
H. harbour; helicopter; hardness; meridian altitude
h hour; henry; hard *(pencil lead)*; hardness; heat content; height; total heat; head; heater; hecto; hexode; heptode; horizontal force
h. hour; altitude; hardness; *(ship's log)* hail
HA *(Am)* helix angle; heavy artillery; high altitude
H.A. Highway Act; hour angle
H/A Le Havre-Antwerp range of ports
ha, ha. hectare
HAA, H.A.A. heavy anti-aircraft [artillery]
h. & c. hot and cold [water]
H. and M. hull and machinery; hull and materials
h. & t. hardened and tempered
HASP high-altitude sounding projectile
HASR high-altitude sounding rocket
h_{ave} height average
HB hard black *(pencil-lead)*
Hb. haemoglobin
H.B.S. herringbone strutting
HC high capacity; heavy current
H.C. Highway Code
h.c. heat coil
h/c hold covered
hcb. hydrocarbon
H.C.F. highest common factor
H.C.P. hexagonal close packed
HD heavy duty; diamond hardness; harmonic distortion
H.D. hourly difference
H/D heavy duty; Le Havre-Dunkirk range of ports
H_D Vickers diamond hardness
Hd, hd head
HDF horizontal distributing frame; high frequency direction finder
hdkf. handkerchief
hdn. [to] harden
HE heavy enamel
H.E. height of eye; high explosive
He *(symbol)* helium
HEC hydroxyethylcellulose
hectog. hectogram[me]
hectol. hectolitre
hectom. hectometre
HED unit skin dose
herf high energy rate forming
HETP height equivalent to a theoretical plate
HETS height equivalent to a theoretical stage
HEW [Department of] Health, Education, and Welfare
Hex., hex, hexag. hexagonal
HF, H.F. high frequency
Hf *(symbol)* hafnium
hf, h-f high frequency
hf bd half bound
HFC high frequency current; high frequency choke
hf cf half-calf
HFDF high frequency direction finding
HFI high-frequency input [power]
H.F.O. hole full of oil
HFRDF, H.F.R.D.F. high frequency repeater distribution frame; high frequency radio direction finding
hfs. hyperfine structure
H.F.W. hole full of water
H.G. hand-ginned
Hg *(symbol)* mercury
hg. hectogram[me]
HGR head gear receiver
HH double hard *(pencil lead)*
H/H Le Havre-Hamburg range of ports
hhd hogshead *(63 gal)*
h.h.d.w. heavy handy deadweight scrap
HHH treble hard *(pencil lead)*
hhv *(Am)* high heat value

HI Hydraulic Institute
H.I. horizontal interval
H.I.B. hoop-iron bond
HIFAM high-fidelity amplitude modulation
Hi-Fi high fidelity
hi.imp. high impedance
Hi.Q. high quality
hiran high precision shoran
HiVo high voltage
h.j. hot junction
H_l Ludwig hardness
hl heavy loaded; hectolitre
hl. hectolitre
HLTTL high level transistor transistor logic
hm hectometre
h_{max} height maximum
H.M.D. hydraulic mean depth
HMG heavy machine gun
H.M.S. heavy media separation; Her (His) Majesty's Ship
HMT hand micro telephone
HN horn
HNC Higher National Certificate
HND hand [operation]; Higher National Diploma
H.N.W. head, nut, and washer
H.O. hydrographic office
Ho *(symbol)* holmium
hor., horiz. horizon
horl horizontal
how howitzer
HP high power; horsepower
H.P. horsepower; high-pressure; hot-pressed *(paper)*; hire purchase; horizontal parallax
hp horsepower; hire purchase; half pay; high pass
h-p high-pressure
hpa Hewlett Packard Associates
HPC hard-processing channel
HPF high-pass filter
HP-G high pressure gelatine
hp.hr., hp-hr horsepower-hour
hp per cu in. per min. horsepower per cubic inch per minute
H.Q., Hq headquarters
HR Rockwell hardness
H.R. hot rolled; high resistance
hr hour; highly resistive
hr. hour
HRB Rockwell B hardness
HRC Rockwell C hardness; high rupturing capacity *(fuses)*
Hrd. and Gr. harden and grind
h_{rms} height root mean square
HRS hydrant refuelling system
hrs. hours
HS handset; hospital ship
H.S. harbour service; hospital ship; heating surface; horizontal stripes
h.s. high-speed

HSR harbour surveillance radar
HSS high-speed steel
H.T. high tide; high tensile; high tension; heat-treated
ht high-tension
ht, ht. height; heat
h.t. high tension; high torque; heat-treated
h-t heat-treated
HTB high-tension battery
htd. heated
HTDC high-tension direct current
htg. heating
HTP high-test peroxide
HTR heater
H.T.S. high tensile steel; high tension separation
HTU height of a transfer unit
HTV hypersonic test vehicle
H.U.C. highest useful compression
H.U.C.R. highest useful compression ratio
H.V., hv, h.v. high voltage
HVAR high velocity aircraft rocket
hvdc high-voltage direct current
H.V.I. high-viscosity index
HVL, H.V.L. half-value layer
HVT, H.V.T. half-value thickness
HW howler
H.W. high water
H.W.F. and C high water full and change
HW-level high water level
H.W.M. high water mark
H.W.O.S.T. high water ordinary spring tides
HWR heavy-water reactor
hy henry
Hyd. hydrographic
hyd. hydrous; hydrated
hydrol. hydrolysis
hydrold. hydrolyzed
hydrolg. hydrolyzing
hydx. hydroxide
hygr[os]. hygroscopic
hyp.log. hyperbolic logarithm
Hz hertz

I

I intensity; inertia; indicator; interphone; interference; Imperial; *(symbol)* iodine; island, isle
i [current] intensity
i. inactive; insoluble
i- iso-; optically inactive
IA infra-audible
IA, I.A. international Ångstrom; international agreement
I_a anode current
IAAF International Amateur Athletic Federation
IAC Indian Airlines Corporation

IACS *(Am)* International Annealed Copper Standard
I.A.D.B. Inter-American Development Bank
IAE Institute of Aeronautical Engineers
IAEA International Atomic Energy Agency
IAF International Aeronautical Federation
IAGC instantaneous automatic gain control
IARU International Amateur Radio Union
IAS indicated air speed; Institute of Aeronautical Sciences
I.A.S. International Association of Sedimentologists
I.A.T.A. International Air Transport Association
IAVC instantaneous automatic volume control
IB incendiary bomb
I.B. inner bottom; in board
I.b. in bond
i/B in bulk
ib. ibidem
IBA International Bowling Association
IBF International Badminton Federation
ibid. ibidem
IBM International Business Machines
IBn identification beacon
$\mathbf{I}_{BO}$ breakover current
I.B.P. initial boiling point
IBU International Boxing Union
IC interior communication; internal connection; integrated circuit
I.C. internal combustion; standard thickness of 0.0213 inches; iron-constantan; index correction
i/c in charge
ICA ignition control additive
ICAO International Civil Aviation Organization
ICBM intercontinental ballistic missile
ICC International Computation Centre; Imperial Communication Committee
ICCF International Correspondence Chess Federation
I.C.D. initial chlorine demand
ICE *(Am)* Institute of Civil Engineers
I.C.E. Institution of Chemical Engineers
ICF International Canoe Federation
ICGM intercontinental guided missile
ICI Imperial Chemical Industries Ltd.
I.C.I. International Commission on Illumination
ICM intercontinental missile
I.C.P. international candle power
I.C.R.P. International Commission on Radiological Protection
ICS interphone control station
ICSC International Committee of Silent Chess
ICSU International Council of Scientific Unions
I.C.T. *(Am)* International Critical Tables
I.C.W. interrupted continuous wave
ID *(Am)* internal (inside) diameter; intermittent duty
I.D. inside diameter; induced draught
i.d. inside diameter
IDC instantaneous deviation control
I.D.C. Industrial Development Corporation of South Africa, Ltd.
ident. identical; identity
identif. identify
identifd. identified
identifn. identification
I.D.F. intermediate distribution frame; indicating direction finder
ID's interior diameters
I.E. index error
I.E.C. International Electrotechnical Commission
I.E.E. Institution of Electrical Engineers
IES Illuminating Engineering Society
I.E.S.S. Institution of Engineers and Shipbuilders in Scotland
IF image frequency
I.F., if, i.f., i-f intermediate frequency
$\mathbf{i}_F$ forward current
IFAC International Federation of Automatic Control
IFF identification of friend or foe
IFL induction field locator
IFR in-flight refuelling; instrument flight rules
IFRB International Frequency Registration Board
I.F.S. independent front suspension
IFT intermediate frequency transformer
i.f.t. intermediate frequency transformation
IFWHA International Federation of Women's Hockey Associations
I.G. Imperial gallon; inert gas welding
$\mathbf{i}_G$ gate current
i.g. involute gear
I.gal. Imperial gallon
$\mathbf{I}_{GF}$ gate firing current
IGN ignition
ign. ignites
ig.p. ignition point
IGT Institute of Gas Technology
IGY international geophysical year
IH indirectly heated
$\mathbf{I}_h$ heater current
ih indirectly heated
I.H.P., ihp, i.h.p. indicated horsepower
ihp-hr indicated horsepower-hour
IIHF International Ice Hockey Federation
IJF International Judo Federation
IJJU intentional-jitter jamming unit
IK Inkrom process
$\mathbf{I}_{k(max)}$ maximum cathode current
IL indicating lamp
I/L Import Licence
Il *(symbol)* illinium
ILA International Language for Aviation

ILAS instrument landing approach system
IL/C irrevocable letter of credit
ILS, i.l.s. instrument landing system
IM interceptor missile; intermodulation
Im ideal modulation
I.M.C. International Minerals and Chemical Corporation Ltd.
ImC intensity millicurie; Sievert unit
IME Institute of Mechanical Engineers
I.M.E. Institution of Mining Engineers
I.M.E.P., imep, i.m.e.p. indicated mean effective pressure
IMF internal magnetic focus
IMk identification mark
IMM Institution of Mining and Metallurgy
IMP impulse
Imp *(Am)* Imperial
imp. impulse
IMPACT implementation, planning, and control technique
IMPC International Mineral Processing Congress
Imp. gal. Imperial gallon
impt. important
I.M.S. area of immersed midship section
IMT impulse-modulated telemetering system
IN input; ampere-turn
In *(symbol)* indium
in, in. inch[es]
INA international normal atmosphere; Indian National Airways
I.N.A. international normal atmosphere; Institute of Naval Architects
inact. inactive
Inc. *(Am)* incorporated; Incorporated Company
inc. incorporated
incl. inclusive; including
inc.l. incoming line
IND indicator
Ind. industry
indef. indefinite
Ind.Engng.Chem. industrial and engineering chemistry
Indic. indicate-
Ind MD Indu. .rial Minerals Division
Ind.Min. industrial minerals
ind.reg. induction regulator
inel. inelastic
inf. infra; infinitive; information
INIT initial
in-lb inch-pound
in^3/min/HP $inches^3$ per minute per horsepower
inorg. inorganic
in.per min. inch[es] per minute
in.per rev. inch[es] per revolution
in per sec. inch[es] per second
INPT input
in/rev. inch[es] per revolution
INS insulated; insulation
Ins. insurance
ins. inches
$in.^3$/sec. cubic inch[es] per second
insol. insoluble
insoly. insolubility
in.-sq. square inch[es]
Inst instantaneous; instrument
inst. instant
inst., instr. instrument
in/stroke inch[es] per stroke
INT interior, internal; intersected; integrating
Int. international; interest
int *(Am)* internal; interior
int.comb. internal combustion
intercom. intercommunication
INTLK interlock
INV inverse
Inv. invoice
Invis. invisible
I.O. international octal base
Io inching operation; *(symbol)* ionium
I.O.C. International Olympic Committee
I.O.E. intake opposite exhaust
I.O.U. I owe you
$I_{out(max)}$ maximum D.C. output current
I.O.W. Institute of Welding
IP induced polarization [method]; Institute of Petroleum
I.P. induced polarization; initial production; intermediate pressure
i.p. instalment paid
i-p *(Am)* intermediate pressure
IPA isopropyl alcohol; intermediate power amplifier
i.p.a. image power amplifier
IPBM interplanetary ballistic missile
IPC information processing centre
IPCEA Insulated Power Cable Engineers Association
i.p.f. intaken piled fathom
IPM, Ipm inches penetration per month; interruptions per minute
ipm, i.p.m. inch[es] per minute
IPS interruptions per second; inch[es] per second; internal plate screen
I.P.S. iron pipe size
ips, ips., i.p.s. inch[es] per second
I.P.T. Institute of Petroleum Technologists
ipt, i.p.t. inch[es] per tooth
IPY, Ipy inch[es] penetration per year *(corrosion rate)*
IQ instrument quality *(measuring accuracy)*
IR infrared; instantaneous relay; internal resistance; interrogator/responder; insoluble residue
I.R. infrared; insoluble residue
Ir *(symbol)* iridium
i_R reverse current

i.r. infrared
Iraser infrared amplification by stimulated emission of radiation *(laser)*
IRBM intermediate-range ballistic missile
IRC International Radiological Congress
IRE Institute of Radio Engineers
irreg. irregular
i.r.s. independent rear suspension
IRSC International Radium Standard Commission
IRT interrogator/responder transponder unit
IS internal shield
I.S. Indian summer freeboard mark
I_S sustaining current
ISA international standard atmosphere; Instrument Society of America; International Federation of the National Standardizing Associations
ISB, i.s.b. independent sideband
Iscor South African Iron and Steel Industrial Corporation
I.S.I. impact strength index
Isl. island
ISO International Standardization Organization
iso. isotropic
isom. isomer; isomerize
isomd. isomerized
isomn. isomerization
ISONE International Standard of Nuclear Electronics
isoth *(Am)* isothermal
ISP International Society for Photogrammetry
ISTM *(Am)* International Society for Testing Materials
ISU International Skating Union
ISWG Imperial Standard Wire Gauge
IT infantry tank; instrument trainer
I.T. isomeric transition
I.T. [curve] isothermal transformation [curve]
it italic [type]
ITA Independent Television Authority
ital italic [type]
I.T.C. International Tin Council
I.T.E. Institute of Traffic Engineers
I.T.M. inch trim moment
ITT International Telephone and Telegraph Corporation
ITTF International Table Tennis Federation
ITU International Telecommunication Union
ITV industrial television
IU interference unit
I.U. international unit
IUC International Union of Crystallography
I.U.C. International Union of Chemistry
IUGG International Union of Geodesy and Geophysics
I.U.P.A.C. International Union of Pure and Applied Chemistry
I.U.P.A.P. International Union of Pure and Applied Physics
I.V. iodine value; insurance value
I/V increased value *(marine insurance)*
i.v. invoice value
I.W. isotopic weight
I/W ratio impulse-weight ratio
I.W.S. Industrial Welfare Society
IYRU International Yacht Racing Union

J

J joule; junction; jack; polar moment of inertia
j joule; jumper
J/A, j.a. joint account
JAL Japan Airlines Company Ltd.
J.Am.Chem.Soc. Journal of the American Chemical Society
JAN [Specification] *(Am)* Joint Army-Navy [Specification]
J. and W.O. jettison and washing overboard
J.Appl.Chem. Journal of Applied Chemistry
JATO jet-assisted take-off
JB junction box
J.Biol.Chem. Journal of Biological Chemistry
JC jack connection
JCEC Joint Communications and Electronics Committee
J.Chem.Soc. Journal of the Chemical Society *(London)*
J.Chromatogr. Journal of Chromatography
JCT junction
JETEC Joint Electron Tube Engineering Council
J.F., J/F, J/f journal folio
J.Indian Chem.Soc. Journal of the Indian Chemical Society
JK jack
jn junction
j.n.d. just noticeable difference
JO jogging operation
J.Org.Chem. Journal of Organic Chemistry
Jour. journal
JP jet pilot: jute protection *(cable)*
jp jet-propelled
J.Pharmac.Sci. Journal of the Pharmaceutical Sciences
J.Physic. Chem. Journal of Physical Chemistry
JPL jet propulsion laboratory
JT joint; junction transistor

K

K Kelvin; *(symbol)* potassium; carat; cathode; key; keel; knot; speed of vessel in knots
°K degree Kelvin

k kilo
KAPL Knolls Atomic Power Laboratory
kB kilo British thermal units
kc kilo-curie *(1000 curies)*; kilocycle
kc. kilocycle
kcal French calorie, kilocalorie, large calorie
K.C.P.S., kcps kilocycles per second
KC/s kilocycles per second
kc/s kilocycles per second
K.E. kinetic energy
kev kilo-electron-volt
kg, kg. kilogram[me]
kg-cal. kilogram[me]-calorie
kgf kilogram[me] force
kgm, kg-m kilogram[me]-metre
kg per cu m kilogrammes per cubic metre
kgps kilogrammes per second
kg.p.sq.cm. kilogrammes per square centimetre
kg.p.sq.in. kilogrammes per square inch
kg-wt. kilogram[me] weight
kip kilopound
kl kilolitre
KM kilomega
km, km. kilometre
km/hr, km.p.h. kilometres per hour
kmps kilometres per second
kn knot
K.O. knock out
k.p.h. kilometres per hour
kp/s kiloperiods per second
Kr *(symbol)* krypton
kt, kt. knot
kv kilovolt
KVA, kva kilovoltampere
kvah kilovoltampere-hour
kvar kilovar *(reactive kilovolt ampere)*
kw kilowatt
kwh[r] kilowatt-hour
kX kilo-X-unit

L

L learner; lead *(gear)*; level; load; length; lime; line; line telegraphy; latent heat per mole; luminance; lambert; left-cut; single acetate covered
L. latitude
L- l[a]evo
l lumen; lightning *(ship's log book)*; line; length; level; load; latent heat per gram[me]
l, l. litre
l- l[a]evo [rotatory]
LA light artillery; *(Am)* lead angle
L.A. low altitude; lightning arrester
L/A Lloyd's Agent
La *(symbol)* lanthanum
LAA light anti-aircraft artillery
LAB low-altitude bombing; laboratory
lab laboratory
LAM laminated
L.A.N. local apparent noon
LANAC laminar navigation and anti-collision system
L. & D. loss and damage
LAQ lacquer
LASER, laser light amplification by stimulated emission of radiation
LASL Los Alamos Scientific Laboratory
L.A.T. local apparent time
lat, lat. latitude
lat. lateral; latent
lat.ht. latent heat
L.B. local battery; lifeboat
L/B ratio of length to breadth
lb, lb. pound
L.B.A. local battery apparatus
lb.C.H.U. pound centigrade heat unit
lbf pound-force
lb-ft. pound-foot
lb/ft.^{2} pounds per square foot
lb.in. pound-inch
lb./in.^{2} pounds per square inch
L.B.M.C.T.X. local battery magneto call telephone exchange
L.B.P. length between perpendiculars
lb. per cu.in. pounds per cubic inch
lb.p.gal. pounds per gallon
L.B.S. local battery system; lifeboat station
lbs, lbs. pounds
lbs per si, lbs. per sq.in. pounds per square inch
L.B.T.S. local battery telephone set
L.B.T.X. local battery telephone exchange
LC level crossing; light-case; lower case; line construction; light current; lead covered; single acetate single cotton
L.C. light cruiser; London clause
L.C., L/C letter of credit
L.C.L., l.c.l. less than carload
L.C.M., lcm *(Am)* lowest (least) common multiple
l.cr. letter of credit
l.c.r. low compression ratio
L.C.T. local civil time
LCZR localizer
LD leak detector; Linz-Donawitz; lethal dose
L.D. long-distance; lethal dose; landing distance; load draught
Ld. limited; lead
L.D.C., l.d.c. lower dead centre
L.D.D. later drilled deeper
ldg landing
ldg. loading
Ldg Gr landing gear
Ldg Pl. landing place

L.D.L. landing direction light
LDO light Diesel oil
LE light equipment; leading edge
L.E. leading edge
LEM lunar excursion module
LESS least cost estimating and scheduling
LET linear energy transfer
LF low frequency; line finder
L.F., lf low frequency
lf load factor
LFC low-frequency choke
LFDF low-frequency direction finder
LG leg
L.G. landing gear; landing ground
lg large; length; logarithm
lgr. ligroin
LH, L.H. left-hand
L.H. latent heat; lighthouse
L.H.A.R. London–Hamburg–Antwerp–Rotterdam *(navigation)*
l-hr lumen-hour
l.h.s. left-hand side
Li *(symbol)* lithium
lim. limit value [of an expression]
lin *(Am)* linear
lin.ft *(Am)* linear foot (feet)
liq. liquid
lit litre
Lkg. and Bkg., lkg. & bkg. leakage and breakage
L.L. live load
ll lightly loaded *(coil)*
L.L.C. Load Line Committee
LLR load limiting resistor; line of least resistance
LLT landline teletype
L.M. low middling
lm lumen
lmb. lumber
L.M.C. Lloyd's machinery certificate
L.M.E. London Metal Exchange
LMG light machine gun
L.M.S. level measuring set
L.M.T. local mean time
LMTD logarithmic mean temperature difference
ln *(Am)* Napierian (natural) logarithm
LNG liquefied natural gas
LO local oscillator
LO_2 liquid oxygen
L.O. lubricating oil
L.O., lo. longitude
L.O.A. length overall
Loc localizer
log [common] logarithm
log-dec. logarithmic decrement
Loktal locked octal tube
long. longitude
longl., long'l, longt. longitudinal
LOP line of position; lookout post
L.O.P. lookout post
LOR low frequency omni range
Lorac long-range accuracy radar system
loran long-range navigational system
LOX liquid oxygen
LOZ liquid ozone
LP low power; long playing; low pass; long play record
L.P. *(Am)* liquid petroleum; low pressure
l.p. low pass
l-p, l.-p. low-pressure
LPG, L.P.G., Lp.-gas liquefied petroleum gas
L.P.G. tanker liquefied petroleum gas tanker
L.P.T.B. London Passenger Transport Board
lpw *(Am)* lumen per watt
lq. liquid
LR level recorder; load ratio; low-resistive; low-resistance; low resistor
L.R. line relay; liquid rocket; Lloyd's Register
lr low-resistive
LRC level recorder controller; load ratio control
LRN long range navigation
LS limit switch; lobe switching; loudspeaker; low speed
L.S. light ship; long stick *(36½ in lap)*
l.s. lump sum, lumpsum
L.S.A. life-saving apparatus
L.S.A. certificate life-saving certificate
LSB lower sideband
Ls/C lines of communications
LSD least significant digit
L.S.S. life saving station
L.S.T. local sidereal time
LT light; line telegraphy; level transmitter; low tension
L.T. local time
L/T line telegraphy; line telephony
lt low tension
lt, l.t. long ton
LTB last trunk busy
L.T.B. low-tension battery
Ltd limited
ltg lighterage
LtHo, Lt.-H. lighthouse
l.tn. long ton
ltng arr. lightning arrester
LTPD lot tolerance percent defective
ltu long ton unit
Lt.-V. light vessel
L.T.V.-process long tube vertical multiple effect distillation process
l.t.w. low tension winding
Lu *(symbol)* lutecium
lub. lubrication
luboil lubricating oil
LUF lowest usable frequency

LUHF lowest usable high frequency
Lun.Int. lunitidal interval
LV low voltage
L.V. light vessel
LVDT linear variable differential transformer
L.V.I. low-viscosity index
LVR low voltage release
lvs. leaves
L.W. low water
Lw *(symbol)* lawrencium
lw, l.w. long wave
l.w. lap welding
L.W.L. load water-line
LW-level low-water level
L.W.M. low-water mark
L.W.O.S.T. low-water ordinary spring tide
lx lux

M

M thousand; metal; medium; microphone; molecular weight; moment; moment of force; moment of resistance; mutual inductance; metallizing; molar; mega; metacentre; minesweeper; mail; Mach number
M. moment; mean; motor; molar [concentration]; magnet; mass; mechanic[al]; meridian; mega; middle; mile; modulation; module; movement; mutual
M- molar
m mile; metre; minute; meridian; melts; melting; mist *(log book)*; million
m. mass; mean; mile; metre; magnet; manual; mark; measure; medium; mine; minimum; minute; moment
m- meta-; meso-
mμ millimicron; millimicro-
μm micron
MA milliammeter; medium artillery; military academy
Ma *(symbol)* masurium
mA, ma milliampere
MAC maximum allowable concentration; machine-sided cognition; multiple-access computer
Mach.Sc. machine screw
macks mast stack
MAC ship merchant aircraft carrier
Madt micro-alloy diffused base-transistor
MAG magnet
MAG. magnetic
Mag. magazine
mag. magnetic
magamp magnetic amplifier
maint. maintenance
MAL Malaysian Airways Ltd.
m.amp. milliampere
mamu millimass unit
M & A management and administration
M & C maintenance and construction
Mang.B. manganese bronze
Mang.S. manganese steel
MANIAC mathematical analyzer, numerical integrator, and computer
man.ring. manual ringer
M.A.P.I. *(Am)* Machinery and Applied Physics Institute
MAR., M.A.R. microanalytical reagent
M.A.R. miscellaneous apparatus rack
mar. maritime
MASER, maser micro-wave amplification by stimulated emission of radiation
MAT micro-alloy transistor; modular assembly technique; mechanical assembly technique
math mathematics; mathematical
MAV maleic anhydride value
max maximum
MAXUPO maximum undistorted power output
M.B. motor-boat; magnetic bearing; main ballast; moulded breadth
mb millibar
MBL main battle line
MBP monobutylphosphate
MBS mutual broadcasting system
MBT mercaptobenzthiazole
M.B.V. [process] modified Bauer Vogel [process]
MC multiple contact; moving coil; motor car; making capacity
M.C. medium-curing
M/C manual control
mc megacycle; motor cycle
m.c. metric carat; machine certificate; manhole cover; metre candle
m/c machine
MCA Ministry of Civil Aviation
MCDR motorcycle dispatch rider
MCF mean carrier frequency
Mcf *(Am)* thousand cubic feet
mCh millicurie/hour
m.c.i. malleable cast-iron
Mcps, mc.p.s. megacycles per second
Mcrit critical Mach number
MCS, MC/s, Mc/sec, mc/s megacycles per second
MCV manual volume control
MCW, M.C.W. modulated continuous wave
m/cy machinery
M.D. months after date; moulded depth
Md *(symbol)* mendelevium
m.d., m/d months after date
M.D.C. Mineral Development Committee; Marine Diamond Corporation
mdd *(Am)* milligrams per square decimeter per day
M.D.F. main distribution frame

M.Dk. main deck
MDL main defence line
MDM microdensitometer
MDS marine distress signals
ME mechanical engineer
M.E. marine engineer
Me methyl
m.e. milliequivalent
MEA Middle East Airlines Comp.
mech. mechanical; mechanics
mech.eff. mechanical efficiency
mech.lub. mechanical lubrication
Medit. Mediterranean
MEG megohm
meg megohm
megc megacycles
megc/s megacycles per second
megger megohm meter
megs megacycles
m.e.h.p. mean effective horsepower
MEK methyl-ethyl ketone
mem. member
MEMB membran
ME-MMA Mining Electro-Mechanical Maintenance Association
MeOH methanol
M.E.P., mep, m.e.p. mean effective pressure
meq. milliequivalent
mer., Merid. meridian
Merid. noon
Merlin medium energy research light water moderated industrial nuclear reactor
met meteorology
meth. methyl
Mev mega-electron volt, million electron volt
MF multifrequency; melamine-formaldehyde; medium frequency
M.F. machine finished; machine finish; middling fair; medium frequency
mf medium frequency; millifarad
MFD microfarad
MFDF medium-frequency direction finder
mfld mine field
MFP magnetic field potential
M.F.P. Ministry of Fuel and Power
Mfrs Std Gage *(Am)* manufacturers' standard gage
MFS multifrequency system
MG magnetic armature; motor generator; margin; marginal relay; machine gun; military government
M.G. machine-glazed; main generator
M-G, M.-G. motor-generator
Mg *(symbol)* magnesium
mg milligram[me]
m.g.d. million gallons per day
M.G.E. machine-gun emplacement
mgh milligram[me]/hour
mgm. milligram[me]
mg.p.l. milligram[me]s per litre
Mgr manager
M.H. main hatch
mh millihenry
mhc *(Am)* mean horizontal candles
mhcp mean horizontal candle-power
MHD magnetohydrodynamic
M.H.H.W. mean higher high water
mhl medium heavy loaded
mho unit of conductance *(Siemens)*
MHVDF medium-, high-, and very high-frequency direction finder
M.H.W.I. mean high water lunitidal interval
M.H.W.S. mean high water springs
M.I. moment of inertia; micro-inch
mi. mile
MIC microphone; multilayer integrated circuit
Mic. microwave
mic.cryst. microcrystalline
M.I.C.E. Member of the Institution of Civil Engineers
Mid. middling
mid. midshipman
Mid.L. middle latitude
Mid.sec. midship section
mi/hr miles per hour
mike micrometer; microphone
mill. million
M.I.M. moving in materials
M.I.M.E. Member of the Institution of Mining Engineers
M.I.Mech.E. Member of the Institution of Mechanical Engineers
min minute
min, min. minimum
Min.Engr. Mining Engineer
min./max. minimum/maximum
M.Inst.C.E. Member of the Institution of Civil Engineers
min width minimum width
M.I.P. marine insurance policy
mip *(Am)*, **m.i.p.** mean indicated pressure
M.I.R.A. Motor Industry Research Association
mirt moving in rotary tools
misc. miscellaneous
MIT, M.I.T. Massachusetts Institute of Technology
mixt. mixture
Mk. mark
mkg., m-kg metre-kilogram[me]
MKS, M.-K.-S., mks metre-kilogram[me]-second
ML mine layer
M.L. motor launch
M_l longitudinal metacentre
mL millilambert
ml millilitre

m.l. metal-lined
m/l more or less
MLB multilayer board
MLD, M.L.D. median lethal dose
mlhc *(Am)* mean lower horizontal candles
m.l.h.c.p. mean lower hemispherical candle-power
M.L.L.W. mean lower low water
MLR main line of resistance
MLSS mixed liquor suspended solids
MLT mean length per turn; micro layer transistor; median lethal time
M.L.W.I. mean low water lunitidal interval
M.L.W.S. mean low water springtide
MM middle marker; measuring mark
M.M. main motor; mercantile marine
mM. millimol
mm millimetre
M.M.A.B. *(Am)* Minerals and Metals Advisory Board of the National Research Council
M.M.B. Miners Medical Bureau
mmf micromicrofarad; picofarad
mmf, m.m.f. magnetomotive force
mmfd micromicrofarad
MMG medium machine gun
MM taper milling machine taper
m mu millimicron; millimass unit
ML main [line]
M.N. Merchant Navy
Mn *(symbol)* manganese
M.N.R. mean neap rise
M/N ratio ion-pair yield
MO motor
M.O. master oscillator
Mo *(symbol)* molybdenum
mO mass of electron at low velocity
MOD modulator; modulation
mod. modification
MODEM modulator-demodulator
modifn. modification
m.of i. planar moment of inertia
M.O.H. medical officer of health
M.O.I.V. mechanically operated inlet valve
MOL manned orbiting laboratory
mol gram-molecule, mole
mol. molecular; molecule
molectronics molecular electronics
mol wt, mol.wt. molecular weight
MON monitor; monitoring
monocl. monoclinic
monohyd. monohydrate
MOPA, M.O.P.A., m.o.p.a. master oscillator power amplifier
MOPAFD master oscillator power amplifier frequency doubler
mor medium frequency omni range
mort moving out rotary tools
MOS metal-oxide semiconductor
M.O.S. Ministry of Supply
MOT motor
M.O.T. milling on tools; Ministry of Transport
Motocross motor-cycle cross-country race
mot op motor operated
MOUSE, mouse minimum orbital unmanned satellite
M.O.W.B. Ministry of Works and Buildings
MP multipole; metallized paper
M.P. melting point; months after payment; middle point
mp, m.p. medium pressure; melting point
M.P.A. Metal Powder Association; maximum permissible amount
MPC maximum permissible concentration; medium processing channel
MPD maximum permissible dose
m.p.e. multiple protective earthing
m.p.g. miles per gallon
mph, m.p.h. miles per hour
m.p.l. maximum permissible level
m.p.m. metres per minute
M.P.N. most probable number
M.P.S. mean piston speed
Mp/s megaperiods per second
m.p.s. metres per second
mpy mil per year
M.Q. metol-quinol [developer], metol-hydroquinone [developer]
M.Q.B. Mining Qualifications Board
M.R. moment of resistance; molar refraction; mate's receipt
M.R.A. medical room assistant
MRBM medium-range ballistic missile
MRCP mobile radar control post
m.rd macadam road
M.R.E. mining research establishment
MRFL master radio frequency list
MRR mains restoration relay
MS. manuscript
M.S. machinery survey; mail steamer; maximum stress; merchant shipping; metric system; mild steel; months after sight; motorship; machine (machinery) steel; manshift; Mine Superintendent
M/S mine sweeper; months after sight
ms- meso
MSA mine safety appliances
MSB mine-sweeping boat
M.S.Blk. mild steel, black finish
M.S.Brt. mild steel, bright finish
M.S.C. mile of standard cable
M.Sc. Master of Science
msc *(Am)* mean spherical candles
m.s.c. mile of standard cable
MSD most significant digit
m-sec millisecond
M.sh. merchant ship

M.S.L. mean sea level
MSR, msr mean square root
M.S.R. mean spring rise
MSS *(Am)* Manufacturers Standardization Society of the Valve and Fittings Industry
MSS. manuscripts
MST measurement
M.S.T. manshifts per 1 000 tons
MT multiple twin; measurement; metric ton; medium thermal
M.T. Morse taper; mechanical transport; motor transport; mean time; measurement tonnage; measurement tons; metric ton; motor tanker
M_T transverse metacentre
Mt. mount[ain]
mt. measurement
M.T.B. motor torpedo-boat
mtbf mean time between failures
MTD mean temperature difference
mtg. mounting
MTI moving-target indicator
M.T.I. *(Am)* Metal Treating Institute; moment to alter trim one inch
MT_1 moment to change trim one inch
M.T.L. mean tidal level
mtl. material
M.T.M. methods time measurement
MTN Mineral Trade Notes
m ton metric ton
M.T.P.I. Member of the Town Planning Institute
MTR multiple track range radar; material testing reactor
MTV missile test vehicle
mu micron
mu a, mua microampere
MUF, M.U.F., m.u.f. maximum usable frequency
mu f microfarad
muh microhenry
mu in. microinch
mu-in.rms microinches square root of mean square
mu mu micro-micro *(piko)*; micromicron
mu v microvolt
mu w microwatt
MUX multiplex
MV motor vehicle; multivibrator
M.V., mv motor vessel; muzzle velocity
mv millivolt
mv., MVB multivibrator
m.v.c. manual volume control
MVDF medium and very high-frequency direction finder
MW modulated wave; megawatt
M.W. molecular weight
mw, m.w. medium wave
M.W.B. Metropolitan Water Board
MW-level mean water level
M.wt. molecular weight
mx maxwell; multiplex
M.Y. motor yacht

N

N north; normal [concentration; solution]; normalized; number; *(symbol)* nitrogen; neper; noise [meter]; new; national
N. naval; navigator; north
n number; normal; nano-; naval; navigator; north; revolutions per minute
n. north
n- normal
N.A. nautical almanac; numerical aperture; neutral axis
N/A non-acceptance
Na *(symbol)* sodium
n.a. numerical aperture; neutral axis; not above
n.a., n/a not available
NAA National Aeronautical Association
n.a.a. not always afloat
NAB naval air base; navigational aid to bombing
NAC New-Zealand National Airways Corporation
NACA National Advisory Committee on Aeronautics
NACE *(Am)* National Association of Corrosion Engineers
N.A.C.M. National Association of Colliery Managers
N.A.C.O.D.S. National Association of Colliery Overmen, Deputies, and Shotfirers
NAE National Academy of Engineering
NAM *(Am)* National Association of Manufacturers
N.A.M.E. National Association of Marine Engine Builders; Society of Naval Architects and Marine Engineers
NAR *(Am)* negative axial rake
NAS national aircraft standard
NASA National Aeronautics and Space Administration
nat. natural
nat.freq. natural frequency
nat log natural logarithm
naut. nautical
naut.mi. nautical mile; sea mile
naut.mi.p.hr. nautical miles per hour; sea miles per hour
nav. naval; navigable
nav., navig. navigation
N.A.Y.R.U. North American Yacht Racing Union
NB no bias; nitrobenzene
N.B. naval base; nota bene

N.B. pipe size nominal bore pipe size
Nb *(symbol)* niobium; number
n.b. nota bene; note well
N.B.C. National Broadcasting Corporation
N.b.E. north by east
NBFM narrow-band frequency modulation
N.B.F.U. National Board of Fire Underwriters
N.Br. naval brass
NBS *(Am)* National Bureau of Standards
N.B.S. new British standard gauge
N.b.W. north by west
N by E, N. by E. north by east
N by W, N. by W. north by west
NC numerical control; nose cone; no connection; national [standard] coarse thread; national centre; national coarse [series of the American standard]
N.C. naval construction; neutralization capasitor
N/C new charter, new crop
nc, n.c. normally closed
NCA National Coal Association
NCB National Coal Board
N.C.B.Inf.Bull. National Coal Board Information Bulletin
N.C.O. non-commissioned officer
NCP *(Am)* normal circular pitch
NCS network control station
NCSA National Crushed Stone Association
N.C.V., n.c.v. no commercial value
ND no delay; non-directional
N.D. natural draught
Nd *(symbol)* neodymium
nd. needles
NDB non-direction beacon
NDHA *(Am)* National District Heating Association
NDP *(Am)* normal diametral pitch
NE nuclear energy
NE, N.E. north-east
Ne *(symbol)* neon
n.e. north-east; not exceeding
NEC *(Am)* National Electric Code
N.E.C.I.E.S. North East Coast Institution of Engineers and Shipbuilders
N.E.D.C. National Economic Development Council
need. needles
NEF American national extra fine [thread]; national extra fine [thread]
neg. negative
NEL National Engineering Laboratory
NELA National Electric Light Association
NEMA National Electrical Manufacturers' Association
NEP, n.e.p. noise equivalent power
N.E.P.A. Nuclear Energy for the Propulsion of Aircraft
NET network
net terms free of charterer's commission
NF national fine [thread series]; noise figure
N.F. night fighter; national standard fine thread; no funds
n/f night frequency
NFM narrow band frequency modulation
NFPA National Fire Protection Association
NFQ, NFq night frequency
NG nitroglycerine
NHP, n.h.p. nominal horsepower
NHS National Horological Society
NI number of ampere turns
N.I. non-inductive
Ni *(symbol)* nickel
Ni.Cr.S. nickel chromium steel
NIF noise improvement factor
Ni.S. nickel steel
NL noise limiter
N.L. north latitude
N.Lat. north latitude
NLR noise loading ratio
NLRB National Labor Relations Board
Nly. northerly
NM magnetic north; noise meter
N/m no mark *(bill of lading)*
Nm^3 standard cubic metre
n.m. nautical mile
NME *(Am)* National Military Establishment Specifications
N.M.P. navigational microfilm projector
NMR, n-m-r nuclear magnetic resonance
NMTBA National Machine Tool Builders' Association
NNE north-northeast
NNW north-northwest
NO naval officer; normally open
N.O. navigating officer
No *(symbol)* nobelium
No. number
no, n.o. normally opened
NOHC *(Am)* National Open Hearth Committee
NOL *(Am)* Naval Ordnance Laboratory
nom. nominal
nomag nonmagnetic
non-vol. non-volatile
n.o.p. not otherwise provided
norm. normalized
Nos., N^{os}, nos numbers
NOT, not number of turns
NOTAM notes to airmen
NP no pin; napalm; north pole; neap
NP. nitroparaffin
N-P nominal horsepower
Np *(symbol)* neptunium; number of primary turns
np neap
n.p. normal pressure
NPA normal pressure angle; National Petroleum

Association; *(Am)* National Production Authority
N.P.D. north polar distance
N.P.I., n.p.i. needles per inch
NPL new programming language
NPL, N.P.L. National Physical Laboratory
npn, n-p-n negative-positive-negative
NPR noise power ratio
NPS American standard (national) straight pipe thread
NPSC American standard (national) straight pipe coupling internal thread
NPSL American standard (national) straight pipe thread for lock-nut connections
NPSM American standard (national) straight pipe thread for free-fitting mechanical joints
NPT American standard (national) taper pipe thread, American national [Briggs] pipe thread; normal pressure and temperature
npt, n.p.t. normal pressure and temperature
NR Reynolds' number
nr near
n.r. no risk
NRC naval radio compass
N.R.C. *(Am)* National Research Council
NRD naval radio direction finder
Nrd. northward
NRI *(Am)* National Radio Institute
N.R.M.E. notched, returned, and mitred ends
Nrn. northern
NRR *(Am)* negative radial rake
NRS national radio station
N.R.T. net register ton; net register tonnage
NRTS National Reactor Testing Station
NRZ, nrz non-return-to-zero
NS special thread of American national form, American national form of special pitch
N.S. new style; naval service; nuclear ship; north-south
Ns number of secondary turns
ns nanosecond; not specified
n.s. not sufficient; not specified
NSC noise suppression circuit; National Safety Council
NST National Standard Taper
N.-sub. nuclear submarine
N.T. normal temperature; non-tight
N/t new terms
NTC, ntc negative temperature coefficient
N.T.I. noise transmission impairment
NTP, N.T.P., n.t.p. normal temperature and pressure; number of theoretical plates
N.T.R. *(Am)* negative true rake
NTS Nevada Test Site; not to scale
N.T.S. not to scale
NTSC National Television System Committee
NTU number of transfer units
nt wt net weight
NU nucleonics
N.U. number unobtainable
nucl. nuclear
N.U.M. National Union of Mineworkers
N.V.M. non-volatile matter
NVR no-voltage release
nvt neutron density velocity time
NW naked wire; north-west
N.W. narrow widths
N.W., n.w. north-west
NWG National Wire Gauge
NWL. normal water level
NWSA *(Am)* National Welding Supply Association
N.W.T. non-watertight
n.wt. net weight
NYA New York Airways Inc.
NYSEM New York Society of Electron Microscopists
N.Z.S.I. New Zealand Standards Institution

O

O old; officer; order; oscillator; *(symbol)* oxygen; observation; ordinary
o. open; over; optimum; optimal
o- ortho-
OA operating assembly
O.A. over all
o.a., o/a on account of; overall
o.a.d. overall dimensions
O.A.S. open hearth acid steel
o.a.t. outer atmospheric temperature
O.B. outboard; outside broadcast
o.b. oil base
o/b on board
OBI omnibearing indicator
O.B.M. ordnance bench mark
O.B.S. open hearth basic steel
Obs. observation; observed
obs. observed
obsn fl. observation flight
obstn. obstruction
OC operating characteristic; open circuit
O.C. oxygen consumed; operations commence
O/C open charter
Oc. ocean
oc, o.c. open circuit
o.-c. open-closed
OCB, O.C.B. oil circuit breaker
Oc.B/L ocean bill of lading
Occ.Lt. occulting light
o-cm ohm-centimetre
OCR oil circuit recloser; *(Am)* Office of Coal Research
oct. octavo
OD optical density; outside diameter

O.D. olive drab; outside diameter
o.d. outside diameter
o/d on demand
o./d., O.Dia. outside diameter
O.Dk. observation deck
ODR omnidirectional range
O.E. operational efficiency
Oe oersted
OECD Organization for Economic Co-operation and Development
OEEC Organization of European Economic Cooperation
O.E.P. Office of Emergency Planning
OF oscillator frequency
OF, O.F. oil fuel
off. officer
O.F.H.C. oxygen-free high-conductivity
OG oxygen-converter gas-recovery
O.G. ogee; ordinary goods
OGO orbiting geophysical observatory
OGR outgoing repeater
O.G.S. Overseas Geological Survey
O.G.T.C. outgoing toll circuit
O.H. open-hearth; oil hardened
Oh, oh ohm
OHA *(Am)* outside helix angle
O.H.C. overhead camshaft
OhD one-hour duty
OHIA Oil Heat Institute of America
OHM ohmmeter
O.H.V. overhead valve
O.I.H. oil in hole
OIL orange indicating lamp
oilt. oiltight
OL overload
O.L. overhead line
ol. olive
o/l our letter
OM outer marker; *(sl)* „old man"
O.M. organic matter; old measurement; overturning moment
O.M.S. output per man shift
ON, O.N. octane number
ONC Ordinary National Certificate
OND Ordinary National Diploma
ONR Office of Naval Research
o/o our order
O.O.O. out of order
OP operator; operated; observation post; outer panel
O.P. open policy
o.p. out of print; over proof
opp. opposite
O-P process Oppenheimer-Phillips-process
OPR operator
ops operations
opt optical; optics; optimum
optim. optimum
opt.min. optical mineralogy
Opt.S. optical sight
OPV, opv ohms per volt
O.Q. oil quenched
O.R. operational research; owner's risk; octane requirement
or. orange
o.r. owner's risk
ORB omnidirectional radio beacon
o.r.b. owner's risk of breakage
ord owner's risk of damage; ordinary
orf owner's risk of fire (freezing)
org. organic
orig. original
ORNL Oak Ridge National Laboratory
O.R.R. Overseas Research Reports
O-R unit Owens-Rutherford-unit
orw owner's risk of wetting
O.S. ordinary seaman; outsize; Ordnance Survey; old style; one side; oil sand
O/S on sale; out of stock; overseas; outsize
O.S. & W. oak, sunk, and weathered
OSC oscillator; own ship's course
OSCG oscillograph string
OT oil tempered; oil-tight
O.T. overland telegraph; oil tanker; oil-tight
O.T., O/T old terms
O/t. on truck
o.t. oil-tight
O.T.C. originating toll centre
O.T.D. old total depth
OTS Office of Technical Services
O.U. Oxford University
OUF optimal usual frequency
OURAD our radiogram
OUT outgoing; output
OV, O.V., o.v. over-voltage
Ov.L. overhead line
OVV, ovv over-voltage
OW order wire circuit; open wire
O/W oil in water
O.W.D.D. old well drilling deeper
O.W.F., o.w.f. optimum working frequency
O.W.L. over water-line
O.W.P.B. old well plugging back
OWS ocean weather station
O.W.S. ocean weather ship
ox oxidizer
oxid. oxidize
oxidg. oxidizing
oxidn. oxidation
Oz, oz, oz. ounce
oz.ap. apothecaries' ounce
oz.av. ounce avoirdupois
oz.ff. fluid ounce
oz-ft ounce-foot
oz-in ounce-inch
oz.t. ounce troy

P

P poise; period; permanence; permeance; [diametral] pitch; pitch of propeller; polarization; power; pressure; *(symbol)* phosphorus; paper; plate; plug; pole; propagation; [car] park; pedestrian crossing; pickled; pair; patent
P. port, portside
p prime; pitch [ratio]; pressure [intensity]; polar distance; period; proton
p. page; part; pass; per; pipe; piece
p- para-; primary
p pressure per unit of area
PA *(Am)* pressure angle; phase angle; pulse amplifier; public address system
P.A. position approximate; preliminary amplifier; power amplifier; public address; Punta Arenas *(wool)*
P.A., P/A particular average
Pa *(symbol)* protactinium
p.a. particular average; per annum
PAA Pan American Airways
PAB para-amino-benzoic acid
P.A.B.X. private automatic branch exchange
PAC pilotless aircraft
Pad.Str. paddle steamer
PADT post alloy diffused transistor
PAL phase alternation line; Philippine Air Lines Inc.
PAM, pam, p.a.m. pulse amplitude modulation
PAN polyvinyl cyanide
PANAGRA Pan American-Grace Airways Inc.
P & A plug and abandon
P. and C. parge and core *(lime and hair mortar mixed with cow dung)*
P & I protection and indemnity
P. & L. profit and loss
P. & O. Co. Peninsular and Oriental Steam Navigation Co.
P & R parallax and refraction
P. and S. planking and strutting
P. and S., P. & S. port and starboard
PAR precision approach radar
Par. parallax
par. parallel
para parachute; parachutist
para. paragraph; paramagnetic
paratps parachute troops
PAS para-amino-salicylic acid
pat. patent
pat., patd. patented
P.A.X. private automatic exchange
payt. payment
PB playback; push button
P.B. permanent bunkers; plugging back
Pb *(symbol)* lead
PBL parachute-braked landing
P.B.P. pulling big pipe
P.B.X. private branch exchange
PC printed circuit; pulsating current; single paper single cotton *(cable)*
PC. post-chlorinated
P.C. prime cost; prices current; Portland cement; pressure controller; pulsating current; per compass; pitch circle
Pc. piece
pc parsec; price
p.c. per cent; pitch circle
PCB printed circuit board
PCC programme controlled computer
PCD, pcd. pitch circle diameter
P.C.E. pyrometric cone equivalent
pcf, p.c.f. pounds per cubic foot
PCM pulse code modulation
P.C.M. punch-card machine
pcm, p.c.m. pulse code modulation
PCO picture control oscilloscope
PCP chamber parallel circular plate chamber
pcs. pieces
p.ct. per cent
PCU, Pcu pound centigrade unit
P.C.V. pressure control valve
PD pitch [circle] diameter; potential difference; periodic duty; pressure distillate; periscope depth
P.D. port dues; polar distance; position doubtful; pitch diameter
P/D proof of delivery
Pd *(symbol)* palladium
pd potential difference; period; painted
pd, pd. paid
p.d. pitch diameter; potential difference
p.d.a. ratio post deflection acceleration ratio
PDB. p-dichloro benzene
PDC single paper double cotton *(cable)*
P.D.C. pure direct current
PDI pilot direction indicator
P.Dk. poop deck
pdl poundal
PDM pulse-duration modulation
P.E. probable error; electric power; photoelectric
Pe Péclet number
pe photoelectron
peb. pebble
PEC Pilot error correction
P.E.C., pec photoelectric cell
PeCe [chlorinated] polyvinyl chloride
ped pedestrian
PEI *(Am)* Porcelain Enamel Institute
P.E.L. proportional [elastic] limit
PEM photoelectromagnetic effect
Pen. peninsula
PENT pentode
PEP pulse effective power

P.E.P. peak-envelope power
P.E.R.A. Production Engineering Research Association
per hr. per hour
Perm. permanent, permanently
PERT, P.E.R.T. programme evaluation and review technique
p.e.s. photoelectric scanner
pet. petroleum
P.E.T.C. polyethylene tetrachloride
pet. eth. petroleum ether
PETN penta-erythritol tetra-nitrate
petro. liquid petrolatum
petrol. petroleum
PF phenol-formaldehyde; permanent force; pre-flight
P.F. plain face; penetration fracture; power factor; pulse frequency; prix fixe
pF, pf pico-farad
pf, p.f. power factor; pulse frequency
p.f., p-f power factor
Pfd. paraffined
P.F.I. & R. part fill in and ram
P.F.M. pulse frequency modulation
P.F. Powder phenol-formaldehyde moulding powder
PFRT preliminary flight rating test
p.f.u. prepared for use
p-f winding pole-face winding
PG *(Am)* precision-ground
PGED Professional Group on Electron Devices
PH phase[meter]
Ph phenyl
pH pico-henry; log of reciprocal of hydrogen ion concentration
ph pico-henry; phon; phot; phantom
PHA pulse height analyzer
PHONE telephone
Phos.B. phosphor bronze
p.hr. per hour
physiol. physiological
phys.sci. physical sciences
PI performance index; position indicator; penetration index
P.I. paper-insulated
pi power input
PIA Pakistan International Airlines Corporation
P.I.B. polyisobut[yl]ene
P.I.C. Plastic Igniter Cord
PIT parameter input tape
PIV peak inverse voltage
P.I.V. positive infinitely variable
p.i.v. peak inverse voltage
P.I.V.$_{(max)}$ maximum peak inverse voltage
PK peak
pk peck *(8 quarts)*
PL plain language; plug; power line; photo-conductor lamp
P.L. partial loss; proportional[ity] limit; pipe line
Pl. plasma; pole; plan
Pl., pl. plate
p/l profit and loss
P.L.A. Port of London Authority
PLI pilot location indicator
PLM pulse length modulation
PLN plan; plain
Pln Msg plain message
P.L.O. pipeline oil
PLOP pressure line of position
PLSTC plastic
PM preventive maintenance; phase modulation; pulse modulation
P.M. post meridiem; photo multiplier; purpose-made
pm. premium
p.m. post meridiem; permanent magnet
P.M.B.X. private manual branch exchange
P.M.C. plaster-moulded cornice
PME photomagnetoelectric
p.m.h. production per man hour
PMI *(Am)* Pressed Metal Institute
PMR Pacific Missile Range
PMST professor of military science and tactics
P.M.X. private manual exchange
P/N. promissory note
P-N, p-n positive-negative
pneu. pneumatic
PNM pulse-numbered modulation
P-N-P, p-n-p positive-negative-positive
pnr pioneer
PO power oscillator; pulse oscillator
P.O. post office; petty officer
P/O pilot officer
Po *(symbol)* polonium
po. pole
P.O.B.A. *(Am)* Patent Office Board of Appeals
P.O.C., p.o.c. port of call
POD Podbielnak [analysis]
P.O.D., p.o.d. pay on delivery
POL polarized
polym. polymer
polymd. polymerized
polymn. polymerization
P.O.O. post office order
P.O.P. printing-out paper
P.O.R. pay on return
P.O.R., p.o.r. port of refuge
porc. porcelain
port. portable
POS pressure operated switch; position
pos. positive
P.O.S.B. Post Office Saving Bank
POSN position
pot. potentiometer
potass. potassium

powd. powder
PP pilotless plane
P.P. proportional parts; parcel post; pulled pipe
pp push pull; pages; peak to peak
pp. pages; push-push
p.p. picked ports; post paid
p-p push-pull; peak-to-peak
PPA push-pull amplifier
P.P.C. plain plaster cornice
p.p.c.c. parts per cubic centimetre
ppd. prepaid
p.ph.m. pulse phase modulation
P.P.I. plan position indicator; policy proof of interest
PPM pulse-phase modulation; parts per million
P.P.M. pulse position modulation
ppm, p.p.m. parts per million
P.P.P.A. push-pull power amplifier
PPPI precision plan position indicator
PPS, pps, p.p.s. pulses per second
pps, p.p.s. periods per second; pounds per square inch
ppt. precipitate
p.p.t. prompt; prompt loading
pptd. precipitated
pptg. precipitating
pptn. precipitation
PR pressure recorder; periodic-reverse; pure rubber; public relations; primary radar
P.R. ply rating of tyres
Pr parachutist; *(symbol)* praseodymium; propyl; Prandtl number
Pr. proceedings
pr pair
pr. prism[s]; pressure
p.r. payment received; pure rubber; pitch [ratio]
pract. practically
PRC periodic reverse current
prchst parachutist
PRE, preamp. preamplifier, preliminary amplifier
prep. prepare; preparation
prepd. prepared
prepg. preparing
prepn. preparation
press pressure
Prev. previous, previously
PRF, p.r.f. pulse recurrence frequency
pri., prim. primary
pris.coeff. prismatic coefficient
Priv.X. private exchange
Proc. proceedings
proc. procedure; process
prod. produce
Provis. provisional, provisionally
prox. proximo
PRR positive radial rake angle; pulse-repetition (recurrence) rate
p.r.r. pulse-repetition rate
PRT pulse-recurrence time
PRTR plutonium recycle test reactor
PS power supply; physical society; panel-sealed
P.S. paddle steamer; passenger steamer; post scriptum; proof stress; pipe size
P/S. public sale
P./S. portside
p.s.c. per standard compass
psf pound[s] per square foot
psi pound[s] per square inch
p.s.i.a. pound[s] per square inch absolute [pressure]
psig *(Am)* pound[s] per square inch gage
PSM pulse slope modulation
PSN, psn. position
P.S.P. peak sideband power
P.S.R. plow steel rope
PSSC Physical Sciency Study Committee
p.s.t. pressure-still tar
PSU power supply unit
PSWR power standing wave ratio
PT pulse transformer; pressure transmitter; post office with telephone
P.T. potential transformer
Pt *(symbol)* platinum
pt, pt. part; pint; port
p.t. per ton; private terms
p.t.a. phase-to-amplitude modulation
PTBL portable
PTC postal telegraph cable; positive temperature coefficient
ptd patented
Ptd.A. pointed arch
P.T.F.E. polytetrafluoroethylene
P.T.G. prepare town gas
P.T.M. pulse-time modulation
P.T.O., p.t.o. power take-off; please turn over
PTR *(Am)* positive true rake
pts. parts
PTT post, telephone, and telegraph
PU, P.U. power unit
Pu *(symbol)* plutonium
p.u. pick-up
PUD pick-up and delivery
pun. puncheon
pup pick up pump
purif. purify
purifd. purified
purifg. purifying
purifn. purification
PV patrol vessel
PVA, P.V.A. polyvinyl alcohol
PVC, P.V.C., p.v.c. polyvinyl chloride
P.V.D.C. polyvinylidene chloride
PVM polyvinyl methyl ether
P.V.P. polyvinylpyrrolidone
PW pulse width; pilot wire

P.W.D. Public Works Department
pwdr powder
p.w.m. pulse width modulation
PWR pressurized water reactor
PWR, pwr power
pwt. pennyweight
PX phantom coil
PX, P.X. private exchange
Py pyridine nucleus
pyr. pyridine; pyrometer
Pyra. pyramidal
p-z peak-to-zero
P.Z.T. photographic zenith tube

Q

Q quality; quantity [of water discharged]; quartile *(equation)*; quarter deck
q quantity; quintal
qagc quiet automatic gain control
QAVC quenched (quiescent, quiet) automatic volume control
QC quality control; quartz crystal
q.c. quantity at captain's option
Q.C.D. quality control division
q.c.o. quantity at captain's option
Q.E.D., q.e.d. quod erat demonstrandum
QF quick firing
Q.I.T. Quebec Iron and Titanium Corp.
ql quintal
qm metric quintal
Q.M.B. quick make and break
Q.O. quick opening
QP quality product
Q.P. quartered partition
Q.P.P. quiescent push-pull
qr, qr. quarter; quart
Q.R.C. quick ram change
Q.S.M. quadruple screw motorship
Q.S.S. quadruple screw ship; quadruple screw steamer
qual. qualitative
quant. quantitative
quart. quarternary; quarter
QWA quarter-wave aerial

R

R Rankine; Reaumur; resistance; reluctance; rectifier; radio; reception; receiver; *(Am)* right-cut; Rockwell hardness number; Reynolds' number; radical; radius; radial dimension; rubber [bond]; radioactive mineral; readability; recovery
R. reaming; radius; range
°**R** degree Rankine; degree Reaumur
R″ Redwood second
r, r. railway; road; right; radius; roentgen; refraction; rubber
r. river
r- racemic
RA rectifier; regular army
R.A. reduction of area; right ascension; right angle
R/A. refer to acceptor
Ra *(symbol)* radium; air resistance
r.a. radioactive
RAA radar aircraft altitude calculator
RAB Radio Advertising Bureau
RaB ring and ball [method]
RAC radio service code; rules of the air traffic control; rectified alternating current
R.A.C. Royal Automobile Club
rac. racemic
r.a.c. rectified alternating current
RACES Radio Amateur Civil Emergency Service
RACON radar beacon
RAD radiator
rad radian; radius; radio; rad unit
rad. radius; radiant; radix
Radar radio detecting and ranging
RADCM radio countermeasures and deception
radn. radian
radn.p.s. radians per second
Rad Opr radio operator
R.A.E. Royal Aircraft Establishment
RaEm radium emanation
RAeS Royal Aeronautical Society
RAF Royal Air Force
R.A.M.S. right ascension mean sun
R and D., R & D, R. & D. research and development
R. & M. repair and maintenance
R. & T. rods and tubing
R & W raised and weighed
RAOB radiosonde observation
RAP resistance to aqueous penetration
RAPCON radar approach control centre
RAPLOT radar plotting
RAS rectified air speed
RATO rocket assisted take-off
RATOG rocket assisted take-off gear
RATT radio teletypewriter
R.A.X. rural automatic exchange
R.B. radio beacon
Rb *(symbol)* rubidium
RBE, R.B.E. relative biological effectiveness
RBS radio beacon station
RBT radio beam tube
RC resistor-capacitor; resistance/capacitance; reaction coupling; regional centre; radio control; rifle caliber; rubber covered
R.C. rapid curing; recording controller; remote control; ringing circuit; radio compass station; reinforced concrete

R-C remote control
Rc radar control; relay computer
r.c. rapid-curing asphalt; remote control
RCA Radio Corporation of America
RCB rubber covered braided wire; radio-controlled boat
RCBWP rubber covered, braided, and weatherproof *(cable)*
R.C.C.T.L. resistor-capacitor-coupled-transistor-logic
Rcd. received
R.C.D.B., r.c.d.b. rubber-covered double-braided
rcdg recording
rcdr recorder
RCE, r.c.e. ray control electrode
RCG reverberation controlled gain
R.C.M.F. Radio Component Manufacturers' Federation
RCO remote-control office
rcpt. receipt
rcstn radio compass station
rcvd received
rcvg receiving
rcvr receiver
RCWP rubber-covered weatherproof wire; rubber-covered and weatherproof
RD recorder; recording demand meter; rated duty
R/D rate of descent
rd road; rod; Rutherford
rd, rd. round
r.d. running days
RdAc radioactinium
RDC rotating disk contactor
R.D.C. running down clause
R.D.F. radio direction finder (finding); repeater distribution frame
Rd M radarman
RDO radio
Rdo.Serv. radio service
Rdo-Stn radio station
Rdo Trans. radio transmitter
RDR restart delay relay
RdTh radiothorium
RE radio equipment; refrigeration effect
Re *(symbol)* rhenium
REAC reactive; reactor
reactn. reaction
reagt. reagent
reagts. reagents
rearr. rearrange; rearrangement
REC recording
Rec. record
rec. reciprocal; record; recorder
recd. received
RECOGSIG recognition signal
Recpt. receipt
recr receiver
recryst. recrystallize
recrystd. recrystallized
recrystn. recrystallization
RECT rectifier
rect. rectified
RED., Red. reduction
RE-DC reinsertion of direct current
redis. redissolve
reduc. reduce
reducd. reduced
reducn. reduction
Ref. refraction
REFL reflector; reflection
REG regulator; regulation
reg. regular
reg., regd. registered
regen regeneration
Reg.U.S.Pat.Off. registered in the United States Patent Office
REL rate of energy loss; Radio Engineering Laboratories
rel. relative; reluctance
rem roentgen equivalent man unit
REMA Refrigeration Equipment Manufacturers' Association
REP radar evaluation pod
Rep. reported
rep. repetition; repair
reppt. reprecipitate
Rep.Sh. repair shop
Rept. Report
R_{eq} equivalent noise resistance
res. resinoid-bonded; resistance; resistor
resid. residue
resin. resinify
resis. resistance
resp. respectively
RETMA Radio-Electronics-Television Manufacturers' Association
Rev. Review
rev. revolution
Rev. A/C revenue account
rev/min revolutions per minute
RF reactive factor meter; resultant force; regular forces
RF, R.F. radio frequency; range finder
R/F reserve forces
Rf ship resistance; skin resistance
Rf. reef
r.f. radio frequency; range finder
RFA radio frequency amplifier
RFC radio frequency coil; radio frequency choke
R.F.D. raised foredeck
RFI radio frequency interference
RFNA red fuming nitric acid
R.F.S. render, float, and set

RFT radio frequency transformer
R.F.W. reserve feed water
RG rifle grenade; recording unit; radio guide
Rg grid-leak resistance
RH, R.H. *(Am)* right-hand; relative humidity
Rh *(symbol)* rhodium; Rockwell hardness
r.h. relative humidity; right-hand
r/h radiation per hour
RHB radar homing beacon
RHEO rheostat
RHI range-height indicator
RHM roentgen per hour [at one metre]
rhomb. rhombic
r.h.p. rated horsepower
r/hr roentgen per hour
r.h.s. right hand side
RIC Radio Industry Council
RID Radio Intelligence Division *(of the FCC)*
rifi radio interference and field intensity
RIL red indicating lamp
RISAFMONE radio-signal reporting code for rated transmission conditions
riv. riveted
R.J. ramjet
RJTV ramjet test vehicle
Rk cathode resistor
RKT rocket
rkva reactive kilovolt-ampere
RL red lamp
R/L radiolocation
RLBI right-left bearing indicator
RLR reverse locking relay
RLS relays
RM radarman; radioman
Rm. room
RMA Radio Manufacturers' Association
RMI radio magnetic indicator
RMS, R.M.S., rms, r.m.s. root mean square
R.M.S.D. root-mean-square deviation
RMT remote
R.N. Royal Navy; reference noise; Reynolds number
Rn *(symbol)* radon
R.N.A. ribonucleic acid
RNG radio range
R.N.R. Royal Navy Reserve
R.N.S.A. Royal Naval Sailing Association
R.O. range only
ROA rules of the air
r.o.b. remaining on board
R.O.G. rig on ground
R.O.L.F. remotely operated longwall face
ROM run-of-mine
R.O.M.coal run-of-mine coal
R.O.R.C. Royal Ocean Racing Club
rot. rotary; rotation
ROTE range optical tracking equipment
ROTS rotary out-trunk switch
RP rocket propellant
R.P. rock pressure; rocket projectile; reference point
R.P.C. remote position control
R.P.F. radio position finding
rph, r.p.h. revolutions per hour
rpi revolutions per inch
RPM, rpm, r.p.m. revolutions per minute
RPS, rps, r.p.s. revolutions per second
RR range recorder
rr recoilless rifle
RRDE Radar Research and Development Establishment
RRL Radio Research Laboratory
RRS Radio Research Station
RS radio set; range setter; resistor
R.S. rotary switch; rocket station; radio station
R.S.C. rolled steel channel
R.S.G.B. Radio Society of Great Britain
R.Sigs. Royal Corps of Signals
R.S.J. rolled steel joist
RSR remote start relay
rss root sum square value
RST [quality of] readability, signal strength, and transmission
R.Stn. relay station
RT radiotelephony; radiotelegraphy; routine test; routine tested
R.T. wireless telephone station; receiver-transmitter; rye terms
R/T radiotelephony; radiotelegraphy; radiotelegraph; radiotelephone
R.T.A. Road Traffic Act
RTC road tank car
R/TComn. radiotelephone communication
rte. route
RTL resistor transistor logic
RTMA Radio and Television Manufacturers' Association
R.T.P. reference telephonic noise
RTR repeater test rack; radio-teletype receiver
R.Tr., R.Trr rotary transformer
R/T set radiotelegraphy set
RTT radio teletype[writer]
RTTW radio teletypewriter
RTTY radio-teletype
RTU rate of a transfer unit
RTV rocket test vehicle
R.T.Y.C. Royal Thames Yacht-Club
R.T.Z. Rio Tinto Zink Corp.
RU reproducing unit
Ru *(symbol)* ruthenium
rub rubber
R.U.R.T. rigging up rotary tools
R.U.S.T. rigging up standard tools
R.V. relative volume; restriking voltage
rva reactive volt-ampere
R.V.L.R. Road Vehicles Light Regulations

RVP Reid vapour pressure
RVR runway visual range
R/W runway
Rw ship resistance
R.W.D. rear-wheel drive
R.W.P. rain-water pipe
R.W.T.H. raised watertight hatch
RX receiver
Ry, Ry. railway
R.Y.S. Royal Yacht Squadron
Ry Tel. railway telegraph

S

S society; south; steel; surface; silicate [bond]; single silk covered; *(symbol)* sulphur; specific gravity; stratus; summer freebord mark
S. starboard; submarine; south; sailing ship; science; series; sign; surface; system; switch
S″ Saybolt second
s shilling; stere; solid; soluble; symmetrical; slip; service; steel; solubility; [internal] shield; south
s, s. second
s- symmetrically substituted; secondary; syn-
SA single armoured *(cable)*; semiautomatic; speech amplifier; switching assembly; slow operate and slow release
S.A. safe arrival; self-acting
s.a. self actuated
s.a., s/a subject to approval
SAA South African Airways
S.A.A. small arms ammunition
SAE Society of Automotive Engineers; Standard of Automotive Engineers
SAE/AIR Society of Automotive Engineers, Aeronautical Information Reports
SAE/AMS Society of Automotive Engineers, Aeronautical Material Specifications
SAE/ARP Society of Automotive Engineers, Aeronautical Recommended Practices
SAE/AS Society of Automotive Engineers, Aeronautical Standards
SAF Strategic Air Force
sal, salv salvage
saloon satellite launched from a balloon
SAM surface-to-air missile
SAN strong acid number
S. & F.A. shipping and forwarding agent
S. & F. cutter side and face milling cutter
S. & H. sundays & holidays; sundays and holidays excepted
S and O sheared and opened
S. & S., s & s spigot and socket
SAP semi-armour-piercing
sapon. saponify
saponn. saponification
S.A.R. special apparatus rack
SARAH search and rescue and homing equipment
SAS sodium aluminium sulphate
S.A.S.I. South African Standards Institution
S.A.S.M.U.T.A. Sheet and Strip Metal Users' Technical Association
SASOL South African Coal, Oil, and Gas Corporation, Ltd.
SAT, sat saturate
satd. saturated
SATN, satn saturation
SB scout bomber
S.B. smooth bore; simultaneous broadcast; simultaneous broadcasting; switchboard; single braid *(cable)*; steamboat; submarine fog bell
Sb *(symbol)* antimony; switchboard
sb stilb
SBA standard beam approach
S.B.A.C. Society of British Aircraft Constructors, Ltd.
s.b.c. small bayonet cap
SBK Sinclair-Baker-Kellogg *(catalytic reforming)*
SBN strong base number
SBP sliding base plate
SBRC single-braided, rubber-covered
S.B.S. surveyed before shipment
SBT surface barrier transistor
SC superimposed current; scout car; searchlight company; submarine chaser
S.C. supercalendered paper; semiconductor; short circuit; single current; squirrel cage; steel cored; superimpose current; selectional centre; side contact; single contact; silk-covered wire; *(Am)* slow curing; scale; small craft; salvage charges
Sc *(symbol)* scandium
sc. scale
s.c. squirrel cage; short circuit; standard conditions; salvage charges
SCA Subsidiary Communications Authorization
SC ALC steel cored aluminium conductor
SCAM sub-carrier amplitude modulation
scan. scanning
SCAR subcaliber aircraft rocket
s.c.c. single cotton covered *(cable)*
SC CuC steel cored copper conductor
scd scheduled
S.C.D.C. single commutation direct current signalling
SCE single-cotton enamelled [wire]
SCEA *(Am)* side cutting edge angle
SCFM sub-carrier frequency modulation
Sc.H. scleroscope hardness
SCI ship-controlled interception; soft cast iron
SCOOP Scientific Compilation of Optimum Programs of the US Air Force
scp spherical candle-power

SCR silicon[-controlled] rectifier; short-circuit ratio
Scr. screening
scr silicon-controlled rectifier
scr. screwed
s.c.r. silicon-controlled rectifier
SCRN screen
S.C.T. [mould] semi-closed top [mould]
SCTD scattered
SCTY security
S-Cu stratocumulus
SCW silk-covered wire *(cable)*
S.D. soft drawn; shut down; semidiameter; short delivery; sight draft
S/D sea-damaged
S.d. short delivery
S.D.B. sodium dodecylbenzene sulfonate
S.D.B.L. sight draft; bill of lading attached
S.D.D. sodium dinaphthylmethane disulfonate
S.D.F. supergroup distribution frame
S.D.P.L. shut down for pipe line
S.D.W.O. shut down for waiting orders
SE, S.E. secondary emission; secondary electron
S.E. steam emulsion [number]; south-east; stopped end
Se *(symbol)* selenium
S/E A/C single engined aircraft
SEC secondary
sec secant; secondary
sec, sec. second
sech hyperbolic secant
sec. of arc seconds of arc
sect. section
segm segment
sels. selsyn
selsyn self-synchronizing
SEN steam emulsion number
S.-engine starboard engine
sens. sensitivity
sep. separate
sepd. separated
sepn. separation
SEPT silicon epitaxial planar transistor
SESA Society for Experimental Stress Analysis
S.F. shearing force; sunk face; sinking fund
s.f. signal frequency
S.F.B. submarine fog bell
SFC, sfc specific fuel consumption
sfm, sfpm surface feet per minute
SFSA *(Am)* Steel Founders' Society of America
SG signal generator; screen grid; single-groove
S.G. specific gravity; spheroidal graphite *(iron)*
s.g. specific gravity; screw gauge
sgd. signed
SGF smoke generating fuel
S.G. iron spheroidal graphite iron
S.G.M. strict good middling
S.G.O. strict good ordinary
S.G.W. salt-glazed ware
SH shield; shunt
Sh. shoal
sh shilling; ship; shipment; shipwright
sh. sheet; shilling
SHF, S.H.F., shf superhigh frequency
Sh.I. sheet iron
SHLD shield
Shl.Dk. shelter deck
SHM, s.h.m. simple harmonic motion
shoran short range
SHP, S.H.P., Shp, s.h.p. shaft horsepower
Shpg.wt. shipping weight
shpt. shipment
Sh.S. sheet steel
sh.tn. *(Am)* short ton
SI specific impulse
S.I. Système International screw thread; statutory instruments; specific impulse; self-induction
Si *(symbol)* silicon
S.I.C., sic specific inductive capacity
S.I.D. sudden ionospheric disturbance
Sid. sidereal
SIE Society of Industrial Engineers
SIG signal
sig. signature
sigm. signalman
sim. similar; of similar composition
S.I.M.A. Scientific Instrument Manufacturers' Association
sin sine
sinh hyperbolic sine
sintd. sintered
S.I.P. shut in, prorated; shut in pressure
S.I.T. self-ignition temperature
SITREP situation report
S.J. soldered joint
sk. sketch
SL search light; sound locator
S.L. separate-lead type *(cable)*; salvage loss
sl. slightly
s.l. salvage loss; sea level
S.lat. south latitude
SLC specific line capacity; straight line capacity; straight line capacitance capacitor
S/L.C sue and labor clause
s.l.c. single lead covered
Sld., sld. sailed
sld. solid
SLDV separated liquid driven vortex
SLF straight line frequency
S.L.M. strict low middling; steel line measurement
S/LOSS, S/Loss salvage loss
S.L.S. sodium laurylsulphate
sl.s., sl.sol. slightly soluble
SLT standard light source

slt searchlight
SLW straight-line wavelength
slw straight-line wavelength capacitor
Sly southerly
sly. slowly
SM *(Am)* sulfurized mineral oil; sales manager; strategic missile
S.M. strict middling; Simpson multiplier; screw motorship
S/M submarine
Sm *(symbol)* samarium
sm mean specific heat
sm. small
s.m. sea mile; servomotor
s/m submarine
S.M.B. State Mining Bureau
SMBL semi-mobile
SME Society of Mining Engineers
SML *(Am)* sulfurized mineral lard oil
S.M.P.E. Society of Motion Picture Engineers standard for sound tracks
SMPTE Society of Motion Picture and Television Engineers
S.M.R.E. Safety in Mines Research Establishment
S.M.-steel Siemens-Martin steel
SMTS special machine tool standard
SN sodium nitrate; synchronizer
S.N. shipping note; steam navigation
S/N speech/noise; signal/noise; stress/number [of cycles]
Sn *(symbol)* tin
s.n.c. single needle code
SND sound
S.N.D.T. Society for Non-Destructive Testing
SNE single-nylon enamelled *(cable)*
SNR, Snr signal-to-noise ratio
SO slow operating; socket
S.O. senior officer; submarine oscillator
S./O. shipping order
s.o. select order
SOA school of artillery
SOAV solenoid-operated air valve
SOC socket
Soc. Society
SOD small object detector
sod. sodium
Sodar *(Am)* sound radar
SOF *(Am)* sound on film
Sofar *(Am)* sound finding and ranging
S/off sign off
SOL solenoid
S.O.L. shipowner's liability
sol. soluble
solidif. solidification
soln. solution
solv. solvents
soly. solubility
SONAR sound navigation and ranging [equipment]
SOR slow operating relay
SOS, S.O.S. save our souls
SOV sound on vision
SP soil pipe; single-pole; standard pressure; solid propellant; smokeless powder; sound-powered
S.P. self-propelled; stern post; single-pole; sparking plug; stirrup pump
Sp. spirit
sp single-pole; single-phase
sp, sp. specific
s.p. single-pole; single-phase; standard pressure
S.P.A. subject to particular average; sudden phase anomaly
SPAR super precision approach radar
SPC single-paper covered
s.p.c.b. single-pole circuit breaker
S.P.D. steamer pays dues, ship pays dues; south polar distance
Spd. spudding
SPDT single-pole double throw
Spec., spec specification
spec. spectrum
specif *(Am)* specification
SPECS specifications
SPG self-propelled gun
sp gr, sp.gr. specific gravity
SPH, s.-ph. single-phase
Spher. spherical
sp ht, sp.ht. specific heat
spkr. speaker
SPL sound pressure level
S.-P.L. single-party line
SPM self-propelled mount
s.p.m., s.p. min. strokes per minute
S.P.R. simplified practice recommendation
S.P.S. standard pipe size
SPST single pole single throw
SPSW single pole switch
spt. seaport
sp.v. specific volume
Sq, sq, sq. square
SQC statistical quality control
sq.cen., sq.cm, sq.cm. square centimetre
sq.dm. square decimetre
sq ft, sq.ft. square foot (feet)
sq in, sq.in. square inch
sq km, sq.km. square kilometre
sq m, sq.m. square metre
sq.mi. square mile
sq mm, sq.mm. square millimetre
sq mu square micron
sq.r., Sq Rd, sq.rd. square rod
sq rt square root
sq yd, sq.yd. square yard
SR secondary radar; selective ringing; slow

release; slow running; starting relay; saturable reactor; slip ring; send/receive; short range; sound ranging; stateroom; straight run
S.R. specific resistance; straight-reaming; solid rocket; self-rectifying; slip ring
S/R slant range; sending-receiving
Sr *(symbol)* strontium
sr steradian
s.r. specific resistance
S.R. & C.C. strike, riots, and civil commotion
Srd. southward
SRE surveillance radar equipment
S/RF strength of radio frequency
Srn. southern
SRS send/receive switch
SS submarine scout; single silk; steamship
S.S. simplified standard; stainless steel; single signal; single sideband; short stick *(35½ in lap)*; shear strength; screw steamer
S.S., S/S steamship
1 S.S. single screw ship
2 S.S. twin-screw ship, twin-screw steamer
3 S.S. triple screw ship
4 S.S. quadruple screw ship
ss sandstone
S.S. and C. salvage & salv. charges
S.S. & C. same sea and country or coast
SSB, s.s.b. single side band
s.s.c. single-silk covered
SSCW single-silk covered wire
SSD source-skin distance
SSE single-silk enamelled
SSE, S.S.E. south-south-east
SSF supersonic frequency
S.S.F. seconds Saybolt Furol
SSG standard signal generator
SSM surface-to-surface missile
S.S.phone ship-shore telephone
SSR secondary surveillance radar; standby supply relay; single-signal receiver
S.S.S. single-signal superhet; single screw ship
SSSC single sideband suppressed carrier
SST supersonic telegraphy; Supersonic Transport Aircraft
SSU, S.S.U., Ssu seconds Saybolt Universal
SSV ship-to-surface vessel radar
SSW south-southwest
ST sound telegraphy; single-throw
S.T. standard temperature; surface trench; specific heat/temperature
S-T studio to transmitter
St steward
St. stone *(weight)*; staff; station; street
s t short ton
st. stone *(weight)*; standard; steel; steam
s.t. short ton
STA steel-tape armoured; station; stationary
S.T.A. shaft tunnel alley
STALO stable local oscillator
sta.mi. statute mile
starb., Stbd. starboard
STC sensitivity time control
S.T. curve specific heat/temperature curve
STD short-time duty
S.T.D. subscriber trunk dialling
Std., std standard
STDP single-throw double-pole
STG storage
Stg. stringer
STL studio-to-transmitter link
STN, stn station
STOL short take-off and landing
S. to S. station to station
S.T.P., s.t.p. standard temperature and pressure
Str. steamer
str. stroke
strm. storage room
STT short time test
stu short ton unit
Sub. submarine
Sub.Fog Sig. submarine fog signal
subl. sublimate
subln. sublimation
subl.w.m. sublimes without melting
subseq. subsequent
subst. substituted
Sub.Tel. submarine telegraph
Succ. succeeding
SUHL Sylvania universal high-level logic
SUM surface-to-underwater missile
SUP supply; suppressor grid
super. superficial
supers., supersat. supersaturated
suppl. supplement
Supt. superintendent
SUR surface
Surm. surmounted
surv. survey; surveyor
S.U.S. Saybolt-Universal seconds
susp. suspended
S.U.V. Saybolt Universal viscosity
SV single silk varnished
S.V. sluice valve; stop valve; saponification value; sailing vessel; surveying vessel
s.v. specific volume
S-value susceptibility value
SVS supervisory signal
SW switch; special weapons; short wave; standing wave; stationary wave; Seaboard World Airlines Inc.; southwest
S.W. salt water; southwest; shipper's weights; short wave
S-W standing wave
sw. switch
s.w. specific weight; steel wire rope; southwest
s-w short wave

SWBD, swbd. switchboard
S.W.D. stoneware drain
S.W.G. standard wire gauge; *(Am)* steel wire gage
swgr. switchgear
SWI short-wave interference
SWL, S.W.L. short-wave listener
S.W.L. safe working load
SWR, swr standing wave ratio
S.W.R. steel wire rope
SWTL surface-wave transmission line
SWVR standing-wave voltage ratio
SX, sx simplex
SY synchronoscope
S.Y. steam yacht
Sy. supply
s.y. square yard
SYCR synchronize
sym. symmetric[al]
syn. synchronization
sync. synchronization; synchronizing; synchronous
synscp. synchronoscope
syst. system; systematic

T

T teeth *(drawing)*; [number of] teeth; [number of] threads; technician; treated steel; test pressure; time; time constant; [absolute] temperature; torque; tritium; tube; telemetering; teletype; telephone; trunk; terminal; tesla; transformer; transmission; transmitter; transmitting; trimmer; transport number; tropical waters; twisting moment
T. tee iron; true *(direction)*
t ton; metric ton; troy; temperature; time; thickness; triode; fluorescent target
t. temperature; time; thickness; ton; long ton; torpedo
t- trans-; tertiary
t temperature
TA telephone apparatus
T.A. temporarily abandoned
Ta *(symbol)* tantalum
TAA Trans-Australia Airlines
tab. tablet; table
TAC Television Advisory Committee; Trades Advisory Council
Tacan tactical air navigation
TAD tactical air direction net
T.A.D.I.C. telemetry analogue to digital information converter
TAM tactical air missile
TAN total acid number
tan tangent
T & AM theoretical and applied mechanics
t. & b. turned and bored
T. & G., t. & g. tongued and grooved
T. & R. transmitter and receiver
T. and S. touch and stay
t. & t.p. tube and tank process
tang tangent
tanh hyperbolic tangent
TAPPI Technical Association of Pulp and Paper Industry
TAS telegraphy with automatic switching
TAS, T.A.S. true air speed
TASI time assignment speech interpolation
TASO Television Allocations Study Organization
TAT to accompany troops
TB triple braided; time base
T.B. torpedo boat; torpedo bomber
Tb *(symbol)* terbium
T.B.D. torpedo boat destroyer
TBE total binding energy; tetrabromoethane
TBF Tactical Bomber Force
TBM tactical ballistic missile
TBN total base number
t.b.p. true boiling point
T.B.S. talk between ships
TBWP triple-braided weatherproof
TC temperature coefficient; thermocouple; temperature controller; tank circuit; telephone central office; transmitter tuning circuit; top cap; time constant; traffic control
T.C. total carbon; tungsten carbide
Tc *(symbol)* technetium; critical temperature
tc critical temperature
t.c. tungsten carbide; tone control
TCA thrust-chamber; Trans-Canada Airlines
TCC triple cotton covered *(cable)*; Telecommunications Coordinating Committee; Trunk Control Centre
TCI terrain clearance indicator
TCP tricresyl phosphate
TCR transistorized car radio
TCR-tube television cathode-ray tube
TCS traffic control station
TCV troop-carrying vehicle
TCXP tricresylxylenyl phosphate
TD total depth; tractor-drawn; time delay
$\mathbf{t_d}$ delay time
T.D.C., t.d.c. top dead centre
TDF trunk distribution frame
TDMS telegraph distortion measurement set
TDR time delay relay
t.d.w. ton deadweight
TE transverse-electric; totally enclosed; trailing edge
Te *(symbol)* tellurium
te transverse-electric
TEAL Tasman Empire Airways Ltd.
tech. technical

TED teleprinter error detector
TEFC totally enclosed forced cooled
t.e.g. top edges gilt
TEL telegraph; telegraphy; telephone; telephony; telephone station; telecommunication; tetraethyl lead
Tel. & Tel. telephone and telegraph
TELCON telephone conversation
Tele. telegraph
TELECOM. telecommunications
Tele.Com. telephone communication
teleph. telephone
TELERAN television and radar navigation
telg. telegram
teln telephone
telstar television star
tem transverse electromagnetic
temp, temp. temperature; temporary
temp.C. temperature Celsius
temp.F. temperature Fahrenheit
TER teleprinter retransmitting system
term terminal
tert., tert- tertiary
T.E.S. turbo-electric ship
TET tetrode
TF transformer; time fuse
T.F. time factor
t.f. time of flight; time factor
TFO tuning fork oscillator
TG telegram; telegraph; telegraphy; tuned grid
tg tangent
TGC time gain control
tgm. telegram
TGTP tuned-grid tuned-plate
TH telegraph
Th *(symbol)* thorium
THD total harmonic distortion; thread
THDS thousands
theor. theoretical
therm. thermometer
THERMO thermostat
thermo. thermodynamics
Th.H.P., THP, thp thrust horsepower
THR total heat rejection
thro'B/L. through bill of lading
TI tuning indicator; tuning inductance; temperature indicator; traffic identification
Ti *(symbol)* titanium
T.I.B. Technical Information Bureau; trimmed in bunkers
TIF telephone interference (influence) factor
TIMM thermionic integrated micro-modules
TIR tolerance of inside radius
T.J. turbo jet
T_J junction temperature
TK trunk
TL transformation line
T/L total loss
Tl *(symbol)* thallium
TLG telegram; telegraph; telegraphy
TLM telemeter
T.L.O. total loss only
TLV television
TM tactical missile
T.M. twisting moment; time modulation; tuning meter; transverse magnetic
Tm *(symbol)* thulium
t.m. transverse magnetic
TMA terminal control area
T.Max. maximum temperature
TMI tuning meter indicator
TML tetramethyl lead
TMR true motion radar
TMS transmission measuring set; telegraphy with manual switching
TMT tetramethyl thiuramdisulphide
TN tuning
Tn *(symbol)* thoron
tn ton
T.N.A. trinitroaniline
T.N.B. trinitrobenzene
T.N.T. trinitrotoluene; tuned not tuned [oscillator]
T.N.X. trinitroxylene
TO take-off
to. ton
t.of f. time of flight
t_{off} turn-off time
tol. toluene
TOM Technical Oil Mission; telex on multiplex
ton long ton
t_{on} turn-on time
tonf tons force
tonn. tonnage
tons d.w.c. tons deadweight capacity
tons/in^2, tons/sq. in tons per square inch
TOPNS theater of operations
Tor.T. torpedo tube
T.O.S. temporarily out of service
TOT time of transmission
T-O.Wt take-off weight
tox. toxic; toxicology
TP test pilot; test point
T.P. turning point; test point; telephone; teleprinter; tuned plate circuit; tubing pressure
t.p. three-phase; three-pole; three-ply; test pressure
t.p.a. tons per annum
TPC triple paper covered *(cable)*
TPH tons per hour
TPI taper per inch; turns per inch
tpi, t.p.i. turns per inch; threads per inch; teeth per inch; tons per inch
t.p.m. tons per month
TPMS *(Am)* tire pressure maintenance system

TPR teleprinter
TPST triple-pole single-throw [switch]
T-P/T-G, t.p.t.g. tuned-plate tuned-grid [oscillator]
tpy tons per year
TR transformer; transmit-receive tube; temperature recorder; torque-synchro receiver
T.R. tons registered; telegraph repeater
t.r. transmit-receive
t_r rise time
TRACDR tractor-drawn
tradic transistor digital computer
Trafo transformer
Trans. transactions
TRANSAC transistor automatic computer
transceiver transmitter/receiver
Trans.Faraday Soc. Transactions of the Faraday Society
TRE Telecommunication Research Establishment
TRF, t.r.f. tuned radio frequency
Tri. trichloroethylene
tri. trigonal
TRICON triple coincidence navigation
TRK trunk
TRML terminal
Trs transmitter
t.r.s. tough rubber sheathed
Tr.S.M.S. triple-screw motorship
TRSP transport seaplane
Tr.S.S. triple-screw ship
TS telephone set
T.S. Television Society; trial speed; twin-screw; training ship; [ultimate] tensile strength; test solution
ts *(Am)*, **t.s.** tensile strength
TSC two subcarrier system
T.S.D. temporarily shut down
ts.d.w. tons deadweight
TSI true [air] speed indicator
t.s.i. tons per square inch
TSM time sharing multiplex
t.s.m.s. twin-screw motorship
t.s.m.y. twin-screw motor-yacht
TSP transponder
T.SS turbine steamship
T.S.S. twin-screw ship; twin-screw steamer
TT type-test[ed]; teletypewriter; transmitting terminal; temperature transmitter
T.T. torpedo tube
tt telephonist
tt. test tube
T.T.C. telephone toll call
TTS teletype setter
TTT transformation-time-temperature
TTY teletypewriter
T.U. traffic unit; transmission unit; thermal unit
Tu tungsten
TV test voltage; tube voltmeter; television
T.V. tank vessel
T.V., t.v. terminal velocity
T.V.A. Tennessee Valley Authority
TVG time-varied gain
TVI, tvi. television interference
TVIST television information storage tube
T.V.O. tractor vaporizing oil
TW travelling wave
Tw Twaddell
tw twisted; travelling wave
°Tw degree Twaddell
TWA Trans World Airlines Inc.
Twad Twaddell
TWR tower
TWT, twt travelling wave tube
TWX telewrite exchange; teletype service without voice communication; teletypewriter exchange
TX transmitter; television receiver; turbo-synchro transmitter
TYPP [number of] thousands [of] yards per pound [avoirdupois]
typw typewriter

U

U *(symbol)* uranium; Board of Trade Unit; micro
U. unwatched; university
u. ultimate; unit; ultra-
UA ultra audible
U/A underwriting account
UAA United Arab Airlines
UAM underwater-to-air missile
U.A.X. unit automatic exchange
Ubb. Ubbelohde
u.c. uncorrected; usual conditions
u/c unclassified; under construction; under consideration
u.c.b. unless caused by
U.D. uniformly distributed; upper deck
UDC, U.D.C., u.d.c. universal decimal classification; upper dead centre; unidirectional current
U.Dk. upper deck
U.D.T. underdeck tonnage
UET United Engineering Trustees, Inc.
UF microfarad
U.F. urea-formaldehyde [plastic]
UFO unidentified flying object
U.F.Powder urea-formaldehyde moulding powder
UHF, uhf, u.h.f. ultra-high frequency
UJTO unijunction transistor oscillator
UK United Kingdom
U.K.A.E.C. United Kingdom Atomic Energy Commission

U.K.fo. United Kingdom for orders
U.K.S.M. United Kingdom Scientific Mission
UL Underwriters' Laboratories, Inc.; unload
ULF, ulf ultra-low frequency
ult *(Am)* ultimate
UM *(Am)* universal mill [plate]
U.M.M. Underground Mining Machinery Ltd.
UNAEC United Nations Atomic Energy Commission
UNC *(Am)* unified national coarse [thread]
uncl[ass]. unclassified
undec. undecomposed
undislvd. undissolved
UNEF *(Am)* unified national extra fine [thread]
UNF *(Am)* unified national fine [thread]
UNIVAC universal automatic computer
unoxid. unoxidized
uns. unsymmetrical
unsat. unsaturated
unst. unstable
unsym. unsymmetrical
UPO undistorted power output
UPU Universal Postal Union
U.S. uniform system; United States
U/S unassorted
u.s. undistorted signal
USA United States Army
usa ultrashort wave
U.S.A.E.C. United States Atomic Energy Commission
USAF United States Air Force
USB upper sideband
USBM United States Bureau of Mines
U.S.B.S. United States Bureau of Standards
USCGS United States Coast and Geodetic Survey
USCS Unified Soils Classification System
U.S.F. [thread] United States form [thread]
U.S.Fed. United States Federal Specifications Board
U.S.G. United States gallon
Usgpm, U.S.g.p.m. United States gallons per minute
USGS United States Geological Survey
USIS United States Information Service
USJCB United States Joint Communication Board
USM underwater-to-surface missile
U.S.M. United States Mail
U.S.N. United States Navy
USNBS United States National Bureau of Standards
USP United States Pharmacopoeia
USS United States standard
U.S.S. United States ship
USSG United States standard gage
usu. usual[ly]
USW ultrashort wave
U.T. universal time
u.t. usual terms
UTRIP universal triangulation program
UUM underwater-to-underwater missile
U.V., uv ultraviolet
U.W. underwater
u/w underwriter
UXB unexploded bomb

V

V vitrified [bond]; *(symbol)* vanadium; vacuum [tube]; volt; voltmeter; valve; total volume; [linear] velocity; value; [specific] volume; viscosity
V. velocity; vessel
v volume; velocity
v. via; very; volume; viscosity; versus; vide
v- vicinal
VA variometer; volt-ammeter; voltampere
V.A. vice-admiral
va variometer; volt-ammeter; voltampere
VAC volt alternating current
VAC, vac. vacuum; in vacuo
vap. vapour
VAR volt-ampere/reactive
Var. variation
var reactive volt-ampere; visual-aural [radio] range
var. variable; variant; variation
VB vibrator
VC comparison voltage; volt-coulomb; viscosity controller; voice coil; varnished cambric
vc critical volume
VCC voice-controlled carrier
VCG vertical line through centre of gravity
V.C.G. vertical line through centre of gravity
V.C.U. variable correction unit
V.D., v.d. vapour density
vdct direct current test voltage
vdcw direct current working voltage
VDF very-high-frequency direction finder; very-high-frequency direction-finding station
VDH, Vdh Vickers diamond hardness, Vickers hardness number
VDI visual Doppler indicator
v.d.i. variable duration impulse system
VDLV vapour drive liquid vortex
VDR voltage-dependent resistor
ve voltage
ve. velocity
−ve negative
+ve positive
VEB variable elevation beam antenna
vel, vel. velocity
velo unit of velocity
vent. ventilation
Verdan vertical digital analyzer

vers, versin versed sine
Vert., vert. vertex; vertical
VF variable frequency; video frequency
V.F. vulcanized fibre; viscosity factor; voice-frequency [telegraphy]
v-f voice-frequency [telegraphy]
VFCT voice-frequency carrier telegraphy
V.F.Dial. voice-frequency dialling
V.F.Eq't voice-frequency equipment
VFO variable-frequency oscillator
VFR visual flight rules
V.F.Rep. voice-frequency repeater
V.F.T. voice-frequency telegraphy
VFX variable-frequency crystal oscillator
VG voltage gain; Vickers [machine] gun
V.G.C. viscosity-gravity constant
VHF, V.H.F., vhf, v.h.f. very high frequency
VHFDF very-high-frequency direction-finding
Vhn Vickers diamond hardness, Vickers [pyramid] hardness number
VI viscosity index
V.I. vertical interval; volume indicator; viscosity index
Vi *(symbol)* virginium
vib. vibration
vic- vicinal
VID variable intermittent duty
VID, vid. video
VIDF vertical [side] intermediate distributing frame
VIE visual indication equipment
vig. vigorous
viol. violent; violet
ViP voltage impulse protection
Vis. visibility
visc. viscous
vit. vitrified[-bonded]
vitr. vitreous
V.K. vertical keel
V/L vapour-to-liquid ratio
VLF, V.L.F., vlf, v.l.f. very low frequency
VLR very long range
VM velocity modulation; voltmeter
V/m, v/m volt[s] per metre
VMDF vertical [side] main distributing frame
VMG Vickers machine gun
VNL via net loss
VO vacuum-tube oscillator
Vocoder voice codes to recreate
VODAS voice-operated device antisinging
Voder voice operation demonstrator
VOGAD voice-operated gain adjusting device
vol. volume
volat. volatile
vol.lim. voltage limiter
VOM volt-ohm-milliammeter
vom volt-ohmmeter
v.o.p. value as in original policy
VOR visual omnirange *(omnidirectional radiator)*
V.O.S., v.o.s. vision on sound
VOX voice-operated transmitter
V.P. vacuum pressure; vent pipe
v.p. vapour pressure; variable pitch; vent pipe
v-p volume-pressure
V.P.I. vapour-phase inhibitor
Vpm volt[s] per metre; volt[s] per mil
vpm volt[s] per mil
v.p.m. vibrations per minute
V.P.N. Vickers pyramid hardness number
V.P.P. variable-pitch propeller
VR voltage relay; vulcanized rubber; viscosity recorder
vr reduced volume
v.r. voltage regulator
VRD vehicle reception depot; vehicle repair depot
vrs. versed
V.S. volumetric solution; visual signal
V/S visual signalling
vs, vs. versus
v.s. very soluble; very slow
v.s.b. vestigial sideband
VSR very short range
VSW, V.S.W., v.s.w. very short wave
v.s.w.r. voltage standing wave ratio
VT vacuum telegraphy
V.T. viscosity transmitter
V.T., v.t. vacuum tube
VTA vacuum tube amplifier
VTB, V.T.B., v.t.b. voltage time to breakdown
VTD vacuum tube detector; variable temporary duty
VTM voltage-tunable magnetron
VTO vacuum tube oscillator; vertical take-off
VTOL vertical take-off and landing
VTVM vacuum tube voltmeter
VU, V.U., vu, v.u. volume unit
v.v. vice versa

W

W [total] weight; work *(energy)*; *(symbol)* tungsten; watt; wattmeter; wattage; wire
W, W., w west
w watt; weight [per cubic foot]; load for foot run
w. water
W.A. with average
w.a. wetting agent
WAAC West African Airways Corporation
WAC worked all countries *(certificate)*; world aeronautical chart
W & M *(Am)* Washburn & Moen wire gage
W.A.Stn wireless auxiliary station
W.B. water ballast; waybill; wheel-base
Wb weber

w.b. wheel-base
W.B./E.I. West Britain/East Ireland
W.B.S. without benefit of salvage
WC wire cable
W.C. water closet; wireless communication; west-central
w.c. water-cooled; water closet
w/c watt per candle
W.C.S.A. westcoast of South Africa
WCT water-cooled tube
wd. wood
w.d. wiring diagram
wdg. winding
WEC Westinghouse Electric Company
WF wave form
W/F wave frequency
w.f. wrong fount
w.f.d. work function difference
WFNA, W.F.N.A. white fuming nitric acid
WG wire gauge; war government
W.G. water gauge; working group
w.g. water gauge; wire gauge; weight guaranteed
wgt. weight
wh. white; watt-hour
Whit Whitworth
W.H.M. watt-hour meter
whr watt-hour
whr m watt-hour meter
W.I. wrought iron
wibon whether inberth or not
WIL white indicating lamp
wk week
wkg. working
WL white lamp; wave length
W.L. water line
W.L., W/L, w.l., w/l wave length
w.l., w/l water line
w.l.coeff. water-line coefficient
W-less wireless
W.long. west longitude
Wly. westerly
WM wattmeter
W.M. white metal
W.M., W/M weight or measurement
WMO World Meteorological Organization
WNW west-northwest
WO wireless officer
W/O water in oil; wireless operator
w/o without
WP weatherproof
W.P. weather permitting; water pressure
wp waterproof; working pressure
w.p. weather permitting
W.P.A. with particular average
W.P.B. waste paper basket
WPC, wpc, w.p.c. watt[s] per candle; watt[s] per channel
wpm, w.p.m. words per minute
w.p.p. waterproof packing
W.Q. water-quenched
WR wall receptacle
W.R. warehouse receipt; wire rope
W/R warehouse receipt
Wrd. westward
Wrk. wreck
wrls wireless
Wrn. western
W.R.O., w.r.o. war risks only *(insurance)*
W.S. working surface
Ws watt second
w.s. wetted surface; working surface
WSW west-southwest
WT weapon training
W.T. water tank; watertight; wireless; watertime
W.T., W/T wireless telegraphy; wireless telephony
wt, wt. watt; weight
w.t. watertight
W.T.B. watertight bulkhead
W/T DF wireless telegraph direction finder
W./T.D.F. wireless telegraph direction finding station
W.T.D.R. wireless telegraphy direction
W.T.H. watertight hatch
wt/h.p. weight per horsepower
W/T Mge wireless telegraphy message
wt p.ft weight per foot
W/TS wireless telegraphy station
WV working voltage
w/v wind velocity; weight in volume concentration
WVAC working volt alternating current
WVDC working volt direct current
WVL wavelength
WW wire-wound
W.W. weather working days
W.W., W/W warehouse warrant
w/w, w/w. weight in weight concentration
w-w wire-wound
w.w.d. weather working days
W.W.G. Warrington wire gauge
wyec star connection; Y-connection

X

X experimental; *(symbol)* xenon
X- röntgen
XAMP horizontal amplifier
xcvr transceiver
x-d, x div ex dividend
XGAM experimental guided air missile
x.in. ex interest
X lab experimental laboratory
xll extra lightly loaded
XLR experimental liquid rocket

XMIT, XMITT transmitter
XMSN transmission
XMTG transmitting
Xmtr transmitter
XP express paid
xpln. explanation
XRA X-ray assistant
XRCD x-ray crystal density method
XSM experimental strategic missile
XT radiotelephone
Xtal, Xtl crystal
X-tgd cross-tongued
xtr. transmitter
xyl. xylene

Y

Y *(symbol)* yttrium
y yoke
y. year
YAMP vertical amplifier
Y.A.R. York Antwerp Rules
Yb *(symbol)* ytterbium
Y.C. Yacht Club
yd yard
yel. yellow
Y.M. yacht measurement
Y.P. yield point
Y.R. yard repair
yr year
Y.R.A. yacht racing association
Y.S. yield strength
yt. yacht

Z

Z zone; modulus of section
Z. azimuth
ZB zero beat
Z.D. zenith distance
zeta zero energy thermonuclear assembly
ZF zone of fire
Z.F., zf, z.f. zero frequency
ZM impedance measuring [device]; impedance meter
Zn *(symbol)* zinc
ZPI zone position indicator
Zr *(symbol)* zirconium
ZSF zero skip frequency
Z.T. zone time